# THE
# WEATHER
# ALMANAC

# THE
# WEATHER
# ALMANAC

A reference guide to weather, climate,
and related issues in the United States
and its key cities

## TWELFTH EDITION

**Steven L. Horstmeyer**

A JOHN WILEY & SONS, INC., PUBLICATION

Published by John Wiley & Sons, Inc., Hoboken, New Jersey.
Published simultaneously in Canada.

For general information on our other products and services or for technical support, please contact our Customer Care Department within the United States at (800) 762-2974, outside the United States at (317) 572-3993 or fax (317) 572-4002.

Wiley also publishes its books in a variety of electronic formats. Some content that appears in print may not be available in electronic formats. For more information about Wiley products, visit our web site at www.wiley.com

*Library of Congress Cataloging-in-Publication Data*

Horstmeyer, Steven L.
   The weather almanac: a reference guide to weather, climate, and related issues in the United States and its key cities.
Twelfth Edition / Steven L. Horstmeyer
      p. cm.
   Includes bibliographical references and index.
   ISBN 978-0-470-41325-8 (cloth)

Printed in the United States of America

oBook ISBN: 978-1-118-01521-6
ePDF ISBN: 978-1-118-01519-3
ePub ISBN: 978-1-118-01520-9

10  9  8  7  6  5  4  3  2  1

# Contents

# Preface

Whenever an established work like *The Weather Almanac* is updated, the primary concern of the editor is to remain true to the traditions established in earlier editions. The editor is faced with retaining some material, updating other material, and saying goodbye to material that has seen better days.

The 12th edition of *The Weather Almanac* is my best effort at staying true to the hard work of my predecessors, but at the same time move the book in the direction that my 33 years as a broadcast meteorologist, public speaker, and educator tell me is most useful to the user of the book seeking answers to weather questions.

If you are looking at material that contains numbers or maps or graphs, admittedly most of the book, then you are reaping the rewards of the hard work of James Patterson. During the course of this rewrite, James was working full- and part-time jobs, finishing his meteorology degree, buying a house, and starting a family. He always found time to get the job done. He has a great ability to find problem numbers and fix them. Thanks James. James, did I say thanks?

Chapter 1 contains 42 climate maps of the United States and communicate what maps can best the spatial patterns that determine what we call climate. Chapter 2 is all new and a response to a heightened interest in renewable energy. In Chapter 2, you will find detailed information about the solar radiation and wind climates of the United States and many wind roses. They are hard to come by, but there are many here.

I have lumped the many varieties of extreme and record-setting weather into Chapter 3 and dug long and hard to assemble a chapter with some very difficult to find but important weather information. I have stayed with the customary chapters for severe and tropical weather (Chapters 4 and 5, respectively). Chapter 6 covers El Niño/La Niña, perhaps the most important climate discovery of the last half of the 20th century. Chapter 6 is updated with all new easy-to-follow diagrams.

The second entirely new chapter is Chapter 7: Global Warming and Climate Change. As a professional meteorologist deeply concerned with human impact on our environment, I am continually dismayed by the amount of bad information out there about global warming. I have intentionally steered clear of forecasts and hope this chapter clarifies physical principles, research methods, the role of proxy data, and what climate models are and what they are not.

Air Pollution is the topic of Chapter 8 and contains the latest available data that comes primarily from the US Environmental Protection Agency (EPA).

Chapter 9 contains international climate data. I have included more locations (321) and each location has more data. While this does not give a complete look at Earth's climate, most places you can easily go are covered.

In Chapter 10, you will find climate data for 128 US cities in the form of the 2009 *Local Climatological Data Annual Summary*. There is a great deal of information here on each station, and there are 20 more stations than the last edition. We also graphed the daily high temperatures, low temperatures, and precipitation for 2009 so you can see at a glance what 2009 was like in each city.

Chapter 11 is an annotated time line of human interaction with and knowledge of the atmosphere. As of this writing, there is no other popular reference work with such a listing. History buffs should find this interesting.

I can now answer the question my wife put to me about a week ago. Yes honey, you now get your husband back.

*Steven L. Horstmeyer*
*August 23, 2010*
*Cincinnati, OH*

# 1

# Climate Maps of the United States

## • WHAT IS CLIMATE?

Weather is the day-to-day, sometimes minute-to-minute, changes in the atmosphere. Everyone has an intuitive idea what weather is. But when the time period is extended to months, seasons, years, decades, and longer, we talk about climate. Climate is the long-term state of the atmosphere. It is how you expect the atmosphere to behave.

The change of seasons is part of what climate is. In the Midwestern United States, residents expect hot, humid summer weather to gradually yield to autumn, characterized by cool mornings and toasty warm afternoons dominated by blue sky. In southern and central California, residents know that the brown hillsides that dominate the landscape from late spring into late autumn will begin to green as seasonal rains replenish soil moisture and plants begin to grow.

In Hawaii there is hardly any seasonal temperature change at all, but there are subtle differences from summer into winter in wind and rain events.

Climate is much more than seasonal change. It has been called the "average" of all weather, but it is still more. It can also be the daily, weekly, monthly, or annual range of a weather variable. Climate can be the frequency of occurrence of any weather event such as lightning. In addition there are more complex statistical measures, such as standard deviation that measures the variation about the average, that can help define climate.

Climate can be the average relative humidity at a specific hour of the day or the number of days the relative humidity drops below a certain value. The number of days snow fall exceeds a given amount gives you an idea of the frequency of traffic snarls, while the number of hours the average wind exceeds a given value during a year may help decide about the placement of a wind-powered turbine.

Climate can be defined however you need it to be. You decide what weather variables affect your project and develop a climatology that describes what to expect. The average afternoon temperature for a given place may give you an idea of how comfortable the location is but including a humidity variable and wind speed will give you a better idea of the "comfort climate."

If you are projecting the heating cost of locating a new office facility, you would want detailed information about lowest temperatures, how long the temperature is colder than a particular value, how sunny the location is, and how windy it is. Each weather variable is part of the "natural gas for heating" climatology, and each affects the demand for natural gas for heating.

In summer a "residential cooling" climatology would include the same variables as for heating along with a humidity variable to account for electrical power demand.

Think of it this way: weather is a rainy day, while climate is a rainy place. All US cities have rainy days, but Seattle has a rainy climate. Portland, ME, has occasional hot days, but Orlando, FL, has a hot climate.

## • THE CLIMATE MAPS

The 42 maps in this chapter represent a detailed picture, a climatology, of what you can expect over the long term in the lower 48 states.

The data were prepared and quality-controlled by the National Climatic Data Center (NCDC) of National Oceanic and Atmospheric Administration (NOAA) for the *Climate Maps of the United States* (CLIMAPS) database. The maps were redesigned and replotted for grayscale reproduction in this volume.

If you have experience with using or creating contour maps, you may be accustomed to having a fixed data interval between contour lines. That almost never works when creating climate maps because the distribution of climate is not regular and there are many factors that complicate how

*The Weather Almanac: A Reference Guide to Weather, Climate, and Related Issues in the United States and Its Key Cities*, Twelfth Edition. Steve Horstmeyer.
© 2011 John Wiley & Sons, Inc. Published 2011 by John Wiley & Sons, Inc.

quickly values change. The contour intervals used here are the intervals chosen by the NCDC.

Elevation is probably the most difficult complicating factor to deal with. When you examine the maps in this chapter, you will see how small some of the areas can be because of dramatic changes in climate over a short distance in mountainous terrain. For that reason it was decided not to use fill patterns because they can be very confusing when small areas are involved.

For most of the climate maps we opted to use a symmetrical shading scale ranging from white at the minimum value through medium–dark gray back to white at the maximum value. To the unaccustomed this may seem confusing, but it is standard practice in many science publications. There are a few geographical areas where maximum white and minimum white come close together, such as from the Central Valley of California into the Sierra Nevada Mountains. In areas such as this you will need to be careful to interpret the date correctly.

To help guide you in using the climate maps, many have specific values plotted for a contour or an area. Doing this is always problematic because map data can be obscured by the numbers. We carefully considered the placement of each and every number so as to minimize covering fine detail.

Not all the climate maps we prepared are in this chapter; many fit better in chapters on specific topics.

Climate maps covering sunshine, solar radiation, cloud cover, and wind variables are in Chapter 2, covering solar and wind renewable energy-generating technologies.

The climatology of US severe weather is in Chapter 4, with maps of tornado tracks, hail, and lightning occurrence, and the climatic information about Atlantic and eastern Pacific hurricanes is included in Chapter 5. There are maps dealing with specific aspects of past climate (paleoclimate) in Chapter 7. Extensive numerical information about 128 US cities is found in Chapter 10, while weather data for 321 locations outside the United States are found in Chapter 9.

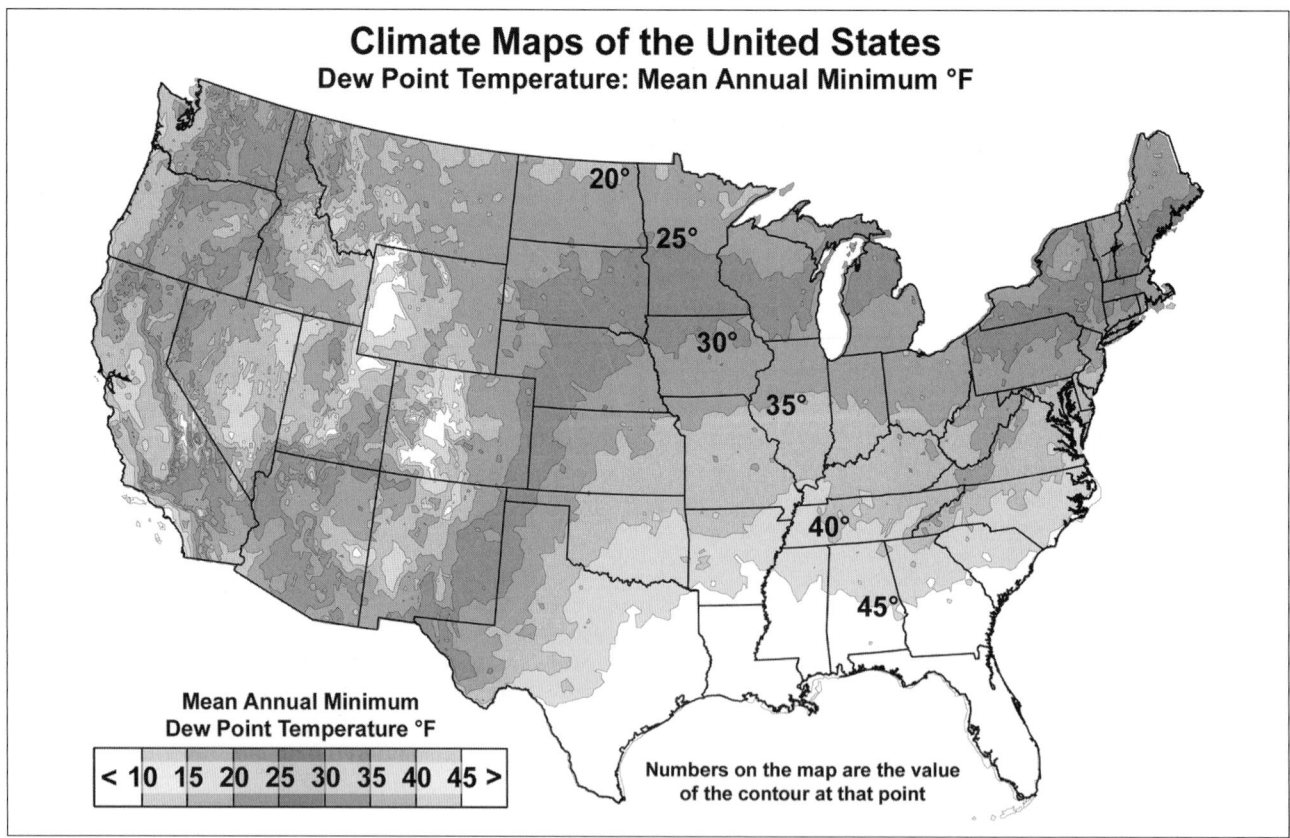

**Figure 1.1**   Mean annual minimum dew point temperature.

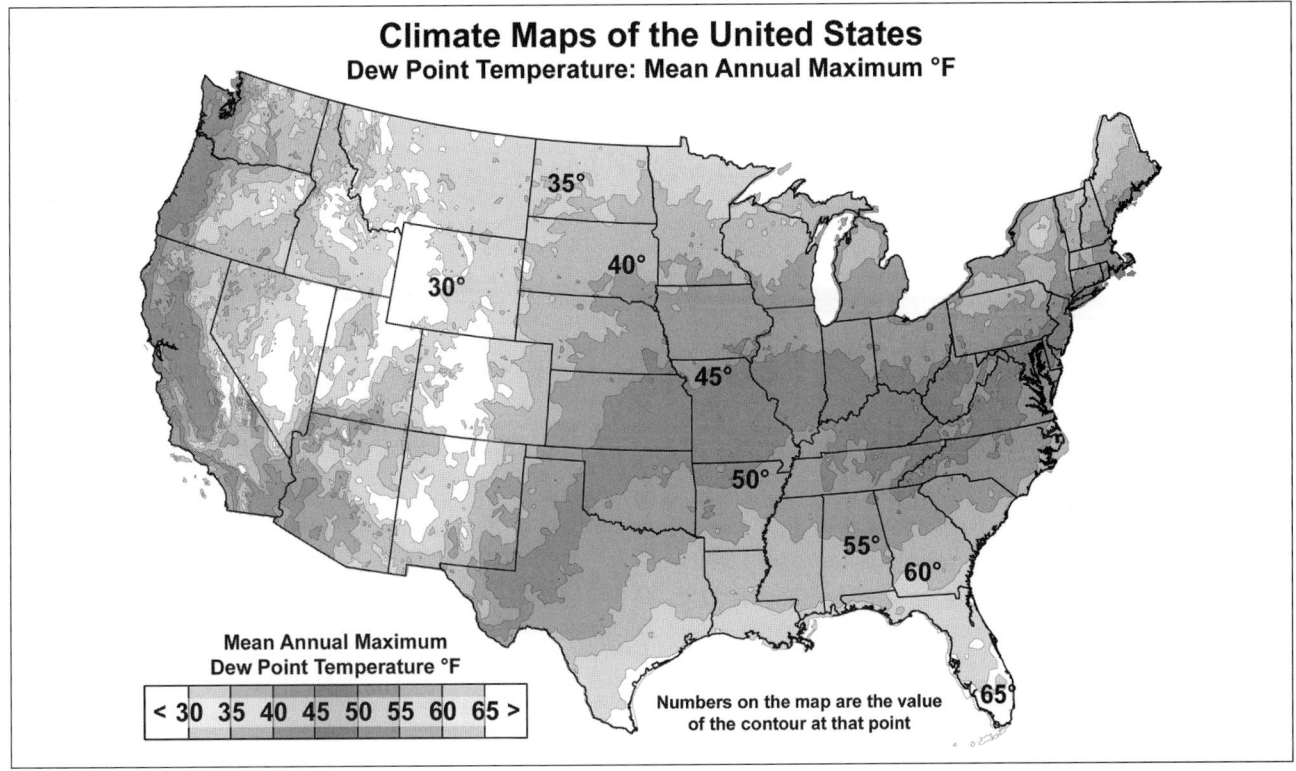

**Figure 1.2**　Mean annual maximum dew point temperature.

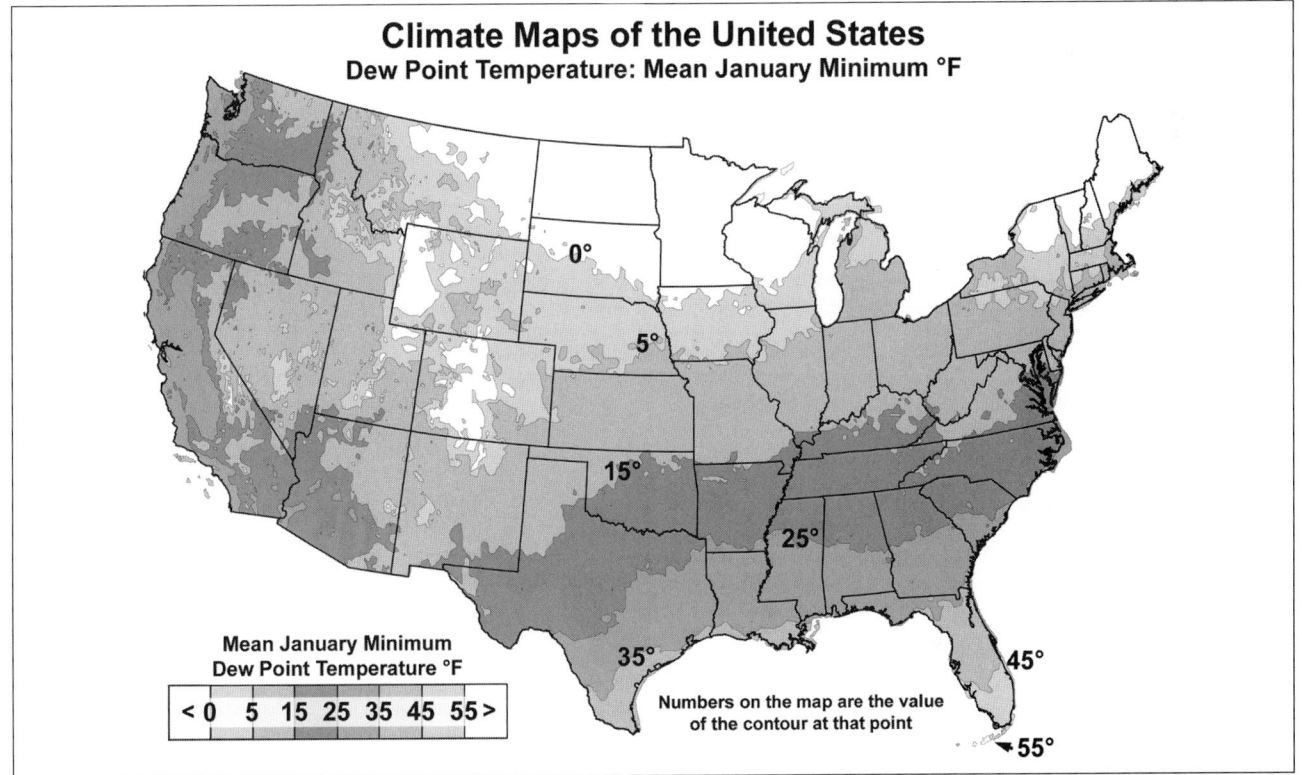

**Figure 1.3**　Mean January minimum dew point temperature.

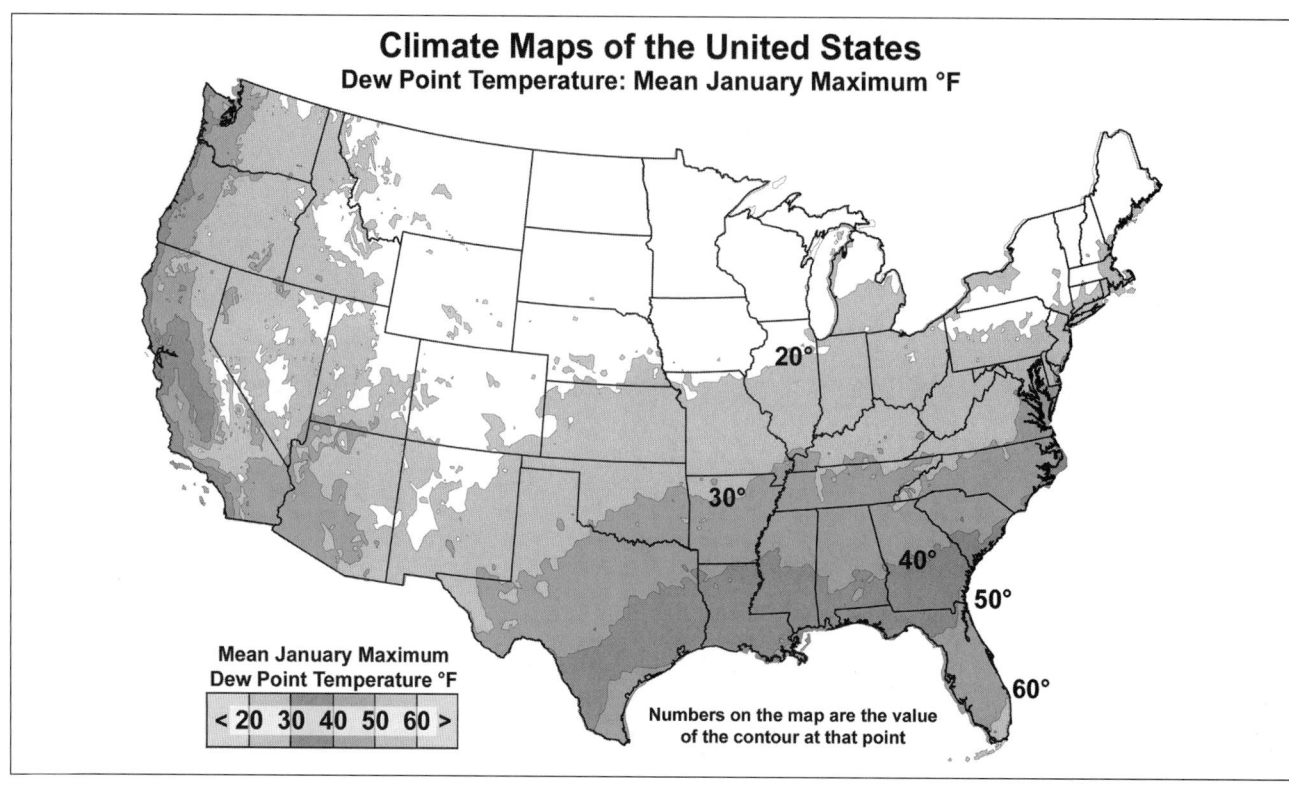

**Figure 1.4** Mean January maximum dew point temperature.

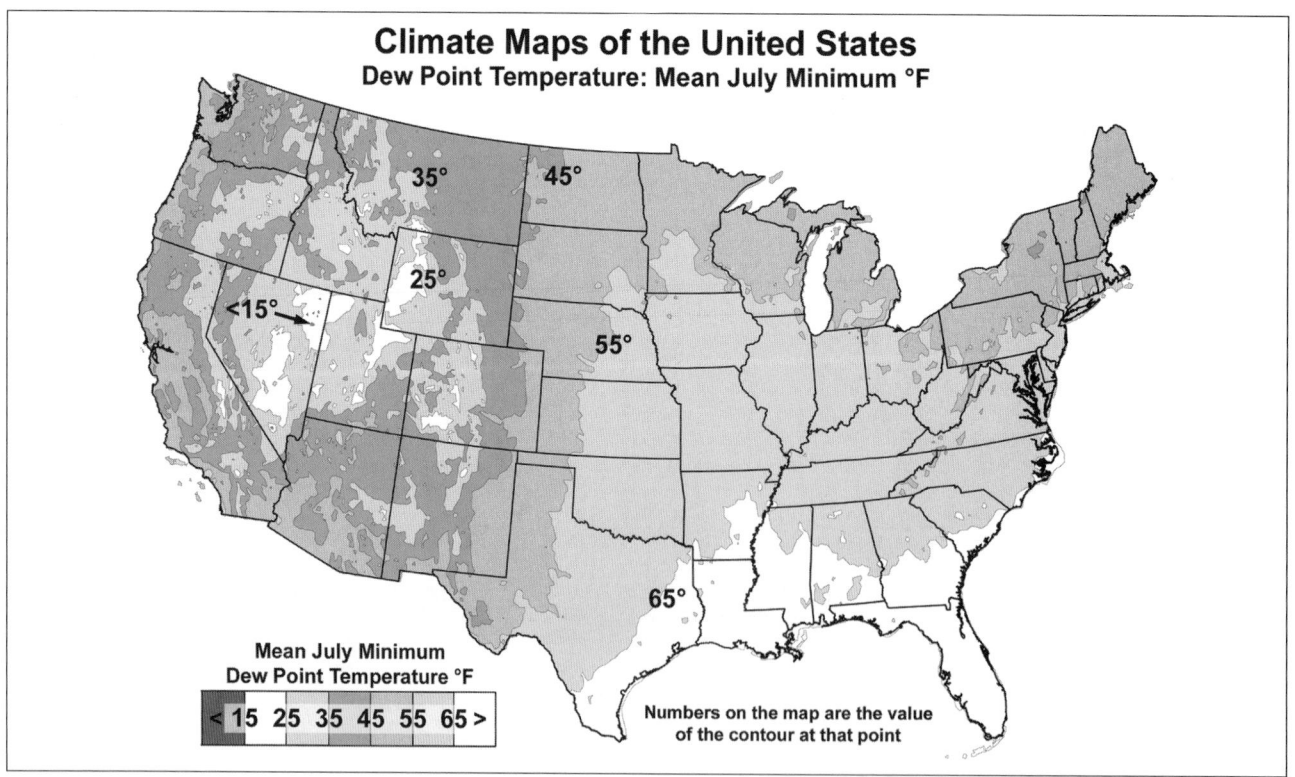

**Figure 1.5** Mean July minimum dew point temperature.

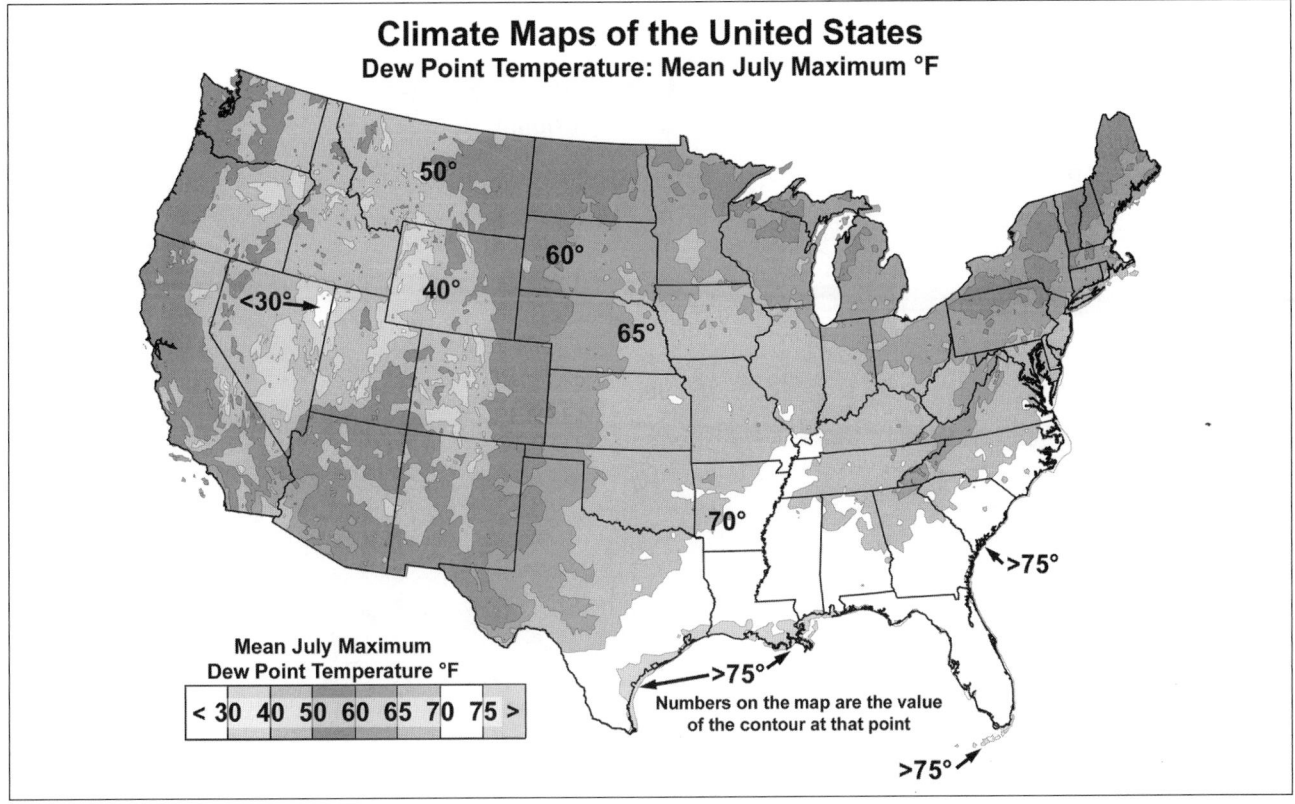

**Figure 1.6**   Mean July maximum dew point temperature.

### Dew Point Temperature

*Notes:* Dew point temperature is one of the many measures of humidity. It is defined as the temperature to which a mass of air must be cooled for condensation to begin or equivalently the temperature at which a relative humidity of 100% occurs when the air is cooled. It is a measure often used by forecasters, and the smaller the difference between dew point temperature and air temperature, the higher the relative humidity. When the dew point temperature equals the ambient air temperature, the relative humidity is 100%.

When the dew point temperature reaches 60°F, nearly everyone feels the humidity, and when the dew point temperature reaches 70°F, nearly everyone says the weather is sticky. Because of that, dew point temperature along with temperature is a primary predictor of energy usage for cooling.

***Energy Use for Heating and Cooling***

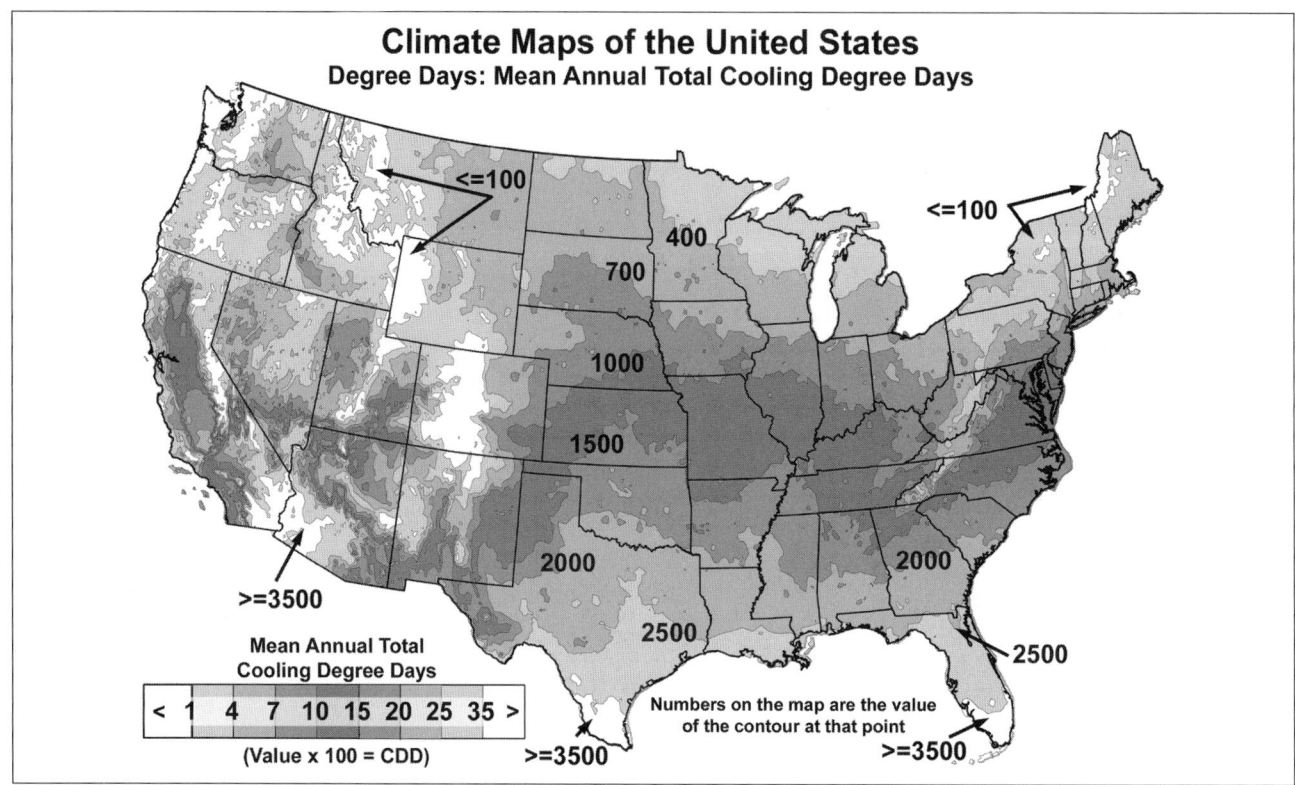

**Figure 1.7** Mean annual number of cooling degree days.

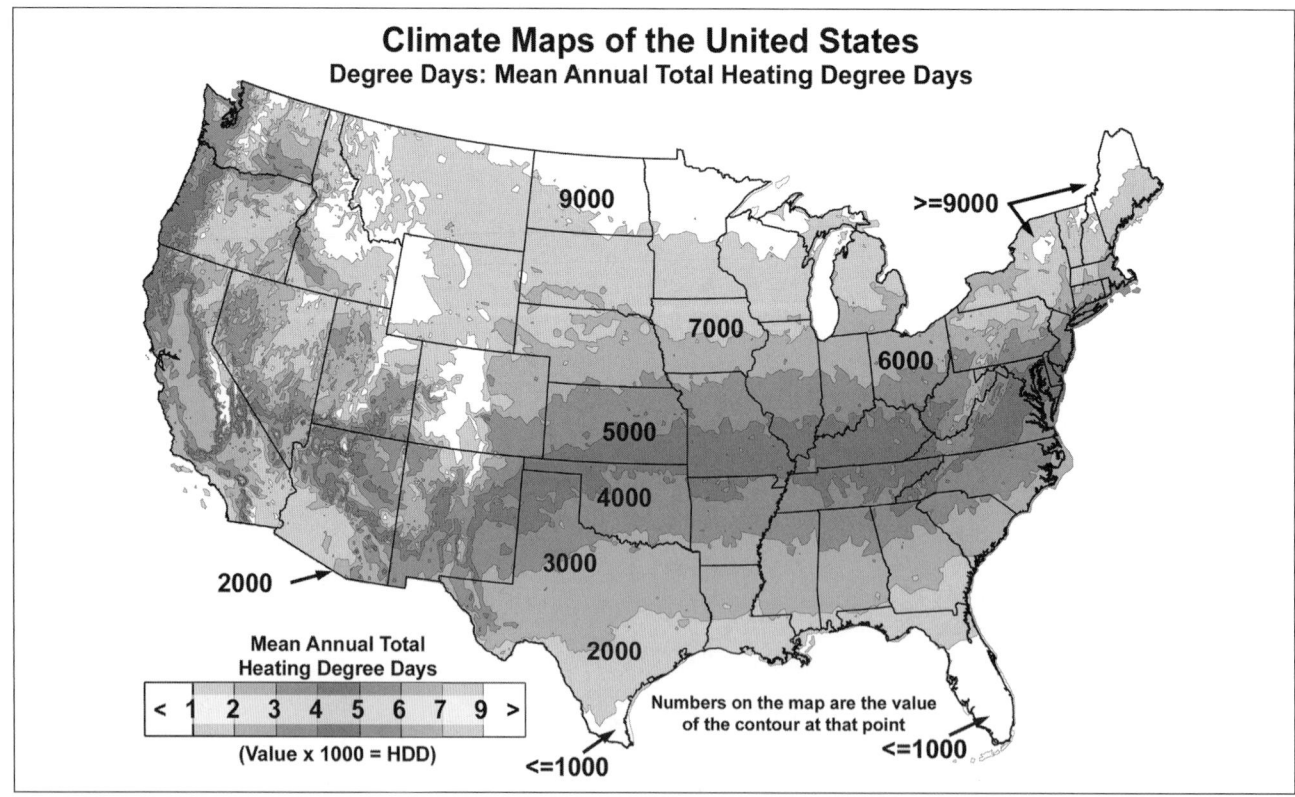

**Figure 1.8** Mean annual number of heating degree days.

### Degree Days Maps

*Notes:* The terms "cooling degree days" and "heating degree days" are confusing and it is best to think of them as cooling units and heating units.

Using a base temperature of 65°F as the dividing point between the need for heating and the need for cooling, both heating and cooling degree days are the difference between the average daily temperature and the base temperature of 65°F. If the average temperature is warmer than 65°F, the difference in °F is the number of cooling degree days. If the average temperature is cooler than 65°F, the difference in °F is equal to the number of heating degree days.

If the high temperature for a day is 80°F and the low is 66°F, the average for the day is 73°F, which is 8°F warmer than the base temperature of 65°F, so the day adds 8 cooling degree days to the running seasonal total.

The way degree days are calculated can lead to a significant error. For example, if at midnight it is 60°F and the temperature drops to 45°F at 1:00 A.M., then for the next 22 hours, the temperature hovers at 40°F; the average temperature for the day using only the high and low is 50°F, but using the 24-hourly temperatures yields an average of 41°F, an 18% difference. Whenever there is a large temperature change early or late in a day, large errors can occur when using only the daily high and daily low to calculate average temperature. For this reason many private utilities use hourly data in forecasting and tracking electrical loads and natural gas demands.

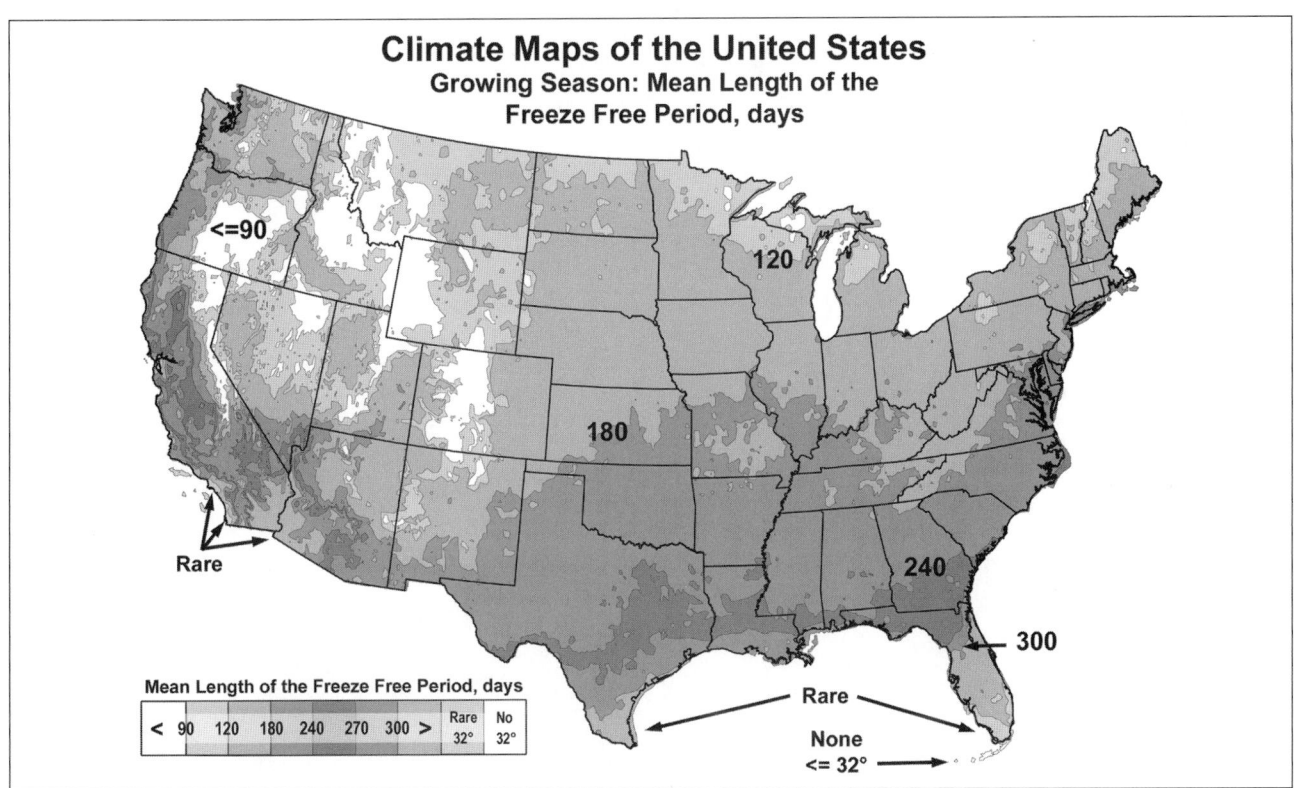

**Figure 1.9**    Mean length of the freeze-free period.

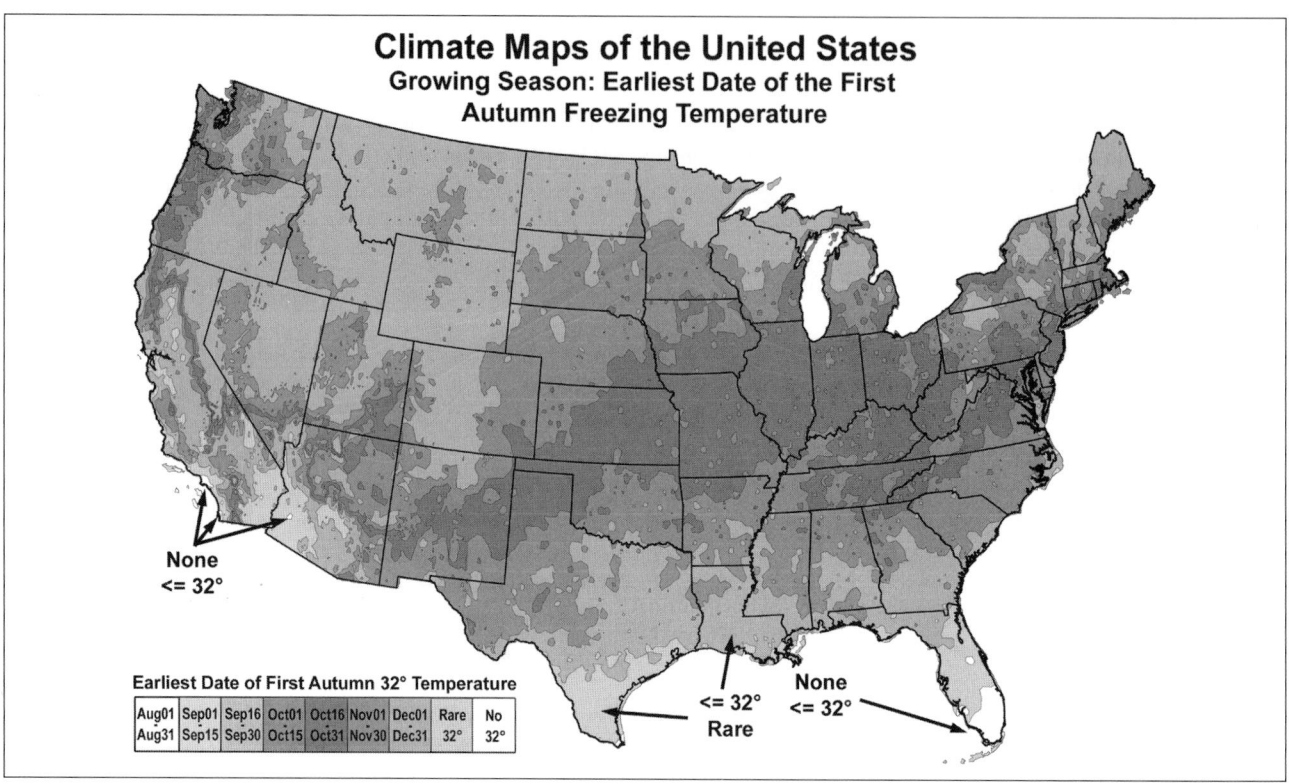

**Figure 1.10**   Earliest date of the first autumn freezing temperature.

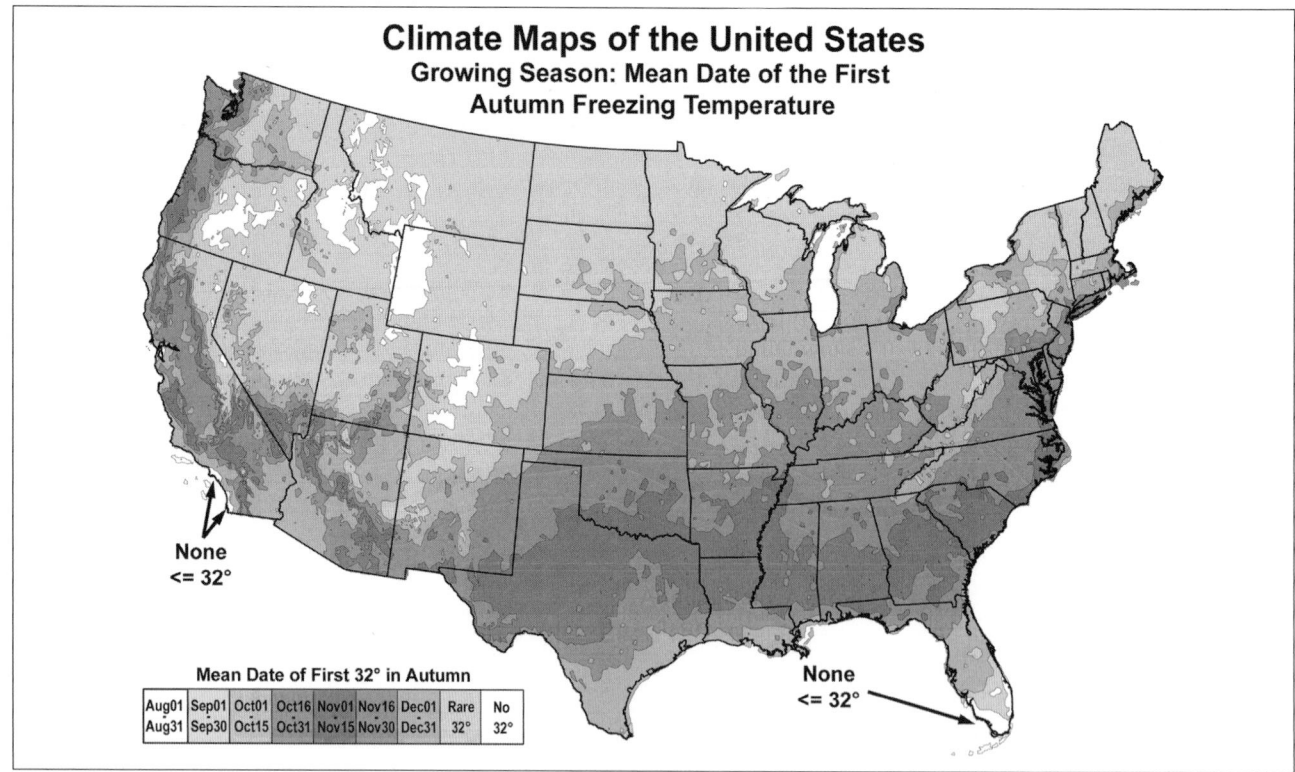

**Figure 1.11**   Mean date of the first autumn freezing temperature.

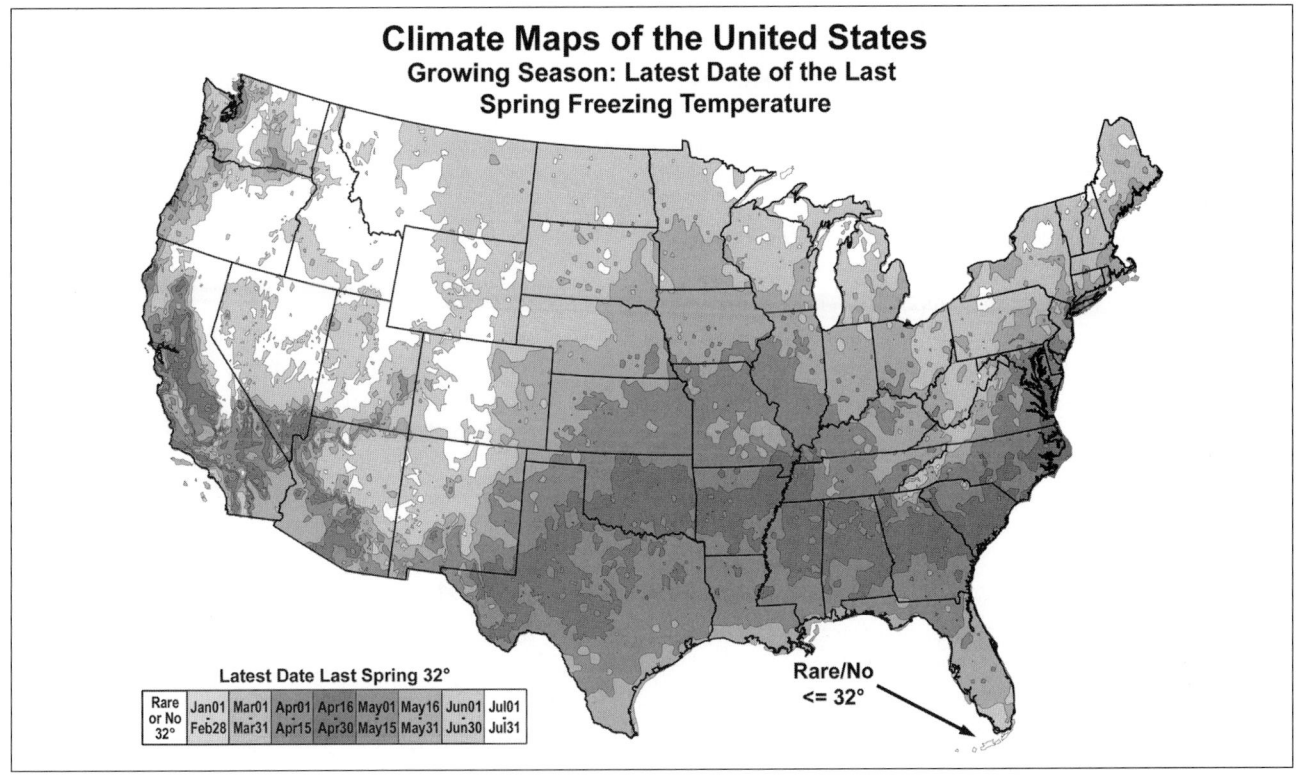

**Figure 1.12**   Latest date of the last spring freezing temperature.

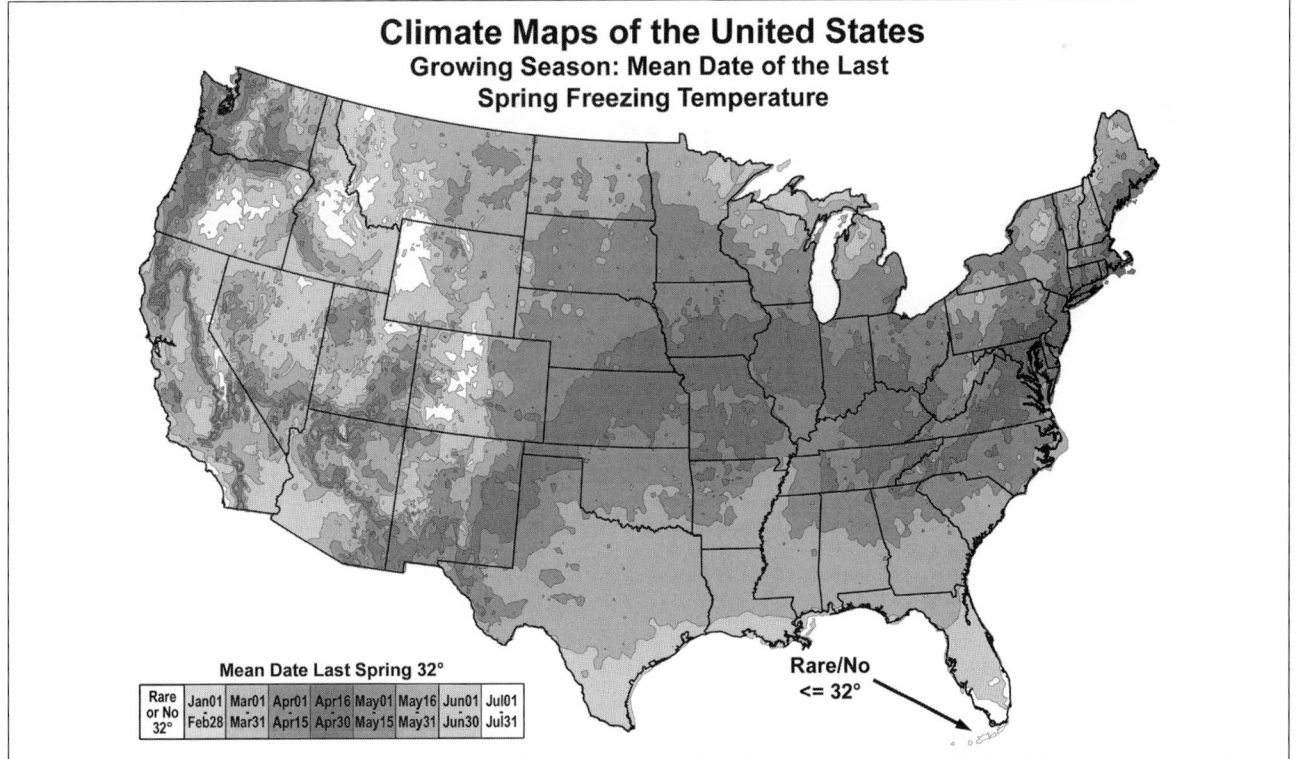

**Figure 1.13**   Mean date of the last spring freezing temperature.

### Growing Season/Freeze–Thaw Maps

*Notes:* Traditionally, the growing season ends with the first-observed frost and begins again after the last frost. Frost can occur when the air temperature at thermometer height is as warm as 35°F–37°F, but because colder air is denser and settles to the ground, the temperature where the frost occurs is freezing or colder.

Instead of the term "growing season," "freeze-free period" is used, and it is defined as the number of days between the last spring freezing temperature and the first autumn freezing temperature. This use prevents confusion between the traditional definition and the one currently in use.

Just because a location reaches 32°F does not mean plant growth has stopped and plant damage has occurred. The length of time the temperature stays at or below a certain temperature threshold is also important. The freeze-free period gives a general indication of the length of the growing season, but crop-specific and site-specific information is required for practical application in agriculture.

### Precipitation Maps

*Notes:* In the United States, measurable precipitation is defined as an amount of 0.01" or more.

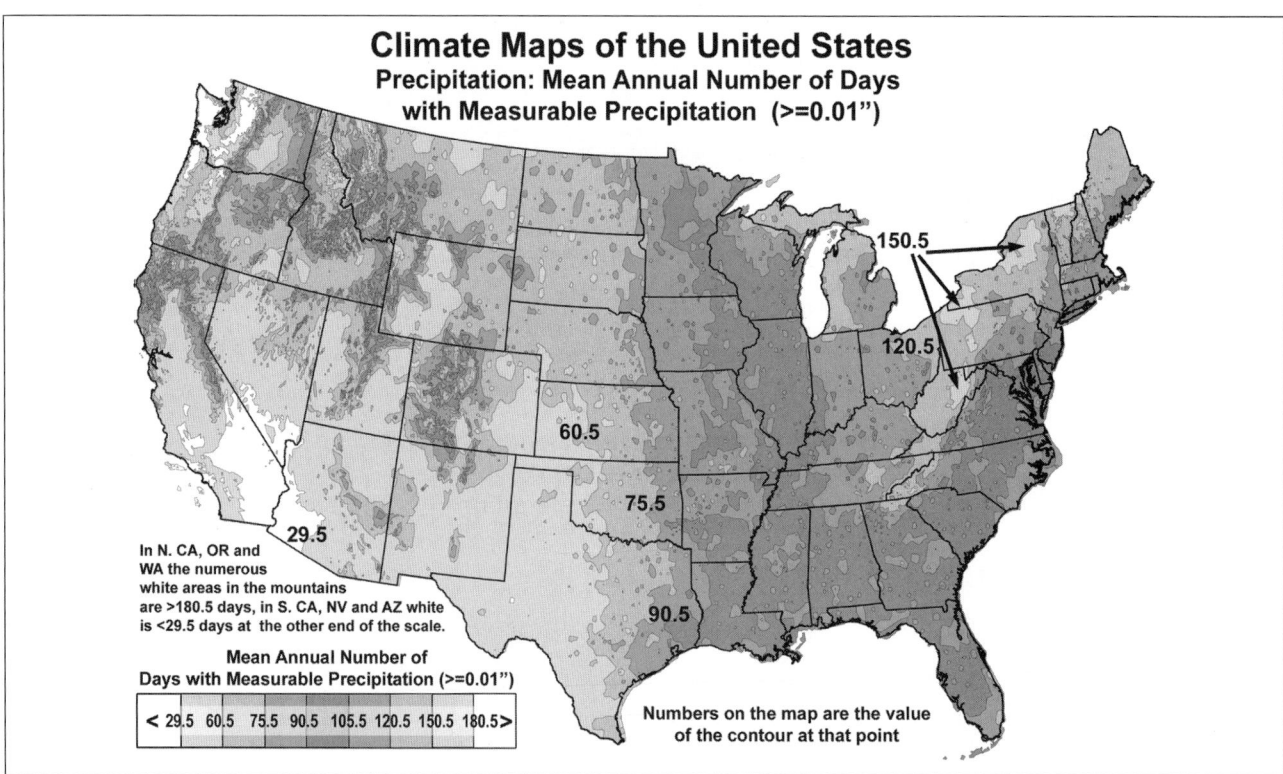

**Figure 1.14**   Mean annual number of days with measurable precipitation.

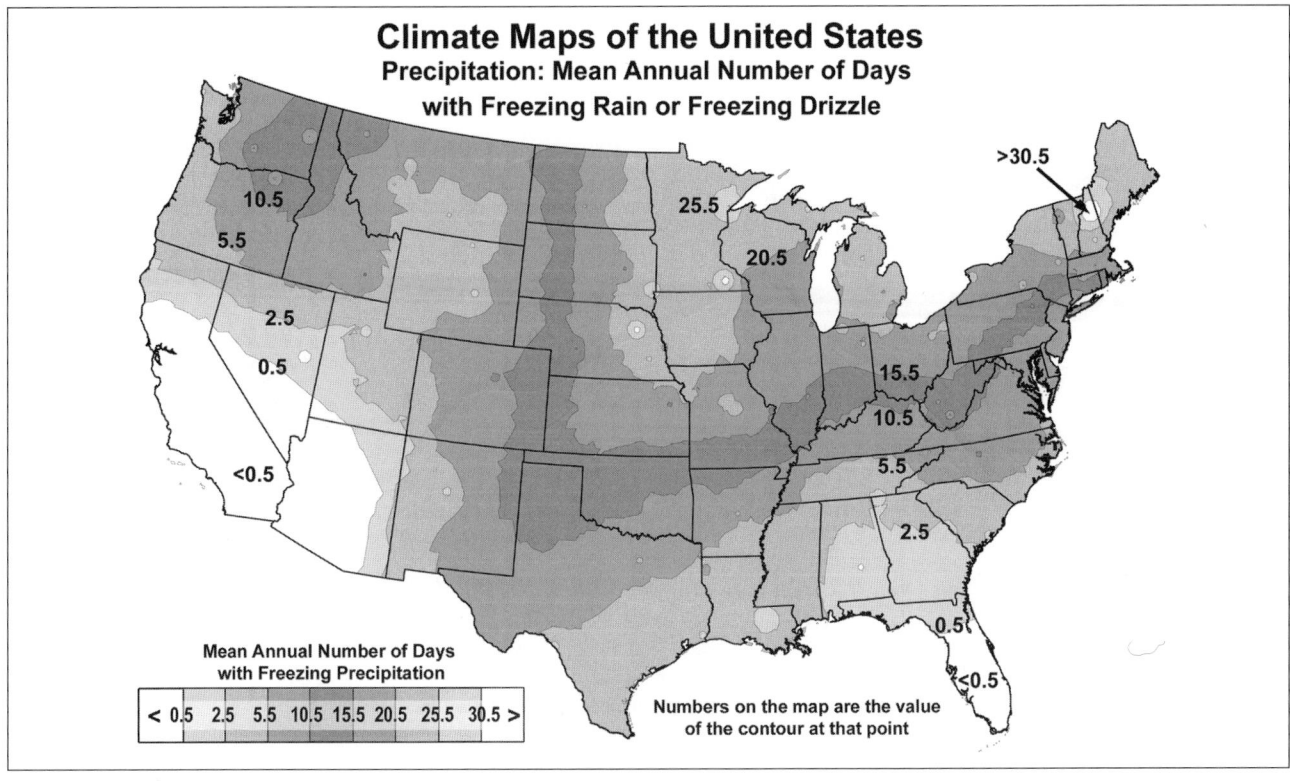

**Figure 1.15**   Mean annual number of days with freezing rain or freezing drizzle.

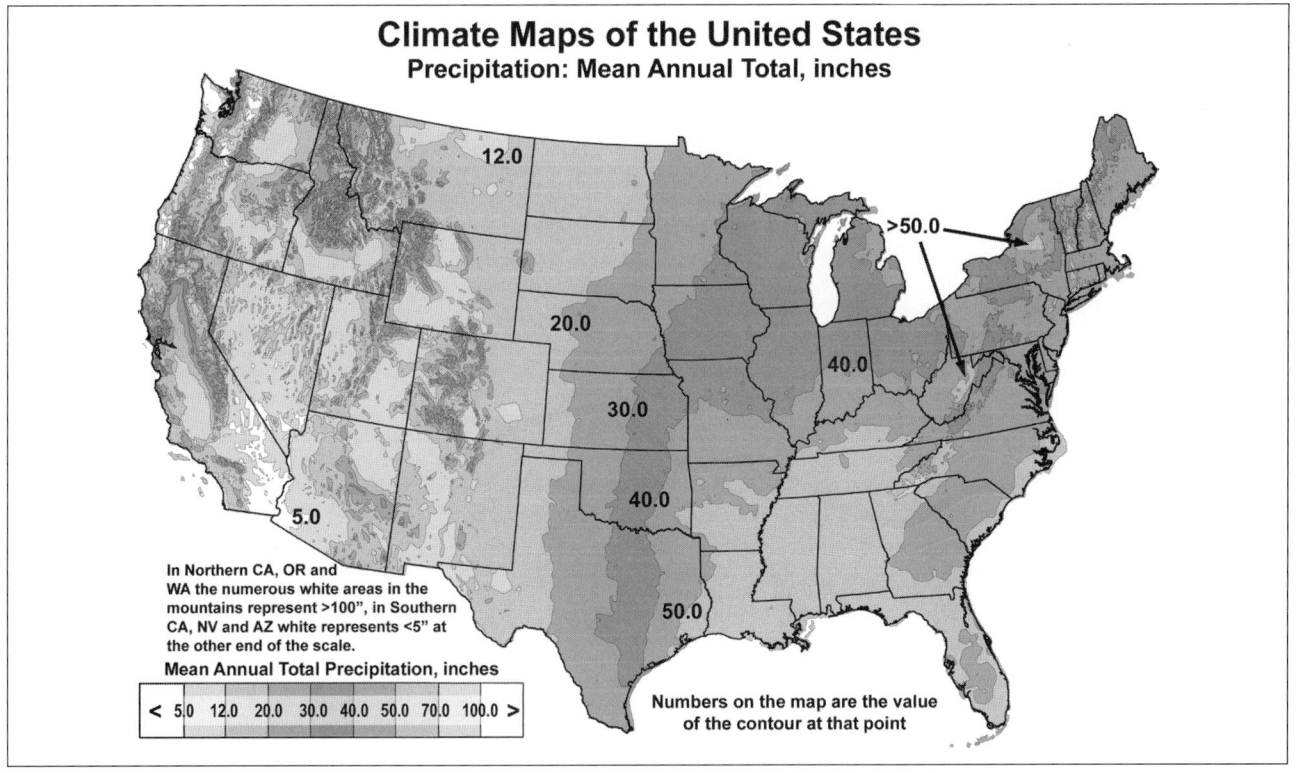

**Figure 1.16**   Mean annual total precipitation.

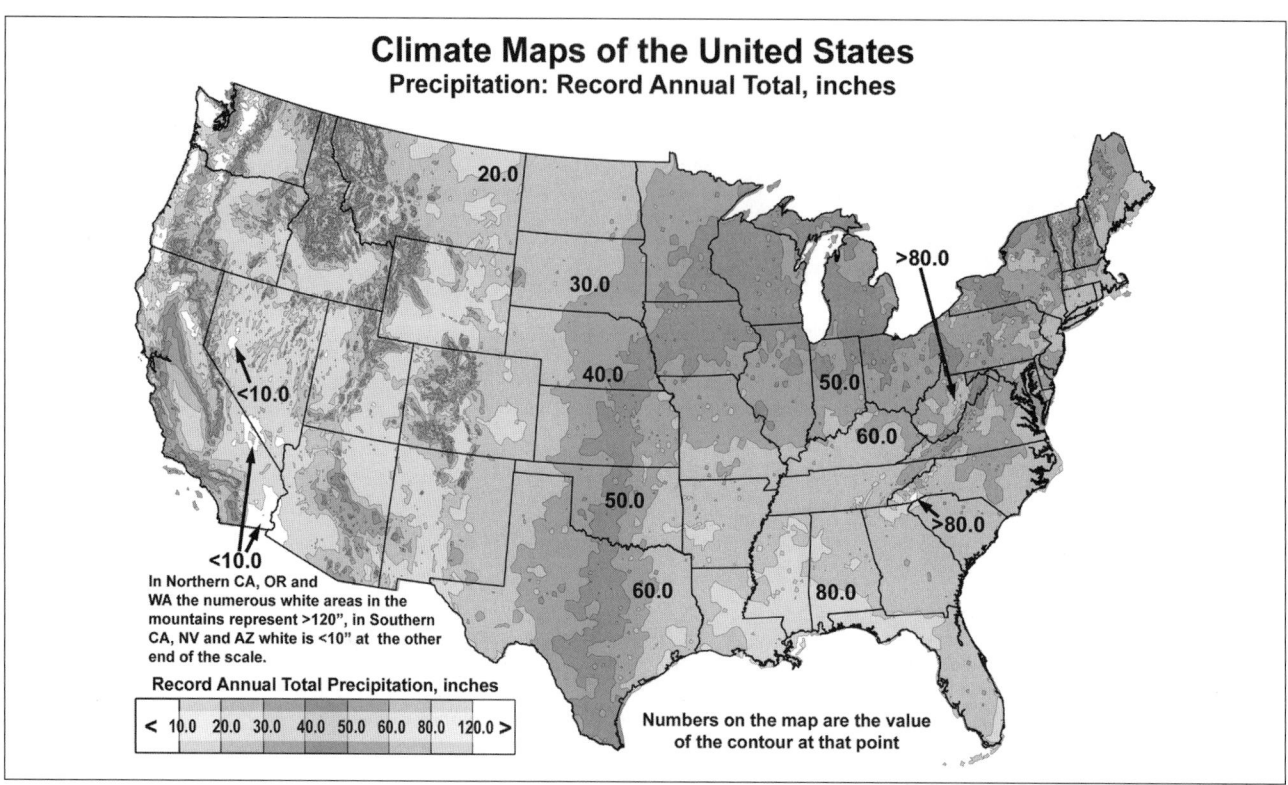

**Figure 1.17**   Record annual total precipitation.

### Atmospheric Pressure (Sea Level)

*Notes:* Atmospheric pressure is always mathematically adjusted to what it would be at sea level the standard reference level in meteorology. Unadjusted pressure values are referred to as "station pressure."

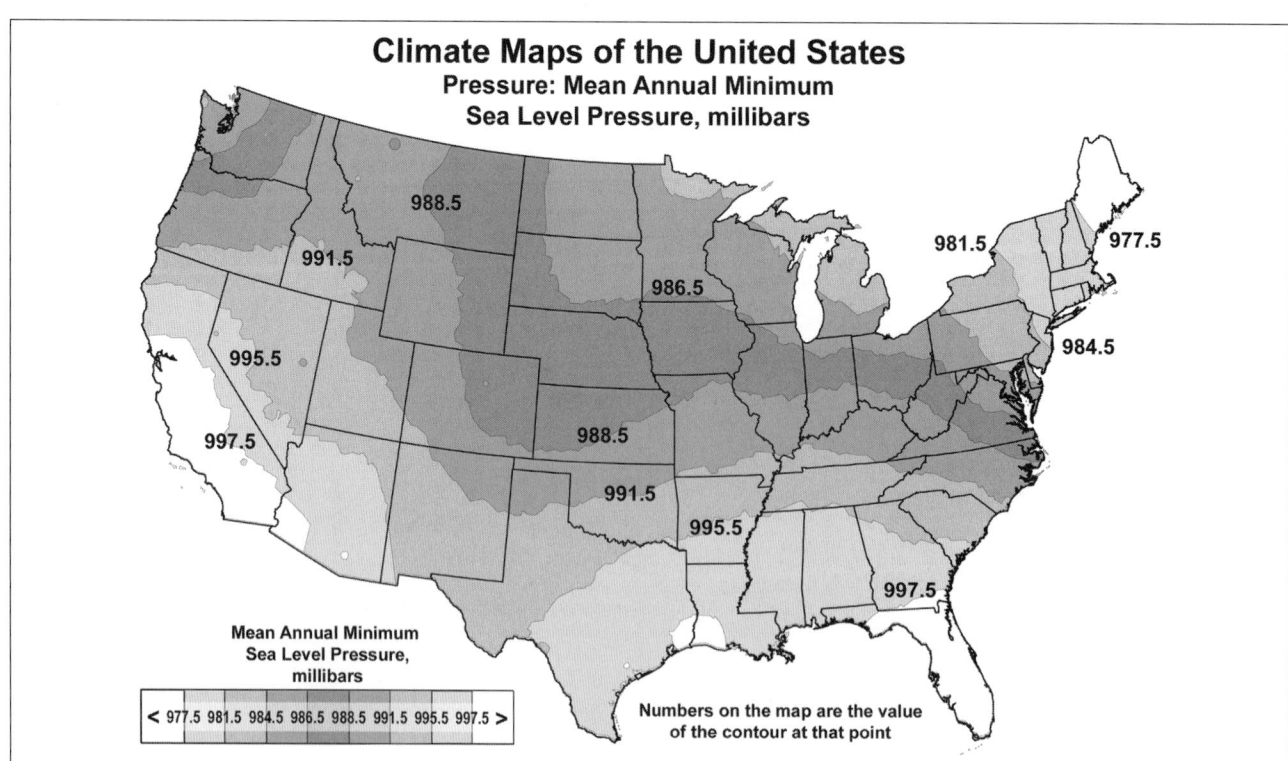

**Figure 1.18**   Mean annual minimum pressure.

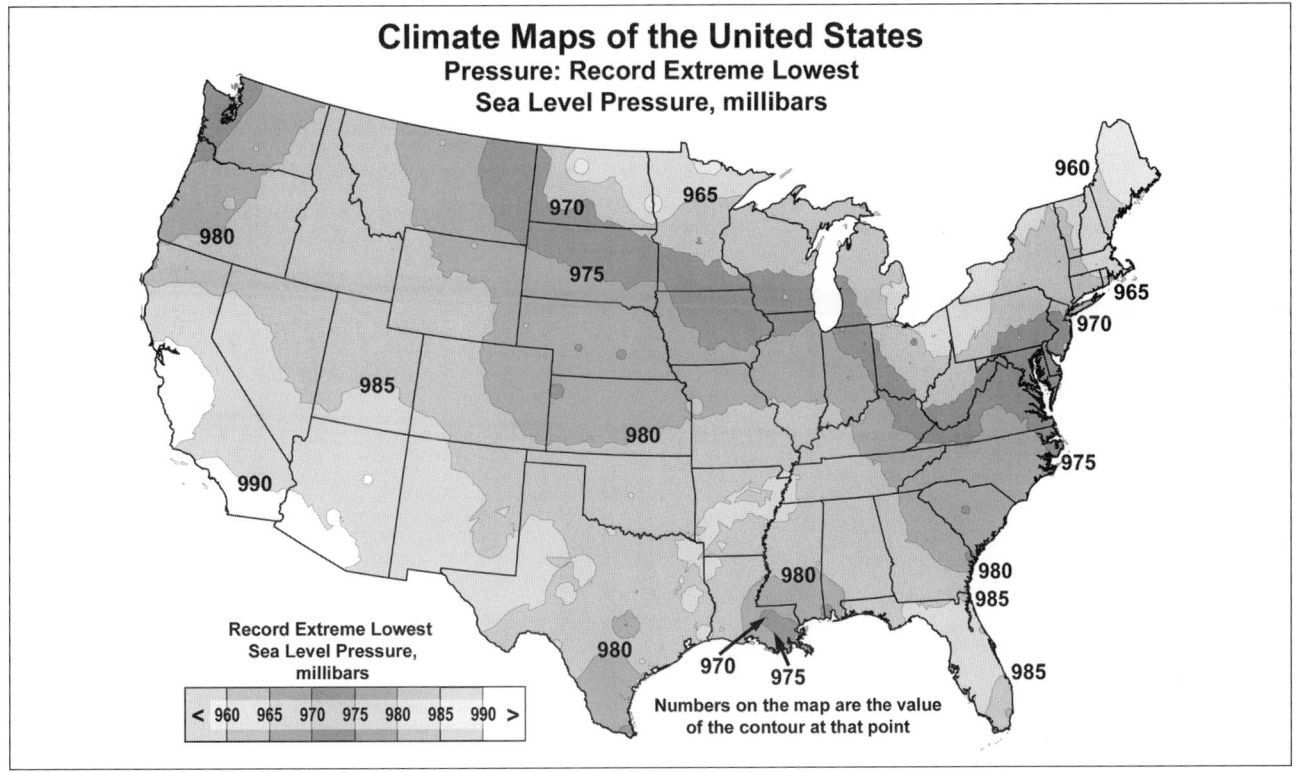

**Figure 1.19** Extreme lowest pressure.

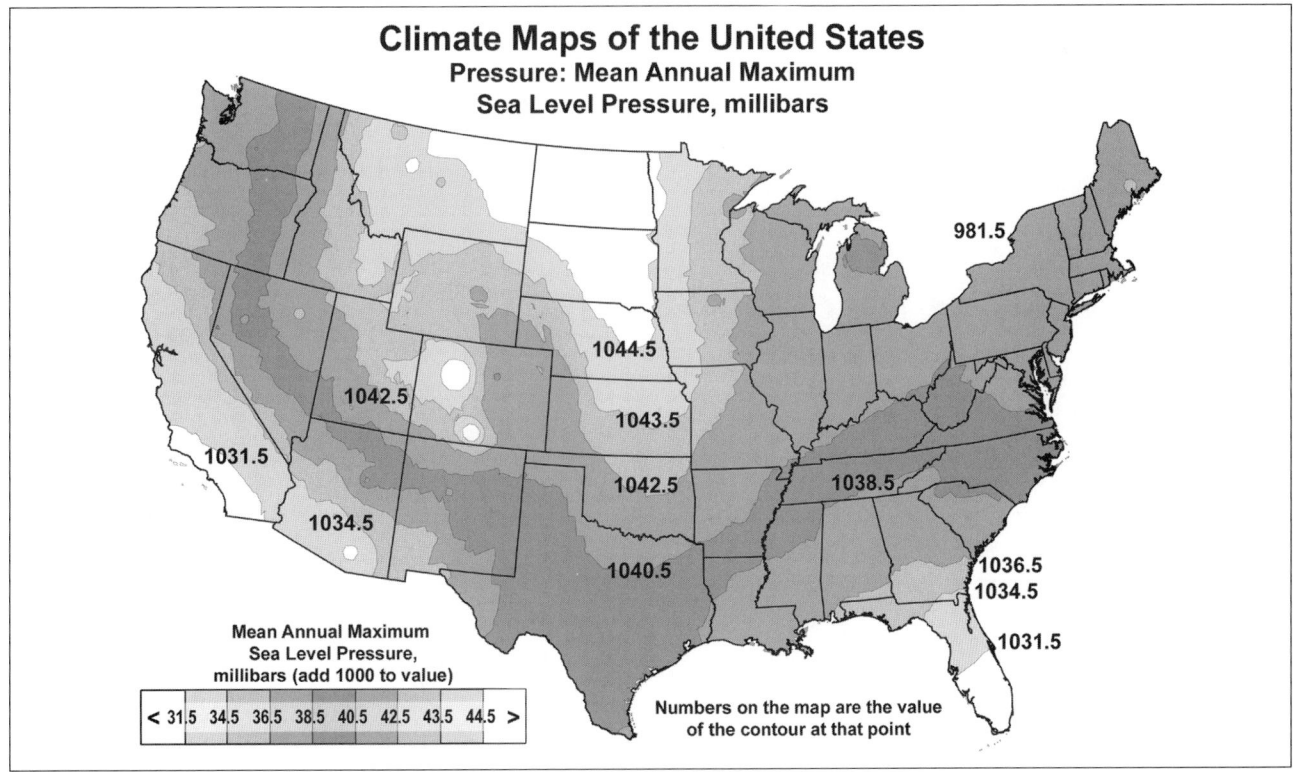

**Figure 1.20** Mean annual maximum pressure.

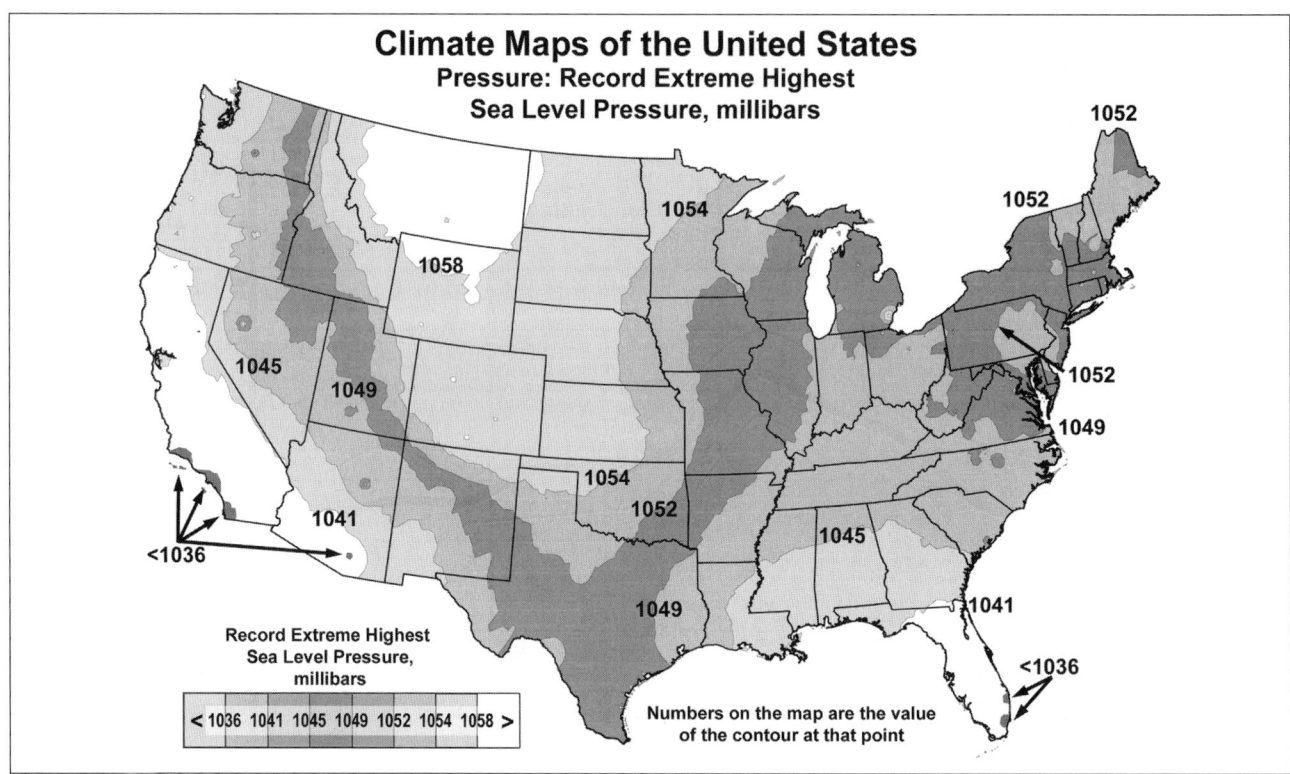

**Figure 1.21**   Extreme highest pressure.

### Snowfall Maps

*Notes:* In the United States, measurable snowfall is defined as a depth of 0.1" or more.

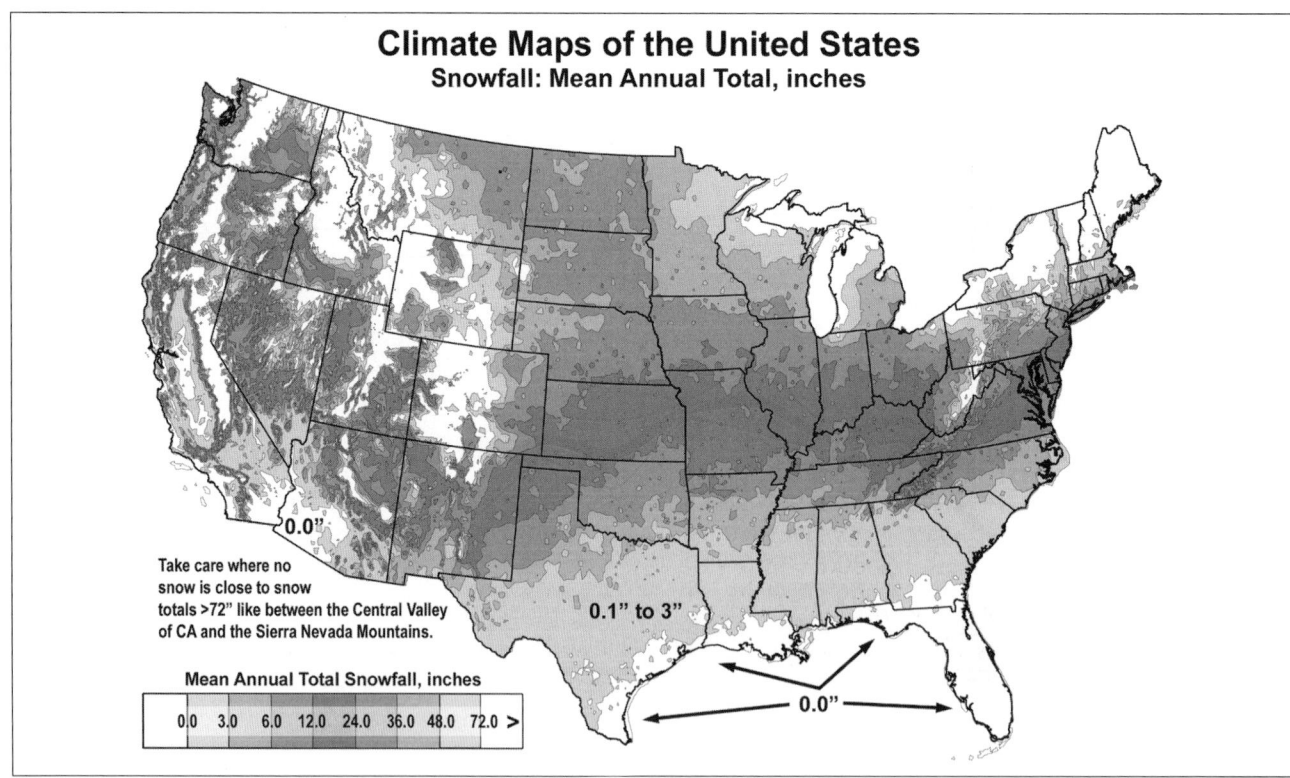

**Figure 1.22**   Mean annual total snowfall.

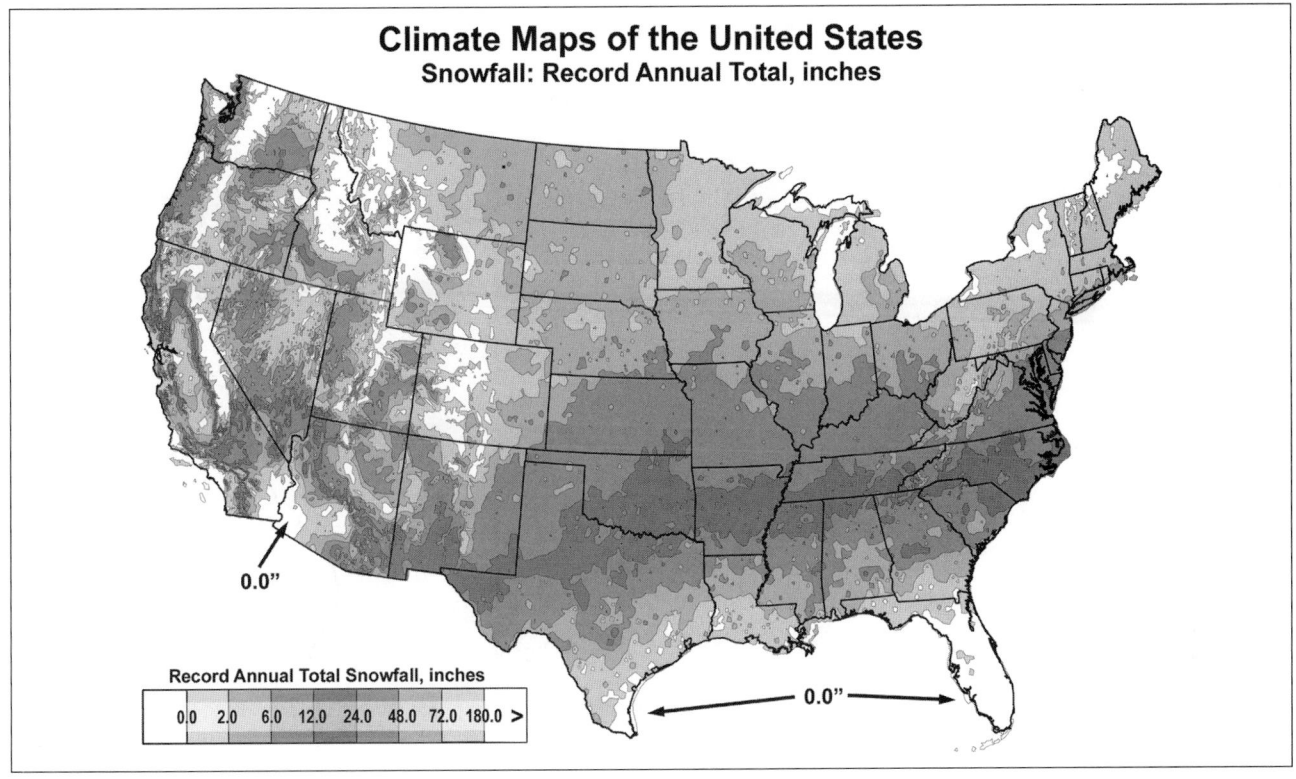

**Figure 1.23**   Record annual total snowfall.

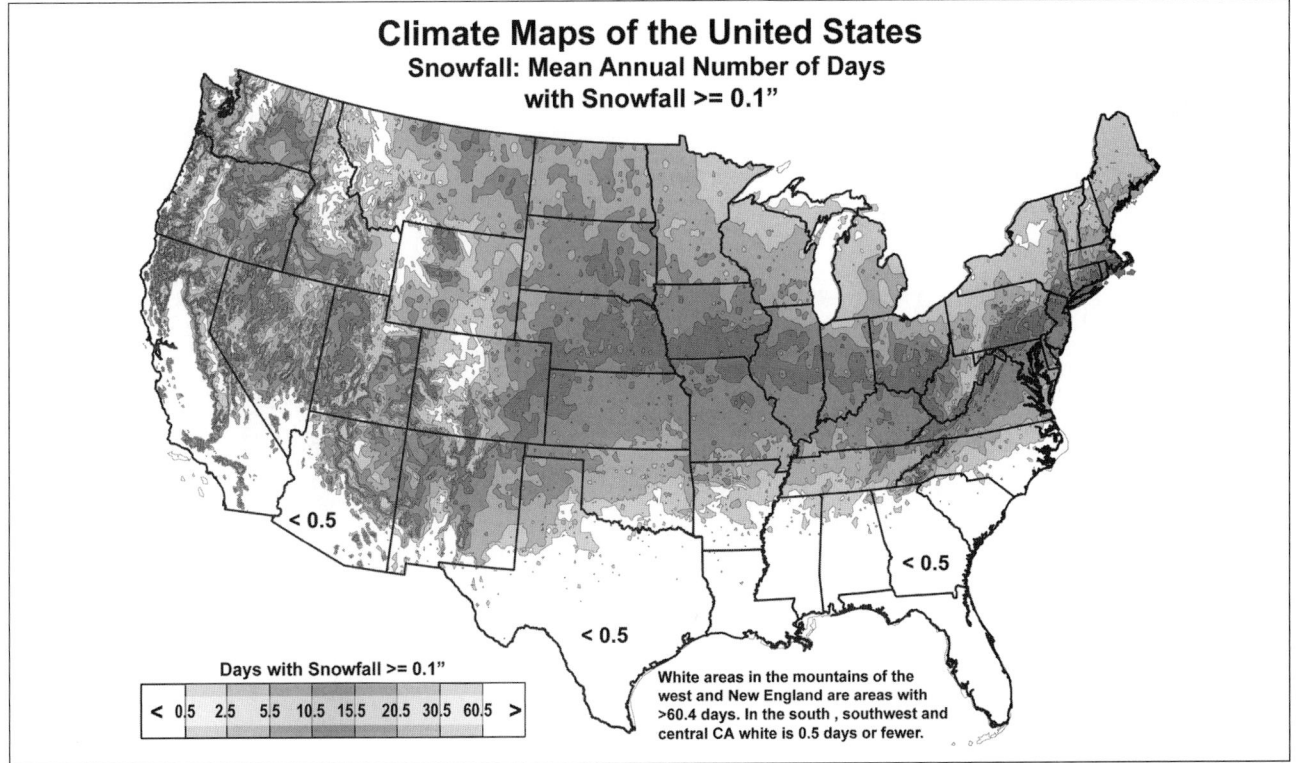

**Figure 1.24**   Mean annual number of days with snowfall of 0.1" or more.

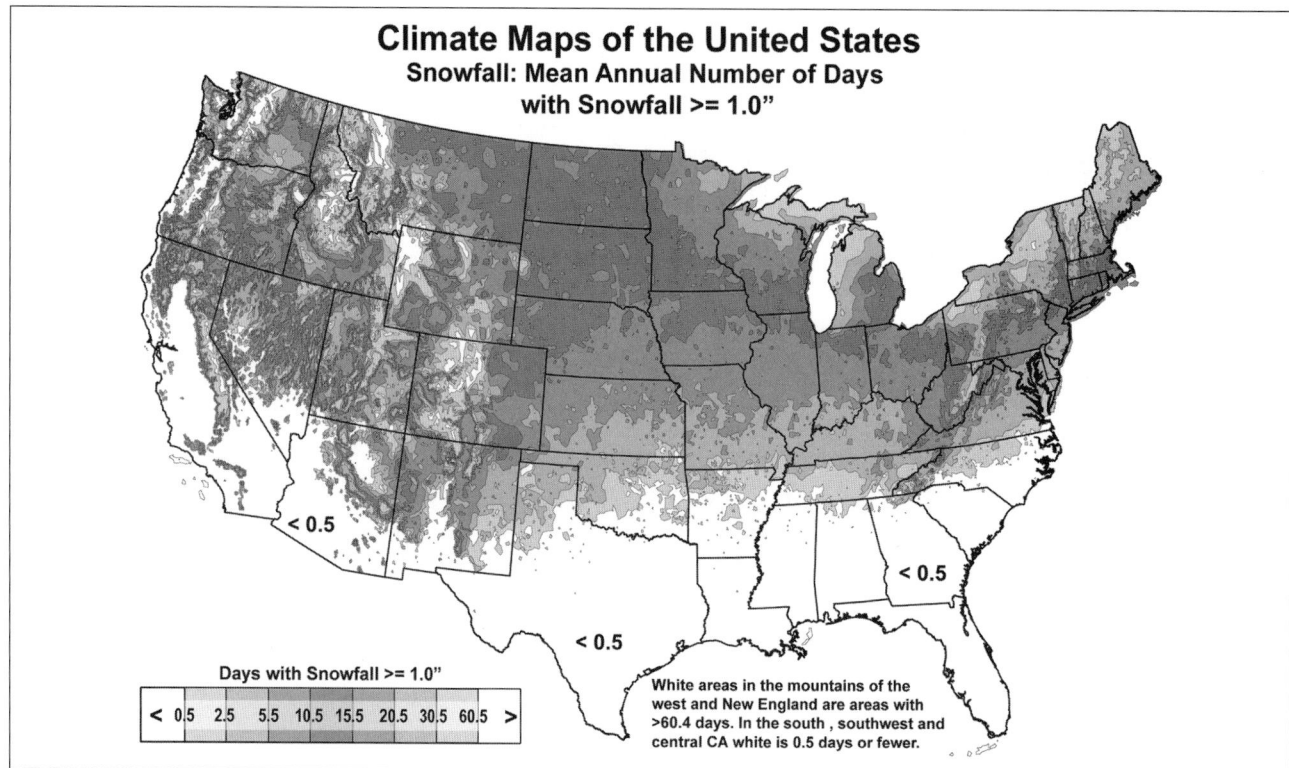

**Figure 1.25**  Mean annual number of days with snowfall of 1" or more.

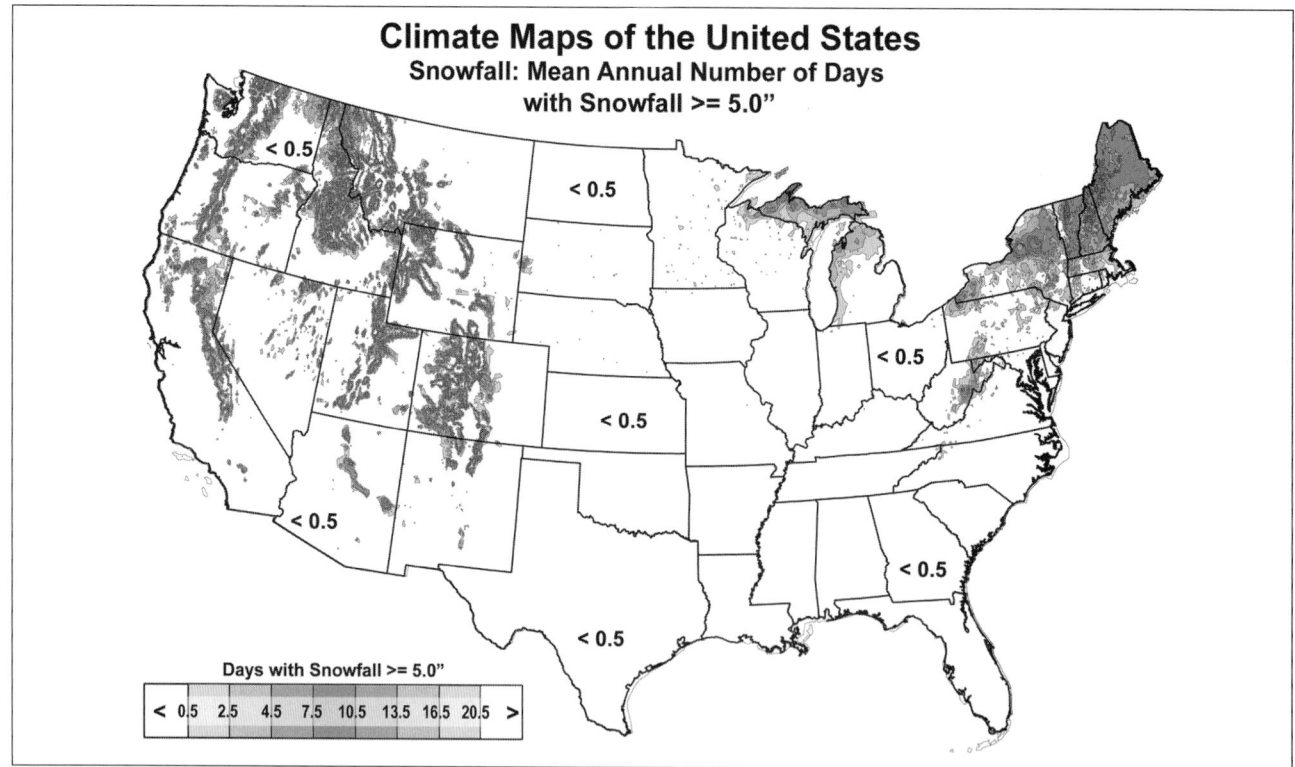

**Figure 1.26**  Mean annual number of days with snowfall of 5" or more.

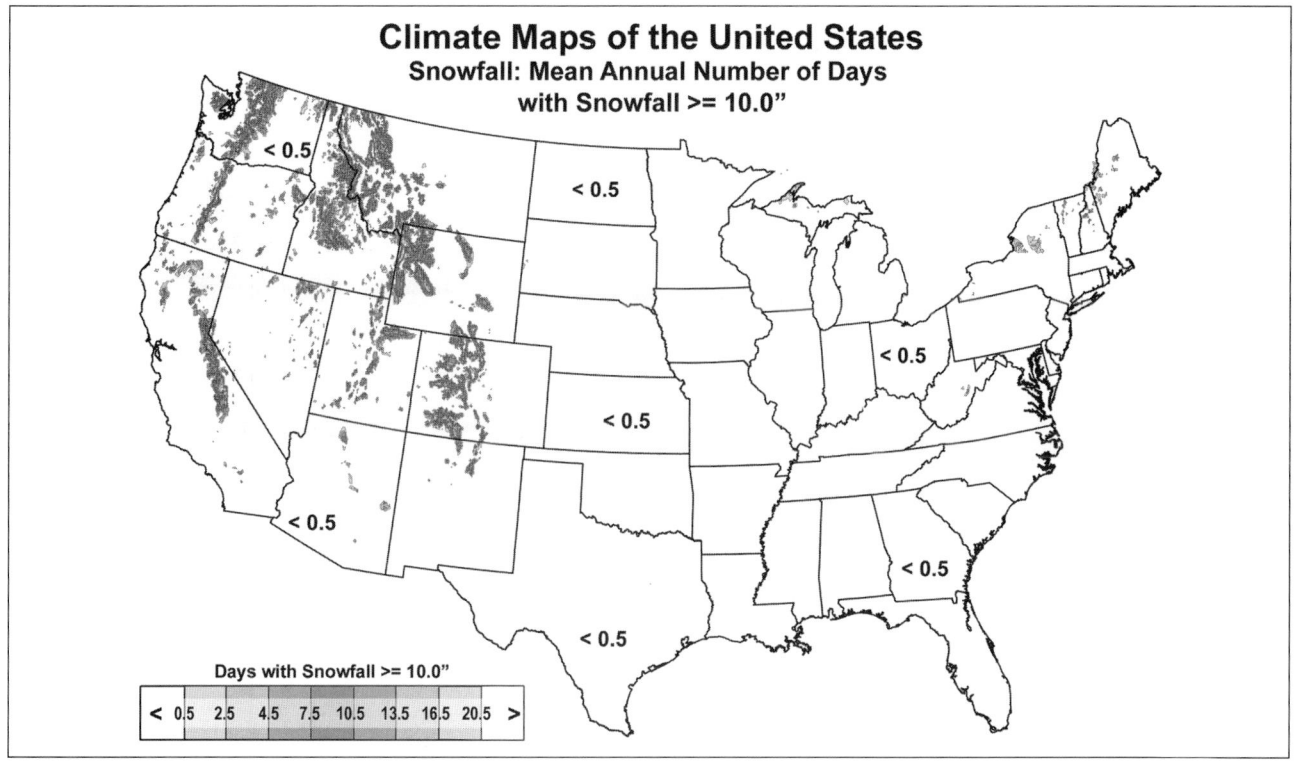

**Figure 1.27**   Mean annual number of days with snowfall of 10" or more.

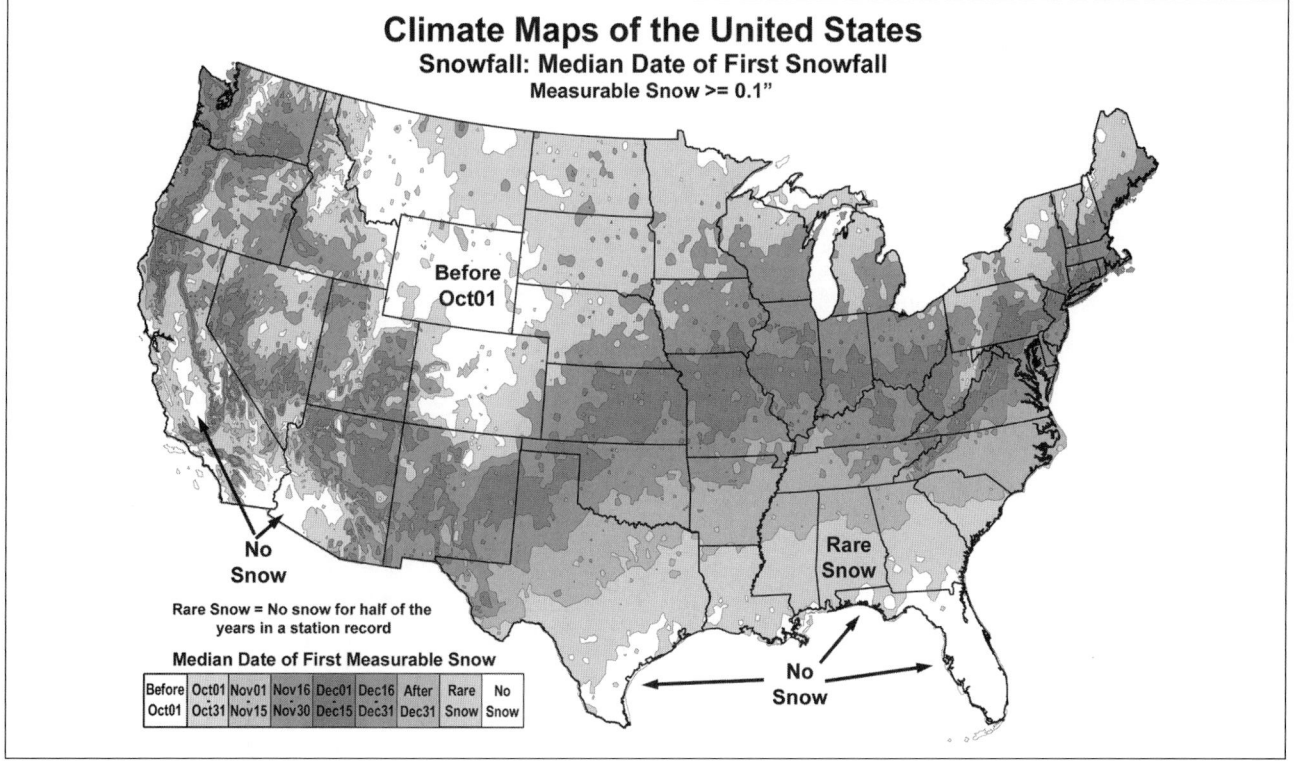

**Figure 1.28**   Median date of first measurable (≥0.1") snowfall.

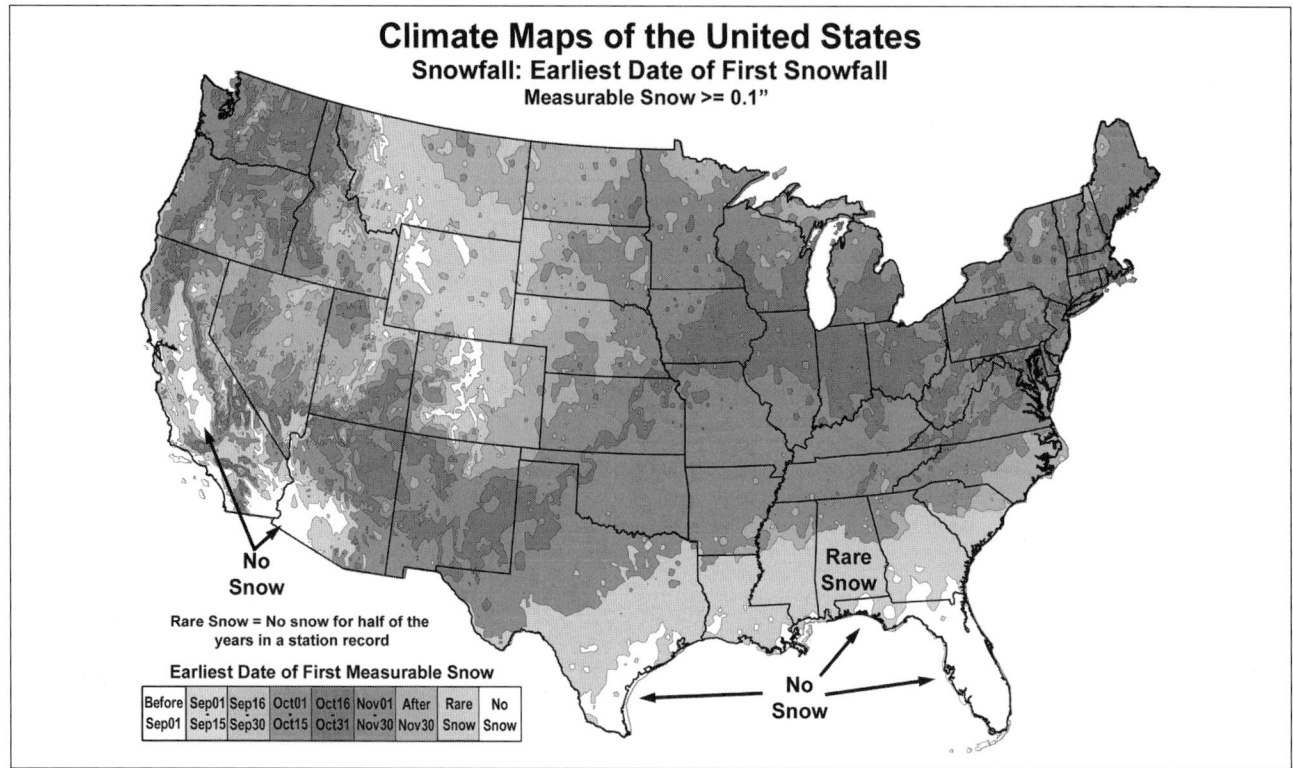

**Figure 1.29**   Extreme first date of first measurable snowfall.

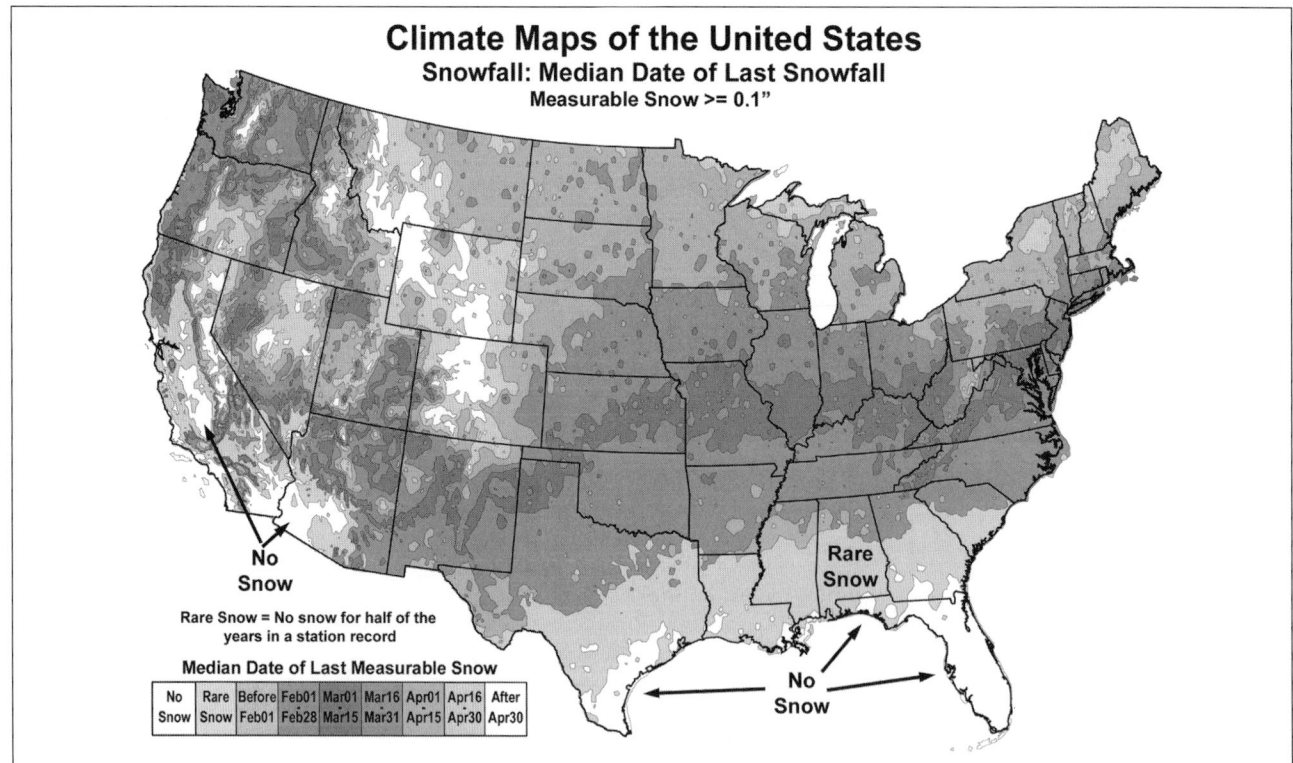

**Figure 1.30**   Median date of last measurable (0.1") snowfall.

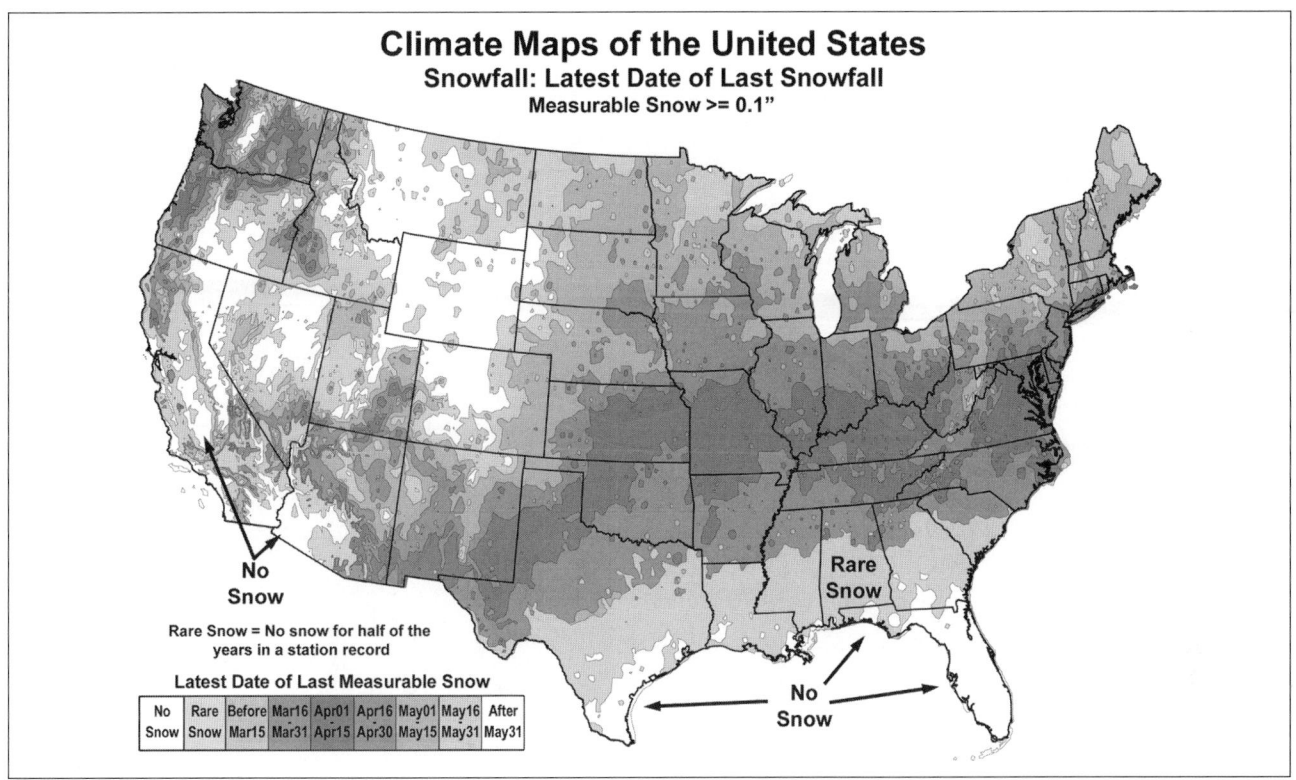

**Figure 1.31**    Extreme last date of last measurable snowfall.

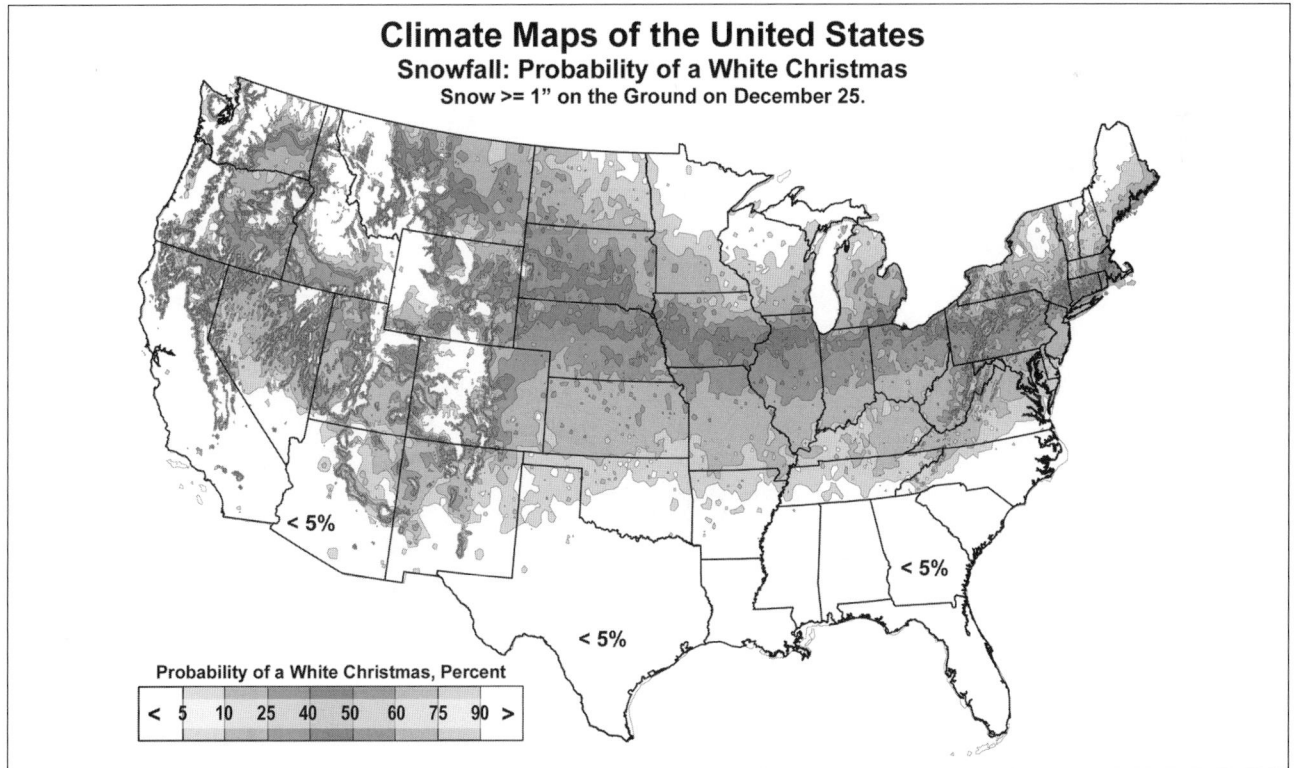

**Figure 1.32**    Probability of a white Christmas.

*Temperature*

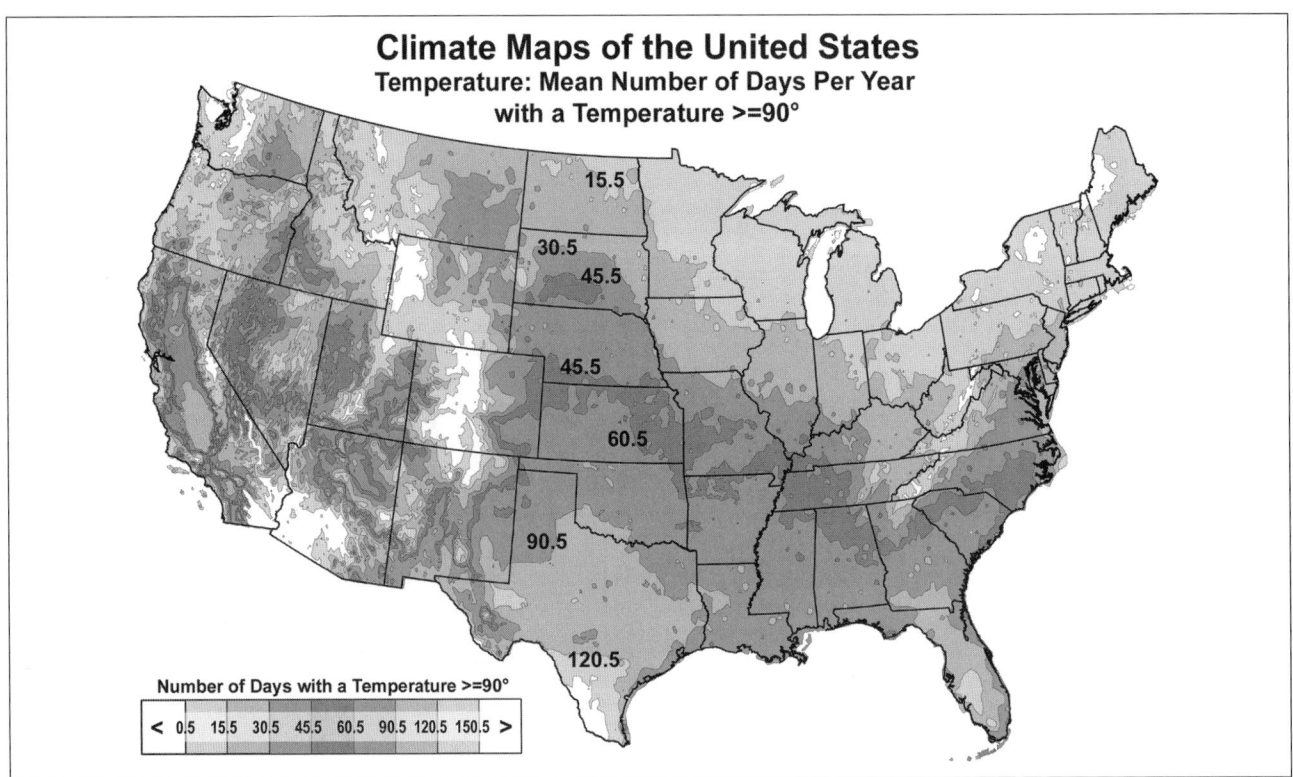

**Figure 1.33**   Mean annual number of days ≥90°.

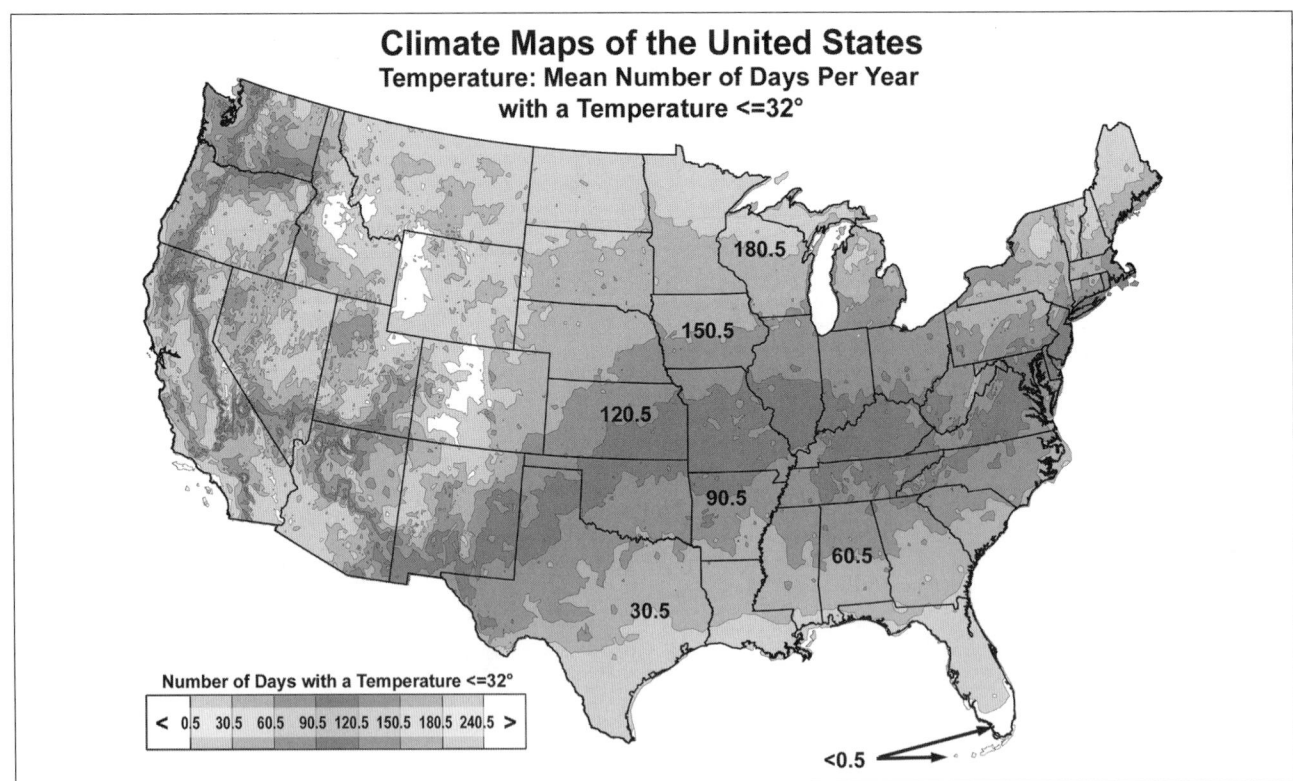

**Figure 1.34**   Mean annual number of days ≤32°.

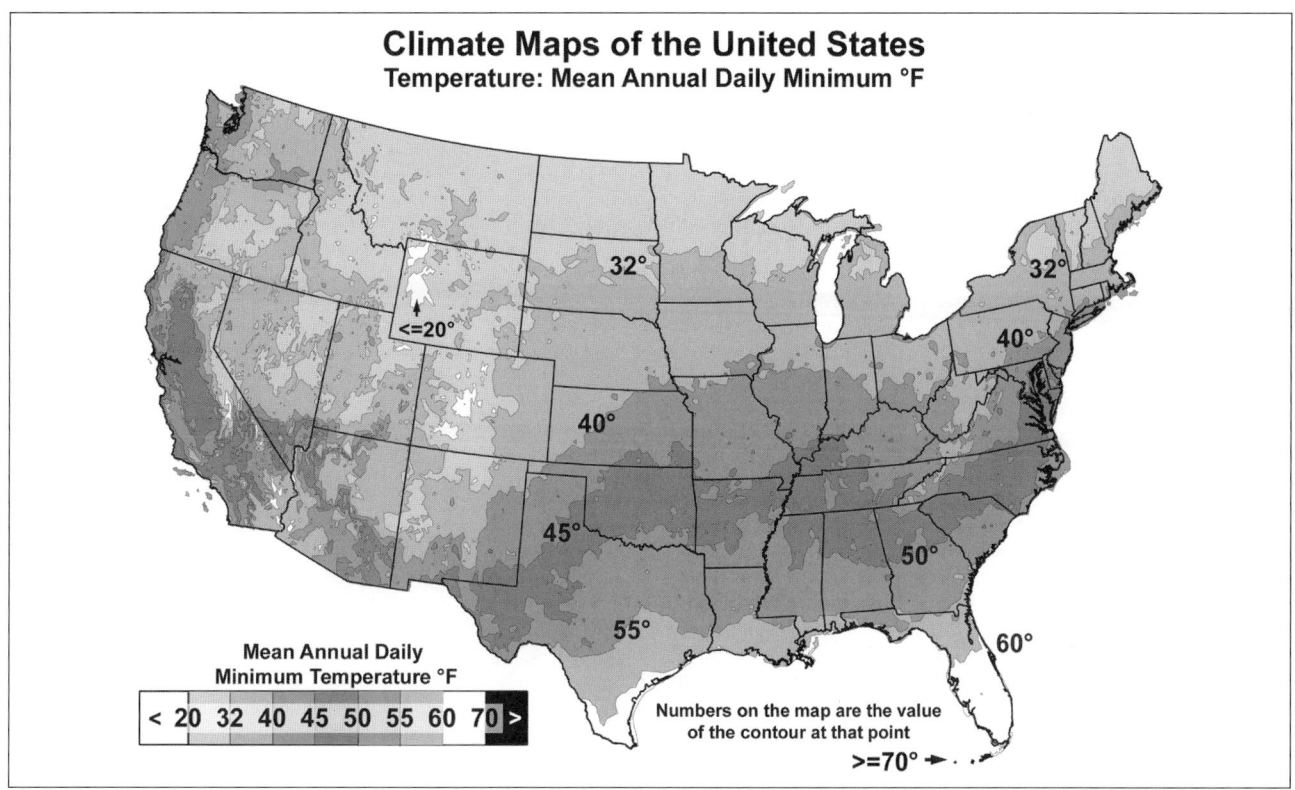

**Figure 1.35**  Mean annual minimum temperature.

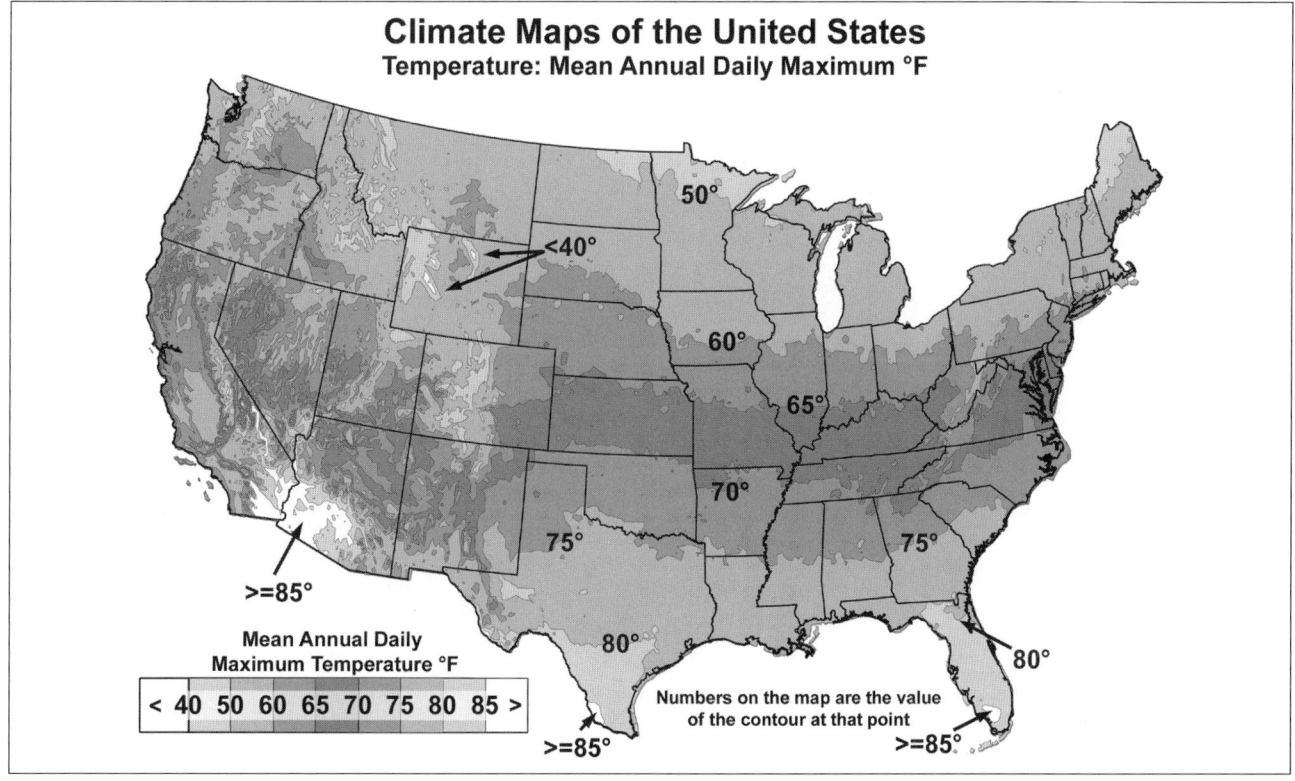

**Figure 1.36**  Mean annual maximum temperature.

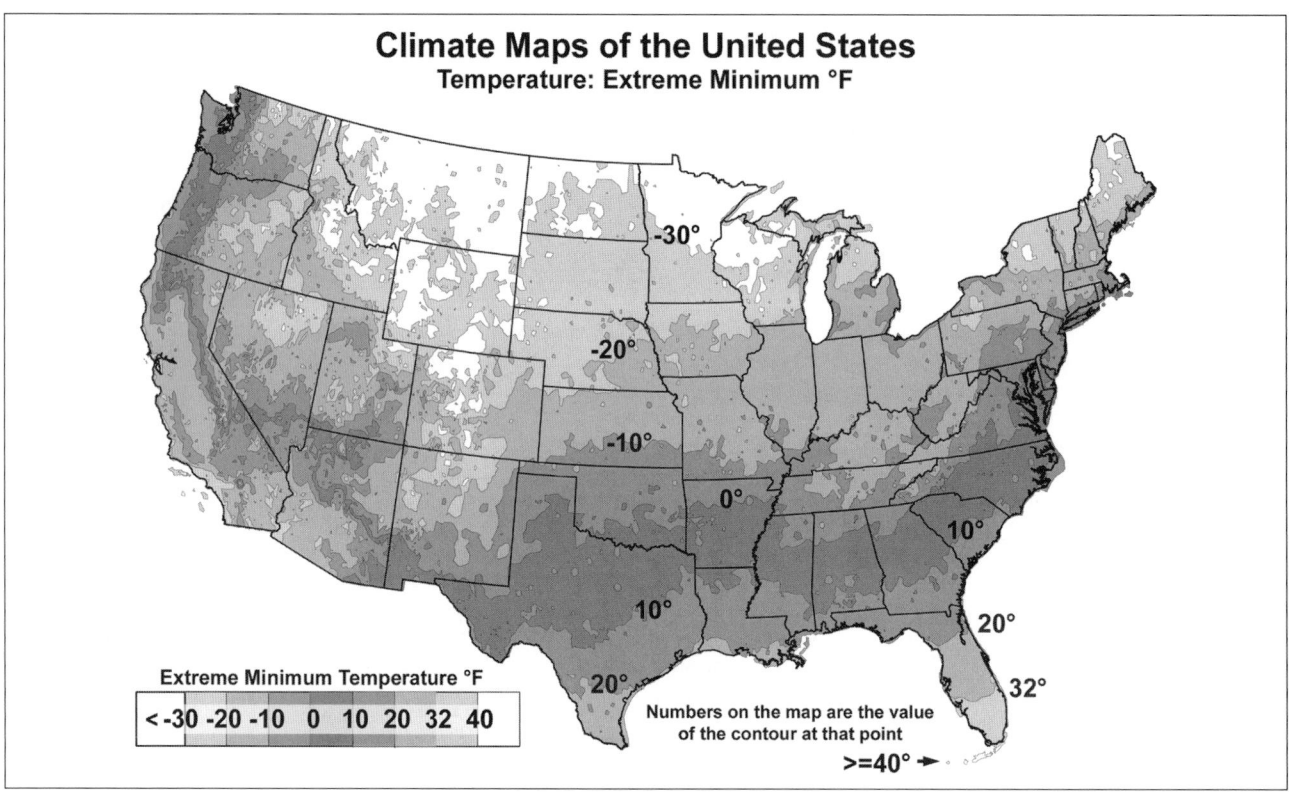

**Figure 1.37** Extreme minimum temperature.

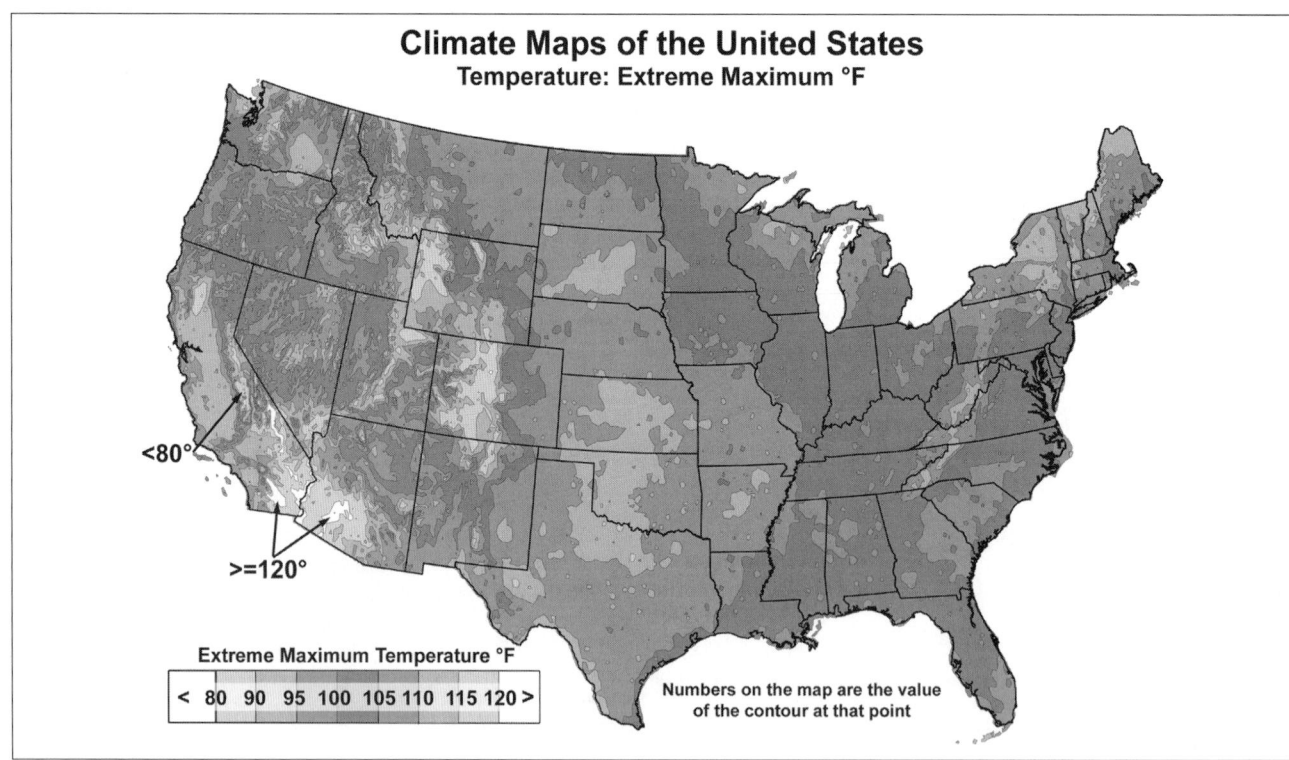

**Figure 1.38** Extreme maximum temperature.

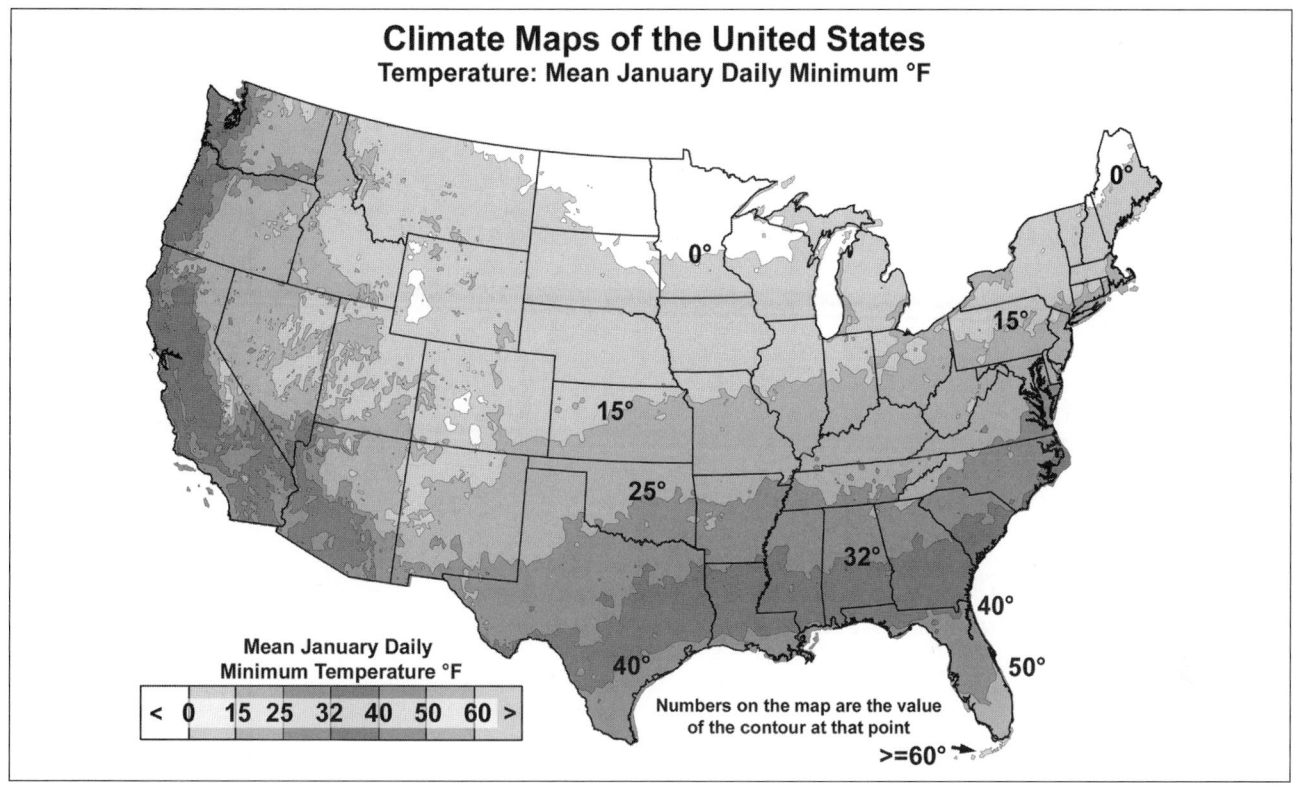

**Figure 1.39**   Mean January minimum temperature.

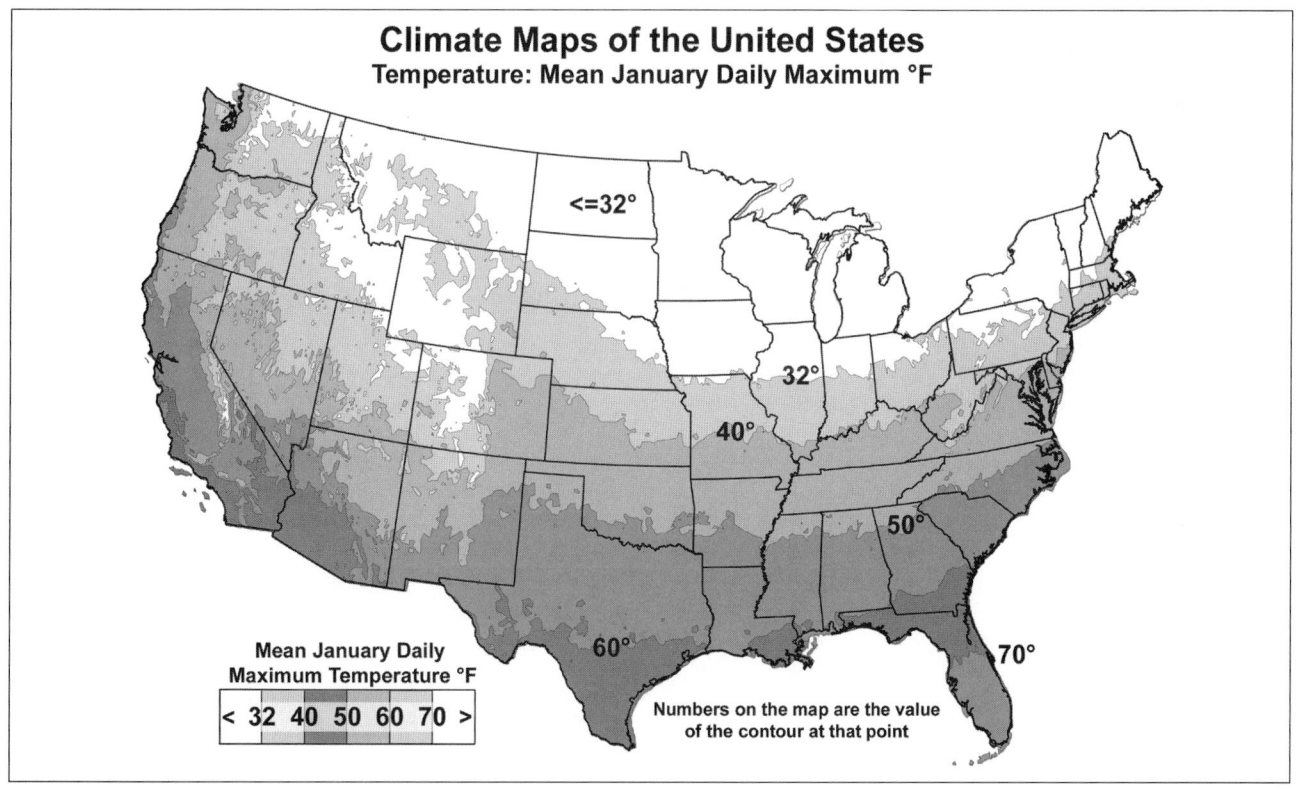

**Figure 1.40**   Mean January maximum temperature.

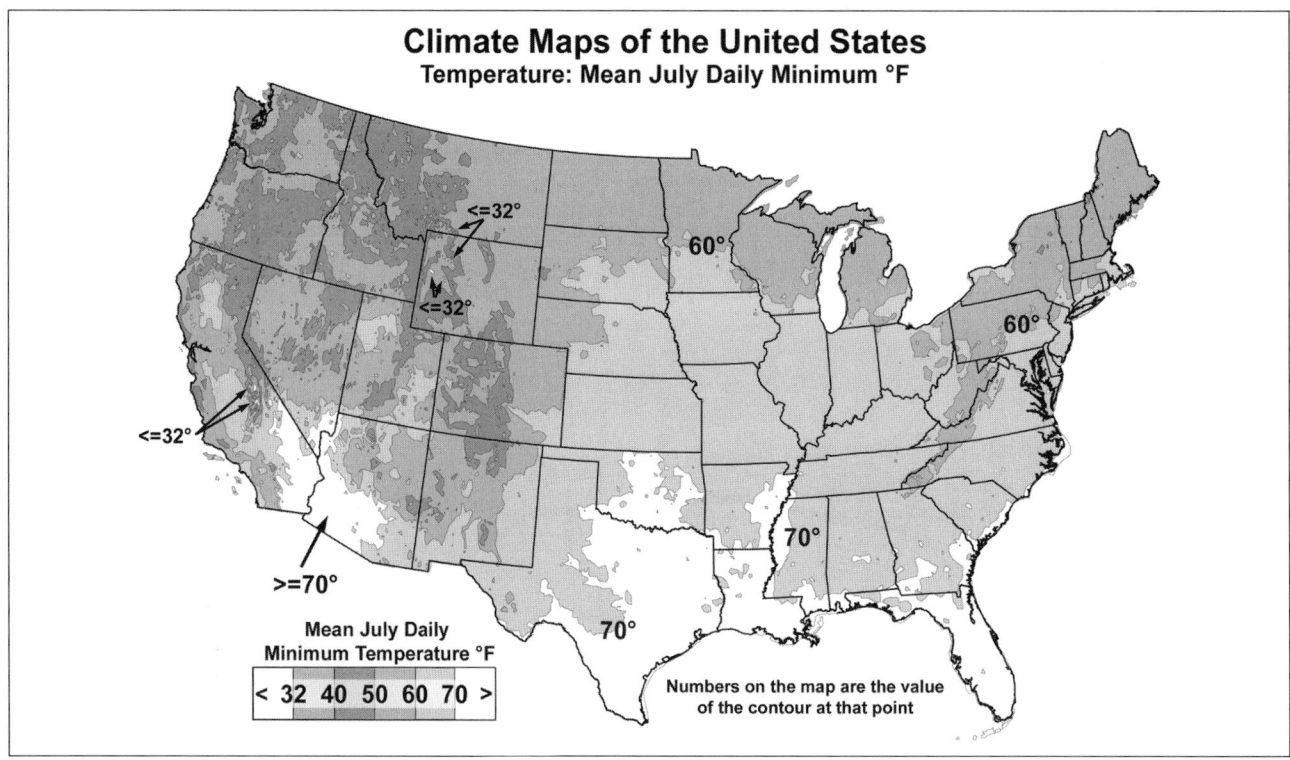

**Figure 1.41** Mean July minimum temperature.

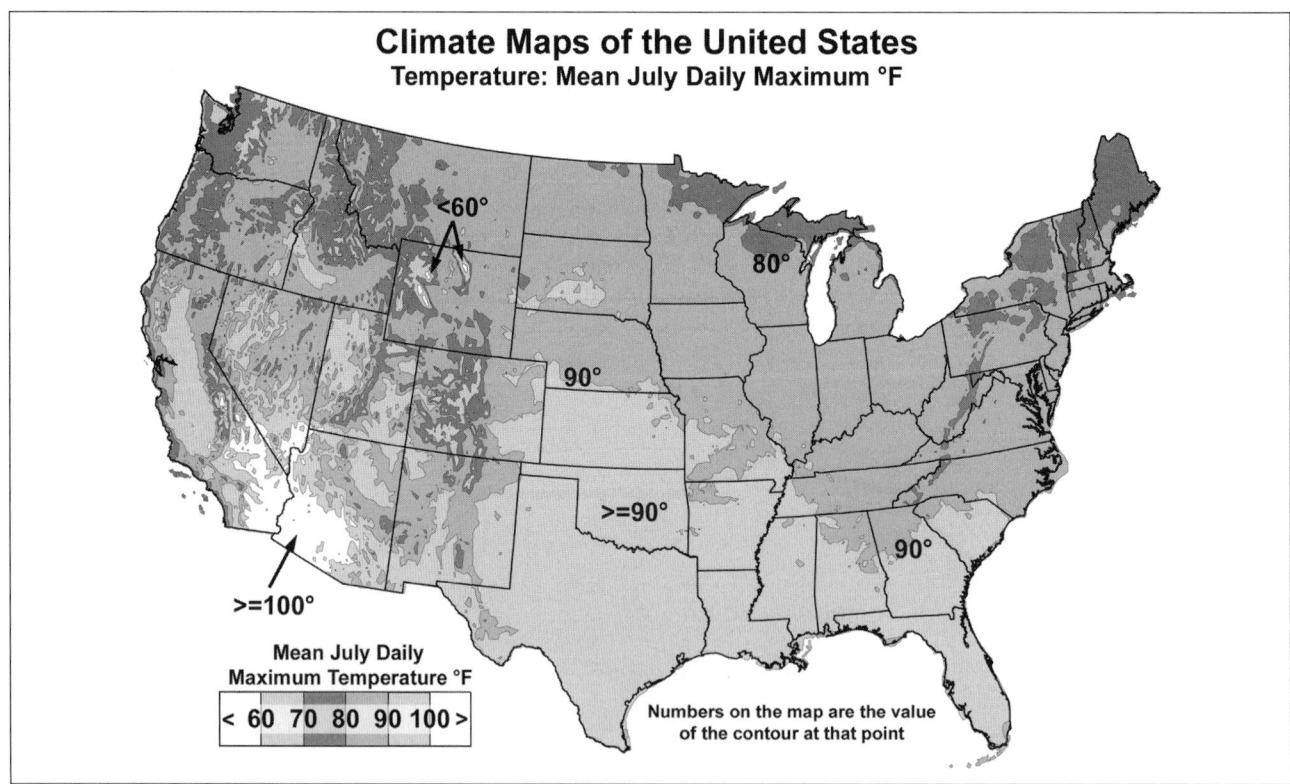

**Figure 1.42** Mean July maximum temperature.

# 2

# Renewable Energy

## • INTRODUCTION

The goal of this chapter is to summarize the "wind power" climate and the "solar energy" climate of the 48 contiguous states and provide the user with data regarding the feasibility of both solar power energy generation and wind power energy generation.

The material in this chapter could easily have been included in Chapter 1, but because of the specific nature of the information I have treated the wind and solar energy climates of the United States separately.

There are 97 figures, a combination of maps, and 132 graphs, in this chapter covering 44 cities. You will find a combination of annual maps covering the lower 48 states, and detailed wind roses, wind speed histograms, and solar energy graphs for specific locations.

## • RENEWABLE ENERGY

Of the renewable energy sources available the most well known are solar energy and wind energy. In addition, the US Department of Energy's National Renewable Energy Laboratory (NREL) lists geothermal, hydrogen (including fuel cells), and biomass as sources of renewable energy. Not listed by the NREL is tidal energy as a potential source in the future.

Weather and climate play a part in the utilization of any energy source for heating and cooling because of varying demand. Only wind and solar renewable energies are directly dependent on weather and climate for supply.

### Definitions

Renewable energy is produced from natural processes that are replenished constantly. In most cases energy is derived directly from the sun, such as wind, solar, and biomass energy sources. Heat generated deep within the earth is also considered renewable, but it is not a product of solar radiation. Geothermal energy is largely the result of radioactive decay deep within the earth. Hydrogen as a fuel source is earth-based, also because water is its source.

Modern biomass energy production, also called bioenergy, involves processing plant materials to a more readily usable form such as a gas or liquid like ethanol. Directly burning wood continues to be the most common form of bioenergy worldwide.

Geothermal energy has been in use for thousands of years. Rome was famous for hot baths and heating small structures using geothermal heat. The modern approach to geothermal energy production involves generating electricity with turbines turned by steam. Geothermal energy production is growing slowly at 3% per year primarily because under present technologies, generation must take place near the sources that are tectonically active regions of earth. The largest geothermal power plant in the world, The Geysers, is located north of San Francisco in Sonoma and Lake Counties and has a generating capacity of 750 MW.

Tidal energy uses the power of incoming and outgoing tidal water to power generators. This source has also been used since Roman times to power mills. Tidal energy falls into two broad categories: the capture of the kinetic energy of water moving horizontally under the influence of tides and the capture of potential energy or water lifted in the tidal bulge. Tidal energy is the only energy source relying on the relative motions of the Earth and moon. Tidal energy production is rare because of infrastructure costs, potential environmental damage, and the lack of suitable sites.

Wind energy is the fastest growing of the renewable sources, with capacity in the United States growing at 30% per year. Worldwide generating capacity in 2008 totaled 121,000 MW. The modern approach to wind energy uses air flow to turn a wind turbine, also called an "aerogenerator." If the mechanical energy is used directly by machinery for grinding grain or pumping water, for example, the term "windmill" is used.

Solar energy is the use of sunlight for power. The NREL lists five categories of solar energy production:

1. **Concentrating solar power** (CSP) technologies use parabolic reflectors to focus sunlight and concentrate the heat to boil water and drive a turbine. CSP technologies

*The Weather Almanac: A Reference Guide to Weather, Climate, and Related Issues in the United States and Its Key Cities*, Twelfth Edition. Steve Horstmeyer.
© 2011 John Wiley & Sons, Inc. Published 2011 by John Wiley & Sons, Inc.

are realized as expensive projects that require large investment and therefore designed as large-scale electrical distribution systems in much the same way as traditional coal- or oil-based electrical power generation. Because of the expense CSP facilities are limited to the sunniest parts of the country.

2. **Photovoltaic** (PV) technology is the most well-known solar energy technology. Sunlight generates electricity using the PV effect, which was discovered in 1954. First-generation solar cells used silicon flat plates and are what most people think of when referring to solar energy production. Second-generation cells involve thin-film technology that can be used to make roof shingles as collectors and thus overcome the esthetic objections to solar panels. Third-generation solar technology uses materials other than silicon alone to more efficiently capture solar energy. Because PV technology is modular and individual modules are relatively inexpensive when compared to CSP technologies, PV technology is well suited for individual residences.

3. **Passive solar technology** involves building designs that allow or promote daytime heating of walls and spaces through exposure to sunlight.

4. **Solar water heating** is simple, water exposed directly or within a collector to sunshine is heated and stored for future use in an insulated tank.

5. **Solar process heating** refers to the use of a variety of technologies in industrial buildings. Some of the technologies are too expensive for residential uses, but solar process heating encompasses the use of PV technology, the primary residential solar energy technology.

## • SOME HISTORY OF WIND MEASUREMENT

In 1805, before anemometers were invented to measure the wind speed, British Admiral Sir Francis Beaufort created a scale named after him. The first version gave a qualitative description of the effect of the wind on the sails on a man of war. There were 13 classes ranging from "just sufficient to give steerage" to "that which no canvas sails could withstand."

In the 1830s, reporting the Beaufort Scale became standard practice in the British Navy. In the 1850s, the scale was adapted for land use, with scale numbers corresponding to anemometer cup rotations. At this time anemometers did not display wind speed.

As steam power supplanted sails the Beaufort Scale was changed in 1906 to reflect the character of the sea, not the state of the sails. In 1923, anemometer cup rotations were standardized and land-based descriptions were added. In 1946, forces 13–17 were added for special cases such as tropical cyclones. Today the Beaufort Scale has been nearly abandoned for measured units of meters per second, kilometers per hour, nautical miles per hour (knots), and miles per hour. The modern Beaufort Scale is found in Table 2.1.

## • WIND ENERGY POTENTIAL IN THE UNITED STATES

Wind energy potential is highly variable from place to place and also highly variable in any one place from time to time. The latter is termed the "intermittency problem" because wind-generated power, even in the most reliable locations, will at times be unavailable. In general, areas with the greatest wind potential are where some factor, often terrain, influences the speed of the wind.

But wind speed is not the only variable that makes a place good or bad for producing wind energy. A place with a reliable, more constant wind may produce more energy than one with higher speeds and less consistency. The suitability of any location for generating wind energy is the product of the complex interaction of many factors that vary at a very small scale. A good location for a wind turbine may not be obvious from casual inspection.

Because wind speed generally increases with height above the surface, wind energy potential is calculated above the surface. Common values used to estimate wind power density are 10 m (33 feet), 30 m (98 feet), and 50 m (164 feet) above ground level.

Because most wind measurements are made near the surface, wind power density estimates apply a rule of thumb called the "wind profile power law" or the "one-seventh power law." This relationship works well over unobstructed ground in stable atmospheric conditions, but over open water/very rough ground or over areas with numerous obstructions to low-level air flow, the error can be substantial.

Better estimates can be made using the log wind profile equation that includes input for surface roughness and stability. When these are missing, which is quite often, the wind profile power law is used.

Table 2.2 shows the wind power density in watts per square meter at 10, 30, and 50 m above ground level. The associated wind speeds in both meters per second and miles per hour calculated (estimated) from the wind profile power law are listed. They are classified using the US Department of Energy's wind potential classification.

Superb potential for wind power generation (800–1600 $W/m^{-2}$) at 50 m above ground level is hard to find. Superb potential is most often found in isolated areas of the Great Plains, coastal areas, and high mountain passes. Electricity from wind can be generated at other areas, but it may not be economically feasible on a large scale. Most of the Great Plains and foothills of the Rocky Mountains are rated as having good or better potential.

The maps and tables in this chapter present a general wind climatology of the contiguous 48 states. Wind roses and wind speed frequency histograms are included for 33 cities along with annual national maps of average wind speed, the occurrence of peak gusts at three limits, and the fastest mile of wind.

The wind speed climatology of a place is also important as input in building and structure design, including towers for capturing wind energy. Modern standards are moving

## Table 2.1   The Modern Beaufort Wind Scale

| Beaufort Force | Description | Wind speed | | | | Wave height | | Sea conditions | Land conditions |
|---|---|---|---|---|---|---|---|---|---|
| | | km/h | mph | kts | m/s | m | ft | | |
| 0 | Calm | < 1 | < 1 | < 1 | < 0.3 | 0 | 0 | Sea is flat. | Calm. Smoke rises vertically. |
| 1 | Light air | 1 – 5 | 1 – 3 | 1 – 2 | 0.3 – 1.5 | 0 – 0.2 | 0 – 1 | Ripples, no crests. | Wind motion just visible in smoke. |
| 2 | Light breeze | 6 – 11 | 4 – 7 | 3 – 6 | 1.5 – 3.3 | 0.2 – 0.5 | 1 – 2 | Small wavelets. Crests glassy look, not breaking. | Wind felt on exposed skin. Leaves rustle. |
| 3 | Gentle breeze | 12 – 19 | 8 – 12 | 7 – 10 | 3.3 – 5.5 | 0.5 – 1 | 2 – 3.5 | Large wavelets. Some crests break; few whitecaps. | Leaves and smaller twigs in constant motion. |
| 4 | Moderate breeze | 20 – 28 | 13 – 17 | 11 – 15 | 5.5 – 8.0 | 1 – 2 | 3.5 – 6 | Small waves with breaking crests. Fairly frequent white horses (breaking waves). | Dust and loose paper raised. Small branches begin to move. |
| 5 | Fresh breeze | 29 – 38 | 18 – 24 | 16 – 20 | 8.0 – 11 | 2 – 3 | 6 – 9 | Moderate waves of some length. Many white horses. Small amounts of spray. | Branches of a moderate size move. Small trees begin to sway. |
| 6 | Strong breeze | 39 – 49 | 25 – 30 | 21 – 26 | 11 – 14 | 3 – 4 | 9 – 13 | Long waves form, white foam crests are frequent, some airborne spray is present. | Large branches in motion. Whistling heard in overhead wires. Umbrella use becomes difficult. Empty plastic garbage cans tip over. |
| 7 | High wind, Moderate gale, Near gale | 50 – 61 | 31 – 38 | 27 – 33 | 14 – 17 | 4 – 5.5 | 13 – 19 | Sea heaps up. Some foam from breaking waves is blown into streaks along wind direction. Moderate amounts of airborne spray. | Whole trees in motion. Effort needed to walk against the wind. Swaying of skyscrapers may be felt, especially by people on upper floors. |
| 8 | Gale Fresh gale | 62 – 74 | 39 – 46 | 34 – 40 | 17 – 20 | 5.5 – 7.5 | 18 – 25 | Moderately high waves with breaking crests forming spindrift. Well-marked streaks of foam are blown along wind direction. Considerable airborne spray. | Some twigs broken from trees. Cars veer on road. Progress on foot is seriously impeded. |
| 9 | Strong gale | 75 – 88 | 47 – 54 | 41 – 47 | 21 – 24 | 7 – 10 | 23 – 32 | High waves whose crests sometimes roll over. Dense foam is blown along wind direction. Large amounts of airborne spray may begin to reduce visibility. | Some branches break off trees, some small trees blow over. Construction signs and barricades blow over. Damage to circus tents and canopies. |
| 10 | Storm Whole gale | 89 – 102 | 55 – 63 | 48 – 55 | 25 – 28 | 9 – 12.5 | 29 – 41 | Very high waves with overhanging crests. Large patches of foam from wave crests give the sea a white appearance. Considerable tumbling of waves with heavy impact. Large amounts of airborne spray reduce visibility. | Trees are broken off or uprooted, saplings bent and deformed. Poorly attached asphalt shingles and shingles in poor condition peel off roofs. |
| 11 | Violent storm | 103 – 117 | 64 – 72 | 56 – 63 | 29 – 32 | 11.5 – 16 | 37 – 52 | Exceptionally high waves. Very large patches of foam, driven before the wind, cover much of the sea surface. Very large amounts of airborne spray severely reduce visibility. | Widespread damage to vegetation. Many roofing surfaces are damaged; asphalt tiles that have curled up and/or fractured due to age may break away completely. |
| 12 | Hurricane | ≥ 118 | ≥ 73 | ≥ 64 | ≥ 33 | ≥ 14 | ≥ 46 | Huge waves. Sea is completely white with foam and spray. Air is filled with driving spray, greatly reducing visibility. | Very widespread damage to vegetation. Some windows may break; mobile homes and poorly constructed sheds and barns are damaged. Debris may be hurled about. |

**Table 2.2  Estimates of Power Density and wind Speed at Three Altitudes Above Ground Level Using the Wind Profile Power Law.**

| U.S. Dept. of Energy Wind Power Class | 10 m (33 ft) Above Ground Level | | 30 m (98 ft) Above Ground Level | | 50 m (164 ft) Above Ground Level | |
|---|---|---|---|---|---|---|
| | Wind power density (W/m²) | Speed m/s (mph) | Wind power density (W/m²) | Speed m/s (mph) | Wind power density (W/m²) | Speed m/s (mph) |
| 1 Poor | 0 - 100 | 0 - 4.4 (0 - 9.8) | 0 - 160 | 0 - 5.1 (0 - 11.4) | 0 - 200 | 0 - 5.6 (0 - 12.5) |
| 2 Marginal | 100 - 150 | 4.4 - 5.1 (9.8 - 11.5) | 160 - 240 | 5.1 - 5.9 (11.4 - 13.2) | 200 - 300 | 5.6 - 6.4 (12.5 - 14.3) |
| 3 Fair | 150 - 200 | 5.1 - 5.6 (11.5 - 12.5) | 240 - 320 | 5.9 - 6.5 (13.2 - 14.6) | 300 - 400 | 6.4 - 7.0 (14.3 - 15.7) |
| 4 Good | 200 - 250 | 5.6 - 6.0 (12.5 - 13.4) | 320 - 400 | 6.5 - 7.0 (14.6 - 15.7) | 400 - 500 | 7.0 - 7.5 (15.7 - 16.8) |
| 5 Excellent | 250 - 300 | 6.0 - 6.4 (13.4 - 14.3) | 400 - 480 | 7.0 - 7.4 (15.7 - 16.6) | 500 - 600 | 7.5 - 8.0 (16.8 - 17.9) |
| 6 Outstanding | 300 - 400 | 6.4 - 7.0 (14.3 - 15.7) | 480 - 640 | 7.4 - 8.2 (16.6 - 18.3) | 600 - 800 | 8.0 - 8.8 (17.9 - 19.7) |
| 7 Superb | 400 - 1000 | 7.0 - 9.4 (15.7 - 21.1) | 640 - 1600 | 8.2 - 11.0 (18.3 - 24.7) | 800 - 2000 | 8.8 - 11.9 (19.7 - 26.6) |

*Source:* The US Department of Energy.

away from the "fastest mile" as a building standard and toward the "3-s wind gust" with a 2% annual occurrence, the same as a 50-year recurrence interval.

The fastest mile of wind is measured in miles per hour and is the fastest average wind speed during the period of 1 min that is observed during the time period required for the air to travel 1 mile past the anemometer. The fastest mile is always slower than the 3-s wind gust.

Table 2.3 provides the approximate conversion from fastest mile to 3-s wind gust according to the International Building Code.

Figure 2.1 shows the wind energy potential at 50 m above ground level of the United States as estimated by the US Department of Energy. Notice how large parts of the country, especially the east and south, have little potential as sites for generation of wind energy.

Figures 2.2–2.6 are maps that show the wind climate of the United States using traditional measures. Figure 2.2 shows the average wind speed in miles per hour; Figures 2.3–2.5 show the occurrences of wind gusts exceeding 30, 40, and 50 mph, respectively; and Figure 2.6 shows the fastest mile of wind for the 48 lower states. Notice the similarities between Figure 2.1 and Figures 2.2–2.6.

Included in this chapter are wind roses and wind speed frequency bar graphs (histograms) for 44 locations in the lower 48 states (Figures 2.7–2.50). It is important to realize

**Table 2.3  Conversion from Fastest-Mile Wind Data to 3-Second Gust Wind Data**

| Fastest - Mile | 3-Second Gust |
|---|---|
| 70 mph 113 kph | 85 mph 137 kph |
| 75 mph 121 kph | 90 mph 45 kph |
| 80 mph 129 kph | 100 mph 161 kph |
| 85 mph 137 kph | 105 mph 169 kph |
| 90 mph 145 kph | 110 mph 177 kph |
| 95 mph 153 kph | 115 mph 185 kph |
| 100 mph 161 kph | 120 mph 193 kph |
| 105 mph 169 kph | 125 mph 201 kph |
| 110 mph 177 kph | 130 mph 209 kph |

*Source:* International Building Code.

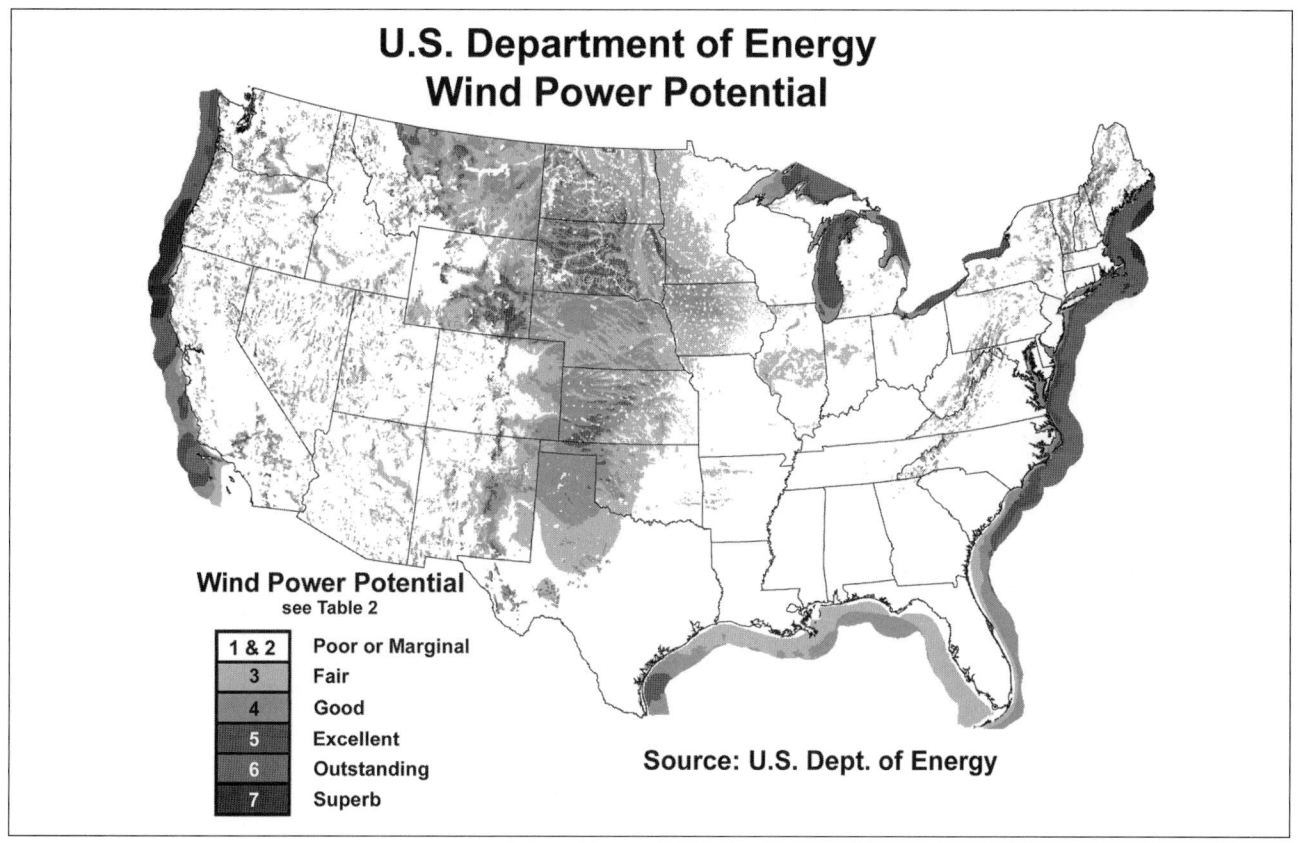

**Figure 2.1**　National wind parameter map.

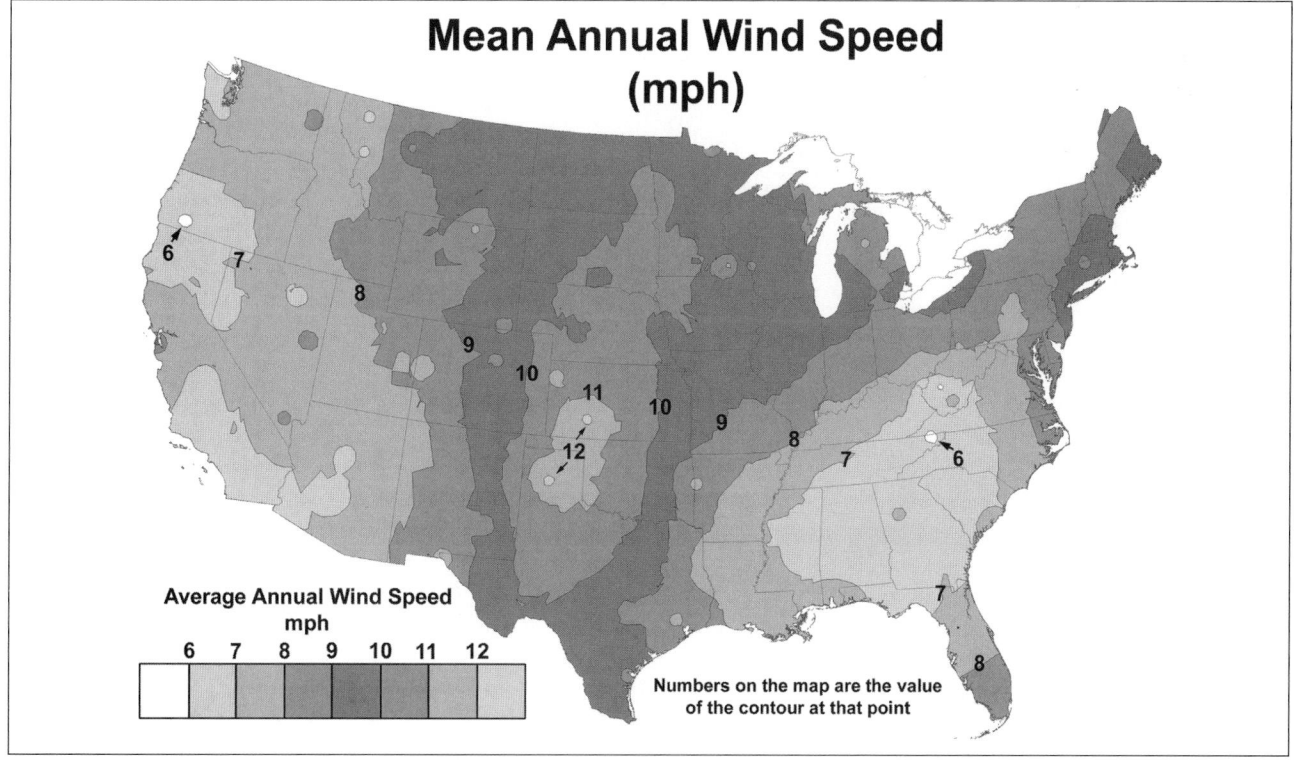

**Figure 2.2**　National wind parameter map.

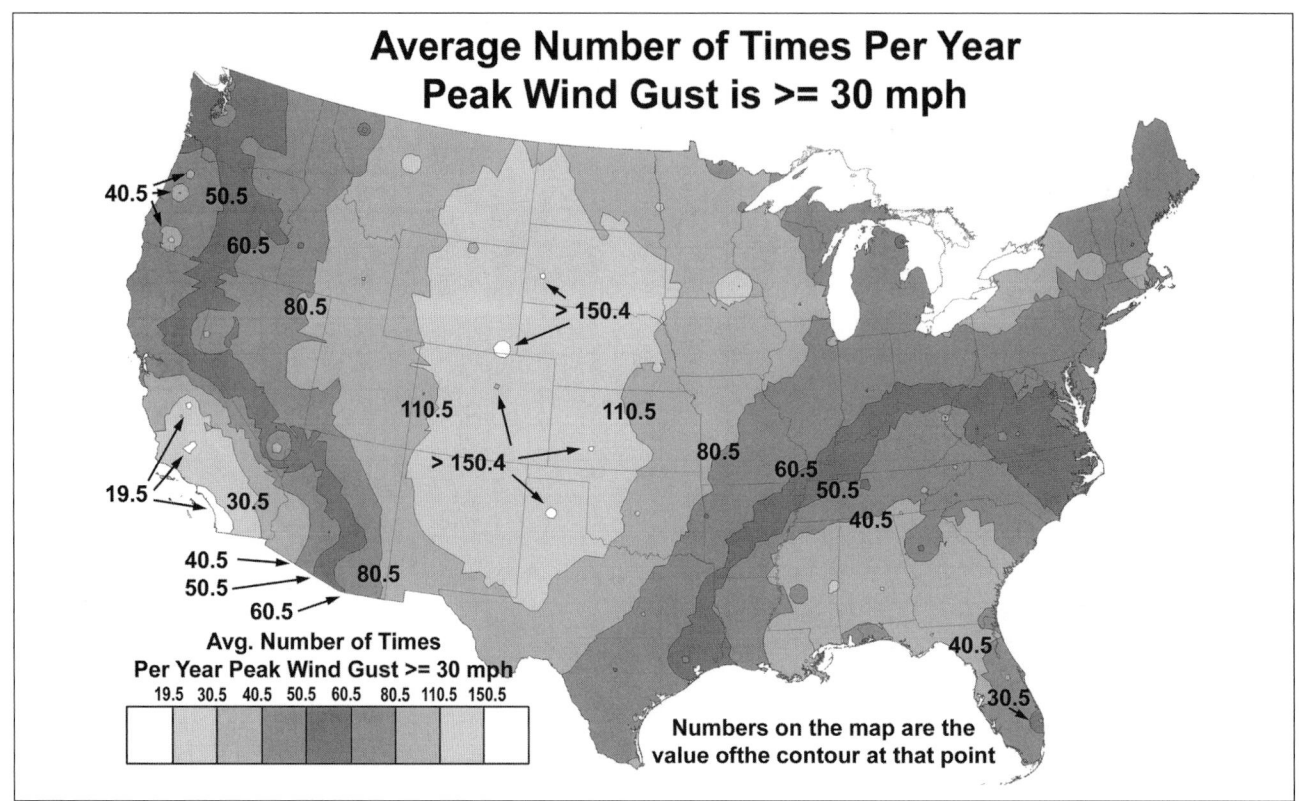

**Figure 2.3** National wind parameter map.

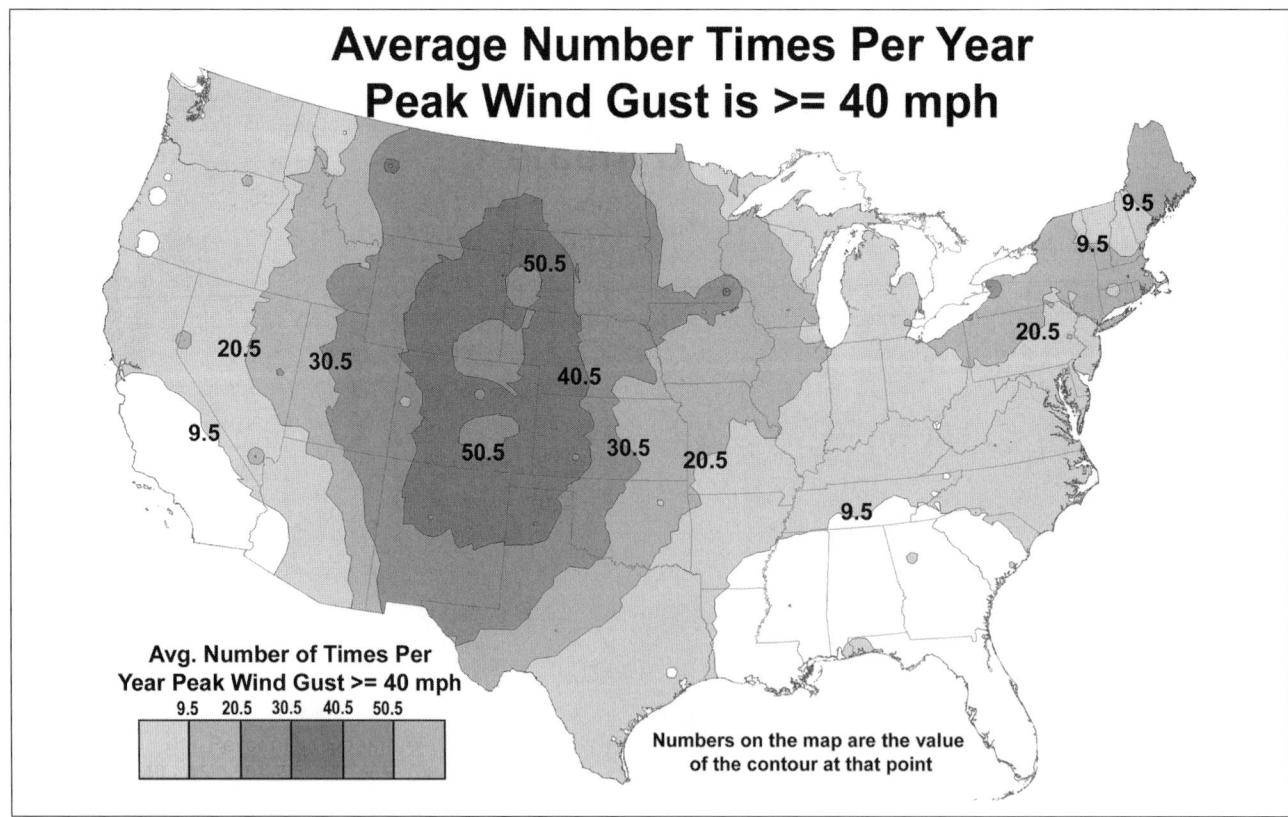

**Figure 2.4** National wind parameter map.

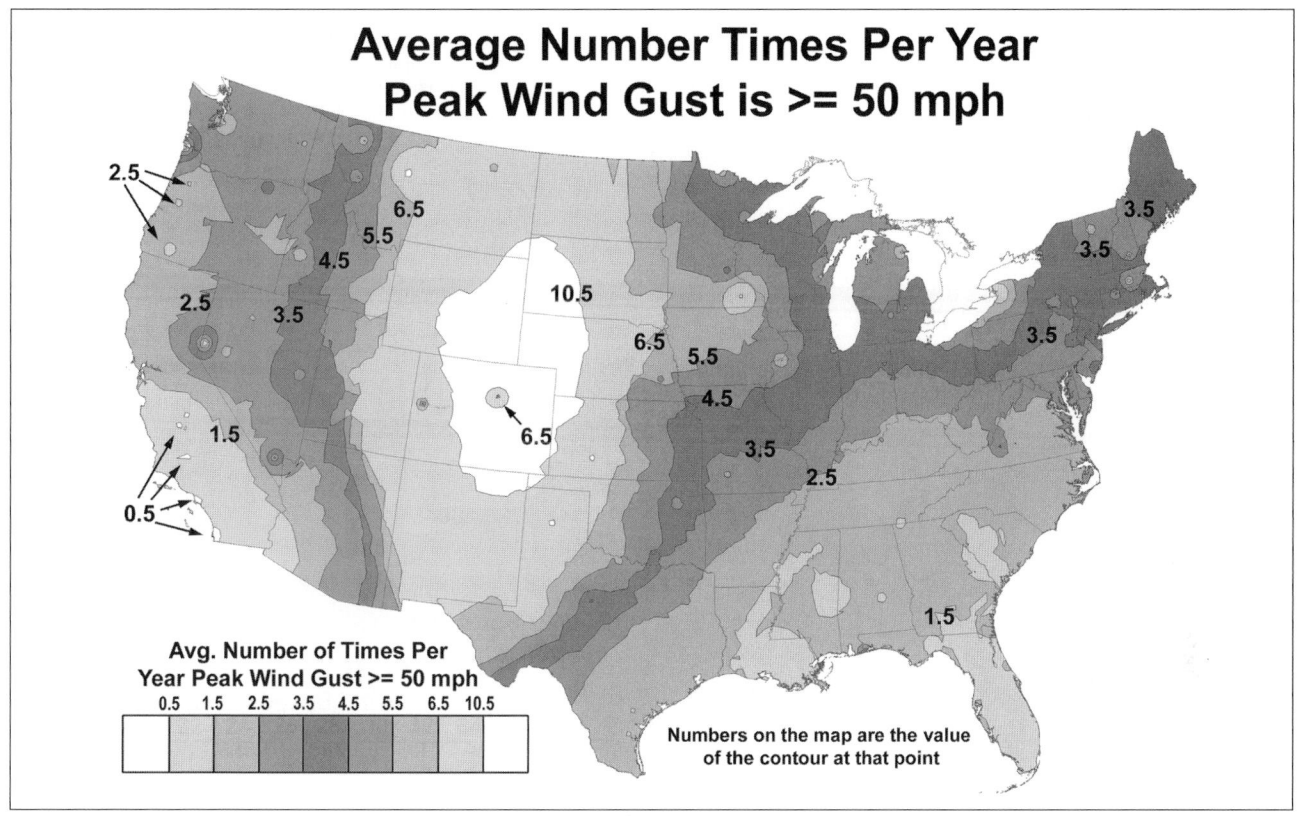

**Figure 2.5**   National wind parameter map.

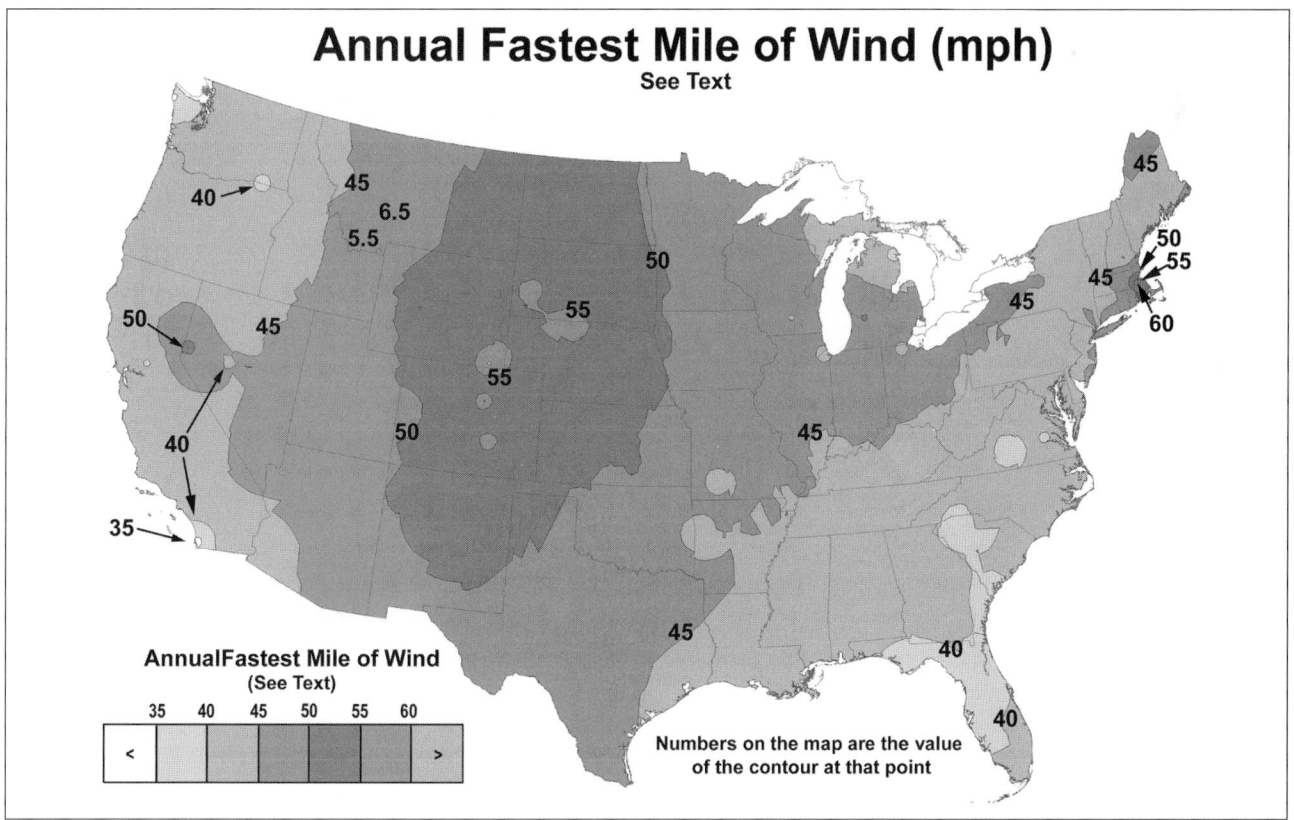

**Figure 2.6**   National wind parameter map.

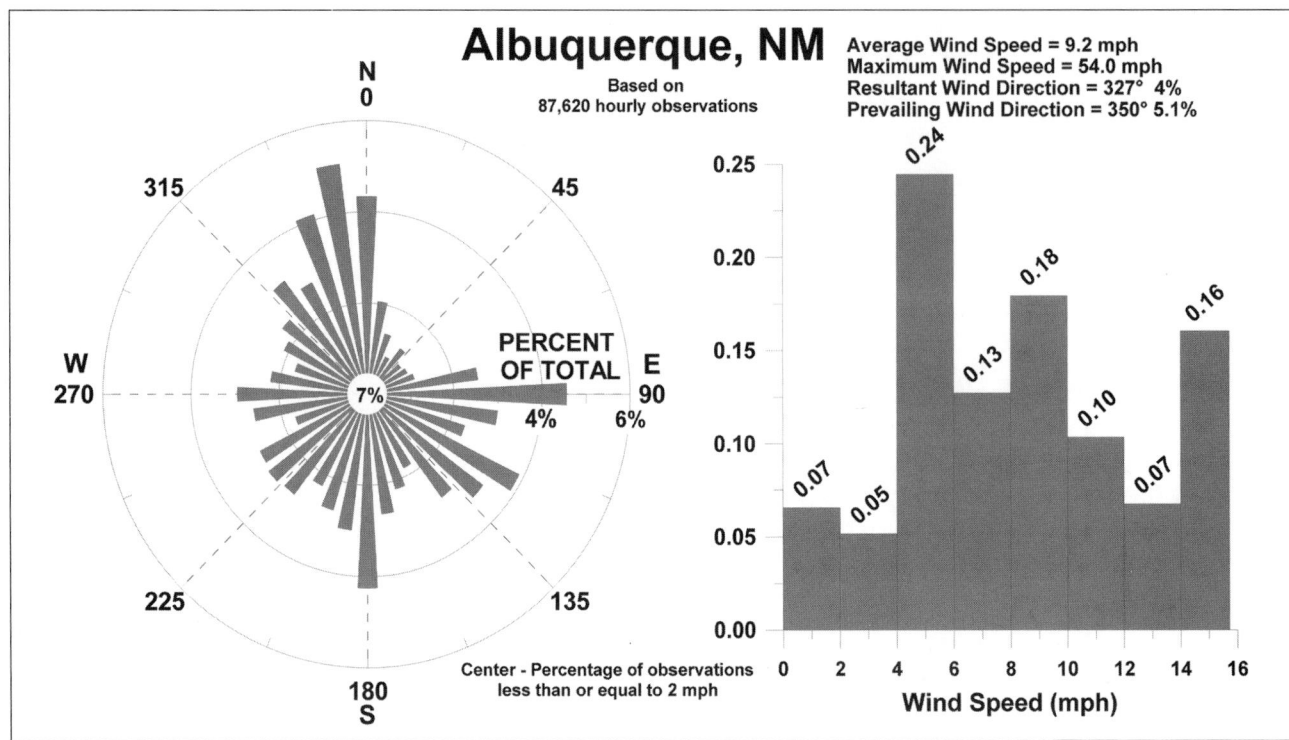

**Figure 2.7** City wind roses and wind speed frequency histogram.

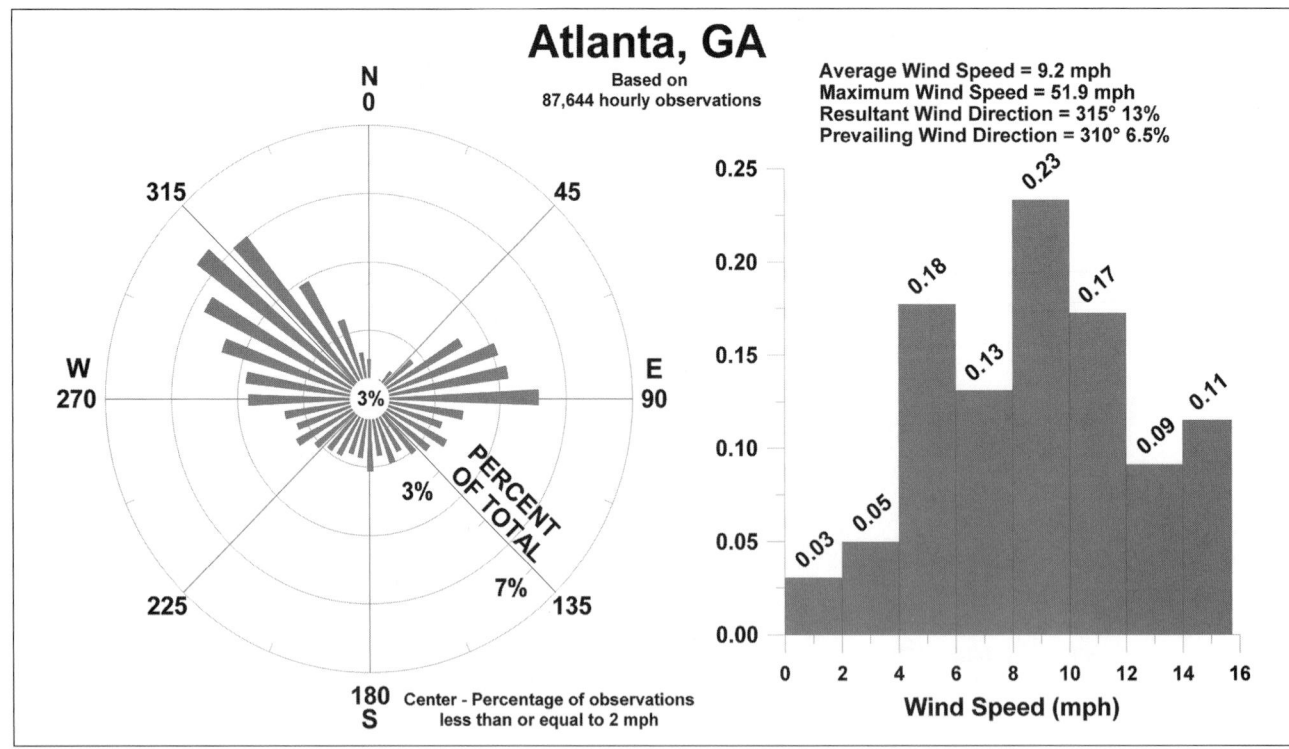

**Figure 2.8** City wind roses and wind speed frequency histogram.

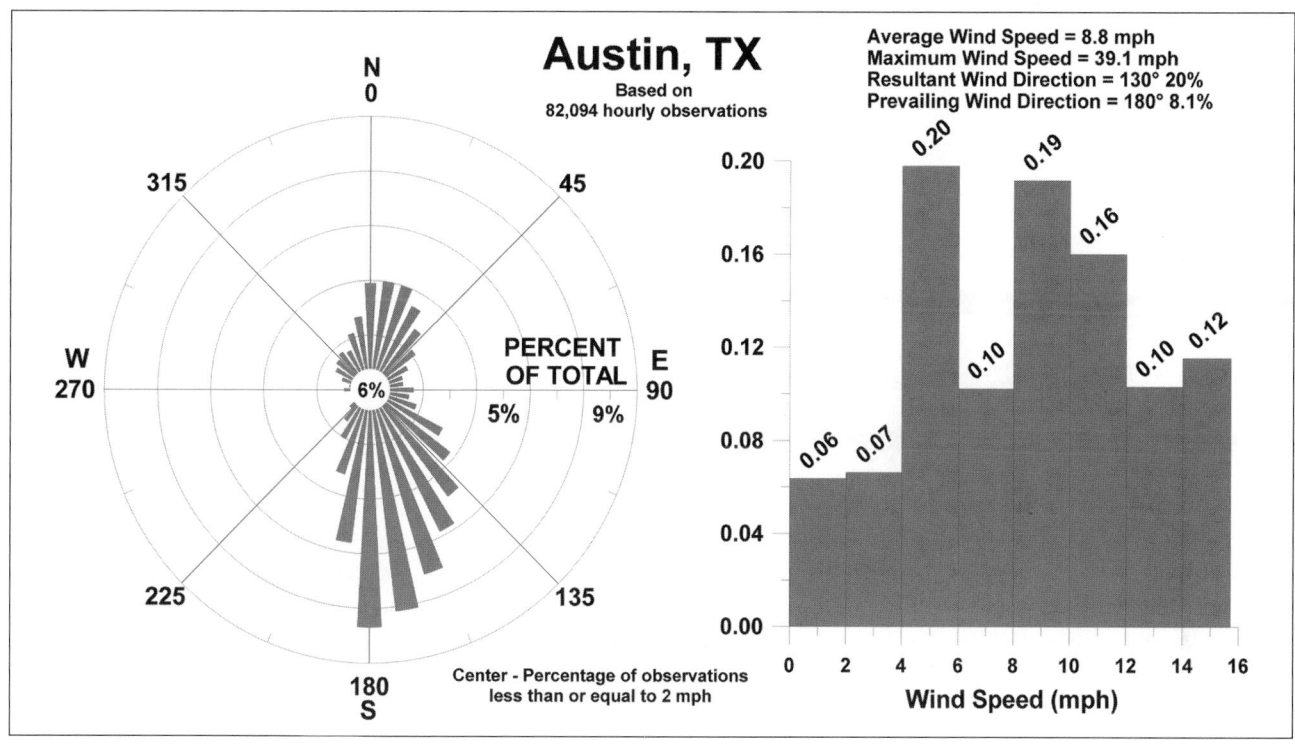

**Figure 2.9**  City wind roses and wind speed frequency histogram.

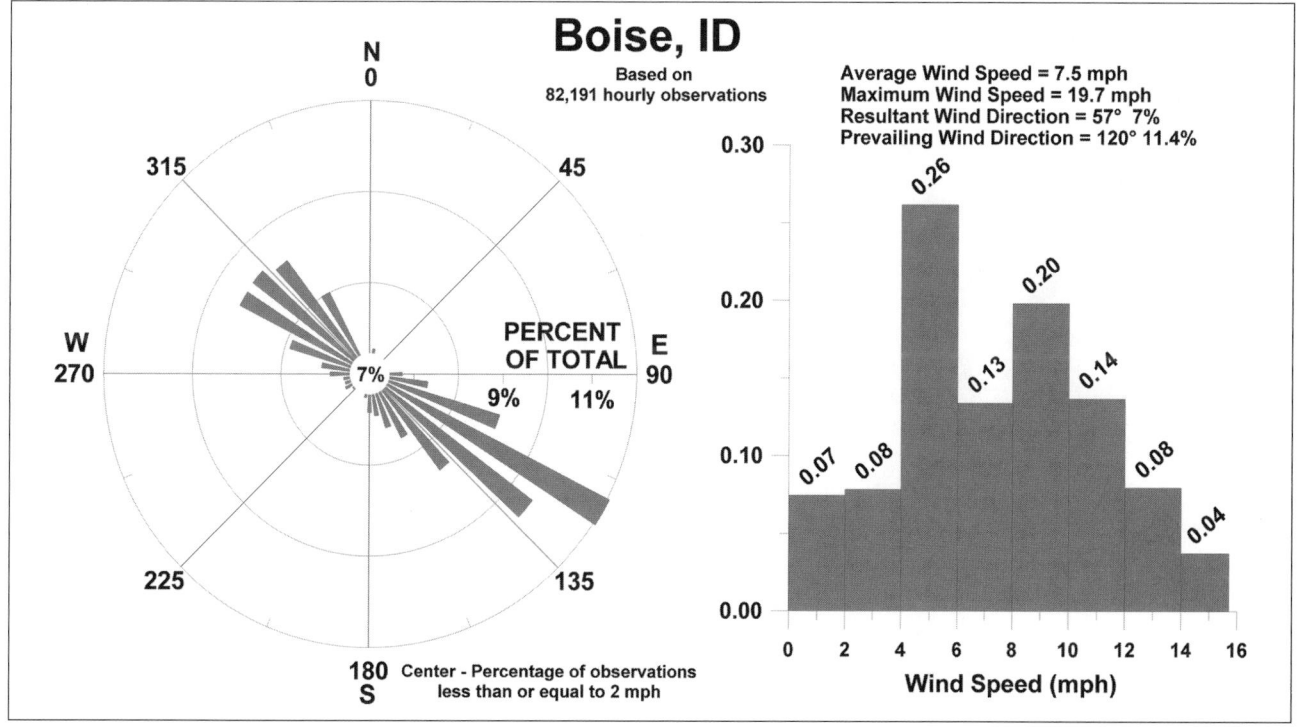

**Figure 2.10**  City wind roses and wind speed frequency histogram.

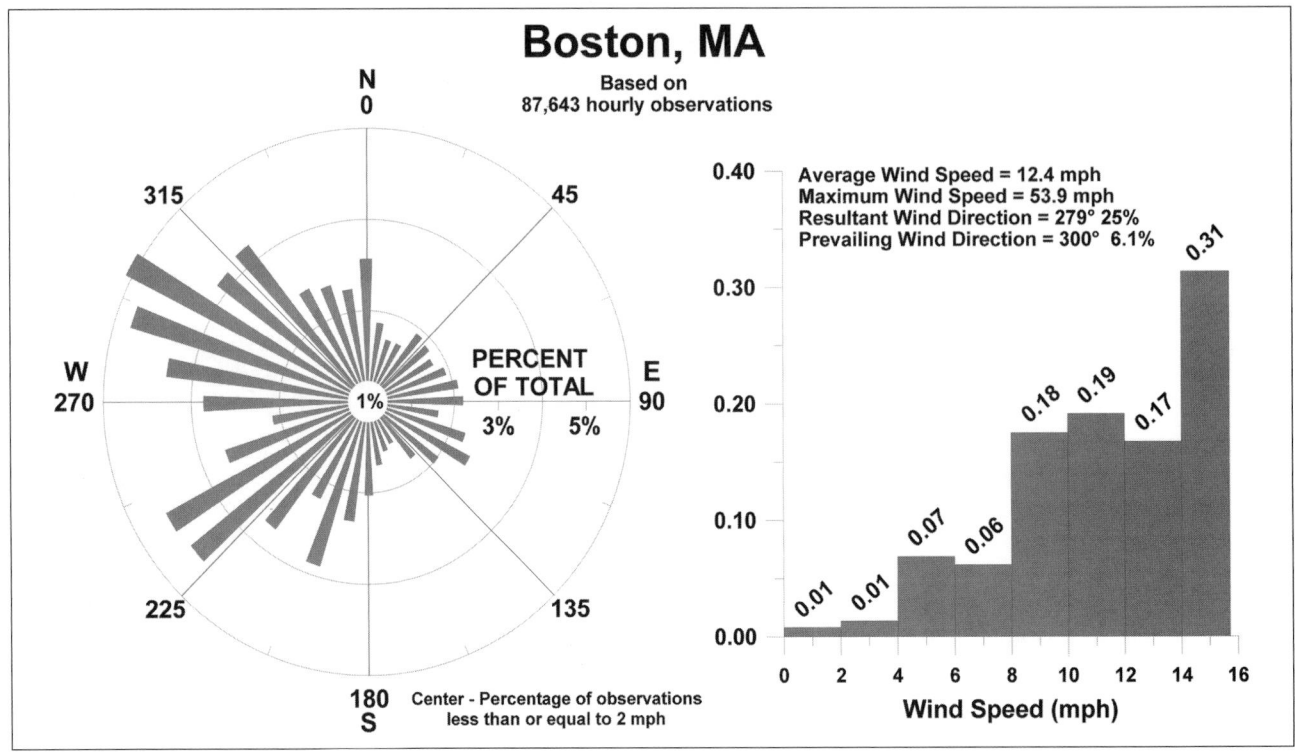

**Figure 2.11**  City wind roses and wind speed frequency histogram.

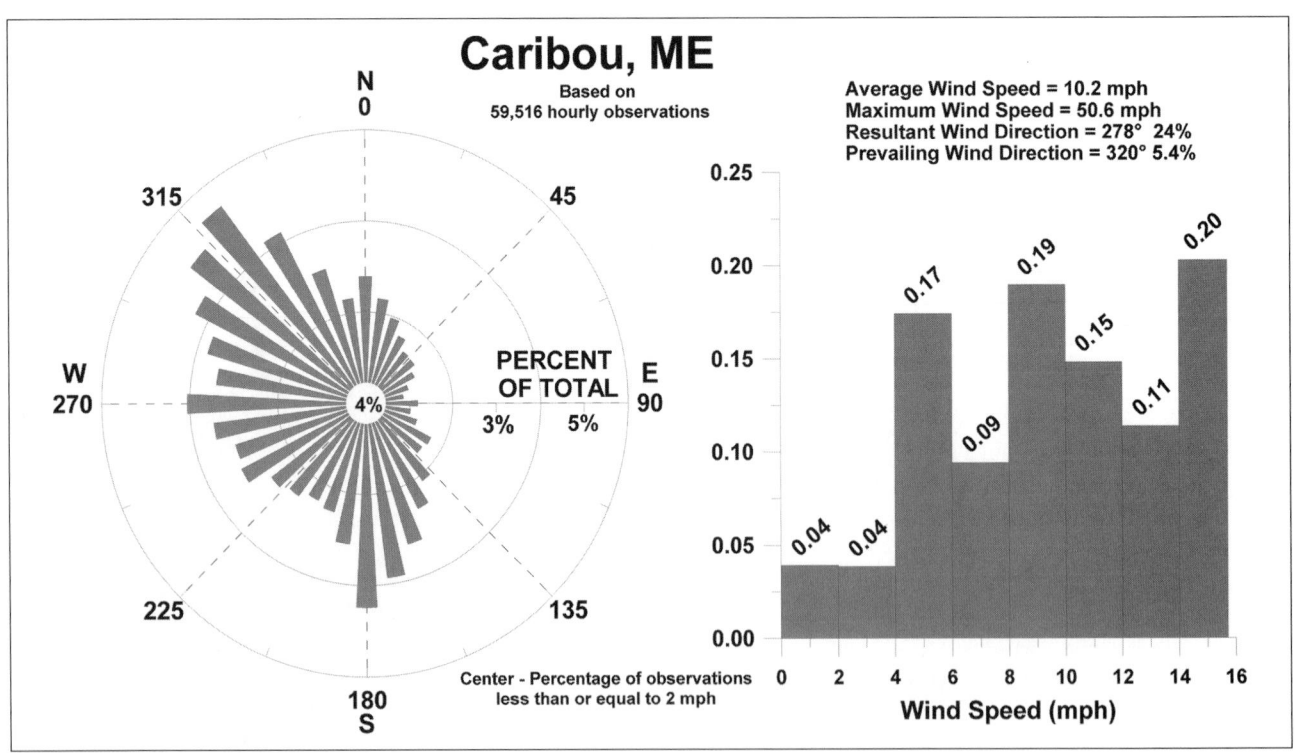

**Figure 2.12**  City wind roses and wind speed frequency histogram.

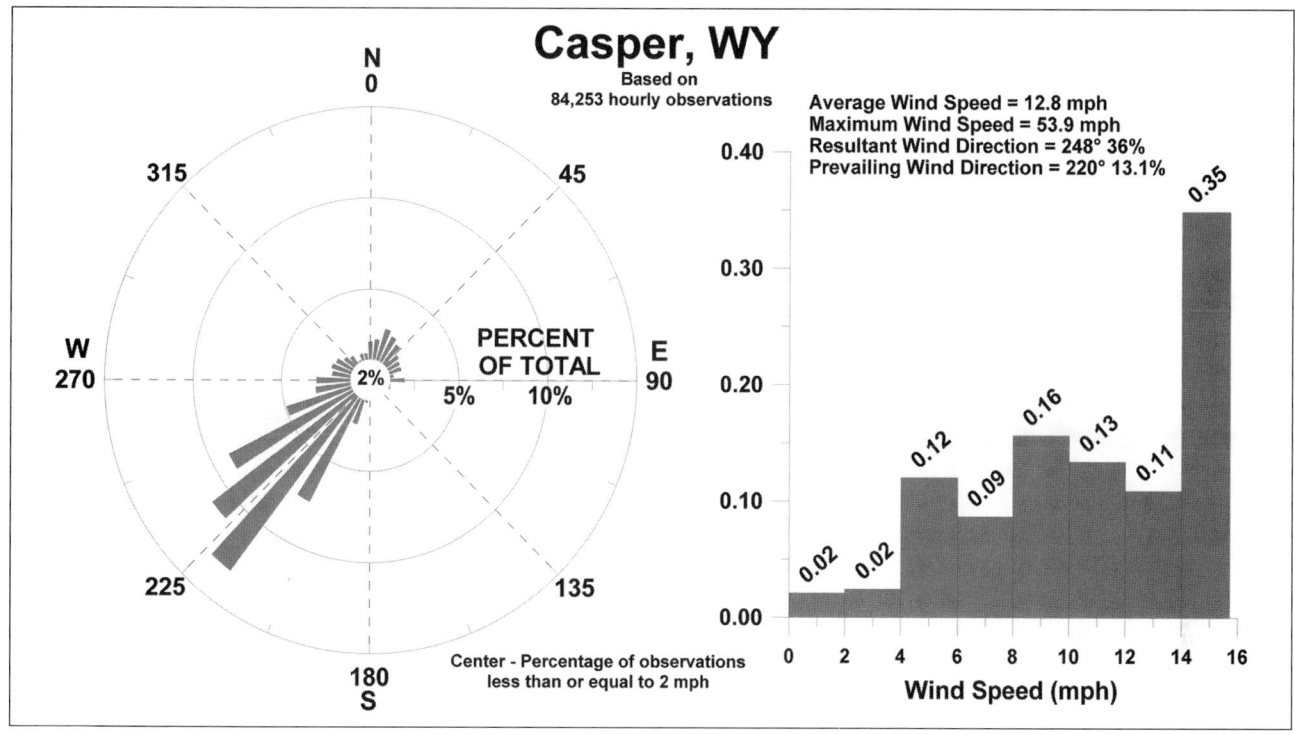

**Figure 2.13**   City wind roses and wind speed frequency histogram.

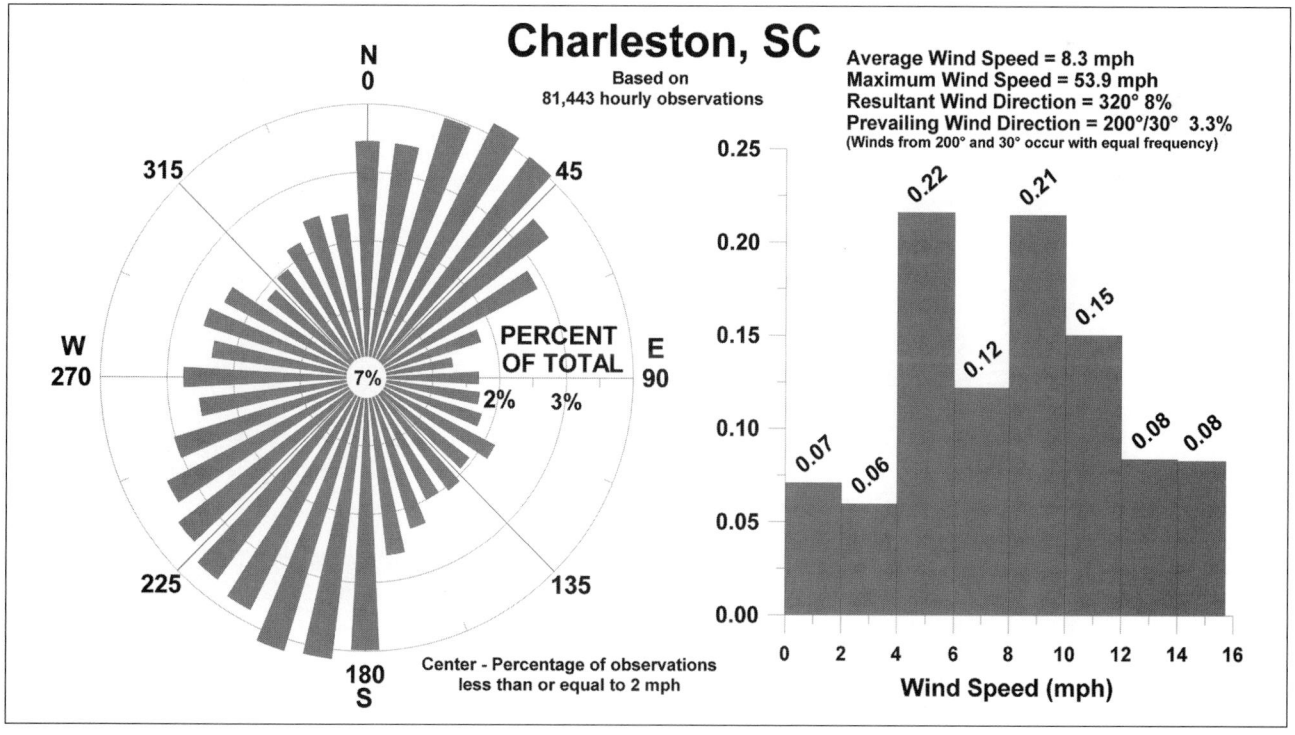

**Figure 2.14**   City wind roses and wind speed frequency histogram.

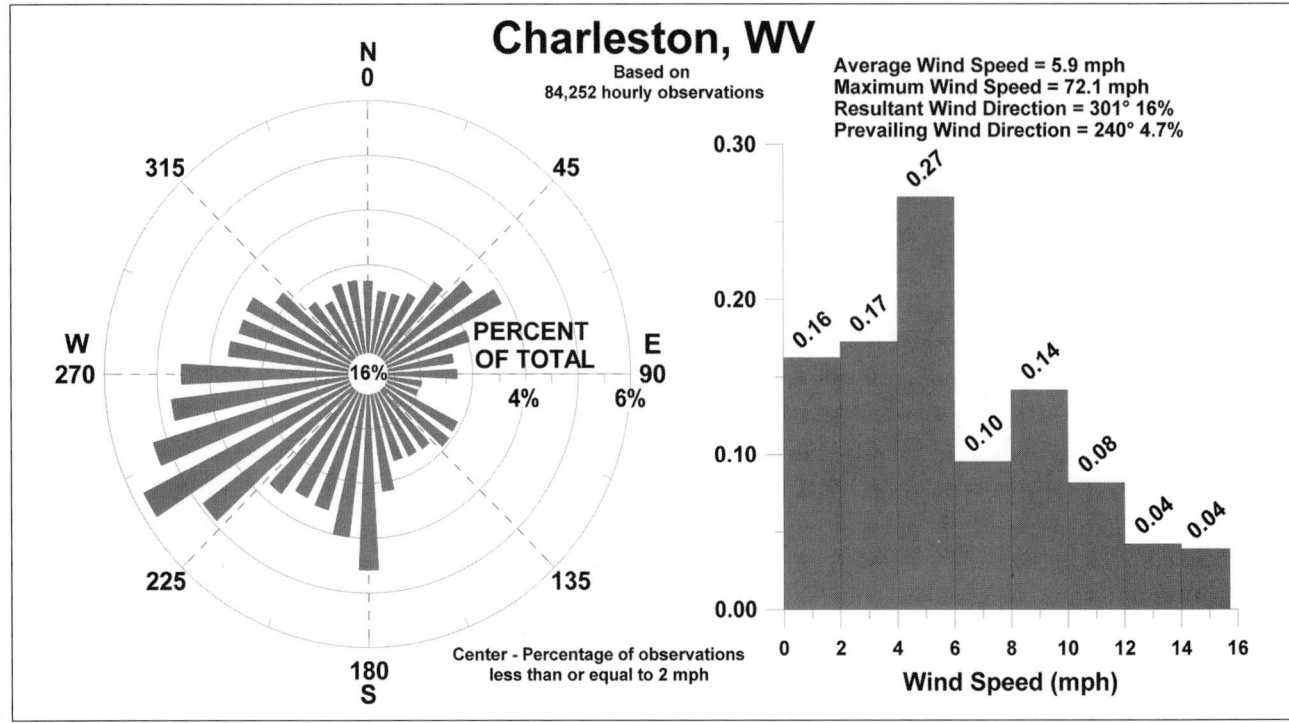

**Figure 2.15**   City wind roses and wind speed frequency histogram.

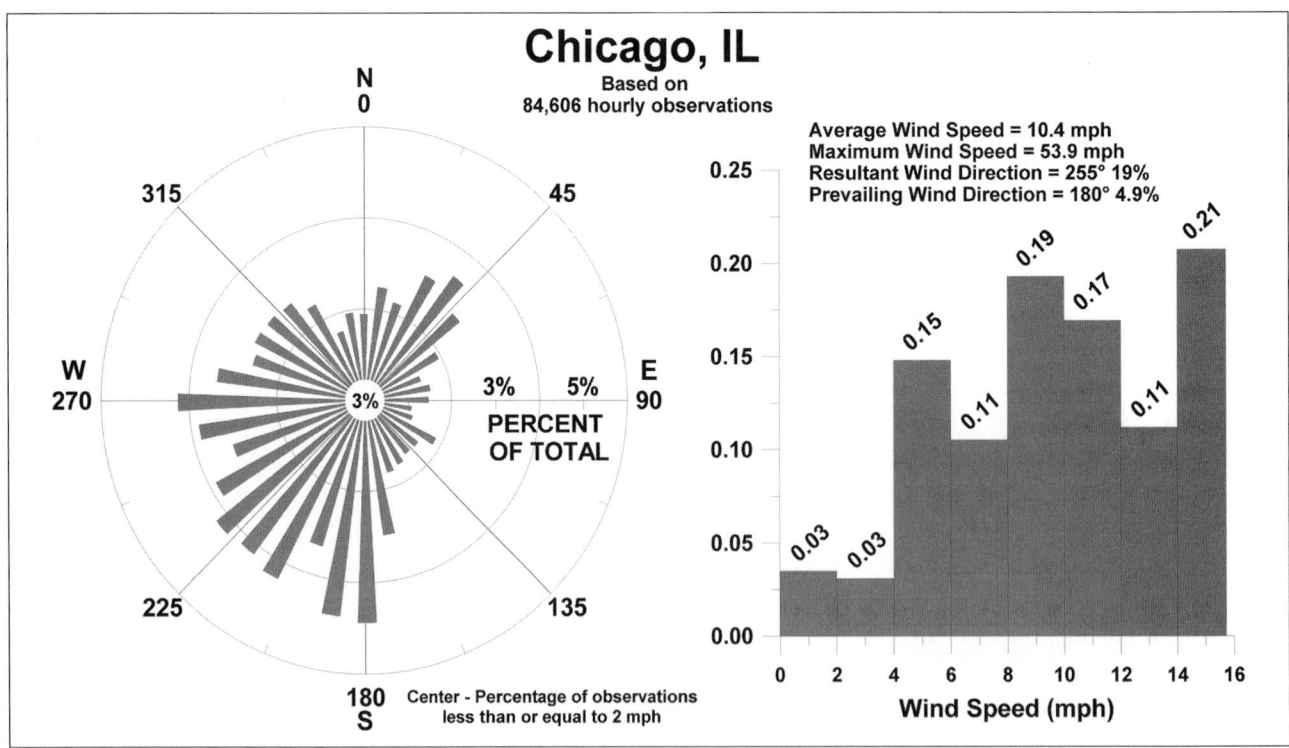

**Figure 2.16**   City wind roses and wind speed frequency histogram.

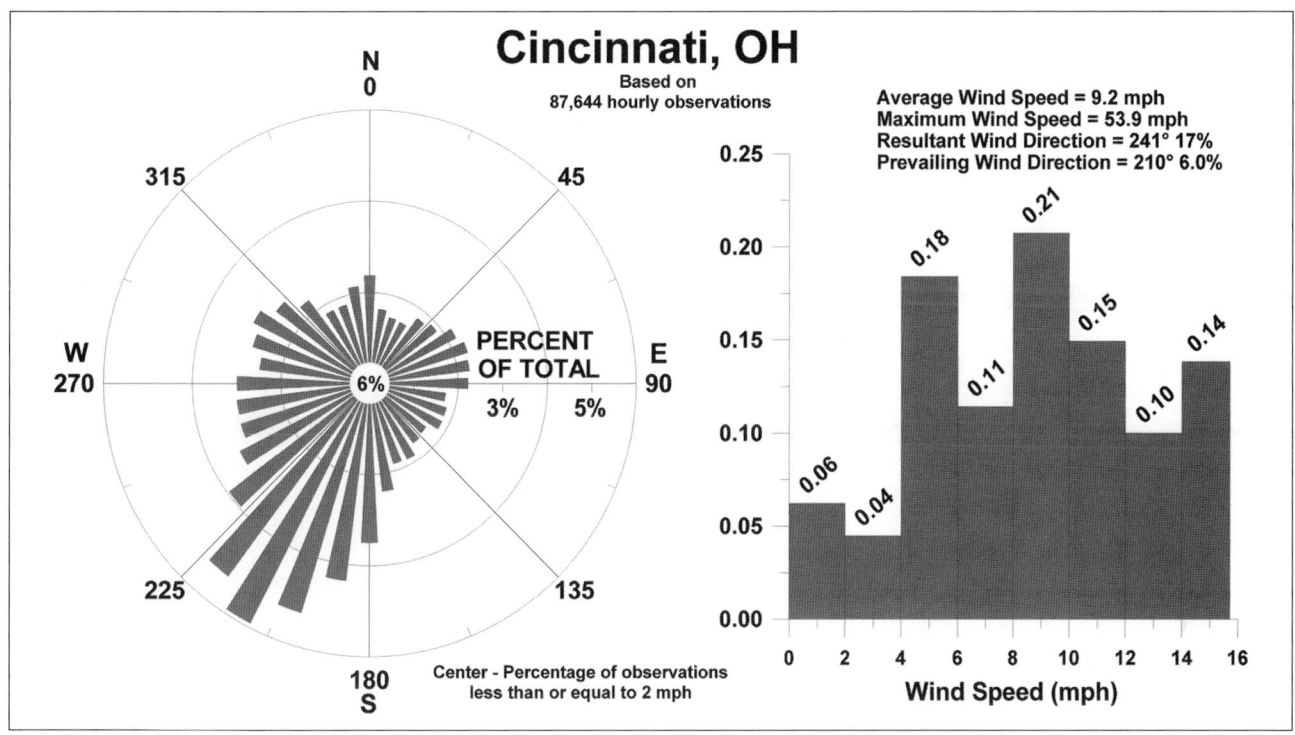

**Figure 2.17**   City wind roses and wind speed frequency histogram.

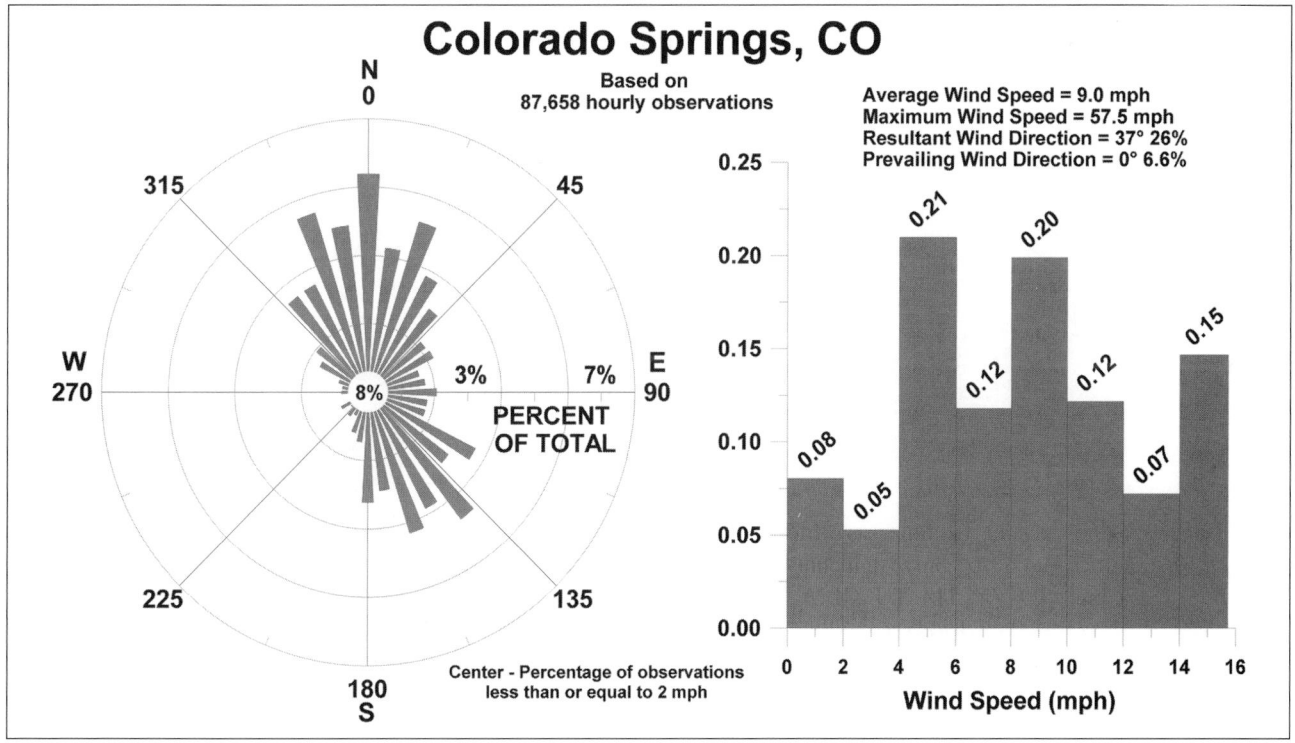

**Figure 2.18**   City wind roses and wind speed frequency histogram.

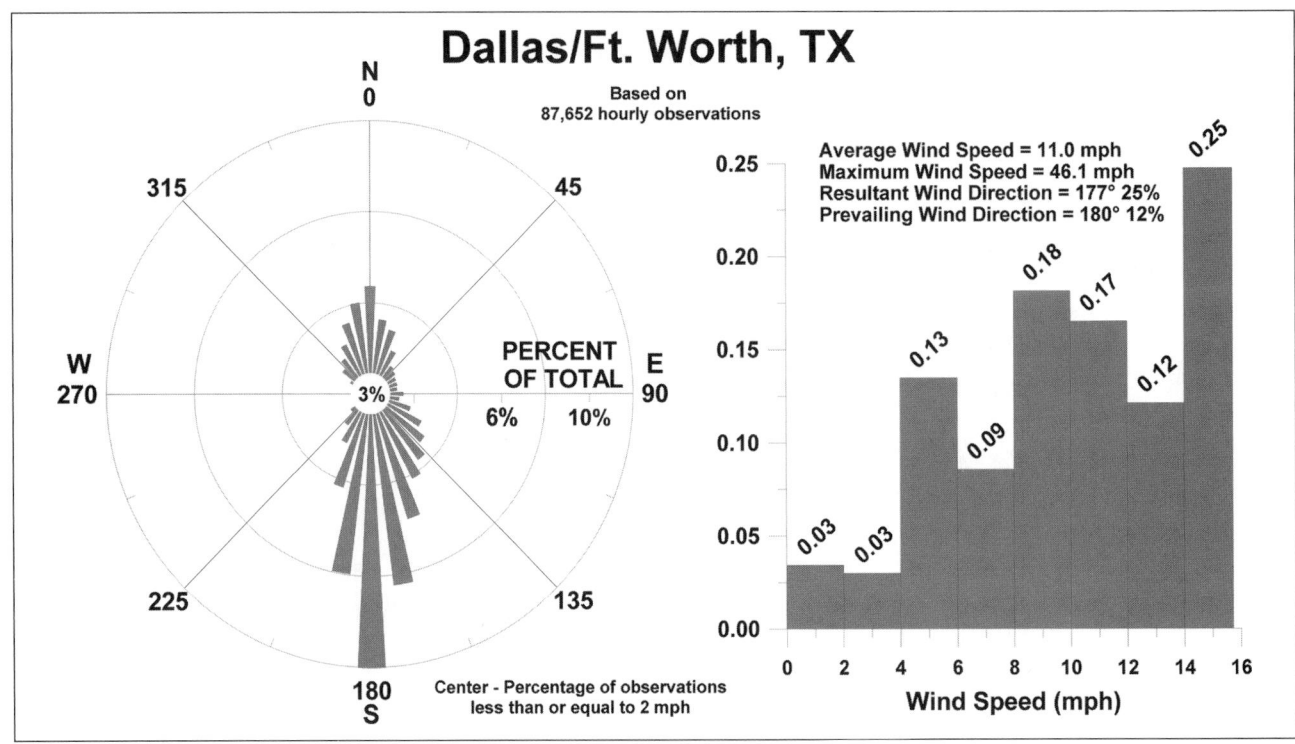

**Figure 2.19** City wind roses and wind speed frequency histogram.

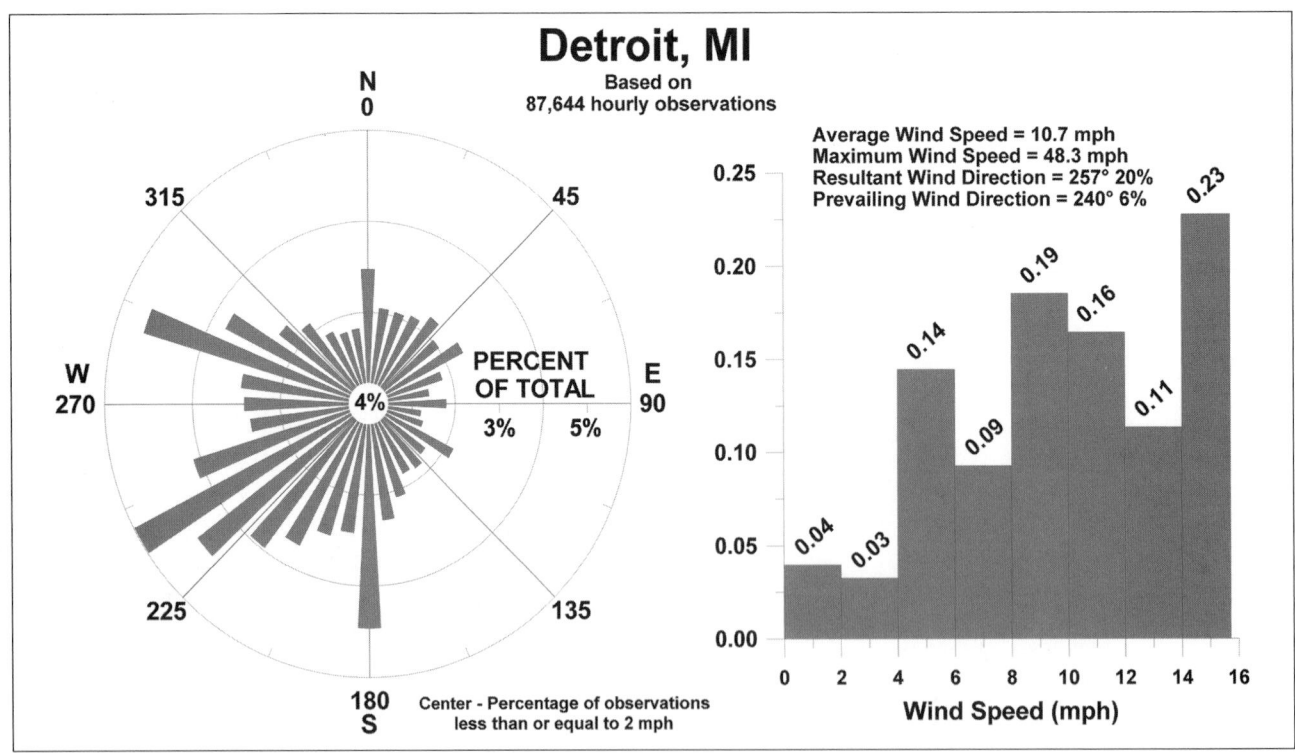

**Figure 2.20** City wind roses and wind speed frequency histogram.

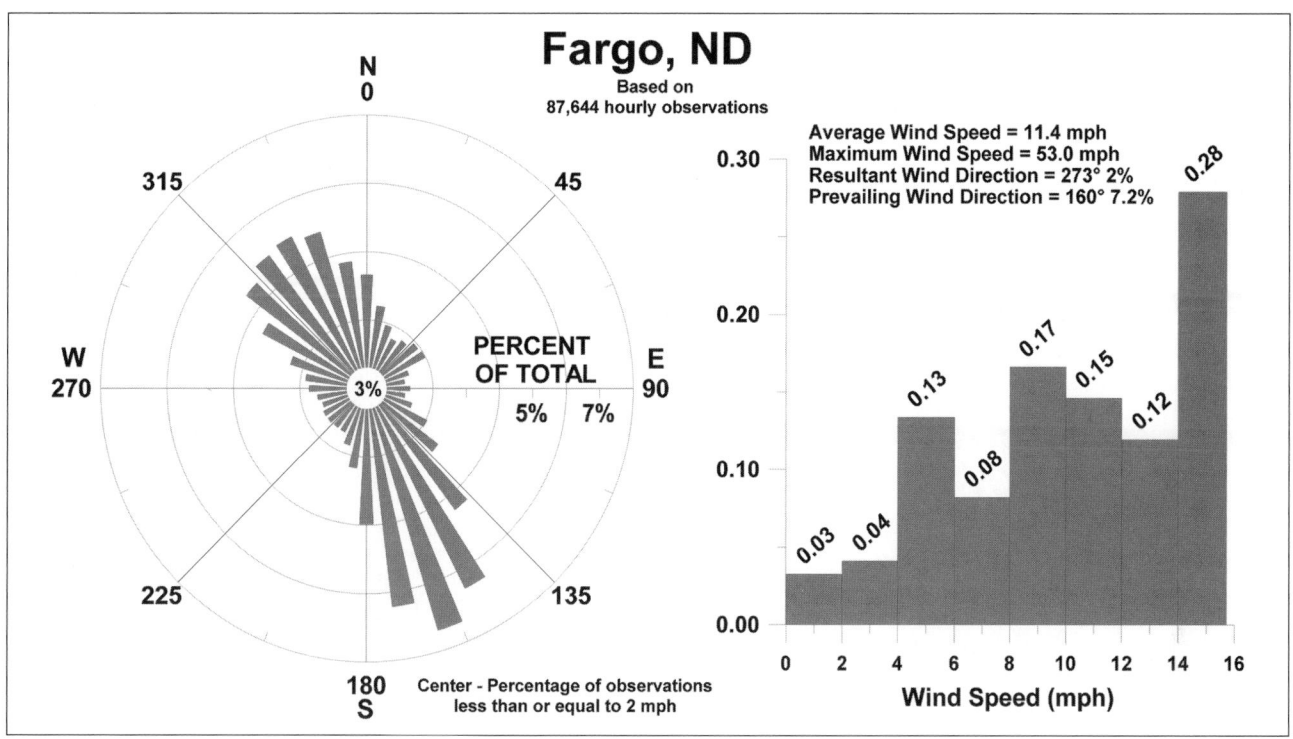

**Figure 2.21**   City wind roses and wind speed frequency histogram.

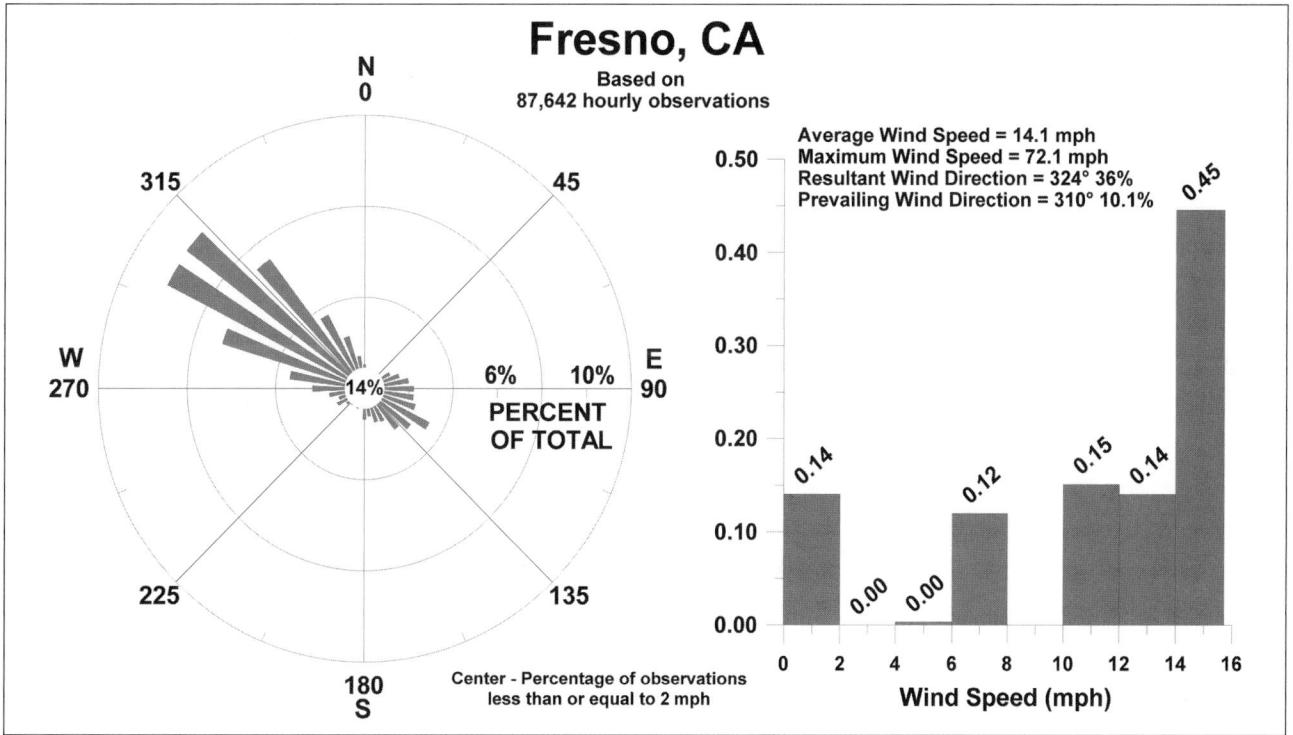

**Figure 2.22**   City wind roses and wind speed frequency histogram.

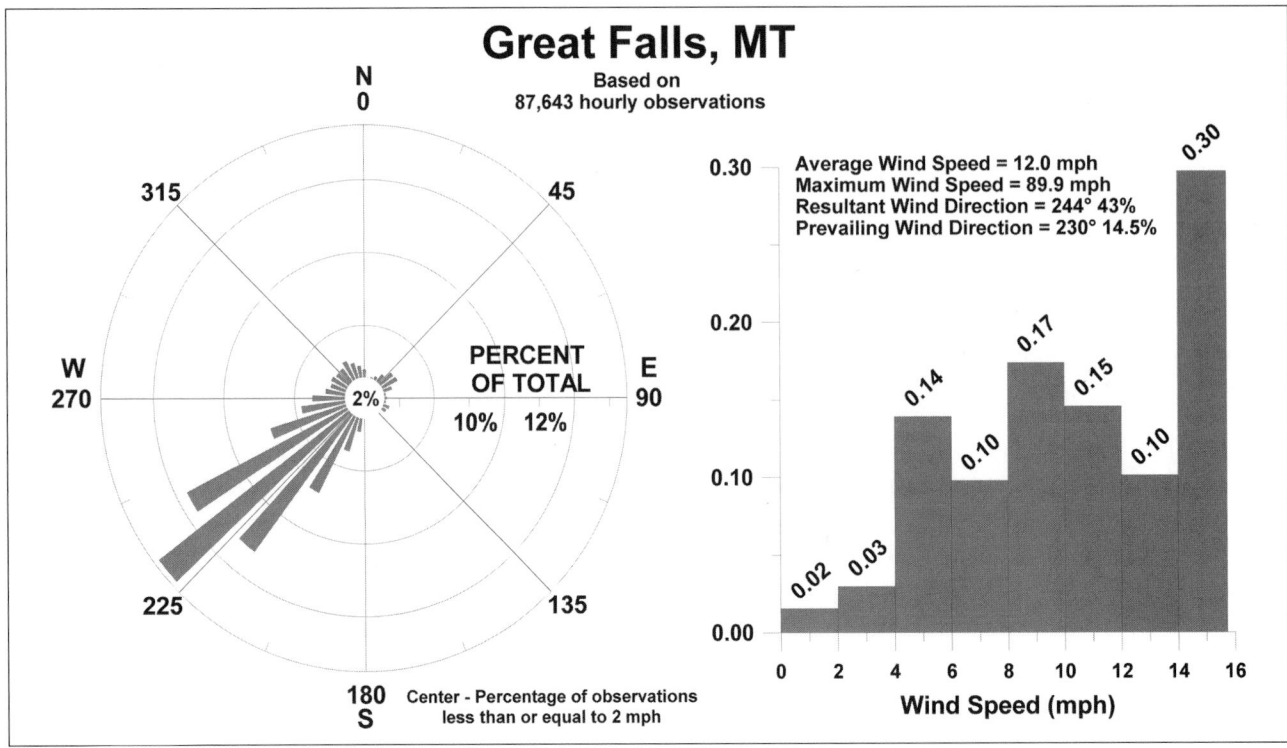

**Figure 2.23**   City wind roses and wind speed frequency histogram.

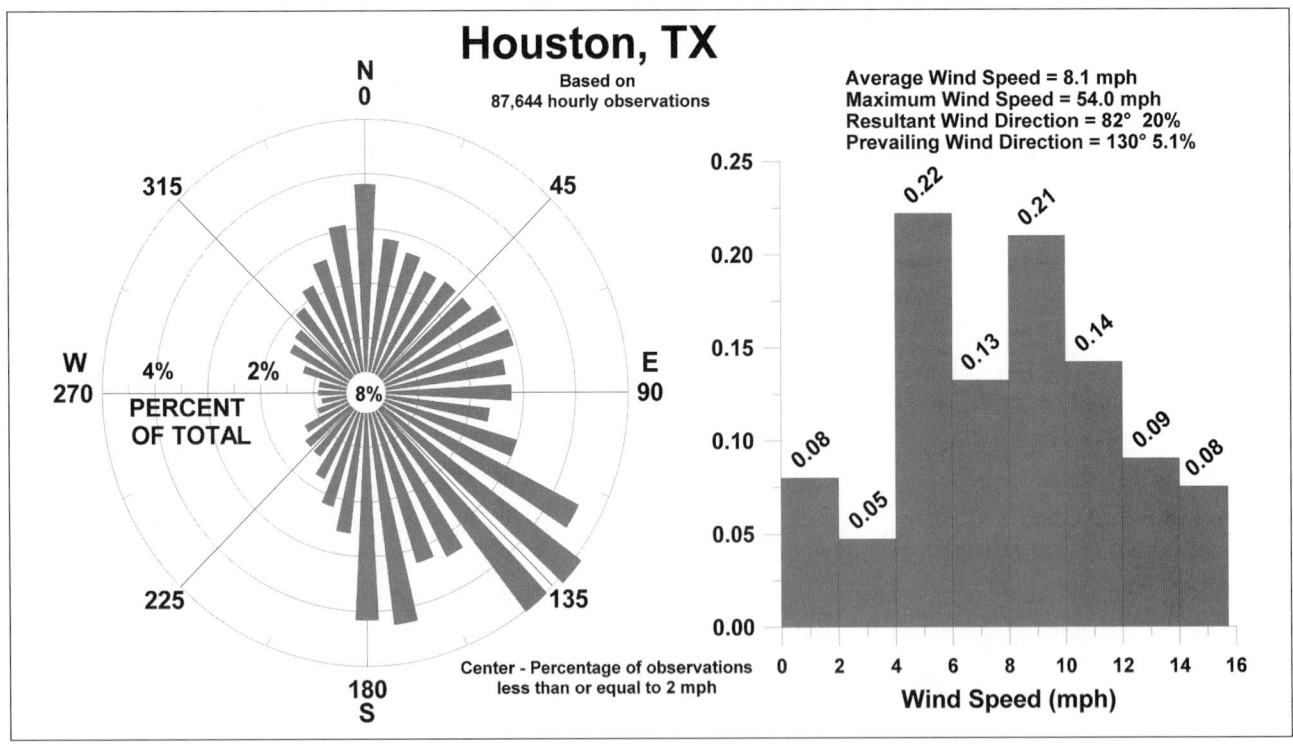

**Figure 2.24**   City wind roses and wind speed frequency histogram.

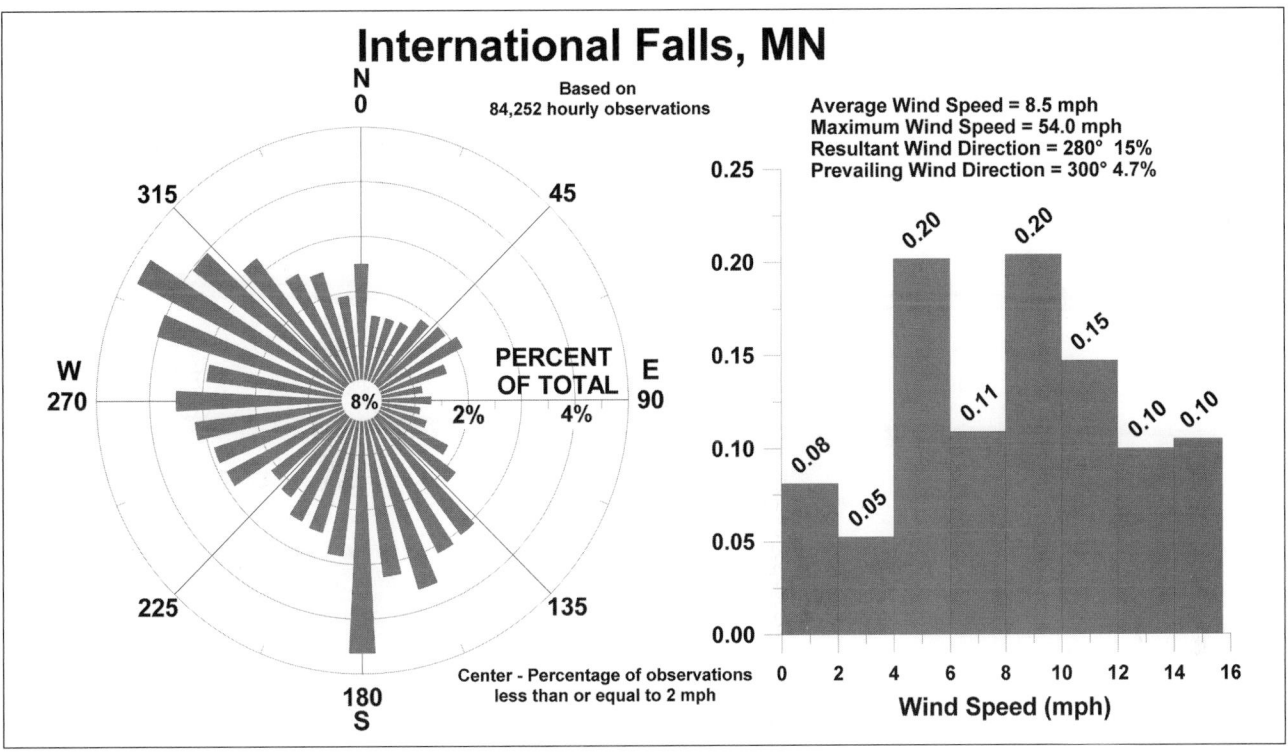

**Figure 2.25**  City wind roses and wind speed frequency histogram.

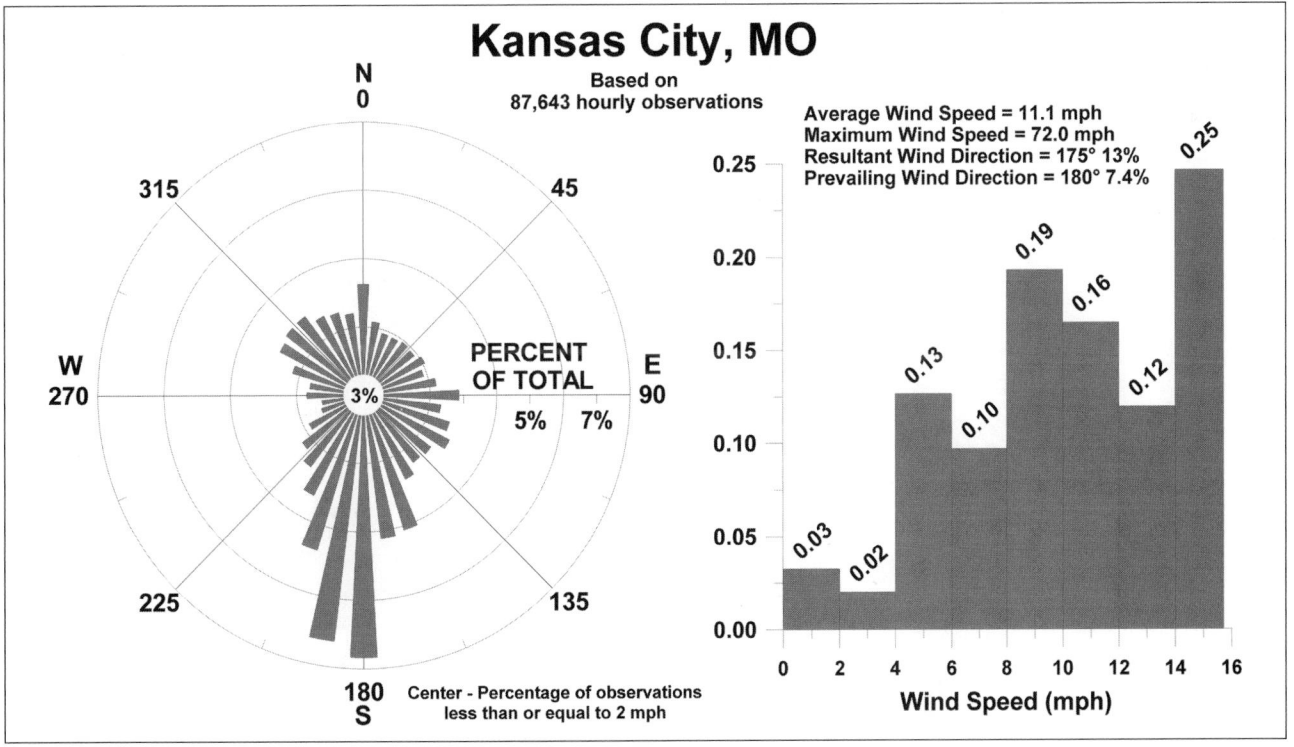

**Figure 2.26**  City wind roses and wind speed frequency histogram.

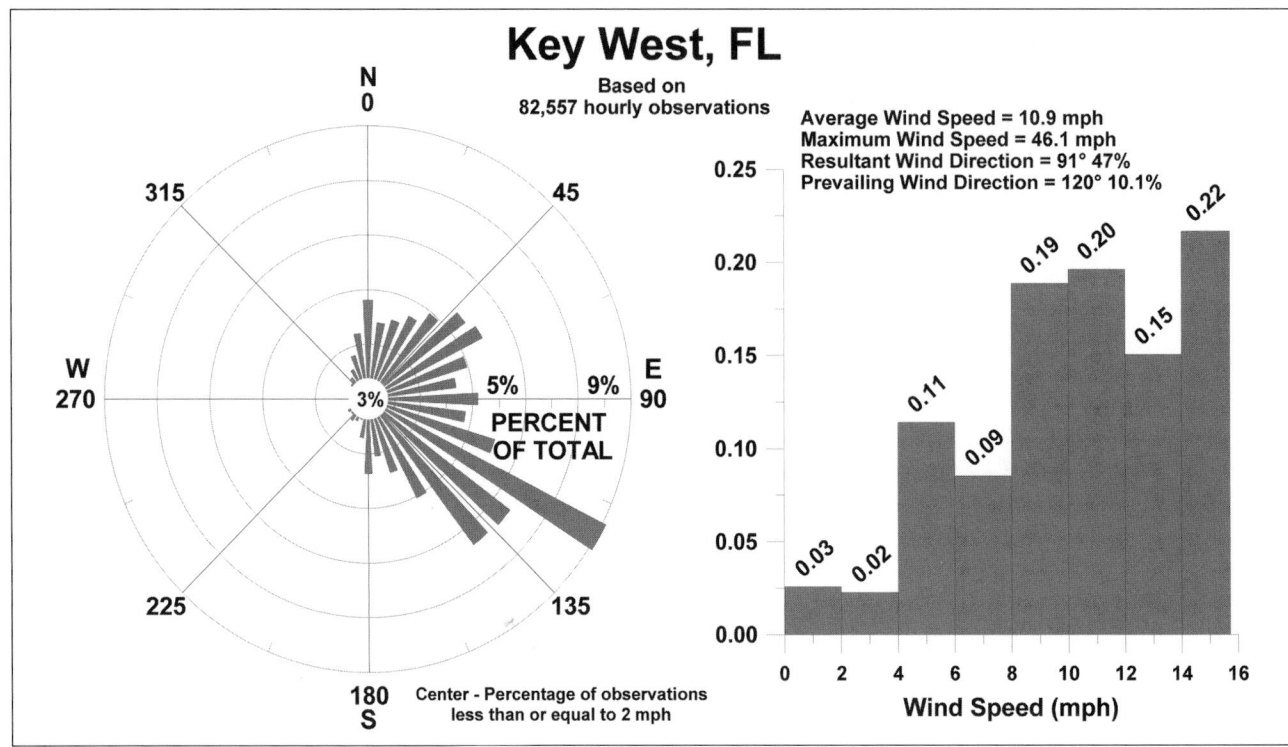

**Figure 2.27** City wind roses and wind speed frequency histogram.

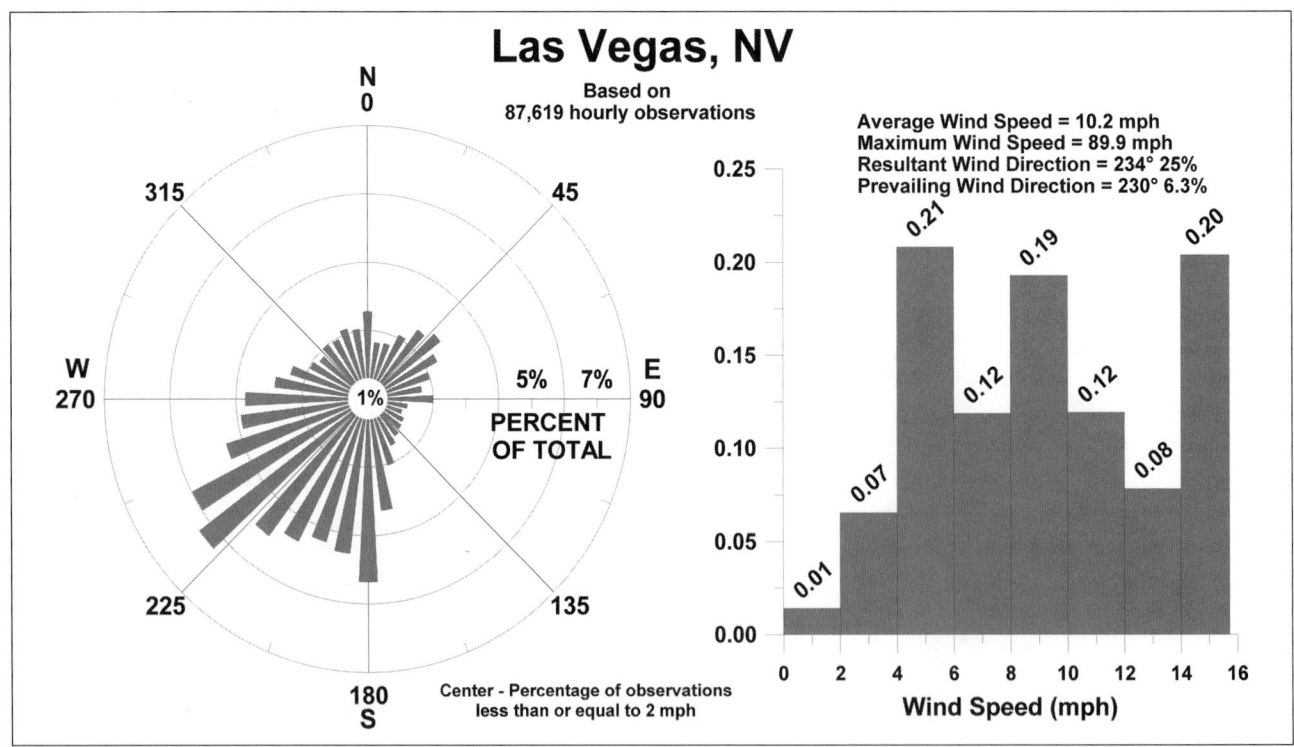

**Figure 2.28** City wind roses and wind speed frequency histogram.

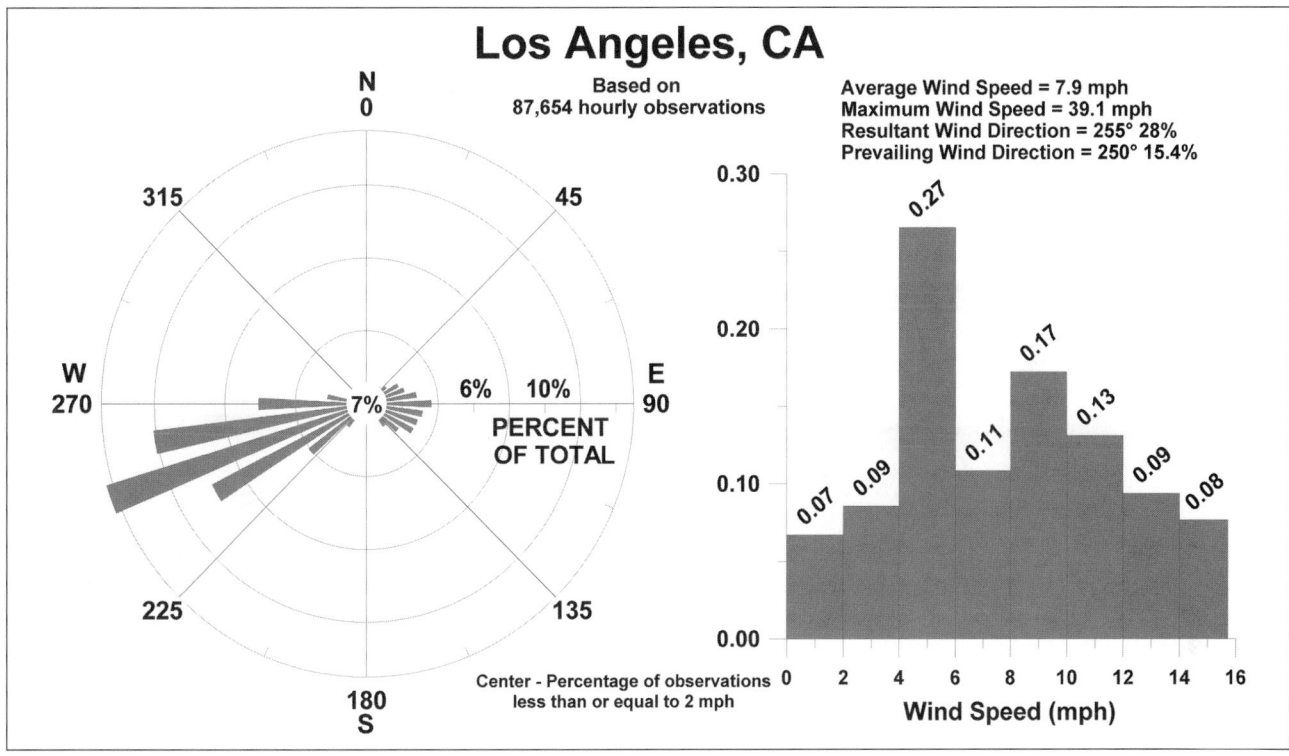

**Figure 2.29**   City wind roses and wind speed frequency histogram.

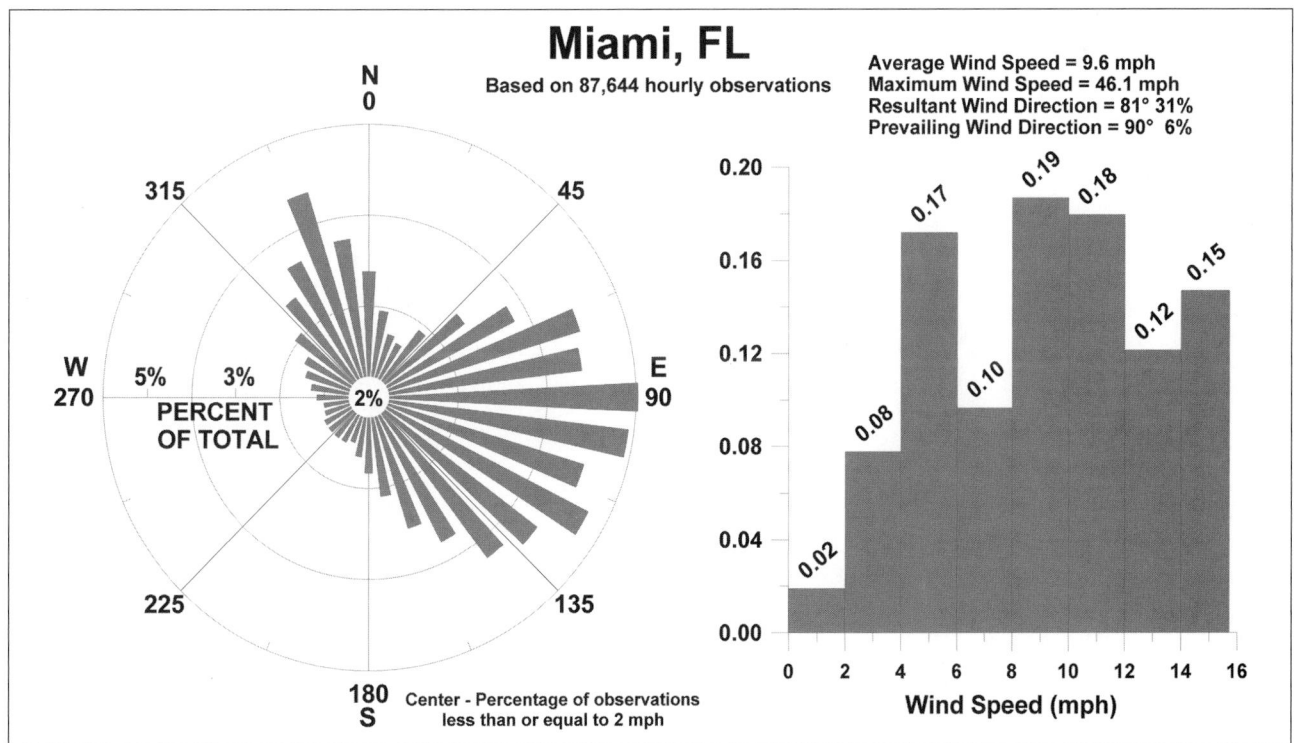

**Figure 2.30**   City wind roses and wind speed frequency histogram.

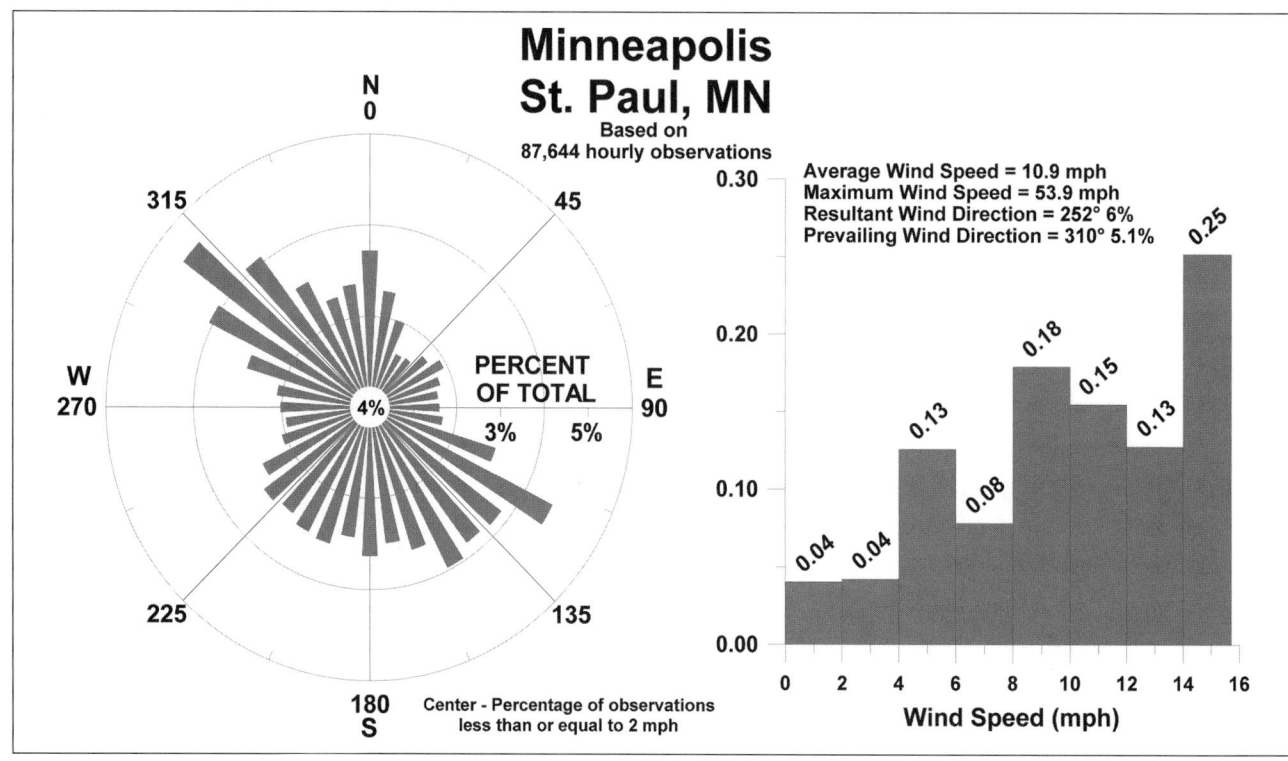

**Figure 2.31** City wind roses and wind speed frequency histogram.

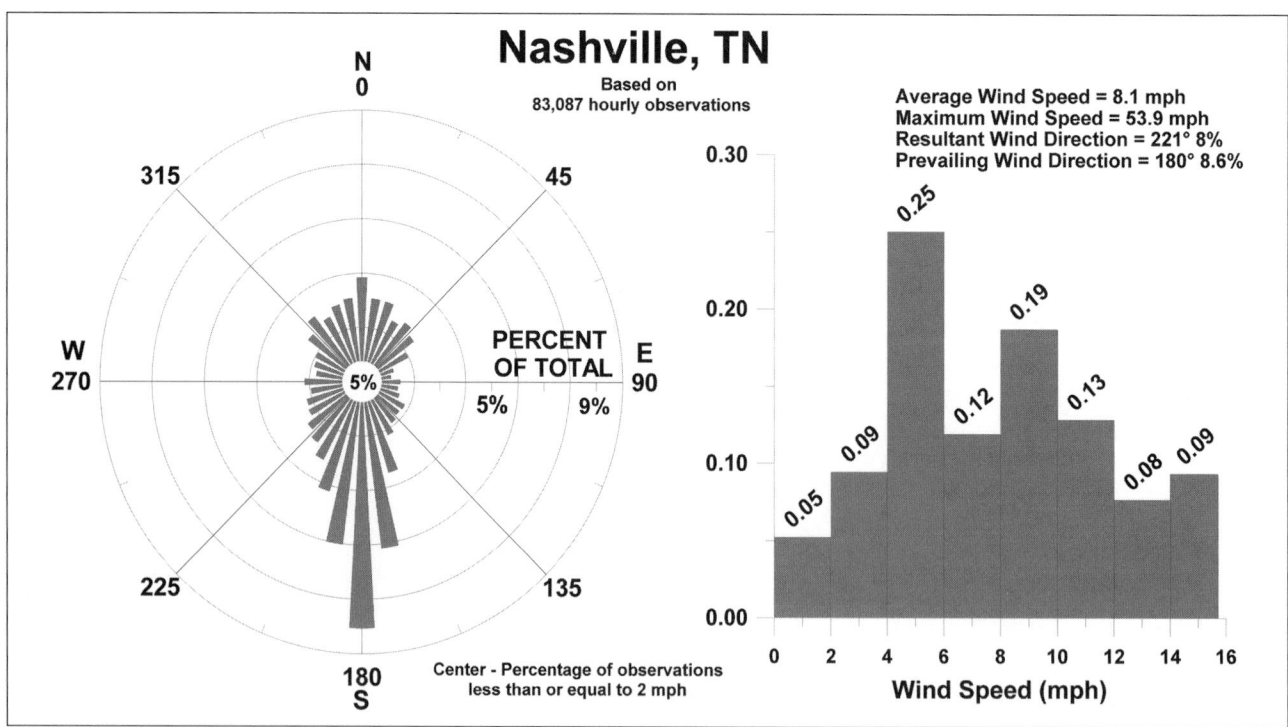

**Figure 2.32** City wind roses and wind speed frequency histogram.

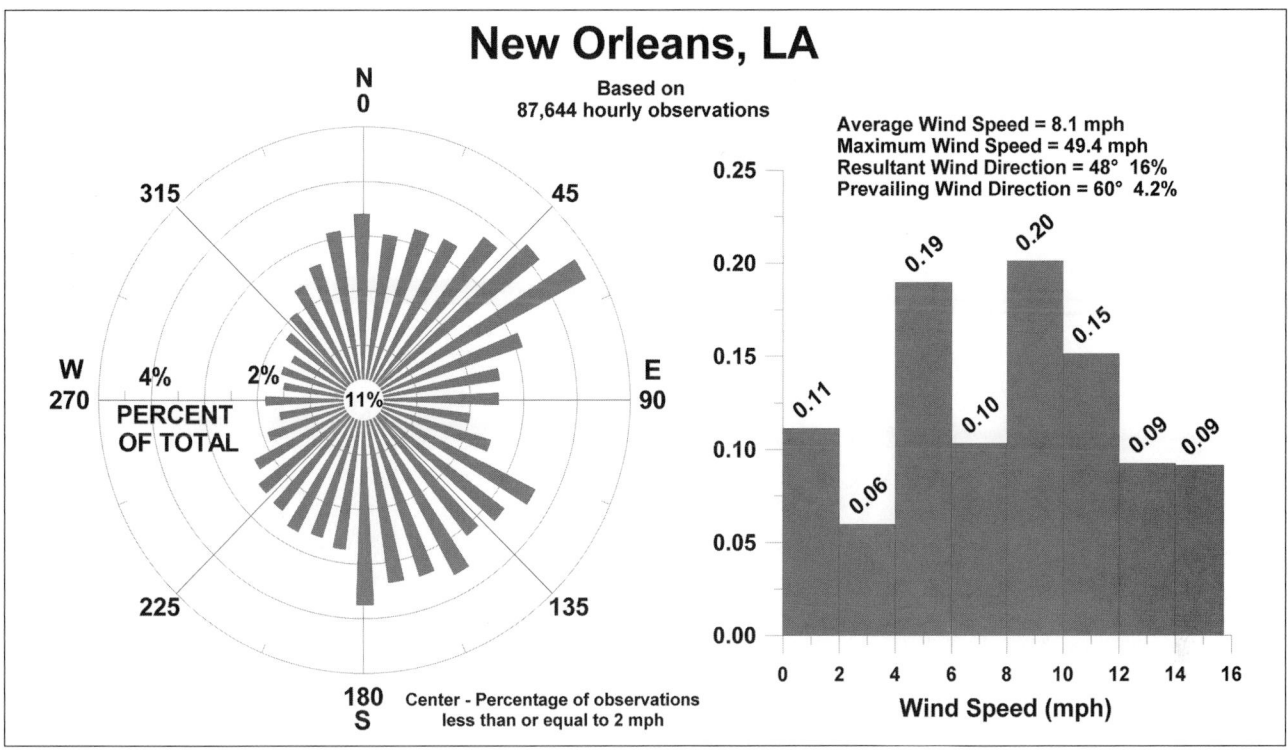

**Figure 2.33**   City wind roses and wind speed frequency histogram.

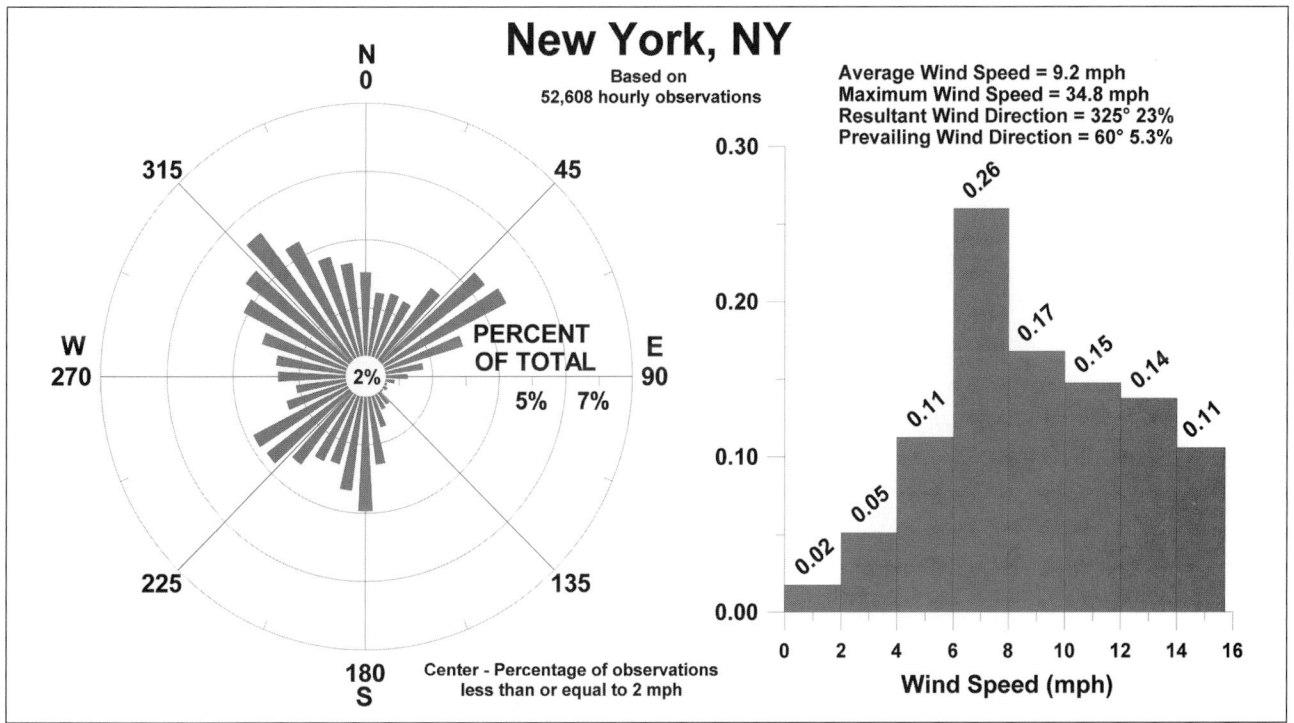

**Figure 2.34**   City wind roses and wind speed frequency histogram.

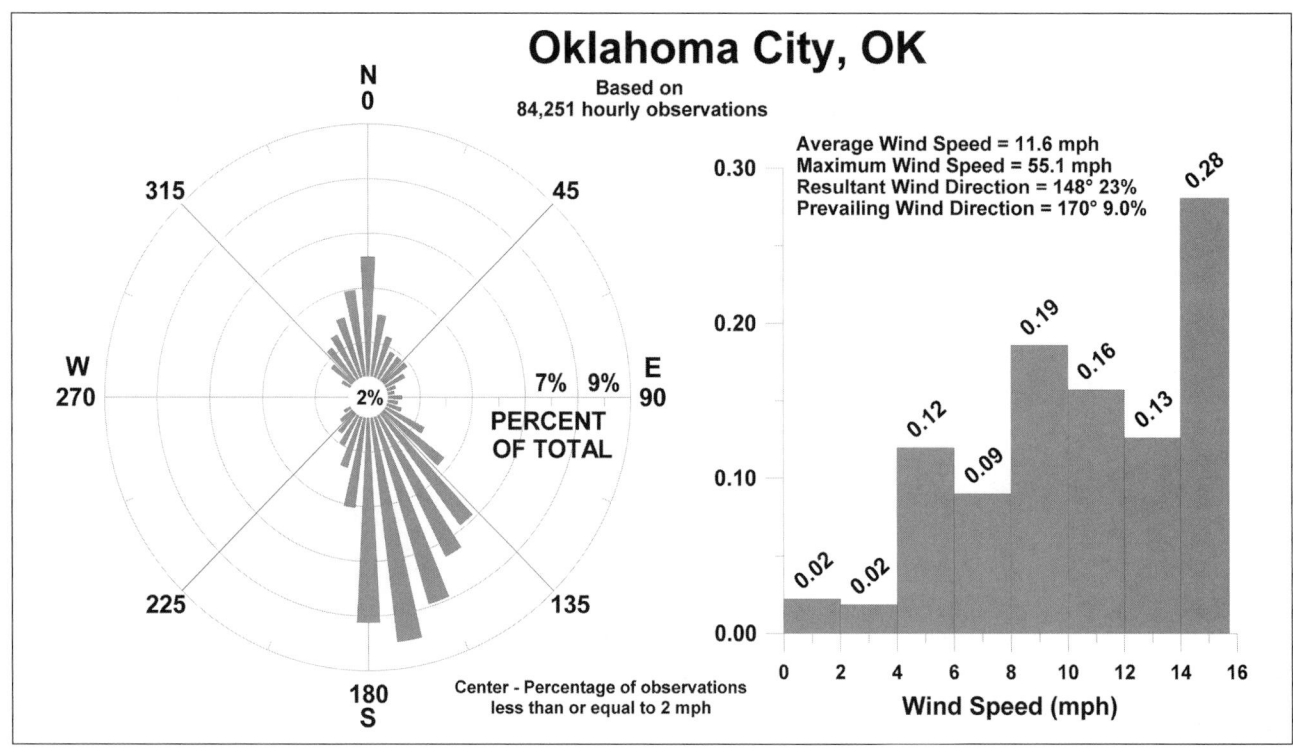

**Figure 2.35** City wind roses and wind speed frequency histogram.

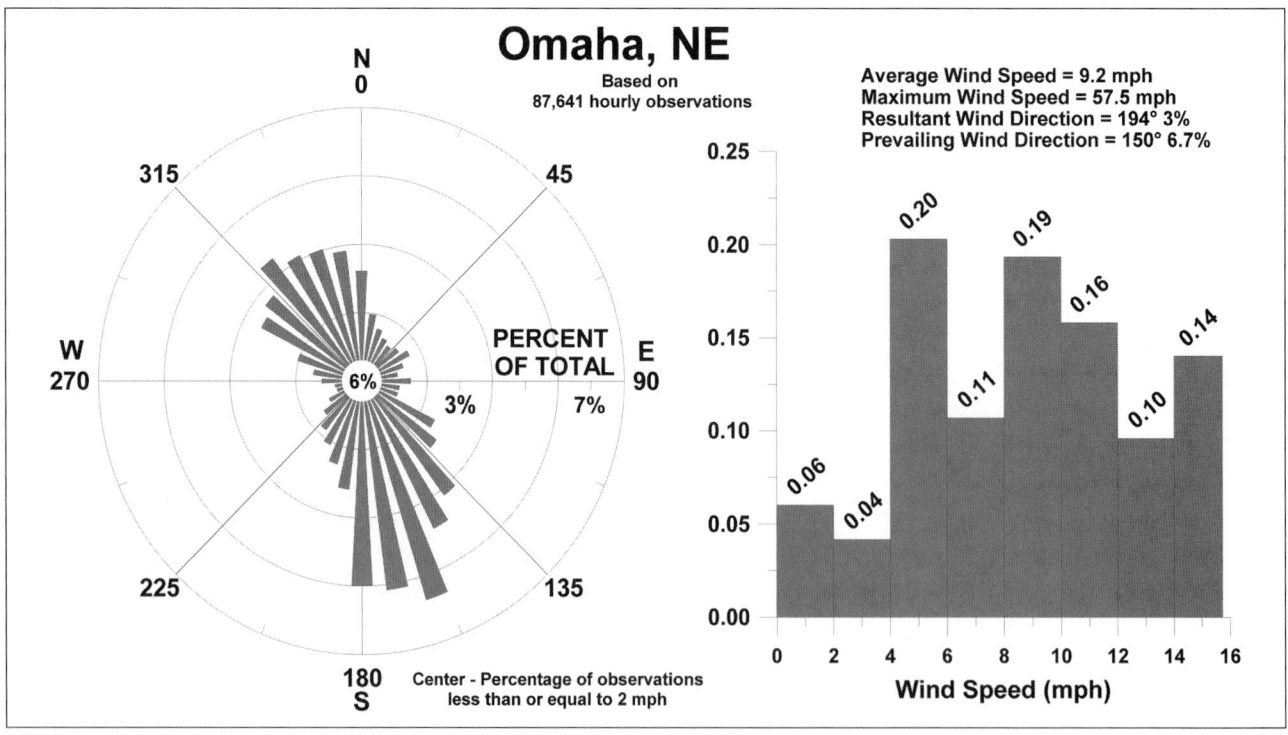

**Figure 2.36** City wind roses and wind speed frequency histogram.

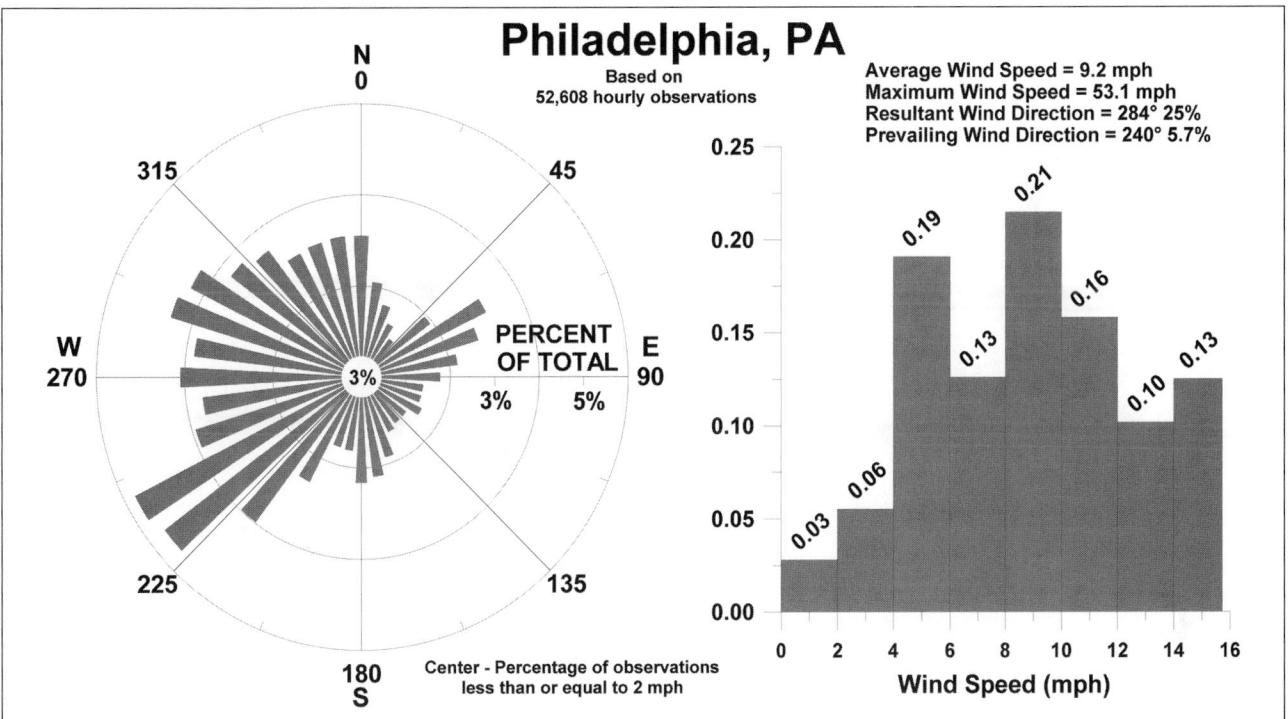

**Figure 2.37**   City wind roses and wind speed frequency histogram.

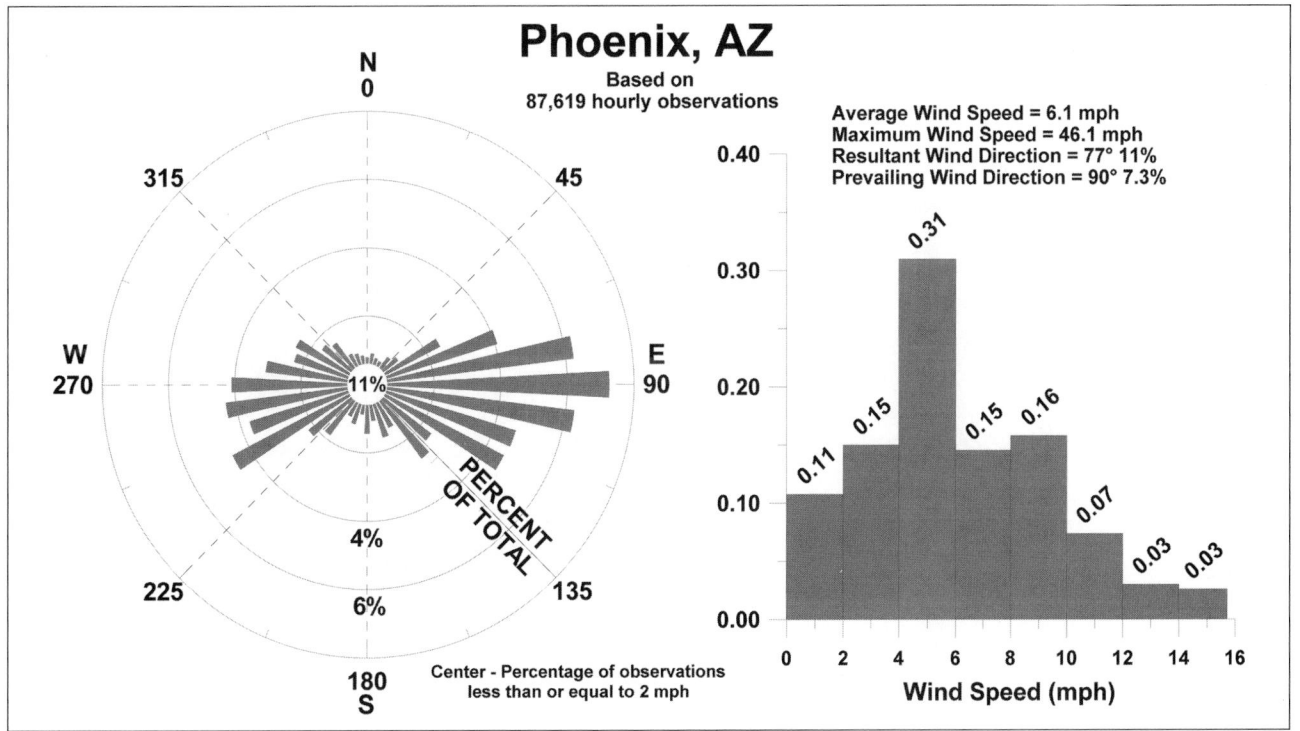

**Figure 2.38**   City wind roses and wind speed frequency histogram.

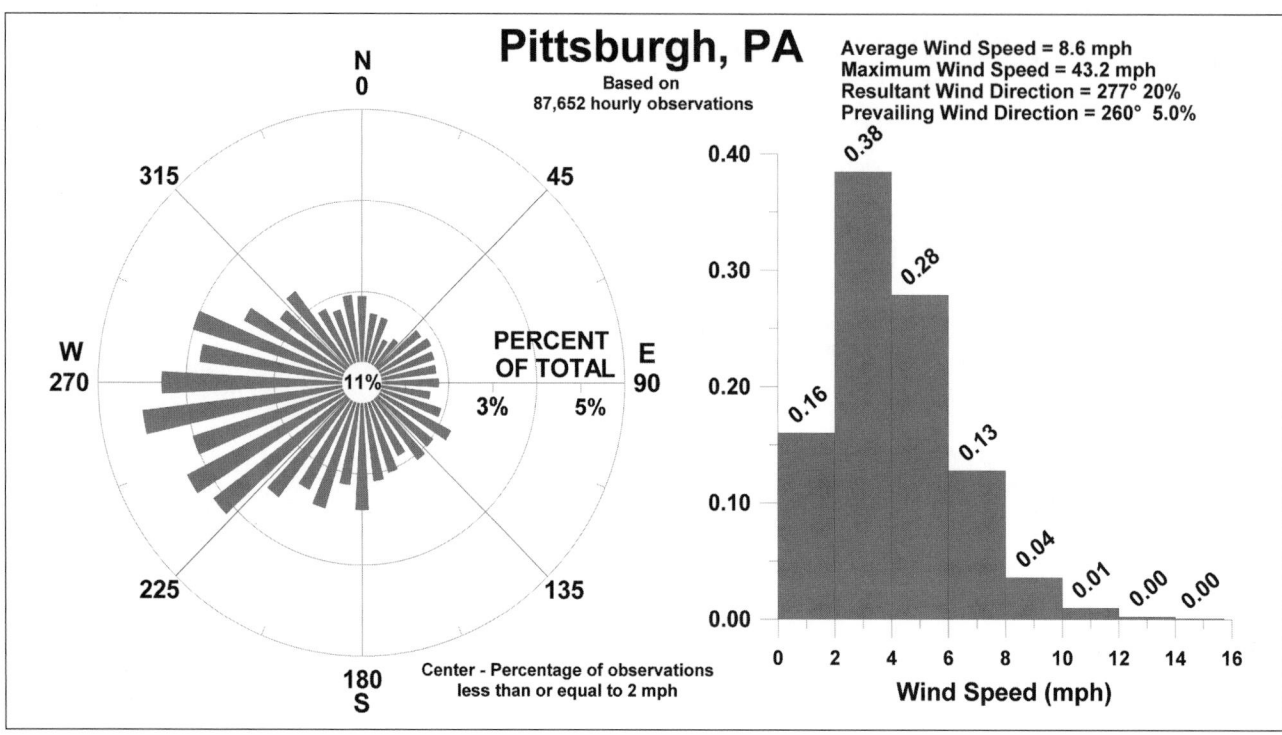

**Figure 2.39** City wind roses and wind speed frequency histogram.

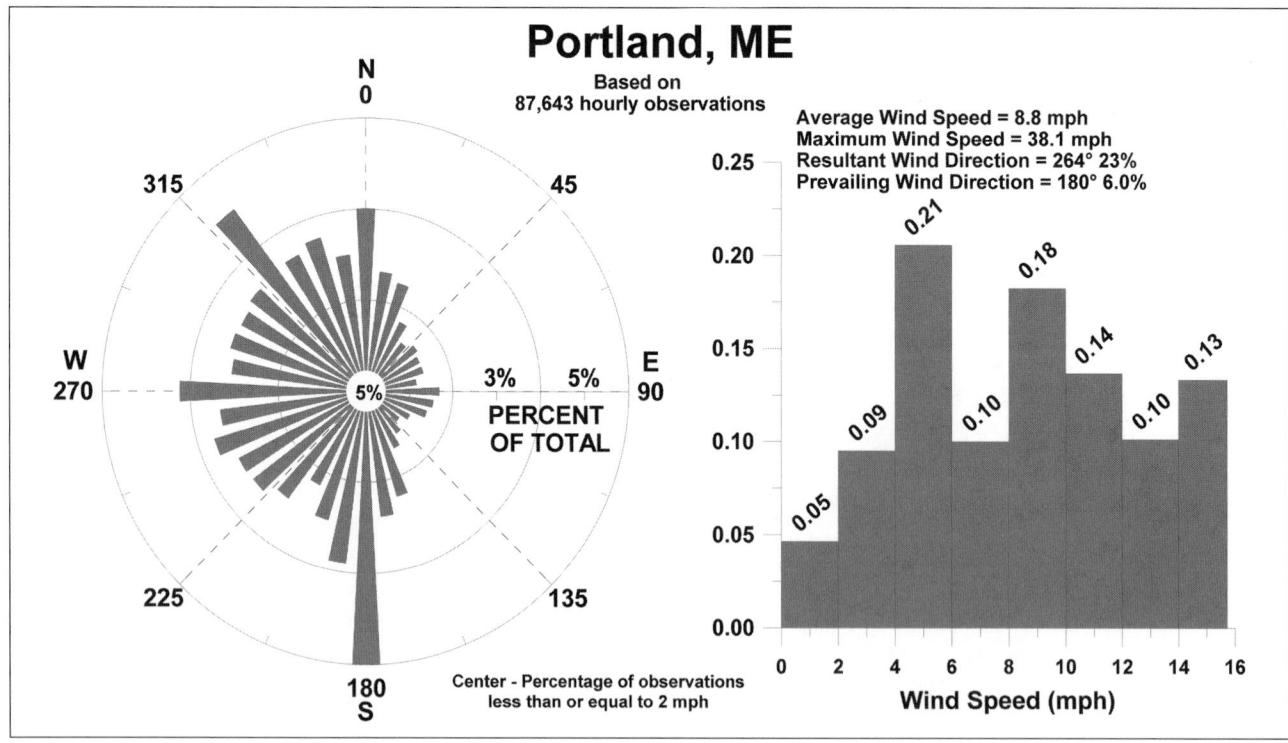

**Figure 2.40** City wind roses and wind speed frequency histogram.

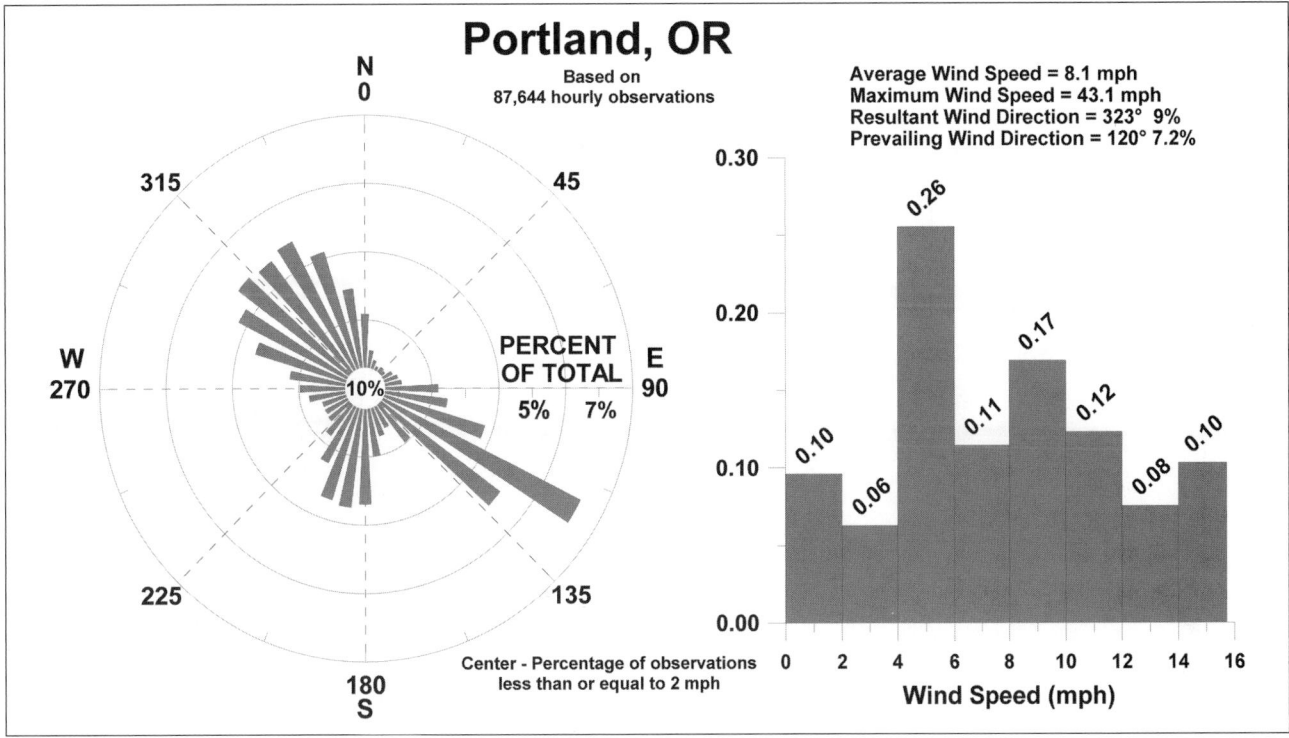

**Figure 2.41**  City wind roses and wind speed frequency histogram.

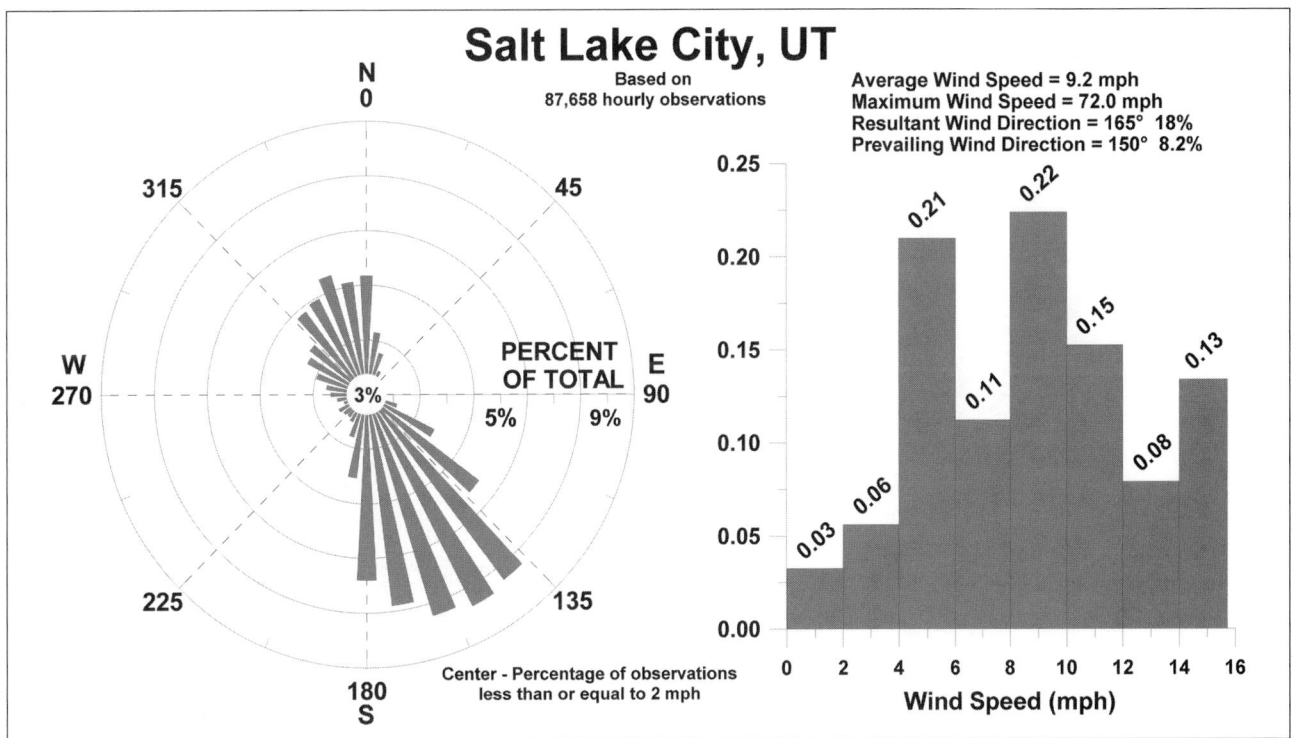

**Figure 2.42**  City wind roses and wind speed frequency histogram.

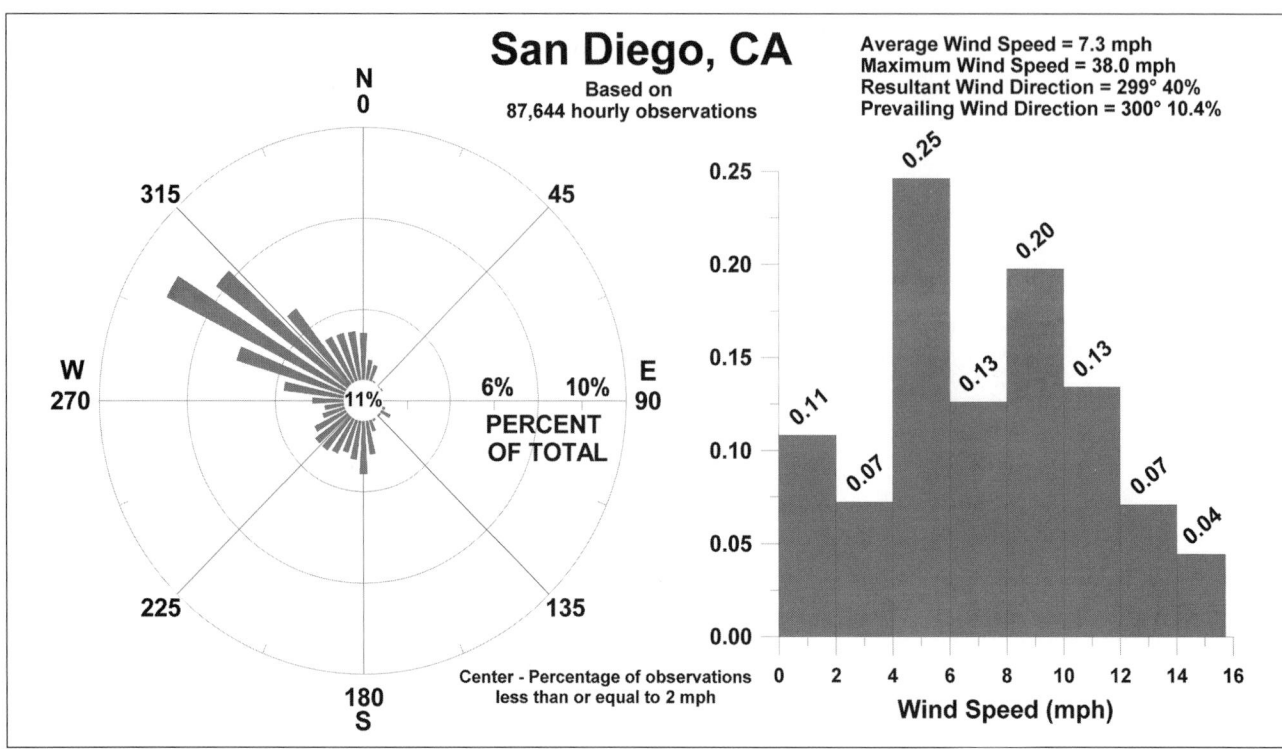

**Figure 2.43** City wind roses and wind speed frequency histogram.

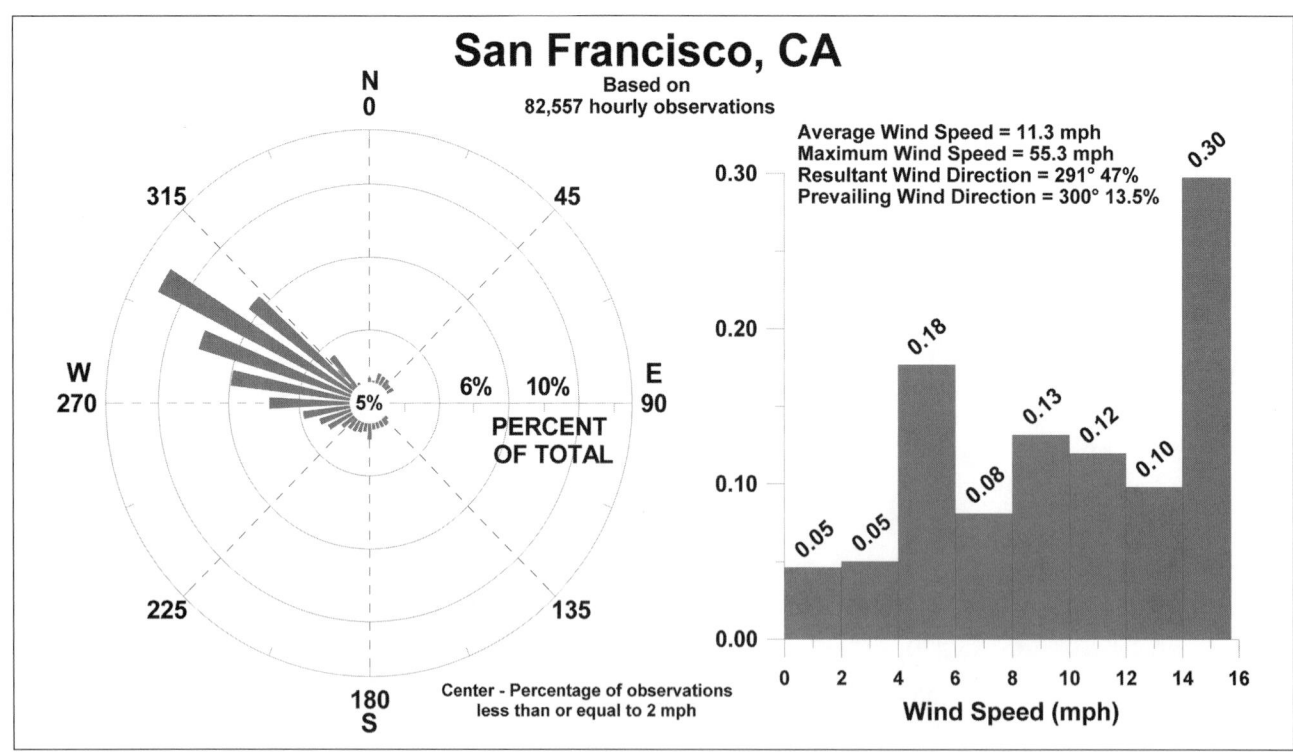

**Figure 2.44** City wind roses and wind speed frequency histogram.

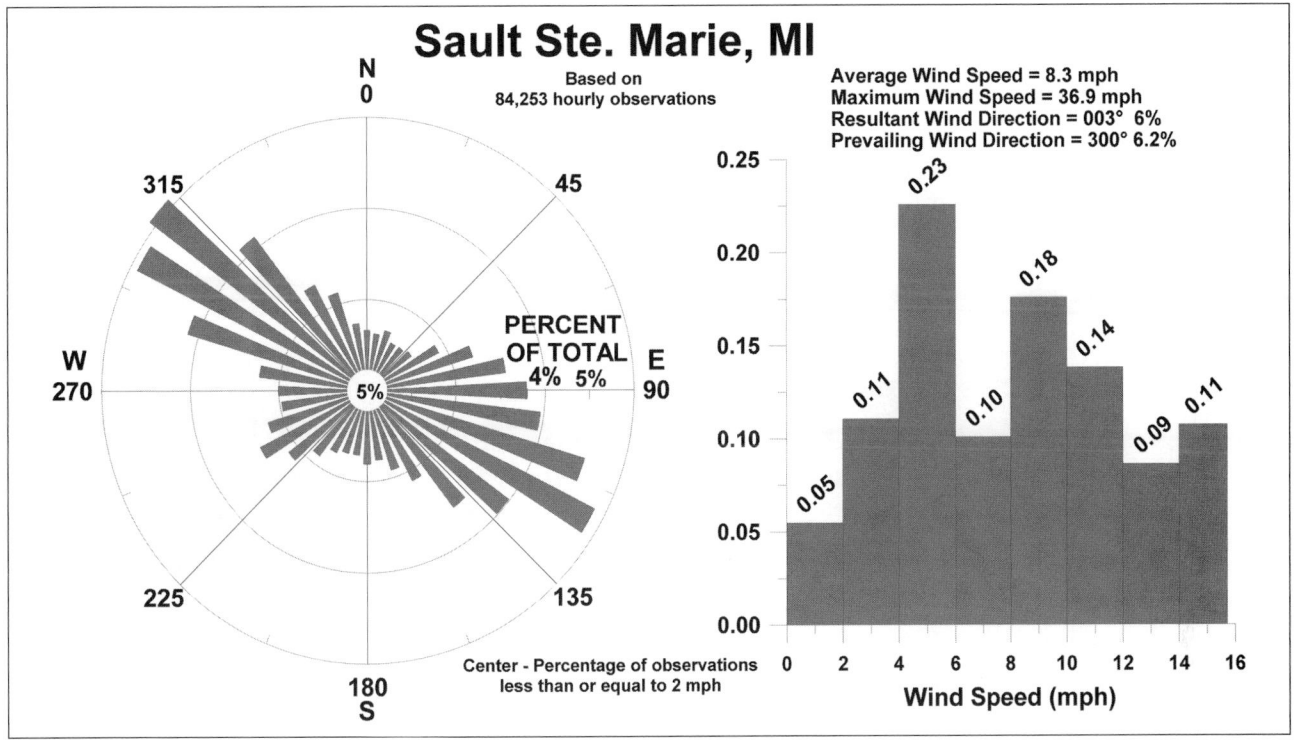

**Figure 2.45**   City wind roses and wind speed frequency histogram.

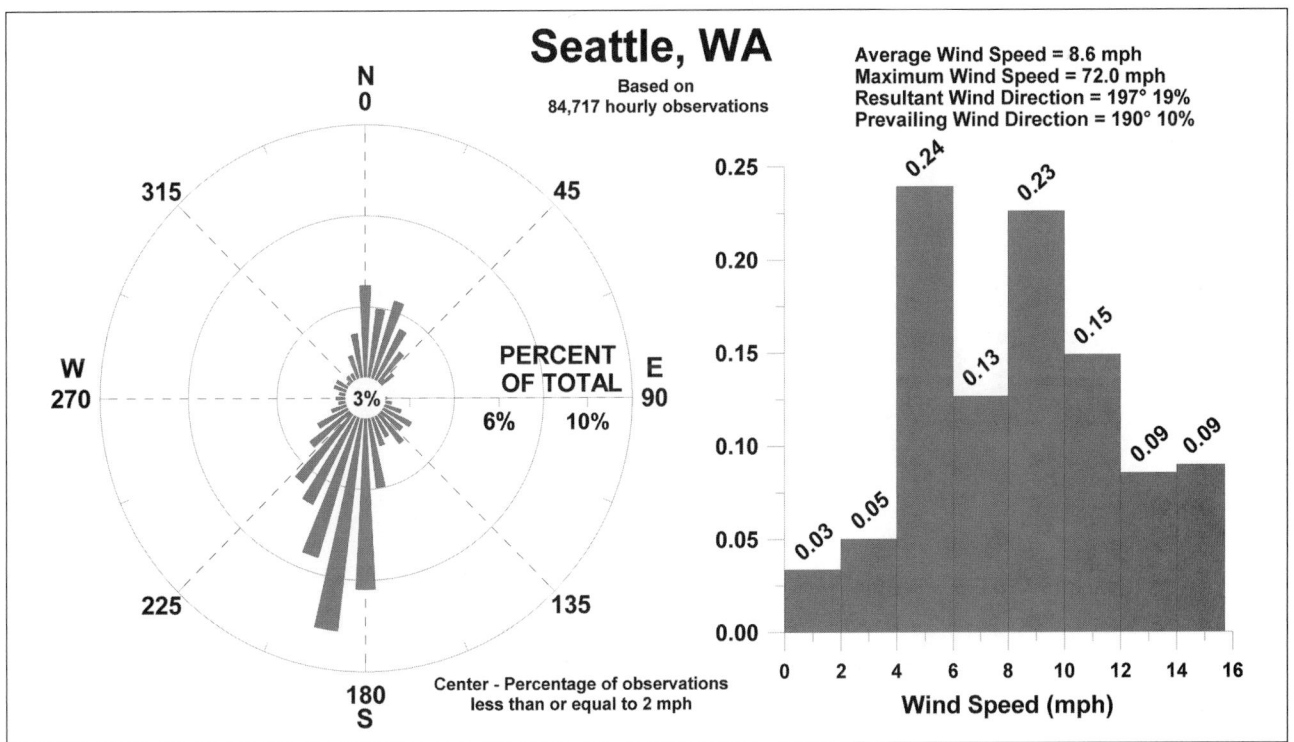

**Figure 2.46**   City wind roses and wind speed frequency histogram.

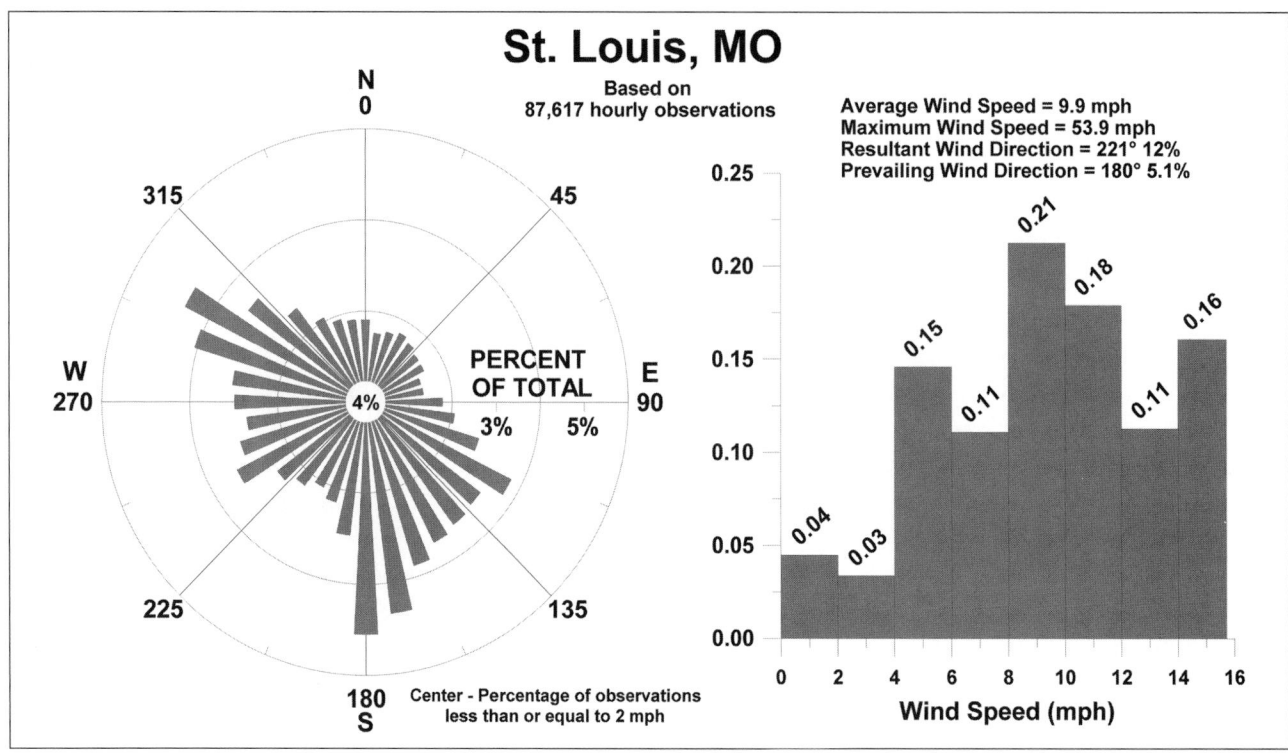

**Figure 2.47**   City wind roses and wind speed frequency histogram.

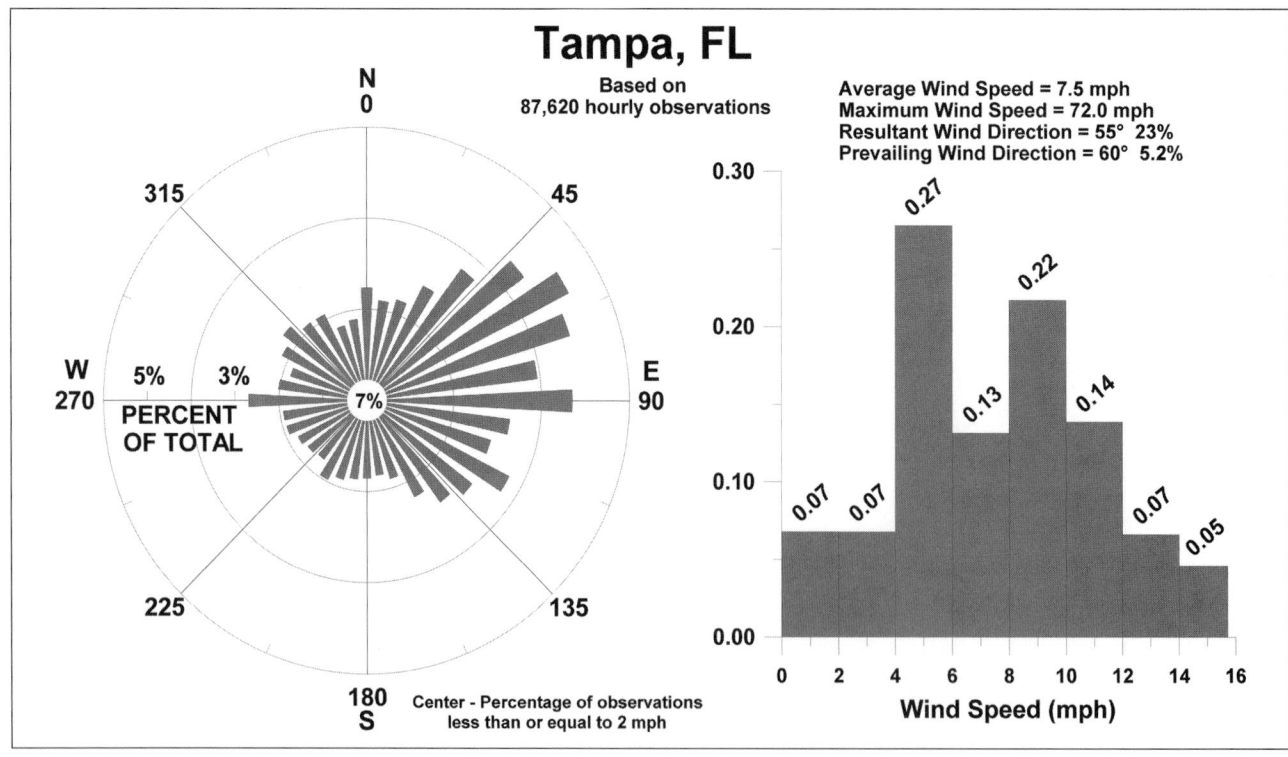

**Figure 2.48**   City wind roses and wind speed frequency histogram.

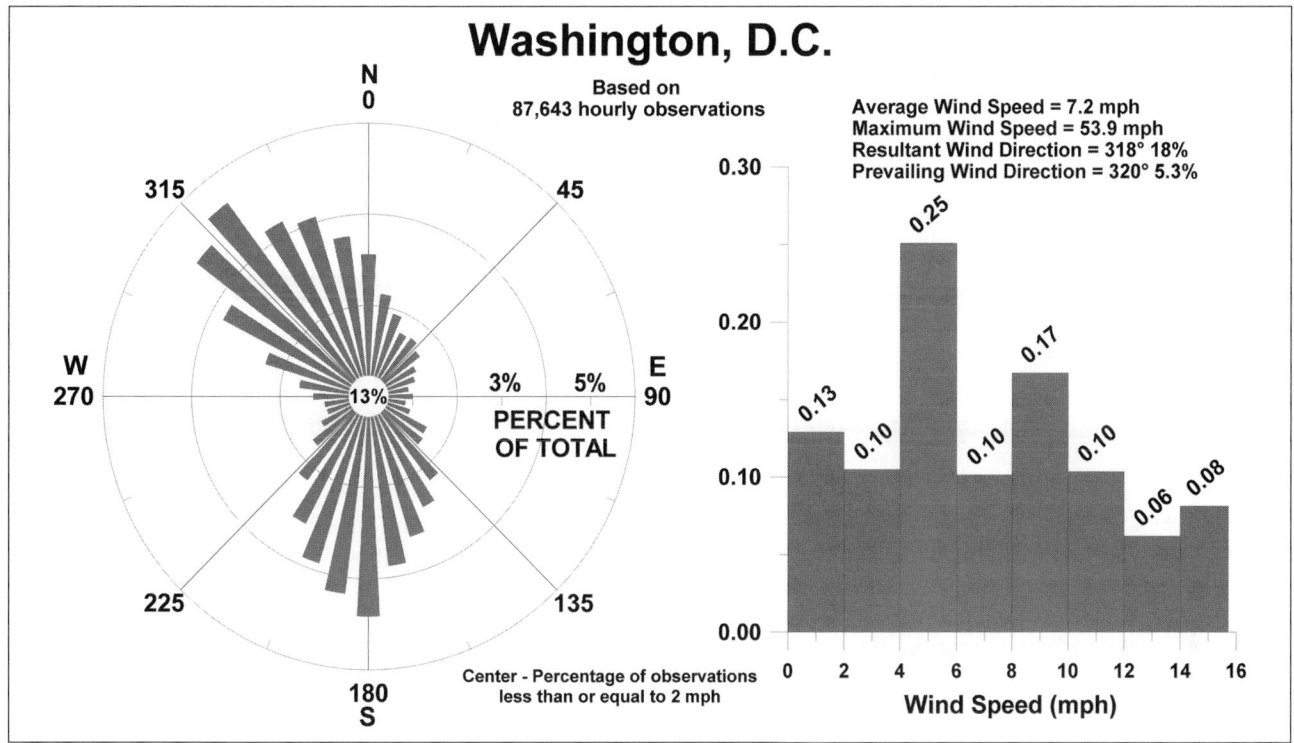

**Figure 2.49**  City wind roses and wind speed frequency histogram.

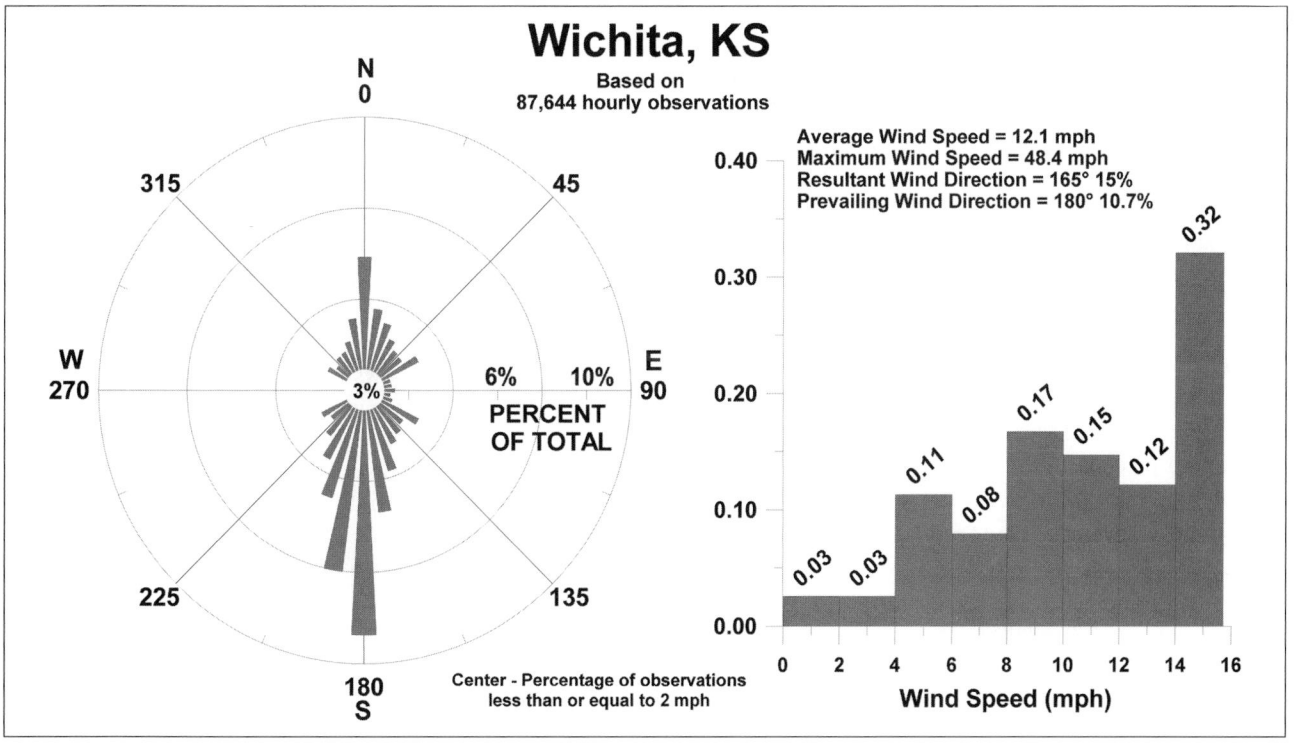

**Figure 2.50**  City wind roses and wind speed frequency histogram.

two factors when using this information: (1) most of these are near-surface velocities, so wind speed at turbine level will be faster, and (2) local variations in exposure and topography can drastically alter the availability of wind power.

Proper estimation of wind power generating potential requires careful application of the wind profile power law. It is also important to remember that detailed field studies are needed to assess the wind power generating potential of a particular location.

## Wind Roses and Wind Speed Frequency Histograms

A wind rose is a circular histogram representing the percentage of winds from all compass directions. Winds are referenced by the direction from which they blow, so when examining a wind rose, the wedge-shaped bars indicate wind blowing toward the center of the circle.

A wind rose with approximately equal-length bars (or petals) from all directions represents a wind environment with great variation of wind direction from day to day. The same goes for the wind speed frequency histogram; the more the bars are close to being equal, the more variable the wind speed.

A wind rose with one or a few long dominant bars represents a location with lower variability, which is usually good for generating wind power if the wind velocity is great enough. Likewise if the wind speed histogram has one or a few dominant bars, the wind speed variation is low. Another way of saying this is that the wind is more constant or reliable.

Each of the 44 wind roses and speed frequency histograms has been calculated from thousands of hourly observations. In most cases, more than 80,000 hourly observations make up the wind roses and wind speed frequency histograms. The data were obtained form NOAA's National Climatic Data Center (NCDC) Solar and Meteorological Surface Observation Network (SAMSON) archive. In this chapter, values from 1981–1990 were used.

Although not a comprehensive atlas of wind direction and wind speed frequency, the 44 sets of wind graphs are widely distributed enough to give a general indication of wind power potential.

### *Using Wind Roses and Wind Speed Frequency Histograms*

At the center of each wind rose, the percentage represents the frequency of calm winds. For the purpose of this chapter, calm is defined as a reported velocity of less than or equal to 2 mph. The histogram represents the frequency of occurrence of wind speeds in the indicated range and is reported as a decimal value above the bar. Two bars may be of slightly different length and have the same decimal value because of rounding.

Above the histogram are four lines of data. Average wind speed is the numerical average of all velocities, including calms. Maximum wind speed is the maximum hourly observation reported in the data set. This is not necessarily the highest wind speed to have occurred at the location; it is merely the highest wind speed observed at the time the hourly observations were made.

Resultant wind direction is the vectorial average of all wind directions and wind speeds reported in the 1981–1990 time period. It was calculated using *WRPLOT View*, © 1998–2008, Lakes Environmental Software. In almost all cases it is different from the prevailing wind direction, which is the most frequently occurring wind category. Neither the resultant wind nor the prevailing wind includes the wind speed only wind direction.

Following both the prevailing wind and the resultant wind is the percentage of time the wind blows from that direction. In the case of the prevailing wind, the percentage represents the percent of time the wind blows from a 10° range centered on the value of the prevailing wind. A prevailing wind entry of 240° 11% means the wind blows between 235° and 245°, 11% of the time.

There are situations when calculating a resultant wind gives almost nonsensical results, for example, if the wind blows with nearly equal frequency from due west and due east. The resultant wind could be from either the north or the south and be of little use. This is caused because the circular nature of the data and the discontinuity as due north is crossed.

Both the prevailing wind direction and the resultant wind direction are less reliable as indicators for wind power as the variability of the wind increases. Prevailing wind may not be a majority wind in highly variable wind direction climates. Resultant wind is a measure of total air movement over a location, and because it is an average, it does not represent an actual wind. Because it is an average, the resultant wind may underrepresent the windiness of an area as winds from opposite directions cancel each other.

An example of this problem is the wind rose for Boise, ID. The wind rose shows a bimodal distribution with two predominant wind directions nearly 180° apart. The resultant wind is from 57°, which is so infrequent that it does not show up on the wind rose. The resultant wind in this situation contains no usable information. The prevailing wind is however obvious from the wind rose, and for wind power generating estimates, prevailing wind direction along with wind speed data stratified by wind direction contains valuable information.

Omaha, NE, is another example of this problem, though not as extreme as Boise.

Los Angeles, CA, is an example where the resultant wind and prevailing wind are close to being the same. The two would be even closer if the wind data were stratified into smaller ranges; in that case the difference could be accounted for by the influence of wind speed in the resultant calculation. Notice that the prevailing wind and resultant wind are in close agreement if there is a single dominant wind direction, in this case from the west-southwest.

Another example like Los Angeles is Dallas/Fort Worth, TX.

A wind rose with a nearly equal distribution of wind directions is not common. New Orleans, LA, comes close. Notice that the resultant wind and prevailing wind differ by 12°. For a wind rose with nearly equal representation of all wind directions prevailing and resultant winds should be nearly the same.

When used together, the prevailing wind and resultant wind can be fairly reliable. The closer the directions and the greater the individual percentages of occurrence, the more reliable the indication of most frequent wind direction.

A very good example of high reliability of the two measures together is the wind rose and histogram of Great Falls, MT. An excellent example of both together being a poor indicator is Fargo, ND. Note how the winds in Great Falls are predominately from the southwest being forced by the terrain to the west. In Fargo, there are two nearly equal dominant directions and the opposites cancel each other to a high degree.

## • SOLAR ENERGY POTENTIAL IN THE UNITED STATES

Solar energy has the potential for producing 20% of the electricity demand in the United States primarily through CSP and PV technologies. Unlike wind power generating potential that is highly dependent on very specific site characteristics, solar energy generating potential relies primarily on exposure. Like wind power generation, solar energy suffers from the intermittency problem. Solar energy is of course not available at night and is severely limited on cloudy days. Haze and other obstructions to visibility can reduce the amount of solar energy available for generating electricity.

CSP technology is almost exclusively a commercial-scale effort. Through the use of concentrating reflectors steam is generated and used to turn a turbine and produce electric current. The cost of such projects puts them well beyond the reach of individual residencies.

PV technology, typically solar panels, is what most people think of when discussing solar energy. It continues to be costly and marginally affordable, but it is expected to benefit from the economies of scale and the modular nature as mass production of solar panels lowers the cost.

In even cloudy northern climates, solar radiation can be a source of electricity using PV technology. The potential for generating electricity is similar over wide geographic areas, being primarily dependent on latitude and cloud cover. Atmospheric transparency is a secondary factor and is affected by anthropogenic pollutants and natural sources of dust. Unlike the potential for generating electricity using wind, local site characteristics do not affect the use of PV technology as drastically.

Maximum solar energy is captured when a PV panel is perpendicular to the sun's incoming rays, which for a specific location varies with time of day and day of the year. Most solar panels are mounted in a fixed position that is calculated to maximize the receipt of solar radiation. This means early and late in the day and at various times of the year the panel will receive less than the maximum possible for that time period.

Solar panel tilt can vary when panels are placed on single- and double-axis mounts using a solar tracker. Single-axis mounts follow the sun through the day to keep the angle at which the solar radiation strikes the panel as near to 90° as possible. Most single-axis mounts have another axis that is adjusted periodically manually to compensate for time of year. Solar panels on double-axis mounts also move, adjusting for time of day and day of year. Single-axis mounts generally output 30% more energy than fixed mounts. Adding a second axis yields an additional 6% beyond a single-axis mount.

Figures 2.51–2.94 are graphs for 44 cities in the contiguous 48 states in this chapter plotted from data provided by the US Department of Energy's NREL. They are a good starting point for evaluating the solar energy potential for most of the lower 48 states.

These graphs are the average daily direct normal solar radiation from 1998 to 2005 in kilowatt-hours (kWh) per square meter per day. One kilowatt-hour is approximately equal to 3412 British thermal units and 860 kilocalories.

Direct solar radiation excludes scattered radiation and is only that which arrives directly from the sun. The word "normal" in the title means that it is measured for a flat plate oriented at 90°, or normal, to the sun's rays. A fixed solar panel will receive less than indicated on the graph because the sun angle varies continuously through the day. A solar-receiving system that tracks the sun will remain perpendicular to the sun's rays through the day and can be expected to receive the indicated amount.

The graphs then represent the maximum energy that can be expected on a given day of the year. Because these are based on 8 years of measured solar energy and not calculated maximum amounts they represent a realistic estimate.

The heavy line is the average daily direct solar radiation, and the upper and lower lines give the values for +1 and −1 standard deviations. Approximately 68% of the time the solar radiation for a day will fall between the upper and lower lines.

In addition to the average normal solar radiation graphs, there are three maps (Figures 2.95–2.97) the 48 contiguous states that summarize the annual solar radiation climate and annual cloudiness.

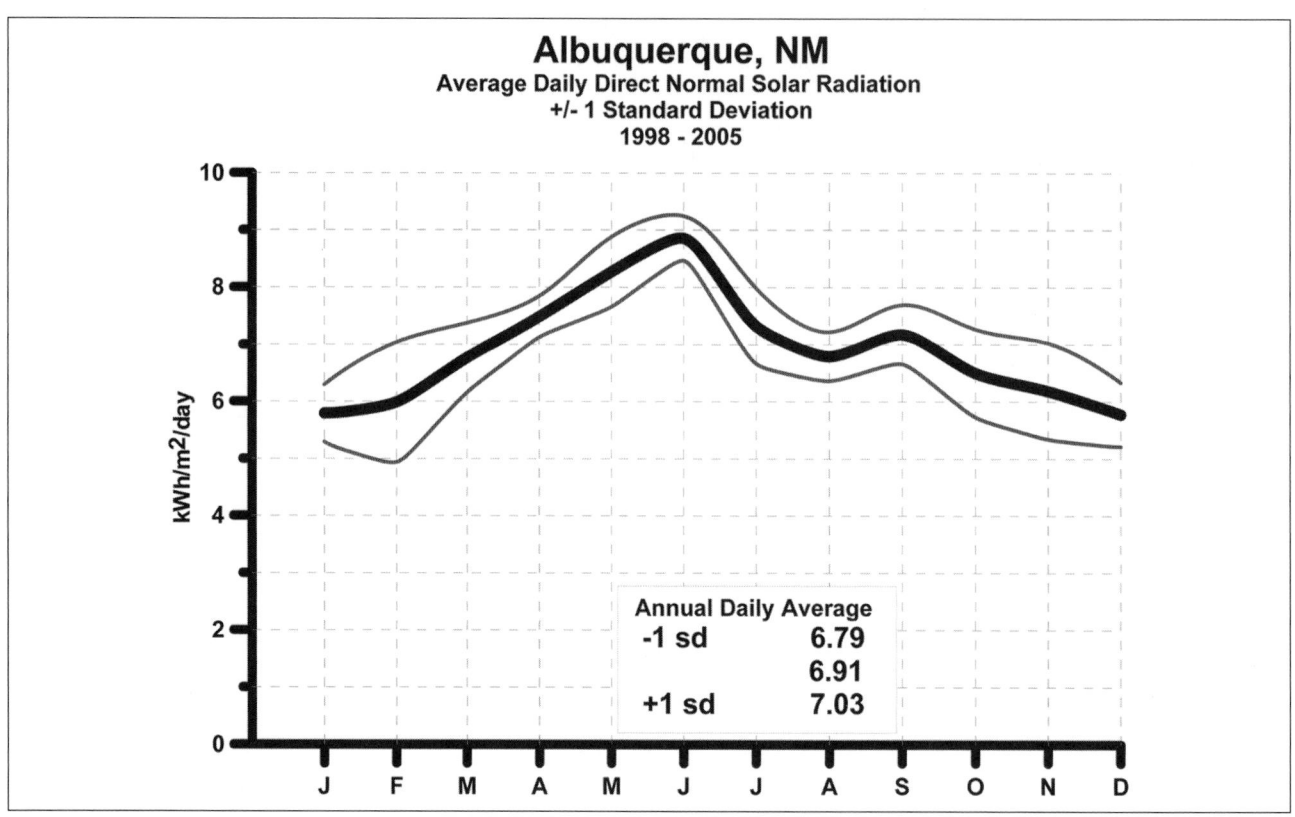

**Figure 2.51**   Solar energy graph.

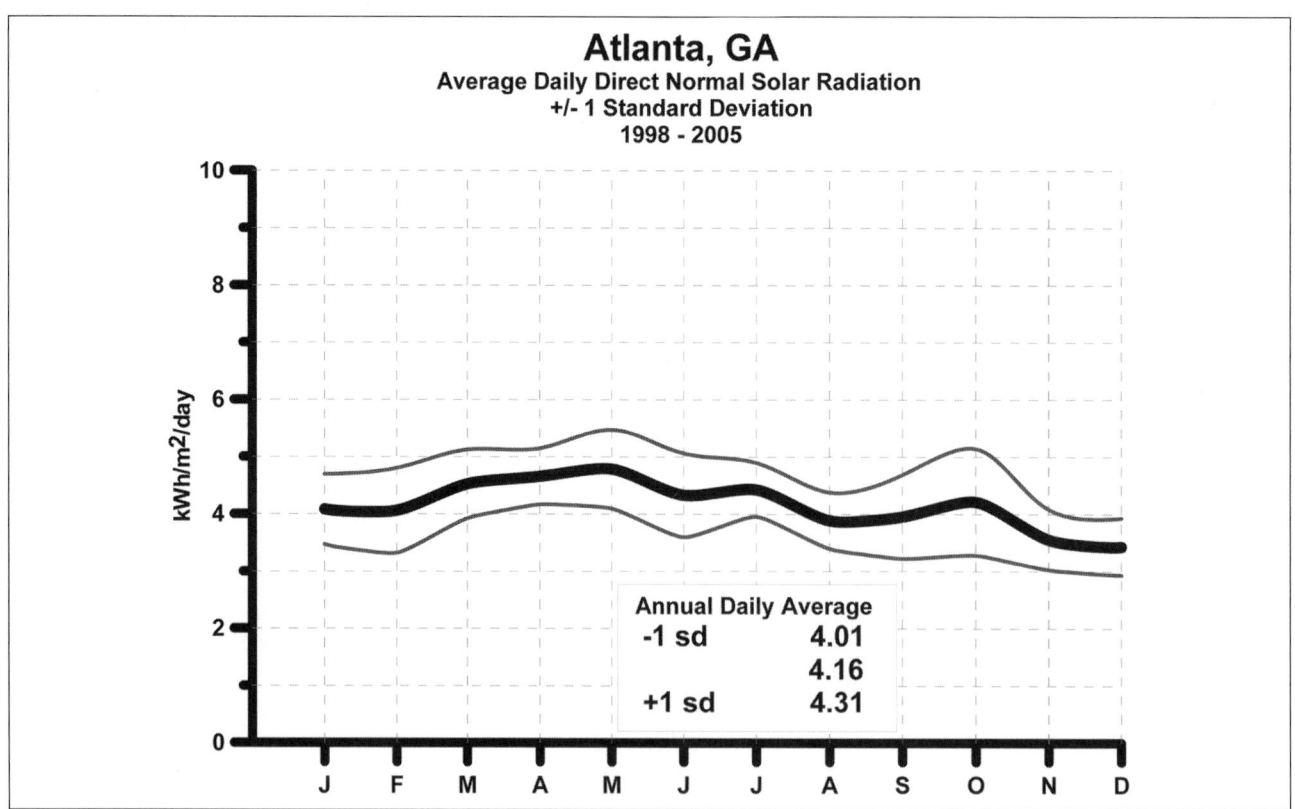

**Figure 2.52**   Solar energy graph.

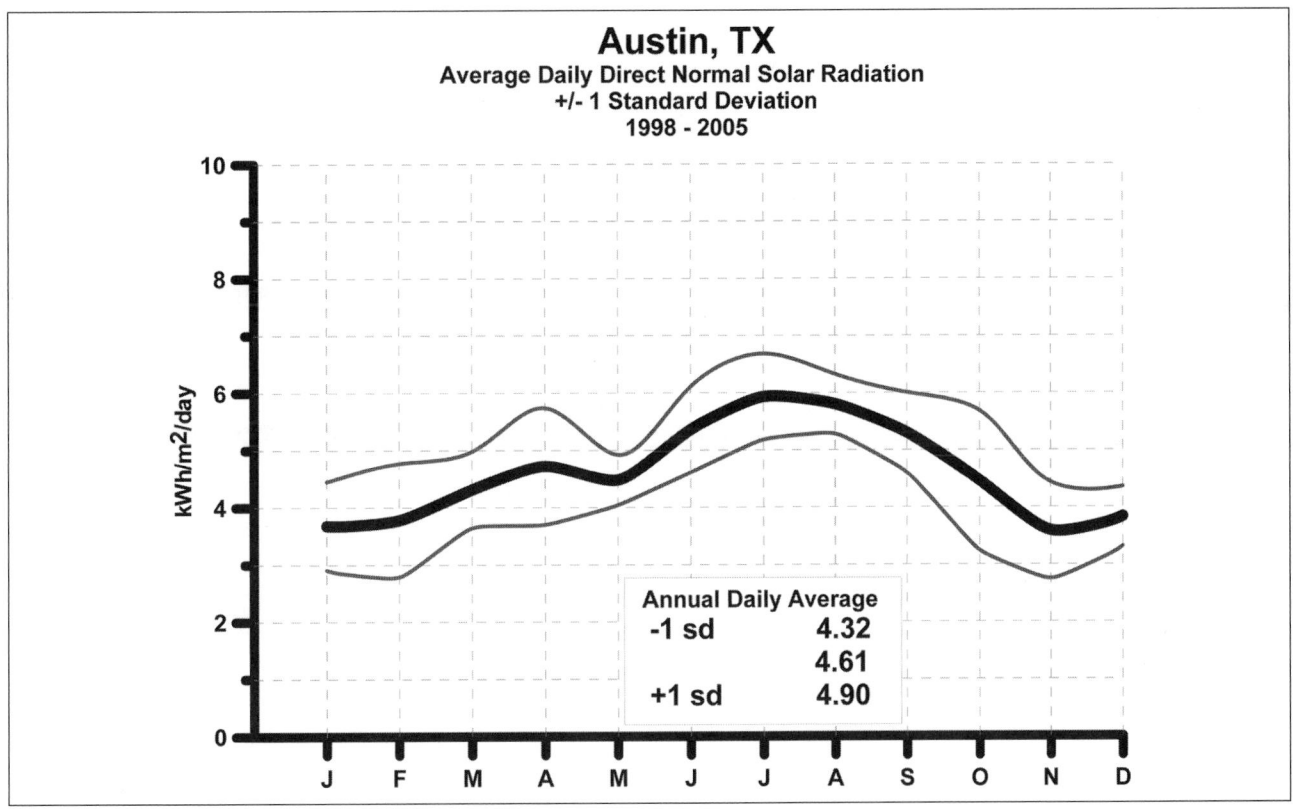

**Figure 2.53** Solar energy graph.

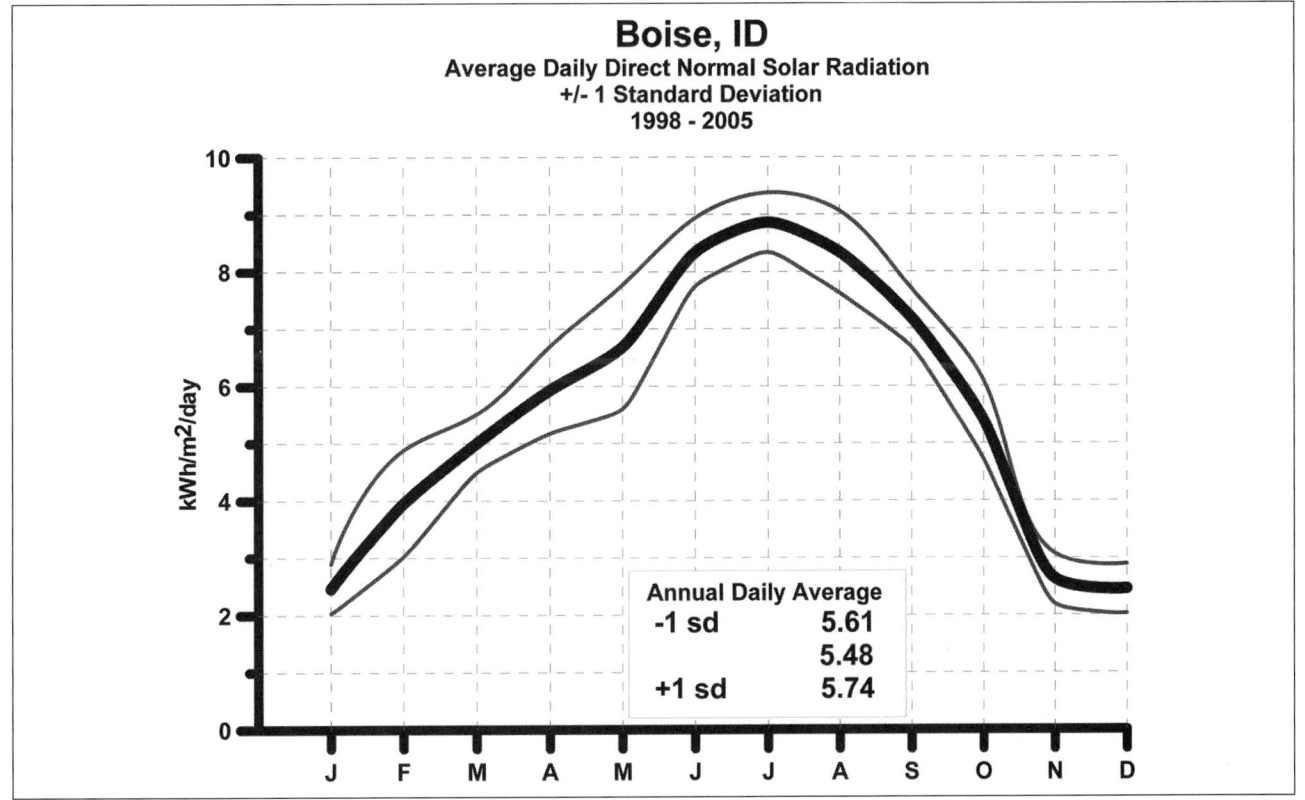

**Figure 2.54** Solar energy graph.

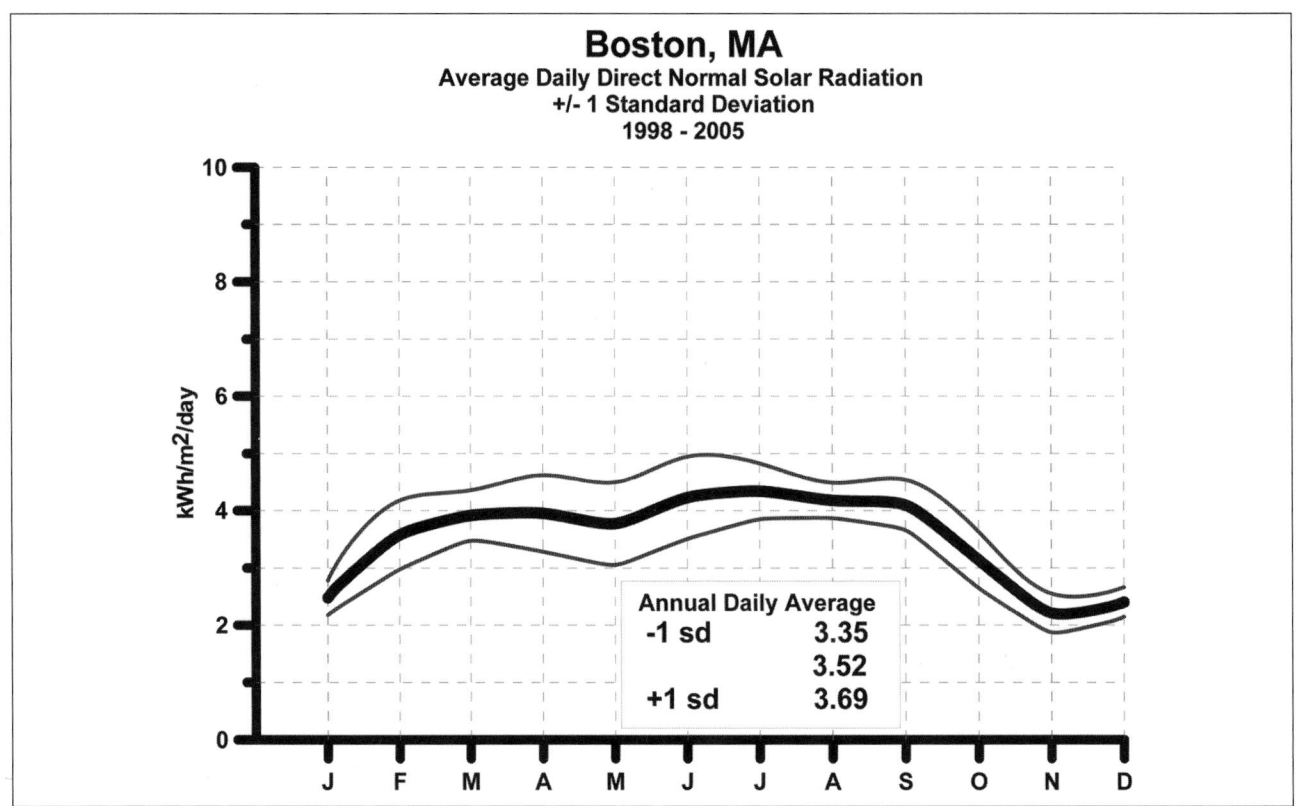

**Figure 2.55**   Solar energy graph.

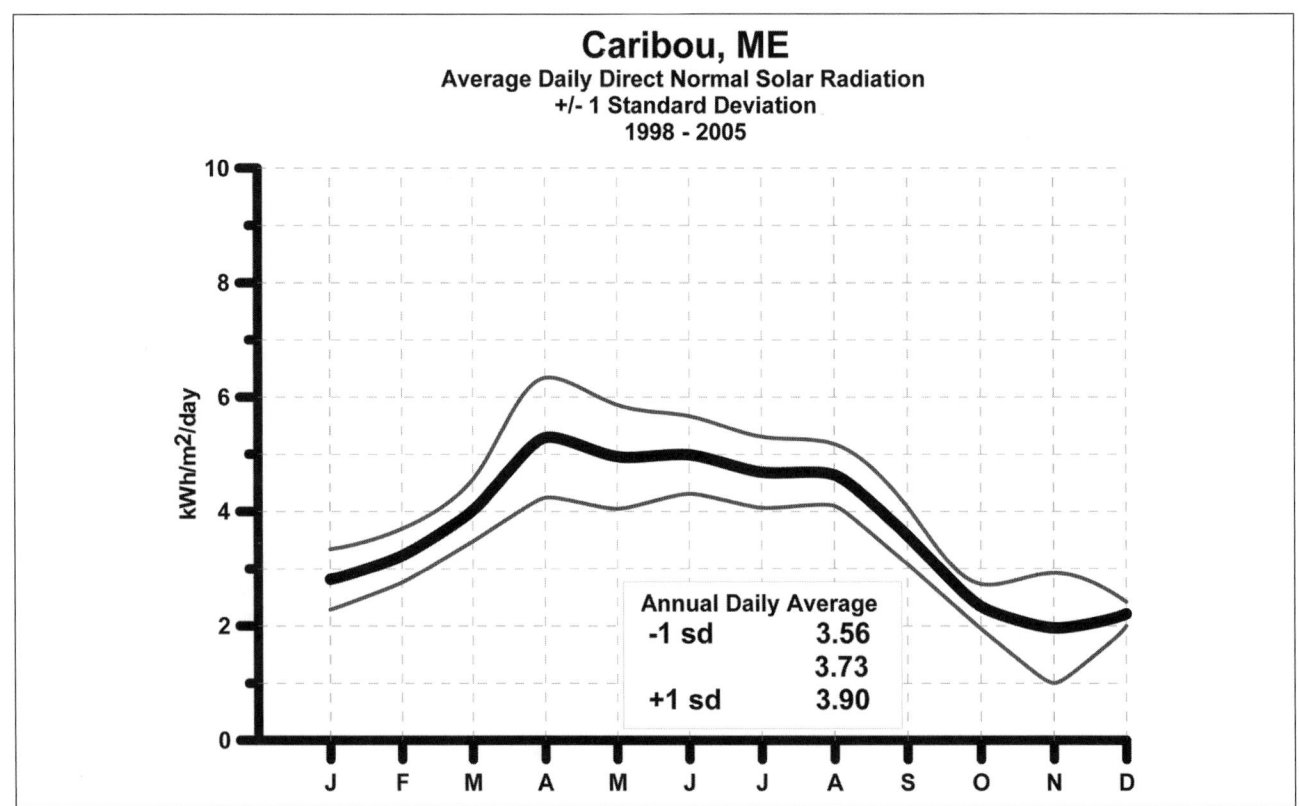

**Figure 2.56**   Solar energy graph.

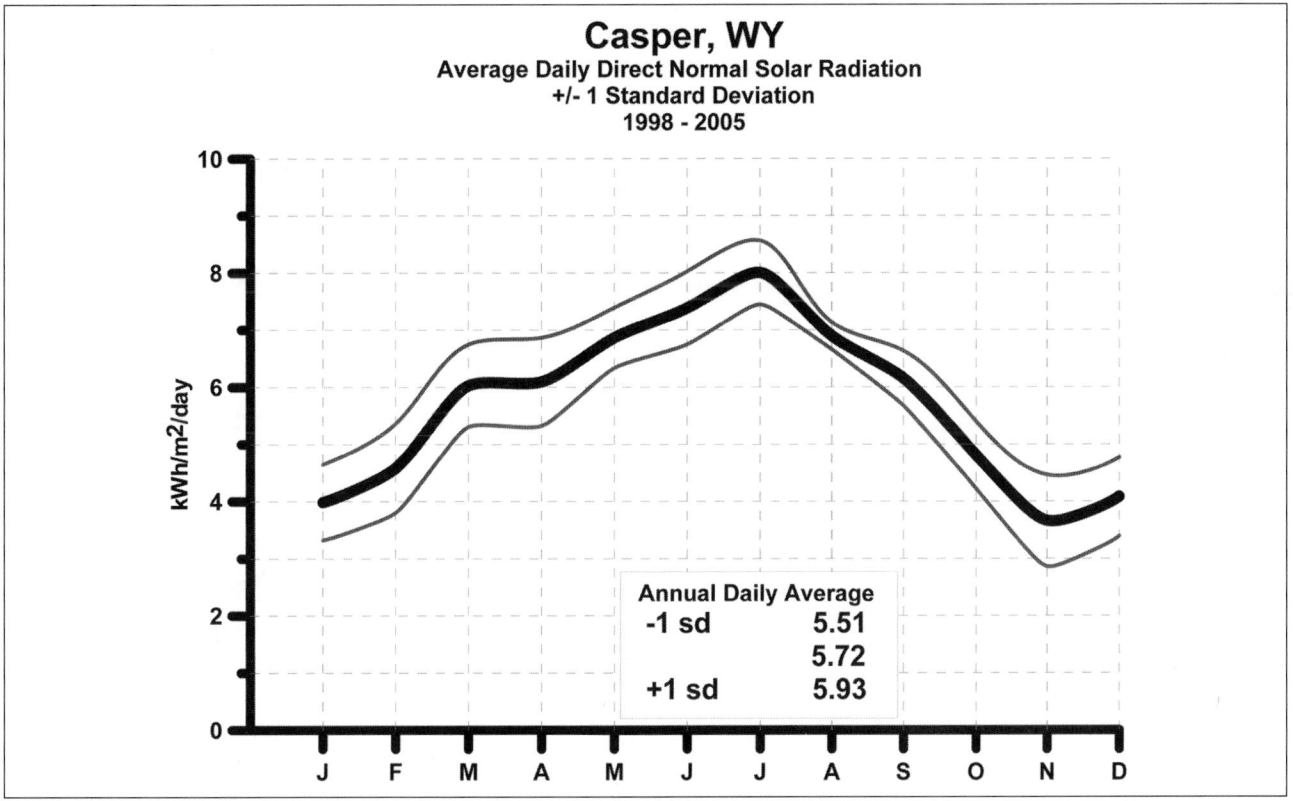

**Figure 2.57**   Solar energy graph.

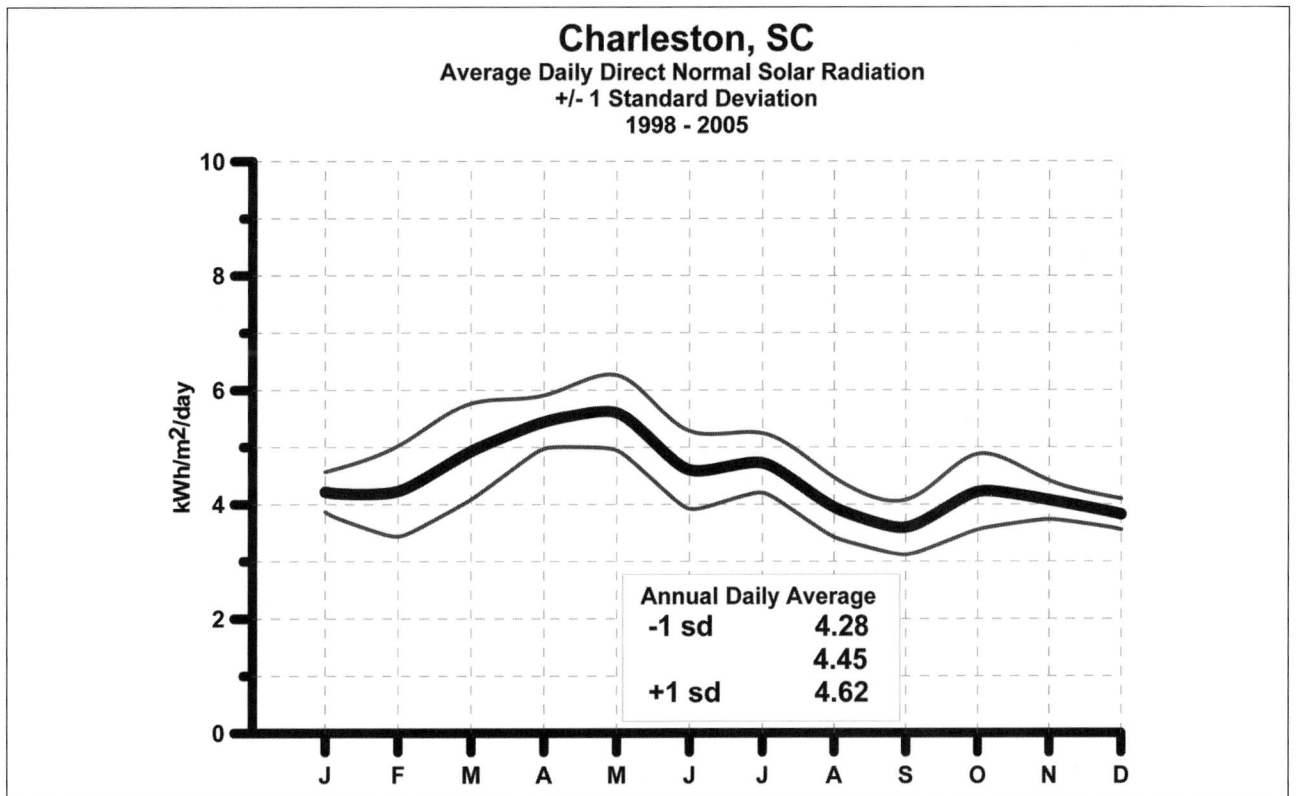

**Figure 2.58**   Solar energy graph.

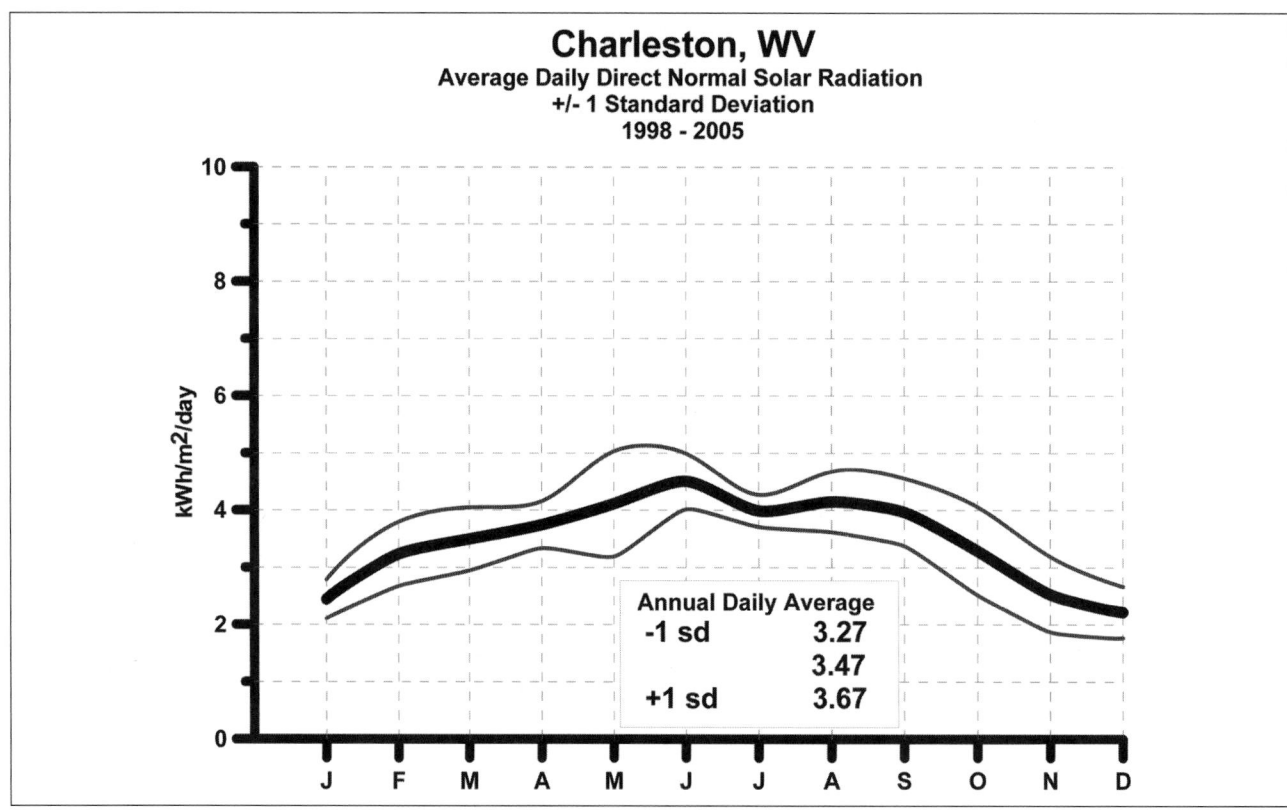

**Figure 2.59** Solar energy graph.

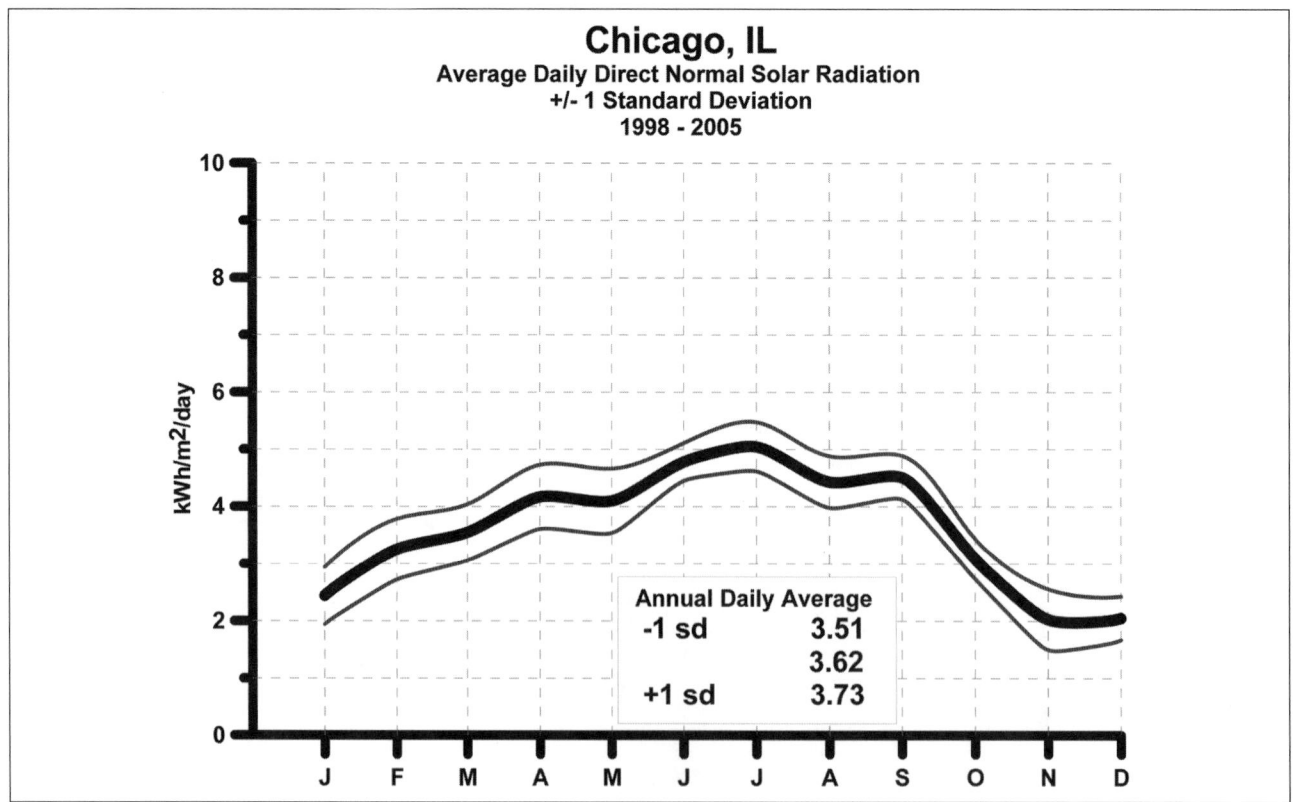

**Figure 2.60** Solar energy graph.

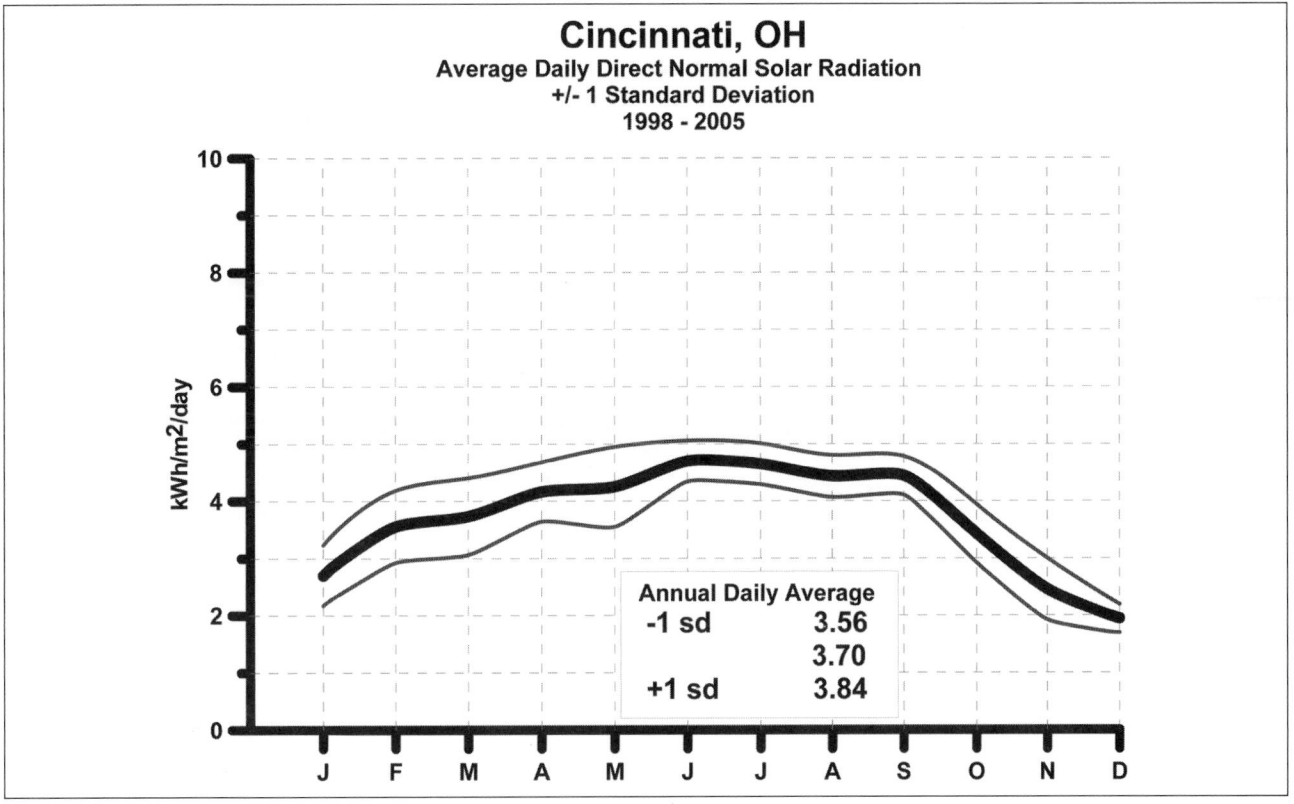

**Figure 2.61**   Solar energy graph.

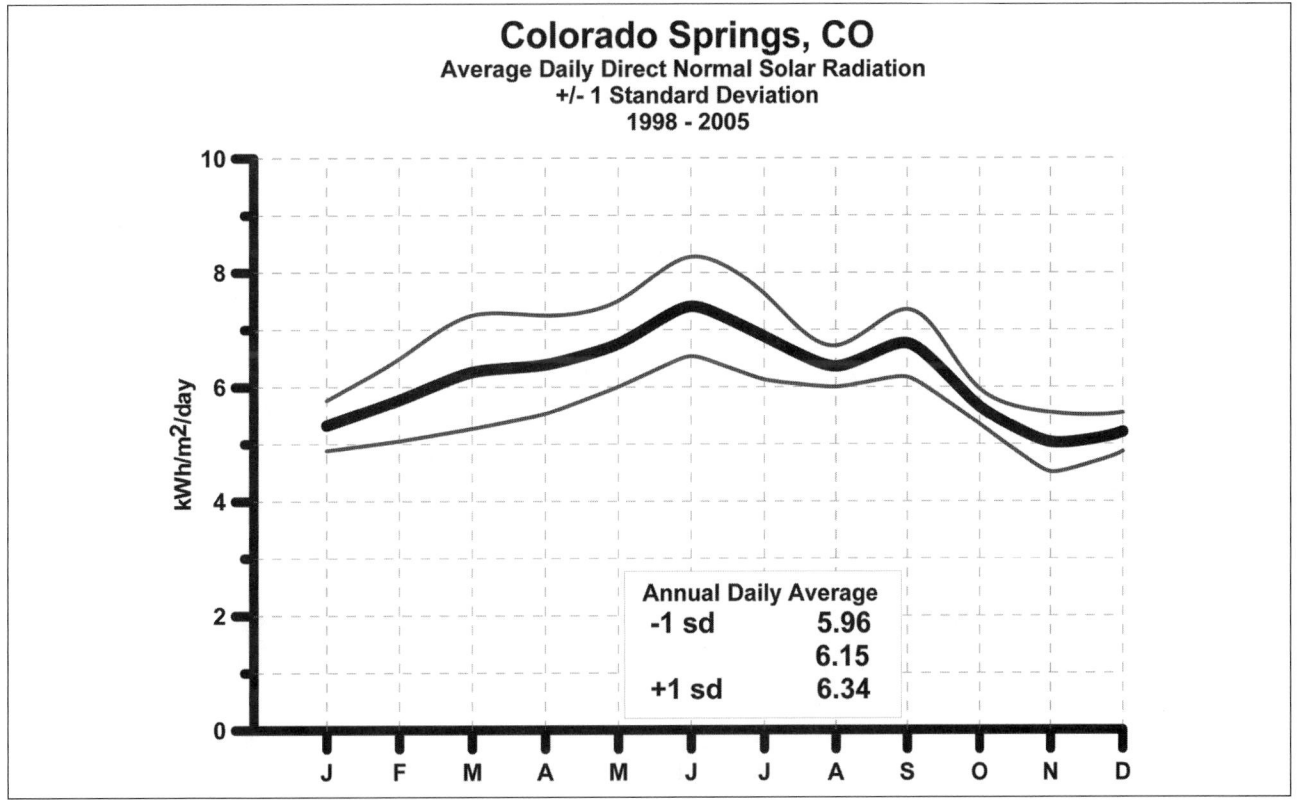

**Figure 2.62**   Solar energy graph.

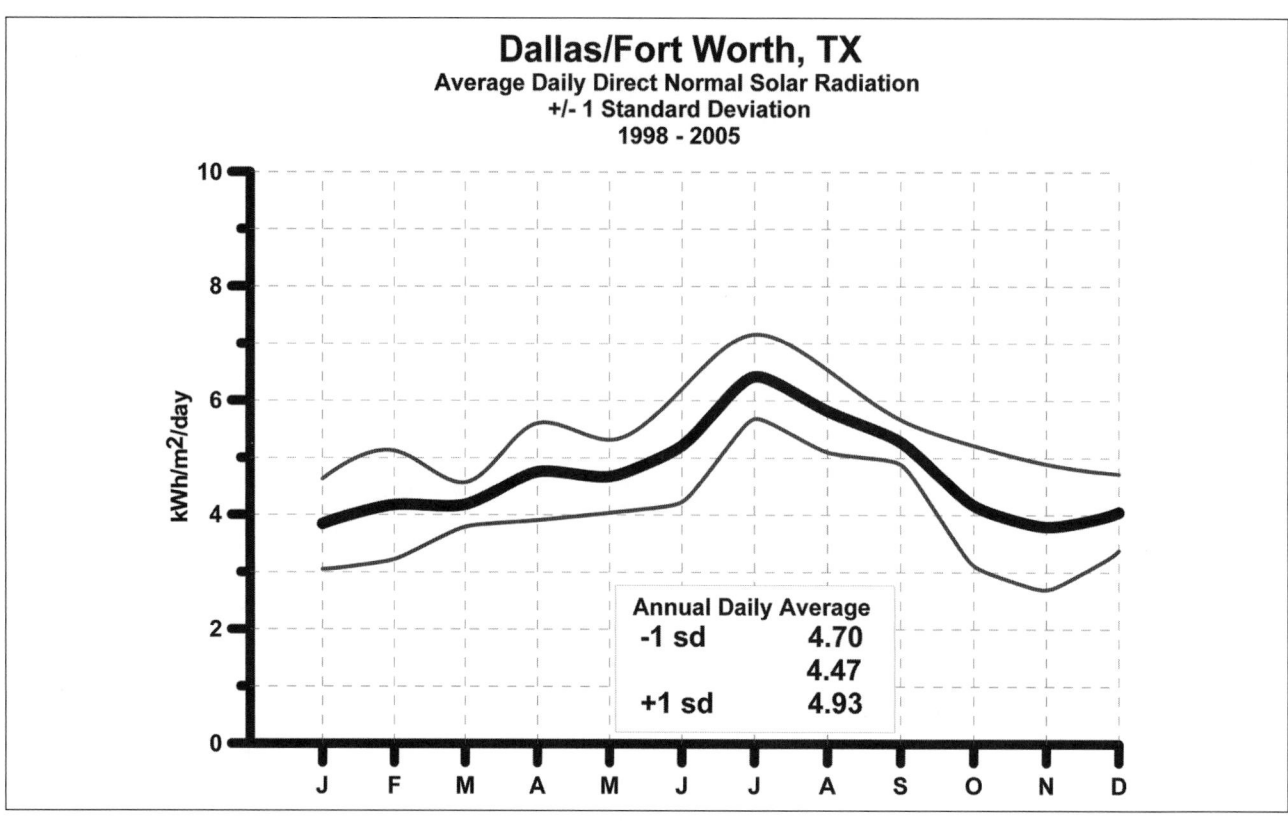

**Figure 2.63**   Solar energy graph.

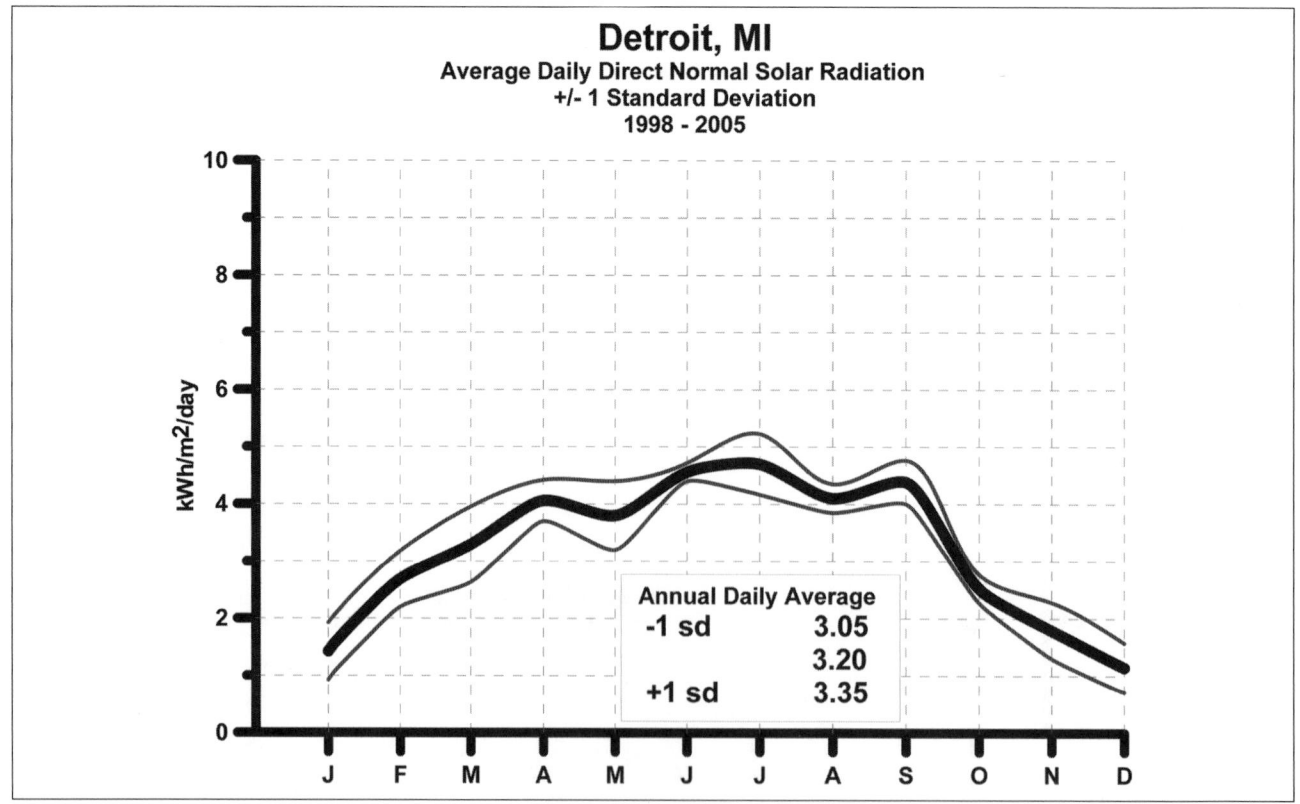

**Figure 2.64**   Solar energy graph.

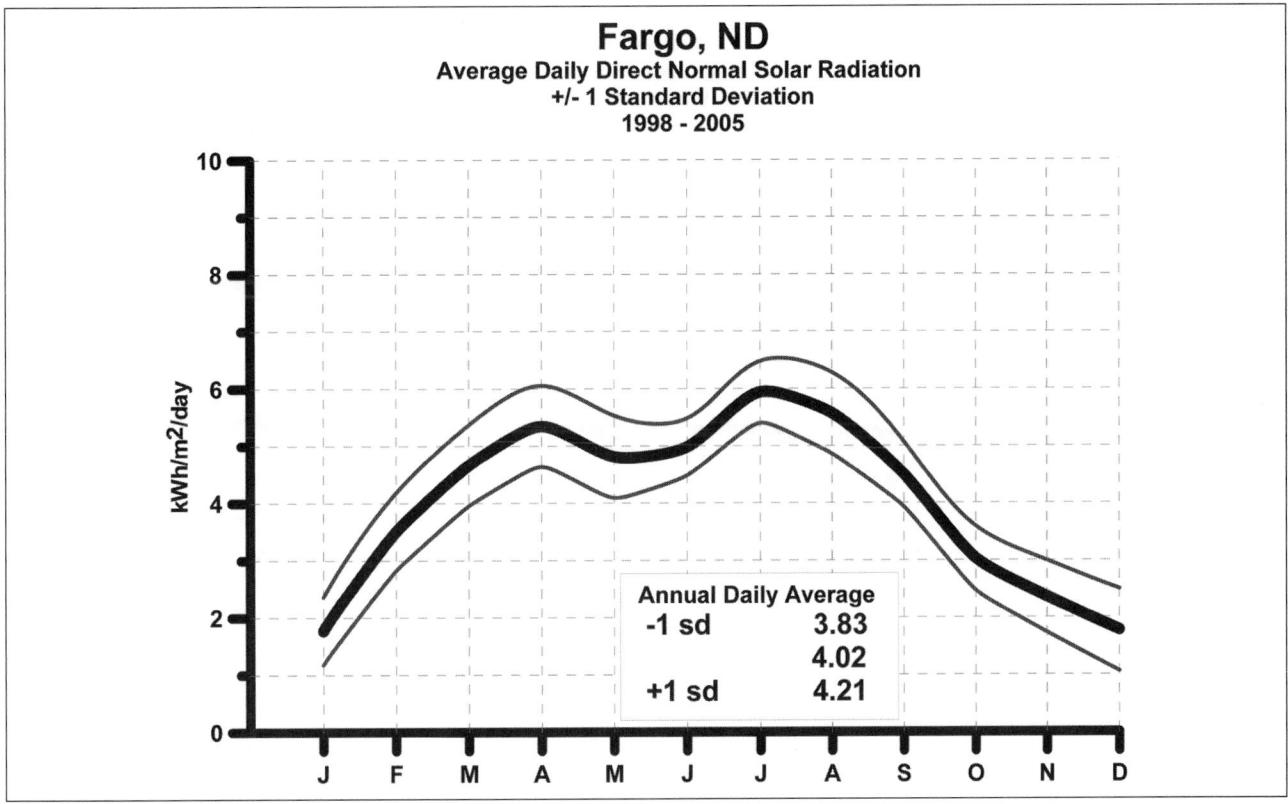

**Figure 2.65**   Solar energy graph.

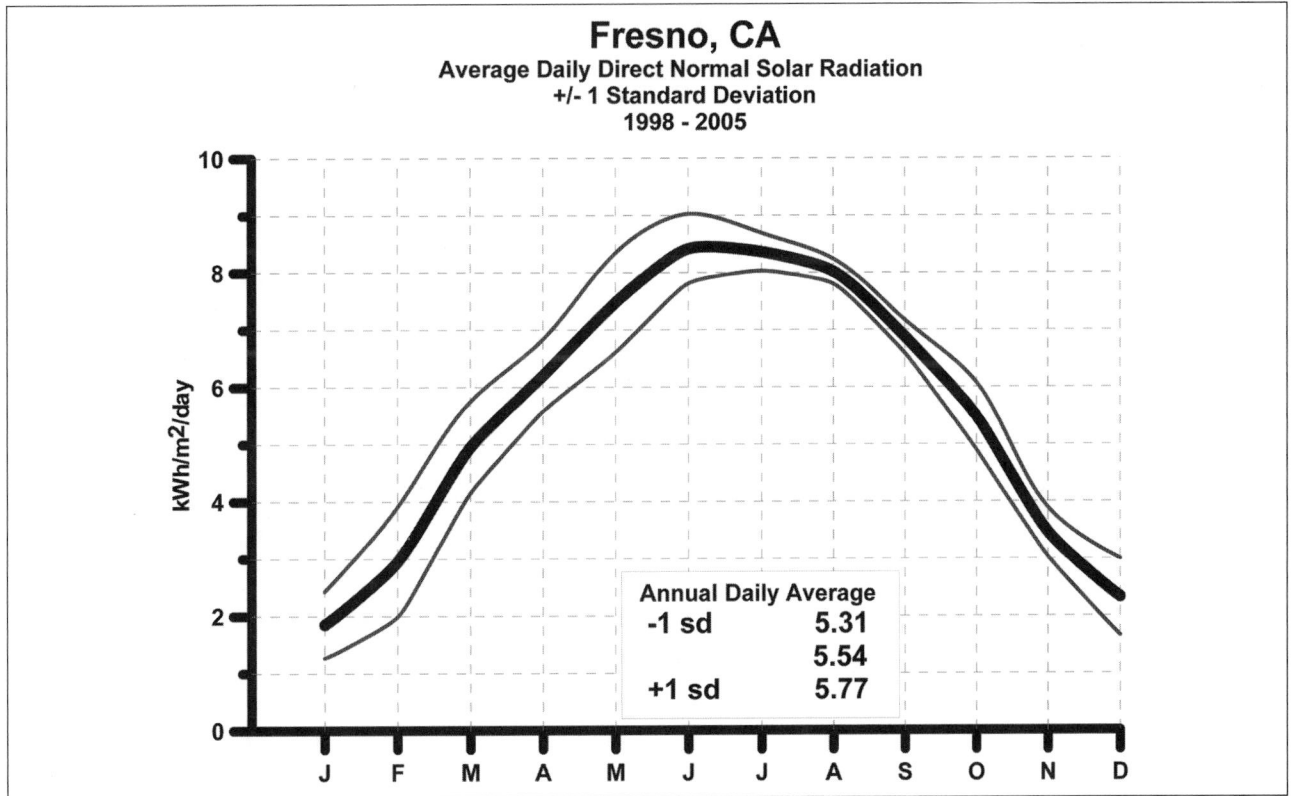

**Figure 2.66**   Solar energy graph.

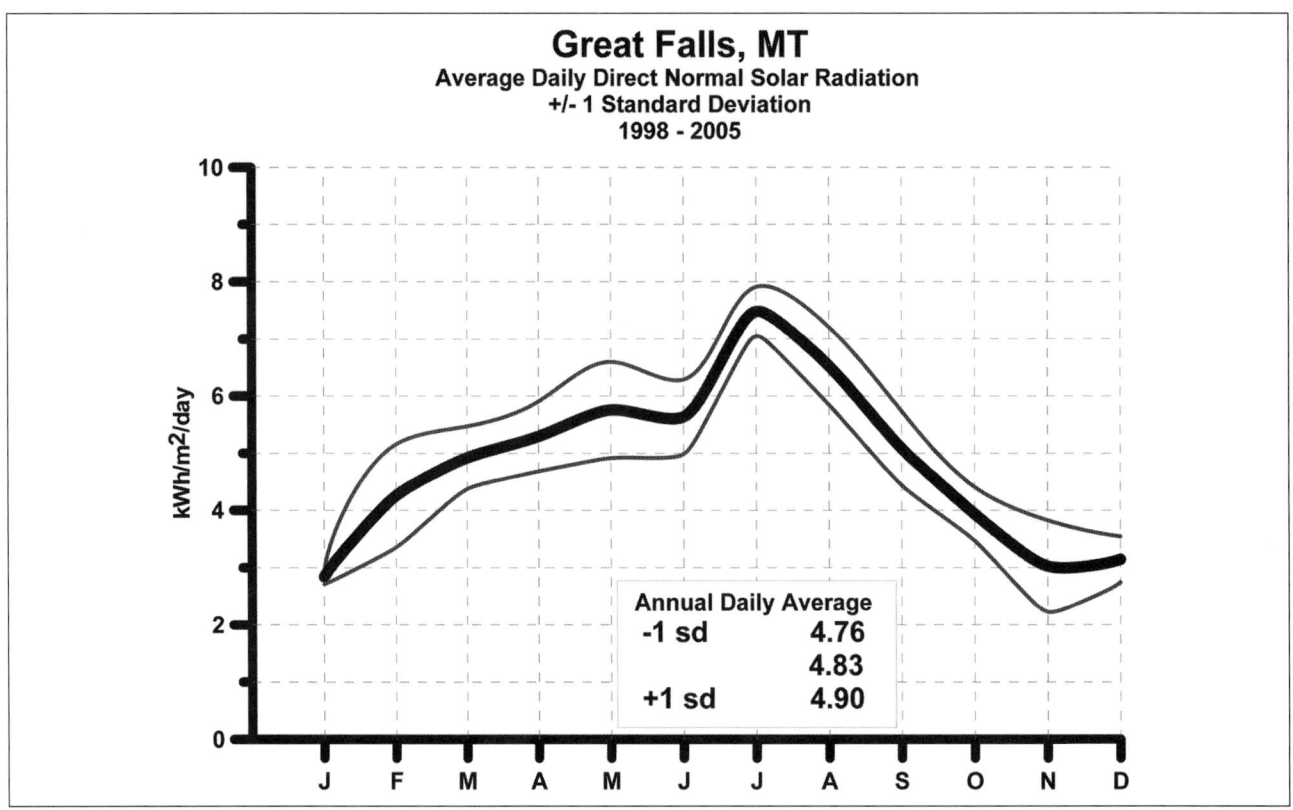

**Figure 2.67**   Solar energy graph.

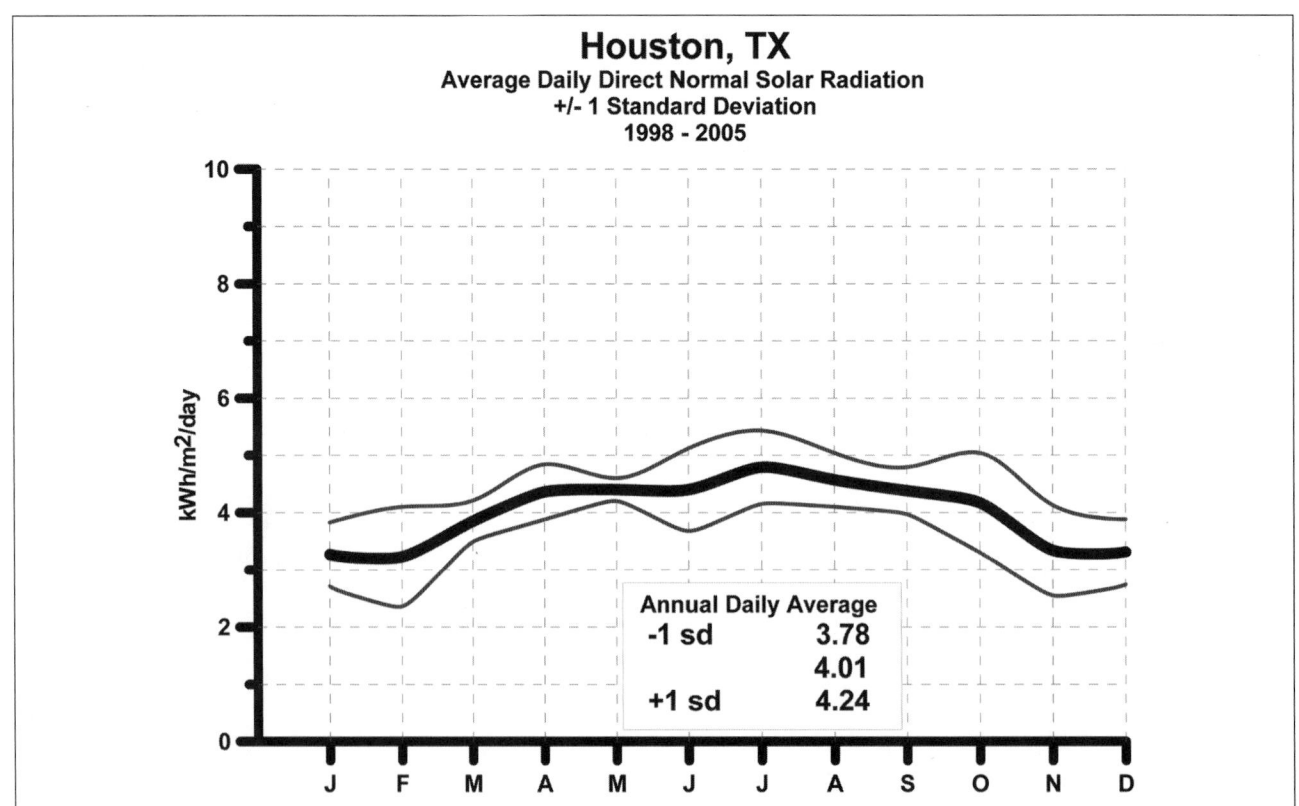

**Figure 2.68**   Solar energy graph.

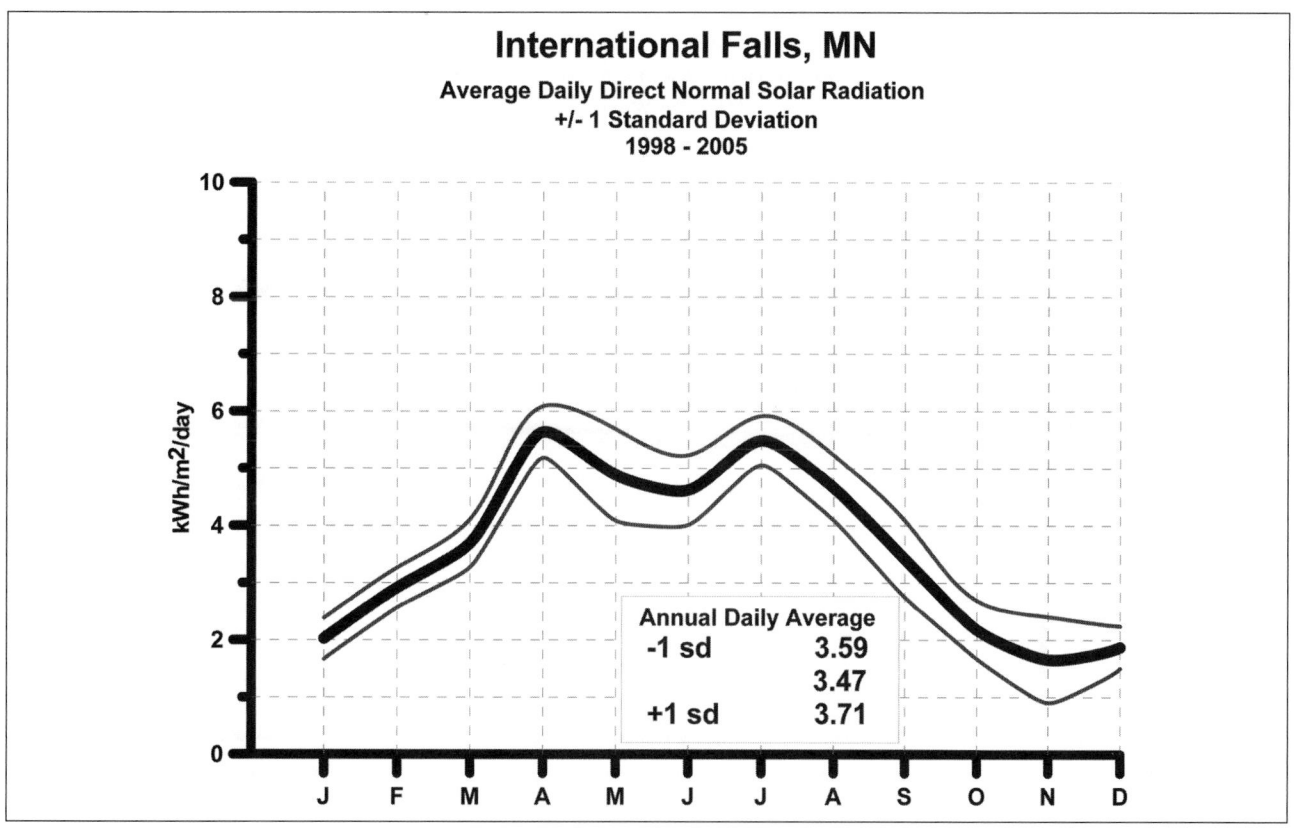

**Figure 2.69**    Solar energy graph.

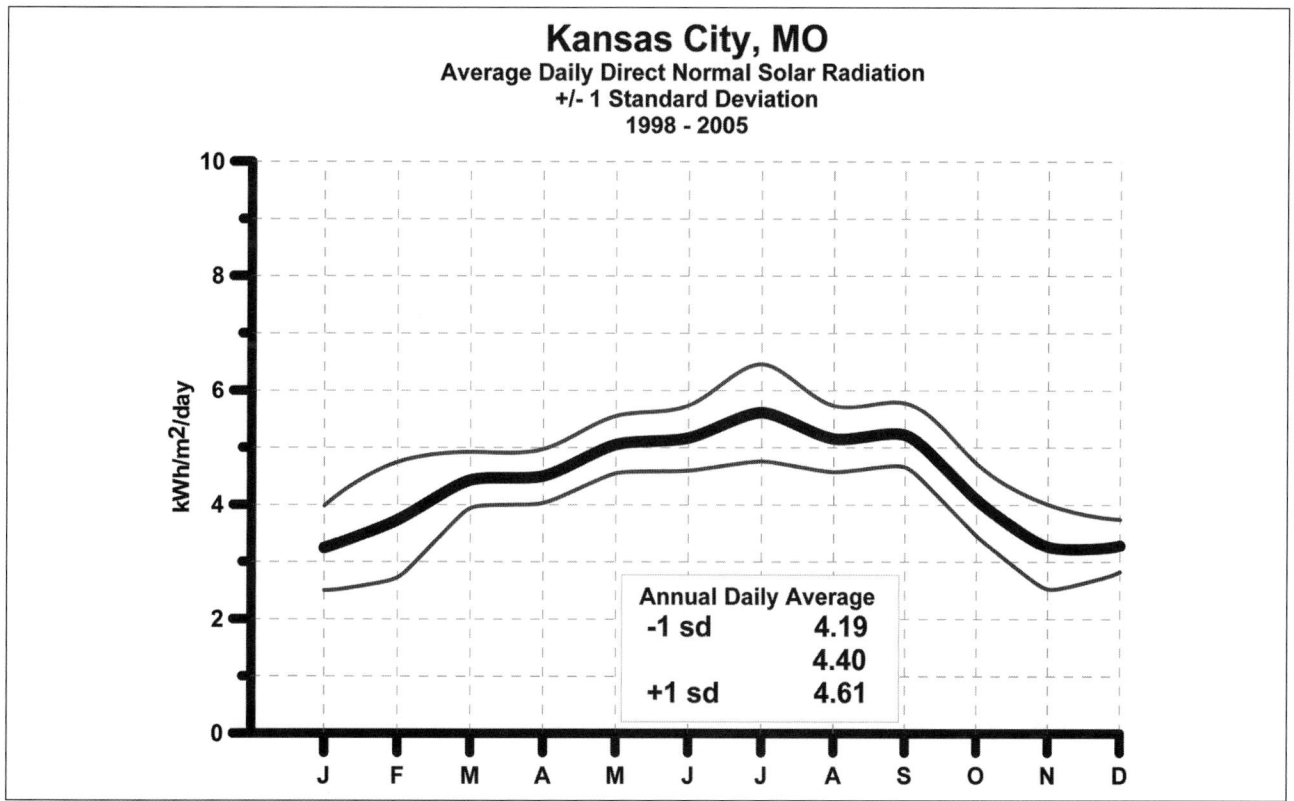

**Figure 2.70**    Solar energy graph.

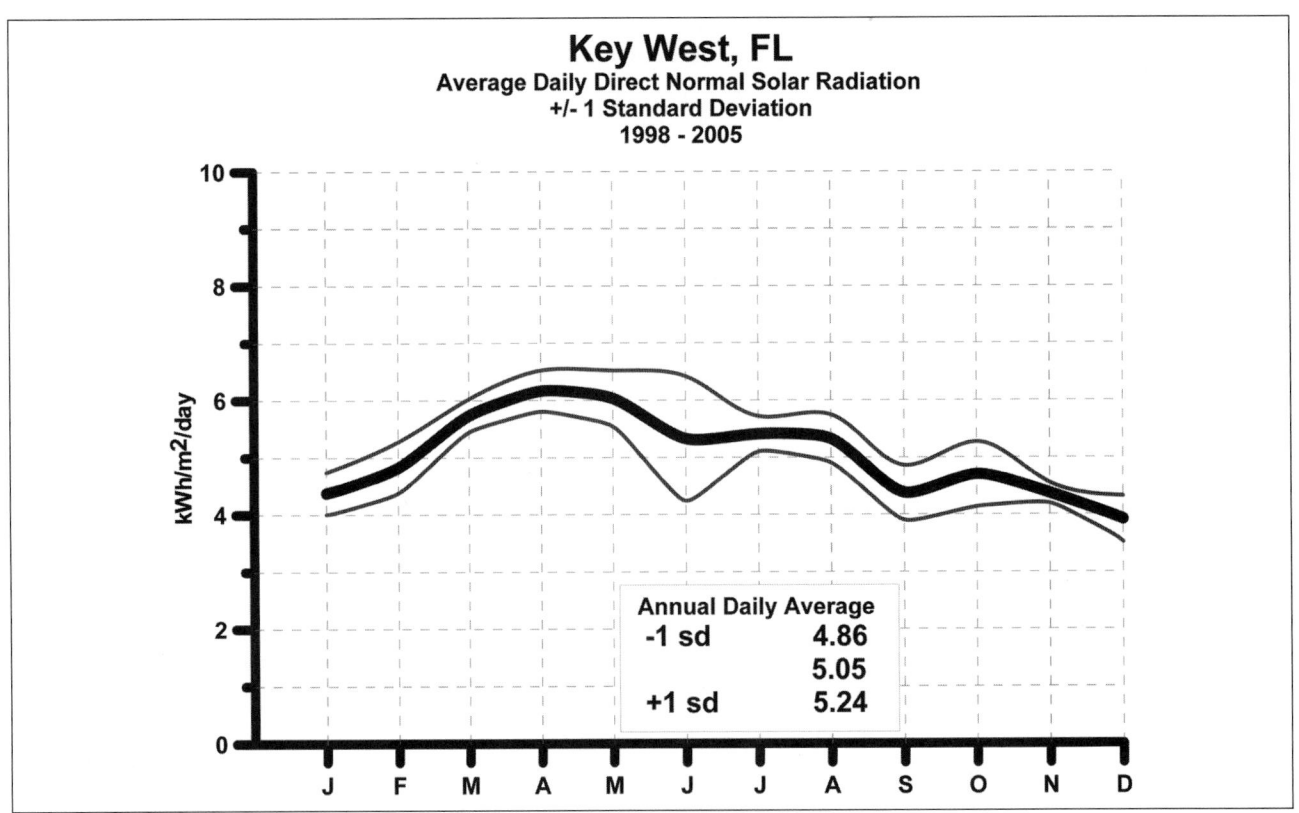

**Figure 2.71** Solar energy graph.

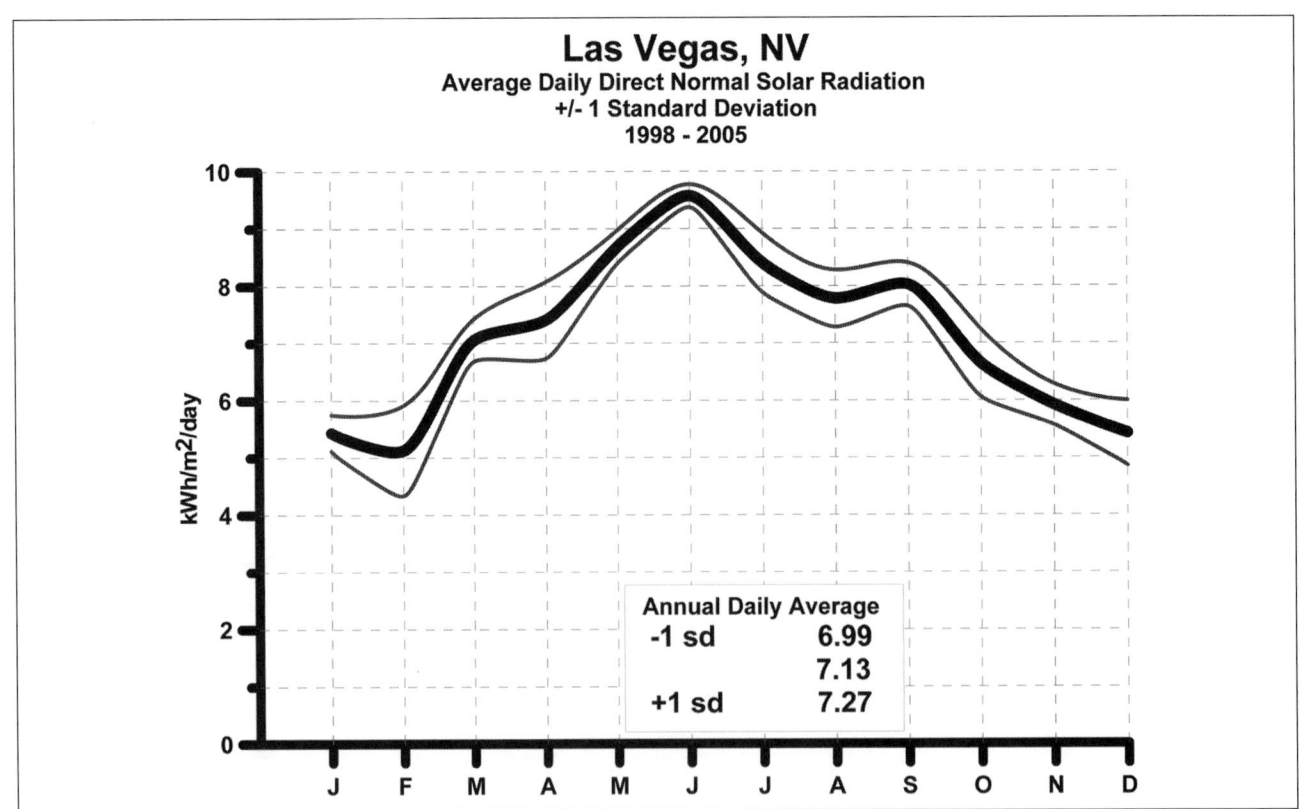

**Figure 2.72** Solar energy graph.

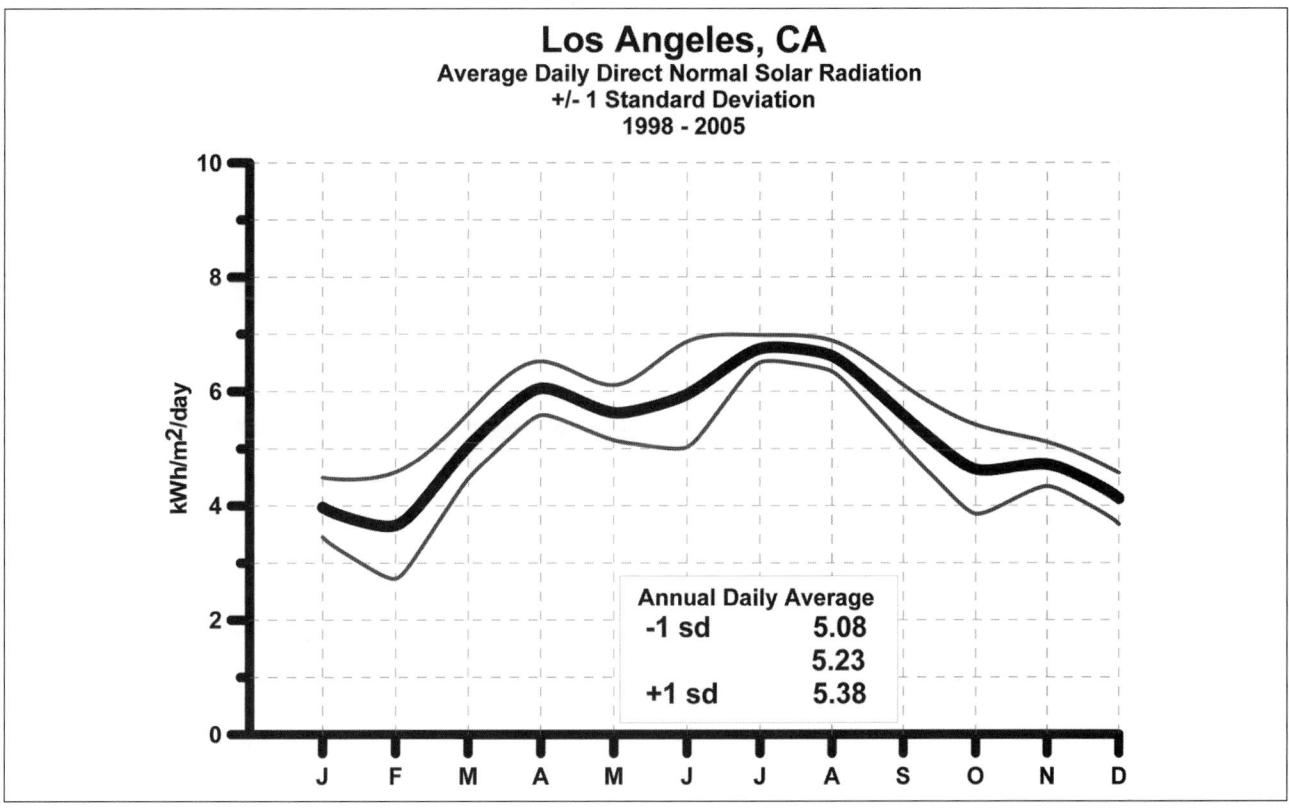

**Figure 2.73**   Solar energy graph.

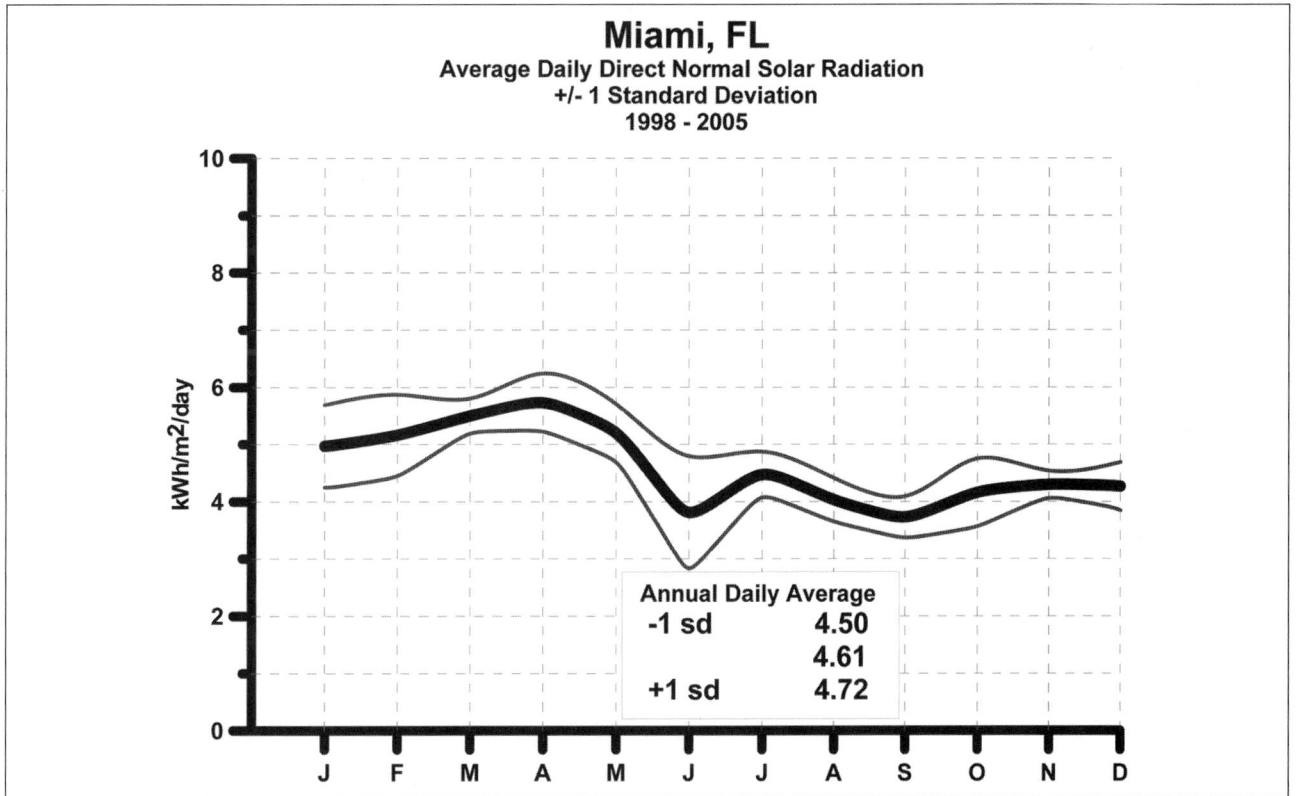

**Figure 2.74**   Solar energy graph.

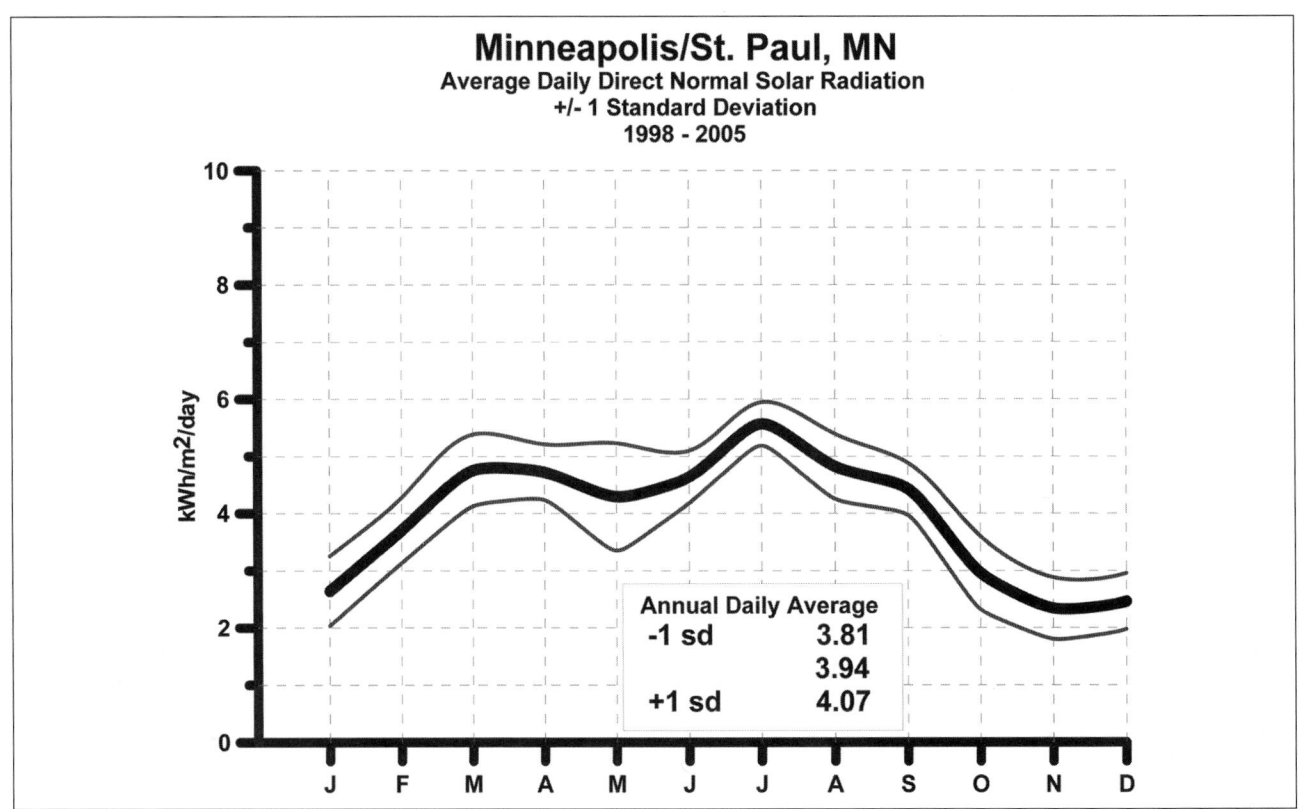

**Figure 2.75**   Solar energy graph.

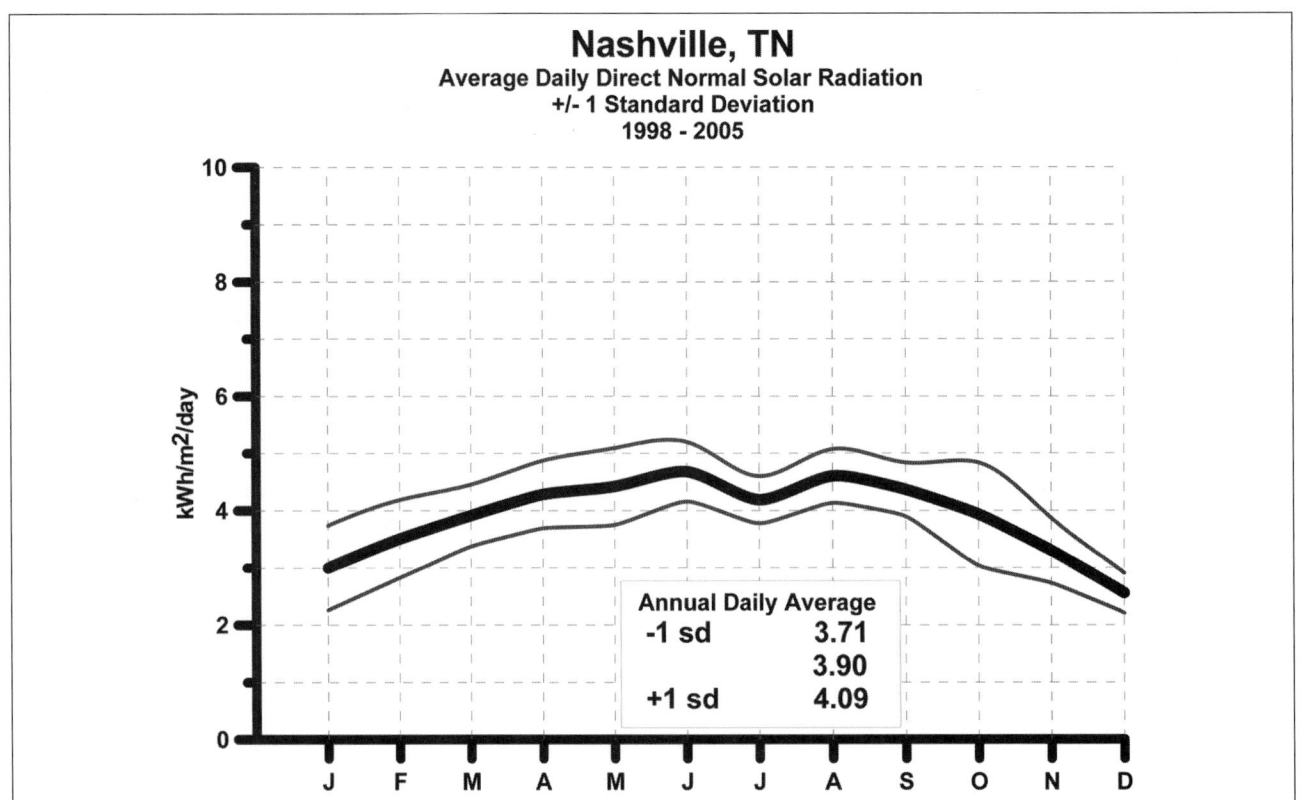

**Figure 2.76**   Solar energy graph.

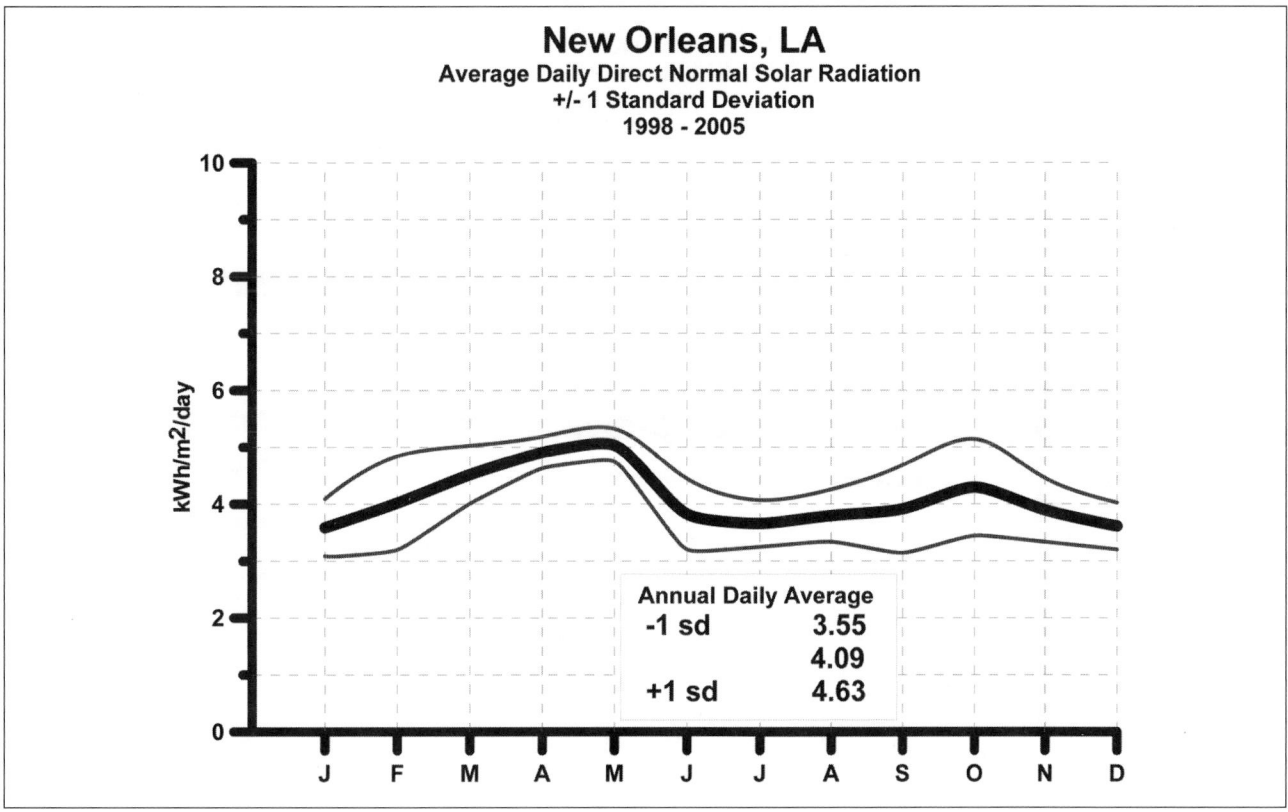

**Figure 2.77**   Solar energy graph.

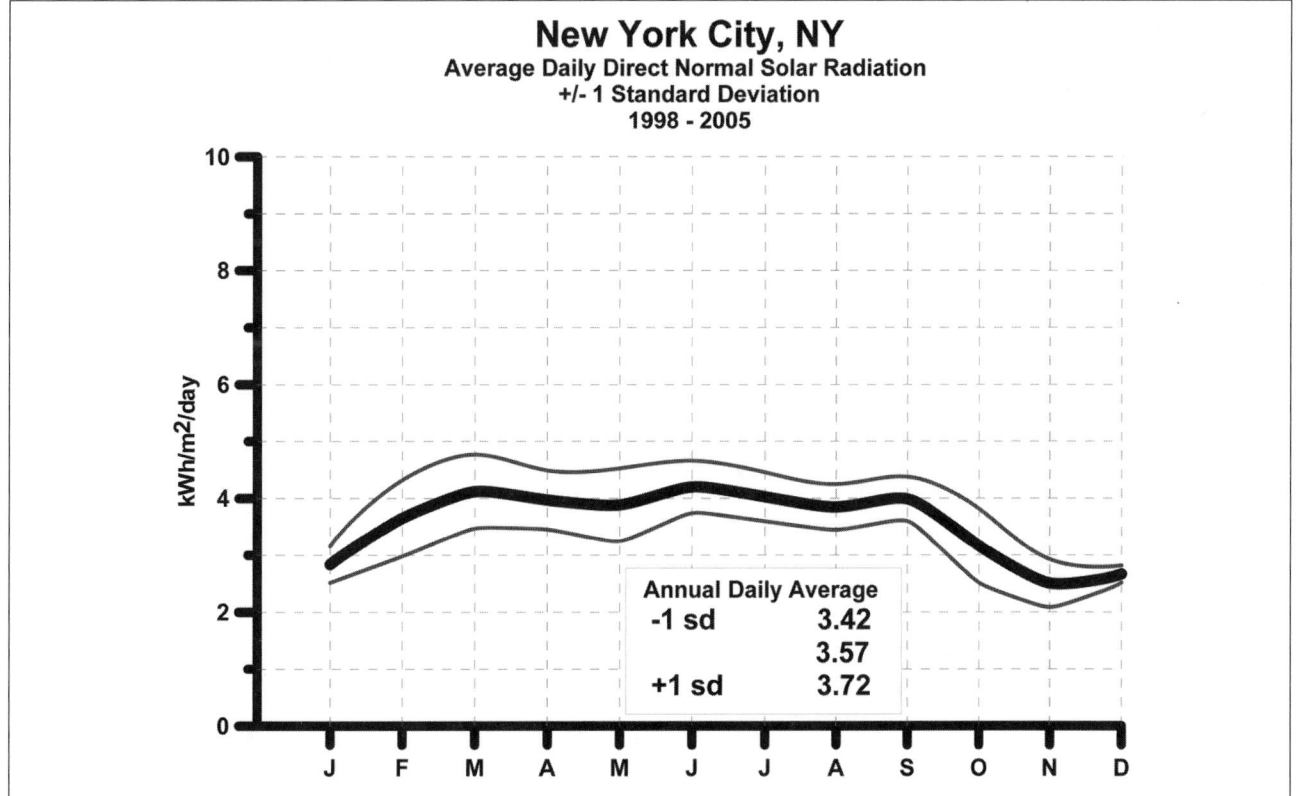

**Figure 2.78**   Solar energy graph.

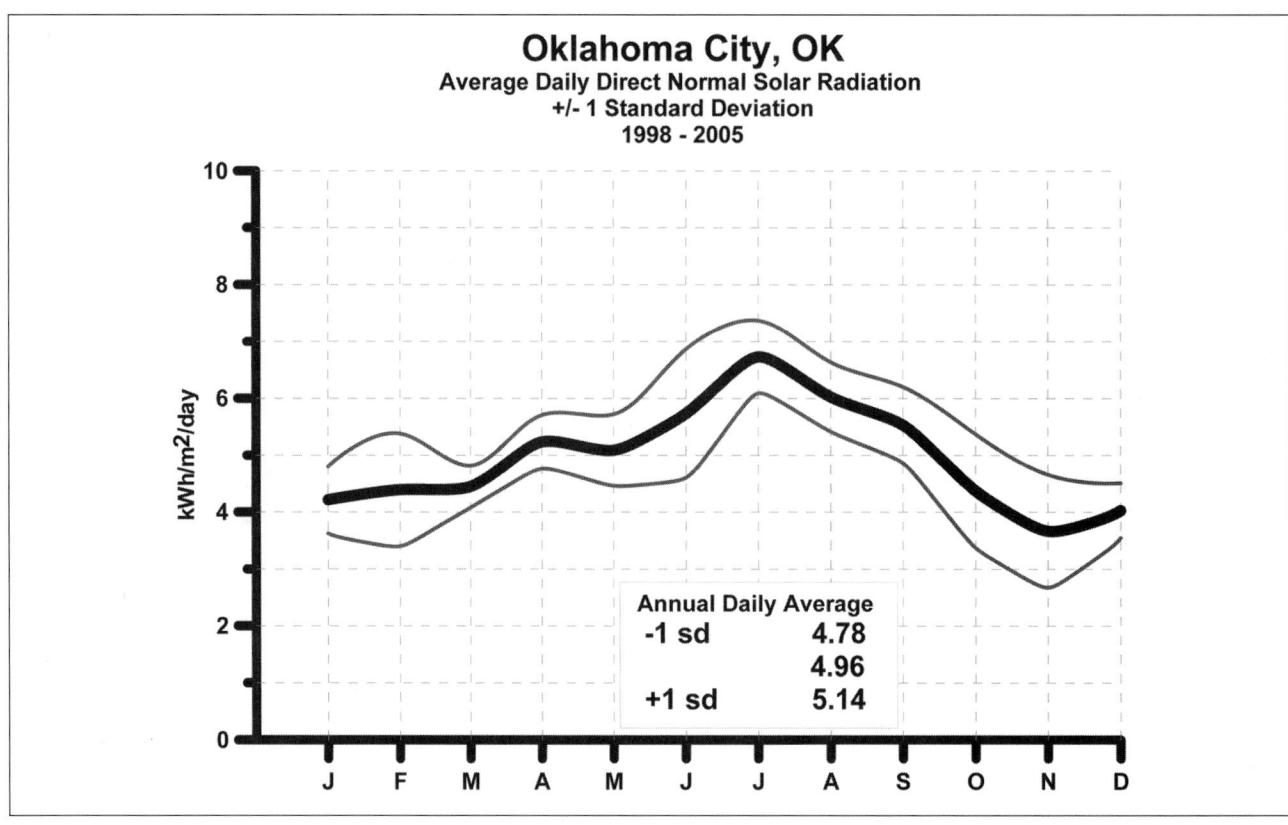

**Figure 2.79**  Solar energy graph.

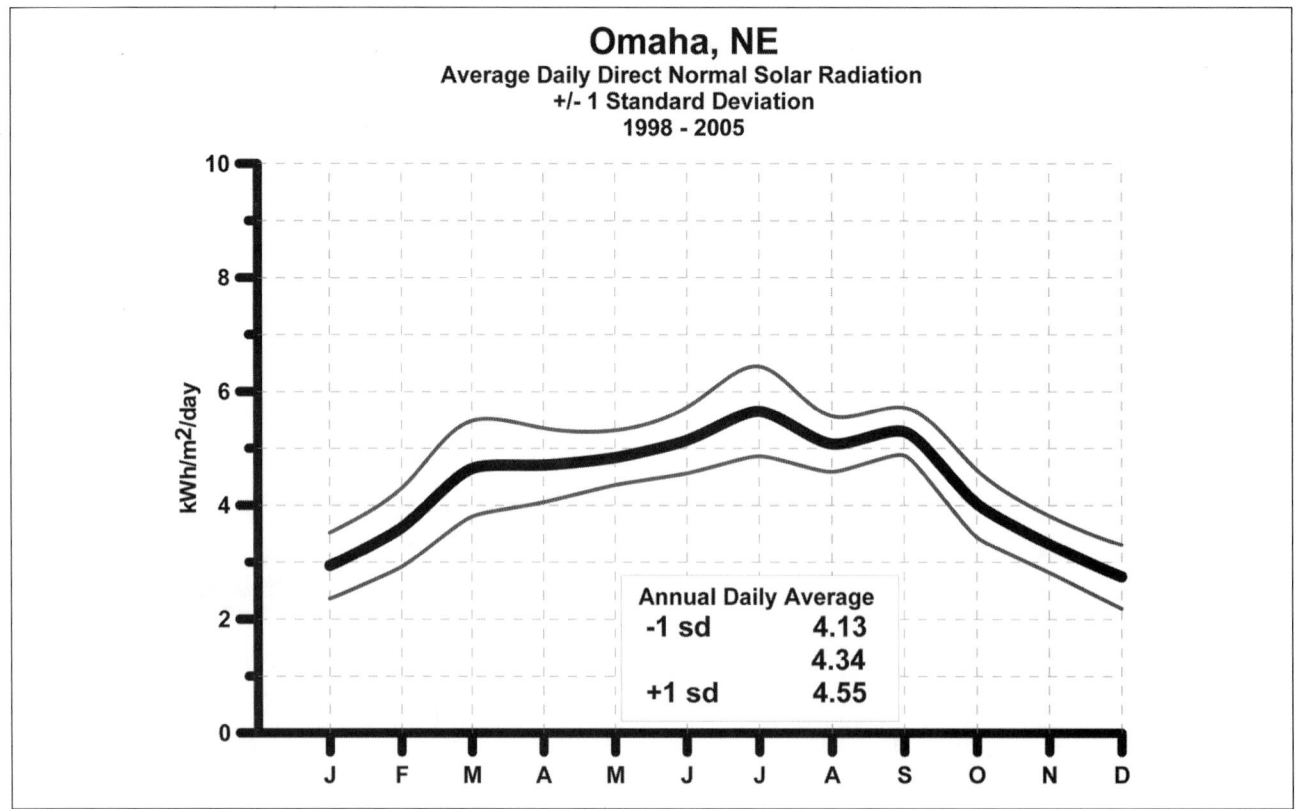

**Figure 2.80**  Solar energy graph.

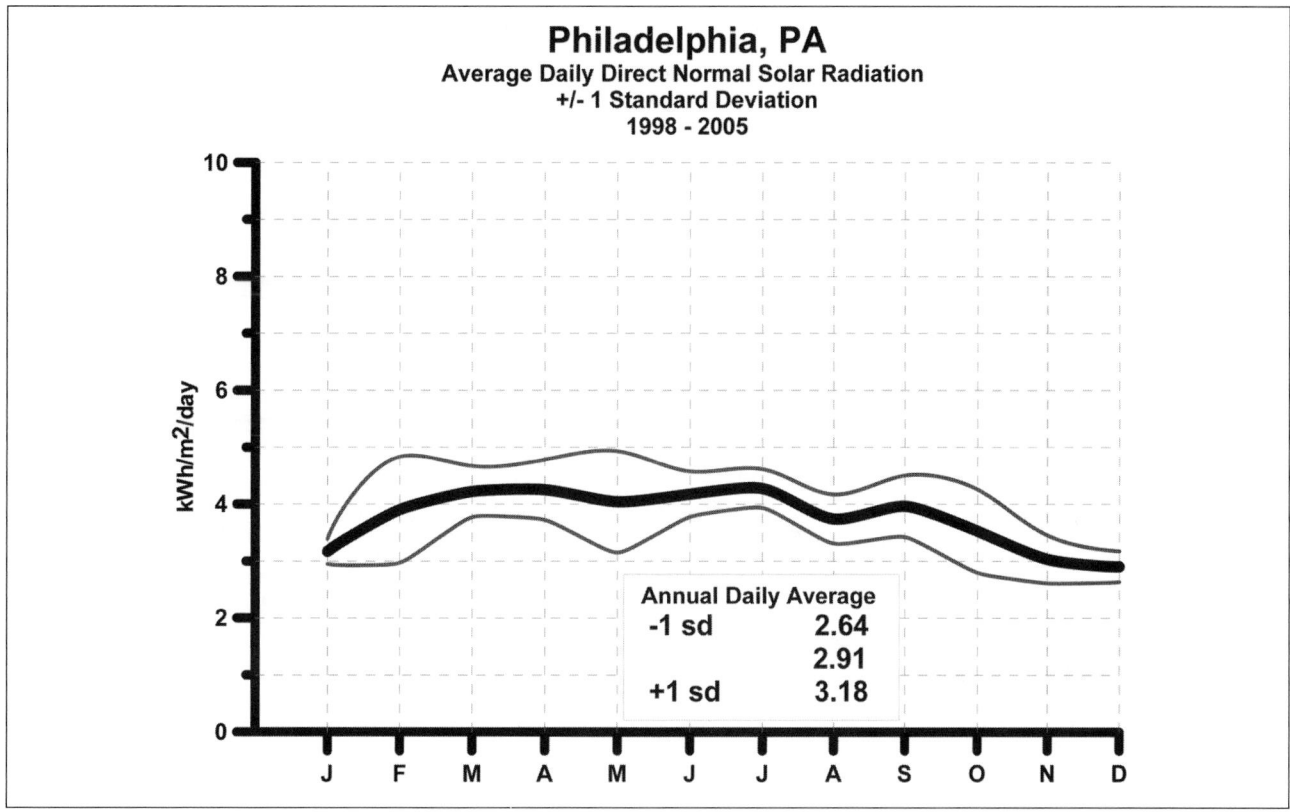

**Figure 2.81**    Solar energy graph.

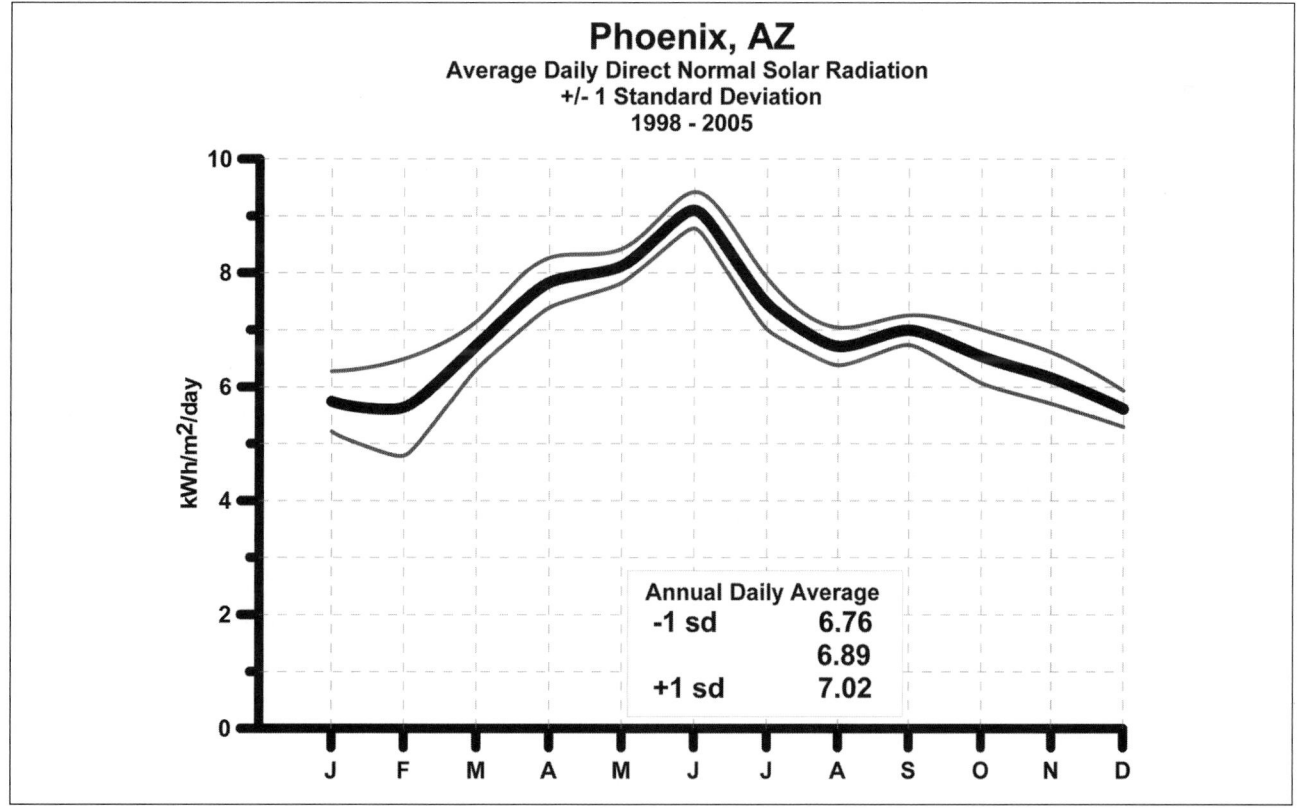

**Figure 2.82**    Solar energy graph.

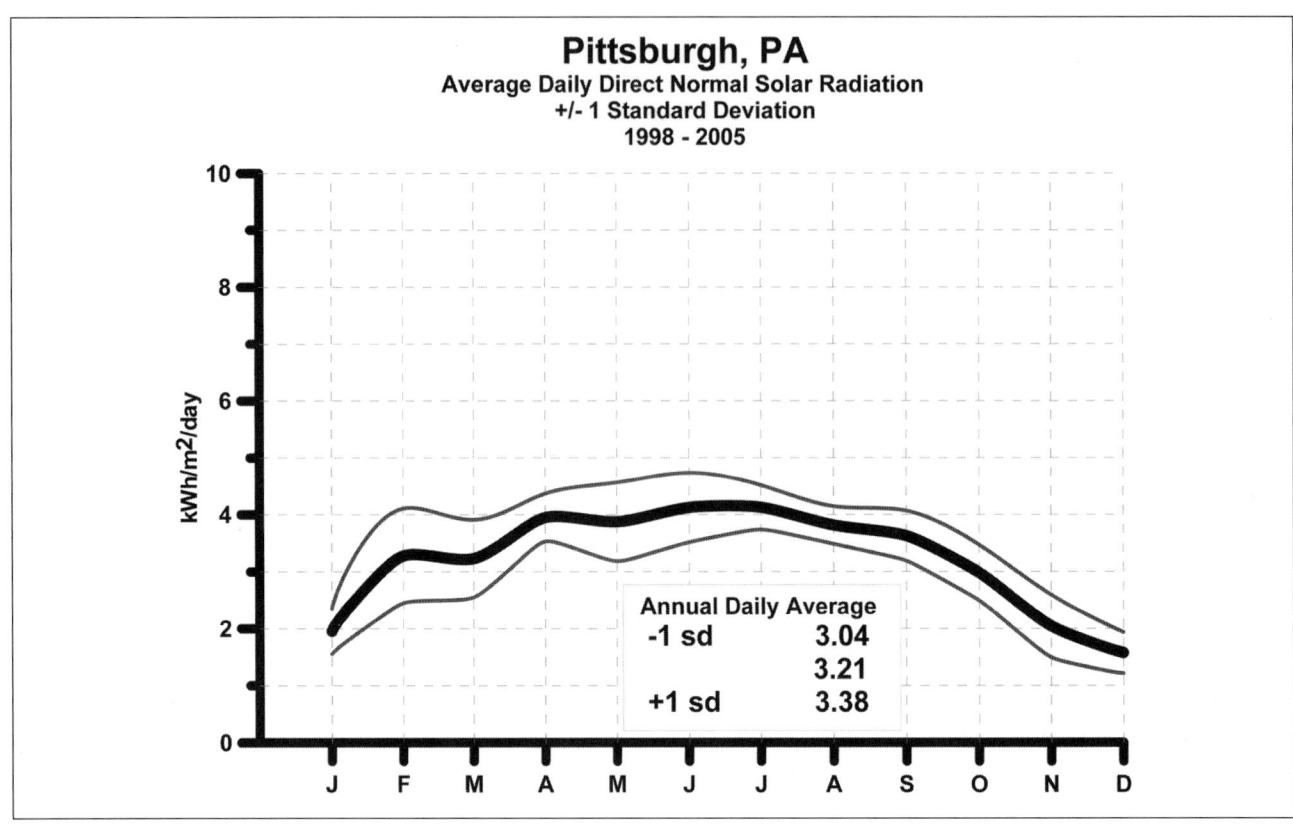

**Figure 2.83** Solar energy graph.

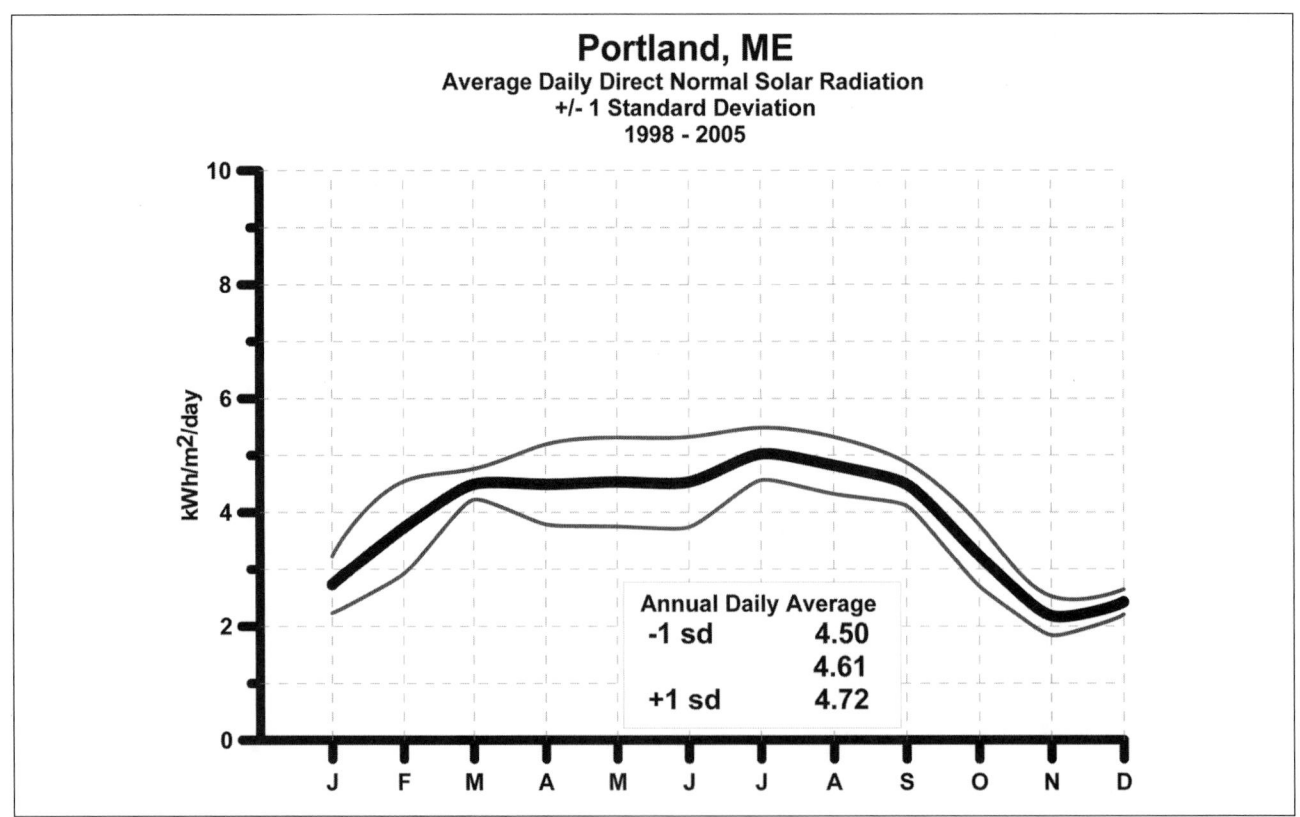

**Figure 2.84** Solar energy graph.

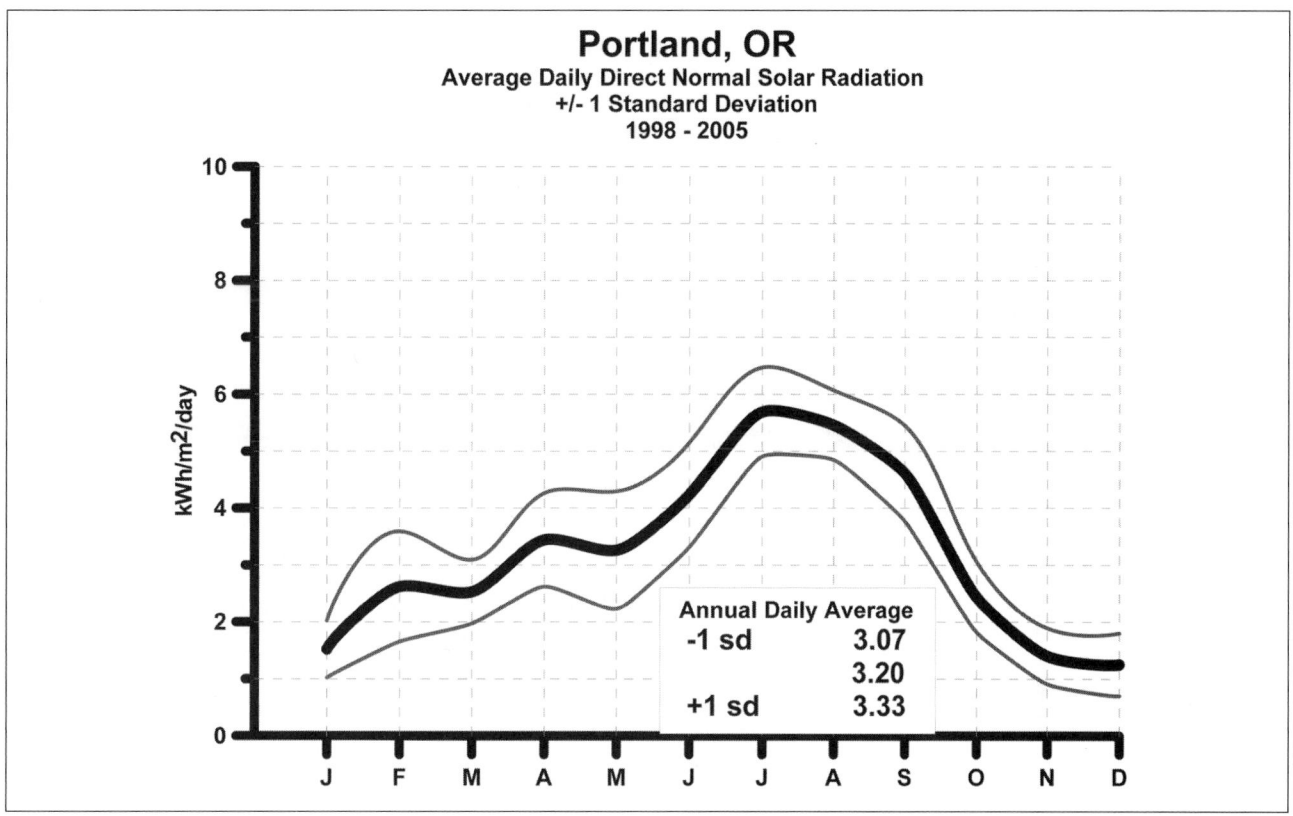

**Figure 2.85**    Solar energy graph.

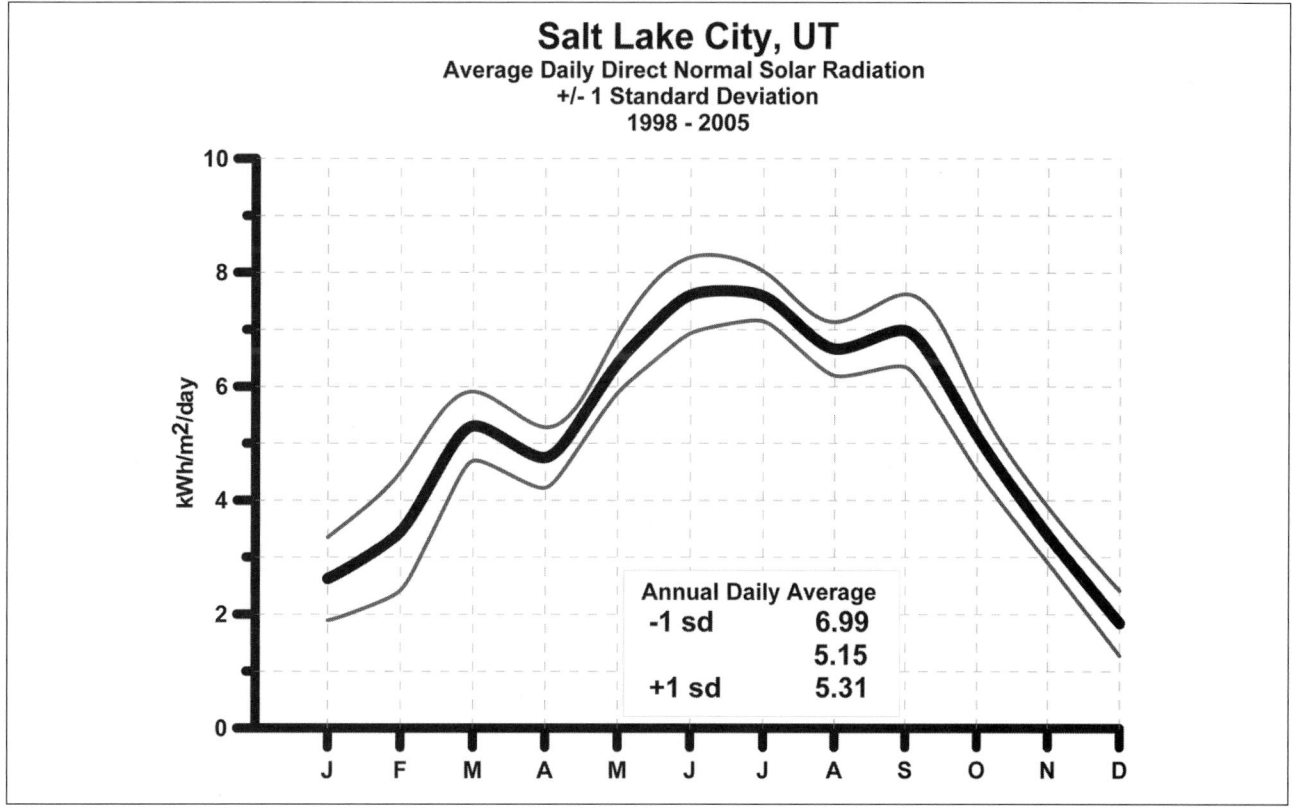

**Figure 2.86**    Solar energy graph.

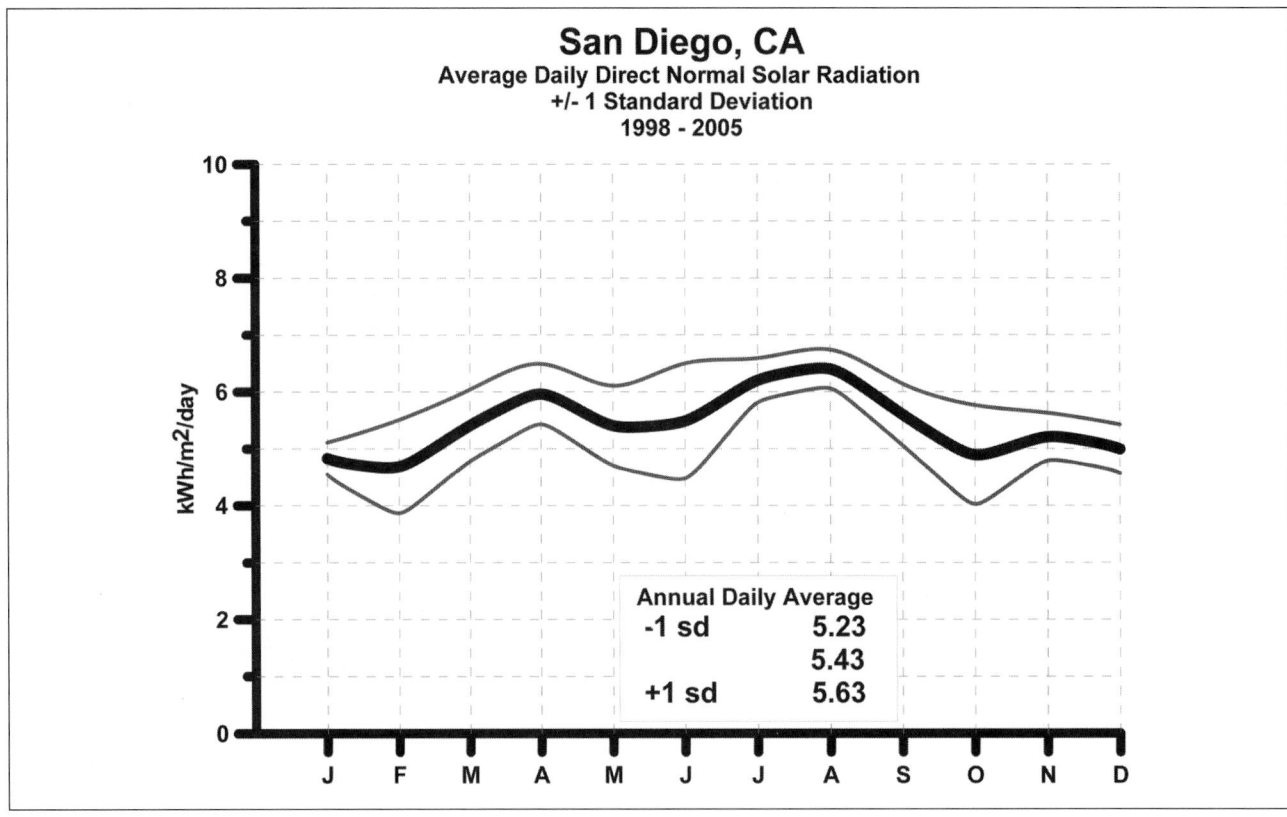

**Figure 2.87**   Solar energy graph.

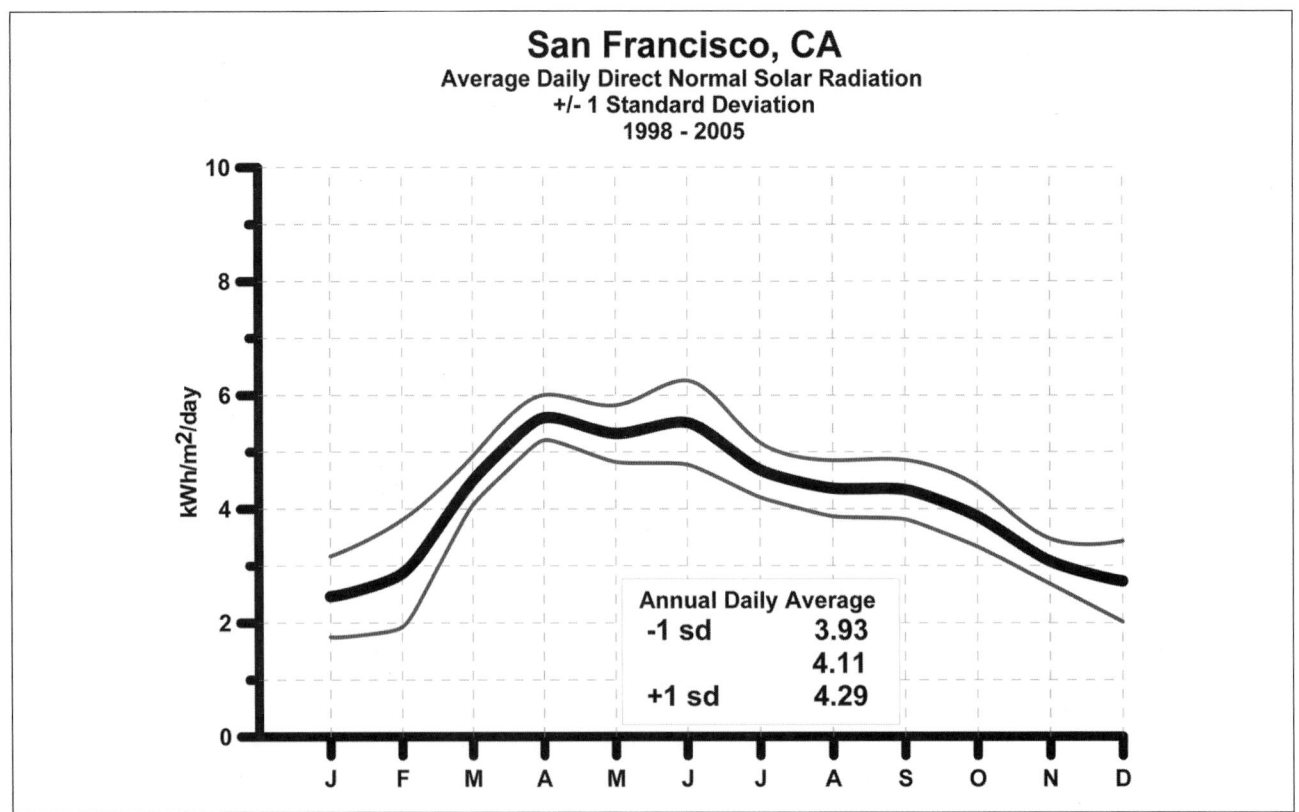

**Figure 2.88**   Solar energy graph.

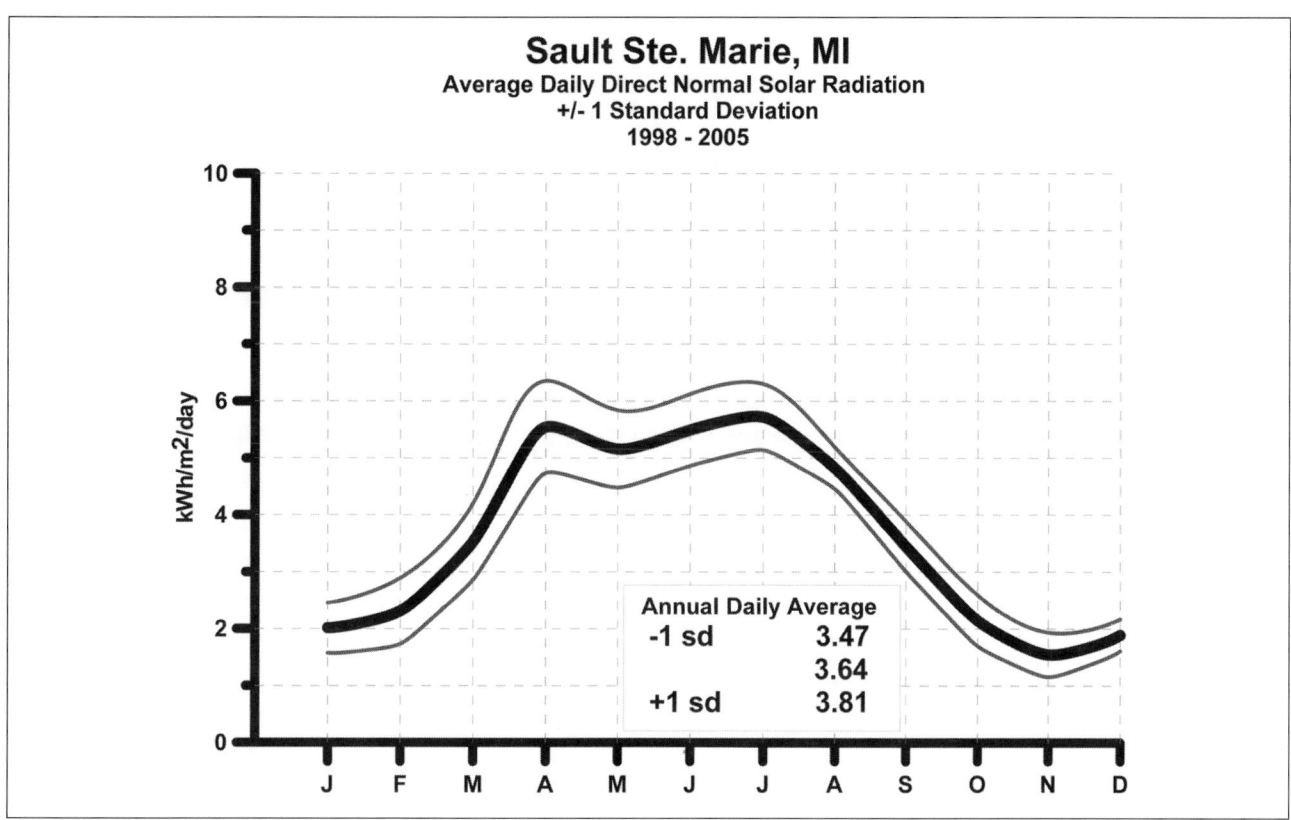

**Figure 2.89**  Solar energy graph.

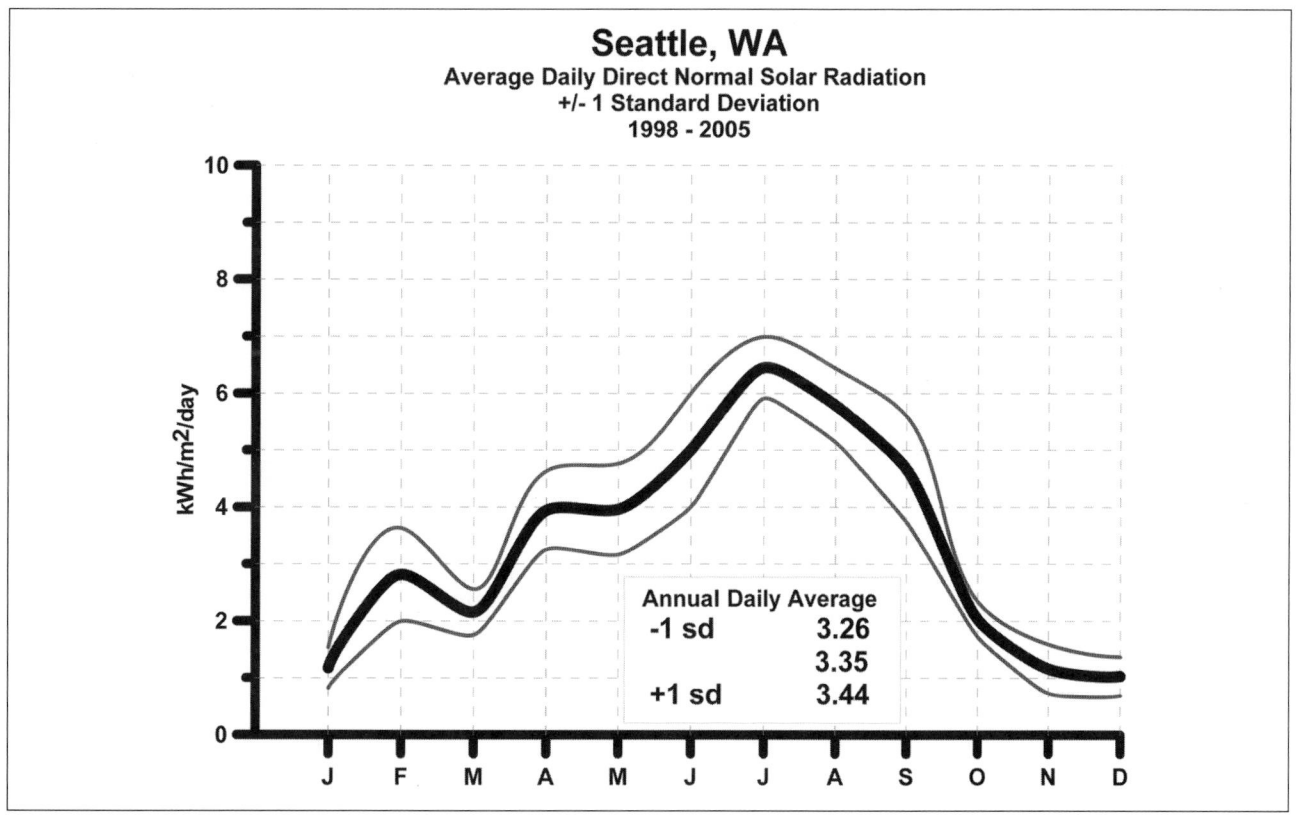

**Figure 2.90**  Solar energy graph.

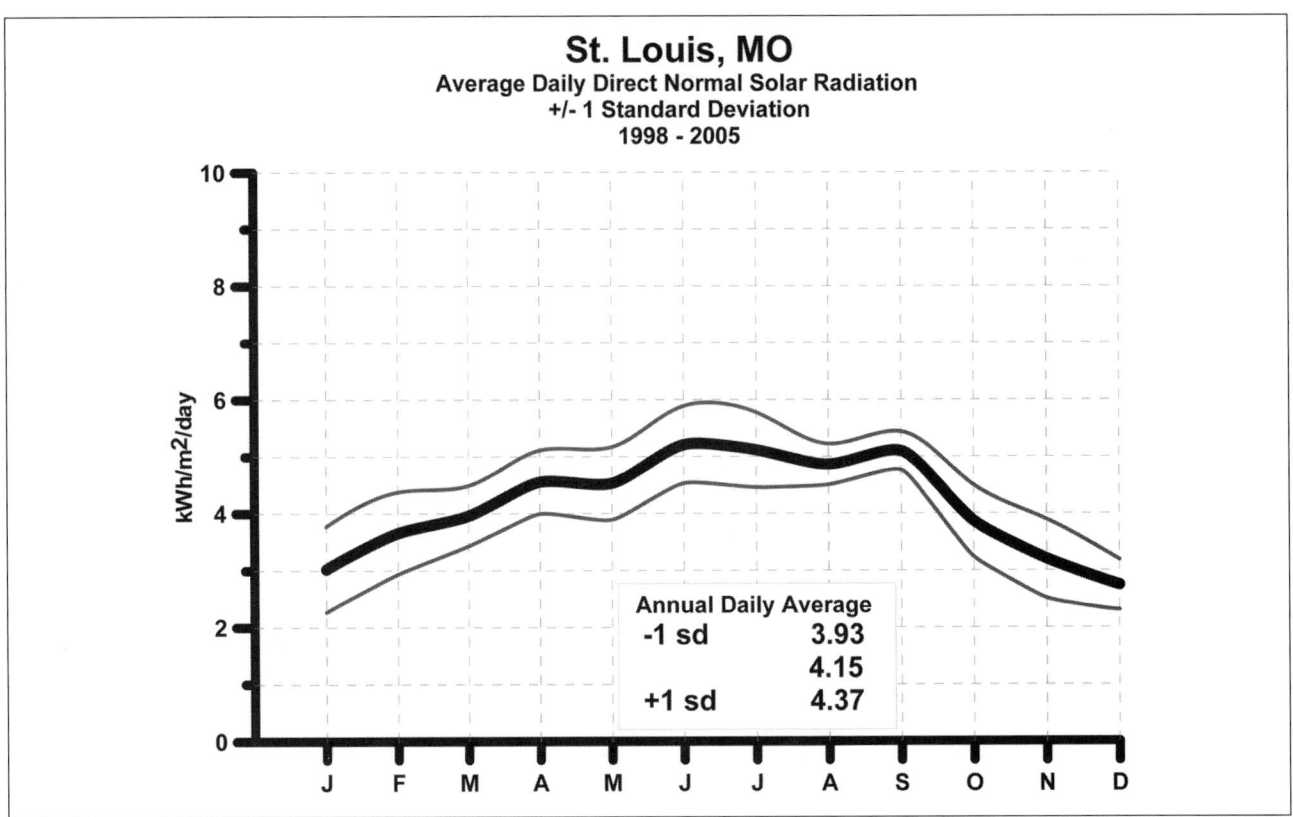

**Figure 2.91**   Solar energy graph.

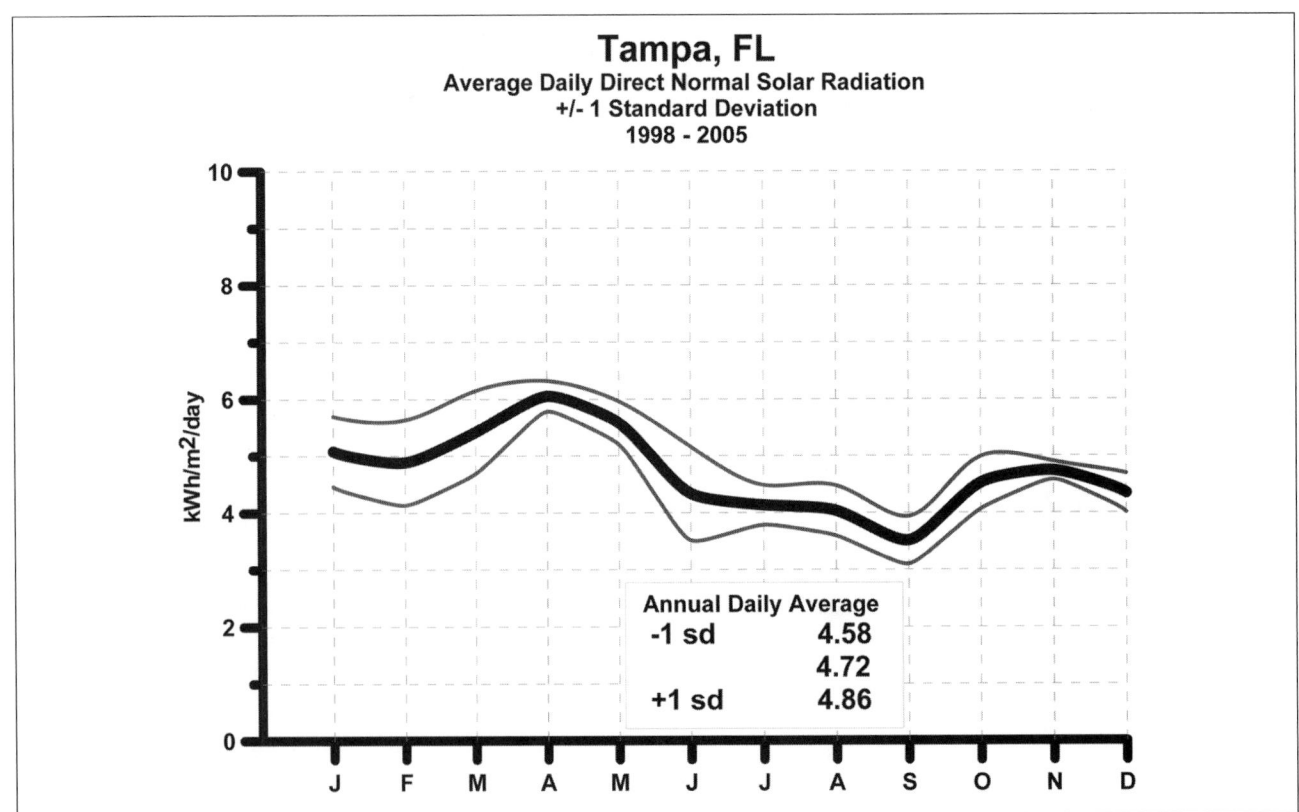

**Figure 2.92**   Solar energy graph.

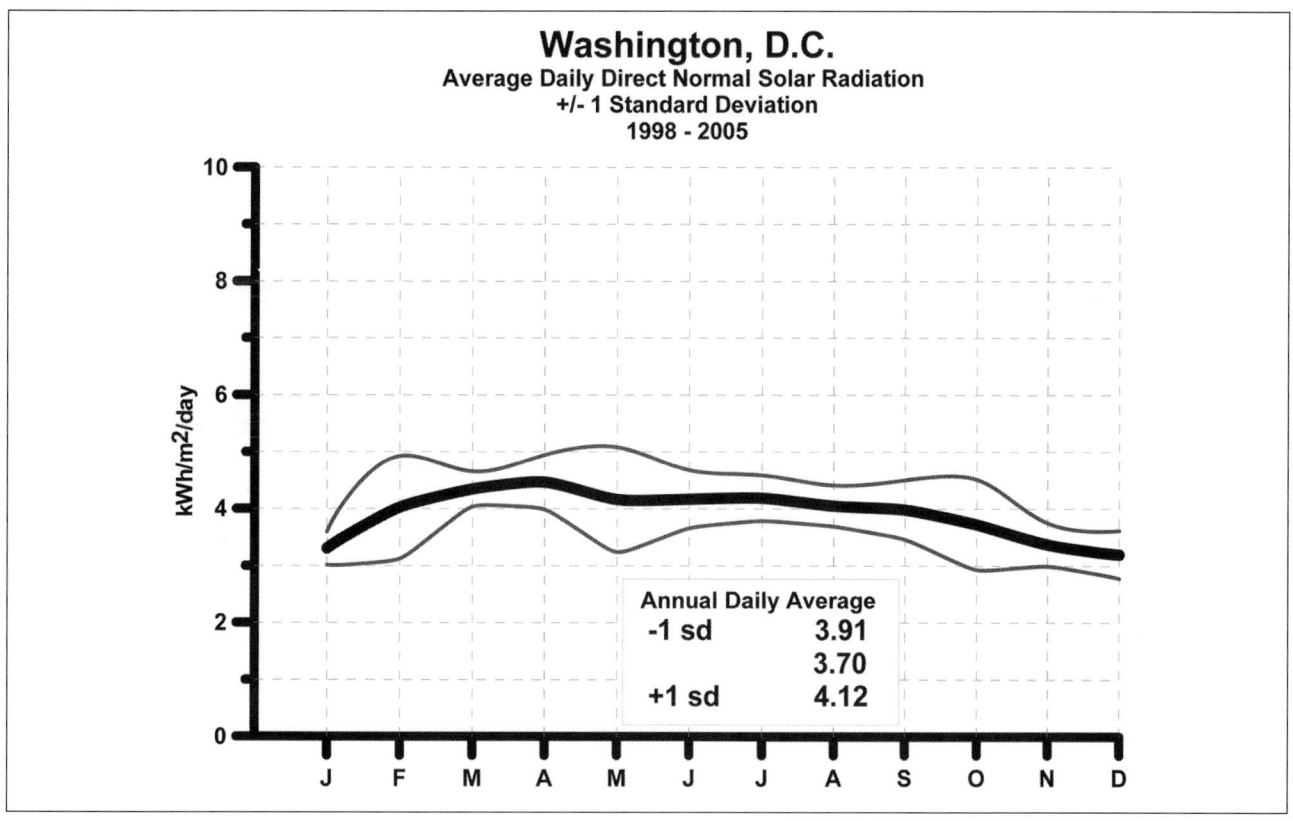

**Figure 2.93**   Solar energy graph.

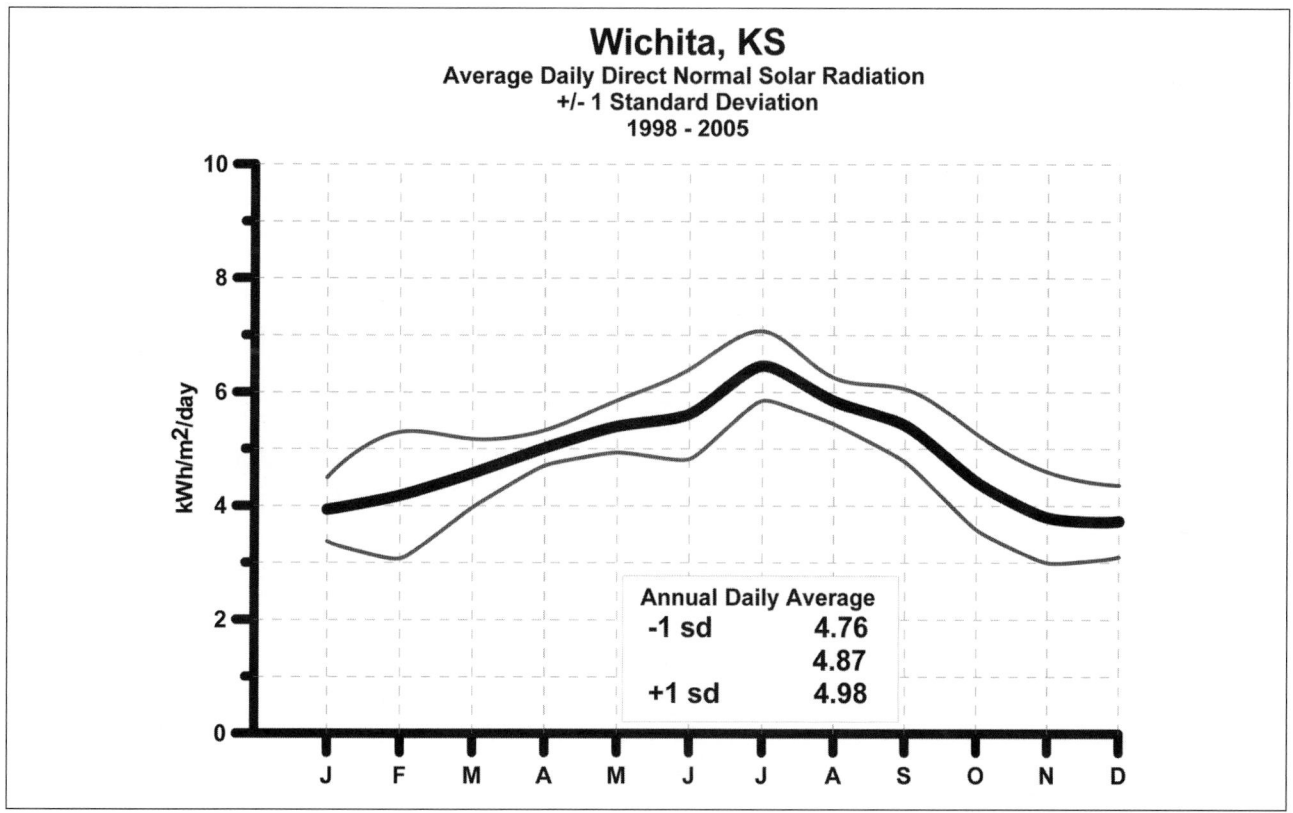

**Figure 2.94**   Solar energy graph.

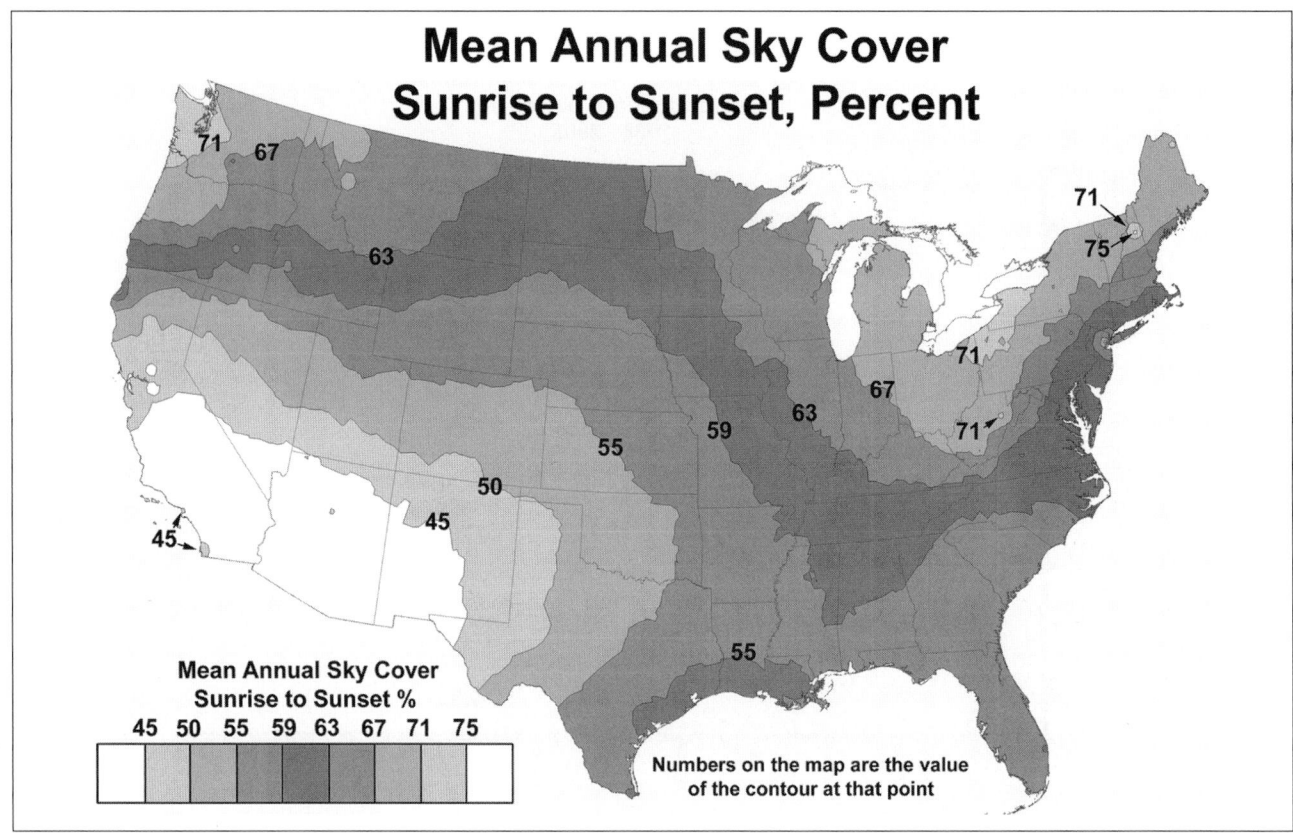

**Figure 2.95** National sunshine/sky cover map.

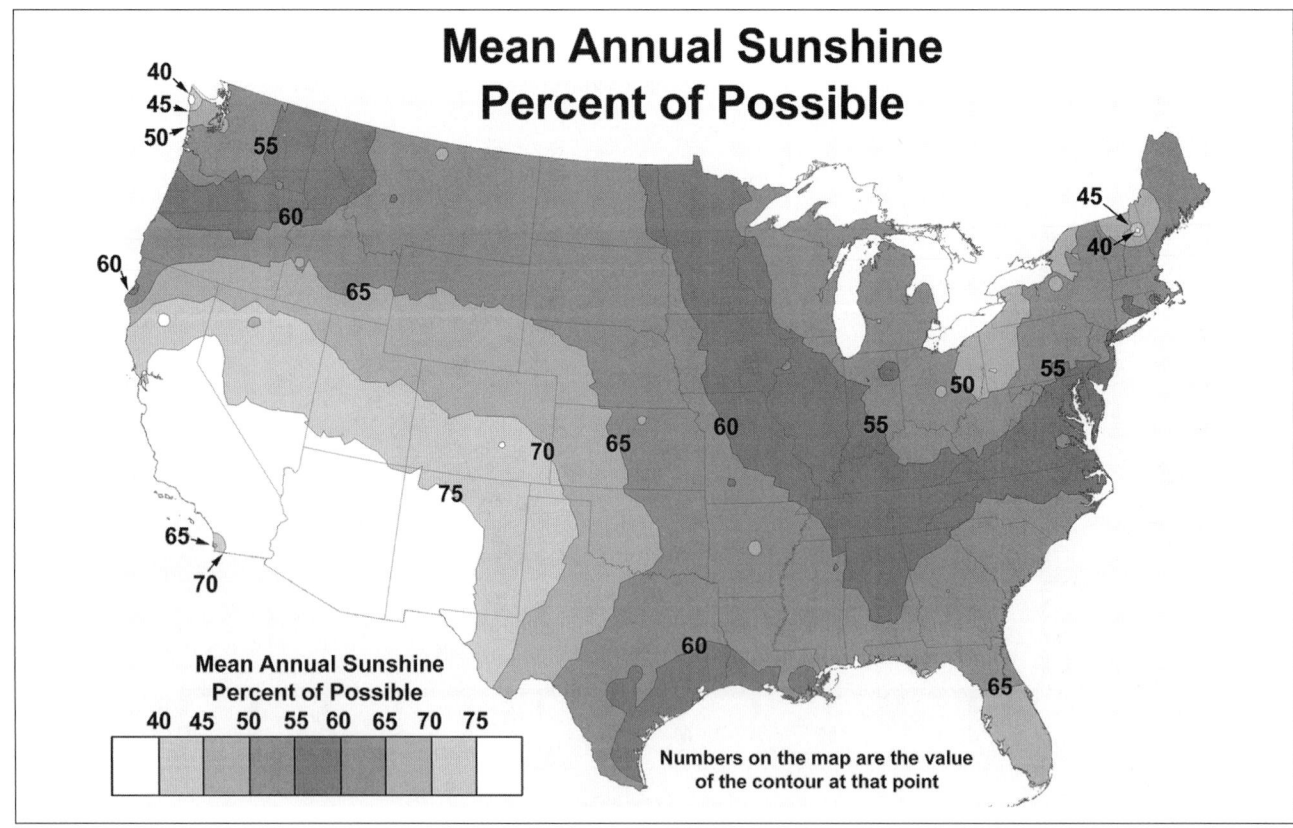

**Figure 2.96** National sunshine/sky cover map.

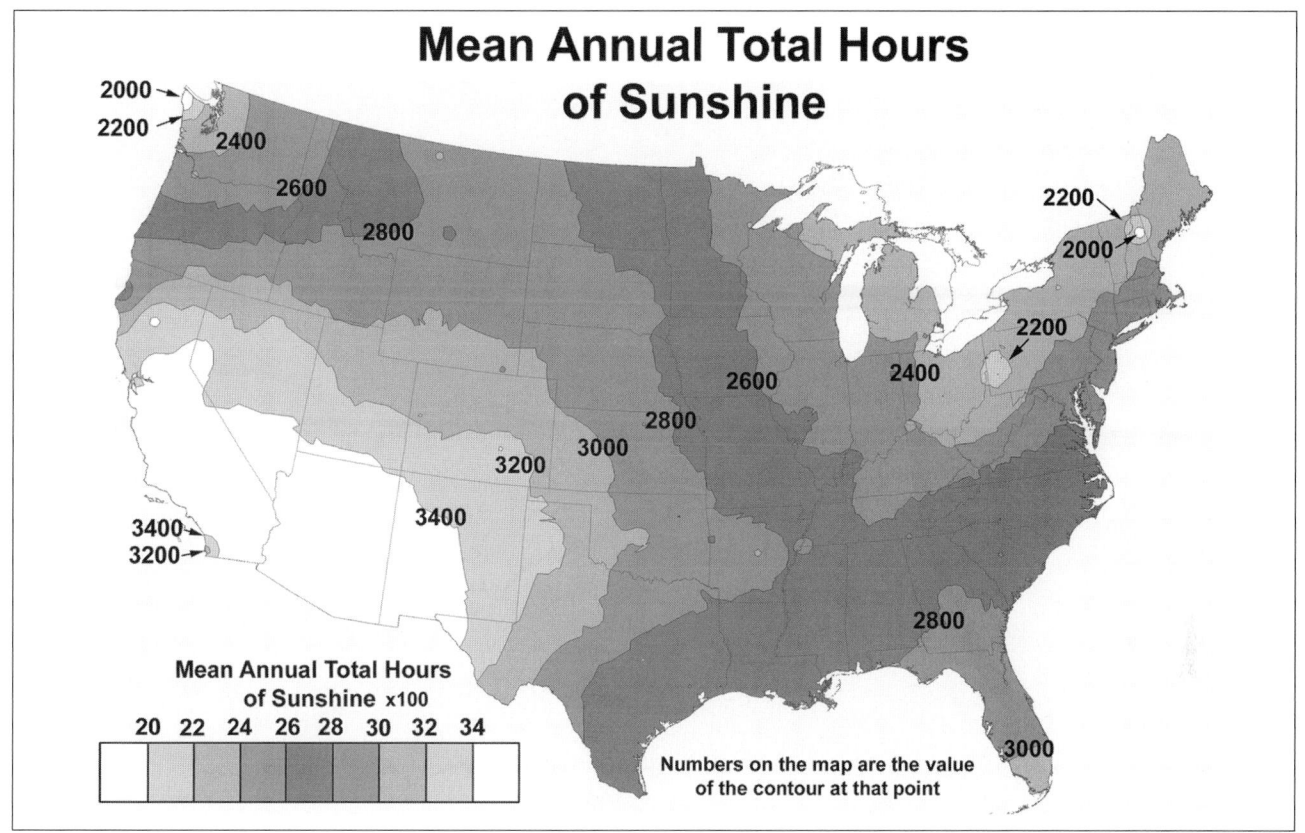

**Figure 2.97** National sunshine/sky cover map.

# 3

# Extreme Weather—Disasters, Records, Floods, Heat, Cold, and Blizzards

In this chapter you will find information about extreme weather. Included are lists of the hottest, coldest, wettest, driest, and snowiest places in the United States and around the world. Record-setting floods, heat waves, cold waves, drought, and blizzards are events that seemingly defy the physics of our atmosphere, but all have a logical explanation and provide a path to a deeper understanding of how weather works.

## • THE COST OF US WEATHER DISASTERS

Between 1955 and 2007, according to the *Extreme Weather Sourcebook*, Societal Impacts Program, National Center for Atmospheric Research, the United States incurred losses totaling more than $620 billion (2007 dollars) that, when adjusted for total national wealth, are in excess of $921 billion. Natural disaster loss figures are estimates, and the estimates of different studies do not necessarily include the same weather elements; therefore, they do not necessarily agree.

The cost estimates from the *Extreme Weather Sourcebook* include only flood, tornado, and hurricane losses.

Table 3.1 ranks the states by total cost of weather disasters from 1955 through 2007 with dollar figures adjusted for inflation to 2007 dollars and adjusted for total national wealth that varies with the economic health of the nation.

## • BILLION DOLLAR DISASTERS IN THE UNITED STATES

In the 29 years from 1980 and 2008, NOAA's National Climatic Data Center (NCDC) has compiled a list of 90 events costing $1 billion or more in dollars adjusted to the year 2007. The total cost of these events, in 2007 dollars, was more than $700 billion.

The list compiled by NOAA's NCDC places the 90 events in 9 categories: nor'easters, ice storms, blizzards, freezes, fires, nontropical floods, heat waves/droughts, severe storms, and tropical storms and hurricanes. This list is more comprehensive than the disasters included for the estimates of Table 3.1. Figure 3.1 is a map summarizing the geographical distribution of the billion dollar disasters and Figure 3.2 is a graph that shows the annual frequency and costs. Both Figures 3.1 and 3.2 were produced by NCDC.

Table 3.2 summarizes the 90 events in chronological order. The numbers are not final because periodically new information becomes available and NCDC updates its files.

Thirty percent of the events (27 events) are tropical in nature and account for more than $367 billion in damages, nearly 52% of the total. This reflects the nature of tropical storms and hurricanes covering wide areas and affecting thousands of square miles per event.

The average tropical billion dollar disaster costs more than $13 billion. With the continued growth of coastal communities this will only continue to increase.

Severe weather that includes tornadoes and severe thunderstorms, along with the wind, hail, and lightning that accompany them, account for nearly 18% of the events and 4.7% of the damage costs. The average billion dollar severe weather disaster costs $2.1 billion.

Of all the disasters to occur from 1980 to, and including, 2008, the most expensive was Hurricane Katrina with damage estimates totaling $133.8 billion, followed by the 1988 heat wave/drought ($71.2 billion) and the 1980 heat wave/drought ($55.4 billion). Table 3.3 lists the 90 events in order of cost from highest to lowest.

*The Weather Almanac: A Reference Guide to Weather, Climate, and Related Issues in the United States and Its Key Cities*, Twelfth Edition. Steve Horstmeyer.
© 2011 John Wiley & Sons, Inc. Published 2011 by John Wiley & Sons, Inc.

**Table 3.1 Total Losses from Weather Disasters, Tornado, Hurricane, and Floods, by State Ranked from Costliest**

| State | Rank | Total Damages | National Wealth Adjusted Damages | State | Rank | Total Damages | National Wealth Adjusted Damages |
|---|---|---|---|---|---|---|---|
| Florida | 1 | $144,074.05 | $183,295.54 | Wisconsin | 26 | $3,965.82 | $6,088.24 |
| Louisiana | 2 | $126,994.87 | $157,489.55 | Arkansas | 27 | $3,790.77 | $6,819.42 |
| Texas | 3 | $55,942.72 | $84,260.21 | Connecticut | 28 | $3,586.73 | $9,266.53 |
| Mississippi | 4 | $44,594.91 | $58,999.45 | West Virginia | 29 | $3,392.61 | $5,588.30 |
| North Carolina | 5 | $30,484.47 | $50,666.96 | Nebraska | 30 | $3,303.20 | $5,913.38 |
| Pennsylvania | 6 | $23,384.44 | $47,385.15 | Massachusetts | 31 | $3,258.61 | $6,691.95 |
| California | 7 | $15,382.96 | $27,587.80 | South Dakota | 32 | $3,048.73 | $5,269.76 |
| Iowa | 8 | $13,110.61 | $20,457.59 | Michigan | 33 | $2,957.54 | $5,642.06 |
| South Carolina | 9 | $12,093.38 | $17,830.80 | Idaho | 34 | $2,665.18 | $5,138.53 |
| Alabama | 10 | $10,737.20 | $15,940.72 | Utah | 35 | $2,474.29 | $3,688.49 |
| New York | 11 | $10,320.53 | $19,119.15 | Tennessee | 36 | $2,176.07 | $3,592.87 |
| Missouri | 12 | $10,113.06 | $17,416.16 | Maryland/D.C. | 37 | $2,149.40 | $3,926.95 |
| Illinois | 13 | $8,533.47 | $14,569.44 | Arizona | 38 | $1,922.59 | $3,266.81 |
| Minnesota | 14 | $7,493.87 | $13,287.45 | Washington | 39 | $1,780.42 | $2,969.84 |
| Oklahoma | 15 | $7,111.81 | $11,624.94 | Nevada | 40 | $1,052.15 | $1,541.41 |
| Indiana | 16 | $6,645.45 | $13,749.34 | Alaska | 41 | $834.17 | $1,714.27 |
| North Dakota | 17 | $6,618.05 | $9,602.73 | Montana | 42 | $635.74 | $1,441.05 |
| Oregon | 18 | $6,231.07 | $10,437.89 | Rhode Island | 43 | $597.22 | $1,217.81 |
| Virginia | 19 | $6,191.61 | $11,094.61 | Vermont | 44 | $481.47 | $850.13 |
| Ohio | 20 | $5,946.58 | $10,277.51 | Maine | 45 | $448.91 | $688.82 |
| Kansas | 21 | $5,645.34 | $10,260.55 | Hawaii | 46 | $395.36 | $576.49 |
| New Jersey | 22 | $4,830.01 | $8,460.18 | New Mexico | 47 | $378.55 | $648.82 |
| Kentucky | 23 | $4,313.05 | $8,173.96 | New Hampshire | 48 | $248.09 | $438.95 |
| Georgia | 24 | $4,005.87 | $6,477.44 | Wyoming | 49 | $212.25 | $364.26 |
| Colorado | 25 | $3,998.00 | $9,158.18 | Delaware | 50 | $117.77 | $141.31 |

*Source:* Extreme Weather Sourcebook.

The deadliest of these expensive disasters was the 1980 heat wave/drought killing an estimated 10,000 people, with the 1988 heat wave/drought (7500 deaths) and Hurricane Katrina (1833 deaths) coming in second and third places. Table 3.4 lists the 90 events in order of death toll from greatest to smallest.

The annotated list in reverse chronological order, along with a brief account of each of the 90 billion dollar plus disaster events follows. For dollar figures, the actual dollars at the time of the disaster is the first figure and the second figure in parentheses is the cost adjusted to 2007 dollars.

## The United States' Billion Dollar Weather Disasters: 1980–2008

### 2008

**Widespread drought.** Entire year, 2008. Severe drought and heat caused agricultural losses in areas of the south and west. Record low lake levels also occurred in areas of the southeast. Includes states of CA, TX, NC, SC, GA, and TN. Estimate of over $2.0 billion in damages/costs.

**Hurricane Ike** (Figure 3.3). September 2008. Category 2 hurricane makes the landfall in Texas as the largest (in size) Atlantic hurricane on record, causing considerable storm surge in coastal TX and significant wind and flooding dam-

age in TX, LA, AR, IL, IN, KY, MO, OH, and PA. Severe gasoline shortages occurred in the southeast US due to damaged oil platforms, storage tanks, pipelines, and off-line refineries. Estimate of over $27.0 billion in damages/costs; 82 deaths; > 100 people missing.

**Hurricane Gustav.** September 2008. Category 2 hurricane makes landfall in Louisiana causing significant wind, storm surge, and flooding damage in AL, AR, LA, and MS. Estimate of at least $5.0 billion in damages/costs; 43 deaths.

**Hurricane Dolly.** July 2008. Category 1 hurricane makes landfall in southern Texas causing considerable wind and flooding damage in TX and NM. Over $1.2 billion in damages/costs; 3 deaths.

**US wildfires.** Summer–fall 2008. Drought conditions across numerous western, central, and southeastern states (AK, AZ, CA, NM, ID, UT, MT, NV, OR, WA, CO, TX, OK, NC, FL) resulted in thousands of wildfires; national acreage burned, exceeding 5.2 million acres (mainly in the west), and over 1000 homes and structures destroyed in California fires alone. Over $2.0 billion in damages/costs; 16 deaths.

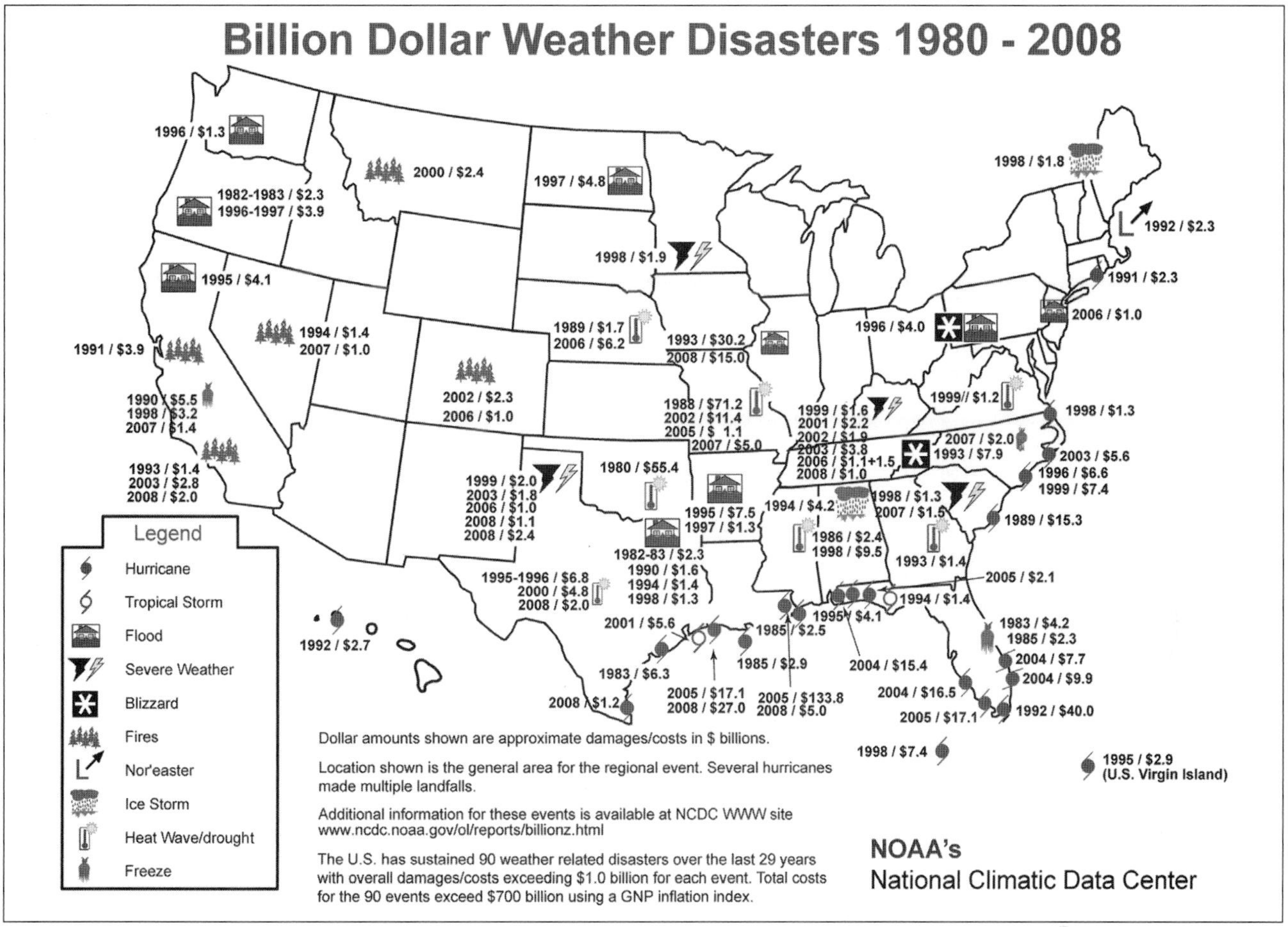

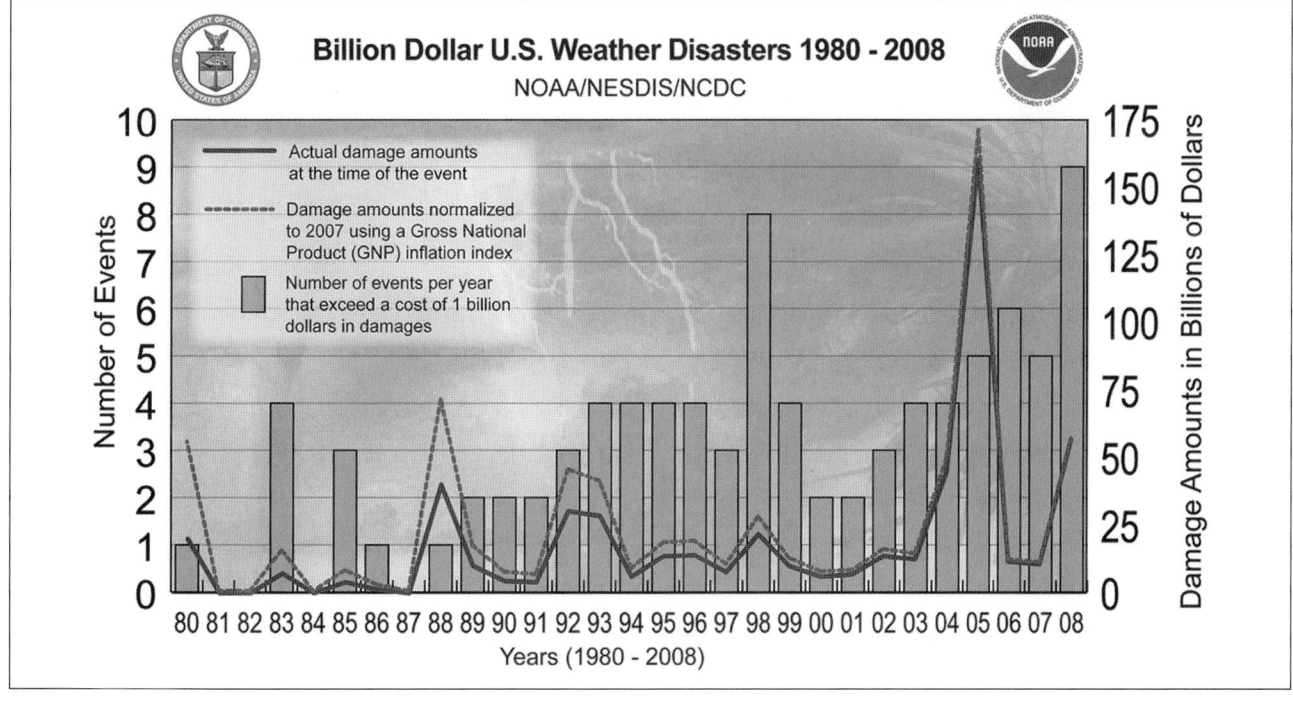

**Figure 3.1**    Billion dollar weather disasters in the United States. (*Source:* NCDC.)

**Figure 3.2**    Billion dollar weather disasters, number of events, and costs. (*Source:* NCDC.)

**Table 3.2    Billion Dollar Weather Disasters—Chronological Listing**

| Year | Event | 2007 Dollars (billions) | Deaths |
|------|-------|------------------------|--------|
| 1980 | Drought/Heat Wave | 55.4 | 10000 |
| 1983 | Hurricane Alicia | 6.3 | 21 |
| 1983 | Freeze Florida | 4.2 | 0 |
| 1983 | Floods/Storms | 2.3 | 50 |
| 1983 | Floods/Storms | 2.3 | 45 |
| 1985 | Freeze FL | 2.3 | 0 |
| 1985 | Hurricane Elena | 2.5 | 4 |
| 1985 | Hurricane Juan | 2.9 | 63 |
| 1986 | Drought/Heat Wave | 2.4 | 100 |
| 1988 | Drought Heat Wave | 71.2 | 7500 |
| 1989 | Hurricane Hugo | 15.3 | 86 |
| 1989 | Drought N Plains | 1.7 | 0 |
| 1990 | Floods S Plains | 1.6 | 13 |
| 1990 | Freeze CA | 5.5 | 0 |
| 1991 | Hurricane Bob | 2.3 | 18 |
| 1991 | Firestorm Oakland | 3.9 | 25 |
| 1992 | Hurricane Andrew | 40 | 61 |
| 1992 | Hurricane Iniki | 2.7 | 7 |
| 1992 | Nor'easter | 2.3 | 19 |
| 1993 | Blizzard East | 7.9 | 270 |
| 1993 | Drought/Heat Wave S'east | 1.4 | 16 |
| 1993 | Floods Midwest | 30.2 | 48 |
| 1993 | WildFires CA | 1.4 | 4 |
| 1994 | Ice Storm S'east | 4.2 | 27 |
| 1994 | TS Alberto | 1.4 | 32 |
| 1994 | Floods TX | 1.4 | 19 |
| 1994 | Wildfire Season | 1.4 | 0 |
| 1995 | Floods CA | 4.1 | 27 |
| 1995 | Severe S'east/S'west | 7.5 | 32 |
| 1995 | Hurricane Marilyn | 2.9 | 13 |
| 1995 | Hurricane Opal | 4.1 | 27 |
| 1996 | Blizzard/Flood | 4.0 | 187 |
| 1996 | Floods Pac NW | 1.3 | 9 |
| 1996 | Drought S Plains | 6.8 | 0 |
| 1996 | Hurricane Fran | 6.6 | 37 |
| 1997 | Flood/Tornadoes Midwest | 1.3 | 67 |
| 1997 | Floods N Plains | 4.8 | 11 |
| 1997 | Floods W Coast | 3.9 | 36 |
| 1998 | Ice Storm New Eng | 1.8 | 16 |
| 1998 | Severe S'east | 1.3 | 132 |
| 1998 | Severe/Hail MN | 1.9 | 1 |
| 1998 | Drought/Heat Wave South | 9.5 | 200 |
| 1998 | Hurricane Bonnie | 1.3 | 3 |
| 1998 | Hurricane Georges | 7.4 | 16 |
| 1998 | Floods TX | 1.3 | 31 |
| 1998 | Freeze CA | 3.2 | 0 |
| 1999 | Tornadoes AR TN | 1.6 | 17 |
| 1999 | Tornadoes OK KS | 2.0 | 55 |
| 1999 | Drought /Heat Wave | 1.2 | 502 |
| 1999 | Hurricane Floyd | 7.4 | 77 |
| 2000 | Drought/Heat Wave | 4.8 | 140 |
| 2000 | Fires West | 2.4 | 0 |
| 2001 | TS Allison | 5.6 | 43 |
| 2001 | Tornadoes/Hail MW OH Valley | 2.2 | 3 |
| 2002 | Drought 30 States | 11.4 | 0 |
| 2002 | Fires West | 2.3 | 21 |
| 2002 | Tornadoes/Severe | 1.9 | 7 |
| 2003 | Severe/Hail | 1.8 | 3 |
| 2003 | Tornadoes/Severe | 3.8 | 51 |
| 2003 | Hurricane Isabel | 5.6 | 55 |
| 2003 | Wildfires S CA | 2.8 | 22 |

**Table 3.2    Billion Dollar Weather Disasters—Chronological Listing [CONTINUED]**

| Year | Event | 2007 Dollars (billions) | Deaths |
|------|-------|------------------------|--------|
| 2004 | Hurricane Charley | 16.5 | 35 |
| 2004 | Hurricane Frances | 9.9 | 48 |
| 2004 | Hurricane Ivan | 15.4 | 57 |
| 2004 | Hurricane Jeanne | 7.7 | 28 |
| 2005 | Hurricane Dennis | 2.1 | 15 |
| 2005 | Hurricane Katrina | 133.8 | 1833 |
| 2005 | Hurricane Rita | 17.1 | 119 |
| 2005 | Drought MW | 1.1 | 0 |
| 2005 | Hurricane Wilma | 17.1 | 35 |
| 2006 | Wildfires | 1.0 | 28 |
| 2006 | Drought | 6.2 | 0 |
| 2006 | Tornadoes/Severe | 1.0 | 10 |
| 2006 | Floods N East | 1.0 | 20 |
| 2006 | Tornadoes Midwest S'east | 1.5 | 10 |
| 2006 | Tornadoes Midwest OH Valley | 1.1 | 27 |
| 2007 | Drought Plains East | 5.0 | 0 |
| 2007 | Wildfires West | 1.0 | 12 |
| 2007 | Freeze Spring | 2.0 | 0 |
| 2007 | Severe East South | 1.5 | 9 |
| 2007 | Freeze CA | 1.4 | 1 |
| 2008 | Tornadoes S'east Midwest | 1.0 | 57 |
| 2008 | Tornadoes/Severe Midwest OH Valley | 2.4 | 13 |
| 2008 | Tornadoes/Severe Midwest Mid Atl | 1.1 | 18 |
| 2008 | Flood Midwest | 15 | 24 |
| 2008 | Wildfires | 2.0 | 16 |
| 2008 | Hurricane Dolly | 1.2 | 3 |
| 2008 | Hurricane Gustav | 5.0 | 43 |
| 2008 | Hurricane Ike | 27 | 100 |
| 2008 | Drought | 2.0 | 0 |

*Source:* NCDC.

**Midwest flood.** June 2008. Heavy rain and flooding caused significant agricultural loss and property damage in IA, IL, IN, MO, MN, NE, and WI, with IA being hardest hit with widespread rainfall totals ranging from 4 inches to over 16 inches. Estimate of over $15 billion in damages/costs; 24 deaths.

**Midwest/Mid-Atlantic severe weather/tornadoes.** June 2008. An outbreak of tornadoes and thunderstorms over the Midwest/Mid-Atlantic states (IA, IL, IN, KS, NE, MI, MN, MO, OK, WI, MD, VA, WV). Over $1.1 billion in damages/ costs; 18 deaths.

**Midwest/Ohio Valley severe weather/tornadoes.** May 2008. Outbreak of tornadoes over the Midwest/Ohio Valley regions (IL, IN, IA, KS, MN, NE, OK, WY, CO) with 235 tornadoes confirmed. Over $2.4 billion in damages/costs; 13 deaths.

**Southeast/Midwest tornadoes.** February 2008. Series of tornadoes and severe thunderstorms across the Southeast and Midwest states (AL, AR, IN, KY, MS, OH, TN, TX) with 87 tornadoes confirmed. Over $1.0 billion in damages/costs; 57 deaths.

*2007*

**Great Plains and eastern drought.** Entire year, 2007. Severe drought with periods of extreme heat over most of the southeast and portions of the Great Plains, Ohio Valley, and Great Lakes area resulting in major reductions in crop yields, along with very low streamflows and lake levels. Includes states of ND, SD, NE, KS, OK, TX, MN, WI, IA, MO, AR, LA, MS, AL, GA, NC, SC, FL, TN, VA, WV, KY, IN, IL, OH, MI, PA, and NY. Preliminary estimate of well over $5.0 billion in damage/costs; some deaths reported due to heat but not beyond typical annual averages.

**Western wildfires.** Summer–fall 2007. Continued drought conditions and high winds over much of the western US (AK, AZ, CA, ID, UT, MT, NV, OR, and WA) resulting in numerous wildfires, with national acreage burned exceeding 8.9 million acres (mainly in the west), and over 3000 homes and structures destroyed in southern California alone. Well over $1.0 billion in damages/costs; at least 12 deaths.

**Spring freeze.** April 2007. Widespread severe freeze over much of the east and midwest (AL, AR, GA, IL, IN, IA, KS, KY, MS, MO, NE, NC, OH, OK, SC, TN, VA, and WV) causing significant losses in fruit crops, field crops

### Table 3.3    Billion Dollar Weather Disasters by Estimated Cost

| Year | Event | 2007 Dollars (billions) | Deaths |
|---|---|---|---|
| 2005 | Hurricane Katrina | 133.8 | 1833 |
| 1988 | Drought Heat Wave | 71.2 | 7500 |
| 1980 | Drought/Heat Wave | 55.4 | 10000 |
| 1992 | Hurricane Andrew | 40 | 61 |
| 1993 | Floods Midwest | 30.2 | 48 |
| 2008 | Hurricane Ike | 27 | 100 |
| 2005 | Hurricane Rita | 17.1 | 119 |
| 2005 | Hurricane Wilma | 17.1 | 35 |
| 2004 | Hurricane Charley | 16.5 | 35 |
| 2004 | Hurricane Ivan | 15.4 | 57 |
| 1989 | Hurricane Hugo | 15.3 | 86 |
| 2008 | Flood Midwest | 15 | 24 |
| 2002 | Drought 30 States | 11.4 | 0 |
| 2004 | Hurricane Frances | 9.9 | 48 |
| 1998 | Drought/Heat Wave South | 9.5 | 200 |
| 1993 | Blizzard East | 7.9 | 270 |
| 2004 | Hurricane Jeanne | 7.7 | 28 |
| 1995 | Severe S'east/S'west | 7.5 | 32 |
| 1998 | Hurricane Georges | 7.4 | 16 |
| 1999 | Hurricane Floyd | 7.4 | 77 |
| 1996 | Drought S Plains | 6.8 | 0 |
| 1996 | Hurricane Fran | 6.6 | 37 |
| 1983 | Hurricane Alicia | 6.3 | 21 |
| 2006 | Drought | 6.2 | 0 |
| 2001 | TS Allison | 5.6 | 43 |
| 2003 | Hurricane Isabel | 5.6 | 55 |
| 1990 | Freeze CA | 5.5 | 0 |
| 2007 | Drought Plains East | 5 | 0 |
| 2008 | Hurricane Gustav | 5 | 43 |
| 1997 | Floods N Plains | 4.8 | 11 |
| 2000 | Drought/Heat Wave | 4.8 | 140 |
| 1983 | Freeze FL | 4.2 | 0 |
| 1994 | Ice Storm S'east | 4.2 | 27 |
| 1995 | Floods CA | 4.1 | 27 |
| 1995 | Hurricane Opal | 4.1 | 27 |
| 1996 | Blizzard/Flood | 4 | 187 |
| 1991 | Firestorm Oakland | 3.9 | 25 |
| 1997 | Floods W Coast | 3.9 | 36 |
| 2003 | Tornadoes/Severe | 3.8 | 51 |
| 1998 | Freeze CA | 3.2 | 0 |
| 1985 | Hurricane Juan | 2.9 | 63 |
| 1995 | Hurricane Marilyn | 2.9 | 13 |
| 2003 | Wildfires S CA | 2.8 | 22 |
| 1992 | Hurricane Iniki | 2.7 | 7 |
| 1985 | Hurricane Elena | 2.5 | 4 |
| 1986 | Drought/Heat Wave | 2.4 | 100 |
| 2000 | Fires West | 2.4 | 0 |
| 2008 | Tornadoes/Severe Ohio Valley | 2.4 | 13 |
| 1983 | Floods/Storms | 2.3 | 50 |
| 1983 | Floods/Storms | 2.3 | 45 |
| 1985 | Freeze FL | 2.3 | 0 |
| 1991 | Hurricane Bob | 2.3 | 18 |
| 1992 | Nor'easter | 2.3 | 19 |
| 2002 | Fires West | 2.3 | 21 |
| 2001 | Tornadoes/Hail Ohio Valley | 2.2 | 3 |
| 2005 | Hurricane Dennis | 2.1 | 15 |
| 1999 | Tornadoes OK KS | 2 | 55 |
| 2007 | Freeze Spring | 2 | 0 |
| 2008 | Wildfires | 2 | 16 |
| 2008 | Drought | 2 | 0 |
| 1998 | Severe/Hail MN | 1.9 | 1 |

**Table 3.3   Billion Dollar Weather Disasters by Estimated Cost** [CONTINUED]

| Year | Event | 2007 Dollars (billions) | Deaths |
|------|-------|-------------------------|--------|
| 2002 | Tornadoes/Severe | 1.9 | 7 |
| 1998 | Ice Storm New Eng | 1.8 | 16 |
| 2003 | Severe/Hail | 1.8 | 3 |
| 1989 | Drought N Plains | 1.7 | 0 |
| 1990 | Floods S Plains | 1.6 | 13 |
| 1999 | Tornadoes AR TN | 1.6 | 17 |
| 2006 | Tornadoes Midwest S'east | 1.5 | 10 |
| 2007 | Severe East South | 1.5 | 9 |
| 1993 | Drought/Heat Wave S'east | 1.4 | 16 |
| 1993 | WildFires CA | 1.4 | 4 |
| 1994 | TS Alberto | 1.4 | 32 |
| 1994 | Floods TX | 1.4 | 19 |
| 1994 | Wildfire Season | 1.4 | 0 |
| 2007 | Freeze CA | 1.4 | 1 |
| 1996 | Floods Pac NW | 1.3 | 9 |
| 1997 | Flood/Tornadoes Midwest | 1.3 | 67 |
| 1998 | Severe S'east | 1.3 | 132 |
| 1998 | Hurricane Bonnie | 1.3 | 3 |
| 1998 | Floods TX | 1.3 | 31 |
| 1999 | Drought /Heat Wave | 1.2 | 502 |
| 2008 | Hurricane Dolly | 1.2 | 3 |
| 2005 | Drought Midwest | 1.1 | 0 |
| 2006 | Tornadoes MW OH Valley | 1.1 | 27 |
| 2008 | Tornadoes/Svr Midwest Mid Atl | 1.1 | 18 |
| 2006 | Wildfires | 1 | 28 |
| 2006 | Tornadoes/Severe | 1 | 10 |
| 2006 | Floods N East | 1 | 20 |
| 2007 | Wildfires West | 1 | 12 |
| 2008 | Tornadoes S'east Midwest | 1 | 57 |

*Source:* NCDC.

(especially wheat), and the ornamental industry. Temperatures in the teens/20s accompanied by rather high winds, nullified typical crop protection systems. Over $2.0 billion in damage/costs; no deaths reported.

**East/south severe weather.** April 2007. Flooding, hail, tornadoes, and severe thunderstorms across numerous states (CT, DE, GA, LA, ME, MD, MA, MS, NH, NJ, NY, NC, PA, RI, SC, TX, VT, VA) in mid-April, including 3 "killer" tornadoes. Over $1.5 billion in damages/costs; 9 deaths.

**California freeze.** January 2007. Widespread agricultural freeze for nearly two weeks in January, overnight temperatures over a good portion of California dipped into the 20s, destroying numerous agricultural crops, with citrus, berry, and vegetable crops most affected. $1.4 billion estimated in damage/costs; 1 fatality reported.

## 2006

**Widespread drought.** Spring–summer 2006. Rather severe drought affected crops, especially during the spring–summer, centered over the Great Plains region with other areas affected across portions of the south and far west, including states of ND, SD, NE, KS, OK, TX, MN, IA, MO,

AR, LA, MS, AL, GA, FL, MT, WY, CO, NM, and CA. Estimate of over $6.0 (6.2) billion in damages/costs; some heat-related deaths but not beyond typical annual averages.

**Northeast flooding.** June 2006. Severe flooding over portions of the northeast due to several weeks of heavy rainfall, affecting the states of NY, PA, DE, MD, NJ, and VA. Over $1.0 billion in damage/costs; at least 20 deaths reported.

**Midwest/Southeast tornadoes** (Figure 3.4). April 2006. Severe weather and numerous tornadoes affecting the states of OK, KS, MO, NE, KY, OH, TN, IN, MS, GA, and AL on April 6–8, with 3 "killer" tornadoes in TN. Over $1.5 billion in damages/costs; 10 deaths.

**Midwest/Ohio Valley tornadoes.** April 2006. Significant outbreak of tornadoes and severe weather affecting the states of IL, IN, IA, AR, MO, KY, and TN on April 2nd with 5 "killer" tornadoes. Approximately $1.1 billion in damages/costs; 27 deaths.

**Severe storms and tornadoes.** March 2006. Outbreak of tornadoes over portions of the midwest and south during a week-long period, affecting the states of AL, AR, KY,

**Table 3.4    Billion Dollar Weather Disasters by Number of Deaths**

| Year | Event | 2007 Dollars (Billions) | Deaths |
|------|-------|------------------------|--------|
| 1980 | Drought/Heat Wave | 55.4 | 10000 |
| 1988 | Drought Heat Wave | 71.2 | 7500 |
| 2005 | Hurricane Katrina | 133.8 | 1833 |
| 1999 | Drought /Heat Wave | 1.2 | 502 |
| 1993 | Blizzard East | 7.9 | 270 |
| 1998 | Drought/Heat WaveSouth | 9.5 | 200 |
| 1996 | Blizzard/Flood | 4 | 187 |
| 2000 | Drought/Heat Wave | 4.8 | 140 |
| 1998 | Severe SE | 1.3 | 132 |
| 2005 | Hurricane Rita | 17.1 | 119 |
| 2008 | Hurricane Ike | 27 | 100 |
| 1986 | Drought/Heat Wave | 2.4 | 100 |
| 1989 | Hurricane Hugo | 15.3 | 86 |
| 1999 | Hurricane Floyd | 7.4 | 77 |
| 1997 | Flood/Tornadoes MW | 1.3 | 67 |
| 1985 | Hurricane Juan | 2.9 | 63 |
| 1992 | Hurricane Andrew | 40 | 61 |
| 2004 | Hurricane Ivan | 15.4 | 57 |
| 2008 | Tornadoes SE MW | 1 | 57 |
| 2003 | Hurricane Isabel | 5.6 | 55 |
| 1999 | Tornadoes OK KS | 2 | 55 |
| 2003 | Tornadoes/Severe | 3.8 | 51 |
| 1983 | Floods/Storms | 2.3 | 50 |
| 1993 | Floods Midwest | 30.2 | 48 |
| 2004 | Hurricane Frances | 9.9 | 48 |
| 1983 | Floods/Storms | 2.3 | 45 |
| 2001 | TS Allison | 5.6 | 43 |
| 2008 | Hurricane Gustav | 5 | 43 |
| 1996 | Hurricane Fran | 6.6 | 37 |
| 1997 | Floods W Coast | 3.9 | 36 |
| 2005 | Hurricane Wilma | 17.1 | 35 |
| 2004 | Hurricane Charley | 16.5 | 35 |
| 1995 | Severe SE/SW | 7.5 | 32 |
| 1994 | TS Alberto | 1.4 | 32 |
| 1998 | Floods TX | 1.3 | 31 |
| 2004 | Hurricane Jeanne | 7.7 | 28 |
| 2006 | Wildfires | 1 | 28 |
| 1994 | Ice Storm SE | 4.2 | 27 |
| 1995 | Floods CA | 4.1 | 27 |
| 1995 | Hurricane Opal | 4.1 | 27 |
| 2006 | Tornadoes MW OH Valley | 1.1 | 27 |
| 1991 | Firestorm Oakland | 3.9 | 25 |
| 2008 | Flood MW | 15 | 24 |
| 2003 | Wildfires S CA | 2.8 | 22 |
| 1983 | Hurricane Alicia | 6.3 | 21 |
| 2002 | Fires West | 2.3 | 21 |
| 2006 | Floods N East | 1 | 20 |
| 1992 | Nor'easter | 2.3 | 19 |
| 1994 | Floods Texas | 1.4 | 19 |
| 1991 | Hurricane Bob | 2.3 | 18 |
| 2008 | Tornadoes/Severe MW Mid Atl | 1.1 | 18 |
| 1999 | Tornadoes AR TN | 1.6 | 17 |
| 1998 | Hurricane Georges | 7.4 | 16 |
| 2008 | Wildfires | 2 | 16 |
| 1998 | Ice Storm New Eng | 1.8 | 16 |
| 1993 | Drought/Heat Wave SE | 1.4 | 16 |
| 2005 | Hurricane Dennis | 2.1 | 15 |

**Table 3.4   Billion Dollar Weather Disasters by Number of Deaths [CONTINUED]**

| Year | Event | 2007 Dollars (Billions) | Deaths |
|------|-------|------------------------|--------|
| 1995 | Hurricane Marilyn | 2.9 | 13 |
| 2008 | Tornadoes/Severe  MW OH Valley | 2.4 | 13 |
| 1990 | Floods S. Plains | 1.6 | 13 |
| 2007 | Wildfires West | 1 | 12 |
| 1997 | Floods N Plains | 4.8 | 11 |
| 2006 | Tornadoes MW SE | 1.5 | 10 |
| 2006 | Tornadoes/Severe | 1 | 10 |
| 2007 | Severe East South | 1.5 | 9 |
| 1996 | Floods Pac NW | 1.3 | 9 |
| 1992 | Hurricane Iniki | 2.7 | 7 |
| 2002 | Tornadoes/Severe | 1.9 | 7 |
| 1985 | Hurricane Elena | 2.5 | 4 |
| 1993 | WildFires CA | 1.4 | 4 |
| 2001 | Tornadoes/Hail MW OH Valley | 2.2 | 3 |
| 2003 | Severe/Hail | 1.8 | 3 |
| 1998 | Hurricane Bonnie | 1.3 | 3 |
| 2008 | Hurricane Dolly | 1.2 | 3 |
| 1998 | Severe/Hail MN | 1.9 | 1 |
| 2007 | Freeze California | 1.4 | 1 |
| 2002 | Drought 30 States | 11.4 | 0 |
| 1996 | Drought S Plains | 6.8 | 0 |
| 2006 | Drought | 6.2 | 0 |
| 1990 | Freeze California | 5.5 | 0 |
| 2007 | Drought G Plains East | 5 | 0 |
| 1983 | Freeze Florida | 4.2 | 0 |
| 1998 | Freeze California | 3.2 | 0 |
| 2000 | Fires  West | 2.4 | 0 |
| 1985 | Freeze Florida | 2.3 | 0 |
| 2007 | Freeze Spring | 2 | 0 |
| 2008 | Drought | 2 | 0 |
| 1989 | Drought N. Plains | 1.7 | 0 |
| 1994 | Wildfire Season | 1.4 | 0 |
| 2005 | Drought MW | 1.1 | 0 |

*Source:* NCDC.

MS, TN, TX, IN, KS, MO, and OK. Over $1.0 billion in damage/costs; at least 10 deaths.

**Numerous wildfires.** Entire year, 2006. Numerous wildfires, mainly over the western half of the country, due to dry weather and high winds, burning nearly 10 million acres (new record for period since 1960), with the most affected states being AK, AZ, CA, CO, FL, ID, MT, NM, NV, OK, OR, TX, WA, and WY. Well over $1.0 billion in overall damages/costs; at least 28 fatalities including 20 firefighters.

## 2005

**Hurricane Wilma.** October 2005. Category 3 hurricane hits southwestern Florida, resulting in strong damaging winds and major flooding across southeastern Florida. Prior to landfall, Wilma, as a Category 5 hurricane, recorded the lowest pressure (882 mb) ever recorded in the Atlantic

basin. Estimate of approximately $16.0 (17.1) billion in damages/costs; estimated 35 deaths.

**Hurricane Rita.** September 2005. Category 3 hurricane hits Texas–Louisiana border coastal region, creating significant storm surge and wind damage along the coast and some inland flooding in the FL panhandle, AL, MS, LA, AR, and TX. Prior to landfall, Rita reached the third lowest pressure (897 mb) ever recorded in the Atlantic basin. Estimate of approximately $16.0 (17.1) billion in damage/costs; 119 deaths reported—most being indirect (many related to evacuations).

**Hurricane Katrina** (Figure 3.5). August 2005. Category 3 hurricane initially impacts the United States as a Category 1 near Miami, FL, then as a strong Category 3 along the eastern LA–western MS coastlines, resulting in severe storm surge damage (maximum surge probably exceeded 25 feet) along the LA–MS–AL coasts, wind damage, and the failure

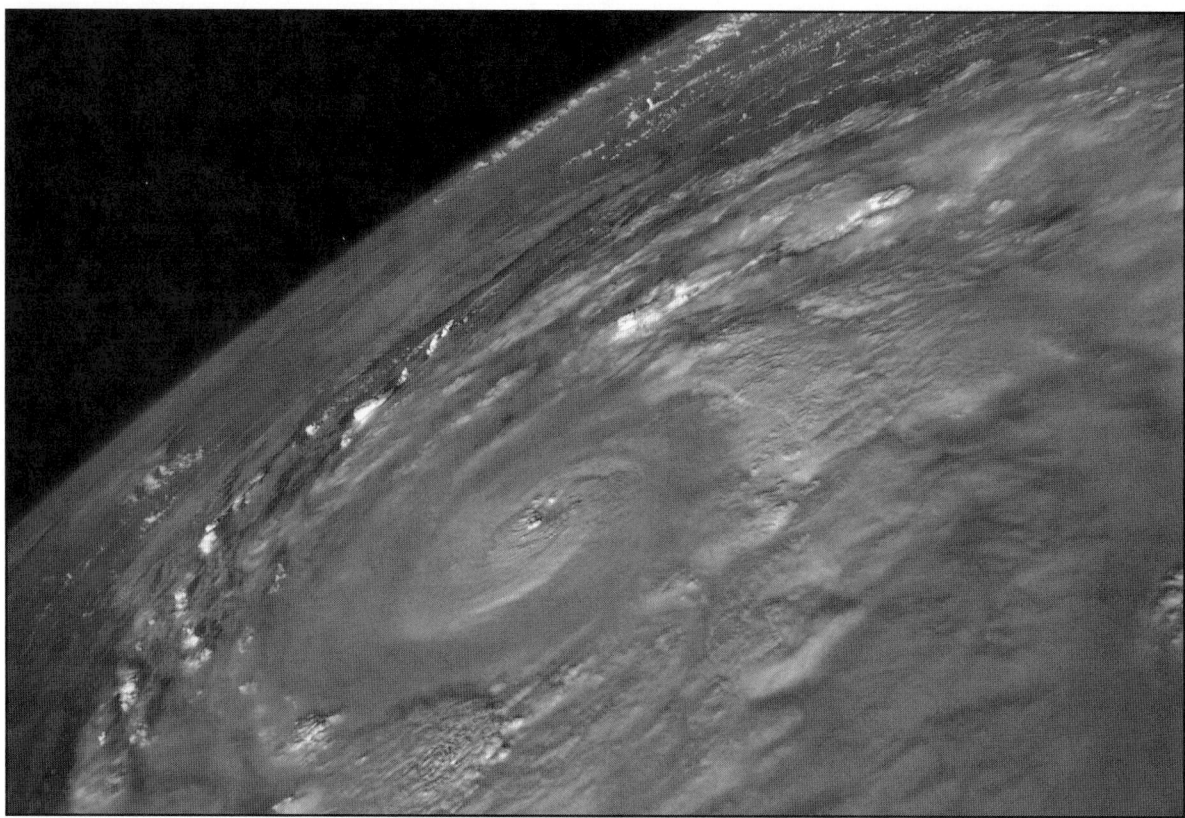

**Figure 3.3**   Hurricane Ike over Cuba, 09.09.2008. (*Source:* NOAA.)

of parts of the levee system in New Orleans. Inland effects included high winds and some flooding in the states of AL, MS, FL, TN, KY, IN, OH, and GA. Estimate of approximately $125 (133.8) billion in damage/costs, making this the most expensive natural disaster in US history; approximately 1833 deaths—the highest US total since the 1928 major hurricane in southern Florida.

**Hurricane Dennis.**   July 2005. Category 3 hurricane makes landfall in western Florida panhandle resulting in storm surge and wind damage along the FL–AL coasts, along with scattered wind and flood damage in GA, MS, and TN. Estimate of over $2.0 (2.1) billion in damage/costs; at least 15 deaths.

**Midwest drought.**   Spring–summer 2005. Rather severe localized drought causes significant crop losses (especially for corn and soybeans) in the states of AR, IL, IN, MO, OH, and WI. Estimate of over $1.0 (1.1) billion in damage/costs; no reported deaths.

*2004*
**Hurricane Jeanne.**   September 2004. Category 3 hurricane makes landfall in east-central Florida, causing considerable wind, storm surge, and flooding damage in FL, with some

flood damage also in the states of GA, SC, NC, VA, MD, DE, NJ, PA, and NY. Puerto Rico also affected. Estimate of over $7.0 (7.7) billion in damage/costs; at least 28 deaths.

**Hurricane Ivan.**   September 2004. Category 3 hurricane makes landfall on Gulf coast of Alabama, with significant wind, storm surge, and flooding damage in coastal AL and FL panhandle, along with wind/flood damage in the states of GA, MS, LA, SC, NC, VA, WV, MD, TN, KY, OH, DE, NJ, PA, and NY. Estimate of over $14.0 (15.4) billion in damage/costs; at least 57 deaths.

**Hurricane Frances.**   September 2004. Category 2 hurricane makes landfall in east-central Florida, causing significant wind, storm surge, and flooding damage in FL, along with considerable flood damage in the states of GA, SC, NC, and NY due to 5–15 inch rains. Estimate of over $9.0 (9.9) billion in damage/costs; at least 48 deaths.

**Hurricane Charley.**   August 2004. Category 4 hurricane makes landfall in southwest Florida, resulting in major wind and some storm surge damage in FL, along with some damage in the states of SC and NC. Estimate of over $15.0 (16.5) billion in damage/costs; at least 35 deaths.

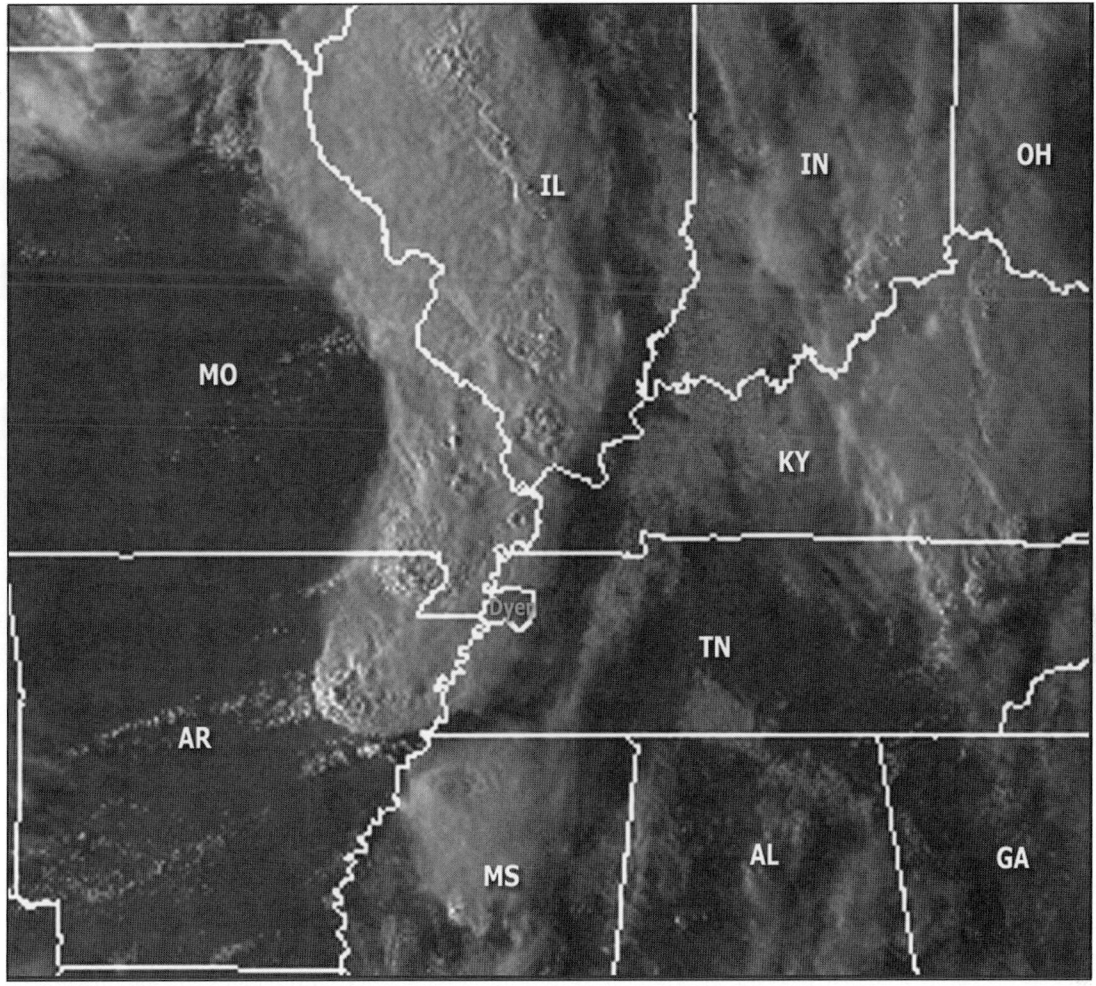

**Figure 3.4**    Severe storms from Arkansas to Illinois, 04.02.2006. (*Source:* NOAA.)

*2003*

**Southern California wildfires.** Late October–early November 2003. Dry weather, high winds, and resulting wildfires in Southern California. More than 743000 acres of brush and timber burned, over 3700 homes destroyed; over $2.5 (2.8) billion in damage/costs; 22 deaths.

**Hurricane Isabel.** September 2003. Category 2 hurricane makes landfall in eastern North Carolina, causing considerable storm surge damage along the coasts of NC, VA, and MD, with wind damage and some flooding due to 4–12 inch rains in NC, VA, MD, DE, WV, NJ, NY, and PA; approximately $5.0 (5.6) billion in damage/costs; 55 deaths.

**Severe storms and tornadoes.** Early May 2003. Numerous tornadoes over the midwest, MS valley, OH/TN valleys, and portions of the southeast, with a modern record one-week total of approximately 400 tornadoes reported; over $3.4 (3.8) billion in damage/costs; 51 deaths.

**Storms and hail.** Early April 2003. Severe storms and large hail over the southern plains and lower MS valley, with Texas being hit hardest, and much of the monetary losses due to hail; over $1.6 (1.8) billion in damage/costs; 3 deaths.

*2002*

**Widespread drought.** Spring through fall 2002. Moderate to extreme drought over large portions of 30 states including the western states, the Great Plains, and much of the eastern United States; estimate of over $10.0 (11.4) billion in damage/costs; no deaths reported.

**Western fire season.** Spring through fall 2002. Major fires over 11 western states from the Rockies to the west coast due to drought and periodic high winds, with over 7.1 million acres burned; over $2.0 (2.3) billion in damage/costs; 21 deaths.

**Figure 3.5** Hurricane Katrina, 08.28.2005. (*Source:* NOAA.)

**Central/Eastern severe weather/tornadoes.** Late April–early May 2002. Numerous tornadoes over the Central and Eastern states (NC, GA, VA, TX, AR, MO, MS, TN, IL, IN, KY, PA, MD, NY, OH, WV, and KS). Over $1.7 billion in damages/costs; 7 deaths.

### 2001

**Tropical Storm Allison.** June 2001. The persistent remnants of Tropical Storm Allison produced rainfall amounts of 30–40 inches in portions of coastal Texas and Louisiana, causing severe flooding, especially in the Houston area, then moved slowly northeastward; fatalities and significant damage reported in TX, LA, MS, FL, VA, and PA; estimate of approximately $5.0 (5.6) billion in damage/costs; at least 43 deaths.

**Midwest and Ohio Valley hail and tornadoes.** April 2001. Storms, tornadoes, and hail in the states of TX, OK, KS, NE, IA, MO, IL, IN, WI, MI, OH, KY, WV, and PA over a 6-day period; over $1.9 (2.2) billion in damage/costs, with the most significant losses due to hail; at least 3 deaths.

### 2000

**Drought/heat wave.** Spring–summer 2000. Severe drought and persistent heat over south-central and southeastern states, causing significant losses to agriculture and related industries;estimate of over $4.0 (4.8) billion in damage/costs; estimated 140 deaths nationwide.

**Western fire season.** Spring–summer 2000. Severe fire season in western states due to drought and frequent winds, with nearly 7 million acres burned; estimate of over $2.0 (2.4) billion in damage/costs (includes fire suppression); no deaths reported.

### 1999

**Hurricane Floyd.** September 1999. Large, Category 2 hurricane makes landfall in eastern NC, causing 10–20 inch rains in 2 days, with severe flooding in NC and some flooding in SC, VA, MD, PA, NY, NJ, DE, RI, CT, MA, NH, and VT; estimate of at least $6.0 (7.4) billion in damage/costs; 77 deaths.

**Eastern drought/heat wave.** Summer 1999. Very dry summer and high temperatures, mainly in eastern United States, with extensive agricultural losses; over $1.0 (1.2) billion in damage/costs; estimated 502 deaths.

**Oklahoma–Kansas tornadoes** (Figures 3.6 and 3.7). May 1999. Outbreak of F4–F5 tornadoes hit the states of Oklahoma and Kansas, along with Texas and Tennessee; Oklahoma City area hardest hit. For the first time, wind speeds over 300 mph were measured in a tornado. At 7:00 P.M. on May 3, near Moore, OK, the Doppler on Wheels program of the University of Oklahoma measured 318 (+/− 10) mph in the F5 tornado. The former record was 257–268 mph

**Figure 3.6**   Tornadic thunderstorms over Oklahoma that caused the Moore, OK F5 tornado, 05.03.1999. (*Source:* NOAA, NASA.)

**Figure 3.7**   Moore, OK tornado hook echo, 05.03.1999. (*Source:* Image by Steve Horstmeyer from NOAA, NCDC data.)

near Red Rock, OK. Damage was over $1.6 (2.0) billion in damage/costs; 55 deaths.

**Arkansas–Tennessee tornadoes.** January 1999. Two outbreaks of tornadoes in 6-day period struck Arkansas and Tennessee; approximately $1.3 (1.6) billion in damage/costs; 17 deaths.

*1998*
**California freeze.** December 1998. A severe freeze damaged fruit and vegetable crops in the Central and Southern San Joaquin Valley. Extended intervals of sub-27°F temperatures occurred over an 8-day period. $2.5 billion estimated damages/costs.

**Texas flooding.** October–November 1998. Severe flooding in southeast Texas from 2 heavy rain events with 10–20 inch rainfall totals; approximately $1.0 (1.3) billion in damage/costs; 31 deaths.

**Hurricane Georges.** September 1998. Category 2 hurricane strikes Puerto Rico, Florida Keys, and Gulf coasts of Louisiana, Mississippi, Alabama, and Florida panhandle, 15–30 inch 2-day rain totals in parts of AL/FL; estimated $5.9 (7.4) billion in damage/costs; 16 deaths.

**Hurricane Bonnie.** August 1998. Category 3 hurricane strikes eastern North Carolina and Virginia; extensive agricultural damage due to winds and flooding with 10-inch rains in 2 days in some locations; approximately $1.0 (1.3) billion in damage/costs; 3 deaths.

**Southern drought/heat wave.** Summer 1998. Severe drought and heat wave from Texas/Oklahoma eastward to the Carolinas; $6.0–$9.0 (7.6–11.3) billion in damage/costs to agriculture and ranching; at least 200 deaths.

**Minnesota severe storms/hail.** May 1998. Very damaging severe thunderstorms with large hail over wide areas of Minnesota; over $1.5 (1.9) billion in damage/costs; 1 death.

**Southeast severe weather.** Winter–spring 1998. Tornadoes and flooding related to El Nino in southeastern states; over $1.0 (1.3) billion in damage/costs; at least 132 deaths.

**Northeast ice storm.** January 1998. Intense ice storm hits Maine, New Hampshire, Vermont, and New York, with extensive forestry losses; over $1.4 (1.8) billion in damage/costs; 16 deaths.

*1997*
**Northern Plains flooding.** April–May 1997. Severe flooding in Dakotas and Minnesota due to heavy spring snowmelt; approximately $3.7 (4.8) billion in damage/costs; 11 deaths.

**Mississippi and Ohio Valleys flooding and tornadoes.** March 1997. Tornadoes and severe flooding hit the states of AR, MO, MS, TN, IL, IN, KY, OH, and WV, with over 10 inches of rain in 24 hours in Louisville; estimated $1.0 (1.3) billion in damage/costs; 67 deaths.

**West Coast flooding.** December 1996–January 1997. Torrential rains (10–40 inches in 2 weeks) and snowmelt produce severe flooding over portions of CA, WA, OR, ID, NV, and MT; approximately $3.0 (3.9) billion in damage/costs; 36 deaths.

*1996*
**Hurricane Fran.** September 1996. Category 3 hurricane strikes North Carolina and Virginia, over 10-inch 24-hour rains in some locations and extensive agricultural and other losses; over $5.0 (6.6) billion in damage/costs; 37 deaths.

**Southern Plains severe drought.** Fall 1995 through summer 1996. Severe drought in agricultural regions of southern plains—Texas and Oklahoma most severely affected; approximately $5.0 (6.8) billion in damage/costs; no deaths.

**Pacific Northwest severe flooding.** February 1996. Very heavy, persistent rains (10–30 inches) and melting snow over OR, WA, ID, and western MT; approximately $1.0 (1.3) billion in damage/costs; 9 deaths.

**Blizzard of 1996 followed by flooding.** January 1996. Very heavy snowstorm (1–4 feet) over Appalachians, Mid-Atlantic, and Northeast; followed by severe flooding in parts of same area due to rain and snowmelt; approximately $3.0 (4.0) billion in damage/costs; 187 deaths.

*1995*
**Hurricane Opal.** October 1995. Category 3 hurricane strikes Florida panhandle, Alabama, western Georgia, eastern Tennessee, and the western Carolinas, causing storm surge, wind, and flooding damage; over $3.0 (4.1) billion in damage/costs; 27 deaths.

**Hurricane Marilyn.** September 1995. Category 2 hurricane devastates US Virgin Islands; estimated $2.1 (2.9) billion in damage/costs; 13 deaths.

**Texas/Oklahoma/Louisiana/Mississippi severe weather and flooding.** May 1995. Torrential rains, hail, and tornadoes across Texas–Oklahoma and southeast Louisiana–southern Mississippi, with Dallas and New Orleans areas (10–25 inch rains in 5 days) hardest hit; $5.0–$6.0 (6.8–8.2) billion in damage/costs; 32 deaths.

**California flooding.** January–March 1995. Frequent winter storms cause 20–70 inch rainfall and periodic flooding across much of California; over $3.0 (4.1) billion in damage/costs; 27 deaths.

### 1994

**Western fire season.** Summer–fall 1994. Severe fire season in western states due to dry weather; approximately $1.0 (1.4) billion in damage/costs; death toll undetermined.

**Texas flooding.** October 1994. Torrential rain (10–25 inches in 5 days) and thunderstorms cause flooding across much of southeast Texas; approximately $1.0 (1.4) billion in damage/costs; 19 deaths.

**Tropical Storm Alberto.** July 1994. Remnants of slow-moving Alberto bring torrential 10–25 inch rains in 3 days, widespread flooding and agricultural damage in parts of Georgia, Alabama, and panhandle of Florida; approximately $1.0 (1.4) billion in damage/costs; 32 deaths.

**Southeast ice storm.** February 1994. Intense ice storm with extensive damage in portions of TX, OK, AR, LA, MS, AL, TN, GA, SC, NC, and VA; approximately $3.0 (4.2) billion in damage/costs; 9 deaths.

### 1993

**California wildfires.** Fall 1993. Dry weather, high winds and wildfires in Southern California; approximately $1.0 (1.4) billion in damage/costs; 4 deaths.

**Midwest flooding.** Summer 1993. Severe, widespread flooding in the central United States due to persistent heavy rains and thunderstorms; approximately $21.0 (30.2) billion in damage/costs; 48 deaths.

**Drought/heat wave.** Summer 1993. Southeastern United States; about $1.0 (1.4) billion damage/costs to agriculture; at least 16 deaths.

**Storm/blizzard.** March 1993. "Storm of the Century" hits entire eastern seaboard with tornadoes (FL), high winds, and heavy snows (2–4 feet); $5.0–$6.0 (7.2–8.6) billion damage/costs; approximately 270 deaths.

### 1992

**Nor'easter of 1992.** December 1992. Slow-moving storm batters northeast US coast, New England hardest hit; $1.0–$2.0 (1.5–3.0) billion in damage/costs; 19 deaths.

**Hurricane Iniki.** September 1992. Category 4 hurricane hits Hawaiian island of Kauai; about $1.8 (2.7) billion in damage/costs; 7 deaths.

**Hurricane Andrew** (Figure 3.8). August 1992. Category 4 hurricane hits Florida and Louisiana, high winds damage or destroy over 125,000 homes; approximately $27.0 (40.0) billion in damage/costs; 61 deaths.

### 1991

**Oakland firestorm.** October 1991. Oakland, California firestorm due to low humidities and high winds; approximately $2.5 (3.9) billion in damage/costs; 25 deaths.

**Hurricane Bob.** August 1991. Category 2 hurricane—mainly coastal North Carolina, Long Island, and New England; $1.5 (2.3) billion in damage/costs; 18 deaths.

### 1990

**California freeze.** December 1990. Severe freeze in the Central and Southern San Joaquin Valley caused the loss of citrus, avocado trees, and other crops in many areas. Several days of subfreezing temperatures occurred, with some valley locations in the teens. $3.4 billion in direct and indirect economic losses, including damage to public buildings, utilities, crops, and residences.

**Texas/Oklahoma/Louisiana/Arkansas flooding.** May 1990. Torrential rains cause flooding along the Trinity, Red, and Arkansas Rivers in TX, OK, LA, and AR; over $1.0 (1.6) billion in damage/costs; 13 deaths.

### 1989

**Hurricane Hugo.** September 1989. Category 4 hurricane devastates South and North Carolina with ~20 foot storm surge and severe wind damage after hitting Puerto Rico and the US Virgin Islands; over $9.0 (15.3) billion in damage/costs (about $7.1 (12.1) billion in Carolinas); 86 deaths (57 in US mainland and 29 in US Islands).

**Northern Plains drought.** Summer 1989. Severe summer drought over much of the northern plains with significant losses to agriculture; at least $1.0 (1.7) billion in damage/costs; no deaths reported.

### 1988

**Drought/heat wave.** Summer 1988. 1988 drought in central and eastern United States with very severe losses to agriculture and related industries; estimated $40.0 (71.2) billion in damage/costs; estimated 5000 to 10000 deaths (includes heat stress-related).

### 1987

None.

### 1986

**Southeast drought/heat wave.** Summer 1986. Severe summer drought in parts of the southeastern United States with severe losses to agriculture; $1.0–$1.5 (1.9–2.8) billion in damage/costs; estimated 100 deaths.

### 1985

**Hurricane Juan.** October–November 1985. Category 1 hurricane—Louisiana and Southeast United States—severe flooding; $1.5 (2.9) billion in damage/costs; 63 deaths.

**Figure 3.8** Hurricane Andrew, after it crossed Florida, 08.24.1992. (*Source:* NOAA.)

**Hurricane Elena.** August–September 1985. Category 3 hurricane—Florida to Louisiana; $1.3 (2.5) billion in damage/costs; 4 deaths.

**Florida freeze.** January 1985. Severe freeze central/northern Florida; about $1.2 (2.3) billion in damage to citrus industry; no deaths.

*1984*
None.

*1983*
**Florida freeze.** December 1983. Severe freeze central/northern Florida; about $2.0 (4.2) billion in damage to citrus industry; no deaths.

**Hurricane Alicia.** August 1983. Category 3 hurricane—Texas; $3.0 (6.3) billion in damage/costs; 21 deaths.

**Western storms and flooding.** 1982–early 1983. Storms and flooding related to El Nino, especially in the states of WA, OR, CA, AZ, NV, ID, UT, and MT; approximately $1.1 (2.3) billion in damage/costs; at least 45 deaths.

**Gulf States storms and flooding.** 1982–early 1983. Storms and flooding related to El Nino, especially in the states of TX, AR, LA, MS, AL, GA, and FL; approximately $1.1 (2.3) billion in damage/costs; at least 50 deaths.

*1982*
None.

*1981*
None.

*1980*
**Drought/heat wave.** June–September 1980. Central and eastern United States; estimated $20.0 ($55.4) billion in

## Table 3.5  State Record Maximum and Minimum Temperature Records by Month

**ALABAMA**

| Month | Max | Max Year/Location | Min | Min Year/Location |
|---|---|---|---|---|
| Jan | 88 | 1898 Mount Willing | -27 | **1966 New Market** |
| Feb | 89 | 1909 Livingston | -18 | 1905 Valley Head |
| Mar | 94 | 1916 Evergreen | 2 | 1993 Birmingham |
| Apr | 98 | 1894 Union | 19 | 1992 Valley Head |
| May | 105 | 1964 Chatom | 29 | 1976 Valley Head |
| Jun | 109 | 1933 Brewton | 35 | 1966 Valley Head |
| Jul | 111 | 1930 Madison | 41 | 1947 Valley Head |
| Aug | 109 | 2000 Centerville | 39 | 1968 Waterloo |
| Sep | 112 | **1925 Centerville** | 29 | 1967 Valley Head |
| Oct | 103 | 1954 Troy | 19 | 1968 Waterloo |
| Nov | 92 | 1922 Selma | -2 | 1950 Valley Head |
| Dec | 88 | 1971 Livingston | -10 | 1983 Heflin |

**ALASKA**

| Month | Max | Max Year/Location | Min | Min Year/Location |
|---|---|---|---|---|
| Jan | 62 | 1981 Petersburg | -80 | **1971 Prospect Creek Camp** |
| Feb | 66 | 1992 Petersburg | -75 | 1947 Tanacross |
| Mar | 69 | 1936 Dutch Harbor | -68 | 1971 Kobuk |
| Apr | 82 | 1976 Annette | -50 | 1986 Umiat |
| May | 92 | 1960 Ladd AFB | -25 | 1992 Chandalar Lake |
| Jun | 100 | **1915 Fort Yukon** | -1 | 1967 Anaktuvuk Pass |
| Jul | 98 | 1976 Haines | 16 | 1969 White Mountain |
| Aug | 99 | 1976 Tenakee Springs | 8 | 1922 Bonanza Mine |
| Sep | 88 | 1940 Tree Point | -13 | 1970 Arctic Village |
| Oct | 74 | 1969 Goose Bay N. | -48 | 1975 Clear Water |
| Nov | 67 | 1970 Annette | -61 | 1935 Fort Yukon |
| Dec | 64 | 1934 Sitka | -72 | 1999 Chicken |

**ARIZONA**

| Month | Max | Max Year/Location | Min | Min Year/Location |
|---|---|---|---|---|
| Jan | 90 | 2000 Organ Pipe NM | -40 | **1971 Hawley Lake** |
| Feb | 97 | 1986 Yuma | -37 | 1985 Flagstaff |
| Mar | 103 | 1896 Fort Mohave | -26 | 1966 Maverick |
| Apr | 113 | 1898 Parker | -16 | 1980 Hawley Lake |
| May | 121 | 1910 Gila Bend | 4 | 1964 Pinion |
| Jun | 128 | **1994 Lake Havasu Cty** | 13 | 1971 Alpine |
| Jul | 127 | 1905 Parker | 25 | 1997 Flagstaff |
| Aug | 124 | 1933 Parker | 20 | 1968 Fort Valley |
| Sep | 123 | 1950 Yuma | 11 | 1934 Williams |
| Oct | 116 | 1917 Sentinel | -9 | 1949 Fort Valley |
| Nov | 100 | 1931 Granite Reef Dm | -30 | 1931 Fort Defiance |
| Dec | 92 | 1958 Bouse | -36 | 1990 Flataff |

**ARKANSAS**

| Month | Max | Max Year/Location | Min | Min Year/Location |
|---|---|---|---|---|
| Jan | 90 | 1928 Dumas | -28 | 1930 Lead Hill |
| Feb | 95 | 1986 Sparkman | -29 | **1905 Pond** |
| Mar | 98 | 1929 Harrison | -14 | 1948 Lead Hill |
| Apr | 99 | 1927 Harrison | 12 | 1957 Lead Hill |
| May | 107 | 1926 Newport | 24 | 1903 Pond |
| Jun | 113 | 1936 Corning | 35 | 1969 Mammoth Spring |
| Jul | 116 | 1954 Ozark | 40 | 1924 Corning |
| Aug | 120 | **1936 Ozark** | 38 | 1986 Yellville |
| Sep | 113 | 2000 Gillham Dam | 28 | 1989 Lead Hill |
| Oct | 105 | 1938 Grannis | 12 | 1917 Dutton |
| Nov | 96 | 1927 Dumas | 0 | 1976 Gilbert |
| Dec | 88 | 1918 Camden | -21 | 1917 Gravette |

**CALIFORNIA**

| Month | Max | Max Year/Location | Min | Min Year/Location |
|---|---|---|---|---|
| Jan | 97 | 1991 Indio | -45 | **1937 Boca** |
| Feb | 100 | 1986 Thermal | -43 | 1989 Boca |
| Mar | 105 | 1900 Ogilby | -35 | 1969 White Mountain |
| Apr | 118 | 1898 Volcano Springs | -30 | 1970 White Mountain |
| May | 122 | 2000 Death Valley | -15 | 1964 White Mountain |
| Jun | 129 | 1902 Volcano Springs | 2 | 1967 White Mountain |
| Jul | 134 | **1913 Greenland Ranch** | 12 | 1997 Bodie |
| Aug | 127 | 1993 Death Valley | 12 | 2000 Bodie |
| Sep | 126 | 1950 Mecca | -5 | 1973 White Mountain |
| Oct | 117 | 980 Mecca | -20 | 1961 White Mountain |
| Nov | 105 | 1906 Craftonville | -28 | 1928 White Mountain |
| Dec | 100 | 1938 La Mesa | -40 | 1972 Termo |

**COLORADO**

| Month | Max | Max Year/Location | Min | Min Year/Location |
|---|---|---|---|---|
| Jan | 84 | 1916 Las Animas | -60 | 1979 Maybell |
| Feb | 90 | 1904 Blaine | -61 | **1985 Maybell** |
| Mar | 96 | 1909 Holly | -46 | 1943 Columbine |
| Apr | 100 | 1992 Las Animas | -35 | 1891 Breckenridge |
| May | 107 | 2000 La Junta 20 mi S | -10 | 1896 Unknown |
| Jun | 113 | 1953 Eversoll Ranch | 3 | 1891 Breckenridge |
| Jul | 114 | **1954 Sedwick** | 10 | 1891 Breckenridge |
| Aug | 112 | 1938 Sedgwick | 5 | 1981 Wolf Creek Ranch |
| Sep | 107 | 1947 Eads | -9 | 1920 Crested Butte |
| Oct | 100 | 1910 Sheridan Lake | -20 | 1917 West Cliffe |
| Nov | 93 | 1915 Sedgwick | -41 | 1931 Garnett |
| Dec | 87 | 1890 Unknown | -53 | 1986 Taylor Park |

## Table 3.5  State Record Maximum and Minimum Temperature Records by Month [CONTINUED]

**CONNECTICUT**

| Month | Max | Max Year/Location | Min | Min Year/Location |
|---|---|---|---|---|
| Jan | 73 | 1950 Waterbury | -32 | **1961 Coventry** |
| Feb | 77 | 1954 Danbury | -32 | **1943 Falls Village** |
| Mar | 92 | 1998 Danbury | -24 | 1967 Coventry |
| Apr | 96 | 1976 Hartford | -7 | 1923 Voluntown |
| May | 99 | 1996 Hartford | 20 | 1970 Falls Village |
| Jun | 103 | 1933 Falls Village | 27 | 1972 Burlington |
| Jul | 106 | **1995 Danbury** | 32 | 1909 Storrs |
| Aug | 104 | 1948 Norwalk | 28 | 1965 Coventry |
| Sep | 103 | 1953 Waterbury | 17 | 1947 Falls Village |
| Oct | 94 | 1927 Waterbury | 11 | 1936 Falls Village |
| Nov | 84 | 1982 Bulls Bridge Dam | -8 | 1938 Salisbury |
| Dec | 77 | 1998 Norwich | -22 | 1942 Salisbury |

**DELAWARE**

| Month | Max | Max Year/Location | Min | Min Year/Location |
|---|---|---|---|---|
| Jan | 78 | 1950 Lewes | -17 | **1893 Millsboro** |
| Feb | 86 | 1999 Lewes | -15 | 1979 Middletown |
| Mar | 91 | 1948 Millsboro | 1 | 1984 Middletown |
| Apr | 99 | 1896 Millsboro | 11 | 1923 Wilmington |
| May | 100 | 1925 Frederick | 25 | 1947 Bridgeville |
| Jun | 104 | 1984 Middletown | 34 | 1897 Kirkwood |
| Jul | 110 | **1930 Millsboro** | 41 | 1988 Newark Univ Farm |
| Aug | 107 | 1918 Wilmington | 41 | 1952 Millsboro |
| Sep | 102 | 1912 Millsboro | 31 | 1947 Newark |
| Oct | 97 | 1941 Bridgeville | 19 | 1969 Milford |
| Nov | 88 | 1950 Lewes | 8 | 1938 Milford |
| Dec | 77 | 1998 Lewes | -12 | 1942 Bridgeville |

**FLORIDA**

| Month | Max | Max Year/Location | Min | Min Year/Location |
|---|---|---|---|---|
| Jan | 92 | 1944 Bushnell | 2 | 1985 Chipley |
| Feb | 94 | 1962 Avon Park | -2 | **1899 Tallahassee** |
| Mar | 100 | 1997 RoyalPalm | 13 | 1980 Smith Creek |
| Apr | 100 | 1971 Pompano Bch | 20 | 1987 Niceville |
| May | 106 | 1962 Monticello | 34 | 1971 Fountain |
| Jun | 109 | **1931 Monticello** | 43 | 1966 De Funiak Springs |
| Jul | 108 | 1930 Cottage Hill | 49 | 1967 Fountain |
| Aug | 105 | 1987 Crescent City | 52 | 1967 Starke |
| Sep | 107 | 1925 Bonifay | 35 | 1967 De Funiak Springs |
| Oct | 100 | 1904 Orange City | 25 | 1917 Bristol |
| Nov | 96 | 1986 Oasis RS | 12 | 1970 Steinhatchee/McCain |
| Dec | 95 | 1978 Tamiami Trail | 5 | 1962 Crestview |

**GEORGIA**

| Month | Max | Max Year/Location | Min | Min Year/Location |
|---|---|---|---|---|
| Jan | 89 | 1975 Lumber City | -17 | **1940 Lafayette** |
| Feb | 90 | 1918 Glennville | -12 | 1899 Diamond |
| Mar | 99 | 1907 Brunswick | -5 | 1993 Blairsville |
| Apr | 99 | 1986 Brunswick | 12 | 1983 Clayton |
| May | 108 | 1953 Bainbridge | 25 | 1963 Blairsville |
| Jun | 110 | 1959 Warrenton | 34 | 1966 Blairsville |
| Jul | 112 | **1952 Louisville** | 40 | 1937 Blairsville |
| Aug | 112 | **1983 Greenville** | 40 | 1968 Clayton |
| Sep | 111 | 1925 Americus | 26 | 1967 Blairsville |
| Oct | 105 | 1954 Fort Gaines | 14 | 1961 Blairsville |
| Nov | 93 | 1961 Brooklet | 0 | 1950 Blairsville |
| Dec | 89 | 1922 St George | -9 | 1962 Blairsville |

**HAWAII**

| Month | Max | Max Year/Location | Min | Min Year/Location |
|---|---|---|---|---|
| Jan | 93 | 1963 Pahala, Hawaii | 14 | 1961 Haleakala Summit, Maui |
| Feb | 94 | 1994 Kahoolawe | 16 | 1981 Mauna Kea, Hawaii |
| Mar | 95 | 1934 Honokaa, Hi | 18 | 1981 Mauna Kea, Hawaii |
| Apr | 100 | **1931 Pahala, Hi** | 18 | 1979 Mauna Kea, Hawaii |
| May | 98 | 1979 Puukohola, Hi | 12 | **1979 Mauna Kea, Hawaii** |
| Jun | 97 | 1919 Mahukona, Hi | 23 | 1981 Mauna Kea, Hawaii |
| Jul | 98 | 1915 Waianae, Oahu | 22 | 1979 Mauna Kea, Hawaii |
| Aug | 98 | 1951 Puunene, Maui | 17 | 1979 Mauna Kea, Hawaii |
| Sep | 98 | 1990 Sea Mtn, Hawaii | 23 | 1979 Mauna Kea, Hawaii |
| Oct | 97 | 1919 Mahukona, Hi | 20 | 1979 Mauna Kea, Hawaii |
| Nov | 96 | 1987 Puukohola, Hi | 23 | 1981 Mauna Kea, Hawaii |
| Dec | 97 | 1913 Mahukona, Hi | 17 | 1981 Mauna Kea, Hawaii |

**IDAHO**

| Month | Max | Max Year/Location | Min | Min Year/Location |
|---|---|---|---|---|
| Jan | 70 | 1953 Grand View | -60 | **1943 Island Park Dam** |
| Feb | 81 | 1995 Hagerman | -57 | 1933 Tetonia |
| Mar | 90 | 1919 Glenns Ferry | -38 | 1960 Idaho Falls |
| Apr | 100 | 1994 Hagerman | -21 | 1936 Alpha |
| May | 107 | 1910 Garnett | 1 | 1944 Landmark |
| Jun | 113 | 1926 Chattin's Flat | 12 | 1923 Stanley |
| Jul | 118 | **1934 Orofino** | 15 | 1968 Stanley |
| Aug | 116 | 1961 Orofino | 10 | 1916 Elk Creek |
| Sep | 110 | 1950 Riggins | -1 | 1926 Obsidian |
| Oct | 98 | 1992 Swan Falls | -16 | 1917 Big Springs |
| Nov | 85 | 1899 American Falls | -38 | 1896 Chesterfield |
| Dec | 82 | 1917 Magic | -54 | 1983 Stanley |

## Table 3.5 State Record Maximum and Minimum Temperature Records by Month [CONTINUED]

**ILLINOIS**

| Month | Max | Year Location | Min | Year Location |
|---|---|---|---|---|
| Jan | 78 | 1986 Cahokia | -36 | **1999 Congerville** |
| Feb | 83 | 2000 Kaskaska Rvr | -35 | 1996 Elizabeth |
| Mar | 94 | 1929 Harrisburg | -21 | 1962 Freeport |
| Apr | 99 | 1989 Kaskaskia | -2 | 1982 Mount Carroll |
| May | 107 | 1934 Sycamore | 21 | 1989 Elizabeth |
| Jun | 108 | 1954 Palestine | 29 | 1972 Marengo |
| Jul | **117** | **1954 East St. Louis** | 35 | 1904 Lanark |
| Aug | 113 | 1934 La Harpe | 31 | 1915 La Grange |
| Sep | 109 | 1925 Harrisburg | 14 | 1899 Lanarck |
| Oct | 98 | 1953 Harrisburg | 1 | 1925 La Harpe |
| Nov | 89 | 1987 Waterloo | -20 | 1947 Rockford |
| Dec | 79 | 1982 Cairo | -29 | 2000 Elizabeth |

**INDIANA**

| Month | Max | Year Location | Min | Year Location |
|---|---|---|---|---|
| Jan | 80 | 1943 Madison | -36 | **1994 New Whiteland** |
| Feb | 83 | 1938 Shoals | -35 | 1951 Greensburg |
| Mar | 91 | 1929 Madison | -19 | 1943 Goshen |
| Apr | 100 | 1930 Winona Lake | 1 | 1982 Goshen College |
| May | 103 | 1934 Collegeville | 18 | 1978 Frankfort Dsp Plt |
| Jun | 111 | 1936 Seymour | 30 | 1918 La Porte |
| Jul | **116** | **1936 Collegeville** | 37 | 1892 Marion |
| Aug | 111 | 1936 Seymour | 33 | 1946 Frankfort |
| Sep | 108 | 1953 Madison | 21 | 1899 Prairie Creek |
| Oct | 100 | 1939 Madison | 8 | 1906 LaPorte |
| Nov | 91 | 1933 Shoals | -10 | 1950 Wheatfield |
| Dec | 78 | 1982 Evansville | -30 | 1924 Rochester |

**IOWA**

| Month | Max | Year Location | Min | Year Location |
|---|---|---|---|---|
| Jan | 73 | 1950 Keokuk | -47 | **1912 Washta** |
| Feb | 82 | 1972 Sidney | -47 | **1996 Elkader** |
| Mar | 95 | 1986 Ida Grove | -35 | 1962 Hampton |
| Apr | 100 | 1980 Waterloo | -9 | 1982 Manchester |
| May | 111 | 1934 Inwood | 10 | 1967 Swea City |
| Jun | 111 | 1934 Lamoni | 27 | 1946 Decorah |
| Jul | **118** | **1934 Keokuk** | 35 | 1984 Elkader |
| Aug | 116 | 1934 Keokuk | 30 | 1950 Sibley |
| Sep | 107 | 1939 Logan | 15 | 1899 Sibley |
| Oct | 97 | 1963 Forest City | -15 | 1925 Inwood |
| Nov | 86 | 1899 Burlington | -25 | 1887 Unknown |
| Dec | 74 | 1939 Thurman | -40 | 1917 Washta |

**KANSAS**

| Month | Max | Year Location | Min | Year Location |
|---|---|---|---|---|
| Jan | 88 | 1967 Kinsley | -35 | 1947 Centralia |
| Feb | 92 | 1981 Aetna | -40 | **1905 Lebanon** |
| Mar | 100 | 1910 Hugoton | -25 | 1948 Oberlin |
| Apr | 107 | 1989 Hayes | -2 | 1935 Dresden |
| May | 108 | 1939 Ellsworth | 14 | 1909 Wallace |
| Jun | 116 | 1911 Clay Center | 30 | 1917 Irene |
| Jul | **121** | **1936 Alton** | 32 | 1880 Unknown |
| Aug | 119 | 1936 Wellington | 33 | 1910 St Francis |
| Sep | 117 | 1947 Lincoln | 15 | 1984 Kirwin Dam |
| Oct | 104 | 1947 St Francis | -3 | 1917 Wallace |
| Nov | 96 | 1909 Kingman | -20 | 1887 Monument |
| Dec | 90 | 1955 Ashland | -34 | 1989 Atwood |

**KENTUCKY**

| Month | Max | Year Location | Min | Year Location |
|---|---|---|---|---|
| Jan | 83 | 1907 Loretto | -37 | **1994 Shelbyville** |
| Feb | 86 | 1890 Princeton | -33 | 1899 Sandyhook |
| Mar | 94 | 1929 Hopkinsville | -14 | 1960 Benton |
| Apr | 98 | 1925 Farmers | 12 | 1964 Blaine |
| May | 106 | 1896 Ashland | 20 | 1966 Falmouth |
| Jun | 110 | 1936 St John | 29 | 1966 Cumberland |
| Jul | **114** | **1930 Greensburg** | 34 | 1988 Ashland |
| Aug | 113 | 1930 St John | 33 | 1986 Ashland |
| Sep | 110 | 1925 Beaver Dam | 24 | 1928 Farmers |
| Oct | 98 | 1953 Frankfort Lk | 10 | 1997 West Liberty |
| Nov | 93 | 2000 Mammoth Cave | -9 | 1929 Shelbyville |
| Dec | 87 | 1982 Pikeville | -24 | 1989 Beaver Dam |

**LOUISIANA**

| Month | Max | Year Location | Min | Year Location |
|---|---|---|---|---|
| Jan | 92 | 1911 Mellville | -8 | 1940 St. Joseph |
| Feb | 92 | 1918 Minden | -16 | **1899 Minden** |
| Mar | 95 | 1929 Ruston | 10 | 1943 Arcadia |
| Apr | 98 | 1918 Reserve | 24 | 1971 Ashland |
| May | 105 | 1911 Abbeville | 30 | 1925 Delhi |
| Jun | 110 | 1936 Dodson | 41 | 1946 Pollock |
| Jul | 111 | 1930 Plain Dealing | 50 | 1967 Converse |
| Aug | **114** | **1936 Plain Dealing** | 45 | 1891 Winnsboro |
| Sep | 111 | 2000 Shreveport Pk | 30 | 1967 Converse |
| Oct | 103 | 1938 Plain Dealing | 21 | 1952 Chatham |
| Nov | 95 | 1913 Reserve | 10 | 1903 Collinston |
| Dec | 90 | 1913 Donaldsonville | -1 | 1929 Plain Dealing |

## Table 3.5 State Record Maximum and Minimum Temperature Records by Month [CONTINUED]

**MAINE**

| Month | Max | Year Location | Min | Year Location |
|---|---|---|---|---|
| Jan | 65 | 1932 Orono | -48 | **1925 Van Buren** |
| Feb | 69 | 1994 Woodland | -44 | 1993 Jackman |
| Mar | 88 | 1998 Sanford | -40 | 1943 Lac Frontiere |
| Apr | 94 | 1990 West Buxton | -14 | 1964 Clayton Lake |
| May | 101 | 1911 Lewiston | 10 | 1966 Clayton Lake |
| Jun | 101 | 1907 Millinocket | 20 | 1909 Van Buren |
| Jul | **105** | **1911 North Bridgeton** | 29 | 1969 Squa Pan Dam |
| Aug | 104 | 1975 Jonesboro | 26 | 1985 Van Buren |
| Sep | 100 | 1895 Farmington | 12 | 1950 Lac Frontiere |
| Oct | 92 | 1930 Bangor | 1 | 1925 Van Buren |
| Nov | 78 | 1938 Orono | -20 | 1933 Van Buren |
| Dec | 75 | 1998 Kennebunkport | -42 | 1914 Van Buren |

**MARYLAND**

| Month | Max | Year Location | Min | Year Location |
|---|---|---|---|---|
| Jan | 83 | 1950 Western Port | -40 | **1912 Oakland** |
| Feb | 88 | 1932 Cumberland | -29 | 1907 Oakland |
| Mar | 97 | 1998 Baltimore | -20 | 1960 Oakland |
| Apr | 102 | 1902 Boettcherville | -3 | 1923 Grantsville |
| May | 102 | 1904 Boettcherville | 16 | 1913 Deer Park |
| Jun | **109** | **1988 Conowingo Dam** | 25 | 1912 Deer Park |
| Jul | **109** | **1936 Cumberland** | 32 | 1909 Deer Park |
| Aug | **109** | **1918 Cumberland** | 30 | 1930 Oakland |
| Sep | 106 | 1932 College Park | 19 | 1942 Oakland |
| Oct | 99 | 1941 Blackwater | 4 | 1895 Deer Park |
| Nov | 89 | 1974 Elkton | -16 | 1930 Oakland |
| Dec | 85 | 1998 BBaltimore | -32 | 1917 Oakland |

**MASSACHUSETTS**

| Month | Max | Year Location | Min | Year Location |
|---|---|---|---|---|
| Jan | 72 | 1950 Arlington | -35 | **1981 Chester** |
| Feb | 76 | 1985 Chester | -31 | 1971 Chester |
| Mar | 92 | 1998 Reading | -22 | 1943 West Cummington |
| Apr | 100 | 1976 Chester | 2 | 1982 Buffumville Lake |
| May | 101 | 1895 Winthrop | 18 | 1981 Borden Brook Res |
| Jun | 103 | 1984 Chester | 24 | 1979 Chester |
| Jul | 106 | 1982 Chester | 32 | 1888 Williamstown |
| Aug | **107** | **1975 New Bedford** | 27 | 1982 Borden Brook Res |
| Sep | 103 | 1953 Lake Cochituate | 18 | 1974 Chester |
| Oct | 92 | 1930 Weston | 6 | 1976 Chester |
| Nov | 85 | 1982 Chester | -10 | 1938 Adams |
| Dec | 78 | 1998 Brockton | -33 | 1980 Chester |

**MICHIGAN**

| Month | Max | Year Location | Min | Year Location |
|---|---|---|---|---|
| Jan | 72 | 1950 Ann Arbor | -48 | 1915 Humboldt |
| Feb | 72 | 1999 Battle Creek | -51 | **1934 Vanderbilt** |
| Mar | 89 | 1910 Lapeer | -45 | 1943 Fife Lake |
| Apr | 96 | 1915 Seney | -34 | 1923 Bergland |
| May | 100 | 1969 Marquette | 8 | 1966 Fife Lake |
| Jun | 107 | 1934 Houghton Lake | 12 | 1897 Humboldt |
| Jul | **112** | **1936 Mio** | 20 | 1903 Wetmore |
| Aug | 108 | 1918 Morenci | 21 | 1915 Baraga |
| Sep | 104 | 1954 Wayne | 9 | 1942 Watersmeet |
| Oct | 94 | 1922 St. Joseph | -3 | 1905 Humboldt |
| Nov | 84 | 1950 Wayne | -23 | 1950 Pellston |
| Dec | 69 | 1998 Dearborn | -41 | 1983 Stambaugh |

**MINNESOTA**

| Month | Max | Year Location | Min | Year Location |
|---|---|---|---|---|
| Jan | 69 | 1981 Montevideo | -57 | 1996 Embarras |
| Feb | 73 | 1896 Plsnt Mounds | -60 | **1996 Tower** |
| Mar | 88 | 1910 Montevideo | -49 | 1897 Pokegama |
| Apr | 101 | 1980 Hawley | -22 | 1982 Tower |
| May | 112 | 1934 Maple Plaine | 4 | 1909 Pine River Dam |
| Jun | 110 | 1931 Canby | 15 | 1964 Big Fork |
| Jul | **114** | **1936 Moorhead** | 24 | 1997 Tower |
| Aug | 110 | 1988 Montevideo | 21 | 1986 Tower |
| Sep | 111 | 1931 Beardsley | 10 | 1930 Big Falls |
| Oct | 98 | 1963 Beardsley | -16 | 1936 Roseau |
| Nov | 84 | 1950 Winona | -45 | 1896 Pokegama |
| Dec | 74 | 1939 Wheaton | -57 | 1898 Pokegama |

**MISSISSIPPI**

| Month | Max | Year Location | Min | Year Location |
|---|---|---|---|---|
| Jan | 89 | 1950 Duckhill | -19 | **1966 Corinth** |
| Feb | 91 | 1918 Yazoo City | -16 | 1951 Batesville |
| Mar | 96 | 1929 Macon | 7 | 1980 University |
| Apr | 97 | 1987 Greenville | 21 | 1987 University |
| May | 104 | 1951 Hattisburg | 30 | 1976 Tupelo |
| Jun | 111 | 1936 Greenwood | 39 | 1990 Hickory Flat |
| Jul | **115** | **1930 Holly Springs** | 47 | 1947 Batesville |
| Aug | 110 | 2000 Oakley/Vicksburg | 42 | 1891 Aberdeen |
| Sep | 111 | 1925 Pontotoc | 31 | 1967 Houston |
| Oct | 100 | 1954 Canton | 16 | 1952 Houston |
| Nov | 92 | 1935 Monticello | 2 | 1950 Booneville |
| Dec | 87 | 1984 Pikayune | -12 | 1963 Holly Springs |

## Table 3.5 State Record Maximum and Minimum Temperature Records by Month [CONTINUED]

**MISSOURI**

| Month | Max | Year | City | Min | Year | City |
|---|---|---|---|---|---|---|
| Jan | 85 | 1950 | Camdenton | -36 | 1912 | Crocker |
| Feb | 90 | 1962 | Berryman | **-40** | **1905** | **Warsaw** |
| Mar | 98 | 1895 | Pickering | -26 | 1960 | Maryville |
| Apr | 100 | 1895 | Darksville | 2 | 1975 | St Joseph |
| May | 110 | 1934 | Maryville | 17 | 1907 | Bethany |
| Jun | 112 | 1936 | Doniphan | 28 | 1988 | Van Buren RS |
| Jul | 118 | 1954 | Warsaw | 38 | 1972 | Cole Camp |
| Aug | 116 | 1936 | Clinton | 35 | 1988 | Mansfield |
| Sep | 110 | 1925 | Jackson | 17 | 1984 | Edgerton |
| Oct | 100 | 1939 | Edgerton | -3 | 1925 | Grant City |
| Nov | 90 | 1955 | Ozark Beach | -19 | 1991 | Lees Summit |
| Dec | 83 | 1918 | Marble Hill | -36 | 1989 | Plattsburg Wtrworks |

**MONTANA**

| Month | Max | Year | City | Min | Year | City |
|---|---|---|---|---|---|---|
| Jan | 79 | 1919 | Choteau | **-70** | **1954** | **Rogers Pass** |
| Feb | 79 | 1932 | Columbus | -61 | 1899 | Fort Logan |
| Mar | 88 | 1910 | Miles City | -45 | 1906 | Fort Logan |
| Apr | 97 | 1980 | Poplar | -30 | 1940 | Summit |
| May | 105 | 1937 | Rock Springs | -5 | 1954 | Polebridge |
| Jun | 112 | 1988 | Wolf Point | 11 | 1943 | Kings Hill |
| Jul | 117 | 1937 | Medicine Lake | 15 | 1919 | Bowen |
| Aug | 112 | 1961 | Iliad | 5 | 1910 | Bowen |
| Sep | 107 | 1983 | Poplar | -7 | 1926 | Pleasant Valley |
| Oct | 99 | 1910 | Springbrook | -30 | 1935 | Summit |
| Nov | 85 | 1975 | Grassrange | -53 | 1959 | Lincoln |
| Dec | 78 | 1939 | Crow Agency | -55 | 1924 | Wheaton |

**NEBRASKA**

| Month | Max | Year | City | Min | Year | City |
|---|---|---|---|---|---|---|
| Jan | 82 | 1894 | Indianola | -45 | 1912 | Walthill |
| Feb | 85 | 1891 | Unknown | **-47** | **1899** | **Camp Clarke** |
| Mar | 101 | 1910 | Grant | -38 | 1891 | Unknown |
| Apr | 106 | 1910 | Ewing | -17 | 1975 | Agate |
| May | 110 | 1895 | Broken Bow | 3 | 1911 | Canton |
| Jun | 114 | 1936 | Franklin | 20 | 1969 | Agate |
| Jul | 118 | 1936 | Minden | 29 | 1971 | Agate |
| Aug | 117 | 1936 | Pawnee City | 23 | 1910 | Canton |
| Sep | 113 | 1897 | Franklin | 1 | 1926 | Gordon |
| Oct | 104 | 1947 | Gothenburg | -15 | 1925 | Hay Springs |
| Nov | 92 | 1893 | Unknown | -32 | 1887 | Unknown |
| Dec | 87 | 1886 | Unknown | **-47** | **1989** | **Oshkosh** |

**NEVADA**

| Month | Max | Year | City | Min | Year | City |
|---|---|---|---|---|---|---|
| Jan | 84 | 1931 | Logandale | **-50** | **1937** | **San Jacinto** |
| Feb | 92 | 1986 | Logandale | -41 | 1894 | Stofiel |
| Mar | 98 | 1940 | Overton | -33 | 1969 | Diamond Valley-Hall |
| Apr | 106 | 1989 | Laughlin | -12 | 1963 | Ruth |
| May | 116 | 2000 | Callville Bay | -7 | 1990 | Pine Valley Bailey Ranch |
| Jun | 125 | 1994 | Laughlin | 8 | 1891 | Belmont |
| Jul | 124 | 1995 | Laughlin | 17 | 1995 | Wildhorse Reservoir |
| Aug | 122 | 1914 | Leeland | 15 | 1964 | Charleston |
| Sep | 117 | 1955 | Indian Springs | 0 | 1895 | Carlin |
| Oct | 111 | 1963 | Overton | -10 | 1970 | Mountain City RS |
| Nov | 98 | 1944 | Mesquite | -23 | 1896 | Wells |
| Dec | 91 | 1910 | Las Vegas | -45 | 1924 | San Jacinto |

**NEW HAMPSHIRE**

| Month | Max | Year | City | Min | Year | City |
|---|---|---|---|---|---|---|
| Jan | 68 | 1950 | Nashua | **-47 1934** | | **Mount Washington** |
| Feb | 72 | 1997 | Greenland | -45 1920 | | Pittsburg |
| Mar | 89 | 1998 | Concord | -38 1950 | | Mount Washington |
| Apr | 95 | 1976 | Concord | -20 1995 | | Mount Washington |
| May | 100 | 1911 | Franklin | -2 1966 | | Mount Washington |
| Jun | 102 | 1991 | North Conway | 13 1980 | | Mount Washington |
| Jul | 106 | 1911 | Nashua | 25 1982 | | Mount Washington |
| Aug | 105 | 1948 | Nashua | 20 1986 | | Mount Washington |
| Sep | 102 | 1953 | Windham | 9 1992 | | Mount Washington |
| Oct | 91 | 1947 | West Lebanon | -2 1976 | | Mount Washington |
| Nov | 84 | 1950 | Windham | -20 1958 | | Mount Washington |
| Dec | 76 | 1998 | Epping | -44 1933 | | Pittsburg |

**NEW JERSEY**

| Month | Max | Year | City | Min | Year | City |
|---|---|---|---|---|---|---|
| Jan | 78 | 1967 | Atlantic City | **-34** | **1904** | **River Vale** |
| Feb | 80 | 1930 | Pleasantville | -26 | 1934 | Canoe Brook |
| Mar | 92 | 1921 | Woodbine | -15 | 1906 | Layton |
| Apr | 98 | 1896 | Passaic | 3 | 1923 | Culvers Lake |
| May | 102 | 1895 | Paterson | 18 | 1947 | Layton |
| Jun | 106 | 1969 | Atlantic City | 29 | 1938 | Layton |
| Jul | 110 | 1936 | Runyon | 33 | 1929 | Layton |
| Aug | 108 | 1918 | Flemington | 32 | 1940 | Charlotteburg |
| Sep | 109 | 1895 | Somerville | 18 | 1947 | Layton |
| Oct | 97 | 1941 | Tuckerton | 9 | 1936 | Layton |
| Nov | 88 | 1950 | Elizabeth | -7 | 1938 | Runyon |
| Dec | 78 | 1998 | Belleplain | -21 | 1917 | Culvers Lake |

## Table 3.5 State Record Maximum and Minimum Temperature Records by Month [CONTINUED]

**NEW MEXICO**

| Month | Max | Year | City | Min | Year | City |
|---|---|---|---|---|---|---|
| Jan | 89 | 1938 | Tularosa | -47 | 1971 | Eagle Nest |
| Feb | 100 | 1904 | Carlsbad | **-50** | **1951** | **Gavilan** |
| Mar | 99 | 1946 | Roswell | -34 | 1948 | Eagle Nest |
| Apr | 104 | 1934 | Artesia | -36 | 1945 | Eagle Nest |
| May | 110 | 2000 | Carlsbad | -2 | 1967 | Eagle Nest |
| Jun | 122 | 1994 | Waste Iso Plt | 10 | 1982 | Lybrook |
| Jul | 116 | 1934 | Orogrande | 19 | 1935 | Therma |
| Aug | 115 | 1994 | Waste Iso Plt | 23 | 1944 | Selsor Ranch |
| Sep | 112 | 1948 | Orogrande | 8 | 1912 | Elizabethtown |
| Oct | 101 | 2000 | Bitter Lks WLR | -15 | 1945 | Red River |
| Nov | 97 | 1903 | Carlsbad | -38 | 1976 | Eagle Nest |
| Dec | 92 | 1933 | Hagerman | -47 | 1961 | Dulce |

**NEW YORK**

| Month | Max | Year | City | Min | Year | City |
|---|---|---|---|---|---|---|
| Jan | 75 | 1950 | Dansville | -46 | 1904 | Paul Smiths |
| Feb | 78 | 1985 | Millbrook | **-52** | **1979** | **Old Forge** |
| Mar | 91 | 1945 | Bedford Hills | -41 | 1938 | Chazy |
| Apr | 98 | 1976 | New Paltz | -24 | 1923 | North Lake |
| May | 102 | 1962 | Poughkeepsie | 10 | 1903 | Paul Smiths |
| Jun | 105 | 1919 | Indian Lake | 21 | 1909 | Nehasane |
| Jul | 108 | 1926 | Troy | 25 | 1963 | Allegheny State Park |
| Aug | 106 | 1948 | Cairo | 20 | 1909 | Indian Lake |
| Sep | 107 | 1953 | Elmira | 15 | 1947 | Roxbury |
| Oct | 99 | 1927 | Addison | 2 | 1925 | Indian Lake |
| Nov | 87 | 1950 | Elmira | -24 | 1933 | Franklinville |
| Dec | 77 | 1998 | Islip | -47 | 1933 | Philadelphia |

**NORTH CAROLINA**

| Month | Max | Year | City | Min | Year | City |
|---|---|---|---|---|---|---|
| Jan | 86 | 1932 | Kinston | **-34** | **1985** | **Mt Mitchell** |
| Feb | 88 | 1962 | New Bern | -23 | 1958 | Mt Mitchell |
| Mar | 100 | 1907 | Southern Pines | -15 | 1964 | Mt Mitchell |
| Apr | 100 | 1990 | Lewiston | 1 | 1944 | Mt Mitchell |
| May | 105 | 1941 | Belhaven | 11 | 1956 | Mt Mitchell |
| Jun | 107 | 1959 | Lake Mitchie | 27 | 1972 | Banner Elk |
| Jul | 109 | 1940 | Albemarle | 32 | 1961 | Celo |
| Aug | 110 | 1983 | Fayetteville | 31 | 1930 | Banner Elk |
| Sep | 109 | 1954 | Weldon | 23 | 1999 | Oconaluftee |
| Oct | 102 | 1954 | Albemarle | 8 | 1952 | Banner Elk |
| Nov | 90 | 1950 | Greenville | -21 | 1929 | Mt Mitchell |
| Dec | 86 | 1998 | Goldsboro | -22 | 1962 | Mt Mitchell |

**NORTH DAKOTA**

| Month | Max | Year | City | Min | Year | City |
|---|---|---|---|---|---|---|
| Jan | 70 | 1908 | Chilcot | -56 | 1916 | Goodall |
| Feb | 72 | 1992 | Fort Yates | **-60** | **1936** | **Parshall** |
| Mar | 90 | 1910 | Edmore | -48 | 1897 | McKinney |
| Apr | 101 | 1980 | Oakes | -24 | 1975 | Powers Lake |
| May | 111 | 1934 | Langdon | -3 | 1967 | Larimore |
| Jun | 112 | 2002 | Brein/Flasher | 18 | 1969 | Belcourt Indian Res |
| Jul | 121 | 1936 | Steele | 23 | 1911 | Manfred |
| Aug | 115 | 1922 | Cando | 19 | 1915 | New Rockford |
| Sep | 109 | 1906 | Larimore | 4 | 1942 | Parshall |
| Oct | 98 | 1963 | Watford City | -18 | 1919 | Zap |
| Nov | 88 | 1909 | Haley | -39 | 1985 | Pembina |
| Dec | 70 | 1939 | New England | -50 | 1983 | Tioga |

**OHIO**

| Month | Max | Year | City | Min | Year | City |
|---|---|---|---|---|---|---|
| Jan | 79 | 1950 | Chesapeake | **-37** | **1994** | **Logan** |
| Feb | 81 | 1930 | Middleport | -39 | 1899 | Milligan |
| Mar | 96 | 1907 | Portsmouth | -21 | 1984 | Fredericktown |
| Apr | 97 | 1925 | Portsmouth | -4 | 1982 | Dorset |
| May | 102 | 1914 | Brilliant | 17 | 1968 | Toledo Sewage |
| Jun | 108 | 1934 | Germantown | 27 | 1972 | Danville |
| Jul | 113 | 1934 | Gallipolis | 34 | 1988 | Cardwell |
| Aug | 111 | 1947 | Napoleon | 27 | 1982 | Canfield |
| Sep | 107 | 1953 | Philo | 23 | 1928 | Peebles |
| Oct | 99 | 1884 | Ironton | 8 | 1895 | Coalton |
| Nov | 89 | 1938 | Gallipolis | -17 | 1958 | Mansfield |
| Dec | 80 | 1982 | Chillicothe | -32 | 1884 | Wauseon |

**OKLAHOMA**

| Month | Max | Year | City | Min | Year | City |
|---|---|---|---|---|---|---|
| Jan | 92 | 1911 | Cloud Chief | **-27** | **1930** | **Watts** |
| Feb | 99 | 1918 | Arapaho | **-27** | **1905** | **Vinita** |
| Mar | 104 | 1971 | Frederick | -18 | 1948 | Kenton |
| Apr | 106 | 1972 | Mangum Research | 6 | 1936 | Boise City |
| May | 114 | 2000 | Weatherford | 19 | 1909 | Hooker |
| Jun | 120 | 1994 | Tipton | 34 | 1919 | Kenton |
| Jul | 120 | 1943 | Tishomingo | 41 | 1915 | Goodwell |
| Aug | 120 | 1936 | Poteau | 38 | 1915 | Bartlesville |
| Sep | 115 | 1947 | Alva | 25 | 1985 | Boise City |
| Oct | 106 | 2000 | Hollis 5E | 6 | 1993 | Kenton |
| Nov | 95 | 1914 | Mutual | -15 | 1976 | Kenton |
| Dec | 92 | 1951 | Ardmore | -19 | 1932 | Goodwell |

## Table 3.5  State Record Maximum and Minimum Temperature Records by Month [CONTINUED]

### OREGON

| Month | Max | Year Location | Min | Year Location |
|---|---|---|---|---|
| Jan | 82 | 1934 Fremont | -52 | 1937 Austin |
| Feb | 89 | 1907 Williams | **-54** | **1933 Ukiah** |
| Mar | 99 | 1900 Merlin | -30 | 1922 Fremont |
| Apr | 102 | 1906 Marble Creek | -23 | 1936 Meacham |
| May | 108 | 1924 Blitzen | 0 | 1968 Juniper Lake |
| Jun | 113 | 1932 Blitzen | 11 | 1952 Crater Lake |
| Jul | 119 | **1898 Prineville** | 14 | 1955 Fremont |
| Aug | 119 | **1898 Pendleton** | 13 | 1937 Seneca |
| Sep | 111 | 1955 Illahe | 2 | 1926 Harney Branch |
| Oct | 104 | 1980 Dora | -9 | 1991 Seneca |
| Nov | 89 | 1936 Mitchell | -32 | 1985 Ukiah |
| Dec | 81 | 1897 Dayville | -53 | 1924 Riverside |

### PENNSYLVANIA

| Month | Max | Year Location | Min | Year Location |
|---|---|---|---|---|
| Jan | 85 | 1906 Freeport | **-42** | **1904 Smethport** |
| Feb | 83 | 1932 Hyndman | -39 | 1899 Lawrenceville |
| Mar | 92 | 1907 Everett | -31 | 1916 West Bingham |
| Apr | 98 | 1976 Norristown | -5 | 1923 Brooksville |
| May | 102 | 1941 Marcus Hook | 10 | 1966 Clermont |
| Jun | 107 | 1933 Sharon | 20 | 1913 Smethport |
| Jul | 111 | **1936 Phoenixville** | 28 | 1963 Clermont |
| Aug | 108 | 1918 Claysville | 23 | 1982 Clermont |
| Sep | 106 | 1953 Stroudsburg | 17 | 1957 Kane |
| Oct | 100 | 1941 Phoenixville | 7 | 1952 Coudersport |
| Nov | 88 | 1961 Claysville | -15 | 1930 Somerset |
| Dec | 82 | 1982 Washington | -29 | 1980 Clermont |

### RHODE ISLAND

| Month | Max | Year Location | Min | Year Location |
|---|---|---|---|---|
| Jan | 69 | 1995 Providence | -23 | 1942 Kingston |
| Feb | 72 | 1985 Providence | **-25** | **1996 Greene** |
| Mar | 90 | 1945 Providence | -10 | 1967 Kingston |
| Apr | 98 | 1976 Providence | 8 | 1923 Kingston |
| May | 95 | 1996 Providence | 25 | 2000 Kingston |
| Jun | 101 | 1952 Providence | 30 | 1945 Kingston |
| Jul | 102 | 1991 Providence | 38 | 1945 Kingston |
| Aug | 104 | **1975 Providence** | 33 | 1965 Kingston |
| Sep | 100 | 1983 Providence | 25 | 1957 Kingston |
| Oct | 90 | 1949 Providence | 13 | 1940 Kingston |
| Nov | 82 | 1950 Greenville | 4 | 1989 North Foster |
| Dec | 77 | 1998 Providence | -17 | 1942 Kingston |

### SOUTH CAROLINA

| Month | Max | Year Location | Min | Year Location |
|---|---|---|---|---|
| Jan | 86 | 1950 Beaufort | **-19** | **1985 Caesars Head** |
| Feb | 89 | 1962 Ridgeland | -1 | 1 1899 Santuc |
| Mar | 99 | 1907 Blackville | -8 | 1980 Cedar Creek |
| Apr | 99 | 1986 Aiken | 17 | 1992 Chesnee |
| May | 106 | 1911 Santuc | 28 | 1989 Ninety-Nine Islands |
| Jun | 111 | **1954 Camden** | 34 | 1961 Ceasars Head |
| Jul | 110 | 1887 Unknown | 45 | 1937 Long Creek |
| Aug | 110 | 1983 Cheraw | 45 | 1889 Brewer Mines |
| Sep | 111 | **1925 Calhoun Falls** | 28 | 1967 Caesars Head |
| Oct | 103 | 1954 Little Mountain | 16 | 1965 Chester |
| Nov | 93 | 1961 Bamberg | -1 | 1950 Caesars Head |
| Dec | 89 | 1906 Blackville | -6 | 1917 Landrum |

### SOUTH DAKOTA

| Month | Max | Year Location | Min | Year Location |
|---|---|---|---|---|
| Jan | 79 | 1921 Spearfish | -57 | 1916 Camp Crook |
| Feb | 79 | 1896 Vermillon | **-58** | **1936 McIntosh** |
| Mar | 96 | 1943 Tyndall | -35 | 1996 Antelope Rge Station |
| Apr | 102 | 1962 Murdo | -17 | 1957 Deerfield |
| May | 113 | 1934 Redfield | -1 | 1943 Ralph |
| Jun | 115 | 1931 La Delle | 14 | 1969 Dearfield |
| Jul | 120 | **1936 Gannvalley** | 24 | 1950 Hardy Ranger Stn |
| Aug | 116 | 1965 Redfield | 20 | 1988 Dearfield |
| Sep | 112 | 1931 Gannvalley | 5 | 1985 Dearfield |
| Oct | 103 | 1947 Murdo | -19 | 1925 Kennebec |
| Nov | 89 | 1999 Kennebec | -34 | 1891 Castlewood |
| Dec | 79 | 1939 Rapid City | -47 | 1990 Rochford |

### TENNESSEE

| Month | Max | Year Location | Min | Year Location |
|---|---|---|---|---|
| Jan | 83 | 1943 Celina | -30 | 1963 Kingston Springs |
| Feb | 85 | 1996 Gatlinburg | -30 | 1899 Erasmus |
| Mar | 94 | 1929 Clarksville | -11 | 1996 Mt Leconte |
| Apr | 98 | 1925 Etowah | 4 | 2000 Mt Leconte |
| May | 102 | 1911 Jackson | 18 | 1989 Mt Leconte |
| Jun | 110 | 1936 Etowah | 28 | 1998 Mt Leconte |
| Jul | 113 | **1930 Perryville** | 36 | 1961 Mountain City |
| Aug | 113 | **1930 Perryville** | 30 | 1989 Mt Leconte |
| Sep | 112 | 1925 Clarksville | 19 | 1990 Mt Leconte |
| Oct | 99 | 1953 Samburg Wldf | 10 | 1989 Mt Leconte |
| Nov | 89 | 1987 Ctrville Wtr Pl | -8 | 1950 Allardt |
| Dec | 82 | 1951 Newport | **-32** | **1917 Mountain City** |

## Table 3.5  State Record Maximum and Minimum Temperature Records by Month [CONTINUED]

### TEXAS

| Month | Max | Year Location | Min | Year Location |
|---|---|---|---|---|
| Jan | 98 | 1997 Zapata | -22 | 1959 Spearman |
| Feb | 104 | 1902 Fort Ringgold | **-23** | **1933 Seminole** |
| Mar | 108 | 1954 Rio Grande Cty | -12 | 1948 Spearman |
| Apr | 113 | 1984 Catarina | 5 | 1936 Romero |
| May | 116 | 1989 Boquillas RS | 15 | 1909 Tulia |
| Jun | **120** | **1994 Monahans** | 32 | 1917 Tulia |
| Jul | 119 | 1910 Tilden | 40 | 1906 Claytonville |
| Aug | **120** | **1936 Seymour** | 39 | 1910 Plemons |
| Sep | 116 | 2000 Columbus | 25 | 1983 Bravo |
| Oct | 109 | 1926 Victoria | 8 | 1993 Dalhart |
| Nov | 102 | 1988 McAllen | -10 | 1976 Stratford |
| Dec | 98 | 1951 Cotulla | -16 | 1983 Lipscomb |

### UTAH

| Month | Max | Year Location | Min | Year Location |
|---|---|---|---|---|
| Jan | 74 | 1996 Lytle Ranch | -50 | 1913 Strawberry East |
| Feb | 84 | 1986 Saint George | **-69** | **1985 Peter's Sink** |
| Mar | 89 | 1986 Saint George | -37 | 1917 East Portal |
| Apr | 98 | 1898 Saint George | -19 | 1917 East Portal |
| May | 108 | 1910 Saint George | 0 | 1965 Silver Lake Brighton |
| Jun | 116 | 1892 Unknown | 10 | 1919 Black's Fork |
| Jul | **117** | **1985 Saint George** | 17 | 1902 Tropic |
| Aug | 113 | 1892 Unknown | 17 | 1992 Randolph |
| Sep | 110 | 1950 Zion Natl Park | 2 | 1926 Woodruff |
| Oct | 99 | 1999 Zion Natl Park | -16 | 1972 Woodruff |
| Nov | 86 | 1924 Saint George | -30 | 1979 Woodruff |
| Dec | 76 | 1906 Rockville | -49 | 1932 Woodruff |

### VERMONT

| Month | Max | Year Location | Min | Year Location |
|---|---|---|---|---|
| Jan | 70 | 1950 Rutland | -44 | 1934 Bloomfield |
| Feb | 68 | 1957 Bennington | -46 | 1943 East Barnet |
| Mar | 88 | 1998 Salisbury | -37 | 1938 St Albans |
| Apr | 97 | 1976 Vernon | -13 | 1982 Mount Mansfield |
| May | 99 | 1911 Cavendish | 5 | 1957 Mount Mansfield |
| Jun | 101 | 1919 St Johnsbury | 19 | 1961 West Burke |
| Jul | **105** | **1911 Vernon** | 24 | 1986 Mount Mansfield |
| Aug | 104 | 1916 Cornwall | 24 | 1919 Somerset |
| Sep | 100 | 1953 Vernon | 15 | 1963 Chelsea |
| Oct | 89 | 1930 St Johnsbury | 2 | 1925 Garfield |
| Nov | 81 | 1950 Bellows Falls | -20 | 1925 Garfield |
| Dec | 72 | 1941 Enosburg Falls | **-50** | **1933 Bloomfield** |

### VIRGINIA

| Month | Max | Year Location | Min | Year Location |
|---|---|---|---|---|
| Jan | 84 | 1950 Clarksville | **-30** | **1985 Mt. Lake Biol Stn** |
| Feb | 87 | 1932 Roanoke | -29 | 1899 Monterey |
| Mar | 96 | 1907 Arvonia | -15 | 1914 Burkes Garden |
| Apr | 100 | 1925 Hopewell | 0 | 1982 Floyd |
| May | 105 | 1970 St Paul | 15 | 1986 Marion |
| Jun | 107 | 1934 Lincoln | 26 | 1930 Burkes Garden |
| Jul | **110** | **1954 Balcony Falls** | 31 | 1926 Burkes Garden |
| Aug | 109 | 1983 Colonial Beach | 31 | 1986 Big Meadows |
| Sep | 108 | 1932 Lincoln | 20 | 1989 Mt Lake Biol Stn |
| Oct | 101 | 1941 Walkerton | 9 | 1952 Burkes Garden |
| Nov | 91 | 1971 Chincoteague WL | -3 | 1970 Partlow |
| Dec | 85 | 1998 Colonial Bch | -27 | 1917 Blacksburg |

### WASHINGTON

| Month | Max | Year Location | Min | Year Location |
|---|---|---|---|---|
| Jan | 74 | 1940 Darrington | -42 | 1937 Deer Park |
| Feb | 83 | 1895 Centerville | -40 | 1933 Deer Park |
| Mar | 88 | 1910 Kennewick | -25 | 1955 Stockdill Ranch |
| Apr | 103 | 1934 Wahluke | -7 | 1936 Lake Keechelus |
| May | 107 | 1986 Dallesport | 11 | 1954 Republic |
| Jun | 112 | 1961 John Day Dam | 20 | 1976 Ranier Paradise RS |
| Jul | **118** | **1928 Wahluke** | 22 | 2000 Ranier Paradise RS |
| Aug | **118** | **1961 Ice Harbor Dam** | 20 | 1896 Cascade Tunnel |
| Sep | 111 | 2000 Mt. Adams RGR | 11 | 1934 Newport |
| Oct | 99 | 1987 Glenoma | -5 | 1935 Bumping Lake |
| Nov | 83 | 1949 Kosmos | -29 | 1896 Ellensburg |
| Dec | 74 | 1939 Sedro Wooley | **-48** | **1968 Mazama** |

### WEST VIRGINIA

| Month | Max | Year Location | Min | Year Location |
|---|---|---|---|---|
| Jan | 83 | 1914 Smithfield | -36 | 1985 Snowshoe |
| Feb | 86 | 1932 Moorefield | -35 | 1899 Dayton |
| Mar | 94 | 1907 Romney | -22 | 1960 Kumbrabow St Forest |
| Apr | 101 | 1925 Martinsburg | -4 | 1923 Cheat Bridge |
| May | 105 | 1895 Nuttallburg | 14 | 1947 Canaan Valley |
| Jun | 109 | 1925 Burlington | 22 | 1945 Pickens' |
| Jul | **112** | **1936 Martinsburg** | 27 | 1988 Canaan Valley |
| Aug | **112** | **1930 Moorefield** | 25 | 1982 Canaan Valley |
| Sep | 107 | 1895 Nuttallburg | 15 | 1942 Bayard |
| Oct | 102 | 1919 Moorefield | 3 | 1962 Buckeye |
| Nov | 89 | 1948 Madison | -14 | 1956 Canaan Valley |
| Dec | 84 | 1951 Hogsett Gal Dm | **-37** | **1917 Lewisburg** |

**Table 3.5   State Record Maximum and Minimum Temperature Records by Month [CONTINUED]**

**WISCONSIN**

| | | | | |
|---|---|---|---|---|
| Jan | 66 | 1897 Prairie DuChien | -54 | 1922 Danbury |
| Feb | 69 | 2000 Afton/Beloit | **-55** | **1996 Couderay** |
| Mar | 86 | 1986 Dodge | -48 | 1962 Couderay |
| Apr | 97 | 1980 Lone Rock | -20 | 1924 Rest Lake |
| May | 109 | 1934 Prairie DuChien | 7 | 1966 Gorden |
| Jun | 106 | 1934 Racine | 20 | 1964 Danbury |
| Jul | **114** | **1936 Wisconsin Dells** | 27 | 1972 Jump River |
| Aug | 108 | 1988 UW- Arboret | 22 | 1950 Coddington Exp Farm |
| Sep | 104 | 1939 Prairie DuChien | 10 | 1949 Coddington Exp Farm |
| Oct | 95 | 1897 Gratiot | -7 | 1925 Long Lake |
| Nov | 84 | 1904 Prairie DuChien | -34 | 1898 Osceola |
| Dec | 67 | 1998 La Crosse | -52 | 1983 Couderay |

**WYOMING**

| | | | | |
|---|---|---|---|---|
| Jan | 72 | 1981 Recluse 14NNW | -58 | 1943 Lamar Ranger Sta. |
| Feb | 79 | 1995 Weston | -66 | 1933 Riverside Ranger Sta. |
| Mar | 86 | 1907 Pine Bluff | -50 | 1906 Snake River Yellowstone NP |
| Apr | 93 | 1948 Basin | -29 | 1920 Lake Yellowstone |
| May | 101 | 1934 Sundance | -8 | 1912 Foxpark |
| Jun | 114 | 1988 Whalen Dam | 5 | 1978 Burgess Junction |
| Jul | 114 | 1900 Basin | 10 | 1911 Lolabana Ranch |
| Aug | **115** | **1983 Basin** | 7 | 1910 Fountain Hotel Yellowstone NP |
| Sep | 108 | 1978 Colony | -15 | 1983 Big Piney |
| Oct | 98 | 1947 Hampshire | -33 | 1917 Soda Butte Yellowstone NP |
| Nov | 86 | 1914 Wheatland | -46 | 1976 Darwin Ranch |
| Dec | 78 | 1939 Sheridan | **-59** | **1924 Riverside Ranger Sta.** Yellowstone NP |

**United States**

| | | | | |
|---|---|---|---|---|
| Jan | 98 | 1997 Zapata, TX | **-80** | **1971 Prospect Creek Camp, AK** |
| Feb | 104 | 1902 Fort Ringgold, TX | -75 | 1947 Tanacross, AK |
| Mar | 108 | 1954 Rio Grande Cty, TX | -68 | 1971 Kobuk, AK |
| Apr | 118 | 1898 Volcano Sprgs., CA | -50 | 1986 Umiat, AK |
| May | 122 | 2000 Death Valley, CA | -25 | 1992 Chandalar Lake, AK |
| Jun | 129 | 1902 Volcano Sprgs, CA | -1 | 1967 Anaktuvuk Pass, AK |
| Jul | **134** | **1913 Greenland Rch, CA** | 12 | 1997 Bodie, CA |
| Aug | 127 | 1993 Death Valley, CA | 5 | 1981 Wolf Creek Rch, CA, 1910 Bowen, MT |
| Sep | 126 | 1950 Mecca, CA | -15 | 1983 Big Piney, WY |
| Oct | 117 | 1906 Mecca, CA | -48 | 1975 Clear Water, AK |
| Nov | 105 | 1906 Craftonville, CA | -61 | 1935 Fort Yukon, AK |
| Dec | 100 | 1938 La Mesa, CA | -72 | 1999 Chicken, AK |

*Source:* NCDC.

damage/costs to agriculture and related industries; estimated 10000 deaths (includes heat stress-related).

## • TABLES OF RECORD-SETTING WEATHER: TABLES 3.5–3.9

### Record Weather in the United States

In this section you will find tables listing temperature, rainfall, and snowfall extremes for all 50 states. Hawaii has no official records of snowfall, but it does snow there.

In the United States, officially accepted weather records are investigated by NOAA's NCDC. However, the process of quality control can only go so far. For records many years old, there may be scant evidence of site characteristics that potentially could alter observations. In addition, research is ongoing and new data sources may be discovered, or problems with accepted data may be uncovered. The listings here are subject to change in the future.

The characteristics of weather itself complicate fine-tuning of records. Rainfall and snowfall accumulations are spotty by their very nature, while air temperature is highly variable due to microclimatic factors. A thermometer on a hilltop will often be warmer at sunrise than a valley thermometer because of cold air drainage and pooling of the cold air in the valley. A large expanse of rain or snow will not be uniform. Not only do large-scale storm characteristics affect

precipitation totals, but so too does terrain and the amount of convective activity in the storm.

Figure 3.9, a 180° panorama by astronomer Dr. Jean-Charles Cuillandre, shows snow on Mauna Kea (White Mountain in the Hawaiian language) around the astronomical observatories, with cloud tops in the background and in the distance, on the left, the summit of Mauna Loa. Hawaii is not included in the tables of snow records because snow is not measured officially. However, it does snow in Hawaii each winter, high on the slopes of the volcanoes Mauna Kea and Mauna Loa on the Big Island and on Haleakala on the island of Maui. It is not unusual to see snow skiers near the observatories each winter. During the last glacial advance, Mauna Kea was covered by a large, permanent ice cap that melted approximately 11,000 years ago. There is geological evidence of 4 glaciations in the last 200,000 years.

### Record Weather Around the World

Tables 3.10–3.14 list record-setting maximum and minimum temperatures and maximum and minimum rainfall for the world and the known values for each continent. The five tables here summarize extreme weather around the world.

Table 3.14 lists extreme rainfall events by duration of the event. You should notice the great influence of tropical cyclones on the numbers in this table.

Following Table 3.14 is a listing of miscellaneous world or regional records or record events thought to be world extremes.

More detailed information on weather around the world is contained in Chapter 9. There you will find average weather tables for 321 locations around the world, with locater maps to help your search. In Chapter 9, the coverage of Antarctica has been increased and there is now coverage of Arctic Ocean Islands.

## • MISCELLANEOUS WEATHER RECORDS AND EXTREMES

### Atmospheric Pressure Adjusted to Sea Level (Excluding Tornadoes)

*Source: World Meteorological Organization (WMO) and US Army Topographic Engineering Center*

The highest atmospheric pressure readings occur in bitterly cold, dense Arctic air masses. The lowest surface pressures occur in tornadoes, but direct observation of tornado central pressure is difficult because tornadoes cover a very small area and, of course, the destructive effect of high winds often guarantee that the instrument will not survive.

**Table 3.6    State 24-Hour Maximum Rainfall Records**

| State | Precip. Inches | Date | Station | Elevation Feet |
|---|---|---|---|---|
| Alabama | 32.52 | Jul. 19-20, 1997 | Dauphin Is. Sea Lab | 8 |
| Alaska | 15.20 | Oct. 12, 1982 | Angoon | 15 |
| Arizona | 11.40 | Sep. 4-5, 1970 | Workman Creek | 6,970 |
| Arkansas | 14.06 | Dec. 3, 1982 | Big Fork | 1,100 |
| California | 26.12 | Jan. 22-23, 1943 | Hoegees Camp | 2,760 |
| Colorado | 11.08 | Jun. 17, 1965 | Holly | 3,390 |
| Connecticut | 12.77 | Aug. 19, 1955 | Burlington | 460 |
| Delaware | 8.50 | Jul. 13, 1975 | Dover | 30 |
| Florida | 38.70 | Sep. 5, 1950 | Yankeetown | 5 |
| Georgia | 21.10 | Jul. 6, 1994 | Americus | 490 |
| Hawaii | 38.00 | Jan. 24-25, 1956 | Kilauea Plantation | 180 |
| Idaho | 7.17 | Nov. 23, 1909 | Rattlesnake Creek | 4,000 |
| Illinois | 16.91 | Jul. 18, 1996 | Aurora | 640 |
| Indiana | 10.50 | Aug. 6, 1905 | Princeton | 480 |
| Iowa | 16.70 | Aug. 5-6, 1959 | Decatur Co. | 1,110 |
| Kansas | 12.59 | May 31-Jun. 1, 1941 | Burlington | 1,010 |
| Kentucky | 10.40 | Jun. 28, 1960 | Dunmor | 610 |
| Louisiana | 22.00 | Aug. 28-29, 1962 | Hackberry | 10 |
| Maine | 13.32 | Oct. 20-21, 1996 | Portland | 45 |
| Maryland | 14.75 | Jul. 26-27, 1897 | Jewell | 165 |
| Massachusetts | 18.15 | Aug. 18-19, 1955 | Westfield | 220 |
| Michigan | 9.78 | Aug. 31-Sep. 1, 1914 | Bloomingdale | 750 |
| Minnesota | 10.84 | Jul. 21-22, 1972 | Fort Ripley | 1,140 |
| Mississippi | 15.68 | Jul. 9, 1968 | Columbus | 190 |
| Missouri | 18.18 | Jul. 20, 1965 | Edgarton | 856 |
| Montana | 11.50 | Jun. 20, 1921 | Circle | 2,440 |
| Nebraska | 13.15 | Jul. 8-9, 1950 | York | 1,610 |
| Nevada | 7.13 | Jan. 31, 1963 | Mt. Rose Hwy. Stn. | 7,360 |
| New Hampshire | 10.38 | Feb. 10-11, 1970 | Mount Washington | 6,262 |
| New Jersey | 14.81 | Aug. 19, 1939 | Tuckerton | 20 |
| New Mexico | 11.28 | May 18-19, 1955 | Lake Maloya | 7,400 |
| New York | 11.17 | Oct. 9, 1903 | NYC Central Park | 130 |
| North Carolina | 22.22 | Jul. 15-16, 1916 | Altapass | 2,600 |
| North Dakota | 8.10 | Jun. 29, 1975 | Litchville | 1,470 |
| Ohio | 10.75 | Aug. 7-8, 1995 | Lockington Dam | 950 |
| Oklahoma | 15.68 | Oct. 11, 1973 | Enid | 1245 |
| Oregon | 11.65 | Nov. 19, 1996 | Port Orford | 150 |
| Pennsylvania | 34.50* | Jul. 17, 1942 | Smethport | 1,510 |
| Rhode Island | 12.13 | Sep. 16-17, 1932 | Westerly | 40 |
| South Carolina | 17.00 | Aug. 27, 1995 | Antreville | 700 |
| South Dakota | 8.00 | Sep. 10, 1900 | Elk Point | 1,127 |
| Tennessee | 11.00 | Mar. 28, 1902 | McMinnville | 900 |
| Texas | 43.00* | Jul. 25-26, 1979 | Alvin | 50 |
| Utah | 6.00* | Sep. 5, 1970 | Bug Point | 6,600 |
| Vermont | 8.77 | Nov. 3-4, 1927 | Somerset | 2,080 |
| Virginia | 27.00* | Aug. 20, 1969 | Nelson Co. | est 500 |
| Washington | 14.26 | Nov. 23-24, 1986 | Mt. Mitchell #2 | 3,600 |
| West Virginia | 19.00* | Jul. 18, 1889 | Rockport | 700 |
| Wisconsin | 11.72 | Jun. 24, 1946 | Mellen | 1,150 |
| Wyoming | 6.06 | Aug. 1, 1985 | Cheyenne | 6,126 |

*Source:* NCDC.

**Table 3.7   State Maximum Annual Rainfall Records**

| State | Precip. Inches | Year | Station | Elevation |
|-------|------|------|---------|-----------|
| Alabama | 98.22 | 1961 | Citronelle | 331 |
| Alaska | 332.29 | 1976 | MacLeod Harbor | 40 |
| Arizona | 58.92 | 1978 | Hawley Lake | 8,180 |
| Arkansas | 98.55 | 1957 | Newhope | 850 |
| California | 153.54 | 1909 | Monumental | 2,420 |
| Colorado | 92.84 | 1897 | Ruby | est.10,000 |
| Connecticut | 78.53 | 1955 | Burlington | 460 |
| Delaware | 72.75 | 1948 | Lewes | 10 |
| Florida | 112.43 | 1966 | Wewahitchka | 50 |
| Georgia | 122.16 | 1959 | Flat Top | est. 3,600 |
| Hawaii | 704.83 | 1982 | Kukui | 5,788 |
| Idaho | 81.05 | 1933 | Roland | 4,150 |
| Illinois | 74.58 | 1950 | New Burnside | 560 |
| Indiana | 97.38 | 1890 | Marengo | 570 |
| Iowa | 74.50 | 1851 | Muscatine | 680 |
| Kansas | 68.55 | 1993 | Blaine | 1,530 |
| Kentucky | 79.68 | 1950 | Russelville | 590 |
| Louisiana | 113.74 | 1991 | New Orleans | 6 |
| Maine | 75.64 | 1845 | Brunswick | 70 |
| Maryland | 76.52 | 1971 | Towson | 390 |
| Massachusetts | 76.49 | 1996 | New Salem | 845 |
| Michigan | 64.01 | 1881 | Adrian | 770 |
| Minnesota | 52.36 | 1993 | Fairmont | 1,187 |
| Mississippi | 104.36 | 1991 | Waveland | 8 |
| Missouri | 92.77 | 1957 | Portageville | 280 |
| Montana | 55.51 | 1953 | Summit | 5,210 |
| Nebraska | 64.52 | 1869 | Omaha | 980 |
| Nevada | 59.03 | 1969 | Mt. Rose Resort | est. 7,300 |
| New Hampshire | 130.14 | 1969 | Mount Washington | 6,260 |
| New Jersey | 85.99 | 1882 | Paterson | 100 |
| New Mexico | 62.45 | 1941 | White Tail | 7,450 |
| New York | 90.97 | 1996 | Slide Mountain | 2,649 |
| North Carolina | 129.60 | 1964 | Rosman | 2,220 |
| North Dakota | 37.98 | 1944 | Milnor | 2,600 |
| Ohio | 70.82 | 1870 | Little Mountain | 1,187 |
| Oklahoma | 84.47 | 1957 | Kiamichi Tower | 2,350 |
| Oregon | 204.04 | 1996 | Laurel Mountain | 3,590 |
| Pennsylvania | 81.64 | 1952 | Mt. Pocono | 1,910 |
| Rhode Island | 70.21 | 1983 | Kingston | 100 |
| South Carolina | 110.79 | 1994 | Jocassee | 2,500 |
| South Dakota | 48.42 | 1946 | Deadwood | 4,550 |
| Tennessee | 114.88 | 1957 | Haw Knob | 4,900 |
| Texas | 109.38 | 1873 | Clarksville | 440 |
| Utah | 108.54 | 1983 | Alta | 8,760 |
| Vermont | 100.96 | 1996 | Mt. Mansfield | 3,950 |
| Virginia | 83.70 | 1996 | Philpott Dam | 1,123 |
| Washington | 184.56 | 1931 | Wynoochee Oxbow | 670 |
| West Virginia | 89.01 | 1926 | Bayard | 2,381 |
| Wisconsin | 62.07 | 1884 | Embarrass | 808 |
| Wyoming | 55.46 | 1945 | Grassy Lake Dam | 7,240 |

*Source:* NCDC.

## Table 3.8 State Minimum Annual Rainfall Records

| State | Precip. Inches | Date | Station | Elevation Feet |
|---|---|---|---|---|
| Alabama | 22.00 | 1954 | Primrose Farm | 180 |
| Alaska | 1.61 | 1935 | Barrow | 31 |
| Arizona | 0.07 | 1956 | Davis Dam | 660 |
| Arkansas | 19.11 | 1936 | Index | 300 |
| California | 0.00 | 1929 | Death Valley | -282 |
| Colorado | 1.69 | 1939 | Buena Vista | 7,980 |
| Connecticut | 23.60 | 1965 | Baltic | 140 |
| Delaware | 21.38 | 1965 | Dover | 30 |
| Florida | 21.16 | 1989 | Conch Key | 6 |
| Georgia | 17.14 | 1954 | Swainsboro | 320 |
| Hawaii | 0.19 | 1953 | Kawaihae | est. 75 |
| Idaho | 2.09 | 1947 | Grand View | 2,360 |
| Illinois | 16.59 | 1956 | Keithsburg | 540 |
| Indiana | 18.67 | 1934 | Brooksville | 630 |
| Iowa | 12.11 | 1958 | Cherokee | 1,360 |
| Kansas | 4.77 | 1956 | Johnson | 3,270 |
| Kentucky | 14.51 | 1968 | Jeremiah | 1,160 |
| Louisiana | 26.44 | 1936 | Shreveport | 170 |
| Maine | 23.06 | 1930 | Machias | 30 |
| Maryland | 17.76 | 1930 | Picardy | 1,030 |
| Massachusetts | 21.76 | 1965 | Chatham L.S. | 20 |
| Michigan | 15.64 | 1936 | Croswell | 730 |
| Minnesota | 7.81 | 1936 | Angus | 870 |
| Mississippi | 25.97 | 1936 | Yazoo City | 120 |
| Missouri | 16.14 | 1956 | La Belle | 770 |
| Montana | 2.97 | 1960 | Belfry | 4,040 |
| Nebraska | 6.30 | 1931 | Hull | 4,400 |
| Nevada | T | 1898 | Hot Springs | 4,072 |
| New Hampshire | 22.31 | 1930 | Bethlehem | 1,440 |
| New Jersey | 19.85 | 1965 | Canton | 20 |
| New Mexico | 1.00 | 1910 | Hermanas | 4,540 |
| New York | 17.64 | 1941 | Lewiston | 320 |
| North Carolina | 22.69 | 1930 | Mount Airy | 1,070 |
| North Dakota | 4.02 | 1934 | Parshall | 1,930 |
| Ohio | 16.96 | 1963 | Elyria | 730 |
| Oklahoma | 6.53 | 1956 | Regnier | 4,280 |
| Oregon | 3.33 | 1939 | Warm Springs Res. | 3,330 |
| Pennsylvania | 15.71 | 1965 | Breezewood | 1,350 |
| Rhode Island | 24.08 | 1965 | Block Island | 40 |
| South Carolina | 20.73 | 1954 | Rock Hill | 667 |
| South Dakota | 2.89 | 1936 | Ludlow | 2,850 |
| Tennessee | 25.23 | 1941 | Halls | 310 |
| Texas | 1.64 | 1956 | Presidio | 2,580 |
| Utah | 1.34 | 1974 | Myton | 5,080 |
| Vermont | 22.98 | 1941 | Burlington | 330 |
| Virginia | 12.52 | 1941 | Moores Creek Dam | 1,950 |
| Washington | 2.61 | 1930 | Wahluke | 416 |
| West Virginia | 9.50 | 1930 | Upper Tract | 1,540 |
| Wisconsin | 12.00 | 1937 | Plum Is. | 590 |
| Wyoming | 1.28 | 1960 | Lysite | 5,260 |

*Source:* NCDC.

## Table 3.9   State Snowfall Records and National Summary

| Alabama | | | |
|---|---|---|---|
| Greatest Daily | 18.5 | REFORM | 1/23/1940 |
| Max 2-Day | 22 | REFORM | 1/24/1940 |
| Max 3-Day | 13.2 | SAINT BERNARD | 1/25/1940 |
| Max 4-Day | 8 | RED BAY | 2/1/1966 |
| Max 5-Day | 8 | RED BAY | 2/2/1966 |
| Max 6-Day | 6.1 | VALLEY HEAD | 1/11/1970 |
| Max 7-Day | 6.5 | VALLEY HEAD | 1/12/1970 |
| Greatest Monthly | 22.7 | REFORM | Jan-40 |
| Greatest Aug-July | 22.6 | FLORENCE | 1968 |
| Greatest Daily Depth | 10 | BIRMINGHAM FAA ARPT | 1/23/1940 |
| **Alaska** | | | |
| Greatest Daily | 62 | THOMPSON PASS | 12/29/1955 |
| Max 2-Day | 120.6 | THOMPSON PASS | 12/30/1955 |
| Max 3-Day | 147 | THOMPSON PASS | 12/30/1955 |
| Max 4-Day | 163 | THOMPSON PASS | 12/30/1955 |
| Max 5-Day | 175.4 | THOMPSON PASS | 12/31/1955 |
| Max 6-Day | 172.6 | THOMPSON PASS | 2/24/1953 |
| Max 7-Day | 186.9 | THOMPSON PASS | 2/25/1953 |
| Greatest Monthly | 297.9 | THOMPSON PASS | Feb-53 |
| Greatest Aug-July | 974.1 | THOMPSON PASS | 1953 |
| Greatest Daily Depth | 133 | WHITTIER | 3/26/1985 |
| **Arizona** | | | |
| Greatest Daily | 38 | HEBER RANGER STN | 12/14/1967 |
| Max 2-Day | 56 | HEBER RANGER STN | 12/14/1967 |
| Max 3-Day | 64 | FLAGSTAFF AP | 01/17/1895 |
| Max 4-Day | 78 | FLAGSTAFF AP | 01/18/1895 |
| Max 5-Day | 92 | FLAGSTAFF AP | 01/19/1895 |
| Max 6-Day | 98 | FLAGSTAFF AP | 01/19/1895 |
| Max 7-Day | 81 | HEBER RANGER STN | 12/19/1967 |
| Greatest Monthly | 126 | FLAGSTAFF AP | 01/1895 |
| Greatest Aug-July | 344 | SUNRISE MOUNTAIN | 1975 |
| Greatest Daily Depth | 86 | PALISADE RANGER STN | 2/11/1966 |
| **Arkansas** | | | |
| Greatest Daily | 18 | BEE BRANCH | 2/19/1921 |
| Max 2-Day | 22 | HARRISON FAA AIRPORT | 12/23/1966 |
| Max 3-Day | 19 | JONESBORO 4 N | 3/5/1965 |
| Max 4-Day | 21 | JONESBORO 4 N | 3/6/1965 |
| Max 5-Day | 15.9 | GREENBRIER | 1/6/1988 |
| Max 6-Day | 14 | MARKED TREE | 12/16/1932 |
| Max 7-Day | 15 | MARKED TREE | 12/16/1932 |
| Greatest Monthly | 48 | CALICO ROCK 2 WSW | Jan-18 |
| Greatest Aug-July | 36.8 | YELLVILLE 2 SSE | 1970 |
| Greatest Daily Depth | 23 | FAYETTEVILLE EXP STN | 1/21/1918 |
| **California** | | | |
| Greatest Daily | 60 | GIANT FOREST | 1/19/1933 |
| Max 2-Day | 86 | MOUNT SHASTA SKI BOWL | 1/21/1964 |
| Max 3-Day | 122 | MOUNT SHASTA SKI BOWL | 1/22/1964 |
| Max 4-Day | 145 | ECHO SUMT SIERRA AT TAHOE | 4/1/1982 |
| Max 5-Day | 154 | ECHO SUMT SIERRA AT TAHOE | 4/2/1982 |
| Max 6-Day | 161 | ECHO SUMT SIERRA AT TAHOE | 4/3/1982 |
| Max 7-Day | 185 | ECHO SUMT SIERRA AT TAHOE | 4/4/1982 |
| Greatest Monthly | 313 | TAMARACK | Mar-07 |
| Greatest Aug-July | 746.5 | ECHO SUMT SIERRA AT TAHOE | 1983 |
| Greatest Daily Depth | 210 | ECHO SUMT SIERRA AT TAHOE | 4/7/1982 |

**Table 3.9   State Snowfall Records and National Summary [CONTINUED]**

| Colorado | | | |
|---|---|---|---|
| Greatest Daily | 63 | GEORGETOWN | 12/04/13 |
| Max 2-Day | 90 | WOLF CREEK PASS 4 W | 01/28/56 |
| Max 3-Day | 104 | WOLF CREEK PASS 4 W | 01/28/56 |
| Max 4-Day | 115 | WOLF CREEK PASS 4 W | 01/29/56 |
| Max 5-Day | 134 | WOLF CREEK PASS 1 E | 01/01/65 |
| Max 6-Day | 142 | WOLF CREEK PASS 1 E | 01/01/65 |
| Max 7-Day | 143 | WOLF CREEK PASS 1 E | 01/01/65 |
| Greatest Monthly | 217 | PAGOSA SPRINGS | 01/01/37 |
| Greatest Aug-July | 520 | WOLF CREEK PASS 4 W | 1948 |
| Greatest Daily Depth | 116 | PAGOSA SPRINGS | 02/26/37 |
| **Connecticut** | | | |
| Greatest Daily | 28 | MIDDLETOWN 4 W | 01/28/1897 |
| Max 2-Day | 30 | FALLS VILLAGE | 2/6/1920 |
| Max 3-Day | 34 | FALLS VILLAGE | 2/7/1920 |
| Max 4-Day | 32.7 | NORFOLK 2 SW | 12/8/1996 |
| Max 5-Day | 32.7 | NORFOLK 2 SW | 12/8/1996 |
| Max 6-Day | 26.4 | NORFOLK 2 SW | 12/17/1970 |
| Max 7-Day | 27.6 | NORFOLK 2 SW | 12/18/1970 |
| Greatest Monthly | 73.6 | NORFOLK 2 SW | Mar-56 |
| Greatest Aug-July | 152.5 | NORFOLK 2 SW | 1967 |
| Greatest Daily Depth | 55 | NORFOLK 2 SW | 2/5/1961 |
| **Delaware** | | | |
| Greatest Daily | 25 | DOVER | 2/19/1979 |
| Max 2-Day | 27 | MIDDLETOWN 3 E | 3/21/1958 |
| Max 3-Day | 25.5 | MILFORD 4 SE | 12/27/1909 |
| Max 4-Day | 22.2 | WILMINGTON WSO ARPT | 1/10/1996 |
| Max 5-Day | 12 | NEWARK UNIVERSITY FARM | 1/21/1978 |
| Max 6-Day | 9 | SMYRNA 1 W | 1/28/2004 |
| Max 7-Day | 4.5 | WILMINGTON WSO ARPT | 2/5/1958 |
| Greatest Monthly | 36.5 | DOVER | Feb-79 |
| Greatest Aug-July | 49.1 | GEORGETOWN 5 SW | 1996 |
| Greatest Daily Depth | 25 | BRIDGEVILLE 1 NW | 2/19/1979 |
| **Florida** | | | |
| Greatest Daily | 4 | MILTON EXPERIMENT STN | 3/6/1954 |
| Max 2-Day | 3 | STARKE | 1/19/1977 |
| Max 3-Day | -8.8 | MILTON EXPERIMENT STN | 1/12/1973 |
| Max 4-Day | 0 | ADAMS BEACH | Jan-57 |
| Max 5-Day | 0 | ADAMS BEACH | Jan-57 |
| Max 6-Day | 0 | ADAMS BEACH | Jan-57 |
| Max 7-Day | 0 | ADAMS BEACH | Jan-57 |
| Greatest Monthly | 4 | MILTON EXPERIMENT STN | Jan-77 |
| Greatest Aug-July | 4 | MILTON EXPERIMENT STN | 1954 |
| Greatest Daily Depth | 3 | MILTON EXPERIMENT STN | 2/10/1973 |
| **Georgia** | | | |
| Greatest Daily | 19.3 | CEDARTOWN 3 NE | 03/03/42 |
| Max 2-Day | 16 | ELLIJAY | 03/14/93 |
| Max 3-Day | 12 | CLAYTON 1 SSW | 01/31/30 |
| Max 4-Day | 11.5 | GAINESVILLE | 03/12/60 |
| Max 5-Day | 7.5 | WALESKA | 03/12/60 |
| Max 6-Day | 7.5 | WALESKA | 03/12/60 |
| Max 7-Day | 7.5 | WALESKA | 03/12/60 |
| Greatest Monthly | 22 | CLAYTON 1 SSW | 03/01/60 |
| Greatest Aug-July | 21.9 | BLAIRSVILLE EXP STA | 1936 |
| Greatest Daily Depth | 10 | ATHENS | 01/24/40 |

**Table 3.9   State Snowfall Records and National Summary [CONTINUED]**

| Idaho | | | |
|---|---|---|---|
| Greatest Daily | 30 | HEADQUARTERS | 12/28/1968 |
| Max 2-Day | 38.5 | ROLAND WEST PORTAL | 2/27/1936 |
| Max 3-Day | 49 | DEADWOOD DAM | 12/29/1931 |
| Max 4-Day | 62 | ISLAND PARK | 1/1/2004 |
| Max 5-Day | 68 | ISLAND PARK | 1/1/2004 |
| Max 6-Day | 64 | ROLAND WEST PORTAL | 2/24/1933 |
| Max 7-Day | 72 | ROLAND WEST PORTAL | 2/10/1932 |
| Greatest Monthly | 129 | ROLAND WEST PORTAL | Dec-48 |
| Greatest Aug-July | 380.1 | ROLAND WEST PORTAL | 1946 |
| Greatest Daily Depth | 108 | DEADWOOD DAM 15 N | 3/31/1983 |
| **Illinois** | | | |
| Greatest Daily | 24 | COATSBURG | 2/27/1900 |
| Max 2-Day | 26 | COATSBURG | 2/27/1900 |
| Max 3-Day | 24 | BARRINGTON 3 SW | 1/14/1979 |
| Max 4-Day | 27 | MONTICELLO NO 2 | 1/3/1999 |
| Max 5-Day | 27 | MONTICELLO NO 2 | 1/4/1999 |
| Max 6-Day | 27 | MONTICELLO NO 2 | 1/4/1999 |
| Max 7-Day | 28 | NEW BURNSIDE | 1/15/1918 |
| Greatest Monthly | 52 | ELGIN | Jan-79 |
| Greatest Aug-July | 103.3 | ANTIOCH | 1979 |
| Greatest Daily Depth | 39 | ANTIOCH | 1/16/1979 |
| **Indiana** | | | |
| Greatest Daily | 22 | CANNELTON | 12/23/04 |
| Max 2-Day | 29 | SOUTH BEND WSO AP | 01/31/09 |
| Max 3-Day | 32 | SOUTH BEND WSO AP | 01/31/09 |
| Max 4-Day | 33 | LA PORTE | 02/18/58 |
| Max 5-Day | 36 | LA PORTE | 02/19/58 |
| Max 6-Day | 37 | LA PORTE | 02/19/58 |
| Max 7-Day | 37.5 | LA PORTE | 12/21/51 |
| Greatest Monthly | 59.5 | LA PORTE | 02/01/58 |
| Greatest Aug-July | 125.6 | LA PORTE | 1978 |
| Greatest Daily Depth | 38 | LA PORTE | 02/18/58 |
| **Iowa** | | | |
| Greatest Daily | 21 | FAYETTE | 3/6/1959 |
| Max 2-Day | 32.5 | ZEARING | 3/14/1905 |
| Max 3-Day | 29 | PERRY 3 N | 3/10/1909 |
| Max 4-Day | 28 | LAKE PARK | 12/25/1968 |
| Max 5-Day | 30.7 | LAKE PARK | 12/23/1968 |
| Max 6-Day | 32.7 | LAKE PARK | 12/24/1968 |
| Max 7-Day | 34.7 | LAKE PARK | 12/25/1968 |
| Greatest Monthly | 43.7 | LAKE PARK | Dec-68 |
| Greatest Aug-July | 85.6 | CRESCO 1 NE | 1951 |
| Greatest Daily Depth | 52 | LAKE PARK | 3/1/1969 |
| **Kansas** | | | |
| Greatest Daily | 25 | COLUMBUS 6 NNW | 03/14/1896 |
| Max 2-Day | 37 | OLATHE 3 E | 3/24/1912 |
| Max 3-Day | 33 | ELKHART 6 NNE | 2/26/1903 |
| Max 4-Day | 30.5 | JOHNSON | 12/20/1918 |
| Max 5-Day | 29 | JOHNSON | 11/6/1946 |
| Max 6-Day | 30 | MC DONALD | 4/2/1980 |
| Max 7-Day | 35 | MC DONALD | 4/2/1980 |
| Greatest Monthly | 55.9 | OLATHE 3 E | Mar-12 |
| Greatest Aug-July | 103.6 | MC DONALD | 1984 |
| Greatest Daily Depth | 31 | HOLTON 4 NE | 3/16/1960 |

**Table 3.9 State Snowfall Records and National Summary** [CONTINUED]

| Kentucky | | | |
|---|---|---|---|
| Greatest Daily | 20 | LEWISPORT 4 S | 12/23/04 |
| Max 2-Day | 26.5 | PHELPS 3 S | 01/08/96 |
| Max 3-Day | 28.3 | PHELPS 3 S | 01/08/96 |
| Max 4-Day | 28.3 | PHELPS 3 S | 01/08/96 |
| Max 5-Day | 28 | LAGRANGE | 01/20/78 |
| Max 6-Day | 28.4 | PINE MOUNTAIN 3 NW | 01/12/96 |
| Max 7-Day | 29.5 | PINE MOUNTAIN 3 NW | 01/13/96 |
| Greatest Monthly | 46.5 | BENHAM | 03/01/60 |
| Greatest Aug-July | 108.2 | BENHAM | 1960 |
| Greatest Daily Depth | 31 | LAGRANGE | 01/20/78 |
| **Louisiana** | | | |
| Greatest Daily | 16 | SHELL BEACH | 02/14/1895 |
| Max 2-Day | 14 | LOGANSPORT | 12/22/29 |
| Max 3-Day | 16.4 | GEORGETOWN | 02/05/05 |
| Max 4-Day | 6.5 | ST JOSEPH 3 N | 02/01/51 |
| Max 5-Day | 6.3 | HOSSTON | 01/22/78 |
| Max 6-Day | 6.3 | HOSSTON | 01/23/78 |
| Max 7-Day | 0 | ABBEVILLE | 01/01/95 |
| Greatest Monthly | 16 | SHELL BEACH | 02/1895 |
| Greatest Aug-July | 16 | FARMERVILLE | 1899 |
| Greatest Daily Depth | 10 | CHATHAM | 01/31/49 |
| **Maine** | | | |
| Greatest Daily | 35 | MIDDLE DAM | 11/23/1943 |
| Max 2-Day | 49 | HARRIS STATION | 2/26/1969 |
| Max 3-Day | 57 | HARRIS STATION | 2/27/1969 |
| Max 4-Day | 57 | HARRIS STATION | 2/27/1969 |
| Max 5-Day | 56 | LONG FALLS DAM | 2/28/1969 |
| Max 6-Day | 36.1 | CARIBOU WSO AIRPORT | 12/30/2005 |
| Max 7-Day | 38.2 | CARIBOU WSO AIRPORT | 12/30/2005 |
| Greatest Monthly | 89 | HARRIS STATION | Feb-69 |
| Greatest Aug-July | 237.6 | LONG FALLS DAM | 1969 |
| Greatest Daily Depth | 84 | FARMINGTON | 2/28/1969 |
| **Maryland** | | | |
| Greatest Daily | 31 | CLEAR SPRING 1 ENE | 3/29/1942 |
| Max 2-Day | 36 | EDGEMONT | 3/30/1942 |
| Max 3-Day | 38 | FROSTBURG 2 | 3/15/1993 |
| Max 4-Day | 43 | FROSTBURG 2 | 3/14/1993 |
| Max 5-Day | 47 | FROSTBURG 2 | 3/15/1993 |
| Max 6-Day | 51.5 | MC HENRY 2 NW | 3/14/1993 |
| Max 7-Day | 51.5 | MC HENRY 2 NW | 3/15/1993 |
| Greatest Monthly | 68 | FROSTBURG 2 | Mar-93 |
| Greatest Aug-July | 169.5 | MC HENRY 2 NW | 1993 |
| Greatest Daily Depth | 54 | FROSTBURG 2 | 3/15/1993 |
| **Massachusetts** | | | |
| Greatest Daily | 29 | NATICK | 4/1/1997 |
| Max 2-Day | 43.7 | IPSWICH | 2/8/1978 |
| Max 3-Day | 43.9 | IPSWICH | 2/8/1978 |
| Max 4-Day | 47.3 | PERU | 3/4/1947 |
| Max 5-Day | 47.3 | PERU | 3/4/1947 |
| Max 6-Day | 48.3 | PERU | 3/4/1947 |
| Max 7-Day | 49.8 | PERU | 3/4/1947 |
| Greatest Monthly | 66 | WASHINGTON 2 | Dec-69 |
| Greatest Aug-July | 156.1 | ASHFIELD | 1996 |
| Greatest Daily Depth | 62 | GREAT BARRINGTON ARPT | 1/13/1996 |

**Table 3.9   State Snowfall Records and National Summary [CONTINUED]**

| Michigan | | | |
|---|---|---|---|
| Greatest Daily | 30 | HERMAN | 12/19/1996 |
| Max 2-Day | 46 | SAULT STE MARIE WSO | 12/10/1995 |
| Max 3-Day | 60 | PETOSKEY | 12/27/2001 |
| Max 4-Day | 66 | PETOSKEY | 12/27/2001 |
| Max 5-Day | 75 | PETOSKEY | 12/27/2001 |
| Max 6-Day | 79 | PETOSKEY | 12/28/2001 |
| Max 7-Day | 85 | PETOSKEY | 12/29/2001 |
| Greatest Monthly | 129.5 | COPPER HARBOR 3 WNW | Jan-82 |
| Greatest Aug-July | 263.5 | COPPER HARBOR 3 WNW | 1982 |
| Greatest Daily Depth | 70 | DETOUR VILLAGE | 2/26/1962 |
| **Minnesota** | | | |
| Greatest Daily | 36 | WOLF RIDGE E L C | 1/7/1994 |
| Max 2-Day | 45 | WOLF RIDGE E L C | 1/7/1994 |
| Max 3-Day | 46.5 | WOLF RIDGE E L C | 1/8/1994 |
| Max 4-Day | 36 | TWO HARBORS | 11/3/1991 |
| Max 5-Day | 35.3 | DULUTH WSO AP | 12/9/1950 |
| Max 6-Day | 36.8 | DULUTH WSO AP | 12/7/1950 |
| Max 7-Day | 38.9 | DULUTH WSO AP | 12/8/1950 |
| Greatest Monthly | 66.4 | COLLEGEVILLE ST JOHN | Mar-65 |
| Greatest Aug-July | 154 | PIGEON RIVER BRIDGE | 1934 |
| Greatest Daily Depth | 88 | MEADOWLANDS 1 NNW | 2/21/1969 |
| **Mississippi \*** | | | |
| Greatest Daily | 14 | BOONEVILLE | 01/17/93 |
| Max 2-Day | 18 | KIPLING & YAZOO CITY | 1/1/1964 & 1/23/1935 |
| Max 3-Day | 15 | YAZOO CITY | 1/24/1935 |
| Max 4-Day | 11 | COLUMBUS & LOUISVILLE | 1/6/1919 & 12/21/1929 |
| Max 5-Day | 11 | COLUMBUS & LOUISVILLE | 1/7/1919 |
| Max 6-Day | 1.2 | CLEVELAND | 1/11/1970 |
| Max 7-Day | TRACE | CLINTON EXP. STA. | 2/12/1960 |
| Greatest Monthly | 23 | CLEVELAND | Feb-66 |
| Greatest Aug-July | 25.2 | SENATOBIA | 1968 |
| Greatest Daily Depth | 10 | COLUMBUS | 02/14/60 |
| **Missouri** | | | |
| Greatest Daily | 24 | BRUNSWICK | 12/5/1925 |
| Max 2-Day | 27.6 | NEOSHO | 3/17/1970 |
| Max 3-Day | 26.6 | GROVESPRING 3 N | 12/14/2000 |
| Max 4-Day | 28.5 | ST. CATHARINE | 4/2/1926 |
| Max 5-Day | 28.7 | ST. CATHARINE | 4/3/1926 |
| Max 6-Day | 24 | STEFFENVILLE | 2/27/1900 |
| Max 7-Day | 25 | STEFFENVILLE | 2/27/1900 |
| Greatest Monthly | 47.5 | POPLAR BLUFF | Jan-18 |
| Greatest Aug-July | 73.9 | EDINA | 1978 |
| Greatest Daily Depth | 36 | UNION | 3/20/1960 |
| **Montana** | | | |
| Greatest Daily | 48 | MILLEGAN 14 SE | 12/27/2003 |
| Max 2-Day | 66 | SHONKIN 7 S | 5/29/1982 |
| Max 3-Day | 73 | RED LODGE 2 N | 4/27/1984 |
| Max 4-Day | 66 | SUMMIT | 1/22/1972 |
| Max 5-Day | 75 | SUMMIT | 1/22/1972 |
| Max 6-Day | 77.5 | SUMMIT | 1/22/1972 |
| Max 7-Day | 76.5 | HAUGAN | 1/18/1913 |
| Greatest Monthly | 131.1 | SUMMIT | Jan-72 |
| Greatest Aug-July | 368.9 | SUMMIT | 1943 |
| Greatest Daily Depth | 147 | SUMMIT | 2/18/1975 |

**Table 3.9   State Snowfall Records and National Summary** [CONTINUED]

| Nebraska | | | |
|---|---|---|---|
| Greatest Daily | 24 | ARTHUR | 3/27/1939 |
| Max 2-Day | 35 | HAY SPRINGS | 4/14/1927 |
| Max 3-Day | 43.5 | HAY SPRINGS | 4/14/1927 |
| Max 4-Day | 49.5 | HAY SPRINGS | 4/15/1927 |
| Max 5-Day | 49.5 | HAY SPRINGS | 4/15/1927 |
| Max 6-Day | 29 | CHADRON 1 SSW | 4/22/1920 |
| Max 7-Day | 29.5 | BLOOMFIELD | 1/2/1960 |
| Greatest Monthly | 50.7 | HAY SPRINGS | Apr-27 |
| Greatest Aug-July | 100.6 | MULLEN 21 NW | 1995 |
| Greatest Daily Depth | 35 | HARTINGTON | 3/13/1962 |
| **Nevada** | | | |
| Greatest Daily | 26 | RENO NWSFO | 12/30/2004 |
| Max 2-Day | 37 | MARLETTE LAKE | 12/28/1931 |
| Max 3-Day | 51 | DAGGET PASS | 2/9/1999 |
| Max 4-Day | 60 | SPOONERS STN | 1/13/1911 |
| Max 5-Day | 74 | MARLETTE LAKE | 2/7/1942 |
| Max 6-Day | 80 | MARLETTE LAKE | 2/7/1942 |
| Max 7-Day | 82 | MARLETTE LAKE | 2/7/1942 |
| Greatest Monthly | 132.9 | SPOONERS STN | Jan-11 |
| Greatest Aug-July | 178.5 | GLENBROOK | 1949 |
| Greatest Daily Depth | 108 | MARLETTE LAKE | 2/27/1917 |
| **New Hampshire** | | | |
| Greatest Daily | 41 | CANNON MOUNTAIN | 12/4/1963 |
| Max 2-Day | 65 | CANNON MOUNTAIN | 12/4/1963 |
| Max 3-Day | 72.5 | PINKHAM NOTCH | 2/27/1969 |
| Max 4-Day | 77 | PINKHAM NOTCH | 2/28/1969 |
| Max 5-Day | 77 | PINKHAM NOTCH | 2/28/1969 |
| Max 6-Day | 64.2 | MOUNT WASHINGTON | 12/27/1969 |
| Max 7-Day | 64.2 | MOUNT WASHINGTON | 12/27/1969 |
| Greatest Monthly | 130 | PINKHAM NOTCH | Feb-69 |
| Greatest Aug-July | 331.1 | MOUNT WASHINGTON | 2006 |
| Greatest Daily Depth | 147 | CANNON MOUNTAIN | 3/10/1969 |
| **New Jersey** | | | |
| Greatest Daily | 33 | ELIZABETH | 02/14/1899 |
| Max 2-Day | 34 | OAK RIDGE RESERVOIR | 12/27/1947 |
| Max 3-Day | 30 | OAK RIDGE RESERVOIR | 2/8/1978 |
| Max 4-Day | 28 | SOMERVILLE 3 NW | 12/29/1933 |
| Max 5-Day | 27 | NEWTON ST PAULS ABBEY | 1/11/1996 |
| Max 6-Day | 23.3 | ESTELL MANOR | 2/20/2003 |
| Max 7-Day | 20.6 | CULVERS LAKE | 3/8/1916 |
| Greatest Monthly | 50 | BELVIDERE | Jan-25 |
| Greatest Aug-July | 104.5 | CANISTEAR RESERVOIR | 1961 |
| Greatest Daily Depth | 52 | CANISTEAR RESERVOIR | 2/5/1961 |
| **New Mexico** | | | |
| Greatest Daily | 36 | ABBOTT 1 SE | 11/24/1940 |
| Max 2-Day | 61.5 | CHACON | 3/22/1919 |
| Max 3-Day | 59 | ANCHOR MINE | 12/4/1913 |
| Max 4-Day | 70 | ANCHOR MINE | 12/4/1913 |
| Max 5-Day | 78 | ANCHOR MINE | 12/5/1913 |
| Max 6-Day | 55 | TAJIQUE NEAR | 1/20/1916 |
| Max 7-Day | 54 | LAKE MALOYA | 2/27/1997 |
| Greatest Monthly | 144 | ANCHOR MINE | Mar-12 |
| Greatest Aug-July | 281 | ANCHOR MINE | 1919 |
| Greatest Daily Depth | 73 | LAKE MALOYA | 4/9/1973 |

**Table 3.9   State Snowfall Records and National Summary [CONTINUED]**

| New York | | | |
|---|---|---|---|
| Greatest Daily | 45 | WATERTOWN | 11/15/1900 |
| Max 2-Day | 63.8 | HIGHMARKET | 1/1/2002 |
| Max 3-Day | 84.2 | HIGHMARKET | 1/1/2002 |
| Max 4-Day | 97 | HIGHMARKET | 1/1/2002 |
| Max 5-Day | 89.5 | MALLORY | 2/1/1966 |
| Max 6-Day | 89.5 | MALLORY | 2/1/1966 |
| Max 7-Day | 89 | BENNETTS BRIDGE | 3/5/1947 |
| Greatest Monthly | 182 | HOOKER 12 NNW | Jan-78 |
| Greatest Aug-July | 379.5 | HOOKER 12 NNW | 1979 |
| Greatest Daily Depth | 84 | BARNES CORNERS | 1/25/1987 |
| **North Carolina** | | | |
| Greatest Daily | 29 | MOUNT MITCHELL | 3/21/2001 |
| Max 2-Day | 31 | MOUNT MITCHELL | 1/28/1998 |
| Max 3-Day | 32 | MOUNT MITCHELL | 3/22/2001 |
| Max 4-Day | 30.5 | MOUNT MITCHELL | 12/20/2003 |
| Max 5-Day | 33 | MOUNT MITCHELL | 12/31/1997 |
| Max 6-Day | 37 | MOUNT MITCHELL | 1/1/1998 |
| Max 7-Day | 28 | GRANDFATHER MOUNTAIN | 3/23/1981 |
| Greatest Monthly | 56.5 | BOONE | Mar-60 |
| Greatest Aug-July | 103.4 | BOONE | 1960 |
| Greatest Daily Depth | 44 | BOONE | 3/14/1960 |
| **North Dakota** | | | |
| Greatest Daily | 24 | AMIDON | 2/28/1998 |
| Max 2-Day | 31 | GRAND FORKS FAA AP | 3/5/1966 |
| Max 3-Day | 37.8 | GRAND FORKS FAA AP | 3/5/1966 |
| Max 4-Day | 40.4 | GRAND FORKS FAA AP | 3/5/1966 |
| Max 5-Day | 36.5 | ELLENDALE 8 NNW | 11/27/1993 |
| Max 6-Day | 36.5 | ELLENDALE 8 NNW | 11/27/1993 |
| Max 7-Day | 38.5 | TAGUS | 4/19/1970 |
| Greatest Monthly | 48 | LANGDON EXP FARM | Feb-08 |
| Greatest Aug-July | 103 | STEELE | 1970 |
| Greatest Daily Depth | 58 | FORBES 10 NW | 2/28/1969 |
| **Ohio** | | | |
| Greatest Daily | 22 | WOODSFIELD 2 N | 2/17/2003 |
| Max 2-Day | 33 | PAINESVILLE 4 NW | 11/25/1950 |
| Max 3-Day | 39 | KIRTLAND-HOLDEN 2 | 11/12/1996 |
| Max 4-Day | 48.2 | KIRTLAND-HOLDEN 2 | 11/13/1996 |
| Max 5-Day | 56.2 | KIRTLAND-HOLDEN 2 | 11/14/1996 |
| Max 6-Day | 56.7 | KIRTLAND-HOLDEN 2 | 11/15/1996 |
| Max 7-Day | 56.7 | KIRTLAND-HOLDEN 2 | 11/15/1996 |
| Greatest Monthly | 69.5 | CHARDON | Dec-62 |
| Greatest Aug-July | 147.5 | CHARDON | 1963 |
| Greatest Daily Depth | 35 | ASHTABULA | 2/8/1977 |
| **Oklahoma** | | | |
| Greatest Daily | 23 | BUFFALO | 02/21/71 |
| Max 2-Day | 36 | BUFFALO | 02/22/71 |
| Max 3-Day | 35 | STILLWATER 3 NNW | 11/21/06 |
| Max 4-Day | 34 | KENTON | 02/26/03 |
| Max 5-Day | 19 | ZOE 1 S | 12/29/75 |
| Max 6-Day | 14.5 | GUYMON LEE RANCH | 03/12/58 |
| Max 7-Day | 16.5 | KENTON | 12/23/18 |
| Greatest Monthly | 39.5 | BUFFALO | 02/01/71 |
| Greatest Aug-July | 56.9 | HOOKER | 1993 |
| Greatest Daily Depth | 25 | EL RENO 1 N | 01/11/05 |

**Table 3.9    State Snowfall Records and National Summary [CONTINUED]**

| Oregon | | | |
|---|---|---|---|
| Greatest Daily | 37 | CHEMULT | 2/6/1949 |
| Max 2-Day | 59 | CRATER LAKE NATL PARK HQ | 2/28/1971 |
| Max 3-Day | 81.8 | SANTIAM PASS | 1/19/1964 |
| Max 4-Day | 98.8 | SANTIAM PASS | 1/20/1964 |
| Max 5-Day | 117 | CRATER LAKE NATL PARK HQ | 2/7/1949 |
| Max 6-Day | 133 | CRATER LAKE NATL PARK HQ | 2/8/1949 |
| Max 7-Day | 143 | CRATER LAKE NATL PARK HQ | 2/9/1949 |
| Greatest Monthly | 248.5 | CRATER LAKE NATL PARK HQ | Feb-49 |
| Greatest Aug-July | 822 | CRATER LAKE NATL PARK HQ | 1949 |
| Greatest Daily Depth | 239 | CRATER LAKE NATL PARK HQ | 3/31/1983 |
| **Pennsylvania** | | | |
| Greatest Daily | 38 | MORGANTOWN | 03/20/58 |
| Max 2-Day | 45.3 | COATESVILLE 1 SW | 02/14/99 |
| Max 3-Day | 52.4 | COATESVILLE 1 SW | 02/14/99 |
| Max 4-Day | 53 | COATESVILLE 1 SW | 02/14/99 |
| Max 5-Day | 43 | EMPORIUM 1 E | 12/29/44 |
| Max 6-Day | 50 | EMPORIUM 1 E | 12/29/44 |
| Max 7-Day | 57 | EMPORIUM 1 E | 12/29/44 |
| Greatest Monthly | 84.4 | EBENSBURG SEWAGE PLANT | 01/01/78 |
| Greatest Aug-July | 186.1 | EBENSBURG SEWAGE PLANT | 1978 |
| Greatest Daily Depth | 55 | BEAVERTOWN 1 NE | 01/13/96 |
| **Rhode Island** | | | |
| Greatest Daily | 30 | WOONSOCKET | 2/7/1978 |
| Max 2-Day | 38 | WOONSOCKET | 2/8/1978 |
| Max 3-Day | 28.4 | WOONSOCKET | 2/27/1969 |
| Max 4-Day | 30.4 | NORTH FOSTER 1 E | 4/10/1996 |
| Max 5-Day | 31.4 | NORTH FOSTER 1 E | 4/11/1996 |
| Max 6-Day | 19.2 | NORTH FOSTER 1 E | 2/13/1994 |
| Max 7-Day | 19.5 | NORTH FOSTER 1 E | 2/14/1994 |
| Greatest Monthly | 52.1 | WOONSOCKET | Feb-69 |
| Greatest Aug-July | 129.1 | NORTH FOSTER 1 E | 1996 |
| Greatest Daily Depth | 42 | NORTH FOSTER 1 E | 2/7/1978 |
| **South Carolina** | | | |
| Greatest Daily | 18 | SOCIETY HILL | 2/25/1914 |
| Max 2-Day | 18.1 | RAINBOW LAKE | 3/3/1942 |
| Max 3-Day | 14.7 | STATEBURG | 02/13/1899 |
| Max 4-Day | 15.6 | CAESARS HEAD | 1/25/1987 |
| Max 5-Day | 15.6 | CAESARS HEAD | 1/26/1987 |
| Max 6-Day | 5 | MCCOLL 3 NNW | 1/30/1966 |
| Max 7-Day | 5 | MCCOLL 3 NNW | 1/31/1966 |
| Greatest Monthly | 28.2 | LANDRUM 1 NE | Mar-60 |
| Greatest Aug-July | 27.3 | CATAWBA | 1936 |
| Greatest Daily Depth | 17 | WINTHROP UNIVERSITY | 2/27/2004 |
| **South Dakota** | | | |
| Greatest Daily | 47 | DEADWOOD | 3/14/1973 |
| Max 2-Day | 59.4 | LEAD | 4/19/2006 |
| Max 3-Day | 74.1 | LEAD | 2/27/1998 |
| Max 4-Day | 97 | LEAD | 3/1/1998 |
| Max 5-Day | 112.4 | LEAD | 3/1/1998 |
| Max 6-Day | 112.6 | LEAD | 3/1/1998 |
| Max 7-Day | 112.7 | LEAD | 3/1/1998 |
| Greatest Monthly | 94 | DUMONT 2 ENE | Mar-50 |
| Greatest Aug-July | 324 | LEAD | 1997 |
| Greatest Daily Depth | 73 | LEAD | 3/1/1998 |

**Table 3.9   State Snowfall Records and National Summary [CONTINUED]**

| Tennessee | | | |
|---|---|---|---|
| Greatest Daily | 20.8 | ELIZABETHTON | 3/18/1936 |
| Max 2-Day | 30 | ELKMONT | 3/3/1942 |
| Max 3-Day | 21 | EMBREEVILLE | 3/19/1936 |
| Max 4-Day | 23 | MT LECONTE | 12/31/1997 |
| Max 5-Day | 29 | MT LECONTE | 12/31/1997 |
| Max 6-Day | 18.7 | BRISTOL WSO AIRPORT | 12/27/1969 |
| Max 7-Day | 18.7 | BRISTOL WSO AIRPORT | 12/27/1969 |
| Greatest Monthly | 39.5 | CROSSVILLE EXP STN | Jan-18 |
| Greatest Aug-July | 75.5 | MOUNTAIN CITY 2 | 1960 |
| Greatest Daily Depth | 30 | MT LECONTE | 1/6/1994 |
| **Texas** | | | |
| Greatest Daily | 24 | CLIFTON 9 E | 12/21/1929 |
| Max 2-Day | 26 | HILLSBORO | 12/22/1929 |
| Max 3-Day | 29.5 | PLAINVIEW | 2/4/1956 |
| Max 4-Day | 30 | PLAINVIEW | 2/5/1956 |
| Max 5-Day | 30 | PLAINVIEW | 2/5/1956 |
| Max 6-Day | 30 | DIVIDE SCHOOL | 1/7/1947 |
| Max 7-Day | 34 | DIVIDE SCHOOL | 1/7/1947 |
| Greatest Monthly | 34 | DIVIDE SCHOOL | Jan-47 |
| Greatest Aug-July | 58.7 | BORGER | 1983 |
| Greatest Daily Depth | 26 | HILLSBORO | 12/22/1929 |
| **Utah** | | | |
| Greatest Daily | 35 | SILVER LAKE BRIGHTON | 1/22/1964 |
| Max 2-Day | 54 | PINE VIEW DAM | 1/20/1955 |
| Max 3-Day | 72 | PINE VIEW DAM | 1/21/1955 |
| Max 4-Day | 79 | ALTA | 3/26/1983 |
| Max 5-Day | 92 | ALTA | 1/29/1965 |
| Max 6-Day | 97 | ALTA | 1/30/1965 |
| Max 7-Day | 99 | ALTA | 1/30/1965 |
| Greatest Monthly | 244.5 | ALTA | Dec-83 |
| Greatest Aug-July | 810.5 | ALTA | 1984 |
| Greatest Daily Depth | 160 | ALTA | 3/27/1975 |
| **Vermont** | | | |
| Greatest Daily | 33 | SAINT JOHNSBURY | 02/25/69 |
| Max 2-day | 37 | JAY PEAK | 03/07/01 |
| Max 3-day | 46 | PERU | 03/05/47 |
| Max 4-day | 56 | PERU | 03/05/47 |
| Max 5-Day | 59 | PERU | 03/05/47 |
| Max 6-day | 60 | PERU | 03/05/47 |
| Max 7-day | 51 | SOMERSET | 03/06/47 |
| Greatest Monthly | 95 | WEST WARDSBORO | 03/01/01 |
| Greatest Jul-Aug | 236.8 | MOUNT MANSFIELD | 1968 |
| Greatest Daily Depth | 78 | SOMERSET | 03/05/47 |
| **Virginia** | | | |
| Greatest Daily | 33.5 | LURAY 5 E | 3/3/1994 |
| Max 2-Day | 48 | BIG MEADOWS | 1/7/1996 |
| Max 3-Day | 40 | LURAY 5 E | 3/4/1994 |
| Max 4-Day | 36 | MOUNT WEATHER | 1/10/1996 |
| Max 5-Day | 33.2 | BLACKSBURG 3 SE | 1/10/1996 |
| Max 6-Day | 36.1 | BLACKSBURG 3 SE | 1/12/1996 |
| Max 7-Day | 41 | THE PLAINS 2 NNE | 1/13/1996 |
| Greatest Monthly | 54 | WARRENTON 5 NE | 02/1899 |
| Greatest Aug-July | 124.2 | WISE 3 E | 1996 |
| Greatest Daily Depth | 47 | BIG MEADOWS | 1/7/1996 |

**Table 3.9   State Snowfall Records and National Summary [CONTINUED]**

| Washington | | | |
|---|---|---|---|
| Greatest Daily | 48 | GUNN'S RANCH | 1/21/1935 |
| Max 2-Day | 71 | SCENIC | 3/27/1936 |
| Max 3-Day | 88 | RAINIER PARADISE RNGER S | 1/25/1971 |
| Max 4-Day | 117 | CASCADE TUNNEL | 02/28/1899 |
| Max 5-Day | 128 | CASCADE TUNNEL | 02/28/1899 |
| Max 6-Day | 136 | RAINIER PARADISE RNGER S | 1/26/1971 |
| Max 7-Day | 143 | RAINIER PARADISE RNGER S | 1/26/1971 |
| Greatest Monthly | 249.5 | RAINIER PARADISE RNGER S | Jan-71 |
| Greatest Aug-July | 1140 | MT. BAKER SKI RESPORT | 1999 |
| Greatest Daily Depth | 293 | RAINIER PARADISE RNGER S | 4/12/1974 |
| **West Virginia** | | | |
| Greatest Daily | 33 | ALPENA | 12/29/67 |
| Max 2-Day | 41 | POLK CREEK | 11/25/50 |
| Max 3-Day | 57.2 | COBURN CREEK | 11/26/50 |
| Max 4-Day | 59.5 | COBURN CREEK | 11/27/50 |
| Max 5-Day | 60.7 | COBURN CREEK | 11/28/50 |
| Max 6-Day | 62 | COBURN CREEK | 11/29/50 |
| Max 7-Day | 69 | PICKENS 2 | 02/27/47 |
| Greatest Monthly | 104 | TERRA ALTA NO 1 | 01/01/77 |
| Greatest Aug-July | 241.3 | DAVIS 3 SE | 2003 |
| Greatest Daily Depth | 56 | CANAAN VALLEY | 02/14/79 |
| **Wisconsin** | | | |
| Greatest Daily | 25 | TREMPEALEAU DAM 6 | 1/20/1952 |
| Max 2-Day | 33 | HURLEY | 1/6/1997 |
| Max 3-Day | 39 | HURLEY | 11/3/1989 |
| Max 4-Day | 40 | HURLEY | 11/3/1989 |
| Max 5-Day | 49 | HURLEY | 12/28/2001 |
| Max 6-Day | 59.5 | HURLEY | 12/29/2001 |
| Max 7-Day | 63 | HURLEY | 12/30/2001 |
| Greatest Monthly | 82 | BRULE R S | Nov-91 |
| Greatest Aug-July | 161.5 | LAC VIEUX DESERT | 1971 |
| Greatest Daily Depth | 64 | BRULE ISLAND | 3/18/1939 |
| **Wyoming** | | | |
| Greatest Daily | 34 | BECHLER RIVER RS | 1/28/1933 |
| Max 2-Day | 46.2 | LANDER AP | 4/23/1999 |
| Max 3-Day | 52 | BURGESS JUNCTION | 4/29/1963 |
| Max 4-Day | 60 | BECHLER RIVER RS | 1/25/1933 |
| Max 5-Day | 64 | BECHLER RIVER RS | 1/26/1933 |
| Max 6-Day | 73 | BECHLER RIVER RS | 1/25/1933 |
| Max 7-Day | 84.5 | BECHLER RIVER RS | 1/25/1933 |
| Greatest Monthly | 188.5 | BECHLER RIVER RS | Jan-33 |
| Greatest Aug-July | 382.4 | SNAKE RIVER | 1976 |
| Greatest Daily Depth | 92 | GRASSY LAKE DAM | 3/24/1946 |
| **Nation** | | | |
| Greatest Daily | 63 | GEORGETOWN, CO | 12/4/1913 |
| Max 2-Day | 120.6 | THOMPSON PASS, AK | 12/30/1955 |
| Max 3-Day | 147 | THOMPSON PASS, AK | 12/30/1955 |
| Max 4-Day | 163 | THOMPSON PASS, AK | 12/30/1955 |
| Max 5-Day | 175.4 | THOMPSON PASS, AK | 12/31/1955 |
| Max 6-Day | 172.6 | THOMPSON PASS, AK | 2/24/1953 |
| Max 7-Day | 186.9 | THOMPSON PASS, AK | 2/25/1953 |
| Greatest Monthly | 313 | TAMARACK, CA | Mar-07 |
| Greatest Aug-July | 1140 | MT. BAKER SKI RESORT | 1999 |
| Greatest Daily Depth | 293 | RAINIER PARADISE RNGR STA, WA | 4/12/1974 |

*Source:* NCDC.

**Figure 3.9**   Snow on Mauna Kea, HI, 180° panorama. (Copyright Dr. Jean-Charles Cuillandre, used with permission.)

### Table 3.10   World Maximum Temperatures by Continent

| Continent | Maximum °F | Place | Elevation Feet | Date |
|---|---|---|---|---|
| Africa | 136 | El Azizia, Libya | 367 | 09/13/22 |
| North America | 134 | Death Valley, CA | -178 | 07/10/13 |
| Asia | 129 | Tirat Tsvi, Israel | -722 | 06/22/42 |
| Australia | 128* | Cloncurry, Queensland | 622 | 01/16/89 |
| Europe | 122 | Seville, Spain | 26 | 08/04/81 |
| South America | 120 | Rivadavia, Argentina | 676 | 12/11/05 |
| Oceania | 108 | Tuguegarao, Philippines | 72 | 04/29/12 |
| Antarctica | 59 | Vanda Station, Scott Coast | 49 | 01/05/74 |

*This temperature was measured using the techniques available at the time of recording which are different to the standard techniques currently used in Australia. The most likely record for Australia's high-temperature using today's standards 50.7°C (123°F) recorded at Oodnadatta  2/1/1960 which is accepted by the World Meteorological Organization.

*Source:* NCDC.

### Table 3.11   World Minimum Temperatures by Continent

| Continent | Minimum °F | Place | Elevation Feet | Date |
|---|---|---|---|---|
| Antarctica | -129 | Vostok | 11220 | 07/21/83 |
| Asia | -90 | Oimekon, Russia | 2625 | 02/06/33 |
| Asia | -90 | Verkhoyansk, Russia | 350 | 02/07/92 |
| Greenland | -87 | Northice | 7687 | 01/09/54 |
| North America | -81.4 | Snag, Yukon, Canada | 2120 | 02/03/47 |
| Europe | -72.6 | Ust'Shchugor, Russia | 279 | 12/31/78 |
| South Amercia | -27 | Sarmiento, Argentina | 879 | 06/01/07 |
| Africa | -11 | Ifrane, Morocco | 5364 | 02/11/35 |
| Australia | -9.4 | Charlotte Pass, NSW | 5758 | 06/29/94 |
| Oceania | 12 | Mauna Kea Observatory ,HI | 13780 | 05/17/79 |

*Source:* NCDC.

**Table 3.12  World Maximum Rainfall Records by Continent**

| Continent | Greatest Average inches | Place | Elevation Feet | Years of Data |
|---|---|---|---|---|
| South America | 523.6*^ | Lloro, Colombia | 520 | 29 |
| Asia | 467.4* | Mawsynram, India | 4597 | 38 |
| Oceania | 460.0* | Mt. Waialeale, Kauai, HI | 5148 | 30 |
| Africa | 405.0 | Debundscha, Cameroon | 30 | 32 |
| South America | 354.0 ^ | Quibdo, Colombia | 120 | 16 |
| Australia | 316.3 | Bellenden Ker, Queensland | 5102 | 9 |
| North America | 276.0 | Henderson Lake, British Colombia | 12 | 14 |
| Europe | 183.0 | Crkvica, Bosnia-Hercegovina | 3337 | 22 |

\* Depending on measurement practices each of these may be the world's highest total.
^ Quibdo, Columbia is the accepted offical record for South America, the amount for Lloro, Columbia is estimated at a location 14 miles from Quibdo.

*Source:* NCDC.

Very low pressure also occurs in tropical cyclones. These too are destructive, but because they are much larger, they pass over numerous observation points and pressure measurements are more numerous. In addition, the Hurricane Hunters fly through many tropical cyclones and release dropsondes. A dropsonde is an instrument package that radios back conditions as it sinks through a tropical cyclone. Pressure is extrapolated to the ground and it was using this method that the record low surface pressure of Typhoon Tip was measured.

For atmospheric pressure to be compared between stations at different altitudes, the station pressure readings are adjusted for altitude, temperature, humidity, and instrument error to mean sea level as a reference. Station pressure is unadjusted and is highest in the lowest places on Earth, areas where the surface is below sea level. Death Valley in the United States is 280 feet below sea level and in Egypt the Qattara Depression is as low as 436 feet below sea level, while in the Xinjiang Autonomous Region in far western China the Turfan Depression bottoms out at 505 feet below sea level. The lowest point on Earth and the highest station pressures are found along the shores of the Dead Sea at 1286 feet below sea level.

**Table 3.13  World Minimum Rainfall Records by Continent**

| Continent | Lowest Average inches | Place | Elevation Feet | Years of Data |
|---|---|---|---|---|
| South America | 0.03 | Arica, Chile | 95 | 59 |
| Africa | < 0.1 | Wadi Halfa, Sudan | 410 | 39 |
| Antarctica | 0.8 * | Amundsen-Scott South Pole Sta. | 9186 | 10 |
| North America | 1.2 | Batagues, Mexico | 16 | 14 |
| Asia | 1.8 | Aden, Yemen | 22 | 50 |
| Australia | 4.05 | Mulka (Troudaninna) | 160 | 42 |
| Europe | 6.4 | Astrakhan, Russia | 45 | 25 |
| Oceania | 7.41 | Mauna Kea Ob. Hawaii, USA | | 10 |

\* The amount listed is snow accumulation for one year as indicated by snow markers. The liquid equivalent has not been determined. Antarctica's interior is probably the world's driest desert.

*Source:* NCDC.

**Table 3.14   World Extreme Rainfall Events by Duration**

| Duration | Amount [1] Inches | Location | Date |
|---|---|---|---|
| 1 min | 1.50 | Barot, Guadeloupe, West Indies | 26 Nov 1970 |
| 8 | 4.96 | Fussen, Bavaria, Germany | 25 May 1920 |
| 15 | 7.80 | Plumb Point, Jamaica | 12 May 1916 |
| 20 | 10.24 | Curtea-de-Arges, Romania | 07 Jul  1889 |
| 42 | 12.00 | Holt, MO, USA | 22 Jun  1947 |
| 60 | 15.79 | Shangdi, Nei Mongol, China | 03 Jul  1975 |
| 2 h 10 m | 19.02 | Rockport, WV, USA | 18 Jul 1889 |
| 2 h 45 m | 22.01 | D'Hannis, TX, USA | 31 May 1935 |
| 4 h 30 m | 30.79 | Smethport, PA, USA | 18 Jul 1942 |
| 6 h | 33.07 | Muduocaidang, China | 01 Aug 1977 |
| 9 h | 42.80 | Belouve, La Réunion | 28 Feb 1964 |
| 10 h | 55.12 | Muduocaidang, China | 01 Aug 1977 |
| 18 h | 37.99 | Thrall, TX, USA (US) | 09 Sep 1921 |
| 18.5 h | 66.50 | Belouve, La Réunion | 28-89 Feb 1964 |
| 24 h | 42.91 | Alvin, TX, USA (US) | 25-26 Jul 1979 |
| 24 h | 71.85 | Foc Foc, La Réunion | 07-08 Jan 1966 |
| 2 days | 91.13 | Aurere, La Réunion | 07-09 Apr 1958 |
| 3 | 154.70 | Cratère Commerson, La Réunion | 24-26 Feb 2007 |
| 4 | 191.70 | Cratère Commerson, La Réunion | 24-27 Feb 2007 |
| 5 | 169.33 | Commerson, La Réunion [3] | 23-27 Jan 1980 |
| 6 | 183.19 | Commerson, La Réunion | 22-27 Jan 1980 |
| 7 | 196.97 | Commerson, La Réunion | 21-27 Jan 1980 |
| 8 | 208.11 | Commerson, La Réunion | 20-27 Jan 1980 |
| 9 | 224.09 | Commerson, La Réunion | 19-27 Jan 1980 |
| 10 | 237.32 | Commerson, La Réunion | 18-27 Jan 1980 |
| 11 | 247.99 | Commerson, La Réunion | 17-27 Jan 1980 |
| 12 | 252.01 | Commerson, La Réunion | 16-27 Jan 1980 |
| 13 | 252.83 | Commerson, La Réunion | 15-27 Jan 1980 |
| 14 | 253.23 | Commerson, La Réunion | 15-28 Jan 1980 |
| 15 | 253.27 | Commerson, La Réunion | 14-28 Jan 1980 |
| 31 | 366.14 | Cherrapunji, India [2] | 01-31 Jul 1861 |
| 2 months | 502.64 | Cherrapunji, India | Jun-Jul 1861 |
| 3 | 644.45 | Cherrapunji, India | May-Jul 1861 |
| 4 | 737.72 | Cherrapunji, India | Apr-Jul 1861 |
| 5 | 803.62 | Cherrapunji, India | Apr-Aug 1861 |
| 6 | 884.02 | Cherrapunji, India | Apr-Sep 1861 |
| 11 | 905.12 | Cherrapunji, India | Jan-Nov 1861 |
| 12 | 1041.77 | Cherrapunji, India | Aug 1860 - Jul 1861 |
| 2 years | 1605.04 | Cherrapunji, India | 1860 - 1861 |

Source: World Meteorological Organization
1. Table 12 lists maximum average rainfall, this table lists individual events.
2. Cherripunji is in the Khasi Hills north of the Bay of Bengal. Extreme rainfall there is related to the Indian/Asian
Monsoon.  The location of Cherripunji means the tropical cyclones and monsoon rain system get additional lift as
The moisture laden winds rise to an ground elevation of 4872 feet, the additional lift releases more rain.
3. La Réunion is an Indian Ocean island east of Madagascar, extreme rainfall there is related to tropical cyclones in the
months of January, February and April. Because La Réunion is in the southern hemisphere those months are peak tropical
cyclone season for the western Indian Ocean.

**Figure 3.10**   Record-setting Aurora, NE hailstone with tape measure showing the 7" diameter in inches. (*Source:* NOAA Central Library.)

**World's Highest Mean Sea Level Pressure**

1085.6 mb 32.06" 12/18/2001, Tosontsengel, Mongolia

1083.3 mb 32.01" 12/31/1968, Agata, Evenhiyskiy, Russia

1079.0 mb 31.87" 1/23/1900, Barnaul, Russia

**US Highest Mean Sea Level Pressure**

1078.6 mb, 31.85", Northway, AK, 31 January 1989

**World's Lowest Mean Sea Level Pressure (excluding tornadoes)**

870 mb 25.69" 10/12/1979 Eye of Typhoon Tip (16°44' N, 137°46' E)

**US Lowest Mean Sea Level Pressure**

892.3 mb, 26.35" Matecumbe Key, FL, 2 September 1935.

**World's Greatest Pressure Drop**

100 mb, .339", Manchester, SD, 24 June 2003, during a tornado

**World's Lowest Tornado Surface Pressure**

850 mb, 25.01", Manchester, SD, 24 June 2003, tornado

**World's Highest Station Pressure**

1080.47 mb, 31.91" Dead Sea, Jordan/Israel, 21 February 1961.

## Hailstones

*Source: WMO*

Large hail is the product of large thunderstorms and they occur in many parts of the world. Hail is rare in tropical set-tings because of the warm atmosphere, but in areas farther poleward it is very common. Most of the largest hailstones that fall are masses of individual smaller hailstones that freeze together as they fall and have a rough, irregular shape (see Figures 3.10 and 3.11). Large hailstones can fall at over 100 mph.

**World's Heaviest Hailstone**

2.25 pounds 4/14/1986 Gopalganj District, Bangladesh

**US Largest Hailstone** (Figures 3.10 and 3.11)

Circumference,18.75", diameter, 7", Aurora, NE 22 June 2003.

**Former US Record Holder**

Circumference, 17.5", diameter, 5.7", Coffeyville, KS, 3 September 1970.

Figures 3.10 and 3.11 show the record-setting Aurora, Nebraska hailstone of June 22, 2003.

## Maximum Wind Gust

*Source: WMO*

**Surface Record and Tropical Cyclone Record Barrow Island, Australia**

253 mph 4/10/1996 1055 UTC, Tropical Cyclone Olivia

This bumps the wind gust at Mt. Washington, NH, (231 mph, 4/12/1934, 1:21 P.M. EST) from its status as the world record holder. It also bumps the 211.3 mph wind gust, 8/30/2008,

**Figure 3.11**   Record-setting Aurora, NE hailstone showing the 19" circumference in inches. (*Source:* NOAA Central Library.)

at Paso Real de San Diego meteorological station in Pinar de Rio, Cuba during Hurricane Gustav from its status as the world record tropical cyclone wind gust.

### Greatest Snowfall

*Source: NCDC*

Snowfall measurement is very tricky, and measurement techniques vary a great deal around the world. In addition, snowfall accumulation varies greatly depending on surface temperature, slope, wind, moisture content, and other factors. In the United States, measurement is standardized, but there are still issues with accuracy. The snowfall records here are all from the United States and represent, to the best of our knowledge, world extremes.

Heavy snow occurs in many situations, and an accumulation of 3 feet is not unheard of in the large metropolitan areas of the eastern United States when a strong nor'easter moves along the east coast. Extreme snow is also associated with mountains or lake effect events.

Mountains provide an extra boost of lift to the snowstorm as air flows upslope, and high in the mountains the temperature is cold, so little snow melts on the way down. In addition, mountain snows are often quite "dry" or powdery, meaning they have little moisture content, snowflakes do not

stick together, and the snow blows and drifts easily. In a cold mountain environment a little moisture goes a long way, and alpine snows may have a snow to liquid equivalent ratio of 30:1 or greater. This means 1" of water will make 30 inches of snow. Lower elevations can get snow this dry on very cold days.

The National Weather Service (NWS) uses a table of melt water equivalents to estimate new snow accumulations where snow measurements are not made. The table cannot be used for old snow, only newly fallen snow. Melt water equivalents of freshly fallen snow are indicated in Table 3.15. The standard conversion when the air temperature is near freezing is 10:1, meaning 10 inches of snow will result from a melt water equivalent of 1" of water.

The column that covers the temperature range from 34°F to 28°F, an average temperature of 31°F, uses the standard 10:1 ratio. At the warm end of this range, the snow to melt water ratio is likely to be as low as 7:1—a very wet snow—while at the cold end, the ratio could be as high 16:1—a very dry snow, not good for snowballs and snowmen. The most accurate results using this chart alone require interpolation.

In reality, the chart is a crude estimate because snow characteristics are often determined by the temperature and available moisture aloft where the snow forms; then the characteristics are modified as the snow falls through the lowest layers of air. In addition, the strength of the upward motion in the precipitation zone can create large differences. The most accurate snow estimates are achieved when all factors in the three-dimensional atmosphere and the change of these factors through time are taken into account.

Old snow undergoes compaction and freeze–thaw cycles that decrease the depth and increase the water content of snow on the ground. Until recently, water content could only be estimated by direct measurement of old snow. Now remote sensing of snow depth and snow water content is being done by the National Operational Hydrologic Remote Sensing Center (NOHRSC), which is part of the NWS. The method relies on optical and microwave satellite data and cross-checking with ground-based observations and snow model output (Table 3.15).

The Great Lakes provide a nearly infinite supply of moisture for arctic winds to tap when they blow across the relatively warm water. As the air is warmed by contact with the water, it rises, taking with it moisture evaporated from the lake surface creating snow squalls. When the squalls reach the shore, the snow can be intense and accumulate rapidly. As the winds rise out of the lake basin, vertical motion is enhanced, and lake effect snow showers can be followed all the way, for example, from Lake Erie to the high mountains of West Virginia.

**Greatest Season**
1140" 1998–1999 Mt. Baker Ski Resort, WA

**Greatest Storm**
189.0" February 13–19, 1959, Mt. Shasta, CA

**Greatest 24-Hour Snowfall**
76" 14–15 April, 1921, Silver Lake, CO

**Greatest 24-Hour Snowfall (Unofficial)**
77", Montague Township, Lewis Co. NY, 11–12 January 1997

**Greatest 1-Month Snowfall**
390" January 1911, Tamarack, CA.

**Greatest Depth on the Ground**
451" 11 March 1911, Tamarack, CA.

## Temperature Changes

*Source: US Army Topographic Engineering Center*

Rapid temperature changes are associated with fronts and mountainous terrain on the eastern slopes of the Rocky Mountains.

Extreme cold fronts generally account for greater temperature drops than warm fronts account for temperature rises. In mountain areas, chinook winds (warm downslope winds warmed by compression) can warm an area quickly. When an arctic cold front arrives, the chinook ceases and the cold air replaces the abnormally warm air. The result can be large drops in temperature like the examples from Browning and Fairfield, MT and Rapid City, SD.

In addition, along the eastern slopes of the Rocky Mountains cold air can slosh up the mountain slopes, then back down, and up again resulting in rapid fluctuations. Eastward from the Rockies, the plains slope downward for more than 500 miles. Gravity can pull the cold air downslope, but meteorological conditions can cause it to surge westward and up the slopes again. Couple this with the complex terrain and downslope winds and temperature fluctuations can be incredible like the examples from Spearfish, SD and the second example from Rapid City, SD.

**Temperature Drop, Browning, MT, January 23–24, 1916**
In 24 hours a total of 100°F, from 44°F to −56°F.

**Temperature Drop, Fairfield, MT, December 24, 1924**
Drop: In 12 hours a total of 84°F from 63°F at noon to −21°F at midnight

**Temperature Drop, Rapid City, SD, January 12, 1911**
Drop: In 2 hours a total of 62°F, from 49°F at 6:00 A.M. to −13°F at 8:00 A.M.

**Rapid Extreme Temperature Fluctuations**
**Spearfish, SD, January 22, 1943**
Rise: In 2 minutes a total of 49°F, from −4°F at 7:30 A.M. to 45°F at 7:32 A.M.

Drop: In 4 hours 43 minutes a total of 29°F, from 45°F at 7:32 A.M. to 16°F at 12:15 P.M.

Rise: In 25 minutes a total of 40°F, from 16°F at 12:15 P.M.to 56°F at 12:40 P.M.

**Rapid Extreme Temperature Fluctuations**
**Rapid City, SD, January 22, 1943**
**Total Event Time 3 hours 20 minutes starting at 9:20 A.M.**
Rise: In 20 minutes a total of 49°F, from 5°F at 9:20 A.M. to 54°F at 9:40 A.M.

Drop: In 50 minutes a total of 43°F from 54°F at 9:40 A.M. to 11°F at 10:30 A.M.

Rise: In 15 minutes a total of 44°F from 11°F at 10:30 A.M. to 55°F at 10:45 A.M.

Drop: In 45 minutes a total of 45°F from 55°F at 10:45 A.M. to 10°F at 11:30 A.M.

Rise: In 20 minutes a total of 24°F from 10°F at 11:30 A.M. to 34°F at 11:50 A.M.

Drop: In 25 minutes a total of 18°F from 34°F at 11:50 A.M. to 16°F at 12:15 P.M.

Rise: In 25 minutes a total of 40°F from 16°F at 12:15 P.M. to 56°F at 12:40 P.M.

Figure 3.12 is a graph of the temperature change from 3:00 A.M. to 1:00 P.M. local time for Rapid City, SD, Alliance, NE to the south and Glasgow, MT to the north, showing the temperature changes at each location. Notice that Glasgow is well in the arctic air and shows a slight daytime warming. The temperature graph at Alliance indicates the arrival of warm air from the southwest and the initiation of chinook winds during the morning. Rapid City, near the edge of the

**Table 3.15   Snow—Melt Water Equivalents for Newly Fallen Snow**

## Air Temperature °F

| Melt Water Equivalent, inches | 34° ↓ 28° | 27° ↓ 20° | 19° ↓ 15° | 14° ↓ 10° | 9° ↓ 0° | -1° ↓ -20° | -21° ↓ -40° |
|---|---|---|---|---|---|---|---|
| | **Estimated New Snowfall, inches** | | | | | | |
| Trace | Trace | 0.1 | 0.2 | 0.3 | 0.4 | 0.5 | 1.0 |
| 0.01 | 0.10 | 0.2 | 0.2 | 0.3 | 0.4 | 0.5 | 1.0 |
| 0.02 | 0.20 | 0.3 | 0.4 | 0.6 | 0.8 | 1.0 | 2.0 |
| 0.03 | 0.30 | 0.5 | 0.6 | 0.9 | 1.2 | 1.5 | 3.0 |
| 0.04 | 0.40 | 0.6 | 0.8 | 1.2 | 1.6 | 2.0 | 4.0 |
| 0.05 | 0.50 | 0.8 | 1.0 | 1.5 | 2.0 | 2.5 | 5.0 |
| 0.06 | 0.60 | 0.9 | 1.2 | 1.8 | 2.4 | 3.0 | 6.0 |
| 0.07 | 0.70 | 1.1 | 1.4 | 2.1 | 2.8 | 3.5 | 7.0 |
| 0.08 | 0.80 | 1.2 | 1.6 | 2.4 | 3.2 | 4.0 | 8.0 |
| 0.09 | 0.90 | 1.4 | 1.8 | 2.7 | 3.6 | 4.5 | 9.0 |
| 0.10 | 1.00 | 1.5 | 2.0 | 3.0 | 4.0 | 5.0 | 10.0 |
| 0.11 | 1.10 | 1.7 | 2.2 | 3.3 | 4.4 | 5.5 | 11.0 |
| 0.12 | 1.20 | 1.8 | 2.4 | 3.6 | 4.8 | 6.0 | 12.0 |
| 0.13 | 1.30 | 2.0 | 2.6 | 3.9 | 5.2 | 6.5 | 13.0 |
| 0.14 | 1.40 | 2.1 | 2.8 | 4.2 | 5.6 | 7.0 | 14.0 |
| 0.15 | 1.50 | 2.3 | 3.0 | 4.5 | 6.0 | 7.5 | 15.0 |
| 0.16 | 1.60 | 2.4 | 3.2 | 4.8 | 6.4 | 8.0 | 16.0 |
| 0.17 | 1.70 | 2.6 | 3.4 | 5.1 | 6.8 | 8.5 | 17.0 |
| 0.18 | 1.80 | 2.7 | 3.6 | 5.4 | 7.2 | 9.0 | 18.0 |
| 0.19 | 1.90 | 2.9 | 3.8 | 5.7 | 7.6 | 9.5 | 19.0 |
| 0.20 | 2.00 | 3.0 | 4.0 | 6.0 | 8.0 | 10.0 | 20.0 |
| 0.21 | 2.10 | 3.1 | 4.2 | 6.3 | 8.4 | 10.5 | 21.0 |
| 0.22 | 2.20 | 3.3 | 4.4 | 6.6 | 8.8 | 11.0 | 22.0 |
| 0.23 | 2.30 | 3.4 | 4.6 | 6.9 | 9.2 | 11.5 | 23.0 |
| 0.24 | 2.40 | 3.6 | 4.8 | 7.2 | 9.6 | 12.0 | 24.0 |
| 0.25 | 2.50 | 3.8 | 5.0 | 7.5 | 10.0 | 12.5 | 25.0 |
| 0.30 | 3.00 | 4.5 | 6.0 | 9.0 | 12.0 | 15.0 | 30.0 |
| 0.35 | 3.50 | 5.3 | 7.0 | 10.5 | 14.0 | 17.5 | 35.0 |
| 0.40 | 4.00 | 6.0 | 8.0 | 12.0 | 16.0 | 20.0 | 40.0 |
| 0.45 | 4.50 | 6.8 | 9.0 | 13.5 | 18.0 | 22.5 | 45.0 |
| 0.50 | 5.00 | 7.5 | 10.0 | 15.0 | 20.0 | 25.0 | 50.0 |
| 0.60 | 6.00 | 9.0 | 12.0 | 18.0 | 24.0 | 30.0 | 60.0 |
| 0.70 | 7.00 | 10.5 | 14.0 | 21.0 | 28.0 | 35.0 | 70.0 |
| 0.80 | 8.00 | 12.0 | 16.0 | 24.0 | 32.0 | 40.0 | 80.0 |
| 0.90 | 9.00 | 13.5 | 18.0 | 27.0 | 36.0 | 45.0 | 90.0 |
| 1.00 | 10.00 | 15.0 | 20.0 | 30.0 | 40.0 | 50.0 | 100.0 |
| 2.00 | 20.00 | 30.0 | 40.0 | 60.0 | 80.0 | 100.0 | 200.0 |
| 3.00 | 30.00 | 45.0 | 60.0 | 90.0 | 120.0 | 150.0 | 300.0 |

*Source:* NWS.

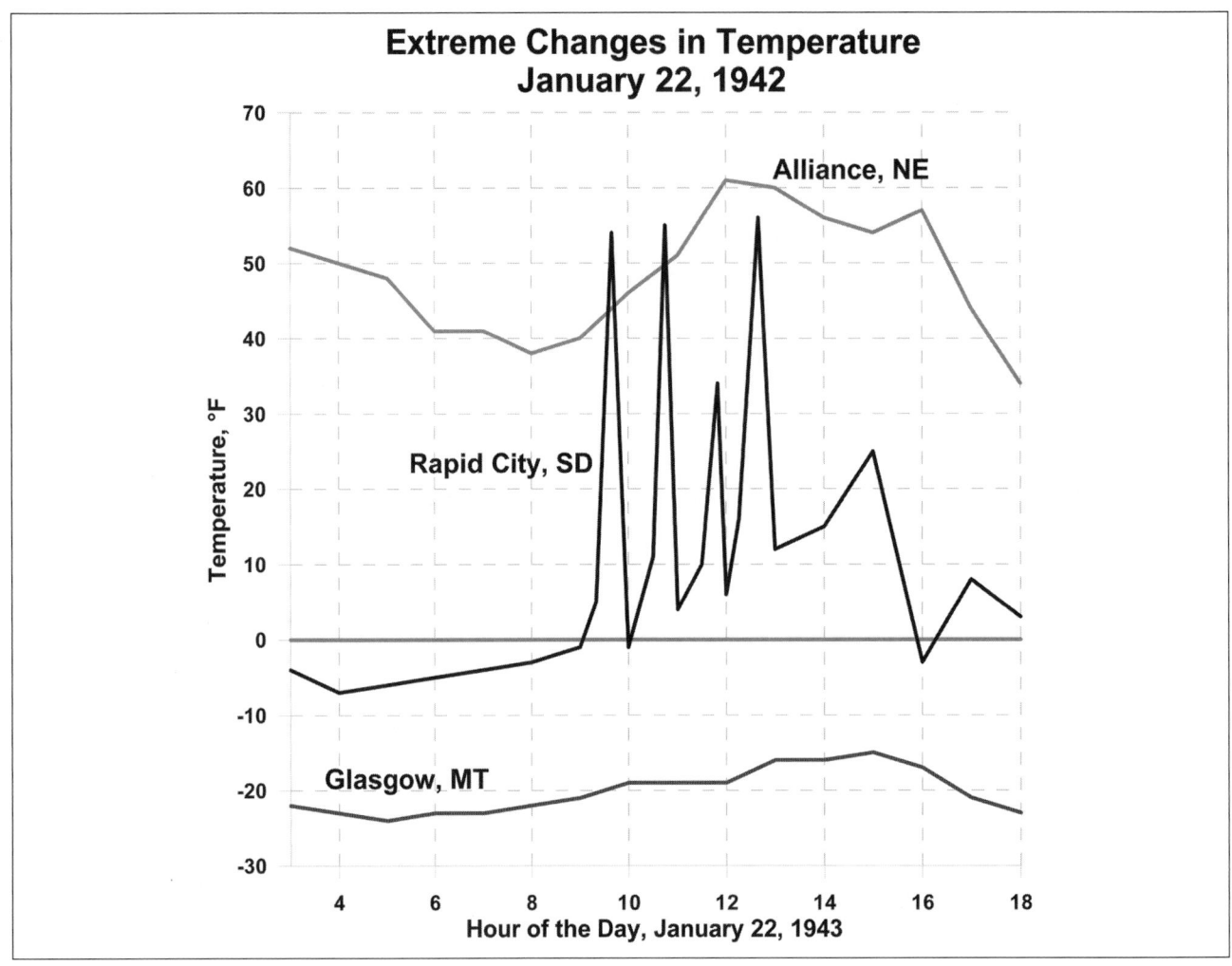

**Figure 3.12** Extreme temperature fluctuations in Rapid City, SD, 01.22.1943. (*Source:* Steve Horstmeyer from NWS data supplied by Weather Graphics Technologies.)

arctic air, fluctuates as the air "sloshes" and chinook winds start and stop.

## Miscellaneous Temperatures

*Source: NCDC*

**Coldest US Winter**
Barrow, AK, Average Temperature −16°F

**Coolest US Summer**
Barrow, AK, Average Temperature 36°F

**Lowest Annual Average**
Barrow, AK Average Temperature 9°F

**Lowest Temperature for the United States Outside Alaska**
Rogers Pass, MT, −69.7°F January 20, 1954

## Rainfall

Extreme rainfall can come from powerful and slow-moving, large-scale middle latitude storms, thunderstorms, tropical systems, stalled fronts, and upslope flow caused by mountains. In short, there are many ways rainfall can be extreme, but one thing these heavy rainfall events must have is moisture, either in place or transported by winds from moisture sources like the oceans or Great Lakes.

Thunderstorms can become stationary along mountain slopes and dump rain hour after hour into the headwaters of a single canyon stream, or they can "train," that is, follow the same path as preceding storms did, like train cars on tracks. Thunderstorms can also develop over the same area for several days in a row, leading to flash floods. In arid areas, scant vegetation does not intercept as much water as in wetter climates and flash flooding can occur quickly.

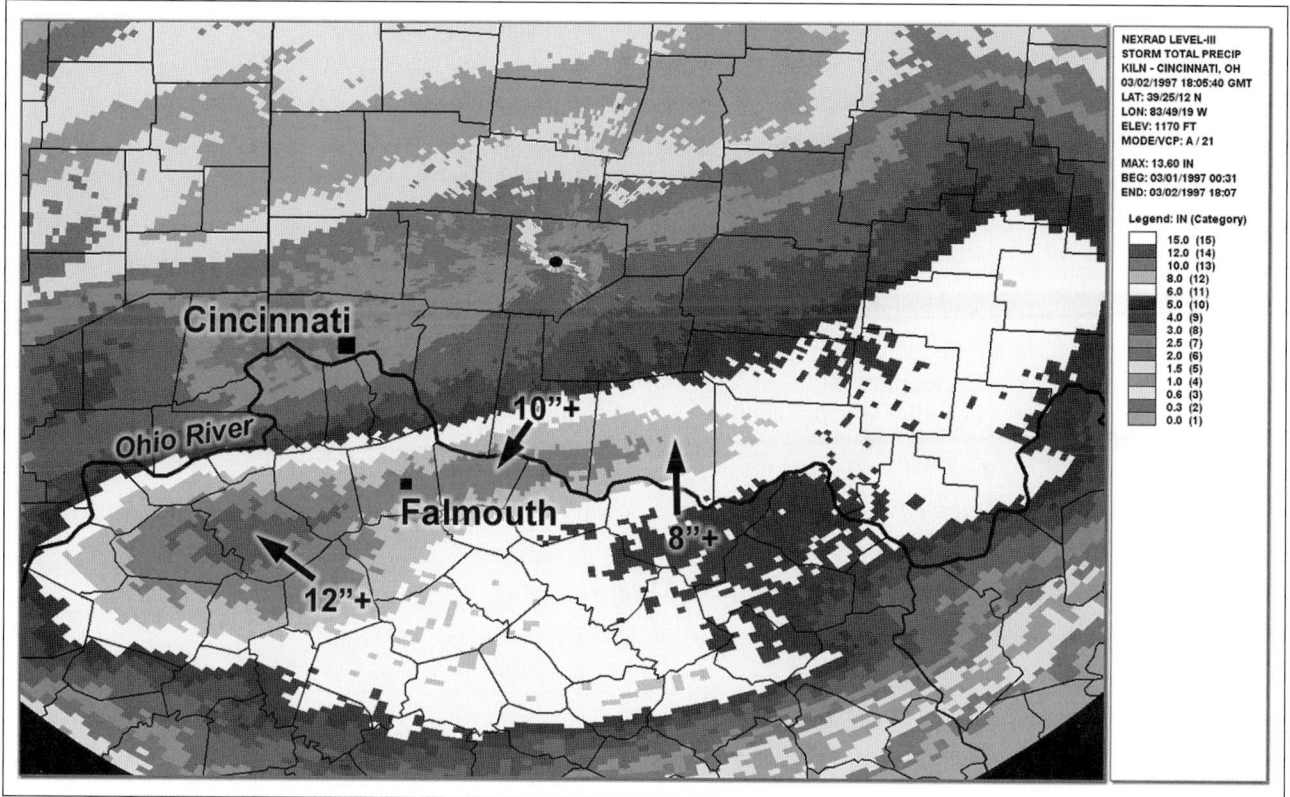

**Figure 3.13**   NEXRAD Doppler radar estimated storm total precipitation, the Falmouth, KY flash flood, 03.02.1997. (*Source:* Steve Horstmeyer from NOAA, NCDC data.)

On July 31/August 1, 1976, a massive stationary thunderstorm dumped 7.5" of rain in an hour and a total of 12" in a few hours on the high elevations of the Big Thompson River in the Front Range of Colorado. The resulting flood killed 144 and destroyed 418 homes and 52 businesses.

Twenty-six died on June 14, 1990 when thunderstorms trained along an outflow boundary left from thunderstorms the day before and thunderstorm dropped 4" of rain in under an hour on the saturated steep valley sides of Wegee and Pipe Creeks upstream from Shadyside in southeast Ohio. A wall of water from 10 to 30 feet tall destroyed 80 homes, caused major damage to 79 and minor damage to 172 other homes.

An example of flooding from a stalled front occurred on March 2–3, 1997 in Falmouth, Kentucky when 80% of the town was under water, hundreds were trapped in the town, 5 died. The Licking river crested higher than ever before. A strong jet stream pulled warm, moisture-laden tropical air up and over a stationary front just south of the town and 10" to 15" of rain fell in a single night. Figure 3.13 is the radar estimated storm total rainfall from the Wilmington, Ohio National Weather Service NEXRAD Doppler Radar.

### US Greatest 24-Hour

43" Alvin, TX, 25–26, July 1979, Tropical Storm Claudette

### US Greatest 12-Month

739", Kukui, Maui, HI, December 1981–December 1982

### US Greatest Annual Average

460", Mount Waialeale, Kauai, HI

## Humidity and Dew Point

Humidity is water vapor in the air and it is highest where there is plenty of water to evaporate and where there is plenty of energy to do the work of evaporation. Cities like Bangkok, Thailand, Kuala Lumpur, Malaysia and Singapore are all well known as cities with a stifling combination of heat and humidity. In the United States cities in Florida and along the Gulf Coast is where the humidity is highest.

Relative humidity is perhaps the best known and most misunderstood measure of humidity because it varies both with the amount of moisture in the air and the air temperature. A warm and muggy summer dawn in the Midwest with a temperature of 72°F and a relative humidity of 93% can turn into a hot sticky afternoon with a temperature of 95°F and a relative humidity of 45%.

The relative humidity is most easily thought of as measuring the proportion of the energy that is available to do the work of evaporation that has been used. In other words a relative humidity of 75% means that 75% of the available energy has been used. In the example the relative humidity

| Dew Point Temp. °F | Human Perception | Relative Humidity Air Temp 90°F |
|---|---|---|
| **Table 3.16** | **Human Perception of Dew Point Temperature** | |
| **75°+** | Extremely uncomfortable, oppressive | 62% |
| **70° - 74°** | Very Humid, quite uncomfortable | 52% - 60% |
| **65° - 69°** | Somewhat uncomfortable for most people at the upper limit | 44% - 52% |
| **60° - 64°** | OK for most, but everyone perceives the humidity at upper limit | 37% - 46% |
| **55° - 59°** | Comfortable | 31% - 41% |
| **50° - 54°** | Very comfortable, evaporative cooling noticeable after swimming. | 31% - 37% |
| **49° or lower** | Feels like the western U.S. A bit dry to some after swimming many feel chilly due to evaporation. | 30% |

has dropped because there is more available energy as the day warms and percentage used decreases.

As air cools, the amount of energy available for evaporation decreases and the relative humidity increases. When the relative humidity reaches 100% most often condensation occurs and the temperature at which condensation first appears is the dew point temperature. If the relative humidity reaches 100% at a temperature below freezing technically it is called the frost point temperature but in practice meteorologists do not distinguish between the two.

Foggy autumn mornings are a common occurrence in valleys and along bodies of water. Air cooling during longer autumn nights often reaches the dew point temperature and fog forms. During the shorter nights of mid–summer, this does not happen as often.

Dew point is often mentioned during TV weather segments in the humid east when the sticky "dog days" of summer arrive. In the less humid west dew point is rarely a factor in discomfort. In fact, the low dew point temperatures in the west lead to rapid evaporation of perspiration and a cooler perception of the temperature.

Table 3.16 lists descriptions of human perception at various dew point temperatures.

Air can also be cooled by evaporating moisture into it. As rain falls, some of it evaporates beneath cloud base using energy from the air that lowers the air temperature. As more and more evaporation occurs the relative humidity increases When the relative humidity reaches 100% solely from cooling caused by evaporation of water into it, the temperature at that point is called the wet bulb temperature.

Wet bulb is often measured by placing a wet "sock" on a thermometer bulb and whirling in the air until the temperature drops no further. The wet bulb temperature is lower than the actual temperature because of evaporative cooling and it is the lowest temperature attainable by evaporative cooling. The difference between the dry bulb temperature and the wet bulb temperature is a measure of relative humidity. In practice the wet bulb temperature tells the meteorologist the coolest temperature that can be reached when evaporation is occurring. In the case of rain evaporating below cloud base if the wet bulb temperature is below freezing the air can be cooled enough to change the falling raindrops to ice pellets or to stop snowflakes from melting to raindrops (Tables 3.16–3.18).

## Fog

Fog is a stratus cloud on the ground. What is a cloud in the air from the vantage point of a valley is fog from the mountain side at the same elevation as the cloud. In this fog discussion mountain tops and mountain passes are excluded because they are special cases of air cooling to 100% relative humidity from being lifted over the terrain. Terrain lifting, also called orographic lifting is an important fog maker, Mt. Washington, NH, for example, averages 308 days per year with fog.

Fog is not water vapor but condensed liquid or solid water. If a fog is composed of small ice crystals it is termed frozen fog. If fog freezes on contact with surfaces it is called freezing fog. To get fog you must have sufficient moisture and a way to cool the air.

Table 3.17   Relative Humidity from Dew Point Temperature

## Relative Humidity, % from Dew Point Temperature, °F and Temperature °F

| Temperature °F \ Dew Point Temperature °F | -20 | -15 | -10 | -5 | 0 | 5 | 10 | 15 | 20 | 25 | 30 | 35 | 40 | 45 | 50 | 55 | 60 | 65 | 70 | 75 | 80 | 85 | 90 |
|---|---|---|---|---|---|---|---|---|---|---|---|---|---|---|---|---|---|---|---|---|---|---|---|
| -20 | 100 | | | | | | | | | | | | | | | | | | | | | | |
| -15 | 86 | 100 | | | | | | | | | | | | | | | | | | | | | |
| -10 | 67 | 78 | 100 | | | | | | | | | | | | | | | | | | | | |
| -5 | 50 | 58 | 75 | 100 | | | | | | | | | | | | | | | | | | | |
| 0 | 40 | 47 | 60 | 80 | 100 | | | | | | | | | | | | | | | | | | |
| 5 | 32 | 37 | 47 | 63 | 79 | 100 | | | | | | | | | | | | | | | | | |
| 10 | 25 | 29 | 38 | 50 | 63 | 79 | 100 | | | | | | | | | | | | | | | | |
| 15 | 20 | 23 | 30 | 40 | 50 | 63 | 79 | 100 | | | | | | | | | | | | | | | |
| 20 | 16 | 19 | 24 | 32 | 41 | 51 | 63 | 81 | 100 | | | | | | | | | | | | | | |
| 25 | 13 | 15 | 20 | 26 | 33 | 41 | 51 | 65 | 80 | 100 | | | | | | | | | | | | | |
| 30 | 11 | 13 | 16 | 21 | 27 | 34 | 41 | 54 | 66 | 82 | 100 | | | | | | | | | | | | |
| 35 | 8.7 | 10 | 13 | 17 | 22 | 28 | 34 | 43 | 54 | 67 | 81 | 100 | | | | | | | | | | | |
| 40 | 7.1 | 8.3 | 11 | 14 | 18 | 23 | 28 | 36 | 44 | 55 | 67 | 82 | 100 | | | | | | | | | | |
| 45 | 5.9 | 6.9 | 8.8 | 12 | 15 | 19 | 23 | 29 | 36 | 45 | 55 | 68 | 82 | 100 | | | | | | | | | |
| 50 | 4.9 | 5.7 | 7.3 | 9.8 | 12 | 15 | 19 | 24 | 30 | 37 | 46 | 56 | 68 | 83 | 100 | | | | | | | | |
| 55 | 4.1 | 4.7 | 6.1 | 8.1 | 10 | 13 | 15 | 20 | 25 | 31 | 38 | 47 | 57 | 69 | 83 | 100 | | | | | | | |
| 60 | 3.4 | 4.0 | 5.1 | 6.8 | 8.5 | 11 | 13 | 17 | 21 | 26 | 32 | 39 | 47 | 58 | 69 | 84 | 100 | | | | | | |
| 65 | 2.8 | 3.3 | 4.3 | 5.7 | 7.1 | 9.0 | 11 | 14 | 18 | 22 | 27 | 33 | 40 | 48 | 58 | 70 | 84 | 100 | | | | | |
| 70 | 2.4 | 2.8 | 3.6 | 4.8 | 6.0 | 7.6 | 9.0 | 12 | 15 | 18 | 22 | 28 | 34 | 41 | 49 | 59 | 71 | 84 | 100 | | | | |
| 75 | 2.0 | 2.4 | 3.0 | 4.1 | 5.1 | 6.4 | 7.6 | 10 | 13 | 15 | 19 | 23 | 28 | 34 | 42 | 50 | 60 | 71 | 85 | 100 | | | |
| 80 | 1.7 | 2.0 | 2.6 | 3.4 | 4.3 | 5.4 | 6.4 | 8.6 | 11 | 13 | 16 | 20 | 24 | 29 | 35 | 42 | 51 | 60 | 72 | 85 | 100 | | |
| 85 | 1.5 | 1.7 | 2.2 | 2.9 | 3.6 | 4.6 | 5.4 | 7.3 | 9.0 | 11 | 14 | 17 | 20 | 25 | 30 | 36 | 43 | 51 | 61 | 72 | 85 | 100 | |
| 90 | 1.2 | 1.5 | 1.9 | 2.5 | 3.1 | 3.9 | 4.6 | 6.2 | 7.7 | 9.5 | 12 | 14 | 17 | 21 | 26 | 31 | 37 | 44 | 52 | 61 | 73 | 85 | 100 |
| 95 | 1.1 | 1.2 | 1.6 | 2.1 | 2.7 | 3.4 | 3.9 | 5.3 | 6.6 | 8.2 | 9.9 | 12 | 15 | 18 | 22 | 27 | 32 | 37 | 44 | 53 | 62 | 73 | 86 |
| 100 | 0.9 | 1.1 | 1.4 | 1.8 | 2.3 | 2.9 | 3.4 | 4.6 | 5.6 | 7.0 | 8.5 | 11 | 13 | 16 | 19 | 23 | 28 | 33 | 39 | 45 | 53 | 63 | 73 |
| 105 | 0.8 | 0.9 | 1.2 | 1.6 | 2.0 | 2.5 | 2.9 | 3.9 | 4.9 | 6.0 | 7.3 | 9.1 | 11 | 13 | 16 | 19 | 23 | 28 | 33 | 38 | 45 | 53 | 63 |
| 110 | 0.7 | 0.8 | 1.0 | 1.4 | 1.7 | 2.2 | 2.5 | 3.4 | 4.2 | 5.2 | 6.3 | 7.8 | 9.5 | 12 | 14 | 17 | 20 | 24 | 28 | 34 | 40 | 47 | 55 |
| 115 | 0.6 | 0.7 | 0.9 | 1.2 | 1.5 | 1.9 | 2.2 | 3.0 | 3.7 | 4.6 | 5.5 | 6.8 | 8.3 | 10 | 12 | 15 | 18 | 21 | 25 | 29 | 35 | 41 | 48 |
| 120 | 0.5 | 0.6 | 0.8 | 1.0 | 1.3 | 1.6 | 1.9 | 2.6 | 3.2 | 4.0 | 4.8 | 5.9 | 7.2 | 8.8 | 11 | 13 | 15 | 18 | 22 | 25 | 30 | 35 | 41 |

**Table 3.18    Relative Humidity from Wet Blub Temperature**

| Air Temperature °F | Depression of the Wet-Bulb Thermometer $(T - T_{wb})°F$ | | | | | | | | | | | | | |
|---|---|---|---|---|---|---|---|---|---|---|---|---|---|---|
| | 1 | 2 | 3 | 4 | 6 | 8 | 10 | 12 | 14 | 16 | 18 | 20 | 25 | 30 |
| 0 | 67 | 33 | 18 | | | | | | | | | | | |
| 5 | 73 | 46 | 20 | | | | | | | | | | | |
| 10 | 78 | 56 | 34 | 13 | | | | | | | | | | |
| 15 | 82 | 64 | 46 | 29 | | | | | | | | | | |
| 20 | 85 | 70 | 55 | 40 | 12 | | | | | | | | | |
| 25 | 87 | 74 | 62 | 49 | 25 | 10 | | | | | | | | |
| 30 | 89 | 78 | 67 | 56 | 36 | 16 | | | | | | | | |
| 35 | 91 | 81 | 72 | 63 | 45 | 27 | 10 | | | | | | | |
| 40 | 92 | 83 | 75 | 68 | 52 | 37 | 22 | 10 | | | | | | |
| 45 | 93 | 86 | 78 | 71 | 57 | 44 | 31 | 18 | 8 | | | | | |
| 50 | 93 | 87 | 80 | 74 | 61 | 49 | 38 | 27 | 16 | 5 | | | | |
| 55 | 94 | 88 | 82 | 76 | 65 | 54 | 43 | 33 | 23 | 14 | 5 | | | |
| 60 | 94 | 89 | 83 | 78 | 68 | 58 | 48 | 39 | 30 | 21 | 13 | 5 | | |
| 65 | 95 | 90 | 85 | 80 | 70 | 61 | 52 | 44 | 35 | 27 | 20 | 12 | | |
| 70 | 95 | 90 | 86 | 81 | 72 | 64 | 55 | 48 | 40 | 33 | 25 | 19 | 4 | |
| 75 | 96 | 91 | 86 | 82 | 74 | 66 | 58 | 51 | 44 | 37 | 30 | 24 | 9 | |
| 80 | 96 | 91 | 87 | 83 | 75 | 68 | 61 | 54 | 47 | 41 | 35 | 29 | 15 | 4 |
| 85 | 96 | 92 | 88 | 84 | 76 | 70 | 63 | 56 | 50 | 44 | 38 | 32 | 20 | 8 |
| 90 | 96 | 92 | 89 | 85 | 78 | 71 | 65 | 58 | 52 | 47 | 41 | 36 | 24 | 13 |
| 95 | 96 | 93 | 89 | 86 | 79 | 72 | 66 | 60 | 54 | 49 | 44 | 38 | 27 | 17 |
| 100 | 96 | 93 | 89 | 86 | 80 | 73 | 68 | 62 | 56 | 51 | 46 | 41 | 30 | 21 |

Fog is most common where water is plentiful and there is a cool surface to chill the air. All along the west coast and along the northern past of the east coast cold ocean currents cool the air to the dew point temperature.

The California current flows southward along the west coast and a combination of cool water from the North Pacific and upwelling create dense fog 40 days per year in San Francisco. Along the coasts of Washington and Oregon the effect is the greatest and at Cape Disappointment at the mouth of the Columbia River is the foggiest spot in the United States being fogged in 29% of the time or 2552 of 8760 hours each year.

Along the east coast the prize goes to Nantucket Island, with 85 days reporting fog with the coasts of the New England States not far behind, especially Maine where most of the coast has 55 foggy days per year. North of this may be the foggiest place on Earth along the coast of Newfoundland, Canada where the cold Labrador current chills the moist ocean air and at Cape Race causing fog 158 days each year.

Away from coasts it is the valleys of the Appalachian Mountain and nearby foothills where cool dense air pools and Gulf of Mexico moisture is trapped. An additional fog-forming factor in the valleys of the Appalachian Mountains is the presence of numerous condensation nuclei, small bits of dust and pine tree terpenes upon which water vapor readily condenses.

Turpenes are volatile organic compounds (VOCs) produced by pine trees that give your Christmas tree that great smell and are the reason the Smoky Mountains are smoky and the Blue Ridge Mountains have that hazy bluish cast. There is more on VOCs as air pollutants in Chapter 7.

Often in winter fog becomes a problem in California's central valley. The combination of warm air above cooler surface air (called an inversion) traps moisture and the presence of large amounts of dust and the residue of agricultural fertilizer provide plenty of condensation nuclei for fog drop formation.

**Average Annual Fog Occurrence**

**Cape Disappointment, WA**
2552 Hours

**Moose Peak Lighthouse, Mistake Island, ME**
1580 Hours

**Nantucket Island**
85 Days with dense fog

## • FLOODS

When you hear the word "flood" you know that means too much water. In the natural world there are many reasons for floods and the United States Geological Survey (USGS) classifies floods as follows:

### Regional Floods

These floods occur in large or regional river basins. Often the floods are seasonal and occur when snow melt combined with rains increase runoff. The soil may be frozen so water does not readily infiltrate the soil.

Other causes of regional floods include slow moving but strong fronts and low pressure systems, and decaying tropical cyclones. Wet weather patterns often precede regional floods setting the stage for widespread runoff by saturating the soil.

### Flash Floods

Water that rises rapidly over a period of a few seconds to a few hours is called a flash flood. Flash floods are the deadliest of the flood types because they catch people with little or no warning. Flash floods are caused by intense rains of short duration often falling from stalled weather systems. Contributing factors include the steepness of the slopes in a basin, basin size, basin shape and the density of vegetation. Basin shapes that approximate a narrow rectangle are more susceptible to flash flooding that similar basins of the same areal extent but more nearly circular.

Urban areas are susceptible to flash floods because much of the surface is covered with nonporous materials and water is prevented from soaking into the ground leading to rapid runoff.

In contrast mature forests slow runoff in several ways. Vegetation intercepts rain drops before they hit the ground. The drops are broken into smaller drops and strike the ground with lower velocity and less force making it more likely the rain will soak into the soil. Some of the water remains in the foliage and either evaporates or slowly falls to the ground later. Intertwined root systems stabilize the soil, which minimizes erosion and further increases infiltration.

### Ice-jam Floods

This type of flood obviously occurs in northern latitudes where rivers freeze either annually or because of a long pe-

riod of time having numerous outbreaks of arctic air. Rivers with the most frequent occurrence of ice-jam flooding are rivers that flow to the north. Southern parts of the basin warm sooner as winter ebbs and water flowing northward encounters still frozen parts of the drainage basin.

When the ice does not completely melt, ice flows can pile up and dam a river backing up the water. When the ice jam breaks, and this can be suddenly, the water rushes downstream and acts like a flash flood.

Ice-jam floods have two additional hazards. The ice flows can move rapidly downstream and inflict great damage when colliding with structures. People caught in the frigid waters of ice-jam floods can quickly develop hypothermia.

### Flood Victims and Cold Water Immersion

Cold water removes heat from the body 25 times faster than cold air of the same temperature. Victims of cold season flooding have an additional set of problems to overcome to survive. The information here is also applicable to individuals who take part in water sports in cold seasons or cold water bodies.

*Cold shock.* Sudden immersion in water colder than 59°F (<15°C) can trigger a debilitating, short duration reflex response called cold shock, which is a combination of respiratory and cardiovascular effects. Both can be life threatening. Two respiratory effects, the involuntary gasp reflex and uncontrollable rapid breathing can lead to drowning through aspiration of water. Near-surface blood vessels, in the cardiovascular effect, constrict quickly leading to a blood pressure surge and rapid heart rate that could result in stroke or a heart attack. The effects of cold shock diminish after approximately three minutes.

*Hypothermia.* In individuals who do not drown, core body temperature begins to drop steadily, a condition called hypothermia. As heat is lost to the cold water the body pulls blood towards the core to slow the loss of heat. Muscles and joints in exposed limbs stiffen and shivering begins making swimming or wading through shallow water difficult. At this point individuals not wearing a life jacket may drown. For those wearing a flotation device drowning may be prevented, but hypothermia will eventually lead to unconsciousness and cardiac arrest.

How long it takes for an individual to succumb to hypothermia is dependent on water temperature, insulation from clothing and body fat, water velocity, and time to rescue. When the core temperature drops to 90°F most people lose consciousness and few survive when the core temperature drops to 85°F (Table 3.19).

### Storm-surge Floods

When storm winds push water onto otherwise dry land it is called a storm-surge flood. Storm-surge floods most frequently occur with the onshore winds of tropical cyclones

**Table 3.19   Still, Cold Water Survival Times**

## Still, Cold Water Survival Times
## Without Protective Clothing*

| Water Temperature °F | °C | Loss of Muscle Control in Hands and Fingers (without protective clothing) | Loss of Large Muscle Control and Possibly Onset of Unconsciousness | Survival Time |
|---|---|---|---|---|
| < 32 | < 0 | < 2 min. | < 12 min. | < 15 min. |
| 32 - 40 | 0 to 4.5 | < 3 min. | 15 to 30 min. | 30 to 90 min. |
| 40 - 50 | 4.5 to 10 | < 5 min. | 30 to 60 min. | 1 to 3 hrs. |
| 50 - 60 | 10 to 15.5 | 10 to 15 min. | 1 to 2 hrs. | 1 to 6 hrs. |
| 60 - 70 | 15.5 to 21 | 30 to 40 min. | 2 to 7 hrs. | 2 to 40 hrs. |
| 70 - 80 | 21 to 26.5 | 1 to 2 hrs. | 2 to 12 hrs. | 3 hrs. to indefinite |
| > 80 | > 26.5 | 2 to 12 hrs. | Indefinite | Indefinite |

*Still water removes heat from the body approximately 25 times faster than air of the same temperature. Rushing water will remove heat even faster this table therefore over estimates survival time in rushing cold water. There is great variation in survival time among individuals because of conditioning, body fat and other factors.

*Source:* US Rescue Task Force, Minnesota Sea Grant, State of Oregon, The Engineering Toolbox.

but large, powerful frontal lows can cause significant storm surge flooding too.

The storm surge is the most dangerous part of a tropical cyclone. Worldwide 9 out of 10 tropical cyclone fatalities are caused by storm surge. Storm-surge floods are at their worst when occurring during high tides. Additionally flash flooding from intense tropical cyclone rains is made worse when the storm surge flows against flooding streams and forces them to back up.

### Dam-failure and Levee-failure Floods

Dams and levees are designed to contain water levels with a certain probability of occurrence. If the water level rises above the design level, the dam or levee will be topped, which can lead to failure of the structures. The water behind the dam or levee rushes through and past the failed structure and a flash flood ensues.

### Debris, Landslide, and Mudflow Floods

Debris floods occur when mud, rocks, logs and other debris block a channel and water is impounded behind the temporary dam. Landslides can also block a channel or create huge waves that cause failure of dams and levees. The most common cause of mudflows is the rapid melting of snow and ice by volcanic activity. The liquefied mud carries rocks, trees and other debris great distances into and down a stream valley. The term lahar is used for mudflows of volcanic origin.

### Flood Forecasting, Advisories, Watches, and Warnings

Three federal organizations are involved with monitoring rivers and river basins and providing river forecasts, advisories, watches and warnings.

The USGS investigates the occurrence, quantity, quality, distribution, and movement of surface and underground wa-

ters and disseminates the data to end users. The USGS network of nearly 9000 stream level gauges continuously record stream levels and the data from this network is used to monitor water resources and forecast stream levels and floods. Additional stream gauges and water data and monitoring is done by the US Army Corps of Engineers.

The NWS is responsible for forecasting river levels and streamflow. Table 3.20 lists the 13 NWS River Forecast Centers (RFCs). The numbers correspond to the numbers in Figure 3.14 locating the areas of responsibility of each RFC.

RFCs use precipitation and other weather forecasts, along with historic and real-time hydrologic data obtained from the USGS gauge network as input to river forecast models (Figures 3.14 and 3.15).

The RFCs produce forecasts that are used by the NWS Weather Forecast Offices (WFOs). It is the responsibility of meteorologists at WFOs in consultation with hydrologists at RFCs to disseminate advisories, watches and warnings for floods and flash floods.

Figure 3.15 shows the major watersheds of the 48 contiguous states and Table 3.21 lists them along with area in square miles and the bodies of water to which the watersheds drain. The numbers in Table 3.33 correspond the numbers on the map identifying the watersheds.

### Significant Floods of the 20th Century in the United States

The United States Geological Society has compiled what are considered to be excellent examples of the types of floods listed above. Figure 3.16 shows the geographical distribution of the floods and Table 3.22 lists the floods with additional information.

**Table 3.20   NWS RFCs**

| Map # | River Forecast Center | Area of Responsibility mi$^2$ | Location |
|:---:|---|:---:|---|
| - | Alaska/Pacific | - | Anchorage, AK |
| 1 | Arkansas-Red Basin | 358,296 | Tulsa, OK |
| 2 | California-Nevada | 426,151 | Sacramento, CA |
| 3 | Colorado Basin | 510,934 | Salt Lake City, UT |
| 4 | Lower Mississippi | 527,909 | Slidell, LA |
| 5 | Middle Atlantic | 137,955 | State College PA |
| 6 | Missouri River Basin | 1,084,322 | Pleasant Hill, MO |
| 7 | North Central | 968,182 | Chanhassen, MN |
| 8 | Northeast | 306,150 | Tauton, MA |
| 9 | Northwest | 613,275 | Portland, OR |
| 10 | Ohio | 300,044 | Wilmington, OH |
| 11 | Southeast | 694,020 | Peachtree City, GA |
| 12 | West Gulf | 892,480 | Fort Worth, TX |

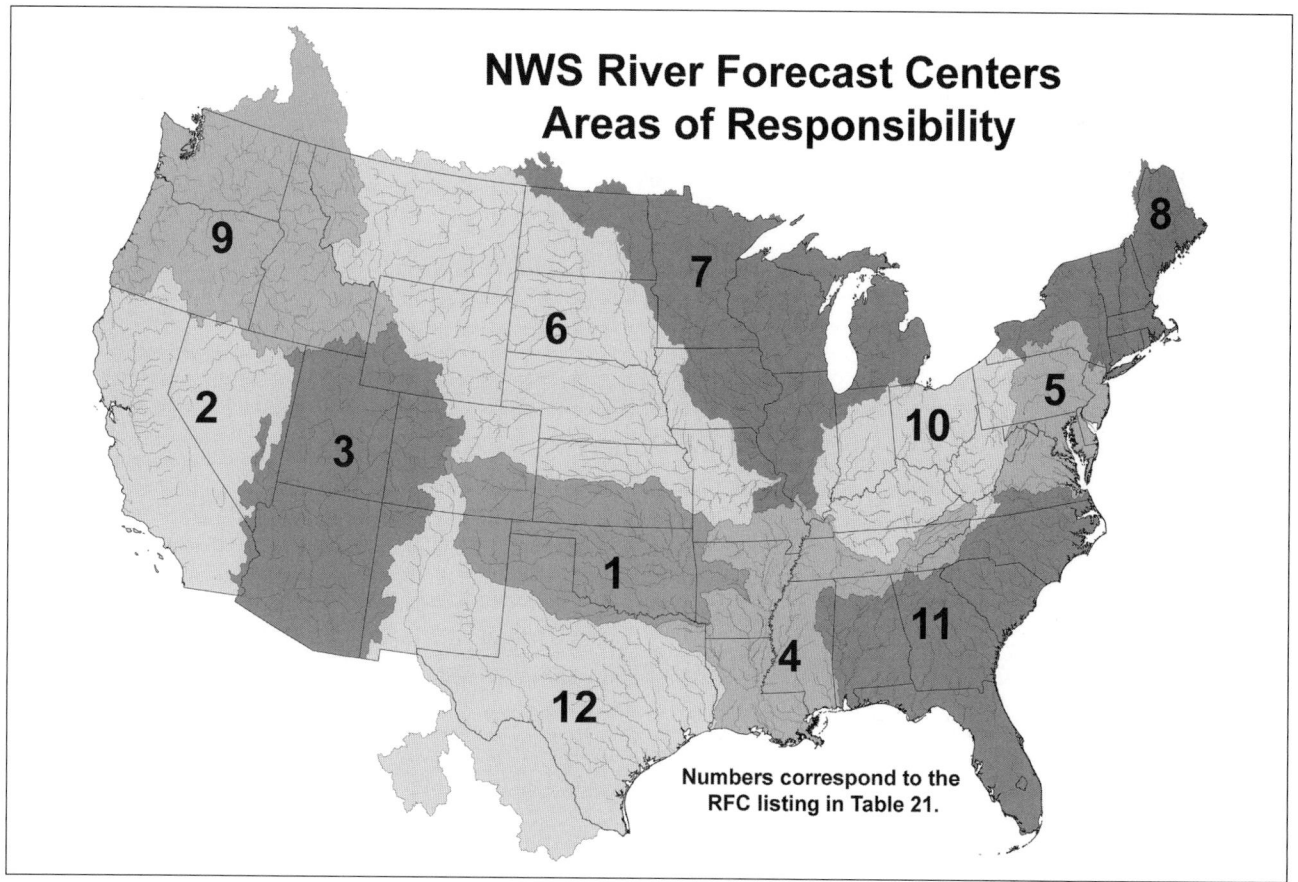

## NWS River Forecast Centers
## Areas of Responsibility

Numbers correspond to the
RFC listing in Table 21.

**Figure 3.14**   NOAA's NWS RFC areas of responsibility. (*Source:* Steve Horstmeyer from NOAA, NWS data.)

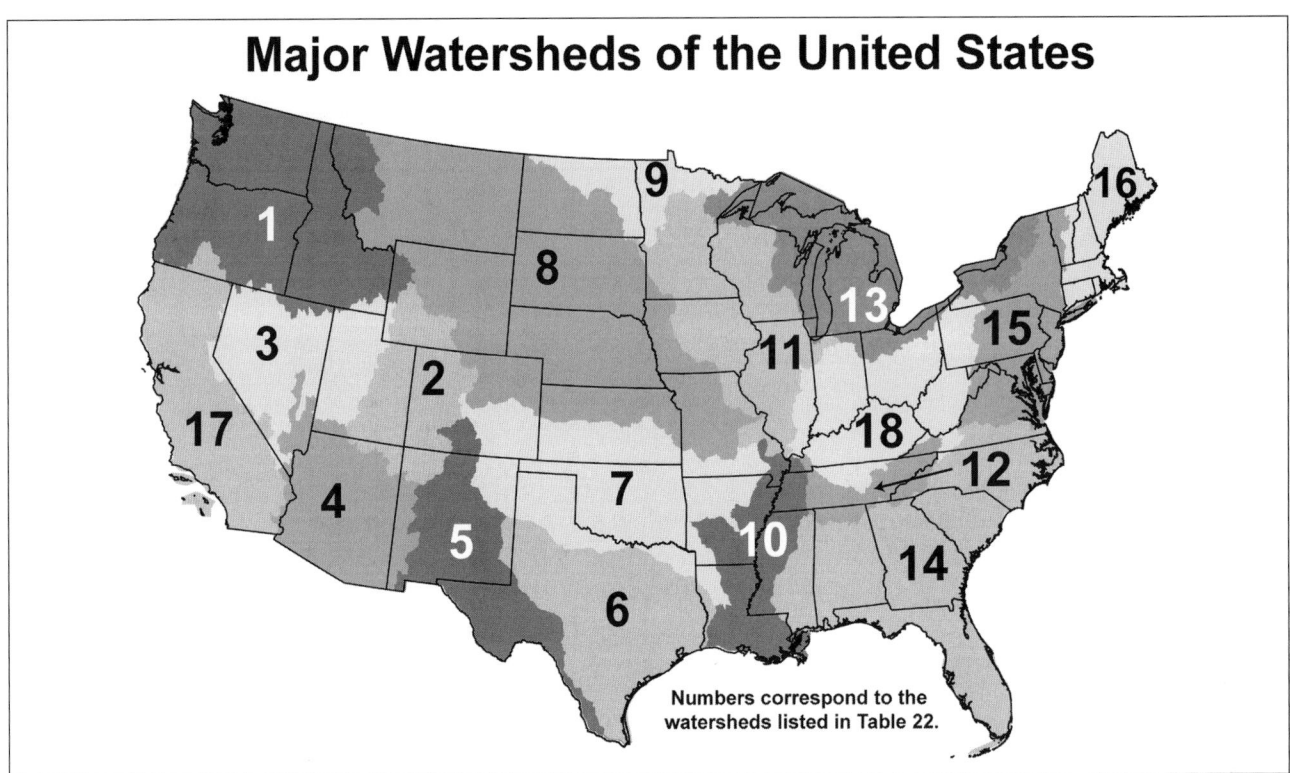

**Figure 3.15**    Major watersheds of the contiguous 48 states. (*Source:* Steve Horstmeyer from USGS data.)

**Table 3.21    Major Watersheds of the Continental United States**

| Map # | Watershed | Area mi² | Drains to |
|-------|-----------|----------|-----------|
| 1 | Pacific Northwest | 454,184 | Pacific Ocean |
| 2 | Upper Colorado River | 212,798 | Gulf of California via #4 |
| 3 | Great Basin | 281,986 | Internal drainage, evaporation |
| 4 | Lower Colorado River | 259,802 | Gulf of California |
| 5 | Rio Grande River | 609,343 | Gulf of Mexico |
| 6 | Texas Gulf | 387,313 | Gulf of Mexico |
| 7 | Arkansas, White, Red Rivers | 527,041 | Gulf of Mexico via #10 |
| 8 | Missouri River | 1,068,195 | Atlantic Ocean via #11 |
| 9 | Souris, Red, Rainey Rivers | 165,147 | Arctic Ocean via Assinaboine R., Nelson R. |
| 10 | Lower Mississippi River | 208,587 | Gulf of Mexico, Atlantic Ocean |
| 11 | Upper Mississippi River | 431,879 | Gulf of Mexico, Atlantic Ocean via #10 |
| 12 | Tennessee River | 100,354 | Gulf of Mexico, Atlantic Ocean via #10 |
| 13 | Great Lakes | 499,571 | St. Lawrence River, Atlantic Ocean |
| 14 | Southeast Atlantic-Gulf | 754,330 | Atlantic Ocean or Gulf of Mexico |
| 15 | Mid Atlantic | 246817 | Atlantic Ocean |
| 16 | Northeast | 135,529 | Atlantic Ocean |
| 17 | California | 332,710 | Pacific Ocean |
| 18 | Ohio River | 304,451 | Gulf of Mexico via #10 |

Notes: The Gulf of Mexico is part of the Atlantic Ocean and the Gulf of California is part of the Pacific Ocean. These are given as drainage destinations for clarity. The Great Basin has no water outlet it is blocked entirely by mountains water is removed by evaporation.

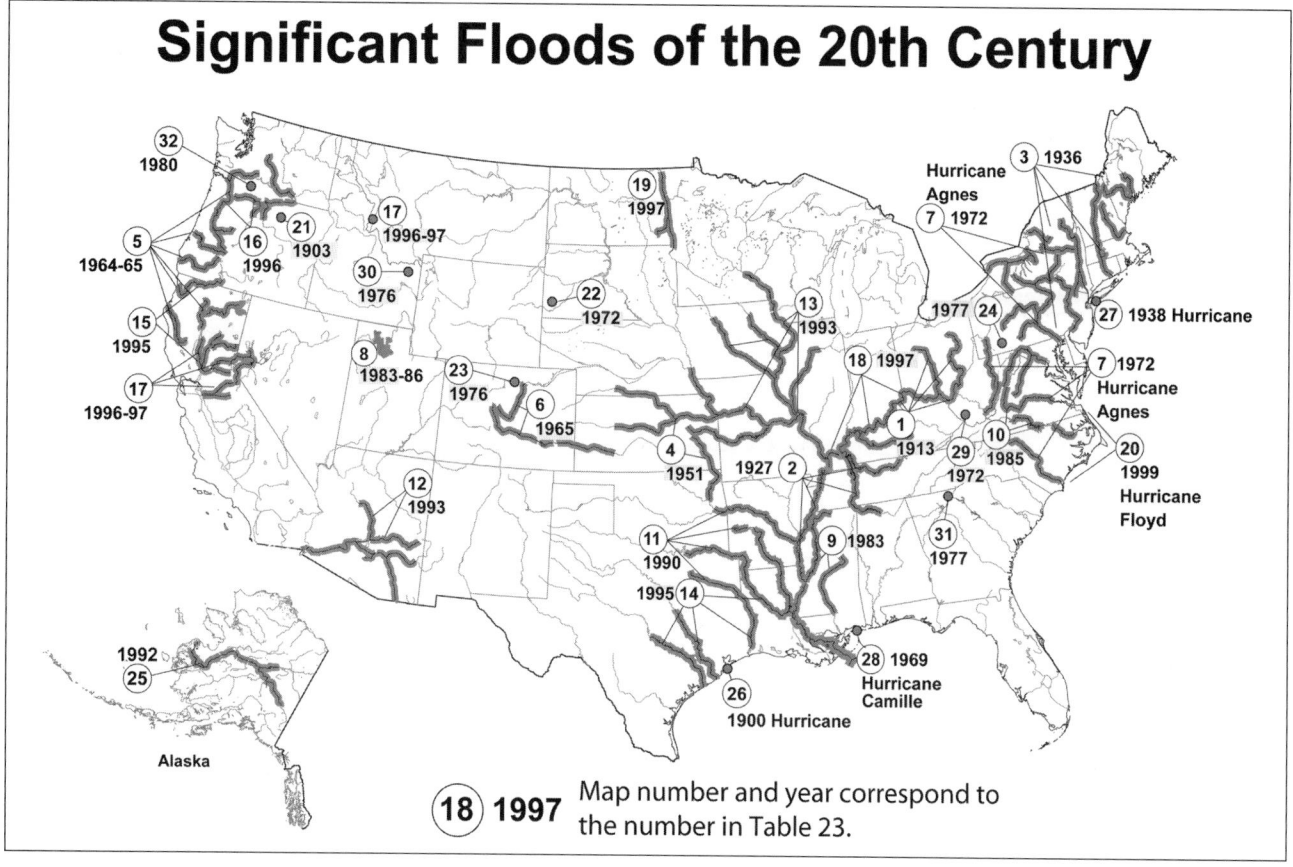

**Figure 3.16**     Significant floods of 20th century. Numbers correspond to numbers in Table 3.23. (*Source:* USGS.)

## • HEAT WAVES

There is no universal definition of heat wave. In general a heat wave is a prolonged period of abnormally hot weather that may or may not be accompanied by high humidity. What qualifies as a heat wave in a northern city may not be as serious in the south because heat waves are relative to the average conditions in any location.

As summer approaches the pool of cold air straddling the northern polar regions contracts and the polar jet stream systems are found farther and farther north. The seasonal invasion of tropical air replaces the retreating polar air that is warmed even more by increasingly strong sunshine. In the span of a few weeks teeth chattering chill is replaced by the heat of summer.

The jet streams and cool air do not necessarily have an uninterrupted stay well north of their winter positions. Occasionally an incursion of polar air along with the jet stream push southward and temporarily replace the warm and sticky tropical air with slightly cooler and noticeably less humid conditions.

When the jet stream stays to the north for a prolonged period and a warm ridge of high pressure builds over a region a heat wave may be on the way.

A high pressure ridge means air is sinking and as sinking air is warmed by compression the relative humidity falls. This means the sky is clear and unrelenting sunshine can heat the ground, which in turn will heat the air close to the surface. In addition with little or no rain the soil dries and energy that would go to evaporating soil moisture is available to heat the air. Heat waves are frequently also droughts.

The combination of heating by compression aloft and solar heating near the surface creates a deep hot air mass that strengthens each day as long as the conditions persist.

Because air sinking from aloft stagnates and within the hot ridge there is little air flow pollutants become trapped and build day after day adding to the stress of the heat and humidity.

In urban areas the large expanses area of pavement and many buildings create an "urban heat island" that is hotter than rural and suburban areas by as much as 7°F–15°F.

The effect of the urban heat island is made even worse in neighborhoods with many old buildings having no air conditioning. Even in cities as far north as Chicago and New York City apartment temperatures can stay warmer than 90°F throughout the night after peaking well above 100°F during the day.

**Table 3.22   Significant Floods of the 20th Century**

| Map # | Dates | Location/Stream | Deaths | Cost $ | Cause |
|---|---|---|---|---|---|
| 1 | March – April 1913 | Ohio | 467 | 143M | Heavy regional rain |
| 2 | April – May 1927 | Mississippi River MO to LA | ??? | 230M | Heavy rain central Mississippi River basin |
| 3 | March 1936 | New England | 150+ | 300M | Heavy rain on deep snow pack |
| 4 | July 1951 | Kansas and Neosho Rivers in Kansas | 15 | 800M | Heavy regional rain, 8" – 13" in 4 days |
| 5 | Dec. 1964 – Jan. 1965 | Pacific Northwest | 47 | 430M | Rapid melt of deep Cascade Mountains. Snow and up to 10" of heavy rain |
| 6 | June 1965 | S. Platte and Arkansas Rivers, Colorado | 24 | 570M | 14 inches of rain in several hours |
| 7 | June 1972 | Northeastern U.S. | 117 | 3.2B | Decaying Hurricane Agnes, rain 10" to 19" |
| 8 | April – June, 1983 | Great Salt Lake | ??? | 621M | Record high lake level from snow melt after the 1982-83 El Niño |
| 9 | May 1983 | Central and northeast MS | 1 | 500M | Heavy regional rain |
| 10 | November 1985 | Shenandoah, James, Roanoke Rivers, VA and WV | 69 | 1.25B | Hurricane Juan followed by heavy regional rain, up to 20". |
| 11 | April 1990 | Trinity, Arkansas, Red Rivers in TX, AR, OK | 17 | 1B | Heavy thunderstorm rain on saturated ground. |
| 12 | January 1993 | Gil, Salt and Santa Cruz Rivers, AZ | ??? | 400M | Persistent heavy rain due El Niño pattern |
| 13 | May – Sep. 1993 | Mississippi River | 48 | 20B | Long period heavy rain |
| 14 | May 1995 | South central U., S. | 32 | 5-6B | Recurring thunderstorm rains |
| 15 | January – March 1995 | California | 27 | 3B | Frequent storms due to El Niño pattern |
| 16 | February 1996 | Pacific Northwest, western Montana | 9 | 1B | Heavy rain followed by heavy snow followed by more heavy rain |
| 17 | Dec 1996 – Jan 1997 | Pacific Northwest and Montana | 36 | 2-3B | Torrential rain on snow |
| 18 | March 1997 | Ohio River and tributary Licking River | 50 | 500M | Torrential rain, stalled frontal system |
| 19 | April – May 1997 | Red River of the North, ND, MN | 8 | 2B | Very rapid melt of thick snow cover |
| 20 | September 1999 | Eastern North Carolina | 42 | 6B | Hurricane Floyd, slow movement |

Table 3.22    Significant Floods of the 20th Century [CONTINUED]

| Map # | Dates | Location/Stream | Deaths | Cost $ | Cause |
|---|---|---|---|---|---|
| 21 | June 14, 1903 | Willow Creek in Oregon | 225 | 1M | Heppner, OR destroyed, cloud burst, debris dam collapse, deadliest natural disaster in OR |
| 22 | June 9-10, 1972 | Rapid City, SD | 237 | 160M | Stationary thunderstorms, 15 inches of rain in 5 hours. |
| 23 | July 31, 1976 | Big Thompson and Cache la Poudre Rivers, CO | 144 | 39M | Stationary thunderstorms over mountains dumped 12" in under 4 hours. |
| 24 | July 19 – 20, 1977 | Conemaugh River, PA | 78 | 300M | 12" of rain in 6 to 8 hours and failure of 5 debris dams  when they were topped. |

## A Significant Ice-Jam Flood of the 20th Century

| Map # | Dates | Location/Stream | Deaths | Cost $ | Cause |
|---|---|---|---|---|---|
| 25 | May 1992 | Yukon River , AK | 0 | ??? | 100-year flood |

## Significant Storm Surge Floods of the 20th Century

| Map # | Dates | Location/Stream | Deaths | Cost $ | Cause |
|---|---|---|---|---|---|
| 26 | September 8, 1900 | Galveston, TX | 6000 | ??? | Great Galveston Hurricane |
| 27 | September 20 - 21, 1938 | New England | 494 | 306M | Great New England Hurricane of 1938 |
| 28 | August 16-17, 1969 | Gulf Coast, MS, LA | 259 | 1.4B | Hurricane Camille, storm surge up to 16' and flood reached 20 miles inland |

## Significant Dam-Failure Floods of the 20th Century

| Map # | Dates | Location/Stream | Deaths | Cost $ | Cause |
|---|---|---|---|---|---|
| 29 | February 26, 1972 | Buffalo Creek, WV | 125 | 60M | Days of heavy rain and at 8AM water reached the crest of a mine slurry impoundment dam which collapsed and destroyed two additional dams. |
| 30 | June 5, 1976 | Teton river, ID | 11 | 400B | Teton Dam, 310 ft. high, 3000 ft. across failed. |
| 31 | November 8, 1977 | Toccoa Creek, GA | 39 | 2.8M | Kelly Barnes Dam failed at 1:30 A.M. After two days of torrential rain. |

## A Significant Mudflow Flood of the 20th Century

| Map # | Dates | Location/Stream | Deaths | Cost $ | Cause |
|---|---|---|---|---|---|
| 32 | May 18, 1980 | Toutle and Cowlitz Rivers, WA | 60 | ??? | Eruption of Mt. St. Helens |

*Source:* USGS.

133

Heat is likely the greatest killer of all weather disasters. Hot weather kills by taxing the body's ability to shed excess heat. Because the human body loses more heat through the evaporation of perspiration than in any other way, when heat is combined with high humidity the danger is especially high.

The American Red Cross has developed a three-part approach to heat wave safety:

1. How to prepare for excessive heat.
2. What to do during a heat wave.
3. Recognize and care for heat related emergencies.

To prepare they recommend discussing heat wave safety with your family, knowing where the elderly and nonmobile live in your neighborhood so you can watch for signs of problems and learning first aid.

The second part includes eating smaller meals, reducing physical exertion, wearing light weight clothing, using a buddy system when working in hot weather and frequently checking on your animals.

In the third part the American Red Cross categorizes illnesses caused by excessive heat.

- **Heat cramps** are often the earliest sign of serious problems to come. Exposure to heat causes excessive sweating that dehydrates the body and depletes electrolytes. Heat cramps are pains or spasms that occur most frequently in the legs and abdomen.
- **Heat exhaustion** is the next stage as a person's body accumulates heat. Pale or flushed skin, excessive sweating, dizziness, lack of awareness and weakness typically occur when a person is suffering from heat exhaustion. Give the victim cool fluids or electrolyte replacement drinks in frequent but small doses and get the victim to a cool area or shade. If the victim loses consciousness call 911.
- **Heat stroke** (also called **sunstroke**) is the most serious heat related illness. The victim's heat control system has stopped functioning and the victim's life is threatened. Try to cool the victim immediately and call 911.

To inform the public of dangerous heat and humidity combinations the NWS uses a measure of the combined effect of called the Heat Index. Measures like the heat index and wind chill are often called effective temperatures. The heat index gives a reliable measure of how hot it feels while in winter the wind chill tells you how cold it feels.

The heat index is the result of many detailed biometeorological studies and it is calculated using a very complicated equation derived using the statistical technique of multiple linear regression. The heat index is designed so that if the dew point temperature is 57°F the heat index is the same as the ambient air temperature. If the dew point temperature is higher than 57°F the heat index is higher than the actual air temperature.

With a dew point temperature of 57°F the relative humidity is 45% when the air temperature is 80°F, 33% at 90°F, and 24% at 100°F. Relative humidity is not a reliable guide for the inexperienced as to the combined effect of heat and humidity.

Heat Index values were devised for shady, light wind conditions, If you are exposed to full sunshine the heat index underestimates the severity of the heat. Full sun exposure can increase the heat index by as much as 15°F.

When the air temperature is warmer than your skin temperature, approximately 90°F, the air in contact with your skin conducts heat to your body. Even though the evaporation of perspiration increases on windy days thus removing additional heat from your body, brisk winds also deliver more hot air to your skin and therefore transfer more heat to your body increasing the danger of heat caused distress (Tables 3.23–3.25).

## Selected Historic Heat Waves in the United States

*Source: David Ludlum, Monthly Weather Review, Midwestern Regional Climate Center, NOAA NCDC*

Heat waves in the United States are born of two sets of conditions. The first and most common east of the Rocky Mountains is a stationary upper-level anticyclone, which is also called a high pressure system.

With the jet stream far to the north and therefore the storm track rain does not fall and because air is sinking in the high pressure system the sky is clear. The sunshine heats the ground, evaporates soil moisture and day after day the region "cooks." Because there is little air flow and because of what is called a subsidence inversion air is trapped under the high and this type of heat wave is often also an episode of unhealthy air pollution levels.

Heat can also be transported from hot areas like the deserts of Southern California, Arizona and northern Mexico into the central and eastern states to create a heat wave. Farmers of the western Great Plains often experience what is called "warm air advection" from the desert southwest, which increases water usage by crops. It is not unusual for the hot air to be transported to the Applachian mountains but heat waves of this type are not as long lasting.

**1876**  San Francisco, Portland, OR.
**1991**  July 25, Los Angeles, California, 109 degrees peak of heat wave.
**1896**  New York City, August 4–14, 420 heat related deaths (possibly as many as 1500), hundreds of horses died, average temperature during the period 90.8°F, high humidity, warm nights and elevated air pollution levels. T. Roosevelt, Pres. Board of Police Commissioners, "The heated term was the worst and most fatal we have ever known."

Pennsylvania, *Stroudsburg Times*, [the summer of 1896], "was the most disagreeable summer since 1876 ... The daily list of deaths and prostrations is appaling in large cities..."

**Table 3.23   Heat Index from Dew Point Temperature**

| Dew Point (°F) | Temperature (°F) | | | | | | | | | | | | | | | |
|---|---|---|---|---|---|---|---|---|---|---|---|---|---|---|---|---|
| | 90 | 91 | 92 | 93 | 94 | 95 | 96 | 97 | 98 | 99 | 100 | 101 | 102 | 103 | 104 | 105 |
| 65 | 94 | 95 | 96 | 97 | 98 | 100 | 101 | 102 | 103 | 104 | 106 | 107 | 108 | 109 | 110 | 112 |
| 66 | 94 | 95 | 97 | 98 | 99 | 100 | 101 | 103 | 104 | 105 | 106 | 108 | 109 | 110 | 111 | 112 |
| 67 | 95 | 96 | 97 | 98 | 100 | 101 | 102 | 103 | 105 | 106 | 107 | 108 | 110 | 111 | 112 | 113 |
| 68 | 95 | 97 | 98 | 99 | 100 | 102 | 103 | 104 | 105 | 107 | 108 | 109 | 110 | 112 | 113 | 114 |
| 69 | 96 | 97 | 99 | 100 | 101 | 103 | 104 | 105 | 106 | 108 | 109 | 110 | 111 | 113 | 114 | 115 |
| 70 | 97 | 98 | 99 | 101 | 102 | 103 | 105 | 106 | 107 | 109 | 110 | 111 | 112 | 114 | 115 | 116 |
| 71 | 98 | 99 | 100 | 102 | 103 | 104 | 106 | 107 | 108 | 109 | 111 | 112 | 113 | 115 | 116 | 117 |
| 72 | 98 | 100 | 101 | 103 | 104 | 105 | 107 | 108 | 109 | 111 | 112 | 113 | 114 | 116 | 117 | 118 |
| 73 | 99 | 101 | 102 | 103 | 105 | 106 | 108 | 109 | 110 | 112 | 113 | 114 | 116 | 117 | 118 | 119 |
| 74 | 100 | 102 | 103 | 104 | 106 | 107 | 109 | 110 | 111 | 113 | 114 | 115 | 117 | 118 | 119 | 121 |
| 75 | 101 | 103 | 104 | 106 | 107 | 108 | 110 | 111 | 113 | 114 | 115 | 117 | 118 | 119 | 121 | 122 |
| 76 | 102 | 104 | 105 | 107 | 108 | 110 | 111 | 112 | 114 | 115 | 117 | 118 | 119 | 121 | 122 | 123 |
| 77 | 103 | 105 | 106 | 108 | 109 | 111 | 112 | 114 | 115 | 117 | 118 | 119 | 121 | 122 | 124 | 125 |
| 78 | 105 | 106 | 108 | 109 | 111 | 112 | 114 | 115 | 117 | 118 | 119 | 121 | 122 | 124 | 125 | 126 |
| 79 | 106 | 107 | 109 | 111 | 112 | 114 | 115 | 117 | 118 | 120 | 121 | 122 | 124 | 125 | 127 | 128 |
| 80 | 107 | 109 | 110 | 112 | 114 | 115 | 117 | 118 | 120 | 121 | 123 | 124 | 126 | 127 | 128 | 130 |
| 81 | 109 | 110 | 112 | 114 | 115 | 117 | 118 | 120 | 121 | 123 | 124 | 126 | 127 | 129 | 130 | 132 |
| 82 | 110 | 112 | 114 | 115 | 117 | 118 | 120 | 122 | 123 | 125 | 126 | 128 | 129 | 131 | 132 | 133 |

**Exposure to full sunshine can increase HI values by up to 15° F**

**1901**  Midwest and east, Bowling Green, KY every day in July 90°F or hotter, 16 days 100°F or hotter 9500 heat related deaths nationwide, 724 in New York City.

**1905**  New England.

**1911**  Kansas, driest and hottest growing season since 1887, Clay Center and Hugoton, KS, both 116°F, state all-time record high temperature. New York City, first two weeks of July, 1000 heat related deaths. New England,105 degrees in Vermont.

### *1930s: The Dust Bowl Years in the United States*

**1930**  July 28, Kentucky, 18 locations set all-time record high temperatures, 41 locations hotter than 100°F. The heat continued into August when many more all-time record high temperatures were set.

**1934**  Great Plains, Ohio Valley, July, hottest month, 21st, Columbus, OH, 106°F, Cincinnati, OH, 109°F,

Hamilton, OH, 111°F, Gallipolis, OH, 113°F, hottest ever in Ohio. Oppressive warm nights and hot days left 160 dead in Ohio from 20th–26th.

St. Louis, MO, 23 days 100°F or hotter, July 24, 110°F. Chicago, 12 days 100°F or hotter.

**1936**  Lincoln, NE, 35 days in-a-row 90°F or hotter, July 25th, high 115°F, low 91°F.

Gann Valley, SD, July 5th, 120°F, Steele, ND, July 6th, 121°F, the hottest temperature ever not in the Desert Southwest. New York City, Central Park, July 9, 106°F, Martinsburg, WV, July 10th, 114°F. July 14th, Iowa, the average high temperature of those observed at 113 observing sites was 108.7°F. Cincinnati, OH, July 8–15, each day hotter than 100°F, 8-day average high temperature was 104°F. Fifteen states set all-time highest temperature records.

**1939**  Los Angeles, September, 546 deaths.

**Table 3.24   Heat Index from Relative Humidity**

| RH (%) | \multicolumn{16}{c}{Temperature (° F)} |
|---|---|
| | 90 | 91 | 92 | 93 | 94 | 95 | 96 | 97 | 98 | 99 | 100 | 101 | 102 | 103 | 104 | 105 |
| 30 | 89 | 90 | 92 | 93 | 95 | 96 | 98 | 99 | 101 | 102 | 104 | 106 | 108 | 110 | 112 | 114 |
| 35 | 91 | 92 | 94 | 95 | 97 | 98 | 100 | 102 | 104 | 106 | 107 | 109 | 112 | 114 | 116 | 118 |
| 40 | 92 | 94 | 96 | 97 | 99 | 101 | 103 | 105 | 107 | 109 | 111 | 113 | 116 | 118 | 121 | 123 |
| 45 | 94 | 96 | 98 | 100 | 102 | 104 | 106 | 108 | 110 | 113 | 115 | 118 | 120 | 123 | 126 | 129 |
| 50 | 96 | 98 | 100 | 102 | 104 | 107 | 109 | 112 | 114 | 117 | 119 | 122 | 125 | 128 | 131 | 135 |
| 55 | 98 | 100 | 103 | 105 | 107 | 110 | 113 | 115 | 118 | 121 | 124 | 127 | 131 | 134 | | |
| 60 | 100 | 103 | 105 | 108 | 111 | 114 | 116 | 120 | 123 | 126 | 129 | 133 | | | | |
| 65 | 103 | 106 | 108 | 111 | 114 | 117 | 121 | 124 | 127 | 131 | 135 | | | | | |
| 70 | 106 | 109 | 112 | 115 | 118 | 122 | 125 | 129 | 133 | | | | | | | |
| 75 | 109 | 112 | 115 | 119 | 122 | 126 | 130 | 134 | | | | | | | | |
| 80 | 112 | 115 | 119 | 123 | 127 | 131 | 135 | | | | | | | | | |

**Exposure to full sunshine can increase HI values by up to 15° F**

**Table 3.25   Possible Health Effects of Excessive Heat and Humidity**

| Heat Index | Possible Health Effects |
|---|---|
| 130° or Higher | Heatstroke/sunstroke like with continued exposure even without physical activity. |
| 105°- 130° | Heat cramps or heat exhaustion is likely and heatstroke is possible with prolonged exposure And/or physical activity |
| 90°- 105° | Heat cramps and heat exhaustion is possible with prolonged exposure and/or physical activity.. |
| 80° - 90° | Fatigue is possible with prolonged exposure and/or physical activity. |

When the heat index is above 105°F increasingly severe heat disorders with continued exposure and/or physical activity can be expected.

*Source:* NWS.

**1941** Washington State, Seattle 100°F, all-time record high temperature (broken in 2009).

**1944** Many cities recorded high temperatures of 100°F or hotter.

**1953** Nationwide, August, Hagerstown, MD 107°F, Fredricksburg, VA 106°F, Newark, NJ 105°F, Louisville, KY 103°F, New York City, Evansville, IN, Cincinnati, OH, Pine River, WI, Baltimore, MD, all 102°F, Washington, D.C. 12 consecutive days 90°F or hotter.

**1955** From the continental divide to the Atlantic Ocean north of the Deep South July and August had one of the most prolonged and persistent heat waves on record with no extreme heat just a persistence of hot days. In the west Orange County, 15-day heat wave starts on opening day for Disneyland July 17. Los Angeles, August 31–Sep 7, >=100°F each day, 110°F Sep 1. Southern California, 946 deaths.

**1959** Southern California, Bakersfield had its last 100°F high temperature on October 17.

**1963** September, Southern California, 580 deaths.

**1966** St. Louis, June 22nd–July 20th, hotter than 90° each day except July 7. Six consecutive days >= 100°F, 246 deaths.

**1980** Beginning in June much of the United States experienced record heat and drought. Kansas City, 17 straight days hotter than 100°F, Dallas, 42 straight days and 69 total days hotter than 100°F, Memphis, 15 straight days hotter than 100°F and an all-time record high temperature of 108°F. Wichita Falls, TX, all-time record high, 117°F.

**1988** Corn belt drought/heat wave, crop losses total millions of dollars, rainfall less than any other summer on record in parts of Illinois. Death estimates range from 5000–10000, crop losses total nearly $40 billion.

**1990** Los Angeles, 112°F, June 26, all-time record high temperature.

**1991** Kentucky, 3rd warmest year on record, 3rd warmest summer on record, 2nd warmest spring on record. Portland, OR, 107°F, all-time record high temperature.

**1998** Southern United States severe heat and drought spread from Texas and Oklahoma to North Carolina and South Carolina, 200 deaths.

**1999** Eastern United States, the worst drought on record for Maryland, Delaware, New Jersey, and Rhode Island. West Virginia was declared a disaster area after 3.81 million acres burned in forest fires. Record heat nationwide caused 502 deaths.

**2002** April, Northeast and New England, July 6–14 western United States.

**2005** Rocky Mountains and western Great Plains, Denver 105°F. July 13th heat starting to appear in spots, July 19–23 many areas with highs each day 100°F or hotter. Denver, July 2nd hottest month of all time. Grand Junction, CO 106°F, all-time record high. Rocky Ford, CO, 108°F.

**2006** July, California, central valley as hot as 115°F, almost 150 deaths.

**2007** Kentucky's warmest year, Louisville, August 26 days 90°F or hotter, 12 daily high temperature records, August 8°F warmer than normal.

**2009** July 27–30, Seattle/Tacoma (SeaTac International Airport), 4 consecutive days hotter than 90°F, only the 5th time this happened, numerous record high and record warm low temperatures. The warmest or nearly the warmest July on record for many locations. Seattle 103°F, all-time record high, Portland, OR, 106°F, 1 degree shy of all-time record, Bellingham, WA 96°F, all-time record high temperature.

## • COLD WAVES

A cold wave occurs when frigid polar or arctic air pushes southward from northern Canada. The coldest of the cold waves occurs when steering currents in the atmosphere flow northward over Alaska, into Siberia and guide extremely cold air across the North Pole. When the "Siberian Express" plunges into the United States temperatures can reach well below 0°F into the Deep South.

A cold wave refers to a single invasion of cold air that drops temperatures rapidly. Record cold seasons are made up of a number of successive cold waves. Popularly, a cold wave is a period of frigid weather of any length.

When the weather is cold the second most important factor in how quickly the human body loses heat is wind speed. Brisk winds remove heat efficiently and to measure how much more quickly heat is lost the wind chill temperature was developed (Table 3.26).

### Selected Historic Cold Waves in the United States

*Source: Ludlum, David, The American Weather Book, 1982, American Meteorological Society, Monthly Weather Review, Midwestern Regional Climate Center, NCDC*

*Notes:* Records are referenced as of the date of occurrence. If a month is reported as the coldest ever, it means coldest ever to that date and the record may have been eclipsed at a later date.

**February 15, 1732** New England, ". . . coldest day in the memory of the oldest living man."

**February 22, 1733** New England, below zero all day.

**February 5, 1788** New England, "Cold Tuesday," extreme wind chill, below 0°F all day Portsmouth, NH 41°F evening before, −13°F dawn.

**February 6, 1807** Ohio, "Cold Friday," below zero all day.

**Table 3.26   Wind Chill**

| mph | Air Temperature °F | | | | | | | | | | | | | | | | | | |
|---|---|---|---|---|---|---|---|---|---|---|---|---|---|---|---|---|---|---|---|
| Calm | 40 | 35 | 30 | 25 | 20 | 15 | 10 | 5 | 0 | -5 | -10 | -15 | -20 | -25 | -30 | -35 | -40 | -45 |
| 5 | 36 | 31 | 25 | 19 | 13 | 7 | 1 | -5 | -11 | -16 | -22 | -28 | -34 | -40 | -46 | -52 | -57 | -63 |
| 10 | 34 | 27 | 21 | 15 | 9 | 3 | -4 | -10 | -16 | -22 | -28 | -35 | -41 | -47 | -53 | -59 | -66 | -72 |
| 15 | 32 | 25 | 19 | 13 | 6 | 0 | -7 | -13 | -19 | -26 | -32 | -39 | -45 | -51 | -58 | -64 | -71 | -77 |
| 20 | 30 | 24 | 17 | 11 | 4 | -2 | -9 | -15 | -22 | -29 | -35 | -42 | -48 | -55 | -61 | -68 | -74 | -81 |
| 25 | 29 | 23 | 16 | 9 | 3 | -4 | -11 | -17 | -24 | -31 | -37 | -44 | -51 | -58 | -64 | -71 | -78 | -84 |
| 30 | 28 | 22 | 15 | 8 | 1 | -5 | -12 | -19 | -26 | -33 | -39 | -46 | -53 | -60 | -67 | -73 | -80 | -87 |
| 35 | 28 | 21 | 14 | 7 | 0 | -7 | -14 | -21 | -27 | -34 | -41 | -48 | -55 | -62 | -69 | -76 | -82 | -89 |
| 40 | 27 | 20 | 13 | 6 | -1 | -8 | -15 | -22 | -29 | -36 | -43 | -50 | -57 | -64 | -71 | -78 | -84 | -91 |
| 45 | 26 | 19 | 12 | 5 | -2 | -9 | -16 | -23 | -30 | -37 | -44 | -51 | -58 | -65 | -72 | -79 | -86 | -93 |
| 50 | 26 | 19 | 12 | 4 | -3 | -10 | -17 | -24 | -31 | -38 | -45 | -52 | -60 | -67 | -74 | -81 | -88 | -95 |
| 55 | 25 | 18 | 11 | 4 | -3 | -11 | -18 | -25 | -32 | -39 | -46 | -54 | -61 | -68 | -75 | -82 | -89 | -97 |
| 60 | 25 | 17 | 10 | 3 | -4 | -11 | -19 | -26 | -33 | -40 | -48 | -55 | -62 | -69 | -76 | -84 | -91 | -99 |

*Source:* NWS.

**January 1835**  Northeast United States, Mercury thermometers froze at Bangor, ME, Bath, ME, Montpelier, VT and White River, VT. Freezing temperature of mercury is −40°F. Hartford, Ct −27°F, New Haven, CT, −23°F, Pittsfield, MA −32°F, Williamstown, MA, −30°F.

**February 1835**  Savanah, GA, 0°F, Charleston, SC, 2°F. On February 8, St. Augustine, FL Low 7°F, High 21°F, Jacksonville, FL, 8°F.

**January 18, 1857**  Fort Ripley, MN, −50°F, Muscatine, IA, −30°F.

**January 23, 1857**  Bath, ME, −52°F, Craftsbury, VT. average daily temperature −28°F Boston, MA, afternoon temperature −5.5°F, coldest ever.

**January 10, 1859**  Montreal, −43.6°F, Burlington, VT, −31.5°F and at 2:00 P.M. −26°F, Woodstock, VT, −45°F, Nantucket Island, −12°F, Brooklyn Heights, NYC, Noon, −9°F.

**January 1, 1864**  Louisville, KY, −19.5°F.

**December 1874–February 1875**  The coldest meteorological winter on record in Iowa.

**February 11–14, 1899**  "Great Arctic Outbreak of 1899"

Tallahassee, FL, −2°F, coldest ever in Florida. New Orleans, LA, 6.8°F, coldest ever

February 8–11, all stations in Minnesota, average temperature −24°F.

**December 1911–February 1912**  Minneapolis-St. Paul, MN

December 31–January 8, below 0°F for 186 consecutive hours.

January 9–13, below 0°F for 113 consecutive hours.

**January 1912**  Coldest month ever in Iowa, 12th, coldest day ever in most of Iowa.

**February 1934**  Northeast United States, coldest month ever.

Average temperature at many locations near 0°F.

Stillwater Reservoir, NY average temperature −1.8°F.

February 9, Vanderbilt, MI −51°F, Stillwater Reservoir, NY −52°F.

**January 18–February 19, 1936**  Omaha, NE, 33 days average temperature −2.8°F.

**January 1940**  St. Bernard, AL −16°F 28th, Leakesville, MS, −14°F, 27th, Lafayette,

GA, −17°F, 27th, Fraser, CO, −47°F, 19th, Gordon, NE, −32°F, 18th, Mason, FL +8°F, 27th.

**January 1949**    Given the title of "worst we ever had" by US Weather Bureau. After a blizzard the first days of the month record cold set in. Coldest month on record for Missoula, MT where opn 22 days the temperature was below zero and the ground froze at a depth of 7' and many water pipes burst. Also the coldest for Boise, ID, Fresno, CA, Bakersfield, CA and San Diego, CA.

**December 1957–February 1958**    Coldest on record in West Palm Bach, FL, Fort Myers, FL, Miami, FL, Lakeland, FL, Mobile, AL, Montgomery, AL, Columbus, GA. Tampa, FL was more than 1°F colder than any other winter since 1825. Crops frozen in FL. Warm west of the Continental Divide.

**January 1963**    January 18–24, Minneapolis, MN below 0°F for 157 consecutive hours.

**January 1977**    The coldest month ever of record in the eastern United States. Snow flurries sighted in Miami, FL and Homestead, FL and as far south as the northern Florida Keys. The Ohio River at Cincinnati froze sufficiently for hundreds to walk across it. This has happened in only 14 of 134 years with data.

**January 1982**    International Falls, MN, coldest January 2nd. Coldest January in Concord, NH, Duluth, MN and Fargo, ND. 3rd coldest January in St. Cloud, MN. 4th coldest January in Minneapolis, MN, Rochester, NY (tie), Milwaukee, WI (tie), Waterloo, IA. 5th coldest January in Rochester, MN, Great Falls, MT and Williston, ND. Omaha, NE 2nd greatest number of days with low temperature below zero, 18 days.

January 10, "Freezer Bowl" Cincinnati Bengals versus San Diego Chargers game-time temperature +9°F, wind chill −59°F.

**December 1983**    Many locations coldest Christmas Day on record.

**January 1985**    21st Ronald Reagan's inauguration held in the Capitol Rotunda due to cold. Frost reported in Miami, FL.

**December 1989**    Coldest December of the century in the eastern United States. Coldest December on record in Cincinnati, OH and the only time in Cincinnati his-

tory the temperature fell to −20°F outside the month of January. Great Lakes snowbelt from northeast Ohio into New York snowfall was 200% of normal in many locations. Miami, FL, 24th, average daily temperature 38°F coldest day ever, 25th average daily temperature 43°F, third coldest day ever and the coldest Christmas Day ever in Miami. By many measures it was the most extreme cold wave on record for Florida.

**January 1994**    Washington, D.C., 19th, first time below 0°F since 1985, high temperature 8°F, coldest high temperature of the century (tie). Much of eastern United States and southern Canada, oldest month since the late 1970s. Flurries in Miami, FL. Extensive damage to the central Florida citrus crop. Indiana, January 18–21, all stations average temperature −5.4°F and east-central, unofficial temperatures to −40°F.

**January–February 1996**    One of the coldest arctic outbreaks of the 20th century, February 2nd, −60°F, Tower, MN
February 2nd, −60°F, Tower, MN. Average temperature for all available stations: in Minnesota,
January 31–February 3, −23.8°F, in Wisconsin, February 1–February 4, −18.3°F. Dallas, TX, 8°F, February 4.

**January 1997**    January 14–17, Dallas, low temperatures below freezing each day, Northern Great Plains numerous temperatures as cold as −40°F with wind chills to −80°F.

**January 2004**    Pacific Northwest, Seattle, WA and Portland, OR both reported unusual snow accumulations and cold. Boston, MA one of the coldest months in 114 years. Virginia Beach, VA many days with low temperatures colder than 32°F.

**January–February 2008**    Tok, AK, −72°F, February 7, February 3–9 Fairbanks, AK, low temperatures equal to or colder than −40°F each day. Fairbanks, AK during winter had only two days warmer than freezing.

## • DROUGHT

### Introduction

A drought is a prolonged period of abnormally dry weather that causes a serious depletion of soil moisture and is apparent by the crop damage, water-supply shortages, lowered lake levels and the streamflow reductions it causes. Droughts can be characterized by areal extent, severity and duration. In general the term drought is reserved for longer, more serious dry spells.

There is no single acceptable comprehensive definition of drought. If less precipitation falls than normal and there is a negative impact on the established local economy then a drought is considered to be in progress.

Annual costs of drought range from $6–$8 billion according to the US Federal Emergency Management Agency (FEMA). The 1988 drought, centered in the corn belt cost $74 billion (2008 dollars) making it the second most costly natural disaster in the United States. Over the past several decades the western United States has experienced longer and more severe droughts and based on tree ring proxy data may be in one of the driest periods in the last 500 years.

As we learn more through the use of proxy data the historical impact of drought is becoming more apparent. In the list of historic droughts at the end of this section you will find that through the history of North America, from the ancient Mayan and Anasazi civilizations to missionary settlements and the earliest European settlements on the east coast drought may have played a major role in the history of each.

## Types of Drought

Drought can be classified as agricultural, meteorological, hydrological or socioeconomic. Agricultural drought primarily refers to soil moisture while hydrological drought deals with streamflow and runoff. Meteorological drought balances precipitation and evapotranspiration, which is the combination of evaporation and transpiration by plants. Socioeconomic drought is concerned with supply and demand issues in the water supply.

Agricultural drought negatively impacts farm production while hydrological drought refers to reduction in streamflow and levels of streams, lakes and reservoirs. Both are meteorological in nature but involve the complexities of plant science, soil physics, ground water flow and many other specialized disciplines.

Meteorological drought is defined in terms of how much moisture is available relative to what is expected during average conditions. Meteorological drought involves preexisting weather conditions and soil moisture is a robust measure of conditions that preceded the period of time under investigation.

Meteorologists first estimate the supply of soil moisture then they determine how it will change if the average amount of rain falls. If the degree of dryness is slight normal rainfall may eliminate the dry conditions. If it is very dry much more rain than normal may be needed. Even if rainfall exceeds normal, drought conditions may still exist until the supply of water eliminates the deficit.

## Moisture Patterns in the United States

The United States is a country that is roughly divided into a moist east and a larger dry west. Except for the mountain slopes facing the Pacific Ocean and other isolated locations, from the coastal ranges in the west to the Rocky Mountains, except the Pacific Northwest, the United States is very dry.

A vast desert exists east of the Sierra Nevada because of the "rain shadow" effect. Plenty of rain falls on the west facing slopes of the mountains and little of the Pacific moisture remains by the time the winds crest the mountains. The dry air flows downslope, is warmed by compression and desiccating winds blow into a vast desert that extends from Mexico into the northern United States. Farther east in the Great Plains rain is more plentiful because of the effect of southerly winds off the Gulf of Mexico.

In dry years the margins of the southwestern deserts expand and in moist years the margins contract. Much of the United States is subject to recurrent drought as the dry desert margins expand eastward when rain is scarce for extended periods. The Great Plains are considered to be desert margin environments and are very susceptible to drought.

Even though the western deserts are normally quite dry they too can experience drought, but drought in very dry Phoenix is different than drought in normally wet Seattle. Until 1965 there was no measure of drought that made the degree of dryness directly comparable across climate types.

## The Palmer Drought Severity Index (PDSI)

To measure the extent, duration and severity of drought in 1965 Wayne Palmer of the Office of Climatology, US Weather Bureau (now NWS) used a supply and demand model to develop an index indicating the degree of dryness and the degree of wetness for the United States.

The Palmer Drought Severity Index (PDSI) uses precipitation, air temperature and local soil moisture along with earlier data as input. PDSI values generally range from −4.0, extremely dry to +4.0, extremely wet and have been standardized so comparisons can be made between regions with very different rainfall climates. If the PDSI is 0 the location under consideration has conditions about equal to the climatological normal. PDSI values can be lower −4.00 in extreme droughts and when the weather has been extremely wet the PDSI can exceed +4.00. In the graphs showing PDSI values reconstructed from tree ring time series (Figures 3.17, 3.18, 3.20–3.22, and 3.24) many years exceed the extreme limits of +/− 4.00.

If both Phoenix and Seattle have PDSI values of −3 they are equally dry relative to their individual average conditions.

The PDSI does not do well in high mountains because the PDSI formula treats all precipitation as rain and in high mountains a significant part of the annual precipitation total falls as snow and is gradually released as melt water

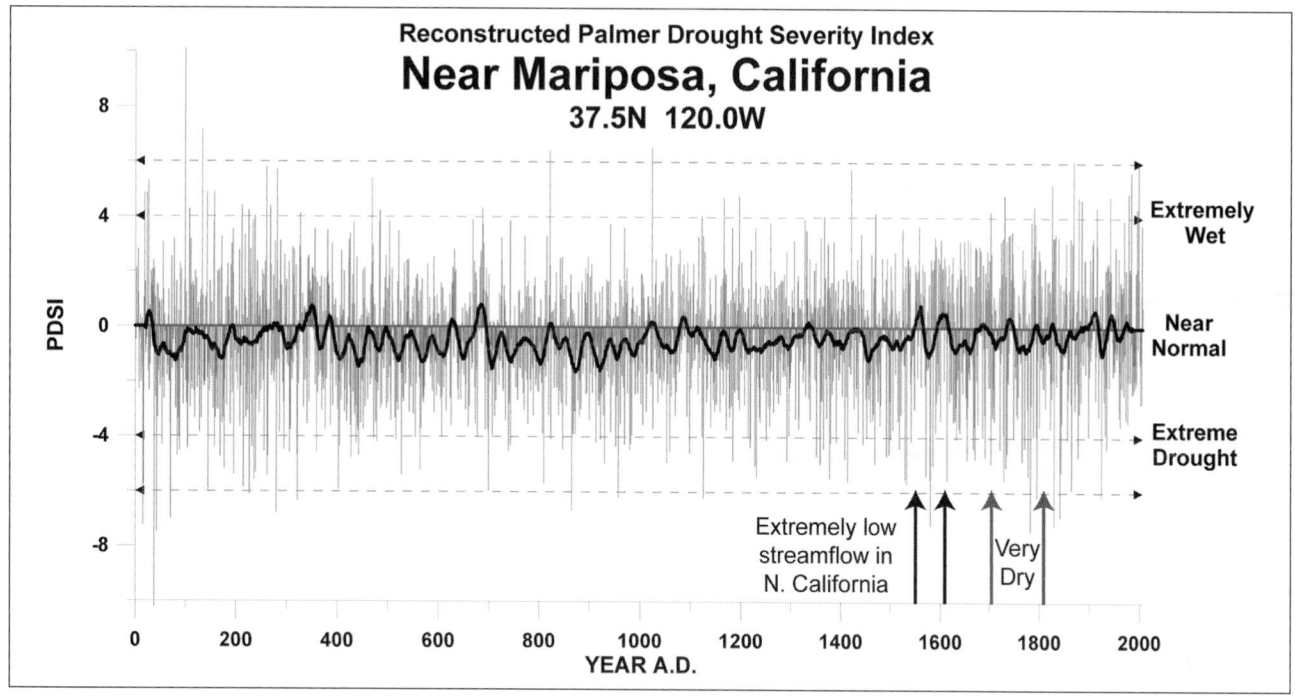

**Figure 3.17**  Reconstructed summer PDSI values near Mariposa, CA. (*Source:* Graph by Steve Horstmeyer from data by Ed Cook, Lamont–Doherty Earth Observatory, Columbia University available at NCDC.)

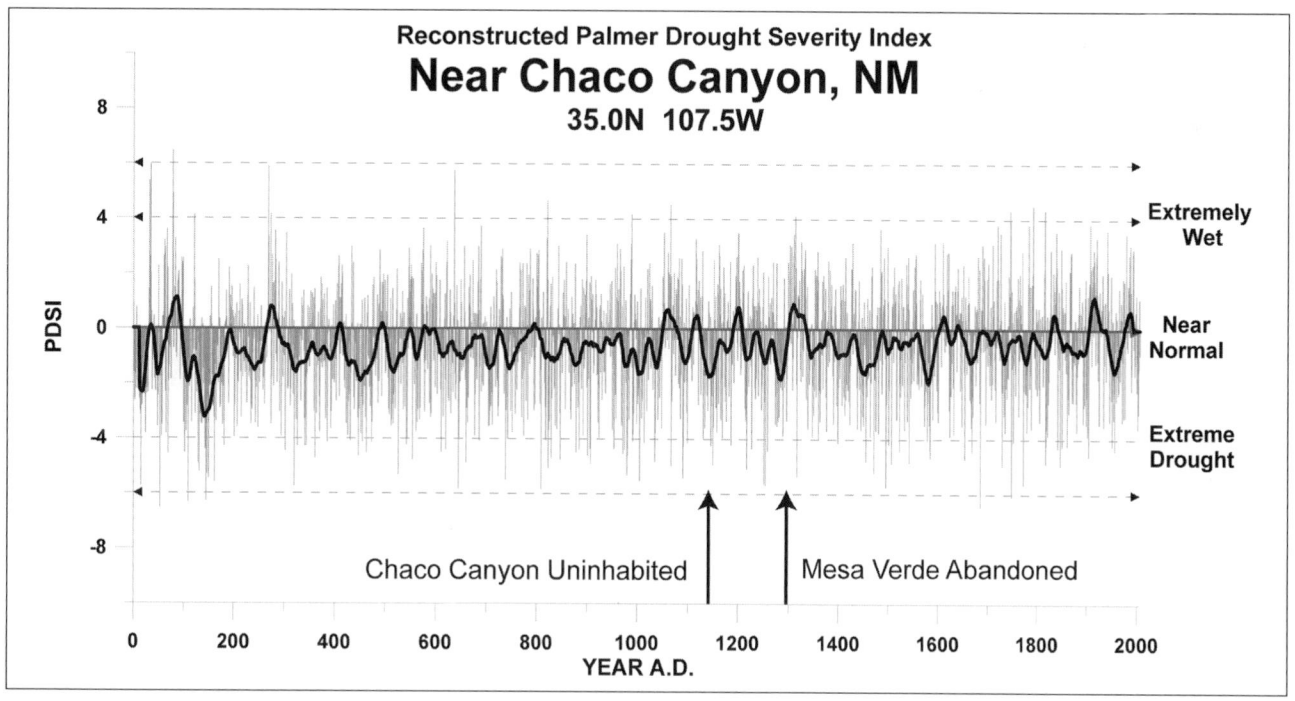

**Figure 3.18**  Reconstructed summer PDSI values near Chaco Canyon, NM, site of an Anasazi settlement. (*Source:* Graph by Steve Horstmeyer from data by Ed Cook, Lamont–Doherty Earth Observatory, Columbia University available at NCDC.)

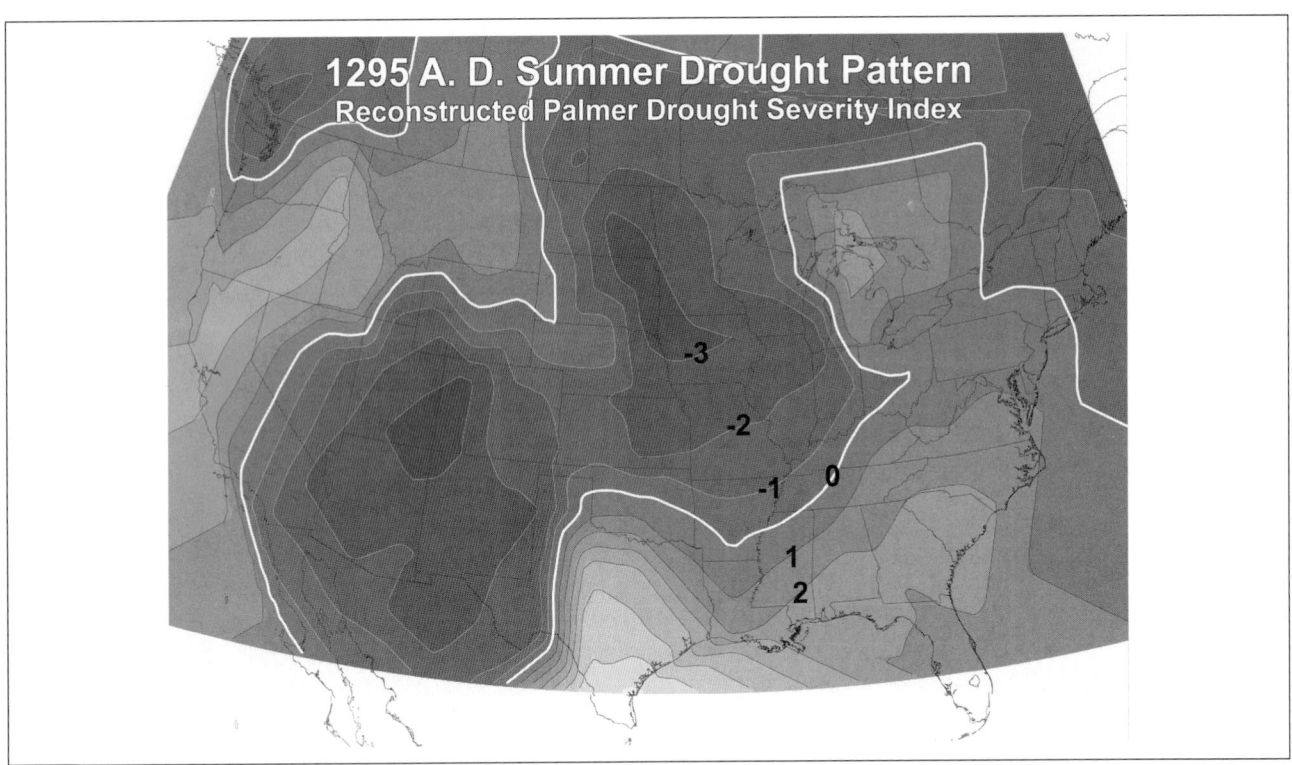

**Figure 3.19** Summer 1295 A.D. drought pattern. Extreme drought in the Four Corners area may have hastened the abandonment of Mesa Verde by the Anasazi. (*Source:* Map by Steve Horstmeyer from data by Ed Cook available from the North American Drought Atlas website.)

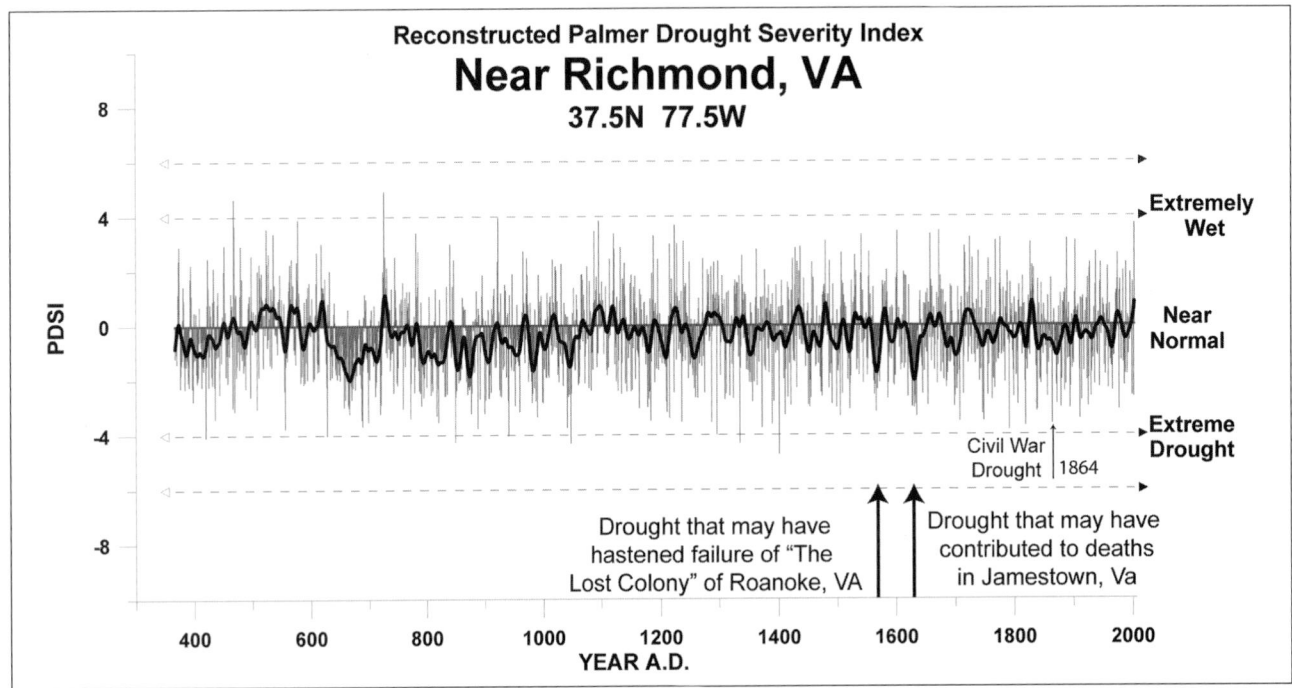

**Figure 3.20** Reconstructed summer PDSI values near Richmond, VA. (*Source:* Graph by Steve Horstmeyer from data by Ed Cook, Lamont–Doherty Earth Observatory, Columbia University available at NCDC.)

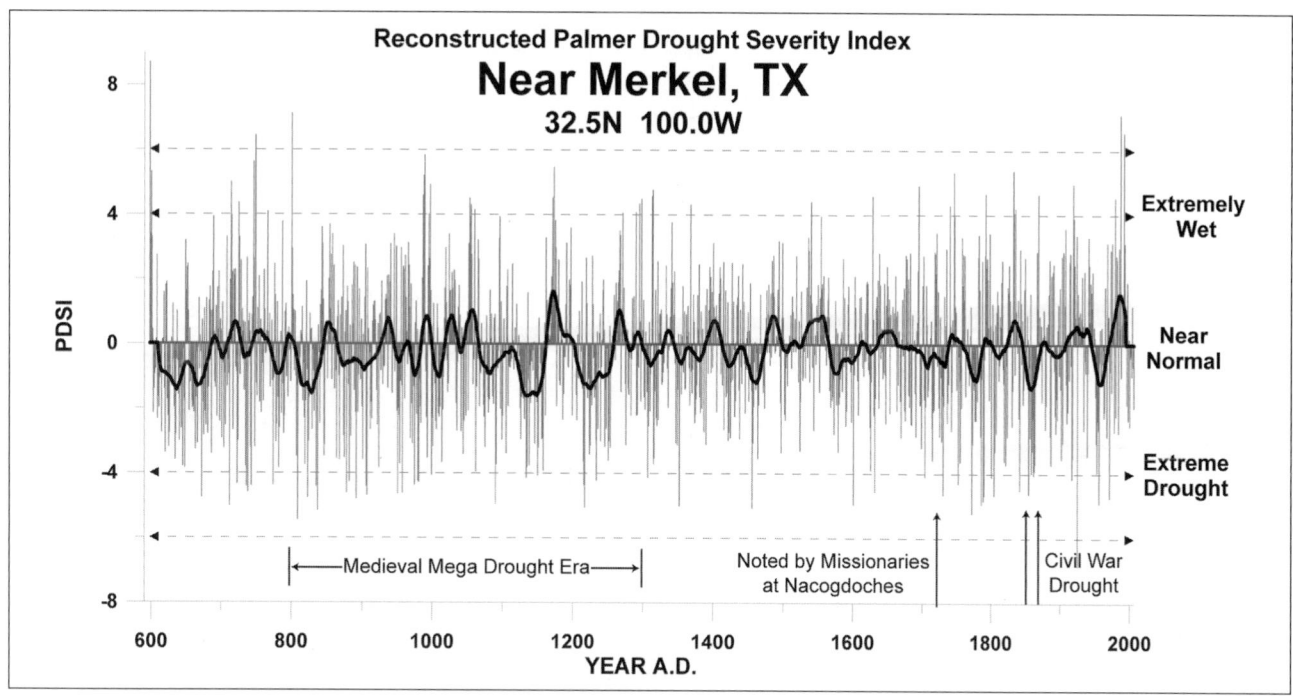

**Figure 3.21**   Reconstructed summer PDSI values near Merkel, TX. (*Source:* Graph by Steve Horstmeyer from data by Ed Cook, Lamont–Doherty Earth Observatory, Columbia University available at NCDC.)

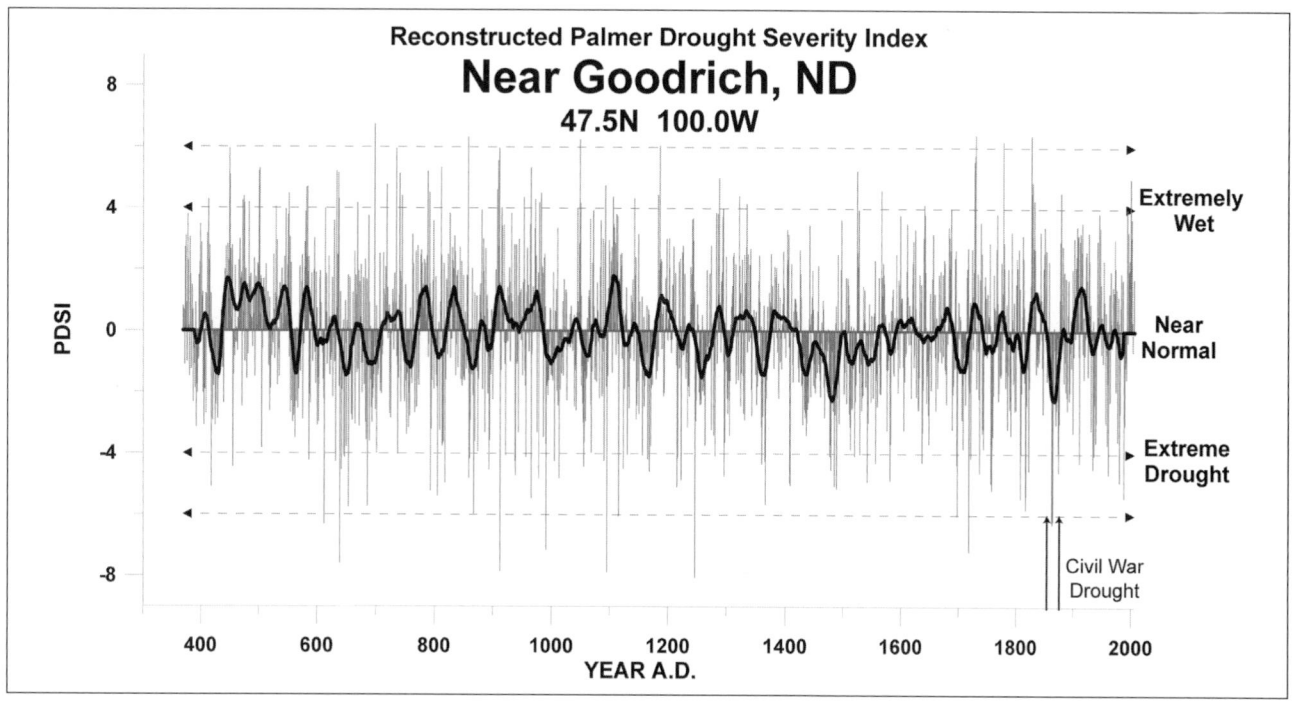

**Figure 3.22**   Reconstructed summer PDSI values near Goodrich, ND. (*Source:* Graph by Steve Horstmeyer from data by Ed Cook, Lamont–Doherty Earth Observatory, Columbia University available at NCDC.)

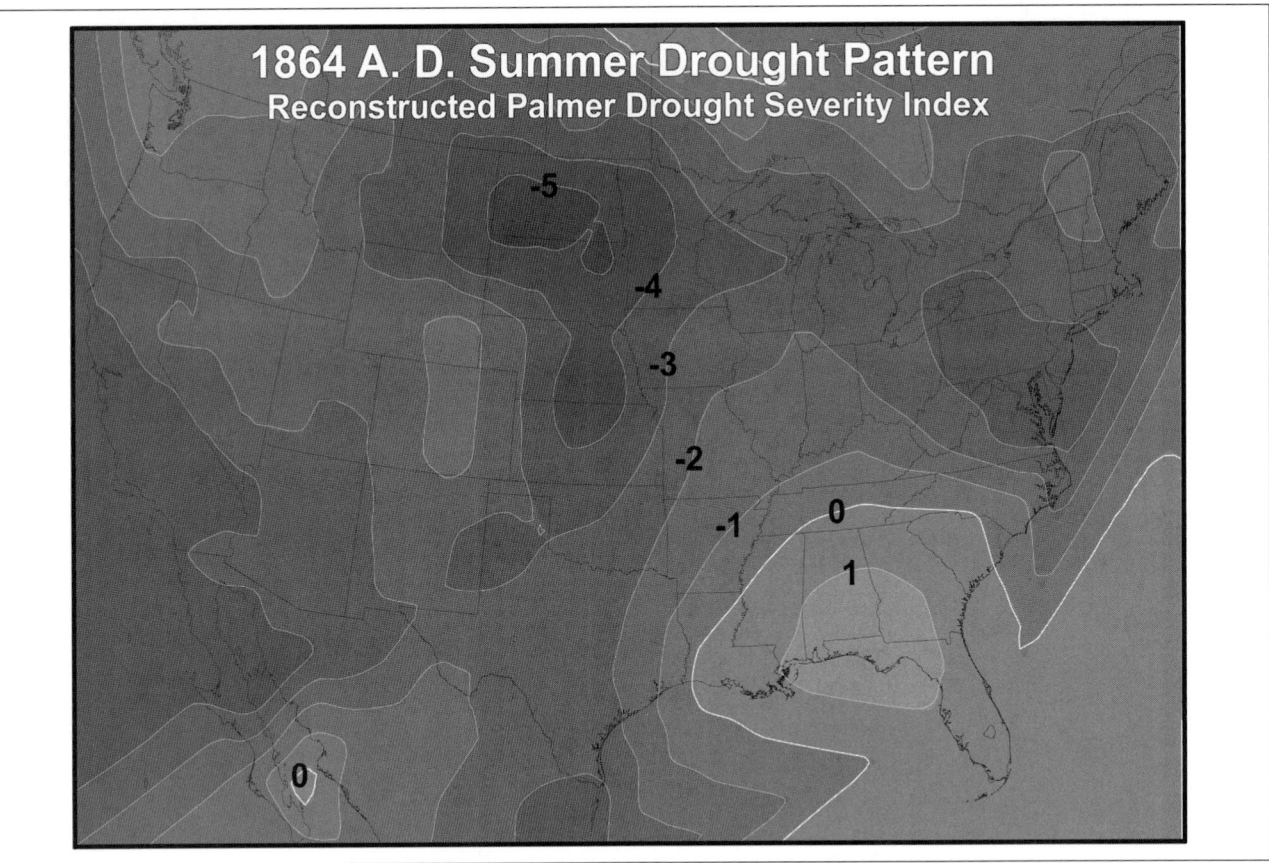

**Figure 3.23** Summer 1864 drought pattern the height of the Civil War Drought. (*Source:* Map by Steve Horstmeyer from data by Ed Cook available from the North American Drought Atlas website.)

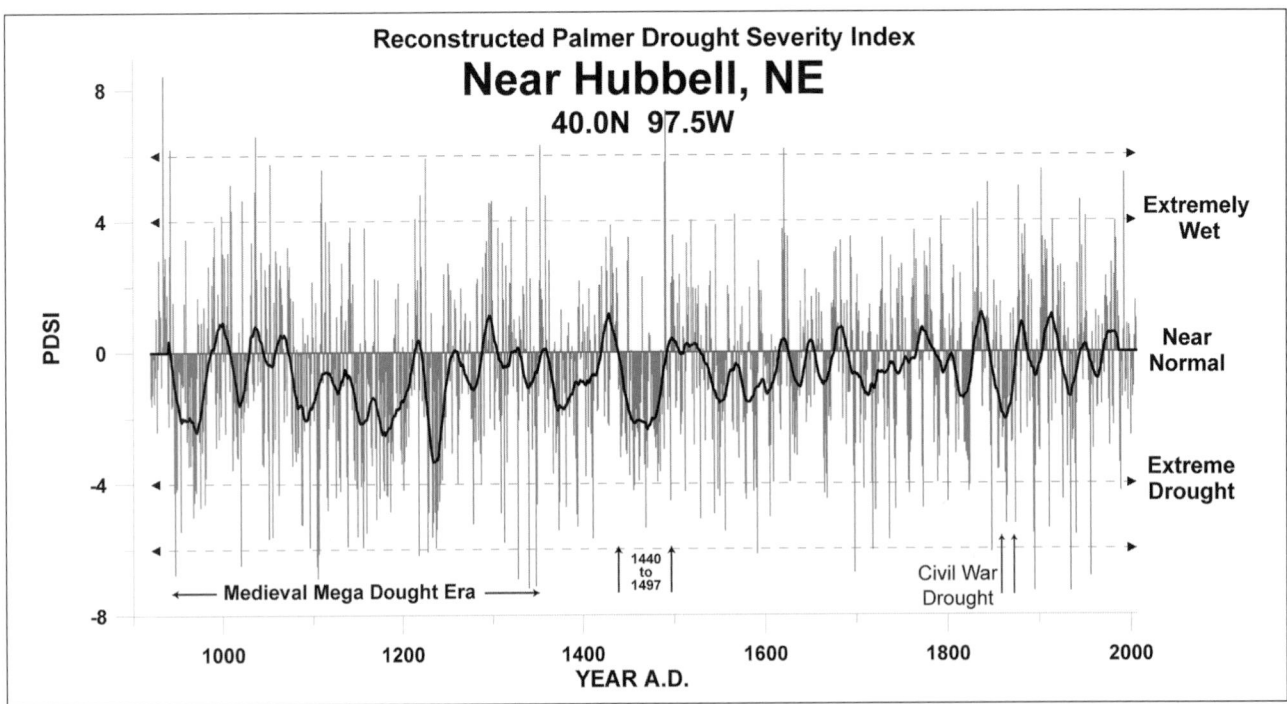

**Figure 3.24** Reconstructed summer PDSI values near Hubbell, NE. (*Source:* Graph by Steve Horstmeyer from data by Ed Cook, Lamont–Doherty Earth Observatory, Columbia University available at NCDC.)

to the soil. The time scale of the PDSI makes it ideal for agricultural purposes, but for measuring longer term effects of drought, like streamflow, it is not generally suitable.

Wayne Palmer's work has been updated and modified so that the following drought indices are available from the NCDC (Table 3.27).

### Table 3.27   Palmer Drought Indices

| Drought Index | Use and Characteristics |
|---|---|
| **Palmer Z Index** | Measures short term drought on a monthly scale |
| **Palmer Drought Severity Index (PDSI)** | Measures long term cumulative intensity of drought Responds quickly to changing weather conditions |
| **Palmer Modified Drought Index (PMDI)** | Operational version of PDSI |
| **Palmer Hydrological Drought Index (PHDI)** | Measures effects of drought on stream flow, groundwater and lake and reservoir levels Responds more slowly thant the PDSI. |

*Source:* NCDC.

### Table 3.28   PDSI Values and Meaning

| PDSI Value | Class |
|---|---|
| >= 4.00 | Extremely Wet |
| 3.00 to 3.99 | Very Wet |
| 2.00 to 2.99 | Moderately Wet |
| 1.00 to 1.99 | Slightly Wet |
| .50 to .99 | Incipient Wet Spell |
| +.49 to -.49 | Near Normal |
| -.50 to -.99 | Incipient Drought |
| -1.00 to -1.99 | Mild Drought |
| -2.00 to -2.99 | Moderate Drought |
| -3.00 to -3.99 | Severe Drought |
| <= -4.00 | Extreme Drought |

*Source:* National Drought Mitigation Center, University of Nebraska, Lincoln.

Drought information is available from the NCDC, US Department of Agriculture's Weekly Weather and Crop Bulletin and the US Army Corps of Engineers.

Many meteorologists, hydrologists and agricultural scientists are moving away from an index approach to tracking drought and, like weather forecasters, are using numerical models to monitor and forecast drought conditions.

## Droughts: Past and Future

Future droughts have become a great concern to scientists and agricultural interests. As more is learned about the rainfall climate of the past in the United States it is becoming more obvious that drought conditions will be more widespread and frequent in the future. The story goes back long before Europeans found the New World.

During much of the Middle Ages the northern hemisphere was dominated by warm climatic conditions originally thought to be benign. But as paleoclimatic research progressed proxy data sources like tree ring time series and lake and ocean sediments scientists have learned that what they have been calling "The Climatic Optimum" or "The Medieval Warm Period" may more accurately be described as "The Era of Mega-Droughts." This warm, dry era lasted from about 800 A.D. to 1300 A.D. and may have been responsible for major population shifts in the ancient Native American cultures.

Numerical climate models of the global atmosphere show wind belts on a warmer Earth are shifted poleward. The arid semitropics, now centered around 30° latitude north and south where the world's great deserts are found, are greatly expanded increasing the amount of land in desert or desert margin climates.

Much of the United States, which is now marginal agricultural land and very productive under irrigation, during the Medieval Warm Period was much drier. If the forecast for global warming into the next century verifies much of the United States will be significantly drier than it is now.

The most severe drought era of the 20th century in the United States is known as the Dust Bowl. Lasting roughly from 1930 to 1940 many of the months during the dust bowl were dry and hot. The heat accelerated soil moisture evaporation and the drought conditions were made worse by poor agricultural practices. The 1950s were nearly as severe and known as "The Little Dust Bowl." Neither drought period is unique there are longer and more severe droughts yet to be written about in prehistory of America (Table 3.29).

## Selected Prehistoric and Historic Droughts in North America

Using tree ring proxy data paleoclimatologist Ed Cook of Lamont–Doherty Earth Observatory has reconstructed the PDSI for North American summers back to the year zero.

The farther we proceed back in time the less proxy data there is so the reconstructions cover less area and are less reliable. The reconstructed PDSI values correspond well to those calculated from meteorological data in the 20th century and scientists are confident they represent past drought patterns reliably.

Figures 3.17, 3.18, 3.20–3.22, and 3.24 are graphs of the PDSI values for 6 locations, the foothills of the Sierra Nevada Mountains near Mariposa, CA, northwest New Mexico, near Chaco Canyon home to the Anasazi Tribe before it was abandoned, eastern Virginia near Richmond, central Texas near Merkel, central North Dakota near Goodrich, There is a reference to each in the droughts list.

**236–377** One of the driest periods in the past 2000 years with frequent and severe drought as indicated by Giant Sequoia tree ring sequences (see Figure 3.17). This dry era is also evident in North Dakota from lake sediment sequences and the Four Corners area (CO, UT, AZ, NM) from tree ring data.

**ca. 250** Mayans abandon a number major cities, probably drought- and population-related.

**ca. 750, ca. 810, ca. 860, ca. 910** A series of droughts and population pressures that stressed the natural environment lead to the progressive collapse of Mayan civilization. Farther north in the western United States, drought frequently occurred from 700–850 A.D.

**900–1300** Medieval mega drought era in the western United States. Peak years and percent of western United States in drought, 936 A.D. (45%), 1034 A.D. (45%), 1150 A.D. (60%), and 1253 A.D. (50%).

**1100–1200** Anasazi civilization was at its height in northwest New Mexico at Chaco Canyon. In the year 1131 A.D. the last log used for construction at the settlement was cut. By 1150 A.D. the settlement had collapsed and the occupants had moved elsewhere. This corresponds to a severe drought indicated on Figure 3.18 that peaked in 1150 A.D.

**1243–1258** Great Plains, longest sustained drought in full record, 16 years in a row with negative PDSI values inw Montana.

**1273–1309** "The Great Drought" in the southwest probably led to the abandonment of Mesa Verde in 1299 A.D., which was built by the Anasazi after the fall of the settlement at Chaco Canyon. This drought is indicated on Figure 3.18 and the map of reconstructed drought conditions for the year 1295 (Figure 3.19) shows a dry pattern that was common preceding the move from Mesa Verde. Rains returned in 1298 and for several years in the early 1300s as the drought eased.

**Table 3.29   20 Severest Drought-Prone Months since 1895 Averaged over the Central Great Plains, 1895–1996**

| Month | Extent of Drought and Notes |
|---|---|
| June 1933 | LA to MT, 2" to 4" below normal |
| June 1988 | Except for New England, FL, AZ and NM severely dry coast to coast |
| July 1936 | Mid Atlantic states to the Rocky Mtns., 2" below normal many areas |
| August 1983 | Eastern MT to the east coast severely to extremely dry. |
| July 1934 | Most of the country 2" - 4" drier than normal. |
| July 1901 | Most of the west and Great Plains, Ohio and Tennessee Valleys |
| June 1931 | From western OH to the Rocky Mountains, very dry |
| August 1947 | Great Plains and Corn Belt, New England |
| July 1930 | Most of the country dry. |
| June 1936 | Appalachian Mountains to the Rocky Mountains, very dry MO, KS, IL |
| July 1954 | Extremely hot and very dry nationwide. |
| August 1936 | Nearly no rain from southern IL to west Texas, Arkansas very dry |
| June 1956 | Only small parts of the Great Plains had normal rain, most received < 50% |
| July 1974 | Hot, dry from ND to OK, Plains rain <40% or normal just as dry from MO to OH |
| June 1918 | West Coast into the Plains where there were spotty showers. |
| June 1911 | Little or no rain from east TX and LA north to IA and west to the Pacific Coast. |
| August 1937 | Appalachian Mountains to West Coast except for spots, below normal rain. |
| August 1913 | Except for parts of AZ and NM <0.5" from the West Coast to Ohio Valley. |
| June 1952 | Eastern MT to the Rio Grande R., east to Georgia very dry with a few exceptions. |
| August 1909 | MO to southern NE south to central TX and West Coast, N. Rocky Mtns., MT, WY |

*Source:* Journal of Climate, v 14, 2296–2316.

**1440–1497** Central Great Plains, in the 57-year period there were only 10 summers with positive PDSI values and 47 of 57 years were dry. Included in this period were streaks of consecutive dry years of 10, 11, and 11 years with negative PDSI values (Figure 3.24).

**1559–1569** Peaking in the late 1560s this drought may have increased mortality in the Spanish settlement of Santa Elena on Parris Island, SC (Figure 3.20). During the 1570s wet weather was the rule but drought returned in the 1580s (see 1587–1589). Documents from Santa Elena refer to the lack of rain for "many months" a "great lack of food." The Juan Pardo expedition of 1567–1568 was ordered to bring food from the interior to alleviate the shortages at Santa Elena.

**1576–1600** Severe drought in the southwestern United States (Figure 3.18). Reconstructed Colorado River flow is lower in the 1579–1598 time period than any other known time period. Extremely low streamflows in northern California also. See the entry for years 2000–2003 and Figure 3.17.

**1580** Hydrological drought northern Great Plains and southern Canadian Plains.

**1587–1589** A multiyear drought that was the worst in 800 years peaked in 1587, the third driest year of the sixteenth century. That year the Spanish abandoned Santa Elena on Parris Island, now in SC. The drought may have been a contributing factor in the decision to leave. The year 1587 was also the year settlers founded Roanoke, Va. destined to be remembered as the "Lost Colony." Drought is thought to have played a role in this unsuccessful colony (Figure 3.20).

**1606–1635** Starting with a seven-year drought, dry weather continued sporadically but often with much drier conditions until 1635. This may have contributed to the deaths of many of the 2600 settlers who died in Jamestown, VA between 1608 and 1624. Evidence indicates that many died of malnutrition. The drought extended northward into Plymouth colony where the Governor called for a special day of religious thanksgiving on July 23, 1623 for rain that fell and was thought to have ended the drought (Figure 3.20).

**1631** Hydrological drought northern Great Plains and southern Canadian Plains.

**1664–1673** Severe drought southwestern United States. Eastward into Texas and north into Iowa and Illinois (Figure 3.21).

**1696–1699 and 1703–1712** Northern Great Plains and extended in various directions during the duration (Figure 3.22).

**1715–1716** New France, St. Lawrence River valley.

**1717–1725** Pacific northwest into the northern Great Plains and southern Canadian plains with intrusions into the southern Great Plains. Missionary Fray Isidro Feliz de Espinoza, in east Texas noted severe drought in 1717–1718 and wrote that missions gave food to Native Americans because of the failed bean harvest. Drought was also noted by missionaries at Nuestra Señora de Guadalupe mission in Nacogdoches, TX, where drought continued into the 1730s (Figure 3.21).

**1735–1744** National drought, driest area was the southwest with PDSI values −4.5 to −6. From the Great Plains to the Ohio and Mississippi River Valleys PDSI values −2.5 to −3.5. It expanded into the northern Rocky Mountains and Pacific Northwest between 1741 and 1744 (Figure 3.22).

**1752–1757** National drought, at the start extreme in the southwest and moderate from the Mississippi River to the east coast. 1754 and 1756 were break years across the northern United States and Ohio Valley into the northeast.

**1759** Hydrological drought Great Plains and southern Canadian Plains.

**1777–1781** Southern California, PDSI as low as −5.5.

**1782** Extreme drought California, Nevada, Utah, Arizona, PDSI as low as −6.0.

**1808–1809** California Mission Buena Ventura received no rain from November 25, 1808 into April 1809. A completely dry wet season means extreme drought with the onset of the dry season in California. PDSI as low as −5.2 (Figure 3.17).

**1816–1826** National drought. Beginning as early as 1799, there were patches of drought, which became a nearly nationwide drought by 1812. In 1814 and 1815 the drought subsided but came back in 1816 in the eastern United States, then spread to a nationwide drought.

**1843–1848** Sierra Nevada very low streamflows, northeast and central Kansas, worst year in California was 1847 when the drought extended over the entire southwest.

**1855–1864 The "Civil War Drought"** was a national drought that peaked in 1864 (Figure 3.23) and 1865 and was one of the most severe ever in Texas maybe even drier than the Dust Bowl (Figure 3.21). In 1860, in northeast and central Kansas, farmers told stories to new settlers of 13 months without a drop of rain in Wyandotte County (Kansas City, KS) (Figure 3.24). This is the culmination of dry weather beginning as early as 1854. April 3, 1860 a great dust storm reported at Fort Scott, KS and on July 10, 1860 at Leavenworth, KS the temperature reached 108°F with a strong desiccating wind like it "blew from the mouth of a blazing furnace."

**1870–1874** Nearly nationwide this drought peaked in 1874. Extreme drought in Kansas in 1874 gave the state the nickname "Droughty Kansas." Dry weather along with swarms of locust in 1875 severely damaged crops.

**1886–1896** After a series of very cold winters two summers of drought along with overgrazing contributed to the demise of the Free-Range cattle industry in the United States, especially in the northern Great Plains. Population pressures, both human and cattle, caused deterioration of natural grass and great stress on the cattle industry.

**1901** Most intense in eastern Texas and Louisiana, extended north to Canada and northeast into Kentucky.

**1910–1913** Texas to Canada and east to the Atlantic Ocean, most intense in Kansas and Nebraska.

**1914** Most of the eastern United States driest from southern Illinois east to western Virginia and south into Tennessee and Alabama.

**1917** Most of Texas, started in 1916, most intense around the Gulf Coast.

**1919** Montana, Idaho and north into Canada.

**1924** California and the Pacific Northwest, Idaho, Nevada.

**1925** Central and southern Great Plains east to the Atlantic Coast.

**1930–1940 The Dust Bowl years.** Some accounts limit the Dust Bowl to the years 1933–1940 however in 1930 most of the Great Plains were near normal with a moderate drought in the east. The Dust Bowl was actually made up of 4 distinct drought events: 1930–1931, 1934, 1936, 1939–1940. During

**Figure 3.25**   Dust storm approaching Spearman, TX, 04.14.1935. (*Source:* NOAA Central Library.)

1931 the east remained dry and moderate to severe drought extended from the Pacific Northwest, across the northern Great Plains into the Western Great Lakes region. The year 1932 was like 1930 with most of the plains states near normal, by 1933 drought set in and intensified. The years 1931, 1934, 1936, 1939 and 1940 had the most extensive drought conditions.

**1930** C. F. Marvin, Chief US Weather Bureau (now NWS), "the central and eastern portions of the country . . . the most severe drought in the climatological history of the United States . . . July and August combined brought the driest weather of record to Pennsylvania, Maryland, Virginia, West Virginia, Kentucky, Missouri and Arkansas with an average for these seven states of only 41 percent of normal."

**1934** The US Department of Agriculture called crop yields a national calamity. On June 4, 1934 President Franklin D. Roosevelt asked Congress for an appropriation of $52.5 million ($836 million in 2008 dollars) for drought relief. By the

end of the Dust Bowl, the appropriations would exceed $1 billion (in 1939 dollars which is $15 billion in 2008 dollars).

**1936** Extreme drought combined with extreme heat. By 1936 21% of rural families in the Great Plains region were receiving federal emergency relief.

**1947** Iowa, most intense central part of the state.

**1950s Little Dust Bowl or "Mid-Century Drought."** In 1951 and 1952 drought began to build in Texas, and in 1953, from northern Texas through Kansas, Missouri and into Kentucky and Ohio, drought was moderate. In 1954, the pattern grew, including most of the United States, east of the Sierra Nevada Mountains and south of the Great Lakes. In 1956 drought eased in the east except or the southeast and continued in the plains. Dry weather peaked in 1956 when parts of Nebraska reached a PDSI of −7 and most of Iowa, Kansas, Oklahoma and Texas were experiencing moderate to extreme drought.

**Figure 3.26**   Tractor nearly buried in drifting dust in the Great Plains, ca. 1935. (*Source:* NOAA Central Library.)

**1959–1966** Pacific Northwest, Northern Rocky Mountains and Northern Plains and later in addition North Dakota into the northeastern United States.

**1963–1964** Great Plains and Rockies.

**1971** Texas, New Mexico, Arizona.

**1972** California.

**1974** Great Plains and the Southwest.

**1977** Most of the United States except Texas.

**1980–1981** Southern Great Plains and the southeastern United States.

**1987–1990** Most of the United States at various times. First in the west then spread to the Ohio Valley states with great agricultural losses in the corn belt during 1988.

**1994–1996** Western United States.

**1998–1999** Northeast United States.

**2000–2003** First the southern United States the most of the west. Colorado River, which supplies water to much of the southwest, flows about half of what it was during the Dust Bowl.

**2006** Texas, Oklahoma, extreme drought.

**2007–2008** Southern California, and the west, extreme drought.

**2009** California, Texas severe to extreme drought.

## • WINTER STORMS

### *Regional Names for Winter Storms*

Winter storms go by many names. Local meteorologists may speak of the Alberta Clipper, the Colorado Lee Slopes Low, the Four Corners Low, the Chattanooga Cho-Cho or the nor'easter. Each is a middle latitude cyclone, a normal low-pressure system you see on a TV weather map with a special name.

The primary source of energy for the middle latitude cyclone is the polar front jet stream. Warm tropical air does not readily mix with cold polar air, where they meet the polar front jet stream forms. The jet stream is a fast moving current of air powered by density differences between the warmer air and the rotation of Earth. When the jet stream is disturbed, mainly by mountains (but there are other ways) the air flowing through the disturbed jet can give birth to a middle latitude cyclone.

The Alberta Clipper forms east of the Canadian Rockies, in or around Alberta, Canada and because this low moves quickly southeastward following the winds in the jet stream aloft, it was compared to the swift clipper ships of sailing days. The clipper is here and gone leaving behind, in most cases a slight covering of fluffy light snow.

The nor'easter is an intense low pressure system that moves northward along the east coast of the United States while tapping the warm moist water of the Gulf Stream as an energy source. Nor'easters may or may not reach blizzard status but because of the contrast between the cold air on the continent and the warm moist air off shore they are frequently intense storms and big snow makers.

The winter of 2009–2010 saw a number of nor'easters dump heavy snow on much of the east coast especially Washington, D.C. As the storm heads north along the coast the cyclonic winds around the storm blow onshore, out of the northeast bringing tremendous amounts of moisture from the ocean. Heavy rain, ice and snow fall depending on temperature.

As the nor'easter heads north along the coast the cyclonic winds around the storm blow onshore to the north of the storm center, out of the northeast bringing tremendous amounts of moisture from the ocean. Heavy rain, ice and snow fall depending on temperature.

A Colorado Lee Slopes Low forms just east of the Rocky Mountains in eastern Colorado and western Kansas and Nebraska while the Four Corners Low forms near the common border point of Colorado, Arizona, New Mexico and Utah. Both can bring big problems because they are close to the Gulf of Mexico and can easily draw in immense amounts of warm, moist Gulf air northward bringing heavy rain and thunderstorms on their south sides and deep, heavy wet snow and ice on their north.

A Chattanooga Choo Choo intensifies over eastern Tennessee and the southern Appalachian Mountains and because of backing winds aloft moves abruptly north-northeastward along the spine of the Appalachian Mountains. This name for the middle latitude cyclone is used regionally and may not be very well known outside of the Ohio Valley states.

### Freezing Rain and Sleet

If conditions are right, on the cold side of the rain band, freezing rain can glaze the surface with a slippery and tough coat of ice. North of the glazed over area, where the air is colder, sleet may fall. These small ice pellets may even bounce several feet into the air if frozen solid enough.

Both freezing rain and sleet originate in warm air upstairs gliding northward above the frigid air at the surface. If raindrops are chilled to about freezing while falling through the lower cold air and strike the surface as liquid they freeze on contact with a surface colder than 32°F. The result is the thick, slippery and heavy glaze of an ice storm. Power lines fall and tree limbs break under the weight of the clear load of ice. In the United States the term freezing rain is used to de-

scribe raindrops freezing on impact, internationally the term glaze is used.

If the raindrops completely freeze into pellets of solid ice in the United States it is called sleet. Sleet does not adhere to surfaces like freezing rain so sleet is not known for the demise of power lines and trees, but it too will cause problems, mainly for transportation. In Great Britain the term sleet means a mix of rain and snow.

### Classification of Winter Storms

Winter storms can be roughly classified as snow storms, ice storms and blizzards. Any of the specially named storms listed above can become intense enough to reach blizzard status and each can cause significant icing when the three dimensional temperature distribution of the atmosphere permits it. The NWS issues advisories, watches and warnings for snow, heavy snow, freezing rain, sleet, ice storms and blizzards.

A blizzard is defined as an intense middle latitude cyclone with wind speeds of 35 mph or greater and visibility reduced frequently to $\frac{1}{4}$ mile or less by blowing snow, falling snow or both. When the NWS issues a blizzard warning these conditions are expected to continue for at least three hours. Often the temperature is below 20°F. During a severe blizzard winds exceed 45 mph and the temperature is below or will drop below 10°F.

## Winter Storm Safety

FEMA deals with managing large disasters in the United States. To make the tasks of rescue and aid during a disaster and recovery after a disaster FEMA has prepared many guides for the public. The following advice on winter storm safety is from FEMA.

### Before Winter Storms and Extreme Cold Arrive

1. **Add the following supplies to your disaster supplies kit:**
   Rock salt to melt ice on walkways.
   Sand to improve traction.
   Snow shovels and other snow removal equipment.
2. **Prepare your home and family:**
   Prepare for possible isolation in your home by having sufficient heating fuel; regular fuel sources may be cut off. For example, store a good supply of dry, seasoned wood for your fireplace or wood-burning stove.
   Winterize your home to extend the life of your fuel supply by insulating walls and attics, caulking and weatherstripping doors and windows, and installing storm windows or covering windows with plastic.
   Winterize your house, barn, shed or any other structure that may provide shelter for your family, neighbors, livestock or equipment. Clear rain gutters; repair roof leaks and cut away tree branches that could fall on a house or other structure during a storm.

Insulate pipes with insulation or newspapers and plastic and allow faucets to drip a little during cold weather to avoid freezing.

Keep fire extinguishers on hand, and make sure everyone in your house knows how to use them. House fires pose an additional risk, as more people turn to alternate heating sources without taking the necessary safety precautions.

Learn how to shut off water valves (in case a pipe bursts).

Know ahead of time what you should do to help elderly or disabled friends, neighbors or employees.

Hire a contractor to check the structural ability of the roof to sustain unusually heavy weight from the accumulation of snow or water, if drains on flat roofs do not work.

3. **Prepare your car:**

*Check or have a mechanic check the following items on your car:*

Antifreeze levels—ensure they are sufficient to avoid freezing.

Battery and ignition system should be in top condition and battery terminals should be clean.

Brakes—check for wear and fluid levels.

Exhaust system—check for leaks and crimped pipes and repair or replace as necessary. Carbon monoxide is deadly and usually gives no warning.

Fuel and air filters—replace and keep water out of the system by using additives and maintaining a full tank of gas.

Heater and defroster—ensure they work properly.

Lights and flashing hazard lights—check for serviceability.

Oil—check for level and weight. Heavier oils congeal more at low temperatures and do not lubricate as well.

Thermostat—ensure it works properly.

Windshield wiper equipment—repair any problems and maintain proper washer fluid level.

*Install good winter tires.* Make sure the tires have adequate tread. All-weather radials are usually adequate for most winter conditions. However, some jurisdictions require that to drive on their roads, vehicles must be equipped with chains or snow tires with studs.

*Maintain at least a half tank of gas during the winter season.*

*Place a winter emergency kit in each car that includes:*

A shovel.
Windshield scraper and small broom.
Flashlight.
Battery-powered radio.
Extra batteries.
Water.
Snack food.
Matches.
Extra hats, socks and mittens.

First aid kit with pocket knife.
Necessary medications.
Blanket(s).
Tow chain or rope.
Road salt and sand.
Booster cables.
Emergency flares.
Fluorescent distress flag.

4. **Dress for the weather:**

Wear several layers of loose fitting, lightweight, warm clothing rather than one layer of heavy clothing. The outer garments should be tightly woven and water repellent.

Wear mittens, which are warmer than gloves.

Wear a hat.

Cover your mouth with a scarf to protect your lungs.

### *During Winter Storms and Extreme Cold*

1. **If you are indoors:**

Listen to your radio, television, or NOAA Weather Radio for weather reports and emergency information.

Eat regularly and drink ample fluids, but avoid caffeine and alcohol.

Conserve fuel, if necessary, by keeping your residence cooler than normal. Temporarily close off heat to some rooms.

If the pipes freeze, remove any insulation or layers of newspapers and wrap pipes in rags. Completely open all faucets and pour hot water over the pipes, starting where they were most exposed to the cold (or where the cold was most likely to penetrate).

Maintain ventilation when using kerosene heaters to avoid build-up of toxic fumes. Refuel kerosene heaters outside and keep them at least three feet from flammable objects.

2. **If you are outdoors:**

Avoid overexertion when shoveling snow. Overexertion can bring on a heart attack—a major cause of death in the winter. If you must shovel snow, stretch before going outside.

Cover your mouth. Protect your lungs from extremely cold air by covering your mouth when outdoors. Try not to speak unless absolutely necessary.

Keep dry. Change wet clothing frequently to prevent a loss of body heat. Wet clothing loses all of its insulating value and transmits heat rapidly.

Watch for signs of frostbite. These include loss of feeling and white or pale appearance in extremities such as fingers, toes, ear lobes, and the tip of the nose. If symptoms are detected, get medical help immediately.

Watch for signs of hypothermia. These include uncontrollable shivering, memory loss, disorientation, incoherence, slurred speech, drowsiness, and apparent exhaustion.

If symptoms of hypothermia are detected:

Get the victim to a warm location.

Remove wet clothing.

Put the person in dry clothing and wrap their entire body in a blanket.

Warm the center of the body first.

Give warm, nonalcoholic or noncaffeinated beverages if the victim is conscious.

Get medical help as soon as possible.

3. **If you are driving:**

*Drive only if it is absolutely necessary. If you must drive, consider the following:*

Travel in the day, don't travel alone, and keep others informed of your schedule.

Stay on main roads; avoid back road shortcuts.

*If a blizzard traps you in the car:*

Pull off the highway. Turn on hazard lights and hang a distress flag from the radio antenna or window.

Remain in your vehicle where rescuers are most likely to find you. Do not set out on foot unless you can see a building close by where you know you can take shelter. Be careful; distances are distorted by blowing snow. A building may seem close, but be too far to walk to in deep snow.

Run the engine and heater about 10 minutes each hour to keep warm. When the engine is running, open a downwind window slightly for ventilation and periodically clear snow from the exhaust pipe. This will protect you from possible carbon monoxide poisoning.

Exercise to maintain body heat, but avoid overexertion. In extreme cold, use road maps, seat covers, and floor mats for insulation. Huddle with passengers and use your coat for a blanket.

Take turns sleeping. One person should be awake at all times to look for rescue crews.

Drink fluids to avoid dehydration.

Be careful not to waste battery power. Balance electrical energy needs—the use of lights, heat, and radio—with supply.

Turn on the inside light at night so work crews or rescuers can see you.

If stranded in a remote area, stomp large block letters in an open area spelling out HELP or SOS and line with rocks or tree limbs to attract the attention of rescue personnel who may be surveying the area by airplane.

Leave the car and proceed on foot—if necessary—once the blizzard passes.

## Selected Historic Winter Storms of the United States

### February and March 1717

"The Great Snow of 1717," New England a series of four storms buried much of New England under nearly four feet on the ground with drifts up to 25 feet high.

### January 1772

Both George Washington and Thomas Jefferson were snowed in at their homes giving "The Washington and Jefferson Snowstorm" its name. The snow was up to three feet deep throughout Maryland and Virginia.

### December 1778

"The Hessian Storm" was named after German mercenaries who fought for Great Britian during the Revolutionary War. Nine froze to death at their posts as the temperature plummeted during the storm and 50 others died around Newport, RI from cold or the results of frostbite. This storm is also called "The Magee Storm" after the captain of the foundered ship *Arnold*.

### December 1779–January 1780

Three nor'easters in succession, 28–29 December, 2–3 January and 5–7 January left 52" of snow in New Haven, CT and 42" to 48" across southern Connecticut with drifts 6' to 10' deep. George Washington wrote from Morristown, NJ, "The snow which in general is eighteen inches deep is much drifted – roads impassable."

### February 1791

One of the first storms to be described west of the Applachians know to have caused widespread deep snow. At Fort Washington, now Cincinnati, OH, General Harmar reported an 11 inch snowfall on the 11th and in Marietta, OH the next day 12 inches fell.

### November 1798

"The Long Storm" started on Saturday November 17 and cleared Wednesday the 21st. From Maryland to Maine, up to a foot-and-a-half of snow fell and the wind caused drifts much higher. Many vessels along the coast were destroyed and at the bodies of least 25 crew members washed ashore.

### February 1802

After a mild January and early February in New England on the 21st a mild morning yielded to a fierce wind out of the northeast and blinding snowfall that lasted for almost a week. Several feet of snow with a hard crust of ice from frozen sleet and freezing rain was strong enough to support carriages.

### May 1803

May 5th to 7th crops were ruined in western Kentucky, central Indiana southeastern Ohio and snow fell and accumulated in those locations to between 4" and 6". The storm dumped 4–6 inches in western and central Pennsylvania as it moved east.

### November 1803

28th: "About eight o'clock last evening it began to snow and continued till [sic] daybreak . . ."

29th: ". . . the snow which fell yesterday and last night is thirteen inches in depth . . ."

From the diary of Meriwether Lewis, Fort Mandan, Dakota, Louisiana Territory.

### December 1811

"The Day Before Christmas Storm" was a true eastern blizzard with temperatures dropping rapidly to near 0°F, gale-force winds and blinding snow. It was first noted gaining strength off Cape Hatteras, NC and struck New York City, Long Island, and southern New England with winds and destructive tides that severely damaged many ships and harbors.

### January 1817

Heavy snow fell with thunder and lightning in many locations including Brunswick, ME, Andover, VT, Philadelphia, PA, Boston, MA, Schenectady, NY. In places 6" accumulated in one hour. There was much speculation about the appearance of lightning and thunder in January.

### January 1857

A severe blizzard called "The Cold Storm" along much of the eastern seaboard caused temperatures to fall to −9°F accompanied by fierce gales and snow accumulations between one and two feet.

### January 1888

The "Schoolhouse Blizzard" or "School Children's Blizzard" or "Children's Blizzard" of January 12th was named for three students of Lois Royce in Plainview, NE who became trapped in the one-room schoolhouse as the blizzard struck. When they ran out of fuel they attempted to go 82 yards to a boarding house. Visibility was so low they lost their way and the students died.

### March 1888

The "Blizzard of 1888," also known as "The Great White Hurricane," produced as much as 50" of snow with hurricane force winds from the Mid-Atlantic states into New England for three days from the 11th to the 14th. New York City was particularly hard hit with damage to its harbor and many overhead electrical wires snapped by the combination of wind and cold. This storm is credited with the installation of underground wires in the city. There were 400 deaths.

### November 1898

The "Portland Gale" was named after the steamship *S.S. Portland*, that left Boston with 192 passengers bound for Portland, ME and sank off the coast of Cape Cod on November 26th. All the passengers perished.

### January 1922

The "Knickerbocker Storm" was named after the Knickerbocker Theatre in Washington, D.C. where almost 100 people died when the roof collapsed under the weight of more than two feet of heavy, wet snow.

### November 1940

"The Armistice Day Storm," November 11th. The Great Lakes and upper Midwest, Manitoba, Minnesota, Wisconsin, and western Ontario were battered by this severe blizzard. In Minnesota 49 died, on Lake 59 sailors were lost and 17" of snow fell in Iowa.

### December 1947

On December 26th a storm began at 5:25 A.M. and dropped 25.8" of snow before dawn the next day snow on New York City stranding thousands on trains, in subways and in the city. The storm affected much of the northeast and 8 people died.

### January 1967

"The Great Chicago Snow of 1967." Snow fell for 29 hours and the official total was 23.0″, Chicago's greatest snow storm ever. Fifty thousand abandoned cars blocked roads and drifts were as tall as 10 feet.

### February 1969

A quick accumulation of 18″ of unexpected snow buried snow plows parked in storage lots. New York City and surrounding communities hired thousands to shovel snow off streets. Thousands were stranded in closed airports, cars and schools.

### February 1976

"The Groundhog Day Gale" brought 115 mph wind gusts to coastal Maine and a tidal surge flood in Bangor, ME. Burlington, VT received 6.5" of snow and post-storm Alake effect snow buried Oswego, NY under 56 inches.

### January 1978

"The Great Blizzard of 1978." January 25th and 26th produced the lowest atmospheric pressure ever recorded in the lower 48 states for a nontropical storm when the pressure at Cleveland, OH dropped to 28.28 inches (958 mb). In Ohio 51 died and the 50000 Ohio National Guard troops were activated into for rescues and relief efforts. Around Chicago snow totals were as great as 58" in lake effect areas and South Bend, IN received 36".

### February 1978

On the 6th and 7th, 50" or more of snow fell in many places in the northeast. Boston received 27.1" and was shut down for a week. In Philadelphia 14.1" fell and New York City received 17.7" of snow.

### February 1979

"The Presidents' Day Storm" was a slow moving coastal storm that produced record snowfall in the Middle Atlantic States, at Washington, D.C., 18.7" in 18.5 hours and a total depth of 23"; Baltimore, MD, airport had 20".

### March 1993

"The Storm of the Century," also known as "The Great Blizzard," dropped 13" of snow on Birmingham, AL, 4" in the Florida panhandle and three feet on Syracuse, NY. Ahead of the storm many tornadoes touched down in the Deep South and more than 200 people died. Hurricane force winds produced storm surges along the Gulf of Mexico coast of Florida of up to 12 feet at Pine Island. In Cuba winds gusted to between 100 And 130 mph, killing 10 and causing over $1 billion for Cuba's most intense nontropical squall.

### January 1996

"The Blizzard of 1996" was an immense blizzard that paralyzed the Mid-Atlantic and Northeast States. Snow accumulations of 1 to 2 feet were common and 100 people died. Most airports in the northeast corridor between Washington, D.C. and Boston were closed for two days.

### February 2003

"President's Day Storm II" or "The President's Day Storm of 2003" broke many snowfall records between February 14th and 19th. Every major metropolitan area from Washington, D.C. to Boston received 15" to 30" of snow. For Baltimore and Boston this was the greatest single storm snowfall ever. Boston received 27.5" and Baltimore received 28.2" of snow.

### February 2006

A nor'easter began February 11 dumped heavy snow from Virginia to Maine through the evening of February 12. All major northeast cities from Baltimore to Boston received at least a foot of snow and in New York the accumulation was an all-time record amount of 26.9".

### February 2008

A record-setting accumulation in Columbus, OH of 20.4" with 15.4" of that in 24 hours, another record, paralyzed central Ohio and stranded thousands in cars along I-70.

### December 2009

A Gulf of Mexico storm that produced record rainfall in Texas moved to the east coast and became a blizzard, dumping 16"–20" of snow and caused whiteout conditions December 19 and 20. In Washington, D.C. a record 24-hour accumulation of 24" was recorded. Philadelphia received 23.2" in the largest December single-storm accumulation. Days later the "2009 Christmas Day Storm" produced 27 tornadoes in the south and central United States and 9" of snow in Texas. The storm was so wound up that warm air wrapped around it and arrived in Rochester, MN from the northeast bringing rain and 35°F while Dallas, TX got snow on Christmas for the first time since 1929.

### • THE COST OF EXTREME WEATHER TO THE UNITED STATES

In the 10 years from 2000 through 2009 the total cost of extreme weather in the United States was over $227 billion. The costliest of the categories was tropical storms and hurricanes, which accounted for nearly 58% of the total.

In the same 10 years 5754 people died from weather related causes. Extreme heat narrowly edged out tropical storms and hurricanes with each accounting for about 20% of the total.

Injuries amounted to more than 27,000 with tornadoes (30%), extreme heat (20%) and lightning (9%) being the top three categories.

Figures 3.27–3.29 show the cost, fatality and injury statistics graphically and Tables 3.30–3.40 list the individual categories by year.

Table 3.30   Weather-related Fatalities, Injuries, and Costs for the United States, 2000. Tables for Individual Years and a 10-Year Total Table

| 2000 Weather Event | Deaths | Injuries | Property | Damage Crop | Total |
|---|---|---|---|---|---|
| | | | Millions of Dollars | | |
| Lightning | 51 | 364 | $39.8 | $0.1 | $39.9 |
| Tornado | 41 | 882 | 423.6 | 6.9 | 430.5 |
| Thunderstorm Wind | 25 | 296 | 245.2 | 58.8 | 304.0 |
| Hail | 2 | 57 | 446.0 | 124.9 | 570.9 |
| Extreme Cold | 26 | 0 | 0.6 | 8.8 | 9.2 |
| Extreme Heat | 158 | 469 | 0.0 | 0.0 | 0.0 |
| Flash Flood | 30 | 36 | 785.7 | 73.6 | 859.3 |
| River Flood | 3 | 2 | 467.5 | 605.7 | 1,073.2 |
| Coastal Storm | 29 | 23 | 0.4 | 0.0 | 0.4 |
| Tsunami | 0 | 0 | 0.0 | 0.0 | 0.0 |
| Rip Current | XX | XX | XX | XX | XX |
| Tropical Storm/Hurricane | 0 | 1 | 8.1 | 0.1 | 8.2 |
| Winter Storm | 41 | 182 | 1,035.3 | 0.0 | 1,035.3 |
| Ice | 0 | 0 | 0.3 | 0.0 | 0.3 |
| Avalanche | 16 | 17 | 0.8 | 0.0 | 0.8 |
| Drought | 0 | 0 | 0.7 | 2,438.1 | 2,438.8 |
| Dust Storm | 1 | 29 | 0.2 | 0.0 | 0.2 |
| Dust Devil | 0 | 0 | 0.0 | 0.0 | 0.0 |
| Rain | 4 | 38 | 4.4 | 4.1 | 8.5 |
| Fog | 10 | 118 | 1.8 | 0.0 | 1.8 |
| High Wind | 26 | 162 | 48.9 | 1.2 | 50.1 |
| Waterspout | 2 | 0 | 0.1 | 0.0 | 0.1 |
| Fire Weather | 3 | 100 | 2,109.2 | 7.1 | 2,116.3 |
| Mud Slide | 0 | 0 | 0.5 | 0.0 | 0.5 |
| Volcanic Ash | 0 | 0 | 0.0 | 0.0 | 0.0 |
| Miscellaneous | 3 | 11 | 0.1 | 0.0 | 0.1 |
| Total | 476 | 2,796 | 5,621.00 | 3,329.20 | 8,950.30 |

*Source:* NOAA.

Table 3.31   Weather-related Fatalities, Injuries, and Costs for the United States, 2001. Tables for Individual Years and a 10-Year Total Table

| 2001 Weather Event | Deaths | Injuries | Damage Property | Crop | Total |
|---|---|---|---|---|---|
| | | | Millions of Dollars | | |
| Lightning | 44 | 371 | 43.60 | 2.00 | 45.60 |
| Tornado | 40 | 743 | 630.10 | 7.40 | 637.50 |
| Thunderstorm Wind | 17 | 341 | 317.80 | 61.00 | 378.80 |
| Hail | 0 | 32 | 2,368.30 | 270.40 | 2,638.70 |
| Extreme Cold | 4 | 0 | 0.00 | 132.20 | 132.20 |
| Extreme Heat | 166 | 445 | 0.00 | 0.00 | 0.00 |
| Flash Flood | 35 | 265 | 856.70 | 41.00 | 897.70 |
| River Flood | 11 | 6 | 362.00 | 1.90 | 363.90 |
| Coastal Storm | 53 | 96 | 17.70 | 0.00 | 17.70 |
| Tsunami | 0 | 0 | 0.00 | 0.00 | 0.00 |
| Rip Current | XX | XX | XX | XX | XX |
| Tropical Storm/Hurricane | 24 | 7 | 5,187.80 | 2.70 | 5,190.50 |
| Winter Storm | 18 | 173 | 103.60 | 0.10 | 103.70 |
| Ice | 0 | 0 | 0.40 | 0.00 | 0.40 |
| Avalanche | 15 | 6 | 0.00 | 0.00 | 0.00 |
| Drought | 0 | 0 | 0.00 | 1,273.90 | 1,273.90 |
| Dust Storm | 0 | 5 | 0.20 | 0.00 | 0.20 |
| Dust Devil | 0 | 1 | 0.00 | 0.00 | 0.00 |
| Rain | 3 | 1 | 20.70 | 21.50 | 42.20 |
| Fog | 7 | 67 | 1.30 | 0.00 | 1.30 |
| High Wind | 14 | 98 | 63.80 | 2.20 | 66.00 |
| Waterspout | 0 | 0 | 0.00 | 0.00 | 0.00 |
| Fire Weather | 5 | 46 | 45.20 | 0.00 | 45.20 |
| Mud Slide | 0 | 0 | 1.30 | 0.00 | 1.30 |
| Volcanic Ash | 0 | 0 | 0.50 | 0.00 | 0.50 |
| Miscellaneous | 6 | 9 | 0.20 | 0.00 | 0.20 |
| Total | 464 | 2,718 | 10,022.80 | 1,816.40 | 11,839.20 |

*Source:* NOAA.

**Table 3.32  Weather-related Fatalities, Injuries, and Costs for the United States, 2002. Tables for Individual Years and a 10-Year Total Table**

| 2002 Weather Event | Deaths | Injuries | Damage Property | Damage Crop | Damage Total |
|---|---|---|---|---|---|
| | | | Millions of Dollars | | |
| Lightning | 51 | 256 | 43.50 | 0.00 | 43.50 |
| Tornado | 55 | 968 | 801.30 | 0.80 | 802.10 |
| Thunderstorm Wind | 17 | 287 | 289.80 | 54.70 | 344.50 |
| Hail | 0 | 125 | 325.20 | 153.80 | 479.00 |
| Extreme Cold | 11 | 0 | 0.30 | 70.40 | 70.70 |
| Extreme Heat | 167 | 378 | 0.00 | 0.00 | 0.00 |
| Flash Flood | 38 | 74 | 329.60 | 44.80 | 374.40 |
| River Flood | 7 | 5 | 322.80 | 37.70 | 360.50 |
| Coastal Storm | 11 | 8 | 0.30 | 0.00 | 0.30 |
| Tsunami | 0 | 0 | 0.00 | 0.00 | 0.00 |
| Rip Current | 43 | 60 | 0.10 | 0.00 | 0.10 |
| Tropical Storm/Hurricane | 51 | 346 | 1,104.40 | 278.00 | 1,382.40 |
| Winter Storm | 17 | 105 | 752.00 | 0.00 | 752.00 |
| Ice | 0 | 0 | 0.10 | 0.00 | 0.10 |
| Avalanche | 15 | 11 | 0.00 | 0.00 | 0.00 |
| Drought | 0 | 0 | 0.00 | 737.60 | 737.60 |
| Dust Storm | 2 | 45 | 0.40 | 0.00 | 0.40 |
| Dust Devil | 0 | 1 | 0.00 | 0.00 | 0.00 |
| Rain | 1 | 1 | 0.80 | 1.10 | 1.90 |
| Fog | 19 | 143 | 3.50 | 0.00 | 3.50 |
| High Wind | 28 | 129 | 82.90 | 33.50 | 116.40 |
| Waterspout | 0 | 1 | 0.00 | 0.00 | 0.00 |
| Fire Weather | 1 | 138 | 200.80 | 2.00 | 202.80 |
| Mud Slide | 0 | 0 | 0.00 | 0.00 | 0.00 |
| Volcanic Ash | 0 | 0 | 0.00 | 0.00 | 0.00 |
| Miscellaneous | 2 | 0 | 0.00 | 0.00 | 0.00 |
| Total | 540 | 3,090 | 4,260.40 | 1,414.40 | 5,674.80 |

*Source:* NOAA.

**Table 3.33   Weather-related Fatalities, Injuries, and Costs for the United States, 2003. Tables for Individual Years and a 10-Year Total Table**

| 2003 Weather Event | Deaths | Injuries | Damage Property | Damage Crop | Damage Total |
|---|---|---|---|---|---|
| | | | Millions of Dollars | | |
| Lightning | 43 | 236 | 25.70 | 0.00 | 25.70 |
| Tornado | 54 | 1,087 | 1,265.60 | 15.90 | 1,281.50 |
| Thunderstorm Wind | 19 | 226 | 355.80 | 66.50 | 422.30 |
| Hail | 0 | 121 | 522.50 | 110.00 | 632.50 |
| Extreme Cold | 20 | 14 | 0.90 | 28.20 | 29.10 |
| Extreme Heat | 36 | 174 | 0.00 | 0.00 | 0.00 |
| Flash Flood | 67 | 61 | 2,123.90 | 18.30 | 2,142.20 |
| River Flood | 18 | 5 | 418.70 | 139.80 | 558.50 |
| Coastal Storm | 14 | 6 | 11.40 | 0.00 | 11.40 |
| Tsunami | 0 | 0 | 0.00 | 0.00 | 0.00 |
| Rip Current | 41 | 47 | 0.00 | 0.00 | 0.00 |
| Tropical Storm/Hurricane | 14 | 233 | 1,879.50 | 40.80 | 1,920.30 |
| Winter Storm | 28 | 112 | 499.40 | 8.60 | 508.00 |
| Ice | 0 | 0 | 0.00 | 0.00 | 0.00 |
| Avalanche | 9 | 7 | 0.00 | 0.00 | 0.00 |
| Drought | 0 | 0 | 645.20 | 572.50 | 1,217.70 |
| Dust Storm | 2 | 91 | 0.30 | 0.00 | 0.30 |
| Dust Devil | 1 | 3 | 0.10 | 0.00 | 0.10 |
| Rain | 5 | 34 | 8.40 | 69.80 | 78.20 |
| Fog | 0 | 0 | 0.00 | 0.00 | 0.00 |
| High Wind | 24 | 156 | 166.20 | 41.00 | 207.20 |
| Waterspout | 0 | 0 | 0.00 | 0.00 | 0.00 |
| Fire Weather | 0 | 2 | 2,331.60 | 0.00 | 2,331.60 |
| Mud Slide | 15 | 16 | 5.60 | 0.00 | 5.60 |
| Volcanic Ash | 0 | 0 | 0.00 | 0.00 | 0.00 |
| Miscellaneous | 27 | 289 | 7.80 | 31.20 | 39.00 |
| **Total** | **438** | **2,924** | **10,269.10** | **1,142.60** | **11,411.70** |

*Source:* NOAA.

**Table 3.34  Weather-related Fatalities, Injuries, and Costs for the United States, 2004. Tables for Individual Years and a 10-Year Total Table**

| 2004 Weather Event | Deaths | Injuries | Damage Property | Damage Crop | Damage Total |
|---|---|---|---|---|---|
| | | | Millions of Dollars | | |
| Lightning | 32 | 280 | 26.10 | 0.00 | 26.10 |
| Tornado | 35 | 396 | 537.10 | 12.10 | 549.20 |
| Thunderstorm Wind | 16 | 244 | 167.30 | 15.60 | 182.90 |
| Hail | 0 | 4 | 463.00 | 73.00 | 536.00 |
| Extreme Cold | 27 | 2 | 0.70 | 0.00 | 0.70 |
| Extreme Heat | 6 | 74 | 0.00 | 0.00 | 0.00 |
| Flash Flood | 58 | 29 | 856.70 | 12.20 | 868.90 |
| River Flood | 24 | 99 | 839.50 | 329.20 | 1,168.70 |
| Coastal Storm | 30 | 29 | 0.50 | 0.00 | 0.50 |
| Tsunami | 0 | 0 | 0.00 | 0.00 | 0.00 |
| Rip Current | 32 | 23 | 0.00 | 0.00 | 0.00 |
| Tropical Storm/Hurricane | 34 | 840 | 18,901.80 | 667.30 | 19,569.10 |
| Winter Storm | 28 | 190 | 183.50 | 0.20 | 183.70 |
| Ice | 0 | 0 | 0.00 | 0.00 | 0.00 |
| Avalanche | 12 | 6 | 0.10 | 0.00 | 0.10 |
| Drought | 0 | 0 | 0.00 | 1.20 | 1.20 |
| Dust Storm | 0 | 11 | 0.10 | 0.00 | 0.10 |
| Dust Devil | 0 | 5 | 0.00 | 0.00 | 0.00 |
| Rain | 4 | 9 | 3.60 | 0.50 | 4.10 |
| Fog | 1 | 45 | 1.60 | 0.00 | 1.60 |
| High Wind | 26 | 68 | 3,312.80 | 340.50 | 3,653.30 |
| Waterspout | 0 | 0 | 0.00 | 0.00 | 0.00 |
| Fire Weather | 0 | 0 | 16.90 | 0.00 | 16.90 |
| Mud Slide | 4 | 11 | 32.30 | 0.00 | 32.30 |
| Volcanic Ash | 0 | 0 | 0.00 | 0.00 | 0.00 |
| Miscellaneous | 0 | 63 | 2.80 | 0.30 | 3.10 |
| Total | 369 | 2,428 | 25,346.40 | 1,452.10 | 26,798.50 |

*Source:* NOAA.

**Table 3.35   Weather-related Fatalities, Injuries, and Costs for the United States, 2005. Tables for Individual Years and a 10-Year Total Table**

| 2005 Weather Event | Deaths | Injuries | Damage Property | Damage Crop | Damage Total |
|---|---|---|---|---|---|
| | | | Millions of Dollars | | |
| Lightning | 38 | 309 | 52.40 | 0.40 | 52.80 |
| Tornado | 38 | 537 | 421.80 | 82.10 | 503.90 |
| Thunderstorm Wind | 16 | 185 | 398.70 | 11.90 | 410.60 |
| Hail | 1 | 8 | 480.50 | 56.10 | 536.60 |
| Extreme Cold | 24 | 1 | 0.10 | 139.20 | 139.30 |
| Extreme Heat | 158 | 298 | 3.00 | 0.00 | 3.00 |
| Flash Flood | 28 | 22 | 293.20 | 35.30 | 328.50 |
| River Flood | 15 | 16 | 1,244.50 | 68.90 | 1,313.40 |
| Coastal Storm | 0 | 0 | 0.00 | 0.00 | 0.00 |
| Tsunami | 6 | 12 | 2.20 | 0.00 | 2.20 |
| Rip Current | 0 | 0 | 0.00 | 0.00 | 0.00 |
| Tropical Storm/Hurricane | 1,016 | 130 | 93,064.40 | 2,075.20 | 95,139.60 |
| Winter Storm | 34 | 72 | 293.80 | 0.10 | 293.90 |
| Ice | 0 | 0 | 0.00 | 0.00 | 0.00 |
| Avalanche | 11 | 16 | 0.20 | 0.00 | 0.20 |
| Drought | 0 | 0 | 77.40 | 1,311.10 | 1,388.50 |
| Dust Storm | 0 | 32 | 0.10 | 0.00 | 0.10 |
| Dust Devil | 0 | 0 | 0.00 | 0.00 | 0.00 |
| Rain | 8 | 39 | 264.20 | 228.60 | 492.80 |
| Fog | 0 | 0 | 0.00 | 0.00 | 0.00 |
| High Wind | 7 | 43 | 58.50 | 21.90 | 80.40 |
| Waterspout | 0 | 0 | 0.00 | 0.00 | 0.00 |
| Fire Weather | 0 | 0 | 40.70 | 0.00 | 40.70 |
| Mud Slide | 11 | 11 | 91.50 | 0.00 | 91.50 |
| Volcanic Ash | 0 | 0 | 0.00 | 0.00 | 0.00 |
| Miscellaneous | 5 | 94 | 1.60 | 0.20 | 1.80 |
| Total | 1,451 | 1,834 | 96,788.80 | 4,031.00 | 100,819.80 |

*Source:* NOAA.

Table 3.36  Weather-related Fatalities, Injuries, and Costs for the United States, 2006. Tables for Individual Years and a 10-Year Total Table

| 2006 Weather Event | Deaths | Injuries | Damage Property (Millions of Dollars) | Damage Crop (Millions of Dollars) | Damage Total (Millions of Dollars) |
|---|---|---|---|---|---|
| Lightning | 48 | 246 | 63.80 | 0.00 | 63.80 |
| Tornado | 67 | 990 | 752.30 | 6.70 | 759.00 |
| Thunderstorm Wind | 14 | 249 | 408.00 | 31.10 | 439.10 |
| Hail | 0 | 18 | 1,569.40 | 132.90 | 1,702.30 |
| Extreme Cold | 2 | 5 | 0.00 | 11.90 | 11.90 |
| Extreme Heat | 253 | 1,513 | 0.20 | 492.50 | 492.70 |
| Flash Flood | 59 | 18 | 2,136.60 | 104.90 | 2,241.50 |
| River Flood | 17 | 5 | 1,631.10 | 95.20 | 1,726.30 |
| Coastal Storm | 7 | 16 | 55.10 | 0.00 | 55.10 |
| Tsunami | 0 | 0 | 0.00 | 0.00 | 0.00 |
| Rip Current | 23 | 24 | 0.00 | 0.00 | 0.00 |
| Tropical Storm/Hurricane | 0 | 1 | 2.40 | 43.30 | 45.70 |
| Winter Storm | 17 | 109 | 571.00 | 0.00 | 571.00 |
| Ice | 0 | 0 | 0.00 | 0.00 | 0.00 |
| Avalanche | 11 | 6 | 0.00 | 0.00 | 0.00 |
| Drought | 0 | 4 | 138.00 | 2,498.10 | 2,636.10 |
| Dust Storm | 2 | 22 | 0.70 | 2.30 | 3.00 |
| Dust Devil | 0 | 1 | 0.00 | 0.00 | 0.00 |
| Rain | 1 | 4 | 52.40 | 49.70 | 102.10 |
| Fog | 0 | 0 | 0.00 | 0.00 | 0.00 |
| High Wind | 26 | 133 | 195.00 | 15.20 | 210.20 |
| Waterspout | 0 | 0 | 0.00 | 0.00 | 0.00 |
| Fire Weather | 0 | 0 | 192.40 | 0.00 | 192.40 |
| Mud Slide | 3 | 2 | 58.90 | 20.00 | 78.90 |
| Volcanic Ash | 0 | 0 | 0.00 | 0.00 | 0.00 |
| Miscellaneous | 17 | 126 | 157.30 | 232.70 | 390.00 |
| Total | 567 | 3,492 | 7,984.60 | 3,736,5 | 11,721.10 |

*Source:* NOAA.

**Table 3.37   Weather-related Fatalities, Injuries, and Costs for the United States, 2007. Tables for Individual Years and a 10-Year Total Table**

| 2007 Weather Event | Deaths | Injuries | Damage Property | Damage Crop | Damage Total |
|---|---|---|---|---|---|
| | | | Millions of Dollars | | |
| Lightning | 45 | 138 | 82.06 | 0.06 | 82.12 |
| Tornado | 81 | 659 | 1,400.76 | 6.77 | 1,407.52 |
| Thunderstorm Wind | 18 | 256 | 238.17 | 191.42 | 429.60 |
| Hail | 0 | 25 | 558.76 | 154.67 | 713.42 |
| Extreme Cold | 47 | 174 | 3.04 | 2,386.59 | 2,389.64 |
| Extreme Heat | 105 | 1,886 | 0.00 | 0.00 | 0.00 |
| Flash Flood | 70 | 51 | 1,227.77 | 49.89 | 1,277.66 |
| River Flood | 17 | 8 | 512.37 | 518.41 | 1,030.78 |
| Coastal Storm | 0 | 6 | 93.37 | 0.00 | 93.37 |
| Tsunami | 0 | 0 | 0.00 | 0.00 | 0.00 |
| Rip Current | 57 | 16 | 7.57 | 0.00 | 7.57 |
| Tropical Storm/Hurricane | 1 | 13 | 38.80 | 0.01 | 38.80 |
| Winter Storm | 9 | 159 | 101.00 | 0.23 | 101.23 |
| Ice | 7 | 11 | 1,379.81 | 0.00 | 1,379.81 |
| Avalanche | 18 | 12 | 0.00 | 0.00 | 0.00 |
| Drought | 0 | 0 | 3.43 | 1,528.66 | 1,532.08 |
| Dust Storm | 0 | 4 | 0.95 | 0.00 | 0.95 |
| Dust Devil | 0 | 0 | 0.00 | 0.00 | 0.00 |
| Rain | 2 | 8 | 78.57 | 2.05 | 80.62 |
| Fog | 0 | 3 | 0.45 | 0.00 | 0.45 |
| High Wind | 16 | 76 | 257.42 | 1.22 | 258.64 |
| Waterspout | 0 | 0 | 0.07 | 0.00 | 0.07 |
| Fire Weather | 19 | 225 | 1,370.19 | 43.61 | 1,413.80 |
| Mud Slide | 3 | 3 | 51.15 | 0.00 | 51.15 |
| Volcanic Ash | 0 | 0 | 0.00 | 0.00 | 0.00 |
| Miscellaneous | 0 | 0 | 0.00 | 0.00 | 0.00 |
| Total | 515 | 3,733 | 7,405.70 | 4,883.58 | 12,289.28 |

*Source:* NOAA.

Table 3.38    Weather-related Fatalities, Injuries, and Costs for the United States, 2008. Tables for Individual Years and a 10-Year Total Table

| 2008 Weather Event | Deaths | Injuries | Damage Property | Damage Crop | Damage Total |
|---|---|---|---|---|---|
| | | | Millions of Dollars | | |
| Lightning | 27 | 216 | 60.11 | 0.10 | 60.21 |
| Tornado | 126 | 1,714 | 1,843.80 | 20.00 | 1,863.80 |
| Thunderstorm Wind | 28 | 271 | 1,261.50 | 29.66 | 1,291.16 |
| Hail | 1 | 13 | 464.34 | 173.61 | 637.95 |
| Extreme Cold | 44 | 0 | 0.04 | 140.99 | 141.03 |
| Extreme Heat | 71 | 217 | 0.05 | 0.50 | 0.55 |
| Flash Flood | 58 | 30 | 1,289.13 | 912.95 | 2,202.08 |
| River Flood | 24 | 16 | 2,116.59 | 1,264.95 | 3,381.54 |
| Coastal Storm | 15 | 0 | 9,926.62 | 0.85 | 9,927.47 |
| Tsunami | 0 | 0 | 0.00 | 0.00 | 0.00 |
| Rip Current | 67 | 79 | 10.63 | 0.00 | 10.63 |
| Tropical Storm/Hurricane | 12 | 24 | 7,619.12 | 473.66 | 8,092.77 |
| Winter Storm | 21 | 121 | 931.89 | 19.70 | 951.59 |
| Ice | 0 | 0 | 104.14 | 0.00 | 104.14 |
| Avalanche | 26 | 15 | 1.13 | 0.00 | 1.13 |
| Drought | 0 | 0 | 0.10 | 1.59 | 1.69 |
| Dust Storm | 0 | 0 | 0.00 | 0.00 | 0.00 |
| Dust Devil | 1 | 14 | 0.06 | 0.00 | 0.06 |
| Rain | 2 | 16 | 7.15 | 4.50 | 11.65 |
| Fog | 0 | 0 | 6.64 | 0.00 | 6.64 |
| High Wind | 42 | 122 | 1,222.79 | 172.18 | 1,394.97 |
| Waterspout | 0 | 0 | 0.01 | 0.00 | 0.01 |
| Fire Weather | 3 | 35 | 236.59 | 2.06 | 238.65 |
| Mud Slide | 0 | 0 | 2.43 | 0.00 | 2.43 |
| Volcanic Ash | 0 | 0 | 0.00 | 0.00 | 0.00 |
| Miscellaneous | 0 | 0 | 0.18 | 0.00 | 0.18 |
| Total | 568 | 2,903 | 27,105.02 | 3,217.30 | 30,322.32 |

*Source:* NOAA.

**Table 3.39   Weather-related Fatalities, Injuries, and Costs for the United States, 2009. Tables for Individual Years and a 10-Year Total Table**

| 2009 Weather Event | Deaths | Injuries | Damage Property | Crop | Total |
|---|---|---|---|---|---|
| | | | Millions of Dollars | | |
| Lightning | 34 | 201 | 43.83 | 0.01 | 43.84 |
| Tornado | 21 | 351 | 566.37 | 18.48 | 584.85 |
| Thunderstorm Wind | 22 | 189 | 1,397.50 | 32.57 | 1,430.07 |
| Hail | 0 | 69 | 1,278.79 | 349.67 | 1,628.46 |
| Extreme Cold | 33 | 4 | 0.09 | 189.05 | 189.14 |
| Extreme Heat | 45 | 204 | 4.06 | 0.00 | 4.06 |
| Flash Flood | 32 | 17 | 438.50 | 9.05 | 447.55 |
| River Flood | 21 | 9 | 607.57 | 29.23 | 636.80 |
| Coastal Storm | 0 | 0 | 281.69 | 0.00 | 281.69 |
| Tsunami | 32 | 129 | 81.00 | 0.02 | 81.02 |
| Rip Current | 54 | 53 | 25.86 | 0.00 | 25.86 |
| Tropical Storm/Hurricane | 2 | 1 | 0.92 | 0.01 | 0.93 |
| Winter Storm | 21 | 394 | 339.56 | 0.50 | 340.06 |
| Ice | 7 | 5 | 1,170.96 | 0.00 | 1,170.96 |
| Avalanche | 7 | 3 | 1.00 | 0.00 | 1.00 |
| Drought | 0 | 0 | 0.05 | 49.58 | 49.63 |
| Dust Storm | 2 | 0 | 0.66 | 5.75 | 6.41 |
| Dust Devil | 0 | 3 | 0.23 | 0.00 | 0.23 |
| Rain | 6 | 9 | 5.89 | 0.16 | 6.05 |
| Fog | 0 | 0 | 2.76 | 0.00 | 2.76 |
| High Wind | 25 | 68 | 198.23 | 0.03 | 198.26 |
| Waterspout | 0 | 0 | 0.01 | 0.00 | 0.01 |
| Fire Weather | 2 | 109 | 110.82 | 1.38 | 112.20 |
| Mud Slide | 0 | 9 | 51.91 | 0.00 | 51.91 |
| Volcanic Ash | 0 | 0 | 0.00 | 0.00 | 0.00 |
| Miscellaneous | 0 | 0 | 0.06 | 0.00 | 0.06 |
| **Total** | **366** | **1,827** | **6,608.30** | **685.49** | **7,293.79** |

*Source:* NOAA.

Table 3.40   Weather-related Fatalities, Injuries, and Costs for the United States, 2000–2009. Tables for Individual Years and a 10-Year Total Table

| Totals 2000 - 2009 Weather Event | Deaths | Injuries | Damage Property | Damage Crop | Damage Total |
|---|---|---|---|---|---|
| | | | Millions of Dollars | | |
| Lightning | 413 | 2,617 | 480.9 | 2.67 | 483.57 |
| Tornado | 558 | 8,327 | 8,642.7 | 177.15 | 8,819.87 |
| Thunderstorm Wind | 192 | 413 | 5,079.8 | 553.25 | 5,633.03 |
| Hail | 4 | 472 | 8,476.8 | 1,599.05 | 10,075.83 |
| Extreme Cold | 238 | 200 | 5.8 | 3,107.33 | 3,112.91 |
| Extreme Heat | 1,165 | 5,658 | 7.3 | 493.00 | 500.31 |
| Flash Flood | 475 | 603 | 10,337.8 | 1,301.99 | 11,639.79 |
| River Flood | 157 | 171 | 8,522.6 | 3,090.99 | 11,613.62 |
| Coastal Storm | 159 | 184 | 10,387.1 | 0.85 | 10,387.93 |
| Tsunami | 38 | 141 | 83.2 | 0.02 | 83.22 |
| Rip Current | 317 | 302 | 44.2 | 0.00 | 44.16 |
| Tropical Storm/Hurricane | 1,154 | 1,596 | 127,807.2 | 3,581.08 | 131,388.30 |
| Winter Storm | 234 | 1,617 | 4,811.1 | 29.43 | 4,840.48 |
| Ice | 14 | 16 | 2,655.7 | 0.00 | 2,655.71 |
| Avalanche | 140 | 99 | 3.2 | 0.00 | 3.23 |
| Drought | 0 | 4 | 864.9 | 10,412.33 | 11,277.20 |
| Dust Storm | 9 | 239 | 3.6 | 8.05 | 11.66 |
| Dust Devil | 2 | 28 | 0.4 | 0.00 | 0.39 |
| Rain | 36 | 159 | 446.1 | 382.01 | 828.12 |
| Fog | 37 | 376 | 18.1 | 0.00 | 18.05 |
| High Wind | 234 | 1,055 | 5,606.5 | 628.93 | 6,235.47 |
| Waterspout | 2 | 1 | 0.2 | 0.00 | 0.19 |
| Fire Weather | 33 | 655 | 6,654.4 | 56.15 | 6,710.55 |
| Mud Slide | 36 | 52 | 295.6 | 20.00 | 315.59 |
| Volcanic Ash | 0 | 0 | 0.5 | 0.00 | 0.50 |
| Miscellaneous | 60 | 592 | 170.0 | 264.40 | 434.44 |
| Total | 5,754 | 27,745 | 201,412.1 | 21,972.07 | 227,120.79 |

*Source:* NOAA.

# 4

# Thunderstorms, Lightning, Hail, and Tornadoes

## • INTRODUCTION

From the dawn of civilization mankind has had a relationship with thunder and lightning based on fear, respect, and superstition. As long ago as 4000 B.C. carvings from Babylon depict a lightning god and the gods of many civilizations; such as the Norse god Thor held the power of lightning and thunder.

During the Middle Ages the Latin phrase *Fulgura Frango* ("I break up the lightning flashes") was cast into many church bells in the belief that ringing the bells would dispel thunderbolts. Hundreds of bell ringers died when the steeples in which they were ringing the church bells were struck by lightning.

Many churches burned, and a number of times, the gunpowder stored within obliterated large parts of the surrounding neighborhood until Benjamin Franklin invented the lightning rod that safely conducted the electric current to ground. One example of the value of Franklin's insight is St. Mark's Basilica in Venice, Italy. It was struck by lightning and damaged in 1388, 1417, 1489, 1548, 1565, 1653, 1745, 1761, and 1762. Since lightning rods were installed in 1766, it has been struck many times, but there has been no more damage.

Thunderstorms are not as mysterious today as they were back then, but there is still much to learn. Since the invention of radar, Doppler radar, and weather satellites, our state of knowledge has increased to the point that we can distinguish storms likely to cause tornadoes from storms containing only straight-line winds and days that may give birth to lines of thunderstorms from days that will see the parent storm of the tornado, the supercell scattered across the landscape.

## • THUNDERSTORM BASICS

The key to understanding thunderstorms is to know the degree of interaction between the storm and the environment. Particularly important are the buoyancy of the environment in which the thunderstorm forms and the change in wind direction and speed, or wind shear, upward from the surface in that environment.

Buoyancy refers to how much less dense a rising parcel of air, lifted in the thunderstorm updraft, is than the air surrounding the storm. Buoyancy results when surface air is heated during the day, and when water vapor the air contains condenses and releases, heat as the air is lifted.

Meteorologists use the term "instability" and measure it in a number of ways to determine the degree of buoyancy. When the atmosphere is unstable, the thunderstorm updraft will be strong and there is a better chance of heavy rain, damaging wind gusts and large hail.

Vertical wind shear refers to the change in both direction and velocity of the wind from the surface upward (Figure 4.1). Both speed shear and directional shear play a part in determining what kind of thunderstorm event develops. At one extreme, if both speed shear and directional shear are weak, isolated convective cells will form. If both shear values are strong supercell thunderstorms, the parent storms of tornadoes are likely.

When all conditions are right, a severe thunderstorm may form. In the United States to be severe, a thunderstorm must have winds of 50 knots (58 mph, 25.8 m/s) and/or hail of 1" (25.4 mm) in diameter. A thunderstorm that spawns a tornado is by definition severe. If none of these criteria are observed, a thunderstorm can be considered severe based on a damage survey conducted after the storm.

*The Weather Almanac: A Reference Guide to Weather, Climate, and Related Issues in the United States and Its Key Cities*, Twelfth Edition. Steve Horstmeyer.
© 2011 John Wiley & Sons, Inc. Published 2011 by John Wiley & Sons, Inc.

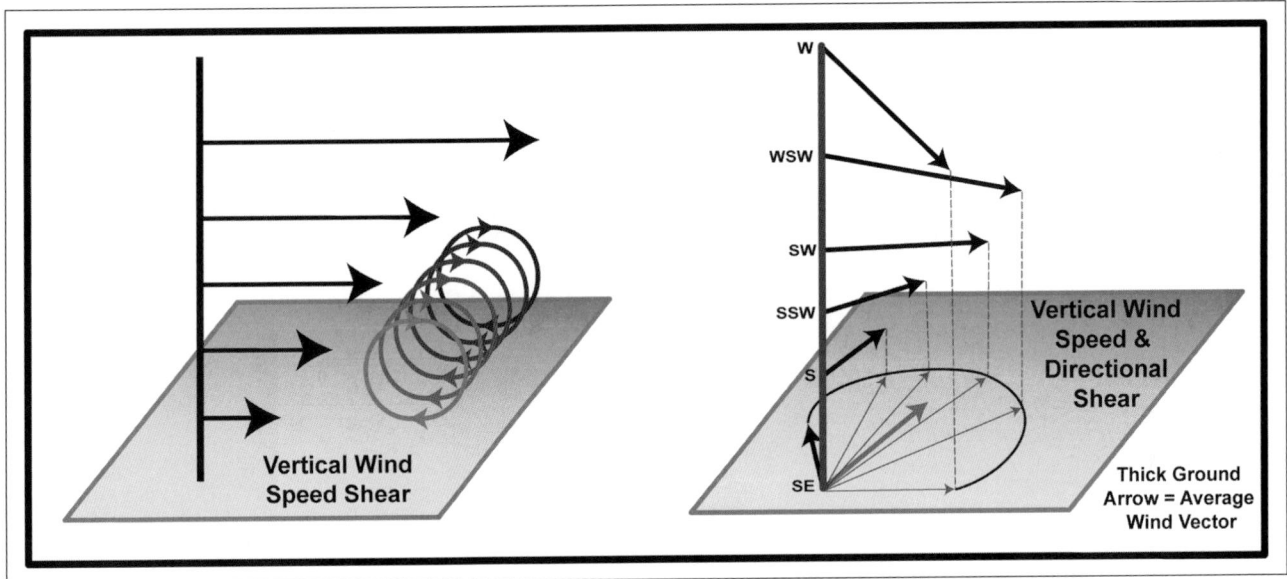

**Figure 4.1** Speed shear, change of wind velocity with increasing altitude, is illustrated on the left and the combination of wind speed change and wind direction change, called directional shear, is on the right. How a thunderstorm interacts with shear in the environment determines the type of storm that develops.

Even if a thunderstorm does not meet severe criteria, it can be very dangerous and a killer because all thunderstorms contain lightning. In the 10 years from 2000 to 2009, 413 people were killed by lightning the fifth deadliest weather phenomenon in the United States.

### • ISOLATED CONVECTIVE CELLS

Isolated convective cells are often called "air mass thunderstorms" because they are the typical thunderstorms that develop on warm, humid summer afternoons and are often powered solely by surface heating.

Thunderstorms at one time were thought to be either "frontal" storms, those that form along a front, or "air mass" storms, those that pop up seemingly randomly within an air mass. This concept is obsolete and no longer used as frequently as it was in the past as it fades from popular usage.

In fact there are many lifting mechanisms that can lead to thunderstorm formation if there is sufficient instability. Lifting mechanisms include upslope air flow, converging surface winds, and dynamic lifting by disturbances in the jet stream in addition to fronts and surface heating. Often several mechanisms work together.

Individual convective cells that form in an environment with weak shear, and sometimes just enough buoyancy for a small thunderstorm, will go through a life cycle as shown in Figure 4.2. As buoyancy increases even isolated convective cells can become severe. When wind speed shear is weak, individual convective cells often drift in the direction of the average tropospheric wind.

The cumulus stage (also called the towering cumulus stage) is dominated by updrafts and vertical cloud growth as the cumulus cloud grows into a towering cumulus. Late in this stage precipitation begins to fall, but there is no lightning and thunder.

In the mature stage, updrafts and downdrafts are present and so are lightning and thunder. The convective cell has grown into a cumulonimbus cloud that is also popularly called a thunderhead. If the buoyancy is great enough, a strong updraft core develops that can punch through the tropopause into the stratosphere. Overshooting tops can be observed on satellite images and are a sign of strong-to-severe thunderstorms.

Figure 4.3 shows a large thunderstorm photographed from the International Space Station off the coast of Chile. The anvil, flattened up against the tropopause, is clearly visible and extends farther downwind (right) of the thunderstorm core. The dome of the overshooting top extends above the anvil penetrating into the stratosphere.

As billions upon billions of raindrops fall through the thunderstorm, each pushes a small amount of air toward the surface. Some of the raindrops evaporate, cooling the surrounding air and increasing the density, thereby resulting in even greater downward motion. The combined effect creates the cool rush of air, called the downdraft, often felt flowing out of a thunderstorm just before rain arrives. The leading edge of the downdraft is called a gust front.

New thunderstorms can form along the gust front because it lifts the warm, moist air flowing into the storm. Thunderstorm outflow can travel hundreds of miles, and outflow boundaries left over from a previous day's thunderstorms can serve to generate thunderstorms the next day.

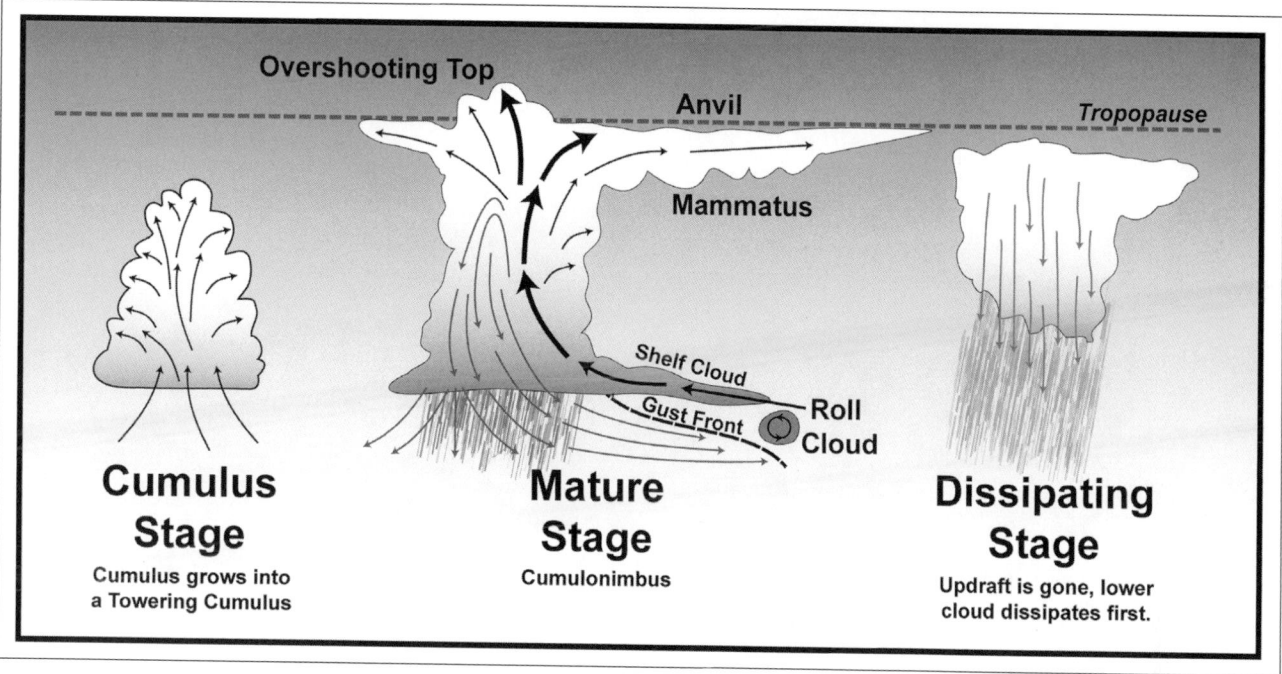

**Figure 4.2**    The life cycle of an isolated convective cell.

**Figure 4.3**    A large thunderstorm off the coast of Chile from the International Space Station. (*Source:* NASA.)

**Figure 4.4** An ominous shelf cloud on June 15, 2010, in Hebron, Kentucky. (*Source:* Stephanie West, used with permission.)

The interaction of the warm, moist inflow immediately above the cool outflow often causes a horizontal extension of the thunderstorm called a shelf cloud. A shelf cloud can have an ominous otherworldly look, dwarfing the landscape as it approaches, such as the spectacular shelf cloud shown in Figure 4.4 from Hebron, KY, on June 15, 2010. Take note of the size of the church in the lower right corner.

A tube of air forming a rolling arc around the leading edge of a thunderstorm, but detached from the main cloud, is an arc cloud. It forms in the same way as the shelf cloud and both are shown in Figure 4.2.

The dissipating stage of an isolated convective cell is dominated by downdrafts, ragged debris clouds, and heavy downpours that weaken as time goes on. Thunder and lightning may be present, but as the dissipating stage progresses, both end. The lowest part of the cloud dissipates first because moisture is no longer supplied because inflow has ended.

In certain cases if the updraft ceases abruptly, the downdraft accelerates and forms a downburst. If the damage area is 2.5 miles in diameter or less, the term "microburst" is used. Sometimes larger downbursts are called "macrobursts." Winds can exceed 150 mph and cause great damage. Microbursts are especially dangerous to aircraft because of the abrupt change of wind flow over the wing and potential loss of lift.

## • MESOSCALE CONVECTIVE SYSTEMS

A mesoscale convective system (MCS) is any collection of thunderstorms in a line or cluster that travel together and interact, both cooperating and competing for moist air during the lifetime of the system.

Buoyancy and wind shear interact in many complex ways to govern the evolution of MCSs. In addition, the MCS is a player in its own evolution; as it grows the environment is modified, which can favor further growth or serve to limit development.

## • SQUALL LINES

Linear MCSs are generally called squall lines, which typically form along or ahead of and nearly parallel to a cold front. Squall lines can form more than 100 miles in advance of a front, and when conditions are favorable, multiple squall lines can form (Figure 4.5).

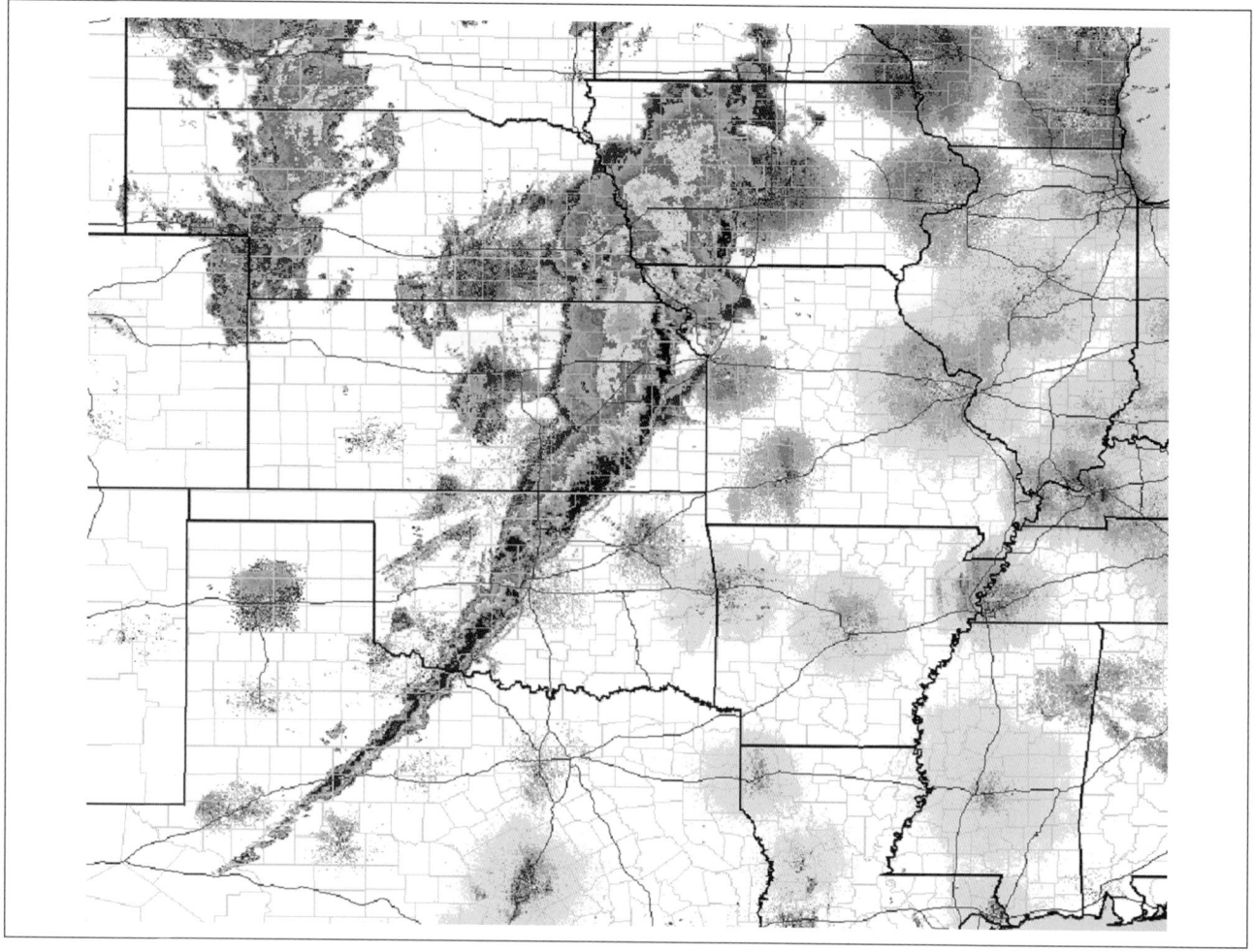

**Figure 4.5**   A squall line on June 5, 2008, as seen on a mosaic of NEXRAD Doppler Radar data. (*Source:* NOAA NCDC.)

Early in the twentieth century the term "squall line" was synonymous with a cold front, but in modern usage, it refers to a line of thunderstorms that is often strong to severe whether a front is present or not.

## • BOW ECHOES

A bow echo, or bowline, is a squall line in the form of an archer's bow when seen on a weather radar display. The center of the line bows out and is associated with damaging straight line winds. Small short-lived tornadoes are also possible with a bow echo. The term dates to 1978 and an investigation by Dr. Theodore Fujita who coined the term. Fujita is famous for his tornado damage or "F" scale.

The bow echo forms from a squall line when speed shear is moderate to strong, directional shear is weak, and the atmosphere is very unstable, that is, has large buoyancy values. The wind velocity above the surface is greatest where the line bows and this is where damaging straight lines often occur at the surface (Figure 4.6).

At either end of the bow echo are "book-end vortices" (also called line-end vortices). The northern vortex eventually dominates as the bow echo begins to dissipate leading to a "coma stage."

Large bow echoes that persist for several hours to a couple of days, are 280 miles or more in length, and travel great distances are called "derechoes." Derecho is a Spanish-language word meaning straight, which refers to the straight-line movement of the system. Derechoes can cause damage all along their length and for hundreds of miles in the direction of the travel.

## • LINE ECHO WAVE PATTERN

Like the bow echo, if the speed shear is strong and the directional shear is weak, a line echo wave pattern (LEWP, pronounced "loop") may form. A LEWP is composed of a bowed line of thunderstorms and an eastward extension called the warm advection band (Figure 4.7). Two or more bow echoes may also combine to form a LEWP.

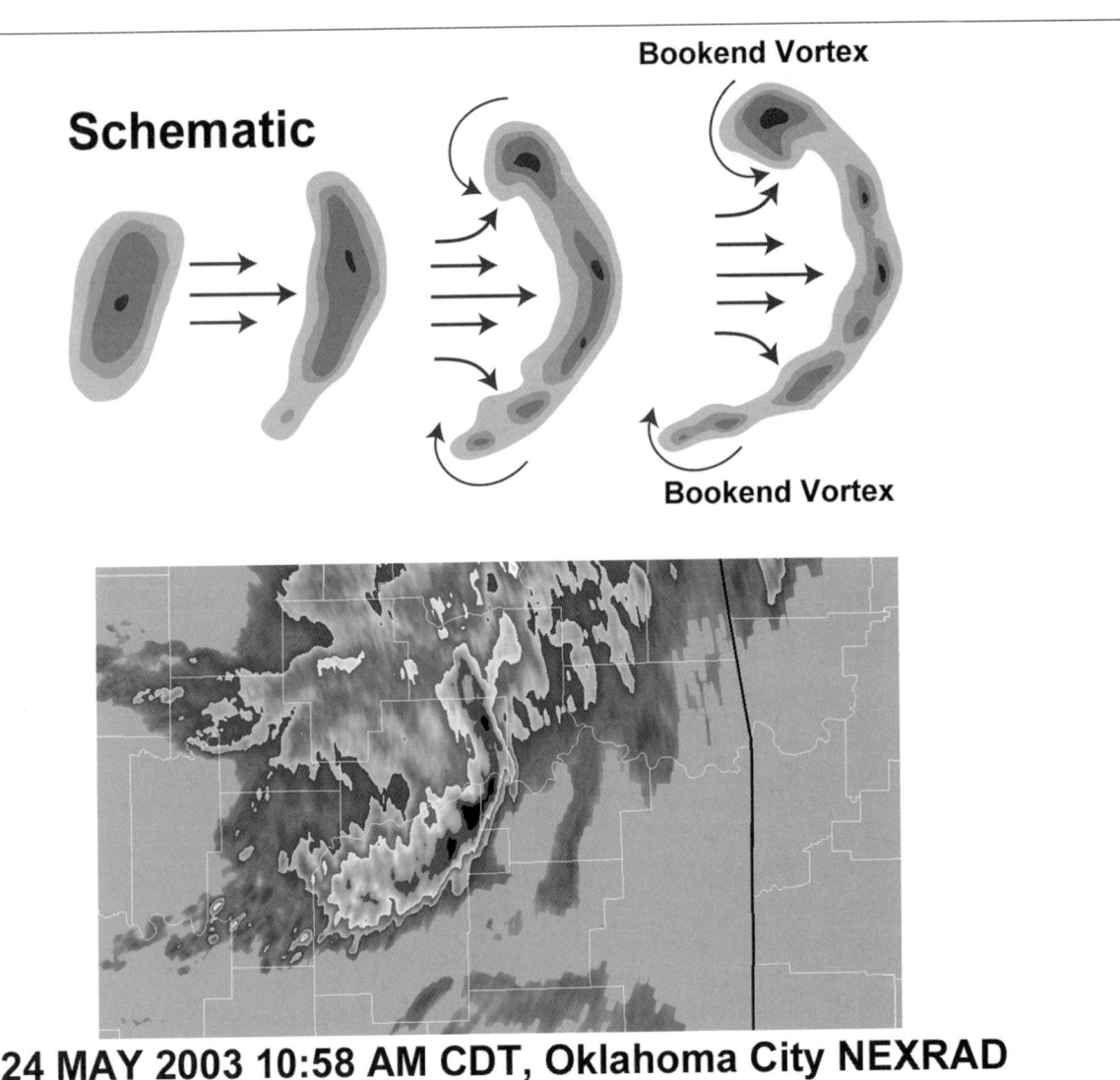

**Figure 4.6**  A bow echo, or bowline, that often forms with little directional wind shear and strong velocity shear. (Top) a schematic based on an original by Dr. T. Fujita. (Bottom) a bowline May 24, 2003, as seen on the Oklahoma City Nexrad Doppler Radar. (*Source:* Map by Steve Horstmeyer from NOAA NCDC data.)

A medium- to small-scale low-pressure system is centered at the apex and this may be the locus for supercells and tornadoes to form. Farther south straight-line winds and gust front vortices (also called gustnadoes) may cause damage.

## • THE MESOSCALE CONVECTIVE COMPLEX

When seen on a satellite image, a mesoscale convective complex (MCC) is a cluster of thunderstorms that can range from nearly circular to elliptical. MCCs typically move to the east, can travel many miles, and have a lifetime of 12 hours or more.

Figure 4.8 is an infrared satellite image of an MCC moving through southern Indiana on May 12, 2010, at 9:40 A.M.

EDT. Frequent, intense lightning occurred with this MCS along with large hail and flooding rains. Across northernmost Kentucky, nearly 5" of rain fell, leading to flash flooding on many small streams.

Nearly every MCC causes some severe weather in the form of one or all of the following: flash flooding, large hail, damaging winds, tornadoes, and frequent, intense lightning, and approximately one MCC in four causes a fatality.

Between 30% and 50% of MCCs begin as individual afternoon thunderstorms on the east slopes of the Rocky Mountains that move east over warm moist air in the Great Plains and then rapidly merge into a large mass of cooperating thunderstorms that travel eastward through the night and dissipate in the morning.

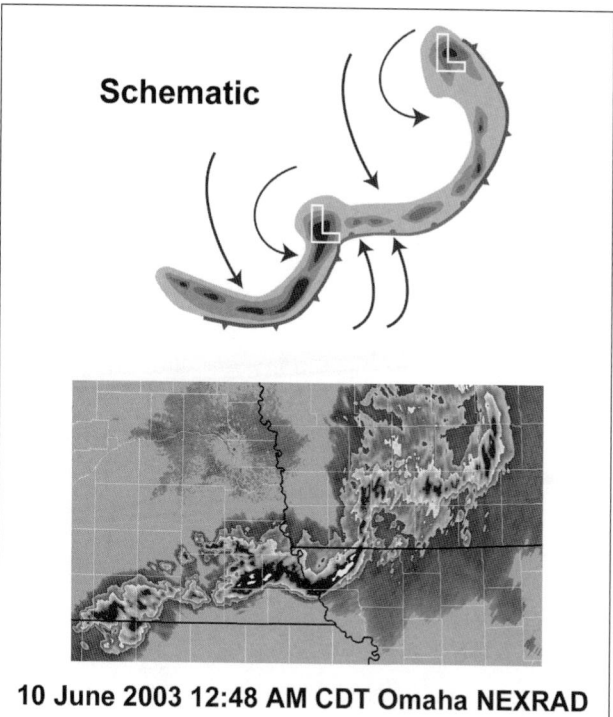

**Schematic**

**10 June 2003 12:48 AM CDT Omaha NEXRAD**

**Figure 4.7** A line echo wave pattern or (LEWP). (Top) schematic; (bottom) a LEWP June 10, 2003, from the Omaha, NE, NEXRAD Doppler Radar. (*Source:* Map by Steve Horstmeyer from NOAA NCDC data.)

Most of the remainder of MCCs form where large quantities of warm, moist air are being pushed northward by southerly winds. An ideal location for this is east of a surface low along a warm front. The MCC tracks eastward parallel to the front. When the MCC enters an environment that is not favorable to sustain thunderstorms, it begins to dissipate. Many MCCs take an abrupt turn to the south near the end of their lifetimes and then dissipate quickly.

### • SUPERCELL THUNDERSTORMS

A supercell is a thunderstorm with a persistent, deep rotating updraft called a "mesocyclone." If conditions are right, supercell storm may "spin up" and a tornado may touchdown. Figure 4.9 shows a typical supercell, characteristic features, and terminology used in discussing the storms.

Supercell thunderstorms can travel many miles and be active for hours. The average supercell has a lifetime of 1–2 hours, but some live 4 hours or more and in that time can travel a great distance. Less than 1% of thunderstorms in the United States are supercells, so they are fairly rare.

Both vertical wind directional shear and vertical wind speed shear (see Figure 4.1) are strong in an environment favorable to supercell formation.

The large-scale environment first sets the stage for rotating updrafts when the wind direction changes in a clockwise

manner and increases in speed with altitude (see Figure 4.1). Most thunderstorms will follow or travel just to the right of the average wind direction in the troposphere and push large amounts of air ahead of them in the direction of travel.

Wind speed shear causes the air being pushed ahead of the storm to form horizontal rotating tubes approximately at right angles to the direction of travel (see Figure 4.1).

When the low-level wind flowing toward the storm is strong, the air spirals toward the storm (Figure 4.10). As the spirals converge on the updraft, the rotational energy is tilted to nearly vertical as it is pulled into the storm's updraft core (Figure 4.11). In the updraft the spirals are stretched becoming longer and narrower, which causes the rate of rotation to increase.

Because a large supercell pulls air from a wide area, small amounts of rotation in individual spirals add up to enough rotation to explain the amount needed for a tornado.

As the supercell travels across the countryside the conversion of rotation around the horizontal axes of the spirals to rotation around the vertical axis of the updraft takes place continuously and can maintain the rotating updraft for several hours. The conversion of rotation from horizontal to vertical is a key to understanding the formation of long-track tornadoes.

The tilt of the supercell updraft is caused by vertical wind speed shear and is another key to the long life of a supercell. The tilt separates the updraft from the downdraft, so heavy rain falls ahead of the updraft (relative to storm motion) near the forward flank downdraft. Because of the tilted updraft in a supercell updraft energy is not being used to lift air against falling raindrops, or as one meteorologist put it, the rain cannot "snuff out the updraft."

A second source of supercell rotation is the pool of rain-cooled air beneath the storm. Here too rotating horizontal tubes of air form and are tilted to the vertical as they are drawn into the thunderstorm supplying rotation (Figure 4.12).

On a radar display the most distinctive feature of a well-developed, potentially tornadic supercell is the hook echo. The hook echo in Figure 4.13 was detected on December 23, 2009, at 4:39 P.M. CST by the Shreveport, LA NEXRAD, and was associated with an EF2 tornado.

Figure 3.7 is the hook echo of the EF5 Moore, OK tornado, one of the strongest tornadoes on record. The mesocyclone shows up as a hook where the cloud and precipitation wrap around the rotating updraft and where unsaturated air from outside the cloud flows into and spirals upward and around the updraft (Figure 4.14).

Doppler radar technology allows the meteorologist to measure the rate of rotation in the updraft, and trained National Weather Service meteorologists issue "Doppler radar indicated" tornado warnings, giving the public more time to take shelter.

Most tornado warnings are now issued before a tornado is ever sighted, which is based on the spin-up in a supercell

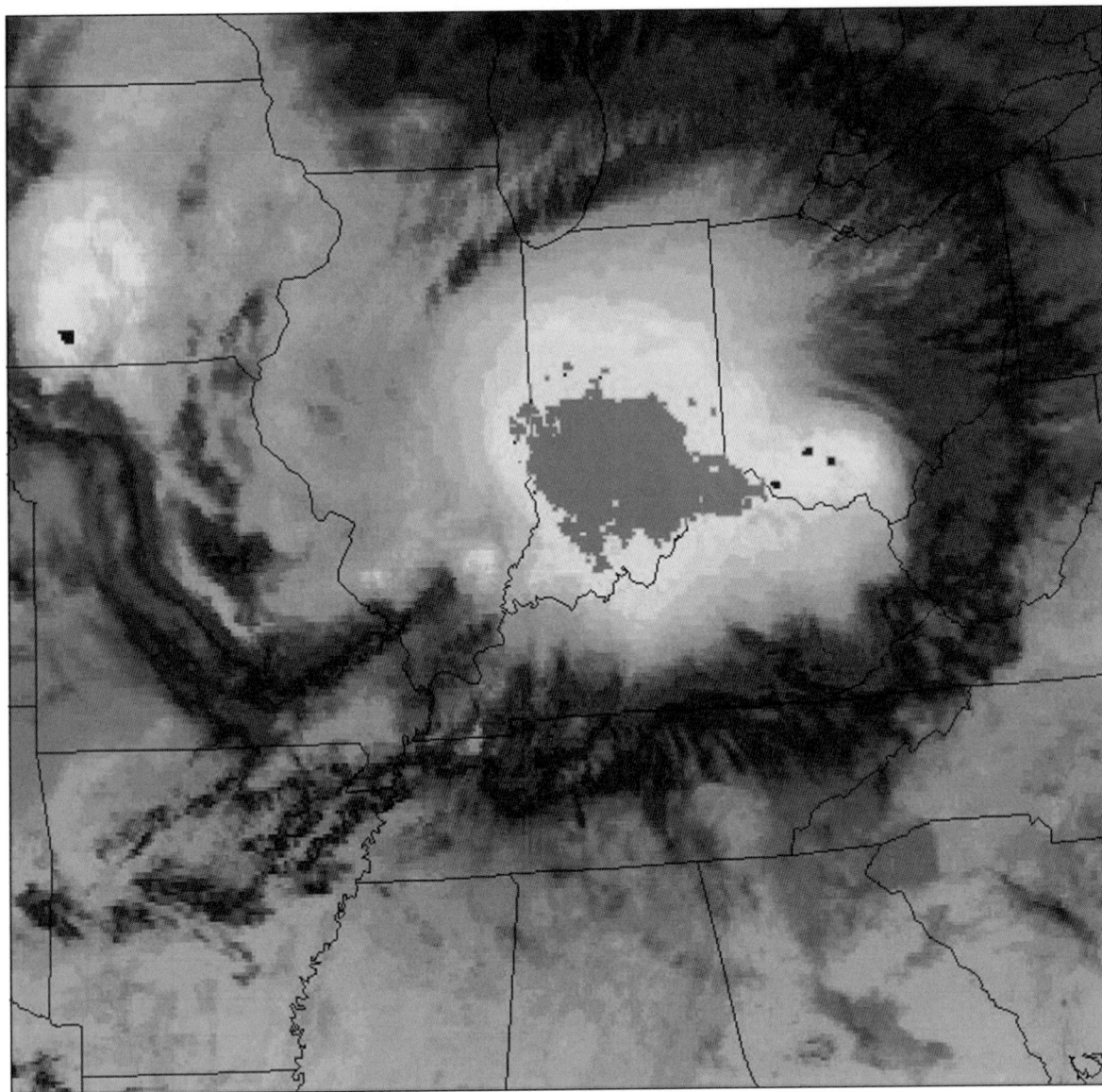

**Figure 4.8**  A mesoscale convective complex, MCC, seen by weather satellite on May 12, 2010. (*Source:* NOAA.)

measured by Doppler radar. The downside of this is that it leads to many false alarms, and nationwide, about 75% of tornado warnings issued by the National Weather Service are false alarms.

## • TORNADO FORMATION

Once the updraft of a thunderstorm begins to rotate, the rotation can be concentrated by stretching the inflowing spiraling winds in the updraft. Just like an ice skater finishing a performance, the rate of spin increases as the air is stretched and narrow. With enough spin-up a tornado will form. Only when a funnel touches the ground is it a tornado. A funnel that does not touch the ground is termed a "funnel aloft."

Most funnels are visible because of the condensation of water vapor. The condensation funnel is often mixed with dust and other debris to produce what an observer sees. Occasionally a tornado can only be detected by dust and debris lifted from the ground because there is insufficient moisture to condense into a funnel. Funnels that cannot be seen can be powerful and deadly.

At the other extreme are "rain-wrapped" tornadoes that form in supercells with large amounts of widespread torrential rain that hides the tornado. These too can be very dangerous because they can look just like heavy rain with little or no indication to the untrained eye of a tornado hidden with the rain shaft.

Not all tornadoes are spawned by mesocyclones. Less intense, small tornadoes that have a short lifetime, but can still

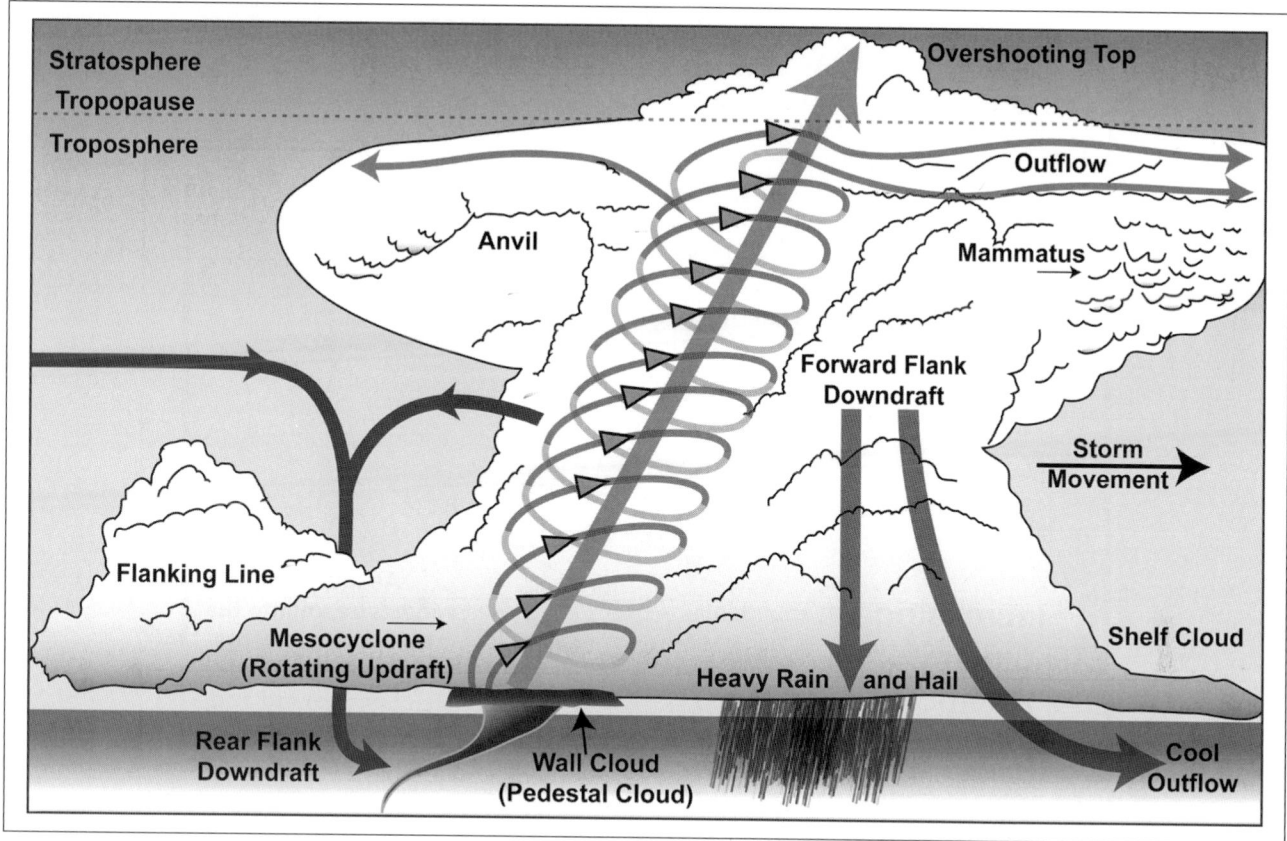

**Figure 4.9**   A tornadic supercell thunderstorm showing the relationships between the various parts of the storm. (*Source:* Steve Horstmeyer, cloud outline based on an original from NOAA, all other parts original.)

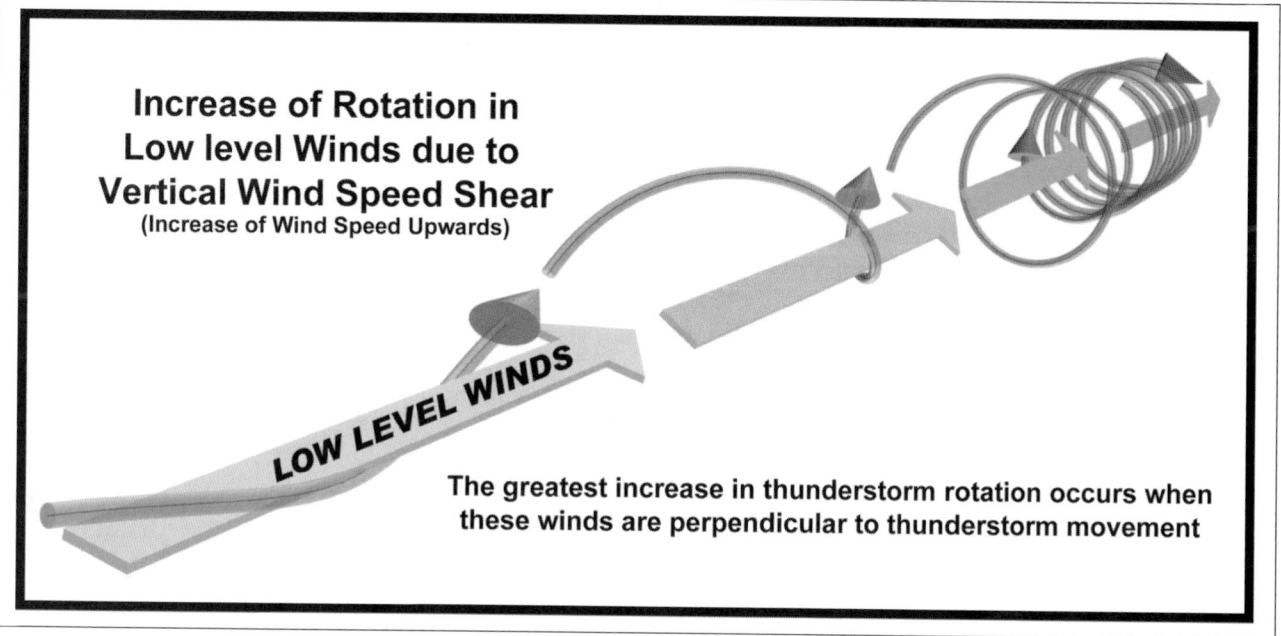

**Figure 4.10**   The increasing rotation of horizontal winds due to wind shear is indicated by the tightening spiral.

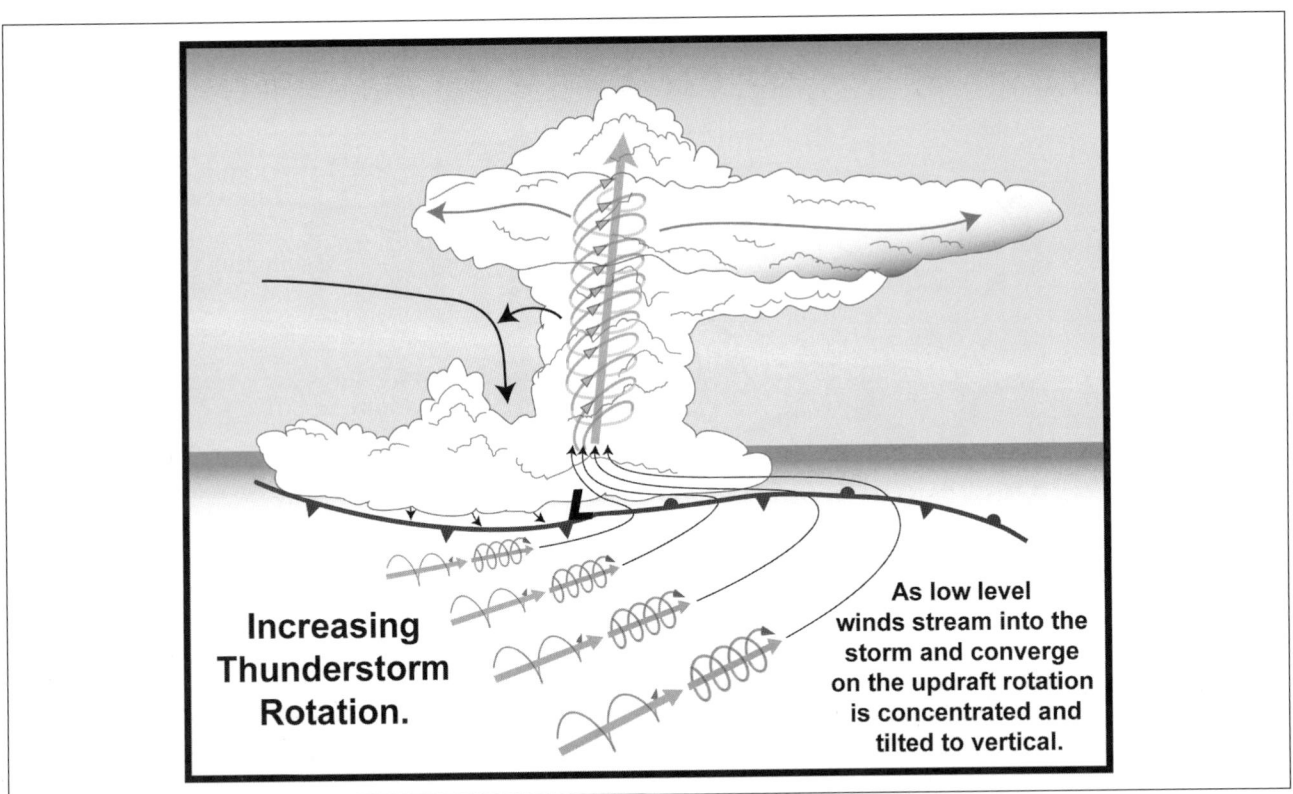

**Figure 4.11** Spiraling winds blowing toward a supercell thunderstorm, being pulled into the updraft, and contributing to the rotation of the updraft.

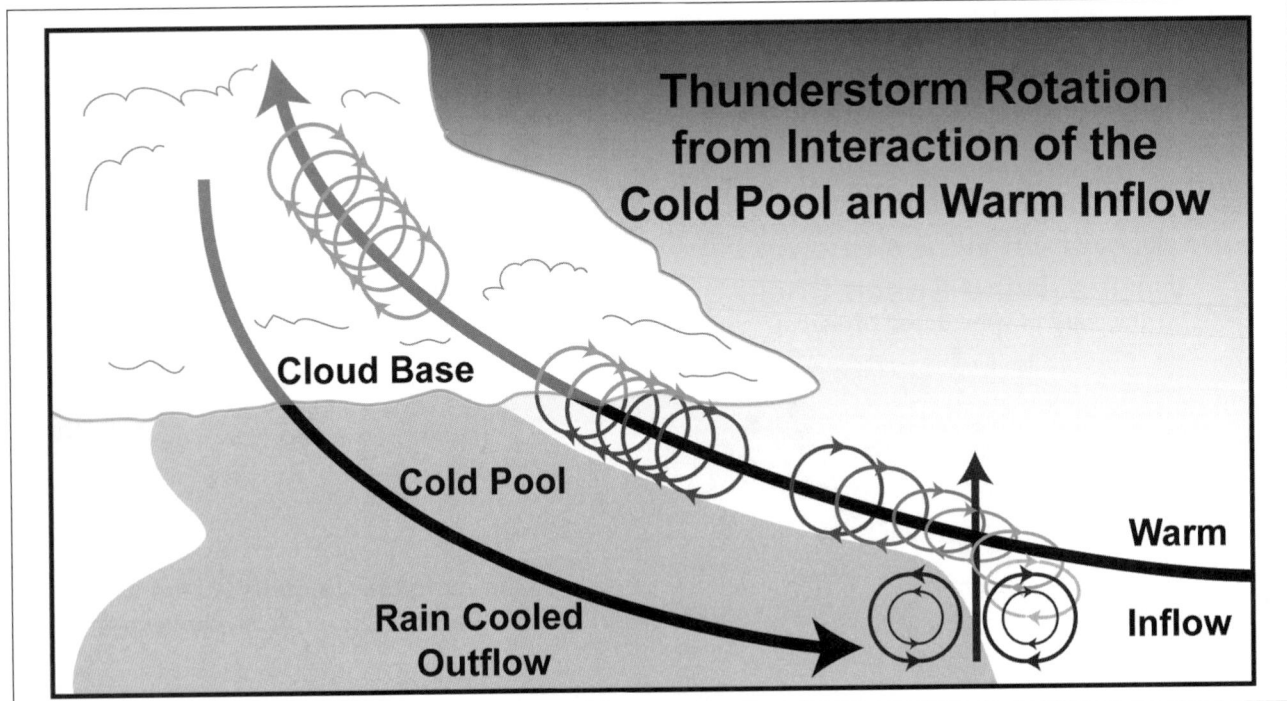

**Figure 4.12** Rotation caused by the interaction of the cool outflow with the warm air in the storm environment and then contributing to the updraft rotation as it is tilted and pulled into the updraft.

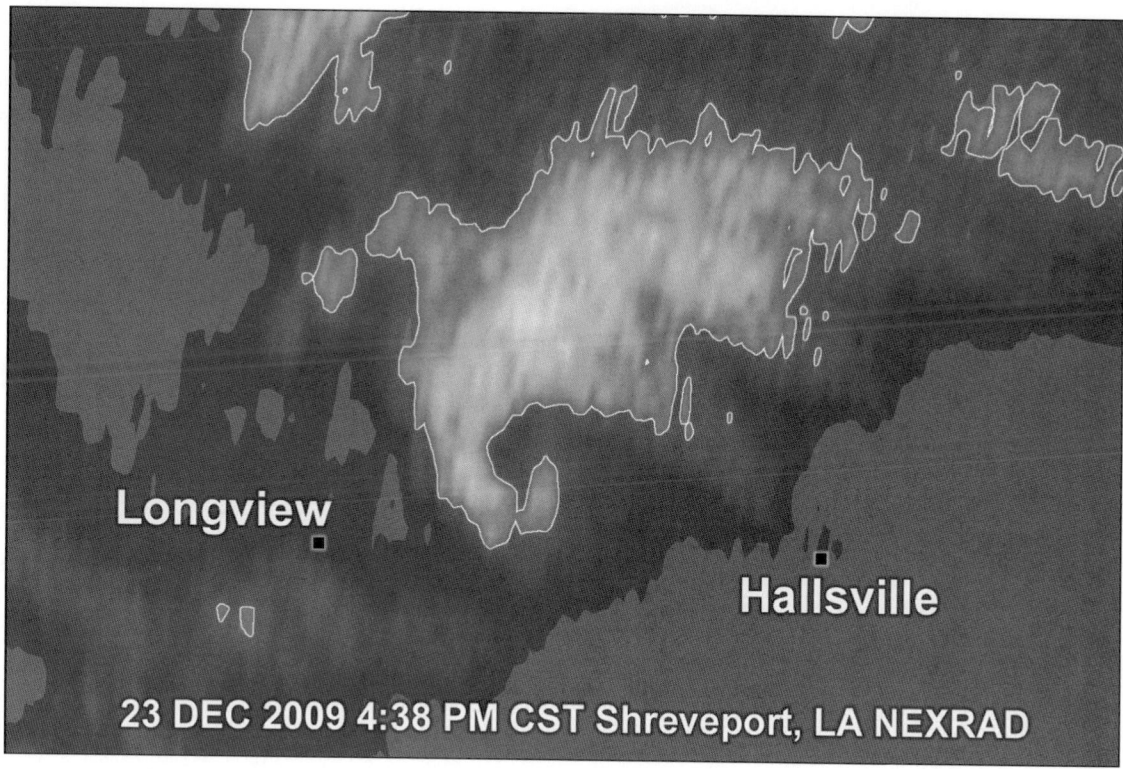

**Figure 4.13**  Hook echo of December 23, 2009, over eastern Texas on the Shreveport, LA, NEXRAD Doppler Radar. (*Source:* Map by Steve Horstmeyer from NOAA NCDC data.)

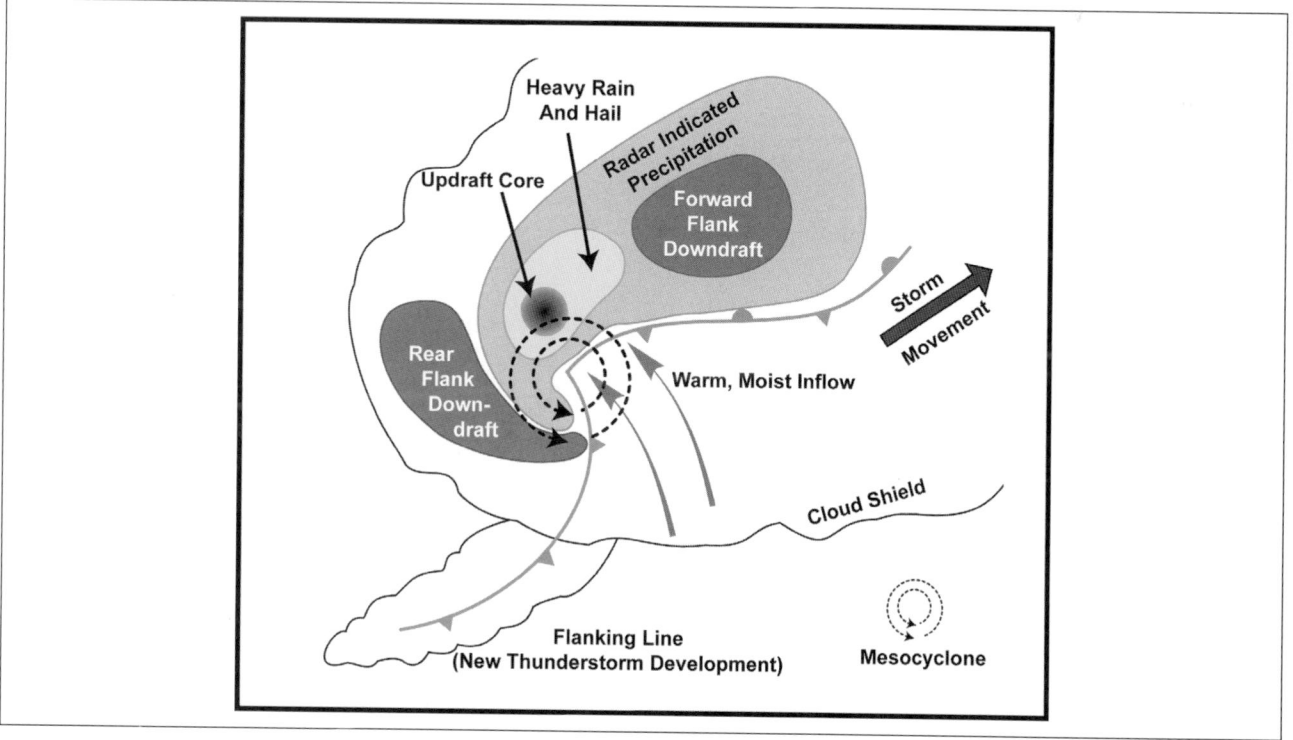

**Figure 4.14**  Schematic of a radar view of a supercell thunderstorm showing the various parts and locations of hail, heavy rain, the hook echo, and its relationship to the hook echo.

cause damage, sometimes go by the name "gustnado," short for gust front tornado.

Gustnadoes form as the cool air flowing out of the storm and warm, moist air just above flowing into the storm cause a nearly horizontal tube of air to roll along ahead of it in the same way a roll cloud is formed. If the tube is pulled toward the updraft and tilted to nearly vertical then stretched vertically a small tornado with a short lifetime may result. Because the updraft rotation was not deep and long term and occurred only when the rotating tube was tilted, a gustnado is not considered to be a mesocyclone tornado.

Even though it seems straightforward, there is no clear-cut boundary between a tornado formed in a mesocyclone and a vortex formed other ways in association with a thunderstorm such as a gustnado. Some meteorologists feel the term tornado should only refer to funnels that are caused by mesocyclones. In practice, any funnel that contacts the ground and is spawned by any thunderstorm is a tornado.

In addition to the rotation mechanisms discussed above, the rear flank downdraft (RFD) may enhance concentration of rotation near the surface. As it sinks to the ground at the rear of the storm, the RFD (Figure 4.9) effectively blocks air from escaping from beneath the mesocyclone and may even get wrapped around the tornado and be lifted into the storm. Researchers are actively investigating the role of the RFD in tornadogenesis.

## • THE FUJITA SCALE

The Fujita Scale or "F-Scale" was developed by Dr. Ted Fujita and first published in 1971. At that time little was known about wind damage to structures, so Fujita was in unknown scientific territory.

The scale was intended to classify tornado damage and damage that allowed estimates of tornado wind speeds. Fujita's initial work was based mostly on frame houses. In this way a tornado database could be developed to further understanding of these violent storms.

As a scientist Fujita wanted to bridge the gap in wind classification between the Beaufort Scale (see Table 2.1) and the Mach Number Scale used to describe velocity relative to the speed of sound for supersonic flight.

Fujita's original scale ranges from F0, which is equal to Force 11 on the Beaufort Scale and just below minimum hurricane velocity, to F12, which is equivalent to Mach 1 or the speed of sound in air at $-3°C$ (26.6°F), which is 738 mph (Figure 4.15). For classifying tornadoes, only F0 through F5 (Table 4.1) was used. But being a wise scientist and knowing the unexpected could occur, just in case Fujita described F6 as an "inconceivable tornado."

## • THE ENHANCED FUJITA SCALE

In his memoirs, *Mystery of Severe Storms* (1992), Fujita realizing the original scale needed to be refined published an update that took into account the type of construction in evaluating wind speeds.

Later research showed that the original Fujita Tornado Damage Scale overestimated wind speeds especially in the range from F3 to F5. To correct this research was conducted by meteorologists and wind and building engineers to update the scale using an approach similar to what Fujita proposed in 1992.

The original Fujita Scale was decommissioned in the United States on February 1, 2007, and superseded by the Enhanced Fujita (EF) Scale. The original F-Scale is still in use in Canada and France.

The EF-Scale was designed by specifically investigating the degree of damage on 23 types of building structures, two types of towers, various poles, and hard- or softwood trees. From this came the 28 damage indicators (DIs) provided in Table 4.2.

Because each of the 28 structure types or 2 tree types receives characteristic damage at given wind speeds, the enhanced scale is more accurate than the original scale, but it is still an estimate of wind speeds. The EF-Scale uses the 3-s maximum wind gust speed. A 3-s wind gust is the maximum wind speed sustained for a period of 3 s.

To determine the EF-Scale of a tornado investigators in the field first categorize the DI for the scene and then access the degree of damage (DOD). The third step is to apply the DOD table for the particular DI and estimate the 3-s wind speed for the location they are studying.

The number of degrees of damage vary from as many as 12 in the double-wide manufactured homes DI category to as few as 3 for the towers and poles DI category.

Table 4.3 shows the DOD scale for the one- and two-family residences of DI category, labeled FR12 in Table 4.2. The wind speed is listed as the expected wind with upper and lower bounds on the wind speed listed. Each of the 28 DI categories has a table like this.

If investigators determine a damage scene that consists of one- and two-family residences and the degree of damage is consistent with Category 6, the expected wind speed is 122 mph with a lower bound of 104 mph and an upper bound of 142 mph. The expected wind speed is then used to assign the tornado an EF2-Scale number as provided in Table 4.4.

## • SOME TORNADO HISTORY

Antarctica stands alone as the only continent where a tornado has never been observed. The undisputed world leader in tornado occurrence is North America, and the country that

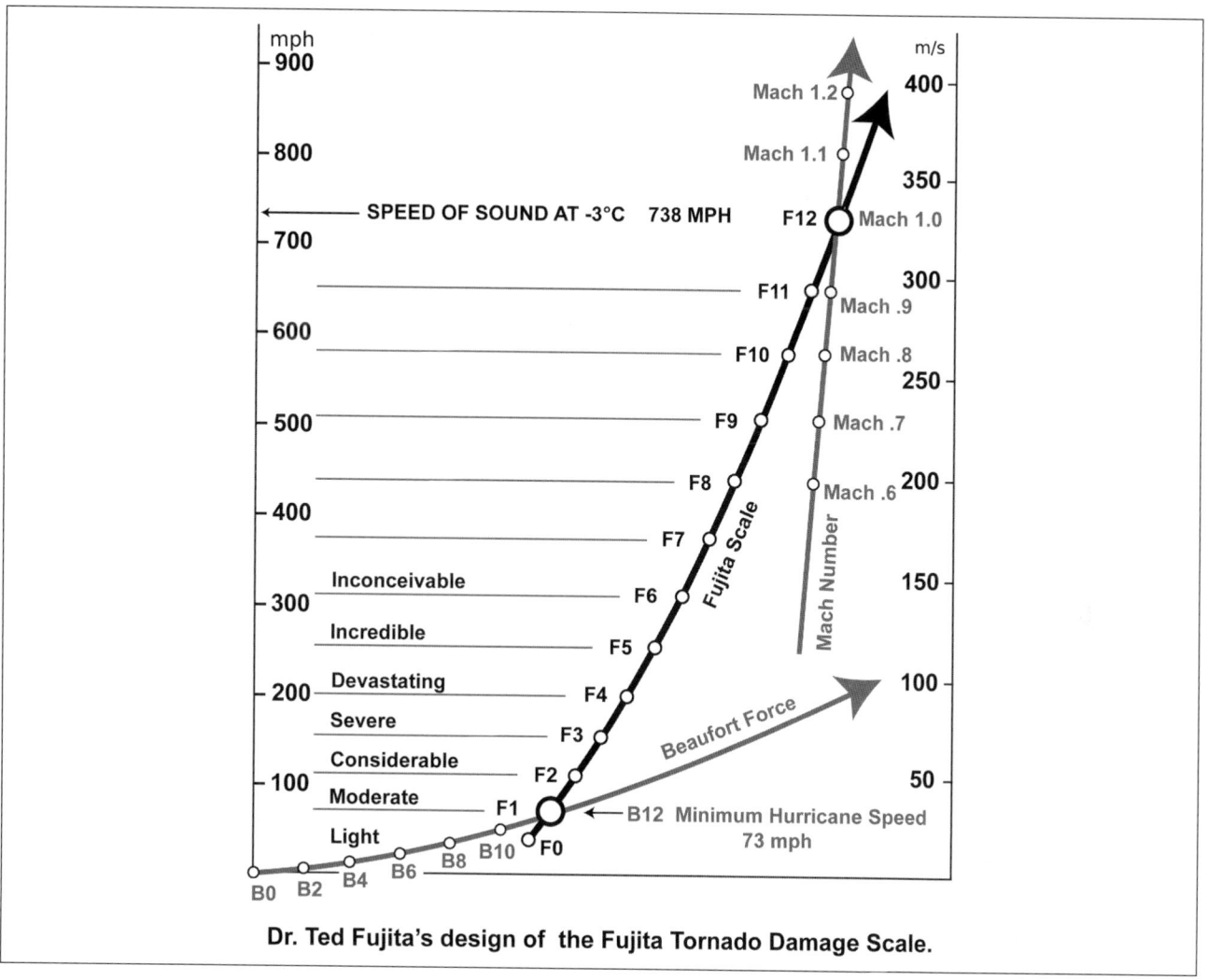

**Dr. Ted Fujita's design of the Fujita Tornado Damage Scale.**

**Figure 4.15** The original concept of the Fujita Tornado Damage Scale as conceptualized by Dr. Theodore Fujita. The F-Scale was designed to bridge the gap between winds commonly experienced in weather systems as categorized in the Beaufort Wind Scale and the Mach Number Scale developed for studying supersonic flight. Only wind speeds F0 through F5 were used to describe tornado damage. (*Source:* Redrawn and modified from the original by Dr. T. Fujita.)

leads all others is the United States, with an average of 1200 touchdowns per year.

That number however is unreliable because of changes in tornado reporting, increased interest by the public, larger population densities in rural areas, errors, and misclassifications. Even today there are rural areas where a tornado may go unreported. In areas like the high Great Plains without many structures and trees to yield evidence of high wind speeds and damaging wind gusts may not be reported.

Despite problems such as these in observing tornadoes and collecting data, we have learned much since the very first modern forecast of tornadoes on March 25, 1948, by Major Ernest J. Fawbush and Captain Robert C. Miller at Tinker Air Force Base in Oklahoma.

Tornadoes had been mentioned in even earlier forecasts. In fact on May 27, 1896, the forecast from the US Weather Bureau called for "destructive local storms." That day a killer tornado touched down in St. Louis (Figure 4.16). The death toll in metropolitan St. Louis from both sides of the Mississippi was 255. Twenty-seven others were killed in Illinois during the outbreak.

Even earlier than that, in 1886, the US Army Signal Corps, the official forecasting agency of the US Government at the time, banned the use of the word "tornado," because "the harm done by a [tornado] prediction would eventually be greater than that which results from the tornado itself."

The ban lasted until March 17, 1952, when the first tornado watch was issued by the newly created Severe Local Storm Warning Center in Norman, OK.

The earliest work on tornadoes was done by John P. Finley, who in 1882 wrote *The Character of Six Hundred Tornadoes*. In 1883 he had 120 "tornado reporters" and that

**Table 4.1   The Original Fujita Tornado Damage Scale**

| F Scale Number | Tornado Description | Wind Speed mph m/s | Typical Damage |
|---|---|---|---|
| F0 | Minor | Up to 72 Up to 32 | Minor or none |
| F1 | Moderate | 73 – 112 33 - 50 | Roof shingles peeled back |
| F2 | Significant | 113 – 157 51 - 70 | Roofs torn from large frame houses |
| F3 | Severe | 158 – 206 72 - 94 | Roofs and walls torn from well constructed buildings |
| F4 | Devastating | 207 – 260 95 - 118 | Well constructed houses leveled |
| F5 | Incredible | 261 – 318 119 - 143 | Strong frame houses lifted from foundation Ground swept clean of debris |
| F6 | Inconceivable | 319 – 379 145 - 172 | Unknown because of F4/5 damage there is little or no evidence. |

grew to 2403 in 1887. He used his research to describe tornado, producing weather patterns that could be used to forecast the storms.

Finley's detailed studies were the first effort to understand and forecast severe weather. Figure 4.17 is Finley's diagram of the tornado that struck Irving, KS, on May 29–30, 1882, and Figure 4.18 is a map he produced of tornado touchdowns from 1760 to 1885.

Figure 4.19 from August 28, 1884, near Howard, Dakota Territory, is widely reported to be the first photograph of a tornado. Finley reproduced it as a lithograph in his book. He also included a lithograph of another tornado photograph taken on April 26, 1884, near Garnett, KS (Figure 4.20). The photograph is in the possession of the Kansas State Historical Society and is the earliest known photograph of a tornado.

## • LIGHTNING AND THUNDER

### Introduction

By definition, there must be thunder to have a thunderstorm, and if there is thunder, there must be lightning because thunder is caused as lightning bolts superheat the air surrounding the bolt's channel to temperatures as high as 54,000°F. The superheated air expands explosively and the resulting shock wave travels outward and is heard as the sound of thunder. Because the shock wave can travel in a curved path because of conditions in the atmosphere, you may see lightning but not hear the thunder.

Temperature generally decreases with height above the surface and this causes the sound waves of thunder to bend upward. At 15 miles from the source, a clap of thunder that starts at 2.5 miles above the ground begins to curve upward. This means that thunder originating at an altitude lower that 2.5 miles, which is most thunder, will not be heard beyond 15 miles.

If the temperature increases with height, meteorologists call this an inversion thunder that may be heard much farther from the source.

In addition the sound of thunder can curve up or down because of the increase of wind speed with altitude. The direction of curvature changes as the direction of the wind relative to the sound wave changes. Considering both the temperature effect and the wind shear effect, it is obvious that very complex patterns to thunder audibility can develop.

The sound of thunder itself becomes lower in pitch as it travels through the air. Higher frequencies (i.e., higher pitch sounds) are more readily scattered by turbulence, absorbed by rain, and absorbed by air molecules, so the farther thunder travels, the more the higher pitch sounds are removed. In addition, the wavelength of the sound wave increases with travel time, which lowers the pitch. The result is that the farther away the listener is from a thunderstorm, the lower the volume and the lower the pitch of the sound.

When lightning strikes, the light reaches your eye so quickly that it is virtually instantaneous. Sound is slower, taking about 5 s to travel a mile. The distance to a thunderstorm can be estimated by counting the seconds between when you see the flash of lightning and when you hear the

**Table 4.2   The Enhanced Fujita Scale Damage Indicator (DI) Categories. The Investigator First Determines What Type of Construction Is Involved in the Damaged Area**

| Damage Indicator Number | Enhanced Fujita Scale Damage Indicator (DI) |
|---|---|
| 1 | Small Barns or Farm Outbuildings (SBO) |
| 2 | One- or Two-Family Residences (FR12) |
| 3 | Manufactured Home – Single Wide (MHSW) |
| 4 | Manufactured Home – Double Wide (MHDW) |
| 5 | Apartments, Condos, Townhouses [3 stories or less] (ACT) |
| 6 | Motel (M) |
| 7 | Masonry Apartment or Motel Building (MAM) |
| 8 | Small Retail Building [Fast Food Restaurants] (SRB) |
| 9 | Small Professional Building [Doctor's Office, Branch Banks] (SPB) |
| 10 | Strip Mall (SM) |
| 11 | Large Shopping Mall (LSM) |
| 12 | Large, Isolated Retail Building [K-Mart, Wal-Mart] (LIRB) |
| 13 | Automobile Showroom (ASR) |
| 14 | Automobile Service Building (ASB) |
| 15 | Elementary School [Single Story; Interior or Exterior Hallways] (ES) |
| 16 | Junior or Senior High School (JHSH) |
| 17 | Low-Rise Building [1-4 Stories] (LRB) |
| 18 | Mid-Rise Building [5-20 Stories] (MRB) |
| 19 | High-Rise Building [More than 20 Stories] (HRB) |
| 20 | Institutional Building [Hospital, Government or University Building] (IB) |
| 21 | Metal Building System (MBS) |
| 22 | Service Station Canopy (SSC) |
| 23 | Warehouse Building [Tilt-up Walls or Heavy-Timber Construction](WHB) |
| 24 | Transmission Line Towers (TLT) |
| 25 | Free-Standing Towers (FST) |
| 26 | Free-Standing Light Poles, Luminary Poles, Flag Poles (FSP) |
| 27 | Trees: Hardwood (TH) |
| 28 | Trees: Softwood (TS) |

*Source:* NOAA.

thunder. Divide the seconds by 5 and you know the approximate distance in miles to the storm.

Lightning is a powerful electrical discharge or spark very much like the shock you may receive when walking over a carpet on a winter day and reaching for a door knob. For both the tiny spark and the giant lightning bolt to occur, opposite electrical charges must accumulate across a gap. When the charge difference across the gap exceeds the insulation ability of the air, an electrical discharge can jump the gap.

Just how charge separation occurs in a thunderstorm is still poorly understood and there are many mechanisms proposed to explain just how this happens. The most accepted, but still incomplete, theory states that when liquid cloud drops and ice crystals collide or ice crystals and water drops break up in a cloud, negatively charged electrons reside on the larger particles that settle toward the base of the cloud. The leftover positive charges on the smaller particles are carried by the updrafts to accumulate high in the cloud.

## Types of Lightning and Lightning Names

Lightning discharges can occur from cloud to ground (CG), cloud to cloud (CC), cloud to air (CA), and within an individual thunderstorm or intracloud (IC). These are the names used to describe observed lightning flashes by operational meteorologists. Most lightning follows a jagged path and often forks on its way and is popularly called "forked

**Table 4.3 The Enhanced Fujita Scale Degree of Damage (DOD) Scale for One- and Two-Family Residences. There Is a DOD Scale for Each DI Category Listed in Table 4.2**

| Degree of Damage | Damage Description for One- and Two- Family Residences EF Scale Damage Indicators Category (FR12) | Expected Wind Speed | Lower Bound | Upper Bound |
|---|---|---|---|---|
| | | miles per hour | | |
| 1 | Threshold of visible damage | 65 | 53 | 80 |
| 2 | Loss of roof covering material (<20%), gutters and/or awning; loss of vinyl or metal siding | 79 | 63 | 97 |
| 3 | Broken glass in doors and windows | 96 | 79 | 114 |
| 4 | Uplift of roof deck and loss of significant roof covering material (>20%); collapse of chimney; garage doors collapse inward or outward; failure of porch or carport | 97 | 81 | 116 |
| 5 | Entire house shifts off foundation | 121 | 103 | 141 |
| 6 | Large sections of roof structure removed; most walls remain standing | 122 | 104 | 142 |
| 7 | Top floor exterior walls collapsed | 132 | 113 | 153 |
| 8 | Most interior walls of top story collapsed | 148 | 128 | 173 |
| 9 | Most walls collapsed in bottom floor, except small interior rooms | 152 | 127 | 178 |
| 10 | Total destruction of entire building | 170 | 142 | 198 |

*Source:* NOAA.

lightning." Other names given to lightning include heat lightning, sheet lightning, rocket lightning, ribbon lightning, bead lightning, and ball lightning.

The name heat lightning comes from lightning observed in the distance on warm summer nights. Because the thunderstorm is a distance too far away for thunder to be heard, it is popularly thought to be a specific type of lightning discharge. Heat lightning however is generally IC or CC lightning that illuminates high clouds and can be seen a great distance away.

Sheet lightning is also IC lightning that illuminates a large portion of a thunderstorm and thus appears to be a sheet of light across the sky.

Rocket lightning refers to nearly horizontal CA discharges that appear to move slowly outward from the cloud.

Figure 4.21 shows ribbon lightning striking the Empire State Building. It occurs when the wind speed is fast enough to move the lightning discharge channel downwind, and a series of lightning bolts appear to be in separate channels parallel to each other.

Bead lightning is not a type of lightning but a stage in the life of a lightning channel. As the channel cools, the luminosity decreases and breaks down into segments and the disconnected bright spots look like a string of beads. Nearly all lightning channels "bead out," but because the scale is small and lasts only a tiny fraction of a second, the beading is rarely observed.

**Table 4.4 The Enhanced Fujita Tornado Damage Scale; the EF Category Is Determined by Determining the DI Category and Then Applying the DOD Scale to Estimate the Wind Speed and Assign an EF Category**

| EF Scale Number | Damage Description | Wind Speed mph m/s |
|---|---|---|
| EF0 | Minor | 65 - 85 29 - 38 |
| EF1 | Moderate | 86 – 110 38 - 49 |
| EF2 | Considerable | 111 - 135 50 - 60 |
| EF3 | Severe | 136 - 165 61 - 73 |
| EF4 | Extreme | 166 - 200 74 - 89 |
| EF5 | Total Destruction | > 200 > 89 |

*Source:* NOAA.

**Figure 4.16** Damage from the 1896 St. Louis, MO, killer tornado. (*Source:* NOAA.)

Ball lightning is still a matter of dispute. Some doubt its existence entirely, but there are so many historical reports of ball lightning that most scientists accept it as a real phenomenon. Ball lightning is generally seen as an orange-to basketball-size sphere with a glow that is not intense. It has been observed both indoors and outdoors and generally moves slowly horizontally and lasts on average about 1 s. There is no consensus on how it forms, but ball lightning is associated with CG lightning strikes.

Other forms of electrical discharge occur from the tops of or well above thunderstorms. Sprites, elves, and blue jets are the names given to the three known types of electrical discharge and go by the collective name of transient luminous events.

Sprites occur high above a thunderstorm in the high atmosphere 50–90 miles up in the mesosphere and are associated with positive lightning discharges (see below). They are mostly seen as red-orange electrical flashes, but a sprite is not lightning because it is a cold electrical discharge and dim when compared to lightning flashes in the thunderstorm below. A sprite is about the brightness of a moderate aurora.

Blue jets originate in the top of the thunderstorm and shoot upward to an altitude of 25–30 miles through the stratosphere. Unlike sprites, blue jets do not appear to be closely associated with lightning below but are much more common during thunderstorms containing a great deal of hail. As research continues, blue jets appear to occur in many sizes from small "blue starters" to "gigantic jets."

Elves are often seen as a flat expanding disk with a red hue round 60 miles above Earth near the top of the mesosphere. Elves are associated with thunderstorms and probably owe their glow to collisions of energized electrons exciting nitrogen molecules.

### Negative Lightning and Positive Lightning

When lightning is generated by a thunderstorm it is either a negative or positive bolt. The name refers to the dominant electrical charge in the portion of the cloud where the discharge originates. About 80% of all CG lightning is negative, while the remaining 20% is positive.

The sequence of events that brings a negative CG stroke to the surface is given in Table 4.5.

Preliminary breakdown is the term used for the initial steps leading to the transport of negative charge to the ground in a lightning stroke. It is not well understood.

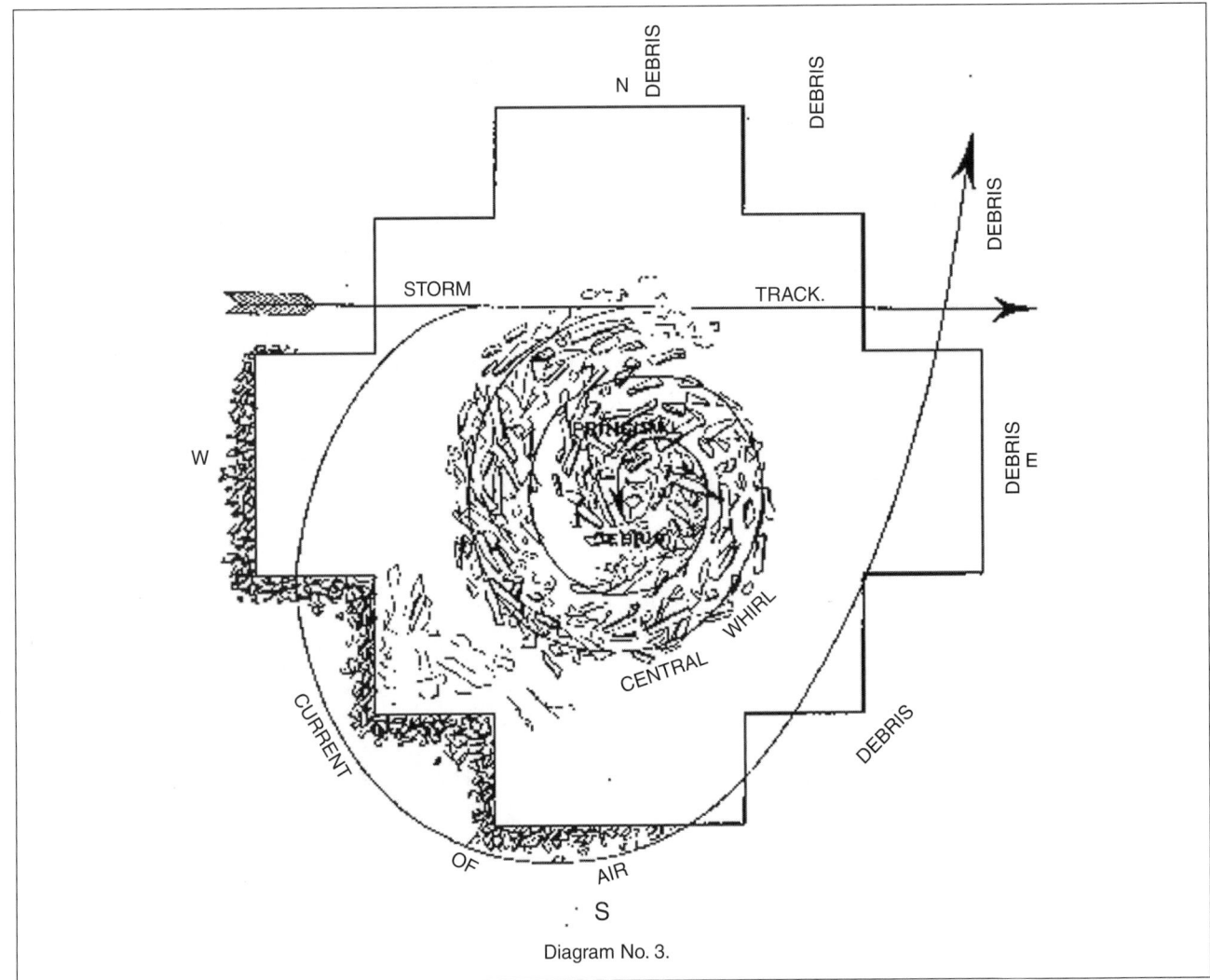

Diagram No. 3.

**Figure 4.17** Diagram of the Irving, KS, tornado May 29–30, 1882, by John P. Finley. (*Source:* NOAA.)

In 20 ms (ms, thousandths of a second) the stepped leader surges downward, establishing a jagged path toward the surface, and is met usually within 100 feet of the ground by the upward-moving ground discharge in what is called the attachment process.

This establishes a channel from the surface back to the cloud, which is followed by the first return stroke. Unlike the stepped leader, the return stroke is a continuous flow of current. It is the fast return stroke that rapidly heats the air surrounding the lightning channel. The rapid expansion of the heated air causes a shock wave that is heard as thunder. This is the end of a single-stroke flash.

If enough current is available, strokes called dart leaders can travel to the surface, each followed by additional return strokes and more thunder. A multiple-stroke flash can have many of these cycles.

Figure 4.22 shows the charge distribution in a typical thunderstorm. Negative charges collect near the base of the

thunderstorm, while higher in the cloud is a region of positive charge. If a lightning bolt originates near cloud base, it is a negative stroke; if a bolt originates from the positive region, it is a positive stroke.

Most lightning strokes are negative CG strokes between the cloud base and the positively charged "shadow" induced on the surface below the thunderstorm. CG, CA, and IC strokes are either positive or negative, but mostly negative.

A positive CG stroke must be much more powerful than a negative CG stroke because it has to jump a much larger gap between high in the thunderstorm and the negatively charged ground sometimes 10–15 miles away. This is also the reason positive CG strokes are less frequent than negative CG strikes.

The term "a bolt from the blue" refers to a sudden unexpected event and is derived from the occurrence of positive CG strikes. A positive strike surges from the side of a

**Figure 4.18**   John P. Finley's map of all US tornadoes he could find from 1760 to 1885. (*Source:* By NOAA.)

thunderstorm high above the surface, travels nearly horizontally, and then goes to ground and strikes often in a location where the blue sky is cloud free. "Bolts out of the blue" are deadly, and each year a few people are killed by lightning when standing under a clear blue sky.

## Lightning Safety

Lightning is very dangerous, and an average of 65 people per year are killed by lightning in the United States according to official statistics. There is no safe location when outdoors during a lightning event. In the center of a city with many high-rise office buildings, strikes that hit the ground are much less numerous than in residential and rural neighborhoods or in open country.

Trees are **never** safe shelter from lightning, and if caught outdoors, it is best to get away from trees and stoop down in the baseball catcher's position until danger passes. When lightning strikes a tree or other object, the charge radiates outward along the ground from the strike location and is eventually dissipated. The farther you are from the strike point and the less your body is in contact with the ground, the better your chances of survival.

Cars are better than being in the open, but the rubber tires do not insulate sufficiently to stop a lightning bolt from traveling to ground through a car. However the metal body may help direct the charge around you, but if you are in contact with or close to a conductor while in the car, you will likely be struck.

In your home, taking a shower or bath and doing dishes are not safe activities because metal plumbing readily conducts the electrical charge. The same holds for land-line telephones because the electrical charge can travel a long distance through telephone wires. Cordless and cell phones are safe to use.

Recent research supports the view that 80% of lightning fatalities occur during nonsevere storms perhaps because there is no specific warning procedure for lightning and lightning awareness is low, especially during routine convective events.

By careful analysis of all available data, including official lightning reports, newspaper accounts, and Centers for Disease Control data, Walker Ashley and Christopher Gilson at Northern Illinois University found that official statistics seriously underestimate the number of lightning fatalities in the

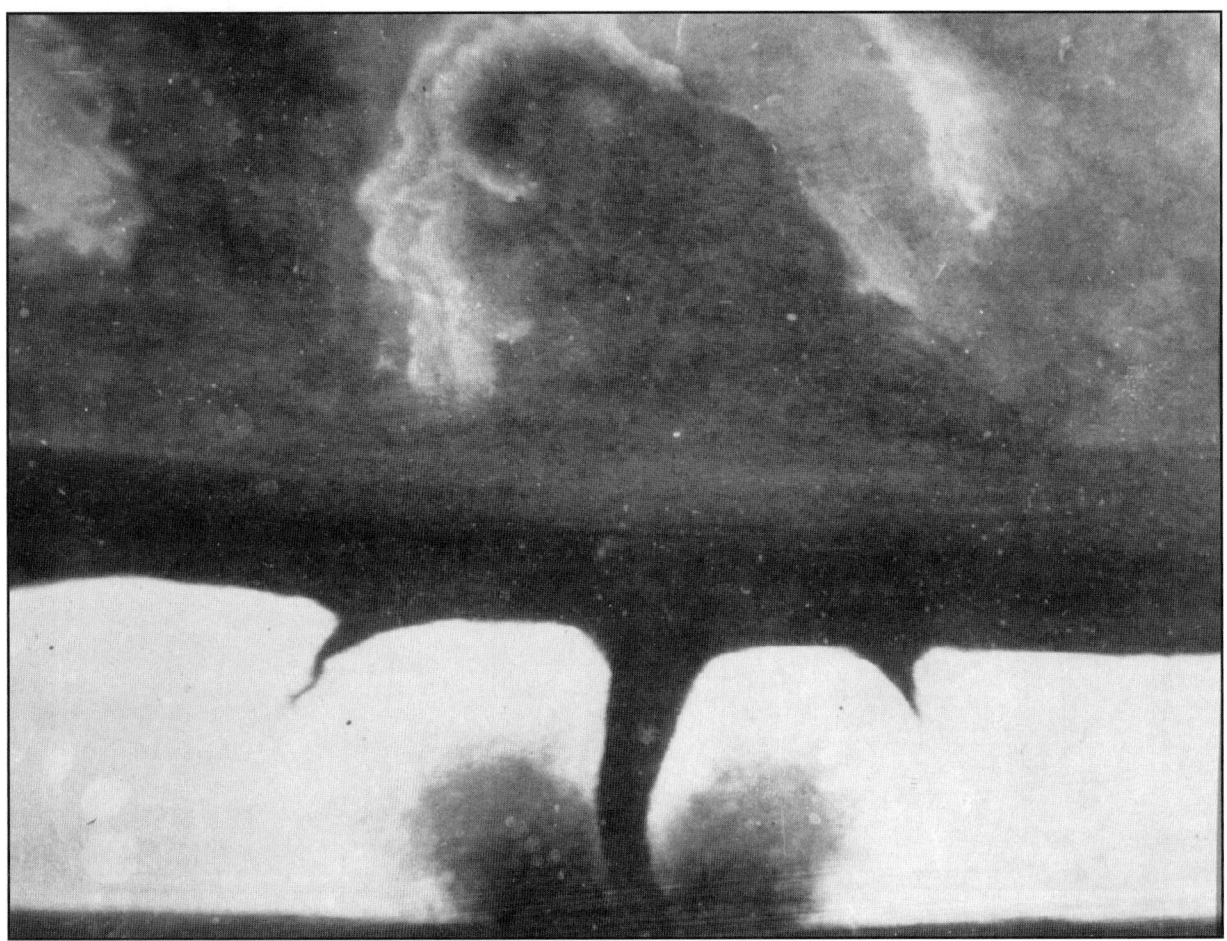

**Figure 4.19**　What is probably the second photograph taken of a tornado, August 28, 1884, near Howard, Dakota Territory. (*Source:* NOAA.)

United States. Their conclusion is that lightning is likely the deadliest component of thunderstorms activity.

## • HAIL

Beginning January 5, 2010, a thunderstorm is considered to be severe if hail 1" in diameter (quarter size) or larger falls from the storm. The former severe storm hail diameter threshold was 0.75", which is described as dime or penny size. Research indicated that most significant damage is done by hail of diameter 1" or greater, so the severe threshold was updated. Table 4.6 lists terminology that describes hail size.

Hail can be used as a measure of the severity of thunderstorms because of what is called terminal velocity. If a hailstone fell through a vacuum, it would obey the laws of gravity precisely. But as hail falls through the atmosphere, "air resistance," also called "drag," slows the fall speed by exerting an upward force on the hailstone.

When the drag equals the force of gravity, the hailstone stops accelerating and falls at a constant speed, called the terminal velocity. The larger the hailstone, the greater the terminal velocity.

Table 4.6 also provides terminal velocities for various-size hailstones. The velocities are based on smooth, spherical hail. When hail is oblong or of any odd shape, it can tumble and that decreases the terminal velocity. In addition, some of the largest hailstones result when smaller stones freeze together or coalesce on the way down and have a very rough surface. The rough surface, because of greater friction, also decreases the terminal velocity. For a hailstone that both tumbles and has a rough surface, the drag is even greater and the terminal velocity correspondingly slower.

For a hailstone to stay aloft, the thunderstorm updraft must be at least equal to the terminal velocity of the hailstone. This is an important concept in determining how hail is formed.

For an internal hailstone structure to develop like the Coffeyville, KS, hailstone of September 3, 1970

**Figure 4.20**  Lithograph from the work of John P. Finley of a tornado April 26, 1884, near Garnett, KS. The original photograph is in the possession of the Kansas State Historical Society and is likely the oldest surviving photograph of a tornado. (*Source:* NOAA.)

(Figure 4.23), with alternating bands of clear and milky ice, the stone has to alternate between traveling through air warmer than freezing and air colder than freezing.

The traditional explanation for hail formation is illustrated in Figure 4.24. Only two trips above the freezing level are shown, but a hailstone is thought to get caught in updrafts and downdrafts traveling above the freezing level (<32°F) where small ice particles freeze on its surface into a milky ice layer and below the freezing level (>32°F) where water coats the surfaces and then freezes to a clear glaze when the hailstone is pushed back above the freezing level.

For the Coffeyville, KS, hailstone, using the traditional explanation there were eight to ten such trips estimated by counting the ice layers. But by the time the hailstone grew to the second layer in from the surface the updraft velocity to push it back up above the freezing level would have been far greater than observed in most thunderstorms.

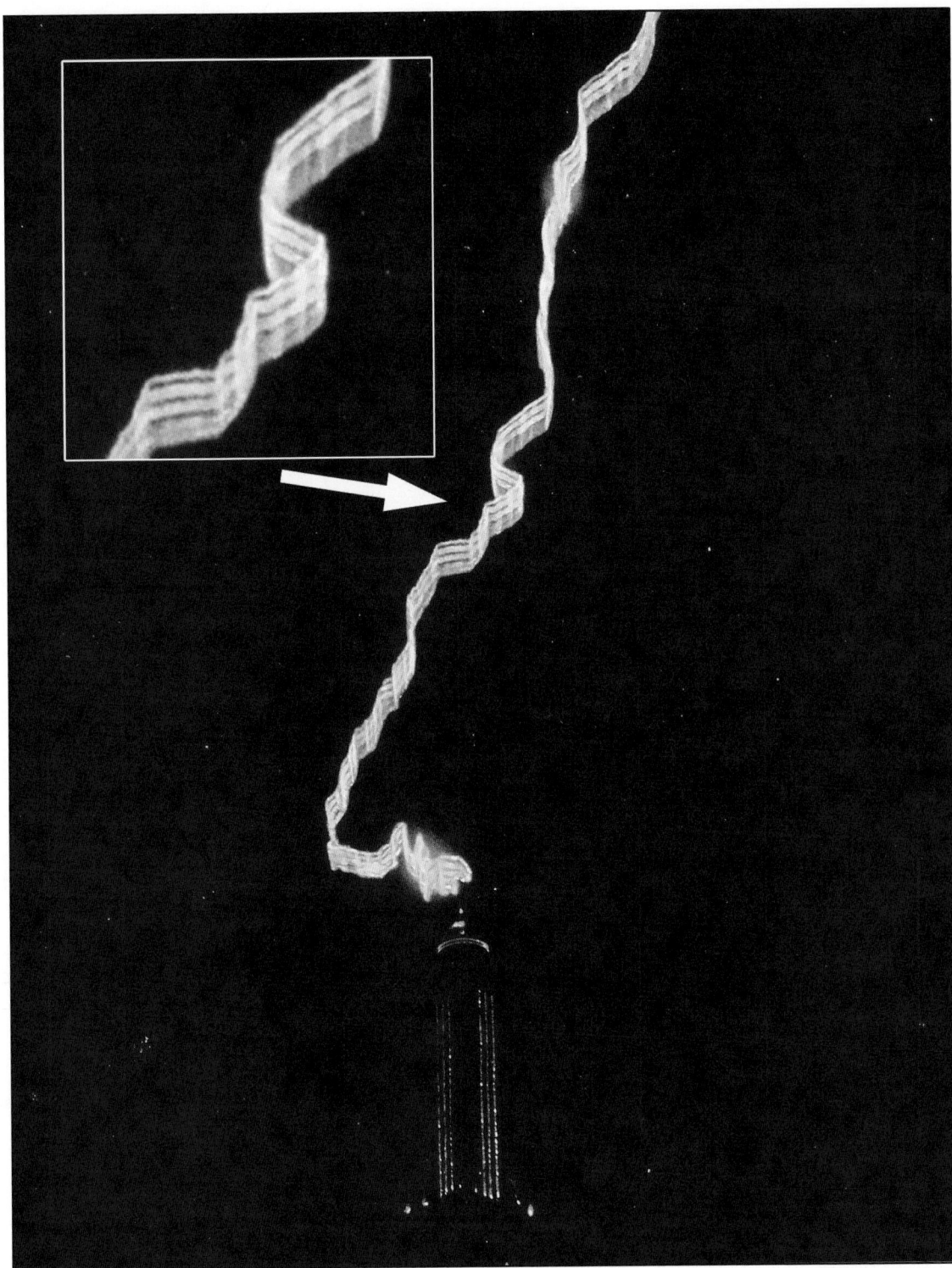

**Figure 4.21**  Ribbon lightning striking the Empire State Building. Multiple strokes are moved downwind, giving the appearance of a flat ribbon of lightning as seen in the enlarged section. (*Source:* NOAA.)

**Table 4.5   Events in the Life of a Lightning Strike**

| Event | Duration | | Direction |
|---|---|---|---|
| **Charge Distribution Established** | --- | --- | --- |
| **Preliminary Breakdown** | --- | --- | --- |
| **Stepped Leader** Surges towards the ground in jagged steps | Start @ 0 milliseconds | End @ 20 milliseconds | Down |
| **Ground Discharge** From one or more places | Start @ 19 milliseconds | End @ 20 milliseconds | Up |
| **Attachment Process** The stepped leader joins with ground discharges from a few feet to 100 feet above the ground. | At about 20 milliseconds | | --- |
| **First Return Stroke** – a  fast, continuous current that flows back towards the cloud and rapidly heats the channel. The shock wave from the expansion causes thunder. | Duration of  50 – 100 microseconds | | Up |
| **End of a single stroke flash** | | | |
| **Additional available electric charge may result in a multiple stroke flash following an inactive interval of 20 to 80 microseconds after the First Return Stroke** | | | |
| **Dart Leader and subsequent dart leaders** A continuous current stroke | Duration of 2 milliseconds | | Down |
| **Return stroke and subsequent return strokes** | Duration of  50 – 100 microseconds | | Up |

*Source:* Data from Martin Uman.

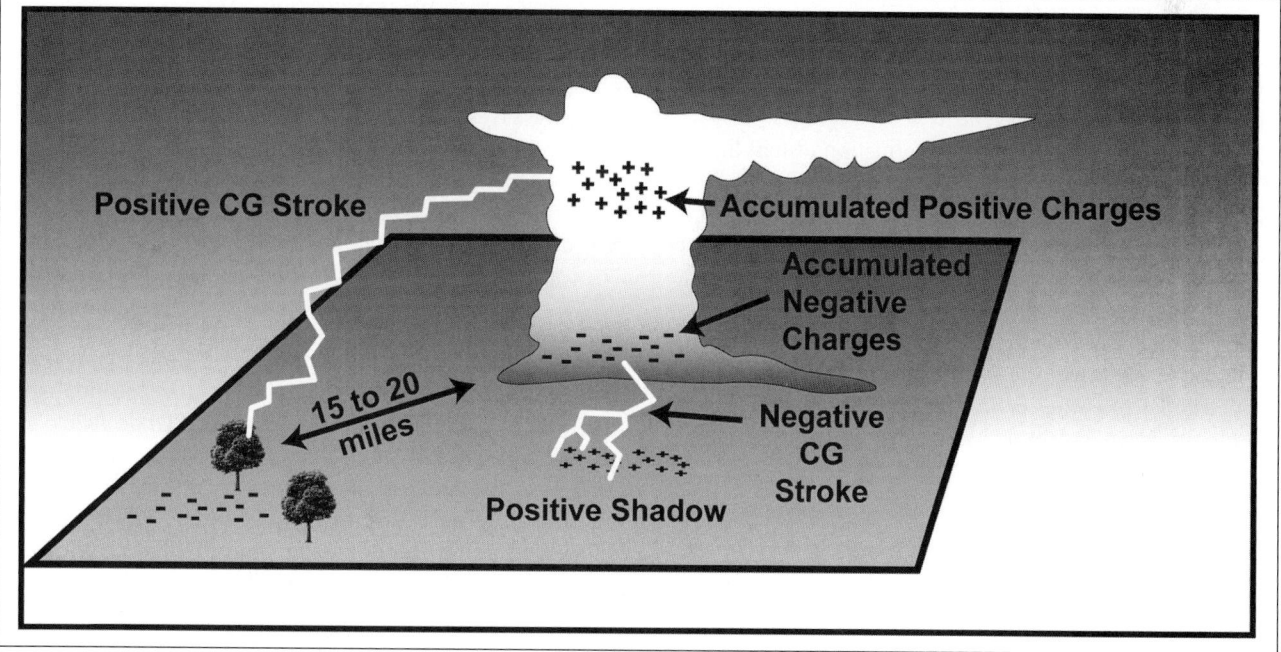

**Figure 4.22**   Charge distribution and positive and negative lightning strokes in a thunderstorm.

**Table 4.6  Hail-Size Terminology and Terminal Velocities or Maximum Speed of Descent**

| Hail Size Description | Hail Size inches | Terminal Velocity mph |
|---|---|---|
| Pea | 0.25" | 34.4 |
| Marble/Mothball | 0.50" | 48.7 |
| Dime/Penny | 0.75" | 59.6 |
| Quarter | 1.00" | 68.8 |
| Ping Pong Ball | 1.50" | 84.3 |
| Golf Ball | 1.75" | 91.1 |
| Tennis Ball | 2.50" | 108.8 |
| Baseball | 2.75" | 114.2 |
| Tea Cup | 3.00" | 119.8 |
| Grapefruit | 4.00" | 136.1 |
| Softball | 4.50" | 146.8 |
| Giant | 5.00"+ | 148.7 |

Terminal velocity calculations based on spherical hailstones with a smooth surface. The terminal velocity of non-spherical hailstones and those with a rough surface or having both characteristics will fall significantly more slowly. Calcuations were made using the calculator found at http://hyperphysics.phy-astr.gsu.edu.

For hailstones, the updraft  velocity required to repeatedly slow the fall and reverse the direction of travel is greater than the terminal velocity, and for large hailstones, the velocity to push the hailstone back upward is just too great to be responsible for the outer layers.

An alternate possibility for the formation of hail that requires only a single trajectory through the thunderstorm is illustrated in Figure 4.25. Updrafts transport warm, moist air upward and downdrafts transport cooled air downward, resulting in a contorted "freezing level."

If the trajectory of the hailstone takes it through alternating warm, moist updrafts (>32°) and cold downdrafts (<32°), the alternating layers of milky and clear ice will form without requiring updraft velocities greater than meteorologists have observed.

## • REFERENCE TABLES AND MAPS

### Tornadoes

**Tornado tracks.** Maps of all tornado tracks from 1950 to 2009 by F/EF-Scale (see Figures 4.26–4.31). Notice how the dominant direction of travel is from the southwest to northeast.

**F-EF5 tornadoes.** The location of all confirmed US F-EF5 tornadoes from 1900 through July 2010 is shown in Figure 4.32. Each F-EF5 tornado is provided in Table 4.7 in reverse chronological order using the locater numbers on the map. Table 4.8 gives other tornadoes that could be F-EF5 strength. For these, either more investigation is needed or there are insufficient data to make a final determination. Some sources list tornado #31 in Table 4.7, the Mt. Hope Tornado, tornado #96 of the 1974 Super Outbreak as an F4.

### Tornado Alley, Tornado Season, Time of Day, and Tornado Trends

Tornado alley refers to the portion of the Great Plains where more tornadoes touch down than in any other location in the world. It is very obvious in Figure 4.33. "Dixie Alley" extends from southern Mississippi into northern Alabama, and in Illinois and Indiana, there is a tornado touchdown maximum; both of these also show on the map. Florida too has an area where tornadoes are numerous; this too is shown in Figure 4.33. The maxima near Denver and Houston are likely due to greater detection of tornadoes as a result of population density.

The "Ozark minimum" in western Arkansas and Missouri and the minimum over the Appalachian Mountains show the effect of elevated terrain on tornado frequency. Elevated

**Figure 4.23**   The Coffeyville, KS, hailstone that fell on September 3, 1970, has a maximum circumference of 17.5" and a maximum diameter of 5.7". This hailstone surpassed the Potter, NE, hailstone on July 6, 1928, and was superseded as largest US hailstone by the Aurora, NE, hailstone that fell on June 22, 2003, which had a maximum diameter of 7" and a maximum circumference of 18.75". (*Source:* NOAA.)

terrain disrupts large-scale air flow, blocks moisture transport, and creates cooler atmospheric temperatures over elevated landmasses. Each of these factors inhibits tornado formation, but tornadoes do occur in mountains.

Figure 4.34 shows the months when tornado occurrence peaks for each of the 48 contiguous states and is complimented by Table 4.9, providing the number of tornadoes for each state and the number per 10,000 square miles from 1953 through 2009.

Figures 4.35–4.38 summarize the occurrence of US tornadoes from 1950 to 2009. Figures 4.35 and 4.36 make it look like that tornadoes are on the increase, but these graphs do not take into account the increase of population, the increased public awareness, and technological and social changes that make tornado detection much more likely now than 60 years ago. Determining if there is a trend is a very difficult problem scientists have not fully solved.

Figures 4.37 and 4.38 show when tornado occurrence peaks during the year and during the day. Clearly April, May, and June, when humid tropical air is returning to the

continent and the jet stream system is still vigorous before weakening to its typical summer strength is when tornado formation is preferred.

In Figure 4.38, the role of the sun's energy in tornado formation is clear, with the peak occurrence of tornadoes during the warmest part of a typical day.

Data for all these tables and figures are from NOAA's Storm Prediction Center and NOAA's National Climatic Data Center, and all maps and graphs are by Steve Horstmeyer.

### *Tornado Outbreaks, Fatalities, and Costs*
There is no official or agreed-upon definition of tornado outbreak, tornado swarm, and tornado outbreak sequence. When tornado research was expanding in the 1970s, an outbreak was the occurrence of ten or more tornadoes, but today it is often considered to have a minimum of six.

An outbreak is generally caused by a single weather system spawning multiple tornadoes in a 24- to 48-hour time period in a single region. The region can be large and the tornadoes spread over a large area.

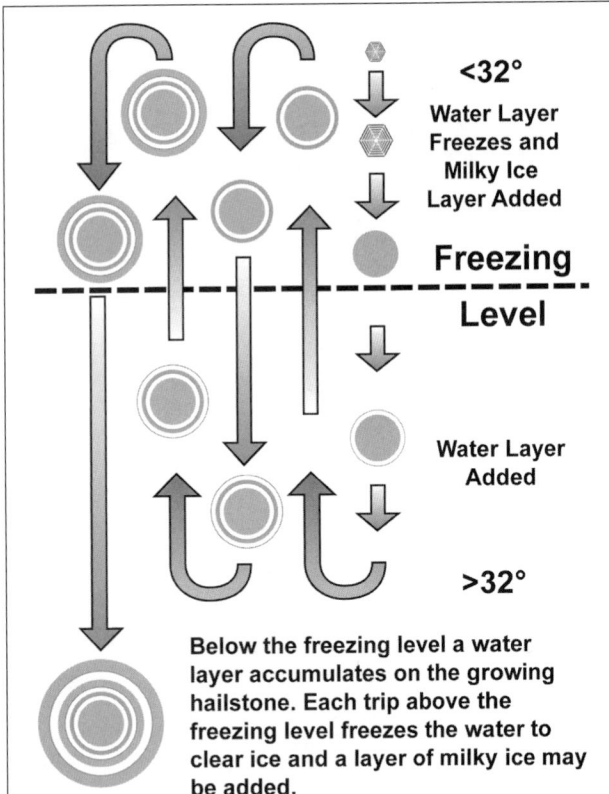

Below the freezing level a water layer accumulates on the growing hailstone. Each trip above the freezing level freezes the water to clear ice and a layer of milky ice may be added.

**Figure 4.24** The traditional explanation for hailstone growth is that updrafts and downdrafts transport the growing hailstone above and below the freezing level. Layers of water are added in the lower, warmer air and then freeze to clear ice layers in the cold air above the freezing level. Milky ice layers are also added in the subfreezing air.

When many tornadoes are spawned over multiple days it is called a tornado outbreak sequence or extended tornado outbreak. "Tornado outbreak sequence" is not a strictly defined term and can consist of nearly continuous tornado activity or a sequence can have decreased nocturnal activity with a daytime peak over several days.

The term "tornado swarm" is mostly used as a descriptive term to characterize time periods with several to many tornadoes.

Table 4.10 is a chronological listing of all known tornado outbreaks since 1860 with 50 or more fatalities. For many there is no designated name and I have used names that describe the location and often month or time of year.

Table 4.11 compliments Table 4.10 and provides tornado outbreaks or outbreak sequences with 30 or more tornadoes going back to 1884. Outbreaks are listed in order of decreasing number of tornadoes.

Tables 4.12 and 4.13 give the 25 deadliest and 10 costliest tornadoes in US history.

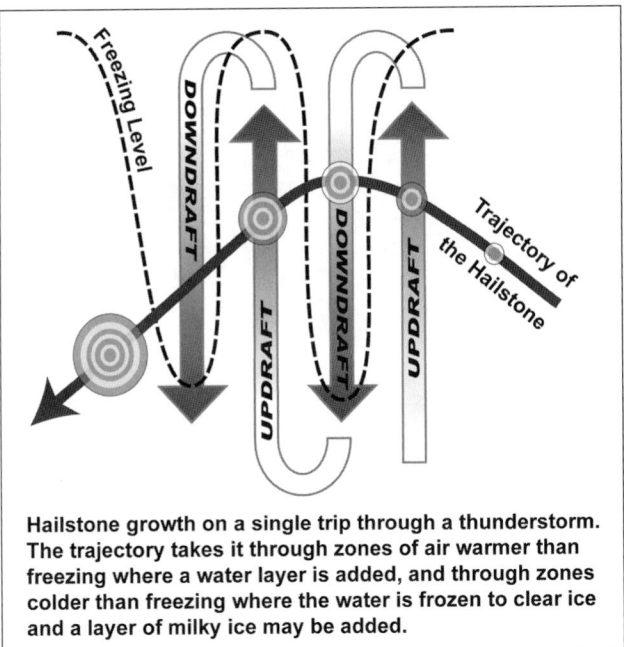

Hailstone growth on a single trip through a thunderstorm. The trajectory takes it through zones of air warmer than freezing where a water layer is added, and through zones colder than freezing where the water is frozen to clear ice and a layer of milky ice may be added.

**Figure 4.25** Another possible explanation for hail formation is that the freezing level is not level but higher in updrafts and lower in downdrafts. A single trajectory through a storm could account for alternating layers of clear ice and milky ice.

Tables 4.14a–c contain fatality statistics by state for the years 2000–2009. Only states with killer tornadoes are listed and fatalities are given by state, by F/EF-Scale and location.

## Thunderstorms through the Year

The annual cycle of thunderstorm occurrence is illustrated in Figures 4.39–4.43, which show the average number of days thunder is detected in the United States. By comparing maps you can see how thunderstorm occurrence spreads northward and northwestward from the Gulf of Mexico as the humid tropical air returns to the continent in spring and summer. All data are from NOAA's NCDC and maps are by Steve Horstmeyer and James Patterson.

## Lightning Seasons, Fatalities, Number of Strikes, and Most Dangerous Metropolitan Areas

The annual cycles of lightning for the United States and the world are mapped in Figures 4.44a–c and 4.45a–c. Data and maps are from NASA's Global Hydrology and Climate Center at the Marshall Space Flight Center. The maps have been modified for flash rates in units of square miles for this book.

Data for Tables 4.15a–j and 4.16–4.18 are from the National Climatic Data Center, and data for Table 4.19 are from the National Lightning Detection Network and published by Vaisala, Inc., who owns and operates the network.

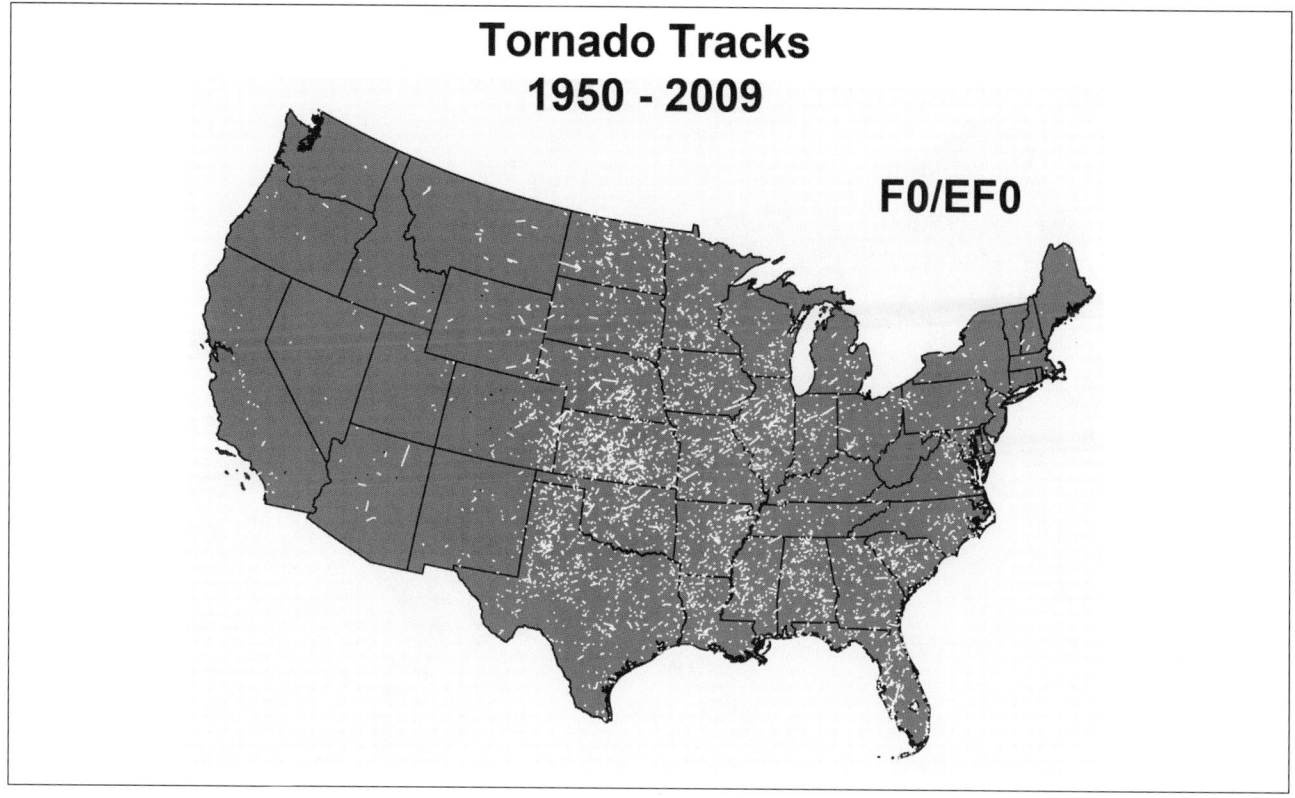

**Figure 4.26**    Tracks of F0/EF0 tornadoes, 1950–2009. (*Source:* Map by Steve Horstmeyer from NOAA SPC data.)

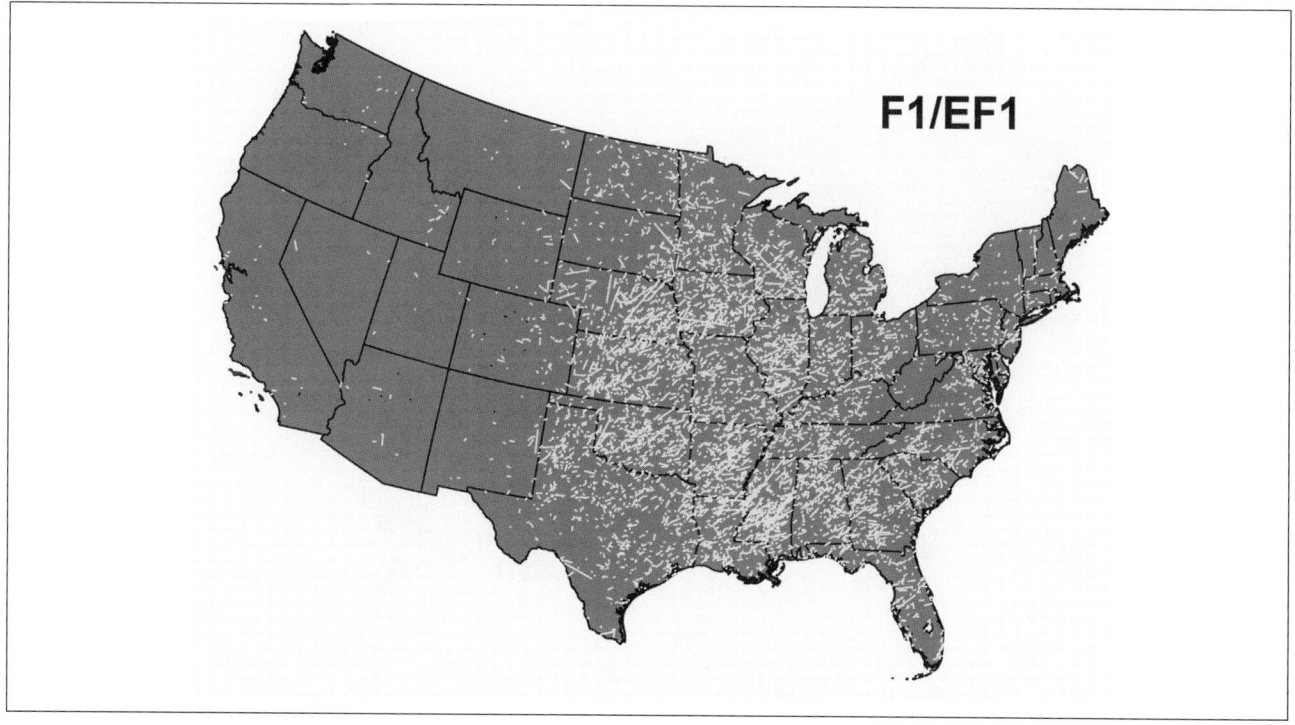

**Figure 4.27**    Tracks of F1/EF1 tornadoes, 1950–2009. (*Source:* Map by Steve Horstmeyer from NOAA SPC data.)

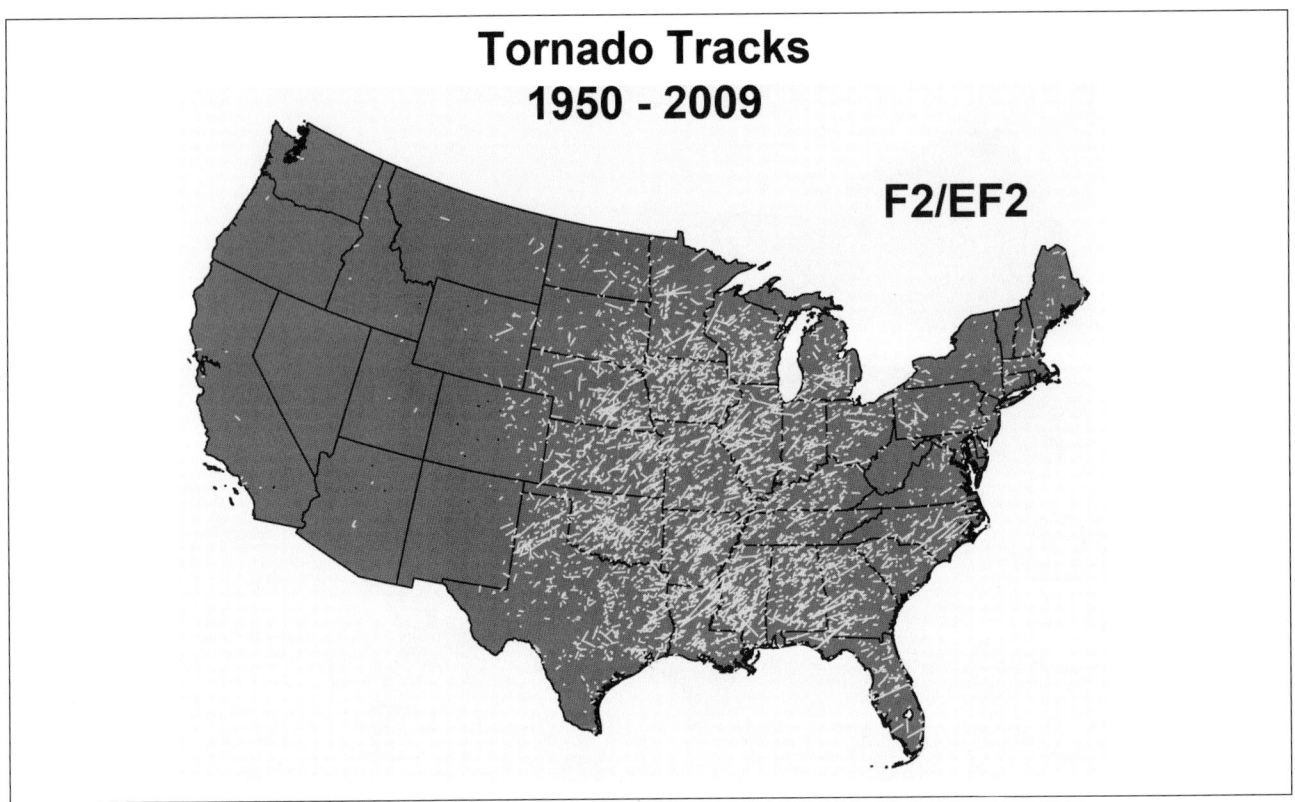

**Figure 4.28** Tracks of F2/EF2 tornadoes, 1950–2009. (*Source:* Map by Steve Horstmeyer from NOAA SPC data.)

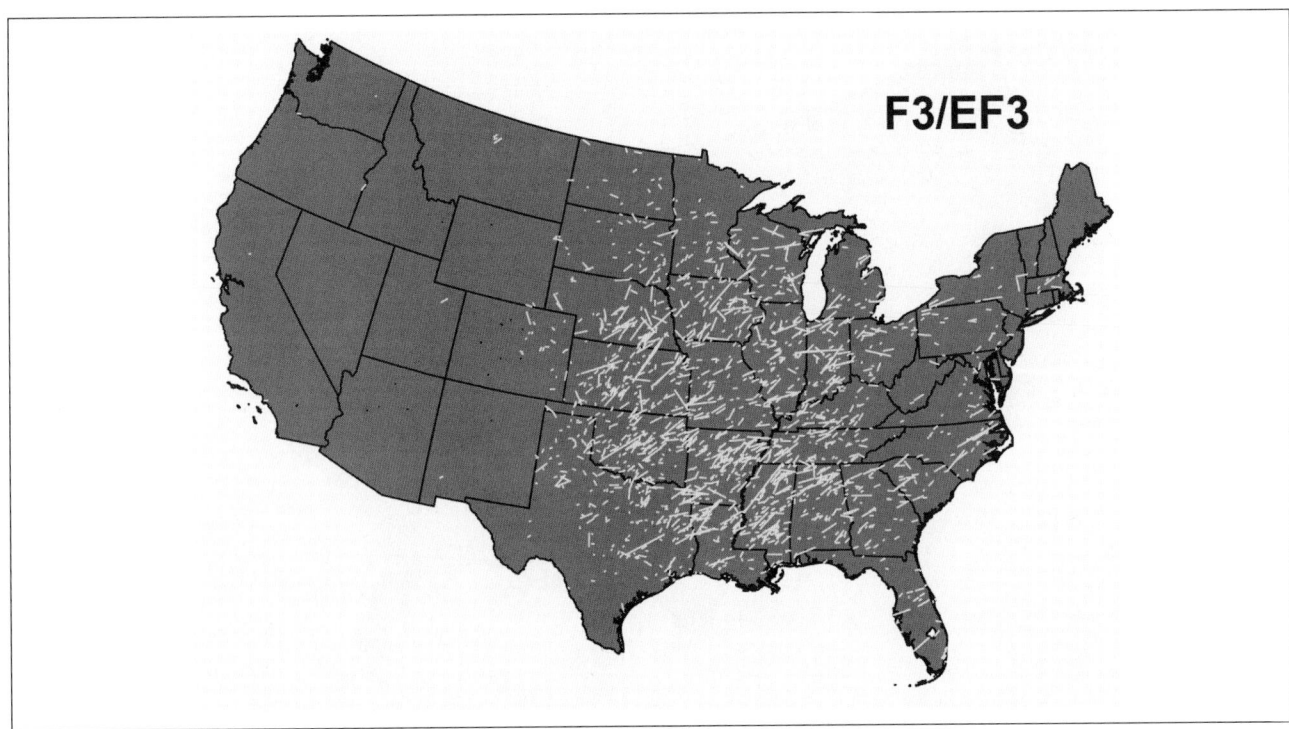

**Figure 4.29** Tracks of F3/EF3 tornadoes, 1950–2009. (*Source:* Map by Steve Horstmeyer from NOAA SPC data.)

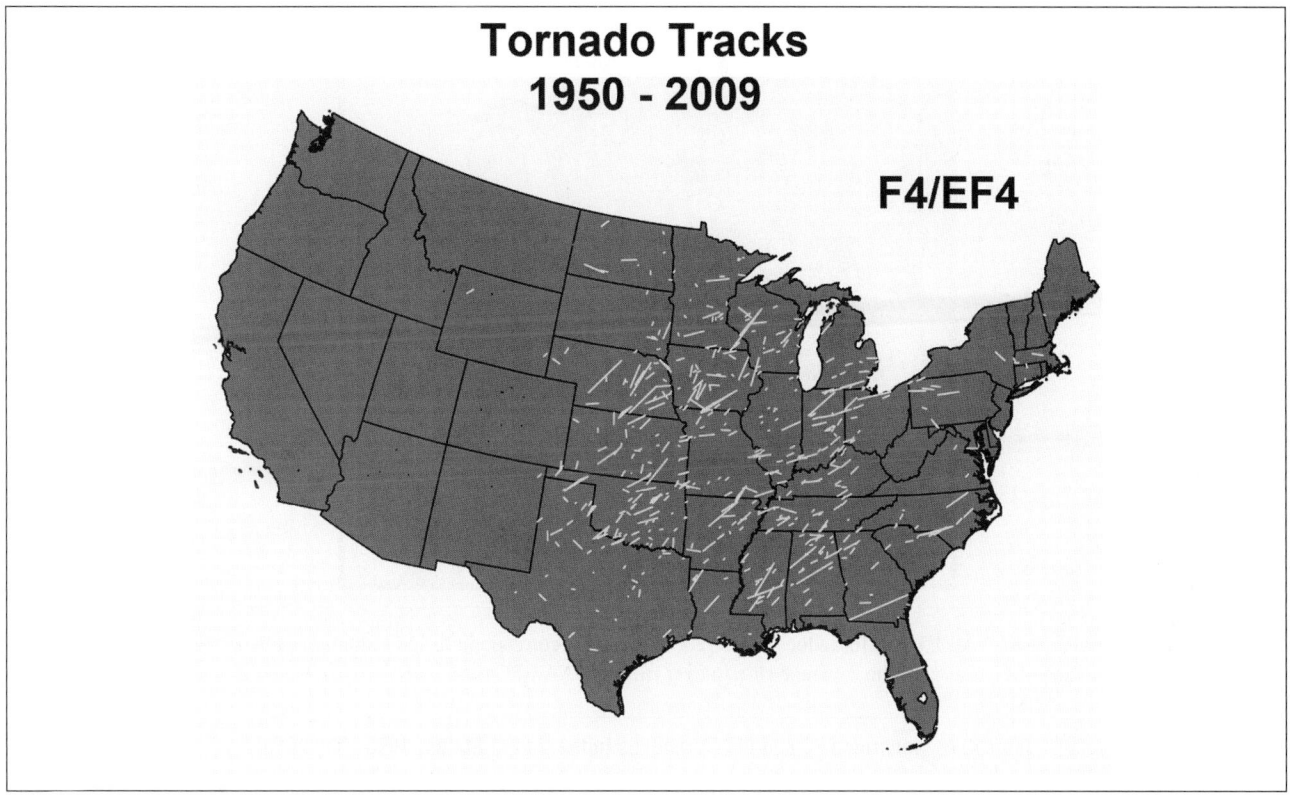

**Figure 4.30**  Tracks of F4/EF4 tornadoes, 1950–2009. (*Source:* Map by Steve Horstmeyer from NOAA SPC data.)

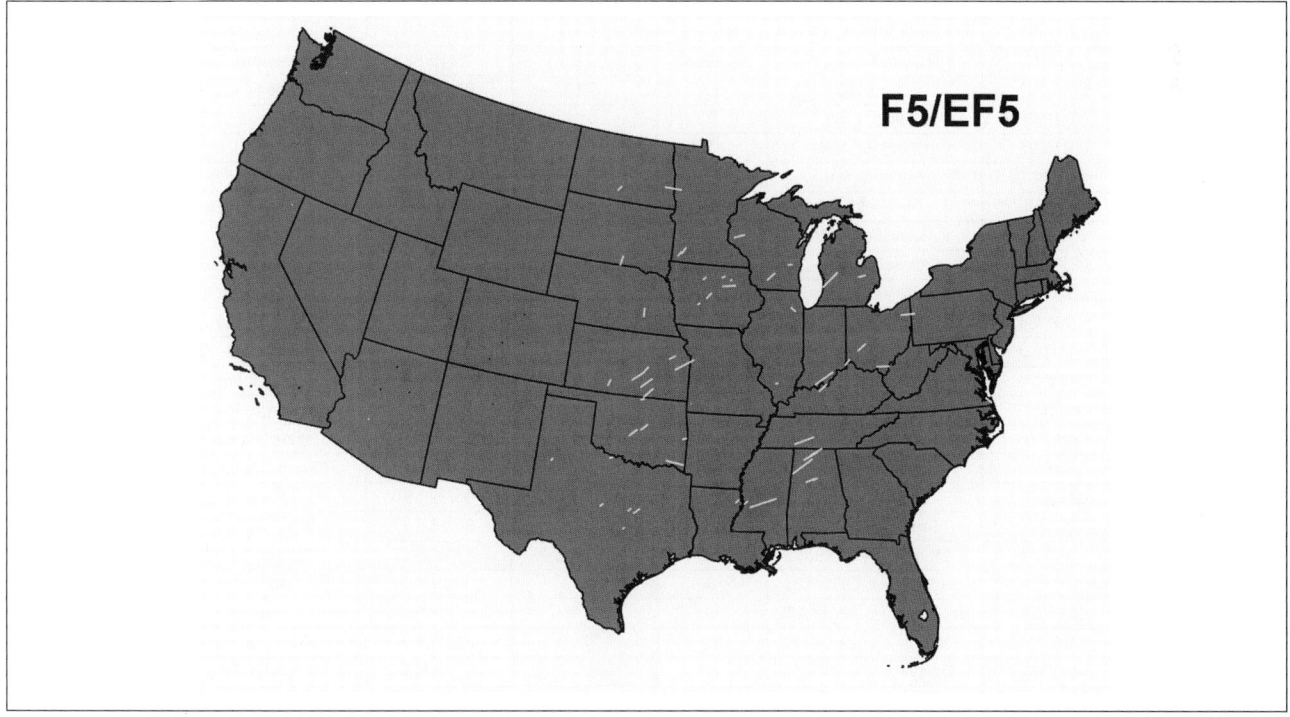

**Figure 4.31**  Tracks of F5/EF5 tornadoes, 1950–2009. (*Source:* Map by Steve Horstmeyer from NOAA SPC data.)

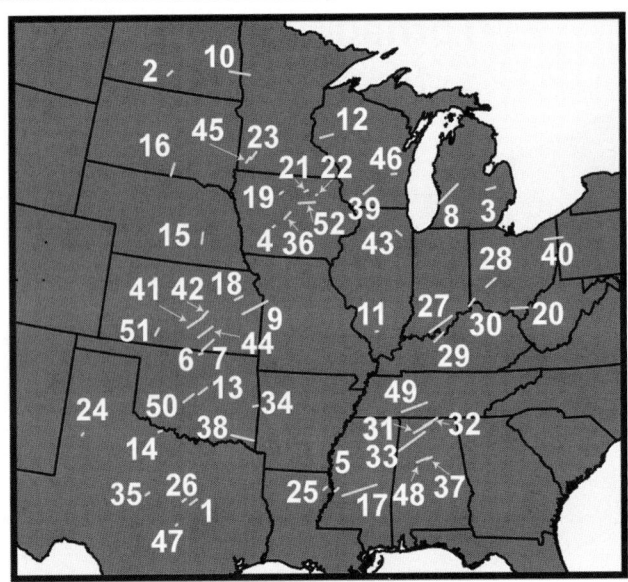

**Figure 4.32** All F5/EF5 tornadoes, 1950–2009; numbers correspond to individual tornado entries in Table 4.7. (*Source:* Map by Steve Horstmeyer from NOAA SPC data.)

### Table 4.7   Confirmed F5 and EF5 Tornadoes in the United States, 1950–July 2010

| Map # | Date | Communities | Deaths | Map # | Date | Communities | Deaths |
|-------|------|-------------|--------|-------|------|-------------|--------|
| 52 | 05/25/08 | Parkersburg, IA | 7 | 26 | 05/06/73 | Valley Mills, TX | 0 |
| 51 | 05/04/07 | Greensburg, KS | 12 | 25 | 02/21/71 | Delhi ,LA | 46 |
| 50 | 05/03/99 | Bridge Creek/Moore, OK | 36 | 24 | 05/11/70 | Lubbock ,TX | 28 |
| 49 | 04/16/98 | Waynesboro, TN | 3 | 23 | 06/13/68 | Tracy, MN | 9 |
| 48 | 04/08/98 | Pleasant Grove, AL | 32 | 22 | 05/15/68 | Maynard, IA | 1 |
| 47 | 05/27/97 | Jarrell, TX | 27 | 21 | 05/15/68 | Charles City, IA | 13 |
| 46 | 07/18/96 | Oakfield, WI | 0 | 20 | 04/23/68 | Gallipolis, OH | 7 |
| 45 | 06/16/92 | Chandler, MN | 1 | 19 | 10/14/66 | Belmond, IA | 16 |
| 44 | 04/26/91 | Andover, KS | 17 | 18 | 06/08/66 | Topeka, KS | 16 |
| 43 | 08/28/90 | Plainfield, IL | 29 | 17 | 03/03/66 | Jackson, MS | 57 |
| 42 | 03/13/90 | Goessel, KS | 1 | 16 | 05/08/65 | Gregory, SD | 0 |
| 41 | 03/13/90 | Hesston, KS | 1 | 15 | 05/05/64 | Bradshaw, NE | 2 |
| 40 | 05/31/85 | Niles, OH | 18 | 14 | 04/03/64 | Wichita Falls, TX | 7 |
| 39 | 06/07/84 | Barneveld, WI | 9 | 13 | 05/05/60 | Prague, OK | 5 |
| 38 | 04/02/82 | Broken Bow, OK | 0 | 12 | 06/04/58 | Menomonie, WI | 20 |
| 37 | 04/04/77 | Birmingham, AL | 22 | 11 | 12/18/57 | Murphysboro, IL | 1 |
| 36 | 06/13/76 | Jordan, IA | 0 | 10 | 06/20/57 | Fargo, ND | 10 |
| 35 | 04/19/76 | Brownwood, TX | 0 | 9 | 05/20/57 | Ruskin Heights, MO | 44 |
| 34 | 03/26/76 | Spiro, OK | 2 | 8 | 04/03/56 | Grand Rapids, MI | 18 |
| 33 | 04/03/74 | Guin, AL (#101) | 30 | 7 | 05/25/55 | Udall, KS | 82 |
| 32 | 04/03/74 | Tanner, AL (#98) | 22 | 6 | 05/25/55 | Blackwell, OK | 20 |
| 31 | 04/03/74 | Mt. Hope, AL (#96) | 28 | 5 | 12/05/53 | Vicksburg, MS | 38 |
| 30 | 04/03/74 | Sayler Park, OH (#43) | 3 | 4 | 06/27/53 | Adair, IA | 1 |
| 29 | 04/03/74 | Brandenburg, KY (# 47) | 31 | 3 | 06/08/53 | Flint, MI | 116 |
| 28 | 04/03/74 | Xenia, OH  (# 37) | 32 | 2 | 05/29/53 | Ft. Rice, ND | 2 |
| 27 | 04/03/74 | Daisy Hill, IN  (# 40) | 6 | 1 | 05/11/53 | Waco, TX | 114 |

Note: The Bridge Creek/Moore, OK tornado was the last F5 evaluated using the original Fujita Scale. Greensburg, KS is the first EF5 tornado. Some sources list #31 the Mt. Hope tornado, # 96 in the 1974 Super Outbreak as an F4.

*Source:* NOAA SPC and NCDC.

**Table 4.8   Possible F5 Tornadoes in the United States, 1880–1949**

| Date | Location | Deaths | Date | Location | Deaths |
|---|---|---|---|---|---|
| 04/24/1880 | Christian Co. IL | 6 – 11 | 06/15/12 | Creighton, Missouri | 5 |
| 06/12/1881 | Nodaway Co., MO | 2 | 06/11/15 | Mullinville, Kansas | 0 |
| 06/17/1882 | Grinell, IA | 68 | 05/25/17 | Sedgwick, Kansas | 23 |
| 08/21/1883 | Rochester, MN | 37 | 05/21/18 | Dennison, Iowa | 4 |
| 04/01/1884 | Oakville, IN | 8 | 06/22/19 | Fergus Falls, Minnesota | 57 |
| 08/28/1884 | Howard, SD | 4 | 03/11/23 | Pinson, Tennessee | 20 |
| 06/15/1892 | Faribault-Freeborn-Steel Co. MN | 12 | 05/14/23 | Big Spring, Texas | 23 |
| 05/22/1893 | Darlington, WI | 3 | 03/18/25 | Missouri-Illinois-Indiana | 695 |
| 07/06/1893 | Cherokee - Buena Vista | 71 | 04/12/27 | Rock Springs, Texas | 74 |
| 09/21/1894 | Northern IA, southern MN | 14 | 05/07/27 | McPherson, Kansas | 10 |
| 05/01/1895 | Sedgwick Co., Kansas | 8 – 11 | 04/10/29 | Sneed, Arkansas | 23 |
| 05/03/1895 | Sioux Co., IA | 9 – 15 | 05/22/33 | Tryon, Nebraska | 8 |
| 05/15/1896 | Sherman, TX | 73 | 04/05/36 | Tupelo, Mississippi | 216 |
| 05/17/1896 | Nemaha, NE | 25 | 04/26/38 | Oshkosh, Nebraska | 3 |
| 05/25/1896 | Oakland Co. MI | 47 | 06/10/38 | Clyde, Texas | 14 |
| 05/18/1898 | Marathon Co., WI | 12 | 04/14/39 | Vici, Oklahoma-Kiowa, Kansas | 7 |
| 06/12/1899 | St. Croix Co. - New Richmond, WI | 117 | 03/16/42 | Lacon, Illinois | 7 |
| 05/10/05 | Snyder, Oklahoma | 97 | 04/29/42 | Oberlin, Kansas | 15 |
| 06/05/05 | Colling, Michigan | 5 | 06/17/44 | Summit, South Dakota | 8 |
| 04/23/08 | Pender, Nebraska | 3 | 04/12/45 | Antlers, Oklahoma | 69 |
| 06/05/08 | Carleton, Nebraska | 11 | 04/09/47 | Woodward, Oklahoma | 181 |
| 06/15/12 | Creighton, Missouri | 5 | 05/31/47 | Leedey, Oklahoma | 7 |

Notes: Each of these needs to be investigated and are in a separate table because they occurred before the earliest date in the tornado database which begins in 1950.

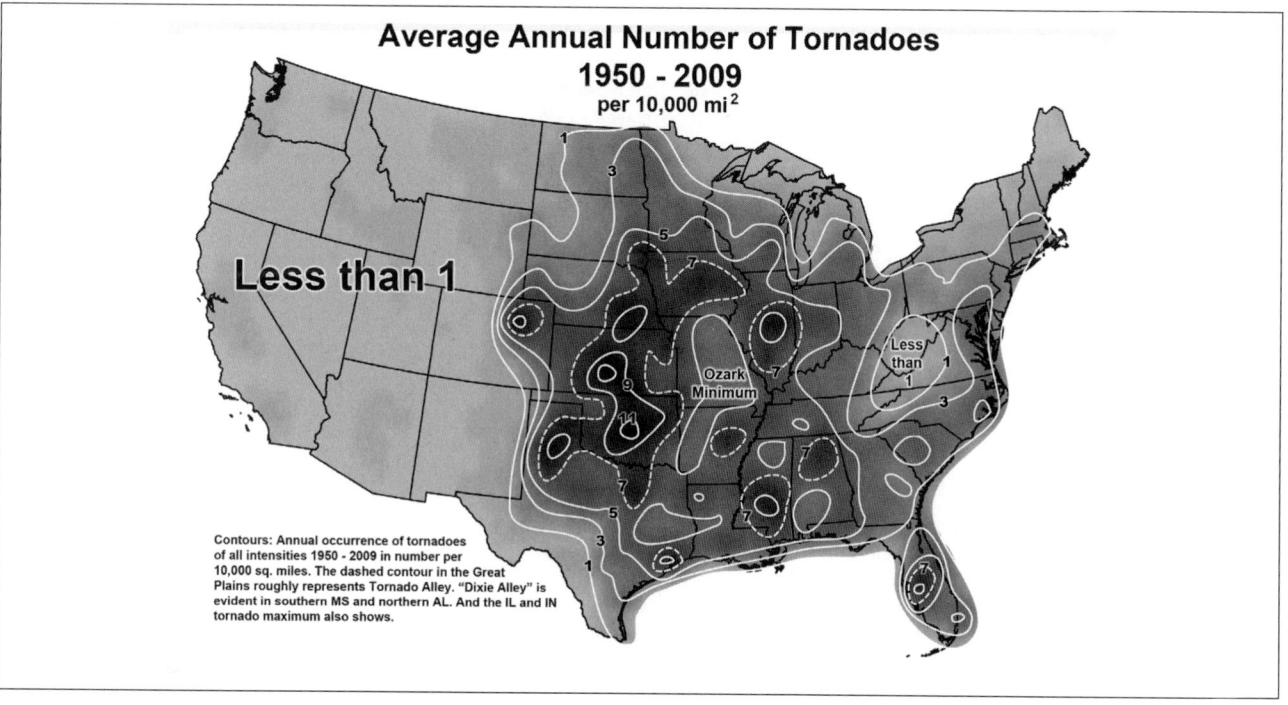

**Figure 4.33**   Annual tornado occurrence in the United States in number of confirmed tornadoes 1950–2009 per 10,000 square miles. (*Source:* Steve Horstmeyer from NOAA SPC data.)

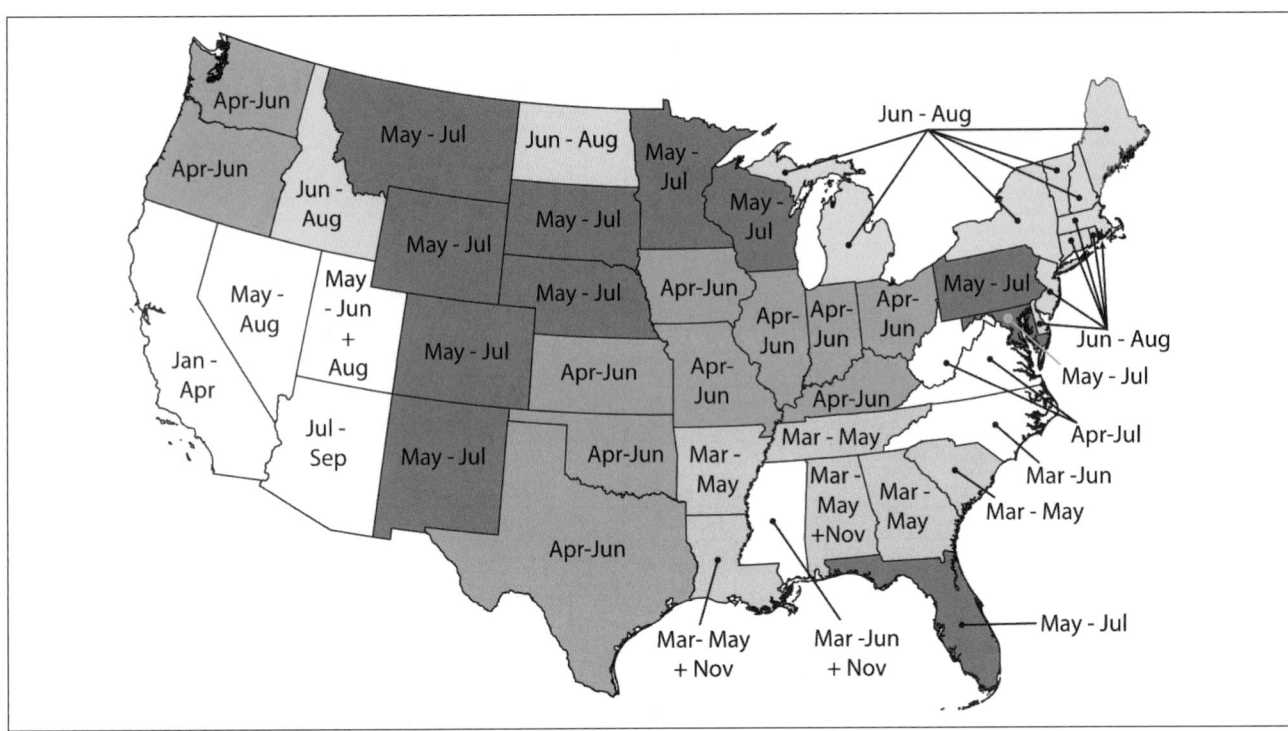

**Figure 4.34** Peak time of year of tornado occurrence by state. (*Source:* Steve Horstmeyer from NOAA SPC data.)

**Table 4.9  Average Number of US Tornadoes by State and per 10,000 Square Miles per State, 1953–2009**

| | Annual # | Rank | Area mi². | Annual # per 10,000 mi². | Rank | | Annual # | Rank | Area mi². | Annual # per 10,000 mi². | Rank |
|---|---|---|---|---|---|---|---|---|---|---|---|
| Alabama | 28 | 12 | 51718 | 5.4 | 11 | Montana | 7 | 30 | 147047 | 0.5 | 41 |
| Alaska | 0 | 50 | 47720 | 0.0 | 50 | Nebraska | 45 | 5 | 77359 | 5.8 | 9 |
| Arizona | 4 | 33 | 114007 | 0.4 | 42 | Nevada | 1 | 44 | 110567 | 0.1 | 49 |
| Arkansas | 27 | 13 | 53183 | 5.1 | 13 | New Hampshire | 2 | 39 | 9283 | 2.2 | 32 |
| California | 6 | 32 | 158648 | 0.3 | 45 | New Jersey | 3 | 34 | 7790 | 3.9 | 18 |
| Colorado | 23 | 16 | 104100 | 2.2 | 31 | New Mexico | 9 | 29 | 121599 | 0.7 | 39 |
| Connecticut | 1 | 44 | 5006 | 2.0 | 33 | New York | 7 | 30 | 49112 | 1.4 | 35 |
| Delaware | 1 | 44 | 2026 | 4.9 | 14 | North Carolina | 20 | 20 | 52672 | 3.8 | 19 |
| Florida | 54 | 4 | 58681 | 9.2 | 1 | North Dakota | 23 | 16 | 70704 | 3.3 | 25 |
| Georgia | 24 | 15 | 58390 | 4.1 | 17 | Ohio | 15 | 23 | 41328 | 3.6 | 24 |
| Hawaii | 1 | 44 | 42146 | 0.2 | 46 | Oklahoma | 56 | 3 | 69903 | 8.0 | 2 |
| Idaho | 3 | 34 | 83574 | 0.4 | 44 | Oregon | 2 | 39 | 97052 | 0.2 | 47 |
| Illinois | 37 | 7 | 56343 | 6.6 | 5 | Pennsylvania | 12 | 26 | 45310 | 2.6 | 30 |
| Indiana | 22 | 18 | 36185 | 6.1 | 7 | Rhode Island | 0.2 | 49 | 1213 | 1.6 | 34 |
| Iowa | 38 | 6 | 56276 | 6.8 | 4 | South Carolina | 15 | 23 | 31117 | 4.8 | 15 |
| Kansas | 62 | 2 | 82282 | 7.5 | 3 | South Dakota | 29 | 10 | 77122 | 3.8 | 20 |
| Kentucky | 13 | 25 | 40411 | 3.2 | 26 | Tennessee | 16 | 22 | 42146 | 3.8 | 21 |
| Louisiana | 29 | 10 | 47720 | 6.0 | 8 | Texas | 139 | 1 | 266874 | 5.2 | 12 |
| Maine | 2 | 39 | 33128 | 0.6 | 40 | Utah | 2 | 39 | 84905 | 0.2 | 48 |
| Maryland | 6 | 32 | 10455 | 5.7 | 10 | Vermont | 1 | 44 | 9615 | 1.0 | 37 |
| Massachusetts | 3 | 34 | 8262 | 3.6 | 23 | Virginia | 11 | 27 | 40598 | 2.7 | 29 |
| Michigan | 17 | 21 | 58513 | 2.9 | 28 | Washington | 3 | 34 | 68126 | 0.4 | 43 |
| Minnesota | 26 | 14 | 84397 | 3.1 | 27 | West Virginia | 2 | 39 | 24231 | 0.8 | 38 |
| Mississippi | 30 | 9 | 47695 | 6.3 | 6 | Wisconsin | 21 | 19 | 56145 | 3.7 | 22 |
| Missouri | 33 | 8 | 69709 | 4.7 | 16 | Wyoming | 11 | 27 | 97818 | 1.1 | 36 |

*Source:* NOAA NCDC.

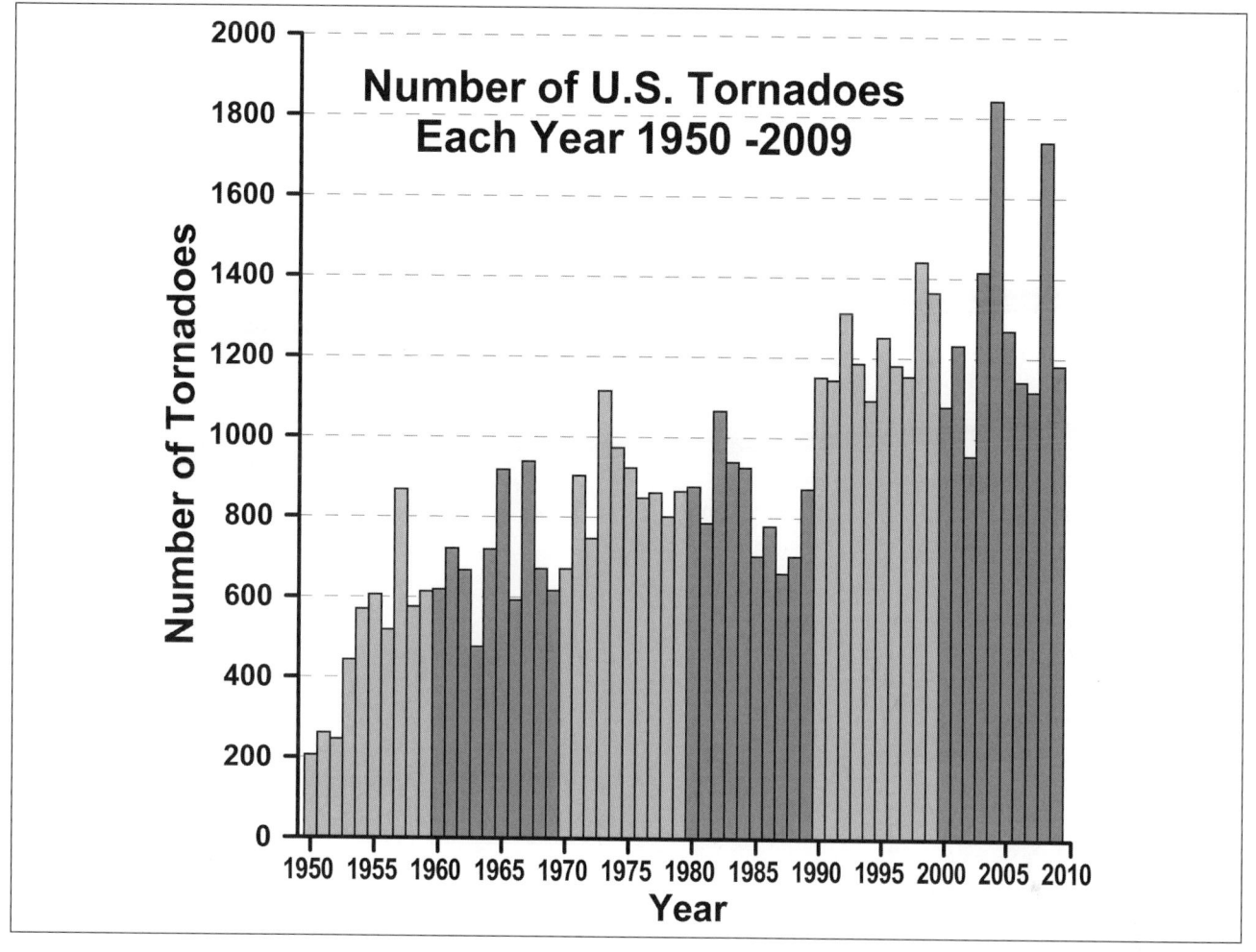

**Figure 4.35**   Number of tornadoes each year, 1950–2009. The apparent trend may be a false one and due to greater population density and greater public awareness and interest. (*Source:* Steve Horstmeyer from NOAA SPC data.)

The data for Tables 4.20 and 4.21 are the results of research conducted by Walker Ashley and Christopher Gilson at Northern Illinois University. They conducted an exhaustive study and found that official lightning fatality statistics seriously underestimate the true death toll.

Their survey included not only official statistics but Centers for Disease Control information and local newspaper accounts to arrive at a better estimate of lightning deaths. Table 4.20 is from their study and Table 4.21 has been modified to list the metropolitan areas using square miles.

### Hail Occurrence

Data for Figures 4.46a,b is from NOAA's Storm Prediction Center and maps were created by Steve Horstmeyer.

The maps depict the annual number of reports of hail per square mile from 1955 through 2009. Figure 4.46a shows the distribution of hail of any size, and Figure 4.46b shows hail with a diameter of 2" or greater.

Because 2" or greater diameter hail is rare in Figure 4.46b, the contours represent the occurrence of hail once in the number of years indicated by the contour label. For example, a large part of Oklahoma will receive 2" or greater hail once in 2 years, while the dashed contour represents one occurrence in 4 years and the outermost contour once in 16 years.

Notice the influence of dense urban populations on both figures and the resultant increase of observations. There may also be an urban heat island effect.

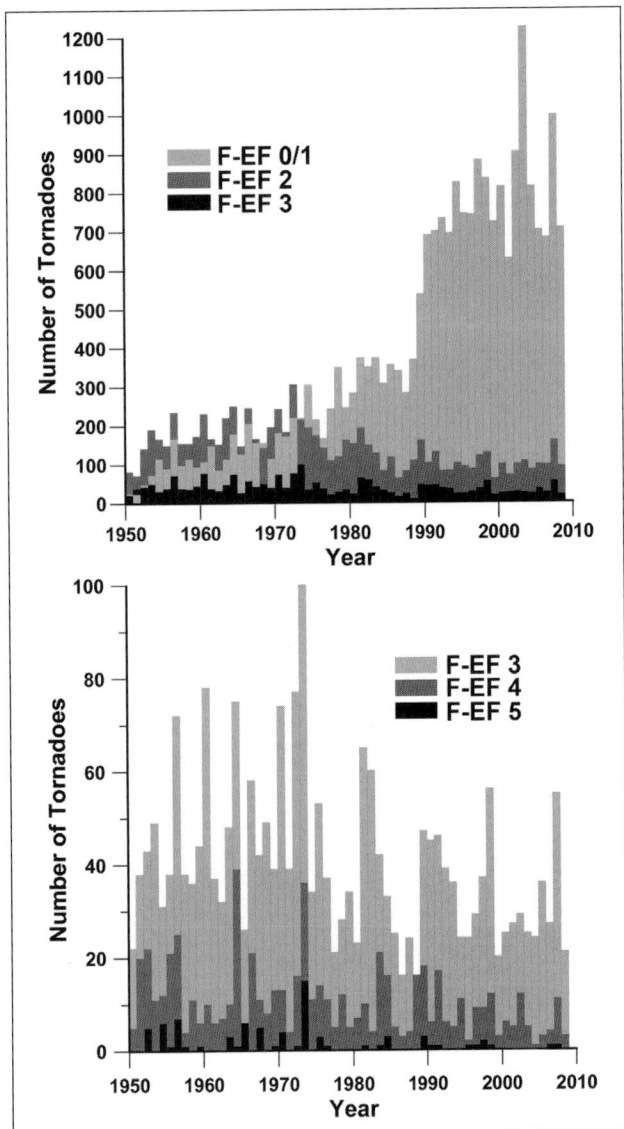

**Figure 4.36** Annual tornado count by F/EF-Scale. (*Source:* Steve Horstmeyer from NOAA SPC data.)

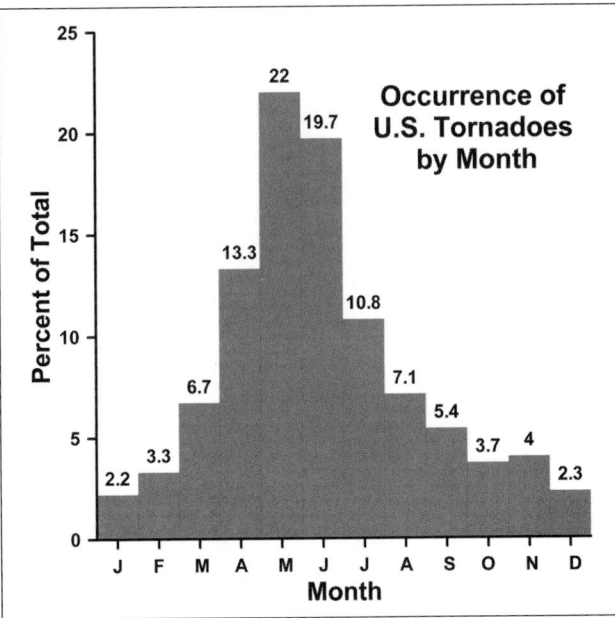

**Figure 4.37** Monthly occurrence of US tornadoes. (*Source:* Steve Horstmeyer from NOAA SPC data.)

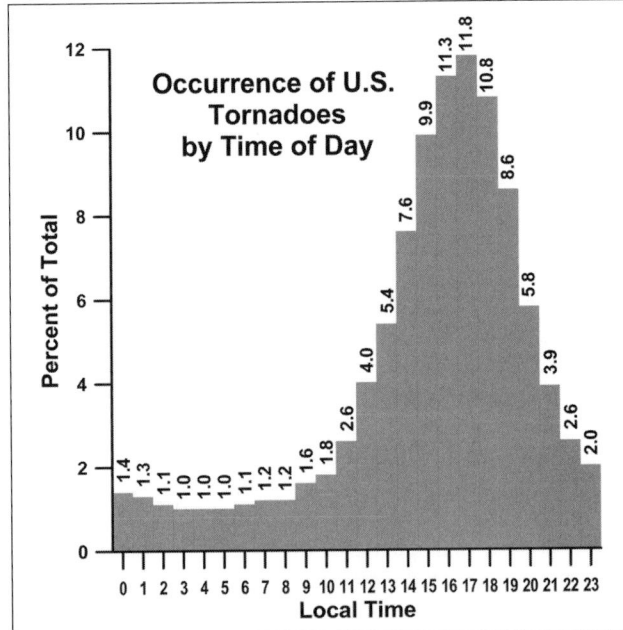

**Figure 4.38** Occurrence of US tornadoes by time of day. (*Source:* Steve Horstmeyer from NOAA SPC data.)

**Table 4.10   US Tornado Outbreaks with 50 or More Fatalities, 1860–July 2010**

| Tornado or Outbreak | Date(s) | Fatalities and Injuries | Notes |
|---|---|---|---|
| Mid-Mississippi Valley Outbreak of 1860 | June 3, 1860 | 148 fatalities 409 injuries | |
| Southeast Outbreak of March 1875 | March 20, 1875 | 93 fatalities 367 injuries | |
| Southeast Outbreak of May 1875 | May 1, 1875 | 58 fatalities 195 injuries | |
| Mississippi Valley Outbreak of April 1880 | April 18, 1880 | 165 fatalities 511 injuries | |
| Southeast Outbreak of April 1883 | April 22–23, 1883 | 109 fatalities 755 injuries | |
| Mississippi Valley Outbreak of 1883 | May 18, 1883 | 64 fatalities 386 injuries | |
| 1884 Enigma Outbreak | February 19–20, 1884 | 178 fatalities 1056 injuries | 41 or more tornadoes |
| 1886 Sauk Rapids, MN Tornado | April 14, 1886 | 72 fatalities 200 injuries | Deadliest tornado in Minnesota history |
| March 1890 Louisville, KY Killer Tornado and Outbreak | March 27, 1890 | 146 fatalities 847 injuries | 76 killed in downtown Louisville. Kentucky |
| September 1894 Upper Mississippi Valley Outbreak | September 21–22, 1894 | 63 fatalities 253 injuries | |
| 1896 Sherman, Texas Outbreak | May 15, 1896 | 85 fatalities 291 injuries | First Outbreak in a series of three  during May 1896 |
| 1896 Des Moines, IA Killer Tornado and Great Lakes Outbreak. | May 24–25, 1896 | 79 fatalities 215 injuries | |
| 1896 St. Louis-East St. Louis Tornado | May 27–28, 1896 | 305 fatalities 1236 injuries | 3rd deadliest tornado in U.S. history |
| January 1898 Arkansas Outbreak | January 11, 1898 | 56 fatalities 119 injuries | |
| May 1898 Mississippi Valley Outbreak | May 17–18, 1898 | 55 fatalities 380 injuries | |
| 1899 Great Natchez, MS Tornado | May 7, 1840 | 317 fatalities 109 injuries | 2nd deadliest tornado in U.S. history |
| 1899 New Richmond , WI Tornado | June 11–12, 1899 | 124 fatalities 203 injuries | |
| 1902 Goliad, Texas Outbreak | May 18, 1902 | 114 fatalities 279 injuries | Tied with Waco tornado as deadliest in Texas history |

**Table 4.10   US Tornado Outbreaks with 50 or More Fatalities, 1860–July 2010** [CONTINUED]

| Tornado or Outbreak | Date(s) | Fatalities and Injuries | Notes |
|---|---|---|---|
| 1905 Snyder, Oklahoma Tornado | May 10, 1905 | 97 fatalities 150 injuries | F5 largely destroyed Snyder |
| 1908 Southeast Outbreak | April 23–25, 1908 | 324 fatalities 1720 injuries | |
| 1909 Mississippi - Tennessee Valleys Outbreak | April 29 – May 1, 1909 | 165 fatalities 696 injuries | |
| April 1912 Great Plains to Southeast U.S. Outbreak Sequence | April 20–29, 1912 | 104 fatalities 630 injuries | |
| Mid-March 1913 Outbreak in the Southeast and Mississippi Valley. | March 13–14, 1913 | 78 fatalities 492 injuries | |
| June 1916 Southern U.S. Outbreak | June 5–6, 1916 | 112 fatalities 741 injuries | |
| May–June 1917 Outbreak Sequence in the South and Southeast | May 25 – June 1, 1917 | 383 fatalities | 8-day nearly continuous Sequence of outbreaks F5 Sedgwick, KS |
| March 1919 Outbreak in the Central U.S. | March 14–16, 1919 | 53 fatalities 219 injuries | |
| April 1919 Southern Great Plains Outbreak | April 6–9, 1919 | 92 fatalities 412 injuries | |
| 1919 Fergus Falls, MN Tornado | June 22, 1919 | 57 fatalities 200 injuries | Likely an F5 |
| 1920 Palm Sunday Outbreak | March 28, 1920 | 380 fatalities 1215 injuries | |
| 1920 Mississippi-Alabama Outbreaks | April 19–21, 1920 | 243 fatalities 1374 injuries | |
| April 1921 Texas - Arkansas Outbreak | April 13–16, 1921 | 90 fatalities 676 injuries | |
| 1924 Horrell Hill, SC Killer Tornado and Outbreak | April 30, 1924 | 110 fatalities 1133 injuries | 7 killed at school in Horrell Hill, South Carolina |
| 1924 Sandusky-Lorain, Ohio Outbreak | June 28, 1924 | 90 fatalities 349 injuries | |
| 1925 Tri-State Tornado | March 18, 1925 | 747 fatalities 2298 injuries | Deadliest , 3rd costliest U.S.and Longest path (219 – 234 miles) and duration (3.5 hours) tornado in the world . Probably an F5 Tornado |
| November 1926 Southern U.S.Outbreak | November 25–26, 1926 | 107 fatalities 451 injuries | |

**Table 4.10   US Tornado Outbreaks with 50 or More Fatalities, 1860–July 2010** [CONTINUED]

| Tornado or Outbreak | Date(s) | Fatalities and Injuries | Notes |
|---|---|---|---|
| 1927 Southern Plains-Midwest Outbreak | April 18–19, 1927 | 146 fatalities 235 injuries | |
| May 1927 Kansas Outbreak | May 7–9, 1927 | 217 fatalities 1156 injuries | One F5 with 10 fatalities and 300 injuries in McPherson, Kansas on May 7 |
| 1927 St. Louis Outbreak | September 29, 1927 | 82 fatalities 620 injuries | Second costliest tornado in history |
| 1929 Slocum, Texas to Statesboro, Georgia Outbreak | April 24–25, 1929 | 63 fatalities 567 injuries | |
| May 1930 Frost, TX killer tornado and Outbreak Sequence | May 1–9, 1930 | 110 fatalities 520 injuries | |
| 1932 Deep South Outbreak | March 21–22, 1932 | 330 fatalities 2145 injuries | |
| Southeastern U.S. Outbreak of March 1933 | March 30–31, 1933 | 87 fatalities 620 injuries | |
| 1936 Tupelo-Gainesville Outbreak | April 5–6, 1936 | 454 fatalities 2498 injuries | F5 in Tupelo the 4th deadliest in U.S. Gainesville F4 – 5th deadliest in U.S. |
| April 1939 Vici, OK to Kiowa, KS F5 and Outbreak Sequence | April 14–17, 1939 | 57 fatalities 316 injuries | F5 Vici, Alva and Capron, OK |
| March1942 Greenwood, MS Killer Tornadoand Outbreak | March 16, 1942 | 148 fatalities 1284 injuries | School buses with children smashed by the storm Lacon, IL F5 |
| April-May 1942 Great Plains Outbreak Sequence | April 27–30 & May 2, 1942 | 123 fatalities 839 injuries | F5 Oberlin, KS |
| 1944 Appalachians Outbreak | June 22–23, 1944 | 163 fatalities 1044 injuries | |
| April 1945 Antlers, OK to Quincy, IL  Outbreak | April 12, 1945 | 128 fatalities 1001 injuries | F5 Antlers, OK killed 67 immediately (19 others died later of injuries) One-half mile wide possible a multiple vortex. |
| 1947 Pampa, TX to Kingman Co., KS Outbreak | April 9–10, 1947 | 181 fatalities 980 injuries | F5 Glazier-Higgins-Woodward, OK Tornado, 6th deadliest in U.S. Clean up hampered by snow. |
| 1949 Warren, Arkansas Outbreak | January 3, 1949 | 60 fatalities 504 injuries | |
| May 1949 Great Plains to West Virginia Outbreak Sequence | May 20–21, 1949 | 66 fatalities 552 injuries | Perhaps 2nd most intense known Outbreak F2-F5 =74 F4-F5 = 9 |
| 1952 Arkansas-Tennessee Outbreak | March 21–22, 1952 | 208 fatalities | |

**Table 4.10   US Tornado Outbreaks with 50 or More Fatalities, 1860–July 2010 [CONTINUED]**

| Tornado or Outbreak | Date(s) | Fatalities and Injuries | Notes |
|---|---|---|---|
| 1953 Waco, TX Tornado and Texas Outbreak | May 9–11, 1953 | 144 fatalities 903 injuries | Waco Tornado, May 11, occurred during Texas Outbreak and is tied for deadliest in Texas history with 114 deaths and 10th deadliest in the United States |
| Sedgwick, CO to Flint, MI to Worcester, MA Outbreak Sequence | June 7–9, 1953 | 247 fatalities | June 8 – F5 -Flint-Beecher tornado last 100-fatality (116) single tornado in US history . June 9 – F4 – Worcester, 94 deaths |
| 1955 Great Plains Outbreak | May 25–26, 1955 | 102 fatalities | May 25 – F5 – Blackwell, OK and F5 Udall, KS the last >= 75 fatality single tornado in the U.S with 80 deaths. |
| May 1957 Central Plains Outbreak | May 19–21, 1957 | 59 fatalities | F5 – Ruskin Heights, MO, 71 mile track from Williamsburg, KS to Little Blue, MO, F5 damage 3-4 blocks wide, maximum width 1 mile. |
| 1965 Palm Sunday Outbreak | April 11–12, 1965 | 256 fatalities | 11 April – F5 – Dunlap, IN, 36 deaths. 11 April – F5 Strongsville, OH, 18 deaths. Among most intense recorded outbreaks |
| Candlestick Park Tornado, Jackson, MS | March 3, 1966 | 58 dead 508 injuries | 202.5 mile path F5 Jackson, MS |
| Belvidere - Oak Lawn, IL Outbreak | April 21, 1967 | 58 fatalities | Large school and traffic death tolls |
| 1968 Arkansas and Kansas to NE OH Outbreak | May 15–16, 1968 | 74 fatalities 1203 injuries | May 15 – F5 – Charles City IA and F5 Olewin, IA 39 tornadoes |
| February 1971 Texas to Mississippi Valley Outbreak | February 21, 1971 | 119 fatalities | F5 – Delhi, LA, only F5 in Louisiana Earliest F5 in the year, deadliest F5 (47). |
| Super Outbreak | April 3–4, 1974 | 315 to 330 fatalities | Largest and most intense Outbreak 148 Tornadoes F5 – 7 (see Table 7), F4 – 23, F3 – 34, F2 – 32, F1 – 33 F0 - 19 |
| 1979 Wichita Falls, TX Tornado and Red River Valley Outbreak | April 10–11, 1979 | 56 fatalities | Wichita Falls Tornado on 10th |
| 1984 Georgia and Carolinas Outbreak | March 28, 1984 | 57 fatalities 1200 injuries | 2 F4's left damage paths 2+ miles wide |
| 1985 United States-Canadian Outbreak | May 31, 1985 | 88 fatalities | F5 – Ravenna Arsenal, Newton Falls, (F3/4) Lordstown (F3/4), Warren, Niles, (F5) OH to Wheatland (F5), Mercer, PA. Half-mile wide, 300 mph winds estimated, 47 mile path length. |
| 1999 Bridge Creek - Moore, OK F5 Tornado and Outbreak | May 3, 1999 | 50 fatalities 665 injuries | First tornado to incur $1 billion in damages. 3 deaths in overpass shelters. Last F5 using original Fujita scale. Doppler velocity 301 mph. |
| 2008 Super Tuesday Outbreak and Memphis, Nashville and Jackson TN Tornadoes. | February 5–6, 2008 | 57 fatalities 200 injuries | Deadliest U.S. Outbreak since 1985 United States-Canadian Outbreak on May 31, 1985 |

*Source:* NOAA.

**Table 4.11   Tornado Outbreaks and Outbreak Sequences Ranked by the Number of Tornadoes, 1884–July 2010**

| Rank | Outbreak or Sequence | Date(s) | Number of Tornadoes Max Strength | Fatalities Injuries | Damages $ Original $ 2009 |
|---|---|---|---|---|---|
| 1 | **Super Outbreak** | April 3-4, 1974 | 148 in 18 hours 30 F4/F5 6 or 7 F5 – see note | 315 – 330 6142 | 600 million 2.9 billion |
| 2 | **May 1999 Oklahoma Outbreak** | May 3-6, 1999 | 140 tornadoes Moore, OK EF5 | 5 2000 | 1.9 billion 2.4 billion |
| 3 | **May 2007 Central Plains Outbreak** | May 4-6, 2007 | 120 tornadoes EF5 Greensburg, KS, first EF5 using new scale. | 14 123 | 268 million 342 million |
| 4 | **November 1992 Outbreak** | November 21-23, 1992 | 94 tornadoes F4 – 6 confirmed | 26 641 | 291 million 371 million |
| 5 | **May 2003 Outbreak Sequence** | April 30 – May 11, 2003 | 94 tornadoes during May 4 outbreak 401 in 12 days EF4 – 6 confirmed | 48 Unknown | 952 million 1.1 billion |
| 6 | **Super Tuesday Outbreak** | February 5-6, 2008 | 87 tornadoes EF4 – 5 confirmed | 57 Unknown | >1 billion >1 billion |
| 7 | **April 2009 Outbreak** | April 9-11, 2009 | 85 tornadoes EF4 – Murfreesboro, TN | 35 200 | 160 million |
| 8 | **Enigma Outbreak** | February 19-20, 1884 | 60 tornadoes F4 – 4 confirmed | 182 to 1200 Unknown | 4 million+ 94 million |
| 9 | **April-May 2010 Outbreak** | April 30 – May 2, 2010 | 58 tornadoes EF3 – 3 confirmed | 5 +24 indirect Unknown | Unknown at time of publication |
| 10 | **March 2009 Outbreak Sequence** | March 23 – 29, 2009 | 56 tornadoes EF3 – 2 confirmed | No fatalities 41 injuries | 14.2 million |
| 11 | **April 2010 Outbreak** | April 22 – 24, 2010 | 54 tornadoes EF4 – 2 confirmed | 10 + 2 indirect Unknown | Unknown at time of publication |
| 12 | **April 1991 Andover, KS Outbreak** | April 26, 1991 | 55 tornadoes EF5 – 1 confirmed, Andover, KS | 21 308 | 277+ million 431+million |
| 13 | **1965 Palm Sunday Outbreak** | April 11-12, 1965 | 47 tornadoes F4 – 17 confirmed | 271 3148 | 200+ million 1.4 billion |
| 14 | **June 2010 Upper Midwest-Great Lakes Outbreak** | June 5-6, 2010 | 46 tornadoes EF4 – Millbury, OH | 7 Unknown | Unknown at time of publication |

**Table 4.11  Tornado Outbreaks and Outbreak Sequences Ranked by the Number of Tornadoes, 1884–July 2010 [CONTINUED]**

| Rank | Outbreak or Sequence | Date(s) | Number of Tornadoes Max Strength | Fatalities Injuries | Damages $ Original $ 2009 |
|------|----------------------|---------|----------------------------------|---------------------|---------------------------|
| 14 | 1955 Great Plains Outbreak | May 25–26, 1955 | 46 tornadoes F5 – 2 confirmed Blackwell, OK and Udall, KS | 102 unknown | Unknown |
| 16 | April 1967 Oak Lawn-Belvidere, IL Outbreak | April 21, 1967 | 45 tornadoes F4 – 5 confirmed | 58 1109 | 100+ million 635+ million |
| 17 | 1984 Upper Midwest Outbreak | June 7-8, 1984 | 42 tornadoes F5 – Barneveld, WI | 13 319 | 100+ million 204+ million |
| 18 | May 2010 Oklahoma Outbreak | May 10-13, 2010 | 39 tornadoes EF4 – 2 confirmed | 3 Unknown | Unknown at time of publication |
| 19 | May 2009 Derecho | May 8, 2009 | 39 tornadoes EF3 – 2 confirmed | 6 +2 indirect Unknown | 500 million |
| 20 | 1932 Deep South Outbreak | March 21-22, 1932 | 33 tornadoes F4 – 10 confirmed | 334 Unknown | 5 million 78 million |
| 21 | 1920 Palm Sunday Outbreak | March 28, 1920 | 30 tornadoes F4 – 13 confirmed | 380 Unknown | damages unknown |
| 22 | January 1967 St. Louis Outbreak | January 24, 1967 | 30 tornadoes F4 – 2 confirmed | 7 268 | 100+ million 636+ million |
| 23 | 2004 Utica, IN Outbreak | April 20, 2004 | 30 tornadoes F3 – Utica, IN | 30 Unknown | damages unknown |
| | Preliminary Info. Mid June 2010 Outbreak | June 16-17, 2010 | 26 confirmed 76+ reported investigation in progress EF4 – 4 confirmed | 3 Unknown | Unknown at time of publication |

Note: Six F5 tornadoes are generally accepted as the number occurring during the 1974 Super Outbreak.   Some investigators consider the Mt. Hope, AL tornado (also called the Tanner, AL #2) to be an F5.   Because Tanner #1 was an F5 tornado, damage done by the Mt. Hope tornado as it followed a nearly identical path is indistinguishable from the damage done by Tanner #1 so damage evidence is inconclusive.

*Source:* NOAA.

**Table 4.12    The 25 Deadliest US Tornadoes**

|      | Date       | Communities              | Deaths |
|------|------------|--------------------------|--------|
| 1    | 3/18/1925  | Tri-State (MO/IL/IN)     | 695    |
| 2    | 5/6/1840   | Natchez MS               | 317    |
| 3    | 5/26/1896  | St. Louis MO             | 255    |
| 4    | 4/5/1936   | Tupelo MS                | 216    |
| 5    | 4/6/1936   | Gainesville GA           | 203    |
| 6    | 4/9/1947   | Woodward OK              | 181    |
| 7    | 4/24/1908  | Amite LA, Purvis MS      | 143    |
| 8    | 12Jun1899  | New Richmond WI          | 117    |
| 9    | 6/8/1953   | Flint MI                 | 115    |
| 10   | 5/11/1953  | Waco TX                  | 114    |
| 10   | 5/18/1902  | Goliad TX                | 114    |
| 12   | 3/23/1913  | Omaha NE                 | 103    |
| 13   | 5/26/1917  | Mattoon IL               | 101    |
| 14   | 6/23/1944  | Shinnston WV             | 100    |
| 15   | 4/18/1880  | Marshfield MO            | 99     |
| 16   | 6/1/1903   | Gainesville, Holland GA  | 98     |
| 16   | 5/9/1927   | Poplar Bluff MO          | 98     |
| 18   | 5/10/1905  | Snyder OK                | 97     |
| 19   | 4/24/1908  | Natchez MS               | 91     |
| 20   | 6/9/1953   | Worcester MA             | 90     |
| 21   | 4/20/1920  | Starkville MS, Waco AL   | 88     |
| 22   | 6/28/1924  | Lorain, Sandusky OH      | 85     |
| 23   | 5/25/1955  | Udall KS                 | 80     |
| 24   | 9/29/1927  | St. Louis MO             | 79     |
| 25   | 3/27/1890  | Louisville KY            | 76     |

*Source:* NOAA SPC, NCDC, and tornadoproject.com.

**Table 4.13    The Ten Costliest US Tornadoes**

|      | Date       | Communities      | Actual Dollars  | 2009 Dollars   |
|------|------------|------------------|-----------------|----------------|
| 1    | 6/8/1966   | Topeka KS        | 250,000,000     | 1,634,800,000  |
| 2    | 5/11/1970  | Lubbock TX       | 250,000,000     | 1,365,980,000  |
| 3    | 5/3/1999   | Oklahoma City OK | 1,000,000,000   | 1,275,510,000  |
| 4    | 4/3/1974   | Xenia OH         | 250,000,000     | 1,075,500,000  |
| 5    | 5/6/1975   | Omaha NE         | 250,603,000     | 988,174,260    |
| 6    | 4/10/1979  | Wichita Falls TX | 277,841,000     | 810,440,600    |
| 7    | 6/3/1980   | Grand Island NE  | 285,050,000     | 732,571,000    |
| 8    | 10/3/1979  | Windsor Locks CT | 250,000,000     | 729,230,000    |
| 9    | 5/8/2003   | Oklahoma City OK | 370,000,000     | 430,520,000    |
| 10   | 6/9/1953   | Worcester MA     | 52,000,000      | 412,380,000    |

*Source:* NOAA, NCDC, and tornadoproject.com.

**Table 4.14(a)  US Tornado Deaths by State, F/EF-Scale and Circumstance, 2002–2003. Only States with Fatalities Are Listed**

## Tornado Fatality Statistics 2000 - 2003

| State | Killer Tornadoes | Deaths | F Scale | Killer Tornadoes | Deaths | Circumstance | Deaths |
|-------|-----------------|--------|---------|-----------------|--------|--------------|--------|
| **2000** | | | | | | | |
| GA | 4 | 19 | F0 | 0 | 0 | Mobile Home | 28 |
| AL | 3 | 13 | F1 | 2 | 2 | Permanent Home | 7 |
| TX | 2 | 4 | F2 | 4 | 5 | Vehicle | 4 |
| TN | 1 | 1 | F3 | 5 | 21 | Outside/Open | 2 |
| IA | 1 | 1 | F4 | 3 | 13 | TOTAL | 41 |
| MN | 1 | 1 | F5 | 0 | 0 | | |
| OH | 1 | 1 | F? | 0 | 0 | | |
| SC | 1 | 1 | TOTAL | 14 | 41 | | |
| TOTAL | 14 | 41 | | | | | |
| **2001** | | | | | | | |
| MS | 5 | 12 | F0 | 1 | 1 | Mobile Home | 17 |
| AL | 3 | 6 | F1 | 3 | 4 | Permanent Home | 15 |
| AR | 3 | 5 | F2 | 11 | 14 | Vehicle | 3 |
| WI | 1 | 3 | F3 | 7 | 18 | Business | 3 |
| IA | 1 | 2 | F4 | 2 | 3 | Outside/Open | 2 |
| TN | 2 | 2 | F5 | 0 | 0 | TOTAL | 40 |
| FL | 2 | 2 | F? | 0 | 0 | | |
| MD | 1 | 2 | TOTAL | 24 | 40 | | |
| IN | 2 | 2 | | | | | |
| LA | 1 | 1 | | | | | |
| MO | 1 | 1 | | | | | |
| KS | 1 | 1 | | | | | |
| OK | 1 | 1 | | | | | |
| TOTAL | 24 | 40 | | | | | |
| **2002** | | | | | | | |
| TN | 5 | 17 | F0 | 0 | 0 | Mobile Home | 32 |
| AL | 5 | 13 | F1 | 4 | 6 | Permanent Home | 15 |
| OH | 3 | 5 | F2 | 11 | 17 | Vehicle | 4 |
| MO | 3 | 4 | F3 | 11 | 29 | Business | 1 |
| IL | 3 | 4 | F4 | 2 | 3 | Outside/Open | 3 |
| MD | 2 | 3 | F5 | 0 | 0 | TOTAL | 55 |
| TX | 2 | 3 | F? | 0 | 0 | | |
| LA | 1 | 2 | TOTAL | 28 | 55 | | |
| KY | 1 | 1 | | | | | |
| PA | 1 | 1 | | | | | |
| MS | 1 | 1 | | | | | |
| AR | 1 | 1 | | | | | |
| TOTAL | 28 | 55 | | | | | |
| **2003** | | | | | | | |
| MO | 8 | 19 | F0 | 1 | 1 | Mobile Home | 25 |
| TN | 2 | 12 | F1 | 2 | 3 | Permanent Home | 24 |
| KS | 3 | 8 | F2 | 4 | 5 | Business | 1 |
| GA | 2 | 6 | F3 | 9 | 25 | Outside/Open | 3 |
| KY | 2 | 3 | F4 | 7 | 20 | Other/Unknown | 1 |
| IL | 2 | 2 | F5 | 0 | 0 | TOTAL | 54 |
| NE | 2 | 2 | F? | 0 | 0 | | |
| FL | 1 | 1 | TOTAL | 23 | 54 | | |
| NJ | 1 | 1 | | | | | |
| TOTAL | 23 | 54 | | | | | |

*Source:* NCDC.

**Table 4.14(b)**   US Tornado Deaths by State, F/EF-Scale and Circumstance, 2004–2006. Only States with Fatalities Are Listed

## Tornado Fatality Statistics 2004 - 2006

| State | Killer Tornadoes | Deaths | F Scale | Killer Torornadoes | Deaths | Circumstance | Deaths |
|---|---|---|---|---|---|---|---|
| | | | | 2004 | | | |
| IL | 2 | 9 | F0 | 1 | 1 | Mobile Home | 11 |
| MO | 3 | 7 | F1 | 5 | 5 | Permanent Home | 21 |
| FL | 4 | 7 | F2 | 4 | 8 | Vehicle | 2 |
| NC | 1 | 3 | F3 | 5 | 12 | Business | 2 |
| TX | 2 | 2 | F4 | 2 | 4 | TOTAL | 36 |
| WI | 1 | 1 | F5 | 0 | 0 | | |
| IN | 1 | 1 | F? | 3 | 6 | | |
| NE | 1 | 1 | TOTAL | 20 | 36 | | |
| GA | 1 | 1 | | | | | |
| SC | 1 | 1 | | | | | |
| LA | 1 | 1 | | | | | |
| MS | 1 | 1 | | | | | |
| AL | 1 | 1 | | | | | |
| TOTAL | 20 | 36 | | | | | |
| | | | | 2005 | | | |
| IN | 1 | 25 | F0 | 0 | 0 | Mobile Home | 34 |
| GA | 3 | 4 | F1 | 2 | 3 | Permanent Home | 3 |
| AR | 2 | 3 | F2 | 4 | 5 | Vehicle | 1 |
| WY | 1 | 2 | F3 | 6 | 31 | Business | 1 |
| MS | 1 | 1 | F4 | 0 | 0 | TOTAL | 39 |
| WI | 1 | 1 | F5 | 0 | 0 | | |
| IA | 1 | 1 | F? | 0 | 0 | | |
| KY | 1 | 1 | TOTAL | 12 | 39 | | |
| MO | 1 | 1 | | | | | |
| TOTAL | 12 | 39 | | | | | |
| | | | | 2006 | | | |
| TN | 6 | 34 | F0 | 0 | 0 | Mobile Home | 27 |
| MO | 8 | 13 | F1 | 7 | 8 | Permanent Home | 31 |
| NC | 2 | 9 | F2 | 7 | 7 | Vehicle | 7 |
| TX | 2 | 4 | F3 | 11 | 52 | Business | 2 |
| MN | 2 | 2 | F4 | 0 | 0 | TOTAL | 67 |
| AL | 1 | 1 | F5 | 0 | 0 | | |
| IA | 1 | 1 | F? | 0 | 0 | | |
| LA | 1 | 1 | TOTAL | 25 | 67 | | |
| IL | 1 | 1 | | | | | |
| PA | 1 | 1 | | | | | |
| TOTAL | 25 | 67 | | | | | |

*Source:* NCDC.

**Table 4.14(c)  US Tornado Deaths by State, F/EF-Scale and Circumstance, 2007–2009. Only States with Fatalities Are Listed**

## Tornado Fatality Statistics 2007 - 2009

| State | Killer Tornadoes | Deaths | F Scale | Killer Tornadoes | Deaths | Circumstance | Deaths |
|---|---|---|---|---|---|---|---|
| **2007** | | | | | | | |
| FL | 2 | 21 | F0 | 0 | 0 | Mobile Home | 52 |
| KS | 4 | 14 | F1 | 3 | 4 | Permanent Home | 16 |
| AL | 2 | 10 | F2 | 10 | 20 | Vehicle | 2 |
| GA | 4 | 10 | F3 | 9 | 35 | Business | 10 |
| TX | 3 | 9 | F4 | 3 | 11 | Outside/Open | 1 |
| MO | 2 | 3 | F5 | 1 | 11 | TOTAL | 81 |
| LA | 2 | 3 | F? | 0 | 0 | | |
| MI | 2 | 3 | TOTAL | 26 | 81 | | |
| NM | 1 | 2 | | | | | |
| OK | 1 | 2 | | | | | |
| CO | 1 | 2 | | | | | |
| ND | 1 | 1 | | | | | |
| SC | 1 | 1 | | | | | |
| TOTAL | 26 | 81 | | | | | |
| **2008** | | | | | | | |
| TN | 4 | 31 | F0 | 0 | 0 | Mobile Home | 56 |
| AR | 5 | 21 | F1 | 4 | 4 | Permanent Home | 43 |
| MO | 3 | 19 | F2 | 13 | 18 | Vehicle | 14 |
| IA | 2 | 13 | F3 | 14 | 53 | Business | 10 |
| KY | 2 | 7 | F4 | 6 | 42 | Outside/Open | 3 |
| AL | 3 | 6 | F5 | 1 | 9 | TOTAL | 126 |
| OK | 1 | 6 | F? | 0 | 0 | | |
| GA | 3 | 5 | TOTAL | 38 | 126 | | |
| KS | 3 | 4 | | | | | |
| IN | 3 | 4 | | | | | |
| LA | 2 | 3 | | | | | |
| NC | 3 | 3 | | | | | |
| FL | 1 | 1 | | | | | |
| CO | 1 | 1 | | | | | |
| MN | 1 | 1 | | | | | |
| NH | 1 | 1 | | | | | |
| TOTAL | 38 | 126 | | | | | |
| **2009** | | | | | | | |
| OK | 1 | 8 | F0 | 0 | 0 | Mobile Home | 12 |
| AR | 1 | 3 | F1 | 3 | 3 | Permanent Home | 7 |
| MO | 2 | 3 | F2 | 1 | 2 | Vehicle | 1 |
| TN | 1 | 2 | F3 | 3 | 6 | Business | 1 |
| KY | 1 | 2 | F4 | 2 | 10 | TOTAL | 21 |
| AL | 1 | 1 | F5 | 0 | 0 | | |
| GA | 1 | 1 | F? | 0 | 0 | | |
| MS | 1 | 1 | TOTAL | 9 | 21 | | |
| TOTAL | 9 | 21 | | | | | |

*Source:* NCDC.

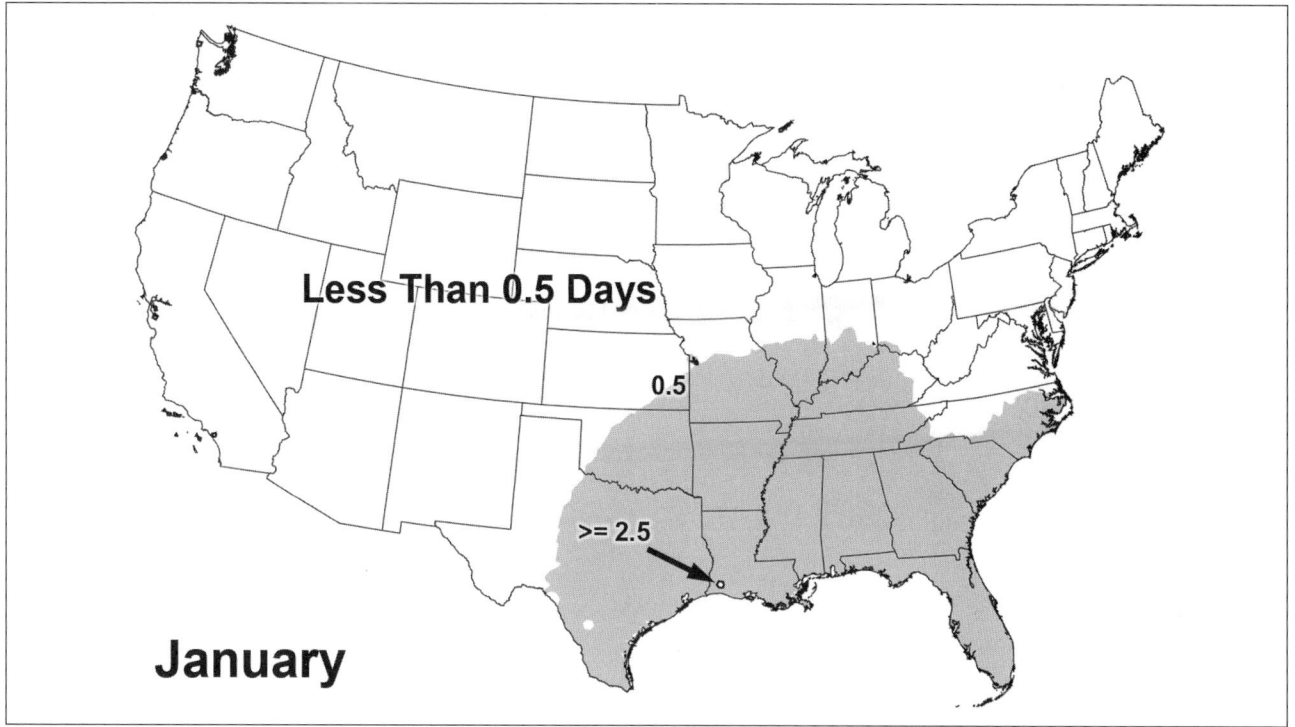

**Figure 4.39**   Average number of days with thunder detected in January. (*Source:* Data by NOAA NCDC.)

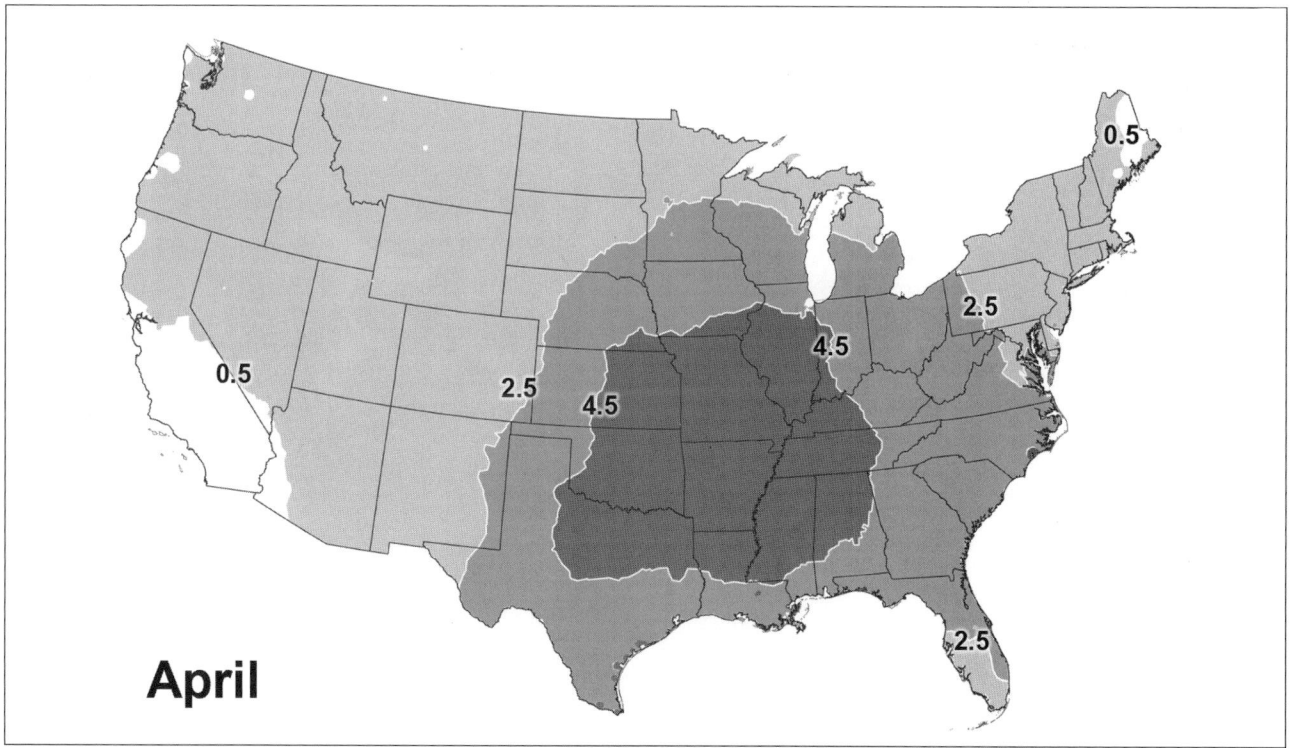

**Figure 4.40**   Average number of days with thunder detected in April. (*Source:* Data by NOAA NCDC.)

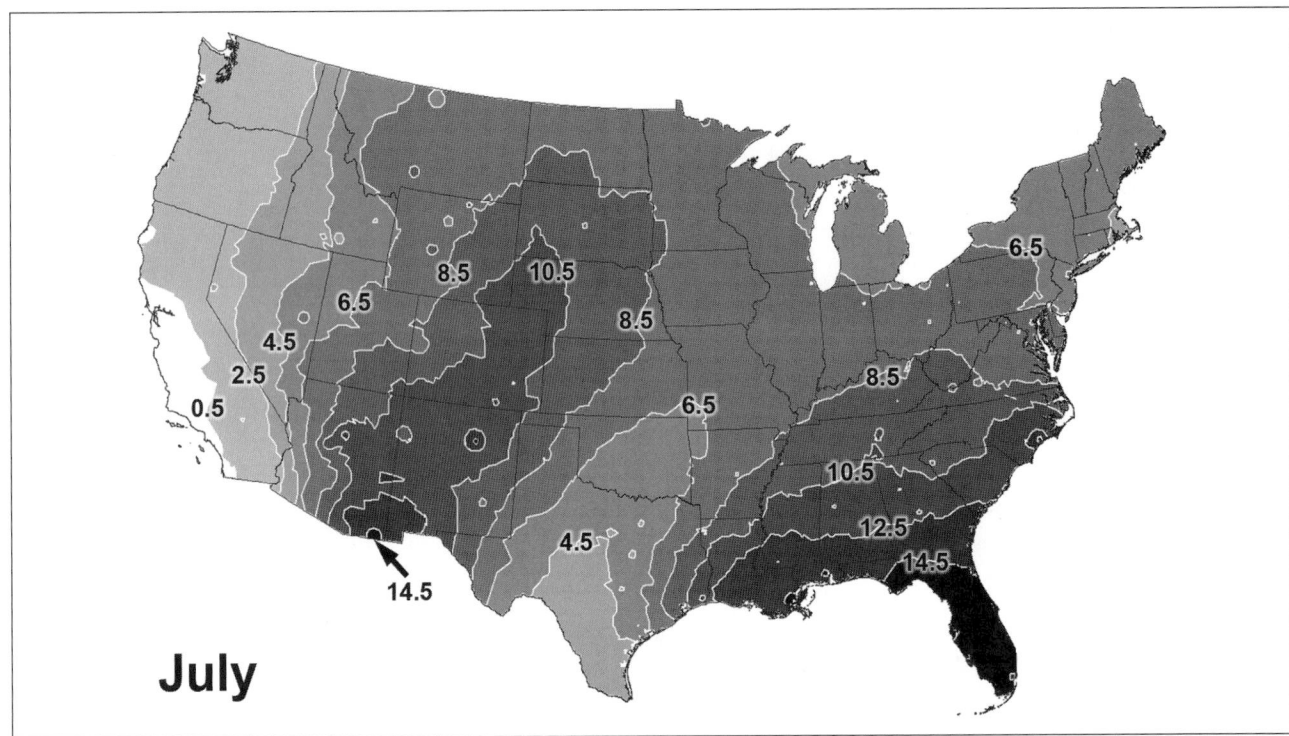

**Figure 4.41** Average number of days with thunder detected in July. (*Source:* Data by NOAA NCDC.)

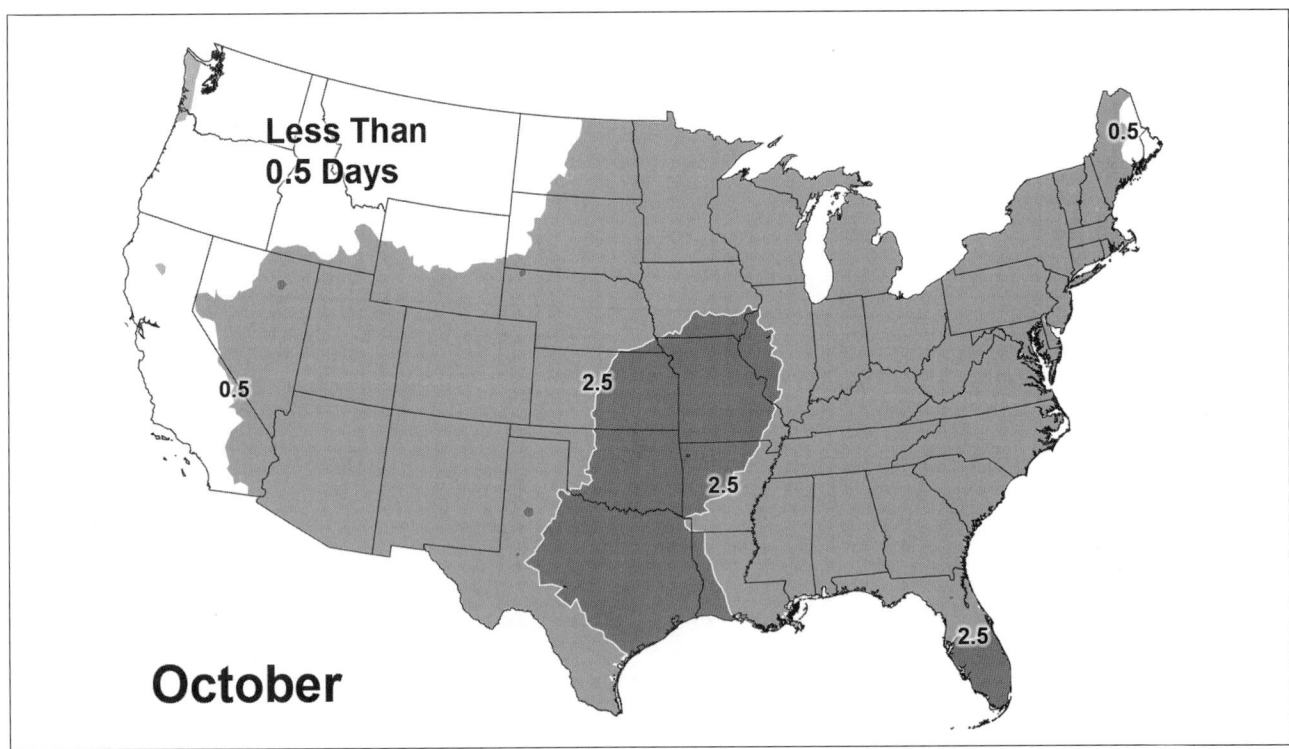

**Figure 4.42** Average number of days with thunder detected in October. (*Source:* Data by NOAA NCDC.)

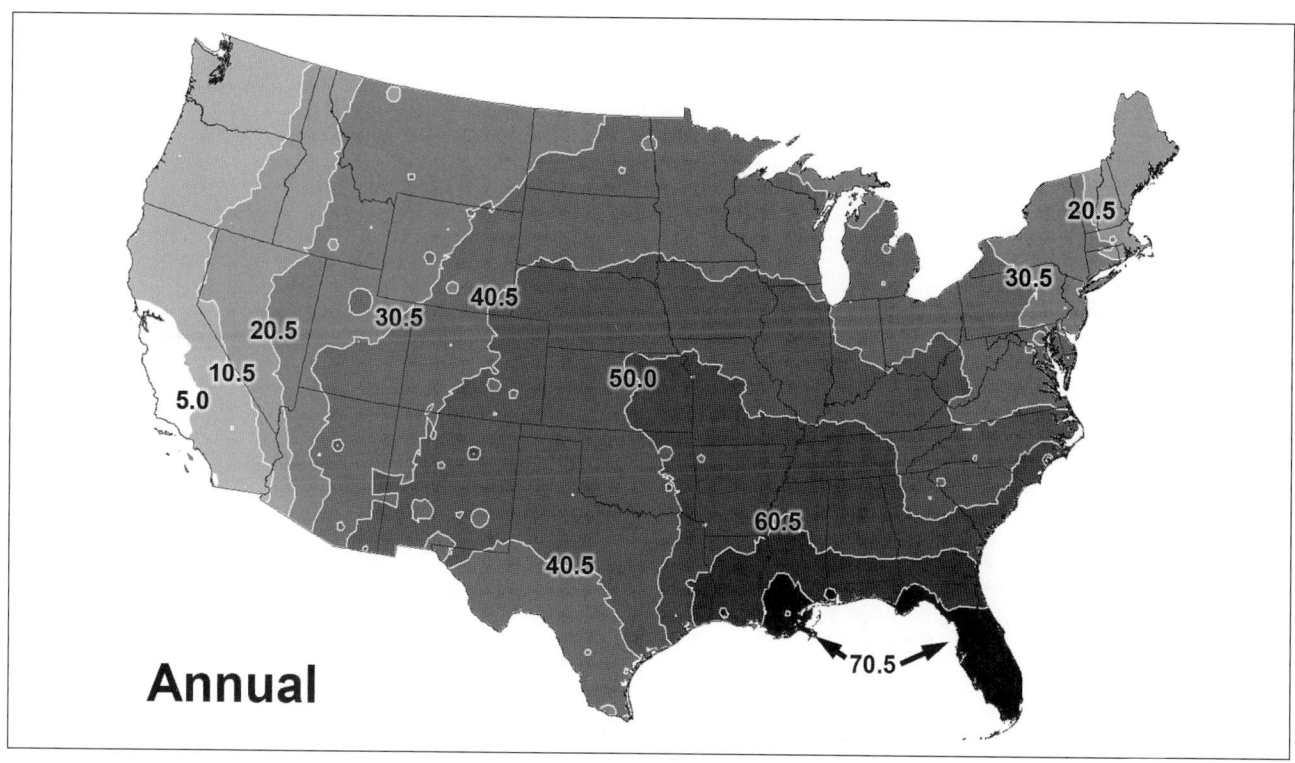

**Figure 4.43**   Average annual number of days with thunder detected. (*Source:* Data by NOAA NCDC.)

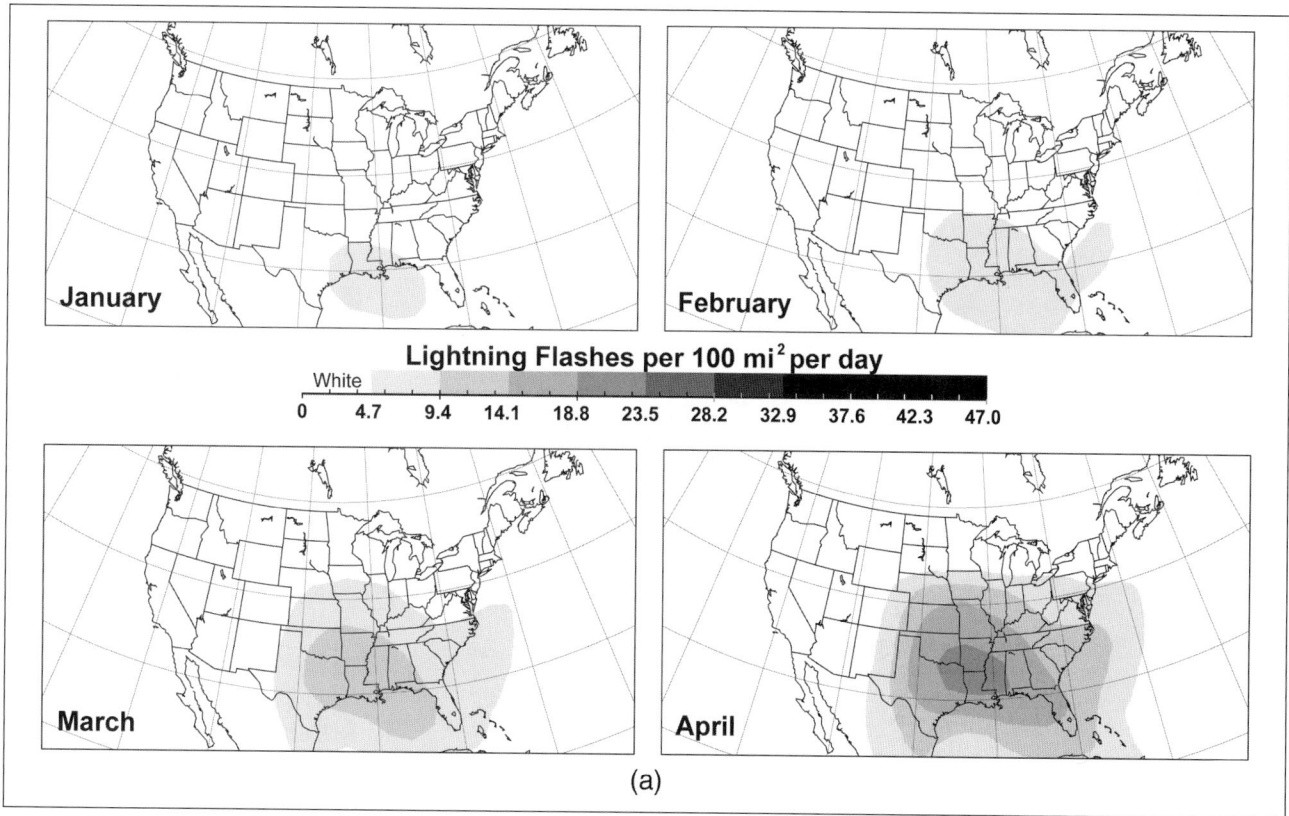

(a)

**Figure 4.44 a–c**   Lightning flashes per 100 square miles per day for each month in the United States as observed by NASA satellites. (*Source:* Modified from original maps by NASA, Marshall Space Flight Center, Global Hydrology and Climate Center.)

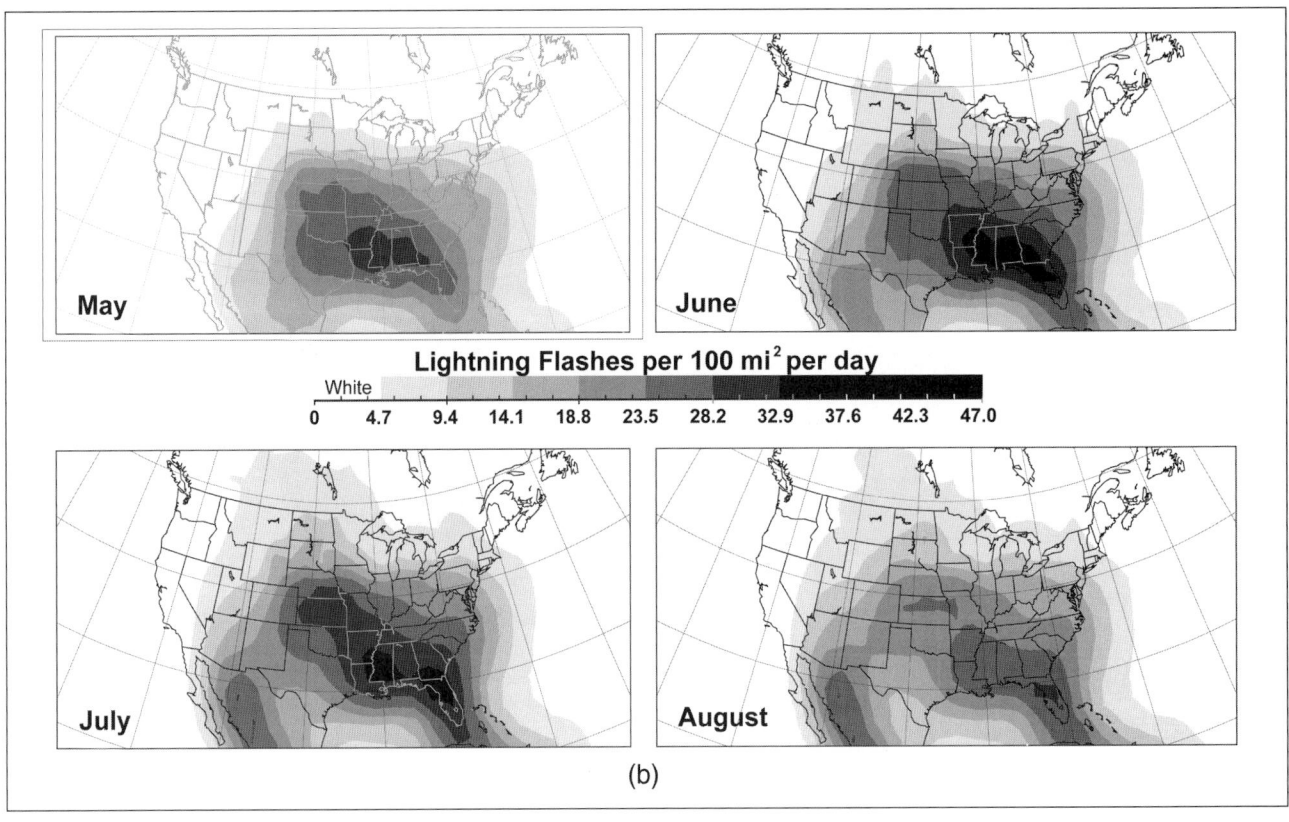

(b)

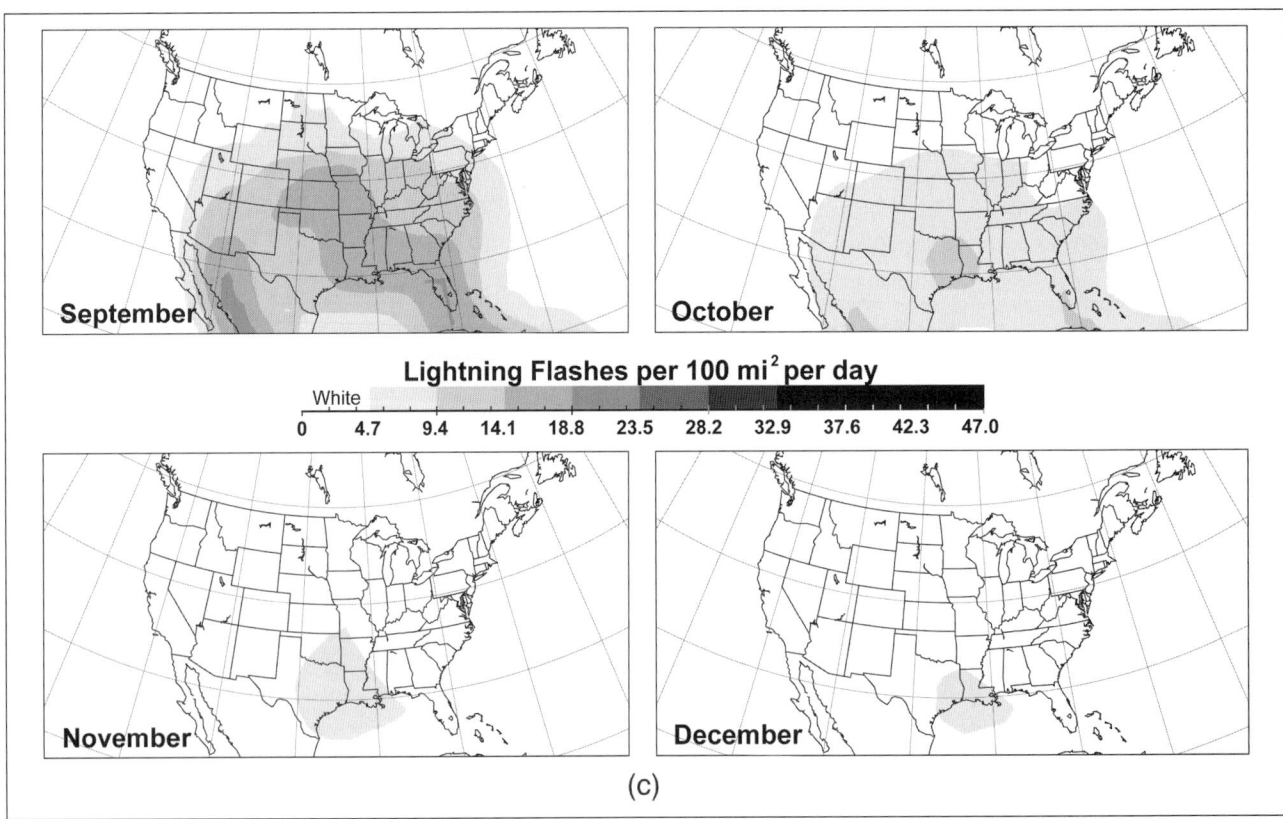

(c)

**Figure 4.44** *(Continued)*

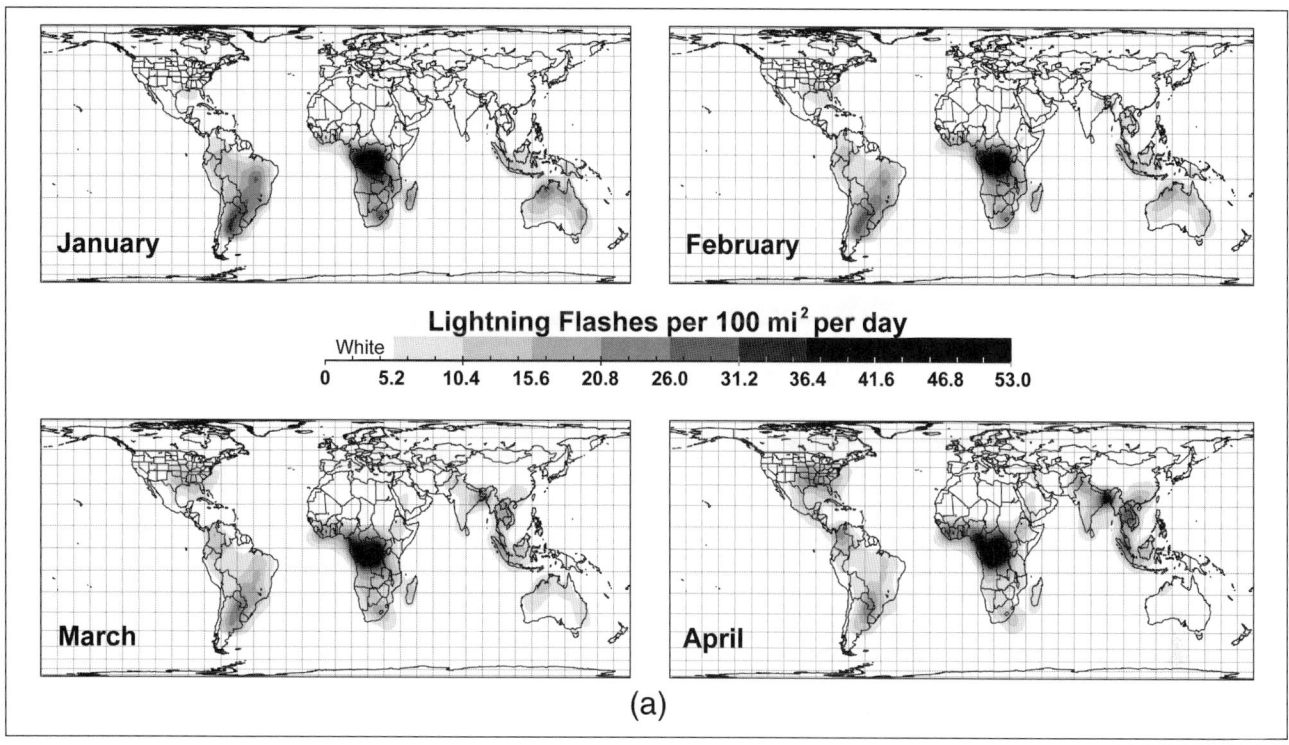

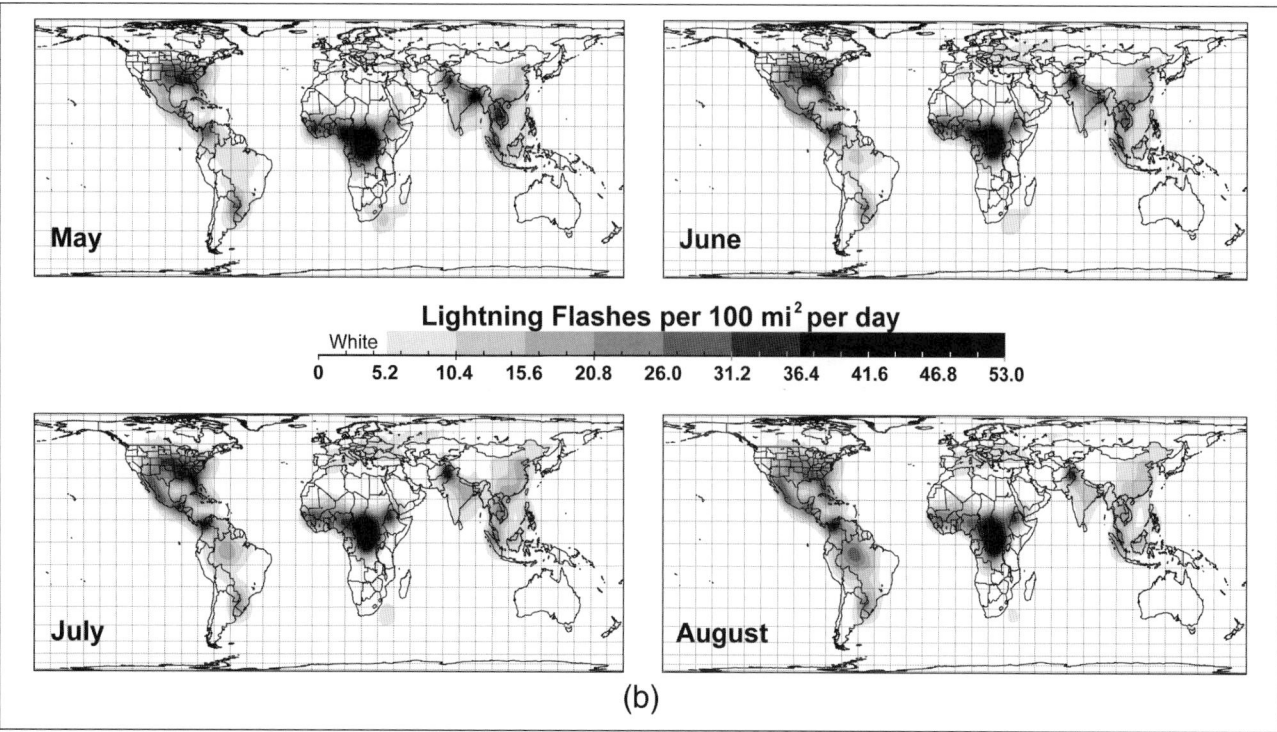

**Figure 4.45 a–c**    Lightning flashes per 100 square miles per day for each month for the entire globe as observed by NASA satellites. (*Source:* Modified from original maps by NASA, Marshall Space Flight Center, Global Hydrology and Climate Center.)

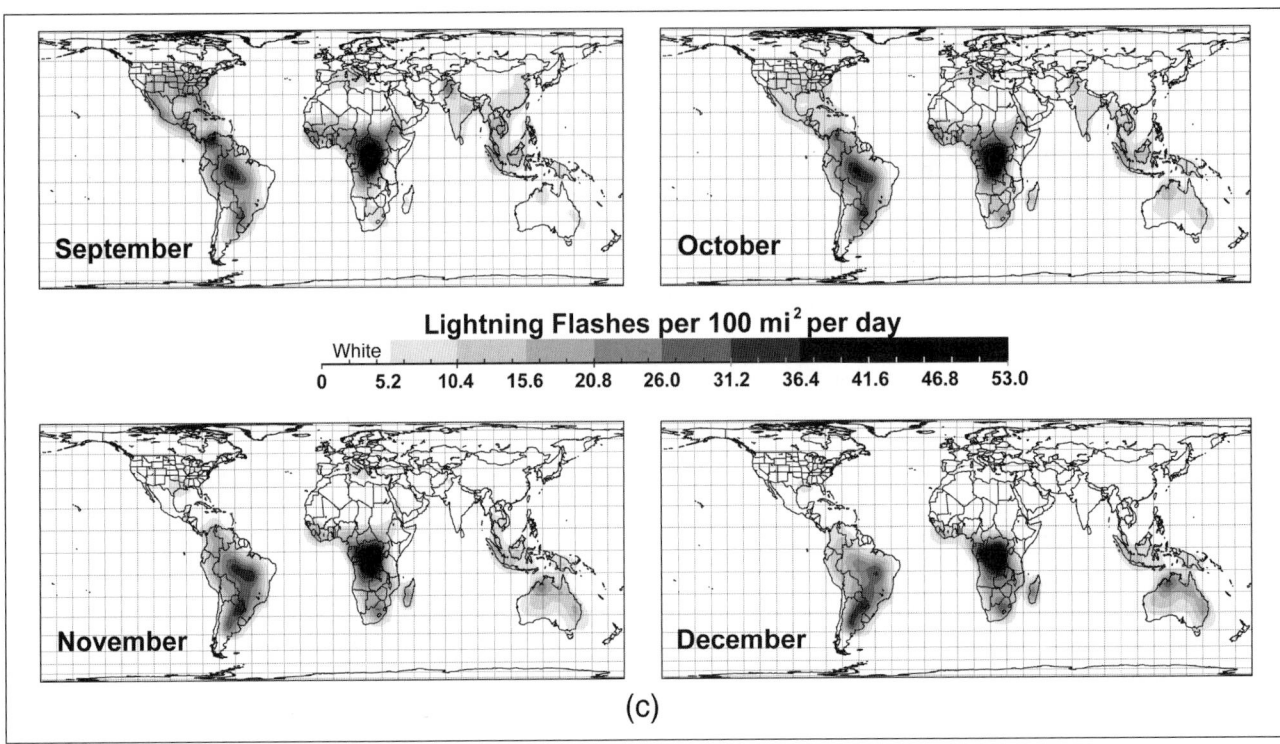

(c)

**Figure 4.45**  (*Continued*)

**Table 4.15(a)   US Lightning Deaths by State and Activity/Location, 2000. Only States with Fatalities Are Listed**

## U.S. Lightning Deaths 2000

| State | Ball Field | Boat | Camping | Near Heavy Equip. | Golf | In Water | Not Known | In the Open | Using Phone | Under Tree | Total |
|---|---|---|---|---|---|---|---|---|---|---|---|
| AZ | 0 | 0 | 0 | 0 | 0 | 0 | 1 | 1 | 0 | 0 | 2 |
| CO | 0 | 0 | 0 | 0 | 0 | 0 | 0 | 2 | 0 | 0 | 2 |
| CT | 0 | 0 | 0 | 0 | 0 | 0 | 1 | 0 | 0 | 0 | 1 |
| FL | 0 | 3 | 0 | 0 | 1 | 2 | 0 | 2 | 0 | 0 | 8 |
| GA | 0 | 0 | 0 | 0 | 0 | 0 | 1 | 2 | 0 | 1 | 4 |
| IA | 0 | 0 | 0 | 0 | 0 | 0 | 1 | 0 | 0 | 0 | 1 |
| KY | 0 | 0 | 0 | 0 | 0 | 0 | 2 | 2 | 0 | 0 | 4 |
| LA | 0 | 0 | 0 | 0 | 2 | 0 | 1 | 0 | 0 | 0 | 3 |
| MA | 0 | 0 | 0 | 0 | 0 | 0 | 0 | 1 | 0 | 0 | 1 |
| MD | 0 | 0 | 1 | 0 | 0 | 0 | 0 | 0 | 0 | 2 | 3 |
| ME | 0 | 0 | 0 | 0 | 0 | 0 | 0 | 1 | 0 | 0 | 1 |
| MI | 0 | 0 | 0 | 0 | 0 | 0 | 0 | 1 | 0 | 0 | 1 |
| MS | 0 | 0 | 0 | 0 | 0 | 0 | 0 | 1 | 0 | 0 | 1 |
| NC | 0 | 0 | 0 | 0 | 0 | 0 | 0 | 1 | 0 | 0 | 1 |
| NY | 0 | 0 | 0 | 0 | 0 | 0 | 1 | 0 | 0 | 0 | 1 |
| OH | 0 | 0 | 0 | 0 | 1 | 0 | 0 | 0 | 0 | 1 | 2 |
| PA | 0 | 0 | 0 | 0 | 0 | 0 | 0 | 2 | 0 | 0 | 2 |
| PR | 0 | 0 | 0 | 0 | 0 | 0 | 0 | 1 | 0 | 0 | 1 |
| SC | 0 | 0 | 0 | 0 | 0 | 0 | 0 | 0 | 0 | 2 | 2 |
| TX | 0 | 0 | 0 | 0 | 0 | 0 | 1 | 3 | 0 | 1 | 5 |
| UT | 0 | 0 | 0 | 0 | 0 | 0 | 0 | 2 | 0 | 0 | 2 |
| VA | 0 | 0 | 0 | 0 | 0 | 0 | 0 | 0 | 0 | 2 | 2 |
| WI | 0 | 0 | 0 | 0 | 0 | 0 | 0 | 1 | 0 | 0 | 1 |
| Total | 0 | 3 | 1 | 0 | 4 | 2 | 9 | 23 | 0 | 9 | 51 |
| Percent | 0 | 6 | 2 | 0 | 8 | 4 | 18 | 45 | 0 | 18 | |

**Table 4.15(b)   US Lightning Deaths by State and Activity/Location, 2001. Only States with Fatalities Are Listed**

## U.S. Lightning Deaths 2001

| State | Ball Field | Boat | Camping | Near Heavy Equip. | Golf | In Water | Not Known | In the Open | Using Phone | Under Tree | Total |
|-------|-----------|------|---------|-------------------|------|----------|-----------|-------------|-------------|------------|-------|
| AL | 1 | 0 | 0 | 0 | 0 | 0 | 0 | 0 | 0 | 1 | 2 |
| CA | 0 | 0 | 0 | 0 | 0 | 0 | 0 | 2 | 0 | 0 | 2 |
| CO | 0 | 0 | 0 | 0 | 0 | 0 | 0 | 3 | 0 | 0 | 3 |
| FL | 0 | 0 | 0 | 0 | 0 | 0 | 2 | 6 | 0 | 0 | 8 |
| GA | 0 | 0 | 0 | 0 | 0 | 0 | 0 | 1 | 0 | 0 | 1 |
| IL | 0 | 0 | 0 | 0 | 1 | 0 | 1 | 3 | 0 | 0 | 5 |
| IN | 0 | 0 | 0 | 0 | 1 | 0 | 0 | 0 | 0 | 0 | 1 |
| LA | 0 | 0 | 0 | 0 | 0 | 0 | 0 | 1 | 0 | 0 | 1 |
| MA | 0 | 0 | 0 | 0 | 0 | 0 | 0 | 1 | 0 | 0 | 1 |
| MI | 0 | 0 | 0 | 0 | 0 | 0 | 0 | 1 | 0 | 0 | 1 |
| MS | 0 | 0 | 0 | 0 | 0 | 0 | 0 | 1 | 0 | 0 | 1 |
| NE | 0 | 0 | 0 | 0 | 0 | 0 | 0 | 1 | 0 | 0 | 1 |
| NJ | 0 | 1 | 0 | 0 | 0 | 0 | 0 | 1 | 0 | 0 | 2 |
| OH | 0 | 0 | 0 | 0 | 0 | 0 | 1 | 0 | 0 | 1 | 2 |
| PA | 1 | 0 | 0 | 0 | 0 | 0 | 0 | 0 | 0 | 1 | 2 |
| PR | 0 | 0 | 0 | 0 | 0 | 0 | 0 | 2 | 0 | 0 | 2 |
| SC | 0 | 0 | 0 | 0 | 0 | 0 | 0 | 1 | 0 | 0 | 1 |
| SD | 0 | 0 | 0 | 0 | 0 | 0 | 0 | 2 | 0 | 0 | 2 |
| TN | 0 | 0 | 0 | 0 | 0 | 0 | 0 | 0 | 0 | 1 | 1 |
| TX | 0 | 0 | 0 | 0 | 0 | 0 | 0 | 1 | 0 | 0 | 1 |
| VA | 0 | 1 | 0 | 0 | 0 | 0 | 2 | 0 | 0 | 0 | 3 |
| WI | 0 | 0 | 0 | 0 | 0 | 0 | 0 | 1 | 0 | 0 | 1 |
| Total | 2 | 2 | 0 | 0 | 2 | 0 | 6 | 28 | 0 | 4 | 44 |
| Percent | 5 | 5 | 0 | 0 | 5 | 0 | 14 | 64 | 0 | 9 | |

Table 4.15(c)    US Lightning Deaths by State and Activity/Location, 2002. Only States with Fatalities Are Listed

## U.S. Lightning Deaths 2002

| State | Ball Field | Boat | Camping | Near Heavy Equip. | Golf | In Water | Not Known | In the Open | Using Phone | Under Tree | Total |
|-------|-----------|------|---------|-------------------|------|----------|-----------|-------------|-------------|------------|-------|
| AL | 0 | 0 | 0 | 0 | 0 | 0 | 0 | 1 | 0 | 0 | 1 |
| AR | 0 | 0 | 0 | 0 | 0 | 0 | 0 | 1 | 0 | 0 | 1 |
| FL | 0 | 1 | 0 | 0 | 0 | 0 | 1 | 6 | 0 | 1 | 9 |
| GA | 0 | 0 | 0 | 0 | 0 | 0 | 0 | 1 | 0 | 1 | 2 |
| ID | 0 | 0 | 0 | 0 | 0 | 0 | 0 | 1 | 0 | 0 | 1 |
| IL | 1 | 0 | 0 | 0 | 0 | 0 | 0 | 0 | 0 | 0 | 1 |
| LA | 0 | 0 | 0 | 0 | 0 | 0 | 0 | 1 | 0 | 1 | 2 |
| MD | 0 | 0 | 0 | 0 | 0 | 0 | 0 | 1 | 0 | 0 | 1 |
| MI | 0 | 0 | 0 | 0 | 0 | 0 | 1 | 0 | 0 | 0 | 1 |
| MO | 0 | 0 | 0 | 0 | 0 | 0 | 0 | 1 | 0 | 4 | 5 |
| MS | 0 | 0 | 1 | 0 | 0 | 0 | 0 | 0 | 0 | 1 | 2 |
| NC | 0 | 0 | 0 | 0 | 0 | 0 | 0 | 1 | 0 | 1 | 2 |
| NJ | 0 | 0 | 0 | 0 | 0 | 0 | 0 | 1 | 0 | 0 | 1 |
| NY | 0 | 0 | 0 | 0 | 0 | 0 | 1 | 1 | 0 | 0 | 2 |
| OH | 0 | 0 | 0 | 0 | 0 | 0 | 1 | 0 | 0 | 0 | 1 |
| OK | 0 | 0 | 0 | 0 | 0 | 0 | 0 | 0 | 0 | 1 | 1 |
| PA | 0 | 0 | 1 | 0 | 0 | 0 | 0 | 0 | 0 | 0 | 1 |
| SC | 0 | 1 | 0 | 0 | 0 | 0 | 0 | 1 | 0 | 0 | 2 |
| TN | 0 | 0 | 0 | 0 | 0 | 0 | 0 | 3 | 0 | 0 | 3 |
| TX | 0 | 0 | 0 | 0 | 0 | 1 | 0 | 1 | 0 | 0 | 2 |
| UT | 0 | 0 | 0 | 0 | 0 | 0 | 0 | 2 | 0 | 0 | 2 |
| VT | 0 | 0 | 1 | 0 | 0 | 0 | 0 | 0 | 0 | 0 | 1 |
| WI | 0 | 0 | 1 | 0 | 0 | 0 | 0 | 0 | 0 | 0 | 1 |
| WY | 0 | 0 | 0 | 0 | 0 | 0 | 0 | 0 | 0 | 1 | 1 |
| Total | 1 | 2 | 4 | 0 | 0 | 1 | 8 | 24 | 0 | 11 | 51 |
| Percent | 2 | 4 | 8 | 0 | 0 | 2 | 16 | 47 | 0 | 22 | |

**Table 4.15(d)   US Lightning Deaths by State and Activity/Location, 2003. Only States with Fatalities Are Listed**

## U.S. Lightning Deaths 2003

| State | Ball Field | Boat | Camping | Near Heavy Equip. | Golf | In Water | Not Known | In the Open | Using Phone | Under Tree | Total |
|-------|-----------|------|---------|-------------------|------|----------|-----------|-------------|-------------|-----------|-------|
| CA | 0 | 0 | 0 | 1 | 0 | 0 | 0 | 0 | 0 | 0 | 1 |
| CO | 0 | 0 | 0 | 0 | 1 | 0 | 0 | 2 | 0 | 3 | 6 |
| FL | 2 | 0 | 0 | 0 | 0 | 1 | 1 | 5 | 0 | 1 | 10 |
| GA | 0 | 0 | 0 | 0 | 0 | 0 | 0 | 2 | 0 | 0 | 2 |
| GU | 0 | 0 | 0 | 0 | 0 | 1 | 0 | 0 | 0 | 0 | 1 |
| IA | 0 | 0 | 1 | 0 | 0 | 0 | 0 | 0 | 0 | 0 | 1 |
| MI | 0 | 0 | 0 | 0 | 0 | 0 | 0 | 1 | 0 | 0 | 1 |
| NC | 0 | 1 | 0 | 0 | 0 | 0 | 3 | 1 | 0 | 0 | 5 |
| NM | 0 | 0 | 0 | 0 | 0 | 0 | 0 | 1 | 0 | 0 | 1 |
| OH | 0 | 0 | 0 | 0 | 0 | 0 | 1 | 0 | 0 | 0 | 1 |
| OK | 0 | 0 | 0 | 0 | 0 | 0 | 0 | 2 | 0 | 0 | 2 |
| SD | 0 | 0 | 0 | 0 | 0 | 1 | 0 | 0 | 0 | 0 | 1 |
| TX | 0 | 0 | 0 | 0 | 0 | 0 | 1 | 1 | 0 | 2 | 4 |
| UT | 0 | 0 | 0 | 0 | 0 | 1 | 0 | 0 | 0 | 2 | 3 |
| VA | 0 | 0 | 0 | 0 | 0 | 0 | 0 | 1 | 0 | 0 | 1 |
| VI | 0 | 0 | 0 | 0 | 0 | 1 | 0 | 0 | 0 | 0 | 1 |
| WY | 0 | 0 | 0 | 0 | 0 | 0 | 0 | 2 | 0 | 0 | 2 |
| Total | 2 | 1 | 1 | 1 | 1 | 5 | 6 | 18 | 0 | 8 | 43 |
| Percent | 5 | 2 | 2 | 2 | 2 | 12 | 14 | 42 | 0 | 19 | |

**Table 4.15(e)  US Lightning Deaths by State and Activity/Location, 2004. Only States with Fatalities Are Listed**

## U.S. Lightning Deaths 2004

| State | Ball Field | Boat | Camping | Near Heavy Equip. | Golf | In Water | Not Known | In the Open | Using Phone | Under Tree | Total |
|---|---|---|---|---|---|---|---|---|---|---|---|
| AL | 0 | 0 | 0 | 0 | 0 | 0 | 1 | 0 | 0 | 0 | 1 |
| AR | 0 | 0 | 0 | 0 | 0 | 0 | 0 | 1 | 0 | 0 | 1 |
| CA | 0 | 0 | 0 | 0 | 0 | 0 | 0 | 1 | 0 | 0 | 1 |
| CO | 0 | 0 | 0 | 0 | 1 | 0 | 0 | 2 | 0 | 0 | 3 |
| FL | 1 | 0 | 0 | 0 | 1 | 0 | 0 | 2 | 0 | 1 | 5 |
| GA | 0 | 0 | 0 | 1 | 0 | 0 | 0 | 1 | 0 | 3 | 5 |
| IN | 1 | 0 | 0 | 0 | 0 | 0 | 0 | 0 | 0 | 0 | 1 |
| LA | 0 | 1 | 0 | 1 | 0 | 0 | 0 | 0 | 0 | 0 | 2 |
| MS | 0 | 1 | 0 | 0 | 0 | 0 | 0 | 0 | 0 | 0 | 1 |
| NC | 0 | 0 | 0 | 0 | 0 | 0 | 0 | 1 | 0 | 0 | 1 |
| NY | 0 | 0 | 0 | 0 | 0 | 0 | 0 | 0 | 0 | 1 | 1 |
| OH | 0 | 0 | 0 | 0 | 0 | 1 | 0 | 0 | 0 | 0 | 1 |
| RI | 0 | 0 | 0 | 0 | 0 | 0 | 0 | 1 | 0 | 0 | 1 |
| SC | 0 | 1 | 0 | 0 | 0 | 0 | 0 | 1 | 0 | 0 | 2 |
| TN | 0 | 0 | 0 | 0 | 0 | 0 | 0 | 1 | 0 | 0 | 1 |
| TX | 0 | 0 | 0 | 0 | 0 | 0 | 0 | 2 | 0 | 1 | 3 |
| VA | 0 | 0 | 0 | 0 | 0 | 0 | 1 | 0 | 0 | 0 | 1 |
| VT | 0 | 0 | 0 | 0 | 1 | 0 | 0 | 0 | 0 | 0 | 1 |
| Total | 2 | 3 | 0 | 2 | 3 | 1 | 2 | 13 | 0 | 6 | 32 |
| Percent | 6 | 9 | 0 | 6 | 9 | 3 | 6 | 41 | 0 | 19 | |

Table 4.15(f)   US Lightning Deaths by State and Activity/Location, 2005. Only States with Fatalities Are Listed

## U.S. Lightning Deaths 2005

| State | Ball Field | Boat | Camping | Near Heavy Equip. | Golf | In Water | Not Known | In the Open | Using Phone | Under Tree | Total |
|---|---|---|---|---|---|---|---|---|---|---|---|
| AL | 0 | 0 | 0 | 0 | 0 | 0 | 0 | 2 | 0 | 0 | 2 |
| AZ | 0 | 0 | 0 | 0 | 0 | 0 | 0 | 0 | 0 | 1 | 1 |
| CA | 0 | 0 | 2 | 0 | 0 | 0 | 0 | 0 | 0 | 0 | 2 |
| CO | 0 | 0 | 0 | 0 | 0 | 0 | 0 | 1 | 0 | 0 | 1 |
| FL | 1 | 0 | 0 | 0 | 0 | 0 | 0 | 4 | 0 | 0 | 5 |
| GA | 0 | 0 | 0 | 0 | 0 | 0 | 0 | 0 | 0 | 2 | 2 |
| IA | 0 | 0 | 0 | 0 | 1 | 0 | 0 | 0 | 0 | 0 | 1 |
| IL | 0 | 0 | 0 | 0 | 0 | 0 | 0 | 1 | 0 | 0 | 1 |
| KY | 0 | 0 | 0 | 0 | 0 | 0 | 0 | 2 | 0 | 0 | 2 |
| LA | 0 | 0 | 0 | 0 | 0 | 0 | 0 | 1 | 0 | 0 | 1 |
| ME | 0 | 1 | 0 | 0 | 0 | 0 | 0 | 0 | 0 | 0 | 1 |
| MO | 0 | 0 | 0 | 0 | 0 | 0 | 0 | 2 | 0 | 0 | 2 |
| MS | 0 | 0 | 0 | 0 | 0 | 0 | 1 | 1 | 0 | 0 | 2 |
| NC | 0 | 0 | 0 | 0 | 0 | 0 | 0 | 0 | 0 | 1 | 1 |
| OH | 0 | 0 | 0 | 0 | 1 | 0 | 0 | 1 | 0 | 1 | 3 |
| OK | 0 | 0 | 0 | 0 | 0 | 0 | 0 | 0 | 0 | 1 | 1 |
| PA | 0 | 0 | 0 | 0 | 1 | 0 | 0 | 2 | 0 | 1 | 4 |
| SC | 0 | 0 | 0 | 0 | 0 | 0 | 0 | 1 | 0 | 0 | 1 |
| TN | 0 | 0 | 0 | 0 | 0 | 0 | 0 | 1 | 0 | 1 | 2 |
| TX | 0 | 0 | 0 | 0 | 0 | 0 | 0 | 1 | 0 | 0 | 1 |
| UT | 0 | 0 | 1 | 0 | 0 | 0 | 0 | 0 | 0 | 0 | 1 |
| WV | 0 | 0 | 0 | 0 | 0 | 0 | 0 | 1 | 0 | 0 | 1 |
| Total | 1 | 1 | 3 | 0 | 3 | 0 | 1 | 21 | 0 | 8 | 38 |
| Percent | 3 | 3 | 8 | 0 | 8 | 0 | 3 | 55 | 0 | 21 | |

Table 4.15(g)   US Lightning Deaths by State and Activity/Location, 2006. Only States with Fatalities Are Listed

## U.S. Lightning Deaths 2006

| State | Ball Field | Boat | Camping | Near Heavy Equip. | Golf | In Water | Not Known | In the Open | Using Phone | Under Tree | Total |
|-------|-----------|------|---------|-------------------|------|----------|-----------|-------------|-------------|------------|-------|
| AL | 0 | 0 | 0 | 0 | 0 | 0 | 0 | 1 | 0 | 0 | 1 |
| AR | 0 | 1 | 0 | 0 | 0 | 0 | 1 | 1 | 0 | 0 | 3 |
| AS | 1 | 0 | 0 | 0 | 0 | 0 | 0 | 0 | 0 | 0 | 1 |
| AZ | 0 | 1 | 0 | 0 | 0 | 0 | 0 | 0 | 0 | 1 | 2 |
| CO | 1 | 0 | 0 | 0 | 0 | 0 | 0 | 3 | 0 | 1 | 5 |
| FL | 0 | 0 | 0 | 0 | 0 | 0 | 2 | 2 | 0 | 1 | 5 |
| GA | 0 | 0 | 0 | 0 | 0 | 0 | 0 | 1 | 0 | 0 | 1 |
| IL | 0 | 0 | 0 | 0 | 1 | 0 | 0 | 0 | 0 | 0 | 1 |
| IN | 0 | 0 | 0 | 0 | 0 | 0 | 0 | 0 | 0 | 1 | 1 |
| KY | 0 | 0 | 0 | 0 | 0 | 0 | 0 | 1 | 0 | 0 | 1 |
| LA | 1 | 0 | 0 | 0 | 0 | 0 | 1 | 0 | 0 | 0 | 2 |
| MI | 0 | 0 | 0 | 0 | 0 | 0 | 0 | 0 | 0 | 3 | 3 |
| MS | 0 | 0 | 0 | 0 | 0 | 0 | 0 | 0 | 1 | 1 | 2 |
| MT | 0 | 0 | 0 | 0 | 0 | 0 | 0 | 1 | 0 | 0 | 1 |
| NC | 0 | 0 | 0 | 0 | 0 | 0 | 1 | 2 | 0 | 0 | 3 |
| NE | 0 | 0 | 0 | 0 | 0 | 0 | 0 | 1 | 0 | 0 | 1 |
| NJ | 2 | 0 | 0 | 0 | 0 | 0 | 0 | 0 | 0 | 1 | 3 |
| NV | 0 | 0 | 0 | 0 | 0 | 0 | 0 | 1 | 0 | 0 | 1 |
| NY | 0 | 0 | 0 | 0 | 1 | 0 | 0 | 0 | 0 | 0 | 1 |
| PA | 0 | 0 | 0 | 0 | 0 | 0 | 0 | 0 | 0 | 1 | 1 |
| PR | 0 | 0 | 0 | 0 | 0 | 0 | 1 | 1 | 0 | 0 | 2 |
| SC | 0 | 0 | 0 | 0 | 0 | 0 | 0 | 1 | 0 | 1 | 2 |
| TN | 0 | 0 | 0 | 0 | 0 | 0 | 1 | 1 | 0 | 1 | 3 |
| TX | 0 | 0 | 0 | 0 | 0 | 0 | 1 | 0 | 0 | 0 | 1 |
| WI | 0 | 0 | 0 | 0 | 0 | 0 | 0 | 0 | 0 | 1 | 1 |
| Total | 5 | 2 | 0 | 0 | 2 | 0 | 8 | 16 | 1 | 13 | 47 |
| Percent | 11 | 4 | 0 | 0 | 4 | 0 | 17 | 34 | 2 | 28 | |

**Table 4.15(h)  US Lightning Deaths by State and Activity/Location, 2007. Only States with Fatalities Are Listed**

## U.S. Lightning Deaths 2007

| State | Ball Field | Boat | Camping | Near Heavy Equip. | Golf | In Water | Not Known | In the Open | Using Phone | Under Tree | Total |
|---|---|---|---|---|---|---|---|---|---|---|---|
| AL | 0 | 0 | 0 | 0 | 0 | 0 | 0 | 0 | 0 | 1 | 1 |
| AZ | 0 | 0 | 0 | 0 | 0 | 0 | 0 | 1 | 0 | 0 | 1 |
| CO | 0 | 0 | 1 | 0 | 0 | 0 | 0 | 1 | 0 | 0 | 2 |
| FL | 1 | 0 | 0 | 0 | 0 | 1 | 1 | 7 | 0 | 1 | 11 |
| GA | 0 | 0 | 0 | 0 | 0 | 0 | 1 | 1 | 0 | 1 | 3 |
| IL | 0 | 0 | 0 | 0 | 0 | 0 | 0 | 1 | 0 | 0 | 1 |
| KS | 0 | 0 | 0 | 0 | 0 | 0 | 0 | 1 | 0 | 0 | 1 |
| KY | 0 | 0 | 0 | 0 | 0 | 0 | 0 | 1 | 0 | 0 | 1 |
| MD | 0 | 0 | 0 | 0 | 0 | 0 | 0 | 0 | 0 | 1 | 1 |
| ME | 0 | 0 | 1 | 0 | 0 | 0 | 0 | 0 | 0 | 0 | 1 |
| MI | 0 | 0 | 0 | 0 | 0 | 0 | 0 | 0 | 0 | 1 | 1 |
| MN | 0 | 0 | 0 | 0 | 0 | 0 | 0 | 1 | 0 | 0 | 1 |
| MO | 0 | 0 | 0 | 0 | 0 | 0 | 0 | 2 | 0 | 0 | 2 |
| MT | 0 | 1 | 0 | 0 | 0 | 0 | 0 | 0 | 0 | 0 | 1 |
| NJ | 0 | 0 | 0 | 0 | 0 | 1 | 1 | 0 | 0 | 0 | 2 |
| NY | 0 | 0 | 0 | 0 | 0 | 0 | 0 | 1 | 0 | 0 | 1 |
| OH | 0 | 0 | 0 | 0 | 0 | 0 | 0 | 1 | 0 | 0 | 1 |
| PA | 0 | 0 | 0 | 0 | 0 | 0 | 0 | 1 | 0 | 0 | 1 |
| SC | 1 | 0 | 0 | 0 | 0 | 0 | 0 | 1 | 0 | 0 | 2 |
| TN | 0 | 0 | 0 | 0 | 0 | 0 | 0 | 1 | 0 | 0 | 1 |
| TX | 0 | 1 | 0 | 0 | 0 | 0 | 0 | 2 | 0 | 3 | 6 |
| UT | 0 | 0 | 1 | 0 | 0 | 0 | 0 | 0 | 0 | 0 | 1 |
| VA | 0 | 0 | 0 | 0 | 0 | 0 | 0 | 1 | 0 | 0 | 1 |
| WI | 0 | 0 | 0 | 0 | 1 | 0 | 0 | 0 | 0 | 0 | 1 |
| Total | 2 | 2 | 3 | 0 | 1 | 2 | 3 | 24 | 0 | 8 | 45 |
| Percent | 4.44 | 4.44 | 6.67 | 0.00 | 2.22 | 4.44 | 6.67 | 53.33 | 0.00 | 17.78 | |

Table 4.15(i)   US Lightning Deaths by State and Activity/Location, 2008. Only States with Fatalities Are Listed

## U.S. Lightning Deaths 2008

| State | Ball Field | Boat | Camping | Near Heavy Equip. | Golf | In Water | Not Known | In the Open | Using Phone | Under Tree | Total |
|---|---|---|---|---|---|---|---|---|---|---|---|
| AR | 0 | 0 | 0 | 0 | 0 | 0 | 0 | 0 | 0 | 1 | 1 |
| CO | 0 | 0 | 0 | 0 | 0 | 0 | 0 | 2 | 0 | 2 | 4 |
| CT | 0 | 0 | 0 | 0 | 0 | 0 | 1 | 0 | 0 | 0 | 1 |
| FL | 0 | 2 | 1 | 0 | 0 | 0 | 0 | 1 | 0 | 0 | 4 |
| IA | 0 | 0 | 0 | 0 | 0 | 0 | 0 | 1 | 0 | 0 | 1 |
| KS | 0 | 0 | 0 | 0 | 0 | 0 | 0 | 0 | 0 | 1 | 1 |
| MA | 0 | 0 | 0 | 0 | 0 | 0 | 0 | 0 | 0 | 1 | 1 |
| ME | 0 | 0 | 0 | 0 | 0 | 0 | 0 | 0 | 0 | 2 | 2 |
| MS | 0 | 0 | 0 | 0 | 0 | 0 | 0 | 0 | 0 | 1 | 1 |
| NC | 0 | 0 | 0 | 0 | 0 | 0 | 0 | 1 | 0 | 0 | 1 |
| NJ | 0 | 0 | 0 | 0 | 0 | 0 | 0 | 1 | 0 | 0 | 1 |
| OH | 0 | 0 | 0 | 0 | 0 | 0 | 0 | 2 | 0 | 0 | 2 |
| PA | 0 | 0 | 0 | 0 | 0 | 1 | 0 | 0 | 0 | 0 | 1 |
| RI | 0 | 0 | 0 | 0 | 0 | 0 | 0 | 1 | 0 | 0 | 1 |
| SC | 0 | 1 | 0 | 0 | 0 | 0 | 1 | 0 | 0 | 0 | 2 |
| TX | 0 | 0 | 0 | 0 | 0 | 0 | 0 | 1 | 0 | 0 | 1 |
| VA | 0 | 0 | 0 | 0 | 0 | 0 | 0 | 1 | 0 | 0 | 1 |
| WI | 0 | 0 | 0 | 0 | 0 | 0 | 0 | 1 | 0 | 0 | 1 |
| Total | 0 | 3 | 1 | 0 | 0 | 1 | 2 | 12 | 0 | 8 | 27 |
| Percent | 0.00 | 11.11 | 3.70 | 0.00 | 0.00 | 3.70 | 7.41 | 44.44 | 0.00 | 29.63 | |

**Table 4.15(j)   US Lightning Deaths by State and Activity/Location, 2009. Only States with Fatalities Are Listed**

## U.S. Lightning Deaths 2009

| State | Ball Field | Boat | Camping | Near Heavy Equip. | Golf | In Water | Not Known | In the Open | Using Phone | Under Tree | Total |
|-------|-----------|------|---------|-------------------|------|----------|-----------|-------------|-------------|-----------|-------|
| AL | 0 | 0 | 0 | 0 | 0 | 0 | 1 | 0 | 0 | 0 | 1 |
| AZ | 0 | 0 | 0 | 0 | 0 | 0 | 0 | 1 | 0 | 0 | 1 |
| CA | 0 | 0 | 0 | 0 | 0 | 0 | 0 | 1 | 0 | 1 | 2 |
| CO | 0 | 0 | 0 | 0 | 0 | 0 | 0 | 1 | 0 | 0 | 1 |
| FL | 1 | 1 | 0 | 0 | 1 | 0 | 0 | 2 | 0 | 0 | 5 |
| IN | 0 | 0 | 1 | 0 | 0 | 0 | 0 | 0 | 0 | 0 | 1 |
| KS | 0 | 0 | 0 | 0 | 0 | 0 | 1 | 0 | 0 | 0 | 1 |
| KY | 0 | 0 | 0 | 0 | 0 | 0 | 0 | 0 | 0 | 1 | 1 |
| LA | 0 | 1 | 0 | 0 | 0 | 0 | 0 | 0 | 0 | 0 | 1 |
| MA | 0 | 0 | 0 | 0 | 0 | 0 | 0 | 1 | 0 | 0 | 1 |
| MN | 0 | 0 | 0 | 0 | 0 | 0 | 0 | 1 | 0 | 1 | 2 |
| MO | 0 | 0 | 0 | 0 | 0 | 0 | 0 | 2 | 0 | 0 | 2 |
| MS | 0 | 0 | 0 | 0 | 0 | 0 | 0 | 1 | 0 | 0 | 1 |
| MT | 0 | 0 | 0 | 0 | 0 | 0 | 0 | 1 | 0 | 0 | 1 |
| NC | 0 | 0 | 0 | 0 | 0 | 1 | 0 | 2 | 0 | 0 | 3 |
| NJ | 0 | 0 | 0 | 0 | 0 | 0 | 0 | 0 | 0 | 1 | 1 |
| NM | 0 | 0 | 0 | 0 | 0 | 0 | 0 | 1 | 0 | 1 | 2 |
| NY | 0 | 0 | 0 | 0 | 0 | 0 | 0 | 1 | 0 | 0 | 1 |
| PA | 0 | 0 | 0 | 0 | 0 | 0 | 0 | 1 | 0 | 0 | 1 |
| PR | 0 | 0 | 0 | 0 | 0 | 0 | 0 | 1 | 0 | 0 | 1 |
| TX | 0 | 0 | 0 | 0 | 0 | 0 | 0 | 3 | 0 | 0 | 3 |
| VA | 1 | 0 | 0 | 0 | 0 | 0 | 0 | 0 | 0 | 0 | 1 |
| Total | 2 | 2 | 1 | 0 | 1 | 1 | 2 | 20 | 0 | 5 | 34 |
| Percent | 5.88 | 5.88 | 2.94 | 0.00 | 2.94 | 2.94 | 5.88 | 58.82 | 0.00 | 14.71 | |

*Source:* NOAA NCDC.

**Table 4.16　US Lightning Deaths by State and Ranking, 1959–2009**

| | Deaths | Rank | | Deaths | Rank |
|---|---|---|---|---|---|
| Alabama | 107 | 12 | Nebraska | 45 | 31 |
| Alaska | 0 | 51 | Nevada | 7 | 48 |
| Arizona | 71 | 24 | New Hampshire | 8 | 45 |
| Arkansas | 122 | 11 | New Jersey | 68 | 25 |
| California | 30 | 33 | New Mexico | 90 | 21 |
| Colorado | 139 | 6 | New York | 138 | 7 |
| Connecticut | 15 | 41 | North Carolina | 190 | 3 |
| Delaware | 15 | 41 | North Dakota | 12 | 44 |
| D.C. | 5 | 49 | Ohio | 143 | 4 |
| Florida | 460 | 1 | Oklahoma | 99 | 17 |
| Georgia | 106 | 13 | Oregon | 8 | 45 |
| Hawaii | 0 | 51 | Pennsylvania | 129 | 9 |
| Idaho | 26 | 38 | Puerto Rico | 34 | 32 |
| Illinois | 100 | 16 | Rhode Island | 8 | 45 |
| Indiana | 88 | 22 | South Carolina | 98 | 18 |
| Iowa | 72 | 23 | South Dakota | 23 | 40 |
| Kansas | 64 | 27 | Tennessee | 140 | 5 |
| Kentucky | 92 | 19 | Texas | 210 | 2 |
| Louisiana | 137 | 8 | Utah | 49 | 30 |
| Maine | 27 | 34 | Vermont | 15 | 41 |
| Maryland | 124 | 10 | Virginia | 66 | 26 |
| Massachusetts | 28 | 35 | Washington | 5 | 49 |
| Michigan | 105 | 14 | West Virginia | 26 | 38 |
| Minnesota | 62 | 28 | Wisconsin | 59 | 29 |
| Mississippi | 104 | 15 | Wyoming | 27 | 37 |
| Missouri | 93 | 20 | | | |
| Montana | 28 | 35 | **United States** | 3,919 | |

*Source:* NOAA NCDC.

**Table 4.17   US Lightning Death Rates per Million Population and State Rankings, 1959–2009**

| | Death Rate | Rank | | Death Rate | Rank |
|---|---|---|---|---|---|
| New Mexico | 1.33 | 1 | Georgia | 0.37 | 27 |
| Wyoming | 1.30 | 2 | Missouri | 0.37 | 28 |
| Arkansas | 1.11 | 3 | Indiana | 0.32 | 29 |
| Colorado | 0.96 | 4 | Wisconsin | 0.29 | 30 |
| Florida | 0.90 | 5 | Minnesota | 0.29 | 31 |
| Mississippi | 0.84 | 6 | Texas | 0.28 | 32 |
| Montana | 0.70 | 7 | West Virginia | 0.28 | 33 |
| Louisiana | 0.69 | 8 | Ohio | 0.27 | 34 |
| Oklahoma | 0.68 | 9 | Puerto Rico | 0.25 | 35 |
| Utah | 0.67 | 10 | Virginia | 0.24 | 36 |
| South Dakota | 0.66 | 11 | Michigan | 0.23 | 37 |
| South Carolina | 0.63 | 12 | Pennsylvania | 0.22 | 38 |
| North Carolina | 0.62 | 13 | New Jersey | 0.18 | 39 |
| Tennessee | 0.62 | 14 | Illinois | 0.18 | 40 |
| Vermont | 0.60 | 15 | New Hampshire | 0.17 | 41 |
| Maryland | 0.58 | 16 | Rhode Island | 0.17 | 42 |
| Nebraska | 0.57 | 17 | New York | 0.15 | 43 |
| Idaho | 0.56 | 18 | D.C. | 0.15 | 44 |
| Alabama | 0.56 | 19 | Nevada | 0.15 | 45 |
| Kansas | 0.53 | 20 | Connecticut | 0.10 | 46 |
| Kentucky | 0.52 | 21 | Massachusetts | 0.09 | 47 |
| Iowa | 0.51 | 22 | Oregon | 0.06 | 48 |
| Delaware | 0.49 | 23 | Washington | 0.02 | 49 |
| Maine | 0.48 | 24 | California | 0.02 | 50 |
| Arizona | 0.45 | 25 | Alaska | 0 | 51 |
| North Dakota | 0.38 | 26 | Hawaii | 0 | 52 |

*Source:* NOAA NCDC.

**Table 4.18　US Lightning Death Rates by State per 1000 Square Mile and State Rankings, 1959–2009**

| | Deaths | Deaths per 1000 sq. mi. | Rank | | Deaths | Deaths per 1000 sq. mi. | Rank |
|---|---|---|---|---|---|---|---|
| Alabama | 107 | 2.060 | 12 | Nebraska | 45 | 0.582 | 31 |
| Alaska | 0 | 0.000 | 51 | Nevada | 7 | 0.063 | 48 |
| Arizona | 71 | 0.614 | 24 | New Hampshire | 8 | 0.862 | 45 |
| Arkansas | 122 | 2.294 | 11 | New Jersey | 68 | 8.601 | 25 |
| California | 30 | 0.176 | 33 | New Mexico | 90 | 0.724 | 21 |
| Colorado | 139 | 1.326 | 6 | New York | 138 | 2.790 | 7 |
| Connecticut | 15 | 2.996 | 42 | North Carolina | 190 | 3.550 | 3 |
| Delaware | 15 | 7.404 | 42 | North Dakota | 12 | 0.170 | 44 |
| D.C. | 5 | 81.967 | 49 | Ohio | 143 | 3.460 | 4 |
| Florida | 460 | 7.754 | 1 | Oklahoma | 99 | 1.416 | 17 |
| Georgia | 106 | 1.815 | 13 | Oregon | 8 | 0.082 | 46 |
| Hawaii | 0 | 0.000 | 52 | Pennsylvania | 129 | 2.825 | 9 |
| Idaho | 26 | 0.311 | 38 | Puerto Rico | 34 | 9.540 | 32 |
| Illinois | 100 | 1.775 | 16 | Rhode Island | 8 | 6.595 | 47 |
| Indiana | 88 | 2.404 | 22 | South Carolina | 98 | 3.149 | 18 |
| Iowa | 72 | 1.279 | 23 | South Dakota | 23 | 0.298 | 40 |
| Kansas | 64 | 0.766 | 27 | Tennessee | 140 | 3.322 | 5 |
| Kentucky | 92 | 2.252 | 19 | Texas | 210 | 0.776 | 2 |
| Louisiana | 137 | 2.850 | 8 | Utah | 49 | 0.577 | 30 |
| Maine | 27 | 0.815 | 34 | Vermont | 15 | 1.560 | 43 |
| Maryland | 124 | 11.860 | 10 | Virginia | 66 | 1.601 | 26 |
| Massachusetts | 28 | 3.268 | 35 | Washington | 5 | 0.073 | 50 |
| Michigan | 105 | 1.794 | 14 | West Virginia | 26 | 1.073 | 39 |
| Minnesota | 62 | 0.711 | 28 | Wisconsin | 59 | 1.051 | 29 |
| Mississippi | 104 | 2.160 | 15 | Wyoming | 27 | 0.276 | 37 |
| Missouri | 93 | 1.305 | 20 | | | | |
| Montana | 28 | 0.184 | 36 | **United States** | **3,919** | | |

*Source:* NOAA SPC.

**Table 4.19   Average Annual Number of Lightning Strikes per State and Density per Square Mile by State (1996–2000)**

| | Avg. Strikes per year | Area sq. mi. | Strikes per sq. mi. | | Avg. Strikes per year | Area sq. mi. | Strikes per sq. mi. |
|---|---|---|---|---|---|---|---|
| Florida | 1,447,914 | 58681 | 24.67 | Nebraska | 522,933 | 77359 | 6.76 |
| Louisiana | 942,128 | 47720 | 19.74 | New Jersey | 45,106 | 7790 | 5.79 |
| Mississippi | 856,384 | 47695 | 17.96 | Arizona | 638,590 | 114007 | 5.60 |
| Alabama | 824,171 | 51718 | 15.94 | Wisconsin | 297,126 | 56145 | 5.29 |
| Oklahoma | 1,017,989 | 69903 | 14.56 | Michigan | 301,462 | 58513 | 5.15 |
| South Carolina | 451,841 | 31117 | 14.52 | Colorado | 529,243 | 104100 | 5.08 |
| Missouri | 995,744 | 69709 | 14.28 | South Dakota | 386,322 | 77122 | 5.01 |
| Indiana | 511,482 | 36185 | 14.14 | New York | 225,839 | 49112 | 4.60 |
| Arkansas | 750,747 | 53183 | 14.12 | Minnesota | 386,131 | 84397 | 4.58 |
| Georgia | 815,836 | 58390 | 13.97 | North Dakota | 291,131 | 70704 | 4.12 |
| Illinois | 780,602 | 56343 | 13.85 | Connecticut | 19,314 | 5006 | 3.86 |
| Tennessee | 579,751 | 42146 | 13.76 | Wyoming | 302,531 | 97818 | 3.09 |
| Kentucky | 541,493 | 40411 | 13.40 | Massachusetts | 25,191 | 8262 | 3.05 |
| D.C. | 793 | 61 | 13.00 | Utah | 253,564 | 84905 | 2.99 |
| Texas | 2,937,283 | 266874 | 11.01 | Vermont | 27,432 | 9615 | 2.85 |
| Kansas | 899,136 | 82282 | 10.93 | New Hampshire | 23,996 | 9283 | 2.58 |
| Iowa | 609,474 | 56276 | 10.83 | Montana | 346,912 | 147047 | 2.36 |
| Ohio | 446,757 | 41328 | 10.81 | Rhode Island | 2,386 | 1213 | 1.97 |
| North Carolina | 546,603 | 52672 | 10.38 | Nevada | 164,405 | 110567 | 1.49 |
| West Virginia | 215,216 | 24231 | 8.88 | Maine | 47,215 | 33128 | 1.43 |
| Virginia | 353,070 | 40598 | 8.70 | Idaho | 81,633 | 83574 | 0.98 |
| Maryland | 88,278 | 10455 | 8.44 | Oregon | 53,167 | 97052 | 0.55 |
| Delaware | 15,745 | 2026 | 7.77 | California | 86,100 | 158648 | 0.54 |
| New Mexico | 919,554 | 121599 | 7.56 | Washington | 19,756 | 68126 | 0.29 |
| Pennsylvania | 316,932 | 45310 | 6.99 | United States | | | 7.60 |

*Source:* Data by NLDN, Vaisala, Inc.

**Table 4.20    Top 25 Lightning Fatality Counts by Metropolitan Area, 1959–2006**

| Rank | Metro Area | Fatalities |
|------|------------|------------|
| 1 | Miami-Fort Lauderdale-Miami Beach, FL | 107 |
| 2 | New York-N. New Jersey-Long Is., NY, NJ, PA | 89 |
| 3 | Chicago-Naperville-Joliet, IL, IN, WI | 70 |
| 4 | Tampa-St. Petersburg-Clearwater, FL | 69 |
| 5 | Houston-Baytown-Sugar Land, TX | 57 |
| 6 | Denver-Aurora, CO | 44 |
| 7 | Orlando, FL | 43 |
| 7 | New Orleans-Metairie-Kenner, LA | 43 |
| 9 | Philadelphia-Camden-Wilmington, PA, NJ, DE, MD | 42 |
| 10 | Wash. D.C.-Arlington-Alexandria, DC, VA, MD, WV | 36 |
| 10 | Dallas-Fort Worth-Arlington, TX | 36 |
| 12 | Jacksonville, FL | 35 |
| 13 | Detroit-Warren-Livonia, MI | 33 |
| 13 | Atlanta-Sandy Springs-Marietta, GA | 33 |
| 15 | Pittsburgh, PA | 31 |
| 15 | St. Louis, MO, IL | 31 |
| 17 | Cincinnati-Middletown, OH, KY, IN | 29 |
| 18 | Minneapolis-St. Paul-Bloomington, MN, WI | 27 |
| 19 | Palm Bay-Melbourne-Titusville, FL | 26 |
| 19 | Nashville-Davidson--Murfreesboro, TN | 26 |
| 21 | Lakeland, FL | 25 |
| 21 | Cleveland-Elyria-Mentor, OH | 25 |
| 23 | Raleigh-Cary, NC | 21 |
| 23 | Baltimore-Towson, MD | 21 |
| 25 | Pensacola-Ferry Pass-Brent, FL | 20 |
| 25 | Colorado Springs, CO | 20 |
| 25 | Columbus, OH | 20 |

*Source:* Modified from Reassessment of Lightning Mortality in the U.S.: Analyses of Contrasting Datasets, Spatial Distributions, and Storm Morphologies, Walker Ashley, Christopher Gilson, and David Keith, Meteorology Program, Department of Geography, Northern Illinois University. Presented at the 17th Conference on Applied Climatology, American Meteorological Society, August 2008 and updated in Bulletin of the American Meteorological Society, October 2009.

**Table 4.21   Top 25 Lightning Fatality Counts by Metropolitan Area Ranked by Fatalities per 100 Square Miles, 1959–2006**

| Rank | Metro Area | Fatalities | Fatalities per 100 mi$^2$ |
|------|------------|------------|---------------------------|
| 1 | Tampa–St. Petersburg–Clearwater, FL | 69 | 2.62 |
| 2 | Muncie, IN | 10 | 2.53 |
| 3 | Palm Bay–Melbourne–Titusville, FL | 26 | 2.46 |
| 4 | Cape Coral–Fort Myers, FL | 19 | 2.34 |
| 5 | Miami–Fort Lauderdale–Miami Beach, FL | 107 | 1.97 |
| 6 | Fort Walton Beach–Crestview-Destin, FL | 13 | 1.38 |
| 7 | Boulder, CO | 10 | 1.35 |
| 8 | New York–N, NJ–Long Is., NY–NJ–PA | 89 | 1.29 |
| 9 | Mobile, AL | 16 | 1.27 |
| 10 | Lakeland, FL | 25 | 1.24* |
| 11 | Cleveland–Elyria–Mentor, OH | 25 | 1.24* |
| 12 | Pensacola–Ferry Pass–Brent, FL | 20 | 1.18 |
| 13 | Deltona–Daytona Beach–Ormond Beach, FL | 14 | 1.16 |
| 14 | Orlando, FL | 43 | 1.07 |
| 15 | Sarasota–Bradenton–Venice, FL | 14 | 1.04 |
| 16 | Jacksonville, FL | 35 | 1.03 |
| 17 | Winston–Salem, NC | 15 | 1.02* |
| 18 | New Orleans–Metairie–Kenner, LA | 43 | 1.02* |
| 19 | Raleigh–Cary, NC | 21 | 0.98 |
| 20 | Chicago–Naperville–Joliet, IL–IN–WI | 70 | 0.96 |
| 21 | Philadelphia–Camden–Wilmington, PA–NJ–DE–MD | 42 | 0.88 |
| 22 | Dayton, OH | 15 | 0.87 |
| 23 | Gulfport–Biloxi, MS | 13 | 0.86 |
| 24 | Gainesville, FL | 11 | 0.83* |
| 25 | Detroit–Warren–Livonia, MI | 33 | 0.83* |

* In converting figures from the original metric to square miles, rounding made these metro areas tied. In the original table using fatalities per 1000 km$^2$ they were not tied and the original order has been preserved.

*Source:* Modified from Reassessment of Lightning Mortality in the U.S.: Analyses of Contrasting Datasets, Spatial Distributions, and Storm Morphologies, Walker Ashley, Christopher Gilson, and David Keith, Meteorology Program, Department of Geography, Northern Illinois University. Presented at the 17th Conference on Applied Climatology, American Meteorological Society, August 2008 and updated in Bulletin of the American Meteorological Society, October 2009.

**Table 4.22**

| Total 2000 - 2009 Weather Event | Deaths | Injuries | Damage Property | Crop | Total |
|---|---|---|---|---|---|
| | | | Millions of Dollars | | |
| Lightning | 413 | 2,617 | 480.9 | 2.67 | 483.57 |
| Tornado | 558 | 8,327 | 8,642.7 | 177.15 | 8,819.87 |
| Thunderstorm Wind | 192 | 413 | 5,079.8 | 553.25 | 5,633.03 |
| Hail | 4 | 472 | 8,476.8 | 1,599.05 | 10,075.83 |
| Extreme Cold | 238 | 200 | 5.8 | 3,107.33 | 3,112.91 |
| Extreme Heat | 1,165 | 5,658 | 7.3 | 493.00 | 500.31 |
| Flash Flood | 475 | 603 | 10,337.8 | 1,301.99 | 11,639.79 |
| River Flood | 157 | 171 | 8,522.6 | 3,090.99 | 11,613.62 |
| Coastal Storm | 159 | 184 | 10,387.1 | 0.85 | 10,387.93 |
| Tsunami | 38 | 141 | 83.2 | 0.02 | 83.22 |
| Rip Current | 317 | 302 | 44.2 | 0.00 | 44.16 |
| Trop. Storm/Hurricane | 1,154 | 1,596 | 127,807.2 | 3,581.08 | 131,388.30 |
| Winter Storm | 234 | 1,617 | 4,811.1 | 29.43 | 4,840.48 |
| Ice | 14 | 16 | 2,655.7 | 0.00 | 2,655.71 |
| Avalanche | 140 | 99 | 3.2 | 0.00 | 3.23 |
| Drought | 0 | 4 | 864.9 | 10,412.33 | 11,277.20 |
| Dust Storm | 9 | 239 | 3.6 | 8.05 | 11.66 |
| Dust Devil | 2 | 28 | 0.4 | 0.00 | 0.39 |
| Rain | 36 | 159 | 446.1 | 382.01 | 828.12 |
| Fog | 37 | 376 | 18.1 | 0.00 | 18.05 |
| High Wind | 234 | 1,055 | 5,606.5 | 628.93 | 6,235.47 |
| Waterspout | 2 | 1 | 0.2 | 0.00 | 0.19 |
| Fire Weather | 33 | 655 | 6,654.4 | 56.15 | 6,710.55 |
| Mud Slide | 36 | 52 | 295.6 | 20.00 | 315.59 |
| Volcanic Ash | 0 | 0 | 0.5 | 0.00 | 0.50 |
| Miscellaneous | 60 | 592 | 170.0 | 264.40 | 434.44 |
| Total | 5,754 | 27,745 | 201,412.1 | 21,972.07 | 227,120.79 |

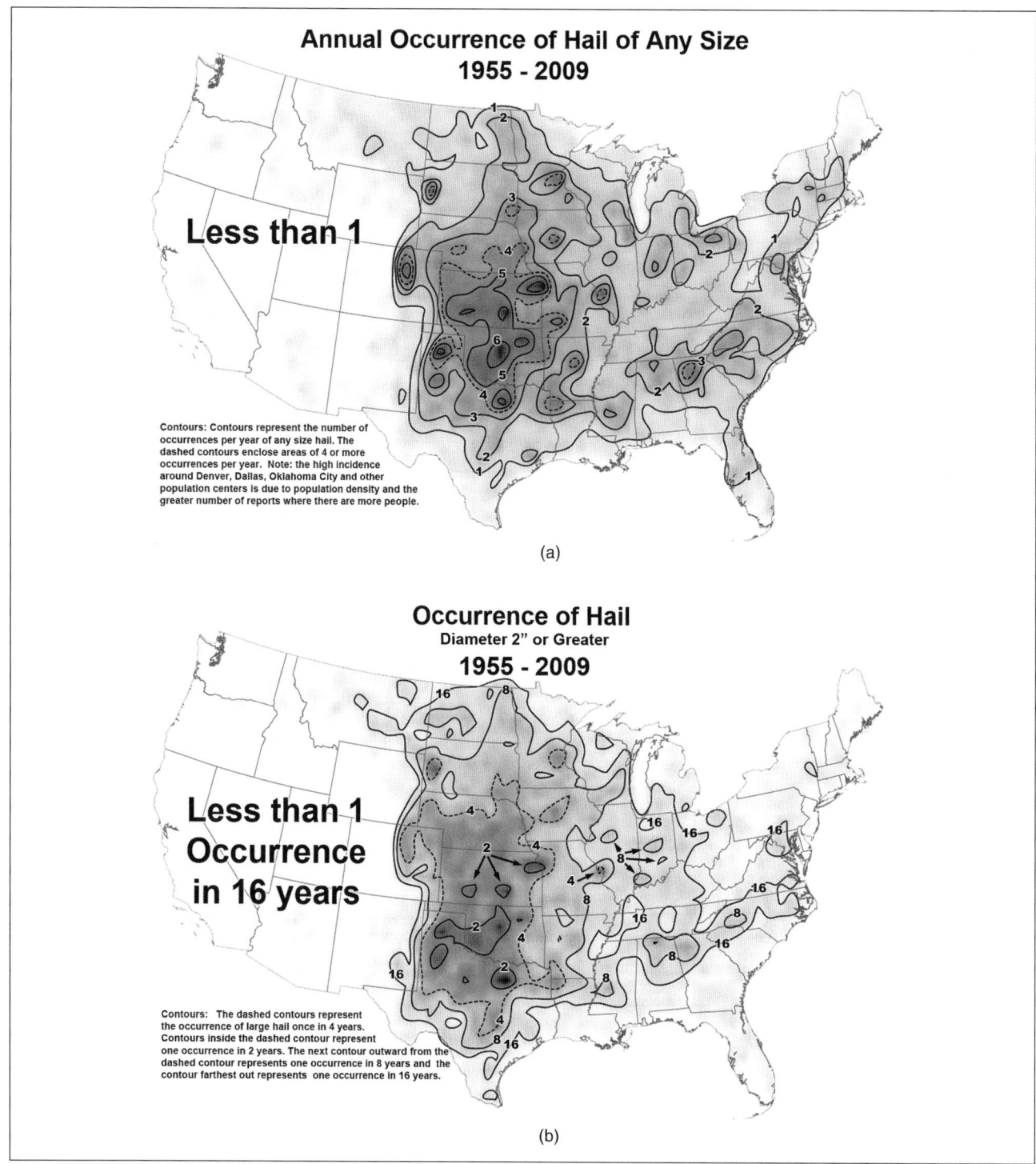

**Figure 4.46 a–b**   Annual occurrence of hail: (a, top) any size hail and (b, bottom) hail of diameter ≥2″ per square mile, based on observed occurrences 1955–2009. Contours in (a) are in units of occurrences per year; contours in (b) are in units of years per occurrence, or then number of years expected between occurrences. (*Source:* Steve Horstmeyer from NOAA SPC data.)

# 5

# Tropical Cyclones

## • INTRODUCTION

The largest, most powerful storms on this planet are tropical cyclones. The cloud shield from a big one can cover thousands of square miles, wind speeds can exceed 200 mph, and the storm surge can push many miles inland along low-lying coasts. The energy released by a medium-sized tropical cyclone in 1 day can exceed the daily total use of energy on the entire planet by 200 times.

As terrifying and deadly as they are, from space a mature tropical cyclone is a natural beauty. Gently undulating clouds flow outward in a gradual spiral from the central eye, an oasis of calm surrounded by screeching winds, and torrential downpours. Figure 5.1 shows the graceful beauty of Hurricane Felix as photographed from the International Space Station.

The beauty of the topside view misrepresents the fury below generated by a storm that may lift 500 trillion (500,000,000,000,000) pounds of water in a 10-day lifetime.

## • GEOGRAPHY AND FORMATION

Whether a tropical cyclone is called hurricane, typhoon, severe tropical cyclone, severe cyclonic storm, or just tropical storm, they are all almost the same. Each requires warm tropical water, an atmosphere with little vertical wind shear, a source of rotation, and a preexisting disturbance aloft.

During July and August in the Atlantic Ocean, the preexisting disturbance aloft is often invisible crossing North Africa until it reaches the sticky air near the coast. Satellite images then show scattered thunderstorms beginning to develop, like those on August 10, 2009 (Figure 5.2). Once the scattered storms reach the ocean, they begin to grow and increase in number and the scattered cells become a cluster of thunderstorms that begin to develop together.

Figure 5.3 shows a thunderstorm of a tropical wave on August 17, 2009. Three days later Hurricane Bill, a Cape Verde-type hurricane, would form from these storms. Cape Verde hurricanes are named for the cape and islands along the coast of Africa near which these storm form.

The origin of the disturbances aloft is tied to the seasonal wind changes that initiate the wet monsoon over India and southern Asia. As the tropical easterly jet (flowing westward from India) forms and interacts with the Himalaya, disturbances propagate toward Africa and give birth to the thunderstorm clusters. There are probably other sources for the preexisting disturbances, but this is a topic not yet well understood by meteorologists.

As the thunderstorms increase in number, they become better organized, but if the cluster forms closer to the equator than $10°$ of latitude, there is not sufficient Earth rotation for the thunderstorm complex to begin to rotate.

If vertical wind shear is too strong, the thunderstorms cannot develop into tall cells with strong updrafts. When the wind increase with height is strong enough, the storms tilt and the well-developed, chimney-like updraft fails to evolve (Figure 5.4).

In weak shear environments, deep convection through 50,000 or more feet of the atmosphere will lift millions of tons of moist air, and as the water vapor condenses, the heat originally used to evaporate the moisture is released back into the atmosphere and drives the thunderstorm complex. Heat released in this way is called "latent heat of condensation."

Poleward of $10°$ north and south of the equator where the water is warm enough and vertical wind shear weak enough is where tropical cyclones are born.

There are other locations where tropical cyclones originate in the Atlantic Ocean, and Figures 5.5a–f show the regions where hurricanes are most likely to originate for each month during hurricane season in the North Atlantic Ocean.

*The Weather Almanac: A Reference Guide to Weather, Climate, and Related Issues in the United States and Its Key Cities*, Twelfth Edition. Steve Horstmeyer.
© 2011 John Wiley & Sons, Inc. Published 2011 by John Wiley & Sons, Inc.

**Figure 5.1** Hurricane Felix, September 3, 2007, 11:38z. (*Source:* NASA, International Space Station.)

In June because of the warm water tropical cyclones are most likely to develop over the Caribbean Sea, Gulf of Mexico, and the Atlantic Ocean off the southeast coast of the United States.

By August the Cape Verde track has become established with tropical cyclone development peaking in September. As the water cools through October and November and the hemispheric circulation moves toward a winter pattern, tropical cyclone development fades.

In the southern hemisphere, tropical cyclone rotation is clockwise, and north of the equator, it is counterclockwise. The rotation is imparted by the rotation of the Earth and tropical cyclones rotate in the same direction as Earth's rotation. This is called the "Coriolis effect," after the French mathematician Gaspard-Gustave de Coriolis who first correctly formalized the mathematics of how the effect works.

A strong tropical cyclone can make it less likely a tropical cyclone; following it on a similar track will strengthen into a major storm by cooling surface waters. There are three ways that this can happen. Cloud cover from the tropical cyclone decreases warming of surface waters by the sun, rain that falls cools the surface waters, and the fierce surface winds churn the waters so much that cool water is brought upward from below the warm surface layer.

A dramatic example of this is illustrated in Figure 5.6. Unforgettable Hurricane Katrina was followed 1 month later by Hurricane Rita, the most intense tropical cyclone to ever cross the Gulf of Mexico.

Katrina cooled the entire Gulf of Mexico an average of 1°C (1.8°F) and in spots along the track as much as 4°C–5°C (7.2°F–9°F), which is a large amount of energy no longer available to fuel the next tropical cyclone.

By the time Rita moved into the Gulf of Mexico the water temperatures partially recovered because the warm layer was so warm and deep. Rita was large and powerful, so the three cooling mechanisms cooled the Gulf of Mexico even more.

As oceanographers and meteorologist study tropical cyclone more, it is becoming apparent that not only it is the surface water temperature that is important, but the depth of the warm layer also plays an important role as shown by Katrina and Rita.

Tropical oceans have the right combination of conditions for the birth of tropical cyclones. Occasionally waters farther poleward warmed by seasonal sunshine or warm

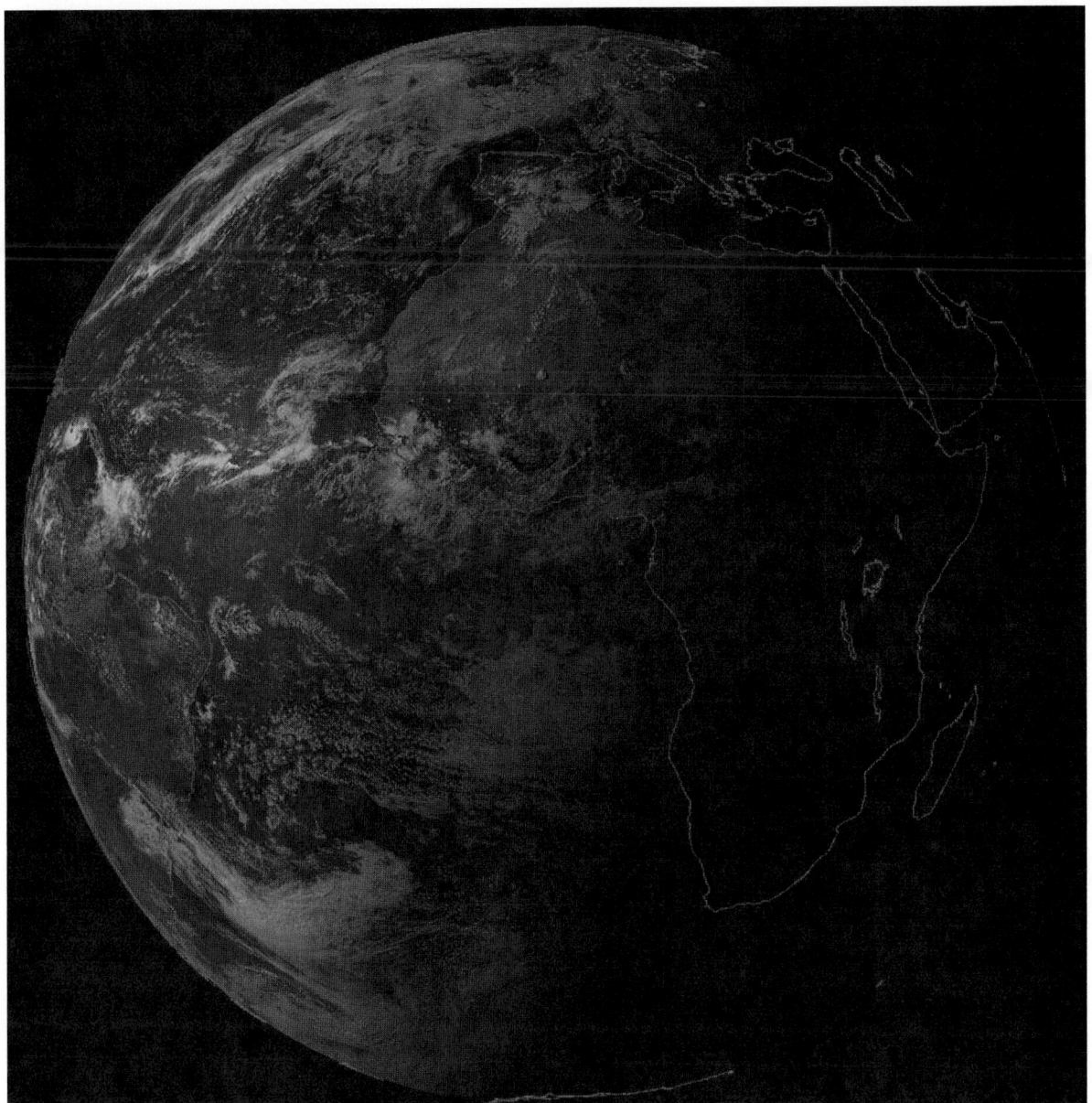

**Figure 5.2** Summer thunderstorms developing over West Africa, August 10, 2009. (*Source:* University of Wisconsin – Madison, Space Science and Engineering Center.)

ocean currents have enough energy to support tropical cyclone formation, but most tropical cyclones form within the tropics.

Sea-surface temperatures of 80°F (26.7°C) are generally thought to be the minimum required for tropical cyclone formation, but where other factors are strongly in favor of development, tropical cyclones can form over waters a couple degrees cooler.

Once a complex of thunderstorms begins to rotate and the central pressure drops, if conditions are right, the winds increase and the pressure continues to fall. In the Atlantic Ocean the complex becomes a tropical depression; after additional strengthening the depression becomes a tropical storm and then a hurricane.

Because of the release of latent heat by condensing water vapor within the storm and warming of sinking air by compression in the eye tropical cyclones are classified as warm-core disturbances.

Figure 5.7 shows all tropical cyclone occurrences on Earth from 1985 to 2005 along with the name used to commonly refer to hurricane-strength storms. Table 5.1 gives the regional names at various levels of organization and development, while Table 5.2 indicates the seasons and average storm numbers.

**Figure 5.3** The tropical wave that would become Hurricane Bill, August 14, 2009, 18z. (*Source:* University of Wisconsin – Madison, Space Science and Engineering Center.)

The tropical South Atlantic is a special case. Only one tropical cyclone is known to have ever reached tropical storm or hurricane strength there (Figure 5.8), while a total of only six tropical cyclones are known. Two additional storms classified as subtropical storms, which have both tropical (deep convective thunderstorms) and middle-latitude characteristics (cool air pool aloft), have formed in the South Atlantic. In Figure 5.8 note that the circulation of Catarina is the opposite of a northern hemisphere tropical cyclone. Just before this chapter was finished, the second known tropical storm strength cyclone was spotted by satellite off

the coast of South America. Storm 90Q is shown in Figure 5.9. The arrow points to 90Q in the visible satellite image. No eye is visible, but 15 hours later, an infrared image from another satellite (inset) shows big thunderstorms near the storm's center and the evidence of an eyelike feature.

Many speculate that between South America and Africa the water is too cool for tropical cyclone formation. That may be true at times as the cool water that wells up off the west coast of Africa flows toward South America, south of the equator. But the controlling factors are that vertical wind

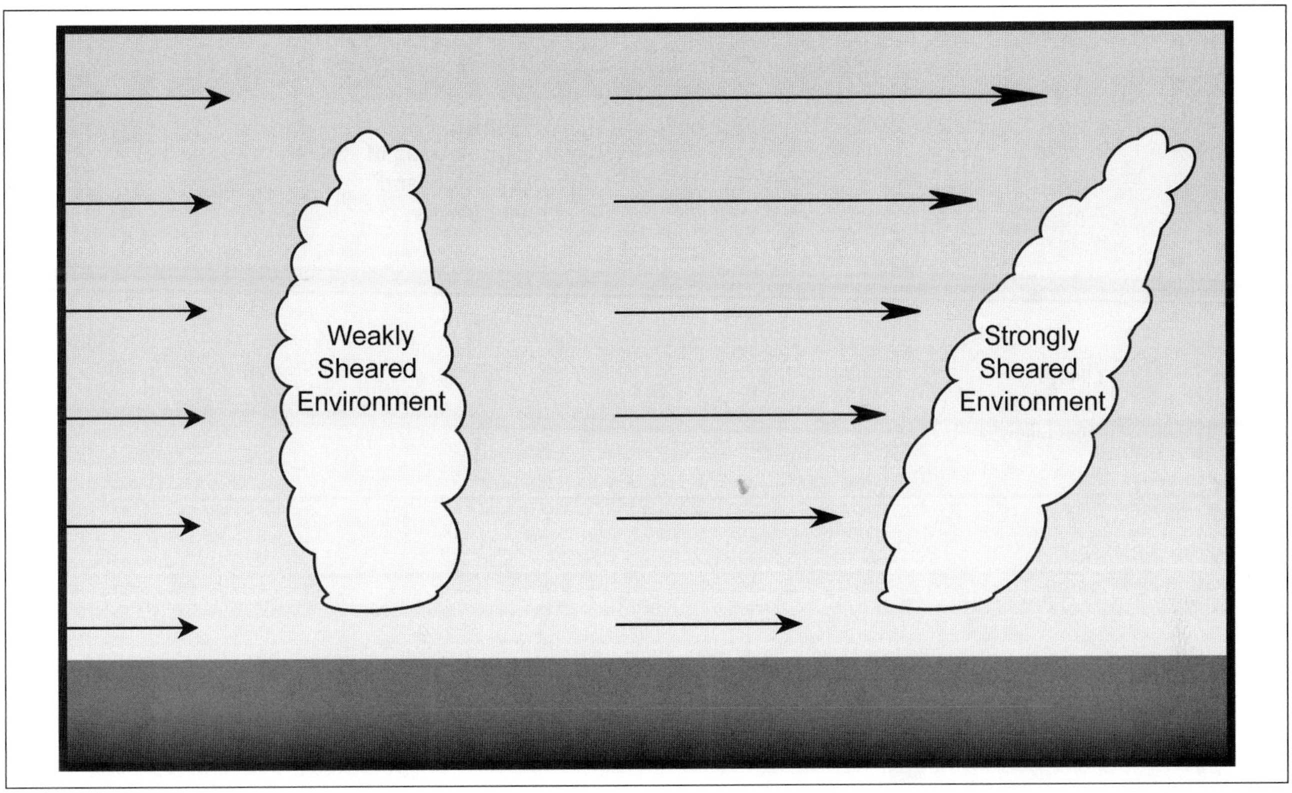

**Figure 5.4**   The effect of wind shear on hurricane development.

shear through the atmosphere is too strong in the region and there are few disturbances there to initiate tropical cyclone formation.

Tropical systems are driven by a completely different energy source than middle-latitude systems that are cold-core disturbances. Tropical cyclones derive their energy from condensing water vapor, while the primary energy source for middle-latitude systems is the jet stream. Water-vapor condensation though is a secondary source of energy for nontropical systems.

## ● TROPICAL CYCLONE STRUCTURE

Tropical cyclones are large rotating storms that are collections of cooperating thunderstorms that create the largest most powerful storms on Earth. From above the feeder bands of thunderstorms are easily visible as they spiral toward the eye of Hurricane Floyd in 1999 (Figure 5.10).

In cross section the thunderstorms are larger and more powerful, closer to the eye peaking in intensity in the towering eye wall. Figure 5.11 shows the cross section of an ideal tropical cyclone with violently rising air in the core of each thunderstorm and sinking air around the periphery of the storm and at the center in the eye.

The sinking air in the eye warms as it is compressed, and for this reason, in the eye the sky often clears. When flying through an intense hurricane with a distinct eye, hurricane hunters often photograph the "stadium effect." Descending air in the eye means the sky is clear but the surrounding eye wall clouds tower above the aircraft as they rotate at high velocity around the eye. It has the look of being in a sports stadium. Figures 5.12 and 5.13 show the stadium effect in the eye of Katrina from a hurricane hunter aircraft.

The eye of the hurricane can be a good indicator of storm strength, with an average diameter of 20–40 miles. Generally a small eye is an indication of a powerful storm and eyes of small diameter are often replaced as an outer concentric eye wall of thunderstorms forms and then shrinks, cutting off the old eye from the inflow of moist tropical air that powers the storm. This is called an "eye wall replacement cycle."

Other possible reasons for the eye wall replacement cycles include the extreme turbulence caused by the high winds tearing the inner (older) eye wall apart, and in a small eye, the thunderstorms cannot remain organized because they interfere with one another. The concentric eye walls of Typhoon Amber (2007) are shown in Figure 5.14.

Hurricane Wilma (2005), the most intense Atlantic hurricane ever, had the smallest eye known with a diameter of 2 miles. Figure 5.15, taken from the International Space Station 222 miles above the storm, shows the eye of Wilma

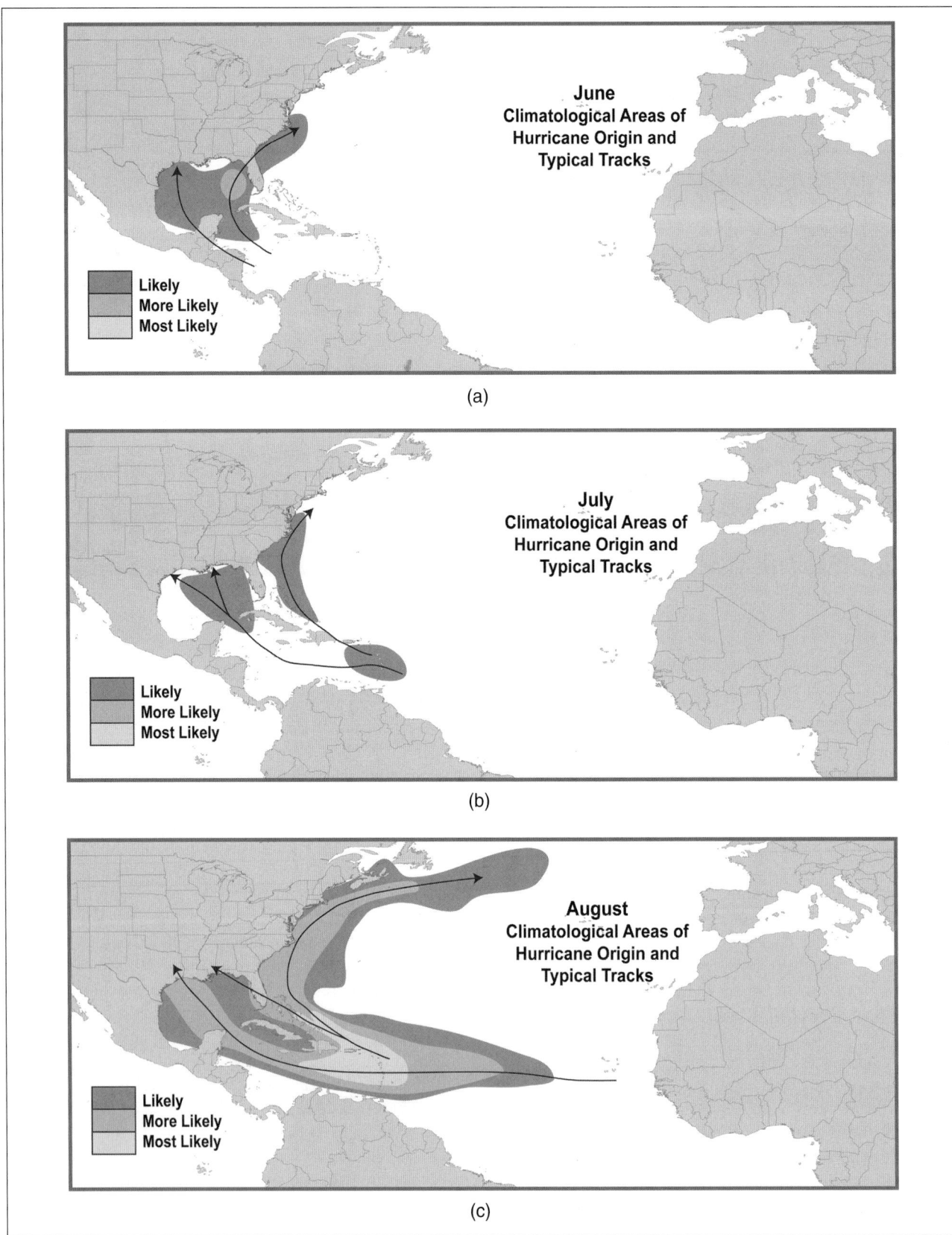

**Figure 5.5a–f** Climatological areas favorable for hurricane formation for each month of the Atlantic hurricane season. (*Source:* Steve Horstmeyer from NASA original.)

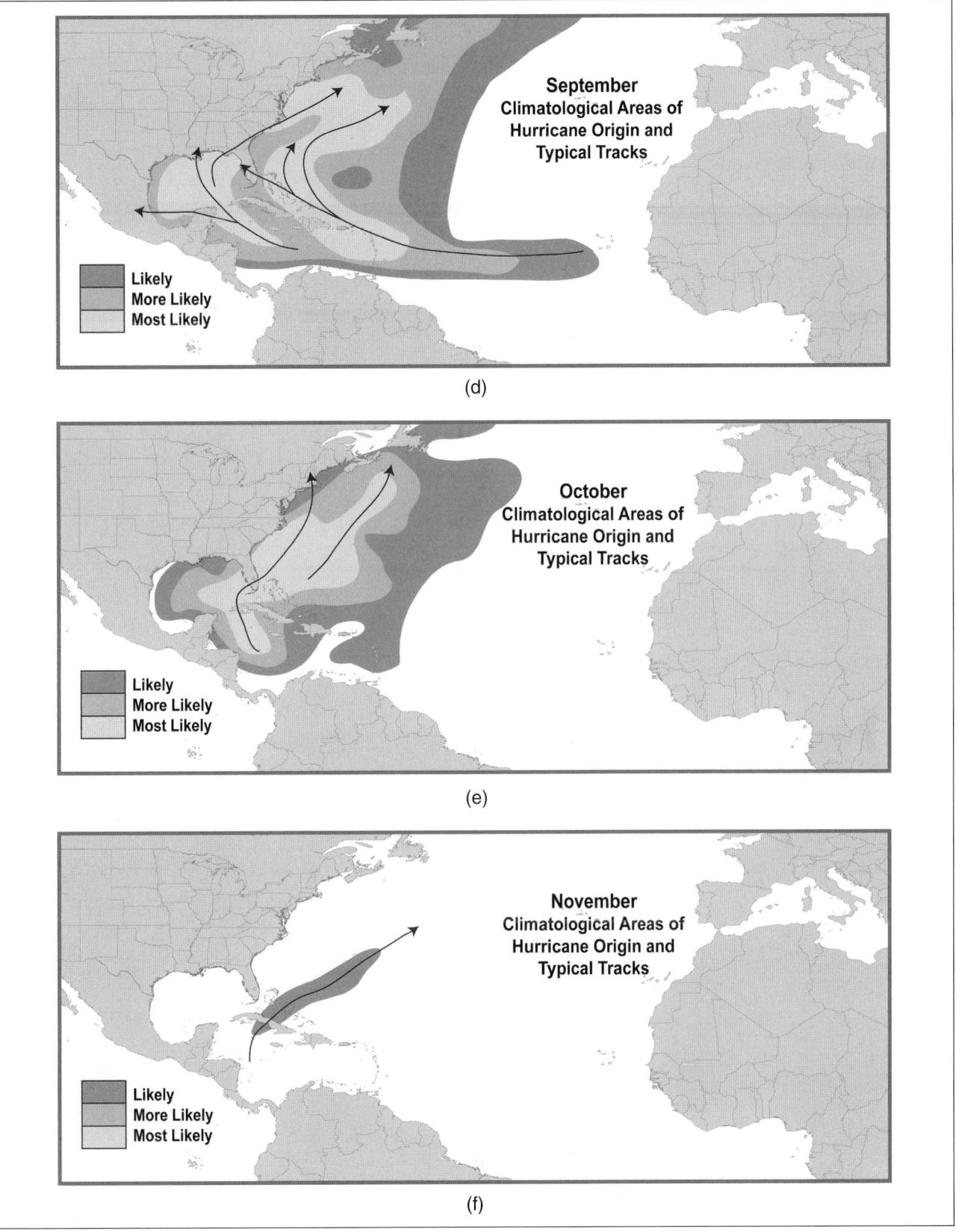

**Figure 5.5** *(Continued)*.

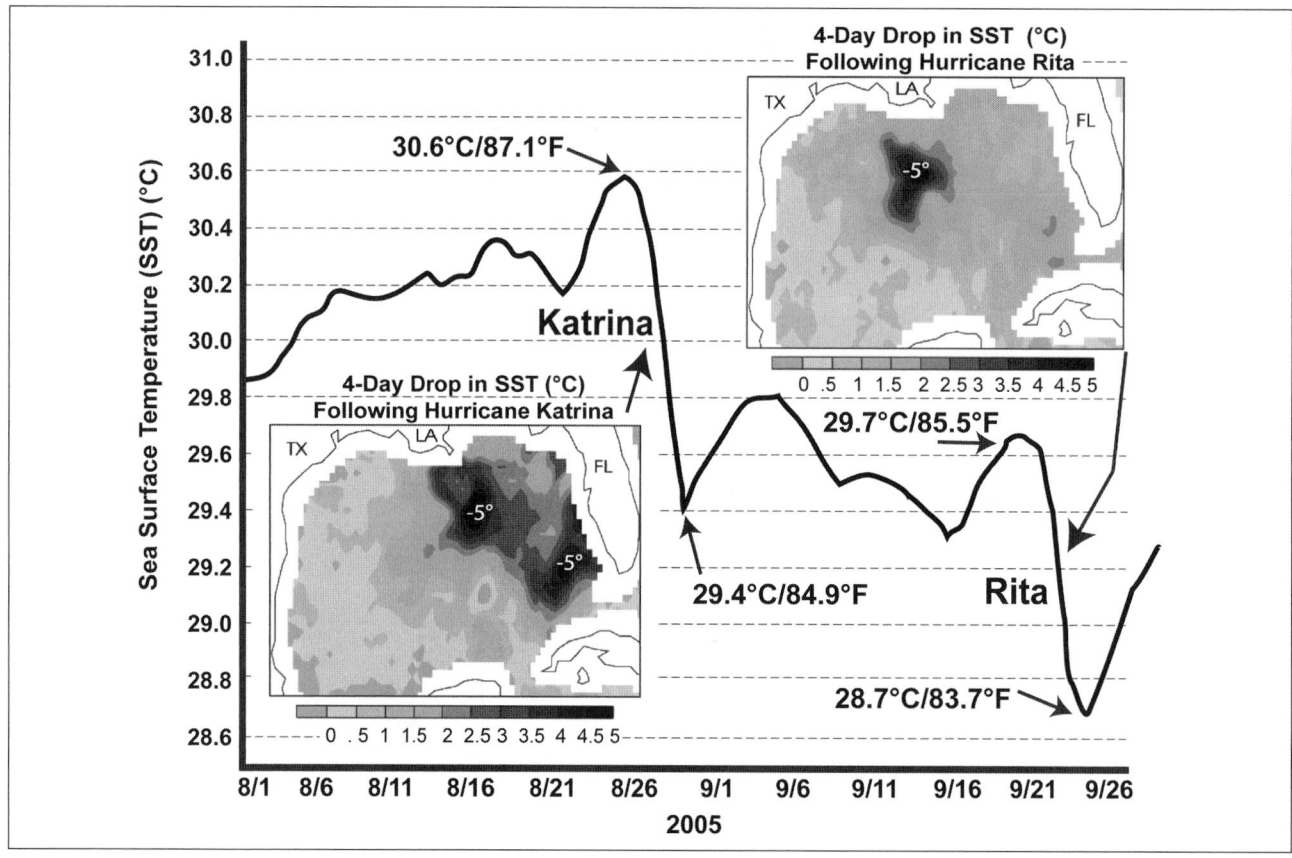

**Figure 5.6**  Cooling of the Gulf of Mexico by hurricanes Katrina and Rita. (*Source:* Steve Horstmeyer from NASA original.)

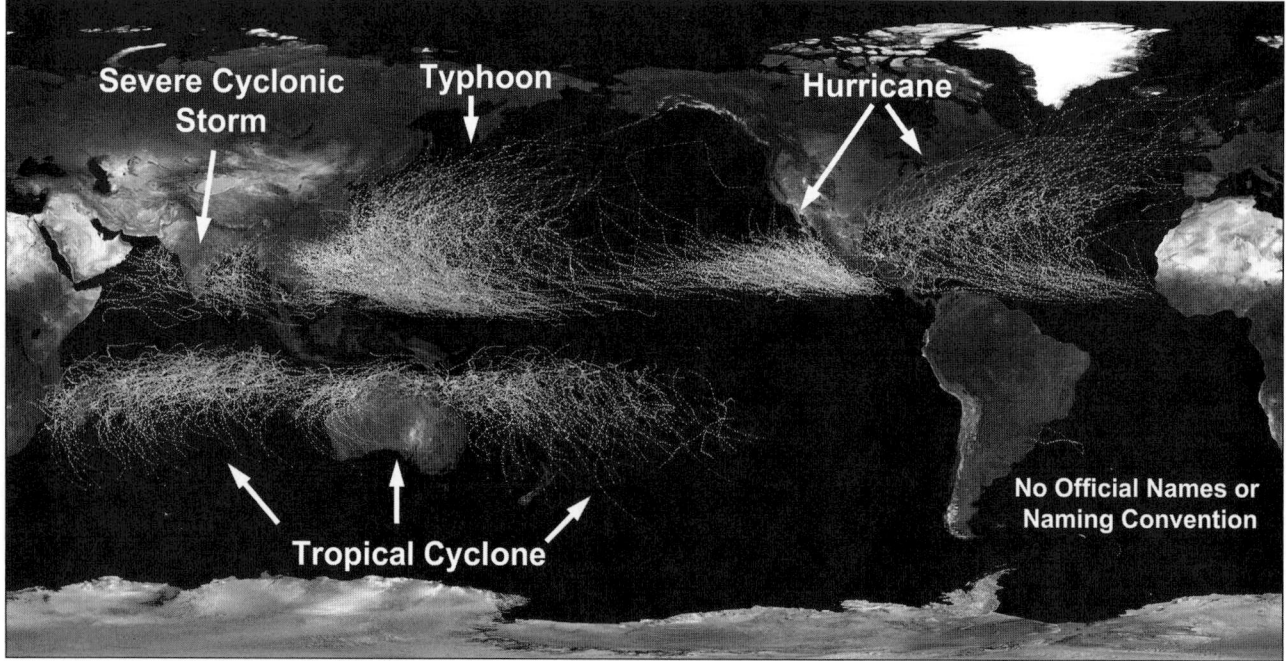

**Figure 5.7**  Worldwide distribution of tropical cyclones with regional names. (*Source:* NASA.)

**Table 5.1 International Tropical System Classification**

| Beaufort Number | 10-minute sustained winds mph | Northern Indian Ocean (Indian Meteorological Department) | Southwest Indian Ocean (Météo France) | Australia Region (Bureau of Meteorology) | Southwest Pacific (Fiji Meteorological Service) | Northwest Pacific (Japan Meteorological Agency) | Northwest Pacific (Joint Typhoon Warning Center) | Northeast Pacific & North Atlantic (National Hurricane Center and Central Pacific Hurricane Center) |
|---|---|---|---|---|---|---|---|---|
| 0 – 6 | <32 | Depression | Tropical Disturbance | Tropical Low | Tropical Depression | Tropical Depression | Tropical Depression | Tropical Depression |
| 7 | 32 – 33 | Depression | Tropical Depression | Tropical Low | Tropical Depression | Tropical Depression | Tropical Depression | Tropical Depression |
| 7 | 34 – 38 | Deep Depression | Tropical Depression | Tropical Low | Tropical Depression | Tropical Depression | Tropical Depression | Tropical Depression |
| 8 – 9 | 39 – 54 | Cyclonic Storm | Moderate Tropical Storm | Tropical Cyclone | Tropical Cyclone | Tropical Storm | Tropical Storm | Tropical Storm |
| 10 | 55 – 63 | Severe Cyclonic Storm | Severe Tropical Storm | Tropical Cyclone | Tropical Cyclone | Severe Tropical Storm | Tropical Storm | Tropical Storm |
| 11 | 64 – 72 | Severe Cyclonic Storm | Severe Tropical Storm | Tropical Cyclone | Tropical Cyclone | Severe Tropical Storm | Tropical Storm | Hurricane |
| 12 | 73 – 83 | | Tropical Cyclone | Severe Tropical Cyclone | Severe Tropical Cyclone | Typhoon | Typhoon | Hurricane |
| 13 | 84 – 98 | | Tropical Cyclone | Severe Tropical Cyclone | Severe Tropical Cyclone | Typhoon | Typhoon | Major Hurricane |
| 14 | 99 – 102 | Very Severe Cyclonic Storm | Intense Tropical Cyclone | Severe Tropical Cyclone | Severe Tropical Cyclone | Typhoon | Typhoon | Major Hurricane |
| 15 | 103 – 114 | Very Severe Cyclonic Storm | Intense Tropical Cyclone | Severe Tropical Cyclone | Severe Tropical Cyclone | Typhoon | Typhoon | Major Hurricane |
| 16 | 115 – 123 | Very Severe Cyclonic Storm | Intense Tropical Cyclone | Severe Tropical Cyclone | Severe Tropical Cyclone | Typhoon | Typhoon | Major Hurricane |
| | 124 – 131 | Very Severe Cyclonic Storm | Very Intense Tropical Cyclone | Severe Tropical Cyclone | Severe Tropical Cyclone | Typhoon | Typhoon | Major Hurricane |
| 17 | 132 – 137 | | Very Intense Tropical Cyclone | Severe Tropical Cyclone | Severe Tropical Cyclone | Typhoon | Super Typhoon | Major Hurricane |
| 17 | >138 | Super Cyclonic Storm | Very Intense Tropical Cyclone | Severe Tropical Cyclone | Severe Tropical Cyclone | Typhoon | Super Typhoon | Major Hurricane |

Note: The South Atlantic tropical cyclone of March 24 – 28, 2005 has unofficially called Cyclone Catarina by Brazilian meteorologists. All tropical cyclone categories and names are agreed upon internationally and proposed by national meteorological agencies affiliated with the World Meteorological Organization. Because tropical cyclones are almost unknown in the South Atlantic Ocean there is no categorical naming convention and no storm names list. To be consistent with other southern hemisphere schemes Catarina is referred to as a cyclone.

*Source: AMOL.*

### Table 5.2 Worldwide Tropical Cyclone Seasons and Occurrences

| Basin | Start of Season | End of Season | Average Number Per Year | | |
|---|---|---|---|---|---|
| | | | > 39 mph | >73 mph | >110 mph |
| **Northern Hemisphere** | | | | | |
| Northwest Pacific | April | January | 26.7 | 16.9 | 8.5 |
| Northeast Pacific | May | November | 16.3 | 9.0 | 4.1 |
| North Atlantic | June | November | 10.6 | 5.9 | 2.0 |
| North Indian | April | December | 5.4 | 2.2 | 0.4 |
| **Southern Hemisphere** | | | | | |
| South Indian | November | April | 20.6 | 10.3 | 4.3 |
| South Atlantic* | December | May | - | - | - |
| Australia Southwest Pacific | November | April | 9 | 4.8 | 1.9 |

*In the South Atlantic tropical cyclones are rare so there are no average numbers. The season is assumed to be the opposite of the North Atlantic season.

*Source:* AOML.

**Figure 5.8** Cyclone Catarina, the only known hurricane-strength tropical cyclone in the South Atlantic. (*Source:* NASA.)

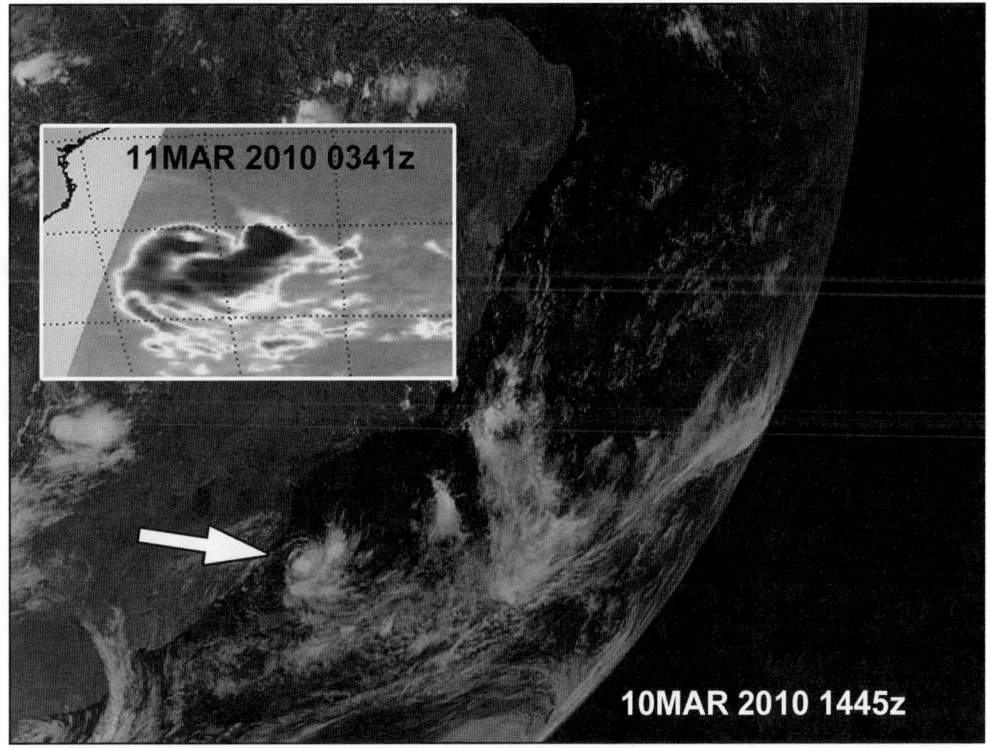

**Figure 5.9**  Tropical Cyclone 90Q, March 10, 2010. Only the second-known tropical cyclone to surpass tropical storm strength winds in the South Atlantic. (*Source:* NASA, GOES Project.)

**Figure 5.10**  Hurricane Floyd, September 14, 1999. (*Source:* NASA.)

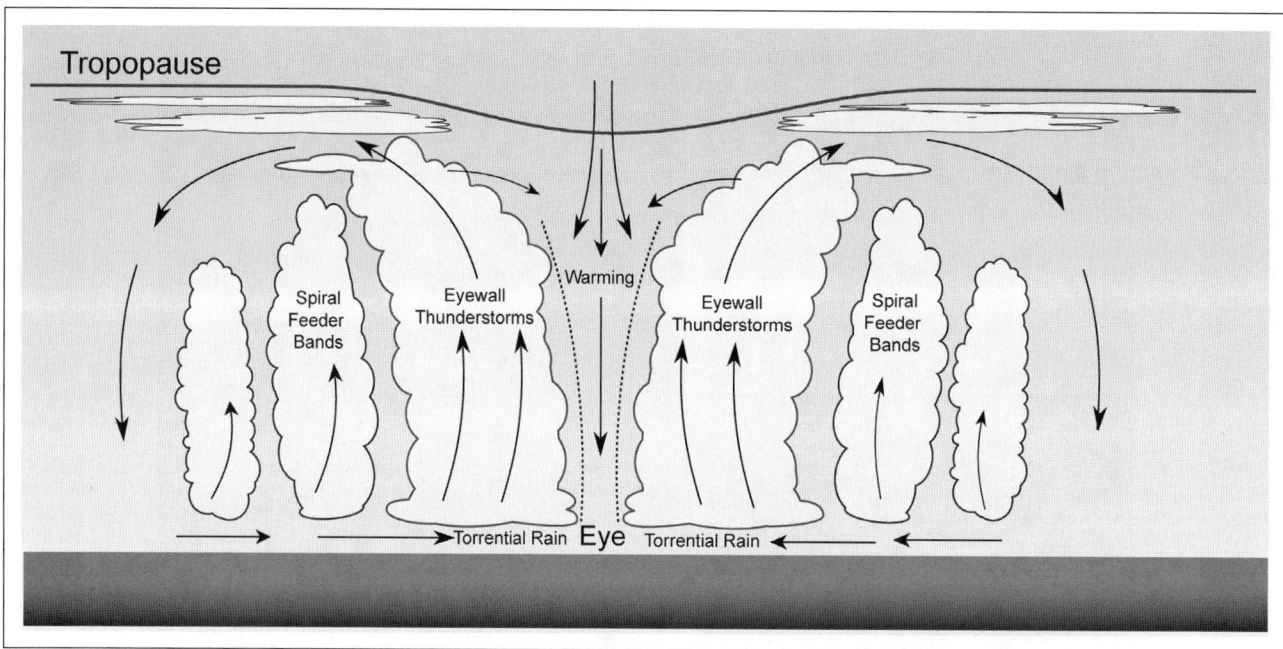

**Figure 5.11** Cross section of a hurricane. (*Source:* Steve Horstmeyer from NASA original.)

on October 19, 2005, at the time. Figure 5.15 was taken when Wilma's winds were 175 mph and the central pressure of 882 mbar (26.05 in. of Hg) was the lowest ever in an Atlantic hurricane.

Typhoon Carmen (August 20, 1960) and Typhoon Winnie (1997) tied for the largest eye with a diameter of 200 miles. Carmen was measured by weather radar on Okinawa (Figure 5.16) and the large eye of Winnie is shown in Figure 5.17. Winnie and probably Carmen belong to a category called "annular tropical cyclones." The nearly rain-free area between the outer and inner eye walls of Winnie is called the "moat."

To forecast the strength and position of a hurricane, scientists must have data and much of that are now derived from satellite imagery. The well-known visible and infrared images we see on TV are used, but there are many sophisticated ways to estimate wind speed, rainfall rates, and intensity changes from the perspective of a weather satellite.

Despite advances in satellite technology, hurricane hunter flights continue to provide a critical detailed look at the inside of storms. Traveling from outside the tropical cyclone toward the eye of the storm, wind increases, atmospheric pressure falls, and rainfall rate varies, depending on the location relative to thunderstorm bands.

Figure 5.18 shows the results of a hurricane hunter flight through Floyd on September 13, 1999. Winds at both the flight level and the surface vary but steadily increase as the eye wall gets closer and then drop abruptly in the eye. Rainfall rates show when the aircraft is passing through thunderstorm bands; in addition wind speeds tend to increase at the same time.

Figure 5.19 is a similar graph showing the results of a flight through Hurricane Hugo, and Figure 5.20 shows observations made at Charleston International Airport and at the McEntire Air National Guard Weather Facility about 30 miles to the west. Figure 5.19 shows more detail because the observations were made at 30-s intervals, while the data in Figure 5.20 were made mostly at hourly intervals. Figures 5.18–5.20 give a detailed picture of the structure of a tropical cyclone, and Figure 5.21 shows the major damage a major hurricane like Hugo can inflict.

## • HURRICANE CLASSIFICATION

Hurricanes and tropical storms are categorized by strength, and the scale used until recently is called the Saffir–Simpson Hurricane Scale (SSHS), after engineer Herbert Saffir and former director of the National Hurricane Center, Robert Simpson, who together introduced it in 1973 (Table 5.3).

Because coastal flooding and storm surge are so dependent on local coastal conditions a generalized storm surge scale like the one in the SSHS can be dangerously misleading.

A recent example of this is Hurricane Ike in 2008. The storm made landfall with Category 2 winds. However, the storm surge at Galveston was equivalent to what is currently defined for the Category 4–5 storm range. Reports from emergency managers indicated that many residents would not evacuate because the storm was only an SSHS Category 2 or 3.

**Figure 5.12** The view from the eye of Hurricane Katrina on August 28, 2005, from a NOAA hurricane hunter aircraft. (*Source:* NOAA.)

**Figure 5.13**    The view from the eye of Hurricane Katrina on August 28, 2005, from a NOAA hurricane hunter aircraft. (*Source:* NOAA.)

### Table 5.3   Saffir–Simpson Hurricane Scale

| Category | Wind speed | | Storm Surge | |
|---|---|---|---|---|
| | mph | km/h | feet | meters |
| Five | ≥156 | ≥250 | >18 | >5.5 |
| Four | 131–155 | 210–249 | 13–18 | 4.0–5.5 |
| Three | 111–130 | 178–209 | 9–12 | 2.7–3.7 |
| Two | 96–110 | 154–177 | 6–8 | 1.8–2.4 |
| One | 74–95 | 119–153 | 4–5 | 1.2–1.5 |
| **Less Intense Storms** | | | | |
| Tropical Storm | 39–73 | 63–117 | 0–3 | 0–0.9 |
| Tropical Depression | 0–38 | 0–62 | 0 | 0 |

**Figure 5.14**    The concentric eye walls in Typhoon Amber, August 27, 2007. (*Source:* University of Wisconsin, CIMSS.)

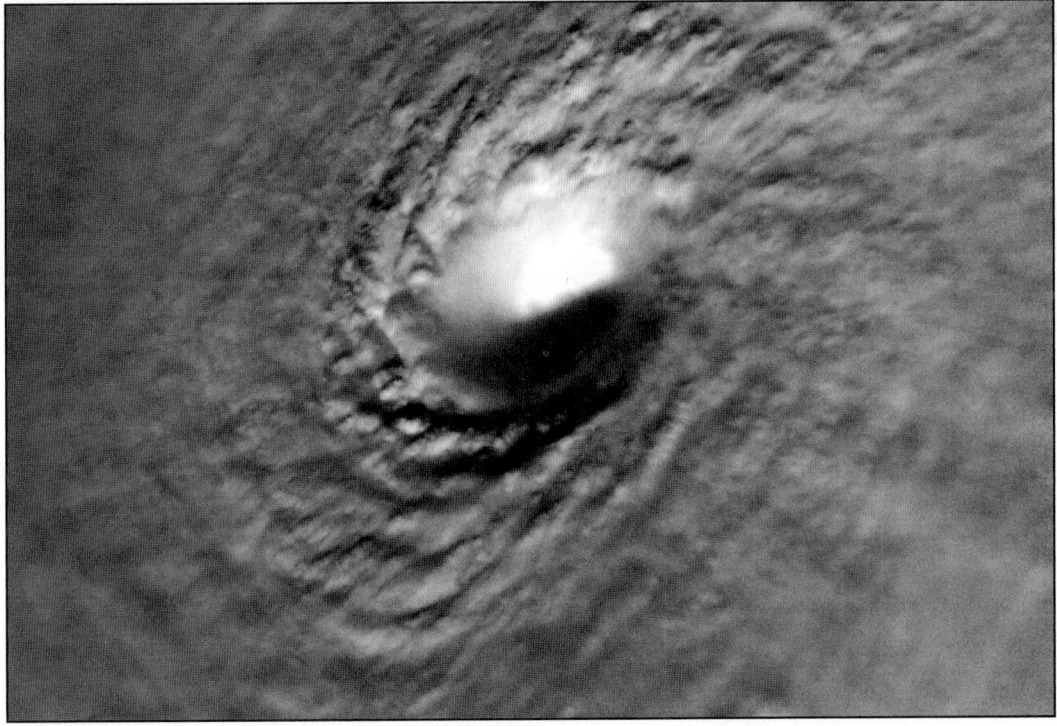

**Figure 5.15**    The eye of Hurricane Wilma, 2 miles in diameter, the smallest ever measured. At the time the winds were 175 mph and atmospheric pressure 882 mbar (26.05 in. Hg) the most intense hurricane on record in the Atlantic Ocean. (*Source:* NASA, ISS.)

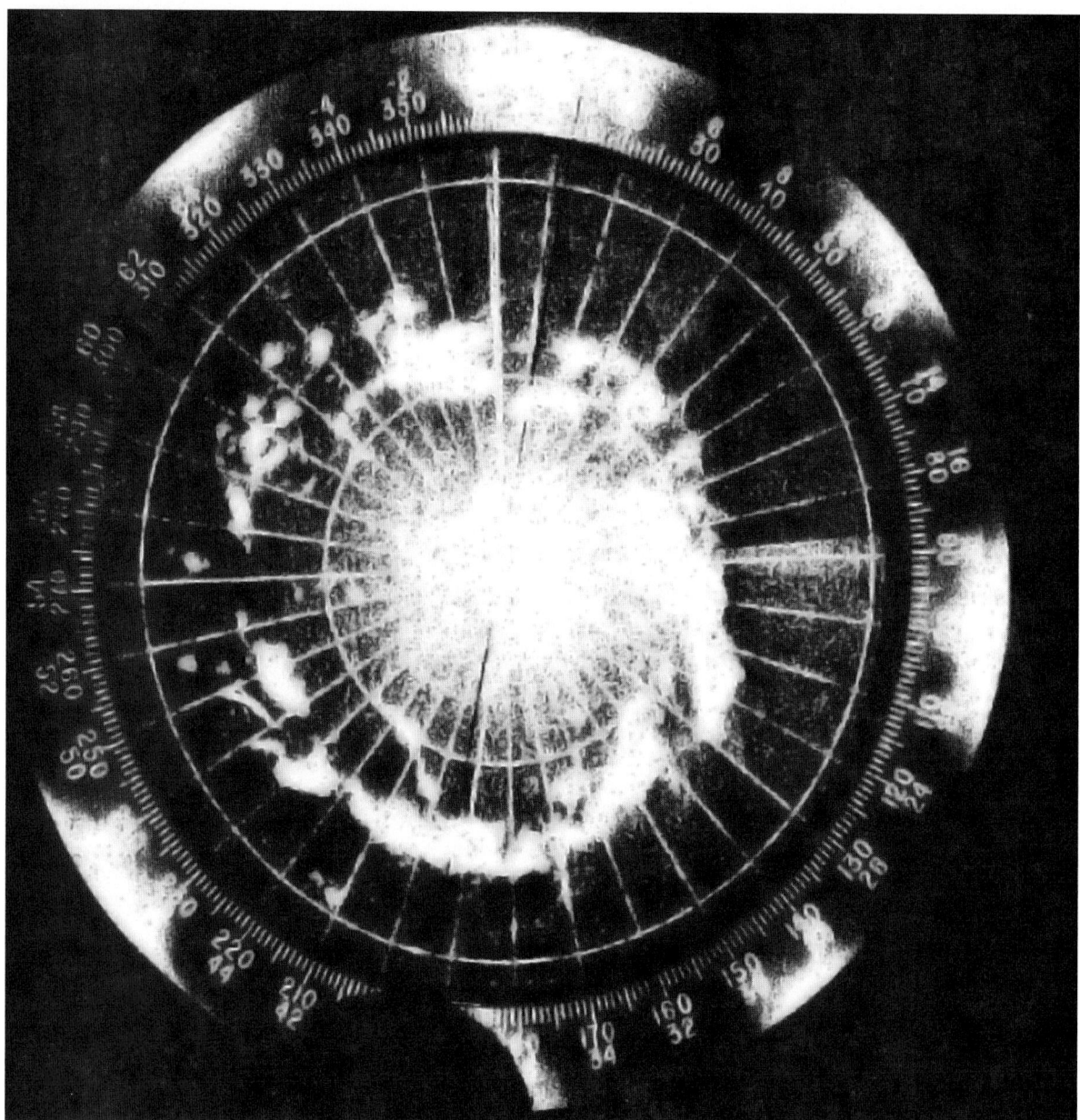

**Figure 5.16** A radar view of Typhoon Carmen, August 20, 1960, from Kadena Air Base, Okinawa. This was the largest eye measured with a diameter of 200 miles until Typhoon Winnie tied Carmen. (*Source:* Joint Typhoon Warning Center.)

In 2004, Hurricane Charley—a Category 4 storm at landfall on the SSHS for winds only—created a storm surge equivalent to the Category 2 storm range. Hurricane Katrina in 2005 was a Category 3 storm at landfall for winds but with storm surge equivalent to the Category 5 storm range.

During the 2009 hurricane season, the Saffir–Simpson Hurricane Scale (SSHS) was replaced with the experimental Saffir–Simpson Hurricane Wind Scale (SSHWS; Tables 5.4a–e).

The difference between the SSHS and the SSHWS is that storm surge estimates and pressure have been removed and the new scale is based on wind speed only. Very general storm surge estimates, like the ones in the SSHS, can be drastically different from the true storm surge because of varying coastal configurations, tide schedules, and strike angles, and can mislead anyone not familiar with the limitations of the SSHS. Figure 5.22 illustrates the storm surge estimates of the SSHS based only on hurricane category that are no longer in use.

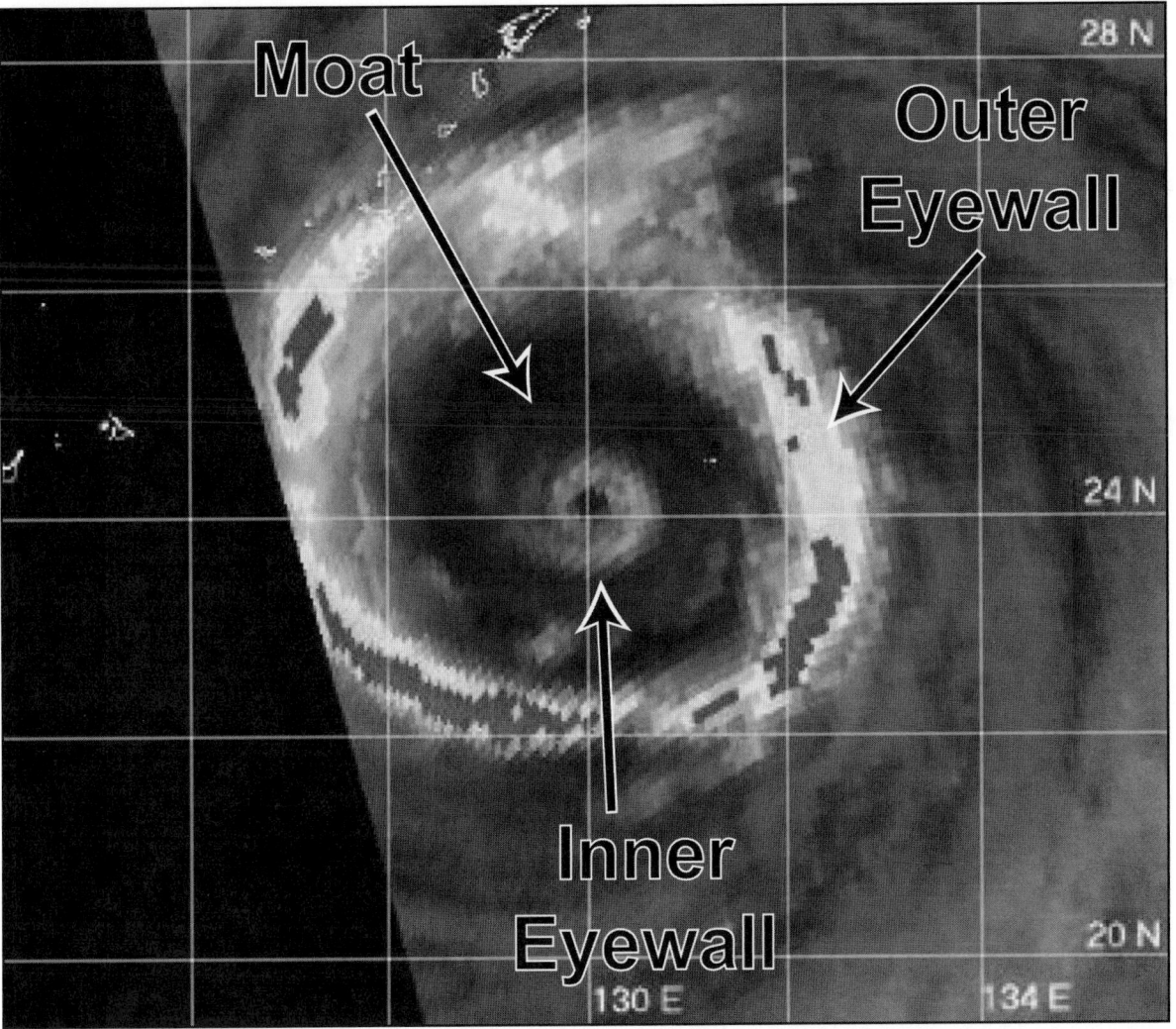

**Figure 5.17** Satellite view of Typhoon Winnie using a microwave of a military weather satellite showing the double eye wall structure and the moat. Winnie is tied with Typhoon Carmen for the largest known eye diameter. (*Source:* Joint Typhoon Warning Center.)

Conditions that may increase the storm surge above the estimates stated in the SSHS are a shallow slope of the sea floor that leads to greater surge; a harbor or funnel-shaped embayment will amplify the surge and the direction in which the wind crosses the coast and interacts with other factors to increase the surge.

Storm surge refers to the rush of water on shore with the approach of a tropical cyclone. The primary cause is the force of the wind pushing surface water. The four secondary causes include (a) The decrease of central pressure with the approach of the storm allowing water to rise. This is often negligible when compared to the wind effect (Figure 5.23). (b) Cessation of downwelling in shallow water. Figure 5.23 also indicates the downwelling of water as it converges on the central part of the storm. When the water is shallow it cannot sink far enough for downwelling to compensate for

the convergence of water in the center. (c) The specific features of a given coastline. (d) In estuaries and other low areas rainfall can be great enough to contribute to the total height of the water.

The total depth of water on land is the storm tide (Figure 5.24) and it is indicated by high water marks (HWM) left by water stains, stranded debris, and deposited sediment. The National Oceanic and Atmospheric Administration (NOAA) conducts a detailed survey of HWMs once the water has subsided.

Determining storm surge is not straightforward because the water that is found onshore is the sum of (1) the astronomical tide, (2) storm surge, (3) wave run-up, and (4) fresh water flooding.

Because both HWM and local tides are referenced to sea level, the predicted tide—where it is known—is subtracted

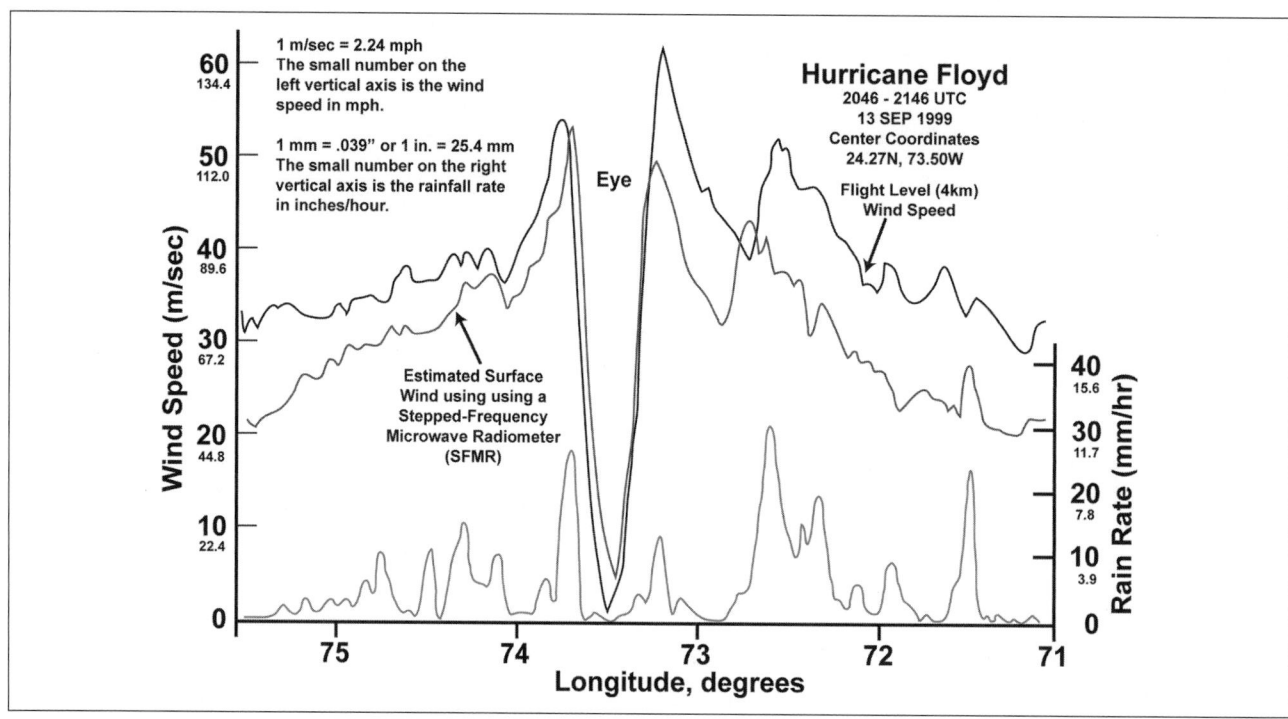

**Figure 5.18** A cross section of Hurricane Floyd measured during a hurricane reconnaissance flight on September 13, 1999, showing the flight-level winds, surface winds, and precipitation rate. (*Source:* Modified by Steve Horstmeyer from NOAA original.)

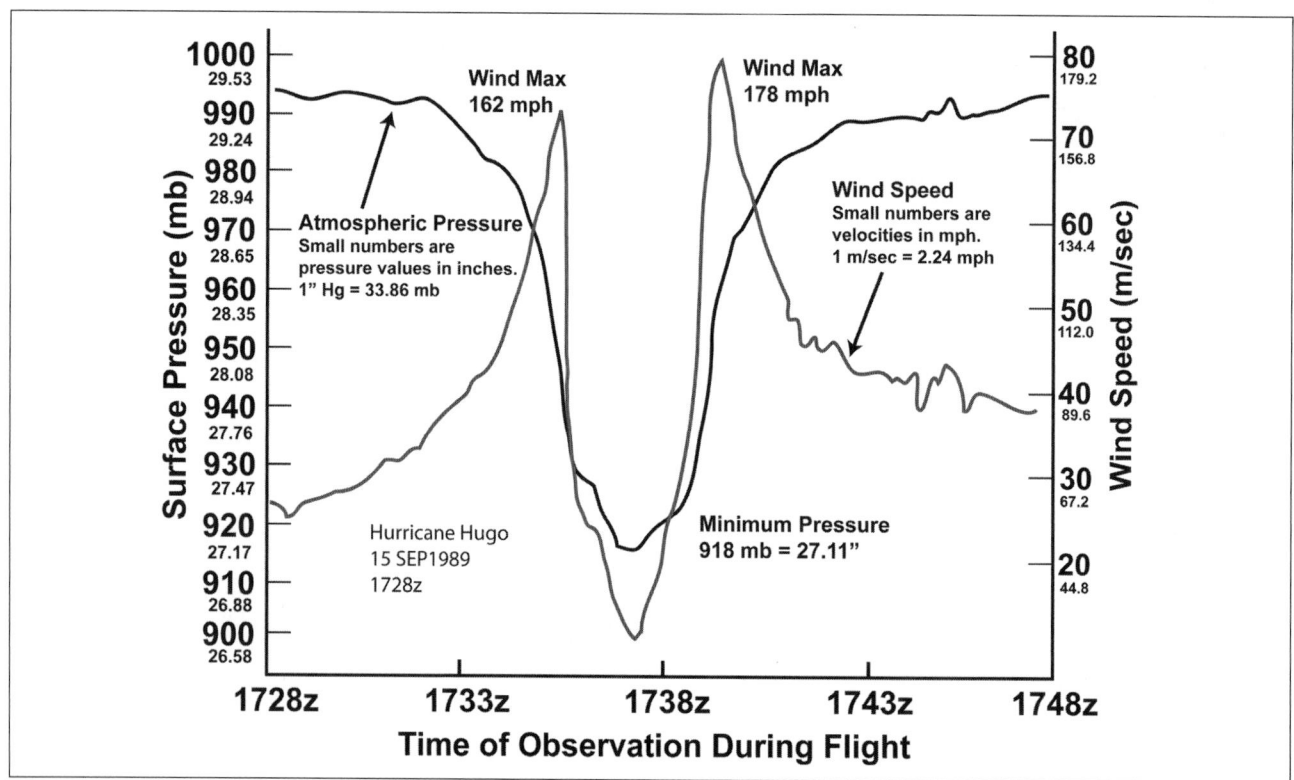

**Figure 5.19** A cross section of Hurricane Hugo measured by a hurricane hunter flight on September 15, 1989, showing the atmospheric pressure and wind speeds when crossing the storm. (*Source:* Modified by Steve Horstmeyer from NOAA original.)

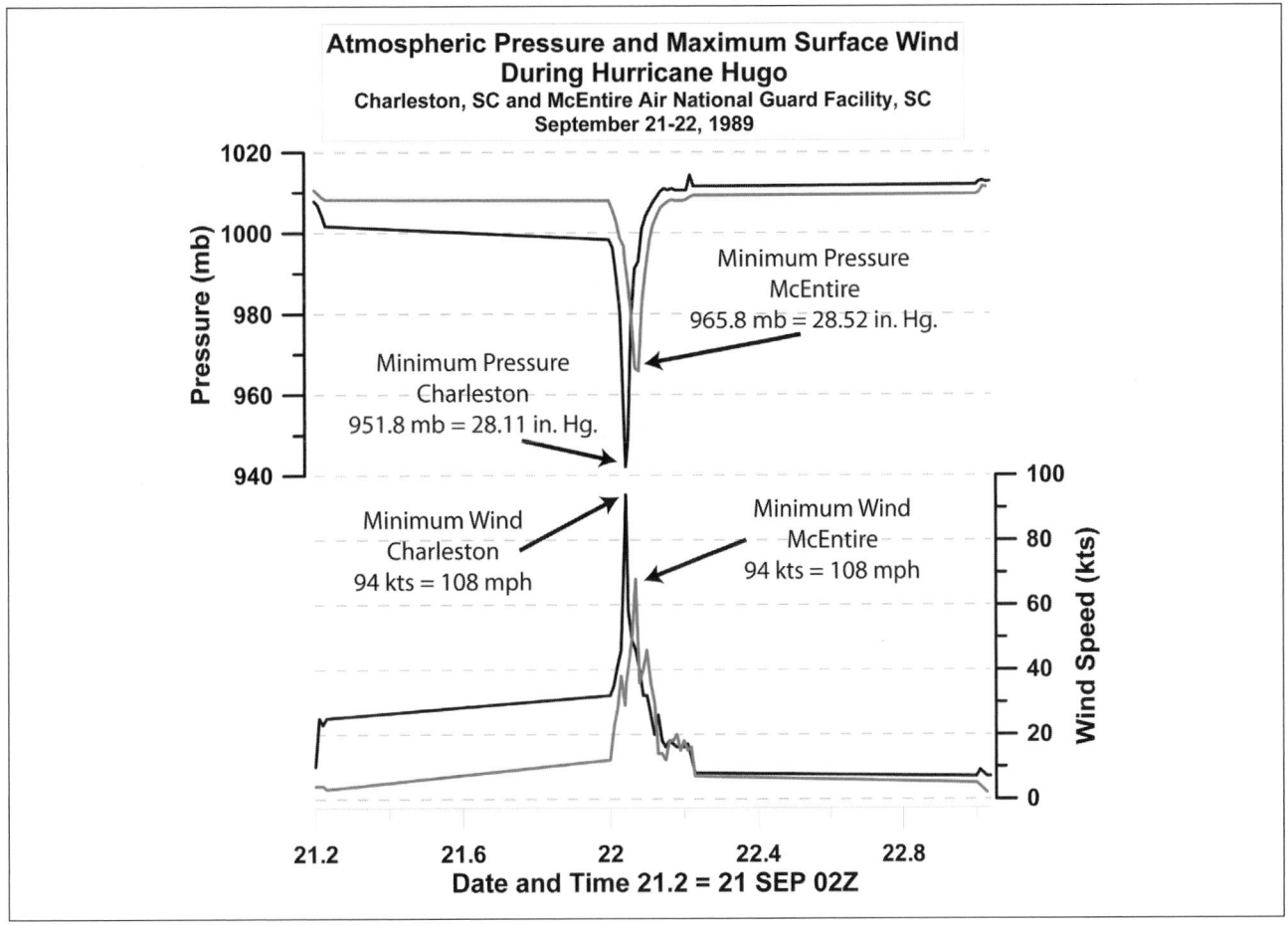

**Figure 5.20**    The change and sea-level pressure and wind speed with the passage of Hurricane Hugo at Charleston, SC, and McEntire Air National Guard Facility, SC, on September 21–22, 1989.

from the HWM, giving the storm surge. If it is impossible to determine the astronomical tide, the storm-caused coastal flooding is reported as the storm tide.

Wave run-up occurs when the surging waters encounter a steep slope and waves break up the slope to a height that can be significantly higher than the HWM would be without wave run-up. Fresh water flooding can be difficult to estimate because there is little evidence left of the degree of flooding before the arrival of the storm surge. If there is evidence for either of the two, the NOAA evaluates the individual contributions to total water depth so that the storm surge can be measured as accurately as possible.

Once the astronomical tide, wave run-up, and fresh water flooding are subtracted from the HWM, what is left is storm surge.

The greatest storm surge potential occurs where the movement of the tropical cyclone and the wind are in the same direction, increasing the wind as it approaches a coast. In the northern hemisphere, this is on the  right side of

the storm track, and in the southern hemisphere, on the left side.

The plus sign in Figures 5.25 and 5.26 indicates this additive effect of the surface winds (curved arrows) with storm movement (dashed arrow). On the opposite side of the storm track, the wind blows in the opposite direction from storm movement and the difference decreases the surge potential.

The storm surge potential varies with the variables discussed above, and with the location of the eye to the coast, the direction of storm movement, and the direction the coast runs. On December 24, 2008, Cyclone Billy (Figure 5.25) was moving parallel to the northern coast of Australia and the additive effect was blowing either offshore or parallel to the shore. At the same time the greatest storm potential existed in Broome where storm movement and surface winds were in opposite directions because of the direction of the coastline.

On October 19, 2005, the situation with Wilma (Figure 5.26) was similar. The greatest potential for storm surge is

Table 5.4(a)  Saffir–Simpson Hurricane Wind Scale

| Category 1  74-95 mph  119-153 km/h Very dangerous winds will produce some damage. | Example Storm: Dolly (2008) South Padre Island, Texas |
|---|---|
| People Livestock Pets | When struck by flying or falling debris could be injured or killed. |
| Mobile Homes | Older (pre-1994) mobile homes could be destroyed, especially if they are not anchored properly. Newer mobile homes that are anchored properly can sustain damage to shingles, metal roofs and loss of vinyl siding, as well as damage to carports, sunrooms, or lanais. |
| Frame Homes | Poorly constructed frame homes  experience major damage, involving loss of the roof covering and damage to gable ends as well as the removal of porch coverings and awnings. Unprotected windows may break. Masonry chimneys can be toppled. Well-constructed frame homes could have damage to roof shingles, vinyl siding, soffit panels, and gutters. |
| Apartments Shopping Centers Industrial Buildings | Some apartment building and shopping center roof coverings could be partially removed. Industrial buildings can lose roofing and siding especially from windward corners, rakes, and eaves. Failures to overhead doors and unprotected windows will be common. |
| High-Rise Windows and Glass | Windows in high-rise buildings can be broken by flying debris. Falling and broken glass will pose a significant danger even after the storm. |
| Signage, Fences Canopies | There will be occasional damage to commercial signage, fences, and canopies. |
| Trees | Large branches of trees will snap and shallow rooted trees can be toppled. |
| Power and Water | Extensive damage to power lines and poles will likely result in power outages that could last a few to several days. |

Table 5.4(b)   Saffir–Simpson Hurricane Wind Scale

| Category 2   96-110 mph   154-177 km/h Extremely dangerous winds with extensive damage. | Example Storm: Francis (2004) Port St. Lucie, Florida |
|---|---|
| **People Livestock Pets** | There is a substantial risk of injury or death due to flying and falling debris |
| **Mobile Homes** | Older (pre-1994) mobile homes, a high chance of being destroyed and flying debris can shred all mobile homes, including newer models. |
| **Frame Homes** | Poorly constructed frame homes have a high chance of having their roof structures removed especially if they are not anchored properly. Unprotected windows will have a high probability of being broken by flying debris. Well-constructed frame homes could sustain major roof and siding damage. Failure of aluminum, screened-in, swimming pool enclosures will be common. |
| **Apartments Shopping Centers Industrial Buildings** | There will be a substantial percentage of roof and siding damage to apartment buildings and industrial buildings. Unreinforced masonry walls can collapse. |
| **High-Rise Windows and Glass** | Windows in high-rise buildings can be broken by flying debris. Falling and broken glass will pose a significant danger even after the storm. |
| **Signage, Fences Canopies** | Commercial signage, fences, and canopies will be damaged and often destroyed. |
| **Trees** | Many shallowly rooted trees will be snapped or uprooted and block numerous roads. |
| **Power and Water** | Near-total power loss is expected with outages that could last from several days to weeks. Potable water could become scarce as filtration systems begin to fail. |

**Table 5.4(c)   Saffir–Simpson Hurricane Wind Scale**

| Category 3   111-130 mph   178-209 km/h Widespread devastating damage | Example Storm: Ivan (2004) Gulf Shores, Alabama |
|---|---|
| People Livestock Pets | There is a high risk of injury or death to people, livestock, and pets due to flying and falling debris. |
| Mobile Homes | Nearly all older (pre-1994) mobile homes will be destroyed. Most newer mobile homes will sustain severe damage with potential for complete roof failure and wall collapse. |
| Frame Homes | Poorly constructed frame homes can be destroyed by the removal of the roof and exterior walls. Unprotected windows will be broken by flying debris. Well-built frame homes can experience major damage involving the removal of roof decking and gable ends. |
| Apartments Shopping Centers Industrial Buildings | There will be a high percentage of roof covering and siding damage to apartment buildings and industrial buildings. Isolated structural damage to wood or steel framing can occur. Complete failure of older metal buildings is possible, and older unreinforced masonry buildings can collapse. |
| High-Rise Windows and Glass | Numerous windows will be blown out of high-rise buildings resulting in falling glass, which will pose a threat for days to weeks after the storm. |
| Signage, Fences Canopies | Most commercial signage, fences, and canopies will be destroyed. |
| Trees | Many trees will be snapped or uprooted, blocking numerous roads. |
| Power and Water | Electricity and water will be unavailable for several days to a few weeks after the storm passes. |

Table 5.4(d)   Saffir–Simpson Hurricane Wind Scale

| Category 4   131-155 mph   210-249 km/h<br>Catastrophic damage will occur | Example Storm: Charley (2004)<br>Punta Gorda, Florida |
|---|---|
| People<br>Livestock<br>Pets | There is a very high risk of injury or death to people, livestock, and pets due to flying and falling debris. |
| Mobile Homes | Nearly all older (pre-1994) mobile homes will be destroyed. A high percentage of newer mobile homes also will be destroyed. |
| Frame<br>Homes | Poorly constructed homes can sustain complete collapse of all walls as well as the loss of the roof structure. Well-built homes also can sustain severe damage with loss of most of the roof structure and/or some exterior walls. Extensive damage to roof coverings, windows, and doors will occur. Large amounts of windborne debris will be lofted into the air. Windborne debris damage will break most unprotected windows and penetrate some protected windows. |
| Apartments<br>Shopping<br>Centers<br>Industrial<br>Buildings | There will be a high percentage of structural damage to the top floors of apartment buildings. Steel frames in older industrial buildings can collapse. There will be a high percentage of collapse to older unreinforced masonry buildings. |
| High-Rise<br>Windows and<br>Glass | Most windows will be blown out of high-rise buildings resulting in falling glass, which will pose a threat for days to weeks after the storm. |
| Signage,<br>Fences<br>Canopies | Nearly all commercial signage, fences, and canopies will be destroyed. |
| Trees | Most trees will be snapped or uprooted and power poles downed. Fallen trees and power poles will isolate residential areas. |
| Power and<br>Water | Power outages will last for weeks to possibly months. Long-term water shortages will increase human suffering. Most of the area will be uninhabitable for weeks or months. |

**Table 5.4(e)** Saffir–Simpson Hurricane Wind Scale

| Category 5  >155 mph  >249 km/h<br>Widespread catastrophic damage will occur | Example Storm: Andrew (1992)<br>Cutler Ridge, Florida |
|---|---|
| **People Livestock Pets** | People, livestock, and pets are at very high risk of injury or death from flying or falling debris, even if indoors in mobile homes or framed homes. |
| **Mobile Homes** | Almost complete destruction of all mobile homes will occur, regardless of age or construction. |
| **Frame Homes** | A high percentage of frame homes will be destroyed, with total roof failure and wall collapse. Extensive damage to roof covers, windows, and doors will occur. Large amounts of windborne debris will be lofted into the air. Windborne debris damage will occur to nearly all unprotected windows and many protected windows. |
| **Apartments Shopping Centers Industrial Buildings** | Significant damage to wood roof commercial buildings will occur due to loss of roof sheathing. Complete collapse of many older metal buildings can occur. Most unreinforced masonry walls will fail which can lead to the collapse of the buildings. A high percentage of industrial buildings and low-rise apartment buildings will be destroyed. |
| **High-Rise Windows and Glass** | Nearly all windows will be blown out of high-rise buildings resulting in falling glass, which will pose a threat for days to weeks after the storm. |
| **Signage, Fences Canopies** | Nearly all trees will be snapped or uprooted and power poles downed. Fallen trees and power poles will isolate residential areas. |
| **Trees** | Nearly all trees will be snapped or uprooted and power poles downed. Fallen trees and power poles will isolate residential areas. |
| **Power and Water** | Power outages will last for weeks to possibly months. Long-term water shortages will increase human suffering. Most of the area will be uninhabitable for weeks or months. |

**Figure 5.21** The Ben Sawyer Bridge to Sullivan Island after Hurricane Hugo. (*Source:* NOAA.)

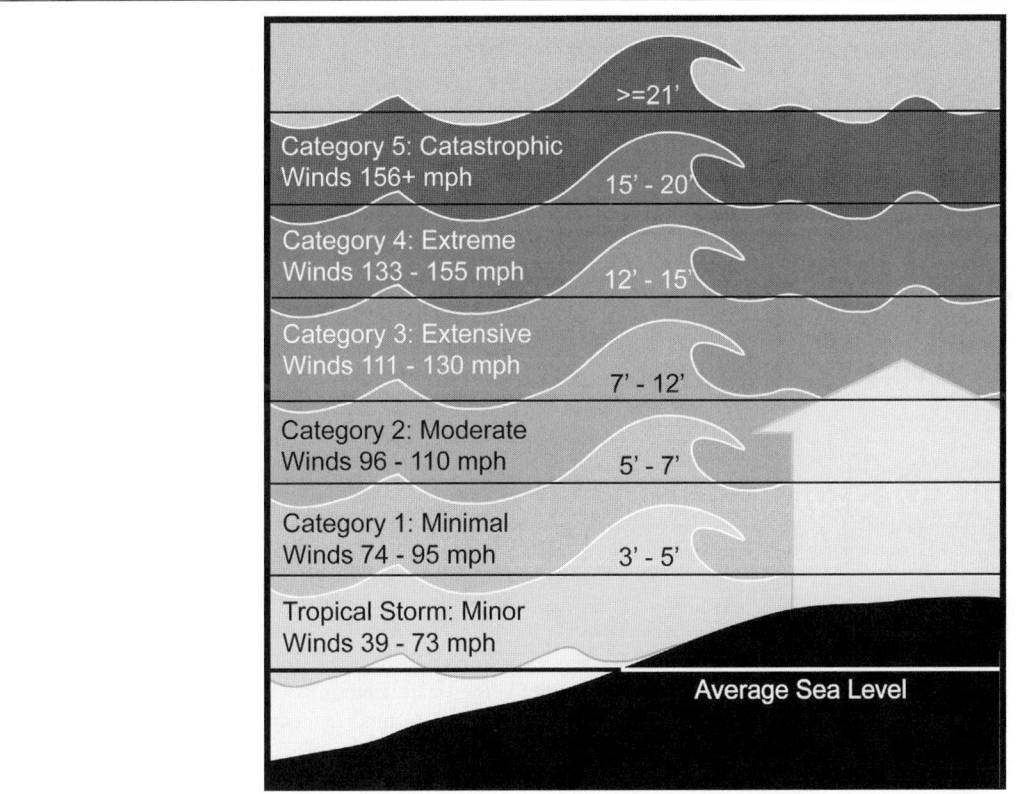

**Figure 5.22** The Saffir–Simpson Hurricane Scale (SSHS). The storm surge estimates can be misleading because surge is so dependent on local conditions. The SSHS has been replaced by the Saffir–Simpson Hurricane Wind Scale. (*Source:* Modified by Steve Horstmeyer from NASA original.)

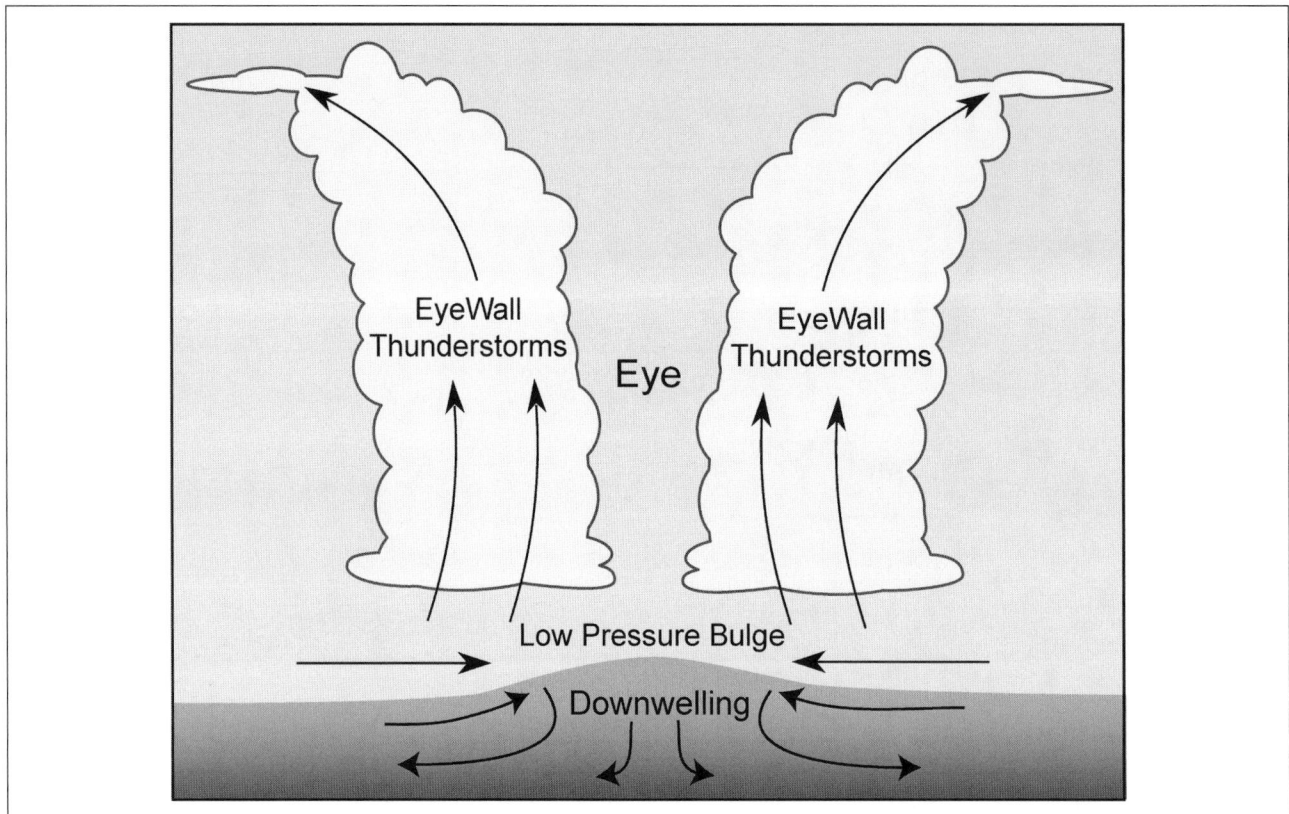

**Figure 5.23** The water convergence, low-pressure bulge, and downwelling of water near the eye of a tropical cyclone. In shallow water downwelling is not possible and the effect is magnified.

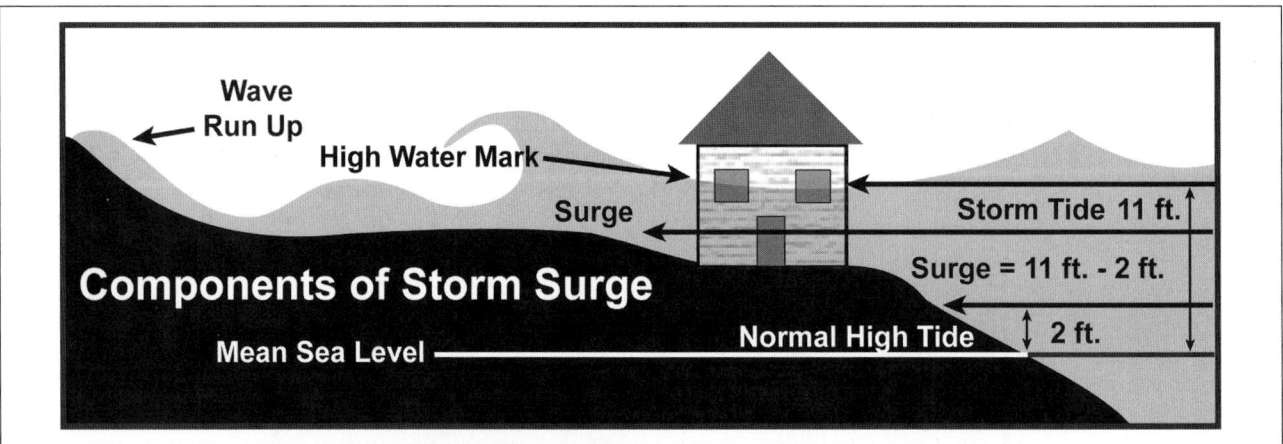

**Figure 5.24** The components of a storm surge. Storm surge is the water depth in addition of the local high tide and does not include wave run-up. There are additional factors (see text). (*Source:* Modified by Steve Horstmeyer from NASA original.)

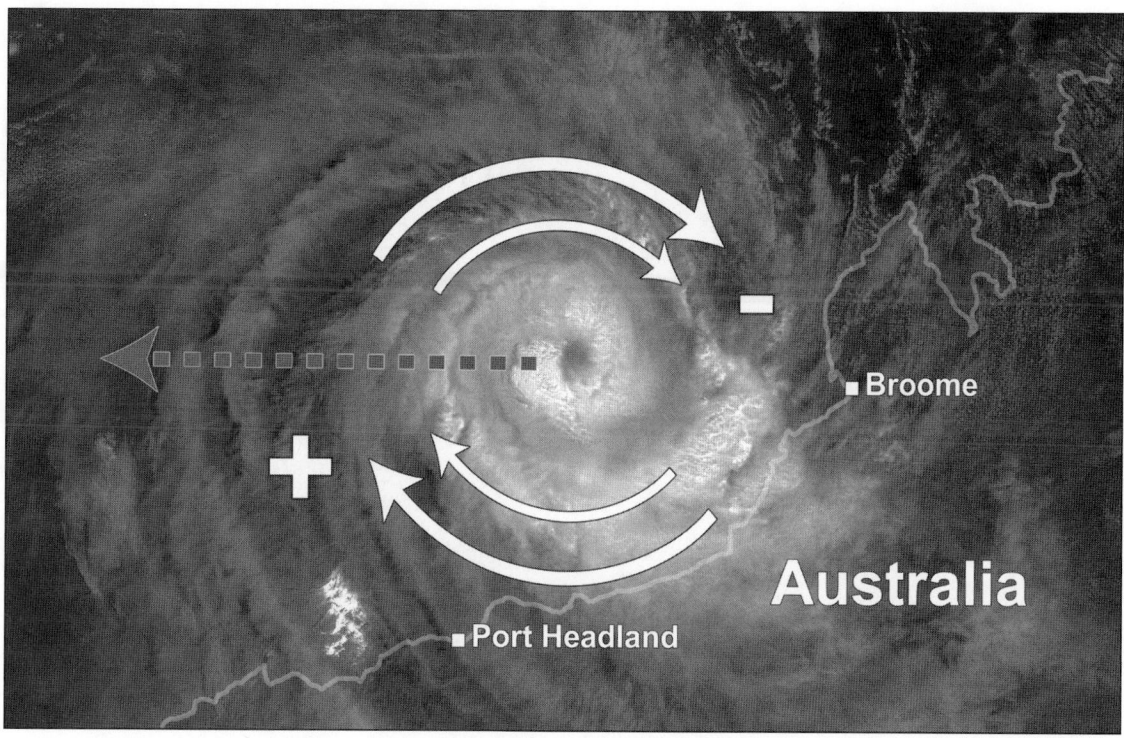

**Figure 5.25**  Cyclone Billy along the northern coast of Australia on December 24, 2008. In the southern hemisphere a cyclone's circulation is clockwise. The additive effect of storm movement and storm winds is indicated by the + sign. The complex relationship that determines storm surge is a combination of the direction the coast runs, storm position, and storm movement. (*Source:* NASA.)

**Figure 5.26**  Hurricane Wilma north of Nicaragua on October 19, 2005. In the northern hemisphere a cyclone's circulation is clockwise. The additive effect of storm movement and storm winds is indicated by the + sign. The complex relationship that determines storm surge is a combination of the direction the coast runs, storm position, and storm movement. (*Source:* NASA.)

**Figure 5.27**  The Richelieu Manor Apartments in Pass Christian, MS, before Hurricane Camille. (*Source:* NOAA.)

where the storm movement and winds are in opposite directions. The additive effect of wind and hurricane movement is out to sea.

Few storm surge records are available, but the greatest is generally accepted to have occurred along the northeast coast of Australia on March 5, 1899, when Tropical Cyclone Mahina generated a surge of 42 feet (13 m) in Bathurst Bay.

In the Bay of Bengal where Bangladesh reports 53% of the world's tropical cyclone deaths despite receiving only 1% of the world's tropical cyclones, the record storm surge was from the Great Backerganj Cyclone of October 29–November 1, 1876, at 40 feet (12 m).

Because of the shape of the Bay of Bengal and the shallow water along the coast of Bangladesh storm surges there are deadly. In 1970, 500,000 people were killed by a storm surge of 34.8 feet (10.6 m) that occurred during the highest tides of the year when the Great Bhola Cyclone made landfall. Of the 30 deadliest tropical cyclones, 22 affected Bangladesh.

The greatest storm surge for the western North Pacific was generated by Typhoon Joe on July 22, 1980, at Nandu Station, China. The surge there was 19.5 feet (5.94 m).

In the Atlantic Ocean, Hurricane Camille generated a storm surge of 22.8' at Pass Christian, MS. Figures 5.27 and 5.28 show the Richelieu Manor Apartments there before and after the storm surge. Despite popular legend there was not

a hurricane party that night in the apartments but 23 people rode the storm out and 8 died.

The water Katrina pushed onshore in southern Mississippi was 5 feet deeper than Camille's surge at Pass Christian for the greatest known Atlantic storm surge at 27.8 feet.

The water crossed north of Interstate 10 in several locations in Mississippi and penetrated as far as 10 miles inland.

Figure 5.29 is a radar view of Katrina just minutes before landfall near Buras, LA. Unlike the examples in Figures 5.25 and 5.26, Katrina was heading directly onshore at an estimated 13 knots (15 mph). The combination of forward velocity of the storm, the storm winds blowing inland, and the shallow water along the Mississippi coast combined to generate an extreme storm surge.

A borderline Category 2/Category 3 hurricane like Katrina at the time should generate a storm surge of around 7 feet based on the SSHS. Katrina taught meteorologists another lesson, this one about storm surge. As a result, the SSHS has been replaced with the SSHWS that does not include a storm surge category.

## • THE DVORAK SATELLITE TECHNIQUE

Just looking at a globe will convince you that it is impossible for hurricane hunters to fly through every developing

**Figure 5.28**  The Richelieu Manor Apartments after the 22.8' storm surge of Category 5 Hurricane Camille on August 17, 1969. Of 23 people who remained in the building 8 died. (*Source:* NOAA.)

**Figure 5.29**  Radar view of Hurricane Katrina 10-min before landfall at Buras, LA, on August 29, 2005. The location of Pass Christian, MS, is indicated. The + and − signs indicate where storm movement works with or against storm winds. (*Source:* Steve Horstmeyer from NWS NEXRAD data at NCDC.)

### Table 5.5  The Dvorak Satellite Technique T-Numbers

| T-Number | Wind (knots) | Estimated Minimum Pressure mb  (in. Hg) | |
|---|---|---|---|
| | | Atlantic | NW Pacific |
| 1.0 - 1.5 | 25 | ---- | ---- |
| 2.0 | 30 | 1009 (29.80) | 1000 (29.53) |
| 2.5 | 35 | 1005 (29.68) | 997 (29.44) |
| 3.0 | 45 | 1000 (29.53) | 991 (29.27) |
| 3.5 | 55 | 994 (29.36) | 984  (29.06) |
| 4.0 | 65 | 987 (29.15) | 976  (28.82) |
| 4.5 | 77 | 979 (28.91) | 966  (28.53) |
| 5.0 | 90 | 970 (28.65) | 954  (28.17) |
| 5.5 | 102 | 960 (28.35) | 941  (28.82) |
| 6.0 | 115 | 948 (28.00) | 927  (27.79) |
| 6.5 | 127 | 935 (27.61) | 914  (26.99) |
| 7.0 | 140 | 921 (27.20) | 898  (26.52) |
| 7.5 | 155 | 906 (26.76) | 879  (25.96) |
| 8.0 | 170 | 890 (26.28) | 858  (25.34) |

tropical cyclone in the Atlantic Ocean simply because of the great distances, and in the much larger Pacific Ocean and isolated south Indian Oceans, frequent flights are beyond consideration.

Vernon Dvorak was a meteorologist with the NOAA's National Environmental Satellite Data and Information Service (NESDIS) Satellite Analysis Branch in 1974 when he developed a technique to estimate developing tropical cyclone intensity visually using satellite imagery. It was now possible to track and forecast tropical cyclones even in the remotest parts of Earth's ocean basins. His technique, though modified, is still in use today.

As tropical cyclones develop, they have similar characteristics when intensities are nearly the same. As a tropical cyclone strengthens, the features change predictably and are reliable indicators of the strength. The visual patterns have been compared to observed data for which wind speeds were measured, and the Dvorak Technique allows a fairly accurate estimate.

The satellite meteorologist looks for visual patterns that are typical of a tropical cyclone within a range of intensity values and assigns a value from 1 to 8, called the "T-number" (tropical number), which indicates storm intensity (Table 5.5). By following changes in the T-number the evolution of tropical cyclones can be tracked and forecasts issued for remote islands and ships at sea.

Using infrared imagery, when the eye is visible its temperature is compared to the temperature of the tops of the

towering eye wall thunderstorms. The greater the difference in temperature, the stronger the storm. This is called the "enhanced infrared (EIR) pattern." Updated versions of the Dvorak technique utilize the advanced capabilities of weather satellites along with the traditional visual patterns and computer algorithms to remove subjectivity.

Some of the patterns meteorologists look for include the curved band pattern (T1.0–4.5), the sheared pattern (T1.5–3.5), the central dense overcast (CDO) pattern (T2.5–5.0), the banding eye pattern (T4.0–4.5), and the eye pattern (T4.5–8.0). Figures 5.30–5.34 show some of the patterns meteorologists look for in applying the Dvorak technique.

A well-developed, strong hurricane often has a distinct eye like the eye of Typhoon Tip shown in Figure 5.30. The smaller and more distinct the eye, the stronger the hurricane. The satellite meteorologist can use the "eye pattern" to estimate intensity. Tip was the most intense tropical cyclone on record. The pressure dropped to 870 mbar (25.69″) for the lowest sea-level pressure ever with winds to 190 mph. Tip was also the largest tropical cyclone ever with a diameter of 1380 miles.

Figure 5.31 shows two views of Wilma 55 hours apart at the same magnification, illustrating how the Dvorak eye pattern can be used to determine intensity change: on the left at peak intensity (175 mph, 882 mbar) and on the right after an eye wall replacement cycle on October 21 (150 mph). The small size of the pinhole eye of Wilma is the smallest tropical

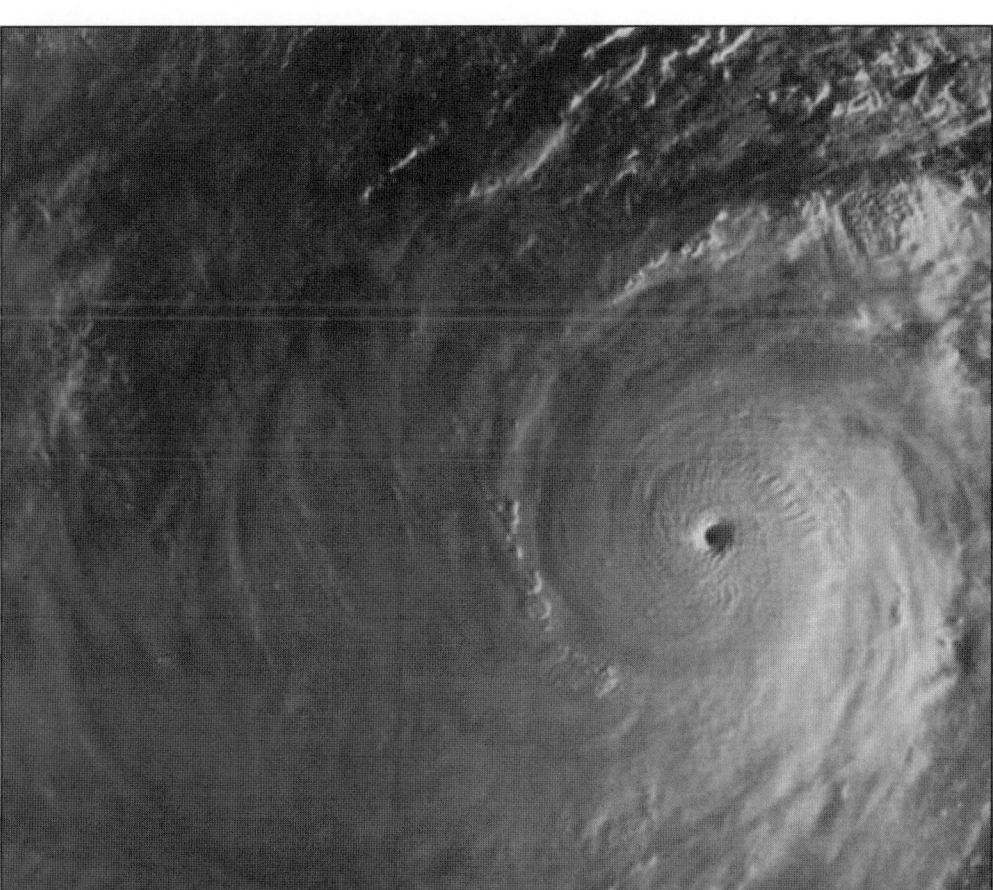

**Figure 5.30**   The compact eye of Typhoon Tip, the most intense tropical cyclone on record. Satellite meteorologists use the "eye pattern" to determine tropical cyclone strength. (*Source:* NOAA.)

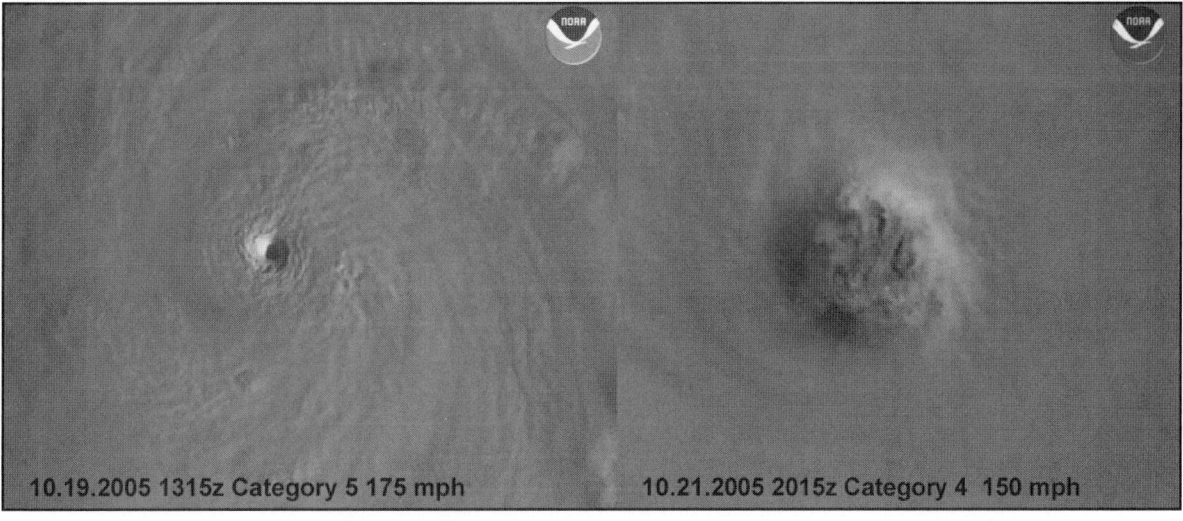

10.19.2005 1315z Category 5 175 mph                    10.21.2005 2015z Category 4  150 mph

**Figure 5.31**   Two views of the eye of Hurricane Wilma. (Left) at peak intensity and (right) after an eye wall replacement cycle. The Dvorak technique tells the meteorologist the hurricane had weakened. (*Source:* NASA.)

**Figure 5.32** A wide view of Hurricane Wilma at peak intensity, October 20, 2005. (*Source:* NASA.)

**Figure 5.33** The central dense overcast (CDO) of developing Hurricane Dennis on July 7, 2005, on the left and the well-developed, visible eye on July 9, 2005. The change of the CDO pattern is evidence of the strengthening storm. (*Source:* NASA.)

**Figure 5.34**   Tropical Cyclone Edzani in the southern Indian Ocean, January 6–7, 2010. On the left the curved bands are just beginning to wrap around the center. On the right the bands spiral all the way to the eye wall. The curved band pattern indicates a well-developed, intense tropical cyclone.

cyclone eye known and shows much better in the context of the wide view (Figure 5.32).

As a tropical cyclone develops it often has a CDO that obscures the eye. This pattern is made of cirrus clouds from the top of thunderstorms. As the tropical cyclone intensifies, the CDO grows, and when sinking air near the center, it creates a hole in the CDO and an eye is visible; it is at hurricane strength. If the CDO is asymmetrical, it can indicate a sheared environment that is not favorable to storm development.

Figure 5.33 shows developing Hurricane Dennis on July 7, 2005, with a CDO (left) and on July 9, 2005, (right) as a Category 3 hurricane with a distinct eye no longer obscured by the high clouds.

Tracking the "curved band pattern" allows the meteorologist to follow changes in intensity and degree of maturity. Figure 5.34 shows Tropical Cyclone Edzani in the southern Indian Ocean. The image on the left is from January 6, 2010, 0450z, and about 27 hours later (January 7, 0825z), there has been a great amount of change to the curved bands. In the left image the developing eye is visible but the curved bands do not wrap completely around the storm center. On the right the bands are more distinct and wrap around the center several times.

Edzani became the first very intense tropical cyclone (see Table 5.1) in the southwest Indian Ocean since Juliet in 2005. Edzani's winds peaked at 140 mph (10 min average) with a minimum pressure of 905 mbar (26.73 in. Hg).

## • REFERENCE MATERIALS

The following reference pages include maps, tables, graphs, and photos of tropical cyclones.

### Reference Maps (Figures 5.35–5.40)

### Reference Tables

#### *Top 100*
Tables 5.6–5.9 are "top 100" lists of Atlantic and Pacific hurricanes. There are 1352 Atlantic storms starting in 1851 through the 2009 Atlantic Hurricane Season. Tropical depressions and subtropical storms are excluded. In the Pacific, the data started in 1949 and continued through the 2009 East Pacific Hurricane Season. In that time there were 854 storms.

The storms are separately ranked by minimum pressure and maximum wind. Care must be used with these data because most minimum-pressure values are missing for Atlantic storms well into the 1960s and for East Pacific storms into the late 1980s. Wind values are much better represented with only a few missing in both data sets through the entire period of record.

### Tropical Cyclone Names, Retired Names, Renaming Storms

Table 5.10 is a list of retired Atlantic and Pacific hurricane names. Storm names are retired when a tropical cyclone is

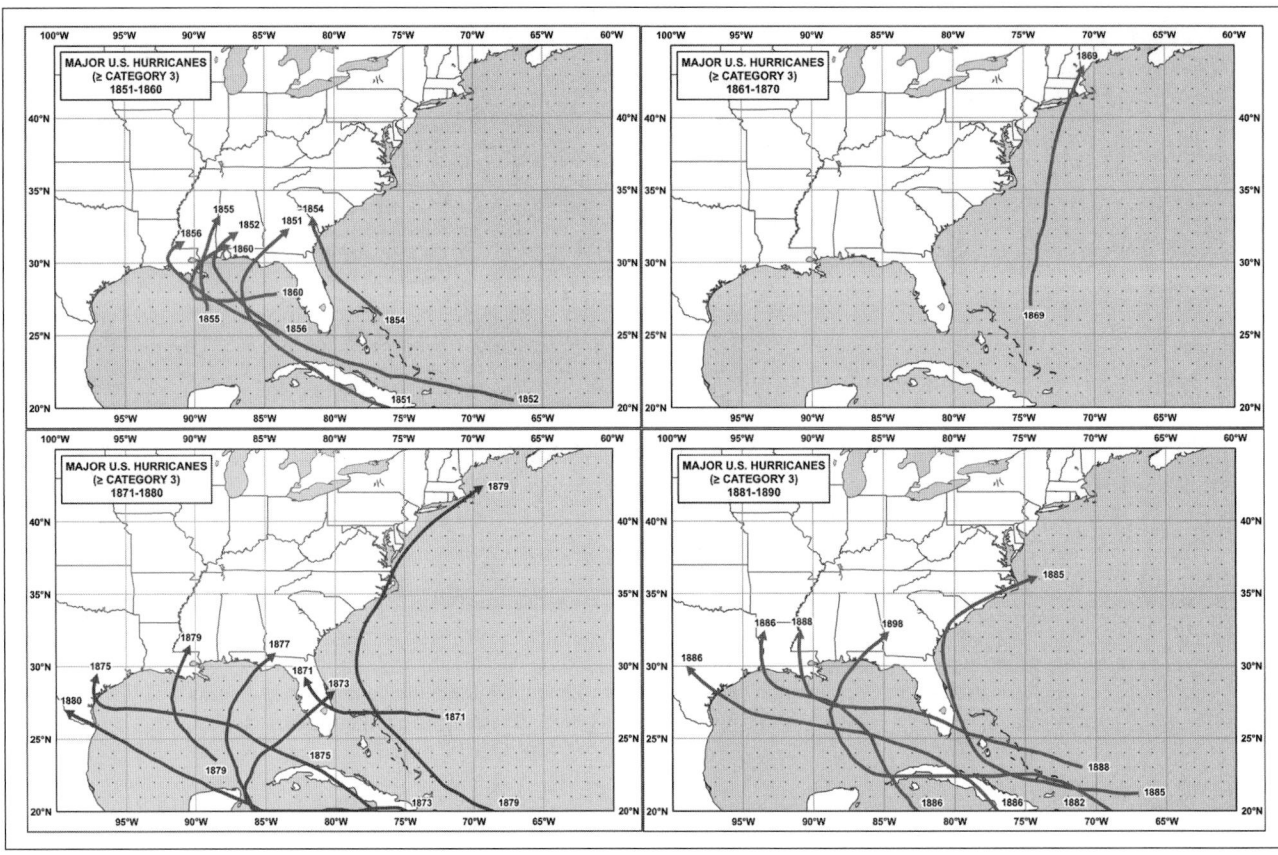

**Figure 5.35a** Landfalling major hurricanes by decade from 1851 through 1890. (*Source:* NOAA, NHC.)

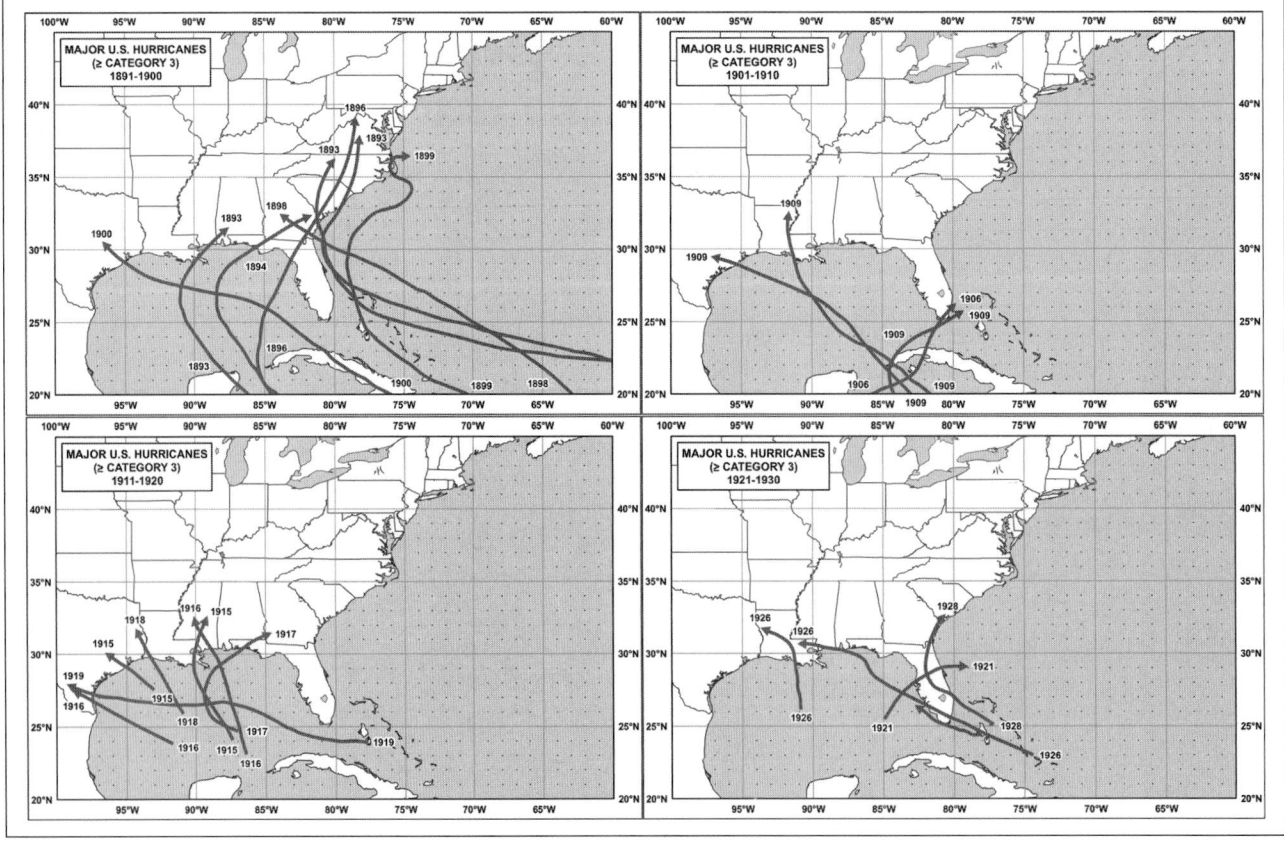

**Figure 5.35b** Landfalling major hurricanes by decade from 1891 through 1930. (*Source:* NOAA, NHC.)

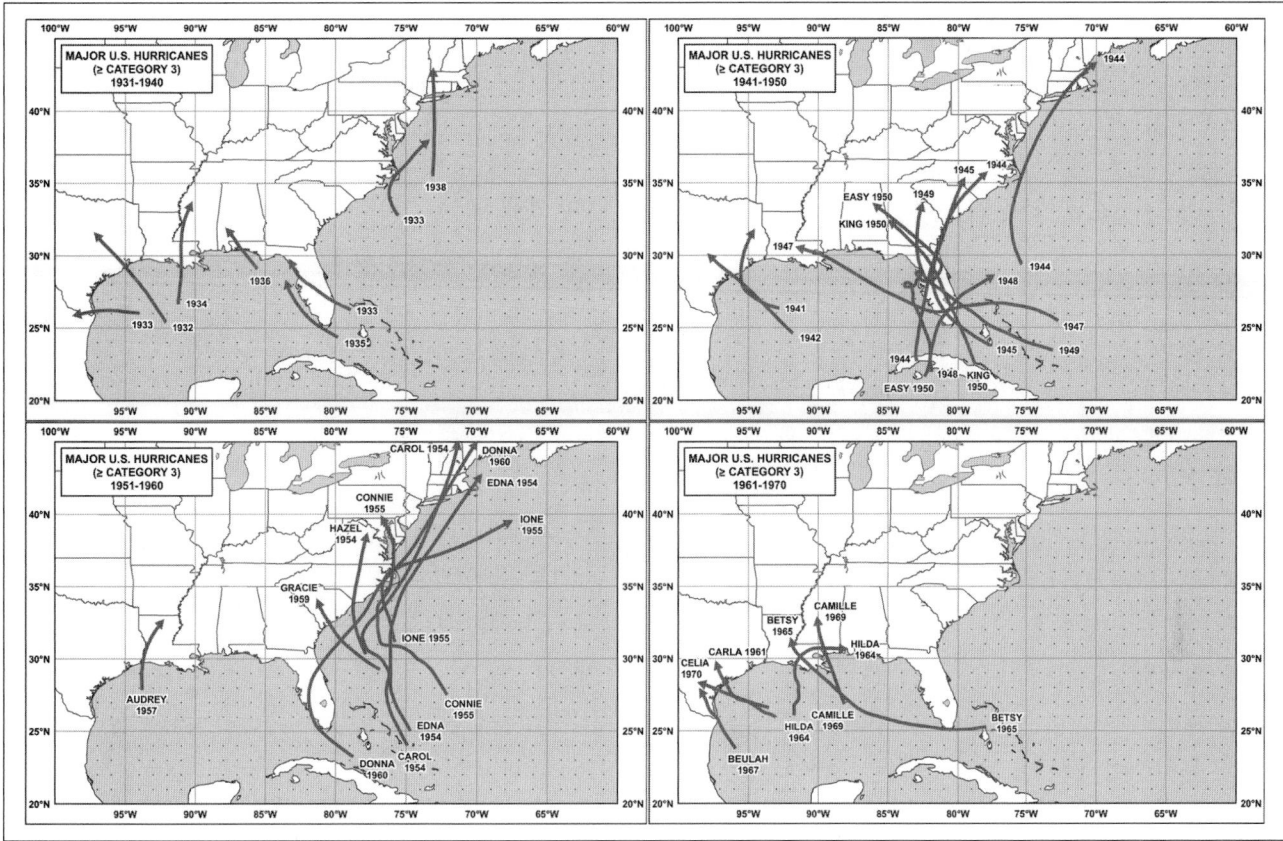

**Figure 5.35c**   Landfalling major hurricanes by decade from 1931 through 1970. (*Source:* NOAA, NHC.)

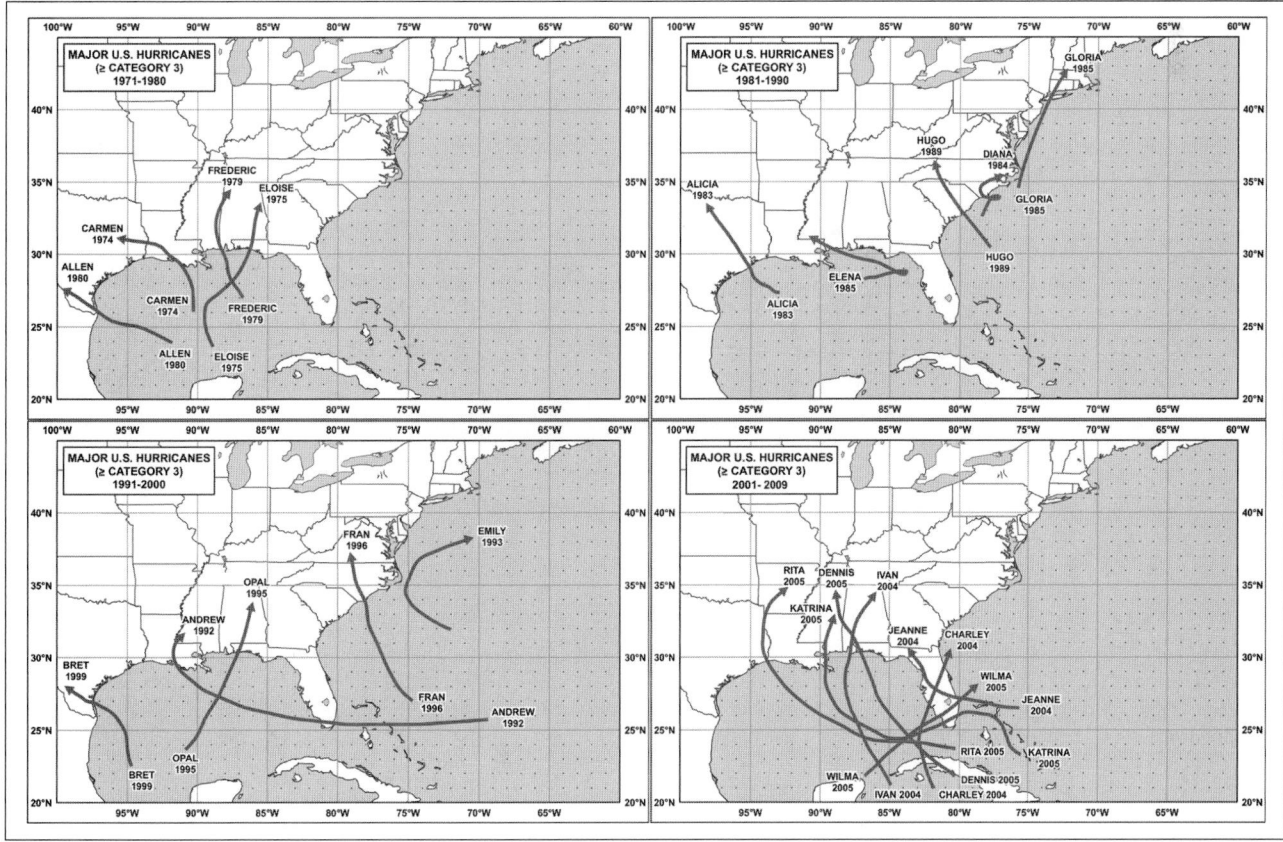

**Figure 5.35d**   Landfalling major hurricanes by decade from 1971 through 2009. (*Source:* NOAA, NHC.)

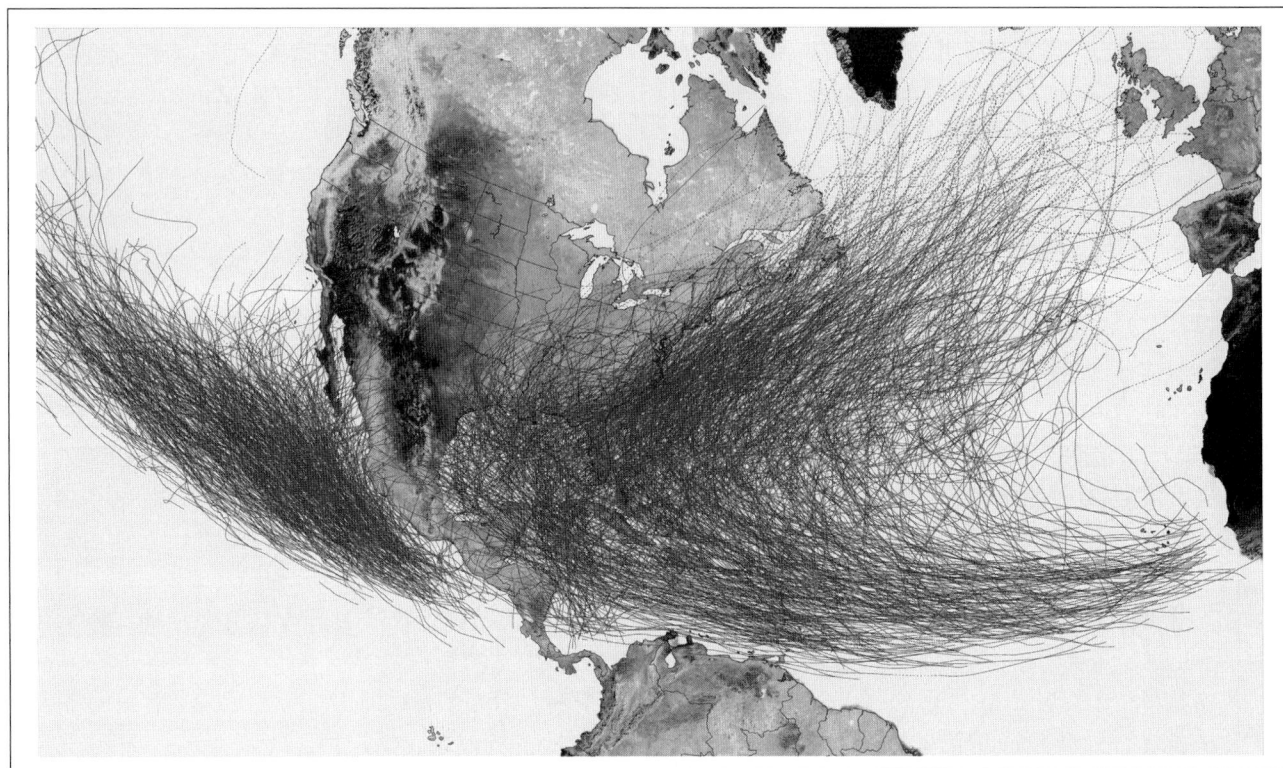

**Figure 5.36** The tracks of all known tropical cyclones in the Eastern North Pacific (1949–2008) and Atlantic (1851–2008) oceans. (*Source:* NOAA, NHC.)

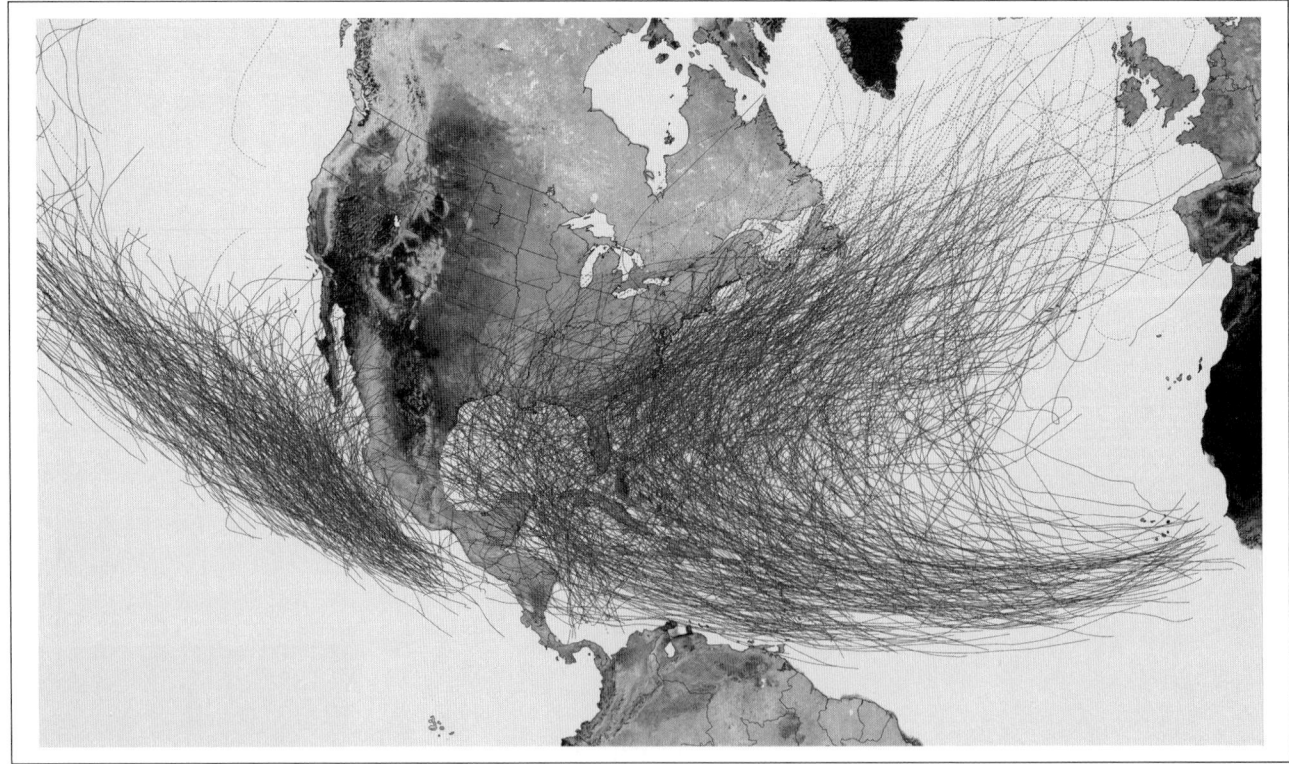

**Figure 5.37** The tracks of all known hurricanes in the Eastern North Pacific (1949–2008) and Atlantic (1851–2008) oceans. (*Source:* NOAA, NHC.)

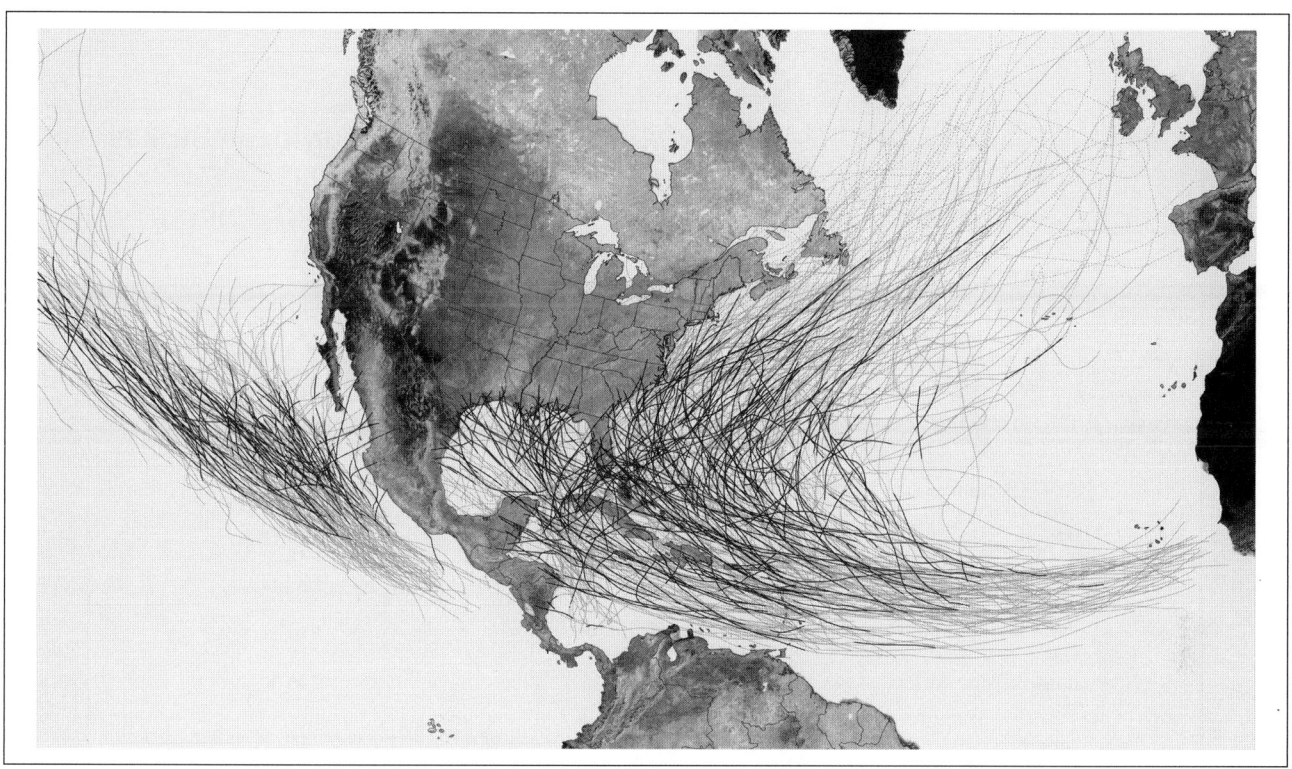

**Figure 5.38**   The tracks of all known major hurricanes (≥Category 3) in the Eastern North Pacific (1949–2008) and Atlantic (1851–2008) oceans. (*Source:* NOAA, NHC.)

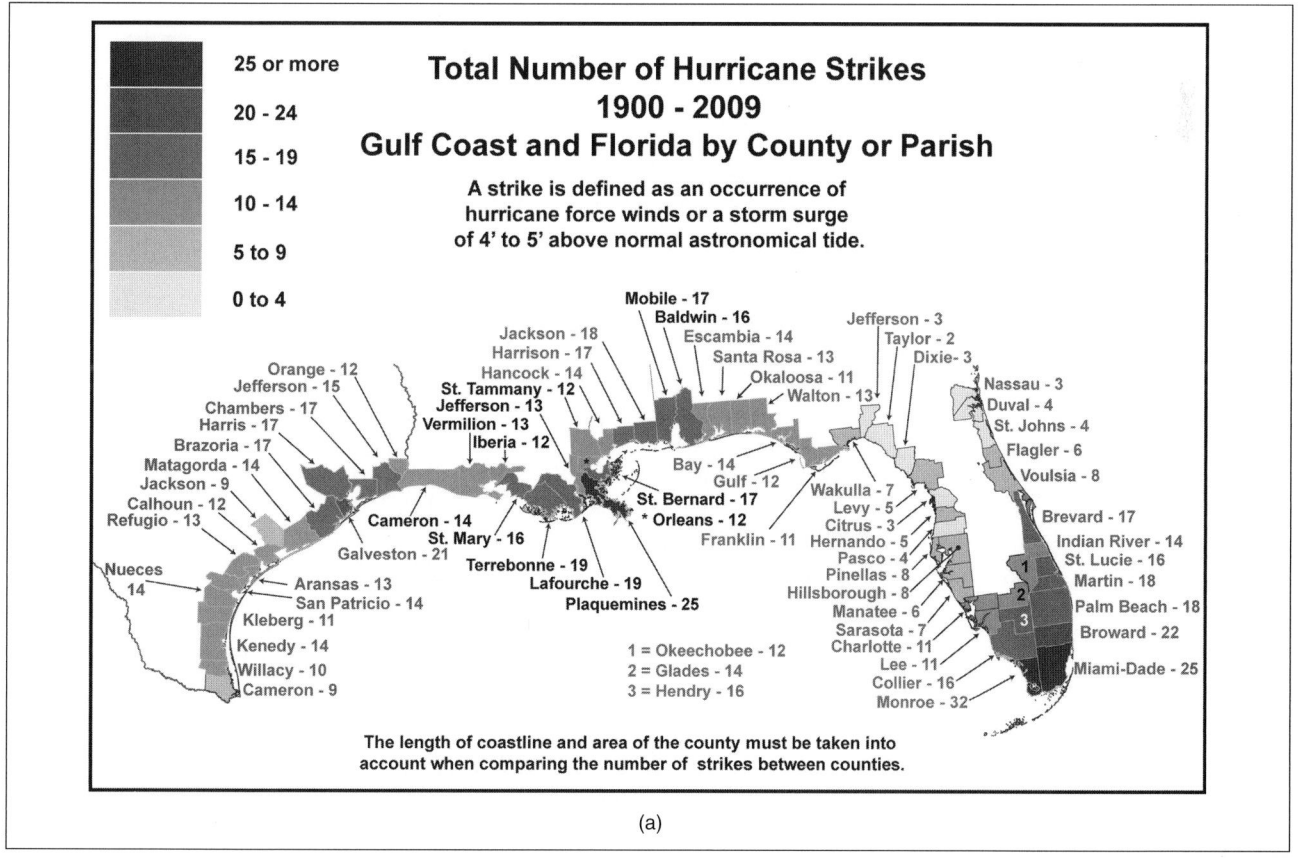

(a)

**Figure 5.39 a–d**   All Atlantic hurricane strikes for each county or parish, 1900–2009. (*Source:* Steve Horstmeyer from NOAA, NHC data.)

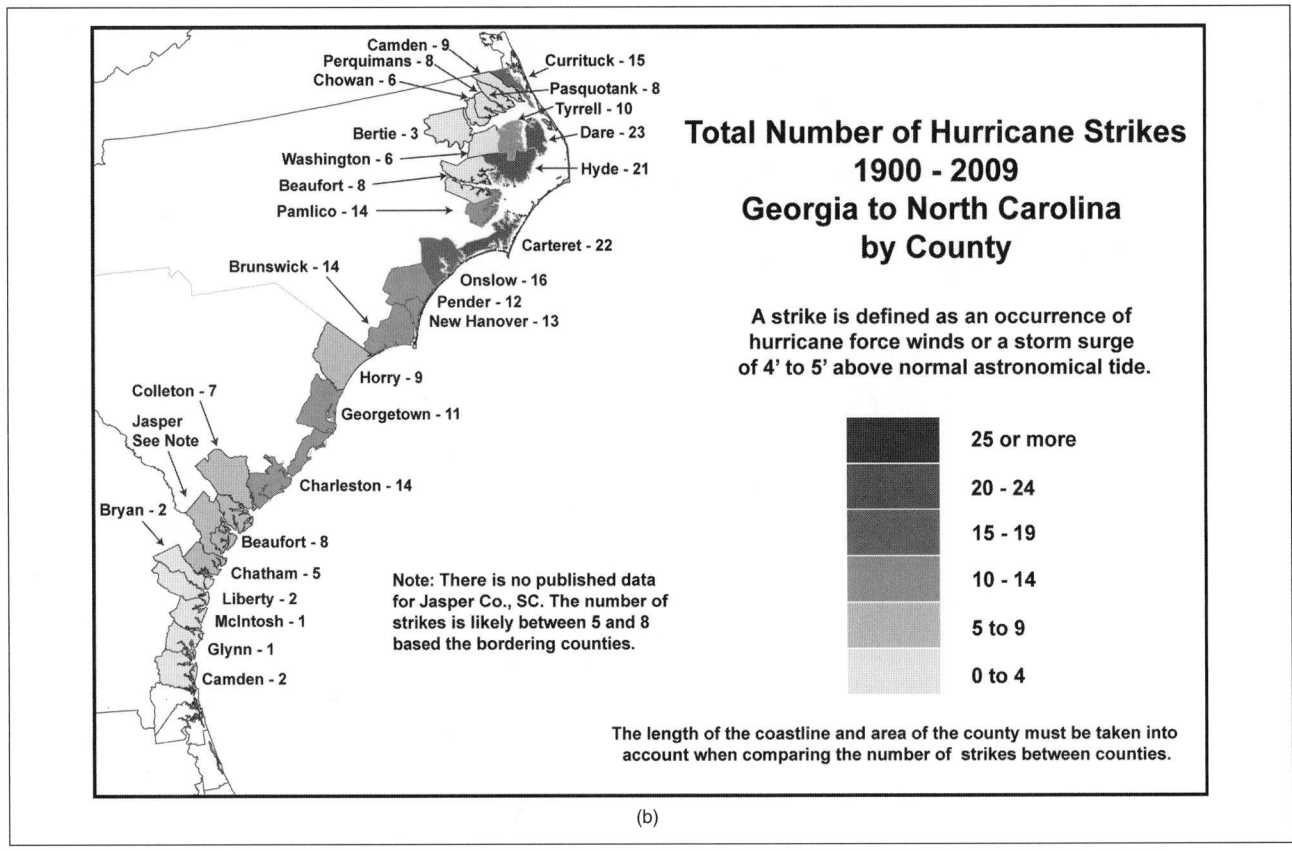

(b)

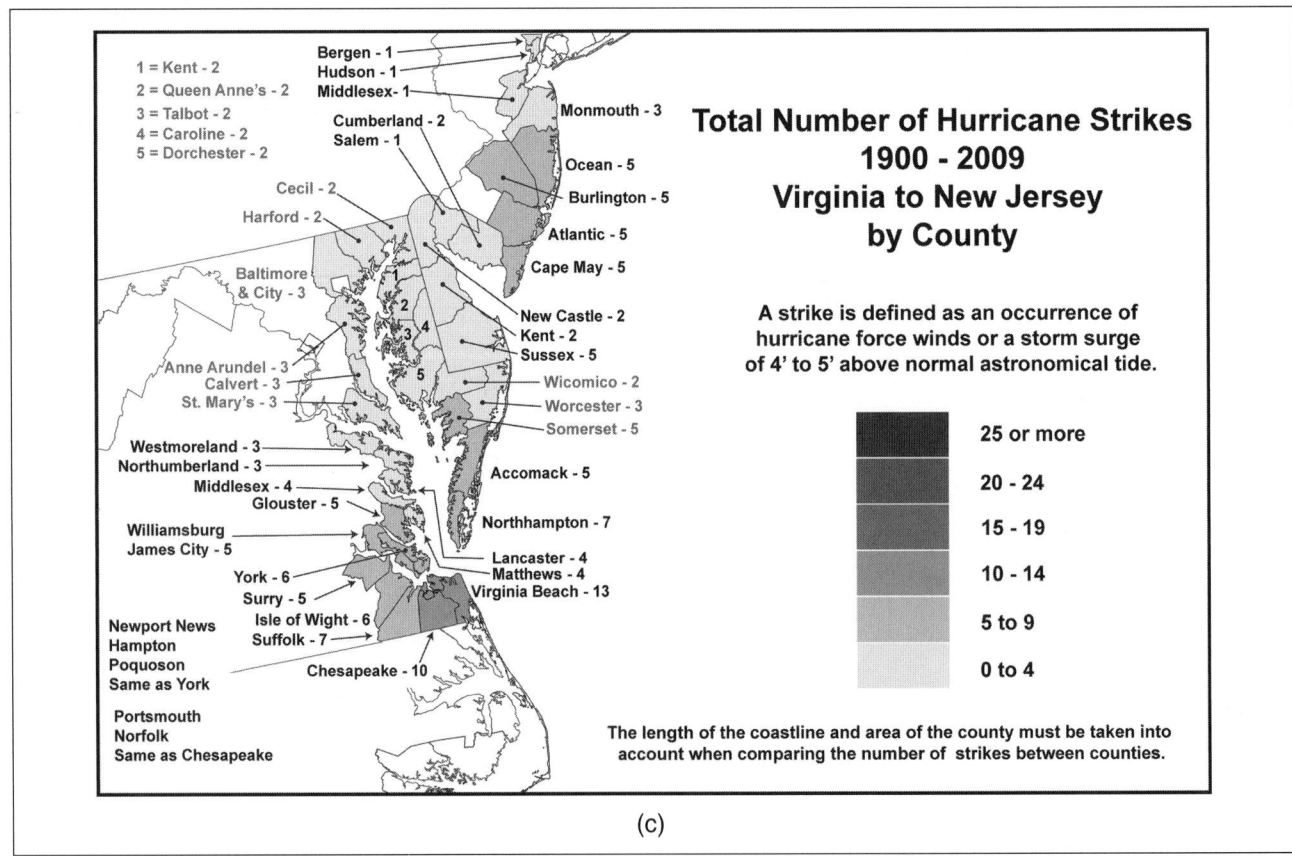

(c)

**Figure 5.39** (*Continued*)

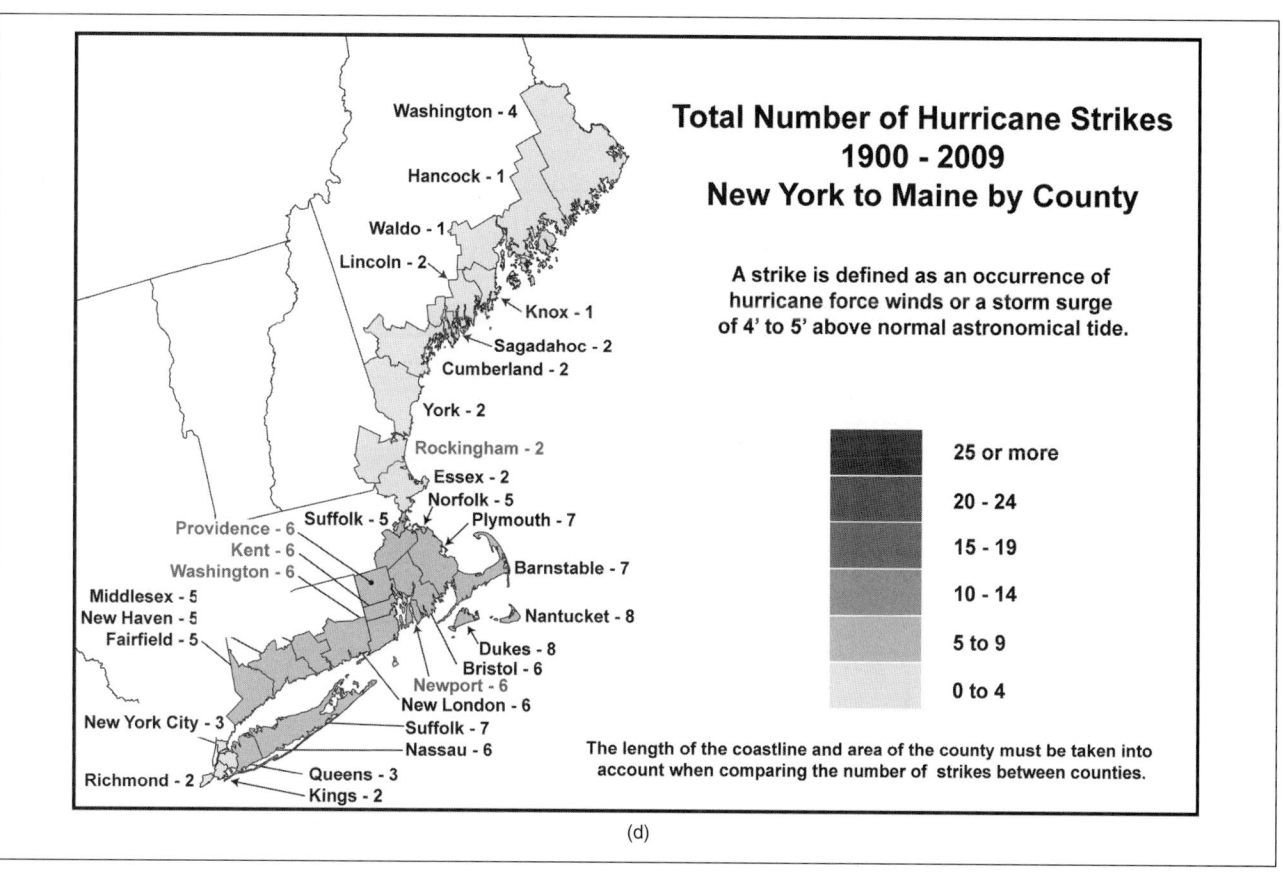

(d)

**Figure 5.39** (*Continued*).

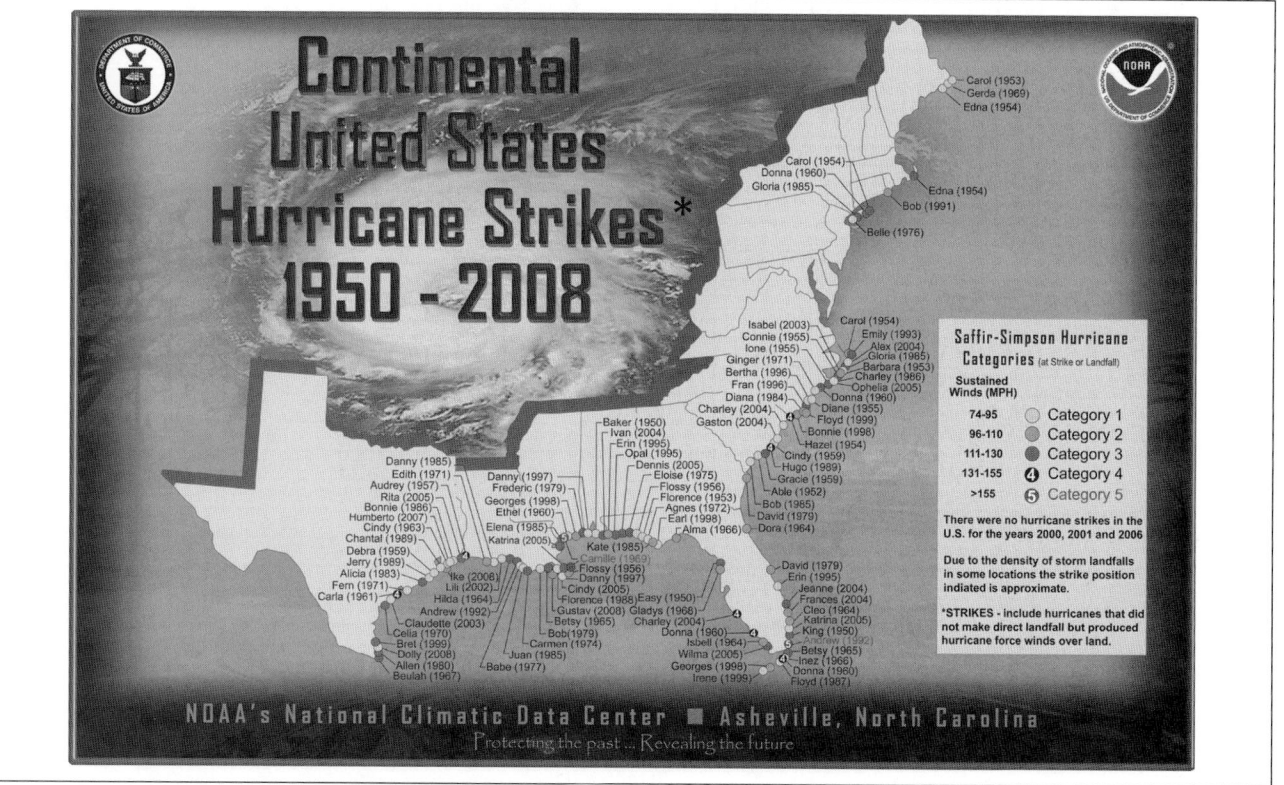

**Figure 5.40** Hurricane strikes on the continental US coast, 1950–2008, by category. (*Source:* Modified by Steve Horstmeyer from NOAA, NCDC data.)

**Table 5.6   Top 100 Atlantic Hurricanes by Maximum Wind Speed**

| | Storm | Year | Month | Day | Hour | Max Wind kts | Press at Max Wind | Lat at Max ° N | Lon at Max ° W |
|---|---|---|---|---|---|---|---|---|---|
| 1 | CAMILLE | 1969 | 8 | 18 | 0 | 165 | 909 | 29.4 | 89.1 |
| 2 | ALLEN | 1980 | 8 | 8 | 18 | 165 | 899 | 21.8 | 86.4 |
| 3 | DOG | 1950 | 9 | 6 | 6 | 160 | - | 26.7 | 68.4 |
| 4 | GILBERT | 1988 | 9 | 14 | 0 | 160 | 888 | 19.7 | 83.8 |
| 5 | WILMA | 2005 | 10 | 19 | 12 | 160 | 882 | 17.3 | 82.8 |
| 6 | MITCH | 1998 | 10 | 27 | 0 | 155 | 910 | 17.2 | 83.8 |
| 7 | RITA | 2005 | 9 | 22 | 3 | 155 | 895 | 24.7 | 87.3 |
| 8 | JANET | 1955 | 9 | 28 | 0 | 150 | - | 18 | 86.1 |
| 9 | CARLA | 1961 | 9 | 11 | 6 | 150 | 936 | 27.2 | 95.7 |
| 10 | ANITA | 1977 | 9 | 2 | 6 | 150 | 926 | 24.2 | 97.1 |
| 11 | DAVID | 1979 | 9 | 1 | 18 | 150 | 926 | 17.9 | 69.7 |
| 12 | ANDREW | 1992 | 8 | 24 | 18 | 150 | 922 | 25.4 | 75.8 |
| 13 | KATRINA | 2005 | 8 | 28 | 18 | 150 | 902 | 26.3 | 88.6 |
| 14 | DEAN | 2007 | 8 | 21 | 8 | 150 | 905 | 18.7 | 87.7 |
| 15 | FELIX | 2007 | 9 | 3 | 7 | 150 | 929 | 14 | 75.3 |
| 16 | ISABEL | 2003 | 9 | 11 | 18 | 145 | 915 | 21.5 | 54.8 |
| 17 | IVAN | 2004 | 9 | 12 | 0 | 145 | 910 | 18.2 | 79.6 |
| 18 | UNNAMED | 1928 | 9 | 14 | 18 | 140 | 931 | 17.9 | 65.8 |
| 19 | UNNAMED | 1932 | 9 | 6 | 18 | 140 | - | 25.7 | 77 |
| 20 | UNNAMED | 1935 | 9 | 3 | 0 | 140 | 892 | 24.5 | 80.1 |
| 21 | UNNAMED | 1938 | 9 | 20 | 18 | 140 | - | 24.1 | 71.6 |
| 22 | UNNAMED | 1947 | 9 | 16 | 6 | 140 | - | 26.5 | 75.4 |
| 23 | EASY | 1951 | 9 | 7 | 12 | 140 | - | 24 | 66 |
| 24 | CLEO | 1958 | 8 | 16 | 0 | 140 | 948 | 19.6 | 49.8 |
| 25 | DONNA | 1960 | 9 | 4 | 12 | 140 | 952 | 16.8 | 59.5 |
| 26 | ETHEL | 1960 | 9 | 15 | 6 | 140 | - | 28.1 | 88.9 |
| 27 | HATTIE | 1961 | 10 | 31 | 0 | 140 | 920 | 17.9 | 86.1 |
| 28 | BEULAH | 1967 | 9 | 20 | 6 | 140 | 931 | 25.1 | 96.8 |
| 29 | EDITH | 1971 | 9 | 10 | 18 | 140 | 943 | 14.8 | 83.2 |
| 30 | HUGO | 1989 | 9 | 16 | 18 | 140 | 918 | 14.6 | 54.6 |
| 31 | EMILY | 2005 | 7 | 17 | 0 | 140 | 929 | 17.1 | 79.5 |
| 32 | CARRIE | 1957 | 9 | 9 | 12 | 135 | 975 | 18.3 | 48.2 |
| 33 | CLEO | 1964 | 8 | 24 | 0 | 135 | 950 | 16.7 | 69.5 |
| 34 | BETSY | 1965 | 9 | 10 | 0 | 135 | 941 | 28.3 | 89.2 |
| 35 | GEORGES | 1998 | 9 | 20 | 6 | 135 | 937 | 16 | 56.3 |
| 36 | FLOYD | 1999 | 9 | 13 | 12 | 135 | 921 | 23.9 | 71.4 |
| 37 | LENNY | 1999 | 11 | 18 | 18 | 135 | 933 | 17.4 | 64.8 |
| 38 | GUSTAV | 2008 | 8 | 30 | 22 | 135 | 941 | 22.4 | 83.1 |
| 39 | UNNAMED | 1853 | 9 | 3 | 12 | 130 | 924 | 19.7 | 56.2 |
| 40 | UNNAMED | 1856 | 8 | 11 | 18 | 130 | 934 | 29.2 | 91.1 |
| 41 | UNNAMED | 1880 | 8 | 13 | 0 | 130 | 931 | 25.7 | 96.9 |
| 42 | UNNAMED | 1912 | 11 | 18 | 0 | 130 | - | 17.6 | 78.7 |
| 43 | UNNAMED | 1922 | 9 | 21 | 18 | 130 | - | 29.9 | 65.5 |
| 44 | UNNAMED | 1926 | 9 | 17 | 18 | 130 | - | 21.8 | 71.2 |
| 45 | UNNAMED | 1930 | 9 | 4 | 18 | 130 | 933 | 18.4 | 70 |
| 46 | UNNAMED | 1933 | 10 | 6 | 0 | 130 | - | 27.2 | 76.1 |
| 47 | UNNAMED | 1949 | 8 | 27 | 0 | 130 | 954 | 26.8 | 80.1 |
| 48 | FOX | 1952 | 10 | 25 | 18 | 130 | 934 | 21.8 | 81 |
| 49 | CAROL | 1953 | 9 | 4 | 18 | 130 | 929 | 19.9 | 60 |
| 50 | HILDA | 1964 | 10 | 2 | 0 | 130 | 942 | 25.2 | 91.4 |
| 51 | INEZ | 1966 | 9 | 29 | 0 | 130 | 929 | 17.1 | 68.5 |
| 52 | CARMEN | 1974 | 9 | 8 | 0 | 130 | 937 | 28.7 | 90.8 |
| 53 | OPAL | 1995 | 10 | 4 | 12 | 130 | 919 | 27.3 | 88.5 |

**Table 5.6   Top 100 Atlantic Hurricanes by Maximum Wind Speed** [CONTINUED]

| | Storm | Year | Month | Day | Hour | Max Wind kts | Press at Max Wind | Lat at Max ° N | Lon at Max ° W |
|---|---|---|---|---|---|---|---|---|---|
| 54 | GERT | 1999 | 9 | 16 | 6 | 130 | 933 | 18 | 51.7 |
| 55 | CHARLEY | 2004 | 9 | 13 | 19 | 130 | 941 | 26.6 | 82.2 |
| 56 | DENNIS | 2005 | 7 | 8 | 12 | 130 | 938 | 20.9 | 79.5 |
| 57 | UNNAMED | 1906 | 9 | 16 | 12 | 125 | - | 32.1 | 71.1 |
| 58 | UNNAMED | 1906 | 9 | 27 | 18 | 125 | - | 28.4 | 87.7 |
| 59 | UNNAMED | 1932 | 8 | 14 | 0 | 125 | 942 | 28.9 | 94.7 |
| 60 | CONNIE | 1955 | 8 | 10 | 12 | 125 | 970 | 30.8 | 75.3 |
| 61 | AUDREY | 1957 | 6 | 27 | 12 | 125 | 946 | 29.3 | 93.8 |
| 62 | ESTHER | 1961 | 9 | 20 | 18 | 125 | 950 | 32 | 72.6 |
| 63 | FLORA | 1963 | 10 | 4 | 0 | 125 | 944 | 18 | 73.1 |
| 64 | GLADYS | 1964 | 9 | 18 | 6 | 125 | 951 | 25.8 | 66.5 |
| 65 | GLORIA | 1985 | 9 | 25 | 6 | 125 | 920 | 25.1 | 70.9 |
| 66 | HELENE | 1988 | 9 | 24 | 18 | 125 | 938 | 15.3 | 46.1 |
| 67 | JOAN | 1988 | 10 | 22 | 6 | 125 | 932 | 11.9 | 83.2 |
| 68 | GABRIELLE | 1989 | 9 | 6 | 0 | 125 | 944 | 22.2 | 58.3 |
| 69 | EDOUARD | 1996 | 8 | 28 | 0 | 125 | 944 | 20.9 | 60.4 |
| 70 | BRET | 1999 | 8 | 22 | 12 | 125 | 944 | 26.2 | 96.1 |
| 71 | IRIS | 2001 | 10 | 9 | 0 | 125 | 948 | 16.5 | 88 |
| 72 | LILI | 2002 | 10 | 3 | 0 | 125 | 940 | 26.7 | 90.3 |
| 73 | KARL | 2004 | 9 | 21 | 6 | 125 | 938 | 19.6 | 47.3 |
| 74 | IKE | 2008 | 9 | 4 | 6 | 125 | 935 | 22.4 | 55 |
| 75 | PALOMA | 2008 | 11 | 8 | 12 | 125 | 944 | 19.8 | 79.6 |
| 76 | UNNAMED | 1866 | 10 | 2 | 0 | 120 | 938 | 24.5 | 77.1 |
| 77 | UNNAMED | 1867 | 10 | 30 | 18 | 120 | 952 | 18.2 | 64.8 |
| 78 | UNNAMED | 1878 | 10 | 3 | 0 | 120 | - | 29.3 | 72.6 |
| 79 | UNNAMED | 1880 | 10 | 2 | 12 | 120 | 928 | 28.3 | 63.7 |
| 80 | UNNAMED | 1909 | 7 | 21 | 18 | 120 | - | 26.9 | 91.3 |
| 81 | UNNAMED | 1909 | 9 | 20 | 18 | 120 | - | 26.2 | 87.8 |
| 82 | UNNAMED | 1915 | 8 | 16 | 12 | 120 | - | 26.2 | 91.1 |
| 83 | UNNAMED | 1919 | 9 | 10 | 12 | 120 | - | 24.6 | 82.7 |
| 84 | UNNAMED | 1921 | 10 | 24 | 12 | 120 | - | 23.4 | 86 |
| 85 | UNNAMED | 1926 | 7 | 26 | 0 | 120 | - | 23.5 | 75.1 |
| 86 | UNNAMED | 1926 | 9 | 14 | 12 | 120 | - | 32 | 70 |
| 87 | UNNAMED | 1929 | 9 | 27 | 0 | 120 | 936 | 24.4 | 78.2 |
| 88 | UNNAMED | 1933 | 9 | 3 | 18 | 120 | - | 23.6 | 72.4 |
| 89 | UNNAMED | 1943 | 8 | 24 | 6 | 120 | - | 28.3 | 66.8 |
| 90 | UNNAMED | 1944 | 9 | 13 | 18 | 120 | - | 26.3 | 72.3 |
| 91 | UNNAMED | 1945 | 8 | 28 | 0 | 120 | 968 | 29.1 | 96 |
| 92 | UNNAMED | 1945 | 9 | 16 | 18 | 120 | - | 24.9 | 79.6 |
| 93 | ABLE | 1950 | 8 | 18 | 0 | 120 | - | 26.1 | 73.8 |
| 94 | FOX | 1950 | 9 | 15 | 18 | 120 | - | 24.6 | 59.4 |
| 95 | HAZEL | 1954 | 10 | 15 | 6 | 120 | - | 30.2 | 77.8 |
| 96 | GRETA | 1956 | 11 | 5 | 12 | 120 | 970 | 25.3 | 61 |
| 97 | GRACIE | 1959 | 9 | 29 | 12 | 120 | 950 | 31.3 | 79.6 |
| 98 | BETSY | 1961 | 9 | 6 | 12 | 120 | 945 | 30.9 | 56.1 |
| 99 | GLADYS | 1975 | 10 | 3 | 18 | 120 | 939 | 37.8 | 67 |
| 100 | ELLA | 1978 | 9 | 4 | 12 | 120 | 956 | 40 | 63 |
| 101 | FELIX | 1995 | 8 | 13 | 18 | 120 | 929 | 24.3 | 61 |
| 102 | LUIS | 1995 | 9 | 5 | 6 | 120 | 939 | 17.3 | 61 |
| 103 | HORTENSE | 1996 | 9 | 13 | 0 | 120 | 935 | 25.9 | 71.5 |
| 104 | CINDY | 1999 | 8 | 29 | 0 | 120 | 944 | 31.5 | 58.4 |
| 105 | ISAAC | 2000 | 9 | 29 | 18 | 120 | 943 | 26.6 | 54.2 |
| 106 | KEITH | 2000 | 10 | 1 | 6 | 120 | 941 | 17.9 | 87.2 |
| 107 | MICHELLE | 2001 | 11 | 5 | 18 | 120 | 949 | 21.5 | 81.8 |
| 108 | FRANCES | 2004 | 9 | 1 | 7 | 120 | 935 | 2.1 | 68.1 |

**Table 5.7  Top 100 Atlantic Hurricanes by Minimum Pressure**

| | Storm | Year | Month | Day | Hour | Wind at Min Press | Min Press mb | Lat at Min ° N | Lon at Min ° W |
|---|---|---|---|---|---|---|---|---|---|
| 1 | WILMA | 2005 | 10 | 19 | 12 | 160 | 882 | 17.3 | 82.8 |
| 2 | GILBERT | 1988 | 9 | 14 | 0 | 160 | 888 | 19.7 | 83.8 |
| 3 | UNNAMED | 1935 | 9 | 3 | 0 | 140 | 892 | 24.5 | 80.1 |
| 4 | RITA | 2005 | 9 | 22 | 3 | 155 | 895 | 24.7 | 87.3 |
| 5 | ALLEN | 1980 | 8 | 8 | 18 | 165 | 899 | 21.8 | 86.4 |
| 6 | KATRINA | 2005 | 8 | 28 | 18 | 150 | 902 | 26.3 | 88.6 |
| 7 | DEAN | 2007 | 8 | 21 | 8 | 150 | 905 | 18.7 | 87.7 |
| 8 | CAMILLE | 1969 | 8 | 18 | 0 | 165 | 909 | 29.4 | 89.1 |
| 9 | MITCH | 1998 | 10 | 27 | 0 | 155 | 910 | 17.2 | 83.8 |
| 10 | IVAN | 2004 | 9 | 12 | 0 | 145 | 910 | 18.2 | 79.6 |
| 11 | ISABEL | 2003 | 9 | 11 | 18 | 145 | 915 | 21.5 | 54.8 |
| 12 | HUGO | 1989 | 9 | 16 | 18 | 140 | 918 | 14.6 | 54.6 |
| 13 | OPAL | 1995 | 10 | 4 | 12 | 130 | 919 | 27.3 | 88.5 |
| 14 | HATTIE | 1961 | 10 | 31 | 0 | 140 | 920 | 17.9 | 86.1 |
| 15 | GLORIA | 1985 | 9 | 25 | 6 | 125 | 920 | 25.1 | 70.9 |
| 16 | FLOYD | 1999 | 9 | 13 | 12 | 135 | 921 | 23.9 | 71.4 |
| 17 | ANDREW | 1992 | 8 | 24 | 18 | 150 | 922 | 25.4 | 75.8 |
| 18 | UNNAMED | 1853 | 9 | 3 | 12 | 130 | 924 | 19.7 | 56.2 |
| 19 | ANITA | 1977 | 9 | 2 | 6 | 150 | 926 | 24.2 | 97.1 |
| 20 | DAVID | 1979 | 9 | 1 | 18 | 150 | 926 | 17.9 | 69.7 |
| 21 | UNNAMED | 1880 | 10 | 2 | 12 | 120 | 928 | 28.3 | 63.7 |
| 22 | CAROL | 1953 | 9 | 4 | 18 | 130 | 929 | 19.9 | 60 |
| 23 | INEZ | 1966 | 9 | 29 | 0 | 130 | 929 | 17.1 | 68.5 |
| 24 | FELIX | 1995 | 8 | 13 | 18 | 120 | 929 | 24.3 | 61 |
| 25 | EMILY | 2005 | 7 | 17 | 0 | 140 | 929 | 17.1 | 79.5 |
| 26 | FELIX | 2007 | 9 | 3 | 7 | 150 | 929 | 14 | 75.3 |
| 27 | UNNAMED | 1880 | 8 | 13 | 0 | 130 | 931 | 25.7 | 96.9 |
| 28 | UNNAMED | 1928 | 9 | 14 | 18 | 140 | 931 | 17.9 | 65.8 |
| 29 | BEULAH | 1967 | 9 | 20 | 6 | 140 | 931 | 25.1 | 96.8 |
| 30 | JOAN | 1988 | 10 | 22 | 6 | 125 | 932 | 11.9 | 83.2 |
| 31 | UNNAMED | 1930 | 9 | 4 | 18 | 130 | 933 | 18.4 | 70 |
| 32 | GERT | 1999 | 9 | 16 | 6 | 130 | 933 | 18 | 51.7 |
| 33 | LENNY | 1999 | 11 | 18 | 18 | 135 | 933 | 17.4 | 64.8 |
| 34 | UNNAMED | 1856 | 8 | 11 | 18 | 130 | 934 | 29.2 | 91.1 |
| 35 | FOX | 1952 | 10 | 25 | 18 | 130 | 934 | 21.8 | 81 |
| 36 | UNNAMED | 1915 | 9 | 29 | 12 | 115 | 935 | 27.8 | 89.7 |
| 37 | HORTENSE | 1996 | 9 | 13 | 0 | 120 | 935 | 25.9 | 71.5 |
| 38 | ISIDORE | 2002 | 9 | 23 | 18 | 110 | 935 | 21.6 | 88.9 |
| 39 | FRANCES | 2004 | 9 | 1 | 7 | 120 | 935 | 2.1 | 68.1 |
| 40 | IKE | 2008 | 9 | 4 | 6 | 125 | 935 | 22.4 | 55 |
| 41 | UNNAMED | 1929 | 9 | 27 | 0 | 120 | 936 | 24.4 | 78.2 |
| 42 | CARLA | 1961 | 9 | 11 | 6 | 150 | 936 | 27.2 | 95.7 |
| 43 | CARMEN | 1974 | 9 | 8 | 0 | 130 | 937 | 28.7 | 90.8 |
| 44 | GEORGES | 1998 | 9 | 20 | 6 | 135 | 937 | 16 | 56.3 |
| 45 | UNNAMED | 1854 | 9 | 7 | 12 | 110 | 938 | 28 | 78.6 |
| 46 | UNNAMED | 1859 | 10 | 6 | 12 | 110 | 938 | 42.5 | 64 |
| 47 | UNNAMED | 1866 | 10 | 2 | 0 | 120 | 938 | 24.5 | 77.1 |
| 48 | HELENE | 1988 | 9 | 24 | 18 | 125 | 938 | 15.3 | 46.1 |
| 49 | KARL | 2004 | 9 | 21 | 6 | 125 | 938 | 19.6 | 47.3 |
| 50 | DENNIS | 2005 | 7 | 8 | 12 | 130 | 938 | 20.9 | 79.5 |
| 51 | GLADYS | 1975 | 10 | 3 | 18 | 120 | 939 | 37.8 | 67 |
| 52 | LUIS | 1995 | 9 | 5 | 6 | 120 | 939 | 17.3 | 61 |
| 53 | FABIAN | 2003 | 9 | 4 | 0 | 115 | 939 | 23.2 | 65.1 |

**Table 5.7   Top 100 Atlantic Hurricanes by Minimum Pressure [CONTINUED]**

| | Storm | Year | Month | Day | Hour | Wind at Min Press | Min Press mb | Lat at Min ° N | Lon at Min ° W |
|---|---|---|---|---|---|---|---|---|---|
| 54 | LILI | 2002 | 10 | 3 | 0 | 125 | 940 | 26.7 | 90.3 |
| 55 | BETSY | 1965 | 9 | 10 | 0 | 135 | 941 | 28.3 | 89.2 |
| 56 | KEITH | 2000 | 10 | 1 | 6 | 120 | 941 | 17.9 | 87.2 |
| 57 | CHARLEY | 2004 | 9 | 13 | 19 | 130 | 941 | 26.6 | 82.2 |
| 58 | GUSTAV | 2008 | 8 | 30 | 22 | 135 | 941 | 22.4 | 83.1 |
| 59 | UNNAMED | 1932 | 8 | 14 | 0 | 125 | 942 | 28.9 | 94.7 |
| 60 | DORA | 1964 | 9 | 6 | 6 | 115 | 942 | 26.1 | 64.4 |
| 61 | HILDA | 1964 | 10 | 2 | 0 | 130 | 942 | 25.2 | 91.4 |
| 62 | HELENE | 1958 | 9 | 28 | 18 | 115 | 943 | 33.9 | 77.5 |
| 63 | EDITH | 1971 | 9 | 10 | 18 | 140 | 943 | 14.8 | 83.2 |
| 64 | ISAAC | 2000 | 9 | 29 | 18 | 120 | 943 | 26.6 | 54.2 |
| 65 | BILL | 2009 | 8 | 21 | 0 | 110 | 943 | 24.1 | 63.7 |
| 66 | FLORA | 1963 | 10 | 4 | 0 | 125 | 944 | 18 | 73.1 |
| 67 | CHARLIE | 1972 | 9 | 22 | 18 | 60 | 944 | 52 | 36 |
| 68 | GABRIELLE | 1989 | 9 | 6 | 0 | 125 | 944 | 22.2 | 58.3 |
| 69 | EDOUARD | 1996 | 8 | 28 | 0 | 125 | 944 | 20.9 | 60.4 |
| 70 | BRET | 1999 | 8 | 22 | 12 | 125 | 944 | 26.2 | 96.1 |
| 71 | CINDY | 1999 | 8 | 29 | 0 | 120 | 944 | 31.5 | 58.4 |
| 72 | PALOMA | 2008 | 11 | 8 | 12 | 125 | 944 | 19.8 | 79.6 |
| 73 | BETSY | 1961 | 9 | 6 | 12 | 120 | 945 | 30.9 | 56.1 |
| 74 | CELIA | 1970 | 8 | 4 | 18 | 110 | 945 | 27.5 | 96.3 |
| 75 | AUDREY | 1957 | 6 | 27 | 12 | 125 | 946 | 29.3 | 93.8 |
| 76 | FREDERIC | 1979 | 9 | 13 | 0 | 115 | 946 | 29.7 | 88 |
| 77 | HARVEY | 1981 | 9 | 15 | 0 | 115 | 946 | 28.4 | 62.6 |
| 78 | CLAUDETTE | 1991 | 9 | 7 | 12 | 115 | 946 | 27.2 | 61.7 |
| 79 | GRETA | 1978 | 9 | 18 | 6 | 115 | 947 | 15.8 | 84.3 |
| 80 | UNNAMED | 1916 | 8 | 18 | 12 | 110 | 948 | 25.3 | 94.7 |
| 81 | CLEO | 1958 | 8 | 16 | 0 | 140 | 948 | 19.6 | 49.8 |
| 82 | FRANCES | 1961 | 10 | 7 | 0 | 110 | 948 | 32.9 | 66.3 |
| 83 | IRIS | 2001 | 10 | 9 | 0 | 125 | 948 | 16.5 | 88 |
| 84 | UNNAMED | 1882 | 9 | 10 | 0 | 100 | 949 | 30 | 87.1 |
| 85 | DIANA | 1984 | 9 | 12 | 0 | 115 | 949 | 33.9 | 77.7 |
| 86 | MICHELLE | 2001 | 11 | 5 | 18 | 120 | 949 | 21.5 | 81.8 |
| 87 | UNNAMED | 1869 | 9 | 9 | 18 | 100 | 950 | 38.8 | 72.6 |
| 88 | GRACIE | 1959 | 9 | 29 | 12 | 120 | 950 | 31.3 | 79.6 |
| 89 | ESTHER | 1961 | 9 | 20 | 18 | 125 | 950 | 32 | 72.6 |
| 90 | CLEO | 1964 | 8 | 24 | 0 | 135 | 950 | 16.7 | 69.5 |
| 91 | DEBBY | 1982 | 9 | 18 | 0 | 115 | 950 | 38.8 | 62.3 |
| 92 | KEITH | 1988 | 10 | 20 | 18 | 65 | 950 | 52 | 46 |
| 93 | BOB | 1991 | 8 | 19 | 6 | 100 | 950 | 36.5 | 74.5 |
| 94 | MARILYN | 1995 | 9 | 17 | 0 | 100 | 950 | 20.4 | 67 |
| 95 | JEANNE | 2004 | 9 | 26 | 4 | 105 | 950 | 27.2 | 80.3 |
| 96 | UNNAMED | 1878 | 10 | 19 | 18 | 100 | 951 | 31 | 56.8 |
| 97 | GLADYS | 1964 | 9 | 18 | 6 | 125 | 951 | 25.8 | 66.5 |
| 98 | ERIKA | 1997 | 9 | 10 | 18 | 110 | 951 | 27.9 | 60.2 |
| 99 | UNNAMED | 1867 | 10 | 30 | 18 | 120 | 952 | 18.2 | 64.8 |
| 100 | DONNA | 1960 | 9 | 4 | 12 | 140 | 952 | 16.8 | 59.5 |
| 101 | FRAN | 1996 | 9 | 5 | 6 | 105 | 952 | 29.8 | 76.7 |
| 102 | KATE | 2003 | 10 | 4 | 18 | 110 | 952 | 30.2 | 54 |
| 103 | BERTHA | 2008 | 7 | 7 | 21 | 110 | 952 | 20.3 | 51.9 |

### Table 5.8   Top 100 East Pacific Hurricanes by Maximum Wind Speed

| | Storm | Year | Month | Day | Hour | Max Wind knots | Press (mb) at Mx Wind | Lat °N | Lon °W |
|---|---|---|---|---|---|---|---|---|---|
| 1 | LINDA | 1997 | 9 | 12 | 6 | 160 | 902 | 17.1 | 109.6 |
| 2 | RICK | 2009 | 10 | 18 | 6-12 | 155 | 906 | 15.2 | 106.6 |
| 3 | KENNA | 2002 | 10 | 25 | 6 | 145 | 915 | 19.3 | 107.5 |
| 4 | GILMA | 1994 | 7 | 25 | 18 | 140 | 920 | 12.1 | 143.3 |
| 5 | GUILLERMO | 1997 | 8 | 5 | 12 | 140 | 921 | 14.4 | 120.9 |
| 6 | ELIDA | 2002 | 7 | 25 | 6 | 140 | 921 | 14.3 | 109.5 |
| 7 | HERNAN | 2002 | 9 | 2 | 18 | 140 | 921 | 17.7 | 112.2 |
| 8 | AVA | 1973 | 6 | 8 | 18 | 140 | 928 | 12.6 | 113.3 |
| 9 | NOT NAMED | 1959 | 10 | 27 | 12 | 140 | 958 | 19.7 | 104.4 |
| 10 | TRUDY | 1990 | 10 | 20 | 12 | 135 | 924 | 15.5 | 111.0 |
| 11 | HERNAN | 1990 | 7 | 24 | 18 | 135 | 928 | 17.5 | 117.9 |
| 12 | JIMENA | 2009 | 8 | 31 | 19 | 135 | 931 | 18.3 | 109.0 |
| 13 | CARLOTTA | 2000 | 6 | 21 | 12 | 135 | 934 | 15.3 | 104.5 |
| 14 | EMILIA | 1994 | 7 | 21 | 18 | 135 | 935 | 13.0 | 153.4 |
| 15 | OLIVIA | 1994 | 9 | 25 | 12 | 130 | 923 | 17.8 | 119.8 |
| 16 | LIDIA | 1993 | 9 | 11 | 6 | 130 | 930 | 16.7 | 105.4 |
| 17 | JAVIER | 2004 | 9 | 14 | 0 | 130 | 930 | 15.9 | 106.8 |
| 18 | KENNETH | 1993 | 9 | 11 | 6 | 130 | 932 | 17.2 | 119.6 |
| 19 | HOWARD | 1998 | 8 | 23 | 0 | 130 | 932 | 14.4 | 108.0 |
| 20 | SARAH | 1967 | 9 | 16 | 12 | 130 | 933 | 19.3 | 193.9 |
| 21 | JULIETTE | 1995 | 9 | 21 | 18 | 130 | 933 | 17.8 | 114.3 |
| 22 | DANIEL | 2006 | 7 | 22 | 0 | 130 | 933 | 13.5 | 128.0 |
| 23 | TINA | 1992 | 10 | 1 | 6 | 130 | 934 | 17.5 | 121.1 |
| 24 | DOT | 1959 | 8 | 3 | 0 | 130 | 952 | 15.3 | 145.7 |
| 25 | JULIETTE | 2001 | 9 | 26 | 0 | 125 | 925 | 16.8 | 107.9 |
| 26 | HECTOR | 1988 | 8 | 3 | 12 | 125 | 935 | 16.7 | 116.2 |
| 27 | RAYMOND | 1989 | 10 | 1 | 6 | 125 | 935 | 17.4 | 114.3 |
| 28 | ODILE | 1990 | 9 | 27 | 12 | 125 | 935 | 17.7 | 126.6 |
| 29 | KEVIN | 1991 | 10 | 1 | 0 | 125 | 935 | 17.6 | 112.0 |
| 30 | FRANK | 1992 | 7 | 18 | 12 | 125 | 935 | 16.1 | 121.8 |
| 31 | FELICIA | 2009 | 8 | 6 | 0 | 125 | 935 | 14.7 | 130.1 |
| 32 | CELIA | 1992 | 6 | 28 | 18 | 125 | 936 | 15.2 | 112.7 |
| 33 | FERNANDA | 1993 | 8 | 12 | 12 | 125 | 936 | 15.5 | 127.3 |
| 34 | FAUSTO | 2002 | 8 | 25 | 18 | 125 | 936 | 16.1 | 122.9 |
| 35 | ORLENE | 1992 | 9 | 7 | 0 | 125 | 937 | 15.9 | 126.9 |
| 36 | INIKI | 1992 | 9 | 12 | 18 | 125 | 938 | 19.5 | 160.0 |
| 37 | ADOLPH | 2001 | 5 | 29 | 12 | 125 | 940 | 15.5 | 104.2 |
| 38 | BUD | 2006 | 7 | 13/17 | 6/18 | 125 | 953 | 17.2 | 119.8 |
| 39 | BARBARA | 1995 | 7 | 14 | 6 | 120 | 941 | 14.4 | 127.1 |
| 40 | FABIO | 1988 | 8 | 3 | 12 | 120 | 943 | 14.7 | 143.8 |
| 41 | ESTELLE | 1992 | 7 | 14 | 18 | 120 | 943 | 16.8 | 122.3 |
| 42 | BLAS | 1998 | 6 | 25 | 6 | 120 | 943 | 15.3 | 106.8 |
| 43 | DORA | 1999 | 8 | 13 | 6 | 120 | 943 | 15.5 | 134.0 |
| 44 | HOWARD | 2004 | 9 | 2 | 12 | 120 | 943 | 17.8 | 113.3 |
| 45 | MARIE | 1990 | 9 | 11 | 12 | 120 | 946 | 14.8 | 133.5 |
| 46 | FLOSSIE | 2007 | 8 | 12 | 0 | 120 | 949 | 13.2 | 141.7 |
| 47 | LIZA | 1976 | 10 | 1 | 18 | 120 | 971 | 22.6 | 108.9 |
| 48 | DOREEN | 1973 | 7 | 21 | 18 | 120 | 972 | 14.0 | 121.2 |
| 49 | CELESTE | 1972 | 8 | 14 | 12 | 115 | 944 | 12.8 | 150.0 |
| 50 | NORBERT | 2008 | 10 | 8 | 18 | 115 | 945 | 16.4 | 110.9 |
| 51 | KENNETH | 2005 | 9 | 18 | 12 | 115 | 947 | 14.2 | 129.2 |
| 52 | OCTAVE | 1989 | 9 | 13 | 0 | 115 | 948 | 20.9 | 113.8 |
| 53 | VIRGIL | 1992 | 10 | 3 | 0 | 115 | 948 | 15.9 | 102.6 |

Table 5.8    Top 100 East Pacific Hurricanes by Maximum Wind Speed [CONTINUED]

| | Storm | Year | Month | Day | Hour | Max Wind knots | Press (mb) at Mx Wind | Lat °N | Lon °W |
|---|---|---|---|---|---|---|---|---|---|
| 54 | JOVA | 1993 | 9 | 1 | 12 | 115 | 948 | 17.8 | 106.6 |
| 55 | ADOLPH | 1995 | 6 | 18 | 6 | 115 | 948 | 17.1 | 107.8 |
| 56 | JIMENA | 1997 | 8 | 29 | 18 | 115 | 948 | 18.6 | 137.7 |
| 57 | PAULINE | 1997 | 10 | 9 | 18 | 115 | 948 | 15.3 | 96.3 |
| 58 | ESTELLE | 1998 | 8 | 2 | 6 | 115 | 948 | 16.8 | 117.5 |
| 59 | JOHN | 2006 | 8 | 30 | 18 | 115 | 948 | 16.9 | 102.7 |
| 60 | GREG | 1993 | 8 | 20 | 12 | 115 | 949 | 18.4 | 119.8 |
| 61 | LANE | 1994 | 9 | 7 | 6 | 115 | 949 | 19.0 | 126.5 |
| 62 | JIMENA | 1991 | 9 | 26 | 0 | 115 | 950 | 13.7 | 117.7 |
| 63 | DORA | 1993 | 7 | 17 | 12 | 115 | 950 | 15.4 | 130.2 |
| 64 | DOUGLAS | 1996 | 8 | 2 | 6 | 115 | 950 | 20.4 | 113.9 |
| 65 | NORA | 1997 | 9 | 21 | 12 | 115 | 950 | 16.6 | 108.0 |
| 66 | FELICIA | 1997 | 7 | 20 | 18 | 115 | 951 | 15.8 | 129.5 |
| 67 | RUBY | 1972 | 11 | 16 | 12 | 110 | 944 | 15.2 | 189.3 |
| 68 | EUGENE | 1993 | 7 | 20 | 18 | 110 | 948 | 15.4 | 123.6 |
| 69 | PRISCILLA | 1971 | 10 | 11 | 18 | 110 | 951 | 16.7 | 104.6 |
| 70 | JOVA | 2005 | 9 | 20 | 0 | 110 | 951 | 16.0 | 143.3 |
| 71 | DANIEL | 2000 | 7 | 26 | 0 | 110 | 954 | 14.5 | 122.5 |
| 72 | LANE | 2006 | 9 | 16 | 19 | 110 | 954 | 24.1 | 107.2 |
| 73 | GUILLERMO | 2009 | 8 | 15 | 12 | 110 | 954 | 18.7 | 132.6 |
| 74 | HYACINTH | 1972 | 9 | 1 | 18 | 110 | 972 | 15.5 | 110.8 |
| 75 | ISMAEL | 1989 | 8 | 20 | 18 | 105 | 955 | 18.7 | 116.8 |
| 76 | KIKO | 1989 | 8 | 27 | 0 | 105 | 955 | 23.5 | 109.3 |
| 77 | CARLOS | 1991 | 6 | 24 | 6 | 105 | 955 | 15.3 | 127.0 |
| 78 | FAUSTO | 1996 | 9 | 12 | 12 | 105 | 955 | 20.5 | 110.0 |
| 79 | ILEANA | 2006 | 8 | 23 | 12 | 105 | 955 | 17.8 | 111.0 |
| 80 | BEATRIZ | 1999 | 7 | 13 | 12 | 105 | 956 | 15.1 | 122.2 |
| 81 | HERNAN | 2008 | 8 | 9 | 12 | 105 | 956 | 15.5 | 122.8 |
| 82 | LINDA | 1991 | 10 | 6 | 18 | 105 | 957 | 16.7 | 108.6 |
| 83 | HILARY | 1993 | 8 | 22 | 18 | 105 | 957 | 18.9 | 108.6 |
| 84 | DARBY | 2004 | 7 | 29 | 6 | 105 | 957 | 16.8 | 126.2 |
| 85 | ISELLE | 1990 | 7 | 26 | 0 | 105 | 958 | 20.2 | 113.6 |
| 86 | FEFA | 1991 | 8 | 2 | 6 | 105 | 960 | 16.9 | 123.6 |
| 87 | DARBY | 1992 | 7 | 6 | 6 | 105 | 968 | 19.9 | 113.0 |
| 88 | OLIVIA | 1971 | 9 | 27 | 18 | 100 | 948 | 19.6 | 109.9 |
| 89 | WINIFRED | 1992 | 10 | 9 | 12 | 100 | 960 | 17.5 | 104.3 |
| 90 | ENRIQUE | 1997 | 7 | 15 | 18 | 100 | 960 | 15.5 | 124.5 |
| 91 | DARBY | 1998 | 7 | 28 | 0 | 100 | 960 | 17.1 | 133.1 |
| 92 | JULIO | 1990 | 8 | 22 | 0 | 100 | 961 | 16.9 | 123.0 |
| 93 | GEORGETTE | 1998 | 8 | 15 | 18 | 100 | 961 | 20.9 | 121.1 |
| 94 | ALMA | 2002 | 5 | 30 | 12 | 100 | 962 | 16.1 | 115.3 |
| 95 | LESTER | 1998 | 10 | 23 | 18 | 100 | 966 | 16.7 | 108.9 |
| 96 | EUGENE | 1999 | 8 | 10 | 0 | 95 | 965 | 14.0 | 133.5 |
| 97 | SERGIO | 2006 | 11 | 15 | 18 | 95 | 965 | 12.0 | 103.7 |
| 98 | CALVIN | 1993 | 7 | 7 | 12 | 95 | 966 | 18.2 | 104.1 |
| 99 | HECTOR | 2006 | 8 | 18 | 6 | 95 | 966 | 15.6 | 124.0 |
| 100 | GEORGETTE | 1992 | 7 | 21 | 12 | 95 | 968 | 17.5 | 120.5 |
| 101 | PATRICIA | 1970 | 10 | 10 | 18 | 95 | 980 | 18.0 | 124.8 |

**Table 5.9   Top 100 East Pacific Hurricanes by Minimum Pressure**

| | Storm | Year | Month | Day | Hour | Wind (kts) at Mn Press | Min Press mb | Lat ° N | Lon ° W |
|---|---|---|---|---|---|---|---|---|---|
| 1 | LINDA | 1997 | 9 | 12 | 6 | 160 | 902 | 17.1 | 109.6 |
| 2 | RICK | 2009 | 10 | 18 | 6-12 | 155 | 906 | 15.2 | 106.6 |
| 3 | KENNA | 2002 | 10 | 25 | 6 | 145 | 915 | 19.3 | 107.5 |
| 4 | GILMA | 1994 | 7 | 25 | 18 | 140 | 920 | 12.1 | 143.3 |
| 5 | GUILLERMO | 1997 | 8 | 5 | 12 | 140 | 921 | 14.4 | 120.9 |
| 6 | ELIDA | 2002 | 7 | 25 | 6 | 140 | 921 | 14.3 | 109.5 |
| 7 | HERNAN | 2002 | 9 | 2 | 18 | 140 | 921 | 17.7 | 112.2 |
| 8 | OLIVIA | 1994 | 9 | 25 | 12 | 130 | 923 | 17.8 | 119.8 |
| 9 | TRUDY | 1990 | 10 | 20 | 12 | 135 | 924 | 15.5 | 111.0 |
| 10 | JULIETTE | 2001 | 9 | 26 | 0 | 125 | 925 | 16.8 | 107.9 |
| 11 | AVA | 1973 | 6 | 8 | 18 | 140 | 928 | 12.6 | 113.3 |
| 12 | HERNAN | 1990 | 7 | 24 | 18 | 135 | 928 | 17.5 | 117.9 |
| 13 | LIDIA | 1993 | 9 | 11 | 6 | 130 | 930 | 16.7 | 105.4 |
| 14 | JAVIER | 2004 | 9 | 14 | 0 | 130 | 930 | 15.9 | 106.8 |
| 15 | JIMENA | 2009 | 8 | 31 | 19 | 135 | 931 | 18.3 | 109.0 |
| 16 | KENNETH | 1993 | 9 | 11 | 6 | 130 | 932 | 17.2 | 119.6 |
| 17 | HOWARD | 1998 | 8 | 23 | 0 | 130 | 932 | 14.4 | 108.0 |
| 18 | SARAH | 1967 | 9 | 16 | 12 | 130 | 933 | 19.3 | 193.9 |
| 19 | JULIETTE | 1995 | 9 | 21 | 18 | 130 | 933 | 17.8 | 114.3 |
| 20 | DANIEL | 2006 | 7 | 22 | 0 | 130 | 933 | 13.5 | 128.0 |
| 21 | CARLOTTA | 2000 | 6 | 21 | 12 | 135 | 934 | 15.3 | 104.5 |
| 22 | TINA | 1992 | 10 | 1 | 6 | 130 | 934 | 17.5 | 121.1 |
| 23 | EMILIA | 1994 | 7 | 21 | 18 | 135 | 935 | 13.0 | 153.4 |
| 24 | HECTOR | 1988 | 8 | 3 | 12 | 125 | 935 | 16.7 | 116.2 |
| 25 | RAYMOND | 1989 | 10 | 1 | 6 | 125 | 935 | 17.4 | 114.3 |
| 26 | ODILE | 1990 | 9 | 27 | 12 | 125 | 935 | 17.7 | 126.6 |
| 27 | KEVIN | 1991 | 10 | 1 | 0 | 125 | 935 | 17.6 | 112.0 |
| 28 | FRANK | 1992 | 7 | 18 | 12 | 125 | 935 | 16.1 | 121.8 |
| 29 | FELICIA | 2009 | 8 | 6 | 0 | 125 | 935 | 14.7 | 130.1 |
| 30 | CELIA | 1992 | 6 | 28 | 18 | 125 | 936 | 15.2 | 112.7 |
| 31 | FERNANDA | 1993 | 8 | 12 | 12 | 125 | 936 | 15.5 | 127.3 |
| 32 | FAUSTO | 2002 | 8 | 25 | 18 | 125 | 936 | 16.1 | 122.9 |
| 33 | ORLENE | 1992 | 9 | 7 | 0 | 125 | 937 | 15.9 | 126.9 |
| 34 | IGNACIO | 1979 | 10 | 28 | 18 | 125 | 938 | 17.0 | 107.3 |
| 35 | INIKI | 1992 | 9 | 12 | 18 | 125 | 938 | 19.5 | 160.0 |
| 36 | ADOLPH | 2001 | 5 | 29 | 12 | 125 | 940 | 15.5 | 104.2 |
| 37 | BARBARA | 1995 | 7 | 14 | 6 | 120 | 941 | 14.4 | 127.1 |
| 38 | FABIO | 1988 | 8 | 3 | 12 | 120 | 943 | 14.7 | 143.8 |
| 39 | ESTELLE | 1992 | 7 | 14 | 18 | 120 | 943 | 16.8 | 122.3 |
| 40 | BLAS | 1998 | 6 | 25 | 6 | 120 | 943 | 15.3 | 106.8 |
| 41 | DORA | 1999 | 8 | 13 | 6 | 120 | 943 | 15.5 | 134.0 |
| 42 | HOWARD | 2004 | 9 | 2 | 12 | 120 | 943 | 17.8 | 113.3 |
| 43 | CELESTE | 1972 | 8 | 14 | 12 | 115 | 944 | 12.8 | 150.0 |
| 44 | RUBY | 1972 | 11 | 16 | 12 | 110 | 944 | 15.2 | 189.3 |
| 45 | NORBERT | 2008 | 10 | 8 | 18 | 115 | 945 | 16.4 | 110.9 |
| 46 | MARIE | 1990 | 9 | 11 | 12 | 120 | 946 | 14.8 | 133.5 |
| 47 | KENNETH | 2005 | 9 | 18 | 12 | 115 | 947 | 14.2 | 129.2 |
| 48 | OCTAVE | 1989 | 9 | 13 | 0 | 115 | 948 | 20.9 | 113.8 |
| 49 | VIRGIL | 1992 | 10 | 3 | 0 | 115 | 948 | 15.9 | 102.6 |
| 50 | JOVA | 1993 | 9 | 1 | 12 | 115 | 948 | 17.8 | 106.6 |
| 51 | ADOLPH | 1995 | 6 | 18 | 6 | 115 | 948 | 17.1 | 107.8 |
| 52 | JIMENA | 1997 | 8 | 29 | 18 | 115 | 948 | 18.6 | 137.7 |
| 53 | PAULINE | 1997 | 10 | 9 | 18 | 115 | 948 | 15.3 | 96.3 |
| 54 | ESTELLE | 1998 | 8 | 2 | 6 | 115 | 948 | 16.8 | 117.5 |
| 55 | JOHN | 2006 | 8 | 30 | 18 | 115 | 948 | 16.9 | 102.7 |

**Table 5.9   Top 100 East Pacific Hurricanes by Minimum Pressure [CONTINUED]**

| | Storm | Year | Month | Day | Hour | Wind (kts) at Mn Press | Min Press mb | Lat °N | Lon °W |
|---|---|---|---|---|---|---|---|---|---|
| 56 | EUGENE | 1993 | 7 | 20 | 18 | 110 | 948 | 15.4 | 123.6 |
| 57 | OLIVIA | 1971 | 9 | 27 | 18 | 100 | 948 | 19.6 | 109.9 |
| 58 | FLOSSIE | 2007 | 8 | 12 | 0 | 120 | 949 | 13.2 | 141.7 |
| 59 | GREG | 1993 | 8 | 20 | 12 | 115 | 949 | 18.4 | 119.8 |
| 60 | LANE | 1994 | 9 | 7 | 6 | 115 | 949 | 19.0 | 126.5 |
| 61 | JIMENA | 1991 | 9 | 26 | 0 | 115 | 950 | 13.7 | 117.7 |
| 62 | DORA | 1993 | 7 | 17 | 12 | 115 | 950 | 15.4 | 130.2 |
| 63 | DOUGLAS | 1996 | 8 | 2 | 6 | 115 | 950 | 20.4 | 113.9 |
| 64 | NORA | 1997 | 9 | 21 | 12 | 115 | 950 | 16.6 | 108.0 |
| 65 | FELICIA | 1997 | 7 | 20 | 18 | 115 | 951 | 15.8 | 129.5 |
| 66 | PRISCILLA | 1971 | 10 | 11 | 18 | 110 | 951 | 16.7 | 104.6 |
| 67 | JOVA | 2005 | 9 | 20 | 0 | 110 | 951 | 16.0 | 143.3 |
| 68 | DOT | 1959 | 8 | 3 | 0 | 130 | 952 | 15.3 | 145.7 |
| 69 | BUD | 2006 | 7 | 13/17 | 6/18 | 125 | 953 | 17.2 | 119.8 |
| 70 | DANIEL | 2000 | 7 | 26 | 0 | 110 | 954 | 14.5 | 122.5 |
| 71 | LANE | 2006 | 9 | 16 | 19 | 110 | 954 | 24.1 | 107.2 |
| 72 | GUILLERMO | 2009 | 8 | 15 | 12 | 110 | 954 | 18.7 | 132.6 |
| 73 | ISMAEL | 1989 | 8 | 20 | 18 | 105 | 955 | 18.7 | 116.8 |
| 74 | KIKO | 1989 | 8 | 27 | 0 | 105 | 955 | 23.5 | 109.3 |
| 75 | CARLOS | 1991 | 6 | 24 | 6 | 105 | 955 | 15.3 | 127.0 |
| 76 | FAUSTO | 1996 | 9 | 12 | 12 | 105 | 955 | 20.5 | 110.0 |
| 77 | ILEANA | 2006 | 8 | 23 | 12 | 105 | 955 | 17.8 | 111.0 |
| 78 | BEATRIZ | 1999 | 7 | 13 | 12 | 105 | 956 | 15.1 | 122.2 |
| 79 | HERNAN | 2008 | 8 | 9 | 12 | 105 | 956 | 15.5 | 122.8 |
| 80 | LINDA | 1991 | 10 | 6 | 18 | 105 | 957 | 16.7 | 108.6 |
| 81 | HILARY | 1993 | 8 | 22 | 18 | 105 | 957 | 18.9 | 108.6 |
| 82 | DARBY | 2004 | 7 | 29 | 6 | 105 | 957 | 16.8 | 126.2 |
| 83 | NOT NAMED | 1959 | 10 | 27 | 12 | 140 | 958 | 19.7 | 104.4 |
| 84 | ISELLE | 1990 | 7 | 26 | 0 | 105 | 958 | 20.2 | 113.6 |
| 85 | FEFA | 1991 | 8 | 2 | 6 | 105 | 960 | 16.9 | 123.6 |
| 86 | WINIFRED | 1992 | 10 | 9 | 12 | 100 | 960 | 17.5 | 104.3 |
| 87 | ENRIQUE | 1997 | 7 | 15 | 18 | 100 | 960 | 15.5 | 124.5 |
| 88 | DARBY | 1998 | 7 | 28 | 0 | 100 | 960 | 17.1 | 133.1 |
| 89 | JULIO | 1990 | 8 | 22 | 0 | 100 | 961 | 16.9 | 123.0 |
| 90 | GEORGETTE | 1998 | 8 | 15 | 18 | 100 | 961 | 20.9 | 121.1 |
| 91 | ALMA | 2002 | 5 | 30 | 12 | 100 | 962 | 16.1 | 115.3 |
| 92 | LORRAINE | 1970 | 8 | 23 | 18 | 85 | 963 | 16.9 | 124.3 |
| 93 | PATRICIA | 1974 | 10 | 10 | 0 | 80 | 964 | 18.0 | 119.6 |
| 94 | EUGENE | 1999 | 8 | 10 | 0 | 95 | 965 | 14.0 | 133.5 |
| 95 | SERGIO | 2006 | 11 | 15 | 18 | 95 | 965 | 12.0 | 103.7 |
| 96 | LESTER | 1998 | 10 | 23 | 18 | 100 | 966 | 16.7 | 108.9 |
| 97 | CALVIN | 1993 | 7 | 7 | 12 | 95 | 966 | 18.2 | 104.1 |
| 98 | HECTOR | 2006 | 8 | 18 | 6 | 95 | 966 | 15.6 | 124.0 |
| 99 | DARBY | 1992 | 7 | 6 | 6 | 105 | 968 | 19.9 | 113.0 |
| 100 | GEORGETTE | 1992 | 7 | 21 | 12 | 95 | 968 | 17.5 | 120.5 |

**Table 5.10 Retired Atlantic and Eastern North Pacific Hurricane Names**

**Atlantic**

| | |
|---|---|
| Carol | 1954 |
| Hazel | 1954 |
| Connie | 1955 |
| Diane | 1955 |
| Ione | 1955 |
| Janet | 1955 |
| Audrey | 1957 |
| Donna | 1960 |
| Carla | 1961 |
| Hattie | 1961 |
| Flora | 1963 |
| Cleo | 1964 |
| Dora | 1964 |
| Hilda | 1964 |
| Betsy | 1965 |
| Inez | 1966 |
| Beulah | 1967 |
| Edna | 1968 |
| Camille | 1969 |
| Celia | 1970 |
| Agnes | 1972 |
| Carmen | 1974 |
| Fifi | 1974 |
| Eloise | 1975 |
| Anita | 1977 |
| David | 1979 |
| Frederic | 1979 |
| Allen | 1980 |
| Alicia | 1983 |
| Elena | 1985 |
| Gloria | 1985 |
| Gilbert | 1988 |
| Joan | 1988 |
| Hugo | 1989 |
| Diana | 1990 |
| Klaus | 1990 |
| Bob | 1991 |

| | |
|---|---|
| Andrew | 1992 |
| Luis | 1995 |
| Marilyn | 1995 |
| Opal | 1995 |
| Roxanne | 1995 |
| Cesar | 1996 |
| Fran | 1996 |
| Hortense | 1996 |
| Georges | 1998 |
| Mitch | 1998 |
| Floyd | 1999 |
| Lenny | 1999 |
| Keith | 2000 |
| Allison | 2001 |
| Iris | 2001 |
| Michelle | 2001 |
| Isidore | 2002 |
| Lili | 2002 |
| Fabian | 2003 |
| Isabel | 2003 |
| Juan | 2003 |
| Charley | 2004 |
| Frances | 2004 |
| Ivan | 2004 |
| Jeanne | 2004 |
| Dennis | 2005 |
| Katrina | 2005 |
| Rita | 2005 |
| Stan | 2005 |
| Wilma | 2005 |
| Dean | 2007 |
| Felix | 2007 |
| Noel | 2007 |
| Gustav | 2008 |
| Ike | 2008 |
| Paloma | 2008 |

**Pacific**

| | |
|---|---|
| Iwa | 1982 |
| Iniki | 1992 |
| Ismael | 1995 |
| Pauline | 1997 |
| Paka | 1997 |
| Kenna | 2002 |
| Ioke | 2006 |
| Alma | 2008 |

extremely destructive or deadly. To retire a storm name, a request is made to the World Meteorological Organization (WMO) by one or more of the countries affected.

Some names, like Adolf, have been retired for political reasons and others for reasons unknown now.

Naming conventions for tropical cyclones vary from region to region. The names are agreed upon and submitted by the WMO members and assigned by the agency responsible for cyclone forecasting and tracking in a given region. Lists consist of names commonly found in the various countries and languages of a region.

The name of a tropical cyclone can be changed for the following reasons:

1. A tropical cyclone crosses from the Atlantic basin to the Pacific basin or the reverse across Central America if the low-level circulation dissipates. If the low-level circulation is maintained, it retains the original name. Before 2001 the policy of the National Hurricane Center was than any storm crossing into the other basin was renamed. In 1988 Hurricane Joan became East Pacific Hurricane Miriam. Since 2001 no storm has been renamed using the current policy.
2. A tropical cyclone enters the southwestern Indian Ocean from east of 90°E into the area of responsibility of Météo-France. If it has already been named, the new name will be a hyphenated combination of (original name)-(new name), like very intense tropical cyclone Adeline-Juliet of April 1–12, 2005.
3. When the remnants of a tropical cyclone redevelop, if the continuity of the system is in question, or if interaction with another system brings about the redevelopment. In 2005 the remnants of a Cape Verde TD10 interacted with

another disturbance to become TD12, which developed into Katrina.

The history of naming tropical cyclones goes back well before World War II. Early European settlers named storms for the saint whose feast day it was when the storm struck. In the late 1800s British meteorologist Clement Lindley Wragge used people's names to refer to storms.

To enhance communication, tropical cyclones were named during World War II; using female names only after a 1941 novel *Storm* by George R. Stewart popularized the idea.

From 1950 through 1952, names from the phonetic alphabet of the US Army and Navy were used by the military for North Atlantic storms, and starting in 1953, the US National Hurricane Center started systematically naming storms.

In 1979 the WMO began using masculine names, alternating them with the feminine. In 2002 subtropical storms of sufficient strength were given names from the lists.

Names lists starting with the 2010 seasons can be found in Tables 5.11a–h and the strongest storms by basin in Table 5.12.

## Reference Graphs

Figures 5.41–5.44 are graphs reproduced from originals, with a few modifications by the National Hurricane Center. Figure 5.40 breaks down each Atlantic season from 1951 through 2005 by storm intensity. Figure 5.41 shows the progression of hurricane season based on the number of storms in 100 years. The final two graphs show the progression of the Atlantic and Eastern North Pacific seasons using the cumulative frequency of storm incidence through an average season.

**Table 5.11(a)  Tropical Cyclone Names Around the World (North Atlantic)**

| North Atlantic Tropical Cyclone Names | | | | | |
|---|---|---|---|---|---|
| U.S. National Hurricane Center | | | | | |
| 2010 | 2011 | 2012 | 2013 | 2014 | 2015 |
| Alex | Arlene | Alberto | Andrea | Arthur | Ana |
| Bonnie | Bret | Beryl | Barry | Bertha | Bill |
| Colin | Cindy | Chris | Chantal | Cristobal | Claudette |
| Danielle | Don | Debby | Dorian | Dolly | Danny |
| Earl | Emily | Ernesto | Erin | Edouard | Erika |
| Fiona | Franklin | Florence | Fernand | Fay | Fred |
| Gaston | Gert | Gordon | Gabrielle | Gonzalo | Grace |
| Hermine | Harvey | Helene | Humberto | Hanna | Henri |
| Igor | Irene | Isaac | Ingrid | Isaias | Ida |
| Julia | Jose | Joyce | Jerry | Josephine | Joaquin |
| Karl | Katia | Kirk | Karen | Kyle | Kate |
| Lisa | Lee | Leslie | Lorenzo | Laura | Larry |
| Matthew | Maria | Michael | Melissa | Marco | Mindy |
| Nicole | Nate | Nadine | Nestor | Nana | Nicholas |
| Otto | Ophelia | Oscar | Olga | Omar | Odette |
| Paula | Philippe | Patty | Pablo | Paulette | Peter |
| Richard | Rina | Rafael | Rebekah | Rene | Rose |
| Shary | Sean | Sandy | Sebastien | Sally | Sam |
| Tomas | Tammy | Tony | Tanya | Teddy | Teresa |
| Virginie | Vince | Valerie | Van | Vicky | Victor |
| Walter | Whitney | William | Wendy | Wilfred | Wanda |

If the tropical cyclone name list is exhausted for a given year new storms are named in order using letters of the Greek alphabet.

**Warning Area**
**U.S. NHC**

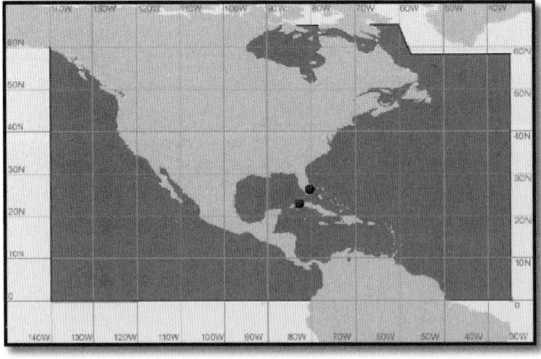

**Table 5.11(b)   Tropical Cyclone Names Around the World (Eastern North Pacific)**

| Eastern North Pacific (East of 140°W) U.S. National Hurricane Center | | | | | |
|---|---|---|---|---|---|
| **2010** | **2011** | **2012** | **2013** | **2014** | **2015** |
| Agatha | Adrian | Aletta | Alvin | Amanda | Andres |
| Blas | Beatriz | Bud | Barbara | Boris | Blanca |
| Celia | Calvin | Carlotta | Cosme | Cristina | Carlos |
| Darby | Dora | Daniel | Dalila | Douglas | Dolores |
| Estelle | Eugene | Emilia | Erick | Elida | Enrique |
| Frank | Fernanda | Fabio | Flossie | Fausto | Felicia |
| Georgette | Greg | Gilma | Gil | Genevieve | Guillermo |
| Howard | Hilary | Hector | Henriette | Hernan | Hilda |
| Isis | Irwin | Ileana | Ivo | Iselle | Ignacio |
| Javier | Jova | John | Juliette | Julio | Jimena |
| Kay | Kenneth | Kristy | Kiko | Karina | Kevin |
| Lester | Lidia | Lane | Lorena | Lowell | Linda |
| Madeline | Max | Miriam | Manuel | Marie | Marty |
| Newton | Norma | Norman | Narda | Norbert | Nora |
| Orlene | Otis | Olivia | Octave | Odile | Olaf |
| Paine | Pilar | Paul | Priscilla | Polo | Patricia |
| Roslyn | Ramon | Rosa | Raymond | Rachel | Rick |
| Seymour | Selma | Sergio | Sonia | Simon | Sandra |
| Tina | Todd | Tara | Tico | Trudy | Terry |
| Virgil | Veronica | Vicente | Velma | Vance | Vivian |
| Winifred | Wiley | Willa | Wallis | Winnie | Waldo |
| Xavier | Xina | Xavier | Xina | Xavier | Xina |
| Yolanda | York | Yolanda | York | Yolanda | York |
| Zeke | Zelda | Zeke | Zelda | Zeke | Zelda |

If the tropical cyclone name list is exhausted for a given year new storms are named in order using leffers of the Greek alphabet.

**Warning Area**
**U.S. NHC**

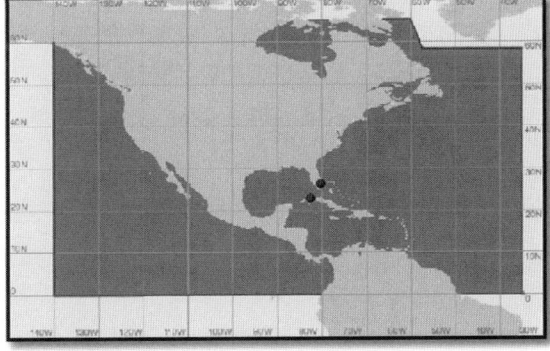

**Table 5.11(c)  Tropical Cyclone Names Around the World (Central North Pacific)**

| Central North Pacific (Dateline to 140°W) Central Pacific Hurricane Center, Honolulu | | | |
|---|---|---|---|
| List 1 | List 2 | List 3 | List 4 |
| Akoni | Aka | Alika | Ana |
| Ema | Ekeka | Ele | Ela |
| Hone | Hene | Huko | Halola |
| Iona | Iolana | Iopa | Iune |
| Keli | Keoni | Kika | Kilo |
| Lala | Lino | Lana | Loke |
| Moke | Mele | Maka | Malia |
| Nolo | Nona | Neki | Niala |
| Olana | Oliwa | Omeka | Oho |
| Pena | Pama | Pewa | Pali |
| Ulana | Upana | Unala | Ulika |
| Wale | Wene | Wali | Walaka |

A season begins where the previous season left off.

**Warning Area
U.S. CPHC Honolulu**

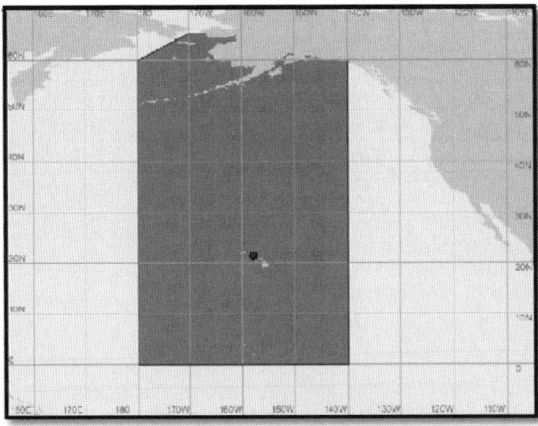

**Table 5.11(d)   Tropical Cyclone Names Around the World (Western North Pacific)**

| Western North Pacific (International Date Line to 100°E) | | | | |
|---|---|---|---|---|
| Regional Specialized Meteorological Center, Tokyo, Japan | | | | |
| **Contributing** | **List I** | **List II** | **List III** | **List IV** | **List V** |
| **Nation** | Damrey | Kong-rey | Nakri | Krovanh | Sarika |
| **Cambodia** | Haikui | Yutu | Fengshen | Dujuan | Haima |
| **China** | Kirogi | Toraji | Kalmaegi | Mujigae | Meari |
| **DPR Korea** | Kai-tak | Man-yi | Fung-wong | Choi-wan | Ma-on |
| **Hong Kong** | Tembin | Usagi | Kammuri | Koppu | Tokage |
| **Japan** | Bolaven | Pabuk | Phanfone | Melor | Nock-ten |
| **Laos** | Sanba | Wutip | Vongfong | Nepartak | Muifa |
| **Macau** | Jelawat | Sepat | Nuri | Lupit | Merbok |
| **Malaysia** | Ewiniar | Fitow | Sinlaku | Mirinae | Nanmadol |
| **Micronesia** | Maliksi | Danas | Hagupit | Nida | Talas |
| **Philippines** | Gaemi | Nari | Jangmi | Omais | Noru |
| **South Korea** | Prapiroon | Wipha | Mekkhala | Conson | Kulap |
| **Thailand** | Maria | Francisco | Higos | Chanthu | Roke |
| **USA** | Son-Tinh | Lekima | Bavi | Dianmu | Sonca |
| **Vietnam** | Bopha | Krosa | Maysak | Mindulle | Nesat |
| **Cambodia** | Wukong | Haiyan | Haishen | Lionrock | Haitang |
| **China** | Sonamu | Podul | Noul | Kompasu | Nalgae |
| **DPR Korea** | Shanshan | Lingling | Dolphin | Namtheun | Banyan |
| **Hong Kong** | Yagi | Kajiki | Kujira | Malou | Washi |
| **Japan** | Leepi | Faxai | Chan-hom | Meranti | Pakhar |
| **Laos** | Bebinca | Peipah | Linfa | Fanapi | Sanvu |
| **Macau** | Rumbia | Tapah | Nangka | Malakas | Mawar |
| **Malaysia** | Soulik | Mitag | Soudelor | Megi | Guchol |
| **Micronesia** | Cimaron | Hagibis | Molave | Chaba | Talim |
| **Philippines** | Jebi | Neoguri | Goni | Aere | Doksuri |
| **South Korea** | Mangkhut | Rammasun | Etau | Songda | Khanun |
| **Thailand** | Utor | Matmo | Vamco | | Vicente |
| **USA** | Trami | Halong | | | Saola |
| **Vietnam** | | | | | |

Names are mostly flowers, animals, birds and trees and descriptive adjectives. Names are not used in
alphabetical order  the contributing countries are. A new season starts where the previous left off.

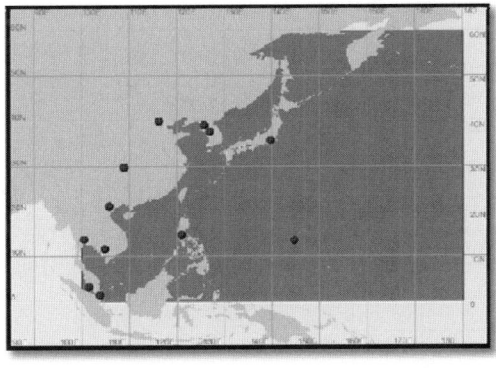

**Warning Area**

**RSMC Tokyo**

**Table 5.11(e)  Tropical Cyclone Names Around the World (Philippine Area)**

| 115°E to 135°E and 5°N to 25°N  PAGASA* Philippines | | | |
|---|---|---|---|
| 2010 | 2011 | 2012 | 2013 |
| Agaton | Amang | Ambo | Auring |
| Basyang | Bebeng | Butchoy | Bising |
| Caloy | Chedeng | Cosme | Crising |
| Domeng | Dodong | Dindo | Dante |
| Ester | Egay | Enteng | Emong |
| Florita | Falcon | Frank | Feria |
| Glenda | Goring | Gener | Gorio |
| Henry | Hanna | Helen | Huaning |
| Inday | Ineng | Igme | Isang |
| Juan | Juaning | Julian | Jolina |
| Katring | Kabayan | Karen | Kiko |
| Luis | Lando | Lawin | Labuyo |
| Milenyo | Mina | Marce | Maring |
| Neneng | Nonoy | Nina | Nando |
| Ompong | Onyok | Ofel | Ondoy |
| Paeng | Pedring | Pablo | Pepeng |
| Queenie | Quiel | Quinta | Quedan |
| Reming | Ramon | Rolly | Ramil |
| Seniang | Sendong | Siony | Santi |
| Tomas | Tisoy | Tonyo | Tino |
| Usman | Ursula | Ulysses | Urduja |
| Venus | Viring | Vicky | Vinta |
| Waldo | Weng | Warren | Wilma |
| Yayang | Yoyoy | Yoyong | Yolanda |
| Zeny | Zigzag | Zosimo | Zoraida |
| **Additional storm names if needed** | | | |
| Agila | Abe | Alakdan | Alamid |
| Bagwis | Berto | Baldo | Bruno |
| Chito | Charo | Clara | Conching |
| Diego | Dado | Dencio | Dolor |
| Elena | Estoy | Estong | Ernie |
| Felino | Felion | Felipe | Florante |
| Gunding | Gening | Gardo | Gerardo |
| Harriet | Herman | Heling | Hernan |
| Indang | Irma | Ismael | Isko |
| Jessa | Jaime | Julio | Jerome |

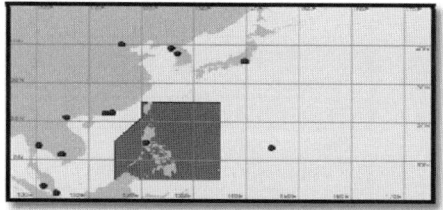

**Warning Area**
**PAGASA**

For local use only within the Philippines.

*Philippine Atmospheric, Geophysical and Astronomical Services Administration

**Table 5.11(f)   Tropical Cyclone Names Around the World (Northern Indian Ocean)**

| Northern Indian Ocean India Meteorological Department, Delhi, India | | | |
|---|---|---|---|
| **Nations Providing Names** | **List 1** | **List 2** | **List 3** | **List 4** |

| **Nations Providing Names** | **List 1** | **List 2** | **List 3** | **List 4** |
|---|---|---|---|---|
| | Onil | Ogni | Nisha | Giri |
| | Agni | Akash | Bijli | Jal |
| **Bangladesh** | Hibaru | Gonu | Aila | Keila |
| **India** | Pyarr | Yemyin | Phyan | Thane |
| **Maldives** | Baaz | Sidr | Ward | Mujan |
| **Myanmar** | Fanoos | Nargis | Laila | Nilam |
| **Oman** | Mala | Rashmi | Bandu | Mahasen |
| **Pakistan** | Mukda | Khai-Muk | Phet | Phailin |
| **Sri Lanka** | | | | |
| **Thailand** | **List 5** | **List 6** | **List 7** | **List 8** |
| | Helen | Chapala | Ockhi | Fani |
| | Leher | Megh | Sagar | Vayu |
| | Madi | Roanu | Makunu | Hikaa |
| | Nanauk | Kyant | Daye | Kyarr |
| | Hudhud | Nada | Luban | Maha |
| | Nilofar | Vardah | Titli | Bulbul |
| | Priya | Asiri | Gigum | Soba |
| | Komen | Mora | Phethai | Amphan |

**Warning Area**

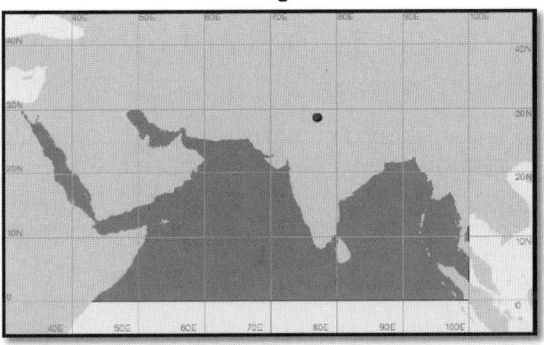

**India Meteorological Department**

**Table 5.11(g–h)  Tropical Cyclone Names Around the World (Southern Indian Ocean)**

| 90° E – 30° Southwest Indian Ocean<br>Regional Specialized Meteorological Center, La Reunion | | | |
|---|---|---|---|
| **2009 - 2010** | | **West of  55°** |
| Anja | Nigel | Name is given by: |
| Bongani | Olympe | The Sub-Regional Tropical |
| Cleo | Pamela | Cyclone Advisory Center |
| David | Quentin | **Madagascar** |
| Edzani | Rahim | |
| Fami | Savana | **55°E to 90°E** |
| Gelane | Themba | Name is given by: |
| Hubert | Uyapo | The Sub-Regional Tropical |
| Imani | Viviane | Cyclone Advisory Center |
| Joel | Walter | **Mauritius** |
| Kanja | Xangy | |
| Lunda | Yemurai | A new name list is compiled |
| Mohono | Zanele | for each tropical cyclone season |

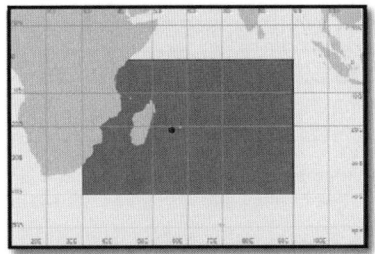

**Warning Area**

**RSMC La Reunion**

| Equator to 10°S and 90°E to 120°E<br>Tropical Cyclone Warning Center Jakarta | | |
|---|---|---|
| List A | Standby | **Warning Area** |
| Anggrek | Anggur | |
| Bakung | Belimbing | |
| Cempaka | Duku | |
| Dahlia | Jambu | |
| Flamboyan | Lengkeng | |
| Kenanga | Mangga | |
| Lili | Nangka | |
| Mawar | Pisang | |
| Seroja | Rambuta | |
| Teratai | Sawo | |

### Table 5.11(i)   Tropical Cyclone Names Around the World (Australia Region)

| Australia Region | | | | |
|---|---|---|---|---|
| **Southeast Indian Ocean, Arafura Sea, Gulf of Carpentaria, Coral Sea and Gulf of Papua** | | | | |
| List 1 | List 2 | List 3 | List 4 | List 5 |
| Anika | Anthony | Alessia | Alfred | Ann |
| Billy | Bianca | Bruce | Blanche | Blake |
| Charlotte | Carlos | *Cathy* | Caleb | Claudia |
| Dominic | Dianne | Dylan | Debbie | Damien |
| Ellie | Errol | Edna | Ernie | Esther |
| Freddy | Fina | Fletcher | Frances | Ferdinand |
| Gabrielle | Grant | Gillian | Greg | Gretel |
| *Hamish* | Heidi | Hadi | Hilda | Harold |
| Ilsa | Iggy | Ita | Ira | Imogen |
| Jasper | Jasmine | Jack | Joyce | Joshua |
| Kirrily | Koji | Kate | Kelvin | Kimi |
| Laurence | Lua | Lam | Linda | Lucas |
| Magda | Mitchell | Marcia | Marcus | Marian |
| Neville | Narelle | Nathan | Nora | Noah |
| Olga | Oswald | Olwyn | Owen | Odette |
| Paul | Peta | Quang | Penny | Paddy |
| Robyn | Rusty | Raquel | Riley | Ruby |
| Sean | Sandra | Stan | Savannah | Seth |
| Tasha | Tim | Tatjana | Trevor | Tiffany |
| Vince | Victoria | Uriah | Veronica | Verdun |
| Zelia | Zane | Yvette | Wallace | |

Used in rotating order the new season picks up where the former season left off.

Names are given when gale force (39 mph) winds are present.

Tropical cyclones entering from other regions retain their names.

Retirement of the names Hamish and Cathy has been requested.

### Warning Areas

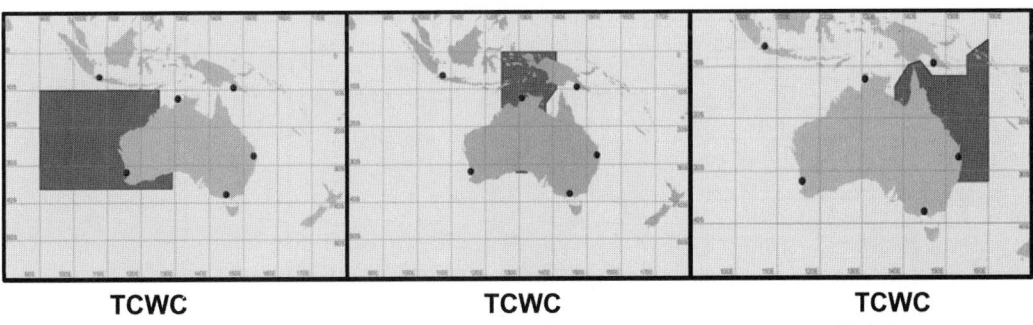

| TCWC | TCWC | TCWC |
|---|---|---|
| **Perth** | **Darwin** | **Brisbane** |

**Table 5.11(j)  Tropical Cyclone Names Around the World (Southern Indian)**

| Equator to 25°S and 160°E to 120°W |||||
|---|---|---|---|---|
| Regional Specialized Meteorological Center Nadi, Fiji |||||
| List A | List B | List C | List D | Standby |
| Ana | Arthur | Atu | Amos | Alvin |
| Bina | Becky | Bune | Bart | Bela |
| Cody | Chip | Cyril | Colin | Cook |
| Dovi | Denia | Daphne | Donna | Dean |
| Eva | Elisa | Evan | Ella | Eden |
| Fili | Fotu | Freda | Frank | Florin |
| Gina | Glen | Garry | Gita | Garth |
| Hagar | Hettie | Heley | Hali | Hart |
| Irene | Innis | Ian | Iris | Isa |
| Judy | Joni | June | Jo | Julie |
| Kerry | Ken | Kofi | Kala | Kevin |
| Lola | Lin | Lusi | Leo | Louise |
| Mal | Mick | Mike | Mona | Moses |
| Nat | Nisha | Nute | Neil | Niko |
| Olof | Oli | Odile | Oma | Opeti |
| Pita | Pat | Pam | Pami | Pearl |
| Rae | Rene | Reuben | Rita | Rex |
| Sheila | Sarah | Solo | Sarai | Suki |
| Tam | Tomas | Tuni | Tino | Troy |
| Urmil | Ului | Ula |  | Vanessa |
| Vaianu | Vania | Victor | Vicky | Wano |
| Wati | Wilma | Winston | Wiki | Yvonne |
| Xavier |  |  |  | Zidane |
| Yani | Yasi | Yalo | Yolande | Saga |
| Zita | Zaka | Zena | Zazu | Lea |
|  |  |  |  | Kamu |
|  |  |  |  | Pena |
|  |  |  |  | Elia |

**Warning area
Nadi, Fiji**

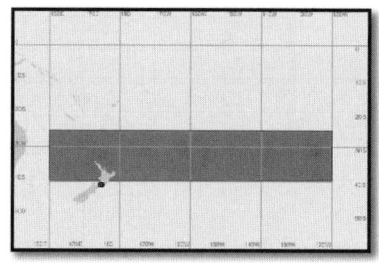

**Warning Area
Wellington, New Zealand**

| North of 10°S and 140°E to 160°E ||
|---|---|
| TCWC Port Moseby, Papua, New Guinea ||
| List A | List B |
| Alu | Nou |
| Buri | Obaha |
| Dodo | Paia |
| Emau | Ranu |
| Fere | Sabi |
| Hibu | Tau |
| Ila | Ume |
| Kama | Vali |
| Lobu | Wau |
| Maila | Auram |

**Warning Area**

**Table 5.12   Strongest Tropical Cyclones by Basin**

| | |
|---|---|
| **Northern Indian Ocean - Arabian Sea** (Strongest Tropical Cyclone that formed there) | **Super Cyclonic Storm Gonu** <br> SSHS:Category 5 <br> June 1-7, 2007 <br> 920 mb, 27.17" Hg <br> 270 km/h 185 mph, 1-minute sustained <br> 235 km/h 145 mph, 3-minute sustained <br> 78 fatalities |
| **Northern Indian Ocean – Bay of Bengal** (Strongest Tropical Cyclone that formed there) | **Super Cyclonic Storm Sidr** <br> SSHS: Category 5 <br> Nov 11 - 16, 2007 <br> 944mb, 27.88" Hg <br> 215 km/h 130 mph 1-minute sustained <br> 260 km/h 160 mph 3-minute sustained <br> Storm surge: 5m (16 ft) <br> 10,000 fatalities |
| **Northern Indian Ocean** | **BOB 06** <br> **(1999  Orissa Cyclone, Paradip Cyclone)** <br> SSHS: Category 5 <br> October 25 – November 3, 1999 <br> 260 km/h 160 mph,  3 and 1-minute sustained <br> <912 mb, 26.93" Hg, <br> 26 foot storm surge in Orissa, India <br> 15,000 fatalities |
| **Southern  Indian Ocean** | **Cyclone Daryl/Agnielle** <br> SSHS: Category 2 <br> November 17 – 25, 1995 <br> 885 mb, 26.14"  (Joint Typhoon Warning Center estimate, 979mb, 28.91" Hg, Australia) - disputed <br> 100 km/hr, 10-minute sustained <br> Stayed over open ocean |
| **Australia Region (Indian and Pacific)** | **Severe Tropical Cyclone Monica** <br> SSHS: Category 5 <br> April 17  – 26, 2006 <br> 285 km/h, 180 mph, 1-minute sustained <br> 360 km/h 225 mph, gusts <br> 879 mb, 25.96" Hg (by the Joint Typhoon Warning Center which is disputed and possibly as low as 869 mb, 25.66" Hg by the Dvorak Technique) which would make Monica the most intense tropical cyclone surpassing Typhoon Tip. <br> Storm surge: 5 – 6 m (16.4 – 19.6 ft), Junction Bay, Australia. <br> No fatalities |
| **Southwest Indian Ocean** | **Very Intense Tropical Cyclone Gafilo** <br> SSHS: Category 5 <br> March 3 – 11, 2004 <br> 230 km/hr 160 mph, 1-minute sustained <br> 895mb, 26.43" Hg |

**Table 5.12   Strongest Tropical Cyclones by Basin [CONTINUED]**

| | |
|---|---|
| **Southern Pacific** | **Category 5 Cyclone Zoe**<br>**(Australian Scale)**<br>SSHS: Category 5<br>December 23, 2002 – January 1, 2003<br>285 km/hr, 180 mph, 1-minute sustained<br>890 mb, 26.28" Hg<br>Storm surge: 5 – 10 m (5.5 – 11 ft), island of Tikopia<br>No Fatalities |
| **Western Pacific** | **Super Typhoon Tip**<br>SSHS: Category 5<br>October 4-19, 1979<br>305 km/hr 190 mph, 1-minute sustained<br>260 km/hr 160mph, 10-minute sustained<br>870 mb, 25.69" Hg, Lowest recorded surface pressure in the world.<br>Largest diameter of tropical storm force winds for any tropical cyclone, 1380 mile (2,220 km)<br>Routine reconnaissance flights ended in August 1987 and so Angela (1995) and Gay (1992) may have been stronger but data is lacking. |
| **Central Pacific** | **Super Typhoon/Hurricane Ioke**<br>SSHS: Category 5<br>August 20 – September 6, 2006<br>260 km/hr, 160 mph, 1-minute sustained<br>195 km/hr, 120 mph, 3-minute sustained<br>915 mb, 27.02" Hg<br>Storm surge: extratropical remnants of Ioke created a 9.1m (30 ft) surge in southeast Alaska.<br>No fatalities |
| **Eastern North Pacific** | **Major Hurricane Linda**<br>SSHS: Category 5<br>September 9-17, 1997<br>295 km/hr, 195 mph, 1-minute sustained<br>902 mb, 26.64" Hg<br>No fatalities |
| **North Atlantic** | **Major Hurricane Wilma**<br>SSHS: Category 5<br>October 15 – 26, 2005<br>295 km/hr, 185 mph, 1-minute sustained<br>882 mb, 26.05" Hg<br>23 fatalities |
| **South Atlantic** | **Cyclone Catarina**<br>SSHS: Category 2<br>March 24 – 28, 2004<br>155 km/hr, 100mph, 1-minute sustained<br>972 mb, 28.71" Hg<br>3 – 10 fatalities |

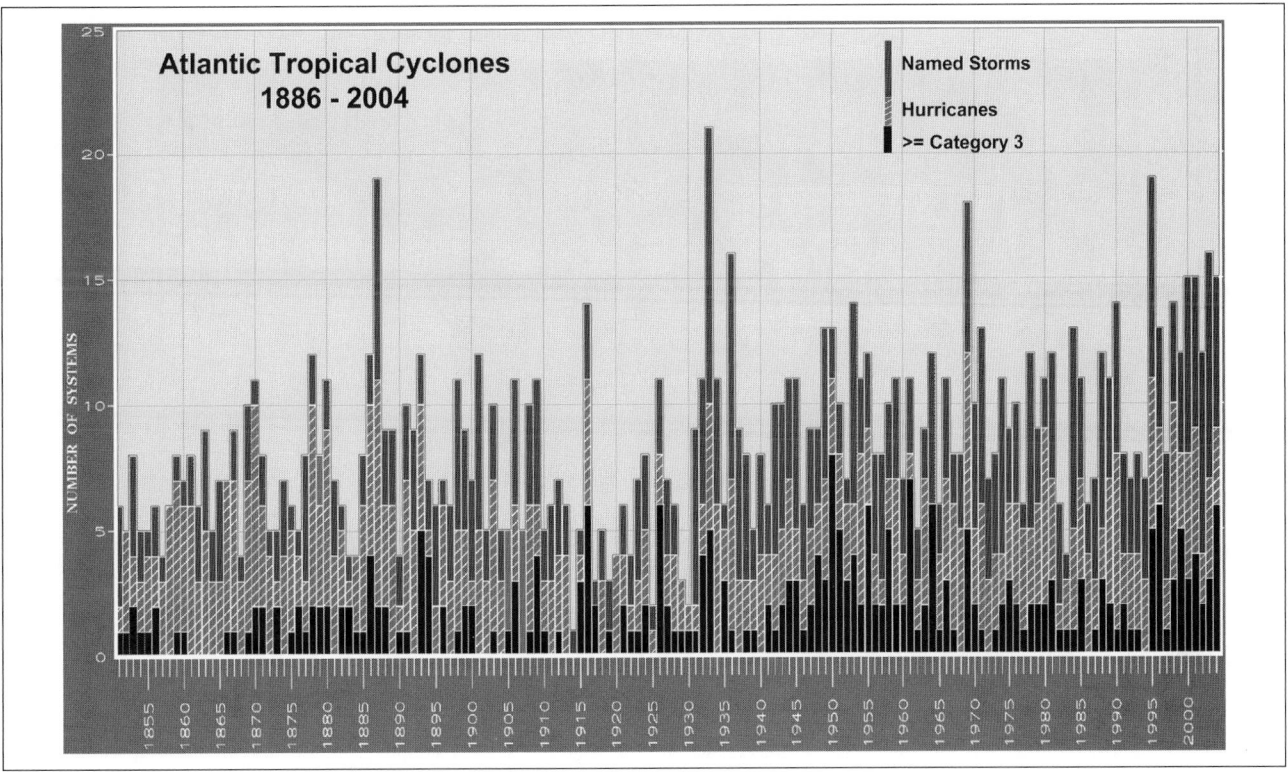

**Figure 5.41**   Number of systems each season, 1851–2005.

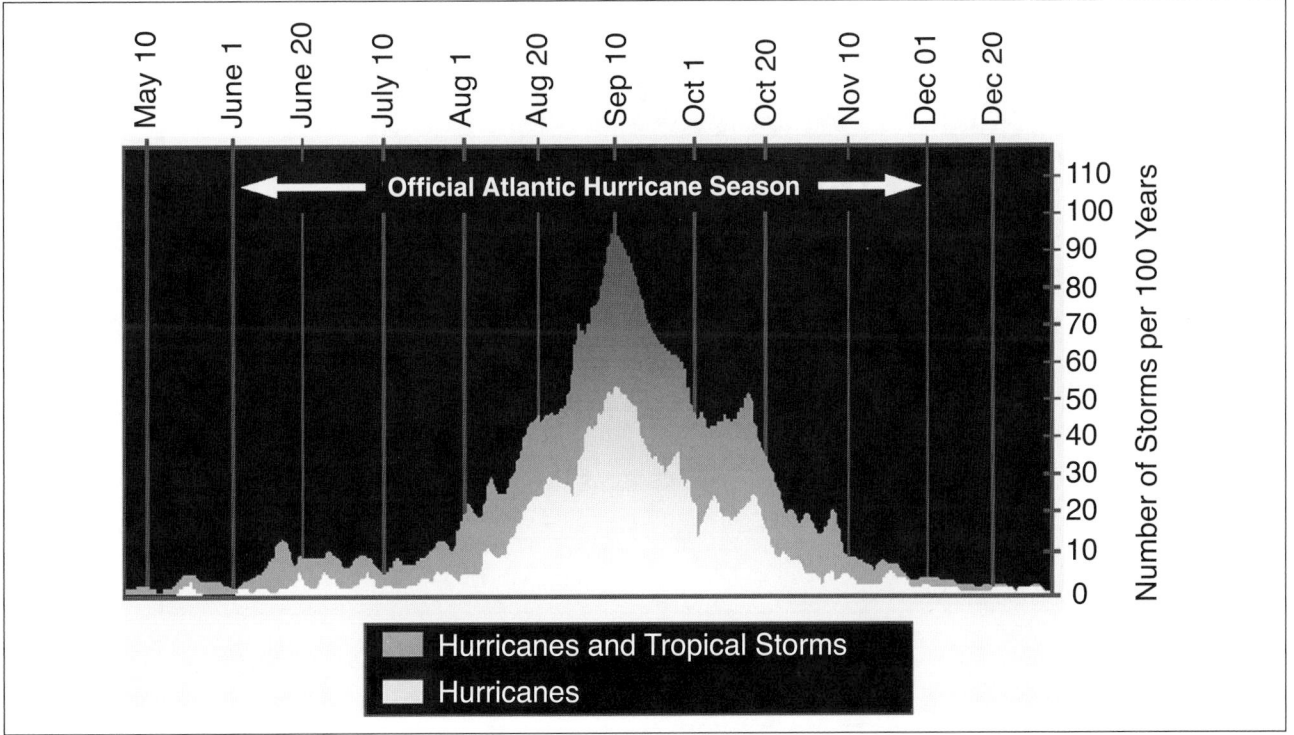

**Figure 5.42**   Seasonal storm distribution based on the average number of storms in a 100-year period.

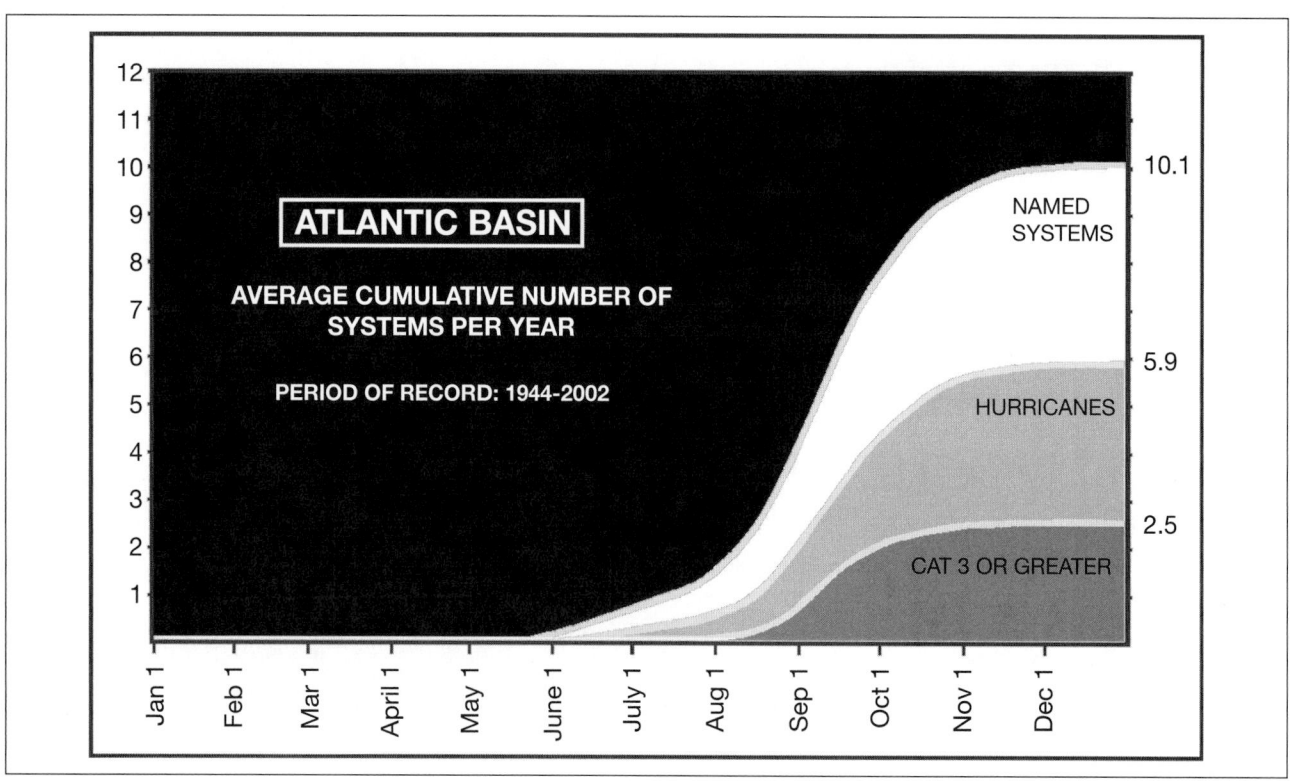

**Figure 5.43**    Average cumulative storm total for the Atlantic season.

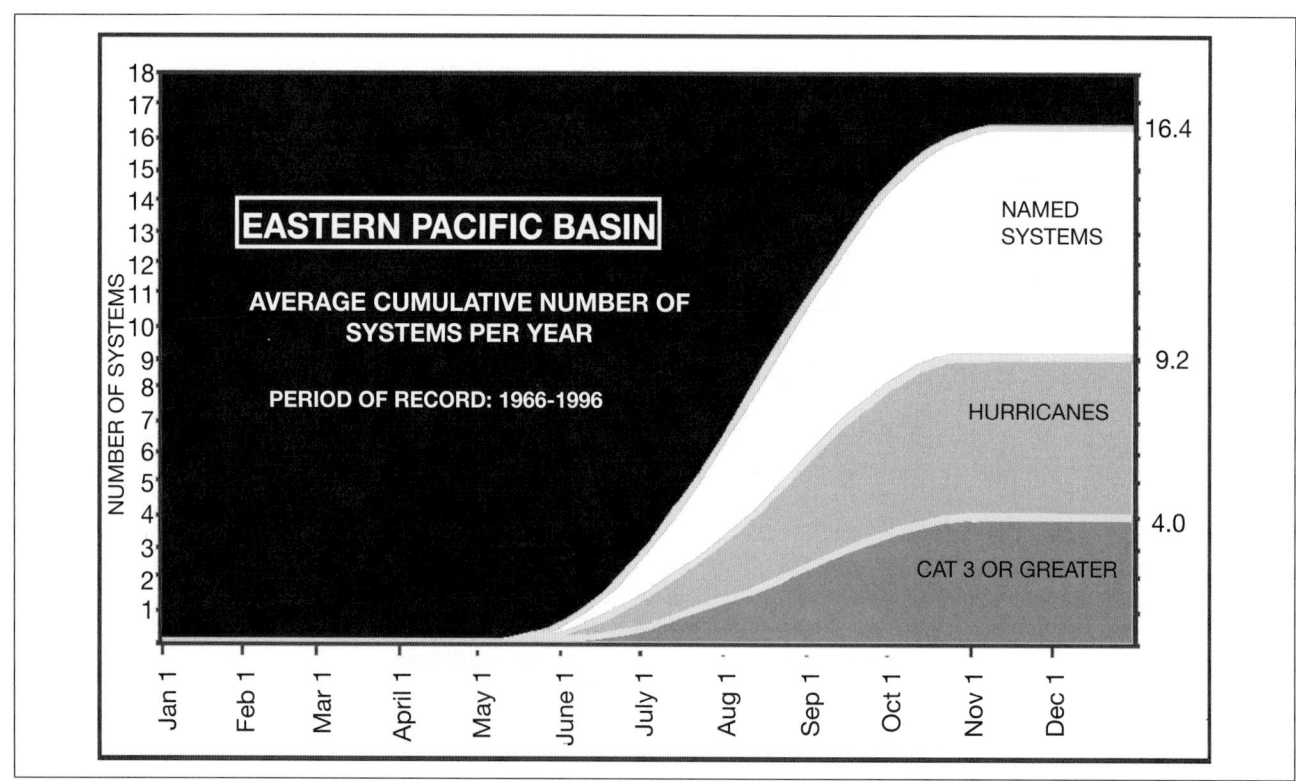

**Figure 5.44**    Average cumulative storm total for the Eastern North Pacific season.

# 6

# El Niño, La Niña, and the Southern Oscillation

## • INTRODUCTION

As I put the finishing touches to this chapter in November 2009, the sea surface temperatures (SSTs) along the equator in the Pacific Ocean are averaging around 2°C warmer than normal. Because of this, the winter ahead in the Midwest of the United States is expected to be fairly dry and temperatures not far from normal. The Gulf Coast should be cool and wet, California wet and stormy, and southern Canada warm. The reason why forecasts like these have been issued is that each time the warm water appears, weather conditions like those forecast have occurred because of a global weather pattern called El Niño.

El Niño, also called El Niño/Southern Oscillation, or ENSO for short, is an irregularly occurring fluctuation in the coupled atmosphere–ocean system that results in either warmer or colder than average water along the equator in the Pacific Ocean. It occurs every 2–8 years, varies in intensity, and has worldwide effects.

NOAA's Climate Prediction Center, the Meteorological Service of Canada, and the National Meteorological Service of Mexico consider an El Niño, or the warm phase of ENSO, to be in progress if sea surface temperatures in the Niño 3.4 region along the equator in the Pacific Ocean is 0.5°C (0.9°F) warmer than average for 3 or more consecutive months. If the water is 0.5°C (0.9°F) colder than average in the same region for the same length of time, a La Niña (or ENSO cold phase) is said to be in progress. Figure 6.1 is a map of the regions defined for monitoring the progress of ENSO events.

Much is known about the global effects of ENSO, but why an El Niño or La Niña starts or ends is a mystery and a topic of much scientific investigation and debate. Rainfall, wind, and air pressure patterns over the equatorial Pacific Ocean are most strongly linked to the underlying sea surface temperatures, from December through April. El Niño and La Niña conditions are often strongest during these months, and this is also when the greatest impact on US weather occurs.

## • SOME EL NIÑO/LA NIÑA HISTORY

The parched Pacific coast of South America from Ecuador into northern Peru is a desert partly because of the cold water that wells up and chills the air near the coast. Each year as the seasonal shift of wind belts takes place and southern summer approaches, a warm water current from the north invades these coastal areas and displaces the cold water that normally flows from the south and wells up in the region.

The arrival of the warmer water often coincides with a brief wet season that, in a few weeks, brings almost all the rain that coastal communities of Ecuador and northern Peru will get in a year.

Fishermen learned to adapt to the changing ocean currents as far back as the 1500s, and in the predominately Catholic coastal communities, farmers depended on the reliable but short rainy season. Because the change in direction of the ocean current coincided with or fell just after Christmas, the annual event was named in honor of the Christ Child. In the region, the proper noun El Niño refers to the Boy Jesus. The rains were considered a gift from the Christ Child, and through the years the brief wet season became known as "El Niño."

Occasionally, the rains would be intense for weeks, causing devastating floods. During these *años de abundancia* (years of abundance) rainfall in the short span of a few weeks would fall that was many times the average annual total. Intense rain on the unconsolidated desert soil caused great destruction and loss of life.

Scientists began to study these events, and both the annual wet season and the *años de abundancia* were treated

*The Weather Almanac: A Reference Guide to Weather, Climate, and Related Issues in the United States and Its Key Cities*, Twelfth Edition. Steve Horstmeyer.
© 2011 John Wiley & Sons, Inc. Published 2011 by John Wiley & Sons, Inc.

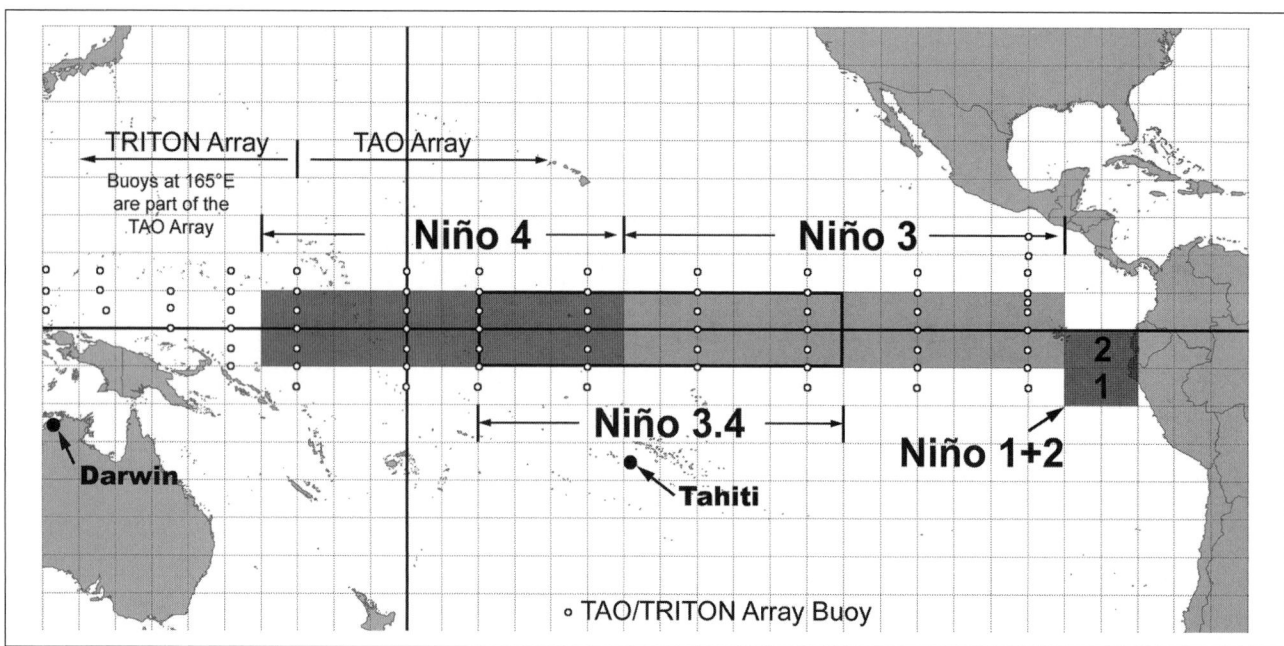

**Figure 6.1** Niño regions and the TAO/TRITON Array used for monitoring the progress of El Niño and La Niña episodes. (*Source:* Steve Horstmeyer from NOAA, NCEP, CPC data.)

as different intensities of the same phenomenon under the name El Niño. By the time it was known that the extremely wet years had a different cause than the annual wet season, the term El Niño was entrenched in popular and scientific literature and the name stuck.

The counterpart to the warm phase of the ENSO became known as La Niña or young girl, the opposite of a male child.

## • SIR GILBERT WALKER—THE INDIAN MONSOON AND ENSO

In 1904 Sir Gilbert Walker a British mathematician, was appointed Director General of Observatories for the Indian Meteorological Service. Because of widespread famine in 1877 and 1899 following the failure of summer monsoon rains, Gilbert was asked to begin a search for the causes of monsoon failure. Once again in 1918, before he completed his work, the monsoon rains failed and famine threatened to spread across India.

Walker published his research in 1928 and noted that, when the average monthly atmospheric pressure difference between Tahiti and Darwin, Australia, was great, due to a stronger than average South Pacific Subtropical High Pressure System (SPHPS), the monsoon rains in India were heavy. When the pressure difference was small because the SPHPS was weak, the rains failed and drought was the result. The locations of Darwin and Tahiti are shown in Figure 6.1.

He called the seesaw variation between the two regions the Southern Oscillation (SO) and measured the intensity

with the Southern Oscillation Index (SOI). Walker's research showed that when the SPHPS was weak, drought conditions also hit Australia, Indonesia, and parts of sub-Saharan Africa, while at the same time Canada had mild winters. Walker was unable to discover the underlying cause.

Little additional work was done in the 1930s, 1940s, and 1950s until 1958 when the strong El Niño of 1957–1958 caught the attention of scientists trying to learn as much about the planet as possible during the International Geophysical Year.

Research during and after the 1958 El Niño revealed links between SSTs and the atmosphere, but it was in 1969 that one of the pioneers of modern meteorology, Jacob Bjerknes, was the first to present a physical explanation for the link between SSTs, trade winds, and the El Niño/La Niña cycle. In his paper, he synthesized much earlier work and unified it to explain the physical causes.

It is now known that during the last 130 years of instrumental record, all failures of monsoon rains occurred during the warm phase of ENSO or El Niño. However, it is also known that not every El Niño resulted in monsoon failure. There are four El Niños that are exceptions, as shown in Figure 6.2, a graph of Indian rainfall from 1871 to 2000. The bars show departure from average annual rainfall for the period 1871–2000. El Niño episodes are indicated by black arrows pointing up and La Niña episodes by grey arrows pointing down.

Nearly one-half of the world's population relies on monsoon rains in the area from East Africa across southern Asia to China. Many people in this region subsist on crops they grow and make some money in years when yield is

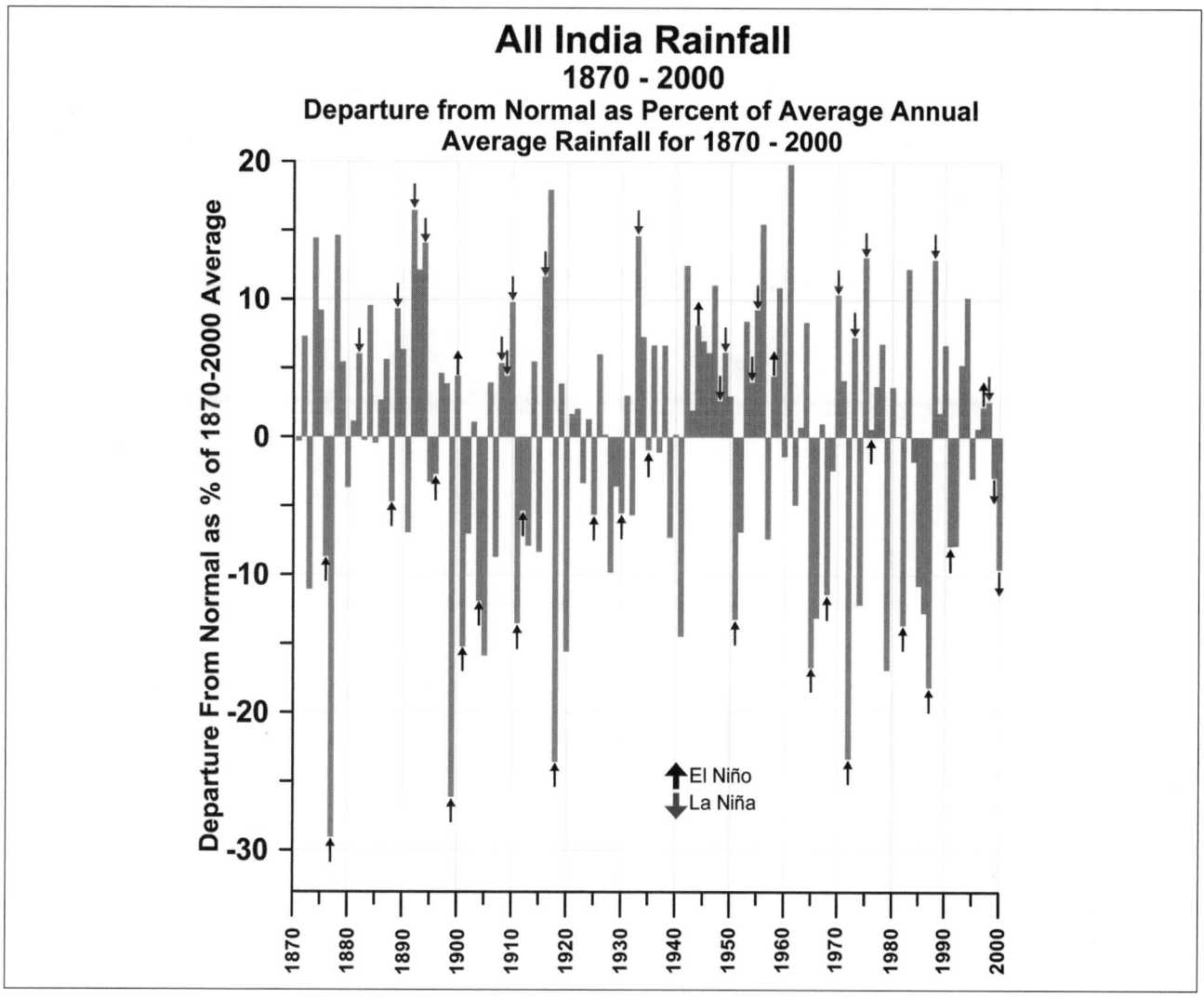

**Figure 6.2** All India Rainfall 1870–2000 and El Niño and La Niña episodes. (*Source:* Steve Horstmeyer from University of Washington, Joint Institute for the Study of the Atmosphere and Ocean data. *Original data source:* Dr. B. Parthasarathy, Indian Institute of Tropical Meteorology.)

sufficient. The importance of a forecast of meager rainfall, so that action can be taken to avoid famine, is obvious.

Preliminary results of current research (2009) suggest that the relationship between ENSO and monsoon rains may be changing. This appears to be related to changes in the location and intensity of the warm water pool in the Pacific. This is a research area of great importance and will undoubtedly be the focus of many future headlines.

### • THE MECHANICS OF THE SOUTHERN OSCILLATION AND EL NIÑO/LA NIÑA

The SO refers to the strengthening and subsequent weakening of the (SPHPS) that drives the El Niño/La Niña cycle.

Figures 6.3a,b are schematic maps that illustrate the warm and cold phases of the ENSO cycle.

La Niña (cold phase) occurs when the pressure is higher than average in the SPHPS resulting in southeast trade winds that are stronger than normal. Along the equator and the west coast of South America, upwelling due to the coriolis effect and Ekman pumping of cold, deep water is intense. The cold water brought to the surface creates a cold tongue that extends approximately to the International Date Line in the central equatorial Pacific. In addition, the stronger than average air flow westward across the Pacific Ocean causes water to pile up in the western Pacific from northern Australia to Papua New Guinea and northward toward Guam and the South China Sea. Sea level there is slightly higher than that in the eastern Pacific. During strong

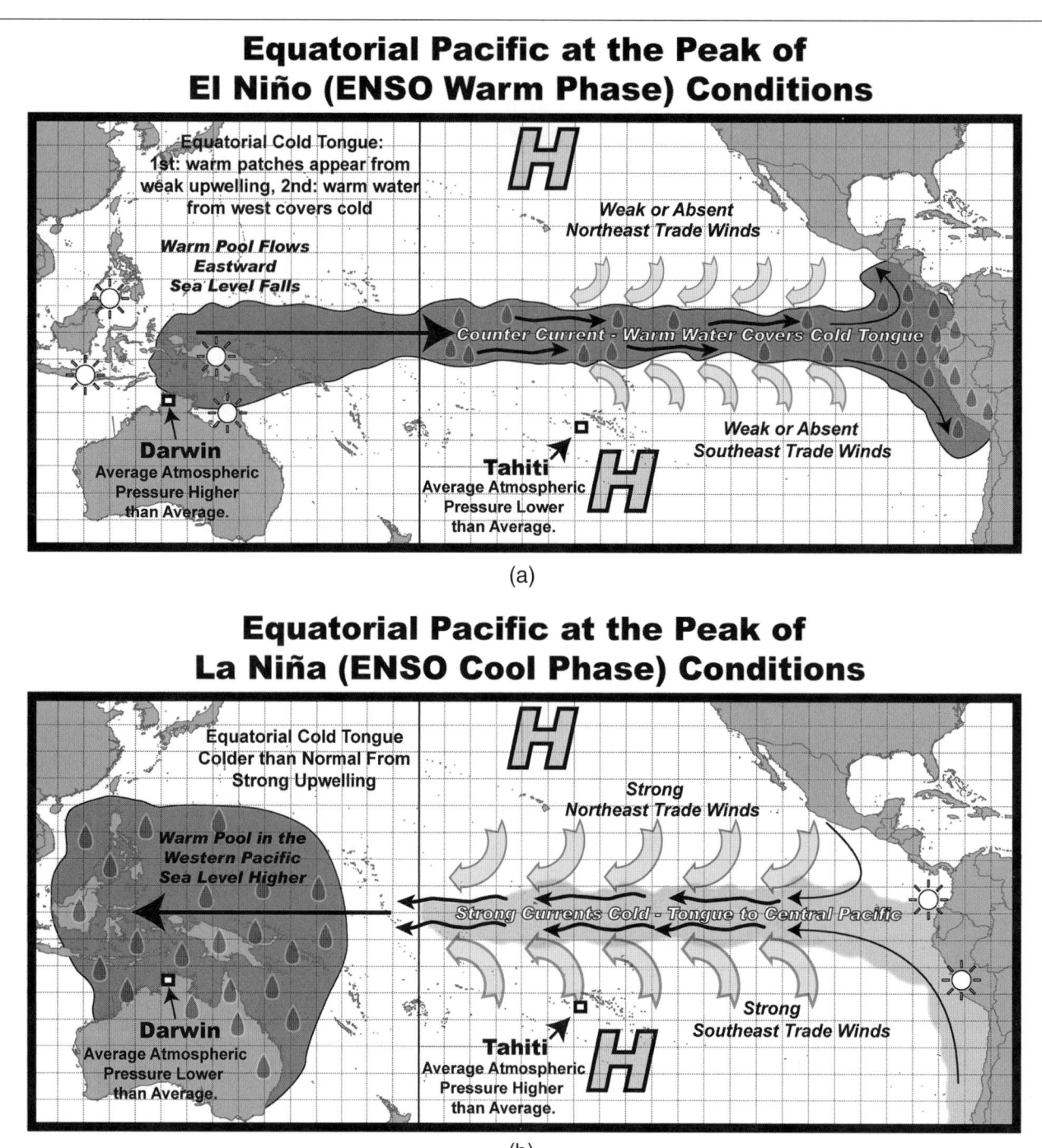

**Figure 6.3 a,b** Schematic maps of ocean and atmosphere changes during El Niño and La Niña episodes.

La Niña conditions, sea level in the western Pacific may be 13 inches or more than that in the east. Figure 6.4 is a schematic cross section along the equator in the Pacific illustrating La Niña conditions in both the ocean and atmosphere.

Rainfall is extensive in the western Pacific where the water is warm and atmospheric pressure is low, and as

Figure 6.4 shows, air is rising. In the eastern Pacific, cold waters chill the surface air from below, making it less likely to rise, while air aloft is generally sinking, creating a dry zone. The east–west aligned circulation cell connecting the rising air in the western Pacific with the sinking air in the eastern Pacific is called the Walker Circulation in honor of Sir Gilbert Walker.

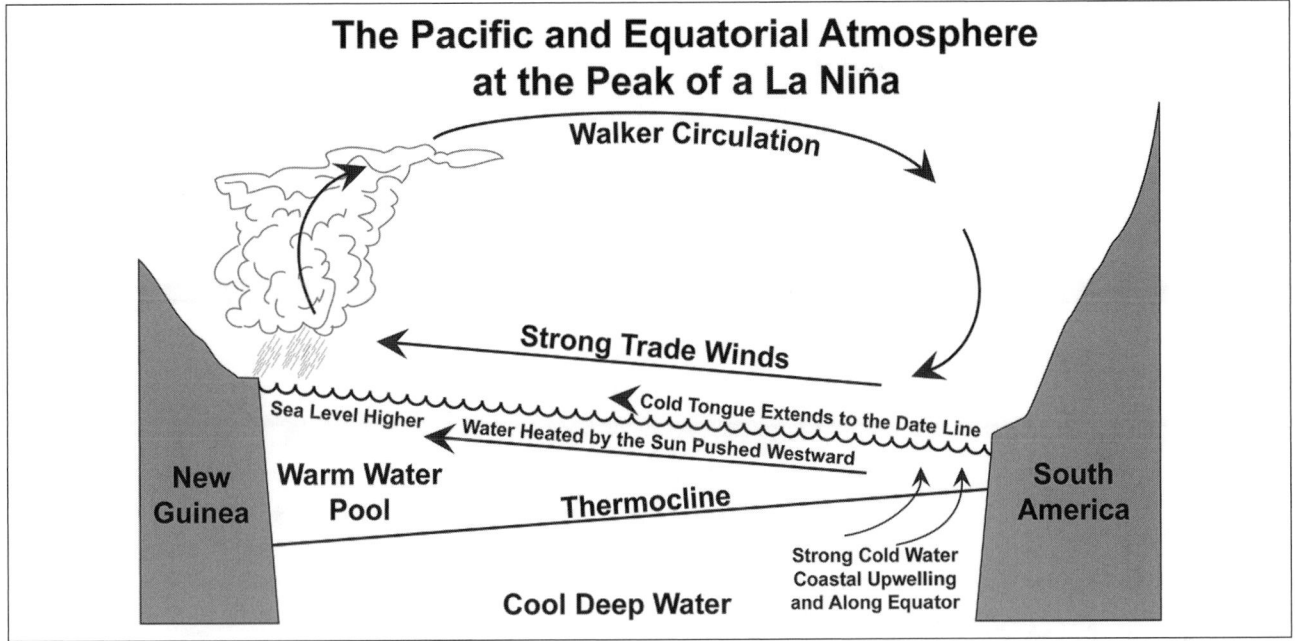

**Figure 6.4**   Schematic cross section along the equator in the Pacific illustrating the ocean and atmosphere during a La Niña episode (ENSO cold phase). (*Source:* Modified by Steve Horstmeyer based on a diagram by W. S. Kessler, NOAA, PMEL.)

Figure 6.5 shows a well-developed cold tongue across the equatorial Pacific Ocean on July 15,1988 during the strong La Niña of 1988–1989. Two factors extend the cold tongue to the International Dateline: (1) stronger than normal trade winds pushing the cold tongue westward, and (2) strong trade winds creating strong surface currents that lead to strong upwelling of cold, deep water along the equator.

Data for Figure 6.5 are from the Advanced Very High Resolution Radiometer (AVHRR) onboard the NOAA polar orbiting satellites. The AVHRR consists of six sensors or "channels" onboard each satellite that are designed to record information from a different part of the electromagnetic spectrum ranging from 0.58 μm (micrometers, millionths of a meter) to 12.5 μm that extends from the

near infrared (NIR) to long-wavelength infrared (LWIR) based on the International Commission on Illumination (CIE) classification.

The highest ground resolution that can be obtained from the AVHRR is 1.09 km (0.7 miles) when the satellite is directly overhead. AVHRR data have been collected continuously since 1981 and archived by the National Climatic Data Center.

When the pressure in the subtropical anticyclone starts to fall, the trade winds weaken, upwelling slows, and patches of warm water begin to appear along the equator (Figure 6.6a). In Figures 6.6b–d, the warm patches are consolidated and a strong warm tongue forms. The Walker Circulation begins to change as the warm water that accumulated in the western Pacific and held there by the strong trade winds during

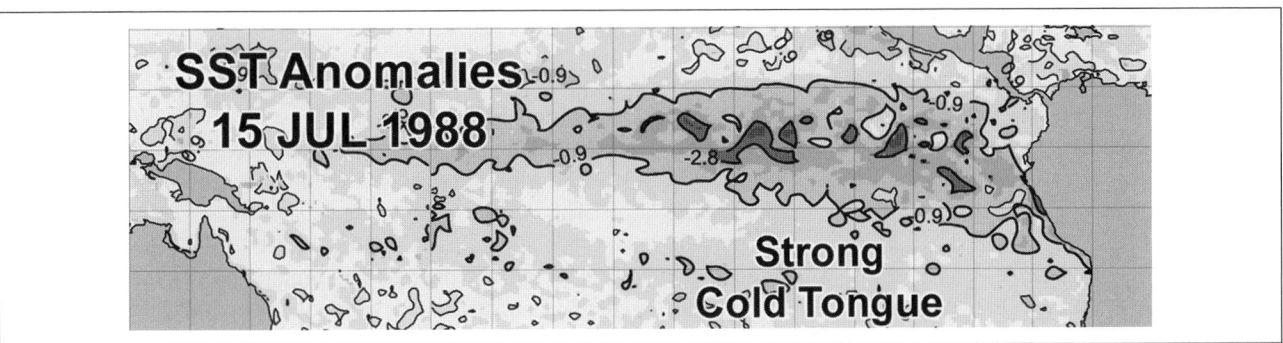

**Figure 6.5**   Well-developed cold tongue in the Equatorial Pacific Ocean, July 15, 1988. (*Source:* Steve Horstmeyer from NOAA, NCDC data using Panoply software by Robert Schmunk, NASA, GISS.)

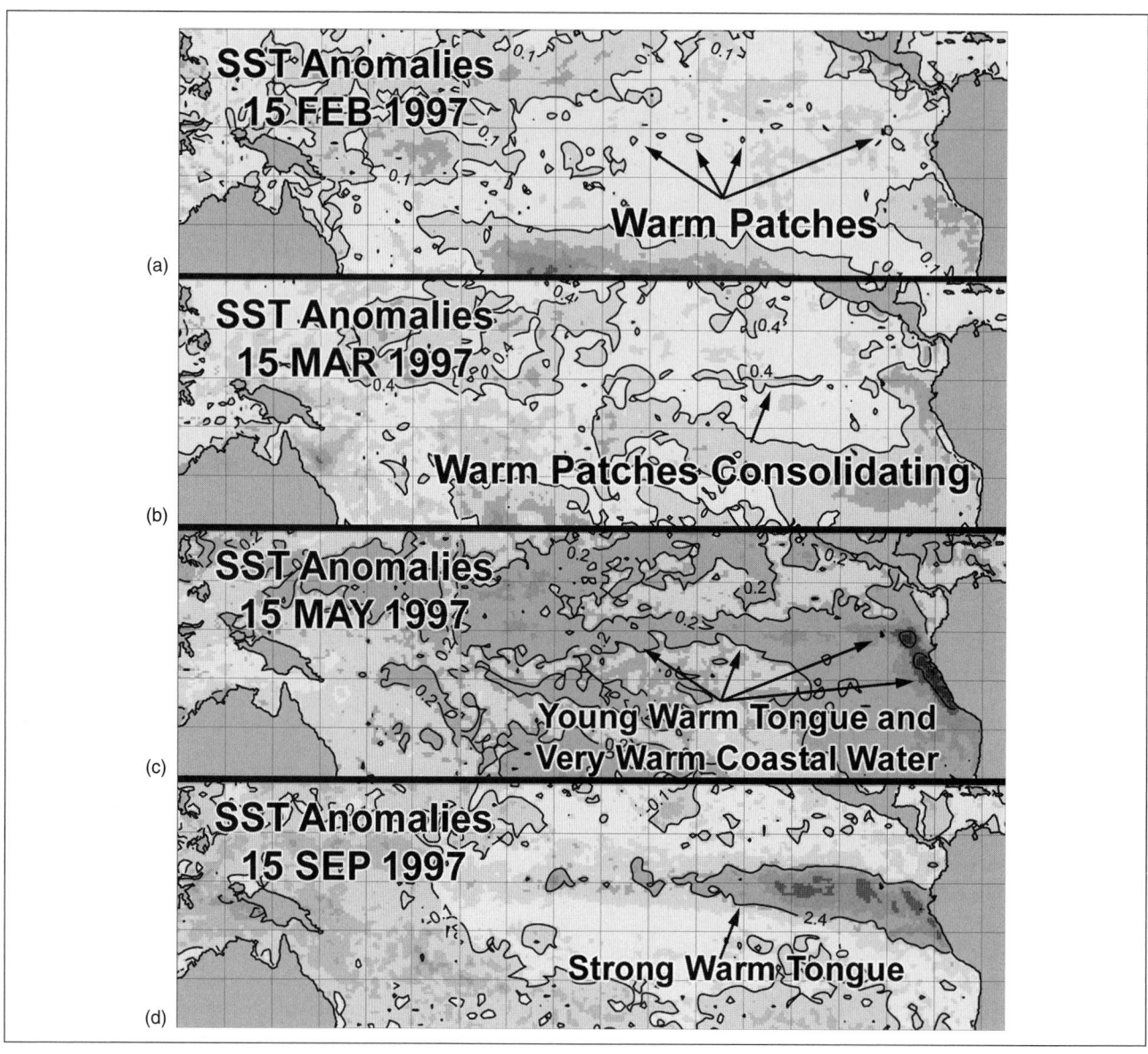

**Figure 6.6** Development of warm water along the equator in 1997. (a) February 15, 1997, spotty warm patches beginning to appear as upwelling weakens with slowing trade winds. (b) March 15, 1997, warm patches consolidating. (c) May 15, 1997, the young warm tongue stretches into the Central Pacific. (d) September 15,1997, a strong warm tongue with a large area of SST anomalies warmer than +2.4°C. (*Source:* Steve Horstmeyer from NOAA, NCDC data using Panoply software by Robert Schmunk, NASA, GISS.)

the ENSO cold phase starts to flow eastward. Figure 6.7 illustrates the changes taking place in the ocean and atmosphere at the onset of the ENSO warm phase. Sea level is falling in the western Pacific as the warm water pool heads eastward.

As the warm phase peaks, warm surface water covers the cool water all the way to the coast of South America and, depending on the strength of the El Niño, some distance along the coast, north and south of th equator. Upwelling either stops or is severely restricted (Figure 6.8). The warmer than normal SSTs heat the atmosphere from below, causing extensive and sometimes torrential rains along the equator in the zone where rain is normally scarce. Figure 6.9 shows the extent of the warmer than normal water on Christmas Day in 1997 during the strongest El Niño on record.

The warm water has fewer nutrients, and less dissolved oxygen and carbon dioxide, than the cold water below. Phytoplankton, the microscopic floating green plants, die, small fish like anchovies that feed on the phytoplankton in turn die, and in strong El Niños, millions of coastal birds die because their primary food source, the anchovy, is scarce.

ENSO is now known to be a rapid response of the ocean to changing winds along the equator, but it is still not known what drives the SO.

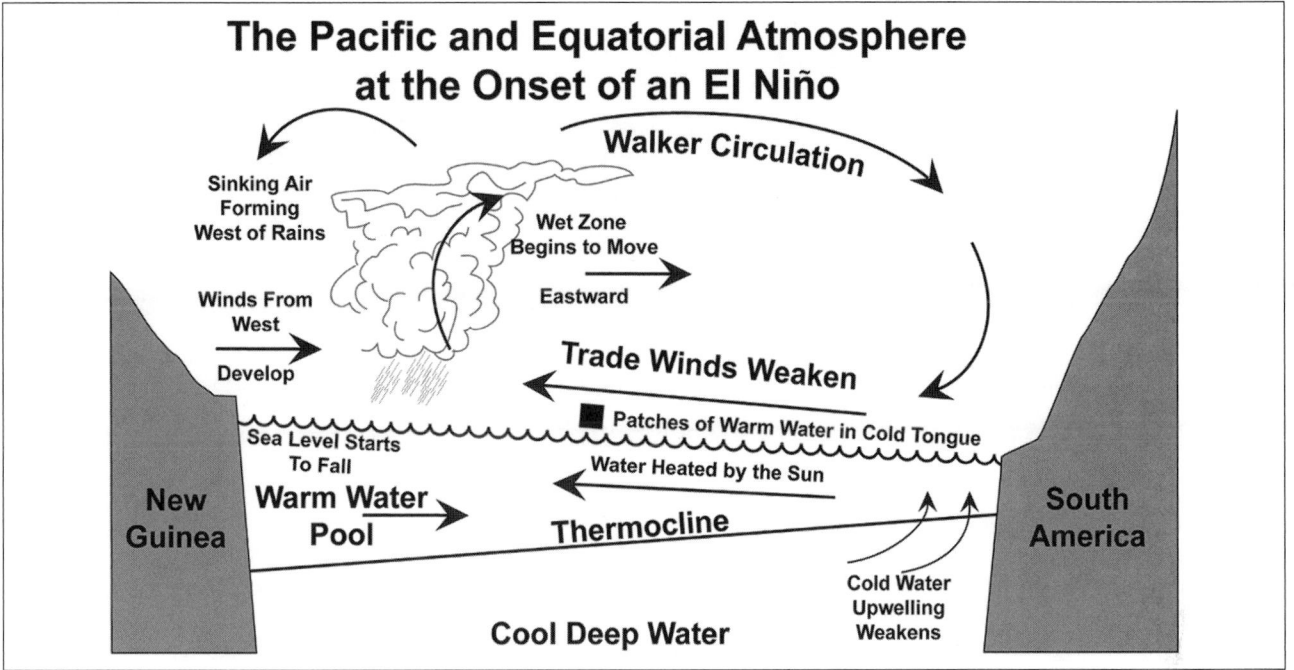

**Figure 6.7** Schematic cross section along the equator in the Pacific illustrating the ocean and atmosphere during the onset of an El Niño episode (ENSO warm phase). (*Source:* Modified by Steve Horstmeyer based on a diagram by W. S. Kessler, NOAA, PMEL.)

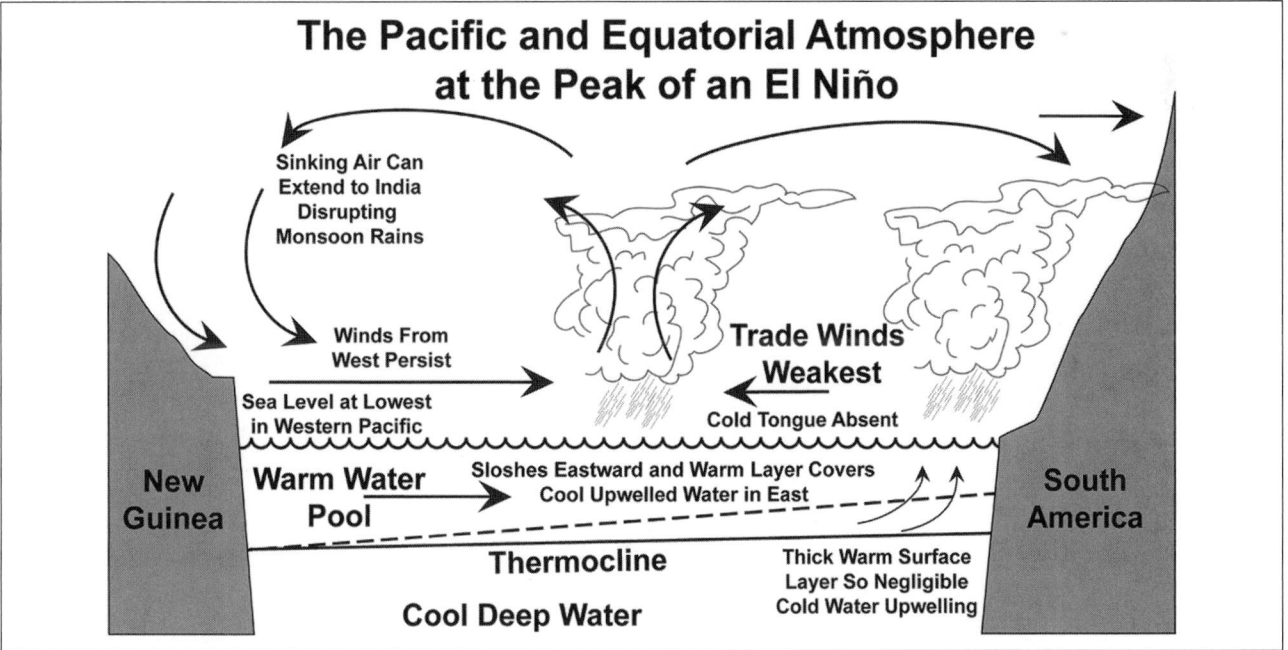

**Figure 6.8** Schematic cross section along the equator in the Pacific illustrating the ocean and atmosphere during the peak of an El Niño episode (ENSO warm phase). (*Source:* Modified by Steve Horstmeyer based on a diagram by W. S. Kessler, NOAA, PMEL.)

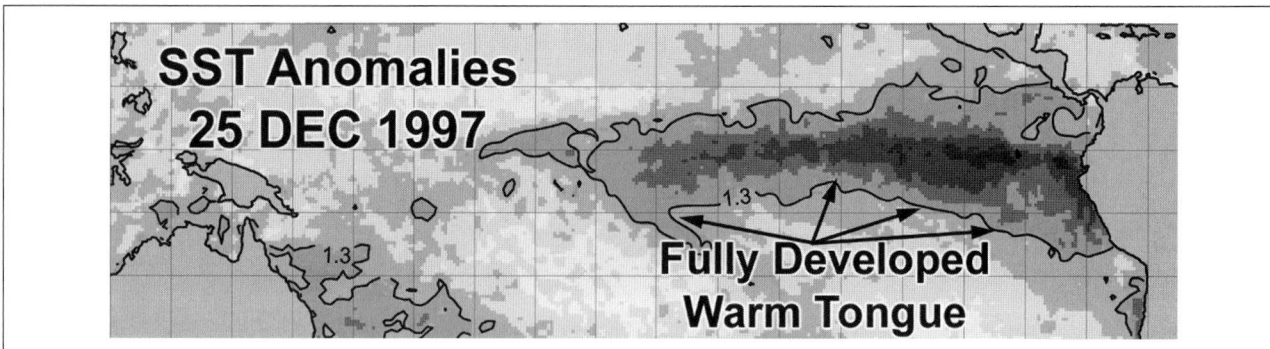

**Figure 6.9** The warm tongue Christmas Day of 1997 during the strongest El Niño on record. (*Source:* Steve Horstmeyer from NOAA, NCDC data using Panoply software by Robert Schmunk, NASA, GISS.)

## • THE SOUTHERN OSCILLATION INDEX AND SSTs

The SOI is a measure of the strength of this "sloshing" of atmospheric pressure in the South Pacific and is obtained by subtracting the average monthly pressure at Darwin, Australia from the average at Tahiti that is near the center of the SPHPS. Both locations are mapped in Figures 6.1 and 6.3a,b. When the index is negative (Darwin pressure > Tahiti pressure), the Equatorial Pacific is in a warm phase because the SPHPS is weaker than average and an El Niño may form. A positive SOI (Tahiti pressure > Darwin pressure) indicates a strong SPHPC and strong trade winds resulting in strong upwelling and cooler water at the surface.

The SOI determines whether the SO is in a warm phase or a cold phase, but it is the resulting SSTs that determine if the ENSO phase makes it to the status of El Niño or La Niña. Figure 6.10 shows the SOI from 1866 into 2009. From the graph we know that the SO is irregular and varies in strength. More is involved in the development of an El Niño or La Niña than just the SOI. SSTs in Niño Region 3.4 must be warm enough or cold enough for a sufficiently long period of time to be classified as an El Niño or La Niña.

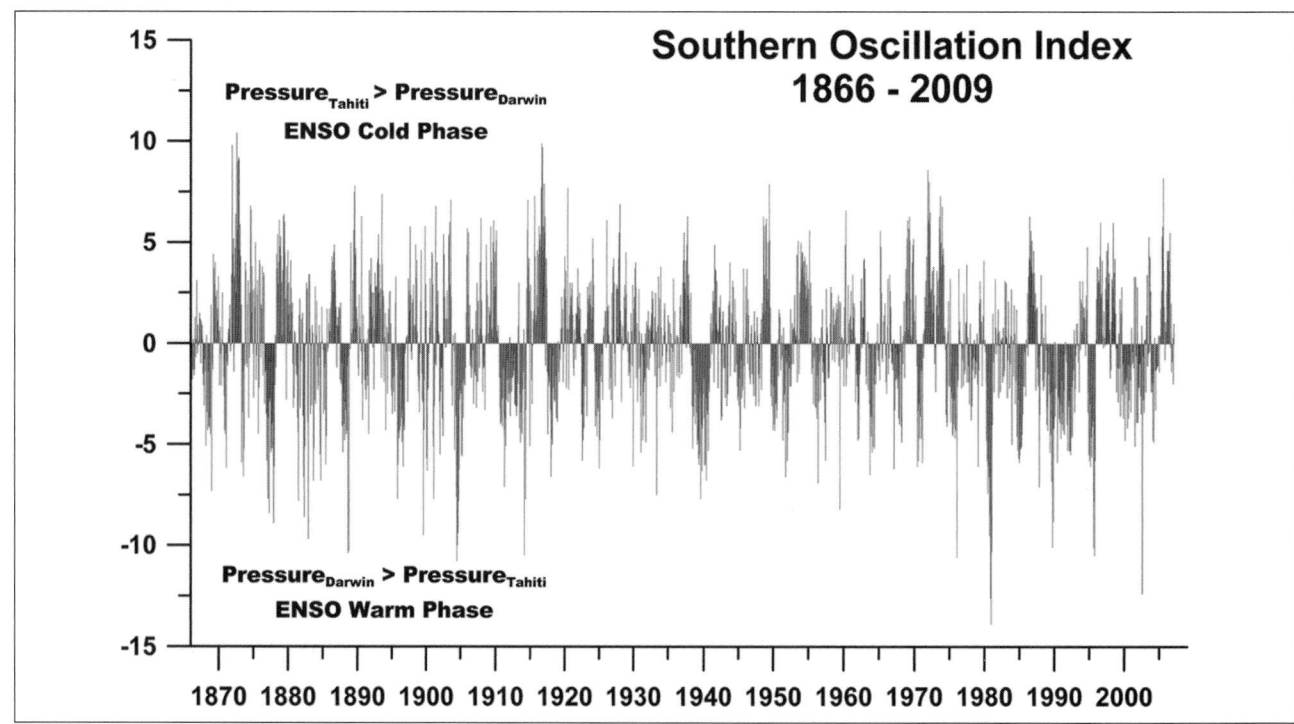

**Figure 6.10** The SOI from 1866 to October 2009. (*Source:* Steve Horstmeyer from NOAA, NCEP, CPC data and The Climate Research Unit, School of Environmental Sciences, University of East Anglia.)

**Table 6.1   Characteristics of Version 3 of the AVHRR Aboard NOAA Polar Orbiting Satellites**

| Channel | Wavelength (μm) | CIE Classification | Typical Use |
|---------|-----------------|--------------------|--------------|
| 1 | 0.58 - 0.68 | Visible Light (yellow to red) | Daytime cloud and surface mapping |
| 2 | 0.725 - 1.00 | Near Infrared | Land-water boundaries |
| 3A | 1.58 - 1.64 | Short Wavelength Infrared | Snow and ice detection |
| 3B | 3.55 - 3.93 | Mid Wavelength Infrared | Night cloud mapping, sea surface temperature |
| 4 | 10.30 - 11.30 | Long Wavelength Infrared | Night cloud mapping, sea surface temperature |
| 5 | 11.50 - 12.50 | Long Wavelength Infrared | Sea surface temperature |

*Source:* NOAA, Satellite Information Service.

The warm waters of El Niño heat the atmosphere from below, and tremendous quantities of warm, moist air are lifted into the atmosphere, creating wide areas with large rainfall totals, areas that are otherwise quite dry. In the Eastern Pacific, much of the energy flux is northward, high in the troposphere. This energy flow strengthens the subtropical jet stream and energizes storms over the Pacific Ocean that crash into the west coast of the United States. Farther east, the Gulf Coast of the United States is wet and cooler than normal due to thick cloud cover and heavy rain. The global effects of El Niño are illustrated in Figures 6.11a,b.

The cool waters of La Niña have the opposite effect by chilling the lower atmosphere. During El Niño from the coast of South America to the International Date Line, rainfall is generally well above normal. During La Niña, the sky is clear and rainfall is scarce in the same region. La Niña is also felt around the globe, and Figures 6.12a,b show the effects.

### • MONITORING THE EQUATORIAL PACIFIC

The Equatorial Pacific has been divided into five regions for research and reporting of oceanic and atmospheric variables associated with ENSO. Table 6.2 is a list of the coordinates of the regions that are mapped in Figure 6.1. Niño Region 3.4

**Table 6.2   Coordinates of Niño Regions in the Equatorial Pacific Ocean and Area**

| Region and Boundaries | | Area (mi²) |
|---|---|---|
| Niño 1+2 | 0° - 10° South | 80° - 90°West | 474,966 |
| Niño 3 | 5°North - 5°South | 150°West - 90°West | 2,860,682 |
| Niño 4 | 5°North - 5°South | 160°East - 150°West | 2,383,901 |
| Niño 3.4 | 5°North - 5°South | 170° - 120°West | 2,383,901 |

*Source:* NOAA, Pacific Marine Environmental Laboratory.

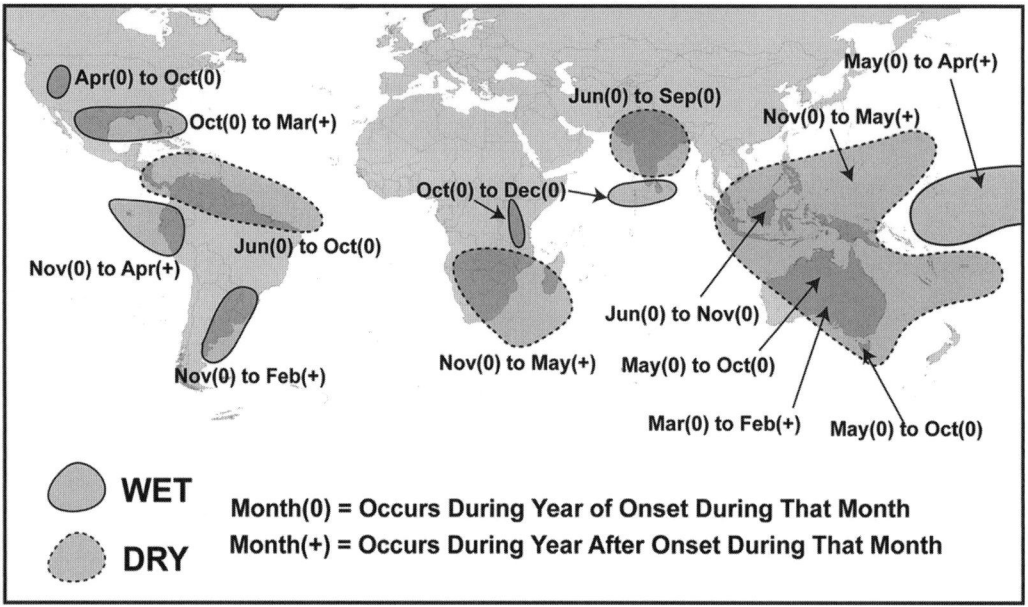

(a)

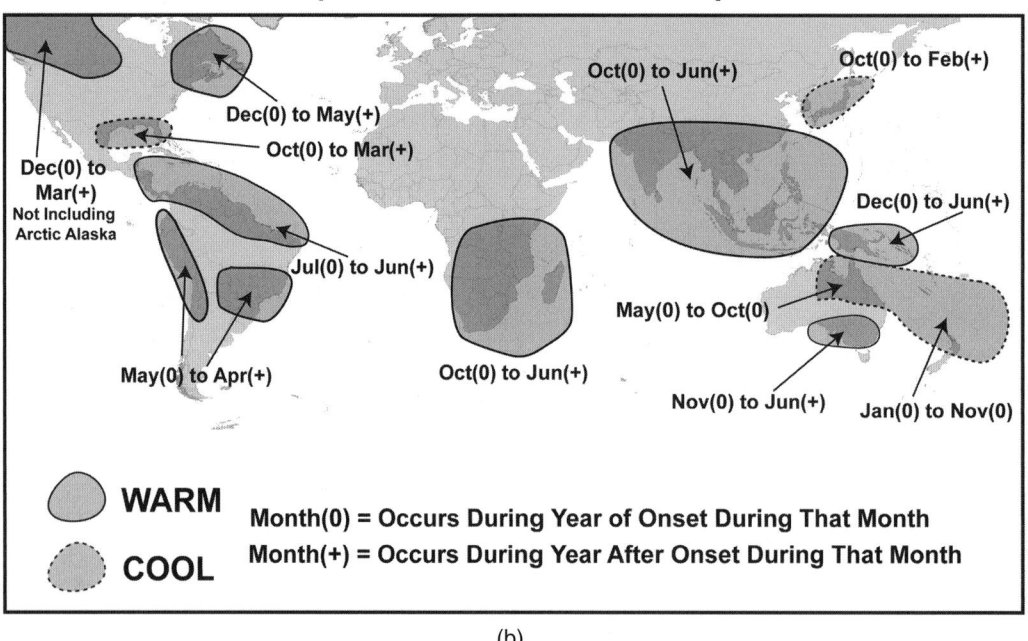

(b)

**Figure 6.11 a,b** Potential and observed global effects of El Niño episodes. (*Source:* Steve Horstmeyer based on diagrams from NOAA, NCEP, CPC.)

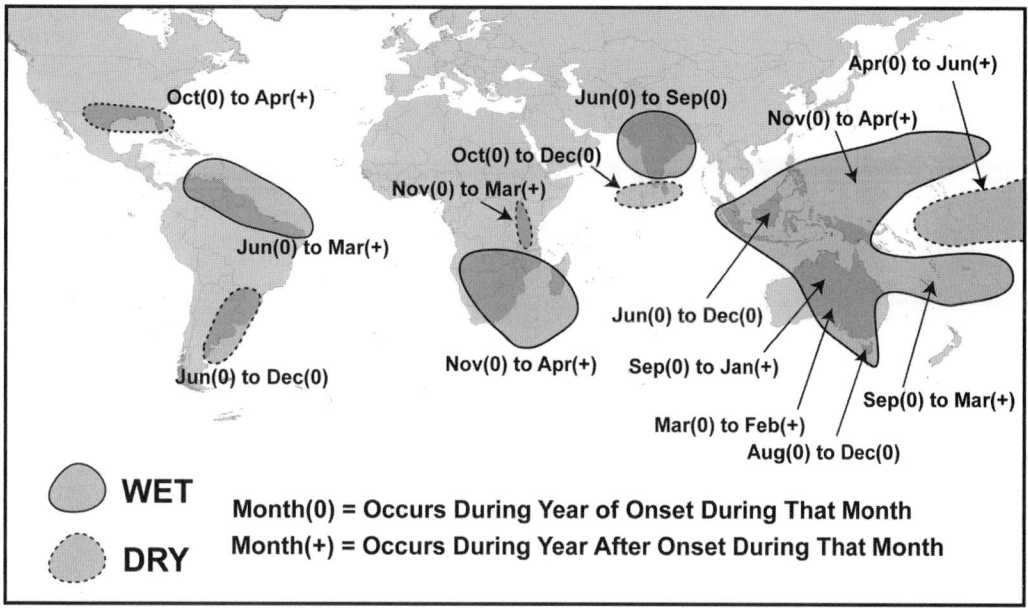

(a)

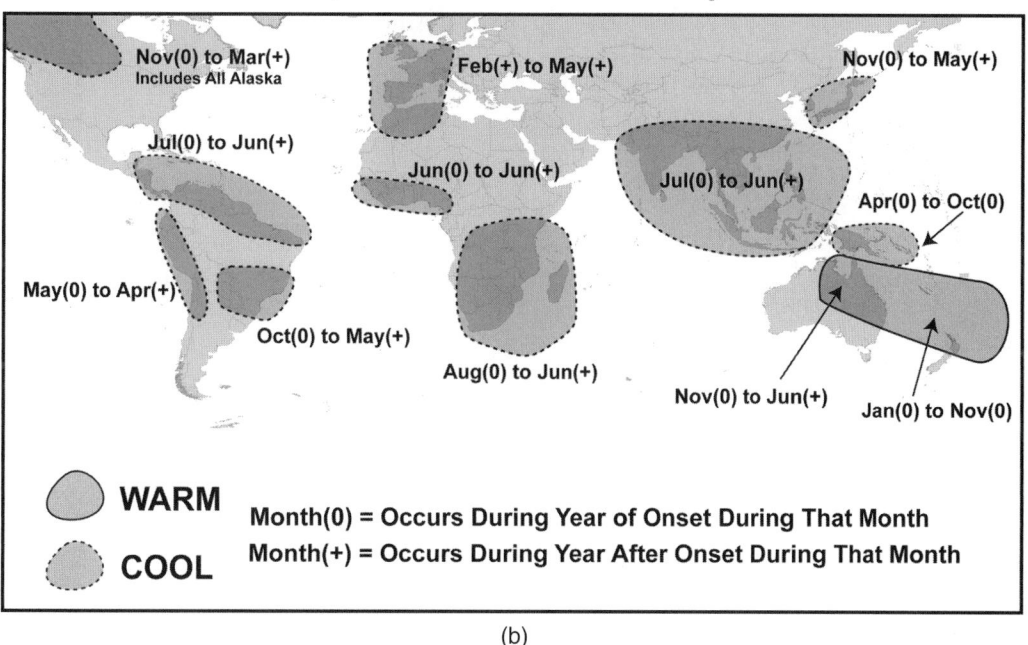

(b)

**Figure 6.12 a,b**  Potential and observed global effects of La Niña episodes. (*Source:* Steve Horstmeyer based on diagrams from NOAA, NCEP, CPC.)

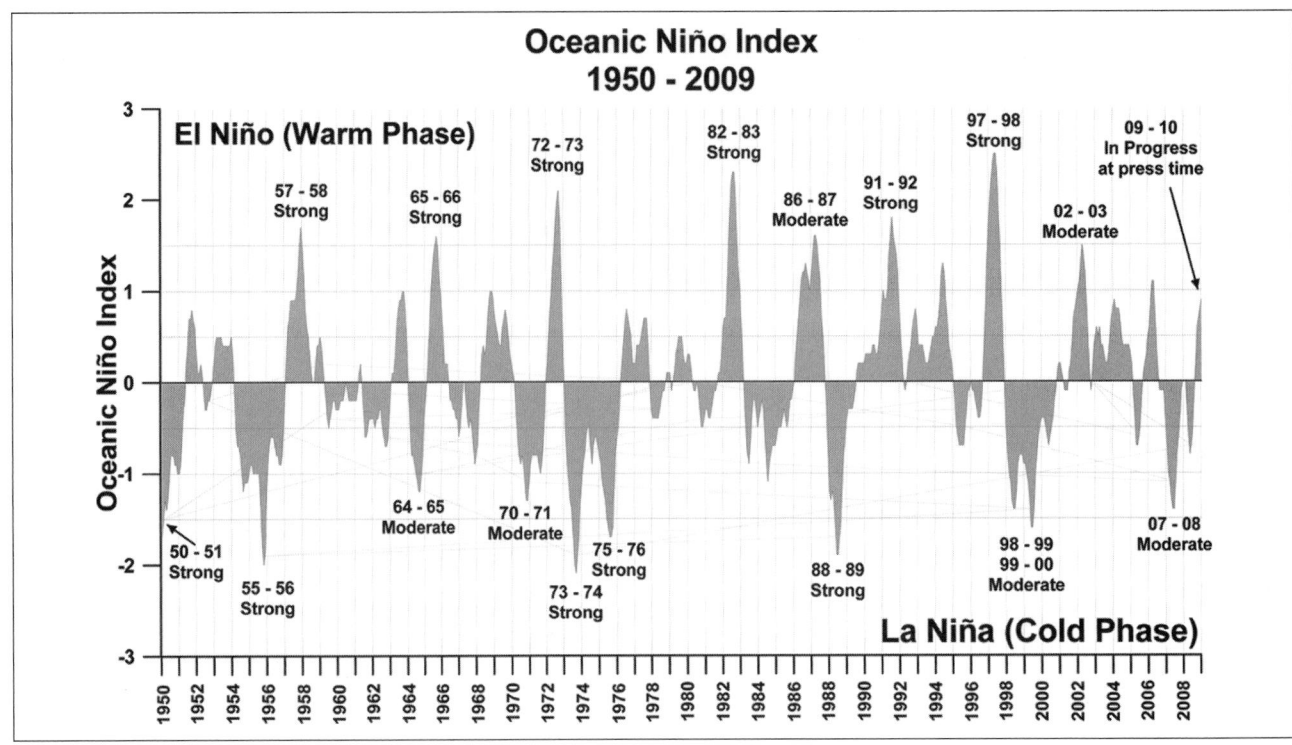

**Figure 6.13**   ONI in Niño Region 3.4, 1950–2009, with strong and moderate El Niño and La Niña episodes indicated. (*Source:* Steve Horstmeyer from NOAA, NCEP, CPC data.)

was added in 1996 so that scientists could easily look at the critical overlap region between Regions 3 and 4. Regions 1 and 2 are now usually considered as a single region. By following developments in all five regions, forecasters can track the development and demise of both an El Niño and La Niña. NOAA's Climate Prediction Center bases the Oceanic Niño Index (ONI) on Region 3.4.

After the strong El Niño of 1982–1983 that was neither forecast nor detected until near peak scientists realized there was a need for a reliable and continuous stream of oceanographic and atmospheric data from along the equator in the Pacific Ocean.

NOAA's Pacific Marine Environmental Laboratory (PMEL) developed the Tropical Atmosphere/Ocean (TAO) Array, consisting of 70 anchored buoys that span the Equatorial Pacific gathering weather data, sea surface temperatures, and surface and subsurface current data. Each type of data is important in forecasting and monitoring changes in the Niño regions. The data are transmitted via satellite and available almost immediately.

In October 1999, the Japan Agency for Marine-Earth Science and Technology (JAMSTEC) assumed responsibility for moorings along and to the west of 156° E longitude. JAMSTEC deployed the Triangle Trans-Ocean Buoy Network (TRITON) buoys at those locations. Data from both the TAO Array and TRITON moorings are merged in a common database and available from both PMEL and JAMSTEC. The

TAO Array was subsequently renamed the TAO/TRITON Array. The locations of individual TAO/TRITON Array buoys are mapped in Figure 6.1.

## • ENERGY AND EL NIÑO/LA NIÑA

Two thresholds must be surpassed for an El Niño or La Niña to occur. The first is the average SST for the Niño 3.4 region. If the average temperature is 0.5°C (0.9°F), or more, warmer than average and the anomaly persists for three overlapping 3-month periods or longer, an El Niño is in progress. If the average SST is 0.5°C, or more, colder than average for three overlapping 3-month periods or longer, a La Niña is in progress.

Figures 6.2 and 6.10 both illustrate ways scientists determine what is happening with ENSO. Figure 6.13 shows a time series of SST's anomalies in Niño Region 3.4 as the ONI, which is used by NOAA's Climate Prediction Center. Niño Region 3.4 is about 76% as large as the lower 48 states, which have an area of 3,119,884.6 square miles.

Figure 6.14 shows the "Cold Tongue Index" through time. It covers a much larger area than the ONI, from 6° S to 6° N and one-quarter of Earth's equatorial circumference from 90° W to 180°. This region has an area of 5.14 million square miles and is larger than the land area of the United States (all 50 states), which encompasses 3.79 million square miles.

**Table 6.3   The Number of Named Atlantic and Pacific Tropical Cyclones for El Niño Years**

|  | Atlantic | | Eastern Pacific | |
|---|---|---|---|---|
|  | Average | El Niño Avg. | Average | El Niño Avg. |
| **Named storms** | 9.4 | 7.1 | 16.7 | 17.6 |
| **Hurricanes** | 5.8 | 4.0 | 9.8 | 10.0 |
| **Intense Hurricanes** | 2.5 | 1.5 | 4.8 | 5.5 |

*Source:* NOAA, TPC, NHC.

The Niño regions cover millions of square miles, and in the case of the Cold Tongue Index the area is larger than the area of all 50 states. Even though the regions look small compared to the Pacific Ocean in Figure 6.1, immense amounts of heat can be transferred from the ocean to the atmosphere over that amount of warm water. In fact, so much energy is transferred that global weather patterns are altered.

**• ENSO AND HURRICANES**

One way global weather is changed during El Niño occurs to the east of the El Niño regions in the Atlantic Ocean basin. It is well known (Table 6.3 and Figures 6.15a, b) that when an El Niño is in progress, the incidence of tropical cyclones in the Atlantic Ocean is less than average. During a La Niña, hurricanes are more numerous than average. The stronger

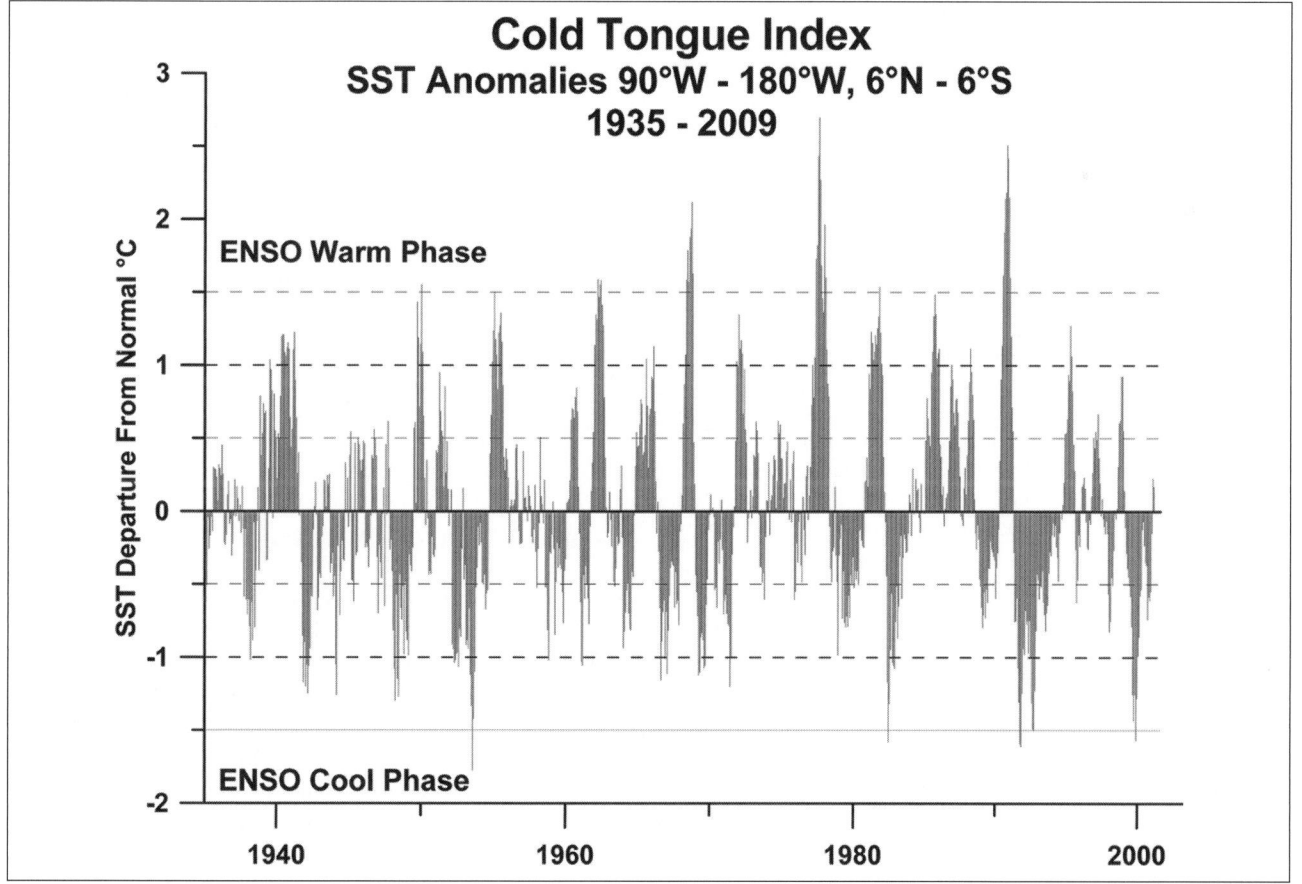

**Figure 6.14**   Cold Tongue Index, 1935–2009, calculated for 5.14 million square miles of the Equatorial Eastern Pacific Ocean. (*Source:* Steve Horstmeyer from University of Washington, Joint Institute for the Study of the Atmosphere and Ocean data.)

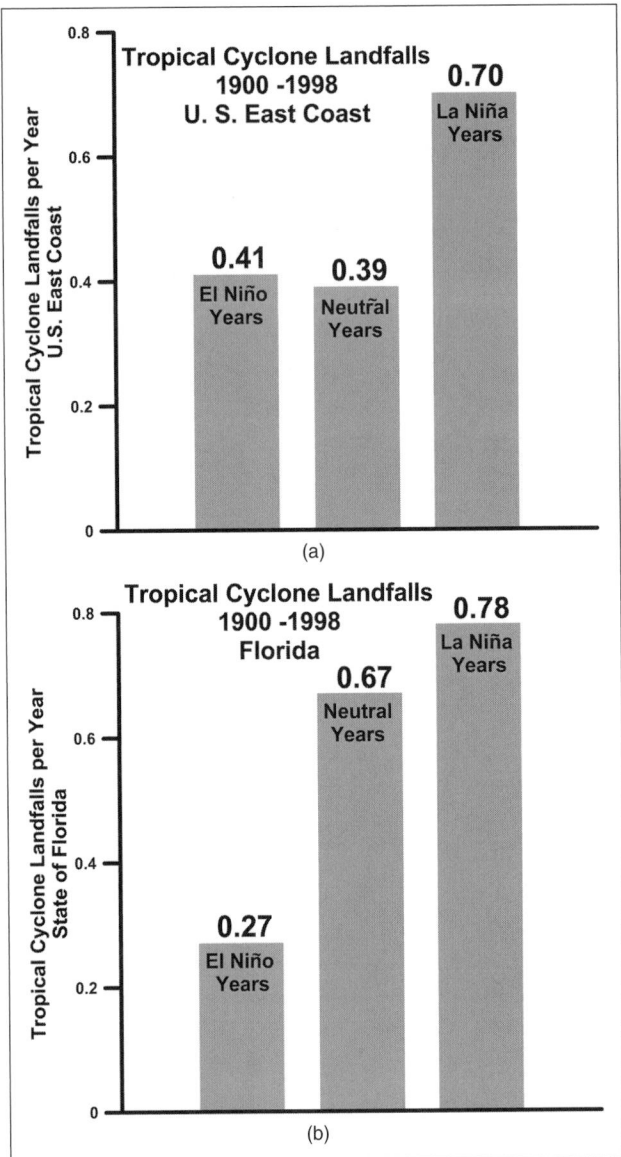

**Figure 6.15 a,b**  The number of tropical cyclones making landfall per year, 1900–1998, during El Niño, neutral, and La Niña years. (*Source:* Florida State University, Center for Ocean-Atmospheric Prediction Studies, Technical Report 2002–2005.)

the El Niño, the smaller the number of tropical storms and hurricanes.

There are three possible causes for this:

1. **Vertical Wind Shear.** Tropical cyclones require warm water and warm water and small wind speed increases upward (i.e.. little vertical wind shear) through the atmosphere. When the wind velocity increases significantly above the surface, the energy needed for hurricane formation cannot become concentrated because it is blown away down wind. During an El Niño, wind shear is large over the Atlantic Ocean and tropical storm and hurricane incidence is small.

2. **The Saharan Air Layer.** During El Niño years, dry air flowing from North Africa transports more dust than average over the Atlantic Ocean. There are four possible ways the Saharan Air Layer could affect tropical cyclone formation:
   a. The dust could block enough sunshine that surface waters are a bit cooler.
   b. The dust may also interfere with condensation and the formation of rain drops that release energy that powers tropical cyclones. There are other microphysical possibilities within clouds that could be caused by the dust.
   c. The dust may have little or no effect on tropical cyclone formation, but when the dry air is entrained into the tropical system, development is stopped because of lack of condensation.
   d. The dust is just a consequence of the airflow, and it is the large-scale atmospheric circulation that inhibits cyclone formation.

3. **Monsoon Disturbances and the Tropical Easterly Jet.** Many Atlantic hurricanes begin as disturbances that originate in the south Asian monsoon system and are energized and steered westward across North Africa to the Atlantic Ocean by the mid-level jet that forms each season. When the monsoon is weak:
   a. Fewer disturbances form, and the disturbances are weaker than average.
   b. The jet itself is weaker and less able to energize the disturbances.
   c. The jet is displaced southward, so there is less rotation available for disturbances to begin to rotate. This is rotation of the Earth, and it is well known that few tropical cyclones form south of 10° N. During late summer and autumn of 1997, as the 1997–1998 El Niño developed, the Tropical Easterly Jet was displaced well south of average and few tropical cyclones formed.

## • ENSO AND MONSOON RAINFALL

Sir Gilbert Walker did not discover how the SO exerted control on monsoon rainfall. Today, it is thought that the displaced Walker Circulation holds the key. The position of the Walker Circulation through an El Niño–La Niña cycle is illustrated in Figures 6.4, 6.7, and 6.8.

While moderate to strong La Niña conditions are always associated with abundant monsoon rainfall, moderate to strong El Niño conditions do not guarantee drought. Figure 6.8 shows the center of rising air at the peak of the warm phase shifted into the central Pacific Ocean. In simple terms, rising air in one location must be balanced by sinking air in another, and numerical models of the atmosphere confirm that air sinks west of the center of rising air over India and southern Asia. Because rising air is required for rainfall, this large-scale result of a shifted Walker Circulation explains why during some warm episodes drought results.

Recent research indicates that the key may lie in the fine details of the warm phase, and Indian rainfall may rely both on how warm the water is and how far it extends into the Pacific. In years of failed monsoon rainfall, the water in the central Pacific is warmer than in years of near normal monsoon rainfall.

## • FOUR HISTORICALLY IMPORTANT EL NIÑOS OF THE 20TH CENTURY

### The El Niño of 1957–1958

Every scientific effort has to have a platform from which to begin, and the unified international effort to unravel the mysteries of weather changes in the Pacific basin began during this El Niño episode.

The International Geophysical Year (IGY) was an extensive series of cooperative investigations involving 67 countries of Earth. The IGY started on July 1, 1957 and concluded on December 31, 1958. During IGY, the United States and the Soviet Union launched their first artificial satellites, the Van Allen Radiation Belts were discovered, and the key to plate tectonics and continental drift, the mid ocean ridge system, was discovered.

In addition, a strong El Niño occurred during this period, and because of some very unusual weather in the Pacific basin, interest in finding the cause increased. In Hawaii the hurricane Nina became the first tropical cyclone on record to affect the island chain, causing the highest wind gust ever recorded in Honolulu at 85 mph.

After the collapse of the California sardine industry in the 1950s due to overfishing, the Peruvian government saw a chance for economic development in large-scale processing and exporting of anchovies as livestock feed. The primary food source of sea bird populations is the anchovy, and sea birds died by the thousands along the coast of Peru as the warm El Niño waters covered the nutrient-rich cold water that usually wells up there and threatened the newly developing fish meal industry.

In 1958, near San Diego, CA, a group of oceanographers and meteorologists met to discuss the changes in the Pacific Ocean basin during the 1957–1958 ENSO warm phase. Here, for the first time, was an attempt to synthesize what was known about the links between the meteorology and oceanography of what we now know as El Niño.

### The El Niño of 1971–1972

This is the first ENSO warm phase where the combination of human activity and an ENSO warm phase led to an ecological catastrophe felt across the globe.

The upwelling of cold nutrient-rich water along the west coast of South America creates one of the most productive marine environments on Earth. Anchovies and other small fish feed on tiny floating green plants called phytoplankton, and millions of sea birds feed on the small fish. The large number of sea birds creates another profitable industry, the mining of bird guano to be processed into fertilizer.

Because of the successful large-scale commercialization of the fish meal industry in Peru, the annual anchovy catch rose from 9.2 thousand metric tons (a metric ton is 2204.6 pounds, or 1000 kg) in 1952 to 10.5 million metric tons in 1961, a 1140% increase. Catches at that level were maintained through 1971.

The strong El Niño of 1971–1972 brought an end to the bountiful harvests. The combination of years of fishing beyond the ability of the fish population to replace harvested individuals (called overfishing) and the cessation of upwelling resulted in a collapse of the anchovy industry, the death of millions of sea birds, and the demise of the guano industry.

Because there was no immediate substitute for fish meal in cattle and chicken feed, prices increased by up to 400% in the early 1970s in countries like the United States that were highly dependent on the anchovy industry.

Despite the catastrophic decrease of the anchovy population, the fishing industry tried to harvest at levels they had come to expect, and in 1973 the catch was less than half of what it was before the El Niño.

The industry substituted sardines for a while, and the anchovy population began to recover when the strongest El Niño to that time developed in 1982.

### The El Niño of 1982–1983

The 1982–1983 El Niño was the strongest and most devastating of the century and, until the late 1990s, thought to be the strongest in historical times. It is blamed for almost 2000 deaths and more than $28 billion (2008 dollars) in damage to property and economic losses worldwide.

Weather-related disasters occurred on all continents except Antarctica. In Australia, Africa, and Indonesia, there were droughts, dust storms, and brush fires. The heaviest rainfall in recorded history fell in Peru. Areas where 6 inches was the average annual total received more than 120 inches in 6 weeks. The heavy rain on unconsolidated desert soil created unprecedented mudslides. Floods and landslides were numerous and beach erosion widespread in California because of extreme rainfall and storminess there.

During that period, the trade winds were weaker than ever recorded and at times reversed direction. The cold deep water that welled up along the coast of South America at times was buried under 500 feet of warm water, completely stopping upwelling then as the warm pool from the western Pacific Ocean arrived along the coast of Peru; SSTs rose 10°C (18°F) in a couple of weeks.

This El Niño motivated the development of the TAO/TRITON Array of moored buoys along the equator in the Pacific Ocean, which is the technology that has made predicting and monitoring ENSO changes possible on a real-time basis.

### The El Niño of 1997–1998

While the 1982–1983 El Niño was unpredicted and almost unnoticed until it peaked, 15 years later, the greatest El Niño to the date of this writing could be followed on the Internet and watched on news programs around the world as it happened.

Because the El Niño of 1982–1983 was a devastating surprise, the TAO/TRITON Array of oceanographic and meteorological buoys had been deployed and was completely operational before the spring of 1997. By July 1997 it was clear that it would be of extraordinary strength, and forecasts of possible catastrophic weather were being issued by August. Television and print journalists were soon informing millions of people as they were following the development of the 1997–1998 El Niño.

Planning and preparing for weather events can be very costly, but when done well, savings can be great. The 1982–1983 El Niño cost California $2.2 billion in storm losses, but storm losses during the stronger 1997–1998 El Niño were $1.1 billion. Much of the difference is attributed to better preparedness due to the TAO/TRITON Array.

In India the monsoon rains were slightly above the 1870–2000 average. The significance of the strongest El Niño on record, being one of four El Niños since 1870 not accompanied by drought in India (Figure 6.2), is that scientists began looking closely at the SST patterns. Preliminary results indicate that monsoon rains are sensitive to where the warm pool is centered and how much the central Pacific Ocean warms during an El Niño and how that affects the position of the Walker Circulation. Researchers are now also looking in more detail at how the monsoon rains are distributed in individual regions, rather than just the "all India" rainfall used in Figure 6.1.

### • EL NIÑO–LA NIÑA OCCURRENCES

Table 6.4 lists all El Niño and La Niña occurrences since 1950 and Table 6.5 lists the strongest by season based on the ONI. An El Niño is in progress if the 3-month running mean SST anomalies in the Niño 3.4 region are warmer than normal by at least 0.5°C for 5 months in a row or longer. A La Niña is in progress if the temperature is at least 0.5°C colder than normal for the same length of time. Extreme ENSO events are listed in Tables 6.6a,b.

### • OTHER OSCILLATIONS AND TELECONNECTION PATTERNS

Table 6.7 lists four characteristic time scales at and typical weather occurrences that operate within those time frames. We see daily changes in the atmosphere as the familiar weather systems like fronts, highs, and lows that traverse the weather map. There are many other systems that operate in the atmosphere across a wide range of time and geographic scales and, in some cases, over enormous geographic distances.

Teleconnection patterns are natural long-distance relationships that occur frequently in our very complex atmosphere–ocean system. They occur primarily because of interactions between different parts of the atmosphere–ocean system, and they influence temperature and rainfall patterns because of large-scale changes in the location and strength of the atmosphere's jet streams around the globe. Teleconnections are the reason abnormal weather patterns occur simultaneously over vast distances.

Each teleconnection pattern consists of centers where atmospheric pressure is higher or lower than normal and corresponding centers with anomalies of opposite sign. Just like ENSO, the pressure differences oscillate, causing corresponding adjustments in other parts of the system.

Figures 6.11a,b and 6.12a,b illustrate the worldwide teleconnection patterns associated with El Niño and La Niña. There are many others in the atmosphere.

### A Few Examples of Teleconnections

The winter of 1995–1996 was very cold and snowy over much of eastern North America, while northern Europe and Scandinavia were cold and southern Europe/northern Africa experienced very wet and stormy conditions. These conditions were all partly related to the teleconnection pattern called the North Atlantic Oscillation.

The winter of 1982–1983 was stormy and destructive in California as storm after storm pounded the west coast. This was a direct result of the very strong El Niño that winter and the global teleconnections associated with it. At the same time, the coastal deserts of Ecuador and Peru received 100 inches or more of rainfall, drought in Australia and Indonesia caused massive wildfires, and the monsoon rains failed in India.

Occasionally during winter, a surge of moisture makes it to the west coast of the United States and flooding rains occur because of what is known as the "Pineapple Express." The moisture originates over the warm water pool of tropical western Pacific Ocean and is carried by a strong subtropical jet stream past Hawaii and to North America. The cause of the Pineapple Express is related to a teleconnection pattern called the Madden–Julian oscillation.

**Table 6.4  El Niño and La Niña Occurrences for 3-Month Periods Based on the ONI**

| Year | DJF | JFM | FMA | MAM | AMJ | MJJ | JJA | JAS | ASO | SON | OND | NDJ |
|------|-----|-----|-----|-----|-----|-----|-----|-----|-----|-----|-----|-----|
| 1950 | -1.7 | -1.5 | -1.3 | -1.4 | -1.3 | -1.1 | -0.8 | -0.8 | -0.8 | -0.9 | -0.9 | -1.0 |
| 1951 | -1.0 | -0.9 | -0.6 | -0.3 | -0.2 | 0.2 | 0.4 | 0.7 | 0.7 | 0.8 | 0.7 | 0.6 |
| 1952 | 0.3 | 0.1 | 0.1 | 0.2 | 0.1 | -0.1 | -0.3 | -0.3 | -0.2 | -0.2 | -0.1 | 0.0 |
| 1953 | 0.2 | 0.4 | 0.5 | 0.5 | 0.5 | 0.5 | 0.4 | 0.4 | 0.4 | 0.4 | 0.4 | 0.4 |
| 1954 | 0.5 | 0.3 | -0.1 | -0.5 | -0.7 | -0.7 | -0.8 | -1.0 | -1.2 | -1.1 | -1.1 | -1.1 |
| 1955 | -1.0 | -0.9 | -0.9 | -1.0 | -1.0 | -1.0 | -1.0 | -1.0 | -1.4 | -1.8 | -2.0 | -1.9 |
| 1956 | -1.3 | -0.9 | -0.7 | -0.6 | -0.6 | -0.6 | -0.7 | -0.8 | -0.8 | -0.9 | -0.9 | -0.8 |
| 1957 | -0.5 | -0.1 | 0.3 | 0.6 | 0.7 | 0.9 | 0.9 | 0.9 | 0.9 | 1.0 | 1.2 | 1.5 |
| 1958 | 1.7 | 1.5 | 1.2 | 0.8 | 0.6 | 0.5 | 0.3 | 0.1 | 0.0 | 0.0 | 0.2 | 0.4 |
| 1959 | 0.4 | 0.5 | 0.4 | 0.2 | 0.0 | -0.2 | -0.4 | -0.5 | -0.4 | -0.3 | -0.2 | -0.2 |
| 1960 | -0.3 | -0.3 | -0.3 | -0.2 | -0.2 | -0.2 | -0.1 | 0.0 | -0.1 | -0.2 | -0.2 | -0.2 |
| 1961 | -0.2 | -0.2 | -0.2 | -0.1 | 0.1 | 0.2 | 0.0 | -0.3 | -0.6 | -0.6 | -0.5 | -0.4 |
| 1962 | -0.4 | -0.4 | -0.4 | -0.5 | -0.4 | -0.4 | -0.3 | -0.3 | -0.5 | -0.6 | -0.7 | -0.7 |
| 1963 | -0.6 | -0.3 | 0.0 | 0.1 | 0.1 | 0.3 | 0.6 | 0.8 | 0.9 | 0.9 | 1.0 | 1.0 |
| 1964 | 0.8 | 0.4 | -0.1 | -0.5 | -0.8 | -0.8 | -0.9 | -1.0 | -1.1 | -1.2 | -1.2 | -1.0 |
| 1965 | -0.8 | -0.4 | -0.2 | 0.0 | 0.3 | 0.6 | 1.0 | 1.2 | 1.4 | 1.5 | 1.6 | 1.5 |
| 1966 | 1.2 | 1.0 | 0.8 | 0.5 | 0.2 | 0.2 | 0.2 | 0.0 | -0.2 | -0.2 | -0.3 | -0.3 |
| 1967 | -0.4 | -0.4 | -0.6 | -0.5 | -0.3 | 0.0 | 0.0 | -0.2 | -0.4 | -0.5 | -0.4 | -0.5 |
| 1968 | -0.7 | -0.9 | -0.8 | -0.7 | -0.3 | 0.0 | 0.3 | 0.4 | 0.3 | 0.4 | 0.7 | 0.9 |
| 1969 | 1.0 | 1.0 | 0.9 | 0.7 | 0.6 | 0.5 | 0.4 | 0.4 | 0.6 | 0.7 | 0.8 | 0.7 |
| 1970 | 0.5 | 0.3 | 0.2 | 0.1 | 0.0 | -0.3 | -0.6 | -0.8 | -0.9 | -0.8 | -0.9 | -1.1 |
| 1971 | -1.3 | -1.3 | -1.1 | -0.9 | -0.8 | -0.8 | -0.8 | -0.8 | -0.8 | -0.9 | -1.0 | -0.9 |
| 1972 | -0.7 | -0.4 | 0.0 | 0.2 | 0.5 | 0.8 | 1.0 | 1.3 | 1.5 | 1.8 | 2.0 | 2.1 |
| 1973 | 1.8 | 1.2 | 0.5 | -0.1 | -0.6 | -0.9 | -1.1 | -1.3 | -1.4 | -1.7 | -2.0 | -2.1 |
| 1974 | -1.9 | -1.7 | -1.3 | -1.1 | -0.9 | -0.8 | -0.6 | -0.5 | -0.5 | -0.7 | -0.9 | -0.7 |
| 1975 | -0.6 | -0.6 | -0.7 | -0.8 | -0.9 | -1.1 | -1.2 | -1.3 | -1.5 | -1.6 | -1.7 | -1.7 |
| 1976 | -1.6 | -1.2 | -0.8 | -0.6 | -0.5 | -0.2 | 0.1 | 0.3 | 0.5 | 0.7 | 0.8 | 0.7 |
| 1977 | 0.6 | 0.5 | 0.2 | 0.2 | 0.2 | 0.4 | 0.4 | 0.4 | 0.5 | 0.6 | 0.7 | 0.7 |
| 1978 | 0.7 | 0.4 | 0.0 | -0.3 | -0.4 | -0.4 | -0.4 | -0.4 | -0.4 | -0.3 | -0.2 | -0.1 |
| 1979 | -0.1 | 0.0 | 0.1 | 0.1 | 0.1 | -0.1 | 0.0 | 0.1 | 0.3 | 0.4 | 0.5 | 0.5 |
| 1980 | 0.5 | 0.3 | 0.2 | 0.2 | 0.3 | 0.3 | 0.2 | 0.0 | -0.1 | -0.1 | 0.0 | -0.1 |
| 1981 | -0.3 | -0.5 | -0.5 | -0.4 | -0.3 | -0.3 | -0.4 | -0.4 | -0.3 | -0.2 | -0.1 | -0.1 |
| 1982 | 0.0 | 0.1 | 0.1 | 0.3 | 0.6 | 0.7 | 0.7 | 1.0 | 1.5 | 1.9 | 2.2 | 2.3 |
| 1983 | 2.3 | 2.0 | 1.5 | 1.2 | 1.0 | 0.6 | 0.2 | -0.2 | -0.6 | -0.8 | -0.9 | -0.7 |
| 1984 | -0.4 | -0.2 | -0.2 | -0.3 | -0.5 | -0.4 | -0.3 | -0.2 | -0.3 | -0.6 | -0.9 | -1.1 |
| 1985 | -0.9 | -0.8 | -0.7 | -0.7 | -0.7 | -0.6 | -0.5 | -0.5 | -0.5 | -0.4 | -0.3 | -0.4 |
| 1986 | -0.5 | -0.4 | -0.2 | -0.2 | -0.1 | 0.0 | 0.3 | 0.5 | 0.7 | 0.9 | 1.1 | 1.2 |
| 1987 | 1.2 | 1.3 | 1.2 | 1.1 | 1.0 | 1.2 | 1.4 | 1.6 | 1.6 | 1.5 | 1.3 | 1.1 |
| 1988 | 0.7 | 0.5 | 0.1 | -0.2 | -0.7 | -1.2 | -1.3 | -1.2 | -1.3 | -1.6 | -1.9 | -1.9 |
| 1989 | -1.7 | -1.5 | -1.1 | -0.8 | -0.6 | -0.4 | -0.3 | -0.3 | -0.3 | -0.3 | -0.2 | -0.1 |
| 1990 | 0.1 | 0.2 | 0.2 | 0.2 | 0.2 | 0.2 | 0.3 | 0.3 | 0.3 | 0.3 | 0.3 | 0.4 |
| 1991 | 0.4 | 0.3 | 0.3 | 0.4 | 0.6 | 0.8 | 1.0 | 0.9 | 0.9 | 1.0 | 1.4 | 1.6 |
| 1992 | 1.8 | 1.6 | 1.5 | 1.4 | 1.2 | 0.8 | 0.5 | 0.2 | 0.0 | -0.1 | 0.0 | 0.2 |
| 1993 | 0.3 | 0.4 | 0.6 | 0.7 | 0.8 | 0.7 | 0.4 | 0.4 | 0.4 | 0.4 | 0.3 | 0.2 |
| 1994 | 0.2 | 0.2 | 0.3 | 0.4 | 0.5 | 0.5 | 0.6 | 0.6 | 0.7 | 0.9 | 1.2 | 1.3 |
| 1995 | 1.2 | 0.9 | 0.7 | 0.4 | 0.3 | 0.2 | 0.0 | -0.2 | -0.5 | -0.6 | -0.7 | -0.7 |
| 1996 | -0.7 | -0.7 | -0.5 | -0.3 | -0.1 | -0.1 | 0.0 | -0.1 | -0.1 | -0.2 | -0.3 | -0.4 |
| 1997 | -0.4 | -0.3 | 0.0 | 0.4 | 0.8 | 1.3 | 1.7 | 2.0 | 2.2 | 2.4 | 2.5 | 2.5 |

**Table 6.4    El Niño and La Niña Occurrences for 3-Month Periods Based on the ONI**
[CONTINUED]

| Year | DJF | JFM | FMA | MAM | AMJ | MJJ | JJA | JAS | ASO | SON | OND | NDJ |
|------|-----|-----|-----|-----|-----|-----|-----|-----|-----|-----|-----|-----|
| 1998 | 2.3 | 1.9 | 1.5 | 1.0 | 0.5 | 0.0 | -0.5 | -0.8 | -1.0 | -1.1 | -1.3 | -1.4 |
| 1999 | -1.4 | -1.2 | -0.9 | -0.8 | -0.8 | -0.8 | -0.9 | -0.9 | -1.0 | -1.1 | -1.3 | -1.6 |
| 2000 | -1.6 | -1.4 | -1.0 | -0.8 | -0.6 | -0.5 | -0.4 | -0.4 | -0.4 | -0.5 | -0.6 | -0.7 |
| 2001 | -0.6 | -0.5 | -0.4 | -0.2 | -0.1 | 0.1 | 0.2 | 0.2 | 0.1 | 0.0 | -0.1 | -0.1 |
| 2002 | -0.1 | 0.1 | 0.2 | 0.4 | 0.7 | 0.8 | 0.9 | 1.0 | 1.1 | 1.3 | 1.5 | 1.4 |
| 2003 | 1.2 | 0.9 | 0.5 | 0.1 | -0.1 | 0.1 | 0.4 | 0.5 | 0.6 | 0.5 | 0.6 | 0.4 |
| 2004 | 0.4 | 0.3 | 0.2 | 0.2 | 0.3 | 0.5 | 0.7 | 0.8 | 0.9 | 0.8 | 0.8 | 0.8 |
| 2005 | 0.7 | 0.5 | 0.4 | 0.4 | 0.4 | 0.4 | 0.4 | 0.3 | 0.2 | -0.1 | -0.4 | -0.7 |
| 2006 | -0.7 | -0.6 | -0.4 | -0.1 | 0.1 | 0.2 | 0.3 | 0.5 | 0.6 | 0.9 | 1.1 | 1.1 |
| 2007 | 0.8 | 0.4 | 0.1 | -0.1 | -0.1 | -0.1 | -0.1 | -0.4 | -0.7 | -1.0 | -1.1 | -1.3 |
| 2008 | -1.4 | -1.4 | -1.1 | -0.8 | -0.6 | -0.4 | -0.1 | 0.0 | 0.0 | 0.0 | -0.3 | -0.6 |
| 2009 | -0.8 | -0.7 | -0.5 | -0.1 | 0.2 | 0.6 | 0.7 | 0.8 | 0.9 | | | |

*Source:* NOAA, NCEP, CPC.

**Table 6.5    El Niño and La Niña Episodes from Table 6.4 Classified by Strength of the SST Anomalies (ONI) in the Niño 3.4 Region as Weak ($\pm$ 0.5–0.9), Moderate ($\pm$ 1.0–1.4), and Strong ($\geq \pm$ 1.5)**

| El Niño | | | La Niña | | |
|---------|-----|--------|---------|-----|--------|
| Weak | Mod | Strong | Weak | Mod | Strong |
| 1951 | 1986 | 1957 | 1950 | 1954 | 1955 |
| 1963 | 1987 | 1965 | 1956 | 1964 | 1973 |
| 1968 | 1994 | 1972 | 1962 | 1970 | 1975 |
| 1969 | 2002 | 1982 | 1967 | 1998 | 1988 |
| 1976 | | 1991 | 1971 | 1999 | |
| 1977 | | 1997 | 1974 | 2007 | |
| 2004 | | | 1984 | | |
| 2006 | | | 1995 | | |
| | | | 2000 | | |

*Source:* NOAA, NCEP, CPC.

**Table 6.6 (a)   Ranking of the Strength of El Niño Episodes, 1896–2009, by Strength and Season**

| El Niño Extreme Years Based on the Southern Oscillation Index | | | | |
|---|---|---|---|---|
| **Rank** | **DJF** | **MAM** | **JJA** | **SON** |
| 1 | 1983 | 1905 | 1905 | 1982 |
| 2 | 1998 | 1987 | 1982 | 1997 |
| 3 | 1992 | 1998 | 1987 | 1940 |
| 4 | 1941 | 1912 | 1940 | 1901 |
| 5 | 1978 | 1994 | 1941 | 1941 |
| 6 | 1919 | 1983 | 1994 | 1994 |
| 7 | 1987 | 1992 | 1965 | 1914 |
| 8 | 1897 | 1900 | 1896 | 1991 |
| 9 | 1905 | 1897 | 1972 | 1965 |
| 10 | 1912 | 1993 | 1977 | 1977 |
| 11 | 1958 | 1991 | 1914 | 1900 |
| 12 | 1942 | 1940 | 1993 | 1951 |
| 13 | 1926 | 1941 | 1911 | 1939 |
| 14 | 1959 | 1953 | 1923 | 1963 |
| 15 | 1973 | 1977 | 1925 | 1923 |
| 16 | 1906 | 1972 | 2002 | 1896 |
| 17 | 1915 | 1966 | 1946 | 1972 |
| 18 | 1993 | 1926 | 1951 | 1925 |
| 19 | 1952 | 1980 | 1976 | 1913 |
| 20 | 1990 | 1995 | 1919 | 1905 |

**Table 6.6 (b)   Ranking of the Strength of La Niña Episodes, 1896–2009, by Strength and Season**

| La Niña Extreme Years Based on the Southern Oscillation Index | | | | |
|---|---|---|---|---|
| **Rank** | **DJF** | **MAM** | **JJA** | **SON** |
| 1 | 1974 | 1917 | 1917 | 1917 |
| 2 | 2008 | 1904 | 1975 | 1975 |
| 3 | 1918 | 1971 | 1950 | 1988 |
| 4 | 1929 | 1950 | 1916 | 1973 |
| 5 | 1904 | 1989 | 1910 | 1906 |
| 6 | 1976 | 1974 | 1938 | 2008 |
| 7 | 1951 | 1956 | 1955 | 1955 |
| 8 | 1939 | 1903 | 1909 | 2000 |
| 9 | 1989 | 1975 | 1956 | 1910 |
| 10 | 1999 | 2000 | 1900 | 1971 |
| 11 | 1971 | 1925 | 1931 | 1950 |
| 12 | 2000 | 1999 | 1973 | 1970 |
| 13 | 1956 | 1918 | 1988 | 1964 |
| 14 | 1917 | 1927 | 1981 | 1956 |
| 15 | 1943 | 1898 | 1964 | 1938 |
| 16 | 1962 | 1921 | 1924 | 1924 |
| 17 | 1930 | 1931 | 1974 | 1908 |
| 18 | 1950 | 1902 | 1901 | 1999 |
| 19 | 1921 | 1910 | 1906 | 1916 |
| 20 | 1925 | 1928 | 1945 | 1928 |

*Source:* NOAA, Environmental Sciences Research Laboratory, Physical Sciences Division.
Updated by Steve Horstmeyer.

**Table 6.7    Some Common Weather Phenomena and the Time Scales at Which They Operate**

| Characteristic Time Scales and Weather Phenomena | |
|---|---|
| **A Few Days** | Normal storm systems and frontal passages |
| **A Few Weeks** | A mid-winter warm-up or a mid-summer wet period |
| **A Few Months** | Particularly cold winters or hot summers |
| **Several Years** | Abnormal winters for several years in a row |

*Source:* NOAA, NCEP, CPC.

**Table 6.8    Five Common Teleconnection Patterns, the Time of Year They Operate, and Some Effects of the Connection**

| Teleconnection | Centers | Time of year | Region and Effects |
|---|---|---|---|
| **North Atlantic Oscillation (NAO)** | Greenland vs. Central N. Atlantic | All Year | Eastern N. America into W. Europe. Temperatures through position, amplitude and strength of polar jet stream. |
| **Arctic Oscillation (AO)** | Arctic vs. Subtropical Atlantic | All Year, Strongest Northern Winter | West European Temperature and Precipitation through strength of the polar jet stream. |
| **Antarctic Oscillation (AAO)** | Antarctic vs. Subtropics | All Year, Strongest Southern Winter | Precipitation, winds and storm tracks south of 40°S latitude. |
| **Pacific North American Pattern (PNA)** | Aleutian Islands and Southeast U.S. vs. Hawaii and Western U.S. | All Year | North American temperature and precipitation through position of East Asian jet stream and adjustments to jet farther east across the Pacific. |
| **Madden-Julian Oscillation (MJO)** | Eastward moving wave in the tropics | All Year | 30 – 60 day periods of rainy then dry weather around the tropics |

*Source:* NOAA, NCEP, CPC.

# 7

# Global Warming and Climate Change

## • INTRODUCTION

Global climate change is a topic often in the headlines. Whether on talk radio, the print media, or the subject of a number of blogs, there is a great deal of misinformation and much misunderstanding.

This chapter is designed to be a reference for the reader to fill in the blanks. In it you will find explanations of processes commonly discussed in the popular media that are poorly understood.

I have purposely avoided discussions of specific forecasts of future climate. The focus of this chapter is how climate changes, how we know, and the terminology used to describe changes in the climate system. Consider this chapter as a basic guide to climate science.

## • INTRODUCTION TO CLIMATE CHANGE

Climate is often described as the long-term state of the atmosphere. It is what you expect annually and over long periods of time. Climate can be thought of as the average of day-to-day weather over several years or longer. But climate is far too complex for a single, simple definition.

Components of the climate system can be studied alone for a specific purpose. We may speak of the "rainfall climate" of the Great Plains of the United States when studying agriculture or the "cloud cover climate" of another location when evaluating that location's suitability for capturing solar energy.

Climate science involves much more than just the study of the atmosphere. Anything that can take part in maintaining or changing the state of Earth's climate system falls under the realm of climate science. Because this encompasses the atmosphere, the oceans (hydrosphere), Earth's ice (chryosphere), the solid Earth (lithosphere), Earth's life (biosphere) and numerous complex interactions between all

the "spheres," the essence of climate science is interdisciplinary.

Each of the "spheres" plays an important role in the climate of Earth. The ocean stores much heat and supplies moisture, mountains block the movement of the moisture and disturb the jet stream, ice reflects sunshine making the air above it cool and the atmosphere moves all the moisture and energy around the Earth.

Multiple interactions between these components make solving the climate change puzzle difficult because those interactions can accelerate or slow a change of climate.

For example a complete description of the fate of the greenhouse gas carbon dioxide cannot be written without understanding, at a minimum, numerous interactions between the following:

a. Chemical reactions in the oceans.
b. The biology of floating green oceanic plants called phytoplankton.
c. Mountain building, which increases erosion, which in turn increases the amount of carbon in minerals buried as sediment on the sea floor.
d. Growth and decay of vast continental ice sheets.
e. The exchange of $CO_2$ between the atmosphere and oceans.
f. Biological processes in soil
g. Human activities.

Changing climate is the norm and that climates are in a constant state of change on our planet is not disputed by scientists. Earth's climates today are very different than in the distant past and they will be very different again in the future.

When climate changes climate scientists say the change has been "forced" by a mechanism. Table 7.1 summarizes a number of climate forcing mechanisms. Changes in climate are complex and result from the interaction of many forcing mechanisms that operate over a wide range of time scales.

*The Weather Almanac: A Reference Guide to Weather, Climate, and Related Issues in the United States and Its Key Cities*, Twelfth Edition. Steve Horstmeyer.
© 2011 John Wiley & Sons, Inc. Published 2011 by John Wiley & Sons, Inc.

**Table 7.1  Climate Forcing Mechanisms**

| "Sphere" of the Forcing Mechanism | Forcing Mechanism | Time Scale (Years) | Notes |
|---|---|---|---|
| **Astronomical** | **Solar evolution**<br>**Solar variation**<br>**Galactic Dust**<br>**Milankovitch Cycles**<br>-eccentricity<br>-obliquity<br>-precession | Billions<br>10 – 200<br>200-500 million<br><br>413k, 125k, 96k<br>41k<br>23k, 19k | All forcings are external radiative. Galactic dust has not been widely accepted as a forcing mechanism |
| **Lithosphere** | **Plate tectonics**<br>-continental drift<br>-mountain building<br>-ocean circulation<br>-atmospheric circulation<br>-chemical weathering<br><br>**Isostatic Sea Level Changes**<br><br>**Volcanism** | 10 million to 500 million<br><br><br><br><br><br>10k to 100 million<br><br>1 year - millions | Can be considered external to the climate system.<br><br>Chemical weathering removes carbon during formation of limestone and related rocks. |
| **Hydrosphere** | **Sea Level Change**<br>From land ice changes<br>**Circulation**<br>-deep water<br>-shallow water<br>**$CO_2$ content** | 10k – 100k<br><br>1000<br>100 – 1000<br>100 years | Deep water circulation is the thermohaline circulation and is often referred to as the Global Conveyor Belt. |
| **Cryosphere** | **Ice sheets**<br>growth and decay<br><br>**Black carbon deposition** | 100 – 10k<br><br>1 - 100 | Black carbon is also called soot and increases the absorption of solar radiation which increases melting. |
| **Atmosphere** | **Greenhouse gases**<br>-long term levels<br>-glacial vs. interglacial<br>**Circulation** | 1 million – 10 million<br>1000 – 10k<br>months – 10 | Long term refers to the evolution of Earth. Glacial vs. interglacial refers to natural greenhouse gas changes. |
| **Biosphere** | **Forest Growth**<br>**Swamps and Peat Bogs** | 100 – 1000 | Swamps and bogs produce methane. |
| **Human Forcings** | **Fossil Fuels**<br>**Land Use** | 10 – 100<br>10 – 100 | Deforestation reduces carbon uptake by photosynthetic plants |

While some climate forcings can take hundreds of millions of years, other variations take place in a few.

At the long end of the time scale moving continents can radically change the climates of Earth. Change is so slow that ecosystems evolve adaptations. At the other end of the time scale several cycles of regional drought can come and go during a human lifetime and cause widespread episodes of famine. Adaptation to fast climate changes is nearly impossible and often the only successful survival technique is migration.

## • THE GREENHOUSE EFFECT

The greenhouse effect involves radiant energy at thermal infrared wavelengths emitted by Earth that is trapped within the atmosphere by greenhouse gases. The energy is then re-radiated in all directions and some escapes to space, some is reabsorbed within the atmosphere and some returns to the surface. The more energy retained by the atmosphere, the warmer the climate will be.

It is a basic fact that an atmosphere with an increasing concentration of greenhouse gases will trap more of Earth's re-radiated infrared energy than an atmosphere with a lower concentration of greenhouse gases. The problem of global warming that Earth faces today is a matter of an enhanced greenhouse effect.

Without the greenhouse effect life as we know it on Earth would be impossible and Earth would be a bitterly cold, inhospitable planet. Because of the radiant energy retained due to the greenhouse effect Earth's average surface temperature is 14°C (57°F). Without the greenhouse effect Earth's average temperature would be −18°C (0.4°F).

Cloud cover is also important in trapping Earth's radiation but because it is a nongaseous component of the climate system it is treated separately from the effect of greenhouse gases.

On a cloudless day short wavelength solar energy penetrates to the surface with almost no absorption. On average 30% of solar radiation is reflected directly back to space and takes no part in warming Earth. Reflected solar radiation is not changed and when reflected it propagates in a different direction and often passes back out to space without having a role in the climate system. The remaining 70% of the sun's energy is absorbed and drives Earth's climate system.

Solar radiation that is absorbed is changed when re-radiated. Because Earth's temperature is cooler than the sun's, the Earth radiates at longer wavelengths, primarily in what is called the infrared band. Infrared radiation is intercepted by greenhouse gases and for Earth to maintain a nearly constant average temperature the incoming solar energy that is absorbed must be balanced by Earth radiation escaping to space.

The term "greenhouse effect" is an inaccurate comparison of the mechanisms by which Earth and actual greenhouse stay warm. Both a greenhouse and Earth are warmed as the sun's rays pass through the envelope of either glass or atmospheric gases and absorbed by everything inside.

Because the sun is hotter than Earth, solar output peaks at shorter wavelengths than the radiation given off by everything inside the greenhouse. Glass is transparent to short wavelength solar radiation but opaque to almost all the longer wavelength thermal infrared energy radiated by everything within the greenhouse.

The re-radiated energy is trapped by the glass and heat accumulates. The warmed air within the greenhouse is isolated by the glass from surrounding air so the temperature within the greenhouse rises. The glass acts as a passive external boundary to movement of air and redistribution of the radiant energy.

In the atmosphere greenhouse gases travel vast distances, are moved vertically and the trapped radiant energy is distributed by global wind belts. In addition infrared radiation is absorbed and emitted multiple times within the atmosphere and between the atmosphere and other components of the climate system. Unlike the glass of a greenhouse the atmosphere is an active agent in the process.

Since the great continental ice sheets began to melt between 12,000 and 15,000 years ago, Earth has been warming, indicating that Earth's energy budget has been out of balance from that time to the present. Earth presently retains more energy than it loses to space and Earth's average temperature continues to rise.

Since the industrial revolution the concentrations of greenhouse gases have been increasing (Figures 7.1–7.4) and the warming of Earth has accelerated (Figures 7.5 and 7.6). Carbon dioxide, methane and nitrous oxide each have anthropogenic and natural sources. Separating the natural and human emissions of these gases is not clear cut and a source of uncertainty in emissions and climate predictions The only source of the greenhouse gases in Figure 7.4 is modern industry through chemical engineering.

Figure 7.1 shows the increase of carbon dioxide. The human contribution is primarily the result of fossil fuel combustion. Figure 7.2 indicates the same for methane. Anthropogenic sources of methane include rice paddies, livestock and the gas and coal industries. Increases of atmospheric nitrous oxide (Figure 7.3) that can be attributed to humans come from agricultural soil management, management of animal manure, waste water management, fossil fuel combustion and production of industrial acids.

The increase of 5 man-made greenhouse gases shown in Figure 7.4. The chlorofluorocarbons (CFCs) are potent greenhouse gases and are ozone depleting chemicals. CFCs were used in fire extinguishers and as refrigerants and propellants and have been replaced by hydrochlorofluorocarbons (HCFCs). HCFCs are only an intermediate replacement for CFCs because they also deplete stratospheric ozone but

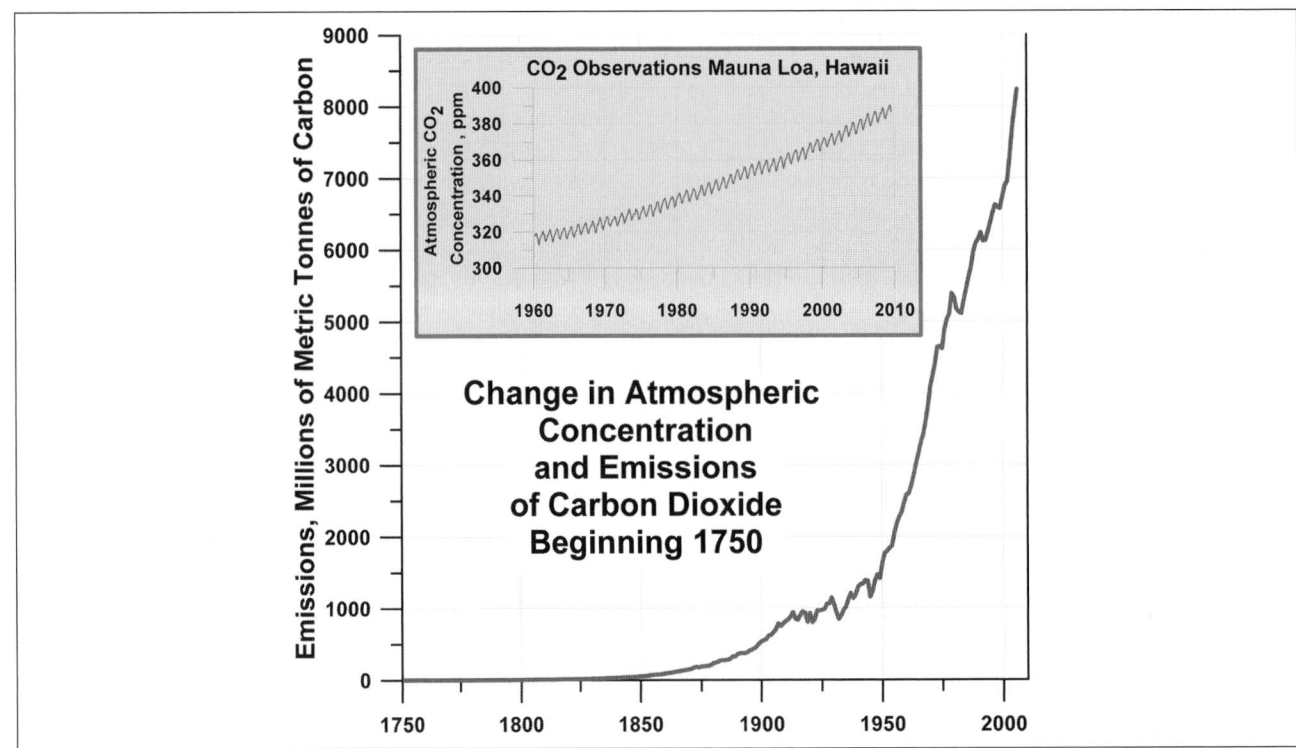

**Figure 7.1** Increase of atmospheric $CO_2$ since 1750 and the "Keeling Curve" showing seasonal variations and long-term trend of $CO_2$ at Mauna Loa Observatory, Hawaii. (*Source:* US EPA and NOAA (inset).)

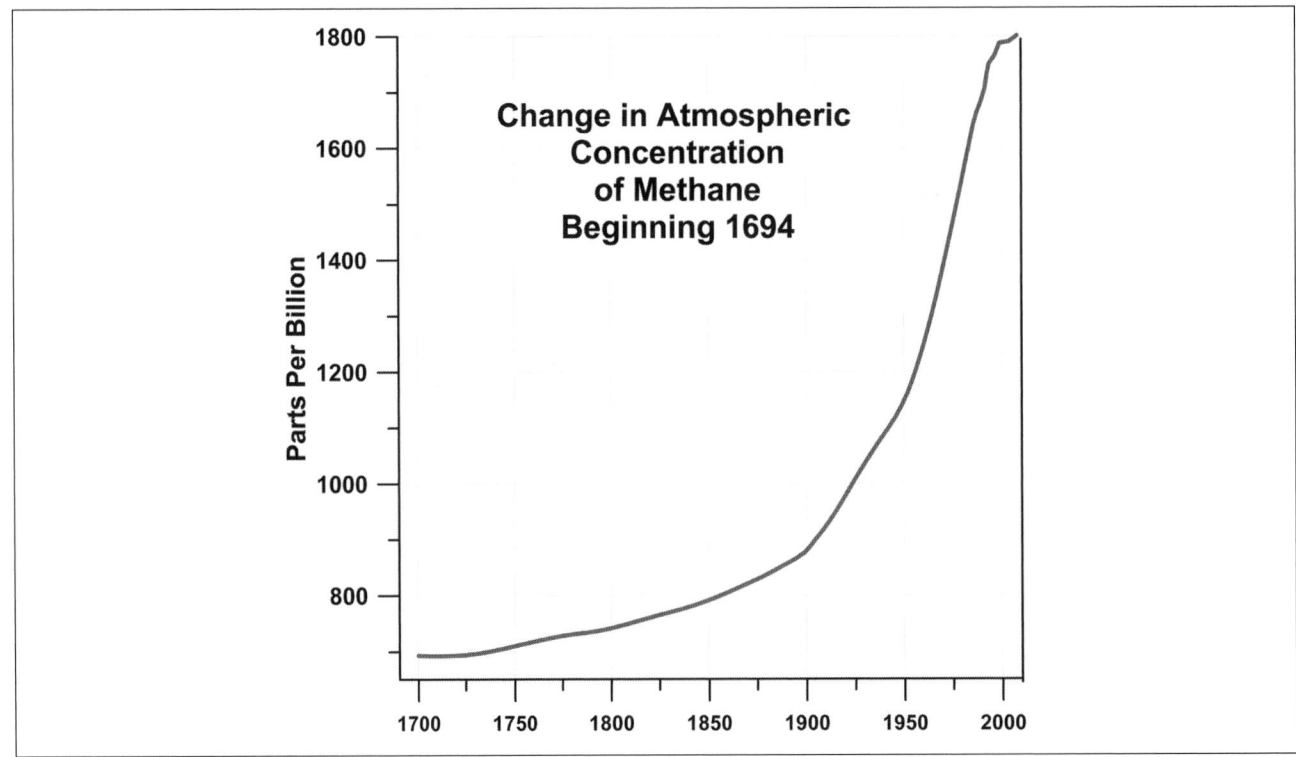

**Figure 7.2** Increase of atmospheric $CH_4$ (methane) since 1694. (*Source:* US EPA.)

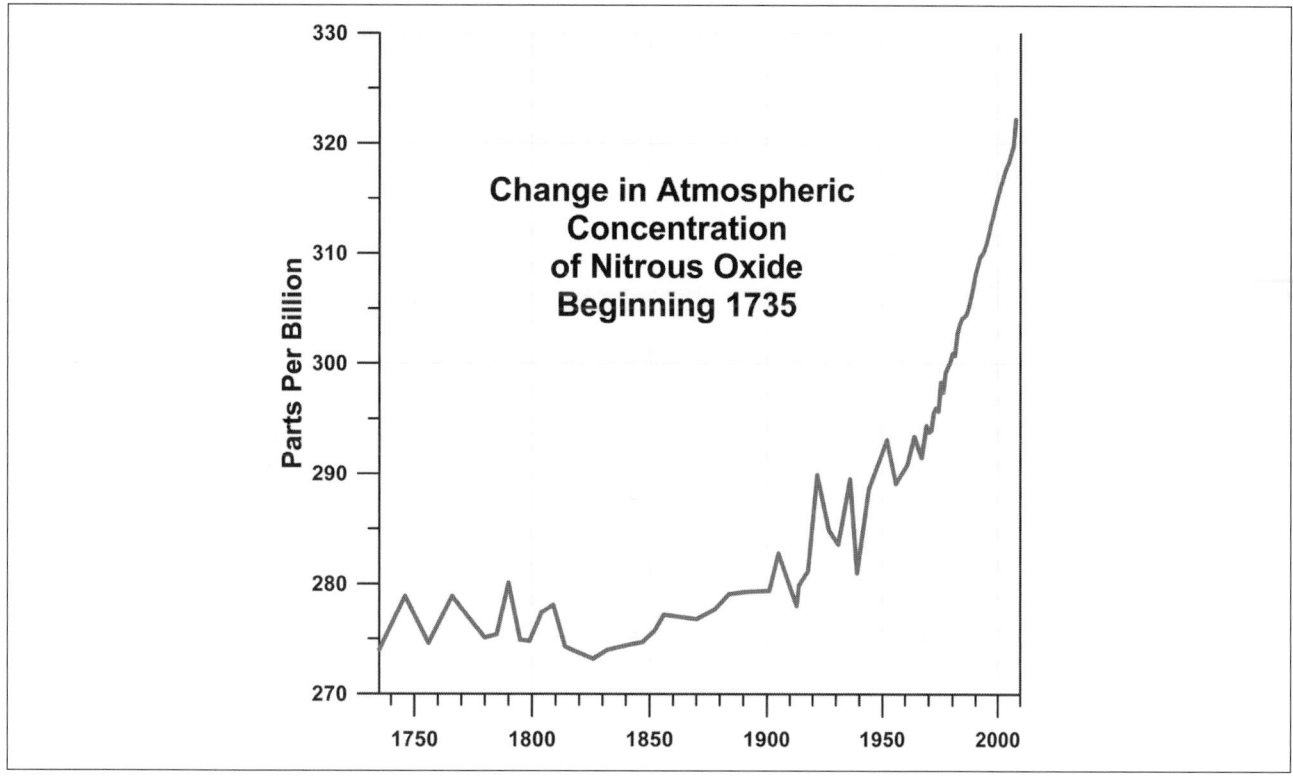

**Figure 7.3**    Increase of atmospheric N₂O (nitrous oxide) since 1735. (*Source:* US EPA.)

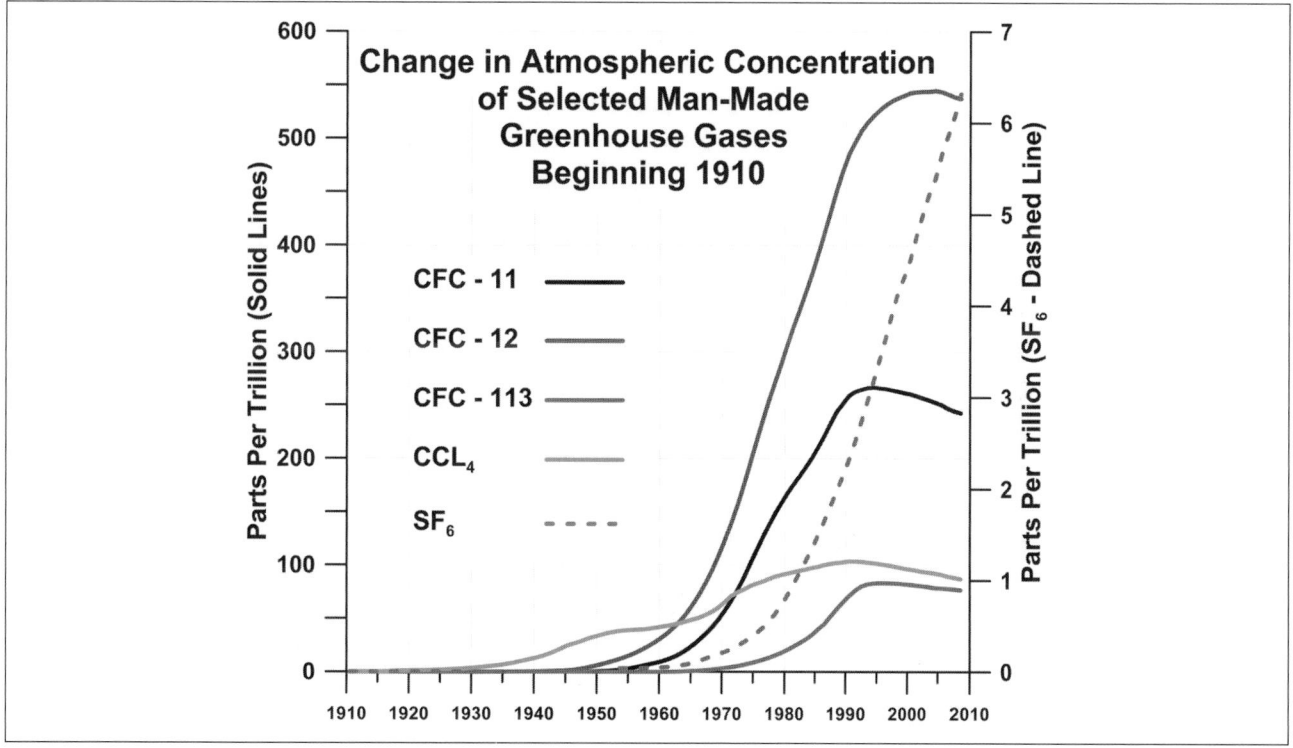

**Figure 7.4**    Change of 5 anthropogenic (man-made) gases since 1910. (*Source:* US EPA.)

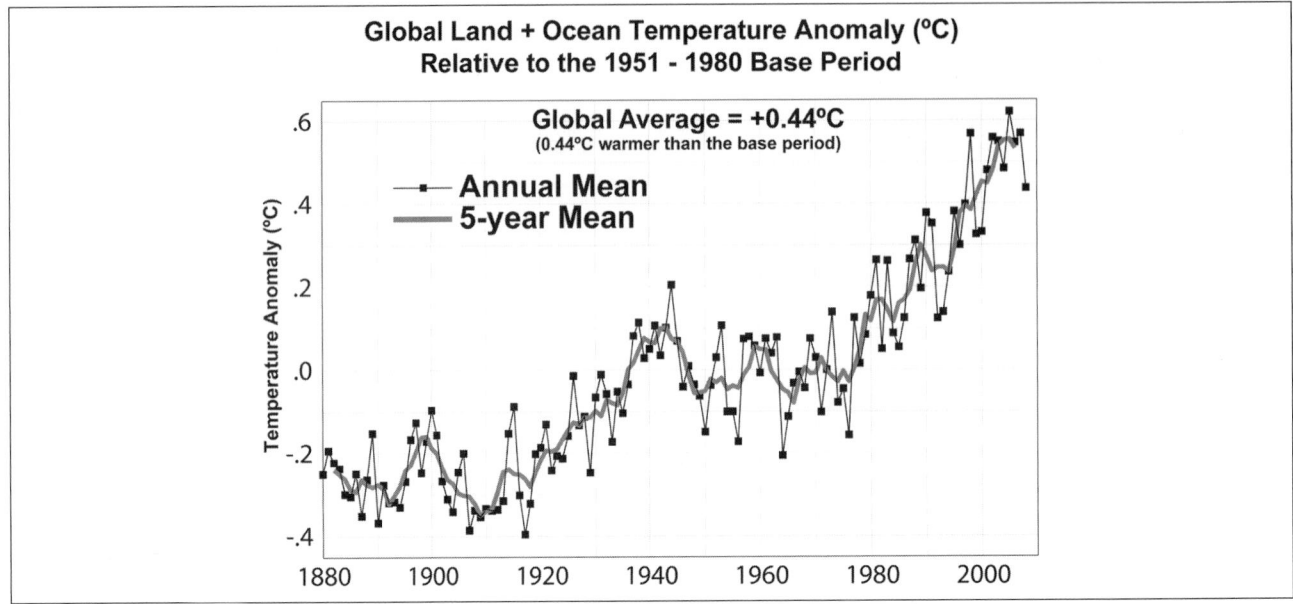

**Figure 7.5**   Global temperature trend using land and ocean data starting in 1880. (*Source:* NOAA.)

to a lesser extent than CFCs because many molecules break down before rising into the stratosphere.

Hydrofluorocarbons (HFCs) will replace all HCFCs by international agreement by 2020 except for developing nations which have until 2030. HFCs have an even shorter lifetime than HCFCs and largely do not make it to the stratosphere. In addition they lack the chlorine atom which catalyzes stratospheric ozone depleting reactions.

Also graphed in Figure 7.4 are carbon tetrachloride or "carbon tet" ($CCl_4$) and sulfur hexafluoride ($SF_6$). $CCl_4$ was used for dry cleaning, in fire extinguishers and as a pesticide for stored grain. It was also used in producing CFC

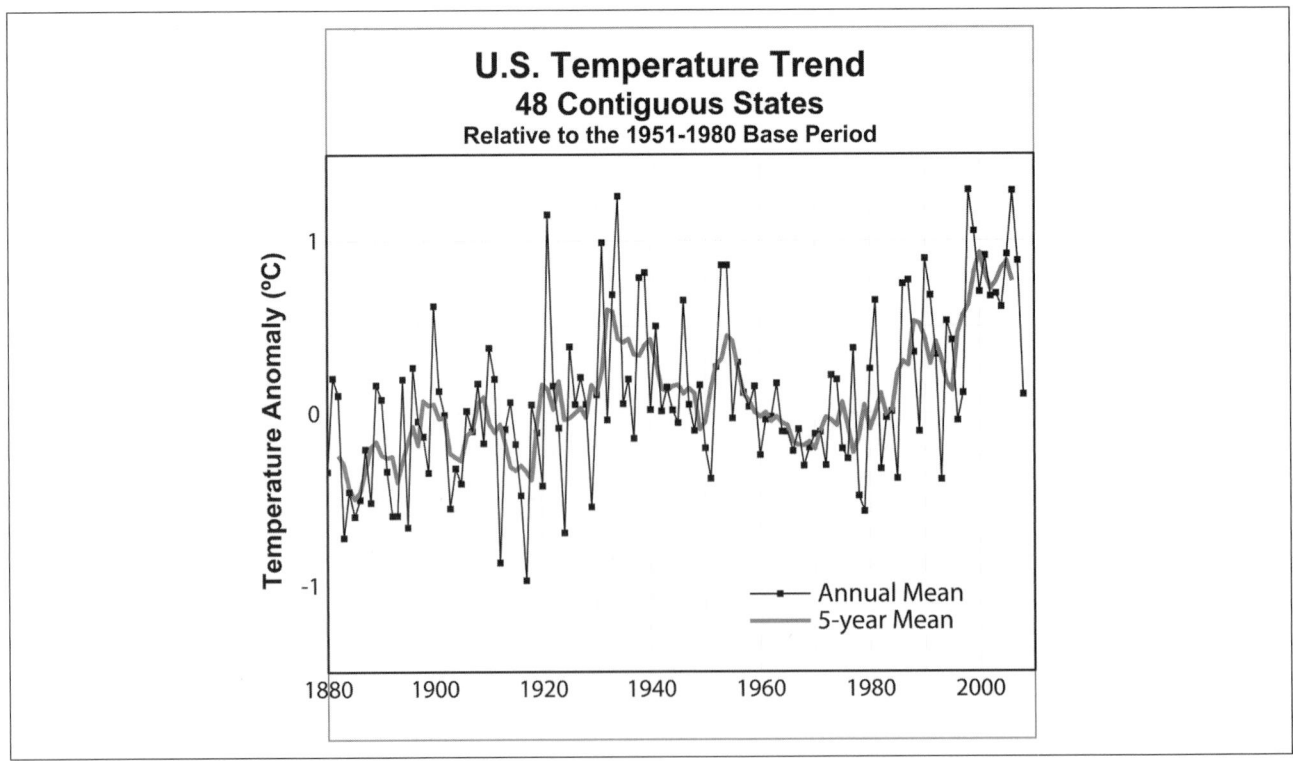

**Figure 7.6**   US temperature trend for the lower 48 states beginning 1880. (*Source:* NOAA.)

**Table 7.2   GWP of Greenhouse Gases**

| Greenhouse Gas | Atmospheric Lifetime years | 20-year GWP | 100-year GWP | 500-year GWP |
|---|---|---|---|---|
| Carbon Dioxide ($CO_2$) | 50 - 200 | 1 | 1 | 1 |
| Methane ($CH_4$) | 12– | 62 | 23 | 7 |
| Nitrous Oxide ($N_2O$) | 114 | 275 | 296 | 156 |
| HFC-23 | 260 | 9400 | 12000 | 10000 |
| HFC-125 | 29 | 5900 | 3400 | 1100 |
| HFC-134a | 13.8 | 3300 | 1300 | 400 |
| HFC-143a | 52 | 5500 | 4300 | 1600 |
| HFC-152a | 1.4 | 410 | 120 | 37 |
| HFC-227ea | 33 | 5600 | 3500 | 1100 |
| HFC-236fa | 220 | 7500 | 9400 | 7100 |
| HFC-43-10mee | 15 | 3700 | 1500 | 470 |
| $CF_4$ (tetrafluoromethane) | 50,000 | 3900 | 5700 | 8900 |
| $C_2F_6$ (hexafluoroethane) | 10,000 | 8000 | 11900 | 18000 |
| $C_4F_{10}$ (perfluorobutane) | 2,600 | 7,000 | 4,800 | 10,100 |
| $C_6F_{14}$ (perfluorohexan) | 3,200 | 5900 | 8600 | 12400 |
| $SF_6$ (sulfur hexaflouride) | 3,200 | 15100 | 22200 | 32400 |

*Note: HFC-23 through HFC-43-10mee are grouped under the label "Hydroflurocarbons".*

*Source:* IPCC, Fourth Assessment Report, 2007.

refrigerants. Because of the high toxicity it has been replaced with less toxic chemicals in many processes. $CCL_4$ is a powerful greenhouse gas.

Sulfur hexafluoride is the most powerful greenhouse gas known with a 100-year global warming potential (GWP) more than 16,000 times the GWP of $CO_2$. It is used as a gaseous insulator in the electric power industry, as an inert gas in casting magnesium and as an inert filler in energy efficient windows.

Scientists measure the potential for future warming using the GWP for individual greenhouse gases. The GWP also provides a simple comparison of greenhouse gases by combining many factors into one number. Because the contribution by an individual gas to global warming is a function of how long the gas remains in the atmosphere and the specific molecular structure of the gas the GWP is not always straightforward. Typically the GWP is listed for 20-year, 100-year and 500-year time spans so that the effects of both longevity of the compound and radiation absorption efficiency can be compared. By definition $CO_2$ has a GWP of 1 and all other gases are compared to it. Table 7.2 lists the GWP of some common greenhouse gases.

The warming Earth is now experiencing is more rapid than any time in the past one million years and is caused by a combination of forcing from natural effects and forcing by the release of greenhouse gases due to human activities.

The greatest single contributor to the greenhouse effect is water vapor. Because there is no human activity that increases water vapor globally the GWP of water vapor is not calculated.

The concentration of water vapor in the atmosphere and therefore its contribution to the greenhouse effect is controlled by Earth's temperature. The warming of Earth increases global evaporation and increases the amount of water vapor in the atmosphere.

Water vapor increases in turn enhance the greenhouse effect and further warm the planet causing evaporation to again increase. This is called positive feedback. Water vapor increases can also cause cooling when highly reflective clouds are formed. This is an example of negative feedback. Feedback mechanisms are discussed later. As of this writing climate model simulations support the position that water vapor's warming effect is larger than the cooling effect.

The greenhouse effect is a natural part of our climate system. Global average temperature today is the result of the combination of three effects, the natural greenhouse effect, post glacial warming which started about 12,000 years ago and is driven by the cyclically changing geometrical relationship between Earth and sun, and the increase of greenhouse gases due to human activity.

### A Brief History of the Greenhouse Effect

In 1824 Jean Baptiste Joseph Fourier (1768–1830), best known as a mathematician, discovered what would later be called the greenhouse effect. He established the principle of planetary energy balance recognizing that Earth not only received radiant energy from the sun but also lost "dark heat," now called infrared radiation, to space. Eventually a balance between the two would be reached and a planet's atmosphere would retain some of the "dark heat" causing the planet to be warmer than if it had no atmosphere.

John Tyndall (1820–1893), an Irish physicist, discovered that certain gases are opaque to infrared radiation and in 1862 he compared the atmosphere to a dam that causes water to accumulate upstream. He established that invisible infrared radiation is radiation like visible light and was the first to prove that water vapor, carbon dioxide and ozone absorb Earth's radiation and trap heat.

The role of fossil fuel combustion in warming Earth was first proposed by Svante Arrhenius (1859–1927) in 1896 when he was studying the causes of ice ages. He suggested that doubling the $CO_2$ concentration of the atmosphere would lead to a 5°C rise in average global temperature. Forty-two years later Guy Stewart Callendar (1898–1964), published a study after painstakingly plotting thousands of observations from around the world by hand. His conclusion was that his graphs showed that carbon dioxide concentrations were on the rise and the planet was warming because of burning of fossil fuels.

The starting point of modern efforts to understand greenhouse gases and temperature variations on Earth goes back to 1957 when oceanographer Roger Revelle (1909–1991) at the Scripps Institute of Oceanography and chemist Hans Suess (1909–1993) from the University of Chicago published a study that indicated global warming may be occurring because of the increase of carbon dioxide.

Dr. Revelle then hired David Keeling (1928–2005), a geochemist, who in 1960 published a paper linking carbon dioxide increases from human sources to warming of the planet. The graph he produced showing the increase of $CO_2$ is known as the "Keeling Curve." The inset panel of Figure 7.1 shows the extension of his work into 2010. Keeling started continuous measurements of $CO_2$ at the South Pole and Mauna Loa Observatory, Hawaii in 1958. His South Pole program ended due to budget cuts in the 1960's but observations continue at Mauna Loa.

Important to the global warming story because of its political and public relations ramifications is that before he retired Roger Revelle taught at Harvard University. One of his students was future Vice President, Academy Award winner, Nobel Peace Prize recipient and environmental activist Al Gore.

### Composition of the Atmosphere

The atmosphere is composed primarily of constant amounts of nitrogen ($N_2$, 78.08%), oxygen ($O_2$, 20.95%) and trace amounts of other gases. Because the concentrations of these gases do not vary they are called the fixed gases. Atmospheric gases present in smaller and variable amounts are termed the variable gases. Despite their low concentrations trace gases are very important.

Listed in order of abundance the variable gases are water vapor, carbon dioxide, methane, nitrous oxide, ozone and CFCs and related compounds. Each of these variable gases is a greenhouse gas.

A greenhouse gas both absorbs and emits radiation in the wavelengths of thermal infrared electromagnetic radiation, the wavelength at which Earth's radiation peaks. Without greenhouse gases all Earth's radiation at these wavelengths would escape to space. Figures 7.1–7.4 show the increase of most important greenhouse gases that humans directly influence. There is more water vapor in the atmosphere than any other greenhouse gas but humans have no way of increasing the global concentration. Water vapor concentration is controlled by Earth's average temperature.

### Variation of Water Vapor

In the short term the water vapor content of the atmosphere varies with location, season, prevailing wind direction and migratory weather systems and altitude.

In the long term which is especially important to global warming, because water vapor concentration is controlled by Earth's temperature, any change in the global amount of this

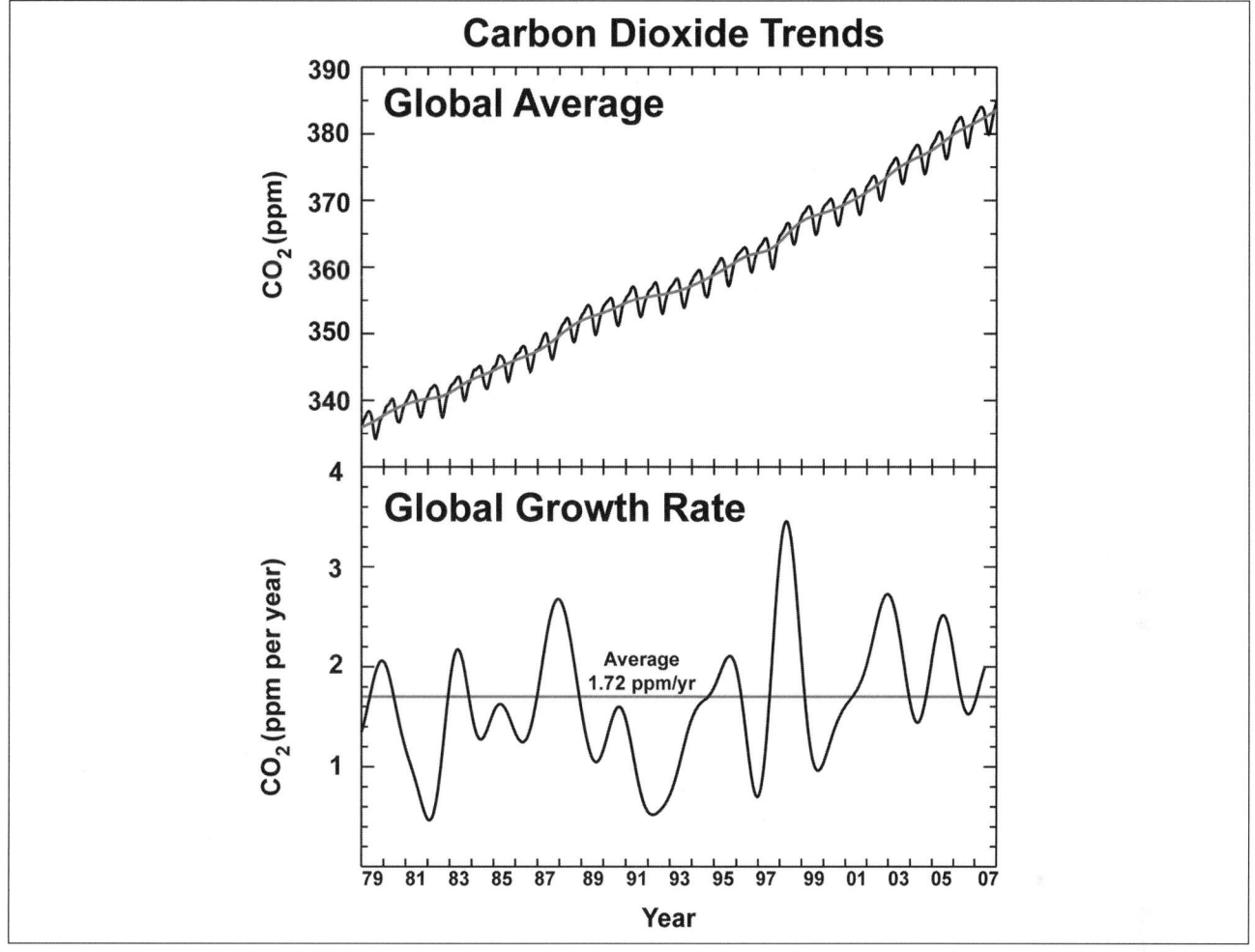

**Figure 7.7**   The trends of atmospheric carbon dioxide concentration and growth rate since 1979. (*Source:* Redrawn from NOAA, ESRL.)

greenhouse gas in the atmosphere is a reaction to global temperature change.

As our planet warms there is more energy available to evaporate water and ultimately more water vapor in the atmosphere means more trapped infrared radiation. On a warming Earth the additional water vapor reinforces the original warming. Whenever the result of a process enhances the original change it is called positive feedback. Feedback is discussed later.

Increases of water vapor may also cause additional low-level cloud cover and the surface is cooled because solar radiation is reflected back to space, which limits the original warming. This is called negative feedback.

### Variation of Carbon Dioxide

Carbon dioxide varies seasonally and is at a minimum each year during northern hemisphere summer when the photosynthesis peaks on the large northern hemisphere landmasses. During northern winter when less $CO_2$ is used for photosynthesis the concentration in the atmosphere in-

creases. The resulting graph looks like the edge of a saw (Figures 7.7 and 7.1, inset).

Longer term variations in atmospheric $CO_2$ are associated with evolution of the climate system, increasing and decreasing global erosion rates and glacial advances and glacial retreats that are paced by changes of Earth's orbital geometry called Milankovitch Cycles.

As Earth warms $CO_2$ levels increase and as Earth cools $CO_2$ levels decrease. Figure 7.8 shows the variation of temperature and carbon dioxide back 800,000 years recovered from air bubbles trapped in Antarctic ice cores. Notice how the temperature varies almost exactly as $CO_2$ varies. There is a small lag with $CO_2$ following the start warming meaning that $CO_2$ is not the initial cause of a warming event.

This has been incorrectly cited as evidence that $CO_2$ is not important in global warming. Scientists know glacial and interglacial cycles are paced by orbital variations of Earth. As Earth warms, $CO_2$ increases in response and warming is accelerated. The role of orbital variations, called Milankovitch

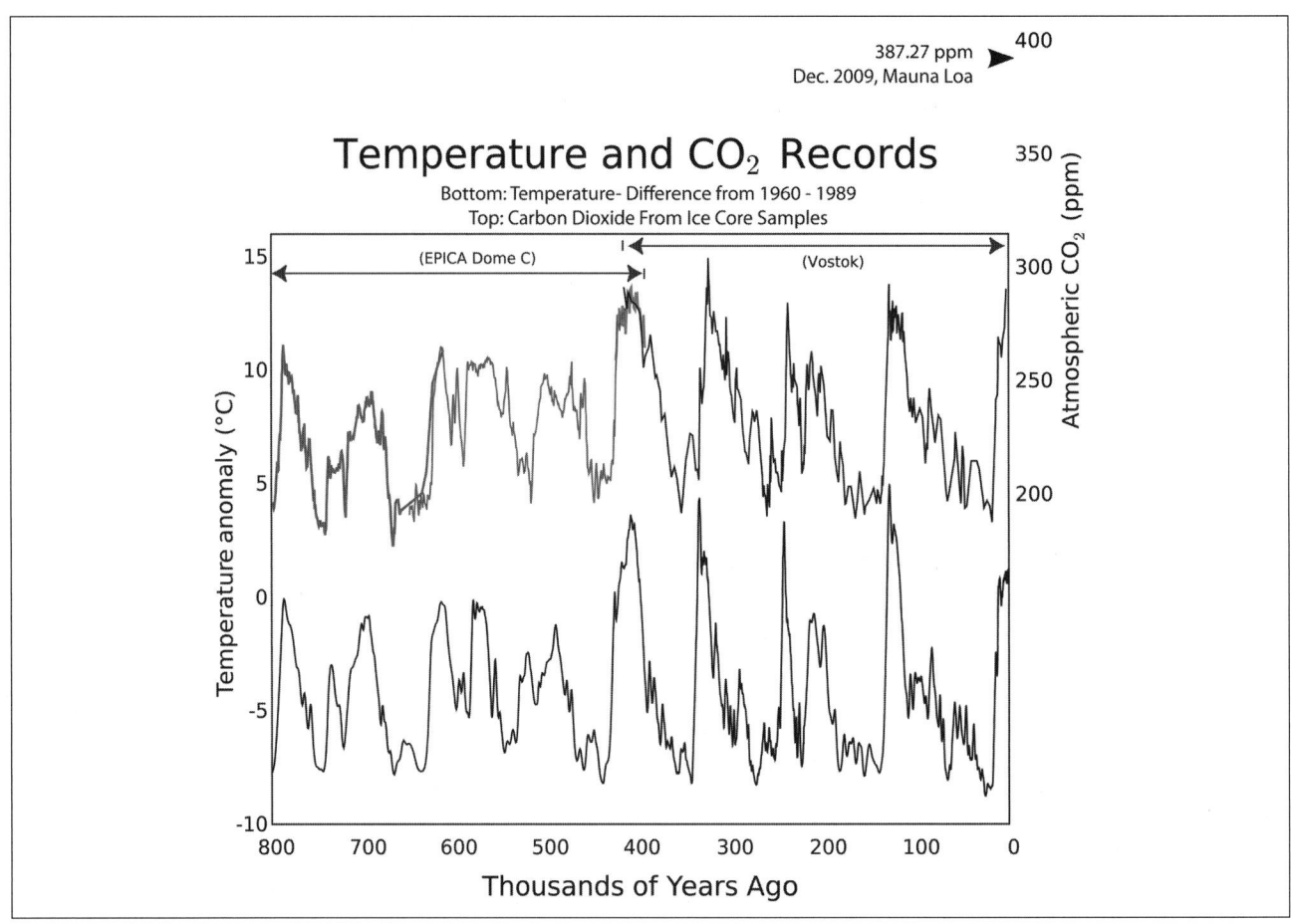

**Figure 7.8** The synchronous change of temperature and $CO_2$ for the last 800,000 years. Note that warming leads $CO_2$ by a few years, which is evidence that $CO_2$ increase is a feedback during warming caused by orbital cycle changes. (*Source:* World Data Center, A Paleoclimatology and NOAA Paleoclimatology Program. Temperature data: J. Jouzel et al. (2007); $CO_2$ data: D. Lüthi et al. (2008).)

Cycles, is discussed in "Milankovitch Cycles" of the "Forcing Mechanisms and Feedback section."

The most likely sources of additional $CO_2$ during warming events are increased decomposition of plant and animal matter in soils and slower uptake of $CO_2$ by oceans as ocean water warms.

Earth's oceans contain approximately 50 times more $CO_2$ than the atmosphere. As ocean waters warm dissolved $CO_2$ escapes to the atmosphere more readily and when the waters cool additional $CO_2$ is absorbed from the atmosphere. Both cooling and warming of ocean waters reinforce those cooling or warming trends and are therefore positive feedback loops.

Since the beginning of the Industrial Revolution the burning of fossil fuels and deforestation have increased the carbon dioxide content of the atmosphere faster than the natural increases during times of glacial retreat and global warming in the past.

Current atmospheric $CO_2$ levels are higher than at any time in the past million years and evidence is conclusive that most of the modern increase in $CO_2$ is the result of human activity. Because of this the majority of climate scientists are convinced the warming that began 10,000–12,000 years ago as the massive continental ice sheets melted is now accelerating.

The majority of the $CO_2$ increase in the atmosphere is anthropogenic. Carbon-14 ($^{14}C$) is a form of the element made radioactive by the bombardment of atmospheric nitrogen by cosmic rays. The half-life of $^{14}C$ is 5730 years, after that amount of time half of the radioactive carbon has decayed back to nonradioactive nitrogen.

Every living thing on Earth has $^{14}C$ within its tissues. When an organism dies the amount of $^{14}C$ begins to decrease and after 5730 years only half remains. Organisms that have been dead for millions of years and isolated from cosmic ray bombardment by thick layers of rock, like the organisms that formed coal and oil deposits, are void of $^{14}C$.

With the increased consumption of fossil fuels and the resultant increase of atmospheric $CO_2$ the relative abundance of $^{14}C$ has been decreasing in the atmosphere. The only known mechanism for the decrease in the concentration of $^{14}C$ is by the release of "dead" carbon from fossil fuels. The

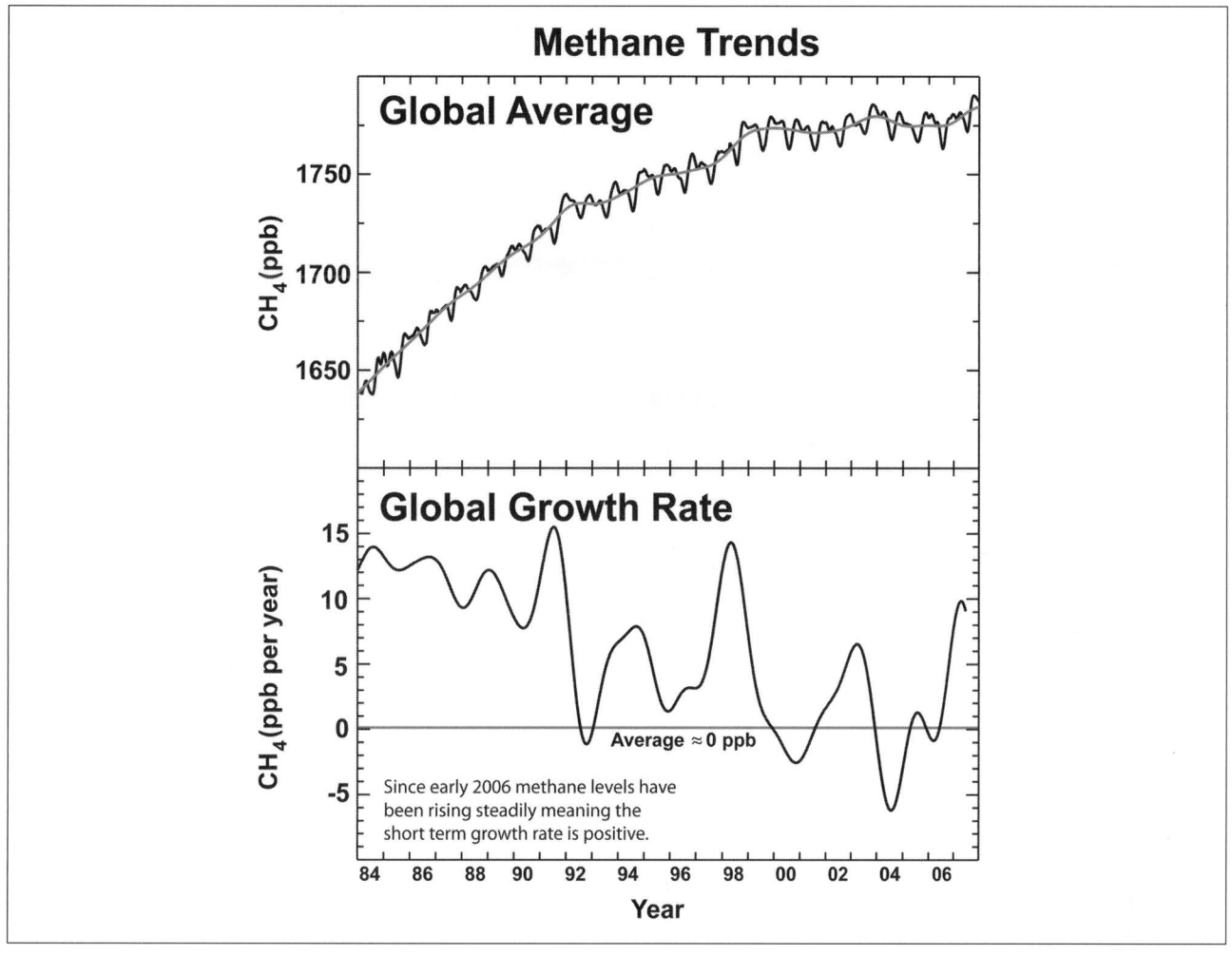

**Figure 7.9**  The trends of atmospheric methane concentration and growth rate since 1979. (*Source:* Redrawn from NOAA, ESRL.)

term "dead carbon" refers to the lack of radioactive $^{14}C$ in anything containing carbon.

### Variation of Methane

Methane ($CH_4$) is the principal compound in natural gas making up about 87% of it by volume. It is a potent greenhouse gas and the concentration of methane in the atmosphere is increasing (Figures 7.2 and 7.9). There are both natural and anthropogenic sources of methane (Figure 7.10).

Like $CO_2$, methane has a seasonal cycle with the concentration being lowest during northern summer when photochemical reactions oxidize it leaving behind $CO_2$ and water vapor. This is an important aspect of methane's role as a greenhouse gas. Methane oxidation in the atmosphere leaves behind two greenhouse gases.

Most atmospheric methane is found in the northern hemisphere where the majority of its sources reside. The concentration of methane has increased in the atmosphere from 700 ppbv (parts per billion by volume) in the pre-industrial world to 1745 ppbv in 1998 an increase of about 150%.

### Nitrous Oxide

The atmospheric concentration of nitrous oxide ($N_2O$) is increasing (Figure 7.3). It is produced both naturally and by human activities. Primary anthropogenic sources include agricultural fertilizer, animal manure management, sewage treatment, combustion of fossil fuel, adipic acid production (for nylon manufacturing and fertilizer production) and nitric acid production.

Nitrous oxide is also produced naturally by a wide variety of biological processes in soil and water. Particularly important is microbial action in wet tropical forests. There is also evidence that nitrification of water courses and wetlands may increase natural production of nitrous oxide. Nitrification occurs when runoff water carries excess nitrogen fertilizer into streams.

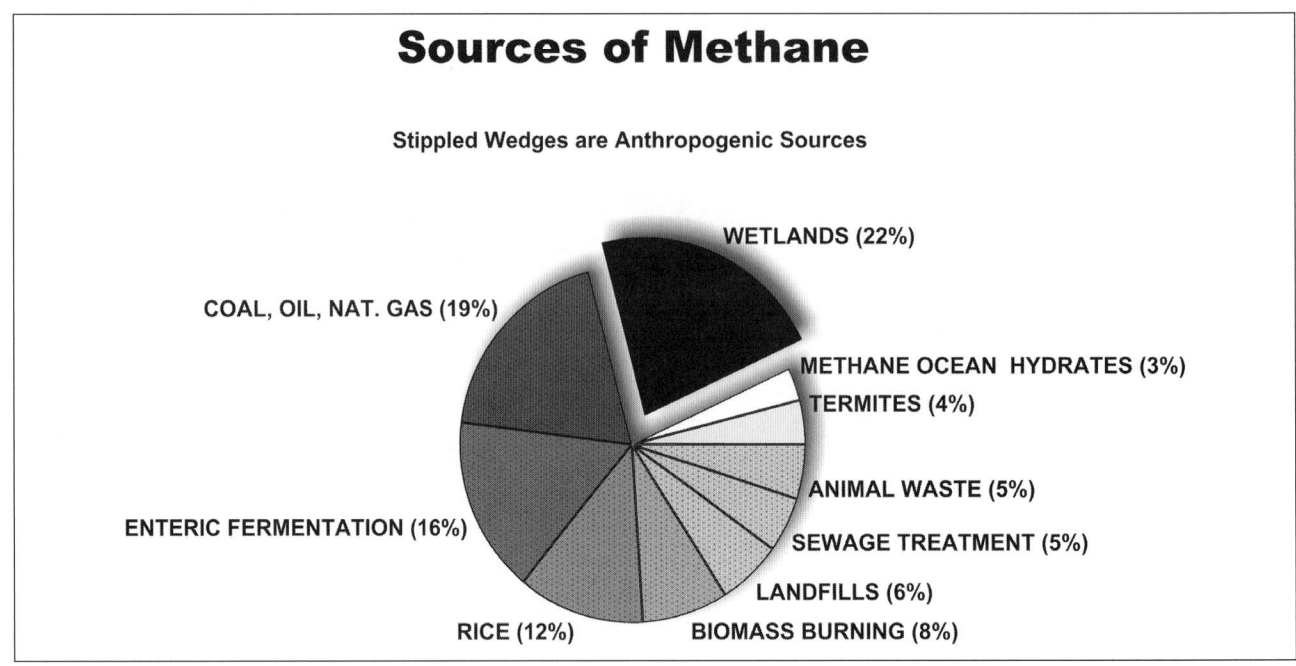

**Figure 7.10**   Natural and anthropogenic sources of methane. (*Source:* US EPA.)

$N_2O$ is a powerful greenhouse gas and will absorb 298 times more infrared radiation in a 100 year period than will and equivalent weight of $CO_2$.

### Tropospheric Ozone

Ozone in the troposphere is both naturally occurring and the product of photochemical reactions between nitrogen oxides, volatile organic compounds and carbon monoxide. The reactants are products of automobile exhaust, industrial emissions and chemical solvents. The reactions are powered by sunlight.

Ozone is a molecule that consists of three oxygen atoms ($O_3$) and is a powerful irritant to the lungs and eyes and a strong oxidizer. In sufficiently high concentrations ozone is toxic.

Stratospheric ozone on the other hand is essential for life on land. Ozone efficiently absorbs deadly ultraviolet solar radiation. The ozone layer though is very diffuse, if brought to the surface the higher pressure there would compress it to a thickness of 3 mm, about the thickness of two dimes.

Ozone has been known as a distinct compound since 1840 and it is named after the Greek verb "to smell" referring to the odor detected after lightning strikes and electrical sparks.

Ozone in the upper atmosphere is a greenhouse gas with approximately 25% of the strength of $CO_2$.

### CFCs

CFCs, HCFCs, and HFCs along with many related substances are chemical compounds with a variety of uses including use as refrigerants and in fire extinguishers. There are no naturally occurring CFCs, HCFCs or HFCs, all are

man-made. The chemistry of this group is beyond the scope of this article but some common gases of these groups are included in Tables 7.2 and 7.3.

CFCs are powerful greenhouse gases and in addition many also deplete stratospheric ozone. Ozone in the stratosphere absorbs much of the cancer-causing ultraviolet radiation and is necessary for most life on Earth.

### Summary

The greenhouse gases, water vapor, carbon dioxide, methane, nitrous oxide, CFC-11, CFC-12 and tropospheric ozone together account for about 97 of greenhouse gas forcing.

## • FORCING MECHANISMS AND FEEDBACK LOOPS

### Forcing Mechanisms

The agents driving Earth's climate system are called forcing mechanisms. The term originated with scientists who simulate past and future climates using complex computer models.

Strictly speaking a forcing is an external input, like the solar energy that drives the climate system. Plate tectonics which pushes the continents across the face of the globe is also an external forcing mechanism. Plate tectonics is not external to Earth but external to the components that make up the climate system and when continents move, climates change. Likewise the release of anthropogenic carbon dioxide can be considered an external

**Table 7.3   Common Greenhouse Gases and Increases of Concentration since 1998**

| Greenhouse Gas | Atmospheric Concentration 2005 | Change 1998 - 2005 | Radiative Forcing 2005 (W m$^{-2}$) |
|---|---|---|---|
| $CO_2$ (carbon dioxide) | 379 ± 0.65 ppm | +13 ppm | 1.66 |
| $CH_4$ (methane) | 1,774 ± 1.8 ppb | +11 ppb | 0.48 |
| $N_2O$ (nitrous oxide) | 319 ± 0.12 ppb | +5 ppb | 0.16 |
| CFC-11 | 251 ± 0.36 ppt | −13 ppt | 0.063 |
| CFC-12 | 538 ± 0.18 ppt | +4 ppt | 0.17 |
| CFC-113 | 79 ± 0.064 ppt | −4 ppt | 0.024 |
| HCFC-22 | 169 ± 1.0 ppt | +38 ppt | 0.033 |
| HCFC-141b | 18 ± 0.068 ppt | +9 ppt | 0.0025 |
| HCFC-142b | 15 ± 0.13 ppt | +6 ppt | 0.0031 |
| $CH_3CCl_3$ (methyl chloroform) | 19 ± 0.47 ppt | −47 ppt | 0.0011 |
| $CCl_4$ (carbon tetrachloride) | 93 ± 0.17 ppt | −7 ppt | 0.012 |
| HFC-125 | 3.7 ± 0.10 ppt | +2.6 ppt | 0.0009 |
| HFC-134a | 35 ± 0.73 ppt | +27 ppt | 0.0055 |
| HFC-152a | 3.9 ± 0.11 ppt | +2.4 ppt | 0.0004 |
| HFC-23 | 18 ± 0.12 ppt | +4 ppt | 0.0033 |
| $SF_6$ (sulfur hexafluoride) | 5.6 ± 0.038 ppt | +1.5 ppt | 0.0029 |
| $CF_4$ (PFC-14) | 74 ± 1.6 ppt | - | 0.0034 |
| $C_2F_6$ (PFC-116) | 2.9 ± 0.025 ppt | +0.5 ppt | 0.0008 |

*Source:* IPCC, Fourth Assessment Report, 2007.

forcing mechanism. Climate forcing mechanisms are listed in Table 7.1.

Forcing can be negative and push Earth toward a cooler state. In periods of frequent volcanic activity, dust and sulfate aerosols are injected into the stratosphere where they may stay for several years. Dust absorbs solar radiation warming the atmosphere while sulfates (Figure 7.11) react with water vapor to form a sulfuric acid haze layer which effectively reflects sunlight. In 1991 Mt. Pinatubo in the Philippine Islands cooled Earth 0.5C for nearly a year because of the sulfates injected into the stratosphere.

The "human volcano" can also serve as a forcing mechanism warming the atmosphere. In the year 2000 more than 6 teragrams (6 trillion grams, which is more than 13 billion pounds) of black carbon, popularly known as soot, was re-

leased into the atmosphere because of incomplete combustion of fossil fuels.

Black carbon absorbs sunlight in the air and reduces the albedo (reflectivity) of surfaces like ice caps by darkening them. The millions of tons of dust and soot released each year warm the planet and Figure 7.12 shows the release of black carbon by source while Figure 7.13 shows the fossil fuel contribution to black carbon starting in 1875.

Climate scientists use the concept of forcing in climate models and as a convenient way of communicating the effects of changes in the atmosphere. Sulfates from volcanoes can be said to be a forcing mechanism that cools Earth, while black carbon from fossil fuel combustion is a forcing mechanism that warms climate. By using the concept of forcing scientists have a convenient, simple and efficient way to communicate the results of their work.

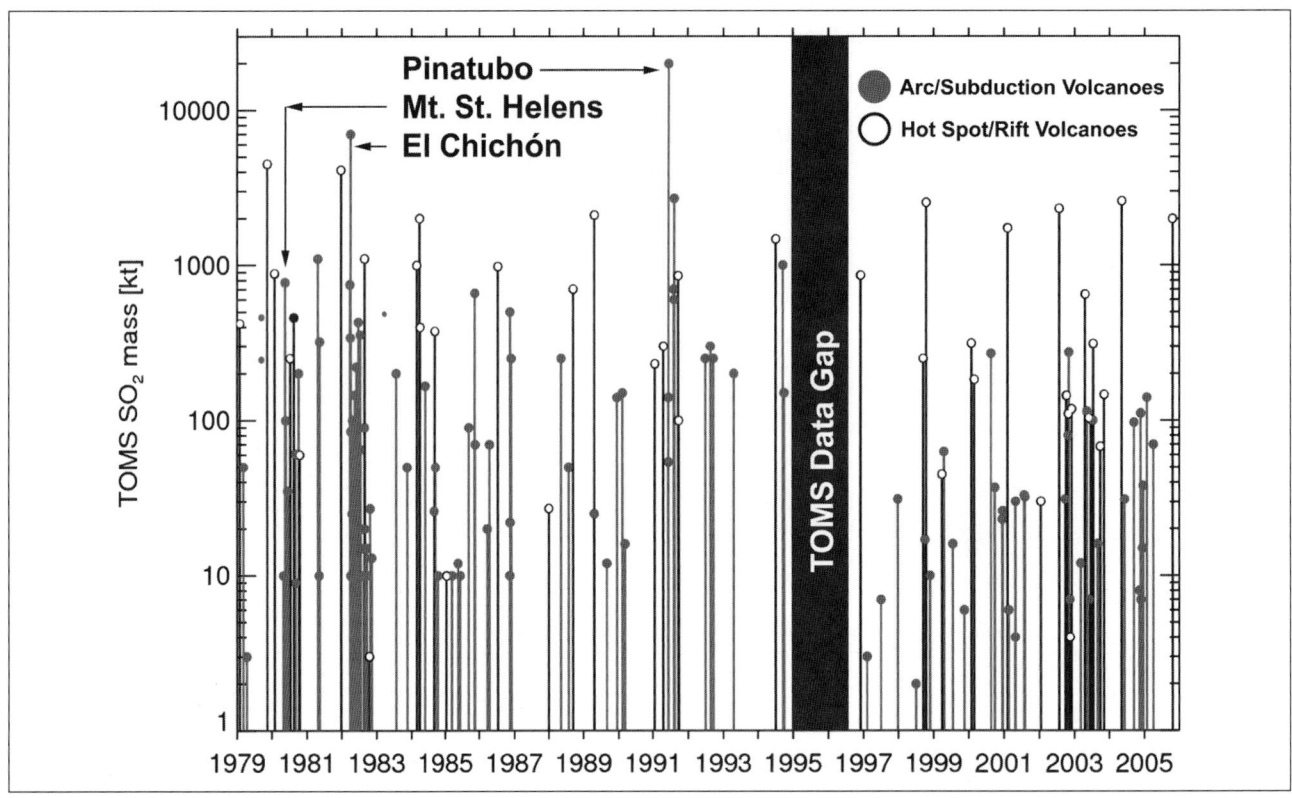

**Figure 7.11**  Injection of sulfates into the atmosphere by volcanoes, 1979–2006 in thousands of metric tonnes. (*Source:* Redrawn and modified from NASA, Goddard Space Flight Center, TOMS Group.)

### Feedback Loops

Earth is warming and the warming is the result of more than simple forcing by both natural and anthropogenic mechanisms. As Earth's climate system retains additional heat, ice caps and glaciers melt exposing bare ground. The dark ground absorbs more solar radiation than highly reflective ice and snow cover further warming Earth and accelerating

the melting of ice. This is called the ice-albedo feedback (albedo is the term used by climate scientists for reflectivity). This feedback is termed "positive" because it works in the same direction as the forcing and accelerates the warming (Figure 7.14).

The ice–albedo feedback also works as a positive feedback as Earth enters a glacial stage. Cooler climates mean

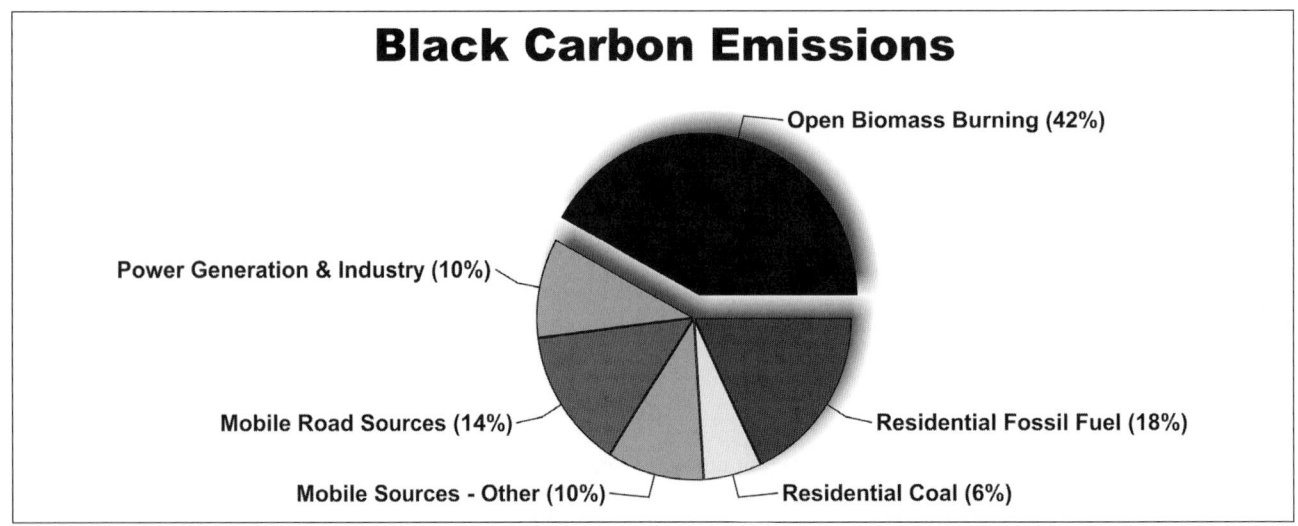

**Figure 7.12**  Release of black carbon by source. (*Source:* US EPA.)

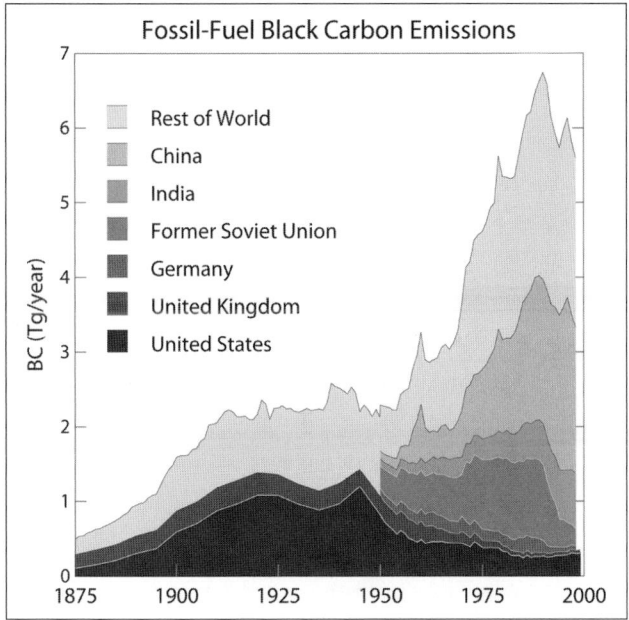

**Figure 7.13**  Black carbon emissions from the burning of fossil fuels by country from 1875. (*Source:* Redrawn from NASA, Goddard Institute for Space Studies.)

more snow and ice which results in more reflection therefore increasing the cooling.

As ocean waters warm less and less $CO_2$ will be absorbed from the atmosphere allowing carbon dioxide in the atmosphere to accumulate more quickly which will accelerate global warming. This too is a positive feedback and is illustrated in Figure 7.15.

Greater evaporation is also likely from the warmer oceans which also speeds warming by increasing the water vapor content of the atmosphere. This is another positive feedback (Figure 7.16).

Each of the mechanisms listed above warms the climate system beyond what was initially forced. Any mechanism that works to increase the original forcing is called positive feedback. In these examples there is a positive ice–albedo feedback loop, a positive carbon dioxide feedback loop and a positive water vapor feedback loop. Positive feedback always increases the effect of the initial forcing.

Feedback that works against the initial forcing is termed negative feedback.

For example, it is reasonable to assume that an Earth with more water vapor will have greater low-level cloud cover which reflects a greater amount of solar radiation back to space than an atmosphere lacking the cloud cover.

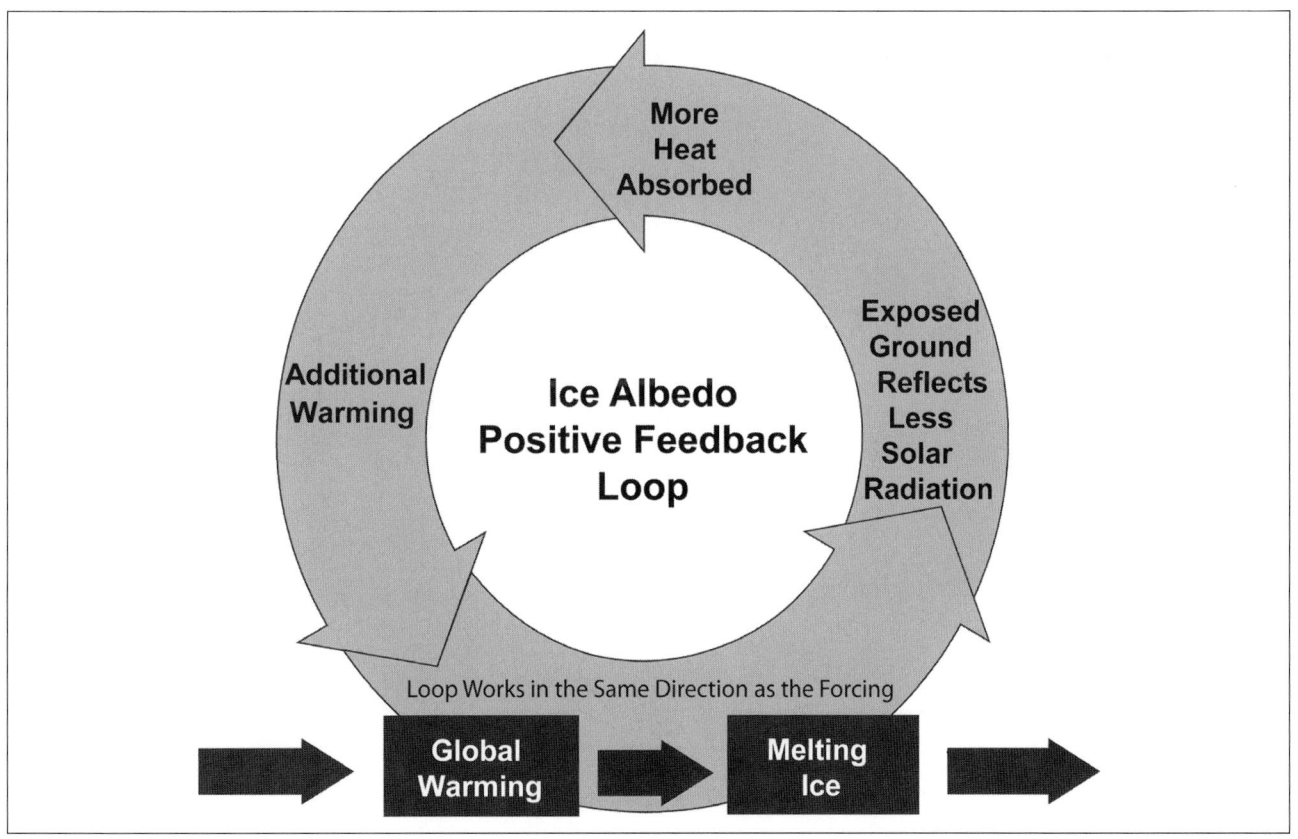

**Figure 7.14**  Ice–albedo positive feedback loop.

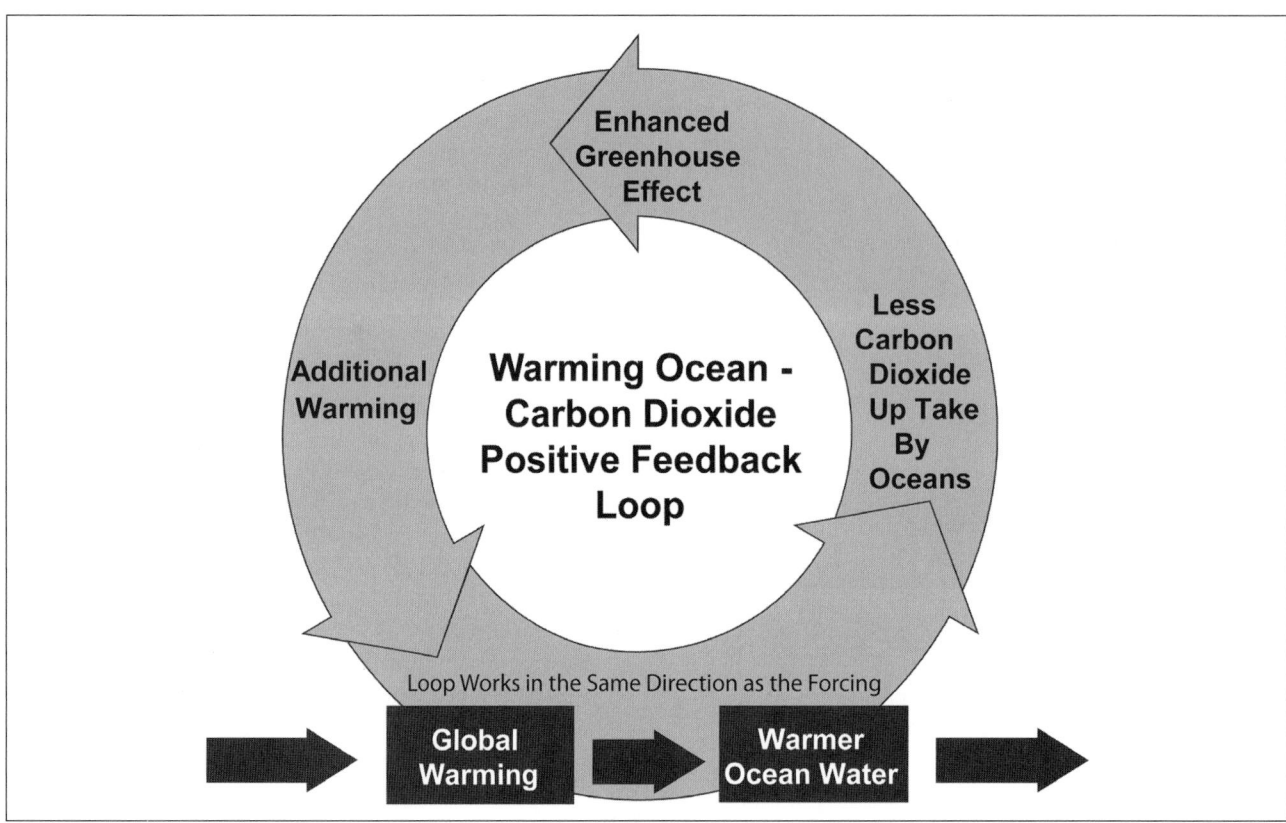

**Figure 7.15**  Warming ocean—carbon dioxide positive feedback loop.

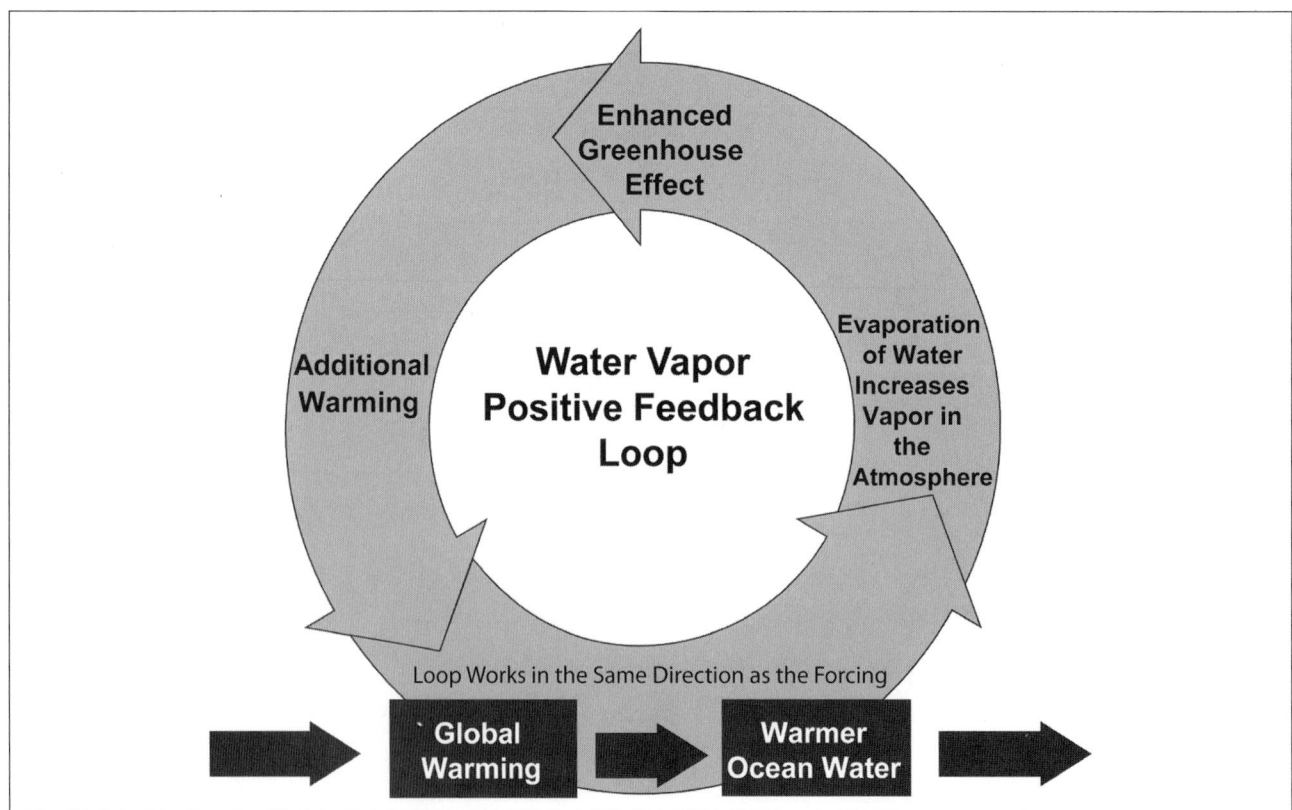

**Figure 7.16**  Global warming—water vapor positive feedback loop.

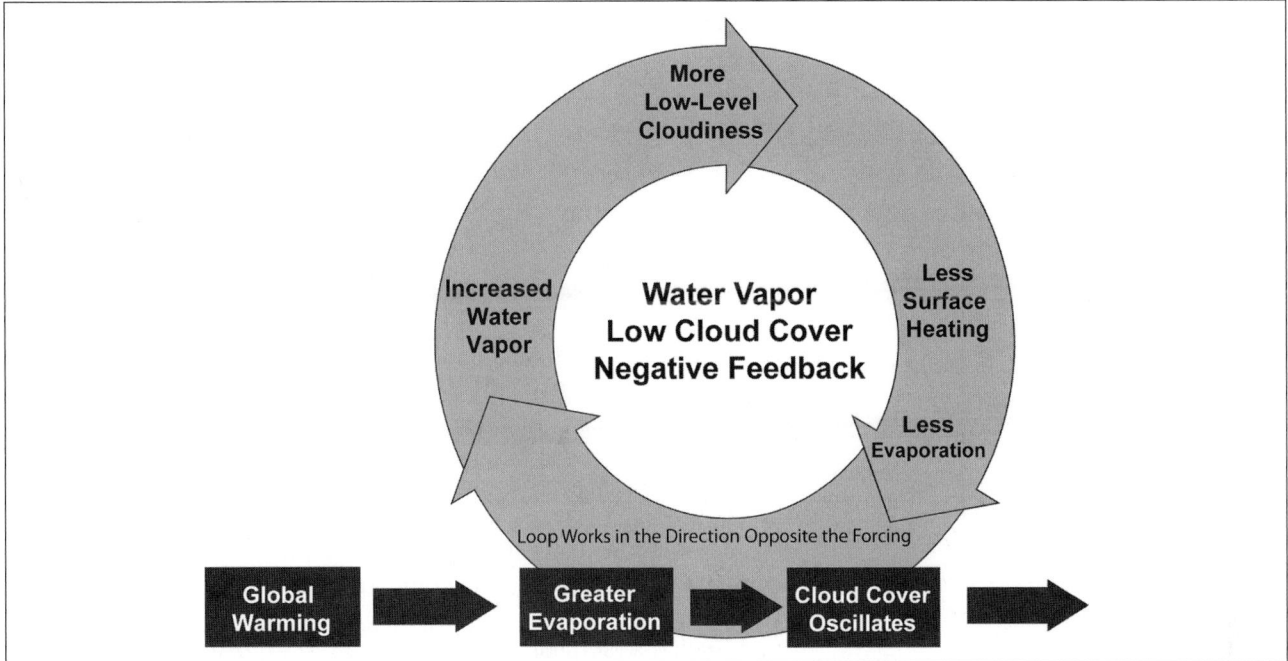

**Figure 7.17**   Water vapor–low cloud cover negative feedback loop.

In response Earth would cool and evaporation would decrease leading to less low-level cloud cover. More sun would then reach the surface and warming would begin again. The negative water vapor feedback loop is diagrammed in Figure 7.17.

Negative feedback puts the brakes on forcing and acts to limit change. The climate system tends return to the initial state. Negative feedback is self-limiting and over time the state of the climate system with the water vapor–low cloud cover negative feedback loop would oscillate between warming and cooling.

Climate scientists are studying the role of water vapor in increasing the greenhouse effect, in changing the water budget at many locations across the Earth and in increasing global average cloud cover. As of this writing cloud feedback is one of the most poorly understood reactions to a warming Earth. Table 7.4 lists a variety of feedback mechanisms which may affect future climate.

The feedback examples presented here illustrate a problem for predicting future climates. Because warming can lead to feedback that both warms and cools the planet's climate scientists must carefully study the many possible complex interactions to determine the resultant temperature change.

Feedback is internal to the climate system and may be relatively simple. Feedback can also be extremely complicated when it results from the interaction of multiple factors over a number of time scales. Climate modelers have worked for years to quantify the interactions within the climate system so predictions of future climates can be as accurate as possible.

### Radiative versus Nonradiative Forcing

A forcing mechanism that directly alters the balance between incoming solar radiation and outgoing Earth radiation is categorized as a radiative forcing mechanism.

Changes in the output of the sun are classified as radiative forcing, so too are changes in the orbital geometry of Earth which are called Milankovitch Cycles. Orbital variations qualify as radiative forcing mechanisms because they affect the receipt of solar energy at the surface and change Earth's energy budget.

Some processes that take place at the surface can be classified as radiative forcing because they modify Earth's energy balance. Over geologic time continents have been bunched together in tropical latitudes. In this situation the energy balance of Earth is quite different than if the continents massed at polar latitudes.

Plate tectonics forces climate changes at time scales of millions of years and some of those forcings are radiative. Because elevated landmasses loose more heat via radiation to space than those at lower elevations, when a large expanse of a land, like the Tibetan Plateau for example, is lifted it has a cooler climate. The Tibetan Plateau is about one million square miles at an average elevation of 16,000 feet above sea level that was lifted as India collided with Asia and created the Himalaya Mountains. The climate on the elevated

Table 7.4   Feedback Mechanisms on a Warming Earth

## Example Feedbacks in Earth's Climate System

| | | |
|---|---|---|
| **Water Vapor** | More evaporation on a warmer Earth leads to more water vapor in the atmosphere | |
| | **Positive** | Increase of atmospheric water vapor leads to a greater greenhouse effect. |
| **Carbon Dioxide** | Less $CO_2$ absorption or $CO_2$ expelled by warmer ocean water results in more accumulating in the atmosphere | |
| | **Positive** | $CO_2$ increases in the atmosphere increasing the greenhouse effect. |
| **Clouds** | More water vapor on a warmer Earth leads to more cloud cover. | |
| | **Negative** | Increase of low clouds that efficiently reflect sunshine and  result in a cooler the surface. |
| | **Positive** | Increase of high clouds that efficiently absorbs infrared radiation warming the high troposphere. |
| **Ice - Albedo** | Ice melts on a warming Earth exposing darker ground that absorbs more solar radiation. | |
| | **Positive** | Greater absorption of solar radiation by exposed surface accelerates warming. |
| | Melt water dilutes dense salty water in the southern Arctic/North Atlantic Oceans. | |
| | **Negative** | Slowing of the oceanic conveyor circulation (thermohaline circulation) because the fresh water inflow lower the density of the water and less sinks driving the circulation resulting in less heat transport poleward and cooling of the far north. |
| **Permafrost** | Melting permafrost leads to increasing decomposition in the soil and release of $CO_2$ and $CH_)$. Methane is released by anaerobic decomposition (decomposition without oxygen). | |
| | **Positive** | Atmospheric carbon dioxide and methane concentrations increase accelerating warming. |
| **Northern Boreal Forest** | Warming climates in Siberia and northern Canada result in the expansion of the vast boreal forests and a longer growing season. | |
| | **Positive** | More forest growth due to greater area covered by trees and a longer growing season leads to removal of carbon from the atmosphere through photosynthesis. This is the sink for what was called the "missing $CO_2$". |
| **Soil Processes** | Mychorrhizal plants, plants having roots with a symbiotic relationship with mychorrhizae (a fungus) increase carbon uptake in warmer soil. | |
| | **Negative** | Symbiotic relationship leads to greater plant growth and greater storage of carbon in the soil, decreasing the amount of carbon in the atmosphere. |
| | Decomposition of organic matter by micro-organisms increases, leading to an increase of atmospheric $CO_2$ (and methane where decomposition is anaerobic). | |
| | **Positive** | Warmer soils lead to greater rates of decomposition and release of $CO_2$ to the atmosphere. |

landmass is much colder than those at the same latitude that are lower.

If the forcing agent does not directly change Earth's energy budget it is termed nonradiative. Typically a nonradiative mechanism works at very long time scales and forces changes in how the ocean and atmosphere redistribute energy across the face of the globe.

Plate tectonics can cause new mountain ranges to grow that can block air from moving into a region thus changing the climate. The growth of the Himalaya cut off Indian Ocean moisture from the interior of Asia. The climate there is much drier than before the mountains existed. In addition the Himalayas and Tibetan Plateau modify the global wind belt called the westerlies. Many winter storms that hit the west coast of the United States get started as the westerly winter jet stream is perturbed by the elevated landmass.

When ocean currents are rerouted or blocked by moving continents the movement of heat from tropical latitudes can be drastically altered. This type of forcing changes the transport of excess tropical heat poleward and can have a profound effect on climate.

## Milankovitch Cycles: A Brief History

As Earth proceeds along its orbital path it interacts with the gravitational fields of the sun, moon and other planets. The varying gravitational tugs by members of the solar system disturb both Earth's orbit and its rotation. These changes in orbital geometry operate on a number of time sales.

The Egyptians were first to realize that Earth's orbit is not perfect and the first computations of orbital variations were done by Joseph-Louis Lagrange (1736–1813) in 1781. When Louis Agassiz (1807–1873) published evidence of ice ages in 1840 there was great interest in tying the glaciations to orbital variations. Then Joseph Adhémar (1797–1862) published *Revolutions of the Sea* in 1842. He proposed the precession (wobble) of the axis of rotation, combined with the eccentricity of Earth's orbit as the factors that caused ice ages. An immediate objection to his theory was that total annual insolation did not change, only the seasonal distribution did.

After years of mathematical refinement Urbain Le Verrier (1811–1877) developed equations for planetary orbits which allowed him to predict the existence of an unknown planet that disturbed the orbit of Uranus. In 1846 the planet Neptune was discovered within $1°$ of arc of the position he predicted. Le Verrier also calculated the changing tilt of Earth's orbit.

Self-educated James Croll (1821–1890) used Leverrier's equations and published, *On the Physical Cause of the Change of Climate During Geological Epochs* in 1864. He worked on improving his theory until his death in 1890.

Croll suggested that the eccentricity of Earth's orbit was sufficient to explain cycles of glacial advance and glacial retreat. But his work included the interaction of precession and

eccentricity. He hypothesized that while the southern hemisphere was in a cold phase and glaciers were growing there the northern hemisphere was the warm.

Most importantly Croll linked orbital changes and climate feedbacks. When the northern hemisphere was in a cold phase, ice accumulated and increased the reflection of sunlight back to space. This, he wrote, "increased the degree of cold." He also discussed cloud feedbacks and ocean current changes which are important to climate change.

In 1904 Ludwig Pilgrim calculated the obliquity (tilt) of the axis and other orbital parameters with a higher degree of accuracy than earlier computations. Between 1920 and 1941 Milutin Milankovitch (1879–1958) refined his proposal that variations of three orbital parameters, precession (axial wobble), obliquity (axial tilt) and orbital eccentricity (orbital deviation from circular) of Earth paced ice ages. Today the collective effect of changes in the three orbital parameters are called Milankovitch Cycles or the Milankovitch Mechanism.

Milankovitch improved the work of those that went before him giving a firm mathematical basis to orbital and rotational changes. The key to ice age initiation according to his theory is the amount of insolation received at high northern latitudes during summer. Milankovitch reasoned that reduced summer insolation would allow snow from the previous winter to go unmelted and over hundreds of years ice sheets would grow on the large continental landmasses. The theory did not find wide acceptance during Milankovitch's lifetime.

In 1976 James Hays, John Imbrie and Sir Nicholas Shackleton published *Variations in the Earth's Orbit: Pacemaker of the Ice Ages*. They used deep sea sediment cores and demonstrated that climate changes matched and were likely paced by Milankovitch Cycles.

### *Orbital Eccentricity*

It refers to, at one extreme, how nearly circular an orbit is and at the other extreme how elongated the orbit is. A circle has an eccentricity of $\varepsilon = 0$. Earth's orbit varies from $\varepsilon = 0.005$ which is almost a circle to $\varepsilon = 0.028$ which is slightly elliptical. Figure 7.18 is a highly exaggerated diagram illustrating orbital eccentricity.

When an orbit is elliptical the sun is at one focus. As Earth moves about the sun is gets closer then farther away. Kepler's Laws tell us as Earth moves farther from the sun it slows, then it speeds up as its orbital path takes it nearer and nearer the sun.

Orbital eccentricity affects climate in two ways. First seasons that occur when Earth is far from the sun are longer because Earth is moving slower. Northern summer today is 4.66 days longer than northern winter and spring is 2.9 days longer than fall. Eccentricity is currently decreasing the difference in length of the seasons as the orbit becomes more nearly circular.

In 30,000 years when eccentricity is at its minimum (less than 0.3%) the difference between northern summer (90.9

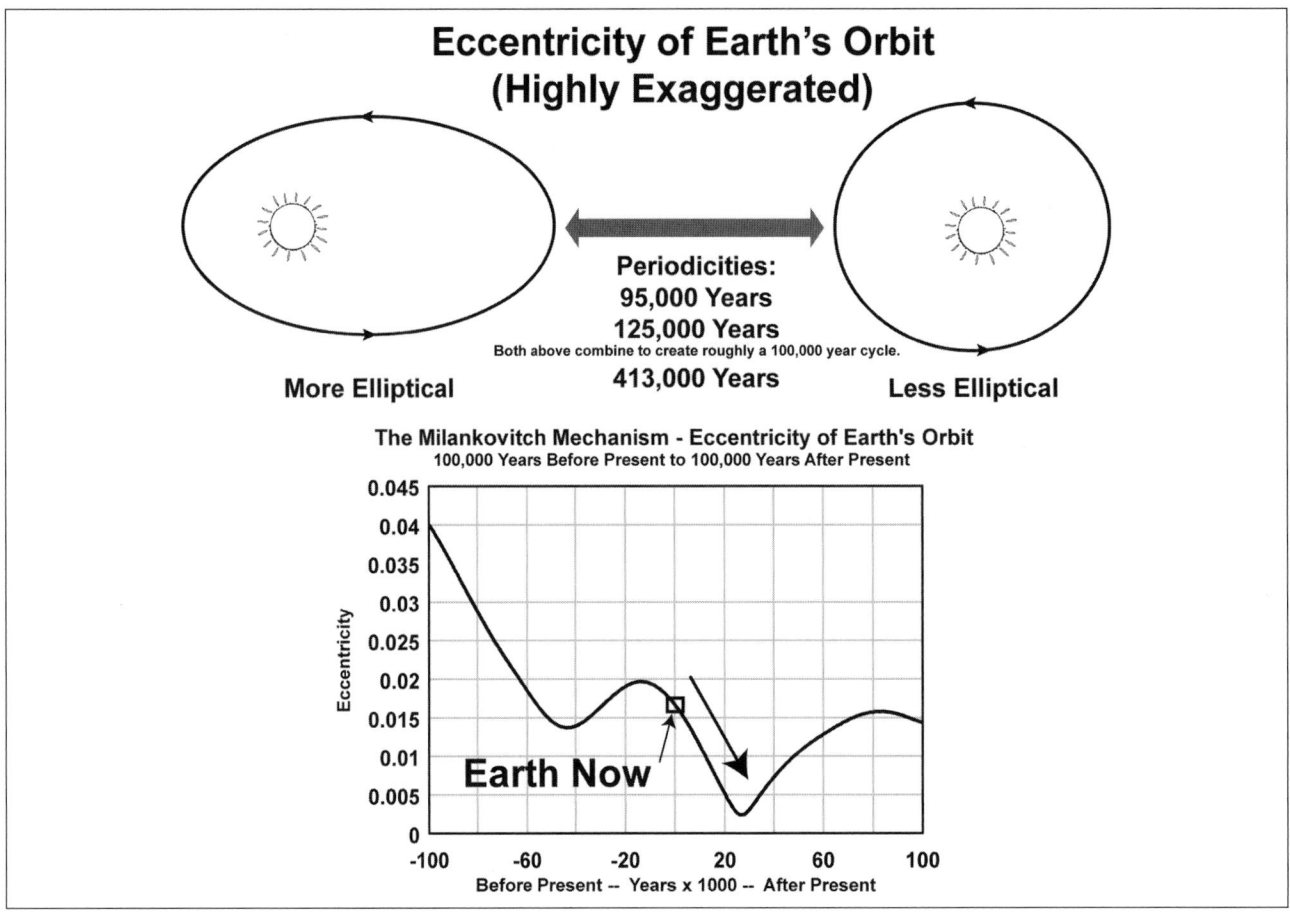

**Figure 7.18** The Milankovitch Mechanism—orbital eccentricity. (*Source:* Steve Horstmeyer. Eccentricity data: World Data Center A—Paleoclimatology, A. Berger and M. Loutre (1991).)

days) and northern winter (91.7 days) will be 0.8 days. One-hundred thousand years ago with an eccentricity of about 4%, the difference between northern summer and northern winter was more that 13 days.

The second climatic effect of orbital eccentricity is distance from the sun to Earth. Because radiation intensity decreases with the inverse square of distance the greater the eccentricity the greater the difference in solar radiation between aphelion (greatest distance) and perihelion (closest distance) of Earth from the sun.

Today with eccentricity small and getting smaller there is a difference of about 6% in received solar radiation between aphelion (closest approach) and perihelion (farthest approach). At time of maximum eccentricity the difference is as much as 30%.

### Precession (Wobble) of the Axis of Rotation

This means that the star called the north star (now Polaris) used for navigation will gradually change as the north end of the axis points in varying directions. This happens because Earth has an equatorial bulge and is not a perfect sphere so the gravitational pull of the sun and moon cause Earth to wobble.

The climatic effect of precession is to change the position in Earth's orbit that seasons occur. Figure 7.19 shows how this occurs. About 11,500 years from now northern winter will occur at perihelion, decreasing winter insolation and northern summer will be during aphelion increasing summer insolation. The difference between summer heat and winter cold will be greater than present because now northern summer occurs when Earth is farthest from the sun.

### Obliquity (Tilt) of Earth's Axis

It changes the intensity of solar radiation along latitude circles. The axial tilt varies from 22.1° to 24.5° and is currently 23.44° and decreasing.

Currently Omaha, NE experiences a decrease in insolation of 59.8% from the summer solstice to the winter solstice. At minimum obliquity (21.4°) the decrease is 52.0% and at maximum obliquity insolation decreases 62.4%.

When Earth's axial tilt is at maximum differences in seasonal temperatures are increased and at minimum axial tilt seasonal differences are decreased. Figure 7.20 illustrates the obliquity of Earth's orbit.

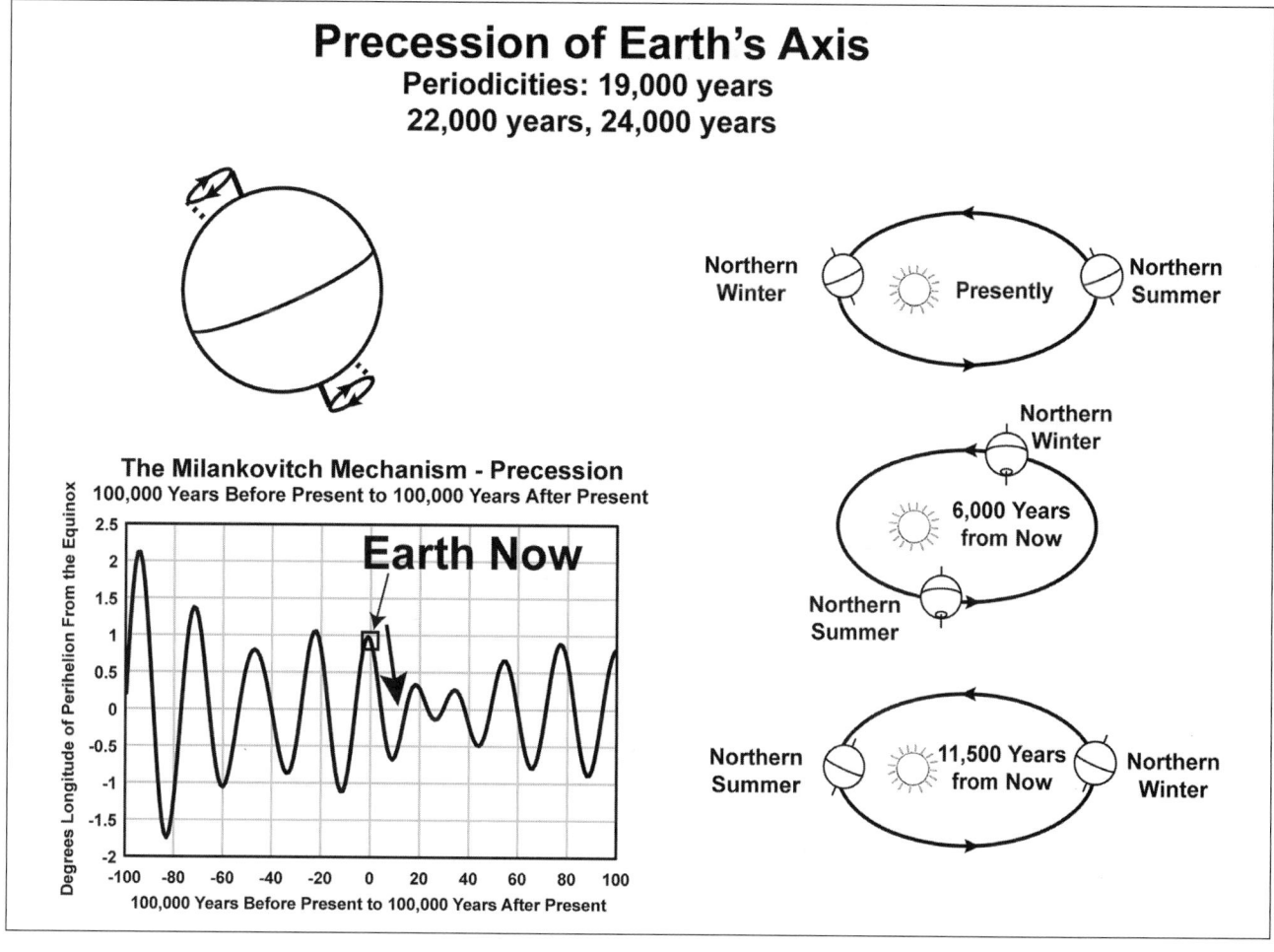

**Figure 7.19** The Milankovitch Mechanism—precession or axial wobble. (*Source:* Steve Horstmeyer. Precession data: World Data Center A—Paleoclimatology, A. Berger and M. Loutre (1991).)

## Combining the Orbital Parameters

It is best done by looking at variations in insolation over time. Each of the illustrations for this discussion indicate Earth is heading into a period of warmer climate. Obliquity is decreasing, eccentricity is decreasing and in about 6000 years the solstices will occur between the aphelion and perihelion meaning decreasing differences between summer and winter.

Figure 7.21 indicates the combined forcing of these three orbital elements will increase the insolation from about 3 W m$^{-2}$ today to almost 5.5 W m$^{-2}$ 12,000 years from now. This is calculated for 65° N latitude as specified by the Milankovitch Mechanism.

Climate scientists rarely find a relationship as strikingly clear as the one between the Milankovitch Mechanism and proxy data shown in Figure 7.22.

Graph A is the solar forcing computed from Milankovitch cycle data and Graph B shows the departure of the deuterium content (heavy hydrogen) from standard mean ocean water (SMOW) obtained from the ice core at Dome C Antarc-

tica. The graph is, inverted because greater concentrations go with colder temperatures. Also from the Dome C ice core is the $CO_2$ concentration in air bubbles trapped in the ice. Graph D shows from an internationally accepted calcium carbonate based standard measure of the isotope oxygen-18 ($^{18}O$). This data set was obtained from an ocean sediment core drilled on the Agulhas Ridge in the south Atlantic Ocean southwest of Africa.

The temperature departures in Graph E are reconstructed from deuterium and oxygen isotope ratios. Oxygen isotope ratios and deuterium concentrations as proxy data are discussed later.

## Problems with the Milankovitch Mechanism

There are six major problems with the Milankovitch Mechanism. Some may be resolved as future research clarifies the role of feedbacks more clearly.

1. Climate extremes seem to exceed Milankovitch forcing so climate scientists must rely on feedbacks within

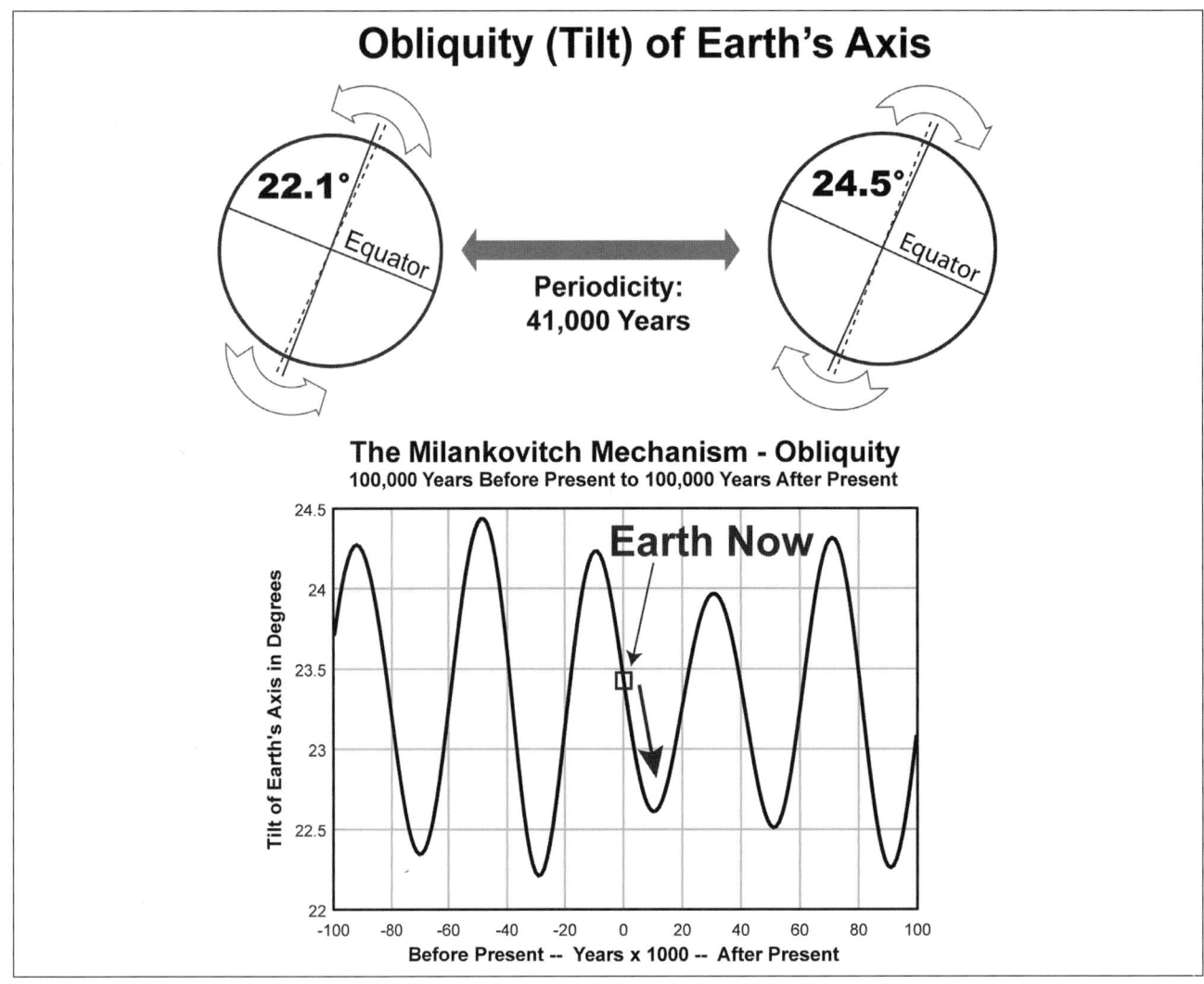

**Figure 7.20** The Milankovitch Mechanism—obliquity or axial tilt. (*Source:* Steve Horstmeyer. Obliquity data: World Data Center A—Paleoclimatology, A. Berger and M. Loutre (1991).)

the climate system to explain the discrepancy. This is obvious when you look at Figure 7.22 and compare the solar forcing to the results.

2. The 100,000 year problem refers to the fact that during the last 1 million years the strongest climate signal is the 100,000 year eccentricity signal despite it being weaker than precession and obliquity. Again feedbacks may explain the discrepancy because a few climate models can reproduce the 100,000 year cycle.

3. The 400,000 year problem refers to the absence of a strong eccentricity signal at 400,000 year periods. This signal is stronger than the 100,000 year signal and may be absent because the response time of the climate system is too slow to show it.

4. Eccentricity has a well-established double periodicity at 95,000 and 125,000 years, but most proxy data chronologies show only a single peak at 100,000 years. It is possible that the proxy data may not have the resolution to

show both peaks. This is called the single peak problem or the unsplit peak problem.

5. About 1 million years ago the frequency mode of climate variations changed. From 1 million years before present back to 3 million years before present the dominant periodicity matched the 41,000 year change in obliquity. Since 1 million years ago it has matched the 100,000 year eccentricity cycle. No one knows why this transition has occurred. This is termed the transition problem.

6. The causality problem refers to the termination of the glaciation just before the most recent glacial stage at about 135,000 years ago which, from available proxy data seems to have ended about 10,000 years before the forcing that is supposed to have ended it. Proxy data that support this scenario include speleothem data from Devil's Hole cave in Nevada, coral terrace data in Papua New Guinea, sediment data from the Bahamas and fossil coral

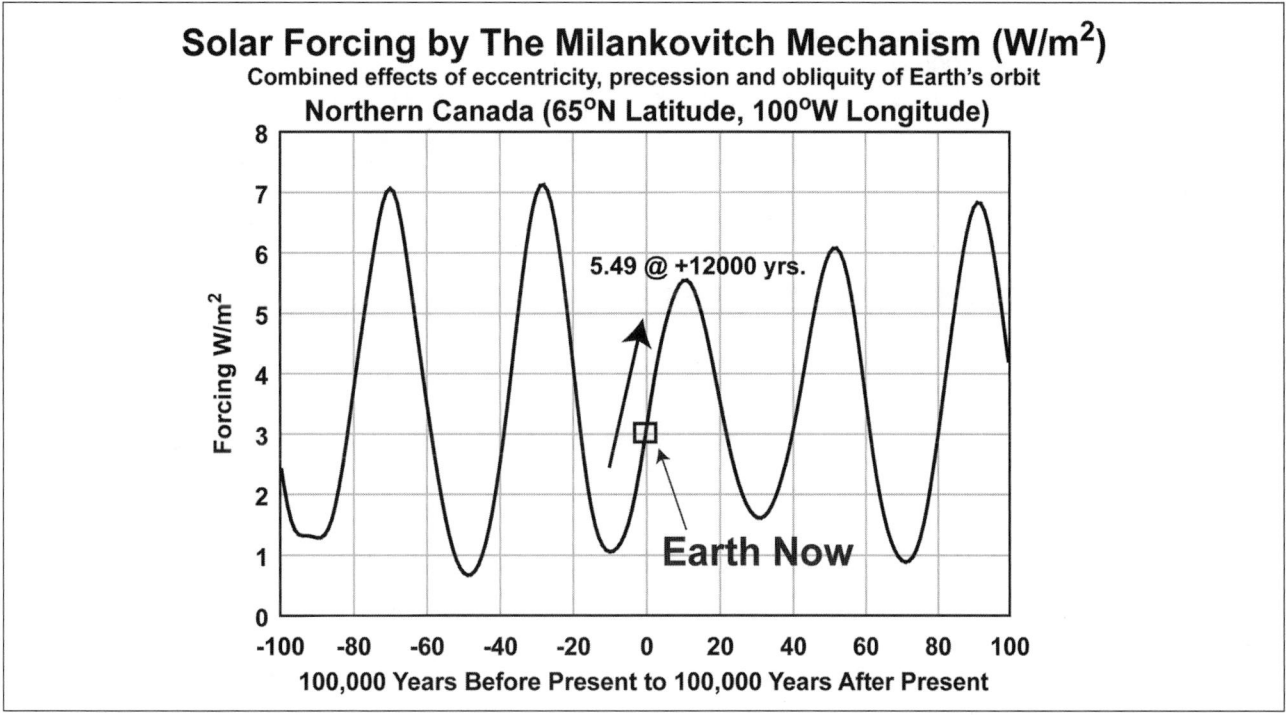

**Figure 7.21**   The Milankovitch Mechanism—total solar forcing. (*Source:* Steve Horstmeyer. Eccentricity data: World Data Center A—Paleoclimatology, A. Berger and M. Loutre (1992).)

data from Barbados. Climate scientists are trying to resolve this issue.

## Summary

As of the publication of this work climate scientists have learned that there are two dominant controls over the advance and retreat of large ice sheets and glaciers.

Ice ages are forced by variations in the orbital geometry of Earth. The key is summer sunshine intensity and possibly total warm season solar radiation around 65° N latitude. Solar radiation intensity varies in a complex manner due to variations in orbital eccentricity, axial obliquity (tilt) and axial precession (wobble) of Earth.

The Milankovitch Mechanism is enhanced or diminished by secondary forcing factors including the concentration of atmospheric carbon dioxide and the arrangement of continents which controls the shape of ocean basins and thus the flow of ocean currents and the transport of heat poleward.

When solar radiation drops below a certain value snows of winter fail to melt completely and eventually nucleation of a large continental ice sheet is said to occur. Because ice sheets require land to grow the large landmasses of the high latitudes of the northern hemisphere are the key to ice age initiation. In the southern hemisphere Antarctica is isolated by ocean waters so the ice sheets cannot grow equatorward.

As feedback mechanisms come into play, in particular the ice albedo feedback, cooling accelerates and atmospheric carbon dioxide levels fall because cooler ocean waters ab-sorb more atmospheric $CO_2$. In addition there is less decomposition in the colder soils. Additional potential sinks of $CO_2$ include an increase productivity of phytoplankton in colder waters and sinking water rich in $CO_2$ in high latitudes where seasonal ice forms.

## •   CLIMATE RESEARCH

### Introduction

Climate research roughly falls into three categories; research to reconstruct past climates, research to understand present climates and research to predict future climates.

Much of the research involving present climate is concentrated on monitoring changes in the composition of the atmosphere especially greenhouse gases, sulfate aerosols and sun-absorbing particles. Figure 7.23 shows the National Oceanic and Atmospheric Administration's worldwide network of monitoring stations.

### Reconstructing Past Climates

Weather instruments are fairly recent inventions and dense, well-distributed networks of instrumented stations have mostly been around less than one hundred years. Some parts of the globe still lack adequate instrumentation.

Because the study of past climates extends back to the formation of Earth, climate reconstructions rely on proxy data. Proxy data are naturally occurring indicators of past

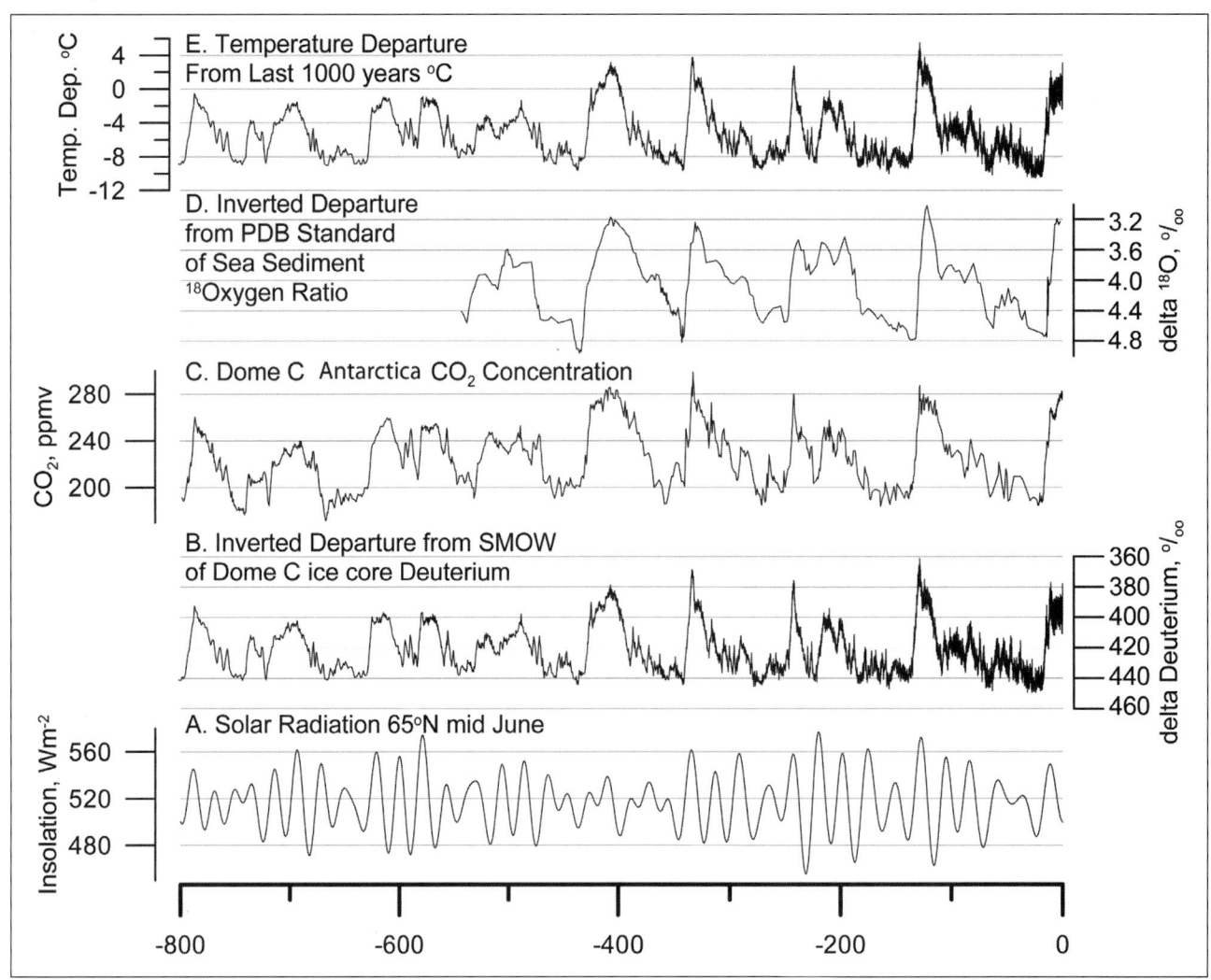

**Figure 7.22**  Matching the Milankovitch Mechanism to proxy data. (*Source:* Steve Horstmeyer. World Data Center A—Paleoclimatology. Carbon dioxide, D. Lüthi et al. (2008); deuterium and temperature, J. Jouzel et al. (2007); solar forcing, A. Berger and M. Loutre (1991), Alfred Wegner Institute for Polar and Marine Research; Oxygen isotope, S. Becquey and R. Gersonde (2003).)

climate like tree rings, ice cores, sea floor sediments, fossil pollen, ancient coral reefs and deep loess (rock dust) deposits.

Using recent proxy data and calibrating against modern instrumental observations allows scientists to develop mathematical functions that yield estimates of past conditions. There are many methods to extrapolate past climate from these data

### Tree Ring Proxy Data—Dendroclimatology

The width of tree rings indicates how favorable the climate was for tree growth the year the ring formed. Using mathematical–statistical functions developed when modern tree ring widths are calibrated against instrumental weather records, scientists reconstruct past climatic conditions. In addition to ring widths, carbon, hydrogen and oxygen iso-

tope analysis of the wood samples in a given ring can also be used to reconstruct climatic conditions during the year the ring formed. Absolute dating of a series of tree rings can be accomplished using well known carbon-14 method.

Tree ring chronologies can be cross referenced to extend the proxy record backward in time. Cross referencing has been accomplished by comparing ring width patterns between trees located in similar climate types and relatively close to one another or ring series in wood beams from ancient dwellings. Carbon isotope ratio chronologies can be constructed using wood removed from annual tree rings. Cross referenced carbon isotope chronologies also enables scientists to extend tree ring proxy data back in time. Isotope studies are covered in more detail when discussing ice core data. Figure 7.24 illustrates how scientists cross reference tree ring data. This method is also used in ice cores,

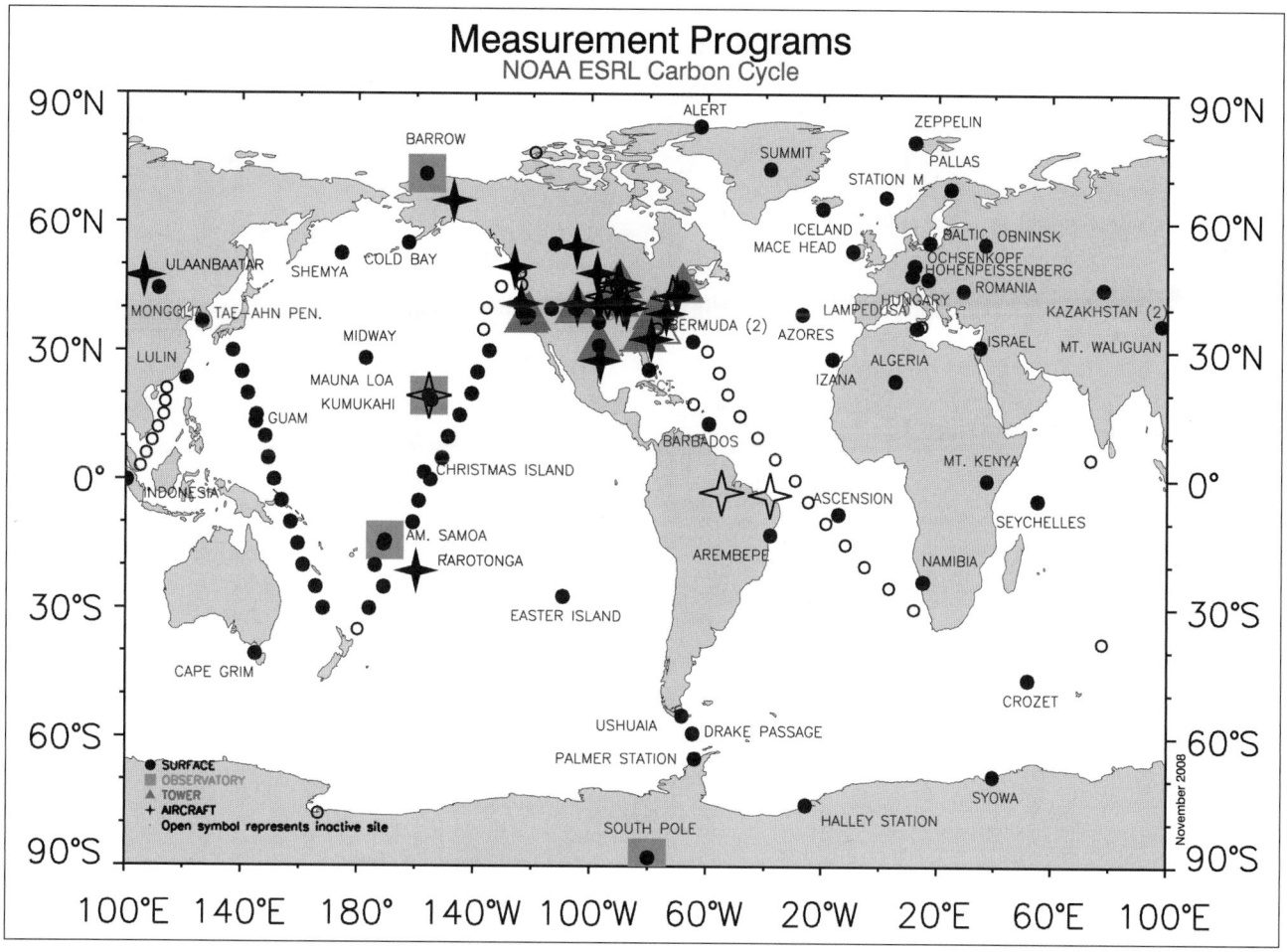

**Figure 7.23**   Carbon cycle monitoring programs of the National Oceanic and Atmospheric Administration. (*Source:* NOAA, ESRL.)

sediment cores, coral cross sections and speleothem deposits (cave formations).

A River Oak chronology from southern Germany, the world's longest, goes back more than 10,000 years and a Bristlecone Pine chronology from the White Mountains of California, the oldest living organisms on Earth, extends back 8500 years.

Absolute dates for tree ring series can be determined using the $^{14}$C technique. Once a tree ring is formed there is no further addition or removal of carbon. Because radioactive $^{14}$C decays at a known rate—the half-life is 5730 years—the ratio of nonradioactive carbon to $^{14}$C indicates the age of the wood.

Figure 7.25 shows a reconstruction of northern hemisphere temperatures as deviation from the 1961–1990 mean based on 66 sites in both North America and Eurasia. All sites are at either the latitudinal treeline or the altitude treeline.

### Ice Core Proxy Data

Ice cores contain annual layers of snow accumulation but it is not always possible to detect them in regions of slight snow accumulation like the high plateau of Antarctica and for very old ice from deep within an ice sheet where the immense pressure of overlying ice eliminates boundaries between adjacent layers. Ice core chronologies can be calibrated using dust, volcanic ash layers and oxygen isotope ratios. By cross referencing with other ice core data and sea floor sediment cores the chronologies can be extended far back in time.

The deeper the ice from which part of an ice core is extracted the greater the pressure and the less distinct individual layers become. The net effect is to decrease the time resolution of the proxy data. In some cases the finest time resolution possible may be hundreds to thousands of years.

**Ancient air in ice cores.**  Ice core layers include small bubbles of ancient air than can be analyzed to measure atmospheric composition of the past. The greenhouse gasses carbon dioxide, methane and nitrous oxide are important in determining how effectively the atmosphere trapped infrared radiation (heat) emitted by Earth. Knowing the concentration of the gases helps climate scientists estimate ancient temperatures on Earth.

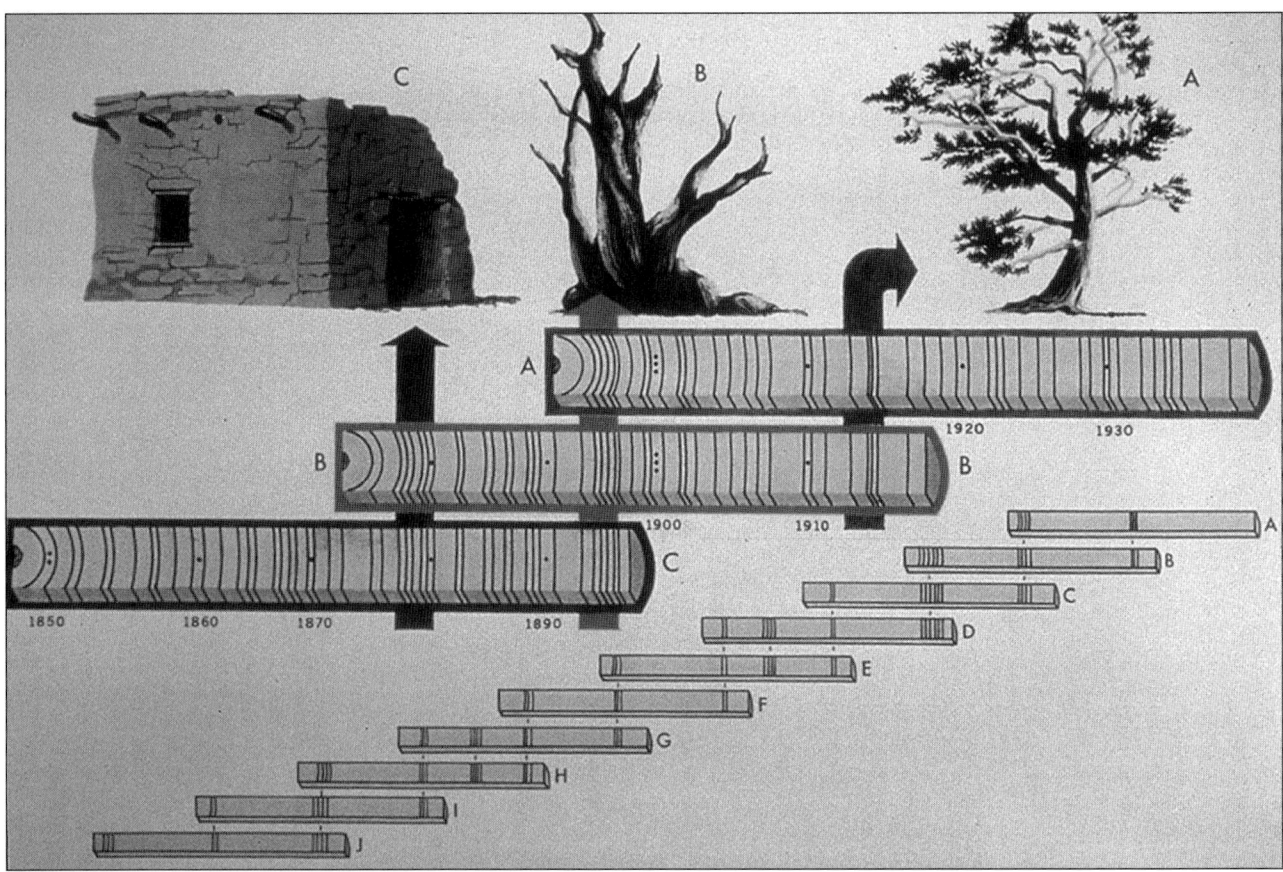

**Figure 7.24** Cross-referencing tree ring sequences. (*Source:* NOAA Central Library.)

Figure 7.26 shows the results of analyzing ancient air in the Vostok, Antactica ice core back to more than 400,000 years before present. Included on the graph is the forcing by changes in Earth's orbital geometry. The match between Milankovitch cycles and climate is not perfect, there is a time shift in some cases but the results are a strong indication that orbital geometry changes pace ice ages on our planet.

**Isotopic fractionation of oxygen and hydrogen.** The ice of a core was once primarily ocean water. It evaporated, condensed and subsequently fell as snow. Ocean water is made mostly of "common" hydrogen ($^{1}H$) and "common" oxygen ($^{16}O$). There are however small amounts of heavier isotopes of both in all water.

A heavy isotope has a greater number of neutrons in the atomic nucleus. For climate research the useful hydrogen isotope, $^{2}H$, has one additional neutron. For isotopic oxygen, $^{18}O$, there are two. Both elements have isotopes in addition to these.

Isotopes of the same substance have different masses and slightly different bonding characteristics. Water molecules containing lighter isotopes more readily evaporate and less readily condense to become precipitation resulting in glacial ice having a greater proportion of light isotopes than the water from which it evaporated.

The isotope $^{2}H$ is named deuterium or heavy hydrogen and has almost twice the mass of common hydrogen because it has a neutron in the nucleus. Because of the additional mass and altered bonding characteristics a water molecule with heavy hydrogen requires more energy to evaporate than "common water." At a given temperature air has a smaller ratio of heavy water to common water than the ocean water from which it evaporated.

Using the same reasoning in reverse, as evaporated moisture is carried by winds, heavy water falls as precipitation more readily than common water. By the time the air is over the high plateaus of the Antarctic or Greenland ice sheets it is even more depleted of water molecules containing deuterium than it was immediately after the water evaporated.

The global climate system naturally fractionates deuterium from common hydrogen depleting the snow that falls on ice sheets of molecules containing deuterium. The greater the depletion of heavy hydrogen in the ice, the colder the air was when the moisture fell as snow.

By the same natural processes water molecules containing "heavy oxygen," $^{18}O$, are fractionated from water molecules containing lighter $^{16}O$. The degree of natural

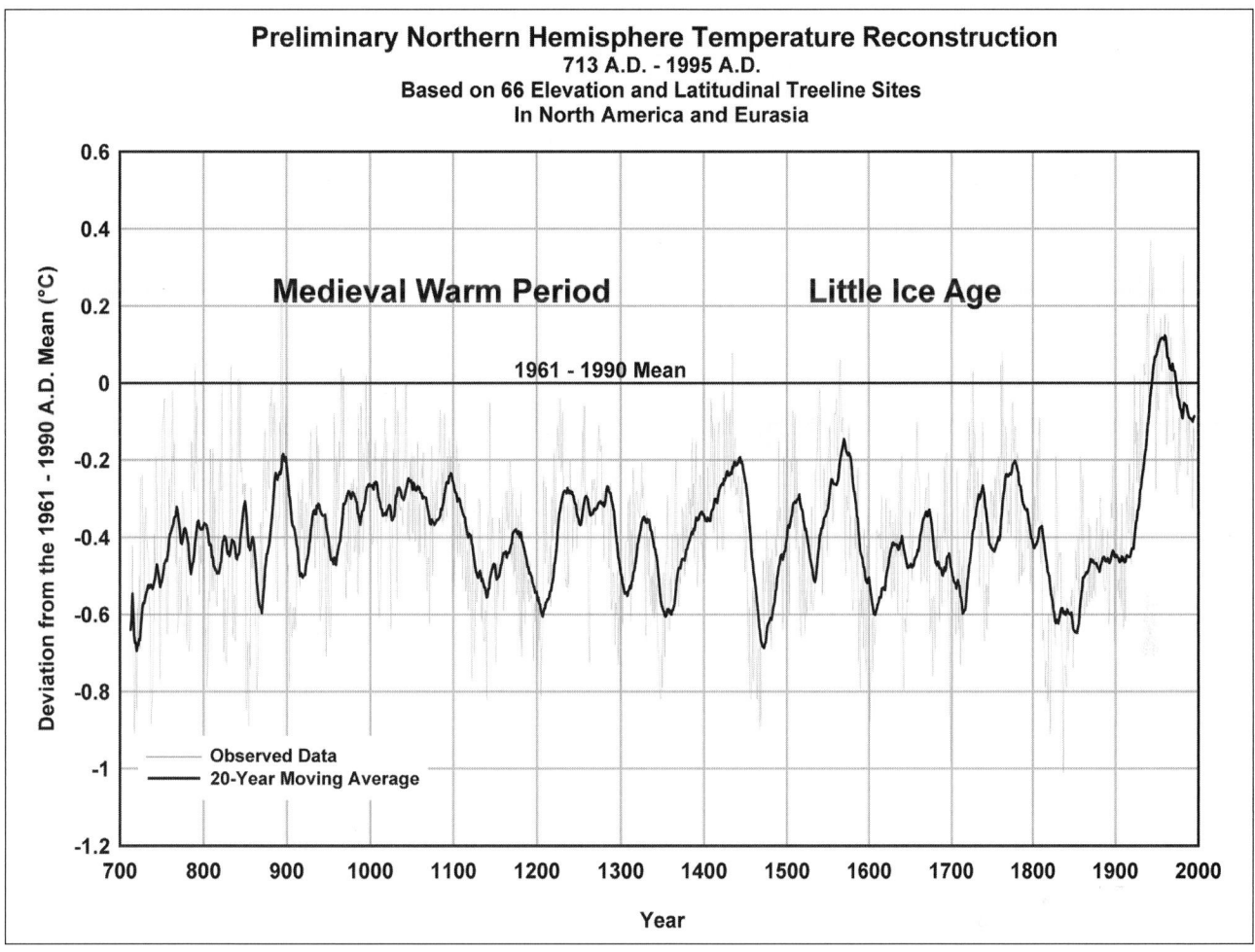

**Figure 7.25**    An example of temperature reconstruction using tree ring data form 66 sites. (*Source:* Steve Horstmeyer. World Data Center A— Paleoclimatology, R. D'Arrigo et al. (2006).)

fractionation can also be measured in wood samples extracted from tree rings, sea sediment cores and the calcium carbonate shells of marine animals.

Chronologies from ice cores extend much farther back in time than the tree ring record. The longest proxy climate record from ice is a combined chronology from two locations in Antarctica, Vostok (elevation 11,444', latitude 78.47° S, longitude 106.87° E) and Dome C (elevation 10,607', latitude 75.10° S, longitude 123.35° E). The chronology extends back 798,512 years at an ice depth of 3190 meters (10,466 feet).

Vostok, now a Russian, French, and American joint research station is where the coldest temperature on Earth was measured (−128.6°F, −89.2°C) but higher on the ice sheet temperatures have probably been lower. The highest ice sheet elevation in Antarctica is Dome A at 13,428 feet above sea level.

*Sea Floor Sediment Cores*

Like tree rings and ice cores sediments on the floor of the ocean can form annual layers. The floor of an ocean basin can be a very energetic environment stirring up sediments that settle there. But if a location is picked carefully so there is little turbulence, the annual layers can yield a high resolution chronology for millions of years into the past.

Throughout Earth's oceans there is a continuous fall of sediment toward the sea floor. The "marine snow" consists of materials eroded from continents, materials rafted and dropped by floating ice and remains of marine plants and animals. Of great interest to paleoclimatologists are the calcium carbonate ($CaCO_3$) shells that accumulate on the sea floor. Oxygen atoms included in the calcium carbonate are excellent paleotemperature indicators. The ratio of "heavy" oxygen ($^{18}O$) to "common" oxygen ($^{16}O$) in the shells is the same as the ratio in the sea water at the time the shell grew.

In 1946 Harold Urey, the chemist who won the Nobel Prize in chemistry for his discovery of deuterium, developed the oxygen isotope ratio technique of reconstructing sea surface temperatures. By raising mollusks in water with a known ratio of heavy to light oxygen, Urey found that oxygen isotope ratios in shells of mollusks depended on two variables, the original oxygen isotope ratio of the

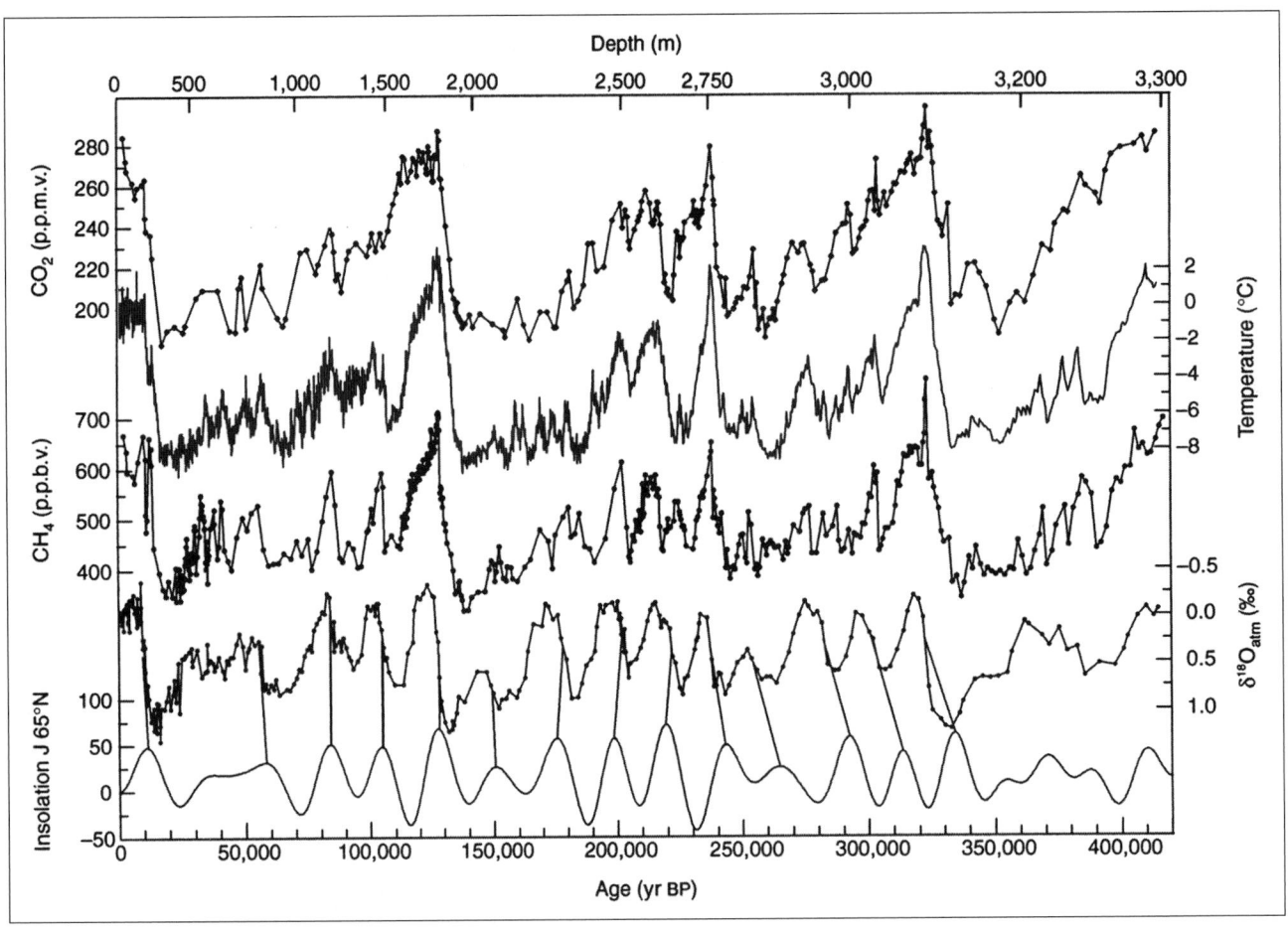

**Figure 7.26** 420,000 years of ice core data from Vostok, Antarctica and Milankovitch solar forcing for reference. (*Source:* US Global Change Research Program.)

water and the temperature of the water. Colder water contained a greater proportion of $^{18}O$ and by carefully calibrating his results Urey discovered that for every 1°C change in water temperature the proportion of heavy oxygen changed by 0.2 parts per thousand (0.2‰).

Oxygen isotope proxy data in sediment cores can be used to indicate both local sea surface temperature and the global volume of land ice. By analyzing shells of bottom dwelling or benthic organisms separately from floating or planktonic organisms found in the same core the "ice volume effect" can be measured.

Antarctic bottom water (AABW) is produced along the Antarctic coast as seasonal ice forms leaving behind cold and salty water which is the most dense ocean water in the world. It sinks to the ocean floor surrounding Antarctica and spreads northward in the Atlantic, Pacific and Indian Oceans with a nearly constant temperature of 1.5°C. Even during ice ages the temperature of bottom water was nearly the same temperature as it is today.

Because the temperature of bottom water does not change any variation in the oxygen isotope ratio measured in the shells of benthic organisms must be due to changes in the

original composition of the water. Isotopic fractionation controls the isotopic composition, as discussed in the previous section, and as more and more ice forms ocean water is increasingly richer in the heavy oxygen isotope $^{18}O$. This is called the "ice volume effect."

Planktonic organisms on the other hand contain both climatic signals. Not only does the oxygen isotope ratio reflect the ice volume but it also reflects the temperature of the water. By comparing the oxygen isotope ratio chronologies of benthic organisms to planktonic organisms both water temperature and ice volume can be estimated.

Sea water isotope proxy data goes back more than 2 billion years. Data that old is not from sediment cores but from ancient carbonate rocks. Figure 7.27 illustrates the variation of $^{18}O$ for the past 550 million years.

### Other Proxy Data

**Coral reefs.** Coral reefs have annual layers defined by oxygen isotope ratios. Salinity also effects oxygen isotope ratios. Coral reefs form in shallow water so ancient reefs in deep water indicate sea level has risen since the coral grew.

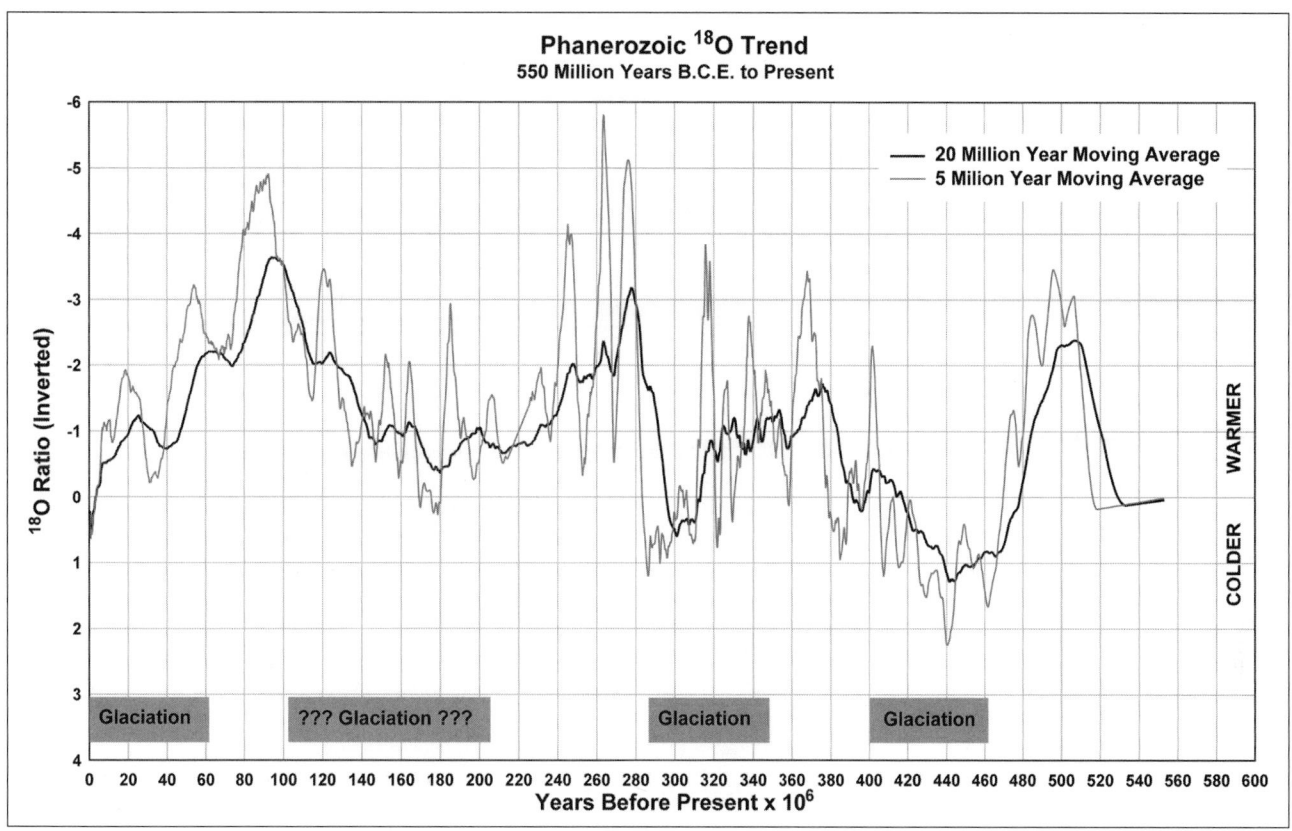

**Figure 7.27**  553 million years of oxygen isotope data from sea floor sediment cores and possible ice ages. *Source:* Steve Horstmeyer from Dr. Jan Veizer, Phanerozioc and Precambrian Isotope Databases, University of Ottawa.)

Data from corals off the mouth of the tropical Burdekin River in northeast Australia have been used to reconstruct annual river flows. When the discharge volume is great, much more humic acid and fulvic acid is washed into the ocean and subsequently incorporated into the calcium carbonate structure of the coral reefs.

These acids are a product of plant and soil microorganism metabolism and when present in coral skeletons they fluoresce, that is, give off yellow-green light when exposed to invisible ultraviolet light. The intensity of the light can be used to statistically reconstruct river flow.

Figure 7.28 shows the reconstruction and how the corals are cross referenced. Figure 7.29 is a graph of reconstructed runoff from the Burdekin River basin from 1644 through 1980. Runoff is calculated as the average depth of water over the basin in milimeters.

**Fossil pollen—palynology.** All flowering plants produce pollen grains. Because of their distinctive shapes pollen grains can be used to identify the plant species from which they came.

Pollen grains are well preserved in the sediment layers so the analysis of the pollen in a sediment core from a lake or ocean tells us what kinds of plants were growing at the time the sediment was deposited. Past climates can then be reconstructed from the pollen proxy data.

**Loess deposits.** Loess is windblown silt, pale yellow or buff in color. It originates as glacier ice grinds surface rock to a fine dust and occurs in deposits hundreds of feet thick and over thousands of square miles. The most famous deposits of loess used in climate research are found in western and northern China which may provide data that extends back 22 million years. In the United States loess is common from Iowa into Nebraska.

Grain size variations in loess layers tell climatologists about wind speeds and directions while layer thickness, chemical alteration of minerals in layers and soil development between layers indicates aridity or lack of it. Organic matter yields data on plant types, oxygen isotope ratios and carbon dating.

**Speleothems (cave deposits).** Stalactites that hang from a cave ceiling, stalagmites that grow from the cave floor and flowstone that covers larger areas are all calcium carbonate and yield good oxygen isotope proxy data.

In regions with strong seasonality in rainfall microscopically thin annual layers are present that allow absolute dating and ring width serves as an indicator of how wet the climate was when the ring formed. Annual layers also promise a long proxy record by cross referencing deposits from nearby caves.

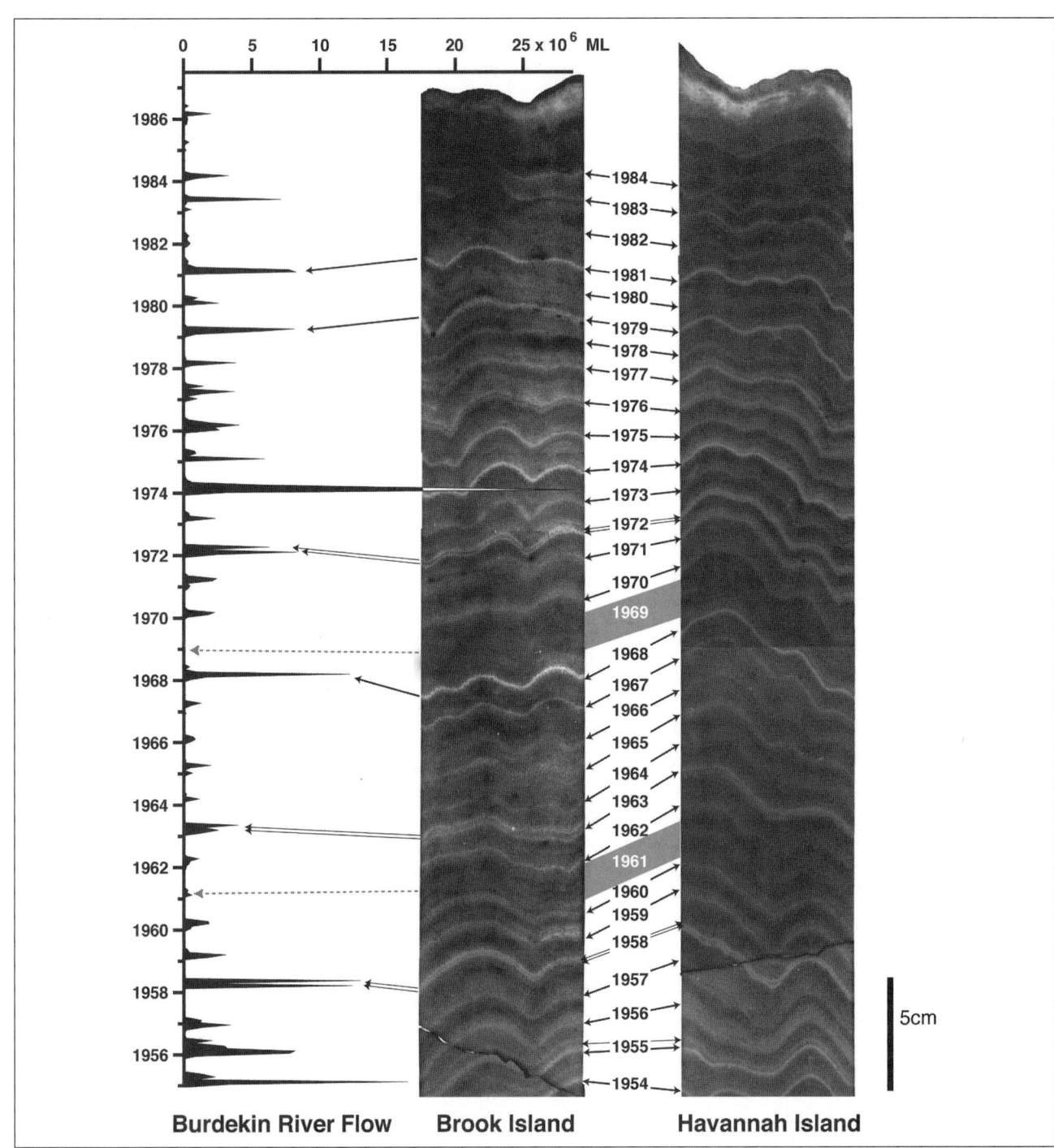

**Figure 7.28** Two coral cross sections showing annual layers and cross referenced. Light colored layers show fluorescence caused by organic acids washed out to sea during high river flow years and incorporated in the coral skeleton. (*Source:* World Data Center A—Paleoclimatology, P. J. Isdale et al. (2001).)

In Figure 7.30, using three separate stalagmites (which grow up from the floor of a cave) from Scotland, Italy, and China, researchers have reconstructed northern hemisphere temperatures from 1500 A.D. to 2000 A.D.

Speleothem studies like this are relatively new in climate research and the climate scientists who did this research urge

caution, their results show much promise but much work needs to be done.

**Borehole data.** Direct temperature measurements from holes bored into the ground yield information on past climate. Heat flows from the surface or to the surface

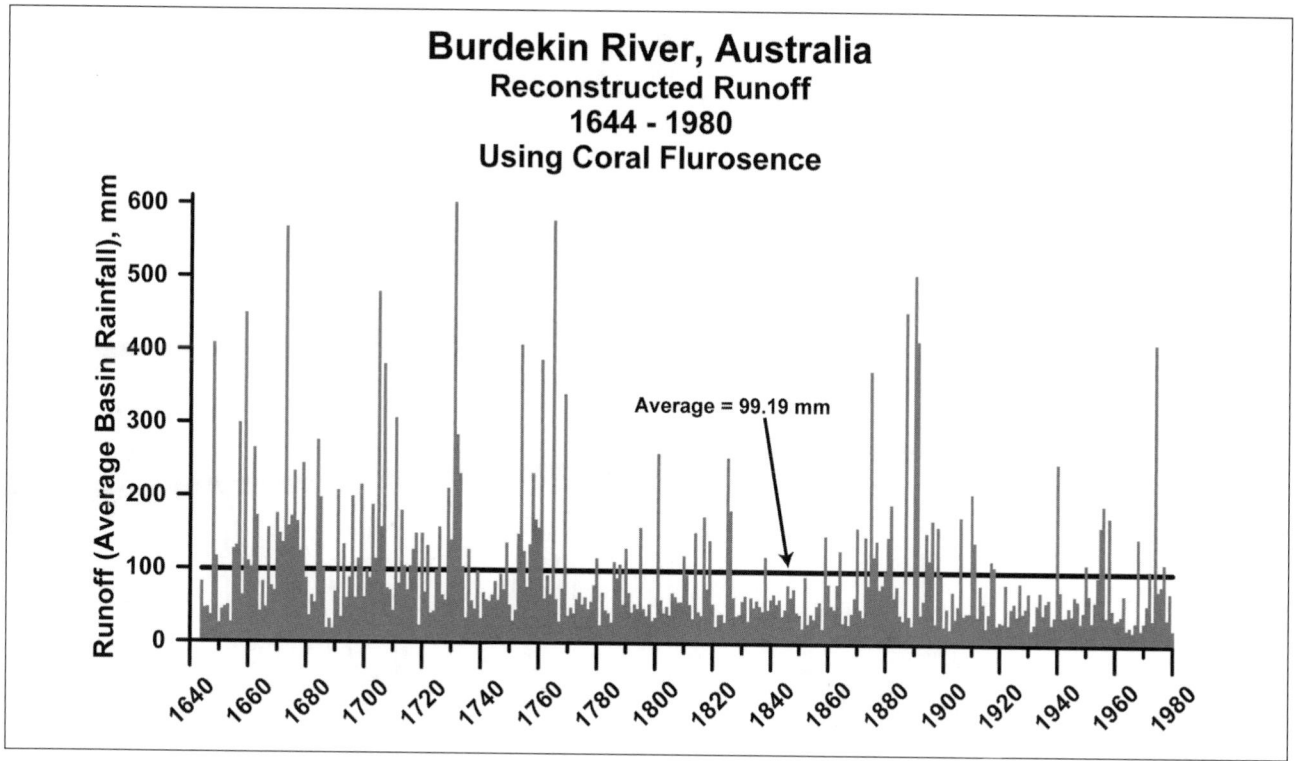

**Figure 7.29**   Reconstructed annual runoff of the Burdekin River, Australia using coral fluorescence. (*Source:* World Data Center A—Paleoclimatology, P. J. Isdale et al. (2001).)

depending on whether the air or ground is warmer. Analysis of the temperature profile can reveal temperature trends in the atmosphere.

**Lake level data.**   This reveals runoff, evaporation and water table information.

**Insect data.**   Both fossil and unfossilized soil remains yield information about past climate by revealing ancient environments.

**Lake and stream sediment cores.**   They are used in paleolimnology to reconstruct former inland water environments.

**Paleoforest fire data.**   Using tree fire scars and other proxies forest fire occurrence can be mapped and followed through time.

**Phytolith data.**   A phytolith is a microscopic hard part produced by many plants, which is made of silica and therefore long lasting. Phytoliths are used to track vegetational changes and identify paleoenvironments. Data goes back in some cases 350 million years. Phytoliths from grasses have been found in dinosaur dung in India.

**Fossil animal data.**   Like plants fossil animals reveal past environments and fossilized animal dung can reveal what the animal was feeding on evealing past environments.

**Pack rat middens.**   Middens are masses of plant material and animal remains encased in crystallized pack rat urine.

They are found in caves and rock shelters in southwest and other arid regions of the world. and reveal much about paleoenvironments. Figure 7.31 shows a pack rat midden the coin is a dime for scale.

## • PREDICTING FUTURE CLIMATES

A climate model is a set of complex equations that expresses the relationship between parts of the climate system and how they interact. Super computers or computer networks solve the equations over a period of future or past time and the results constitute a climate simulation. Climate models are the primary tool climate scientists use to understand how climate may change in the future.

### Fundamental Processes

There are four fundamental types of processes in computer models of Earth's climate system:

1. *Radiative processes*—the transfer of radiation, both solar and terrestrial, into, through and out of the climate system.
2. *Dynamic processes*—the transfer of energy both horizontal and vertical through the climate system. This is done primarily by global winds and ocean currents.

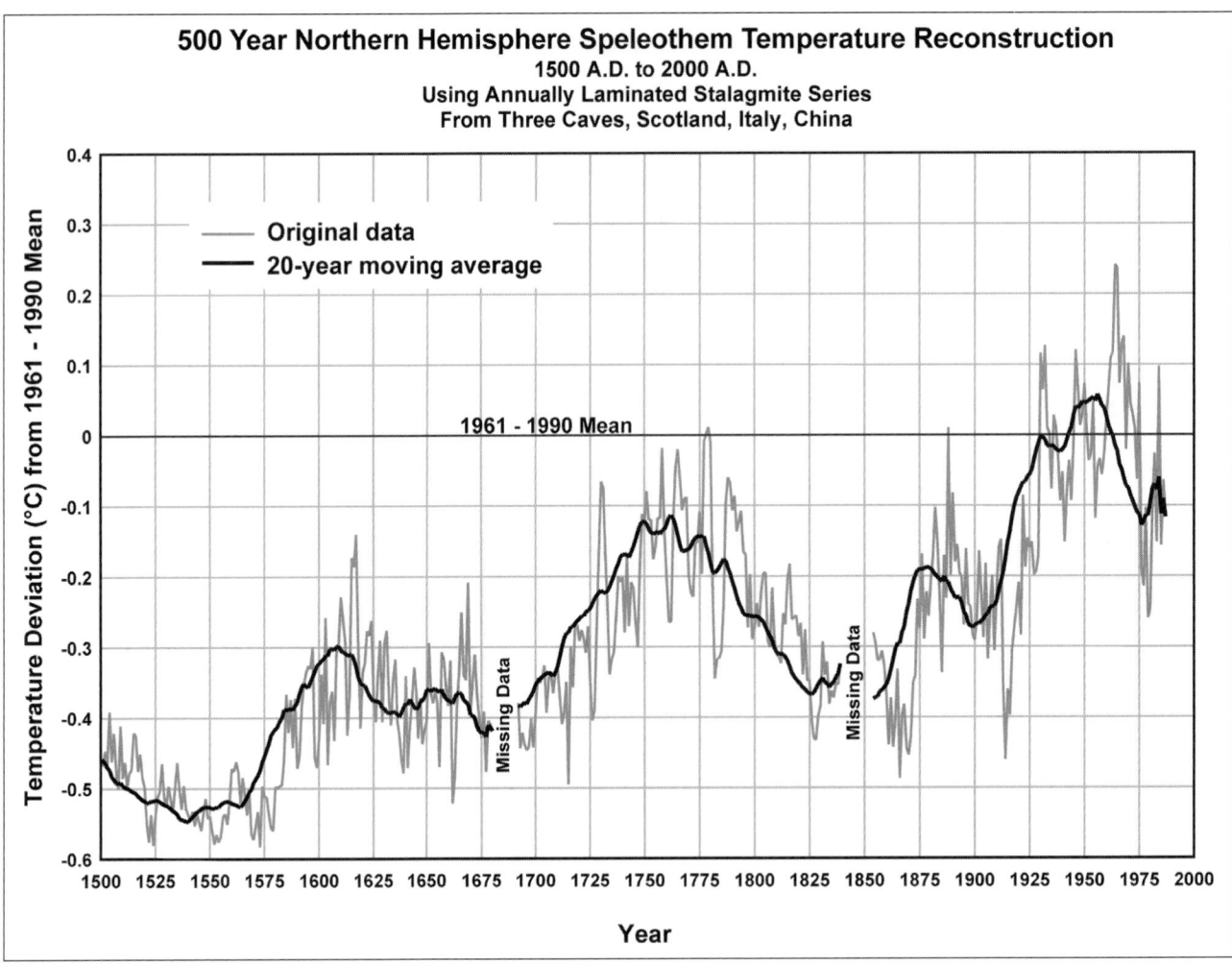

**Figure 7.30** Northern hemisphere temperature reconstruction from speleothem (cave deposits) oxygen isotope ratios, 1500 A.D.–2000 A.D. (*Source:* World Data Center A—Paleoclimatology, C.S. Smith et al. (2006).)

3. *Surface processes*—changes of surface characteristics (soil moisture, snow cover, sea ice, sea surface temperature, vegetation and more) and how these changes effect the interaction of the surface with the climate system.
4. *Chemical processes*—the chemical composition of the atmosphere, ocean, and other parts of the climate system along with the myriad of chemical reactions that take place between the components.

## Climate Models—Types

Generally climate models fall into four categories. In order of increasing complexity and increasing computational requirements they are:

1. Energy balance models (EBMs)
2. Radiative–convective models (RCMs)
3. Statistical–dynamical models (SDMs)
4. General circulation models (GCMs: AGCM, OGCM, CGCM)

### EBMs

These are the simplest and oldest of climate models. An EBM attempts to account for all energy coming into and going out of Earth's climate system and the equator to pole transport of heat energy in an effort to balance Earth's radiation budget. EBMs do not consider the nature of the atmosphere in the vertical only the incoming and outgoing energy. Typically EBMs operate over varying latitudes.

The basic variables of an EBM are: shortwave radiation received from the sun (insolation), shortwave solar radiation reflected by Earth (albedo), infrared radiation emitted by Earth and energy stored within the atmosphere by greenhouse gases.

The balance calculations of an EBM are relatively simple which makes EBMs good teaching tools and good for investigating individual climate forcing mechanisms like Milankovitch cycles, albedo on a glaciated Earth, cloud cover changes in isolation from the numerous complicating interactions and feedbacks.

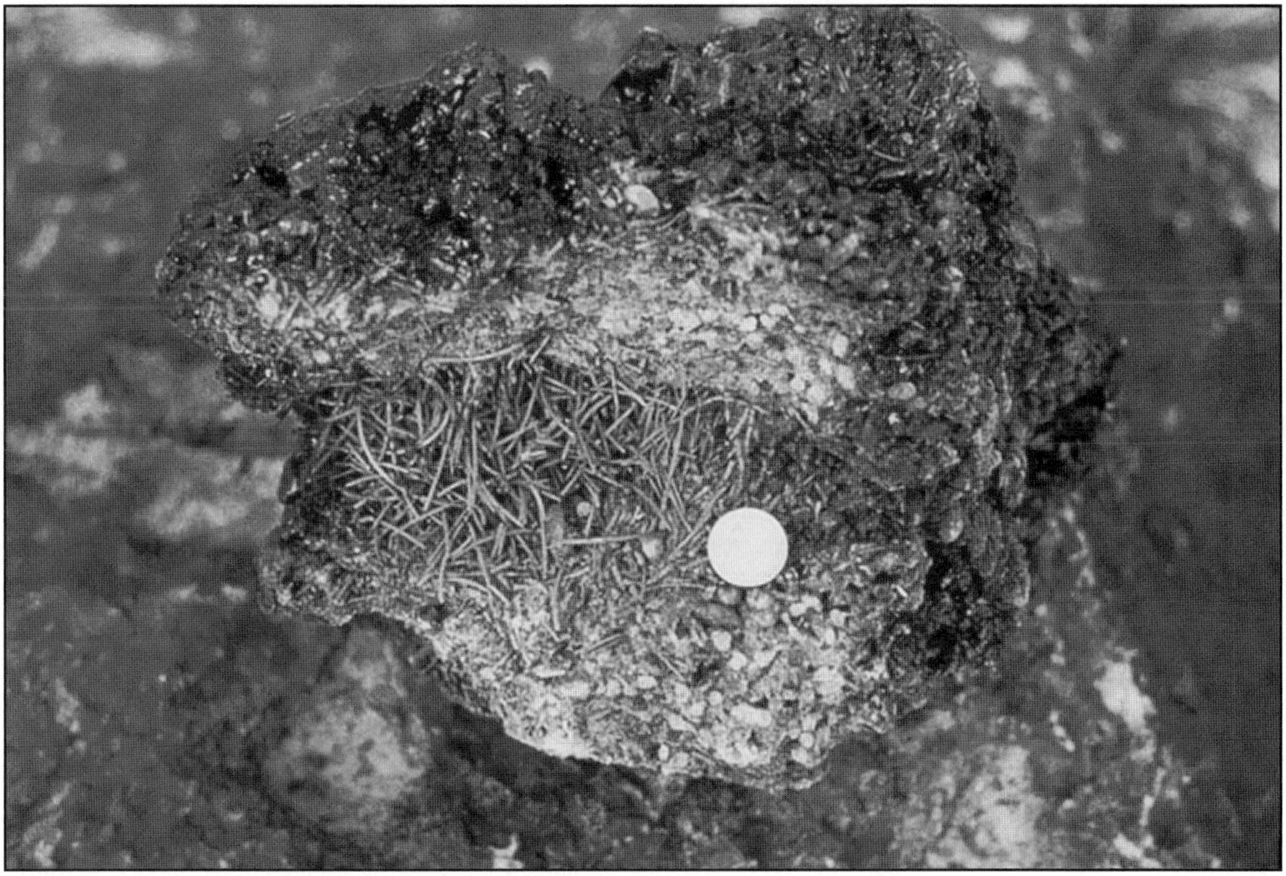

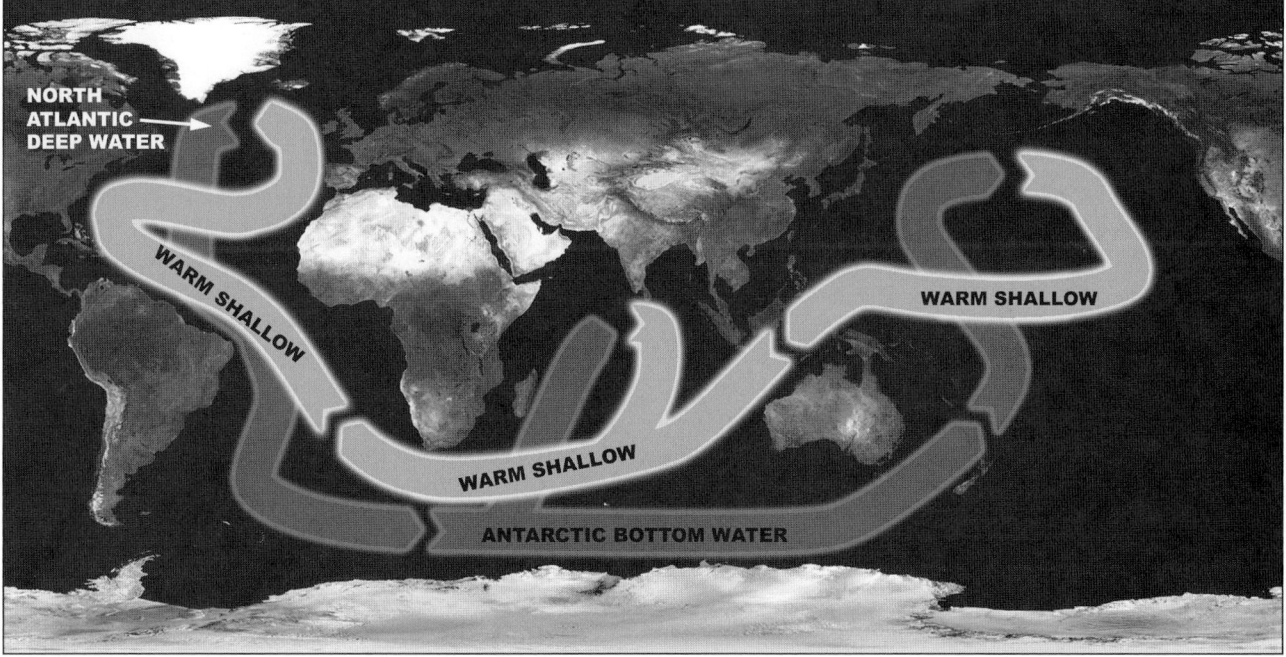

**Figure 7.31**　A pack rat midden is a mass of plant and animal material held together by crystalized pack rat urine many contain valuable paleoclimatic. (*Source:* NOAA Paleoclimatology Program.)

**Figure 7.32**　The global oceanic conveyor belt, also called the thermohaline circulation, is driven by cold, salty, and therefore dense water sinking as winter pack ice forms in both the Arctic and Antarctic.

EBMs can also be used to develop new ways to parameterize (represent) fundamental processes in more complicated models.

### RCMs

They operate only in the vertical dimension. RCMs take into account the upward and downward transport of heat in the atmosphere both radiative and convective. Vertical transport of parcels of air including heat content is called convection.

In the vertical a RCM can estimate surface temperature and the temperature at varying levels of the atmosphere. RCMs can also estimate the greenhouse effect more accurately than EBMs because they can account for varying greenhouse gas concentrations at different altitudes and the interactions between those layers.

The vertical column is often a global average at a single grid point of a GCM, the most complex climate models. RCMs are ideal to study the sensitivity of the climate system to humidity, clouds, greenhouse gas concentrations and temperature lapse rate from surface upward through the atmosphere in the GCM.

### SDMs

These are two-dimensional and they represent the vertical and latitudinal (north–south) dimensions. A model run simulates a 2D slice of the atmosphere and the vertical and north–south transport of water vapor, heat and kinetic energy. SDMs use atmospheric dynamics, i.e. the equations of motion, and statistical representations of winds to estimate horizontal transports. Statistical representations reduce the number of computations required during a climate simulation.

SDMs are particularly useful at simulating the effects of small changes in sun-diminishing tropospheric aerosols, stratospheric ozone and volcanically derived stratospheric sulfate aerosols because a SDM simulation can run as much as 1000 times faster than in a GCM.

### GCMs

They are the most complex and sophisticated models used to simulate past and future climate. GCMs try to represent every component of the climate system, surface vegetation, ocean ice, land use changes, sea surface temperatures and the interactions between the ocean and atmosphere.

When global climate change is in the headlines and the term "climate model" appears it is most often a GCM that is being discussed. GCM in the popular literature is often said to mean "global climate model" instead of "global circulation model." There is little or no controversy in using the terms interchangeably.

GCMs come in several varieties depending on what is being simulated. An OGCM is an Ocean GCM that simulates the general circulation of Earth's oceans while AGCM stands for Atmospheric GCM. If an atmospheric GCM is combined with an oceanic GCM the term coupled GCM (CGCM) may be used.

Essentially modern GCMs consist of individual modules designed to simulate one part of the global climate system. The modules representing land, atmosphere, ocean and sea ice, for example are linked by a coupler program that transfers radiant energy, kinetic energy and mass (water vapor, carbon dioxide, etc) between the modules.

AGCMs are designed to solve for radiant energy, near-surface atmospheric conditions, surface conditions, convective precipitation and clouds and large-scale precipitation at each time-step and each Earth coordinate. Earth coordinates cover the entire globe over a number of vertical layers.

OGCMs are designed to accurately represent wind driven surface (or shallow water) currents, the density driven deep circulation and the interactions of the two. Both require accurate representations of air–sea interaction. The role of the atmosphere is obvious for shallow water currents.

Deep water circulation is also called the thermohaline circulation or meridional overturning circulation. In general the thermohaline circulation originates in both the North Atlantic in the vicinity of Greenland and around the entire Antarctic continent. Each southern winter as sea ice forms around Antarctica the salinity of unfrozen water increases and the temperature cools. Both increasing salinity and cooling temperature increase water density and the water sinks to the sea floor and spreads northward along the floor of the world's oceans. During northern winter the same happens, at a smaller scale, in the vicinity of Greenland.

The sinking Antarctic waters are the most dense open ocean waters on Earth and the resulting water masses are called AABW. The north Atlantic counterpart is called North Atlantic deep water (NADW) and it flows at intermediate depths between the shallow water currents above and AABW below.

The two dense water masses are connected in a global circulation called the Global Ocean Conveyor Belt (Figure 7.32). There is evidence that if sea ice formation decreases the conveyor belt wsill slow resulting in a decrease of the transport of heat poleward, especially in the northern hemisphere.

Paleoclimatic proxy evidence also indicates the possibility of the rapid onset of an ice age with the shutdown of the thermohaline circulation. As global warming proceeds, ice formation decreases and the melting of land ice increases resulting in greater fresh water inflow to the ocean. Each subseqent winter there will be less water dense enough to sink. The reduction of dense water, which drives the global ocean conveyor belt may cause the circulation to slow and the transport of heat poleward to decrease, cooling, in particular, the large landmasses of the northern hemisphere.

Accurately simulating the state of Earth's climate system, past, present and future, requires an in-depth knowledge of all the climate system's components. Early climate modeling concentrated on the atmosphere at times ignoring the ocean and other components. Often this was a consequence of limited computational power. As climate science

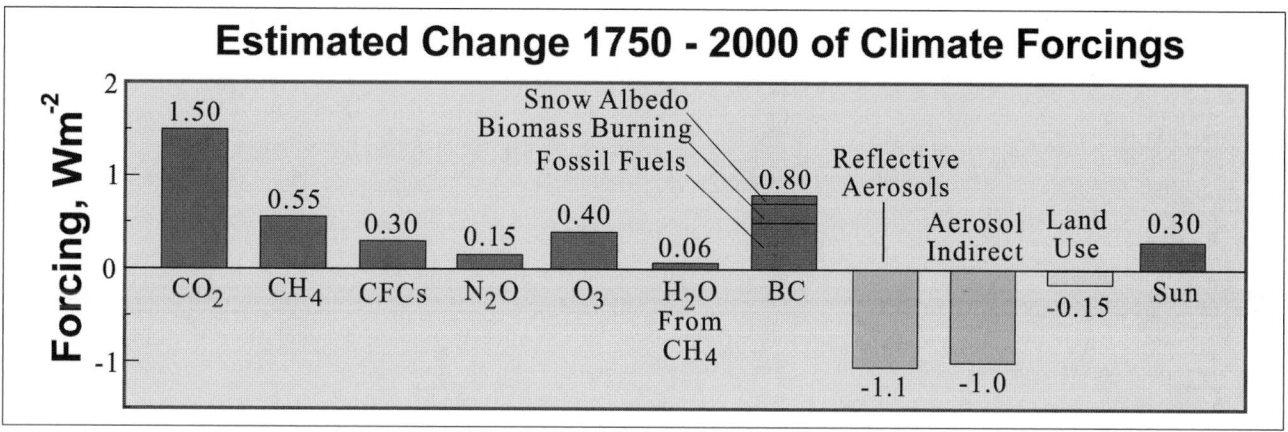

**Figure 7.33**  Estimated change in radiative forcings, 1750–2000. (*Source:* Steve Horstmeyer, Redrawn from NASA, Goddard Institute for Space Studies.)

advanced climate modelers realized that all components of Earth's climate system, lithosphere, biosphere, chryosphere, atmosphere and hydrosphere play an important role in determining the state of Earth's climates making the very basis of climate science interdisciplinary.

### Climate Sensitivity

A primary task of climate science is to determine the sensitivity of Earth's climate system to forcings. And to do so they must answer this question: "How much does the average global surface temperature change, when the energy budget is again in balance after being forced?"

But two forcings that seem to be equal may turn out to be very different after feedbacks are taken into account. Climate scientists then calculate adjusted measures of forcing. An in depth discussion of the many measures of climate sensitivity is beyond the scope of this chapter. Figure 7.33 shows the estimated change from 1750–2000 in forcings used in NASA's Goddard Institute for Space Studies General Circulation Model to estimate climate sensitivity. Figure 7.34 shows greenhouse gas and black carbon forcings as a percentage of the total.

Ultimately the sensitivity of a climate model to a forcing should exactly match the sensitivity of the climate

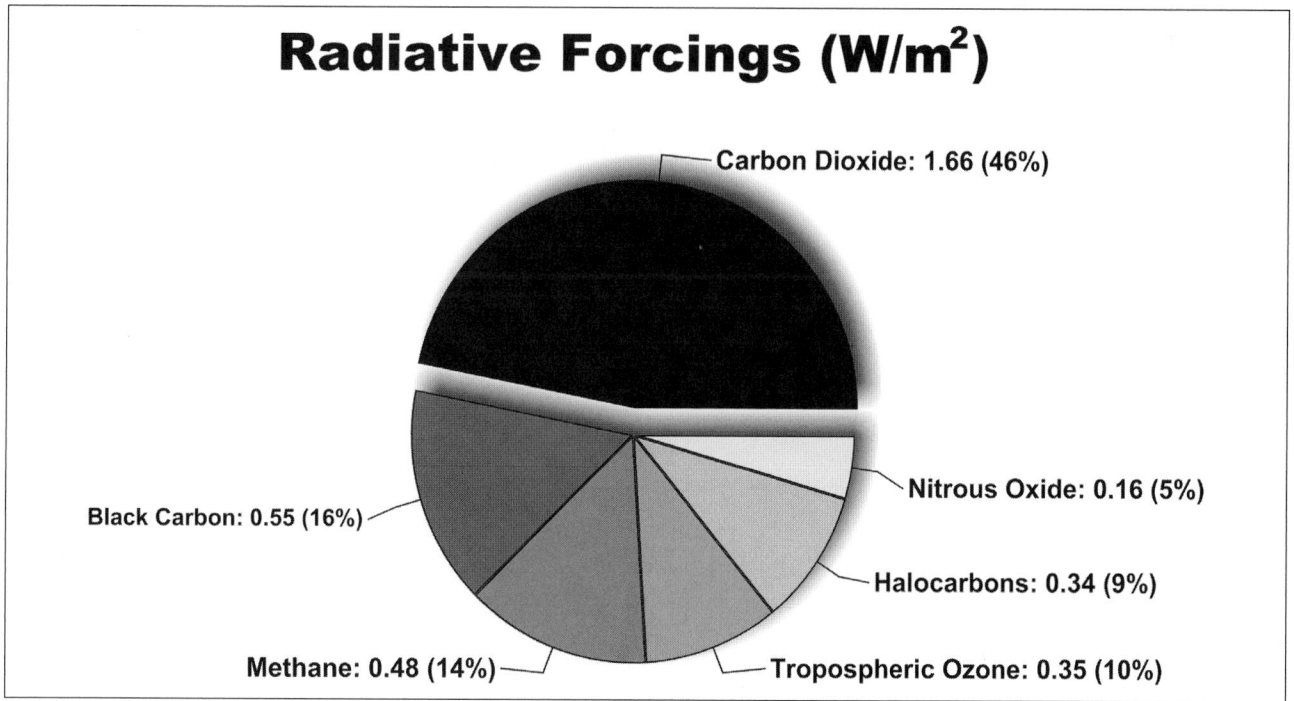

**Figure 7.34**  Individual contribution of greenhouse gases and black carbon to global warming. (*Source:* Steve Horstmeyer. Redrawn from NASA, Goddard Institute for Space Studies.)

system. Because carbon dioxide plays such an important role in global warming many climate modelers have focused on the sensitivity (model and real climate) to changes in $CO_2$ and a commonly quoted measure is the change of global average surface temperature when $CO_2$ is doubled.

Early in the history of climate models (1979) two efforts at modeling determined climate sensitivities for a doubling of atmospheric $CO_2$ to be 2°C and 4°C with a margin of error of ±0.5°C. The range of sensitivity most often quoted today, 1.5°C to 4.5°C is from that early study.

Using observations of present climate and increasingly accurate model simulations the estimated range from 1979 has neither changed much nor has the range been narrowed much. State of the art models are now beginning to restrict the range of climate sensitivity to between 2.6°C and 4.1°C. Notice this is at the higher end of the original estimate which means early studies man have underestimated the warming potential of increasing greenhouse gas concentrations.

Climate sensitivity is not the same as the predicted surface temperature change. Forecast surface temperature changes range from 1.4°C to 5.8°C by the year 2100 for a doubling of $CO_2$. Because the estimates include uncertainties in future greenhouse gas emissions the range is greater than the climate sensitivity.

# 8

# Air Pollution

## • INTRODUCTION

Dirty air is not a recent problem, and the attempts to regulate air pollution in the United States did not begin in the 1970s with the birth of the US Environmental Protection Agency, but as long as 100 years ago with a few scattered efforts. The widespread realization that air pollution is a serious health threat and that the federal government should play a leading role is however new. Out of these concerns and initiatives has come the creation of the most comprehensive air pollution control laws in the history of the United States.

## • SOME AIR POLLUTION HISTORY

In 61 A.D. Lucius Seneca complained about the quality of the air in Rome, the world's first city to reach 1 million residents. In 1306 Edward I of England ordered that the burning of charcoal (sea coal) in craftsman's furnaces be prohibited because of the foul-smelling fumes.

In the sixteenth century, Elizabeth I banned the burning of coal in London while Parliament was in session because of the fumes and soot.

In the United States during the 1840s, Louisa Taft—mother of future President William Howard Taft—complained about coal soot landing on her clothes when shopping in the fastest-growing American city of the time Cincinnati. About 50 years later, in 1896, Dr. Julius Cohen published the first comprehensive book on air pollution, *The Air of Towns.*

Nine years later, Dr. Henry Antoine Des Voeux presented a paper titled *Fog and Smoke* at the Public Health Congress of 1905. On July 26, he was quoted by the London newspaper *Daily Graphic.*

"He said it required no science to see that there was something produced in great cities which was not found in the country, and that was smoky fog, or what was known as 'smog'."

The *Daily Graphic* claimed he coined the term, but "smog" appears as early as January 19, 1893, in the *Los Angeles Times.*

Dr. Des Voeux may not have coined it, but it is almost certain that he introduced the term into public health and air pollution control circles. The term smog came into popular use after Dr. Des Voeux's 1911 report to the Manchester Conference of the Smoke Abatement League regarding the deadly 1909 air pollution episode in both Glasgow and Edinburgh, Scotland.

Air pollution initially was thought to be more a nuisance than a threat to human health. Even where emissions were excessive if the wind dispersed the pollutants over a large region, the concentrations remained low enough that health effects were not acute. But as pollution increased occasionally, weather conditions trapped the dirty air and the health effects became obvious. After five famous and deadly smog episodes in the twentieth century, attitudes toward air pollution began to change.

After World War II the increasing popularity of the automobile and the suburbanization of America meant that factories and residential coal consumption would be replaced as the number one source of airborne pollution. The automobile not only introduced new pollutants, but also a new type of pollution source, the mobile source, and the term "smog" was no longer sufficient to characterize air pollution.

"Sulfurous smog" was used to describe air pollution that results in the episodes provided in Table 8.1. It results from the burning of fuels with a high concentration of sulfur. Often dampness or fog and a high concentration of particulate matter would aggravate the episode. Typically found in eastern cities where coal was the primary fuel for residential uses, sulfurous smog was made worse in cities in valleys where air could easily be trapped by temperature inversions.

Sulfurous smog is now relatively rare in the United States because of air pollution abatement efforts and the change to natural gas and other low-sulfur, petroleum-based fuels. Photochemical smog is now the main type of urban air

*The Weather Almanac: A Reference Guide to Weather, Climate, and Related Issues in the United States and Its Key Cities*, Twelfth Edition. Steve Horstmeyer.
© 2011 John Wiley & Sons, Inc. Published 2011 by John Wiley & Sons, Inc.

**Table 8.1  Five Deadly Air Pollution Episodes of the Twentieth Century**

| Location | Date | Cause | Consequences |
|---|---|---|---|
| Glasgow and Edinburgh, Scotland | 1909 | Smoke, sulfurous gases and fog | An estimated 1000 deaths |
| Meuse River Valley, Belgium | 1930 | High concentrations of sulfur dioxide during a temperature inversion. | 63 deaths, 1000 sick |
| Donora, Pennsylvania | 1948 | High concentrations of sulfur dioxide, temperature inversion, foggy weather. | 20 deaths due to cardiac and respiratory disease, 6,000 complained of cough, respiratory tract irritation, chest pain, headaches, nausea, and vomiting. |
| Poza Rica, Mexico | 1950 | Natural gas plant released hydrogen sulfide coupled with temperature inversion and foggy weather. | 22 deaths 320 hospitalized |
| London, England "Great London Fog" | 1952 | A five day temperature inversion trapped deadly acid aerosols | > 4000 deaths from respiratory and cardiac disease. 8000 more may have died later. |

pollution. But even with the large growth in automobile traffic, pollution control efforts have lowered the concentrations of photochemical compounds from the levels of the middle and late 1960s just before the *Clean Air Act* became law. More often than not the term "smog" is used to describe all types of air pollution, but it usually refers to photochemical pollution because the vast majority of air pollution in the United States is related to automobile exhaust.

The automobile emits nitrogen oxides and hydrocarbons, which undergo transformation when sunlight energizes chemical reactions to produce ozone and other compounds that are oxidants and therefore corrosive. The term "photochemical smog" is used to describe this type of air pollution in which smoke and fog play no part.

### The Clean Air Act

In 1955 Congress passed the *Air Pollution Control Act*, which was the first federal air pollution legislation. The law funded research into the problem of dirty air.

The next step was taken in 1963 with the authorization of a national program to address pollution issues, and funding was approved to develop ways of minimizing air pollution.

Research was expanded in 1967 when new legislation tackled the problem of interstate transportation of air pollutants.

Fifteen years after the first federal air pollution statute, the *Clean Air Act* of 1970 became law. This legislation

1. Authorized the establishment of National Ambient Air Quality Standards (NAAQS)
2. Established requirements for state implementation plans to achieve the NAAQS
3. Authorized the establishment of new source performance standards for new and modified stationary sources
4. Authorized the establishment of National Emission Standards for Hazardous Air Pollutants
5. Increased enforcement authority
6. Authorized requirements for control of motor vehicle emissions.

On December 2, 1970, the same year as the *Clean Air Act* became law, the US Environmental Protection Agency (EPA) opened for business. The EPA's responsibilities include monitoring pollution, setting pollutant standards, and enforcing pollution laws in addition to doing research to protect human health and safeguard the natural environment.

The *Clean Air Act* was amended in 1977, with provisions to prevent areas that attained NAAQS from slipping back into noncompliance. In 1990 additional amendments increased the authority of the federal government in controlling air pollution and created a new program to control acid deposition (acid rain and related phenomena). NAAQS were included for toxic air pollutants, and provisions for stratospheric ozone were included.

## The Role of the EPA

The EPA states that its mission is "to protect human health and the environment." Congress writes environmental laws and the EPA is the government agency that implements those laws.

The EPA accomplishes its mission by

1. Developing and enforcing regulations, developing and setting standards that are enforced by states and tribes through their own regulations, helping states and tribes meet standards, enforcing regulations, and helping companies understand requirements
2. Studying and solving environmental issues at laboratories around the nation and sharing information with private organizations
3. Awarding grants to help states with environmental programs and other institutions with educational and research programs
4. Sponsoring partnerships with businesses, nonprofits, and state and local governments for programs that help solve environmental issues
5. Teaching people about the environment
6. Publishing information.

There are many areas that in the public's mind count as environmental protection but are not directly under the authority of the EPA, though the EPA often works with the responsible agencies to solve a specific problem. Here are some examples of areas the EPA does not regulate or enforce:

1. The *Endangered Species Act* is managed by the US Fish and Wildlife Service.
2. Wetlands issues fall under the authority of the US Army Corps of Engineers.
3. Workplace environmental issues are handled by the Occupational Safety and Health Administration.
4. The Consumer Product Safety Commission deals with the safety of products people buy on a daily basis.
5. The Food and Drug Administration is responsible for any substance applied to the human body, while pesticides and other substances used to control pests are under the EPA.
6. Noise pollution is a matter of local regulations and no longer regulated by the EPA.
7. Road dust and landfill issues are also local matters.

For administrative purposes, the EPA has divided the United States into ten regions, as shown in Figure 8.1. Regional offices are responsible for implementing and monitoring EPA programs for the states in that region. By dividing the country into smaller units, scientists in the regional offices can investigate environmental problems in greater detail than if everything was covered at the national level.

Air pollutants are categorized by the EPA into three general types: (1) criteria pollutants; (2) hazardous air pollutants (HAPs), also called toxic air pollutants; and (3) volatile organic compounds (VOCs).

### *Pollutant Inventories—Criteria Pollutants*

Criteria pollutants are common pollutants found in every community. They are called "criteria pollutants" because regulations, standards, and limits are developed through scientific research into human health criteria and environmental criteria.

There are six air pollutants the EPA classifies as "criteria pollutants," which are used as a ruler to measure the air quality of a region; they are particulate matter, ozone, carbon monoxide, lead, nitrogen oxides, and sulfur dioxide. Exposure to these pollutants is associated with numerous effects on human health, including increased respiratory symptoms, hospitalization for heart or lung diseases, and even premature death.

The EPA measures the concentration of each criteria pollutant at monitoring sites across the country and estimates the total amount of each criteria pollutant released using engineering methods and computer models.

The EPA uses the six "criteria pollutants" as a ruler to measure the overall air quality of a region. These "yardstick" concentrations are called the National Ambient Air Quality Standards. The NAAQS are summarized in Table 8.2.

**Ozone** ($O_3$) is the major component of smog. $O_3$ in the stratosphere is beneficial to life by absorbing harmful ultraviolet radiation from the sun. At ground level, high concentrations of $O_3$ are a major health and environmental concern. $O_3$ is formed through complex chemical reactions between volatile organic compounds (VOCs) and oxides of nitrogen ($NO_x$) energized by sunlight. Peak $O_3$ levels occur typically during the warmer times of the year because the photochemical reactions are temperature dependent. Both VOCs and $NO_x$ are emitted by transportation and industrial sources. VOCs are emitted from sources as diverse as automobiles, chemical manufacturing, dry cleaners, paint shops, and other industries using solvents.

$O_3$ causes health problems because it damages lung tissue, reduces lung function, and makes the lungs vulnerable to other irritants. Scientific evidence indicates that $O_3$ affects not only people with impaired respiratory systems, but healthy adults and children too.

**Carbon monoxide** (CO) is a colorless, odorless, poisonous gas produced by incomplete combustion of carbon in fuels.

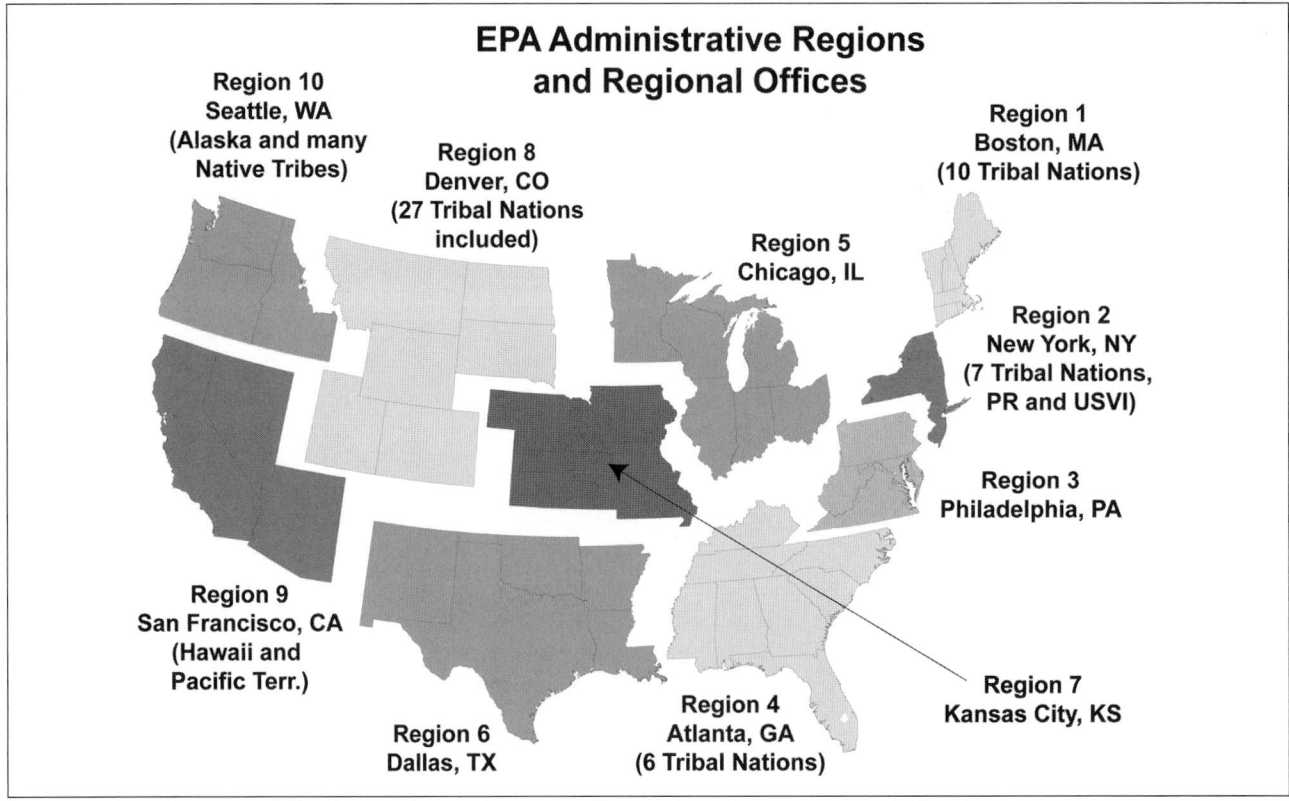

**Figure 8.1** Regions of the Environmental Protection Agency. (*Source:* US Environmental Protection Agency.)

The largest contribution of CO is from highway motor vehicles and so CO monitoring has been focused on urban areas where traffic is concentrated. Other major CO sources are wood-burning stoves, incinerators, and industrial sources.

When CO enters the bloodstream, the delivery of oxygen to the body's organs and tissues is reduced. The most seriously affected individuals are those who suffer from cardiovascular disease, particularly angina and peripheral vascular disease.

**Sulfur dioxide** ($SO_2$) is a colorless, pungent gas primarily from stationary sources such as coal and oil-burning power plants, refineries, steel mills, and paper mills. It affects breathing and may aggravate existing respiratory and cardiovascular disease. Sensitive individuals include asthmatics, those with bronchitis or emphysema, children, and the elderly. $SO_2$ is also a primary contributor to acid deposition (acid rain and related phenomena), which causes acidification of lakes and streams and can damage trees, crops, historic buildings, and statues. Sulfur compounds in the air also contribute to the reduction of visibility in large parts of the country. This is especially noticeable in national parks.

**Nitrogen dioxide** ($NO_2$) is a brownish, highly reactive gas that is present in all urban atmospheres. Nitrogen oxides ($NO_x$) form when fuel is burned at high temperatures and

there are two major emissions sources, transportation and stationary fuel combustion sources such as electric utility and industrial boilers. $NO_2$ can irritate the lungs, cause bronchitis and pneumonia, and lower resistance to respiratory infections. $NO_x$ are important in the formation of ozone ($O_3$) and acid rain, which affects both terrestrial and aquatic ecosystems.

**Particulates** or **particulate matter** is a class of air pollutants that include dust, dirt, soot, smoke, and liquid droplets directly emitted into the air by sources such as factories, power plants, cars, construction activity, fires, and natural windblown dust. Particles formed in the atmosphere by condensation or the transformation of emitted gases such as $SO_2$ and VOCs are also considered particulate matter.

These particles come in many shapes and sizes and are made of hundreds of different substances. The EPA classifies particle pollution as "inhalable coarse particles," with diameters larger than 2.5 μm ($PM_{10}$) and smaller than 10 μm, and "fine particles," with diameters that are 2.5 μm and smaller ($PM_{2.5}$). A 2.5-μm-diameter particle is 36 times smaller than a typical grain of fine sand that is 90 μm in diameter and 28 times smaller than a human hair that is about 70 μm in diameter.

Primary particles are emitted directly from a source, such as construction sites, unpaved roads, fields, smokestacks,

## Table 8.2  National Ambient Air Quality Standards

| Pollutant | Primary Standards | | Secondary Standards | |
|---|---|---|---|---|
| | Level | Averaging Time | Level | Averaging Time |
| Carbon Monoxide | 9 ppm (10 mg/m$^3$) | 8-hour(1) | None | |
| | 35 ppm (40 mg/m$^3$) | 1-hour(1) | | |
| Lead | 0.15 µg/m$^3$(1) | Rolling 3-Month Average | Same as Primary | |
| | 1.5 µg/m$^3$ | Quarterly Average | Same as Primary | |
| Nitrogen Dioxide | 0.053 ppm (100 µg/m$^3$) | Annual (Arithmetic Mean) | Same as Primary | |
| Particulate Matter (PM$_{10}$) | 150 µg/m$^3$ | 24-hour (3) | Same as Primary | |
| Particulate Matter (PM$_{2.5}$) | 15.0 µg/m$^3$ | Annual (4) (Arithmetic Mean) | Same as Primary | |
| | 35 µg/m$^3$ | 24-hour (5) | Same as Primary | |
| Ozone | 0.075 ppm (2008 std) | 8-hour (6) | Same as Primary | |
| | 0.08 ppm (1997 std) | 8-hour (7) | Same as Primary | |
| | 0.12 ppm | 1-hour (8) | Same as Primary | |
| Sulfur Dioxide | 0.03 ppm | Annual (Arithmetic Mean) | 0.5 ppm (1300 µg/m$^3$) | 3-hour (1) |
| | 0.14 ppm | 24-hour (1) | | |

(1) Not to be exceeded more than once per year.

(2) Final rule signed October 15, 2008.

(3) Not to be exceeded more than once per year on average over 3 years.

(4) To attain this standard, the 3-year average of the weighted annual mean PM2.5 concentrations from single or multiple community-oriented monitors must not exceed 15.0 µg/m3.

(5) To attain this standard, the 3-year average of the 98th percentile of 24-hour concentrations at each population-oriented monitor within an area must not exceed 35 µg/m3 (effective December 17, 2006).

(6) To attain this standard, the 3-year average of the fourth-highest daily maximum 8-hour average ozone concentrations measured at each monitor within an area over each year must not exceed 0.075 ppm. (effective May 27, 2008)

(7) (a) To attain this standard, the 3-year average of the fourth-highest daily maximum 8-hour average ozone concentrations measured at each monitor within an area over each year must not exceed 0.08 ppm.

   (b) The 1997 standard—and the implementation rules for that standard—will remain in place for implementation purposes as EPA undertakes rule making to address the transition from the 1997 ozone standard to the 2008 ozone standard.

(8) (a) The standard is attained when the expected number of days per calendar year with maximum hourly average concentrations above 0.12 ppm is $\leq$ 1.

   (b) As of June 15, 2005 EPA has revoked the 1-hour ozone standard in all areas except the fourteen 8-hour ozone non-attainment Early Action Compact (EAC) Areas. For one of the 14 EAC areas (Denver, CO), the 1-hour standard was revoked on November 20, 2008. For the other 13 EAC areas, the 1-hour standard was revoked on April 15, 2009.

*Source:* US Environmental Protection Agency.

or fires. Secondary particles result from complex reactions between chemicals such as sulfur dioxides and nitrogen oxides that are emitted from power plants, industries, and automobiles.

Particulates are a major concern for human health, including effects on breathing, aggravation of existing respiratory and cardiovascular disease, alterations in the body's defense systems against foreign materials, damage to lungs, formation of cancer, and premature death. The elderly, children, and individuals with chronic obstructive pulmonary disease, cardiovascular disease, influenza, and asthma are especially sensitive to airborne particulate matter. Particulates are also responsible for reduction of visibility.

**Lead** (Pb) can be ingested in a number of ways, including inhalation of air with lead particulates and ingestion in food, water, soil, or dust. Blood levels of lead have been reduced dramatically since 1975 mainly because of the introduction of unleaded gasoline. In 1984, unleaded gasoline accounted for 60% of sales; by 1993, unleaded gasoline sales had risen to 99% of the total. The elimination of lead in soldered cans is an additional factor in lowering Pb blood levels.

Infants and young children are especially susceptible to low doses of Pb, and this age group still shows the highest blood levels. Low doses of Pb can lead to central nervous system damage.

### Pollutant Inventories—Hazardous Air Pollutants

Hazardous air pollutants, also known as toxic air pollutants, are chemicals and substances that are known or suspected to cause cancer or other serious health effects or adverse environmental effects. The EPA is working with state, local, and tribal governments to reduce air toxic releases of 188 HAPs to the environment. Examples of toxic air pollutants include benzene, which is found in gasoline; perchlorethylene, which is emitted from some dry-cleaning facilities; and methylene chloride, which is used as a solvent and paint stripper by a number of industries. Examples of other listed air toxics include dioxin, asbestos, toluene, and metals such as cadmium, mercury, chromium, and lead compounds.

There are currently 188 HAPs regulated by the *Clean Air Act* and thousands of additional other potentially toxic substances that are being emitted into the environment. Some sources may be of greater importance in your state because of amount emitted and more concern to local officials, so not all parts of the United States treat these pollutants in the same way.

### Pollutant Inventories—Volatile Organic Compounds

In addition to criteria pollutants and HAPs, the EPA also regulates emissions of VOCs under the criteria pollutant program. VOCs are organic compounds that easily vaporize in the atmosphere. Exposure can be through inhalation of vapor, skin contact with or swallowing of a liquid, and occasionally through contact with a solid form. VOCs can be hazardous when inhaled and should be avoided by pregnant women.

There are both natural and synthetic VOCs, and the EPA estimates that 72% are from natural sources and 28% are anthropogenic. Methane is technically a VOC but is excluded from the VOC lists and treated separately because of its role in global warming.

Gasoline, paints, pesticides, adhesives, cleaning solvents, paint strippers, photocopiers, and building materials (plywood for example) may emit VOCs.

Natural sources of VOCs include pine and citrus trees. The strong scent of pine needles is due to a group of VOCs chemists call monoterpenes. The "smoke" or blue haze that gives the Smoky Mountains their name is thought to be caused by a VOC group called sesquiterpenes, which is emitted by the pine trees. Limonene is a VOC that gives oranges their smell and is found in all citrus fruits.

One of the best-known effects of VOC exposure is "sick building syndrome." Many paints, adhesives, and building materials emit the VOC formaldehyde, the most common culprit in "sick building syndrome," which irritates mucous membranes and can lead to general irritability.

VOCs are ozone precursors, which means they can lead to increased surface ozone concentrations when they react with nitrogen oxides in the atmosphere. VOCs are emitted from motor vehicles, fuel distribution, chemical manufacturing, and a wide variety of industrial, commercial, and consumer solvent uses.

## • NATIONAL AMBIENT AIR QUALITY STANDARDS AND PROGRESS IN CLEANING AMERICA'S AIR

Tables 8.2–8.5 tell the story of cleaning America's air. The NAAQS are listed in Table 8.2. Each pollutant has a maximum concentration and averaging time, and sulfur dioxide has a secondary standard. The six pollutants ($PM_{2.5}$ and $PM_{10}$ count as one) are the criteria pollutants.

The *Clean Air Act* established two types of national air quality standards. Primary standards are designed to protect public health, including the health of "sensitive" populations such as asthmatics, children, and the elderly, while secondary standards set pollutant limits to protect public welfare, including protection against visibility impairment, damage to animals, crops, vegetation, and buildings. The *Clean Air Act* requires periodic review of the science upon which the standards are based and the standards themselves.

Tables 8.3–8.5 indicate across-the-board improvement for all criteria pollutants from 1980 through 2008, but there is still work to do because Figure 8.2 shows that millions continue to live in areas that violate the NAAQS.

**Table 8.3    Percent Change in Air Quality**

| (Negative numbers indicate less pollution) | | |
|---|---|---|
| | **1980 to 2008** | **1990 to 2008** |
| **Carbon Monoxide (CO)** | -79 | -68 |
| **Ozone ($O_3$) (8-hr)** | -25 | -14 |
| **Lead (Pb)** | -92 | -78 |
| **Nitrogen Dioxide ($NO_2$)** | -46 | -35 |
| **$PM_{10}$ (24-hr)** | --- | -31 |
| **$PM_{2.5}$ (annual)** | --- | -19 |
| **$PM_{2.5}$ (24-hr)** | --- | -20 |
| **Sulfur Dioxide ($SO_2$)** | -71 | -59 |

Notes:
1. --- Trend data not available
2. PM2.5 air quality based on data since 2000
3. Negative numbers indicate improvements in air quality

*Source:* US Environmental Protection Agency.

**Table 8.4    Percent Change in Emissions**

| | **1980 to 2008** | **1990 to 2008** |
|---|---|---|
| **Carbon Monoxide (CO)** | -56 | -46 |
| **Lead (Pb)** | -99 | -79 |
| **Nitrogen Oxides ($NO_X$)** | -40 | -35 |
| **Volatile Organic Compounds (VOC)** | -47 | -31 |
| **Direct $PM_{10}$** | -68 | -39 |
| **Direct $PM_{2.5}$** | --- | -58 |
| **Sulfur Dioxide ($SO_2$)** | -56 | -51 |

Notes:
1. --- Trend data not available
2. Direct PM10 emissions for 1980 are based on data since 1985
3. Negative numbers indicate reductions in emissions

*Source:* US Environmental Protection Agency.

**Table 8.5   National Emissions Estimates**

| National Emissions Estimates (Fires and Dust Excluded) For Common Pollutants and their Precursors | | | | | | | |
|---|---|---|---|---|---|---|---|
| **Pollutant** | **Millions of Tons Per Year** | | | | | | |
| | **1980** | **1985** | **1990** | **1995** | **2000** | **2005** | **2008** |
| Carbon Monoxide (CO) | 178 | 170 | 144 | 120 | 102 | 93 | 78 |
| Lead | 0.074 | 0.023 | 0.005 | 0.004 | 0.003 | 0.002 | 0.002 |
| Nitrogen Oxides (NOx) | 27 | 26 | 25 | 25 | 22 | 19 | 16 |
| Volatile Organic Compounds (VOC) | 30 | 27 | 23 | 22 | 17 | 18 | 16 |
| Particulate Matter - $PM_{10}$ | 6 | 4 | 3 | 3 | 2 | 2 | 2 |
| Particulate Matter - $PM_{2.5}$ | NA | NA | 2 | 2 | 2 | 1 | 1 |
| Sulfur Dioxide ($SO_2$) | 26 | 23 | 23 | 19 | 16 | 15 | 11 |
| Totals | 267 | 250 | 218 | 189 | 159 | 147 | 123 |

Notes:
In 1985 and 1996 EPA refined its methods for estimating emissions. Between 1970 and 1975, EPA revised its methods for estimating PM emissions.

The estimates for 2005 and beyond are from the final version 2 of the 2005 NEI.

For CO, NOx, SO2 and VOC emissions, fires are excluded because they are highly variable; for direct PM emissions both fires and dust are excluded.

PM estimates do not include condensable PM.

EPA has not estimated PM2.5 emissions prior to 1990.

The 1999 estimate for lead is used for 2000, and the 2002 estimate for lead is used for 2005 and 2008. PM2.5 emissions are not added when calculating the total because they are included in the PM10 estimate.

*Source:* US Environmental Protection Agency.

Figures 8.3–8.9 are graphs showing that the concentration of each of the criteria pollutants has decreased beginning in 1980 and continuing through 2008. The upper line represents the 90th percentile or the concentration that 90% of the monitoring stations fall below. The lower line is the 10th percentile; only 10% of the stations have pollutant concentrations less than this level.

Ozone continues to be a problem throughout most of the United States, and though lead levels have fallen dramatically since 1980, the concentrations of lead are still near the NAAQS and not significantly below it.

Figures 8.10–8.19 show the emissions of VOCs for each EPA region, and Figure 8.20 shows emissions of VOCs by source for 1990 and for 1996–2002. Each graph is based on the latest data available from the EPA.

The average emissions of toxic air pollutants by region for the years 1990–1993 and the 2002 value are illustrated in Figures 8.21 and 8.22.

## • INFORMING THE PUBLIC—AIR QUALITY INDEX

Table 8.6 is the Air Quality Index (AQI) which tells you how clean or polluted the air is and the health effects that might be a concern for you. Though there could be long-term health effects from prolonged exposure of certain pollutants, the AQI focuses on health effects you may experience within a few hours or days after breathing polluted air.

The AQI takes into account the concentrations of five major air pollutants regulated by the *Clean Air Act*:

## Table 8.6   The Air Quality Index

| Air Quality Index Levels of Health Concern | Numerical Value | Meaning |
|---|---|---|
| **Good** (Green) | 0 to 50 | Air quality is considered satisfactory, and air pollution poses little or no risk |
| **Moderate** (Yellow) | 51 to 100 | Air quality is acceptable; however, for some pollutants there may be a moderate health concern for a very small number of people who are unusually sensitive to air pollution. |
| **Unhealthy for Sensitive Groups** (Orange) | 101 to 150 | Members of sensitive groups may experience health effects. The general public is not likely to be affected. |
| **Unhealthy** (Red) | 151 to 200 | Everyone may begin to experience health effects; members of sensitive groups may experience more serious health effects. |
| **Very Unhealthy** (Purple) | 201 to 300 | Health alert: everyone may experience more serious health effects |
| **Hazardous** (Maroon) | 301 to 500 | Health warnings of emergency conditions. The entire population is more likely to be affected. |

*Source:* US Environmental Protection Agency.

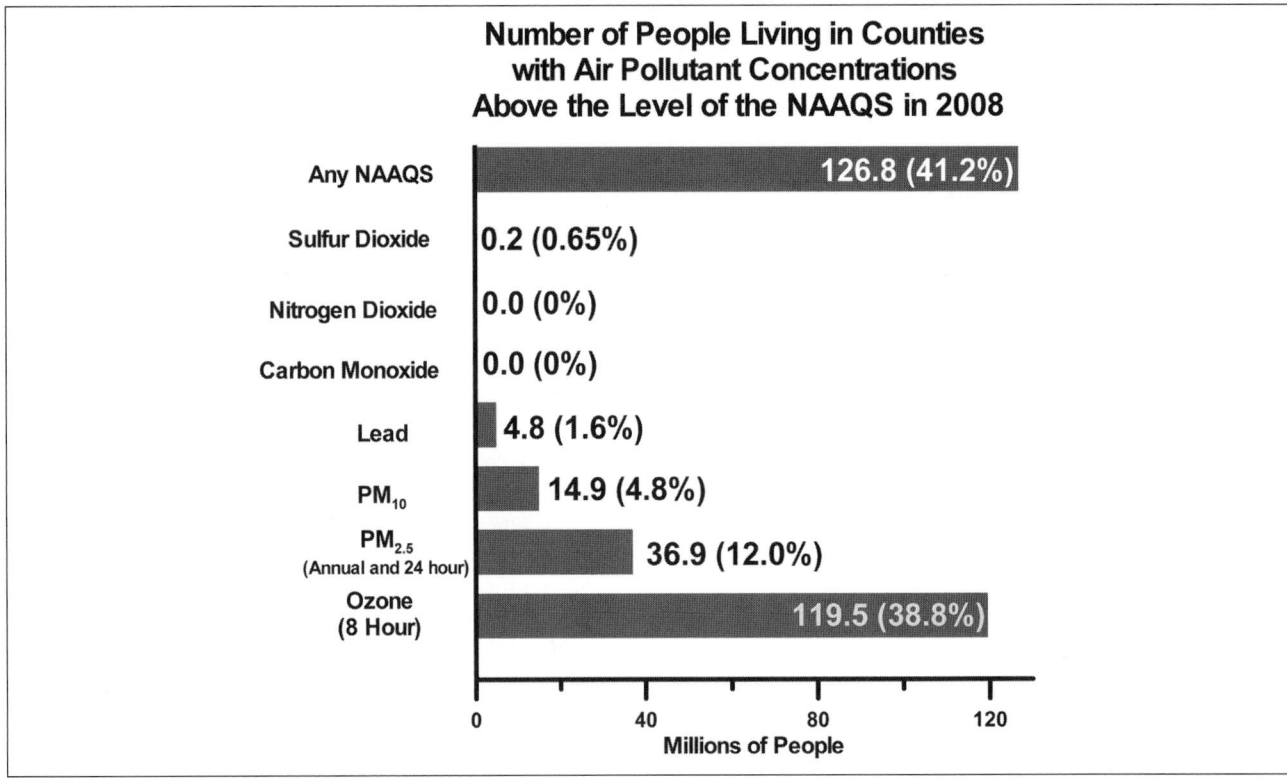

**Figure 8.2** Number of people living in counties with air pollutant concentrations above the level of the National Ambient Air Quality Standards in 2008. (*Source:* Steve Horstmeyer from US Environmental Protection Agency data.)

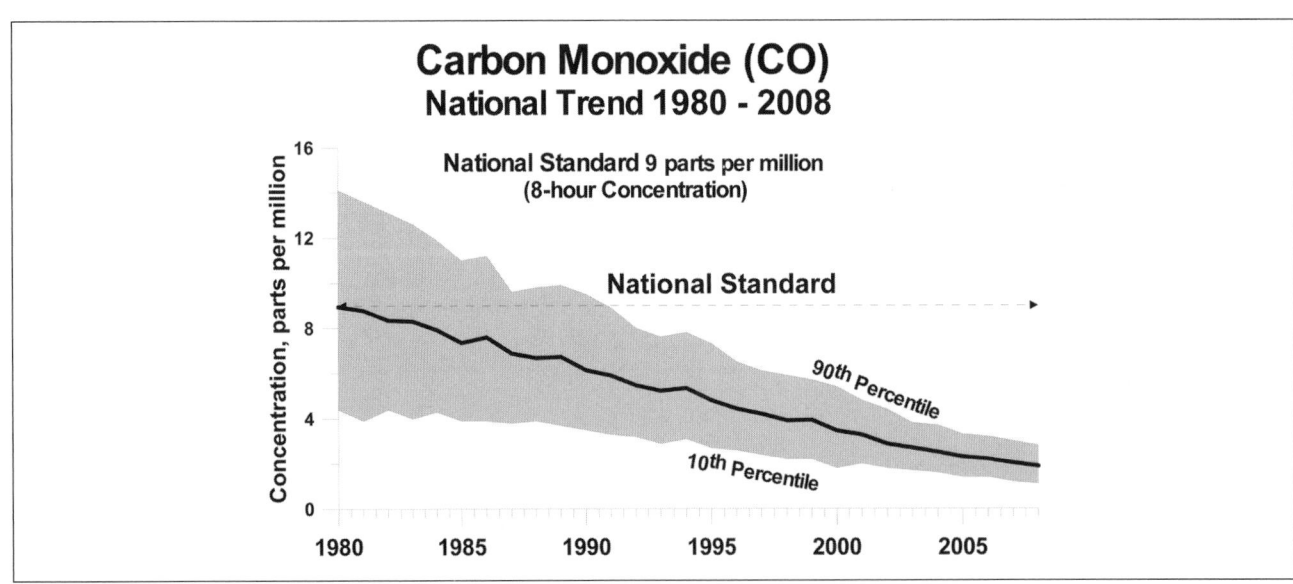

**Figure 8.3** Trends in the concentrations of the individual criteria pollutants, 1990–2008, except PM₂.₅, 2000–2008. (*Source:* Steve Horstmeyer from US Environmental Protection Agency data.)

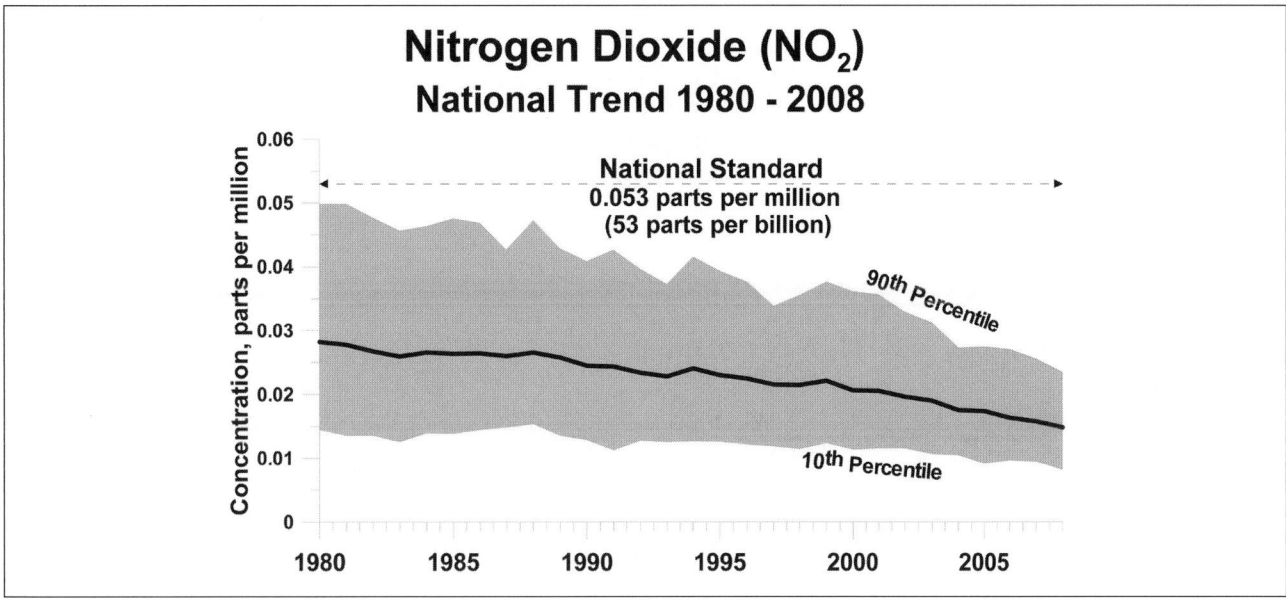

**Figure 8.4**  Trends in the concentrations of the individual criteria pollutants, 1990–2008, except PM$_{2.5}$, 2000–2008. (*Source:* Steve Horstmeyer from US Environmental Protection Agency data.)

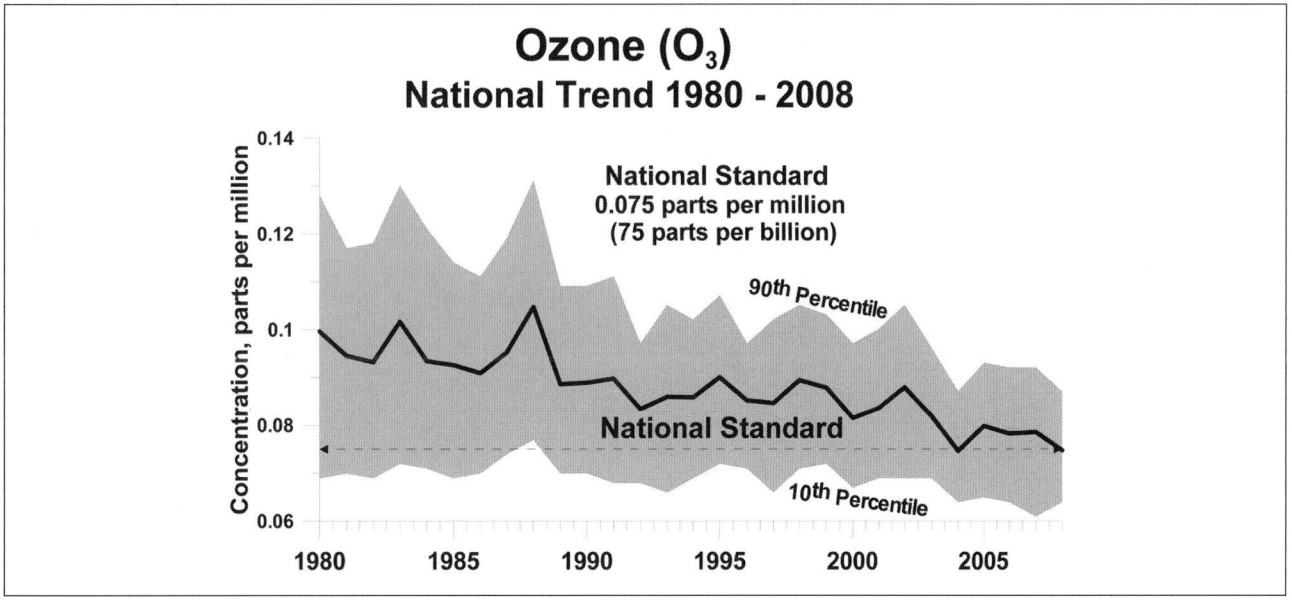

**Figure 8.5**  Trends in the concentrations of the individual criteria pollutants, 1990–2008, except PM$_{2.5}$, 2000–2008. (*Source:* Steve Horstmeyer from US Environmental Protection Agency data.)

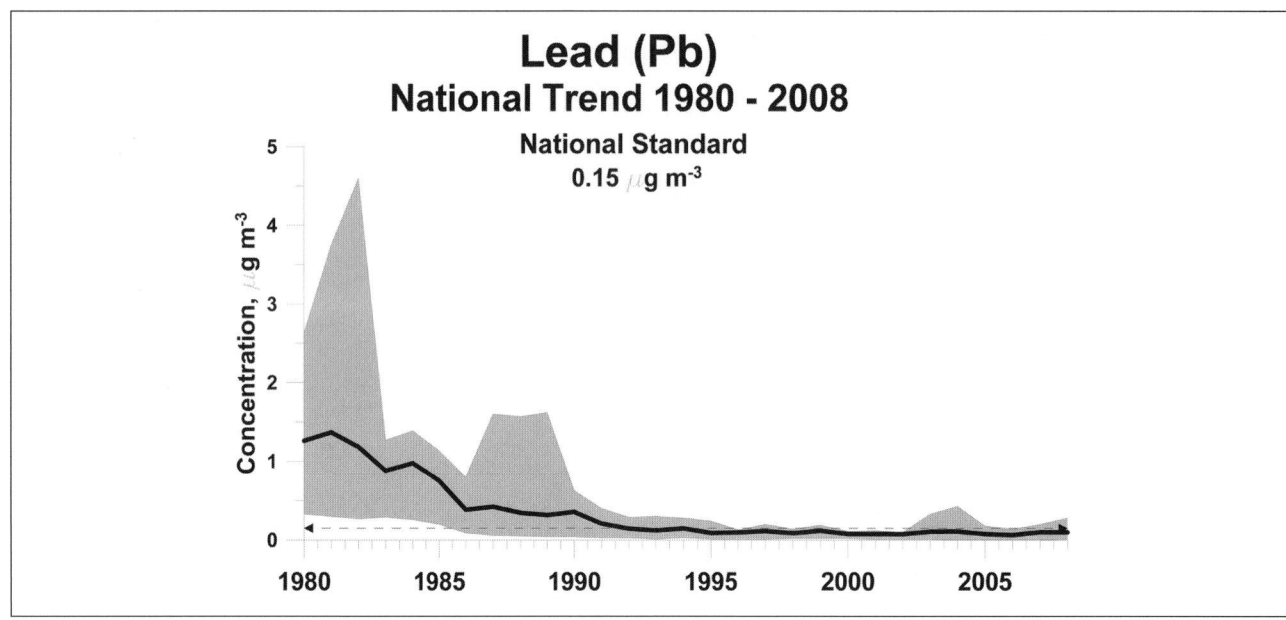

**Figure 8.6** Trends in the concentrations of the individual criteria pollutants, 1990–2008, except PM₂.₅, 2000–2008. (*Source:* Steve Horstmeyer from US Environmental Protection Agency data.)

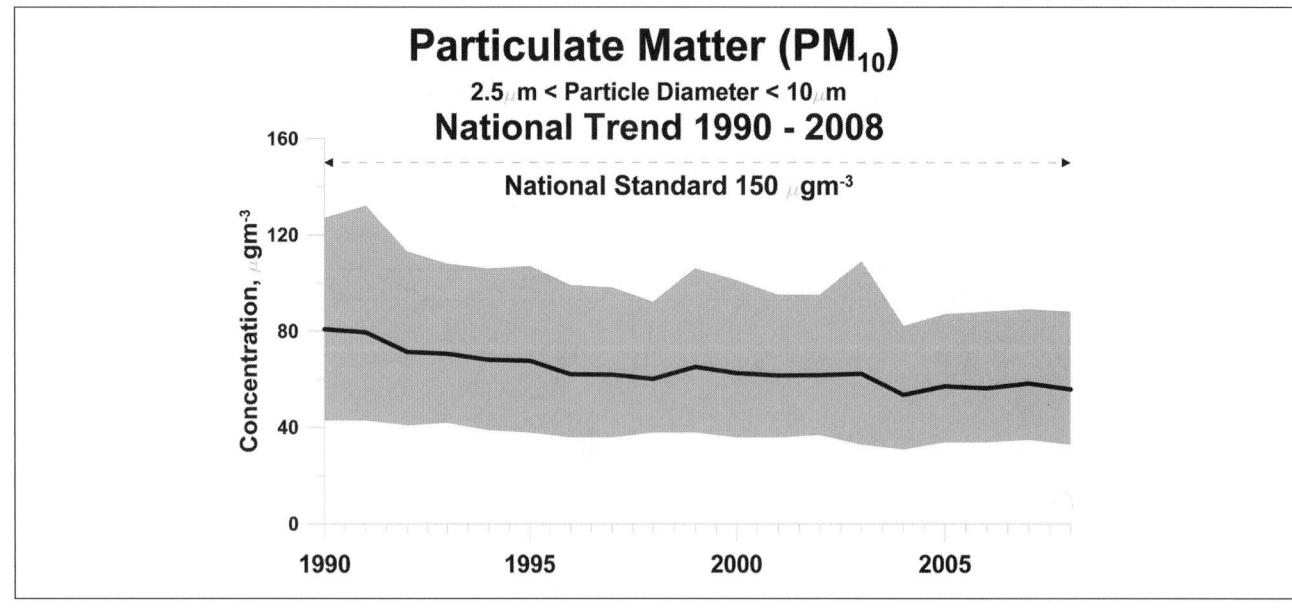

**Figure 8.7** Trends in the concentrations of the individual criteria pollutants, 1990–2008, except PM₂.₅, 2000–2008. (*Source:* Steve Horstmeyer from US Environmental Protection Agency data.)

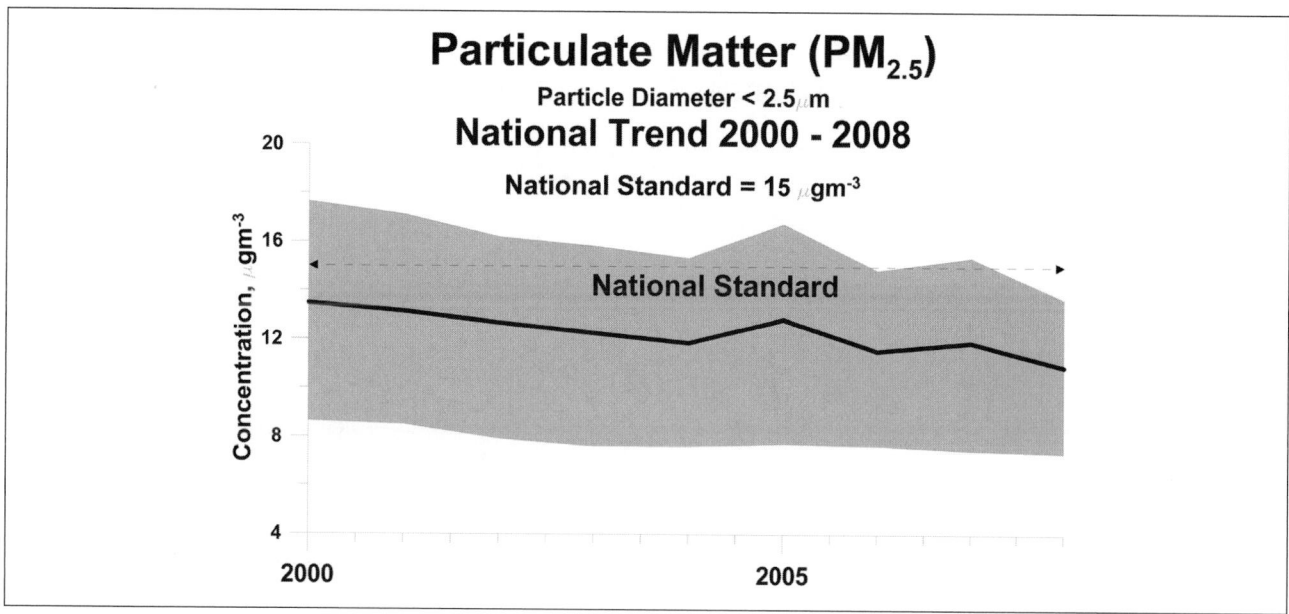

**Figure 8.8**   Trends in the concentrations of the individual criteria pollutants, 1990–2008, except PM₂.₅, 2000–2008. (*Source:* Steve Horstmeyer from US Environmental Protection Agency data.)

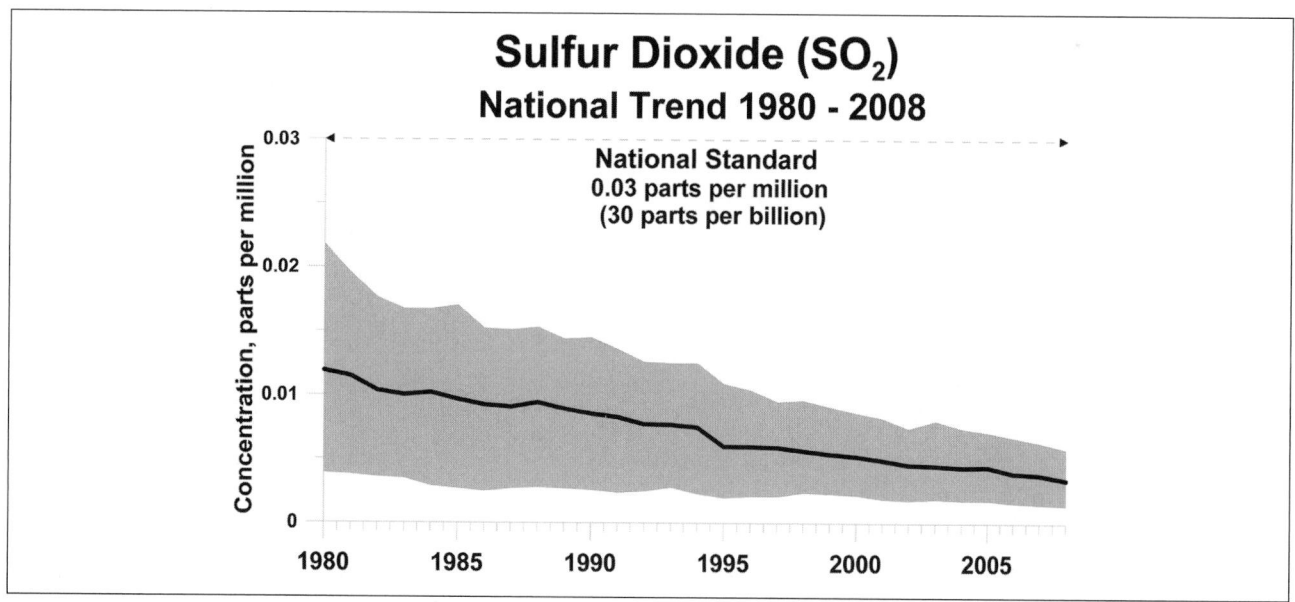

**Figure 8.9**   Trends in the concentrations of the individual criteria pollutants, 1990–2008, except PM₂.₅, 2000–2008. (*Source:* Steve Horstmeyer from US Environmental Protection Agency data.)

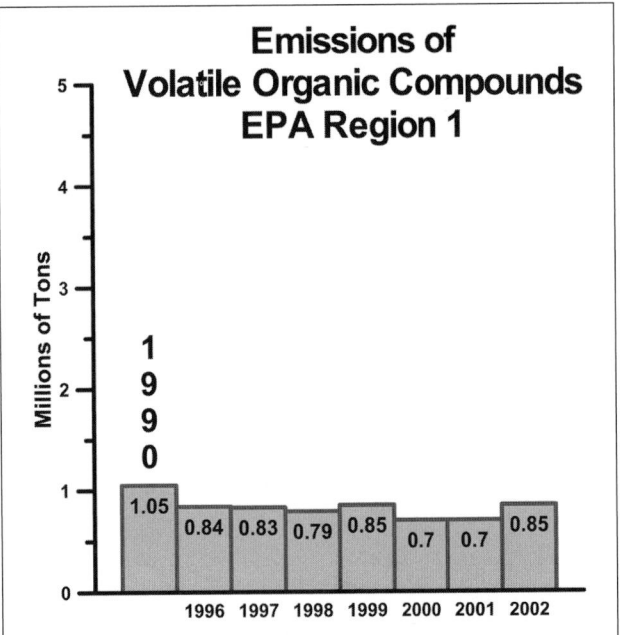

**Figure 8.10** Volatile organic compound emissions for Environmental Protection Agency Region 1. (*Source:* Steve Horstmeyer from US Environmental Protection Agency data.)

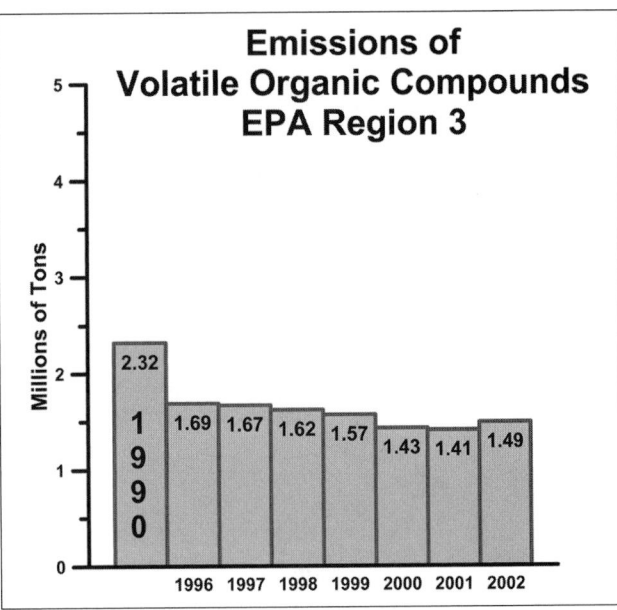

**Figure 8.12** Volatile organic compound emissions for Environmental Protection Agency Region 3. (*Source:* Steve Horstmeyer from US Environmental Protection Agency data.)

ground-level ozone, particulate matter, carbon monoxide, sulfur dioxide, and nitrogen dioxide. The EPA has established NAAQS for each of these to protect public health. Of the five, ground-level ozone and airborne particles pose the greatest threats to human health in this country.

## Understanding the AQI

The AQI is a pollution scale that runs from 0 to 500. The higher the AQI on this yardstick, the greater the level of air pollution and the greater the possible negative effect on health.

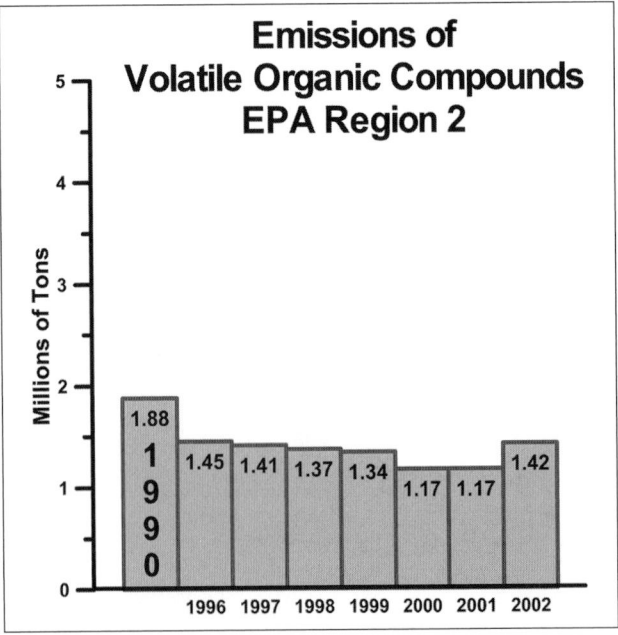

**Figure 8.11** Volatile organic compound emissions for Environmental Protection Agency Region 2. (*Source:* Steve Horstmeyer from US Environmental Protection Agency data.)

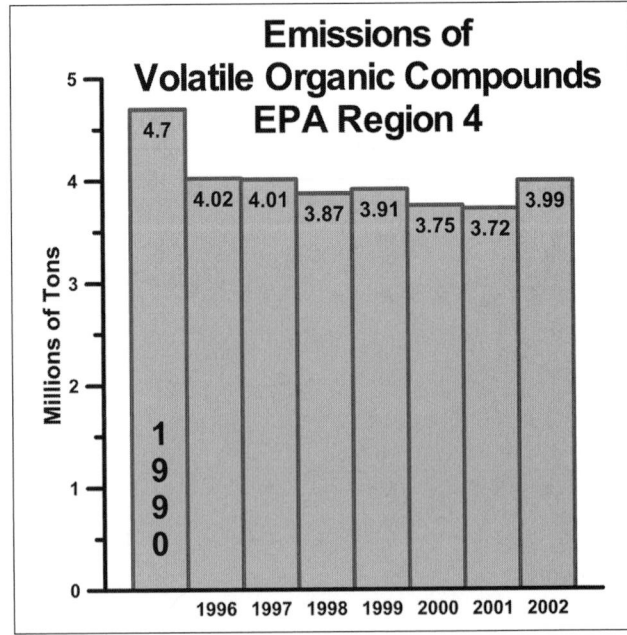

**Figure 8.13** Volatile organic compound emissions for Environmental Protection Agency Region 4. (*Source:* Steve Horstmeyer from US Environmental Protection Agency data.)

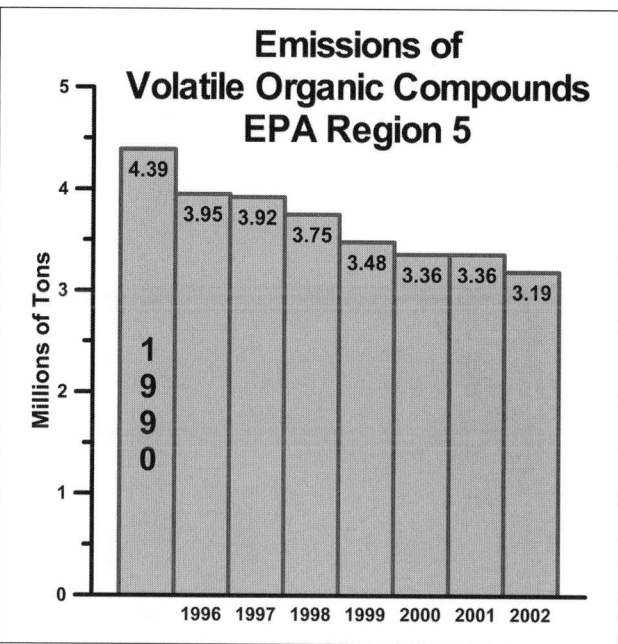

**Figure 8.14** Volatile organic compound emissions for Environmental Protection Agency Region 5. (*Source:* Steve Horstmeyer from US Environmental Protection Agency data.)

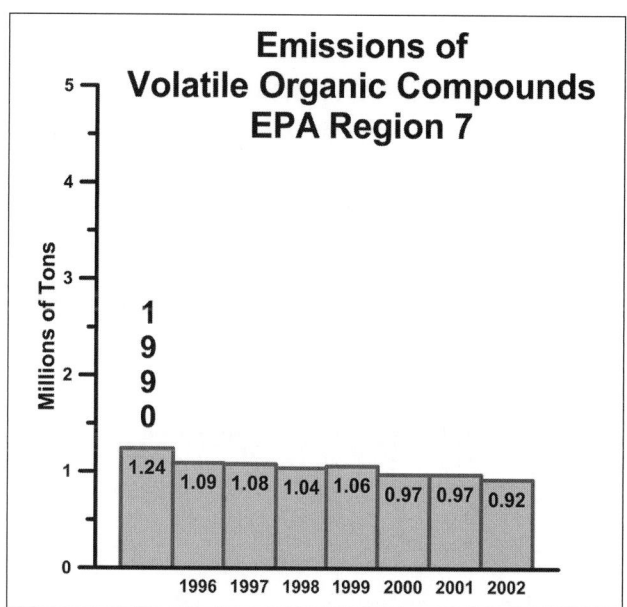

**Figure 8.16** Volatile organic compound emissions for Environmental Protection Agency Region 7. (*Source:* Steve Horstmeyer from US Environmental Protection Agency data.)

Each of the five pollutants has a different concentration set by the EPA as the NAAQS based on research into the potential effects on the health of the public. To avoid the confusion of unfamiliar units, multiple concentrations, and standards that vary from pollutant to pollutant, the EPA developed the AQI.

An AQI value of 100 corresponds to the national air quality standard for a pollutant, which is the level the EPA has

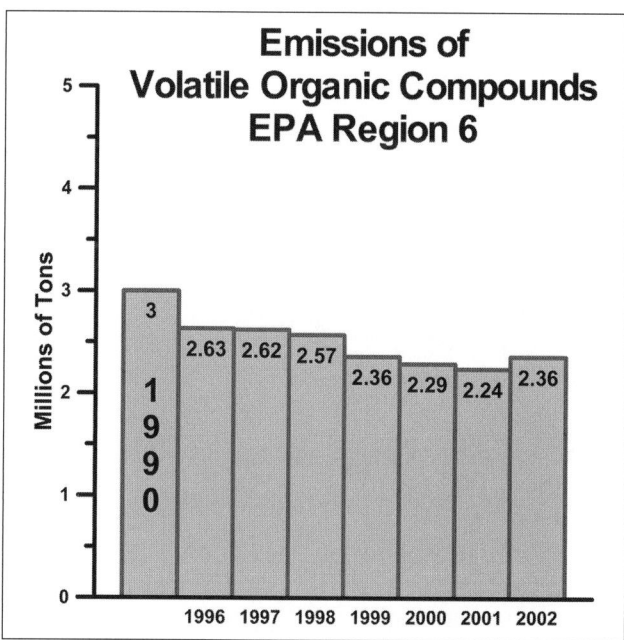

**Figure 8.15** Volatile organic compound emissions for Environmental Protection Agency Region 6. (*Source:* Steve Horstmeyer from US Environmental Protection Agency data.)

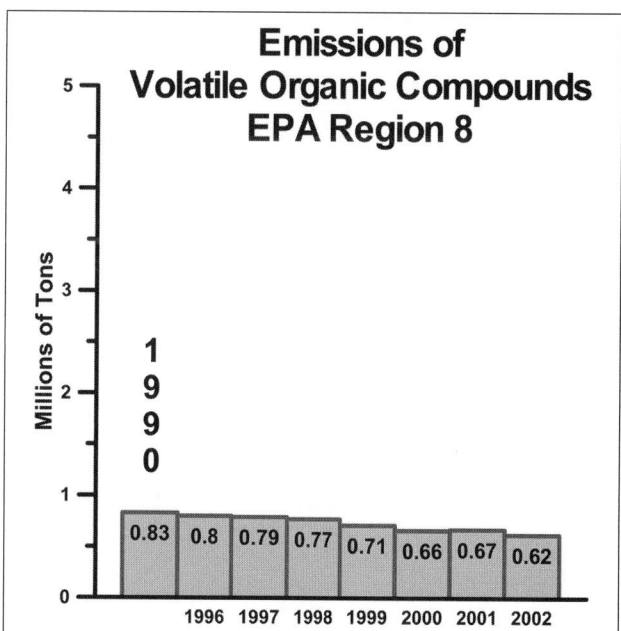

**Figure 8.17** Volatile organic compound emissions for Environmental Protection Agency Region 8. (*Source:* Steve Horstmeyer from US Environmental Protection Agency data.)

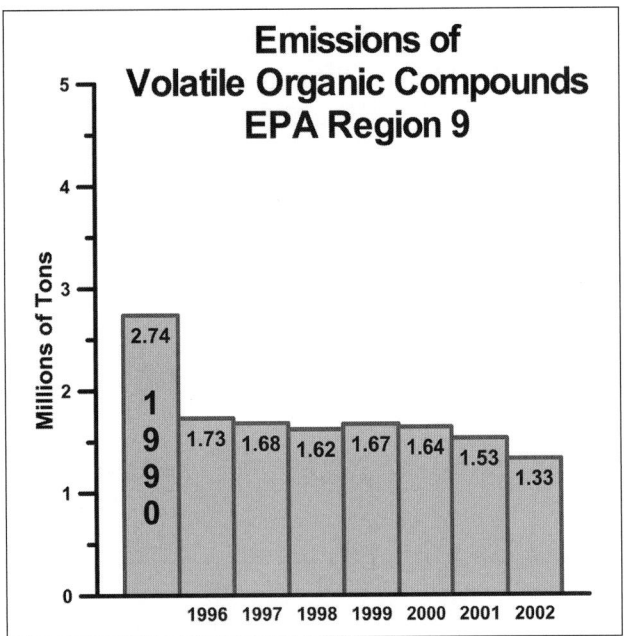

**Figure 8.18** Volatile organic compound emissions for Environmental Protection Agency Region 9. (*Source:* Steve Horstmeyer from US Environmental Protection Agency data.)

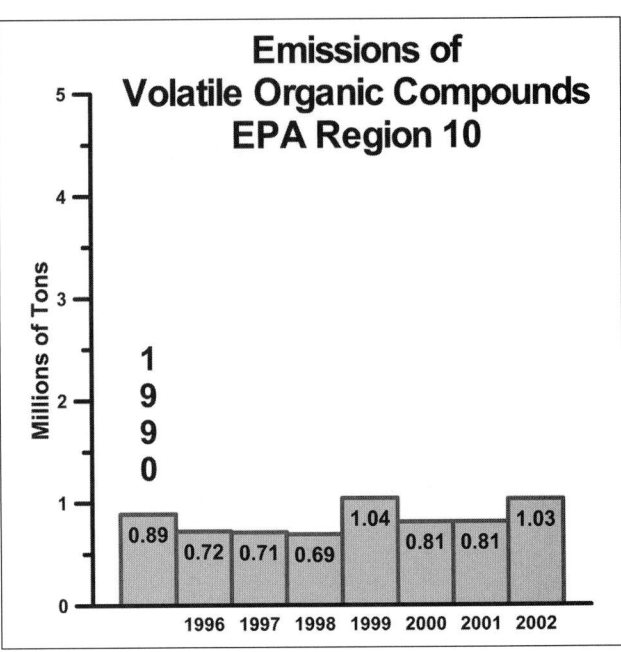

**Figure 8.19** Volatile organic compound emissions for Environmental Protection Agency Region 10. (*Source:* Steve Horstmeyer from US Environmental Protection Agency data.)

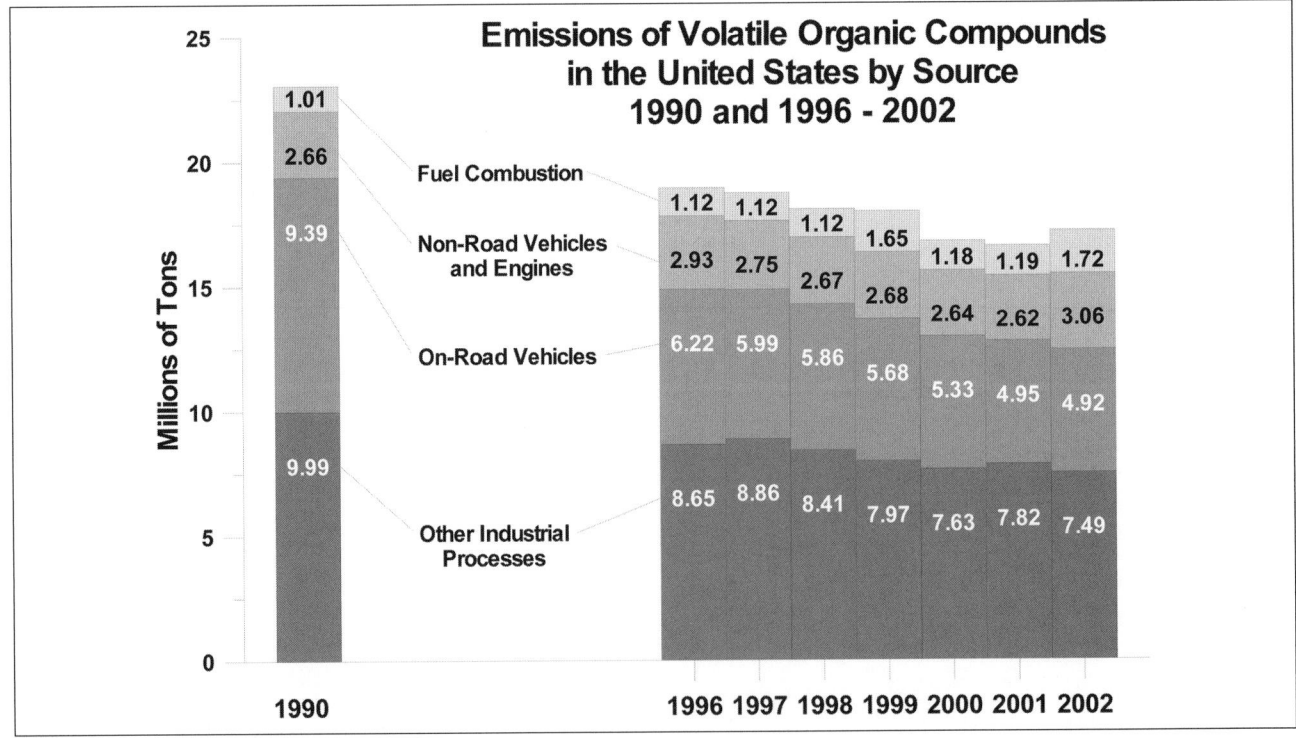

**Figure 8.20** Volatile organic compound emissions by source for 1990 and 1996–2002. (*Source:* Steve Horstmeyer from US Environmental Protection Agency data.)

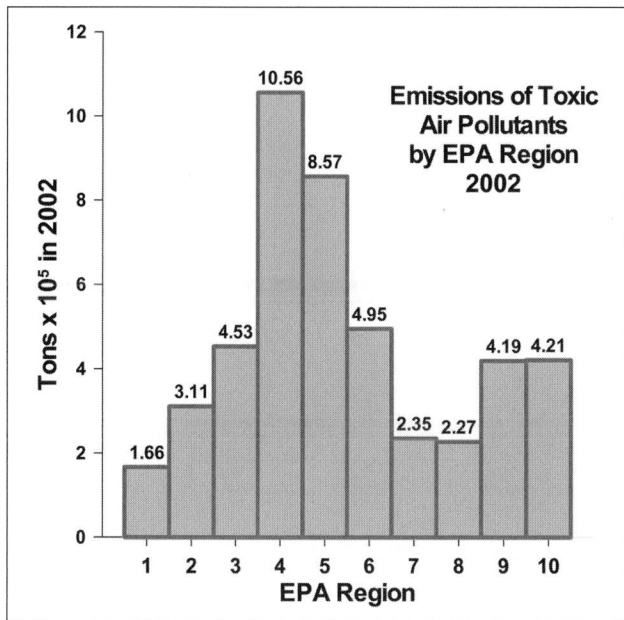

**Figure 8.21**   Toxic air pollutant emissions by region for 2002. (*Source:* Steve Horstmeyer from US Environmental Protection Agency data.)

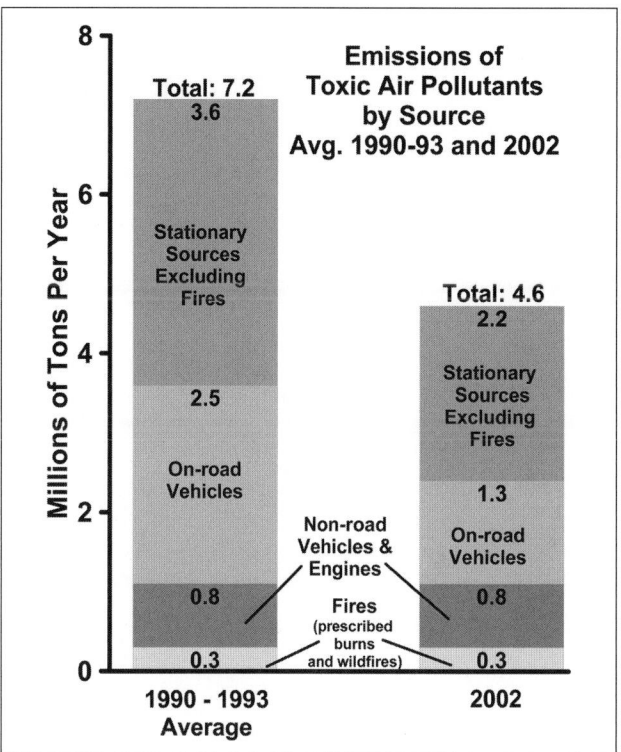

**Figure 8.22**   Toxic air pollutant emissions by source, average 1990–1993 and 2002. (*Source:* Steve Horstmeyer from US Environmental Protection Agency data.)

set to protect public health. AQI values below 100 are generally thought of as satisfactory. When AQI values are above 100, air quality is considered to be unhealthy. First to be affected by polluted air are people sensitive to elevated pollution levels because of age (the elderly and infants) or medical condition. As the AQI increases, more are affected, and

if the concentration of pollutants is great enough, everyone should reduce or try to reduce exposure.

# 9

# Climate Data from Around the World

## • INTRODUCTION

Earth is home to an astounding variety of climates, and while examining numbers can never be as impressive as the stifling humidity after an Amazon River Basin thunderstorm or the penetrating cold and wind-whipped snow of a Great Plains blizzard, numbers can help you prepare for your journey, gain insight into cultural activities, and learn about weather processes.

In this edition of *The Weather Almanac*, I have expanded worldwide weather data to 321 stations and increased the amount of information for each station. The data are presented in a user-friendly format, with a lookup table in nested alphabetical order and index maps showing all stations.

## • CLIMATE CONTROLS

It is beyond the scope of *The Weather Almanac* to provide a detailed explanation of Earth's climates, but a short explanation will provide some perspective for using the data tables.

Climate controls can be broken down into causes at the following scales: global, continental, and local. These are not formally defined scales but will illustrate how the climate of a place can come to be.

### Global Scale Controls

At the global scale the primary control of climate is latitude and that determines which atmospheric circulation cell affects your location and the intensity of solar radiation received.

Intensity of solar radiation varies from north to south across latitude circles. At the northern summer solstice, the maximum solar radiation intensity is at the Tropic of Cancer, 23.5° north, while the South Pole is in the dark. Six months later the Tropic of Capricorn at 23.5° south receives maximum solar radiation intensity and the North Pole is in the dark.

For a location near 40° latitude, Omaha, Kansas City, Cincinnati, Denver, Indianapolis for example, winter solar radiation intensity is half of what it is during summer, which easily explains the seasons.

If you put the latitudinal variation of solar radiation through the year on a rotating Earth, atmospheric circulation cells such as those in Figure 9.1 develop. Figure 9.1 is simplified, and if in addition you throw in continents with high mountains, the pattern becomes much more complicated.

The circulation cells shift northward and southward following the zone of maximum solar radiation by 5°–12° of latitude. In the monsoon area, roughly covering the Indian Ocean the shift is dramatic and can be as much at 40° of latitude.

If you are near the equator, air is generally rising and rainfall is support for the growth of tropical rainforests. A close examination though shows there are complications. For example, maximum rainfall has a seasonal change that can be explained by the northward and southward shift of the global wind belts.

Near 30° north and south, you find the world's great deserts that owe their existence to sinking air at those latitudes. Poleward of the deserts of both hemispheres are the middle latitudes where lows, highs, and fronts signal the clash of warm and cold air and bring alternating wet and dry weather.

Even the deserts get some rain, which again can be explained by the annual migration of the circulation cells, and in the middle latitudes, fronts and storms are stronger in winter when the difference between the cold air and the warm air is greatest.

### Continental Scale Controls

At the continental scale altitude, distance from large water bodies, especially the oceans, and the location of mountain ranges are the primary controls of climate. These can greatly modify the climate of a place from what is expected, using only the global atmospheric circulation cells as a guide.

*The Weather Almanac: A Reference Guide to Weather, Climate, and Related Issues in the United States and Its Key Cities*, Twelfth Edition. Steve Horstmeyer.
© 2011 John Wiley & Sons, Inc. Published 2011 by John Wiley & Sons, Inc.

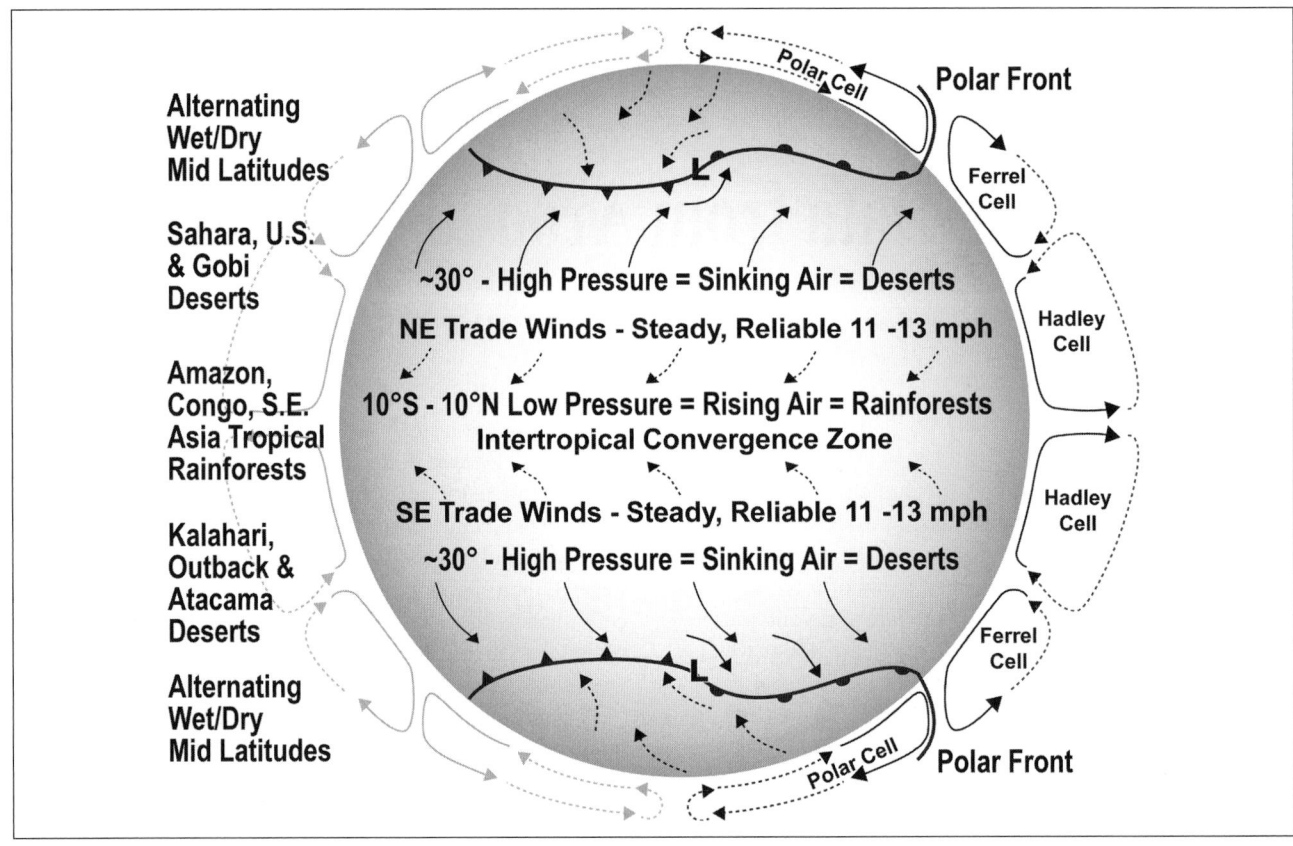

**Figure 9.1** Global atmospheric circulation.

Altitude causes cooler weather, and for every 1000 feet of altitude increase, the average temperature decreases about 3.5°F. To get the same decrease by moving poleward would require that you move about 300 miles. There are many complications to this rule of thumb, but it illustrates how important altitude is.

Locations near the center of a continent get colder and hotter than locations along the coast because ocean water acts to slow the change because of its great heat capacity. West coasts in middle latitudes have less temperature variation than east coasts because air there arrives from the ocean. Along east coasts in middle latitudes the air blows from the continent.

Long mountain ranges block moisture. Western Oregon and Washington are very dry despite being located in the middle latitudes where fronts should bring plenty of rainfall. The Cascade Mountain Range lifts air from the Pacific, causing rain to fall on west-facing slopes. When the air arrives in western Washington and Oregon, there is little moisture left for rain. This is called the "rain shadow effect."

If you live in a belt from Cleveland, OH, through Buffalo, NY, and farther northeast through most of Upstate New York, you know well what positioning a city on the lee side (downwind side) of a big lake can mean. The "lake-effect snow belt" is famous for local- to regional-scale immense snowfalls as bitter arctic winds blow across the warmer waters of the Great Lakes, evaporate moisture, and dump paralyzing depths of snow in whiteouts that can last for hours.

The western shore of Lower Michigan and nearly all of the Upper Peninsula of Michigan also receive lake-effect snow.

**Local Controls**

Anyone who has ridden a bicycle or motorcycle as the sun is setting has felt the cooler air in even a small dip in the road. Like water, denser cool air flows downhill and pools in low spots, as a result valley bottoms are often much cooler than higher-up valley sides. Terrain variation is a primary local control of climate.

Pooling of cool air, disruption of wind flow, transport of warmer or cooler air, and moisture and change of sunshine are the primary reasons for local climate variation. Any factor that modifies these is a local climate control.

Downwind of a large urban area is often warmer than areas upwind because of the plume of warm air from the city. Within urban areas the onset of the first killing frost may be delayed a couple months, but a short distance away within the city, a low spot in a large park may collect cool air and experience frost at times not too different from those out of the city.

On a winter backpacking trip to Yosemite National Park we were hiking on the north wall of the valley in shorts and tee shirts comfortable in the 75°F air and sunshine. Across the valley in the perpetual shade of the steep walls, temperatures were below freezing and several inches of snow was on the ground. Over the distance of a mile sun exposure differences caused two distinct microclimates and a temperature difference of more than 40°F.

## • INTERNATIONAL WEATHER DATA

Getting accurate international data is always a challenge. Standards, time of day, equipment, site characteristics, and frequency of observation all vary from country to country. Every effort has been made to assure the accuracy of these data tables, and they will work well for all popular purposes, travel, school reports, or just curiosity about the world.

### Table Organization and Use

The world data tables are arranged alphabetically by continent, then alphabetically by country within each continent, and finally alphabetically by location within each country. Following the continents, island locations are listed alphabetically by basin and then alphabetically by location within each basin. These are the major divisions of the World Weather Data Tables:

Africa
Antarctica
Asia
Australia
Europe
North America
South America
Arctic Ocean Islands
Atlantic Ocean/Caribbean Sea Islands
Indian Ocean Islands
Mediterranean Sea Islands
Pacific Ocean Islands
Southern Ocean Islands

Table 9.1 provides all locations using the scheme described above. Each location is also assigned a number from 1 to 321 from first to last in order of the nested alphabetical scheme. The numbers are used to locate a station on the index maps (Figures 9.2–9.15).

Note that in some cases for ease of use, I have taken some liberties. For example, all Russian locations are included in data for Asia.

The Southern Ocean is unlike any other because it is not confined to a basin between continents. The Southern Ocean surrounds Antarctica and for political reasons is loosely defined as extending outward to 60°S latitude by the International Hydrographic Organization. Using this definition the Southern Ocean is the fourth largest of Earth's oceans by surface area.

Oceanographers often define the Southern Ocean using the physical characteristics of different water masses which in part help determine chemical and biological properties of the water. The outer limit of the Southern Ocean is often considered to be where the cold, salty water flowing northward from Antarctica meets warmer water at what is called the Antarctic Convergence (AAC). The distance of the AAC from Antarctica varies seasonally.

Some geographers and oceanographers do not recognize the existence of the Southern Ocean and consider it the southern extremes of the Pacific, Atlantic, and Indian Oceans.

## • USING WEATHER AND CLIMATE DATA

Interpolation is very handy and helps fill in gaps. If you need an estimate of the average maximum temperature on the first of the month, find the average of the current month's average high temperature and the previous month's average high temperature for example.

Be careful while using interpolation geographically. There are two geographical settings, in particular, that lead to inaccurate results. The first involves elevation. The average high temperature of a location decreases about 3.5°F for every 1000 feet of elevation increase. That is approximately equivalent to a poleward move in the middle latitudes of 300 miles. This is an approximation because it does not take into account cloud cover and other factors such as sun exposure.

Poleward of the Tropics of Cancer and Capricorn southern hemisphere, south-facing mountain slopes and in the northern hemisphere north-facing mountain slopes receive little or no direct solar radiation. Snow will stay on the ground longer and temperatures will be dramatically cooler than the opposite sun-exposed slopes.

If two stations are in close proximity, but of different elevation, a better result may be achieved by accounting for elevation change in your interpolation by subtracting 3.5°F for every 1000 feet of elevation increase.

The second major problem in using interpolation is proximity to large bodies of water. If the body of water happens to be cold and a station is downwind of the water body, two factors will make interpolation to an inland location inaccurate. First, the wind will carry cool air over the location, keeping it cooler than inland locations, and cloud cover and precipitation may also be a factor.

The best example of this in the United States is along the west coast. San Diego, Los Angeles, and San Francisco are dramatically cooler than inland stations. In Los Angeles it is not uncommon for the international airport that is right on the coast to have a high temperature in the upper 70s (°F) during summer, while Downtown Los Angeles and the San Fernando Valley are in the low 100s. San Francisco is famous for a chilly breeze in midsummer.

**Table 9.1   World Weather Data Station List**

# Africa

| Map # | Country | City or Location | Latitude ddmm | Longitude ddmm | Elevation feet |
|---|---|---|---|---|---|
| 1 | ALGERIA | ALGIERS | 3642N | 00313E | 82 |
| 2 | ANGOLA | LUANDA | 0851S | 01314E | 243 |
| 3 | BENIN | COTONOU | 0621N | 00223E | 16 |
| 4 | BOTSWANA | MAUN | 1959S | 02325E | 3101 |
| 5 | BURKINA FASO | OUAGADOUGOU | 1221N | 00131W | 1037 |
| 6 | BURUNDI | BUJUMBURA | 0320S | 02919E | 2566 |
| 7 | CAMEROON | YAOUNDE | 0349N | 01131E | 2470 |
| 8 | CEN. AFRICAN REP. | BANGUI | 0423N | 01833E | 1208 |
| 9 | CHAD | N'DJAMENA | 1208N | 01502E | 975 |
| 10 | CONGO, DEM. REP | KINSHASA | 0423S | 01526E | 1014 |
| 11 | CONGO, REP OF | BRAZZAVILLE | 0415S | 01515E | 1047 |
| 12 | CÔTE D'IVOIRE | ABIDJANUARY | 0515N | 00355W | 20 |
| 13 | DJIBOUTI | DJIBOUTI | 1133N | 04309E | 30 |
| 14 | EGYPT | ASWAN DAM | 2357N | 03249E | 656 |
| 15 | EGYPT | CAIRO | 3008N | 03124E | 366 |
| 16 | EQUATORIAL GUINEA | MALABO | 0345N | 00845E | 164 |
| 17 | ERITREA | ASMARCHA | 1518N | 03855E | 7628 |
| 18 | ETHIOPIA | ADDIS ABABA | 0859N | 03847E | 7642 |
| 19 | GABON | LIBREVILLE | 0028N | 00926E | 39 |
| 20 | GAMBIA | YUNDUM | 1321N | 01640W | 85 |
| 21 | GUINEA | CONAKRY | 0934N | 01337W | 85 |
| 22 | GUINEA-BISSAU | BISSAU | 1153N | 01539W | 128 |
| 23 | KENYA | NAIROBI | 0119S | 03655E | 5327 |
| 24 | LIBERIA | MONROVIA | 0614N | 01021W | 31 |
| 25 | LIBYA | KURFA | 2413N | 02318E | 1254 |
| 26 | LIBYA | TRIPOLI | 3240N | 01309E | 263 |
| 27 | MADAGASCAR | ANTANANARIVO | 1854S | 04732E | 4702 |
| 28 | MALAWI | LILONGWE | 1358S | 03342E | 3717 |
| 29 | MALI | BAMAKO | 1237N | 00801W | 1079 |
| 30 | MAURITANIA | NOUAKCHOTT | 1805N | 01557W | 7 |
| 31 | MOROCCO | CASABLANCA | 3333N | 00739W | 203 |
| 32 | MOROCCO | MIDELT | 3241N | 00444W | 4987 |
| 33 | MOZAMBIQUE | MAPUTO | 2555S | 03234E | 128 |
| 34 | NAMIBIA | WINDHOEK | 2237S | 01705E | 5578 |
| 35 | NIGER | NIAMEY | 1328N | 00210E | 732 |
| 36 | NIGERIA | LAGOS | 0635N | 00320E | 132 |
| 37 | RWANDA | KAMEMBE | 0227S | 02854E | 5190 |
| 38 | SENEGAL | DAKAR | 1444N | 01730W | 89 |
| 39 | SIERRA LEONE | FREETOWN | 0837N | 01312W | 82 |
| 40 | SOMALIA | MOGADISHU | 0201N | 04519E | 26 |
| 41 | SOUTH AFRICA | CAPETOWN | 3354S | 01829E | 36 |
| 42 | SOUTH AFRICA | PRETORIA | 2545S | 02814E | 4491 |
| 43 | SUDAN | KHARTOUM | 1536N | 03235E | 1256 |
| 44 | TANZANIA | DAR ES SALAAM | 0652S | 03912E | 181 |
| 45 | TOGO | LOME | 0610N | 00115E | 72 |
| 46 | TUNISIA | TUNIS | 3651N | 01014E | 17 |
| 47 | UGANDA | ENTEBBE | 0002N | 03227E | 3789 |
| 48 | WESTERN SAHARA | EL AAIUN | 2709N | 01312W | 207 |
| 49 | ZAMBIA | LUSAKA | 1525S | 02820E | 4208 |
| 50 | ZIMBABWE | HARARE | 1755S | 03106E | 4904 |

**Table 9.1  World Weather Data Station List** [CONTINUED]

# Antarctica

| Map # | Country | City or Location | Latitude ddmm | Longitude ddmm | Elevation feet |
|-------|---------|------------------|---------------|----------------|----------------|
| 51 | ANTARCTICA | MCMURDO STATION | 7753S | 16648W | 8 |
| 52 | ANTARCTICA | SOUTH POLE STATION | 8959S | 00000W | 9186 |
| 53 | ANTARCTICA | VOSTOK 2 | 7827S | 10652E | 11220 |

# Asia

| Map # | Country | City or Location | Latitude ddmm | Longitude ddmm | Elevation feet |
|-------|---------|------------------|---------------|----------------|----------------|
| 54 | AFGHANISTAN | KABUL | 3433N | 06912E | 5909 |
| 55 | ARMENIA | YEREVAN | 4008N | 04428E | 2976 |
| 56 | AZERBAIJAN | BAKU | 3844N | 04850E | -36 |
| 57 | BAHRAIN | BAHRAIN | 2616N | 05037E | 6 |
| 58 | BANGLADESH | DHAKA | 2346N | 09023E | 24 |
| 59 | BHUTAN | THIMPHU | 2728N | 08938E | 7656 |
| 60 | BRUNEI | BANDAR SERI BEGAWAN | 0455N | 11455E | 10 |
| 61 | CAMBODIA | PHNOM PENH | 1133N | 10451E | 39 |
| 62 | CHINA | BEIJING | 3956N | 11620E | 167 |
| 63 | CHINA | CHITING-HSILIN | 3357N | 09237E | 16564 |
| 64 | CHINA | CHONGQING | 2930N | 10633E | 855 |
| 65 | CHINA | HAIKOU | 2000N | 11025E | 46 |
| 66 | CHINA | HARBIN | 4545N | 12638E | 476 |
| 67 | CHINA | HONG KONG | 2219N | 11411E | 15 |
| 68 | CHINA | LHASA | 2943N | 09102E | 11796 |
| 69 | CHINA | SHANG-HAI | 3112N | 12126E | 16 |
| 70 | CHINA | WUSU | 4423N | 08430E | 1345 |
| 71 | CHINA | XI'AN | 3415N | 10855E | 1312 |
| 72 | CHINA, REPUBLIC OF | TAIPEI | 2504N | 12132E | 21 |
| 73 | GEORGIA | TBILISI | 4141N | 04457E | 1608 |
| 74 | INDIA | BANGALORE | 1257N | 07740E | 2937 |
| 75 | INDIA | KOLKATA (CALCUTTA) | 2239N | 08826E | 13 |
| 76 | INDIA | CHENNAI (MADRAS) | 1259N | 08010E | 34 |
| 77 | INDIA | MUMBAI (BOMBAY) | 1905N | 07250E | 9 |
| 78 | INDIA | NEW DELHI | 2834N | 07706E | 755 |
| 79 | INDIA (JAMMU AND KASHMIR) | LEH | 3409N | 07734E | 11529 |
| 80 | INDONESIA | JAKARTA | 0609S | 10650E | 16 |
| 81 | INDONESIA | SUMBAWA-BESAR | 0829S | 11725E | 28 |
| 82 | INDONESIA | TJIBEUREUM | 0721S | 10815E | 1200 |
| 83 | IRAN | TEHERAN | 3541N | 05119E | 3937 |
| 84 | IRAQ | BAGHDAD | 3319N | 04421E | 112 |
| 85 | ISRAEL | TEL AVIV | 3206N | 03446E | 33 |
| 86 | JAPAN | TOKYO | 3533N | 13945E | 14 |
| 87 | JORDAN | AMMAN | 3158N | 03559E | 2547 |
| 88 | KAZAKHSTAN | TARAZ | 4251N | 07123E | 2106 |
| 89 | KOREA, NORTH | PYONGYANG | 3901N | 12549E | 94 |
| 90 | KOREA, SOUTH | SEOUL | 3731N | 12655E | 34 |
| 91 | KUWAIT | KUWAIT | 2914N | 04758E | 182 |
| 92 | KYRGYZSTAN | BISHKEK | 5252N | 07436E | 1288 |
| 93 | LAOS | LUANG-PRABANG | 1953N | 10208E | 997 |
| 94 | LEBANON | BEIRUT | 3348N | 03529E | 86 |
| 95 | MALAYSIA | KOTA KINABALU | 0556N | 11603E | 9 |
| 96 | MALAYSIA | KUALA LUMPUR | 0308N | 10133E | 89 |
| 97 | MALAYASIA | KUCHING | 0129N | 11020E | 85 |
| 98 | MONGOLIA | ALTAI | 4624N | 09615E | 7044 |

**Table 9.1  World Weather Data Station List [CONTINUED]**

# Asia, continued

| Map # | Country | City or Location | Latitude ddmm | Longitude ddmm | Elevation feet |
|---|---|---|---|---|---|
| 99 | MONGOLIA | ULAANBAATAR | 4751N | 10645E | 4157 |
| 100 | MYANMAR (BURMA) | NAYPYIDAW | 1943N | 09613E | 340 |
| 101 | NEPAL | KATMANDU | 2742N | 08522E | 4423 |
| 102 | OMAN | MUSCAT | 2345N | 05835E | 20 |
| 103 | PAKISTAN | KARACHI | 2454N | 06709E | 95 |
| 104 | PHILIPPINES, REP. OF | BAGUIO | 1625N | 12035E | 4962 |
| 105 | PHILIPPINES, REP. OF | MANILA | 1431N | 12100E | 49 |
| 106 | QATAR | DOHA | 2516N | 05133E | 33 |
| 107 | RUSSIA | ASTRAKHAN | 4616N | 04802E | 59 |
| 108 | RUSSIA | DIKSON | 7330N | 08014E | 66 |
| 109 | RUSSIA | IRKUTSK | 5216N | 10421E | 1591 |
| 110 | RUSSIA | KAMCHATKA | 5906N | 15959E | 89 |
| 111 | RUSSIA | KHANTY-MANSIYSK | 6058N | 06904E | 131 |
| 112 | RUSSIA | MOSCOW | 5545N | 03734E | 512 |
| 113 | RUSSIA | MURMANSK | 6858N | 03303E | 151 |
| 114 | RUSSIA | OKHOTSKIY-PEREVO | 6153N | 13533E | 479 |
| 115 | RUSSIA | OKHOTSK | 5922N | 14312E | 20 |
| 116 | RUSSIA | OYMYAKON NEAR | 6316N | 14309E | 2382 |
| 117 | RUSSIA | SAINT PETERSBURG | 5958N | 03018E | 13 |
| 118 | RUSSIA | TIKSI | 7135N | 12855E | 26 |
| 119 | RUSSIA | VERKHOYANSK | 6733N | 13323E | 449 |
| 120 | RUSSIA | VLADIVOSTOK | 4307N | 13154E | 453 |
| 121 | RUSSIA | YAKUTSK | 6205N | 12945E | 338 |
| 122 | SAUDI ARABIA | RIYADH | 2443N | 04643E | 1922 |
| 123 | SINGAPORE, REP. OF | SINGAPORE | 0122N | 10348E | 59 |
| 124 | SRI LANKA | COLOMBO | 0654N | 07952E | 22 |
| 125 | SYRIA | DAMASCUS | 3328N | 03613E | 2605 |
| 126 | TAJIKISTAN | DUSHANBE | 3835N | 06847E | 2703 |
| 127 | THAILAND | BANGKOK | 1344N | 10030E | 53 |
| 128 | TURKEY | ANKARA | 4007N | 03259E | 3122 |
| 129 | TURKMENISTAN | ASHGABAT | 3758N | 05820E | 775 |
| 130 | UNITED ARAB EMIRATES | DUBAI | 2515N | 05520E | 5 |
| 131 | UZBEKISTAN | TASHKENT | 4116N | 06916E | 1404 |
| 132 | VIETNAM | HANOI | 2103N | 10552E | 20 |
| 133 | YEMEN | ADEN | 1249N | 04501E | 10 |

**Australia**

| Map # | Country | City or Location | Latitude ddmm | Longitude ddmm | Elevation feet |
|---|---|---|---|---|---|
| 134 | AUSTRALIA | ADELAIDE | 3457S | 13832E | 20 |
| 135 | AUSTRALIA | ALICE SPRINGS | 2348S | 13353E | 1791 |
| 136 | AUSTRALIA | BRISBANE | 2725S | 15305E | 17 |
| 137 | AUSTRALIA | CAIRNS | 1653S | 14545E | 7 |
| 138 | AUSTRALIA | DARWIN | 1225S | 13052E | 104 |
| 139 | AUSTRALIA | HOBART | 4250S | 14733E | 9 |
| 140 | AUSTRALIA | PERTH | 3156S | 11558E | 64 |
| 141 | AUSTRALIA | SIDNEY | 3357S | 15101E | 16 |
| 142 | AUSTRALIA | WYNDHAM | 1531S | 12809E | 14 |

## Table 9.1   World Weather Data Station List [CONTINUED]

# Europe

| Map # | Country | City or Location | Latitude ddmm | Longitude ddmm | Elevation feet |
|-------|---------|------------------|---------------|----------------|----------------|
| 143 | ALBANIA | KORÇË | 4036N | 02046E | 2948 |
| 144 | ALBANIA | TIRANA | 4120N | 01947E | 241 |
| 145 | AUSTRIA | INNSBRUCK | 4715N | 01120E | 1906 |
| 146 | AUSTRIA | VIENNA | 4815N | 01622E | 666 |
| 147 | BELARUS | MINSK | 5352N | 02732E | 768 |
| 148 | BELGIUM | BRUSSELS | 5054N | 00429E | 180 |
| 149 | BOSNIA AND HERZEGOVINA | SARAJEVO | 4352N | 01818E | 1620 |
| 150 | BULGARIA | SOFIA | 4249N | 02323E | 1929 |
| 151 | CROATIA | ZAGREB | 4549N | 01558E | 535 |
| 152 | CZECH REPUBLIC | PRAGUE | 5006N | 01417E | 1247 |
| 153 | DENMARK | COPENHAGEN | 5538N | 01240E | 16 |
| 154 | FINLAND | HELSINKI | 6019N | 02457E | 167 |
| 155 | FINLAND | IVALO | 6836N | 02725E | 459 |
| 156 | FRANCE | NICE-CÔTE D' AZUR | 4339N | 00712E | 13 |
| 157 | FRANCE | PARIS | 4843N | 00224E | 292 |
| 158 | FRANCE | PIC DU MIDI OBSERVATORY | 4256N | 00008E | 9378 |
| 159 | GERMANY | BERLIN | 5233N | 01318E | 115 |
| 160 | GERMANY | MUNICH | 4808N | 01142E | 1728 |
| 161 | GIBRALTAR U.K. | NORTH FRONT | 3609N | 00521W | 8 |
| 162 | GREECE | ATHENS | 3753N | 02343E | 90 |
| 163 | HUNGARY | BUDAPEST | 4731N | 01902E | 425 |
| 164 | ICELAND | REYKJAVIK | 6407N | 02156W | 45 |
| 165 | IRELAND | DUBLIN | 5322N | 00621W | 155 |
| 166 | ITALY | ROME | 4148N | 01236E | 430 |
| 167 | KOSOVO | PRISTINA | 4240N | 02110E | 2139 |
| 168 | LATVIA | RIGA | 5658N | 02404E | 10 |
| 169 | LITHUANIA | VILNIUS | 5453N | 02353E | 246 |
| 170 | LUXEMBOURG | LUXEMBOURG CITY | 4937N | 00603E | 1083 |
| 171 | MACEDONIA | SKOPJE | 4159N | 02128E | 784 |
| 172 | MOLDOVA | CHISINAU | 4700N | 02854E | 115 |
| 173 | MONTENEGRO | PODGORICA | 4221N | 01915E | 121 |
| 174 | NETHERLANDS | AMSTERDAM | 5219N | 00447E | -10 |
| 175 | NORWAY | OSLO | 5954N | 01037E | 56 |
| 176 | NORWAY | VARDO | 7022N | 03106E | 43 |
| 177 | POLAND | WARSAW | 5211N | 02058E | 351 |
| 178 | PORTUGAL | LISBON | 3846N | 00908W | 361 |
| 179 | ROMANIA | BUCHAREST | 4430N | 02606E | 302 |
| 180 | SERBIA | BELGRADE | 4448N | 02028E | 433 |
| 181 | SLOVAK REPUBLIC | BRATISLAVA | 4810N | 01713E | 433 |
| 182 | SLOVENIA | LJUBLJANA | 4603N | 01433E | 951 |
| 183 | SPAIN | MADRID | 4028N | 00334W | 1972 |
| 184 | SWEDEN | STOCKHOLM | 5939N | 01755E | 122 |
| 185 | SWITZERLAND | ZURICH | 4723N | 00833E | 1617 |
| 186 | TURKEY | ANKARA | 4007N | 03259E | 3122 |
| 187 | UKRAINE | KIEV | 5024N | 03027E | 587 |
| 188 | U..K. GREAT BRITIAN | LONDON | 5128N | 00027W | 80 |
| 189 | U.K. NORTHERN IRELAND | BELFAST | 5436N | 00552W | 24 |
| 190 | U.K. SCOTLAND | EDINBURGH | 5556N | 00320W | 135 |
| 191 | U.K.WALES | CARDIFF | 5124N | 00321W | 220 |

**Table 9.1   World Weather Data Station List [CONTINUED]**

# North America

| Map # | Country | City or Location | Latitude ddmm | Longitude ddmm | Elevation feet |
|---|---|---|---|---|---|
| 192 | BELIZE | BELIZE CITY | 1732N | 08818W | 16 |
| 193 | CANADA | ALERT | 8231N | 06220W | 95 |
| 194 | CANADA | QIKIQTARJUAQ (BROUGHTON) | 6733N | 06402W | 25 |
| 195 | CANADA | CALGARY | 5106N | 11401W | 3557 |
| 196 | CANADA | CAMBRIDGE BAY | 6906N | 10508W | 90 |
| 197 | CANADA | CHURCHILL | 5845N | 09404W | 94 |
| 198 | CANADA | DAWSON | 6404N | 13929W | 1062 |
| 199 | CANADA | FORT NELSON | 5850N | 12235W | 1253 |
| 200 | CANADA | HALIFAX | 4453N | 06331W | 476 |
| 201 | CANADA | INUVIK | 6818N | 13329W | 223 |
| 202 | CANADA | MONTREAL | 4528N | 07345W | 117 |
| 203 | CANADA | MOOSONEE | 5116N | 08039W | 34 |
| 204 | CANADA | QUEBEC | 4648N | 07123W | 239 |
| 205 | CANADA | REGINA | 5026N | 10440W | 1894 |
| 206 | CANADA | QAUSUITTUQ (RESOLUTE) | 7443N | 09459W | 220 |
| 207 | CANADA | SAGLEK | 5829N | 06239W | 269 |
| 208 | CANADA | SAINT JOHN NB | 4519N | 06553W | 356 |
| 209 | CANADA | TORONTO | 4341N | 07937W | 569 |
| 210 | CANADA | VANCOUVER | 4911N | 12310W | 9 |
| 211 | CANADA | WINNIPEG | 4954N | 09714W | 783 |
| 212 | CANADA | YELLOWKNIFE | 6228N | 11427W | 674 |
| 213 | COSTA RICA | SAN JOSE | 0958N | 08449W | 55 |
| 214 | EL SALVADOR | SAN SALVADOR | 1340N | 08905W | 2014 |
| 215 | GREENLAND | EISMITTE | 7053N | 04042W | 9843 |
| 216 | GREENLAND | NUUK (GODTHAB) | 6410N | 05143W | 66 |
| 217 | GREENLAND | THULE AIR BASE | 7631N | 06844W | 251 |
| 218 | GUATEMALA | GUATEMALA CITY | 1435N | 09032W | 4885 |
| 219 | HONDURAS | TEGUCIGALPA | 1403N | 08713W | 3294 |
| 220 | MEXICO | CAMPECHE | 1950N | 09030W | 30 |
| 221 | MEXICO | CIUDAD JUAREZ | 3140N | 10626W | 3830 |
| 222 | MEXICO | MERIDA | 2056N | 08939W | 30 |
| 223 | MEXICO | MEXICO CITY | 1926N | 09904W | 7340 |
| 224 | NICARAGUA | MANAGUA | 1208N | 08610W | 180 |
| 225 | PANAMA | PANAMA CITY | 0858N | 07930W | 30 |

# South America

| Map # | Country | City or Location | Latitude ddmm | Longitude ddmm | Elevation feet |
|---|---|---|---|---|---|
| 225 | ARGENTINA | BUENOS AIRES | 3435S | 05829W | 82 |
| 226 | ARGENTINA | RESISTENCIA | 2728S | 05859W | 164 |
| 227 | ARGENTINA | USHUAIA | 5449S | 06819W | 10 |
| 228 | BOLIVIA | EL ALTO | 1630S | 06811W | 13354 |
| 229 | BOLIVIA | SUCRE | 1903S | 06516W | 9351 |
| 230 | BOLIVIA | TRINIDAD | 1445S | 06448W | 774 |
| 231 | BRAZIL | BELEM | 0128S | 04827W | 33 |
| 232 | BRAZIL | BRASILIA | 1551S | 04756W | 3481 |
| 233 | BRAZIL | MANAUS | 0308S | 06001W | 144 |
| 234 | BRAZIL | SAO PAULO | 2337S | 04639W | 2628 |
| 235 | BRAZIL | SAO SALVADOR | 1255S | 03820W | 20 |
| 236 | CHILE | ARICA | 1830S | 07019W | 328 |

**Table 9.1  World Weather Data Station List [CONTINUED]**

# South America, continued

| Map # | Country | City or Location | Latitude ddmm | Longitude ddmm | Elevation feet |
|-------|---------|------------------|---------------|----------------|----------------|
| 237 | CHILE | PUNTA ARENAS | 5308S | 07053W | 6 |
| 238 | CHILE | ANTOFAGASTA | 2327S | 07028W | 442 |
| 239 | CHILE | PUERTO MONTT | 4128S | 07256W | 43 |
| 240 | CHILE | SANTIAGO | 3327S | 07042W | 1706 |
| 241 | COLOMBIA | BOGOTA | 0444N | 07417W | 8379 |
| 242 | COLOMBIA | VILLAVICENCIO | 0409N | 07334W | 1414 |
| 243 | ECUADOR | GUYAQUIL | 0210S | 07952W | 13 |
| 244 | ECUADOR | QUITO | 0008S | 07829W | 9222 |
| 245 | FRENCH GUIANA | CAYENNE | 0449N | 05222W | 26 |
| 246 | GUYANA | GEORGETOWN | 0630N | 05815W | 96 |
| 247 | PARAGUAY | ASUNCION | 2516S | 05738W | 210 |
| 248 | PERU | CUZCO | 1333S | 07159W | 10866 |
| 249 | PERU | LIMA | 1201S | 07708W | 105 |
| 250 | SURINAME | PARAMARIBO | 0527N | 05512W | 54 |
| 251 | URUGUAY | MONTEVIDEO | 3452S | 05612W | 72 |
| 252 | VENEZUELA | CARACAS | 1029N | 06650W | 2760 |

# Arctic Ocean Islands

| Map # | Country | City or Location | Latitude ddmm | Longitude ddmm | Elevation feet |
|-------|---------|------------------|---------------|----------------|----------------|
| 253 | ALEXANDRA LAND | RUSSIA | 8040N | 04358E | 89 |
| 254 | BARTER ISLAND | USA | 7008N | 14335W | 8 |
| 255 | CAPE GOLOMJANNYI | RUSSIA | 7933N | 09037E | 10 |
| 256 | CAPE ZHELANIYA (SEVERNY IS.) | RUSSIA | 7657N | 06835E | 26 |
| 257 | DRUZHNYY ISLAND | RUSSIA | 8037N | 05803E | 66 |
| 258 | ISACHSEN | CANADA | 7847N | 10332W | 175 |
| 259 | KOTELNYY ISLAND | RUSSIA | 7600N | 13754E | 33 |
| 260 | MOULD BAY | CANADA | 7614N | 11918W | 40 |
| 261 | RUSSKAYA GAVAN | RUSSIA | 7611N | 06234E | 30 |
| 262 | WRANGEL ISLAND | RUSSIA | 7058N | 17832W | 10 |

# Atlantic/Caribbean Islands

| Map # | Country | City or Location | Latitude ddmm | Longitude ddmm | Elevation feet |
|-------|---------|------------------|---------------|----------------|----------------|
| 263 | ANTIGUA AND BARBUDA | ST. JOHN'S | 1707N | 06147W | 62 |
| 264 | AZORES ISLANDS, PORTUGAL | PONTA DELGADA | 3745N | 02540W | 118 |
| 265 | BAHAMAS, COMMONWEALTH OF | NASSAU | 2502N | 07728W | 7 |
| 266 | BARBADOS | BRIDGETOWN | 1308N | 05936W | 181 |
| 267 | BERMUDA | HAMILTON | 3221N | 06441W | 12 |
| 268 | CANARY ISLANDS SPAIN | SANTA CRUZ | 2827N | 01615W | 151 |
| 269 | CAPE VERDE, REP. OF | PRAIA | 1455N | 02331W | 115 |
| 270 | CUBA | HAVANA | 2301N | 08224W | 225 |
| 271 | DOMINICAN REPUBLIC | SANTO DOMINGO | 1826N | 06940W | 57 |
| 272 | HAITI | PETIONVILLE | 1831N | 07217W | 1312 |
| 273 | JAMAICA | KINGSTON | 1754N | 07717W | 18 |
| 274 | NETHERLANDS ANTILLES | WILLEMSTAD | 1211N | 06857W | 27 |
| 275 | MADEIRA ISLANDS | FUNCHAL | 3241N | 01646W | 190 |
| 276 | PUERTO RICO | SAN JUAN | 1828N | 06607W | 82 |
| 277 | ST. HELENA | ASCENSION | 0758S | 01424W | 272 |
| 278 | ST. LUCIA | CASTRIES | 1344N | 06057W | 10 |
| 279 | SAINT MAARTEN | JULIANA APT | 1802N | 06306W | 0 |
| 280 | SAO TOME AND PRINCIPE | SAO TOME | 0023N | 00643E | 26 |
| 281 | TRINDAD AND TOBAGO | CHAGUARAMAS | 1041N | 06137W | 42 |
| 282 | VIRGIN ISLAND, U.S. | ALEXANDER HAMILTON ARPT. | 1742N | 06447W | 60 |

**Table 9.1  World Weather Data Station List [CONTINUED]**

# Indian Ocean Islands

| Map # | Country | City or Location | Latitude ddmm | Longitude ddmm | Elevation feet |
|---|---|---|---|---|---|
| 283 | ANDAMAN/NICOBAR ISLANDS | PORT BLAIR | 1140N | 09243E | 259 |
| 284 | COCOS-KEELING Is. | COCOS ISLAND | 1211S | 09650E | 11 |
| 285 | KIRIBATI, REP. OF | CHRISTMAS IS | 0159N | 15720W | 5 |
| 286 | MALDIVES, REP. OF | SEENU ATOLL | 0041S | 07309E | 6 |
| 287 | MASCARENE ISLAND GROUP | DIEGO GARCIA | 0714S | 07226E | 7 |
| 288 | MAURITIUS | PORT LOUIS | 2006S | 05732E | 181 |
| 289 | REUNION | ST DENIS | 2035S | 05531E | 66 |
| 290 | SEYCHELLES | VICTORIA | 0437S | 05527E | 8 |

# Mediterranean Islands

| Map # | Country | City or Location | Latitude ddmm | Longitude ddmm | Elevation feet |
|---|---|---|---|---|---|
| 291 | BALEARIC ISLANDS, SPAIN | PALMA, MALLORCA | 3933N | 00243E | 13 |
| 292 | CORSICA,FRANCE | AJACCIO | 4155N | 00848E | 20 |
| 293 | CRETE,GREECE | HERAKLION | 3520N | 02511E | 115 |
| 294 | CYPRUS, REP. OF | NICOSIA | 3509N | 03316E | 732 |
| 295 | SARDINIA, ITLAY | CAGLIARI | 3914N | 00903E | 12 |
| 296 | SICILY, ITALY | PALERMO | 3806N | 01318E | 345 |
| 297 | MALTA | VALLETTA | 3553N | 01425E | 340 |

# Pacific Ocean Islands

| Map # | Country | City or Location | Latitude ddmm | Longitude ddmm | Elevation feet |
|---|---|---|---|---|---|
| 298 | AMERICAN SAMOA | PAGO PAGO INTL | 1419S | 17043W | 29 |
| 299 | OGASAWARA IS., JAPAN | IWO JIMA | 2447N | 14119E | 353 |
| 300 | CAROLINE ISLANDS | TRUK | 0727N | 15150E | 6 |
| 301 | CHILE | RAPA NUI (EASTER IS.) | 2712S | 10924W | 134 |
| 302 | FIJI ISLANDS, REP. OF | SUVA | 1808S | 17826E | 20 |
| 303 | GUAM | ANDERSEN AFB | 1334N | 14455E | 605 |
| 304 | NANSEI/RYUKYU IS, JAPAN | NAHA (OKINAWA) | 2611N | 12738E | 14 |
| 305 | JOHNSTON ISLAND, USA | JOHNSTON ISLAND | 1644N | 16931W | 7 |
| 306 | MARSHALL ISLANDS, REP. OF | ENIWETOK | 1120N | 16219E | 13 |
| 307 | MICRONESIA, FED. STATES OF | WENO | 0727N | 15150E | 6 |
| 308 | MIDWAY ISLANDS, USA | MIDWAY NS | 2811N | 17722W | 13 |
| 309 | NEW CALEDONIA, FRANCE | NOUMEA | 2216S | 16627E | 246 |
| 310 | NEW ZEALAND | QUEENSTOWN | 4501S | 16844E | 1167 |
| 311 | NEW ZEALAND | WELLINGTON | 4119S | 17448E | 38 |
| 312 | SAMOA | APIA | 1348S | 17147W | 7 |
| 313 | SOCIETY ISLANDS (TAHITI) | PAPEETE | 1733S | 14936W | 7 |
| 314 | SOLOMON ISLANDS | HONIARA | 0925S | 16002E | 10 |
| 315 | VANUATU | PORT VILA | 1744S | 16819E | 66 |
| 316 | WAKE ISLAND | WAKE ISLAND | 1916N | 16638E | 12 |

# Southern Ocean Islands

| Map # | Country | City or Location | Latitude ddmm | Longitude ddmm | Elevation feet |
|---|---|---|---|---|---|
| 317 | DECEPTION ISLAND | DECEPTION ISLAND | 6259S | 06034W | 26 |
| 318 | HEARD IS., AUST. | HEARD IS. | 5306S | 07331E | 65 |
| 319 | MACQUARIE IS., AUST. | MACQUARIE IS. | 5430S | 15857E | 20 |
| 320 | MARION IS., S. AFRICA | PRINCE EDWARD IS. | 4653S | 03752S | 72 |
| 321 | S. SANDWICH IS., GR. BR. | SOUTH THULE IS | 5927S | 02719W | 268 |

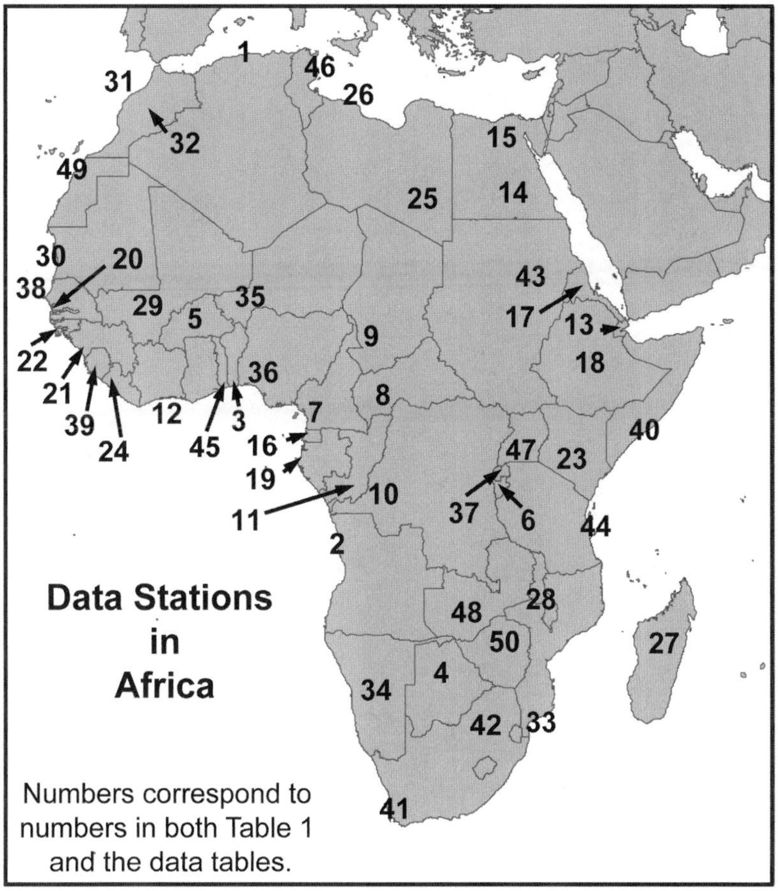

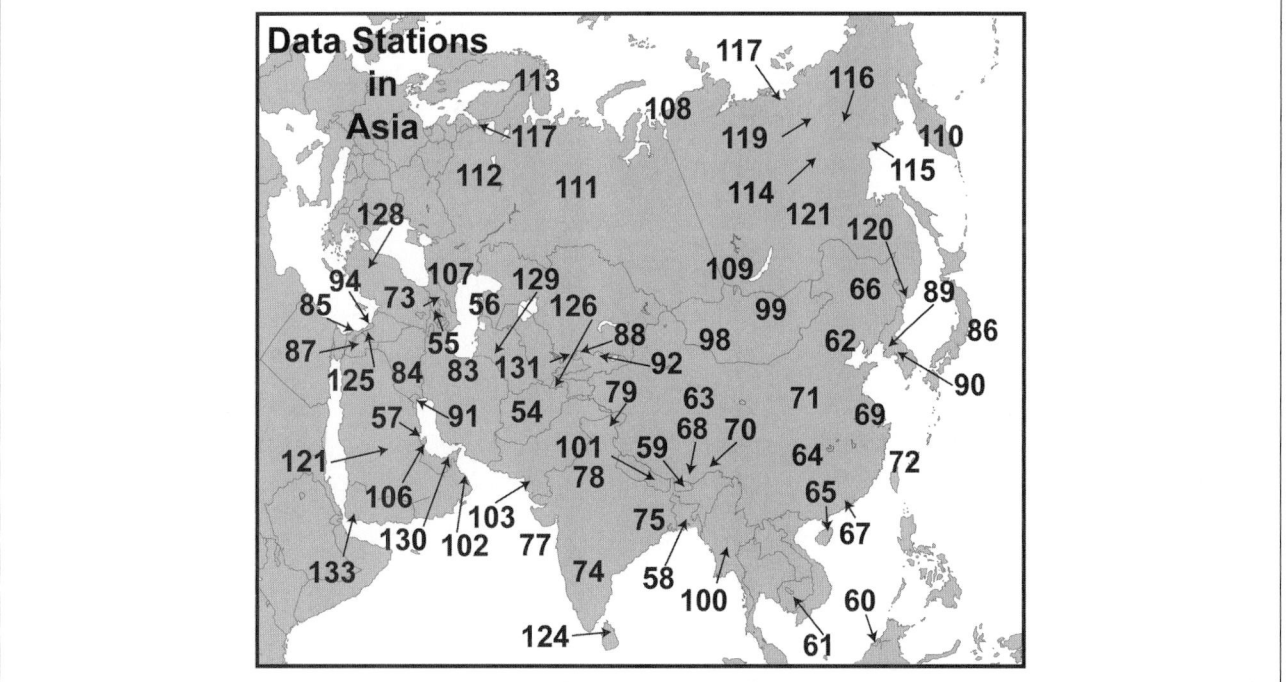

**Figure 9.2**   Locater map for weather data station.

**Figure 9.3**   Locater map for weather data station.

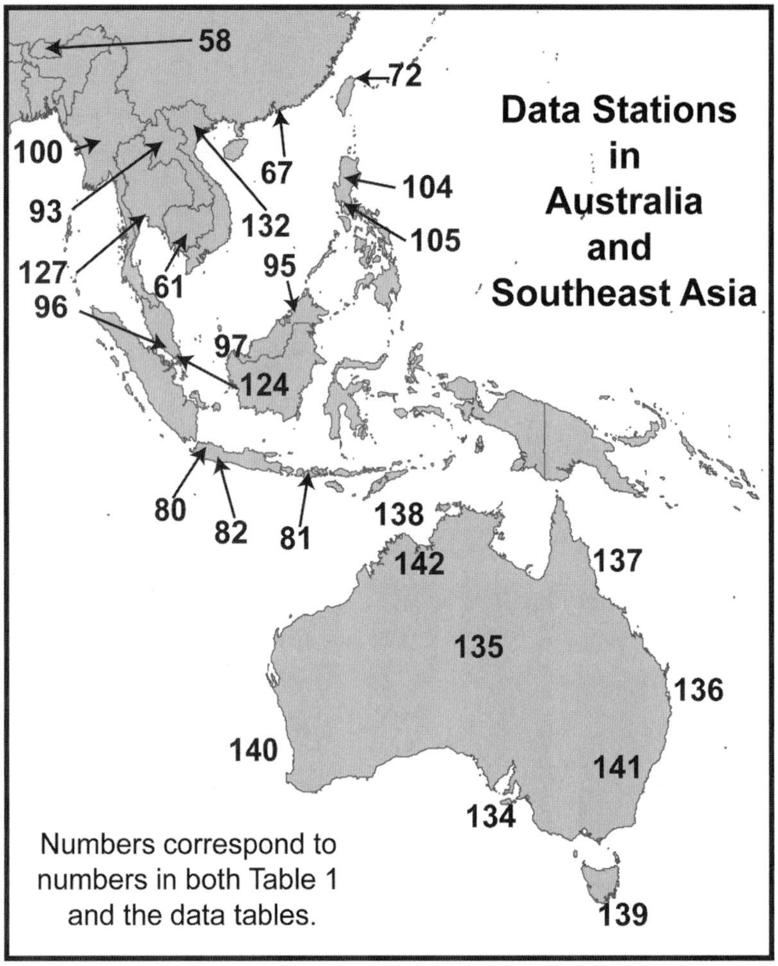

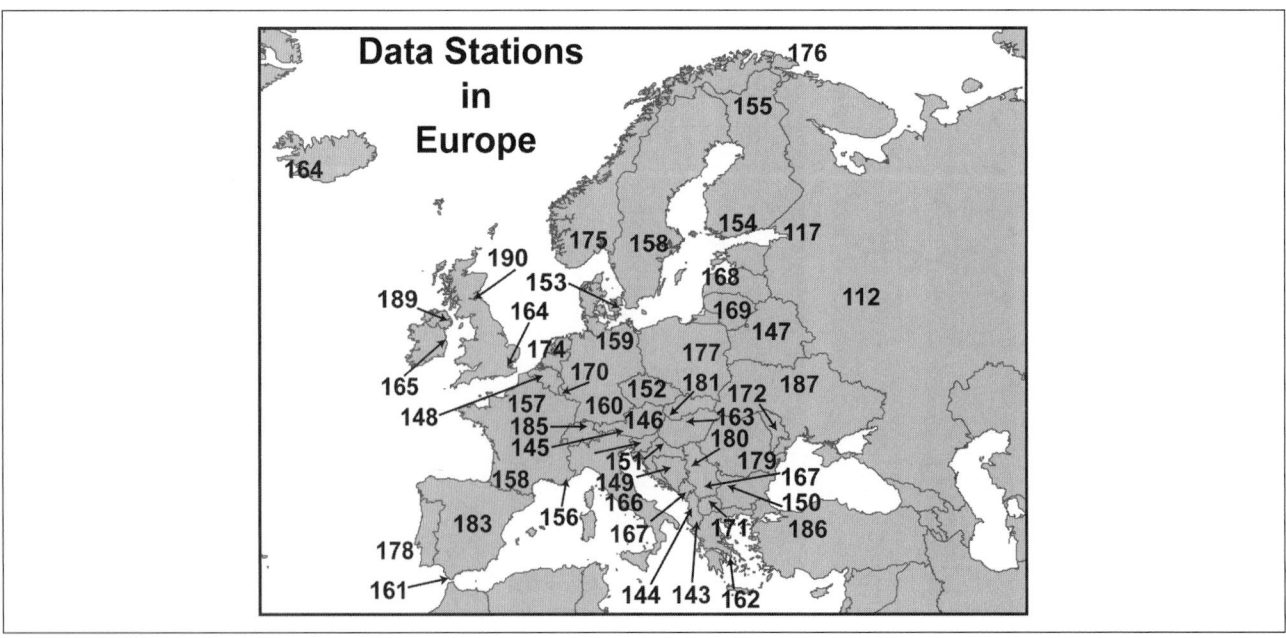

**Figure 9.4** Locater map for weather data station.

**Figure 9.5** Locater map for weather data station.

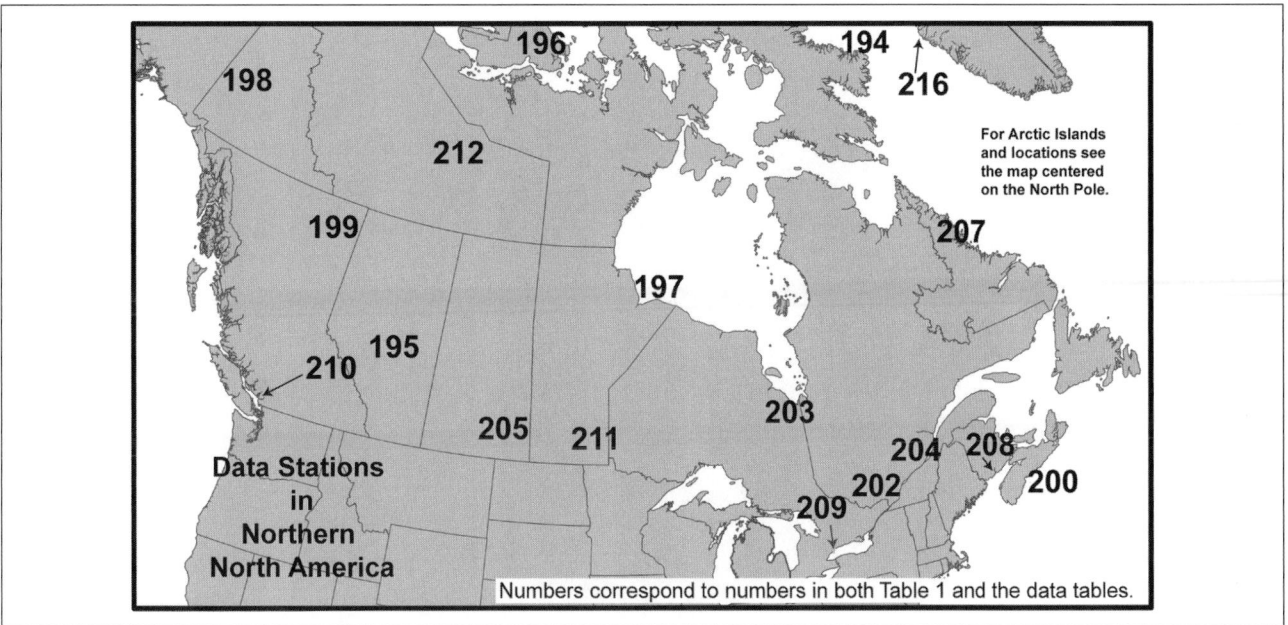

**Figure 9.6**  Locater map for weather data station.

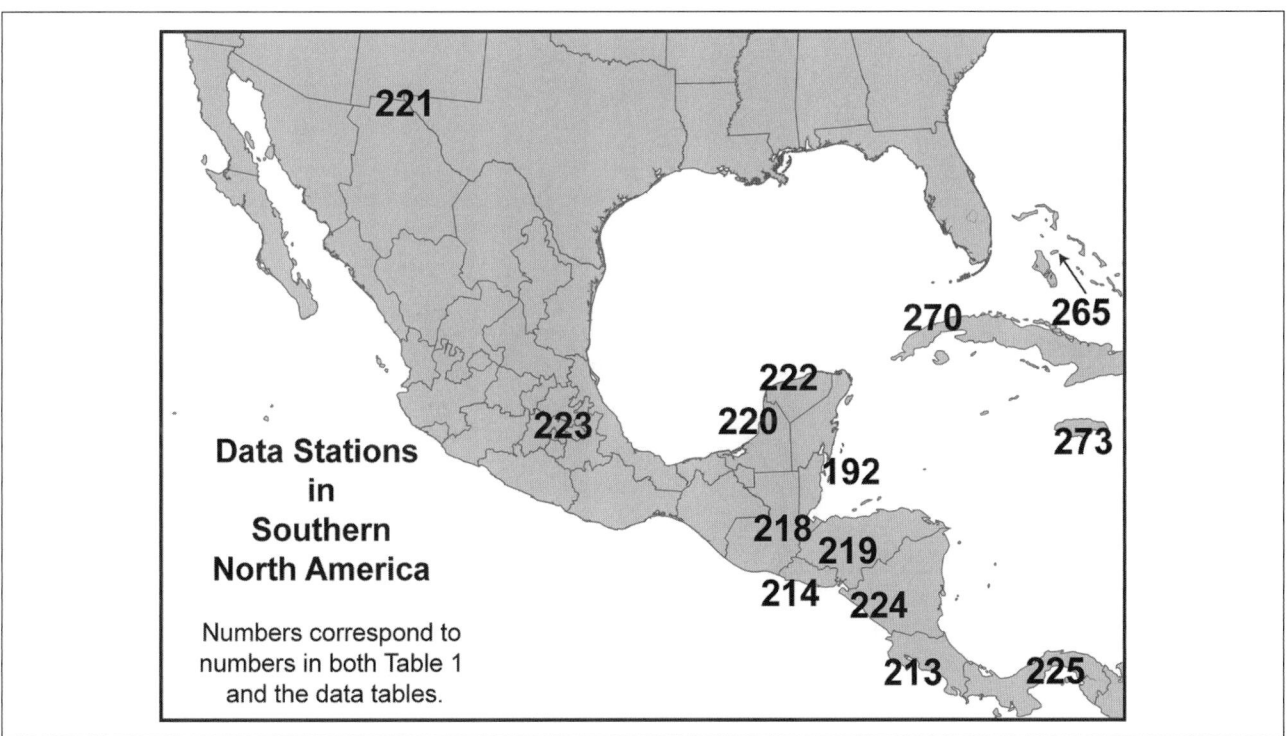

**Figure 9.7**  Locater map for weather data station.

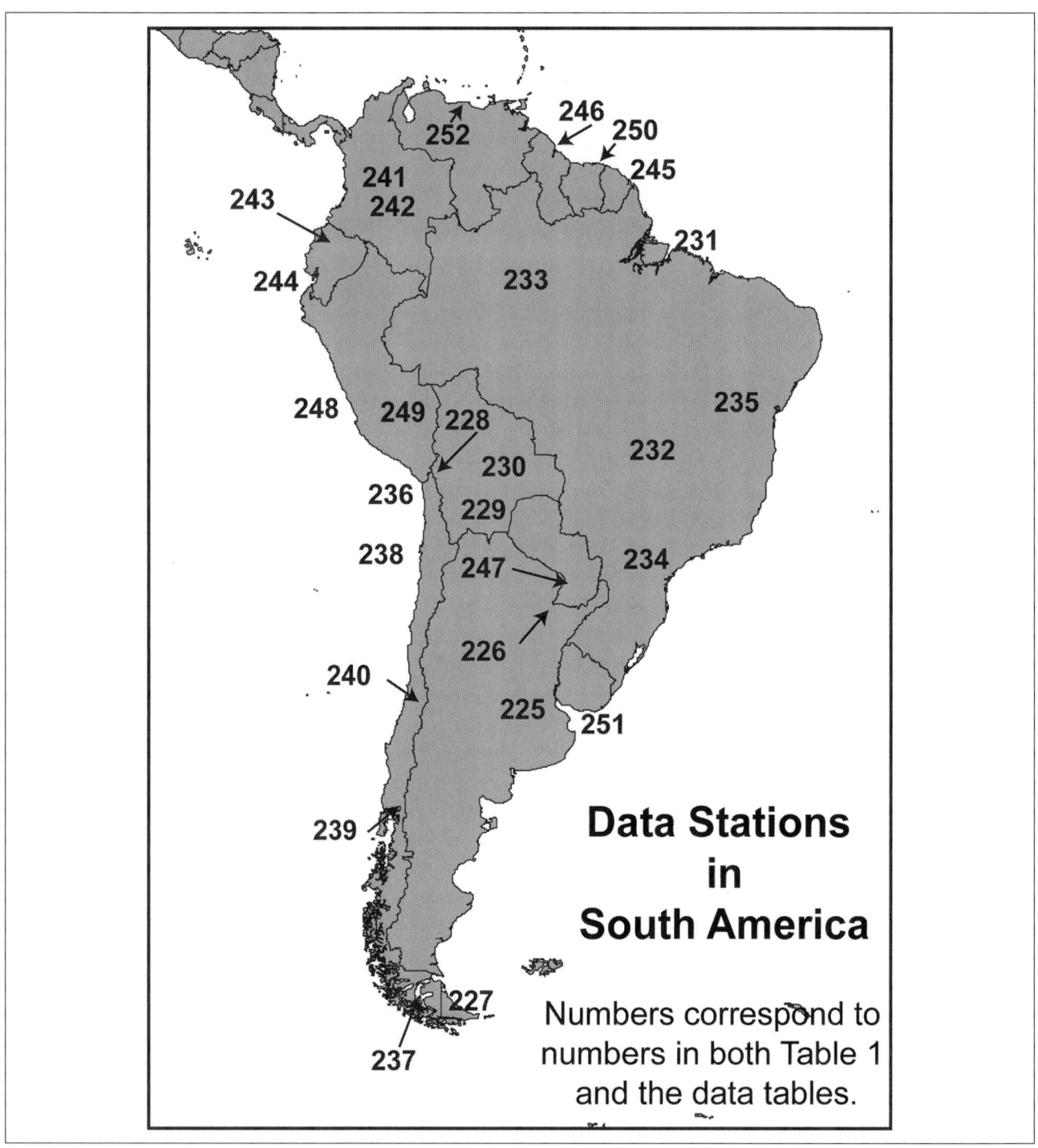

**Figure 9.8** Locater map for weather data station.

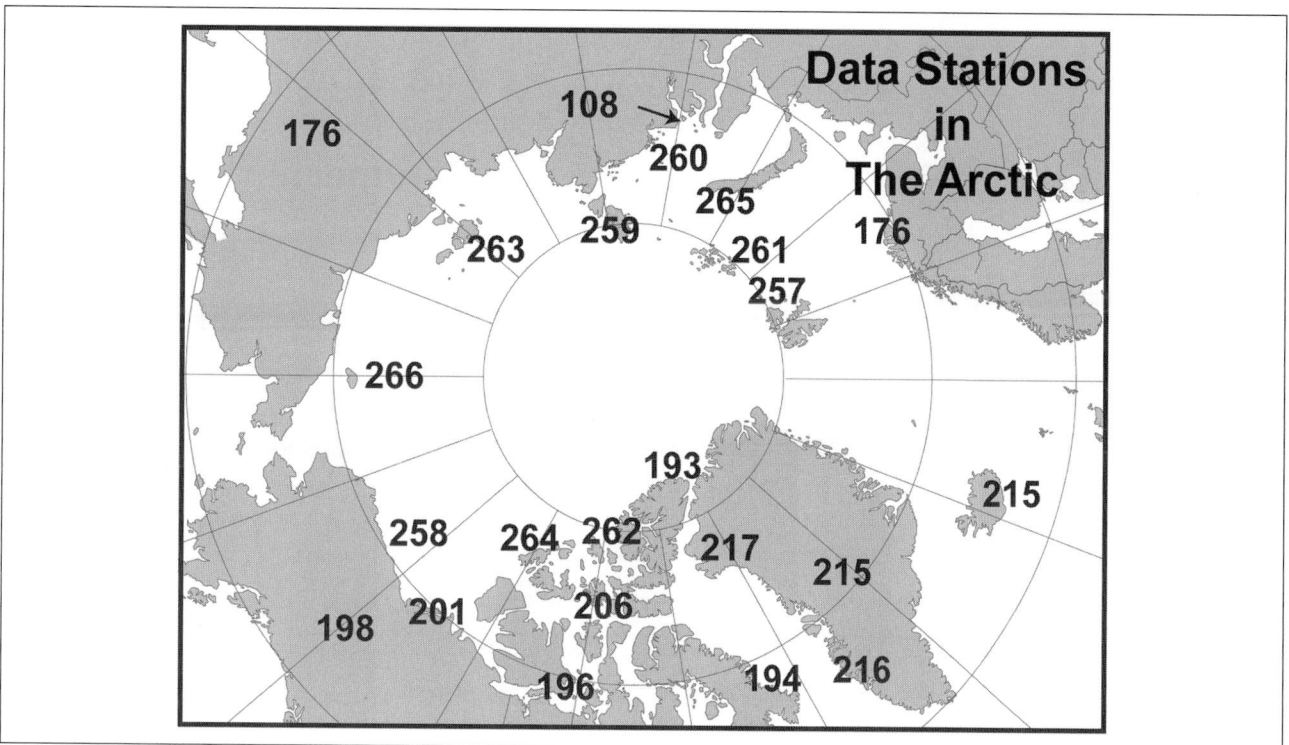

**Figure 9.9**  Locater map for weather data station.

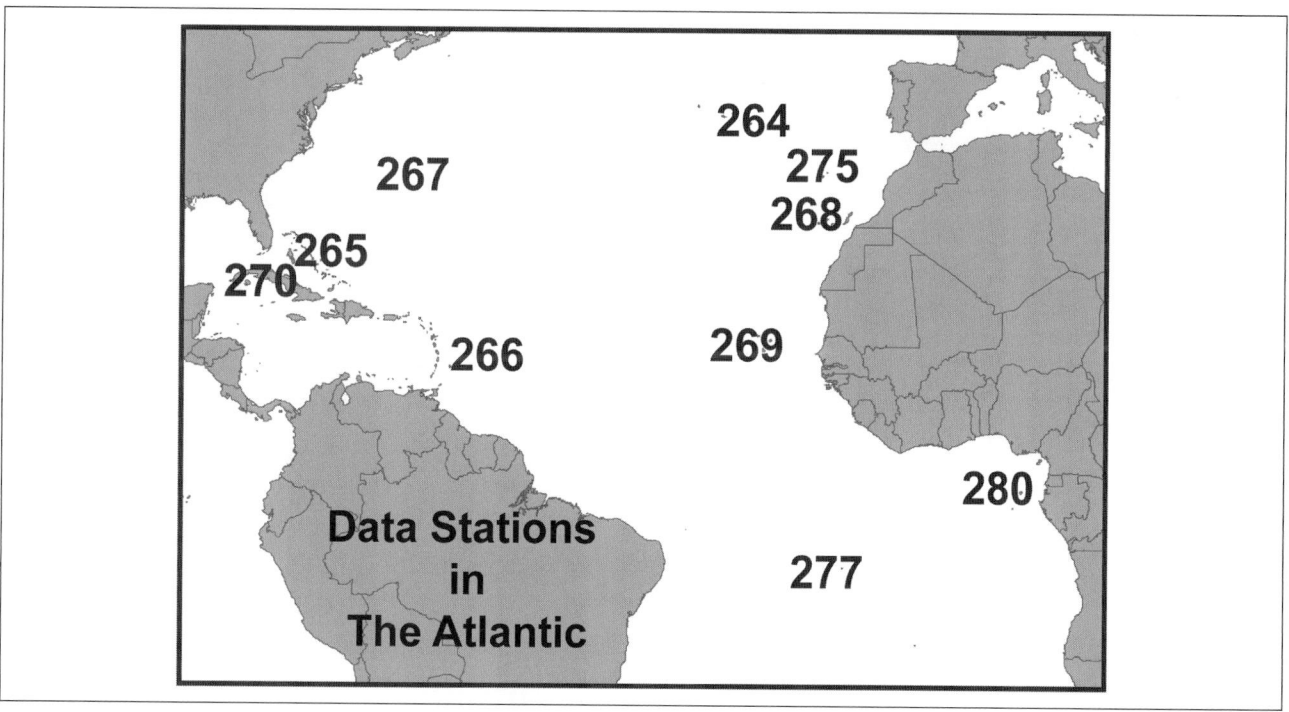

**Figure 9.10**  Locater map for weather data station.

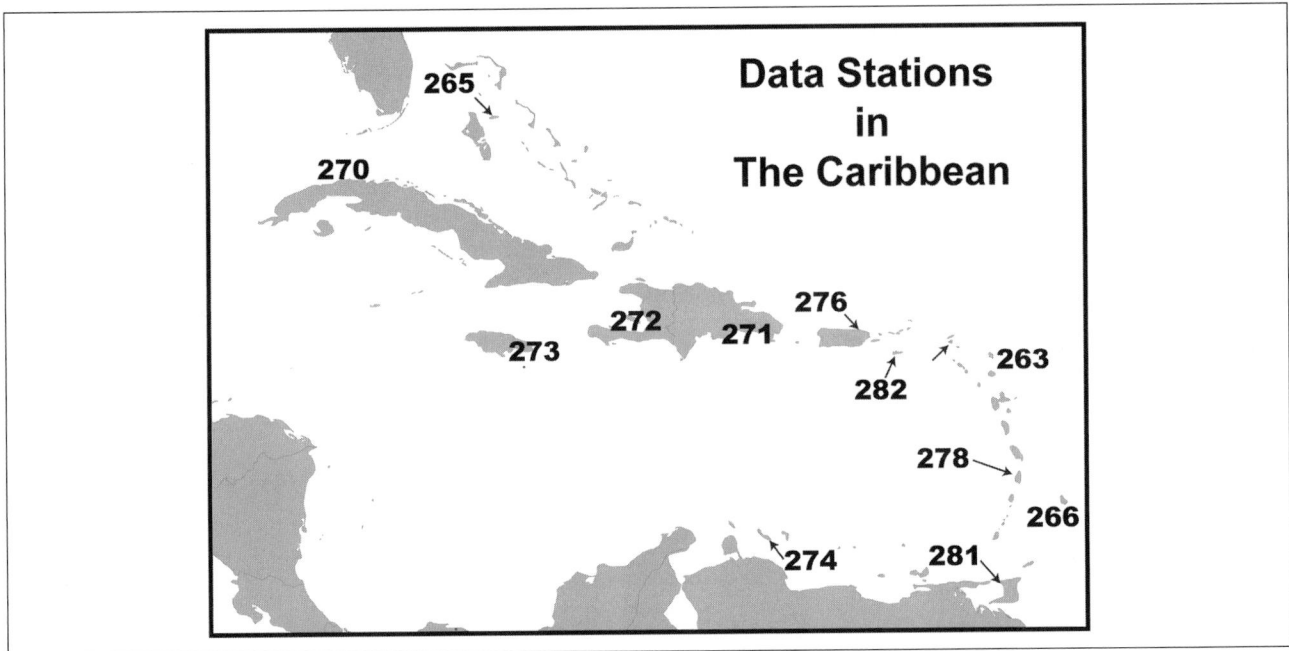

**Figure 9.11**  Locater map for weather data station.

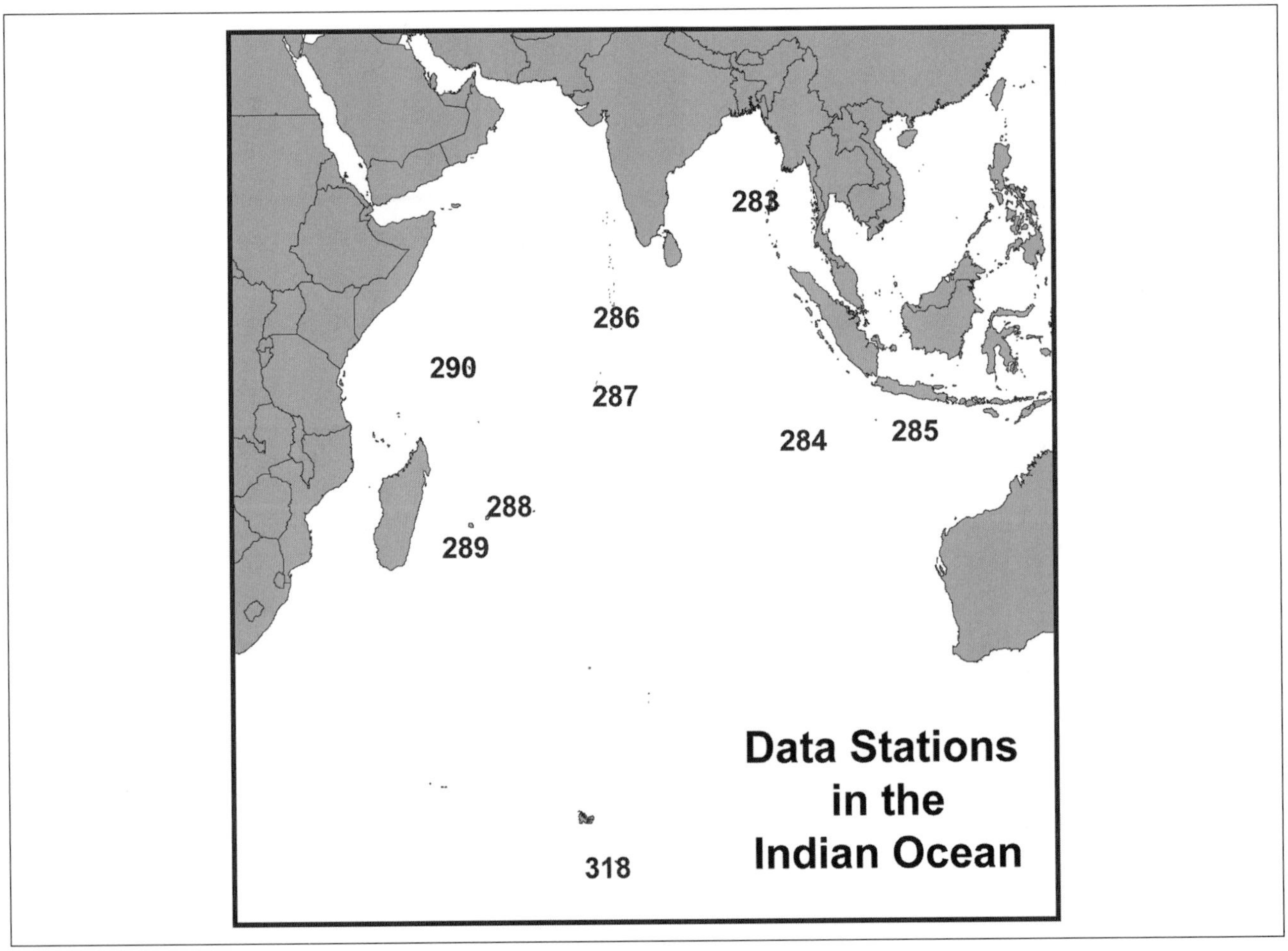

**Figure 9.12**  Locater map for weather data station.

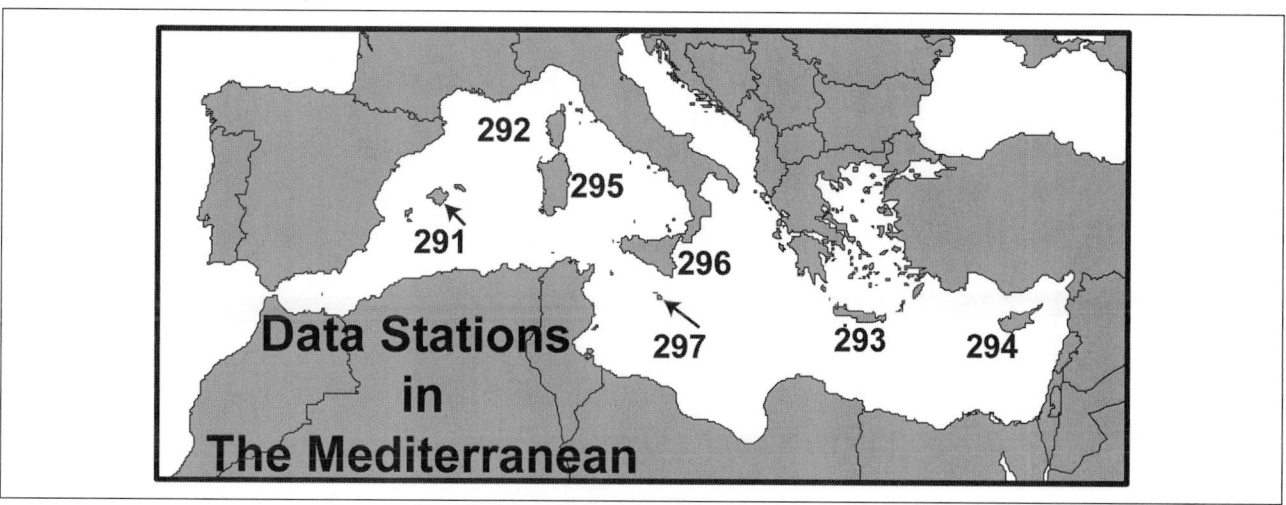

**Figure 9.13** Locater map for weather data station.

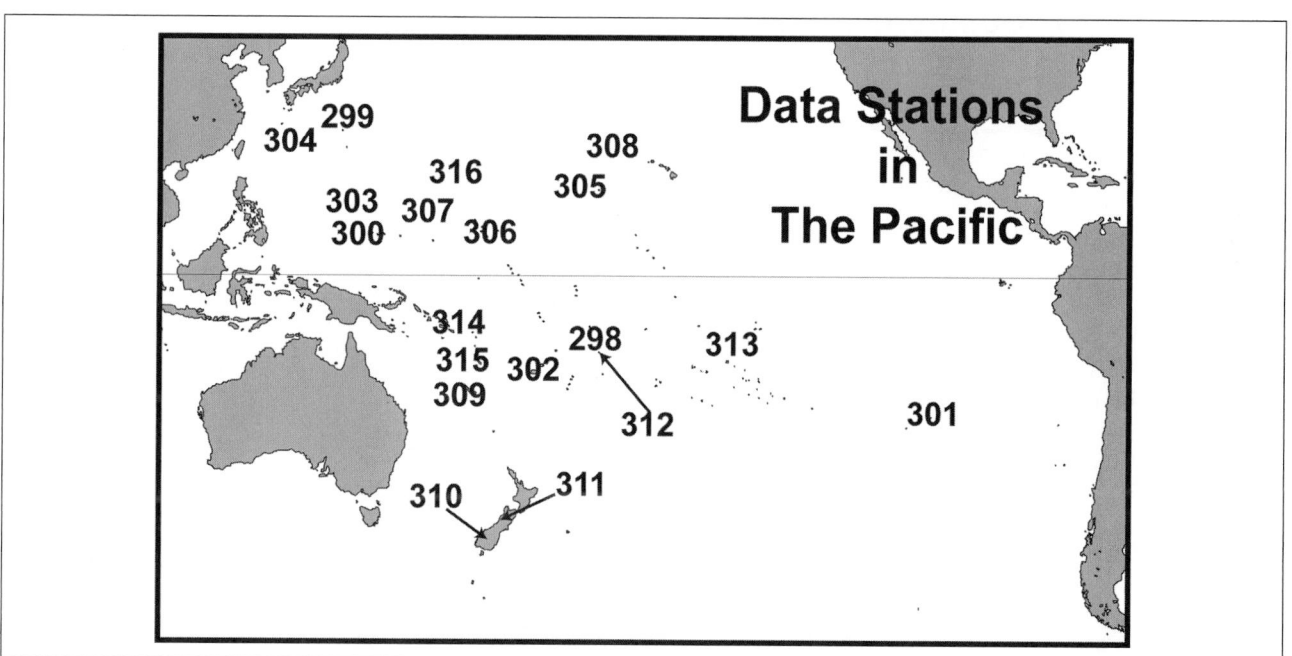

**Figure 9.14** Locater map for weather data station.

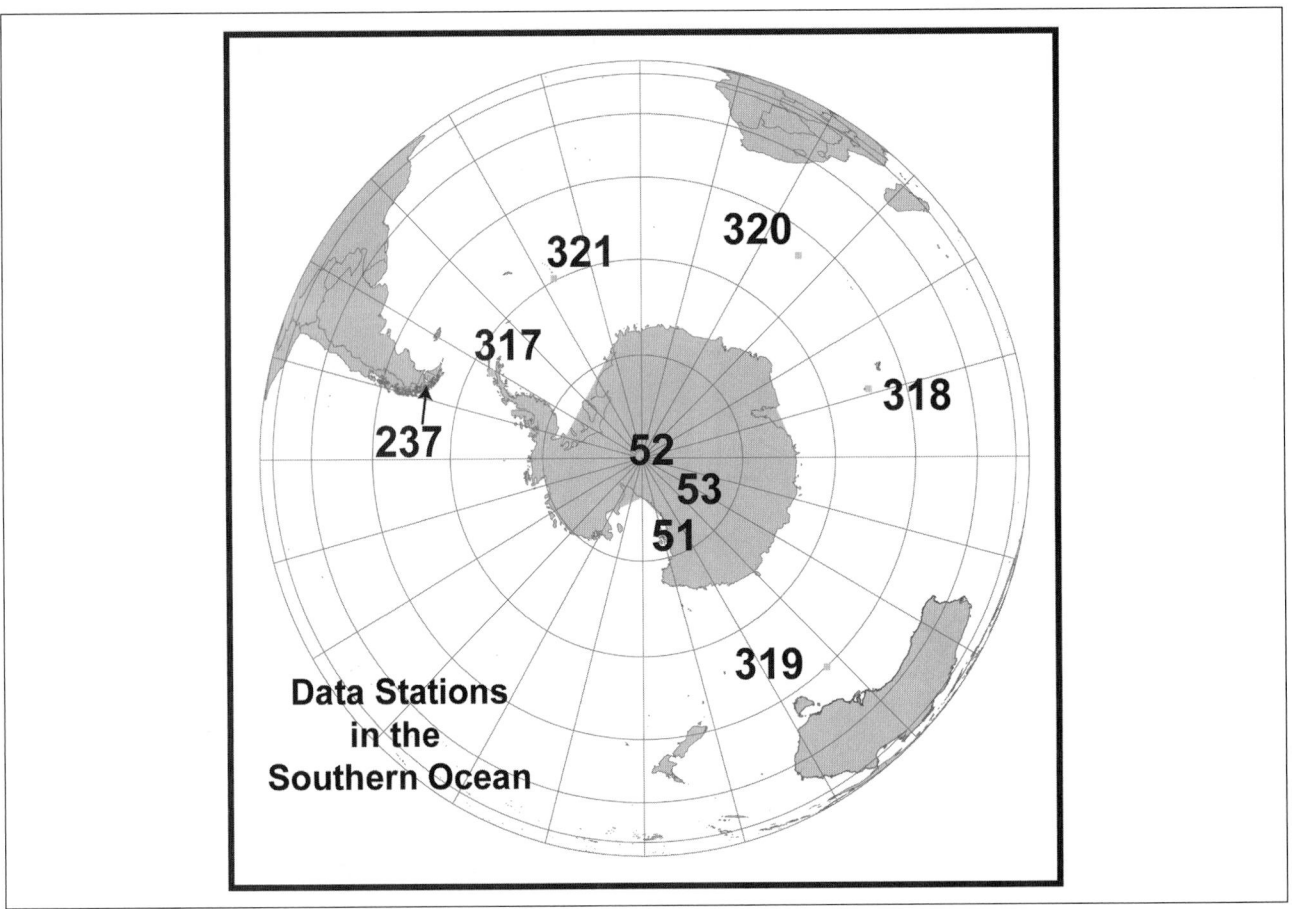

**Figure 9.15** Locater map for weather data station.

- **DATA KEY**

Figure 9.16 is the key to station data (Tables 9.2), and for the reader's convenience, it is repeated at eight locations within the tables where space allows.

## Data Table Key

| Map Locater Number | MONTH | | |
|---|---|---|---|
| **Location** | Average Max T °F | | Record Max T °F |
| | Average Min T °F | | Record Min T °F |
| Lat - Lon ddmm Elev ft | Avg. Precip in. (Rain Days) (Thunder Days) | | Avg. Snow in. (Snow Days) |

| | |
|---|---|
| **\*\*\* or Blank** | No data, missing data or data of insufficient reliability |
| **Rain Days** | Number of Days with > = 0.01" of precipitation falling |
| **Snow Days** | Number of Days with > = 1.5" of snow falling |

**Figure 9.16** Weather data key.

**Table 9.2  World Weather Data**

# Africa, Algeria to Burkina Faso                                   Map # 1 to 5

### Map Location 1 — Algeria, Algiers  3642N 00313E 82

| | JAN | FEB | MAR | APR | MAY | JUN | JUL | AUG | SEP | OCT | NOV | DEC | ANNUAL |
|---|---|---|---|---|---|---|---|---|---|---|---|---|---|
| Record High | 76 | 86 | 84 | 99 | 102 | 107 | 106 | 117 | 112 | 100 | 88 | 117 | 117 |
| Normal High | 59 | 61 | 63 | 68 | 73 | 78 | 83 | 85 | 81 | 74 | 66 | 60 | 71 |
| Normal Low | 49 | 52 | 52 | 55 | 59 | 65 | 70 | 71 | 69 | 63 | 56 | 51 | 59 |
| Record Low | 31 | 31 | 33 | 39 | 39 | 49 | 56 | 57 | 53 | 45 | 35 | 31 | 31 |
| Precip | 4.57 (9) (2) | 2.99 (6.4) (2) | 2.24 (4.8) (2) | 2.56 (5.4) (1) | 1.42 (3.3) (2) | 0.55 (1.2) (2) | 0.08 (0.3) (1) | 0.16 (0.4) (1) | 1.06 (2.2) (3) | 3.31 (5.9) (2) | 3.66 (6.4) (2) | 4.65 (9.1) (2) | 27.3 (54.4) (22) |
| Snow | (0) (0) | (0) (0) | (0) (0) | (0) (0) | (0) (0) | (0) (0) | (0) (0) | (0) (0) | (0) (0) | (0) (0) | (0) (0) | (0) (0) | (0) (0) |

### Map Location 2 — Angola, Luanda  0851S 01314E 243

| | JAN | FEB | MAR | APR | MAY | JUN | JUL | AUG | SEP | OCT | NOV | DEC | ANNUAL |
|---|---|---|---|---|---|---|---|---|---|---|---|---|---|
| Record High | 91 | 95 | 95 | 94 | 97 | 89 | 85 | 83 | 84 | 89 | 94 | 98 | 98 |
| Normal High | 83 | 85 | 86 | 85 | 82 | 77 | 74 | 74 | 76 | 79 | 82 | 83 | 81 |
| Normal Low | 74 | 75 | 75 | 75 | 73 | 68 | 64 | 64 | 67 | 71 | 73 | 74 | 71 |
| Record Low | 69 | 70 | 70 | 70 | 64 | 59 | 58 | 58 | 62 | 65 | 68 | 67 | 58 |
| Precip | 1 (2.3) (3) | 1.4 (3.2) (4) | 3 (6) (9) | 4.6 (7.9) (11) | 0.5 (1.3) (2) | 0.05 (0.2) (0) | 0 (0) (0) | 0.05 (0.2) (0.1) | 0.1 (0.4) (0.4) | 0.2 (0.6) (0.8) | 1.1 (2.2) (2) | 0.8 (1.9) (2) | 12.8 (26.2) (34.3) |
| Snow | (0) (0) | (0) (0) | (0) (0) | (0) (0) | (0) (0) | (0) (0) | (0) (0) | (0) (0) | (0) (0) | (0) (0) | (0) (0) | (0) (0) | (0) (0) |

### Map Location 3 — Benin, Cotonou  0621N 00223E 16

| | JAN | FEB | MAR | APR | MAY | JUN | JUL | AUG | SEP | OCT | NOV | DEC | ANNUAL |
|---|---|---|---|---|---|---|---|---|---|---|---|---|---|
| Record High | 90 | 93 | 94 | 95 | 94 | 91 | 89 | 89 | 89 | 91 | 92 | 91 | 95 |
| Normal High | 80 | 82 | 83 | 83 | 81 | 78 | 78 | 77 | 78 | 80 | 82 | 81 | 80 |
| Normal Low | 74 | 77 | 79 | 78 | 76 | 74 | 74 | 73 | 74 | 75 | 77 | 76 | 76 |
| Record Low | 66 | 70 | 70 | 70 | 70 | 65 | 68 | 69 | 68 | 71 | 71 | 74 | 65 |
| Precip | 1.3 (2.9) (2) | 1.3 (2.9) (3) | 4.6 (7.9) (8) | 4.9 (8.2) (12) | 1 (2.5) (13) | 1.44 (2.9) (13) | 3.5 (6.3) (4) | 2.6 (4.8) (4) | 5.3 (8.7) (11) | 0.2 (0.6) (0.8) | 1 (1.3) (2) | 0.5 (1.2) (2) | 30.2 (55.6) (83) |
| Snow | (0) (0) | (0) (0) | (0) (0) | (0) (0) | (0) (0) | (0) (0) | (0) (0) | (0) (0) | (0) (0) | (0) (0) | (0) (0) | (0) (0) | (0) (0) |

### Map Location 4 — Botswana, Maun  1959S 02325E 3101

| | JAN | FEB | MAR | APR | MAY | JUN | JUL | AUG | SEP | OCT | NOV | DEC | ANNUAL |
|---|---|---|---|---|---|---|---|---|---|---|---|---|---|
| Record High | 104 | 103 | 102 | 99 | 96 | 92 | 97 | 101 | 108 | 109 | 110 | 108 | 110 |
| Normal High | 90 | 89 | 88 | 87 | 82 | 77 | 83 | 91 | 97 | 95 | 94 | 91 | 87 |
| Normal Low | 66 | 66 | 63 | 58 | 49 | 42 | 47 | 55 | 64 | 64 | 66 | 66 | 57 |
| Record Low | 48 | 48 | 50 | 38 | 29 | 25 | 24 | 31 | 31 | 43 | 49 | 52 | 24 |
| Precip | 4.3 (8.6) (***) | 3.8 (7.9) (***) | 3.5 (6.7) (***) | 1.1 (2.7) (***) | 0.2 (0.6) (***) | 0.05 (0.2) (***) | 0 (***) | 0.05 (0.3) (***) | 0.5 (1.1) (***) | 0.24 (0.7) (1) | 1.9 (3.6) (***) | 2.8 (6.1) (***) | 18.2 (37.8) (***) |
| Snow | (0) (0) | (0) (0) | (0) (0) | (0) (0) | (0) (0) | (0) (0) | (0) (0) | (0) (0) | (0) (0) | (0) (0) | (0) (0) | (0) (0) | (0) (0) |

### Map Location 5 — Burkina Faso, Ouagadougou  1221N 00131W 1037

| | JAN | FEB | MAR | APR | MAY | JUN | JUL | AUG | SEP | OCT | NOV | DEC | ANNUAL |
|---|---|---|---|---|---|---|---|---|---|---|---|---|---|
| Record High | 113 | 113 | 113 | 116 | 118 | 111 | 106 | 101 | 102 | 106 | 107 | 113 | 118 |
| Normal High | 92 | 98 | 104 | 103 | 101 | 96 | 87 | 87 | 89 | 95 | 96 | 95 | 96 |
| Normal Low | 60 | 68 | 73 | 79 | 78 | 76 | 72 | 71 | 73 | 74 | 71 | 62 | 72 |
| Record Low | 48 | 54 | 59 | 59 | 66 | 63 | 65 | 58 | 66 | 64 | 60 | 51 | 48 |
| Precip | 0 (0) (0.3) | 0.08 (0.3) (0.3) | 0.24 (0.7) (1) | 0.71 (1.8) (6) | 3.39 (6.6) (13) | 4.49 (7.7) (16) | 7.6 (11.2) (19) | 10.35 (13.4) (18) | 6.02 (9.6) (16) | 1.61 (3.1) (13) | 0.04 (0.3) (1) | 0 (0) (0) | 34.5 (54.7) (103.6) |
| Snow | (0) (0) | (0) (0) | (0) (0) | (0) (0) | (0) (0) | (0) (0) | (0) (0) | (0) (0) | (0) (0) | (0) (0) | (0) (0) | (0) (0) | (0) (0) |

**Table 9.2   World Weather Data** [CONTINUED]

# Africa, Burundi to Congo, Dem. Rep. of

## Map # 6 to 10

**Map Location 6 — Burundi, Bujumbura — 0320S 02919E 2566**

| | JAN | FEB | MAR | APR | MAY | JUN | JUL | AUG | SEP | OCT | NOV | DEC | ANNUAL |
|---|---|---|---|---|---|---|---|---|---|---|---|---|---|
| High | 90 | 94 | 92 | 93 | 90 | 89 | 89 | 87 | 91 | 92 | 89 | 90 | 97 |
| Low | 60 | 60 | 60 | 60 | 60 | 58 | 56 | 59 | 59 | 61 | 59 | 60 | 54 |
| Snow | (0)(0) | (0)(0) | (0)(0) | (0)(0) | (0)(0) | (0)(0) | (0)(0) | (0)(0) | (0)(0) | (0)(0) | (0)(0) | (0)(0) | (0)(0) |
| High | 82 | 83 | 84 | 83 | 83 | 83 | 83 | 86 | 86 | 85 | 81 | 84 | 84 |
| Low | 66 | 65 | 66 | 66 | 66 | 63 | 62 | 64 | 66 | 66 | 66 | 65 | 65 |
| Precip | 3.69 (7.7) (22.2) | 4.04 (8.2) (17.9) | 4.71 (8) (20.8) | 6.51 (9.2) (13.3) | 2.41 (5.1) (8) | 0.23 (0.6) (2.5) | 0.13 (0.4) (0.4) | 0.11 (0.3) (3.8) | 1.23 (2.5) (8.4) | 2.77 (5) (14.8) | 3.84 (7.9) (16.5) | 3.89 (6.7) (16.3) | 33.6 (61.6) (144.9) |

**Map Location 7 — Cameroon, Yaounde — 0349N 01131E 2470**

| | JAN | FEB | MAR | APR | MAY | JUN | JUL | AUG | SEP | OCT | NOV | DEC | ANNUAL |
|---|---|---|---|---|---|---|---|---|---|---|---|---|---|
| High | 91 | 92 | 91 | 96 | 94 | 90 | 87 | 93 | 88 | 91 | 89 | 90 | 96 |
| Low | 54 | 59 | 58 | 59 | 60 | 59 | 59 | 60 | 59 | 59 | 61 | 60 | 54 |
| Snow | (0)(0) | (0)(0) | (0)(0) | (0)(0) | (0)(0) | (0)(0) | (0)(0) | (0)(0) | (0)(0) | (0)(0) | (0)(0) | (0)(0) | (0)(0) |
| High | 85 | 85 | 85 | 85 | 83 | 81 | 80 | 80 | 81 | 81 | 83 | 83 | 83 |
| Low | 67 | 67 | 67 | 66 | 67 | 66 | 65 | 65 | 65 | 65 | 66 | 66 | 66 |
| Precip | 0.87 (2) (3) | 2.48 (5.4) (3) | 5.75 (8.8) (13) | 7.17 (9.5) (17) | 8.03 (9.8) (16) | 5.95 (9.5) (13) | 2.21 (4.2) (4) | 2.91 (5.4) (4) | 7.95 (11.8) (14) | 11.81 (15.2) (23) | 5 (8.3) (22) | 0.79 (1.8) (7) | 60.9 (91.7) (139) |

**Map Location 8 — Cen. African Rep., Bangui — 0423N 01833E 1208**

| | JAN | FEB | MAR | APR | MAY | JUN | JUL | AUG | SEP | OCT | NOV | DEC | ANNUAL |
|---|---|---|---|---|---|---|---|---|---|---|---|---|---|
| High | 99 | 102 | 103 | 100 | 103 | 96 | 94 | 94 | 95 | 96 | 98 | 97 | 103 |
| Low | 55 | 56 | 61 | 64 | 61 | 62 | 63 | 63 | 63 | 63 | 62 | 57 | 55 |
| Snow | (0)(0) | (0)(0) | (0)(0) | (0)(0) | (0)(0) | (0)(0) | (0)(0) | (0)(0) | (0)(0) | (0)(0) | (0)(0) | (0)(0) | (0)(0) |
| High | 91 | 93 | 92 | 91 | 89 | 88 | 86 | 87 | 87 | 87 | 89 | 89 | 89 |
| Low | 67 | 68 | 70 | 71 | 70 | 69 | 70 | 68 | 68 | 68 | 68 | 67 | 68 |
| Precip | 0.83 (1.9) (4) | 1.85 (4.1) (7) | 4.88 (8.1) (11) | 5.04 (8.3) (13) | 6.81 (9.4) (14) | 5.32 (8.7) (15) | 7.28 (10.9) (15) | 8.86 (12.3) (16) | 7.28 (11.1) (17) | 7.95 (11.8) (19) | 3.98 (6.9) (13) | 1.34 (3) (7) | 61.4 (96.5) (151) |

**Map Location 9 — Chad, N'Djamena — 1208N 01502E 975**

| | JAN | FEB | MAR | APR | MAY | JUN | JUL | AUG | SEP | OCT | NOV | DEC | ANNUAL |
|---|---|---|---|---|---|---|---|---|---|---|---|---|---|
| High | 113 | 113 | 118 | 117 | 116 | 113 | 113 | 102 | 107 | 110 | 109 | 110 | 118 |
| Low | 47 | 50 | 55 | 57 | 60 | 64 | 62 | 62 | 61 | 57 | 52 | 47 | 46 |
| Snow | (0)(0) | (0)(0) | (0)(0) | (0)(0) | (0)(0) | (0)(0) | (0)(0) | (0)(0) | (0)(0) | (0)(0) | (0)(0) | (0)(0) | (0)(0) |
| High | 93 | 96 | 103 | 107 | 104 | 100 | 88 | 88 | 92 | 98 | 98 | 95 | 100 |
| Low | 57 | 60 | 67 | 74 | 77 | 75 | 72 | 72 | 72 | 71 | 63 | 66 | 68 |
| Precip | 0 (0) (0.3) | 0 (0) (0.3) | 0 (0) (0.3) | 0.2 (0.6) (2) | 1.42 (3.3) (9) | 2.6 (4.9) (10) | 6.14 (9.7) (14) | 10.12 (13.3) (14) | 4.09 (7) (13) | 0.91 (1.9) (7) | 0.04 (0.3) (0.3) | 0 (0) (0.3) | 25.5 (41) (70.5) |

**Map Location 10 — Dem. Rep. Congo, Kinshasa — 0423S 01526E 1014**

| | JAN | FEB | MAR | APR | MAY | JUN | JUL | AUG | SEP | OCT | NOV | DEC | ANNUAL |
|---|---|---|---|---|---|---|---|---|---|---|---|---|---|
| High | 97 | 98 | 102 | 99 | 100 | 100 | 90 | 93 | 95 | 95 | 98 | 94 | 102 |
| Low | 68 | 67 | 69 | 68 | 67 | 59 | 56 | 57 | 63 | 66 | 67 | 67 | 56 |
| Snow | (0)(0) | (0)(0) | (0)(0) | (0)(0) | (0)(0) | (0)(0) | (0)(0) | (0)(0) | (0)(0) | (0)(0) | (0)(0) | (0)(0) | (0)(0) |
| High | 86 | 87 | 88 | 88 | 84 | 80 | 80 | 82 | 84 | 85 | 85 | 85 | 85 |
| Low | 73 | 72 | 72 | 72 | 72 | 67 | 64 | 66 | 70 | 71 | 72 | 72 | 70 |
| Precip | 4.96 (9.4) (13.3) | 5.55 (10) (12.8) | 6.65 (9.3) (14.7) | 8.35 (9.9) (22.5) | 5.39 (8.5) (12.7) | 0.2 (0.5) (1.2) | 0.04 (0.2) (0) | 0.12 (0.4) (0.5) | 1.3 (2.6) (4.7) | 5.47 (8.9) (12.6) | 9.25 (13.1) (17.8) | 6.69 (10.4) (15.2) | 54 (83.2) (128) |

Table 9.2  World Weather Data [CONTINUED]

# Africa, Congo, Rep. of to Egypt

## Map # 11 to 15

### Map Location 11 — Congo, Rep. of, Brazzaville — 0415S 01515E 1047

| Month | Rec Hi | Norm Hi | Norm Lo | Rec Lo | Snow (days) | Precip in (rec)(days) |
|---|---|---|---|---|---|---|
| JANUARY | 99 | 87 | 70 | 64 | 0 (0) | 5.8 (10.1) (11) |
| FEBRUARY | 102 | 88 | 70 | 64 | 0 (0) | 5.7 (10.1) (12) |
| MARCH | 101 | 89 | 70 | 64 | 0 (0) | 7 (9.5) (15) |
| APRIL | 96 | 89 | 70 | 62 | 0 (0) | 8.5 (9.9) (18) |
| MAY | 97 | 87 | 70 | 60 | 0 (0) | 4.8 (8.1) (11) |
| JUNE | 95 | 83 | 65 | 55 | 0 (0) | 0.3 (0.7) (1) |
| JULY | 93 | 80 | 61 | 53 | 0 (0) | 0.1 (0.3) (0) |
| AUGUST | 97 | 82 | 64 | 55 | 0 (0) | 0.2 (0.5) (1) |
| SEPTEMBER | 99 | 86 | 67 | 59 | 0 (0) | 1.3 (2.6) (3) |
| OCTOBER | 99 | 87 | 70 | 63 | 0 (0) | 5.8 (9.3) (11) |
| NOVEMBER | 97 | 87 | 70 | 64 | 0 (0) | 8.1 (12) (15) |
| DECEMBER | 96 | 86 | 70 | 59 | 0 (0) | 6.5 (10.4) (16) |
| ANNUAL | 102 | | | 53 | 0 (0) | 54.1 (83.5) (114) |

### Map Location 12 — Côte d'Ivoire, Abidjan — 0515N 00355W 20

| Month | Rec Hi | Norm Hi | Norm Lo | Rec Lo | Snow (days) | Precip in (rec)(days) |
|---|---|---|---|---|---|---|
| JANUARY | 94 | 88 | 73 | 59 | 0 (0) | 1.22 (2.8) (5) |
| FEBRUARY | 95 | 90 | 75 | 64 | 0 (0) | 2.17 (4.8) (9) |
| MARCH | 96 | 90 | 75 | 67 | 0 (0) | 4.61 (7.9) (15) |
| APRIL | 95 | 90 | 75 | 68 | 0 (0) | 4.84 (8.1*) (18) |
| MAY | 94 | 88 | 75 | 68 | 0 (0) | 14.13 (***) (16) |
| JUNE | 93 | 85 | 73 | 68 | 0 (0) | 21.73 (16.9) (14) |
| JULY | 91 | 83 | 71 | 64 | 0 (0) | 8.39 (11.9) (1) |
| AUGUST | 88 | 82 | 71 | 63 | 0 (0) | 1.77 (3.5) (0.3) |
| SEPTEMBER | 90 | 83 | 73 | 65 | 0 (0) | 2.8 (5.1) (3) |
| OCTOBER | 92 | 85 | 74 | 67 | 0 (0) | 8.54 (12.4) (13) |
| NOVEMBER | 94 | 87 | 74 | 67 | 0 (0) | 7.8 (11.7) (20) |
| DECEMBER | 95 | 88 | 74 | 62 | 0 (0) | 3.46 (7.3) (14) |
| ANNUAL | 96 | 87 | 74 | 59 | 0 (0) | 81.25 (***) (128.3) |

### Map Location 13 — Djibouti, Djibouti — 1133N 04309E 30

| Month | Rec Hi | Norm Hi | Norm Lo | Rec Lo | Snow (days) | Precip in (rec)(days) |
|---|---|---|---|---|---|---|
| JANUARY | 93 | 84 | 73 | 66 | 0 (0) | 0.4 (1) (0) |
| FEBRUARY | 93 | 84 | 75 | 65 | 0 (0) | 0.48 (1.2) (0) |
| MARCH | 98 | 87 | 77 | 69 | 0 (0) | 0.87 (2.2) (0.3) |
| APRIL | 101 | 90 | 79 | 70 | 0 (0) | 0.49 (1.3) (1) |
| MAY | 112 | 93 | 82 | 70 | 0 (0) | 0.2 (0.6) (1) |
| JUNE | 117 | 99 | 86 | 73 | 0 (0) | 0.02 (0.2) (1) |
| JULY | 117 | 106 | 87 | 72 | 0 (0) | 0.21 (0.5) (4) |
| AUGUST | 116 | 103 | 85 | 72 | 0 (0) | 0.34 (0.8) (4) |
| SEPTEMBER | 112 | 96 | 85 | 73 | 0 (0) | 0.24 (0.6) (3) |
| OCTOBER | 102 | 92 | 80 | 70 | 0 (0) | 0.43 (1) (2) |
| NOVEMBER | 96 | 88 | 77 | 65 | 0 (0) | 0.95 (2) (0.3) |
| DECEMBER | 95 | 85 | 73 | 62 | 0 (0) | 0.52 (1.3) (0.3) |
| ANNUAL | 117 | 92 | 80 | 63 | 0 (0) | 5.1 (12.7) (16.9) |

### Map Location 14 — Egypt, Aswan Dam — 2357N 03249E 656

| Month | Rec Hi | Norm Hi | Norm Lo | Rec Lo | Snow (days) | Precip in (rec)(days) |
|---|---|---|---|---|---|---|
| JANUARY | 100 | 74 | 50 | 38 | 0 (***) | 0 (***) |
| FEBRUARY | 102 | 78 | 52 | 35 | 0 (0) | 0 (***) |
| MARCH | 111 | 87 | 58 | 43 | 0 (0) | 0.01 (***) |
| APRIL | 115 | 96 | 66 | 49 | 0 (0) | 0.02 (0.1) (***) |
| MAY | 118 | 103 | 74 | 52 | 0 (0) | 0.04 (0.1) (***) |
| JUNE | 124 | 107 | 78 | 68 | 0 (0) | 0 (0) (***) |
| JULY | 124 | 106 | 79 | 70 | 0 (0) | 0 (***) |
| AUGUST | 120 | 106 | 79 | 67 | 0 (0) | 0 (***) |
| SEPTEMBER | 117 | 103 | 75 | 63 | 0 (0) | 0 (***) |
| OCTOBER | 113 | 98 | 71 | 57 | 0 (0) | 0 (***) |
| NOVEMBER | 108 | 87 | 62 | 43 | 0 (0) | 0 (***) |
| DECEMBER | 99 | 77 | 53 | 39 | 0 (0) | 0 (***) |
| ANNUAL | 124 | 94 | 66 | 35 | 0 (0) | 0.1 (0.3) (***) |

### Map Location 15 — Egypt, Cairo — 3008N 03124E 366

| Month | Rec Hi | Norm Hi | Norm Lo | Rec Lo | Snow (days) | Precip in (rec)(days) |
|---|---|---|---|---|---|---|
| JANUARY | 78 | 65 | 48 | 41 | 0 (0) | 0.05 (0) (0) |
| FEBRUARY | 88 | 67 | 48 | 41 | 0 (0) | 0.07 (0.3) (0) |
| MARCH | 95 | 73 | 51 | 43 | 0 (0) | 0.09 (0.3) (0) |
| APRIL | 105 | 82 | 56 | 45 | 0 (0) | 0.07 (0.2) (0) |
| MAY | 113 | 90 | 64 | 52 | 0 (0) | 0 (0) (0) |
| JUNE | 110 | 95 | 69 | 59 | 0 (0) | 0 (0) (0.3) |
| JULY | 108 | 96 | 71 | 66 | 0 (0) | 0 (0) (0) |
| AUGUST | 103 | 96 | 72 | 68 | 0 (0) | 0 (0) (0) |
| SEPTEMBER | 99 | 90 | 70 | 63 | 0 (0) | 0 (0) (0) |
| OCTOBER | 100 | 85 | 64 | 55 | 0 (0) | 0.07 (0.2) (0) |
| NOVEMBER | 81 | 69 | 51 | 43 | 0 (0) | 0.47 (0.7) (0) |
| DECEMBER | | | | | 0 (0) | 0.52 (0.7) (1.6) |
| ANNUAL | 113 | 82 | 60 | 41 | 0 (0) | 1.3 (2.9) (1.2) |

Table 9.2 World Weather Data [CONTINUED]

# Africa, Equatorial Guinea to Gambia

## Map # 16 to 20

**Map Location 16 — Eq. Guinea, Malabo — 0345E 00845N 164**

| | JAN | FEB | MAR | APR | MAY | JUN | JUL | AUG | SEP | OCT | NOV | DEC | ANNUAL |
|---|---|---|---|---|---|---|---|---|---|---|---|---|---|
| Avg High | 87 | 89 | 88 | 89 | 87 | 85 | 84 | 85 | 86 | 86 | 87 | 87 | |
| Avg Low | 67 | 69 | 69 | 70 | 71 | 69 | 69 | 69 | 69 | 70 | 70 | 67 | |
| Record High | 89 | 91 | 90 | 90 | 89 | 87 | 88 | | | | | | |
| Record Low | 64 | 66 | 66 | 67 | 66 | 65 | 31 | | | | | | |
| Precip in (days) | 1 (2.3) | 2.46 (5.4) | 6.16 (9) | 5.27 (8.4) | 6.51 (9.2) | 8.56 (12.1) | 8.38 (11.9) | 7.48 (11.1) | 9.25 (13.1) | 9.17 (13) | 2.22 (4.9) | 4.08 (7) | 70.5 (107.4) (***) |

**Map Location 17 — Eritrea, Asmarcha — 1518N 03855E 7628**

| | JAN | FEB | MAR | APR | MAY | JUN | JUL | AUG | SEP | OCT | NOV | DEC | ANNUAL |
|---|---|---|---|---|---|---|---|---|---|---|---|---|---|
| Avg High | 74 | 76 | 77 | 78 | 82 | 78 | 71 | 71 | 74 | 72 | 71 | 71 | |
| Avg Low | 44 | 46 | 48 | 51 | 45 | 53 | 53 | 53 | 51 | 53 | 49 | 50 | |
| Record High | 81 | 84 | 85 | 88 | 85 | 85 | 85 | 81 | 82 | 79 | 78 | 82 | |
| Record Low | 36 | 37 | 37 | 40 | 31 | 41 | 41 | 47 | 42 | 38 | 37 | 31 | |
| Precip in (days) | 0.05 (0.3) | 0.4 (1.1) | 0.4 (0.9) | 1.5 (3.4) | 1.5 (3.5) | 1.3 (2.6) | 6.7 (10.3) | 5 (8.3) | 0.3 (0.8) | 1.3 (2.6) | 0.4 (1.1) | 0.05 (0.3) | 18.5 (34.5) (62.8) |

**Map Location 18 — Ethiopia, Addis Ababa — 0859N 03847E 7642**

| | JAN | FEB | MAR | APR | MAY | JUN | JUL | AUG | SEP | OCT | NOV | DEC | ANNUAL |
|---|---|---|---|---|---|---|---|---|---|---|---|---|---|
| Avg High | 75 | 76 | 77 | 77 | 77 | 74 | 69 | 69 | 72 | 75 | 73 | 73 | |
| Avg Low | 43 | 47 | 49 | 50 | 50 | 49 | 50 | 50 | 49 | 45 | 43 | 41 | |
| Record High | 82 | 86 | 84 | 88 | 97 | 95 | | 84 | 82 | 81 | 82 | 82 | |
| Record Low | 30 | 36 | 37 | 40 | 36 | 36 | | 41 | 41 | 32 | 32 | 32 | |
| Precip in (days) | 0.5 (1.2) (2) | 1.5 (3.4) (1.1) | 2.6 (5.4) (5) | 3.4 (6.6) (2.1) | 3.4 (6.6) (0) | 5.4 (8.8) (8.3) | 11 (13.8) (11.6) | 11.8 (14.3) (10.6) | 7.5 (11.3) (7) | 0.8 (1.7) (2.6) | 0.6 (1.3) (0) | 0.2 (0.6) (0.6) | 48.7 (75) (50.9) |

**Map Location 19 — Gabon, Libreville — 0028N 00926E 39**

| | JAN | FEB | MAR | APR | MAY | JUN | JUL | AUG | SEP | OCT | NOV | DEC | ANNUAL |
|---|---|---|---|---|---|---|---|---|---|---|---|---|---|
| Avg High | 87 | 88 | 89 | 89 | 88 | 85 | 83 | 84 | 85 | 86 | 86 | 87 | |
| Avg Low | 73 | 72 | 73 | 73 | 72 | 70 | 68 | 69 | 71 | 71 | 71 | 72 | |
| Record High | 94 | 94 | 96 | 95 | 99 | 95 | 92 | 92 | 92 | 92 | 95 | 92 | |
| Record Low | 63 | 63 | 63 | 64 | 64 | 63 | 62 | 63 | 66 | 67 | 67 | 66 | |
| Precip in (days) | 9.8 (14) | 9.3 (16) | 13.2 (19) | 13.4 (20) | 9.6 (10.2) (12) | 0.5 (1.1) (2) | 0.1 (0.3) | 0.7 (1.5) | 4.1 (7.1) (3) | 13.6 (16.5) (12) | 14.7 (17.1) (18) | 9.38 (18) | 98.8 (134) |

**Map Location 20 — Gambia, Yundum — 1321N 01640W 85**

| | JAN | FEB | MAR | APR | MAY | JUN | JUL | AUG | SEP | OCT | NOV | DEC | ANNUAL |
|---|---|---|---|---|---|---|---|---|---|---|---|---|---|
| Avg High | 88 | 90 | 94 | 91 | 89 | 89 | 86 | 85 | 87 | 89 | 89 | 88 | |
| Avg Low | 59 | 61 | 63 | 65 | 67 | 73 | 74 | 73 | 73 | 72 | 65 | 61 | |
| Record High | 99 | 102 | 104 | 106 | 106 | 100 | 93 | 92 | 94 | 99 | 96 | 96 | |
| Record Low | 45 | 50 | 53 | 54 | 57 | 65 | 69 | 68 | 63 | 61 | 54 | 48 | |
| Precip in (days) | 0.1 (0.4) (0.3) | 0.1 (0.4) (0.3) | 0.05 (0.2) (0) | 0.05 (0.2) (0) | 0.4 (1.1) (0.3) | 2.3 (4.4) (10) | 11.1 (13.9) (18) | 19.7 (16.7) (16) | 12.2 (15.5) (18) | 4.3 (7.3) (13) | 0.7 (1.5) (1) | 0.1 (0.4) (0.3) | 51.1 (62) (77.2) |

Table 9.2  World Weather Data [CONTINUED]

# Africa, Guinea to Libya, Kurfa

## Map # 21 to 25

### Map Location 21 — Guinea, Conakry — 0934N 01337W 85

| Month | Rec High | Avg High | Avg Low | Rec Low | Precipitation |
|---|---|---|---|---|---|
| JANUARY | 104 | 88 | 72 | 64 | 0.04 (0.2) (1) |
| FEBRUARY | 94 | 88 | 73 | 63 | 0.08 (0.3) (0.3) |
| MARCH | 96 | 89 | 73 | 69 | 0.16 (0.5) (2) |
| APRIL | 95 | 90 | 73 | 68 | 0.91 (2.3) (3) |
| MAY | 95 | 89 | 75 | 66 | 6.46 (9.2) (11) |
| JUNE | 92 | 86 | 73 | 65 | 23.15 (17.1*) (24) |
| JULY | 89 | 83 | 72 | 67 | 52.64 (***) (17) |
| AUGUST | 87 | 82 | 72 | 68 | 42.24 (***) (6) |
| SEPTEMBER | 90 | 85 | 73 | 66 | 27.56 (21.6) (18) |
| OCTOBER | 91 | 87 | 73 | 64 | 12.76 (15.9) (24) |
| NOVEMBER | 91 | 87 | 75 | 69 | 4.76 (8) (13) |
| DECEMBER | 104 | 88 | 74 | 66 | 0.55 (1.3) (3) |
| ANNUAL | 104 | 87 | 73 | 63 | 171.3 (***) (122.3) |

### Map Location 22 — Guinea-Bissau, Bissau — 1153N 01539W 128

| Month | Rec High | Avg High | Avg Low | Rec Low | Precipitation |
|---|---|---|---|---|---|
| JANUARY | 98 | 88 | 64 | 54 | 0.02 (0.2) (0) |
| FEBRUARY | 101 | 91 | 65 | 56 | 0.03 (0.2) (0) |
| MARCH | 102 | 93 | 67 | 60 | 0.02 (0.1) (0) |
| APRIL | 106 | 92 | 69 | 62 | 0.03 (0.1) (0) |
| MAY | 103 | 91 | 72 | 63 | 0.68 (1.7) (2) |
| JUNE | 96 | 88 | 73 | 67 | 6.88 (10.5) (12) |
| JULY | 92 | 85 | 73 | 67 | 18.6 (16.5) (12) |
| AUGUST | 91 | 86 | 73 | 67 | 26.87 (17.7) (8) |
| SEPTEMBER | 93 | 88 | 73 | 67 | 17.12 (18.3) (13) |
| OCTOBER | 94 | 88 | 73 | 68 | 7.67 (11.5) (12) |
| NOVEMBER | 95 | 89 | 72 | 59 | 1.63 (3.2) (3) |
| DECEMBER | 96 | 87 | 66 | 55 | 0.08 (0.3) (0) |
| ANNUAL | 106 | 89 | 70 | 54 | 79.6 (80.3) (62) |

### Map Location 23 — Kenya, Nairobi — 0119S 03655E 5327

| Month | Rec High | Avg High | Avg Low | Rec Low | Precipitation |
|---|---|---|---|---|---|
| JANUARY | 87 | 77 | 54 | 47 | 1.39 (3.1) (1) |
| FEBRUARY | 88 | 79 | 55 | 48 | 1.57 (3.5) (2) |
| MARCH | 88 | 77 | 57 | 49 | 4.48 (7.8) (4) |
| APRIL | 84 | 75 | 58 | 52 | 8.3 (9.9) (5) |
| MAY | 82 | 72 | 56 | 48 | 5.13 (8.3) (3) |
| JUNE | 83 | 70 | 53 | 45 | 1.76 (3.5) (2) |
| JULY | 79 | 69 | 51 | 43 | 0.54 (1.2) (1) |
| AUGUST | 79 | 70 | 52 | 44 | 0.82 (1.7) (1) |
| SEPTEMBER | 84 | 75 | 52 | 41 | 0.98 (2) (1) |
| OCTOBER | 86 | 76 | 55 | 45 | 2.11 (4) (1) |
| NOVEMBER | 85 | 74 | 56 | 43 | 4.51 (7.6) (1) |
| DECEMBER | 84 | 74 | 55 | 47 | 2.67 (5.8) (3) |
| ANNUAL | 88 | 74 | 55 | 41 | 34.3 (58.4) (25) |

### Map Location 24 — Liberia, Monrovia — 0614N 01021W 31

| Month | Rec High | Avg High | Avg Low | Rec Low | Precipitation |
|---|---|---|---|---|---|
| JANUARY | 94 | 89 | 70 | 60 | 1.25 (1.4) (4.9) |
| FEBRUARY | 97 | 91 | 72 | 62 | 1.58 (2.1) (5.3) |
| MARCH | 96 | 91 | 72 | 67 | 4.01 (6.5) (12.1) |
| APRIL | 100 | 91 | 72 | 68 | 5.86 (9.8) (17.3) |
| MAY | 94 | 89 | 73 | 69 | 7.89 (15.5) (18.7) |
| JUNE | 92 | 86 | 72 | 68 | 21.9 (20) (7.7) |
| JULY | 88 | 81 | 72 | 64 | 27.52 (17.8) (0.5) |
| AUGUST | 86 | 80 | 71 | 65 | 18.21 (20.5) (0) |
| SEPTEMBER | 90 | 83 | 72 | 68 | 26.83 (23.7) (6.2) |
| OCTOBER | 92 | 86 | 72 | 69 | 16.35 (20.2) (14) |
| NOVEMBER | 92 | 89 | 72 | 69 | 7.19 (12.6) (12) |
| DECEMBER | 92 | 88 | 71 | 62 | 3.61 (5.6) (9.1) |
| ANNUAL | 100 | 87 | 72 | 60 | 142.2 (155.7) (107.8) |

### Map Location 25 — Libya, Kurfa — 2413N 02318E 1254

| Month | Rec High | Avg High | Avg Low | Rec Low | Precipitation |
|---|---|---|---|---|---|
| JANUARY | 90 | 69 | 41 | 26 | 0.06 (0.1) (0.1) |
| FEBRUARY | 99 | 74 | 44 | 29 | 0 (0) (0.4) |
| MARCH | 104 | 82 | 50 | 33 | 0 (0) (0.2) |
| APRIL | 111 | 91 | 59 | 45 | 0 (0) (0.1) |
| MAY | 118 | 97 | 67 | 48 | 0.02 (0.1) (0.5) |
| JUNE | 122 | 102 | 72 | 60 | 0 (0) (0) |
| JULY | 111 | 100 | 73 | 62 | 0 (0) (0) |
| AUGUST | 115 | 101 | 74 | 63 | 0 (0) (0) |
| SEPTEMBER | 108 | 96 | 69 | 56 | 0 (0) (0) |
| OCTOBER | 106 | 90 | 62 | 39 | 0 (0) (0) |
| NOVEMBER | 101 | 81 | 53 | 37 | 0 (0) (0) |
| DECEMBER | 91 | 73 | 45 | 30 | 0 (0) (0) |
| ANNUAL | 122 | 88 | 59 | 26 | 0.1 (0.2) (1.3) |

**Table 9.2 World Weather Data [CONTINUED]**

# Africa, Libya, Tripoli to Mauritania

## Map # 26 to 30

For each month the values are listed as: Record High / Average High / Average Low / Record Low; Precipitation mean (greatest) ( ); Snow mean (days).

### Map Location 26 — Libya, Tripoli — 3240N 01309E 263

| Month | Rec Hi | Avg Hi | Avg Lo | Rec Lo | Precip | Snow |
|---|---|---|---|---|---|---|
| JANUARY | 84 | 64 | 42 | 25 | 2.5 (5.5) (1.7) | *** (0) |
| FEBRUARY | 93 | 67 | 44 | 32 | 2 (4.4) (1.2) | 0 (0) |
| MARCH | 113 | 73 | 46 | 33 | 1.1 (2.7) (0.4) | 0 (0) |
| APRIL | 113 | 82 | 52 | 35 | 0.5 (1.3) (2) | 0 (0) |
| MAY | 119 | 87 | 58 | 37 | 0.2 (0.6) (0.6) | 0 (0) |
| JUNE | 121 | 94 | 64 | 45 | 0.1 (0.3) (0.5) | 0 (0) |
| JULY | 124 | 98 | 66 | 54 | 0.05 (0.2) (0) | 0 (0) |
| AUGUST | 125 | 98 | 67 | 51 | 0.05 (0.2) (0) | 0 (0) |
| SEPTEMBER | 121 | 93 | 65 | 49 | 0.4 (0.9) (2.7) | 0 (0) |
| OCTOBER | 109 | 87 | 61 | 44 | 1 (2.1) (3.3) | 0 (0) |
| NOVEMBER | 107 | 76 | 53 | 38 | 1.4 (2.8) (2.6) | 0 (0) |
| DECEMBER | 85 | 66 | 45 | 32 | 3 (6.4) (0.8) | 0 (0) |
| ANNUAL | 125 | 82 | 55 | 25 | 12.3 (27.4) (15.8) | 0 (0) |

### Map Location 27 — Madagascar, Antananarivo — 1854S 04732E 4702

| Month | Rec Hi | Avg Hi | Avg Lo | Rec Lo | Precip | Snow |
|---|---|---|---|---|---|---|
| JANUARY | 91 | 79 | 61 | 53 | 11.8 (***) (22) | 0 (0) |
| FEBRUARY | 90 | 78 | 61 | 52 | 11 (***) (17) | 0 (0) |
| MARCH | 87 | 79 | 60 | 51 | 7 (9.5) (19) | 0 (0) |
| APRIL | 87 | 76 | 58 | 45 | 2.1 (4.6) (12) | 0 (0) |
| MAY | 85 | 73 | 54 | 40 | 0.7 (1.8) (3) | 0 (0) |
| JUNE | 80 | 69 | 50 | 34 | 0.3 (0.7) (1) | 0 (0) |
| JULY | 80 | 68 | 48 | 37 | 0.3 (0.7) (1) | 0 (0) |
| AUGUST | 85 | 70 | 48 | 35 | 0.4 (0.9) (2) | 0 (0) |
| SEPTEMBER | 92 | 74 | 51 | 35 | 0.7 (1.5) (2) | 0 (0) |
| OCTOBER | 92 | 80 | 54 | 38 | 2.4 (4.4) (14) | 0 (0) |
| NOVEMBER | 94 | 81 | 58 | 42 | 5.3 (8.7) (18) | 0 (0) |
| DECEMBER | 91 | 80 | 60 | 52 | 11.3 (***) (21) | 0 (0) |
| ANNUAL | 95 | 76 | 55 | 34 | 53.3 (***) (132) | 0 (0) |

### Map Location 28 — Malawi, Lilongwe — 1358S 03342E 3717

| Month | Rec Hi | Avg Hi | Avg Lo | Rec Lo | Precip | Snow |
|---|---|---|---|---|---|---|
| JANUARY | 90 | 80 | 63 | 55 | 8.03 (***) (***) | 0 (0) |
| FEBRUARY | 88 | 80 | 63 | 53 | 8.27 (***) (***) | 0 (0) |
| MARCH | 89 | 80 | 62 | 50 | 5.32 (8.5) (***) | 0 (0) |
| APRIL | 86 | 80 | 57 | 40 | 1.46 (3.4) (***) | 0 (0) |
| MAY | 88 | 78 | 50 | 37 | 0.2 (0.6) (***) | 0 (0) |
| JUNE | 85 | 75 | 45 | 30 | 0.08 (0.3) (***) | 0 (0) |
| JULY | 84 | 75 | 43 | 26 | 0 (0) (***) | 0 (0) |
| AUGUST | 78 | 78 | 46 | 29 | 0.08 (0.3) (***) | 0 (0) |
| SEPTEMBER | 92 | 82 | 52 | 35 | 0.16 (0.5) (***) | 0 (0) |
| OCTOBER | 96 | 86 | 58 | 48 | 0.2 (0.6) (***) | 0 (0) |
| NOVEMBER | 97 | 86 | 62 | 53 | 2.95 (5.3) (***) | 0 (0) |
| DECEMBER | 92 | 82 | 64 | 51 | 6.73 (10.4) (***) | 0 (0) |
| ANNUAL | 97 | 80 | 55 | 26 | 33.5 (***) (***) | 0 (0) |

### Map Location 29 — Mali, Bamako — 1237N 00801W 1079

| Month | Rec Hi | Avg Hi | Avg Lo | Rec Lo | Precip | Snow |
|---|---|---|---|---|---|---|
| JANUARY | 107 | 91 | 61 | 48 | 0.04 (0.2) (1) | 0 (0) |
| FEBRUARY | 117 | 97 | 66 | 51 | 0 (0) (0.3) | 0 (0) |
| MARCH | 109 | 102 | 71 | 58 | 0.16 (0.5) (2) | 0 (0) |
| APRIL | 111 | 103 | 76 | 65 | 0.67 (1.7) (5) | 0 (0) |
| MAY | 115 | 102 | 76 | 66 | 2.72 (5.6) (10) | 0 (0) |
| JUNE | 105 | 94 | 73 | 64 | 5.39 (8.8) (17) | 0 (0) |
| JULY | 102 | 89 | 71 | 64 | 9.09 (12.5) (20) | 0 (0) |
| AUGUST | 102 | 87 | 71 | 64 | 13.19 (15) (20) | 0 (0) |
| SEPTEMBER | 97 | 89 | 71 | 63 | 8.23 (12.1) (18) | 0 (0) |
| OCTOBER | 104 | 93 | 71 | 59 | 2.44 (4.5) (13) | 0 (0) |
| NOVEMBER | 110 | 94 | 65 | 65 | 3.9 (6.8) (1) | 0 (0) |
| DECEMBER | 104 | 92 | 62 | 53 | 0 (0) (0) | 0 (0) |
| ANNUAL | 117 | 94 | 70 | 47 | 45.8 (67.7) (107.3) | 0 (0) |

### Map Location 30 — Mauritania, Nouakchott — 1805N 01557W 7

| Month | Rec Hi | Avg Hi | Avg Lo | Rec Lo | Precip | Snow |
|---|---|---|---|---|---|---|
| JANUARY | 96 | 85 | 57 | 45 | 0 (0) (0.3) | 0 (0) |
| FEBRUARY | 103 | 87 | 59 | 49 | 0.08 (0.3) (0.3) | 0 (0) |
| MARCH | 106 | 89 | 63 | 51 | 0.04 (0.1) (0.3) | 0 (0) |
| APRIL | 109 | 90 | 64 | 54 | 0.04 (0.1) (0.3) | 0 (0) |
| MAY | 115 | 93 | 69 | 58 | 0.04 (0.1) (0) | 0 (0) |
| JUNE | 114 | 92 | 73 | 64 | 0.04 (0.2) (1) | 0 (0) |
| JULY | 109 | 89 | 74 | 70 | 0.47 (1.1) (1) | 0 (0) |
| AUGUST | 108 | 90 | 75 | 68 | 2.28 (4.3) (5) | 0 (0) |
| SEPTEMBER | 111 | 93 | 75 | 71 | 1.54 (3) (4) | 0 (0) |
| OCTOBER | 109 | 91 | 71 | 63 | 0.39 (0.9) (2) | 0 (0) |
| NOVEMBER | 107 | 89 | 65 | 56 | 0.08 (0.3) (0.3) | 0 (0) |
| DECEMBER | 83 | 83 | 56 | 56 | 0.12 (0.4) (0.3) | 0 (0) |
| ANNUAL | 115 | 89 | 67 | 44 | 5.1 (10.8) (14.8) | 0 (0) |

## Africa, Morocco to Niger

**Table 9.2 World Weather Data [CONTINUED]**

## Map # 31 to 35

### Map Location 31 — Morocco, Casablanca — 3333N 00739W 203

| | Jan | Feb | Mar | Apr | May | Jun | Jul | Aug | Sep | Oct | Nov | Dec | Annual |
|---|---|---|---|---|---|---|---|---|---|---|---|---|---|
| Record High | 86 | 97 | 98 | 100 | 100 | 97 | 108 | 110 | 110 | 107 | 95 | | 110 |
| Record Low | 31 | 32 | 35 | 41 | 45 | 46 | 54 | 53 | 50 | 46 | 30 | | 30 |
| Normal High | 63 | 64 | 67 | 69 | 72 | 76 | 79 | 81 | 79 | 76 | 69 | 65 | |
| Normal Low | 45 | 46 | 49 | 52 | 56 | 61 | 65 | 66 | 63 | 58 | 52 | 47 | |
| Precip | 2.1 (4.6) (1) | 1.9 (4.2) (1) | 2.2 (4.8) (1) | 1.4 (3.3) (1) | 0.9 (2.2) (0.3) | 0.2 (0.5) (0.3) | 0 (0) (0.3) | 0.05 (0.2) (0.3) | 0.3 (0.8) (0.3) | 1.5 (2.9) (1) | 2.6 (4.8) (1) | 2.8 (6.1) (0.3) | 15.9 (34.4) (7.8) |

### Map Location 32 — Morocco, Midelt — 3241N 00444W 4987

| | Jan | Feb | Mar | Apr | May | Jun | Jul | Aug | Sep | Oct | Nov | Dec | Annual |
|---|---|---|---|---|---|---|---|---|---|---|---|---|---|
| Record High | 71 | 77 | 84 | 86 | 94 | 97 | 104 | 101 | 99 | 87 | 77 | 69 | 104 |
| Record Low | 19 | 14 | 20 | 29 | 32 | 40 | 48 | 49 | 41 | 34 | 26 | 20 | 14 |
| Normal High | 53 | 56 | 62 | 68 | 74 | 83 | 93 | 92 | 82 | 70 | 62 | 55 | |
| Normal Low | 32 | 34 | 38 | 43 | 47 | 55 | 61 | 61 | 55 | 47 | 40 | 35 | |
| Precip | 0.4 (1) (***) | 0.8 (1.9) (***) | 0.9 (2.2) (***) | 1.3 (3.1) (***) | 1 (2.5) (***) | 0.5 (1.1) (***) | 0.3 (0.7) (***) | 0.3 (0.7) (***) | 0.9 (1.9) (***) | 0.9 (1.9) (***) | 0.9 (1.9) (***) | 0.8 (1.9) (***) | 9 (20.8) (***) |

### Map Location 33 — Mozambique, Maputo — 2555S 03234E 28

| | Jan | Feb | Mar | Apr | May | Jun | Jul | Aug | Sep | Oct | Nov | Dec | Annual |
|---|---|---|---|---|---|---|---|---|---|---|---|---|---|
| Record High | 110 | 103 | 104 | 102 | 101 | 94 | 96 | 100 | 114 | 113 | 112 | | 114 |
| Record Low | 61 | 62 | 60 | 52 | 46 | 46 | 45 | 47 | 49 | 53 | 52 | | 45 |
| Normal High | 86 | 87 | 85 | 83 | 80 | 77 | 76 | 78 | 80 | 82 | 83 | 85 | |
| Normal Low | 71 | 71 | 69 | 66 | 60 | 56 | 55 | 57 | 61 | 64 | 67 | 69 | |
| Precip | 5.1 (9.6) (4.5) | 4.9 (9.4) (3.4) | 4.9 (8.2) (3) | 2.1 (4.6) (2.5) | 1.1 (2.7) (1.7) | 0.8 (1.7) (0.6) | 0.5 (1.1) (0.6) | 0.5 (1.1) (0.6) | 1.1 (2.2) (2.6) | 1.9 (3.6) (4.1) | 3.2 (5.7) (4.8) | 3.8 (7.9) (5.6) | 29.9 (57.8) (3.9) |

### Map Location 34 — Namibia, Windhoek — 2237S 01705E 5578

| | Jan | Feb | Mar | Apr | May | Jun | Jul | Aug | Sep | Oct | Nov | Dec | Annual |
|---|---|---|---|---|---|---|---|---|---|---|---|---|---|
| Record High | 97 | 94 | 94 | 87 | 89 | 79 | 77 | 85 | 91 | 93 | 93 | | 97 |
| Record Low | 49 | 44 | 39 | 36 | 29 | 27 | 27 | 25 | 31 | 35 | 33 | | 27 |
| Normal High | 85 | 83 | 80 | 77 | 72 | 68 | 68 | 73 | 77 | 84 | 84 | 86 | |
| Normal Low | 63 | 61 | 59 | 55 | 48 | 44 | 43 | 47 | 53 | 59 | 59 | 62 | |
| Precip | 3.03 (6.5) (8) | 2.87 (6.2) (7) | 3.11 (6.2) (6.7) | 1.61 (3.7) (3.1) | 0.28 (0.8) (0.7) | 0.04 (0.2) (0.4) | 0.04 (0.2) (0.2) | 0.08 (0.3) (1.1) | 0.08 (0.3) (1) | 0.39 (0.9) (3.6) | 0.91 (1.9) (4.3) | 1.85 (4.1) (7) | 14.3 (31.2) (42.3) |

### Map Location 35 — Niger, Niamey — 1328N 00210E 732

| | Jan | Feb | Mar | Apr | May | Jun | Jul | Aug | Sep | Oct | Nov | Dec | Annual |
|---|---|---|---|---|---|---|---|---|---|---|---|---|---|
| Record High | 104 | 109 | 112 | 118 | 120 | 114 | 108 | 106 | 110 | 109 | 106 | | 120 |
| Record Low | 47 | 50 | 51 | 62 | 67 | 67 | 64 | 63 | 67 | 61 | 53 | | 47 |
| Normal High | 93 | 98 | 105 | 108 | 106 | 101 | 94 | 89 | 93 | 101 | 101 | 94 | |
| Normal Low | 58 | 63 | 71 | 77 | 80 | 77 | 74 | 73 | 73 | 74 | 65 | 59 | |
| Precip | 0 (0) (1) | 0 (0) (1) | 0.08 (0.3) (2) | 0.28 (0.8) (3) | 1.46 (3.4) (7) | 3.15 (5.7) (13) | 5.31 (8.7) (14) | 8.62 (12.1) (16) | 3.66 (6.4) (12) | 0.79 (1.7) (5) | 0.04 (0.3) (0.3) | 0 (0) (0) | 23.4 (39.4) (74.3) |

**Table 9.2  World Weather Data [CONTINUED]**

# Africa, Nigeria to Somalia

## Map # 36 to 40

**Map Location 36 — Nigeria, Lagos — 0635N 00320E 132**

| Month | Rec High | Rec Low | Avg High | Avg Low | Snow | Precip |
|---|---|---|---|---|---|---|
| JANUARY | 95 | 63 | 88 | 74 | 0 (0) | 1.1 (2.5) (2) |
| FEBRUARY | 96 | 66 | 89 | 77 | 0 (0) | 1.8 (4) (9) |
| MARCH | 99 | 60 | 89 | 78 | 0 (0) | 3.9 (7.2) (13) |
| APRIL | 99 | 69 | 89 | 77 | 0 (0) | 5.7 (8.8) (16) |
| MAY | 104 | 62 | 87 | 76 | 0 (0) | 10.9 (10.4) (18) |
| JUNE | 93 | 69 | 85 | 74 | 0 (0) | 17.7 (16.3) (16) |
| JULY | 93 | 68 | 83 | 74 | 0 (0) | 10.5 (13.5) (6) |
| AUGUST | 93 | 65 | 82 | 73 | 0 (0) | 2.6 (4.9) (2) |
| SEPTEMBER | 94 | 68 | 85 | 74 | 0 (0) | 5.6 (9.1) (9) |
| OCTOBER | 96 | 69 | 88 | 75 | 0 (0) | 8.2 (12.1) (19) |
| NOVEMBER | 99 | 69 | 88 | 75 | 0 (0) | 2.7 (4.9) (15) |
| DECEMBER | 95 | 68 | 86 | 75 | 0 (0) | 1 (2.3) (4) |
| ANNUAL | 104 | 60 | 86 | 75 | 0 (0) | 71.7 (96) (129) |

**Map Location 37 — Rwanda, Kamembe — 0227S 02854E 5190**

| Month | Rec High | Rec Low | Avg High | Avg Low | Snow | Precip |
|---|---|---|---|---|---|---|
| JANUARY | 88 | 56 | 78 | 59 | 0 (0) | 5.43 (9.9) (23) |
| FEBRUARY | 85 | 56 | 78 | 59 | 0 (0) | 6.14 (10.3) (22) |
| MARCH | 85 | 54 | 78 | 60 | 0 (0) | 6.38 (9.2) (23) |
| APRIL | 85 | 56 | 78 | 60 | 0 (0) | 5.55 (8.7) (26) |
| MAY | 84 | 56 | 79 | 59 | 0 (0) | 3.39 (6.6) (20) |
| JUNE | 85 | 53 | 78 | 58 | 0 (0) | 0.67 (1.5) (7) |
| JULY | 87 | 52 | 80 | 57 | 0 (0) | 0.55 (1.2) (5) |
| AUGUST | 90 | 50 | 82 | 58 | 0 (0) | 1.42 (2.8) (5) |
| SEPTEMBER | 93 | 55 | 81 | 59 | 0 (0) | 3.94 (6.8) (16) |
| OCTOBER | 91 | 56 | 80 | 60 | 0 (0) | 5.75 (9.3) (24) |
| NOVEMBER | 85 | 56 | 79 | 59 | 0 (0) | 6.61 (10.3) (25) |
| DECEMBER | 86 | 59 | 78 | 59 | 0 (0) | 5.91 (10.2) (25) |
| ANNUAL | 93 | 50 | 79 | 59 | 0 (0) | 51.7 (86.8) (221) |

**Map Location 38 — Senegal, Dakar — 1444N 01730W 89**

| Month | Rec High | Rec Low | Avg High | Avg Low | Snow | Precip |
|---|---|---|---|---|---|---|
| JANUARY | 102 | 56 | 79 | 64 | 0 (0) | 0 (0) (0) |
| FEBRUARY | 100 | 58 | 80 | 64 | 0 (0) | 0.08 (0.3) (0.3) |
| MARCH | 109 | 59 | 80 | 64 | 0 (0) | 0 (0) (0) |
| APRIL | 101 | 61 | 81 | 65 | 0 (0) | 0 (0) (0) |
| MAY | 100 | 61 | 84 | 68 | 0 (0) | 0.04 (0.1) (0.3) |
| JUNE | 100 | 65 | 88 | 73 | 0 (0) | 0.59 (1.3) (1) |
| JULY | 99 | 69 | 88 | 76 | 0 (0) | 3.47 (6.2) (5) |
| AUGUST | 99 | 69 | 87 | 76 | 0 (0) | 9.8 (13) (11) |
| SEPTEMBER | 100 | 69 | 89 | 76 | 0 (0) | 6.42 (10.1) (12) |
| OCTOBER | 101 | 70 | 89 | 76 | 0 (0) | 1.93 (3.7) (5) |
| NOVEMBER | 99 | 64 | 86 | 73 | 0 (0) | 0.2 (0.6) (0.3) |
| DECEMBER | 95 | 53 | 81 | 67 | 0 (0) | 0.24 (0.7) (0) |
| ANNUAL | 109 | 53 | 84 | 70 | 0 (0) | 22.8 (36) (35.2) |

**Map Location 39 — Sierra Leone, Freetown — 0837N 01312W 82**

| Month | Rec High | Rec Low | Avg High | Avg Low | Snow | Precip |
|---|---|---|---|---|---|---|
| JANUARY | 97 | 62 | 90 | 73 | 0 (0) | 0.41 (1) (0) |
| FEBRUARY | 100 | 61 | 91 | 74 | 0 (0) | 0.3 (0.8) (0.3) |
| MARCH | 99 | 68 | 91 | 75 | 0 (0) | 1.16 (2.8) (2) |
| APRIL | 99 | 67 | 91 | 75 | 0 (0) | 4.06 (7.3) (4) |
| MAY | 107 | 65 | 90 | 74 | 0 (0) | 11.47 (10.5) (17) |
| JUNE | 100 | 65 | 88 | 73 | 0 (0) | 20.04 (16.7) (18) |
| JULY | 100 | 65 | 86 | 73 | 0 (0) | 35.58 (22.5) (3) |
| AUGUST | 104 | 62 | 84 | 72 | 0 (0) | 36.57 (22.5) (3) |
| SEPTEMBER | 96 | 62 | 86 | 72 | 0 (0) | 28.48 (21.8) (8) |
| OCTOBER | 94 | 64 | 88 | 72 | 0 (0) | 12.62 (15.8) (21) |
| NOVEMBER | 99 | 63 | 89 | 73 | 0 (0) | 5.12 (8.5) (13) |
| DECEMBER | 94 | 64 | 89 | 74 | 0 (0) | 1.42 (3.2) (6) |
| ANNUAL | 107 | 61 | 89 | 73 | 0 (0) | 157.2 (132.6) (99.3) |

**Map Location 40 — Somalia, Mogadishu — 0201N 04519E 26**

| Month | Rec High | Rec Low | Avg High | Avg Low | Snow | Precip |
|---|---|---|---|---|---|---|
| JANUARY | 100 | 64 | 89 | 74 | 0 (0) | 0.05 (0.3) (0) |
| FEBRUARY | 98 | 65 | 89 | 78 | 0 (0) | 0.05 (0.3) (1) |
| MARCH | 93 | 68 | 89 | 76 | 0 (0) | 0.05 (0.2) (0) |
| APRIL | 108 | 68 | 91 | 77 | 0 (0) | 2.3 (5) (1.7) |
| MAY | 94 | 65 | 90 | 76 | 0 (0) | 2.3 (5) (0) |
| JUNE | 90 | 68 | 84 | 75 | 0 (0) | 3.8 (6.7) (0) |
| JULY | 89 | 59 | 84 | 74 | 0 (0) | 2.5 (4.7) (0) |
| AUGUST | 89 | 59 | 85 | 73 | 0 (0) | 1.9 (3.7) (0.7) |
| SEPTEMBER | 93 | 60 | 87 | 73 | 0 (0) | 1 (2.1) (0) |
| OCTOBER | 94 | 65 | 87 | 75 | 0 (0) | 0.9 (1.9) (0) |
| NOVEMBER | 98 | 69 | 88 | 76 | 0 (0) | 1.6 (3.1) (0) |
| DECEMBER | 96 | 68 | 88 | 74 | 0 (0) | 0.5 (1.2) (0) |
| ANNUAL | 108 | 59 | 87 | 75 | 0 (0) | 16.9 (34.2) (3.4) |

**Table 9.2　World Weather Data [CONTINUED]**

# Africa, South Africa to Togo

## Map # 41 to 45

### Map Location 41 — South Africa, Capetown — 3354S 01829E 36

| | JANUARY | FEBRUARY | MARCH | APRIL | MAY | JUNE |
|---|---|---|---|---|---|---|
| Normal High | 78 | 79 | 77 | 72 | 67 | 65 |
| Record High | 99 | 100 | 103 | 102 | 95 | 85 |
| Normal Low | 60 | 60 | 58 | 53 | 49 | 46 |
| Record Low | 44 | 41 | 42 | 38 | 31 | 29 |
| Precipitation | 0.67 (1.6) (0.3) | 0.59 (1.4) (0.3) | 0.87 (2.2) (0.3) | 1.93 (4.3) (1) | 3.7 (6.9) (0.3) | 4.29 (7.4) (1) |
| Snow | 0 (0) | 0 (0) | 0 (0) | 0 (0) | 0 (0) | 0 (0) |

| | JULY | AUGUST | SEPTEMBER | OCTOBER | NOVEMBER | DECEMBER | ANNUAL |
|---|---|---|---|---|---|---|---|
| Normal High | 63 | 64 | 65 | 70 | 73 | 76 | 71 |
| Record High | 84 | 89 | 93 | 90 | 93 | 100 | 103 |
| Normal Low | 46 | 46 | 49 | 52 | 55 | 58 | 53 |
| Record Low | 28 | 31 | 33 | 34 | 40 | 41 | 28 |
| Precipitation | 3.7 (6.6) (0.3) | 3.27 (5.9) (0.3) | 2.28 (4.3) (0.3) | 1.57 (3.1) (0.3) | 1.02 (2.1) (0.3) | 0.79 (1.8) (0.3) | 24.7 (47.6) (5) |
| Snow | *** (***) | 0 (0) | 0 (0) | 0 (0) | 0 (0) | 0 (0) | *** (***) |

### Map Location 42 — South Africa, Pretoria — 2545S 02814E 4491

| | JANUARY | FEBRUARY | MARCH | APRIL | MAY | JUNE |
|---|---|---|---|---|---|---|
| Normal High | 81 | 81 | 78 | 75 | 70 | 66 |
| Record High | 95 | 104 | 91 | 83 | 79 | 77 |
| Normal Low | 60 | 60 | 57 | 50 | 42 | 37 |
| Record Low | 49 | 41 | 43 | 33 | 26 | 24 |
| Precipitation | 4.92 (9.4) (14) | 4.25 (8.5) (11) | 3.54 (6.7) (8) | 1.61 (3.7) (4) | 0.98 (2.4) (2) | 0.39 (0.9) (1) |
| Snow | 0 (0) | 0 (0) | 0 (0) | 0 (0) | 0 (0) | *** (***) |

| | JULY | AUGUST | SEPTEMBER | OCTOBER | NOVEMBER | DECEMBER | ANNUAL |
|---|---|---|---|---|---|---|---|
| Normal High | 66 | 71 | 77 | 80 | 80 | 82 | 76 |
| Record High | 75 | 83 | 89 | 92 | 96 | 95 | 96 |
| Normal Low | 37 | 42 | 49 | 55 | 57 | 59 | 50 |
| Record Low | 24 | 28 | 30 | 37 | 41 | 43 | 24 |
| Precipitation | 0.39 (0.9) (1) | 0.2 (0.5) (1) | 0.83 (1.7) (3) | 2.21 (4.1) (8) | 4.96 (8.2) (12) | 5.08 (9.6) (11) | 29.4 (56.6) (76) |
| Snow | *** (***) | *** (***) | *** (***) | 0 (0) | 0 (0) | 0 (0) | *** (***) |

### Map Location 43 — Sudan, Khartoum — 1536N 03235E 1256

| | JANUARY | FEBRUARY | MARCH | APRIL | MAY | JUNE |
|---|---|---|---|---|---|---|
| Normal High | 90 | 93 | 100 | 105 | 107 | 106 |
| Record High | 104 | 111 | 113 | 117 | 117 | 118 |
| Normal Low | 59 | 61 | 66 | 72 | 77 | 79 |
| Record Low | 41 | 44 | 49 | 53 | 61 | 67 |
| Precipitation | 0.05 (0.3) (0) | 0.025 (0.3) (0) | 0.05 (0.2) (0) | 0.05 (0.2) (0) | 0.1 (0.3) (3) | 0.3 (0.7) (1) |
| Snow | 0 (0) | 0 (0) | 0 (0) | 0 (0) | 0 (0) | 0 (0) |

| | JULY | AUGUST | SEPTEMBER | OCTOBER | NOVEMBER | DECEMBER | ANNUAL |
|---|---|---|---|---|---|---|---|
| Normal High | 101 | 98 | 102 | 104 | 97 | 92 | 100 |
| Record High | 117 | 109 | 113 | 113 | 107 | 95 | 118 |
| Normal Low | 77 | 76 | 77 | 75 | 68 | 62 | 71 |
| Record Low | 65 | 64 | 61 | 62 | 54 | 45 | 41 |
| Precipitation | 2.1 (4) (5) | 2.8 (5.2) (6) | 0.7 (1.5) (3) | 0.2 (0.6) (2) | 0.05 (0.3) (0.3) | 0 (0) (0) | 6.4 (13.6) (20.3) |
| Snow | 0 (0) | 0 (0) | 0 (0) | 0 (0) | 0 (0) | 0 (0) | 0 (0) |

### Map Location 44 — Tanzania, Dar es Salaam — 0652S 03912E 181

| | JANUARY | FEBRUARY | MARCH | APRIL | MAY | JUNE |
|---|---|---|---|---|---|---|
| Normal High | 87 | 88 | 88 | 86 | 85 | 84 |
| Record High | 95 | 95 | 96 | 95 | 91 | 90 |
| Normal Low | 77 | 77 | 75 | 73 | 71 | 68 |
| Record Low | 69 | 68 | 69 | 66 | 64 | 60 |
| Precipitation | 2.62 (5.7) (3) | 2.26 (5.4) (4) | 4.78 (8) (10) | 10.95 (10.4) (7) | 7.82 (9.7) (1) | 1.31 (2.7) (0) |
| Snow | 0 (0) | 0 (0) | 0 (0) | 0 (0) | 0 (0) | 0 (0) |

| | JULY | AUGUST | SEPTEMBER | OCTOBER | NOVEMBER | DECEMBER | ANNUAL |
|---|---|---|---|---|---|---|---|
| Normal High | 83 | 83 | 83 | 85 | 86 | 87 | 85 |
| Record High | 90 | 89 | 91 | 92 | 94 | 95 | 96 |
| Normal Low | 66 | 66 | 67 | 69 | 72 | 75 | 71 |
| Record Low | 60 | 59 | 61 | 62 | 66 | 69 | 59 |
| Precipitation | 1.18 (2.4) (0) | 0.97 (2) (0) | 1.13 (2.3) (0) | 1.84 (3.5) (0) | 2.94 (5.3) (2) | 3.52 (7.4) (5) | 41.3 (64.4) (32) |
| Snow | 0 (0) | 0 (0) | 0 (0) | 0 (0) | 0 (0) | 0 (0) | 0 (0) |

### Map Location 45 — Togo, Lome — 0610N 00115E 72

| | JANUARY | FEBRUARY | MARCH | APRIL | MAY | JUNE |
|---|---|---|---|---|---|---|
| Normal High | 85 | 87 | 87 | 81 | 86 | 83 |
| Record High | 100 | 103 | 102 | 102 | 104 | 102 |
| Normal Low | 71 | 74 | 74 | 74 | 74 | 73 |
| Record Low | 58 | 65 | 68 | 68 | 69 | 68 |
| Precipitation | 0.64 (1.5) (1) | 0.93 (2.1) (2) | 1.59 (3.7) (4) | 4.43 (7.7) (5) | 5.37 (8.5) (7) | 8.72 (12.2) (5) |
| Snow | 0 (0) | 0 (0) | 0 (0) | 0 (0) | 0 (0) | 0 (0) |

| | JULY | AUGUST | SEPTEMBER | OCTOBER | NOVEMBER | DECEMBER | ANNUAL |
|---|---|---|---|---|---|---|---|
| Normal High | 80 | 79 | 80 | 83 | 86 | 85 | 84 |
| Record High | 100 | 90 | 97 | 98 | 99 | 98 | 104 |
| Normal Low | 71 | 71 | 72 | 72 | 73 | 72 | 73 |
| Record Low | 67 | 66 | 65 | 68 | 68 | 66 | 58 |
| Precipitation | 2.57 (4.8) (2) | 0.37 (0.9) (1) | 0.99 (2) (2) | 2.22 (4.2) (6) | 1.02 (2.1) (4) | 0.48 (1.2) (2) | 29.3 (50.9) (41) |
| Snow | 0 (0) | 0 (0) | 0 (0) | 0 (0) | 0 (0) | 0 (0) | 0 (0) |

**Table 9.2  World Weather Data [CONTINUED]**

# Africa, Tunisia to Zimbabwe

## Map # 46 to 50

### Map Location 46 — Tunisia, Tunis — 3651N 01014E 17

| | JANUARY | FEBRUARY | MARCH | APRIL | MAY | JUNE |
|---|---|---|---|---|---|---|
| Avg High / Rec High | 58 / 77 | 61 / 84 | 65 / 91 | 70 / 104 | 76 / 104 | 84 / 109 |
| Avg Low / Rec Low | 43 / 30 | 44 / 32 | 47 / 34 | 51 / 37 | 56 / 43 | 63 / 48 |
| Precip | 2.64 (5.7) (1) | 1.93 (4.3) (1) | 1.5 (3.5) (0.3) | 1.22 (2.9) (2) | 0.75 (1.9) (2) | 0.24 (0.6) (2) |
| Snow | 0 (0) | 0 (0) | 0 (0) | 0 (0) | 0 (0) | 0 (0) |

| | JULY | AUGUST | SEPTEMBER | OCTOBER | NOVEMBER | DECEMBER | ANNUAL |
|---|---|---|---|---|---|---|---|
| Avg High / Rec High | 90 / 118 | 91 / 117 | 87 / 111 | 77 / 104 | 68 / 90 | 60 / 81 | 74 / 118 |
| Avg Low / Rec Low | 68 / 50 | 69 / 52 | 66 / 51 | 59 / 45 | 51 / 34 | 44 / 30 | 55 / 30 |
| Precip | 0.04 (0.2) (1) | 0.35 (0.8) (3) | 1.3 (2.6) (3) | 1.65 (3.2) (4) | 1.89 (3.6) (1) | 2.83 (6.1) (1) | 16.3 (35.4) (21.3) |
| Snow | 0 (0) | 0 (0) | 0 (0) | 0 (0) | 0 (0) | 0 (0) | 0 (0) |

### Map Location 47 — Uganda, Entebbe — 0002N 03227E 3789

| | JANUARY | FEBRUARY | MARCH | APRIL | MAY | JUNE |
|---|---|---|---|---|---|---|
| Avg High / Rec High | 80 / 88 | 80 / 92 | 80 / 88 | 78 / 89 | 78 / 89 | 77 / 82 |
| Avg Low / Rec Low | 63 / 55 | 63 / 51 | 64 / 54 | 64 / 51 | 64 / 52 | 63 / 57 |
| Precip | 2.61 (5.7) (14) | 3.45 (7.3) (13) | 6.16 (9) (15) | 10.13 (10.2) (17) | 9.68 (10.2) (18) | 4.61 (7.8) (18) |
| Snow | 0 (0) | 0 (0) | 0 (0) | 0 (0) | 0 (0) | 0 (0) |

| | JULY | AUGUST | SEPTEMBER | OCTOBER | NOVEMBER | DECEMBER | ANNUAL |
|---|---|---|---|---|---|---|---|
| Avg High / Rec High | 77 / 85 | 77 / 84 | 77 / 87 | 78 / 88 | 79 / 90 | 79 / 90 | 78 / 92 |
| Avg Low / Rec Low | 61 / 53 | 62 / 54 | 62 / 54 | 63 / 52 | 63 / 56 | 63 / 53 | 63 / 51 |
| Precip | 3.03 (5.5) (19) | 2.97 (5.5) (19) | 3.09 (5.5) (19) | 3.66 (6.4) (20) | 5.07 (8.4) (18) | 4.47 (8.9) (16) | 58.9 (90.4) (206) |
| Snow | 0 (0) | 0 (0) | 0 (0) | 0 (0) | 0 (0) | 0 (0) | 0 (0) |

### Map Location 49 — Western Sahara, El Aaiun — 2709N 01312W 207

| | JANUARY | FEBRUARY | MARCH | APRIL | MAY | JUNE |
|---|---|---|---|---|---|---|
| Avg High / Rec High | 67 / 89 | 67 / 84 | 68 / 88 | 70 / 103 | 70 / 80 | 72 / 83 |
| Avg Low / Rec Low | 56 / 41 | 57 / 43 | 58 / 49 | 60 / 48 | 62 / 55 | 64 / 55 |
| Precip | 0.29 (0.8) (***) | 0.2 (0.6) (***) | 0.11 (0.3) (***) | 0.02 (0.1) (***) | 0 (0) (***) | 0 (0) (***) |
| Snow | 0 (0) | 0 (0) | 0 (0) | 0 (0) | 0 (0) | 0 (0) |

| | JULY | AUGUST | SEPTEMBER | OCTOBER | NOVEMBER | DECEMBER | ANNUAL |
|---|---|---|---|---|---|---|---|
| Avg High / Rec High | 73 / 97 | 74 / 86 | 74 / 94 | 74 / 101 | 73 / 95 | 69 / 84 | 71 / 103 |
| Avg Low / Rec Low | 65 / 59 | 65 / 57 | 65 / 57 | 64 / 53 | 61 / 50 | 57 / 44 | 61 / 41 |
| Precip | 0 (0) (***) | 0.01 (0.1) (***) | 0.18 (0.5) (***) | 0.1 (0.4) (***) | 0.47 (1.1) (***) | 0.33 (0.8) (***) | 1.7 (4.7) (***) |
| Snow | 0 (0) | 0 (0) | 0 (0) | 0 (0) | 0 (0) | 0 (0) | 0 (0) |

### Map Location 48 — Zambia, Lusaka — 1525S 02820E 4208

| | JANUARY | FEBRUARY | MARCH | APRIL | MAY | JUNE |
|---|---|---|---|---|---|---|
| Avg High / Rec High | 78 / 88 | 79 / 87 | 78 / 86 | 79 / 87 | 77 / 85 | 73 / 83 |
| Avg Low / Rec Low | 63 / 58 | 63 / 56 | 62 / 55 | 59 / 50 | 54 / 47 | 50 / 39 |
| Precip | 9.06 (***) (18) | 7.45 (10.1) (15) | 5.61 (8.7) (12) | 0.68 (1.7) (3) | 0.06 (0.2) (1) | 0.04 (0.2) (0.3) |
| Snow | 0 (0) | 0 (0) | 0 (0) | 0 (0) | 0 (0) | 0 (0) |

| | JULY | AUGUST | SEPTEMBER | OCTOBER | NOVEMBER | DECEMBER | ANNUAL |
|---|---|---|---|---|---|---|---|
| Avg High / Rec High | 73 / 83 | 77 / 84 | 84 / 95 | 88 / 100 | 84 / 98 | 80 / 93 | 79 / 100 |
| Avg Low / Rec Low | 49 / 40 | 53 / 43 | 59 / 46 | 64 / 54 | 64 / 55 | 63 / 57 | 59 / 39 |
| Precip | 0.01 (0.1) (0) | 0 (0) (0) | 0.03 (0.2) (1) | 0.42 (1) (4) | 3.61 (6.3) (13) | 5.91 (10.2) (16) | 39.2 (***) (83.3) |
| Snow | 0 (0) | 0 (0) | 0 (0) | 0 (0) | 0 (0) | 0 (0) | 0 (0) |

### Map Location 50 — Zimbabwe, Harare — 1755S 03106E 4904

| | JANUARY | FEBRUARY | MARCH | APRIL | MAY | JUNE |
|---|---|---|---|---|---|---|
| Avg High / Rec High | 80 / 97 | 79 / 94 | 80 / 94 | 78 / 91 | 74 / 87 | 71 / 85 |
| Avg Low / Rec Low | 59 / 48 | 59 / 48 | 58 / 46 | 54 / 37 | 47 / 34 | 42 / 31 |
| Precip | 7.7 (9.9) (14) | 7 (10.3) (11) | 4.6 (7.9) (7) | 1.1 (2.7) (3) | 0.5 (1.3) (1) | 0.1 (0.3) (0.3) |
| Snow | 0 (0) | 0 (0) | 0 (0) | 0 (0) | 0 (0) | 0 (0) |

| | JULY | AUGUST | SEPTEMBER | OCTOBER | NOVEMBER | DECEMBER | ANNUAL |
|---|---|---|---|---|---|---|---|
| Avg High / Rec High | 71 / 82 | 75 / 82 | 82 / 87 | 86 / 99 | 85 / 100 | 82 / 102 | 79 / 102 |
| Avg Low / Rec Low | 42 / 25 | 45 / 31 | 51 / 31 | 55 / 42 | 58 / 46 | 59 / 48 | 52 / 25 |
| Precip | 0.05 (0.2) (0.3) | 0.1 (0.3) (0.3) | 0.2 (0.6) (1) | 1.1 (2.2) (5) | 3.8 (6.6) (13) | 6.4 (10.4) (12) | 32.6 (52.7) (67.8) |
| Snow | *** (***) | 0 (0) | 0 (0) | 0 (0) | 0 (0) | 0 (0) | 0 (0) |

**Table 9.2  World Weather Data** [CONTINUED]

# Antarctica

## Map # 51 to 53

### Map Location 51 — Antarctica, Mcmurdo Station — 7753S 16648W 8

| Month | Avg Max T °F | Avg Min T °F | Avg. Precip in. (Rain Days)(Thunder Days) | Record Max T °F | Record Min T °F | Avg. Snow in. (Snow Days) |
|---|---|---|---|---|---|---|
| JANUARY | 30 | 21 | 0.48 (1.3) (0) | 42 | 4 | 3.8 (0.8) |
| FEBRUARY | 20 | 11 | 0.65 (1.6) (0) | 39 | -7 | 4.4 (1.1) |
| MARCH | 3 | -8 | 0.35 (1.1) (0) | 26 | -30 | 3.8 (0.7) |
| APRIL | -1 | -13 | 0.38 (1.1) (0) | 23 | -39 | 3.9 (0.7) |
| MAY | -4 | -17 | 0.4 (0.8) (0) | 19 | -45 | 3.9 (0.7) |
| JUNE | -4 | -17 | 0.32 (1.1) (0) | 18 | -39 | 3.2 (0.3) |
| JULY | -9 | -24 | 0.23 (0.8) (0) | 24 | -59 | 2.3 (0.2) |
| AUGUST | -12 | -28 | 0.33 (1) (0) | 29 | -57 | 3.1 (0.5) |
| SEPTEMBER | -4 | -19 | 0.39 (1.4) (0) | 18 | -42 | 3.9 (0.7) |
| OCTOBER | 2 | -12 | 0.23 (0.8) (0) | 24 | -39 | 2.1 (0.3) |
| NOVEMBER | 21 | 10 | 0.2 (1) (0) | 37 | -18 | 1.8 (0.2) |
| DECEMBER | 29 | 21 | 0.3 (1.3) (0) | 42 | 21 | 3.4 (1.2) |
| ANNUAL | 6 | -5 | 4.3 (13.3) (0) | 42 | -59 | 39.6 (7.4) |

### Map Location 52 — Antarctica, South Pole Station — 8959S 00000W 9186

| Month | Avg Max T °F | Avg Min T °F | Avg. Precip in. (Rain Days)(Thunder Days) | Record Max T °F | Record Min T °F | Avg. Snow in. (Snow Days) |
|---|---|---|---|---|---|---|
| JANUARY | -16 | -23 | 0.03 (0) (0) | 6 | -35 | 0.2 (0) |
| FEBRUARY | -34 | -42 | 0.05 (0) (0) | -7 | -69 | 0.2 (0) |
| MARCH | -62 | -73 | 0 (0) (0) | -35 | -95 | 0 (0) |
| APRIL | -66 | -79 | 0 (0) (0) | -26 | -99 | 0 (0) |
| MAY | -62 | -76 | 0 (0) (0) | -30 | -98 | 0.1 (0) |
| JUNE | -65 | -79 | 0 (0) (0) | -30 | -102 | 0 (0) |
| JULY | -67 | -81 | 0 (0) (0) | -31 | -102 | 0 (0) |
| AUGUST | -69 | -82 | 0 (0) (0) | -30 | -107 | 0.1 (0) |
| SEPTEMBER | -68 | -81 | 0 (0) (0) | -35 | -107 | 0 (0) |
| OCTOBER | -55 | -64 | 0.01 (0) (0) | -21 | -86 | 0.1 (0) |
| NOVEMBER | -34 | -41 | 0 (0) (0) | -34 | -41 | 0.1 (0) |
| DECEMBER | -16 | -21 | 0.02 (0) (0) | -2 | -37 | 0.2 (0) |
| ANNUAL | -50 | -61 | 0.01 (0) (0) | 6 | -107 | 1 (0) |

### Map Location 53 — Antarctica, Vostok — 7827S 10652E 11220

| Month | Avg Max T °F | Avg Min T °F | Avg. Precip in. (Rain Days)(Thunder Days) | Record Max T °F | Record Min T °F | Avg. Snow in. (Snow Days) |
|---|---|---|---|---|---|---|
| JANUARY | -22 | -40 | 0.02 (***) (***) | -8 | -54 | *** (***) |
| FEBRUARY | -39 | -58 | 0.04 (***) (***) | -13 | -83 | *** (***) |
| MARCH | -60 | -74 | 0.28 (***) (***) | -36 | -103 | *** (***) |
| APRIL | -75 | -88 | 0.17 (***) (***) | -46 | -100 | *** (***) |
| MAY | -76 | -89 | 0.33 (***) (***) | -45 | -109 | *** (***) |
| JUNE | -83 | -95 | 0.49 (***) (***) | -53 | -114 | *** (***) |
| JULY | -82 | -96 | 0.23 (***) (***) | -47 | -117 | *** (***) |
| AUGUST | -88 | -103 | 0.21 (***) (***) | -62 | -127 | *** (***) |
| SEPTEMBER | -82 | -98 | 0.19 (***) (***) | -48 | -116 | *** (***) |
| OCTOBER | -66 | -84 | 0.07 (***) (***) | -41 | -104 | *** (***) |
| NOVEMBER | -39 | -58 | 0.02 (***) (***) | -28 | -74 | *** (***) |
| DECEMBER | -19 | -37 | 0.03 (***) (***) | -14 | -54 | *** (***) |
| ANNUAL | -60 | -76 | 2.1 (***) (***) | -8 | -127 | *** (***) |

## Data Key

|  |  |
|---|---|
| Map Locater Number | MONTH |
| Location | Average Max T °F — Record Max T °F |
| Lat - Lon ddmm Elev ft | Average Min T °F — Record Min T °F |
| | Avg. Precip in. (Rain Days) (Thunder Days) — Avg. Snow in. (Snow Days) |

| | |
|---|---|
| *** or Blank | No data, missing data or data of insufficient reliability |
| Rain Days | Number of Days with > = 0.01" of precipitation falling |
| Snow Days | Number of Days with > = 1.5" of snow falling |

**Table 9.2  World Weather Data [CONTINUED]**

# Asia, Afghanistan to Bangladesh

**Map # 54 to 58**

## Map Location 54 — Afghanistan, Kabul 3433N 06912E 5909

| Month | Avg High | Rec High | Avg Low | Rec Low | Precip (days)(snow) | |
|---|---|---|---|---|---|---|
| JANUARY | 42 | 66 | 19 | -14 | 0.98 (2.2)(0) | (***)(***) |
| FEBRUARY | 44 | 65 | 23 | -7 | 2.44 (4.8)(0) | (***)(***) |
| MARCH | 55 | 73 | 34 | 12 | 2.87 (5)(0) | (***)(***) |
| APRIL | 64 | 80 | 42 | 28 | 4.36 (6.2)(1.5) | (***)(***) |
| MAY | 74 | 90 | 47 | 30 | 1.09 (2.4)(0) | (0)(0) |
| JUNE | 86 | 95 | 53 | 36 | 0.06 (0.2)(0) | (0)(0) |
| JULY | 88 | 98 | 58 | 39 | 0.26 (0.5)(0) | (0)(0) |
| AUGUST | 89 | 97 | 56 | 35 | 0.05 (0.2)(1.7) | (0)(0) |
| SEPTEMBER | 84 | 93 | 48 | 33 | 0.02 (0.2)(1) | (0)(0) |
| OCTOBER | 73 | 89 | 39 | 25 | 0.06 (0.3)(0) | (***)(0) |
| NOVEMBER | 58 | 75 | 29 | 15 | 0.96 (2)(0) | (***)(***) |
| DECEMBER | 46 | 66 | 23 | -2 | 0.98 (2.2)(0) | (***)(***) |
| ANNUAL | 67 | 98 | 39 | -14 | 14.1 (26.2)(4.2) | (***)(***) |

## Map Location 55 — Armenia, Yerevan 4008N 04428E 2976

| Month | Avg High | Rec High | Avg Low | Rec Low | Precip (days)(snow) | |
|---|---|---|---|---|---|---|
| JANUARY | 36 | 68 | 22 | -8 | 0.75 (5)(0.2) | (***)(***) |
| FEBRUARY | 42 | 64 | 25 | 3 | 0.95 (5.6)(0.4) | (***)(***) |
| MARCH | 53 | 72 | 34 | 5 | 0.65 (4.6)(0.6) | (***)(***) |
| APRIL | 65 | 82 | 44 | 21 | 1.36 (7.1)(5.6) | (***)(***) |
| MAY | 76 | 95 | 53 | 32 | 1.78 (8.2)(10.9) | (0)(0) |
| JUNE | 85 | 102 | 58 | 37 | 0.86 (5.7)(11.6) | (0)(0) |
| JULY | 91 | 104 | 66 | 54 | 0.63 (2.5)(7.3) | (0)(0) |
| AUGUST | 90 | 104 | 64 | 54 | 0.31 (2.6)(4.6) | (0)(0) |
| SEPTEMBER | 82 | 95 | 56 | 41 | 0.81 (3.2)(4.1) | (0)(0) |
| OCTOBER | 69 | 88 | 46 | 25 | 0.95 (5.8)(1.8) | (***)(***) |
| NOVEMBER | 56 | 72 | 37 | 21 | 0.76 (3.4)(0.1) | (***)(***) |
| DECEMBER | 42 | 57 | 28 | 10 | 0.48 (4.3)(0) | (***)(***) |
| ANNUAL | 66 | 108 | 44 | -8 | 10.3 (58)(47.2) | (***)(***) |

## Map Location 56 — Azerbaijan, Baku 3844N 04850E -36

| Month | Avg High | Rec High | Avg Low | Rec Low | Precip (days)(snow) | |
|---|---|---|---|---|---|---|
| JANUARY | 44 | 79 | 33 | 5 | 2.68 (7.8)(0) | (***)(***) |
| FEBRUARY | 46 | 79 | 35 | 10 | 2.07 (7.6)(0) | (***)(***) |
| MARCH | 51 | 84 | 40 | 19 | 4.58 (11.5)(0.4) | (***)(***) |
| APRIL | 60 | 84 | 47 | 28 | 2.52 (10.5)(1.3) | (***)(***) |
| MAY | 72 | 93 | 57 | 41 | 0.77 (6)(3.8) | (0)(0) |
| JUNE | 81 | 97 | 64 | 46 | 3.38 (4)(3.5) | (0)(0) |
| JULY | 86 | 100 | 69 | 50 | 1.06 (2.9)(0.4) | (0)(0) |
| AUGUST | 85 | 100 | 68 | 54 | 2.1 (3.8)(1.3) | (0)(0) |
| SEPTEMBER | 78 | 93 | 63 | 46 | 7.79 (10.8)(2.6) | (0)(0) |
| OCTOBER | 68 | 90 | 55 | 34 | 9.4 (11.6)(2.2) | (0)(0) |
| NOVEMBER | 56 | 90 | 46 | 25 | 5.7 (9.6)(0.4) | (***)(***) |
| DECEMBER | 49 | 77 | 38 | 7 | 2.6 (7)(0.6) | (***)(***) |
| ANNUAL | 65 | 100 | 51 | 5 | 44.6 (93.1)(16.5) | (***)(***) |

## Map Location 57 — Bahrain, Bahrain 2616N 05037E 6

| Month | Avg High | Rec High | Avg Low | Rec Low | Precip (days)(snow) | |
|---|---|---|---|---|---|---|
| JANUARY | 68 | 84 | 58 | 37 | 1.14 (2.5)(1) | (0)(0) |
| FEBRUARY | 70 | 90 | 60 | 48 | 0.52 (1.3)(0.5) | (0)(0) |
| MARCH | 76 | 93 | 65 | 46 | 0.37 (0.9)(1.8) | (0)(0) |
| APRIL | 84 | 100 | 71 | 59 | 0.36 (0.9)(2) | (0)(0) |
| MAY | 91 | 111 | 79 | 66 | 0.09 (0.3)(0.9) | (0)(0) |
| JUNE | 96 | 113 | 84 | 73 | 0 (0)(0) | (0)(0) |
| JULY | 99 | 111 | 87 | 77 | 0 (0)(0.1) | (0)(0) |
| AUGUST | 100 | 113 | 88 | 81 | 0 (0)(0) | (0)(0) |
| SEPTEMBER | 96 | 108 | 84 | 72 | 0 (0)(0) | (0)(0) |
| OCTOBER | 90 | 106 | 78 | 63 | 0.03 (0.2)(0) | (0)(0) |
| NOVEMBER | 81 | 95 | 71 | 55 | 0.11 (0.4)(0.7) | (0)(0) |
| DECEMBER | 71 | 85 | 62 | 46 | 0.55 (1.3)(0.8) | (0)(0) |
| ANNUAL | 85 | 113 | 74 | 37 | 3.2 (7.8)(7.8) | (0)(0) |

## Map Location 58 — Bangladesh, Dhaka 2346N 09023E 24

| Month | Avg High | Rec High | Avg Low | Rec Low | Precip (days)(snow) | |
|---|---|---|---|---|---|---|
| JANUARY | 77 | 88 | 56 | 43 | 0.3 (0.8)(1) | (0)(0) |
| FEBRUARY | 82 | 94 | 59 | 44 | 1.2 (2)(2.6) | (0)(0) |
| MARCH | 90 | 105 | 68 | 51 | 2.4 (3)(5.2) | (0)(0) |
| APRIL | 92 | 108 | 74 | 62 | 5.4 (7)(11.5) | (0)(0) |
| MAY | 91 | 106 | 76 | 58 | 9.6 (11)(13.5) | (0)(0) |
| JUNE | 89 | 98 | 78 | 67 | 12.4 (15)(12.6) | (0)(0) |
| JULY | 88 | 95 | 79 | 70 | 13 (18)(5.6) | (0)(0) |
| AUGUST | 88 | 97 | 79 | 71 | 13.3 (18)(6.8) | (0)(0) |
| SEPTEMBER | 89 | 98 | 79 | 70 | 9.8 (12)(13.5) | (0)(0) |
| OCTOBER | 88 | 99 | 75 | 63 | 5.3 (6)(6.8) | (0)(0) |
| NOVEMBER | 84 | 94 | 66 | 52 | 1 (1)(0) | (0)(0) |
| DECEMBER | 79 | 87 | 58 | 45 | 0.2 (0.4)(0) | (0)(0) |
| ANNUAL | 86 | 108 | 71 | 43 | 73.9 (94.2)(79.1) | (0)(0) |

**Table 9.2  World Weather Data [CONTINUED]**

# Asia, Bhutan to China, Chiting

## Map # 59 to 63

### Map Location 59 — Bhutan, Thimphu — 2728N 08938E 7656

| | JANUARY | FEBRUARY | MARCH | APRIL | MAY | JUNE |
|---|---|---|---|---|---|---|
| | 54 | 58 | 62 | 68 | 74 | 81 |
| | 27 | 33 | 39 | 45 | 51 | 56 |
| | *** | *** | *** | *** | *** | *** |
| | *** | *** | *** | *** | *** | *** |
| | 0.03 (***) (***) | 0.4 (**) (***) | 1.3 (***) (***) | 3.1 (***) (***) | 7.1 (***) (***) | 16 (***) (***) |

| | JULY | AUGUST | SEPTEMBER | OCTOBER | NOVEMBER | DECEMBER | ANNUAL |
|---|---|---|---|---|---|---|---|
| | 79 | 75 | 74 | 71 | 64 | 58 | 68 |
| | 59 | 58 | 53 | 51 | 41 | 30 | 45 |
| | *** | *** | *** | *** | *** | *** | *** |
| | *** | *** | *** | *** | *** | *** | *** |
| | 20 (***) (***) | 14.8 (***) (***) | 9.3 (***) (***) | 3.1 (***) (***) | 0.4 (***) (***) | 0.2 (***) (***) | 76.1 (***) (***) |

### Map Location 60 — Brunei, Bandar Seri Begawa — 0455N 11455E 10

| | JANUARY | FEBRUARY | MARCH | APRIL | MAY | JUNE |
|---|---|---|---|---|---|---|
| | 86 | 86 | 87 | 88 | 88 | 88 |
| | 74 | 74 | 74 | 75 | 75 | 75 |
| | 91 | 90 | 93 | 92 | 93 | 95 |
| | 69 | 69 | 67 | 69 | 71 | 70 |
| | 16.75 (***) (4..9) | 6.52 (7.5) (2.9) | 5.53 (5.5) (7.3) | 4.37 (5.5) (9.2) | 8.21 (5.2) (11.2) | 12.03 (16) (8.5) |

| | JULY | AUGUST | SEPTEMBER | OCTOBER | NOVEMBER | DECEMBER | ANNUAL |
|---|---|---|---|---|---|---|---|
| | 88 | 88 | 87 | 87 | 87 | 87 | 87 |
| | 74 | 74 | 74 | 74 | 74 | 74 | 74 |
| | 93 | 94 | 93 | 92 | 92 | 92 | 95 |
| | 70 | 70 | 69 | 71 | 69 | 69 | 67 |
| | 8.49 (13) (8.5) | 8.43 (13) (7.2) | 11.79 (14) (8.9) | 11.73 (14) (9) | 14.45 (15.6) (6.3) | 11.27 (***) (6.7) | 119.6 (***) (90.6) |

### Map Location 61 — Cambodia, Phnom Penh — 1133N 10451E 39

| | JANUARY | FEBRUARY | MARCH | APRIL | MAY | JUNE |
|---|---|---|---|---|---|---|
| | 88 | 91 | 93 | 95 | 93 | 91 |
| | 71 | 72 | 74 | 76 | 76 | 76 |
| | 97 | 98 | 102 | 105 | 102 | 101 |
| | 55 | 59 | 66 | 64 | 69 | 70 |
| | 0.3 (0.7) (0) | 0.4 (0.9) (0.7) | 1.4 (3.1) (0.6) | 3.1 (5) (3.3) | 5.7 (5.5) (1.3) | 5.8 (10.4) (3) |

| | JULY | AUGUST | SEPTEMBER | OCTOBER | NOVEMBER | DECEMBER | ANNUAL |
|---|---|---|---|---|---|---|---|
| | 90 | 90 | 88 | 87 | 86 | 86 | 90 |
| | 76 | 76 | 76 | 76 | 74 | 71 | 75 |
| | 99 | 100 | 96 | 97 | 94 | 94 | 105 |
| | 68 | 68 | 70 | 63 | 62 | 62 | 55 |
| | 6 (10.6) (5.1) | 6.1 (10.7) (2.5) | 8.9 (11.9) (4.6) | 9.9 (12.7) (5) | 5.5 (8.8) (2.1) | 1.7 (3.3) (0.8) | 54.8 (83.6) (29) |

### Map Location 62 — China, Beijing — 3956N 11620E 167

| | JANUARY | FEBRUARY | MARCH | APRIL | MAY | JUNE |
|---|---|---|---|---|---|---|
| | 34 | 43 | 54 | 69 | 81 | 88 |
| | 15 | 21 | 32 | 44 | 56 | 65 |
| | 50 | 66 | 75 | 86 | 99 | 106 |
| | -2 | 5 | 12 | 28 | 41 | 52 |
| | 0.15 (1.9) (0) | 0.18 (2.1) (0) | 0.34 (1.8) (0) | 0.62 (3) (0.9) | 1.3 (5.4) (3.4) | 3.26 (7.9) (8.5) |

| | JULY | AUGUST | SEPTEMBER | OCTOBER | NOVEMBER | DECEMBER | ANNUAL |
|---|---|---|---|---|---|---|---|
| | 89 | 86 | 79 | 66 | 50 | 39 | 65 |
| | 72 | 69 | 58 | 44 | 31 | 20 | 44 |
| | 100 | 97 | 97 | 84 | 68 | 57 | 106 |
| | 61 | 55 | 41 | 27 | 10 | 7 | -2 |
| | 9.82 (14.7) (11.8) | 5.75 (11.4) (9.7) | 2.28 (7.8) (4.4) | 0.67 (2.2) (1.1) | 0.23 (0.9) (0) | 0.09 (1.6) (0.1) | 24.7 (60.7) (39.9) |

### Map Location 63 — China, Chiting-Hsilin — 3357N 09237E 16564

| | JANUARY | FEBRUARY | MARCH | APRIL | MAY | JUNE |
|---|---|---|---|---|---|---|
| | 18 | 22 | 30 | 42 | 47 | 52 |
| | -12 | -12 | -1 | 10 | 20 | 31 |
| | 32 | 32 | 45 | 55 | 61 | 63 |
| | -35 | -24 | -18 | -8 | 7 | 25 |
| | *** (***) (0) | *** (***) (0.4) | *** (***) (0) | *** (***) (0.5) | *** (***) (13.1) | *** (***) (14.6) |

| | JULY | AUGUST | SEPTEMBER | OCTOBER | NOVEMBER | DECEMBER | ANNUAL |
|---|---|---|---|---|---|---|---|
| | 56 | 56 | 52 | 38 | 28 | 20 | 38 |
| | 36 | 34 | 30 | 13 | -3 | -12 | 11 |
| | 64 | 64 | 63 | 59 | 39 | 37 | 64 |
| | 27 | 25 | 19 | -4 | -15 | -26 | -35 |
| | *** (***) (19.5) | *** (***) (12.3) | *** (***) (12.7) | *** (***) (0.4) | *** (***) (0) | *** (***) (0.3) | *** (***) (73.8) |

## Asia, China, Chongqing to Lhasa

Table 9.2  World Weather Data [CONTINUED]

### Map # 64 to 68

**Map Location 64 — China, Chongqing — 2930N 10633E 855**

| | JANUARY | FEBRUARY | MARCH | APRIL | MAY | JUNE |
|---|---|---|---|---|---|---|
| Record high | 63 | 72 | 86 | 95 | 99 | 102 |
| Average high | 51 | 55 | 67 | 74 | 78 | 86 |
| Average low | 41 | 45 | 53 | 60 | 65 | 72 |
| Record low | 28 | 34 | 37 | 46 | 54 | 61 |
| Precipitation | 0.64 (4.1)(0.3) | 0.88 (5)(0.8) | 1.48 (6)(2.8) | 3.84 (11.4)(4.4) | 5.63 (13.4)(5.8) | 7.08 (12.8)(4.3) |
| Snow | 0 (0) | 0 (0) | 0 (0) | 0 (0) | 0 (0) | 0 (0) |

| | JULY | AUGUST | SEPTEMBER | OCTOBER | NOVEMBER | DECEMBER | ANNUAL |
|---|---|---|---|---|---|---|---|
| Record high | 104 | 104 | 99 | 91 | 79 | 66 | 104 |
| Average high | 93 | 93 | 85 | 71 | 62 | 54 | 72 |
| Average low | 77 | 76 | 70 | 61 | 54 | 46 | 60 |
| Record low | 68 | 66 | 57 | 48 | 41 | 32 | 28 |
| Precipitation | 5.5 (11.1)(10.6) | 4.57 (9.9*)(9.1) | 5.75 (***)(2.7) | 4.26 (***)(1) | 2 (6.8)(1.3) | 0.79 (4.7)(0) | 42.4 (***)(43.1) |
| Snow | 0 (0) | 0 (0) | 0 (0) | 0 (0) | 0 (0) | 0 (0) | *** (***) |

**Map Location 65 — China, Haikou — 2000N 11025E 46**

| | JANUARY | FEBRUARY | MARCH | APRIL | MAY | JUNE |
|---|---|---|---|---|---|---|
| Record high | 86 | 95 | 100 | 100 | 102 | 100 |
| Average high | 69 | 71 | 80 | 86 | 91 | 91 |
| Average low | 57 | 59 | 66 | 71 | 76 | 77 |
| Record low | 37 | 45 | 48 | 57 | 66 | 72 |
| Precipitation | 0.95 (5.2)(0.2) | 1.24 (6.1)(1.4) | 1.84 (7.1)(3.8) | 3.8 (11.3)(6.9) | 6.16 (13.8)(22.1) | 8.09 (13.6)(20.4) |
| Snow | 0 (0) | 0 (0) | 0 (0) | 0 (0) | 0 (0) | 0 (0) |

| | JULY | AUGUST | SEPTEMBER | OCTOBER | NOVEMBER | DECEMBER | ANNUAL |
|---|---|---|---|---|---|---|---|
| Record high | 100 | 97 | 95 | 91 | 91 | 90 | 102 |
| Average high | 91 | 90 | 88 | 83 | 79 | 74 | 83 |
| Average low | 77 | 77 | 76 | 71 | 67 | 61 | 70 |
| Record low | 73 | 70 | 70 | 57 | 50 | 46 | 37 |
| Precipitation | 8.53 (13.9)(20.9) | 7.35 (13.0*)(22.5) | 10.1 (***)(15.2) | 6.66 (***)(3.5) | 3.16 (10.2)(1.6) | 1.86 (8)(0.1) | 59.7 (***)(118.6) |
| Snow | 0 (0) | 0 (0) | 0 (0) | 0 (0) | 0 (0) | 0 (0) | 0 (0) |

**Map Location 66 — China, Harbin — 4545N 12638E 476**

| | JANUARY | FEBRUARY | MARCH | APRIL | MAY | JUNE |
|---|---|---|---|---|---|---|
| Record high | 30 | 52 | 61 | 82 | 95 | 93 |
| Average high | 8 | 19 | 35 | 55 | 69 | 80 |
| Average low | -14 | -5 | 13 | 31 | 45 | 57 |
| Record low | -35 | -24 | -20 | 9 | 27 | 41 |
| Precipitation | 0.2 (2.2)(0) | 0.2 (2.2)(0) | 0.4 (2.1)(0) | 0.9 (4)(0.6) | 1.7 (6.7)(2.7) | 4.1 (9.2)(7.1) |
| Snow | *** (***) | *** (***) | *** (***) | *** (***) | *** (***) | *** (***) |

| | JULY | AUGUST | SEPTEMBER | OCTOBER | NOVEMBER | DECEMBER | ANNUAL |
|---|---|---|---|---|---|---|---|
| Record high | 95 | 93 | 90 | 73 | 59 | 43 | 95 |
| Average high | 83 | 79 | 70 | 54 | 33 | 16 | 50 |
| Average low | 61 | 61 | 50 | 32 | 13 | -4 | 29 |
| Record low | 52 | 46 | 32 | 10 | -9 | -29 | -35 |
| Precipitation | 5.8 (11.4)(8.4) | 4.2 (9.3)(7.1) | 2.2 (7.5)(5.2) | 1.2 (4)(0.5) | 0.4 (1.4)(0.1) | 0.2 (2.2)(0) | 21.5 (62.2)(31.7) |
| Snow | *** (***) | *** (***) | *** (***) | *** (***) | *** (***) | *** (***) | *** (***) |

**Map Location 67 — China, Hong Kong — 2219N 11411E 15**

| | JANUARY | FEBRUARY | MARCH | APRIL | MAY | JUNE |
|---|---|---|---|---|---|---|
| Record high | 79 | 82 | 84 | 92 | 92 | 94 |
| Average high | 64 | 63 | 67 | 75 | 82 | 85 |
| Average low | 56 | 55 | 60 | 67 | 74 | 78 |
| Record low | 32 | 38 | 44 | 52 | 60 | 67 |
| Precipitation | 1.3 (3.1)(0.3) | 1.8 (4.1)(0.3) | 2.9 (5.3)(2) | 5.4 (8.1)(4) | 11.5 (11.9)(5) | 15.5 (13.2)(7) |
| Snow | 0 (0) | 0 (0) | 0 (0) | 0 (0) | 0 (0) | 0 (0) |

| | JULY | AUGUST | SEPTEMBER | OCTOBER | NOVEMBER | DECEMBER | ANNUAL |
|---|---|---|---|---|---|---|---|
| Record high | 96 | 97 | 94 | 94 | 87 | 85 | 97 |
| Average high | 87 | 87 | 85 | 81 | 74 | 68 | 77 |
| Average low | 78 | 78 | 77 | 73 | 65 | 59 | 68 |
| Record low | 71 | 71 | 65 | 57 | 44 | 41 | 32 |
| Precipitation | 14.2 (12.8)(7) | 15 (13)(6) | 10.1 (12.7)(4) | 4.5 (6.6)(2) | 1.7 (3.1)(0.3) | 1.2 (2.9)(0.3) | 85.1 (96.8)(38.2) |
| Snow | 0 (0) | 0 (0) | 0 (0) | 0 (0) | 0 (0) | 0 (0) | 0 (0) |

**Map Location 68 — China (Tibet), Lhasa — 2943N 09102E 11796**

| | JANUARY | FEBRUARY | MARCH | APRIL | MAY | JUNE |
|---|---|---|---|---|---|---|
| Record high | 61 | 70 | 70 | 77 | 81 | 81 |
| Average high | 44 | 48 | 51 | 61 | 66 | 72 |
| Average low | 13 | 19 | 25 | 34 | 40 | 48 |
| Record low | 3 | 3 | 10 | 19 | 25 | 36 |
| Precipitation | 0.1 (1.6)(0) | 0.5 (3.6)(0) | 0.3 (1.7)(0.2) | 0.2 (1.2)(1.6) | 1 (4.4)(7.4) | 2.5 (6.5)(15.7) |
| Snow | *** (***) | *** (***) | *** (***) | *** (***) | *** (***) | *** (***) |

| | JULY | AUGUST | SEPTEMBER | OCTOBER | NOVEMBER | DECEMBER | ANNUAL |
|---|---|---|---|---|---|---|---|
| Record high | 82 | 82 | 72 | 72 | 66 | 63 | 82 |
| Average high | 71 | 69 | 67 | 62 | 53 | 46 | 59 |
| Average low | 50 | 49 | 45 | 34 | 21 | 15 | 33 |
| Record low | 43 | 34 | 32 | 19 | 5 | -4 | -4 |
| Precipitation | 4.8 (10.2)(20.7) | 3.5 (8.3)(20.4) | 2.6 (8.8)(11.4) | 0.5 (1.7)(1.6) | 0.1 (0.5)(0) | 0 (0.2)(0) | 16.1 (48.5)(79.2) |
| Snow | *** (***) | *** (***) | *** (***) | *** (***) | *** (***) | *** (***) | *** (***) |

**Table 9.2  World Weather Data** [CONTINUED]

# Asia, China, Shang-Hai to Georgia

# Map # 69 to 73

*Each month cell lists: Normal High, Record High, Normal Low, Record Low, and Precipitation as inches (mean days) (snow).*

## Map Location 69 — China, Shang-Hai 3112N 12126E 16

| Month | Normal High | Record High | Normal Low | Record Low | Precip (in) (days) (snow) |
|---|---|---|---|---|---|
| JANUARY | 45 | 64 | 31 | 14 | 1.94 (9.6) (0) |
| FEBRUARY | 50 | 75 | 34 | 18 | 2.39 (10) (0.3) |
| MARCH | 58 | 81 | 43 | 23 | 3.32 (12.2) (0.9) |
| APRIL | 66 | 82 | 51 | 32 | 3.55 (12.5) (3.6) |
| MAY | 74 | 97 | 60 | 45 | 3.77 (12) (3.3) |
| JUNE | 81 | 95 | 69 | 55 | 7.02 (13.9) (5.5) |
| JULY | 91 | 99 | 78 | 68 | 5.81 (11) (10.2) |
| AUGUST | 89 | 99 | 77 | 68 | 5.5 (10.7) (9.5) |
| SEPTEMBER | 81 | 95 | 70 | 55 | 5.16 (11.8) (4.3) |
| OCTOBER | 72 | 86 | 57 | 36 | 2.92 (8.8) (0.5) |
| NOVEMBER | 63 | 82 | 48 | 25 | 2.09 (8.1) (0.1) |
| DECEMBER | 52 | 68 | 38 | 21 | 1.48 (7.6) (0) |
| ANNUAL | 69 | 99 | 55 | 14 | 44.9 (128.2) (38.2) |

## Map Location 70 — China, Wusu 4423N 08430E 1345

| Month | Normal High | Record High | Normal Low | Record Low | Precip (in) (days) (snow) |
|---|---|---|---|---|---|
| JANUARY | 11 | 28 | -8 | -27 | *** (***) (0) |
| FEBRUARY | 21 | 41 | 2 | -24 | *** (***) (0) |
| MARCH | 38 | 70 | 21 | -11 | *** (***) (0) |
| APRIL | 63 | 86 | 43 | 27 | *** (***) (0.5) |
| MAY | 73 | 95 | 51 | 32 | *** (***) (1.6) |
| JUNE | 84 | 99 | 62 | 46 | *** (***) (8.4) |
| JULY | 87 | 99 | 66 | 55 | *** (***) (9.7) |
| AUGUST | 86 | 100 | 65 | 48 | *** (***) (6.7) |
| SEPTEMBER | 76 | 97 | 56 | 30 | *** (***) (1.2) |
| OCTOBER | 58 | 95 | 38 | 19 | *** (***) (0) |
| NOVEMBER | 35 | 55 | 21 | -15 | *** (***) (0.2) |
| DECEMBER | 19 | 41 | 6 | -26 | *** (***) (0) |
| ANNUAL | 54 | 100 | 35 | -27 | *** (***) (28.3) |

## Map Location 71 — China, Xi'an 3415N 10855E 1312

| Month | Normal High | Record High | Normal Low | Record Low | Precip (in) (days) (snow) |
|---|---|---|---|---|---|
| JANUARY | 41 | 61 | 23 | 9 | 0.17 (2) (0) |
| FEBRUARY | 49 | 70 | 28 | 12 | 0.35 (2.9) (0) |
| MARCH | 59 | 82 | 39 | 23 | 0.82 (3.7) (0) |
| APRIL | 69 | 88 | 48 | 25 | 1.5 (6.1) (1.7) |
| MAY | 77 | 97 | 56 | 39 | 2.27 (8.2) (1.7) |
| JUNE | 90 | 106 | 67 | 50 | 2.06 (5.7) (4.2) |
| JULY | 92 | 106 | 73 | 64 | 3.28 (7.9) (7.1) |
| AUGUST | 88 | 100 | 70 | 54 | 4.08 (9.2*) (6.7) |
| SEPTEMBER | 78 | 93 | 60 | 43 | 4.73 (***) (1.7) |
| OCTOBER | 66 | 90 | 49 | 32 | 2.05 (7) (0.4) |
| NOVEMBER | 53 | 72 | 38 | 21 | 0.78 (2.6) (0) |
| DECEMBER | 44 | 64 | 28 | 14 | 0.23 (2.3) (0.1) |
| ANNUAL | 67 | 106 | 48 | 9 | 22.3 (***) (23.6) |

## Map Location 72 — China, Republic of, Taipei 2504N 12132E 21

| Month | Normal High | Record High | Normal Low | Record Low | Precip (in) (days) (snow) |
|---|---|---|---|---|---|
| JANUARY | 66 | 84 | 53 | 34 | 3.75 (6.7) (0) |
| FEBRUARY | 67 | 86 | 56 | 37 | 5.34 (8.6) (0.1) |
| MARCH | 71 | 89 | 58 | 45 | 4.26 (8.6) (0) |
| APRIL | 77 | 95 | 64 | 50 | 5.25 (9) (0.4) |
| MAY | 84 | 98 | 69 | 57 | 6.92 (8.2) (0.5) |
| JUNE | 88 | 98 | 73 | 61 | 8.77 (10.6) (0.9) |
| JULY | 92 | 99 | 76 | 70 | 8.77 (9.7) (1.6) |
| AUGUST | 91 | 99 | 74 | 60 | 8.18 (7.3) (0.6) |
| SEPTEMBER | 88 | 97 | 74 | 60 | 5.47 (7.5) (0) |
| OCTOBER | 80 | 91 | 68 | 54 | 4.19 (7.9) (0) |
| NOVEMBER | 75 | 90 | 63 | 48 | 2.86 (7) (0) |
| DECEMBER | 69 | 86 | 58 | 33 | 2.21 (9.2) (2.1) |
| ANNUAL | 79 | 99 | 66 | 33 | 72.5 (100.6) (5.3) |

## Map Location 73 — Georgia, Tbilisi 4141N 04457E 1608

| Month | Normal High | Record High | Normal Low | Record Low | Precip (in) (days) (snow) |
|---|---|---|---|---|---|
| JANUARY | 45 | 63 | 30 | 7 | 0.56 (3.9) (0.1) |
| FEBRUARY | 47 | 72 | 36 | 7 | 0.41 (3.2) (0.1) |
| MARCH | 53 | 73 | 36 | 19 | 0.72 (4.4) (0.3) |
| APRIL | 61 | 84 | 44 | 21 | 2.21 (9.2) (2.1) |
| MAY | 74 | 91 | 54 | 39 | 2.59 (9.1) (9.3) |
| JUNE | 79 | 95 | 60 | 45 | 1.93 (7.4) (10.8) |
| JULY | 85 | 100 | 65 | 32 | 2.55 (6.9) (6.9) |
| AUGUST | 84 | 99 | 65 | 52 | 1.09 (6.1) (6.7) |
| SEPTEMBER | 76 | 90 | 58 | 43 | 1.8 (5.5) (4.7) |
| OCTOBER | 65 | 82 | 48 | 25 | 1.98 (5.7) (1.4) |
| NOVEMBER | 55 | 70 | 44 | 21 | 1.05 (4.7) (0) |
| DECEMBER | 46 | 70 | 33 | 10 | 0.51 (4.1) (0) |
| ANNUAL | 64 | 100 | 47 | 7 | 17.4 (70.2) (42.4) |

## Asia, India, Bangalore to New Delhi    Map # 74 to 78

Table 9.2  World Weather Data [CONTINUED]

### Map Location 74 — India, Bangalore — 1257N 07740E 2937

| | JAN | FEB | MAR | APR | MAY | JUN | JUL | AUG | SEP | OCT | NOV | DEC | ANNUAL |
|---|---|---|---|---|---|---|---|---|---|---|---|---|---|
| Record High | 100 | 100 | 100 | 101 | 102 | 100 | 94 | 92 | 96 | 97 | 97 | 97 | |
| Record Low | 46 | 49 | 52 | 57 | 61 | 57 | 61 | 58 | 59 | 56 | 51 | 51 | |
| Normal High | 80 | 86 | 90 | 93 | 91 | 84 | 81 | 81 | 82 | 82 | 79 | 78 | |
| Normal Low | 57 | 60 | 65 | 69 | 69 | 67 | 66 | 66 | 65 | 64 | 62 | 58 | |
| Precip (in)(Y)(days) | 0.24 (0.7) (0) | 0.26 (0.7) (1) | 0.4 (1) (2) | 1.61 (3.3) (9) | 4.16 (6.1) (13) | 2.86 (4.7) (4) | 3.93 (6.2) (3) | 4.98 (7.6) (4) | 6.66 (7.3) (5) | 5.87 (6.9) (6) | 2.67 (4.5) (1) | 0.44 (1.1) (0) | 34.1 (50.1) (48) |

### Map Location 75 — India, Kolkata (Calcutta) — 2239N 08826E 13

| | JAN | FEB | MAR | APR | MAY | JUN | JUL | AUG | SEP | OCT | NOV | DEC | ANNUAL |
|---|---|---|---|---|---|---|---|---|---|---|---|---|---|
| Record High | 90 | 99 | 106 | 110 | 109 | 113 | 96 | 95 | 97 | 97 | 91 | 89 | |
| Record Low | 47 | 48 | 57 | 67 | 68 | 73 | 73 | 73 | 72 | 63 | 57 | 47 | |
| Normal High | 80 | 86 | 94 | 98 | 97 | 93 | 90 | 90 | 90 | 89 | 85 | 81 | |
| Normal Low | 57 | 62 | 71 | 78 | 80 | 80 | 79 | 79 | 79 | 72 | 65 | 58 | |
| Precip (in)(Y)(days) | 0.68 (1) (0.4) | 1.6 (1.7) (2.6) | 0.54 (1.6) (1.2) | 3.05 (2.9) (7.1) | 3.78 (5.8) (6.5) | 10.46 (13) (14.6) | 11.16 (18.1) (8.2) | 12.95 (16.5) (13.3) | 8.9 (9.3) (5.9) | 12.44 (17.5) (8.5) | 0.58 (0.7) (0) | 0.16 (0.3) (0) | 66.3 (88.4) (68.3) |

### Map Location 76 — India, Chennai (Madras) — 1259N 08010E 34

| | JAN | FEB | MAR | APR | MAY | JUN | JUL | AUG | SEP | OCT | NOV | DEC | ANNUAL |
|---|---|---|---|---|---|---|---|---|---|---|---|---|---|
| Record High | 93 | 98 | 102 | 109 | 113 | 110 | 106 | 104 | 102 | 97 | 94 | 91 | |
| Record Low | 57 | 56 | 62 | 68 | 70 | 69 | 70 | 69 | 69 | 63 | 59 | 57 | |
| Normal High | 85 | 88 | 91 | 96 | 101 | 100 | 96 | 95 | 94 | 90 | 85 | 84 | |
| Normal Low | 67 | 68 | 72 | 78 | 82 | 81 | 79 | 78 | 77 | 75 | 72 | 69 | |
| Precip (in)(Y)(days) | 1.41 (2) (0.3) | 0.41 (0.7) (0.3) | 0.29 (0.4) (1) | 0.61 (0.9) (2) | 1.03 (1) (3) | 1.86 (4) (6) | 3.6 (7) (5) | 4.58 (8) (7) | 4.68 (7) (10) | 12.04 (11) (9) | 13.96 (11) (4) | 5.45 (5) (1) | 49.9 (58) (48.6) |

### Map Location 77 — India, Mumbai (Bombay) — 1905N 07250E 9

| | JAN | FEB | MAR | APR | MAY | JUN | JUL | AUG | SEP | OCT | NOV | DEC | ANNUAL |
|---|---|---|---|---|---|---|---|---|---|---|---|---|---|
| Record High | 96 | 96 | 110 | 105 | 102 | 102 | 96 | 96 | 98 | 102 | 98 | 98 | |
| Record Low | 46 | 46 | 55 | 66 | 68 | 70 | 71 | 69 | 70 | 67 | 60 | 60 | |
| Normal High | 88 | 89 | 92 | 93 | 93 | 93 | 87 | 88 | 90 | 93 | 92 | 90 | |
| Normal Low | 62 | 63 | 68 | 74 | 78 | 75 | 76 | 76 | 74 | 73 | 70 | 65 | |
| Precip (in)(Y)(days) | 0.14 (0.5) (0) | 0.08 (0.3) (0) | 0.05 (0.2) (0) | 0.03 (0.1) (1) | 0.65 (1.6) (3.2) | 19.06 (18.2) (4.7) | 24.27 (20.1) (2.2) | 13.39 (15.3) (1.6) | 10.39 (9) (0.4) | 2.54 (4.3) (2.1) | 0.53 (1.2) (0) | 0.08 (0.3) (0.3) | 71.2 (71.1) (15.4) |

### Map Location 78 — India, New Delhi — 2834N 07706E 755

| | JAN | FEB | MAR | APR | MAY | JUN | JUL | AUG | SEP | OCT | NOV | DEC | ANNUAL |
|---|---|---|---|---|---|---|---|---|---|---|---|---|---|
| Record High | 84 | 92 | 103 | 111 | 115 | 115 | 113 | 104 | 105 | 101 | 92 | 83 | |
| Record Low | 31 | 35 | 45 | 53 | 64 | 63 | 70 | 69 | 63 | 49 | 39 | 39 | |
| Normal High | 64 | 75 | 85 | 97 | 105 | 102 | 95 | 93 | 93 | 93 | 83 | 74 | |
| Normal Low | 43 | 49 | 57 | 68 | 79 | 83 | 80 | 78 | 76 | 64 | 52 | 45 | |
| Precip (in)(Y)(days) | 0.9 (2) (1) | 0.7 (2) (3) | 0.5 (1) (2) | 0.3 (1) (3) | 0.5 (2) (5) | 2.9 (4) (6) | 7.1 (8) (7) | 6.8 (8) (6) | 4.6 (4) (3) | 0.4 (1) (1) | 0.1 (0.2) (0.3) | 0.4 (1) (1) | 25.2 (34.2) (38.3) |

## Asia, India, Leh to Iran

Map # 79 to 83

**Table 9.2  World Weather Data [CONTINUED]**

**Map Location 79 — India, Leh (Jammu And Kashmir)  3409N 07734E 11529**

| Month | Avg Max | Avg Min | Rec High | Rec Low | Precip |
|---|---|---|---|---|---|
| JANUARY | 29 | 8 | 47 | -19 | 0.38 (1) (0) |
| FEBRUARY | 33 | 10 | 55 | -13 | 0.31 (0.8) (0) |
| MARCH | 44 | 21 | 67 | -3 | 0.28 (0.7) (0.3) |
| APRIL | 55 | 30 | 74 | 0 | 0.23 (0.6) (0.2) |
| MAY | 64 | 37 | 84 | 18 | 0.22 (0.6) (0.5) |
| JUNE | 71 | 44 | 93 | 30 | 0.18 (0.4) (0.3) |
| JULY | 76 | 50 | 92 | 33 | 0.47 (0.9) (0.6) |
| AUGUST | 76 | 50 | 90 | 38 | 0.59 (1.1) (0.6) |
| SEPTEMBER | 70 | 42 | 87 | 20 | 0.27 (0.7) (0) |
| OCTOBER | 58 | 31 | 86 | 10 | 0.1 (0.3) (0) |
| NOVEMBER | 47 | 20 | 68 | 0 | 0.4 (1) (0) |
| DECEMBER | 35 | 13 | 57 | -14 | 0.19 (0.6) (0) |
| ANNUAL | | | 93 | -19 | 3.6 (8.7) (2.5) |

**Map Location 80 — Indonesia, Jakarta  0609S 10650E 16**

| Month | Avg Max | Avg Min | Rec High | Rec Low | Precip |
|---|---|---|---|---|---|
| JANUARY | 84 | 74 | 92 | 69 | 11.8 (***) (13) |
| FEBRUARY | 84 | 74 | 91 | 69 | 11.8 (***) (12) |
| MARCH | 85 | 74 | 92 | 69 | 8.3 (5.2) (14) |
| APRIL | 86 | 75 | 86 | 75 | 5.8 (5.5) (14) |
| MAY | 87 | 75 | 92 | 70 | 5.2 (5.6) (12) |
| JUNE | 86 | 74 | 92 | 67 | 3.8 (8.3) (8) |
| JULY | 86 | 73 | 92 | 67 | 2.5 (6.8) (5) |
| AUGUST | 87 | 73 | 94 | 67 | 1.7 (5.8) (5) |
| SEPTEMBER | 88 | 73 | 96 | 66 | 2.6 (5.6) (7) |
| OCTOBER | 88 | 74 | 96 | 69 | 4.4 (7.6) (13) |
| NOVEMBER | 87 | 74 | 96 | 68 | 5.6 (8.9) (16) |
| DECEMBER | 85 | 74 | 93 | 67 | 8 (***) (13) |
| ANNUAL | | | 96 | 66 | 71.5 (***) (132) |

**Map Location 81 — Indonesia, Sumbawa-Besar  0829S 11725E 28**

| Month | Avg Max | Avg Min | Rec High | Rec Low | Precip |
|---|---|---|---|---|---|
| JANUARY | 88 | *** | *** | *** | 7.32 (***) (6.7) |
| FEBRUARY | 88 | *** | *** | *** | 19.82 (***) (11.2) |
| MARCH | 88 | *** | *** | *** | 9.63 (5.5) (13.1) |
| APRIL | 88 | *** | *** | *** | 2.42 (4.4) (5.4) |
| MAY | 88 | *** | *** | *** | 3.71 (5.3) (1) |
| JUNE | 87 | *** | *** | *** | 2.46 (6.7) (0.7) |
| JULY | 87 | *** | *** | *** | 0 (0) (0.3) |
| AUGUST | 87 | *** | *** | *** | 0 (0) (0.3) |
| SEPTEMBER | 89 | *** | *** | *** | 0.1 (2.4) (0.3) |
| OCTOBER | 90 | *** | *** | *** | 1.88 (4.7) (4.1) |
| NOVEMBER | 90 | *** | *** | *** | 7.05 (10.3) (6.6) |
| DECEMBER | 88 | *** | *** | *** | 3.31 (5.4) (13.7) |
| ANNUAL | | | *** | *** | 57.7 (***) (63.1) |

**Map Location 82 — Indonesia, Tjibeureum  0721S 10815E 1200**

| Month | Avg Max | Avg Min | Rec High | Rec Low | Precip |
|---|---|---|---|---|---|
| JANUARY | *** | *** | *** | *** | 17.44 (***) (***) |
| FEBRUARY | *** | *** | *** | *** | 12.22 (***) (***) |
| MARCH | *** | *** | *** | *** | 16.82 (***) (***) |
| APRIL | *** | *** | *** | *** | 10.53 (6) (***) |
| MAY | *** | *** | *** | *** | 6.78 (5.4) (***) |
| JUNE | *** | *** | *** | *** | 4.49 (9) (***) |
| JULY | *** | *** | *** | *** | 17.98 (19.8) (***) |
| AUGUST | *** | *** | *** | *** | 18.45 (20.1) (***) |
| SEPTEMBER | *** | *** | *** | *** | 6.16 (9.5) (***) |
| OCTOBER | *** | *** | *** | *** | 16.92 (***) (***) |
| NOVEMBER | *** | *** | *** | *** | 14.25 (15.5) (***) |
| DECEMBER | *** | *** | *** | *** | 6.08 (9.4) (***) |
| ANNUAL | | | *** | *** | 148.1 (***) (***) |

**Map Location 83 — Iran, Teheran  3541N 05119E 3937**

| Month | Avg Max | Avg Min | Rec High | Rec Low | Precip |
|---|---|---|---|---|---|
| JANUARY | 45 | 27 | 65 | -5 | 1.8 (3.7) (0.2) |
| FEBRUARY | 50 | 32 | 73 | 4 | 1.5 (3.2) (0.4) |
| MARCH | 59 | 39 | 85 | 16 | 1.8 (3.6) (0.9) |
| APRIL | 71 | 49 | 91 | 28 | 1.4 (3) (2.9) |
| MAY | 82 | 58 | 99 | 37 | 0.5 (1.2) (3.6) |
| JUNE | 93 | 66 | 107 | 46 | 0.1 (0.3) (0.5) |
| JULY | 99 | 72 | 109 | 57 | 0.1 (0.3) (1) |
| AUGUST | 97 | 71 | 109 | 52 | 0.1 (0.3) (0.7) |
| SEPTEMBER | 90 | 64 | 102 | 46 | 0.1 (0.3) (0.2) |
| OCTOBER | 76 | 53 | 93 | 38 | 0.3 (0.8) (0.1) |
| NOVEMBER | 63 | 43 | 84 | 19 | 0.8 (1.7) (0.1) |
| DECEMBER | 51 | 33 | 68 | 10 | 1.2 (2.6) (0.2) |
| ANNUAL | | | 109 | -5 | 9.7 (21) (10.8) |

Table 9.2  World Weather Data [CONTINUED]

# Asia, Iraq to Kazakhstan

# Map # 84 to 88

**Map Location 84 — Iraq, Baghdad — 3319N 04421E 112**

| Month | Normal Max | Normal Min | Record Max | Record Min | Greatest Precip | Normal Precip in (cm) (days) |
|---|---|---|---|---|---|---|
| JANUARY | 61 | 39 | 73 | 21 | *** (***) | 1.04 (2.3) (0.9) |
| FEBRUARY | 65 | 43 | 82 | 23 | *** (***) | 1.31 (2.8) (1) |
| MARCH | 73 | 49 | 90 | 27 | *** (***) | 0.85 (2) (2.1) |
| APRIL | 85 | 59 | 108 | 39 | 0 (0) | 0.27 (0.7) (3) |
| MAY | 98 | 68 | 113 | 50 | 0 (0) | 0.05 (0.2) (2.8) |
| JUNE | 106 | 74 | 119 | 61 | 0 (0) | 0 (0) (0.2) |
| JULY | 111 | 78 | 120 | 66 | 0 (0) | 0 (0) (0) |
| AUGUST | 110 | 77 | 120 | 66 | 0 (0) | 0 (0) (0) |
| SEPTEMBER | 104 | 71 | 114 | 58 | 0 (0) | 0.01 (0.1) (0.3) |
| OCTOBER | 93 | 62 | 106 | 44 | 0 (0) | 0.03 (0.2) (0.9) |
| NOVEMBER | 75 | 51 | 95 | 27 | *** (***) | 1.1 (2.2) (1.7) |
| DECEMBER | 63 | 43 | 77 | 25 | *** (***) | 1.02 (2.3) (0.9) |
| ANNUAL | 87 | 60 | 120 | 21 | *** (***) | 5.7 (12.8) (13.8) |

**Map Location 85 — Israel, Tel Aviv — 3206N 03446E 33**

| Month | Normal Max | Normal Min | Record Max | Record Min | Greatest Precip | Normal Precip in (cm) (days) |
|---|---|---|---|---|---|---|
| JANUARY | 64 | 50 | 79 | 36 | 0 (0) | 4.88 (8.1) (3.1) |
| FEBRUARY | 64 | 50 | 88 | 29 | 0 (0) | 2.36 (4.7) (2.5) |
| MARCH | 66 | 52 | 95 | 39 | 0 (0) | 2.16 (4.2) (2.6) |
| APRIL | 71 | 57 | 104 | 43 | 0 (0) | 0.9 (2.1) (0.7) |
| MAY | 74 | 62 | 116 | 50 | *** (***) | 0.14 (0.4) (0.1) |
| JUNE | 79 | 68 | 102 | 57 | 0 (0) | 0 (0) (0) |
| JULY | 82 | 72 | 99 | 64 | 0 (0) | 0 (0) (0) |
| AUGUST | 84 | 73 | 94 | 68 | 0 (0) | 0 (0) (0) |
| SEPTEMBER | 82 | 71 | 96 | 61 | 0 (0) | 0.1 (0.3) (0.1) |
| OCTOBER | 79 | 65 | 101 | 54 | 0 (0) | 0.47 (1.1) (0.7) |
| NOVEMBER | 74 | 59 | 96 | 41 | 0 (0) | 0.45 (6.1) (3.5) |
| DECEMBER | 67 | 53 | 88 | 39 | 0 (0) | 7.19 (9.9) (4.2) |
| ANNUAL | 74 | 61 | 116 | 29 | *** (***) | 22.6 (36.9) (17.5) |

**Map Location 86 — Japan, Tokyo — 3533N 13945E 14**

| Month | Normal Max | Normal Min | Record Max | Record Min | Greatest Precip | Normal Precip in (cm) (days) |
|---|---|---|---|---|---|---|
| JANUARY | 47 | 29 | 72 | 17 | 1.4 (0.2) | 1.9 (4.3) (0.2) |
| FEBRUARY | 48 | 31 | 77 | 18 | 5.4 (0.8) | 2.9 (6.2) (0) |
| MARCH | 54 | 36 | 77 | 22 | 4.2 (6.9) (0) | 2.9 (5.5) (0.3) |
| APRIL | 63 | 46 | 85 | 30 | 5.3 (8.3) (0.3) | 1.2 (2.6) (0.5) |
| MAY | 71 | 54 | 90 | 36 | 5.8 (8.4) (0) | 1.4 (0.2) |
| JUNE | 76 | 63 | 93 | 47 | 6.5 (8.6) (0.1) | 2.9 (6.2) (0) |
| JULY | 83 | 70 | 99 | 55 | 5.6 (7.8) (1.1) | 6 (8.2) (0.8) |
| AUGUST | 86 | 72 | 101 | 60 | 6 (8.2) (0.8) | 9.2 (11.8) (0.2) |
| SEPTEMBER | 79 | 66 | 96 | 51 | 9.2 (11.8) (0.2) | 8.2 (10.8) (0) |
| OCTOBER | 69 | 55 | 90 | 36 | 8.2 (10.8) (0) | 3.8 (5.7) (0) |
| NOVEMBER | 60 | 43 | 81 | 26 | 3.8 (5.7) (0) | 2.2 (4.9) (0) |
| DECEMBER | 52 | 33 | 74 | 20 | 2.2 (4.9) (0) | 61.6 (91.6) (2.5) |
| ANNUAL | 66 | 50 | 101 | 17 | 8.3 (1.2) | 61.6 (91.6) (2.5) |

**Map Location 87 — Jordan, Amman — 3158N 03559E 2547**

| Month | Normal Max | Normal Min | Record Max | Record Min | Greatest Precip | Normal Precip in (cm) (days) |
|---|---|---|---|---|---|---|
| JANUARY | 54 | 39 | 76 | 21 | *** (***) | 2.7 (5.2) (0.4) |
| FEBRUARY | 56 | 40 | 85 | 23 | *** (***) | 2.9 (5.5) (0.3) |
| MARCH | 60 | 43 | 90 | 26 | *** (***) | 1.2 (2.6) (0.5) |
| APRIL | 73 | 49 | 103 | 34 | 0 (0) | 0.6 (1.4) (0.7) |
| MAY | 83 | 57 | 105 | 39 | 0 (0) | 0.2 (0.5) (0.3) |
| JUNE | 87 | 61 | 109 | 46 | 0 (0) | 0 (0) (0) |
| JULY | 89 | 65 | 106 | 54 | 0 (0) | 0 (0) (0) |
| AUGUST | 90 | 65 | 109 | 55 | 0 (0) | 0 (0) (0) |
| SEPTEMBER | 88 | 62 | 103 | 52 | 0 (0) | 0.03 (0.2) (0.2) |
| OCTOBER | 81 | 57 | 99 | 44 | 0 (0) | 0.2 (0.5) (0.9) |
| NOVEMBER | 70 | 50 | 91 | 27 | *** (***) | 1.3 (2.6) (0.9) |
| DECEMBER | 59 | 42 | 77 | 25 | *** (***) | 1.8 (3.7) (0.5) |
| ANNUAL | 74 | 53 | 109 | 21 | *** (***) | 10.9 (22.2) (4.7) |

**Map Location 88 — Kazakhstan, Taraz — 4251N 07123E 2106**

| Month | Normal Max | Normal Min | Record Max | Record Min | Greatest Precip | Normal Precip in (cm) (days) |
|---|---|---|---|---|---|---|
| JANUARY | 32 | 13 | 70 | -42 | *** (***) | 1.17 (6.5) (0.1) |
| FEBRUARY | 37 | 17 | 79 | -40 | *** (***) | 1.82 (9.6) (0.1) |
| MARCH | 49 | 29 | 93 | -15 | *** (***) | 1.44 (7.9) (0.4) |
| APRIL | 64 | 40 | 97 | 16 | *** (***) | 1.87 (7.9) (1.6) |
| MAY | 75 | 49 | 108 | 27 | *** (***) | 1.85 (7.5) (5.2) |
| JUNE | 83 | 55 | 108 | 34 | *** (***) | 0.83 (3.7) (6.5) |
| JULY | 89 | 58 | 111 | 41 | *** (***) | 0.42 (3.8) (4.1) |
| AUGUST | 86 | 54 | 106 | 34 | *** (***) | 0.48 (1.5) (1.2) |
| SEPTEMBER | 76 | 44 | 104 | 23 | *** (***) | 0.18 (2) (0.5) |
| OCTOBER | 63 | 35 | 97 | 3 | *** (***) | 0.74 (4.6) (0.4) |
| NOVEMBER | 47 | 24 | 82 | -35 | *** (***) | 2.19 (6.7) (0.1) |
| DECEMBER | 36 | 17 | 79 | -42 | *** (***) | 1.89 (8.8) (0) |
| ANNUAL | 61 | 36 | 111 | -42 | *** (***) | 14.9 (70.5) (20.2) |

Table 9.2  World Weather Data [CONTINUED]

# Asia, Korea, North to Laos

## Map # 89 to 93

### Map Location 89 — Korea, North, Pyongyang 3901N

| | JAN | FEB | MAR | APR | MAY | JUNE | JULY | AUG | SEP | OCT | NOV | DEC | ANNUAL |
|---|---|---|---|---|---|---|---|---|---|---|---|---|---|
| Avg High | 28 | 35 | 45 | 63 | 75 | 79 | 83 | 84 | 77 | 65 | 49 | 34 | 60 |
| Rec High | 48 | 55 | 66 | 82 | 90 | 93 | 93 | 100 | 90 | 82 | 68 | 59 | 100 |
| Avg Low | 10 | 16 | 27 | 40 | 52 | 61 | 70 | 70 | 58 | 44 | 32 | 19 | 42 |
| Rec Low | -15 | -9 | 3 | 23 | 36 | 50 | 54 | 37 | 39 | 27 | 7 | -9 | -15 |
| Precip | 0.58 (1.8) (0) | 0.51 (1.5) (0) | 1.15 (2.9) (0.3) | 2.74 (3.9) (0.8) | 1.95 (3.3) (1.6) | 4.19 (5.7) (4.1) | 14.68 (11.6) (5.6) | 10.21 (9.4) (5.8) | 4.47 (4.6) (2) | 1.48 (3.2) (1.4) | 1.43 (3.7) (0.8) | 0.8 (2.2) (0.1) | 44.2 (53.8) (22.5) *** (***) |

### Map Location 90 — Korea, South, Seoul 3731N 12655E 34

| | JAN | FEB | MAR | APR | MAY | JUNE | JULY | AUG | SEP | OCT | NOV | DEC | ANNUAL |
|---|---|---|---|---|---|---|---|---|---|---|---|---|---|
| Avg High | 32 | 37 | 47 | 62 | 72 | 80 | 84 | 87 | 78 | 67 | 51 | 37 | 61 |
| Rec High | 54 | 61 | 73 | 85 | 90 | 99 | 98 | 99 | 91 | 86 | 74 | 67 | 99 |
| Avg Low | 15 | 20 | 29 | 41 | 51 | 61 | 70 | 71 | 59 | 45 | 32 | 20 | 43 |
| Rec Low | -8 | -3 | 5 | 24 | 36 | 49 | 55 | 58 | 38 | 25 | 11 | -12 | -12 |
| Precip | 1.2 (2.9) (0) | 0.8 (2.1) (0.1) | 1.5 (3) (0) | 3 (5.5) (0.7) | 3.2 (5.7) (0.7) | 5.1 (7.4) (1.5) | 14.8 (13) (2.9) | 10.5 (11.3) (1.6) | 4.7 (6.8) (0.8) | 1.6 (2.9) (0.6) | 1.8 (3.2) (0.6) | 1 (2.5) (0) | 49.2 (66.3) (9.5) 10.8 (2.7) |

### Map Location 91 — Kuwait, Kuwait City 2914N 04758E 182

| | JAN | FEB | MAR | APR | MAY | JUNE | JULY | AUG | SEP | OCT | NOV | DEC | ANNUAL |
|---|---|---|---|---|---|---|---|---|---|---|---|---|---|
| Avg High | 67 | 68 | 80 | 87 | 102 | 111 | 113 | 113 | 107 | 96 | 82 | 71 | 91 |
| Rec High | 81 | 77 | 97 | 104 | 113 | 120 | 120 | 120 | 117 | 108 | 95 | 79 | 120 |
| Avg Low | 46 | 51 | 57 | 65 | 75 | 82 | 85 | 84 | 79 | 69 | 59 | 48 | 67 |
| Rec Low | 27 | 37 | 48 | 50 | 63 | 75 | 79 | 75 | 68 | 54 | 37 | 30 | 27 |
| Precip | 0.58 (1.4) (0) | 1.19 (2.6) (0.9) | 1.26 (2.7) (0.2) | 1.31 (2.8) (1.6) | 0.5 (1.2) (1) | 0 (0) (0) | 0 (0) (0.2) | 0 (0) (0) | 0 (0) (0) | 0 (0) (0.2) | 0.36 (0.9) (1.3) | 0.15 (0.5) (0.8) | 5.3 (12.1) (6.2) *** (***) |

### Map Location 92 — Kyrgyzstan, Bishkek 5252N 07436E 1288

| | JAN | FEB | MAR | APR | MAY | JUNE | JULY | AUG | SEP | OCT | NOV | DEC | ANNUAL |
|---|---|---|---|---|---|---|---|---|---|---|---|---|---|
| Avg High | 34 | 37 | 46 | 59 | 70 | 76 | 80 | 78 | 68 | 58 | 44 | 35 | *** |
| Rec High | *** | *** | *** | *** | *** | *** | *** | *** | *** | *** | *** | *** | *** |
| Avg Low | 22 | 23 | 30 | 39 | 48 | 55 | 60 | 58 | 49 | 39 | 33 | 23 | *** |
| Rec Low | *** | *** | *** | *** | *** | *** | *** | *** | *** | *** | *** | *** | *** |
| Precip | *** (***) (***) | *** (***) (***) | *** (***) (***) | *** (***) (***) | *** (***) (***) | *** (***) (***) | *** (***) (***) | *** (***) (***) | *** (***) (***) | *** (***) (***) | *** (***) (***) | *** (***) (***) | *** (***) (***) |

### Map Location 93 — Laos, Luang-Prabang 1953N 10208E 997

| | JAN | FEB | MAR | APR | MAY | JUNE | JULY | AUG | SEP | OCT | NOV | DEC | ANNUAL |
|---|---|---|---|---|---|---|---|---|---|---|---|---|---|
| Avg High | 82 | 89 | 93 | 96 | 95 | 93 | 90 | 90 | 91 | 89 | 85 | 81 | 90 |
| Rec High | 103 | 102 | 106 | 113 | 111 | 104 | 102 | 104 | 102 | 101 | 97 | 91 | 113 |
| Avg Low | 56 | 58 | 63 | 69 | 73 | 74 | 74 | 74 | 73 | 69 | 64 | 59 | 67 |
| Rec Low | 32 | 45 | 48 | 57 | 54 | 57 | 59 | 57 | 51 | 54 | 43 | 34 | 32 |
| Precip | 0.6 (1.3) (0.3) | 0.7 (1.5) (1) | 1.2 (2.7) (2) | 4.3 (5.5) (4) | 6.4 (5.4) (9) | 6.1 (10.7) (5) | 9.1 (13.6) (5) | 11.8 (15.8) (5) | 6.5 (9.8) (4) | 3.1 (6.2) (1) | 1.2 (3.9) (0.3) | 0.5 (1.1) (0.3) | 51.5 (77.5) (36.9) 0 (0) |

**Asia, Lebanon to Mongolia, Altai**

Table 9.2   World Weather Data [CONTINUED]

**Map # 94 to 98**

**Map Location 94 — Lebanon, Beirut — 3348N 03529E 86**

| | JANUARY | FEBRUARY | MARCH | APRIL | MAY | JUNE | ANNUAL |
|---|---|---|---|---|---|---|---|
| Record High | 77 | 87 | 97 | 99 | 107 | 104 | 107 |
| Normal High | 62 | 63 | 66 | 72 | 78 | 83 | |
| Normal Low | 51 | 51 | 54 | 58 | 64 | 69 | |
| Record Low | 31 | 30 | 36 | 43 | 32 | 50 | 30 |
| Precip | 7.5 (10) (9.4) | 6.2 (9.3) (7) | 3.7 (5.8) (6.5) | 2.2 (4.2) (3.2) | 0.7 (1.7) (2) | 0.1 (0.3) (0.3) | 35.2 (52.1) (53.7) |
| Snow | 0 (0) | 0 (0) | 0 (0) | 0 (0) | 0 (0) | 0 (0) | 0 (0) |

| | JULY | AUGUST | SEPTEMBER | OCTOBER | NOVEMBER | DECEMBER |
|---|---|---|---|---|---|---|
| Record High | 98 | 99 | 99 | 101 | 91 | 84 |
| Normal High | 87 | 89 | 86 | 81 | 73 | 65 |
| Normal Low | 73 | 74 | 73 | 69 | 61 | 55 |
| Record Low | 59 | 62 | 60 | 46 | 34 | 30 |
| Precip | 0.03 (0.1) (0) | 0.03 (0.1) (0) | 0.2 (0.5) (1.3) | 2 (3.6) (5.4) | 5.2 (6.6) (9.6) | 7.3 (9.9) (9) |
| Snow | 0 (0) | 0 (0) | 0 (0) | 0 (0) | 0 (0) | 0 (0) |

**Map Location 95 — Malaysia, Kota Kinabalu — 0556N 11603E 9**

| | JANUARY | FEBRUARY | MARCH | APRIL | MAY | JUNE | ANNUAL |
|---|---|---|---|---|---|---|---|
| Record High | 90 | 89 | 92 | 91 | 92 | 94 | 94 |
| Normal High | 85 | 85 | 87 | 88 | 88 | 88 | |
| Normal Low | 73 | 73 | 73 | 75 | 75 | 75 | |
| Record Low | 67 | 67 | 66 | 70 | 72 | 71 | 66 |
| Precip | 4.77 (6.07) (1.8) | 2.28 (4.1) (2.7) | 3.39 (5.1) (4.9) | 5.88 (5.5) (11.3) | 9.27 (5.3) (14.6) | 12 (15.9) (8.3) | 107.6 (***) (78.1) |
| Snow | 0 (0) | 0 (0) | 0 (0) | 0 (0) | 0 (0) | 0 (0) | 0 (0) |

| | JULY | AUGUST | SEPTEMBER | OCTOBER | NOVEMBER | DECEMBER |
|---|---|---|---|---|---|---|
| Record High | 91 | 92 | 90 | 89 | 95 | 89 |
| Normal High | 87 | 87 | 87 | 86 | 86 | 86 |
| Normal Low | 74 | 74 | 74 | 74 | 74 | 74 |
| Record Low | 70 | 69 | 71 | 71 | 69 | 71 |
| Precip | 10.01 (14.3) (4.6) | 13.05 (14.8) (6.5) | 14.17 (15.4) (6.4) | 11.19 (13.6) (14.5) | 12.05 (14.2) (4.7) | 10.2 (***) (3.9) |
| Snow | 0 (0) | 0 (0) | 0 (0) | 0 (0) | 0 (0) | 0 (0) |

**Map Location 96 — Malaysia, Kuala Lumpur — 0308N**

| | JANUARY | FEBRUARY | MARCH | APRIL | MAY | JUNE | ANNUAL |
|---|---|---|---|---|---|---|---|
| Record High | 96 | 99 | 97 | 95 | 97 | 97 | 99 |
| Normal High | 90 | 91 | 92 | 92 | 91 | 91 | |
| Normal Low | 73 | 72 | 73 | 74 | 75 | 73 | |
| Record Low | 66 | 66 | 66 | 71 | 70 | 69 | 66 |
| Precip | 6.76 (7.5) (9.1) | 4.78 (6.7) (6.6) | 10.43 (5.9) (16.2) | 11.52 (7) (21.9) | 9.02 (5.3) (12.9) | 5.35 (10) (11.7) | 105.7 (***) (154.4) |
| Snow | 0 (0) | 0 (0) | 0 (0) | 0 (0) | 0 (0) | 0 (0) | 0 (0) |

| | JULY | AUGUST | SEPTEMBER | OCTOBER | NOVEMBER | DECEMBER |
|---|---|---|---|---|---|---|
| Record High | 95 | 95 | 94 | 93 | 93 | 94 |
| Normal High | 90 | 90 | 89 | 89 | 89 | 88 |
| Normal Low | 70 | 73 | 73 | 73 | 73 | 72 |
| Record Low | 69 | 68 | 70 | 65 | 64 | 64 |
| Precip | 6.03 (11.9) (9.7) | 8.2 (11.3) (13.3) | 13.89 (15.3) (15.9) | 11.32 (***) (12.8) | 9.7 (5.5) (1.4) | 13.1 (14.8) (13.1) |
| Snow | 0 (0) | 0 (0) | 0 (0) | 0 (0) | 0 (0) | 0 (0) |

**Map Location 97 — Malaysia, Kuching — 0129N 11020E 85**

| | JANUARY | FEBRUARY | MARCH | APRIL | MAY | JUNE | ANNUAL |
|---|---|---|---|---|---|---|---|
| Record High | 93 | 93 | 93 | 96 | 96 | 96 | 97 |
| Normal High | 88 | 88 | 89 | 90 | 91 | 91 | |
| Normal Low | 72 | 72 | 72 | 72 | 72 | 73 | |
| Record Low | 65 | 67 | 65 | 69 | 66 | 66 | 64 |
| Precip | 27.1 (***) (6.4) | 19.7 (***) (7.2) | 14.2 (***) (10) | 13.1 (14.8) (13.1) | 9.7 (5.5) (1.4) | 8.5 (13) (10.5) | 159.2 (***) (140) |
| Snow | 0 (0) | 0 (0) | 0 (0) | 0 (0) | 0 (0) | 0 (0) | 0 (0) |

| | JULY | AUGUST | SEPTEMBER | OCTOBER | NOVEMBER | DECEMBER |
|---|---|---|---|---|---|---|
| Record High | 97 | 99 | 94 | 93 | 93 | 97 |
| Normal High | 92 | 92 | 90 | 90 | 88 | 90 |
| Normal Low | 72 | 72 | 72 | 72 | 72 | 72 |
| Record Low | 66 | 64 | 68 | 64 | 64 | 64 |
| Precip | 6.9 (11.5) (10.5) | 8.8 (13.3) (12.2) | 9.5 (12.4) (15.7) | 12.6 (14.5) (16.9) | 20.1 (***) (9) | 9.27 (5.3) (14.5) |
| Snow | 0 (0) | 0 (0) | 0 (0) | 0 (0) | 0 (0) | 0 (0) |

**Map Location 98 — Mongolia, Altai — 4624N 09615E 7044**

| | JANUARY | FEBRUARY | MARCH | APRIL | MAY | JUNE | ANNUAL |
|---|---|---|---|---|---|---|---|
| Record High | 27 | 46 | 50 | 72 | 73 | 81 | 84 |
| Normal High | 8 | 15 | 30 | 42 | 54 | 65 | |
| Normal Low | -14 | -11 | 5 | -2 | 32 | 43 | |
| Record Low | -35 | -38 | -20 | -31 | 7 | 21 | -38 |
| Precip | 0 (0) (0) | 0.31 (0.8) (0) | 0.1 (0) (0) | 0 (0) (0.2) | 0.75 (3) (0.5) | 0.7 (5.2) (1.4) | 6.9 (19.9) (11.3) |
| Snow | *** (***) | *** (***) | *** (***) | *** (***) | *** (***) | *** (***) | *** (***) |

| | JULY | AUGUST | SEPTEMBER | OCTOBER | NOVEMBER | DECEMBER |
|---|---|---|---|---|---|---|
| Record High | 84 | 84 | 72 | 63 | 48 | 17 |
| Normal High | 65 | 65 | 56 | 39 | 25 | 17 |
| Normal Low | 47 | 44 | 34 | 18 | 3 | -6 |
| Record Low | 32 | 30 | 18 | -18 | -31 | -33 |
| Precip | 0.16 (0) (6.4) | 3.92 (8.2) (2.6) | 0.31 (0.8) (0) | 0.33 (1) (0.2) | 0.36 (0.9) (0) | 0 (0) (0) |
| Snow | 0 (0) | 0 (0) | 0* (0*) | *** (***) | *** (***) | *** (***) |

**Table 9.2 World Weather Data [CONTINUED]**

# Asia, Mongolia, Ulan Bator to Pakistan

## Map # 99 to 103

### Map Location 99 — Mongolia, Ulaanbaatar — 4751N 10645E 4157

| | JANUARY | FEBRUARY | MARCH | APRIL | MAY | JUNE |
|---|---|---|---|---|---|---|
| Record High | 25 | 39 | 61 | 75 | 86 | 93 |
| Normal High | 0 | 12 | 30 | 47 | 61 | 70 |
| Normal Low | -25 | -18 | 1 | 19 | 31 | 44 |
| Record Low | -47 | -44 | -38 | -8 | 3 | 25 |
| Precip | 0.03 (0) (0) | 0.16 (0.5) (0) | 0.22 (0.2) (0) | 0.17 (0.5) (0) | 0.46 (1.2) (0.6) | 2.06 (3.4) (3.8) |

| | JULY | AUGUST | SEPTEMBER | OCTOBER | NOVEMBER | DECEMBER | ANNUAL |
|---|---|---|---|---|---|---|---|
| Record High | 91 | 99 | 81 | 70 | 52 | 31 | 99 |
| Normal High | 71 | 70 | 60 | 45 | 20 | 6 | 41 |
| Normal Low | 50 | 46 | 32 | 17 | -5 | -20 | 14 |
| Record Low | 37 | 27 | 10 | -4 | -31 | -44 | -47 |
| Precip | 3.65 (8.2) (4.1) | 1.44 (4.2) (4.9) | 0.85 (2.4) (1.3) | 1.7 (1.1) (0) | 0.26 (1) (0) | 0.19 (0.2) (0) | 11.2 (22.9) (14.7) |

### Map Location 100 — Myanmar (Burma), Naypyidaw — 1943N 09613E 340

| | JANUARY | FEBRUARY | MARCH | APRIL | MAY | JUNE |
|---|---|---|---|---|---|---|
| Record High | 93 | 100 | 104 | 106 | 109 | 102 |
| Normal High | 87 | 91 | 97 | 100 | 96 | 88 |
| Normal Low | 59 | 62 | 68 | 76 | 77 | 75 |
| Record Low | 48 | 52 | 55 | 68 | 70 | 72 |
| Precip | 0.08 (0.7) (***) | 0.03 (0) (***) | 0 (0) (***) | 0.49 (1.4) (***) | 3.81 (8.5) (***) | 13.73 (17.2) (***) |

| | JULY | AUGUST | SEPTEMBER | OCTOBER | NOVEMBER | DECEMBER | ANNUAL |
|---|---|---|---|---|---|---|---|
| Record High | 99 | 99 | 97 | 99 | 93 | 91 | 109 |
| Normal High | 87 | 87 | 88 | 89 | 88 | 86 | 90 |
| Normal Low | 74 | 74 | 75 | 74 | 68 | 60 | 70 |
| Record Low | 68 | 68 | 72 | 68 | 57 | 50 | 48 |
| Precip | 7.51 (13.4) (***) | 12.97 (17.6) (***) | 12.67 (15.9) (***) | 4.12 (7.6) (***) | 0.52 (1) (***) | 0.05 (0) (***) | 56 (83.3) (***) |

### Map Location 101 — Nepal, Katmandu — 2742N 08522E 4423

| | JANUARY | FEBRUARY | MARCH | APRIL | MAY | JUNE |
|---|---|---|---|---|---|---|
| Record High | 76 | 83 | 92 | 95 | 97 | 99 |
| Normal High | 65 | 68 | 77 | 84 | 85 | 85 |
| Normal Low | 36 | 40 | 45 | 53 | 60 | 67 |
| Record Low | 27 | 30 | 35 | 40 | 49 | 57 |
| Precip | 0.6 (1) (***) | 1.6 (5) (***) | 0.9 (2) (***) | 2.3 (6) (***) | 4.8 (10) (***) | 9.7 (15) (***) |

| | JULY | AUGUST | SEPTEMBER | OCTOBER | NOVEMBER | DECEMBER | ANNUAL |
|---|---|---|---|---|---|---|---|
| Record High | 91 | 92 | 92 | 92 | 85 | 73 | 99 |
| Normal High | 84 | 83 | 83 | 80 | 73 | 66 | 78 |
| Normal Low | 69 | 68 | 66 | 56 | 46 | 38 | 54 |
| Record Low | 61 | 61 | 56 | 42 | 33 | 29 | 27 |
| Precip | 14.7 (21) (***) | 13.6 (20) (***) | 6.1 (12) (***) | 1.5 (4) (***) | 0.3 (1) (***) | 0.1 (0.2) (***) | 56.2 (97.2) (***) |

### Map Location 102 — Oman, Muscat — 2345N 05835E 20

| | JANUARY | FEBRUARY | MARCH | APRIL | MAY | JUNE |
|---|---|---|---|---|---|---|
| Record High | 89 | 80 | 94 | 104 | 110 | 110 |
| Normal High | 73 | 74 | 78 | 88 | 96 | 95 |
| Normal Low | 66 | 67 | 73 | 79 | 85 | 88 |
| Record Low | 60 | 59 | 63 | 64 | 73 | 73 |
| Precip | 0.96 (1) (***) | 0.5 (1.2) (***) | 0.34 (1.2) (***) | 0 (0) (***) | 0.07 (0.3) (***) | 0 (0) (***) |

| | JULY | AUGUST | SEPTEMBER | OCTOBER | NOVEMBER | DECEMBER | ANNUAL |
|---|---|---|---|---|---|---|---|
| Record High | 105 | 101 | 99 | 100 | 110 | 110 | 110 |
| Normal High | 93 | 90 | 87 | 87 | 82 | 76 | 85 |
| Normal Low | 85 | 84 | 82 | 79 | 74 | 69 | 78 |
| Record Low | 70 | 71 | 75 | 70 | 64 | 60 | 59 |
| Precip | 0.02 (0) (***) | 0 (0) (***) | 0 (0) (***) | 0 (0) (***) | 0 (0) (***) | 0.91 (0.9) (***) | 2.8 (4.6) (***) |

### Map Location 103 — Pakistan, Karachi — 2454N 06709E 95

| | JANUARY | FEBRUARY | MARCH | APRIL | MAY | JUNE |
|---|---|---|---|---|---|---|
| Record High | 89 | 99 | 107 | 111 | 118 | 115 |
| Normal High | 77 | 79 | 85 | 90 | 93 | 93 |
| Normal Low | 55 | 58 | 67 | 73 | 79 | 82 |
| Record Low | 39 | 41 | 47 | 56 | 63 | 68 |
| Precip | 0.5 (1.2) (0.2) | 0.4 (1) (0.5) | 0.1 (0.3) (0.1) | 0.1 (0.3) (0) | 0.1 (0.3) (0) | 0.7 (1.3) (0.7) |

| | JULY | AUGUST | SEPTEMBER | OCTOBER | NOVEMBER | DECEMBER | ANNUAL |
|---|---|---|---|---|---|---|---|
| Record High | 110 | 102 | 108 | 108 | 102 | 94 | 118 |
| Normal High | 91 | 88 | 91 | 91 | 87 | 80 | 87 |
| Normal Low | 81 | 79 | 77 | 72 | 64 | 57 | 70 |
| Record Low | 72 | 72 | 68 | 55 | 48 | 39 | 39 |
| Precip | 3.2 (5.2) (3.1) | 1.6 (2.8) (1.8) | 0.5 (1.1) (1.4) | 0.1 (0.3) (0.2) | 0.1 (0.3) (0.1) | 0.2 (0.6) (0.4) | 7.8 (15.2) (9.3) |

**Table 9.2   World Weather Data [CONTINUED]**

# Asia, Philippines to Russia, Dikson

## Map # 104 to 108

*(For each month the cell group shows, in order: record high / record low / snow, then mean high / mean low / precipitation in (mm) (days). Values read from the rotated table.)*

### Map Location 104 — Philippines, Baguio · 1625N 12035E 4962

| Month | Rec High | Rec Low | Snow | Mean High | Mean Low | Precipitation |
|---|---|---|---|---|---|---|
| JANUARY | 79 | 47 | 0 (0) | 72 | 55 | 0.9 (2.3) (0) |
| FEBRUARY | 81 | 47 | 0 (0) | 73 | 56 | 0.9 (2.3) (0.3) |
| MARCH | 82 | 52 | 0 (0) | 76 | 58 | 1.7 (3.4) (1) |
| APRIL | 84 | 50 | 0 (0) | 77 | 60 | 4.3 (7) (3) |
| MAY | 81 | 57 | 0 (0) | 76 | 61 | 15.8 (16) (6) |
| JUNE | 80 | 53 | 0 (0) | 75 | 61 | 17.2 (13.6*) (4) |
| JULY | 80 | 54 | 0 (0) | 71 | 60 | 42.3 (***) (3) |
| AUGUST | 80 | 55 | 0 (0) | 71 | 60 | 45.7 (***) (2) |
| SEPTEMBER | 79 | 57 | 0 (0) | 71 | 60 | 28.1 (***) (2) |
| OCTOBER | 81 | 52 | 0 (0) | 73 | 60 | 15 (16.2) (1) |
| NOVEMBER | 80 | 49 | 0 (0) | 74 | 59 | 4.9 (7) (1) |
| DECEMBER | 80 | 46 | 0 (0) | 74 | 59 | 2 (4.5) (0.3) |
| ANNUAL | 84 | 46 | 0 (0) | 75 | 59 | 178.8 (***) (23.6) |

### Map Location 105 — Philippines, Manila · 1431N 12100E 49

| Month | Rec High | Rec Low | Snow | Mean High | Mean Low | Precipitation |
|---|---|---|---|---|---|---|
| JANUARY | 96 | 58 | 0 (0) | 86 | 69 | 0.9 (2.3) (0.2) |
| FEBRUARY | 96 | 58 | 0 (0) | 88 | 69 | 0.5 (1.4) (0.1) |
| MARCH | 98 | 61 | 0 (0) | 91 | 71 | 0.7 (1.4) (0.2) |
| APRIL | 100 | 62 | 0 (0) | 93 | 73 | 1.3 (2.7) (0.5) |
| MAY | 101 | 67 | 0 (0) | 93 | 75 | 5.1 (7.8) (4.8) |
| JUNE | 101 | 68 | 0 (0) | 91 | 75 | 10 (11) (8.5) |
| JULY | 97 | 67 | 0 (0) | 88 | 75 | 16.6 (13.5) (8.8) |
| AUGUST | 95 | 68 | 0 (0) | 87 | 75 | 16.6 (13.5) (8.8) |
| SEPTEMBER | 95 | 66 | 0 (0) | 88 | 74 | 14 (15.7) (7.9) |
| OCTOBER | 95 | 65 | 0 (0) | 88 | 72 | 7.6 (10.2) (6.4) |
| NOVEMBER | 93 | 62 | 0 (0) | 87 | 72 | 5.7 (8) (1.2) |
| DECEMBER | 94 | 60 | 0 (0) | 86 | 70 | 2.6 (5.6) (0.4) |
| ANNUAL | 101 | 58 | 0 (0) | 89 | 73 | 82 (93.2) (49.1) |

### Map Location 106 — Qatar, Doha · 2516N 05133E 33

| Month | Rec High | Rec Low | Snow | Mean High | Mean Low | Precipitation |
|---|---|---|---|---|---|---|
| JANUARY | 96 | 36 | 0 (0) | 70 | 53 | 0.85 (1.9) (0.1) |
| FEBRUARY | 92 | 34 | 0 (0) | 72 | 54 | 0.48 (1.2) (0.1) |
| MARCH | 100 | 43 | 0 (0) | 79 | 60 | 0.46 (1.1) (0.3) |
| APRIL | 110 | 52 | 0 (0) | 90 | 68 | 0.13 (0.4) (0.1) |
| MAY | 116 | 63 | 0 (0) | 99 | 77 | 0.09 (0.3) (0.1) |
| JUNE | 119 | 67 | 0 (0) | 106 | 82 | 0 (0) (0) |
| JULY | 119 | 70 | 0 (0) | 108 | 85 | 0 (0) (0.1) |
| AUGUST | 120 | 72 | 0 (0) | 108 | 84 | 0 (0) (0) |
| SEPTEMBER | 116 | 70 | 0 (0) | 103 | 79 | 0 (0) (0) |
| OCTOBER | 112 | 54 | 0 (0) | 95 | 72 | 0 (0) (0) |
| NOVEMBER | 99 | 48 | 0 (0) | 84 | 65 | 0.18 (0.5) (0.1) |
| DECEMBER | 87 | 40 | 0 (0) | 73 | 56 | 0.82 (1.9) (0.1) |
| ANNUAL | 120 | 34 | 0 (0) | 91 | 70 | 3 (7.3) (1) |

### Map Location 107 — Russia, Astrakhan · 4616N 04802E 59

| Month | Rec High | Rec Low | Snow | Mean High | Mean Low | Precipitation |
|---|---|---|---|---|---|---|
| JANUARY | 55 | -24 | *** (***) | 26 | 13 | 0.54 (6.3) (0) |
| FEBRUARY | 63 | -27 | *** (***) | 29 | 15 | 0.5 (3.9) (0) |
| MARCH | 75 | -17 | *** (***) | 41 | 24 | 0.37 (4.5) (0) |
| APRIL | 90 | 16 | *** (***) | 60 | 40 | 0.35 (2.3) (0.4) |
| MAY | 99 | 30 | 0 (0) | 74 | 54 | 0.53 (2.8) (0.8) |
| JUNE | 102 | 43 | 0 (0) | 83 | 64 | 0.68 (3.1) (2.7) |
| JULY | 104 | 52 | 0 (0*) | 88 | 68 | 0.41 (2.6) (2.7) |
| AUGUST | 102 | 43 | 0* (0*) | 85 | 64 | 1.07 (3) (3.3) |
| SEPTEMBER | 97 | 28 | *** (***) | 74 | 54 | 0.79 (2.9) (1) |
| OCTOBER | 86 | 16 | *** (***) | 59 | 41 | 0.88 (3.1) (0.1) |
| NOVEMBER | 70 | -15 | *** (***) | 43 | 30 | 0.71 (4.6) (0.1) |
| DECEMBER | 61 | -29 | *** (***) | 31 | 21 | 0.93 (7.7) (0) |
| ANNUAL | 104 | -29 | *** (***) | 58 | 41 | 7.8 (46.8) (11.1) |

### Map Location 108 — Russia, Dikson · 7330N 08014E 66

| Month | Rec High | Rec Low | Snow | Mean High | Mean Low | Precipitation |
|---|---|---|---|---|---|---|
| JANUARY | 23 | -51 | *** (***) | -14 | -23 | 1.65 (14.2) (0) |
| FEBRUARY | 25 | -51 | *** (***) | -12 | -23 | 1.57 (12.9) (0) |
| MARCH | 28 | -45 | *** (***) | -5 | -15 | 1.26 (14.4) (0) |
| APRIL | 32 | -36 | *** (***) | 5 | -6 | 1.12 (11.4) (0) |
| MAY | 43 | -20 | *** (***) | 18 | 9 | 0.85 (11.7) (0) |
| JUNE | 54 | 3 | *** (***) | 34 | 27 | 1.3 (13.2) (0) |
| JULY | 81 | 27 | *** (***) | 46 | 37 | 1.56 (8.4) (0.8) |
| AUGUST | 75 | 28 | *** (***) | 45 | 38 | 2.23 (12.8) (0.2) |
| SEPTEMBER | 63 | 19 | *** (***) | 37 | 32 | 2.72 (15.5) (0) |
| OCTOBER | 39 | -17 | *** (***) | 20 | 14 | 1.42 (17.4) (0) |
| NOVEMBER | 32 | -45 | *** (***) | 1 | -8 | 1.07 (13.2) (0) |
| DECEMBER | 14 | -45 | *** (***) | -5 | -14 | 1.2 (14.4) (0) |
| ANNUAL | 81 | -51 | *** (***) | 14 | 6 | 17.9 (159.9) (1) |

**Table 9.2  World Weather Data [CONTINUED]**

# Asia, Russia, Irktusk to Murmansk

## Map # 109 to 113

Each month cell: Extreme High / Extreme Low; Normal High / Normal Low; Precipitation (days) (snow)

### Map Location 109 — Russia, Irktusk — 5216N 10421E 1591

| Month | Ext High / Low | Norm High / Low | Precip (days) (snow) |
|---|---|---|---|
| JANUARY | 30 / -47 | 8 / -10 | 0.48 (8.2) (0) |
| FEBRUARY | 43 / -36 | 15 / -8 | 0.47 (6) (0) |
| MARCH | 55 / -31 | 30 / 6 | 0.5 (5.1) (0) |
| APRIL | 73 / -18 | 44 / 23 | 0.76 (5.5) (0) |
| MAY | 90 / 19 | 61 / 35 | 1.22 (6.8) (0.7) |
| JUNE | 90 / 27 | 70 / 46 | 2.46 (9.5) (3.1) |
| JULY | 93 / 39 | 74 / 52 | 4.64 (11.3) (8.1) |
| AUGUST | 90 / 32 | 69 / 49 | 4.16 (10.1) (4) |
| SEPTEMBER | 81 / 18 | 58 / 36 | 1.57 (7.2) (0.8) |
| OCTOBER | 70 / -4 | 43 / 24 | 1.19 (6.7) (0) |
| NOVEMBER | 54 / -38 | 22 / 5 | 0.72 (9.1) (0) |
| DECEMBER | 36 / -45 | 8 / -8 | 0.86 (11.7) (0) |
| ANNUAL | 93 / -47 | 42 / 21 | 19 (97.2) (16.7) |

### Map Location 110 — Russia, Kamchatka — 5906N 15959E 89

| Month | Ext High / Low | Norm High / Low | Precip (days) (snow) |
|---|---|---|---|
| JANUARY | 43 / -51 | 11 / -6 | 1.54 (11.8) (0) |
| FEBRUARY | 43 / -45 | 8 / -6 | 0.89 (6.9) (0.1) |
| MARCH | 46 / -38 | 21 / 1 | 1.31 (8.3) (0.1) |
| APRIL | 54 / -17 | 31 / 14 | 1.55 (9.7) (0.6) |
| MAY | 63 / 0 | 45 / 30 | 1.3 (8) (0) |
| JUNE | 77 / 23 | 55 / 37 | 0.98 (4.6) (0.1) |
| JULY | 82 / 28 | 62 / 43 | 2.93 (9.8) (0.3) |
| AUGUST | 79 / 21 | 59 / 43 | 2.82 (9.7) (0.1) |
| SEPTEMBER | 68 / 12 | 52 / 34 | 2.46 (9.8) (0.1) |
| OCTOBER | 57 / 10 | 38 / 26 | 3.56 (15) (0.3) |
| NOVEMBER | 50 / -27 | 23 / 10 | 1.55 (9.7) (0.6) |
| DECEMBER | 50 / -40 | 15 / 1 | 2.06 (12.3) (0.1) |
| ANNUAL | 82 / -51 | 35 / 19 | 24.9 (123) (2.1) |

### Map Location 111 — Russia, Khanty-Mansiysk — 6058N 06904E 131

| Month | Ext High / Low | Norm High / Low | Precip (days) (snow) |
|---|---|---|---|
| JANUARY | 37 / -51 | 3 / -11 | 1.47 (13.1) (0) |
| FEBRUARY | 46 / -51 | 8 / -7 | 1.16 (10.1) (0) |
| MARCH | 55 / -38 | 20 / 2 | 1.02 (9.5) (0) |
| APRIL | 73 / -22 | 39 / 23 | 0.69 (7.8) (0.2) |
| MAY | 93 / -2 | 53 / 36 | 1.37 (7.6) (1.6) |
| JUNE | 93 / 27 | 67 / 50 | 2.03 (8.8) (4.1) |
| JULY | 99 / 34 | 71 / 56 | 2.88 (8.8) (6.4) |
| AUGUST | 88 / 30 | 67 / 52 | 3.06 (10.8) (3.4) |
| SEPTEMBER | 82 / 19 | 54 / 41 | 2.76 (11.4) (0.8) |
| OCTOBER | 68 / -22 | 36 / 26 | 2.33 (14.3) (0.3) |
| NOVEMBER | 46 / -44 | 19 / 6 | 1.79 (13) (0) |
| DECEMBER | 39 / -60 | 6 / -8 | 1.21 (11.9) (0) |
| ANNUAL | 99 / -60 | 37 / 22 | 21.8 (127.1) (16.8) |

### Map Location 112 — Russia, Moscow — 5545N 03734E 512

| Month | Ext High / Low | Norm High / Low | Precip (days) (snow) |
|---|---|---|---|
| JANUARY | 48 / -44 | 18 / 9 | 1.98 (14.8) (0) |
| FEBRUARY | 47 / -37 | 23 / 13 | 1.55 (12.2) (0) |
| MARCH | 63 / -26 | 32 / 21 | 1.43 (10.3) (0.1) |
| APRIL | 82 / -6 | 50 / 35 | 1.45 (8.2) (1.1) |
| MAY | 92 / 19 | 66 / 48 | 2.02 (8.6) (4) |
| JUNE | 95 / 37 | 72 / 53 | 1.95 (8.4) (6.2) |
| JULY | 101 / 45 | 74 / 57 | 3.47 (11.3) (7.6) |
| AUGUST | 99 / 32 | 71 / 55 | 3.01 (10.5) (6.9) |
| SEPTEMBER | 90 / 27 | 61 / 46 | 1.97 (8.7) (1.5) |
| OCTOBER | 75 / 10 | 47 / 37 | 1.7 (10.2) (0.4) |
| NOVEMBER | 54 / -27 | 32 / 25 | 1.51 (10.4) (0) |
| DECEMBER | 41 / -38 | 23 / 18 | 2.13 (16.4) (0) |
| ANNUAL | 101 / -44 | 47 / 35 | 24.2 (130) (27.8) |

### Map Location 113 — Russia, Murmansk — 6858N 03303E 151

| Month | Ext High / Low | Norm High / Low | Precip (days) (snow) |
|---|---|---|---|
| JANUARY | 43 / -29 | 16 / 8 | 1.49 (15.1) (0.1) |
| FEBRUARY | 41 / -36 | 15 / 6 | 1.35 (12.1) (0.2) |
| MARCH | 43 / -26 | 24 / 13 | 0.81 (12.6) (0.2) |
| APRIL | 61 / 0 | 35 / 24 | 0.81 (11.5) (0) |
| MAY | 77 / 16 | 45 / 34 | 1.28 (10.3) (0.2) |
| JUNE | 82 / 28 | 56 / 43 | 2.28 (12.4) (2.8) |
| JULY | 90 / 37 | 62 / 49 | 2.19 (11.5) (3.3) |
| AUGUST | 88 / 36 | 58 / 47 | 2.88 (14.1) (1.7) |
| SEPTEMBER | 68 / 25 | 48 / 40 | 2.09 (13.6) (0.2) |
| OCTOBER | 54 / -6 | 36 / 30 | 2.24 (14.9) (0.2) |
| NOVEMBER | 48 / -13 | 28 / 23 | 1.47 (13.1) (0) |
| DECEMBER | 39 / -22 | 20 / 12 | 1.58 (14.6) (0.3) |
| ANNUAL | 90 / -36 | 37 / 27 | 20.5 (155.8) (9.2) |

**Table 9.2  World Weather Data [CONTINUED]**

# Asia, Russia, Okhotskiy Perevo to Tiksi

# Map # 114 to 118

### Map Location 114 — Russia, Okhotskiy-Perevo — 6153N 13533E 479

| Month | | | | *** (***) | Precip |
|---|---|---|---|---|---|
| JANUARY | -43 | -52 | 0 | -71 | *** (***) | 0.34 (9.3) (0) |
| FEBRUARY | -26 | -45 | 7 | -69 | *** (***) | 0.32 (8.7) (0) |
| MARCH | 6 | -26 | 45 | -63 | *** (***) | 0.6 (8.4) (0) |
| APRIL | 31 | 2 | 61 | -44 | *** (***) | 0.57 (5.5) (0) |
| MAY | 50 | 30 | 77 | 3 | *** (***) | 1.62 (8.6) (0.8) |
| JUNE | 69 | 44 | 91 | 28 | *** (***) | 1.68 (7.9) (3.5) |
| JULY | 75 | 49 | 95 | 25 | *** (***) | 1.76 (7.6) (3.5) |
| AUGUST | 69 | 45 | 91 | 23 | *** (***) | 1.98 (8.8) (3) |
| SEPTEMBER | 52 | 32 | 77 | 7 | *** (***) | 1.53 (8.2) (1.1) |
| OCTOBER | 23 | 9 | 54 | -38 | *** (***) | 1.69 (13.2) (0) |
| NOVEMBER | -15 | -27 | 37 | -63 | *** (***) | 0.96 (12.2) (0) |
| DECEMBER | -38 | -46 | 10 | -71 | *** (***) | 0.47 (10.3) (0) |
| ANNUAL | 21 | 1 | 95 | -71 | *** (***) | 13.5 (108.7) (13.2) |

### Map Location 115 — Russia, Okhotsk — 5922N 14312E 20

| Month | | | | *** (***) | Precip |
|---|---|---|---|---|---|
| JANUARY | -4 | -12 | 30 | -33 | *** (***) | 0.42 (4) (0) |
| FEBRUARY | 5 | -8 | 34 | -29 | *** (***) | 0.35 (3.5) (0) |
| MARCH | 17 | -2 | 39 | -33 | *** (***) | 0.51 (5.8) (0) |
| APRIL | 30 | 13 | 57 | -6 | *** (***) | 1.29 (7.4) (0.1) |
| MAY | 40 | 30 | 64 | 14 | *** (***) | 1.96 (8.7) (0.1) |
| JUNE | 49 | 40 | 70 | 28 | *** (***) | 3.08 (9.7) (0.2) |
| JULY | 61 | 51 | 82 | 37 | *** (***) | 3.12 (9.4) (2) |
| AUGUST | 61 | 51 | 86 | 39 | *** (***) | 3.2 (9.3) (0.4) |
| SEPTEMBER | 54 | 41 | 70 | 23 | *** (***) | 3.19 (8.7) (0.4) |
| OCTOBER | 34 | 22 | 57 | -2 | *** (***) | 1.9 (6.3) (0) |
| NOVEMBER | 9 | 1 | 36 | -26 | *** (***) | 0.76 (4.8) (0) |
| DECEMBER | -1 | -9 | 30 | -35 | *** (***) | 0.82 (5.1) (0) |
| ANNUAL | 29 | 18 | 86 | -35 | *** (***) | 20.6 (82.7) (3.2) |

### Map Location 116 — Russia, Oymyakon (near) — 6316N 14309E 2382

| Month | | | | *** (***) | Precip |
|---|---|---|---|---|---|
| JANUARY | -49 | -60 | 2 | -80 | *** (***) | 0.57 (7.3) (0) |
| FEBRUARY | -35 | -56 | 3 | -90 | *** (***) | 0.51 (7.7) (0) |
| MARCH | -7 | -40 | 30 | -71 | *** (***) | 0.34 (7.2) (0.1) |
| APRIL | 25 | -15 | 53 | -44 | *** (***) | 0.29 (4.5) (0) |
| MAY | 45 | 21 | 72 | -11 | *** (***) | 0.56 (5.8) (0.1) |
| JUNE | 65 | 36 | 88 | 21 | *** (***) | 1.24 (8.4) (3.5) |
| JULY | 69 | 40 | 91 | 16 | *** (***) | 1.78 (10) (4.8) |
| AUGUST | 65 | 34 | 86 | 12 | *** (***) | 1.54 (7.5) (2.3) |
| SEPTEMBER | 48 | 23 | 72 | -2 | *** (***) | 0.82 (5.9) (0.3) |
| OCTOBER | 13 | -8 | 45 | -49 | *** (***) | 0.77 (8.4) (0) |
| NOVEMBER | -26 | -42 | 28 | -72 | *** (***) | 0.56 (10.2) (0) |
| DECEMBER | -43 | -55 | 14 | -78 | *** (***) | 0.52 (9.9) (0) |
| ANNUAL | 14 | -9 | 91 | -90 | *** (***) | 9.5 (92.8) (11.1) |

### Map Location 117 — Russia, St. Petersburg — 5958N 03018E 13

| Month | | | | *** (***) | Precip |
|---|---|---|---|---|---|
| JANUARY | 18 | 9 | 37 | -26 | *** (***) | 1.12 (14.1) (0) |
| FEBRUARY | 21 | 11 | 39 | -35 | *** (***) | 0.93 (11.3) (0) |
| MARCH | 31 | 18 | 54 | -22 | *** (***) | 1.05 (9.6) (0) |
| APRIL | 46 | 30 | 70 | -4 | *** (***) | 1.38 (8) (0.1) |
| MAY | 60 | 40 | 82 | 21 | *** (***) | 1.35 (8.2) (3.1) |
| JUNE | 68 | 48 | 86 | 28 | *** (***) | 2.31 (8.5) (5.3) |
| JULY | 70 | 52 | 93 | 36 | *** (***) | 2.9 (10) (5.9) |
| AUGUST | 68 | 51 | 88 | 34 | *** (***) | 2.8 (10.7) (4.9) |
| SEPTEMBER | 59 | 44 | 84 | 25 | *** (***) | 1.88 (10.2) (1.9) |
| OCTOBER | 47 | 37 | 66 | 12 | *** (***) | 2.51 (12.1) (0) |
| NOVEMBER | 33 | 27 | 54 | -11 | *** (***) | 1.31 (11.9) (0) |
| DECEMBER | 25 | 18 | 41 | -17 | *** (***) | 1.74 (14.8) (0) |
| ANNUAL | 46 | 32 | 93 | -35 | *** (***) | 29.3 (129.4) (21.4) |

### Map Location 118 — Russia, Tiksi — 7135N 12855E 26

| Month | | | | *** (***) | Precip |
|---|---|---|---|---|---|
| JANUARY | -21 | -30 | 9 | -53 | *** (***) | 1.45 (10.9) (0) |
| FEBRUARY | -17 | -26 | 14 | -58 | *** (***) | 1.63 (9.9) (0) |
| MARCH | -10 | -22 | 27 | -51 | *** (***) | 1.06 (11.3) (0) |
| APRIL | 6 | -10 | 41 | -47 | *** (***) | 0.54 (7.4) (0) |
| MAY | 25 | 13 | 54 | -13 | *** (***) | 0.84 (9.8) (0) |
| JUNE | 44 | 32 | 86 | 14 | *** (***) | 1.55 (8.5) (0.1) |
| JULY | 52 | 40 | 81 | 28 | *** (***) | 1.78 (10.5) (0.6) |
| AUGUST | 49 | 39 | 86 | 28 | *** (***) | 2.08 (11.1) (0.3) |
| SEPTEMBER | 39 | 30 | 61 | 9 | *** (***) | 1.41 (11.8) (0.2) |
| OCTOBER | 17 | 9 | 41 | -31 | *** (***) | 1.15 (12.5) (0) |
| NOVEMBER | -6 | -14 | 25 | -42 | *** (***) | 1.76 (12.6) (0) |
| DECEMBER | -16 | -25 | 19 | -44 | *** (***) | 1.55 (10.8) (0) |
| ANNUAL | 14 | 3 | 86 | -58 | *** (***) | 16.8 (127.1) (1.2) |

**Table 9.2   World Weather Data [CONTINUED]**

# Asia, Russia, Verkhoyansk to Singapore

## Map # 119 to 123

### Map Location 119 — Russia, Verkhoyansk 6733N 13323E 449

| | JANUARY | FEBRUARY | MARCH | APRIL | MAY | JUNE |
|---|---|---|---|---|---|---|
| | -17 | 16 | 34 | 50 | 75 | 93 |
| | -52 | -37 | -6 | 23 | 46 | 67 |
| | -59 | -53 | -38 | -11 | 23 | 43 |
| | -76 | -90 | -72 | -54 | -20 | 23 |
| | 0.19 (4.9) (0) | 0.32 (5.4) (0) | 0.25 (4.5) (0) | 0.28 (4.2) (0) | 0.49 (5.1) (0) | 0.86 (5.6) (1.8) |

| | JULY | AUGUST | SEPTEMBER | OCTOBER | NOVEMBER | DECEMBER | ANNUAL |
|---|---|---|---|---|---|---|---|
| | 93 | 91 | 72 | 52 | 28 | 7 | 93 |
| | 69 | 63 | 45 | 14 | -28 | -46 | 13 |
| | 45 | 39 | 28 | -4 | -38 | -54 | -6 |
| | 27 | 14 | 1 | -47 | -71 | -74 | -78 |
| | 1.15 (7.2) (1.9) | 0.81 (5.8) (0.6) | 0.97 (6.8) (0) | 0.64 (8.2) (0) | 0.62 (9.6) (0) | 0.55 (7.3) (0) | 7.1 (74.6) (4.3) |

### Map Location 120 — Russia, Vladivostok 4307N 13154E 453

| | JANUARY | FEBRUARY | MARCH | APRIL | MAY | JUNE |
|---|---|---|---|---|---|---|
| | 41 | 52 | 61 | 73 | 84 | 91 |
| | 17 | 23 | 35 | 49 | 58 | 64 |
| | 1 | 6 | 20 | 34 | 44 | 52 |
| | -22 | -18 | -8 | 18 | 32 | 41 |
| | 0.47 (2.7) (0) | 0.61 (2.6) (0) | 1.07 (5.9) (0) | 1.81 (7.4) (0) | 2.78 (10.2) (0.5) | 5.18 (16.1) (2) |

| | JULY | AUGUST | SEPTEMBER | OCTOBER | NOVEMBER | DECEMBER | ANNUAL |
|---|---|---|---|---|---|---|---|
| | 93 | 91 | 88 | 77 | 68 | 50 | 99 |
| | 77 | 70 | 70 | 58 | 39 | 23 | 49 |
| | 65 | 56 | 52 | 43 | 25 | 9 | 35 |
| | 50 | 36 | 34 | 19 | 0 | -17 | -22 |
| | 5.36 (16.3) (1.2) | 4.47 (8) (2.6) | 1.97 (6.4) (0.6) | 1.55 (5.2) (0.3) | 0.83 (4) (0) | *** | 33.8 (99.9) (8.5) |

### Map Location 121 — Russia, Yakutsk 6205N 12945E 338

| | JANUARY | FEBRUARY | MARCH | APRIL | MAY | JUNE |
|---|---|---|---|---|---|---|
| | 9 | 10 | 36 | 61 | 75 | 95 |
| | -39 | -25 | 4 | 30 | 52 | 71 |
| | -47 | -39 | -20 | 7 | 31 | 47 |
| | -72 | -69 | -56 | -42 | 10 | 27 |
| | 0.58 (10.3) (0) | 0.39 (8.7) (0) | 0.38 (6.4) (0) | 0.38 (4.8) (0) | 0.78 (5.8) (0.5) | 1.26 (6.4) (2.7) |

| | JULY | AUGUST | SEPTEMBER | OCTOBER | NOVEMBER | DECEMBER | ANNUAL |
|---|---|---|---|---|---|---|---|
| | 95 | 95 | 73 | 55 | 28 | 12 | 95 |
| | 76 | 70 | 52 | 25 | -15 | -37 | 22 |
| | 53 | 48 | 34 | 12 | -25 | -45 | 5 |
| | 34 | 25 | 10 | -31 | -65 | -74 | -74 |
| | 1.59 (6.7) (2.8) | 1.29 (6.3) (2.1) | 1.01 (6.6) (0.3) | 0.82 (9.6) (0.1) | 0.76 (13.3) (0) | 0.5 (10.1) (0) | 9.7 (95) (8.5) |

### Map Location 122 — Saudi Arabia, Riyadh 2443N 04643E 1922

| | JANUARY | FEBRUARY | MARCH | APRIL | MAY | JUNE |
|---|---|---|---|---|---|---|
| | 84 | 95 | 99 | 107 | 110 | 117 |
| | 70 | 73 | 82 | 91 | 100 | 106 |
| | 48 | 50 | 57 | 65 | 73 | 77 |
| | 28 | 34 | 32 | 50 | 61 | 68 |
| | 0.29 (0.8) (0) | 0.05 (0.3) (0) | 0.43 (1.1) (1) | 0.13 (0.4) (3.6) | 0 (0) (0.7) | 0 (0) (0) |

| | JULY | AUGUST | SEPTEMBER | OCTOBER | NOVEMBER | DECEMBER | ANNUAL |
|---|---|---|---|---|---|---|---|
| | 120 | 114 | 112 | 102 | 91 | 86 | 120 |
| | 107 | 102 | 93 | 93 | 79 | 70 | 90 |
| | 78 | 72 | 63 | 63 | 57 | 48 | 64 |
| | 67 | 68 | 62 | 50 | 41 | 31 | 28 |
| | 0 (0) (0) | 0 (0) (0) | 0 (0) (0) | 0 (0) (0) | 0.44 (1) (0) | 0.23 (0.7) (0) | *** (***) (***) |

### Map Location 123 — Singapore, Rep. of 0122N 10348E 59

| | JANUARY | FEBRUARY | MARCH | APRIL | MAY | JUNE |
|---|---|---|---|---|---|---|
| | *** | *** | *** | *** | *** | *** |
| | 85 | 87 | 88 | 89 | 88 | 88 |
| | 73 | 75 | 75 | 75 | 77 | 75 |
| | *** | *** | *** | *** | *** | *** |
| | 9.4 (***) (4) | 6.5 (***) (6) | 6.8 (***) (13) | 6.6 (***) (20) | 6.7 (***) (20) | 6.4 (***) (16) |

| | JULY | AUGUST | SEPTEMBER | OCTOBER | NOVEMBER | DECEMBER | ANNUAL |
|---|---|---|---|---|---|---|---|
| | *** | *** | *** | *** | *** | *** | *** |
| | 87 | 87 | 88 | 87 | 87 | 85 | 87 |
| | 75 | 75 | 75 | 75 | 75 | 73 | 75 |
| | *** | *** | *** | *** | *** | *** | *** |
| | 5.9 (***) (14) | 6.7 (***) (14) | 6.7 (***) (15) | 7.6 (***) (17) | 9.8 (***) (20) | 10.6 (***) (13) | 89.7 (***) (172) |

# Asia, Sri Lanka to Turkey

Table 9.2   World Weather Data [CONTINUED]

## Map # 124 to 128

### Map Location 124 — Sri Lanka, Colombo — 0654N 07952E 22

| | JANUARY | FEBRUARY | MARCH | APRIL | MAY | JUNE |
|---|---|---|---|---|---|---|
| Record high | 94 | 96 | 96 | 99 | 91 | 96 |
| Record low | 59 | 61 | 64 | 70 | 69 | 72 |
| Snow | 0 (0) | 0 (0) | 0 (0) | 0 (0) | 0 (0) | 0 (0) |
| Normal high | 86 | 87 | 84 | 88 | 87 | 85 |
| Normal low | 72 | 72 | 74 | 76 | 78 | 77 |
| Precip | 3.5 (6.4) (4) | 2.7 (5.2) (4) | 5.8 (6.8) (8) | 9.1 (7.4) (12) | 14.6 (13.2) (8) | 8.8 (11.7) (2) |

| | JULY | AUGUST | SEPTEMBER | OCTOBER | NOVEMBER | DECEMBER | ANNUAL |
|---|---|---|---|---|---|---|---|
| Record high | 89 | 88 | 89 | 89 | 90 | 91 | 99 |
| Record low | 71 | 70 | 71 | 69 | 66 | 63 | 59 |
| Snow | 0 (0) | 0 (0) | 0 (0) | 0 (0) | 0 (0) | 0 (0) | 0 (0) |
| Normal high | 85 | 85 | 85 | 85 | 85 | 85 | 86 |
| Normal low | 77 | 77 | 77 | 75 | 73 | 72 | 75 |
| Precip | 5.3 (8) (1) | 4.3 (6.7) (1) | 6.3 (7.1) (2) | 13.7 (12.1) (6) | 12.4 (10.5) (8) | 5.8 (9) (6) | 92.3 (104.1) (62) |

### Map Location 125 — Syria, Damascus — 3328N 03613E 2605

| | JANUARY | FEBRUARY | MARCH | APRIL | MAY | JUNE |
|---|---|---|---|---|---|---|
| Record high | 72 | 73 | 88 | 95 | 99 | 102 |
| Record low | 23 | 23 | 27 | 36 | 43 | 48 |
| Snow | *** (***) | *** (***) | *** (***) | 0 (0) | 0 (0) | 0 (0) |
| Normal high | 54 | 58 | 64 | 75 | 85 | 93 |
| Normal low | 37 | 38 | 42 | 48 | 57 | 62 |
| Precip | 2.32 (4.6) (0.6) | 1.25 (2.7) (0.7) | 0.7 (1.7) (0) | 0.29 (0.7) (1.1) | 0.62 (1.5) (1.8) | 0 (0) (0) |

| | JULY | AUGUST | SEPTEMBER | OCTOBER | NOVEMBER | DECEMBER | ANNUAL |
|---|---|---|---|---|---|---|---|
| Record high | 109 | 111 | 102 | 99 | 82 | 70 | 111 |
| Record low | 52 | 54 | 48 | 41 | 27 | 30 | 23 |
| Snow | 0 (0) | 0 (0) | 0 (0) | *** (***) | *** (***) | 0 (0) | *** (***) |
| Normal high | 97 | 98 | 90 | 82 | 68 | 59 | 77 |
| Normal low | 63 | 64 | 59 | 55 | 47 | 41 | 51 |
| Precip | 0 (0) (0) | 0 (0) (0) | 0 (0) (0.3) | 0.23 (0.6) (0.3) | 0.85 (1.8) (0.7) | 1.97 (4) (0.8) | 8.2 (17.6) (6.3) |

### Map Location 126 — Tajikistan, Dushanbe — 3835N 06847E 2703

| | JANUARY | FEBRUARY | MARCH | APRIL | MAY | JUNE |
|---|---|---|---|---|---|---|
| Record high | 70 | 73 | 82 | 90 | 100 | 104 |
| Record low | 0 | 7 | 14 | 32 | 43 | 52 |
| Snow | *** (***) | *** (***) | *** (***) | 0 (0) | 0 (0) | 0 (0) |
| Normal high | 49 | 51 | 60 | 69 | 79 | 91 |
| Normal low | 31 | 35 | 42 | 49 | 56 | 64 |
| Precip | 0.85 (3.3) (0) | 2.59 (8.4) (0.3) | 4.46 (11.4) (1.3) | 4.57 (10) (3.9) | 2.58 (8.7) (10.2) | 0.47 (2.6) (4.5) |

| | JULY | AUGUST | SEPTEMBER | OCTOBER | NOVEMBER | DECEMBER | ANNUAL |
|---|---|---|---|---|---|---|---|
| Record high | 108 | 104 | 99 | 91 | 77 | 68 | 108 |
| Record low | 57 | 50 | 39 | 28 | 21 | 14 | 0 |
| Snow | 0 (0) | 0 (0) | 0 (0*) | *** (***) | *** (***) | *** (***) | *** (***) |
| Normal high | 96 | 94 | 86 | 74 | 60 | 51 | 77 |
| Normal low | 67 | 63 | 55 | 46 | 38 | 33 | 48 |
| Precip | 0.23 (1.1) (1.5) | 0.01 (0.5) (0.5) | 0.18 (0.9) (0.1) | 1.29 (3.8) (1.2) | 1.94 (5.4) (0.4) | 2 (6.3) (0.3) | 21.2 (62.4) (24.2) |

### Map Location 127 — Thailand, Bangkok — 1344N 10030E 53

| | JANUARY | FEBRUARY | MARCH | APRIL | MAY | JUNE |
|---|---|---|---|---|---|---|
| Record high | 98 | 103 | 104 | 104 | 104 | 100 |
| Record low | 50 | 61 | 61 | 65 | 70 | 70 |
| Snow | 0 (0) | 0 (0) | 0 (0) | 0 (0) | 0 (0) | 0 (0) |
| Normal high | 89 | 91 | 93 | 95 | 93 | 91 |
| Normal low | 68 | 73 | 76 | 78 | 77 | 77 |
| Precip | 0.5 (1.1) (0) | 0.9 (1.9) (2) | 1.5 (3.2) (5) | 2.3 (5.2) (11) | 6.2 (5.5) (17) | 6 (10.6) (11) |

| | JULY | AUGUST | SEPTEMBER | OCTOBER | NOVEMBER | DECEMBER | ANNUAL |
|---|---|---|---|---|---|---|---|
| Record high | 97 | 96 | 97 | 96 | 96 | 97 | 114 |
| Record low | 68 | 70 | 66 | 68 | 60 | 66 | 50 |
| Snow | 0 (0) | 0 (0) | 0 (0) | 0 (0) | 0 (0) | 0 (0) | 0 (0) |
| Normal high | 90 | 89 | 89 | 88 | 88 | 88 | 90 |
| Normal low | 76 | 76 | 75 | 75 | 73 | 69 | 75 |
| Precip | 6.6 (11.2) (11) | 6.8 (11.4) (11) | 11.8 (14) (13) | 9.2 (12.2) (11) | 3.6 (5.2) (11) | 0.4 (0.9) (0) | 55.8 (82.4) (96) |

### Map Location 128 — Turkey, Ankara — 4007N 03259E 3122

| | JANUARY | FEBRUARY | MARCH | APRIL | MAY | JUNE |
|---|---|---|---|---|---|---|
| Record high | 61 | 64 | 84 | 88 | 88 | 93 |
| Record low | -11 | -11 | 9 | 23 | 28 | 36 |
| Snow | *** (***) | *** (***) | *** (***) | *** (***) | *** (***) | 0 (0) |
| Normal high | 40 | 42 | 52 | 65 | 72 | 79 |
| Normal low | 24 | 26 | 31 | 40 | 47 | 52 |
| Precip | 1.63 (3.4) (0.2) | 1.58 (3.3) (0) | 1.68 (3.5) (1.2) | 1.78 (3.6) (3) | 2.25 (4.3) (5.8) | 1.24 (2.2) (3.9) |

| | JULY | AUGUST | SEPTEMBER | OCTOBER | NOVEMBER | DECEMBER | ANNUAL |
|---|---|---|---|---|---|---|---|
| Record high | 100 | 104 | 97 | 91 | 75 | 61 | 104 |
| Record low | 43 | 37 | 34 | 18 | 9 | 3 | -11 |
| Snow | 0 (0) | 0 (0) | 0 (0) | *** (***) | *** (***) | *** (***) | *** (***) |
| Normal high | 86 | 88 | 81 | 68 | 57 | 45 | 65 |
| Normal low | 57 | 58 | 51 | 41 | 35 | 30 | 41 |
| Precip | 0.68 (1.3) (3.1) | 0.33 (0.7) (1.2) | 0.33 (0.8) (1.7) | 1.01 (2.1) (0.6) | 1.22 (2.4) (0.4) | 1.19 (2.6) (0) | 14.9 (30.2) (21.1) |

**Table 9.2   World Weather Data [CONTINUED]**

# Asia, Turkmenistan to Yemen

## Map # 129 to 133

### Map Location 129 — Turkmenistan, Ashgabat — 3758N 05820E 775

| Month | Avg High | Avg Low | Rec High | Rec Low | Precip (max)(24h) | Snow |
|---|---|---|---|---|---|---|
| JANUARY | 50 | 31 | 82 | 10 | 0.33 (2.8) (0) | *** (***) |
| FEBRUARY | 53 | 34 | 81 | 12 | 0.65 (3.1) (0.1) | *** (***) |
| MARCH | 61 | 41 | 90 | 21 | 1.41 (7.6) (0.4) | *** (***) |
| APRIL | 71 | 51 | 100 | 32 | 1.81 (8) (2.4) | *** (***) |
| MAY | 87 | 63 | 111 | 48 | 0.85 (3.4) (2.9) | 0 (0) |
| JUNE | 95 | 69 | 111 | 46 | 0.18 (1.2) (2) | 0 (0) |
| JULY | 100 | 75 | 109 | 61 | 0.18 (0.6) (0.7) | 0 (0) |
| AUGUST | 98 | 70 | 113 | 57 | 0.01 (0.1) (0.4) | 0 (0) |
| SEPTEMBER | 87 | 60 | 104 | 34 | 0.08 (0.7) (0.5) | 0 (0) |
| OCTOBER | 73 | 49 | 93 | 28 | 0.8 (3.3) (0.9) | 0 (0) |
| NOVEMBER | 60 | 39 | 88 | 10 | 0.75 (3.5) (0.1) | 0 (0) |
| DECEMBER | 49 | 33 | 81 | 3 | 0.56 (3.7) (0.1) | 0 (0) |
| ANNUAL | 74 | 51 | 113 | 3 | 7.6 (38) (10.5) | *** (***) |

### Map Location 130 — United Arab Emirates, Dubai — 2515N 05520E 5

| Month | Avg High | Avg Low | Rec High | Rec Low | Precip (max)(24h) | Snow |
|---|---|---|---|---|---|---|
| JANUARY | 74 | 55 | 85 | 45 | 1.52 (3.2) (0.6) | 0 (0) |
| FEBRUARY | 75 | 57 | 90 | 46 | 0.24 (0.7) (0.5) | 0 (0) |
| MARCH | 82 | 62 | 101 | 51 | 0.4 (1) (1.1) | 0 (0) |
| APRIL | 87 | 66 | 105 | 54 | 1.13 (2.5) (1) | 0 (0) |
| MAY | 93 | 71 | 112 | 59 | 0.13 (0.4) (0.6) | 0 (0) |
| JUNE | 97 | 76 | 111 | 66 | 0 (0) (0) | 0 (0) |
| JULY | 101 | 83 | 113 | 73 | 0.11 (0.3) (0.3) | 0 (0) |
| AUGUST | 102 | 82 | 113 | 70 | 0 (0) (0) | 0 (0) |
| SEPTEMBER | 98 | 78 | 113 | 64 | 0.03 (0.2) (0.3) | 0 (0) |
| OCTOBER | 93 | 70 | 103 | 59 | 0 (0) (0.2) | 0 (0) |
| NOVEMBER | 85 | 64 | 95 | 50 | 0.68 (1.5) (0.7) | 0 (0) |
| DECEMBER | 77 | 57 | 90 | 45 | 0.9 (2) (0.8) | 0 (0) |
| ANNUAL | 89 | 68 | 113 | 45 | 5.1 (11.8) (6.1) | 0 (0) |

### Map Location 131 — Uzbekistan, Tashkent — 4116N 06916E 1404

| Month | Avg High | Avg Low | Rec High | Rec Low | Precip (max)(24h) | Snow |
|---|---|---|---|---|---|---|
| JANUARY | 40 | 24 | 72 | -18 | 0.68 (3.1) (0) | *** (***) |
| FEBRUARY | 45 | 28 | 79 | -15 | 0.87 (4.1) (0.1) | *** (***) |
| MARCH | 56 | 37 | 91 | -4 | 1.81 (5) (0) | *** (***) |
| APRIL | 69 | 47 | 95 | 21 | 3.17 (6.2) (0.7) | *** (***) |
| MAY | 81 | 56 | 108 | 32 | 0.44 (4) (2.9) | 0 (0) |
| JUNE | 91 | 62 | 111 | 39 | 0.35 (2.2) (1.4) | 0 (0) |
| JULY | 96 | 65 | 111 | 46 | 0.23 (1.1) (0.2) | 0 (0) |
| AUGUST | 93 | 61 | 109 | 45 | 0.25 (0.9) (0.4) | 0 (0) |
| SEPTEMBER | 83 | 52 | 104 | 32 | 0.05 (0.8) (0.1) | 0 (0) |
| OCTOBER | 69 | 43 | 100 | 12 | 0.65 (3.6) (0.4) | 0 (0) |
| NOVEMBER | 56 | 34 | 88 | -8 | 1.08 (5.1) (0.4) | 0 (0) |
| DECEMBER | 44 | 28 | 75 | -22 | 1.8 (7.5) (0) | 0 (0) |
| ANNUAL | 69 | 45 | 111 | -22 | 11.4 (43.6) (6.6) | *** (***) |

### Map Location 132 — Vietnam, Hanoi — 2103N 10552E 20

| Month | Avg High | Avg Low | Rec High | Rec Low | Precip (max)(24h) | Snow |
|---|---|---|---|---|---|---|
| JANUARY | 69 | 57 | 92 | 41 | 0.7 (1.5) (0.3) | 0 (0) |
| FEBRUARY | 69 | 58 | 95 | 43 | 1.1 (2.2) (0.3) | 0 (0) |
| MARCH | 74 | 63 | 98 | 46 | 1.5 (3.2) (2) | 0 (0) |
| APRIL | 81 | 69 | 101 | 50 | 3.2 (5) (4) | 0 (0) |
| MAY | 89 | 75 | 109 | 60 | 7.7 (5.3) (8) | 0 (0) |
| JUNE | 92 | 78 | 108 | 64 | 9.4 (13.8) (9) | 0 (0) |
| JULY | 91 | 78 | 104 | 63 | 12.7 (16.5) (10) | 0 (0) |
| AUGUST | 90 | 78 | 101 | 63 | 13.5 (17) (11) | 0 (0) |
| SEPTEMBER | 88 | 76 | 99 | 63 | 10 (12.8) (6) | 0 (0) |
| OCTOBER | 84 | 71 | 96 | 54 | 3.9 (7.1) (2) | 0 (0) |
| NOVEMBER | 78 | 65 | 94 | 44 | 1.7 (4.5) (1) | 0 (0) |
| DECEMBER | 72 | 59 | 89 | 41 | 0.8 (1.7) (0) | 0 (0) |
| ANNUAL | 81 | 69 | 109 | 41 | 66.2 (90.6) (53.6) | 0 (0) |

### Map Location 133 — Yemen, Aden — 1249N 04501E 10

| Month | Avg High | Avg Low | Rec High | Rec Low | Precip (max)(24h) | Snow |
|---|---|---|---|---|---|---|
| JANUARY | 82 | 73 | 88 | 61 | 0.32 (0.8) (0) | 0 (0) |
| FEBRUARY | 83 | 73 | 87 | 63 | 0.11 (0.4) (0) | 0 (0) |
| MARCH | 85 | 75 | 93 | 61 | 0.13 (0.4) (0.1) | 0 (0) |
| APRIL | 89 | 79 | 99 | 64 | 0.02 (0.1) (0.1) | 0 (0) |
| MAY | 93 | 81 | 108 | 64 | 0.19 (0.5) (0.5) | 0 (0) |
| JUNE | 98 | 84 | 110 | 77 | 0 (0) (0.2) | 0 (0) |
| JULY | 97 | 83 | 104 | 75 | 0.15 (0.4) (1.9) | 0 (0) |
| AUGUST | 96 | 82 | 104 | 72 | 0.08 (0.2) (1.4) | 0 (0) |
| SEPTEMBER | 96 | 83 | 103 | 75 | 0.09 (0.3) (0.8) | 0 (0) |
| OCTOBER | 91 | 76 | 102 | 66 | 0.05 (0.2) (0.3) | 0 (0) |
| NOVEMBER | 86 | 75 | 95 | 64 | 0.33 (0.8) (0) | 0 (0) |
| DECEMBER | 83 | 73 | 88 | 63 | 0.15 (0.5) (0) | 0 (0) |
| ANNUAL | 90 | 78 | 110 | 61 | 1.6 (4.6) (5.4) | 0 (0) |

**Table 9.2  World Weather Data [CONTINUED]**

# Australia, Adelaide to Darwin

## Map # 134 to 138

### Map Location 134 — Australia, Adelaide — 3457S 13832E 20

| Month | Normal High | Normal Low | Precip | Record High | Record Low | Snow |
|---|---|---|---|---|---|---|
| JANUARY | 86 | 61 | 0.8 (2.8) (2) | 118 | 45 | 0 (0) |
| FEBRUARY | 86 | 62 | 0.7 (2.5) (2) | 114 | 45 | 0 (0) |
| MARCH | 81 | 59 | 1 (8.5) (1) | 111 | 44 | 0 (0) |
| APRIL | 73 | 55 | 1.8 (9.4) (1) | 99 | 40 | 0 (0) |
| MAY | 66 | 50 | 2.7 (10.3) (1) | 89 | 36 | 0 (0) |
| JUNE | 61 | 47 | 3 (8.4) (1) | 77 | 32 | 0 (0) |
| JULY | 59 | 45 | 2.6 (7.6) (1) | 74 | 32 | 0 (0) |
| AUGUST | 62 | 46 | 2.6 (7.6) (1) | 85 | 32 | 0 (0) |
| SEPTEMBER | 66 | 48 | 2.1 (6) (2) | 91 | 33 | 0 (0) |
| OCTOBER | 73 | 51 | 1.7 (4.9) (3) | 103 | 36 | 0 (0) |
| NOVEMBER | 79 | 55 | 1.1 (3.2) (1) | 113 | 41 | 0 (0) |
| DECEMBER | 83 | 59 | 1 (3.2) (2) | 115 | 43 | 0 (0) |
| ANNUAL | 73 | 53 | 21.1 (74.4) (18) | 118 | 32 | 0 (0) |

### Map Location 135 — Australia, Alice Springs — 2348S 13353E 1791

| Month | Normal High | Normal Low | Precip | Record High | Record Low | Snow |
|---|---|---|---|---|---|---|
| JANUARY | 97 | 70 | 1.7 (4.8) (1.5) | 113 | 51 | 0 (0) |
| FEBRUARY | 95 | 69 | 1.3 (3.9) (1.6) | 113 | 47 | 0 (0) |
| MARCH | 90 | 63 | 1.1 (8.6) (0.4) | 108 | 45 | 0 (0) |
| APRIL | 81 | 54 | 0.4 (7.8) (0.6) | 104 | 36 | 0* (0*) |
| MAY | 73 | 46 | 0.6 (8) (0.4) | 96 | 28 | *** (***) |
| JUNE | 67 | 41 | 0.5 (2.6) (0.1) | 89 | 22 | *** (***) |
| JULY | 67 | 39 | 0.3 (2.1) (0.1) | 89 | 19 | *** (***) |
| AUGUST | 73 | 43 | 0.3 (2.1) (0) | 96 | 25 | *** (***) |
| SEPTEMBER | 81 | 49 | 0.3 (0.6) (0.4) | 102 | 31 | 0 (0) |
| OCTOBER | 88 | 58 | 0.7 (1.9) (2.2) | 107 | 39 | 0 (0) |
| NOVEMBER | 93 | 64 | 1.2 (3.5) (1.8) | 109 | 42 | 0 (0) |
| DECEMBER | 96 | 68 | 1.5 (4.4) (3.8) | 112 | 50 | 0 (0) |
| ANNUAL | 83 | 55 | 9.9 (50.3) (12.9) | 113 | 19 | 0 (0) |

### Map Location 136 — Australia, Brisbane — 2725S 15305E 17

| Month | Normal High | Normal Low | Precip | Record High | Record Low | Snow |
|---|---|---|---|---|---|---|
| JANUARY | 85 | 69 | 6.4 (12.4) (2.7) | 110 | 59 | 0 (0) |
| FEBRUARY | 85 | 68 | 6.3 (12.3) (2.4) | 106 | 58 | 0 (0) |
| MARCH | 82 | 66 | 5.7 (12.8) (1.1) | 99 | 52 | 0 (0) |
| APRIL | 79 | 61 | 3.7 (11.2) (0.9) | 95 | 44 | 0 (0) |
| MAY | 74 | 56 | 2.8 (10.4) (0.4) | 90 | 41 | 0 (0) |
| JUNE | 69 | 51 | 2.6 (7.6) (0.2) | 89 | 37 | 0 (0) |
| JULY | 68 | 49 | 2.2 (6.8) (0.1) | 83 | 35 | 0 (0) |
| AUGUST | 71 | 50 | 1.9 (6.1) (0.5) | 88 | 37 | 0 (0) |
| SEPTEMBER | 76 | 55 | 1.9 (5.4) (1.4) | 95 | 41 | 0 (0) |
| OCTOBER | 80 | 60 | 2.5 (7) (3.4) | 102 | 43 | 0 (0) |
| NOVEMBER | 82 | 64 | 3.7 (9.6) (3.5) | 106 | 48 | 0 (0) |
| DECEMBER | 85 | 67 | 5 (10.6) (4.4) | 106 | 54 | 0 (0) |
| ANNUAL | 78 | 60 | 44.7 (112.2) (21) | 110 | 35 | 0 (0) |

### Map Location 137 — Australia, Cairns — 1653S 14545E 7

| Month | Normal High | Normal Low | Precip | Record High | Record Low | Snow |
|---|---|---|---|---|---|---|
| JANUARY | 90 | 74 | 16.6 (18.5) (2.1) | 110 | 67 | 0 (0) |
| FEBRUARY | 89 | 74 | 15.7 (18.2) (2.8) | 108 | 64 | 0 (0) |
| MARCH | 87 | 73 | 18.1 (19.7) (1) | 100 | 62 | 0 (0) |
| APRIL | 85 | 70 | 11.3 (16.1) (0.1) | 95 | 57 | 0 (0) |
| MAY | 81 | 66 | 4.4 (11.8) (0.1) | 90 | 50 | 0 (0) |
| JUNE | 79 | 64 | 2.9 (8.2) (0) | 88 | 43 | 0 (0) |
| JULY | 78 | 61 | 1.6 (5.4) (0) | 86 | 43 | 0 (0) |
| AUGUST | 80 | 62 | 1.7 (5.7) (0) | 88 | 43 | 0 (0) |
| SEPTEMBER | 83 | 64 | 1.7 (4.9) (0) | 92 | 48 | 0 (0) |
| OCTOBER | 86 | 68 | 2.1 (6) (1) | 96 | 54 | 0 (0) |
| NOVEMBER | 88 | 70 | 3.9 (10) (1.1) | 98 | 58 | 0 (0) |
| DECEMBER | 90 | 73 | 8.7 (14.7) (1.7) | 103 | 44 | 0 (0) |
| ANNUAL | 85 | 68 | 88.7 (139.2) (9.9) | 110 | 43 | 0 (0) |

### Map Location 138 — Australia, Darwin — 1225S 13052E 104

| Month | Normal High | Normal Low | Precip | Record High | Record Low | Snow |
|---|---|---|---|---|---|---|
| JANUARY | 90 | 77 | 15.2 (18) (10.1) | 100 | 68 | 0 (0) |
| FEBRUARY | 90 | 77 | 12.3 (16.9) (8.8) | 101 | 63 | 0 (0) |
| MARCH | 91 | 77 | 10 (15.5) (10.6) | 102 | 68 | 0 (0) |
| APRIL | 92 | 76 | 3.8 (11.3) (4.2) | 104 | 64 | 0 (0) |
| MAY | 91 | 73 | 0.6 (8) (0) | 102 | 59 | 0 (0) |
| JUNE | 88 | 69 | 0.1 (1.5) (0) | 99 | 55 | 0 (0) |
| JULY | 87 | 67 | 0.03 (1.3) (0.1) | 98 | 56 | 0 (0) |
| AUGUST | 89 | 70 | 0.1 (1.5) (0) | 98 | 57 | 0 (0) |
| SEPTEMBER | 91 | 74 | 0.5 (1.2) (0) | 102 | 63 | 0 (0) |
| OCTOBER | 93 | 77 | 2 (5.7) (4.1) | 105 | 69 | 0 (0) |
| NOVEMBER | 94 | 78 | 4.7 (11.4) (11.8) | 103 | 67 | 0 (0) |
| DECEMBER | 92 | 78 | 9.4 (15.2) (12.2) | 102 | 69 | 0 (0) |
| ANNUAL | 91 | 74 | 58.7 (107.5) (61.9) | 105 | 55 | 0 (0) |

**Table 9.2  World Weather Data [CONTINUED]**

# Australia, Hobart to Wyndham

## Map # 139 to 142

### Map Location 139 — Australia, Hobart — 4250S 14733E 91

| | JANUARY | FEBRUARY | MARCH | APRIL | MAY | JUNE | JULY | AUGUST | SEPTEMBER | OCTOBER | NOVEMBER | DECEMBER | ANNUAL |
|---|---|---|---|---|---|---|---|---|---|---|---|---|---|
| Record Max T°F | 105 | 104 | 99 | 87 | 78 | 69 | 66 | 72 | 82 | 92 | 98 | 105 | 105 |
| Average Max T°F | 71 | 71 | 68 | 63 | 58 | 53 | 52 | 55 | 59 | 63 | 66 | 69 | 62 |
| Average Min T°F | 53 | 53 | 51 | 48 | 44 | 41 | 40 | 41 | 43 | 46 | 48 | 51 | 47 |
| Record Min T°F | 40 | 39 | 35 | 33 | 29 | 29 | 28 | 30 | 30 | 32 | 35 | 38 | 28 |
| Avg. Precip in. (Rain)(Thunder) | 1.9 (5.2)(1) | 1.5 (4.4)(1) | 1.8 (9.4)(1) | 1.9 (9.4)(1) | 1.8 (9.4)(0) | 2.2 (6.8)(0.3) | 2.1 (6.6)(0) | 1.9 (6.1)(0) | 2.1 (6)(0) | 2.3 (6.5)(1) | 2.4 (6.7)(1) | 2.1 (5.6)(1) | 24 (82.2)(6.6) |
| Avg. Snow in. (Snow) | 0 (0) | 0 (0) | 0 (0) | 0 (0) | 0 (0) | 0* (0*) | *** (***) | 0 (0) | 0 (0) | 0 (0) | 0 (0) | 0 (0) | 0 (0) |

### Map Location 140 — Australia, Perth — 3156S 11558E 64

| | JANUARY | FEBRUARY | MARCH | APRIL | MAY | JUNE | JULY | AUGUST | SEPTEMBER | OCTOBER | NOVEMBER | DECEMBER | ANNUAL |
|---|---|---|---|---|---|---|---|---|---|---|---|---|---|
| Record Max T°F | 111 | 112 | 106 | 100 | 90 | 82 | 76 | 83 | 91 | 95 | 105 | 108 | 112 |
| Average Max T°F | 85 | 85 | 81 | 76 | 69 | 64 | 63 | 64 | 67 | 70 | 76 | 81 | 73 |
| Average Min T°F | 63 | 63 | 61 | 57 | 53 | 50 | 48 | 48 | 50 | 53 | 57 | 61 | 55 |
| Record Min T°F | 43 | 46 | 44 | 36 | 34 | 31 | 32 | 34 | 32 | 39 | 40 | 41 | 31 |
| Avg. Precip in. (Rain)(Thunder) | 0.3 (1.5)(0.4) | 0.4 (1.8)(0.5) | 0.8 (8.3)(0.2) | 1.7 (9.3)(0.5) | 5.1 (12.4)(1.5) | 7.1 (13.8)(1.2) | 6.7 (13.4)(1.3) | 5.7 (12.4)(0.8) | 3.4 (9)(0.2) | 2.2 (6.2)(0.8) | 0.8 (2.2)(0.8) | 0.5 (2)(0.6) | 34.7 (92.3)(8.8) |
| Avg. Snow in. (Snow) | 0 (0) | 0 (0) | 0 (0) | 0 (0) | 0 (0) | 0 (0) | 0 (0) | 0 (0) | 0 (0) | 0 (0) | 0 (0) | 0 (0) | 0 (0) |

### Map Location 141 — Australia, Sidney — 3357S 15101E 16

| | JANUARY | FEBRUARY | MARCH | APRIL | MAY | JUNE | JULY | AUGUST | SEPTEMBER | OCTOBER | NOVEMBER | DECEMBER | ANNUAL |
|---|---|---|---|---|---|---|---|---|---|---|---|---|---|
| Record Max T°F | 114 | 108 | 103 | 91 | 86 | 80 | 78 | 88 | 92 | 92 | 105 | 110 | 114 |
| Average Max T°F | 78 | 78 | 76 | 71 | 66 | 61 | 60 | 63 | 67 | 71 | 74 | 77 | 70 |
| Average Min T°F | 65 | 65 | 63 | 58 | 52 | 48 | 46 | 48 | 51 | 56 | 60 | 63 | 56 |
| Record Min T°F | 49 | 49 | 47 | 44 | 40 | 36 | 35 | 35 | 39 | 39 | 43 | 48 | 35 |
| Avg. Precip in. (Rain)(Thunder) | 3.5 (8.3)(1.5) | 4 (9.1)(1.2) | 5 (12.3)(0.8) | 5.3 (12.5)(0.8) | 5 (12.3)(0.6) | 4.6 (11)(0.1) | 4.6 (11)(0.2) | 3 (8.4)(0.3) | 2.9 (7.9)(0.7) | 2.8 (7.7)(1.8) | 2.9 (7.2)(3.1) | 2.9 (7.2)(3.1) | 46.5 (115.6)(13) |
| Avg. Snow in. (Snow) | 0 (0) | 0 (0) | 0 (0) | 0 (0) | 0 (0) | 0 (0) | 0 (0) | 0 (0) | 0 (0) | 0 (0) | 0 (0) | 0 (0) | 0 (0) |

### Map Location 142 — Australia, Wyndham — 1531S 12809E 14

| | JANUARY | FEBRUARY | MARCH | APRIL | MAY | JUNE | JULY | AUGUST | SEPTEMBER | OCTOBER | NOVEMBER | DECEMBER | ANNUAL |
|---|---|---|---|---|---|---|---|---|---|---|---|---|---|
| Record Max T°F | 112 | 108 | 106 | 105 | 100 | 98 | 95 | 99 | 104 | 110 | 112 | 110 | 112 |
| Average Max T°F | 96 | 96 | 95 | 95 | 90 | 86 | 89 | 89 | 94 | 97 | 98 | 97 | 93 |
| Average Min T°F | 80 | 80 | 79 | 77 | 72 | 68 | 70 | 75 | 75 | 80 | 81 | 81 | 76 |
| Record Min T°F | 67 | 64 | 67 | 64 | 55 | 54 | 50 | 56 | 60 | 68 | 69 | 68 | 50 |
| Avg. Precip in. (Rain)(Thunder) | 7.3 (13.4)(16) | 6 (12)(11) | 4.6 (12)(4) | 0.8 (8.3)(1) | 0.2 (7.6)(0.3) | 0.2 (1.8)(0) | 0.2 (1.8)(0) | 0.03 (1.3)(0.3) | 0.1 (0)(0.3) | 0.4 (0.9)(4) | 2 (5.7)(8) | 4.1 (9.3)(17) | 25.9 (74.1)(61.9) |
| Avg. Snow in. (Snow) | 0 (0) | 0 (0) | 0 (0) | 0 (0) | 0 (0) | 0 (0) | 0 (0) | 0 (0) | 0 (0) | 0 (0) | 0 (0) | 0 (0) | 0 (0) |

## Data Key

|  | |
|---|---|
| Map Locater Number | MONTH |
| Location | Average Max T°F |
| Lat • Lon ddmm Elev ft | Average Min T°F |
| | Avg. Precip in. (Rain Days) (Thunder Days) |
| | Record Max T°F |
| | Record Min T°F |
| | Avg. Snow in. (Snow Days) |

| | |
|---|---|
| *** or Blank | No data, missing data or data of insufficient reliability |
| Rain Days | Number of Days with >= 0.01" of precipitation falling |
| Snow Days | Number of Days with >= 1.5" of snow falling |

**Table 9.2  World Weather Data [CONTINUED]**

# Europe, Albania to Belarus

## Map # 143 to 147

### Map Location 143 — Albania, Korçë 4120N 01947E 241

| Month | Avg Max | Avg Min | Rec Max | Rec Min | Precip (Days)(Snow) |
|---|---|---|---|---|---|
| JANUARY | 40 | 27 | 55 | 5 | 2.29(7)(*) |
| FEBRUARY | 45 | 29 | 62 | 1 | 3.48(7.2)(*) |
| MARCH | 46 | 30 | 72 | 15 | 2.02(5.7)(*) |
| APRIL | 57 | 39 | 73 | 13 | 2.14(5.7)(*) |
| MAY | 66 | 48 | 83 | 32 | 3.85(8.1)(*) |
| JUNE | 75 | 53 | 93 | 42 | 1.1(3.0)(*) |
| JULY | 82 | 57 | 94 | 50 | 0.7(1.7)(*) |
| AUGUST | 84 | 59 | 98 | 50 | 0.44(1.7)(*) |
| SEPTEMBER | 72 | 51 | 90 | 40 | 2(3.7)(*) |
| OCTOBER | 62 | 44 | 77 | 31 | 3.17(6.0)(*) |
| NOVEMBER | 50 | 38 | 65 | 29 | 5.15(11.0)(*) |
| DECEMBER | 43 | 31 | 61 | 8 | 2.06(5.7)(*) |
| ANNUAL | 60 | 42 | 98 | 1 | 28.3(66.5)(*) |

### Map Location 144 — Albania, Tirana 4120N 01947E 241

| Month | Avg Max | Avg Min | Rec Max | Rec Min | Precip (Days)(Snow) |
|---|---|---|---|---|---|
| JANUARY | 53 | 36 | 65 | 22 | 5.08(8.2)(***) |
| FEBRUARY | 55 | 37 | 71 | 20 | 5.81(7.9)(***) |
| MARCH | 57 | 38 | 78 | 25 | 4.2(7.8)(***) |
| APRIL | 65 | 46 | 81 | 31 | 5.84(8)(***) |
| MAY | 73 | 54 | 88 | 37 | 5.74(9.3)(***) |
| JUNE | 82 | 60 | 99 | 51 | 3.06(3)(***) |
| JULY | 88 | 63 | 101 | 55 | 1.51(3.3)(***) |
| AUGUST | 90 | 63 | 105 | 55 | 0.45(1.3)(***) |
| SEPTEMBER | 82 | 57 | 95 | 46 | 1.64(3)(***) |
| OCTOBER | 73 | 50 | 86 | 35 | 4.93(7.2)(***) |
| NOVEMBER | 62 | 46 | 78 | 33 | 5.97(9.7)(***) |
| DECEMBER | 56 | 38 | 70 | 20 | 5.34(9.3)(***) |
| ANNUAL | 70 | 49 | 105 | 20 | 49.6(78)(***) |

### Map Location 145 — Austria, Innsbruck 4715N 01120E 1906

| Month | Avg Max | Avg Min | Rec Max | Rec Min | Precip (Days)(Snow) |
|---|---|---|---|---|---|
| JANUARY | 34 | 20 | 61 | -15 | 2.09(6.5)(0.3) |
| FEBRUARY | 40 | 24 | 60 | -22 | 1.58(5.2)(0) |
| MARCH | 51 | 31 | 77 | 2 | 1.65(5.3)(0.3) |
| APRIL | 60 | 39 | 83 | 19 | 2.24(6.3)(0.3) |
| MAY | 69 | 46 | 91 | 18 | 2.95(7.0)(3.0) |
| JUNE | 75 | 52 | 97 | 31 | 4.09(8.8)(5.0) |
| JULY | 78 | 55 | 95 | 38 | 4.76(9.3)(7.0) |
| AUGUST | 76 | 54 | 97 | 34 | 4.57(9.2)(5.0) |
| SEPTEMBER | 69 | 49 | 87 | 30 | 3.03(7.4)(2) |
| OCTOBER | 58 | 40 | 76 | 18 | 2.4(6.4)(0.3) |
| NOVEMBER | 46 | 32 | 73 | -4 | 2.24(6.1)(0.3) |
| DECEMBER | 36 | 24 | 66 | -13 | 2.09(6.5)(0) |
| ANNUAL | 58 | 39 | 97 | -22 | 33.7(84)(23.5) |

### Map Location 146 — Austria, Vienna 4815N 01622E 666

| Month | Avg Max | Avg Min | Rec Max | Rec Min | Precip (Days)(Snow) |
|---|---|---|---|---|---|
| JANUARY | 34 | 26 | 62 | -8 | 1.5(4.9)(0) |
| FEBRUARY | 38 | 28 | 67 | -14 | 1.4(4.6)(0.3) |
| MARCH | 47 | 34 | 77 | 3 | 1.8(5.6)(0.3) |
| APRIL | 57 | 41 | 83 | 18 | 2(6)(1) |
| MAY | 66 | 50 | 92 | 27 | 2.8(6.9)(5) |
| JUNE | 71 | 56 | 97 | 39 | 2.7(7)(6) |
| JULY | 75 | 59 | 98 | 45 | 3(7.4)(7) |
| AUGUST | 73 | 58 | 97 | 42 | 2.7(7)(6) |
| SEPTEMBER | 66 | 52 | 90 | 31 | 2(5.7)(2) |
| OCTOBER | 55 | 44 | 82 | 16 | 1.8(5.5)(0.3) |
| NOVEMBER | 44 | 36 | 71 | 6 | 1.9(5.5)(0.3) |
| DECEMBER | 37 | 30 | 66 | -4 | 1.8(5.8)(0.3) |
| ANNUAL | 55 | 43 | 98 | -14 | 25.6(72.1)(28.5) |

### Map Location 147 — Belarus, Minsk 5352N 02732E 768

| Month | Avg Max | Avg Min | Rec Max | Rec Min | Precip (Days)(Snow) |
|---|---|---|---|---|---|
| JANUARY | 21 | 14 | 37 | -20 | 1.35(14.1)(0) |
| FEBRUARY | 25 | 16 | 39 | -13 | 1.51(11.9)(0) |
| MARCH | 34 | 23 | 66 | -17 | 1.69(11.3)(0.3) |
| APRIL | 51 | 36 | 73 | 14 | 1.38(8.6)(0.8) |
| MAY | 64 | 47 | 81 | 28 | 2.64(10.9)(6.2) |
| JUNE | 71 | 53 | 90 | 32 | 2.69(8.6)(5.9) |
| JULY | 72 | 55 | 91 | 45 | 2.26(9.6)(6.4) |
| AUGUST | 70 | 54 | 90 | 37 | 3.4(10.1)(4.9) |
| SEPTEMBER | 62 | 47 | 82 | 28 | 1.86(8.4)(1.5) |
| OCTOBER | 50 | 40 | 73 | 19 | 1.79(9.6)(0.2) |
| NOVEMBER | 36 | 30 | 55 | 0 | 2.26(13.2)(0) |
| DECEMBER | 27 | 22 | 50 | -8 | 1.63(15)(0) |
| ANNUAL | 49 | 36 | 91 | -20 | 24.5(131.3)(26.2) |

**Table 9.2  World Weather Data [CONTINUED]**

# Europe, Belgium to Czech Republic

# Map # 148 to 152

## Map Location 148 — Belgium, Brussels — 5054N 00429E 180

| Month | Normal High | Normal Low | Precip in (mm) (days) | Record High | Record Low |
|---|---|---|---|---|---|
| January | 40 | 33 | 2.7 (7.9) (0) | 57 | 10 |
| February | 41 | 31 | 1.9 (6.1) (0.3) | 64 | 1 |
| March | 49 | 36 | 2 (6) (1) | 70 | 14 |
| April | 57 | 40 | 2.5 (6.6) (1) | 86 | 28 |
| May | 64 | 47 | 2.7 (6.8) (4) | 88 | 32 |
| June | 69 | 52 | 1.7 (4.9) (3) | 88 | 37 |
| July | 73 | 56 | 3.1 (7.6) (4) | 97 | 43 |
| August | 73 | 56 | 2.7 (7) (3) | 97 | 43 |
| September | 67 | 52 | 2.4 (6.4) (3) | 93 | 37 |
| October | 59 | 45 | 1.7 (5.1) (1) | 75 | 23 |
| November | 48 | 40 | 3.2 (7.7) (0.3) | 68 | 23 |
| December | 43 | 36 | 3.1 (8.6) (0.3) | 63 | 3 |
| Annual | 57 | 44 | 29.7 (80.7) (20.9) | 97 | 1 |

## Map Location 149 — Bosnia and Herzegovina, Sarajevo — 4352N 01818E 1620

| Month | Normal High | Normal Low | Precip in (mm) (days) | Record High | Record Low |
|---|---|---|---|---|---|
| January | 36 | 23 | 2.04 (5.9) (0) | 64 | -13 |
| February | 41 | 22 | 3.06 (6.2) (0.1) | 68 | -13 |
| March | 49 | 30 | 1.95 (6.3) (0.2) | 79 | 3 |
| April | 60 | 39 | 2.45 (6.5) (0.8) | 84 | 21 |
| May | 68 | 45 | 3.48 (8.4) (2.4) | 90 | 27 |
| June | 75 | 51 | 3 (7.3) (4.3) | 93 | 34 |
| July | 80 | 54 | 2.55 (5.6) (4.7) | 97 | 41 |
| August | 82 | 53 | 2.08 (4.6) (3.9) | 100 | 39 |
| September | 73 | 46 | 2.82 (4.9) (1.2) | 95 | 30 |
| October | 62 | 42 | 3.17 (7) (1) | 82 | 27 |
| November | 48 | 36 | 4.28 (8.6) (1.4) | 70 | 10 |
| December | 40 | 29 | 3.79 (8.6) (0.2) | 66 | -9 |
| Annual | 60 | 39 | 34.7 (79.9) (20.2) | 100 | -13 |

## Map Location 150 — Bulgaria, Sofia — 4249N 02323E 1929

| Month | Normal High | Normal Low | Precip in (mm) (days) | Record High | Record Low |
|---|---|---|---|---|---|
| January | 34 | 23 | 1.55 (4.9) (0.2) | 61 | -11 |
| February | 40 | 26 | 1.15 (4.4) (0.2) | 70 | -11 |
| March | 48 | 31 | 1.67 (4.9) (0.3) | 77 | -4 |
| April | 60 | 40 | 2.28 (6.7) (2) | 84 | 25 |
| May | 68 | 48 | 2.91 (7.9) (4.6) | 88 | 32 |
| June | 75 | 54 | 3.23 (7) (7.6) | 97 | 36 |
| July | 80 | 57 | 2.23 (5.5) (6.7) | 97 | 41 |
| August | 81 | 57 | 1.18 (2.8) (4.2) | 102 | 43 |
| September | 74 | 51 | 1.16 (2.9) (1.8) | 97 | 34 |
| October | 63 | 43 | 1.59 (4.2) (0.8) | 93 | 25 |
| November | 50 | 36 | 2.03 (6.1) (0.5) | 72 | 7 |
| December | 39 | 28 | 1.73 (5.3) (0.1) | 68 | 3 |
| Annual | 61 | 44 | 22.7 (62.6) (29) | 102 | -11 |

## Map Location 151 — Croatia, Zagreb — 4549N 01558E 535

| Month | Normal High | Normal Low | Precip in (mm) (days) | Record High | Record Low |
|---|---|---|---|---|---|
| January | 37 | 28 | 1.8 (5.8) (0) | 63 | 1 |
| February | 42 | 30 | 1.9 (6.1) (0) | 66 | -7 |
| March | 51 | 37 | 2.2 (6.3) (1) | 74 | 11 |
| April | 62 | 46 | 2.8 (6.9) (2) | 81 | 29 |
| May | 70 | 53 | 3.1 (7.1) (7) | 87 | 34 |
| June | 77 | 59 | 3.9 (8.6) (8) | 99 | 40 |
| July | 82 | 62 | 3.2 (7.7) (7) | 100 | 48 |
| August | 79 | 61 | 3.2 (7.7) (6) | 95 | 45 |
| September | 71 | 55 | 3.4 (7.9) (3) | 88 | 37 |
| October | 60 | 47 | 3.9 (8.5) (2) | 81 | 26 |
| November | 50 | 41 | 3.1 (7.5) (1) | 73 | 15 |
| December | 38 | 31 | 2.4 (7.3) (0.3) | 62 | -2 |
| Annual | 60 | 46 | 34.9 (87.4) (37.3) | 100 | -7 |

## Map Location 152 — Czech Rep., Prague — 5006N 01417E 1247

| Month | Normal High | Normal Low | Precip in (mm) (days) | Record High | Record Low |
|---|---|---|---|---|---|
| January | 32 | 23 | 0.76 (2.2) (0) | 52 | -8 |
| February | 33 | 23 | 0.9 (2.5) (0.1) | 64 | -18 |
| March | 43 | 30 | 1.05 (3.9) (0.2) | 68 | 3 |
| April | 56 | 39 | 1.43 (4.1) (1.9) | 82 | 16 |
| May | 63 | 45 | 2.29 (6) (4.4) | 84 | 25 |
| June | 70 | 52 | 2.93 (7.4) (6.5) | 88 | 34 |
| July | 72 | 55 | 3.47 (6.5) (5.7) | 93 | 41 |
| August | 72 | 54 | 2.27 (5.9) (6.4) | 95 | 41 |
| September | 66 | 49 | 1.48 (4.8) (1.9) | 86 | 32 |
| October | 55 | 41 | 1.3 (4.1) (0.2) | 77 | 25 |
| November | 42 | 34 | 1 (2.8) (0) | 63 | 14 |
| December | 34 | 27 | 1.05 (3.2) (0.2) | 84 | 0 |
| Annual | 53 | 39 | 19.9 (53.4) (27.5) | 95 | -18 |

**Table 9.2  World Weather Data [CONTINUED]**

# Europe, Denmark to France, Paris

## Map # 153 to 157

### Map Location 153 — Denmark, Copenhagen — 5538N 01240E 16

| Month | High | Low | Record High | Record Low | Precip (in)(cm)(days) |
|---|---|---|---|---|---|
| JANUARY | 35 | 28 | | | 1.49 (4.9) (0.1) |
| FEBRUARY | 36 | 27 | | | 1.22 (4) (0) |
| MARCH | 40 | 30 | 62 | -1 | 1.46 (4.8) (0) |
| APRIL | 50 | 36 | 79 | 20 | 1.61 (5.2) (0.5) |
| MAY | 61 | 44 | 85 | 26 | 1.61 (5.2) (1.1) |
| JUNE | 69 | 51 | 90 | 31 | 1.85 (5.2) (1.9) |
| JULY | 72 | 54 | 91 | 39 | 2.4 (6.4) (4.2) |
| AUGUST | 69 | 53 | 89 | 33 | 2.99 (7.4) (3.2) |
| SEPTEMBER | 63 | 48 | 86 | 26 | 1.97 (5.6) (1.1) |
| OCTOBER | 53 | 42 | 74 | 19 | 2.21 (6.1) (0.5) |
| NOVEMBER | 43 | 35 | 57 | 5 | 1.85 (5.4) (0.2) |
| DECEMBER | 38 | 31 | 54 | -1 | 2.13 (6.6) (0.2) |
| ANNUAL | 52 | 40 | 91 | -13 | 22.8 (66.8) (13) |

### Map Location 154 — Finland, Helsinki — 6019N 02457E 167

| Month | High | Low | Record High | Record Low | Precip (in)(cm)(days) |
|---|---|---|---|---|---|
| JANUARY | 27 | 17 | 47 | -27 | 1.7 (5.5) (0) |
| FEBRUARY | 26 | 14 | 49 | -29 | 1.5 (4.9) (0) |
| MARCH | 30 | 19 | 54 | -18 | 1.4 (4.6) (0) |
| APRIL | 41 | 30 | 70 | 0 | 1.8 (5.6) (2) |
| MAY | 54 | 40 | 81 | 21 | |
| JUNE | 64 | 50 | 88 | 32 | 1.8 (5.1) (2) |
| JULY | 67 | 55 | 89 | 39 | 2.2 (6) (3) |
| AUGUST | 65 | 53 | | | 2.9 (7.3) (2) |
| SEPTEMBER | 56 | 45 | 77 | 25 | 2.5 (6.6) (1) |
| OCTOBER | 46 | 38 | 63 | 14 | 2.6 (6.7) (0.3) |
| NOVEMBER | 36 | 29 | 51 | -4 | 2.5 (6.6) (0) |
| DECEMBER | 35 | 21 | 47 | -22 | 2 (6.3) (0) |
| ANNUAL | 46 | 34 | 89 | -29 | 24.3 (69.8) (10.6) |

### Map Location 155 — Finland, Ivalo — 6836N 02725E 459

| Month | High | Low | Record High | Record Low | Precip (in)(cm)(days) |
|---|---|---|---|---|---|
| JANUARY | 13 | 2 | 39 | -42 | 1.32 (4.1) (0) |
| FEBRUARY | 15 | 1 | 41 | -44 | 0.93 (2.6) (0) |
| MARCH | 25 | 7 | 48 | -35 | 0.65 (2.2) (0) |
| APRIL | 36 | 17 | 57 | -24 | 0.94 (3) (0) |
| MAY | 48 | 33 | 79 | 10 | 1.48 (5.1) (0.3) |
| JUNE | 59 | 43 | 84 | 27 | 1.92 (6.5) (1.7) |
| JULY | 65 | 48 | 90 | 28 | 2.6 (6.4) (2.2) |
| AUGUST | 59 | 44 | | | 2.55 (7.8) (0.6) |
| SEPTEMBER | 49 | 37 | 73 | 18 | 1.93 (5.3) (0.3) |
| OCTOBER | 36 | 27 | 54 | -6 | 1.4 (4.5) (0) |
| NOVEMBER | 24 | 16 | 45 | -36 | 1.24 (4) (0) |
| DECEMBER | 15 | 5 | 39 | -42 | 1.03 (3.1) (0) |
| ANNUAL | 37 | 23 | 90 | -44 | 18 (54.6) (5.1) |

### Map Location 156 — France, Nice-Côte d'Azur — 4339N 00712E 13

| Month | High | Low | Record High | Record Low | Precip (in)(cm)(days) |
|---|---|---|---|---|---|
| JANUARY | 55 | 40 | 70 | 18 | 2.52 (7.5) (0.6) |
| FEBRUARY | 56 | 40 | 72 | 20 | 2.24 (6.9) (0.6) |
| MARCH | 59 | 44 | 72 | 23 | 2.68 (6.8) (1.3) |
| APRIL | 63 | 48 | 81 | 29 | 2.24 (6.3) (1.7) |
| MAY | 69 | 54 | 87 | 36 | 2.09 (6.1) (3.3) |
| JUNE | 75 | 61 | 92 | 36 | 1.61 (4.7) (5.4) |
| JULY | 80 | 65 | 100 | | 0.59 (1.7) (3.8) |
| AUGUST | 80 | 64 | | | 0.94 (2.8) (3.6) |
| SEPTEMBER | 77 | 61 | 92 | 40 | 2.36 (6.3) (4) |
| OCTOBER | 69 | 54 | 82 | 34 | 5.83 (9.7) (3.1) |
| NOVEMBER | 62 | 46 | 77 | 28 | 4.37 (9) (1.7) |
| DECEMBER | 56 | 41 | 77 | 21 | 3.07 (8.5) (0.8) |
| ANNUAL | 67 | 53 | 100 | 18 | 30.9 (76.8) (29.6) |

### Map Location 157 — France, Paris — 4843N 00224E 292

| Month | High | Low | Record High | Record Low | Precip (in)(cm)(days) |
|---|---|---|---|---|---|
| JANUARY | 43 | 33 | 58 | 9 | 2.17 (6.7) (0) |
| FEBRUARY | 45 | 34 | 63 | 5 | 1.77 (5.7) (0.2) |
| MARCH | 53 | 37 | 76 | 23 | 1.38 (4.6) (0.7) |
| APRIL | 59 | 42 | 86 | 29 | 1.65 (5.3) (1.1) |
| MAY | 66 | 48 | 89 | 32 | |
| JUNE | 72 | 53 | 99 | 39 | 2.17 (5.9) (3.1) |
| JULY | 76 | 57 | 103 | 46 | 2.17 (5.9) (3.7) |
| AUGUST | 75 | 56 | | | 2.56 (6.7) (3.2) |
| SEPTEMBER | 69 | 52 | 91 | 38 | 2.17 (6) (1.7) |
| OCTOBER | 60 | 46 | 82 | 25 | 1.97 (5.6) (0.2) |
| NOVEMBER | 48 | 39 | 68 | 21 | 2.13 (5.9) (0.2) |
| DECEMBER | 42 | 34 | 61 | 9 | 2.01 (6.3) (0) |
| ANNUAL | 59 | 44 | 103 | 5 | 24.2 (70.7) (17.4) |

Table 9.2   World Weather Data [CONTINUED]

# Europe, France, Pic du Midi Observatory to Greece                Map # 158 to 162

For each month: two temperature columns (mean / record) and two moisture columns (precipitation / snowfall). Precipitation and snowfall given as mean (greatest)(days). Temperatures in °F.

## Map Location 158 — France, Pic du Midi — 4256N 00008E 9378

| Month | Mean Max | Mean Min | Rec Max | Rec Min | Precip | Snow |
|---|---|---|---|---|---|---|
| JANUARY | 24 | 14 | 50 | -23 | 7.1 (11.9)(0) | *** (***) |
| FEBRUARY | 23 | 13 | 50 | -27 | 17.8 (9.3)(0) | *** (***) |
| MARCH | 28 | 17 | 55 | -8 | 5.9 (9.7)(0) | *** (***) |
| APRIL | 31 | 19 | 52 | -5 | 2.8 (9.3)(1) | *** (***) |
| MAY | 37 | 25 | 57 | 0 | 5.8 (12.2)(2) | *** (***) |
| JUNE | 45 | 33 | 72 | 10 | 1.7 (4.6)(5) | *** (***) |
| JULY | 51 | 39 | 68 | 16 | 0.2 (1.9)(5) | *** (***) |
| AUGUST | 50 | 38 | 70 | 18 | 0.3 (0)(5) | *** (***) |
| SEPTEMBER | 45 | 34 | 62 | 7 | 6.5 (12.0)(4) | *** (***) |
| OCTOBER | 36 | 27 | 62 | -4 | 6.9 (15.5)(1) | *** (***) |
| NOVEMBER | 30 | 20 | 49 | -7 | 5.5 (6.5)(0) | *** (***) |
| DECEMBER | 24 | 15 | 45 | -13 | 14.2 (16.2)(0) | *** (***) |
| ANNUAL | 35 | 25 | 72 | -27 | 74.5 (109.1)(23) | *** (***) |

## Map Location 159 — Germany, Berlin — 5233N 01318E 115

| Month | Mean Max | Mean Min | Rec Max | Rec Min | Precip | Snow |
|---|---|---|---|---|---|---|
| JANUARY | 35 | 26 | 55 | -1 | 1.9 (6.1)(0) | 5.3 (1.4) |
| FEBRUARY | 38 | 27 | 62 | -15 | 1.3 (4.3)(0) | 6.2 (1.3) |
| MARCH | 46 | 32 | 73 | 7 | 1.5 (4.9)(1) | 1.2 (0.3) |
| APRIL | 55 | 38 | 85 | 20 | 1.7 (5.4)(1) | 0.1 (0) |
| MAY | 65 | 46 | 92 | 28 | 1.9 (5.8)(4) | 0 (0) |
| JUNE | 70 | 51 | 94 | 35 | 2.3 (6.2)(5) | 0 (0) |
| JULY | 74 | 55 | 101 | 43 | 3.1 (7.6)(5) | 0 (0) |
| AUGUST | 72 | 54 | 94 | 43 | 2.2 (6)(4) | 0 (0) |
| SEPTEMBER | 66 | 48 | 94 | 31 | 1.9 (5.5)(1) | 0 (0) |
| OCTOBER | 55 | 41 | 77 | 15 | 1.7 (5.1)(0.3) | 0 (0) |
| NOVEMBER | 43 | 33 | 64 | 8 | 1.7 (5.1)(0.3) | 1.7 (0.3) |
| DECEMBER | 37 | 29 | 60 | -3 | 1.9 (6.1)(0.3) | 1.4 (0.2) |
| ANNUAL | 55 | 40 | 101 | -15 | 23.1 (68.1)(21.9) | 15.9 (3.5) |

## Map Location 160 — Germany, Munich — 4808N 01142E 1728

| Month | Mean Max | Mean Min | Rec Max | Rec Min | Precip | Snow |
|---|---|---|---|---|---|---|
| JANUARY | 36 | 22 | 62 | -14 | 2.72 (7.2) | 12.4 (3) |
| FEBRUARY | 36 | 21 | 64 | -29 | 2 (5.7) | 11 (2.5) |
| MARCH | 48 | 29 | 75 | 1 | 1.97 (5.6) | 3.1 (0.6) |
| APRIL | 56 | 36 | 78 | 10 | 2.31 (6.5) | 2.4 (0.3) |
| MAY | 64 | 42 | 86 | 23 | 4.3 (9.6) | 0.6 (0.2) |
| JUNE | 69 | 49 | 93 | 34 | 4.89 (10.8) | 0 (0) |
| JULY | 73 | 53 | 96 | 22 | 6.02 (10.8) | 0 (0) |
| AUGUST | 72 | 51 | 95 | 32 | 3.79 (9.1) | 0 (0) |
| SEPTEMBER | 67 | 46 | 89 | 31 | 3.02 (7.4) | 0 (0) |
| OCTOBER | 56 | 37 | 78 | 21 | 1.67 (5) | 1.9 (0.5) |
| NOVEMBER | 44 | 31 | 68 | -2 | 2.29 (5.6) | 3.4 (0.7) |
| DECEMBER | 38 | 26 | 60 | 0 | 1.87 (6.4) | 9.5 (1.8) |
| ANNUAL | 55 | 37 | 96 | -29 | 36.8 (89.7) | 44.3 (9.6) |

## Map Location 161 — Gibraltar, U.K. — 3609N 00521W 8

| Month | Mean Max | Mean Min | Rec Max | Rec Min | Precip | Snow |
|---|---|---|---|---|---|---|
| JANUARY | 60 | 50 | 74 | 36 | 6.06 (11)(2) | 0 (0) |
| FEBRUARY | 62 | 51 | 75 | 33 | 3.98 (9.8)(1) | 0 (0) |
| MARCH | 65 | 54 | 81 | 38 | 4.69 (7.7)(1) | 0 (0) |
| APRIL | 69 | 56 | 82 | 45 | 2.44 (6.6)(1) | 0 (0) |
| MAY | 73 | 60 | 87 | 47 | 1.06 (3.6)(1) | 0 (0) |
| JUNE | 79 | 65 | 91 | 57 | 0.2 (0.3)(1) | 0 (0) |
| JULY | 83 | 68 | 101 | 58 | 0.04 (0)(1) | 0 (0) |
| AUGUST | 84 | 70 | 99 | 57 | 0.12 (0.1)(1) | 0 (0) |
| SEPTEMBER | 80 | 66 | 92 | 57 | 0.39 (2.1)(1) | 0 (0) |
| OCTOBER | 74 | 63 | 92 | 50 | 2.52 (6.6)(1) | 0 (0) |
| NOVEMBER | 67 | 57 | 84 | 46 | 5.67 (9.7)(1) | 0 (0) |
| DECEMBER | 62 | 53 | 73 | 42 | 4.92 (10.5)(1) | 0 (0) |
| ANNUAL | 72 | 59 | 101 | 33 | 32.1 (68)(12.3) | 0 (0) |

## Map Location 162 — Greece, Athens — 3753N 02343E 90

| Month | Mean Max | Mean Min | Rec Max | Rec Min | Precip | Snow |
|---|---|---|---|---|---|---|
| JANUARY | 54 | 42 | 72 | 20 | 2.2 (6.8)(1) | *** (***) |
| FEBRUARY | 55 | 43 | 73 | 21 | 1.6 (5.2)(1) | *** (***) |
| MARCH | 60 | 46 | 95 | 44 | 1.4 (4.6)(1) | *** (***) |
| APRIL | 67 | 52 | 91 | 35 | 0.8 (2.7)(2) | 0* (0*) |
| MAY | 77 | 60 | 104 | 42 | 0.8 (2.7)(2) | *** (***) |
| JUNE | 85 | 67 | 109 | 54 | 0.6 (1.7)(2) | *** (***) |
| JULY | 90 | 72 | 111 | 59 | 0.2 (0.3)(1) | *** (***) |
| AUGUST | 90 | 72 | 106 | 58 | 0.4 (1.1)(1) | *** (***) |
| SEPTEMBER | 83 | 66 | 103 | 48 | 0.6 (2.6)(1) | *** (***) |
| OCTOBER | 74 | 60 | 94 | 44 | 1.7 (5.1)(2) | *** (***) |
| NOVEMBER | 64 | 52 | 87 | 30 | 2.8 (7.1)(2) | *** (***) |
| DECEMBER | 57 | 46 | 87 | 24 | 2.8 (8.1)(1) | *** (***) |
| ANNUAL | 71 | 57 | 111 | 20 | 15.9 (48)(16) | *** (***) |

**Table 9.2  World Weather Data [CONTINUED]**

# Europe, Hungary to Kosovo

## Map # 163 to 167

*Temperature columns: Normal High / Record High / Normal Low / Record Low (°F). Precipitation: mean (greatest) (least) inches.*

### Map Location 163 — Hungary, Budapest — 4731N 01902E 425

| Month | Norm High | Rec High | Norm Low | Rec Low | Precipitation |
|---|---|---|---|---|---|
| JANUARY | 34 | 54 | 27 | -2 | 1.67 (5.2) (0.2) |
| FEBRUARY | 35 | 50 | 26 | 4 | 1.68 (4.6) (0.2) |
| MARCH | 47 | 72 | 34 | 14 | 1.04 (3.1) (0) |
| APRIL | 61 | 82 | 43 | 28 | 1.69 (5.1) (1.3) |
| MAY | 70 | 90 | 51 | 36 | 2.29 (6.1) (4.6) |
| JUNE | 78 | 90 | 60 | 46 | 2.28 (5.3) (6.3) |
| JULY | 81 | 97 | 62 | 48 | 1.72 (3.7) (3.4) — 0 (0) |
| AUGUST | 81 | 102 | 61 | 52 | 2.16 (4.7) (4.4) — 0 (0) |
| SEPTEMBER | 73 | 90 | 54 | 41 | 2.16 (2.7) (1) — 0 (0*) |
| OCTOBER | 61 | 81 | 45 | 28 | 2.5 (4.6) (0.4) — *** (***) |
| NOVEMBER | 45 | 64 | 37 | 19 | 2.02 (4.6) (0) — *** (***) |
| DECEMBER | 40 | 59 | 34 | 14 | 2.52 (6.9) (0) — *** (***) |
| ANNUAL | | 102 | | -2 | 23.7 (56.6) (21.8) — *** (***) |

### Map Location 164 — Iceland, Reykjavik — 6407N 02156W 45

| Month | Norm High | Rec High | Norm Low | Rec Low | Precipitation |
|---|---|---|---|---|---|
| JANUARY | 36 | 50 | 28 | 0 | 4 (11.1) (0.2) |
| FEBRUARY | 37 | 50 | 28 | 4 | 3.1 (9.8) (0.4) |
| MARCH | 39 | 58 | 30 | 6 | 3 (10.2) (0) |
| APRIL | 43 | 59 | 33 | 10 | 2.1 (8.6) (0) |
| MAY | 50 | 64 | 39 | 16 | 1.6 (5.5) (0) |
| JUNE | 55 | 70 | 44 | 31 | 1.7 (3.3) (0) |
| JULY | 58 | 71 | 47 | 30 | 2.6 (8.7) (0) |
| AUGUST | 57 | 74 | 47 | 39 | 2 (5.4) (0.1) |
| SEPTEMBER | 51 | 68 | 42 | 27 | 3.1 (10) (0.2) |
| OCTOBER | 44 | 59 | 36 | 14 | 3.4 (10.1) (0.1) |
| NOVEMBER | 39 | 53 | 32 | 10 | 3.6 (16.7) (0.3) |
| DECEMBER | 38 | 53 | 30 | 3 | 3.7 (12.2) (0) |
| ANNUAL | | 74 | | 0 | 33.9 (111.6) (1.3) — *** (***) |

### Map Location 165 — Ireland, Dublin — 5322N 00621W 155

| Month | Norm High | Rec High | Norm Low | Rec Low | Precipitation |
|---|---|---|---|---|---|
| JANUARY | 46 | 62 | 35 | 4 | 2.7 (7.9) (0.3) |
| FEBRUARY | 47 | 65 | 34 | 8 | 2.19 (6.8) (0.3) |
| MARCH | 49 | 72 | 35 | 15 | 2.01 (6) (0.3) |
| APRIL | 53 | 72 | 37 | 19 | 1.86 (5.7) (0.3) |
| MAY | 58 | 80 | 42 | 22 | 2.34 (6.5) (1) |
| JUNE | 64 | 84 | 47 | 31 | 1.95 (5.5) (1) |
| JULY | 66 | 85 | 51 | 33 | 2.84 (7.2) (2) |
| AUGUST | 65 | 86 | 50 | 35 | 2.98 (7.4) (1) |
| SEPTEMBER | 62 | 82 | 46 | 29 | 2.76 (7) (1) |
| OCTOBER | 55 | 73 | 41 | 22 | 2.73 (6.9) (0.3) |
| NOVEMBER | 50 | 67 | 38 | 15 | 2.65 (6.8) (0.3) |
| DECEMBER | 47 | 63 | 35 | 7 | 2.64 (7.8) (0) |
| ANNUAL | | 86 | | 4 | 29.6 (81.5) (7.8) — *** (***) |

### Map Location 166 — Italy, Rome — 4148N 01236E 430

| Month | Norm High | Rec High | Norm Low | Rec Low | Precipitation |
|---|---|---|---|---|---|
| JANUARY | 52 | 65 | 37 | 23 | 3.23 (8.8) (1) |
| FEBRUARY | 54 | 67 | 38 | 20 | 2.68 (7.8) (1.4) |
| MARCH | 61 | 77 | 43 | 23 | 2.87 (7) (0.9) |
| APRIL | 67 | 79 | 48 | 28 | 2.6 (6.7) (2.1) |
| MAY | 75 | 92 | 55 | 42 | 2.17 (6.2) (2.4) |
| JUNE | 82 | 99 | 61 | 49 | 1.58 (4.6) (1.4) |
| JULY | 88 | 103 | 65 | 54 | 0.67 (2) (2.6) |
| AUGUST | 87 | 105 | 65 | 55 | |
| SEPTEMBER | 82 | 103 | 61 | 46 | 2.56 (6.7) (2.9) |
| OCTOBER | 72 | 87 | 53 | 36 | 5.04 (9.4) (3.8) |
| NOVEMBER | 62 | 74 | 45 | 28 | 4.41 (9) (3) |
| DECEMBER | 56 | 70 | 41 | 29 | 3.86 (9.6) (1.3) |
| ANNUAL | | 105 | | 20 | 32.7 (80.9) (24.9) — 0.1 (0) |

### Map Location 167 — Kosovo, Pristina — 4240N 02110E 2139

| Month | Norm High | Rec High | Norm Low | Rec Low | Precipitation |
|---|---|---|---|---|---|
| JANUARY | 39 | 58 | 25 | -8 | 2.28 (7) (0.3) |
| FEBRUARY | 43 | 63 | 25 | -3 | 2.6 (7.7) (0.3) |
| MARCH | 53 | 79 | 33 | 3 | 2.52 (6.7) (0.3) |
| APRIL | 64 | 85 | 41 | 20 | 2.91 (7) (0.3) |
| MAY | 71 | 87 | 49 | 31 | 3.54 (7.3) (1) |
| JUNE | 79 | 95 | 55 | 42 | 3.7 (8.4) (3) |
| JULY | 84 | 103 | 60 | 41 | 1.81 (5.1) (4) |
| AUGUST | 83 | 103 | 58 | 44 | 1.73 (5) (2) |
| SEPTEMBER | 77 | 93 | 53 | 32 | 2.21 (6.1) (2) |
| OCTOBER | 65 | 87 | 45 | 23 | 4.49 (9.1) (1) |
| NOVEMBER | 55 | 73 | 37 | 12 | 3.58 (8.1) (0.3) |
| DECEMBER | 40 | 64 | 26 | -2 | 4.57 (10.3) (0.3) |
| ANNUAL | | 103 | | -8 | 35.9 (87.8) (14.8) — *** (***) |

**Europe, Latvia to Moldova**

Table 9.2  World Weather Data [CONTINUED]

# Map # 168 to 172

## Map Location 168 — Latvia, Riga — 5658N 02404E 10

| | JAN | FEB | MAR | APR | MAY | JUN | JUL | AUG | SEP | OCT | NOV | DEC | ANNUAL |
|---|---|---|---|---|---|---|---|---|---|---|---|---|---|
| Normal Max | 28 | 29 | 36 | 49 | 61 | 67 | 71 | 69 | 61 | 49 | 39 | 31 | 49 |
| Record Max | 45 | 52 | 68 | 77 | 86 | 90 | 93 | 91 | 84 | 75 | 64 | 54 | 93 |
| Normal Min | 17 | 18 | 23 | 33 | 41 | 48 | 54 | 52 | 46 | 38 | 31 | 24 | 35 |
| Record Min | -26 | -31 | -22 | 9 | 19 | 28 | 37 | 36 | 23 | 12 | -8 | -15 | -31 |
| Precip (in)(days)(snow) | 1.3 (14) (0) | 1.15 (11) (0.1) | 1.33 (10.7) (0.2) | 1.04 (7.6) (0.6) | 1.51 (8.8) (3.3) | 2.04 (8.3) (4.9) | 2.66 (10.4) (5.6) | 2.83 (11) (4) | 3.1 (11.6) (2.2) | 2.2 (10.6) (0.4) | 2.14 (11.2) (0.5) | 2.05 (12.9) (0) | 23.3 (128.1) (21.8) |

## Map Location 169 — Lithuania, Vilnius — 5453N 02353E 246

| | JAN | FEB | MAR | APR | MAY | JUN | JUL | AUG | SEP | OCT | NOV | DEC | ANNUAL |
|---|---|---|---|---|---|---|---|---|---|---|---|---|---|
| Normal Max | 25 | 28 | 36 | 52 | 63 | 71 | 72 | 70 | 63 | 52 | 39 | 30 | 50 |
| Record Max | 41 | 46 | 70 | 75 | 82 | 88 | 91 | 91 | 84 | 75 | 63 | 52 | 91 |
| Normal Min | 18 | 18 | 24 | 36 | 45 | 52 | 54 | 53 | 47 | 41 | 32 | 23 | 37 |
| Record Min | -15 | -13 | -11 | 19 | 28 | 39 | 36 | 32 | 28 | 18 | 7 | -11 | -15 |
| Precip (in)(days)(snow) | 1.61 (12.4) (0) | 1.31 (12.1) (0.1) | 1.34 (10) (0.2) | 1.19 (7.5) (0.3) | 3.01 (11.5) (3.9) | 2.65 (8.9) (4.8) | 1.58 (8.1) (5) | 2.72 (11.7) (4.4) | 2.12 (9.1) (2.5) | 1.72 (9.4) (0.3) | 2.11 (11.2) (0) | 1.73 (12.5) (0) | 23.1 (124.4) (21.5) |

## Map Location 170 — Luxembourg, Luxembourg City — 4937N 00603E 1083

| | JAN | FEB | MAR | APR | MAY | JUN | JUL | AUG | SEP | OCT | NOV | DEC | ANNUAL |
|---|---|---|---|---|---|---|---|---|---|---|---|---|---|
| Normal Max | 36 | 40 | 49 | 58 | 65 | 71 | 74 | 73 | 65 | 56 | 45 | 39 | 56 |
| Record Max | 57 | 63 | 73 | 85 | 91 | 98 | 99 | 97 | 91 | 80 | 66 | 58 | 99 |
| Normal Min | 29 | 30 | 33 | 40 | 46 | 52 | 55 | 55 | 50 | 43 | 37 | 32 | 42 |
| Record Min | -10 | -7 | 7 | 21 | 26 | 29 | 38 | 33 | 23 | 19 | 13 | -8 | -10 |
| Precip (in)(days)(snow) | 2.3 (7) (0.3) | 2 (6.3) (0.3) | 1.9 (5.8) (0.3) | 2.1 (6.1) (2) | 2.4 (6.5) (3) | 2.5 (6.6) (5) | 2.8 (7.1) (5) | 2.6 (6.8) (5) | 2.4 (6.4) (2) | 2.7 (6.9) (0.3) | 2.7 (6.9) (0.3) | 2.8 (8.1) (0) | 29.2 (80.5) (23.5) |

## Map Location 171 — Macedonia, Skopje — 4159N 02128E 784

| | JAN | FEB | MAR | APR | MAY | JUN | JUL | AUG | SEP | OCT | NOV | DEC | ANNUAL |
|---|---|---|---|---|---|---|---|---|---|---|---|---|---|
| Normal Max | 40 | 44 | 55 | 66 | 74 | 83 | 89 | 87 | 80 | 69 | 56 | 42 | 65 |
| Record Max | 68 | 75 | 94 | 86 | 97 | 102 | 106 | 106 | 103 | 94 | 86 | 70 | 106 |
| Normal Min | 26 | 27 | 34 | 42 | 50 | 56 | 60 | 58 | 52 | 45 | 38 | 29 | 43 |
| Record Min | -14 | -11 | -1 | 24 | 28 | 37 | 42 | 42 | 26 | 24 | 12 | -7 | -14 |
| Precip (in)(days)(snow) | 1.81 (5.8) (0) | 1.61 (5.3) (0) | 1.5 (4.9) (0.3) | 1.34 (4.5) (0.3) | 2.05 (6) (4) | 1.93 (5.4) (5) | 1.38 (4.1) (4) | 1.47 (4.3) (4) | 1.65 (5) (2) | 2.28 (6.2) (1) | 2.8 (7.1) (0.3) | 1.69 (5.5) (0.3) | 21.5 (64.1) (21.2) |

## Map Location 172 — Moldova, Chisinau — 4700N 02854E 115

| | JAN | FEB | MAR | APR | MAY | JUN | JUL | AUG | SEP | OCT | NOV | DEC | ANNUAL |
|---|---|---|---|---|---|---|---|---|---|---|---|---|---|
| Normal Max | 32 | 34 | 45 | 60 | 71 | 77 | 79 | 79 | 72 | 60 | 46 | 37 | 58 |
| Record Max | *** | *** | *** | *** | *** | *** | *** | *** | *** | *** | *** | *** | *** |
| Normal Min | 21 | 24 | 31 | 42 | 52 | 58 | 61 | 60 | 52 | 43 | 35 | 27 | 42 |
| Record Min | *** | *** | *** | *** | *** | *** | *** | *** | *** | *** | *** | *** | *** |
| Precip (in)(days)(snow) | *** (***) (***) | *** (***) (***) | *** (***) (***) | *** (***) (***) | *** (***) (***) | *** (***) (***) | *** (***) (***) | *** (***) (***) | *** (***) (***) | *** (***) (***) | *** (***) (***) | *** (***) (***) | *** (***) (***) |

**Table 9.2 World Weather Data [CONTINUED]**

# Europe, Montenegro to Poland

# Map # 173 to 177

**Map Location 173 — Montenegro, Podgorica — 4221N 01915E 121**

| Month | Rec High | Avg High | Avg Low | Rec Low | Precip (Snow) (Days) |
|---|---|---|---|---|---|
| JANUARY | 61 | 48 | 37 | 14 | 7.05 (11.2) (2) |
| FEBRUARY | 68 | 52 | 39 | 19 | 7.68 (11.3) (2) |
| MARCH | 79 | 58 | 43 | 25 | 5.32 (8.2) (2) |
| APRIL | 88 | 69 | 50 | 36 | 3.86 (7.4) (2) |
| MAY | 90 | 74 | 55 | 41 | 4.13 (7.5) (4) |
| JUNE | 99 | 85 | 65 | 52 | 2.36 (6.3) (5) |
| JULY | 104 | 90 | 69 | 57 | 1.58 (4.6) (3) |
| AUGUST | 106 | 93 | 71 | 52 | 2.48 (6.6) (2) |
| SEPTEMBER | 102 | 82 | 63 | 52 | 4.45 (9*) (3) |
| OCTOBER | 88 | 71 | 55 | 41 | 7.95 (***) (4) |
| NOVEMBER | 79 | 60 | 46 | 27 | 8.39 (***) (3) |
| DECEMBER | 68 | 52 | 41 | 19 | 9.02 (11.7) (2) |
| ANNUAL | 106 | | | 14 | 64.3 (72.7) (34) |

**Map Location 174 — Netherlands, Amsterdam — 5219N 00447E -10**

| Month | Rec High | Avg High | Avg Low | Rec Low | Precip (Snow) (Days) |
|---|---|---|---|---|---|
| JANUARY | 55 | 40 | 34 | 3 | 2 (6.3) (0.3) |
| FEBRUARY | 61 | 41 | 34 | -4 | 1.4 (4.6) (0.3) |
| MARCH | 66 | 46 | 37 | 15 | 1.3 (4.4) (1) |
| APRIL | 77 | 52 | 43 | 27 | 1.6 (5.2) (2) |
| MAY | 88 | 60 | 50 | 30 | 1.8 (5.6) (3) |
| JUNE | 95 | 65 | 55 | 39 | 1.8 (5.1) (5) |
| JULY | 91 | 69 | 59 | 43 | 2.6 (6.8) (5) |
| AUGUST | 90 | 68 | 56 | 43 | 2.8 (7.1) (4) |
| SEPTEMBER | 77 | 64 | 45 | 26 | 2.5 (6.6) (1) |
| OCTOBER | 70 | 56 | 48 | 27 | 2.9 (7.2) (0.3) |
| NOVEMBER | 63 | 47 | 41 | 16 | 2.6 (6.7) (1) |
| DECEMBER | 57 | 41 | 35 | 6 | 2.2 (6.8) (1) |
| ANNUAL | 95 | | | -4 | 25.6 (72.7) (29.6) |

**Map Location 175 — Norway, Oslo — 5954N 01037E 56**

| Month | Rec High | Avg High | Avg Low | Rec Low | Precip (Snow) (Days) |
|---|---|---|---|---|---|
| JANUARY | 53 | 30 | 20 | -21 | 1.7 (5.5) (0) |
| FEBRUARY | 57 | 32 | 20 | -18 | 1.3 (4.3) (0) |
| MARCH | 63 | 40 | 25 | -10 | 1.4 (4.6) (0) |
| APRIL | 75 | 50 | 34 | 5 | 1.6 (5.2) (0) |
| MAY | 84 | 62 | 43 | 26 | 1.8 (5.6) (1) |
| JUNE | 93 | 69 | 51 | 33 | 2.4 (6.4) (3) |
| JULY | 91 | 73 | 56 | 42 | 2.9 (7.3) (6) |
| AUGUST | 88 | 69 | 53 | 37 | 3.8 (8.5) (4) |
| SEPTEMBER | 77 | 60 | 45 | 26 | 2.5 (6.6) (1) |
| OCTOBER | 70 | 49 | 37 | 12 | 2.9 (7.2) (0.3) |
| NOVEMBER | 57 | 37 | 29 | 2 | 2.3 (7) (0.3) |
| DECEMBER | 54 | 31 | 24 | -10 | 2.3 (7) (0.3) |
| ANNUAL | 93 | | | -21 | 26.9 (74.4) (15.9) |

**Map Location 176 — Norway, Vardo — 7022N 03106E 43**

| Month | Rec High | Avg High | Avg Low | Rec Low | Precip (Snow) (Days) |
|---|---|---|---|---|---|
| JANUARY | 42 | 27 | 19 | -9 | 2.5 (7.5) (0) |
| FEBRUARY | 43 | 26 | 18 | -11 | 2.5 (7.5) (0.3) |
| MARCH | 47 | 29 | 20 | -3 | 2.3 (6.4) (0) |
| APRIL | 56 | 34 | 26 | 6 | 1.5 (4.9) (0) |
| MAY | 69 | 40 | 32 | 14 | 1.3 (4.4) (0) |
| JUNE | 78 | 47 | 38 | 25 | 1.3 (3.8) (1) |
| JULY | 80 | 53 | 44 | 30 | 1.5 (4.4) (1) |
| AUGUST | 77 | 53 | 44 | 30 | 1.7 (4.9) (0.3) |
| SEPTEMBER | 68 | 47 | 40 | 23 | 1.9 (5.5) (0.3) |
| OCTOBER | 56 | 38 | 32 | 8 | 2.1 (5.9) (0) |
| NOVEMBER | 49 | 33 | 26 | 5 | 2.4 (7.3) (0) |
| DECEMBER | 44 | 30 | 22 | -3 | 2.4 (7.3) (0) |
| ANNUAL | 80 | | | -11 | 23.5 (69.1) (2.9) |

**Map Location 177 — Poland, Warsaw — 5211N 02058E 351**

| Month | Rec High | Avg High | Avg Low | Rec Low | Precip (Snow) (Days) |
|---|---|---|---|---|---|
| JANUARY | 52 | 30 | 22 | -18 | 0.97 (3) (0.1) |
| FEBRUARY | 54 | 31 | 21 | -17 | 1.39 (4) (0.2) |
| MARCH | 64 | 40 | 27 | -8 | 0.86 (3.1) (0.1) |
| APRIL | 82 | 55 | 37 | 19 | 1.28 (4) (1.7) |
| MAY | 86 | 64 | 46 | 27 | 2.08 (5.7) (5) |
| JUNE | 91 | 72 | 52 | 36 | 2.7 (6.3) (6.5) |
| JULY | 95 | 74 | 56 | 41 | 3.08 (7.3) (7) |
| AUGUST | 95 | 73 | 54 | 41 | 2.03 (5) (4.8) |
| SEPTEMBER | 88 | 66 | 47 | 32 | 1.75 (5.1) (2.2) |
| OCTOBER | 77 | 55 | 40 | 18 | 1.19 (4.4) (0.2) |
| NOVEMBER | 63 | 42 | 34 | 14 | 1.67 (5.3) (0.1) |
| DECEMBER | 59 | 33 | 27 | -9 | 1.36 (4.4) (0) |
| ANNUAL | 95 | | | -18 | 20.4 (57.6) (27.9) |

Table 9.2  World Weather Data [CONTINUED]

# Europe, Portugal to Slovenia

## Map # 178 to 182

### Map Location 178 — Portugal, Lisbon — 3846N 00908W 361

| | JAN | FEB | MAR | APR | MAY | JUN | JUL | AUG | SEP | OCT | NOV | DEC | ANNUAL |
|---|---|---|---|---|---|---|---|---|---|---|---|---|---|
| Record High | 68 | 77 | 83 | 87 | 94 | 99 | 103 | 102 | 99 | 93 | 77 | 66 | 103 |
| Normal Max | 56 | 58 | 61 | 64 | 69 | 75 | 79 | 80 | 76 | 69 | 62 | 57 | 67 |
| Normal Min | 46 | 47 | 49 | 52 | 56 | 60 | 63 | 64 | 62 | 57 | 52 | 47 | 55 |
| Record Low | 30 | 28 | 34 | 37 | 42 | 49 | 52 | 52 | 51 | 43 | 34 | 31 | 28 |
| Precipitation | 3.3 (8.9) (0.4) | 3.2 (8.8) (0.6) | 3.1 (7.1) (0.9) | 2.4 (6.5) (0.9) | 1.7 (5.4) (0.9) | 0.7 (2.1) (0.5) | 0.2 (0.3) (0.3) | 0.2 (0.3) (0.2) | 1.4 (4.4) (0.9) | 3.1 (7.5) (0.5) | 4.2 (8.8) (0.6) | 3.6 (9.3) (0.6) | 27.1 (69.4) (7.3) |
| Snow | 0 (0) | 0 (0) | 0 (0) | 0 (0) | 0 (0) | 0 (0) | 0 (0) | 0 (0) | 0 (0) | 0 (0) | 0 (0) | 0 (0) | 0 (0) |

### Map Location 179 — Romania, Bucharest — 4430N 02606E 302

| | JAN | FEB | MAR | APR | MAY | JUN | JUL | AUG | SEP | OCT | NOV | DEC | ANNUAL |
|---|---|---|---|---|---|---|---|---|---|---|---|---|---|
| Record High | 57 | 63 | 77 | 90 | 91 | 99 | 104 | 104 | 99 | 95 | 75 | 64 | 104 |
| Normal Max | 34 | 38 | 48 | 63 | 72 | 81 | 85 | 86 | 77 | 66 | 50 | 38 | 62 |
| Normal Min | 22 | 24 | 31 | 42 | 51 | 58 | 61 | 60 | 53 | 45 | 36 | 28 | 43 |
| Record Low | -11 | -15 | 3 | 27 | 34 | 43 | 50 | 46 | 34 | 27 | 10 | 3 | -15 |
| Precipitation | 1.81 (4.8) (0) | 1.51 (4.4) (0) | 1.3 (3.8) (0.5) | 1.94 (4.6) (1.1) | 3.36 (8) (7) | 2.92 (5.8) (8.4) | 2.32 (4.4) (6.2) | 1.76 (3.6) (5.4) | 1.52 (3.4) (1.7) | 1.44 (3) (0.5) | 2.55 (5.6) (0.4) | 1.85 (5.4) (0) | 24.3 (56.8) (31.2) |
| Snow | *** (***) | *** (***) | *** (***) | *** (***) | *** (***) | *** (***) | *** (***) | *** (***) | *** (***) | *** (***) | *** (***) | *** (***) | *** (***) |

### Map Location 180 — Serbia, Belgrade — 4448N 02028E 433

| | JAN | FEB | MAR | APR | MAY | JUN | JUL | AUG | SEP | OCT | NOV | DEC | ANNUAL |
|---|---|---|---|---|---|---|---|---|---|---|---|---|---|
| Record High | 62 | 68 | 80 | 88 | 92 | 98 | 103 | 107 | 95 | 94 | 85 | 64 | 107 |
| Normal Max | 37 | 41 | 53 | 64 | 74 | 79 | 84 | 83 | 76 | 65 | 52 | 40 | 62 |
| Normal Min | 27 | 27 | 35 | 45 | 53 | 58 | 61 | 60 | 55 | 47 | 39 | 30 | 45 |
| Record Low | -2 | -14 | 6 | 21 | 29 | 41 | 49 | 45 | 35 | 9 | 12 | -3 | -14 |
| Precipitation | 1.89 (6) (0) | 1.81 (5.8) (0) | 1.81 (5.6) (0.3) | 2.13 (6.2) (3) | 2.95 (7) (6) | 3.78 (8.5) (7) | 2.36 (6.3) (7) | 2.17 (5.9) (4) | 1.97 (5.6) (3) | 2.17 (6) (1) | 2.4 (6.4) (0.3) | 2.17 (6.7) (0) | 27.6 (76) (31.6) |
| Snow | *** (***) | *** (***) | *** (***) | *** (***) | *** (***) | *** (***) | *** (***) | *** (***) | *** (***) | *** (***) | *** (***) | *** (***) | *** (***) |

### Map Location 181 — Slovak Republic, Bratislava — 4810N 01713E 433

| | JAN | FEB | MAR | APR | MAY | JUN | JUL | AUG | SEP | OCT | NOV | DEC | ANNUAL |
|---|---|---|---|---|---|---|---|---|---|---|---|---|---|
| Record High | 59 | 62 | 72 | 82 | 90 | 97 | 101 | 100 | 94 | 83 | 69 | 58 | 101 |
| Normal Max | 33 | 38 | 49 | 60 | 69 | 75 | 80 | 78 | 72 | 59 | 47 | 37 | 58 |
| Normal Min | 23 | 26 | 33 | 41 | 49 | 55 | 58 | 57 | 51 | 43 | 37 | 28 | 42 |
| Record Low | -15 | -25 | 0 | 18 | 29 | 36 | 44 | 42 | 30 | 19 | 13 | -9 | -25 |
| Precipitation | 1.81 (5.8) (0.1) | 1.54 (5) (0) | 1.58 (4.9) (0.3) | 2.09 (6) (1.3) | 2.56 (7.1) (5.5) | 2.01 (5.1) (6.8) | 2.76 (6.8) (7.4) | 2.52 (6.3) (4.6) | 1.97 (5.4) (1.5) | 2.13 (5.7) (0.3) | 2.72 (6.5) (0.1) | 2.21 (7) (0) | 25.9 (71.6) (27.9) |
| Snow | *** (***) | *** (***) | *** (***) | *** (***) | *** (***) | *** (***) | *** (***) | *** (***) | *** (***) | *** (***) | *** (***) | *** (***) | *** (***) |

### Map Location 182 — Slovenia, Ljubljana — 4603N 01433E 951

| | JAN | FEB | MAR | APR | MAY | JUN | JUL | AUG | SEP | OCT | NOV | DEC | ANNUAL |
|---|---|---|---|---|---|---|---|---|---|---|---|---|---|
| Record High | 57 | 66 | 74 | 81 | 90 | 100 | 102 | 95 | 91 | 80 | 69 | 60 | 102 |
| Normal Max | 35 | 41 | 50 | 59 | 68 | 75 | 80 | 78 | 71 | 59 | 47 | 38 | 58 |
| Normal Min | 24 | 25 | 32 | 40 | 48 | 54 | 57 | 56 | 51 | 42 | 36 | 28 | 41 |
| Record Low | -14 | -14 | -2 | 19 | 25 | 34 | 41 | 38 | 31 | 20 | 4 | -4 | -14 |
| Precipitation | 3.47 (9.1) (0.3) | 2.76 (8) (0.3) | 4.41 (7.6) (1) | 4.72 (7.7) (3) | 5.83 (8.9) (6) | 5.67 (9.7) (8) | 4.53 (9.2) (9) | 6.02 (9.8*) (7) | 7.56 (***) (3) | 7.76 (***) (3) | 6.38 (9.7) (1) | 4.61 (10.3) (0.3) | 63.7 (***) (41.9) |
| Snow | *** (***) | *** (***) | *** (***) | *** (***) | *** (***) | *** (***) | *** (***) | *** (***) | *** (***) | *** (***) | *** (***) | *** (***) | *** (***) |

## Europe, Spain to Ukraine

### Map # 183 to 187

Table 9.2  World Weather Data [CONTINUED]

Each monthly cell lists: Normal High / Normal Low, Record High / Record Low, and Precipitation (…) (…).

**Map Location 183 — Spain, Madrid — 4028N 00334W 1972**

| Month | Norm High | Norm Low | Rec High | Rec Low | Precip |
|---|---|---|---|---|---|
| JANUARY | 49 | 34 | 66 | 14 | 1.73 (5) (0) |
| FEBRUARY | 54 | 35 | 77 | 14 | 1.66 (4.2) (0) |
| MARCH | 62 | 40 | 82 | 25 | 2.02 (5.2) (0.4) |
| APRIL | 66 | 43 | 82 | 30 | 1.55 (4.7) (1.2) |
| MAY | 74 | 50 | 95 | 32 | 2.01 (5.1) (1.8) |
| JUNE | 82 | 57 | 104 | 37 | 1.12 (2.9) (1.8) |
| JULY | 91 | 63 | 104 | 48 | 0.47 (1.5) (3.4) |
| AUGUST | 89 | 62 | 104 | 48 | 0.66 (1.6) (1.2) |
| SEPTEMBER | 82 | 57 | 99 | 43 | 1.2 (2.5) (1) |
| OCTOBER | 68 | 49 | 84 | 30 | 2.59 (5.9) (0.4) |
| NOVEMBER | 58 | 41 | 75 | 27 | 2.13 (4.3) (0.6) |
| DECEMBER | 51 | 36 | 70 | 19 | 2.45 (5.2) (0) |
| ANNUAL | 69 | 47 | 104 | 14 | 19.6 (48.1) (11.8) |

**Map Location 184 — Sweden, Stockholm — 5939N 01755E 122**

| Month | Norm High | Norm Low | Rec High | Rec Low | Precip |
|---|---|---|---|---|---|
| JANUARY | 31 | 23 | 51 | -26 | 1.5 (4.9) (0) |
| FEBRUARY | 31 | 22 | 54 | -22 | 1.1 (3.6) (0) |
| MARCH | 37 | 26 | 59 | -14 | 1.1 (3.7) (0) |
| APRIL | 45 | 32 | 77 | -8 | 1.5 (4.9) (0.3) |
| MAY | 57 | 41 | 84 | 19 | 1.6 (5.2) (1) |
| JUNE | 65 | 49 | 91 | 32 | 1.9 (5.3) (2) |
| JULY | 70 | 55 | 97 | 40 | 2.8 (7.1) (4) |
| AUGUST | 66 | 53 | 91 | 36 | 3.1 (7.6) (2) |
| SEPTEMBER | 58 | 46 | 84 | 23 | 2.1 (5.9) (1) |
| OCTOBER | 48 | 39 | 68 | 16 | 2.1 (5.9) (0.3) |
| NOVEMBER | 38 | 31 | 57 | 0 | 1.9 (5.5) (0) |
| DECEMBER | 33 | 26 | 52 | -11 | 1.9 (6.1) (0) |
| ANNUAL | 48 | 37 | 97 | -26 | 22.6 (65.7) (10.6) |

**Map Location 185 — Switzerland, Zurich — 4723N 00833E 1617**

| Month | Norm High | Norm Low | Rec High | Rec Low | Precip |
|---|---|---|---|---|---|
| JANUARY | 36 | 26 | 62 | 2 | 1.93 (6.1) (0.3) |
| FEBRUARY | 41 | 29 | 66 | 4 | 2.24 (6.9) (0) |
| MARCH | 52 | 34 | 70 | 12 | 2.95 (7) (0.3) |
| APRIL | 60 | 41 | 84 | 25 | 3.78 (7.4) (1) |
| MAY | 67 | 47 | 87 | 29 | 4.49 (7.6) (3) |
| JUNE | 73 | 53 | 97 | 39 | 5.28 (9.6) (4) |
| JULY | 77 | 56 | 100 | 41 | 5.16 (9.6) (4) |
| AUGUST | 76 | 56 | 97 | 41 | 5.24 (9.6) (4) |
| SEPTEMBER | 70 | 52 | 89 | 34 | 4.33 (8.9) (2) |
| OCTOBER | 57 | 43 | 82 | 29 | 4.06 (8.7) (0.3) |
| NOVEMBER | 46 | 36 | 67 | 21 | 2.8 (7.1) (0) |
| DECEMBER | 36 | 29 | 52 | 9 | 2.91 (8.3) (0.3) |
| ANNUAL | 58 | 42 | 100 | 2 | 45.2 (96.8) (20.2) |

**Map Location 186 — Turkey, Ankara — 4007N 03259E 3122**

| Month | Norm High | Norm Low | Rec High | Rec Low | Precip |
|---|---|---|---|---|---|
| JANUARY | 40 | 24 | 61 | -11 | 1.63 (3.4) (0.2) |
| FEBRUARY | 42 | 26 | 64 | -11 | 1.58 (3.3) (0) |
| MARCH | 52 | 31 | 84 | 9 | 1.68 (3.5) (1.2) |
| APRIL | 65 | 40 | 88 | 23 | 1.78 (3.6) (3) |
| MAY | 72 | 47 | 88 | 28 | 2.25 (4.3) (5.8) |
| JUNE | 79 | 52 | 93 | 36 | 1.24 (2.2) (3.9) |
| JULY | 86 | 57 | 100 | 43 | 0.68 (1.3) (0) |
| AUGUST | 88 | 58 | 104 | 37 | 0.33 (0.7) (1.2) |
| SEPTEMBER | 81 | 51 | 97 | 34 | 0.33 (0.8) (1.7) |
| OCTOBER | 68 | 41 | 82 | 18 | 1.01 (2.1) (0.6) |
| NOVEMBER | 57 | 35 | 75 | 9 | 1.22 (2.4) (0.4) |
| DECEMBER | 45 | 30 | 61 | 3 | 1.19 (2.6) (0) |
| ANNUAL | 65 | 41 | 104 | -11 | 14.9 (30.2) (21.1) |

**Map Location 187 — Ukraine, Kiev — 5024N 03027E 587**

| Month | Norm High | Norm Low | Rec High | Rec Low | Precip |
|---|---|---|---|---|---|
| JANUARY | 24 | 16 | 45 | -18 | 2.06 (12.4) (0) |
| FEBRUARY | 28 | 18 | 46 | -18 | 1.21 (10.2) (0) |
| MARCH | 36 | 26 | 70 | -18 | 1.38 (10.6) (0.2) |
| APRIL | 55 | 40 | 82 | 18 | 1.16 (6) (1.6) |
| MAY | 68 | 50 | 86 | 30 | 1.9 (8.8) (4.3) |
| JUNE | 74 | 55 | 93 | 37 | 2 (7.7) (5.2) |
| JULY | 76 | 57 | 91 | 46 | 2.98 (8) (6.3) |
| AUGUST | 74 | 56 | 93 | 36 | 3.12 (8.3) (5) |
| SEPTEMBER | 66 | 48 | 86 | 32 | 1.26 (5.3) (0.7) |
| OCTOBER | 54 | 41 | 79 | 19 | 1.57 (6.3) (0.4) |
| NOVEMBER | 39 | 32 | 63 | -4 | 2.21 (10.8) (0) |
| DECEMBER | 30 | 23 | 54 | -8 | 1.91 (12.4) (0) |
| ANNUAL | 52 | 39 | 93 | -18 | 22.8 (106.8) (23.7) |

Table 9.2 World Weather Data [CONTINUED]

# Europe, U.K., Great Britian to Wales

# Maps 188 to 191

## Map Location 188 — U.K. Great Britian, London — 5128N

| Month | Avg Max °F | Avg Min °F | Precip (Rain)(Thunder) | Record Max °F | Record Min °F | Snow (Snow Days) |
|---|---|---|---|---|---|---|
| JANUARY | 45 | 36 | 1.8 (5.8) (0.2) | 58 | 9 | *** (***) |
| FEBRUARY | 46 | 36 | 1.5 (4.9) (0.1) | 63 | 11 | *** (***) |
| MARCH | 49 | 37 | 1.7 (5.4) (0.8) | 70 | 17 | *** (***) |
| APRIL | 55 | 40 | 1.5 (4.9) (1) | 82 | 26 | *** (***) |
| MAY | 63 | 46 | 1.7 (5.4) (3) | 89 | 30 | 0 (0) |
| JUNE | 68 | 51 | 2.1 (5.8) (2) | 94 | 37 | 0 (0) |
| JULY | 71 | 55 | 2.2 (6) (3) | 93 | 42 | 0 (0) |
| AUGUST | 70 | 54 | 2.2 (6) (3) | 94 | 41 | 0 (0) |
| SEPTEMBER | 65 | 50 | 1.9 (5.5) (1) | 92 | 31 | 0 (0*) |
| OCTOBER | 57 | 45 | 2.7 (6.9) (0.4) | 83 | 25 | *** (***) |
| NOVEMBER | 49 | 39 | 2.2 (6) (0.2) | 63 | 20 | *** (***) |
| DECEMBER | 46 | 37 | 2.3 (7) (0.2) | 60 | 11 | *** (***) |
| ANNUAL | 57 | 44 | 23.8 (69.6) (14.9) | 94 | 9 | *** (***) |

## Map Location 189 — U.K. N.Ireland, Belfast — 5436N 00552W 24

| Month | Avg Max °F | Avg Min °F | Precip (Rain)(Thunder) | Record Max °F | Record Min °F | Snow (Snow Days) |
|---|---|---|---|---|---|---|
| JANUARY | 43 | 35 | 4.13 (9.9) (***) | 56 | 9 | *** (***) |
| FEBRUARY | 44 | 35 | 2.83 (8.1) (***) | 57 | 11 | *** (***) |
| MARCH | 49 | 37 | 2.37 (6.5) (***) | 67 | 10 | *** (***) |
| APRIL | 53 | 39 | 2.39 (6.5) (***) | 69 | 24 | *** (***) |
| MAY | 59 | 43 | 2.53 (6.7) (***) | 77 | 26 | *** (***) |
| JUNE | 63 | 48 | 2.41 (6.4) (***) | 83 | 30 | 0 (0) |
| JULY | 65 | 52 | 3.37 (8) (***) | 85 | 36 | 0 (0) |
| AUGUST | 65 | 51 | 3.45 (8.1) (***) | 82 | 34 | 0* (0*) |
| SEPTEMBER | 61 | 49 | 3.29 (7.8) (***) | 76 | 28 | *** (***) |
| OCTOBER | 55 | 44 | 3.94 (8.6) (***) | 69 | 24 | *** (***) |
| NOVEMBER | 48 | 39 | 3.61 (8.2) (***) | 61 | 21 | *** (***) |
| DECEMBER | 45 | 37 | 3.94 (9.7) (***) | 58 | 13 | *** (***) |
| ANNUAL | 54 | 42 | 38.3 (94.5) (***) | 85 | 9 | *** (***) |

## Map Location 190 — U.K. Scotland, Edinburgh — 5556N 00320W 135

| Month | Avg Max °F | Avg Min °F | Precip (Rain)(Thunder) | Record Max °F | Record Min °F | Snow (Snow Days) |
|---|---|---|---|---|---|---|
| JANUARY | 44 | 38 | 1.73 (5.6) (0.1) | 58 | 24 | *** (***) |
| FEBRUARY | 44 | 37 | 1.56 (5.1) (0) | 58 | 20 | *** (***) |
| MARCH | 46 | 37 | 1.77 (5.5) (0.1) | 62 | 23 | *** (***) |
| APRIL | 49 | 39 | 1.38 (4.6) (5) | 63 | 30 | *** (***) |
| MAY | 54 | 44 | 1.89 (5.8) (1) | 73 | 32 | 0 (0) |
| JUNE | 59 | 49 | 1.85 (5.2) (1) | 79 | 40 | 0 (0) |
| JULY | 63 | 53 | 2.64 (6.8) (3) | 81 | 43 | 0 (0) |
| AUGUST | 62 | 53 | 3.11 (7.6) (2) | 79 | 43 | 0 (0) |
| SEPTEMBER | 59 | 50 | 1.89 (5.5) (8) | 79 | 39 | 0 (0) |
| OCTOBER | 53 | 45 | 2.6 (6.7) (3) | 69 | 31 | *** (***) |
| NOVEMBER | 47 | 41 | 2.17 (6) (0) | 61 | 29 | *** (***) |
| DECEMBER | 44 | 39 | 2.13 (6.6) (0) | 57 | 26 | *** (***) |
| ANNUAL | 52 | 44 | 24.7 (71) (23.2) | 81 | 20 | *** (***) |

## Map Location 191 — U.K. Wales, Cardiff — 5124N 00321W 220

| Month | Avg Max °F | Avg Min °F | Precip (Rain)(Thunder) | Record Max °F | Record Min °F | Snow (Snow Days) |
|---|---|---|---|---|---|---|
| JANUARY | 45 | 36 | 3.7 (9.4) (0.4) | 59 | 2 | *** (***) |
| FEBRUARY | 46 | 36 | 2.9 (8.3) (0.1) | 59 | 12 | *** (***) |
| MARCH | 50 | 37 | 3.1 (7.1) (0.4) | 70 | 18 | *** (***) |
| APRIL | 55 | 41 | 2.5 (6.6) (0.5) | 75 | 27 | *** (***) |
| MAY | 61 | 45 | 2.5 (6.6) (1) | 84 | 31 | 0 (0) |
| JUNE | 66 | 51 | 3.1 (7.6) (2) | 87 | 39 | 0 (0) |
| JULY | 69 | 54 | 4.2 (8.9) (1) | 90 | 45 | 0 (0) |
| AUGUST | 68 | 54 | 3.1 (7.5) (0.5) | 91 | 41 | 0 (0) |
| SEPTEMBER | 64 | 51 | 4.7 (9.2) (0.5) | 83 | 36 | 0 (0*) |
| OCTOBER | 57 | 45 | 4.7 (9.2) (0.5) | 76 | 26 | *** (***) |
| NOVEMBER | 50 | 40 | 4.1 (8.7) (0.2) | 65 | 23 | *** (***) |
| DECEMBER | 46 | 37 | 5 (10.6) (0.8) | 57 | 19 | *** (***) |
| ANNUAL | 56 | 44 | 41.4 (97.1) (7.9) | 91 | 2 | *** (***) |

## Data Key

|  |  |
|---|---|
| Map Locater Number | |
| Location | |
| Lat - Lon ddmm Elev ft | |
| | Average Max T °F |
| | Average Min T °F |
| | Avg. Precip in. (Rain Days) (Thunder Days) |
| MONTH | |
| | Record Max T °F |
| | Record Min T °F |
| | Avg. Snow in. (Snow Days) |

|  |  |
|---|---|
| *** or Blank | No data, missing data or data of insufficient reliability |
| Rain Days | Number of Days with >= 0.01" of precipitation falling |
| Snow Days | Number of Days with >= 1.5" of snow falling |

**Table 9.2   World Weather Data [CONTINUED]**

# North America, Belize to Canada, Cambridge Bay

# Maps 192 - 196

*Each month: Record High / Normal High / Normal Low / Record Low, and Precipitation (mean, greatest, days/snow).*

## Map Location 192 — Belize, Belize City — 1732N 08818W 16

| Month | Rec High | Norm High | Norm Low | Rec Low | Precipitation |
|---|---|---|---|---|---|
| JANUARY | 90 | 81 | 67 | 49 | 6.6 (23.6) (0) |
| FEBRUARY | 93 | 82 | 69 | 49 | 2.8 (5.7) (1) |
| MARCH | 95 | 84 | 71 | 54 | 1.8 (4.2) (0) |
| APRIL | 97 | 86 | 74 | 59 | 2.8 (6) (0) |
| MAY | 96 | 87 | 75 | 60 | 5.8 (9.4) (1) |
| JUNE | 97 | 87 | 75 | 64 | 9.8 (14.7) (1) |
| JULY | 95 | 87 | 75 | 62 | 7.7 (13.3) (1) |
| AUGUST | 96 | 88 | 75 | 60 | 7.8 (13.3) (1) |
| SEPTEMBER | 97 | 86 | 74 | 60 | 9.2 (14.3) (0) |
| OCTOBER | 96 | 86 | 72 | 58 | 13.2 (19.3) (1) |
| NOVEMBER | 95 | 83 | 68 | 52 | 10.5 (16.3) (0) |
| DECEMBER | 92 | 81 | 68 | 49 | 7.7 (28.6) (0) |
| ANNUAL | 97 | 85 | 72 | 49 | 85.7 (168.7) (6) |

## Map Location 193 — Canada, Alert — 8231N 06220W 95

| Month | Rec High | Norm High | Norm Low | Rec Low | Precipitation |
|---|---|---|---|---|---|
| JANUARY | 32 | -19 | -29 | -48 | 0.21 (0.4) (0) |
| FEBRUARY | 37 | -21 | -32 | -49 | 0.31 (0.8) (0) |
| MARCH | 26 | -22 | -33 | -53 | 0.28 (0.8) (0) |
| APRIL | 18 | -8 | -18 | -45 | 0.29 (0.8) (0) |
| MAY | 33 | 13 | 6 | -17 | 0.45 (1.4) (0) |
| JUNE | 62 | 34 | 28 | 12 | 0.64 (2.2) (0) |
| JULY | 67 | 44 | 36 | 26 | 0.48 (1.8) (0) |
| AUGUST | 64 | 36 | 30 | 14 | 1.14 (3.4) (0) |
| SEPTEMBER | 41 | 18 | 10 | -13 | 1.03 (3.2) (0) |
| OCTOBER | 33 | 2 | -7 | -28 | 0.94 (3.1) (0) |
| NOVEMBER | 28 | -9 | -18 | -39 | 0.21 (1.6) (0) |
| DECEMBER | 40 | -16 | -27 | -45 | 0.36 (1) (0) |
| ANNUAL | 67 | 4 | -4 | -53 | 6.3 (20.5) (0) |

## Map Location 194 — Canada, Qikiqtarjuaq (Broughton) — 6733N 06402W 25

| Month | Rec High | Norm High | Norm Low | Rec Low | Precipitation |
|---|---|---|---|---|---|
| JANUARY | 31 | -10 | -21 | -49 | 0.42 (1.3) (0) |
| FEBRUARY | 19 | -14 | -25 | -46 | 0.25 (0.5) (0) |
| MARCH | 27 | -5 | -17 | -45 | 0.15 (0.4) (0) |
| APRIL | 43 | 11 | -3 | -27 | 0.14 (0.1) (0) |
| MAY | 48 | 29 | 16 | -14 | 0.35 (0.9) (0) |
| JUNE | 56 | 41 | 30 | 20 | 0.31 (1) (0) |
| JULY | 73 | 49 | 35 | 25 | 0.55 (1.4) (0) |
| AUGUST | 61 | 47 | 35 | 14 | 1.28 (4.7) (0) |
| SEPTEMBER | 60 | 39 | 30 | 7 | 1.28 (5) (0) |
| OCTOBER | 53 | 27 | 19 | -17 | 0.69 (2.6) (0) |
| NOVEMBER | 44 | 15 | 5 | -22 | 0.46 (1) (0) |
| DECEMBER | 32 | 0 | -10 | -40 | 0.33 (0.6) (0) |
| ANNUAL | 73 | 19 | 8 | -49 | 6.2 (18.6) (0.1) |

## Map Location 195 — Canada, Calgary — 5106N 11401W 3557

| Month | Rec High | Norm High | Norm Low | Rec Low | Precipitation |
|---|---|---|---|---|---|
| JANUARY | 61 | 24 | 2 | -48 | 0.51 (1.6) (0) |
| FEBRUARY | 76 | 28 | 6 | -49 | 0.55 (1.7) (0) |
| MARCH | 75 | 37 | 14 | -35 | 0.84 (2.7) (0) |
| APRIL | 85 | 53 | 27 | -22 | 0.99 (3.1) (0) |
| MAY | 90 | 63 | 36 | 2 | 2.34 (6.6) (1) |
| JUNE | 95 | 69 | 43 | 12 | 3.14 (6.9) (4) |
| JULY | 97 | 76 | 47 | 31 | 2.51 (6) (9) |
| AUGUST | 96 | 74 | 45 | 28 | 2.29 (5.7) (5) |
| SEPTEMBER | 90 | 64 | 37 | 8 | 1.5 (4.1) (1) |
| OCTOBER | 85 | 54 | 29 | -8 | 0.69 (2.6) (0) |
| NOVEMBER | 71 | 38 | 17 | -31 | 0.72 (2.6) (0) |
| DECEMBER | 67 | 29 | 9 | -45 | 0.57 (1.8) (0) |
| ANNUAL | 97 | 51 | 26 | -49 | 16.6 (45.4) (20) |

## Map Location 196 — Canada, Cambridge Bay — 6906N 10508W 90

| Month | Rec High | Norm High | Norm Low | Rec Low | Precipitation |
|---|---|---|---|---|---|
| JANUARY | 21 | -20 | -34 | -63 | 0.35 (1) (0.8) |
| FEBRUARY | 11 | -24 | -36 | -59 | 0.16 (1.1) (0.3) |
| MARCH | 23 | -14 | -29 | -52 | 0.2 (0.8) (0.7) |
| APRIL | 43 | 2 | -15 | -42 | 0.14 (0.4) (0.4) |
| MAY | 52 | 22 | 8 | -31 | 0.22 (1.1) (0.4) |
| JUNE | 72 | 40 | 30 | 6 | 0.21 (0.6) (0) |
| JULY | 75 | 54 | 40 | 30 | 0.93 (3.1) (0) |
| AUGUST | 76 | 50 | 39 | 16 | 1.97 (3.7) (0) |
| SEPTEMBER | 60 | 35 | 28 | 7 | 1.11 (3.7) (0.8) |
| OCTOBER | 39 | 18 | 6 | -25 | 0.57 (2.9) (1.1) |
| NOVEMBER | 27 | -5 | -17 | -44 | 0.54 (2.4) (0.9) |
| DECEMBER | 36 | -16 | -28 | -57 | 0.38 (1.8) (1.3) |
| ANNUAL | 76 | 12 | 0 | -63 | 6.8 (22.6) (6.7) |

## North America, Canada, Churchill to Inuvik

**Maps 197 – 201**

**Table 9.2  World Weather Data [CONTINUED]**

Each monthly cell lists: Average Max / Average Min / Record Max / Record Min (°F) and Precipitation (Snow)(—).

### Map Location 197 — Canada, Churchill — 5845N 09404W 94

| Month | Avg Max | Avg Min | Rec Max | Rec Min | Precip (Snow)(—) |
|---|---|---|---|---|---|
| JANUARY | -11 | -27 | 39 | -57 | 0.48 (1.5)(0) |
| FEBRUARY | -8 | -25 | 34 | -52 | 0.61 (1.9)(0) |
| MARCH | 4 | -16 | 41 | -52 | 0.87 (2.8)(0.1) |
| APRIL | 24 | -26 | 64 | -26 | 0.89 (2.8)(0.1) |
| MAY | 38 | 22 | 87 | -14 | 0.93 (3)(0.3) |
| JUNE | 52 | 34 | 88 | 13 | 1.85 (4.9)(0.8) |
| JULY | 62 | 43 | 96 | 22 | 2.19 (5.5)(1.4) |
| AUGUST | 60 | 44 | 90 | 25 | 2.69 (6.3)(1.9) |
| SEPTEMBER | 49 | 34 | 79 | 8 | 2.33 (5.6)(0.1) |
| OCTOBER | 34 | 20 | 69 | -17 | 1.43 (4)(0) |
| NOVEMBER | 13 | -2 | 45 | -53 | 1.03 (3.2)(0) |
| DECEMBER | -3 | -19 | 35 | -47 | 0.66 (2.1)(0) |
| ANNUAL | 27 | 9 | 96 | -57 | 16 (43.6)(4.6) |

### Map Location 198 — Canada, Dawson — 6404N 13929W 1062

| Month | Avg Max | Avg Min | Rec Max | Rec Min | Precip (Snow)(—) |
|---|---|---|---|---|---|
| JANUARY | -14 | -28 | 47 | -68 | 0.9 (2.9)(0) |
| FEBRUARY | -4 | -20 | 48 | -73 | 0.7 (2.2)(0) |
| MARCH | 16 | -8 | 52 | -54 | 0.5 (1.6)(0) |
| APRIL | 41 | 16 | 69 | -41 | 0.5 (1.6)(0) |
| MAY | 59 | 34 | 86 | 4 | 0.5 (0) |
| JUNE | 70 | 43 | 95 | 25 | 1.2 (3.6)(2) |
| JULY | 73 | 46 | 95 | 29 | 1.5 (4.2)(4) |
| AUGUST | 67 | 42 | 88 | 17 | 1.5 (4.2)(1.8) |
| SEPTEMBER | 52 | 32 | 79 | 8 | 1.4 (3.9)(0) |
| OCTOBER | 33 | 19 | 68 | -23 | 1.2 (3.6)(0) |
| NOVEMBER | 7 | -4 | 52 | -50 | 1.1 (3.4)(0) |
| DECEMBER | -8 | -20 | 55 | -66 | 1 (3.2)(0) |
| ANNUAL | 33 | 13 | 95 | -73 | 12.5 (37.6)(8) |

### Map Location 199 — Canada, Fort Nelson — 5850N 12235W 1253

| Month | Avg Max | Avg Min | Rec Max | Rec Min | Precip (Snow)(—) |
|---|---|---|---|---|---|
| JANUARY | 1 | -15 | 45 | -61 | 0.86 (2.8)(0) |
| FEBRUARY | 10 | -6 | 57 | -55 | 1.16 (3.7)(0) |
| MARCH | 26 | 3 | 62 | -39 | 0.71 (2.3)(0) |
| APRIL | 47 | 25 | 76 | -30 | 0.84 (2.7)(0) |
| MAY | 60 | 37 | 89 | 5 | 1.44 (4.4)(1) |
| JUNE | 68 | 46 | 93 | 30 | 2.53 (6)(4) |
| JULY | 74 | 51 | 98 | 34 | 2.36 (5.8)(5) |
| AUGUST | 70 | 48 | 98 | 29 | 1.52 (4.2)(2) |
| SEPTEMBER | 59 | 38 | 91 | 12 | 1.32 (3.8)(0) |
| OCTOBER | 43 | 25 | 78 | -18 | 1.03 (3.2)(0) |
| NOVEMBER | 18 | 3 | 57 | -42 | 1.37 (3.9)(0) |
| DECEMBER | 3 | -11 | 49 | -54 | 1.23 (3.9)(0) |
| ANNUAL | 40 | 20 | 98 | -61 | 16.4 (46.7)(12) |

### Map Location 200 — Canada, Halifax — 4453N 06331W 476

| Month | Avg Max | Avg Min | Rec Max | Rec Min | Precip (Snow)(—) |
|---|---|---|---|---|---|
| JANUARY | 32 | 15 | 57 | -17 | 5.4 (11.7)(0.2) |
| FEBRUARY | 31 | 15 | 54 | -21 | 4.35 (10.5)(0) |
| MARCH | 38 | 23 | 70 | -10 | 4.85 (10.9)(0.1) |
| APRIL | 47 | 31 | 87 | 7 | 4.54 (10.5)(1.1) |
| MAY | 59 | 40 | 90 | 22 | 4.14 (9.9)(0.8) |
| JUNE | 68 | 48 | 94 | 32 | 4.04 (8)(1.5) |
| JULY | 74 | 55 | 99 | 40 | 3.79 (7.7)(1.4) |
| AUGUST | 74 | 56 | 94 | 39 | 4.38 (8.3)(1.8) |
| SEPTEMBER | 67 | 50 | 94 | 29 | 4.13 (8.7)(1) |
| OCTOBER | 57 | 41 | 94 | 29 | 5.42 (10.7)(0.3) |
| NOVEMBER | 46 | 32 | 75 | 4 | 5.31 (10.5)(0.3) |
| DECEMBER | 35 | 21 | 62 | -14 | 5.39 (11.7)(0.2) |
| ANNUAL | 52 | 36 | 99 | -21 | 55.7 (119.1)(8.7) |

### Map Location 201 — Canada, Inuvik — 6818N 13329W 223

| Month | Avg Max | Avg Min | Rec Max | Rec Min | Precip (Snow)(—) |
|---|---|---|---|---|---|
| JANUARY | -10 | -26 | 49 | -59 | 0.55 (1.7)(0.2) |
| FEBRUARY | -9 | -24 | 49 | -62 | 0.49 (1.5)(0) |
| MARCH | 0 | -17 | 49 | -56 | 0.38 (1.1)(0.1) |
| APRIL | 19 | -2 | 57 | -44 | 0.5 (1.6)(0) |
| MAY | 40 | 22 | 77 | -14 | 0.49 (1.5)(0) |
| JUNE | 58 | 40 | 86 | 20 | 0.8 (2.6)(0) |
| JULY | 66 | 47 | 93 | 30 | 1.39 (4)(0.5) |
| AUGUST | 58 | 42 | 93 | 25 | 1.42 (4)(0.4) |
| SEPTEMBER | 44 | 32 | 76 | 7 | 0.91 (3)(0) |
| OCTOBER | 25 | 15 | 55 | -22 | 0.86 (2.9)(0) |
| NOVEMBER | 3 | -9 | 44 | -50 | 0.76 (2.7)(0.1) |
| DECEMBER | -8 | -24 | 50 | -57 | 0.42 (1.2)(0) |
| ANNUAL | 24 | 8 | 93 | -62 | 9 (27.8)(1.3) |

Table 9.2  World Weather Data [CONTINUED]

# North America, Canada, Monreal to Resolute     Map # 202 -206

**Map Location 202 — Canada, Montreal — 4528N 07345W 117**

| Month | Normal Max | Record High | Normal Min | Record Low | Precip (Snow)(Days) | Snow |
|---|---|---|---|---|---|---|
| JANUARY | 21 | 56 | 6 | -35 | 3.8 (9.7) (0) | 27.7 (5.7) |
| FEBRUARY | 23 | 52 | 8 | -28 | 3 (8.3) (0) | 23.3 (4.9) |
| MARCH | 33 | 70 | 19 | -21 | 3.5 (8.8) (0.1) | 20.1 (4.7) |
| APRIL | 50 | 83 | 33 | 2 | 2.6 (7.2) (0.3) | 5.5 (1.1) |
| MAY | 64 | 94 | 47 | 23 | 3.1 (8.1) (1.1) | 0.1 (0) |
| JUNE | 74 | 94 | 57 | 37 | 3.4 (7.2) (2.6) | 0 (0) |
| JULY | 78 | 97 | 61 | 45 | 3.7 (7.6) (3.2) | 0 (0) |
| AUGUST | 75 | 96 | 59 | 41 | 3.5 (7.3) (3.3) | 0 (0) |
| SEPTEMBER | 67 | 90 | 51 | 28 | 3.7 (8) (2) | 0 (0) |
| OCTOBER | 54 | 80 | 40 | 20 | 3.4 (7.5) (0.5) | 0.9 (0.1) |
| NOVEMBER | 39 | 70 | 27 | -18 | 3.5 (7.6) (0.1) | 10.9 (2.4) |
| DECEMBER | 26 | 60 | 13 | -29 | 3.6 (9.4) (0) | 23.8 (5) |
| ANNUAL | 50 | 97 | 35 | -35 | 40.8 (96.7) (13.2) | 112.3 (23.9) |

**Map Location 203 — Canada, Moosonee — 5116N 08039W 34**

| Month | Normal Max | Record High | Normal Min | Record Low | Precip (Snow)(Days) | Snow |
|---|---|---|---|---|---|---|
| JANUARY | 5 | 45 | -16 | -52 | 1.93 (5.9) (0) | 18.8 (4) |
| FEBRUARY | 10 | 51 | -14 | -50 | 1.91 (5.9) (0) | 19 (4.1) |
| MARCH | 22 | 60 | -2 | -43 | 1.64 (5) (0) | 13 (2.8) |
| APRIL | 36 | 80 | 16 | -25 | 1.63 (4.9) (0) | 10.3 (2.1) |
| MAY | 50 | 92 | 31 | 1 | 2.74 (7.4) (1) | 3.8 (0.7) |
| JUNE | 64 | 94 | 42 | 21 | 3.58 (7.4) (3) | 0.3 (0) |
| JULY | 71 | 96 | 48 | 29 | 2.94 (6.6) (4) | 0 (0) |
| AUGUST | 69 | 95 | 48 | 30 | 3.51 (7.4) (4) | 0 (0) |
| SEPTEMBER | 60 | 89 | 41 | 21 | 3.22 (7.2) (1) | 0.2 (0) |
| OCTOBER | 46 | 80 | 31 | 2 | 2.83 (6.5) (0) | 6.7 (1.3) |
| NOVEMBER | 28 | 66 | 14 | -30 | 2.53 (6) (0) | 16.6 (4.1) |
| DECEMBER | 14 | 51 | -5 | -44 | 2.31 (6.9) (0) | 21.3 (4.6) |
| ANNUAL | 40 | 96 | 20 | -52 | 30.8 (77.1) (13) | 110 (23.7) |

**Map Location 204 — Canada, Quebec — 4648N 07123W 239**

| Month | Normal Max | Record High | Normal Min | Record Low | Precip (Snow)(Days) | Snow |
|---|---|---|---|---|---|---|
| JANUARY | 18 | 52 | 2 | -34 | 3.45 (9.2) (0) | 29.3 (6) |
| FEBRUARY | 20 | 53 | 4 | -33 | 2.74 (7.8) (0) | 23.1 (4.9) |
| MARCH | 31 | 64 | 15 | -23 | 3.02 (8) (0) | 20.8 (4.9) |
| APRIL | 44 | 76 | 29 | -2 | 2.35 (6.6) (1) | 8.7 (1.8) |
| MAY | 61 | 91 | 41 | 18 | 3.15 (8.2) (2) | 0.5 (0) |
| JUNE | 72 | 94 | 52 | 31 | 3.68 (7.6) (4) | 0 (0) |
| JULY | 76 | 97 | 54 | 39 | 4.02 (7.9) (5) | 0 (0) |
| AUGUST | 73 | 96 | 54 | 37 | 3.98 (7.9) (3) | 0 (0) |
| SEPTEMBER | 64 | 89 | 48 | 22 | 3.6 (7.8) (1) | 0 (0) |
| OCTOBER | 51 | 82 | 37 | 14 | 3.41 (7.5) (0) | 1.8 (0.3) |
| NOVEMBER | 36 | 71 | 24 | -14 | 3.23 (7.2) (0) | 14.4 (3.5) |
| DECEMBER | 22 | 59 | 9 | -32 | 3.22 (8.7) (0) | 25.1 (5.3) |
| ANNUAL | 47 | 97 | 31 | -34 | 39.8 (94.4) (16) | 123.7 (26.7) |

**Map Location 205 — Canada, Regina — 5026N 10440W 1894**

| Month | Normal Max | Record High | Normal Min | Record Low | Precip (Snow)(Days) | Snow |
|---|---|---|---|---|---|---|
| JANUARY | 10 | 48 | -11 | -54 | 0.51 (1.6) (0) | 4.7 (0.8) |
| FEBRUARY | 13 | 60 | -9 | -56 | 0.35 (1) (0) | 3.4 (0.5) |
| MARCH | 27 | 76 | 6 | -44 | 0.67 (2.1) (0) | 5.4 (1.1) |
| APRIL | 50 | 91 | 26 | -20 | 0.74 (2.4) (1) | 3 (0.6) |
| MAY | 65 | 99 | 37 | 7 | 1.84 (5.5) (2) | 0.6 (0) |
| JUNE | 73 | 103 | 47 | 23 | 3.25 (7) (5) | 0.1 (0) |
| JULY | 79 | 110 | 51 | 28 | 2.38 (5.8) (7) | 0 (0) |
| AUGUST | 77 | 105 | 48 | 23 | 1.76 (4.7) (6) | 0 (0) |
| SEPTEMBER | 65 | 99 | 38 | 3 | 1.32 (3.8) (1) | 0.6 (0.1) |
| OCTOBER | 52 | 88 | 27 | -15 | 0.86 (2.9) (0) | 2.3 (0.4) |
| NOVEMBER | 32 | 73 | 11 | -47 | 0.6 (2.4) (0) | 4.8 (0.9) |
| DECEMBER | 16 | 59 | -1 | -55 | 0.42 (1.2) (0) | 3.9 (0.6) |
| ANNUAL | 47 | 110 | 23 | -56 | 14.7 (40.4) (22) | 28.8 (5) |

**Map Location 206 — Canada, Qausuittuq (Resolute) — 7443N 09459W 220**

| Month | Normal Max | Record High | Normal Min | Record Low | Precip (Snow)(Days) | Snow |
|---|---|---|---|---|---|---|
| JANUARY | -20 | 23 | -33 | -53 | 0.08 (0.2) (0) | 0.9 (0) |
| FEBRUARY | -23 | 7 | -36 | -57 | 0.09 (0) (0) | 1 (0) |
| MARCH | -18 | 20 | -31 | -61 | 0.16 (0.2) (0) | 1.7 (0) |
| APRIL | -1 | 30 | -16 | -40 | 0.2 (0.2) (0) | 2.1 (0.2) |
| MAY | 19 | 40 | 8 | -20 | 0.51 (1) (0) | 5.4 (0.5) |
| JUNE | 37 | 57 | 29 | 8 | 0.75 (1.8) (0) | 2.8 (0.3) |
| JULY | 45 | 61 | 35 | 28 | 0.92 (2.7) (0) | 0.5 (0) |
| AUGUST | 42 | 59 | 33 | 17 | 1.11 (3.5) (0) | 2.1 (0.2) |
| SEPTEMBER | 28 | 48 | 20 | 0 | 0.78 (3) (0) | 5.7 (0.5) |
| OCTOBER | 11 | 32 | 0 | -30 | 0.52 (1.5) (0) | 6.6 (0.8) |
| NOVEMBER | -5 | 27 | -18 | -43 | 0.24 (0.4) (0) | 2.6 (0.3) |
| DECEMBER | -14 | 17 | -27 | -51 | 0.13 (0) (0) | 1.4 (0) |
| ANNUAL | 8 | 61 | -2 | -61 | 5.5 (14.5) (0) | 32.8 (2.8) |

## Table 9.2  World Weather Data [CONTINUED]

## North America, Canada, Saglek to Winnipeg　　　　　Map # 207 -211

### Map Location 207 — Canada, Saglek — 5829N 06239W 269

| | JANUARY | FEBRUARY | MARCH | APRIL | MAY | JUNE |
|---|---|---|---|---|---|---|
| | 40 | 44 | 39 | 46 | 55 | 83 |
| | 11 | 13 | 17 | 25 | 36 | 45 |
| | 1 | 0 | 3 | 12 | 27 | 34 |
| | -32 | -26 | -25 | -15 | 7 | 21 |
| | 3.1(7.2)(0) | 2.9(6.1)(0) | 2.4(5.3)(0) | 1.9(5.1)(0) | 2.5(4.8)(0) | 2.4(4.8)(0) |
| | 26(5.5) | 27.1(4.3) | 24.1(4.2) | 19(4.0) | 11.3(2.4) | 5.8(1.8) |

| | JULY | AUGUST | SEPTEMBER | OCTOBER | NOVEMBER | DECEMBER | ANNUAL |
|---|---|---|---|---|---|---|---|
| | 79 | 73 | 73 | 60 | 52 | 36 | 83 |
| | 54 | 52 | 48 | 38 | 28 | 18 | 32 |
| | 41 | 42 | 38 | 31 | 21 | 9 | 22 |
| | 31 | 33 | 25 | 16 | -3 | -17 | -32 |
| | 3.3(6.7)(0.2) | 3.4(7.3)(0) | 4.4(7.4)(0) | 2.3(5.6)(0) | 3.8(8.3)(0) | 3.6(7.8)(0) | 35.9(76.4)(0.2) |
| | 1.2(0.2) | 0(0) | 5.1(0.8) | 8.4(2.3) | 34.1(6.7) | 34.9(6.0) | 197(38.2) |

### Map Location 208 — Canada, Saint John NB — 4519N

| | JANUARY | FEBRUARY | MARCH | APRIL | MAY | JUNE |
|---|---|---|---|---|---|---|
| | 55 | 51 | 62 | 75 | 87 | 89 |
| | 28 | 28 | 36 | 43 | 57 | 64 |
| | 11 | 12 | 22 | 32 | 41 | 49 |
| | -24 | -22 | -12 | 1 | 24 | 32 |
| | 4.1(10.2)(0) | 3.1(8.5)(0) | 3.7(9.2)(0) | 3.2(8.3)(1) | 3.1(8.1)(1) | 3.2(7)(2) |
| | 18.8(4) | 17.2(3.7) | 11.5(2.4) | 5.4(1.1) | 0.1(0) | 0(0) |

| | JULY | AUGUST | SEPTEMBER | OCTOBER | NOVEMBER | DECEMBER | ANNUAL |
|---|---|---|---|---|---|---|---|
| | 91 | 90 | 93 | 84 | 71 | 59 | 93 |
| | 69 | 69 | 63 | 54 | 43 | 32 | 49 |
| | 54 | 54 | 49 | 41 | 30 | 17 | 34 |
| | 38 | 37 | 29 | 16 | -11 | -21 | -24 |
| | 3.1(6.8)(3) | 3.6(7.5)(2) | 3.7(8)(1) | 4.1(8.6)(1) | 3.9(8.3)(0) | 3.8(9.7)(0) | 42.6(100.2)(11) |
| | 0(0) | 0(0) | 0(0) | 0.2(0) | 5.1(0.9) | 12.8(2.7) | 71.1(14.8) |

### Map Location 209 — Canada, Toronto — 4341N 07937W 569

| | JANUARY | FEBRUARY | MARCH | APRIL | MAY | JUNE |
|---|---|---|---|---|---|---|
| | 61 | 55 | 80 | 90 | 93 | 97 |
| | 30 | 30 | 37 | 50 | 63 | 73 |
| | 16 | 15 | 23 | 34 | 44 | 54 |
| | -26 | -25 | -17 | 5 | 25 | 28 |
| | 2.71(7.7)(0) | 2.43(7.1)(0) | 2.58(7.1)(0.5) | 2.48(6.9)(1.7) | 2.91(7.8)(1.9) | 2.67(6.2)(3.4) |
| | 16(3.4) | 15.3(3.3) | 10.7(2.2) | 2.8(0.5) | 0.1(0) | 0(0) |

| | JULY | AUGUST | SEPTEMBER | OCTOBER | NOVEMBER | DECEMBER | ANNUAL |
|---|---|---|---|---|---|---|---|
| | 105 | 102 | 98 | 86 | 77 | 61 | 105 |
| | 79 | 77 | 69 | 56 | 43 | 33 | 53 |
| | 59 | 58 | 51 | 40 | 31 | 21 | 37 |
| | 39 | 40 | 28 | 16 | -5 | -22 | -26 |
| | 2.95(6.6)(3.8) | 2.73(6.3)(3) | 2.9(6.6)(2.1) | 2.43(5.8)(0.8) | 2.76(6.4)(0.4) | 2.63(7.6)(0) | 32.2(82.1)(17.6) |
| | 0(0) | 0(0) | 0.4(0) | 2.1(0.3) | 4.2(0.7) | 12.4(2.6) | 61.9(12.7) |

### Map Location 210 — Canada, Vancouver — 4911N 12310W 9

| | JANUARY | FEBRUARY | MARCH | APRIL | MAY | JUNE |
|---|---|---|---|---|---|---|
| | 59 | 61 | 68 | 79 | 84 | 92 |
| | 41 | 44 | 50 | 58 | 64 | 69 |
| | 32 | 34 | 37 | 40 | 46 | 52 |
| | 2 | 8 | 15 | 27 | 33 | 35 |
| | 8.57(13.4)(0) | 5.79(12)(0) | 5.03(11.2)(0) | 3.34(8.6)(0) | 2.84(7.6)(1) | 2.45(5.9)(1) |
| | 11.5(2.4) | 6.4(1.2) | 4.0(0.5) | 0.3(0) | 0(0) | 0(0) |

| | JULY | AUGUST | SEPTEMBER | OCTOBER | NOVEMBER | DECEMBER | ANNUAL |
|---|---|---|---|---|---|---|---|
| | 91 | 92 | 85 | 77 | 74 | 57 | 92 |
| | 73 | 74 | 65 | 57 | 48 | 43 | 57 |
| | 54 | 60 | 49 | 44 | 39 | 35 | 43 |
| | 39 | 37 | 30 | 21 | 10 | 8 | 2 |
| | 1.69(4.6)(1) | 1.22(3.6)(1) | 3.63(7.8)(0) | 5.78(11.2)(0) | 8.28(14.9)(0) | 8.76(13.5)(0) | 57.4(114.3)(4) |
| | 0(0) | 0(0) | 0.3(0) | 2.1(0.3) | 5.7(1) | | 28.8(5.4) |

### Map Location 211 — Canada, Winnipeg — 4954N 09714W 783

| | JANUARY | FEBRUARY | MARCH | APRIL | MAY | JUNE |
|---|---|---|---|---|---|---|
| | 46 | 53 | 74 | 93 | 100 | 101 |
| | 7 | 12 | 27 | 48 | 65 | 74 |
| | -13 | -9 | 5 | 27 | 40 | 50 |
| | -48 | -48 | -38 | -18 | 11 | 21 |
| | 0.92(3)(0) | 0.86(2.8)(0) | 1.19(3.7)(0) | 1.37(4.2)(1) | 2.26(6.4)(2) | 3.15(5.9)(1) |
| | 9.1(1.9) | 8.4(1.7) | 10(2.1) | 3.9(0.7) | 1.1(0.1) | 0(0) |

| | JULY | AUGUST | SEPTEMBER | OCTOBER | NOVEMBER | DECEMBER | ANNUAL |
|---|---|---|---|---|---|---|---|
| | 108 | 105 | 99 | 86 | 71 | 53 | 108 |
| | 79 | 76 | 65 | 51 | 30 | 15 | 46 |
| | 55 | 52 | 43 | 31 | 14 | -3 | 24 |
| | 35 | 30 | 17 | -5 | -34 | -54 | -54 |
| | 3.08(6.8)(6) | 2.45(5.9)(6) | 2.35(5.7)(2) | 1.49(4.1)(1) | 1.12(3.4)(0) | 0.95(3.1)(0) | 21.2(56)(23) |
| | 0(0) | | 0.1(0) | 2.9(0.5) | 9(1.9) | 9.1(1.9) | 53.6(10.8) |

Table 9.2 World Weather Data [CONTINUED]

# North America, Canada, Yellowknife to Greenland, Nuuk    Map # 212 -216

## Map Location 212 — Canada, Yellowknife — 6228N 11427W 674

| | JANUARY | FEBRUARY | MARCH | APRIL | MAY | JUNE |
|---|---|---|---|---|---|---|
| | 37 | 43 | 43 | 60 | 79 | 85 |
| | -8 | -6 | 10 | 29 | 46 | 63 |
| | -23 | -23 | -10 | 9 | 28 | 45 |
| | -60 | -60 | -47 | -38 | -9 | 28 |
| | 0.84 (2.2) (0) | 0.57 (2) (0) | 0.66 (1.9) (0) | 0.42 (1.2) (0) | 0.67 (1.8) (0) | 0.59 (1.4) (0) |
| | 8.2 (2) | 5.6 (1.2) | 5.6 (1) | 3 (0.6) | 0.2 (0) | 0 (0) |

| | JULY | AUGUST | SEPTEMBER | OCTOBER | NOVEMBER | DECEMBER | ANNUAL |
|---|---|---|---|---|---|---|---|
| | 90 | 86 | 79 | 65 | 46 | 37 | 90 |
| | 69 | 65 | 50 | 36 | 13 | -3 | 30 |
| | 52 | 50 | 38 | 26 | -1 | -18 | 14 |
| | 33 | 34 | 18 | -9 | -43 | -55 | -60 |
| | 1.47 (4.2) (0.9) | 1.41 (3.5) (0.7) | 0.95 (2.5) (0.1) | 1.34 (4.3) (0) | 1.02 (3.8) (0) | 0.81 (2.9) (0) | 10.8 (31.7) (1.7) |
| | 0 (0) | 0 (0) | 0.2 (0) | 2.8 (0.8) | 7.5 (1.3) | 6.2 (1.2) | 39.3 (8.1) |

## Map Location 213 — Costa Rica, San Jose — 0958N 08449W 55

| | JANUARY | FEBRUARY | MARCH | APRIL | MAY | JUNE |
|---|---|---|---|---|---|---|
| | 87 | 88 | 91 | 89 | 88 | 92 |
| | 75 | 76 | 79 | 79 | 80 | 79 |
| | 58 | 58 | 59 | 62 | 62 | 62 |
| | 49 | 51 | 50 | 53 | 54 | 56 |
| | 0.6 (1.3) (0.3) | 0.2 (1.5) (0.3) | 0.8 (2.2) (2.0) | 1.8 (4.2) (2.0) | 9.0 (14.5) (8.0) | 9.5 (14.5) (8.0) |
| | 0 (0) | 0 (0) | 0 (0) | 0 (0) | 0 (0) | 0 (0) |

| | JULY | AUGUST | SEPTEMBER | OCTOBER | NOVEMBER | DECEMBER | ANNUAL |
|---|---|---|---|---|---|---|---|
| | 84 | 85 | 86 | 85 | 84 | 87 | 92 |
| | 77 | 78 | 78 | 77 | 77 | 75 | 78 |
| | 62 | 61 | 61 | 60 | 60 | 58 | 60 |
| | 54 | 56 | 56 | 55 | 52 | 49 | 49 |
| | 8.3 (13.7) (5.0) | 9.5 (14.5) (5.0) | 12.0 (18.1) (10.0) | 11.8 (17.9) (6.0) | 5.7 (8.6) (2.0) | 1.6 (2.3) (0.3) | 70.8 (108.2) (46.9) |
| | 0 (0) | 0 (0) | 0 (0) | 0 (0) | 0 (0) | 0 (0) | 0 (0) |

## Map Location 214 — El Salvador, San Salvador — 1340N

| | JANUARY | FEBRUARY | MARCH | APRIL | MAY | JUNE |
|---|---|---|---|---|---|---|
| | 101 | 103 | 105 | 104 | 103 | 98 |
| | 90 | 92 | 94 | 93 | 91 | 87 |
| | 60 | 60 | 62 | 65 | 67 | 66 |
| | 45 | 49 | 45 | 54 | 58 | 56 |
| | 0.3 (1.4) (0.3) | 0.2 (1.5) (0.3) | 0.4 (1.3) (2) | 1.7 (4) (7) | 7.7 (10.1) (17) | 12.9 (16.7) (17) |
| | 0 (0) | 0 (0) | 0 (0) | 0 (0) | 0 (0) | 0 (0) |

| | JULY | AUGUST | SEPTEMBER | OCTOBER | NOVEMBER | DECEMBER | ANNUAL |
|---|---|---|---|---|---|---|---|
| | 98 | 99 | 99 | 101 | 102 | 101 | 105 |
| | 89 | 89 | 89 | 87 | 87 | 89 | 89 |
| | 65 | 66 | 66 | 63 | 63 | 61 | 64 |
| | 58 | 60 | 53 | 54 | 49 | 47 | 45 |
| | 11.5 (15.7) (18) | 11.7 (15.9) (18) | 12.1 (18.3) (18) | 9.5 (14.8) (10) | 1.6 (3.2) (5) | 0.4 (1.4) (2) | 70 (104.3) (114.6) |
| | 0 (0) | 0 (0) | 0 (0) | 0 (0) | 0 (0) | 0 (0) | 0 (0) |

## Map Location 215 — Greenland, Eismitte (Mid Ice) — 7053N 04042W 9843

| | JANUARY | FEBRUARY | MARCH | APRIL | MAY | JUNE |
|---|---|---|---|---|---|---|
| | 5 | 9 | 4 | 10 | 16 | 23 |
| | -33 | -42 | -29 | -14 | 6 | 13 |
| | -53 | -64 | -51 | -37 | -18 | -9 |
| | -84 | -84 | -85 | -74 | -50 | -22 |
| | 0.6 (***) (***) | 0.2 (***) (***) | 0.3 (***) (***) | 0.2 (***) (***) | 0.1 (***) (***) | 0.1 (***) (***) |
| | *** (***) | *** (***) | *** (***) | *** (***) | *** (***) | *** (***) |

| | JULY | AUGUST | SEPTEMBER | OCTOBER | NOVEMBER | DECEMBER | ANNUAL |
|---|---|---|---|---|---|---|---|
| | 27 | 22 | 17 | 8 | -2 | 8 | 27 |
| | 19 | 11 | 4 | -23 | -33 | -28 | -14 |
| | 1 | -13 | -20 | -42 | -57 | -46 | -33 |
| | -19 | -31 | -38 | -69 | -73 | -73 | -85 |
| | 0.1 (***) (***) | 0.4 (***) (***) | 0.3 (***) (***) | 0.5 (***) (***) | 0.5 (***) (***) | 1 (***) (***) | 4.3 (***) (***) |
| | *** (***) | *** (***) | *** (***) | *** (***) | *** (***) | *** (***) | *** (***) |

## Map Location 216 — Greenland, Nuuk — 6410N 05143W 66

| | JANUARY | FEBRUARY | MARCH | APRIL | MAY | JUNE |
|---|---|---|---|---|---|---|
| | 52 | 51 | 53 | 56 | 61 | 74 |
| | 19 | 20 | 24 | 31 | 40 | 47 |
| | 10 | 9 | 13 | 20 | 29 | 34 |
| | -20 | -17 | -19 | -6 | 11 | 22 |
| | 1.4 (***) (0) | 1.7 (***) (0) | 1.6 (***) (0) | 1.2 (***) (0) | 1.7 (***) (0.1) | 1.4 (***) (0) |
| | 3.1 (***) (0.1) | 3.3 (***) (0) | 2.5 (***) (0.1) | 1.9 (***) (0.1) | 1.5 (***) (0) | *** (***) |

| | JULY | AUGUST | SEPTEMBER | OCTOBER | NOVEMBER | DECEMBER | ANNUAL |
|---|---|---|---|---|---|---|---|
| | 76 | 71 | 62 | 65 | 58 | 59 | 76 |
| | 52 | 51 | 43 | 35 | 28 | 23 | 34 |
| | 38 | 38 | 34 | 26 | 19 | 14 | 24 |
| | 29 | 27 | 18 | 6 | -1 | -14 | -20 |
| | 2.2 (***) (0.1) | 1.4 (***) (0) | 3.3 (***) (0) | 1.6 (***) (0) | 1.2 (***) (0) | 1.5 (***) (0) | 23.5 (***) (0.5) |
| | *** (***) | *** (***) | *** (***) | *** (***) | *** (***) | *** (***) | *** (***) |

Table 9.2  World Weather Data [CONTINUED]

# North America, Greenland, Thule AFB to Mexico, Juarez   Map # 217 -221

## Map Location 217 — Greenland, Thule Air Base — 7631N 06844W 251

| | JAN | FEB | MAR | APR | MAY | JUN | JUL | AUG | SEP | OCT | NOV | DEC | ANNUAL |
|---|---|---|---|---|---|---|---|---|---|---|---|---|---|
| Record High | 37 | 32 | 31 | 41 | 47 | 63 | 62 | 63 | 49 | 40 | 35 | 36 | 63 |
| Normal High | -4 | -5 | -8 | 10 | 28 | 41 | 46 | 44 | 32 | 19 | 6 | -1 | 17 |
| Normal Low | -17 | -20 | -22 | -7 | 16 | 32 | 38 | 36 | 23 | 8 | -6 | -14 | 6 |
| Record Low | -39 | -42 | -44 | -32 | -11 | 22 | 29 | 23 | 0 | -12 | -28 | -33 | -44 |
| Precipitation | 0.36 (0.7) (0) | 0.34 (1.1) (0) | 0.2 (0.4) (0) | 0.16 (0.3) (0) | 0.25 (0.6) (0) | 0.24 (0.4) (0) | 0.67 (1.6) (0) | 0.6 (1.7) (0) | 0.63 (2.1) (0) | 0.65 (2) (0) | 0.5 (1.5) (0) | 0.22 (0.2) (0) | 4.8 (12.6) (0) |
| Snowfall | 3.3 (0.3) | 3.5 (0.8) | 2.1 (0.2) | 1.7 (0.2) | 2.8 (0.4) | 0.3 (0) | 0 (0) | 0.3 (0) | 4 (0.5) | 7 (1.5) | 5.1 (0.8) | 3.2 (0.2) | 33.3 (4.9) |

## Map Location 218 — Guatemala, Guatemala City — 1435N 09032W 4885

| | JAN | FEB | MAR | APR | MAY | JUN | JUL | AUG | SEP | OCT | NOV | DEC | ANNUAL |
|---|---|---|---|---|---|---|---|---|---|---|---|---|---|
| Record High | 80 | 82 | 84 | 85 | 87 | 82 | 83 | 88 | 81 | 80 | 81 | 88 | 88 |
| Normal High | 72 | 74 | 77 | 80 | 79 | 76 | 77 | 77 | 75 | 72 | 71 | 75 | 75 |
| Normal Low | 52 | 53 | 55 | 58 | 60 | 59 | 59 | 59 | 59 | 57 | 54 | 57 | 57 |
| Record Low | 43 | 47 | 48 | 51 | 52 | 55 | 55 | 53 | 55 | 49 | 41 | 41 | 41 |
| Precipitation | 0.08 (0) (0) | 0.01 (0) (0) | 0.67 (2.2) (0.9) | 0.79 (1.3) (2.9) | 4.49 (7.8) (9.3) | 10.02 (14.3) (10.7) | 11.03 (11) (7.3) | 8.96 (16.5) (17.4) | 10.36 (16) (19) | 4.11 (4.2) (3.8) | 0.57 (1) (1.5) | 0.59 (1) (2.4) | 70 (104.3) (114.6) |

## Map Location 219 — Honduras, Tegucigalpa — 1403N 08713W 3294

| | JAN | FEB | MAR | APR | MAY | JUN | JUL | AUG | SEP | OCT | NOV | DEC | ANNUAL |
|---|---|---|---|---|---|---|---|---|---|---|---|---|---|
| Record High | 87 | 90 | 92 | 92 | 91 | 90 | 87 | 88 | 89 | 86 | 86 | 84 | 92 |
| Normal High | 78 | 81 | 82 | 88 | 91 | 85 | 85 | 86 | 87 | 83 | 84 | 84 | 85 |
| Normal Low | 43 | 40 | 41 | 50 | 54 | 58 | 52 | 51 | 53 | 53 | 48 | 48 | 48 |
| Record Low | 39 | | | | | | | | | | | | 35 |
| Precipitation | 0.55 (1.3) (0) | 0.2 (1.5) (0) | 0.43 (1.4) (0) | 1.1 (2.8) (0.2) | 6.14 (9.6) (1.6) | 6.54 (12.2) (2) | 3.62 (8) (0.2) | 4.37 (9.3) (0.1) | 6.93 (10.6) (0.3) | 5.55 (8.4) (0.2) | 1.58 (3.2) (0) | 0.59 (1.3) (0.1) | 37.6 (69.6) (4.7) |

## Map Location 220 — Mexico, Campeche — 1950N 09030W 30.

| | JAN | FEB | MAR | APR | MAY | JUN | JUL | AUG | SEP | OCT | NOV | DEC | ANNUAL |
|---|---|---|---|---|---|---|---|---|---|---|---|---|---|
| Record High | 90 | 92 | 95 | 99 | 98 | 95 | 94 | 91 | 91 | 91 | 90 | 90 | 99 |
| Normal High | 78 | 81 | 84 | 85 | 87 | 86 | 86 | 86 | 85 | 83 | 80 | 78 | 83 |
| Normal Low | 68 | 69 | 74 | 74 | 79 | 76 | 76 | 76 | 76 | 74 | 70 | 66 | 73 |
| Record Low | 54 | 56 | 56 | 62 | 62 | 67 | 70 | 70 | 69 | 64 | 60 | 59 | 54 |
| Precipitation | 1.1 (1.6) (0) | 0.3 (1.4) (0) | 0.8 (2.2) (0.2) | 0.04 (0.5) (0) | 3.2 (6.6) (1.3) | 7.2 (12.8) (3) | 6.4 (12) (4) | 5.1 (10.4) (4.2) | 5.2 (7.8) (1.6) | 3.1 (4.8) (0.4) | 1.4 (3) (0) | 1.2 (1.7) (0) | 35 (64.8) (14.7) |

## Map Location 221 — Mexico, Ciudad Juarez — 3140N 10626W 3830

| | JAN | FEB | MAR | APR | MAY | JUN | JUL | AUG | SEP | OCT | NOV | DEC | ANNUAL |
|---|---|---|---|---|---|---|---|---|---|---|---|---|---|
| Record High | 77 | 86 | 93 | 95 | 102 | 109 | 109 | 103 | 101 | 94 | 85 | 77 | 109 |
| Normal High | 57 | 62 | 69 | 77 | 86 | 94 | 93 | 91 | 86 | 77 | 66 | 57 | 76 |
| Normal Low | 32 | 37 | 42 | 50 | 58 | 67 | 70 | 68 | 63 | 52 | 40 | 33 | 51 |
| Record Low | -8 | 5 | 14 | 26 | 36 | 46 | 56 | 52 | 41 | 26 | 11 | -5 | -8 |
| Precipitation | 0.4 (1.4) (0.2) | 0.5 (1.3) (0.2) | 0.3 (1.1) (0.2) | 0.2 (0.9) (1) | 0.3 (1.1) (2) | 0.6 (0.3) (4) | 1.8 (3.9) (8) | 1.6 (3.3) (8) | 1.3 (2.9) (3) | 0.7 (2.5) (2) | 0.5 (2.3) (0.2) | 0.5 (1.3) (0.2) | 8.7 (22.3) (29) |
| Snowfall | 0 (0) | 1.6 (0.4) | 0.8 (0.2) | *** (0) | 0 (0) | 0 (0) | 0 (0) | 0 (0) | 0 (0) | *** (0) | 1.3 (0.2) | 1.2 (0.4) | *** (1.3) |

**Table 9.2  World Weather Data [CONTINUED]**

# North America, Mexico, Merida to Panama

## Map # 222 -224

### Map Location 222 — Mexico, Merida — 2056N 08939W 30

| | JANUARY | FEBRUARY | MARCH | APRIL | MAY | JUNE |
|---|---|---|---|---|---|---|
| Record Max T °F | 92 | 95 | 98 | 106 | 104 | 103 |
| Average Max T °F | 83 | 85 | 89 | 92 | 94 | 92 |
| Average Min T °F | 62 | 63 | 66 | 69 | 72 | 73 |
| Record Min T °F | 53 | 51 | 52 | 58 | 63 | 69 |
| Avg. Precip in. (Rain Days)(Thunder Days) | 1.06 (1.6) (0) | 0.67 (1.3) (0.1) | 1.02 (2.7) (0.1) | 0.87 (2.3) (0.3) | 2.64 (5.7) (0.8) | 7.05 (12.7) (1.5) |
| Avg. Snow in. (Snow Days) | 0 (0) | 0 (0) | 0 (0) | 0 (0) | 0 (0) | 0 (0) |

| | JULY | AUGUST | SEPTEMBER | OCTOBER | NOVEMBER | DECEMBER | ANNUAL |
|---|---|---|---|---|---|---|---|
| Record Max T °F | 97 | 100 | 96 | 94 | 91 | 92 | 106 |
| Average Max T °F | 92 | 91 | 90 | 87 | 85 | 82 | 89 |
| Average Min T °F | 73 | 73 | 73 | 71 | 67 | 64 | 69 |
| Record Min T °F | 64 | 67 | 68 | 63 | 56 | 55 | 51 |
| Avg. Precip in. (Rain Days)(Thunder Days) | 4.53 (9.6) (2.3) | 5.39 (10.8) (2.5) | 5.32 (8) (0.9) | 3.31 (5.1) (0.3) | 1.57 (3.2) (0) | 1.1 (1.6) (0.1) | 34.5 (64.6) (8.9) |
| Avg. Snow in. (Snow Days) | 0 (0) | 0 (0) | 0 (0) | 0 (0) | 0 (0) | 0 (0) | 0 (0) |

### Map Location 223 — Mexico, Mexico City — 1926N 09904W 7340

| | JANUARY | FEBRUARY | MARCH | APRIL | MAY | JUNE |
|---|---|---|---|---|---|---|
| Record Max T °F | 74 | 81 | 85 | 89 | 92 | 87 |
| Average Max T °F | 66 | 70 | 75 | 78 | 79 | 76 |
| Average Min T °F | 42 | 44 | 48 | 52 | 54 | 55 |
| Record Min T °F | 27 | 24 | 32 | 40 | 40 | 43 |
| Avg. Precip in. (Rain Days)(Thunder Days) | 0.24 (1.4) (0.2) | 0.28 (1.4) (0.5) | 0.47 (1.5) (1.2) | 0.71 (2) (1.8) | 1.89 (4.4) (3) | 4.06 (8.8) (3.3) |
| Avg. Snow in. (Snow Days) | *** (***) | *** (***) | 0 (0) | 0 (0) | 0 (0) | 0 (0) |

| | JULY | AUGUST | SEPTEMBER | OCTOBER | NOVEMBER | DECEMBER | ANNUAL |
|---|---|---|---|---|---|---|---|
| Record Max T °F | 84 | 84 | 78 | 80 | 80 | 84 | 92 |
| Average Max T °F | 74 | 74 | 72 | 70 | 68 | 66 | 72 |
| Average Min T °F | 54 | 54 | 54 | 50 | 47 | 43 | 50 |
| Record Min T °F | 47 | 46 | 34 | 36 | 30 | 29 | 24 |
| Avg. Precip in. (Rain Days)(Thunder Days) | 4.49 (9.5) (5.5) | 4.29 (9.2) (5.2) | 4.06 (6.1) (3.5) | 1.57 (3.2) (1.3) | 0.47 (2.3) (0.8) | 0.28 (1.4) (0.2) | 22.8 (51.2) (26.5) |
| Avg. Snow in. (Snow Days) | 0 (0) | 0 (0) | 0 (0) | 0 (0) | 0 (0) | 0 (0) | 0 (0) |

### Map Location 224 — Nicaragua, Managua — 1208N 08610W 180

| | JANUARY | FEBRUARY | MARCH | APRIL | MAY | JUNE |
|---|---|---|---|---|---|---|
| Record Max T °F | 92 | 93 | 94 | 94 | 98 | 95 |
| Average Max T °F | 88 | 89 | 91 | 94 | 93 | 88 |
| Average Min T °F | 69 | 70 | 72 | 73 | 74 | 73 |
| Record Min T °F | 62 | 63 | 67 | 68 | 65 | 69 |
| Avg. Precip in. (Rain Days)(Thunder Days) | 0.08 (0.2) (0) | 0.05 (0.3) (0) | 0.02 (0) (0.5) | 0.13 (0.5) (0) | 3.06 (5.2) (5) | 8.23 (12.7) (12.5) |
| Avg. Snow in. (Snow Days) | 0 (0) | 0 (0) | 0 (0) | 0 (0) | 0 (0) | 0 (0) |

| | JULY | AUGUST | SEPTEMBER | OCTOBER | NOVEMBER | DECEMBER | ANNUAL |
|---|---|---|---|---|---|---|---|
| Record Max T °F | 88 | 92 | 94 | 94 | 92 | 91 | 98 |
| Average Max T °F | 88 | 89 | 89 | 88 | 88 | 87 | 89 |
| Average Min T °F | 73 | 73 | 73 | 74 | 71 | 70 | 72 |
| Record Min T °F | 70 | 70 | 69 | 66 | 64 | 59 | 59 |
| Avg. Precip in. (Rain Days)(Thunder Days) | 3.65 (11) (12.5) | 5.15 (10) (10) | 6.81 (10.7) (14) | 6.32 (9.5) (6.7) | 1.24 (3) (1) | 0.33 (0.7) (0.3) | 35.1 (63.8) (62.5) |
| Avg. Snow in. (Snow Days) | 0 (0) | 0 (0) | 0 (0) | 0 (0) | 0 (0) | 0 (0) | 0 (0) |

### Map Location 225 — Panama, Panama City — 0858N 07930W 30

| | JANUARY | FEBRUARY | MARCH | APRIL | MAY | JUNE |
|---|---|---|---|---|---|---|
| Record Max T °F | 94 | 94 | 94 | 95 | 97 | 94 |
| Average Max T °F | 89 | 90 | 91 | 92 | 89 | 88 |
| Average Min T °F | 73 | 74 | 74 | 75 | 76 | 75 |
| Record Min T °F | 65 | 67 | 70 | 68 | 72 | 70 |
| Avg. Precip in. (Rain Days)(Thunder Days) | 1.98 (3.2) (1.2) | 0.85 (1.6) (0.6) | 0.29 (0.6) (0.1) | 2.05 (3.2) (3.8) | 9.31 (11.7) (15.7) | 8.16 (13.2) (16.8) |
| Avg. Snow in. (Snow Days) | 0 (0) | 0 (0) | 0 (0) | 0 (0) | 0 (0) | 0 (0) |

| | JULY | AUGUST | SEPTEMBER | OCTOBER | NOVEMBER | DECEMBER | ANNUAL |
|---|---|---|---|---|---|---|---|
| Record Max T °F | 94 | 94 | 94 | 94 | 95 | 94 | 97 |
| Average Max T °F | 88 | 88 | 87 | 87 | 87 | 88 | 88 |
| Average Min T °F | 75 | 75 | 74 | 74 | 74 | 74 | 74 |
| Record Min T °F | 70 | 70 | 70 | 70 | 70 | 69 | 65 |
| Avg. Precip in. (Rain Days)(Thunder Days) | 8.65 (13.6) (21) | 8.53 (13.9) (21.5) | 7.91 (10.5) (19.6) | 12.18 (13.5) (16.2) | 10.92 (16.2) (14) | 6.55 (10.4) (5.1) | 77.4 (111.6) (135.6) |
| Avg. Snow in. (Snow Days) | 0 (0) | 0 (0) | 0 (0) | 0 (0) | 0 (0) | 0 (0) | 0 (0) |

## Data Key

| Map Locator Number | | |
|---|---|---|
| Location | Average Max T °F | Record Max T °F |
| Lat - Lon ddmm Elev ft | Average Min T °F | Record Min T °F |
| | MONTH | |
| Avg. Precip in. (Rain Days) (Thunder Days) | | Avg. Snow in. (Snow Days) |

| *** or Blank | No data, missing data or data of insufficient reliability |
|---|---|
| Rain Days | Number of Days with >= 0.01" of precipitation falling |
| Snow Days | Number of Days with >= 1.5" of snow falling |

**Table 9.2  World Weather Data [CONTINUED]**

# South America, Argentina to Boliva, Sucre

## Map # 225 -229

**Map Location 225 — Argentina, Buenos Aires — 3435S 05829W 82**

| Month | Record High | Normal High | Normal Low | Record Low | Precip. mean (greatest)(days) | Snow mean (greatest) |
|---|---|---|---|---|---|---|
| JANUARY | 104 | 85 | 63 | 43 | 3.1 (5.3)(4) | 0 (0) |
| FEBRUARY | 103 | 83 | 63 | 40 | 2.8 (4.9)(3) | 0 (0) |
| MARCH | 99 | 79 | 60 | 39 | 4.3 (6.5)(4) | 0 (0) |
| APRIL | 97 | 72 | 53 | 28 | 3.5 (5.5)(3) | 0 (0) |
| MAY | 84 | 64 | 47 | 25 | 3 (4.8)(3) | 0 (0) |
| JUNE | 77 | 57 | 41 | 23 | 2.4 (4.3)(2) | 0 (0) |
| JULY | 84 | 57 | 42 | 22 | 2.2 (4)(2) | *** (***) |
| AUGUST | 87 | 60 | 43 | 27 | 2.4 (4.3)(3) | *** (***) |
| SEPTEMBER | 93 | 64 | 46 | 28 | 3.1 (5.3)(3) | *** (***) |
| OCTOBER | 91 | 69 | 50 | 34 | 3.4 (5.7)(3) | *** (***) |
| NOVEMBER | 95 | 76 | 56 | 36 | 3.3 (5.6)(4) | *** (***) |
| DECEMBER | 102 | 82 | 61 | 39 | 3.9 (6.4)(5) | *** (***) |
| ANNUAL | 104 | 71 | 52 | 22 | 37.4 (62.6)(39) | *** (***) |

**Map Location 226 — Argentina, Resistencia — 2728S 05859W 164**

| Month | Record High | Normal High | Normal Low | Record Low | Precip. mean (greatest)(days) | Snow mean (greatest) |
|---|---|---|---|---|---|---|
| JANUARY | 108 | 93 | 71 | 54 | 7.14 (6)(3) | 0 (0) |
| FEBRUARY | 106 | 91 | 70 | 54 | 4.23 (4.9)(8) | 0 (0) |
| MARCH | 99 | 86 | 67 | 50 | 6.41 (6.6)(2) | 0 (0) |
| APRIL | 97 | 80 | 61 | 43 | 4.6 (5.1)(5) | 0 (0) |
| MAY | 90 | 75 | 59 | 34 | 3.9 (5.2)(1) | 0 (0) |
| JUNE | 88 | 71 | 55 | 30 | 2.45 (4.9)(1) | 0 (0) |
| JULY | 90 | 71 | 53 | 30 | 2.12 (3.8)(1.5) | 0 (0) |
| AUGUST | 90 | 74 | 54 | 34 | 2.19 (3.2)(2.5) | 0 (0) |
| SEPTEMBER | 102 | 76 | 57 | 34 | 2.95 (5.9)(5.1) | 0 (0) |
| OCTOBER | 104 | 82 | 62 | 43 | 4.15 (5.8)(3) | 0 (0) |
| NOVEMBER | 102 | 85 | 64 | 48 | 5.68 (5.9)(7) | 0 (0) |
| DECEMBER | 108 | 90 | 69 | 54 | 5.17 (6.1)(10) | 0 (0) |
| ANNUAL | 108 | 81 | 62 | 30 | 51 (63.4)(49.1) | 0 (0) |

**Map Location 227 — Argentina, Ushuaia — 5449S 06819W 10**

| Month | Record High | Normal High | Normal Low | Record Low | Precip. mean (greatest)(days) | Snow mean (greatest) |
|---|---|---|---|---|---|---|
| JANUARY | 85 | 57 | 41 | 28 | 2 (3.8)(0.3) | *** (***) |
| FEBRUARY | 79 | 58 | 41 | 27 | 2.6 (4.6)(0.3) | *** (***) |
| MARCH | 79 | 55 | 38 | 27 | 1.9 (3.4)(0.3) | *** (***) |
| APRIL | 68 | 48 | 33 | 18 | 2.1 (3.7)(0.3) | *** (***) |
| MAY | 66 | 43 | 29 | -4 | 1.5 (2.9)(0) | *** (***) |
| JUNE | 60 | 39 | 26 | -1 | 1.2 (2.3)(0.3) | *** (***) |
| JULY | 63 | 39 | 25 | -6 | 1.2 (2.3)(0.3) | *** (***) |
| AUGUST | 57 | 42 | 27 | 6 | 1.1 (2.1)(0) | *** (***) |
| SEPTEMBER | 66 | 46 | 31 | 17 | 1.3 (2.8)(0) | *** (***) |
| OCTOBER | 79 | 52 | 35 | 24 | 1.6 (3.3)(0.3) | *** (***) |
| NOVEMBER | 75 | 54 | 36 | 21 | 1.5 (3.1)(0.3) | *** (***) |
| DECEMBER | 80 | 56 | 39 | 27 | 1.9 (3.6)(0.3) | *** (***) |
| ANNUAL | 85 | 49 | 33 | -6 | 19.9 (37.9)(2.7) | *** (***) |

**Map Location 228 — Bolivia, El Alto — 1630S 06811W 13354**

| Month | Record High | Normal High | Normal Low | Record Low | Precip. mean (greatest)(days) | Snow mean (greatest) |
|---|---|---|---|---|---|---|
| JANUARY | 77 | 63 | 43 | 26 | 4.5 (7.1)(2) | *** (***) |
| FEBRUARY | 76 | 63 | 43 | 26 | 4.2 (6.7)(2) | *** (***) |
| MARCH | 76 | 64 | 42 | 27 | 2.6 (4.3)(3) | *** (***) |
| APRIL | 75 | 65 | 40 | 24 | 1.3 (2.6)(1) | *** (***) |
| MAY | 72 | 64 | 37 | 14 | 0.5 (1.5)(0.3) | *** (***) |
| JUNE | 70 | 62 | 34 | 12 | 0.3 (0.8)(1) | *** (***) |
| JULY | 71 | 62 | 33 | 5 | 0.4 (1)(0) | *** (***) |
| AUGUST | 72 | 63 | 35 | 9 | 0.5 (1.1)(0) | *** (***) |
| SEPTEMBER | 80 | 64 | 38 | 18 | 1.1 (2.5)(2) | *** (***) |
| OCTOBER | 76 | 66 | 40 | 16 | 1.6 (3.3)(1) | *** (***) |
| NOVEMBER | 77 | 67 | 42 | 23 | 1.9 (3.7)(0.3) | *** (***) |
| DECEMBER | 76 | 65 | 42 | 22 | 3.7 (6.1)(2) | *** (***) |
| ANNUAL | 80 | 64 | 39 | 5 | 22.6 (40.7)(13.6) | *** (***) |

**Map Location 229 — Bolivia, Sucre — 1903S 06516W 9351**

| Month | Record High | Normal High | Normal Low | Record Low | Precip. mean (greatest)(days) | Snow mean (greatest) |
|---|---|---|---|---|---|---|
| JANUARY | 88 | 63 | 48 | 40 | 7.3 (10.2)(2) | *** (***) |
| FEBRUARY | 85 | 62 | 48 | 41 | 4.9 (7.6)(2) | *** (***) |
| MARCH | 87 | 64 | 48 | 38 | 3.7 (5.7)(1) | *** (***) |
| APRIL | 84 | 63 | 45 | 37 | 1.6 (3)(1) | *** (***) |
| MAY | 81 | 62 | 40 | 29 | 0.2 (1.1)(0.3) | 0* (0) |
| JUNE | 81 | 61 | 38 | 27 | 0.1 (0.5)(2) | *** (***) |
| JULY | 78 | 61 | 37 | 25 | 0.2 (0.6)(1) | *** (***) |
| AUGUST | 82 | 65 | 40 | 30 | 0.3 (0.8)(1) | *** (***) |
| SEPTEMBER | 81 | 67 | 44 | 29 | 1 (2.3)(2) | *** (***) |
| OCTOBER | 83 | 65 | 46 | 33 | 1.6 (3.3)(1) | *** (***) |
| NOVEMBER | 83 | 68 | 48 | 40 | 2.6 (4.7)(2) | *** (***) |
| DECEMBER | 84 | 66 | 49 | 41 | 4.3 (6.9)(1) | *** (***) |
| ANNUAL | 88 | 64 | 44 | 25 | 27.8 (46.7)(16.3) | *** (***) |

**Table 9.2  World Weather Data [CONTINUED]**

# South America, Boliva, Trinidad to Brazil, Sao Paulo    Map # 230 -234

**Map Location 230 — Bolivia, Trinidad — 1445S 06448W 774**

| Month | Max | Min | (snow) | Max | Min | Precip |
|---|---|---|---|---|---|---|
| JANUARY | 105 | 65 | 0 (0) | 89 | 74 | 11.21 (12.6) (***) |
| FEBRUARY | 104 | 68 | 0 (0) | 89 | 74 | 9.25 (10.8) (***) |
| MARCH | 96 | 65 | 0 (0) | 88 | 73 | 7.56 (8.4) (***) |
| APRIL | 92 | 61 | 0 (0) | 85 | 71 | 5.83 (5.9) (***) |
| MAY | 92 | 49 | 0 (0) | 82 | 65 | 4.15 (4.8) (***) |
| JUNE | 93 | 50 | 0 (0) | 85 | 64 | 4.08 (3.2) (***) |
| JULY | 98 | 47 | 0 (0) | 87 | 63 | 2.82 (2.3) (***) |
| AUGUST | 103 | 46 | 0 (0) | 85 | 61 | 1.81 (2) (***) |
| SEPTEMBER | 110 | 49 | 0 (0) | 91 | 66 | 4.21 (3.8) (***) |
| OCTOBER | 104 | 57 | 0 (0) | 90 | 70 | 6.09 (7.1) (***) |
| NOVEMBER | 104 | 56 | 0 (0) | 89 | 70 | 7.81 (7.9) (***) |
| DECEMBER | 106 | 61 | 0 (0) | 90 | 73 | 8.97 (9.9) (***) |
| ANNUAL | 110 | 46 | 0 (0) | | | 73.8 (78.7) |

**Map Location 231 — Brazil, Belem — 0128S 04827W 33**

| Month | Max | Min | (snow) | Max | Min | Precip |
|---|---|---|---|---|---|---|
| JANUARY | 95 | 61 | 0 (0) | 87 | 72 | 12.5 (14.5) (8) |
| FEBRUARY | 94 | 68 | 0 (0) | 86 | 72 | 14.1 (15.5) (6) |
| MARCH | 95 | 66 | 0 (0) | 87 | 73 | 14.1 (***) (7) |
| APRIL | 95 | 69 | 0 (0) | 87 | 73 | 12.6 (16.3) (7) |
| MAY | 94 | 68 | 0 (0) | 88 | 73 | 10.2 (13.6) (7) |
| JUNE | 94 | 68 | 0 (0) | 88 | 72 | 6.7 (11.3) (7) |
| JULY | 95 | 64 | 0 (0) | 88 | 71 | 5.9 (10.1) (7) |
| AUGUST | 95 | 67 | 0 (0) | 88 | 71 | 4.4 (7.7) (7) |
| SEPTEMBER | 96 | 65 | 0 (0) | 89 | 71 | 3.5 (5.8) (7) |
| OCTOBER | 98 | 63 | 0 (0) | 89 | 71 | 3.3 (5.6) (5) |
| NOVEMBER | 97 | 67 | 0 (0) | 90 | 71 | 2.6 (4.7) (5) |
| DECEMBER | 97 | 66 | 0 (0) | 89 | 72 | 6.1 (9) (7) |
| ANNUAL | 98 | 61 | 0 (0) | | | 96 (***) (80) |

**Map Location 232 — Brazil, Brasilia — 1551S 04756W 3481**

| Month | Max | Min | (snow) | Max | Min | Precip |
|---|---|---|---|---|---|---|
| JANUARY | 85 | 61 | 0 (0) | 80 | 65 | 9 (11.8) (1.2) |
| FEBRUARY | 86 | 61 | 0 (0) | 81 | 64 | 7.8 (10.7) (1.2) |
| MARCH | 89 | 58 | 0 (0) | 82 | 64 | 4.8 (7.1) (1.4) |
| APRIL | 87 | 57 | 0 (0) | 82 | 62 | 3.4 (5.4) (0.8) |
| MAY | 84 | 49 | 0 (0) | 79 | 56 | 1.4 (2.7) (0.3) |
| JUNE | 82 | 47 | 0 (0) | 77 | 52 | 0.01 (0.3) (0.2) |
| JULY | 81 | 46 | 0 (0) | 78 | 51 | 0 (0) (0.1) |
| AUGUST | 87 | 48 | 0 (0) | 82 | 55 | 0.04 (0.4) (0.3) |
| SEPTEMBER | 93 | 55 | 0 (0) | 87 | 60 | 1.3 (2.8) (0.6) |
| OCTOBER | 92 | 59 | 0 (0) | 82 | 64 | 4.9 (7.2) (0.6) |
| NOVEMBER | 89 | 59 | 0 (0) | 82 | 66 | 9.7 (11.1) (1.2) |
| DECEMBER | 86 | 61 | 0 (0) | 78 | 64 | 11.7 (14) (1.2) |
| ANNUAL | 93 | 46 | 0 (0) | | | 54 (73.5) (9.1) |

**Map Location 233 — Brazil, Manaus — 0308S 06001W 144**

| Month | Max | Min | (snow) | Max | Min | Precip |
|---|---|---|---|---|---|---|
| JANUARY | 99 | 65 | 0 (0) | 87 | 74 | 9.8 (12.5) (4) |
| FEBRUARY | 100 | 68 | 0 (0) | 86 | 74 | 9.1 (11.9) (3) |
| MARCH | 97 | 66 | 0 (0) | 87 | 74 | 10.3 (13.8) (4) |
| APRIL | 97 | 66 | 0 (0) | 86 | 74 | 8.7 (11.9) (5) |
| MAY | 95 | 68 | 0 (0) | 88 | 75 | 6.7 (9.5) (3) |
| JUNE | 95 | 65 | 0 (0) | 88 | 74 | 3.3 (5.8) (4) |
| JULY | 95 | 64 | 0 (0) | 89 | 74 | 2.3 (4.2) (3) |
| AUGUST | 98 | 66 | 0 (0) | 90 | 74 | 1.5 (2.8) (4) |
| SEPTEMBER | 99 | 64 | 0 (0) | 91 | 75 | 1.8 (3.6) (5) |
| OCTOBER | 100 | 68 | 0 (0) | 91 | 75 | 4.2 (6.6) (7) |
| NOVEMBER | 99 | 67 | 0 (0) | 90 | 75 | 5.6 (7.9) (7) |
| DECEMBER | 101 | 63 | 0 (0) | 89 | 75 | 8 (10.9) (6) |
| ANNUAL | 101 | 63 | 0 (0) | | | 71.3 (101.4) (55) |

**Map Location 234 — Brazil, Sao Paulo — 2337S 04639W 2628**

| Month | Max | Min | (snow) | Max | Min | Precip |
|---|---|---|---|---|---|---|
| JANUARY | 97 | 50 | 0 (0) | 77 | 63 | 8.8 (11.7) (8) |
| FEBRUARY | 100 | 52 | 0 (0) | 79 | 64 | 7.81 (10.7) (7) |
| MARCH | 91 | 50 | 0 (0) | 76 | 62 | 6.05 (8.7) (5) |
| APRIL | 91 | 41 | 0 (0) | 73 | 59 | 2.25 (3.9) (1) |
| MAY | 95 | 36 | 0 (0) | 68 | 54 | 2.95 (4.8) (0) |
| JUNE | 86 | 34 | 0 (0) | 66 | 54 | 2.36 (4.3) (0) |
| JULY | 88 | 32 | 0 (0) | 66 | 53 | 1.51 (2.8) (0) |
| AUGUST | 90 | 36 | 0 (0) | 67 | 53 | 2.14 (3.9) (0) |
| SEPTEMBER | 90 | 32 | 0 (0) | 68 | 57 | 3.45 (5.7) (0) |
| OCTOBER | 95 | 39 | 0 (0) | 72 | 57 | 4.59 (7) (3) |
| NOVEMBER | 95 | 45 | 0 (0) | 72 | 59 | 6.01 (8.2) (3) |
| DECEMBER | 99 | 48 | 0 (0) | 75 | 62 | 9.37 (12.2) (6) |
| ANNUAL | 100 | 32 | 0 (0) | | | 57.3 (83.9) (33) |

**Table 9.2  World Weather Data [CONTINUED]**

# South America, Brazil, Sao Salvador to Chile, Puerto Montt  Map # 235 -299

## Map Location 235 — Brazil, Sao Salvador — 1255S 03820W 20

| | JANUARY | FEBRUARY | MARCH | APRIL | MAY | JUNE |
|---|---|---|---|---|---|---|
| Record High | 94 | 100 | 100 | 94 | 97 | 90 |
| Record Low | 68 | 52 | 63 | 68 | 64 | 50 |
| Snow | 0 (0) | 0 (0) | 0 (0) | 0 (0) | 0 (0) | 0 (0) |
| Normal High | 86 | 86 | 86 | 84 | 82 | 80 |
| Normal Low | 74 | 74 | 74 | 74 | 72 | 71 |
| Precip | 2.6 (4.6) (3) | 5.3 (8.1) (4) | 6.1 (8.8) (4) | 11.2 (14.8) (2) | 10.8 (14.3) (1) | 9.4 (14.7) (0) |

| | JULY | AUGUST | SEPTEMBER | OCTOBER | NOVEMBER | DECEMBER | ANNUAL |
|---|---|---|---|---|---|---|---|
| Record High | 85 | 90 | 90 | 99 | 97 | 92 | 100 |
| Record Low | 54 | 54 | 54 | 59 | 55 | 66 | 50 |
| Snow | 0 (0) | 0 (0) | 0 (0) | 0 (0) | 0 (0) | 0 (0) | 0 (0) |
| Normal High | 79 | 79 | 81 | 83 | 84 | 84 | 83 |
| Normal Low | 69 | 69 | 70 | 71 | 72 | 73 | 72 |
| Precip | 7.2 (12) (0) | 4.8 (8.3) (0) | 3.3 (5.6) (0) | 4 (6.4) (1) | 4.5 (6.9) (2) | 5.6 (8.4) (3) | 74.8 (112.9) (20) |

## Map Location 236 — Chile, Arica — 1830S 07019W 328

| | JANUARY | FEBRUARY | MARCH | APRIL | MAY | JUNE |
|---|---|---|---|---|---|---|
| Record High | 91 | 93 | 93 | 86 | 81 | 79 |
| Record Low | 50 | 50 | 52 | 43 | 43 | 39 |
| Snow | 0 (0) | 0 (0) | 0 (0) | 0 (0) | 0 (0) | 0 (0) |
| Normal High | 78 | 79 | 77 | 74 | 70 | 67 |
| Normal Low | 64 | 65 | 63 | 60 | 58 | 57 |
| Precip | 0.05 (0.7) (0) | 0 (0) (0) | 0 (0) (0) | 0 (0) (0) | 0 (0) (0) | 0 (0) (0) |

| | JULY | AUGUST | SEPTEMBER | OCTOBER | NOVEMBER | DECEMBER | ANNUAL |
|---|---|---|---|---|---|---|---|
| Record High | 75 | 77 | 81 | 79 | 82 | 86 | 93 |
| Record Low | 39 | 41 | 43 | 41 | 46 | 50 | 39 |
| Snow | 0 (0) | 0 (0) | 0 (0) | 0 (0) | 0 (0) | 0 (0) | 0 (0) |
| Normal High | 66 | 65 | 67 | 69 | 72 | 75 | 72 |
| Normal Low | 54 | 55 | 56 | 58 | 60 | 62 | 59 |
| Precip | 0 (0) (0) | 0.05 (0.4) (0) | 0 (0) (0) | 0 (0) (0) | 0 (0) (0) | 0.05 (0.7) (0) | 0.1 (1.8) (0) |

## Map Location 237 — Chile, Punta Arenas — 5308S 07053W 6

| | JANUARY | FEBRUARY | MARCH | APRIL | MAY | JUNE |
|---|---|---|---|---|---|---|
| Record High | 86 | 79 | 75 | 69 | 63 | 52 |
| Record Low | 26 | 28 | 23 | 23 | 16 | 11 |
| Snow | 0 (***) | 0 (***) | 0 (***) | 0 (***) | 0 (***) | 1.6 (***) |
| Normal High | 59 | 58 | 55 | 49 | 43 | 40 |
| Normal Low | 45 | 44 | 43 | 39 | 35 | 33 |
| Precip | 1.2 (2.6) (0.3) | 1.2 (2.6) (0.3) | 1.6 (3) (0.3) | 2.1 (3.7) (0.3) | 2.2 (3.8) (0) | 2 (3.6) (0) |

| | JULY | AUGUST | SEPTEMBER | OCTOBER | NOVEMBER | DECEMBER | ANNUAL |
|---|---|---|---|---|---|---|---|
| Record High | 55 | 55 | 61 | 67 | 74 | 81 | 86 |
| Record Low | 12 | 15 | 19 | 25 | 23 | 25 | 11 |
| Snow | 2 (***) | 3.9 (***) | 3.2 (***) | 0 (***) | 0 (***) | 0 (***) | 10.7 (***) |
| Normal High | 38 | 40 | 45 | 50 | 55 | 57 | 49 |
| Normal Low | 33 | 33 | 35 | 38 | 43 | 43 | 39 |
| Precip | 1.8 (3.3) (0) | 1.9 (3.5) (0) | 1.6 (3.3) (0) | 1 (2.3) (0.3) | 1.3 (2.8) (0) | 1.3 (2.7) (0.3) | 19.2 (37.2) (1.8) |

## Map Location 238 — Chile, Antofagasta — 2327S 07028W 442

| | JANUARY | FEBRUARY | MARCH | APRIL | MAY | JUNE |
|---|---|---|---|---|---|---|
| Record High | 86 | 86 | 83 | 81 | 75 | 78 |
| Record Low | 50 | 46 | 48 | 39 | 46 | 41 |
| Snow | 0 (0) | 0 (0) | 0 (0) | 0 (0) | 0 (0) | 0 (0) |
| Normal High | 76 | 76 | 74 | 70 | 67 | 65 |
| Normal Low | 63 | 63 | 61 | 58 | 55 | 52 |
| Precip | 0 (0) (0) | 0 (0) (0) | 0 (0) (0) | 0.05 (0.9) (0) | 0.05 (0.9) (0) | 0.1 (0.5) (0) |

| | JULY | AUGUST | SEPTEMBER | OCTOBER | NOVEMBER | DECEMBER | ANNUAL |
|---|---|---|---|---|---|---|---|
| Record High | 63 | 75 | 75 | 79 | 81 | 72 | 86 |
| Record Low | 37 | 41 | 42 | 44 | 46 | 46 | 34 |
| Snow | 0 (0) | 0 (0) | 0 (0) | 0 (0) | 0 (0) | 0 (0) | 0 (0) |
| Normal High | 63 | 62 | 64 | 66 | 69 | 72 | 68 |
| Normal Low | 51 | 52 | 53 | 55 | 58 | 60 | 57 |
| Precip | 0.2 (0.6) (0) | 0.1 (0.5) (0) | 0.05 (0.6) (0) | 0.1 (0.7) (0) | 0.05 (0.6) (0) | 0 (0) (0) | 0.65 (1.7) (0) |

## Map Location 239 — Chile, Puerto Montt — 4128S 07256W 43

| | JANUARY | FEBRUARY | MARCH | APRIL | MAY | JUNE |
|---|---|---|---|---|---|---|
| Record High | 82 | 84 | 77 | 72 | 66 | 64 |
| Record Low | 41 | 39 | 36 | 34 | 29 | 28 |
| Snow | 0 (0) | *** (***) | *** (***) | 0* (0*) | 0* (0*) | *** (***) |
| Normal High | 68 | 68 | 64 | 60 | 56 | 52 |
| Normal Low | 52 | 52 | 49 | 47 | 44 | 43 |
| Precip | 4.66 (7.3) (0.3) | 4.3 (6.9) (0.6) | 6.06 (8.7) (0.7) | 7.49 (10.5) (0.6) | 10.88 (14.4) (0.6) | 9.78 (15.1) (0.8) |

| | JULY | AUGUST | SEPTEMBER | OCTOBER | NOVEMBER | DECEMBER | ANNUAL |
|---|---|---|---|---|---|---|---|
| Record High | 63 | 67 | 78 | 71 | 80 | 65 | 84 |
| Record Low | 27 | 28 | 32 | 32 | 34 | 39 | 27 |
| Snow | *** (***) | *** (***) | *** (***) | *** (***) | *** (***) | *** (***) | *** (***) |
| Normal High | 51 | 53 | 56 | 59 | 62 | 65 | 60 |
| Normal Low | 41 | 41 | 42 | 44 | 47 | 50 | 46 |
| Precip | 10.77 (16) (0.5) | 9.35 (14.7) (0.3) | 6.29 (8.5) (0.5) | 5.41 (7.7) (0.5) | 5.5 (7.8) (0.3) | 5.49 (8.3) (0.2) | 86 (125.9) (5.9) |

**Table 9.2  World Weather Data [CONTINUED]**

# South America, Chile, Santiago to Ecuador          Map # 240 -244

*(For each month: Record High, Average High, Average Low, Record Low, Precipitation mean (max) (min), Snow mean (max))*

### Map Location 240 — Chile, Santiago  3327S 07042W 1706

| Month | Rec. Hi | Avg. Hi | Avg. Lo | Rec. Lo | Precip. | Snow |
|---|---|---|---|---|---|---|
| JANUARY | 96 | 85 | 53 | 43 | 0.1 (0.8) (0.2) | 0 (0) |
| FEBRUARY | 98 | 84 | 52 | 43 | 0.1 (0.8) (0.1) | 0 (0) |
| MARCH | 94 | 80 | 49 | 38 | 0.2 (1.1) (0.3) | 0 (0) |
| APRIL | 90 | 74 | 45 | 33 | 0.5 (1.5) (0.3) | 0 (0*) |
| MAY | 87 | 65 | 41 | 27 | 2.5 (4.2) (0.3) | 0 (***) |
| JUNE | 80 | 58 | 37 | 26 | 3.3 (5.8) (0.2) | 2 (***) |
| JULY | 81 | 59 | 37 | 24 | 3 (5.3) (0.1) | 2 (***) |
| AUGUST | 83 | 62 | 39 | 26 | 2.2 (4) (0.1) | 5 (***) |
| SEPTEMBER | 88 | 66 | 42 | 31 | 1.2 (2.7) (0.4) | 5 (***) |
| OCTOBER | 92 | 72 | 45 | 32 | 0.6 (1.6) (0.1) | 0 (0) |
| NOVEMBER | 97 | 78 | 48 | 37 | 0.3 (1.1) (0.2) | 0 (0) |
| DECEMBER | 99 | 83 | 51 | 36 | 0.2 (1) (0) | 0 (0) |
| ANNUAL | 99 | 72 | 45 | 24 | 14.2 (29.9) (2.3) | 9 (***) |

### Map Location 241 — Colombia, Bogota  0444N 07417W 8379

| Month | Rec. Hi | Avg. Hi | Avg. Lo | Rec. Lo | Precip. | Snow |
|---|---|---|---|---|---|---|
| JANUARY | 73 | 67 | 48 | 36 | 2.28 (4.2) (0.2) | 0 (0) |
| FEBRUARY | 75 | 68 | 49 | 30 | 2.6 (4.6) (1.5) | 0 (0) |
| MARCH | 75 | 67 | 50 | 36 | 3.98 (6.1) (4.8) | 0 (0) |
| APRIL | 75 | 67 | 51 | 32 | 5.75 (8.3) (4.5) | 0 (0) |
| MAY | 74 | 66 | 51 | 36 | 4.45 (6.7) (3.8) | 0 (0) |
| JUNE | 72 | 65 | 51 | 32 | 2.44 (4.4) (0.2) | 0 (0) |
| JULY | 72 | 64 | 50 | 41 | 2.01 (3.7) (0.2) | 0 (0) |
| AUGUST | 72 | 65 | 50 | 34 | 2.2 (4) (0.5) | 0 (0) |
| SEPTEMBER | 73 | 66 | 49 | 37 | 2.44 (4.5) (1.5) | 0 (0) |
| OCTOBER | 73 | 66 | 50 | 34 | 6.3 (8.5) (8.2) | 0 (0) |
| NOVEMBER | 73 | 66 | 50 | 36 | 4.68 (7) (5.8) | 0 (0) |
| DECEMBER | 75 | 66 | 49 | 36 | 2.6 (4.6) (5.2) | 0 (0) |
| ANNUAL | 75 | 66 | 50 | 30 | 41.7 (66.6) (36.4) | 0 (0) |

### Map Location 242 — Colombia, Villavicencio  0409N 07334W 1414

| Month | Rec. Hi | Avg. Hi | Avg. Lo | Rec. Lo | Precip. | Snow |
|---|---|---|---|---|---|---|
| JANUARY | 99 | 91 | 71 | 59 | 1.6 (1.5) (0) | 0 (0) |
| FEBRUARY | 101 | 90 | 73 | 59 | 3.9 (1.5) (1) | 0 (0) |
| MARCH | 100 | 89 | 73 | 59 | 5.7 (2.5) (1) | 0 (0) |
| APRIL | 99 | 89 | 70 | 57 | 19.8 (14.7) (3) | 0 (0) |
| MAY | 98 | 86 | 71 | 51 | 20 (11) (2) | 0 (0) |
| JUNE | 102 | 87 | 70 | 59 | 20.4 (19) (1) | 0 (0) |
| JULY | 93 | 86 | 69 | 57 | 17.9 (20.8) (2) | 0 (0) |
| AUGUST | 100 | 87 | 70 | 48 | 15.2 (14) (2) | 0 (0) |
| SEPTEMBER | 99 | 88 | 69 | 48 | 15.9 (13) (4) | 0 (0) |
| OCTOBER | 106 | 91 | 70 | 48 | 15.2 (15.5) (6) | 0 (0) |
| NOVEMBER | 104 | 89 | 69 | 51 | 13.9 (18) (5) | 0 (0) |
| DECEMBER | 104 | 90 | 69 | 51 | 6.4 (9) (2) | 0 (0) |
| ANNUAL | 106 | 89 | 70 | 48 | 155.9 (140.5) (29) | 0 (0) |

### Map Location 243 — Ecuador, Guayaquil  0210S 07952W 13

| Month | Rec. Hi | Avg. Hi | Avg. Lo | Rec. Lo | Precip. | Snow |
|---|---|---|---|---|---|---|
| JANUARY | 93 | 87 | 72 | 63 | 0.52 (1.5) (0.6) | 0 (0) |
| FEBRUARY | 93 | 87 | 72 | 64 | 1.86 (4) (2.2) | 0 (0) |
| MARCH | 93 | 87 | 72 | 64 | 2.22 (3.7) (3.6) | 0 (0) |
| APRIL | 88 | 88 | 72 | 64 | 0.4 (1.5) (2.3) | 0 (0) |
| MAY | 87 | 87 | 70 | 64 | 0.08 (0.2) (0.2) | 0 (0) |
| JUNE | 84 | 84 | 68 | 61 | 0.09 (0.2) (0.3) | 0 (0) |
| JULY | 90 | 84 | 67 | 57 | 0 (0) (0) | 0 (0) |
| AUGUST | 90 | 85 | 66 | 59 | 0 (0) (0) | 0 (0) |
| SEPTEMBER | 93 | 86 | 67 | 52 | 0 (0) (0.1) | 0 (0) |
| OCTOBER | 93 | 86 | 68 | 52 | 0 (0) (0.1) | 0 (0) |
| NOVEMBER | 93 | 87 | 68 | 63 | 0.04 (0.3) (0) | 0 (0) |
| DECEMBER | 95 | 89 | 70 | 64 | 0.12 (0.2) (0.3) | 0 (0) |
| ANNUAL | 95 | 86 | 69 | 52 | 5.3 (11.6) (9.6) | 0 (0) |

### Map Location 244 — Ecuador, Quito  0008S 07829W 9222

| Month | Rec. Hi | Avg. Hi | Avg. Lo | Rec. Lo | Precip. | Snow |
|---|---|---|---|---|---|---|
| JANUARY | 81 | 67 | 46 | 34 | 5 (7.7) (4) | 0 (0) |
| FEBRUARY | 79 | 67 | 47 | 32 | 5.3 (8.1) (4) | 0 (0) |
| MARCH | 80 | 68 | 47 | 32 | 6.1 (8.8) (6) | 0 (0) |
| APRIL | 78 | 69 | 47 | 36 | 6.8 (9.6) (7) | 0 (0) |
| MAY | 81 | 69 | 47 | 32 | 5.4 (7.9) (6) | 0 (0) |
| JUNE | 78 | 69 | 45 | 32 | 1.8 (3.3) (3) | 0 (0) |
| JULY | 79 | 71 | 44 | 25 | 0.7 (1.5) (1) | *** (***) |
| AUGUST | 72 | 72 | 44 | 34 | 0.8 (1.6) (2) | 0* (0*) |
| SEPTEMBER | 82 | 72 | 45 | 28 | 1.4 (3) (6) | *** (***) |
| OCTOBER | 86 | 71 | 46 | 32 | 5.2 (7.5) (8) | 0 (0) |
| NOVEMBER | 81 | 70 | 45 | 32 | 4.3 (6.7) (4) | 0 (0) |
| DECEMBER | 81 | 70 | 46 | 32 | 4.3 (6.9) (3) | 0 (0) |
| ANNUAL | 86 | 70 | 46 | 25 | 47.1 (72.6) (54) | *** (***) |

Table 9.2 World Weather Data [CONTINUED]

# South America, French Guiana to Peru

## Map # 245 -249

### Map Location 245 — French Guiana, Cayenne — 0449N 05222W 26

| | JAN | FEB | MAR | APR | MAY | JUN | JUL | AUG | SEP | OCT | NOV | DEC | ANNUAL |
|---|---|---|---|---|---|---|---|---|---|---|---|---|---|
| Normal Max | 84 | 85 | 85 | 86 | 85 | 87 | 88 | 90 | 91 | 89 | 89 | 86 | 87 |
| Normal Min | 74 | 74 | 74 | 75 | 74 | 73 | 73 | 73 | 74 | 74 | 74 | 74 | 74 |
| Extreme Max | 91 | 93 | 92 | 92 | 92 | 93 | 93 | 96 | 97 | 95 | 95 | 93 | 97 |
| Extreme Min | 67 | 68 | 66 | 65 | 68 | 68 | 66 | 68 | 66 | 66 | 66 | 68 | 65 |
| Precipitation | 14.4 (15.7) (4) | 12.3 (14.4*) (1) | 15.8 (***) (1) | 18.9 (***) (2) | 21.7 (***) (8) | 15.5 (***) (11) | 6.9 (11.6) (15) | 2.8 (5) (7) | 1.2 (2.7) (10) | 1.3 (2.8) (7) | 4.6 (7) (8) | 10.7 (13.2) (7) | 126.1 (***) (81) |
| Snow | 0 (0) | 0 (0) | 0 (0) | 0 (0) | 0 (0) | 0 (0) | 0 (0) | 0 (0) | 0 (0) | 0 (0) | 0 (0) | 0 (0) | 0 (0) |

### Map Location 246 — Guyana, Georgetown — 0630N 05815W 96

| | JAN | FEB | MAR | APR | MAY | JUN | JUL | AUG | SEP | OCT | NOV | DEC | ANNUAL |
|---|---|---|---|---|---|---|---|---|---|---|---|---|---|
| Normal Max | 85 | 85 | 86 | 87 | 87 | 86 | 88 | 90 | 91 | 90 | 89 | 85 | 87 |
| Normal Min | 69 | 69 | 70 | 71 | 72 | 71 | 71 | 71 | 70 | 70 | 71 | 70 | 70 |
| Extreme Max | 91 | 94 | 92 | 96 | 94 | 92 | 94 | 98 | 99 | 95 | 95 | 92 | 99 |
| Extreme Min | 64 | 63 | 64 | 64 | 66 | 67 | 68 | 65 | 62 | 66 | 65 | 63 | 62 |
| Precipitation | 7.55 (12.2) (0.4) | 5.81 (9.9) (0.5) | 4.92 (9.4) (0.4) | 7.9 (10.8) (0.3) | 12.25 (18.3) (2.2) | 15.01 (22) (2.9) | 12.24 (20.1) (7) | 7.53 (14.9) (7.5) | 4.56 (9.1) (6) | 4.41 (9.1) (6) | 6.2 (10.6) (3.2) | 9.98 (16.4) (1.1) | 98.4 (162.2) (37.6) |
| Snow | 0 (0) | 0 (0) | 0 (0) | 0 (0) | 0 (0) | 0 (0) | 0 (0) | 0 (0) | 0 (0) | 0 (0) | 0 (0) | 0 (0) | 0 (0) |

### Map Location 247 — Paraguay, Asuncion — 2516S 05738W 210

| | JAN | FEB | MAR | APR | MAY | JUN | JUL | AUG | SEP | OCT | NOV | DEC | ANNUAL |
|---|---|---|---|---|---|---|---|---|---|---|---|---|---|
| Normal Max | 92 | 92 | 88 | 80 | 74 | 75 | 73 | 76 | 77 | 84 | 87 | 91 | 82 |
| Normal Min | 75 | 73 | 72 | 65 | 64 | 60 | 58 | 62 | 63 | 66 | 71 | 73 | 67 |
| Extreme Max | 109 | 109 | 105 | 104 | 99 | 98 | 103 | 101 | 105 | 106 | 108 | 110 | 110 |
| Extreme Min | 54 | 52 | 49 | 42 | 34 | 29 | 29 | 30 | 37 | 38 | 45 | 47 | 29 |
| Precipitation | 5.5 (8.3) (4.2) | 5.1 (7.8) (3.5) | 4.3 (6.5) (4.2) | 5.2 (7.6) (1.2) | 4.6 (6.9) (1.1) | 2.7 (4.8) (1.1) | 2.2 (4) (0.2) | 1.5 (2.8) (1.1) | 3.1 (5.3) (2.6) | 5.5 (7.8) (2.4) | 5.9 (8.1) (3.4) | 6.2 (9.1) (2.7) | 51.8 (79) (27.7) |
| Snow | 0 (0) | 0 (0) | 0 (0) | 0 (0) | 0 (0) | 0 (0) | ***(***) | ***(***) | ***(***) | ***(***) | ***(***) | ***(***) | ***(***) |

### Map Location 248 — Peru, Cuzco — 1333S 07159W 10866

| | JAN | FEB | MAR | APR | MAY | JUN | JUL | AUG | SEP | OCT | NOV | DEC | ANNUAL |
|---|---|---|---|---|---|---|---|---|---|---|---|---|---|
| Normal Max | 68 | 69 | 70 | 71 | 70 | 69 | 70 | 72 | 72 | 72 | 72 | 71 | 71 |
| Normal Min | 45 | 45 | 44 | 40 | 35 | 33 | 31 | 34 | 40 | 43 | 43 | 44 | 39 |
| Extreme Max | 77 | 81 | 79 | 79 | 84 | 77 | 77 | 82 | 81 | 84 | 83 | 82 | 86 |
| Extreme Min | 34 | 36 | 35 | 25 | 24 | 23 | 16 | 30 | 30 | 30 | 30 | 32 | 16 |
| Precipitation | 6.4 (9.3) (***) | 5.9 (8.8) (***) | 4.3 (6.5) (***) | 2 (3.5) (***) | 0.6 (1.7) (***) | 0.2 (0.6) (***) | 0.2 (0.6) (***) | 0.4 (1) (***) | 1 (2.3) (***) | 2.6 (4.7) (***) | 3 (5.2) (***) | 5.4 (8.2) (***) | 32 (52.4) (***) |
| Snow | ***(***) | ***(***) | ***(***) | ***(***) | ***(***) | ***(***) | ***(***) | ***(***) | ***(***) | ***(***) | ***(***) | ***(***) | ***(***) |

### Map Location 249 — Peru, Lima — 1201S 07708W 105

| | JAN | FEB | MAR | APR | MAY | JUN | JUL | AUG | SEP | OCT | NOV | DEC | ANNUAL |
|---|---|---|---|---|---|---|---|---|---|---|---|---|---|
| Normal Max | 76 | 78 | 78 | 74 | 69 | 64 | 63 | 62 | 63 | 65 | 68 | 72 | 69 |
| Normal Min | 66 | 67 | 67 | 64 | 61 | 59 | 57 | 57 | 57 | 58 | 60 | 63 | 61 |
| Extreme Max | 82 | 84 | 83 | 82 | 81 | 74 | 74 | 70 | 71 | 72 | 80 | 80 | 84 |
| Extreme Min | 60 | 55 | 59 | 57 | 54 | 52 | 52 | 52 | 52 | 50 | 54 | 54 | 50 |
| Precipitation | 0.1 (0.8) (0) | 0.03 (0.7) (0) | 0.03 (0.9) (0) | 0.1 (0.7) (0) | 0.2 (0.6) (0) | 0.2 (0.6) (0) | 0.3 (0.8) (0) | 0.3 (0.8) (0) | 0.3 (1.1) (0) | 0.1 (0.7) (0) | 0.03 (0.9) (0) | 0.03 (0.7) (0) | 1.7 (9.8) (0) |
| Snow | 0 (0) | 0 (0) | 0 (0) | 0 (0) | 0 (0) | 0 (0) | 0 (0) | 0 (0) | 0 (0) | 0 (0) | 0 (0) | 0 (0) | 0 (0) |

Table 9.2   World Weather Data [CONTINUED]

# South America, Suriname to Venezuela

## Map # 250 to 252

### Map Location 250 — Suriname, Paramaribo — 0527N 05512W 54

| Month | Avg Max | Rec Max | Avg Min | Rec Min | Avg Precip (Rain Days)(Thunder Days) | Snow (Snow Days) |
|---|---|---|---|---|---|---|
| JANUARY | 87 | 91 | 71 | 61 | 11.11 (16.8) (6.4) | 0 (0) |
| FEBRUARY | 86 | 93 | 70 | 63 | 8.81 (16.6) (4.1) | 0 (0) |
| MARCH | 86 | 93 | 71 | 64 | 9.22 (14.6) (3.5) | 0 (0) |
| APRIL | 88 | 93 | 71 | 66 | 11.89 (15.9) (6.3) | 0 (0) |
| MAY | 87 | 97 | 72 | 64 | 13.09 (22.1) (12.7) | 0 (0) |
| JUNE | 88 | 93 | 72 | 66 | 15.48 (20.1) (18) | 0 (0) |
| JULY | 89 | 93 | 71 | 59 | 9.98 (16.9) (20.6) | 0 (0) |
| AUGUST | 91 | 97 | 71 | 66 | 7.73 (12.1) (19.5) | 0 (0) |
| SEPTEMBER | 92 | 95 | 71 | 66 | 3.96 (9) (17.2) | 0 (0) |
| OCTOBER | 92 | 100 | 72 | 68 | 2.6 (7.4) (13.7) | 0 (0) |
| NOVEMBER | 91 | 97 | 72 | 68 | 4.68 (8.1) (12.2) | 0 (0) |
| DECEMBER | 89 | 95 | 71 | 64 | 6.23 (14.6) (9.4) | 0 (0) |
| ANNUAL | 89 | 100 | 71 | 59 | 104.8 (174.2) (143.6) | 0 (0) |

### Map Location 251 — Uruguay, Montevideo — 3452S 05612W 72

| Month | Avg Max | Rec Max | Avg Min | Rec Min | Avg Precip (Rain Days)(Thunder Days) | Snow (Snow Days) |
|---|---|---|---|---|---|---|
| JANUARY | 83 | 109 | 62 | 46 | 2.9 (5) (4) | 0 (0) |
| FEBRUARY | 82 | 105 | 61 | 46 | 2.6 (4.6) (3) | 0 (0) |
| MARCH | 78 | 101 | 59 | 40 | 3.9 (6) (3) | 0 (0) |
| APRIL | 71 | 98 | 53 | 36 | 3.9 (6) (3) | 0 (0) |
| MAY | 64 | 87 | 48 | 29 | 3.3 (5.2) (3) | 0* (0*) |
| JUNE | 59 | 81 | 43 | 28 | 3.2 (5.7) (2) | *** (***) |
| JULY | 58 | 83 | 43 | 26 | 2.9 (5.2) (2) | *** (***) |
| AUGUST | 59 | 79 | 43 | 25 | 3.1 (5.5) (3) | *** (***) |
| SEPTEMBER | 63 | 86 | 46 | 29 | 3 (5.2) (3) | 0 (0) |
| OCTOBER | 68 | 94 | 49 | 29 | 2.6 (4.7) (2) | 0 (0) |
| NOVEMBER | 74 | 98 | 54 | 38 | 2.9 (5.1) (4) | 0 (0) |
| DECEMBER | 79 | 102 | 59 | 41 | 3.1 (5.3) (5) | 0 (0) |
| ANNUAL | 70 | 109 | 52 | 25 | 37.4 (63.5) (37) | *** (***) |

### Map Location 252 — Venezuela, Caracas — 1029N 06650W 2760

| Month | Avg Max | Rec Max | Avg Min | Rec Min | Avg Precip (Rain Days)(Thunder Days) | Snow (Snow Days) |
|---|---|---|---|---|---|---|
| JANUARY | 75 | 83 | 56 | 47 | 0.9 (2.1) (0.4) | 0 (0) |
| FEBRUARY | 77 | 88 | 56 | 46 | 0.4 (1.3) (0) | 0 (0) |
| MARCH | 79 | 91 | 58 | 45 | 0.6 (1.7) (0) | 0 (0) |
| APRIL | 81 | 89 | 60 | 51 | 1.3 (2.6) (0.8) | 0 (0) |
| MAY | 80 | 89 | 62 | 52 | 3.1 (5) (3.3) | 0 (0) |
| JUNE | 78 | 86 | 62 | 53 | 4 (7) (4.2) | 0 (0) |
| JULY | 78 | 84 | 61 | 52 | 4.3 (7.5) (5) | 0 (0) |
| AUGUST | 79 | 84 | 61 | 53 | 4.3 (7.5) (4.8) | 0 (0) |
| SEPTEMBER | 80 | 85 | 61 | 53 | 4.2 (6.6) (6.8) | 0 (0) |
| OCTOBER | 79 | 86 | 61 | 54 | 4.3 (6.7) (6.2) | 0 (0) |
| NOVEMBER | 77 | 84 | 60 | 51 | 3.7 (6) (3.3) | 0 (0) |
| DECEMBER | 78 | 83 | 58 | 47 | 1.8 (3.5) (0.8) | 0 (0) |
| ANNUAL | 78 | 91 | 60 | 45 | 32.9 (57.5) (35.6) | 0 (0) |

## Data Key

| | |
|---|---|
| Map Locater Number | |
| Location | |
| Lat - Lon ddmm Elev ft | MONTH |
| Average Max T °F | Record Max T °F |
| Average Min T °F | Record Min T °F |
| Avg. Precip in. (Rain Days) (Thunder Days) | Avg. Snow in. (Snow Days) |

| | |
|---|---|
| *** or Blank | No data, missing data or data of insufficient reliability |
| Rain Days | Number of Days with >= 0.01" of precipitation falling |
| Snow Days | Number of Days with >= 1.5" of snow falling |

# Arctic Ocean Islands

Table 9.2 World Weather Data [CONTINUED]

## Map # 253 - 257

### Map Location 253 — Alexandra Land, Russia — 8040N 04358E 89

| Month | Avg High | Rec High | Avg Low | Rec Low | Precip (days) |
|---|---|---|---|---|---|
| JANUARY | -10 | 30 | -23 | -51 | 1 (12) (0) |
| FEBRUARY | -13 | 25 | -24 | -58 | 1.17 (9.9) (0) |
| MARCH | -7 | 32 | -20 | -47 | 1.14 (11.2) (0) |
| APRIL | 4 | 34 | -8 | -40 | 1 (11) (0) |
| MAY | 16 | 39 | 7 | -18 | 0.88 (13.9) (0) |
| JUNE | 32 | 45 | 25 | 7 | 0.98 (11.1) (0) |
| JULY | 36 | 52 | 31 | 23 | 1 (9) (0) |
| AUGUST | 35 | 50 | 30 | 19 | 1.19 (10.1) (0) |
| SEPTEMBER | 28 | 41 | 21 | -6 | 1.33 (12.8) (0) |
| OCTOBER | 13 | 41 | 3 | -33 | 1.21 (15.6) (0) |
| NOVEMBER | 5 | 37 | -6 | -45 | 1.61 (11) (0) |
| DECEMBER | 0 | 37 | -11 | -47 | 1.76 (12.1) (0) |
| ANNUAL | 12 | 52 | 2 | -58 | 14.3 (139) (0) |

### Map Location 254 — Barter Island, USA — 7008N 14335W 8

| Month | Avg High | Rec High | Avg Low | Rec Low | Precip (days) |
|---|---|---|---|---|---|
| JANUARY | -10 | 37 | -23 | -50 | 0.54 (1.4) (0) |
| FEBRUARY | -15 | 34 | -28 | -59 | 0.32 (1.1) (0) |
| MARCH | -8 | 30 | -22 | -50 | 0.23 (0.5) (0) |
| APRIL | 7 | 43 | -8 | -37 | 0.25 (0.5) (0) |
| MAY | 26 | 48 | 16 | -11 | 0.33 (0.7) (0) |
| JUNE | 38 | 67 | 30 | 18 | 0.69 (2.3) (0.2) |
| JULY | 46 | 71 | 35 | 24 | 1.18 (3.6) (0) |
| AUGUST | 44 | 72 | 35 | 28 | 1.06 (3.6) (0) |
| SEPTEMBER | 35 | 64 | 28 | 7 | 1.18 (3.1) (0) |
| OCTOBER | 22 | 43 | 11 | -16 | 1.05 (2.9) (0) |
| NOVEMBER | 6 | 37 | -6 | -36 | 0.47 (1.3) (0) |
| DECEMBER | -8 | 30 | -20 | -51 | 0.21 (0.4) (0) |
| ANNUAL | 15 | 72 | 4 | -59 | 7.5 (21.4) (0.5) |

### Map Location 255 — Cape Golomjannyi, Russia — 7933N 09037E 10

| Month | Avg High | Rec High | Avg Low | Rec Low | Precip (days) |
|---|---|---|---|---|---|
| JANUARY | -17 | 25 | -27 | -54 | 0.67 (10.3) (0) |
| FEBRUARY | -17 | 25 | -27 | -56 | 0.64 (9.2) (0) |
| MARCH | -14 | 14 | -24 | -51 | 0.57 (9.3) (0) |
| APRIL | -1 | 28 | -11 | -33 | 0.5 (8.2) (0) |
| MAY | 14 | 36 | 5 | -22 | 0.48 (13.5) (0) |
| JUNE | 31 | 46 | 26 | 5 | 0.67 (10.3) (0) |
| JULY | 37 | 55 | 32 | 27 | 1.99 (10.6) (0.4) |
| AUGUST | 35 | 52 | 30 | 18 | 1.42 (13.1) (0.2) |
| SEPTEMBER | 29 | 41 | 23 | -4 | 1.26 (14.1) (0.1) |
| OCTOBER | 12 | 36 | 3 | -22 | 1.29 (13.2) (0) |
| NOVEMBER | -2 | 30 | -11 | -36 | 0.88 (9.6) (0) |
| DECEMBER | -6 | 30 | -16 | -51 | 1.21 (13.5) (0) |
| ANNUAL | 8 | 55 | 0 | -56 | 11.6 (134.9) (0.7) |

### Map Location 256 — Cape Zhelaniya (Severny Is.), Russia — 7657N 06835E 26

| Month | Avg High | Rec High | Avg Low | Rec Low | Precip (days) |
|---|---|---|---|---|---|
| JANUARY | 4 | 36 | -9 | -51 | 1.98 (13) (0) |
| FEBRUARY | 5 | 34 | -9 | -47 | 2.15 (11.1) (0) |
| MARCH | 2 | 34 | -11 | -56 | 2.18 (12.3) (0) |
| APRIL | 9 | 41 | -4 | -36 | 2.16 (13.9) (0) |
| MAY | 22 | 55 | 13 | -24 | 1.7 (16.3) (0) |
| JUNE | 31 | 59 | 26 | -4 | 1.19 (9.9) (0.1) |
| JULY | 40 | 70 | 32 | 9 | 1.11 (9.8) (0) |
| AUGUST | 40 | 70 | 33 | 19 | 1.85 (12.3) (0.1) |
| SEPTEMBER | 35 | 59 | 29 | 3 | 1.89 (14.6) (0) |
| OCTOBER | 26 | 46 | 19 | -17 | 1.61 (17.1) (0) |
| NOVEMBER | 14 | 41 | 3 | -36 | 1.15 (12.4) (0) |
| DECEMBER | 7 | 37 | -5 | -42 | 2 (13) (0) |
| ANNUAL | 20 | 70 | 10 | -56 | 20.8 (155.5) (0.2) |

### Map Location 257 — Druzhnyy Island, Russia — 8037N 05803E 66

| Month | Avg High | Rec High | Avg Low | Rec Low | Precip (days) |
|---|---|---|---|---|---|
| JANUARY | -14 | 30 | -23 | -44 | 1.74 (11.3) (0) |
| FEBRUARY | -13 | 32 | -23 | -47 | 1.71 (11.3) (0) |
| MARCH | -11 | 27 | -21 | -42 | 1.95 (15) (0) |
| APRIL | 2 | 32 | -8 | -36 | 1.19 (9.3) (0) |
| MAY | 15 | 36 | 7 | -18 | 0.59 (11.2) (0) |
| JUNE | 32 | 41 | 26 | 9 | 1 (9) (0) |
| JULY | 36 | 45 | 31 | 23 | 1.52 (8.2) (0) |
| AUGUST | 34 | 45 | 30 | 21 | 1.57 (10.1) (0) |
| SEPTEMBER | 29 | 39 | 24 | 1 | 1.54 (13.6) (0) |
| OCTOBER | 12 | 37 | 5 | -24 | 1.49 (11.6) (0) |
| NOVEMBER | 2 | 36 | -7 | -36 | 2 (11) (0) |
| DECEMBER | -3 | 34 | -11 | -44 | 2.19 (13.4) (0) |
| ANNUAL | 10 | 45 | 3 | -47 | 18.1 (135.7) (0) |

**Table 9.2  World Weather Data [CONTINUED]**

# Arctic Ocean Islands

## Map # 258 - 262

### Map Location 258 — Isachsen Island, Canada — 7847N 10332W 175

| | JAN | FEB | MAR | APR | MAY | JUN | JUL | AUG | SEP | OCT | NOV | DEC | ANNUAL |
|---|---|---|---|---|---|---|---|---|---|---|---|---|---|
| Record High | 25 | -5 | 17 | 30 | 36 | 62 | 66 | 58 | 37 | 29 | 25 | 15 | 66 |
| Avg High | -24 | -25 | -25 | -5 | 17 | 35 | 43 | 39 | 22 | 5 | -12 | -20 | 4 |
| Avg Low | -37 | -40 | -37 | -18 | 6 | 28 | 34 | 31 | 12 | -8 | -25 | -32 | -6 |
| Record Low | -63 | -60 | -65 | -44 | -21 | 6 | 26 | 8 | -17 | -35 | -50 | -60 | -65 |
| Precip | 0.06 (0) (0) | 0.06 (0) (0) | 0 (0) | 0.17 (0.4) (0) | 0.3 (0.9) (0) | 1 (0) | 0.87 (2.8) (0.1) | 0.91 (2.9) (0) | 0.71 (2.6) (0) | 0.4 (2) (0) | 0.14 (1.5) (0) | 0.06 (0) (0) | 3.8 (13.9) (0.1) |

### Map Location 259 — Kotelnyy Island, Russia — 7600N 13754E 33

| | JAN | FEB | MAR | APR | MAY | JUN | JUL | AUG | SEP | OCT | NOV | DEC | ANNUAL |
|---|---|---|---|---|---|---|---|---|---|---|---|---|---|
| Record High | 12 | 14 | 12 | 32 | 37 | 64 | 68 | 57 | 45 | 32 | 25 | 18 | 68 |
| Avg High | -20 | -19 | -16 | 0 | 18 | 35 | 41 | 38 | 30 | 14 | -2 | -14 | 9 |
| Avg Low | -29 | -28 | -26 | -12 | 8 | 28 | 33 | 30 | 24 | 6 | -11 | -22 | 0 |
| Record Low | -45 | -49 | -49 | -51 | -18 | 10 | 25 | 21 | 0 | -33 | -36 | -38 | -51 |
| Precip | 0.41 (6.9) (0) | 0.56 (7.7) (0) | 1 (8) (0) | 0.49 (6.4) (0) | 1.06 (10.7) (0) | 0 (6) (0) | 1.32 (9.3) (0.1) | 1.42 (10.9) (0.2) | 0.62 (9.9) (0) | 0.81 (13.9) (0) | 0.53 (9.9) (0) | 0.73 (8.2) (0) | 9.1 (108.1) (0.3) |

### Map Location 260 — Mould Bay, Canada — 7614N 11918W 40

| | JAN | FEB | MAR | APR | MAY | JUN | JUL | AUG | SEP | OCT | NOV | DEC | ANNUAL |
|---|---|---|---|---|---|---|---|---|---|---|---|---|---|
| Record High | 15 | 13 | 13 | 29 | 35 | 56 | 60 | 57 | 46 | 32 | 25 | 38 | 60 |
| Avg High | -21 | -25 | -19 | -2 | 18 | 36 | 43 | 39 | 25 | 7 | -9 | -19 | 6 |
| Avg Low | -36 | -38 | -33 | -17 | 7 | 27 | 34 | 31 | 16 | -6 | -22 | -31 | -5 |
| Record Low | -55 | -58 | -56 | -43 | -20 | 8 | 25 | 14 | -13 | -33 | -46 | -78 | -78 |
| Precip | 0.08 (0) (0) | 0.07 (0.2) (0) | 0 (0) (0) | 0.13 (0.2) (0) | 0.29 (0.5) (0) | 0 (0) (0) | 0.48 (1.3) (0.1) | 0.37 (1.2) (0) | 0.45 (1.5) (0) | 0.16 (0) (0) | 0.11 (0) (0) | 0.12 (0) (0) | 2.7 (6.3) (0.1) |

### Map Location 261 — Russkaya Gavan, Russia — 7611N 06234E 30

| | JAN | FEB | MAR | APR | MAY | JUN | JUL | AUG | SEP | OCT | NOV | DEC | ANNUAL |
|---|---|---|---|---|---|---|---|---|---|---|---|---|---|
| Record High | 37 | 39 | 36 | 45 | 61 | 66 | 73 | 64 | 48 | 41 | 43 | 37 | 73 |
| Avg High | 10 | 9 | 7 | 15 | 26 | 37 | 44 | 43 | 28 | 22 | 19 | 13 | 24 |
| Avg Low | -4 | -4 | -7 | 0 | 16 | 29 | 35 | 35 | 20 | 15 | 7 | 0 | 13 |
| Record Low | -45 | -44 | -51 | -33 | -22 | -2 | 10 | 23 | 3 | -11 | -33 | -40 | -51 |
| Precip | 1.95 (12.9) (0) | 1.7 (12) (0) | 1 (11) (0) | 1.85 (12.9) (0) | 1.38 (15.8) (0) | 1 (12) (0) | 1.64 (10.9) (0) | 1.53 (11.2) (0.3) | 1.77 (13.2) (0) | 1.68 (15.5) (0.2) | 1.33 (12.7) (0) | 1.51 (12.5) (0) | 19 (152.6) (0.6) |

### Map Location 262 — Wrangel Island, Russia — 7058N 17832W 10

| | JAN | FEB | MAR | APR | MAY | JUN | JUL | AUG | SEP | OCT | NOV | DEC | ANNUAL |
|---|---|---|---|---|---|---|---|---|---|---|---|---|---|
| Record High | 30 | 32 | 32 | 32 | 48 | 52 | 57 | 52 | 46 | 41 | 32 | 28 | 57 |
| Avg High | -3 | -9 | -5 | 6 | 25 | 37 | 41 | 39 | 33 | 22 | 11 | -1 | 16 |
| Avg Low | -13 | -20 | -16 | -8 | 15 | 29 | 33 | 33 | 28 | 15 | 4 | -10 | 8 |
| Record Low | -44 | -47 | -42 | -27 | -26 | 14 | 27 | 28 | 12 | -11 | -22 | -33 | -47 |
| Precip | 1.14 (11.7) (0) | 0.63 (7.8) (0) | 1 (8) (0) | 0.41 (5.3) (0) | 0.81 (10.3) (0.1) | 0 (5) (0) | 1.39 (8.3) (0.2) | 1.38 (10.3) (0.4) | 1.4 (11.2) (0.2) | 1.3 (12.1) (0.3) | 0.9 (11) (0) | 0.91 (8.2) (0) | 11.8 (109.4) (1.2) |

Table 9.2  World Weather Data [CONTINUED]

# Atlantic Ocean and Caribbean Islands

# Map # 263 - 267

### Map Location 263 — Antigua And Barbuda, St. John's 1707N

| | JAN | FEB | MAR | APR | MAY | JUN | JUL | AUG | SEP | OCT | NOV | DEC | ANNUAL |
|---|---|---|---|---|---|---|---|---|---|---|---|---|---|
| Record Hi / Lo | 86 / 61 | 86 / 64 | 86 / 65 | 88 / 67 | 92 / 66 | 90 / 70 | 89 / 70 | 90 / 71 | 90 / 68 | 91 / 65 | 90 / 69 | 87 / 66 | 92 / 61 |
| (snow) | 0 (0) | 0 (0) | 0 (0) | 0 (0) | 0 (0) | 0 (0) | 0 (0) | 0 (0) | 0 (0) | 0 (0) | 0 (0) | 0 (0) | 0 (0) |
| Normal Hi / Lo | 81 / 72 | 81 / 72 | 82 / 72 | 83 / 74 | 84 / 75 | 85 / 77 | 86 / 77 | 86 / 77 | 86 / 76 | 86 / 75 | 84 / 75 | 83 / 73 | 84 / 75 |
| Precip | 2.36 (7.2)(0) | 1.31 (3.8)(0) | 1.34 (2.9)(0) | 1.8 (3.2)(0.1) | 3.48 (5.6)(0.4) | 3 (6.9)(1.2) | 2.48 (6.6)(1) | 3.22 (8.8)(1) | 4.88 (7.3)(2.4) | 4.1 (9.3)(1) | 4.58 (8.1)(0.7) | 2.82 (6.9)(0.4) | 35.4 (76.6)(8.2) |

### Map Location 264 — Azores Islands, Ponta Delgada 3745N 02540W 118

| | JAN | FEB | MAR | APR | MAY | JUN | JUL | AUG | SEP | OCT | NOV | DEC | ANNUAL |
|---|---|---|---|---|---|---|---|---|---|---|---|---|---|
| Record Hi / Lo | 72 / 37 | 70 / 39 | 71 / 40 | 74 / 43 | 82 / 45 | 84 / 49 | 85 / 52 | 85 / 52 | 88 / 50 | 82 / 48 | 79 / 46 | 82 / 43 | 88 / 37 |
| (snow) | 0 (0) | 0 (0) | 0 (0) | 0 (0) | 0 (0) | 0 (0) | 0 (0) | 0 (0) | 0 (0) | 0 (0) | 0 (0) | 0 (0) | 0 (0) |
| Normal Hi / Lo | 62 / 54 | 62 / 53 | 63 / 53 | 64 / 55 | 67 / 56 | 71 / 60 | 76 / 64 | 78 / 65 | 75 / 64 | 71 / 61 | 67 / 58 | 64 / 55 | 68 / 58 |
| Precip | 4 (8.2)(1) | 3.5 (7.4)(1) | 3.5 (6.7)(1) | 2.5 (5.3)(0.3) | 2.3 (5)(0.3) | 1.4 (2.8)(0.3) | 1 (2.1)(0.3) | 1.2 (2.4)(1) | 2.9 (5.2)(1) | 3.6 (6.3)(1) | 3.7 (6.5)(1) | 3 (6.4)(1) | 32.6 (64.3)(9.2) |

### Map Location 265 — Bahamas, Nassau 2502N 07728W 7

| | JAN | FEB | MAR | APR | MAY | JUN | JUL | AUG | SEP | OCT | NOV | DEC | ANNUAL |
|---|---|---|---|---|---|---|---|---|---|---|---|---|---|
| Record Hi / Lo | 85 / 41 | 86 / 43 | 88 / 46 | 91 / 53 | 92 / 53 | 94 / 62 | 94 / 67 | 94 / 67 | 92 / 65 | 92 / 54 | 89 / 49 | 86 / 45 | 94 / 41 |
| (snow) | 0 (0) | 0 (0) | 0 (0) | 0 (0) | 0 (0) | 0 (0) | 0 (0) | 0 (0) | 0 (0) | 0 (0) | 0 (0) | 0 (0) | 0 (0) |
| Normal Hi / Lo | 77 / 65 | 77 / 64 | 79 / 66 | 81 / 69 | 84 / 71 | 87 / 74 | 88 / 75 | 89 / 76 | 88 / 75 | 85 / 73 | 81 / 70 | 79 / 67 | 83 / 70 |
| Precip | 1.4 (2)(0.9) | 1.5 (2.2)(0.6) | 1.4 (3.4)(0.9) | 2.5 (5.5)(1) | 4.6 (8.4)(3) | 6.4 (12)(4.7) | 5.8 (11.4)(5.1) | 5.3 (10.7)(4.7) | 6.9 (10.6)(3.3) | 6.5 (9.9)(1) | 2.8 (4.5)(0.2) | 1.3 (1.8)(0.1) | 46.4 (82.4)(25.5) |

### Map Location 266 — Barbados, Bridgetown 1308N 05936W 181

| | JAN | FEB | MAR | APR | MAY | JUN | JUL | AUG | SEP | OCT | NOV | DEC | ANNUAL |
|---|---|---|---|---|---|---|---|---|---|---|---|---|---|
| Record Hi / Lo | 87 / 61 | 87 / 61 | 89 / 62 | 89 / 64 | 91 / 66 | 90 / 67 | 90 / 68 | 95 / 69 | 92 / 67 | 89 / 62 | 89 / 66 | 88 / 64 | 95 / 61 |
| (snow) | 0 (0) | 0 (0) | 0 (0) | 0 (0) | 0 (0) | 0 (0) | 0 (0) | 0 (0) | 0 (0) | 0 (0) | 0 (0) | 0 (0) | 0 (0) |
| Normal Hi / Lo | 83 / 70 | 83 / 69 | 85 / 70 | 86 / 72 | 87 / 73 | 87 / 74 | 86 / 74 | 87 / 76 | 86 / 73 | 86 / 73 | 85 / 73 | 83 / 71 | 85 / 72 |
| Precip | 2.6 (5)(0) | 1.1 (1.6)(1) | 1.3 (3.2)(0) | 1.4 (3.4)(0) | 3.31 (5.1)(1) | 4.4 (9.4)(1) | 5.8 (11.4)(1) | 6.7 (10.3)(0) | 6.7 (10.3)(0) | 7 (10.7)(1) | 8.1 (12.6)(0) | 3.8 (9.7)(0) | 50.3 (93.8)(6) |

### Map Location 267 — Bermuda, Hamilton 3221N 06441W 12

| | JAN | FEB | MAR | APR | MAY | JUN | JUL | AUG | SEP | OCT | NOV | DEC | ANNUAL |
|---|---|---|---|---|---|---|---|---|---|---|---|---|---|
| Record Hi / Lo | 73 / 46 | 74 / 48 | 75 / 50 | 78 / 52 | 83 / 60 | 87 / 64 | 91 / 68 | 91 / 68 | 90 / 68 | 87 / 63 | 81 / 55 | 74 / 46 | 91 / 46 |
| (snow) | 0 (0) | 0 (0) | 0 (0) | 0 (0) | 0 (0) | 0 (0) | 0 (0) | 0 (0) | 0 (0) | 0 (0) | 0 (0) | 0 (0) | 0 (0) |
| Normal Hi / Lo | 66 / 60 | 66 / 60 | 67 / 60 | 69 / 62 | 74 / 67 | 79 / 72 | 83 / 76 | 84 / 77 | 83 / 76 | 78 / 72 | 73 / 67 | 69 / 63 | 74 / 68 |
| Precip | 5.28 (10.9)(2.3) | 4.07 (8.2)(1) | 4.53 (9.3)(2.2) | 3.31 (5.5)(1.5) | 3.55 (5.7)(1.6) | 5.77 (8.3)(2.7) | 4.49 (8.4)(4.3) | 5.99 (8.3)(4.6) | 5.7 (9.7)(2.6) | 7.07 (9.8)(2.1) | 5.15 (8)(1.4) | 5.12 (10.1)(0.5) | 60 (102.2)(26.8) |

## Atlantic Ocean and Caribbean Islands

### Map # 268 - 272

Table 9.2　World Weather Data [CONTINUED]

**Map Location 268 — Canary Is., Spain, Santa Cruz — 2827N 01615W**

| Month | Record High | Record Low | Normal High | Normal Low | Precip in (mm) (snow) |
|---|---|---|---|---|---|
| JANUARY | 81 | 48 | 69 | 58 | 1.41 (3.2) (***) |
| FEBRUARY | 84 | 47 | 69 | 58 | 1.54 (3.5) (***) |
| MARCH | 96 | 49 | 71 | 59 | 1.06 (2.6) (***) |
| APRIL | 94 | 49 | 73 | 60 | 0.51 (1.3) (***) |
| MAY | 103 | 55 | 75 | 62 | 0.27 (0.7) (***) |
| JUNE | 98 | 55 | 79 | 66 | 0 (0) |
| JULY | 109 | 61 | 83 | 69 | 0 (0) (***) |
| AUGUST | 105 | 63 | 85 | 70 | 0 (0) (***) |
| SEPTEMBER | 100 | 63 | 82 | 70 | 0.12 (0.4) (***) |
| OCTOBER | | | 79 | 67 | 1.22 (2.4) (***) |
| NOVEMBER | 87 | 50 | 74 | 63 | 1.77 (3.4) (***) |
| DECEMBER | 82 | 49 | 71 | 60 | 2.01 (4.5) (***) |
| ANNUAL | 109 | 47 | 76 | 64 | 9.9 (22) (***) |

**Map Location 269 — Cape Verde, Rep. of, Praia — 1455N 02331W**

| Month | Record High | Record Low | Normal High | Normal Low | Precip in (mm) (snow) |
|---|---|---|---|---|---|
| JANUARY | 90 | 63 | 77 | 68 | 0.1 (0.4) (0.3) |
| FEBRUARY | 90 | 56 | 77 | 67 | 0.05 (0.3) (0.3) |
| MARCH | 91 | 62 | 78 | 68 | 0.04 (0.1) (0) |
| APRIL | 93 | 63 | 79 | 69 | 0.09 (0.3) (0) |
| MAY | 99 | 61 | 81 | 70 | 0 (0) (0) |
| JUNE | 95 | 66 | 82 | 70 | 0.04 (0.2) (0) |
| JULY | 91 | 66 | 83 | 75 | 0.52 (1.2) (0) |
| AUGUST | 91 | 66 | 83 | 76 | 3.91 (6.9) (1) |
| SEPTEMBER | 97 | 70 | 84 | 77 | 1.14 (2.3) (1) |
| OCTOBER | 95 | 68 | 85 | 76 | 1.87 (3.6) (1) |
| NOVEMBER | 90 | 68 | 82 | 74 | 0.31 (0.8) (0.3) |
| DECEMBER | 87 | 64 | 79 | 71 | 0.18 (0.5) (0) |
| ANNUAL | 99 | 56 | 81 | 72 | 8.3 (16.6) (3.9) |

**Map Location 270 — Cuba, Havana — 2301N 08224W — 225**

| Month | Record High | Record Low | Normal High | Normal Low | Precip in (mm) (snow) |
|---|---|---|---|---|---|
| JANUARY | 93 | 43 | 79 | 65 | 2.8 (5.7) (0.2) |
| FEBRUARY | 93 | 46 | 79 | 65 | 1.8 (2.8) (0.8) |
| MARCH | 104 | 46 | 81 | 67 | 1.8 (4.2) (0.8) |
| APRIL | 104 | 43 | 84 | 69 | 2.3 (5.1) (2.4) |
| MAY | 97 | 57 | 86 | 72 | 4.7 (8.5) (3.8) |
| JUNE | 99 | 63 | 88 | 74 | 6.5 (12.1) (6.3) |
| JULY | 97 | 64 | 89 | 75 | 4.9 (10.2) (14.4) |
| AUGUST | 97 | 64 | 89 | 75 | 5.3 (10.7) (12.8) |
| SEPTEMBER | 100 | 63 | 88 | 75 | 5.9 (8.9) (7.8) |
| OCTOBER | 100 | 52 | 85 | 73 | 6.8 (10.4) (4.2) |
| NOVEMBER | 93 | 45 | 81 | 69 | 3.1 (4.8) (0.4) |
| DECEMBER | 90 | 45 | 79 | 67 | 2.3 (4) (0.2) |
| ANNUAL | 104 | 43 | 84 | 71 | 48.2 (87.4) (54.1) |

**Map Location 271 — Dominican Republic, Santo Domingo — 1826N 06940W — 57**

| Month | Record High | Record Low | Normal High | Normal Low | Precip in (mm) (snow) |
|---|---|---|---|---|---|
| JANUARY | 94 | 58 | 84 | 66 | 2.4 (4.3) (0) |
| FEBRUARY | 93 | 60 | 85 | 66 | 1.4 (2) (0) |
| MARCH | 94 | 60 | 84 | 67 | 1.9 (4.4) (0) |
| APRIL | 95 | 62 | 85 | 69 | 3.9 (7.6) (0) |
| MAY | 94 | 62 | 86 | 71 | 6.8 (9.9) (9) |
| JUNE | 96 | 67 | 87 | 72 | 6.2 (11.8) (8) |
| JULY | 96 | 64 | 88 | 72 | 6.4 (12) (15) |
| AUGUST | 98 | 68 | 88 | 73 | 6.3 (11.9) (14) |
| SEPTEMBER | 98 | 64 | 88 | 72 | 7.3 (11.2) (17) |
| OCTOBER | 95 | 66 | 87 | 72 | 6 (9.1) (8) |
| NOVEMBER | 97 | 61 | 86 | 70 | 4.8 (7.2) (3) |
| DECEMBER | 95 | 55 | 85 | 67 | 2.4 (4.3) (1) |
| ANNUAL | 98 | 55 | 86 | 70 | 55.8 (95.7) (75) |

**Map Location 272 — Haiti, Petionville — 1831N 07217W — 1312**

| Month | Record High | Record Low | Normal High | Normal Low | Precip in (mm) (snow) |
|---|---|---|---|---|---|
| JANUARY | 90 | 57 | 84 | 63 | 1 (1.5) (0) |
| FEBRUARY | 90 | 54 | 84 | 64 | 1.9 (3) (0) |
| MARCH | 92 | 58 | 86 | 65 | 3.5 (7) (0) |
| APRIL | 92 | 58 | 86 | 66 | 7.3 (10.1*) (0) |
| MAY | 93 | 59 | 86 | 68 | 10 (***) (0.2) |
| JUNE | 93 | 58 | 88 | 68 | 5 (10.3) (0.2) |
| JULY | 96 | 64 | 90 | 69 | 3.6 (8) (0.3) |
| AUGUST | 89 | 69 | 89 | 69 | 5.9 (11.5) (0.6) |
| SEPTEMBER | 97 | 64 | 88 | 69 | 7.5 (11.6) (0.6) |
| OCTOBER | 95 | 58 | 86 | 68 | 7.1 (10.9) (0.1) |
| NOVEMBER | 91 | 59 | 84 | 67 | 3.2 (4.9) (0.1) |
| DECEMBER | 90 | 56 | 83 | 64 | 1.3 (1.8) (0.1) |
| ANNUAL | 97 | 54 | 86 | 67 | 57.3 (***) (2.2) |

**Table 9.2  World Weather Data [CONTINUED]**

# Atlantic Ocean and Caribbean Islands

## Map # 273 - 277

### Map Location 273 — Jamaica, Kingston — 1754N 07717W 18

| Month | Record High | Normal High | Normal Low | Record Low | Precipitation | Snow |
|---|---|---|---|---|---|---|
| January | 95 | 86 | 67 | 57 | 0.9 (1.4) (0.1) | 0 (0) |
| February | 94 | 86 | 67 | 56 | 0.6 (1.3) (0.1) | 0 (0) |
| March | 94 | 86 | 68 | 58 | 0.9 (2.4) (0.3) | 0 (0) |
| April | 94 | 87 | 70 | 62 | 1.2 (3) (1.1) | 0 (0) |
| May | 95 | 87 | 72 | 60 | 4 (7.7) (5.5) | 0 (0) |
| June | 95 | 89 | 74 | 66 | 3.5 (7.8) (4) | 0 (0) |
| July | 96 | 90 | 73 | 66 | 1.5 (3) (7.6) | 0 (0) |
| August | 98 | 90 | 73 | 55 | 3.6 (8) (9.1) | 0 (0) |
| September | 96 | 89 | 73 | 67 | 3.9 (5.9) (11.8) | 0 (0) |
| October | 96 | 88 | 73 | 65 | 7.1 (10.9) (8.4) | 0 (0) |
| November | 96 | 87 | 71 | 57 | 2.9 (4.6) (3.3) | 0 (0) |
| December | 96 | 87 | 69 | 57 | 1.4 (2) (0.8) | 0 (0) |
| ANNUAL | 97 | 88 | 71 | 56 | 31.5 (58) (52.1) | 0 (0) |

### Map Location 274 — Netherlands Antilles, Willemstad — 1211N 06857W 27

| Month | Record High | Normal High | Normal Low | Record Low | Precipitation | Snow |
|---|---|---|---|---|---|---|
| January | 88 | 84 | 75 | 68 | 3.15 (4.5) (0.2) | 0 (0) |
| February | 87 | 84 | 75 | 69 | 0.44 (1.3) (0) | 0 (0) |
| March | 88 | 85 | 76 | 68 | 0.46 (1.2) (0.2) | 0 (0) |
| April | 89 | 86 | 77 | 71 | 0.58 (0.7) (0.3) | 0 (0) |
| May | 91 | 86 | 78 | 70 | 1.51 (2.8) (0.9) | 0 (0) |
| June | 91 | 87 | 79 | 70 | 0.68 (2) (1.1) | 0 (0) |
| July | 94 | 88 | 79 | 71 | 1.19 (2.8) (1.6) | 0 (0) |
| August | 94 | 88 | 79 | 72 | 2.43 (4.5) (2) | 0 (0) |
| September | 94 | 89 | 80 | 72 | 0.86 (1.1) (2.2) | 0 (0) |
| October | 94 | 88 | 79 | 69 | 2.88 (4.9) (4.7) | 0 (0) |
| November | 91 | 86 | 77 | 70 | 4.18 (7.7) (4.6) | 0 (0) |
| December | 90 | 85 | 76 | 70 | 2.9 (6.4) (1.3) | 0 (0) |
| ANNUAL | 94 | 86 | 78 | 68 | 21.3 (39.9) (19.1) | 0 (0) |

### Map Location 275 — Madeira Islands, Funchal — 3241N 01646W 190

| Month | Record High | Normal High | Normal Low | Record Low | Precipitation | Snow |
|---|---|---|---|---|---|---|
| January | 79 | 66 | 56 | 42 | 2.47 (5.4) (0.3) | 0 (0) |
| February | 82 | 65 | 56 | 40 | 2.92 (6.3) (1) | 0 (0) |
| March | 82 | 66 | 56 | 44 | 3.09 (6.2) (1) | 0 (0) |
| April | 84 | 67 | 58 | 44 | 1.34 (3.2) (0.3) | 0 (0) |
| May | 88 | 69 | 60 | 48 | 0.75 (1.9) (0.3) | 0 (0) |
| June | 96 | 72 | 63 | 48 | 0.23 (0.6) (0.3) | 0 (0) |
| July | 103 | 75 | 66 | 52 | 0.02 (0.2) (0) | 0 (0) |
| August | 103 | 76 | 67 | 52 | 0.06 (0.2) (0.3) | 0 (0) |
| September | 95 | 76 | 67 | 52 | 1.04 (2.1) (0.3) | 0 (0) |
| October | 92 | 74 | 65 | 47 | 3.04 (5.5) (0.3) | 0 (0) |
| November | 87 | 71 | 61 | 45 | 3.48 (6.1) (1) | 0 (0) |
| December | 82 | 67 | 58 | 41 | 3.33 (7.1) (0.3) | 0 (0) |
| ANNUAL | 103 | 70 | 61 | 40 | 21.8 (44.8) (5.4) | 0 (0) |

### Map Location 276 — Puerto Rico, San Juan — 1828N 06607W 82

| Month | Record High | Normal High | Normal Low | Record Low | Precipitation | Snow |
|---|---|---|---|---|---|---|
| January | 88 | 81 | 70 | 63 | 4.01 (9.7) (0.3) | 0 (0) |
| February | 88 | 81 | 70 | 64 | 3.44 (6.3) (0.1) | 0 (0) |
| March | 93 | 82 | 71 | 64 | 2.23 (5.4) (0.3) | 0 (0) |
| April | 91 | 84 | 72 | 66 | 4.74 (8.7) (1.8) | 0 (0) |
| May | 92 | 85 | 74 | 68 | 7.33 (11.8) (6.4) | 0 (0) |
| June | 92 | 86 | 75 | 70 | 6.24 (10.7) (7.1) | 0 (0) |
| July | 92 | 87 | 76 | 72 | 6.94 (13.1) (8.6) | 0 (0) |
| August | 88 | 87 | 76 | 72 | 7.42 (12.9) (9.3) | 0 (0) |
| September | 92 | 88 | 75 | 72 | 6.33 (12.3) (11.8) | 0 (0) |
| October | 91 | 86 | 74 | 69 | 6.51 (11.8) (9.3) | 0 (0) |
| November | 89 | 84 | 73 | 65 | 6.81 (11.5) (3.7) | 0 (0) |
| December | 88 | 82 | 71 | 65 | 5.97 (11.7) (0.7) | 0 (0) |
| ANNUAL | 93 | 84 | 73 | 63 | 68 (125.9) (59.4) | 0 (0) |

### Map Location 277 — St. Helena, Ascension — 0758S 01424W 272

| Month | Record High | Normal High | Normal Low | Record Low | Precipitation | Snow |
|---|---|---|---|---|---|---|
| January | 89 | 85 | 73 | 67 | 1.4 (2.9) (0) | 0 (0) |
| February | 88 | 87 | 74 | 66 | 0.9 (2.1) (0) | 0 (0) |
| March | 88 | 88 | 75 | 65 | 0.9 (2.1) (1) | 0 (0) |
| April | 88 | 88 | 75 | 66 | 0.5 (1.5) (0) | 0 (0) |
| May | 92 | 87 | 74 | 67 | 0.2 (1.1) (0) | 0 (0) |
| June | 90 | 85 | 73 | 65 | 0.03 (0.4) (0) | 0 (0) |
| July | 88 | 84 | 72 | 65 | 0.03 (0.4) (0) | 0 (0) |
| August | 88 | 83 | 71 | 65 | 0.03 (0.4) (0) | 0 (0) |
| September | 88 | 82 | 71 | 65 | 0.2 (0.9) (0) | 0 (0) |
| October | 88 | 83 | 71 | 65 | 1.1 (2.5) (0) | 0 (0) |
| November | 89 | 84 | 72 | 67 | 2.1 (4) (0) | 0 (0) |
| December | 95 | 84 | 72 | 67 | 1.6 (3.2) (0) | 0 (0) |
| ANNUAL | 95 | 85 | 73 | 65 | 9 (21.5) (1) | 0 (0) |

**Table 9.2  World Weather Data [CONTINUED]**

# Atlantic Ocean and Caribbean Islands

# Map # 278 - 281

## Map Location 278 — St. Lucia, Castries — 1344N 06057W 10

| | JANUARY | FEBRUARY | MARCH | APRIL | MAY | JUNE |
|---|---|---|---|---|---|---|
| Extreme Hi/Lo | 88 / 63 | 89 / 62 | 90 / 66 | 92 / 62 | 94 / 70 | 91 / 70 |
| | 0 (0) | 0 (0) | 0 (0) | 0 (0) | 0 (0) | 0 (0) |
| Mean Hi/Lo | 84 / 73 | 85 / 73 | 85 / 74 | 86 / 74 | 87 / 76 | 87 / 76 |
| Precip | 2.75 (8) (0) | 1.96 (6.2) (0) | 1.83 (4.7) (0) | 1.98 (5.1) (0.2) | 3.97 (10) (0.2) | 5.73 (13.1) 90.7 |

| | JULY | AUGUST | SEPTEMBER | OCTOBER | NOVEMBER | DECEMBER | ANNUAL |
|---|---|---|---|---|---|---|---|
| Extreme Hi/Lo | 91 / 70 | 87 / 57 | 90 / 70 | 92 / 70 | 91 / 70 | 89 / 67 | 94 / 62 |
| | 0 (0) | 0 (0) | 0 (0) | 0 (0) | 0 (0) | 0 (0) | 0 (0) |
| Mean Hi/Lo | 86 / 76 | 87 / 77 | 88 / 77 | 88 / 76 | 87 / 76 | 85 / 75 | 86 / 75 |
| Precip | 8.65 (16.6) (1.6) | 5.55 (11.9) (1.8) | 8.56 (13.5) (4) | 8.06 (12.4) (3.6) | 8.26 (10.3) (1.1) | 3.81 (10) (0.3) | 61.1 (121.8) (13.5) |

## Map Location 279 — Saint Maarten, Juliana Intn'l — 1802N 06306W 0

| | JANUARY | FEBRUARY | MARCH | APRIL | MAY | JUNE |
|---|---|---|---|---|---|---|
| Extreme Hi/Lo | *** / *** | *** / *** | *** / *** | *** / *** | *** / *** | *** / *** |
| | 0 (0) | 0 (0) | 0 (0) | 0 (0) | 0 (0) | 0 (0) |
| Mean Hi/Lo | 83 / 73 | 82 / 72 | 84 / 73 | 85 / 74 | 86 / 74 | 87 / 77 |
| Precip | 2.56 (8) (***) | 2.11 (5) (***) | 1.87 (3) (***) | 2.52 (3) (***) | 3.66 (3) (***) | 2.48 (3) (***) |

| | JULY | AUGUST | SEPTEMBER | OCTOBER | NOVEMBER | DECEMBER | ANNUAL |
|---|---|---|---|---|---|---|---|
| Extreme Hi/Lo | *** / *** | *** / *** | *** / *** | *** / *** | *** / *** | *** / *** | *** / *** |
| | 0 (0) | 0 (0) | 0 (0) | 0 (0) | 0 (0) | 0 (0) | 0 (0) |
| Mean Hi/Lo | 88 / 77 | 88 / 78 | 88 / 77 | 88 / 77 | 86 / 75 | 84 / 74 | 86 / 75 |
| Precip | 2.87 (6) (***) | 3.95 (5) (***) | 5.56 (5) (***) | 4.45 (8) (***) | 5.96 (10) (***) | 3.63 (12) (***) | 41.62 (71) (***) |

## Map Location 280 — São Tomé/Principe, São Tomé — 0023N 00643 E26

| | JANUARY | FEBRUARY | MARCH | APRIL | MAY | JUNE |
|---|---|---|---|---|---|---|
| Extreme Hi/Lo | 90 / 66 | 92 / 67 | 92 / 67 | 92 / 67 | 93 / 65 | 82 / 57 |
| | 0 (0) | 0 (0) | 0 (0) | 0 (0) | 0 (0) | 0 (0) |
| Mean Hi/Lo | 85 / 72 | 86 / 73 | 86 / 73 | 86 / 73 | 85 / 73 | 82 / 71 |
| Precip | 3.18 (6.8) (10) | 3.3 (7) (9) | 5.16 (8.4) (13) | 4.79 (8.1) (21) | 4.44 (7.7) (6) | 0.75 (1.6) (0) |

| | JULY | AUGUST | SEPTEMBER | OCTOBER | NOVEMBER | DECEMBER | ANNUAL |
|---|---|---|---|---|---|---|---|
| Extreme Hi/Lo | 87 / 57 | 88 / 56 | 89 / 61 | 89 / 65 | 89 / 65 | *** / *** | 93 / 56 |
| | 0 (0) | 0 (0) | 0 (0) | 0 (0) | 0 (0) | 0 (0) | 0 (0) |
| Mean Hi/Lo | 81 / 69 | 82 / 69 | 83 / 70 | 84 / 71 | 84 / 68 | 84 / 72 | 84 / 71 |
| Precip | 0 (0) (0) | 0.04 (0.2) (0) | 0.67 (1.4) (0) | 4.32 (7.4) (3) | 3.9 (6.8) (8) | 4.26 (8.6) (11) | 34.8 (64) (81) |

## Map Location 281 — Trinidad And Tobago, Chaguaramas — 1041N 06137W 42

| | JANUARY | FEBRUARY | MARCH | APRIL | MAY | JUNE |
|---|---|---|---|---|---|---|
| Extreme Hi/Lo | 92 / 65 | 92 / 68 | 96 / 68 | 100 / 69 | 96 / 70 | 95 / 70 |
| | 0 (0) | 0 (0) | 0 (0) | 0 (0) | 0 (0) | 0 (0) |
| Mean Hi/Lo | 86 / 72 | 87 / 72 | 87 / 73 | 88 / 73 | 89 / 74 | 87 / 74 |
| Precip | 2.6 (6.9) (0.2) | 2.83 (5.8) (0.8) | 1.61 (4.1) (0) | 1.94 (4.6) (0.1) | 3.05 (6.4) (1.5) | 8.64 (13.6) (4.1) |

| | JULY | AUGUST | SEPTEMBER | OCTOBER | NOVEMBER | DECEMBER | ANNUAL |
|---|---|---|---|---|---|---|---|
| Extreme Hi/Lo | 94 / 70 | 95 / 67 | 96 / 68 | 93 / 67 | 94 / 70 | 94 / 69 | 100 / 65 |
| | 0 (0) | 0 (0) | 0 (0) | 0 (0) | 0 (0) | 0 (0) | 0 (0) |
| Mean Hi/Lo | 88 / 74 | 89 / 74 | 90 / 75 | 90 / 74 | 89 / 74 | 87 / 73 | 88 / 74 |
| Precip | 7.86 (17) (9.6) | 9.22 (15.5) (13.5) | 6.99 (13.6) (14) | 5.79 (11.9) (11.9) | 5.29 (11.9) (6.3) | 5.21 (10.1) (1.2) | 61 (121.4) (63.2) |

## Map Location 282 — Virgin Islands, U.S., Alexander Hamilton — 1742N 06447W 60

| | JANUARY | FEBRUARY | MARCH | APRIL | MAY | JUNE |
|---|---|---|---|---|---|---|
| Extreme Hi/Lo | 88 / 61 | 89 / 66 | 90 / 62 | 93 / 63 | 92 / 68 | 92 / 67 |
| | 0 (0) | 0 (0) | 0 (0) | 0 (0) | 0 (0) | 0 (0) |
| Mean Hi/Lo | 84 / 71 | 84 / 71 | 84 / 71 | 86 / 73 | 87 / 74 | 88 / 76 |
| Precip | 2.05 (5.5) (0) | 1.71 (4.5) (0.1) | 1.45 (4) (0.2) | 2.82 (5.5) (0.7) | 3.81 (7.3) (2.1) | 3.03 (7) (2.9) |

| | JULY | AUGUST | SEPTEMBER | OCTOBER | NOVEMBER | DECEMBER | ANNUAL |
|---|---|---|---|---|---|---|---|
| Extreme Hi/Lo | 91 / 68 | 93 / 70 | 92 / 67 | 93 / 67 | 93 / 64 | 88 / 65 | 93 / 61 |
| | 0 (0) | 0 (0) | 0 (0) | 0 (0) | 0 (0) | 0 (0) | 0 (0) |
| Mean Hi/Lo | 88 / 76 | 89 / 76 | 88 / 75 | 88 / 75 | 86 / 73 | 87 / 72 | 86 / 74 |
| Precip | 4.04 (9.4) (3.8) | 4.68 (9.4) (3.7) | 6.08 (10.6) (5.1) | 4.77 (8.5) (3.6) | 5.03 (8.5) (2.5) | 3.9 (8.3) (0.8) | 43.4 (88.5) (25.5) |

# Indian Ocean Islands

# Map # 283 - 287

**Table 9.2   World Weather Data [CONTINUED]**

## Map Location 283 — Andaman/Nicobar Is., Port Blair — 1140N 09243E 259

| | JANUARY | FEBRUARY | MARCH | APRIL | MAY | JUNE |
|---|---|---|---|---|---|---|
| Avg Max | 84 | 86 | 88 | 89 | 87 | 84 |
| Avg Min | 72 | 71 | 72 | 75 | 75 | 75 |
| Ext Max | 91 | 93 | 98 | 99 | 97 | 96 |
| Ext Min | 62 | 62 | 64 | 67 | 66 | 67 |
| Precip | 1.79 (3.7) (0.7) | 1.11 (2.4) (0.6) | 1.12 (2.5) (3) | 2.36 (4.4) (8) | 15.13 (14.4) (8) | 21.75 (19.2) (7) |
| Snow | 0 (0) | 0 (0) | 0 (0) | 0 (0) | 0 (0) | 0 (0) |

| | JULY | AUGUST | SEPTEMBER | OCTOBER | NOVEMBER | DECEMBER | ANNUAL |
|---|---|---|---|---|---|---|---|
| Avg Max | 84 | 83 | 83 | 84 | 84 | 84 | 85 |
| Avg Min | 75 | 75 | 74 | 74 | 74 | 73 | 74 |
| Ext Max | 90 | 89 | 89 | 96 | 90 | 91 | 99 |
| Ext Min | 68 | 68 | 65 | 64 | 67 | 68 | 62 |
| Precip | 15.43 (16.5) (4) | 16.29 (16.9*) (3) | 17.4 (***) (3) | 12.51 (10.7) (3) | 10.52 (9) (3) | 7.92 (10.1) (1.5) | 123.3 (***) (44.8) |
| Snow | 0 (0) | 0 (0) | 0 (0) | 0 (0) | 0 (0) | 0 (0) | 0 (0) |

## Map Location 284 — Cocos-Keeling Is., Cocos Island — 1211S 09650E 11

| | JANUARY | FEBRUARY | MARCH | APRIL | MAY | JUNE |
|---|---|---|---|---|---|---|
| Avg Max | 86 | 87 | 86 | 85 | 84 | 83 |
| Avg Min | 77 | 77 | 78 | 78 | 77 | 76 |
| Ext Max | 93 | 93 | 91 | 94 | 90 | 90 |
| Ext Min | 70 | 69 | 70 | 70 | 71 | 69 |
| Precip | 5.4 (11.2) (0) | 7.7 (13.8) (0.6) | 8.5 (14.6) (0) | 10.4 (15.7) (0) | 7.9 (14.3) (0.6) | 9 (15.1) (0) |
| Snow | 0 (0) | 0 (0) | 0 (0) | 0 (0) | 0 (0) | 0 (0) |

| | JULY | AUGUST | SEPTEMBER | OCTOBER | NOVEMBER | DECEMBER | ANNUAL |
|---|---|---|---|---|---|---|---|
| Avg Max | 82 | 83 | 83 | 84 | 85 | 85 | 84 |
| Avg Min | 76 | 75 | 75 | 76 | 76 | 77 | 77 |
| Ext Max | 89 | 89 | 90 | 90 | 91 | 91 | 94 |
| Ext Min | 69 | 65 | 69 | 69 | 70 | 68 | 65 |
| Precip | 8.7 (14.9) (0) | 4.8 (11.3) (0) | 3.7 (9.6) (0.8) | 3.3 (8.8) (0.7) | 4.2 (10.6) (0) | 4.6 (10.1) (0) | 78.2 (150) (2.7) |
| Snow | 0 (0) | 0 (0) | 0 (0) | 0 (0) | 0 (0) | 0 (0) | 0 (0) |

## Map Location 285 — Kiribati, Rep. of, Christmas Is. — 0159N 15720W 5

| | JANUARY | FEBRUARY | MARCH | APRIL | MAY | JUNE |
|---|---|---|---|---|---|---|
| Avg Max | 85 | 85 | 86 | 86 | 87 | 87 |
| Avg Min | 75 | 75 | 76 | 76 | 76 | 76 |
| Ext Max | 89 | 91 | 92 | 92 | 93 | 90 |
| Ext Min | 66 | 71 | 71 | 70 | 73 | 68 |
| Precip | 1.02 (2.4) (0) | 2.83 (4.6) (0) | 2.49 (6) (0) | 8.05 (13.8) (0.2) | 3.54 (6.8) (0) | 3.23 (6.1) (0.4) |
| Snow | 0 (0) | 0 (0) | 0 (0) | 0 (0) | 0 (0) | 0 (0) |

| | JULY | AUGUST | SEPTEMBER | OCTOBER | NOVEMBER | DECEMBER | ANNUAL |
|---|---|---|---|---|---|---|---|
| Avg Max | 86 | 87 | 87 | 87 | 86 | 86 | 86 |
| Avg Min | 76 | 77 | 76 | 75 | 76 | 75 | 76 |
| Ext Max | 90 | 91 | 91 | 92 | 92 | 91 | 93 |
| Ext Min | 72 | 71 | 69 | 68 | 67 | 69 | 66 |
| Precip | 2 (3) (0) | 0.56 (1.8) (0) | 0.1 (0.1) (0) | 0.1 (0.3) (0) | 0.34 (0.7) (0) | 0.63 (1.7) (0) | 24.9 (47.3) (0.6) |
| Snow | 0 (0) | 0 (0) | 0 (0) | 0 (0) | 0 (0) | 0 (0) | 0 (0) |

## Map Location 286 — Maldives, Rep. of, Seenu Atoll — 0041S 07309E 6

| | JANUARY | FEBRUARY | MARCH | APRIL | MAY | JUNE |
|---|---|---|---|---|---|---|
| Avg Max | 86 | 87 | 87 | 88 | 88 | 87 |
| Avg Min | 77 | 77 | 77 | 78 | 78 | 77 |
| Ext Max | *** | *** | *** | *** | *** | *** |
| Ext Min | *** | *** | *** | *** | *** | *** |
| Precip | 11.5 (***) (***) | 2.3 (4.6) (***) | 12.8 (10) (***) | 3.9 (5.9) (***) | 6.5 (6.9) (***) | 10.9 (13.5) (***) |
| Snow | (***) (***) | (***) (***) | (***) (***) | (***) (***) | (***) (***) | (***) (***) |

| | JULY | AUGUST | SEPTEMBER | OCTOBER | NOVEMBER | DECEMBER | ANNUAL |
|---|---|---|---|---|---|---|---|
| Avg Max | 88 | 86 | 86 | 87 | 86 | 88 | 87 |
| Avg Min | 77 | 76 | 76 | 76 | 76 | 75 | 77 |
| Ext Max | *** | *** | *** | *** | *** | *** | *** |
| Ext Min | *** | *** | *** | *** | *** | *** | *** |
| Precip | 4.8 (7.4) (***) | 6.5 (9.4) (***) | 8.4 (8) (***) | 13 (11.2) (***) | 6.8 (9.7) (***) | 8.4 (8) (***) | 95.8 (***) (***) |
| Snow | (***) (***) | (***) (***) | (***) (***) | (***) (***) | (***) (***) | (***) (***) | (***) (***) |

## Map Location 287 — Mascarene Islands, Diego Garcia — 0714S 07226E 7

| | JANUARY | FEBRUARY | MARCH | APRIL | MAY | JUNE |
|---|---|---|---|---|---|---|
| Avg Max | 86 | 86 | 88 | 87 | 85 | 83 |
| Avg Min | 77 | 77 | 78 | 78 | 77 | 76 |
| Ext Max | 91 | 90 | 93 | 91 | 90 | 88 |
| Ext Min | 73 | 73 | 72 | 73 | 72 | 72 |
| Precip | 12.84 (***) (***) | 8.82 (12.7) (***) | 12.87 (16) (***) | 5.87 (9.4) (***) | 6.58 (9.3) (***) | 5.24 (8.6) (***) |
| Snow | 0 (0) | 0 (0) | 0 (0) | 0 (0) | 0 (0) | 0 (0) |

| | JULY | AUGUST | SEPTEMBER | OCTOBER | NOVEMBER | DECEMBER | ANNUAL |
|---|---|---|---|---|---|---|---|
| Avg Max | 82 | 82 | 82 | 83 | 85 | 85 | 85 |
| Avg Min | 75 | 75 | 75 | 76 | 77 | 77 | 77 |
| Ext Max | 86 | 86 | 85 | 88 | 87 | 88 | 93 |
| Ext Min | 72 | 72 | 72 | 72 | 72 | 72 | 72 |
| Precip | 5.16 (8.5) (***) | 9.13 (12.5) (***) | 8.9 (***) (***) | 6.5 (9.2) (***) | 5.95 (8.9) (***) | 6.69 (10.4) (***) | 94.5 (***) (***) |
| Snow | 0 (0) | 0 (0) | 0 (0) | 0 (0) | 0 (0) | 0 (0) | 0 (0) |

Table 9.2 World Weather Data [CONTINUED]

# Indian Ocean Islands, continued

## Map # 288 - 290

### Map Location 288 — Mauritius, Port Louis — 2006S 05732E 181

| Month | Avg Max | Record Max | Avg Min | Record Min | Avg Precip in. (Rain Days) (Thunder Days) |
|---|---|---|---|---|---|
| JANUARY | 86 | 95 | 73 | 63 | 8.5 (***) (0.3) |
| FEBRUARY | 85 | 91 | 73 | 64 | 7.8 (9.8) (1) |
| MARCH | 84 | 90 | 72 | 63 | 8.7 (10) (1) |
| APRIL | 82 | 88 | 70 | 58 | 5 (8.2) (1) |
| MAY | 79 | 85 | 66 | 55 | 3.8 (7.1) (0.3) |
| JUNE | 76 | 83 | 63 | 51 | 2.6 (4.9) (0) |
| JULY | 75 | 80 | 62 | 51 | 2.3 (4.4) (0) |
| AUGUST | 75 | 80 | 62 | 50 | 2.5 (4.7) (0) |
| SEPTEMBER | 77 | 83 | 63 | 51 | 1.4 (2.8) (0.3) |
| OCTOBER | 80 | 88 | 64 | 55 | 1.6 (3.1) (0.3) |
| NOVEMBER | 83 | 91 | 67 | 57 | 1.8 (3.5) (0.3) |
| DECEMBER | 85 | 95 | 71 | 62 | 4.6 (9) (0.3) |
| ANNUAL | 81 | 95 | 67 | 50 | 50.6 (***) (4.8) |

### Map Location 289 — Reunion, St Denis — 2035S 05531E 66

| Month | Avg Max | Record Max | Avg Min | Record Min | Avg Precip in. (Rain Days) (Thunder Days) |
|---|---|---|---|---|---|
| JANUARY | 88 | *** | 74 | *** | 11.06 (***) (1) |
| FEBRUARY | 88 | *** | 75 | *** | 7.36 (10.1*) (3) |
| MARCH | 87 | *** | 74 | *** | 13.98 (***) (3) |
| APRIL | 86 | *** | 73 | *** | 5.87 (8.9) (1) |
| MAY | 82 | *** | 70 | *** | 2.21 (4.8) (1) |
| JUNE | 79 | *** | 66 | *** | 1.97 (3.8) (0) |
| JULY | 78 | *** | 65 | *** | 1.54 (3.1) (0) |
| AUGUST | 79 | *** | 65 | *** | 1.3 (2.6) (0) |
| SEPTEMBER | 81 | *** | 66 | *** | 1.42 (2.8) (0) |
| OCTOBER | 83 | *** | 68 | *** | 1.18 (2.4) (0) |
| NOVEMBER | 87 | *** | 71 | *** | 2.76 (5) (1) |
| DECEMBER | 86 | *** | 72 | *** | 5.24 (9.7) (1) |
| ANNUAL | 84 | *** | 70 | *** | 55.9 (***) (11) |

### Map Location 290 — Seychelles, Victoria — 0437S 05527E 8

| Month | Avg Max | Record Max | Avg Min | Record Min | Avg Precip in. (Rain Days) (Thunder Days) |
|---|---|---|---|---|---|
| JANUARY | 83 | 88 | 76 | 69 | 15.2 (***) (3) |
| FEBRUARY | 84 | 89 | 77 | 71 | 10.5 (***) (3) |
| MARCH | 85 | 90 | 77 | 69 | 9.2 (10.1) (3) |
| APRIL | 86 | 92 | 77 | 71 | 7.2 (9.5) (2) |
| MAY | 85 | 91 | 77 | 69 | 6.7 (9.3) (2) |
| JUNE | 83 | 89 | 77 | 67 | 4 (7) (1) |
| JULY | 81 | 86 | 75 | 67 | 3.3 (6) (0.3) |
| AUGUST | 81 | 86 | 75 | 67 | 2.7 (5) (0.3) |
| SEPTEMBER | 82 | 88 | 76 | 68 | 5.1 (8.4) (1) |
| OCTOBER | 83 | 89 | 75 | 68 | 6.1 (9.7) (0.3) |
| NOVEMBER | 84 | 89 | 75 | 68 | 9.1 (13) (0.3) |
| DECEMBER | 83 | 91 | 75 | 69 | 13.4 (***) (0.3) |
| ANNUAL | 83 | 92 | 76 | 67 | 92.5 (***) (16.5) |

## Data Key

| Map Locater Number | |
|---|---|
| Location | MONTH |
| Lat - Lon ddmm Elev ft | |
| Average Max T °F | Record Max T °F |
| Average Min T °F | Record Min T °F |
| Avg. Precip in. (Rain Days) (Thunder Days) | Avg. Snow in. (Snow Days) |

| | |
|---|---|
| *** or Blank | No data, missing data or data of insufficient reliability |
| Rain Days | Number of Days with >= 0.01" of precipitation falling |
| Snow Days | Number of Days with >= 1.5" of snow falling |

# Mediterranean Islands

**Table 9.2  World Weather Data [CONTINUED]**

## Map # 291 - 295

Each monthly cell is given as: *normal daily high / normal daily low / precipitation (max) (max 24-hr) / record high / record low / snow (snow)*.

### Map Location 291 — Mallorca, Spain, Palma — 3933N 00243E 13

| | JANUARY | FEBRUARY | MARCH | APRIL | MAY | JUNE |
|---|---|---|---|---|---|---|
| | 57 / 42 / 1.4 (4.6) (3.3) / 75 / 28 / *** (***) | 59 / 43 / 1.6 (5.2) (1.8) / 77 / 28 / *** (***) | 62 / 45 / 1.5 (4.9) (0.7) / 79 / 30 / 0 (0) | 66 / 49 / 1.3 (4.4) (0.8) / 84 / 33 / 0 (0) | 73 / 55 / 91 / 41 / 0 (0) | 80 / 61 / 1 (3) (2.4) / 100 / 48 / 0 (0) |

| | JULY | AUGUST | SEPTEMBER | OCTOBER | NOVEMBER | DECEMBER | ANNUAL |
|---|---|---|---|---|---|---|---|
| | 84 / 66 / 0.2 (0.3) (1.5) / 101 / 54 / 0 (0) | 86 / 67 / 0.8 (2.4) (1.8) / 102 / 56 / 0 (0) | 81 / 64 / 2.5 (6.6) (2) / 94 / 48 / 0 (0) | 74 / 57 / 2.8 (7.1) (4.2) / 89 / 35 / 0 (0) | 65 / 50 / 2.8 (7.1) (0.8) / 78 / 33 / 0 (0) | 59 / 44 / 2.2 (6.8) (1.7) / 73 / 27 / *** (***) | 71 / 54 / 19.4 (56.8) (23) / 102 / 27 / *** (***) |

### Map Location 292 — Corsica, France, Ajaccio — 4155N 00848E 20

| | JANUARY | FEBRUARY | MARCH | APRIL | MAY | JUNE |
|---|---|---|---|---|---|---|
| | 56 / 40 / 2.99 (8.4) (1) / 72 / 25 / *** (***) | 58 / 42 / 2.56 (7.6) (1) / 72 / 23 / *** (***) | 62 / 44 / 2.09 (6.1) (0.3) / 82 / 28 / *** (***) | 66 / 48 / 1.89 (5.8) (2) / 88 / 36 / 0 (0) | 72 / 53 / 1.97 (5.9) (1) / 99 / 38 / 0 (0) | 79 / 60 / 0.83 (2.5) (2) / 97 / 45 / 0 (0) |

| | JULY | AUGUST | SEPTEMBER | OCTOBER | NOVEMBER | DECEMBER | ANNUAL |
|---|---|---|---|---|---|---|---|
| | 85 / 64 / 0.39 (1) (1) / 101 / 47 / 0 (0) | 85 / 64 / 0.63 (1.8) (1) / 103 / 50 / 0 (0) | 81 / 61 / 1.97 (5.6) (2) / 97 / 45 / 0 (0) | 72 / 55 / 3.47 (8) (2) / 88 / 39 / 0 (0) | 64 / 49 / 3.82 (8.4) (3) / 82 / 31 / 0 (0) | 59 / 44 / 3.86 (9.6) (1) / 73 / 26 / *** (***) | 70 / 52 / 26.5 (70.7) (17.3) / 103 / 23 / *** (***) |

### Map Location 293 — Crete, Greece, Heraklion — 3520N 02511E 115

| | JANUARY | FEBRUARY | MARCH | APRIL | MAY | JUNE |
|---|---|---|---|---|---|---|
| | 60 / 48 / 5.08 (10.6) (1) / 76 / 34 / 0 (0) | 60 / 48 / 3.94 (9.7) (1) / 82 / 30 / 0 (0) | 64 / 50 / 2.64 (6.8) (1) / 91 / 35 / 0 (0) | 70 / 54 / 1.14 (3.9) (0.3) / 98 / 41 / 0 (0) | 76 / 60 / 0.59 (1.8) (1) / 100 / 48 / 0 (0) | 82 / 67 / 0.08 (0) (0.3) / 114 / 50 / 0 (0) |

| | JULY | AUGUST | SEPTEMBER | OCTOBER | NOVEMBER | DECEMBER | ANNUAL |
|---|---|---|---|---|---|---|---|
| | 85 / 72 / 0.04 (0) (0) / 106 / 61 / 0 (0) | 85 / 71 / 0.12 (0.1) (0) / 104 / 56 / 0 (0) | 82 / 68 / 1.26 (4.1) (0.3) / 102 / 54 / 0 (0) | 77 / 62 / 1.5 (4.7) (1) / 96 / 45 / 0 (0) | 71 / 56 / 4.8 (9.3) (1) / 87 / 37 / 0 (0) | 64 / 51 / 6.65 (11.1) (1) / 80 / 37 / 0 (0) | 73 / 59 / 27.8 (62.1) (7.9) / 114 / 30 / 0 (0) |

### Map Location 294 — Cyprus, Nicosia — 3509N 03316E 732

| | JANUARY | FEBRUARY | MARCH | APRIL | MAY | JUNE |
|---|---|---|---|---|---|---|
| | 58 / 42 / 2.9 (5.5) (3.2) / 70 / 25 / *** (***) | 59 / 42 / 2 (4.1) (2.7) / 76 / 23 / *** (***) | 65 / 44 / 1.3 (2.8) (2.9) / 88 / 27 / *** (***) | 74 / 50 / 0.8 (1.9) (2.4) / 95 / 32 / 0 (0) | 83 / 60 / 1.1 (2.5) (3.7) / 109 / 40 / 0 (0) | 91 / 65 / 0.4 (0.8) (1.2) / 105 / 49 / 0 (0) |

| | JULY | AUGUST | SEPTEMBER | OCTOBER | NOVEMBER | DECEMBER | ANNUAL |
|---|---|---|---|---|---|---|---|
| | 97 / 69 / 0.03 (0.1) (0.4) / 116 / 52 / 0 (0) | 97 / 69 / 0.03 (0.1) (0.4) / 108 / 57 / 0 (0) | 91 / 65 / 0.2 (0.5) (1.4) / 106 / 49 / 0 (0) | 81 / 58 / 0.9 (1.9) (4.4) / 105 / 40 / 0 (0) | 72 / 51 / 1.7 (3.2) (3.4) / 95 / 26 / *** (***) | 62 / 45 / 3 (5.7) (3.6) / 76 / 29 / 0 (0) | 78 / 55 / 14.4 (29.1) (29.7) / 116 / 23 / *** (***) |

### Map Location 295 — Sardinia, Cagliari — 3914N 00903E 12

| | JANUARY | FEBRUARY | MARCH | APRIL | MAY | JUNE |
|---|---|---|---|---|---|---|
| | 57 / 42 / 1.65 (5.4) (0.8) / 68 / 26 / *** (***) | 58 / 43 / 1.42 (4.7) (1.8) / 69 / 26 / *** (***) | 62 / 45 / 2.05 (6) (1) / 81 / 28 / *** (***) | 65 / 49 / 1.61 (5.2) (1.8) / 82 / 35 / 0 (0) | 72 / 55 / 0.98 (3.3) (1.8) / 85 / 44 / 0 (0) | 80 / 56 / 0.91 (2.7) (1.5) / 100 / 47 / 0 (0) |

| | JULY | AUGUST | SEPTEMBER | OCTOBER | NOVEMBER | DECEMBER | ANNUAL |
|---|---|---|---|---|---|---|---|
| | 86 / 56 / 0.08 (0) (1) / 100 / 55 / 0 (0) | 85 / 67 / 0.24 (0.5) (1.5) / 101 / 54 / 0 (0) | 80 / 64 / 1.14 (3.9) (3.4) / 102 / 51 / 0 (0) | 74 / 57 / 2.01 (5.7) (1.8) / 88 / 43 / 0 (0) | 66 / 50 / 3.98 (8.6) (1.9) / 77 / 33 / 0 (0) | 62 / 45 / 2.72 (7.9) (0.7) / 71 / 32 / 0 (0) | 71 / 52 / 18.8 (53.9) (19) / 102 / 26 / *** (***) |

**Table 9.2  World Weather Data** [CONTINUED]

## Mediterranean Islands, continued

### Map # 296 - 297

### Map Location 296 — Sicily, Italy, Palermo — 3806N 01318E 345

| | JANUARY | FEBRUARY | MARCH | APRIL | MAY | JUNE |
|---|---|---|---|---|---|---|
| Avg Max / Rec Max | 58 / 71 | 60 / 78 | 62 / 91 | 67 / 86 | 83 / 98 | 82 / 104 |
| Avg Min / Rec Min | 47 / 34 | 47 / 32 | 49 / 31 | 53 / 42 | 59 / 48 | 66 / 52 |
| Precip (Rain)(Thunder) / Snow (Snow) | 4.41 (10.2) (2.5) / 0 (0) | 3.82 (9.6) (3.4) / 0 (0) | 2.6 (6.7) (1) / 0 (0) | 2.09 (6.1) (0) / 0 (0) | 1.42 (4.7) (1) / 0 (0) | 0.59 (1.7) (0) / 0 (0) |

| | JULY | AUGUST | SEPTEMBER | OCTOBER | NOVEMBER | DECEMBER | ANNUAL |
|---|---|---|---|---|---|---|---|
| Avg Max / Rec Max | 86 / 108 | 87 / 113 | 83 / 106 | 75 / 90 | 67 / 84 | 61 / 73 | 73 / 113 |
| Avg Min / Rec Min | 71 / 59 | 72 / 61 | 69 / 51 | 62 / 48 | 55 / 41 | 50 / 58 | 58 / 31 |
| Precip (Rain)(Thunder) / Snow (Snow) | 0.32 (0.8) (0) / 0 (0) | 0.59 (1.7) (0.8) / 0 (0) | 1.81 (5.3) (1.4) / 0 (0) | 3.7 (8.3) (2.7) / 0 (0) | 4.21 (8.8) (1.4) / 0 (0) | 5.2 (10.7) (2.5) / 0 (0) | 30.8 (74.6) (16.7) / 0 (0) |

### Map Location 297 — Malta, Valletta — 3553N 01425E 340

| | JANUARY | FEBRUARY | MARCH | APRIL | MAY | JUNE |
|---|---|---|---|---|---|---|
| Avg Max / Rec Max | 59 / 76 | 59 / 76 | 62 / 83 | 66 / 93 | 71 / 92 | 79 / 99 |
| Avg Min / Rec Min | 51 / 39 | 51 / 34 | 52 / 37 | 56 / 44 | 61 / 49 | 67 / 57 |
| Precip (Rain)(Thunder) / Snow (Snow) | 3.3 (8.9) (2) / 0 (0) | 2.3 (7) (1) / 0 (0) | 1.5 (4.9) (2) / 0 (0) | 0.8 (2.7) (1) / 0 (0) | 0.4 (0.9) (1) / 0 (0) | 0.1 (0) (1) / 0 (0) |

| | JULY | AUGUST | SEPTEMBER | OCTOBER | NOVEMBER | DECEMBER | ANNUAL |
|---|---|---|---|---|---|---|---|
| Avg Max / Rec Max | 84 / 104 | 85 / 105 | 81 / 100 | 76 / 94 | 68 / 82 | 62 / 75 | 71 / 105 |
| Avg Min / Rec Min | 72 / 62 | 73 / 62 | 71 / 57 | 66 / 45 | 59 / 42 | 54 / 39 | 61 / 34 |
| Precip (Rain)(Thunder) / Snow (Snow) | 0.05 (0) (0.3) / 0 (0) | 0.2 (0.3) (1) / 0 (0) | 1.3 (4.2) (2) / 0 (0) | 2.7 (6.9) (6) / 0 (0) | 3.6 (8.2) (3) / 0 (0) | 3.9 (9.7) (4) / 0 (0) | 20.1 (53.7) (24.3) / 0 (0) |

## Data Key

| Map Locater Number | MONTH | |
|---|---|---|
| Location | Average Max T °F | Record Max T °F |
| Lat.·Lon ddmm Elev ft | Average Min T °F | Record Min T °F |
| | Avg. Precip in. (Rain Days) (Thunder Days) | Avg. Snow in. (Snow Days) |

| | |
|---|---|
| *** or Blank | No data, missing data or data of insufficient reliability |
| Rain Days | Number of Days with > = 0.01" of precipitation falling |
| Snow Days | Number of Days with > = 1.5" of snow falling |

# Pacific Ocean Islands

## Map # 298 - 302

Table 9.2　World Weather Data [CONTINUED]

### Map Location 298 — American Samoa, Pago Pago Intl — 1419S 17043W 29

| Month | Rec High | Rec Low | Norm High | Norm Low | Precip | Snow |
|---|---|---|---|---|---|---|
| JANUARY | 96 | 72 | 87 | 75 | 24.5 (23.2) (7) | 0 (0) |
| FEBRUARY | 94 | 72 | 86 | 74 | 20.5 (20.2) (28) | 0 (0) |
| MARCH | 94 | 72 | 87 | 75 | 19.2 (20.4) (31) | 0 (0) |
| APRIL | 92 | 72 | 87 | 76 | 16.5 (18.7) (30) | 0 (0) |
| MAY | 97 | 71 | 85 | 76 | 15.4 (18.2) (31) | 0 (0) |
| JUNE | 96 | 70 | 85 | 78 | 12.3 (16.3) (***) | 0 (0) |
| JULY | 92 | 67 | 83 | 74 | 10 (15.6) (31) | 0 (0) |
| AUGUST | 98 | 70 | 83 | 75 | 8.2 (14.6) (2.5) | 0 (0) |
| SEPTEMBER | 91 | 71 | 85 | 75 | 13.1 (17.1) (2) | 0 (0) |
| OCTOBER | 90 | 70 | 85 | 75 | 14.9 (17.3) (8) | 0 (0) |
| NOVEMBER | 92 | 70 | 85 | 74 | 19.2 (18.9) (4) | 0 (0) |
| DECEMBER | 92 | 70 | 88 | 75 | 19.8 (19.8) (6) | 0 (0) |
| ANNUAL | 98 | 67 | 85 | 75 | 193.6 (220.3) (***) | 0 (0) |

### Map Location 299 — Ogasawara Is., Japan, Iwo Jima — 2447N 14119E 353

| Month | Rec High | Rec Low | Norm High | Norm Low | Precip | Snow |
|---|---|---|---|---|---|---|
| JANUARY | 80 | 57 | 71 | 63 | 3.03 (5) (0.2) | 0 (0) |
| FEBRUARY | 80 | 55 | 71 | 63 | 2.95 (4.5) (0.1) | 0 (0) |
| MARCH | 81 | 55 | 73 | 65 | 1.8 (3.3) (0.2) | 0 (0) |
| APRIL | 85 | 61 | 78 | 69 | 4.18 (6.1) (0.7) | 0 (0) |
| MAY | 94 | 64 | 82 | 74 | 4.35 (6.4) (0.4) | 0 (0) |
| JUNE | 90 | 66 | 85 | 77 | 3.92 (6.7) (0.3) | 0 (0) |
| JULY | 91 | 73 | 86 | 78 | 7.08 (9.3) (0.9) | 0 (0) |
| AUGUST | 91 | 71 | 86 | 78 | 6.63 (9.6) (1.2) | 0 (0) |
| SEPTEMBER | 90 | 72 | 86 | 78 | 4.37 (7.2) (1.2) | 0 (0) |
| OCTOBER | 88 | 69 | 84 | 76 | 6.63 (8.6) (0.5) | 0 (0) |
| NOVEMBER | 87 | 62 | 80 | 73 | 4.93 (8.3) (0.5) | 0 (0) |
| DECEMBER | 84 | 60 | 75 | 68 | 4.51 (7.7) (0.2) | 0 (0) |
| ANNUAL | 94 | 55 | 80 | 72 | 54.4 (82.7) (6.4) | 0 (0) |

### Map Location 300 — Caroline Is., Truk — 0727N 15150E 6

| Month | Rec High | Rec Low | Norm High | Norm Low | Precip | Snow |
|---|---|---|---|---|---|---|
| JANUARY | 89 | 71 | 85 | 78 | 9.57 (11) (1.3) | 0 (0) |
| FEBRUARY | 91 | 71 | 86 | 78 | 6.72 (10.4) (0.3) | 0 (0) |
| MARCH | 94 | 72 | 86 | 77 | 7.47 (11.4) (1) | 0 (0) |
| APRIL | 92 | 70 | 86 | 77 | 15.28 (16.2) (1) | 0 (0) |
| MAY | 94 | 72 | 86 | 77 | 17.12 (18.3) (1.2) | 0 (0) |
| JUNE | 93 | 71 | 87 | 77 | 12.52 (18.9) (1.2) | 0 (0) |
| JULY | 94 | 70 | 87 | 76 | 13.93 (16.9) (2.3) | 0 (0) |
| AUGUST | 96 | 71 | 87 | 76 | 12.48 (18.3) (1.5) | 0 (0) |
| SEPTEMBER | 94 | 70 | 87 | 76 | 13.4 (17.1) (1.6) | 0 (0) |
| OCTOBER | 97 | 70 | 87 | 76 | 12.98 (17) (2.7) | 0 (0) |
| NOVEMBER | 93 | 71 | 87 | 77 | 12.03 (16.4) (1.8) | 0 (0) |
| DECEMBER | 91 | 72 | 86 | 77 | 14.79 (16.1) (1.6) | 0 (0) |
| ANNUAL | 97 | 70 | 86 | 77 | 148.3 (188) (17.5) | 0 (0) |

### Map Location 301 — Chile, Rapa Nui (Easter Is.) — 2712S 10924W 134

| Month | Rec High | Rec Low | Norm High | Norm Low | Precip | Snow |
|---|---|---|---|---|---|---|
| JANUARY | 80 | 59 | 67 | 62 | 2.68 (*) (*) | * (*)(*) |
| FEBRUARY | 81 | 62 | 67 | 62 | 3.27 (*) (*) | * (*)(*) |
| MARCH | 80 | 46 | 67 | 59 | 3.54 (*) (*) | * (*)(*) |
| APRIL | 95 | 39 | 77 | 61 | 4.37 (*) (*) | * (*)(*) |
| MAY | 86 | 50 | 75 | 62 | 5.75 (*) (*) | * (*)(*) |
| JUNE | 95 | 45 | 71 | 50 | 4.02 (*) (*) | * (*)(*) |
| JULY | 88 | 45 | 70 | 59 | 3.54 (*) | * (*)(*) |
| AUGUST | 88 | 45 | 70 | 49 | 3.5 (*) (*) | * (*)(*) |
| SEPTEMBER | 86 | 50 | 71 | 57 | 3.35 (*) | * (*)(*) |
| OCTOBER | 86 | 55 | 73 | 64 | 2.6 (*) (*) | * (*)(*) |
| NOVEMBER | 98 | 46 | 75 | 64 | 2.83 (*) (*) | * (*)(*) |
| DECEMBER | 98 | 55 | 78 | 64 | 3.5 (*) (*) | * (*)(*) |
| ANNUAL | 98 | 39 | 75 | 63 | 42.95 (*) (*) | * (*)(*) |

### Map Location 302 — Fiji Islands, Suva — 1808S 17826E 20

| Month | Rec High | Rec Low | Norm High | Norm Low | Precip | Snow |
|---|---|---|---|---|---|---|
| JANUARY | 95 | 67 | 86 | 74 | 11.4 (16.4) (4) | 0 (0) |
| FEBRUARY | 97 | 67 | 86 | 74 | 10.7 (16) (4) | 0 (0) |
| MARCH | 98 | 66 | 86 | 74 | 14.5 (17.7) (6) | 0 (0) |
| APRIL | 94 | 61 | 84 | 73 | 12.2 (16.6) (3) | 0 (0) |
| MAY | 93 | 61 | 82 | 71 | 10.1 (15.5) (1) | 0 (0) |
| JUNE | 90 | 58 | 80 | 69 | 6.7 (13.4) (91) | 0 (0) |
| JULY | 90 | 55 | 79 | 68 | 4.9 (11.4) (1) | 0 (0) |
| AUGUST | 90 | 57 | 79 | 68 | 8.3 (14.7) (91) | 0 (0) |
| SEPTEMBER | 90 | 57 | 80 | 69 | 7.7 (15) (1) | 0 (0) |
| OCTOBER | 93 | 57 | 81 | 70 | 8.3 (15.4) (1) | 0 (0) |
| NOVEMBER | 93 | 57 | 83 | 71 | 9.8 (16.2) (2) | 0 (0) |
| DECEMBER | 97 | 62 | 85 | 73 | 12.5 (17) (3) | 0 (0) |
| ANNUAL | 98 | 55 | 83 | 71 | 117.1 (185.3) (28) | 0 (0) |

**Table 9.2   World Weather Data [CONTINUED]**

# Pacific Ocean Islands, continued

## Map # 303 - 307

*For each month the data are given as: Extreme High / Extreme Low (upper), Normal High / Normal Low (lower), and Precipitation in. (mm) (days), Snow in. (in).*

### Map Location 303 — Guam, Andersen AFB — 1334N 14455E 605

| Month | Ext Hi/Lo | Nrm Hi/Lo | Precip in. (mm) (days) | Snow |
|---|---|---|---|---|
| JANUARY | 86 / 66 | 81 / 75 | 4.15 (8.8) (0.2) | 0 (0) |
| FEBRUARY | 86 / 70 | 81 / 74 | 4.57 (7.7) (0.1) | 0 (0) |
| MARCH | 85 / 69 | 81 / 75 | 2.38 (6) (0.1) | 0 (0) |
| APRIL | 94 / 69 | 83 / 76 | 4.9 (7.7) (0.8) | 0 (0) |
| MAY | 88 / 71 | 83 / 76 | 5.72 (8.8) (0.4) | 0 (0) |
| JUNE | 89 / 70 | 84 / 77 | 4.56 (10.7) (1.7) | 0 (0) |
| JULY | 89 / 70 | 84 / 76 | 8.02 (14.1) (4.1) | 0 (0) |
| AUGUST | 90 / 70 | 84 / 76 | 11.73 (16.4) (3.5) | 0 (0) |
| SEPTEMBER | 88 / 71 | 83 / 75 | 13.5 (18.5) (6.4) | 0 (0) |
| OCTOBER | 87 / 71 | 83 / 76 | 15.37 (17.8) (5.3) | 0 (0) |
| NOVEMBER | 89 / 71 | 83 / 77 | 6.84 (13.4) (1.7) | 0 (0) |
| DECEMBER | 87 / 68 | 82 / 76 | 6.74 (11.7) (0.7) | 0 (0) |
| ANNUAL | 94 / 66 | 83 / 76 | 88.5 (141.6) (25) | 0 (0) |

### Map Location 304 — Nansei/Ryukyu Is, Japan, Naha (Okinawa) — 2611N 12738E 14

| Month | Ext Hi/Lo | Nrm Hi/Lo | Precip in. (mm) (days) | Snow |
|---|---|---|---|---|
| JANUARY | 80 / 43 | 67 / 56 | 5.3 (10.1) (0.2) | 0 (0) |
| FEBRUARY | 81 / 41 | 67 / 55 | 5.4 (10.2) (0.8) | 0 (0) |
| MARCH | 82 / 43 | 70 / 59 | 6.1 (8.7) (2.2) | 0 (0) |
| APRIL | 87 / 48 | 76 / 64 | 6.1 (8.7) (2.2) | 0 (0) |
| MAY | 91 / 52 | 80 / 68 | 8.9 (10.4) (2.5) | 0 (0) |
| JUNE | 94 / 59 | 85 / 75 | 10 (11) (3) | 0 (0) |
| JULY | 96 / 69 | 89 / 77 | 7.1 (9.1) (2.6) | 0 (0) |
| AUGUST | 95 / 71 | 88 / 77 | 10 (11) (2.7) | 0 (0) |
| SEPTEMBER | 93 / 63 | 87 / 75 | 7.1 (9.1) (2.6) | 0 (0) |
| OCTOBER | 92 / 59 | 81 / 69 | 6.6 (9) (0.2) | 0 (0) |
| NOVEMBER | 89 / 47 | 75 / 64 | 5.9 (8.2) (0.4) | 0 (0) |
| DECEMBER | 82 / 44 | 70 / 58 | 4.3 (8.5) (0.5) | 0 (0) |
| ANNUAL | 96 / 41 | 78 / 66 | 82.8 (114.5) (18.6) | 0 (0) |

### Map Location 305 — Johnston Is., USA — 1644N 16931W 7

| Month | Ext Hi/Lo | Nrm Hi/Lo | Precip in. (mm) (days) | Snow |
|---|---|---|---|---|
| JANUARY | 86 / 63 | 80 / 73 | 5.62 (5.5) (0.7) | 0 (0) |
| FEBRUARY | 85 / 67 | 80 / 73 | 2.06 (3.5) (0.1) | 0 (0) |
| MARCH | 85 / 68 | 80 / 73 | 2.34 (3.7) (0.4) | 0 (0) |
| APRIL | 86 / 68 | 81 / 74 | 2.63 (5.2) (0.4) | 0 (0) |
| MAY | 88 / 69 | 82 / 75 | 1.09 (3.1) (0) | 0 (0) |
| JUNE | 88 / 70 | 84 / 76 | 0.86 (2.8) (0.1) | 0 (0) |
| JULY | 93 / 71 | 85 / 77 | 1.31 (4) (0.1) | 0 (0) |
| AUGUST | 93 / 71 | 86 / 78 | 1.59 (4.2) (0) | 0 (0) |
| SEPTEMBER | 94 / 71 | 86 / 78 | 2.13 (4.4) (0.1) | 0 (0) |
| OCTOBER | 90 / 68 | 85 / 77 | 2.71 (5.3) (0.2) | 0 (0) |
| NOVEMBER | 90 / 69 | 83 / 76 | 1.9 (4) (0.1) | 0 (0) |
| DECEMBER | 89 / 68 | 81 / 75 | 3.44 (6.4) (0.4) | 0 (0) |
| ANNUAL | 94 / 63 | 83 / 75 | 27.7 (52.1) (2.6) | 0 (0) |

### Map Location 306 — Marshall Is., Eniwetok — 1120N 16219E 13

| Month | Ext Hi/Lo | Nrm Hi/Lo | Precip in. (mm) (days) | Snow |
|---|---|---|---|---|
| JANUARY | 88 / 71 | 85 / 78 | 1.03 (3.4) (0) | 0 (0) |
| FEBRUARY | 88 / 73 | 85 / 78 | 0.85 (2.5) (0) | 0 (0) |
| MARCH | 89 / 73 | 86 / 78 | 1.73 (4.3) (0) | 0 (0) |
| APRIL | 90 / 72 | 86 / 78 | 2.47 (5.3) (0) | 0 (0) |
| MAY | 91 / 73 | 86 / 78 | 5.65 (9.2) (0.2) | 0 (0) |
| JUNE | 90 / 73 | 87 / 79 | 4.06 (9.6) (0.4) | 0 (0) |
| JULY | 92 / 71 | 87 / 79 | 7.12 (13) (1.3) | 0 (0) |
| AUGUST | 92 / 71 | 87 / 79 | 6.67 (13.1) (0) | 0 (0) |
| SEPTEMBER | 93 / 72 | 88 / 79 | 6.76 (13.1) (1.4) | 0 (0) |
| OCTOBER | 91 / 71 | 87 / 79 | 9.76 (14.6) (1.9) | 0 (0) |
| NOVEMBER | 94 / 72 | 86 / 79 | 7.26 (13) (0.5) | 0 (0) |
| DECEMBER | 88 / 71 | 86 / 78 | 3.61 (7.2) (0.2) | 0 (0) |
| ANNUAL | 94 / 71 | 86 / 79 | 57 (108.3) (6.8) | 0 (0) |

### Map Location 307 — Micronesia, Weno — 0727N 15150E 6

| Month | Ext Hi/Lo | Nrm Hi/Lo | Precip in. (mm) (days) | Snow |
|---|---|---|---|---|
| JANUARY | 89 / 71 | 85 / 78 | 9.57 (11) (1.3) | 0 (0) |
| FEBRUARY | 91 / 71 | 86 / 78 | 6.72 (10.4) (0.3) | 0 (0) |
| MARCH | 94 / 72 | 86 / 77 | 7.47 (11.4) (1) | 0 (0) |
| APRIL | 92 / 70 | 86 / 77 | 15.28 (16.2) (1) | 0 (0) |
| MAY | 94 / 72 | 86 / 77 | 17.12 (18.3) (1.2) | 0 (0) |
| JUNE | 93 / 71 | 87 / 77 | 12.52 (18.9) (1.2) | 0 (0) |
| JULY | 94 / 70 | 87 / 76 | 13.93 (16.9) (2.3) | 0 (0) |
| AUGUST | 96 / 71 | 87 / 76 | 12.48 (17) (2.7) | 0 (0) |
| SEPTEMBER | 94 / 70 | 87 / 76 | 13.4 (17.1) (1.6) | 0 (0) |
| OCTOBER | 97 / 70 | 87 / 76 | 12.98 (17) (2.7) | 0 (0) |
| NOVEMBER | 93 / 71 | 87 / 77 | 12.03 (16.4) (1.8) | 0 (0) |
| DECEMBER | 91 / 72 | 86 / 77 | 14.79 (16.1) (1.6) | 0 (0) |
| ANNUAL | 97 / 70 | 86 / 77 | 148.3 (188) (17.5) | 0 (0) |

Table 9.2   World Weather Data [CONTINUED]

# Pacific Ocean Islands, continued

## Map # 308 - 312

Columns per month: Avg High / Avg Low / Record High / Record Low / Snow (days) / Precipitation total (max) (days)

### Map Location 308 — Midway Is., USA; Midway Naval Station; 2811N 17722W 13

| Month | Avg High | Avg Low | Rec High | Rec Low | Snow | Precipitation |
|---|---|---|---|---|---|---|
| JANUARY | 69 | 62 | 76 | 54 | 0 (0) | 5.25 (9.2) (0.6) |
| FEBRUARY | 69 | 62 | 76 | 53 | 0 (0) | 4.13 (6.5) (0.1) |
| MARCH | 69 | 63 | 77 | 52 | 0 (0) | 2.23 (4.6) (0.1) |
| APRIL | 70 | 64 | 78 | 53 | 0 (0) | 2.98 (4.3) (0.5) |
| MAY | 74 | 68 | 81 | 60 | 0 (0) | 2.23 (3) (0.2) |
| JUNE | 79 | 72 | 85 | 62 | 0 (0) | 3.4 (3.6) (0.4) |
| JULY | 81 | 74 | 88 | 67 | 0 (0) | 2.91 (6.3) (0.7) |
| AUGUST | 82 | 75 | 89 | 68 | 0 (0) | 4.79 (7.2) (1.3) |
| SEPTEMBER | 82 | 75 | 88 | 67 | 0 (0) | 3.47 (6.9) (0.8) |
| OCTOBER | 78 | 72 | 87 | 60 | 0 (0) | 4.58 (6.7) (0.7) |
| NOVEMBER | 75 | 69 | 82 | 59 | 0 (0) | 2.42 (4.3) (0.3) |
| DECEMBER | 72 | 65 | 75 | 56 | 0 (0) | 3.41 (8) (0.2) |
| ANNUAL | 75 | 68 | 89 | 52 | 0 (0) | 41.8 (70.6) (5.9) |

### Map Location 309 — New Caledonia, France, Noumea; 2216S 16627E 246

| Month | Avg High | Avg Low | Rec High | Rec Low | Snow | Precipitation |
|---|---|---|---|---|---|---|
| JANUARY | 86 | 72 | 97 | 64 | 0 (0) | 3.7 (8.6) (3) |
| FEBRUARY | 85 | 73 | 99 | 64 | 0 (0) | 5.1 (10.8) (2) |
| MARCH | 85 | 72 | 95 | 63 | 0 (0) | 5.7 (12.8) (1) |
| APRIL | 83 | 70 | 96 | 61 | 0 (0) | 5.2 (12.4) (1) |
| MAY | 79 | 66 | 91 | 56 | 0 (0) | 4.4 (11.8) (1) |
| JUNE | 77 | 64 | 89 | 55 | 0 (0) | 3.7 (9.6) (0.3) |
| JULY | 76 | 62 | 87 | 52 | 0 (0) | 3.6 (9.5) (1) |
| AUGUST | 76 | 61 | 85 | 54 | 0 (0) | 2.6 (7.6) (0.3) |
| SEPTEMBER | 78 | 63 | 90 | 55 | 0 (0) | 2.5 (7) (0.3) |
| OCTOBER | 80 | 65 | 93 | 56 | 0 (0) | 2 (5.7) (0.3) |
| NOVEMBER | 83 | 68 | 94 | 60 | 0 (0) | 2.4 (6.7) (0) |
| DECEMBER | 86 | 70 | 98 | 63 | 0 (0) | 2.6 (6.6) (3) |
| ANNUAL | 81 | 67 | 99 | 52 | 0 (0) | 43.5 (109.1) (13.2) |

### Map Location 310 — New Zealand, Queenstown; 4501S 16844E 1167

| Month | Avg High | Avg Low | Rec High | Rec Low | Snow | Precipitation |
|---|---|---|---|---|---|---|
| JANUARY | 71 | 49 | 93 | 33 | 0 (0) | 3.2 (7.8) (1) |
| FEBRUARY | 70 | 50 | 90 | 34 | 0 (0) | 2.9 (7.2) (1) |
| MARCH | 67 | 47 | 85 | 32 | 0 (0) | 3 (10.6) (0.3) |
| APRIL | 60 | 43 | 73 | 29 | 0* (0*) | 2.9 (10.5) (0.3) |
| MAY | 52 | 36 | 70 | 23 | *** (***) | 2.6 (10.2) (0.3) |
| JUNE | 47 | 33 | 68 | 21 | *** (***) | 2.2 (6.8) (0.3) |
| JULY | 46 | 31 | 65 | 21 | *** (***) | 2.2 (6.8) (0.3) |
| AUGUST | 50 | 34 | 69 | 25 | *** (***) | 2.6 (7.2) (1) |
| SEPTEMBER | 56 | 37 | 78 | 28 | *** (***) | 3 (8.2) (1) |
| OCTOBER | 60 | 41 | 84 | 29 | *** (***) | 2.5 (7) (1) |
| NOVEMBER | 64 | 44 | — | — | 0 (0) | — |
| DECEMBER | 69 | 48 | — | — | 0 (0) | — |
| ANNUAL | 59 | 41 | 94 | 19 | *** (***) | 32 (96) (8.5) |

### Map Location 311 — New Zealand, Wellington; 4119S 17448E 38

| Month | Avg High | Avg Low | Rec High | Rec Low | Snow | Precipitation |
|---|---|---|---|---|---|---|
| JANUARY | 69 | 56 | 85 | 39 | 0 (0) | 3.2 (7.8) (1) |
| FEBRUARY | 69 | 56 | 88 | 41 | 0 (0) | 3.2 (7.8) (0.3) |
| MARCH | 67 | 54 | 81 | 39 | 0 (0) | 3.2 (10.8) (0.3) |
| APRIL | 63 | 51 | 74 | 36 | 0 (0) | 3.8 (11.3) (0.3) |
| MAY | 58 | 47 | 71 | 32 | 0 (0) | 4.6 (12) (0.3) |
| JUNE | 55 | 44 | 69 | 30 | 0 (0) | 4.6 (11) (0.3) |
| JULY | 53 | 42 | 66 | 29 | 0 (0) | 5.4 (12.1) (1) |
| AUGUST | 54 | 43 | 66 | 29 | 0 (0) | 4.6 (11) (0.3) |
| SEPTEMBER | 57 | 46 | 69 | 31 | 0 (0) | 3.8 (9.8) (1) |
| OCTOBER | 60 | 48 | 75 | 34 | 0 (0) | 3.2 (10.8) (0.3) |
| NOVEMBER | 63 | 50 | 81 | 36 | 0 (0) | 3.5 (9.2) (1) |
| DECEMBER | 67 | 54 | 83 | 38 | 0 (0) | 3.5 (8.3) (1) |
| ANNUAL | 61 | 49 | 88 | 29 | 0 (0) | 47.4 (121.3) (7.1) |

### Map Location 312 — Samoa, Apia; 1348S 17147W 7

| Month | Avg High | Avg Low | Rec High | Rec Low | Snow | Precipitation |
|---|---|---|---|---|---|---|
| JANUARY | 86 | 75 | 91 | 69 | 0 (0) | 17.9 (19) (4) |
| FEBRUARY | 85 | 76 | 92 | 70 | 0 (0) | 15.2 (18) (4) |
| MARCH | 86 | 74 | 91 | 70 | 0 (0) | 14.1 (17.5) (4) |
| APRIL | 86 | 75 | 91 | 69 | 0 (0) | 10 (15.5) (4) |
| MAY | 85 | 74 | 90 | 67 | 0 (0) | 6.3 (13.2) (3) |
| JUNE | 85 | 74 | 90 | 67 | 0 (0) | 5.1 (11.7) (2) |
| JULY | 85 | 74 | 91 | 63 | 0 (0) | 3.2 (8.7) (1) |
| AUGUST | 84 | 74 | 90 | 65 | 0 (0) | 3.5 (9.3) (1) |
| SEPTEMBER | 84 | 74 | 90 | 65 | 0 (0) | 5.2 (12.2) (2) |
| OCTOBER | 85 | 75 | 93 | 66 | 0 (0) | 6.7 (14.1) (3) |
| NOVEMBER | 86 | 74 | 92 | 69 | 0 (0) | 10.5 (16.5) (5) |
| DECEMBER | 85 | 74 | 92 | 69 | 0 (0) | 14.6 (17.8) (5) |
| ANNUAL | 85 | 75 | 93 | 63 | 0 (0) | 112.3 (173.5) (38) |

## Table 9.2 World Weather Data [CONTINUED]

# Pacific Ocean Islands, continued

# Map # 313 - 316

**Map Location 313 — Society Islands, (Tahiti), Papeete — 1733S 14936W 7**

| | JANUARY | FEBRUARY | MARCH | APRIL | MAY | JUNE |
|---|---|---|---|---|---|---|
| Avg Max / Min T | 89 / 72 | 89 / 72 | 89 / 72 | 89 / 72 | 87 / 70 | 86 / 69 |
| Avg Precip (Rain Days)(Thunder Days) | 13.15 (17.2) (5) | 11.5 (16.5) (4) | 6.5 (13.4) (5) | 6.77 (13.6) (2) | 4.92 (13.6) (2) | 3.15 (8.7) (1) |
| Record Max / Min T | 92 / 67 | 92 / 67 | 93 / 67 | 92 / 67 | 90 / 65 | 90 / 61 |
| Avg Snow (Snow Days) | 0 (0) | 0 (0) | 0 (0) | 0 (0) | 0 (0) | 0 (0) |

| | JULY | AUGUST | SEPTEMBER | OCTOBER | NOVEMBER | DECEMBER | ANNUAL |
|---|---|---|---|---|---|---|---|
| Avg Max / Min T | 86 / 68 | 86 / 68 | 86 / 69 | 87 / 70 | 88 / 71 | 88 / 72 | 88 / 70 |
| Avg Precip (Rain Days)(Thunder Days) | 2.6 (7.6) (1) | 1.85 (6) (0.3) | 2.28 (6.4) (1) | 3.35 (8.9) (1) | 6.46 (13.8) (2) | 11.85 (16.7) (3) | 74.4 (141) (27.3) |
| Record Max / Min T | 89 / 61 | 88 / 61 | 91 / 62 | 91 / 62 | 93 / 64 | 93 / 66 | 93 / 61 |
| Avg Snow (Snow Days) | 0 (0) | 0 (0) | 0 (0) | 0 (0) | 0 (0) | 0 (0) | 0 (0) |

**Map Location 314 — Solomon Is., Honiara — 0925S 16002E 10**

| | JANUARY | FEBRUARY | MARCH | APRIL | MAY | JUNE |
|---|---|---|---|---|---|---|
| Avg Max / Min T | 88 / 74 | 88 / 74 | 87 / 73 | 88 / 73 | 88 / 73 | 87 / 72 |
| Avg Precip (Rain Days)(Thunder Days) | 14.06 (16.3) (7.3) | 13.26 (16.1) (7.6) | 16.71 (20.5) (8.2) | 10.57 (14) (5.3) | 8.09 (12.5) (4.4) | 6.72 (10.8) (2.5) |
| Record Max / Min T | 94 / 68 | 95 / 68 | 93 / 69 | 93 / 69 | 93 / 68 | 92 / 67 |
| Avg Snow (Snow Days) | 0 (0) | 0 (0) | 0 (0) | 0 (0) | 0 (0) | 0 (0) |

| | JULY | AUGUST | SEPTEMBER | OCTOBER | NOVEMBER | DECEMBER | ANNUAL |
|---|---|---|---|---|---|---|---|
| Avg Max / Min T | 86 / 72 | 87 / 72 | 88 / 72 | 88 / 72 | 88 / 73 | 88 / 73 | 88 / 73 |
| Avg Precip (Rain Days)(Thunder Days) | 5.95 (10.4) (3.4) | 4.36 (9.2) (1.8) | 4.57 (10.2) (4.2) | 7.65 (10.9) (7.1) | 7.65 (12.7) (7.8) | 9.49 (14.2) (9.8) | 109.1 (157.8) (69.7) |
| Record Max / Min T | 92 / 67 | 94 / 66 | 95 / 65 | 93 / 67 | 94 / 65 | 94 / 65 | 95 / 65 |
| Avg Snow (Snow Days) | 0 (0) | 0 (0) | 0 (0) | 0 (0) | 0 (0) | 0 (0) | 0 (0) |

**Map Location 315 — Vanuatu, Port Vila — 1744S 16819E 66**

| | JANUARY | FEBRUARY | MARCH | APRIL | MAY | JUNE |
|---|---|---|---|---|---|---|
| Avg Max / Min T | 85 / 73 | 86 / 74 | 85 / 73 | 83 / 71 | 81 / 70 | 80 / 68 |
| Avg Precip (Rain Days)(Thunder Days) | 10.2 (15.7) (4.7) | 11.2 (16.3) (2.7) | 11.7 (16.3) (2.7) | 9.6 (15.3) (2) | 5.6 (12.7) (1.3) | 4.9 (11.4) (0.7) |
| Record Max / Min T | 90 / 66 | 91 / 66 | 91 / 66 | 89 / 65 | 90 / 60 | 87 / 56 |
| Avg Snow (Snow Days) | 0 (0) | 0 (0) | 0 (0) | 0 (0) | 0 (0) | 0 (0) |

| | JULY | AUGUST | SEPTEMBER | OCTOBER | NOVEMBER | DECEMBER | ANNUAL |
|---|---|---|---|---|---|---|---|
| Avg Max / Min T | 78 / 66 | 79 / 66 | 79 / 68 | 82 / 69 | 84 / 71 | 85 / 72 | 82 / 70 |
| Avg Precip (Rain Days)(Thunder Days) | 3.8 (9.8) (0.2) | 3.5 (9.3) (0.8) | 3.8 (9.8) (0.7) | 4.8 (11.6) (0.9) | 6.6 (14) (2.4) | 7.1 (13.2) (4.2) | 82.8 (155.4) (24.9) |
| Record Max / Min T | 85 / 55 | 87 / 55 | 88 / 58 | 90 / 60 | 91 / 62 | 90 / 64 | 91 / 55 |
| Avg Snow (Snow Days) | 0 (0) | 0 (0) | 0 (0) | 0 (0) | 0 (0) | 0 (0) | 0 (0) |

**Map Location 316 — Wake Island — 1916N 16638E 12**

| | JANUARY | FEBRUARY | MARCH | APRIL | MAY | JUNE |
|---|---|---|---|---|---|---|
| Avg Max / Min T | 82 / 73 | 82 / 72 | 83 / 73 | 83 / 73 | 84 / 75 | 87 / 77 |
| Avg Precip (Rain Days)(Thunder Days) | 0.98 (3.2) (0) | 1.14 (3.1) (0) | 1.79 (4.2) (0) | 1.92 (5.9) (0.2) | 1.54 (4.4) (0.1) | 2.71 (5.7) (0.9) |
| Record Max / Min T | 87 / 65 | 86 / 65 | 88 / 67 | 87 / 68 | 89 / 70 | 90 / 71 |
| Avg Snow (Snow Days) | 0 (0) | 0 (0) | 0 (0) | 0 (0) | 0 (0) | 0 (0) |

| | JULY | AUGUST | SEPTEMBER | OCTOBER | NOVEMBER | DECEMBER | ANNUAL |
|---|---|---|---|---|---|---|---|
| Avg Max / Min T | 87 / 77 | 88 / 77 | 88 / 78 | 86 / 77 | 85 / 76 | 83 / 74 | 85 / 75 |
| Avg Precip (Rain Days)(Thunder Days) | 3.78 (7.9) (0.9) | 5.61 (10.2) (1.4) | 4.66 (9.9) (1.5) | 5.54 (10.6) (2) | 3.14 (5.8) (0.2) | 1.66 (3.7) (0.2) | 34.5 (74.6) (7.4) |
| Record Max / Min T | 90 / 69 | 92 / 68 | 91 / 70 | 91 / 70 | 88 / 68 | 88 / 68 | 92 / 64 |
| Avg Snow (Snow Days) | 0 (0) | 0 (0) | 0 (0) | 0 (0) | 0 (0) | 0 (0) | 0 (0) |

# Data Key

| Map Locater Number | | MONTH | |
|---|---|---|---|
| Location | Average Max T °F | | Record Max T °F |
| Lat - Lon ddmm Elev ft | Average Min T °F | | Record Min T °F |
| | Avg. Precip in. (Rain Days) (Thunder Days) | | Avg. Snow in. (Snow Days) |

| *** or Blank | No data, missing data or data of insufficient reliability |
|---|---|
| Rain Days | Number of Days with > = 0.01" of precipitation falling |
| Snow Days | Number of Days with > = 1.5" of snow falling |

**Table 9.2  World Weather Data [CONTINUED]**

# Southern Ocean Islands

# Map # 317 - 321

## Map Location 317 — Deception Island — 6259S 06034W 26

| Month | Avg High | Rec High | Avg Low | Rec Low | Precip |
|---|---|---|---|---|---|
| JANUARY | 37 | 51 | 31 | 19 | 2.3 (***0) (***) |
| FEBRUARY | 37 | 48 | 31 | 20 | 2.1 (***0) (***) |
| MARCH | 35 | 46 | 29 | 12 | 2.7 (***) (***) |
| APRIL | 31 | 47 | 24 | 2 | 2 (***) (***) |
| MAY | 27 | 42 | 19 | -6 | 0.2 (***) (***) |
| JUNE | 23 | 39 | 14 | -16 | 0.3 (***) (***) |
| JULY | 21 | 40 | 11 | -13 | 0.6 (***) (***) |
| AUGUST | 22 | 41 | 12 | -18 | 1 (***) (***) |
| SEPTEMBER | 26 | 39 | 17 | -10 | 0.9 (***) (***) |
| OCTOBER | 31 | 45 | 23 | 2 | 4.3 (***) (***) |
| NOVEMBER | 32 | 42 | 25 | 7 | 3.8 (***) (***) |
| DECEMBER | 36 | 45 | 30 | 18 | 2 (***) (***) |
| ANNUAL | 30 | 51 | 22 | -18 | 22.2 (***) (***) |

## Map Location 318 — Heard Is., Australia — 5306S 07331E 65

| Month | Avg High | Rec High | Avg Low | Rec Low | Precip |
|---|---|---|---|---|---|
| JANUARY | 41 | 54 | 33 | 30 | 5.67 (***) (***) |
| FEBRUARY | 41 | 56 | 33 | 29 | 5.83 (***) (***) |
| MARCH | 40 | 55 | 32 | 27 | 5.51 (***) (***) |
| APRIL | 39 | 55 | 32 | 25 | 5.98 (***) (***) |
| MAY | 37 | 51 | 31 | 23 | 5.84 (***) (***) |
| JUNE | 35 | 45 | 27 | 19 | 3.82 (***) (***) |
| JULY | 34 | 52 | 26 | 18 | 3.54 (***) (***) |
| AUGUST | 34 | 52 | 27 | 14 | 2.17 (***) (***) |
| SEPTEMBER | 33 | 43 | 26 | 17 | 2.44 (***) (***) |
| OCTOBER | 34 | 45 | 28 | 18 | 3.66 (***) (***) |
| NOVEMBER | 36 | 44 | 30 | 24 | 3.94 (***) (***) |
| DECEMBER | 39 | 53 | 33 | 29 | 4.96 (***) (***) |
| ANNUAL | 37 | 52 | 29 | 17 | 53.4 (***) (***) |

## Map Location 319 — Macquarie Is., Aust. — 5430S 15857E 20

| Month | Avg High | Rec High | Avg Low | Rec Low | Precip |
|---|---|---|---|---|---|
| JANUARY | 47 | 52 | 40 | 35 | 2.95 (***) (***) |
| FEBRUARY | 47 | 51 | 41 | 33 | 3.84 (***) (***) |
| MARCH | 46 | 50 | 40 | 30 | 3.42 (***) (***) |
| APRIL | 45 | 49 | 39 | 31 | 3.01 (***) (***) |
| MAY | 42 | 50 | 36 | 20 | 2.46 (***) (***) |
| JUNE | 42 | 47 | 35 | 24 | 4.11 (***) (***) |
| JULY | 41 | 47 | 34 | 17 | 3.26 (***) (***) |
| AUGUST | 41 | 45 | 35 | 25 | 3.13 (***) (***) |
| SEPTEMBER | 42 | 46 | 35 | 22 | 2.66 (***) (***) |
| OCTOBER | 43 | 44 | 36 | 29 | 1.78 (***) (***) |
| NOVEMBER | 43 | 48 | 37 | 28 | 3.17 (***) (***) |
| DECEMBER | 46 | 52 | 39 | 17 | 1.67 (***) (***) |
| ANNUAL | 44 | 52 | 37 | 17 | 35.5 (***) (***) |

## Map Location 320 — Prince Edward Is., South Africa — 4653S 03752E 72

| Month | Avg High | Rec High | Avg Low | Rec Low | Precip |
|---|---|---|---|---|---|
| JANUARY | 51 | 62 | 41 | 34 | 8.6 (***) (***) |
| FEBRUARY | 52 | 62 | 42 | 34 | 7.7 (***) (***) |
| MARCH | 51 | 63 | 41 | 34 | 8.5 (***) (***) |
| APRIL | 49 | 59 | 39 | 32 | 8.1 (***) (***) |
| MAY | 46 | 56 | 37 | 30 | 9.1 (***) (***) |
| JUNE | 45 | 54 | 38 | 28 | 8.0 (***) (***) |
| JULY | 44 | 52 | 35 | 26 | 7.4 (***) (***) |
| AUGUST | 43 | 52 | 34 | 26 | 7.2 (***) (***) |
| SEPTEMBER | 44 | 53 | 35 | 26 | 6.7 (***) (***) |
| OCTOBER | 46 | 56 | 36 | 28 | 6.7 (***) (***) |
| NOVEMBER | 48 | 59 | 37 | 30 | 6.7 (***) (***) |
| DECEMBER | 50 | 60 | 39 | 32 | 8.0 (***) (***) |
| ANNUAL | 47 | 63 | 38 | 26 | 94.4 (***) (***) |

## Map Location 321 — South Sandwich Is., Gr. Br., South Thule Is. — 5927S 02719W 268

| Month | Avg High | Rec High | Avg Low | Rec Low | Precip |
|---|---|---|---|---|---|
| JANUARY | 38 | *** | 31 | *** | 4.45 (***) (***) |
| FEBRUARY | 36 | *** | 31 | *** | 4.76 (***) (***) |
| MARCH | 32 | *** | 28 | *** | 4.02 (***) (***) |
| APRIL | 31 | *** | 27 | *** | 3.50 (***) (***) |
| MAY | 25 | *** | 20 | *** | 3.62 (***) (***) |
| JUNE | 24 | *** | 11 | *** | 4.69 (***) (***) |
| JULY | 23 | *** | 15 | *** | 4.68 (***) (***) |
| AUGUST | 25 | *** | 12 | *** | 4.53 (***) (***) |
| SEPTEMBER | 26 | *** | 14 | *** | 5.47 (***) (***) |
| OCTOBER | 27 | *** | 17 | *** | 4.96 (***) (***) |
| NOVEMBER | 29 | *** | 37 | *** | *** (***) (***) |
| DECEMBER | 30 | *** | 22 | *** | 4.84 (***) (***) |
| ANNUAL | 44 | *** | 37 | *** | 56.37 (***) (***) |

**Table 9.2  World Weather Data [CONTINUED]**

# Southern Ocean Islands

# Map # 317 - 321

**Map Location 317 — Deception Island, 6259S 06034W 26**

| | JAN | FEB | MAR | APR | MAY | JUN | JUL | AUG | SEP | OCT | NOV | DEC | ANNUAL |
|---|---|---|---|---|---|---|---|---|---|---|---|---|---|
| Normal High | 37 | 37 | 35 | 31 | 27 | 23 | 21 | 22 | 26 | 31 | 32 | 36 | 30 |
| Normal Low | 31 | 31 | 29 | 24 | 19 | 14 | 11 | 12 | 17 | 23 | 25 | 30 | 22 |
| Record High | 51 | 48 | 46 | 47 | 42 | 39 | 40 | 41 | 39 | 45 | 42 | 45 | 51 |
| Record Low | 19 | 20 | 12 | 2 | -6 | -16 | -13 | -18 | -10 | 2 | 7 | 18 | -18 |
| Precip | 2.3 (***) (***) | 2.1 (***) (***) | 2.7 (***) (***) | 2 (***) (***) | 0.2 (***) (***) | 0.3 (***) (***) | 0.6 (***) (***) | 1 (***) (***) | 0.9 (***) (***) | 4.3 (***) (***) | 3.8 (***) (***) | 2 (***) (***) | 22.2 (***) (***) |

**Map Location 318 — Heard Is., Australia, 5306S 07331E 65**

| | JAN | FEB | MAR | APR | MAY | JUN | JUL | AUG | SEP | OCT | NOV | DEC | ANNUAL |
|---|---|---|---|---|---|---|---|---|---|---|---|---|---|
| Normal High | 41 | 41 | 40 | 39 | 37 | 35 | 34 | 34 | 33 | 34 | 36 | 39 | 44 |
| Normal Low | 33 | 33 | 32 | 32 | 31 | 27 | 26 | 27 | 26 | 28 | 30 | 33 | 37 |
| Record High | 54 | 56 | 55 | 55 | 51 | 45 | 48 | 52 | 43 | 45 | 44 | 53 | 52 |
| Record Low | 30 | 29 | 27 | 25 | 23 | 19 | 18 | 14 | 17 | 18 | 24 | 29 | 17 |
| Precip | 5.67 (***) (***) | 5.83 (***) (***) | 5.51 (***) (***) | 5.98 (***) (***) | 5.84 (***) (***) | 3.82 (***) (***) | 3.54 (***) (***) | 2.17 (***) (***) | 2.44 (***) (***) | 3.66 (***) (***) | 3.94 (***) (***) | 4.96 (***) (***) | 53.4 (***) (***) |

**Map Location 319 — Macquarie Is., Aust., 5430S 15857E 20**

| | JAN | FEB | MAR | APR | MAY | JUN | JUL | AUG | SEP | OCT | NOV | DEC | ANNUAL |
|---|---|---|---|---|---|---|---|---|---|---|---|---|---|
| Normal High | 47 | 47 | 46 | 45 | 42 | 42 | 41 | 41 | 42 | 43 | 43 | 46 | 44 |
| Normal Low | 40 | 41 | 40 | 39 | 36 | 35 | 35 | 35 | 35 | 36 | 37 | 39 | 37 |
| Record High | 52 | 51 | 50 | 49 | 50 | 47 | 47 | 45 | 46 | 44 | 48 | 44 | 52 |
| Record Low | 35 | 33 | 30 | 31 | 20 | 24 | 17 | 25 | 22 | 29 | 28 | 17 | 17 |
| Precip | 2.95 (***) (***) | 3.84 (***) (***) | 3.42 (***) (***) | 3.01 (***) (***) | 2.46 (***) (***) | 4.11 (***) (***) | 3.26 (***) (***) | 3.13 (***) (***) | 2.66 (***) (***) | 1.78 (***) (***) | 3.17 (***) (***) | 1.67 (***) (***) | 35.5 (***) (***) |

**Map Location 320 — Prince Edward Is., South Africa, 4653S 03752E 72**

| | JAN | FEB | MAR | APR | MAY | JUN | JUL | AUG | SEP | OCT | NOV | DEC | ANNUAL |
|---|---|---|---|---|---|---|---|---|---|---|---|---|---|
| Normal High | 51 | 52 | 51 | 49 | 46 | 45 | 44 | 43 | 44 | 46 | 48 | 50 | 47 |
| Normal Low | 41 | 42 | 41 | 39 | 37 | 38 | 35 | 34 | 35 | 36 | 37 | 39 | 38 |
| Record High | 62 | 62 | 63 | 59 | 56 | 54 | 52 | 53 | 53 | 56 | 59 | 60 | 60 |
| Record Low | 34 | 34 | 34 | 32 | 30 | 28 | 27 | 26 | 26 | 28 | 30 | 32 | 26 |
| Precip | 8.6 (***) (***) | 7.7 (***) (***) | 8.5 (***) (***) | 8.1 (***) (***) | 9.1 (***) (***) | 8.0 (***) (***) | 7.4 (***) (***) | 7.2 (***) (***) | 7.2 (***) (***) | 6.7 (***) (***) | 6.7 (***) (***) | 8.0 (***) (***) | 94.4 (***) (***) |

**Map Location 321 — South Sandwich Is., Gr. Br., South Thule Is., 5927S 02719W 268**

| | JAN | FEB | MAR | APR | MAY | JUN | JUL | AUG | SEP | OCT | NOV | DEC | ANNUAL |
|---|---|---|---|---|---|---|---|---|---|---|---|---|---|
| Normal High | 38 | 36 | 32 | 31 | 25 | 24 | 23 | 25 | 26 | 27 | 29 | 30 | 44 |
| Normal Low | 31 | 31 | 28 | 27 | 20 | 11 | 15 | 12 | 14 | 17 | 37 | 22 | 37 |
| Record High | *** | *** | *** | *** | *** | *** | *** | *** | *** | *** | *** | *** | *** |
| Record Low | *** | *** | *** | *** | *** | *** | *** | *** | *** | *** | *** | *** | *** |
| Precip | 4.45 (***) (***) | 4.76 (***) (***) | 4.02 (***) (***) | 3.50 (***) (***) | 3.62 (***) (***) | 4.69 (***) (***) | 4.68 (***) (***) | 4.53 (***) (***) | 5.47 (***) (***) | 4.96 (***) (***) | 6.85 (***) (***) | 4.84 (***) (***) | 56.37 (***) (***) |

# 10

# Local Climatological Data Annual Summaries 2009

## • INTRODUCTION

Local climatological data (LCD) are observed at principal meteorological stations by trained observers or automated instruments that go by the name automated surface observing systems (ASOS), the nation's primary surface weather-observing network.

Observing stations are located worldwide and operated by the National Weather Service (NWS), the US Air Force/Air Weather Service (AWS), the US Navy/National Oceanographic Command (NAVOCFANCOM), and the Federal Aviation Administration (FAA).

The majority of stations collect hourly observations and special (between-hour) observations when weather changes rapidly, is extreme, or is especially dangerous for aviation interests. The primary way most users now get their weather data is via the Internet. Mobile device access is becoming more and more important.

ASOS observe, format, and transmit the following basic weather elements:

- Sky conditions such as cloud height and cloud amount up to 12,000 feet
- Surface visibility up to at least 10 statute miles
- Basic current weather such as the type and intensity for rain, snow, and freezing rain
- Obstructions to visibility such as fog, haze, and/or dust
- Sea-level pressure and altimeter settings
- Air and dew point temperatures
- Wind direction, speed, and character (gusts and squalls)
- Precipitation accumulation
- Selected remarks including variable cloud height, variable visibility, precipitation beginning/ending times, rapid pressure changes, pressure change tendency, wind shift, and peak wind.

ASOS see only directly overhead and will miss nearby fog or a thunderstorm close by because their eyes see only directly overhead.

ASOS are not designed to report clouds above 12,000 feet, virga (precipitation evaporating before it gets to the surface), tornadoes, or funnel clouds. In addition, ice crystals, snow pellets, ice pellets, drizzle, and freezing drizzle are often just reported as generic precipitation and the data user have to determine from temperature and other conditions and reports what is falling from the sky.

Blowing obstructions to visibility such as snow, dust (or sand), snowfall, and snow depth are not reported. Many of the ASOS stations, with staffed air traffic control towers, are monitored, and human observers can edit or augment the automated observations. New sensors are being developed to measure some of the unreported weather elements.

## • READING LCD REPORTS FOR 128 US CITIES

LCD are collected, quality-controlled, and archived by the National Oceanic and Atmospheric Administration's National Climatic Data Center. In this chapter there are data for 128 stations. There is at least one station from each state, Washington, D.C. and Puerto Rico. The stations were chosen to achieve a nearly even distribution, represent the great climatic diversity of the United States, highlight popular destinations, and illustrate large differences over short distances, such as mountain stations versus low-elevation stations.

If a location you are looking for is not present, as long as elevation differences are not too great a reliable picture of your location's weather can be estimated by interpolating between nearby stations. When elevation differences are a

*The Weather Almanac: A Reference Guide to Weather, Climate, and Related Issues in the United States and Its Key Cities*, Twelfth Edition. Steve Horstmeyer.
© 2011 John Wiley & Sons, Inc. Published 2011 by John Wiley & Sons, Inc.

factor, you can estimate an average temperature for an elevated station by taking 3.5°F off the observation for every 1000 feet of elevation increase. This is not exact and can be greatly affected by local site characteristics, but it will give you an idea of the temperature. Differences are usually due to site characteristics, such as a vegetation-covered site versus a bare rocky site.

Stations are presented alphabetically by state and alphabetically by station name within each state where there are multiple stations. The data for each location consist of five pages: (1) a climatological narrative report and a state location index map; (2) a table of normals, means, and extremes; (3) tables of monthly and annual precipitation and average temperature for up to the last 30 years; (4) tables of heating and cooling degree days, which are the monthly deviations of average temperature from 65°F for up to the last 30 years; and (5) a table of snowfall for the last 30 years followed by two graphs for 2009: one of daily maximum, minimum, and average temperatures, and normal maximum and minimum temperatures; and a second of daily liquid-equivalent precipitation.

### Narrative Report

The narrative report is prepared by a local climatologist or National Weather Service meteorologist, and it covers local-area climate. It describes the large-scale setting and local terrain and how both affect temperature, rainfall, snowfall, and other weather factors.

For example, an airport in a valley near a major river may report fog many mornings, while 5 miles away at an airport in an upland setting, visibility is unobstructed. Mountains, swamps, and even plowed fields influence air masses as they move toward a city.

### Normals, Means, and Extremes Page

Climate statistics are numbers representing the probable weather over a sufficiently long period of time. While these statistics contain no information about how much the weather can vary on a daily basis, the *Normals, Means, and Extremes* page paints a picture of the annual cycle of temperature and precipitation.

*Page contents.*
    Station name and four-letter airport code
    Station location
    Precise latitude, longitude, and altitude

WBAN. Throughout each station's data pages, you will see "WBAN" followed by a five-digit number. WBAN is an acronym for **W**eather-**B**ureau-**A**rmy-**N**avy and was the first organization scheme for weather-reporting stations. It was developed in the 1950s as aircraft use rapidly increased and there was a need to better organize and access data. The WBAN system also included standardized forms for data. The WBAN identifier system has been superseded by the World Meteorological Organization station identifier numbers.

Both the International Civil Aviation Organization (ICAO) and the International Air Transport Association (IATA) also have station codes. The ICAO identifier is a four-letter sequence, while the IATA is a three-letter sequence. In addition, the US FAA uses three-letter identifiers that in almost all cases are identical to the IATA identifier.

The ICAO identifier for Los Angeles International Airport is KLAX. The "K" indicates LAX is in the contiguous United States. The IATA identifier is "LAX." CYYC is the ICAO identifier for Calgary, Alberta, Canada. The "C" indicates Canada and the IATA identifier is "YYC."

### Temperature Data

"Mean" temperature data are simply the arithmetic average of the observations, while "normal" temperature data have been fit to a curve called a spline. "Mean" data are for the entire length of time weather observations have been made at a particular station. Normal data are based on a 30-year period of record (POR). Currently the POR for climatic normals is the 30 years from 1971 through 2000.

In a statistical sense, 30 observations is a small sample and a single extreme value could severely skew the seasonal trend. For example, an extremely cold March day could reverse the seasonal warming trend in spring. To prevent this, sophisticated curve fitting adjusts "normal" temperatures so that there is a consistent seasonal progression to warmer temperatures into the summer and to colder temperatures into the winter.

Most of the temperature data are self-explanatory. The "mean of extreme maximums" is the average of the warmest monthly high temperatures, and the "mean of extreme minimums" is the average of the coldest monthly low temperatures for the POR. Note the POR is usually the number of years observations have been made at a location, not the 30-year POR used for calculation of climatic normals.

Dry-bulb temperature is simply the ambient air temperature, while the wet-bulb temperature is a measure of humidity.

Before electronic instruments, observation locations included two thermometers. The bulb, or fluid reservoir of the second, was covered by a "sock" wet with pure distilled water. Evaporation of distilled water from the sock would cool the "wet bulb," and how much cooler the wet-bulb temperature was than the dry-bulb temperature indicated the humidity.

Wet bulb is the temperature achieved by evaporating moisture into the air until the relative humidity is 100%. It is calculated from measured temperature and measured dew point temperature. At ASOS stations dew point is measured with a chilled mirror device that optically senses when condensation has taken place on the cooled mirror.

### Heating and Cooling Degree Days Data

Think of degree days as a measure of the demand for heating or cooling. The terms are confusing unless you think of degree days as "units" of heating or cooling. The normal number of heating degree days for each month is the sum total of the daily deviation of normal temperature from 65°F for the given month and year for the period 1971–2000. A normal daily temperature of 75°F would yield 10 cooling degree days, while a normal daily temperature of 30°F would yield 35 heating degree days.

To prevent cooling and heating degree days from canceling each other out, they are summed separately. If the average temperature is below 65°F, the cooling degree days are recorded as zero, and if warmer than 65°F, the heating degree days are recorded as zero.

### Relative Humidity Data

Given in percent the monthly normal and the 1:00 A.M. LST, 7:00 A.M. LST, 1:00 P.M. LST, and 7:00 P.M. LST values.

### Sunshine Data

The average percentage of daytime hours subject to direct radiation from the sun at the present site. The percentage is given without regard for the intensity of sunshine. That is, thin clouds, light haze, or other minor obstructions to direct solar rays may be present, but would not mitigate the full counting of an hour. Blanks indicate that a given station does not record sunshine data.

### Thunderstorm and Fog Days

*Cloud data.* ASOS stations—cloudiness is based on time-averaged ceilometer data for clouds at or below 12,000 feet and on satellite data for clouds above 12,000 feet. Cloud data from human observers are based on visual sighting by the observer. Average amount of daytime sky and 24-hour sky obscured by any type of cover expressed in oktas (eighths of the sky covered).

Daytime sky coverage expressed as clear (0–2 oktas of the sky covered by clouds), partly cloudy (3–6 oktas of the sky covered by clouds), and cloudy (7–8 oktas of the sky covered by clouds).

*Pressure data.* The average station pressure (observed pressure at a station's altitude) and average pressure adjusted to sea level in inches of mercury.

*Wind data.* Prevailing wind direction is the direction that is observed most often.

Maximum 2-minute wind speed is a measure of gustiness, the velocity sustained for a 2-minute period.

Maximum 5-second wind speed is also a measure of gustiness but of a shorter duration, the velocity sustained for a 5-second period.

Note the 5-second value is considerably higher than the 2-minute value, and both exceed the mean.

Wind direction is expressed as degrees from true north, numerically 09 = 90° (wind from the east), 18 = 180°

(wind from the south), 27 = 270° (wind from the west), and 36 = 360° (wind from the north). Calm is recorded as "00."

*Precipitation data.* Precipitation in inches of water equivalent (frozen forms are melted and measured) for each month and year during the period 1971–2000.

### Snowfall

Values include sleet and all other frozen forms of precipitation and are similar to those for total precipitation. The values are expressed in inches of actual snow or ice fall.

## Special Symbols and Conventions

Symbols that appear on many of the individual summaries: ( ) "blank" indicates missing or unreported data; (+) indicates the value occurs also on earlier dates; (* or 0.*) the value or mean-days-with is between 0.00 and 0.05; (T) trace, an amount too small to measure; (−) temperatures below zero are preceded by a minus sign.

## • OTHER DATA PAGES

The data are observations of average monthly and annual precipitation, heating and cooling degree days, and snowfall. All are self-explanatory.

Figure 10.1 shows the key to the temperature graphs for 2009 for each station; precipitation graphs are

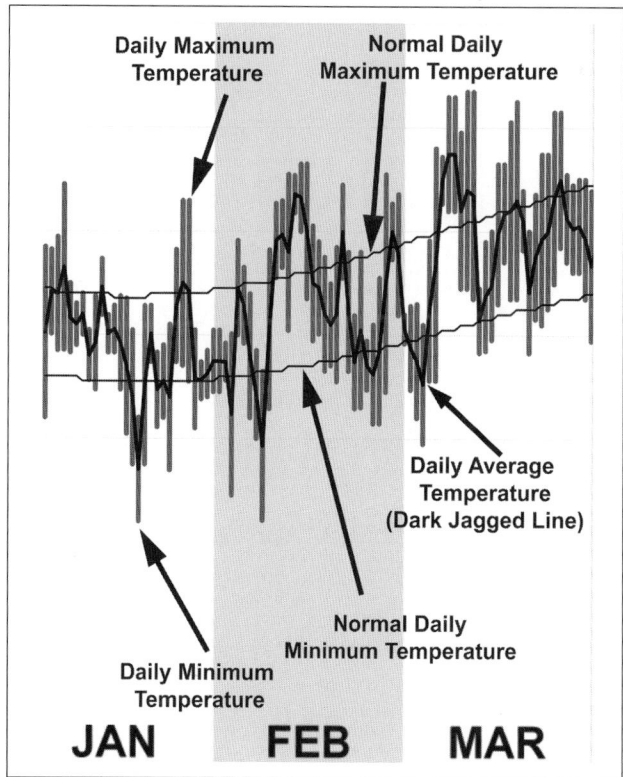

**Figure 10.1**   Key to graphs of 2009 weather data.

self-explanatory. By careful use of the temperature graphs the user can estimate the daily high temperature, daily low temperature, daily temperature range, average daily temperature, normal high temperature, normal low temperature, and deviation of high, low, and average temperatures from normal for each day of 2009.

Note that the graphs, with a few exceptions to allow for extreme temperatures, are plotted on the same scale (from −30°F to 110°F) so that they can be compared directly. For example, compare Key West, FL, to International Falls, MN. Not only is the entire graph for Key West higher on the scale, but the range is much smaller than International Falls.

You can directly compare the graphs of Barrow, AK, and Key West, FL, but the graph of Fairbanks, AK, uses a different scale (−40°F to 100°F) because of lower winter temperatures and so direct comparison cannot be made.

# 2009
# BIRMINGHAM (MUNICIPAL AIRPORT)
# ALABAMA (KBHM)

Birmingham is located in a hilly area of north-central Alabama in the foothills of the Appalachians about 300 miles inland from the Gulf of Mexico. There is a series of southwest to northeast valleys and ridges in the area.

The city is far enough inland to be protected from destructive tropical hurricanes, yet close enough that the Gulf has a pronounced modifying effect on the climate.

Although summers are long and hot, they are not generally excessively hot. On a typical mid-summer day, the temperature will be nearly 70 degrees at daybreak, approach 90 degrees at mid-day, and level off in the low 90s during the afternoon. It is not unusual for the temperature to remain below 100 degrees for several years in a row. However, every few years an extended heat wave will bring temperatures over 100 degrees. July is normally the hottest month but there is little difference from mid-June to mid-August. Rather persistent high humidity adds to the summer discomfort.

January is normally the coldest month but there is not much difference from mid-December to mid-February. Overall, winters are relatively mild. Even in cold spells, it is unusual for the temperature to remain below freezing all day. Sub-zero cold is extremely rare, occurring only a very few times this century. Extremely low temperatures almost always occur under clear skies after a snowfall.

Snowfall is erratic. Sometimes there is a two- or three-year span with no measurable snow. On rare occasions, there may be a 2 to 4 inch snowstorm. The snow usually melts quickly. Even 1 or 2 inches of snow can effectively shut down this sunbelt city because of the hilly terrain, the wetness of the snow and the unfamiliarity of motorists driving on snow and ice.

Birmingham is blessed with abundant rainfall. It is fairly well distributed throughout the year. However, some of the wetter winter months, plus March and July, have twice the rainfall of October, the driest month. Summer rainfall is almost entirely from scattered afternoon and early evening thunderstorms. Serious droughts are rare and most dry spells are not severe.

The stormiest time of the year with the greatest risk of severe thunderstorms and tornadoes is in spring, especially in March and April.

In a normal year, the last 32 degree minimum temperature in the spring is in mid to late March and the first in autumn is in early November.

# NORMALS, MEANS, AND EXTREMES
## BIRMINGHAM (KBHM)

LATITUDE: 33° 33'N   LONGITUDE: -86° 45'W   ELEVATION (FT): GRND: 615   BARO: 630   TIME ZONE: CENTRAL (UTC -6)   WBAN: 13876

| | ELEMENT | POR | JAN | FEB | MAR | APR | MAY | JUN | JUL | AUG | SEP | OCT | NOV | DEC | YEAR |
|---|---|---|---|---|---|---|---|---|---|---|---|---|---|---|---|
| **TEMPERATURE °F** | NORMAL DAILY MAXIMUM | 30 | 52.8 | 58.3 | 66.5 | 74.1 | 81.0 | 87.5 | 90.6 | 90.2 | 84.6 | 74.9 | 64.5 | 56.0 | 73.4 |
| | MEAN DAILY MAXIMUM | 80 | 53.9 | 56.9 | 65.8 | 74.0 | 81.7 | 87.5 | 90.4 | 90.1 | 84.4 | 75.5 | 63.9 | 55.9 | 73.3 |
| | HIGHEST DAILY MAXIMUM | 66 | 81 | 83 | 89 | 92 | 99 | 102 | 106 | 105 | 100 | 94 | 85 | 80 | 106 |
| | YEAR OF OCCURRENCE | | 1949 | 1996 | 1982 | 1987 | 1962 | 1954 | 1980 | 2007 | 1990 | 1954 | 2003 | 1951 | JUL 1980 |
| | MEAN OF EXTREME MAXS. | 80 | 71.2 | 74.9 | 81.6 | 86.2 | 90.5 | 95.2 | 97.2 | 96.3 | 93.8 | 86.4 | 78.7 | 72.1 | 85.3 |
| | NORMAL DAILY MINIMUM | 30 | 32.3 | 35.4 | 42.4 | 48.4 | 57.6 | 65.4 | 69.7 | 68.9 | 63.0 | 50.9 | 41.8 | 35.2 | 50.9 |
| | MEAN DAILY MINIMUM | 80 | 34.3 | 35.9 | 43.1 | 50.3 | 59.1 | 66.2 | 70.3 | 69.6 | 63.4 | 52.1 | 41.7 | 36.1 | 51.8 |
| | LOWEST DAILY MINIMUM | 66 | -6 | 3 | 2 | 26 | 35 | 42 | 51 | 51 | 37 | 27 | 5 | 1 | -6 |
| | YEAR OF OCCURRENCE | | 1985 | 1958 | 1993 | 1973 | 1944 | 1966 | 1967 | 1946 | 1967 | 1956 | 1950 | 1989 | JAN 1985 |
| | MEAN OF EXTREME MINS. | 80 | 14.3 | 18.3 | 24.9 | 33.1 | 43.5 | 54.5 | 62.0 | 60.7 | 48.5 | 34.5 | 24.6 | 17.8 | 36.4 |
| | NORMAL DRY BULB | 30 | 42.6 | 46.8 | 54.5 | 61.3 | 69.3 | 76.4 | 80.2 | 79.6 | 73.8 | 62.9 | 53.1 | 45.6 | 62.2 |
| | MEAN DRY BULB | 80 | 44.2 | 46.4 | 54.5 | 62.2 | 70.4 | 76.9 | 80.4 | 79.8 | 73.9 | 63.8 | 52.8 | 46.1 | 62.6 |
| | MEAN WET BULB | 26 | 38.5 | 41.3 | 47.1 | 54.0 | 62.6 | 68.8 | 72.0 | 71.2 | 66.1 | 56.4 | 47.9 | 40.8 | 55.6 |
| | MEAN DEW POINT | 26 | 34.6 | 37.4 | 42.6 | 50.1 | 59.9 | 66.9 | 70.3 | 69.3 | 63.5 | 53.4 | 44.5 | 37.0 | 52.5 |
| | NORMAL NO. DAYS WITH: | | | | | | | | | | | | | | |
| | MAXIMUM >= 90 | 30 | 0.0 | 0.0 | 0.0 | 0.1 | 1.9 | 11.3 | 19.0 | 17.7 | 7.4 | 0.1 | 0.0 | 0.0 | 57.5 |
| | MAXIMUM <= 32 | 30 | 1.4 | 0.6 | 0.1 | 0.0 | 0.0 | 0.0 | 0.0 | 0.0 | 0.0 | 0.0 | 0.0 | 0.6 | 2.7 |
| | MINIMUM <= 32 | 30 | 16.4 | 11.5 | 5.3 | 1.2 | 0.0 | 0.0 | 0.0 | 0.0 | 0.0 | 0.0 | 6.3 | 14.0 | 55.0 |
| | MINIMUM <= 0 | 30 | 0.1 | 0.0 | 0.0 | 0.0 | 0.0 | 0.0 | 0.0 | 0.0 | 0.0 | 0.0 | 0.0 | 0.0 | 0.1 |
| **H/C** | NORMAL HEATING DEG. DAYS | 30 | 691 | 514 | 339 | 154 | 31 | 1 | 0 | 0 | 11 | 133 | 359 | 590 | 2823 |
| | NORMAL COOLING DEG. DAYS | 30 | 1 | 3 | 16 | 51 | 167 | 351 | 476 | 455 | 280 | 69 | 9 | 3 | 1881 |
| **RH** | NORMAL (PERCENT) | 30 | 70 | 68 | 64 | 66 | 71 | 73 | 74 | 73 | 73 | 71 | 72 | 71 | 71 |
| | HOUR 00 LST | 30 | 77 | 74 | 73 | 78 | 84 | 85 | 86 | 86 | 84 | 83 | 80 | 77 | 81 |
| | HOUR 06 LST | 30 | 81 | 80 | 80 | 84 | 87 | 87 | 87 | 89 | 90 | 88 | 87 | 84 | 85 |
| | HOUR 12 LST | 30 | 61 | 56 | 53 | 50 | 56 | 57 | 59 | 57 | 57 | 54 | 57 | 60 | 56 |
| | HOUR 18 LST | 30 | 66 | 59 | 53 | 51 | 59 | 61 | 64 | 64 | 67 | 70 | 69 | 68 | 63 |
| **S** | PERCENT POSSIBLE SUNSHINE | 34 | 42 | 50 | 55 | 63 | 66 | 65 | 59 | 63 | 61 | 66 | 55 | 46 | 58 |
| **W/O** | MEAN NO. DAYS WITH: | | | | | | | | | | | | | | |
| | HEAVY FOG(VISBY <= 1/4 MI) | 46 | 1.0 | 0.6 | 0.7 | 0.4 | 0.3 | 0.5 | 0.4 | 0.4 | 0.4 | 0.5 | 0.8 | 1.0 | 7.0 |
| | THUNDERSTORMS | 62 | 1.8 | 2.4 | 4.5 | 5.0 | 7.1 | 8.9 | 11.8 | 8.8 | 4.0 | 1.4 | 2.0 | 1.3 | 59.0 |
| **CLOUDNESS** | MEAN: | | | | | | | | | | | | | | |
| | SUNRISE-SUNSET (OKTAS) | | | | | | | | | | | | | | |
| | MIDNIGHT-MIDNIGHT (OKTAS) | | | | | | | | | | | | | | |
| | MEAN NO. DAYS WITH: | | | | | | | | | | | | | | |
| | CLEAR | | | | | | | | | | | | | | |
| | PARTLY CLOUDY | | | | | | | | | | | | | | |
| | CLOUDY | | | | | | | | | | | | | | |
| **PR** | MEAN STATION PRESSURE(IN) | 26 | 29.48 | 29.45 | 29.39 | 29.36 | 29.35 | 29.35 | 29.37 | 29.36 | 29.38 | 29.43 | 29.45 | 29.48 | 29.40 |
| | MEAN SEA-LEVEL PRES. (IN) | 26 | 30.17 | 30.13 | 30.07 | 30.02 | 30.01 | 30.00 | 30.03 | 30.02 | 30.03 | 30.10 | 30.14 | 30.17 | 30.07 |
| **WINDS** | MEAN SPEED (MPH) | 26 | 7.0 | 7.4 | 7.7 | 7.3 | 6.3 | 5.5 | 5.3 | 5.0 | 5.7 | 5.6 | 6.1 | 6.6 | 6.3 |
| | PREVAIL.DIR(TENS OF DEGS) | 32 | 36 | 36 | 36 | 19 | 24 | 16 | 28 | 10 | 09 | 36 | 36 | 36 | 09 |
| | MAXIMUM 2-MINUTE: | | | | | | | | | | | | | | |
| | SPEED (MPH) | 11 | 44 | 51 | 41 | 39 | 35 | 40 | 53 | 38 | 39 | 30 | 38 | 20 | 53 |
| | DIR. (TENS OF DEGS) | | 28 | 24 | 28 | 29 | 03 | 01 | 23 | 03 | 07 | 15 | 26 | 09 | 23 |
| | YEAR OF OCCURRENCE | | 2008 | 2001 | 2006 | 2006 | 2005 | 2006 | 2008 | 2000 | 2004 | 2006 | 2003 | 2009 | JUL 2008 |
| | MAXIMUM 3-SECOND | | | | | | | | | | | | | | |
| | SPEED (MPH) | 11 | 55 | 68 | 55 | 54 | 46 | 56 | 72 | 53 | 54 | 41 | 49 | 53 | 72 |
| | DIR. (TENS OF DEGS) | | 29 | 24 | 23 | 29 | 28 | 31 | 26 | 32 | 07 | 14 | 28 | 19 | 26 |
| | YEAR OF OCCURRENCE | | 2008 | 2001 | 2009 | 2006 | 2002 | 2009 | 2008 | 2007 | 2004 | 2006 | 2001 | 2009 | JUL 2008 |
| **PRECIPITATION** | NORMAL (IN) | 30 | 5.45 | 4.21 | 6.10 | 4.67 | 4.83 | 3.78 | 5.09 | 3.48 | 4.05 | 3.23 | 4.63 | 4.47 | 53.99 |
| | MAXIMUM MONTHLY (IN) | 66 | 11.00 | 17.67 | 15.80 | 13.75 | 17.22 | 9.04 | 13.70 | 10.85 | 10.96 | 11.90 | 15.25 | 13.98 | 17.67 |
| | YEAR OF OCCURRENCE | | 1949 | 1961 | 1980 | 1979 | 2003 | 1999 | 1950 | 1967 | 2004 | 1995 | 1948 | 1961 | FEB 1961 |
| | MINIMUM MONTHLY (IN) | 66 | 1.09 | 1.20 | 1.02 | 0.42 | 0.88 | 0.67 | 0.30 | 0.38 | T | 0.07 | 0.42 | 0.81 | T |
| | YEAR OF OCCURRENCE | | 1981 | 1968 | 2007 | 1986 | 2000 | 1968 | 1983 | 1989 | 1955 | 1991 | 1949 | 1980 | SEP 1955 |
| | MAXIMUM IN 24 HOURS (IN) | 66 | 5.81 | 6.57 | 7.05 | 5.08 | 5.78 | 3.85 | 5.47 | 5.13 | 9.75 | 6.94 | 4.87 | 5.29 | 9.75 |
| | YEAR OF OCCURRENCE | | 1949 | 1961 | 1970 | 1966 | 2003 | 1957 | 1985 | 1952 | 2004 | 1995 | 1948 | 1961 | SEP 2004 |
| | NORMAL NO. DAYS WITH: | | | | | | | | | | | | | | |
| | PRECIPITATION >= 0.01 | 30 | 11.4 | 9.5 | 11.0 | 9.1 | 10.6 | 10.4 | 12.1 | 9.3 | 8.0 | 6.5 | 9.4 | 10.5 | 117.8 |
| | PRECIPITATION >= 1.00 | 30 | 1.6 | 1.2 | 1.9 | 1.5 | 1.7 | 0.9 | 1.4 | 1.0 | 1.5 | 0.9 | 1.6 | 1.3 | 16.5 |
| **SNOWFALL** | NORMAL (IN) | 30 | 0.7 | 0.1 | 0.6 | 0.2 | 0.0 | 0.0 | 0.0 | 0.0 | 0.0 | 0.* | 0.* | 0.1 | 1.7 |
| | MAXIMUM MONTHLY (IN) | 66 | 6.6 | 2.3 | 13.0 | 5.0 | T | T | T | 0.0 | 0.0 | T | 1.4 | 8.0 | 13.0 |
| | YEAR OF OCCURRENCE | | 1982 | 1960 | 1993 | 1987 | 1996 | 2006 | 1994 | | | 1992 | 1993 | 1950 | 1963 | MAR 1993 |
| | MAXIMUM IN 24 HOURS (IN) | 62 | 4.5 | 2.3 | 13.0 | 5.0 | T | T | T | 0.0 | 0.0 | T | 1.4 | 8.4 | 13.0 |
| | YEAR OF OCCURRENCE' | | 1948 | 1960 | 1993 | 1987 | 1996 | 1992 | 1994 | | | 1992 | 1993 | 1950 | 1963 | MAR 1993 |
| | MAXIMUM SNOW DEPTH (IN) | 62 | 8 | 2 | 13 | 5 | 0 | 0 | 0 | 0 | 0 | 0 | 1 | 11 | 13 |
| | YEAR OF OCCURRENCE | | 1964 | 1960 | 1993 | 1987 | | | | | | | 1950 | 1958 | MAR 1993 |
| | NORMAL NO. DAYS WITH: | | | | | | | | | | | | | | |
| | SNOWFALL >= 1.0 | 30 | 0.3 | 0.0 | 0.1 | 0.0 | 0.0 | 0.0 | 0.0 | 0.0 | 0.0 | 0.0 | 0.0 | 0.0 | 0.4 |

## PRECIPITATION (inches) 2009 BIRMINGHAM (KBHM)

| YEAR | JAN | FEB | MAR | APR | MAY | JUN | JUL | AUG | SEP | OCT | NOV | DEC | ANNUAL |
|------|------|------|------|------|------|------|------|------|------|------|------|------|--------|
| 1980 | 6.63 | 2.36 | 15.80 | 9.10 | 7.30 | 3.01 | 2.11 | 2.84 | 5.26 | 3.21 | 3.04 | 0.81 | 61.47 |
| 1981 | 1.09 | 4.87 | 7.23 | 2.45 | 2.81 | 2.49 | 3.88 | 5.30 | 0.93 | 3.34 | 1.67 | 5.82 | 41.88 |
| 1982 | 5.19 | 6.29 | 2.71 | 7.86 | 3.19 | 5.39 | 3.53 | 2.68 | 0.66 | 3.73 | 7.11 | 9.51 | 57.85 |
| 1983 | 3.26 | 6.42 | 5.06 | 8.28 | 9.57 | 3.81 | 0.30 | 0.99 | 2.83 | 3.67 | 9.14 | 12.63 | 65.96 |
| 1984 | 3.96 | 2.79 | 4.07 | 8.61 | 6.07 | 1.41 | 5.06 | 3.86 | 0.16 | 3.91 | 5.42 | 2.30 | 47.62 |
| 1985 | 5.22 | 5.79 | 1.71 | 2.86 | 4.36 | 5.34 | 10.07 | 4.07 | 1.97 | 4.12 | 2.62 | 2.54 | 50.67 |
| 1986 | 1.21 | 1.79 | 2.45 | 0.42 | 3.66 | 3.87 | 1.61 | 5.56 | 2.52 | 5.24 | 9.66 | 3.08 | 41.07 |
| 1987 | 5.89 | 5.82 | 4.77 | 1.03 | 6.03 | 4.59 | 2.30 | 3.96 | 3.52 | 1.16 | 3.17 | 3.08 | 45.32 |
| 1988 | 5.55 | 2.52 | 3.18 | 3.18 | 1.22 | 0.79 | 2.95 | 3.43 | 8.57 | 3.41 | 6.33 | 2.84 | 43.97 |
| 1989 | 4.76 | 4.31 | 5.70 | 3.40 | 3.82 | 8.00 | 6.42 | 0.38 | 7.38 | 1.52 | 4.63 | 3.39 | 53.71 |
| 1990 | 7.38 | 7.43 | 5.81 | 2.38 | 4.12 | 2.08 | 3.16 | 0.59 | 2.04 | 2.98 | 4.02 | 5.47 | 47.46 |
| 1991 | 3.19 | 4.27 | 5.42 | 4.87 | 8.90 | 7.52 | 4.00 | 4.59 | 3.02 | 0.07 | 4.00 | 3.64 | 53.49 |
| 1992 | 3.22 | 3.96 | 3.36 | 2.61 | 1.18 | 5.34 | 7.41 | 7.43 | 5.70 | 2.18 | 7.94 | 5.27 | 55.60 |
| 1993 | 6.11 | 2.35 | 4.40 | 2.99 | 3.93 | 1.47 | 1.16 | 2.72 | 4.62 | 3.22 | 2.22 | 4.01 | 39.20 |
| 1994 | 5.10 | 4.78 | 7.56 | 3.77 | 3.74 | 5.41 | 7.75 | 4.05 | 4.67 | 5.39 | 3.93 | 4.10 | 60.25 |
| 1995 | 3.85 | 4.37 | 3.63 | 4.42 | 3.15 | 3.07 | 1.81 | 1.51 | 5.53 | 11.90 | 6.97 | 4.91 | 55.12 |
| 1996 | 9.59 | 3.09 | 10.59 | 2.70 | 5.16 | 2.17 | 8.50 | 4.45 | 5.83 | 3.55 | 4.00 | 3.28 | 62.91 |
| 1997 | 6.31 | 4.82 | 3.23 | 5.08 | 4.41 | 5.55 | 5.99 | 2.74 | 4.00 | 5.49 | 3.74 | 4.13 | 55.49 |
| 1998 | 8.06 | 8.52 | 6.36 | 7.99 | 4.03 | 3.25 | 7.75 | 8.98 | 0.52 | 1.17 | 4.37 | 6.27 | 67.27 |
| 1999 | 8.63 | 2.34 | 6.75 | 2.03 | 5.37 | 9.04 | 3.13 | 0.81 | 0.65 | 4.16 | 2.73 | 3.13 | 48.77 |
| 2000 | 5.72 | 2.17 | 10.67 | 8.19 | 0.88 | 2.89 | 4.62 | 2.20 | 1.66 | 1.26 | 8.14 | 1.84 | 50.24 |
| 2001 | 5.23 | 4.35 | 8.43 | 7.30 | 5.25 | 7.53 | 3.61 | 7.38 | 6.26 | 2.43 | 4.20 | 4.76 | 66.73 |
| 2002 | 6.02 | 2.88 | 6.47 | 3.02 | 4.15 | 5.27 | 7.72 | 1.60 | 9.94 | 5.46 | 4.65 | 7.23 | 64.41 |
| 2003 | 2.22 | 5.80 | 4.29 | 4.39 | 17.22 | 6.63 | 5.89 | 9.48 | 2.65 | 0.49 | 3.26 | 3.26 | 65.58 |
| 2004 | 2.77 | 5.92 | 3.14 | 3.24 | 5.04 | 7.01 | 3.29 | 2.66 | 10.96 | 2.61 | 11.13 | 3.55 | 61.32 |
| 2005 | 1.93 | 4.06 | 5.84 | 5.23 | 5.60 | 4.91 | 9.49 | 1.68 | 1.73 | 0.49 | 3.62 | 4.62 | 49.20 |
| 2006 | 5.97 | 8.70 | 4.72 | 7.81 | 2.53 | 4.77 | 5.03 | 2.69 | 3.74 | 4.97 | 2.60 | 3.03 | 56.56 |
| 2007 | 3.00 | 2.56 | 1.02 | 2.62 | 1.08 | 1.63 | 3.43 | 4.88 | 3.30 | 1.71 | 1.69 | 1.94 | 28.86 |
| 2008 | 4.47 | 4.66 | 4.59 | 5.24 | 7.98 | 3.97 | 5.05 | 7.90 | 0.40 | 1.94 | 2.49 | 6.40 | 55.09 |
| 2009 | 6.28 | 5.07 | 6.82 | 2.12 | 6.26 | 3.03 | 7.59 | 4.49 | 10.69 | 7.88 | 5.33 | 6.10 | 71.66 |
| POR= 80 YRS | 5.08 | 4.85 | 6.02 | 4.61 | 4.38 | 3.95 | 5.19 | 4.18 | 3.86 | 3.00 | 4.11 | 4.81 | 54.04 |

WBAN : 13876

## AVERAGE TEMPERATURE (°F) 2009 BIRMINGHAM (KBHM)

| YEAR | JAN | FEB | MAR | APR | MAY | JUN | JUL | AUG | SEP | OCT | NOV | DEC | ANNUAL |
|------|------|------|------|------|------|------|------|------|------|------|------|------|--------|
| 1980 | 45.2 | 42.3 | 51.4 | 60.9 | 70.0 | 77.6 | 84.4 | 83.0 | 78.4 | 60.3 | 52.4 | 44.0 | 62.5 |
| 1981 | 39.2 | 48.2 | 51.3 | 67.1 | 67.8 | 80.6 | 82.2 | 79.2 | 71.9 | 61.8 | 54.6 | 42.7 | 62.2 |
| 1982 | 41.9 | 47.3 | 58.9 | 59.1 | 72.6 | 75.7 | 80.7 | 79.6 | 73.1 | 64.2 | 54.1 | 51.2 | 63.2 |
| 1983 | 40.6 | 44.8 | 50.8 | 56.6 | 67.3 | 74.1 | 80.3 | 81.7 | 71.7 | 63.0 | 51.9 | 39.8 | 60.2 |
| 1984 | 38.5 | 46.8 | 52.4 | 59.0 | 67.5 | 77.1 | 78.2 | 77.5 | 72.1 | 71.1 | 51.4 | 54.3 | 62.2 |
| 1985 | 35.5 | 43.2 | 56.7 | 63.7 | 69.5 | 76.2 | 78.1 | 78.2 | 72.0 | 67.3 | 61.0 | 40.2 | 61.8 |
| 1986 | 41.9 | 49.9 | 55.5 | 61.8 | 71.2 | 78.6 | 82.5 | 77.8 | 76.9 | 63.9 | 57.4 | 44.0 | 63.5 |
| 1987 | 42.1 | 46.9 | 54.6 | 59.5 | 73.8 | 76.7 | 80.5 | 82.0 | 73.1 | 56.8 | 54.5 | 49.1 | 62.5 |
| 1988 | 39.7 | 43.5 | 54.1 | 61.2 | 67.6 | 77.5 | 79.8 | 81.5 | 74.4 | 57.7 | 55.2 | 45.8 | 61.5 |
| 1989 | 48.9 | 45.7 | 56.4 | 60.3 | 67.7 | 75.5 | 79.2 | 79.5 | 72.5 | 61.9 | 52.9 | 38.0 | 61.5 |
| 1990 | 49.0 | 54.5 | 57.0 | 60.7 | 68.9 | 77.8 | 79.8 | 82.3 | 77.5 | 63.9 | 55.7 | 49.7 | 64.7 |
| 1991 | 44.8 | 49.7 | 55.8 | 65.7 | 73.1 | 76.9 | 81.4 | 80.8 | 75.4 | 65.1 | 50.0 | 49.3 | 64.0 |
| 1992 | 43.4 | 50.6 | 53.8 | 61.0 | 67.4 | 74.7 | 80.4 | 75.6 | 73.4 | 61.5 | 51.7 | 45.8 | 61.6 |
| 1993 | 47.0 | 44.9 | 50.4 | 58.8 | 68.7 | 78.5 | 83.8 | 81.2 | 73.2 | 61.9 | 51.7 | 44.4 | 62.0 |
| 1994 | 39.6 | 49.3 | 54.6 | 65.3 | 67.5 | 78.5 | 78.0 | 78.0 | 72.1 | 64.1 | 57.0 | 49.2 | 62.8 |
| 1995 | 44.4 | 46.2 | 57.4 | 63.7 | 72.0 | 75.6 | 83.2 | 84.8 | 74.5 | 62.5 | 48.0 | 43.6 | 63.0 |
| 1996 | 42.6 | 46.0 | 50.9 | 59.1 | 73.3 | 77.2 | 79.9 | 78.6 | 71.8 | 62.6 | 52.0 | 48.7 | 61.9 |
| 1997 | 44.7 | 50.5 | 59.5 | 58.2 | 66.0 | 74.4 | 80.3 | 77.9 | 74.9 | 62.7 | 49.7 | 43.0 | 61.8 |
| 1998 | 46.7 | 48.0 | 52.7 | 60.9 | 73.4 | 80.3 | 82.2 | 80.4 | 78.2 | 67.6 | 57.3 | 50.7 | 64.9 |
| 1999 | 49.3 | 51.2 | 51.6 | 67.2 | 69.9 | 77.1 | 81.7 | 83.6 | 74.8 | 64.0 | 55.5 | 46.7 | 64.4 |
| 2000 | 44.5 | 52.2 | 58.0 | 59.4 | 74.3 | 77.6 | 82.6 | 81.6 | 74.2 | 65.8 | 52.0 | 38.1 | 63.4 |
| 2001 | 40.1 | 51.0 | 49.7 | 64.5 | 70.5 | 75.3 | 80.1 | 78.5 | 71.6 | 60.6 | 58.6 | 49.3 | 62.5 |
| 2002 | 46.9 | 44.5 | 54.5 | 65.7 | 70.2 | 76.9 | 80.6 | 81.1 | 77.6 | 66.9 | 50.5 | 44.4 | 63.3 |
| 2003 | 38.4 | 47.1 | 56.9 | 64.1 | 70.5 | 75.0 | 79.0 | 79.9 | 73.5 | 63.5 | 58.6 | 43.8 | 62.5 |
| 2004 | 43.3 | 45.2 | 59.1 | 62.2 | 73.7 | 76.4 | 79.5 | 76.6 | 74.2 | 69.0 | 57.6 | 44.9 | 63.5 |
| 2005 | 48.1 | 50.4 | 53.6 | 61.4 | 67.8 | 76.8 | 80.5 | 81.3 | 77.7 | 63.6 | 56.4 | 42.4 | 63.3 |
| 2006 | 50.7 | 45.0 | 56.1 | 67.8 | 71.2 | 79.2 | 83.5 | 83.7 | 73.4 | 62.5 | 53.9 | 49.9 | 64.7 |
| 2007 | 45.9 | 44.6 | 61.4 | 60.5 | 73.1 | 81.0 | 80.2 | 86.9 | 77.3 | 67.8 | 53.9 | 51.7 | 65.4 |
| 2008 | 43.1 | 49.1 | 56.0 | 63.5 | 70.4 | 79.2 | 81.6 | 79.7 | 75.4 | 63.0 | 51.2 | 49.2 | 63.5 |
| 2009 | 44.7 | 48.1 | 57.5 | 62.3 | 71.2 | 79.4 | 78.5 | 78.6 | 74.7 | 61.8 | 53.6 | 44.3 | 62.9 |
| POR= 80 YRS | 44.2 | 46.4 | 54.5 | 62.2 | 70.4 | 76.9 | 80.4 | 79.8 | 73.9 | 63.8 | 52.8 | 46.1 | 62.6 |

## HEATING DEGREE DAYS (base 65°F) 2009 BIRMINGHAM (KBHM)

| YEAR | JUL | AUG | SEP | OCT | NOV | DEC | JAN | FEB | MAR | APR | MAY | JUN | TOTAL |
|------|-----|-----|-----|-----|-----|-----|-----|-----|-----|-----|-----|-----|-------|
| 1980-81 | 0 | 0 | 7 | 181 | 372 | 642 | 795 | 464 | 428 | 46 | 51 | 0 | 2986 |
| 1981-82 | 0 | 0 | 19 | 138 | 314 | 682 | 711 | 490 | 250 | 199 | 4 | 0 | 2807 |
| 1982-83 | 0 | 0 | 16 | 134 | 331 | 449 | 751 | 558 | 437 | 262 | 39 | 0 | 2977 |
| 1983-84 | 0 | 0 | 26 | 108 | 388 | 774 | 817 | 519 | 392 | 202 | 68 | 2 | 3296 |
| 1984-85 | 0 | 0 | 10 | 27 | 410 | 330 | 907 | 604 | 278 | 123 | 16 | 3 | 2708 |
| 1985-86 | 0 | 0 | 14 | 62 | 165 | 761 | 711 | 415 | 305 | 132 | 21 | 0 | 2586 |
| 1986-87 | 0 | 1 | 0 | 112 | 239 | 644 | 705 | 500 | 320 | 201 | 0 | 0 | 2722 |
| 1987-88 | 0 | 0 | 5 | 248 | 315 | 488 | 777 | 616 | 347 | 134 | 18 | 0 | 2948 |
| 1988-89 | 0 | 0 | 0 | 234 | 289 | 589 | 492 | 545 | 289 | 200 | 65 | 0 | 2703 |
| 1989-90 | 0 | 0 | 27 | 140 | 363 | 834 | 491 | 299 | 265 | 172 | 32 | 0 | 2623 |
| 1990-91 | 0 | 0 | 16 | 138 | 282 | 474 | 620 | 421 | 304 | 46 | 7 | 0 | 2308 |
| 1991-92 | 0 | 0 | 12 | 97 | 455 | 492 | 663 | 409 | 344 | 173 | 46 | 0 | 2691 |
| 1992-93 | 0 | 0 | 6 | 116 | 398 | 587 | 550 | 559 | 452 | 196 | 20 | 0 | 2884 |
| 1993-94 | 0 | 0 | 16 | 153 | 411 | 633 | 780 | 433 | 328 | 90 | 31 | 0 | 2875 |
| 1994-95 | 0 | 0 | 3 | 82 | 241 | 483 | 632 | 520 | 250 | 101 | 24 | 0 | 2336 |
| 1995-96 | 0 | 0 | 12 | 137 | 509 | 662 | 686 | 560 | 440 | 203 | 12 | 0 | 3221 |
| 1996-97 | 0 | 0 | 10 | 121 | 390 | 498 | 628 | 399 | 188 | 209 | 46 | 0 | 2489 |
| 1997-98 | 0 | 0 | 1 | 162 | 450 | 676 | 562 | 469 | 400 | 150 | 13 | 5 | 2888 |
| 1998-99 | 0 | 0 | 0 | 47 | 231 | 462 | 483 | 383 | 408 | 73 | 8 | 0 | 2095 |
| 1999-00 | 0 | 0 | 12 | 107 | 279 | 559 | 634 | 370 | 219 | 175 | 1 | 0 | 2356 |
| 2000-01 | 0 | 0 | 10 | 84 | 419 | 828 | 766 | 392 | 466 | 101 | 3 | 0 | 3069 |
| 2001-02 | 0 | 0 | 32 | 183 | 200 | 478 | 564 | 568 | 355 | 84 | 42 | 0 | 2506 |
| 2002-03 | 0 | 0 | 0 | 68 | 437 | 630 | 818 | 497 | 264 | 89 | 4 | 0 | 2807 |
| 2003-04 | 0 | 0 | 17 | 88 | 229 | 651 | 673 | 571 | 205 | 137 | 27 | 0 | 2598 |
| 2004-05 | 0 | 0 | 0 | 31 | 239 | 615 | 516 | 406 | 360 | 128 | 45 | 0 | 2340 |
| 2005-06 | 0 | 0 | 0 | 146 | 296 | 692 | 436 | 555 | 302 | 48 | 23 | 0 | 2498 |
| 2006-07 | 0 | 0 | 5 | 160 | 335 | 460 | 587 | 565 | 168 | 185 | 4 | 0 | 2469 |
| 2007-08 | 0 | 0 | 81 | 327 | 415 | 671 | 462 | 282 | 110 | 20 | 0 | | 2368 |
| 2008-09 | 0 | 0 | 0 | 141 | 407 | 492 | 623 | 467 | 246 | 147 | 17 | 0 | 2540 |
| 2009- | 0 | 0 | 7 | 140 | 336 | 637 | | | | | | | |

WBAN : 13876

## COOLING DEGREE DAYS (base 65°F) 2009 BIRMINGHAM (KBHM)

| YEAR | JAN | FEB | MAR | APR | MAY | JUN | JUL | AUG | SEP | OCT | NOV | DEC | TOTAL |
|------|-----|-----|-----|-----|-----|-----|-----|-----|-----|-----|-----|-----|-------|
| 1980 | 0 | 4 | 6 | 25 | 176 | 386 | 607 | 563 | 413 | 43 | 2 | 0 | 2225 |
| 1981 | 0 | 1 | 9 | 113 | 145 | 475 | 539 | 448 | 231 | 46 | 9 | 0 | 2016 |
| 1982 | 1 | 0 | 65 | 28 | 247 | 327 | 495 | 459 | 265 | 118 | 12 | 27 | 2044 |
| 1983 | 0 | 0 | 7 | 17 | 115 | 281 | 481 | 528 | 233 | 53 | 2 | 0 | 1717 |
| 1984 | 0 | 0 | 11 | 30 | 151 | 373 | 421 | 395 | 230 | 221 | 6 | 6 | 1844 |
| 1985 | 0 | 1 | 27 | 90 | 163 | 349 | 413 | 417 | 233 | 139 | 51 | 1 | 1884 |
| 1986 | 0 | 0 | 16 | 43 | 222 | 413 | 547 | 406 | 364 | 84 | 18 | 0 | 2113 |
| 1987 | 0 | 0 | 5 | 46 | 280 | 357 | 487 | 532 | 254 | 1 | 7 | 3 | 1972 |
| 1988 | 0 | 0 | 13 | 26 | 105 | 382 | 467 | 516 | 292 | 14 | 1 | 0 | 1816 |
| 1989 | 0 | 14 | 27 | 65 | 155 | 323 | 447 | 459 | 257 | 51 | 3 | 0 | 1801 |
| 1990 | 1 | 14 | 21 | 49 | 160 | 393 | 466 | 543 | 398 | 110 | 9 | 4 | 2168 |
| 1991 | 0 | 0 | 28 | 74 | 266 | 364 | 515 | 496 | 333 | 106 | 9 | 10 | 2201 |
| 1992 | 0 | 0 | 2 | 59 | 129 | 298 | 484 | 333 | 265 | 14 | 5 | 0 | 1589 |
| 1993 | 0 | 0 | 5 | 19 | 140 | 410 | 591 | 511 | 271 | 63 | 18 | 0 | 2028 |
| 1994 | 0 | 0 | 10 | 109 | 115 | 408 | 410 | 411 | 221 | 58 | 8 | 0 | 1750 |
| 1995 | 2 | 0 | 23 | 67 | 247 | 323 | 571 | 620 | 301 | 67 | 3 | 3 | 2227 |
| 1996 | 0 | 16 | 10 | 34 | 280 | 374 | 470 | 428 | 219 | 54 | 8 | 1 | 1894 |
| 1997 | 6 | 1 | 23 | 14 | 82 | 287 | 481 | 404 | 303 | 99 | 0 | 0 | 1700 |
| 1998 | 1 | 0 | 27 | 34 | 277 | 470 | 538 | 484 | 403 | 134 | 6 | 26 | 2400 |
| 1999 | 7 | 5 | 0 | 146 | 168 | 371 | 524 | 580 | 312 | 83 | 0 | 0 | 2196 |
| 2000 | 3 | 8 | 9 | 12 | 298 | 385 | 553 | 521 | 294 | 117 | 37 | 0 | 2237 |
| 2001 | 0 | 6 | 0 | 95 | 182 | 314 | 473 | 423 | 234 | 53 | 17 | 0 | 1797 |
| 2002 | 12 | 0 | 36 | 112 | 210 | 363 | 491 | 507 | 386 | 134 | 9 | 0 | 2260 |
| 2003 | 0 | 3 | 18 | 70 | 184 | 307 | 440 | 469 | 282 | 48 | 45 | 0 | 1866 |
| 2004 | 6 | 0 | 31 | 62 | 304 | 350 | 459 | 364 | 284 | 162 | 25 | 0 | 2047 |
| 2005 | 1 | 3 | 12 | 29 | 140 | 360 | 488 | 513 | 390 | 111 | 44 | 0 | 2091 |
| 2006 | 2 | 0 | 35 | 141 | 222 | 434 | 579 | 587 | 262 | 87 | 8 | 0 | 2357 |
| 2007 | 0 | 0 | 63 | 57 | 259 | 489 | 480 | 686 | 380 | 174 | 2 | 9 | 2599 |
| 2008 | 2 | 5 | 10 | 72 | 192 | 430 | 524 | 464 | 315 | 87 | 1 | 9 | 2111 |
| 2009 | 0 | 0 | 22 | 77 | 216 | 437 | 424 | 429 | 305 | 49 | 0 | | 1959 |

## SNOWFALL (inches)  2009  BIRMINGHAM (KBHM)

| YEAR | JUL | AUG | SEP | OCT | NOV | DEC | JAN | FEB | MAR | APR | MAY | JUN | TOTAL |
|------|-----|-----|-----|-----|-----|-----|-----|-----|-----|-----|-----|-----|-------|
| 1980-81 | 0.0 | 0.0 | 0.0 | 0.0 | T | T | 0.0 | T | 0.0 | 0.0 | 0.0 | 0.0 | T |
| 1981-82 | 0.0 | 0.0 | 0.0 | 0.0 | 0.0 | T | 2.1 | 0.0 | T | 0.0 | 0.0 | 0.0 | 2.1 |
| 1982-83 | 0.0 | 0.0 | 0.0 | 0.0 | 0.0 | T | 1.0 | T | 1.5 | 0.0 | 0.0 | 0.0 | 2.5 |
| 1983-84 | 0.0 | 0.0 | 0.0 | 0.0 | 0.0 | T | T | T | 2.0 | 0.0 | 0.0 | 0.0 | 2.0 |
| 1984-85 | 0.0 | 0.0 | 0.0 | 0.0 | 0.0 | T | T | 0.3 | 0.0 | 0.0 | 0.0 | 0.0 | 0.3 |
| 1985-86 | 0.0 | 0.0 | 0.0 | 0.0 | 0.0 | T | T | T | 0.0 | 0.0 | 0.0 | 0.0 | T |
| 1986-87 | 0.0 | 0.0 | 0.0 | 0.0 | 0.0 | T | 2.6 | 0.0 | T | 5.0 | 0.0 | 0.0 | 7.6 |
| 1987-88 | 0.0 | 0.0 | 0.0 | 0.0 | 0.0 | 0.0 | 1.0 | T | 0.0 | 0.0 | 0.0 | 0.0 | 1.0 |
| 1988-89 | 0.0 | 0.0 | 0.0 | 0.0 | 0.0 | T | 0.0 | T | T | T | 0.0 | 0.0 | T |
| 1989-90 | 0.0 | 0.0 | 0.0 | 0.0 | T | 0.4 | 0.0 | T | 0.0 | 0.0 | T | T | 0.4 |
| 1990-91 | T | 0.0 | 0.0 | 0.0 | 0.0 | T | T | T | T | T | 0.0 | T | T |
| 1991-92 | 0.0 | 0.0 | 0.0 | 0.0 | T | 0.0 | 4.4 | 0.0 | T | 0.0 | T | T | 4.4 |
| 1992-93 | 0.0 | 0.0 | T | T | 0.0 | 0.0 | 0.0 | T | 13.0 | 0.0 | T | 0.0 | 13.0 |
| 1993-94 | 0.0 | 0.0 | 0.0 | T | 0.0 | T | T | 0.0 | T | 0.0 | 0.0 | 0.0 | T |
| 1994-95 | T | 0.0 | 0.0 | 0.0 | 0.0 | 0.0 | T | 1.0 | T | 0.0 | T | 0.0 | 1.0 |
| 1995-96 | 0.0 | 0.0 | 0.0 | 0.0 | T | T | 0.3 | 1.2 | T | 0.0 | T | 0.0 | 1.5 |
| 1996-97 | 0.0 | 0.0 | 0.0 | 0.0 | 0.0 | T | T | T | 0.0 | 0.0 | T | 0.0 | T |
| 1997-98 | 0.0 | 0.0 | 0.0 | 0.0 | T | 1.7 | T | T | T | T | 0.0 | T | 1.7 |
| 1998-99 | T | 0.0 |  |  | 0.0 | 0.0 | T | T | 0.0 | T | 0.0 | T |  |
| 1999-00 | 0.0 | 0.0 | 0.0 | 0.0 | 0.0 | 0.0 | 3.0 | T | T | T | 0.0 | 0.0 | 3.0 |
| 2000-01 | T | 0.0 | 0.0 | 0.0 | T | T | T | T | T | T | 0.0 | T | T |
| 2001-02 | 0.0 | 0.0 | 0.0 | 0.0 | 0.0 | 0.0 | T | T | 0.0 | 0.0 | 0.0 | 0.0 | T |
| 2002-03 | 0.0 | 0.0 | 0.0 | 0.0 | 0.0 | 0.0 | T | T | 0.0 | 0.0 | T | 0.0 | T |
| 2003-04 | T | 0.0 | 0.0 | 0.0 | 0.0 | 0.0 | 0.0 | T | T | 0.0 | 0.0 | 0.0 | T |
| 2004-05 | 0.0 | 0.0 | 0.0 | T | 0.0 | 0.0 | 0.0 | 0.0 | 0.0 | T | 0.0 | 0.0 | T |
| 2005-06 | 0.0 | 0.0 | 0.0 | 0.0 | 0.0 | 0.0 | T | T | 0.0 | T | 0.0 | T | T |
| 2006-07 | 0.0 | 0.0 | 0.0 | 0.0 | 0.0 | 0.0 | T | T | 0.0 | 0.0 | 0.0 | 0.0 | T |
| 2007-08 | 0.0 | 0.0 | 0.0 | 0.0 | 0.0 | 0.0 | T | T | 0.1 | 0.0 | 0.0 | 0.0 | 0.1 |
| 2008-09 | 0.0 | 0.0 | 0.0 | 0.0 | 0.0 | 0.0 | 0.0 | 0.0 | 2.0 | T | 0.0 | 0.0 | 2.0 |
| 2009- | 0.0 | 0.0 | 0.0 | 0.0 | 0.0 | T |  |  |  |  |  |  |  |
| POR=<br>80 YRS | T | 0.0 | T | T | T | 0.3 | 0.8 | 0.2 | 0.3 | 0.1 | T | T | 1.7 |

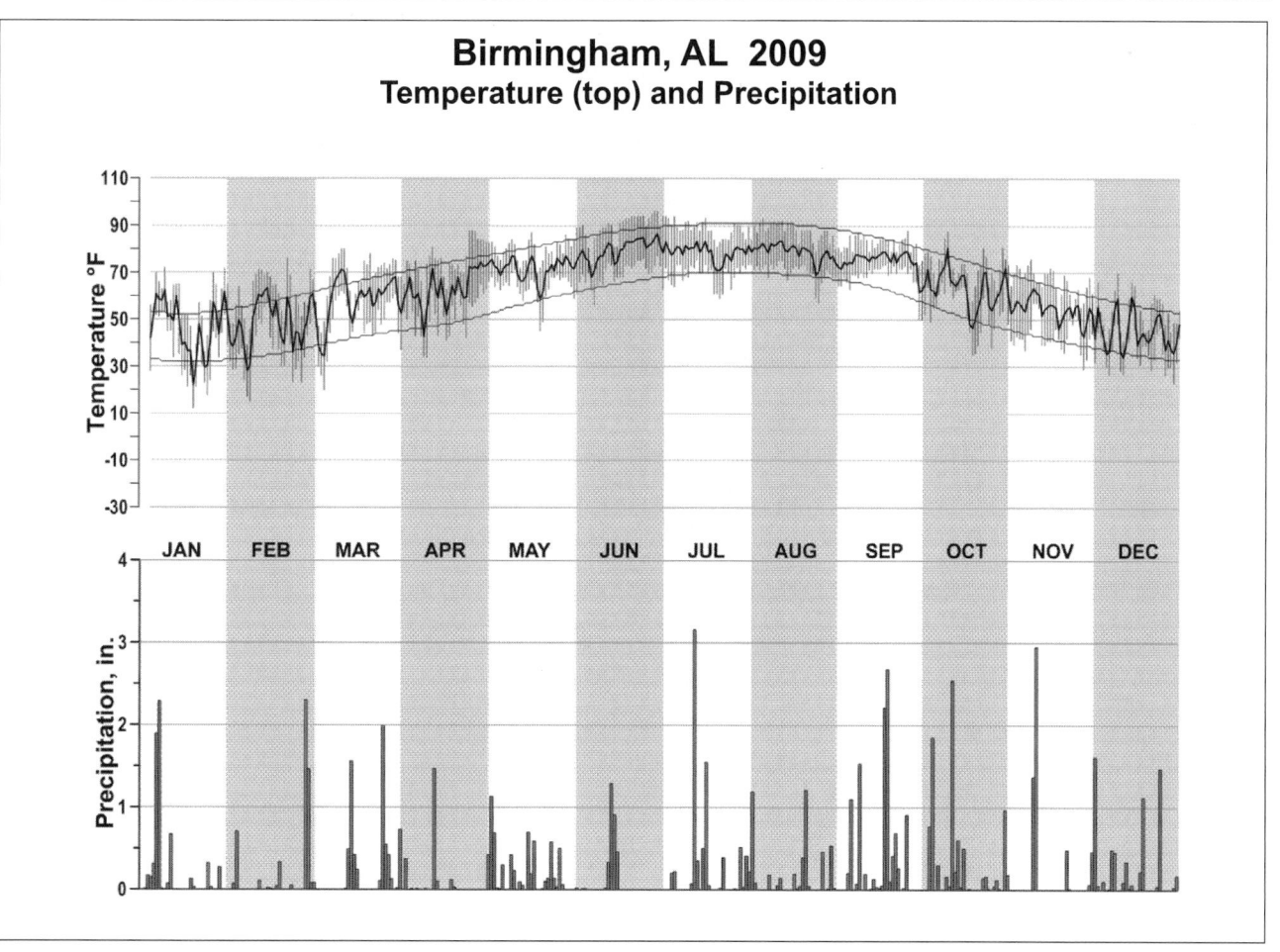

# Birmingham, AL  2009
## Temperature (top) and Precipitation

# 2009
# MOBILE
# ALABAMA (KMOB)

Mobile is located at the head of Mobile Bay and approximately 30 miles from the Gulf of Mexico. Its weather is influenced to a considerable extent by the Gulf.

The summers are consistently warm, but temperatures are seldom as high as they are at inland stations. Normally, in summer, the day begins in the low 70s and the temperature rises rapidly before noon to the high 80s or low 90s, when it is checked by the onset of the sea breeze. On the rare occasions when northerly winds prevail throughout the day, temperatures may reach the high 90s or rise slightly above 100 degrees.

Winter weather is usually mild except for occasional invasions of cold air that last about three days. January is the coldest month in the year. Unusual winters may produce readings that require extensive protective measures as some citrus fruit is grown in the area and outdoor nurseries are numerous.

Based on the 1951-1980 period, the average first occurrence of 32 degrees Fahrenheit in the fall is November 26 and the average last occurrence in the spring is February 27.

The yearly rainfall is among the highest in the United States. It is fairly evenly distributed throughout the year with a slight maximum at the height of the summer thunderstorm season and a slight minimum during the late fall. Rainfall is usually of the shower type and long periods of continuous rain are rare.

Frontal thunderstorms may occur in any month of the year. There may be a thunderstorm every other day in July and August. The summer storms are usually not too violent and seldom produce hail.

The area is subject to hurricanes from the West Indies, the western Caribbean, and the Gulf of Mexico.

# NORMALS, MEANS, AND EXTREMES
## MOBILE (KMOB)

| LATITUDE: | LONGITUDE: | ELEVATION (FT): | TIME ZONE: | WBAN: 13894 |
|---|---|---|---|---|
| 30 ° 41'N | -88 ° 14'W | GRND: 215    BARO: 212 | CENTRAL    (UTC -6) | |

| | ELEMENT | POR | JAN | FEB | MAR | APR | MAY | JUN | JUL | AUG | SEP | OCT | NOV | DEC | YEAR |
|---|---|---|---|---|---|---|---|---|---|---|---|---|---|---|---|
| **TEMPERATURE °F** | NORMAL DAILY MAXIMUM | 30 | 60.7 | 64.5 | 71.2 | 77.4 | 84.2 | 89.4 | 91.2 | 90.8 | 86.8 | 79.2 | 70.1 | 62.9 | 77.4 |
| | MEAN DAILY MAXIMUM | 62 | 61.0 | 64.5 | 69.7 | 77.8 | 84.7 | 89.6 | 91.0 | 90.7 | 86.8 | 79.3 | 69.9 | 63.0 | 77.3 |
| | HIGHEST DAILY MAXIMUM | 68 | 84 | 82 | 90 | 94 | 100 | 102 | 104 | 105 | 99 | 93 | 87 | 81 | 105 |
| | YEAR OF OCCURRENCE | | 1949 | 1989 | 1946 | 1987 | 1953 | 1952 | 1952 | 2000 | 1990 | 1963 | 2003 | 1998 | AUG 2000 |
| | MEAN OF EXTREME MAXS. | 62 | 75.1 | 77.4 | 82.2 | 86.6 | 92.3 | 96.0 | 96.9 | 96.3 | 93.6 | 88.6 | 81.8 | 76.6 | 87.0 |
| | NORMAL DAILY MINIMUM | 30 | 39.5 | 42.4 | 49.2 | 54.8 | 62.8 | 69.2 | 71.8 | 71.7 | 67.6 | 56.3 | 47.8 | 41.6 | 56.2 |
| | MEAN DAILY MINIMUM | 62 | 40.8 | 43.6 | 48.9 | 56.5 | 64.3 | 70.4 | 72.9 | 72.6 | 68.6 | 57.7 | 48.1 | 42.8 | 57.3 |
| | LOWEST DAILY MINIMUM | 68 | 3 | 11 | 21 | 32 | 43 | 49 | 60 | 59 | 42 | 30 | 22 | 8 | 3 |
| | YEAR OF OCCURRENCE | | 1985 | 1996 | 1993 | 1987 | 1960 | 1984 | 1947 | 1956 | 1967 | 1993 | 1950 | 1983 | JAN 1985 |
| | MEAN OF EXTREME MINS. | 62 | 22.3 | 25.8 | 32.2 | 41.1 | 51.9 | 62.3 | 68.5 | 67.3 | 56.7 | 40.9 | 31.3 | 25.0 | 43.8 |
| | NORMAL DRY BULB | 30 | 50.1 | 53.5 | 60.2 | 66.1 | 73.5 | 79.3 | 81.5 | 81.3 | 77.2 | 67.7 | 58.9 | 52.3 | 66.8 |
| | MEAN DRY BULB | 62 | 50.9 | 54.1 | 59.3 | 67.2 | 74.5 | 80.1 | 81.9 | 81.7 | 77.7 | 68.5 | 58.9 | 52.9 | 67.3 |
| | MEAN WET BULB | 26 | 45.7 | 48.3 | 53.5 | 59.2 | 66.7 | 72.0 | 74.3 | 74.1 | 70.1 | 61.3 | 53.5 | 47.7 | 60.5 |
| | MEAN DEW POINT | 26 | 42.3 | 45.0 | 50.1 | 56.0 | 64.2 | 70.3 | 72.7 | 72.4 | 68.1 | 58.4 | 50.6 | 44.5 | 57.9 |
| | NORMAL NO. DAYS WITH: | | | | | | | | | | | | | | |
| | MAXIMUM >= 90 | 30 | 0.0 | 0.0 | 0.0 | 0.2 | 3.0 | 15.5 | 22.2 | 21.3 | 10.4 | 0.7 | 0.0 | 0.0 | 73.3 |
| | MAXIMUM <= 32 | 30 | 0.1 | * | 0.0 | 0.0 | 0.0 | 0.0 | 0.0 | 0.0 | 0.0 | 0.0 | 0.0 | 0.1 | 0.2 |
| | MINIMUM <= 32 | 30 | 8.0 | 4.9 | 1.3 | * | 0.0 | 0.0 | 0.0 | 0.0 | 0.0 | * | 1.2 | 6.2 | 21.6 |
| | MINIMUM <= 0 | 30 | 0.0 | 0.0 | 0.0 | 0.0 | 0.0 | 0.0 | 0.0 | 0.0 | 0.0 | 0.0 | 0.0 | 0.0 | 0.0 |
| **H/C** | NORMAL HEATING DEG. DAYS | 30 | 455 | 326 | 182 | 57 | 4 | 0 | 0 | 0 | 2 | 51 | 208 | 396 | 1681 |
| | NORMAL COOLING DEG. DAYS | 30 | 9 | 11 | 45 | 106 | 282 | 445 | 527 | 520 | 384 | 151 | 41 | 18 | 2539 |
| **RH** | NORMAL (PERCENT) | 30 | 74 | 72 | 72 | 71 | 74 | 76 | 78 | 78 | 77 | 73 | 75 | 75 | 75 |
| | HOUR 00 LST | 30 | 81 | 81 | 83 | 84 | 86 | 87 | 89 | 89 | 87 | 85 | 84 | 82 | 85 |
| | HOUR 06 LST | 30 | 83 | 85 | 86 | 88 | 89 | 89 | 91 | 92 | 91 | 88 | 87 | 85 | 88 |
| | HOUR 12 LST | 30 | 61 | 57 | 55 | 53 | 55 | 58 | 61 | 61 | 60 | 54 | 58 | 62 | 58 |
| | HOUR 18 LST | 30 | 70 | 65 | 65 | 62 | 65 | 68 | 73 | 74 | 73 | 69 | 73 | 73 | 69 |
| **S** | PERCENT POSSIBLE SUNSHINE | | | | | | | | | | | | | | |
| **W/O** | MEAN NO. DAYS WITH: | | | | | | | | | | | | | | |
| | HEAVY FOG(VISBY <= 1/4 MI) | 46 | 6.0 | 4.2 | 6.3 | 4.5 | 3.2 | 1.3 | 1.3 | 1.8 | 1.8 | 3.3 | 4.6 | 5.0 | 43.3 |
| | THUNDERSTORMS | 62 | 2.3 | 2.5 | 4.7 | 4.6 | 7.2 | 12.1 | 17.8 | 14.8 | 7.3 | 2.3 | 2.3 | 2.4 | 80.3 |
| **CLOUDNESS** | MEAN: | | | | | | | | | | | | | | |
| | SUNRISE-SUNSET (OKTAS) | 48 | 5.3 | 5.0 | 4.9 | 4.6 | 4.6 | 4.6 | 5.1 | 4.6 | 4.4 | 3.5 | 4.2 | 4.9 | 4.6 |
| | MIDNIGHT-MIDNIGHT (OKTAS) | 32 | 5.0 | 4.6 | 4.7 | 4.2 | 4.2 | 4.0 | 4.5 | 4.2 | 3.9 | 3.3 | 4.0 | 4.7 | 4.3 |
| | MEAN NO. DAYS WITH: | | | | | | | | | | | | | | |
| | CLEAR | 48 | 7.9 | 7.6 | 8.7 | 9.2 | 8.6 | 7.0 | 4.1 | 6.3 | 9.0 | 14.2 | 11.0 | 8.8 | 102.4 |
| | PARTLY CLOUDY | 48 | 6.2 | 6.6 | 7.8 | 8.7 | 11.3 | 13.8 | 15.0 | 14.8 | 10.1 | 7.6 | 7.2 | 6.4 | 115.5 |
| | CLOUDY | 48 | 16.9 | 14.0 | 14.5 | 12.0 | 11.1 | 9.2 | 11.9 | 9.9 | 11.0 | 9.1 | 11.8 | 15.9 | 147.3 |
| **PR** | MEAN STATION PRESSURE(IN) | 26 | 29.92 | 29.88 | 29.83 | 29.79 | 29.77 | 29.76 | 29.80 | 29.77 | 29.76 | 29.82 | 29.88 | 29.92 | 29.83 |
| | MEAN SEA-LEVEL PRES. (IN) | 26 | 30.16 | 30.12 | 30.06 | 30.02 | 30.00 | 29.99 | 30.03 | 30.00 | 29.99 | 30.06 | 30.11 | 30.15 | 30.06 |
| **WINDS** | MEAN SPEED (MPH) | 26 | 8.8 | 9.2 | 9.2 | 8.8 | 7.8 | 6.5 | 6.0 | 5.8 | 7.1 | 7.4 | 7.9 | 8.3 | 7.7 |
| | PREVAIL.DIR(TENS OF DEGS) | 35 | 36 | 36 | 15 | 15 | 19 | 19 | 21 | 05 | 05 | 01 | 36 | 36 | 36 |
| | MAXIMUM 2-MINUTE: | | | | | | | | | | | | | | |
| | SPEED (MPH) | 13 | 43 | 36 | 39 | 41 | 33 | 45 | 41 | 66 | 59 | 40 | 41 | 43 | 66 |
| | DIR. (TENS OF DEGS) | | 23 | 22 | 18 | 09 | 36 | 16 | 08 | 13 | 01 | 22 | 22 | 22 | 13 |
| | YEAR OF OCCURRENCE | | 2001 | 2006 | 2001 | 2005 | 2006 | 2001 | 2008 | 2005 | 2004 | 2001 | 2006 | 2002 | AUG 2005 |
| | MAXIMUM 3-SECOND | | | | | | | | | | | | | | |
| | SPEED (MPH) | 13 | 52 | 47 | 53 | 47 | 44 | 57 | 58 | 83 | 75 | 54 | 52 | 63 | 83 |
| | DIR. (TENS OF DEGS) | | 22 | 23 | 24 | 09 | 21 | 34 | 10 | 12 | 02 | 21 | 21 | 22 | 12 |
| | YEAR OF OCCURRENCE | | 2001 | 2008 | 1999 | 2005 | 2008 | 1998 | 2008 | 2005 | 2004 | 2001 | 2006 | 2002 | AUG 2005 |
| **PRECIPITATION** | NORMAL (IN) | 30 | 5.75 | 5.10 | 7.20 | 5.06 | 6.10 | 5.01 | 6.54 | 6.20 | 6.01 | 3.25 | 5.41 | 4.66 | 66.29 |
| | MAXIMUM MONTHLY (IN) | 68 | 16.92 | 11.89 | 15.58 | 17.69 | 15.08 | 20.66 | 19.29 | 15.19 | 24.13 | 13.20 | 13.65 | 15.37 | 24.13 |
| | YEAR OF OCCURRENCE | | 1998 | 1983 | 1946 | 1955 | 1980 | 2003 | 1949 | 1984 | 1998 | 1985 | 1948 | 2009 | SEP 1998 |
| | MINIMUM MONTHLY (IN) | 68 | 0.55 | 1.09 | 0.24 | 0.08 | 0.36 | 1.19 | 1.72 | 1.04 | 0.58 | T | 0.25 | 1.29 | T |
| | YEAR OF OCCURRENCE | | 2003 | 1999 | 2006 | 1999 | 1996 | 1966 | 1983 | 1997 | 1963 | 1978 | 1960 | 1980 | OCT 1978 |
| | MAXIMUM IN 24 HOURS (IN) | 68 | 8.34 | 5.37 | 10.57 | 13.36 | 8.86 | 7.38 | 10.07 | 6.62 | 10.06 | 5.65 | 7.02 | 7.46 | 13.36 |
| | YEAR OF OCCURRENCE | | 1965 | 1981 | 1990 | 1955 | 1995 | 1961 | 1997 | 1969 | 1998 | 1985 | 1975 | 1995 | APR 1955 |
| | NORMAL NO. DAYS WITH: | | | | | | | | | | | | | | |
| | PRECIPITATION >= 0.01 | 30 | 10.9 | 8.8 | 9.6 | 7.5 | 9.0 | 11.1 | 15.1 | 13.6 | 10.4 | 5.4 | 8.6 | 9.5 | 119.5 |
| | PRECIPITATION >= 1.00 | 30 | 1.7 | 1.6 | 2.5 | 1.7 | 1.8 | 1.6 | 2.0 | 2.2 | 1.6 | 1.1 | 1.6 | 1.5 | 20.9 |
| **SNOWFALL** | NORMAL (IN) | 30 | 0.1 | 0.2 | 0.1 | 0.* | 0.0 | 0.0 | 0.0 | 0.0 | 0.0 | 0.0 | 0.0 | 0.* | 0.4 |
| | MAXIMUM MONTHLY (IN) | 66 | 3.5 | 3.6 | 2.7 | T | T | T | T | 0.0 | 0.0 | 0.0 | T | 3.0 | 3.6 |
| | YEAR OF OCCURRENCE | | 1955 | 1973 | 1993 | 2007 | 1991 | 2006 | 1995 | | | | 1966 | 1963 | FEB 1973 |
| | MAXIMUM IN 24 HOURS (IN) | 66 | 3.5 | 3.6 | 2.7 | T | T | T | T | 0.0 | 0.0 | 0.0 | T | 3.0 | 3.6 |
| | YEAR OF OCCURRENCE' | | 1955 | 1973 | 1993 | 1988 | 1991 | 2006 | 1995 | | | | 1966 | 1963 | FEB 1973 |
| | MAXIMUM SNOW DEPTH (IN) | 59 | 22 | 3 | 2 | 0 | 0 | 0 | 0 | 0 | 0 | 0 | 0 | 0 | 22 |
| | YEAR OF OCCURRENCE | | 1964 | 1973 | 1993 | | | | | | | | | | JAN 1964 |
| | NORMAL NO. DAYS WITH: | | | | | | | | | | | | | | |
| | SNOWFALL >= 1.0 | 30 | 0.0 | 0.1 | 0.0 | 0.0 | 0.0 | 0.0 | 0.0 | 0.0 | 0.0 | 0.0 | 0.0 | 0.0 | 0.1 |

## PRECIPITATION (inches) 2009 MOBILE (KMOB)

| YEAR | JAN | FEB | MAR | APR | MAY | JUN | JUL | AUG | SEP | OCT | NOV | DEC | ANNUAL |
|------|------|------|------|------|------|------|------|------|------|------|------|------|--------|
| 1980 | 4.95 | 1.58 | 13.46 | 15.43 | 15.08 | 2.57 | 6.54 | 6.42 | 5.94 | 2.90 | 1.96 | 1.29 | 78.12 |
| 1981 | 1.23 | 8.75 | 3.00 | 0.96 | 12.51 | 6.50 | 6.53 | 5.06 | 3.97 | 0.94 | 0.85 | 6.82 | 57.12 |
| 1982 | 3.56 | 7.42 | 6.81 | 4.48 | 2.84 | 11.00 | 13.14 | 7.00 | 4.68 | 1.61 | 3.66 | 8.24 | 74.44 |
| 1983 | 5.82 | 11.89 | 6.89 | 12.53 | 1.53 | 8.09 | 1.72 | 11.57 | 5.97 | 3.81 | 5.30 | 8.34 | 83.46 |
| 1984 | 6.13 | 4.79 | 4.75 | 3.53 | 4.28 | 2.07 | 2.36 | 15.19 | 0.74 | 6.19 | 1.67 | 2.12 | 53.82 |
| 1985 | 5.06 | 6.39 | 5.49 | 1.22 | 5.77 | 3.94 | 8.31 | 4.32 | 10.32 | 13.20 | 1.61 | 4.34 | 69.97 |
| 1986 | 2.67 | 4.17 | 4.53 | 2.16 | 4.18 | 7.53 | 7.13 | 5.60 | 4.41 | 4.83 | 8.45 | 3.68 | 59.34 |
| 1987 | 5.81 | 8.64 | 6.18 | 0.83 | 10.69 | 7.68 | 4.18 | 10.33 | 3.62 | 0.02 | 5.54 | 3.60 | 67.12 |
| 1988 | 4.64 | 6.26 | 7.80 | 4.19 | 0.58 | 2.34 | 6.04 | 10.43 | 14.04 | 1.83 | 2.30 | 1.80 | 62.25 |
| 1989 | 2.13 | 1.47 | 5.57 | 3.55 | 6.47 | 9.82 | 7.16 | 3.82 | 4.55 | 0.90 | 11.33 | 7.23 | 64.00 |
| 1990 | 7.35 | 8.13 | 12.24 | 4.51 | 6.47 | 2.68 | 2.28 | 1.46 | 2.65 | 1.23 | 1.77 | 5.20 | 55.97 |
| 1991 | 16.07 | 3.35 | 4.83 | 10.43 | 15.03 | 7.04 | 8.15 | 5.17 | 1.87 | 0.86 | 5.78 | 3.09 | 81.67 |
| 1992 | 9.89 | 9.94 | 3.61 | 1.90 | 1.58 | 5.24 | 7.77 | 8.27 | 3.26 | 1.03 | 12.70 | 5.27 | 70.46 |
| 1993 | 8.03 | 4.48 | 10.45 | 4.80 | 5.24 | 3.01 | 5.64 | 3.52 | 3.35 | 5.32 | 2.87 | 3.69 | 60.40 |
| 1994 | 6.57 | 1.32 | 4.34 | 5.48 | 4.59 | 5.09 | 10.72 | 2.05 | 1.76 | 5.91 | 4.05 | 3.04 | 54.92 |
| 1995 | 7.09 | 3.01 | 8.53 | 7.46 | 12.08 | 3.32 | 4.63 | 7.60 | 1.16 | 8.70 | 8.05 | 8.86 | 80.49 |
| 1996 | 4.65 | 6.23 | 8.97 | 11.93 | 0.36 | 5.18 | 4.95 | 8.29 | 5.01 | 1.72 | 2.52 | 6.92 | 66.73 |
| 1997 | 5.67 | 5.98 | 5.42 | 5.66 | 8.50 | 8.49 | 18.52 | 1.04 | 1.98 | 5.83 | 8.68 | 4.37 | 80.14 |
| 1998 | 16.92 | 4.29 | 10.92 | 5.19 | 1.78 | 2.87 | 6.90 | 5.02 | 24.13 | 1.35 | 4.95 | 2.20 | 86.52 |
| 1999 | 5.19 | 1.09 | 9.44 | 0.08 | 4.71 | 7.58 | 4.30 | 2.87 | 4.41 | 4.75 | 3.05 | 3.43 | 50.90 |
| 2000 | 2.67 | 1.30 | 6.94 | 2.43 | 2.38 | 3.21 | 3.90 | 3.67 | 3.43 | 0.47 | 11.54 | 3.80 | 45.74 |
| 2001 | 4.08 | 2.70 | 11.04 | 0.88 | 1.52 | 6.20 | 8.55 | 9.49 | 2.59 | 3.53 | 1.24 | 2.83 | 54.65 |
| 2002 | 3.52 | 2.87 | 6.08 | 1.74 | 4.45 | 4.24 | 9.38 | 5.02 | 12.94 | 8.35 | 4.92 | 8.97 | 72.48 |
| 2003 | 0.55 | 5.57 | 4.30 | 3.59 | 9.51 | 20.66 | 9.48 | 5.17 | 1.57 | 3.04 | 3.70 | 3.79 | 70.93 |
| 2004 | 3.43 | 10.78 | 0.42 | 3.64 | 2.06 | 16.07 | 6.90 | 12.53 | 6.06 | 2.42 | 8.48 | 3.37 | 76.16 |
| 2005 | 2.40 | 3.85 | 5.28 | 16.62 | 2.75 | 7.33 | 16.23 | 11.01 | 2.34 | 0.21 | 2.22 | 3.59 | 73.83 |
| 2006 | 3.15 | 3.94 | 0.24 | 3.37 | 1.28 | 2.91 | 4.76 | 7.88 | 3.13 | 5.63 | 9.10 | 3.96 | 49.35 |
| 2007 | 3.60 | 1.56 | 0.99 | 6.61 | 2.05 | 2.27 | 11.81 | 6.52 | 3.28 | 4.95 | 3.55 | 8.06 | 55.25 |
| 2008 | 6.86 | 3.95 | 2.65 | 10.47 | 4.55 | 4.86 | 6.11 | 14.56 | 4.05 | 3.05 | 3.62 | 4.37 | 69.10 |
| 2009 | 3.54 | 3.81 | 12.34 | 1.73 | 5.29 | 2.45 | 5.68 | 10.18 | 6.69 | 4.91 | 4.48 | 15.37 | 76.47 |
| POR= 62 YRS | 4.96 | 4.93 | 6.41 | 5.26 | 5.17 | 5.66 | 7.82 | 6.90 | 5.98 | 3.12 | 4.40 | 5.37 | 65.98 |

WBAN : 13894

## AVERAGE TEMPERATURE (°F) 2009 MOBILE (KMOB)

| YEAR | JAN | FEB | MAR | APR | MAY | JUN | JUL | AUG | SEP | OCT | NOV | DEC | ANNUAL |
|------|------|------|------|------|------|------|------|------|------|------|------|------|--------|
| 1980 | 56.2 | 49.7 | 61.6 | 65.9 | 74.6 | 81.2 | 84.6 | 83.2 | 81.8 | 66.3 | 57.8 | 51.6 | 67.9 |
| 1981 | 46.0 | 53.4 | 59.5 | 71.0 | 71.6 | 82.6 | 84.0 | 82.7 | 76.8 | 69.4 | 63.0 | 49.9 | 67.5 |
| 1982 | 49.7 | 53.3 | 63.5 | 67.1 | 74.4 | 80.7 | 81.0 | 81.0 | 76.0 | 68.5 | 60.4 | 56.9 | 67.7 |
| 1983 | 47.1 | 51.1 | 55.0 | 61.8 | 72.1 | 76.4 | 81.8 | 81.4 | 73.6 | 67.4 | 57.7 | 47.7 | 64.4 |
| 1984 | 46.1 | 52.4 | 58.6 | 65.7 | 72.5 | 78.1 | 79.9 | 78.8 | 75.6 | 72.7 | 56.3 | 60.4 | 66.4 |
| 1985 | 43.5 | 50.7 | 64.5 | 67.7 | 73.7 | 79.9 | 79.7 | 81.0 | 76.0 | 71.4 | 66.1 | 48.8 | 66.9 |
| 1986 | 49.7 | 56.4 | 59.7 | 66.1 | 74.4 | 81.0 | 83.3 | 80.5 | 79.5 | 68.5 | 64.3 | 51.1 | 67.9 |
| 1987 | 48.8 | 53.6 | 59.2 | 64.6 | 75.7 | 78.9 | 81.8 | 82.2 | 77.2 | 62.5 | 59.2 | 56.5 | 66.7 |
| 1988 | 46.8 | 50.6 | 58.6 | 67.3 | 72.8 | 79.5 | 81.8 | 81.3 | 78.3 | 65.3 | 62.7 | 53.8 | 66.6 |
| 1989 | 57.1 | 54.0 | 61.7 | 65.9 | 74.1 | 79.0 | 81.0 | 81.6 | 76.7 | 66.1 | 59.1 | 44.4 | 66.7 |
| 1990 | 54.6 | 59.0 | 61.7 | 65.0 | 73.4 | 81.1 | 81.7 | 83.1 | 78.6 | 68.0 | 60.9 | 56.7 | 68.7 |
| 1991 | 52.4 | 55.9 | 61.8 | 69.3 | 76.0 | 79.7 | 81.9 | 81.3 | 77.1 | 69.4 | 54.1 | 54.5 | 67.8 |
| 1992 | 49.4 | 57.0 | 60.3 | 65.4 | 72.2 | 79.3 | 82.4 | 79.4 | 77.2 | 67.4 | 56.4 | 55.6 | 66.8 |
| 1993 | 55.1 | 52.7 | 57.0 | 62.1 | 71.9 | 79.9 | 82.6 | 82.8 | 78.3 | 68.1 | 56.8 | 49.8 | 66.4 |
| 1994 | 47.2 | 56.1 | 60.0 | 68.6 | 73.4 | 79.6 | 79.9 | 80.4 | 76.6 | 68.6 | 63.8 | 55.2 | 67.5 |
| 1995 | 51.0 | 55.1 | 61.8 | 67.1 | 75.6 | 78.6 | 82.5 | 83.2 | 78.6 | 68.5 | 56.7 | 51.6 | 67.5 |
| 1996 | 49.0 | 52.4 | 55.4 | 63.8 | 75.6 | 78.8 | 81.7 | 80.0 | 76.4 | 67.3 | 59.2 | 54.3 | 66.2 |
| 1997 | 51.0 | 54.9 | 66.3 | 63.5 | 72.2 | 77.4 | 81.5 | 81.4 | 79.2 | 67.2 | 55.1 | 49.4 | 66.6 |
| 1998 | 52.2 | 53.7 | 57.1 | 65.2 | 76.4 | 82.5 | 83.5 | 83.2 | 78.5 | 70.3 | 63.3 | 57.2 | 68.6 |
| 1999 | 55.3 | 57.3 | 58.6 | 70.8 | 73.7 | 79.1 | 81.4 | 84.4 | 76.0 | 68.6 | 60.0 | 51.6 | 68.1 |
| 2000 | 53.6 | 57.6 | 63.6 | 64.7 | 76.9 | 79.6 | 84.3 | 83.9 | 77.3 | 68.2 | 57.0 | 46.1 | 67.7 |
| 2001 | 47.2 | 58.3 | 56.9 | 69.6 | 74.5 | 79.1 | 81.6 | 80.7 | 76.0 | 65.1 | 63.9 | 55.1 | 67.3 |
| 2002 | 52.5 | 50.3 | 59.1 | 69.8 | 73.7 | 79.6 | 81.3 | 80.6 | 79.3 | 71.6 | 56.2 | 51.1 | 67.1 |
| 2003 | 46.2 | 53.7 | 62.0 | 67.9 | 77.3 | 80.0 | 81.1 | 82.0 | 77.6 | 69.2 | 63.3 | 49.9 | 67.5 |
| 2004 | 51.0 | 51.2 | 64.2 | 65.8 | 75.1 | 80.2 | 82.7 | 80.2 | 78.5 | 74.2 | 62.7 | 50.7 | 68.0 |
| 2005 | 55.1 | 56.7 | 58.9 | 64.5 | 73.1 | 79.2 | 82.2 | 82.1 | 81.0 | 68.0 | 61.0 | 50.6 | 67.7 |
| 2006 | 56.0 | 52.8 | 62.4 | 70.9 | 74.1 | 80.7 | 81.8 | 82.8 | 76.5 | 67.7 | 57.0 | 53.8 | 68.0 |
| 2007 | 51.5 | 51.4 | 62.9 | 64.5 | 73.1 | 80.1 | 81.0 | 84.2 | 78.6 | 70.0 | 58.9 | 57.1 | 67.8 |
| 2008 | 49.2 | 55.6 | 59.0 | 66.0 | 73.9 | 80.0 | 81.7 | 80.2 | 77.1 | 66.1 | 56.7 | 56.5 | 66.8 |
| 2009 | 52.4 | 54.5 | 62.6 | 66.2 | 75.4 | 82.5 | 82.1 | 80.0 | 79.0 | 69.0 | 57.6 | 50.4 | 67.6 |
| POR= 62 YRS | 50.9 | 54.1 | 59.3 | 67.2 | 74.5 | 80.1 | 82.0 | 81.7 | 77.7 | 68.5 | 59.0 | 52.9 | 67.3 |

## HEATING DEGREE DAYS (base 65°F) 2009  MOBILE (KMOB)

| YEAR | JUL | AUG | SEP | OCT | NOV | DEC | JAN | FEB | MAR | APR | MAY | JUN | TOTAL |
|------|-----|-----|-----|-----|-----|-----|-----|-----|-----|-----|-----|-----|-------|
| 1980-81 | 0 | 0 | 0 | 63 | 235 | 420 | 581 | 323 | 180 | 9 | 4 | 0 | 1815 |
| 1981-82 | 0 | 0 | 5 | 42 | 113 | 463 | 485 | 323 | 161 | 48 | 0 | 0 | 1640 |
| 1982-83 | 0 | 0 | 5 | 63 | 185 | 296 | 545 | 383 | 306 | 119 | 4 | 0 | 1906 |
| 1983-84 | 0 | 0 | 7 | 60 | 243 | 529 | 582 | 363 | 218 | 69 | 10 | 0 | 2081 |
| 1984-85 | 0 | 0 | 2 | 17 | 270 | 172 | 665 | 397 | 84 | 50 | 1 | 0 | 1658 |
| 1985-86 | 0 | 0 | 1 | 22 | 69 | 503 | 469 | 251 | 188 | 40 | 0 | 0 | 1543 |
| 1986-87 | 0 | 0 | 0 | 43 | 105 | 433 | 500 | 311 | 193 | 106 | 0 | 0 | 1691 |
| 1987-88 | 0 | 0 | 0 | 108 | 194 | 283 | 565 | 421 | 214 | 33 | 0 | 0 | 1818 |
| 1988-89 | 0 | 0 | 0 | 53 | 122 | 363 | 254 | 335 | 170 | 81 | 2 | 0 | 1380 |
| 1989-90 | 0 | 0 | 6 | 71 | 204 | 630 | 326 | 186 | 128 | 71 | 0 | 0 | 1622 |
| 1990-91 | 0 | 0 | 2 | 58 | 145 | 287 | 382 | 258 | 156 | 13 | 0 | 0 | 1301 |
| 1991-92 | 0 | 0 | 3 | 32 | 344 | 342 | 477 | 236 | 165 | 77 | 16 | 0 | 1692 |
| 1992-93 | 0 | 0 | 0 | 18 | 276 | 297 | 307 | 340 | 260 | 110 | 3 | 0 | 1611 |
| 1993-94 | 0 | 0 | 0 | 64 | 278 | 464 | 543 | 260 | 182 | 46 | 0 | 0 | 1837 |
| 1994-95 | 0 | 0 | 0 | 35 | 93 | 304 | 426 | 281 | 148 | 41 | 3 | 0 | 1331 |
| 1995-96 | 0 | 0 | 0 | 29 | 263 | 445 | 489 | 381 | 323 | 97 | 3 | 0 | 2030 |
| 1996-97 | 0 | 0 | 0 | 48 | 212 | 346 | 440 | 293 | 63 | 80 | 5 | 0 | 1487 |
| 1997-98 | 0 | 0 | 0 | 89 | 296 | 475 | 390 | 313 | 266 | 60 | 0 | 0 | 1889 |
| 1998-99 | 0 | 0 | 0 | 24 | 94 | 287 | 319 | 234 | 204 | 30 | 3 | 0 | 1195 |
| 1999-00 | 0 | 0 | 2 | 52 | 162 | 414 | 363 | 242 | 85 | 68 | 0 | 0 | 1388 |
| 2000-01 | 0 | 0 | 6 | 45 | 302 | 579 | 542 | 208 | 262 | 36 | 0 | 0 | 1980 |
| 2001-02 | 0 | 0 | 4 | 95 | 89 | 312 | 404 | 405 | 238 | 24 | 10 | 0 | 1581 |
| 2002-03 | 0 | 0 | 0 | 18 | 279 | 424 | 576 | 314 | 113 | 48 | 0 | 0 | 1772 |
| 2003-04 | 0 | 0 | 1 | 16 | 143 | 459 | 444 | 395 | 87 | 66 | 9 | 0 | 1620 |
| 2004-05 | 0 | 0 | 0 | 10 | 124 | 449 | 312 | 241 | 205 | 54 | 5 | 0 | 1400 |
| 2005-06 | 0 | 0 | 0 | 70 | 180 | 448 | 285 | 339 | 145 | 1 | 2 | 0 | 1470 |
| 2006-07 | 0 | 0 | 0 | 63 | 254 | 341 | 419 | 377 | 121 | 99 | 0 | 0 | 1674 |
| 2007-08 | 0 | 0 | 0 | 50 | 202 | 280 | 495 | 283 | 202 | 67 | 0 | 0 | 1579 |
| 2008-09 | 0 | 0 | 0 | 87 | 265 | 302 | 395 | 295 | 113 | 60 | 1 | 0 | 1518 |
| 2009- | 0 | 0 | 0 | 81 | 217 | 446 | | | | | | | |

WBAN : 13894

## COOLING DEGREE DAYS (base 65°F) 2009  MOBILE (KMOB)

| YEAR | JAN | FEB | MAR | APR | MAY | JUN | JUL | AUG | SEP | OCT | NOV | DEC | TOTAL |
|------|-----|-----|-----|-----|-----|-----|-----|-----|-----|-----|-----|-----|-------|
| 1980 | 2 | 17 | 61 | 89 | 305 | 495 | 618 | 571 | 512 | 110 | 25 | 9 | 2814 |
| 1981 | 0 | 4 | 15 | 195 | 217 | 538 | 595 | 557 | 366 | 185 | 60 | 4 | 2736 |
| 1982 | 17 | 2 | 119 | 118 | 296 | 475 | 502 | 503 | 342 | 177 | 54 | 53 | 2658 |
| 1983 | 0 | 0 | 4 | 27 | 229 | 351 | 532 | 514 | 271 | 140 | 29 | 2 | 2099 |
| 1984 | 0 | 3 | 24 | 96 | 249 | 401 | 469 | 435 | 327 | 261 | 16 | 34 | 2315 |
| 1985 | 4 | 2 | 77 | 133 | 275 | 452 | 461 | 504 | 339 | 227 | 108 | 7 | 2589 |
| 1986 | 0 | 15 | 31 | 79 | 295 | 486 | 573 | 489 | 444 | 158 | 89 | 8 | 2667 |
| 1987 | 4 | 0 | 21 | 99 | 341 | 422 | 527 | 544 | 373 | 41 | 27 | 28 | 2427 |
| 1988 | 7 | 10 | 21 | 108 | 251 | 442 | 529 | 513 | 404 | 67 | 61 | 20 | 2433 |
| 1989 | 16 | 33 | 74 | 116 | 289 | 431 | 498 | 523 | 366 | 114 | 31 | 0 | 2491 |
| 1990 | 7 | 23 | 34 | 80 | 267 | 493 | 525 | 569 | 417 | 155 | 29 | 36 | 2635 |
| 1991 | 1 | 6 | 66 | 149 | 348 | 448 | 528 | 513 | 371 | 176 | 23 | 22 | 2651 |
| 1992 | 0 | 9 | 31 | 97 | 249 | 437 | 548 | 454 | 373 | 102 | 20 | 13 | 2333 |
| 1993 | 5 | 2 | 21 | 33 | 224 | 455 | 554 | 559 | 408 | 165 | 37 | 0 | 2463 |
| 1994 | 0 | 18 | 33 | 161 | 268 | 443 | 469 | 485 | 357 | 155 | 66 | 6 | 2461 |
| 1995 | 0 | 12 | 55 | 106 | 338 | 416 | 554 | 571 | 417 | 145 | 20 | 35 | 2669 |
| 1996 | 1 | 23 | 30 | 69 | 335 | 419 | 526 | 473 | 349 | 125 | 44 | 17 | 2411 |
| 1997 | 15 | 18 | 109 | 42 | 236 | 377 | 521 | 519 | 435 | 163 | 7 | 0 | 2442 |
| 1998 | 0 | 0 | 31 | 73 | 359 | 531 | 581 | 569 | 412 | 196 | 49 | 52 | 2853 |
| 1999 | 24 | 25 | 11 | 212 | 280 | 432 | 516 | 611 | 340 | 173 | 19 | 3 | 2646 |
| 2000 | 16 | 32 | 48 | 69 | 377 | 446 | 607 | 592 | 383 | 151 | 67 | 0 | 2788 |
| 2001 | 1 | 27 | 16 | 177 | 303 | 427 | 524 | 492 | 338 | 104 | 62 | 12 | 2483 |
| 2002 | 26 | 0 | 60 | 176 | 285 | 444 | 513 | 488 | 435 | 230 | 22 | 0 | 2679 |
| 2003 | 0 | 5 | 23 | 143 | 386 | 459 | 504 | 536 | 388 | 153 | 100 | 0 | 2697 |
| 2004 | 18 | 0 | 69 | 97 | 329 | 460 | 556 | 479 | 414 | 306 | 62 | 10 | 2800 |
| 2005 | 10 | 16 | 24 | 46 | 265 | 432 | 538 | 536 | 486 | 172 | 66 | 8 | 2599 |
| 2006 | 15 | 6 | 70 | 187 | 293 | 479 | 530 | 558 | 350 | 151 | 22 | 1 | 2662 |
| 2007 | 9 | 0 | 62 | 91 | 258 | 462 | 504 | 602 | 414 | 213 | 24 | 41 | 2680 |
| 2008 | 8 | 16 | 25 | 101 | 283 | 459 | 523 | 480 | 371 | 128 | 21 | 47 | 2462 |
| 2009 | 11 | 8 | 43 | 101 | 333 | 530 | 536 | 473 | 426 | 212 | 3 | 4 | 2680 |

**SNOWFALL (inches) 2009 MOBILE (KMOB)**

| YEAR | JUL | AUG | SEP | OCT | NOV | DEC | JAN | FEB | MAR | APR | MAY | JUN | TOTAL |
|---|---|---|---|---|---|---|---|---|---|---|---|---|---|
| 1980-81 | 0.0 | 0.0 | 0.0 | 0.0 | 0.0 | 0.0 | T | T | 0.0 | 0.0 | 0.0 | 0.0 | T |
| 1981-82 | 0.0 | 0.0 | 0.0 | 0.0 | 0.0 | 0.0 | T | 0.0 | 0.0 | 0.0 | 0.0 | 0.0 | T |
| 1982-83 | 0.0 | 0.0 | 0.0 | 0.0 | 0.0 | 0.0 | 0.0 | 0.0 | 0.0 | 0.0 | 0.0 | 0.0 | 0.0 |
| 1983-84 | 0.0 | 0.0 | 0.0 | 0.0 | 0.0 | T | 0.0 | T | 0.0 | 0.0 | 0.0 | 0.0 | T |
| 1984-85 | 0.0 | 0.0 | 0.0 | 0.0 | 0.0 | 0.0 | T | T | 0.0 | 0.0 | 0.0 | 0.0 | T |
| 1985-86 | 0.0 | 0.0 | 0.0 | 0.0 | 0.0 | 0.0 | 0.0 | 0.0 | 0.0 | 0.0 | 0.0 | 0.0 | 0.0 |
| 1986-87 | 0.0 | 0.0 | 0.0 | 0.0 | 0.0 | 0.0 | T | 0.0 | 0.0 | T | 0.0 | 0.0 | T |
| 1987-88 | 0.0 | 0.0 | 0.0 | 0.0 | 0.0 | 0.0 | 0.0 | 1.7 | 0.0 | T | 0.0 | 0.0 | 1.7 |
| 1988-89 | 0.0 | 0.0 | 0.0 | 0.0 | 0.0 | 0.0 | 0.0 | 0.0 | T | 0.0 | 0.0 | 0.0 | T |
| 1989-90 | T | 0.0 | 0.0 | 0.0 | 0.0 | T | 0.0 | 0.0 | 0.0 | 0.0 | 0.0 | 0.0 | T |
| 1990-91 | 0.0 | 0.0 | 0.0 | 0.0 | 0.0 | 0.0 | 0.0 | 0.0 | 0.0 | 0.0 | T | 0.0 | T |
| 1991-92 | 0.0 | 0.0 | 0.0 | 0.0 | 0.0 | 0.0 | 0.0 | 0.0 | 0.0 | 0.0 | 0.0 | 0.0 | 0.0 |
| 1992-93 | T | 0.0 | 0.0 | 0.0 | 0.0 | 0.0 | 0.0 | 0.0 | 2.7 | 0.0 | 0.0 | 0.0 | 2.7 |
| 1993-94 | 0.0 | 0.0 | 0.0 | 0.0 | 0.0 | T | 0.0 | T | 0.0 | 0.0 | 0.0 | 0.0 | T |
| 1994-95 | 0.0 | 0.0 | 0.0 | 0.0 | 0.0 | 0.0 | T | T | 0.0 | 0.0 | 0.0 | 0.0 | T |
| 1995-96 | T | 0.0 | 0.0 | 0.0 | 0.0 | T | 0.0 | | | | | | |
| 1996-97 | | | | | | 1.0 | | | | | | | |
| 1997-98 | | | | | | | | | | | | | |
| 1998-99 | | | | 0.0 | 0.0 | | 0.0 | 0.0 | 0.0 | 0.0 | 0.0 | 0.0 | |
| 1999-00 | 0.0 | 0.0 | 0.0 | 0.0 | 0.0 | 0.0 | 0.0 | 0.0 | 0.0 | T | 0.0 | 0.0 | T |
| 2000-01 | 0.0 | 0.0 | 0.0 | 0.0 | 0.0 | 0.0 | T | 0.0 | 0.0 | 0.0 | 0.0 | 0.0 | T |
| 2001-02 | 0.0 | 0.0 | 0.0 | 0.0 | 0.0 | 0.0 | T | 0.0 | T | 0.0 | 0.0 | 0.0 | T |
| 2002-03 | 0.0 | 0.0 | 0.0 | 0.0 | 0.0 | 0.0 | T | 0.0 | 0.0 | 0.0 | 0.0 | 0.0 | T |
| 2003-04 | 0.0 | 0.0 | 0.0 | 0.0 | 0.0 | 0.0 | 0.0 | T | 0.0 | 0.0 | 0.0 | 0.0 | T |
| 2004-05 | 0.0 | 0.0 | 0.0 | 0.0 | 0.0 | T | 0.0 | 0.0 | 0.0 | T | 0.0 | 0.0 | T |
| 2005-06 | 0.0 | 0.0 | 0.0 | 0.0 | 0.0 | 0.0 | 0.0 | 0.0 | 0.0 | 0.0 | 0.0 | T | T |
| 2006-07 | 0.0 | 0.0 | 0.0 | 0.0 | 0.0 | 0.0 | 0.0 | 0.0 | 0.0 | T | 0.0 | 0.0 | T |
| 2007-08 | 0.0 | 0.0 | 0.0 | 0.0 | 0.0 | 0.0 | T | 0.0 | 0.0 | 0.0 | 0.0 | 0.0 | T |
| 2008-09 | 0.0 | 0.0 | 0.0 | 0.0 | 0.0 | 0.0 | 0.0 | 0.0 | T | 0.0 | 0.0 | 0.0 | T |
| 2009- | 0.0 | 0.0 | 0.0 | 0.0 | 0.0 | 0.0 | | | | | | | |
| POR= 59 YRS | T | 0.0 | 0.0 | 0.0 | 0.0 | 0.1 | 0.1 | 0.1 | 0.1 | T | T | T | 0.4 |

# Mobile, AL 2009
## Temperature and Precipitation

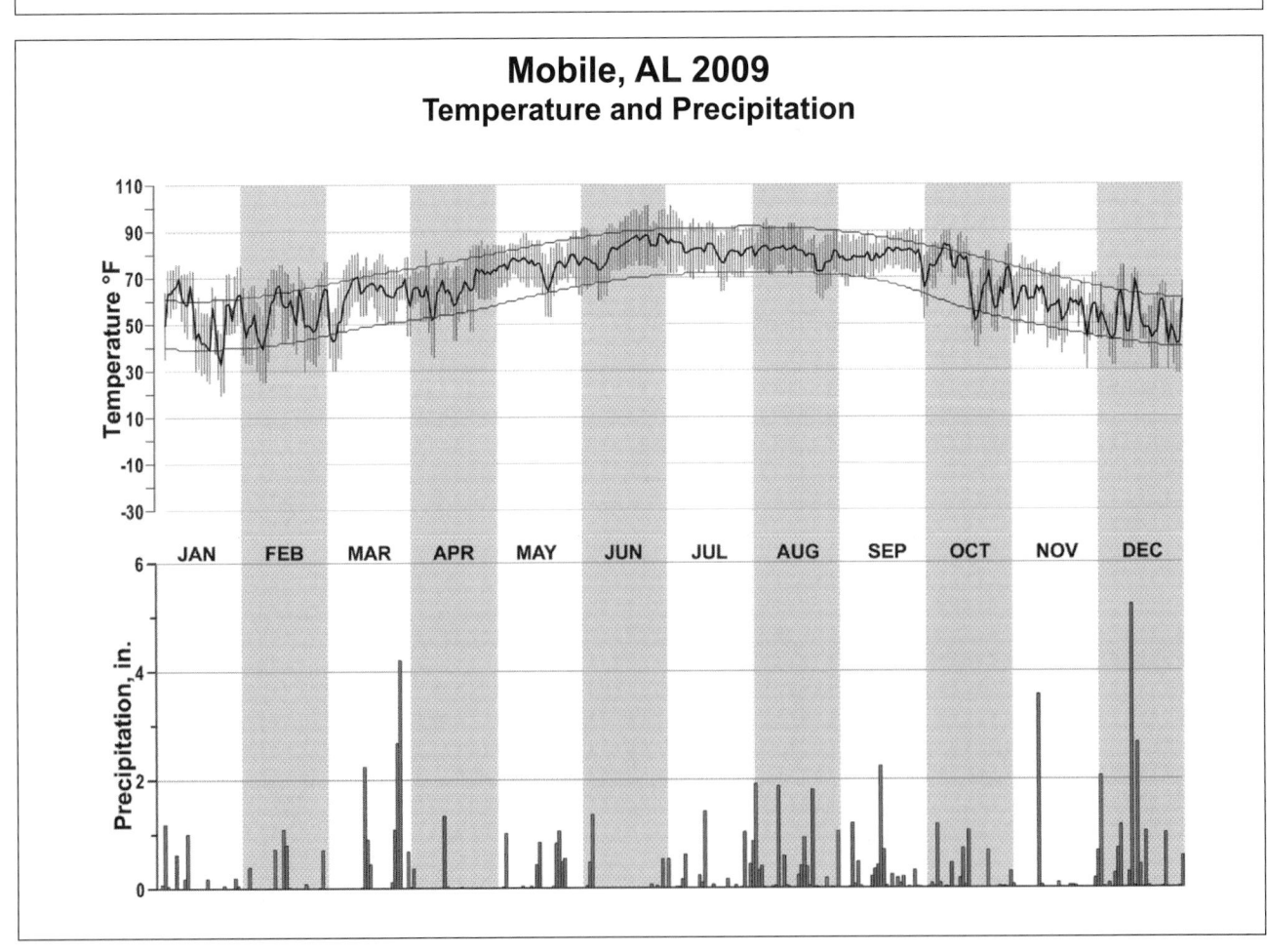

# 2009
# ANCHORAGE
# ALASKA (PANC)

Anchorage is in a broad valley with adjacent narrow bodies of water. Cook Inlet, including Knik Arm and Turnagain Arm, lies approximately 2 miles to the west, north, and south. The terrain rises gradually to the east for about 10 miles, with marshes interspersed with glacial moraines, shallow depressions, small streams, and knolls. Beyond this area, the Chugach Mountains rise abruptly into a range oriented north-northeast to south-southwest, with average elevation 4,000 to 5,000 feet and some peaks to 8,000 or 10,000 feet. The Chugach Range acts as a barrier to the influx of warm, moist air from the Gulf of Alaska, so the average annual precipitation is only 10 to 15 percent of that at stations located on the Gulf of Alaska side of the Chugach Range. The Alaska Mountain Range lies in a long arc from southwest, through northwest, to northeast, approximately 100 miles distant from Anchorage. During the winter, this range is an effective barrier to the influx of very cold air from the north side of the range.

The four seasons are well marked in Anchorage. In the summer, high temperatures average about 60 degrees and low temperatures nearly 50 degrees. Temperatures in the 70s are considered very warm. On summer days, temperatures on the east side of Anchorage may be about 10 degrees warmer than the official airport readings. Rain increases after mid-June. About two-thirds of the days in July and August are cloudy and one-third have rain.

Autumn is brief, beginning in early September and ending in mid-October. Temperatures begin to fall in September with snow becoming more frequent in October.

Winter can be considered as mid-October to early April when streams and lakes are frozen. Temperatures steadily decrease into January when the highs are near 20 degrees and lows near 5 degrees. The coldest weather is normally in January, when very cold days have high temperatures below zero. Cold days generally have clear skies and calm wind. Mild days do occur with temperatures in the 30s. On cold winter nights, temperatures on the east side of Anchorage may be 10-20 degrees lower than airport readings on the west side. Most winter precipitation is snow, but rain may occur on a few days.

Annual snowfall varies from about 70 inches on the west side to about 90 inches on the east side of Anchorage at low elevations. Along the Chugach Mountains, snow totals increase steadily with increasing elevations and winter arrives a month earlier and stays a month longer at the 1,000 to 2,000 foot level. Most snow is light or dry, i.e., low in water content. Freezing rain is extremely rare. Fog, made of water droplets, occurs on about fifteen days. In general, ice-fog does not occur in Anchorage.

Spring begins in late April and May when days are warm and sunny, nights are cool, and precipitation is exceedingly small. Foliage turns green by late May.

The wind in Anchorage is generally light. However, on several days each winter, strong northerly winds, up to 90 mph, affect the entire Anchorage area. Also during the winter there are about eight occurrences of very strong southeast winds which affect only the east side of Anchorage and the slopes of the Chugach Mountains. These winds occur more often above the 800 feet elevation in the Chugach where winds are funneled thru creek canyons. On the east side of Anchorage, damaging winds of over 100 mph have been recorded.

The average occurrence of the first snow is mid-October, but has occurred as early as mid-September. The average date of the last snow is mid-April, but has occurred as late as early May. The growing season is about 125 days. Average occurrence of the last temperature of 32 degrees in spring is mid-May and the first in fall is mid-September. Daylight varies from about 19 hours in late June to 6 hours in late December with 12 hours of daylight occurring in late September and late March.

# NORMALS, MEANS, AND EXTREMES
## ANCHORAGE (PANC)

| LATITUDE: 61° 10'N | LONGITUDE: -149° 59'W | ELEVATION (FT): GRND: 120  BARO: 222 | TIME ZONE: ALASKA (UTC -9) | WBAN: 26451 |

| ELEMENT | POR | JAN | FEB | MAR | APR | MAY | JUN | JUL | AUG | SEP | OCT | NOV | DEC | YEAR |
|---|---|---|---|---|---|---|---|---|---|---|---|---|---|---|
| **TEMPERATURE °F** | | | | | | | | | | | | | | |
| NORMAL DAILY MAXIMUM | 30 | 22.2 | 25.8 | 33.6 | 43.9 | 54.9 | 62.3 | 65.3 | 63.3 | 55.0 | 40.0 | 27.7 | 23.7 | 43.1 |
| MEAN DAILY MAXIMUM | 56 | 21.6 | 25.9 | 32.7 | 43.3 | 54.7 | 62.1 | 65.3 | 63.2 | 54.9 | 40.5 | 27.9 | 22.8 | 42.9 |
| HIGHEST DAILY MAXIMUM | 56 | 50 | 48 | 51 | 69 | 77 | 85 | 84 | 82 | 73 | 64 | 54 | 48 | 85 |
| YEAR OF OCCURRENCE | | 1961 | 1991 | 1984 | 2005 | 2002 | 1969 | 2003 | 1978 | 1957 | 2006 | 2002 | 2005 | JUN 1969 |
| MEAN OF EXTREME MAXS. | 56 | 40.4 | 41.9 | 44.1 | 54.7 | 67.0 | 73.3 | 75.6 | 73.1 | 64.5 | 53.7 | 42.4 | 40.9 | 56.0 |
| NORMAL DAILY MINIMUM | 30 | 9.3 | 11.7 | 18.2 | 28.7 | 38.9 | 47.0 | 51.5 | 49.4 | 41.4 | 28.3 | 15.9 | 11.4 | 29.3 |
| MEAN DAILY MINIMUM | 56 | 8.9 | 12.2 | 17.5 | 28.5 | 38.9 | 47.0 | 51.6 | 49.6 | 41.4 | 28.6 | 16.1 | 10.5 | 29.2 |
| LOWEST DAILY MINIMUM | 56 | -34 | -28 | -24 | -4 | 17 | 33 | 38 | 31 | 19 | -5 | -21 | -30 | -34 |
| YEAR OF OCCURRENCE | | 1975 | 1999 | 1971 | 1985 | 1964 | 1961 | 1964 | 1984 | 1992 | 1956 | 1956 | 1964 | JAN 1975 |
| MEAN OF EXTREME MINS. | 56 | -12.2 | -7.2 | -0.0 | 16.3 | 29.9 | 38.9 | 44.6 | 40.2 | 29.6 | 12.3 | -2.0 | -10.7 | 15.0 |
| NORMAL DRY BULB | 30 | 15.8 | 18.7 | 25.9 | 36.3 | 46.9 | 54.7 | 58.4 | 56.4 | 48.2 | 34.1 | 21.8 | 17.5 | 36.2 |
| MEAN DRY BULB | 56 | 15.3 | 19.1 | 25.1 | 35.9 | 46.8 | 54.8 | 58.5 | 56.4 | 48.1 | 34.6 | 22.0 | 16.6 | 36.1 |
| MEAN WET BULB | 26 | 15.4 | 17.7 | 22.0 | 31.6 | 40.8 | 48.6 | 53.3 | 52.1 | 44.7 | 31.7 | 20.1 | 17.9 | 33.0 |
| MEAN DEW POINT | 26 | 12.5 | 14.7 | 17.5 | 26.4 | 35.6 | 44.5 | 50.5 | 49.5 | 41.7 | 28.6 | 17.4 | 15.6 | 29.5 |
| NORMAL NO. DAYS WITH: | | | | | | | | | | | | | | |
| MAXIMUM >= 70 | 30 | 0.0 | 0.0 | 0.0 | 0.0 | 0.6 | 3.4 | 6.5 | 3.4 | 0.1 | 0.0 | 0.0 | 0.0 | 14.0 |
| MAXIMUM <= 32 | 30 | 24.0 | 19.6 | 11.9 | 2.1 | 0.0 | 0.0 | 0.0 | 0.0 | 0.0 | 5.0 | 20.9 | 24.8 | 108.3 |
| MINIMUM <= 32 | 30 | 30.5 | 27.4 | 29.3 | 20.5 | 2.8 | 0.0 | 0.0 | 0.1 | 3.3 | 20.5 | 28.5 | 30.3 | 193.2 |
| MINIMUM <= 0 | 30 | 8.6 | 6.6 | 2.1 | * | 0.0 | 0.0 | 0.0 | 0.0 | 0.0 | 0.1 | 2.8 | 6.4 | 26.6 |
| **H/C** | | | | | | | | | | | | | | |
| NORMAL HEATING DEG. DAYS | 30 | 1526 | 1295 | 1212 | 861 | 560 | 311 | 206 | 268 | 505 | 957 | 1297 | 1472 | 10470 |
| NORMAL COOLING DEG. DAYS | 30 | 0 | 0 | 0 | 0 | 0 | 0 | 3 | 0 | 0 | 0 | 0 | 0 | 3 |
| **RH** | | | | | | | | | | | | | | |
| NORMAL (PERCENT) | 30 | 77 | 73 | 69 | 65 | 62 | 66 | 73 | 76 | 77 | 76 | 79 | 80 | 73 |
| HOUR 03 LST | 30 | 77 | 76 | 74 | 75 | 74 | 77 | 82 | 85 | 83 | 79 | 81 | 80 | 79 |
| HOUR 09 LST | 30 | 77 | 76 | 72 | 66 | 63 | 67 | 73 | 78 | 79 | 79 | 80 | 80 | 74 |
| HOUR 15 LST | 30 | 75 | 68 | 59 | 54 | 51 | 56 | 63 | 65 | 65 | 68 | 76 | 79 | 65 |
| HOUR 21 LST | 30 | 77 | 74 | 71 | 67 | 61 | 65 | 72 | 78 | 79 | 78 | 80 | 80 | 74 |
| **S** PERCENT POSSIBLE SUNSHINE | 40 | 34 | 42 | 50 | 50 | 50 | 46 | 42 | 38 | 38 | 35 | 31 | 26 | 40 |
| **W/O** MEAN NO. DAYS WITH: | | | | | | | | | | | | | | |
| HEAVY FOG(VISBY <= 1/4 MI) | 46 | 4.7 | 3.7 | 1.3 | 0.7 | 0.2 | 0.0 | 0.1 | 0.6 | 0.9 | 2.2 | 3.7 | 4.5 | 22.6 |
| THUNDERSTORMS | 56 | 0.0 | 0.0 | 0.0 | 0.0 | 0.1 | 0.2 | 0.4 | 0.4 | 0.2 | 0.0 | 0.0 | 0.0 | 1.3 |
| **CLOUDNESS** MEAN: | | | | | | | | | | | | | | |
| SUNRISE-SUNSET (OKTAS) | 44 | 5.6 | 5.7 | 5.4 | 5.8 | 6.1 | 6.3 | 6.3 | 6.3 | 6.3 | 6.1 | 5.8 | 6.0 | 6.0 |
| MIDNIGHT-MIDNIGHT (OKTAS) | 33 | 5.5 | 5.4 | 5.3 | 5.6 | 6.1 | 6.4 | 6.4 | 6.2 | 6.0 | 5.8 | 5.6 | 5.9 | 5.9 |
| MEAN NO. DAYS WITH: | | | | | | | | | | | | | | |
| CLEAR | 45 | 7.1 | 6.3 | 7.7 | 5.7 | 3.9 | 2.8 | 3.2 | 3.2 | 3.6 | 5.0 | 5.5 | 5.6 | 59.6 |
| PARTLY CLOUDY | 45 | 4.8 | 3.7 | 5.6 | 6.1 | 6.6 | 6.9 | 5.8 | 6.1 | 5.3 | 4.6 | 4.3 | 3.9 | 64.0 |
| CLOUDY | 45 | 19.3 | 18.1 | 17.6 | 18.3 | 20.4 | 20.2 | 21.2 | 21.1 | 20.5 | 20.6 | 19.3 | 20.9 | 237.5 |
| **PR** | | | | | | | | | | | | | | |
| MEAN STATION PRESSURE(IN) | 26 | 29.51 | 29.59 | 29.60 | 29.65 | 29.73 | 29.77 | 29.81 | 29.74 | 29.62 | 29.51 | 29.47 | 29.45 | 29.62 |
| MEAN SEA-LEVEL PRES. (IN) | 26 | 29.67 | 29.76 | 29.75 | 29.80 | 29.88 | 29.91 | 29.95 | 29.88 | 29.77 | 29.66 | 29.62 | 29.60 | 29.77 |
| **WINDS** | | | | | | | | | | | | | | |
| MEAN SPEED (MPH) | 26 | 6.6 | 6.7 | 7.4 | 7.3 | 8.5 | 8.3 | 7.4 | 6.8 | 7.1 | 6.7 | 6.7 | 6.4 | 7.2 |
| PREVAIL.DIR(TENS OF DEGS) | 46 | 01 | 36 | 36 | 17 | 17 | 17 | 17 | 17 | 17 | 36 | 01 | 01 | 17 |
| MAXIMUM 2-MINUTE: | | | | | | | | | | | | | | |
| SPEED (MPH) | 11 | 35 | 41 | 54 | 37 | 29 | 33 | 30 | 33 | 33 | 32 | 36 | 38 | 54 |
| DIR. (TENS OF DEGS) | | 14 | 15 | 02 | 17 | 15 | 18 | 17 | 14 | 15 | 15 | 14 | 16 | 02 |
| YEAR OF OCCURRENCE | | 2009 | 2000 | 2003 | 2005 | 2008 | 1999 | 2001 | 2005 | 2000 | 2006 | 2002 | 1999 | MAR 2003 |
| MAXIMUM 3-SECOND | | | | | | | | | | | | | | |
| SPEED (MPH) | 11 | 54 | 58 | 71 | 51 | 43 | 46 | 41 | 45 | 47 | 47 | 47 | 49 | 71 |
| DIR. (TENS OF DEGS) | | 16 | 12 | 01 | 17 | 15 | 16 | 14 | 14 | 15 | 13 | 15 | 16 | 01 |
| YEAR OF OCCURRENCE | | 2009 | 2000 | 2003 | 2005 | 2008 | 2007 | 2009 | 2005 | 2000 | 2006 | 2009 | 1999 | MAR 2003 |
| **PRECIPITATION** | | | | | | | | | | | | | | |
| NORMAL (IN) | 30 | 0.68 | 0.74 | 0.65 | 0.52 | 0.70 | 1.06 | 1.70 | 2.93 | 2.87 | 2.09 | 1.09 | 1.05 | 16.08 |
| MAXIMUM MONTHLY (IN) | 56 | 2.13 | 3.07 | 2.76 | 2.32 | 1.93 | 3.40 | 4.49 | 9.77 | 7.35 | 4.28 | 2.84 | 2.67 | 9.77 |
| YEAR OF OCCURRENCE | | 1949 | 1955 | 1979 | 2008 | 1989 | 1962 | 2001 | 1989 | 2004 | 2002 | 1976 | 1955 | AUG 1989 |
| MINIMUM MONTHLY (IN) | 56 | 0.02 | 0.07 | T | T | 0.02 | 0.17 | 0.42 | 0.33 | 0.72 | 0.35 | 0.04 | 0.09 | T |
| YEAR OF OCCURRENCE | | 1982 | 1958 | 1983 | 1969 | 1957 | 1993 | 1972 | 1969 | 1998 | 1960 | 2006 | 1995 | MAR 1983 |
| MAXIMUM IN 24 HOURS (IN) | 56 | 1.19 | 1.16 | 1.25 | 1.32 | 1.18 | 1.84 | 2.37 | 4.12 | 1.92 | 1.62 | 1.66 | 1.62 | 4.12 |
| YEAR OF OCCURRENCE | | 1961 | 1956 | 1986 | 2008 | 1980 | 1962 | 2001 | 1989 | 1961 | 2003 | 1964 | 1955 | AUG 1989 |
| NORMAL NO. DAYS WITH: | | | | | | | | | | | | | | |
| PRECIPITATION >= 0.01 | 30 | 8.1 | 7.3 | 6.8 | 5.5 | 7.0 | 8.2 | 11.3 | 13.8 | 14.5 | 12.3 | 9.3 | 11.1 | 115.2 |
| PRECIPITATION >= 1.00 | 30 | * | 0.0 | * | 0.0 | 0.0 | * | 0.1 | 0.3 | 0.2 | 0.1 | 0.0 | 0.0 | 0.7 |
| **SNOWFALL** | | | | | | | | | | | | | | |
| NORMAL (IN) | 30 | 9.0 | 11.0 | 10.3 | 4.1 | 0.1 | 0.0 | 0.0 | 0.* | 0.2 | 8.5 | 11.4 | 14.9 | 69.5 |
| MAXIMUM MONTHLY (IN) | 56 | 29.3 | 52.1 | 31.0 | 30.8 | 6.1 | 0.0 | 0.0 | T | 6.3 | 28.1 | 38.8 | 41.6 | 52.1 |
| YEAR OF OCCURRENCE | | 2007 | 1996 | 1979 | 2008 | 2001 | | | 2008 | 2004 | 1996 | 1994 | 1955 | FEB 1996 |
| MAXIMUM IN 24 HOURS (IN) | 56 | 10.5 | 13.9 | 22.0 | 9.1 | 5.0 | 0.0 | 0.0 | T | 6.0 | 14.6 | 16.4 | 17.7 | 22.0 |
| YEAR OF OCCURRENCE' | | 1955 | 1996 | 2002 | 1955 | 2001 | | | 1997 | 2004 | 1996 | 1964 | 1955 | MAR 2002 |
| MAXIMUM SNOW DEPTH (IN) | 55 | 833 | 840 | 906 | 356 | 17 | 0 | 0 | 0 | 6 | 105 | 416 | 715 | 906 |
| YEAR OF OCCURRENCE | | 1956 | 1956 | 1959 | 1955 | 1955 | | | | 2004 | 1991 | 1994 | 1994 | MAR 1959 |
| NORMAL NO. DAYS WITH: | | | | | | | | | | | | | | |
| SNOWFALL >= 1.0 | 30 | 3.0 | 2.8 | 2.6 | 1.3 | 0.0 | 0.0 | 0.0 | 0.0 | 0.1 | 2.4 | 3.5 | 3.9 | 19.6 |

## PRECIPITATION (inches) 2009 ANCHORAGE (PANC)

| YEAR | JAN | FEB | MAR | APR | MAY | JUN | JUL | AUG | SEP | OCT | NOV | DEC | ANNUAL |
|---|---|---|---|---|---|---|---|---|---|---|---|---|---|
| 1980 | 1.28 | 1.18 | 0.30 | 0.19 | 1.68 | 2.73 | 2.27 | 3.06 | 2.53 | 3.05 | 0.49 | 0.41 | 19.17 |
| 1981 | 0.93 | 0.97 | 0.41 | 0.19 | 0.81 | 0.83 | 4.39 | 4.96 | 2.15 | 3.49 | 1.85 | 0.36 | 21.34 |
| 1982 | 0.02 | 0.69 | 0.42 | 0.27 | 0.54 | 1.56 | 2.41 | 2.33 | 4.66 | 2.95 | 1.72 | 0.11 | 17.68 |
| 1983 | 0.21 | 0.23 | T | 1.36 | 0.59 | 0.66 | 0.55 | 2.89 | 2.29 | 2.67 | 0.23 | 0.48 | 12.16 |
| 1984 | 1.30 | 1.08 | 0.08 | 0.93 | 0.96 | 1.10 | 1.11 | 3.21 | 2.59 | 1.38 | 0.15 | 1.08 | 14.97 |
| 1985 | 0.70 | 0.67 | 0.86 | 0.50 | 1.45 | 1.01 | 0.99 | 3.54 | 3.17 | 1.07 | 0.08 | 1.47 | 15.51 |
| 1986 | 0.20 | 0.55 | 1.70 | 0.42 | 0.50 | 0.33 | 2.02 | 3.62 | 2.85 | 4.11 | 1.23 | 1.42 | 18.95 |
| 1987 | 1.72 | 0.20 | 0.17 | 0.24 | 0.67 | 1.09 | 1.89 | 0.43 | 1.91 | 2.60 | 1.90 | 1.12 | 13.94 |
| 1988 | 0.38 | 0.32 | 0.65 | 0.37 | 0.56 | 0.79 | 0.64 | 3.77 | 1.26 | 2.96 | 1.11 | 1.51 | 14.32 |
| 1989 | 0.26 | 0.17 | 0.22 | 0.98 | 1.93 | 1.14 | 2.89 | 9.77 | 3.92 | 3.63 | 1.01 | 1.63 | 27.55 |
| 1990 | 1.42 | 1.46 | 0.46 | 0.27 | 0.71 | 1.52 | 0.81 | 1.90 | 6.64 | 0.73 | 1.31 | 1.78 | 19.01 |
| 1991 | 0.62 | 0.42 | 0.65 | 0.23 | 0.12 | 0.18 | 2.82 | 3.54 | 3.41 | 1.93 | 1.57 | 1.82 | 17.31 |
| 1992 | 1.17 | 1.04 | 0.31 | 0.08 | 0.58 | 1.21 | 0.79 | 2.49 | 2.83 | 2.08 | 1.17 | 0.69 | 14.44 |
| 1993 | 0.94 | 1.17 | 0.29 | 0.09 | 1.17 | 0.17 | 0.57 | 4.02 | 4.27 | 1.90 | 2.00 | 0.30 | 16.89 |
| 1994 | 0.59 | 0.28 | 1.51 | 0.45 | 0.51 | 1.34 | 0.57 | 1.02 | 1.66 | 1.21 | 2.47 | 1.51 | 13.12 |
| 1995 | 0.52 | 1.00 | 0.88 | 0.08 | 1.11 | 0.91 | 3.01 | 2.19 | 2.93 | 0.95 | 0.09 | 0.09 | 13.76 |
| 1996 | 0.11 | 2.40 | 0.42 | 0.08 | 0.20 | 0.50 | 2.04 | 2.53 | 1.93 | 2.63 | 1.38 | .24 | 14.46 |
| 1997 | 0.12 | 0.52 | 0.01 | 0.25 | 1.12 | 0.60 | 1.36 | 8.37 | 2.53 | 1.93 | 0.87 | 1.80 | 19.48 |
| 1998 | 0.45 | 0.24 | 0.07 | 0.39 | 0.63 | 2.70 | 1.01 | 3.25 | 0.72 | 0.54 | 0.18 | 1.47 | 11.65 |
| 1999 | 0.37 | 0.28 | 0.61 | 0.29 | 1.30 | 1.10 | 2.15 | 4.62 | 3.17 | 2.63 | 0.35 | 1.43 | 18.30 |
| 2000 | 1.04 | 0.54 | 0.48 | 0.39 | 0.69 | 1.43 | 2.58 | 1.68 | 3.24 | 0.59 | 1.13 | 0.58 | 14.37 |
| 2001 | 1.10 | 0.85 | 0.88 | 0.34 | 0.48 | 0.24 | 4.49 | 0.97 | 1.14 | 1.57 | 0.26 | 0.20 | 12.52 |
| 2002 | 0.72 | 0.35 | 1.61 | 0.29 | 0.27 | 1.01 | 1.46 | 3.51 | 3.36 | 4.28 | 0.27 | 1.66 | 18.79 |
| 2003 | 0.39 | 0.90 | 0.34 | 0.17 | 0.67 | 0.95 | 1.23 | 2.34 | 1.96 | 3.06 | 2.57 | 2.10 | 16.68 |
| 2004 | 0.49 | 0.73 | 0.86 | 0.77 | 1.02 | 0.95 | 0.88 | 1.17 | 7.35 | 1.18 | 2.40 | 1.73 | 19.53 |
| 2005 | 0.61 | 1.29 | 1.04 | 0.16 | 0.29 | 0.81 | 1.03 | 3.44 | 4.57 | 0.78 | 0.99 | 0.90 | 15.91 |
| 2006 | 0.37 | 0.71 | 0.73 | 0.49 | 0.55 | 1.41 | 1.47 | 6.60 | 3.56 | 2.02 | 0.04 | 2.38 | 20.33 |
| 2007 | 1.34 | 0.14 | 0.18 | 0.17 | 0.66 | 1.10 | 1.81 | 2.09 | 4.30 | 1.69 | 1.29 | 0.62 | 15.39 |
| 2008 | 1.16 | 0.85 | 0.41 | 2.32 | 0.40 | 0.63 | 3.25 | 0.92 | 3.22 | 1.77 | 1.11 | 0.99 | 17.03 |
| 2009 | 0.97 | 0.45 | 1.10 | 0.13 | 0.76 | 0.57 | 1.40 | 2.89 | 1.17 | 2.20 | 1.22 | 0.78 | 13.64 |
| POR= 56 YRS | 0.75 | 0.81 | 0.64 | 0.55 | 0.66 | 1.00 | 1.88 | 2.69 | 2.75 | 1.92 | 1.10 | 1.11 | 15.86 |

WBAN : 26451

## AVERAGE TEMPERATURE (°F) 2009 ANCHORAGE (PANC)

| YEAR | JAN | FEB | MAR | APR | MAY | JUN | JUL | AUG | SEP | OCT | NOV | DEC | ANNUAL |
|---|---|---|---|---|---|---|---|---|---|---|---|---|---|
| 1980 | 14.3 | 27.4 | 27.2 | 39.3 | 45.8 | 53.2 | 57.0 | 54.4 | 46.7 | 37.2 | 27.6 | 0.8 | 35.9 |
| 1981 | 31.5 | 24.8 | 34.4 | 36.0 | 50.7 | 53.8 | 57.4 | 54.8 | 47.9 | 36.0 | 21.8 | 15.9 | 38.8 |
| 1982 | 6.4 | 15.5 | 26.3 | 33.1 | 44.5 | 52.9 | 56.2 | 54.7 | 47.5 | 26.6 | 21.0 | 21.5 | 33.9 |
| 1983 | 16.2 | 21.4 | 28.7 | 37.4 | 48.7 | 55.9 | 58.5 | 56.1 | 45.3 | 34.4 | 24.9 | 16.7 | 37.0 |
| 1984 | 18.8 | 19.4 | 36.4 | 38.8 | 49.6 | 58.8 | 60.8 | 56.6 | 49.3 | 35.5 | 19.8 | 18.9 | 38.6 |
| 1985 | 30.3 | 13.5 | 26.7 | 28.4 | 45.1 | 51.9 | 58.5 | 55.2 | 47.6 | 30.3 | 14.0 | 27.5 | 35.8 |
| 1986 | 25.6 | 21.8 | 24.1 | 31.0 | 46.6 | 54.6 | 58.0 | 54.3 | 48.6 | 39.0 | 25.0 | 28.3 | 38.1 |
| 1987 | 22.8 | 25.3 | 26.8 | 37.9 | 47.2 | 51.9 | 57.1 | 57.3 | 48.0 | 38.9 | 26.9 | 18.2 | 38.2 |
| 1988 | 18.0 | 22.7 | 31.3 | 37.1 | 48.5 | 55.2 | 58.8 | 56.0 | 48.0 | 33.3 | 20.5 | 22.0 | 37.6 |
| 1989 | 3.5 | 17.6 | 23.6 | 39.3 | 46.3 | 55.3 | 59.4 | 59.0 | 50.6 | 34.0 | 17.2 | 24.2 | 35.8 |
| 1990 | 15.5 | 3.8 | 28.6 | 39.9 | 49.9 | 57.1 | 58.6 | 57.8 | 49.6 | 32.3 | 9.9 | 14.8 | 34.8 |
| 1991 | 15.9 | 19.6 | 23.7 | 37.7 | 46.6 | 55.7 | 57.5 | 55.5 | 51.0 | 33.0 | 25.0 | 20.5 | 36.8 |
| 1992 | 20.3 | 15.0 | 24.9 | 35.2 | 46.1 | 55.9 | 59.5 | 55.8 | 40.3 | 31.2 | 27.1 | 15.0 | 35.5 |
| 1993 | 14.5 | 21.0 | 28.8 | 40.6 | 50.7 | 56.3 | 61.1 | 58.8 | 48.8 | 38.7 | 25.2 | 24.0 | 39.0 |
| 1994 | 21.7 | 17.1 | 25.7 | 38.6 | 47.1 | 56.7 | 58.8 | 58.8 | 48.6 | 33.5 | 15.3 | 15.7 | 36.5 |
| 1995 | 15.6 | 20.7 | 18.6 | 40.4 | 48.8 | 56.0 | 59.2 | 57.9 | 53.7 | 38.1 | 21.0 | 19.0 | 37.4 |
| 1996 | 6.1 | 15.8 | 29.2 | 38.6 | 50.1 | 56.9 | 59.9 | 56.7 | 46.5 | 25.4 | 19.0 | 13.0 | 34.8 |
| 1997 | 15.9 | 30.7 | 24.6 | 38.2 | 48.0 | 56.7 | 60.8 | 58.1 | 50.4 | 29.6 | 28.0 | 16.8 | 38.2 |
| 1998 | 15.5 | 25.8 | 30.1 | 40.2 | 47.3 | 54.7 | 57.3 | 53.8 | 49.0 | 35.9 | 23.5 | 14.3 | 37.3 |
| 1999 | 11.9 | 7.2 | 24.1 | 34.5 | 45.7 | 55.3 | 58.4 | 56.9 | 48.9 | 34.0 | 19.7 | 14.9 | 34.3 |
| 2000 | 14.8 | 25.9 | 29.6 | 37.4 | 46.1 | 55.1 | 56.8 | 54.8 | 47.1 | 34.8 | 29.7 | 25.1 | 38.1 |
| 2001 | 27.5 | 23.1 | 28.8 | 37.7 | 44.8 | 58.1 | 57.7 | 58.4 | 49.2 | 30.1 | 19.6 | 10.8 | 37.2 |
| 2002 | 23.7 | 21.3 | 22.0 | 30.8 | 48.4 | 53.4 | 59.8 | 57.0 | 50.1 | 41.2 | 35.2 | 24.0 | 38.9 |
| 2003 | 22.1 | 31.3 | 26.8 | 38.4 | 48.1 | 55.8 | 62.3 | 57.9 | 49.0 | 40.1 | 21.5 | 21.9 | 39.2 |
| 2004 | 9.6 | 26.4 | 24.2 | 37.3 | 50.2 | 57.5 | 61.9 | 61.2 | 45.0 | 38.2 | 29.0 | 21.9 | 38.5 |
| 2005 | 18.8 | 20.5 | 32.1 | 40.1 | 50.6 | 56.9 | 61.4 | 58.1 | 51.4 | 36.7 | 16.8 | 25.4 | 39.1 |
| 2006 | 10.6 | 21.9 | 23.7 | 35.8 | 48.6 | 54.4 | 58.2 | 54.9 | 49.5 | 39.0 | 11.5 | 21.6 | 35.8 |
| 2007 | 16.7 | 17.3 | 14.4 | 38.4 | 47.3 | 54.5 | 58.2 | 58.2 | 50.5 | 35.6 | 30.8 | 19.5 | 36.8 |
| 2008 | 13.4 | 16.8 | 30.3 | 33.4 | 45.8 | 51.6 | 55.8 | 55.6 | 48.5 | 29.3 | 21.2 | 14.2 | 34.7 |
| 2009 | 13.0 | 17.5 | 21.8 | 35.3 | 48.5 | 54.2 | 59.4 | 56.2 | 49.0 | 40.7 | 20.6 | 20.3 | 36.4 |
| POR= 56 YRS | 15.3 | 19.1 | 25.1 | 35.9 | 46.8 | 54.8 | 58.5 | 56.4 | 48.2 | 34.6 | 22.0 | 16.6 | 36.1 |

## HEATING DEGREE DAYS (base 65°F) 2009  ANCHORAGE (PANC)

| YEAR | JUL | AUG | SEP | OCT | NOV | DEC | JAN | FEB | MAR | APR | MAY | JUN | TOTAL |
|---|---|---|---|---|---|---|---|---|---|---|---|---|---|
| 1980-81 | 243 | 320 | 542 | 855 | 1115 | 1990 | 1032 | 1122 | 943 | 863 | 438 | 329 | 9792 |
| 1981-82 | 230 | 307 | 507 | 893 | 1290 | 1516 | 1813 | 1382 | 1191 | 949 | 625 | 356 | 11059 |
| 1982-83 | 261 | 313 | 520 | 1184 | 1315 | 1342 | 1507 | 1216 | 1117 | 821 | 500 | 267 | 10363 |
| 1983-84 | 194 | 269 | 585 | 945 | 1194 | 1491 | 1425 | 1319 | 880 | 778 | 471 | 179 | 9730 |
| 1984-85 | 129 | 254 | 464 | 906 | 1350 | 1423 | 1070 | 1437 | 1182 | 1091 | 610 | 388 | 10304 |
| 1985-86 | 193 | 298 | 516 | 1065 | 1523 | 1155 | 1215 | 1206 | 1260 | 1013 | 564 | 307 | 10315 |
| 1986-87 | 215 | 325 | 486 | 800 | 1194 | 1133 | 1303 | 1104 | 1176 | 805 | 543 | 386 | 9470 |
| 1987-88 | 243 | 232 | 506 | 801 | 1136 | 1444 | 1450 | 1221 | 1037 | 830 | 504 | 285 | 9689 |
| 1988-89 | 184 | 270 | 503 | 975 | 1331 | 1326 | 1908 | 1322 | 1277 | 765 | 573 | 286 | 10720 |
| 1989-90 | 173 | 181 | 423 | 956 | 1428 | 1255 | 1533 | 1715 | 1121 | 746 | 465 | 237 | 10233 |
| 1990-91 | 191 | 222 | 457 | 1006 | 1648 | 1552 | 1518 | 1265 | 1273 | 813 | 563 | 273 | 10781 |
| 1991-92 | 226 | 287 | 414 | 988 | 1193 | 1373 | 1380 | 1444 | 1240 | 891 | 579 | 268 | 10283 |
| 1992-93 | 161 | 280 | 735 | 1039 | 1131 | 1543 | 1563 | 1226 | 1117 | 725 | 436 | 252 | 10208 |
| 1993-94 | 125 | 187 | 477 | 808 | 1191 | 1267 | 1334 | 1335 | 1212 | 785 | 548 | 243 | 9512 |
| 1994-95 | 183 | 190 | 485 | 968 | 1488 | 1523 | 1526 | 1239 | 1433 | 734 | 496 | 265 | 10530 |
| 1995-96 | 172 | 214 | 335 | 826 | 1314 | 1425 | 1827 | 1423 | 1102 | 783 | 456 | 239 | 10116 |
| 1996-97 | 151 | 251 | 549 | 1220 | 1375 | 1608 | 1516 | 956 | 1246 | 796 | 520 | 249 | 10437 |
| 1997-98 | 123 | 207 | 432 | 1090 | 1103 | 1486 | 1530 | 1093 | 1073 | 739 | 540 | 302 | 9718 |
| 1998-99 | 232 | 340 | 475 | 895 | 1240 | 1566 | 1638 | 1611 | 1262 | 908 | 592 | 286 | 11045 |
| 1999-00 | 204 | 248 | 478 | 953 | 1352 | 1544 | 1548 | 1126 | 1090 | 821 | 579 | 291 | 10234 |
| 2000-01 | 245 | 312 | 534 | 931 | 1052 | 1230 | 1154 | 1168 | 1113 | 812 | 618 | 199 | 9368 |
| 2001-02 | 220 | 200 | 466 | 1072 | 1356 | 1673 | 1272 | 1216 | 1327 | 1017 | 506 | 341 | 10666 |
| 2002-03 | 156 | 243 | 442 | 731 | 889 | 1264 | 1323 | 937 | 1177 | 791 | 519 | 268 | 8740 |
| 2003-04 | 97 | 220 | 474 | 763 | 1299 | 1478 | 1708 | 1114 | 1256 | 826 | 451 | 222 | 9908 |
| 2004-05 | 102 | 115 | 593 | 820 | 1073 | 1330 | 1424 | 1239 | 1012 | 742 | 438 | 236 | 9124 |
| 2005-06 | 112 | 207 | 401 | 872 | 1436 | 1220 | 1678 | 1199 | 1272 | 869 | 498 | 309 | 10073 |
| 2006-07 | 202 | 306 | 458 | 799 | 1598 | 1341 | 1491 | 1327 | 1565 | 792 | 539 | 313 | 10731 |
| 2007-08 | 203 | 202 | 430 | 903 | 1021 | 1402 | 1594 | 1391 | 1070 | 939 | 591 | 394 | 10140 |
| 2008-09 | 276 | 283 | 489 | 1097 | 1307 | 1569 | 1603 | 1322 | 1332 | 883 | 503 | 315 | 10979 |
| 2009- | 171 | 266 | 474 | 744 | 1326 | 1379 | | | | | | | |

WBAN : 26451

## COOLING DEGREE DAYS (base 65°F) 2009  ANCHORAGE (PANC)

| YEAR | JAN | FEB | MAR | APR | MAY | JUN | JUL | AUG | SEP | OCT | NOV | DEC | TOTAL |
|---|---|---|---|---|---|---|---|---|---|---|---|---|---|
| 1980 | 0 | 0 | 0 | 0 | 0 | 0 | 0 | 0 | 0 | 0 | 0 | 0 | 0 |
| 1981 | 0 | 0 | 0 | 0 | 0 | 0 | 0 | 0 | 0 | 0 | 0 | 0 | 0 |
| 1982 | 0 | 0 | 0 | 0 | 0 | 0 | 0 | 0 | 0 | 0 | 0 | 0 | 0 |
| 1983 | 0 | 0 | 0 | 0 | 0 | 0 | 0 | 0 | 0 | 0 | 0 | 0 | 0 |
| 1984 | 0 | 0 | 0 | 0 | 0 | 0 | 5 | 1 | 0 | 0 | 0 | 0 | 6 |
| 1985 | 0 | 0 | 0 | 0 | 0 | 0 | 0 | 0 | 0 | 0 | 0 | 0 | 0 |
| 1986 | 0 | 0 | 0 | 0 | 0 | 0 | 4 | 0 | 0 | 0 | 0 | 0 | 4 |
| 1987 | 0 | 0 | 0 | 0 | 0 | 0 | 2 | 0 | 0 | 0 | 0 | 0 | 2 |
| 1988 | 0 | 0 | 0 | 0 | 0 | 0 | 0 | 0 | 0 | 0 | 0 | 0 | 0 |
| 1989 | 0 | 0 | 0 | 0 | 0 | 0 | 5 | 2 | 0 | 0 | 0 | 0 | 7 |
| 1990 | 0 | 0 | 0 | 0 | 0 | 3 | 1 | 2 | 0 | 0 | 0 | 0 | 6 |
| 1991 | 0 | 0 | 0 | 0 | 0 | 0 | 0 | 0 | 0 | 0 | 0 | 0 | 0 |
| 1992 | 0 | 0 | 0 | 0 | 0 | 0 | 0 | 0 | 0 | 0 | 0 | 0 | 0 |
| 1993 | 0 | 0 | 0 | 0 | 0 | 0 | 11 | 0 | 0 | 0 | 0 | 0 | 11 |
| 1994 | 0 | 0 | 0 | 0 | 0 | 0 | 0 | 2 | 0 | 0 | 0 | 0 | 2 |
| 1995 | 0 | 0 | 0 | 0 | 0 | 1 | 0 | 0 | 0 | 0 | 0 | 0 | 1 |
| 1996 | 0 | 0 | 0 | 0 | 0 | 0 | 0 | 0 | 0 | 0 | 0 | 0 | 0 |
| 1997 | 0 | 0 | 0 | 0 | 0 | 5 | 0 | 1 | 0 | 0 | 0 | 0 | 6 |
| 1998 | 0 | 0 | 0 | 0 | 0 | 0 | 0 | 0 | 0 | 0 | 0 | 0 | 0 |
| 1999 | 0 | 0 | 0 | 0 | 0 | 0 | 7 | 0 | 0 | 0 | 0 | 0 | 7 |
| 2000 | 0 | 0 | 0 | 0 | 0 | 0 | 0 | 0 | 0 | 0 | 0 | 0 | 0 |
| 2001 | 0 | 0 | 0 | 0 | 0 | 0 | 0 | 0 | 0 | 0 | 0 | 0 | 0 |
| 2002 | 0 | 0 | 0 | 0 | 0 | 0 | 1 | 0 | 0 | 0 | 0 | 0 | 1 |
| 2003 | 0 | 0 | 0 | 0 | 0 | 0 | 19 | 4 | 0 | 0 | 0 | 0 | 23 |
| 2004 | 0 | 0 | 0 | 0 | 0 | 2 | 11 | 4 | 0 | 0 | 0 | 0 | 17 |
| 2005 | 0 | 0 | 0 | 0 | 0 | 0 | 7 | 1 | 0 | 0 | 0 | 0 | 8 |
| 2006 | 0 | 0 | 0 | 0 | 0 | 0 | 0 | 0 | 0 | 0 | 0 | 0 | 0 |
| 2007 | 0 | 0 | 0 | 0 | 0 | 3 | 0 | 0 | 0 | 0 | 0 | 0 | 3 |
| 2008 | 0 | 0 | 0 | 0 | 0 | 0 | 0 | 0 | 0 | 0 | 0 | 0 | 0 |
| 2009 | 0 | 0 | 0 | 0 | 0 | 0 | 2 | 0 | 0 | 0 | 0 | 0 | 2 |

## SNOWFALL (inches)  2009  ANCHORAGE (PANC)

| YEAR | JUL | AUG | SEP | OCT | NOV | DEC | JAN | FEB | MAR | APR | MAY | JUN | TOTAL |
|------|-----|-----|-----|-----|-----|-----|-----|-----|-----|-----|-----|-----|-------|
| 1980-81 | 0.0 | 0.0 | 0.0 | 10.2 | 4.2 | 1.4 | 5.0 | 6.6 | 4.4 | 1.1 | T | 0.0 | 32.9 |
| 1981-82 | 0.0 | 0.0 | 1.5 | 6.3 | 20.0 | 7.6 | 0.5 | 0.6 | 5.6 | 3.5 | 0.7 | 0.0 | 46.3 |
| 1982-83 | 0.0 | 0.0 | 0.0 | 27.1 | 23.4 | 1.9 | 3.7 | 4.3 | T | 11.0 | 0.0 | 0.0 | 71.4 |
| 1983-84 | 0.0 | 0.0 | T | 23.7 | 2.1 | 10.5 | 15.0 | 18.9 | 0.2 | 9.8 | 0.0 | 0.0 | 80.2 |
| 1984-85 | 0.0 | 0.0 | 0.0 | 3.3 | 1.8 | 18.0 | 9.7 | 7.9 | 12.8 | 7.3 | 1.3 | 0.0 | 62.1 |
| 1985-86 | 0.0 | 0.0 | 0.0 | 0.8 | 1.5 | 6.1 | 5.1 | 6.1 | 21.0 | 5.4 | 0.1 | 0.0 | 46.1 |
| 1986-87 | 0.0 | 0.0 | 0.0 | T | 3.8 | 10.1 | 18.5 | 2.2 | 2.5 | 1.6 | 0.0 | 0.0 | 38.7 |
| 1987-88 | 0.0 | 0.0 | 0.0 | T | 29.2 | 26.3 | 4.7 | 9.2 | 8.5 | 2.0 | 0.0 | 0.0 | 79.9 |
| 1988-89 | 0.0 | 0.0 | 0.0 | 12.0 | 15.3 | 18.6 | 10.1 | 2.3 | 5.1 | T | 0.2 | 0.0 | 63.6 |
| 1989-90 | 0.0 | 0.0 | 0.0 | 16.3 | 10.1 | 20.0 | 27.5 | 23.0 | 4.7 | 0.8 | T | 0.0 | 102.4 |
| 1990-91 | 0.0 | 0.0 | 0.0 | 1.6 | 16.9 | 21.4 | 7.7 | 5.4 | 12.7 | T | 0.0 | 0.0 | 65.7 |
| 1991-92 | 0.0 | 0.0 | 0.0 | 11.6 | 19.3 | 26.2 | 21.4 | 18.3 | 2.7 | T | 0.2 | 0.0 | 99.7 |
| 1992-93 | 0.0 | 0.0 | 3.0 | 13.0 | 9.1 | 12.1 | 13.7 | 18.3 | 5.7 | 0.0 | 0.0 | 0.0 | 74.9 |
| 1993-94 | 0.0 | 0.0 | T | 4.4 | 11.9 | 5.1 | 7.5 | 1.7 | 29.9 | 6.0 | 0.0 | 0.0 | 66.5 |
| 1994-95 | 0.0 | 0.0 | 0.0 | 9.1 | 38.8 | 29.0 | 12.6 | 15.3 | 16.7 | 0.0 | 0.0 | 0.0 | 121.5 |
| 1995-96 | 0.0 | 0.0 | 0.0 | 4.0 | 0.9 | 2.5 | 2.5 | 52.1 | 6.1 | 0.9 | 0.0 | 0.0 | 69.0 |
| 1996-97 | 0.0 | 0.0 | .1 | 28.1 | 25.7 | 4.7 | 3.1 | 5.1 | 0.8 | 0.2 | 0.0 | 0.0 | 67.8 |
| 1997-98 | 0.0 | T | 0.0 | 11.6 | 6.4 | 26.6 | 6.8 | 3.1 | 1.2 | 2.9 | 0.0 | | 79.3 |
| 1998-99 | 0.0 | T | 0.0 | 0.4 | 9.1 | 34.5 | 8.3 | 6.6 | 17.4 | 3.0 | T | 0.0 | 79.3 |
| 1999-00 | 0.0 | 0.0 | 0.0 | 5.7 | 8.5 | 18.6 | 28.6 | 4.6 | 7.4 | 2.8 | 0.0 | 0.0 | 76.2 |
| 2000-01 | 0.0 | 0.0 | T | 2.1 | 3.2 | 4.2 | 11.0 | 19.7 | 15.9 | 1.3 | 6.1 | 0.0 | 63.5 |
| 2001-02 | 0.0 | 0.0 | 0.0 | 20.6 | 6.6 | 7.9 | 7.1 | 9.0 | 29.5 | 0.1 | 0.7 | 0.0 | 81.5 |
| 2002-03 | 0.0 | 0.0 | 0.0 | T | 2.1 | 23.5 | 3.0 | 0.3 | 7.5 | 0.4 | 0.0 | 0.0 | 36.8 |
| 2003-04 | 0.0 | 0.0 | 0.0 | 0.0 | 28.8 | 37.6 | 9.4 | 10.4 | 22.3 | 5.4 | 0.0 | 0.0 | 113.9 |
| 2004-05 | 0.0 | 0.0 | 6.3 | 3.4 | 16.0 | 21.5 | 2.2 | 15.4 | 10.7 | 1.1 | 0.0 | 0.0 | 76.6 |
| 2005-06 | 0.0 | 0.0 | 0.0 | 1.5 | 16.7 | 14.7 | 10.7 | 8.5 | 12.3 | 5.1 | 0.3 | 0.0 | 69.8 |
| 2006-07 | 0.0 | T | 0.0 | 8.7 | 1.0 | 36.9 | 29.3 | 4.4 | 4.0 | T | 0.0 | 0.0 | 84.3 |
| 2007-08 | 0.0 | 0.0 | 0.0 | 3.1 | 14.8 | 13.8 | 27.2 | 17.4 | 2.0 | 30.8 | 0.0 | 0.0 | 109.1 |
| 2008-09 | 0.0 | T | 0.0 | 13.1 | 21.7 | 17.8 | 10.8 | 13.6 | 14.8 | 1.6 | 0.0 | 0.0 | 93.4 |
| 2009- | 0.0 | 0.0 | T | T | 13.9 | 11.4 | | | | | | | |
| POR=<br>57 YRS | 0.0 | T | 0.3 | 7.6 | 11.8 | 15.8 | 10.1 | 11.9 | 9.9 | 4.9 | 0.3 | 0.1 | 72.7 |

# Anchorage, AK  2009
## Temperature and Precipitation

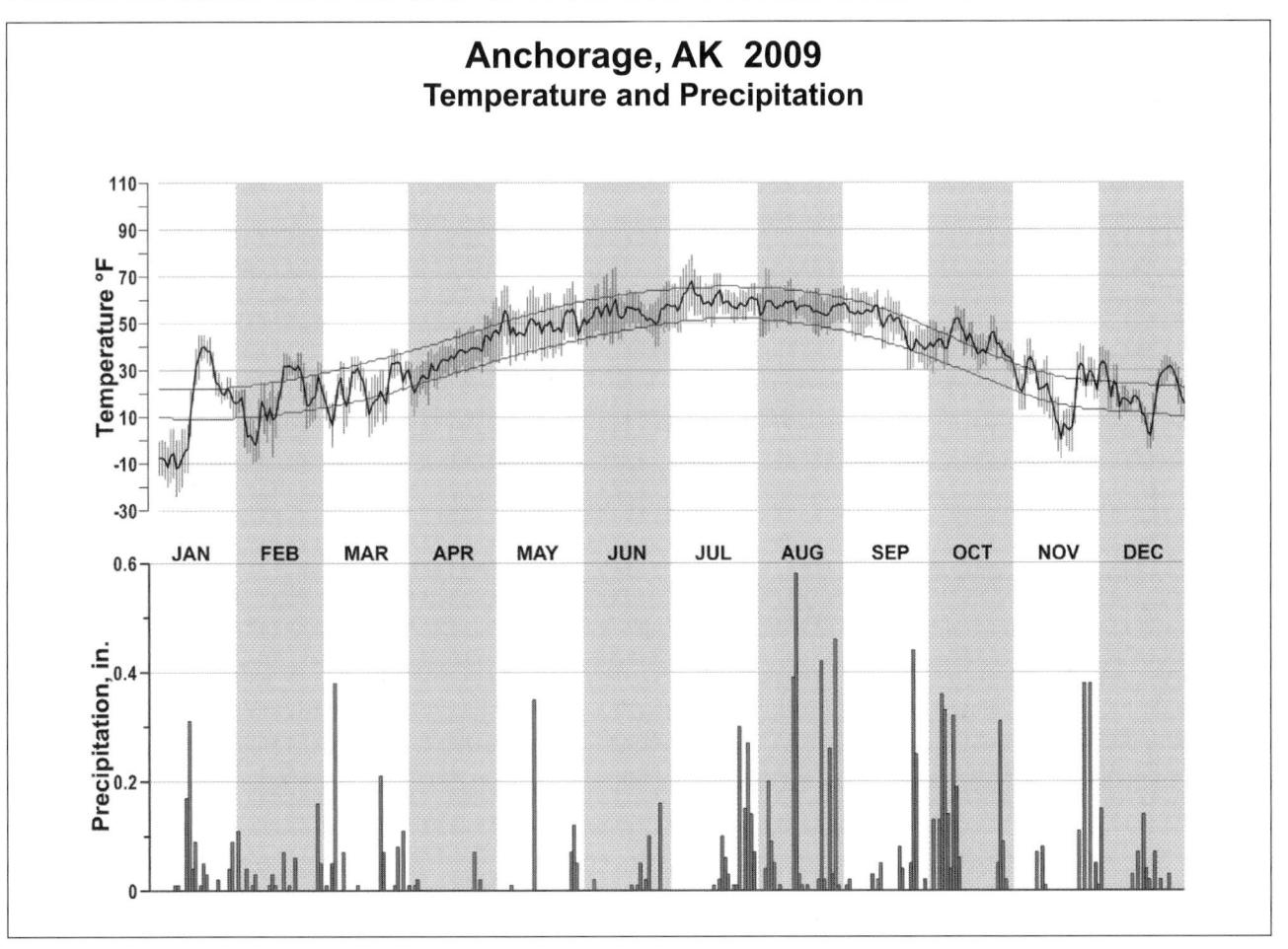

# 2009
# BARROW
# ALASKA (PABR)

Barrow is the most northerly First-Order station operated by the National Weather Service. Although this station generally records one of the lowest mean temperatures for the winter months, the surrounding topography prevents the establishment of the lowest minima for the state. With the Arctic Ocean to the north, east, and west, and level tundra stretching 200 miles to the south, there are no natural wind barriers to assist in stilling the wind, permitting the lowering of temperatures by radiation, and no downslope drainage area to aid the flow of cold air to lower levels. Consequently, temperature inversions in the lower levels of the atmosphere are not as marked as those observed at stations in the central interior.

Temperatures at this northern station remain below the freezing point through most of the year, with the daily maxima reaching higher than 32 degrees on an average of only 109 days a year. Freezing temperatures have been observed every month of the year. February is generally the coldest month and March temperatures are but little higher than those observed in the winter months. In April, temperatures begin a general upward trend, with May becoming the definite transitional period from winter to the summer season. July is the warmest month of the year and the frequency of minimum temperatures of 32 degrees or less are about one day out of two for July and August. During late July or early August, the Arctic Ocean is usually ice-free for the first time in summer. The end of the short summer is reached in September. By November about half of the daily mean temperatures are zero or below, and Barrow definitely returns to the clutches of winter cold.

At 1250 p.m. on November 18, the sun dips below the horizon and is not seen again until 1151 a.m. on January 24. Then the amount of possible sunshine each day increases by never less than 9 minutes per day. By 106 a.m. on May 10th the possible sunshine has increased to 24 hours per day. The sun remains visible from that time to August 2, when it again sets for 1 hour and 25 minutes. The decrease in hours of sunshine is as rapid as the increase.

The amount of sunshine appears to have a direct relationship to the occurrence of cloudiness, precipitation, and heavy fog. All three build up to a maximum along with the hours of sunshine. Maximum cloudiness does continue into the fall months, although the amount of sunshine, precipitation, and fog are on the decrease. Since an accurate estimate of cloudiness cannot be made under conditions of darkness, the record of cloudiness for that time is not summarized. However, average cloudiness probably approximates that observed during late winter and spring months.

Variation of wind speed during the year is small, with the fall months being windiest. Extreme winds in the upper 40s and low 50s have been recorded for all months.

# NORMALS, MEANS, AND EXTREMES
## BARROW (PABR)

| LATITUDE: 71° 17'N | LONGITUDE: -156° 45'W | ELEVATION (FT): GRND: 40 BARO: 38 | TIME ZONE: ALASKA (UTC -9) | WBAN: 27502 |
|---|---|---|---|---|

| | ELEMENT | POR | JAN | FEB | MAR | APR | MAY | JUN | JUL | AUG | SEP | OCT | NOV | DEC | YEAR |
|---|---|---|---|---|---|---|---|---|---|---|---|---|---|---|---|
| **TEMPERATURE °F** | NORMAL DAILY MAXIMUM | 30 | -7.7 | -9.8 | -7.4 | 6.3 | 24.9 | 39.5 | 46.5 | 43.6 | 34.8 | 19.3 | 4.6 | -4.7 | 15.8 |
| | MEAN DAILY MAXIMUM | 91 | -7.5 | -9.7 | -7.3 | 6.9 | 24.8 | 38.3 | 45.9 | 43.2 | 34.2 | 20.9 | 6.1 | -4.0 | 16.0 |
| | HIGHEST DAILY MAXIMUM | 89 | 36 | 36 | 34 | 42 | 47 | 72 | 79 | 76 | 62 | 43 | 39 | 34 | 79 |
| | YEAR OF OCCURRENCE | | 1974 | 1982 | 1998 | 1936 | 1996 | 1996 | 1993 | 1968 | 1995 | 1954 | 1937 | 1932 | JUL 1993 |
| | MEAN OF EXTREME MAXS. | 91 | 18.7 | 14.0 | 14.0 | 26.4 | 36.9 | 55.3 | 64.6 | 60.8 | 48.9 | 33.9 | 24.8 | 19.0 | 34.8 |
| | NORMAL DAILY MINIMUM | 30 | -19.6 | -22.0 | -20.0 | -7.3 | 15.3 | 30.4 | 34.3 | 33.8 | 27.5 | 9.8 | -6.4 | -16.4 | 5.0 |
| | MEAN DAILY MINIMUM | 91 | -19.8 | -21.4 | -20.1 | -6.8 | 14.6 | 29.0 | 33.8 | 33.4 | 27.2 | 11.5 | -4.5 | -15.3 | 5.1 |
| | LOWEST DAILY MINIMUM | 89 | -53 | -56 | -52 | -42 | -19 | 4 | 22 | 20 | 1 | -32 | -40 | -55 | -56 |
| | YEAR OF OCCURRENCE | | 1975 | 1924 | 1971 | 1924 | 1984 | 1969 | 1936 | 1925 | 1975 | 1970 | 1948 | 1924 | FEB 1924 |
| | MEAN OF EXTREME MINS. | 91 | -37.5 | -38.8 | -35.5 | -24.7 | -1.7 | 21.6 | 28.5 | 27.4 | 17.2 | -9.1 | -20.9 | -31.7 | -8.7 |
| | NORMAL DRY BULB | 30 | -13.7 | -15.9 | -13.7 | -.5 | 20.1 | 35.0 | 40.4 | 38.7 | 31.2 | 14.6 | -.9 | -10.6 | 10.4 |
| | MEAN DRY BULB | 91 | -13.5 | -15.4 | -13.7 | 0.1 | 19.7 | 33.9 | 39.9 | 38.4 | 30.7 | 16.2 | 0.8 | -9.6 | 10.6 |
| | MEAN WET BULB | 23 | -11.9 | -13.4 | -11.4 | 1.4 | 20.6 | 33.8 | 38.7 | 37.6 | 31.3 | 17.1 | 0.1 | -7.8 | 11.3 |
| | MEAN DEW POINT | 23 | -15.9 | -17.9 | -15.5 | -1.1 | 19.0 | 32.2 | 37.3 | 36.5 | 29.9 | 15.2 | -2.5 | -10.7 | 8.9 |
| | NORMAL NO. DAYS WITH: | | | | | | | | | | | | | | |
| | MAXIMUM >= 70 | 30 | 0.0 | 0.0 | 0.0 | 0.0 | 0.0 | 0.1 | 0.7 | 0.3 | 0.0 | 0.0 | 0.0 | 0.0 | 1.1 |
| | MAXIMUM <= 32 | 30 | 31.0 | 28.2 | 29.9 | 29.5 | 25.0 | 3.0 | 0.0 | 1.3 | 11.6 | 28.6 | 29.9 | 31.0 | 249.0 |
| | MINIMUM <= 32 | 30 | 31.0 | 28.2 | 30.0 | 30.0 | 30.6 | 22.6 | 12.1 | 14.6 | 23.5 | 29.7 | 30.0 | 31.0 | 313.3 |
| | MINIMUM <= 0 | 30 | 29.5 | 26.6 | 29.1 | 22.6 | 2.3 | 0.0 | 0.0 | 0.0 | 0.0 | 6.9 | 21.1 | 28.6 | 166.7 |
| **H/C** | NORMAL HEATING DEG. DAYS | 30 | 2440 | 2267 | 2423 | 1967 | 1391 | 903 | 763 | 815 | 1016 | 1564 | 1978 | 2346 | 19873 |
| | NORMAL COOLING DEG. DAYS | 30 | 0 | 0 | 0 | 0 | 0 | 0 | 0 | 0 | 0 | 0 | 0 | 0 | 0 |
| **RH** | NORMAL (PERCENT) | 30 | 75 | 75 | 75 | 80 | 87 | 87 | 87 | 90 | 90 | 86 | 81 | 77 | 83 |
| | HOUR 03 LST | 30 | 75 | 75 | 75 | 81 | 90 | 92 | 94 | 94 | 91 | 86 | 81 | 78 | 84 |
| | HOUR 09 LST | 30 | 75 | 75 | 75 | 81 | 87 | 86 | 87 | 91 | 90 | 86 | 81 | 78 | 83 |
| | HOUR 15 LST | 30 | 76 | 76 | 75 | 78 | 83 | 82 | 82 | 85 | 86 | 86 | 81 | 78 | 81 |
| | HOUR 21 LST | 30 | 75 | 75 | 75 | 81 | 87 | 87 | 87 | 90 | 90 | 86 | 81 | 77 | 83 |
| **S** | PERCENT POSSIBLE SUNSHINE | | | | | | | | | | | | | | |
| **W/O** | MEAN NO. DAYS WITH: | | | | | | | | | | | | | | |
| | HEAVY FOG(VISBY <= 1/4 MI) | 46 | 1.4 | 1.1 | 1.2 | 2.2 | 6.6 | 8.2 | 10.9 | 11.1 | 5.2 | 2.7 | 1.8 | 1.0 | 53.4 |
| | THUNDERSTORMS | 61 | 0.0 | 0.1 | 0.0 | 0.0 | 0.0 | 0.0 | 0.1 | 0.0 | 0.0 | 0.0 | 0.0 | 0.0 | 0.2 |
| **CLOUDNESS** | MEAN: | | | | | | | | | | | | | | |
| | SUNRISE-SUNSET (OKTAS) | 8 | 1.4 | 4.2 | 3.8 | 4.9 | 6.6 | 6.5 | 6.3 | 7.2 | 7.6 | 7.0 | 2.8 | 0.0 | 4.9 |
| | MIDNIGHT-MIDNIGHT (OKTAS) | 34 | 3.9 | 3.7 | 3.6 | 4.4 | 6.6 | 6.3 | 6.1 | 7.1 | 7.4 | 6.7 | 5.0 | 4.1 | 5.4 |
| | MEAN NO. DAYS WITH: | | | | | | | | | | | | | | |
| | CLEAR | 54 | 6.2 | 11.7 | 13.5 | 10.4 | 3.3 | 3.5 | 3.4 | 1.4 | 1.2 | 2.3 | 5.5 | 3.5 | 65.9 |
| | PARTLY CLOUDY | 54 | 1.5 | 5.6 | 7.1 | 7.0 | 5.1 | 5.8 | 7.0 | 3.8 | 2.6 | 3.8 | 2.6 | 0.0 | 51.9 |
| | CLOUDY | 54 | 2.7 | 10.8 | 10.4 | 12.5 | 22.7 | 20.7 | 20.2 | 25.3 | 25.9 | 24.6 | 10.5 | 0.0 | 186.3 |
| **PR** | MEAN STATION PRESSURE(IN) | 26 | 30.03 | 30.08 | 30.16 | 30.07 | 30.04 | 29.95 | 29.92 | 29.88 | 29.86 | 29.90 | 29.95 | 29.96 | 29.98 |
| | MEAN SEA-LEVEL PRES. (IN) | 26 | 30.04 | 30.10 | 30.17 | 30.09 | 30.06 | 29.97 | 29.94 | 29.90 | 29.88 | 29.92 | 29.97 | 29.98 | 30.00 |
| **WINDS** | MEAN SPEED (MPH) | 26 | 12.6 | 12.7 | 11.7 | 11.9 | 12.6 | 11.9 | 12.4 | 12.4 | 13.2 | 13.8 | 13.2 | 13.0 | 12.6 |
| | PREVAIL.DIR(TENS OF DEGS) | 29 | 07 | 06 | 06 | 08 | 08 | 09 | 09 | 10 | 09 | 08 | 08 | 07 | 07 |
| | MAXIMUM 2-MINUTE: | | | | | | | | | | | | | | |
| | SPEED (MPH) | 11 | 53 | 58 | 43 | 51 | 38 | 36 | 46 | 55 | 44 | 55 | 49 | 46 | 58 |
| | DIR. (TENS OF DEGS) | | 07 | 24 | 09 | 26 | 09 | 14 | 27 | 27 | 03 | 09 | 07 | 09 | 24 |
| | YEAR OF OCCURRENCE | | 2001 | 2006 | 2009 | 2002 | 2002 | 2008 | 2003 | 2000 | 2004 | 2006 | 2000 | 2003 | FEB 2006 |
| | MAXIMUM 3-SECOND | | | | | | | | | | | | | | |
| | SPEED (MPH) | 11 | 60 | 70 | 51 | 60 | 41 | 41 | 58 | 64 | 54 | 67 | 55 | 55 | 70 |
| | DIR. (TENS OF DEGS) | | 07 | 24 | 09 | 25 | 09 | 14 | 27 | 27 | 04 | 09 | 35 | 10 | 24 |
| | YEAR OF OCCURRENCE | | 2001 | 2006 | 2009 | 2002 | 2002 | 2008 | 2003 | 2000 | 2004 | 2006 | 2006 | 2003 | FEB 2006 |
| **PRECIPITATION** | NORMAL (IN) | 30 | 0.12 | 0.12 | 0.09 | 0.12 | 0.12 | 0.32 | 0.87 | 1.04 | 0.69 | 0.39 | 0.16 | 0.12 | 4.16 |
| | MAXIMUM MONTHLY (IN) | 89 | 1.04 | 0.81 | 1.49 | 1.36 | 0.81 | 1.15 | 3.19 | 2.81 | 1.88 | 1.65 | 1.15 | 0.76 | 3.19 |
| | YEAR OF OCCURRENCE | | 1962 | 1959 | 1963 | 1963 | 1933 | 1955 | 1989 | 1963 | 2002 | 1925 | 1965 | 1967 | JUL 1989 |
| | MINIMUM MONTHLY (IN) | 89 | 0.00 | 0.00 | 0.00 | 0.00 | T | T | T | T | 0.01 | 0.12 | T | 0.00 | 0.00 |
| | YEAR OF OCCURRENCE | | 1939 | 1936 | 1928 | 1938 | 1995 | 1937 | 1937 | 1934 | 1969 | 1936 | 1989 | 1936 | JAN 1939 |
| | MAXIMUM IN 24 HOURS (IN) | 89 | 0.70 | 0.36 | 0.71 | 0.42 | 0.30 | 0.82 | 1.32 | .89 | 0.62 | 1.00 | 0.41 | 0.34 | 1.32 |
| | YEAR OF OCCURRENCE | | 1937 | 1959 | 1963 | 1963 | 1969 | 1955 | 1987 | 2005 | 2003 | 1926 | 1925 | 2009 | JUL 1987 |
| | NORMAL NO. DAYS WITH: | | | | | | | | | | | | | | |
| | PRECIPITATION >= 0.01 | 30 | 4.9 | 4.1 | 4.0 | 4.3 | 4.8 | 6.0 | 9.1 | 12.1 | 13.1 | 11.9 | 6.0 | 5.1 | 85.4 |
| | PRECIPITATION >= 1.00 | 30 | 0.0 | 0.0 | 0.0 | 0.0 | 0.0 | 0.0 | 0.1 | 0.0 | 0.0 | 0.0 | 0.0 | 0.0 | 0.1 |
| **SNOWFALL** | NORMAL (IN) | 30 | 2.0 | 1.9 | 1.7 | 2.1 | 1.7 | 0.8 | 0.2 | 0.9 | 5.0 | 7.4 | 3.1 | 2.2 | 29.0 |
| | MAXIMUM MONTHLY (IN) | 89 | 11.9 | 11.4 | 15.8 | 15.4 | 12.9 | 6.6 | 9.0 | 4.0 | 16.2 | 22.7 | 19.0 | 12.1 | 22.7 |
| | YEAR OF OCCURRENCE | | 1962 | 2009 | 1963 | 1963 | 1933 | 1933 | 1922 | 1969 | 1987 | 2008 | 1925 | 1925 | OCT 2008 |
| | MAXIMUM IN 24 HOURS (IN) | 89 | 7.3 | 3.6 | 7.1 | 4.2 | 5.0 | 3.2 | 6.0 | 2.8 | 5.1 | 15.0 | 6.3 | 5.0 | 15.0 |
| | YEAR OF OCCURRENCE' | | 2001 | 1959 | 1963 | 1963 | 1996 | 1981 | 1922 | 2003 | 1987 | 1926 | 2001 | 1922 | OCT 1926 |
| | MAXIMUM SNOW DEPTH (IN) | 60 | 22 | 29 | 30 | 30 | 25 | 14 | 1 | 3 | 20 | 12 | 18 | 17 | 30 |
| | YEAR OF OCCURRENCE | | 1962 | 1962 | 1962 | 1962 | 1963 | 1950 | 1963 | 2003 | 1962 | 2008 | 2007 | 2005 | APR 1962 |
| | NORMAL NO. DAYS WITH: | | | | | | | | | | | | | | |
| | SNOWFALL >= 1.0 | 30 | 0.4 | 0.4 | 0.2 | 0.5 | 0.4 | 0.3 | 0.0 | 0.2 | 1.7 | 1.6 | 0.7 | 0.4 | 6.8 |

## PRECIPITATION (inches) 2009 BARROW (PABR)

| YEAR | JAN | FEB | MAR | APR | MAY | JUN | JUL | AUG | SEP | OCT | NOV | DEC | ANNUAL |
|------|-----|-----|-----|-----|-----|-----|-----|-----|-----|-----|-----|-----|--------|
| 1980 | 0.15 | 0.15 | 0.02 | 0.06 | 0.01 | 0.56 | 0.77 | 1.41 | 0.73 | 0.36 | 0.12 | 0.08 | 4.42 |
| 1981 | 0.22 | 0.06 | 0.03 | 0.19 | 0.06 | 0.51 | 1.77 | 0.56 | 0.52 | 0.32 | 0.06 | 0.09 | 4.39 |
| 1982 | 0.17 | 0.43 | 0.24 | 0.34 | 0.39 | 0.21 | 0.78 | 0.86 | 0.59 | 0.56 | 0.02 | 0.13 | 4.72 |
| 1983 | 0.03 | 0.09 | T | 0.20 | 0.07 | 0.11 | 0.10 | 1.04 | 0.93 | 0.36 | 0.25 | 0.05 | 3.23 |
| 1984 | 0.19 | 0.16 | 0.11 | 0.27 | 0.07 | 0.03 | 0.83 | 1.64 | 0.15 | 0.33 | 0.12 | 0.08 | 3.98 |
| 1985 | 0.05 | 0.10 | 0.15 | 0.05 | 0.25 | 0.64 | 0.61 | 0.51 | 0.58 | 0.45 | 0.25 | 0.16 | 3.80 |
| 1986 | 0.16 | 0.14 | 0.09 | 0.03 | 0.07 | 0.07 | 0.79 | 0.69 | 1.45 | 0.43 | 0.14 | 0.09 | 4.15 |
| 1987 | 0.13 | 0.08 | 0.03 | T | 0.13 | 0.06 | 1.94 | 1.00 | 1.37 | 0.17 | 0.05 | 0.18 | 5.14 |
| 1988 | 0.02 | 0.04 | 0.10 | 0.03 | 0.02 | 0.15 | 0.74 | 1.57 | 0.41 | 0.24 | 0.01 | 0.26 | 3.59 |
| 1989 | 0.01 | 0.29 | | 0.42 | 0.02 | 0.36 | 3.19 | 1.69 | 0.69 | 0.20 | T | 0.20 | |
| 1990 | 0.03 | 0.06 | 0.13 | 0.08 | 0.13 | 0.38 | 1.35 | 1.19 | 0.55 | 0.42 | 0.17 | 0.12 | 4.61 |
| 1991 | 0.07 | 0.08 | 0.02 | 0.07 | 0.17 | 0.08 | 0.22 | 0.20 | 0.35 | 0.39 | 0.07 | 0.03 | 1.75 |
| 1992 | 0.04 | 0.12 | 0.13 | 0.11 | 0.08 | 0.16 | 0.26 | 0.66 | 0.47 | 0.21 | 0.22 | 0.24 | 2.70 |
| 1993 | 0.45 | 0.17 | 0.11 | 0.03 | 0.11 | 0.44 | 0.67 | 0.98 | 1.50 | 0.50 | 0.25 | 0.12 | 5.33 |
| 1994 | 0.05 | 0.02 | 0.13 | 0.02 | 0.34 | 0.15 | 0.56 | 2.02 | 0.44 | 0.33 | 0.07 | 0.15 | 4.28 |
| 1995 | 0.06 | 0.03 | 0.23 | 0.13 | T | 0.36 | 1.09 | 0.31 | 0.16 | 0.14 | 0.19 | 0.05 | 2.75 |
| 1996 | T | 0.11 | 0.06 | 0.11 | 0.39 | 0.07 | 1.04 | 0.57 | 0.81 | 0.17 | 0.12 | 0.03 | 3.48 |
| 1997 | 0.08 | T | 0.03 | 0.11 | 0.03 | 0.63 | 0.48 | 2.64 | 0.59 | 0.41 | 0.08 | 0.06 | 5.14 |
| 1998 | T | 0.02 | 0.04 | 0.14 | 0.19 | 0.20 | 0.56 | 1.55 | 0.90 | 0.65 | 0.09 | 0.36 | 4.70 |
| 1999 | 0.09 | 0.08 | 0.09 | 0.07 | 0.03 | 0.31 | 0.67 | 1.38 | 0.37 | 0.23 | 0.23 | 0.13 | 3.68 |
| 2000 | 0.36 | 0.08 | 0.05 | 0.03 | 0.29 | 0.75 | 2.06 | 1.05 | 0.62 | 0.45 | 0.06 | 0.02 | 5.82 |
| 2001 | 0.63 | 0.09 | 0.01 | 0.12 | 0.15 | 0.20 | 1.55 | 1.25 | 0.53 | 0.20 | 0.46 | 0.06 | 5.25 |
| 2002 | 0.06 | T | 0.04 | 0.25 | 0.12 | 0.62 | 0.06 | 1.04 | 1.88 | 0.55 | 0.11 | 0.02 | 4.75 |
| 2003 | 0.08 | 0.13 | 0.01 | 0.15 | 0.23 | 0.12 | 0.97 | 0.88 | 1.27 | 1.49 | 0.36 | 0.11 | 5.80 |
| 2004 | 0.09 | 0.06 | 0.20 | 0.01 | 0.22 | 0.93 | 1.60 | 0.81 | 1.28 | 0.43 | 0.21 | 0.17 | 6.01 |
| 2005 | 0.04 | 0.14 | 0.04 | 0.03 | 0.33 | 0.35 | 0.73 | 1.64 | 0.86 | 0.22 | 0.25 | 0.23 | 4.86 |
| 2006 | 0.13 | 0.36 | 0.13 | 0.20 | 0.03 | 0.38 | 1.06 | 0.64 | 0.37 | 0.33 | 0.34 | 0.20 | 4.17 |
| 2007 | 0.20 | 0.04 | 0.11 | 0.11 | 0.26 | 0.01 | 0.04 | 0.37 | 0.34 | 0.34 | 0.47 | 0.06 | 2.35 |
| 2008 | 0.10 | 0.22 | T | 0.73 | 0.25 | 0.50 | 1.33 | 0.30 | 0.07 | 0.86 | 0.32 | 0.14 | 4.82 |
| 2009 | 0.30 | 0.44 | 0.10 | 0.19 | 0.45 | 0.18 | 0.75 | 2.13 | 0.67 | 0.44 | 0.18 | 0.34 | 6.17 |
| POR=<br>90 YRS | 0.16 | 0.25 | 0.13 | 0.15 | 0.15 | 0.32 | 0.90 | 0.92 | 0.60 | 0.49 | 0.25 | 0.17 | 4.49 |

WBAN : 27502

## AVERAGE TEMPERATURE (°F) 2009 BARROW (PABR)

| YEAR | JAN | FEB | MAR | APR | MAY | JUN | JUL | AUG | SEP | OCT | NOV | DEC | ANNUAL |
|------|-----|-----|-----|-----|-----|-----|-----|-----|-----|-----|-----|-----|--------|
| 1980 | -13.9 | -10.3 | -11.5 | -3.8 | 17.0 | 37.0 | 35.7 | 33.9 | 25.1 | 14.5 | -5.0 | -15.6 | 8.6 |
| 1981 | -0.9 | -15.7 | -10.9 | 1.1 | 23.6 | 34.6 | 39.7 | 33.6 | 25.5 | 14.2 | -0.6 | -8.1 | 11.3 |
| 1982 | -11.1 | -6.5 | -12.4 | -1.0 | 16.6 | 33.9 | 38.0 | 36.7 | 29.6 | 6.9 | -10.1 | -9.3 | 9.3 |
| 1983 | -19.2 | -15.4 | -13.4 | 2.7 | 16.9 | 34.6 | 38.1 | 34.3 | 23.9 | 7.0 | 1.3 | 0.6 | 9.3 |
| 1984 | -15.3 | -33.0 | -16.5 | -10.6 | 16.6 | 37.9 | 41.0 | 38.4 | 34.2 | 17.2 | -8.1 | -13.6 | 7.4 |
| 1985 | -7.3 | -17.2 | -12.9 | -6.6 | 22.7 | 35.8 | 39.0 | 38.6 | 28.9 | 10.4 | 2.7 | -6.9 | 10.6 |
| 1986 | -15.0 | -8.8 | -17.7 | -7.7 | 20.1 | 34.5 | 42.0 | 39.5 | 36.9 | 16.4 | 0.4 | -6.4 | 11.2 |
| 1987 | -13.0 | -20.0 | -11.7 | -4.7 | 20.2 | 34.1 | 38.9 | 38.9 | 28.3 | 22.9 | -5.4 | -8.9 | 10.0 |
| 1988 | -10.5 | -14.4 | -12.7 | 1.4 | 19.8 | 33.6 | 38.9 | 35.7 | 28.2 | 2.0 | -13.6 | -9.3 | 8.3 |
| 1989 | -24.0 | 9.3 | | 6.0 | 17.4 | 36.5 | 45.5 | 46.8 | 35.5 | 18.0 | -12.6 | -9.5 | |
| 1990 | -23.0 | -23.2 | -11.4 | 7.5 | 26.5 | 37.7 | 42.3 | 37.3 | 31.5 | 16.9 | -6.9 | -15.7 | 10.0 |
| 1991 | -13.4 | -18.2 | -17.7 | 3.3 | 28.0 | 36.9 | 37.9 | 36.1 | 29.9 | 17.5 | -8.0 | -15.8 | 9.7 |
| 1992 | -18.6 | -20.3 | -9.2 | 0.8 | 20.7 | 35.9 | 39.7 | 39.0 | 25.0 | 12.5 | -1.1 | -8.7 | 9.6 |
| 1993 | -12.6 | -11.1 | -12.6 | 6.5 | 23.0 | 36.4 | 45.2 | 36.8 | 32.4 | 23.1 | 5.2 | -8.1 | 13.7 |
| 1994 | -9.0 | -8.9 | -18.2 | -0.1 | 18.6 | 32.6 | 41.6 | 43.1 | 27.2 | 8.4 | -8.6 | -16.0 | 9.2 |
| 1995 | -12.0 | -16.2 | -16.7 | 7.6 | 26.4 | 36.6 | 40.4 | 36.5 | 35.6 | 18.5 | 4.0 | -11.9 | 12.4 |
| 1996 | -7.1 | -16.3 | -5.6 | 0.0 | 25.9 | 37.8 | 42.2 | 35.9 | 27.9 | 5.5 | 9.4 | -4.3 | 12.6 |
| 1997 | -17.5 | -14.2 | -13.1 | 3.7 | 19.7 | 34.2 | 40.9 | 42.0 | 35.0 | 19.7 | 11.1 | -11.6 | 12.5 |
| 1998 | -14.6 | -14.6 | -2.7 | 11.2 | 23.8 | 37.8 | 44.6 | 42.4 | 37.7 | 24.7 | 15.1 | -1.8 | 17.0 |
| 1999 | -18.7 | -15.2 | -15.9 | -2.3 | 22.8 | 35.4 | 42.8 | 42.1 | 33.5 | 17.7 | -.5 | -16.8 | 10.4 |
| 2000 | -13.8 | -15.7 | -12.0 | -.8 | 13.9 | 37.1 | 38.4 | 38.5 | 31.6 | 17.2 | .5 | -5.1 | 10.8 |
| 2001 | -10.6 | -2.7 | -15.7 | 1.4 | 11.1 | 35.0 | 38.6 | 35.3 | 32.0 | 9.6 | .8 | -8.5 | 10.5 |
| 2002 | -16.8 | -17.9 | -2.1 | 1.6 | 24.8 | 33.6 | 39.2 | 36.7 | 36.0 | 23.8 | 6.7 | -2.1 | 13.6 |
| 2003 | -10.5 | -15.7 | -10.6 | 8.4 | 21.5 | 34.3 | 40.9 | 35.7 | 32.0 | 24.0 | 2.8 | -7.8 | 12.9 |
| 2004 | -8.3 | -22.8 | -15.2 | 1.0 | 23.4 | 38.8 | 42.6 | 44.2 | 33.0 | 21.4 | 2.3 | -8.8 | 12.6 |
| 2005 | -8.0 | -12.7 | -7.9 | 2.2 | 22.5 | 34.1 | 39.1 | 42.3 | 33.6 | 21.9 | -1.1 | -4.1 | 13.5 |
| 2006 | -11.6 | -8.0 | -17.9 | -3.9 | 23.9 | 37.7 | 39.1 | 36.4 | 36.2 | 25.5 | 6.0 | -2.7 | 13.4 |
| 2007 | -15.6 | -10.2 | -15.1 | 8.3 | 16.9 | 36.0 | 44.2 | 44.8 | 37.7 | 23.4 | 13.9 | -0.3 | 15.3 |
| 2008 | -15.6 | -13.8 | -16.0 | 7.5 | 22.2 | 37.2 | 40.0 | 37.6 | 34.0 | 22.6 | 4.0 | 0.5 | 13.4 |
| 2009 | -12.7 | -15.8 | -15.5 | 2.3 | 25.6 | 35.2 | 43.8 | 41.7 | 34.9 | 24.9 | 0.6 | -3.2 | 13.5 |
| POR=<br>91 YRS | -13.5 | -15.4 | -13.7 | 0.1 | 19.7 | 33.9 | 39.9 | 38.4 | 30.7 | 16.2 | 0.8 | -9.6 | 10.6 |

## HEATING DEGREE DAYS (base 65°F) 2009 BARROW (PABR)

| YEAR | JUL | AUG | SEP | OCT | NOV | DEC | JAN | FEB | MAR | APR | MAY | JUN | TOTAL |
|---|---|---|---|---|---|---|---|---|---|---|---|---|---|
| 1980-81 | 902 | 961 | 1191 | 1558 | 2102 | 2501 | 2043 | 2265 | 2360 | 1917 | 1275 | 906 | 19981 |
| 1981-82 | 777 | 968 | 1176 | 1569 | 1966 | 2269 | 2363 | 2002 | 2403 | 1983 | 1495 | 925 | 19896 |
| 1982-83 | 833 | 871 | 1057 | 1796 | 2261 | 2307 | 2617 | 2255 | 2435 | 1871 | 1485 | 906 | 20694 |
| 1983-84 | 830 | 941 | 1225 | 1798 | 1907 | 1998 | 2495 | 2848 | 2532 | 2273 | 1498 | 806 | 21151 |
| 1984-85 | 738 | 816 | 917 | 1479 | 2197 | 2197 | 2442 | 2248 | 2304 | 2423 | 2153 | 1303 | 870 | 19890 |
| 1985-86 | 800 | 810 | 1078 | 1691 | 1867 | 2233 | 2485 | 2066 | 2573 | 2183 | 1384 | 909 | 20079 |
| 1986-87 | 707 | 784 | 840 | 1500 | 1941 | 2216 | 2424 | 2388 | 2382 | 2093 | 1380 | 920 | 19575 |
| 1987-88 | 801 | 802 | 1095 | 1297 | 2114 | 2295 | 2344 | 2305 | 2415 | 1906 | 1394 | 938 | 19706 |
| 1988-89 | 803 | 898 | 1097 | 1956 | 2363 | 2307 | 2763 | 1556 |  | 1767 | 1468 | 851 |  |
| 1989-90 | 595 | 557 | 879 | 1449 | 2335 | 2311 | 2730 | 2477 | 2372 | 1720 | 1184 | 816 | 19425 |
| 1990-91 | 696 | 850 | 997 | 1487 | 2161 | 2511 | 2433 | 2336 | 2568 | 1849 | 1138 | 834 | 19860 |
| 1991-92 | 837 | 887 | 1046 | 1469 | 2194 | 2513 | 2598 | 2481 | 2307 | 1930 | 1365 | 866 | 20493 |
| 1992-93 | 778 | 800 | 1193 | 1619 | 1982 | 2285 | 2407 | 2130 | 2410 | 1751 | 1294 | 850 | 19499 |
| 1993-94 | 609 | 868 | 970 | 1291 | 1795 | 2267 | 2300 | 2073 | 2587 | 1951 | 1431 | 968 | 19110 |
| 1994-95 | 721 | 671 | 1126 | 1753 | 2212 | 2515 | 2388 | 2278 | 2534 | 1719 | 1189 | 846 | 19952 |
| 1995-96 | 760 | 876 | 877 | 1432 | 1829 | 2388 | 2239 | 2361 | 2189 | 1951 | 1207 | 808 | 18917 |
| 1996-97 | 698 | 894 | 1108 | 1843 | 1665 | 2137 | 2551 | 2209 | 2414 | 1834 | 1398 | 915 | 19666 |
| 1997-98 | 741 | 707 | 894 | 1399 | 1610 | 2368 | 2460 | 2224 | 2093 | 1609 | 1270 | 810 | 18185 |
| 1998-99 | 626 | 693 | 810 | 1240 | 1492 | 2069 | 2589 | 2241 | 2501 | 2011 | 1299 | 881 | 18452 |
| 1999-00 | 681 | 704 | 938 | 1458 | 1958 | 2528 | 2435 | 2332 | 2380 | 1967 | 1577 | 828 | 19786 |
| 2000-01 | 820 | 814 | 994 | 1473 | 1927 | 2167 | 2335 | 1887 | 2493 | 1901 | 1663 | 894 | 19368 |
| 2001-02 | 811 | 916 | 982 | 1709 | 1917 | 2273 | 2528 | 2317 | 2071 | 1894 | 1237 | 935 | 19590 |
| 2002-03 | 792 | 868 | 863 | 1271 | 1740 | 2074 | 2332 | 2255 | 2338 | 1693 | 1342 | 913 | 18481 |
| 2003-04 | 738 | 901 | 985 | 1263 | 1860 | 2249 | 2265 | 2541 | 2479 | 1912 | 1284 | 781 | 19258 |
| 2004-05 | 688 | 634 | 953 | 1346 | 1875 | 2281 | 2258 | 2169 | 2251 | 1877 | 1315 | 919 | 18566 |
| 2005-06 | 799 | 695 | 933 | 1332 | 1975 | 2135 | 2373 | 2037 | 2568 | 2060 | 1269 | 811 | 18987 |
| 2006-07 | 795 | 880 | 858 | 1218 | 1761 | 2097 | 2493 | 2102 | 2480 | 1693 | 1483 | 862 | 18722 |
| 2007-08 | 637 | 619 | 814 | 1281 | 1527 | 2017 | 2492 | 2284 | 2506 | 1716 | 1320 | 828 | 18041 |
| 2008-09 | 770 | 844 | 922 | 1309 | 1824 | 1995 | 2408 | 2260 | 2494 | 1875 | 1219 | 888 | 18808 |
| 2009- | 652 | 715 | 896 | 1234 | 1926 | 2113 |  |  |  |  |  |  |  |

WBAN : 27502

## COOLING DEGREE DAYS (base 65°F) 2009 BARROW (PABR)

| YEAR | JAN | FEB | MAR | APR | MAY | JUN | JUL | AUG | SEP | OCT | NOV | DEC | TOTAL |
|---|---|---|---|---|---|---|---|---|---|---|---|---|---|
| 1980 | 0 | 0 | 0 | 0 | 0 | 0 | 0 | 0 | 0 | 0 | 0 | 0 | 0 |
| 1981 | 0 | 0 | 0 | 0 | 0 | 0 | 0 | 0 | 0 | 0 | 0 | 0 | 0 |
| 1982 | 0 | 0 | 0 | 0 | 0 | 0 | 0 | 0 | 0 | 0 | 0 | 0 | 0 |
| 1983 | 0 | 0 | 0 | 0 | 0 | 0 | 0 | 0 | 0 | 0 | 0 | 0 | 0 |
| 1984 | 0 | 0 | 0 | 0 | 0 | 0 | 0 | 0 | 0 | 0 | 0 | 0 | 0 |
| 1985 | 0 | 0 | 0 | 0 | 0 | 0 | 0 | 0 | 0 | 0 | 0 | 0 | 0 |
| 1986 | 0 | 0 | 0 | 0 | 0 | 0 | 0 | 0 | 0 | 0 | 0 | 0 | 0 |
| 1987 | 0 | 0 | 0 | 0 | 0 | 0 | 0 | 0 | 0 | 0 | 0 | 0 | 0 |
| 1988 | 0 | 0 | 0 | 0 | 0 | 0 | 0 | 0 | 0 | 0 | 0 | 0 | 0 |
| 1989 | 0 | 0 | 0 | 0 | 0 | 0 | 0 | 0 | 0 | 0 | 0 | 0 | 0 |
| 1990 | 0 | 0 | 0 | 0 | 0 | 0 | 0 | 0 | 0 | 0 | 0 | 0 | 0 |
| 1991 | 0 | 0 | 0 | 0 | 0 | 0 | 0 | 0 | 0 | 0 | 0 | 0 | 0 |
| 1992 | 0 | 0 | 0 | 0 | 0 | 0 | 0 | 0 | 0 | 0 | 0 | 0 | 0 |
| 1993 | 0 | 0 | 0 | 0 | 0 | 0 | 0 | 0 | 0 | 0 | 0 | 0 | 0 |
| 1994 | 0 | 0 | 0 | 0 | 0 | 0 | 0 | 0 | 0 | 0 | 0 | 0 | 0 |
| 1995 | 0 | 0 | 0 | 0 | 0 | 0 | 0 | 0 | 0 | 0 | 0 | 0 | 0 |
| 1996 | 0 | 0 | 0 | 0 | 0 | 0 | 0 | 0 | 0 | 0 | 0 | 0 | 0 |
| 1997 | 0 | 0 | 0 | 0 | 0 | 0 | 0 | 0 | 0 | 0 | 0 | 0 | 0 |
| 1998 | 0 | 0 | 0 | 0 | 0 | 0 | 0 | 0 | 0 | 0 | 0 | 0 | 0 |
| 1999 | 0 | 0 | 0 | 0 | 0 | 0 | 0 | 0 | 0 | 0 | 0 | 0 | 0 |
| 2000 | 0 | 0 | 0 | 0 | 0 | 0 | 0 | 0 | 0 | 0 | 0 | 0 | 0 |
| 2001 | 0 | 0 | 0 | 0 | 0 | 0 | 0 | 0 | 0 | 0 | 0 | 0 | 0 |
| 2002 | 0 | 0 | 0 | 0 | 0 | 0 | 0 | 0 | 0 | 0 | 0 | 0 | 0 |
| 2003 | 0 | 0 | 0 | 0 | 0 | 0 | 0 | 0 | 0 | 0 | 0 | 0 | 0 |
| 2004 | 0 | 0 | 0 | 0 | 0 | 0 | 0 | 0 | 0 | 0 | 0 | 0 | 0 |
| 2005 | 0 | 0 | 0 | 0 | 0 | 0 | 0 | 0 | 0 | 0 | 0 | 0 | 0 |
| 2006 | 0 | 0 | 0 | 0 | 0 | 0 | 0 | 0 | 0 | 0 | 0 | 0 | 0 |
| 2007 | 0 | 0 | 0 | 0 | 0 | 0 | 0 | 0 | 0 | 0 | 0 | 0 | 0 |
| 2008 | 0 | 0 | 0 | 0 | 0 | 0 | 0 | 0 | 0 | 0 | 0 | 0 | 0 |
| 2009 | 0 | 0 | 0 | 0 | 0 | 0 | 0 | 0 | 0 | 0 | 0 | 0 | 0 |

## SNOWFALL (inches) 2009 BARROW (PABR)

| YEAR | JUL | AUG | SEP | OCT | NOV | DEC | JAN | FEB | MAR | APR | MAY | JUN | TOTAL |
|------|-----|-----|-----|-----|-----|-----|-----|-----|-----|-----|-----|-----|-------|
| 1980-81 | T | 1.0 | 2.4 | 7.9 | 2.4 | 1.6 | 4.5 | 1.2 | 0.4 | 2.8 | 0.7 | 3.7 | 28.6 |
| 1981-82 | T | 0.6 | 5.5 | 5.7 | 0.6 | 0.9 | 1.7 | 4.3 | 2.4 | 3.4 | 4.4 | 0.2 | 29.7 |
| 1982-83 | 1.7 | 0.4 | 5.1 | 5.8 | 0.2 | 1.3 | 0.3 | 0.9 | T | 1.5 | 0.7 | 0.7 | 18.6 |
| 1983-84 | 0.4 | 1.8 | 6.9 | 3.4 | 3.1 | 0.5 | 1.9 | 1.6 | 1.1 | 2.7 | 0.7 | T | 24.1 |
| 1984-85 | T | 2.4 | 0.7 | 3.7 | 1.2 | 0.8 | 0.5 | 1.1 | 1.6 | 0.6 | 2.4 | 0.4 | 15.4 |
| 1985-86 | 0.0 | 0.7 | 6.1 | 4.7 | 2.7 | 1.7 | 1.6 | 1.4 | 0.9 | 0.3 | 0.9 | T | 21.0 |
| 1986-87 | T | 0.9 | 4.2 | 9.4 | 3.4 | 3.0 | 2.9 | 2.8 | 0.7 | 0.4 | 1.4 | 0.8 | 29.9 |
| 1987-88 | T | 0.4 | 16.2 | 2.7 | 0.5 | 2.7 | 0.5 | 1.3 | 1.7 | 0.4 | 1.3 | 0.6 | 28.3 |
| 1988-89 | T | 2.5 | 4.1 | 2.7 | 0.1 | 2.6 | 0.1 | 3.3 | | 4.2 | 0.6 | 0.2 | |
| 1989-90 | T | 0.0 | 0.4 | 6.6 | 0.4 | 4.0 | 0.7 | 0.8 | 3.0 | 1.9 | 1.1 | 1.2 | 20.1 |
| 1990-91 | 0.4 | T | 4.7 | 11.0 | 5.6 | 2.4 | 1.9 | 2.0 | 0.6 | 2.0 | 2.7 | 1.1 | 34.4 |
| 1991-92 | T | 1.1 | 4.6 | 7.9 | 1.5 | 1.2 | 1.1 | 2.7 | 4.6 | 3.1 | 1.3 | 0.5 | 29.6 |
| 1992-93 | T | 2.1 | 8.9 | 7.5 | 6.8 | 4.7 | 8.6 | 3.2 | 4.2 | 3.4 | 1.4 | 0.1 | 50.9 |
| 1993-94 | 0.0 | 1.5 | 11.8 | 8.5 | 5.5 | 3.1 | 1.4 | 0.4 | 4.3 | 1.3 | 3.8 | 2.5 | 44.1 |
| 1994-95 | T | 1.1 | 3.4 | 11.1 | 2.7 | 3.0 | 1.8 | 0.7 | 5.2 | 2.7 | 0.1 | T | 31.8 |
| 1995-96 | T | 1.3 | 1.4 | 5.4 | 3.7 | 1.7 | 0.8 | 4.2 | 1.9 | 3.4 | 6.0 | 0.4 | 30.2 |
| 1996-97 | T | .2 | 7.2 | 6.3 | 4.5 | .8 | 4.8 | T | 1.7 | 3.0 | 2.1 | T | 30.6 |
| 1997-98 | T | T | 3.8 | 19.1 | 5.0 | 3.0 | 0.5 | 1.5 | 1.3 | 3.9 | 2.2 | T | 40.3 |
| 1998-99 | 0.0 | T | 0.8 | 12.6 | 3.8 | 7.1 | 1.0 | 1.6 | 1.9 | 1.7 | 0.7 | 1.5 | 32.7 |
| 1999-00 | T | 0.7 | 4.1 | 11.5 | 9.9 | 2.9 | 7.0 | 2.6 | 1.2 | 2.2 | 4.9 | T | 47.0 |
| 2000-01 | 0.7 | 2.5 | 9.9 | 11.3 | 3.5 | 1.5 | 8.1 | 1.9 | 0.8 | 2.4 | 3.0 | 0.1 | 45.7 |
| 2001-02 | 0.2 | 0.2 | 1.6 | 6.8 | 17.5 | 4.9 | 2.6 | 0.1 | 1.5 | 7.4 | 1.7 | 0.4 | 44.9 |
| 2002-03 | 0.8 | 1.1 | 2.8 | 12.5 | 6.7 | 4.2 | 5.4 | 6.6 | 1.0 | 5.6 | 4.7 | T | 51.4 |
| 2003-04 | 0.2 | 3.5 | 5.5 | 15.6 | 8.4 | 3.7 | 1.9 | 2.9 | 3.9 | 0.5 | 0.6 | T | 46.7 |
| 2004-05 | 0.2 | T | 7.5 | 6.3 | 6.6 | 4.4 | 1.2 | 2.9 | 0.9 | 1.3 | 5.5 | 1.6 | 38.4 |
| 2005-06 | T | 0.2 | 0.1 | 5.9 | 10.2 | 12.1 | 2.8 | 6.4 | 2.9 | 4.9 | 0.9 | 1.8 | 48.2 |
| 2006-07 | T | 0.3 | 2.4 | 10.9 | 11.7 | 3.7 | 5.4 | 1.2 | 2.3 | 3.6 | 5.4 | T | 46.9 |
| 2007-08 | 0.0 | 0.0 | 0.4 | 10.2 | 14.4 | 4.9 | 3.1 | 4.9 | T | 11.1 | 4.6 | 0.2 | 53.8 |
| 2008-09 | 0.9 | T | 0.6 | 22.7 | 12.3 | 5.1 | 7.0 | 11.4 | 2.8 | 4.5 | 9.3 | T | 76.6 |
| 2009- | 0.0 | 0.1 | 2.7 | 8.8 | 7.6 | 5.7 | | | | | | | |
| POR= 83 YRS | 0.5 | 0.7 | 3.5 | 7.4 | 4.3 | 2.9 | 2.4 | 3.4 | 2.0 | 2.6 | 2.1 | 0.6 | 32.4 |

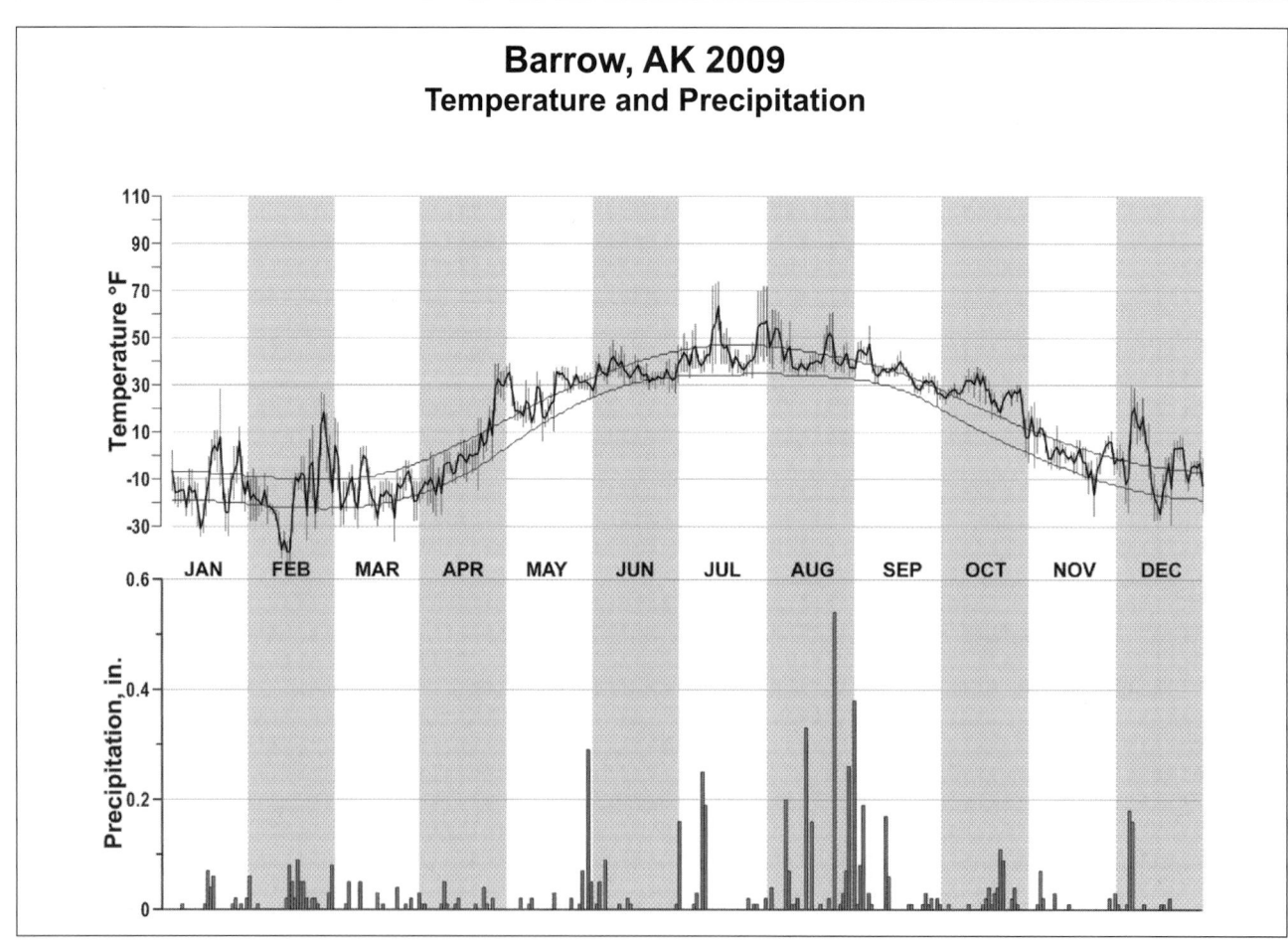

# Barrow, AK 2009
## Temperature and Precipitation

# 2009
# FAIRBANKS
# ALASKA (PAFA)

Fairbanks is located in the Tanana Valley, in the interior of Alaska. It has a distinctly continental climate, with large variation of temperature from winter to summer.

The climate in Fairbanks is conditioned mainly by the response of the land mass to large changes in solar heat received by the area during the year. The sun is above the horizon from 18 to 21 hours during June and July. During this period, daily average maximum temperatures reach the lower 70s. Temperatures of 80 degrees or higher occur on about 10 days each summer. In contrast, from November to early March, when the period of daylight ranges from 10 to less than 4 hours per day, the lowest temperature readings normally fall below zero quite regularly. Low temperatures of -40 degrees or colder occur each winter. The range of temperatures in summer is comparatively low, from the lower 30s to the mid 90s. In winter, this range is larger, from about 65 below to 45 degrees above. This large winter range of temperature reflects the great difference between frigid weather associated with dry northerly airflow from the Arctic to mild temperatures associated with southerly airflow from the Gulf of Alaska, accompanied by chinook winds off the Alaska Range, 80 miles to the south of Fairbanks.

Snow cover is persistent in Fairbanks, without interruption, from October through April. Snowfalls of 4 inches or more in a day occur only three times during winter. Blizzard conditions are almost never seen, as winds in Fairbanks are above 20 miles an hour less than 1 percent of the time. Precipitation normally reaches a minimum in spring, and a maximum in August, when rainfall is common. During summer, thunderstorms occur in Fairbanks on an average of about eight days. Thunderstorms are about three times more frequent over the hills to the north and east of Fairbanks. Damaging hail or wind rarely accompany thunderstorms around Fairbanks.

There are rolling hills reaching elevations up to 2,000 feet above Fairbanks to the north and east of the city. During winter, the uplands are often warmer than Fairbanks, as cold air settles into the valley. In some months, temperatures in the uplands will average more than 10 degrees warmer than Fairbanks. During summer, the uplands are a few degrees cooler than the city. Precipitation in the uplands around Fairbanks is heavier than it is in the city by roughly 20 to 50 percent. Fairbanks exhibits an urban heat island, especially during winter. Low lying areas nearby, such as the community of North Pole, are often colder than the city, sometimes by as much as 15 degrees.

During winter, with temperatures of -20 degrees or colder, ice fog frequently forms in the city. Cold snaps accompanied by ice fog generally last about a week, but can last three weeks in unusual situations. The fog is almost always less than 300 feet deep, so that the surrounding uplands are usually in the clear, with warmer temperatures. Visibility in the ice fog is sometimes quite low, and this can hinder aircraft operations for as much as a day in severe cases. Aside from the low visibility in winter ice fog, flying weather in Fairbanks is quite favorable, especially from February through May, when crystal clear weather is common and the length of daylight is rapidly increasing.

Hardy vegetables and grains grow luxuriantly. Freezing of local rivers normally begins in the first week of October. The date when ice will normally support a persons weight is October 27. Rivers remain frozen and safe for travel until early April. Breakup of the river ice usually occurs in the first week of May.

# NORMALS, MEANS, AND EXTREMES
## FAIRBANKS (PAFA)

| LATITUDE: 64 ° 49'N | LONGITUDE: -147° 51'W | | ELEVATION (FT): GRND: 432 BARO: 464 | | | | TIME ZONE: ALASKA (UTC -9) | | | WBAN: 26411 | | |

| | ELEMENT | POR | JAN | FEB | MAR | APR | MAY | JUN | JUL | AUG | SEP | OCT | NOV | DEC | YEAR |
|---|---|---|---|---|---|---|---|---|---|---|---|---|---|---|---|
| **TEMPERATURE °F** | NORMAL DAILY MAXIMUM | 30 | -.3 | 8.0 | 25.0 | 43.6 | 60.6 | 70.9 | 73.0 | 66.3 | 54.3 | 31.4 | 11.2 | 3.3 | 37.3 |
| | MEAN DAILY MAXIMUM | 61 | -0.2 | 8.8 | 23.9 | 42.4 | 59.9 | 70.4 | 72.2 | 66.2 | 54.5 | 32.2 | 11.7 | 1.8 | 37.0 |
| | HIGHEST DAILY MAXIMUM | 58 | 52 | 47 | 56 | 76 | 89 | 96 | 94 | 93 | 84 | 72 | 49 | 45 | 96 |
| | YEAR OF OCCURRENCE | | 2009 | 1987 | 1994 | 2009 | 1960 | 1969 | 1975 | 1994 | 1957 | 2003 | 1997 | 1999 | JUN 1969 |
| | MEAN OF EXTREME MAXS. | 61 | 29.0 | 33.5 | 44.2 | 60.1 | 74.8 | 83.5 | 85.2 | 80.6 | 68.8 | 52.0 | 33.8 | 30.0 | 56.3 |
| | NORMAL DAILY MINIMUM | 30 | -19.0 | -15.6 | -2.7 | 19.8 | 36.9 | 48.5 | 51.9 | 46.2 | 34.7 | 15.6 | -6.6 | -15.2 | 16.2 |
| | MEAN DAILY MINIMUM | 61 | -17.1 | -12.4 | -2.5 | 20.1 | 37.7 | 48.9 | 51.9 | 46.6 | 35.5 | 17.4 | -3.9 | -14.0 | 17.4 |
| | LOWEST DAILY MINIMUM | 58 | -61 | -58 | -49 | -24 | -1 | 29 | 35 | 27 | 3 | -27 | -46 | -62 | -62 |
| | YEAR OF OCCURRENCE | | 1969 | 1993 | 1956 | 1986 | 1964 | 2006 | 1959 | 1987 | 1992 | 1992 | 1990 | 1961 | DEC 1961 |
| | MEAN OF EXTREME MINS. | 61 | -40.5 | -34.9 | -25.2 | -3.0 | 26.2 | 38.2 | 43.0 | 34.8 | 22.6 | -4.6 | -24.9 | -36.8 | -0.3 |
| | NORMAL DRY BULB | 30 | -9.7 | -3.8 | 11.1 | 31.7 | 48.8 | 59.7 | 62.4 | 56.2 | 44.5 | 23.5 | 2.3 | -5.9 | 26.7 |
| | MEAN DRY BULB | 61 | -8.6 | -1.7 | 10.7 | 31.3 | 48.8 | 59.9 | 62.1 | 56.4 | 45.0 | 24.8 | 3.9 | -6.0 | 27.2 |
| | MEAN WET BULB | 22 | -2.3 | -0.5 | 9.2 | 26.6 | 40.3 | 51.3 | 54.8 | 50.8 | 40.1 | 22.0 | 1.3 | -2.4 | 24.3 |
| | MEAN DEW POINT | 22 | -6.4 | -5.1 | 3.5 | 19.6 | 32.6 | 45.7 | 50.8 | 47.4 | 36.7 | 19.2 | -1.1 | -5.0 | 19.8 |
| | NORMAL NO. DAYS WITH: | | | | | | | | | | | | | | |
| | MAXIMUM >= 70 | 30 | 0.0 | 0.0 | 0.0 | 0.1 | 4.2 | 17.0 | 21.5 | 10.9 | 1.5 | 0.0 | 0.0 | 0.0 | 55.2 |
| | MAXIMUM <= 32 | 30 | 29.8 | 26.2 | 20.9 | 5.1 | 0.1 | 0.0 | 0.0 | 0.0 | 0.4 | 16.3 | 28.1 | 29.9 | 156.8 |
| | MINIMUM <= 32 | 30 | 31.0 | 28.3 | 31.0 | 26.1 | 5.5 | 0.0 | 0.0 | 0.4 | 9.1 | 27.8 | 30.0 | 31.0 | 220.2 |
| | MINIMUM <= 0 | 30 | 25.3 | 22.2 | 16.0 | 2.2 | 0.0 | 0.0 | 0.0 | 0.0 | 0.0 | 3.2 | 19.3 | 25.0 | 113.2 |
| **H/C** | NORMAL HEATING DEG. DAYS | 30 | 2315 | 1926 | 1670 | 999 | 504 | 179 | 121 | 283 | 615 | 1287 | 1882 | 2199 | 13980 |
| | NORMAL COOLING DEG. DAYS | 30 | 0 | 0 | 0 | 0 | 0 | 20 | 42 | 11 | 1 | 0 | 0 | 0 | 74 |
| **RH** | NORMAL (PERCENT) | 30 | 72 | 70 | 63 | 54 | 49 | 57 | 65 | 72 | 71 | 76 | 75 | 74 | 67 |
| | HOUR 03 LST | 30 | 71 | 71 | 70 | 68 | 67 | 76 | 82 | 86 | 82 | 79 | 75 | 74 | 75 |
| | HOUR 09 LST | 30 | 71 | 71 | 68 | 56 | 50 | 59 | 68 | 76 | 76 | 79 | 76 | 74 | 69 |
| | HOUR 15 LST | 30 | 71 | 66 | 52 | 41 | 35 | 42 | 49 | 55 | 55 | 66 | 74 | 75 | 57 |
| | HOUR 21 LST | 30 | 71 | 71 | 65 | 53 | 46 | 52 | 61 | 73 | 74 | 77 | 76 | 75 | 66 |
| **S** | PERCENT POSSIBLE SUNSHINE | | | | | | | | | | | | | | |
| **W/O** | MEAN NO. DAYS WITH: | | | | | | | | | | | | | | |
| | HEAVY FOG(VISBY <= 1/4 MI) | 46 | 3.6 | 1.5 | 0.3 | 0.3 | 0.2 | 0.2 | 0.7 | 1.5 | 1.0 | 1.6 | 0.6 | 1.9 | 13.4 |
| | THUNDERSTORMS | 61 | 0.4 | 0.3 | 0.0 | 0.0 | 0.4 | 3.1 | 2.9 | 0.9 | 0.1 | 0.0 | 0.1 | 0.1 | 8.3 |
| **CLOUDNESS** | MEAN: | | | | | | | | | | | | | | |
| | SUNRISE-SUNSET (OKTAS) | 46 | 5.0 | 5.0 | 4.7 | 5.3 | 5.5 | 5.8 | 5.9 | 6.2 | 6.1 | 6.3 | 5.5 | 5.6 | 5.6 |
| | MIDNIGHT-MIDNIGHT (OKTAS) | 33 | 4.8 | 4.5 | 4.4 | 5.0 | 5.4 | 5.8 | 5.9 | 6.0 | 5.8 | 6.0 | 5.2 | 5.2 | 5.3 |
| | MEAN NO. DAYS WITH: | | | | | | | | | | | | | | |
| | CLEAR | 46 | 9.1 | 8.1 | 9.9 | 6.7 | 4.6 | 3.0 | 3.3 | 2.8 | 4.1 | 3.8 | 6.8 | 6.7 | 68.9 |
| | PARTLY CLOUDY | 46 | 5.9 | 6.0 | 7.0 | 7.9 | 11.0 | 10.4 | 8.9 | 6.8 | 6.1 | 4.9 | 5.0 | 5.7 | 85.6 |
| | CLOUDY | 46 | 16.0 | 14.2 | 14.1 | 15.4 | 15.4 | 16.6 | 18.2 | 20.9 | 19.3 | 21.5 | 17.6 | 18.0 | 207.2 |
| **PR** | MEAN STATION PRESSURE(IN) | 26 | 29.33 | 29.39 | 29.39 | 29.34 | 29.37 | 29.36 | 29.41 | 29.38 | 29.30 | 29.27 | 29.28 | 29.25 | 29.34 |
| | MEAN SEA-LEVEL PRES. (IN) | 26 | 29.86 | 29.93 | 29.92 | 29.85 | 29.87 | 29.86 | 29.90 | 29.88 | 29.80 | 29.78 | 29.81 | 29.79 | 29.85 |
| **WINDS** | MEAN SPEED (MPH) | 26 | 2.6 | 3.2 | 4.9 | 5.8 | 6.7 | 6.2 | 5.7 | 5.5 | 5.1 | 4.3 | 2.8 | 2.3 | 4.6 |
| | PREVAIL.DIR(TENS OF DEGS) | 37 | 03 | 04 | 36 | 36 | 36 | 24 | 24 | 36 | 36 | 36 | 04 | 04 | 36 |
| | MAXIMUM 2-MINUTE: | 12 | | | | | | | | | | | | | |
| | SPEED (MPH) | | 29 | 31 | 32 | 31 | 30 | 31 | 31 | 31 | 23 | 25 | 22 | 26 | 32 |
| | DIR. (TENS OF DEGS) | | 18 | 05 | 05 | 25 | 36 | 09 | 01 | 26 | 04 | 26 | 05 | 19 | 05 |
| | YEAR OF OCCURRENCE | | 2005 | 2008 | 2005 | 2002 | 2004 | 2005 | 2002 | 2003 | 2009 | 2002 | 1999 | 2004 | MAR 2005 |
| | MAXIMUM 3-SECOND | 12 | | | | | | | | | | | | | |
| | SPEED (MPH) | | 40 | 39 | 40 | 59 | 37 | 38 | 40 | 37 | 30 | 36 | 28 | 36 | 59 |
| | DIR. (TENS OF DEGS) | | 19 | 25 | 28 | 05 | 11 | 04 | 27 | 26 | 19 | 26 | 25 | 21 | 05 |
| | YEAR OF OCCURRENCE | | 2005 | 2006 | 2009 | 2003 | 1998 | 2008 | 1999 | 2003 | 2004 | 2002 | 2003 | 2004 | APR 2003 |
| **PRECIPITATION** | NORMAL (IN) | 30 | 0.56 | 0.36 | 0.28 | 0.21 | 0.60 | 1.40 | 1.73 | 1.74 | 1.12 | 0.92 | 0.68 | 0.74 | 10.34 |
| | MAXIMUM MONTHLY (IN) | 58 | 2.40 | 1.75 | 2.24 | 3.06 | 1.96 | 3.52 | 5.96 | 6.20 | 3.05 | 2.19 | 3.32 | 3.23 | 6.20 |
| | YEAR OF OCCURRENCE | | 1993 | 1966 | 1991 | 2002 | 2004 | 1955 | 2003 | 1967 | 1960 | 1983 | 1970 | 1984 | AUG 1967 |
| | MINIMUM MONTHLY (IN) | 58 | 0.01 | 0.01 | T | T | 0.05 | 0.19 | 0.06 | .24 | 0.15 | 0.08 | T | T | T |
| | YEAR OF OCCURRENCE | | 1966 | 2000 | 1987 | 1991 | 2009 | 1966 | 2009 | 2005 | 1968 | 1954 | 1953 | 1969 | APR 1991 |
| | MAXIMUM IN 24 HOURS (IN) | 58 | 0.75 | 0.97 | 1.17 | 1.06 | 0.88 | 1.52 | 2.56 | 3.42 | 1.21 | 2.22 | 0.84 | 1.25 | 3.42 |
| | YEAR OF OCCURRENCE | | 1993 | 1966 | 1991 | 2002 | 1955 | 1955 | 2003 | 1967 | 1954 | 1976 | 1970 | 1968 | AUG 1967 |
| | NORMAL NO. DAYS WITH: | | | | | | | | | | | | | | |
| | PRECIPITATION >= 0.01 | 30 | 7.8 | 5.9 | 5.1 | 3.7 | 7.4 | 11.0 | 12.5 | 12.7 | 10.4 | 12.0 | 10.5 | 9.8 | 108.8 |
| | PRECIPITATION >= 1.00 | 30 | 0.0 | 0.0 | 0.0 | 0.0 | 0.0 | 0.1 | 0.1 | * | 0.0 | 0.0 | 0.0 | 0.0 | 0.2 |
| **SNOWFALL** | NORMAL (IN) | 30 | 10.6 | 7.1 | 5.2 | 2.4 | 0.6 | 0.0 | 0.0 | 0.0 | 2.2 | 12.0 | 13.8 | 13.5 | 67.4 |
| | MAXIMUM MONTHLY (IN) | 58 | 40.2 | 43.1 | 30.4 | 15.4 | 14.1 | T | T | 0.1 | 24.4 | 25.9 | 54.0 | 50.7 | 54.0 |
| | YEAR OF OCCURRENCE | | 1993 | 1966 | 1991 | 2002 | 1992 | 2006 | 1990 | 1995 | 1992 | 1982 | 1970 | 1984 | NOV 1970 |
| | MAXIMUM IN 24 HOURS (IN) | 58 | 10.1 | 20.1 | 12.6 | 6.3 | 9.4 | T | T | 0.1 | 9.0 | 10.4 | 14.6 | 14.7 | 20.1 |
| | YEAR OF OCCURRENCE' | | 1993 | 1966 | 1963 | 1992 | 1992 | 1993 | 1990 | 1995 | 1992 | 1974 | 1970 | 1968 | FEB 1966 |
| | MAXIMUM SNOW DEPTH (IN) | 60 | 46 | 52 | 54 | 49 | 14 | 0 | 0 | 0 | 12 | 16 | 42 | 46 | 54 |
| | YEAR OF OCCURRENCE | | 1993 | 1966 | 1991 | 1991 | 1991 | | | | 1992 | 1982 | 1970 | 1990 | MAR 1991 |
| | NORMAL NO. DAYS WITH: | | | | | | | | | | | | | | |
| | SNOWFALL >= 1.0 | 30 | 3.4 | 2.2 | 1.7 | 0.8 | 0.1 | 0.0 | 0.0 | 0.0 | 0.7 | 4.7 | 4.8 | 4.2 | 22.6 |

## PRECIPITATION (inches) 2009  FAIRBANKS (PAFA)

| YEAR | JAN | FEB | MAR | APR | MAY | JUN | JUL | AUG | SEP | OCT | NOV | DEC | ANNUAL |
|------|-----|-----|-----|-----|-----|-----|-----|-----|-----|-----|-----|-----|--------|
| 1980 | 0.52 | 0.22 | 0.13 | 0.10 | 0.31 | 1.38 | 1.37 | 1.68 | 0.76 | 0.39 | 0.70 | 0.32 | 7.88 |
| 1981 | 0.31 | 0.78 | 0.07 | 0.32 | 0.73 | 1.91 | 2.41 | 1.35 | 0.80 | 0.91 | 0.91 | 0.58 | 11.08 |
| 1982 | 0.34 | 0.38 | 0.39 | 0.93 | 0.96 | 1.96 | 2.33 | 1.67 | 0.77 | 1.48 | 1.49 | 0.23 | 12.93 |
| 1983 | 0.24 | 0.18 | 0.09 | 0.27 | 0.14 | 0.57 | 1.71 | 3.33 | 0.92 | 2.19 | 0.08 | 0.65 | 10.37 |
| 1984 | 0.89 | 0.64 | 0.03 | 0.47 | 1.17 | 0.48 | 2.95 | 1.15 | 0.22 | 0.70 | 0.42 | 3.23 | 12.35 |
| 1985 | 0.52 | 0.48 | 0.57 | 0.36 | 0.41 | 1.80 | 1.13 | 1.88 | 2.59 | 1.00 | 0.90 | 0.08 | 11.72 |
| 1986 | 0.13 | 0.19 | 0.32 | 0.07 | 0.54 | 0.87 | 2.12 | 2.36 | 0.65 | 1.79 | 0.48 | 0.34 | 9.86 |
| 1987 | 0.68 | 0.10 | T | 0.05 | 0.21 | 1.02 | 1.70 | 0.56 | 0.57 | 0.39 | 0.64 | 0.51 | 6.43 |
| 1988 | 0.32 | 0.13 | 0.13 | 0.21 | 1.51 | 2.26 | 1.02 | 1.95 | 0.73 | 1.07 | 0.68 | 0.46 | 10.47 |
| 1989 | 0.52 | 0.98 | 0.13 | 0.05 | 0.99 | 2.53 | 0.91 | 0.78 | 0.72 | 1.28 | 0.97 | 0.57 | 10.43 |
| 1990 | 0.52 | 0.72 | 0.11 | 0.07 | 0.40 | 1.73 | 4.87 | 3.60 | 1.74 | 0.31 | 1.51 | 2.94 | 18.52 |
| 1991 | 1.17 | 0.17 | 2.24 | T | 0.10 | 0.36 | 0.81 | 1.18 | 1.16 | 0.71 | 0.48 | 1.02 | 9.40 |
| 1992 | 0.85 | 0.66 | 0.07 | 0.47 | 1.23 | 2.15 | 2.32 | 0.59 | 1.34 | 0.91 | 0.93 | 1.21 | 12.73 |
| 1993 | 2.40 | 0.31 | 0.26 | 0.03 | 0.63 | 1.24 | 0.35 | 1.58 | 2.63 | 0.61 | 0.86 | 0.43 | 11.33 |
| 1994 | 0.47 | 0.32 | 0.17 | 0.07 | 0.22 | 2.41 | 1.11 | 1.38 | 0.61 | 0.82 | 1.67 | 0.48 | 9.73 |
| 1995 | 0.25 | 0.29 | 0.18 | 0.17 | 0.73 | 1.91 | 1.32 | 2.10 | 1.33 | 0.31 | 0.20 | 0.06 | 8.85 |
| 1996 | 0.32 | 1.40 | 0.53 | 0.05 | 0.14 | 1.56 | 1.07 | 2.83 | 1.06 | 1.13 | .81 | .46 | 11.36 |
| 1997 | 0.25 | 0.34 | 0.06 | 0.01 | 0.07 | 1.03 | 1.08 | 1.70 | 0.48 | 0.94 | 0.26 | 0.52 | 6.74 |
| 1998 | 0.08 | 0.11 | T | 0.05 | 0.41 | 1.33 | 3.35 | 3.18 | 1.19 | 0.27 | 0.21 | 0.56 | 10.74 |
| 1999 | 0.22 | 0.21 | 0.19 | 0.04 | 0.31 | 1.30 | 2.11 | 1.85 | 1.83 | 0.90 | 0.58 | 0.73 | 10.27 |
| 2000 | 1.97 | T | 0.10 | T | 0.74 | 0.72 | 1.29 | 3.04 | 1.48 | 0.91 | 0.37 | 0.16 | 10.78 |
| 2001 | 0.40 | 0.66 | 0.43 | 0.16 | 0.62 | 0.65 | 2.45 | 2.01 | 0.25 | 0.55 | 0.06 | 0.24 | 8.48 |
| 2002 | 0.54 | 0.18 | 0.14 | 3.06 | 0.38 | 0.78 | 2.63 | 3.02 | 1.09 | 1.05 | 0.05 | 0.41 | 13.33 |
| 2003 | 0.28 | 0.89 | 0.02 | 0.05 | 0.27 | 0.61 | 5.96 | 1.89 | 1.27 | 0.34 | 1.68 | 0.59 | 13.85 |
| 2004 | 0.33 | 0.33 | 0.29 | 0.03 | 1.96 | 0.31 | 1.13 | 0.37 | 1.34 | 1.12 | 0.63 | 0.75 | 8.59 |
| 2005 | 1.17 | 0.25 | 0.26 | 0.18 | 1.26 | 1.93 | 3.44 | 0.24 | 1.72 | 0.45 | 0.75 | 0.15 | 11.80 |
| 2006 | 0.22 | 0.76 | 0.22 | 0.40 | 0.18 | 0.71 | 2.24 | 2.16 | 0.56 | 0.56 | 0.09 | 0.48 | 8.58 |
| 2007 | 0.50 | 0.12 | 0.20 | 0.10 | 0.86 | 1.88 | 3.67 | 1.52 | 1.58 | 0.51 | 0.11 | 0.31 | 11.36 |
| 2008 | 0.93 | 0.18 | 0.10 | 1.27 | 0.50 | 2.08 | 4.12 | 2.66 | 0.64 | 0.76 | 0.28 | 0.50 | 14.02 |
| 2009 | 0.52 | 0.59 | 1.09 | 0.09 | 0.05 | 1.55 | 0.06 | 2.72 | 0.53 | 0.50 | 0.31 | 0.36 | 8.37 |
| POR=<br>61 YRS | 0.59 | 0.43 | 0.35 | 0.28 | 0.59 | 1.37 | 2.03 | 1.88 | 1.06 | 0.77 | 0.65 | 0.70 | 10.70 |

WBAN : 26411

## AVERAGE TEMPERATURE (°F) 2009  FAIRBANKS (PAFA)

| YEAR | JAN | FEB | MAR | APR | MAY | JUN | JUL | AUG | SEP | OCT | NOV | DEC | ANNUAL |
|------|-----|-----|-----|-----|-----|-----|-----|-----|-----|-----|-----|-----|--------|
| 1980 | -9.5 | 16.0 | 17.3 | 35.9 | 50.7 | 56.5 | 60.9 | 53.6 | 43.0 | 33.0 | 11.5 | -24.0 | 28.7 |
| 1981 | 18.1 | 5.1 | 27.1 | 31.4 | 51.3 | 58.9 | 56.5 | 53.6 | 44.1 | 29.5 | 12.3 | -4.2 | 32.0 |
| 1982 | -18.0 | -3.9 | 13.1 | 27.7 | 46.8 | 58.5 | 63.1 | 56.6 | 49.3 | 18.5 | 4.4 | 2.2 | 26.5 |
| 1983 | -11.0 | 3.4 | 13.8 | 37.4 | 50.4 | 62.3 | 64.2 | 53.5 | 41.2 | 23.6 | 8.6 | -3.7 | 28.6 |
| 1984 | -5.9 | -13.3 | 21.6 | 30.3 | 47.2 | 61.6 | 60.9 | 53.8 | 46.8 | 25.9 | 0.1 | -3.1 | 27.2 |
| 1985 | 11.1 | -9.4 | 14.6 | 20.8 | 46.7 | 57.8 | 63.1 | 56.3 | 42.9 | 18.8 | -4.7 | 7.7 | 27.1 |
| 1986 | -2.1 | 4.7 | 6.0 | 24.0 | 47.8 | 62.6 | 63.6 | 54.7 | 46.3 | 27.1 | 0.2 | 7.3 | 28.5 |
| 1987 | 0.7 | 1.5 | 13.5 | 35.0 | 50.9 | 61.9 | 64.2 | 57.7 | 44.1 | 33.0 | 6.0 | -3.2 | 30.4 |
| 1988 | -5.3 | 3.9 | 17.8 | 33.6 | 52.8 | 62.9 | 65.8 | 58.4 | 44.4 | 17.4 | -3.1 | 4.2 | 29.4 |
| 1989 | -21.3 | 3.4 | 6.7 | 36.2 | 47.8 | 60.1 | 64.6 | 60.8 | 48.6 | 26.2 | -7.1 | 4.5 | 27.5 |
| 1990 | -12.9 | -21.7 | 18.5 | 38.1 | 55.1 | 61.6 | 65.3 | 60.0 | 44.8 | 24.1 | -4.9 | -6.3 | 26.8 |
| 1991 | -4.7 | -1.2 | 11.9 | 35.4 | 51.2 | 63.8 | 60.6 | 54.4 | 48.0 | 24.9 | 0.6 | -2.9 | 28.5 |
| 1992 | -4.7 | -8.7 | 14.0 | 26.1 | 41.8 | 60.0 | 64.1 | 56.5 | 31.7 | 17.5 | 10.4 | -7.6 | 25.1 |
| 1993 | -3.9 | 2.4 | 17.2 | 41.1 | 53.7 | 62.0 | 65.6 | 56.1 | 44.1 | 29.3 | 8.4 | 0.6 | 31.4 |
| 1994 | -1.5 | -6.4 | 10.5 | 34.7 | 51.6 | 58.4 | 64.7 | 59.6 | 43.7 | 21.0 | 0.3 | -8.5 | 27.3 |
| 1995 | -9.4 | -0.2 | 3.8 | 40.1 | 53.5 | 60.9 | 63.1 | 57.3 | 52.8 | 28.0 | -2.7 | -9.0 | 28.2 |
| 1996 | -16.8 | -2.6 | 15.5 | 33.3 | 49.1 | 59.5 | 63.4 | 53.4 | 42.0 | 13.2 | -1.9 | -13.4 | 24.6 |
| 1997 | -16.2 | 13.0 | 4.7 | 35.4 | 49.0 | 63.1 | 64.9 | 58.7 | 49.7 | 17.5 | 10.6 | -7.0 | 28.6 |
| 1998 | -13.6 | 2.1 | 18.4 | 38.6 | 50.0 | 58.8 | 62.7 | 53.0 | 46.1 | 26.1 | 5.0 | -5.8 | 28.5 |
| 1999 | -17.0 | -16.7 | 7.2 | 32.7 | 46.9 | 61.4 | 61.5 | 58.0 | 45.2 | 19.5 | -4.7 | -12.8 | 23.4 |
| 2000 | -9.9 | 6.7 | 17.6 | 31.9 | 44.4 | 61.5 | 59.6 | 51.7 | 41.7 | 22.3 | 8.7 | .3 | 28.0 |
| 2001 | 7.8 | 7.0 | 9.9 | 33.3 | 44.5 | 61.1 | 60.0 | 57.3 | 47.5 | 22.0 | .6 | -10.5 | 28.4 |
| 2002 | 3.5 | -.5 | 9.9 | 23.2 | 50.4 | 58.2 | 61.8 | 54.6 | 47.4 | 31.7 | 19.2 | 4.9 | 30.4 |
| 2003 | -3.3 | 11.0 | 8.6 | 33.1 | 47.2 | 60.2 | 60.7 | 56.2 | 41.8 | 32.0 | 9.2 | -9.1 | 29.0 |
| 2004 | -15.9 | 1.1 | 6.0 | 35.1 | 52.3 | 66.9 | 64.5 | 62.2 | 38.8 | 29.7 | 7.1 | -3.3 | 28.7 |
| 2005 | -9.1 | -2.3 | 19.6 | 32.2 | 55.6 | 61.5 | 62.2 | 57.3 | 46.3 | 27.4 | -5.1 | .9 | 28.9 |
| 2006 | -22.0 | 5.6 | 3.0 | 29.5 | 49.7 | 58.1 | 61.3 | 54.4 | 49.5 | 31.7 | -9.8 | -1.9 | 25.8 |
| 2007 | -6.5 | -6.6 | -6.4 | 37.6 | 51.1 | 61.6 | 64.4 | 60.5 | 47.1 | 21.2 | 11.5 | -3.2 | 27.7 |
| 2008 | -9.1 | -5.8 | 15.4 | 29.5 | 49.9 | 60.2 | 60.6 | 55.0 | 46.7 | 15.1 | -1.3 | -7.7 | 25.7 |
| 2009 | -11.8 | -1.4 | 5.6 | 31.2 | 51.6 | 60.3 | 66.5 | 54.5 | 48.5 | 30.7 | -1.1 | -2.7 | 27.7 |
| POR=<br>61 YRS | -8.6 | -1.7 | 10.7 | 31.3 | 48.8 | 59.9 | 62.1 | 56.4 | 45.0 | 24.8 | 3.9 | -6.0 | 27.2 |

## HEATING DEGREE DAYS (base 65°F) 2009  FAIRBANKS (PAFA)

| YEAR | JUL | AUG | SEP | OCT | NOV | DEC | JAN | FEB | MAR | APR | MAY | JUN | TOTAL |
|------|-----|-----|-----|-----|-----|-----|-----|-----|-----|-----|-----|-----|-------|
| 1980-81 | 127 | 351 | 654 | 985 | 1599 | 2766 | 1447 | 1676 | 1168 | 999 | 418 | 188 | 12378 |
| 1981-82 | 255 | 347 | 622 | 1094 | 1573 | 2150 | 2581 | 1929 | 1602 | 1113 | 555 | 216 | 14037 |
| 1982-83 | 86 | 252 | 465 | 1434 | 1816 | 1946 | 2356 | 1725 | 1581 | 823 | 451 | 133 | 13068 |
| 1983-84 | 62 | 351 | 705 | 1280 | 1688 | 2126 | 2201 | 2277 | 1338 | 1035 | 549 | 120 | 13732 |
| 1984-85 | 140 | 344 | 538 | 1205 | 1950 | 2111 | 1666 | 2086 | 1558 | 1321 | 558 | 215 | 13692 |
| 1985-86 | 72 | 267 | 654 | 1430 | 2095 | 1776 | 2079 | 1686 | 1825 | 1224 | 527 | 113 | 13748 |
| 1986-87 | 110 | 312 | 559 | 1169 | 1943 | 1787 | 1994 | 1776 | 1594 | 893 | 428 | 128 | 12693 |
| 1987-88 | 61 | 218 | 620 | 987 | 1768 | 2111 | 2185 | 1768 | 1455 | 934 | 371 | 96 | 12574 |
| 1988-89 | 39 | 202 | 611 | 1469 | 2045 | 1883 | 2676 | 1722 | 1804 | 859 | 529 | 149 | 13988 |
| 1989-90 | 73 | 134 | 484 | 1195 | 2164 | 1875 | 2420 | 2433 | 1431 | 798 | 310 | 127 | 13444 |
| 1990-91 | 74 | 178 | 600 | 1261 | 2097 | 2212 | 2161 | 1849 | 1640 | 877 | 421 | 130 | 13500 |
| 1991-92 | 143 | 321 | 504 | 1234 | 1935 | 2105 | 2163 | 2140 | 1577 | 1160 | 711 | 157 | 14150 |
| 1992-93 | 45 | 259 | 995 | 1463 | 1636 | 2253 | 2134 | 1751 | 1478 | 711 | 343 | 103 | 13171 |
| 1993-94 | 43 | 273 | 620 | 1099 | 1694 | 1995 | 2061 | 2002 | 1689 | 900 | 408 | 210 | 12994 |
| 1994-95 | 56 | 217 | 634 | 1357 | 1941 | 2280 | 2310 | 1827 | 1896 | 739 | 354 | 143 | 13754 |
| 1995-96 | 92 | 235 | 365 | 1141 | 2033 | 2299 | 2541 | 1962 | 1528 | 946 | 486 | 171 | 13799 |
| 1996-97 | 71 | 353 | 681 | 1600 | 2010 | 2424 | 2509 | 1450 | 1862 | 880 | 485 | 106 | 14431 |
| 1997-98 | 40 | 204 | 453 | 1467 | 1624 | 2224 | 2429 | 1755 | 1437 | 786 | 460 | 199 | 13078 |
| 1998-99 | 103 | 365 | 563 | 1200 | 1793 | 2188 | 2535 | 2282 | 1784 | 963 | 553 | 131 | 14460 |
| 1999-00 | 147 | 221 | 584 | 1403 | 2084 | 2403 | 2315 | 1680 | 1467 | 987 | 631 | 112 | 14034 |
| 2000-01 | 165 | 407 | 693 | 1318 | 1682 | 2001 | 1764 | 1616 | 1701 | 940 | 629 | 126 | 13042 |
| 2001-02 | 164 | 231 | 521 | 1326 | 1926 | 2333 | 1899 | 1828 | 1700 | 1246 | 453 | 197 | 13824 |
| 2002-03 | 121 | 324 | 524 | 1023 | 1366 | 1857 | 2110 | 1505 | 1742 | 950 | 547 | 141 | 12210 |
| 2003-04 | 154 | 265 | 689 | 1015 | 1667 | 2289 | 2502 | 1847 | 1823 | 889 | 388 | 44 | 13572 |
| 2004-05 | 49 | 125 | 780 | 1085 | 1730 | 2112 | 2290 | 1876 | 1401 | 977 | 283 | 127 | 12835 |
| 2005-06 | 81 | 248 | 554 | 1158 | 2093 | 1981 | 2692 | 1658 | 1914 | 1058 | 464 | 202 | 14103 |
| 2006-07 | 118 | 323 | 458 | 1026 | 2241 | 2067 | 2215 | 2003 | 2208 | 816 | 427 | 113 | 14015 |
| 2007-08 | 44 | 144 | 533 | 1351 | 1597 | 2111 | 2293 | 2050 | 1529 | 1057 | 460 | 146 | 13315 |
| 2008-09 | 158 | 305 | 544 | 1538 | 1986 | 2248 | 2380 | 1856 | 1832 | 1008 | 409 | 144 | 14408 |
| 2009- | 27 | 323 | 488 | 1055 | 1979 | 2096 | | | | | | | |

WBAN : 26411

## COOLING DEGREE DAYS (base 65°F) 2009  FAIRBANKS (PAFA)

| YEAR | JAN | FEB | MAR | APR | MAY | JUN | JUL | AUG | SEP | OCT | NOV | DEC | TOTAL |
|------|-----|-----|-----|-----|-----|-----|-----|-----|-----|-----|-----|-----|-------|
| 1980 | 0 | 0 | 0 | 0 | 0 | 0 | 8 | 6 | 0 | 0 | 0 | 0 | 14 |
| 1981 | 0 | 0 | 0 | 0 | 1 | 11 | 0 | 2 | 0 | 0 | 0 | 0 | 14 |
| 1982 | 0 | 0 | 0 | 0 | 0 | 27 | 36 | 0 | 0 | 0 | 0 | 0 | 63 |
| 1983 | 0 | 0 | 0 | 0 | 5 | 61 | 40 | 0 | 0 | 0 | 0 | 0 | 106 |
| 1984 | 0 | 0 | 0 | 0 | 0 | 22 | 21 | 2 | 0 | 0 | 0 | 0 | 45 |
| 1985 | 0 | 0 | 0 | 0 | 0 | 8 | 20 | 4 | 0 | 0 | 0 | 0 | 32 |
| 1986 | 0 | 0 | 0 | 0 | 0 | 46 | 74 | 0 | 0 | 0 | 0 | 0 | 120 |
| 1987 | 0 | 0 | 0 | 0 | 0 | 42 | 42 | 0 | 0 | 0 | 0 | 0 | 84 |
| 1988 | 0 | 0 | 0 | 0 | 0 | 41 | 72 | 2 | 0 | 0 | 0 | 0 | 115 |
| 1989 | 0 | 0 | 0 | 0 | 0 | 10 | 67 | 11 | 0 | 0 | 0 | 0 | 88 |
| 1990 | 0 | 0 | 0 | 0 | 11 | 32 | 91 | 35 | 0 | 0 | 0 | 0 | 169 |
| 1991 | 0 | 0 | 0 | 0 | 0 | 100 | 13 | 0 | 0 | 0 | 0 | 0 | 113 |
| 1992 | 0 | 0 | 0 | 0 | 0 | 17 | 21 | 4 | 0 | 0 | 0 | 0 | 42 |
| 1993 | 0 | 0 | 0 | 0 | 0 | 18 | 70 | 3 | 0 | 0 | 0 | 0 | 91 |
| 1994 | 0 | 0 | 0 | 0 | 0 | 19 | 55 | 55 | 0 | 0 | 0 | 0 | 129 |
| 1995 | 0 | 0 | 0 | 0 | 5 | 24 | 41 | 0 | 4 | 0 | 0 | 0 | 74 |
| 1996 | 0 | 0 | 0 | 0 | 0 | 11 | 29 | 0 | 0 | 0 | 0 | 0 | 40 |
| 1997 | 0 | 0 | 0 | 0 | 0 | 56 | 46 | 15 | 0 | 0 | 0 | 0 | 117 |
| 1998 | 0 | 0 | 0 | 0 | 0 | 18 | 37 | 1 | 0 | 0 | 0 | 0 | 56 |
| 1999 | 0 | 0 | 0 | 0 | 0 | 27 | 45 | 11 | 0 | 0 | 0 | 0 | 83 |
| 2000 | 0 | 0 | 0 | 0 | 0 | 16 | 5 | 0 | 0 | 0 | 0 | 0 | 21 |
| 2001 | 0 | 0 | 0 | 0 | 0 | 15 | 14 | 0 | 0 | 0 | 0 | 0 | 29 |
| 2002 | 0 | 0 | 0 | 0 | 8 | 0 | 30 | 7 | 0 | 0 | 0 | 0 | 45 |
| 2003 | 0 | 0 | 0 | 0 | 0 | 3 | 27 | 1 | 0 | 0 | 0 | 0 | 31 |
| 2004 | 0 | 0 | 0 | 0 | 0 | 108 | 40 | 45 | 0 | 0 | 0 | 0 | 193 |
| 2005 | 0 | 0 | 0 | 0 | 0 | 27 | 5 | 18 | 0 | 0 | 0 | 0 | 50 |
| 2006 | 0 | 0 | 0 | 0 | 0 | 1 | 11 | 0 | 0 | 0 | 0 | 0 | 12 |
| 2007 | 0 | 0 | 0 | 0 | 3 | 19 | 32 | 12 | 0 | 0 | 0 | 0 | 66 |
| 2008 | 0 | 0 | 0 | 0 | 0 | 10 | 30 | 0 | 0 | 0 | 0 | 0 | 40 |
| 2009 | 0 | 0 | 0 | 0 | 0 | 6 | 79 | 3 | 0 | 0 | 0 | 0 | 88 |

## SNOWFALL (inches) 2009 FAIRBANKS (PAFA)

| YEAR | JUL | AUG | SEP | OCT | NOV | DEC | JAN | FEB | MAR | APR | MAY | JUN | TOTAL |
|------|-----|-----|-----|-----|-----|-----|-----|-----|-----|-----|-----|-----|-------|
| 1980-81 | 0.0 | 0.0 | 3.4 | 5.1 | 10.6 | 4.8 | 6.2 | 10.2 | 0.9 | 1.9 | 0.0 | 0.0 | 43.1 |
| 1981-82 | 0.0 | 0.0 | 0.3 | 7.8 | 16.2 | 10.6 | 7.6 | 7.0 | 7.2 | 11.4 | 0.3 | 0.0 | 68.4 |
| 1982-83 | 0.0 | 0.0 | 0.6 | 25.9 | 27.8 | 3.8 | 4.8 | 3.6 | 2.1 | 1.9 | 0.4 | 0.0 | 70.9 |
| 1983-84 | 0.0 | 0.0 | 0.4 | 17.3 | 2.4 | 14.4 | 13.8 | 11.1 | 0.8 | 6.7 | T | 0.0 | 66.9 |
| 1984-85 | 0.0 | 0.0 | 0.0 | 11.3 | 8.5 | 50.7 | 8.1 | 8.0 | 7.4 | 5.9 | 1.0 | 0.0 | 100.9 |
| 1985-86 | 0.0 | 0.0 | 2.1 | 5.4 | 14.7 | 1.8 | 2.3 | 2.6 | 3.7 | 1.0 | 0.0 | 0.0 | 33.6 |
| 1986-87 | 0.0 | 0.0 | T | 11.8 | 7.4 | 6.0 | 12.0 | 1.5 | T | 1.0 | T | 0.0 | 39.7 |
| 1987-88 | 0.0 | 0.0 | T | 3.5 | 14.1 | 10.4 | 6.3 | 2.5 | 2.6 | 0.2 | 0.3 | 0.0 | 39.9 |
| 1988-89 | 0.0 | 0.0 | 0.2 | 12.9 | 15.2 | 10.5 | 10.8 | 13.4 | 3.9 | 0.4 | 0.8 | 0.0 | 68.1 |
| 1989-90 | 0.0 | 0.0 | 0.5 | 19.7 | 18.1 | 11.1 | 10.4 | 16.0 | 2.6 | T | 0.0 | T | 78.4 |
| 1990-91 | T | 0.0 | 1.6 | 6.9 | 37.3 | 47.5 | 20.6 | 3.0 | 30.4 | T | 0.0 | 0.0 | 147.3 |
| 1991-92 | 0.0 | 0.0 | 0.0 | 12.2 | 9.7 | 18.9 | 15.5 | 15.0 | 1.8 | 11.5 | 14.1 | 0.0 | 98.7 |
| 1992-93 | 0.0 | 0.0 | 24.4 | 16.7 | 18.7 | 28.5 | 40.2 | 5.4 | 5.2 | T | 0.0 | T | 139.1 |
| 1993-94 | 0.0 | 0.0 | 7.5 | 7.3 | 16.3 | 8.5 | 11.2 | 8.1 | 4.5 | 0.8 | 0.0 | 0.0 | 64.2 |
| 1994-95 | 0.0 | 0.0 | 0.0 | 15.3 | 35.3 | 13.1 | 5.9 | 7.4 | 3.1 | 1.3 | 0.0 | 0.0 | 81.4 |
| 1995-96 | 0.0 | 0.1 | 0.0 | 7.1 | 3.2 | 0.8 | 5.4 | 29.3 | 9.6 | 0.7 | 0.0 | 0.0 | 56.2 |
| 1996-97 | 0.0 | T | 7.1 | 16.2 | 18.4 | 9.4 | 4.9 | 8.8 | 2.3 | 0.4 | 0.0 | T | 67.5 |
| 1997-98 | 0.0 | 0.0 | 0.0 | 19.3 | 7.6 | 13.1 | 1.3 | 3.2 | 0.9 | 0.5 | 0.1 | 0.0 | 46.0 |
| 1998-99 | 0.0 | 0.0 | T | 6.2 | 4.6 | 8.5 | 4.2 | 2.7 | 3.5 | 0.9 | 0.4 | 0.0 | 31.0 |
| 1999-00 | 0.0 | 0.0 | 3.0 | 11.1 | 11.2 | 13.8 | 27.9 | 0.2 | 3.0 | T | T | 0.0 | 70.2 |
| 2000-01 | 0.0 | 0.0 | 1.4 | 9.8 | 7.5 | 2.7 | 6.7 | 10.7 | 8.6 | 2.2 | 7.0 | 0.0 | 56.6 |
| 2001-02 | 0.0 | 0.0 | 0.3 | 6.4 | 2.8 | 4.6 | 11.4 | 4.2 | 2.2 | 15.4 | 1.8 | 0.0 | 49.1 |
| 2002-03 | 0.0 | T | T | 14.7 | 1.6 | 9.6 | 6.9 | 7.8 | 0.8 | T | T | 0.0 | 41.4 |
| 2003-04 | 0.0 | 0.0 | 0.1 | 2.6 | 27.6 | 9.6 | 8.5 | 7.8 | 5.2 | 0.2 | T | 0.0 | 61.6 |
| 2004-05 | 0.0 | 0.0 | 1.8 | 17.0 | 11.9 | 15.0 | 17.8 | 4.5 | 5.2 | 4.5 | 0.0 | 0.0 | 77.7 |
| 2005-06 | 0.0 | T | 0.3 | 6.4 | 14.3 | 4.2 | 4.0 | 13.2 | 4.7 | 3.3 | 0.0 | T | 50.4 |
| 2006-07 | 0.0 | 0.0 | 0.4 | 4.6 | 1.4 | 7.8 | 8.9 | 1.6 | 3.3 | T | 0.0 | 0.0 | 28.0 |
| 2007-08 | 0.0 | 0.0 | 0.0 | 9.5 | 4.2 | 6.0 | 14.4 | 5.4 | 1.4 | 14.7 | 0.0 | 0.0 | 55.6 |
| 2008-09 | 0.0 | T | 1.8 | 12.2 | 6.5 | 12.2 | 9.2 | 14.1 | 15.5 | T | 0.0 | 0.0 | 71.5 |
| 2009- | 0.0 | 0.0 | 0.2 | 5.3 | 7.4 | 6.8 | | | | | | | |
| POR= 62 YRS | T | T | 1.4 | 10.5 | 12.6 | 12.5 | 10.8 | 8.5 | 6.0 | 3.2 | 0.7 | 0.1 | 66.3 |

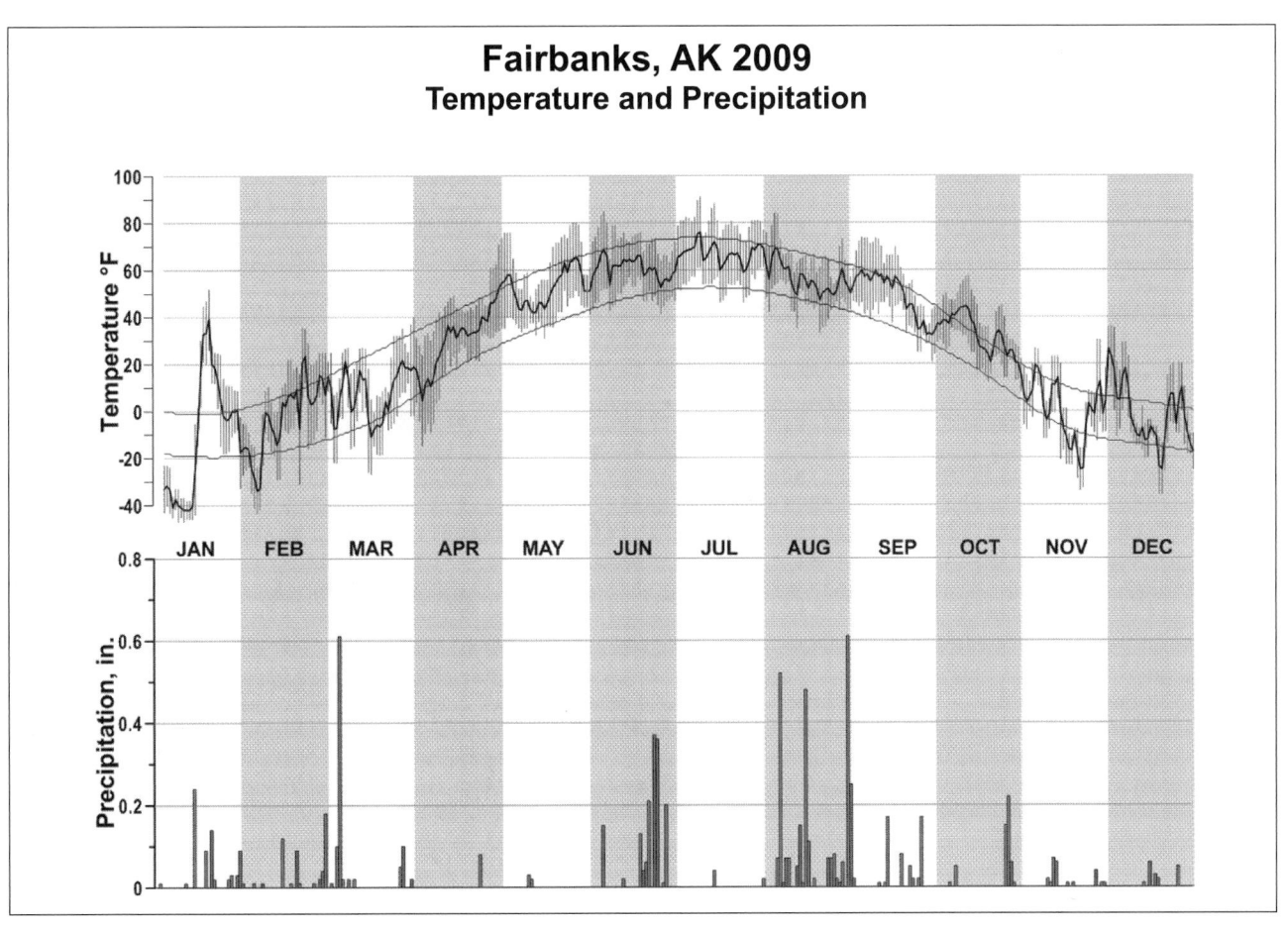

## Fairbanks, AK 2009
### Temperature and Precipitation

# 2009
# JUNEAU
# ALASKA (PAJN)

Juneau lies well within the area of maritime influences which prevail over the coastal areas of southeastern Alaska, and is in the path of most storms that cross the Gulf of Alaska. Consequently, the area has little sunshine, generally moderate temperatures, and abundant precipitation. In contrast with the characteristic lack of sunshine there are greatly appreciated intervals, sometimes lasting for several days at a stretch, during which clear skies prevail. The rugged terrain exerts a fundamental influence upon local temperatures and the distribution of precipitation, creating considerable variations in both weather elements within relatively short distances.

Temperature variations, both daily and seasonal, are usually confined to relatively narrow limits by the dominant maritime influences. There are, however, periods of comparatively severe cold, which usually start with strong northerly winds, and are most often caused by the flow of cold air from northwestern Canada through nearby mountain passes and over the Juneau ice field. These are generally of brief duration. During such periods strong, gusty winds, known locally as Taku Winds, often occur especially in downtown Juneau, Douglas, and other local areas, but generally they are not felt in the Mendenhall Valley. At times these are strong enough to cause considerable damage. During periods of calm or light winds, temperature differences within short distances are frequently very pronounced. Variations in local sunlight and air drainage patterns produce wide differences in temperatures particularly between upland or sloping areas and areas of low, flat terrain. Juneau International Airport, located on low, flat terrain formed by the Mendenhall River delta, and in the path of drainage air from the Mendenhall Glacier, averages about 10 days a year with minimum readings below zero. Downtown Juneau, located on a sloping portion of a rugged mountain area, experiences on the average only about one day each year with minimum readings below zero. At the airport the growing season averages 146 days, from May 4 to September 28, while the downtown average is 181 days, from April 22 to October 21.

The months of February to June mark the period of lightest precipitation, with monthly averages of about 3 inches. After June the monthly amounts increase gradually, reaching an average of 7.71 inches in October. Due to the rugged topography, precipitation throughout the year tends to vary greatly within short distances. At the Juneau Airport, yearly precipitation is 53 inches while downtown, only 8 miles away, it is 93 inches. The maximum yearly amount received in the city is almost double the maximum received at the airport.

Although a trace of snow has fallen as early as September 9, first falls usually occur in the latter part of October, and sometimes not until the first part of December. On the average there is very little accumulation on the ground at low levels until the last of November, although at higher elevations, and particularly on mountain tops, a cover is usually established in early October. Snow accumulation usually reaches its greatest depth during the middle of February. Individual storms may produce heavy falls as late as the first half of May. However, snow cover is usually gone before the middle of April. Ice accumulations due to alternating thawing and freezing of snow or due to freezing precipitation are frequent problems in the Juneau area during the winter months.

# NORMALS, MEANS, AND EXTREMES
## JUNEAU (PAJN)

LATITUDE: 58 ° 21'N   LONGITUDE: -134° 34'W   ELEVATION (FT): GRND: 22   BARO: 40   TIME ZONE: ALASKA   (UTC -9)   WBAN: 25309

| | ELEMENT | POR | JAN | FEB | MAR | APR | MAY | JUN | JUL | AUG | SEP | OCT | NOV | DEC | YEAR |
|---|---|---|---|---|---|---|---|---|---|---|---|---|---|---|---|
| **TEMPERATURE °F** | NORMAL DAILY MAXIMUM | 30 | 30.6 | 34.3 | 39.5 | 48.1 | 55.7 | 61.6 | 64.3 | 63.1 | 56.1 | 46.9 | 37.6 | 33.0 | 47.6 |
| | MEAN DAILY MAXIMUM | 65 | 30.0 | 34.0 | 38.6 | 47.2 | 55.5 | 61.4 | 63.7 | 62.7 | 54.9 | 47.0 | 37.7 | 32.6 | 47.1 |
| | HIGHEST DAILY MAXIMUM | 65 | 57 | 57 | 61 | 74 | 82 | 86 | 90 | 84 | 73 | 61 | 56 | 54 | 90 |
| | YEAR OF OCCURRENCE | | 1958 | 1992 | 1998 | 2003 | 1947 | 1969 | 1975 | 2004 | 1996 | 2003 | 1949 | 1999 | JUL 1975 |
| | MEAN OF EXTREME MAXS. | 66 | 42.9 | 44.0 | 47.8 | 59.9 | 69.6 | 75.9 | 78.0 | 75.6 | 65.6 | 55.6 | 48.0 | 44.1 | 58.9 |
| | NORMAL DAILY MINIMUM | 30 | 20.7 | 23.5 | 27.8 | 33.4 | 40.1 | 46.1 | 49.2 | 48.3 | 43.8 | 37.7 | 28.9 | 24.4 | 35.3 |
| | MEAN DAILY MINIMUM | 64 | 19.4 | 22.9 | 26.4 | 32.1 | 39.2 | 45.1 | 48.4 | 47.6 | 42.6 | 37.1 | 28.5 | 23.4 | 34.4 |
| | LOWEST DAILY MINIMUM | 65 | -22 | -22 | -15 | 6 | 25 | 31 | 36 | 27 | 23 | 11 | -5 | -21 | -22 |
| | YEAR OF OCCURRENCE | | 1972 | 1968 | 1972 | 1963 | 1972 | 1971 | 1950 | 1948 | 1972 | 1984 | 2006 | 1949 | JAN 1972 |
| | MEAN OF EXTREME MINS. | 66 | -0.0 | 4.8 | 10.0 | 22.8 | 30.0 | 37.1 | 41.7 | 39.2 | 31.8 | 24.4 | 12.9 | 3.9 | 21.6 |
| | NORMAL DRY BULB | 30 | 25.7 | 28.9 | 33.7 | 40.8 | 47.9 | 53.9 | 56.8 | 55.7 | 50.0 | 42.3 | 33.3 | 28.7 | 41.5 |
| | MEAN DRY BULB | 64 | 24.7 | 28.5 | 32.5 | 39.6 | 47.4 | 53.5 | 56.1 | 55.2 | 48.9 | 42.1 | 33.1 | 28.0 | 40.8 |
| | MEAN WET BULB | 25 | 29.0 | 29.5 | 31.5 | 37.7 | 44.0 | 49.8 | 52.8 | 52.4 | 48.1 | 41.5 | 34.0 | 31.4 | 40.1 |
| | MEAN DEW POINT | 25 | 27.0 | 27.2 | 28.5 | 34.4 | 40.9 | 47.3 | 50.9 | 50.4 | 46.8 | 40.1 | 32.3 | 29.8 | 38.0 |
| | NORMAL NO. DAYS WITH: | | | | | | | | | | | | | | |
| | MAXIMUM >= 70 | 30 | 0.0 | 0.0 | 0.0 | 0.1 | 1.4 | 5.1 | 7.2 | 5.7 | 0.2 | 0.0 | 0.0 | 0.0 | 19.7 |
| | MAXIMUM <= 32 | 30 | 14.8 | 8.7 | 3.0 | * | 0.0 | 0.0 | 0.0 | 0.0 | 0.0 | 0.3 | 5.3 | 11.5 | 43.6 |
| | MINIMUM <= 32 | 30 | 24.1 | 21.6 | 20.5 | 11.7 | 1.8 | * | 0.0 | 0.0 | 1.2 | 6.6 | 17.1 | 22.1 | 126.7 |
| | MINIMUM <= 0 | 30 | 3.0 | 1.1 | 0.4 | 0.0 | 0.0 | 0.0 | 0.0 | 0.0 | 0.0 | 0.0 | * | 1.3 | 5.8 |
| **H/C** | NORMAL HEATING DEG. DAYS | 30 | 1219 | 1010 | 973 | 728 | 529 | 335 | 257 | 288 | 453 | 704 | 953 | 1125 | 8574 |
| | NORMAL COOLING DEG. DAYS | 30 | 0 | 0 | 0 | 0 | 0 | 0 | 0 | 0 | 0 | 0 | 0 | 0 | 0 |
| **RH** | NORMAL (PERCENT) | 30 | 84 | 82 | 80 | 78 | 76 | 77 | 81 | 84 | 89 | 88 | 86 | 86 | 83 |
| | HOUR 03 LST | 30 | 85 | 86 | 87 | 89 | 90 | 89 | 89 | 93 | 95 | 92 | 89 | 88 | 89 |
| | HOUR 09 LST | 30 | 86 | 85 | 83 | 78 | 76 | 78 | 83 | 86 | 90 | 90 | 88 | 88 | 84 |
| | HOUR 15 LST | 30 | 82 | 76 | 69 | 63 | 62 | 65 | 71 | 74 | 80 | 83 | 83 | 86 | 75 |
| | HOUR 21 LST | 30 | 85 | 85 | 83 | 81 | 77 | 76 | 80 | 84 | 92 | 91 | 88 | 87 | 84 |
| **S** | PERCENT POSSIBLE SUNSHINE | 33 | 32 | 32 | 37 | 39 | 39 | 34 | 31 | 32 | 26 | 19 | 23 | 20 | 30 |
| **W/O** | MEAN NO. DAYS WITH: | | | | | | | | | | | | | | |
| | HEAVY FOG(VISBY <= 1/4 MI) | 45 | 2.1 | 2.6 | 2.0 | 1.1 | 0.6 | 0.3 | 0.2 | 1.0 | 2.4 | 3.4 | 3.2 | 2.6 | 21.5 |
| | THUNDERSTORMS | 60 | 0.0 | 0.0 | 0.0 | 0.0 | 0.0 | 0.1 | 0.1 | 0.0 | 0.0 | 0.0 | 0.0 | 0.0 | 0.2 |
| **CLOUDNESS** | MEAN: | | | | | | | | | | | | | | |
| | SUNRISE-SUNSET (OKTAS) | 48 | 6.2 | 6.4 | 6.4 | 6.4 | 6.3 | 6.5 | 6.6 | 6.3 | 6.8 | 7.0 | 6.7 | 6.7 | 6.5 |
| | MIDNIGHT-MIDNIGHT (OKTAS) | 16 | 5.2 | 5.7 | 6.1 | 5.9 | 6.3 | 6.1 | 6.2 | 5.9 | 6.4 | 6.6 | 6.0 | 6.1 | 6.0 |
| | MEAN NO. DAYS WITH: | | | | | | | | | | | | | | |
| | CLEAR | 48 | 5.7 | 4.3 | 4.4 | 3.7 | 3.8 | 3.4 | 2.8 | 3.9 | 2.7 | 2.4 | 3.3 | 3.2 | 43.6 |
| | PARTLY CLOUDY | 48 | 2.6 | 3.0 | 3.2 | 4.2 | 4.4 | 4.2 | 4.5 | 4.7 | 3.3 | 2.1 | 2.3 | 2.0 | 40.5 |
| | CLOUDY | 48 | 22.5 | 21.0 | 23.4 | 22.1 | 22.6 | 22.4 | 23.0 | 21.9 | 23.4 | 26.0 | 24.0 | 25.2 | 277.5 |
| **PR** | MEAN STATION PRESSURE(IN) | 25 | 30.39 | 30.46 | 30.40 | 30.44 | 30.46 | 30.48 | 30.49 | 30.46 | 30.43 | 30.38 | 30.35 | 30.37 | 30.43 |
| | MEAN SEA-LEVEL PRES. (IN) | 25 | 29.84 | 29.86 | 29.85 | 29.89 | 29.91 | 29.92 | 29.94 | 29.91 | 29.88 | 29.83 | 29.79 | 29.81 | 29.87 |
| **WINDS** | MEAN SPEED (MPH) | 25 | 7.3 | 7.3 | 7.8 | 7.7 | 7.6 | 6.9 | 6.8 | 6.9 | 7.9 | 8.5 | 7.9 | 8.0 | 7.6 |
| | PREVAIL.DIR(TENS OF DEGS) | 28 | 09 | 09 | 11 | 12 | 12 | 11 | 09 | 09 | 09 | 11 | 12 | 12 | 11 |
| | MAXIMUM 2-MINUTE: | | | | | | | | | | | | | | |
| | SPEED (MPH) | 11 | 48 | 46 | 40 | 37 | 36 | 26 | 30 | 38 | 41 | 47 | 48 | 48 | 48 |
| | DIR. (TENS OF DEGS) | | 11 | 12 | 12 | 11 | 12 | 11 | 12 | 12 | 12 | 11 | 12 | 11 | 12 |
| | YEAR OF OCCURRENCE | | 2008 | 2002 | 2004 | 2005 | 2001 | 2006 | 2003 | 1999 | 2001 | 2004 | 2009 | 2004 | NOV 2009 |
| | MAXIMUM 3-SECOND: | | | | | | | | | | | | | | |
| | SPEED (MPH) | 11 | 66 | 54 | 49 | 49 | 47 | 35 | 39 | 48 | 53 | 61 | 62 | 60 | 66 |
| | DIR. (TENS OF DEGS) | | 11 | 12 | 12 | 11 | 15 | 11 | 10 | 13 | 20 | 11 | 11 | 11 | 11 |
| | YEAR OF OCCURRENCE | | 2008 | 2002 | 2000 | 2005 | 1999 | 2009 | 2008 | 1999 | 2000 | 2007 | 2009 | 2004 | JAN 2008 |
| **PRECIPITATION** | NORMAL (IN) | 30 | 4.81 | 4.02 | 3.51 | 2.96 | 3.48 | 3.36 | 4.14 | 5.37 | 7.54 | 8.30 | 5.43 | 5.41 | 58.33 |
| | MAXIMUM MONTHLY (IN) | 65 | 9.38 | 8.48 | 6.50 | 7.48 | 9.20 | 6.22 | 10.36 | 12.31 | 15.14 | 15.25 | 13.38 | 13.61 | 15.25 |
| | YEAR OF OCCURRENCE | | 2009 | 1964 | 1994 | 1999 | 1992 | 1996 | 1997 | 1961 | 1991 | 1974 | 2005 | 1997 | OCT 1974 |
| | MINIMUM MONTHLY (IN) | 65 | 0.94 | 0.07 | 0.59 | 0.27 | 0.84 | 1.08 | 1.15 | 0.56 | 2.34 | 2.71 | 1.15 | 0.49 | 0.07 |
| | YEAR OF OCCURRENCE | | 1969 | 1989 | 1983 | 1948 | 2004 | 1950 | 1972 | 1979 | 1965 | 1950 | 1983 | 1983 | FEB 1989 |
| | MAXIMUM IN 24 HOURS (IN) | 65 | 2.74 | 2.71 | 1.81 | 2.05 | 2.30 | 2.26 | 2.46 | 2.62 | 3.35 | 4.66 | 3.57 | 3.56 | 4.66 |
| | YEAR OF OCCURRENCE | | 1948 | 1993 | 1992 | 1997 | 1992 | 1996 | 2001 | 1974 | 1996 | 1946 | 2005 | 1956 | OCT 1946 |
| | NORMAL NO. DAYS WITH: | | | | | | | | | | | | | | |
| | PRECIPITATION >= 0.01 | 30 | 19.4 | 16.3 | 18.8 | 17.5 | 17.1 | 15.9 | 16.7 | 18.1 | 21.1 | 23.8 | 20.3 | 20.8 | 225.8 |
| | PRECIPITATION >= 1.00 | 30 | 0.7 | 0.6 | 0.2 | 0.2 | 0.3 | 0.3 | 0.5 | 0.9 | 1.4 | 1.5 | 0.9 | 0.7 | 8.2 |
| **SNOWFALL** | NORMAL (IN) | 30 | 28.8 | 18.0 | 11.1 | 1.0 | 0.* | 0.0 | 0.0 | 0.0 | 0.* | 1.1 | 13.0 | 20.9 | 93.9 |
| | MAXIMUM MONTHLY (IN) | 65 | 75.2 | 86.3 | 62.7 | 46.3 | 1.2 | T | 0.0 | 0.0 | T | 15.6 | 69.8 | 54.7 | 86.3 |
| | YEAR OF OCCURRENCE | | 2009 | 1965 | 2007 | 1963 | 1964 | 1970 | | | 1974 | 1956 | 1994 | 1964 | FEB 1965 |
| | MAXIMUM IN 24 HOURS (IN) | 65 | 20.1 | 23.7 | 31.0 | 24.2 | 0.7 | T | 0.0 | 0.0 | T | 8.8 | 19.4 | 25.6 | 31.0 |
| | YEAR OF OCCURRENCE' | | 1975 | 1949 | 1948 | 1963 | 1945 | 1970 | | | 1974 | 1956 | 1994 | 1962 | MAR 1948 |
| | MAXIMUM SNOW DEPTH (IN) | 59 | 38 | 41 | 40 | 33 | 0 | 0 | 0 | 0 | 0 | 7 | 28 | 36 | 41 |
| | YEAR OF OCCURRENCE | | 1966 | 1949 | 1972 | 1963 | | | | | | 1956 | 2006 | 1962 | FEB 1949 |
| | NORMAL NO. DAYS WITH: | | | | | | | | | | | | | | |
| | SNOWFALL >= 1.0 | 30 | 7.2 | 4.9 | 2.8 | 0.3 | 0.0 | 0.0 | 0.0 | 0.0 | 0.0 | 0.2 | 3.4 | 4.6 | 23.4 |

## PRECIPITATION (inches) 2009 JUNEAU (PAJN)

| YEAR | JAN | FEB | MAR | APR | MAY | JUN | JUL | AUG | SEP | OCT | NOV | DEC | ANNUAL |
|------|-----|-----|-----|-----|-----|-----|-----|-----|-----|-----|-----|-----|--------|
| 1980 | 3.44 | 2.83 | 2.75 | 5.32 | 2.53 | 4.37 | 6.49 | 5.61 | 7.91 | 11.26 | 7.10 | 2.27 | 61.88 |
| 1981 | 4.66 | 2.57 | 1.88 | 2.11 | 3.27 | 2.44 | 4.25 | 6.19 | 11.61 | 6.18 | 6.93 | 2.24 | 54.33 |
| 1982 | 3.74 | 1.42 | 2.52 | 2.44 | 5.10 | 1.86 | 1.73 | 5.97 | 5.10 | 7.97 | 2.10 | 1.17 | 41.12 |
| 1983 | 4.00 | 1.69 | 0.59 | 2.53 | 5.37 | 2.69 | 3.16 | 9.52 | 6.13 | 4.24 | 1.15 | 0.49 | 41.56 |
| 1984 | 6.06 | 5.40 | 3.75 | 2.11 | 1.84 | 4.17 | 6.92 | 6.26 | 3.39 | 6.69 | | | |
| 1985 | | | | | | | | | | | | 8.33 | |
| 1986 | 7.00 | 3.25 | 6.08 | 2.98 | 2.54 | 2.76 | 2.38 | 6.89 | 2.40 | 12.33 | 5.96 | 6.42 | 60.99 |
| 1987 | 3.99 | 3.13 | 2.12 | 2.08 | 2.60 | 6.02 | 2.54 | 4.54 | 8.92 | 10.36 | 7.17 | 5.32 | 58.79 |
| 1988 | 2.58 | 6.55 | 4.15 | 2.25 | 3.91 | 2.05 | 5.21 | 5.53 | 5.46 | 9.71 | 8.62 | 4.75 | 60.77 |
| 1989 | 6.77 | 0.07 | 1.33 | 0.87 | 3.44 | 1.10 | 3.81 | 2.82 | 7.29 | 6.37 | 6.23 | 6.78 | 46.88 |
| 1990 | 3.72 | 4.54 | 4.86 | 1.06 | 1.72 | 3.32 | 4.65 | 5.35 | 10.63 | 6.59 | 4.89 | 6.03 | 57.36 |
| 1991 | 4.16 | 6.55 | 4.41 | 4.73 | 4.72 | 3.41 | 4.85 | 9.60 | 15.14 | 8.63 | 9.63 | 9.32 | 85.15 |
| 1992 | 8.69 | 7.24 | 6.37 | 3.63 | 9.20 | 2.98 | 5.18 | 5.02 | 11.45 | 5.90 | 7.91 | 5.73 | 79.30 |
| 1993 | 9.11 | 8.09 | 3.50 | 1.94 | 2.19 | 4.92 | 2.25 | 3.20 | 8.44 | 9.00 | 11.06 | 7.89 | 71.59 |
| 1994 | 7.05 | 2.52 | 6.50 | 3.68 | 4.20 | 1.83 | 4.32 | 2.68 | 11.17 | 9.15 | 9.57 | 6.22 | 68.89 |
| 1995 | 1.78 | 2.83 | 3.01 | 2.08 | 2.85 | 3.45 | 4.36 | 5.01 | 7.43 | 6.04 | 2.93 | 4.58 | 46.35 |
| 1996 | 2.26 | 8.43 | 4.12 | 2.19 | 1.80 | 6.22 | 3.16 | 7.91 | 10.68 | 6.20 | 2.75 | 4.73 | 60.45 |
| 1997 | 2.73 | 8.17 | 3.91 | 4.41 | 3.25 | 3.51 | 10.36 | 3.93 | 8.26 | 7.85 | 4.63 | 13.61 | 74.62 |
| 1998 | 2.54 | 1.90 | 3.71 | 3.12 | 2.21 | 2.50 | 4.95 | 6.80 | 6.17 | 12.13 | 1.72 | 5.45 | 53.20 |
| 1999 | 8.14 | 2.66 | 2.58 | 7.48 | 5.69 | 2.69 | 4.10 | 6.77 | 10.62 | 12.19 | 5.77 | 10.30 | 78.99 |
| 2000 | 4.82 | 1.56 | 5.75 | 4.40 | 3.25 | 5.72 | 6.65 | 6.12 | 10.05 | 10.11 | 6.37 | 4.17 | 68.97 |
| 2001 | 7.43 | 4.40 | 3.33 | 2.19 | 5.19 | 1.65 | 7.26 | 3.66 | 8.37 | 7.80 | 3.62 | 4.49 | 59.39 |
| 2002 | 3.28 | 5.62 | 1.33 | 0.47 | 2.37 | 3.40 | 4.72 | 10.50 | 6.08 | 10.69 | 7.95 | 5.86 | 62.27 |
| 2003 | 5.68 | 1.44 | 3.56 | 0.86 | 2.90 | 3.74 | 3.44 | 4.53 | 11.41 | 4.44 | 6.21 | 5.92 | 54.13 |
| 2004 | 5.89 | 5.66 | 5.59 | 4.43 | 0.84 | 1.30 | 3.54 | 2.51 | 9.23 | 7.18 | 8.38 | 10.67 | 65.22 |
| 2005 | 5.90 | 6.12 | 4.18 | 2.94 | 0.89 | 3.00 | 5.25 | 6.58 | 9.92 | 9.12 | 13.38 | 6.74 | 74.02 |
| 2006 | 2.93 | 2.07 | 1.55 | 4.24 | 4.56 | 5.93 | 4.43 | 11.02 | 13.01 | 11.78 | 3.40 | 9.37 | 74.29 |
| 2007 | 6.25 | 3.00 | 4.81 | 2.99 | 4.27 | 1.85 | 6.71 | 2.35 | 9.09 | 11.67 | 3.15 | 3.75 | 59.89 |
| 2008 | 4.91 | 4.92 | 4.00 | 4.79 | 3.87 | 1.68 | 8.25 | 5.33 | 10.84 | 15.05 | 5.89 | 3.92 | 73.45 |
| 2009 | 9.38 | 3.89 | 2.95 | 2.17 | 2.67 | 2.83 | 2.35 | 7.30 | 8.92 | 5.65 | 7.20 | 3.95 | 59.26 |
| POR= 65 YRS | 4.57 | 3.83 | 3.42 | 2.92 | 3.37 | 3.14 | 4.49 | 5.29 | 7.54 | 8.34 | 5.71 | 5.26 | 57.88 |

WBAN : 25309

## AVERAGE TEMPERATURE (°F) 2009 JUNEAU (PAJN)

| YEAR | JAN | FEB | MAR | APR | MAY | JUN | JUL | AUG | SEP | OCT | NOV | DEC | ANNUAL |
|------|-----|-----|-----|-----|-----|-----|-----|-----|-----|-----|-----|-----|--------|
| 1980 | 19.5 | 33.8 | 34.1 | 42.2 | 49.4 | 55.6 | 55.6 | 54.7 | 49.0 | 44.6 | 38.7 | 21.7 | 41.6 |
| 1981 | 37.6 | 32.7 | 39.4 | 39.1 | 52.1 | 54.3 | 56.1 | 55.9 | 49.2 | 42.8 | 36.8 | 26.9 | 43.6 |
| 1982 | 13.8 | 21.3 | 31.8 | 37.1 | 45.4 | 56.3 | 57.7 | 54.8 | 50.3 | 42.2 | 30.6 | 31.6 | 39.4 |
| 1983 | 30.2 | 31.8 | 34.8 | 42.6 | 49.7 | 55.6 | 57.1 | 54.6 | 48.0 | 42.1 | 31.8 | 18.9 | 41.4 |
| 1984 | 32.0 | | 39.6 | 43.1 | 49.1 | 53.5 | 55.5 | 55.3 | 50.0 | 40.5 | | | |
| 1985 | | | | | | | | | | | | 32.5 | |
| 1986 | 34.3 | 28.7 | 35.3 | 37.3 | 46.6 | 54.3 | 56.8 | 54.4 | 50.7 | 45.8 | 30.5 | 36.0 | 42.6 |
| 1987 | 33.1 | 34.7 | 31.8 | 41.6 | 47.8 | 51.6 | 58.6 | 57.5 | 50.1 | 44.4 | 40.0 | 34.3 | 43.8 |
| 1988 | 27.0 | 32.6 | 37.6 | 41.7 | 48.5 | 54.0 | 53.8 | 53.9 | 48.2 | 44.1 | 36.2 | 31.3 | 42.4 |
| 1989 | 25.5 | 23.9 | 29.4 | 42.7 | 49.0 | 55.4 | 60.1 | 58.2 | 52.4 | 41.7 | 32.9 | 36.0 | 42.3 |
| 1990 | 26.4 | 25.1 | 36.4 | 42.7 | 49.7 | 55.0 | 59.3 | 58.1 | 51.2 | 40.7 | 26.4 | 23.7 | 41.2 |
| 1991 | 24.9 | 35.3 | 33.1 | 41.7 | 48.2 | 55.0 | 55.2 | 55.0 | 50.3 | 40.1 | 36.3 | 33.6 | 42.4 |
| 1992 | 35.7 | 32.7 | 36.9 | 41.1 | 46.4 | 55.2 | 57.0 | 56.0 | 46.8 | 40.5 | 38.4 | 25.6 | 42.7 |
| 1993 | 24.5 | 30.2 | 35.8 | 44.4 | 52.1 | 55.8 | 59.6 | 57.1 | 51.2 | 45.8 | 36.8 | 35.9 | 44.1 |
| 1994 | 27.7 | 18.4 | 36.7 | 43.3 | 48.1 | 54.9 | 57.4 | 59.3 | 50.2 | 43.3 | 29.4 | 27.7 | 41.4 |
| 1995 | 27.2 | 29.0 | 30.3 | 43.4 | 51.4 | 55.4 | 57.0 | 54.8 | 53.8 | | 32.4 | 25.2 | |
| 1996 | 16.6 | 29.0 | 32.8 | 40.7 | 48.7 | 54.1 | 57.7 | 54.7 | 48.9 | 40.2 | 31.3 | 26.5 | 40.1 |
| 1997 | 26.1 | 35.3 | 31.5 | 41.5 | 51.5 | 56.6 | 58.2 | 58.8 | 53.3 | 41.4 | 37.3 | 36.7 | 44.0 |
| 1998 | 24.5 | 37.1 | 35.0 | 42.4 | 50.1 | 55.9 | 57.5 | 54.2 | 49.4 | 42.8 | 33.1 | 27.9 | 42.5 |
| 1999 | 26.6 | 28.6 | 33.1 | 39.5 | 44.8 | 53.8 | 57.4 | 56.4 | 49.7 | 43.5 | 36.1 | 35.9 | 42.1 |
| 2000 | 26.5 | 31.7 | 36.3 | 39.7 | 46.9 | 53.0 | 55.3 | 54.4 | 49.0 | 41.9 | 37.7 | 31.2 | 42.0 |
| 2001 | 36.7 | 28.2 | 33.4 | 40.3 | 45.2 | 54.2 | 55.4 | 57.2 | 50.3 | 41.9 | 32.9 | 29.0 | 42.1 |
| 2002 | 30.6 | 30.8 | 28.3 | 36.3 | 47.1 | 54.8 | 54.6 | 54.4 | 49.6 | 44.3 | 40.1 | 31.9 | 41.9 |
| 2003 | 31.6 | 31.7 | 30.5 | 41.7 | 46.8 | 53.7 | 58.0 | 54.9 | 48.8 | 43.3 | 30.6 | 31.8 | 42.0 |
| 2004 | 25.2 | 35.5 | 34.7 | 41.0 | 51.7 | 58.0 | 59.6 | 59.4 | 48.8 | 41.2 | 36.8 | 32.1 | 43.7 |
| 2005 | 25.2 | 30.7 | 37.5 | 44.0 | 52.7 | 56.1 | 56.3 | 57.2 | 51.1 | 41.7 | 35.5 | 33.7 | 43.5 |
| 2006 | 29.5 | 28.5 | 27.8 | 39.0 | 47.8 | 54.5 | 56.3 | 53.1 | 50.0 | 42.3 | 19.4 | 34.3 | 40.2 |
| 2007 | 31.0 | 25.1 | 28.4 | 38.6 | 46.8 | 54.4 | 56.7 | 56.5 | 49.7 | 42.0 | 34.9 | 27.4 | 41.0 |
| 2008 | 26.0 | 27.9 | 35.1 | 38.3 | 48.0 | 50.9 | 53.3 | 53.8 | 49.5 | 41.1 | 35.4 | 22.3 | 40.1 |
| 2009 | 25.8 | 27.1 | 28.5 | 39.2 | 48.1 | 54.9 | 59.9 | 55.9 | 49.8 | 42.6 | 35.0 | 27.9 | 41.2 |
| POR= 64 YRS | 24.7 | 28.5 | 32.5 | 39.6 | 47.4 | 53.5 | 56.1 | 55.2 | 48.9 | 42.1 | 33.1 | 28.0 | 40.8 |

**HEATING DEGREE DAYS (base 65°F) 2009  JUNEAU (PAJN)**

| YEAR | JUL | AUG | SEP | OCT | NOV | DEC | JAN | FEB | MAR | APR | MAY | JUN | TOTAL |
|------|-----|-----|-----|-----|-----|-----|-----|-----|-----|-----|-----|-----|-------|
| 1980-81 | 283 | 308 | 472 | 628 | 783 | 1333 | 843 | 899 | 786 | 772 | 392 | 316 | 7815 |
| 1981-82 | 269 | 275 | 469 | 682 | 841 | 1175 | 1579 | 1214 | 1021 | 830 | 601 | 257 | 9213 |
| 1982-83 | 220 | 310 | 435 | 699 | 1027 | 1029 | 1073 | 924 | 931 | 663 | 470 | 275 | 8056 |
| 1983-84 | 237 | 317 | 502 | 701 | 991 | 1423 | 1014 |  | 780 | 649 | 486 | 338 |  |
| 1984-85 | 286 | 291 | 444 | 754 |  |  |  |  |  |  |  |  |  |
| 1985-86 |  |  |  |  |  | 1001 | 943 | 1011 | 913 | 822 | 564 | 316 |  |
| 1986-87 | 249 | 319 | 423 | 587 | 1028 | 891 | 982 | 841 | 1020 | 697 | 526 | 394 | 7957 |
| 1987-88 | 199 | 222 | 440 | 635 | 744 | 944 | 1169 | 935 | 844 | 691 | 508 | 324 | 7655 |
| 1988-89 | 338 | 338 | 497 | 641 | 855 | 1040 | 1217 | 1144 | 1097 | 663 | 491 | 283 | 8604 |
| 1989-90 | 159 | 210 | 370 | 713 | 959 | 890 | 1191 | 1109 | 879 | 661 | 467 | 295 | 7903 |
| 1990-91 | 180 | 210 | 407 | 748 | 1152 | 1274 | 1238 | 823 | 981 | 694 | 516 | 298 | 8521 |
| 1991-92 | 294 | 303 | 435 | 764 | 855 | 966 | 902 | 930 | 865 | 712 | 571 | 292 | 7889 |
| 1992-93 | 240 | 274 | 540 | 750 | 791 | 1217 | 1250 | 966 | 899 | 612 | 394 | 271 | 8204 |
| 1993-94 | 166 | 240 | 406 | 588 | 838 | 891 | 1147 | 1298 | 869 | 641 | 514 | 299 | 7897 |
| 1994-95 | 229 | 171 | 437 | 668 | 1061 | 1150 | 1165 | 1003 | 1066 | 641 | 416 | 285 | 8292 |
| 1995-96 | 241 | 309 | 329 |  | 971 | 1229 | 1493 | 1038 | 993 | 719 | 498 | 322 |  |
| 1996-97 | 218 | 312 | 476 | 764 | 1004 | 1187 | 1196 | 826 | 1031 | 698 | 414 | 252 | 8378 |
| 1997-98 | 203 | 183 | 345 | 725 | 823 | 869 | 1247 | 776 | 926 | 673 | 457 | 266 | 7493 |
| 1998-99 | 224 | 331 | 463 | 681 | 951 | 1141 | 1181 | 1012 | 978 | 758 | 619 | 329 | 8668 |
| 1999-00 | 230 | 264 | 452 | 660 | 858 | 895 | 1184 | 959 | 882 | 751 | 552 | 354 | 8041 |
| 2000-01 | 295 | 322 | 471 | 710 | 813 | 1040 | 873 | 1026 | 972 | 734 | 605 | 320 | 8181 |
| 2001-02 | 292 | 235 | 434 | 709 | 956 | 1111 | 1060 | 951 | 1131 | 856 | 549 | 299 | 8583 |
| 2002-03 | 316 | 325 | 455 | 636 | 740 | 1019 | 1027 | 927 | 1064 | 690 | 555 | 330 | 8084 |
| 2003-04 | 210 | 306 | 478 | 665 | 1026 | 1020 | 1225 | 849 | 933 | 713 | 406 | 220 | 8051 |
| 2004-05 | 163 | 172 | 478 | 730 | 838 | 1011 | 1222 | 954 | 848 | 622 | 373 | 260 | 7671 |
| 2005-06 | 265 | 240 | 413 | 715 | 877 | 961 | 1094 | 1019 | 1147 | 774 | 526 | 306 | 8337 |
| 2006-07 | 264 | 364 | 444 | 700 | 1361 | 943 | 1048 | 1108 | 1130 | 784 | 559 | 313 | 9018 |
| 2007-08 | 251 | 257 | 452 | 707 | 895 | 1159 | 1200 | 1071 | 919 | 794 | 519 | 415 | 8639 |
| 2008-09 | 357 | 338 | 457 | 732 | 882 | 1317 | 1210 | 1057 | 1123 | 770 | 518 | 296 | 9057 |
| 2009- | 156 | 276 | 450 | 685 | 892 | 1141 |  |  |  |  |  |  |  |

WBAN : 25309

**COOLING DEGREE DAYS (base 65°F) 2009  JUNEAU (PAJN)**

| YEAR | JAN | FEB | MAR | APR | MAY | JUN | JUL | AUG | SEP | OCT | NOV | DEC | TOTAL |
|------|-----|-----|-----|-----|-----|-----|-----|-----|-----|-----|-----|-----|-------|
| 1980 | 0 | 0 | 0 | 0 | 0 | 1 | 0 | 0 | 0 | 0 | 0 | 0 | 1 |
| 1981 | 0 | 0 | 0 | 0 | 0 | 0 | 0 | 0 | 0 | 0 | 0 | 0 | 0 |
| 1982 | 0 | 0 | 0 | 0 | 0 | 2 | 0 | 0 | 0 | 0 | 0 | 0 | 2 |
| 1983 | 0 | 0 | 0 | 0 | 0 | 0 | 0 | 0 | 0 | 0 | 0 | 0 | 0 |
| 1984 | 0 |  | 0 | 0 | 0 | 0 | 0 | 0 | 0 | 0 |  |  |  |
| 1985 |  |  |  |  |  |  |  |  |  |  |  | 0 |  |
| 1986 | 0 | 0 | 0 | 0 | 0 | 2 | 0 | 0 | 0 | 0 | 0 | 0 | 2 |
| 1987 | 0 | 0 | 0 | 0 | 0 | 0 | 5 | 0 | 0 | 0 | 0 | 0 | 5 |
| 1988 | 0 | 0 | 0 | 0 | 0 | 0 | 0 | 0 | 0 | 0 | 0 | 0 | 0 |
| 1989 | 0 | 0 | 0 | 0 | 0 | 0 | 14 | 0 | 0 | 0 | 0 | 0 | 14 |
| 1990 | 0 | 0 | 0 | 0 | 0 | 1 | 8 | 3 | 0 | 0 | 0 | 0 | 12 |
| 1991 | 0 | 0 | 0 | 0 | 0 | 6 | 0 | 0 | 0 | 0 | 0 | 0 | 6 |
| 1992 | 0 | 0 | 0 | 0 | 0 | 6 | 0 | 0 | 0 | 0 | 0 | 0 | 6 |
| 1993 | 0 | 0 | 0 | 0 | 0 | 0 | 5 | 0 | 0 | 0 | 0 | 0 | 5 |
| 1994 | 0 | 0 | 0 | 0 | 0 | 0 | 0 | 2 | 0 | 0 | 0 | 0 | 2 |
| 1995 | 0 | 0 | 0 | 0 | 0 | 1 | 0 | 0 | 0 |  | 0 | 0 |  |
| 1996 | 0 | 0 | 0 | 0 | 0 | 0 | 2 | 0 | 0 | 0 | 0 | 0 | 2 |
| 1997 | 0 | 0 | 0 | 0 | 0 | 6 | 0 | 0 | 0 | 0 | 0 | 0 | 6 |
| 1998 | 0 | 0 | 0 | 0 | 0 | 0 | 0 | 0 | 0 | 0 | 0 | 0 | 0 |
| 1999 | 0 | 0 | 0 | 0 | 0 | 0 | 2 | 3 | 0 | 0 | 0 | 0 | 5 |
| 2000 | 0 | 0 | 0 | 0 | 0 | 0 | 0 | 0 | 0 | 0 | 0 | 0 | 0 |
| 2001 | 0 | 0 | 0 | 0 | 0 | 0 | 0 | 0 | 0 | 0 | 0 | 0 | 0 |
| 2002 | 0 | 0 | 0 | 0 | 0 | 0 | 0 | 0 | 0 | 0 | 0 | 0 | 0 |
| 2003 | 0 | 0 | 0 | 0 | 0 | 0 | 3 | 0 | 0 | 0 | 0 | 0 | 3 |
| 2004 | 0 | 0 | 0 | 0 | 0 | 17 | 2 | 5 | 0 | 0 | 0 | 0 | 24 |
| 2005 | 0 | 0 | 0 | 0 | 0 | 0 | 0 | 2 | 0 | 0 | 0 | 0 | 2 |
| 2006 | 0 | 0 | 0 | 0 | 0 | 0 | 0 | 0 | 0 | 0 | 0 | 0 | 0 |
| 2007 | 0 | 0 | 0 | 0 | 0 | 0 | 0 | 0 | 0 | 0 | 0 | 0 | 0 |
| 2008 | 0 | 0 | 0 | 0 | 0 | 0 | 0 | 0 | 0 | 0 | 0 | 0 | 0 |
| 2009 | 0 | 0 | 0 | 0 | 0 | 0 | 6 | 0 | 0 | 0 | 0 | 0 | 6 |

### SNOWFALL (inches) 2009 JUNEAU (PAJN)

| YEAR | JUL | AUG | SEP | OCT | NOV | DEC | JAN | FEB | MAR | APR | MAY | JUN | TOTAL |
|---|---|---|---|---|---|---|---|---|---|---|---|---|---|
| 1980-81 | 0.0 | 0.0 | 0.0 | 0.0 | 0.5 | 40.5 | 2.4 | 16.4 | 0.5 | 2.2 | 0.0 | 0.0 | 62.5 |
| 1981-82 | 0.0 | 0.0 | 0.0 | 0.0 | 4.0 | 6.0 | 69.2 | 29.6 | 8.4 | 1.1 | T | 0.0 | 118.3 |
| 1982-83 | 0.0 | 0.0 | 0.0 | 2.0 | 0.4 | 10.8 | 40.1 | 15.7 | 0.2 | T | 0.0 | 0.0 | 69.2 |
| 1983-84 | 0.0 | 0.0 | 0.0 | 0.0 | 8.1 | 13.3 | 43.1 | 0.7 | 1.0 | T | T | 0.0 | 66.2 |
| 1984-85 | 0.0 | 0.0 | 0.0 | 0.0 | | | | | | | | | |
| 1985-86 | | | | | | 2.0 | 10.3 | 7.4 | 30.4 | 4.4 | T | 0.0 | |
| 1986-87 | 0.0 | 0.0 | 0.0 | T | 22.1 | 1.4 | 3.3 | 1.4 | 7.3 | T | 0.0 | 0.0 | 35.5 |
| 1987-88 | 0.0 | 0.0 | 0.0 | T | 4.6 | 6.8 | 3.5 | 8.0 | 1.0 | 0.5 | 0.0 | 0.0 | 24.4 |
| 1988-89 | 0.0 | 0.0 | 0.0 | 0.0 | 4.8 | 11.3 | 44.7 | 0.2 | 10.0 | T | 0.0 | 0.0 | 71.0 |
| 1989-90 | 0.0 | 0.0 | 0.0 | 0.6 | 32.5 | 6.4 | 36.5 | 39.4 | 0.6 | 0.0 | 0.0 | 0.0 | 116.0 |
| 1990-91 | 0.0 | 0.0 | 0.0 | 0.0 | 48.8 | 33.2 | 31.8 | 15.5 | 9.4 | 0.8 | T | 0.0 | 139.5 |
| 1991-92 | 0.0 | 0.0 | 0.0 | 5.4 | 7.7 | 49.3 | 14.3 | 12.4 | 4.1 | T | T | 0.0 | 93.2 |
| 1992-93 | 0.0 | 0.0 | 0.0 | 1.1 | 4.4 | 25.0 | 32.5 | 36.8 | 2.6 | T | 0.0 | 0.0 | 102.4 |
| 1993-94 | 0.0 | 0.0 | 0.0 | 0.0 | 4.3 | 10.4 | 61.2 | 32.4 | 22.9 | T | 0.0 | 0.0 | 131.2 |
| 1994-95 | 0.0 | 0.0 | 0.0 | 0.0 | 69.8 | 25.8 | 8.6 | 14.0 | 28.4 | T | 0.0 | 0.0 | 146.6 |
| 1995-96 | 0.0 | 0.0 | 0.0 | 0.0 | 2.9 | 23.0 | 20.8 | 33.1 | 5.5 | T | 0.0 | 0.0 | 85.3 |
| 1996-97 | 0.0 | 0.0 | 0.0 | .5 | 2.5 | 4.6 | 12.3 | 17.3 | 19.2 | 0.4 | 0.0 | 0.0 | 56.8 |
| 1997-98 | 0.0 | 0.0 | 0.0 | 1.9 | 1.2 | 14.9 | 12.9 | 0.5 | | | | | |
| 1998-99 | | | | | | 22.8 | 53.1 | 34.2 | 7.3 | 1.1 | T | 0.0 | |
| 1999-00 | 0.0 | 0.0 | 0.0 | T | 5.1 | 19.9 | 13.6 | 4.6 | T | T | 0.0 | 0.0 | 43.2 |
| 2000-01 | 0.0 | 0.0 | T | 2.3 | 1.0 | 2.3 | 7.6 | 14.9 | 0.6 | T | T | 0.0 | 28.7 |
| 2001-02 | 0.0 | 0.0 | 0.0 | 3.4 | 3.1 | 25.5 | 18.9 | 28.9 | 2.5 | T | 0.0 | 0.0 | 82.3 |
| 2002-03 | 0.0 | 0.0 | 0.0 | 0.0 | 0.0 | 17.7 | 15.3 | 6.2 | 17.7 | T | 0.0 | 0.0 | 56.9 |
| 2003-04 | 0.0 | 0.0 | 0.0 | 0.0 | 24.9 | 11.9 | 34.2 | 2.8 | 18.5 | T | 0.0 | 0.0 | 92.3 |
| 2004-05 | 0.0 | 0.0 | 0.0 | T | 1.9 | 10.8 | 46.7 | 16.6 | 1.0 | 1.6 | 0.0 | 0.0 | 78.6 |
| 2005-06 | 0.0 | 0.0 | 0.0 | 0.0 | 8.3 | 4.6 | 30.8 | 3.2 | 6.9 | 0.5 | T | 0.0 | 54.3 |
| 2006-07 | 0.0 | 0.0 | 0.0 | T | 64.2 | 25.4 | 27.5 | 10.6 | 62.7 | 0.0 | 0.0 | 0.0 | 190.4 |
| 2007-08 | 0.0 | 0.0 | 0.0 | 0.0 | 4.5 | 12.6 | 22.9 | 43.9 | 13.5 | 13.7 | T | 0.0 | 111.1 |
| 2008-09 | 0.0 | 0.0 | 0.0 | 4.6 | 3.8 | 32.7 | 75.2 | 30.4 | 31.3 | 2.2 | 0.0 | 0.0 | 180.2 |
| 2009- | 0.0 | 0.0 | 0.0 | 0.2 | 19.8 | 8.9 | | | | | | | |
| POR= 64 YRS | 0.0 | 0.0 | T | 1.1 | 11.4 | 21.5 | 26.2 | 18.8 | 14.0 | 2.5 | T | 0.0 | 95.5 |

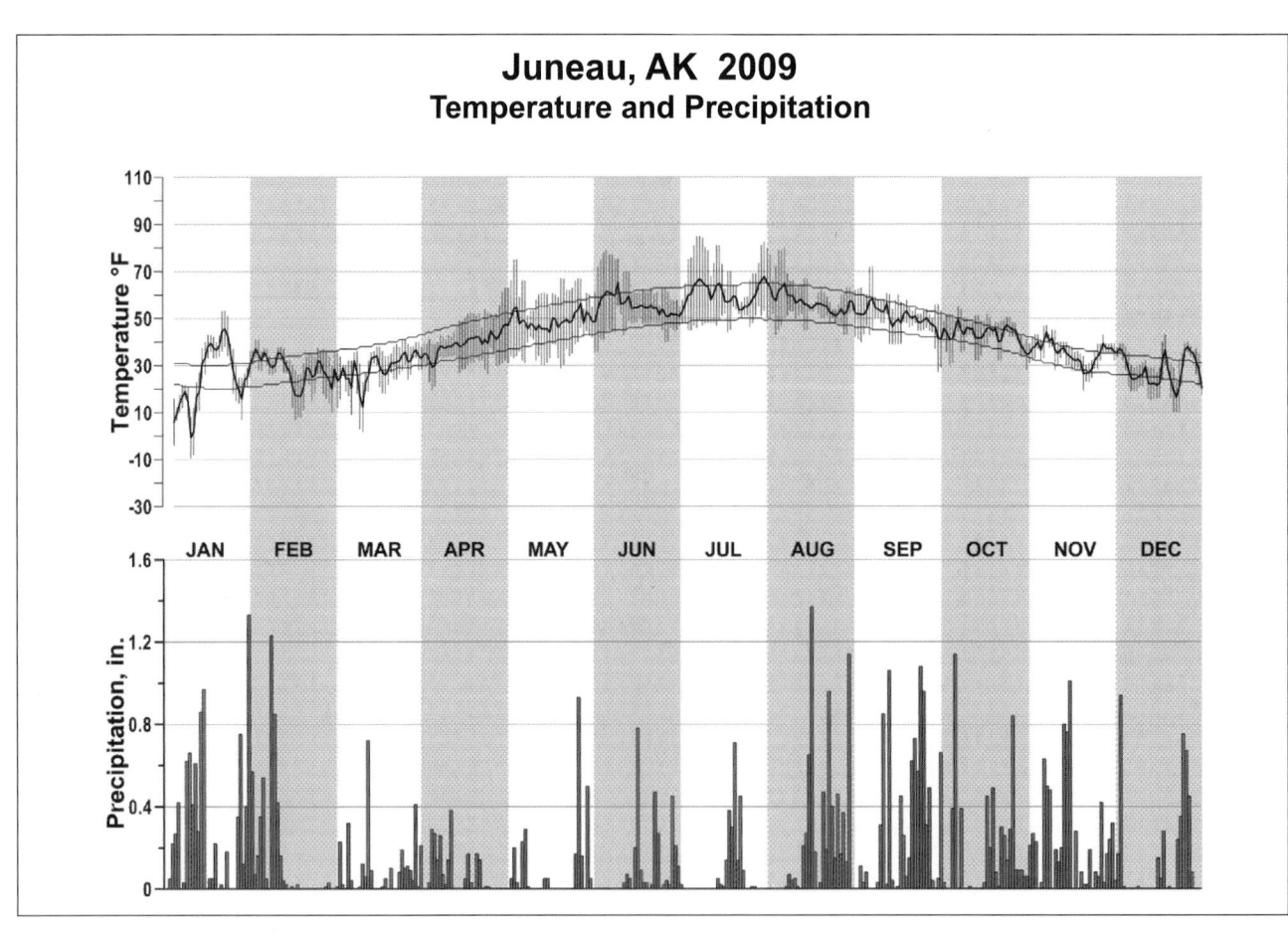

# Juneau, AK 2009
## Temperature and Precipitation

# 2009
# NOME
# ALASKA (PAOM)

The weather station at Nome is located at Nome Field, approximately 1 mile northwest of the city. Low, marshy flats lie between the station and Norton Sound to the south, exposing the station to winds from the southeast through the west. A series of foothills, with heights of 500 to 1,200 feet, extend from northwest through north to east at a distance of from 4 to 8 miles. The terrain increases in ruggedness and height farther north, with the Kigluaik Mountains reaching a height of 5,000 feet at a distance of 30 miles. The ground along the coastal flats is swampy during the summer months, but is permanently frozen below a depth of 2 to 3 feet. Vegetation in the Nome area consists mostly of grass and numerous small flowering plants.

The moderating influence of the open water of Norton Sound is effective only from early June to about the middle of November. Storms moving through this area during these months result in extended periods of cloudiness and rain. There is a nearly continuous cloud cover during July and August. During the summer months the daily temperature range is very slight. The freezing of Norton Sound in November causes a rather abrupt change from a maritime to a continental climate. The majority of low pressure systems during this period take a path south of Nome, resulting in strong easterly winds, accompanied by frequent blizzards, with the winds later becoming northerly and reaching Nome across the colder frozen areas of northern Alaska.

Temperatures generally remain well below freezing from the middle of November to the latter part of April, with January usually the coldest month of the year. Temperatures usually begin to rise near the end of February and continue to rise until they reach a maximum in July.

Precipitation reaches its maximum during the late summer months and drops to a minimum in April and May. Snow begins to fall in September, but usually does not accumulate on the ground until the first part of November. The snow cover decreases rapidly in April and May, and normally disappears by the middle of June. Snow depths in Nome have exceeded 70 inches.

Severe windstorms do occur with winds over 70 mph recorded several times. Strong winds during the winter months when there is snow cover produce blowing snow conditions that severely hinder transportation in the area.

# NORMALS, MEANS, AND EXTREMES
## NOME (PAOM)

| LATITUDE: 64 ° 30'N | LONGITUDE: -165° 26'W | | ELEVATION (FT): GRND: 20   BARO: 18 | | | | TIME ZONE: ALASKA   (UTC -9) | | | WBAN: 26617 | | |

| | ELEMENT | POR | JAN | FEB | MAR | APR | MAY | JUN | JUL | AUG | SEP | OCT | NOV | DEC | YEAR |
|---|---|---|---|---|---|---|---|---|---|---|---|---|---|---|---|
| **TEMPERATURE °F** | NORMAL DAILY MAXIMUM | 30 | 13.4 | 13.6 | 17.7 | 26.8 | 43.0 | 53.9 | 58.6 | 56.0 | 48.6 | 34.0 | 23.0 | 15.8 | 33.7 |
| | MEAN DAILY MAXIMUM | 85 | 12.9 | 13.1 | 16.9 | 26.0 | 41.9 | 52.5 | 56.9 | 55.4 | 47.8 | 33.9 | 22.4 | 13.9 | 32.8 |
| | HIGHEST DAILY MAXIMUM | 63 | 43 | 48 | 43 | 51 | 78 | 83 | 86 | 81 | 71 | 59 | 47 | 43 | 86 |
| | YEAR OF OCCURRENCE | | 1977 | 1986 | 1984 | 1988 | 1981 | 2004 | 1977 | 2005 | 1979 | 1954 | 2002 | 1969 | JUL 1977 |
| | MEAN OF EXTREME MAXS. | 85 | 34.0 | 33.7 | 34.4 | 40.8 | 60.1 | 70.7 | 73.0 | 68.4 | 59.4 | 46.1 | 36.7 | 33.2 | 49.2 |
| | NORMAL DAILY MINIMUM | 30 | -1.8 | -2.3 | 1.0 | 12.4 | 31.1 | 40.6 | 46.6 | 45.2 | 37.2 | 22.9 | 10.8 | .9 | 20.4 |
| | MEAN DAILY MINIMUM | 85 | -1.5 | -1.8 | 0.5 | 10.9 | 29.2 | 38.7 | 44.9 | 43.7 | 35.9 | 22.8 | 10.1 | .9 | 19.5 |
| | LOWEST DAILY MINIMUM | 63 | -54 | -42 | -46 | -30 | -11 | 23 | 30 | 26 | 9 | -10 | -39 | -41 | -54 |
| | YEAR OF OCCURRENCE | | 1989 | 1978 | 1971 | 1968 | 1949 | 1974 | 1994 | 1994 | 1992 | 1966 | 1948 | 1961 | JAN 1989 |
| | MEAN OF EXTREME MINS. | 85 | -26.0 | -25.9 | -23.3 | -11.9 | 14.5 | 29.8 | 36.3 | 32.7 | 23.6 | 5.6 | -11.5 | -23.0 | 1.7 |
| | NORMAL DRY BULB | 30 | 5.8 | 5.7 | 9.4 | 19.6 | 37.1 | 47.3 | 52.6 | 50.6 | 42.9 | 28.5 | 16.9 | 8.4 | 27.1 |
| | MEAN DRY BULB | 85 | 5.7 | 5.6 | 8.7 | 18.5 | 35.6 | 45.8 | 50.9 | 49.6 | 41.9 | 28.4 | 16.2 | 7.1 | 26.2 |
| | MEAN WET BULB | 26 | 4.0 | 7.2 | 8.4 | 18.8 | 32.9 | 43.7 | 48.8 | 47.3 | 40.0 | 26.6 | 15.1 | 8.0 | 25.1 |
| | MEAN DEW POINT | 26 | 0.3 | 2.4 | 4.5 | 15.3 | 29.9 | 40.3 | 46.8 | 45.6 | 37.4 | 23.6 | 11.7 | 4.6 | 21.9 |
| | NORMAL NO. DAYS WITH: | | | | | | | | | | | | | | |
| | MAXIMUM >= 70 | 30 | 0.0 | 0.0 | 0.0 | 0.0 | 0.3 | 2.1 | 3.6 | 1.0 | 0.0 | 0.0 | 0.0 | 0.0 | 7.0 |
| | MAXIMUM <= 32 | 30 | 27.9 | 25.0 | 26.9 | 19.6 | 4.2 | 0.0 | 0.0 | 0.0 | 0.2 | 12.2 | 24.1 | 27.3 | 167.4 |
| | MINIMUM <= 32 | 30 | 30.9 | 28.1 | 30.9 | 28.8 | 18.2 | 3.0 | 0.2 | 1.2 | 9.0 | 25.5 | 28.9 | 30.9 | 235.6 |
| | MINIMUM <= 0 | 30 | 16.6 | 15.8 | 14.9 | 7.4 | 0.2 | 0.0 | 0.0 | 0.0 | 0.0 | 0.6 | 7.5 | 15.4 | 78.4 |
| **H/C** | NORMAL HEATING DEG. DAYS | 30 | 1836 | 1663 | 1710 | 1361 | 867 | 533 | 387 | 446 | 664 | 1134 | 1444 | 1756 | 13801 |
| | NORMAL COOLING DEG. DAYS | 30 | 0 | 0 | 0 | 0 | 0 | 0 | 2 | 0 | 0 | 0 | 0 | 0 | 2 |
| **RH** | NORMAL (PERCENT) | 30 | 74 | 72 | 71 | 74 | 75 | 74 | 80 | 82 | 77 | 76 | 76 | 74 | 75 |
| | HOUR 03 LST | 30 | 74 | 73 | 73 | 77 | 81 | 82 | 87 | 87 | 82 | 79 | 77 | 75 | 79 |
| | HOUR 09 LST | 30 | 74 | 74 | 73 | 74 | 75 | 73 | 80 | 83 | 79 | 80 | 77 | 74 | 76 |
| | HOUR 15 LST | 30 | 74 | 71 | 69 | 72 | 72 | 70 | 76 | 78 | 70 | 70 | 74 | 74 | 73 |
| | HOUR 21 LST | 30 | 74 | 73 | 73 | 75 | 74 | 70 | 77 | 81 | 79 | 78 | 77 | 75 | 76 |
| **S** | PERCENT POSSIBLE SUNSHINE | 40 | 40 | 55 | 54 | 54 | 50 | 43 | 37 | 32 | 36 | 35 | 31 | 34 | 42 |
| **W/O** | MEAN NO. DAYS WITH: | | | | | | | | | | | | | | |
| | HEAVY FOG(VISBY <= 1/4 MI) | 46 | 1.6 | 1.5 | 1.5 | 1.5 | 2.8 | 3.2 | 2.5 | 1.2 | 0.4 | 0.5 | 1.4 | 1.4 | 19.5 |
| | THUNDERSTORMS | 61 | 0.0 | 0.0 | 0.0 | 0.0 | 0.1 | 0.2 | 0.3 | 0.1 | 0.0 | 0.0 | 0.0 | 0.0 | 0.7 |
| **CLOUDNESS** | MEAN: | | | | | | | | | | | | | | |
| | SUNRISE-SUNSET (OKTAS) | 52 | 5.0 | 4.3 | 4.5 | 4.9 | 5.3 | 5.4 | 6.1 | 6.5 | 5.9 | 5.5 | 5.3 | 5.0 | 5.3 |
| | MIDNIGHT-MIDNIGHT (OKTAS) | 34 | 4.5 | 4.0 | 4.1 | 4.6 | 5.2 | 5.4 | 6.0 | 6.2 | 5.7 | 5.0 | 5.1 | 4.6 | 5.0 |
| | MEAN NO. DAYS WITH: | | | | | | | | | | | | | | |
| | CLEAR | 52 | 10.0 | 11.4 | 11.2 | 9.1 | 7.1 | 5.8 | 3.7 | 2.7 | 4.5 | 6.9 | 7.4 | 9.8 | 89.6 |
| | PARTLY CLOUDY | 52 | 4.4 | 3.8 | 5.2 | 6.4 | 7.6 | 9.0 | 6.8 | 5.7 | 5.5 | 5.6 | 4.2 | 3.9 | 68.1 |
| | CLOUDY | 52 | 16.6 | 13.1 | 14.6 | 14.5 | 16.5 | 15.3 | 19.8 | 22.0 | 19.4 | 18.0 | 18.0 | 16.7 | 204.5 |
| **PR** | MEAN STATION PRESSURE(IN) | 26 | 29.74 | 29.79 | 29.86 | 29.84 | 29.83 | 29.86 | 29.86 | 29.81 | 29.71 | 29.70 | 29.67 | 29.63 | 29.78 |
| | MEAN SEA-LEVEL PRES. (IN) | 26 | 29.76 | 29.82 | 29.88 | 29.86 | 29.89 | 29.88 | 29.89 | 29.83 | 29.73 | 29.72 | 29.70 | 29.66 | 29.80 |
| **WINDS** | MEAN SPEED (MPH) | 26 | 9.1 | 10.4 | 9.2 | 9.1 | 9.0 | 8.8 | 9.1 | 9.8 | 10.1 | 9.2 | 10.4 | 10.1 | 9.5 |
| | PREVAIL.DIR(TENS OF DEGS) | 29 | 08 | 08 | 08 | 08 | 26 | 26 | 26 | 26 | 36 | 07 | 08 | 08 | 08 |
| | MAXIMUM 2-MINUTE: | | | | | | | | | | | | | | |
| | SPEED (MPH) | 11 | 49 | 47 | 45 | 43 | 33 | 32 | 35 | 35 | 47 | 48 | 47 | 43 | 49 |
| | DIR. (TENS OF DEGS) | | 28 | 10 | 04 | 01 | 02 | 12 | 18 | 21 | 21 | 21 | 14 | 09 | 28 |
| | YEAR OF OCCURRENCE | | 2008 | 2006 | 2005 | 2001 | 2006 | 2003 | 2009 | 2003 | 2005 | 2004 | 2003 | 2003 | JAN 2008 |
| | MAXIMUM 3-SECOND | | | | | | | | | | | | | | |
| | SPEED (MPH) | 11 | 56 | 56 | 51 | 48 | 40 | 40 | 43 | 40 | 56 | 59 | 58 | 54 | 59 |
| | DIR. (TENS OF DEGS) | | 28 | 09 | 36 | 30 | 36 | 12 | 19 | 20 | 20 | 20 | 15 | 06 | 20 |
| | YEAR OF OCCURRENCE | | 2008 | 2006 | 2007 | 2002 | 2001 | 2003 | 2009 | 2009 | 2005 | 2004 | 2003 | 2007 | OCT 2004 |
| **PRECIPITATION** | NORMAL (IN) | 30 | 0.92 | 0.75 | 0.60 | 0.65 | 0.74 | 1.14 | 2.15 | 3.23 | 2.51 | 1.58 | 1.28 | 1.01 | 16.56 |
| | MAXIMUM MONTHLY (IN) | 63 | 2.43 | 2.11 | 1.95 | 2.15 | 2.95 | 4.15 | 4.78 | 8.58 | 7.46 | 3.84 | 4.39 | 2.16 | 8.58 |
| | YEAR OF OCCURRENCE | | 2000 | 1989 | 1954 | 1961 | 2004 | 1978 | 1999 | 1998 | 1986 | 1972 | 1979 | 1951 | AUG 1998 |
| | MINIMUM MONTHLY (IN) | 63 | T | T | T | 0.01 | 0.04 | 0.04 | 0.25 | 0.40 | 0.06 | T | 0.03 | 0.03 | T |
| | YEAR OF OCCURRENCE | | 1970 | 1979 | 1971 | 1992 | 1994 | 1964 | 1964 | 1971 | 2008 | 1974 | 1962 | 1956 | FEB 1979 |
| | MAXIMUM IN 24 HOURS (IN) | 63 | 1.23 | 0.77 | 0.68 | 0.75 | 1.01 | 2.03 | 1.77 | 2.99 | 1.49 | 2.28 | 1.15 | 1.09 | 2.99 |
| | YEAR OF OCCURRENCE | | 1963 | 1975 | 2009 | 1995 | 1996 | 1953 | 1954 | 1976 | 1978 | 1957 | 1994 | 1951 | AUG 1976 |
| | NORMAL NO. DAYS WITH: | | | | | | | | | | | | | | |
| | PRECIPITATION >= 0.01 | 30 | 11.0 | 7.6 | 8.4 | 7.9 | 8.8 | 9.2 | 11.8 | 16.1 | 14.4 | 11.8 | 12.4 | 11.5 | 130.9 |
| | PRECIPITATION >= 1.00 | 30 | 0.0 | 0.0 | 0.0 | 0.0 | 0.0 | 0.0 | 0.2 | 0.3 | 0.1 | * | 0.0 | 0.0 | 0.6 |
| **SNOWFALL** | NORMAL (IN) | 30 | 11.6 | 9.0 | 7.4 | 7.0 | 2.6 | 0.2 | 0.0 | 0.* | 0.5 | 5.4 | 13.2 | 12.5 | 69.4 |
| | MAXIMUM MONTHLY (IN) | 63 | 39.8 | 35.1 | 29.0 | 23.3 | 10.0 | 2.8 | 0.0 | 0.1 | 3.5 | 20.1 | 30.9 | 29.7 | 39.8 |
| | YEAR OF OCCURRENCE | | 2007 | 2009 | 2009 | 1961 | 1977 | 2008 | | 1965 | 1992 | 1996 | 1994 | 1994 | JAN 2007 |
| | MAXIMUM IN 24 HOURS (IN) | 63 | 10.4 | 9.8 | 15.5 | 7.1 | 5.3 | 1.7 | 0.0 | 0.1 | 2.6 | 6.7 | 8.7 | 14.0 | 15.5 |
| | YEAR OF OCCURRENCE' | | 1999 | 1996 | 2009 | 1997 | 1952 | 1982 | | 1965 | 1948 | 1996 | 1979 | 1997 | MAR 2009 |
| | MAXIMUM SNOW DEPTH (IN) | 60 | 60 | 67 | 78 | 72 | 37 | 3 | 0 | 0 | 2 | 9 | 26 | 46 | 78 |
| | YEAR OF OCCURRENCE | | 1995 | 1949 | 2009 | 2009 | 1949 | 1952 | | | 1993 | 1948 | 1948 | 1994 | MAR 2009 |
| | NORMAL NO. DAYS WITH: | | | | | | | | | | | | | | |
| | SNOWFALL >= 1.0 | 30 | 3.7 | 3.0 | 2.4 | 2.4 | 1.0 | 0.0 | 0.0 | 0.0 | 0.2 | 1.7 | 4.6 | 4.2 | 23.2 |

## PRECIPITATION (inches) 2009 NOME (PAOM)

| YEAR | JAN | FEB | MAR | APR | MAY | JUN | JUL | AUG | SEP | OCT | NOV | DEC | ANNUAL |
|------|-----|-----|-----|-----|-----|-----|-----|-----|-----|-----|-----|-----|--------|
| 1980 | 1.45 | 1.23 | 0.71 | 0.43 | 1.08 | 3.78 | 2.51 | 1.50 | 1.21 | 1.65 | 0.94 | 0.47 | 16.96 |
| 1981 | 1.46 | 1.44 | 0.82 | 1.03 | 0.66 | 0.80 | 3.37 | 2.52 | 0.92 | 1.63 | 0.68 | 1.19 | 16.52 |
| 1982 | 1.51 | 1.59 | 0.84 | 0.69 | 1.23 | 1.55 | 1.62 | 1.72 | 3.72 | 1.39 | 2.20 | 1.19 | 19.25 |
| 1983 | 0.22 | 0.07 | 0.12 | 1.18 | 0.51 | 0.31 | 1.47 | 3.34 | 2.25 | 1.87 | 1.64 | 1.11 | 14.09 |
| 1984 | 0.61 | 0.01 | 0.22 | 0.87 | 0.45 | 0.80 | 3.30 | 4.45 | 1.58 | 0.92 | 0.40 | 1.31 | 14.92 |
| 1985 | 0.61 | 0.56 | 1.55 | 0.07 | 1.12 | 1.38 | 1.12 | 3.90 | 4.81 | 3.71 | 1.12 | 1.29 | 21.24 |
| 1986 | 0.40 | 0.66 | 0.03 | 0.51 | 0.31 | 0.24 | 2.84 | 2.76 | 7.46 | 0.35 | 1.09 | 1.17 | 17.82 |
| 1987 | 0.95 | 0.56 | 0.49 | 0.11 | 0.37 | 0.60 | 0.97 | 1.91 | 2.42 | 1.83 | 0.65 | 1.17 | 12.03 |
| 1988 | 0.45 | 1.41 | 0.86 | 0.29 | 0.97 | 1.41 | 0.32 | 4.12 |  | 0.28 | 0.44 | 1.39 |  |
| 1989 | 0.84 | 2.11 | 0.39 | 1.73 | 2.02 | 0.29 | 4.66 | 3.29 | 2.26 | 1.23 | 1.20 | 0.72 | 20.74 |
| 1990 | 0.89 | 0.52 | 0.70 | 1.09 | 2.00 | 1.17 | 4.66 | 4.30 | 1.86 | 2.35 | 1.56 | 1.28 | 22.38 |
| 1991 | 1.13 | 0.43 | 0.55 | 0.64 | 0.35 | 2.28 | 2.44 | 0.75 | 2.69 | 2.31 | 0.14 | 0.46 | 14.17 |
| 1992 | 0.37 | 0.53 | 1.18 | 0.01 | 0.23 | 0.42 | 0.61 | 3.94 | 1.75 | 1.51 | 0.95 | 1.29 | 12.79 |
| 1993 | 1.14 | 1.14 | 1.68 | 0.18 | 0.52 | 0.39 | 0.92 | 3.15 | 2.57 | 2.78 | 2.83 | 1.88 | 19.18 |
| 1994 | 0.50 | 0.76 | 1.21 | 0.29 | 0.04 | 0.14 | 2.28 | 7.79 | 1.90 | 1.50 | 1.94 | 1.87 | 20.22 |
| 1995 | 1.52 | 1.47 | 0.49 | 1.67 | 0.84 | 0.90 | 1.18 | 1.86 | 1.59 | 2.97 | 0.39 | 0.58 | 15.46 |
| 1996 | 0.89 | 2.01 | 0.40 | 0.07 | 2.32 | 0.75 | 1.89 | 3.46 | .90 | 1.68 | 1.26 | .64 | 16.27 |
| 1997 | 1.14 | 0.75 | 0.14 | 0.85 | 1.17 | 0.06 | 2.25 | 4.46 | 2.09 | 0.98 | 3.33 | 1.16 | 18.38 |
| 1998 | 1.04 | 0.19 | 1.78 | 1.64 | 2.78 | 2.34 | 0.90 | 8.58 | 3.48 | 1.92 | 1.21 | 1.04 | 26.90 |
| 1999 | 1.41 | 0.75 | 0.12 | 0.97 | 0.12 | 0.63 | 4.78 | 1.85 | 3.42 | 0.65 | 0.32 | 0.21 | 15.23 |
| 2000 | 2.43 | 1.17 | 0.21 | 0.11 | 0.56 | 0.19 | 3.39 | 4.36 | 3.06 | 1.40 | 2.14 | 1.44 | 20.46 |
| 2001 | 1.36 | 1.33 | 0.12 | 1.71 | 0.45 | 1.17 | 2.15 | 3.16 | 0.73 | 0.65 | 0.03 | 0.57 | 13.43 |
| 2002 | 1.42 | 1.36 | 0.49 | 1.50 | 0.54 | 0.30 | 1.52 | 0.62 | 2.97 | 1.45 | 0.82 | 0.98 | 13.97 |
| 2003 | 0.45 | 0.95 | 0.55 | 1.10 | 0.42 | 1.52 | 2.04 | 3.98 | 0.88 | 1.51 | 3.01 | 1.21 | 17.62 |
| 2004 | 0.26 | 0.36 | 0.56 | 0.31 | 2.95 | 1.26 | 1.21 | 4.56 | 0.62 | 2.75 | 1.37 | 1.45 | 17.66 |
| 2005 | 0.26 | 1.24 | 0.61 | 0.32 | 1.06 | 0.86 | 1.63 | 2.92 | 4.85 | 1.61 | 0.45 | 0.72 | 16.53 |
| 2006 | 0.18 | 1.54 | 0.78 | 0.39 | 0.44 | 2.08 | 2.43 | 1.48 | 3.57 | 3.24 | 1.09 | 0.27 | 17.49 |
| 2007 | 1.69 | 0.08 | 0.18 | 0.40 | 0.08 | 1.62 | 0.74 | 2.00 | 3.84 | 0.91 | 1.33 | 1.43 | 14.30 |
| 2008 | 1.87 | 0.37 | 0.83 | 0.95 | 0.59 | 1.13 | 3.09 | 0.56 | 0.06 | 0.11 | 0.37 | 1.00 | 10.93 |
| 2009 | 0.79 | 1.79 | 1.26 | 1.20 | 0.39 | 1.66 | 1.32 | 2.41 | 0.97 | 0.91 | 0.97 | 0.94 | 14.61 |
| POR= 83 YRS | 0.95 | 0.76 | 0.69 | 0.66 | 0.71 | 1.15 | 2.35 | 3.23 | 2.52 | 1.38 | 1.05 | 0.92 | 16.37 |

WBAN : 26617

## AVERAGE TEMPERATURE (°F) 2009 NOME (PAOM)

| YEAR | JAN | FEB | MAR | APR | MAY | JUN | JUL | AUG | SEP | OCT | NOV | DEC | ANNUAL |
|------|-----|-----|-----|-----|-----|-----|-----|-----|-----|-----|-----|-----|--------|
| 1980 | 0.9 | 13.1 | 17.9 | 23.8 | 43.2 | 43.8 | 53.8 | 49.7 | 42.6 | 31.8 | 19.9 | 1.4 | 28.5 |
| 1981 | 19.5 | 8.1 | 22.2 | 24.3 | 42.7 | 50.2 | 49.7 | 50.9 | 41.1 | 28.5 | 16.4 | 7.5 | 30.1 |
| 1982 | 11.0 | 11.7 | 9.9 | 12.4 | 33.6 | 49.2 | 55.4 | 53.0 | 45.1 | 25.1 | 19.8 | 13.8 | 28.3 |
| 1983 | 2.7 | 11.1 | 16.9 | 25.0 | 43.8 | 49.2 | 56.2 | 47.6 | 40.3 | 26.7 | 22.0 | 20.9 | 30.2 |
| 1984 | 2.0 | -14.2 | 18.3 | 12.4 | 31.1 | 49.7 | 51.1 | 44.9 | 45.0 | 29.5 | 16.0 | 12.0 | 24.8 |
| 1985 | 24.4 | -0.6 | 5.8 | 1.3 | 31.8 | 41.1 | 51.5 | 50.9 | 40.8 | 25.8 | 20.7 | 24.3 | 26.5 |
| 1986 | 5.2 | 14.5 | 6.3 | 12.1 | 33.6 | 51.9 | 52.6 | 48.6 | 41.9 | 28.0 | 21.0 | 19.4 | 27.9 |
| 1987 | 6.1 | 10.5 | 16.8 | 18.5 | 36.9 | 49.1 | 53.0 | 52.1 | 38.8 | 33.2 | 11.8 | 4.6 | 27.6 |
| 1988 | 13.6 | 7.6 | 6.5 | 23.0 | 43.2 | 47.9 | 53.0 | 51.4 |  | 25.6 | 7.9 | 8.4 |  |
| 1989 | -15.2 | 22.5 | 13.2 | 24.7 | 31.4 | 46.8 | 49.8 | 50.6 | 46.3 | 30.1 | 9.7 | 12.5 | 26.9 |
| 1990 | 2.4 | -17.3 | 10.8 | 26.0 | 40.0 | 47.2 | 56.0 | 54.4 | 40.3 | 30.4 | 14.6 | 3.8 | 25.7 |
| 1991 | 6.9 | 5.9 | 8.6 | 25.0 | 38.8 | 51.1 | 55.1 | 51.6 | 48.0 | 33.3 | 17.3 | 0.1 | 28.5 |
| 1992 | 8.3 | -0.1 | 5.2 | 21.7 | 27.5 | 47.7 | 52.0 | 48.3 | 33.7 | 23.7 | 12.8 | 4.1 | 23.7 |
| 1993 | -4.6 | 14.8 | 13.2 | 27.6 | 40.2 | 51.8 | 56.2 | 49.9 | 38.9 | 32.5 | 21.6 | 15.3 | 29.8 |
| 1994 | 5.0 | 12.3 | 3.3 | 19.5 | 42.3 | 47.8 | 50.6 | 48.6 | 42.4 | 23.8 | 8.5 | 3.7 | 25.7 |
| 1995 | 8.0 | 11.1 | 4.9 | 25.7 | 41.6 | 43.0 | 49.7 | 50.0 | 47.7 | 29.3 | 15.7 | 15.1 | 28.5 |
| 1996 | 4.5 | 3.9 | 17.3 | 21.4 | 38.5 | 44.1 | 52.6 | 47.8 | 39.7 | 21.8 | 18.9 | 8.8 | 26.6 |
| 1997 | 5.2 | 13.4 | 4.7 | 26.7 | 40.2 | 48.3 | 51.1 | 51.8 | 48.4 | 25.6 | 23.2 | .6 | 28.3 |
| 1998 | 2.9 | 8.6 | 21.2 | 26.0 | 35.3 | 47.4 | 53.5 | 46.0 | 41.8 | 31.6 | 20.9 | 5.6 | 28.4 |
| 1999 | -1.9 | -1.9 | 4.2 | 17.1 | 31.3 | 47.6 | 48.8 | 49.2 | 43.8 | 25.1 | 11.6 | -9.4 | 22.1 |
| 2000 | .8 | 21.4 | 16.2 | 21.3 | 31.9 | 48.9 | 50.1 | 47.7 | 39.4 | 29.5 | 23.3 | 23.9 | 29.5 |
| 2001 | 16.5 | 18.5 | 7.6 | 21.9 | 27.4 | 44.7 | 48.3 | 48.1 | 42.3 | 22.9 | 13.7 | -3.5 | 25.7 |
| 2002 | 1.7 | 11.0 | 20.0 | 20.0 | 38.9 | 48.8 | 52.0 | 47.8 | 42.5 | 34.8 | 26.2 | 16.6 | 30.0 |
| 2003 | 13.1 | 11.3 | 8.0 | 26.2 | 35.7 | 52.7 | 49.0 | 50.3 | 41.9 | 32.3 | 21.1 | 6.2 | 29.0 |
| 2004 | 5.0 | 7.9 | 9.7 | 29.0 | 40.5 | 52.2 | 55.7 | 55.7 | 40.0 | 34.7 | 23.2 | 11.3 | 30.4 |
| 2005 | 11.6 | 3.3 | 15.1 | 15.4 | 40.7 | 50.1 | 52.5 | 52.9 | 45.2 | 29.2 | 10.6 | 11.5 | 28.2 |
| 2006 | -7.1 | 11.6 | 7.3 | 12.4 | 35.7 | 42.5 | 49.9 | 48.3 | 45.2 | 36.1 | 20.7 | 5.3 | 25.7 |
| 2007 | 6.3 | 12.3 | 0.2 | 26.9 | 38.5 | 47.8 | 55.5 | 54.5 | 47.8 | 28.8 | 21.6 | 11.4 | 29.3 |
| 2008 | -0.6 | -3.6 | 9.0 | 15.1 | 36.4 | 44.5 | 49.6 | 49.7 | 45.2 | 22.0 | 9.1 | 12.5 | 24.1 |
| 2009 | 0.9 | 1.7 | 5.3 | 16.6 | 34.7 | 47.0 | 54.5 | 49.5 | 42.2 | 30.5 | 8.5 | 15.0 | 25.5 |
| POR= 85 YRS | 5.7 | 5.6 | 8.7 | 18.5 | 35.6 | 45.8 | 50.9 | 49.6 | 41.9 | 28.4 | 16.2 | 7.1 | 26.2 |

## HEATING DEGREE DAYS (base 65°F) 2009 NOME (PAOM)

| YEAR | JUL | AUG | SEP | OCT | NOV | DEC | JAN | FEB | MAR | APR | MAY | JUN | TOTAL |
|---|---|---|---|---|---|---|---|---|---|---|---|---|---|
| 1980-81 | 341 | 468 | 669 | 1022 | 1346 | 1969 | 1404 | 1595 | 1317 | 1215 | 682 | 437 | 12465 |
| 1981-82 | 468 | 428 | 709 | 1123 | 1454 | 1779 | 1669 | 1494 | 1705 | 1575 | 969 | 476 | 13849 |
| 1982-83 | 288 | 367 | 590 | 1233 | 1350 | 1583 | 1930 | 1507 | 1490 | 1194 | 648 | 465 | 12645 |
| 1983-84 | 267 | 533 | 734 | 1179 | 1287 | 1362 | 1956 | 2305 | 1445 | 1575 | 1043 | 455 | 14141 |
| 1984-85 | 422 | 617 | 593 | 1094 | 1463 | 1640 | 1251 | 1840 | 1834 | 1910 | 1023 | 711 | 14398 |
| 1985-86 | 424 | 433 | 717 | 1207 | 1325 | 1254 | 1855 | 1412 | 1816 | 1583 | 963 | 392 | 13381 |
| 1986-87 | 379 | 503 | 687 | 1142 | 1310 | 1410 | 1823 | 1521 | 1490 | 1389 | 863 | 467 | 12984 |
| 1987-88 | 364 | 392 | 780 | 978 | 1592 | 1876 | 1588 | 1665 | 1813 | 1256 | 670 | 506 | 13480 |
| 1988-89 | 363 | 415 | | 1212 | 1711 | 1751 | 2488 | 1183 | 1601 | 1206 | 1030 | 539 | |
| 1989-90 | 466 | 443 | 558 | 1074 | 1654 | 1623 | 1940 | 2308 | 1679 | 1164 | 768 | 527 | 14204 |
| 1990-91 | 284 | 320 | 734 | 1066 | 1508 | 1897 | 1797 | 1654 | 1743 | 1198 | 804 | 410 | 13415 |
| 1991-92 | 301 | 407 | 504 | 975 | 1423 | 2014 | 1753 | 1887 | 1850 | 1295 | 1154 | 513 | 14076 |
| 1992-93 | 398 | 510 | 934 | 1272 | 1560 | 1886 | 2159 | 1404 | 1599 | 1115 | 764 | 390 | 13991 |
| 1993-94 | 284 | 461 | 774 | 999 | 1298 | 1536 | 1860 | 1473 | 1915 | 1358 | 695 | 511 | 13164 |
| 1994-95 | 443 | 503 | 670 | 1270 | 1690 | 1901 | 1764 | 1507 | 1863 | 1173 | 718 | 654 | 14156 |
| 1995-96 | 467 | 459 | 512 | 1101 | 1470 | 1543 | 1876 | 1772 | 1473 | 1304 | 815 | 619 | 13411 |
| 1996-97 | 379 | 526 | 751 | 1333 | 1375 | 1735 | 1850 | 1439 | 1865 | 1139 | 760 | 496 | 13648 |
| 1997-98 | 426 | 401 | 493 | 1215 | 1248 | 1990 | 1917 | 1571 | 1351 | 1160 | 916 | 522 | 13210 |
| 1998-99 | 353 | 580 | 688 | 1025 | 1317 | 1834 | 2069 | 1865 | 1878 | 1427 | 1035 | 517 | 14588 |
| 1999-00 | 494 | 481 | 626 | 1230 | 1598 | 2301 | 1985 | 1259 | 1504 | 1304 | 1018 | 474 | 14274 |
| 2000-01 | 455 | 533 | 761 | 1095 | 1244 | 1267 | 1494 | 1295 | 1770 | 1285 | 1157 | 602 | 12958 |
| 2001-02 | 508 | 517 | 674 | 1295 | 1532 | 2115 | 1957 | 1506 | 1389 | 1343 | 805 | 479 | 14120 |
| 2002-03 | 396 | 524 | 668 | 931 | 1157 | 1493 | 1601 | 1499 | 1759 | 1157 | 901 | 361 | 12447 |
| 2003-04 | 490 | 450 | 688 | 1003 | 1313 | 1814 | 1850 | 1649 | 1706 | 1072 | 751 | 384 | 13170 |
| 2004-05 | 281 | 284 | 744 | 932 | 1247 | 1659 | 1647 | 1725 | 1539 | 1480 | 744 | 441 | 12723 |
| 2005-06 | 384 | 380 | 588 | 1101 | 1625 | 1649 | 2230 | 1488 | 1782 | 1570 | 900 | 668 | 14365 |
| 2006-07 | 462 | 508 | 585 | 890 | 1321 | 1845 | 1814 | 1469 | 2000 | 1134 | 814 | 510 | 13352 |
| 2007-08 | 294 | 320 | 510 | 1112 | 1293 | 1653 | 2031 | 1986 | 1731 | 1490 | 877 | 609 | 13906 |
| 2008-09 | 478 | 468 | 585 | 1326 | 1671 | 1617 | 1979 | 1768 | 1840 | 1449 | 932 | 533 | 14646 |
| 2009- | 320 | 472 | 678 | 1066 | 1690 | 1543 | | | | | | | |

WBAN : 26617

## COOLING DEGREE DAYS (base 65°F) 2009 NOME (PAOM)

| YEAR | JAN | FEB | MAR | APR | MAY | JUN | JUL | AUG | SEP | OCT | NOV | DEC | TOTAL |
|---|---|---|---|---|---|---|---|---|---|---|---|---|---|
| 1980 | 0 | 0 | 0 | 0 | 0 | 0 | 0 | 0 | 0 | 0 | 0 | 0 | 0 |
| 1981 | 0 | 0 | 0 | 0 | 0 | 0 | 0 | 0 | 0 | 0 | 0 | 0 | 0 |
| 1982 | 0 | 0 | 0 | 0 | 0 | 7 | 0 | 0 | 0 | 0 | 0 | 0 | 7 |
| 1983 | 0 | 0 | 0 | 0 | 0 | 0 | 0 | 0 | 0 | 0 | 0 | 0 | 0 |
| 1984 | 0 | 0 | 0 | 0 | 0 | 2 | 0 | 0 | 0 | 0 | 0 | 0 | 2 |
| 1985 | 0 | 0 | 0 | 0 | 0 | 0 | 12 | 0 | 0 | 0 | 0 | 0 | 12 |
| 1986 | 0 | 0 | 0 | 0 | 0 | 4 | 2 | 0 | 0 | 0 | 0 | 0 | 6 |
| 1987 | 0 | 0 | 0 | 0 | 0 | 0 | 0 | 0 | 0 | 0 | 0 | 0 | 0 |
| 1988 | 0 | 0 | 0 | 0 | 0 | 0 | 0 | 0 | | 0 | 0 | 0 | |
| 1989 | 0 | 0 | 0 | 0 | 0 | 0 | 0 | 0 | 0 | 0 | 0 | 0 | 0 |
| 1990 | 0 | 0 | 0 | 0 | 0 | 0 | 7 | 0 | 0 | 0 | 0 | 0 | 7 |
| 1991 | 0 | 0 | 0 | 0 | 0 | 0 | 0 | 0 | 0 | 0 | 0 | 0 | 0 |
| 1992 | 0 | 0 | 0 | 0 | 0 | 0 | 0 | 0 | 0 | 0 | 0 | 0 | 0 |
| 1993 | 0 | 0 | 0 | 0 | 0 | 0 | 19 | 0 | 0 | 0 | 0 | 0 | 19 |
| 1994 | 0 | 0 | 0 | 0 | 0 | 0 | 0 | 0 | 0 | 0 | 0 | 0 | 0 |
| 1995 | 0 | 0 | 0 | 0 | 0 | 0 | 0 | 0 | 0 | 0 | 0 | 0 | 0 |
| 1996 | 0 | 0 | 0 | 0 | 0 | 0 | 2 | 0 | 0 | 0 | 0 | 0 | 2 |
| 1997 | 0 | 0 | 0 | 0 | 0 | 0 | 0 | 0 | 0 | 0 | 0 | 0 | 0 |
| 1998 | 0 | 0 | 0 | 0 | 0 | 0 | 3 | 0 | 0 | 0 | 0 | 0 | 3 |
| 1999 | 0 | 0 | 0 | 0 | 0 | 0 | 0 | 0 | 0 | 0 | 0 | 0 | 0 |
| 2000 | 0 | 0 | 0 | 0 | 0 | 0 | 0 | 0 | 0 | 0 | 0 | 0 | 0 |
| 2001 | 0 | 0 | 0 | 0 | 0 | 0 | 0 | 0 | 0 | 0 | 0 | 0 | 0 |
| 2002 | 0 | 0 | 0 | 0 | 0 | 0 | 1 | 0 | 0 | 0 | 0 | 0 | 1 |
| 2003 | 0 | 0 | 0 | 0 | 0 | 0 | 0 | 0 | 0 | 0 | 0 | 0 | 0 |
| 2004 | 0 | 0 | 0 | 0 | 0 | 7 | 0 | 3 | 0 | 0 | 0 | 0 | 10 |
| 2005 | 0 | 0 | 0 | 0 | 0 | 0 | 0 | 10 | 0 | 0 | 0 | 0 | 10 |
| 2006 | 0 | 0 | 0 | 0 | 0 | 0 | 0 | 0 | 0 | 0 | 0 | 0 | 0 |
| 2007 | 0 | 0 | 0 | 0 | 0 | 0 | 5 | 0 | 0 | 0 | 0 | 0 | 5 |
| 2008 | 0 | 0 | 0 | 0 | 0 | 0 | 5 | 0 | 0 | 0 | 0 | 0 | 5 |
| 2009 | 0 | 0 | 0 | 0 | 0 | 0 | 0 | 0 | 0 | 0 | 0 | 0 | 0 |

## SNOWFALL (inches) 2009 NOME (PAOM)

| YEAR | JUL | AUG | SEP | OCT | NOV | DEC | JAN | FEB | MAR | APR | MAY | JUN | TOTAL |
|------|-----|-----|-----|-----|-----|-----|-----|-----|-----|-----|-----|-----|-------|
| 1980-81 | 0.0 | T | T | 5.2 | 9.6 | 4.9 | 14.7 | 11.6 | 5.9 | 10.4 | 4.6 | 0.0 | 66.9 |
| 1981-82 | 0.0 | 0.0 | 1.4 | 4.0 | 7.8 | 14.8 | 17.3 | 2.6 | 8.0 | 6.8 | 2.7 | 2.5 | 67.9 |
| 1982-83 | 0.0 | 0.0 | T | 8.2 | 23.6 | 13.0 | 2.4 | 1.4 | 1.7 | 10.2 | 1.5 | 0.0 | 62.0 |
| 1983-84 | 0.0 | 0.0 | 0.1 | 1.4 | 14.8 | 8.8 | 6.6 | 0.1 | 1.1 | 9.8 | 4.8 | 0.0 | 47.5 |
| 1984-85 | 0.0 | T | 0.0 | 2.5 | 5.5 | 13.6 | 4.5 | 5.9 | 11.2 | 0.7 | 8.5 | T | 52.4 |
| 1985-86 | 0.0 | 0.0 | T | 1.2 | 10.9 | 9.1 | 4.4 | 6.1 | 0.3 | 4.2 | 1.8 | 0.0 | 38.0 |
| 1986-87 | 0.0 | 0.0 | 1.8 | 2.3 | 4.6 | 10.6 | 9.7 | 5.7 | 4.9 | 1.5 | 0.9 | T | 42.0 |
| 1987-88 | 0.0 | 0.0 | 0.5 | 6.2 | 6.8 | 11.0 | 4.8 | 14.2 | 8.6 | 2.6 | 1.2 | 0.0 | 55.9 |
| 1988-89 | 0.0 | 0.0 | | 1.4 | 5.7 | 15.0 | 10.6 | 23.2 | 5.8 | 16.4 | 6.4 | T | |
| 1989-90 | 0.0 | 0.0 | T | 4.8 | 22.2 | 8.4 | 11.1 | 7.6 | 8.2 | 8.8 | 0.8 | 0.0 | 71.9 |
| 1990-91 | 0.0 | 0.0 | 0.1 | 4.9 | 17.1 | 14.2 | 11.3 | 5.5 | 7.7 | 8.6 | T | T | 69.4 |
| 1991-92 | 0.0 | 0.0 | 0.0 | 1.8 | 1.4 | 6.8 | 5.8 | 7.4 | 19.2 | 0.1 | 3.5 | 0.0 | 46.0 |
| 1992-93 | 0.0 | 0.0 | 3.5 | 11.0 | 10.6 | 14.9 | 12.2 | 11.5 | 19.6 | 0.8 | 1.5 | 0.0 | 85.6 |
| 1993-94 | 0.0 | 0.0 | 2.9 | 3.2 | 21.5 | 28.7 | 8.2 | 4.0 | 12.4 | 3.6 | T | T | 84.5 |
| 1994-95 | 0.0 | 0.0 | 0.5 | 7.1 | 30.9 | 29.7 | 21.5 | 19.7 | 8.8 | 11.2 | T | T | 129.4 |
| 1995-96 | 0.0 | 0.0 | 0.0 | 8.6 | 5.2 | 8.0 | 12.4 | 33.0 | 5.9 | 1.8 | 4.3 | T | 79.2 |
| 1996-97 | 0.0 | 0.0 | 0.1 | 20.1 | 10.5 | 11.5 | 17.9 | 10.9 | 1.4 | 11.9 | 2.5 | 0.0 | 86.8 |
| 1997-98 | 0.0 | 0.0 | 0.0 | 4.1 | 28.0 | 20.2 | 14.1 | 3.0 | 19.2 | 15.0 | 5.9 | 0.0 | 109.5 |
| 1998-99 | 0.0 | 0.0 | 0.0 | 6.4 | 10.4 | 13.8 | 20.9 | 16.6 | 2.4 | 12.9 | 0.3 | 0.3 | 84.0 |
| 1999-00 | 0.0 | 0.0 | 0.8 | 0.7 | 6.9 | 5.5 | 33.2 | 17.1 | 3.2 | 1.1 | 3.3 | 0.0 | 71.8 |
| 2000-01 | 0.0 | 0.0 | T | 11.4 | 15.9 | 19.0 | 18.3 | 18.2 | 2.4 | 19.7 | 5.2 | 1.0 | 111.1 |
| 2001-02 | 0.0 | 0.0 | T | T | 0.9 | 8.0 | 24.6 | 24.1 | 5.2 | 18.3 | 3.6 | 0.0 | 84.7 |
| 2002-03 | 0.0 | 0.0 | 0.0 | 1.1 | 4.8 | 5.7 | 8.8 | 15.8 | 7.9 | 7.4 | 0.2 | 0.0 | 51.7 |
| 2003-04 | 0.0 | 0.0 | 0.0 | 0.5 | 17.2 | 22.2 | 3.4 | 5.6 | 8.8 | 1.5 | 0.7 | 0.0 | 59.9 |
| 2004-05 | 0.0 | 0.0 | 2.8 | 1.9 | 16.6 | 26.3 | 4.0 | 29.3 | 8.5 | 2.1 | 0.4 | 0.0 | 91.9 |
| 2005-06 | 0.0 | 0.0 | 0.0 | 4.9 | 8.4 | 15.8 | 6.1 | 23.0 | 23.6 | 6.2 | 2.2 | 1.5 | 91.7 |
| 2006-07 | 0.0 | 0.0 | 1.1 | 1.4 | 14.8 | 9.0 | 39.8 | 3.9 | 3.3 | 4.3 | T | 0.0 | 77.6 |
| 2007-08 | 0.0 | 0.0 | 0.0 | 5.6 | 8.7 | 24.7 | 34.0 | 9.9 | 10.0 | 15.0 | 1.8 | 2.8 | 112.5 |
| 2008-09 | 0.0 | 0.0 | 0.0 | 2.9 | 9.1 | 10.3 | 17.4 | 35.1 | 29.0 | 9.5 | 0.1 | 0.0 | 113.4 |
| 2009- | 0.0 | 0.0 | 0.5 | 2.5 | 15.3 | 11.7 | | | | | | | |
| POR=<br>68 YRS | 0.0 | T | 0.5 | 4.2 | 10.7 | 10.8 | 11.2 | 9.1 | 8.5 | 6.5 | 1.8 | 0.3 | 63.6 |

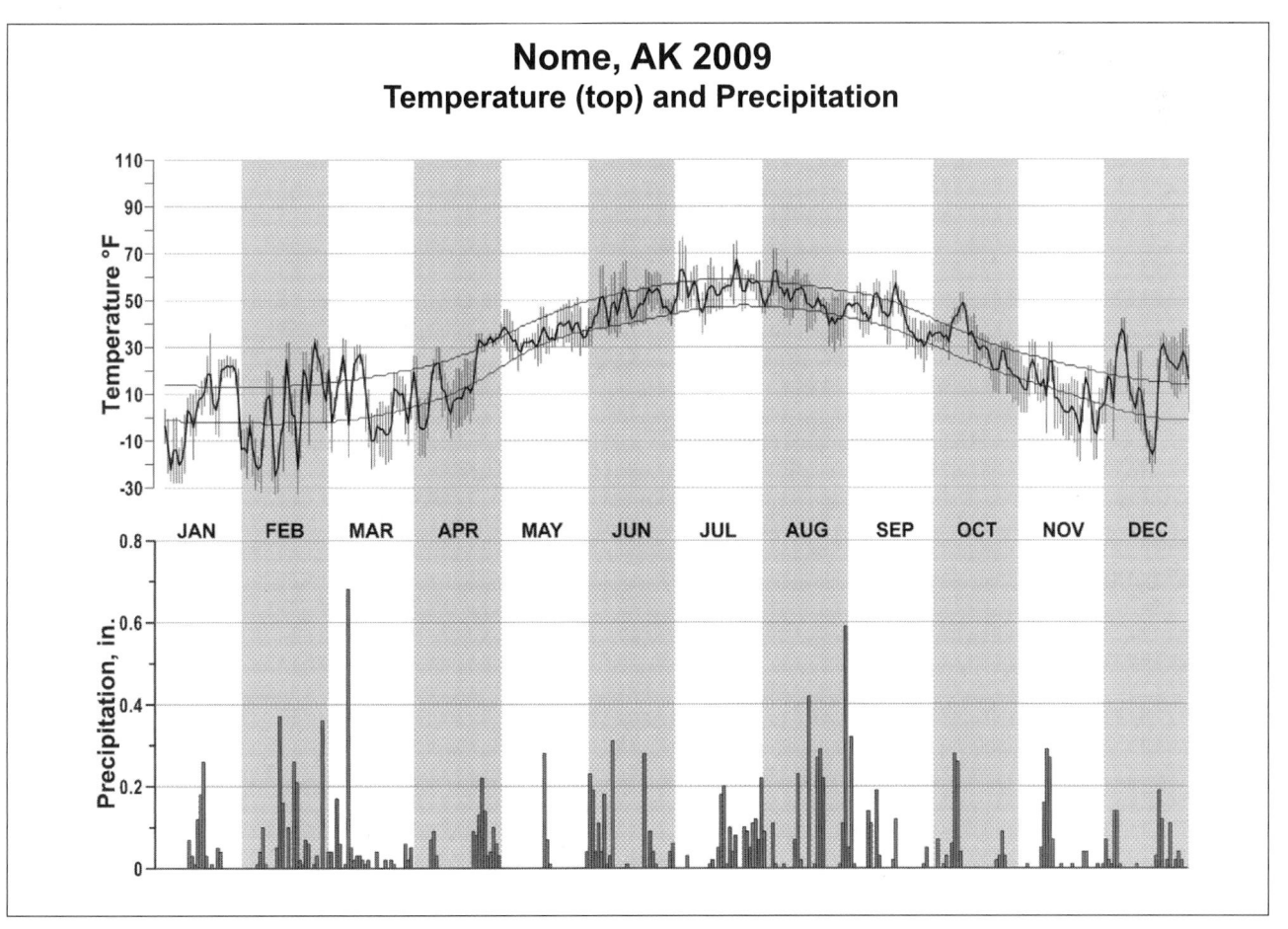

# Nome, AK 2009
## Temperature (top) and Precipitation

# 2009
# FLAGSTAFF
# ARIZONA (KFLG)

Flagstaff, elevation 7,000 feet, is situated on a volcanic plateau at the base of the highest mountains in Arizona. The climate may be classified as vigorous with cold winters, mild, pleasantly cool summers, moderate humidity, and considerable diurnal temperature change. Only limited farming exists due to the short growing season. The stormy months are January, February, March, July, and August.

Based on the 1951-1980 period, the average first occurrence of 32 degrees Fahrenheit in the fall is September 21 and the average last occurrence in the spring is June 13.

Temperatures in Flagstaff are characteristic of high altitude climates. The average daily range of temperature is relatively high, especially in the winter months, October to March, as a result of extensive snow cover and clear skies. Winter minimum temperatures frequently reach zero or below and temperatures of -25 degrees or less have occurred. Summer maximum temperatures are often above 80 degrees and occasionally, temperatures have exceeded 95 degrees.

The Flagstaff area is semi-arid. Several months have recorded little or no precipitation. Over 90 consecutive days without measurable precipitation have occurred. Annual precipitation ranges from less than 10 inches to more than 35 inches. Winter snowfalls can be heavy, exceeding 100 inches during one month and over 200 inches during the winter season. However, accumulations are quite variable from year to year. Some winter months may experience little or no snow and the winter season has produced total snow accumulations of less than 12 inches.

# NORMALS, MEANS, AND EXTREMES
## FLAGSTAFF (KFLG)

| LATITUDE: 35 ° 8 'N | LONGITUDE: -111° 40'W | | ELEVATION (FT): GRND: 7003  BARO: 7003 | | | | TIME ZONE: MOUNTAIN  (UTC -7) | | | WBAN: 03103 | | |

| | ELEMENT | POR | JAN | FEB | MAR | APR | MAY | JUN | JUL | AUG | SEP | OCT | NOV | DEC | YEAR |
|---|---|---|---|---|---|---|---|---|---|---|---|---|---|---|---|
| **TEMPERATURE °F** | NORMAL DAILY MAXIMUM | 30 | 42.9 | 45.6 | 50.3 | 58.4 | 67.6 | 78.7 | 82.2 | 79.7 | 73.8 | 63.1 | 50.8 | 43.7 | 61.4 |
| | MEAN DAILY MAXIMUM | 83 | 42.1 | 43.8 | 50.1 | 57.3 | 68.0 | 77.1 | 81.6 | 79.1 | 72.8 | 63.4 | 51.1 | 43.5 | 60.8 |
| | HIGHEST DAILY MAXIMUM | 60 | 66 | 71 | 73 | 80 | 89 | 96 | 97 | 92 | 90 | 85 | 74 | 68 | 97 |
| | YEAR OF OCCURRENCE | | 1971 | 1986 | 2007 | 1992 | 2002 | 1970 | 1973 | 1978 | 1950 | 1980 | 1977 | 1950 | JUL 1973 |
| | MEAN OF EXTREME MAXS. | 85 | 57.2 | 59.3 | 64.3 | 71.7 | 80.0 | 88.3 | 90.3 | 87.4 | 83.1 | 75.7 | 66.3 | 58.7 | 73.5 |
| | NORMAL DAILY MINIMUM | 30 | 16.5 | 18.8 | 22.8 | 27.3 | 34.0 | 41.4 | 49.9 | 49.1 | 41.7 | 31.1 | 22.1 | 16.6 | 30.9 |
| | MEAN DAILY MINIMUM | 83 | 15.3 | 17.6 | 22.4 | 27.7 | 34.4 | 41.2 | 50.9 | 49.7 | 41.5 | 31.4 | 22.3 | 16.7 | 30.9 |
| | LOWEST DAILY MINIMUM | 60 | -22 | -23 | -16 | -2 | 0 | 22 | 32 | 24 | 23 | -2 | -13 | -23 | -23 |
| | YEAR OF OCCURRENCE | | 1971 | 1985 | 1966 | 1975 | 1996 | 1955 | 1955 | 1968 | 1971 | 1971 | 1958 | 1990 | DEC 1990 |
| | MEAN OF EXTREME MINS. | 85 | -4.4 | -0.6 | 4.8 | 14.6 | 22.6 | 29.4 | 41.1 | 40.4 | 30.0 | 18.5 | 6.1 | -2.4 | 16.7 |
| | NORMAL DRY BULB | 30 | 29.7 | 32.2 | 36.6 | 42.9 | 50.8 | 60.1 | 66.1 | 64.4 | 57.8 | 47.1 | 36.5 | 30.2 | 46.2 |
| | MEAN DRY BULB | 83 | 28.7 | 30.7 | 36.3 | 42.7 | 51.2 | 59.2 | 66.3 | 64.4 | 57.2 | 47.4 | 36.8 | 30.2 | 45.9 |
| | MEAN WET BULB | 26 | 23.7 | 26.3 | 28.8 | 32.5 | 38.7 | 44.0 | 52.1 | 52.8 | 46.0 | 36.3 | 28.1 | 23.4 | 36.1 |
| | MEAN DEW POINT | 26 | 18.3 | 19.7 | 22.2 | 24.2 | 28.6 | 31.3 | 43.9 | 47.7 | 39.6 | 29.6 | 21.9 | 17.1 | 28.7 |
| | NORMAL NO. DAYS WITH: | | | | | | | | | | | | | | |
| | MAXIMUM >= 90 | 30 | 0.0 | 0.0 | 0.0 | 0.0 | 0.0 | 1.7 | 2.1 | 0.6 | 0.0 | 0.0 | 0.0 | 0.0 | 4.4 |
| | MAXIMUM <= 32 | 30 | 4.0 | 2.5 | 1.0 | 0.4 | 0.0 | 0.0 | 0.0 | 0.0 | 0.0 | 0.1 | 1.1 | 4.1 | 13.2 |
| | MINIMUM <= 32 | 30 | 30.5 | 27.7 | 29.5 | 24.6 | 13.0 | 2.7 | 0.0 | 0.0 | 2.3 | 18.2 | 28.0 | 30.0 | 206.5 |
| | MINIMUM <= 0 | 30 | 2.2 | 1.0 | 0.3 | * | 0.0 | 0.0 | 0.0 | 0.0 | 0.0 | * | 0.3 | 1.8 | 5.6 |
| **H/C** | NORMAL HEATING DEG. DAYS | 30 | 1099 | 930 | 880 | 668 | 446 | 174 | 33 | 56 | 224 | 554 | 850 | 1085 | 6999 |
| | NORMAL COOLING DEG. DAYS | 30 | 0 | 0 | 0 | 0 | 0 | 23 | 64 | 36 | 3 | 0 | 0 | 0 | 126 |
| **RH** | NORMAL (PERCENT) | 30 | 64 | 64 | 60 | 48 | 44 | 35 | 49 | 60 | 56 | 54 | 58 | 63 | 55 |
| | HOUR 05 LST | 30 | 75 | 76 | 74 | 67 | 64 | 54 | 68 | 77 | 75 | 71 | 70 | 72 | 70 |
| | HOUR 11 LST | 30 | 53 | 50 | 45 | 35 | 30 | 23 | 34 | 40 | 39 | 39 | 44 | 49 | 40 |
| | HOUR 17 LST | 30 | 51 | 47 | 44 | 33 | 31 | 22 | 36 | 44 | 37 | 36 | 43 | 49 | 39 |
| | HOUR 23 LST | 30 | 71 | 70 | 67 | 56 | 50 | 40 | 57 | 69 | 67 | 64 | 65 | 68 | 62 |
| **S** | PERCENT POSSIBLE SUNSHINE | 15 | 77 | 73 | 76 | 82 | 88 | 86 | 75 | 76 | 81 | 79 | 75 | 73 | 78 |
| **W/O** | MEAN NO. DAYS WITH: | | | | | | | | | | | | | | |
| | HEAVY FOG(VISBY <= 1/4 MI) | 42 | 3.4 | 2.9 | 2.2 | 1.8 | 0.2 | 0.0 | 0.2 | 0.4 | 0.6 | 1.1 | 1.3 | 2.9 | 17.0 |
| | THUNDERSTORMS | 46 | 0.1 | 0.3 | 1.0 | 1.1 | 3.0 | 3.2 | 13.4 | 14.2 | 6.6 | 2.0 | 0.5 | 0.1 | 45.5 |
| **CLOUDINESS** | MEAN: | | | | | | | | | | | | | | |
| | SUNRISE-SUNSET (OKTAS) | | | | | | | | | | | | | | |
| | MIDNIGHT-MIDNIGHT (OKTAS) | | | | | | | | | | | | | | |
| | MEAN NO. DAYS WITH: | | | | | | | | | | | | | | |
| | CLEAR | | | | | | | | | | | | | | |
| | PARTLY CLOUDY | | | | | | | | | | | | | | |
| | CLOUDY | | | | | | | | | | | | | | |
| **PR** | MEAN STATION PRESSURE(IN) | 26 | 23.25 | 23.21 | 23.20 | 23.20 | 23.24 | 23.30 | 23.38 | 23.39 | 23.34 | 23.30 | 23.27 | 23.24 | 23.28 |
| | MEAN SEA-LEVEL PRES. (IN) | 26 | 30.13 | 30.06 | 29.99 | 29.93 | 29.91 | 29.92 | 29.99 | 30.02 | 30.00 | 30.03 | 30.09 | 30.13 | 30.02 |
| **WINDS** | MEAN SPEED (MPH) | 26 | 5.9 | 6.3 | 6.4 | 7.4 | 7.0 | 6.6 | 5.0 | 4.2 | 4.9 | 5.4 | 5.8 | 5.7 | 5.9 |
| | PREVAIL.DIR(TENS OF DEGS) | 14 | 22 | 22 | 22 | 22 | 22 | 22 | 23 | 22 | 22 | 22 | 22 | 22 | 22 |
| | MAXIMUM 2-MINUTE: | | | | | | | | | | | | | | |
| | SPEED (MPH) | 15 | 46 | 40 | 51 | 45 | 41 | 44 | 35 | 38 | 39 | 48 | 37 | 45 | 51 |
| | DIR. (TENS OF DEGS) | | 22 | 20 | 21 | 21 | 22 | 22 | 26 | 07 | 21 | 05 | 06 | 22 | 21 |
| | YEAR OF OCCURRENCE | | 2005 | 2006 | 2009 | 2009 | 2007 | 2007 | 2002 | 2004 | 2004 | 2007 | 2000 | 2008 | MAR 2009 |
| | MAXIMUM 3-SECOND | | | | | | | | | | | | | | |
| | SPEED (MPH) | 15 | 61 | 52 | 64 | 59 | 59 | 64 | 51 | 64 | 56 | 60 | 52 | 60 | 64 |
| | DIR. (TENS OF DEGS) | | 23 | 24 | 20 | 21 | 21 | 23 | 05 | 22 | 20 | 05 | 26 | 22 | 20 |
| | YEAR OF OCCURRENCE | | 2008 | 2008 | 2009 | 2007 | 2007 | 2007 | 2004 | 2005 | 2007 | 2007 | 2001 | 2008 | MAR 2009 |
| **PRECIPITATION** | NORMAL (IN) | 30 | 2.18 | 2.56 | 2.62 | 1.29 | 0.80 | 0.43 | 2.40 | 2.89 | 2.12 | 1.93 | 1.86 | 1.83 | 22.91 |
| | MAXIMUM MONTHLY (IN) | 60 | 9.55 | 10.05 | 6.75 | 5.62 | 4.14 | 2.92 | 6.62 | 8.06 | 6.75 | 9.86 | 6.64 | 7.30 | 10.05 |
| | YEAR OF OCCURRENCE | | 1993 | 1993 | 1970 | 1965 | 1992 | 1955 | 1986 | 1986 | 1983 | 1972 | 1985 | 1967 | FEB 1993 |
| | MINIMUM MONTHLY (IN) | 60 | 0.00 | T | T | T | 0.00 | 0.00 | T | 0.26 | 0.00 | T | 0.00 | T | 0.00 |
| | YEAR OF OCCURRENCE | | 1972 | 1967 | 1972 | 2008 | 2002 | 1971 | 1993 | 1962 | 1995 | 1952 | 1999 | 1958 | MAY 2002 |
| | MAXIMUM IN 24 HOURS (IN) | 60 | 2.71 | 4.48 | 2.96 | 1.79 | 1.11 | 2.79 | 2.55 | 3.04 | 3.43 | 2.73 | 3.69 | 3.75 | 4.48 |
| | YEAR OF OCCURRENCE | | 1993 | 1993 | 1970 | 1985 | 1965 | 1956 | 1964 | 1986 | 1965 | 1972 | 1978 | 2004 | FEB 1993 |
| | NORMAL NO. DAYS WITH: | | | | | | | | | | | | | | |
| | PRECIPITATION >= 0.01 | 30 | 8.0 | 8.1 | 9.4 | 5.8 | 5.4 | 2.7 | 11.3 | 12.8 | 7.7 | 5.7 | 5.1 | 6.6 | 88.6 |
| | PRECIPITATION >= 1.00 | 30 | 0.3 | 0.5 | 0.3 | 0.1 | 0.0 | 0.0 | 0.4 | 0.5 | 0.5 | 0.4 | 0.4 | 0.3 | 3.7 |
| **SNOWFALL** | NORMAL (IN) | 30 | 23.0 | 21.3 | 24.4 | 9.7 | 1.3 | 0.* | 0.0 | 0.0 | 0.* | 2.8 | 13.3 | 14.0 | 109.8 |
| | MAXIMUM MONTHLY (IN) | 52 | 63.4 | 45.5 | 79.4 | 58.3 | 8.7 | T | T | 0.3 | 2.0 | 24.7 | 40.7 | 86.0 | 86.0 |
| | YEAR OF OCCURRENCE | | 1980 | 1990 | 1991 | 1965 | 2008 | 2006 | 1992 | 1992 | 1965 | 1971 | 1985 | 1967 | DEC 1967 |
| | MAXIMUM IN 24 HOURS (IN) | 52 | 23.1 | 23.1 | 26.3 | 17.2 | 6.6 | T | T | 0.3 | 2.0 | 13.5 | 19.6 | 27.3 | 27.3 |
| | YEAR OF OCCURRENCE' | | 1980 | 1987 | 1970 | 1977 | 1965 | 1993 | 1992 | 1992 | 1965 | 1974 | 1991 | 1967 | DEC 1967 |
| | MAXIMUM SNOW DEPTH (IN) | 51 | 34 | 40 | 37 | 31 | 8 | 0 | 0 | 0 | T | 11 | 23 | 66 | 66 |
| | YEAR OF OCCURRENCE | | 1979 | 1979 | 1969 | 1973 | 1954 | | | | 1965 | 1974 | 1975 | 1967 | DEC 1967 |
| | NORMAL NO. DAYS WITH: | | | | | | | | | | | | | | |
| | SNOWFALL >= 1.0 | 30 | 4.8 | 4.7 | 5.7 | 2.3 | 0.5 | 0.0 | 0.0 | 0.0 | 0.0 | 0.6 | 2.7 | 4.0 | 25.3 |

## PRECIPITATION (inches) 2009 FLAGSTAFF (KFLG)

| YEAR | JAN | FEB | MAR | APR | MAY | JUN | JUL | AUG | SEP | OCT | NOV | DEC | ANNUAL |
|------|-----|-----|-----|-----|-----|-----|-----|-----|-----|-----|-----|-----|--------|
| 1980 | 6.52 | 7.81 | 4.16 | 1.21 | 1.79 | 0.25 | 2.49 | 2.19 | 0.65 | 1.08 | T | 1.15 | 29.30 |
| 1981 | 1.31 | 1.16 | 4.04 | 1.50 | 0.72 | 1.09 | 2.87 | 3.73 | 2.54 | 1.81 | 2.43 | 0.17 | 23.37 |
| 1982 | 4.62 | 2.55 | 5.69 | 0.25 | 0.86 | T | 1.89 | 2.32 | 3.17 | 0.71 | 5.36 | 3.67 | 31.09 |
| 1983 | 1.61 | 3.04 | 4.36 | 2.18 | 0.06 | 0.28 | 2.86 | 3.53 | 6.75 | 0.75 | 1.53 | 2.52 | 29.47 |
| 1984 | 0.36 | 0.13 | 0.89 | 0.73 | 0.21 | 0.13 | 4.20 | 3.86 | 1.75 | 1.43 | 1.40 | 5.00 | 20.09 |
| 1985 | 2.38 | 1.67 | 2.54 | 3.39 | 0.26 | 0.09 | 2.36 | 1.07 | 3.68 | 2.44 | 6.64 | 0.15 | 26.67 |
| 1986 | 0.31 | 1.76 | 2.60 | 1.23 | 0.72 | 1.16 | 6.62 | 8.06 | 4.80 | 2.02 | 1.68 | 1.43 | 32.39 |
| 1987 | 2.51 | 2.54 | 1.69 | 0.21 | 0.70 | 0.25 | 1.93 | 3.74 | 1.49 | 4.64 | 2.91 | 1.37 | 23.98 |
| 1988 | 1.64 | 2.30 | 0.14 | 3.83 | 0.14 | 1.86 | 3.48 | 4.77 | 0.15 | 1.23 | 1.14 | 1.00 | 21.68 |
| 1989 | 1.84 | 1.35 | 2.08 | 0.01 | 0.62 | 0.23 | 2.28 | 3.40 | 0.30 | 1.28 | T | 1.05 | 14.44 |
| 1990 | 1.54 | 3.20 | 2.17 | 2.32 | 0.73 | 0.24 | 4.32 | 1.71 | 6.18 | 0.49 | 1.09 | 1.68 | 25.67 |
| 1991 | 1.76 | 2.08 | 6.00 | T | 0.14 | 0.59 | 1.04 | 1.64 | 0.26 | 1.12 | 4.47 | 2.73 | 21.83 |
| 1992 | 2.03 | 3.69 | 4.40 | 0.78 | 4.14 | 0.32 | 2.67 | 5.80 | T | 3.64 | 0.46 | 6.78 | 34.71 |
| 1993 | 9.55 | 10.05 | 1.54 | 0.26 | 0.44 | 0.55 | T | 4.19 | 1.95 | 3.29 | 3.02 | 0.76 | 35.60 |
| 1994 | 0.38 | 2.47 | 3.03 | 2.48 | 1.01 | | 1.70 | 3.61 | 2.75 | 1.12 | 1.91 | 1.43 | |
| 1995 | 2.39 | 3.77 | 3.99 | 1.49 | 0.44 | 0.06 | 0.61 | 2.30 | 0.00 | 0.01 | 0.31 | 0.35 | 15.72 |
| 1996 | 0.19 | 1.36 | 0.54 | 0.07 | T | T | 1.79 | 0.92 | 3.73 | 0.93 | | 0.64 | |
| 1997 | 3.21 | 0.99 | 0.03 | 0.60 | 0.08 | 0.18 | 0.21 | 2.83 | 3.68 | 1.15 | 0.80 | 1.85 | 15.61 |
| 1998 | 1.30 | 2.15 | 3.74 | 1.52 | 1.31 | 0.00 | 4.72 | 2.82 | 4.45 | 3.11 | 1.76 | 0.42 | 27.30 |
| 1999 | 0.28 | 0.48 | 0.53 | 2.80 | 0.42 | 0.95 | 3.27 | 2.45 | 4.54 | T | 0.00 | T | 15.72 |
| 2000 | 0.62 | 1.61 | 3.12 | 0.19 | 0.12 | 1.11 | 0.29 | 2.83 | 0.36 | 3.85 | 1.07 | 0.21 | 15.38 |
| 2001 | 2.60 | 1.68 | 1.28 | 1.40 | 0.82 | 0.03 | 2.80 | 3.46 | 0.68 | 1.21 | 0.43 | 1.16 | 17.55 |
| 2002 | 0.02 | 0.07 | 0.62 | 0.51 | 0.00 | 0.00 | 2.60 | 1.00 | 4.01 | 1.88 | 1.48 | 0.69 | 12.88 |
| 2003 | 0.14 | 2.75 | 1.13 | 0.44 | 0.73 | 0.04 | 3.40 | 3.03 | 2.62 | 0.14 | 2.51 | 0.92 | 17.85 |
| 2004 | 0.76 | 1.06 | 0.74 | 1.81 | 0.00 | 0.02 | 1.47 | 4.71 | 1.76 | 3.51 | 3.10 | 4.67 | 23.61 |
| 2005 | 6.58 | 4.19 | 2.43 | 2.15 | 0.08 | 0.40 | 2.51 | 3.41 | 0.46 | 1.59 | 0.20 | 0.01 | 24.01 |
| 2006 | 0.23 | 0.09 | 2.16 | 0.99 | 0.08 | 0.65 | 4.07 | 2.83 | 1.24 | 2.55 | 0.06 | 0.61 | 15.56 |
| 2007 | 1.20 | 0.81 | 0.51 | 0.38 | 0.14 | 0.01 | 2.86 | 2.56 | 2.90 | 0.38 | 1.38 | 4.33 | 17.46 |
| 2008 | 3.95 | 2.56 | 0.04 | T | 1.17 | 0.02 | 2.35 | 2.40 | 0.69 | 0.21 | 1.29 | 4.17 | 18.85 |
| 2009 | 0.73 | 1.48 | 0.22 | 0.33 | 2.08 | 0.36 | 1.00 | 0.74 | 0.77 | 0.21 | 0.88 | 2.85 | 11.65 |
| POR= 83 YRS | 2.07 | 2.00 | 2.05 | 1.26 | 0.66 | 0.45 | 2.44 | 2.72 | 1.88 | 1.44 | 1.55 | 2.02 | 20.54 |

WBAN : 03103

## AVERAGE TEMPERATURE (°F) 2009 FLAGSTAFF (KFLG)

| YEAR | JAN | FEB | MAR | APR | MAY | JUN | JUL | AUG | SEP | OCT | NOV | DEC | ANNUAL |
|------|-----|-----|-----|-----|-----|-----|-----|-----|-----|-----|-----|-----|--------|
| 1980 | 30.7 | 32.8 | 32.3 | 41.9 | 45.6 | 60.9 | 69.0 | 66.3 | 59.7 | 49.0 | 41.4 | 39.9 | 47.5 |
| 1981 | 36.2 | 36.3 | 36.6 | 48.5 | 52.1 | 66.1 | 68.2 | 65.4 | 58.7 | 46.7 | 41.8 | 36.4 | 49.4 |
| 1982 | 28.3 | 30.5 | 35.3 | 43.5 | 49.9 | 57.1 | 63.8 | 65.6 | 57.6 | 44.0 | 35.2 | 28.1 | 44.9 |
| 1983 | 31.0 | 32.3 | 36.2 | 37.0 | 49.9 | 57.3 | 65.5 | 63.8 | 60.7 | 48.6 | 36.9 | 34.0 | 46.1 |
| 1984 | 31.7 | 32.8 | 38.3 | 40.7 | 56.8 | 58.5 | 65.7 | 63.8 | 59.3 | 42.4 | 35.3 | 29.0 | 46.2 |
| 1985 | 27.5 | 28.9 | 35.5 | 46.2 | 51.5 | 62.6 | 67.2 | 65.3 | 53.1 | 47.7 | 33.8 | 32.4 | 46.0 |
| 1986 | 37.0 | 34.2 | 39.7 | 44.1 | 52.0 | 61.2 | 64.0 | 65.7 | 53.0 | 43.5 | 37.6 | 30.2 | 46.9 |
| 1987 | 27.6 | 31.1 | 32.8 | 45.9 | 50.7 | 61.3 | 62.8 | 62.8 | 56.5 | 50.4 | 36.2 | 27.1 | 45.4 |
| 1988 | 29.1 | 34.2 | 37.4 | 44.0 | 50.8 | 61.5 | 67.4 | 64.3 | 55.9 | 52.5 | 36.9 | 27.9 | 46.8 |
| 1989 | 26.2 | 32.3 | 41.8 | 50.4 | 54.2 | 60.8 | 68.1 | 63.6 | 58.7 | 47.5 | 38.4 | 31.6 | 47.8 |
| 1990 | 28.6 | 29.3 | 38.3 | 46.1 | 50.3 | 64.4 | 66.9 | 62.6 | 59.8 | 48.9 | 37.6 | 25.1 | 46.5 |
| 1991 | 30.5 | 37.2 | 32.2 | 43.1 | 49.4 | 57.3 | 67.1 | 66.5 | 59.6 | 50.8 | 36.9 | 30.2 | 46.7 |
| 1992 | 27.3 | 35.2 | 38.1 | 49.1 | 53.3 | 59.0 | 65.1 | 64.2 | 59.3 | 50.4 | 33.8 | 26.3 | 46.8 |
| 1993 | 32.2 | 32.5 | 40.2 | 46.4 | 53.9 | 60.4 | 65.5 | 64.8 | 57.9 | 47.4 | 35.0 | 30.1 | 47.2 |
| 1994 | 32.8 | 29.8 | 40.6 | 44.3 | 51.7 | | 65.4 | 66.4 | 58.1 | 44.2 | 32.2 | 32.1 | |
| 1995 | 28.4 | 37.2 | 38.2 | 40.4 | | 55.1 | 64.6 | 66.9 | | 47.7 | 42.3 | 31.9 | |
| 1996 | 31.4 | 36.9 | 37.8 | 44.9 | | 61.9 | | | | | | | |
| 1997 | 28.0 | 29.6 | 39.1 | 40.5 | 54.1 | 57.1 | 63.8 | 63.8 | 59.4 | 45.0 | 37.7 | 26.6 | 45.4 |
| 1998 | 31.4 | 27.2 | 33.7 | 37.2 | 46.8 | 55.0 | 66.2 | 64.9 | 57.3 | 44.6 | 37.1 | 31.1 | 44.4 |
| 1999 | 33.9 | 35.1 | 39.0 | 37.5 | 49.4 | 58.0 | 64.2 | 62.1 | 56.1 | 47.7 | 40.5 | 29.3 | 46.1 |
| 2000 | 32.9 | 35.6 | 35.4 | 46.6 | 55.4 | 62.2 | 66.5 | 65.0 | 60.1 | 46.2 | 30.8 | 34.7 | 47.6 |
| 2001 | 27.2 | 29.7 | 39.0 | 43.5 | 55.3 | 62.2 | 65.6 | 64.3 | 59.9 | 50.2 | 38.8 | 27.0 | 46.9 |
| 2002 | 31.1 | 34.0 | 35.9 | 47.5 | 52.1 | 70.0 | 66.0 | 58.1 | 45.5 | 38.3 | 28.3 | 47.5 | |
| 2003 | 37.2 | 33.1 | 37.2 | 40.6 | 53.8 | 60.7 | 69.5 | 64.9 | 59.1 | 51.8 | 35.7 | 30.2 | 47.8 |
| 2004 | 28.9 | 27.2 | 42.7 | 42.7 | 52.6 | 60.0 | 64.6 | 62.4 | 55.9 | 44.5 | 34.7 | 31.4 | 45.6 |
| 2005 | 31.6 | 33.1 | 36.6 | 42.8 | 52.5 | 57.9 | 67.3 | 62.9 | 57.3 | 47.4 | 38.8 | 32.1 | 46.7 |
| 2006 | 30.6 | 34.3 | 32.8 | 44.3 | 54.4 | 64.2 | 67.8 | 62.9 | 54.4 | 44.7 | 38.5 | 30.0 | 46.6 |
| 2007 | 25.4 | 32.5 | 41.0 | 44.6 | 53.1 | 62.7 | 68.6 | 66.3 | 57.7 | 47.7 | 41.6 | 26.0 | 47.3 |
| 2008 | 24.4 | 30.9 | 38.2 | 44.1 | 48.5 | 61.9 | 67.5 | 66.9 | 58.5 | 48.4 | 41.6 | 30.6 | 46.8 |
| 2009 | 32.0 | 31.8 | 39.0 | 42.6 | 56.0 | 57.6 | 68.0 | 64.9 | 59.0 | 44.9 | 39.4 | 24.5 | 46.6 |
| POR= 83 YRS | 28.7 | 30.7 | 36.3 | 42.7 | 51.2 | 59.2 | 66.3 | 64.4 | 57.2 | 47.4 | 36.8 | 30.2 | 45.9 |

## HEATING DEGREE DAYS (base 65°F) 2009  FLAGSTAFF (KFLG)

| YEAR | JUL | AUG | SEP | OCT | NOV | DEC | JAN | FEB | MAR | APR | MAY | JUN | TOTAL |
|---|---|---|---|---|---|---|---|---|---|---|---|---|---|
| 1980-81 | 6 | 43 | 153 | 491 | 703 | 774 | 886 | 794 | 873 | 486 | 398 | 50 | 5657 |
| 1981-82 | 1 | 39 | 182 | 558 | 689 | 880 | 1130 | 963 | 911 | 640 | 458 | 230 | 6681 |
| 1982-83 | 65 | 22 | 218 | 643 | 888 | 1136 | 1046 | 911 | 888 | 835 | 461 | 222 | 7335 |
| 1983-84 | 26 | 64 | 134 | 502 | 837 | 952 | 1025 | 929 | 820 | 722 | 247 | 204 | 6462 |
| 1984-85 | 21 | 51 | 165 | 695 | 884 | 1109 | 1155 | 1005 | 911 | 557 | 411 | 102 | 7066 |
| 1985-86 | 26 | 31 | 351 | 532 | 931 | 1005 | 862 | 855 | 777 | 619 | 392 | 119 | 6500 |
| 1986-87 | 58 | 28 | 353 | 661 | 816 | 1069 | 1150 | 942 | 990 | 566 | 435 | 114 | 7182 |
| 1987-88 | 82 | 86 | 246 | 447 | 859 | 1167 | 1103 | 885 | 848 | 621 | 434 | 124 | 6902 |
| 1988-89 | 9 | 41 | 266 | 381 | 836 | 1141 | 1196 | 910 | 710 | 432 | 330 | 139 | 6391 |
| 1989-90 | 7 | 67 | 184 | 540 | 792 | 1030 | 1124 | 996 | 822 | 558 | 449 | 84 | 6653 |
| 1990-91 | 10 | 93 | 167 | 494 | 816 | 1231 | 1060 | 772 | 1008 | 652 | 476 | 225 | 7004 |
| 1991-92 | 7 | 12 | 159 | 438 | 835 | 1071 | 1159 | 854 | 830 | 470 | 356 | 180 | 6371 |
| 1992-93 | 38 | 80 | 163 | 444 | 932 | 1191 | 1014 | 903 | 761 | 551 | 339 | 158 | 6574 |
| 1993-94 | 29 | 45 | 207 | 538 | 893 | 1073 | 990 | 979 | 747 | 614 | 406 |  |  |
| 1994-95 | 35 | 11 | 203 | 637 | 974 | 1014 | 1128 | 772 | 821 | 731 |  | 290 |  |
| 1995-96 | 60 | 10 |  | 531 | 671 | 1015 | 1035 | 807 | 834 | 597 |  | 108 |  |
| 1996-97 | 2 | 36 | 291 | 620 |  | 966 | 1141 | 984 | 797 | 729 | 330 | 231 |  |
| 1997-98 | 61 | 45 | 161 | 613 | 813 | 1182 | 1037 | 1050 | 964 | 828 | 559 | 300 | 7613 |
| 1998-99 | 15 | 27 | 226 | 624 | 832 | 1044 | 957 | 829 | 798 | 817 | 480 | 220 | 6869 |
| 1999-00 | 45 | 86 | 260 | 529 | 726 | 1099 | 987 | 846 | 911 | 547 | 293 | 98 | 6427 |
| 2000-01 | 26 | 49 | 148 | 577 | 1020 | 933 | 1165 | 983 | 801 | 641 | 295 | 109 | 6747 |
| 2001-02 | 25 | 40 | 149 | 452 | 781 | 1172 | 1047 | 863 | 894 | 521 | 397 | 69 | 6410 |
| 2002-03 | 2 | 38 | 204 | 598 | 792 | 1130 | 856 | 889 | 856 | 726 | 342 | 131 | 6564 |
| 2003-04 | 0 | 30 | 171 | 403 | 872 | 1074 | 1112 | 1091 | 678 | 661 | 377 | 141 | 6610 |
| 2004-05 | 44 | 89 | 264 | 630 | 902 | 1034 | 1029 | 888 | 876 | 657 | 382 | 216 | 7011 |
| 2005-06 | 17 | 62 | 222 | 537 | 781 | 1012 | 1060 | 852 | 990 | 611 | 320 | 60 | 6524 |
| 2006-07 | 14 | 65 | 310 | 623 | 790 | 1080 | 1221 | 901 | 738 | 608 | 359 | 86 | 6795 |
| 2007-08 | 3 | 9 | 215 | 532 | 694 | 1200 | 1252 | 982 | 824 | 620 | 504 | 116 | 6951 |
| 2008-09 | 4 | 7 | 188 | 510 | 695 | 1058 | 1015 | 926 | 800 | 666 | 274 | 225 | 6368 |
| 2009- | 2 | 53 | 176 | 615 | 759 | 1251 |  |  |  |  |  |  |  |

WBAN : 03103

## COOLING DEGREE DAYS (base 65°F) 2009  FLAGSTAFF (KFLG)

| YEAR | JAN | FEB | MAR | APR | MAY | JUN | JUL | AUG | SEP | OCT | NOV | DEC | TOTAL |
|---|---|---|---|---|---|---|---|---|---|---|---|---|---|
| 1980 | 0 | 0 | 0 | 0 | 0 | 43 | 133 | 91 | 2 | 0 | 0 | 0 | 269 |
| 1981 | 0 | 0 | 0 | 0 | 0 | 91 | 108 | 58 | 0 | 0 | 0 | 0 | 257 |
| 1982 | 0 | 0 | 0 | 0 | 0 | 1 | 36 | 46 | 6 | 0 | 0 | 0 | 89 |
| 1983 | 0 | 0 | 0 | 0 | 0 | 0 | 45 | 31 | 12 | 0 | 0 | 0 | 88 |
| 1984 | 0 | 0 | 0 | 0 | 2 | 14 | 49 | 23 | 2 | 0 | 0 | 0 | 90 |
| 1985 | 0 | 0 | 0 | 0 | 0 | 36 | 102 | 47 | 0 | 0 | 0 | 0 | 185 |
| 1986 | 0 | 0 | 0 | 0 | 0 | 11 | 33 | 55 | 0 | 0 | 0 | 0 | 99 |
| 1987 | 0 | 0 | 0 | 0 | 0 | 13 | 21 | 26 | 0 | 0 | 0 | 0 | 60 |
| 1988 | 0 | 0 | 0 | 0 | 0 | 28 | 94 | 28 | 1 | 0 | 0 | 0 | 151 |
| 1989 | 0 | 0 | 0 | 0 | 0 | 22 | 111 | 29 | 0 | 0 | 0 | 0 | 162 |
| 1990 | 0 | 0 | 0 | 0 | 0 | 76 | 76 | 28 | 18 | 0 | 0 | 0 | 198 |
| 1991 | 0 | 0 | 0 | 0 | 0 | 0 | 79 | 65 | 4 | 0 | 0 | 0 | 148 |
| 1992 | 0 | 0 | 0 | 0 | 0 | 6 | 48 | 62 | 0 | 0 | 0 | 0 | 116 |
| 1993 | 0 | 0 | 0 | 0 | 0 | 26 | 50 | 46 | 1 | 0 | 0 | 0 | 123 |
| 1994 | 0 | 0 | 0 | 0 | 0 |  | 53 | 57 | 0 | 0 | 0 | 0 |  |
| 1995 | 0 | 0 | 0 | 0 |  | 0 | 53 | 76 |  | 0 | 0 | 0 |  |
| 1996 | 0 | 0 | 0 | 0 |  | 18 |  |  |  | 0 | 0 | 0 |  |
| 1997 | 0 | 0 | 0 | 0 | 1 | 0 | 29 | 14 | 2 | 0 | 0 | 0 | 46 |
| 1998 | 0 | 0 | 0 | 0 | 0 | 5 | 60 | 28 | 0 | 0 | 0 | 0 | 93 |
| 1999 | 0 | 0 | 0 | 0 | 0 | 15 | 29 | 5 | 0 | 0 | 0 | 0 | 49 |
| 2000 | 0 | 0 | 0 | 0 | 3 | 18 | 81 | 55 | 8 | 0 | 0 | 0 | 165 |
| 2001 | 0 | 0 | 0 | 0 | 0 | 30 | 50 | 26 | 0 | 0 | 0 | 0 | 106 |
| 2002 | 0 | 0 | 0 | 0 | 5 | 19 | 163 | 79 | 5 | 0 | 0 | 0 | 271 |
| 2003 | 0 | 0 | 0 | 0 | 3 | 8 | 146 | 32 | 1 | 0 | 0 | 0 | 190 |
| 2004 | 0 | 0 | 0 | 0 | 0 | 0 | 42 | 16 | 0 | 0 | 0 | 0 | 58 |
| 2005 | 0 | 0 | 0 | 0 | 0 | 8 | 97 | 4 | 0 | 0 | 0 | 0 | 109 |
| 2006 | 0 | 0 | 0 | 0 | 0 | 41 | 106 | 9 | 0 | 0 | 0 | 0 | 156 |
| 2007 | 0 | 0 | 0 | 0 | 0 | 26 | 122 | 56 | 3 | 0 | 0 | 0 | 207 |
| 2008 | 0 | 0 | 0 | 0 | 1 | 29 | 88 | 74 | 0 | 0 | 0 | 0 | 192 |
| 2009 | 0 | 0 | 0 | 0 | 0 | 12 | 103 | 54 | 2 | 0 | 0 | 0 | 171 |

**SNOWFALL (inches) 2009 FLAGSTAFF (KFLG)**

| YEAR | JUL | AUG | SEP | OCT | NOV | DEC | JAN | FEB | MAR | APR | MAY | JUN | TOTAL |
|------|-----|-----|-----|-----|-----|-----|-----|-----|-----|-----|-----|-----|-------|
| 1980-81 | 0.0 | 0.0 | 0.0 | 6.9 | T | 6.8 | 11.7 | 11.9 | 45.6 | 9.5 | 0.0 | 0.0 | 92.4 |
| 1981-82 | 0.0 | 0.0 | | 0.0 | 0.0 | 0.0 | 47.5 | 20.4 | 26.7 | 1.9 | 0.4 | 0.0 | |
| 1982-83 | 0.0 | 0.0 | T | T | 22.6 | 27.1 | 15.3 | 25.0 | 38.5 | 13.6 | 0.5 | 0.0 | 142.6 |
| 1983-84 | 0.0 | 0.0 | 0.0 | 0.0 | 14.3 | 5.8 | 4.3 | 1.8 | 0.3 | 5.5 | 0.0 | 0.0 | 32.0 |
| 1984-85 | 0.0 | 0.0 | 0.0 | 0.6 | 9.4 | 28.7 | 26.2 | 31.3 | 21.3 | 18.5 | 0.0 | 0.0 | 136.0 |
| 1985-86 | 0.0 | 0.0 | 0.0 | 0.0 | 40.7 | 2.6 | 0.4 | 26.9 | 32.8 | 1.6 | 0.4 | 0.0 | 105.4 |
| 1986-87 | 0.0 | 0.0 | 0.9 | 0.6 | 4.8 | 9.5 | 38.6 | 40.7 | 25.0 | 1.5 | 0.0 | 0.0 | 121.6 |
| 1987-88 | 0.0 | 0.0 | 0.0 | 0.0 | 2.9 | 16.7 | 28.9 | 21.0 | 1.5 | 33.1 | 0.4 | 0.0 | 104.5 |
| 1988-89 | 0.0 | 0.0 | 0.0 | 0.0 | 11.9 | 15.0 | 21.7 | 12.0 | 16.6 | T | 0.5 | 0.0 | 77.7 |
| 1989-90 | 0.0 | T | T | T | T | 13.1 | 24.2 | 45.5 | 25.0 | 4.2 | 1.4 | 0.0 | 113.4 |
| 1990-91 | T | T | T | T | 9.6 | 22.3 | 3.1 | 13.5 | 79.4 | T | T | 0.0 | 127.9 |
| 1991-92 | 0.0 | 0.0 | T | 5.9 | 39.5 | 24.0 | 24.7 | 24.9 | 35.9 | 4.0 | T | T | 158.9 |
| 1992-93 | T | 0.3 | 0.0 | 0.5 | 2.1 | 41.7 | 55.7 | 35.2 | 11.9 | 2.3 | 0.3 | T | 150.0 |
| 1993-94 | 0.0 | T | T | 0.0 | 23.0 | 13.4 | 6.9 | 26.1 | 12.3 | 27.8 | T | | |
| 1994-95 | 0.0 | T | 0.0 | | | | | | | | | 0.0 | |
| 1995-96 | 0.0 | 0.0 | 0.0 | 0.0 | 0.0 | | | | | | | | |
| 1996-97 | | | | | | | | | | | | | |
| 1997-98 | | | | | | | | | | | | | |
| 1998-99 | | | | | | | | | | | | | |
| 1999-00 | | | | | | | | | | | | | |
| 2000-01 | | | | | | | | | | | | | |
| 2001-02 | | | | | | | | | | | | | |
| 2002-03 | | | | | | | | | | | | | |
| 2003-04 | | | | | | | 9.8 | 24.5 | 12.5 | T | 0.0 | 0.0 | |
| 2004-05 | 0.0 | 0.0 | 0.0 | 16.4 | 11.8 | 10.9 | 56.3 | 16.8 | 18.4 | 1.1 | T | 0.0 | 131.7 |
| 2005-06 | 0.0 | T | 0.0 | 0.0 | T | T | 1.6 | 0.0 | 40.0 | 3.0 | 0.0 | T | 44.6 |
| 2006-07 | 0.0 | 0.0 | 0.0 | T | 0.6 | 9.5 | 25.4 | 11.0 | 1.2 | 2.3 | 0.4 | 0.0 | 50.4 |
| 2007-08 | 0.0 | 0.0 | 0.0 | 0.0 | 0.0 | 25.4 | 26.9 | 35.9 | 2.5 | 0.1 | 8.7 | 0.0 | 99.5 |
| 2008-09 | 0.0 | 0.0 | 0.0 | 0.0 | 2.5 | 39.1 | 7.8 | 31.3 | 0.3 | 5.0 | 0.0 | 0.0 | 86.0 |
| 2009- | 0.0 | 0.0 | 0.0 | 0.3 | T | 33.0 | | | | | | | |
| POR=<br>70 YRS | T | T | 0.1 | 1.5 | 7.4 | 13.9 | 18.7 | 15.0 | 16.3 | 6.6 | 1.6 | T | 81.1 |

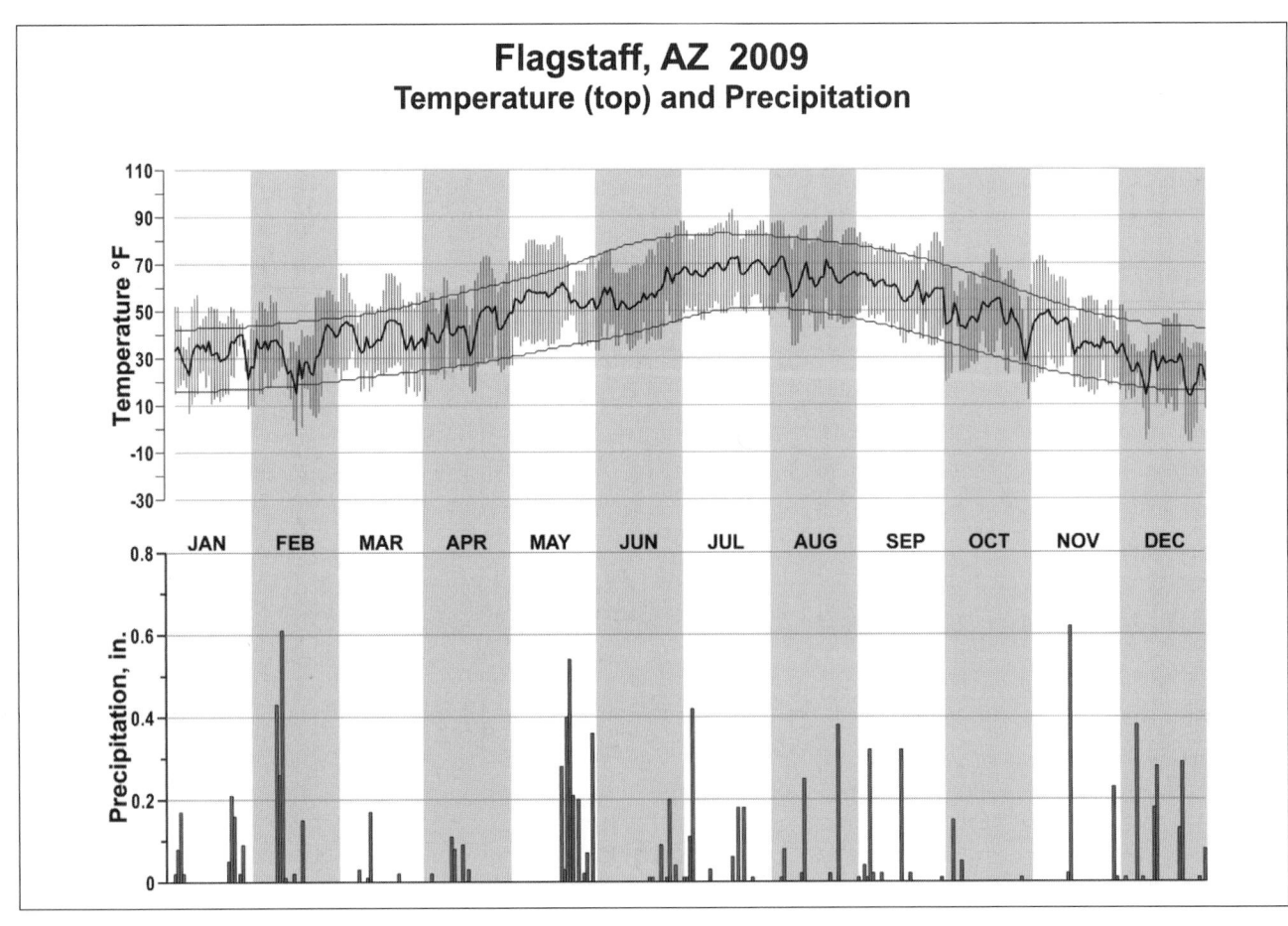

# Flagstaff, AZ  2009
## Temperature (top) and Precipitation

# 2009
# PHOENIX
# ARIZONA (KPHX)

Phoenix is located in the Salt River Valley at an elevation of about 1,100 feet. The valley is oval shaped and flat except for scattered precipitous mountains rising a few hundred to as much as 1,500 feet above the valley floor. Sky Harbor Airport, where the weather observations are taken, is in the southern part of the city. Six miles to the south of the airport are the South Mountains rising to 2,500 feet. Eighteen miles southwest, the Estrella Mountains rise to 4,500 feet, and 30 miles to the west are the White Tank Mountains rising to 4,100 feet. The Superstition Mountains, over 30 miles to the east, rise to as much as 5,000 feet. The valley, though located in the Sonora Desert, supports large acreages of cotton, citrus, and other agriculture along with one of the largest urban populations in the United States. The water supply for this complex desert community is partly from reservoirs on the impounded Salt and Verde Rivers, and partly from a large underground water table.

Temperatures range from very hot in summer to mild in winter. Many winter days reach over 70 degrees and typical high temperatures in the middle of the winter are in the 60s. The climate becomes less attractive in the summer. The normal high temperature is over 90 degrees from early May through early October, and over 100 degrees from early June through early September. Many days each summer will exceed 110 degrees in the afternoon and remain above 85 degrees all night. When temperatures are extremely high, the low humidity does not provide much comfort.

Indeed, the climate is very dry. Annual precipitation is only about 7 inches, and afternoon humidities range from about 30 percent in winter to only about 10 percent in June. Rain comes mostly in two seasons. From about Thanksgiving to early April there are periodic rains from Pacific storms. Moisture from the south and southeast results in a summer thunderstorm peak in July and August. Usually the break from extreme dryness in June to the onset of thunderstorms in early July is very abrupt. Afternoon humidities suddenly double to about 20 percent, which with the great heat, gives a feeling of mugginess. Fog is rare, occurring about once per winter, and is unknown in the other seasons.

The valley is characterized by light winds. High winds associated with thunderstorms occur periodically in the summer. These occasionally create duststorms which move large distances across the deserts. Strong thunderstorm winds occur any month of the year, but are rare outside the summer months. Persistent strong winds of 30 mph or more are rare except for two or three events in an average spring due to Pacific storms. Winter storms rarely bring high winds due to the relatively stable air in the valley during that season.

Based on the 1951-1980 period, the average first occurrence of 32 degrees Fahrenheit in the fall is December 13 and the average last occurrence in the spring is February 7.

# NORMALS, MEANS, AND EXTREMES
## PHOENIX (KPHX)

LATITUDE: 33° 26'N    LONGITUDE: -111° 59'W    ELEVATION (FT): GRND: 1105   BARO: 1106    TIME ZONE: MOUNTAIN (UTC -7)    WBAN: 23183

| | ELEMENT | POR | JAN | FEB | MAR | APR | MAY | JUN | JUL | AUG | SEP | OCT | NOV | DEC | YEAR |
|---|---|---|---|---|---|---|---|---|---|---|---|---|---|---|---|
| **TEMPERATURE °F** | NORMAL DAILY MAXIMUM | 30 | 67.3 | 71.4 | 76.1 | 84.7 | 93.9 | 103. | 106. | 104. | 99.0 | 87.7 | 75.0 | 67.1 | 86.3 |
| | MEAN DAILY MAXIMUM | 71 | 66.1 | 69.1 | 75.8 | 84.1 | 93.7 | 102.4 | 105.6 | 103.4 | 98.6 | 88.2 | 75.0 | 66.5 | 85.7 |
| | HIGHEST DAILY MAXIMUM | 72 | 88 | 92 | 100 | 105 | 113 | 122 | 121 | 116 | 118 | 107 | 96 | 88 | 122 |
| | YEAR OF OCCURRENCE | | 1971 | 1986 | 1988 | 1992 | 1984 | 1990 | 1995 | 2003 | 1950 | 1980 | 2009 | 1950 | JUN 1990 |
| | MEAN OF EXTREME MAXS. | 71 | 77.8 | 82.3 | 88.5 | 97.2 | 105.0 | 112.2 | 113.5 | 111.0 | 107.6 | 99.5 | 87.2 | 77.6 | 96.6 |
| | NORMAL DAILY MINIMUM | 30 | 44.8 | 48.4 | 53.0 | 57.6 | 67.4 | 75.6 | 82.9 | 81.6 | 75.6 | 62.1 | 50.4 | 43.9 | 61.9 |
| | MEAN DAILY MINIMUM | 71 | 41.4 | 43.8 | 48.9 | 55.5 | 64.0 | 72.2 | 80.4 | 79.2 | 72.5 | 60.8 | 48.2 | 41.7 | 59.1 |
| | LOWEST DAILY MINIMUM | 72 | 17 | 22 | 25 | 32 | 40 | 50 | 61 | 60 | 47 | 34 | 25 | 22 | 17 |
| | YEAR OF OCCURRENCE | | 1950 | 1948 | 1966 | 1945 | 1967 | 1944 | 1944 | 1942 | 1965 | 1971 | 1938 | 1948 | JAN 1950 |
| | MEAN OF EXTREME MINS. | 71 | 31.6 | 35.2 | 39.1 | 45.5 | 53.8 | 63.8 | 72.2 | 71.3 | 63.4 | 49.4 | 38.0 | 32.2 | 49.6 |
| | NORMAL DRY BULB | 30 | 56.1 | 59.9 | 64.6 | 71.2 | 80.7 | 89.8 | 94.8 | 93.1 | 87.3 | 74.9 | 62.7 | 55.5 | 74.2 |
| | MEAN DRY BULB | 71 | 53.8 | 56.4 | 62.4 | 69.8 | 78.9 | 87.4 | 93.1 | 91.3 | 85.6 | 74.5 | 61.6 | 54.1 | 72.4 |
| | MEAN WET BULB | 26 | 42.9 | 45.0 | 47.2 | 49.6 | 54.3 | 58.6 | 67.7 | 69.0 | 63.6 | 54.7 | 46.9 | 42.7 | 53.5 |
| | MEAN DEW POINT | 26 | 35.0 | 35.9 | 37.1 | 36.7 | 39.6 | 43.3 | 57.4 | 61.1 | 54.1 | 44.9 | 37.6 | 34.3 | 43.1 |
| | NORMAL NO. DAYS WITH: | | | | | | | | | | | | | | |
| | MAXIMUM >= 90 | 30 | 0.0 | 0.1 | 2.2 | 10.5 | 23.1 | 29.2 | 30.8 | 30.7 | 28.0 | 15.3 | 0.7 | 0.0 | 170.6 |
| | MAXIMUM <= 32 | 30 | 0.0 | * | 0.0 | 0.0 | 0.0 | 0.0 | 0.0 | 0.0 | 0.0 | 0.0 | 0.0 | 0.0 | 0.0 |
| | MINIMUM <= 32 | 30 | 1.3 | 0.4 | 0.1 | 0.0 | 0.0 | 0.0 | 0.0 | 0.0 | 0.0 | 0.0 | 0.1 | 0.9 | 2.8 |
| | MINIMUM <= 0 | 30 | 0.0 | 0.0 | 0.0 | 0.0 | 0.0 | 0.0 | 0.0 | 0.0 | 0.0 | 0.0 | 0.0 | 0.0 | 0.0 |
| **H/C** | NORMAL HEATING DEG. DAYS | 30 | 290 | 169 | 100 | 28 | 1 | 0 | 0 | 0 | 0 | 14 | 121 | 304 | 1027 |
| | NORMAL COOLING DEG. DAYS | 30 | 2 | 15 | 74 | 226 | 478 | 745 | 910 | 859 | 667 | 336 | 51 | 1 | 4364 |
| **RH** | NORMAL (PERCENT) | 30 | 50 | 45 | 40 | 28 | 23 | 19 | 31 | 35 | 34 | 35 | 42 | 49 | 36 |
| | HOUR 05 LST | 30 | 64 | 59 | 55 | 42 | 34 | 29 | 43 | 48 | 47 | 49 | 56 | 64 | 49 |
| | HOUR 11 LST | 30 | 44 | 39 | 34 | 23 | 19 | 16 | 28 | 32 | 31 | 31 | 36 | 44 | 31 |
| | HOUR 17 LST | 30 | 32 | 28 | 25 | 17 | 14 | 12 | 20 | 23 | 23 | 23 | 26 | 32 | 23 |
| | HOUR 23 LST | 30 | 53 | 47 | 41 | 28 | 23 | 19 | 32 | 36 | 36 | 39 | 46 | 54 | 38 |
| **S** | PERCENT POSSIBLE SUNSHINE | 11 | 79 | 82 | 85 | 89 | 94 | 95 | 86 | 86 | 89 | 88 | 83 | 77 | 86 |
| **W/O** | MEAN NO. DAYS WITH: | | | | | | | | | | | | | | |
| | HEAVY FOG(VISBY <= 1/4 MI) | 46 | 0.4 | 0.0 | 0.1 | 0.0 | 0.0 | 0.0 | 0.0 | 0.0 | 0.0 | 0.1 | 0.1 | 0.4 | 1.1 |
| | THUNDERSTORMS | 62 | 0.5 | 0.6 | 1.0 | 0.5 | 0.9 | 0.9 | 5.9 | 6.6 | 3.4 | 1.5 | 0.5 | 0.5 | 22.8 |
| **CLOUDNESS** | MEAN: | | | | | | | | | | | | | | |
| | SUNRISE-SUNSET (OKTAS) | 1 | | 2.4 | 2.0 | 2.4 | 1.2 | 1.2 | 1.6 | 0.8 | | | | 1.6 | |
| | MIDNIGHT-MIDNIGHT (OKTAS) | 1 | | 2.4 | | 2.4 | 1.2 | 1.2 | 1.6 | 0.8 | | 0.0 | | | |
| | MEAN NO. DAYS WITH: | | | | | | | | | | | | | | |
| | CLEAR | 1 | 5.0 | 17.5 | 18.0 | 14.0 | 26.0 | 17.0 | 6.0 | 10.0 | 15.0 | 15.0 | | 13.0 | |
| | PARTLY CLOUDY | 1 | 1.0 | 5.0 | 5.0 | 4.0 | 1.5 | 4.0 | | 1.0 | | 2.0 | | 5.0 | |
| | CLOUDY | | | 6.0 | 4.0 | 1.0 | 1.0 | | | 1.0 | | | | | |
| **PR** | MEAN STATION PRESSURE(IN) | 26 | 28.89 | 28.84 | 28.77 | 28.71 | 28.66 | 28.63 | 28.67 | 28.68 | 28.67 | 28.74 | 28.83 | 28.89 | 28.75 |
| | MEAN SEA-LEVEL PRES. (IN) | 26 | 30.05 | 30.00 | 29.92 | 29.85 | 29.78 | 29.74 | 29.78 | 29.80 | 29.79 | 29.87 | 29.98 | 30.06 | 29.89 |
| **WINDS** | MEAN SPEED (MPH) | 26 | 5.1 | 5.6 | 6.4 | 7.0 | 7.0 | 6.8 | 7.1 | 6.7 | 6.2 | 5.4 | 4.9 | 4.6 | 6.1 |
| | PREVAIL.DIR(TENS OF DEGS) | 38 | 11 | 09 | 11 | 09 | 28 | 28 | 28 | 09 | 09 | 11 | 11 | 09 | 11 |
| | MAXIMUM 2-MINUTE: | | | | | | | | | | | | | | |
| | SPEED (MPH) | 15 | 36 | 35 | 43 | 51 | 38 | 45 | 51 | 54 | 39 | 37 | 32 | 43 | 54 |
| | DIR. (TENS OF DEGS) | | 25 | 27 | 24 | 30 | 27 | 06 | 07 | 10 | 26 | 26 | 27 | 27 | 10 |
| | YEAR OF OCCURRENCE | | 1995 | 2004 | 1998 | 2001 | 2006 | 2006 | 2006 | 2008 | 2008 | 2004 | 2008 | 2009 | AUG 2008 |
| | MAXIMUM 3-SECOND | | | | | | | | | | | | | | |
| | SPEED (MPH) | 15 | 39 | 43 | 53 | 59 | 52 | 54 | 77 | 75 | 51 | 48 | 38 | 55 | 77 |
| | DIR. (TENS OF DEGS) | | 24 | 26 | 36 | 29 | 30 | 09 | 07 | 11 | 26 | 24 | 27 | 27 | 07 |
| | YEAR OF OCCURRENCE | | 1995 | 2004 | 2009 | 2001 | 2009 | 2006 | 2005 | 2008 | 2008 | 1998 | 2008 | 2009 | JUL 2005 |
| **PRECIPITATION** | NORMAL (IN) | 30 | 0.83 | 0.77 | 1.07 | 0.25 | 0.16 | 0.09 | 0.99 | 0.94 | 0.75 | 0.79 | 0.73 | 0.92 | 8.29 |
| | MAXIMUM MONTHLY (IN) | 72 | 5.22 | 3.15 | 4.16 | 2.10 | 1.06 | 1.70 | 5.15 | 5.56 | 4.23 | 4.40 | 3.04 | 3.98 | 5.56 |
| | YEAR OF OCCURRENCE | | 1993 | 2003 | 1941 | 1941 | 1976 | 1972 | 1984 | 1951 | 1939 | 1972 | 1952 | 1967 | AUG 1951 |
| | MINIMUM MONTHLY (IN) | 72 | 0.00 | 0.00 | 0.00 | 0.00 | 0.00 | 0.00 | T | T | 0.00 | 0.00 | 0.00 | 0.00 | 0.00 |
| | YEAR OF OCCURRENCE | | 1972 | 1967 | 1959 | 1962 | 1983 | 1983 | 1995 | 1975 | 1973 | 1973 | 1980 | 1981 | JUN 1983 |
| | MAXIMUM IN 24 HOURS (IN) | 72 | 1.84 | 1.65 | 2.04 | 1.38 | 0.96 | 1.64 | 2.75 | 3.07 | 2.43 | 2.32 | 2.16 | 1.89 | 3.07 |
| | YEAR OF OCCURRENCE | | 1993 | 2003 | 1983 | 1941 | 1976 | 1972 | 1984 | 1943 | 1970 | 1988 | 1993 | 1967 | AUG 1943 |
| | NORMAL NO. DAYS WITH: | | | | | | | | | | | | | | |
| | PRECIPITATION >= 0.01 | 30 | 4.2 | 4.3 | 4.6 | 1.7 | 1.1 | 0.7 | 4.2 | 4.5 | 3.1 | 2.9 | 2.7 | 3.7 | 37.7 |
| | PRECIPITATION >= 1.00 | 30 | * | * | 0.1 | 0.0 | 0.0 | * | 0.2 | 0.2 | 0.2 | 0.1 | 0.1 | 0.1 | 1.0 |
| **SNOWFALL** | NORMAL (IN) | 30 | 0.* | 0.* | 0.* | 0.0 | 0.0 | 0.0 | 0.0 | 0.0 | 0.0 | 0.0 | 0.0 | 0.* | 0.0 |
| | MAXIMUM MONTHLY (IN) | 62 | T | 0.6 | T | T | T | 0.0 | 0.0 | 0.0 | 0.0 | T | 0.0 | 0.4 | 0.6 |
| | YEAR OF OCCURRENCE | | 1993 | 1939 | 1991 | 1949 | 1992 | | | | | 1992 | | 1990 | FEB 1939 |
| | MAXIMUM IN 24 HOURS (IN) | 62 | T | 0.6 | T | T | T | 0.0 | 0.0 | 0.0 | 0.0 | T | 0.0 | 0.4 | 0.6 |
| | YEAR OF OCCURRENCE' | | 1993 | 1939 | 1991 | 1949 | 1992 | | | | | 1992 | | 1990 | FEB 1939 |
| | MAXIMUM SNOW DEPTH (IN) | 48 | 0 | 0 | 0 | 0 | 0 | 0 | 0 | 0 | 0 | 0 | 0 | 0 | 0 |
| | YEAR OF OCCURRENCE | | | | | | | | | | | | | | |
| | NORMAL NO. DAYS WITH: | | | | | | | | | | | | | | |
| | SNOWFALL >= 1.0 | 30 | 0.0 | 0.0 | 0.0 | 0.0 | 0.0 | 0.0 | 0.0 | 0.0 | 0.0 | 0.0 | 0.0 | 0.0 | 0.0 |

## PRECIPITATION (inches) 2009 PHOENIX (KPHX)

| YEAR | JAN | FEB | MAR | APR | MAY | JUN | JUL | AUG | SEP | OCT | NOV | DEC | ANNUAL |
|------|------|------|------|------|------|------|------|------|------|------|------|------|--------|
| 1980 | 1.58 | 2.09 | 0.86 | 0.44 | 0.21 | 0.03 | 0.56 | 0.06 | 0.13 | 0.02 | 0.00 | 0.08 | 6.06 |
| 1981 | 0.71 | 1.08 | 0.98 | 0.20 | 0.03 | T | 1.14 | 0.11 | 0.18 | 1.34 | 0.95 | 0.00 | 6.72 |
| 1982 | 0.81 | 0.67 | 1.30 | T | 0.50 | T | 0.43 | 1.97 | 0.12 | T | 2.50 | 1.64 | 9.94 |
| 1983 | 0.70 | 1.17 | 3.17 | 0.18 | 0.00 | 0.00 | 0.38 | 2.48 | 2.43 | 0.71 | 0.43 | 1.16 | 12.81 |
| 1984 | 0.31 | 0.00 | 0.00 | 0.91 | 0.18 | 0.18 | 5.15 | 0.87 | 3.36 | 0.31 | 0.71 | 2.93 | 14.91 |
| 1985 | 0.95 | 0.18 | 0.46 | 0.17 | T | 0.00 | 0.98 | 0.21 | 1.60 | 0.92 | 1.59 | 0.86 | 7.92 |
| 1986 | 0.07 | 1.19 | 1.58 | 0.01 | T | 0.01 | 1.19 | 1.27 | 0.47 | 0.41 | 0.03 | 1.38 | 7.61 |
| 1987 | 0.67 | 2.06 | 0.28 | 0.09 | 0.06 | 0.01 | 1.08 | 0.45 | 0.57 | 0.47 | 1.04 | 1.62 | 8.40 |
| 1988 | 0.90 | 0.23 | 0.17 | 1.09 | 0.00 | 0.02 | 0.87 | 0.63 | 0.00 | 2.38 | 0.78 | 0.14 | 7.21 |
| 1989 | 1.19 | T | 1.25 | 0.00 | T | 0.00 | 0.13 | 1.11 | 0.47 | 0.46 | 0.14 | 0.19 | 4.94 |
| 1990 | 0.80 | 0.70 | 0.35 | 0.17 | 0.16 | 0.04 | 1.05 | 2.70 | 1.11 | 0.04 | 0.15 | 0.46 | 7.73 |
| 1991 | 0.63 | 0.56 | 2.05 | 0.00 | 0.00 | T | 0.14 | 0.12 | 0.81 | 1.16 | 1.25 | 1.63 | 8.35 |
| 1992 | 1.62 | 0.90 | 2.49 | 0.49 | 1.05 | 0.04 | 2.95 | 1.30 | 0.03 | 0.26 | 0.03 | 3.08 | 14.24 |
| 1993 | 5.22 | 1.72 | 1.62 | 0.00 | 0.08 | 0.01 | T | 0.55 | 0.06 | 1.27 | 2.79 | 0.02 | 13.34 |
| 1994 | 0.13 | 0.54 | 1.36 | 0.09 | 0.39 | T | 0.25 | 0.02 | 1.74 | 0.55 | 0.68 | 3.03 | 8.78 |
| 1995 | 1.41 | 0.34 | 1.04 | 0.29 | 0.09 | T | T | 3.50 | 1.08 | 0.00 | 1.75 | 0.01 | 9.51 |
| 1996 | 0.25 | 1.03 | 0.55 | T | T | T | 1.04 | 0.34 | 0.54 | 0.37 | 0.23 | 0.02 | 4.37 |
| 1997 | 0.90 | 0.55 | 0.01 | 0.24 | T | T | 0.17 | 1.39 | 0.45 | 0.07 | 0.06 | 0.83 | 4.67 |
| 1998 | 0.35 | 2.93 | 1.31 | 0.43 | 0.04 | 0.00 | 1.94 | 1.05 | 0.58 | 1.03 | 0.19 | 0.68 | 10.53 |
| 1999 | 0.01 | 0.17 | 0.11 | 1.13 | 0.00 | T | 2.96 | 0.92 | 1.31 | 0.00 | 0.00 | 0.00 | 6.61 |
| 2000 | 0.01 | T | 2.98 | 0.00 | 0.00 | 0.34 | 0.28 | 0.58 | 0.01 | 3.17 | 0.50 | 0.00 | 7.87 |
| 2001 | 1.77 | 0.86 | 0.77 | 1.06 | 0.02 | 0.01 | 0.67 | 0.46 | 0.00 | 0.02 | 0.20 | 0.88 | 6.72 |
| 2002 | 0.05 | 0.00 | 0.08 | 0.06 | 0.00 | T | 1.18 | T | 0.50 | 0.37 | 0.42 | 0.16 | 2.82 |
| 2003 | 0.57 | 3.15 | 0.51 | 0.17 | 0.00 | 0.00 | 0.61 | 0.51 | 0.25 | 0.21 | 0.65 | 0.19 | 6.82 |
| 2004 | 0.82 | 1.02 | 1.28 | 0.90 | 0.00 | 0.00 | 0.59 | 0.36 | 0.15 | 0.78 | 0.52 | 1.56 | 7.98 |
| 2005 | 1.85 | 3.01 | 0.36 | 0.12 | T | T | 0.16 | 1.21 | 0.16 | 0.17 | 0.00 | 0.00 | 7.04 |
| 2006 | 0.00 | 0.00 | 1.56 | T | T | T | 1.29 | 1.26 | 0.78 | 0.22 | T | 0.34 | 5.45 |
| 2007 | 0.49 | 0.40 | 0.83 | 0.21 | T | 0.00 | 0.36 | 0.31 | 0.07 | 0.04 | 1.25 | 1.09 | 5.05 |
| 2008 | 1.58 | 0.39 | 0.00 | 0.00 | 0.45 | T | 2.15 | 3.55 | T | T | 0.49 | 0.97 | 9.58 |
| 2009 | 0.15 | 1.32 | T | 0.19 | 0.25 | 0.02 | 0.40 | 0.29 | 0.16 | T | 0.01 | 0.47 | 3.26 |
| POR= 71 YRS | 0.77 | 0.77 | 0.84 | 0.29 | 0.14 | 0.09 | 0.83 | 1.03 | 0.70 | 0.59 | 0.56 | 0.90 | 7.51 |

WBAN : 23183

## AVERAGE TEMPERATURE (°F) 2009 PHOENIX (KPHX)

| YEAR | JAN | FEB | MAR | APR | MAY | JUN | JUL | AUG | SEP | OCT | NOV | DEC | ANNUAL |
|------|------|------|------|------|------|------|------|------|------|------|------|------|--------|
| 1980 | 56.6 | 60.6 | 60.7 | 69.8 | 76.0 | 88.9 | 95.6 | 92.2 | 87.3 | 75.6 | 64.1 | 61.3 | 74.1 |
| 1981 | 59.2 | 61.4 | 63.8 | 76.0 | 80.5 | 93.4 | 95.2 | 95.8 | 89.2 | 73.6 | 66.1 | 58.6 | 76.1 |
| 1982 | 53.9 | 60.1 | 62.4 | 72.5 | 80.4 | 88.1 | 93.7 | 93.7 | 86.7 | 73.5 | 61.9 | 54.1 | 73.4 |
| 1983 | 56.0 | 58.4 | 62.2 | 66.6 | 80.6 | 88.6 | 95.5 | 92.6 | 91.0 | 77.2 | 62.4 | 57.2 | 74.0 |
| 1984 | 57.4 | 60.1 | 67.6 | 70.7 | 87.0 | 88.9 | 91.7 | 91.2 | 87.5 | 71.4 | 61.9 | 53.7 | 74.1 |
| 1985 | 54.3 | 57.4 | 62.8 | 75.1 | 84.2 | 92.4 | 94.9 | 94.5 | 82.3 | 75.1 | 61.3 | 55.9 | 74.2 |
| 1986 | 61.4 | 61.0 | 69.3 | 74.2 | 82.3 | 92.8 | 92.3 | 94.5 | 84.1 | 74.7 | 65.0 | 56.4 | 75.7 |
| 1987 | 54.7 | 59.7 | 63.4 | 77.9 | 82.6 | 93.0 | 93.1 | 92.2 | 86.9 | 80.9 | 63.1 | 52.7 | 75.0 |
| 1988 | 55.1 | 62.5 | 66.3 | 73.0 | 81.4 | 93.1 | 96.2 | 93.9 | 87.4 | 82.4 | 64.4 | 55.7 | 76.0 |
| 1989 | 54.4 | 61.9 | 70.1 | 80.1 | 83.1 | 92.1 | 97.4 | 93.7 | 89.9 | 77.3 | 66.4 | 57.0 | 77.0 |
| 1990 | 55.6 | 56.6 | 67.2 | 76.2 | 81.1 | 93.8 | 93.6 | 90.8 | 87.6 | 78.7 | 65.9 | 53.6 | 75.1 |
| 1991 | 55.9 | 66.0 | 60.3 | 72.2 | 79.7 | 87.8 | 95.1 | 94.5 | 88.5 | 80.2 | 63.5 | 57.3 | 75.1 |
| 1992 | 56.4 | 62.1 | 64.7 | 77.0 | 83.1 | 90.1 | 92.8 | 92.3 | 90.5 | 79.8 | 61.5 | 53.8 | 75.3 |
| 1993 | 58.2 | 58.2 | 65.7 | 73.8 | 83.7 | 89.6 | 92.9 | 91.6 | 87.9 | 76.7 | 61.4 | 56.2 | 74.7 |
| 1994 | 56.8 | 58.1 | 65.0 | 71.8 | 78.3 | 92.2 | 93.9 | 95.3 | 86.9 | 73.1 | 58.0 | 55.3 | 73.7 |
| 1995 | 54.4 | 63.1 | 64.3 | 68.4 | 76.2 | 86.2 | 94.5 | 94.6 | 89.2 | 76.2 | 66.6 | 56.9 | 74.2 |
| 1996 | 55.8 | 62.3 | 65.1 | 74.1 | 83.0 | 91.9 | 95.2 | 94.1 | 84.3 | 74.9 | 64.5 | 56.5 | 75.1 |
| 1997 | 55.7 | 57.8 | 69.4 | 70.9 | 86.3 | 87.7 | 93.8 | 92.9 | 90.0 | 74.7 | 64.1 | 52.9 | 74.7 |
| 1998 | 56.6 | 54.1 | 61.9 | 66.7 | 76.2 | 85.4 | 94.5 | 94.7 | 87.9 | 74.1 | 62.7 | 54.4 | 72.4 |
| 1999 | 56.8 | 60.3 | 64.9 | 65.9 | 79.8 | 88.8 | 91.3 | 93.0 | 87.4 | 79.3 | 68.0 | 55.3 | 74.2 |
| 2000 | 58.1 | 60.9 | 63.1 | 75.2 | 84.8 | 91.0 | 95.0 | 92.5 | 90.3 | 72.5 | 56.9 | 57.3 | 74.8 |
| 2001 | 54.1 | 56.9 | 65.6 | 71.7 | 86.9 | 92.0 | 94.4 | 94.7 | 92.2 | 79.4 | 68.4 | 53.8 | 75.8 |
| 2002 | 56.0 | 61.3 | 65.2 | 76.7 | 81.7 | 93.4 | 96.0 | 95.1 | 89.8 | 75.1 | 66.9 | 55.2 | 76.0 |
| 2003 | 62.0 | 59.4 | 64.8 | 70.3 | 83.3 | 91.8 | 97.7 | 94.6 | 90.7 | 82.7 | 61.9 | 56.8 | 76.3 |
| 2004 | 57.5 | 55.8 | 72.3 | 72.7 | 83.4 | 91.2 | 94.5 | 92.7 | 88.1 | 75.1 | 61.0 | 56.0 | 75.0 |
| 2005 | 57.8 | 59.2 | 63.9 | 72.3 | 82.7 | 90.4 | 97.3 | 92.2 | 89.6 | 78.3 | 66.3 | 56.8 | 75.6 |
| 2006 | 57.7 | 61.8 | 62.3 | 72.7 | 84.7 | 94.6 | 96.5 | 92.6 | 85.7 | 76.1 | 67.1 | 54.3 | 75.5 |
| 2007 | 52.9 | 59.9 | 68.7 | 73.7 | 84.6 | 92.8 | 95.8 | 96.2 | 90.4 | 78.2 | 70.1 | 53.2 | 76.4 |
| 2008 | 54.7 | 58.3 | 66.7 | 74.0 | 78.5 | 93.2 | 94.9 | 93.0 | 89.9 | 77.9 | 67.2 | 55.9 | 75.4 |
| 2009 | 58.7 | 60.7 | 67.5 | 71.4 | 86.2 | 88.7 | 98.3 | 95.2 | 90.0 | 74.5 | 67.5 | 53.7 | 76.0 |
| POR= 71 YRS | 53.8 | 56.4 | 62.4 | 69.8 | 78.9 | 87.4 | 93.1 | 91.3 | 85.6 | 74.5 | 61.6 | 54.1 | 72.4 |

## HEATING DEGREE DAYS (base 65°F) 2009 PHOENIX (KPHX)

| YEAR | JUL | AUG | SEP | OCT | NOV | DEC | JAN | FEB | MAR | APR | MAY | JUN | TOTAL |
|------|-----|-----|-----|-----|-----|-----|-----|-----|-----|-----|-----|-----|-------|
| 1980-81 | 0 | 0 | 0 | 12 | 108 | 122 | 181 | 131 | 74 | 8 | 0 | 0 | 636 |
| 1981-82 | 0 | 0 | 0 | 1 | 56 | 196 | 335 | 151 | 99 | 4 | 0 | 0 | 842 |
| 1982-83 | 0 | 0 | 0 | 1 | 103 | 331 | 272 | 181 | 120 | 53 | 0 | 0 | 1061 |
| 1983-84 | 0 | 0 | 0 | 0 | 154 | 236 | 228 | 139 | 16 | 23 | 0 | 0 | 796 |
| 1984-85 | 0 | 0 | 0 | 7 | 126 | 345 | 328 | 222 | 102 | 5 | 0 | 0 | 1135 |
| 1985-86 | 0 | 0 | 0 | 0 | 149 | 274 | 110 | 158 | 66 | 2 | 1 | 0 | 760 |
| 1986-87 | 0 | 0 | 0 | 0 | 43 | 260 | 318 | 172 | 95 | 4 | 0 | 0 | 892 |
| 1987-88 | 0 | 0 | 0 | 0 | 98 | 375 | 311 | 100 | 60 | 20 | 2 | 0 | 966 |
| 1988-89 | 0 | 0 | 0 | 0 | 135 | 284 | 321 | 133 | 46 | 0 | 0 | 0 | 919 |
| 1989-90 | 0 | 0 | 0 | 1 | 36 | 243 | 291 | 253 | 76 | 0 | 0 | 0 | 900 |
| 1990-91 | 0 | 0 | 0 | 0 | 65 | 348 | 275 | 27 | 161 | 5 | 0 | 0 | 881 |
| 1991-92 | 0 | 0 | 0 | 34 | 107 | 233 | 260 | 89 | 56 | 7 | 0 | 0 | 786 |
| 1992-93 | 0 | 0 | 0 | 0 | 127 | 340 | 205 | 184 | 61 | 0 | 0 | 0 | 917 |
| 1993-94 | 0 | 0 | 0 | 0 | 125 | 264 | 246 | 192 | 54 | 13 | 0 | 0 | 894 |
| 1994-95 | 0 | 0 | 0 | 15 | 211 | 293 | 320 | 61 | 74 | 48 | 6 | 0 | 1028 |
| 1995-96 | 0 | 0 | 0 | 0 | 22 | 243 | 278 | 111 | 63 | 0 | 0 | 0 | 717 |
| 1996-97 | 0 | 0 | 0 | 46 | 67 | 260 | 284 | 202 | 39 | 36 | 0 | 0 | 934 |
| 1997-98 | 0 | 0 | 0 | 9 | 83 | 369 | 250 | 299 | 138 | 77 | 2 | 0 | 1227 |
| 1998-99 | 0 | 0 | 0 | 4 | 73 | 322 | 245 | 127 | 65 | 104 | 2 | 0 | 942 |
| 1999-00 | 0 | 0 | 0 | 0 | 53 | 297 | 214 | 121 | 110 | 8 | 0 | 0 | 803 |
| 2000-01 | 0 | 0 | 0 | 19 | 238 | 234 | 330 | 222 | 87 | 45 | 0 | 0 | 1175 |
| 2001-02 | 0 | 0 | 0 | 0 | 68 | 341 | 272 | 115 | 74 | 0 | 0 | 0 | 870 |
| 2002-03 | 0 | 0 | 0 | 1 | 21 | 293 | 95 | 158 | 81 | 15 | 0 | 0 | 664 |
| 2003-04 | 0 | 0 | 0 | 0 | 106 | 251 | 231 | 260 | 46 | 9 | 0 | 0 | 903 |
| 2004-05 | 0 | 0 | 0 | 17 | 134 | 271 | 220 | 156 | 63 | 0 | 0 | 0 | 861 |
| 2005-06 | 0 | 0 | 0 | 0 | 50 | 247 | 227 | 102 | 119 | 4 | 0 | 0 | 749 |
| 2006-07 | 0 | 0 | 0 | 0 | 48 | 324 | 366 | 151 | 53 | 7 | 0 | 0 | 949 |
| 2007-08 | 0 | 0 | 0 | 0 | 30 | 360 | 313 | 201 | 51 | 0 | 3 | 0 | 958 |
| 2008-09 | 0 | 0 | 0 | 5 | 34 | 278 | 198 | 145 | 25 | 23 | 0 | 0 | 708 |
| 2009- | 0 | 0 | 0 | 23 | 44 | 342 | | | | | | | |

WBAN : 23183

## COOLING DEGREE DAYS (base 65°F) 2009 PHOENIX (KPHX)

| YEAR | JAN | FEB | MAR | APR | MAY | JUN | JUL | AUG | SEP | OCT | NOV | DEC | TOTAL |
|------|-----|-----|-----|-----|-----|-----|-----|-----|-----|-----|-----|-----|-------|
| 1980 | 0 | 5 | 2 | 187 | 344 | 724 | 956 | 852 | 675 | 346 | 88 | 13 | 4192 |
| 1981 | 5 | 36 | 40 | 345 | 489 | 857 | 943 | 961 | 731 | 277 | 95 | 5 | 4784 |
| 1982 | 0 | 21 | 24 | 234 | 481 | 697 | 899 | 897 | 658 | 272 | 12 | 0 | 4195 |
| 1983 | 2 | 1 | 38 | 112 | 489 | 715 | 951 | 861 | 787 | 388 | 85 | 0 | 4429 |
| 1984 | 0 | 2 | 107 | 203 | 688 | 724 | 836 | 821 | 681 | 208 | 41 | 0 | 4311 |
| 1985 | 0 | 17 | 40 | 316 | 603 | 826 | 934 | 920 | 525 | 319 | 47 | 0 | 4547 |
| 1986 | 3 | 52 | 209 | 282 | 543 | 844 | 853 | 921 | 582 | 307 | 51 | 1 | 4648 |
| 1987 | 3 | 30 | 51 | 396 | 553 | 846 | 879 | 850 | 665 | 499 | 48 | 0 | 4820 |
| 1988 | 10 | 31 | 108 | 265 | 520 | 851 | 972 | 904 | 678 | 543 | 124 | 3 | 5009 |
| 1989 | 1 | 49 | 210 | 459 | 566 | 820 | 1013 | 897 | 751 | 392 | 87 | 0 | 5245 |
| 1990 | 5 | 25 | 150 | 339 | 506 | 873 | 895 | 806 | 683 | 431 | 101 | 0 | 4814 |
| 1991 | 0 | 61 | 22 | 227 | 465 | 691 | 937 | 920 | 712 | 514 | 70 | 0 | 4619 |
| 1992 | 0 | 13 | 54 | 373 | 567 | 762 | 868 | 851 | 771 | 464 | 30 | 0 | 4753 |
| 1993 | 1 | 1 | 89 | 271 | 585 | 746 | 871 | 832 | 695 | 369 | 24 | 0 | 4484 |
| 1994 | 0 | 3 | 59 | 223 | 421 | 821 | 904 | 943 | 664 | 273 | 10 | 0 | 4321 |
| 1995 | 0 | 19 | 58 | 158 | 365 | 642 | 923 | 927 | 733 | 354 | 79 | 0 | 4258 |
| 1996 | 0 | 41 | 70 | 278 | 564 | 817 | 943 | 908 | 583 | 358 | 60 | 4 | 4626 |
| 1997 | 2 | 8 | 185 | 217 | 665 | 689 | 899 | 873 | 756 | 315 | 62 | 0 | 4671 |
| 1998 | 0 | 0 | 50 | 136 | 361 | 619 | 924 | 928 | 691 | 295 | 13 | 2 | 4019 |
| 1999 | 0 | 7 | 68 | 138 | 468 | 724 | 820 | 875 | 679 | 453 | 151 | 1 | 4384 |
| 2000 | 6 | 7 | 60 | 321 | 624 | 789 | 939 | 860 | 764 | 259 | 0 | 0 | 4629 |
| 2001 | 0 | 0 | 112 | 250 | 685 | 816 | 920 | 928 | 823 | 452 | 177 | 0 | 5163 |
| 2002 | 0 | 19 | 89 | 358 | 525 | 858 | 971 | 940 | 749 | 325 | 82 | 0 | 4916 |
| 2003 | 8 | 6 | 79 | 179 | 576 | 810 | 1023 | 924 | 779 | 556 | 20 | 4 | 4964 |
| 2004 | 6 | 1 | 281 | 249 | 580 | 791 | 920 | 867 | 699 | 341 | 20 | 0 | 4755 |
| 2005 | 4 | 0 | 35 | 227 | 557 | 770 | 1005 | 850 | 745 | 418 | 98 | 0 | 4709 |
| 2006 | 8 | 21 | 46 | 241 | 619 | 898 | 988 | 861 | 626 | 349 | 119 | 0 | 4776 |
| 2007 | 0 | 11 | 177 | 277 | 615 | 839 | 963 | 973 | 768 | 419 | 189 | 0 | 5231 |
| 2008 | 0 | 11 | 108 | 277 | 429 | 852 | 935 | 873 | 752 | 412 | 104 | 0 | 4753 |
| 2009 | 9 | 33 | 108 | 222 | 663 | 718 | 1038 | 946 | 757 | 321 | 125 | 0 | 4940 |

## SNOWFALL (inches) 2009 PHOENIX (KPHX)

| YEAR | JUL | AUG | SEP | OCT | NOV | DEC | JAN | FEB | MAR | APR | MAY | JUN | TOTAL |
|---|---|---|---|---|---|---|---|---|---|---|---|---|---|
| 1977-78 | 0.0 | 0.0 | 0.0 | 0.0 | 0.0 | 0.0 | 0.0 | 0.0 | 0.0 | 0.0 | 0.0 | 0.0 | 0.0 |
| 1978-79 | 0.0 | 0.0 | 0.0 | 0.0 | 0.0 | 0.0 | 0.0 | 0.0 | 0.0 | 0.0 | 0.0 | 0.0 | 0.0 |
| 1979-80 | 0.0 | 0.0 | 0.0 | 0.0 | 0.0 | 0.0 | 0.0 | 0.0 | 0.0 | 0.0 | 0.0 | 0.0 | 0.0 |
| 1980-81 | 0.0 | 0.0 | 0.0 | 0.0 | 0.0 | 0.0 | 0.0 | 0.0 | 0.0 | 0.0 | 0.0 | 0.0 | 0.0 |
| 1981-82 | 0.0 | 0.0 | 0.0 | 0.0 | 0.0 | 0.0 | 0.0 | 0.0 | 0.0 | 0.0 | 0.0 | 0.0 | 0.0 |
| 1982-83 | 0.0 | 0.0 | 0.0 | 0.0 | 0.0 | 0.0 | 0.0 | 0.0 | 0.0 | 0.0 | 0.0 | 0.0 | 0.0 |
| 1983-84 | 0.0 | 0.0 | 0.0 | 0.0 | 0.0 | 0.0 | 0.0 | 0.0 | 0.0 | 0.0 | 0.0 | 0.0 | 0.0 |
| 1984-85 | 0.0 | 0.0 | 0.0 | 0.0 | 0.0 | 0.0 | 0.0 | T | 0.0 | 0.0 | 0.0 | 0.0 | T |
| 1985-86 | 0.0 | 0.0 | 0.0 | 0.0 | 0.0 | 0.1 | 0.0 | 0.0 | 0.0 | 0.0 | 0.0 | 0.0 | 0.1 |
| 1986-87 | 0.0 | 0.0 | 0.0 | 0.0 | 0.0 | 0.0 | T | 0.0 | 0.0 | 0.0 | 0.0 | 0.0 | T |
| 1987-88 | 0.0 | 0.0 | 0.0 | 0.0 | 0.0 | 0.0 | 0.0 | 0.0 | 0.0 | 0.0 | 0.0 | 0.0 | 0.0 |
| 1988-89 | 0.0 | 0.0 | 0.0 | 0.0 | 0.0 | 0.0 | 0.0 | 0.0 | 0.0 | 0.0 | 0.0 | 0.0 | 0.0 |
| 1989-90 | 0.0 | 0.0 | 0.0 | 0.0 | 0.0 | 0.0 | 0.0 | 0.0 | 0.0 | 0.0 | 0.0 | 0.0 | 0.0 |
| 1990-91 | 0.0 | 0.0 | 0.0 | 0.0 | 0.0 | 0.4 | 0.0 | 0.0 | T | 0.0 | 0.0 | 0.0 | 0.4 |
| 1991-92 | 0.0 | 0.0 | 0.0 | 0.0 | 0.0 | 0.0 | 0.0 | 0.0 | 0.0 | 0.0 | T | 0.0 | T |
| 1992-93 | 0.0 | 0.0 | 0.0 | T | 0.0 | 0.0 | T | 0.0 | 0.0 | 0.0 | 0.0 | 0.0 | T |
| 1993-94 | 0.0 | 0.0 | 0.0 | 0.0 | 0.0 | 0.0 | 0.0 | T | 0.0 | 0.0 | 0.0 | 0.0 | T |
| 1994-95 | 0.0 | 0.0 | 0.0 | 0.0 | 0.0 | 0.0 | 0.0 | 0.0 | 0.0 | 0.0 | 0.0 | 0.0 | 0.0 |
| 1995-96 | 0.0 | 0.0 | 0.0 | 0.0 | 0.0 | 0.0 | 0.0 | 0.0 | 0.0 | 0.0 | 0.0 | 0.0 | 0.0 |
| 1996-97 | 0.0 | 0.0 | 0.0 | 0.0 | 0.0 | 0.0 | 0.0 | 0.0 | 0.0 | 0.0 | 0.0 | 0.0 | 0.0 |
| 1997-98 | 0.0 | 0.0 | 0.0 | 0.0 | 0.0 | 0.0 | 0.0 | 0.0 | 0.0 | 0.0 | 0.0 | 0.0 | 0.0 |
| 1998-99 | 0.0 | 0.0 | 0.0 | 0.0 | 0.0 | T | 0.0 | 0.0 | 0.0 | 0.0 | 0.0 | 0.0 | T |
| 1999-00 | 0.0 | 0.0 | 0.0 | 0.0 | 0.0 | 0.0 | | | | | | | |
| 2000-01 | | | | | | | | | | | | | |
| 2001-02 | | | | | | | | | | | | | |
| 2002-03 | | | | | | | | | | | | | |
| 2003-04 | | | | | | | | | | | | | |
| 2004-05 | | | | | | | | | | | | | |
| 2005- | | | | | | | | | | | | | |
| 2006- | | | | | | | | | | | | | |
| POR=<br>54 YRS | 0.0 | 0.0 | 0.0 | T | 0.0 | T | T | T | T | 0.0 | T | 0.0 | T |

# Phoenix, AZ 2009
## Temperature and Precipitation

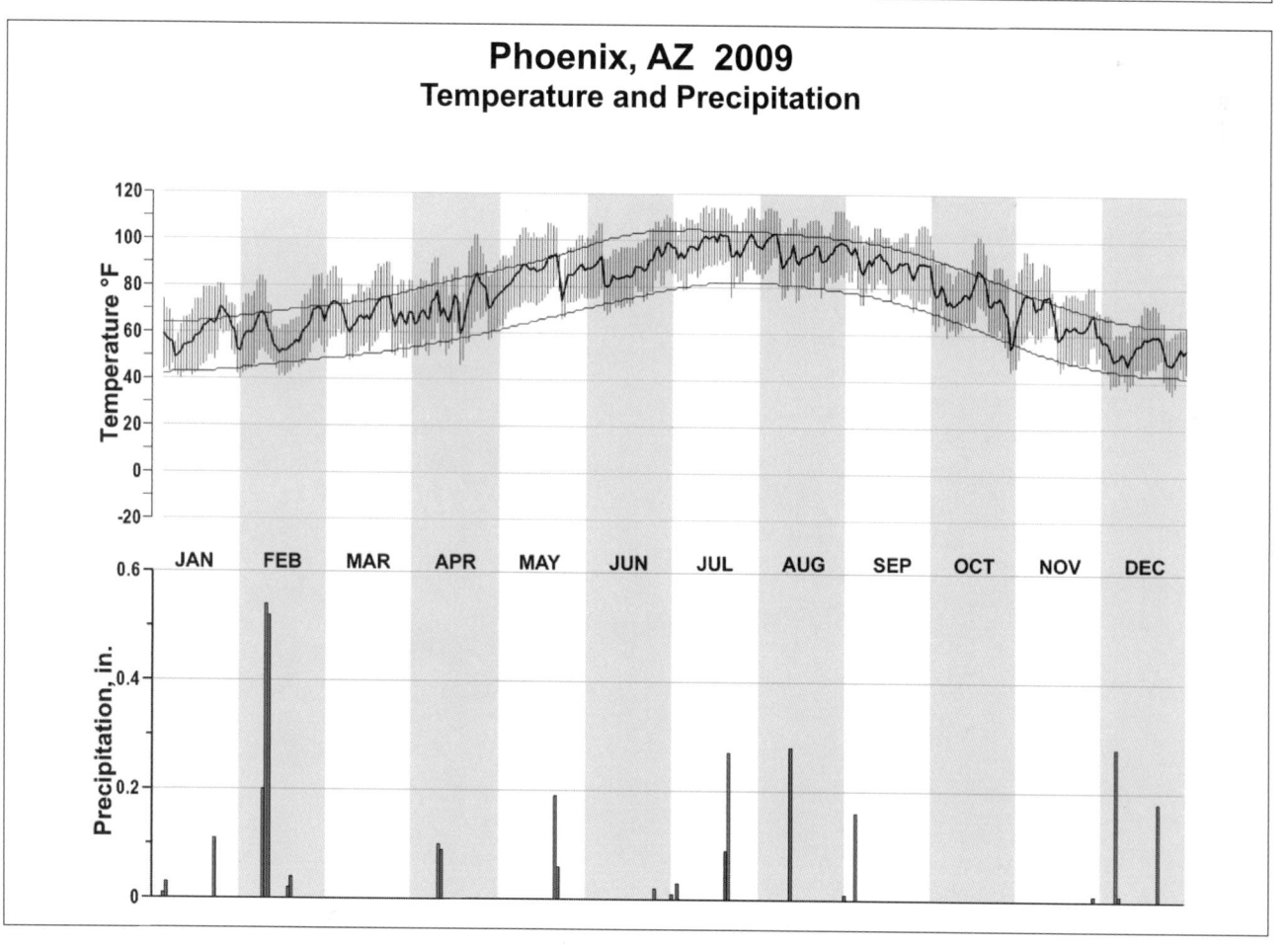

# 2009
# TUCSON
# ARIZONA (KTUS)

Tucson lies at the foot of the Catalina Mountains, north of the airport. The area within about 15 miles of the airport station is flat or gently rolling, with many dry washes. The soil is sandy, and vegetation is mostly brush, cacti, and small trees. Rugged mountains encircle the valley. The mountains to the north, east, and south rise to over 5,000 feet above the airport. The western hills and mountains range from 500 to 4,000 feet.

The climate of Tucson is characterized by a long hot season, from April to October. Temperatures above 90 degrees prevail from May through September. Temperatures of 100 degrees or higher average 41 days annually, including 14 days each for June and July, but these extreme temperatures are moderated by low relative humidities. The temperature range is large, averaging 30 degrees or more a day.

More than 50 percent of the annual precipitation falls between July 1 and September 15, and over 20 percent falls from December through March. During the summer, scattered convective or orographic showers and thunderstorms often fill dry washes to overflowing. On occasion, brief, torrential downpours cause destructive flash floods in the Tucson area. Hail rarely occurs in thunderstorms. The December through March precipitation occurs as prolonged rainstorms that replenish the ground water. During these storms, snow often falls on the higher mountains, but snow in Tucson is infrequent, particularly in accumulations exceeding an inch in depth.

From the first of the year, the humidity decreases steadily until the summer thunderstorm season, when it shows a marked increase. From mid-September, the end of the thunderstorm season, the humidity decreases again until late November. Occasionally during the summer, humidities are high enough to produce discomfort, but only for short periods. During the hot season, humidity values sometimes fall below 5 percent.

Tucson lies in the zone receiving more sunshine than any other section of the United States. Cloudless days are commonplace, and average cloudiness is low.

Surface winds are generally light, with no major seasonal changes in velocity or direction. Occasional duststorms occur in areas where the ground has been disturbed. During the spring, winds may briefly be strong enough to cause some damage to trees and buildings. Wind velocities and directions are influenced by the surrounding mountains, and the general slope of the terrain. Usually local winds tend to be in the southeast quadrant during the night and early morning hours, veering to northwest during the day. Highest velocities usually occur with winds from the southwest and east to south.

While dust and haze are frequently visible, their effect on the general clarity of the atmosphere is not great. Visibility is normally high.

Based on the 1951-1980 period, the average first occurrence of 32 degrees Fahrenheit in the fall is November 29 and the average last occurrence in the spring is February 28.

# NORMALS, MEANS, AND EXTREMES
## TUCSON (KTUS)

LATITUDE: 32 ° 7 'N  LONGITUDE: -110° 57'W  ELEVATION (FT): GRND: 2549  BARO: 2581  TIME ZONE: MOUNTAIN (UTC -7)  WBAN: 23160

| | ELEMENT | POR | JAN | FEB | MAR | APR | MAY | JUN | JUL | AUG | SEP | OCT | NOV | DEC | YEAR |
|---|---|---|---|---|---|---|---|---|---|---|---|---|---|---|---|
| **TEMPERATURE °F** | NORMAL DAILY MAXIMUM | 30 | 64.5 | 68.4 | 73.3 | 81.5 | 90.4 | 100. | 99.6 | 97.4 | 94.0 | 84.0 | 72.3 | 64.6 | 82.5 |
| | MEAN DAILY MAXIMUM | 61 | 64.9 | 68.3 | 73.4 | 81.6 | 90.5 | 99.6 | 99.3 | 97.0 | 94.2 | 84.6 | 73.3 | 65.2 | 82.7 |
| | HIGHEST DAILY MAXIMUM | 69 | 87 | 92 | 99 | 104 | 109 | 117 | 114 | 112 | 107 | 102 | 93 | 84 | 117 |
| | YEAR OF OCCURRENCE | | 1999 | 1957 | 1988 | 1989 | 2005 | 1990 | 1995 | 1993 | 2000 | 1993 | 2009 | 1954 | JUN 1990 |
| | MEAN OF EXTREME MAXS. | 61 | 78.0 | 81.6 | 86.6 | 93.5 | 101.1 | 107.8 | 107.7 | 104.6 | 102.1 | 95.5 | 85.1 | 77.8 | 93.5 |
| | NORMAL DAILY MINIMUM | 30 | 38.9 | 41.6 | 45.1 | 50.5 | 58.6 | 68.0 | 73.4 | 72.4 | 67.7 | 57.0 | 45.1 | 39.2 | 54.8 |
| | MEAN DAILY MINIMUM | 61 | 38.8 | 41.0 | 44.9 | 50.9 | 58.9 | 68.1 | 74.4 | 72.6 | 67.9 | 56.9 | 45.5 | 39.0 | 54.9 |
| | LOWEST DAILY MINIMUM | 69 | 16 | 20 | 20 | 27 | 38 | 47 | 59 | 61 | 44 | 26 | 24 | 16 | 16 |
| | YEAR OF OCCURRENCE | | 1949 | 1955 | 1965 | 1945 | 1950 | 1955 | 1992 | 1956 | 1965 | 1971 | 1979 | 1974 | DEC 1974 |
| | MEAN OF EXTREME MINS. | 61 | 27.0 | 29.6 | 32.5 | 39.4 | 47.0 | 57.2 | 67.0 | 66.6 | 59.0 | 43.6 | 32.2 | 26.8 | 44.0 |
| | NORMAL DRY BULB | 30 | 51.7 | 55.0 | 59.2 | 66.0 | 74.5 | 84.1 | 86.5 | 84.9 | 80.9 | 70.5 | 58.7 | 51.9 | 68.7 |
| | MEAN DRY BULB | 61 | 51.8 | 54.7 | 59.2 | 66.2 | 74.7 | 84.0 | 86.9 | 84.8 | 81.1 | 70.7 | 59.4 | 52.1 | 68.8 |
| | MEAN WET BULB | 26 | 39.5 | 40.8 | 42.6 | 45.1 | 50.5 | 55.9 | 65.6 | 67.1 | 61.4 | 51.5 | 43.3 | 39.6 | 50.2 |
| | MEAN DEW POINT | 26 | 31.2 | 31.8 | 31.9 | 31.8 | 35.4 | 41.0 | 56.7 | 61.0 | 52.8 | 42.2 | 33.9 | 31.0 | 40.1 |
| | NORMAL NO. DAYS WITH: | | | | | | | | | | | | | | |
| | MAXIMUM >= 90 | 30 | 0.0 | 0.0 | 0.8 | 5.4 | 18.3 | 28.5 | 29.2 | 29.0 | 24.3 | 9.4 | 0.2 | 0.0 | 145.1 |
| | MAXIMUM <= 32 | 30 | 0.0 | 0.0 | 0.0 | 0.0 | 0.0 | 0.0 | 0.0 | 0.0 | 0.0 | 0.0 | 0.0 | 0.0 | 0.0 |
| | MINIMUM <= 32 | 30 | 5.0 | 2.1 | 0.5 | 0.0 | 0.0 | 0.0 | 0.0 | 0.0 | 0.0 | * | 1.5 | 4.8 | 13.9 |
| | MINIMUM <= 0 | 30 | 0.0 | 0.0 | 0.0 | 0.0 | 0.0 | 0.0 | 0.0 | 0.0 | 0.0 | 0.0 | 0.0 | 0.0 | 0.0 |
| **H/C** | NORMAL HEATING DEG. DAYS | 30 | 401 | 275 | 194 | 76 | 7 | 0 | 0 | 0 | 0 | 33 | 195 | 397 | 1578 |
| | NORMAL COOLING DEG. DAYS | 30 | 0 | 3 | 25 | 107 | 300 | 577 | 672 | 625 | 477 | 211 | 20 | 0 | 3017 |
| **RH** | NORMAL (PERCENT) | 30 | 49 | 45 | 39 | 28 | 24 | 21 | 40 | 47 | 41 | 39 | 43 | 50 | 39 |
| | HOUR 05 LST | 30 | 62 | 59 | 55 | 43 | 37 | 33 | 55 | 64 | 57 | 53 | 56 | 62 | 53 |
| | HOUR 11 LST | 30 | 41 | 36 | 31 | 21 | 18 | 17 | 32 | 37 | 33 | 32 | 34 | 41 | 31 |
| | HOUR 17 LST | 30 | 31 | 27 | 24 | 16 | 14 | 13 | 27 | 32 | 27 | 25 | 27 | 33 | 25 |
| | HOUR 23 LST | 30 | 57 | 50 | 43 | 31 | 26 | 23 | 44 | 52 | 46 | 44 | 49 | 56 | 43 |
| **S** | PERCENT POSSIBLE SUNSHINE | 52 | 80 | 81 | 86 | 92 | 93 | 92 | 78 | 80 | 87 | 88 | 85 | 79 | 85 |
| **W/O** | MEAN NO. DAYS WITH: | | | | | | | | | | | | | | |
| | HEAVY FOG(VISBY <= 1/4 MI) | 46 | 0.3 | 0.1 | 0.0 | 0.0 | 0.0 | 0.0 | 0.0 | 0.0 | 0.0 | 0.0 | 0.1 | 0.3 | 0.8 |
| | THUNDERSTORMS | 61 | 0.3 | 0.3 | 0.5 | 0.7 | 1.5 | 2.6 | 13.0 | 13.2 | 5.1 | 1.8 | 0.4 | 0.3 | 39.7 |
| **CLOUDNESS** | MEAN: | | | | | | | | | | | | | | |
| | SUNRISE-SUNSET (OKTAS) | | | | 2.4 | | 0.0 | 1.6 | | | | | | | |
| | MIDNIGHT-MIDNIGHT (OKTAS) | | | | | | 0.0 | 1.6 | | | | | | | |
| | MEAN NO. DAYS WITH: | | | | | | | | | | | | | | |
| | CLEAR | 1 | 1.0 | 5.0 | 12.0 | | 27.0 | 15.0 | | | | | | | |
| | PARTLY CLOUDY | 1 | 1.0 | 3.0 | 3.0 | | 1.0 | 2.0 | | | | | | | |
| | CLOUDY | | | 4.0 | 1.0 | | | 1.0 | | | | | | | |
| **PR** | MEAN STATION PRESSURE(IN) | 26 | 27.41 | 27.38 | 27.33 | 27.29 | 27.26 | 27.26 | 27.31 | 27.32 | 27.30 | 27.33 | 27.38 | 27.41 | 27.33 |
| | MEAN SEA-LEVEL PRES. (IN) | 26 | 30.04 | 29.99 | 29.92 | 29.86 | 29.80 | 29.77 | 29.83 | 29.84 | 29.82 | 29.89 | 29.98 | 30.04 | 29.90 |
| **WINDS** | MEAN SPEED (MPH) | 26 | 7.4 | 7.6 | 8.1 | 8.6 | 8.4 | 8.4 | 8.2 | 7.7 | 8.0 | 7.7 | 7.5 | 7.1 | 7.9 |
| | PREVAIL.DIR(TENS OF DEGS) | 30 | 14 | 15 | 15 | 15 | 15 | 15 | 15 | 15 | 15 | 15 | 15 | 15 | 15 |
| | MAXIMUM 2-MINUTE: | | | | | | | | | | | | | | |
| | SPEED (MPH) | 14 | 36 | 37 | 40 | 39 | 36 | 48 | 55 | 48 | 44 | 38 | 33 | 37 | 55 |
| | DIR. (TENS OF DEGS) | | 20 | 20 | 14 | 18 | 21 | 05 | 07 | 15 | 15 | 11 | 13 | 23 | 07 |
| | YEAR OF OCCURRENCE | | 2008 | 1998 | 2004 | 2002 | 2009 | 2006 | 2001 | 2006 | 2002 | 2000 | 2008 | 2009 | JUL 2001 |
| | MAXIMUM 3-SECOND | | | | | | | | | | | | | | |
| | SPEED (MPH) | 14 | 46 | 48 | 46 | 46 | 53 | 56 | 60 | 60 | 52 | 53 | 53 | 48 | 60 |
| | DIR. (TENS OF DEGS) | | 12 | 22 | 14 | 20 | 21 | 05 | 07 | 12 | 14 | 26 | 19 | 23 | 07 |
| | YEAR OF OCCURRENCE | | 2008 | 1998 | 2004 | 2009 | 2009 | 2006 | 2001 | 2000 | 2002 | 1996 | 2004 | 2009 | JUL 2001 |
| **PRECIPITATION** | NORMAL (IN) | 30 | 0.99 | 0.88 | 0.81 | 0.28 | 0.24 | 0.24 | 2.07 | 2.30 | 1.45 | 1.21 | 0.67 | 1.03 | 12.17 |
| | MAXIMUM MONTHLY (IN) | 69 | 4.81 | 3.20 | 2.26 | 1.66 | 1.11 | 1.56 | 6.17 | 7.93 | 5.11 | 4.98 | 1.90 | 5.02 | 7.93 |
| | YEAR OF OCCURRENCE | | 1993 | 1998 | 1952 | 1951 | 1992 | 2000 | 1981 | 1955 | 1964 | 1983 | 1952 | 1965 | AUG 1955 |
| | MINIMUM MONTHLY (IN) | 69 | T | 0.00 | 0.00 | 0.00 | 0.00 | 0.00 | 0.04 | 0.23 | 0.00 | 0.00 | 0.00 | 0.00 | 0.00 |
| | YEAR OF OCCURRENCE | | 2006 | 1972 | 1956 | 1972 | 1974 | 1983 | 1995 | 1976 | 1953 | 1982 | 1980 | 1981 | JUN 1983 |
| | MAXIMUM IN 24 HOURS (IN) | 69 | 1.46 | 1.70 | 1.19 | 1.28 | 0.89 | 1.27 | 3.93 | 2.48 | 3.05 | 3.58 | 1.86 | 2.12 | 3.93 |
| | YEAR OF OCCURRENCE | | 1993 | 2007 | 1952 | 1999 | 1943 | 1954 | 1958 | 1961 | 1964 | 1983 | 1968 | 1994 | JUL 1958 |
| | NORMAL NO. DAYS WITH: | | | | | | | | | | | | | | |
| | PRECIPITATION >= 0.01 | 30 | 4.9 | 4.1 | 4.4 | 2.0 | 2.1 | 2.3 | 9.8 | 9.7 | 5.2 | 3.7 | 3.2 | 4.5 | 55.9 |
| | PRECIPITATION >= 1.00 | 30 | 0.1 | * | 0.0 | * | 0.0 | 0.0 | 0.4 | 0.4 | 0.4 | 0.3 | * | 0.1 | 1.7 |
| **SNOWFALL** | NORMAL (IN) | 30 | 0.3 | 0.2 | 0.2 | 0.1 | 0.0 | 0.0 | 0.0 | 0.0 | 0.0 | 0.0 | 0.* | 0.4 | 1.2 |
| | MAXIMUM MONTHLY (IN) | 69 | 4.7 | 3.9 | 5.7 | 2.0 | 0.0 | T | T | T | T | 0.0 | 6.4 | 6.8 | 6.8 |
| | YEAR OF OCCURRENCE | | 1987 | 1965 | 1964 | 1976 | 2006 | 2007 | 1995 | 2006 | 1996 | 2008 | 1958 | 1971 | DEC 1971 |
| | MAXIMUM IN 24 HOURS (IN) | 68 | 4.3 | 3.9 | 5.7 | 2.0 | T | 0.0 | T | T | T | T | 6.4 | 6.8 | 6.8 |
| | YEAR OF OCCURRENCE' | | 1987 | 1965 | 1964 | 1976 | 1992 | | 1995 | 1995 | 1990 | 1991 | 1958 | 1971 | DEC 1971 |
| | MAXIMUM SNOW DEPTH (IN) | 60 | 1 | 4 | 5 | 0 | 0 | 0 | 0 | 0 | 0 | 0 | 1 | 5 | 5 |
| | YEAR OF OCCURRENCE | | 1987 | 1965 | 1964 | | | | | | | | 1958 | 1971 | DEC 1971 |
| | NORMAL NO. DAYS WITH: | | | | | | | | | | | | | | |
| | SNOWFALL >= 1.0 | 30 | 0.1 | 0.1 | 0.0 | 0.0 | 0.0 | 0.0 | 0.0 | 0.0 | 0.0 | 0.0 | 0.0 | 0.1 | 0.3 |

## PRECIPITATION (inches) 2009 TUCSON (KTUS)

| YEAR | JAN | FEB | MAR | APR | MAY | JUN | JUL | AUG | SEP | OCT | NOV | DEC | ANNUAL |
|------|-----|-----|-----|-----|-----|-----|-----|-----|-----|-----|-----|-----|--------|
| 1980 | 0.73 | 2.90 | 1.22 | 0.08 | T | 0.23 | 1.78 | 1.95 | 2.93 | 0.22 | 0.00 | 0.19 | 12.23 |
| 1981 | 1.29 | 0.71 | 1.98 | 0.56 | 0.26 | 0.16 | 6.17 | 0.80 | 1.10 | 0.06 | 0.61 | 0.00 | 13.70 |
| 1982 | 1.56 | 0.06 | 1.26 | 0.05 | 0.51 | 0.13 | 2.13 | 2.51 | 2.69 | 0.00 | 1.30 | 1.59 | 13.79 |
| 1983 | 1.70 | 0.94 | 1.28 | 0.14 | T | 0.00 | 1.98 | 4.24 | 4.28 | 4.98 | 1.71 | 0.61 | 21.86 |
| 1984 | 0.62 | 0.00 | 0.00 | 0.36 | 0.06 | 1.05 | 2.92 | 4.19 | 1.81 | 0.77 | 0.45 | 3.30 | 15.53 |
| 1985 | 1.71 | 1.08 | 0.20 | 0.45 | T | 0.07 | 3.14 | 1.97 | 1.13 | 2.03 | 0.95 | 0.15 | 12.88 |
| 1986 | 0.98 | 1.13 | 1.30 | T | 0.44 | 0.06 | 1.82 | 3.56 | 0.31 | 0.50 | 0.42 | 1.28 | 11.80 |
| 1987 | 0.59 | 1.64 | 0.83 | 0.80 | 0.74 | 0.16 | 0.37 | 2.79 | 2.30 | 0.34 | 0.44 | 1.50 | 12.50 |
| 1988 | 0.41 | 0.53 | 0.35 | 1.15 | 0.02 | 0.15 | 1.69 | 3.64 | 0.80 | 2.09 | 0.75 | 0.05 | 11.63 |
| 1989 | 0.96 | 0.23 | 0.62 | 0.00 | 0.13 | 0.06 | 1.42 | 0.90 | 0.02 | 1.84 | 0.12 | 0.18 | 6.48 |
| 1990 | 0.96 | 0.71 | 0.38 | 0.10 | 0.03 | 0.64 | 5.45 | 2.70 | 1.63 | 0.58 | 0.23 | 1.54 | 14.95 |
| 1991 | 1.15 | 0.91 | 1.40 | 0.00 | 0.00 | 0.20 | 0.44 | 2.17 | 1.54 | 0.73 | 0.80 | 1.44 | 10.78 |
| 1992 | 1.21 | 1.80 | 2.12 | 0.19 | 1.11 | 0.07 | 0.93 | 4.55 | 0.94 | 0.03 | T | 3.47 | 16.42 |
| 1993 | 4.81 | 1.50 | 0.49 | 0.00 | 0.59 | 0.02 | 0.26 | 4.93 | 0.46 | 0.81 | 0.98 | 0.14 | 14.99 |
| 1994 | 0.02 | 1.03 | 1.14 | 0.04 | 0.52 | 0.26 | 0.41 | 0.45 | 1.46 | 0.76 | 1.83 | 3.71 | 11.63 |
| 1995 | 1.41 | 1.32 | 0.54 | 0.28 | 0.15 | T | 0.04 | 3.71 | 2.29 | 0.36 | 0.86 | 0.22 | 11.18 |
| 1996 | 0.01 | 0.82 | 0.32 | T | 0.00 | T | 1.88 | 1.87 | 3.68 | 1.74 | .19 | T | 10.51 |
| 1997 | 0.93 | 0.67 | 0.02 | 0.47 | 0.44 | 0.02 | 0.51 | 2.32 | 1.43 | 0.38 | 0.49 | 2.88 | 10.56 |
| 1998 | 0.17 | 3.20 | 1.64 | 0.39 | T | 0.00 | 4.06 | 1.70 | 1.10 | 0.24 | 0.67 | 0.45 | 13.62 |
| 1999 | 0.01 | T | T | 1.34 | 0.00 | 0.16 | 4.15 | 3.05 | 0.97 | T | 0.00 | T | 9.68 |
| 2000 | 0.10 | 0.19 | 0.93 | T | 0.00 | 1.56 | 1.59 | 1.70 | 0.02 | 4.98 | 1.36 | T | 12.43 |
| 2001 | 1.24 | 0.46 | 0.88 | 0.84 | 0.24 | 0.54 | 1.09 | 0.85 | 0.33 | 0.69 | 0.05 | 0.60 | 7.81 |
| 2002 | 0.34 | 0.27 | 0.07 | 0.00 | 0.00 | 0.00 | 2.47 | 1.63 | 1.68 | 0.50 | 0.23 | 0.64 | 7.83 |
| 2003 | 0.08 | 1.02 | 0.51 | 0.04 | 0.13 | T | 2.50 | 2.04 | 2.16 | 0.38 | 1.03 | 0.16 | 10.05 |
| 2004 | 0.79 | 0.45 | 1.12 | 1.05 | T | T | 0.86 | 0.95 | 0.61 | 0.62 | 0.44 | 0.71 | 7.60 |
| 2005 | 1.35 | 1.27 | 0.37 | 0.33 | 0.63 | 0.01 | 0.72 | 4.52 | 0.05 | 0.31 | 0.00 | 0.01 | 9.57 |
| 2006 | T | T | 0.41 | T | T | 0.50 | 5.40 | 3.01 | 1.60 | 0.27 | T | 0.62 | 11.81 |
| 2007 | 0.71 | 0.04 | 0.59 | 0.16 | 0.14 | T | 5.22 | 0.90 | 0.45 | 0.02 | 0.80 | 0.76 | 9.79 |
| 2008 | 0.17 | 1.22 | 0.37 | 0.05 | 0.02 | 0.16 | 3.42 | 1.70 | 0.24 | T | 0.27 | 1.00 | 8.62 |
| 2009 | 0.63 | 0.56 | 0.18 | 0.29 | 0.67 | 0.01 | 1.78 | 0.33 | 0.74 | 0.05 | 0.13 | 0.30 | 5.67 |
| POR= 61 YRS | 0.86 | 0.73 | 0.69 | 0.31 | 0.18 | 0.23 | 3.59 | 2.25 | 1.32 | 0.86 | 0.59 | 0.93 | 12.54 |

WBAN : 23160

## AVERAGE TEMPERATURE (°F) 2009 TUCSON (KTUS)

| YEAR | JAN | FEB | MAR | APR | MAY | JUN | JUL | AUG | SEP | OCT | NOV | DEC | ANNUAL |
|------|-----|-----|-----|-----|-----|-----|-----|-----|-----|-----|-----|-----|--------|
| 1980 | 54.3 | 57.9 | 57.5 | 65.6 | 71.5 | 84.9 | 88.6 | 84.6 | 80.5 | 69.6 | 59.5 | 58.1 | 69.4 |
| 1981 | 54.8 | 57.1 | 57.1 | 69.1 | 73.4 | 85.2 | 86.4 | 86.4 | 80.7 | 68.1 | 62.2 | 55.0 | 69.6 |
| 1982 | 50.7 | 54.7 | 57.7 | 66.1 | 72.3 | 80.5 | 84.8 | 83.9 | 79.2 | 67.0 | 57.7 | 50.1 | 67.1 |
| 1983 | 52.9 | 53.8 | 57.3 | 60.4 | 73.8 | 81.6 | 86.9 | 84.0 | 82.2 | 69.5 | 57.4 | 53.5 | 67.8 |
| 1984 | 51.8 | 53.7 | 60.5 | 64.0 | 79.9 | 83.1 | 84.2 | 82.9 | 81.5 | 66.3 | 57.8 | 51.5 | 68.1 |
| 1985 | 50.3 | 53.1 | 58.7 | 68.7 | 75.9 | 85.8 | 87.5 | 86.1 | 77.4 | 70.0 | 58.0 | 52.9 | 68.7 |
| 1986 | 58.7 | 56.9 | 63.8 | 69.0 | 76.8 | 86.6 | 85.5 | 86.0 | 79.0 | 69.6 | 59.8 | 52.3 | 70.3 |
| 1987 | 50.9 | 54.2 | 57.9 | 70.1 | 74.3 | 86.3 | 87.4 | 85.1 | 79.9 | 75.1 | 58.9 | 50.3 | 69.2 |
| 1988 | 53.0 | 59.4 | 61.4 | 68.0 | 76.4 | 86.8 | 87.9 | 85.9 | 80.4 | 75.3 | 59.2 | 51.9 | 70.5 |
| 1989 | 49.9 | 58.2 | 65.0 | 73.8 | 77.4 | 85.4 | 90.0 | 86.6 | 84.5 | 71.1 | 61.7 | 53.0 | 71.4 |
| 1990 | 51.8 | 52.8 | 61.8 | 69.7 | 75.2 | 88.7 | 85.0 | 82.6 | 82.2 | 73.1 | 61.6 | 51.1 | 69.6 |
| 1991 | 52.3 | 59.8 | 55.4 | 65.2 | 73.5 | 81.5 | 87.5 | 86.6 | 80.7 | 74.0 | 58.9 | 54.3 | 69.1 |
| 1992 | 51.6 | 57.3 | 59.4 | 70.8 | 76.7 | 84.5 | 86.8 | 85.1 | 83.6 | 74.2 | 56.1 | 51.4 | 69.8 |
| 1993 | 55.2 | 54.0 | 61.3 | 68.6 | 78.1 | 85.0 | 88.0 | 85.5 | 81.4 | 72.6 | 58.8 | 53.4 | 70.2 |
| 1994 | 53.7 | 55.2 | 62.9 | 68.6 | 75.6 | 89.2 | 90.4 | 90.3 | 84.2 | 70.5 | 56.7 | 53.9 | 70.9 |
| 1995 | 52.6 | 60.7 | 61.2 | 64.8 | 72.6 | 83.3 | 88.4 | 87.3 | 82.9 | 72.4 | 63.1 | 54.0 | 70.3 |
| 1996 | 53.6 | 58.8 | 61.1 | 68.9 | 79.0 | 87.4 | 88.6 | 86.4 | 77.7 | 70.4 | 61.0 | 53.7 | 70.6 |
| 1997 | 52.4 | 53.6 | 64.8 | 65.8 | 79.7 | 83.1 | 88.0 | 85.8 | 84.2 | 70.0 | 60.7 | 48.6 | 69.7 |
| 1998 | 53.2 | 50.8 | 57.9 | 61.4 | 72.9 | 81.8 | 86.5 | 86.5 | 82.6 | 70.7 | 60.5 | 52.0 | 68.1 |
| 1999 | 53.6 | 56.9 | 61.4 | 62.3 | 74.7 | 83.8 | 84.0 | 85.2 | 81.8 | 74.7 | 65.4 | 51.3 | 69.6 |
| 2000 | 55.0 | 57.4 | 58.8 | 70.3 | 80.2 | 84.5 | 88.2 | 84.9 | 84.8 | 68.2 | 52.8 | 54.3 | 70.0 |
| 2001 | 49.7 | 52.8 | 60.2 | 66.6 | 79.3 | 85.6 | 86.2 | 85.9 | 84.4 | 73.2 | 62.9 | 49.4 | 69.7 |
| 2002 | 51.6 | 57.1 | 59.2 | 71.6 | 75.8 | 88.0 | 86.8 | 86.1 | 82.5 | 69.5 | 61.4 | 50.7 | 70.0 |
| 2003 | 58.2 | 54.9 | 59.8 | 65.7 | 78.0 | 84.8 | 89.2 | 86.3 | 82.8 | 75.4 | 59.0 | 52.8 | 70.7 |
| 2004 | 53.0 | 50.8 | 66.7 | 66.3 | 77.8 | 84.8 | 87.0 | 84.9 | 81.0 | 69.9 | 56.2 | 52.7 | 69.3 |
| 2005 | 54.5 | 55.8 | 59.3 | 67.8 | 77.4 | 85.7 | 90.6 | 83.9 | 83.6 | 72.8 | 62.6 | 54.6 | 70.7 |
| 2006 | 54.5 | 58.3 | 58.7 | 68.5 | 79.0 | 88.4 | 88.3 | 83.7 | 78.5 | 70.2 | 63.0 | 51.1 | 70.2 |
| 2007 | 48.6 | 55.4 | 63.1 | 68.6 | 78.1 | 86.0 | 87.9 | 86.9 | 83.0 | 72.9 | 65.7 | 49.2 | 70.5 |
| 2008 | 51.7 | 55.3 | 60.9 | 68.3 | 73.3 | 87.0 | 85.6 | 85.4 | 82.4 | 72.9 | 63.0 | 53.8 | 70.0 |
| 2009 | 55.5 | 57.3 | 63.2 | 67.0 | 80.1 | 82.9 | 90.1 | 88.7 | 83.7 | 69.8 | 64.4 | 50.8 | 71.1 |
| POR= 61 YRS | 51.8 | 54.7 | 59.2 | 66.2 | 74.7 | 84.0 | 86.9 | 84.8 | 81.1 | 70.7 | 59.4 | 52.1 | 68.8 |

## HEATING DEGREE DAYS (base 65°F) 2009 TUCSON (KTUS)

| YEAR | JUL | AUG | SEP | OCT | NOV | DEC | JAN | FEB | MAR | APR | MAY | JUN | TOTAL |
|---|---|---|---|---|---|---|---|---|---|---|---|---|---|
| 1980-81 | 0 | 0 | 0 | 66 | 197 | 210 | 310 | 220 | 244 | 31 | 0 | 0 | 1278 |
| 1981-82 | 0 | 0 | 0 | 34 | 106 | 304 | 437 | 291 | 223 | 46 | 10 | 0 | 1451 |
| 1982-83 | 0 | 0 | 0 | 41 | 211 | 456 | 371 | 309 | 239 | 168 | 6 | 0 | 1801 |
| 1983-84 | 0 | 0 | 0 | 0 | 232 | 348 | 402 | 323 | 140 | 110 | 0 | 0 | 1555 |
| 1984-85 | 0 | 0 | 0 | 49 | 221 | 413 | 448 | 328 | 200 | 41 | 0 | 0 | 1700 |
| 1985-86 | 0 | 0 | 0 | 9 | 217 | 369 | 193 | 244 | 117 | 22 | 6 | 0 | 1177 |
| 1986-87 | 0 | 0 | 0 | 11 | 154 | 387 | 429 | 299 | 225 | 24 | 0 | 0 | 1529 |
| 1987-88 | 0 | 0 | 0 | 0 | 188 | 452 | 366 | 171 | 161 | 46 | 12 | 0 | 1396 |
| 1988-89 | 0 | 0 | 0 | 0 | 220 | 402 | 461 | 199 | 82 | 9 | 4 | 0 | 1377 |
| 1989-90 | 0 | 0 | 0 | 25 | 107 | 361 | 402 | 340 | 156 | 16 | 3 | 0 | 1410 |
| 1990-91 | 0 | 0 | 0 | 5 | 152 | 427 | 384 | 140 | 296 | 47 | 3 | 0 | 1454 |
| 1991-92 | 0 | 0 | 0 | 56 | 195 | 325 | 408 | 215 | 169 | 24 | 0 | 0 | 1392 |
| 1992-93 | 0 | 0 | 0 | 0 | 261 | 418 | 298 | 299 | 129 | 28 | 0 | 0 | 1433 |
| 1993-94 | 0 | 0 | 0 | 5 | 186 | 355 | 345 | 272 | 94 | 42 | 0 | 0 | 1299 |
| 1994-95 | 0 | 0 | 0 | 24 | 255 | 335 | 377 | 123 | 143 | 84 | 17 | 0 | 1358 |
| 1995-96 | 0 | 0 | 0 | 3 | 64 | 332 | 344 | 173 | 135 | 23 | 0 | 0 | 1074 |
| 1996-97 | 0 | 0 | 0 | 91 | 147 | 346 | 386 | 315 | 66 | 77 | 0 | 0 | 1428 |
| 1997-98 | 0 | 0 | 0 | 52 | 146 | 502 | 360 | 392 | 232 | 153 | 3 | 0 | 1840 |
| 1998-99 | 0 | 0 | 0 | 20 | 136 | 398 | 346 | 220 | 126 | 155 | 7 | 0 | 1408 |
| 1999-00 | 0 | 0 | 0 | 0 | 83 | 416 | 308 | 217 | 186 | 26 | 0 | 0 | 1236 |
| 2000-01 | 0 | 0 | 0 | 60 | 357 | 323 | 470 | 335 | 178 | 95 | 1 | 0 | 1819 |
| 2001-02 | 0 | 0 | 0 | 0 | 134 | 479 | 409 | 221 | 197 | 3 | 0 | 0 | 1443 |
| 2002-03 | 0 | 0 | 0 | 22 | 114 | 435 | 204 | 283 | 178 | 67 | 0 | 0 | 1303 |
| 2003-04 | 0 | 0 | 0 | 0 | 186 | 370 | 366 | 405 | 87 | 55 | 0 | 0 | 1469 |
| 2004-05 | 0 | 0 | 0 | 36 | 274 | 375 | 322 | 247 | 183 | 29 | 0 | 0 | 1466 |
| 2005-06 | 0 | 0 | 0 | 0 | 122 | 314 | 321 | 185 | 203 | 18 | 0 | 0 | 1163 |
| 2006-07 | 0 | 0 | 0 | 19 | 97 | 424 | 498 | 262 | 131 | 37 | 0 | 0 | 1468 |
| 2007-08 | 0 | 0 | 0 | 2 | 68 | 480 | 403 | 278 | 155 | 24 | 4 | 0 | 1414 |
| 2008-09 | 0 | 0 | 0 | 24 | 100 | 343 | 296 | 226 | 110 | 62 | 0 | 0 | 1161 |
| 2009- | 0 | 0 | 0 | 47 | 89 | 437 | | | | | | | |

WBAN : 23160

## COOLING DEGREE DAYS (base 65°F) 2009 TUCSON (KTUS)

| YEAR | JAN | FEB | MAR | APR | MAY | JUN | JUL | AUG | SEP | OCT | NOV | DEC | TOTAL |
|---|---|---|---|---|---|---|---|---|---|---|---|---|---|
| 1980 | 0 | 4 | 1 | 109 | 211 | 606 | 742 | 615 | 474 | 216 | 37 | 3 | 3018 |
| 1981 | 0 | 8 | 4 | 159 | 267 | 639 | 633 | 670 | 476 | 137 | 27 | 2 | 3022 |
| 1982 | 0 | 4 | 4 | 82 | 244 | 471 | 622 | 594 | 437 | 112 | 0 | 0 | 2570 |
| 1983 | 0 | 0 | 8 | 36 | 288 | 503 | 688 | 600 | 523 | 145 | 10 | 0 | 2801 |
| 1984 | 0 | 0 | 6 | 87 | 469 | 549 | 601 | 562 | 503 | 96 | 12 | 0 | 2885 |
| 1985 | 0 | 1 | 7 | 159 | 345 | 633 | 704 | 660 | 379 | 173 | 14 | 0 | 3075 |
| 1986 | 2 | 23 | 88 | 150 | 378 | 653 | 643 | 657 | 431 | 158 | 3 | 0 | 3186 |
| 1987 | 0 | 2 | 12 | 184 | 297 | 644 | 702 | 630 | 452 | 325 | 12 | 0 | 3260 |
| 1988 | 2 | 13 | 58 | 142 | 374 | 658 | 716 | 657 | 471 | 327 | 51 | 1 | 3470 |
| 1989 | 0 | 16 | 89 | 281 | 397 | 619 | 780 | 676 | 592 | 221 | 16 | 0 | 3687 |
| 1990 | 0 | 6 | 63 | 164 | 327 | 719 | 625 | 553 | 522 | 262 | 56 | 0 | 3297 |
| 1991 | 0 | 1 | 6 | 58 | 274 | 501 | 703 | 675 | 479 | 345 | 21 | 0 | 3063 |
| 1992 | 0 | 0 | 4 | 204 | 372 | 590 | 683 | 627 | 563 | 291 | 1 | 0 | 3335 |
| 1993 | 1 | 0 | 22 | 142 | 413 | 604 | 721 | 641 | 500 | 250 | 11 | 3 | 3308 |
| 1994 | 0 | 4 | 34 | 156 | 332 | 733 | 794 | 788 | 584 | 203 | 11 | 0 | 3639 |
| 1995 | 0 | 8 | 31 | 85 | 261 | 558 | 733 | 698 | 544 | 241 | 12 | 0 | 3171 |
| 1996 | 0 | 1 | 24 | 148 | 440 | 679 | 740 | 670 | 387 | 266 | 33 | 2 | 3390 |
| 1997 | 0 | 2 | 65 | 105 | 462 | 549 | 718 | 651 | 581 | 212 | 26 | 0 | 3371 |
| 1998 | 1 | 0 | 19 | 54 | 255 | 512 | 672 | 671 | 536 | 203 | 9 | 1 | 2933 |
| 1999 | 0 | 1 | 18 | 83 | 318 | 573 | 598 | 635 | 512 | 306 | 103 | 1 | 3148 |
| 2000 | 3 | 1 | 1 | 194 | 479 | 595 | 725 | 624 | 599 | 164 | 0 | 0 | 3385 |
| 2001 | 0 | 0 | 37 | 150 | 450 | 623 | 667 | 655 | 586 | 265 | 79 | 0 | 3512 |
| 2002 | 0 | 4 | 26 | 208 | 340 | 695 | 683 | 661 | 532 | 169 | 13 | 0 | 3331 |
| 2003 | 0 | 3 | 23 | 96 | 409 | 629 | 756 | 667 | 539 | 326 | 11 | 2 | 3461 |
| 2004 | 0 | 1 | 144 | 97 | 404 | 603 | 691 | 627 | 487 | 193 | 17 | 0 | 3264 |
| 2005 | 2 | 0 | 13 | 120 | 390 | 626 | 801 | 593 | 562 | 248 | 58 | 0 | 3413 |
| 2006 | 2 | 4 | 14 | 131 | 444 | 709 | 731 | 586 | 414 | 188 | 46 | 0 | 3269 |
| 2007 | 0 | 0 | 80 | 149 | 413 | 636 | 714 | 684 | 547 | 250 | 96 | 0 | 3569 |
| 2008 | 0 | 4 | 36 | 129 | 268 | 668 | 646 | 637 | 527 | 276 | 48 | 0 | 3239 |
| 2009 | 7 | 18 | 61 | 131 | 474 | 541 | 785 | 744 | 568 | 203 | 80 | 0 | 3612 |

## SNOWFALL (inches) 2009 TUCSON (KTUS)

| YEAR | JUL | AUG | SEP | OCT | NOV | DEC | JAN | FEB | MAR | APR | MAY | JUN | TOTAL |
|------|-----|-----|-----|-----|-----|-----|-----|-----|-----|-----|-----|-----|-------|
| 1980-81 | 0.0 | 0.0 | 0.0 | 0.0 | 0.0 | 0.0 | 0.0 | 0.0 | T | 0.0 | 0.0 | 0.0 | T |
| 1981-82 | 0.0 | 0.0 | 0.0 | 0.0 | 0.0 | 0.0 | T | 0.0 | T | 0.0 | 0.0 | 0.0 | T |
| 1982-83 | 0.0 | 0.0 | 0.0 | 0.0 | 0.0 | T | 0.0 | 0.0 | 0.0 | 0.0 | 0.0 | 0.0 | T |
| 1983-84 | 0.0 | 0.0 | 0.0 | 0.0 | 0.0 | 0.0 | 0.0 | 0.0 | 0.0 | 0.0 | 0.0 | 0.0 | 0.0 |
| 1984-85 | 0.0 | 0.0 | 0.0 | 0.0 | 0.0 | T | 0.0 | 2.2 | 0.0 | 0.0 | 0.0 | 0.0 | 2.2 |
| 1985-86 | 0.0 | 0.0 | 0.0 | 0.0 | 0.0 | T | 0.0 | T | 0.0 | 0.0 | 0.0 | 0.0 | T |
| 1986-87 | 0.0 | 0.0 | 0.0 | 0.0 | 0.0 | 0.0 | 4.7 | 0.0 | T | 0.0 | 0.0 | 0.0 | 4.7 |
| 1987-88 | 0.0 | 0.0 | 0.0 | 0.0 | 0.0 | 3.6 | 0.0 | 0.0 | 0.0 | 0.0 | 0.0 | 0.0 | 3.6 |
| 1988-89 | 0.0 | 0.0 | 0.0 | 0.0 | 0.0 | T | 0.0 | T | 0.0 | 0.0 | 0.0 | 0.0 | T |
| 1989-90 | 0.0 | 0.0 | 0.0 | 0.0 | 0.0 | 0.0 | 2.7 | 2.3 | 0.0 | T | 0.0 | 0.0 | 5.0 |
| 1990-91 | 0.0 | T | T | 0.0 | 0.0 | 0.6 | 0.0 | T | 0.3 | 0.0 | 0.0 | 0.0 | 0.9 |
| 1991-92 | 0.0 | 0.0 | 0.0 | T | 0.0 | T | 0.0 | 0.0 | T | 0.0 | T | 0.0 | T |
| 1992-93 | 0.0 | T | 0.0 | 0.0 | 0.0 | 0.0 | 0.0 | 0.0 | 0.0 | 0.0 | 0.0 | 0.0 | T |
| 1993-94 | 0.0 | T | 0.0 | 0.0 | 0.0 | 0.0 | T | 0.0 | T | 0.0 | 0.0 | 0.0 | T |
| 1994-95 | 0.0 | 0.0 | 0.0 | 0.0 | T | 0.0 | 0.0 | 0.0 | T | 0.0 | 0.0 | 0.0 | T |
| 1995-96 | T | T | 0.0 | 0.0 | 0.0 | 0.0 | 0.0 | 0.0 | T | 0.0 | 0.0 | 0.0 | T |
| 1996-97 | 0.0 | 0.0 | T | 0.0 | 0.0 | 0.0 | T | 0.0 | 0.0 | T | 0.0 | 0.0 | T |
| 1997-98 | 0.0 | 0.0 | 0.0 | T | 0.0 | 0.0 | 0.0 | 0.0 | T | T | 0.0 | 0.0 | T |
| 1998-99 | 0.0 | 0.0 | 0.0 | 0.0 | 0.0 | T | 0.0 | 0.0 | 0.0 | T | 0.0 | 0.0 | TOTAL |
| 1999-00 | 0.0 | 0.0 | 0.0 | 0.0 | 0.0 | 0.0 | 0.0 | 0.0 | 0.0 | 0.0 | 0.0 | 0.0 | 0.0 |
| 2000-01 | 0.0 | 0.0 | 0.0 | 0.0 | 0.0 | 0.0 | T | T | 0.0 | 0.0 | 0.0 | 0.0 | T |
| 2001-02 | 0.0 | 0.0 | 0.0 | 0.0 | 0.0 | T | 0.6 | 0.0 | 0.0 | 0.0 | 0.0 | 0.0 | 0.6 |
| 2002-03 | T | 0.0 | 0.0 | 0.0 | 0.0 | 0.0 | 0.0 | 0.0 | 0.0 | 0.0 | 0.0 | 0.0 | T |
| 2003-04 | 0.0 | 0.0 | 0.0 | 0.0 | 0.0 | 0.0 | 0.0 | 0.0 | 0.0 | 0.0 | 0.0 | 0.0 | 0.0 |
| 2004-05 | 0.0 | 0.0 | 0.0 | 0.0 | 0.0 | 0.0 | 0.0 | 0.0 | 0.0 | 0.0 | 0.0 | 0.0 | 0.0 |
| 2005-06 | 0.0 | 0.0 | 0.0 | 0.0 | 0.0 | 0.0 | 0.0 | 0.0 | 0.0 | 0.0 | 0.0 | 0.0 | 0.0 |
| 2006-07 | 0.0 | T | 0.0 | 0.0 | 0.0 | 0.0 | T | 0.0 | 0.0 | 0.0 | 0.0 | 0.0 | T |
| 2007-08 | 0.0 | 0.0 | 0.0 | 0.0 | 0.0 | 0.0 | 0.0 | 0.0 | 0.0 | 0.0 | 0.0 | 0.0 | 0.0 |
| 2008-09 | 0.0 | 0.0 | 0.0 | 0.0 | 0.0 | 0.0 | 0.0 | T | 0.0 | 0.0 | 0.0 | 0.0 | T |
| 2009- | 0.0 | 0.0 | 0.0 | 0.0 | 0.0 | 0.0 | | | | | | | |
| POR=<br>62 YRS | T | T | T | T | 0.1 | 0.2 | 0.3 | 0.2 | 0.2 | T | T | 0.0 | 1.0 |

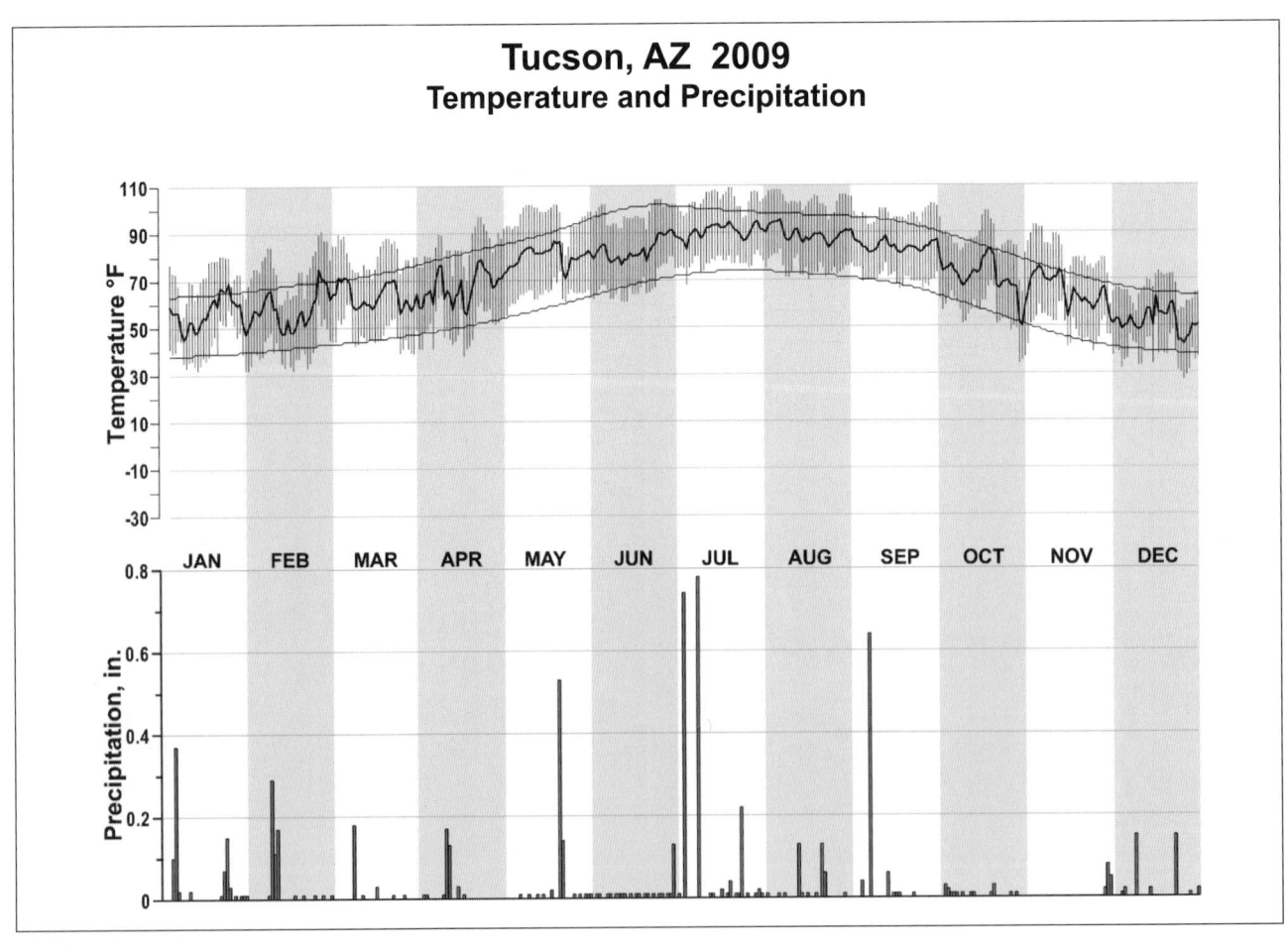

# Tucson, AZ 2009
## Temperature and Precipitation

# 2009
# LITTLE ROCK
# ARKANSAS (KLIT)

Little Rock is located on the Arkansas River near the geographical center of the state. It is situated on the dividing line between the Ouachita Mountains to the west and the flat lowlands comprising the Mississippi River Valley to the east. Elevations range from 222 feet at the river level to 257 feet over much of the flat land, including the airport in the southeast, to near 600 feet in the hilly residential area of the western portions of the city. Two minor temperature variations are observed due to the terrain; somewhat lower minimum temperatures are observed in the airport vicinity and a slight downslope adiabatic heating effect accompanies airflow from the ridges and hills in the west and northwest.

The modified continental climate of Little Rock includes exposure to all of the North American air mass types. However, with its proximity to the Gulf of Mexico, the summer season is marked by prolonged periods of warm and humid weather. The growing season averages 233 days in which 62 percent of the normal precipitation occurs. Winters are mild, but polar and Arctic outbreaks are not uncommon.

Precipitation is fairly well distributed throughout the year. Summer rainfall is almost completely of the convective type. The driest period usually occurs in the late summer and early fall. Snow is almost negligible. Glaze and ice storms, although infrequent, are at times severe. Warm front weather in the winter and early spring, characterized by shallow surface cold air flow from the north under warm moist Gulf air, results in excellent conditions for the production of freezing precipitation.

# NORMALS, MEANS, AND EXTREMES
## LITLE ROCK (KLIT)

LATITUDE: 34 ° 44'N  LONGITUDE: -92 ° 13'W  ELEVATION (FT): GRND: 253  BARO: 292  TIME ZONE: CENTRAL  (UTC -6)  WBAN: 13963

| | ELEMENT | POR | JAN | FEB | MAR | APR | MAY | JUN | JUL | AUG | SEP | OCT | NOV | DEC | YEAR |
|---|---|---|---|---|---|---|---|---|---|---|---|---|---|---|---|
| **TEMPERATURE °F** | NORMAL DAILY MAXIMUM | 30 | 49.5 | 55.6 | 64.2 | 72.9 | 81.0 | 89.0 | 92.8 | 92.1 | 85.1 | 75.1 | 62.0 | 52.5 | 72.7 |
| | MEAN DAILY MAXIMUM | 113 | 50.5 | 52.1 | 63.3 | 71.5 | 80.3 | 86.9 | 91.5 | 91.1 | 83.6 | 74.7 | 61.3 | 52.3 | 71.6 |
| | HIGHEST DAILY MAXIMUM | 68 | 83 | 85 | 91 | 95 | 98 | 105 | 112 | 109 | 106 | 97 | 86 | 80 | 112 |
| | YEAR OF OCCURRENCE | | 1950 | 1986 | 1974 | 1987 | 1998 | 1988 | 1986 | 2000 | 1947 | 1963 | 1955 | 2005 | JUL 1986 |
| | MEAN OF EXTREME MAXS. | 113 | 72.2 | 75.8 | 81.7 | 86.4 | 91.6 | 96.6 | 100.2 | 100.1 | 95.6 | 88.8 | 79.7 | 72.6 | 86.8 |
| | NORMAL DAILY MINIMUM | 30 | 30.8 | 34.8 | 42.6 | 50.0 | 59.2 | 67.8 | 72.0 | 70.5 | 63.6 | 51.5 | 41.5 | 33.9 | 51.5 |
| | MEAN DAILY MINIMUM | 113 | 32.5 | 33.8 | 43.2 | 51.2 | 60.3 | 67.4 | 72.0 | 71.0 | 63.5 | 52.8 | 41.5 | 34.6 | 52.0 |
| | LOWEST DAILY MINIMUM | 68 | -4 | -5 | 11 | 28 | 40 | 46 | 54 | 52 | 37 | 29 | 17 | -1 | -5 |
| | YEAR OF OCCURRENCE | | 1962 | 1951 | 1951 | 2007 | 1971 | 1969 | 1972 | 1986 | 1942 | 1989 | 1976 | 1989 | FEB 1951 |
| | MEAN OF EXTREME MINS. | 113 | 14.3 | 18.9 | 26.2 | 35.8 | 46.6 | 57.0 | 63.7 | 62.0 | 48.9 | 36.6 | 25.2 | 17.8 | 37.8 |
| | NORMAL DRY BULB | 30 | 40.1 | 45.2 | 53.4 | 61.4 | 70.1 | 78.4 | 82.4 | 81.3 | 74.4 | 63.3 | 51.7 | 43.2 | 62.1 |
| | MEAN DRY BULB | 113 | 41.5 | 43.0 | 53.3 | 61.3 | 70.3 | 77.3 | 81.8 | 81.1 | 73.6 | 63.7 | 51.4 | 43.4 | 61.8 |
| | MEAN WET BULB | 26 | 38.9 | 42.1 | 48.6 | 56.2 | 64.3 | 70.5 | 73.1 | 72.4 | 66.6 | 57.5 | 48.8 | 41.0 | 56.7 |
| | MEAN DEW POINT | 26 | 35.3 | 38.2 | 44.3 | 52.6 | 61.7 | 68.2 | 71.0 | 70.1 | 63.8 | 54.2 | 45.4 | 37.4 | 53.5 |
| | NORMAL NO. DAYS WITH: | | | | | | | | | | | | | | |
| | MAXIMUM >= 90 | 3 0 | 0.0 | 0.0 | * | 0.3 | 3.0 | 15.8 | 22.9 | 21.1 | 9.7 | 0.7 | 0.0 | 0.0 | 73.5 |
| | MAXIMUM <= 32 | 3 0 | 2.9 | 1.3 | 0.0 | 0.0 | 0.0 | 0.0 | 0.0 | 0.0 | 0.0 | 0.0 | * | 1.7 | 5.9 |
| | MINIMUM <= 32 | 30 | 18.4 | 11.2 | 4.0 | 0.5 | 0.0 | 0.0 | 0.0 | 0.0 | 0.0 | 0.1 | 5.1 | 14.4 | 53.7 |
| | MINIMUM <= 0 | 30 | * | 0.0 | 0.0 | 0.0 | 0.0 | 0.0 | 0.0 | 0.0 | 0.0 | 0.0 | 0.0 | 0.1 | 0.1 |
| **H/C** | NORMAL HEATING DEG. DAYS | 30 | 775 | 563 | 369 | 150 | 24 | 0 | 0 | 0 | 13 | 124 | 400 | 666 | 3084 |
| | NORMAL COOLING DEG. DAYS | 30 | 1 | 1 | 14 | 52 | 188 | 408 | 542 | 502 | 296 | 72 | 8 | 2 | 2086 |
| **RH** | NORMAL (PERCENT) | 30 | 71 | 68 | 67 | 68 | 73 | 72 | 72 | 71 | 72 | 71 | 72 | 72 | 71 |
| | HOUR 00 LST | 30 | 76 | 74 | 72 | 76 | 84 | 84 | 83 | 83 | 84 | 82 | 79 | 77 | 80 |
| | HOUR 06 LST | 30 | 81 | 80 | 80 | 84 | 88 | 88 | 89 | 88 | 88 | 87 | 84 | 81 | 85 |
| | HOUR 12 LST | 30 | 63 | 59 | 57 | 56 | 59 | 57 | 56 | 55 | 57 | 54 | 60 | 63 | 58 |
| | HOUR 18 LST | 30 | 65 | 59 | 56 | 55 | 61 | 59 | 59 | 59 | 59 | 62 | 66 | 67 | 61 |
| **S** | PERCENT POSSIBLE SUNSHINE | 32 | 46 | 54 | 57 | 62 | 68 | 73 | 71 | 73 | 68 | 69 | 56 | 48 | 62 |
| **W/O** | MEAN NO. DAYS WITH: | | | | | | | | | | | | | | |
| | HEAVY FOG(VISBY <= 1/4 MI) | 46 | 2.5 | 1.6 | 1.3 | 0.7 | 0.7 | 0.5 | 0.4 | 0.5 | 0.8 | 1.5 | 2.1 | 2.8 | 15.4 |
| | THUNDERSTORMS | 62 | 1.8 | 2.4 | 4.8 | 6.3 | 7.3 | 7.7 | 8.9 | 6.7 | 3.5 | 2.7 | 2.7 | 1.9 | 56.7 |
| **CLOUDNESS** | MEAN: | | | | | | | | | | | | | | |
| | SUNRISE-SUNSET (OKTAS) | 35 | 5.2 | 4.8 | 5.0 | 4.9 | 4.8 | 4.3 | 4.3 | 3.9 | 4.0 | 3.5 | 4.3 | 4.8 | 4.5 |
| | MIDNIGHT-MIDNIGHT (OKTAS) | 12 | 4.5 | 4.0 | 4.4 | 4.2 | 3.9 | 3.4 | 3.6 | 3.4 | 3.7 | 3.2 | 4.0 | 4.4 | 3.9 |
| | MEAN NO. DAYS WITH: | | | | | | | | | | | | | | |
| | CLEAR | 35 | 8.6 | 9.1 | 8.6 | 8.7 | 8.0 | 9.5 | 8.8 | 11.6 | 11.2 | 14.4 | 11.0 | 9.2 | 118.7 |
| | PARTLY CLOUDY | 35 | 6.1 | 5.7 | 7.0 | 7.5 | 10.8 | 11.6 | 12.9 | 10.9 | 8.6 | 7.1 | 5.9 | 5.9 | 100.0 |
| | CLOUDY | 35 | 16.3 | 13.5 | 15.4 | 13.8 | 12.3 | 8.9 | 9.3 | 8.5 | 10.1 | 9.5 | 13.1 | 16.0 | 146.7 |
| **PR** | MEAN STATION PRESSURE(IN) | 26 | 29.88 | 29.84 | 29.77 | 29.71 | 29.70 | 29.70 | 29.72 | 29.73 | 29.75 | 29.80 | 29.83 | 29.87 | 29.78 |
| | MEAN SEA-LEVEL PRES. (IN) | 26 | 30.18 | 30.13 | 30.06 | 29.99 | 29.97 | 29.97 | 30.00 | 30.01 | 30.03 | 30.08 | 30.12 | 30.17 | 30.06 |
| **WINDS** | MEAN SPEED (MPH) | 26 | 7.7 | 8.0 | 8.6 | 8.2 | 7.0 | 6.3 | 6.0 | 5.7 | 6.1 | 6.1 | 7.1 | 7.4 | 7.0 |
| | PREVAIL.DIR(TENS OF DEGS) | 35 | 25 | 05 | 19 | 19 | 19 | 19 | 24 | 25 | 05 | 25 | 25 | 25 | 24 |
| | MAXIMUM 2-MINUTE: | | | | | | | | | | | | | | |
| | SPEED (MPH) | 11 | 41 | 47 | 37 | 33 | 40 | 46 | 46 | 41 | 36 | 36 | 35 | 38 | 47 |
| | DIR. (TENS OF DEGS) | | 30 | 25 | 24 | 25 | 31 | 30 | 02 | 26 | 21 | 22 | 32 | 32 | 25 |
| | YEAR OF OCCURRENCE | | 2008 | 2008 | 2006 | 2008 | 2006 | 2001 | 2003 | 2006 | 2008 | 2004 | 2005 | 2007 | FEB 2008 |
| | MAXIMUM 3-SECOND | | | | | | | | | | | | | | |
| | SPEED (MPH) | 11 | 54 | 67 | 52 | 44 | 48 | 87 | 55 | 48 | 48 | 43 | 47 | 44 | 87 |
| | DIR. (TENS OF DEGS) | | 28 | 24 | 24 | 24 | 30 | 32 | 02 | 23 | 10 | 29 | 32 | 25 | 32 |
| | YEAR OF OCCURRENCE | | 2008 | 2008 | 2006 | 2008 | 2006 | 1999 | 2003 | 2003 | 2008 | 2006 | 2005 | 2008 | JUN 1999 |
| **PRECIPITATION** | NORMAL (IN) | 30 | 3.61 | 3.33 | 4.88 | 5.47 | 5.05 | 3.95 | 3.31 | 2.93 | 3.71 | 4.25 | 5.73 | 4.71 | 50.93 |
| | MAXIMUM MONTHLY (IN) | 68 | 12.53 | 11.02 | 10.40 | 14.20 | 13.06 | 7.82 | 11.65 | 14.46 | 10.17 | 16.56 | 13.14 | 16.48 | 16.56 |
| | YEAR OF OCCURRENCE | | 1950 | 1956 | 1990 | 1973 | 2009 | 1974 | 2009 | 1966 | 1978 | 2009 | 1988 | 1987 | OCT 2009 |
| | MINIMUM MONTHLY (IN) | 68 | 0.25 | 0.51 | 0.73 | 0.50 | 0.69 | T | 0.14 | T | 0.28 | 0.01 | 0.26 | .63 | T |
| | YEAR OF OCCURRENCE | | 2003 | 1947 | 1966 | 1987 | 1970 | 1952 | 1986 | 1995 | 1956 | 1944 | 1999 | 2005 | AUG 1995 |
| | MAXIMUM IN 24 HOURS (IN) | 68 | 5.18 | 5.15 | 4.56 | 7.96 | 7.71 | 4.61 | 3.58 | 7.32 | 4.57 | 5.67 | 7.81 | 7.79 | 7.96 |
| | YEAR OF OCCURRENCE | | 1969 | 1950 | 1990 | 1974 | 1955 | 1960 | 1988 | 1966 | 2008 | 1990 | 1988 | 2009 | APR 1974 |
| | NORMAL NO. DAYS WITH: | | | | | | | | | | | | | | |
| | PRECIPITATION >= 0.01 | 30 | 9.6 | 8.3 | 10.0 | 9.7 | 10.6 | 9.2 | 8.3 | 6.5 | 7.9 | 7.5 | 9.0 | 9.7 | 106.3 |
| | PRECIPITATION >= 1.00 | 30 | 0.9 | 1.0 | 1.8 | 1.7 | 1.5 | 1.0 | 0.8 | 1.0 | 1.0 | 1.5 | 2.0 | 1.4 | 15.6 |
| **SNOWFALL** | NORMAL (IN) | 30 | 2.0 | 1.3 | 0.6 | 0.* | 0.0 | 0.0 | 0.0 | 0.0 | 0.0 | 0.* | 0.3 | 0.1 | 4.3 |
| | MAXIMUM MONTHLY (IN) | 59 | 13.6 | 9.8 | 7.0 | T | T | T | 0.0 | 0.0 | 0.0 | T | 4.8 | 9.8 | 13.6 |
| | YEAR OF OCCURRENCE | | 1988 | 1979 | 1971 | 2009 | 1997 | 1998 | | | | 1993 | 1971 | 1963 | JAN 1988 |
| | MAXIMUM IN 24 HOURS (IN) | 53 | 12.1 | 9.6 | 6.7 | T | T | T | 0.0 | 0.0 | 0.0 | T | 4.8 | 9.8 | 12.1 |
| | YEAR OF OCCURRENCE' | | 1988 | 1966 | 1971 | 2009 | 1988 | 1998 | | | | 1993 | 1971 | 1963 | JAN 1988 |
| | MAXIMUM SNOW DEPTH (IN) | 52 | 13 | 5 | 4 | 0 | 0 | 0 | 0 | 0 | 0 | 0 | 3 | 8 | 13 |
| | YEAR OF OCCURRENCE | | 1988 | 1985 | 1984 | | | | | | | | 1971 | 1963 | JAN 1988 |
| | NORMAL NO. DAYS WITH: | | | | | | | | | | | | | | |
| | SNOWFALL >= 1.0 | 30 | 0.7 | 0.4 | 0.2 | 0.0 | 0.0 | 0.0 | 0.0 | 0.0 | 0.0 | 0.0 | 0.1 | 0.0 | 1.4 |

## PRECIPITATION (inches) 2009 LITTLE ROCK (KLIT)

| YEAR | JAN | FEB | MAR | APR | MAY | JUN | JUL | AUG | SEP | OCT | NOV | DEC | ANNUAL |
|------|-----|-----|-----|-----|-----|-----|-----|-----|-----|-----|-----|-----|--------|
| 1980 | 2.73 | 0.89 | 6.60 | 5.85 | 4.57 | 0.53 | 0.99 | 0.19 | 5.09 | 2.64 | 6.28 | 1.86 | 38.22 |
| 1981 | 1.11 | 3.89 | 4.00 | 2.75 | 9.73 | 7.80 | 3.15 | 2.91 | 1.37 | 6.11 | 1.64 | 1.34 | 45.80 |
| 1982 | 8.74 | 3.37 | 2.87 | 9.32 | 5.63 | 4.10 | 1.01 | 4.52 | 1.47 | 2.26 | 9.72 | 8.28 | 61.29 |
| 1983 | 2.25 | 1.49 | 4.19 | 6.72 | 7.58 | 3.34 | 1.07 | 0.79 | 0.41 | 3.73 | 4.47 | 9.07 | 45.11 |
| 1984 | 1.31 | 3.52 | 5.58 | 3.77 | 8.22 | 1.06 | 4.15 | 5.69 | 3.28 | 15.35 | 8.49 | 3.54 | 63.96 |
| 1985 | 3.11 | 2.78 | 5.27 | 8.63 | 2.99 | 2.40 | 3.30 | 3.52 | 4.36 | 3.91 | 5.78 | 2.97 | 49.02 |
| 1986 | 0.50 | 3.45 | 3.68 | 7.33 | 4.07 | 6.42 | 0.14 | 4.56 | 1.94 | 6.05 | 5.67 | 3.86 | 47.67 |
| 1987 | 2.07 | 7.07 | 3.52 | 0.50 | 4.56 | 4.63 | 1.60 | 2.12 | 7.56 | 1.37 | 10.96 | 16.48 | 62.44 |
| 1988 | 3.71 | 3.41 | 3.50 | 3.82 | 2.05 | 1.04 | 7.95 | 2.19 | 2.54 | 1.95 | 13.14 | 2.91 | 48.21 |
| 1989 | 3.01 | 9.55 | 7.64 | 2.57 | 4.04 | 3.95 | 7.87 | 1.21 | 3.57 | 1.70 | 1.95 | 2.19 | 49.25 |
| 1990 | 6.50 | 4.82 | 10.40 | 7.73 | 7.71 | 0.80 | 4.63 | 1.57 | 4.08 | 8.75 | 3.29 | 6.79 | 67.07 |
| 1991 | 6.88 | 3.03 | 3.56 | 12.44 | 2.87 | 2.28 | 2.03 | 6.78 | 3.01 | 7.00 | 5.18 | 4.59 | 59.65 |
| 1992 | 1.75 | 2.05 | 6.48 | 1.86 | 3.67 | 5.07 | 6.76 | 2.14 | 2.90 | 0.67 | 4.71 | 3.85 | 41.91 |
| 1993 | 5.06 | 2.44 | 3.05 | 5.40 | 5.49 | 2.04 | 1.24 | 2.77 | 1.44 | 4.10 | 6.33 | 4.41 | 43.77 |
| 1994 | 4.87 | 3.21 | 5.60 | 5.20 | 3.97 | 5.57 | 4.28 | 4.00 | 2.11 | 3.88 | 6.13 | 4.58 | 53.40 |
| 1995 | 3.94 | 2.40 | 3.74 | 4.95 | 4.56 | 1.85 | 2.99 | T | 1.88 | 5.54 | 2.33 | 2.83 | 37.01 |
| 1996 | 2.60 | 2.14 | 3.57 | 4.24 | 3.98 | 2.82 | 3.56 | 1.23 | 6.39 | 6.35 | 7.41 | 2.81 | 47.10 |
| 1997 | 1.88 | 4.65 | 6.48 | 7.73 | 3.92 | 5.41 | 1.88 | 2.18 | 3.75 | 4.35 | 3.87 | 3.66 | 49.76 |
| 1998 | 4.70 | 4.11 | 4.80 | 3.31 | 2.86 | 2.16 | 2.96 | 3.23 | 3.45 | 3.39 | 2.30 | 4.41 | 41.68 |
| 1999 | 6.11 | 1.10 | 4.85 | 5.33 | 3.40 | 6.10 | 2.41 | 0.91 | 1.56 | 4.97 | 0.26 | 5.25 | 42.25 |
| 2000 | 1.03 | 3.93 | 3.86 | 2.94 | 5.80 | 5.66 | 0.94 | 0.04 | 2.39 | 0.79 | 10.99 | 3.42 | 41.79 |
| 2001 | 2.98 | 8.46 | 3.94 | 1.40 | 4.04 | 2.05 | 1.56 | 1.76 | 2.54 | 5.24 | 5.30 | 8.31 | 47.58 |
| 2002 | 3.41 | 1.93 | 9.51 | 1.76 | 5.44 | 2.64 | 3.47 | 2.89 | 1.49 | 4.51 | 1.84 | 7.70 | 46.59 |
| 2003 | 0.25 | 5.56 | 2.47 | 1.80 | 4.27 | 7.68 | 2.05 | 1.26 | 3.54 | 2.16 | 4.94 | 3.50 | 39.48 |
| 2004 | 3.29 | 4.44 | 4.06 | 6.89 | 3.98 | 4.99 | 3.30 | 3.18 | 0.51 | 9.27 | 9.63 | 2.51 | 56.05 |
| 2005 | 4.88 | 3.00 | 3.36 | 3.05 | 1.06 | 2.80 | 4.00 | 4.00 | 3.71 | 1.02 | 3.04 | 0.63 | 34.55 |
| 2006 | 3.76 | 2.00 | 4.39 | 8.55 | 4.09 | 2.95 | 1.63 | 1.95 | 4.25 | 3.13 | 6.07 | 5.98 | 48.75 |
| 2007 | 9.56 | 1.55 | 1.61 | 4.50 | 3.46 | 1.97 | 3.28 | 0.14 | 4.65 | 6.17 | 2.51 | 5.03 | 44.43 |
| 2008 | 1.40 | 3.87 | 7.60 | 9.67 | 4.76 | 4.21 | 2.15 | 5.80 | 7.51 | 4.94 | 2.56 | 3.69 | 58.16 |
| 2009 | 2.60 | 2.16 | 4.63 | 5.33 | 13.06 | 3.08 | 11.65 | 2.75 | 6.44 | 16.56 | 1.20 | 12.33 | 81.79 |
| POR= 113 YRS | 4.33 | 3.63 | 4.62 | 5.21 | 4.88 | 3.59 | 3.32 | 3.08 | 3.42 | 3.47 | 4.29 | 4.40 | 48.24 |

WBAN : 13963

## AVERAGE TEMPERATURE (°F) 2009 LITTLE ROCK (KLIT)

| YEAR | JAN | FEB | MAR | APR | MAY | JUN | JUL | AUG | SEP | OCT | NOV | DEC | ANNUAL |
|------|-----|-----|-----|-----|-----|-----|-----|-----|-----|-----|-----|-----|--------|
| 1980 | 44.0 | 40.9 | 50.3 | 61.5 | 70.6 | 79.4 | 88.1 | 87.0 | 78.6 | 60.4 | 50.4 | 43.1 | 62.9 |
| 1981 | 39.7 | 44.6 | 52.5 | 67.5 | 67.4 | 80.1 | 83.5 | 79.9 | 75.3 | 61.4 | 55.5 | 43.2 | 62.6 |
| 1982 | 37.5 | 41.3 | 57.1 | 58.0 | 72.7 | 76.6 | 83.1 | 82.1 | 74.2 | 64.7 | 53.2 | 48.3 | 62.4 |
| 1983 | 39.2 | 43.8 | 51.1 | 54.4 | 67.7 | 77.4 | 82.5 | 86.1 | 76.0 | 64.0 | 50.8 | 30.9 | 60.3 |
| 1984 | 36.7 | 46.6 | 50.1 | 59.7 | 68.0 | 79.8 | 79.7 | 78.1 | 71.0 | 65.0 | 49.0 | 52.1 | 61.3 |
| 1985 | 33.7 | 39.3 | 57.6 | 63.0 | 70.0 | 78.2 | 81.2 | 80.8 | 72.5 | 66.2 | 56.1 | 38.1 | 61.4 |
| 1986 | 42.5 | 48.2 | 55.4 | 63.6 | 71.4 | 79.7 | 86.3 | 78.1 | 77.6 | 63.1 | 49.7 | 42.4 | 63.2 |
| 1987 | 40.2 | 47.1 | 53.4 | 62.4 | 76.3 | 79.9 | 82.2 | 84.4 | 74.9 | 59.1 | 53.0 | 45.2 | 63.2 |
| 1988 | 35.6 | 42.9 | 52.2 | 61.5 | 70.4 | 78.8 | 81.7 | 82.4 | 75.8 | 60.4 | 53.3 | 44.4 | 61.6 |
| 1989 | 46.3 | 38.4 | 52.5 | 62.6 | 69.6 | 76.2 | 79.3 | 80.2 | 71.2 | 63.3 | 55.3 | 35.7 | 60.9 |
| 1990 | 48.1 | 51.3 | 55.4 | 62.0 | 68.1 | 80.6 | 83.2 | 82.2 | 77.5 | 61.5 | 56.3 | 43.1 | 64.1 |
| 1991 | 39.0 | 49.2 | 56.2 | 64.4 | 74.1 | 79.5 | 82.8 | 80.0 | 73.7 | 64.3 | 49.2 | 46.8 | 63.3 |
| 1992 | 42.7 | 50.6 | 54.4 | 62.6 | 69.0 | 76.5 | 81.0 | 76.5 | 72.7 | 64.4 | 50.1 | 43.9 | 62.0 |
| 1993 | 40.4 | 43.2 | 51.3 | 58.3 | 68.8 | 78.7 | 86.0 | 83.7 | 73.5 | 61.4 | 48.3 | 44.9 | 61.5 |
| 1994 | 38.3 | 45.2 | 53.9 | 64.4 | 68.3 | 81.8 | 80.0 | 78.9 | 72.7 | 64.3 | 56.1 | 46.3 | 62.5 |
| 1995 | 43.0 | 46.5 | 55.5 | 62.4 | 70.7 | 77.6 | 83.1 | 86.7 | 72.5 | 64.1 | 50.1 | 42.1 | 62.9 |
| 1996 | 40.0 | 46.2 | 48.2 | 60.4 | 74.3 | 79.4 | 81.9 | 80.8 | 73.5 | 63.6 | 48.4 | 46.4 | 61.9 |
| 1997 | 41.0 | 47.2 | 56.7 | 58.2 | 68.3 | 77.1 | 84.0 | 80.4 | 76.4 | 63.9 | 49.9 | 42.6 | 62.1 |
| 1998 | 46.7 | 48.9 | 51.7 | 62.0 | 75.9 | 83.1 | 87.2 | 83.9 | 80.6 | 65.8 | 54.8 | 44.9 | 65.5 |
| 1999 | 43.8 | 51.0 | 50.0 | 65.1 | 70.1 | 78.3 | 83.6 | 83.3 | 74.0 | 64.0 | 56.8 | 46.1 | 63.8 |
| 2000 | 43.0 | 50.7 | 55.9 | 61.1 | 72.3 | 76.5 | 82.7 | 86.5 | 75.4 | 65.7 | 48.0 | 32.0 | 62.5 |
| 2001 | 38.6 | 46.4 | 49.5 | 67.3 | 71.3 | 76.9 | 83.2 | 82.0 | 72.7 | 60.6 | 55.7 | 45.9 | 62.5 |
| 2002 | 44.4 | 42.9 | 49.3 | 64.6 | 67.9 | 77.9 | 81.6 | 81.2 | 76.8 | 61.2 | 49.7 | 43.3 | 61.7 |
| 2003 | 38.0 | 40.5 | 52.3 | 63.7 | 71.9 | 74.7 | 81.7 | 83.1 | 72.5 | 65.4 | 55.6 | 44.1 | 62.0 |
| 2004 | 42.9 | 42.1 | 58.2 | 62.5 | 72.1 | 77.5 | 80.0 | 77.5 | 75.3 | 67.4 | 54.7 | 43.3 | 62.8 |
| 2005 | 45.8 | 49.3 | 53.1 | 62.6 | 70.1 | 80.1 | 82.1 | 84.8 | 78.5 | 64.3 | 55.4 | 43.1 | 64.1 |
| 2006 | 49.8 | 42.7 | 55.5 | 68.2 | 72.4 | 79.0 | 84.1 | 84.4 | 73.0 | 62.2 | 53.3 | 47.1 | 64.3 |
| 2007 | 41.5 | 43.5 | 61.7 | 59.6 | 73.5 | 80.2 | 80.2 | 87.2 | 76.8 | 65.7 | 54.4 | 45.4 | 64.1 |
| 2008 | 40.6 | 45.5 | 53.8 | 60.6 | 70.6 | 79.8 | 83.4 | 80.4 | 73.7 | 62.7 | 50.7 | 42.9 | 62.1 |
| 2009 | 40.6 | 49.0 | 54.8 | 61.9 | 69.7 | 81.2 | 79.0 | 79.1 | 73.8 | 59.4 | 55.4 | 40.2 | 62.0 |
| POR= 113 YRS | 41.5 | 43.0 | 53.3 | 61.3 | 70.3 | 77.3 | 81.8 | 81.1 | 73.6 | 63.7 | 51.4 | 43.4 | 61.8 |

## HEATING DEGREE DAYS (base 65°F) 2009  LITTLE ROCK (KLIT)

| YEAR | JUL | AUG | SEP | OCT | NOV | DEC | JAN | FEB | MAR | APR | MAY | JUN | TOTAL |
|------|-----|-----|-----|-----|-----|-----|-----|-----|-----|-----|-----|-----|-------|
| 1980-81 | 0 | 0 | 16 | 184 | 437 | 673 | 774 | 565 | 388 | 37 | 45 | 0 | 3119 |
| 1981-82 | 0 | 0 | 4 | 186 | 278 | 668 | 847 | 656 | 298 | 223 | 6 | 0 | 3166 |
| 1982-83 | 0 | 0 | 12 | 119 | 369 | 536 | 795 | 587 | 425 | 332 | 33 | 0 | 3208 |
| 1983-84 | 0 | 0 | 19 | 89 | 422 | 1050 | 872 | 530 | 460 | 190 | 24 | 0 | 3656 |
| 1984-85 | 0 | 0 | 44 | 81 | 476 | 408 | 962 | 713 | 251 | 101 | 8 | 0 | 3044 |
| 1985-86 | 0 | 0 | 31 | 82 | 283 | 825 | 691 | 467 | 298 | 91 | 7 | 0 | 2775 |
| 1986-87 | 0 | 1 | 0 | 112 | 454 | 694 | 762 | 496 | 353 | 145 | 0 | 0 | 3017 |
| 1987-88 | 0 | 0 | 0 | 182 | 358 | 609 | 904 | 637 | 388 | 123 | 4 | 0 | 3205 |
| 1988-89 | 0 | 0 | 1 | 163 | 358 | 633 | 573 | 738 | 395 | 156 | 39 | 0 | 3056 |
| 1989-90 | 0 | 0 | 23 | 112 | 313 | 898 | 516 | 380 | 316 | 152 | 31 | 0 | 2741 |
| 1990-91 | 0 | 0 | 8 | 173 | 260 | 675 | 798 | 438 | 294 | 69 | 10 | 0 | 2725 |
| 1991-92 | 0 | 0 | 23 | 97 | 483 | 560 | 682 | 413 | 324 | 134 | 40 | 0 | 2756 |
| 1992-93 | 0 | 0 | 14 | 69 | 441 | 647 | 755 | 606 | 423 | 215 | 15 | 1 | 3186 |
| 1993-94 | 0 | 0 | 13 | 174 | 506 | 613 | 824 | 548 | 339 | 105 | 49 | 0 | 3171 |
| 1994-95 | 0 | 0 | 22 | 111 | 276 | 573 | 678 | 512 | 318 | 119 | 27 | 0 | 2636 |
| 1995-96 | 0 | 0 | 24 | 93 | 445 | 703 | 767 | 540 | 520 | 180 | 7 | 0 | 3279 |
| 1996-97 | 0 | 0 | 11 | 103 | 492 | 576 | 741 | 488 | 261 | 217 | 28 | 0 | 2917 |
| 1997-98 | 0 | 0 | 1 | 149 | 451 | 685 | 562 | 445 | 446 | 123 | 0 | 0 | 2862 |
| 1998-99 | 0 | 0 | 0 | 67 | 301 | 627 | 650 | 387 | 458 | 73 | 1 | 0 | 2564 |
| 1999-00 | 0 | 0 | 4 | 100 | 250 | 581 | 672 | 418 | 288 | 131 | 8 | 0 | 2452 |
| 2000-01 | 0 | 0 | 16 | 97 | 513 | 1014 | 811 | 515 | 477 | 64 | 7 | 0 | 3514 |
| 2001-02 | 0 | 0 | 17 | 165 | 274 | 585 | 640 | 613 | 486 | 111 | 51 | 0 | 2942 |
| 2002-03 | 0 | 0 | 0 | 170 | 463 | 664 | 827 | 679 | 386 | 102 | 0 | 0 | 3291 |
| 2003-04 | 0 | 0 | 8 | 73 | 298 | 640 | 690 | 659 | 232 | 124 | 21 | 0 | 2745 |
| 2004-05 | 0 | 0 | 0 | 44 | 308 | 664 | 597 | 433 | 373 | 98 | 41 | 0 | 2558 |
| 2005-06 | 0 | 0 | 4 | 132 | 309 | 672 | 463 | 619 | 323 | 48 | 13 | 0 | 2583 |
| 2006-07 | 0 | 0 | 9 | 165 | 353 | 549 | 723 | 598 | 163 | 201 | 3 | 0 | 2764 |
| 2007-08 | 0 | 0 | 0 | 109 | 329 | 602 | 754 | 564 | 355 | 161 | 19 | 0 | 2893 |
| 2008-09 | 0 | 0 | 0 | 125 | 422 | 678 | 755 | 444 | 337 | 153 | 25 | 0 | 2939 |
| 2009- | 0 | 0 | 1 | 181 | 283 | 764 | | | | | | | |

WBAN : 13963

## COOLING DEGREE DAYS (base 65°F) 2009  LITTLE ROCK (KLIT)

| YEAR | JAN | FEB | MAR | APR | MAY | JUN | JUL | AUG | SEP | OCT | NOV | DEC | TOTAL |
|------|-----|-----|-----|-----|-----|-----|-----|-----|-----|-----|-----|-----|-------|
| 1980 | 0 | 1 | 1 | 42 | 196 | 439 | 725 | 688 | 432 | 50 | 5 | 0 | 2579 |
| 1981 | 0 | 0 | 9 | 117 | 131 | 458 | 580 | 470 | 318 | 78 | 2 | 0 | 2163 |
| 1982 | 0 | 0 | 57 | 21 | 253 | 355 | 570 | 540 | 294 | 114 | 20 | 24 | 2248 |
| 1983 | 0 | 0 | 0 | 21 | 121 | 381 | 550 | 660 | 355 | 63 | 6 | 0 | 2157 |
| 1984 | 0 | 1 | 4 | 38 | 126 | 451 | 462 | 416 | 234 | 88 | 3 | 17 | 1840 |
| 1985 | 0 | 0 | 31 | 48 | 167 | 404 | 508 | 501 | 265 | 125 | 21 | 0 | 2070 |
| 1986 | 0 | 5 | 7 | 56 | 211 | 446 | 668 | 415 | 385 | 63 | 0 | 0 | 2256 |
| 1987 | 0 | 0 | 3 | 74 | 359 | 456 | 540 | 610 | 304 | 8 | 3 | 0 | 2357 |
| 1988 | 0 | 0 | 2 | 24 | 177 | 423 | 523 | 546 | 332 | 25 | 15 | 0 | 2067 |
| 1989 | 0 | 0 | 13 | 91 | 189 | 345 | 450 | 479 | 213 | 70 | 29 | 0 | 1879 |
| 1990 | 0 | 4 | 26 | 67 | 135 | 475 | 571 | 540 | 392 | 73 | 8 | 0 | 2291 |
| 1991 | 0 | 1 | 30 | 56 | 299 | 445 | 558 | 472 | 290 | 82 | 14 | 2 | 2249 |
| 1992 | 0 | 0 | 4 | 70 | 170 | 352 | 505 | 364 | 249 | 60 | 0 | 0 | 1774 |
| 1993 | 0 | 0 | 4 | 22 | 138 | 419 | 656 | 591 | 276 | 69 | 14 | 0 | 2189 |
| 1994 | 0 | 3 | 3 | 96 | 161 | 510 | 472 | 435 | 256 | 100 | 15 | 0 | 2051 |
| 1995 | 5 | 0 | 31 | 47 | 207 | 383 | 568 | 683 | 254 | 73 | 4 | 0 | 2255 |
| 1996 | 0 | 5 | 6 | 49 | 303 | 437 | 529 | 499 | 273 | 62 | 1 | 6 | 2170 |
| 1997 | 5 | 0 | 13 | 21 | 138 | 369 | 597 | 485 | 351 | 122 | 4 | 0 | 2105 |
| 1998 | 0 | 0 | 40 | 40 | 344 | 548 | 697 | 592 | 473 | 98 | 3 | 11 | 2846 |
| 1999 | 0 | 2 | 0 | 85 | 168 | 406 | 582 | 571 | 280 | 74 | 11 | 1 | 2180 |
| 2000 | 0 | 7 | 10 | 21 | 237 | 353 | 556 | 676 | 334 | 128 | 9 | 0 | 2331 |
| 2001 | 0 | 0 | 0 | 138 | 209 | 366 | 574 | 533 | 254 | 35 | 6 | 0 | 2115 |
| 2002 | 10 | 0 | 7 | 108 | 147 | 396 | 518 | 508 | 361 | 61 | 8 | 0 | 2124 |
| 2003 | 0 | 0 | 0 | 69 | 221 | 300 | 527 | 566 | 241 | 90 | 22 | 0 | 2036 |
| 2004 | 10 | 0 | 26 | 55 | 247 | 382 | 471 | 393 | 319 | 126 | 6 | 0 | 2035 |
| 2005 | 9 | 1 | 9 | 34 | 206 | 459 | 535 | 620 | 416 | 120 | 27 | 0 | 2436 |
| 2006 | 0 | 0 | 36 | 151 | 250 | 418 | 600 | 607 | 253 | 83 | 8 | 0 | 2406 |
| 2007 | 0 | 2 | 69 | 46 | 277 | 464 | 477 | 695 | 365 | 139 | 18 | 0 | 2552 |
| 2008 | 8 | 5 | 16 | 34 | 203 | 450 | 577 | 484 | 265 | 60 | 0 | 0 | 2102 |
| 2009 | 2 | 3 | 24 | 68 | 178 | 491 | 441 | 443 | 272 | 16 | 2 | 0 | 1940 |

## SNOWFALL (inches) 2009 LITTLE ROCK (KLIT)

| YEAR | JUL | AUG | SEP | OCT | NOV | DEC | JAN | FEB | MAR | APR | MAY | JUN | TOTAL |
|------|-----|-----|-----|-----|-----|-----|-----|-----|-----|-----|-----|-----|-------|
| 1980-81 | 0.0 | 0.0 | 0.0 | 0.0 | 1.8 | T | T | 0.0 | 0.0 | 0.0 | 0.0 | 0.0 | 1.8 |
| 1981-82 | 0.0 | 0.0 | 0.0 | 0.0 | 0.0 | 0.0 | 5.0 | 6.3 | T | 0.0 | 0.0 | 0.0 | 11.3 |
| 1982-83 | 0.0 | 0.0 | 0.0 | 0.0 | 0.0 | T | T | T | T | T | 0.0 | 0.0 | T |
| 1983-84 | 0.0 | 0.0 | 0.0 | 0.0 | 0.0 | 0.8 | 1.5 | 0.2 | 4.5 | 0.0 | 0.0 | 0.0 | 7.0 |
| 1984-85 | 0.0 | 0.0 | 0.0 | 0.0 | 0.0 | T | 6.3 | 5.0 | 0.0 | 0.0 | 0.0 | 0.0 | 11.3 |
| 1985-86 | 0.0 | 0.0 | 0.0 | 0.0 | 0.0 | T | 0.0 | 1.5 | 0.0 | 0.0 | 0.0 | 0.0 | 1.5 |
| 1986-87 | 0.0 | 0.0 | 0.0 | 0.0 | T | 0.0 | 1.0 | T | 0.8 | 0.0 | 0.0 | 0.0 | 1.8 |
| 1987-88 | 0.0 | 0.0 | 0.0 | 0.0 | 0.0 | T | 13.6 | 2.5 | T | 0.0 | 0.0 | 0.0 | 16.1 |
| 1988-89 | 0.0 | 0.0 | 0.0 | 0.0 | 0.0 | T | 2.0 | T | 1.0 | 0.0 | 0.0 | 0.0 | 3.0 |
| 1989-90 | 0.0 | 0.0 | 0.0 | 0.0 | 0.0 | T | T | 0.0 | 0.0 | 0.0 | 0.0 | 0.0 | T |
| 1990-91 | 0.0 | 0.0 | 0.0 | 0.0 | 0.0 | T | T | T | T | T | 0.0 | 0.0 | T |
| 1991-92 | 0.0 | 0.0 | 0.0 | 0.0 | T | T | T | 0.0 | 0.0 | 0.0 | 0.0 | 0.0 | T |
| 1992-93 | 0.0 | 0.0 | 0.0 | 0.0 | T | T | 0.0 | T | T | 0.0 | 0.0 | 0.0 | T |
| 1993-94 | 0.0 | 0.0 | 0.0 | T | 0.0 | 0.0 | 0.1 | 0.7 | T | T | 0.0 | 0.0 | 0.8 |
| 1994-95 | 0.0 | 0.0 | 0.0 | 0.0 | 0.0 | T | 7.0 | T | 1.0 | 0.0 | 0.0 | 0.0 | 8.0 |
| 1995-96 | 0.0 | 0.0 | 0.0 | 0.0 | 0.1 | T | 0.7 | 0.6 | T | 0.0 | 0.0 | 0.0 | 1.4 |
| 1996-97 | 0.0 | 0.0 | 0.0 | 0.0 | 0.0 | T | T | 4.7 | 0.0 | T | T | 0.0 | 4.7 |
| 1997-98 | 0.0 | 0.0 | 0.0 | 0.0 | T | T | 0.0 | 0.0 | T | T | 0.0 | T | T |
| 1998-99 | 0.0 | 0.0 | 0.0 | | | | | | | | | | |
| 1999-00 | | | | | | | | | | | | | |
| 2000-01 | | | | | | | | | | | | | |
| 2001-02 | | | | | | | | | | | | | |
| 2002-03 | | | | | | | | | | | | | |
| 2003-04 | | | | | | | | | | | | | |
| 2004-05 | | | | | | | | | | | | | |
| 2005-06 | | | | | | | | | | | | | |
| 2006-07 | | | | | | | | | | | | | |
| 2007-08 | | | | | | T | T | 0.0 | 2.6 | 0.0 | 0.0 | 0.0 | |
| 2008-09 | 0.0 | 0.0 | 0.0 | 0.0 | 0.0 | T | T | 0.2 | T | T | 0.0 | 0.0 | 0.2 |
| 2009- | 0.0 | 0.0 | 0.0 | 0.0 | 0.0 | T | | | | | | | |
| POR= 113 YRS | 0.0 | 0.0 | 0.0 | T | 0.1 | 0.7 | 1.7 | 1.3 | 0.4 | T | T | T | 4.2 |

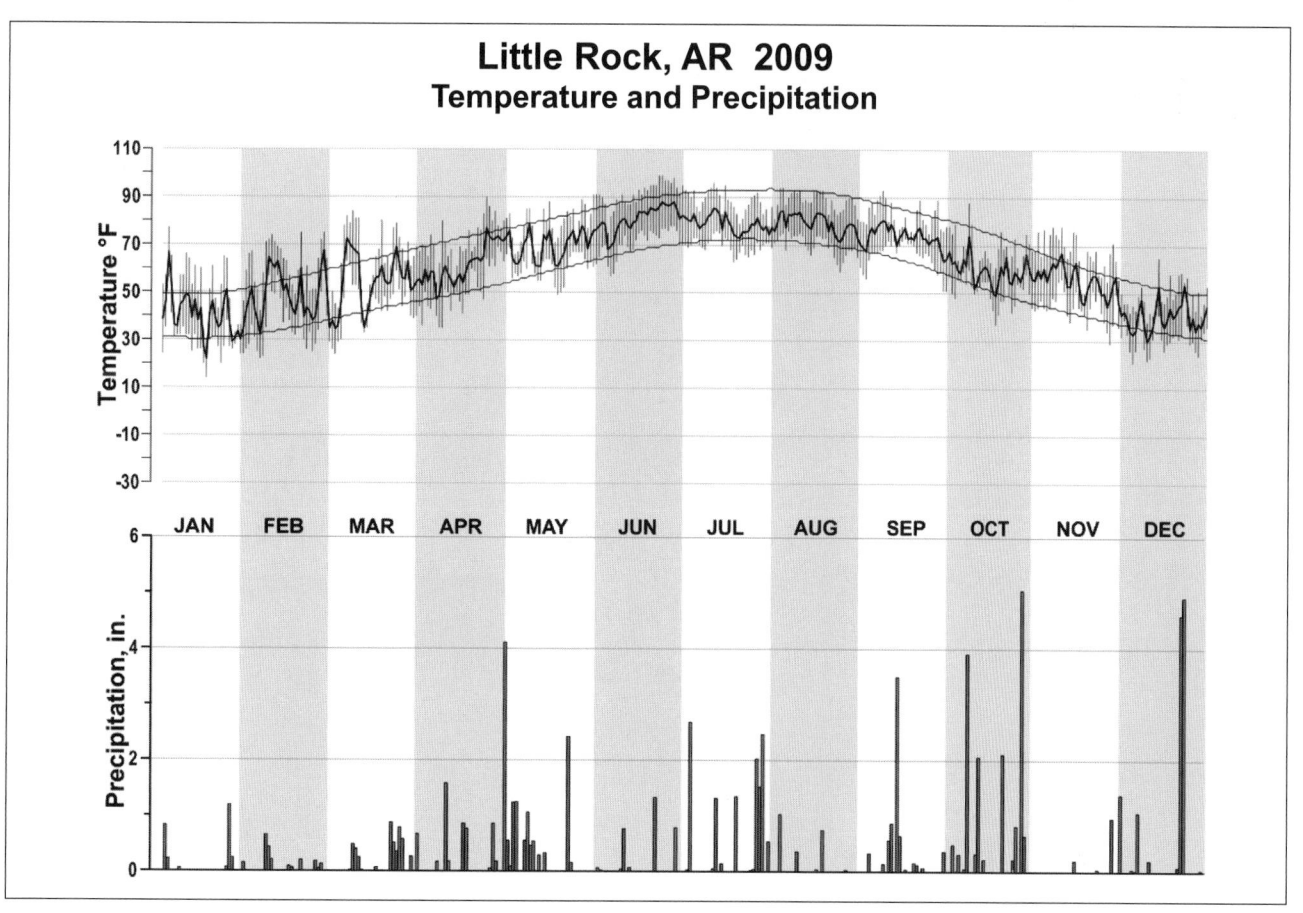

# Little Rock, AR 2009
## Temperature and Precipitation

# 2009
# EUREKA
# CALIFORNIA (KEKA)

Humboldt Bay is one-quarter mile north and one mile west of the station. There are no hills in Eureka of any consequence. The land slopes upward gently from the Bay toward the Coast Range, which begins about 3 miles east of the station and reaches the top of its first ridge approximately 10 miles to the east. The elevation of the ridge is 2,000 feet and extends in a semicircle from a point 20 miles north of Eureka to a point 25 miles south.

The climate of Eureka is completely maritime with high humidity prevailing the entire year. There are definite rainy and dry seasons. The rainy season begins in October and continues through April, accounting for about 90 percent of the annual precipitation. The dry season from May through September is marked by considerable fog or low cloudiness that usually clears in the late morning and sunny weather is generally the case during the early afternoon hours.

Temperatures are moderate the entire year. Although record highs have reached the mid 80s and record lows near 20 degrees, the usual yearly range is from lows in the mid 30s to highs in the mid 70s.

The principal industries are lumbering, fishing, tourism, and dairy farming. There is very little truck farming due to the low temperatures and lack of sunshine, however, the climate is nearly ideal for berries and flowers.

Based on the 1951-1980 period, the average first occurrence of 32 degrees Fahrenheit in the fall is December 10 and the average last occurrence in the spring is February 6.

# NORMALS, MEANS, AND EXTREMES
## EUREKA (KEKA)

| LATITUDE: 40°48'N | LONGITUDE: -124° 9 'W | | ELEVATION (FT): GRND: 19　BARO: 20 | | | | TIME ZONE: PACIFIC　(UTC -8) | | | | WBAN: 24213 | | |

| ELEMENT | | POR | JAN | FEB | MAR | APR | MAY | JUN | JUL | AUG | SEP | OCT | NOV | DEC | YEAR |
|---|---|---|---|---|---|---|---|---|---|---|---|---|---|---|---|
| **TEMPERATURE °F** | NORMAL DAILY MAXIMUM | 30 | 55.0 | 55.9 | 56.1 | 57.4 | 59.6 | 61.8 | 63.3 | 63.9 | 63.6 | 61.3 | 58.0 | 55.1 | 59.3 |
| | MEAN DAILY MAXIMUM | 68 | 54.4 | 55.3 | 55.3 | 56.3 | 58.5 | 60.7 | 62.0 | 62.8 | 62.9 | 61.0 | 58.0 | 54.9 | 58.5 |
| | HIGHEST DAILY MAXIMUM | 99 | 78 | 85 | 78 | 80 | 84 | 85 | 76 | 82 | 86 | 87 | 78 | 77 | 87 |
| | YEAR OF OCCURRENCE | | 1986 | 1930 | 1914 | 1989 | 1939 | 1945 | 1992 | 1991 | 1983 | 1993 | 1987 | 1963 | OCT 1993 |
| | MEAN OF EXTREME MAXS. | 68 | 65.4 | 66.3 | 65.6 | 66.3 | 68.9 | 68.5 | 68.3 | 69.9 | 73.5 | 73.6 | 68.1 | 65.0 | 68.3 |
| | NORMAL DAILY MINIMUM | 30 | 40.8 | 41.8 | 42.2 | 44.0 | 47.6 | 50.7 | 52.8 | 53.4 | 51.2 | 47.7 | 43.9 | 40.6 | 46.4 |
| | MEAN DAILY MINIMUM | 68 | 41.3 | 42.4 | 43.0 | 44.5 | 47.8 | 50.6 | 52.4 | 53.1 | 51.2 | 48.2 | 44.8 | 41.9 | 46.8 |
| | LOWEST DAILY MINIMUM | 99 | 25 | 27 | 29 | 29 | 35 | 40 | 45 | 44 | 41 | 32 | 29 | 21 | 21 |
| | YEAR OF OCCURRENCE | | 2007 | 1990 | 1917 | 2009 | 2002 | 2003 | 1924 | 1935 | 2007 | 1971 | 1994 | 1972 | DEC 1972 |
| | MEAN OF EXTREME MINS. | 68 | 31.5 | 33.3 | 34.5 | 37.1 | 40.9 | 45.3 | 48.3 | 49.0 | 45.3 | 40.1 | 35.2 | 32.1 | 39.4 |
| | NORMAL DRY BULB | 30 | 47.9 | 48.9 | 49.2 | 50.7 | 53.6 | 56.3 | 58.1 | 58.7 | 57.4 | 54.5 | 51.0 | 47.9 | 52.9 |
| | MEAN DRY BULB | 68 | 47.2 | 48.9 | 49.1 | 49.7 | 53.2 | 55.7 | 57.2 | 57.9 | 57.1 | 54.6 | 51.4 | 47.7 | 52.5 |
| | MEAN WET BULB | | | | | | | | | | | | | | |
| | MEAN DEW POINT | | | | | | | | | | | | | | |
| | NORMAL NO. DAYS WITH: | | | | | | | | | | | | | | |
| | MAXIMUM >= 90 | 30 | 0.0 | 0.0 | 0.0 | 0.0 | 0.0 | 0.0 | 0.0 | 0.0 | 0.0 | 0.0 | 0.0 | 0.0 | 0.0 |
| | MAXIMUM <= 32 | 30 | 0.0 | 0.1 | 0.0 | 0.0 | 0.0 | 0.0 | 0.0 | 0.0 | 0.0 | 0.0 | 0.0 | 0.0 | 0.1 |
| | MINIMUM <= 32 | 30 | 1.5 | 1.0 | 0.2 | * | 0.0 | 0.0 | 0.0 | 0.0 | 0.0 | * | 0.6 | 2.0 | 5.3 |
| | MINIMUM <= 0 | 30 | 0.0 | 0.0 | 0.0 | 0.0 | 0.0 | 0.0 | 0.0 | 0.0 | 0.0 | 0.0 | 0.0 | 0.0 | 0.0 |
| **H/C** | NORMAL HEATING DEG. DAYS | 30 | 530 | 451 | 475 | 430 | 354 | 263 | 216 | 198 | 232 | 326 | 423 | 532 | 4430 |
| | NORMAL COOLING DEG. DAYS | 30 | 0 | 0 | 0 | 0 | 0 | 0 | 0 | 2 | 5 | 0 | 0 | 0 | 7 |
| **RH** | NORMAL (PERCENT) | 30 | | | | | | | | | | | | | |
| | HOUR 04 LST | 30 | | | | | | | | | | | | | |
| | HOUR 10 LST | 30 | | | | | | | | | | | | | |
| | HOUR 16 LST | 30 | | | | | | | | | | | | | |
| | HOUR 22 LST | 30 | | | | | | | | | | | | | |
| **S** | PERCENT POSSIBLE SUNSHINE | 84 | 43 | 46 | 52 | 57 | 58 | 59 | 55 | 51 | 55 | 50 | 44 | 41 | 51 |
| **W/O** | MEAN NO. DAYS WITH: | | | | | | | | | | | | | | |
| | HEAVY FOG(VISBY <= 1/4 MI) | 31 | 4.1 | 2.0 | 1.7 | 1.3 | 1.5 | 1.7 | 3.6 | 5.6 | 7.3 | 8.9 | 3.9 | 3.9 | 45.5 |
| | THUNDERSTORMS | 47 | 0.5 | 0.5 | 0.3 | 0.1 | 0.1 | 0.2 | 0.1 | 0.1 | 0.3 | 0.3 | 0.5 | 0.5 | 3.5 |
| **CLOUDNESS** | MEAN: | | | | | | | | | | | | | | |
| | SUNRISE-SUNSET (OKTAS) | 63 | 5.8 | 5.9 | 5.8 | 5.6 | 5.4 | 5.2 | 5.1 | 5.3 | 4.8 | 5.1 | 5.7 | 5.6 | 5.4 |
| | MIDNIGHT-MIDNIGHT (OKTAS) | 2 | 2.6 | 3.2 | 0.0 | 0.0 | 2.6 | 0.0 | 2.5 | 0.0 | 2.8 | 2.4 | 0.0 | 2.4 | 1.5 |
| | MEAN NO. DAYS WITH: | | | | | | | | | | | | | | |
| | CLEAR | 94 | 5.8 | 5.1 | 5.6 | 6.0 | 6.5 | 7.2 | 6.4 | 5.4 | 8.7 | 8.1 | 6.1 | 6.3 | 77.2 |
| | PARTLY CLOUDY | 95 | 6.4 | 5.8 | 8.1 | 8.4 | 10.1 | 9.7 | 11.1 | 10.8 | 8.6 | 8.4 | 6.6 | 6.5 | 100.5 |
| | CLOUDY | 95 | 18.8 | 17.3 | 17.5 | 15.5 | 14.5 | 13.1 | 13.4 | 14.2 | 12.3 | 14.0 | 16.9 | 17.9 | 185.4 |
| **PR** | MEAN STATION PRESSURE(IN) | | | | | | | | | | | | | | |
| | MEAN SEA-LEVEL PRES. (IN) | | | | | | | | | | | | | | |
| **WINDS** | MEAN SPEED (MPH) | 54 | 6.9 | 7.2 | 7.6 | 8.0 | 7.9 | 7.4 | 6.8 | 5.8 | 5.5 | 5.6 | 6.0 | 6.4 | 6.8 |
| | PREVAIL.DIR(TENS OF DEGS) | | | | | | | | | | | | | | |
| | MAXIMUM 2-MINUTE: | | | | | | | | | | | | | | |
| | SPEED (MPH) | 84 | 54 | 48 | 48 | 49 | 40 | 39 | 35 | 34 | 44 | 56 | 55 | 56 | 56 |
| | DIR. (TENS OF DEGS) | | 18 | 22 | 22 | 36 | 32 | 32 | 36 | 36 | 36 | 22 | 18 | 18 | 22 |
| | YEAR OF OCCURRENCE | | 1955 | 1960 | 1953 | 1915 | 1955 | 1949 | 1986 | 1920 | 1941 | 1962 | 1981 | 1931 | OCT 1962 |
| | MAXIMUM 3-SECOND | | | | | | | | | | | | | | |
| | SPEED (MPH) | | | | | | | | | | | | | | |
| | DIR. (TENS OF DEGS) | | | | | | | | | | | | | | |
| | YEAR OF OCCURRENCE | | | | | | | | | | | | | | |
| **PRECIPITATION** | NORMAL (IN) | 30 | 5.97 | 5.51 | 5.55 | 2.91 | 1.62 | 0.65 | 0.16 | 0.38 | 0.86 | 2.36 | 5.78 | 6.35 | 38.10 |
| | MAXIMUM MONTHLY (IN) | 99 | 13.92 | 13.95 | 13.97 | 11.25 | 6.05 | 3.08 | 1.34 | 3.42 | 3.56 | 13.04 | 16.58 | 23.31 | 23.31 |
| | YEAR OF OCCURRENCE | | 1969 | 1998 | 1938 | 2003 | 1960 | 2005 | 1916 | 1983 | 1925 | 1950 | 1973 | 2002 | DEC 2002 |
| | MINIMUM MONTHLY (IN) | 99 | 0.66 | 0.50 | 0.07 | 0.31 | 0.03 | 0.00 | 0.00 | 0.00 | 0.00 | 0.00 | T | 0.52 | 0.00 |
| | YEAR OF OCCURRENCE | | 1985 | 1923 | 1926 | 1956 | 1955 | 1917 | 1967 | 1940 | 1929 | 1917 | 1967 | 1976 | JUL 1967 |
| | MAXIMUM IN 24 HOURS (IN) | 99 | 4.42 | 4.88 | 4.02 | 2.56 | 2.23 | 1.73 | 1.18 | 2.21 | 1.54 | 5.83 | 5.21 | 6.85 | 6.85 |
| | YEAR OF OCCURRENCE | | 1912 | 1959 | 1975 | 1983 | 1943 | 1943 | 1916 | 1983 | 1977 | 1950 | 1998 | 2002 | DEC 2002 |
| | NORMAL NO. DAYS WITH: | | | | | | | | | | | | | | |
| | PRECIPITATION >= 0.01 | 30 | 15.5 | 15.2 | 16.9 | 12.5 | 8.7 | 5.7 | 2.3 | 3.2 | 4.5 | 8.3 | 15.0 | 15.5 | 123.3 |
| | PRECIPITATION >= 1.00 | 30 | 1.4 | 1.1 | 1.0 | 0.4 | * | 0.1 | 0.0 | 0.1 | 0.1 | 0.4 | 1.5 | 1.6 | 7.7 |
| **SNOWFALL** | NORMAL (IN) | 30 | 0.1 | 0.2 | 0.* | 0.* | 0.0 | 0.0 | 0.0 | 0.0 | 0.0 | 0.0 | 0.* | 0.1 | 0.4 |
| | MAXIMUM MONTHLY (IN) | 99 | 3.0 | 3.5 | 1.0 | 0.1 | 0.0 | T | 0.0 | T | 0.0 | 0.0 | 0.1 | 1.9 | 3.5 |
| | YEAR OF OCCURRENCE | | 1935 | 1989 | 1966 | 2003 | | 2008 | | 2006 | | | 1977 | 1972 | FEB 1989 |
| | MAXIMUM IN 24 HOURS (IN) | 99 | 3.0 | 2.0 | 1.0 | 0.1 | 0.0 | T | 0.0 | 0.0 | 0.0 | 0.0 | 0.1 | 1.9 | 3.0 |
| | YEAR OF OCCURRENCE' | | 1935 | 1989 | 1966 | 2003 | | 2008 | | | | | 1977 | 1972 | JAN 1935 |
| | MAXIMUM SNOW DEPTH (IN) | 67 | 0 | 1 | 1 | 0 | 0 | 0 | 0 | 0 | 0 | 0 | 0 | 3 | 3 |
| | YEAR OF OCCURRENCE | | | 1989 | 1999 | | | | | | | | | 2007 | DEC 2007 |
| | NORMAL NO. DAYS WITH: | | | | | | | | | | | | | | |
| | SNOWFALL >= 1.0 | 30 | 0.0 | 0.1 | 0.0 | 0.0 | 0.0 | 0.0 | 0.0 | 0.0 | 0.0 | 0.0 | 0.0 | 0.1 | 0.2 |

## PRECIPITATION (inches) 2009 EUREKA (KEKA)

| YEAR | JAN | FEB | MAR | APR | MAY | JUN | JUL | AUG | SEP | OCT | NOV | DEC | ANNUAL |
|------|-----|-----|-----|-----|-----|-----|-----|-----|-----|-----|-----|-----|--------|
| 1980 | 3.19 | 4.67 | 6.14 | 4.18 | 1.70 | 0.42 | T | 0.07 | 0.14 | 1.38 | 2.49 | 6.10 | 30.48 |
| 1981 | 7.67 | 3.72 | 4.64 | 0.71 | 2.02 | 0.57 | T | 0.01 | 0.97 | 3.71 | 9.39 | 9.88 | 43.29 |
| 1982 | 4.75 | 5.76 | 7.06 | 5.97 | 0.07 | 0.78 | 0.08 | 0.03 | 0.62 | 4.89 | 7.83 | 10.30 | 48.14 |
| 1983 | 8.48 | 9.18 | 10.73 | 5.47 | 1.12 | 0.65 | 0.89 | 3.42 | 0.87 | 1.87 | 10.40 | 14.13 | 67.21 |
| 1984 | 0.76 | 5.18 | 4.70 | 2.76 | 2.51 | 1.07 | 0.03 | 0.05 | 0.55 | 3.67 | 15.15 | 4.27 | 40.70 |
| 1985 | 0.66 | 3.69 | 4.68 | 0.45 | 1.14 | 0.89 | 0.15 | 0.52 | 1.06 | 4.07 | 2.98 | 2.78 | 23.07 |
| 1986 | 7.19 | 10.08 | 6.12 | 1.46 | 2.34 | 0.21 | 0.02 | T | 2.70 | 1.75 | 1.85 | 3.83 | 37.55 |
| 1987 | 6.48 | 3.38 | 6.10 | 1.15 | 0.41 | 0.26 | 0.20 | 0.06 | 0.02 | 1.05 | 4.23 | 10.92 | 34.26 |
| 1988 | 7.13 | 0.54 | 1.18 | 2.06 | 2.70 | 2.22 | 0.05 | T | 0.12 | 0.41 | 8.93 | 6.26 | 31.60 |
| 1989 | 4.71 | 2.88 | 7.63 | 2.01 | 1.67 | 0.21 | 0.08 | 0.13 | 0.85 | 2.90 | 1.60 | 0.80 | 25.47 |
| 1990 | 7.20 | 4.50 | 3.30 | 1.41 | 3.74 | 0.32 | 0.22 | 0.71 | 0.19 | 1.73 | 3.07 | 2.95 | 29.34 |
| 1991 | 1.65 | 2.75 | 6.94 | 2.52 | 2.16 | 0.26 | 1.13 | 0.37 | T | 1.06 | 1.95 | 2.36 | 23.15 |
| 1992 | 3.99 | 3.80 | 3.51 | 2.42 | 0.06 | 1.27 | 0.25 | 0.01 | 0.33 | 2.08 | 2.21 | 9.33 | 29.26 |
| 1993 | 7.15 | 5.93 | 4.72 | 5.94 | 4.44 | 1.23 | 0.37 | 0.54 | 0.03 | 0.56 | 1.35 | 7.12 | 39.38 |
| 1994 | 5.09 | 7.12 | 2.06 | 3.30 | 1.10 | 0.71 | 0.08 | T | 0.06 | 0.54 | 8.21 | 7.00 | 35.27 |
| 1995 | 12.74 | 1.40 | 11.18 | 7.47 | 1.21 | 1.85 | 0.08 | 0.22 | 0.69 | 0.53 | 2.26 | 11.56 | 51.19 |
| 1996 | 10.74 | 8.11 | 3.51 | 4.64 | 2.40 | 0.05 | .03 | T | 1.21 | 3.50 | 5.16 | 21.26 | 60.61 |
| 1997 | 8.81 | 2.55 | 2.73 | 3.06 | 0.90 | 1.25 | T | 0.84 | 2.05 | 2.73 | 7.39 | 4.73 | 37.04 |
| 1998 | 13.42 | 13.95 | 7.83 | 2.23 | 3.12 | 0.33 | 0.16 | 0.01 | 0.08 | 3.06 | 14.09 | 5.40 | 63.68 |
| 1999 | 4.37 | 10.32 | 8.94 | 1.79 | 1.62 | 0.15 | 0.04 | 0.30 | 0.05 | 1.60 | 7.36 | 3.02 | 39.56 |
| 2000 | 9.71 | 7.00 | 2.81 | 2.15 | 1.86 | 0.54 | 0.04 | T | 0.55 | 2.99 | 3.51 | 1.97 | 33.13 |
| 2001 | 3.79 | 3.60 | 2.45 | 2.54 | 0.71 | 0.69 | 0.20 | 0.21 | 0.28 | 1.00 | 7.71 | 11.56 | 34.74 |
| 2002 | 6.37 | 5.76 | 4.32 | 2.42 | 0.55 | 0.28 | 0.03 | 0.01 | 0.06 | 0.06 | 2.66 | 23.31 | 45.83 |
| 2003 | 5.51 | 3.84 | 4.91 | 11.25 | 1.74 | 0.04 | 0.02 | 0.49 | 0.35 | 0.55 | 5.78 | 11.35 | 45.83 |
| 2004 | 6.29 | 8.12 | 2.38 | 1.68 | 1.37 | 0.06 | 0.06 | 0.43 | 0.68 | 5.71 | 1.87 | 9.43 | 38.08 |
| 2005 | 5.91 | 2.41 | 6.24 | 4.70 | 3.90 | 3.08 | 0.05 | 0.07 | 0.08 | 2.40 | 8.52 | 12.72 | 50.08 |
| 2006 | 12.09 | 6.34 | 11.11 | 4.08 | 1.03 | 0.35 | 0.04 | T | 0.09 | 0.58 | 7.41 | 7.09 | 50.21 |
| 2007 | 1.86 | 11.86 | 2.51 | 2.72 | 0.86 | 0.46 | 0.97 | 0.08 | 0.60 | 4.92 | 2.33 | 7.30 | 36.47 |
| 2008 | 9.70 | 2.73 | 3.16 | 2.12 | 0.04 | 0.24 | 0.02 | 0.47 | 0.05 | 0.93 | 4.05 | 6.66 | 30.17 |
| 2009 | 1.58 | 6.20 | 5.45 | 1.58 | 2.93 | 0.18 | 0.06 | 0.02 | 1.03 | 1.95 | 4.15 | 1.58 | 26.71 |
| POR= 68 YRS | 6.52 | 5.35 | 5.14 | 3.00 | 1.79 | 0.63 | 0.15 | 0.31 | 0.71 | 2.70 | 5.69 | 6.94 | 38.93 |

WBAN : 24213

## AVERAGE TEMPERATURE (°F) 2009 EUREKA (KEKA)

| YEAR | JAN | FEB | MAR | APR | MAY | JUN | JUL | AUG | SEP | OCT | NOV | DEC | ANNUAL |
|------|-----|-----|-----|-----|-----|-----|-----|-----|-----|-----|-----|-----|--------|
| 1980 | 48.3 | 53.5 | 48.6 | 51.9 | 52.3 | 55.5 | 57.3 | 55.0 | 56.6 | 54.3 | 51.3 | 51.2 | 53.0 |
| 1981 | 52.5 | 51.3 | 50.2 | 50.6 | 53.3 | 56.5 | 55.2 | 58.1 | 57.9 | 54.3 | 53.6 | 52.1 | 53.8 |
| 1982 | 44.9 | 49.5 | 48.3 | 50.8 | 52.8 | 56.3 | 58.3 | 59.6 | 58.7 | 57.3 | 52.1 | 49.6 | 53.2 |
| 1983 | 51.3 | 53.4 | 53.4 | 51.8 | 54.4 | 58.0 | 60.5 | 61.9 | 60.7 | 57.9 | 53.6 | 50.6 | 55.6 |
| 1984 | 49.4 | 50.0 | 52.8 | 51.3 | 55.0 | 55.2 | 57.6 | 60.2 | 58.3 | 55.3 | 52.1 | 46.8 | 53.7 |
| 1985 | 48.4 | 47.9 | 47.4 | 51.6 | 53.9 | 56.6 | 58.6 | 58.5 | 56.7 | 54.5 | 46.6 | 47.7 | 52.4 |
| 1986 | 54.3 | 52.7 | 53.1 | 50.7 | 53.9 | 59.0 | 57.4 | 57.2 | 57.2 | 55.6 | 53.5 | 51.3 | 54.7 |
| 1987 | 49.1 | 51.4 | 52.7 | 54.2 | 56.6 | 57.7 | 59.5 | 58.5 | 57.3 | 57.8 | 54.8 | 49.7 | 54.9 |
| 1988 | 50.4 | 50.2 | 50.3 | 52.8 | 56.3 | 57.7 | 58.8 | 57.7 | 55.8 | 55.2 | 53.3 | 47.7 | 53.9 |
| 1989 | 46.0 | 45.6 | 51.8 | 54.6 | 55.9 | 57.6 | 59.4 | 59.3 | 56.6 | 54.9 | 52.6 | 49.2 | 53.6 |
| 1990 | 48.3 | 45.5 | 50.2 | 52.8 | 54.0 | 58.2 | 59.8 | 60.4 | 61.6 | 54.3 | 49.7 | 42.8 | 53.1 |
| 1991 | 47.8 | 52.8 | 48.1 | 50.9 | 52.2 | 53.8 | 57.3 | 59.9 | 57.2 | 56.1 | 51.4 | 48.3 | 53.0 |
| 1992 | 50.3 | 53.9 | 54.1 | 56.4 | 56.7 | 58.3 | 60.6 | 59.1 | 57.5 | 58.0 | 51.7 | 46.9 | 55.3 |
| 1993 | 46.4 | 49.4 | 53.4 | 53.6 | 57.9 | 57.9 | 57.4 | 59.5 | 55.8 | 56.9 | 49.9 | 49.5 | 54.0 |
| 1994 | 50.5 | 48.5 | 51.1 | 52.8 | 55.5 | 57.0 | 57.1 | 61.4 | 59.6 | 54.4 | 45.6 | 46.7 | 53.4 |
| 1995 | 52.4 | 51.7 | 50.4 | 51.1 | 53.7 | 56.0 | 60.2 | 58.4 | 60.3 | 54.9 | 53.9 | 52.0 | 54.6 |
| 1996 | 49.5 | 52.3 | 50.7 | 52.9 | 53.7 | 55.5 | 57.4 | 58.1 | 55.8 | 54.5 | 51.3 | 50.9 | 53.6 |
| 1997 | 48.5 | 47.9 | 49.5 | 51.4 | 57.8 | 57.8 | 59.5 | 61.4 | 62.2 | 55.1 | 53.4 | 47.7 | 54.4 |
| 1998 | 52.0 | 50.1 | 50.2 | 50.4 | 53.9 | 56.7 | 58.5 | 58.8 | 57.4 | 55.0 | 51.7 | 44.6 | 53.3 |
| 1999 | 47.4 | 47.3 | 46.8 | 48.6 | 51.3 | 55.0 | 56.5 | 60.1 | 54.9 | 52.5 | 54.7 | 46.0 | 51.8 |
| 2000 | 48.3 | 51.6 | 48.4 | 52.3 | 55.3 | 56.7 | 58.7 | 59.0 | 58.7 | 54.3 | 48.6 | 48.4 | 53.4 |
| 2001 | 46.9 | 47.0 | 50.2 | 49.2 | 54.0 | 56.0 | 57.4 | 59.5 | 56.2 | 54.4 | 51.8 | 49.2 | 52.7 |
| 2002 | 46.6 | 49.0 | 47.0 | 51.4 | 52.3 | 56.3 | 58.6 | 57.0 | 57.3 | 51.9 | 52.3 | 50.6 | 52.5 |
| 2003 | 54.0 | 47.9 | 52.2 | 50.6 | 54.1 | 57.2 | 58.8 | 61.2 | 58.0 | 56.5 | 49.8 | 49.1 | 54.1 |
| 2004 | 49.6 | 49.2 | 51.5 | 52.6 | 56.7 | 58.1 | 60.3 | 61.1 | 58.1 | 54.9 | 48.9 | 48.4 | 54.1 |
| 2005 | 47.6 | 50.2 | 52.0 | 51.7 | 57.2 | 57.4 | 58.7 | 57.0 | 55.1 | 53.8 | 49.6 | 50.4 | 53.4 |
| 2006 | 49.2 | 47.6 | 47.3 | 50.9 | 54.2 | 58.0 | 58.4 | 56.8 | 55.3 | 52.4 | 51.9 | 48.0 | 52.5 |
| 2007 | 42.9 | 48.3 | 50.4 | 50.7 | 52.7 | 56.2 | 61.3 | 59.6 | 57.0 | 53.8 | 50.3 | 44.8 | 52.3 |
| 2008 | 44.9 | 45.9 | 47.3 | 48.1 | 53.6 | 53.7 | 56.9 | 58.7 | 55.2 | 54.1 | 53.0 | 45.2 | 51.4 |
| 2009 | -1.1 | 48.3 | 47.9 | -1.1 | 53.5 | 57.3 | 56.7 | 59.0 | 59.5 | 54.6 | 50.1 | -1.1 | 40.3 |
| POR= 68 YRS | 47.2 | 48.9 | 49.1 | 49.7 | 53.2 | 55.7 | 57.2 | 57.9 | 57.1 | 54.6 | 51.4 | 47.7 | 52.5 |

## HEATING DEGREE DAYS (base 65°F) 2009  EUREKA (KEKA)

| YEAR | JUL | AUG | SEP | OCT | NOV | DEC | JAN | FEB | MAR | APR | MAY | JUN | TOTAL |
|------|-----|-----|-----|-----|-----|-----|-----|-----|-----|-----|-----|-----|-------|
| 1980-81 | 230 | 303 | 246 | 328 | 402 | 422 | 384 | 377 | 451 | 423 | 357 | 249 | 4172 |
| 1981-82 | 299 | 205 | 203 | 324 | 339 | 396 | 616 | 430 | 512 | 419 | 373 | 258 | 4374 |
| 1982-83 | 204 | 158 | 181 | 232 | 381 | 468 | 415 | 317 | 355 | 391 | 320 | 203 | 3625 |
| 1983-84 | 133 | 90 | 129 | 215 | 336 | 443 | 475 | 429 | 369 | 403 | 301 | 285 | 3608 |
| 1984-85 | 222 | 142 | 195 | 295 | 378 | 556 | 507 | 472 | 532 | 396 | 338 | 243 | 4276 |
| 1985-86 | 195 | 194 | 244 | 316 | 546 | 533 | 329 | 337 | 365 | 422 | 338 | 169 | 3988 |
| 1986-87 | 227 | 236 | 227 | 290 | 341 | 417 | 487 | 372 | 377 | 316 | 260 | 213 | 3763 |
| 1987-88 | 163 | 196 | 226 | 221 | 302 | 470 | 446 | 423 | 453 | 358 | 265 | 215 | 3738 |
| 1988-89 | 187 | 218 | 274 | 297 | 345 | 529 | 582 | 535 | 403 | 309 | 278 | 214 | 4171 |
| 1989-90 | 164 | 171 | 243 | 306 | 365 | 482 | 513 | 541 | 453 | 356 | 332 | 198 | 4124 |
| 1990-91 | 154 | 141 | 95 | 325 | 451 | 680 | 528 | 336 | 516 | 418 | 387 | 330 | 4361 |
| 1991-92 | 229 | 164 | 225 | 274 | 401 | 511 | 450 | 316 | 329 | 250 | 248 | 193 | 3590 |
| 1992-93 | 135 | 176 | 220 | 211 | 393 | 556 | 570 | 433 | 354 | 337 | 214 | 203 | 3802 |
| 1993-94 | 229 | 152 | 268 | 253 | 446 | 472 | 442 | 455 | 426 | 357 | 286 | 233 | 4019 |
| 1994-95 | 241 | 102 | 155 | 321 | 576 | 559 | 383 | 368 | 445 | 412 | 341 | 263 | 4166 |
| 1995-96 | 142 | 196 | 133 | 308 | 325 | 394 | 472 | 360 | 435 | 352 | 344 | 278 | 3739 |
| 1996-97 | 231 | 205 | 268 | 323 | 407 | 431 | 501 | 476 | 475 | 401 | 212 | 209 | 4139 |
| 1997-98 | 163 | 106 | 82 | 300 | 341 | 529 | 397 | 409 | 450 | 432 | 337 | 242 | 3788 |
| 1998-99 | 192 | 186 | 223 | 305 | 393 | 624 | 541 | 490 | 556 | 485 | 417 | 296 | 4708 |
| 1999-00 | 254 | 146 | 295 | 382 | 303 | 580 | 511 | 382 | 506 | 378 | 298 | 242 | 4277 |
| 2000-01 | 191 | 179 | 181 | 326 | 484 | 506 | 555 | 500 | 453 | 467 | 337 | 264 | 4443 |
| 2001-02 | 228 | 168 | 257 | 322 | 389 | 481 | 566 | 442 | 554 | 400 | 387 | 253 | 4447 |
| 2002-03 | 190 | 237 | 225 | 402 | 373 | 438 | 333 | 476 | 389 | 427 | 329 | 229 | 4048 |
| 2003-04 | 184 | 113 | 205 | 257 | 449 | 487 | 471 | 449 | 411 | 369 | 251 | 201 | 3847 |
| 2004-05 | 140 | 114 | 200 | 306 | 476 | 506 | 531 | 408 | 392 | 392 | 233 | 218 | 3916 |
| 2005-06 | 191 | 241 | 293 | 341 | 455 | 445 | 486 | 479 | 542 | 418 | 327 | 201 | 4419 |
| 2006-07 | 199 | 244 | 284 | 383 | 386 | 520 | 677 | 461 | 446 | 423 | 375 | 258 | 4656 |
| 2007-08 | 113 | 162 | 234 | 336 | 436 | 620 | 616 | 550 | 541 | 501 | 354 | 330 | 4793 |
| 2008-09 | 246 | 186 | 289 | 330 | 354 | 608 | 559 | 462 | 523 | 559 | 350 | 227 | 4693 |
| 2009- | 252 | 180 | 158 | 316 | 440 | 559 | | | | | | | |

WBAN : 24213

## COOLING DEGREE DAYS (base 65°F) 2009  EUREKA (KEKA)

| YEAR | JAN | FEB | MAR | APR | MAY | JUN | JUL | AUG | SEP | OCT | NOV | DEC | TOTAL |
|------|-----|-----|-----|-----|-----|-----|-----|-----|-----|-----|-----|-----|-------|
| 1980 | 0 | 2 | 0 | 0 | 0 | 0 | 0 | 0 | 0 | 3 | 0 | 0 | 5 |
| 1981 | 4 | 0 | 0 | 0 | 0 | 0 | 0 | 0 | 0 | 0 | 0 | 0 | 4 |
| 1982 | 0 | 0 | 0 | 0 | 0 | 3 | 0 | 0 | 0 | 2 | 0 | 0 | 5 |
| 1983 | 0 | 0 | 0 | 0 | 0 | 0 | 0 | 2 | 7 | 0 | 0 | 0 | 9 |
| 1984 | 0 | 0 | 0 | 0 | 0 | 0 | 0 | 0 | 4 | 1 | 0 | 0 | 5 |
| 1985 | 0 | 0 | 0 | 0 | 0 | 0 | 1 | 0 | 0 | 2 | 0 | 0 | 3 |
| 1986 | 0 | 0 | 0 | 0 | 0 | 0 | 0 | 0 | 0 | 5 | 0 | 0 | 5 |
| 1987 | 0 | 0 | 0 | 0 | 3 | 0 | 0 | 0 | 0 | 5 | 0 | 0 | 8 |
| 1988 | 0 | 0 | 0 | 0 | 0 | 0 | 0 | 0 | 4 | 0 | 0 | 0 | 4 |
| 1989 | 0 | 0 | 0 | 1 | 0 | 0 | 0 | 0 | 0 | 0 | 0 | 0 | 1 |
| 1990 | 0 | 0 | 0 | 0 | 0 | 0 | 0 | 4 | 0 | 0 | 0 | 0 | 4 |
| 1991 | 0 | 0 | 0 | 0 | 0 | 0 | 0 | 10 | 0 | 4 | 0 | 0 | 14 |
| 1992 | 0 | 0 | 0 | 0 | 0 | 0 | 2 | 0 | 3 | 2 | 0 | 0 | 7 |
| 1993 | 0 | 3 | 0 | 0 | 3 | 0 | 0 | 0 | 0 | 8 | 0 | 0 | 14 |
| 1994 | 0 | 0 | 0 | 0 | 0 | 0 | 0 | 0 | 0 | 0 | 0 | 0 | 0 |
| 1995 | 0 | 0 | 0 | 0 | 0 | 0 | 0 | 0 | 0 | 0 | 0 | 0 | 0 |
| 1996 | 0 | 0 | 0 | 0 | 0 | 0 | 0 | 0 | 0 | 1 | 0 | 0 | 1 |
| 1997 | 0 | 0 | 1 | 0 | 4 | 0 | 0 | 3 | 5 | 0 | 0 | 0 | 13 |
| 1998 | 0 | 0 | 0 | 0 | 0 | 0 | 0 | 0 | 0 | 0 | 0 | 0 | 0 |
| 1999 | 0 | 0 | 0 | 0 | 0 | 0 | 0 | 0 | 0 | 0 | 1 | 0 | 1 |
| 2000 | 0 | 0 | 0 | 0 | 0 | 0 | 0 | 0 | 0 | 0 | 0 | 0 | 0 |
| 2001 | 0 | 0 | 0 | 0 | 0 | 0 | 0 | 3 | 0 | 0 | 0 | 0 | 3 |
| 2002 | 0 | 0 | 0 | 0 | 0 | 0 | 0 | 0 | 0 | 0 | 0 | 0 | 0 |
| 2003 | 0 | 0 | 0 | 0 | 0 | 0 | 0 | 1 | 1 | 0 | 0 | 0 | 2 |
| 2004 | 0 | 0 | 0 | 0 | 0 | 0 | 2 | 0 | 0 | 0 | 0 | 0 | 2 |
| 2005 | 0 | 0 | 0 | 0 | 0 | 0 | 1 | 0 | 0 | 0 | 0 | 0 | 1 |
| 2006 | 0 | 0 | 0 | 0 | 0 | 0 | 0 | 0 | 0 | 0 | 0 | 0 | 0 |
| 2007 | 0 | 0 | 0 | 0 | 0 | 0 | 6 | 3 | 0 | 0 | 0 | 0 | 9 |
| 2008 | 0 | 0 | 0 | 0 | 3 | 0 | 0 | 0 | 0 | 0 | 0 | 0 | 3 |
| 2009 | 0 | 0 | 0 | 0 | 0 | 0 | 0 | 0 | 0 | 0 | 0 | 0 | 0 |

## SNOWFALL (inches) 2009 EUREKA (KEKA)

| YEAR | JUL | AUG | SEP | OCT | NOV | DEC | JAN | FEB | MAR | APR | MAY | JUN | TOTAL |
|------|-----|-----|-----|-----|-----|-----|-----|-----|-----|-----|-----|-----|-------|
| 1980-81 | 0.0 | 0.0 | 0.0 | 0.0 | 0.0 | 0.0 | 0.0 | 0.0 | 0.0 | 0.0 | 0.0 | 0.0 | 0.0 |
| 1981-82 | 0.0 | 0.0 | 0.0 | 0.0 | 0.0 | 0.0 | T | 0.0 | T | T | 0.0 | 0.0 | T |
| 1982-83 | 0.0 | 0.0 | 0.0 | 0.0 | 0.0 | T | 0.0 | 0.0 | T | 0.0 | 0.0 | 0.0 | T |
| 1983-84 | 0.0 | 0.0 | 0.0 | 0.0 | T | 1.0 | 0.0 | 0.0 | 0.0 | 0.0 | 0.0 | 0.0 | 1.0 |
| 1984-85 | 0.0 | 0.0 | 0.0 | 0.0 | 0.0 | 0.0 | 0.0 | 0.0 | 0.0 | 0.0 | 0.0 | 0.0 | 0.0 |
| 1985-86 | 0.0 | 0.0 | 0.0 | 0.0 | 0.0 | 0.0 | 0.0 | 0.0 | 0.0 | 0.0 | 0.0 | 0.0 | 0.0 |
| 1986-87 | 0.0 | 0.0 | 0.0 | 0.0 | 0.0 | 0.0 | 0.0 | 0.0 | 0.0 | 0.0 | 0.0 | 0.0 | 0.0 |
| 1987-88 | 0.0 | 0.0 | 0.0 | 0.0 | 0.0 | T | 0.0 | 0.0 | 0.0 | 0.0 | 0.0 | 0.0 | T |
| 1988-89 | 0.0 | 0.0 | 0.0 | 0.0 | 0.0 | T | 0.0 | 3.5 | 0.0 | 0.0 | 0.0 | 0.0 | 3.5 |
| 1989-90 | 0.0 | 0.0 | 0.0 | 0.0 | 0.0 | 0.0 | 0.0 | 0.0 | 1.0 | 0.0 | 0.0 | 0.0 | 1.0 |
| 1990-91 | 0.0 | 0.0 | 0.0 | 0.0 | 0.0 | T | 0.0 | 0.0 | T | 0.0 | 0.0 | 0.0 | T |
| 1991-92 | 0.0 | 0.0 | 0.0 | 0.0 | 0.0 | T | 0.0 | 0.0 | 0.0 | 0.0 | 0.0 | 0.0 | T |
| 1992-93 | 0.0 | 0.0 | 0.0 | 0.0 | 0.0 | T | 0.0 | T | 0.0 | T | 0.0 | 0.0 | T |
| 1993-94 | 0.0 | 0.0 | 0.0 | 0.0 | 0.0 | 0.0 | 0.0 | T | T | 0.0 | 0.0 | 0.0 | T |
| 1994-95 | 0.0 | 0.0 | 0.0 | 0.0 | T | 0.0 | 0.0 | 0.0 | T | T | 0.0 | 0.0 | T |
| 1995-96 | 0.0 | 0.0 | 0.0 | 0.0 | 0.0 | 0.0 | T | T | T | 0.0 | 0.0 | 0.0 | T |
| 1996-97 | 0.0 | 0.0 | 0.0 | 0.0 | 0.0 | T | T | 0.0 | T | 0.0 | 0.0 | 0.0 | T |
| 1997-98 | 0.0 | 0.0 | 0.0 | 0.0 | 0.0 | 0.0 | T | T | 0.0 | 0.0 | 0.0 | 0.0 | T |
| 1998-99 | 0.0 | 0.0 | 0.0 | 0.0 | 0.0 | T | T | T | T | 0.6 | T | 0.0 | 0.6 |
| 1999-00 | 0.0 | 0.0 | 0.0 | 0.0 | 0.0 | T | T | 0.0 | 0.0 | 0.0 | 0.0 | 0.0 | T |
| 2000-01 | 0.0 | 0.0 | 0.0 | 0.0 | 0.0 | 0.0 | 0.0 | 0.0 | T | 0.0 | 0.0 | 0.0 | T |
| 2001-02 | 0.0 | 0.0 | 0.0 | 0.0 | 0.0 | T | 0.5 | 0.0 | T | T | 0.0 | 0.0 | 0.5 |
| 2002-03 | 0.0 | 0.0 | 0.0 | 0.0 | 0.0 | T | 0.0 | 0.0 | 0.0 | 0.1 | 0.0 | 0.0 | 0.1 |
| 2003-04 | 0.0 | 0.0 | 0.0 | 0.0 | 0.0 | T | T | 0.0 | 0.0 | 0.0 | 0.0 | 0.0 | T |
| 2004-05 | 0.0 | 0.0 | 0.0 | 0.0 | 0.0 | T | T | 0.0 | 0.0 | T | 0.0 | 0.0 | T |
| 2005-06 | 0.0 | 0.0 | 0.0 | 0.0 | 0.0 | 0.0 | 0.0 | 0.0 | 0.0 | 0.0 | 0.0 | 0.0 | 0.0 |
| 2006-07 | 0.0 | 0.0 | 0.0 | 0.0 | 0.0 | 0.0 | T | T | T | 0.0 | 0.0 | 0.0 | T |
| 2007-08 | 0.0 | 0.0 | 0.0 | 0.0 | 0.0 | T | T | 0.0 | 0.0 | T | 0.0 | T | T |
| 2008-09 | 0.0 | 0.0 | 0.0 | 0.0 | 0.0 | T | 0.0 | 0.0 | T | 0.0 | 0.0 | 0.0 | T |
| 2009- | 0.0 | 0.0 | 0.0 | 0.0 | 0.0 | 0.0 | | | | | | | |
| POR= 68 YRS | 0.0 | 0.0 | 0.0 | 0.0 | T | T | 0.1 | 0.1 | T | T | 0.0 | T | 0.2 |

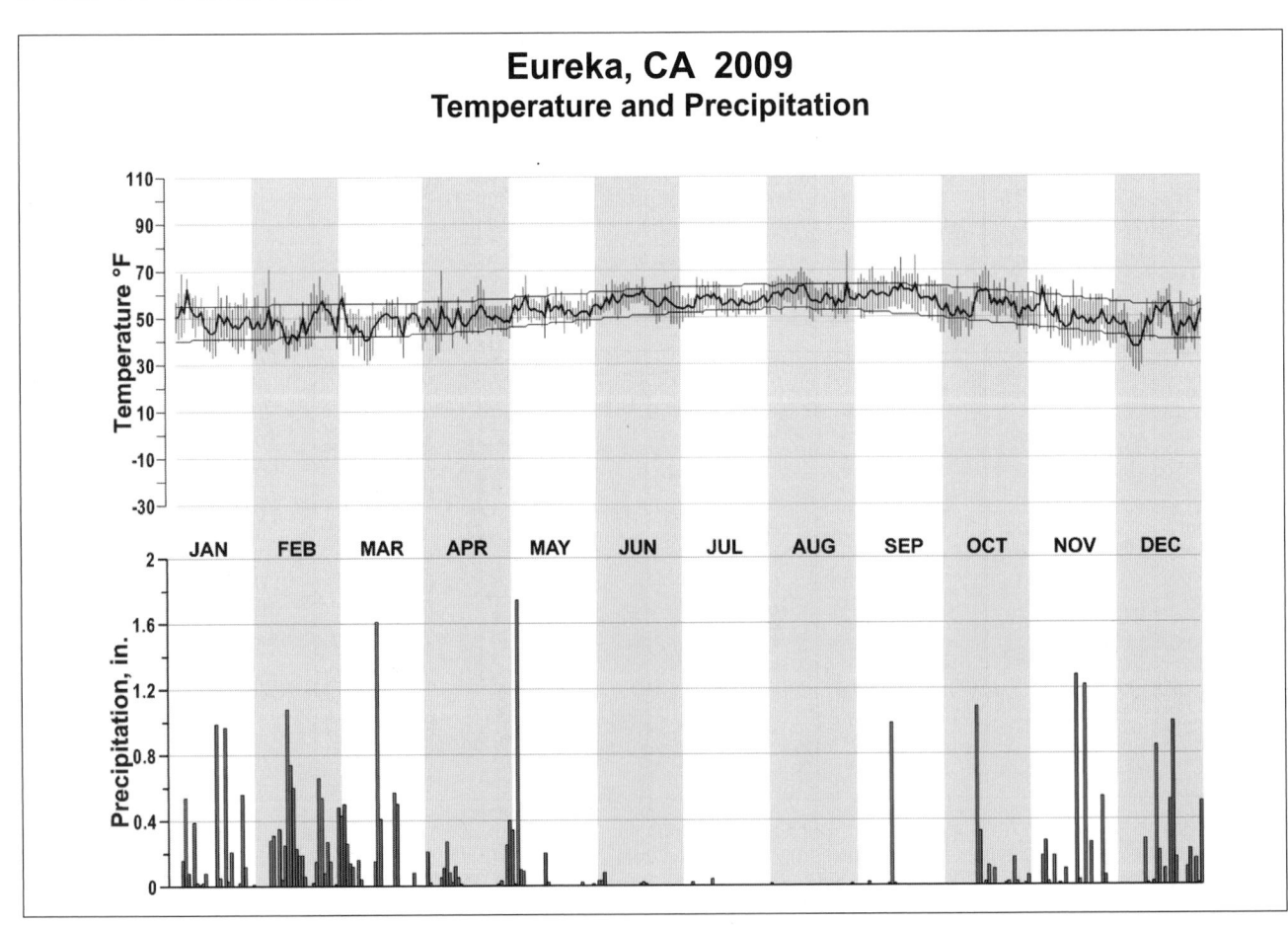

## Eureka, CA 2009
### Temperature and Precipitation

# 2009
# FRESNO
# CALIFORNIA (KFAT)

Fresno is located about midway and toward the eastern edge of the San Joaquin Valley, which is oriented northwest to southeast and has a length of about 225 miles and an average width of 50 miles. The San Joaquin Valley is generally flat. About 15 miles east of Fresno the terrain slopes upward with the foothills of the Sierra Nevada. The Sierra Nevada attain an elevation of more than 14,000 feet 50 miles east of Fresno. West of the city 45 miles lie the foothills of the Coastal Range.

The climate of Fresno is dry and mild in winter and hot in summer. Nearly nine-tenths of the annual precipitation falls in the six months from November to April.

Due to clear skies during the summer and the protection of the San Joaquin Valley from marine effects, the normal daily maximum temperature reaches the high 90s during the latter part of July. The daily maximum temperature during the warmest month has ranged from 76 to 115 degrees. Low relative humidities and some wind movement substantially lower the sensible temperature during periods of high readings. Humidity readings of 15 percent are common on summer afternoons, and readings as low as 8 percent have been recorded. In contrast to this, humidity readings average 90 percent during the morning hours of December and January.

Winds flow with the major axis of the San Joaquin Valley, generally from the northwest. This feature is especially beneficial since, during the warmest months, the northwest winds increase during the evenings. These refreshing breezes and the normally large temperature variation of about 35 degrees between the highest and lowest readings of the day, generally result in comfortable evening and night temperatures.

Winter temperatures are usually mild with infrequent cold spells dropping the readings below freezing. Heavy frost occurs almost every year, and the first frost usually occurs during the last week of November. The last frost in spring is usually in early March, however, one year in five will have the last frost after the first of April. The growing season is 291 days.

Although the heaviest rains recorded at Fresno for short periods have occurred in June, usually any rainfall during the summer is very light. Snow is a rare occurrence in Fresno.

Fresno enjoys a very high percentage of sunshine, receiving more than 80 percent of the possible amounts during all but the four months of November, December, January, and February. Reduction of sunshine during these months is caused by fog and short periods of stormy weather.

During foggy periods, at times lasting nearly two weeks, sunshine is reduced to a minimum. This fog frequently lifts to a few hundred feet above the surface of the valley and presents the appearance of a heavy, solid cloud layer.

Spring and autumn are very enjoyable seasons in Fresno, with clear skies, light rainfall and winds and mild temperatures.

# NORMALS, MEANS, AND EXTREMES
## FRESNO (KFAT)

| LATITUDE: 36 ° 46'N | LONGITUDE: -119° 43'W | ELEVATION (FT): GRND: 333  BARO: 375 | TIME ZONE: PACIFIC  (UTC -8) | WBAN: 93193 |
|---|---|---|---|---|

| | ELEMENT | POR | JAN | FEB | MAR | APR | MAY | JUN | JUL | AUG | SEP | OCT | NOV | DEC | YEAR |
|---|---|---|---|---|---|---|---|---|---|---|---|---|---|---|---|
| TEMPERATURE °F | NORMAL DAILY MAXIMUM | 30 | 53.6 | 61.3 | 66.1 | 74.0 | 82.7 | 90.9 | 96.6 | 94.8 | 88.8 | 78.1 | 63.0 | 53.4 | 75.3 |
| | MEAN DAILY MAXIMUM | 60 | 54.5 | 61.5 | 67.1 | 74.4 | 83.6 | 91.6 | 98.3 | 96.3 | 90.6 | 79.7 | 65.3 | 54.7 | 76.5 |
| | HIGHEST DAILY MAXIMUM | 60 | 78 | 80 | 90 | 100 | 107 | 110 | 113 | 112 | 111 | 102 | 89 | 77 | 113 |
| | YEAR OF OCCURRENCE | | 1986 | 1991 | 1972 | 1981 | 1984 | 2008 | 2006 | 1996 | 1955 | 1980 | 1949 | 2006 | JUL 2006 |
| | MEAN OF EXTREME MAXS. | 60 | 67.5 | 73.0 | 80.2 | 90.1 | 98.8 | 104.7 | 107.1 | 105.4 | 102.1 | 93.6 | 79.5 | 67.0 | 89.1 |
| | NORMAL DAILY MINIMUM | 30 | 38.4 | 41.4 | 44.9 | 48.4 | 54.9 | 61.2 | 66.1 | 64.9 | 60.4 | 51.9 | 42.3 | 37.0 | 51.0 |
| | MEAN DAILY MINIMUM | 60 | 37.7 | 40.7 | 43.8 | 47.9 | 54.4 | 60.4 | 65.8 | 64.0 | 59.5 | 51.0 | 42.5 | 37.3 | 50.4 |
| | LOWEST DAILY MINIMUM | 60 | 19 | 24 | 26 | 32 | 36 | 44 | 50 | 49 | 37 | 27 | 26 | 18 | 18 |
| | YEAR OF OCCURRENCE | | 1963 | 1990 | 1966 | 1982 | 1975 | 1955 | 1955 | 1966 | 1950 | 1972 | 1975 | 1990 | DEC 1990 |
| | MEAN OF EXTREME MINS. | 60 | 27.9 | 31.3 | 34.3 | 38.7 | 44.5 | 51.1 | 57.0 | 56.4 | 50.7 | 41.0 | 32.3 | 27.9 | 41.1 |
| | NORMAL DRY BULB | 30 | 46.0 | 51.4 | 55.5 | 61.2 | 68.8 | 76.1 | 81.4 | 79.9 | 74.6 | 65.0 | 52.7 | 45.2 | 63.2 |
| | MEAN DRY BULB | 60 | 46.1 | 51.1 | 55.4 | 61.2 | 69.0 | 76.0 | 82.0 | 80.2 | 75.1 | 65.4 | 53.9 | 46.0 | 63.5 |
| | MEAN WET BULB | 26 | 43.3 | 46.4 | 49.6 | 51.2 | 55.3 | 59.2 | 63.3 | 62.5 | 59.6 | 54.1 | 47.9 | 42.2 | 52.9 |
| | MEAN DEW POINT | 26 | 41.7 | 43.6 | 46.0 | 45.5 | 48.3 | 51.9 | 55.9 | 55.3 | 53.1 | 48.9 | 44.6 | 40.4 | 47.9 |
| | NORMAL NO. DAYS WITH: | | | | | | | | | | | | | | |
| | MAXIMUM >= 90 | 30 | 0.0 | 0.0 | * | 1.9 | 9.1 | 19.8 | 28.2 | 26.4 | 17.2 | 4.3 | 0.0 | 0.0 | 106.9 |
| | MAXIMUM <= 32 | 30 | 0.0 | 0.0 | 0.0 | 0.0 | 0.0 | 0.0 | 0.0 | 0.0 | 0.0 | 0.0 | 0.0 | 0.1 | 0.1 |
| | MINIMUM <= 32 | 30 | 6.5 | 2.1 | 0.4 | 0.1 | 0.0 | 0.0 | 0.0 | 0.0 | 0.0 | 0.1 | 1.5 | 7.8 | 18.5 |
| | MINIMUM <= 0 | 30 | 0.0 | 0.0 | 0.0 | 0.0 | 0.0 | 0.0 | 0.0 | 0.0 | 0.0 | 0.0 | 0.0 | 0.0 | 0.0 |
| H/C | NORMAL HEATING DEG. DAYS | 30 | 578 | 377 | 283 | 140 | 37 | 4 | 0 | 0 | 3 | 73 | 354 | 598 | 2447 |
| | NORMAL COOLING DEG. DAYS | 30 | 0 | 0 | 3 | 40 | 170 | 351 | 524 | 478 | 307 | 89 | 1 | 0 | 1963 |
| RH | NORMAL (PERCENT) | 30 | 84 | 77 | 70 | 57 | 48 | 43 | 40 | 44 | 49 | 58 | 74 | 83 | 61 |
| | HOUR 04 LST | 30 | 92 | 90 | 87 | 80 | 71 | 65 | 62 | 66 | 71 | 78 | 88 | 92 | 79 |
| | HOUR 10 LST | 30 | 85 | 77 | 66 | 51 | 44 | 39 | 38 | 41 | 45 | 52 | 71 | 83 | 58 |
| | HOUR 16 LST | 30 | 69 | 57 | 49 | 35 | 28 | 24 | 22 | 25 | 28 | 35 | 53 | 67 | 41 |
| | HOUR 22 LST | 30 | 89 | 83 | 76 | 62 | 51 | 44 | 42 | 46 | 51 | 63 | 81 | 88 | 65 |
| S | PERCENT POSSIBLE SUNSHINE | 46 | 47 | 65 | 77 | 85 | 90 | 95 | 97 | 96 | 94 | 88 | 66 | 46 | 79 |
| W/O | MEAN NO. DAYS WITH: | | | | | | | | | | | | | | |
| | HEAVY FOG(VISBY <= 1/4 MI) | 46 | 11.0 | 4.9 | 1.4 | 0.2 | 0.0 | 0.0 | 0.0 | 0.0 | 0.0 | 0.6 | 5.1 | 10.8 | 34.0 |
| | THUNDERSTORMS | 60 | 0.2 | 0.5 | 0.8 | 0.6 | 0.6 | 0.4 | 0.2 | 0.2 | 0.6 | 0.5 | 0.2 | 0.3 | 5.1 |
| CLOUDNESS | MEAN: | | | | | | | | | | | | | | |
| | SUNRISE-SUNSET (OKTAS) | | | | | | | | | | | | | | |
| | MIDNIGHT-MIDNIGHT (OKTAS) | | | | | | | | | | | | | | |
| | MEAN NO. DAYS WITH: | | | | | | | | | | | | | | |
| | CLEAR | | | | | | | | | | | | | | |
| | PARTLY CLOUDY | | | | | | | | | | | | | | |
| | CLOUDY | | | | | | | | | | | | | | |
| PR | MEAN STATION PRESSURE(IN) | 26 | 29.81 | 29.74 | 29.70 | 29.65 | 29.58 | 29.53 | 29.53 | 29.53 | 29.54 | 29.63 | 29.75 | 29.77 | 29.65 |
| | MEAN SEA-LEVEL PRES. (IN) | 26 | 30.16 | 30.09 | 30.05 | 30.00 | 29.93 | 29.87 | 29.87 | 29.87 | 29.88 | 29.98 | 30.10 | 30.16 | 30.00 |
| WINDS | MEAN SPEED (MPH) | 26 | 4.4 | 5.2 | 6.0 | 7.4 | 8.3 | 8.3 | 7.5 | 6.9 | 6.1 | 4.8 | 4.1 | 4.2 | 6.1 |
| | PREVAIL.DIR(TENS OF DEGS) | 34 | 14 | 32 | 32 | 32 | 32 | 31 | 31 | 31 | 31 | 31 | 31 | 14 | 31 |
| | MAXIMUM 2-MINUTE: | | | | | | | | | | | | | | |
| | SPEED (MPH) | 14 | 38 | 36 | 32 | 36 | 32 | 32 | 24 | 25 | 28 | 35 | 30 | 35 | 38 |
| | DIR. (TENS OF DEGS) | | 16 | 13 | 31 | 29 | 32 | 31 | 30 | 30 | 28 | 28 | 28 | 28 | 16 |
| | YEAR OF OCCURRENCE | | 2005 | 1998 | 2007 | 1999 | 1998 | 2008 | 2006 | 1999 | 2005 | 2007 | 2001 | 2008 | JAN 2005 |
| | MAXIMUM 3-SECOND | | | | | | | | | | | | | | |
| | SPEED (MPH) | 14 | 46 | 43 | 41 | 41 | 39 | 40 | 33 | 33 | 33 | 45 | 37 | 41 | 46 |
| | DIR. (TENS OF DEGS) | | 16 | 29 | 18 | 32 | 32 | 31 | 07 | 23 | 16 | 33 | 29 | 29 | 16 |
| | YEAR OF OCCURRENCE | | 2005 | 1999 | 2006 | 2002 | 2008 | 2008 | 2007 | 2007 | 2003 | 2009 | 2001 | 2008 | JAN 2005 |
| PRECIPITATION | NORMAL (IN) | 30 | 2.16 | 2.12 | 2.20 | 0.76 | 0.39 | 0.23 | 0.01 | 0.01 | 0.26 | 0.65 | 1.10 | 1.34 | 11.23 |
| | MAXIMUM MONTHLY (IN) | 60 | 8.56 | 6.12 | 7.24 | 4.41 | 1.65 | 1.93 | 0.22 | 0.25 | 1.19 | 2.45 | 3.50 | 6.73 | 8.56 |
| | YEAR OF OCCURRENCE | | 1969 | 2000 | 1991 | 1967 | 1990 | 1998 | 1992 | 1964 | 1976 | 2000 | 1972 | 1955 | JAN 1969 |
| | MINIMUM MONTHLY (IN) | 60 | 0.04 | T | 0.00 | T | 0.00 | 0.00 | 0.00 | 0.00 | 0.00 | 0.00 | 0.00 | 0.00 | 0.00 |
| | YEAR OF OCCURRENCE | | 1976 | 1964 | 1972 | 2008 | 1982 | 1983 | 1983 | 1981 | 1981 | 1978 | 1959 | 1989 | DEC 1989 |
| | MAXIMUM IN 24 HOURS (IN) | 60 | 2.74 | 1.99 | 2.43 | 1.39 | 1.42 | 1.80 | 0.22 | 0.25 | 0.97 | 1.76 | 1.35 | 1.82 | 2.74 |
| | YEAR OF OCCURRENCE | | 2006 | 1969 | 1995 | 1983 | 1990 | 1998 | 1992 | 1964 | 1978 | 1992 | 1953 | 2007 | JAN 2006 |
| | NORMAL NO. DAYS WITH: | | | | | | | | | | | | | | |
| | PRECIPITATION >= 0.01 | 30 | 8.0 | 8.0 | 8.0 | 3.8 | 1.9 | 0.8 | 0.2 | 0.3 | 1.2 | 2.6 | 5.6 | 6.3 | 46.7 |
| | PRECIPITATION >= 1.00 | 30 | 0.1 | 0.3 | 0.2 | * | * | 0.1 | 0.0 | 0.0 | 0.0 | 0.1 | * | 0.1 | 0.9 |
| SNOWFALL | NORMAL (IN) | 30 | 0.* | 0.* | 0.0 | 0.0 | 0.0 | 0.0 | 0.0 | 0.0 | 0.0 | 0.0 | 0.0 | 0.* | 0.0 |
| | MAXIMUM MONTHLY (IN) | 50 | 2.2 | T | T | T | 0.0 | T | T | T | 0.0 | T | 0.0 | 1.2 | 2.2 |
| | YEAR OF OCCURRENCE | | 1962 | 1994 | 1991 | 2008 | | 1995 | 2007 | 2009 | | 1974 | | 1968 | JAN 1962 |
| | MAXIMUM IN 24 HOURS (IN) | 50 | 1.5 | T | T | 0.0 | 0.0 | T | 0.0 | 0.0 | 0.0 | T | 0.0 | 1.2 | 1.5 |
| | YEAR OF OCCURRENCE' | | 1962 | 1994 | 1991 | | | 1995 | | | | 1974 | | 1968 | JAN 1962 |
| | MAXIMUM SNOW DEPTH (IN) | 49 | 0 | 0 | 0 | 0 | 0 | 0 | 0 | 0 | 0 | 0 | 0 | 1 | 1 |
| | YEAR OF OCCURRENCE | | | | | | | | | | | | | 1968 | DEC 1968 |
| | NORMAL NO. DAYS WITH: | | | | | | | | | | | | | | |
| | SNOWFALL >= 1.0 | 30 | 0.0 | 0.0 | 0.0 | 0.0 | 0.0 | 0.0 | 0.0 | 0.0 | 0.0 | 0.0 | 0.0 | 0.0 | 0.0 |

## PRECIPITATION (inches) 2009  FRESNO (KFAT)

| YEAR | JAN | FEB | MAR | APR | MAY | JUN | JUL | AUG | SEP | OCT | NOV | DEC | ANNUAL |
|------|-----|-----|-----|-----|-----|-----|-----|-----|-----|-----|-----|-----|--------|
| 1980 | 3.83 | 3.30 | 2.05 | 0.25 | 0.18 | T | 0.01 | 0.00 | 0.00 | 0.03 | 0.14 | 0.49 | 10.28 |
| 1981 | 2.67 | 1.29 | 2.59 | 1.01 | T | 0.00 | 0.00 | 0.00 | 0.00 | 0.58 | 1.22 | 0.65 | 10.01 |
| 1982 | 2.11 | 0.58 | 4.76 | 0.89 | 0.00 | 0.31 | 0.00 | T | 1.10 | 1.58 | 3.16 | 1.59 | 16.08 |
| 1983 | 5.14 | 3.70 | 4.53 | 2.76 | 0.01 | 0.00 | 0.00 | 0.09 | 1.03 | 0.09 | 2.51 | 1.75 | 21.61 |
| 1984 | 0.15 | 1.05 | 0.48 | 0.25 | 0.02 | 0.20 | T | T | 0.00 | 0.70 | 1.94 | 1.98 | 6.77 |
| 1985 | 0.43 | 0.71 | 1.73 | 0.12 | 0.00 | 0.33 | 0.04 | 0.02 | 0.43 | 0.85 | 3.02 | 0.72 | 8.40 |
| 1986 | 2.12 | 3.66 | 3.42 | 0.36 | 0.16 | 0.00 | T | 0.00 | 0.38 | 0.00 | 0.01 | 2.30 | 12.41 |
| 1987 | 1.93 | 1.36 | 2.39 | 0.07 | 0.87 | 0.01 | 0.00 | 0.00 | T | 0.85 | 0.52 | 1.19 | 9.19 |
| 1988 | 1.52 | 0.83 | 0.27 | 2.41 | 0.45 | 0.03 | 0.00 | 0.00 | 0.00 | 0.00 | 1.42 | 2.46 | 9.39 |
| 1989 | 0.48 | 1.18 | 2.25 | 0.05 | 0.89 | 0.00 | 0.00 | 0.03 | 1.11 | 0.42 | 0.50 | 0.00 | 6.91 |
| 1990 | 2.82 | 1.33 | 0.67 | 0.92 | 1.65 | 0.00 | T | 0.00 | 0.15 | 0.05 | 0.46 | 0.68 | 8.73 |
| 1991 | 0.13 | 1.01 | 7.24 | 0.02 | 0.03 | T | 0.00 | T | T | 0.80 | 0.04 | 1.22 | 10.49 |
| 1992 | 1.94 | 4.73 | 2.14 | 0.18 | T | T | 0.22 | T | T | 2.19 | T | 2.68 | 14.08 |
| 1993 | 5.18 | 2.44 | 1.76 | 0.20 | 0.25 | 1.61 | 0.00 | 0.00 | 0.00 | 0.12 | 1.16 | 1.03 | 13.75 |
| 1994 | 1.15 | 1.92 | 0.52 | 1.36 | 1.30 | 0.00 | T | 0.00 | 0.20 | 0.77 | 1.57 | 1.33 | 10.12 |
| 1995 | 5.42 | 0.93 | 5.88 | 1.08 | 1.19 | 0.66 | 0.01 | T | 0.00 | 0.00 | T | 2.12 | 17.29 |
| 1996 | 2.07 | 3.57 | 1.52 | 1.17 | 0.38 | 0.08 | T | 0.00 | 0.00 | 1.97 | 1.94 | 4.27 | 16.97 |
| 1997 | 3.53 | 0.17 | 0.10 | T | T | 0.01 | T | 0.00 | 0.15 | 0.07 | 2.66 | 0.99 | 7.68 |
| 1998 | 3.40 | 4.89 | 3.44 | 1.26 | 1.37 | 1.93 | 0.00 | 0.00 | 0.15 | 0.16 | 0.43 | 0.62 | 17.65 |
| 1999 | 2.82 | 1.18 | 0.49 | 0.93 | 0.03 | 0.20 | 0.00 | 0.01 | T | T | 0.48 | 0.03 | 6.17 |
| 2000 | 3.15 | 6.12 | 1.35 | 1.16 | 0.05 | 0.56 | 0.00 | T | 0.32 | 2.45 | 0.01 | 0.07 | 15.24 |
| 2001 | 2.66 | 2.22 | 0.96 | 1.87 | 0.00 | 0.00 | 0.08 | 0.00 | T | 0.29 | 1.99 | 1.95 | 12.02 |
| 2002 | 0.76 | 0.40 | 0.95 | 0.21 | 0.38 | 0.02 | 0.00 | 0.00 | T | 0.00 | 1.78 | 2.25 | 6.75 |
| 2003 | 0.40 | 1.22 | 0.63 | 2.84 | 0.68 | 0.00 | T | 0.04 | T | T | 0.40 | 2.93 | 9.14 |
| 2004 | 0.88 | 1.69 | 1.54 | 0.03 | 0.07 | 0.00 | 0.00 | 0.00 | 0.00 | 2.45 | 0.81 | 3.16 | 10.63 |
| 2005 | 2.42 | 2.30 | 2.51 | 0.56 | 1.62 | 0.01 | 0.00 | T | 0.04 | 0.05 | 0.17 | 2.00 | 11.68 |
| 2006 | 3.40 | 0.54 | 4.73 | 3.27 | 0.36 | 0.00 | T | 0.00 | 0.00 | 0.08 | 0.23 | 1.33 | 13.94 |
| 2007 | 0.59 | 2.29 | 0.97 | 0.49 | 0.05 | 0.00 | T | 0.02 | 0.02 | 0.20 | 0.09 | 2.31 | 7.03 |
| 2008 | 3.32 | 2.12 | 0.02 | T | 0.30 | 0.00 | 0.01 | 0.00 | 0.00 | 0.00 | 0.23 | 1.37 | 1.09 | 8.46 |
| 2009 | 1.02 | 2.43 | 0.24 | 0.72 | 0.46 | 0.20 | 0.00 | T | 0.01 | 1.39 | 0.20 | 2.41 | 9.08 |
| POR= 60 YRS | 2.13 | 1.92 | 1.84 | 1.02 | 0.37 | 0.14 | 0.01 | 0.01 | 0.16 | 0.53 | 1.14 | 1.60 | 10.87 |

WBAN : 93193

## AVERAGE TEMPERATURE (°F) 2009  FRESNO (KFAT)

| YEAR | JAN | FEB | MAR | APR | MAY | JUN | JUL | AUG | SEP | OCT | NOV | DEC | ANNUAL |
|------|-----|-----|-----|-----|-----|-----|-----|-----|-----|-----|-----|-----|--------|
| 1980 | 49.4 | 53.8 | 53.7 | 61.8 | 67.2 | 73.7 | 84.0 | 80.7 | 75.6 | 68.4 | 54.2 | 46.8 | 64.1 |
| 1981 | 47.9 | 52.0 | 54.5 | 63.2 | 70.9 | 82.8 | 84.9 | 82.9 | 76.5 | 61.4 | 55.5 | 47.7 | 65.0 |
| 1982 | 41.7 | 50.5 | 51.4 | 58.0 | 69.3 | 72.9 | 81.0 | 80.4 | 72.3 | 65.0 | 51.1 | 45.4 | 61.6 |
| 1983 | 45.2 | 53.1 | 55.9 | 57.9 | 69.7 | 76.3 | 79.0 | 82.1 | 78.8 | 68.5 | 54.6 | 51.1 | 64.4 |
| 1984 | 47.8 | 50.7 | 58.4 | 60.8 | 74.8 | 77.5 | 87.0 | 83.5 | 81.0 | 62.4 | 53.6 | 46.5 | 65.3 |
| 1985 | 43.3 | 51.3 | 53.1 | 67.2 | 69.4 | 81.8 | 86.0 | 80.5 | 72.3 | 65.0 | 52.5 | 43.8 | 63.9 |
| 1986 | 53.6 | 55.7 | 60.3 | 62.7 | 71.2 | 79.4 | 81.9 | 84.2 | 71.3 | 66.9 | 56.7 | 47.5 | 66.0 |
| 1987 | 45.3 | 52.8 | 55.6 | 66.7 | 71.8 | 78.4 | 77.0 | 80.2 | 75.5 | 70.1 | 52.3 | 44.2 | 64.2 |
| 1988 | 46.0 | 52.2 | 56.8 | 61.6 | 67.0 | 75.6 | 85.5 | 81.2 | 76.4 | 68.7 | 54.3 | 44.5 | 64.2 |
| 1989 | 42.9 | 48.8 | 57.9 | 67.3 | 69.6 | 77.0 | 82.5 | 79.3 | 74.3 | 65.3 | 54.3 | 43.8 | 63.6 |
| 1990 | 45.5 | 48.0 | 57.3 | 65.7 | 68.1 | 76.8 | 84.0 | 80.6 | 75.8 | 67.7 | 52.9 | 41.5 | 63.7 |
| 1991 | 47.0 | 55.8 | 51.5 | 59.5 | 66.1 | 74.7 | 83.8 | 78.6 | 79.9 | 70.5 | 55.8 | 47.0 | 64.2 |
| 1992 | 42.7 | 55.5 | 58.8 | 66.8 | 76.0 | 77.0 | 81.3 | 83.2 | 77.0 | 68.6 | 54.3 | 45.3 | 65.5 |
| 1993 | 47.1 | 51.9 | 60.3 | 61.7 | 69.9 | 75.7 | 80.2 | 79.7 | 75.7 | 67.8 | 53.9 | 45.6 | 64.1 |
| 1994 | 46.9 | 49.9 | 59.3 | 63.2 | 68.5 | 77.7 | 83.3 | 82.3 | 75.4 | 64.8 | 48.1 | 45.3 | 63.7 |
| 1995 | 51.9 | 54.1 | 56.2 | 60.7 | 66.2 | 73.3 | 80.7 | 82.6 | 76.3 | 66.8 | 58.7 | 50.5 | 64.8 |
| 1996 | 48.3 | 54.2 | 57.2 | 63.6 | 69.9 | 77.8 | 85.4 | 83.4 | 74.8 | 64.1 | 53.9 | 49.1 | 65.1 |
| 1997 | 48.7 | 50.3 | 60.0 | 63.5 | 75.3 | 75.8 | 81.3 | 80.6 | 77.3 | 63.8 | 56.9 | 44.7 | 64.9 |
| 1998 | 49.0 | 50.0 | 55.5 | 59.0 | 62.0 | 71.5 | 82.1 | 84.1 | 75.8 | 63.1 | 53.1 | 42.8 | 62.3 |
| 1999 | 44.7 | 49.9 | 53.5 | 58.5 | 68.0 | 75.9 | 80.6 | 78.4 | 77.3 | 68.7 | 56.9 | 47.0 | 63.3 |
| 2000 | 50.2 | 53.8 | 56.5 | 64.2 | 71.0 | 79.8 | 78.8 | 81.2 | 74.5 | 63.9 | 49.2 | 47.8 | 64.2 |
| 2001 | 46.2 | 48.7 | 58.8 | 58.6 | 77.3 | 79.7 | 81.6 | 81.9 | 77.0 | 68.5 | 56.4 | 47.4 | 65.2 |
| 2002 | 45.0 | 52.2 | 55.1 | 62.8 | 69.6 | 78.1 | 84.1 | 80.0 | 77.1 | 65.2 | 56.2 | 49.3 | 64.6 |
| 2003 | 50.6 | 51.1 | 58.1 | 58.6 | 69.5 | 78.4 | 86.5 | 81.4 | 79.2 | 69.8 | 52.2 | 49.3 | 64.7 |
| 2004 | 46.6 | 50.5 | 62.6 | 65.8 | 70.9 | 77.4 | 83.3 | 81.3 | 75.9 | 64.1 | 51.7 | 46.5 | 64.7 |
| 2005 | 47.4 | 54.4 | 57.8 | 59.6 | 69.4 | 73.6 | 86.8 | 84.0 | 73.9 | 65.9 | 57.6 | 51.0 | 65.1 |
| 2006 | 48.7 | 52.4 | 50.1 | 59.7 | 71.9 | 80.7 | 87.9 | 80.2 | 75.8 | 64.0 | 55.4 | 47.1 | 64.5 |
| 2007 | 43.7 | 51.4 | 60.3 | 63.0 | 71.5 | 78.0 | 83.2 | 82.8 | 73.7 | 64.4 | 57.4 | 45.5 | 64.6 |
| 2008 | 47.0 | 51.1 | 57.0 | 61.7 | 70.3 | 79.1 | 83.8 | 84.1 | 78.0 | 67.1 | 57.5 | 44.9 | 65.1 |
| 2009 | 47.7 | 51.5 | 56.0 | 62.0 | 75.3 | 75.7 | 85.0 | 81.8 | 79.7 | 63.7 | 54.1 | 47.2 | 65.0 |
| POR= 60 YRS | 46.1 | 51.1 | 55.4 | 61.2 | 69.0 | 76.0 | 82.0 | 80.2 | 75.1 | 65.4 | 53.9 | 46.0 | 63.4 |

## HEATING DEGREE DAYS (base 65°F) 2009  FRESNO (KFAT)

| YEAR | JUL | AUG | SEP | OCT | NOV | DEC | JAN | FEB | MAR | APR | MAY | JUN | TOTAL |
|---|---|---|---|---|---|---|---|---|---|---|---|---|---|
| 1980-81 | 0 | 0 | 0 | 69 | 318 | 553 | 521 | 359 | 316 | 114 | 9 | 0 | 2259 |
| 1981-82 | 0 | 0 | 0 | 118 | 278 | 530 | 711 | 398 | 412 | 217 | 21 | 4 | 2689 |
| 1982-83 | 0 | 0 | 13 | 62 | 411 | 602 | 607 | 327 | 276 | 206 | 55 | 0 | 2559 |
| 1983-84 | 0 | 0 | 1 | 3 | 304 | 421 | 530 | 408 | 198 | 149 | 6 | 0 | 2020 |
| 1984-85 | 0 | 0 | 0 | 128 | 335 | 566 | 664 | 378 | 361 | 39 | 8 | 3 | 2482 |
| 1985-86 | 0 | 0 | 0 | 63 | 369 | 651 | 345 | 258 | 156 | 98 | 30 | 0 | 1970 |
| 1986-87 | 0 | 0 | 13 | 22 | 242 | 537 | 602 | 337 | 282 | 56 | 26 | 0 | 2117 |
| 1987-88 | 0 | 0 | 0 | 7 | 374 | 636 | 583 | 366 | 251 | 124 | 69 | 12 | 2422 |
| 1988-89 | 0 | 0 | 0 | 20 | 316 | 629 | 679 | 450 | 213 | 52 | 14 | 0 | 2373 |
| 1989-90 | 0 | 0 | 7 | 73 | 310 | 649 | 598 | 470 | 236 | 35 | 19 | 1 | 2398 |
| 1990-91 | 0 | 0 | 0 | 17 | 356 | 722 | 549 | 253 | 412 | 163 | 65 | 0 | 2537 |
| 1991-92 | 0 | 0 | 0 | 81 | 276 | 551 | 683 | 267 | 183 | 25 | 0 | 1 | 2067 |
| 1992-93 | 0 | 0 | 0 | 18 | 316 | 602 | 549 | 359 | 145 | 113 | 9 | 12 | 2123 |
| 1993-94 | 0 | 0 | 0 | 12 | 326 | 595 | 553 | 414 | 168 | 97 | 37 | 0 | 2202 |
| 1994-95 | 0 | 0 | 0 | 58 | 500 | 602 | 398 | 298 | 269 | 146 | 60 | 16 | 2347 |
| 1995-96 | 0 | 0 | 0 | 30 | 184 | 444 | 513 | 304 | 238 | 99 | 8 | 0 | 1820 |
| 1996-97 | 0 | 0 | 0 | 148 | 329 | 486 | 500 | 405 | 169 | 97 | 2 | 0 | 2136 |
| 1997-98 | 0 | 0 | 0 | 92 | 246 | 621 | 490 | 412 | 293 | 226 | 104 | 7 | 2491 |
| 1998-99 | 0 | 0 | 7 | 79 | 351 | 682 | 619 | 418 | 348 | 227 | 35 | 12 | 2778 |
| 1999-00 | 0 | 0 | 0 | 14 | 235 | 550 | 452 | 317 | 259 | 72 | 27 | 3 | 1929 |
| 2000-01 | 0 | 0 | 0 | 103 | 466 | 526 | 577 | 451 | 208 | 222 | 0 | 0 | 2553 |
| 2001-02 | 0 | 0 | 0 | 23 | 251 | 538 | 610 | 352 | 310 | 109 | 30 | 0 | 2223 |
| 2002-03 | 0 | 0 | 0 | 67 | 256 | 477 | 440 | 382 | 216 | 191 | 49 | 0 | 2078 |
| 2003-04 | 0 | 0 | 0 | 24 | 378 | 482 | 565 | 413 | 113 | 64 | 3 | 0 | 2042 |
| 2004-05 | 0 | 0 | 6 | 124 | 391 | 566 | 537 | 291 | 217 | 158 | 30 | 1 | 2321 |
| 2005-06 | 0 | 0 | 0 | 41 | 217 | 424 | 500 | 345 | 456 | 170 | 9 | 0 | 2162 |
| 2006-07 | 0 | 0 | 2 | 56 | 283 | 546 | 654 | 373 | 158 | 117 | 19 | 1 | 2209 |
| 2007-08 | 0 | 0 | 6 | 59 | 223 | 600 | 552 | 396 | 243 | 149 | 20 | 0 | 2248 |
| 2008-09 | 0 | 0 | 0 | 39 | 219 | 616 | 531 | 369 | 274 | 145 | 0 | 0 | 2193 |
| 2009- | 0 | 0 | 2 | 87 | 322 | 544 | | | | | | | |

WBAN : 93193

## COOLING DEGREE DAYS (base 65°F) 2009  FRESNO (KFAT)

| YEAR | JAN | FEB | MAR | APR | MAY | JUN | JUL | AUG | SEP | OCT | NOV | DEC | TOTAL |
|---|---|---|---|---|---|---|---|---|---|---|---|---|---|
| 1980 | 0 | 0 | 0 | 39 | 120 | 265 | 594 | 493 | 326 | 181 | 0 | 0 | 2018 |
| 1981 | 0 | 0 | 0 | 67 | 200 | 545 | 622 | 562 | 352 | 14 | 0 | 0 | 2362 |
| 1982 | 0 | 0 | 0 | 12 | 162 | 251 | 501 | 483 | 240 | 70 | 0 | 0 | 1719 |
| 1983 | 0 | 0 | 0 | 0 | 207 | 343 | 440 | 537 | 422 | 119 | 0 | 0 | 2068 |
| 1984 | 0 | 0 | 1 | 30 | 318 | 382 | 688 | 581 | 487 | 55 | 0 | 0 | 2542 |
| 1985 | 0 | 0 | 0 | 111 | 153 | 516 | 657 | 487 | 227 | 69 | 2 | 0 | 2222 |
| 1986 | 0 | 1 | 18 | 34 | 231 | 440 | 530 | 603 | 206 | 87 | 0 | 0 | 2150 |
| 1987 | 0 | 0 | 0 | 114 | 243 | 409 | 379 | 480 | 323 | 172 | 0 | 0 | 2120 |
| 1988 | 0 | 0 | 3 | 28 | 139 | 338 | 642 | 511 | 349 | 143 | 3 | 0 | 2156 |
| 1989 | 0 | 0 | 4 | 129 | 166 | 366 | 546 | 449 | 291 | 90 | 0 | 0 | 2041 |
| 1990 | 0 | 0 | 2 | 61 | 122 | 360 | 595 | 490 | 333 | 108 | 0 | 0 | 2071 |
| 1991 | 0 | 0 | 0 | 6 | 107 | 298 | 588 | 428 | 454 | 259 | 5 | 0 | 2145 |
| 1992 | 0 | 0 | 0 | 88 | 350 | 366 | 511 | 572 | 365 | 135 | 0 | 0 | 2387 |
| 1993 | 0 | 0 | 3 | 20 | 168 | 342 | 476 | 462 | 331 | 105 | 0 | 0 | 1907 |
| 1994 | 0 | 0 | 1 | 52 | 151 | 389 | 576 | 547 | 318 | 59 | 0 | 0 | 2093 |
| 1995 | 0 | 0 | 0 | 25 | 104 | 273 | 494 | 551 | 347 | 91 | 0 | 0 | 1885 |
| 1996 | 0 | 0 | 4 | 66 | 162 | 389 | 640 | 579 | 300 | 125 | 0 | 0 | 2265 |
| 1997 | 0 | 0 | 18 | 61 | 330 | 334 | 514 | 492 | 373 | 61 | 11 | 0 | 2194 |
| 1998 | 0 | 0 | 6 | 50 | 18 | 210 | 536 | 600 | 338 | 25 | 0 | 0 | 1783 |
| 1999 | 0 | 0 | 0 | 39 | 135 | 348 | 487 | 423 | 373 | 135 | 0 | 0 | 1940 |
| 2000 | 0 | 0 | 0 | 54 | 217 | 454 | 434 | 509 | 291 | 81 | 0 | 0 | 2040 |
| 2001 | 0 | 0 | 20 | 37 | 389 | 447 | 521 | 533 | 365 | 137 | 0 | 0 | 2449 |
| 2002 | 0 | 0 | 9 | 50 | 180 | 400 | 599 | 472 | 372 | 81 | 0 | 0 | 2163 |
| 2003 | 0 | 0 | 7 | 5 | 192 | 406 | 671 | 518 | 431 | 180 | 0 | 0 | 2410 |
| 2004 | 0 | 0 | 45 | 97 | 188 | 376 | 576 | 514 | 341 | 99 | 0 | 0 | 2236 |
| 2005 | 0 | 0 | 4 | 2 | 170 | 266 | 682 | 597 | 271 | 79 | 2 | 0 | 2073 |
| 2006 | 0 | 0 | 0 | 20 | 231 | 478 | 715 | 475 | 337 | 31 | 1 | 0 | 2288 |
| 2007 | 0 | 0 | 20 | 64 | 229 | 396 | 569 | 560 | 274 | 50 | 0 | 0 | 2162 |
| 2008 | 0 | 0 | 0 | 54 | 192 | 431 | 592 | 599 | 394 | 114 | 1 | 0 | 2377 |
| 2009 | 0 | 0 | 1 | 62 | 330 | 328 | 628 | 527 | 451 | 53 | 3 | 0 | 2383 |

## SNOWFALL (inches) 2009 FRESNO (KFAT)

| YEAR | JUL | AUG | SEP | OCT | NOV | DEC | JAN | FEB | MAR | APR | MAY | JUN | TOTAL |
|---|---|---|---|---|---|---|---|---|---|---|---|---|---|
| 1980-81 | 0.0 | 0.0 | 0.0 | 0.0 | 0.0 | 0.0 | 0.0 | 0.0 | 0.0 | 0.0 | 0.0 | 0.0 | 0.0 |
| 1981-82 | 0.0 | 0.0 | 0.0 | 0.0 | 0.0 | 0.0 | 0.0 | 0.0 | 0.0 | 0.0 | 0.0 | 0.0 | 0.0 |
| 1982-83 | 0.0 | 0.0 | 0.0 | 0.0 | 0.0 | 0.0 | 0.0 | 0.0 | 0.0 | 0.0 | 0.0 | 0.0 | 0.0 |
| 1983-84 | 0.0 | 0.0 | 0.0 | 0.0 | 0.0 | 0.0 | 0.0 | 0.0 | 0.0 | 0.0 | 0.0 | 0.0 | 0.0 |
| 1984-85 | 0.0 | 0.0 | 0.0 | 0.0 | 0.0 | 0.0 | 0.0 | 0.0 | 0.0 | 0.0 | 0.0 | 0.0 | 0.0 |
| 1985-86 | 0.0 | 0.0 | 0.0 | 0.0 | 0.0 | 0.0 | 0.0 | 0.0 | 0.0 | 0.0 | 0.0 | 0.0 | 0.0 |
| 1986-87 | 0.0 | 0.0 | 0.0 | 0.0 | 0.0 | 0.0 | 0.0 | 0.0 | 0.0 | 0.0 | 0.0 | 0.0 | 0.0 |
| 1987-88 | 0.0 | 0.0 | 0.0 | 0.0 | 0.0 | 0.0 | 0.0 | 0.0 | 0.0 | 0.0 | 0.0 | 0.0 | 0.0 |
| 1988-89 | 0.0 | 0.0 | 0.0 | 0.0 | 0.0 | 0.0 | 0.0 | T | 0.0 | 0.0 | 0.0 | 0.0 | T |
| 1989-90 | 0.0 | 0.0 | 0.0 | 0.0 | 0.0 | 0.0 | 0.0 | T | 0.0 | 0.0 | 0.0 | 0.0 | T |
| 1990-91 | 0.0 | 0.0 | 0.0 | 0.0 | 0.0 | T | 0.0 | 0.0 | T | 0.0 | 0.0 | 0.0 | T |
| 1991-92 | 0.0 | 0.0 | 0.0 | 0.0 | 0.0 | 0.0 | 0.0 | T | 0.0 | 0.0 | 0.0 | 0.0 | T |
| 1992-93 | 0.0 | 0.0 | 0.0 | 0.0 | 0.0 | 0.0 | 0.0 | 0.0 | 0.0 | 0.0 | 0.0 | 0.0 | 0.0 |
| 1993-94 | 0.0 | 0.0 | 0.0 | 0.0 | 0.0 | 0.0 | 0.0 | T | 0.0 | 0.0 | 0.0 | 0.0 | T |
| 1994-95 | 0.0 | 0.0 | 0.0 | 0.0 | 0.0 | 0.0 | T | 0.0 | 0.0 | 0.0 | 0.0 | T | T |
| 1995-96 | 0.0 | 0.0 | 0.0 | 0.0 | 0.0 | 0.0 | 0.0 | T | 0.0 | 0.0 | 0.0 | 0.0 | T |
| 1996-97 | 0.0 | 0.0 | 0.0 | 0.0 | 0.0 | T | 0.0 | 0.0 | 0.0 | 0.0 | 0.0 | 0.0 | T |
| 1997-98 | 0.0 | 0.0 | 0.0 | 0.0 | 0.0 | 0.0 | 0.0 | T | T | 0.0 | 0.0 | 0.0 | T |
| 1998-99 | 0.0 | 0.0 | 0.0 | 0.0 | 0.0 | 0.5 | T | T | 0.0 | 0.0 | 0.0 | 0.0 | 0.5 |
| 1999-00 | 0.0 | 0.0 | 0.0 | 0.0 | 0.0 | 0.0 | 0.0 | T | 0.0 | 0.0 | 0.0 | 0.0 | T |
| 2000-01 | 0.0 | 0.0 | 0.0 | 0.0 | 0.0 | 0.0 | 0.0 | 0.0 | 0.0 | T | 0.0 | 0.0 | T |
| 2001-02 | 0.0 | 0.0 | 0.0 | 0.0 | 0.0 | 0.0 | 0.0 | 0.0 | 0.0 | 0.0 | 0.0 | 0.0 | 0.0 |
| 2002-03 | 0.0 | 0.0 | 0.0 | 0.0 | 0.0 | 0.0 | 0.0 | 0.0 | 0.0 | 0.0 | 0.0 | 0.0 | 0.0 |
| 2003-04 | 0.0 | 0.0 | 0.0 | 0.0 | 0.0 | 0.0 | 0.0 | 0.0 | 0.0 | 0.0 | 0.0 | 0.0 | 0.0 |
| 2004-05 | 0.0 | 0.0 | 0.0 | T | 0.0 | 0.0 | 0.0 | 0.0 | 0.0 | 0.0 | 0.0 | 0.0 | T |
| 2005- | 0.0 | 0.0 | 0.0 | 0.0 | 0.0 | T | 0.0 | 0.0 | 0.0 | 0.0 | 0.0 | 0.0 | T |
| 2006-07 | 0.0 | 0.0 | 0.0 | 0.0 | 0.0 | 0.0 | 0.0 | 0.0 | 0.0 | 0.0 | 0.0 | 0.0 | 0.0 |
| 2007-08 | 0.0 | 0.0 | 0.0 | 0.0 | 0.0 | 0.0 | 0.0 | 0.0 | 0.0 | 0.0 | 0.0 | 0.0 | 0.0 |
| 2008-09 | 0.0 | 0.0 | 0.0 | 0.0 | 0.0 | T | 0.0 | 0.0 | 0.0 | 0.0 | 0.0 | 0.0 | T |
| 2009- | 0.0 | 0.0 | 0.0 | 0.0 | 0.0 | 0.0 | | | | | | | |
| POR=<br>60 YRS | 0.0 | 0.0 | 0.0 | T | 0.0 | T | T | T | T | T | 0.0 | T | T |

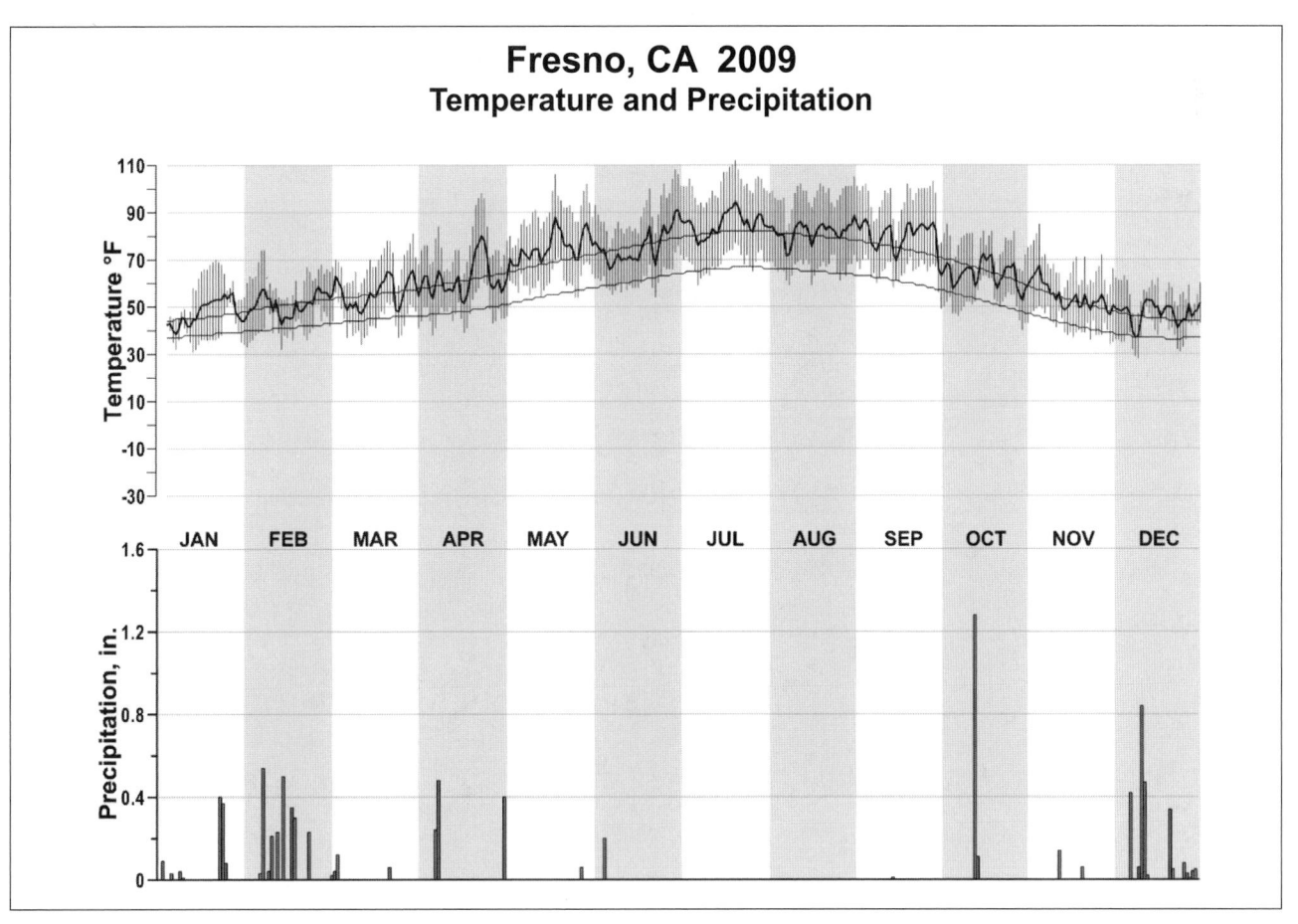

**Fresno, CA 2009**
**Temperature and Precipitation**

# 2009
# LOS ANGELES
# INTERNATIONAL AIRPORT (KLAX)

Predominating influences on the climate of the Los Angeles International Airport are the Pacific Ocean, 3 miles to the west, the southern California coastal mountain ranges which line the inland side of the coastal plain surrounding the airport, and the large scale weather patterns which allow Pacific storm paths to extend as far south as the Los Angeles area only during late fall, winter, and early spring. Marine air covers the coastal plain most of the year but air from the interior reaches the coast at times, especially during the fall and winter months. The coast ranges act as a buffer to the more extreme conditions of the interior. Pronounced differences in temperature, humidity, cloudiness, fog, sunshine, and rain occur over fairly short distances on the coastal plains and the adjoining foothills due to the local topography and the decreased marine effect further inland. In general, temperature ranges are least and humidity highest close to the coast, while precipitation increases with elevation on the foothills.

The most characteristic feature of the climate of the coastal plain around the station is the night and morning low cloudiness and sunny afternoons which prevail during the spring and summer months and occur often during the remainder of the year. The coastal low cloudiness, combined with the westerly sea breeze, produces mild temperatures throughout the year. Daily temperature range is usually less than 15 degrees in spring and summer and about 20 degrees in fall and winter. Hot weather is not frequent at any season along the coast, although readings have exceeded the mid 80s at the airport every month of the year. When high temperatures do occur, the humidity is almost always low so that discomfort is unusual. Nighttime temperatures are generally cool but minimum temperatures below 40 degrees are rare and periods of over 10 years have passed with no readings below freezing at the airport. Prevailing daytime winds are from the west, but night and early morning breezes are usually light and from the east and northeast. Strongest winds observed at the station have been from the west and north following winter storms. During the fall, winter, and spring, gusty dry northeasterly Santa Ana winds blow over southern California mountains and through passes to the coast. These winds rarely reach L.A. International Airport but extremely dry air and dust clouds associated with them can be expected several times each year.

Precipitation occurs mainly in the winter. Measurable rain may fall on about one day in four from late October into early April, but in three years out of four, traces or less are reported for the entire months of July and August. Thunderstorms do not occur often near the coast, but showers and thunderstorms are observed over the coastal ranges at times during the summer when moist air from the south and southeast invades southern California. Annual rainfall at Los Angeles International Airport is somewhat less than that recorded on the Palos Verdes Hills, rising to an elevation of nearly 1,500 feet on a peninsula 12 miles to the south, and on the Hollywood Hills and Santa Monica Mountains which extend east-west 12 miles north of the station with peaks reaching to nearly 2,000 feet. Traces of snow have fallen at Los Angeles International Airport only a few times, melting as they fell.

Visibility at Los Angeles International Airport is frequently restricted by haze, fog, or smoke. Low visibilities are favored by a layer of moist marine air with warm dry air above and light winds but at times a moderate afternoon sea breeze will bring a fog bank ashore and over the airport. Light fog occurs at some time nearly every month, but heavy fog is observed least during the summer and can be expected on about one night or early morning in four during the winter.

# NORMALS, MEANS, AND EXTREMES
## LOS ANGELES (KLAX)

LATITUDE: 33°56'N  LONGITUDE: -118°24'W   ELEVATION (FT): GRND: 112  BARO: 326   TIME ZONE: PACIFIC (UTC -8)   WBAN: 23174

| | ELEMENT | POR | JAN | FEB | MAR | APR | MAY | JUN | JUL | AUG | SEP | OCT | NOV | DEC | YEAR |
|---|---|---|---|---|---|---|---|---|---|---|---|---|---|---|---|
| **TEMPERATURE °F** | NORMAL DAILY MAXIMUM | 30 | 65.6 | 65.8 | 65.3 | 68.0 | 69.3 | 72.6 | 75.3 | 76.8 | 76.5 | 74.3 | 70.4 | 66.7 | 70.6 |
| | MEAN DAILY MAXIMUM | 65 | 64.9 | 65.2 | 65.2 | 67.3 | 69.2 | 71.9 | 75.3 | 76.2 | 75.6 | 73.4 | 70.1 | 65.9 | 70.0 |
| | HIGHEST DAILY MAXIMUM | 74 | 91 | 92 | 95 | 102 | 97 | 104 | 97 | 98 | 110 | 106 | 101 | 94 | 110 |
| | YEAR OF OCCURRENCE | | 2003 | 1963 | 1988 | 1989 | 1979 | 1981 | 1985 | 1955 | 1963 | 1961 | 1966 | 1958 | SEP 1963 |
| | MEAN OF EXTREME MAXS. | 65 | 80.5 | 79.9 | 79.1 | 81.8 | 80.7 | 81.5 | 83.9 | 85.6 | 90.6 | 90.6 | 85.2 | 80.1 | 83.3 |
| | NORMAL DAILY MINIMUM | 30 | 48.6 | 50.1 | 51.3 | 53.6 | 56.9 | 60.1 | 63.3 | 64.5 | 63.6 | 59.4 | 52.7 | 48.5 | 56.1 |
| | MEAN DAILY MINIMUM | 65 | 47.4 | 48.8 | 50.4 | 52.9 | 56.4 | 59.6 | 63.0 | 63.8 | 62.5 | 58.5 | 52.3 | 47.9 | 55.3 |
| | LOWEST DAILY MINIMUM | 74 | 23 | 32 | 34 | 39 | 43 | 48 | 49 | 51 | 47 | 41 | 34 | 32 | 23 |
| | YEAR OF OCCURRENCE | | 1937 | 1942 | 1939 | 1942 | 1938 | 1950 | 1942 | 1948 | 1948 | 1942 | 1939 | 1968 | JAN 1937 |
| | MEAN OF EXTREME MINS. | 65 | 39.1 | 41.9 | 43.4 | 46.7 | 51.2 | 55.1 | 58.9 | 60.0 | 57.5 | 51.8 | 44.2 | 39.8 | 49.1 |
| | NORMAL DRY BULB | 30 | 57.1 | 58.0 | 58.3 | 60.8 | 63.1 | 66.4 | 69.3 | 70.7 | 70.1 | 66.9 | 61.6 | 57.6 | 63.3 |
| | MEAN DRY BULB | 65 | 56.2 | 57.0 | 57.8 | 60.1 | 62.8 | 65.8 | 69.1 | 70.0 | 69.1 | 66.0 | 61.2 | 56.9 | 62.7 |
| | MEAN WET BULB | 26 | 48.7 | 50.4 | 52.2 | 53.9 | 57.3 | 60.1 | 63.4 | 63.8 | 62.8 | 58.5 | 52.9 | 48.3 | 56.0 |
| | MEAN DEW POINT | 26 | 44.4 | 46.3 | 49.3 | 51.0 | 55.1 | 57.9 | 61.4 | 61.8 | 60.6 | 55.9 | 47.8 | 42.6 | 52.8 |
| | NORMAL NO. DAYS WITH: | | | | | | | | | | | | | | |
| | MAXIMUM >= 90 | 30 | 0.0 | * | 0.1 | 0.3 | 0.1 | 0.5 | 0.2 | 0.5 | 1.5 | 1.3 | 0.5 | * | 5.0 |
| | MAXIMUM <= 32 | 30 | 0.0 | 0.0 | 0.0 | 0.0 | 0.0 | 0.0 | 0.0 | 0.0 | 0.0 | 0.0 | 0.0 | 0.0 | 0.0 |
| | MINIMUM <= 32 | 30 | 0.0 | 0.0 | 0.0 | 0.0 | 0.0 | 0.0 | 0.0 | 0.0 | 0.0 | 0.0 | 0.0 | 0.0 | 0.0 |
| | MINIMUM <= 0 | 30 | 0.0 | 0.0 | 0.0 | 0.0 | 0.0 | 0.0 | 0.0 | 0.0 | 0.0 | 0.0 | 0.0 | 0.0 | 0.0 |
| **H/C** | NORMAL HEATING DEG. DAYS | 30 | 252 | 205 | 200 | 141 | 78 | 19 | 1 | 0 | 2 | 21 | 121 | 234 | 1274 |
| | NORMAL COOLING DEG. DAYS | 30 | 4 | 6 | 6 | 15 | 19 | 58 | 135 | 175 | 154 | 81 | 22 | 4 | 679 |
| **RH** | NORMAL (PERCENT) | 30 | 66 | 70 | 73 | 72 | 76 | 76 | 77 | 77 | 76 | 72 | 64 | 62 | 72 |
| | HOUR 04 LST | 30 | 73 | 77 | 81 | 81 | 84 | 86 | 87 | 86 | 85 | 80 | 72 | 68 | 80 |
| | HOUR 10 LST | 30 | 58 | 61 | 63 | 61 | 67 | 68 | 68 | 67 | 65 | 59 | 52 | 52 | 62 |
| | HOUR 16 LST | 30 | 62 | 65 | 68 | 65 | 68 | 68 | 68 | 69 | 68 | 67 | 61 | 58 | 66 |
| | HOUR 22 LST | 30 | 71 | 74 | 77 | 76 | 81 | 83 | 83 | 83 | 81 | 78 | 71 | 67 | 77 |
| **S** | PERCENT POSSIBLE SUNSHINE | | | | | | | | | | | | | | |
| **W/O** | MEAN NO. DAYS WITH: | | | | | | | | | | | | | | |
| | HEAVY FOG(VISBY <= 1/4 MI) | 46 | 2.9 | 2.0 | 1.7 | 1.2 | 0.8 | 0.7 | 1.2 | 1.3 | 2.0 | 2.9 | 3.0 | 2.8 | 22.5 |
| | THUNDERSTORMS | 63 | 0.4 | 0.5 | 0.7 | 0.3 | 0.1 | 0.1 | 0.3 | 0.3 | 0.4 | 0.3 | 0.2 | 0.3 | 3.9 |
| **CLOUDNESS** | MEAN: | | | | | | | | | | | | | | |
| | SUNRISE-SUNSET (OKTAS) | 49 | 4.2 | 4.2 | 4.2 | 3.8 | 4.1 | 3.9 | 3.2 | 2.9 | 3.3 | 3.5 | 3.5 | 3.8 | 3.7 |
| | MIDNIGHT-MIDNIGHT (OKTAS) | 33 | 3.9 | 4.2 | 4.0 | 3.5 | 4.2 | 4.1 | 3.6 | 3.3 | 3.6 | 3.6 | 3.4 | 3.6 | 3.8 |
| | MEAN NO. DAYS WITH: | | | | | | | | | | | | | | |
| | CLEAR | 62 | 12.0 | 11.0 | 11.5 | 11.4 | 10.4 | 9.9 | 12.4 | 13.8 | 13.2 | 13.0 | 14.4 | 12.7 | 145.7 |
| | PARTLY CLOUDY | 62 | 8.0 | 6.5 | 8.7 | 9.2 | 10.8 | 11.3 | 12.9 | 11.7 | 10.3 | 9.8 | 7.7 | 8.3 | 115.2 |
| | CLOUDY | 62 | 10.9 | 10.7 | 10.8 | 9.3 | 9.8 | 8.9 | 5.2 | 5.0 | 6.1 | 7.8 | 7.5 | 9.6 | 101.6 |
| **PR** | MEAN STATION PRESSURE(IN) | 26 | 29.86 | 29.86 | 29.82 | 29.79 | 29.75 | 29.72 | 29.73 | 29.72 | 29.70 | 29.72 | 29.83 | 29.88 | 29.78 |
| | MEAN SEA-LEVEL PRES. (IN) | 26 | 30.10 | 30.05 | 30.02 | 29.98 | 29.94 | 29.91 | 29.93 | 29.91 | 29.89 | 29.95 | 30.03 | 30.08 | 29.98 |
| **WINDS** | MEAN SPEED (MPH) | 26 | 6.6 | 7.8 | 8.1 | 8.7 | 8.3 | 8.2 | 8.1 | 7.8 | 7.5 | 6.9 | 6.5 | 6.6 | 7.6 |
| | PREVAIL.DIR(TENS OF DEGS) | 46 | 26 | 26 | 26 | 26 | 26 | 26 | 26 | 26 | 26 | 26 | 27 | 27 | 26 |
| | MAXIMUM 2-MINUTE: | | | | | | | | | | | | | | |
| | SPEED (MPH) | 12 | 36 | 38 | 43 | 45 | 35 | 28 | 28 | 25 | 32 | 38 | 39 | 43 | 45 |
| | DIR. (TENS OF DEGS) | | 01 | 27 | 27 | 26 | 27 | 25 | 27 | 26 | 26 | 25 | 27 | 26 | 26 |
| | YEAR OF OCCURRENCE | | 2007 | 2009 | 2009 | 1999 | 2003 | 2000 | 2003 | 2009 | 2004 | 2007 | 2008 | 2003 | APR 1999 |
| | MAXIMUM 3-SECOND | | | | | | | | | | | | | | |
| | SPEED (MPH) | 12 | 49 | 45 | 53 | 53 | 47 | 33 | 31 | 32 | 38 | 46 | 46 | 49 | 53 |
| | DIR. (TENS OF DEGS) | | 36 | 29 | 27 | 26 | 27 | 27 | 28 | 26 | 26 | 26 | 33 | 29 | 27 |
| | YEAR OF OCCURRENCE | | 2007 | 2009 | 2009 | 1999 | 2008 | 2007 | 2007 | 2009 | 2007 | 2007 | 1999 | 2006 | MAR 2009 |
| **PRECIPITATION** | NORMAL (IN) | 30 | 2.98 | 3.11 | 2.40 | 0.63 | 0.24 | 0.08 | 0.03 | 0.14 | 0.26 | 0.36 | 1.13 | 1.79 | 13.15 |
| | MAXIMUM MONTHLY (IN) | 74 | 12.71 | 13.79 | 6.37 | 4.52 | 2.55 | 0.74 | 0.32 | 2.47 | 4.39 | 3.78 | 7.92 | 6.57 | 13.79 |
| | YEAR OF OCCURRENCE | | 1995 | 1998 | 1983 | 1965 | 1977 | 1993 | 1992 | 1977 | 1939 | 2004 | 1946 | 1936 | FEB 1998 |
| | MINIMUM MONTHLY (IN) | 74 | 0.00 | T | 0.00 | 0.00 | 0.00 | 0.00 | 0.00 | 0.00 | 0.00 | 0.00 | 0.00 | 0.00 | 0.00 |
| | YEAR OF OCCURRENCE | | 1976 | 1964 | 1959 | 1979 | 1943 | 1978 | 1983 | 1981 | 1968 | 1969 | 1980 | 1990 | DEC 1990 |
| | MAXIMUM IN 24 HOURS (IN) | 74 | 6.19 | 4.16 | 3.54 | 1.88 | 1.72 | 0.74 | 0.28 | 2.40 | 4.20 | 1.77 | 5.60 | 4.72 | 6.19 |
| | YEAR OF OCCURRENCE | | 1956 | 1962 | 1968 | 1960 | 1977 | 1993 | 1992 | 1977 | 1939 | 1972 | 1967 | 2004 | JAN 1956 |
| | NORMAL NO. DAYS WITH: | | | | | | | | | | | | | | |
| | PRECIPITATION >= 0.01 | 30 | 6.4 | 6.3 | 6.5 | 2.6 | 1.3 | 0.5 | 0.4 | 0.5 | 1.2 | 2.0 | 3.1 | 4.7 | 35.5 |
| | PRECIPITATION >= 1.00 | 30 | 1.0 | 0.9 | 0.5 | 0.1 | * | 0.0 | 0.0 | 0.1 | 0.1 | 0.1 | 0.3 | 0.5 | 3.6 |
| **SNOWFALL** | NORMAL (IN) | 30 | 0.0 | 0.0 | 0.0 | 0.0 | 0.0 | 0.0 | 0.0 | 0.0 | 0.0 | 0.0 | 0.0 | 0.0 | 0.0 |
| | MAXIMUM MONTHLY (IN) | 62 | T | T | T | 0.0 | 0.0 | 0.0 | 0.0 | 0.0 | 0.0 | 0.0 | 0.0 | T | T |
| | YEAR OF OCCURRENCE | | 1982 | 1951 | 1991 | | | | | | | | | 1971 | MAR 1991 |
| | MAXIMUM IN 24 HOURS (IN) | 62 | T | T | T | 0.0 | 0.0 | 0.0 | 0.0 | 0.0 | 0.0 | 0.0 | 0.0 | T | T |
| | YEAR OF OCCURRENCE' | | 1982 | 1951 | 1991 | | | | | | | | | 1971 | MAR 1991 |
| | MAXIMUM SNOW DEPTH (IN) | 51 | 0 | 0 | 0 | 0 | 0 | 0 | 0 | 0 | 0 | 0 | 0 | 0 | 0 |
| | YEAR OF OCCURRENCE | | | | | | | | | | | | | | |
| | NORMAL NO. DAYS WITH: | | | | | | | | | | | | | | |
| | SNOWFALL >= 1.0 | 30 | 0.0 | 0.0 | 0.0 | 0.0 | 0.0 | 0.0 | 0.0 | 0.0 | 0.0 | 0.0 | 0.0 | 0.0 | 0.0 |

## PRECIPITATION (inches) 2009 LOS ANGELES (KLAX)

| YEAR | JAN | FEB | MAR | APR | MAY | JUN | JUL | AUG | SEP | OCT | NOV | DEC | ANNUAL |
|---|---|---|---|---|---|---|---|---|---|---|---|---|---|
| 1980 | 6.97 | 9.13 | 3.69 | 0.17 | 0.07 | T | 0.00 | 0.00 | T | T | 0.00 | 1.57 | 21.60 |
| 1981 | 1.51 | 1.58 | 3.24 | 0.46 | T | T | 0.00 | 0.00 | 0.05 | 0.40 | 2.63 | 1.52 | 11.39 |
| 1982 | 2.78 | 0.66 | 3.41 | 1.61 | 0.11 | 0.01 | 0.00 | T | 0.78 | 0.18 | 3.48 | 0.66 | 13.68 |
| 1983 | 5.25 | 5.64 | 6.37 | 3.18 | 0.04 | 0.03 | 0.00 | 1.25 | 1.91 | 0.94 | 2.74 | 2.11 | 29.46 |
| 1984 | 0.39 | 0.01 | 0.14 | 1.16 | T | T | 0.00 | 0.29 | 0.09 | 0.28 | 1.24 | 4.21 | 7.81 |
| 1985 | 0.70 | 1.91 | 0.72 | T | 0.16 | 0.00 | T | 0.00 | 0.28 | 0.36 | 4.75 | 0.44 | 9.32 |
| 1986 | 2.31 | 5.36 | 4.89 | 0.30 | 0.00 | T | 0.09 | T | 1.44 | 0.10 | 1.14 | 0.30 | 15.93 |
| 1987 | 1.27 | 0.64 | 0.92 | 0.02 | T | 0.09 | 0.08 | T | 0.08 | 1.74 | 0.60 | 1.79 | 7.23 |
| 1988 | 1.61 | 1.79 | 0.08 | 1.14 | T | T | 0.00 | 0.02 | 0.07 | T | 0.73 | 2.52 | 7.96 |
| 1989 | 0.59 | 1.72 | 0.86 | T | 0.04 | T | T | T | 0.26 | 0.34 | 0.38 | 0.00 | 4.19 |
| 1990 | 1.18 | 2.60 | 0.14 | 0.34 | 0.83 | T | 0.00 | 0.02 | T | 0.00 | 0.10 | 0.03 | 5.24 |
| 1991 | 1.38 | 2.53 | 3.96 | T | T | T | 0.17 | T | 0.09 | 0.06 | T | 2.86 | 11.05 |
| 1992 | 1.61 | 4.70 | 5.08 | 0.18 | 0.04 | T | 0.32 | 0.00 | 0.00 | 0.50 | 0.00 | 4.16 | 16.59 |
| 1993 | 10.63 | 5.48 | 1.83 | 0.00 | T | 0.74 | T | 0.00 | T | 0.09 | 0.93 | 0.97 | 20.67 |
| 1994 | 0.33 | 4.36 | 1.01 | 0.44 | 0.08 | 0.00 | T | T | 0.00 | 0.14 | 0.66 | 1.05 | 8.07 |
| 1995 | 12.71 | 0.62 | 5.67 | 0.74 | 0.61 | 0.60 | 0.06 | 0.00 | T | 0.01 | 0.10 | 2.16 | 23.28 |
| 1996 | 1.94 | 4.19 | 1.36 | 0.42 | 0.05 | 0.00 | T | .00 | .00 | 1.46 | 1.93 | 4.74 | 16.09 |
| 1997 | 5.12 | 0.05 | T | T | T | 0.00 | T | T | 0.27 | T | 2.66 | 3.93 | 12.03 |
| 1998 | 3.71 | 13.79 | 3.37 | 1.00 | 2.46 | 0.09 | 0.00 | T | 0.01 | T | 1.89 | 0.74 | 27.06 |
| 1999 | 1.19 | 0.50 | 2.12 | 2.23 | T | 0.59 | T | 0.00 | 0.00 | 0.00 | 0.28 | T | 6.91 |
| 2000 | 0.85 | 4.71 | 2.39 | 1.88 | T | 0.00 | 0.00 | 0.03 | 0.03 | 1.12 | 0.00 | T | 11.01 |
| 2001 | 4.68 | 7.30 | 1.29 | 1.10 | 0.01 | T | T | 0.00 | 0.00 | 0.04 | 1.34 | 1.25 | 17.01 |
| 2002 | 0.73 | 0.35 | 0.27 | 0.02 | 0.11 | 0.05 | 0.00 | T | 0.08 | 0.05 | 1.60 | 1.77 | 5.03 |
| 2003 | T | 3.78 | 1.66 | 0.49 | 0.95 | T | 0.02 | T | 0.00 | 0.71 | 0.80 | 1.14 | 9.55 |
| 2004 | 0.49 | 4.61 | 0.77 | 0.03 | 0.04 | T | 0.00 | 0.00 | T | 3.78 | 0.11 | 6.49 | 16.32 |
| 2005 | 6.87 | 6.95 | 1.08 | 0.90 | 0.33 | T | T | 0.00 | 0.25 | 1.01 | 0.47 | 0.95 | 18.81 |
| 2006 | 1.42 | 2.03 | 2.52 | 1.63 | 0.60 | 0.01 | 0.10 | 0.01 | T | T | 0.25 | 0.61 | 9.18 |
| 2007 | 0.39 | 0.82 | 0.09 | 0.36 | 0.00 | T | 0.01 | T | 0.49 | 0.64 | 0.50 | 1.59 | 4.89 |
| 2008 | 4.67 | 2.17 | 0.03 | 0.03 | 0.11 | 0.00 | 0.00 | 0.00 | T | T | 1.50 | 2.51 | 11.02 |
| 2009 | 0.51 | 3.41 | 0.05 | T | T | 0.15 | 0.00 | 0.00 | T | 1.31 | 0.00 | 2.05 | 7.48 |
| POR= 65 YRS | 2.68 | 2.73 | 1.84 | 0.76 | 0.17 | 0.05 | 0.02 | 0.07 | 0.16 | 0.38 | 1.41 | 1.73 | 12.00 |

WBAN : 23174

## AVERAGE TEMPERATURE (°F) 2009 LOS ANGELES (KLAX)

| YEAR | JAN | FEB | MAR | APR | MAY | JUN | JUL | AUG | SEP | OCT | NOV | DEC | ANNUAL |
|---|---|---|---|---|---|---|---|---|---|---|---|---|---|
| 1980 | 59.7 | 61.3 | 58.3 | 60.6 | 60.1 | 65.7 | 68.7 | 70.9 | 67.0 | 66.2 | 62.4 | 60.2 | 63.4 |
| 1981 | 59.5 | 60.6 | 58.1 | 61.2 | 64.8 | 71.9 | 71.7 | 71.3 | 69.6 | 65.3 | 62.1 | 59.4 | 64.6 |
| 1982 | 54.6 | 59.1 | 57.6 | 60.2 | 62.5 | 63.6 | 69.3 | 71.4 | 71.6 | 69.2 | 61.2 | 56.3 | 63.1 |
| 1983 | 58.9 | 57.4 | 57.8 | 59.2 | 62.5 | 65.4 | 69.6 | 73.6 | 72.4 | 69.5 | 61.0 | 57.1 | 63.7 |
| 1984 | 58.2 | 58.5 | 61.3 | 61.4 | 66.1 | 67.1 | 71.3 | 73.1 | 76.5 | 65.6 | 58.7 | 55.2 | 64.4 |
| 1985 | 55.5 | 56.6 | 55.2 | 60.7 | 61.2 | 66.0 | 71.4 | 69.9 | 68.7 | 67.1 | 58.6 | 59.1 | 62.5 |
| 1986 | 62.3 | 58.7 | 59.2 | 61.7 | 63.4 | 66.5 | 68.5 | 68.9 | 65.8 | 66.4 | 65.0 | 58.4 | 63.7 |
| 1987 | 55.0 | 58.3 | 58.9 | 63.1 | 64.2 | 64.7 | 66.3 | 67.7 | 69.6 | 69.0 | 61.9 | 53.9 | 62.7 |
| 1988 | 56.7 | 60.6 | 62.5 | 61.3 | 63.0 | 63.6 | 69.2 | 68.3 | 67.7 | 66.1 | 60.1 | 56.5 | 63.0 |
| 1989 | 55.4 | 54.8 | 58.9 | 64.7 | 62.5 | 65.2 | 69.4 | 67.7 | 68.3 | 65.9 | 65.1 | 60.5 | 63.2 |
| 1990 | 57.1 | 55.0 | 57.5 | 62.5 | 62.7 | 67.5 | 71.4 | 69.7 | 71.0 | 69.6 | 64.2 | 56.8 | 63.8 |
| 1991 | 57.2 | 59.9 | 55.0 | 61.0 | 60.3 | 63.9 | 67.0 | 69.0 | 67.8 | 67.3 | 63.4 | 58.7 | 62.5 |
| 1992 | 58.4 | 60.8 | 59.2 | 65.6 | 66.1 | 66.9 | 71.5 | 72.9 | 70.7 | 67.0 | 63.7 | 55.8 | 64.8 |
| 1993 | 56.2 | 57.0 | 60.5 | 62.8 | 65.1 | 68.3 | 69.5 | 69.5 | 69.0 | 68.4 | 63.6 | 58.4 | 64.0 |
| 1994 | 59.2 | 57.0 | 60.7 | 60.2 | 62.0 | 68.3 | 67.9 | 74.7 | 70.7 | 66.9 | 57.9 | 57.5 | 63.6 |
| 1995 | 56.8 | 61.7 | 60.1 | 60.5 | 60.3 | 64.4 | 68.2 | 68.9 | 70.7 | 67.3 | 62.9 | 58.9 | 63.4 |
| 1996 | 58.2 | 58.3 | 60.1 | 64.2 | 65.9 | 67.3 | 69.5 | 73.3 | 69.1 | 63.9 | 62.1 | 58.2 | 64.2 |
| 1997 | 57.3 | 58.7 | 59.9 | 61.8 | 67.6 | 68.8 | 68.9 | 72.9 | 74.7 | 68.6 | 63.2 | 56.9 | 65.0 |
| 1998 | 56.2 | 55.1 | 58.0 | 57.3 | 61.0 | 64.7 | 69.6 | 72.1 | 70.4 | 65.7 | 59.7 | 55.6 | 62.1 |
| 1999 | 57.2 | 56.5 | 55.7 | 58.4 | 63.4 | 65.0 | 70.4 | 68.8 | 67.8 | 68.1 | 60.6 | 59.2 | 62.6 |
| 2000 | 58.4 | 57.8 | 57.9 | 61.0 | 64.3 | 68.0 | 68.7 | 70.8 | 69.7 | 63.7 | 58.1 | 58.4 | 63.1 |
| 2001 | 54.9 | 54.2 | 57.8 | 57.1 | 64.0 | 67.4 | 67.9 | 68.4 | 67.9 | 65.2 | 60.6 | 55.6 | 61.8 |
| 2002 | 55.4 | 58.9 | 58.0 | 58.9 | 61.8 | 65.7 | 68.5 | 67.6 | 67.6 | 62.8 | 63.8 | 56.3 | 62.1 |
| 2003 | 62.1 | 58.1 | 59.2 | 58.3 | 61.1 | 65.1 | 71.5 | 71.0 | 67.8 | 67.4 | 59.6 | 56.8 | 63.2 |
| 2004 | 56.9 | 55.6 | 61.0 | 62.1 | 66.7 | 66.8 | 69.3 | 69.4 | 72.5 | 63.7 | 58.9 | 57.3 | 63.4 |
| 2005 | 57.4 | 58.2 | 58.6 | 59.9 | 64.0 | 65.0 | 68.6 | 69.7 | 67.6 | 65.0 | 63.8 | 59.1 | 63.1 |
| 2006 | 58.2 | 59.0 | 54.7 | 59.4 | 64.4 | 69.2 | 74.3 | 70.8 | 68.9 | 65.9 | 63.6 | 57.3 | 63.8 |
| 2007 | 54.5 | 57.5 | 60.0 | 60.1 | 62.1 | 64.8 | 69.9 | 71.4 | 68.3 | 66.3 | 61.5 | 55.9 | 62.7 |
| 2008 | 55.9 | 56.8 | 60.1 | 62.7 | 64.1 | 68.4 | 69.9 | 70.0 | 68.7 | 68.9 | 64.1 | 54.9 | 63.7 |
| 2009 | 60.1 | 56.3 | 57.7 | 60.2 | 63.9 | 65.0 | 68.4 | 68.8 | 69.9 | 65.0 | 61.6 | 56.2 | 62.8 |
| POR= 65 YRS | 56.2 | 57.0 | 57.8 | 60.1 | 62.8 | 65.8 | 69.1 | 70.0 | 69.1 | 66.0 | 61.2 | 56.9 | 62.7 |

## HEATING DEGREE DAYS (base 65°F) 2009 LOS ANGELES (KLAX)

| YEAR | JUL | AUG | SEP | OCT | NOV | DEC | JAN | FEB | MAR | APR | MAY | JUN | TOTAL |
|---|---|---|---|---|---|---|---|---|---|---|---|---|---|
| 1980-81 | 1 | 0 | 3 | 27 | 77 | 158 | 164 | 147 | 207 | 134 | 29 | 0 | 947 |
| 1981-82 | 0 | 0 | 0 | 36 | 112 | 172 | 314 | 165 | 225 | 147 | 75 | 45 | 1291 |
| 1982-83 | 0 | 0 | 0 | 3 | 119 | 261 | 203 | 205 | 218 | 167 | 80 | 23 | 1279 |
| 1983-84 | 0 | 0 | 0 | 0 | 129 | 237 | 206 | 182 | 116 | 117 | 22 | 0 | 1009 |
| 1984-85 | 0 | 0 | 0 | 29 | 183 | 299 | 287 | 243 | 296 | 148 | 115 | 14 | 1614 |
| 1985-86 | 0 | 0 | 1 | 17 | 192 | 181 | 107 | 184 | 182 | 114 | 52 | 7 | 1037 |
| 1986-87 | 0 | 0 | 21 | 15 | 34 | 199 | 306 | 189 | 190 | 84 | 44 | 19 | 1101 |
| 1987-88 | 1 | 1 | 0 | 6 | 111 | 337 | 257 | 137 | 127 | 118 | 72 | 61 | 1228 |
| 1988-89 | 0 | 0 | 5 | 16 | 145 | 265 | 294 | 291 | 189 | 80 | 72 | 25 | 1382 |
| 1989-90 | 0 | 2 | 2 | 15 | 44 | 152 | 237 | 275 | 223 | 74 | 82 | 5 | 1111 |
| 1990-91 | 0 | 0 | 0 | 1 | 62 | 261 | 236 | 145 | 303 | 130 | 145 | 41 | 1324 |
| 1991-92 | 0 | 0 | 0 | 29 | 79 | 190 | 202 | 130 | 172 | 25 | 0 | 3 | 830 |
| 1992-93 | 0 | 0 | 0 | 2 | 67 | 277 | 269 | 216 | 147 | 69 | 23 | 10 | 1080 |
| 1993-94 | 0 | 0 | 0 | 3 | 59 | 199 | 181 | 217 | 142 | 144 | 88 | 1 | 1034 |
| 1994-95 | 0 | 0 | 0 | 15 | 207 | 226 | 249 | 125 | 147 | 135 | 138 | 37 | 1279 |
| 1995-96 | 0 | 0 | 0 | 8 | 66 | 182 | 216 | 194 | 149 | 65 | 19 | 4 | 903 |
| 1996-97 | 4 | 0 | 0 | 43 | 119 | 204 | 234 | 174 | 169 | 114 | 1 | 0 | 1062 |
| 1997-98 | 0 | 0 | 0 | 9 | 103 | 246 | 265 | 269 | 212 | 225 | 116 | 20 | 1465 |
| 1998-99 | 0 | 0 | 0 | 18 | 151 | 290 | 235 | 231 | 280 | 210 | 48 | 30 | 1493 |
| 1999-00 | 0 | 0 | 2 | 20 | 131 | 172 | 200 | 207 | 214 | 121 | 39 | 1 | 1107 |
| 2000-01 | 0 | 0 | 0 | 59 | 200 | 198 | 307 | 306 | 217 | 236 | 39 | 0 | 1562 |
| 2001-02 | 0 | 0 | 1 | 22 | 125 | 286 | 288 | 180 | 208 | 176 | 101 | 8 | 1395 |
| 2002-03 | 0 | 0 | 9 | 79 | 78 | 261 | 117 | 188 | 192 | 200 | 112 | 8 | 1244 |
| 2003-04 | 0 | 0 | 0 | 12 | 160 | 249 | 246 | 267 | 143 | 112 | 7 | 0 | 1196 |
| 2004-05 | 0 | 0 | 0 | 51 | 179 | 238 | 239 | 186 | 193 | 153 | 45 | 12 | 1296 |
| 2005-06 | 4 | 0 | 0 | 49 | 76 | 182 | 219 | 180 | 312 | 164 | 31 | 0 | 1217 |
| 2006-07 | 0 | 0 | 0 | 19 | 76 | 232 | 325 | 215 | 153 | 141 | 107 | 28 | 1296 |
| 2007-08 | 0 | 0 | 4 | 43 | 122 | 274 | 276 | 236 | 157 | 137 | 69 | 7 | 1325 |
| 2008-09 | 0 | 0 | 0 | 20 | 77 | 306 | 177 | 244 | 221 | 171 | 40 | 12 | 1268 |
| 2009- | 3 | 1 | 0 | 49 | 104 | 263 | | | | | | | |

WBAN : 23174

## COOLING DEGREE DAYS (base 65°F) 2009 LOS ANGELES (KLAX)

| YEAR | JAN | FEB | MAR | APR | MAY | JUN | JUL | AUG | SEP | OCT | NOV | DEC | TOTAL |
|---|---|---|---|---|---|---|---|---|---|---|---|---|---|
| 1980 | 3 | 9 | 0 | 15 | 0 | 62 | 124 | 190 | 70 | 71 | 9 | 15 | 568 |
| 1981 | 0 | 30 | 0 | 25 | 29 | 212 | 214 | 204 | 145 | 53 | 31 | 4 | 947 |
| 1982 | 0 | 3 | 1 | 10 | 6 | 7 | 144 | 204 | 205 | 140 | 15 | 0 | 735 |
| 1983 | 22 | 0 | 0 | 1 | 11 | 40 | 151 | 274 | 231 | 146 | 17 | 0 | 893 |
| 1984 | 4 | 0 | 8 | 14 | 61 | 69 | 202 | 257 | 352 | 54 | 0 | 1 | 1022 |
| 1985 | 0 | 14 | 0 | 25 | 2 | 51 | 203 | 160 | 118 | 91 | 6 | 6 | 676 |
| 1986 | 29 | 11 | 9 | 24 | 9 | 52 | 115 | 124 | 52 | 64 | 40 | 2 | 531 |
| 1987 | 3 | 11 | 9 | 34 | 27 | 17 | 47 | 94 | 148 | 136 | 22 | 0 | 548 |
| 1988 | 9 | 16 | 56 | 14 | 17 | 26 | 134 | 109 | 91 | 56 | 1 | 8 | 537 |
| 1989 | 5 | 12 | 6 | 80 | 2 | 34 | 145 | 93 | 105 | 48 | 55 | 18 | 603 |
| 1990 | 1 | 1 | 1 | 5 | 16 | 86 | 204 | 151 | 187 | 147 | 43 | 13 | 855 |
| 1991 | 1 | 4 | 0 | 17 | 9 | 13 | 70 | 129 | 92 | 107 | 37 | 2 | 481 |
| 1992 | 5 | 14 | 0 | 50 | 43 | 36 | 207 | 250 | 177 | 73 | 31 | 0 | 886 |
| 1993 | 4 | 0 | 13 | 12 | 32 | 116 | 148 | 147 | 128 | 114 | 23 | 5 | 742 |
| 1994 | 7 | 0 | 18 | 6 | 4 | 105 | 94 | 310 | 176 | 81 | 2 | 2 | 805 |
| 1995 | 3 | 37 | 0 | 7 | 0 | 24 | 108 | 130 | 176 | 87 | 6 | 0 | 578 |
| 1996 | 12 | 9 | 5 | 47 | 53 | 81 | 150 | 265 | 131 | 15 | 39 | 0 | 807 |
| 1997 | 0 | 5 | 16 | 25 | 86 | 121 | 143 | 252 | 298 | 130 | 56 | 2 | 1134 |
| 1998 | 0 | 0 | 1 | 0 | 0 | 18 | 150 | 227 | 171 | 42 | 0 | 3 | 612 |
| 1999 | 0 | 1 | 0 | 17 | 5 | 35 | 176 | 124 | 94 | 121 | 4 | 0 | 577 |
| 2000 | 4 | 3 | 2 | 7 | 23 | 98 | 118 | 182 | 146 | 25 | 2 | 2 | 612 |
| 2001 | 4 | 9 | 0 | 5 | 11 | 80 | 94 | 110 | 95 | 34 | 2 | 2 | 446 |
| 2002 | 0 | 15 | 0 | 0 | 9 | 33 | 115 | 87 | 92 | 18 | 49 | 0 | 418 |
| 2003 | 33 | 1 | 19 | 6 | 1 | 18 | 210 | 193 | 89 | 94 | 6 | 1 | 671 |
| 2004 | 0 | 0 | 29 | 33 | 69 | 60 | 141 | 141 | 232 | 18 | 0 | 5 | 728 |
| 2005 | 9 | 0 | 0 | 6 | 20 | 19 | 122 | 151 | 81 | 55 | 50 | 4 | 517 |
| 2006 | 11 | 20 | 0 | 0 | 19 | 130 | 291 | 182 | 122 | 56 | 51 | 1 | 883 |
| 2007 | 6 | 11 | 17 | 0 | 27 | 26 | 161 | 203 | 111 | 92 | 25 | 0 | 679 |
| 2008 | 2 | 5 | 14 | 76 | 50 | 117 | 156 | 163 | 115 | 148 | 58 | 0 | 904 |
| 2009 | 32 | 4 | 1 | 33 | 11 | 17 | 115 | 127 | 155 | 59 | 9 | 0 | 563 |

**SNOWFALL (inches) 2009 LOS ANGELES (KLAX)**

| YEAR | JUL | AUG | SEP | OCT | NOV | DEC | JAN | FEB | MAR | APR | MAY | JUN | TOTAL |
|------|-----|-----|-----|-----|-----|-----|-----|-----|-----|-----|-----|-----|-------|
| 1976-77 | 0.0 | 0.0 | 0.0 | 0.0 | 0.0 | 0.0 | 0.0 | 0.0 | 0.0 | 0.0 | 0.0 | 0.0 | 0.0 |
| 1977-78 | 0.0 | 0.0 | 0.0 | 0.0 | 0.0 | 0.0 | 0.0 | 0.0 | 0.0 | 0.0 | 0.0 | 0.0 | 0.0 |
| 1978-79 | 0.0 | 0.0 | 0.0 | 0.0 | 0.0 | 0.0 | 0.0 | 0.0 | 0.0 | 0.0 | 0.0 | 0.0 | 0.0 |
| 1979-80 | 0.0 | 0.0 | 0.0 | 0.0 | 0.0 | 0.0 | 0.0 | 0.0 | 0.0 | 0.0 | 0.0 | 0.0 | 0.0 |
| 1980-81 | 0.0 | 0.0 | 0.0 | 0.0 | 0.0 | 0.0 | 0.0 | 0.0 | 0.0 | 0.0 | 0.0 | 0.0 | 0.0 |
| 1981-82 | 0.0 | 0.0 | 0.0 | 0.0 | 0.0 | 0.0 | T | 0.0 | 0.0 | 0.0 | 0.0 | 0.0 | T |
| 1982-83 | 0.0 | 0.0 | 0.0 | 0.0 | 0.0 | 0.0 | 0.0 | 0.0 | 0.0 | 0.0 | 0.0 | 0.0 | 0.0 |
| 1983-84 | 0.0 | 0.0 | 0.0 | 0.0 | 0.0 | 0.0 | 0.0 | 0.0 | 0.0 | 0.0 | 0.0 | 0.0 | 0.0 |
| 1984-85 | 0.0 | 0.0 | 0.0 | 0.0 | 0.0 | 0.0 | 0.0 | 0.0 | 0.0 | 0.0 | 0.0 | 0.0 | 0.0 |
| 1985-86 | 0.0 | 0.0 | 0.0 | 0.0 | 0.0 | 0.0 | 0.0 | 0.0 | 0.0 | 0.0 | 0.0 | 0.0 | 0.0 |
| 1986-87 | 0.0 | 0.0 | 0.0 | 0.0 | 0.0 | 0.0 | 0.0 | 0.0 | 0.0 | 0.0 | 0.0 | 0.0 | 0.0 |
| 1987-88 | 0.0 | 0.0 | 0.0 | 0.0 | 0.0 | 0.0 | 0.0 | 0.0 | 0.0 | 0.0 | 0.0 | 0.0 | 0.0 |
| 1988-89 | 0.0 | 0.0 | 0.0 | 0.0 | 0.0 | 0.0 | 0.0 | 0.0 | 0.0 | 0.0 | 0.0 | 0.0 | 0.0 |
| 1989-90 | 0.0 | 0.0 | 0.0 | 0.0 | 0.0 | 0.0 | 0.0 | 0.0 | T | 0.0 | 0.0 | 0.0 | T |
| 1990-91 | 0.0 | 0.0 | 0.0 | 0.0 | 0.0 | 0.0 | 0.0 | 0.0 | T | 0.0 | 0.0 | 0.0 | T |
| 1991-92 | 0.0 | 0.0 | 0.0 | 0.0 | 0.0 | 0.0 | 0.0 | 0.0 | 0.0 | 0.0 | 0.0 | 0.0 | 0.0 |
| 1992-93 | 0.0 | 0.0 | 0.0 | 0.0 | 0.0 | 0.0 | 0.0 | 0.0 | 0.0 | 0.0 | 0.0 | 0.0 | 0.0 |
| 1993-94 | 0.0 | 0.0 | 0.0 | 0.0 | 0.0 | 0.0 | 0.0 | 0.0 | 0.0 | 0.0 | 0.0 | 0.0 | 0.0 |
| 1994-95 | 0.0 | 0.0 | 0.0 | 0.0 | 0.0 | 0.0 | 0.0 | 0.0 | 0.0 | 0.0 | 0.0 | 0.0 | 0.0 |
| 1995-96 | 0.0 | 0.0 | 0.0 | 0.0 | 0.0 | 0.0 | 0.0 | 0.0 | 0.0 | 0.0 | 0.0 | 0.0 | 0.0 |
| 1996-97 | 0.0 | 0.0 | 0.0 | 0.0 | 0.0 | 0.0 | 0.0 | 0.0 | | | | | |
| 1997-98 | | | | | | | | | | | | | |
| 1998-99 | | | | | | | | | | | | | |
| 1999-00 | | | | | | | | | | | | | |
| 2000-01 | | | | | | | | | | | | | |
| 2001-02 | | | | | | | | | | | | | |
| 2002-03 | | | | | | | | | | | | | |
| 2003-04 | | | | | | | | | | | | | |
| 2004-05 | | | | | | | | | | | | | |
| 2005- | | | | | | | | | | | | | |
| POR= 52 YRS | 0.0 | 0.0 | 0.0 | 0.0 | 0.0 | 0.0 | T | 0.0 | T | 0.0 | 0.0 | 0.0 | T |

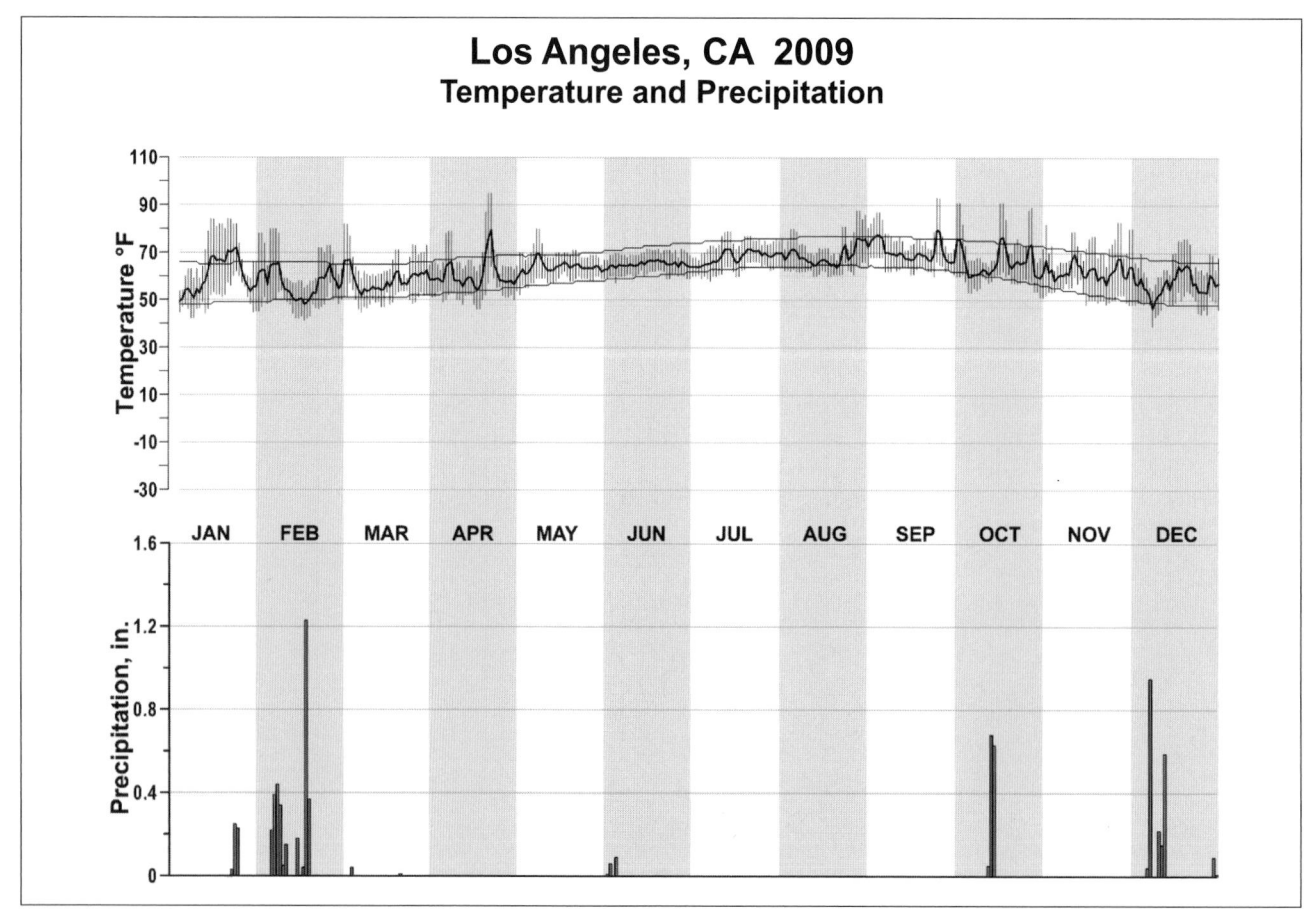

## Los Angeles, CA 2009
### Temperature and Precipitation

# 2009
# REDDING
# CALIFORNIA (KRDD)

Redding, the county seat of Shasta County, is located on the Sacramento River, at the extreme northern in of California's rich Central Valley. It is often referred to as "The Hub City of Northern California", because of it's strategic, central location.

The portion of the Sacramento Valley where Redding lies is bounded in three directions by several mountain ranges. The Trinity Mountains, to the West, are generally 4,000 to 5,000 feet in elevation, but with peaks to near 7,000 feet. To the north, the southern portion of the Cascade Range forms a large part of the very rugged terrain. To the east, the northern slopes of the Sierra Nevada dominate. The elevation of that portion of the Valley where Redding lies is about 500 feet.

The configuration of the mountains on either side of the Sacramento Valley forms a funnel that results in a concentration of precipitation near the upper end of the valley during periods of winter storms with southerly winds.

Winters are cool and wet, with the majority of rain falling during the months from November through April. Normal rainfall for Redding is over 33 inches. Snowfall, while unusal, does occur but seldom remains on the ground for more than a day.

Summers are hot and dry. Daily highs over 100 degrees are common, and 110 degrees is not unusal. What little rainfall occurs during the summers are from the occasional thunderstorm.

Winds are profoundly influenced by the mountains in the area. The general north-south orientation of the major mountain chains reduces east-west air flow to a minimum. Prevailing winds are from the north or northwest, and to a lesser extent, from the south or souheast.

# NORMALS, MEANS, AND EXTREMES
## REDDING (KRDD)

LATITUDE: 40 ° 30'N  LONGITUDE: -122° 18'W  ELEVATION (FT): GRND: 485  BARO: 502  TIME ZONE: PACIFIC  (UTC -8)  WBAN: 24257

| ELEMENT | POR | JAN | FEB | MAR | APR | MAY | JUN | JUL | AUG | SEP | OCT | NOV | DEC | YEAR |
|---|---|---|---|---|---|---|---|---|---|---|---|---|---|---|
| **TEMPERATURE °F** | | | | | | | | | | | | | | |
| NORMAL DAILY MAXIMUM | 30 | 55.4 | 60.1 | 63.9 | 70.6 | 80.7 | 90.7 | 98.5 | 96.9 | 90.2 | 78.4 | 62.4 | 55.6 | 75.3 |
| MEAN DAILY MAXIMUM | 23 | 55.2 | 59.7 | 65.3 | 71.3 | 81.0 | 90.3 | 98.8 | 97.0 | 91.0 | 79.0 | 63.5 | 54.9 | 75.6 |
| HIGHEST DAILY MAXIMUM | 23 | 78 | 83 | 88 | 96 | 106 | 117 | 118 | 115 | 116 | 105 | 88 | 78 | 118 |
| YEAR OF OCCURRENCE | | 2009 | 1992 | 2004 | 2009 | 2008 | 2006 | 1988 | 1990 | 1988 | 1991 | 1995 | 1998 | JUL 1988 |
| MEAN OF EXTREME MAXS. | 23 | 69.7 | 74.0 | 79.9 | 88.3 | 97.6 | 105.5 | 111.1 | 108.3 | 104.5 | 94.9 | 79.1 | 68.6 | 90.1 |
| NORMAL DAILY MINIMUM | 30 | 35.5 | 38.1 | 41.1 | 44.9 | 51.6 | 59.6 | 64.1 | 60.8 | 56.5 | 48.0 | 39.8 | 35.0 | 47.9 |
| MEAN DAILY MINIMUM | 23 | 36.6 | 39.1 | 42.7 | 46.5 | 53.9 | 61.1 | 65.9 | 62.9 | 57.6 | 49.4 | 41.0 | 36.0 | 49.4 |
| LOWEST DAILY MINIMUM | 23 | 19 | 21 | 28 | 28 | 34 | 42 | 53 | 51 | 40 | 33 | 23 | 16 | 16 |
| YEAR OF OCCURRENCE | | 2007 | 1989 | 2006 | 1999 | 2000 | 1990 | 2000 | 1995 | 2007 | 1989 | 1993 | 2009 | DEC 2009 |
| MEAN OF EXTREME MINS. | 23 | 26.8 | 27.8 | 31.0 | 34.9 | 42.0 | 49.9 | 57.3 | 55.1 | 48.3 | 39.0 | 30.3 | 24.8 | 38.9 |
| NORMAL DRY BULB | 30 | 45.5 | 49.1 | 52.5 | 57.8 | 66.2 | 75.2 | 81.3 | 78.9 | 73.4 | 63.2 | 51.1 | 45.3 | 61.6 |
| MEAN DRY BULB | 23 | 45.9 | 49.8 | 54.0 | 58.9 | 67.5 | 75.8 | 82.4 | 79.9 | 74.3 | 64.3 | 52.3 | 45.5 | 62.6 |
| MEAN WET BULB | 23 | 40.2 | 42.7 | 45.5 | 48.4 | 53.4 | 57.4 | 61.4 | 59.2 | 55.6 | 50.5 | 45.0 | 39.9 | 49.9 |
| MEAN DEW POINT | 23 | 37.6 | 38.3 | 40.9 | 43.1 | 47.5 | 50.0 | 53.5 | 51.2 | 47.5 | 43.3 | 40.8 | 36.6 | 44.2 |
| NORMAL NO. DAYS WITH: | | | | | | | | | | | | | | |
| MAXIMUM >= 90 | 30 | 0.0 | 0.0 | 0.0 | 0.9 | 6.1 | 16.0 | 27.4 | 25.1 | 18.6 | 5.5 | 0.0 | 0.0 | 99.6 |
| MAXIMUM <= 32 | 30 | 0.0 | 0.0 | 0.0 | 0.0 | 0.0 | 0.0 | 0.0 | 0.0 | 0.0 | 0.0 | 0.0 | 0.1 | 0.1 |
| MINIMUM <= 32 | 30 | 10.4 | 5.0 | 1.9 | 0.1 | 0.0 | 0.0 | 0.0 | 0.0 | 0.0 | 0.0 | 3.1 | 12.9 | 33.4 |
| MINIMUM <= 0 | 30 | 0.0 | 0.0 | 0.0 | 0.0 | 0.0 | 0.0 | 0.0 | 0.0 | 0.0 | 0.0 | 0.0 | 0.0 | 0.0 |
| **H/C** | | | | | | | | | | | | | | |
| NORMAL HEATING DEG. DAYS | 30 | 606 | 445 | 390 | 239 | 99 | 7 | 0 | 0 | 13 | 131 | 420 | 611 | 2961 |
| NORMAL COOLING DEG. DAYS | 30 | 0 | 0 | 3 | 22 | 133 | 310 | 504 | 430 | 263 | 74 | 2 | 0 | 1741 |
| **RH** | | | | | | | | | | | | | | |
| NORMAL (PERCENT) | 30 | 73 | 60 | 62 | 56 | 49 | 43 | 35 | 39 | 43 | 53 | 68 | 69 | 54 |
| HOUR 04 LST | 30 | 82 | 76 | 77 | 75 | 69 | 63 | 56 | 60 | 62 | 71 | 81 | 80 | 71 |
| HOUR 10 LST | 30 | 73 | 59 | 57 | 49 | 42 | 37 | 31 | 32 | 38 | 44 | 64 | 70 | 50 |
| HOUR 16 LST | 30 | 56 | 39 | 43 | 34 | 31 | 25 | 18 | 19 | 25 | 30 | 48 | 51 | 35 |
| HOUR 22 LST | 30 | 79 | 65 | 66 | 62 | 54 | 46 | 38 | 42 | 47 | 59 | 75 | 75 | 59 |
| **S** PERCENT POSSIBLE SUNSHINE | 10 | 72 | 82 | 85 | 90 | 91 | 94 | 97 | 97 | 94 | 92 | 84 | 73 | 88 |
| **W/O** MEAN NO. DAYS WITH: | | | | | | | | | | | | | | |
| HEAVY FOG(VISBY <= 1/4 MI) | 23 | 5.5 | 1.4 | 0.5 | 0.3 | 0.1 | 0.0 | 0.0 | 0.0 | 0.0 | 0.3 | 2.5 | 4.7 | 15.3 |
| THUNDERSTORMS | 23 | 0.5 | 0.6 | 1.1 | 1.0 | 1.3 | 1.3 | 0.6 | 0.7 | 0.6 | 0.5 | 0.3 | 0.3 | 8.8 |
| **CLOUDNESS** MEAN: | | | | | | | | | | | | | | |
| SUNRISE-SUNSET (OKTAS) | 10 | 5.4 | 5.0 | 5.0 | 5.0 | 4.0 | 2.6 | 1.1 | 1.3 | 1.6 | 2.6 | 3.9 | 4.5 | 3.5 |
| MIDNIGHT-MIDNIGHT (OKTAS) | 10 | 5.1 | 4.5 | 4.6 | 4.4 | 3.8 | 2.5 | 0.9 | 1.1 | 1.2 | 2.3 | 3.5 | 4.1 | 3.2 |
| MEAN NO. DAYS WITH: | | | | | | | | | | | | | | |
| CLEAR | 10 | 8.0 | 8.3 | 8.4 | 8.0 | 11.4 | 17.3 | 25.2 | 24.7 | 22.4 | 17.7 | 11.0 | 9.8 | 172.2 |
| PARTLY CLOUDY | 10 | 4.8 | 5.4 | 6.6 | 8.4 | 9.5 | 8.3 | 4.3 | 4.8 | 4.2 | 6.8 | 7.9 | 5.2 | 76.6 |
| CLOUDY | 10 | 18.2 | 14.2 | 16.0 | 13.6 | 10.1 | 4.4 | 1.3 | 1.6 | 3.4 | 6.5 | 11.1 | 16.0 | 116.4 |
| **PR** | | | | | | | | | | | | | | |
| MEAN STATION PRESSURE(IN) | 23 | 29.59 | 29.51 | 29.49 | 29.46 | 29.37 | 29.34 | 29.33 | 29.34 | 29.35 | 29.45 | 29.56 | 29.59 | 29.45 |
| MEAN SEA-LEVEL PRES. (IN) | 23 | 30.15 | 30.05 | 30.03 | 30.00 | 29.92 | 29.87 | 29.86 | 29.86 | 29.88 | 29.98 | 30.10 | 30.14 | 29.99 |
| **WINDS** | | | | | | | | | | | | | | |
| MEAN SPEED (MPH) | 23 | 5.8 | 6.7 | 7.0 | 6.9 | 7.0 | 7.2 | 6.2 | 5.8 | 5.8 | 5.9 | 5.3 | 6.1 | 6.3 |
| PREVAIL.DIR(TENS OF DEGS) | 24 | 36 | 36 | 36 | 19 | 36 | 35 | 19 | 17 | 36 | 36 | 36 | 36 | 36 |
| MAXIMUM 2-MINUTE: | | | | | | | | | | | | | | |
| SPEED (MPH) | 13 | 55 | 49 | 44 | 39 | 36 | 33 | 32 | 31 | 35 | 40 | 51 | 58 | 58 |
| DIR. (TENS OF DEGS) | | 18 | 17 | 17 | 17 | 17 | 36 | 18 | 02 | 03 | 17 | 17 | 17 | 17 |
| YEAR OF OCCURRENCE | | 2008 | 1999 | 2003 | 1999 | 2002 | 2000 | 2007 | 2008 | 2006 | 2009 | 1998 | 2002 | DEC 2002 |
| MAXIMUM 3-SECOND | | | | | | | | | | | | | | |
| SPEED (MPH) | 13 | 70 | 63 | 55 | 48 | 43 | 44 | 43 | 44 | 43 | 55 | 63 | 71 | 71 |
| DIR. (TENS OF DEGS) | | 17 | 17 | 18 | 17 | 17 | 36 | 18 | 16 | 03 | 18 | 17 | 17 | 17 |
| YEAR OF OCCURRENCE | | 2008 | 2000 | 1999 | 2006 | 2002 | 2000 | 2007 | 1999 | 2006 | 2009 | 1998 | 2002 | DEC 2002 |
| **PRECIPITATION** | | | | | | | | | | | | | | |
| NORMAL (IN) | 30 | 6.50 | 5.49 | 5.15 | 2.40 | 1.66 | 0.69 | 0.05 | 0.22 | 0.48 | 2.18 | 4.03 | 4.67 | 33.52 |
| MAXIMUM MONTHLY (IN) | 23 | 22.93 | 15.80 | 14.78 | 6.09 | 7.67 | 2.28 | 1.15 | 0.83 | 4.83 | 6.26 | 10.11 | 14.72 | 22.93 |
| YEAR OF OCCURRENCE | | 1995 | 1998 | 1995 | 2006 | 1998 | 2009 | 2007 | 1993 | 1989 | 1992 | 1988 | 2002 | JAN 1995 |
| MINIMUM MONTHLY (IN) | 23 | 0.38 | 0.14 | 0.29 | 0.14 | 0.01 | 0.00 | 0.00 | 0.00 | 0.00 | 0.00 | 0.26 | 0.00 | 0.00 |
| YEAR OF OCCURRENCE | | 2007 | 1988 | 2008 | 1987 | 1987 | 2002 | 1990 | 1987 | 1988 | 2002 | 1995 | 1989 | OCT 2002 |
| MAXIMUM IN 24 HOURS (IN) | 23 | 3.96 | 3.17 | 3.18 | 2.33 | 3.79 | 1.46 | 1.15 | 0.63 | 3.15 | 4.09 | 3.23 | 4.45 | 4.45 |
| YEAR OF OCCURRENCE | | 1990 | 2001 | 1995 | 1993 | 1993 | 1997 | 2007 | 1997 | 1989 | 1992 | 1988 | 2004 | DEC 2004 |
| NORMAL NO. DAYS WITH: | | | | | | | | | | | | | | |
| PRECIPITATION >= 0.01 | 30 | 14.2 | 11.4 | 12.1 | 8.1 | 7.6 | 4.0 | 0.7 | 0.8 | 1.5 | 4.3 | 8.5 | 10.1 | 83.3 |
| PRECIPITATION >= 1.00 | 30 | 2.4 | 1.9 | 1.8 | 0.3 | 0.8 | 0.1 | 0.0 | 0.0 | 0.2 | 0.6 | 1.0 | 1.1 | 10.2 |
| **SNOWFALL** | | | | | | | | | | | | | | |
| NORMAL (IN) | 30 | 0.8 | 0.4 | 0.5 | 0.0 | 0.0 | 0.0 | 0.0 | 0.0 | 0.0 | 0.0 | T | 3.5 | 5.2 |
| MAXIMUM MONTHLY (IN) | 10 | 10.9 | 1.4 | 1.8 | T | 1.5 | T | 0.0 | T | 0.0 | 0.0 | T | 17.0 | 17.0 |
| YEAR OF OCCURRENCE | | 1996 | 1990 | 1987 | 1996 | 1990 | 1992 | | 1993 | | | 1988 | 1988 | DEC 1988 |
| MAXIMUM IN 24 HOURS (IN) | 10 | 5.5 | 1.4 | 1.8 | T | 1.5 | T | 0.0 | T | 0.0 | 0.0 | T | 10.0 | 10.0 |
| YEAR OF OCCURRENCE' | | 1996 | 1990 | 1987 | 1996 | 1990 | 1992 | 1987 | 1993 | 1986 | 1986 | 1988 | 1988 | DEC 1988 |
| MAXIMUM SNOW DEPTH (IN) | 9 | 3 | 0 | 0 | 0 | 0 | 0 | 0 | 0 | 0 | 0 | 0 | 7 | 7 |
| YEAR OF OCCURRENCE | | 1989 | | | | | | | | | | | 1988 | DEC 1988 |
| NORMAL NO. DAYS WITH: | | | | | | | | | | | | | | |
| SNOWFALL >= 1.0 | 30 | 0.5 | 0.3 | 0.3 | 0.0 | 0.0 | 0.0 | 0.0 | 0.0 | 0.0 | 0.0 | 0.0 | 0.6 | 1.7 |

## PRECIPITATION (inches) 2009 REDDING (KRDD)

| YEAR | JAN | FEB | MAR | APR | MAY | JUN | JUL | AUG | SEP | OCT | NOV | DEC | ANNUAL |
|---|---|---|---|---|---|---|---|---|---|---|---|---|---|
| 1986 | | | | | | | | | 2.18 | 0.80 | 0.41 | 1.94 | |
| 1987 | 7.01 | 4.97 | 7.00 | 0.14 | 0.01 | T | 0.21 | 0.00 | T | 0.48 | 3.53 | 9.07 | 32.42 |
| 1988 | 7.25 | 0.14 | 0.52 | 3.29 | 3.99 | 1.74 | T | T | 0.00 | 0.11 | 10.11 | 3.68 | 30.83 |
| 1989 | 2.14 | 1.11 | 10.94 | 3.76 | 0.73 | 0.95 | 0.00 | 0.23 | 4.83 | 3.69 | 1.20 | 0.00 | 29.58 |
| 1990 | 8.14 | 1.37 | 2.40 | 0.65 | 6.60 | 0.82 | 0.49 | 1.06 | 1.17 | 0.83 | 0.67 | 0.56 | 24.76 |
| 1991 | 0.89 | 3.97 | 9.67 | 0.52 | 2.13 | 0.11 | 0.03 | 0.00 | 0.00 | 1.93 | 1.27 | 5.09 | 25.61 |
| 1992 | 3.03 | 10.15 | 3.41 | 1.91 | 0.03 | 1.66 | T | T | 0.00 | 6.26 | 0.92 | 10.37 | 37.74 |
| 1993 | 10.38 | 7.52 | 6.34 | 3.66 | 6.72 | 0.65 | 0.00 | 0.99 | 0.00 | 2.92 | 1.52 | 3.16 | 43.86 |
| 1994 | 3.34 | 6.41 | 1.92 | 1.86 | 1.41 | 0.03 | 0.00 | 0.00 | 0.20 | 0.04 | 5.01 | 5.45 | 25.67 |
| 1995 | 22.93 | 1.65 | 14.78 | 4.26 | 0.97 | 1.93 | T | 0.00 | 0.00 | T | 0.26 | 10.81 | 57.59 |
| 1996 | 9.66 | 9.06 | 1.84 | 2.54 | 4.28 | 0.14 | .33 | T | .55 | 1.25 | 3.00 | 8.38 | 41.03 |
| 1997 | 6.76 | 0.72 | 1.76 | 2.43 | 0.43 | 1.91 | 0.01 | 0.63 | 1.62 | 3.36 | 9.06 | 3.30 | 31.99 |
| 1998 | 14.00 | 15.80 | 5.62 | 2.83 | 7.67 | 1.71 | 0.14 | T | 0.06 | 2.30 | 9.29 | 2.17 | 61.59 |
| 1999 | 3.11 | 7.66 | 3.43 | 1.73 | 0.58 | 0.40 | 0.00 | 0.23 | T | 1.02 | 5.51 | 0.63 | 24.30 |
| 2000 | 7.67 | 9.28 | 4.08 | 3.57 | 1.18 | 1.11 | 0.11 | T | 3.08 | 3.74 | 0.98 | 1.89 | 36.69 |
| 2001 | 5.73 | 8.07 | 3.37 | 2.05 | 0.03 | 1.10 | T | 0.00 | 0.49 | 0.83 | 7.38 | 9.31 | 38.36 |
| 2002 | 3.37 | 2.82 | 2.58 | 1.40 | 0.68 | 0.00 | 0.00 | 0.00 | 0.11 | 0.00 | 2.41 | 14.72 | 28.09 |
| 2003 | 6.65 | 2.26 | 3.79 | 4.20 | 0.97 | 0.00 | 0.05 | 0.64 | 0.16 | T | 6.22 | 11.77 | 36.71 |
| 2004 | 2.99 | 10.08 | 1.43 | 1.18 | 1.38 | 0.11 | T | T | 0.30 | 5.76 | 1.72 | 10.82 | 35.77 |
| 2005 | 4.35 | 2.97 | 4.99 | 2.12 | 4.95 | 0.78 | 0.00 | 0.00 | 0.02 | 0.38 | 4.84 | 13.90 | 39.30 |
| 2006 | 7.16 | 4.44 | 7.56 | 6.09 | 0.64 | 0.28 | T | 0.04 | 0.00 | 0.22 | 3.88 | 6.62 | 36.93 |
| 2007 | 0.38 | 7.36 | 0.51 | 2.21 | 1.51 | T | 1.15 | T | 0.17 | 2.96 | 0.48 | 5.02 | 21.75 |
| 2008 | 9.98 | 3.16 | 0.29 | 0.44 | 0.35 | 0.02 | 0.00 | 0.01 | T | 1.30 | 2.82 | 3.33 | 21.70 |
| 2009 | 0.93 | 8.97 | 1.15 | 0.72 | 2.20 | 2.28 | 0.00 | T | 0.12 | 1.99 | 0.67 | 4.03 | 23.06 |
| POR= 23 YRS | 6.43 | 5.65 | 4.32 | 2.33 | 2.15 | 0.77 | 0.11 | 0.17 | 0.63 | 1.76 | 3.47 | 6.08 | 33.87 |

WBAN : 24257

## AVERAGE TEMPERATURE (°F) 2009 REDDING (KRDD)

| YEAR | JAN | FEB | MAR | APR | MAY | JUN | JUL | AUG | SEP | OCT | NOV | DEC | ANNUAL |
|---|---|---|---|---|---|---|---|---|---|---|---|---|---|
| 1986 | | 58.3 | | | | | | | 67.8 | 63.8 | 54.8 | 45.5 | |
| 1987 | 43.3 | 49.3 | 51.6 | 63.3 | 72.1 | 79.3 | 78.1 | 81.2 | 74.1 | 68.2 | 52.2 | 44.9 | 63.1 |
| 1988 | 45.5 | 53.9 | 56.5 | 59.7 | 63.5 | 75.0 | 86.7 | 81.9 | 76.5 | 69.1 | 50.7 | 45.4 | 63.7 |
| 1989 | 44.5 | 45.9 | 52.0 | 63.0 | 66.5 | 76.1 | 80.2 | 77.5 | 71.3 | 61.0 | 53.6 | 46.4 | 61.5 |
| 1990 | 45.0 | 45.9 | 55.2 | 64.6 | 65.4 | 75.0 | 83.7 | 79.6 | 74.9 | 65.4 | 51.6 | 40.3 | 62.2 |
| 1991 | 45.8 | 53.7 | 48.7 | 56.4 | 63.6 | 73.3 | 85.0 | 77.8 | 79.1 | 68.9 | 54.4 | 46.1 | 62.7 |
| 1992 | 44.5 | 53.2 | 55.3 | 61.5 | 75.1 | 76.1 | 80.3 | 81.6 | 74.5 | 65.9 | 52.0 | 43.1 | 63.6 |
| 1993 | 43.4 | 47.2 | 57.0 | 57.4 | 65.1 | 73.5 | 81.1 | 78.5 | 74.0 | 65.2 | 50.7 | 45.5 | 61.6 |
| 1994 | 48.3 | 45.8 | 55.8 | 60.1 | 67.5 | 74.5 | 84.0 | 78.5 | 74.7 | 62.8 | 45.7 | 41.9 | 61.6 |
| 1995 | 48.2 | 53.0 | 51.4 | 56.0 | 64.3 | 72.1 | 80.4 | 79.2 | 75.6 | 65.6 | 57.3 | 48.3 | 62.6 |
| 1996 | 45.6 | 51.7 | 55.2 | 58.1 | 65.5 | 77.0 | 83.8 | 81.8 | 72.4 | 61.9 | 52.9 | 48.5 | 62.9 |
| 1997 | 46.5 | 51.2 | 56.3 | 59.4 | 71.2 | 74.9 | 81.6 | 77.6 | 72.7 | 60.4 | 52.3 | 45.7 | 62.5 |
| 1998 | 46.2 | 46.5 | 52.2 | 55.9 | 57.1 | 69.8 | 81.5 | 81.5 | 75.8 | 61.9 | 50.2 | 44.1 | 60.2 |
| 1999 | 47.5 | 45.5 | 49.0 | 59.0 | 67.3 | 75.0 | 79.3 | 77.9 | 76.5 | 64.8 | 53.4 | 47.4 | 61.9 |
| 2000 | 46.4 | 48.7 | 53.8 | 60.1 | 65.8 | 78.9 | 77.5 | 78.4 | 73.4 | 61.3 | 47.4 | 47.3 | 61.6 |
| 2001 | 44.0 | 46.6 | 57.5 | 55.4 | 74.0 | 75.4 | 80.7 | 80.6 | 75.8 | 67.3 | 53.0 | 46.0 | 63.0 |
| 2002 | 44.7 | 51.7 | 52.2 | 60.3 | 67.5 | 78.5 | 84.1 | 81.1 | 74.5 | 64.7 | 53.9 | 46.2 | 63.3 |
| 2003 | 51.1 | 49.6 | 54.2 | 52.5 | 65.6 | 79.7 | 84.9 | 78.6 | 77.7 | 68.6 | 49.5 | 45.6 | 63.1 |
| 2004 | 43.5 | 48.8 | 60.8 | 62.5 | 68.1 | 78.2 | 83.4 | 80.4 | 73.8 | 62.3 | 52.3 | 48.7 | 63.6 |
| 2005 | 46.7 | 52.7 | 55.5 | 57.1 | 66.1 | 71.1 | 85.4 | 82.2 | 71.3 | 62.7 | 52.9 | 47.7 | 62.6 |
| 2006 | 46.4 | 51.0 | 46.9 | 57.3 | 70.3 | 79.7 | 85.7 | 79.2 | 74.1 | 63.5 | 51.4 | 46.0 | 62.6 |
| 2007 | 45.4 | 48.6 | 58.0 | 59.6 | 69.5 | 77.2 | 81.8 | 79.8 | 71.3 | 60.5 | 55.9 | 45.0 | 62.7 |
| 2008 | 43.9 | 49.4 | 53.7 | 57.4 | 70.0 | 76.8 | 82.5 | 83.2 | 75.5 | 64.3 | 55.7 | 42.9 | 62.9 |
| 2009 | 49.7 | 47.5 | 53.2 | 58.6 | 71.6 | 75.3 | 83.3 | 80.1 | 76.6 | 62.2 | 51.4 | 43.4 | 62.7 |
| POR= 23 YRS | 45.9 | 49.8 | 54.0 | 58.9 | 67.5 | 75.8 | 82.4 | 79.9 | 74.3 | 64.3 | 52.3 | 45.5 | 62.5 |

## HEATING DEGREE DAYS (base 65°F) 2009  REDDING (KRDD)

| YEAR | JUL | AUG | SEP | OCT | NOV | DEC | JAN | FEB | MAR | APR | MAY | JUN | TOTAL |
|------|-----|-----|-----|-----|-----|-----|-----|-----|-----|-----|-----|-----|-------|
| 1985-86 | | | | | | | | | | | | | |
| 1986-87 | | | 87 | 94 | 309 | 598 | 665 | 435 | 408 | 90 | 18 | 0 | |
| 1987-88 | 0 | 0 | 0 | 27 | 381 | 615 | 601 | 315 | 257 | 168 | 124 | 43 | 2531 |
| 1988-89 | 0 | 0 | 0 | 30 | 421 | 602 | 627 | 530 | 397 | 140 | 49 | 4 | 2800 |
| 1989-90 | 0 | 0 | 12 | 135 | 336 | 569 | 611 | 531 | 301 | 52 | 76 | 7 | 2630 |
| 1990-91 | 0 | 1 | 0 | 50 | 396 | 760 | 591 | 310 | 499 | 253 | 106 | 0 | 2966 |
| 1991-92 | 0 | 0 | 0 | 105 | 314 | 579 | 628 | 337 | 295 | 120 | 0 | 15 | 2393 |
| 1992-93 | 0 | 0 | 0 | 57 | 386 | 671 | 663 | 493 | 242 | 242 | 54 | 21 | 2829 |
| 1993-94 | 0 | 0 | 0 | 56 | 426 | 599 | 509 | 532 | 279 | 157 | 55 | 2 | 2615 |
| 1994-95 | 0 | 0 | 0 | 93 | 572 | 710 | 513 | 330 | 415 | 265 | 106 | 24 | 3028 |
| 1995-96 | 0 | 0 | 0 | 52 | 231 | 509 | 597 | 382 | 300 | 221 | 70 | 0 | 2362 |
| 1996-97 | 0 | 0 | 1 | 171 | 354 | 504 | 565 | 378 | 273 | 168 | 21 | 6 | 2441 |
| 1997-98 | 0 | 0 | 8 | 145 | 378 | 590 | 576 | 511 | 393 | 276 | 241 | 8 | 3126 |
| 1998-99 | 0 | 0 | 9 | 121 | 436 | 642 | 535 | 539 | 487 | 213 | 65 | 24 | 3071 |
| 1999-00 | 0 | 0 | 0 | 72 | 342 | 535 | 568 | 467 | 339 | 147 | 116 | 1 | 2587 |
| 2000-01 | 0 | 0 | 10 | 162 | 522 | 544 | 643 | 507 | 236 | 293 | 6 | 3 | 2926 |
| 2001-02 | 0 | 0 | 0 | 45 | 351 | 581 | 620 | 368 | 409 | 164 | 53 | 0 | 2591 |
| 2002-03 | 0 | 0 | 1 | 83 | 325 | 577 | 423 | 429 | 331 | 370 | 102 | 0 | 2641 |
| 2003-04 | 0 | 0 | 0 | 36 | 460 | 594 | 658 | 463 | 159 | 131 | 19 | 0 | 2520 |
| 2004-05 | 0 | 0 | 10 | 192 | 373 | 502 | 558 | 339 | 297 | 229 | 75 | 28 | 2603 |
| 2005-06 | 0 | 0 | 1 | 100 | 358 | 527 | 569 | 386 | 554 | 265 | 12 | 0 | 2772 |
| 2006-07 | 0 | 0 | 3 | 90 | 401 | 583 | 601 | 451 | 220 | 181 | 45 | 4 | 2579 |
| 2007-08 | 0 | 0 | 29 | 142 | 273 | 612 | 650 | 445 | 342 | 246 | 20 | 0 | 2759 |
| 2008-09 | 0 | 0 | 0 | 64 | 282 | 678 | 468 | 483 | 363 | 226 | 34 | 5 | 2603 |
| 2009- | 0 | 0 | 3 | 113 | 400 | 662 | | | | | | | |

WBAN : 24257

## COOLING DEGREE DAYS (base 65°F) 2009  REDDING (KRDD)

| YEAR | JAN | FEB | MAR | APR | MAY | JUN | JUL | AUG | SEP | OCT | NOV | DEC | TOTAL |
|------|-----|-----|-----|-----|-----|-----|-----|-----|-----|-----|-----|-----|-------|
| 1986 | | | | | | | | | 177 | 63 | 9 | 0 | |
| 1987 | 0 | 0 | 0 | 45 | 245 | 439 | 412 | 512 | 279 | 132 | 1 | 0 | 2065 |
| 1988 | 0 | 0 | 4 | 16 | 82 | 349 | 679 | 531 | 350 | 165 | 0 | 1 | 2177 |
| 1989 | 0 | 0 | 0 | 85 | 103 | 344 | 477 | 394 | 210 | 17 | 0 | 0 | 1630 |
| 1990 | 0 | 0 | 1 | 47 | 97 | 314 | 586 | 463 | 306 | 67 | 0 | 0 | 1881 |
| 1991 | 0 | 0 | 0 | 2 | 72 | 257 | 627 | 404 | 430 | 234 | 3 | 0 | 2029 |
| 1992 | 0 | 0 | 0 | 22 | 318 | 355 | 481 | 521 | 291 | 90 | 0 | 0 | 2078 |
| 1993 | 0 | 0 | 0 | 19 | 65 | 282 | 504 | 425 | 279 | 71 | 4 | 0 | 1649 |
| 1994 | 0 | 0 | 1 | 16 | 140 | 296 | 594 | 425 | 299 | 29 | 0 | 0 | 1800 |
| 1995 | 0 | 0 | 0 | 0 | 92 | 244 | 484 | 447 | 325 | 81 | 4 | 0 | 1677 |
| 1996 | 0 | 4 | 1 | 22 | 93 | 367 | 614 | 528 | 229 | 83 | 0 | 0 | 1941 |
| 1997 | 0 | 0 | 9 | 10 | 225 | 312 | 522 | 394 | 244 | 11 | 3 | 0 | 1730 |
| 1998 | 0 | 0 | 0 | 11 | 2 | 160 | 518 | 519 | 340 | 31 | 0 | 0 | 1581 |
| 1999 | 0 | 0 | 0 | 40 | 145 | 330 | 449 | 408 | 350 | 75 | 0 | 0 | 1797 |
| 2000 | 0 | 0 | 0 | 9 | 147 | 425 | 396 | 424 | 269 | 54 | 0 | 0 | 1724 |
| 2001 | 0 | 0 | 11 | 11 | 294 | 321 | 495 | 493 | 331 | 125 | 0 | 0 | 2081 |
| 2002 | 0 | 0 | 20 | 28 | 135 | 415 | 598 | 504 | 292 | 82 | 0 | 0 | 2074 |
| 2003 | 0 | 0 | 2 | 0 | 128 | 449 | 625 | 429 | 387 | 154 | 0 | 0 | 2174 |
| 2004 | 0 | 0 | 34 | 60 | 121 | 403 | 578 | 485 | 278 | 116 | 0 | 0 | 2075 |
| 2005 | 0 | 0 | 8 | 0 | 119 | 219 | 638 | 540 | 200 | 34 | 0 | 0 | 1758 |
| 2006 | 0 | 3 | 0 | 43 | 187 | 447 | 650 | 449 | 285 | 49 | 2 | 0 | 2115 |
| 2007 | 0 | 0 | 9 | 27 | 190 | 378 | 530 | 467 | 225 | 11 | 3 | 0 | 1840 |
| 2008 | 0 | 0 | 0 | 27 | 184 | 360 | 549 | 569 | 320 | 47 | 9 | 0 | 2065 |
| 2009 | 1 | 0 | 4 | 41 | 246 | 324 | 578 | 476 | 359 | 31 | 0 | 0 | 2060 |

**SNOWFALL (inches)  2009  REDDING (KRDD)**

| YEAR | JUL | AUG | SEP | OCT | NOV | DEC | JAN | FEB | MAR | APR | MAY | JUN | TOTAL |
|---|---|---|---|---|---|---|---|---|---|---|---|---|---|
| 1985-86 | | | | | | | | | | | | | |
| 1986-87 | | | 0.0 | 0.0 | 0.0 | 0.0 | 0.0 | 0.0 | 1.8 | 0.0 | 0.0 | 0.0 | |
| 1987-88 | 0.0 | 0.0 | 0.0 | 0.0 | 0.0 | 0.5 | 0.0 | 0.0 | 0.0 | 0.0 | 0.0 | 0.0 | 0.5 |
| 1988-89 | 0.0 | 0.0 | 0.0 | 0.0 | T | 17.0 | 3.2 | T | T | T | 0.0 | 0.0 | 20.2 |
| 1989-90 | 0.0 | 0.0 | 0.0 | 0.0 | 0.0 | 0.0 | 0.0 | 1.4 | T | 0.0 | 1.5 | 0.0 | 2.9 |
| 1990-91 | 0.0 | 0.0 | 0.0 | 0.0 | 0.0 | 0.0 | T | 0.0 | 0.0 | 0.0 | 0.0 | 0.0 | T |
| 1991-92 | 0.0 | 0.0 | 0.0 | 0.0 | 0.0 | 0.0 | T | 0.0 | 0.0 | 0.0 | 0.0 | T | T |
| 1992-93 | 0.0 | 0.0 | 0.0 | 0.0 | 0.0 | T | 0.4 | T | 0.0 | 0.0 | 0.0 | 0.0 | 0.4 |
| 1993-94 | 0.0 | T | 0.0 | 0.0 | 0.0 | 0.0 | 0.0 | 0.9 | T | 0.0 | 0.0 | 0.0 | 0.9 |
| 1994-95 | 0.0 | 0.0 | 0.0 | 0.0 | 0.0 | 0.0 | 0.0 | T | T | T | 0.0 | 0.0 | T |
| 1995-96 | 0.0 | 0.0 | 0.0 | 0.0 | 0.0 | 0.0 | 10.9 | T | T | T | 0.0 | 0.0 | 10.9 |
| 1996-97 | | | | | | | | | | | | | |
| 1997-98 | | | | | | | | | | | | | |
| 1998-99 | | | | | | | | | | | | | |
| 1999-00 | | | | | | | | | | | | | |
| 2000-01 | | | | | | | | | | | | | |
| 2001-02 | | | | | | | | | | | | | |
| 2002-03 | | | | | | | | | | | | | |
| 2003-04 | | | | | | 4.3 | | | | | | | |
| 2004-05 | | | | | | | | | | | | | |
| 2005- | | | | | | | | | | | | | |
| POR= 9 YRS | 0.0 | T | 0.0 | 0.0 | T | 2.0 | 1.3 | 0.2 | 0.2 | T | 0.1 | T | 3.8 |

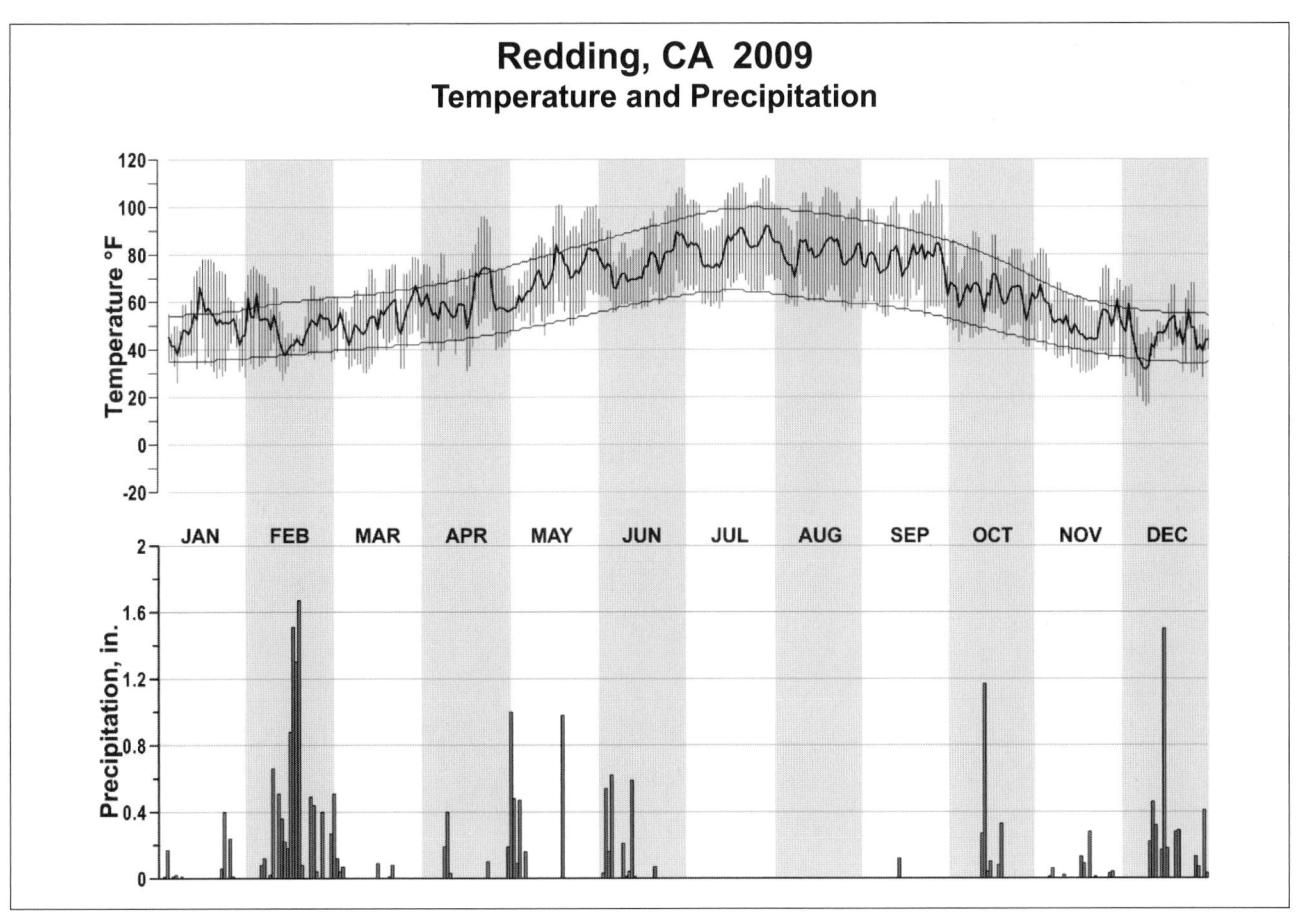

# Redding, CA  2009
## Temperature and Precipitation

# 2009
# SACRAMENTO
# CALIFORNIA (KSAC)

Sacramento, and the lower Sacramento Valley, has a mild climate with abundant sunshine most of the year. A nearly cloud-free sky prevails throughout the summer months, and in much of the spring and fall. The summers are usually dry with warm to hot afternoons and mostly mild nights. The rainy season generally is November through March. About 75 percent of the annual precipitation occurs then, but measurable rain falls only on an average of nine days per month during that period. The shielding effect of mountains to the north, east, and west usually modifies winter storms. The Sierra Nevada snow fields, only 70 miles east of Sacramento, usually provide an adequate water supply during the dry season, and an important recreational area in winter. Heavy snowfall and torrential rains frequently fall on the western Sierra slopes, and may produce flood conditions along the Sacramento River and its tributaries. In the valley, however, excessive rainfall as well as damaging winds are rare.

The prevailing wind at Sacramento is southerly every month but November, when it is northerly. Topographic effects, the north-south alignment of the valley, the coast range, and the Sierra Nevada strongly influence the wind flow in the valley. A sea level gap in the coast range permits cool, oceanic air to flow, occasionally, into the valley during the summer season with a marked lowering of temperature through the Sacramento-San Joaquin River Delta to the capital. In the spring and fall, a large north-to-south pressure gradient develops over the northern part of the state. Air flowing over the Siskiyou mountains to the north warms and dries as it descends to the valley floor. This gusty, blustery north wind is a local variation of the chinook. It apparently carries a form of pollen which may cause allergic responses by susceptible individuals.

As is well known, relative humidity has a marked influence on the reactions of plants and animals to temperature. The extremely low relative humidity that ordinarily accompanies high temperatures in this valley should be considered when comparing temperatures here with those of cities in more humid regions. The extreme hot spells, with temperatures exceeding 100 degrees, are usually caused by air flow from a sub-tropical high pressure area that brings light to nearly calm winds and humidities below 20 percent.

Thunderstorms are few in number, usually mild in character, and occur mainly in the spring. An occasional thunderstorm may drift over the valley from the Sierra Nevada in the summer. Snow falls so rarely, and in such small amounts, that its occurrence may be disregarded as a climatic feature. Heavy fog occurs mostly in midwinter, never in summer, and seldom in spring or autumn. An occasional winter fog, under stagnant atmospheric conditions, may continue for several days. Light and moderate fogs are more frequent, and may come anytime during the wet, cold season. The fog is the radiational cooling type, and is usually confined to the early morning hours.

Sacramento is the geographical center of the great interior valley of California that reaches from Red Bluff in the north to Bakersville in the south. This predominantly agricultural region produces an extremely wide and abundant variety of fruits, grains, and vegetables ranging from the semi-tropical to the hardier varieties.

Based on the 1951-1980 period, the average first occurrence of 32 degrees Fahrenheit in the fall is December 1 and the average last occurrence in the spring is February 14.

# NORMALS, MEANS, AND EXTREMES
## SACRAMENTO (KSAC)

LATITUDE: 38° 30'N   LONGITUDE: -121° 29'W   ELEVATION (FT): GRND: 15   BARO: 41   TIME ZONE: PACIFIC (UTC -8)   WBAN: 23232

| | ELEMENT | POR | JAN | FEB | MAR | APR | MAY | JUN | JUL | AUG | SEP | OCT | NOV | DEC | YEAR |
|---|---|---|---|---|---|---|---|---|---|---|---|---|---|---|---|
| TEMPERATURE °F | NORMAL DAILY MAXIMUM | 30 | 53.8 | 60.5 | 64.7 | 71.4 | 80.0 | 87.4 | 92.4 | 91.4 | 87.5 | 78.2 | 63.7 | 53.9 | 73.7 |
| | MEAN DAILY MAXIMUM | 68 | 53.5 | 59.3 | 64.7 | 71.3 | 79.9 | 87.3 | 92.7 | 91.5 | 87.5 | 77.7 | 63.2 | 53.7 | 73.5 |
| | HIGHEST DAILY MAXIMUM | 59 | 74 | 76 | 88 | 95 | 105 | 115 | 114 | 110 | 108 | 104 | 87 | 72 | 115 |
| | YEAR OF OCCURRENCE | | 2009 | 1992 | 1988 | 2004 | 1984 | 1961 | 1972 | 1996 | 1988 | 2001 | 1960 | 1999 | JUN 1961 |
| | MEAN OF EXTREME MAXS. | 68 | 63.3 | 69.7 | 76.5 | 85.8 | 95.7 | 102.9 | 105.0 | 103.4 | 100.2 | 91.6 | 76.2 | 64.0 | 86.2 |
| | NORMAL DAILY MINIMUM | 30 | 38.8 | 41.9 | 44.2 | 46.3 | 50.9 | 55.5 | 58.3 | 58.1 | 55.8 | 50.6 | 42.8 | 37.7 | 48.4 |
| | MEAN DAILY MINIMUM | 68 | 37.9 | 40.7 | 43.1 | 45.9 | 50.6 | 55.4 | 58.1 | 57.8 | 55.8 | 50.1 | 42.4 | 38.2 | 48.0 |
| | LOWEST DAILY MINIMUM | 59 | 21 | 23 | 26 | 31 | 36 | 41 | 48 | 49 | 42 | 36 | 26 | 18 | 18 |
| | YEAR OF OCCURRENCE | | 2007 | 1989 | 1971 | 1999 | 1974 | 1990 | 1983 | 1978 | 2007 | 2002 | 1993 | 1990 | DEC 1990 |
| | MEAN OF EXTREME MINS. | 68 | 27.9 | 31.3 | 34.1 | 37.7 | 42.2 | 48.1 | 52.5 | 52.3 | 48.7 | 41.3 | 32.3 | 28.3 | 39.7 |
| | NORMAL DRY BULB | 30 | 46.3 | 51.2 | 54.5 | 58.9 | 65.5 | 71.5 | 75.4 | 74.8 | 71.7 | 64.4 | 53.3 | 45.8 | 61.1 |
| | MEAN DRY BULB | 68 | 45.7 | 50.0 | 53.9 | 58.6 | 65.3 | 71.4 | 75.5 | 74.7 | 71.7 | 63.9 | 52.8 | 46.0 | 60.8 |
| | MEAN WET BULB | 26 | 44.3 | 47.3 | 49.7 | 51.2 | 54.8 | 58.3 | 61.0 | 60.0 | 57.9 | 53.6 | 48.1 | 43.6 | 52.5 |
| | MEAN DEW POINT | 26 | 42.4 | 44.2 | 46.3 | 46.9 | 50.0 | 53.0 | 55.9 | 55.2 | 52.9 | 48.8 | 44.7 | 41.3 | 48.5 |
| | NORMAL NO. DAYS WITH: | | | | | | | | | | | | | | |
| | MAXIMUM >= 90 | 30 | 0.0 | 0.0 | 0.0 | 0.5 | 5.8 | 12.1 | 20.7 | 18.6 | 12.7 | 3.3 | 0.0 | 0.0 | 73.7 |
| | MAXIMUM <= 32 | 30 | 0.0 | * | 0.0 | 0.0 | 0.0 | 0.0 | 0.0 | 0.0 | 0.0 | 0.0 | 0.0 | * | 0.0 |
| | MINIMUM <= 32 | 30 | 5.2 | 1.9 | 0.3 | * | 0.0 | 0.0 | 0.0 | 0.0 | 0.0 | 0.0 | 1.4 | 6.8 | 15.6 |
| | MINIMUM <= 0 | 30 | 0.0 | 0.0 | 0.0 | 0.0 | 0.0 | 0.0 | 0.0 | 0.0 | 0.0 | 0.0 | 0.0 | 0.0 | 0.0 |
| H/C | NORMAL HEATING DEG. DAYS | 30 | 580 | 387 | 335 | 208 | 97 | 10 | 0 | 0 | 11 | 84 | 359 | 595 | 2666 |
| | NORMAL COOLING DEG. DAYS | 30 | 0 | 0 | 6 | 24 | 110 | 204 | 320 | 303 | 210 | 66 | 5 | 0 | 1248 |
| RH | NORMAL (PERCENT) | 30 | 84 | 78 | 74 | 66 | 61 | 56 | 55 | 57 | 58 | 64 | 76 | 83 | 68 |
| | HOUR 04 LST | 30 | 90 | 89 | 87 | 82 | 81 | 79 | 78 | 79 | 79 | 81 | 87 | 90 | 84 |
| | HOUR 10 LST | 30 | 85 | 78 | 70 | 57 | 50 | 47 | 48 | 50 | 51 | 56 | 72 | 83 | 62 |
| | HOUR 16 LST | 30 | 70 | 60 | 54 | 43 | 37 | 31 | 30 | 30 | 31 | 37 | 55 | 67 | 45 |
| | HOUR 22 LST | 30 | 87 | 83 | 79 | 73 | 69 | 65 | 63 | 64 | 66 | 71 | 81 | 86 | 74 |
| S | PERCENT POSSIBLE SUNSHINE | 47 | 48 | 65 | 74 | 82 | 90 | 94 | 97 | 96 | 93 | 86 | 66 | 49 | 78 |
| W/O | MEAN NO. DAYS WITH: | | | | | | | | | | | | | | |
| | HEAVY FOG(VISBY <= 1/4 MI) | 46 | 9.9 | 4.7 | 1.6 | 0.3 | 0.1 | 0.0 | 0.0 | 0.0 | 0.1 | 1.0 | 5.0 | 9.3 | 32.0 |
| | THUNDERSTORMS | 62 | 0.4 | 0.5 | 0.7 | 0.5 | 0.4 | 0.3 | 0.2 | 0.2 | 0.4 | 0.3 | 0.2 | 0.2 | 4.3 |
| CLOUDNESS | MEAN: | | | | | | | | | | | | | | |
| | SUNRISE-SUNSET (OKTAS) | 50 | 5.7 | 5.0 | 4.5 | 3.8 | 2.9 | 1.8 | 0.9 | 1.1 | 1.5 | 2.6 | 4.5 | 5.5 | 3.3 |
| | MIDNIGHT-MIDNIGHT (OKTAS) | 31 | 5.3 | 4.6 | 3.9 | 3.3 | 2.3 | 1.7 | 0.8 | 1.0 | 1.2 | 2.1 | 3.9 | 4.8 | 2.9 |
| | MEAN NO. DAYS WITH: | | | | | | | | | | | | | | |
| | CLEAR | 50 | 6.5 | 7.6 | 9.9 | 11.9 | 16.9 | 21.7 | 26.9 | 25.1 | 23.2 | 18.9 | 9.8 | 7.6 | 186.0 |
| | PARTLY CLOUDY | 50 | 5.9 | 7.0 | 8.7 | 9.6 | 8.6 | 5.9 | 3.2 | 4.1 | 4.2 | 6.0 | 7.4 | 5.8 | 76.4 |
| | CLOUDY | 50 | 18.7 | 13.6 | 12.4 | 8.4 | 5.5 | 2.4 | 1.0 | 1.3 | 2.1 | 5.4 | 12.4 | 17.0 | 100.2 |
| PR | MEAN STATION PRESSURE(IN) | 26 | 30.10 | 30.03 | 30.00 | 29.97 | 29.90 | 29.85 | 29.85 | 29.85 | 29.85 | 29.93 | 30.05 | 30.09 | 29.96 |
| | MEAN SEA-LEVEL PRES. (IN) | 26 | 30.12 | 30.06 | 30.02 | 30.00 | 29.93 | 29.88 | 29.88 | 29.88 | 29.88 | 29.96 | 30.08 | 30.12 | 29.98 |
| WINDS | MEAN SPEED (MPH) | 26 | 5.5 | 6.3 | 7.1 | 7.4 | 7.9 | 8.4 | 7.9 | 7.3 | 6.2 | 5.5 | 4.9 | 5.6 | 6.7 |
| | PREVAIL.DIR(TENS OF DEGS) | 29 | 15 | 15 | 23 | 22 | 23 | 22 | 21 | 21 | 21 | 34 | 33 | 15 | 21 |
| | MAXIMUM 2-MINUTE: | 11 | | | | | | | | | | | | | |
| | SPEED (MPH) | | 38 | 37 | 36 | 33 | 38 | 31 | 26 | 26 | 36 | 36 | 38 | 38 | 38 |
| | DIR. (TENS OF DEGS) | | 14 | 33 | 32 | 35 | 33 | 34 | 22 | 22 | 32 | 35 | 15 | 14 | 14 |
| | YEAR OF OCCURRENCE | | 2006 | 2008 | 2000 | 1999 | 2001 | 2008 | 2007 | 2007 | 2006 | 2000 | 2001 | 2003 | JAN 2006 |
| | MAXIMUM 3-SECOND | | | | | | | | | | | | | | |
| | SPEED (MPH) | 11 | 58 | 53 | 44 | 43 | 46 | 39 | 33 | 35 | 48 | 48 | 62 | 53 | 62 |
| | DIR. (TENS OF DEGS) | | 15 | 15 | 16 | 35 | 33 | 32 | 21 | 23 | 32 | 14 | 21 | 14 | 21 |
| | YEAR OF OCCURRENCE | | 2008 | 2006 | 2005 | 1999 | 2001 | 2008 | 2008 | 2007 | 2006 | 2009 | 2007 | 2003 | NOV 2007 |
| PRECIPITATION | NORMAL (IN) | 30 | 3.84 | 3.54 | 2.80 | 1.02 | 0.53 | 0.20 | 0.05 | 0.06 | 0.36 | 0.89 | 2.19 | 2.45 | 17.93 |
| | MAXIMUM MONTHLY (IN) | 70 | 9.69 | 9.95 | 8.13 | 4.76 | 3.13 | 1.26 | 0.79 | 0.65 | 2.78 | 7.51 | 7.41 | 12.64 | 12.64 |
| | YEAR OF OCCURRENCE | | 1995 | 1998 | 1995 | 1941 | 1948 | 1993 | 1974 | 1976 | 1989 | 1962 | 1970 | 1955 | DEC 1955 |
| | MINIMUM MONTHLY (IN) | 70 | 0.05 | 0.15 | 0.05 | 0.00 | T | 0.00 | 0.00 | 0.00 | 0.00 | 0.00 | T | 0.00 | 0.00 |
| | YEAR OF OCCURRENCE | | 2007 | 1964 | 2008 | 1949 | 1992 | 1981 | 1983 | 1982 | 1980 | 1966 | 1995 | 1989 | DEC 1989 |
| | MAXIMUM IN 24 HOURS (IN) | 70 | 3.41 | 3.01 | 2.30 | 2.22 | 1.73 | 1.21 | 0.78 | 0.65 | 1.79 | 5.59 | 2.95 | 3.64 | 5.59 |
| | YEAR OF OCCURRENCE | | 1967 | 1986 | 1982 | 1958 | 2002 | 1993 | 1974 | 1965 | 1989 | 1962 | 1970 | 1955 | OCT 1962 |
| | NORMAL NO. DAYS WITH: | | | | | | | | | | | | | | |
| | PRECIPITATION >= 0.01 | 30 | 10.5 | 9.1 | 9.4 | 4.9 | 2.9 | 1.2 | 0.2 | 0.4 | 1.5 | 3.6 | 7.2 | 8.2 | 59.1 |
| | PRECIPITATION >= 1.00 | 30 | 0.8 | 0.8 | 0.3 | 0.1 | 0.1 | * | 0.0 | 0.0 | * | 0.2 | 0.5 | 0.4 | 3.2 |
| SNOWFALL | NORMAL (IN) | 30 | 0.* | 0.1 | 0.0 | 0.0 | 0.0 | 0.0 | 0.0 | 0.0 | 0.0 | 0.0 | 0.0 | 0.* | 0.1 |
| | MAXIMUM MONTHLY (IN) | 50 | T | 2.0 | T | 0.0 | T | 0.0 | 0.0 | 0.0 | 0.0 | 0.0 | 0.0 | T | 2.0 |
| | YEAR OF OCCURRENCE | | 1974 | 1976 | 1982 | | 1994 | | | | | | | 1995 | FEB 1976 |
| | MAXIMUM IN 24 HOURS (IN) | 50 | T | 2.0 | T | 0.0 | T | 0.0 | 0.0 | 0.0 | 0.0 | 0.0 | 0.0 | T | 2.0 |
| | YEAR OF OCCURRENCE' | | 1974 | 1976 | 1982 | | 1994 | | | | | | | 1995 | FEB 1976 |
| | MAXIMUM SNOW DEPTH (IN) | 48 | 0 | 0 | 0 | 0 | 0 | 0 | 0 | 0 | 0 | 0 | 0 | 0 | 0 |
| | YEAR OF OCCURRENCE | | | | | | | | | | | | | | |
| | NORMAL NO. DAYS WITH: | | | | | | | | | | | | | | |
| | SNOWFALL >= 1.0 | 30 | 0.0 | 0.0 | 0.0 | 0.0 | 0.0 | 0.0 | 0.0 | 0.0 | 0.0 | 0.0 | 0.0 | 0.0 | 0.0 |

## PRECIPITATION (inches) 2009 SACRAMENTO (KSAC)

| YEAR | JAN | FEB | MAR | APR | MAY | JUN | JUL | AUG | SEP | OCT | NOV | DEC | ANNUAL |
|------|-----|-----|-----|-----|-----|-----|-----|-----|-----|-----|-----|-----|--------|
| 1980 | 5.64 | 7.12 | 2.62 | 1.06 | 0.49 | 0.04 | 0.40 | 0.00 | 0.00 | 0.06 | 0.12 | 1.79 | 19.34 |
| 1981 | 4.56 | 0.87 | 3.55 | 0.66 | 0.50 | 0.00 | 0.00 | 0.00 | 0.25 | 2.57 | 6.09 | 3.28 | 22.33 |
| 1982 | 5.50 | 2.35 | 7.12 | 3.07 | T | 0.15 | 0.00 | 0.00 | 1.81 | 2.61 | 5.74 | 3.25 | 31.60 |
| 1983 | 4.92 | 5.56 | 6.75 | 4.21 | 0.25 | 0.40 | 0.00 | 0.11 | 0.66 | 0.40 | 4.91 | 5.26 | 33.43 |
| 1984 | 0.16 | 1.22 | 1.35 | 0.34 | 0.01 | 0.10 | T | 0.01 | 0.07 | 1.39 | 3.61 | 1.23 | 9.49 |
| 1985 | 0.66 | 1.52 | 2.01 | T | 0.01 | 0.15 | T | 0.06 | 0.56 | 0.53 | 3.72 | 2.34 | 11.56 |
| 1986 | 3.67 | 8.60 | 3.20 | 0.91 | 0.07 | 0.00 | 0.00 | 0.00 | 0.60 | 0.19 | 0.14 | 0.76 | 18.14 |
| 1987 | 2.29 | 3.23 | 3.05 | 0.20 | T | T | 0.00 | 0.00 | 0.00 | 1.28 | 2.53 | 3.25 | 15.83 |
| 1988 | 2.96 | 0.99 | 0.17 | 1.58 | 0.89 | 0.19 | 0.00 | 0.00 | 0.00 | 0.19 | 1.68 | 2.73 | 11.38 |
| 1989 | 0.71 | 1.25 | 6.29 | 0.31 | 0.06 | 0.43 | 0.00 | 0.20 | 2.78 | 1.76 | 1.32 | 0.00 | 15.11 |
| 1990 | 4.97 | 2.91 | 0.93 | 0.73 | 2.10 | 0.00 | T | 0.00 | 0.00 | 0.09 | 0.43 | 1.60 | 13.76 |
| 1991 | 0.36 | 3.10 | 6.14 | 0.29 | 0.25 | 0.53 | T | 0.14 | 0.04 | 1.25 | 0.19 | 1.60 | 13.89 |
| 1992 | 1.39 | 5.47 | 2.05 | 0.92 | T | 0.15 | 0.00 | T | 0.00 | 1.31 | 0.28 | 4.94 | 16.51 |
| 1993 | 8.63 | 4.94 | 2.39 | 0.63 | 1.14 | 1.26 | 0.00 | 0.00 | 0.00 | 0.47 | 2.28 | 1.75 | 23.49 |
| 1994 | 2.12 | 3.15 | 0.05 | 0.67 | 1.68 | 0.00 | 0.00 | 0.00 | | | | 2.68 | |
| 1995 | 9.69 | 0.20 | 8.13 | 1.46 | 1.06 | 0.47 | 0.00 | 0.00 | 0.00 | T | T | 5.49 | 26.50 |
| 1996 | 4.16 | 5.49 | 1.73 | 1.25 | 0.79 | 0.00 | 0.00 | T | T | .67 | 1.97 | 6.39 | 22.45 |
| 1997 | 9.05 | 0.28 | 0.34 | 0.18 | 0.35 | 0.59 | 0.00 | 0.32 | 0.16 | 0.82 | 4.56 | 2.91 | 19.56 |
| 1998 | 6.40 | 9.95 | 2.47 | 1.05 | 2.98 | 0.58 | 0.00 | 0.00 | 0.23 | 0.76 | 2.84 | 0.58 | 27.84 |
| 1999 | 2.63 | 4.45 | 1.50 | 0.89 | 0.07 | 0.03 | 0.00 | T | 0.00 | 0.18 | 1.63 | 0.06 | 11.44 |
| 2000 | 6.49 | 8.49 | 2.03 | 1.39 | 1.17 | 0.04 | 0.00 | T | 0.09 | 1.62 | 0.68 | 0.59 | 22.59 |
| 2001 | 3.75 | 4.57 | 2.04 | 1.50 | T | 0.08 | T | 0.00 | 0.50 | 0.36 | 2.43 | 6.27 | 21.50 |
| 2002 | 2.19 | 1.13 | 2.87 | 0.12 | 2.07 | 0.00 | 0.00 | 0.00 | 0.00 | 0.00 | 2.34 | 6.26 | 16.98 |
| 2003 | 1.29 | 1.29 | 1.87 | 2.53 | 1.17 | 0.00 | T | 0.57 | T | 0.04 | 1.52 | 4.23 | 14.51 |
| 2004 | 2.11 | 5.01 | 0.48 | 0.09 | 0.17 | 0.00 | 0.00 | 0.00 | 0.16 | 2.71 | 2.69 | 4.14 | 17.56 |
| 2005 | 3.83 | 2.33 | 3.30 | 0.84 | 1.23 | 0.66 | T | 0.00 | T | 0.15 | 0.85 | 8.98 | 22.17 |
| 2006 | 2.53 | 2.09 | 5.29 | 3.27 | 0.30 | 0.00 | T | 0.00 | 0.00 | 0.16 | 1.12 | 3.01 | 17.77 |
| 2007 | 0.05 | 4.44 | 0.35 | 1.34 | 0.41 | 0.00 | 0.01 | 0.00 | 0.06 | 1.05 | 0.85 | 3.17 | 11.73 |
| 2008 | 6.67 | 1.81 | 0.05 | T | 0.04 | 0.00 | 0.00 | 0.00 | 0.00 | 0.84 | 2.38 | 1.51 | 13.30 |
| 2009 | 1.41 | 5.07 | 2.09 | 1.46 | 1.01 | 0.56 | T | 0.00 | 0.14 | 3.24 | 0.26 | 3.64 | 18.88 |
| POR= 68 YRS | 3.63 | 3.09 | 2.42 | 1.20 | 0.54 | 0.16 | 0.03 | 0.06 | 0.27 | 0.95 | 2.10 | 3.02 | 17.47 |

WBAN : 23232

## AVERAGE TEMPERATURE (°F) 2009 SACRAMENTO (KSAC)

| YEAR | JAN | FEB | MAR | APR | MAY | JUN | JUL | AUG | SEP | OCT | NOV | DEC | ANNUAL |
|------|-----|-----|-----|-----|-----|-----|-----|-----|-----|-----|-----|-----|--------|
| 1980 | 46.9 | 51.9 | 51.6 | 59.6 | 62.7 | 66.8 | 75.0 | 71.4 | 69.4 | 63.7 | 53.5 | 45.4 | 59.8 |
| 1981 | 46.8 | 50.4 | 51.2 | 57.9 | 64.7 | 74.8 | 75.1 | 74.5 | 69.7 | 63.5 | 60.3 | 48.6 | 61.5 |
| 1982 | 42.0 | 50.5 | 50.8 | 55.5 | 64.6 | 66.2 | 72.1 | 71.7 | 68.2 | 61.0 | 46.9 | 43.0 | 57.7 |
| 1983 | 43.1 | 52.2 | 53.4 | 54.7 | 64.2 | 70.8 | 72.2 | 76.6 | 74.9 | 67.5 | 53.7 | 51.0 | 61.2 |
| 1984 | 48.2 | 50.2 | 58.1 | 58.7 | 70.0 | 71.7 | 78.3 | 75.3 | 75.5 | 62.8 | 53.6 | 45.1 | 62.3 |
| 1985 | 42.4 | 51.4 | 50.8 | 61.5 | 63.2 | 75.1 | 77.0 | 72.9 | 68.5 | 63.3 | 49.8 | 42.6 | 59.9 |
| 1986 | 51.4 | 54.7 | 58.8 | 58.4 | 65.5 | 71.6 | 75.0 | 75.2 | 66.2 | 64.8 | 55.5 | 45.7 | 61.9 |
| 1987 | 44.9 | 51.3 | 53.8 | 62.7 | 69.1 | 72.4 | 71.8 | 74.9 | 71.8 | 67.6 | 53.4 | 47.2 | 61.7 |
| 1988 | 48.0 | 54.2 | 58.0 | 60.9 | 64.7 | 72.9 | 80.4 | 75.9 | 72.5 | 66.5 | 53.8 | 46.2 | 62.8 |
| 1989 | 44.1 | 47.1 | 55.6 | 63.2 | 65.8 | 71.7 | 76.2 | 73.8 | 69.6 | 62.4 | 54.3 | 44.3 | 60.7 |
| 1990 | 47.5 | 48.6 | 55.4 | 63.4 | 65.5 | 72.4 | 77.7 | 76.6 | 74.0 | 66.6 | 53.0 | 41.0 | 61.8 |
| 1991 | 47.3 | 55.0 | 51.0 | 58.0 | 63.3 | 70.2 | 77.1 | 73.2 | 74.8 | 68.8 | 55.9 | 46.3 | 61.7 |
| 1992 | 43.6 | 54.1 | 56.2 | 62.1 | 70.6 | 70.9 | 75.3 | 77.0 | 72.4 | 66.6 | 53.4 | 44.1 | 62.2 |
| 1993 | 45.2 | 49.5 | 57.9 | 58.4 | 64.6 | 71.7 | 74.3 | 74.1 | 71.5 | 65.0 | 51.5 | 44.3 | 60.7 |
| 1994 | 47.0 | 48.8 | 56.7 | 60.1 | 65.3 | 71.6 | 74.0 | 75.2 | 71.7 | | 47.6 | 43.7 | |
| 1995 | 51.3 | 52.1 | 53.0 | 57.7 | 63.1 | 69.0 | 74.2 | 75.1 | 72.3 | | 59.3 | 51.1 | |
| 1996 | 48.2 | 54.3 | 56.7 | 61.1 | 67.0 | 73.3 | 78.7 | 78.3 | 71.0 | 63.9 | 55.0 | 51.1 | 63.2 |
| 1997 | 48.3 | 52.7 | 57.9 | 62.4 | 71.9 | 72.9 | 76.5 | 75.9 | 75.1 | 64.1 | 56.9 | 46.2 | 63.4 |
| 1998 | 49.7 | 50.4 | 55.1 | 57.5 | 58.7 | 67.0 | 74.8 | 76.8 | 72.6 | 61.3 | 52.5 | 42.5 | 59.9 |
| 1999 | 44.7 | 48.1 | 50.8 | 57.4 | 62.8 | 69.9 | 72.0 | 73.0 | 72.3 | 65.1 | 54.6 | 46.8 | 59.8 |
| 2000 | 48.8 | 51.4 | 55.5 | 60.7 | 65.7 | 73.2 | 72.1 | 74.0 | 71.2 | 61.9 | 48.6 | 47.0 | 60.8 |
| 2001 | 45.8 | 48.8 | 57.2 | 55.9 | 71.7 | 73.4 | 73.4 | 74.5 | 71.1 | 65.8 | 55.9 | 48.8 | 61.9 |
| 2002 | 44.8 | 50.7 | 52.5 | 58.2 | 64.1 | 72.1 | 75.8 | 73.6 | 72.8 | 62.6 | 54.5 | 49.8 | 61.0 |
| 2003 | 50.7 | 50.2 | 55.8 | 54.3 | 65.2 | 72.5 | 79.3 | 75.1 | 73.7 | 66.5 | 51.3 | 49.1 | 62.0 |
| 2004 | 46.9 | 50.8 | 59.8 | 62.5 | 66.9 | 72.3 | 75.1 | 75.6 | 72.1 | 62.2 | 51.2 | 46.7 | 61.8 |
| 2005 | 45.6 | 52.4 | 56.2 | 57.2 | 65.0 | 69.2 | 78.8 | 76.3 | 68.9 | 63.3 | 54.6 | 49.3 | 61.4 |
| 2006 | 48.1 | 50.9 | 49.7 | 57.4 | 67.1 | 74.2 | 79.1 | 72.9 | 70.0 | 60.8 | 53.4 | 46.3 | 60.8 |
| 2007 | 43.8 | 50.7 | 57.8 | 60.1 | 66.3 | 71.8 | 75.3 | 75.5 | 68.6 | 61.2 | 55.1 | 45.7 | 61.0 |
| 2008 | 45.9 | 49.2 | 54.3 | 57.6 | 67.0 | 72.4 | 75.2 | 76.0 | 72.2 | 64.3 | 55.5 | 44.0 | 61.1 |
| 2009 | 47.4 | 50.9 | 54.4 | 59.5 | 68.7 | 71.2 | 75.4 | 75.0 | 74.9 | 62.2 | 52.8 | 45.3 | 61.5 |
| POR= 68 YRS | 45.7 | 50.0 | 53.9 | 58.6 | 65.3 | 71.4 | 75.5 | 74.7 | 71.7 | 63.9 | 52.8 | 46.0 | 60.8 |

## HEATING DEGREE DAYS (base 65°F) 2009  SACRAMENTO (KSAC)

| YEAR | JUL | AUG | SEP | OCT | NOV | DEC | JAN | FEB | MAR | APR | MAY | JUN | TOTAL |
|------|-----|-----|-----|-----|-----|-----|-----|-----|-----|-----|-----|-----|-------|
| 1980-81 | 2 | 0 | 4 | 134 | 339 | 596 | 557 | 405 | 420 | 229 | 81 | 2 | 2769 |
| 1981-82 | 0 | 0 | 9 | 66 | 145 | 498 | 708 | 398 | 434 | 282 | 70 | 40 | 2650 |
| 1982-83 | 3 | 0 | 31 | 125 | 532 | 675 | 670 | 353 | 354 | 303 | 99 | 4 | 3149 |
| 1983-84 | 3 | 0 | 0 | 7 | 333 | 425 | 514 | 421 | 206 | 191 | 22 | 11 | 2133 |
| 1984-85 | 0 | 0 | 0 | 115 | 335 | 611 | 693 | 377 | 433 | 122 | 89 | 11 | 2786 |
| 1985-86 | 0 | 2 | 15 | 95 | 450 | 689 | 411 | 284 | 192 | 200 | 73 | 0 | 2411 |
| 1986-87 | 0 | 0 | 53 | 47 | 277 | 593 | 614 | 377 | 340 | 95 | 37 | 0 | 2433 |
| 1987-88 | 1 | 0 | 0 | 11 | 339 | 544 | 522 | 307 | 212 | 138 | 94 | 27 | 2195 |
| 1988-89 | 0 | 0 | 3 | 38 | 329 | 576 | 640 | 496 | 285 | 106 | 50 | 3 | 2526 |
| 1989-90 | 0 | 0 | 11 | 107 | 316 | 634 | 536 | 453 | 289 | 71 | 53 | 6 | 2476 |
| 1990-91 | 0 | 0 | 0 | 24 | 356 | 739 | 543 | 274 | 427 | 205 | 104 | 6 | 2678 |
| 1991-92 | 0 | 0 | 0 | 82 | 267 | 572 | 657 | 310 | 265 | 104 | 0 | 9 | 2266 |
| 1992-93 | 0 | 0 | 0 | 24 | 340 | 643 | 605 | 426 | 214 | 202 | 55 | 21 | 2530 |
| 1993-94 | 0 | 0 | 5 | 33 | 399 | 634 | 550 | 449 | 248 | 147 | 48 | 1 | 2514 |
| 1994-95 | 0 | 0 | 0 | 0 | 515 | 654 | 415 | 354 | 364 | 210 | 105 | 26 | |
| 1995-96 | 0 | 0 | 0 | | 166 | 421 | 513 | 302 | 250 | 154 | 21 | 1 | |
| 1996-97 | 0 | 0 | 0 | 121 | 294 | 423 | 511 | 336 | 214 | 104 | 7 | 1 | 2011 |
| 1997-98 | 0 | 0 | 0 | 56 | 248 | 577 | 465 | 404 | 299 | 233 | 190 | 13 | 2485 |
| 1998-99 | 0 | 0 | 8 | 113 | 367 | 689 | 621 | 465 | 431 | 239 | 92 | 29 | 3054 |
| 1999-00 | 1 | 0 | 0 | 57 | 303 | 556 | 496 | 389 | 291 | 138 | 83 | 2 | 2316 |
| 2000-01 | 0 | 4 | 5 | 123 | 484 | 551 | 588 | 445 | 240 | 276 | 4 | 0 | 2720 |
| 2001-02 | 0 | 0 | 0 | 44 | 264 | 496 | 616 | 393 | 380 | 200 | 78 | 0 | 2471 |
| 2002-03 | 0 | 0 | 2 | 114 | 310 | 466 | 435 | 408 | 278 | 314 | 85 | 1 | 2413 |
| 2003-04 | 0 | 0 | 0 | 33 | 402 | 486 | 552 | 404 | 163 | 116 | 16 | 1 | 2173 |
| 2004-05 | 0 | 0 | 13 | 141 | 407 | 563 | 592 | 346 | 266 | 227 | 63 | 20 | 2638 |
| 2005-06 | 0 | 0 | 4 | 75 | 304 | 481 | 518 | 385 | 468 | 234 | 31 | 4 | 2504 |
| 2006-07 | 0 | 0 | 10 | 131 | 338 | 574 | 623 | 395 | 219 | 161 | 43 | 4 | 2498 |
| 2007-08 | 0 | 0 | 28 | 121 | 291 | 587 | 583 | 452 | 326 | 236 | 36 | 0 | 2660 |
| 2008-09 | 0 | 0 | 3 | 49 | 277 | 645 | 537 | 389 | 321 | 193 | 23 | 2 | 2439 |
| 2009- | 0 | 0 | 2 | 101 | 359 | 604 | | | | | | | |

WBAN : 23232

## COOLING DEGREE DAYS (base 65°F) 2009  SACRAMENTO (KSAC)

| YEAR | JAN | FEB | MAR | APR | MAY | JUN | JUL | AUG | SEP | OCT | NOV | DEC | TOTAL |
|------|-----|-----|-----|-----|-----|-----|-----|-----|-----|-----|-----|-----|-------|
| 1980 | 0 | 0 | 0 | 8 | 42 | 91 | 317 | 207 | 145 | 99 | 0 | 0 | 909 |
| 1981 | 0 | 0 | 0 | 26 | 78 | 303 | 318 | 301 | 155 | 28 | 7 | 0 | 1216 |
| 1982 | 0 | 0 | 0 | 2 | 67 | 83 | 230 | 213 | 133 | 9 | 0 | 0 | 737 |
| 1983 | 0 | 0 | 0 | 0 | 81 | 183 | 235 | 368 | 304 | 92 | 0 | 0 | 1263 |
| 1984 | 0 | 0 | 0 | 6 | 183 | 216 | 419 | 327 | 320 | 57 | 0 | 0 | 1528 |
| 1985 | 0 | 0 | 0 | 22 | 41 | 319 | 380 | 254 | 128 | 48 | 0 | 0 | 1192 |
| 1986 | 0 | 0 | 10 | 9 | 95 | 207 | 315 | 321 | 95 | 47 | 0 | 0 | 1099 |
| 1987 | 0 | 0 | 0 | 34 | 171 | 234 | 220 | 314 | 212 | 100 | 0 | 0 | 1285 |
| 1988 | 0 | 0 | 5 | 22 | 88 | 269 | 484 | 346 | 233 | 92 | 0 | 0 | 1539 |
| 1989 | 0 | 0 | 1 | 60 | 83 | 211 | 354 | 280 | 158 | 32 | 0 | 0 | 1179 |
| 1990 | 0 | 0 | 0 | 33 | 75 | 236 | 399 | 367 | 276 | 82 | 0 | 0 | 1468 |
| 1991 | 0 | 0 | 0 | 3 | 54 | 171 | 379 | 261 | 300 | 208 | 0 | 0 | 1376 |
| 1992 | 0 | 0 | 0 | 23 | 180 | 193 | 330 | 381 | 231 | 81 | 0 | 0 | 1419 |
| 1993 | 0 | 0 | 1 | 9 | 49 | 227 | 294 | 291 | 207 | 38 | 0 | 0 | 1116 |
| 1994 | 0 | 0 | 0 | 9 | 67 | 205 | 285 | 320 | 209 | | 0 | 0 | |
| 1995 | 0 | 0 | 0 | 0 | 54 | 152 | 294 | 322 | 228 | | 0 | 0 | |
| 1996 | 0 | 0 | 0 | 42 | 91 | 258 | 430 | 422 | 187 | 96 | 0 | 0 | 1526 |
| 1997 | 0 | 0 | 1 | 31 | 227 | 244 | 362 | 348 | 312 | 33 | 11 | 0 | 1569 |
| 1998 | 0 | 0 | 0 | 12 | 2 | 78 | 311 | 371 | 241 | 8 | 0 | 0 | 1023 |
| 1999 | 0 | 0 | 0 | 17 | 30 | 184 | 225 | 253 | 225 | 69 | 0 | 0 | 1003 |
| 2000 | 0 | 0 | 2 | 15 | 111 | 255 | 227 | 290 | 198 | 32 | 0 | 0 | 1130 |
| 2001 | 0 | 0 | 4 | 9 | 219 | 259 | 266 | 300 | 190 | 75 | 0 | 0 | 1322 |
| 2002 | 0 | 0 | 1 | 4 | 57 | 220 | 341 | 270 | 245 | 48 | 0 | 0 | 1186 |
| 2003 | 0 | 0 | 0 | 0 | 97 | 237 | 451 | 323 | 267 | 86 | 0 | 0 | 1461 |
| 2004 | 0 | 0 | 8 | 47 | 82 | 226 | 320 | 337 | 234 | 61 | 0 | 0 | 1315 |
| 2005 | 0 | 0 | 1 | 0 | 71 | 155 | 435 | 359 | 127 | 29 | 0 | 0 | 1177 |
| 2006 | 0 | 0 | 0 | 11 | 100 | 284 | 445 | 253 | 167 | 7 | 0 | 0 | 1267 |
| 2007 | 0 | 0 | 2 | 20 | 91 | 213 | 324 | 329 | 144 | 8 | 0 | 0 | 1131 |
| 2008 | 0 | 0 | 0 | 21 | 107 | 230 | 322 | 350 | 225 | 33 | 0 | 0 | 1288 |
| 2009 | 0 | 0 | 0 | 36 | 145 | 193 | 329 | 316 | 303 | 22 | 0 | 0 | 1344 |

## SNOWFALL (inches) 2009 SACRAMENTO (KSAC)

| YEAR | JUL | AUG | SEP | OCT | NOV | DEC | JAN | FEB | MAR | APR | MAY | JUN | TOTAL |
|------|-----|-----|-----|-----|-----|-----|-----|-----|-----|-----|-----|-----|-------|
| 1976-77 | 0.0 | 0.0 | 0.0 | 0.0 | 0.0 | 0.0 | 0.0 | 0.0 | 0.0 | 0.0 | 0.0 | 0.0 | 0.0 |
| 1977-78 | 0.0 | 0.0 | 0.0 | 0.0 | 0.0 | 0.0 | 0.0 | 0.0 | 0.0 | 0.0 | 0.0 | 0.0 | 0.0 |
| 1978-79 | 0.0 | 0.0 | 0.0 | 0.0 | 0.0 | 0.0 | 0.0 | 0.0 | 0.0 | 0.0 | 0.0 | 0.0 | 0.0 |
| 1979-80 | 0.0 | 0.0 | 0.0 | 0.0 | 0.0 | 0.0 | 0.0 | 0.0 | 0.0 | 0.0 | 0.0 | 0.0 | 0.0 |
| 1980-81 | 0.0 | 0.0 | 0.0 | 0.0 | 0.0 | 0.0 | 0.0 | 0.0 | 0.0 | 0.0 | 0.0 | 0.0 | 0.0 |
| 1981-82 | 0.0 | 0.0 | 0.0 | 0.0 | 0.0 | 0.0 | 0.0 | 0.0 | T | 0.0 | 0.0 | 0.0 | T |
| 1982-83 | 0.0 | 0.0 | 0.0 | 0.0 | 0.0 | 0.0 | 0.0 | 0.0 | 0.0 | 0.0 | 0.0 | 0.0 | 0.0 |
| 1983-84 | 0.0 | 0.0 | 0.0 | 0.0 | 0.0 | 0.0 | 0.0 | 0.0 | 0.0 | 0.0 | 0.0 | 0.0 | 0.0 |
| 1984-85 | 0.0 | 0.0 | 0.0 | 0.0 | 0.0 | 0.0 | 0.0 | 0.0 | 0.0 | 0.0 | 0.0 | 0.0 | 0.0 |
| 1985-86 | 0.0 | 0.0 | 0.0 | 0.0 | 0.0 | 0.0 | 0.0 | 0.0 | 0.0 | 0.0 | 0.0 | 0.0 | 0.0 |
| 1986-87 | 0.0 | 0.0 | 0.0 | 0.0 | 0.0 | 0.0 | 0.0 | 0.0 | 0.0 | 0.0 | 0.0 | 0.0 | 0.0 |
| 1987-88 | 0.0 | 0.0 | 0.0 | 0.0 | 0.0 | 0.0 | 0.0 | 0.0 | 0.0 | 0.0 | 0.0 | 0.0 | 0.0 |
| 1988-89 | 0.0 | 0.0 | 0.0 | 0.0 | 0.0 | T | 0.0 | 0.0 | 0.0 | 0.0 | 0.0 | 0.0 | T |
| 1989-90 | 0.0 | 0.0 | 0.0 | 0.0 | 0.0 | 0.0 | 0.0 | 0.0 | 0.0 | 0.0 | 0.0 | 0.0 | 0.0 |
| 1990-91 | 0.0 | 0.0 | 0.0 | 0.0 | 0.0 | 0.0 | 0.0 | 0.0 | 0.0 | 0.0 | 0.0 | 0.0 | 0.0 |
| 1991-92 | 0.0 | 0.0 | 0.0 | 0.0 | 0.0 | 0.0 | 0.0 | 0.0 | 0.0 | 0.0 | 0.0 | 0.0 | 0.0 |
| 1992-93 | 0.0 | 0.0 | 0.0 | 0.0 | 0.0 | T | 0.0 | 0.0 | 0.0 | 0.0 | 0.0 | 0.0 | T |
| 1993-94 | 0.0 | 0.0 | 0.0 | 0.0 | 0.0 | T | 0.0 | T | 0.0 | 0.0 | T | 0.0 | T |
| 1994-95 | 0.0 | 0.0 | | 0.0 | | 0.0 | 0.0 | 0.0 | 0.0 | 0.0 | 0.0 | 0.0 | |
| 1995-96 | 0.0 | 0.0 | 0.0 | 0.0 | 0.0 | T | 0.0 | T | 0.0 | 0.0 | 0.0 | 0.0 | T |
| 1996-97 | 0.0 | 0.0 | 0.0 | 0.0 | 0.0 | 0.0 | 0.0 | 0.0 | 0.0 | 0.0 | 0.0 | 0.0 | 0.0 |
| 1997-98 | 0.0 | 0.0 | 0.0 | 0.0 | 0.0 | 0.0 | 0.0 | 0.0 | 0.0 | | | | |
| 1998-99 | | | | | | | | | | | | | |
| 1999-00 | | | | | | | | | | | | | |
| 2000-01 | | | | | | | | | | | | | |
| 2001-02 | | | | | | | | | | | | | |
| 2002-03 | | | | | | | | | | | | | |
| 2003-04 | | | | | | | | | | | | | |
| 2004-05 | | | | | | | | | | | | | |
| 2005- | | | | | | | | | | | | | |
| POR= 50 YRS | 0.0 | 0.0 | 0.0 | 0.0 | 0.0 | T | 0.0 | T | T | 0.0 | T | 0.0 | T |

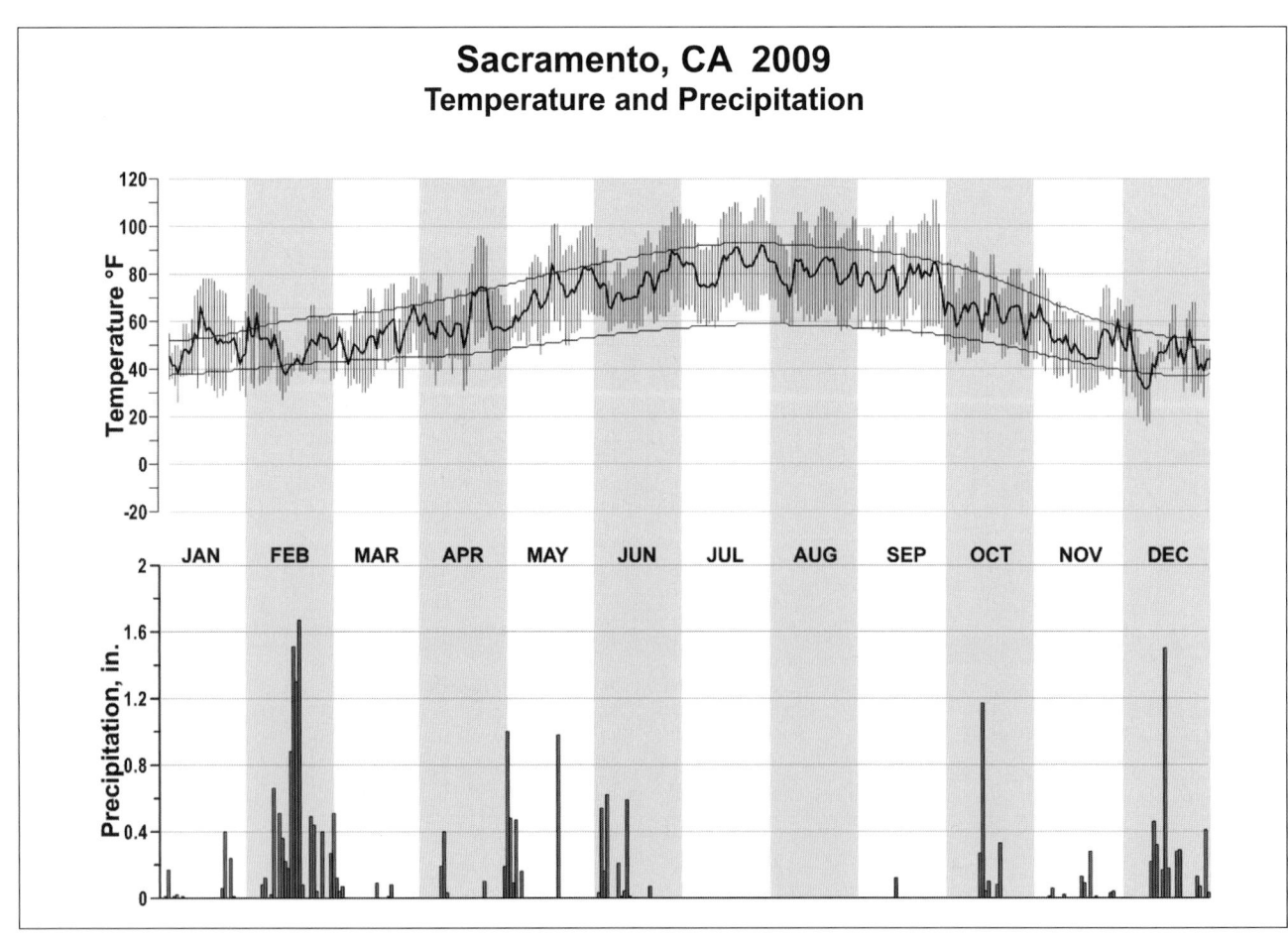

# Sacramento, CA  2009
## Temperature and Precipitation

# 2009
# SAN DIEGO
# CALIFORNIA (KSAN)

The city of San Diego is located on San Diego Bay in the southwest corner of southern California. The prevailing winds and weather are tempered by the Pacific Ocean, with the result that summers are cool and winters warm in comparison with other places along the same general latitude. Temperatures of freezing or below have rarely occurred at the station since the record began in 1871, but hot weather, 90 degrees or above, is more frequent.

Dry easterly winds sometimes blow in the vicinity for several days at a time, bringing temperatures in the 90s and at times even in the 100s in the eastern sections of the city and outlying suburbs. At the National Weather Service station itself, however, there have been relatively few days on which 100 degrees or higher was reached.

As these hot winds are predominant in the fall, highest temperatures occur in the months of September and October. Records show that over 60 percent of the days with 90 degrees or higher have occurred in these two months. High temperatures are almost invariably accompanied by very low relative humidities, which often drop below 20 percent and occasionally below 10 percent.

A marked feature of the climate is the wide variation in temperature within short distances. In nearby valleys daytimes are much warmer in summer and nights noticeably cooler in winter, and freezing occurs much more frequently than in the city. Although records show unusually small daily temperature ranges, only about 15 degrees between the highest and lowest readings, a few miles inland these ranges increase to 30 degrees or more.

Strong winds and gales associated with Pacific, or tropical storms, are infrequent due to the latitude.

The seasonal rainfall is about 10 inches in the city, but increases with elevation and distance from the coast. In the mountains to the north and east the average is between 20 and 40 inches, depending on slope and elevation. Most of the precipitation falls in winter, except in the mountains where there is an occasional thunderstorm. Eighty-five percent of the rainfall occurs from November through March, but wide variations take place in monthly and seasonal totals. Infrequent measurable amounts of hail occur in San Diego, but snow is practically unknown at the Weather Service Office location. In each occurrence of snowfall only a trace was recorded officially, but in some locations amounts up to or slightly exceeding a half-inch fell, and remained on the ground for an hour or more.

As on the rest of the Pacific Coast, a dominant characteristic of spring and summer is the nighttime and early morning cloudiness. Low clouds form regularly and frequently extend inland over the coastal valleys and foothills, but they usually dissipate during the morning and the afternoons are generally clear.

Considerable fog occurs along the coast, but the amount decreases with distance inland. The fall and winter months are usually the foggiest. Thunderstorms are rare, averaging about three a year in the city. Visibilities are good as a rule. The sunshine is plentiful for a marine location, with a marked increase toward the interior.

# NORMALS, MEANS, AND EXTREMES
## SAN DIEGO (KSAN)

| LATITUDE:<br>32 ° 44'N | LONGITUDE:<br>-117° 10'W | | ELEVATION (FT):<br>GRND: 13 BARO: 81 | | | | TIME ZONE:<br>PACIFIC (UTC -8) | | | | WBAN: 23188 | |

| | ELEMENT | POR | JAN | FEB | MAR | APR | MAY | JUN | JUL | AUG | SEP | OCT | NOV | DEC | YEAR |
|---|---|---|---|---|---|---|---|---|---|---|---|---|---|---|---|
| **TEMPERATURE °F** | NORMAL DAILY MAXIMUM | 30 | 65.8 | 66.3 | 66.3 | 68.7 | 69.3 | 72.2 | 75.8 | 77.5 | 77.0 | 74.0 | 69.9 | 66.3 | 70.8 |
| | MEAN DAILY MAXIMUM | 83 | 65.0 | 64.0 | 66.0 | 67.0 | 68.6 | 70.5 | 75.2 | 76.6 | 75.4 | 73.4 | 69.5 | 66.1 | 69.8 |
| | HIGHEST DAILY MAXIMUM | 69 | 88 | 90 | 93 | 98 | 96 | 101 | 99 | 98 | 111 | 107 | 97 | 88 | 111 |
| | YEAR OF OCCURRENCE | | 1953 | 1995 | 1988 | 2009 | 1953 | 1979 | 2006 | 1955 | 1963 | 1961 | 1976 | 1963 | SEP 1963 |
| | MEAN OF EXTREME MAXS. | 83 | 78.4 | 78.3 | 79.2 | 81.3 | 79.3 | 81.1 | 83.6 | 85.5 | 90.1 | 88.0 | 83.1 | 77.8 | 82.1 |
| | NORMAL DAILY MINIMUM | 30 | 49.7 | 51.5 | 53.6 | 56.4 | 59.8 | 62.6 | 65.9 | 67.4 | 66.1 | 61.2 | 53.6 | 48.9 | 58.1 |
| | MEAN DAILY MINIMUM | 83 | 48.2 | 48.7 | 52.0 | 54.4 | 58.0 | 60.5 | 64.6 | 65.9 | 63.6 | 59.6 | 52.6 | 48.7 | 56.4 |
| | LOWEST DAILY MINIMUM | 69 | 29 | 36 | 39 | 41 | 48 | 51 | 55 | 57 | 51 | 43 | 38 | 34 | 29 |
| | YEAR OF OCCURRENCE | | 1949 | 1949 | 1971 | 1945 | 1967 | 1967 | 1948 | 1944 | 1948 | 1971 | 1964 | 1987 | JAN 1949 |
| | MEAN OF EXTREME MINS. | 83 | 40.6 | 43.5 | 45.8 | 49.4 | 53.5 | 57.7 | 61.5 | 62.7 | 59.7 | 53.0 | 45.4 | 40.9 | 51.1 |
| | NORMAL DRY BULB | 30 | 57.8 | 58.9 | 60.0 | 62.6 | 64.6 | 67.4 | 70.9 | 72.5 | 71.6 | 67.6 | 61.8 | 57.6 | 64.4 |
| | MEAN DRY BULB | 83 | 56.6 | 56.4 | 59.0 | 60.7 | 63.3 | 65.6 | 69.9 | 71.3 | 69.5 | 66.5 | 61.1 | 57.4 | 63.1 |
| | MEAN WET BULB | 26 | 49.6 | 51.0 | 52.7 | 54.7 | 57.8 | 60.4 | 64.0 | 65.2 | 64.0 | 59.4 | 53.8 | 49.1 | 56.8 |
| | MEAN DEW POINT | 26 | 45.7 | 47.2 | 49.5 | 51.7 | 55.2 | 58.1 | 62.0 | 63.4 | 61.9 | 57.0 | 49.6 | 44.4 | 53.8 |
| | NORMAL NO. DAYS WITH: | | | | | | | | | | | | | | |
| | MAXIMUM >= 90 | 30 | 0.0 | * | 0.1 | 0.2 | * | 0.5 | 0.2 | 0.3 | 1.3 | 0.7 | 0.2 | 0.0 | 3.5 |
| | MAXIMUM <= 32 | 30 | 0.0 | 0.0 | 0.0 | 0.0 | 0.0 | 0.0 | 0.0 | 0.0 | 0.0 | 0.0 | 0.0 | 0.0 | 0.0 |
| | MINIMUM <= 32 | 30 | 0.0 | 0.0 | 0.0 | 0.0 | 0.0 | 0.0 | 0.0 | 0.0 | 0.0 | 0.0 | 0.0 | 0.0 | 0.0 |
| | MINIMUM <= 0 | 30 | 0.0 | 0.0 | 0.0 | 0.0 | 0.0 | 0.0 | 0.0 | 0.0 | 0.0 | 0.0 | 0.0 | 0.0 | 0.0 |
| **H/C** | NORMAL HEATING DEG. DAYS | 30 | 227 | 176 | 160 | 90 | 47 | 10 | 0 | 0 | 1 | 12 | 109 | 231 | 1063 |
| | NORMAL COOLING DEG. DAYS | 30 | 2 | 4 | 5 | 17 | 32 | 81 | 183 | 230 | 199 | 97 | 15 | 1 | 866 |
| **RH** | NORMAL (PERCENT) | 30 | 67 | 69 | 70 | 69 | 73 | 75 | 76 | 76 | 75 | 72 | 67 | 65 | 71 |
| | HOUR 04 LST | 30 | 73 | 76 | 77 | 77 | 79 | 82 | 83 | 83 | 82 | 78 | 73 | 71 | 78 |
| | HOUR 10 LST | 30 | 58 | 61 | 62 | 61 | 66 | 69 | 70 | 70 | 67 | 63 | 55 | 54 | 63 |
| | HOUR 16 LST | 30 | 59 | 61 | 62 | 60 | 65 | 67 | 67 | 67 | 67 | 66 | 61 | 58 | 63 |
| | HOUR 22 LST | 30 | 73 | 74 | 74 | 74 | 77 | 79 | 81 | 81 | 79 | 77 | 73 | 71 | 76 |
| **S** | PERCENT POSSIBLE SUNSHINE | 56 | 72 | 71 | 70 | 68 | 59 | 58 | 68 | 70 | 69 | 68 | 75 | 73 | 68 |
| **W/O** | MEAN NO. DAYS WITH: | | | | | | | | | | | | | | |
| | HEAVY FOG(VISBY <= 1/4 MI) | 46 | 2.2 | 2.0 | 1.0 | 0.5 | 0.3 | 0.4 | 0.2 | 0.3 | 1.3 | 2.8 | 2.7 | 3.0 | 16.7 |
| | THUNDERSTORMS | 62 | 0.2 | 0.3 | 0.4 | 0.2 | 0.1 | 0.1 | 0.2 | 0.3 | 0.3 | 0.3 | 0.3 | 0.2 | 2.9 |
| **CLOUDNESS** | MEAN: | | | | | | | | | | | | | | |
| | SUNRISE-SUNSET (OKTAS) | 56 | 4.1 | 4.2 | 4.2 | 4.2 | 4.6 | 4.4 | 3.6 | 3.2 | 3.2 | 3.5 | 3.3 | 3.7 | 3.9 |
| | MIDNIGHT-MIDNIGHT (OKTAS) | 32 | 3.9 | 4.4 | 4.5 | 4.1 | 5.1 | 5.0 | 4.3 | 4.0 | 4.0 | 3.8 | 3.4 | 3.5 | 4.2 |
| | MEAN NO. DAYS WITH: | | | | | | | | | | | | | | |
| | CLEAR | 56 | 12.3 | 10.1 | 10.8 | 10.3 | 8.5 | 9.3 | 12.7 | 15.1 | 15.0 | 13.7 | 14.7 | 13.6 | 146.1 |
| | PARTLY CLOUDY | 56 | 7.6 | 7.6 | 9.5 | 10.0 | 11.3 | 11.8 | 12.8 | 11.5 | 9.5 | 9.7 | 8.0 | 7.7 | 117.0 |
| | CLOUDY | 56 | 11.1 | 10.5 | 10.7 | 9.7 | 11.2 | 9.0 | 5.0 | 4.4 | 5.5 | 7.6 | 7.3 | 9.6 | 101.6 |
| **PR** | MEAN STATION PRESSURE(IN) | 26 | 30.06 | 30.03 | 30.00 | 29.97 | 29.93 | 29.89 | 29.90 | 29.89 | 29.87 | 29.93 | 30.01 | 30.05 | 29.96 |
| | MEAN SEA-LEVEL PRES. (IN) | 26 | 30.09 | 30.06 | 30.03 | 30.00 | 29.96 | 29.92 | 29.93 | 29.92 | 29.90 | 29.96 | 30.04 | 30.08 | 29.99 |
| **WINDS** | MEAN SPEED (MPH) | 26 | 5.6 | 6.5 | 7.1 | 7.6 | 7.6 | 7.6 | 7.4 | 7.2 | 6.9 | 6.2 | 5.4 | 5.2 | 6.7 |
| | PREVAIL.DIR(TENS OF DEGS) | 31 | 31 | 31 | 30 | 30 | 31 | 30 | 30 | 30 | 31 | 31 | 31 | 31 | 30 |
| | MAXIMUM 2-MINUTE: | 13 | | | | | | | | | | | | | |
| | SPEED (MPH) | | 31 | 38 | 38 | 32 | 26 | 23 | 21 | 20 | 22 | 30 | 35 | 46 | 46 |
| | DIR. (TENS OF DEGS) | | 16 | 17 | 25 | 26 | 17 | 25 | 30 | 30 | 29 | 27 | 33 | 18 | 18 |
| | YEAR OF OCCURRENCE | | 2008 | 1998 | 2000 | 1999 | 1998 | 2000 | 2006 | 2007 | 2003 | 2004 | 2004 | 2004 | DEC 2004 |
| | MAXIMUM 3-SECOND | | | | | | | | | | | | | | |
| | SPEED (MPH) | 13 | 44 | 46 | 45 | 37 | 32 | 25 | 23 | 26 | 25 | 39 | 45 | 58 | 58 |
| | DIR. (TENS OF DEGS) | | 16 | 14 | 25 | 25 | 18 | 30 | 30 | 30 | 18 | 20 | 31 | 17 | 17 |
| | YEAR OF OCCURRENCE | | 2008 | 2001 | 2000 | 1999 | 1998 | 2009 | 2009 | 2007 | 2007 | 2004 | 2004 | 2004 | DEC 2004 |
| **PRECIPITATION** | NORMAL (IN) | 30 | 2.28 | 2.04 | 2.26 | 0.75 | 0.20 | 0.09 | 0.03 | 0.09 | 0.21 | 0.44 | 1.07 | 1.31 | 10.77 |
| | MAXIMUM MONTHLY (IN) | 69 | 9.09 | 7.65 | 6.96 | 3.71 | 1.79 | 0.87 | 0.24 | 2.13 | 1.90 | 4.98 | 5.82 | 7.60 | 9.09 |
| | YEAR OF OCCURRENCE | | 1993 | 1998 | 1991 | 1988 | 1977 | 1990 | 1991 | 1977 | 1963 | 2004 | 1965 | 1943 | JAN 1993 |
| | MINIMUM MONTHLY (IN) | 69 | T | 0.00 | 0.00 | 0.00 | 0.00 | 0.00 | 0.00 | 0.00 | 0.00 | 0.00 | 0.00 | 0.01 | 0.00 |
| | YEAR OF OCCURRENCE | | 1976 | 1967 | 1997 | 1993 | 1952 | 1981 | 1982 | 1981 | 1979 | 1967 | 1980 | 2000 | MAR 1997 |
| | MAXIMUM IN 24 HOURS (IN) | 69 | 2.65 | 2.61 | 2.40 | 1.98 | 1.50 | 0.82 | 0.23 | 2.13 | 1.00 | 2.71 | 2.44 | 3.07 | 3.07 |
| | YEAR OF OCCURRENCE | | 1978 | 1979 | 1952 | 1988 | 1977 | 1990 | 1991 | 2013 | 1986 | 2004 | 1944 | 1945 | DEC 1945 |
| | NORMAL NO. DAYS WITH: | | | | | | | | | | | | | | |
| | PRECIPITATION >= 0.01 | 30 | 7.2 | 6.6 | 7.2 | 4.1 | 2.0 | 1.1 | 0.6 | 0.6 | 1.5 | 2.8 | 4.0 | 5.2 | 42.9 |
| | PRECIPITATION >= 1.00 | 30 | 0.5 | 0.6 | 0.4 | 0.1 | * | 0.0 | 0.0 | * | 0.0 | 0.0 | 0.2 | 0.2 | 2.0 |
| **SNOWFALL** | NORMAL (IN) | 30 | 0.0 | 0.0 | 0.0 | 0.0 | 0.0 | 0.0 | 0.0 | 0.0 | 0.0 | 0.0 | 0.0 | 0.0 | 0.0 |
| | MAXIMUM MONTHLY (IN) | 59 | T | 0.0 | T | T | 0.0 | 0.0 | 0.0 | 0.0 | 0.0 | 0.0 | T | T | T |
| | YEAR OF OCCURRENCE | | 1949 | | 1985 | 1999 | | | | | | | 1985 | 1967 | APR 1999 |
| | MAXIMUM IN 24 HOURS (IN) | 59 | T | 0.0 | T | T | 0.0 | 0.0 | 0.0 | 0.0 | 0.0 | 0.0 | T | T | T |
| | YEAR OF OCCURRENCE' | | 1949 | | 1985 | 1999 | | | | | | | 1985 | 1967 | APR 1999 |
| | MAXIMUM SNOW DEPTH (IN) | 51 | 0 | 0 | 0 | 0 | 0 | 0 | 0 | 0 | 0 | 0 | 0 | 0 | 0 |
| | YEAR OF OCCURRENCE | | | | | | | | | | | | | | |
| | NORMAL NO. DAYS WITH: | | | | | | | | | | | | | | |
| | SNOWFALL >= 1.0 | 30 | 0.0 | 0.0 | 0.0 | 0.0 | 0.0 | 0.0 | 0.0 | 0.0 | 0.0 | 0.0 | 0.0 | 0.0 | 0.0 |

## PRECIPITATION (inches) 2009 SAN DIEGO (KSAN)

| YEAR | JAN | FEB | MAR | APR | MAY | JUN | JUL | AUG | SEP | OCT | NOV | DEC | ANNUAL |
|---|---|---|---|---|---|---|---|---|---|---|---|---|---|
| 1980 | 5.58 | 4.47 | 2.71 | 1.18 | 0.65 | 0.01 | T | 0.00 | T | 0.05 | 0.00 | 0.31 | 14.96 |
| 1981 | 1.48 | 2.26 | 3.74 | 0.22 | 0.04 | 0.00 | T | 0.00 | 0.03 | 0.14 | 1.79 | 0.54 | 10.24 |
| 1982 | 2.71 | 0.88 | 4.74 | 0.62 | 0.01 | 0.04 | 0.00 | T | 0.38 | 0.05 | 2.10 | 1.43 | 12.96 |
| 1983 | 2.10 | 3.88 | 6.57 | 1.74 | 0.01 | T | 0.01 | 0.39 | 0.21 | 0.40 | 1.94 | 1.53 | 18.78 |
| 1984 | 0.46 | 0.09 | 0.04 | 0.62 | 0.00 | 0.04 | 0.19 | 0.06 | T | 0.29 | 2.37 | 4.55 | 8.71 |
| 1985 | 0.52 | 0.77 | 0.58 | 0.32 | T | T | 0.00 | T | 0.20 | 0.29 | 4.92 | 1.06 | 8.66 |
| 1986 | 0.75 | 2.59 | 3.12 | 1.17 | 0.00 | T | 0.01 | 0.00 | 1.04 | 1.39 | 1.16 | 0.95 | 12.18 |
| 1987 | 1.68 | 1.53 | 1.04 | 0.78 | 0.03 | T | 0.03 | 0.01 | 0.70 | 1.74 | 1.33 | 2.73 | 11.60 |
| 1988 | 0.89 | 1.37 | 0.59 | 3.71 | 0.08 | 0.00 | T | T | T | T | 1.39 | 2.23 | 10.26 |
| 1989 | 0.42 | 0.70 | 0.69 | 0.12 | 0.04 | 0.06 | 0.00 | T | 0.23 | 0.47 | 0.09 | 1.01 | 3.83 |
| 1990 | 2.52 | 1.13 | 0.25 | 0.76 | 0.51 | 0.87 | T | 0.01 | T | T | 0.65 | 0.59 | 7.29 |
| 1991 | 1.06 | 2.46 | 6.96 | 0.05 | 0.01 | T | 0.24 | 0.01 | 0.28 | 0.69 | 0.05 | 1.70 | 13.51 |
| 1992 | 1.81 | 3.34 | 4.42 | 0.28 | 0.07 | 0.04 | 0.03 | 0.05 | 0.00 | 0.18 | 0.03 | 2.56 | 12.81 |
| 1993 | 9.09 | 4.73 | 1.22 | 0.00 | 0.01 | 0.41 | 0.03 | T | T | 0.22 | 0.77 | 0.78 | 17.26 |
| 1994 | 0.70 | 2.75 | 3.67 | 0.93 | 0.07 | T | 0.03 | 0.01 | T | 0.01 | 0.46 | 0.80 | 9.43 |
| 1995 | 8.06 | 1.93 | 3.81 | 0.96 | 0.59 | 0.46 | 0.05 | 0.00 | T | T | 0.30 | 0.88 | 17.04 |
| 1996 | 1.52 | 0.88 | 1.10 | 0.36 | 0.02 | 0.00 | .09 | T | .03 | .94 | 1.70 | .63 | 7.27 |
| 1997 | 3.02 | 0.31 | 0.00 | 0.28 | T | T | T | 0.00 | 0.85 | 0.02 | 1.17 | 1.35 | 7.00 |
| 1998 | 2.68 | 7.65 | 2.21 | 1.11 | 0.64 | 0.10 | 0.20 | T | 0.03 | 0.08 | 0.69 | 0.66 | 16.05 |
| 1999 | 1.54 | 0.70 | 1.09 | 1.62 | 0.06 | 0.04 | T | 0.00 | 0.02 | 0.00 | 0.04 | 0.32 | 5.43 |
| 2000 | 0.17 | 3.67 | 1.00 | 0.54 | T | T | 0.00 | 0.01 | T | 1.24 | 0.26 | 0.01 | 6.90 |
| 2001 | 3.28 | 2.38 | 0.63 | 0.76 | 0.01 | 0.00 | T | 0.00 | 0.00 | 0.00 | 0.95 | 0.46 | 8.47 |
| 2002 | 0.32 | 0.17 | 0.46 | 0.63 | T | T | 0.00 | T | 0.31 | 0.04 | 0.32 | 1.98 | 4.23 |
| 2003 | 0.02 | 4.88 | 1.36 | 1.41 | 0.30 | T | T | 0.00 | T | T | 0.60 | 0.61 | 9.18 |
| 2004 | 0.34 | 2.81 | 0.22 | 0.60 | T | 0.00 | 0.00 | 0.00 | T | 4.98 | 0.33 | 4.01 | 13.29 |
| 2005 | 4.49 | 5.83 | 2.12 | 0.59 | 0.12 | 0.02 | 0.01 | T | 0.10 | 0.46 | 0.12 | 0.25 | 14.11 |
| 2006 | 0.36 | 1.11 | 1.36 | 0.88 | 0.77 | T | 0.04 | 0.01 | 0.00 | 0.76 | 0.15 | 0.71 | 6.15 |
| 2007 | 0.51 | 1.12 | 0.09 | 0.46 | T | 0.00 | 0.00 | 0.00 | 0.05 | 0.37 | 0.97 | 0.80 | 4.37 |
| 2008 | 3.34 | 1.21 | 0.26 | T | 0.23 | 0.02 | 0.00 | T | T | 0.18 | 2.49 | 3.38 | 11.11 |
| 2009 | 0.08 | 2.63 | 0.18 | 0.14 | 0.04 | 0.03 | 0.00 | T | T | T | 0.12 | 2.28 | 5.50 |
| POR= 83 YRS | 1.96 | 2.05 | 1.68 | 0.74 | 0.19 | 0.06 | 0.02 | 0.07 | 0.17 | 0.46 | 1.01 | 1.69 | 10.10 |

WBAN : 23188

## AVERAGE TEMPERATURE (°F) 2009 SAN DIEGO (KSAN)

| YEAR | JAN | FEB | MAR | APR | MAY | JUN | JUL | AUG | SEP | OCT | NOV | DEC | ANNUAL |
|---|---|---|---|---|---|---|---|---|---|---|---|---|---|
| 1980 | 61.1 | 63.5 | 61.5 | 63.9 | 63.8 | 68.5 | 72.9 | 74.2 | 70.4 | 67.3 | 62.7 | 60.8 | 65.9 |
| 1981 | 61.3 | 62.2 | 61.1 | 64.4 | 67.3 | 72.9 | 75.6 | 75.8 | 73.7 | 67.1 | 63.5 | 60.3 | 67.1 |
| 1982 | 56.6 | 60.7 | 60.5 | 63.8 | 65.8 | 66.7 | 71.9 | 73.5 | 73.1 | 70.1 | 62.1 | 57.4 | 65.2 |
| 1983 | 60.7 | 60.9 | 62.0 | 62.4 | 66.2 | 68.1 | 72.6 | 77.4 | 76.8 | 72.2 | 64.4 | 60.6 | 67.0 |
| 1984 | 61.2 | 60.2 | 63.7 | 64.3 | 68.1 | 69.9 | 77.2 | 76.6 | 78.9 | 68.5 | 61.4 | 56.7 | 67.2 |
| 1985 | 57.0 | 57.2 | 58.9 | 63.6 | 64.8 | 69.0 | 75.3 | 72.4 | 69.8 | 67.9 | 60.1 | 58.0 | 64.5 |
| 1986 | 61.0 | 58.9 | 60.5 | 62.8 | 64.6 | 67.4 | 69.6 | 71.8 | 66.9 | 65.5 | 62.8 | 57.6 | 64.1 |
| 1987 | 55.4 | 58.0 | 59.1 | 63.4 | 64.7 | 65.8 | 67.1 | 69.9 | 69.9 | 69.5 | 61.8 | 53.9 | 63.2 |
| 1988 | 56.7 | 59.9 | 61.6 | 62.4 | 63.9 | 64.9 | 70.4 | 71.0 | 70.0 | 66.7 | 60.1 | 56.0 | 63.6 |
| 1989 | 54.7 | 56.7 | 59.8 | 65.6 | 63.7 | 66.0 | 70.1 | 71.0 | 70.4 | 66.3 | 63.1 | 58.7 | 63.8 |
| 1990 | 56.6 | 55.2 | 58.7 | 63.2 | 64.3 | 69.0 | 72.3 | 71.6 | 71.7 | 68.6 | 62.7 | 55.6 | 64.1 |
| 1991 | 57.4 | 59.4 | 56.5 | 61.7 | 62.1 | 64.1 | 67.4 | 68.9 | 69.4 | 68.0 | 62.3 | 57.3 | 62.9 |
| 1992 | 57.4 | 61.1 | 60.4 | 67.0 | 68.0 | 68.1 | 71.8 | 74.9 | 72.4 | 68.2 | 62.6 | 55.3 | 65.6 |
| 1993 | 56.9 | 58.0 | 61.3 | 63.8 | 66.0 | 68.6 | 69.8 | 70.2 | 69.0 | 67.3 | 61.6 | 57.0 | 64.1 |
| 1994 | 57.9 | 56.5 | 60.4 | 61.0 | 62.1 | 68.1 | 69.5 | 74.0 | 72.5 | 66.8 | 56.4 | 55.8 | 63.4 |
| 1995 | 56.9 | 61.4 | 60.4 | 61.5 | 62.0 | 64.8 | 69.0 | 71.9 | 71.5 | 67.1 | 63.2 | 58.3 | 64.0 |
| 1996 | 57.6 | 58.8 | 60.1 | 64.6 | 66.8 | 67.8 | 70.0 | 72.8 | 70.6 | 64.3 | 61.6 | 57.8 | 64.4 |
| 1997 | 58.0 | 58.4 | 61.6 | 62.5 | 68.7 | 67.5 | 69.3 | 72.9 | 75.2 | 68.7 | 64.0 | 57.4 | 65.4 |
| 1998 | 58.2 | 57.3 | 59.3 | 59.6 | 62.8 | 65.7 | 68.8 | 72.8 | 70.3 | 65.7 | 59.7 | 55.4 | 63.0 |
| 1999 | 57.5 | 57.8 | 58.3 | 58.9 | 60.5 | 62.8 | 68.5 | 68.0 | 67.0 | 68.8 | 60.5 | 57.8 | 62.2 |
| 2000 | 58.2 | 59.0 | 58.1 | 62.4 | 64.6 | 68.2 | 69.2 | 72.0 | 70.5 | 65.1 | 58.4 | 58.1 | 63.7 |
| 2001 | 54.7 | 54.9 | 58.9 | 58.5 | 63.7 | 67.6 | 69.0 | 69.6 | 68.2 | 66.3 | 61.4 | 55.5 | 62.4 |
| 2002 | 55.7 | 57.5 | 57.7 | 59.6 | 61.9 | 64.7 | 67.8 | 68.9 | 70.1 | 64.2 | 63.4 | 56.8 | 62.4 |
| 2003 | 61.7 | 59.0 | 60.3 | 59.9 | 62.4 | 64.2 | 70.0 | 72.7 | 69.9 | 67.7 | 60.8 | 56.9 | 63.8 |
| 2004 | 56.9 | 56.7 | 62.6 | 64.4 | 68.5 | 67.8 | 72.1 | 70.8 | 72.6 | 65.4 | 60.1 | 57.8 | 64.6 |
| 2005 | 58.8 | 59.8 | 61.0 | 61.8 | 65.5 | 66.4 | 70.0 | 71.0 | 68.5 | 65.9 | 62.8 | 58.5 | 64.2 |
| 2006 | 57.0 | 58.3 | 57.2 | 61.9 | 64.7 | 70.8 | 76.3 | 73.1 | 70.2 | 65.8 | 63.4 | 56.5 | 64.6 |
| 2007 | 54.1 | 57.5 | 59.9 | 59.9 | 63.0 | 65.3 | 69.8 | 73.3 | 69.6 | 66.7 | 61.2 | 54.2 | 62.9 |
| 2008 | 54.4 | 55.6 | 58.9 | 61.8 | 63.4 | 66.9 | 69.6 | 72.2 | 71.2 | 69.0 | 64.6 | 56.6 | 63.7 |
| 2009 | 59.9 | 57.9 | 59.5 | 62.6 | 64.1 | 66.2 | 70.7 | 72.4 | 72.1 | 65.6 | 61.2 | 56.7 | 64.1 |
| POR= 83 YRS | 56.6 | 56.4 | 59.0 | 60.7 | 63.3 | 65.6 | 69.9 | 71.3 | 69.5 | 66.5 | 61.1 | 57.4 | 63.1 |

## HEATING DEGREE DAYS (base 65°F) 2009 SAN DIEGO (KSAN)

| YEAR | JUL | AUG | SEP | OCT | NOV | DEC | JAN | FEB | MAR | APR | MAY | JUN | TOTAL |
|------|-----|-----|-----|-----|-----|-----|-----|-----|-----|-----|-----|-----|-------|
| 1980-81 | 0 | 0 | 0 | 6 | 75 | 133 | 113 | 101 | 116 | 40 | 1 | 0 | 585 |
| 1981-82 | 0 | 0 | 0 | 9 | 57 | 136 | 258 | 119 | 139 | 64 | 9 | 2 | 793 |
| 1982-83 | 0 | 0 | 0 | 1 | 93 | 228 | 137 | 110 | 88 | 83 | 9 | 0 | 749 |
| 1983-84 | 0 | 0 | 0 | 0 | 66 | 130 | 123 | 134 | 51 | 43 | 4 | 0 | 551 |
| 1984-85 | 0 | 0 | 0 | 4 | 104 | 250 | 238 | 219 | 183 | 60 | 18 | 2 | 1078 |
| 1985-86 | 0 | 0 | 0 | 3 | 145 | 211 | 118 | 173 | 132 | 85 | 29 | 0 | 896 |
| 1986-87 | 0 | 0 | 7 | 10 | 66 | 223 | 291 | 197 | 178 | 72 | 21 | 6 | 1071 |
| 1987-88 | 0 | 0 | 0 | 0 | 98 | 338 | 250 | 147 | 125 | 85 | 53 | 22 | 1118 |
| 1988-89 | 0 | 0 | 0 | 4 | 141 | 275 | 313 | 237 | 158 | 37 | 40 | 14 | 1219 |
| 1989-90 | 0 | 0 | 1 | 13 | 67 | 188 | 252 | 268 | 185 | 52 | 39 | 1 | 1066 |
| 1990-91 | 0 | 0 | 0 | 3 | 88 | 284 | 227 | 152 | 254 | 104 | 96 | 28 | 1236 |
| 1991-92 | 0 | 0 | 0 | 24 | 96 | 231 | 228 | 115 | 136 | 8 | 1 | 0 | 839 |
| 1992-93 | 0 | 0 | 0 | 0 | 84 | 294 | 240 | 191 | 111 | 44 | 7 | 6 | 977 |
| 1993-94 | 0 | 0 | 0 | 4 | 103 | 242 | 214 | 235 | 147 | 116 | 85 | 1 | 1147 |
| 1994-95 | 0 | 0 | 0 | 12 | 249 | 276 | 242 | 106 | 136 | 105 | 87 | 24 | 1237 |
| 1995-96 | 0 | 0 | 0 | 3 | 49 | 200 | 223 | 171 | 146 | 54 | 6 | 0 | 852 |
| 1996-97 | 0 | 0 | 0 | 45 | 115 | 216 | 217 | 185 | 124 | 81 | 0 | 1 | 984 |
| 1997-98 | 0 | 0 | 0 | 2 | 65 | 227 | 203 | 208 | 172 | 155 | 69 | 9 | 1110 |
| 1998-99 | 1 | 0 | 0 | 13 | 152 | 293 | 224 | 198 | 200 | 189 | 132 | 64 | 1466 |
| 1999-00 | 4 | 0 | 3 | 3 | 130 | 217 | 212 | 171 | 206 | 82 | 23 | 0 | 1051 |
| 2000-01 | 0 | 0 | 0 | 27 | 191 | 208 | 314 | 281 | 183 | 191 | 39 | 5 | 1439 |
| 2001-02 | 0 | 0 | 0 | 4 | 107 | 289 | 283 | 208 | 219 | 154 | 101 | 18 | 1383 |
| 2002-03 | 0 | 0 | 1 | 36 | 63 | 248 | 117 | 163 | 151 | 150 | 82 | 33 | 1044 |
| 2003-04 | 0 | 0 | 0 | 5 | 122 | 243 | 247 | 236 | 99 | 43 | 0 | 0 | 995 |
| 2004-05 | 0 | 0 | 0 | 30 | 142 | 218 | 193 | 139 | 118 | 101 | 19 | 0 | 960 |
| 2005-06 | 0 | 0 | 0 | 21 | 79 | 197 | 240 | 183 | 233 | 86 | 19 | 0 | 1058 |
| 2006-07 | 0 | 0 | 0 | 11 | 68 | 255 | 332 | 208 | 159 | 147 | 76 | 25 | 1281 |
| 2007-08 | 0 | 0 | 0 | 22 | 122 | 326 | 321 | 265 | 186 | 142 | 79 | 19 | 1482 |
| 2008-09 | 0 | 0 | 0 | 12 | 55 | 255 | 161 | 198 | 170 | 102 | 38 | 1 | 992 |
| 2009- | 0 | 0 | 0 | 27 | 111 | 251 | | | | | | | |

WBAN : 23188

## COOLING DEGREE DAYS (base 65°F) 2009 SAN DIEGO (KSAN)

| YEAR | JAN | FEB | MAR | APR | MAY | JUN | JUL | AUG | SEP | OCT | NOV | DEC | TOTAL |
|------|-----|-----|-----|-----|-----|-----|-----|-----|-----|-----|-----|-----|-------|
| 1980 | 2 | 13 | 3 | 35 | 15 | 110 | 253 | 289 | 170 | 86 | 15 | 7 | 998 |
| 1981 | 7 | 29 | 0 | 26 | 81 | 244 | 335 | 343 | 265 | 80 | 21 | 0 | 1431 |
| 1982 | 0 | 7 | 6 | 32 | 42 | 58 | 219 | 271 | 250 | 164 | 12 | 0 | 1061 |
| 1983 | 11 | 0 | 1 | 9 | 51 | 99 | 242 | 392 | 364 | 231 | 55 | 0 | 1455 |
| 1984 | 13 | 0 | 15 | 31 | 107 | 156 | 387 | 366 | 422 | 119 | 4 | 0 | 1620 |
| 1985 | 0 | 7 | 0 | 22 | 19 | 128 | 325 | 235 | 153 | 104 | 6 | 0 | 999 |
| 1986 | 2 | 11 | 4 | 29 | 23 | 78 | 152 | 218 | 73 | 31 | 9 | 0 | 630 |
| 1987 | 0 | 6 | 5 | 29 | 17 | 35 | 71 | 158 | 154 | 147 | 10 | 0 | 632 |
| 1988 | 0 | 5 | 28 | 16 | 25 | 26 | 176 | 193 | 161 | 64 | 0 | 5 | 699 |
| 1989 | 0 | 13 | 2 | 63 | 5 | 48 | 165 | 193 | 168 | 58 | 17 | 0 | 732 |
| 1990 | 0 | 0 | 2 | 6 | 21 | 127 | 233 | 211 | 209 | 123 | 25 | 0 | 957 |
| 1991 | 0 | 0 | 0 | 9 | 11 | 10 | 80 | 130 | 138 | 122 | 21 | 0 | 521 |
| 1992 | 0 | 12 | 1 | 72 | 99 | 100 | 216 | 313 | 228 | 106 | 20 | 0 | 1167 |
| 1993 | 0 | 0 | 4 | 12 | 47 | 119 | 155 | 170 | 126 | 81 | 8 | 0 | 722 |
| 1994 | 0 | 0 | 10 | 3 | 1 | 100 | 146 | 289 | 231 | 76 | 0 | 0 | 856 |
| 1995 | 0 | 14 | 1 | 8 | 1 | 27 | 133 | 223 | 199 | 75 | 2 | 0 | 683 |
| 1996 | 0 | 0 | 4 | 53 | 70 | 92 | 164 | 251 | 175 | 30 | 19 | 0 | 858 |
| 1997 | 6 | 4 | 23 | 12 | 119 | 83 | 138 | 252 | 314 | 123 | 40 | 2 | 1116 |
| 1998 | 0 | 0 | 1 | 0 | 6 | 34 | 127 | 250 | 167 | 40 | 0 | 0 | 625 |
| 1999 | 0 | 0 | 0 | 14 | 0 | 5 | 118 | 100 | 69 | 126 | 3 | 0 | 435 |
| 2000 | 8 | 5 | 0 | 8 | 19 | 101 | 137 | 226 | 170 | 39 | 0 | 1 | 714 |
| 2001 | 0 | 3 | 2 | 5 | 4 | 90 | 132 | 149 | 102 | 53 | 4 | 0 | 544 |
| 2002 | 0 | 5 | 0 | 0 | 9 | 16 | 94 | 125 | 157 | 18 | 23 | 0 | 447 |
| 2003 | 20 | 0 | 11 | 3 | 8 | 18 | 161 | 246 | 153 | 98 | 2 | 0 | 720 |
| 2004 | 0 | 0 | 32 | 30 | 115 | 92 | 226 | 186 | 235 | 48 | 2 | 1 | 967 |
| 2005 | 8 | 0 | 3 | 12 | 40 | 47 | 159 | 196 | 112 | 52 | 20 | 3 | 652 |
| 2006 | 2 | 1 | 0 | 0 | 18 | 181 | 357 | 257 | 163 | 46 | 29 | 0 | 1054 |
| 2007 | 0 | 4 | 10 | 0 | 21 | 40 | 159 | 264 | 142 | 83 | 14 | 0 | 737 |
| 2008 | 0 | 0 | 5 | 49 | 39 | 84 | 150 | 229 | 190 | 144 | 48 | 0 | 938 |
| 2009 | 11 | 4 | 7 | 37 | 16 | 41 | 181 | 236 | 219 | 50 | 5 | 0 | 807 |

**SNOWFALL (inches)  2009  SAN DIEGO (KSAN)**

| YEAR | JUL | AUG | SEP | OCT | NOV | DEC | JAN | FEB | MAR | APR | MAY | JUN | TOTAL |
|---|---|---|---|---|---|---|---|---|---|---|---|---|---|
| 1976-77 | 0.0 | 0.0 | 0.0 | 0.0 | 0.0 | 0.0 | 0.0 | 0.0 | 0.0 | 0.0 | 0.0 | 0.0 | 0.0 |
| 1977-78 | 0.0 | 0.0 | 0.0 | 0.0 | 0.0 | 0.0 | 0.0 | 0.0 | 0.0 | 0.0 | 0.0 | 0.0 | 0.0 |
| 1978-79 | 0.0 | 0.0 | 0.0 | 0.0 | 0.0 | 0.0 | 0.0 | 0.0 | 0.0 | 0.0 | 0.0 | 0.0 | 0.0 |
| 1979-80 | 0.0 | 0.0 | 0.0 | 0.0 | 0.0 | 0.0 | 0.0 | 0.0 | 0.0 | 0.0 | 0.0 | 0.0 | 0.0 |
| 1980-81 | 0.0 | 0.0 | 0.0 | 0.0 | 0.0 | 0.0 | 0.0 | 0.0 | 0.0 | 0.0 | 0.0 | 0.0 | 0.0 |
| 1981-82 | 0.0 | 0.0 | 0.0 | 0.0 | 0.0 | 0.0 | 0.0 | 0.0 | 0.0 | 0.0 | 0.0 | 0.0 | 0.0 |
| 1982-83 | 0.0 | 0.0 | 0.0 | 0.0 | 0.0 | 0.0 | 0.0 | 0.0 | 0.0 | 0.0 | 0.0 | 0.0 | 0.0 |
| 1983-84 | 0.0 | 0.0 | 0.0 | 0.0 | 0.0 | 0.0 | 0.0 | 0.0 | 0.0 | 0.0 | 0.0 | 0.0 | 0.0 |
| 1984-85 | 0.0 | 0.0 | 0.0 | 0.0 | 0.0 | 0.0 | 0.0 | 0.0 | T | 0.0 | 0.0 | 0.0 | T |
| 1985-86 | 0.0 | 0.0 | 0.0 | 0.0 | T | 0.0 | 0.0 | 0.0 | 0.0 | 0.0 | 0.0 | 0.0 | T |
| 1986-87 | 0.0 | 0.0 | 0.0 | 0.0 | 0.0 | 0.0 | 0.0 | 0.0 | 0.0 | 0.0 | 0.0 | 0.0 | 0.0 |
| 1987-88 | 0.0 | 0.0 | 0.0 | 0.0 | 0.0 | 0.0 | 0.0 | 0.0 | 0.0 | 0.0 | 0.0 | 0.0 | 0.0 |
| 1988-89 | 0.0 | 0.0 | 0.0 | 0.0 | 0.0 | 0.0 | 0.0 | 0.0 | 0.0 | 0.0 | 0.0 | 0.0 | 0.0 |
| 1989-90 | 0.0 | 0.0 | 0.0 | 0.0 | 0.0 | 0.0 | 0.0 | 0.0 | 0.0 | 0.0 | 0.0 | 0.0 | 0.0 |
| 1990-91 | 0.0 | 0.0 | 0.0 | 0.0 | 0.0 | 0.0 | 0.0 | 0.0 | 0.0 | 0.0 | 0.0 | 0.0 | 0.0 |
| 1991-92 | 0.0 | 0.0 | 0.0 | 0.0 | 0.0 | 0.0 | 0.0 | 0.0 | 0.0 | 0.0 | 0.0 | 0.0 | 0.0 |
| 1992-93 | 0.0 | 0.0 | 0.0 | 0.0 | 0.0 | 0.0 | 0.0 | 0.0 | 0.0 | 0.0 | 0.0 | 0.0 | 0.0 |
| 1993-94 | 0.0 | 0.0 | 0.0 | 0.0 | 0.0 | 0.0 | 0.0 | 0.0 | 0.0 | 0.0 | 0.0 | 0.0 | 0.0 |
| 1994-95 | 0.0 | 0.0 | 0.0 | 0.0 | 0.0 | 0.0 | 0.0 | 0.0 | 0.0 | 0.0 | 0.0 | 0.0 | 0.0 |
| 1995-96 | 0.0 | 0.0 | 0.0 | 0.0 | 0.0 | 0.0 | 0.0 | 0.0 | 0.0 | 0.0 | 0.0 | 0.0 | 0.0 |
| 1996-97 | 0.0 | | | | | | | | | | | | |
| 1997-98 | | | | 0.0 | 0.0 | 0.0 | 0.0 | 0.0 | T | 0.0 | 0.0 | 0.0 | |
| 1998-99 | 0.0 | 0.0 | 0.0 | 0.0 | 0.0 | 0.0 | 0.0 | 0.0 | 0.0 | T | 0.0 | 0.0 | T |
| 1999-00 | 0.0 | 0.0 | 0.0 | 0.0 | 0.0 | 0.0 | 0.0 | 0.0 | 0.0 | 0.0 | 0.0 | | |
| 2000-01 | | | | | | | | | | | | | |
| 2001-02 | | | | | | | | | | | | | |
| 2002-03 | | | | | | | | | | | | | |
| 2003-04 | | | | | | | | | | | | | |
| 2004-05 | | | | | | | | | | | | | |
| 2005- | | | | | | | | | | | | | |
| POR= 72 YRS | 0.0 | 0.1 | 0.0 | 0.0 | T | 0.0 | 0.0 | 0.0 | T | T | 0.0 | 0.0 | 0.1 |

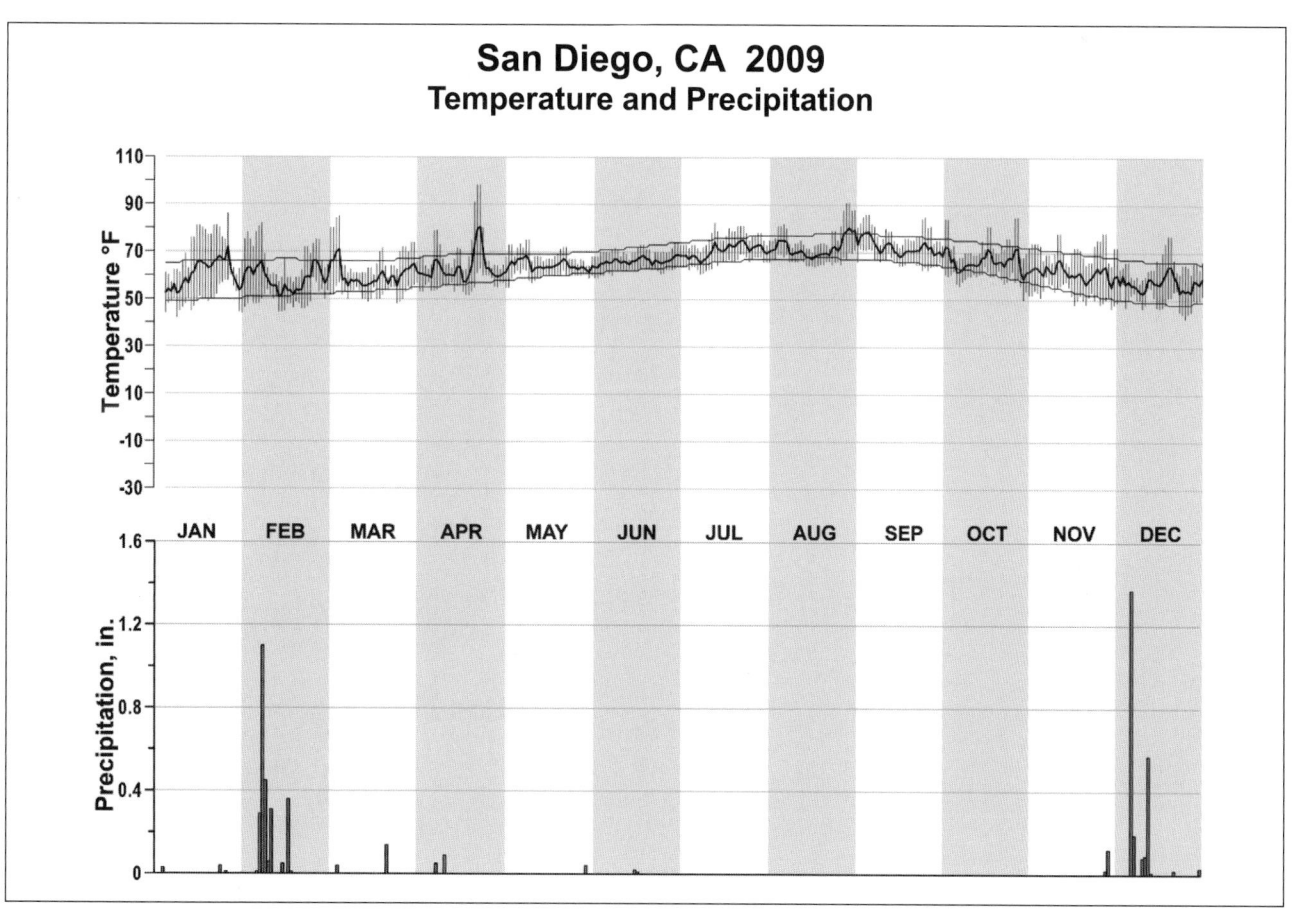

# San Diego, CA  2009
## Temperature and Precipitation

# 2009
# SAN FRANCISCO INTERNATIONAL AIRPORT
# CALIFORNIA (KSFO)

The station is located in the central Terminal Building of the San Francisco International Airport, which is on flat filled tideland on the west shore of San Francisco Bay. The bay borders the airport from the north to the south-southeast. San Bruno Mountain, 5 miles to the north-northwest, rises to 1,300 feet. A north-south trending ridge of coastal mountains, 4 miles to the west, varies in elevation from 700 to 1,900 feet, being highest southward along the peninsula. The Pacific Ocean west of the ridge is 6 miles from the airport. A broad gap to the northwest of the station, between San Bruno Mountain and the coastal mountains, allows a strong flow of marine air over the station and dominate the local climate.

San Francisco Airport enjoys a marine-type climate characterized by mild and moderately wet winters and by dry, cool summers. Winter rains, occurring from November through March, account for over 80 percent of the annual rainfall, and measurable precipitation occurs on an average of 10 days per month during this period. However, there are frequent dry periods lasting well over a week. Severe winter storms with gale winds and heavy rains occur only occasionally. Thunderstorms average two a year and may occur in any month.

The daily and annual range in temperature is small. A few frosty mornings occur during the winter but the temperature seldom drops below freezing. Winter temperatures generally rise to the high 50s in the early afternoon.

The summer weather is dominated by a cool sea breeze resulting in an average summer wind speed of nearly 15 mph. Winds are light in the early morning but normally reach 20 to 25 mph in the afternoon.

A sea fog, arriving over the station during the late evening or night as a low cloud, is another persistent feature of the summer weather. This high fog, occasionally producing drizzle or mist, usually disappears during the late forenoon. Despite the morning overcast, summer days are sunny. On the average a total of only 14 days during the four months from June through September are classified as cloudy.

Daytime temperatures are held down both by the morning low overcast and the afternoon strengthening sea breeze, resulting in daily maximum readings averaging about 70 degrees from May through August. However, during these months occasional hot spells, lasting a few days, are experienced without the usual high fog and sea breeze. September, when the sea breeze becomes less pronounced, is the warmest month with highs in the 70s. Low temperatures during the summer are in the mid-50s.

A strong temperature inversion with its base usually about 1,500 feet persists throughout the summer. Inversions close to the ground are infrequent in summer but rather common in fall and winter. As a consequence of these factors and the continued population and economic growth of the area, atmospheric pollution has become a problem of increasing importance.

# NORMALS, MEANS, AND EXTREMES
## SAN FRANCISCO (KSFO)

**LATITUDE:** 37° 37'N　**LONGITUDE:** -122° 23'W　**ELEVATION (FT):** GRND: 8　BARO: 89　**TIME ZONE:** PACIFIC (UTC -8)　**WBAN: 23234**

| | ELEMENT | POR | JAN | FEB | MAR | APR | MAY | JUN | JUL | AUG | SEP | OCT | NOV | DEC | YEAR |
|---|---|---|---|---|---|---|---|---|---|---|---|---|---|---|---|
| **TEMPERATURE °F** | NORMAL DAILY MAXIMUM | 30 | 55.9 | 59.3 | 61.2 | 64.3 | 66.8 | 69.9 | 71.1 | 71.7 | 72.7 | 69.7 | 62.0 | 56.1 | 65.1 |
| | MEAN DAILY MAXIMUM | 64 | 55.8 | 59.0 | 61.1 | 63.8 | 66.6 | 70.0 | 71.3 | 72.0 | 73.3 | 70.1 | 62.8 | 56.3 | 65.2 |
| | HIGHEST DAILY MAXIMUM | 82 | 72 | 78 | 85 | 92 | 97 | 106 | 105 | 100 | 103 | 99 | 85 | 75 | 106 |
| | YEAR OF OCCURRENCE | | 2009 | 1930 | 1952 | 1989 | 1984 | 1961 | 1988 | 1993 | 1971 | 1987 | 1967 | 1958 | JUN 1961 |
| | MEAN OF EXTREME MAXS. | 64 | 64.4 | 68.3 | 73.1 | 79.0 | 84.3 | 88.4 | 86.6 | 86.6 | 90.1 | 85.6 | 74.0 | 64.5 | 78.7 |
| | NORMAL DAILY MINIMUM | 30 | 42.9 | 45.5 | 46.8 | 48.1 | 50.5 | 52.9 | 54.5 | 55.5 | 55.1 | 52.4 | 47.5 | 43.0 | 49.6 |
| | MEAN DAILY MINIMUM | 64 | 42.3 | 44.8 | 46.1 | 47.6 | 50.2 | 52.7 | 54.1 | 54.9 | 54.7 | 51.9 | 47.3 | 43.2 | 49.2 |
| | LOWEST DAILY MINIMUM | 82 | 24 | 25 | 30 | 31 | 36 | 41 | 43 | 42 | 38 | 34 | 25 | 20 | 20 |
| | YEAR OF OCCURRENCE | | 1928 | 1929 | 1929 | 1929 | 1929 | 1932 | 1928 | 1935 | 1929 | 1929 | 1931 | 1932 | DEC 1932 |
| | MEAN OF EXTREME MINS. | 64 | 33.9 | 37.1 | 39.0 | 41.6 | 45.1 | 48.4 | 50.2 | 51.0 | 49.7 | 45.2 | 39.2 | 34.9 | 42.9 |
| | NORMAL DRY BULB | 30 | 49.4 | 52.4 | 54.0 | 56.2 | 58.7 | 61.4 | 62.8 | 63.9 | 63.9 | 61.0 | 54.7 | 49.5 | 57.3 |
| | MEAN DRY BULB | 64 | 49.1 | 51.9 | 53.6 | 55.7 | 58.4 | 61.3 | 62.7 | 63.5 | 64.0 | 61.0 | 55.1 | 49.8 | 57.2 |
| | MEAN WET BULB | 26 | 46.5 | 48.2 | 49.3 | 50.1 | 52.4 | 54.4 | 56.5 | 57.2 | 56.7 | 54.1 | 50.5 | 46.5 | 51.9 |
| | MEAN DEW POINT | 26 | 44.6 | 45.8 | 46.7 | 47.1 | 49.4 | 51.4 | 53.9 | 54.8 | 54.2 | 51.3 | 47.8 | 44.1 | 49.3 |
| | NORMAL NO. DAYS WITH: | | | | | | | | | | | | | | |
| | MAXIMUM >= 90 | 30 | 0.0 | 0.0 | 0.0 | 0.2 | 0.4 | 1.0 | 0.6 | 0.3 | 1.1 | 0.4 | 0.0 | 0.0 | 4.0 |
| | MAXIMUM <= 32 | 30 | 0.0 | 0.0 | 0.0 | 0.0 | 0.0 | 0.0 | 0.0 | 0.0 | 0.0 | 0.0 | 0.0 | 0.0 | 0.0 |
| | MINIMUM <= 32 | 30 | 0.4 | 0.1 | 0.0 | 0.0 | 0.0 | 0.0 | 0.0 | 0.0 | 0.0 | 0.0 | 0.0 | 0.8 | 1.3 |
| | MINIMUM <= 0 | 30 | 0.0 | 0.0 | 0.0 | 0.0 | 0.0 | 0.0 | 0.0 | 0.0 | 0.0 | 0.0 | 0.0 | 0.0 | 0.0 |
| **H/C** | NORMAL HEATING DEG. DAYS | 30 | 482 | 354 | 339 | 266 | 201 | 120 | 77 | 56 | 62 | 131 | 298 | 476 | 2862 |
| | NORMAL COOLING DEG. DAYS | 30 | 0 | 0 | 0 | 4 | 10 | 21 | 23 | 26 | 38 | 20 | 0 | 0 | 142 |
| **RH** | NORMAL (PERCENT) | 30 | 80 | 78 | 76 | 72 | 72 | 72 | 74 | 75 | 75 | 73 | 76 | 78 | 75 |
| | HOUR 04 LST | 30 | 87 | 86 | 83 | 82 | 84 | 84 | 86 | 87 | 85 | 83 | 84 | 85 | 85 |
| | HOUR 10 LST | 30 | 80 | 77 | 72 | 65 | 64 | 63 | 66 | 68 | 68 | 69 | 72 | 77 | 70 |
| | HOUR 16 LST | 30 | 68 | 66 | 64 | 60 | 60 | 58 | 60 | 61 | 60 | 60 | 63 | 67 | 62 |
| | HOUR 22 LST | 30 | 81 | 80 | 79 | 77 | 79 | 79 | 82 | 83 | 81 | 78 | 78 | 79 | 80 |
| **S** | PERCENT POSSIBLE SUNSHINE | | | | | | | | | | | | | | |
| **W/O** | MEAN NO. DAYS WITH: | | | | | | | | | | | | | | |
| | HEAVY FOG(VISBY <= 1/4 MI) | 46 | 2.5 | 1.6 | 0.3 | 0.1 | 0.0 | 0.0 | 0.0 | 0.1 | 0.2 | 0.5 | 1.1 | 1.7 | 8.1 |
| | THUNDERSTORMS | 64 | 0.4 | 0.5 | 0.4 | 0.2 | 0.1 | 0.0 | 0.1 | 0.1 | 0.3 | 0.2 | 0.2 | 0.3 | 2.8 |
| **CLOUDNESS** | MEAN: | | | | | | | | | | | | | | |
| | SUNRISE-SUNSET (OKTAS) | 55 | 5.0 | 5.0 | 4.6 | 4.2 | 3.6 | 3.0 | 2.4 | 2.6 | 2.6 | 3.1 | 4.2 | 4.7 | 3.8 |
| | MIDNIGHT-MIDNIGHT (OKTAS) | 32 | 4.7 | 4.9 | 4.6 | 3.9 | 3.4 | 3.2 | 2.6 | 2.8 | 2.6 | 3.0 | 4.0 | 4.3 | 3.7 |
| | MEAN NO. DAYS WITH: | | | | | | | | | | | | | | |
| | CLEAR | 69 | 8.5 | 7.8 | 9.5 | 10.8 | 13.5 | 16.2 | 20.3 | 18.8 | 17.9 | 15.5 | 11.1 | 9.4 | 159.3 |
| | PARTLY CLOUDY | 69 | 7.5 | 7.3 | 8.7 | 9.3 | 9.7 | 8.6 | 7.5 | 8.6 | 8.2 | 8.9 | 8.3 | 7.5 | 100.1 |
| | CLOUDY | 69 | 15.0 | 13.1 | 12.8 | 9.9 | 7.8 | 5.2 | 2.8 | 3.3 | 3.6 | 6.5 | 10.6 | 14.1 | 104.7 |
| **PR** | MEAN STATION PRESSURE(IN) | 26 | 30.11 | 30.05 | 30.01 | 30.03 | 29.98 | 29.93 | 29.94 | 29.90 | 29.92 | 29.98 | 30.07 | 30.11 | 30.00 |
| | MEAN SEA-LEVEL PRES. (IN) | 26 | 30.14 | 30.08 | 30.07 | 30.05 | 30.00 | 29.96 | 29.96 | 29.95 | 29.94 | 30.01 | 30.10 | 30.13 | 30.03 |
| **WINDS** | MEAN SPEED (MPH) | 26 | 7.1 | 8.7 | 10.7 | 12.7 | 13.8 | 14.1 | 13.3 | 12.4 | 11.0 | 9.3 | 7.7 | 7.4 | 10.7 |
| | PREVAIL.DIR(TENS OF DEGS) | 35 | 28 | 28 | 28 | 28 | 29 | 29 | 29 | 29 | 29 | 29 | 28 | 28 | 29 |
| | MAXIMUM 2-MINUTE: | | | | | | | | | | | | | | |
| | SPEED (MPH) | 13 | 54 | 60 | 46 | 49 | 46 | 47 | 40 | 38 | 40 | 48 | 51 | 48 | 60 |
| | DIR. (TENS OF DEGS) | | 17 | 18 | 26 | 28 | 26 | 28 | 26 | 27 | 28 | 27 | 18 | 17 | 18 |
| | YEAR OF OCCURRENCE | | 2008 | 2006 | 2000 | 2009 | 1999 | 2005 | 2006 | 2007 | 2009 | 2009 | 2002 | 2002 | FEB 2006 |
| | MAXIMUM 3-SECOND | | | | | | | | | | | | | | |
| | SPEED (MPH) | 13 | 68 | 71 | 52 | 60 | 54 | 58 | 49 | 48 | 52 | 62 | 66 | 64 | 71 |
| | DIR. (TENS OF DEGS) | | 17 | 18 | 19 | 29 | 26 | 28 | 26 | 27 | 27 | 17 | 17 | 18 | 18 |
| | YEAR OF OCCURRENCE | | 2008 | 2006 | 2009 | 2009 | 1999 | 2005 | 2005 | 2007 | 2009 | 2009 | 2002 | 2005 | FEB 2006 |
| **PRECIPITATION** | NORMAL (IN) | 30 | 4.45 | 4.01 | 3.26 | 1.18 | 0.38 | 0.11 | 0.03 | 0.07 | 0.20 | 1.04 | 2.49 | 2.89 | 20.11 |
| | MAXIMUM MONTHLY (IN) | 82 | 11.26 | 13.64 | 9.01 | 6.36 | 3.81 | 0.86 | 0.35 | 0.66 | 2.30 | 7.30 | 7.94 | 12.30 | 13.64 |
| | YEAR OF OCCURRENCE | | 1993 | 1998 | 1958 | 1958 | 1957 | 1967 | 1977 | 1976 | 1959 | 1962 | 1973 | 1955 | FEB 1998 |
| | MINIMUM MONTHLY (IN) | 82 | 0.24 | T | T | T | 0.00 | 0.00 | 0.00 | 0.00 | 0.00 | T | 0.00 | 0.01 | 0.00 |
| | YEAR OF OCCURRENCE | | 1991 | 1953 | 1934 | 1977 | 2001 | 1928 | 1930 | 1996 | 2002 | 1978 | 1929 | 1989 | SEP 2002 |
| | MAXIMUM IN 24 HOURS (IN) | 82 | 5.71 | 3.41 | 2.46 | 2.66 | 1.54 | 0.83 | 0.35 | 0.60 | 2.30 | 3.74 | 2.43 | 3.33 | 5.71 |
| | YEAR OF OCCURRENCE | | 1982 | 1998 | 1982 | 1958 | 1957 | 1967 | 1977 | 1997 | 1959 | 1962 | 1994 | 1955 | JAN 1982 |
| | NORMAL NO. DAYS WITH: | | | | | | | | | | | | | | |
| | PRECIPITATION >= 0.01 | 30 | 11.0 | 10.3 | 10.9 | 5.8 | 3.1 | 1.1 | 0.3 | 0.6 | 1.6 | 3.6 | 7.9 | 9.0 | 65.2 |
| | PRECIPITATION >= 1.00 | 30 | 1.4 | 1.1 | 0.6 | 0.1 | * | 0.0 | 0.0 | 0.0 | 0.0 | 0.3 | 0.5 | 0.6 | 4.6 |
| **SNOWFALL** | NORMAL (IN) | 30 | 0.* | 0.* | 0.0 | 0.0 | 0.0 | 0.0 | 0.0 | 0.0 | 0.0 | 0.0 | 0.0 | 0.* | 0.0 |
| | MAXIMUM MONTHLY (IN) | 69 | 1.5 | T | T | 0.0 | 0.0 | 0.0 | 0.0 | 0.0 | 0.0 | 0.0 | 0.0 | 1.0 | 1.5 |
| | YEAR OF OCCURRENCE | | 1962 | 1996 | 1995 | | | | | | | | | 1932 | JAN 1962 |
| | MAXIMUM IN 24 HOURS (IN) | 69 | 1.5 | T | T | T | 0.0 | 0.0 | 0.0 | 0.0 | 0.0 | 0.0 | 0.0 | 1.0 | 1.5 |
| | YEAR OF OCCURRENCE' | | 1962 | 1996 | 1995 | 1987 | | | | | | | | 1932 | JAN 1962 |
| | MAXIMUM SNOW DEPTH (IN) | 48 | 0 | 0 | 0 | 0 | 0 | 0 | 0 | 0 | 0 | 0 | 0 | 0 | 0 |
| | YEAR OF OCCURRENCE | | | | | | | | | | | | | | |
| | NORMAL NO. DAYS WITH: | | | | | | | | | | | | | | |
| | SNOWFALL >= 1.0 | 30 | 0.0 | 0.0 | 0.0 | 0.0 | 0.0 | 0.0 | 0.0 | 0.0 | 0.0 | 0.0 | 0.0 | 0.0 | 0.0 |

## PRECIPITATION (inches) 2009  SAN FRANCISCO (KSFO)

| YEAR | JAN | FEB | MAR | APR | MAY | JUN | JUL | AUG | SEP | OCT | NOV | DEC | ANNUAL |
|---|---|---|---|---|---|---|---|---|---|---|---|---|---|
| 1980 | 4.85 | 7.62 | 2.65 | 0.90 | 0.24 | 0.03 | 0.10 | T | T | 0.10 | 0.12 | 1.73 | 18.34 |
| 1981 | 5.92 | 2.21 | 3.60 | 0.24 | 0.07 | T | T | T | 0.28 | 2.35 | 4.89 | 3.91 | 23.47 |
| 1982 | 8.81 | 2.82 | 7.63 | 3.25 | T | 0.06 | T | T | 0.96 | 1.95 | 5.34 | 3.99 | 34.81 |
| 1983 | 6.83 | 6.64 | 8.50 | 3.11 | 0.32 | 0.01 | T | T | 0.57 | 0.10 | 6.03 | 6.23 | 38.34 |
| 1984 | 0.46 | 1.47 | 1.36 | 0.68 | T | 0.03 | T | 0.11 | 0.05 | 1.96 | 6.12 | 1.89 | 14.13 |
| 1985 | 0.74 | 2.35 | 3.30 | 0.12 | 0.05 | 0.29 | 0.03 | 0.02 | 0.18 | 0.69 | 3.19 | 1.61 | 12.57 |
| 1986 | 4.04 | 8.09 | 5.84 | 0.39 | 0.15 | T | 0.01 | T | 0.47 | 0.02 | 0.06 | 1.66 | 20.73 |
| 1987 | 2.80 | 3.52 | 1.98 | 0.16 | 0.06 | T | T | T | T | 0.93 | 1.64 | 4.51 | 15.60 |
| 1988 | 3.92 | 0.38 | 0.05 | 2.02 | 0.29 | 0.60 | T | T | 0.03 | 0.42 | 2.31 | 3.65 | 13.67 |
| 1989 | 1.25 | 1.28 | 4.00 | 0.78 | 0.04 | 0.01 | T | T | 1.24 | 1.40 | 1.34 | 0.01 | 11.35 |
| 1990 | 3.06 | 2.28 | 0.79 | 0.20 | 1.55 | T | 0.01 | T | 0.20 | 0.19 | 0.28 | 1.79 | 10.35 |
| 1991 | 0.24 | 3.76 | 6.07 | 0.61 | 0.21 | 0.11 | T | 0.27 | 0.04 | 1.73 | 0.23 | 2.70 | 15.97 |
| 1992 | 2.04 | 6.44 | 4.12 | 0.25 | T | 0.39 | 0.00 | 0.14 | T | 1.12 | 0.15 | 6.04 | 20.69 |
| 1993 | 11.26 | 4.68 | 2.34 | 0.41 | 0.55 | 0.16 | T | T | T | 0.45 | 1.47 | 2.19 | 23.51 |
| 1994 | 2.50 | 5.26 | 0.24 | 1.12 | 1.52 | 0.03 | T | T | 0.10 | 0.33 | 5.73 | 2.49 | 19.32 |
| 1995 | 8.89 | 0.38 | 8.75 | 1.41 | 0.93 | 0.60 | T | .00 | T | 0.03 | 0.02 | 6.41 | 27.42 |
| 1996 | 6.92 | 6.03 | 2.89 | 1.40 | 1.24 | T | T | T | T | .76 | 2.56 | 6.97 | 28.77 |
| 1997 | 7.52 | 0.31 | 0.25 | 0.30 | 0.21 | 0.24 | T | 0.60 | T | 0.68 | 6.41 | 3.87 | 20.39 |
| 1998 | 8.20 | 13.64 | 2.05 | 2.24 | 2.37 | 0.03 | T | 0.00 | 0.09 | 0.62 | 2.43 | 0.96 | 32.63 |
| 1999 | 2.96 | 4.59 | 2.80 | 2.18 | 0.10 | 0.18 | 0.00 | 0.06 | 0.27 | 0.46 | 1.47 | 0.43 | 15.50 |
| 2000 | 5.83 | 8.46 | 1.74 | 1.30 | 0.53 | 0.14 | 0.00 | 0.01 | 0.07 | 2.14 | 0.91 | 0.44 | 21.57 |
| 2001 | 3.87 | 6.12 | 1.02 | 1.56 | 0.00 | 0.10 | 0.00 | 0.00 | 0.11 | 0.31 | 4.51 | 8.54 | 26.14 |
| 2002 | 1.38 | 1.50 | 2.13 | 0.36 | 0.48 | T | 0.00 | 0.00 | 0.00 | T | 2.94 | 10.75 | 19.54 |
| 2003 | 1.43 | 2.45 | 1.17 | 4.42 | 0.63 | 0.00 | T | T | T | T | 2.17 | 6.41 | 18.68 |
| 2004 | 3.02 | 4.57 | 0.67 | 0.10 | 0.07 | T | 0.00 | T | 0.04 | 3.19 | 1.22 | 6.42 | 19.30 |
| 2005 | 4.27 | 5.10 | 3.74 | 1.70 | 1.15 | 0.30 | T | 0.00 | 0.08 | 0.08 | 1.23 | 9.34 | 26.99 |
| 2006 | 2.45 | 2.30 | 6.13 | 4.01 | 0.37 | T | 0.00 | 0.00 | 0.00 | 0.33 | 1.64 | 3.37 | 20.60 |
| 2007 | 0.65 | 4.14 | 0.27 | 1.14 | 0.09 | 0.00 | 0.01 | 0.00 | 0.15 | 1.97 | 0.58 | 2.65 | 11.65 |
| 2008 | 7.61 | 2.04 | 0.23 | 0.03 | T | 0.00 | T | 0.01 | 0.00 | 0.32 | 1.82 | 2.36 | 14.42 |
| 2009 | 0.69 | 6.40 | 2.35 | 0.27 | 0.36 | 0.04 | T | 0.00 | 0.27 | 2.96 | 0.20 | 3.07 | 16.61 |
| POR= 64 YRS | 4.20 | 3.53 | 2.79 | 1.34 | 0.38 | 0.11 | 0.02 | 0.04 | 0.17 | 1.01 | 2.33 | 3.73 | 19.65 |

WBAN : 23234

## AVERAGE TEMPERATURE (°F) 2009  SAN FRANCISCO (KSFO)

| YEAR | JAN | FEB | MAR | APR | MAY | JUN | JUL | AUG | SEP | OCT | NOV | DEC | ANNUAL |
|---|---|---|---|---|---|---|---|---|---|---|---|---|---|
| 1980 | 50.5 | 54.4 | 53.0 | 55.9 | 56.3 | 59.9 | 63.0 | 61.5 | 63.2 | 61.2 | 55.6 | 50.7 | 57.1 |
| 1981 | 51.1 | 54.0 | 53.2 | 56.2 | 59.0 | 65.0 | 61.7 | 63.0 | 62.7 | 58.6 | 56.3 | 52.2 | 57.8 |
| 1982 | 45.1 | 51.7 | 51.3 | 54.6 | 57.4 | 59.7 | 61.7 | 63.3 | 64.0 | 61.1 | 52.3 | 48.9 | 55.9 |
| 1983 | 48.0 | 53.4 | 54.1 | 54.7 | 57.8 | 61.5 | 65.1 | 66.9 | 68.3 | 64.4 | 55.5 | 53.4 | 58.6 |
| 1984 | 51.3 | 52.9 | 57.0 | 56.0 | 61.8 | 61.3 | 65.6 | 64.4 | 69.7 | 60.4 | 53.8 | 47.7 | 58.5 |
| 1985 | 46.4 | 51.6 | 51.4 | 59.0 | 58.6 | 65.2 | 64.8 | 64.0 | 63.2 | 60.7 | 52.0 | 47.1 | 57.0 |
| 1986 | 53.7 | 56.3 | 57.1 | 56.2 | 58.6 | 62.5 | 62.4 | 61.2 | 62.9 | 61.4 | 57.1 | 50.3 | 58.3 |
| 1987 | 49.3 | 53.3 | 54.9 | 59.2 | 61.6 | 62.4 | 63.1 | 65.1 | 64.0 | 63.9 | 57.0 | 50.5 | 58.7 |
| 1988 | 50.6 | 54.5 | 56.5 | 58.1 | 59.5 | 62.5 | 65.3 | 65.0 | 63.1 | 61.4 | 56.5 | 50.4 | 58.6 |
| 1989 | 48.3 | 48.4 | 54.9 | 60.8 | 59.8 | 62.7 | 62.8 | 64.0 | 61.4 | 60.8 | 56.4 | 50.1 | 57.5 |
| 1990 | 49.9 | 49.2 | 53.3 | 58.5 | 59.0 | 62.4 | 64.5 | 66.3 | 66.4 | 63.1 | 56.0 | 46.4 | 57.9 |
| 1991 | 50.1 | 55.3 | 52.2 | 55.7 | 56.9 | 59.4 | 63.7 | 64.4 | 62.9 | 62.8 | 57.5 | 50.8 | 57.6 |
| 1992 | 48.9 | 56.1 | 57.4 | 60.6 | 63.4 | 63.6 | 65.8 | 63.8 | 65.3 | 65.4 | 56.8 | 49.8 | 59.7 |
| 1993 | 49.3 | 52.5 | 57.5 | 58.1 | 62.2 | 64.8 | 64.8 | 67.3 | 63.1 | 63.4 | 56.3 | 50.1 | 59.1 |
| 1994 | 51.5 | 50.6 | 56.0 | 56.8 | 58.5 | 61.5 | 62.1 | 64.5 | 64.1 | 60.1 | 49.9 | 48.0 | 57.0 |
| 1995 | 52.2 | 54.1 | 53.9 | 55.0 | 57.0 | 60.8 | 64.8 | 63.0 | 63.2 | 62.5 | 58.4 | 53.7 | 58.2 |
| 1996 | 51.8 | 54.9 | 56.0 | 59.1 | 60.4 | 61.5 | 63.2 | 62.7 | 63.0 | 60.9 | 56.3 | 53.9 | 58.6 |
| 1997 | 50.9 | 53.4 | 56.1 | 58.2 | 64.4 | 62.3 | 64.0 | 66.9 | 68.3 | 61.8 | 57.7 | 51.1 | 59.6 |
| 1998 | 52.8 | 52.3 | 55.0 | 55.7 | 57.7 | 61.6 | 62.8 | 64.3 | 64.3 | 60.3 | 54.2 | 47.0 | 57.3 |
| 1999 | 49.3 | 50.4 | 51.0 | 54.6 | 55.5 | 59.6 | 61.9 | 63.8 | 63.6 | 62.5 | 57.1 | 51.1 | 56.7 |
| 2000 | 52.4 | 53.5 | 54.4 | 58.2 | 60.2 | 62.9 | 61.4 | 63.6 | 66.2 | 60.0 | 52.6 | 51.6 | 58.1 |
| 2001 | 48.8 | 51.0 | 55.7 | 53.6 | 62.6 | 62.6 | 62.7 | 63.2 | 62.9 | 62.4 | 57.3 | 51.7 | 57.9 |
| 2002 | 49.0 | 53.3 | 53.8 | 56.7 | 58.0 | 61.4 | 63.9 | 64.4 | 64.4 | 61.2 | 57.5 | 52.6 | 58.1 |
| 2003 | 53.8 | 52.2 | 55.9 | 54.3 | 58.9 | 62.8 | 63.0 | 66.4 | 66.9 | 63.2 | 53.9 | 52.0 | 58.6 |
| 2004 | 50.3 | 52.2 | 59.4 | 59.4 | 60.4 | 61.7 | 64.8 | 66.5 | 67.4 | 60.8 | 55.3 | 52.2 | 59.2 |
| 2005 | 49.7 | 55.0 | 57.2 | 56.8 | 61.5 | 61.9 | 64.4 | 63.4 | 62.4 | 61.3 | 57.2 | 53.1 | 58.7 |
| 2006 | 51.6 | 52.9 | 51.1 | 55.6 | 59.8 | 63.3 | 65.8 | 63.5 | 62.4 | 60.5 | 55.5 | 50.3 | 57.7 |
| 2007 | 48.0 | 52.4 | 55.9 | 56.1 | 58.8 | 61.3 | 64.2 | 64.4 | 64.3 | 59.9 | 55.7 | 49.5 | 57.5 |
| 2008 | 48.4 | 51.7 | 53.7 | 55.4 | 58.7 | 61.5 | 62.8 | 64.4 | 65.0 | 63.2 | 57.8 | 48.9 | 57.6 |
| 2009 | 51.3 | 52.1 | 53.8 | 56.4 | 59.4 | 62.9 | 62.3 | 65.6 | 66.2 | 61.8 | 56.6 | 49.2 | 58.1 |
| POR= 64 YRS | 49.1 | 51.9 | 53.6 | 55.7 | 58.4 | 61.3 | 62.7 | 63.5 | 64.0 | 61.0 | 55.1 | 49.8 | 57.2 |

## HEATING DEGREE DAYS (base 65°F) 2009 SAN FRANCISCO (KSFO)

| YEAR | JUL | AUG | SEP | OCT | NOV | DEC | JAN | FEB | MAR | APR | MAY | JUN | TOTAL |
|---|---|---|---|---|---|---|---|---|---|---|---|---|---|
| 1980-81 | 76 | 109 | 74 | 145 | 275 | 436 | 424 | 301 | 358 | 279 | 180 | 53 | 2710 |
| 1981-82 | 112 | 65 | 71 | 197 | 252 | 389 | 611 | 364 | 416 | 307 | 241 | 154 | 3179 |
| 1982-83 | 100 | 63 | 47 | 130 | 376 | 491 | 521 | 322 | 330 | 301 | 225 | 113 | 3019 |
| 1983-84 | 33 | 1 | 18 | 43 | 281 | 354 | 415 | 342 | 242 | 269 | 124 | 115 | 2237 |
| 1984-85 | 43 | 34 | 5 | 147 | 328 | 527 | 570 | 370 | 416 | 180 | 192 | 27 | 2839 |
| 1985-86 | 49 | 51 | 60 | 158 | 382 | 546 | 343 | 236 | 239 | 260 | 191 | 78 | 2593 |
| 1986-87 | 77 | 113 | 62 | 122 | 228 | 447 | 477 | 320 | 309 | 168 | 128 | 85 | 2536 |
| 1987-88 | 60 | 16 | 40 | 70 | 233 | 445 | 440 | 296 | 259 | 212 | 184 | 78 | 2333 |
| 1988-89 | 40 | 29 | 71 | 128 | 246 | 447 | 511 | 455 | 308 | 162 | 160 | 94 | 2651 |
| 1989-90 | 70 | 38 | 103 | 138 | 249 | 454 | 459 | 437 | 356 | 189 | 185 | 94 | 2772 |
| 1990-91 | 33 | 13 | 8 | 77 | 262 | 570 | 454 | 265 | 387 | 273 | 244 | 166 | 2752 |
| 1991-92 | 57 | 44 | 65 | 96 | 223 | 434 | 494 | 252 | 226 | 133 | 52 | 58 | 2134 |
| 1992-93 | 20 | 49 | 34 | 40 | 235 | 466 | 480 | 346 | 225 | 201 | 100 | 54 | 2250 |
| 1993-94 | 33 | 11 | 71 | 67 | 256 | 455 | 408 | 396 | 273 | 238 | 200 | 118 | 2526 |
| 1994-95 | 91 | 43 | 46 | 154 | 446 | 522 | 388 | 301 | 338 | 293 | 242 | 143 | 3007 |
| 1995-96 | 48 | 79 | 73 | 98 | 192 | 344 | 398 | 289 | 270 | 187 | 149 | 128 | 2255 |
| 1996-97 | 75 | 73 | 71 | 155 | 255 | 333 | 431 | 317 | 270 | 194 | 76 | 74 | 2324 |
| 1997-98 | 32 | 8 | 2 | 109 | 222 | 422 | 370 | 348 | 301 | 269 | 218 | 109 | 2410 |
| 1998-99 | 81 | 59 | 57 | 147 | 320 | 550 | 483 | 403 | 428 | 315 | 289 | 170 | 3302 |
| 1999-00 | 104 | 47 | 75 | 92 | 229 | 423 | 381 | 327 | 328 | 207 | 169 | 94 | 2476 |
| 2000-01 | 106 | 56 | 32 | 151 | 367 | 409 | 497 | 384 | 282 | 334 | 107 | 95 | 2820 |
| 2001-02 | 76 | 60 | 74 | 100 | 225 | 408 | 489 | 321 | 340 | 241 | 212 | 120 | 2666 |
| 2002-03 | 48 | 56 | 60 | 138 | 218 | 379 | 339 | 354 | 276 | 313 | 190 | 108 | 2479 |
| 2003-04 | 70 | 16 | 27 | 88 | 325 | 396 | 448 | 365 | 181 | 187 | 142 | 101 | 2346 |
| 2004-05 | 23 | 6 | 27 | 151 | 286 | 390 | 470 | 275 | 237 | 240 | 101 | 90 | 2296 |
| 2005-06 | 34 | 58 | 86 | 115 | 225 | 363 | 410 | 333 | 426 | 276 | 166 | 68 | 2560 |
| 2006-07 | 53 | 59 | 84 | 137 | 278 | 448 | 517 | 346 | 281 | 264 | 205 | 117 | 2789 |
| 2007-08 | 39 | 45 | 48 | 158 | 270 | 474 | 508 | 378 | 344 | 284 | 228 | 138 | 2914 |
| 2008-09 | 90 | 52 | 53 | 79 | 215 | 492 | 419 | 354 | 340 | 277 | 182 | 70 | 2623 |
| 2009- | 91 | 35 | 31 | 104 | 250 | 483 | | | | | | | |

WBAN : 23234

## COOLING DEGREE DAYS (base 65°F) 2009 SAN FRANCISCO (KSFO)

| YEAR | JAN | FEB | MAR | APR | MAY | JUN | JUL | AUG | SEP | OCT | NOV | DEC | TOTAL |
|---|---|---|---|---|---|---|---|---|---|---|---|---|---|
| 1980 | 0 | 0 | 0 | 0 | 0 | 10 | 22 | 7 | 30 | 33 | 1 | 0 | 103 |
| 1981 | 0 | 0 | 0 | 17 | 1 | 61 | 17 | 7 | 7 | 3 | 0 | 0 | 113 |
| 1982 | 0 | 0 | 0 | 1 | 12 | 5 | 7 | 15 | 23 | 12 | 0 | 0 | 75 |
| 1983 | 0 | 0 | 0 | 0 | 7 | 16 | 42 | 66 | 119 | 32 | 0 | 0 | 282 |
| 1984 | 0 | 0 | 0 | 4 | 33 | 10 | 70 | 24 | 152 | 9 | 0 | 0 | 302 |
| 1985 | 0 | 0 | 0 | 8 | 1 | 38 | 50 | 28 | 11 | 33 | 0 | 0 | 169 |
| 1986 | 0 | 0 | 1 | 2 | 0 | 11 | 5 | 0 | 7 | 16 | 0 | 0 | 42 |
| 1987 | 0 | 0 | 0 | 4 | 29 | 15 | 9 | 26 | 17 | 43 | 0 | 0 | 143 |
| 1988 | 0 | 0 | 0 | 11 | 19 | 8 | 55 | 34 | 23 | 24 | 0 | 0 | 174 |
| 1989 | 0 | 0 | 0 | 40 | 6 | 35 | 8 | 15 | 2 | 16 | 0 | 0 | 122 |
| 1990 | 0 | 0 | 0 | 1 | 3 | 23 | 23 | 58 | 58 | 24 | 0 | 0 | 190 |
| 1991 | 0 | 0 | 0 | 0 | 0 | 6 | 21 | 29 | 13 | 33 | 2 | 0 | 104 |
| 1992 | 0 | 0 | 0 | 9 | 10 | 21 | 53 | 18 | 49 | 58 | 0 | 0 | 218 |
| 1993 | 0 | 0 | 0 | 0 | 18 | 58 | 33 | 91 | 19 | 26 | 1 | 0 | 246 |
| 1994 | 0 | 0 | 0 | 0 | 4 | 18 | 4 | 30 | 27 | 8 | 0 | 0 | 91 |
| 1995 | 0 | 0 | 0 | 0 | 0 | 26 | 48 | 26 | 25 | 29 | 0 | 0 | 154 |
| 1996 | 0 | 0 | 0 | 18 | 16 | 30 | 22 | 9 | 21 | 36 | 0 | 0 | 152 |
| 1997 | 0 | 0 | 1 | 0 | 65 | 1 | 5 | 78 | 107 | 15 | 7 | 0 | 279 |
| 1998 | 0 | 0 | 0 | 0 | 0 | 13 | 23 | 44 | 45 | 6 | 0 | 0 | 131 |
| 1999 | 0 | 0 | 0 | 9 | 0 | 15 | 17 | 19 | 41 | 24 | 0 | 0 | 125 |
| 2000 | 0 | 0 | 2 | 10 | 27 | 36 | 2 | 20 | 73 | 1 | 0 | 0 | 171 |
| 2001 | 0 | 0 | 0 | 0 | 38 | 30 | 13 | 10 | 16 | 28 | 0 | 0 | 135 |
| 2002 | 0 | 0 | 0 | 0 | 3 | 18 | 20 | 46 | 59 | 31 | 0 | 0 | 177 |
| 2003 | 0 | 0 | 0 | 0 | 10 | 48 | 16 | 69 | 90 | 38 | 0 | 0 | 271 |
| 2004 | 0 | 0 | 13 | 25 | 4 | 7 | 25 | 63 | 104 | 26 | 0 | 0 | 267 |
| 2005 | 0 | 0 | 5 | 0 | 1 | 3 | 22 | 17 | 14 | 6 | 0 | 0 | 68 |
| 2006 | 0 | 0 | 0 | 0 | 11 | 26 | 82 | 17 | 14 | 6 | 0 | 0 | 156 |
| 2007 | 0 | 0 | 1 | 0 | 21 | 12 | 19 | 32 | 35 | 6 | 0 | 0 | 126 |
| 2008 | 0 | 0 | 0 | 5 | 39 | 40 | 30 | 40 | 62 | 30 | 5 | 0 | 251 |
| 2009 | 0 | 0 | 0 | 23 | 15 | 14 | 13 | 61 | 73 | 12 | 3 | 0 | 214 |

## SNOWFALL (inches) 2009 SAN FRANCISCO (KSFO)

| YEAR | JUL | AUG | SEP | OCT | NOV | DEC | JAN | FEB | MAR | APR | MAY | JUN | TOTAL |
|------|-----|-----|-----|-----|-----|-----|-----|-----|-----|-----|-----|-----|-------|
| 1976-77 | 0.0 | 0.0 | 0.0 | 0.0 | 0.0 | 0.0 | 0.0 | 0.0 | 0.0 | 0.0 | 0.0 | 0.0 | 0.0 |
| 1977-78 | 0.0 | 0.0 | 0.0 | 0.0 | 0.0 | 0.0 | 0.0 | 0.0 | 0.0 | 0.0 | 0.0 | 0.0 | 0.0 |
| 1978-79 | 0.0 | 0.0 | 0.0 | 0.0 | 0.0 | 0.0 | T | 0.0 | 0.0 | 0.0 | 0.0 | 0.0 | T |
| 1979-80 | 0.0 | 0.0 | 0.0 | 0.0 | 0.0 | 0.0 | 0.0 | 0.0 | T | 0.0 | 0.0 | 0.0 | T |
| 1980-81 | 0.0 | 0.0 | 0.0 | 0.0 | 0.0 | 0.0 | 0.0 | T | 0.0 | 0.0 | 0.0 | 0.0 | T |
| 1981-82 | 0.0 | 0.0 | 0.0 | 0.0 | 0.0 | 0.0 | T | 0.0 | T | 0.0 | 0.0 | 0.0 | T |
| 1982-83 | 0.0 | 0.0 | 0.0 | 0.0 | 0.0 | 0.0 | 0.0 | T | T | 0.0 | 0.0 | 0.0 | T |
| 1983-84 | 0.0 | 0.0 | 0.0 | 0.0 | 0.0 | 0.0 | 0.0 | 0.0 | 0.0 | 0.0 | 0.0 | 0.0 | 0.0 |
| 1984-85 | 0.0 | 0.0 | 0.0 | 0.0 | 0.0 | 0.0 | 0.0 | 0.0 | 0.0 | 0.0 | 0.0 | 0.0 | 0.0 |
| 1985-86 | 0.0 | 0.0 | 0.0 | 0.0 | 0.0 | 0.0 | 0.0 | 0.0 | 0.0 | 0.0 | 0.0 | 0.0 | 0.0 |
| 1986-87 | 0.0 | 0.0 | 0.0 | 0.0 | 0.0 | 0.0 | 0.0 | T | 0.0 | 0.0 | 0.0 | 0.0 | T |
| 1987-88 | 0.0 | 0.0 | 0.0 | 0.0 | 0.0 | 0.0 | 0.0 | 0.0 | 0.0 | 0.0 | 0.0 | 0.0 | 0.0 |
| 1988-89 | 0.0 | 0.0 | 0.0 | 0.0 | 0.0 | T | T | T | 0.0 | 0.0 | 0.0 | 0.0 | T |
| 1989-90 | 0.0 | 0.0 | 0.0 | 0.0 | 0.0 | 0.0 | 0.0 | 0.0 | 0.0 | 0.0 | 0.0 | 0.0 | 0.0 |
| 1990-91 | 0.0 | 0.0 | 0.0 | 0.0 | 0.0 | 0.0 | 0.0 | 0.0 | 0.0 | 0.0 | 0.0 | 0.0 | 0.0 |
| 1991-92 | 0.0 | 0.0 | 0.0 | 0.0 | 0.0 | 0.0 | 0.0 | 0.0 | T | 0.0 | 0.0 | 0.0 | T |
| 1992-93 | 0.0 | 0.0 | 0.0 | 0.0 | 0.0 | T | 0.0 | 0.0 | 0.0 | 0.0 | 0.0 | 0.0 | T |
| 1993-94 | 0.0 | 0.0 | 0.0 | 0.0 | 0.0 | 0.0 | 0.0 | T | 0.0 | 0.0 | 0.0 | 0.0 | T |
| 1994-95 | 0.0 | 0.0 | 0.0 | 0.0 | 0.0 | 0.0 | 0.0 | 0.0 | T | 0.0 | 0.0 | 0.0 | T |
| 1995-96 | 0.0 | 0.0 | 0.0 | 0.0 | 0.0 | 0.0 | 0.0 | T | 0.0 | 0.0 | 0.0 | 0.0 | T |
| 1996-97 | 0.0 | 0.0 | 0.0 | | | | | | | | | | |
| 1997-98 | | | | | | | | | | | | | |
| 1998-99 | | | | | | | | | | | | | |
| 1999-00 | | | | | | | | | | | | | |
| 2000-01 | | | | | | | | | | | | | |
| 2001-02 | | | | | | | | | | | | | |
| 2002-03 | | | | | | | | | | | | | |
| 2003-04 | | | | | | | | | | | | | |
| 2004-05 | | | | | | | | | | | | | |
| 2005- | | | | | | | | | | | | | |
| POR= 52 YRS | 0.0 | 0.0 | 0.0 | 0.0 | 0.0 | T | T | T | T | 0.0 | 0.0 | 0.0 | T |

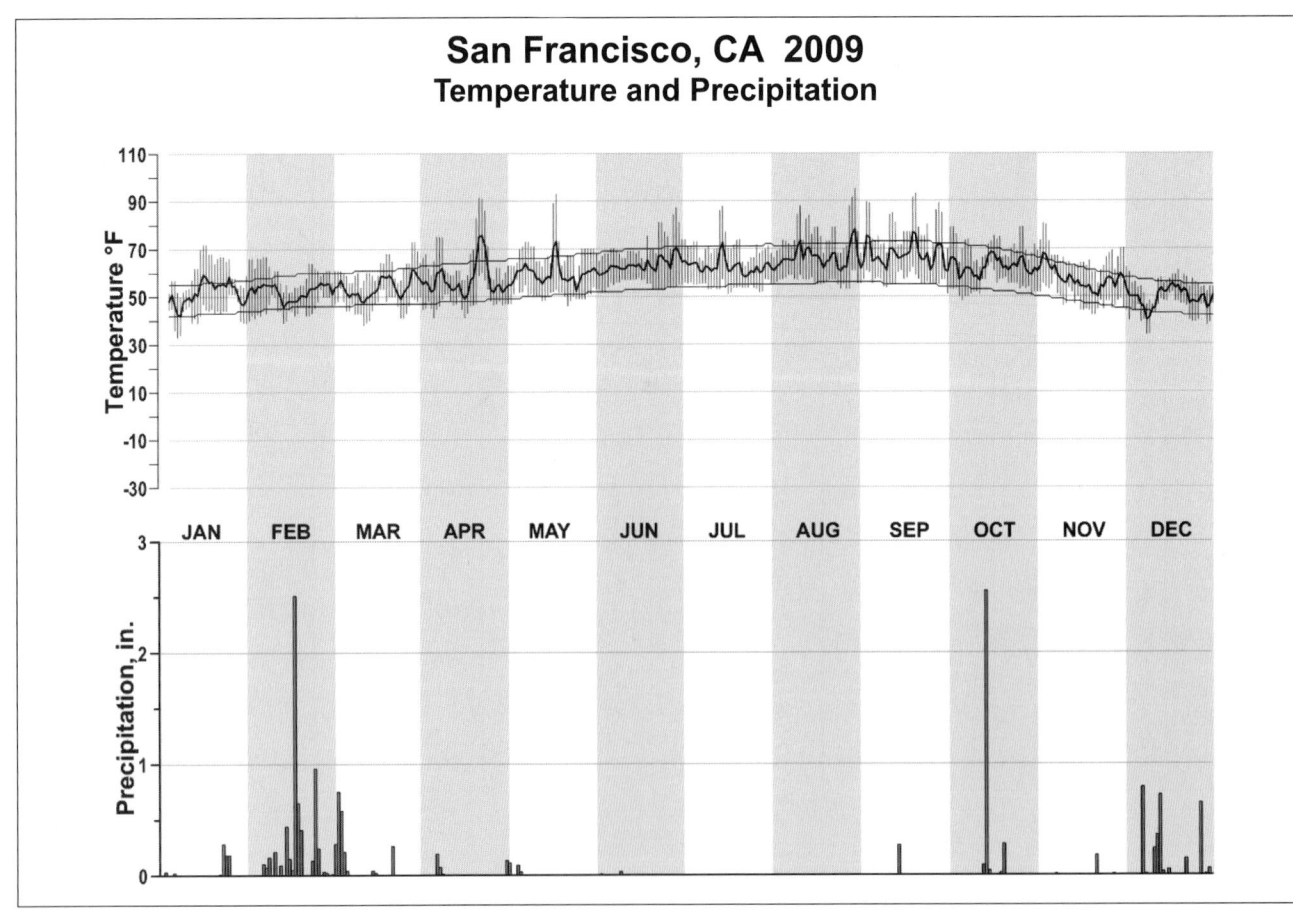

## San Francisco, CA 2009
### Temperature and Precipitation

# 2009
# COLORADO SPRINGS
# COLORADO (KCOS)

At an elevation near 6,200 feet above sea level, Colorado Springs is located in relatively flat semi-arid country on the eastern slope of the Rocky Mountains. Immediately to the west the mountains rise abruptly to heights ranging from 10,000 to 14,000 feet but generally averaging near 11,000 feet. To the east lie gently undulating prairie lands. The land slopes upward to the north, reaching an average height of about 8,000 feet in 20 miles at the top of Palmer Lake Divide.

Colorado Springs is in the Arkansas River drainage basin. The principal tributary feeding the Arkansas from this area is Fountain Creek which rises in the high mountains west of the city and is fed by Monument Creek originating to the north in the Palmer Lake Divide area.

Other topographical features of the area, and particularly its wide range of elevations, help to give Colorado Springs the various and altogether delightful plains and mountain mixture of climate that has established the locality as a highly desirable place to live. The higher elevations immediately to the west and north of the city produce significant differences in temperature and precipitation. Precipitation amounts at these higher elevations are approximately twice those at nearby lower elevations and the number of rainy days is almost triple.

In Colorado Springs itself, precipitation is relatively sparse. Over 80 percent of it falls between April 1 and September 30, mostly as heavy downpours accompanying summer thunderstorms. Temperatures, in view of the station latitude and elevation, are mild. Uncomfortable extremes, in either summer or winter, are comparatively rare and of short duration. Relative humidity is normally low and wind movement moderately high. This is notably true of the west-to-east movement of the chinook winds, that cause rapid rises in winter temperatures and remind us that the Indian meaning of CHINOOK is SNOW EATER.

Colorado Springs is best known as a resort city, but is also important to the high-tech industry and military community. Several military installations, including the United States Air Force Academy and the Space Command are located within or near the city. The surrounding prairie is also important for cattle raising and a considerable amount of grazing land is used for sheep in the summer months. The growing season varies considerably in length but averages from the first week in May to the first week of October.

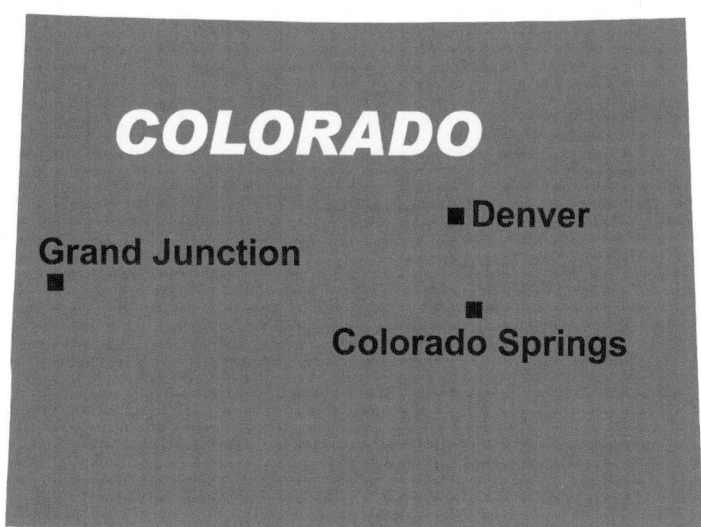

# NORMALS, MEANS, AND EXTREMES
## COLORADO SPRINGS (KCOS)

LATITUDE: 38 ° 48'N  LONGITUDE: -104° 42'W  ELEVATION (FT): GRND: 6181  BARO: 6183  TIME ZONE: MOUNTAIN (UTC -7)  WBAN: 93037

| | ELEMENT | POR | JAN | FEB | MAR | APR | MAY | JUN | JUL | AUG | SEP | OCT | NOV | DEC | YEAR |
|---|---|---|---|---|---|---|---|---|---|---|---|---|---|---|---|
| **TEMPERATURE °F** | NORMAL DAILY MAXIMUM | 30 | 41.7 | 45.4 | 51.6 | 59.2 | 68.4 | 79.2 | 84.4 | 81.6 | 74.1 | 63.4 | 49.8 | 42.4 | 61.8 |
| | MEAN DAILY MAXIMUM | 49 | 42.6 | 43.8 | 50.4 | 59.0 | 69.0 | 77.4 | 85.2 | 82.2 | 74.2 | 64.2 | 50.6 | 43.6 | 61.9 |
| | HIGHEST DAILY MAXIMUM | 60 | 73 | 76 | 81 | 87 | 94 | 100 | 100 | 99 | 94 | 86 | 78 | 77 | 100 |
| | YEAR OF OCCURRENCE | | 1997 | 1963 | 1971 | 1992 | 2000 | 1954 | 2003 | 2008 | 1995 | 1979 | 2006 | 1955 | JUL 2003 |
| | MEAN OF EXTREME MAXS. | 61 | 63.7 | 65.7 | 70.7 | 78.8 | 85.2 | 92.2 | 95.5 | 92.5 | 87.7 | 80.2 | 71.5 | 64.0 | 79.0 |
| | NORMAL DAILY MINIMUM | 30 | 14.5 | 18.0 | 23.9 | 31.4 | 40.7 | 49.5 | 54.8 | 53.6 | 45.4 | 34.3 | 22.6 | 15.6 | 35.2 |
| | MEAN DAILY MINIMUM | 49 | 16.6 | 18.6 | 24.4 | 32.5 | 42.5 | 50.1 | 56.9 | 55.4 | 46.7 | 36.3 | 24.6 | 18.0 | 35.2 |
| | LOWEST DAILY MINIMUM | 60 | -26 | -27 | -11 | -3 | 21 | 32 | 42 | 39 | 22 | 5 | -8 | -24 | -27 |
| | YEAR OF OCCURRENCE | | 1951 | 1951 | 1956 | 1959 | 1954 | 1951 | 1952 | 1992 | 1985 | 1969 | 1976 | 1990 | FEB 1951 |
| | MEAN OF EXTREME MINS. | 61 | -1.2 | 0.3 | 9.1 | 18.9 | 30.1 | 40.6 | 50.4 | 48.5 | 33.9 | 20.6 | 7.5 | -1.3 | 21.5 |
| | NORMAL DRY BULB | 30 | 28.1 | 31.7 | 37.8 | 45.3 | 54.6 | 64.4 | 69.6 | 67.6 | 59.8 | 48.9 | 36.2 | 29.0 | 47.8 |
| | MEAN DRY BULB | 61 | 29.6 | 31.6 | 37.5 | 45.8 | 55.7 | 64.2 | 70.9 | 68.8 | 60.4 | 50.1 | 37.6 | 30.7 | 48.6 |
| | MEAN WET BULB | 26 | 21.8 | 24.1 | 28.8 | 34.7 | 44.1 | 51.0 | 55.9 | 55.9 | 47.9 | 37.7 | 27.9 | 22.2 | 37.7 |
| | MEAN DEW POINT | 26 | 14.0 | 15.5 | 21.0 | 26.9 | 37.0 | 44.5 | 49.6 | 50.6 | 41.0 | 29.7 | 20.3 | 14.2 | 30.4 |
| | NORMAL NO. DAYS WITH: | | | | | | | | | | | | | | |
| | MAXIMUM >= 90 | 30 | 0.0 | 0.0 | 0.0 | 0.0 | 0.3 | 3.7 | 8.4 | 3.2 | 0.7 | 0.0 | 0.0 | 0.0 | 16.3 |
| | MAXIMUM <= 32 | 30 | 7.2 | 5.2 | 2.7 | 0.7 | 0.0 | 0.0 | 0.0 | 0.0 | * | 0.4 | 3.4 | 6.2 | 25.8 |
| | MINIMUM <= 32 | 30 | 29.9 | 26.7 | 25.1 | 13.1 | 2.0 | 0.0 | 0.0 | 0.0 | 1.1 | 9.1 | 24.9 | 29.5 | 161.4 |
| | MINIMUM <= 0 | 30 | 2.4 | 1.3 | 0.1 | 0.0 | 0.0 | 0.0 | 0.0 | 0.0 | 0.0 | 0.0 | 0.3 | 1.9 | 6.0 |
| **H/C** | NORMAL HEATING DEG. DAYS | 30 | 1130 | 917 | 827 | 576 | 312 | 76 | 19 | 20 | 172 | 483 | 848 | 1100 | 6480 |
| | NORMAL COOLING DEG. DAYS | 30 | 0 | 0 | 0 | 1 | 5 | 74 | 176 | 116 | 32 | 0 | 0 | 0 | 404 |
| **RH** | NORMAL (PERCENT) | 30 | 52 | 51 | 51 | 50 | 53 | 50 | 51 | 56 | 52 | 49 | 52 | 53 | 52 |
| | HOUR 05 LST | 30 | 58 | 59 | 62 | 64 | 69 | 67 | 68 | 71 | 67 | 60 | 60 | 58 | 64 |
| | HOUR 11 LST | 30 | 42 | 41 | 40 | 38 | 40 | 37 | 36 | 41 | 39 | 36 | 40 | 43 | 39 |
| | HOUR 17 LST | 30 | 47 | 40 | 38 | 37 | 39 | 36 | 36 | 38 | 43 | 38 | 46 | 49 | 41 |
| | HOUR 23 LST | 30 | 59 | 58 | 58 | 58 | 62 | 58 | 58 | 60 | 66 | 61 | 57 | 60 | 60 |
| **S** | PERCENT POSSIBLE SUNSHINE | | | | | | | | | | | | | | |
| **W/O** | MEAN NO. DAYS WITH: | | | | | | | | | | | | | | |
| | HEAVY FOG(VISBY <= 1/4 MI) | 32 | 2.7 | 3.4 | 3.8 | 2.8 | 1.8 | 1.0 | 0.9 | 1.4 | 1.4 | 2.3 | 3.1 | 2.7 | 27.3 |
| | THUNDERSTORMS | 32 | 0.0 | 0.1 | 0.7 | 2.2 | 7.6 | 10.5 | 12.8 | 13.0 | 4.4 | 0.7 | 0.1 | 0.0 | 52.1 |
| **CLOUDNESS** | MEAN: | | | | | | | | | | | | | | |
| | SUNRISE-SUNSET (OKTAS) | 1 | 3.2 | 3.7 | 5.6 | 6.0 | 4.8 | 3.6 | 3.6 | 5.6 | 3.2 | 4.0 | 3.2 | 2.4 | 4.1 |
| | MIDNIGHT-MIDNIGHT (OKTAS) | 1 | 4.0 | 3.6 | 6.4 | 6.0 | 4.8 | 3.2 | 3.6 | 5.6 | 3.2 | 4.4 | 2.8 | 2.4 | 4.2 |
| | MEAN NO. DAYS WITH: | | | | | | | | | | | | | | |
| | CLEAR | 3 | 5.3 | 8.3 | 8.0 | 5.0 | 1.0 | 3.0 | 7.0 | 6.0 | 11.0 | 10.5 | 6.0 | 11.0 | 82.1 |
| | PARTLY CLOUDY | 3 | 2.0 | 5.0 | 5.7 | 5.5 | 5.0 | 4.0 | 4.0 | 8.5 | 4.0 | 1.0 | 2.0 | 4.5 | 51.2 |
| | CLOUDY | 3 | 1.7 | 6.7 | 7.0 | 8.0 | 2.0 | 2.0 | 1.5 | 7.5 | 6.0 | 5.0 | 1.5 | 1.0 | 49.9 |
| **PR** | MEAN STATION PRESSURE(IN) | 26 | 23.90 | 23.88 | 23.87 | 23.87 | 23.92 | 23.98 | 24.06 | 24.07 | 24.03 | 23.98 | 23.93 | 23.90 | 23.95 |
| | MEAN SEA-LEVEL PRES. (IN) | 26 | 30.10 | 30.05 | 29.97 | 29.91 | 29.89 | 29.88 | 29.94 | 29.98 | 29.99 | 30.02 | 30.06 | 30.10 | 29.99 |
| **WINDS** | MEAN SPEED (MPH) | 26 | 9.0 | 9.5 | 10.5 | 11.4 | 10.6 | 9.9 | 8.9 | 8.6 | 9.1 | 9.5 | 8.9 | 8.9 | 9.6 |
| | PREVAIL.DIR(TENS OF DEGS) | 28 | 36 | 36 | 36 | 36 | 36 | 17 | 36 | 36 | 36 | 36 | 36 | 02 | 36 |
| | MAXIMUM 2-MINUTE: | | | | | | | | | | | | | | |
| | SPEED (MPH) | 16 | 49 | 61 | 56 | 61 | 58 | 51 | 49 | 45 | 45 | 59 | 47 | 53 | 61 |
| | DIR. (TENS OF DEGS) | | 27 | 28 | 28 | 28 | 23 | 27 | 22 | 34 | 24 | 27 | 35 | 29 | 28 |
| | YEAR OF OCCURRENCE | | 1996 | 1999 | 2004 | 1996 | 2006 | 2009 | 2000 | 2000 | 2009 | 2001 | 2004 | 1993 | FEB 1999 |
| | MAXIMUM 3-SECOND | | | | | | | | | | | | | | |
| | SPEED (MPH) | 16 | 55 | 78 | 68 | 68 | 66 | 62 | 60 | 63 | 58 | 70 | 57 | 63 | 78 |
| | DIR. (TENS OF DEGS) | | 26 | 28 | 29 | 30 | 24 | 28 | 23 | 04 | 24 | 30 | 26 | 28 | FEB 1999 |
| | YEAR OF OCCURRENCE | | 1996 | 1999 | 2004 | 1996 | 2006 | 2001 | 2000 | 2009 | 2009 | 1996 | 1995 | 2000 | FEB 1999 |
| **PRECIPITATION** | NORMAL (IN) | 30 | 0.28 | 0.35 | 1.06 | 1.62 | 2.39 | 2.34 | 2.85 | 3.48 | 1.23 | 0.86 | 0.52 | 0.42 | 17.40 |
| | MAXIMUM MONTHLY (IN) | 60 | 1.17 | 2.45 | 2.42 | 7.50 | 5.67 | 8.00 | 5.27 | 7.04 | 4.97 | 5.01 | 2.21 | 1.05 | 8.00 |
| | YEAR OF OCCURRENCE | | 1987 | 1987 | 1998 | 1999 | 1957 | 1965 | 1968 | 1999 | 2008 | 1984 | 1957 | 1988 | JUN 1965 |
| | MINIMUM MONTHLY (IN) | 60 | T | T | 0.01 | 0.01 | 0.33 | 0.13 | 0.28 | 0.15 | T | 0.01 | T | T | T |
| | YEAR OF OCCURRENCE | | 1995 | 1991 | 1966 | 1964 | 2008 | 1990 | 2008 | 1962 | 1953 | 1980 | 1995 | 1995 | DEC 1995 |
| | MAXIMUM IN 24 HOURS (IN) | 60 | 0.79 | 1.49 | 1.63 | 3.30 | 2.57 | 3.09 | 3.66 | 4.11 | 4.97 | 1.60 | 1.45 | 0.69 | 4.97 |
| | YEAR OF OCCURRENCE | | 1987 | 1987 | 1998 | 1999 | 1955 | 1954 | 1997 | 1999 | 2008 | 1960 | 1979 | 1981 | SEP 2008 |
| | NORMAL NO. DAYS WITH: | | | | | | | | | | | | | | |
| | PRECIPITATION >= 0.01 | 30 | 4.7 | 4.4 | 7.5 | 8.7 | 10.7 | 9.9 | 12.2 | 13.4 | 7.5 | 5.1 | 5.0 | 5.0 | 94.1 |
| | PRECIPITATION >= 1.00 | 30 | 0.0 | * | 0.1 | 0.2 | 0.5 | 0.4 | 0.6 | 0.6 | 0.2 | 0.1 | 0.0 | 0.0 | 2.7 |
| **SNOWFALL** | NORMAL (IN) | 30 | 5.4 | 5.1 | 9.4 | 6.3 | 1.3 | 0.* | 0.0 | 0.0 | 0.5 | 3.7 | 6.2 | 6.7 | 44.6 |
| | MAXIMUM MONTHLY (IN) | 61 | 28.7 | 23.2 | 23.2 | 42.7 | 19.4 | 1.1 | T | T | 27.9 | 25.9 | 26.3 | 18.2 | 42.7 |
| | YEAR OF OCCURRENCE | | 1987 | 1987 | 1984 | 1957 | 1978 | 1975 | 2007 | 2007 | 1959 | 1984 | 1991 | 1983 | APR 1957 |
| | MAXIMUM IN 24 HOURS (IN) | 61 | 22.0 | 14.8 | 15.0 | 18.0 | 17.4 | 1.1 | T | T | 17.1 | 19.9 | 14.5 | 9.6 | 22.0 |
| | YEAR OF OCCURRENCE' | | 1987 | 1987 | 1998 | 1957 | 1978 | 1975 | 1992 | 1992 | 1959 | 1997 | 1972 | 1979 | JAN 1987 |
| | MAXIMUM SNOW DEPTH (IN) | 33 | 16 | 12 | 15 | 12 | 11 | 0 | 0 | 0 | 2 | 20 | 11 | 10 | 20 |
| | YEAR OF OCCURRENCE | | 1987 | 1987 | 1998 | 1997 | 1978 | | | | 1985 | 1997 | 1979 | 1979 | OCT 1997 |
| | NORMAL NO. DAYS WITH: | | | | | | | | | | | | | | |
| | SNOWFALL >= 1.0 | 30 | 1.6 | 1.5 | 2.6 | 1.7 | 0.3 | 0.0 | 0.0 | 0.0 | 0.1 | 0.8 | 1.6 | 2.0 | 12.2 |

## PRECIPITATION (inches) 2009  COLORADO SPRINGS (KCOS)

| YEAR | JAN | FEB | MAR | APR | MAY | JUN | JUL | AUG | SEP | OCT | NOV | DEC | ANNUAL |
|------|-----|-----|-----|-----|-----|-----|-----|-----|-----|-----|-----|-----|--------|
| 1980 | 0.25 | 0.54 | 1.30 | 3.64 | 4.99 | 1.60 | 1.69 | 4.59 | 0.65 | 0.01 | 0.35 | 0.05 | 19.66 |
| 1981 | 0.07 | 0.12 | 0.93 | 0.13 | 3.14 | 1.98 | 3.64 | 5.24 | 0.52 | 0.37 | 0.03 | 0.82 | 16.99 |
| 1982 | 0.25 | 0.27 | 0.73 | 0.76 | 3.07 | 3.81 | 3.64 | 5.37 | 3.02 | 0.22 | 0.10 | 0.70 | 21.94 |
| 1983 | 0.43 | 0.09 | 1.79 | 0.97 | 3.08 | 2.41 | 0.99 | 2.59 | 0.37 | 0.28 | 1.09 | 0.70 | 14.79 |
| 1984 | 0.32 | 0.09 | 1.93 | 1.66 | 0.74 | 1.54 | 3.97 | 4.03 | 0.93 | 5.01 | 0.14 | 0.64 | 21.00 |
| 1985 | 0.42 | 0.24 | 1.68 | 2.07 | 3.36 | 0.78 | 4.92 | 1.56 | 1.49 | 0.52 | 0.42 | 0.55 | 18.01 |
| 1986 | 0.01 | 0.30 | 0.31 | 0.65 | 1.89 | 2.47 | 1.63 | 6.06 | 0.61 | 1.41 | 0.64 | 0.28 | 16.26 |
| 1987 | 1.17 | 2.45 | 1.79 | 0.50 | 3.82 | 2.89 | 0.67 | 2.77 | 0.55 | 0.54 | 0.44 | 0.64 | 18.23 |
| 1988 | 0.43 | 0.68 | 0.90 | 0.27 | 1.01 | 1.69 | 2.07 | 2.88 | 1.19 | 0.08 | 0.36 | 1.05 | 12.61 |
| 1989 | 0.23 | 1.23 | 0.49 | 1.06 | 1.11 | 3.42 | 2.26 | 2.63 | 2.30 | 0.28 | 0.02 | 0.41 | 15.44 |
| 1990 | 0.53 | 0.59 | 1.77 | 2.04 | 3.90 | 0.13 | 5.13 | 1.45 | 1.50 | 1.46 | 0.30 | 0.27 | 19.07 |
| 1991 | 0.09 | T | 0.42 | 1.76 | 0.80 | 3.07 | 2.87 | 4.57 | 0.56 | 0.88 | 2.05 | 0.45 | 17.52 |
| 1992 | 0.06 | 0.02 | 2.36 | 0.92 | 2.07 | 3.91 | 0.76 | 3.37 | 0.13 | 0.30 | 0.75 | 0.11 | 14.76 |
| 1993 | 0.52 | 0.21 | 0.79 | 1.02 | 1.60 | 1.27 | 2.38 | 2.17 | 1.44 | 0.91 | 0.97 | 0.11 | 13.39 |
| 1994 | 0.18 | 0.28 | 0.54 | 1.49 | 4.10 | 4.32 | 1.29 | 3.92 | 1.52 | 2.67 | 0.32 | 0.13 | 20.76 |
| 1995 | T | 0.21 | 0.71 | 3.05 | 4.81 | 7.78 | 1.91 | 1.77 | 1.87 | 0.02 | T | T | 22.13 |
| 1996 | 0.16 | 0.34 | 0.82 | 0.39 | 2.22 | 1.58 | 4.46 | 3.46 | 2.04 | 0.89 | 0.17 | 0.04 | 16.57 |
| 1997 | 0.11 | 0.18 | 0.34 | 3.30 | 1.16 | 5.44 | 4.63 | 4.70 | 1.78 | 0.98 | 0.22 | 0.10 | 22.94 |
| 1998 | 0.03 | 0.34 | 2.42 | 1.38 | 0.72 | 1.27 | 5.26 | 2.75 | 0.51 | 0.93 | 0.44 | 0.15 | 16.20 |
| 1999 | 0.12 | 0.05 | 0.41 | 7.50 | 3.57 | 1.36 | 4.70 | 7.04 | 0.52 | 1.10 | 1.01 | 0.20 | 27.58 |
| 2000 | 0.68 | 0.23 | 1.97 | 0.62 | 1.27 | 1.73 | 2.72 | 5.82 | 0.55 | 0.86 | 0.19 | 0.25 | 16.89 |
| 2001 | 0.82 | 0.26 | 1.38 | 0.98 | 3.21 | 2.14 | 3.25 | 1.47 | 1.01 | 0.02 | 0.37 | 0.09 | 15.00 |
| 2002 | 0.25 | 0.11 | 0.29 | 0.02 | 1.12 | 1.17 | 1.62 | 0.43 | 1.31 | 1.33 | 0.09 | 0.11 | 7.85 |
| 2003 | 0.03 | 0.63 | 1.02 | 0.97 | 0.90 | 5.07 | 1.14 | 1.89 | 0.58 | 0.09 | 0.04 | 0.06 | 12.42 |
| 2004 | 0.52 | 0.39 | 0.38 | 2.68 | 0.61 | 6.01 | 4.13 | 4.84 | 0.50 | 0.18 | 0.65 | 0.24 | 21.13 |
| 2005 | 0.78 | 0.04 | 1.03 | 1.08 | 0.73 | 2.10 | 1.91 | 2.65 | 0.68 | 0.48 | 0.08 | 0.30 | 11.86 |
| 2006 | 0.24 | 0.04 | 0.24 | 0.09 | 0.81 | 0.82 | 4.42 | 3.52 | 1.51 | 1.57 | 0.19 | 0.39 | 13.84 |
| 2007 | 0.31 | 0.17 | 0.66 | 1.85 | 2.35 | 0.94 | 1.74 | 2.69 | 0.34 | 0.25 | 0.10 | 0.39 | 11.79 |
| 2008 | 0.46 | 0.19 | 0.96 | 0.39 | 0.33 | 0.51 | 0.28 | 4.30 | 4.97 | 0.14 | 0.25 | 0.15 | 12.93 |
| 2009 | 0.09 | 0.04 | 0.45 | 1.52 | 2.39 | 2.89 | 3.82 | 1.84 | 1.20 | 0.36 | 0.45 | 0.67 | 15.72 |
| POR= 61 YRS | 0.30 | 0.31 | 0.88 | 1.36 | 2.14 | 2.30 | 2.87 | 2.94 | 1.28 | 0.78 | 0.45 | 0.32 | 15.93 |

WBAN : 93037

## AVERAGE TEMPERATURE (°F) 2009  COLORADO SPRINGS (KCOS)

| YEAR | JAN | FEB | MAR | APR | MAY | JUN | JUL | AUG | SEP | OCT | NOV | DEC | ANNUAL |
|------|-----|-----|-----|-----|-----|-----|-----|-----|-----|-----|-----|-----|--------|
| 1980 | 26.7 | 34.3 | 35.7 | 44.3 | 53.4 | 69.2 | 75.3 | 70.4 | 62.3 | 49.9 | 39.5 | 39.8 | 50.1 |
| 1981 | 34.9 | 34.4 | 39.3 | 53.8 | 54.5 | 69.4 | 71.9 | 67.3 | 63.4 | 50.8 | 43.3 | 32.7 | 51.3 |
| 1982 | 29.4 | 29.0 | 38.1 | 45.7 | 52.7 | 60.1 | 70.2 | 68.5 | 59.0 | 47.6 | 35.4 | 29.8 | 47.1 |
| 1983 | 32.5 | 34.4 | 35.7 | 40.1 | 50.5 | 61.1 | 72.3 | 71.9 | 63.8 | 51.3 | 37.7 | 18.4 | 47.5 |
| 1984 | 26.1 | 33.3 | 35.3 | 41.4 | 58.2 | 65.1 | 71.5 | 68.4 | 59.5 | 42.8 | 38.4 | 33.1 | 47.8 |
| 1985 | 25.1 | 26.3 | 37.9 | 48.7 | 57.2 | 65.0 | 70.3 | 69.8 | 58.0 | 49.0 | 32.1 | 27.9 | 47.3 |
| 1986 | 38.2 | 34.7 | 44.3 | 48.5 | 54.6 | 65.8 | 70.4 | 67.6 | 59.1 | 48.0 | 37.7 | 29.8 | 49.9 |
| 1987 | 29.4 | 33.0 | 35.3 | 48.5 | 56.0 | 65.1 | 70.7 | 66.1 | 60.0 | 50.4 | 39.2 | 29.0 | 48.6 |
| 1988 | 24.3 | 31.8 | 36.2 | 48.2 | 56.4 | 68.7 | 70.6 | 70.4 | 60.8 | 53.0 | 39.2 | 29.2 | 49.1 |
| 1989 | 32.8 | 21.8 | 43.7 | 49.0 | 57.6 | 62.1 | 71.8 | 68.4 | 61.0 | 49.7 | 41.5 | 27.3 | 48.9 |
| 1990 | 33.6 | 31.7 | 38.9 | 47.1 | 53.8 | 69.5 | 68.0 | 68.0 | 64.5 | 49.5 | 42.7 | 24.3 | 49.3 |
| 1991 | 28.0 | 38.0 | 39.9 | 45.8 | 58.0 | 67.0 | 69.4 | 68.1 | 60.5 | 50.3 | 33.0 | 31.0 | 49.1 |
| 1992 | 32.6 | 37.6 | 41.6 | 52.1 | 57.8 | 62.5 | 68.4 | 66.6 | 62.9 | 52.4 | 31.9 | 29.3 | 49.6 |
| 1993 | 26.7 | 29.4 | 39.8 | 46.1 | 55.6 | 65.0 | 70.7 | 67.3 | 58.1 | 48.0 | 32.3 | 32.3 | 47.6 |
| 1994 | 31.5 | 31.6 | 40.6 | 45.5 | 57.7 | 69.1 | 69.8 | 70.2 | 62.7 | 49.1 | 37.4 | 33.6 | 49.9 |
| 1995 | 31.4 | 35.8 | 38.7 | 41.3 | 49.4 | 60.6 | 67.4 | 71.9 | 59.0 | 48.9 | 41.4 | 32.1 | 48.2 |
| 1996 | 27.2 | 34.0 | 35.5 | 47.2 | 59.6 | 65.7 | 69.8 | 67.6 | 57.4 | 49.0 | 38.1 | 33.0 | 48.7 |
| 1997 | 27.1 | 30.2 | 40.7 | 39.5 | 53.8 | 64.3 | 70.5 | 67.1 | 62.2 | 48.5 | 33.5 | 30.5 | 47.3 |
| 1998 | 32.1 | 31.6 | 35.4 | 43.0 | 58.9 | 63.8 | 71.0 | 67.6 | 65.4 | 50.0 | 41.7 | 28.3 | 49.1 |
| 1999 | 33.7 | 38.1 | 42.1 | 42.2 | 53.4 | 62.9 | 71.1 | 68.7 | 57.7 | 50.8 | 44.9 | 33.6 | 49.9 |
| 2000 | 32.1 | 38.0 | 39.0 | 48.8 | 59.4 | 64.2 | 71.9 | 71.4 | 62.0 | 49.7 | 30.3 | 28.0 | 49.6 |
| 2001 | 27.1 | 30.3 | 37.4 | 49.4 | 55.1 | 66.5 | 73.5 | 69.8 | 63.1 | 50.9 | 41.0 | 31.6 | 49.6 |
| 2002 | 30.1 | 31.8 | 35.4 | 50.4 | 56.3 | 70.8 | 73.9 | 70.0 | 62.4 | 44.9 | 37.2 | 31.8 | 49.6 |
| 2003 | 35.8 | 26.8 | 40.4 | 49.8 | 57.0 | 61.5 | 75.8 | 71.1 | 58.3 | 53.9 | 37.1 | 33.4 | 50.1 |
| 2004 | 30.7 | 30.2 | 44.7 | 46.3 | 59.2 | 62.7 | 67.3 | 64.5 | 61.0 | 50.8 | 35.9 | 32.7 | 48.8 |
| 2005 | 33.4 | 35.1 | 37.9 | 46.6 | 56.4 | 65.4 | 73.3 | 67.7 | 64.5 | 50.4 | 41.3 | 28.6 | 50.1 |
| 2006 | 34.8 | 30.0 | 38.5 | 51.0 | 59.0 | 69.9 | 71.7 | 68.1 | 56.2 | 47.5 | 40.1 | 31.3 | 49.8 |
| 2007 | 24.6 | 33.0 | 44.5 | 45.1 | 55.4 | 64.7 | 72.2 | 72.6 | 63.4 | 52.6 | 41.3 | 27.3 | 49.7 |
| 2008 | 26.7 | 32.7 | 38.4 | 45.1 | 55.2 | 65.9 | 74.2 | 68.5 | 60.4 | 50.8 | 41.7 | 30.2 | 49.2 |
| 2009 | 34.0 | 36.1 | 41.0 | 45.6 | 57.2 | 64.0 | 68.7 | 67.4 | 59.7 | 42.7 | 41.4 | 23.3 | 48.4 |
| POR= 61 YRS | 29.6 | 31.6 | 37.5 | 45.8 | 55.7 | 64.2 | 70.9 | 68.8 | 60.4 | 50.1 | 37.6 | 30.7 | 48.6 |

## HEATING DEGREE DAYS (base 65°F) 2009  COLORADO SPRINGS (KCOS)

| YEAR | JUL | AUG | SEP | OCT | NOV | DEC | JAN | FEB | MAR | APR | MAY | JUN | TOTAL |
|---|---|---|---|---|---|---|---|---|---|---|---|---|---|
| 1980-81 | 0 | 7 | 113 | 463 | 759 | 776 | 928 | 850 | 789 | 335 | 321 | 38 | 5379 |
| 1981-82 | 5 | 30 | 70 | 433 | 643 | 993 | 1095 | 1001 | 827 | 571 | 374 | 163 | 6205 |
| 1982-83 | 8 | 11 | 198 | 532 | 880 | 1084 | 1001 | 851 | 904 | 742 | 444 | 159 | 6814 |
| 1983-84 | 2 | 0 | 101 | 417 | 811 | 1438 | 1198 | 911 | 912 | 700 | 220 | 58 | 6768 |
| 1984-85 | 0 | 6 | 200 | 684 | 790 | 982 | 1233 | 1077 | 830 | 481 | 242 | 77 | 6602 |
| 1985-86 | 5 | 8 | 253 | 487 | 978 | 1142 | 822 | 840 | 635 | 487 | 315 | 49 | 6021 |
| 1986-87 | 4 | 14 | 174 | 519 | 813 | 1081 | 1096 | 888 | 912 | 491 | 272 | 50 | 6314 |
| 1987-88 | 17 | 74 | 150 | 445 | 767 | 1108 | 1256 | 958 | 886 | 499 | 273 | 25 | 6458 |
| 1988-89 | 7 | 8 | 154 | 366 | 767 | 1099 | 989 | 1207 | 655 | 475 | 247 | 134 | 6108 |
| 1989-90 | 0 | 4 | 172 | 473 | 699 | 1164 | 966 | 928 | 805 | 526 | 345 | 24 | 6106 |
| 1990-91 | 28 | 21 | 83 | 473 | 663 | 1258 | 1142 | 750 | 773 | 568 | 219 | 33 | 6011 |
| 1991-92 | 16 | 16 | 145 | 453 | 954 | 1048 | 998 | 788 | 717 | 383 | 219 | 96 | 5833 |
| 1992-93 | 21 | 53 | 91 | 383 | 990 | 1101 | 1179 | 991 | 776 | 558 | 286 | 84 | 6513 |
| 1993-94 | 0 | 40 | 212 | 519 | 972 | 1008 | 1032 | 926 | 749 | 576 | 223 | 14 | 6271 |
| 1994-95 | 10 | 14 | 98 | 486 | 821 | 969 | 1035 | 811 | 808 | 702 | 477 | 152 | 6383 |
| 1995-96 | 38 | 3 | 231 | 490 | 700 | 1011 | 1162 | 890 | 908 | 527 | 192 | 48 | 6200 |
| 1996-97 | 2 | 17 | 237 | 490 | 800 | 986 | 1167 | 967 | 747 | 758 | 341 | 70 | 6582 |
| 1997-98 | 6 | 28 | 111 | 506 | 937 | 1060 | 1012 | 928 | 911 | 653 | 208 | 116 | 6476 |
| 1998-99 | 3 | 6 | 43 | 458 | 691 | 1129 | 965 | 748 | 702 | 674 | 357 | 101 | 5877 |
| 1999-00 | 0 | 5 | 239 | 434 | 600 | 968 | 1010 | 773 | 798 | 478 | 206 | 83 | 5594 |
| 2000-01 | 0 | 2 | 150 | 473 | 1033 | 1137 | 1169 | 964 | 850 | 462 | 305 | 54 | 6599 |
| 2001-02 | 0 | 6 | 97 | 431 | 714 | 1029 | 1075 | 925 | 909 | 431 | 280 | 22 | 5919 |
| 2002-03 | 0 | 7 | 135 | 622 | 828 | 1023 | 898 | 1062 | 755 | 449 | 265 | 116 | 6160 |
| 2003-04 | 0 | 16 | 200 | 337 | 828 | 972 | 1056 | 1002 | 620 | 554 | 185 | 112 | 5882 |
| 2004-05 | 33 | 65 | 150 | 432 | 868 | 998 | 973 | 830 | 834 | 544 | 280 | 71 | 6078 |
| 2005-06 | 2 | 24 | 72 | 450 | 703 | 1120 | 930 | 974 | 814 | 414 | 216 | 17 | 5736 |
| 2006-07 | 4 | 21 | 257 | 537 | 739 | 1039 | 1247 | 890 | 629 | 593 | 288 | 70 | 6314 |
| 2007-08 | 0 | 0 | 106 | 382 | 704 | 1164 | 1183 | 932 | 819 | 591 | 312 | 57 | 6250 |
| 2008-09 | 0 | 38 | 145 | 432 | 689 | 1073 | 955 | 804 | 736 | 575 | 244 | 94 | 5785 |
| 2009- | 22 | 19 | 175 | 683 | 703 | 1287 | | | | | | | |

WBAN : 93037

## COOLING DEGREE DAYS (base 65°F) 2009  COLORADO SPRINGS (KCOS)

| YEAR | JAN | FEB | MAR | APR | MAY | JUN | JUL | AUG | SEP | OCT | NOV | DEC | TOTAL |
|---|---|---|---|---|---|---|---|---|---|---|---|---|---|
| 1980 | 0 | 0 | 0 | 0 | 0 | 169 | 327 | 180 | 41 | 0 | 0 | 0 | 717 |
| 1981 | 0 | 0 | 0 | 4 | 2 | 176 | 226 | 105 | 27 | 0 | 0 | 0 | 540 |
| 1982 | 0 | 0 | 0 | 0 | 0 | 23 | 176 | 127 | 26 | 0 | 0 | 0 | 352 |
| 1983 | 0 | 0 | 0 | 0 | 1 | 48 | 236 | 219 | 71 | 0 | 0 | 0 | 575 |
| 1984 | 0 | 0 | 0 | 0 | 17 | 68 | 207 | 119 | 42 | 0 | 0 | 0 | 453 |
| 1985 | 0 | 0 | 0 | 0 | 5 | 83 | 179 | 163 | 51 | 0 | 0 | 0 | 481 |
| 1986 | 0 | 0 | 0 | 0 | 1 | 82 | 180 | 102 | 3 | 0 | 0 | 0 | 368 |
| 1987 | 0 | 0 | 0 | 0 | 0 | 62 | 199 | 113 | 6 | 0 | 0 | 0 | 380 |
| 1988 | 0 | 0 | 0 | 0 | 12 | 143 | 190 | 181 | 33 | 0 | 0 | 0 | 559 |
| 1989 | 0 | 0 | 0 | 3 | 25 | 54 | 220 | 117 | 57 | 3 | 0 | 0 | 479 |
| 1990 | 0 | 0 | 0 | 0 | 6 | 168 | 128 | 121 | 73 | 0 | 0 | 0 | 496 |
| 1991 | 0 | 0 | 0 | 0 | 8 | 101 | 161 | 120 | 15 | 4 | 0 | 0 | 409 |
| 1992 | 0 | 0 | 0 | 4 | 3 | 28 | 131 | 106 | 32 | 0 | 0 | 0 | 304 |
| 1993 | 0 | 0 | 0 | 0 | 2 | 89 | 183 | 117 | 11 | 1 | 0 | 0 | 403 |
| 1994 | 0 | 0 | 0 | 0 | 5 | 143 | 165 | 182 | 33 | 0 | 0 | 0 | 528 |
| 1995 | 0 | 0 | 0 | 0 | 0 | 30 | 120 | 226 | 60 | 0 | 0 | 0 | 436 |
| 1996 | 0 | 0 | 0 | 4 | 31 | 77 | 159 | 106 | 16 | 1 | 0 | 0 | 394 |
| 1997 | 0 | 0 | 0 | 0 | 0 | 55 | 182 | 101 | 33 | 0 | 0 | 0 | 371 |
| 1998 | 0 | 0 | 0 | 0 | 23 | 88 | 194 | 95 | 61 | 0 | 0 | 0 | 461 |
| 1999 | 0 | 0 | 0 | 0 | 0 | 42 | 196 | 130 | 26 | 0 | 0 | 0 | 394 |
| 2000 | 0 | 0 | 0 | 0 | 39 | 65 | 220 | 208 | 67 | 5 | 0 | 0 | 604 |
| 2001 | 0 | 0 | 0 | 0 | 5 | 106 | 272 | 164 | 45 | 0 | 0 | 0 | 592 |
| 2002 | 0 | 0 | 0 | 0 | 17 | 204 | 284 | 170 | 61 | 0 | 0 | 0 | 736 |
| 2003 | 0 | 0 | 0 | 0 | 22 | 18 | 342 | 212 | 5 | 0 | 0 | 0 | 599 |
| 2004 | 0 | 0 | 0 | 0 | 13 | 49 | 111 | 55 | 38 | 0 | 0 | 0 | 266 |
| 2005 | 0 | 0 | 0 | 0 | 22 | 90 | 265 | 111 | 63 | 4 | 0 | 0 | 555 |
| 2006 | 0 | 0 | 0 | 0 | 37 | 170 | 219 | 124 | 0 | 2 | 0 | 0 | 552 |
| 2007 | 0 | 0 | 0 | 0 | 0 | 65 | 230 | 246 | 65 | 5 | 0 | 0 | 611 |
| 2008 | 0 | 0 | 0 | 0 | 14 | 93 | 289 | 151 | 12 | 0 | 0 | 0 | 559 |
| 2009 | 0 | 0 | 0 | 0 | 7 | 71 | 141 | 102 | 24 | 0 | 0 | 0 | 345 |

## SNOWFALL (inches) 2009 COLORADO SPRINGS (KCOS)

| YEAR | JUL | AUG | SEP | OCT | NOV | DEC | JAN | FEB | MAR | APR | MAY | JUN | TOTAL |
|------|-----|-----|-----|-----|-----|-----|-----|-----|-----|-----|-----|-----|-------|
| 1980-81 | 0.0 | 0.0 | 0.0 | 0.2 | 4.4 | 1.4 | 1.0 | 1.7 | 9.0 | 0.3 | 0.2 | 0.0 | 18.2 |
| 1981-82 | 0.0 | 0.0 | 0.0 | 0.4 | 0.5 | 9.1 | 3.6 | 6.2 | 8.4 | 2.3 | 3.9 | 0.0 | 34.4 |
| 1982-83 | 0.0 | 0.0 | 0.0 | 0.2 | 0.9 | 8.2 | 4.0 | 1.1 | 16.3 | 4.8 | 0.8 | 0.0 | 36.3 |
| 1983-84 | 0.0 | 0.0 | 0.0 | 0.0 | 10.3 | 18.2 | 7.8 | 1.4 | 23.2 | 9.0 | 0.8 | 0.0 | 70.7 |
| 1984-85 | 0.0 | 0.0 | 0.9 | 25.9 | 2.0 | 10.9 | 8.0 | 4.7 | 22.3 | 0.8 | T | 0.0 | 75.5 |
| 1985-86 | 0.0 | 0.0 | 1.9 | 1.7 | 8.3 | 6.3 | 0.2 | 4.6 | 2.9 | 4.0 | T | 0.0 | 29.9 |
| 1986-87 | 0.0 | 0.0 | 0.0 | 1.4 | 7.3 | 4.4 | 28.7 | 23.2 | 14.9 | 3.3 | T | 0.0 | 83.2 |
| 1987-88 | 0.0 | 0.0 | 0.0 | 0.0 | 4.8 | 9.5 | 4.9 | 11.5 | 12.6 | 1.0 | 0.3 | 0.0 | 44.6 |
| 1988-89 | 0.0 | 0.0 | 0.0 | 0.0 | 1.6 | 13.6 | 3.0 | 18.9 | T | 1.0 | 0.0 | T | 38.1 |
| 1989-90 | T | T | T | 2.1 | 0.2 | 7.5 | 8.7 | 9.3 | 9.7 | 11.3 | 4.2 | 0.0 | 53.0 |
| 1990-91 | 0.0 | 0.0 | 0.0 | 8.2 | 2.7 | 4.1 | 0.9 | T | 5.2 | 5.9 | 1.5 | 0.0 | 28.5 |
| 1991-92 | 0.0 | 0.0 | 0.0 | 7.5 | 26.3 | 4.5 | 1.2 | 0.2 | 3.0 | T | 0.0 | T | 42.7 |
| 1992-93 | T | T | 0.0 | T | 11.4 | 2.6 | 7.5 | 3.3 | 2.3 | 1.2 | T | T | 28.3 |
| 1993-94 | T | 0.0 | 0.1 | 0.5 | 12.3 | 0.5 | 2.2 | 1.5 | 6.8 | 8.7 | T | T | 32.6 |
| 1994-95 | 0.0 | 0.0 | T | 0.2 | 6.1 | 3.9 | 4.5 | 6.5 | 7.7 | 0.0 | 0.0 | T | 28.9 |
| 1995-96 | 0.0 | 0.0 | 0.5 | 0.2 | T | T | 2.9 | 4.7 | 6.9 | 3.2 | | | |
| 1996-97 | | | | 8.7 | | | 9.3 | 3.4 | 2.4 | 12.3 | T | 0.0 | |
| 1997-98 | 0.0 | T | 0.0 | 19.9 | 3.5 | 1.8 | 0.3 | 4.3 | 17.5 | 9.5 | T | T | 56.8 |
| 1998-99 | T | T | 0.0 | 1.1 | 1.7 | 7.3 | 2.9 | 0.3 | 2.7 | 17.6 | 0.3 | T | 33.9 |
| 1999-00 | 0.0 | 0.0 | 0.2 | 3.0 | 10.6 | 2.3 | 7.4 | 2.0 | 12.6 | 2.0 | T | 0.0 | 40.1 |
| 2000-01 | T | 0.0 | 1.5 | 1.5 | 2.9 | 3.7 | 14.6 | 5.9 | 12.0 | 8.1 | 6.5 | T | 56.7 |
| 2001-02 | 0.0 | T | 2.4 | 0.0 | 3.7 | 1.2 | 4.5 | 1.1 | 3.0 | T | T | T | 15.9 |
| 2002-03 | T | 0.0 | 0.0 | 1.0 | 1.7 | 2.7 | 0.3 | 9.3 | 6.2 | 4.0 | 0.4 | T | 25.6 |
| 2003-04 | T | 0.0 | 0.0 | T | 0.7 | 0.8 | 9.3 | 3.7 | 2.2 | 6.6 | T | T | 23.3 |
| 2004-05 | T | T | 0.0 | 0.5 | 6.5 | 4.4 | 9.1 | 0.4 | 7.5 | 2.2 | T | T | 30.6 |
| 2005-06 | T | T | T | T | 0.2 | 10.4 | 2.8 | 1.0 | 2.5 | T | 0.5 | T | 17.4 |
| 2006-07 | 0.0 | 0.0 | 0.0 | 7.6 | 2.7 | 7.4 | 6.2 | 2.2 | T | 3.8 | T | T | 29.9 |
| 2007-08 | T | T | 0.0 | T | 1.4 | 7.2 | 8.0 | 2.7 | 11.7 | 2.5 | 0.5 | 0.0 | 34.0 |
| 2008-09 | 0.0 | 0.0 | 0.0 | T | 2.7 | 5.2 | 3.3 | 0.6 | 4.7 | 1.4 | 0.0 | 0.0 | 17.9 |
| 2009- | 0.0 | 0.0 | T | 1.8 | 1.5 | 9.4 | | | | | | | |
| POR= 47 YRS | T | T | 1.0 | 3.1 | 4.9 | 5.3 | 5.1 | 4.5 | 8.6 | 6.2 | 1.3 | T | 40.0 |

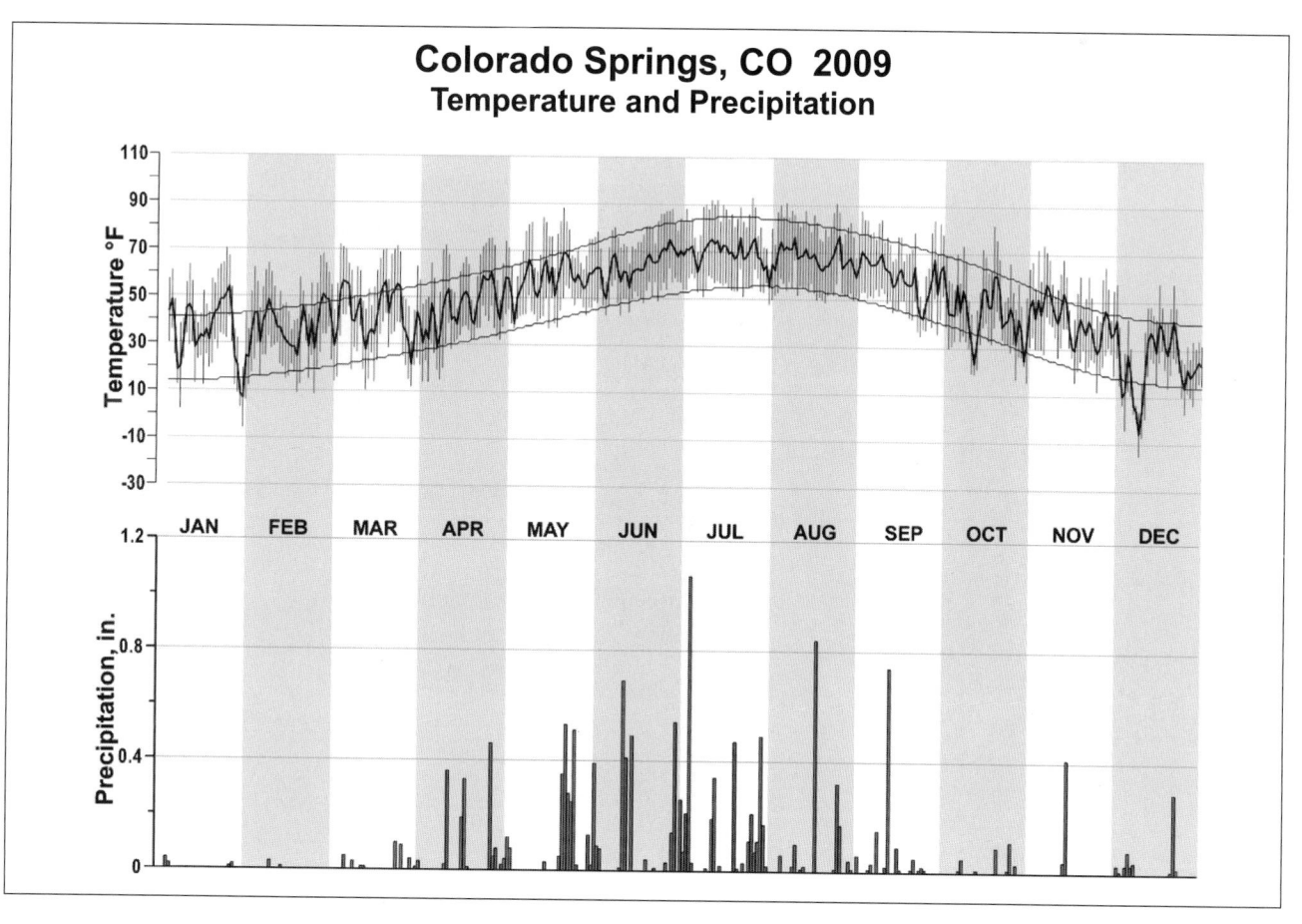

## Colorado Springs, CO 2009
### Temperature and Precipitation

# 2009
# DENVER
# COLORADO (KDEN)

Denver enjoys the invigorating climate that prevails over much of the central Rocky Mountain region, without the extremely cold mornings of the high elevations during winter, or the hot afternoons of summer at lower altitudes. Extremely warm or cold weather in Denver is usually of short duration.

Situated a long distance from any moisture source, and separated from the Pacific Ocean by several high mountain barriers, Denver enjoys low relative humidity, light precipitation, and abundant sunshine.

Air masses from four different sources influence Denver weather. These include arctic air from Canada and Alaska, warm, moist air from the Gulf of Mexico, warm, dry air from Mexico and the southwestern deserts, and Pacific air modified by its passage over mountains to the west.

In winter, the high altitude and mountains to the west combine to moderate temperatures in Denver. Invasions of cold air from the north, intensified by the high altitude, can be abrupt and severe. However, many of the cold air masses that spread southward out of Canada never reach the altitude of Denver, but move off over the lower plains to the east. Surges of air from the west are moderated in their descent down the east face of the Rockies, and reach Denver in the form of chinook winds that often raise temperatures into the 60s, even in midwinter.

In spring, polar air often collides with warm, moist air from the Gulf of Mexico and these collisions result in frequent, rapid and drastic weather changes. Spring is the cloudiest, windiest, and wettest season in the city. Much of the precipitation falls as snow, especially in March and early April. Stormy periods are interspersed with stretches of mild, sunny weather that quickly melt previous snow cover.

Summer precipitation falls mainly from scattered thunderstorms during the afternoon and evening. Mornings are usually clear and sunny, with clouds forming during early afternoon to cut off the sunshine at what would otherwise be the hottest part of the day. Severe thunderstorms, with large hail and heavy rain occasionally occur in the city, but these conditions are more common on the plains to the east.

Autumn is the most pleasant season. Few thunderstorms occur and invasions of cold air are infrequent. As a result, there is more sunshine and less severe weather than at any other time of the year.

Based on the 1951-1980 period, the average first occurrence of 32 degrees Fahrenheit in the fall is October 8 and the average last occurrence in the spring is May 3.

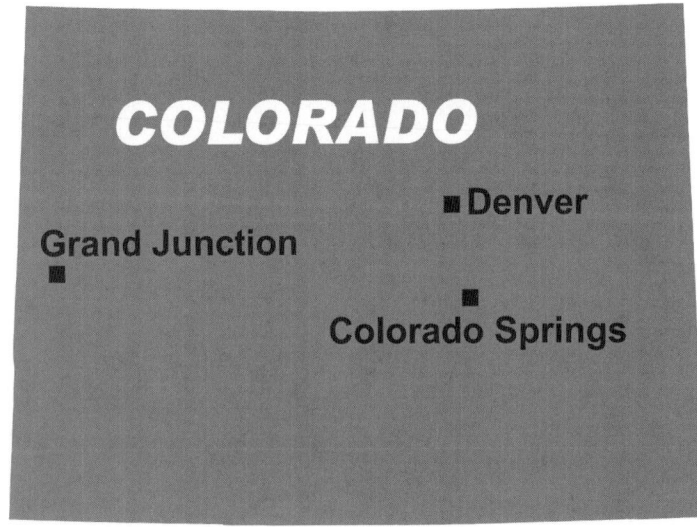

# NORMALS, MEANS, AND EXTREMES
## DENVER (KDEN)

| LATITUDE: 39° 49'N | LONGITUDE: -104° 39'W | ELEVATION (FT): GRND: 5437  BARO: 5382 | TIME ZONE: MOUNTAIN  (UTC -7) | WBAN: 03017 |
|---|---|---|---|---|

| | ELEMENT | POR | JAN | FEB | MAR | APR | MAY | JUN | JUL | AUG | SEP | OCT | NOV | DEC | YEAR |
|---|---|---|---|---|---|---|---|---|---|---|---|---|---|---|---|
| TEMPERATURE °F | NORMAL DAILY MAXIMUM | 30 | 43.2 | 47.2 | 53.7 | 60.9 | 70.5 | 82.1 | 88.0 | 86.0 | 77.4 | 66.0 | 51.5 | 44.1 | 64.2 |
| | MEAN DAILY MAXIMUM | 14 | 44.0 | 46.2 | 54.2 | 60.6 | 71.2 | 80.4 | 89.6 | 86.5 | 77.6 | 64.2 | 53.0 | 43.0 | 64.2 |
| | HIGHEST DAILY MAXIMUM | 14 | 72 | 77 | 79 | 85 | 94 | 102 | 105 | 104 | 97 | 89 | 80 | 72 | 105 |
| | YEAR OF OCCURRENCE | | 1997 | 2006 | 2004 | 2006 | 2003 | 2006 | 2005 | 2008 | 1995 | 2005 | 2006 | 2007 | JUL 2005 |
| | MEAN OF EXTREME MAXS. | 14 | 65.6 | 66.6 | 74.5 | 79.9 | 88.7 | 93.7 | 99.5 | 97.1 | 91.1 | 83.6 | 73.4 | 64.5 | 81.5 |
| | NORMAL DAILY MINIMUM | 30 | 15.2 | 19.1 | 25.4 | 34.2 | 43.8 | 53.0 | 58.7 | 57.4 | 47.3 | 35.9 | 23.5 | 16.4 | 35.8 |
| | MEAN DAILY MINIMUM | 14 | 18.3 | 19.9 | 26.2 | 33.1 | 43.2 | 51.0 | 59.9 | 57.7 | 48.6 | 36.2 | 26.2 | 17.9 | 36.5 |
| | LOWEST DAILY MINIMUM | 14 | -14 | -18 | -4 | 6 | 21 | 31 | 44 | 42 | 25 | 3 | -3 | -19 | -19 |
| | YEAR OF OCCURRENCE | | 1997 | 2007 | 2002 | 1997 | 2008 | 2007 | 1997 | 2004 | 1996 | 1997 | 2003 | 2008 | DEC 2008 |
| | MEAN OF EXTREME MINS. | 14 | -2.1 | -1.6 | 9.0 | 18.6 | 29.3 | 38.9 | 51.0 | 48.3 | 34.5 | 21.2 | 4.9 | -2.8 | 20.8 |
| | NORMAL DRY BULB | 30 | 29.2 | 33.2 | 39.6 | 47.6 | 57.2 | 67.6 | 73.4 | 71.7 | 62.4 | 51.0 | 37.5 | 30.3 | 50.1 |
| | MEAN DRY BULB | 14 | 31.2 | 33.0 | 40.2 | 46.8 | 57.2 | 66.6 | 74.8 | 72.2 | 63.1 | 50.2 | 39.6 | 30.5 | 50.5 |
| | MEAN WET BULB | 14 | 21.8 | 23.4 | 28.5 | 34.5 | 44.3 | 50.7 | 56.3 | 55.8 | 47.8 | 37.5 | 28.5 | 22.5 | 37.6 |
| | MEAN DEW POINT | 14 | 18.1 | 19.1 | 23.8 | 30.2 | 40.6 | 46.9 | 51.4 | 51.4 | 42.7 | 31.8 | 24.0 | 17.3 | 33.1 |
| | NORMAL NO. DAYS WITH: | | | | | | | | | | | | | | |
| | MAXIMUM >= 90 | 30 | 0.0 | 0.0 | 0.0 | * | 0.5 | 7.3 | 15.0 | 9.4 | 2.9 | 0.0 | 0.0 | 0.0 | 35.1 |
| | MAXIMUM <= 32 | 30 | 6.2 | 4.3 | 2.0 | 0.5 | 0.0 | 0.0 | 0.0 | 0.0 | * | 0.4 | 2.7 | 5.3 | 21.4 |
| | MINIMUM <= 32 | 30 | 29.6 | 25.8 | 23.1 | 11.0 | 1.1 | 0.0 | 0.0 | 0.0 | 1.1 | 8.2 | 23.9 | 28.8 | 152.6 |
| | MINIMUM <= 0 | 30 | 3.2 | 1.5 | 0.2 | * | 0.0 | 0.0 | 0.0 | 0.0 | 0.0 | 0.0 | 0.2 | 2.5 | 7.6 |
| H/C | NORMAL HEATING DEG. DAYS | 30 | 1111 | 892 | 788 | 524 | 267 | 60 | 1 | 9 | 136 | 436 | 826 | 1078 | 6128 |
| | NORMAL COOLING DEG. DAYS | 30 | 0 | 0 | 0 | 2 | 23 | 136 | 261 | 217 | 57 | 0 | 0 | 0 | 696 |
| RH | NORMAL (PERCENT) | 30 | | | | | | | | | | | | | |
| | HOUR 05 LST | 30 | | | | | | | | | | | | | |
| | HOUR 11 LST | 30 | | | | | | | | | | | | | |
| | HOUR 17 LST | 30 | | | | | | | | | | | | | |
| | HOUR 23 LST | 30 | | | | | | | | | | | | | |
| S | PERCENT POSSIBLE SUNSHINE | | | | | | | | | | | | | | |
| W/O | MEAN NO. DAYS WITH: | | | | | | | | | | | | | | |
| | HEAVY FOG(VISBY <= 1/4 MI) | 14 | 3.1 | 2.9 | 3.6 | 2.7 | 2.7 | 1.2 | 1.0 | 0.9 | 1.2 | 2.2 | 2.5 | 1.7 | 25.7 |
| | THUNDERSTORMS | 14 | 0.0 | 0.1 | 0.5 | 2.3 | 7.3 | 11.5 | 12.1 | 11.3 | 5.3 | 0.9 | 0.1 | 0.0 | 51.4 |
| CLOUDNESS | MEAN: | | | | | | | | | | | | | | |
| | SUNRISE-SUNSET (OKTAS) | | | | | | | | | | | | | | |
| | MIDNIGHT-MIDNIGHT (OKTAS) | | | | | | | | | | | | | | |
| | MEAN NO. DAYS WITH: | | | | | | | | | | | | | | |
| | CLEAR | 1 | 3.0 | 10.0 | 9.0 | 6.0 | 10.0 | 12.0 | 2.0 | 7.0 | 6.0 | 9.0 | | 13.0 | |
| | PARTLY CLOUDY | 1 | 4.0 | 2.0 | 6.0 | 4.0 | 5.5 | 9.0 | 2.0 | 9.0 | 6.0 | | | 1.0 | |
| | CLOUDY | 1 | 3.0 | 6.0 | 10.0 | 13.0 | 5.5 | 5.0 | 1.0 | 3.0 | 5.0 | 2.0 | | 2.0 | |
| PR | MEAN STATION PRESSURE(IN) | 14 | 24.55 | 24.53 | 24.53 | 24.52 | 24.56 | 24.60 | 24.68 | 24.69 | 24.66 | 24.61 | 24.60 | 24.56 | 24.59 |
| | MEAN SEA-LEVEL PRES. (IN) | 14 | 30.02 | 29.98 | 29.92 | 29.86 | 29.84 | 29.83 | 29.89 | 29.92 | 29.93 | 29.95 | 30.02 | 30.02 | 29.93 |
| WINDS | MEAN SPEED (MPH) | 14 | 9.7 | 9.9 | 10.6 | 11.6 | 10.4 | 10.2 | 9.6 | 9.4 | 9.4 | 9.7 | 9.4 | 9.8 | 10.0 |
| | PREVAIL.DIR(TENS OF DEGS) | 13 | 22 | 22 | 22 | 22 | 23 | 17 | 22 | 23 | 23 | 22 | 22 | 22 | 22 |
| | MAXIMUM 2-MINUTE: | | | | | | | | | | | | | | |
| | SPEED (MPH) | 14 | 43 | 48 | 53 | 53 | 49 | 53 | 54 | 49 | 45 | 46 | 52 | 54 | 54 |
| | DIR. (TENS OF DEGS) | | 35 | 36 | 28 | 33 | 22 | 26 | 13 | 28 | 31 | 29 | 32 | 28 | 28 |
| | YEAR OF OCCURRENCE | | 2009 | 2007 | 1995 | 2001 | 2002 | 2009 | 1999 | 2001 | 2006 | 2001 | 2005 | 2005 | DEC 2005 |
| | MAXIMUM 3-SECOND | | | | | | | | | | | | | | |
| | SPEED (MPH) | 14 | 49 | 60 | 56 | 63 | 61 | 68 | 64 | 61 | 54 | 54 | 61 | 64 | 68 |
| | DIR. (TENS OF DEGS) | | 35 | 36 | 18 | 36 | 18 | 27 | 13 | 29 | 31 | 29 | 32 | 28 | 27 |
| | YEAR OF OCCURRENCE | | 2009 | 2007 | 2007 | 2009 | 2008 | 2009 | 1999 | 2001 | 2006 | 2001 | 2005 | 2005 | JUN 2009 |
| PRECIPITATION | NORMAL (IN) | 30 | 0.51 | 0.49 | 1.28 | 1.93 | 2.32 | 1.56 | 2.16 | 1.82 | 1.14 | 0.99 | 0.98 | 0.63 | 15.81 |
| | MAXIMUM MONTHLY (IN) | 14 | 0.78 | 0.64 | 3.05 | 5.86 | 4.67 | 4.86 | 5.92 | 4.03 | 2.34 | 3.03 | 0.72 | 1.21 | 5.92 |
| | YEAR OF OCCURRENCE | | 2001 | 2001 | 2003 | 1999 | 1995 | 2009 | 1998 | 2008 | 1996 | 2007 | 2001 | 2006 | JUL 1998 |
| | MINIMUM MONTHLY (IN) | 14 | 0.03 | .02 | 0.14 | 0.23 | .71 | 0.12 | 0.24 | 0.56 | .07 | 0.08 | 0.05 | 0.04 | 0.02 |
| | YEAR OF OCCURRENCE | | 2003 | 2005 | 2008 | 2002 | 2005 | 2006 | 2008 | 1996 | 2005 | 2001 | 2003 | 2004 | FEB 2005 |
| | MAXIMUM IN 24 HOURS (IN) | 14 | 0.51 | 0.42 | 1.64 | 2.10 | 2.00 | 2.43 | 3.06 | 1.71 | 1.22 | 2.65 | 0.47 | 0.84 | 3.06 |
| | YEAR OF OCCURRENCE | | 2001 | 2007 | 2003 | 2007 | 2000 | 2003 | 1997 | 2004 | 1996 | 2007 | 1999 | 2006 | JUL 1997 |
| | NORMAL NO. DAYS WITH: | | | | | | | | | | | | | | |
| | PRECIPITATION >= 0.01 | 30 | 5.8 | 5.6 | 8.1 | 8.8 | 11.4 | 8.6 | 9.3 | 9.3 | 7.0 | 5.1 | 6.3 | 5.7 | 91.0 |
| | PRECIPITATION >= 1.00 | 30 | 0.0 | 0.0 | 0.2 | 0.2 | 0.4 | 0.2 | 0.6 | 0.4 | 0.2 | 0.1 | 0.0 | 0.1 | 2.4 |
| SNOWFALL | NORMAL (IN) | 30 | 7.7 | 6.3 | 11.6 | 8.8 | 1.3 | 0.* | 0.0 | 0.0 | 1.9 | 3.9 | 10.6 | 8.9 | 61.0 |
| | MAXIMUM MONTHLY (IN) | 4 | 15.9 | 7.4 | 13.8 | 7.4 | 3.4 | T | T | T | 0.0 | 17.2 | 9.3 | 29.4 | 29.4 |
| | YEAR OF OCCURRENCE | | 2007 | 2007 | 2009 | 2009 | 2008 | 2009 | 2008 | 2009 | | 2009 | 2009 | 2006 | DEC 2006 |
| | MAXIMUM IN 24 HOURS (IN) | 4 | 3.3 | 1.9 | 7.8 | 4.5 | 0.2 | T | T | T | 0.0 | 7.2 | 3.8 | 17.4 | 17.4 |
| | YEAR OF OCCURRENCE' | | 2006 | 2006 | 2009 | 2009 | 2006 | 2009 | 2008 | 2009 | | 2009 | 2009 | 2006 | DEC 2006 |
| | MAXIMUM SNOW DEPTH (IN) | 4 | 14 | 11 | 7 | 4 | T | 0 | 0 | 0 | 0 | 5 | 5 | 21 | 21 |
| | YEAR OF OCCURRENCE | | 2007 | 2007 | 2009 | 2009 | 2006 | | | | | 2009 | 2009 | 2006 | DEC 2006 |
| | NORMAL NO. DAYS WITH: | | | | | | | | | | | | | | |
| | SNOWFALL >= 1.0 | 30 | 2.3 | 2.2 | 3.3 | 2.7 | 0.3 | 0.0 | 0.0 | 0.0 | 0.5 | 1.2 | 3.1 | 2.8 | 18.4 |

## PRECIPITATION (inches) 2009 DENVER (KDEN)

| YEAR | JAN | FEB | MAR | APR | MAY | JUN | JUL | AUG | SEP | OCT | NOV | DEC | ANNUAL |
|---|---|---|---|---|---|---|---|---|---|---|---|---|---|
| 1995 | | | 0.28 | 2.44 | 4.67 | 3.07 | 2.31 | 1.04 | 2.28 | 0.72 | 0.31 | 0.06 | |
| 1996 | 0.29 | 0.09 | 0.77 | 0.33 | 2.40 | 1.77 | 1.01 | 0.56 | 2.34 | 0.39 | 0.38 | 0.06 | 10.39 |
| 1997 | 0.26 | 0.54 | 0.26 | 1.30 | 1.57 | 2.57 | 5.60 | 3.52 | 0.97 | 1.87 | 0.61 | 0.50 | 19.57 |
| 1998 | 0.05 | 0.23 | 0.86 | 2.47 | 1.73 | 0.73 | 5.92 | 1.19 | 0.73 | 1.20 | 0.40 | 0.42 | 15.93 |
| 1999 | 0.38 | 0.15 | 0.19 | 5.86 | 2.37 | 2.52 | 3.84 | 3.37 | 1.20 | 0.31 | 0.47 | 0.29 | 20.95 |
| 2000 | 0.24 | 0.23 | 1.96 | 0.71 | 3.09 | 0.79 | 1.42 | 3.06 | 1.52 | 0.52 | 0.61 | 0.27 | 14.42 |
| 2001 | 0.78 | 0.64 | 1.10 | 1.20 | 3.80 | 1.53 | 4.76 | 0.71 | 1.00 | 0.08 | 0.72 | 0.14 | 16.46 |
| 2002 | 0.48 | 0.32 | 0.53 | 0.23 | 0.94 | 1.45 | 1.39 | 0.78 | 0.58 | 0.49 | 0.24 | 0.05 | 7.48 |
| 2003 | 0.03 | 0.47 | 3.05 | 2.22 | 1.91 | 3.95 | 0.54 | 1.24 | 0.26 | 0.08 | 0.05 | 0.12 | 13.92 |
| 2004 | 0.23 | 0.21 | 0.14 | 1.76 | 1.30 | 2.33 | 2.51 | 2.84 | 1.99 | 0.86 | 0.45 | 0.04 | 14.66 |
| 2005 | 0.37 | 0.02 | 0.59 | 2.45 | 0.71 | 3.99 | 0.27 | 1.33 | 0.07 | 2.16 | 0.48 | 0.35 | 12.79 |
| 2006 | 0.28 | 0.15 | 0.56 | 0.67 | 0.94 | 0.12 | 1.37 | 1.13 | 0.84 | 1.03 | 0.34 | 1.21 | 8.64 |
| 2007 | 0.55 | 0.36 | 0.57 | 2.65 | 1.79 | 0.52 | 0.43 | 2.76 | 0.54 | 3.03 | 0.20 | 0.60 | 14.00 |
| 2008 | 0.08 | 0.18 | 0.14 | 0.32 | 1.56 | 0.73 | 0.24 | 4.03 | 1.04 | 1.44 | 0.18 | 0.24 | 10.18 |
| 2009 | 0.13 | 0.04 | 0.83 | 3.22 | 1.30 | 4.86 | 3.56 | 1.14 | 0.74 | 1.36 | 0.49 | 0.45 | 18.12 |
| POR= 14 YRS | 0.30 | 0.26 | 0.79 | 1.86 | 2.01 | 2.06 | 2.34 | 1.91 | 1.07 | 1.04 | 0.40 | 0.32 | 14.36 |

WBAN : 03017

## AVERAGE TEMPERATURE (°F) 2009 DENVER (KDEN)

| YEAR | JAN | FEB | MAR | APR | MAY | JUN | JUL | AUG | SEP | OCT | NOV | DEC | ANNUAL |
|---|---|---|---|---|---|---|---|---|---|---|---|---|---|
| 1995 | | | 39.3 | 42.9 | 50.0 | 62.2 | 70.9 | 75.3 | 61.7 | 48.5 | 41.8 | 33.0 | |
| 1996 | 27.0 | 33.9 | 36.0 | 48.0 | 58.1 | 68.2 | 73.4 | 71.6 | 60.8 | 50.9 | 37.2 | 33.0 | 49.8 |
| 1997 | 27.9 | 30.0 | 42.1 | 40.5 | 56.6 | 67.8 | 73.1 | 69.7 | 64.3 | 49.7 | 34.8 | 27.9 | 48.7 |
| 1998 | 32.7 | 33.9 | 36.9 | 44.8 | 59.1 | 63.0 | 74.3 | 71.7 | 68.0 | 50.2 | 42.1 | 28.9 | 50.5 |
| 1999 | 33.7 | 38.6 | 43.7 | 42.6 | 54.8 | 64.2 | 73.9 | 71.2 | 59.2 | 52.5 | 47.3 | 33.8 | 51.3 |
| 2000 | 33.0 | 39.2 | 40.4 | 49.8 | 59.2 | 67.0 | 76.7 | 74.5 | 63.6 | 50.5 | 28.9 | 28.3 | 50.9 |
| 2001 | 30.0 | 28.3 | 39.8 | 49.6 | 57.1 | 69.4 | 76.7 | 73.5 | 66.8 | 51.5 | 40.9 | 31.7 | 51.3 |
| 2002 | 29.3 | 33.6 | 33.8 | 50.0 | 56.1 | 71.1 | 76.3 | 71.5 | 63.6 | 44.0 | 37.5 | 33.6 | 50.0 |
| 2003 | 36.9 | 27.6 | 39.9 | 50.4 | 57.5 | 62.1 | 76.9 | 73.7 | 59.3 | 55.1 | 36.2 | 32.5 | 50.7 |
| 2004 | 31.9 | 30.4 | 46.4 | 47.5 | 59.3 | 63.7 | 70.8 | 68.2 | 62.7 | 50.9 | 37.2 | 34.9 | 50.3 |
| 2005 | 32.5 | 35.8 | 39.4 | 46.4 | 57.0 | 66.0 | 77.7 | 71.6 | 67.2 | 52.0 | 42.5 | 30.2 | 51.5 |
| 2006 | 37.4 | 30.9 | 38.4 | 51.5 | 60.4 | 72.8 | 76.2 | 72.8 | 58.9 | 49.5 | 40.5 | 31.7 | 51.8 |
| 2007 | 20.8 | 29.1 | 46.1 | 46.8 | 58.0 | 68.8 | 76.4 | 75.4 | 65.1 | 53.5 | 41.4 | 26.7 | 50.7 |
| 2008 | 27.9 | 33.9 | 39.6 | 46.1 | 55.9 | 67.4 | 77.6 | 71.5 | 61.8 | 52.1 | 43.1 | 26.7 | 50.3 |
| 2009 | 34.9 | 37.3 | 41.8 | 45.9 | 59.0 | 64.4 | 70.3 | 70.3 | 63.5 | 42.9 | 42.6 | 24.2 | 49.8 |
| POR= 14 YRS | 31.2 | 33.0 | 40.2 | 46.8 | 57.2 | 66.6 | 74.8 | 72.2 | 63.1 | 50.2 | 39.6 | 30.5 | 50.5 |

## HEATING DEGREE DAYS (base 65°F) 2009  DENVER (KDEN)

| YEAR | JUL | AUG | SEP | OCT | NOV | DEC | JAN | FEB | MAR | APR | MAY | JUN | TOTAL |
|------|-----|-----|-----|-----|-----|-----|-----|-----|-----|-----|-----|-----|-------|
| 1994-95 | | | | | | | | | 788 | 655 | 457 | 132 | |
| 1995-96 | 26 | 2 | 188 | 505 | 686 | 981 | 1166 | 894 | 893 | 230 | 29 | 0 | 5600 |
| 1996-97 | 0 | 4 | 192 | 444 | 824 | 985 | 1142 | 975 | 704 | 728 | 264 | 35 | 6297 |
| 1997-98 | 2 | 11 | 92 | 475 | 895 | 1142 | 996 | 865 | 865 | 597 | 186 | 137 | 6263 |
| 1998-99 | 1 | 1 | 46 | 453 | 680 | 1113 | 962 | 731 | 654 | 666 | 311 | 85 | 5703 |
| 1999-00 | 1 | 3 | 194 | 383 | 528 | 962 | 984 | 744 | 754 | 446 | 215 | 61 | 5275 |
| 2000-01 | 0 | 5 | 149 | 447 | 1074 | 1131 | 1079 | 1021 | 775 | 455 | 256 | 46 | 6438 |
| 2001-02 | 0 | 4 | 65 | 416 | 717 | 1026 | 1098 | 870 | 961 | 448 | 302 | 26 | 5933 |
| 2002-03 | 0 | 5 | 118 | 643 | 816 | 966 | 865 | 1041 | 770 | 430 | 260 | 107 | 6021 |
| 2003-04 | 0 | 10 | 192 | 312 | 858 | 1001 | 1022 | 998 | 569 | 519 | 192 | 99 | 5772 |
| 2004-05 | 17 | 31 | 131 | 431 | 830 | 926 | 999 | 810 | 786 | 552 | 274 | 84 | 5871 |
| 2005-06 | 3 | 19 | 54 | 412 | 670 | 1075 | 848 | 948 | 817 | 398 | 192 | 4 | 5440 |
| 2006-07 | 7 | 8 | 195 | 487 | 727 | 1028 | 1361 | 1001 | 582 | 544 | 223 | 45 | 6208 |
| 2007-08 | 0 | 0 | 76 | 354 | 701 | 1181 | 1143 | 895 | 781 | 558 | 303 | 64 | 6056 |
| 2008-09 | 0 | 28 | 116 | 393 | 651 | 1179 | 926 | 764 | 711 | 566 | 204 | 78 | 5616 |
| 2009- | 13 | 9 | 117 | 676 | 664 | 1260 | | | | | | | |

WBAN : 03017

## COOLING DEGREE DAYS (base 65°F) 2009  DENVER (KDEN)

| YEAR | JAN | FEB | MAR | APR | MAY | JUN | JUL | AUG | SEP | OCT | NOV | DEC | TOTAL |
|------|-----|-----|-----|-----|-----|-----|-----|-----|-----|-----|-----|-----|-------|
| 1995 | | | 0 | 0 | 0 | 55 | 212 | 327 | 98 | 0 | 0 | 0 | |
| 1996 | 0 | 0 | 0 | 3 | 26 | 133 | 269 | 215 | 71 | 13 | 0 | 0 | 730 |
| 1997 | 0 | 0 | 0 | 0 | 11 | 126 | 260 | 160 | 77 | 8 | 0 | 0 | 642 |
| 1998 | 0 | 0 | 0 | 0 | 13 | 88 | 296 | 215 | 143 | 0 | 0 | 0 | 755 |
| 1999 | 0 | 0 | 0 | 0 | 2 | 69 | 283 | 203 | 30 | 2 | 0 | 0 | 589 |
| 2000 | 0 | 0 | 0 | 0 | 43 | 127 | 368 | 305 | 115 | 5 | 0 | 0 | 963 |
| 2001 | 0 | 0 | 0 | 0 | 18 | 184 | 373 | 274 | 125 | 5 | 0 | 0 | 979 |
| 2002 | 0 | 0 | 0 | 7 | 34 | 218 | 355 | 214 | 84 | 0 | 0 | 0 | 912 |
| 2003 | 0 | 0 | 0 | 0 | 36 | 28 | 376 | 289 | 29 | 13 | 0 | 0 | 771 |
| 2004 | 0 | 0 | 0 | 0 | 23 | 67 | 204 | 138 | 67 | 0 | 0 | 0 | 499 |
| 2005 | 0 | 0 | 0 | 0 | 34 | 119 | 405 | 230 | 125 | 14 | 0 | 0 | 927 |
| 2006 | 0 | 0 | 0 | 0 | 56 | 247 | 363 | 261 | 20 | 14 | 0 | 0 | 961 |
| 2007 | 0 | 0 | 0 | 5 | 15 | 167 | 359 | 329 | 86 | 2 | 0 | 0 | 963 |
| 2008 | 0 | 0 | 0 | 0 | 28 | 144 | 395 | 232 | 26 | 0 | 0 | 0 | 825 |
| 2009 | 0 | 0 | 0 | 0 | 26 | 68 | 181 | 180 | 78 | 0 | 0 | 0 | 533 |

## SNOWFALL (inches) 2009 DENVER (KDEN)

| YEAR | JUL | AUG | SEP | OCT | NOV | DEC | JAN | FEB | MAR | APR | MAY | JUN | TOTAL |
|------|-----|-----|-----|-----|-----|-----|-----|-----|-----|-----|-----|-----|-------|
| 2005-06 | | | | | | | 3.6 | 3.0 | 8.6 | 0.3 | 0.2 | T | 72.6 |
| 2006-07 | 0.0 | T | 0.0 | 9.8 | 4.4 | 29.4 | 15.9 | 7.4 | 4.8 | 0.9 | 0.0 | 0.0 | 44.2 |
| 2007-08 | 0.0 | 0.0 | 0.0 | 3.0 | 2.5 | 20.9 | 1.9 | 5.1 | 4.5 | 2.9 | 3.4 | T | 38.1 |
| 2008-09 | T | T | 0.0 | T | 1.7 | 10.3 | 4.9 | T | 13.8 | 7.4 | 0.0 | T | |
| 2009- | 0.0 | T | 0.0 | 17.2 | 9.3 | 11.1 | | | | | | | |
| POR= 4 YRS | T | T | 0.0 | 7.5 | 4.5 | 17.9 | 6.6 | 3.9 | 7.9 | 2.9 | 0.9 | T | 52.1 |

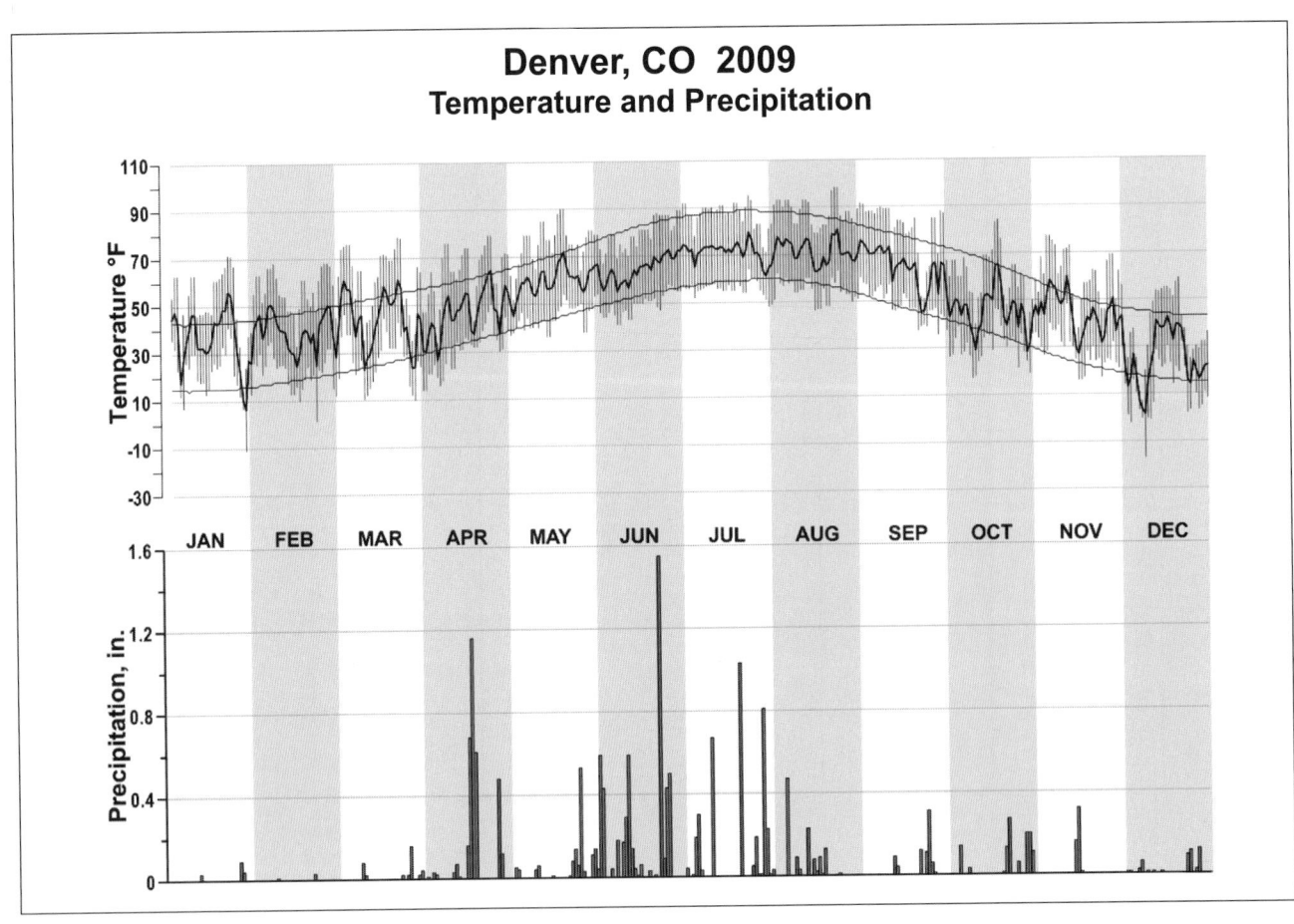

## Denver, CO 2009
### Temperature and Precipitation

# 2009
# GRAND JUNCTION
# COLORADO (KGJT)

Grand Junction is located at the junction of the Colorado and Gunnison Rivers. It is on the west slope of the Rockies, in a large mountain valley. The area has a climate marked by the wide seasonal range usual to interior localities at this latitude. Thanks, however, to the protective topography of the vicinity, sudden and severe weather changes are very infrequent. The valley floor slopes from 4,800 feet near Palisade to 4,400 feet at the west end near Fruita. Mountains are on all sides at distances of from 10 to 60 miles and reach heights of 9,000 to over 12,000 feet.

This mountain valley location, with attendant valley breezes, provides protection from spring and fall frosts. This results in a growing season averaging 191 days in the city. This varies considerably in the outlying districts. It is about the same in the upper valley around Palisade, and 3 to 4 weeks shorter near the river west of Grand Junction. The growing season is sufficiently long to permit commercial growth of almost all fruits except citrus varieties. Summer grazing of cattle and sheep on nearby mountain ranges is extensive.

The interior, continental location, ringed by mountains on all sides, results in quite low precipitation in all seasons. Consequently, agriculture is dependent on irrigation. Adequate supplies of water are available from mountain snows and rains. Summer rains occur chiefly as scattered light showers and thunderstorms which develop over nearby mountains. Winter snows are fairly frequent, but are mostly light and quick to melt. Even the infrequent snows of from 4 to 8 inches seldom remain on the ground for prolonged periods. Blizzard conditions in the valley are extremely rare.

Temperatures above 100 degrees are infrequent, and about one-third of the winters have no readings below zero. Summer days with maximum temperatures in the middle 90s and minimums in the low 60s are common. Relative humidity is very low during the summer, with values similar to other dry locations such as the southern parts of New Mexico and Arizona. Spells of cold winter weather are sometimes prolonged due to cold air becoming trapped in the valley. Winds are usually very light during the coldest weather. Changes in winter are normally gradual, and abrupt changes are much less frequent than in eastern Colorado. Cold waves are rare. Sunny days predominate in all seasons.

The prevailing wind is from the east-southeast due to the valley breeze effect. The strongest winds are associated with thunderstorms or with pre-frontal weather. They usually are from the south or southwest.

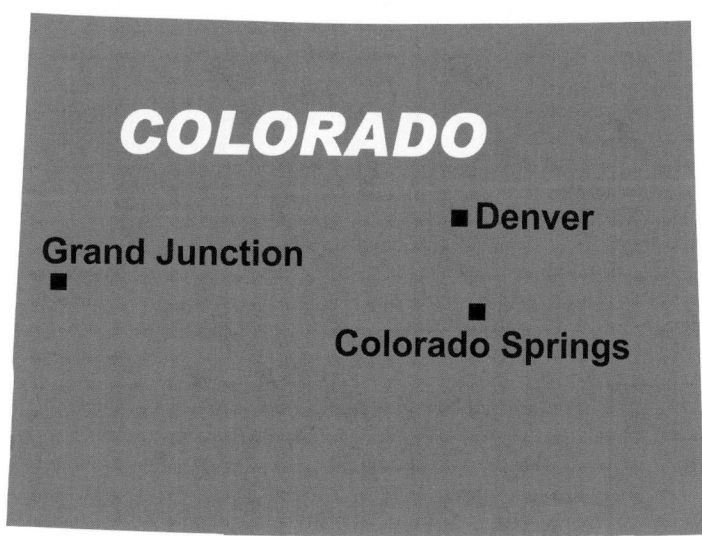

# NORMALS, MEANS, AND EXTREMES
## GRAND JUNCTION (KGJT)

| LATITUDE: 39 ° 8 'N | LONGITUDE: -108° 32'W | | ELEVATION (FT): GRND: 4820  BARO: 4826 | | | | TIME ZONE: MOUNTAIN | (UTC -7) | | | WBAN: 23066 | |
|---|---|---|---|---|---|---|---|---|---|---|---|---|

| ELEMENT | POR | JAN | FEB | MAR | APR | MAY | JUN | JUL | AUG | SEP | OCT | NOV | DEC | YEAR |
|---|---|---|---|---|---|---|---|---|---|---|---|---|---|---|
| **TEMPERATURE °F** | | | | | | | | | | | | | | |
| NORMAL DAILY MAXIMUM | 30 | 36.6 | 45.4 | 55.7 | 64.3 | 74.5 | 86.9 | 92.1 | 89.6 | 80.3 | 66.7 | 49.8 | 38.9 | 65.1 |
| MEAN DAILY MAXIMUM | 110 | 36.6 | 42.9 | 55.1 | 64.2 | 75.6 | 85.5 | 92.8 | 89.4 | 79.4 | 67.1 | 50.4 | 38.8 | 64.8 |
| HIGHEST DAILY MAXIMUM | 63 | 60 | 68 | 81 | 89 | 101 | 105 | 106 | 103 | 100 | 88 | 76 | 64 | 106 |
| YEAR OF OCCURRENCE | | 1971 | 1986 | 2004 | 1992 | 2000 | 1990 | 2005 | 2000 | 1995 | 2003 | 2008 | 1980 | JUL 2005 |
| MEAN OF EXTREME MAXS. | 110 | 49.5 | 58.7 | 71.0 | 80.7 | 89.9 | 98.2 | 101.0 | 98.2 | 92.6 | 81.6 | 65.1 | 52.2 | 78.2 |
| NORMAL DAILY MINIMUM | 30 | 15.6 | 22.7 | 31.0 | 37.5 | 46.4 | 55.3 | 61.4 | 59.7 | 50.4 | 38.6 | 26.3 | 17.5 | 38.5 |
| MEAN DAILY MINIMUM | 110 | 15.9 | 22.4 | 31.2 | 38.7 | 48.2 | 56.3 | 64.1 | 61.9 | 52.2 | 40.9 | 27.8 | 18.6 | 39.9 |
| LOWEST DAILY MINIMUM | 63 | -23 | -18 | 5 | 11 | 26 | 34 | 44 | 43 | 29 | 18 | -4 | -17 | -23 |
| YEAR OF OCCURRENCE | | 1963 | 1989 | 1948 | 1975 | 2002 | 1976 | 1993 | 1968 | 1978 | 1993 | 2004 | 1990 | JAN 1963 |
| MEAN OF EXTREME MINS. | 110 | 1.7 | 8.2 | 17.7 | 25.5 | 34.5 | 44.3 | 55.7 | 52.9 | 39.9 | 28.0 | 14.6 | 4.8 | 27.3 |
| NORMAL DRY BULB | 30 | 26.1 | 34.1 | 43.4 | 50.9 | 60.5 | 71.1 | 76.8 | 74.7 | 65.4 | 52.7 | 38.1 | 28.2 | 51.8 |
| MEAN DRY BULB | 110 | 26.3 | 32.7 | 43.1 | 51.5 | 61.9 | 71.0 | 78.5 | 75.7 | 65.9 | 54.0 | 39.1 | 28.7 | 52.4 |
| MEAN WET BULB | 26 | 23.2 | 28.3 | 33.8 | 38.5 | 45.1 | 49.7 | 56.1 | 55.8 | 49.3 | 40.8 | 31.2 | 24.0 | 39.7 |
| MEAN DEW POINT | 26 | 19.3 | 23.4 | 26.0 | 29.4 | 34.6 | 37.6 | 45.6 | 47.1 | 40.3 | 32.3 | 25.9 | 19.4 | 31.7 |
| NORMAL NO. DAYS WITH: | | | | | | | | | | | | | | |
| MAXIMUM >= 90 | 3 0 | 0.0 | 0.0 | 0.0 | 0.0 | 1.1 | 15.4 | 24.0 | 19.6 | 4.3 | 0.0 | 0.0 | 0.0 | 64.4 |
| MAXIMUM <= 32 | 3 0 | 10.0 | 2.3 | 0.1 | 0.0 | 0.0 | 0.0 | 0.0 | 0.0 | 0.0 | 0.0 | 0.8 | 6.0 | 19.2 |
| MINIMUM <= 32 | 30 | 30.1 | 24.4 | 15.7 | 6.1 | 0.4 | 0.0 | 0.0 | 0.0 | 0.2 | 3.5 | 21.9 | 29.8 | 132.1 |
| MINIMUM <= 0 | 30 | 3.0 | 0.7 | 0.0 | 0.0 | 0.0 | 0.0 | 0.0 | 0.0 | 0.0 | 0.0 | * | 1.0 | 4.7 |
| **H/C** | | | | | | | | | | | | | | |
| NORMAL HEATING DEG. DAYS | 30 | 1194 | 860 | 656 | 409 | 178 | 33 | 12 | 1 | 73 | 367 | 792 | 1125 | 5700 |
| NORMAL COOLING DEG. DAYS | 30 | 0 | 0 | 0 | 2 | 44 | 232 | 394 | 318 | 100 | 1 | 0 | 0 | 1091 |
| **RH** | | | | | | | | | | | | | | |
| NORMAL (PERCENT) | 30 | 72 | 64 | 52 | 44 | 40 | 30 | 34 | 38 | 40 | 47 | 60 | 69 | 49 |
| HOUR 05 LST | 30 | 79 | 74 | 65 | 59 | 56 | 44 | 49 | 52 | 54 | 60 | 71 | 78 | 62 |
| HOUR 11 LST | 30 | 65 | 55 | 43 | 36 | 32 | 24 | 29 | 32 | 34 | 39 | 51 | 61 | 42 |
| HOUR 17 LST | 30 | 63 | 49 | 36 | 29 | 26 | 19 | 22 | 25 | 27 | 33 | 48 | 59 | 36 |
| HOUR 23 LST | 30 | 77 | 69 | 58 | 49 | 45 | 32 | 37 | 41 | 43 | 52 | 66 | 74 | 54 |
| **S** PERCENT POSSIBLE SUNSHINE | 56 | 61 | 65 | 65 | 70 | 73 | 81 | 79 | 77 | 79 | 74 | 63 | 61 | 71 |
| **W/O** MEAN NO. DAYS WITH: | | | | | | | | | | | | | | |
| HEAVY FOG(VISBY <= 1/4 MI) | 46 | 2.8 | 1.6 | 0.6 | 0.2 | 0.0 | 0.0 | 0.0 | 0.0 | 0.0 | 0.2 | 0.9 | 2.1 | 8.4 |
| THUNDERSTORMS | 62 | 0.1 | 0.2 | 0.8 | 1.8 | 4.4 | 4.5 | 7.4 | 7.7 | 5.3 | 1.6 | 0.4 | 0.1 | 34.3 |
| **CLOUDNESS** MEAN: | | | | | | | | | | | | | | |
| SUNRISE-SUNSET (OKTAS) | 50 | 4.9 | 5.0 | 5.0 | 4.8 | 4.4 | 3.2 | 3.3 | 3.4 | 2.9 | 3.4 | 4.2 | 4.5 | 4.1 |
| MIDNIGHT-MIDNIGHT (OKTAS) | 32 | 4.6 | 4.4 | 4.6 | 4.3 | 4.0 | 3.0 | 3.2 | 3.3 | 2.7 | 3.2 | 4.0 | 4.3 | 3.8 |
| MEAN NO. DAYS WITH: | | | | | | | | | | | | | | |
| CLEAR | 50 | 9.1 | 7.6 | 8.0 | 7.9 | 9.6 | 14.9 | 13.7 | 13.1 | 16.0 | 14.7 | 10.4 | 9.6 | 134.6 |
| PARTLY CLOUDY | 50 | 7.0 | 7.4 | 8.6 | 9.4 | 10.7 | 9.4 | 11.2 | 11.2 | 8.1 | 7.6 | 7.4 | 7.6 | 105.6 |
| CLOUDY | 50 | 14.8 | 13.3 | 14.4 | 12.7 | 10.7 | 5.7 | 5.5 | 6.1 | 5.3 | 8.0 | 11.6 | 13.1 | 121.2 |
| **PR** MEAN STATION PRESSURE(IN) | 26 | 25.25 | 25.18 | 25.12 | 25.09 | 25.09 | 25.12 | 25.19 | 25.22 | 25.20 | 25.20 | 25.23 | 25.25 | 25.18 |
| MEAN SEA-LEVEL PRES. (IN) | 26 | 30.23 | 30.13 | 29.96 | 29.87 | 29.82 | 29.81 | 29.86 | 29.90 | 29.93 | 30.01 | 30.12 | 30.22 | 29.99 |
| **WINDS** MEAN SPEED (MPH) | 26 | 5.6 | 6.6 | 8.0 | 8.9 | 9.2 | 9.5 | 9.3 | 8.9 | 8.6 | 7.7 | 6.3 | 5.6 | 7.9 |
| PREVAIL.DIR(TENS OF DEGS) | 4 | 08 | 11 | 11 | 11 | 12 | 12 | 12 | 12 | 12 | 12 | 11 | 08 | 12 |
| MAXIMUM 2-MINUTE: | | | | | | | | | | | | | | |
| SPEED (MPH) | 13 | 40 | 41 | 53 | 49 | 46 | 48 | 45 | 45 | 44 | 51 | 45 | 44 | 53 |
| DIR. (TENS OF DEGS) | | 18 | 20 | 34 | 19 | 18 | 30 | 35 | 31 | 26 | 24 | 33 | 22 | 34 |
| YEAR OF OCCURRENCE | | 2005 | 1998 | 2001 | 2007 | 2008 | 1999 | 2002 | 2002 | 2007 | 2006 | 2005 | 2007 | MAR 2001 |
| MAXIMUM 3-SECOND | | | | | | | | | | | | | | |
| SPEED (MPH) | 13 | 49 | 55 | 66 | 63 | 68 | 55 | 68 | 64 | 59 | 62 | 55 | 54 | 68 |
| DIR. (TENS OF DEGS) | | 18 | 19 | 26 | 20 | 28 | 31 | 36 | 26 | 18 | 25 | 33 | 21 | 28 |
| YEAR OF OCCURRENCE | | 2005 | 1998 | 2009 | 2007 | 2004 | 1999 | 2002 | 1999 | 2009 | 2006 | 2005 | 2007 | MAY 2004 |
| **PRECIPITATION** NORMAL (IN) | 30 | 0.60 | 0.50 | 1.00 | 0.86 | 0.98 | 0.41 | 0.66 | 0.84 | 0.91 | 1.00 | 0.71 | 0.52 | 8.99 |
| MAXIMUM MONTHLY (IN) | 63 | 2.46 | 1.56 | 2.02 | 2.30 | 2.04 | 2.07 | 1.92 | 3.48 | 2.84 | 3.45 | 2.00 | 2.05 | 3.48 |
| YEAR OF OCCURRENCE | | 1957 | 1948 | 1979 | 2004 | 1995 | 1969 | 1983 | 1957 | 1997 | 1972 | 1983 | 2007 | AUG 1957 |
| MINIMUM MONTHLY (IN) | 63 | T | T | 0.02 | 0.06 | T | T | 0.01 | 0.04 | T | 0.00 | T | 0.01 | 0.00 |
| YEAR OF OCCURRENCE | | 1961 | 1972 | 1972 | 1958 | 1970 | 1980 | 1994 | 1956 | 1953 | 1952 | 1989 | 1976 | OCT 1952 |
| MAXIMUM IN 24 HOURS (IN) | 63 | .74 | 0.81 | 1.15 | 1.33 | 1.13 | 1.57 | 1.42 | 1.68 | 1.35 | 1.24 | 0.83 | 1.16 | 1.68 |
| YEAR OF OCCURRENCE | | 2005 | 1996 | 1993 | 1965 | 1983 | 1969 | 1974 | 1997 | 1965 | 1957 | 1983 | 1951 | AUG 1997 |
| NORMAL NO. DAYS WITH: | | | | | | | | | | | | | | |
| PRECIPITATION >= 0.01 | 30 | 6.6 | 5.4 | 7.8 | 7.1 | 7.1 | 4.0 | 5.5 | 6.3 | 6.5 | 6.2 | 6.0 | 5.6 | 74.1 |
| PRECIPITATION >= 1.00 | 30 | 0.0 | 0.0 | * | 0.0 | 0.0 | 0.0 | * | 0.1 | 0.0 | 0.0 | 0.0 | 0.0 | 0.1 |
| **SNOWFALL** NORMAL (IN) | 30 | 6.0 | 3.2 | 2.8 | 1.5 | 0.2 | 0.0 | 0.0 | 0.0 | 0.0 | 0.7 | 2.3 | 4.8 | 21.5 |
| MAXIMUM MONTHLY (IN) | 63 | 33.7 | 18.4 | 14.9 | 14.3 | 5.0 | T | T | T | 3.1 | 6.1 | 12.1 | 19.0 | 33.7 |
| YEAR OF OCCURRENCE | | 1957 | 1948 | 1948 | 1975 | 1979 | 2009 | 2001 | 1993 | 1965 | 1975 | 1964 | 1983 | JAN 1957 |
| MAXIMUM IN 24 HOURS (IN) | 63 | 9.1 | 9.0 | 6.1 | 8.9 | 5.0 | T | T | T | 3.1 | 6.1 | 8.4 | 7.2 | 9.1 |
| YEAR OF OCCURRENCE' | | 1957 | 1989 | 1948 | 1975 | 1979 | 2009 | 2001 | 1993 | 1965 | 1975 | 1954 | 2009 | JAN 1957 |
| MAXIMUM SNOW DEPTH (IN) | 61 | 16 | 12 | 8 | 7 | 1 | 0 | 0 | 0 | 2 | 5 | 8 | 11 | 16 |
| YEAR OF OCCURRENCE | | 1957 | 1957 | 1960 | 1975 | 1979 | | | | 1965 | 1975 | 1954 | 1983 | JAN 1957 |
| NORMAL NO. DAYS WITH: | | | | | | | | | | | | | | |
| SNOWFALL >= 1.0 | 30 | 2.2 | 1.0 | 0.9 | 0.5 | 0.1 | 0.0 | 0.0 | 0.0 | 0.0 | 0.2 | 0.8 | 1.8 | 7.5 |

## PRECIPITATION (inches) 2009  GRAND JUNCTION (KGJT)

| YEAR | JAN | FEB | MAR | APR | MAY | JUN | JUL | AUG | SEP | OCT | NOV | DEC | ANNUAL |
|------|-----|-----|-----|-----|-----|-----|-----|-----|-----|-----|-----|-----|--------|
| 1980 | 0.57 | 1.10 | 1.77 | 0.53 | 1.17 | T | 0.96 | 1.39 | 0.58 | 1.31 | 0.52 | 0.24 | 10.14 |
| 1981 | 0.44 | 0.16 | 1.35 | 0.56 | 1.49 | 0.17 | 0.41 | 0.82 | 0.25 | 2.06 | 0.47 | 0.60 | 8.78 |
| 1982 | 0.29 | 0.41 | 0.79 | 0.09 | 0.75 | 0.21 | 0.35 | 0.94 | 2.81 | 0.83 | 0.48 | 0.27 | 8.22 |
| 1983 | 0.50 | 0.64 | 1.59 | 0.90 | 1.68 | 1.54 | 1.92 | 0.73 | 1.11 | 0.36 | 2.00 | 1.85 | 14.82 |
| 1984 | 0.28 | 0.11 | 1.57 | 1.21 | 0.55 | 1.68 | 0.62 | 1.77 | 0.34 | 2.65 | 0.38 | 0.43 | 11.59 |
| 1985 | 0.51 | 0.26 | 0.92 | 1.78 | 1.09 | 0.39 | 1.21 | 0.24 | 1.67 | 2.32 | 1.10 | 0.73 | 12.22 |
| 1986 | 0.13 | 0.33 | 0.25 | 0.71 | 1.15 | 0.15 | 0.94 | 0.97 | 1.52 | 1.22 | 1.02 | 0.47 | 8.86 |
| 1987 | 0.30 | 1.21 | 1.95 | 0.46 | 1.51 | 0.23 | 1.51 | 0.83 | 0.13 | 0.65 | 1.92 | 0.83 | 11.53 |
| 1988 | 1.07 | 0.21 | 0.72 | 0.99 | 1.10 | 0.21 | 0.18 | 1.37 | 0.76 | 0.02 | 1.02 | 0.20 | 7.85 |
| 1989 | 0.98 | 1.33 | 0.51 | 0.23 | 0.39 | 0.24 | 0.27 | 1.01 | 0.33 | 0.14 | T | 0.08 | 5.51 |
| 1990 | 0.59 | 0.55 | 1.07 | 0.71 | 0.05 | 0.26 | 0.96 | 0.49 | 1.23 | 0.95 | 0.57 | 0.98 | 8.41 |
| 1991 | 0.92 | 0.13 | 0.70 | 0.87 | 0.20 | 0.30 | 0.40 | 0.57 | 2.30 | 1.20 | 1.10 | 0.54 | 9.23 |
| 1992 | 0.24 | 0.35 | 1.71 | 0.15 | 1.81 | 0.17 | 1.03 | 0.84 | 0.33 | 1.45 | 0.76 | 0.35 | 9.19 |
| 1993 | 1.36 | 1.09 | 1.72 | 1.30 | 1.99 | 0.03 | 0.04 | 1.42 | 0.41 | 1.34 | 0.41 | 0.57 | 11.68 |
| 1994 | 0.23 | 0.56 | 0.25 | 1.81 | 0.19 | 0.04 | 0.01 | 0.48 | 1.50 | 0.58 | 0.69 | 0.64 | 6.98 |
| 1995 | 0.62 | 0.52 | 1.74 | 0.96 | 2.04 | 1.32 | 0.87 | 0.47 | 0.66 | 0.24 | 0.20 | 0.55 | 10.19 |
| 1996 | 0.65 | 1.07 | 0.53 | 0.90 | 0.99 | 0.58 | .77 | .15 | 1.53 | 1.35 | 1.01 | .53 | 10.06 |
| 1997 | 0.63 | 0.34 | 0.53 | 2.15 | 1.53 | 0.29 | 0.28 | 2.67 | 2.84 | 1.20 | 0.62 | 0.14 | 13.22 |
| 1998 | 0.47 | 0.48 | 1.36 | 0.75 | 0.21 | 0.55 | 1.21 | 0.61 | 1.44 | 1.36 | 0.42 | 0.26 | 9.12 |
| 1999 | 0.09 | 0.28 | 0.03 | 2.06 | 0.68 | 0.60 | 0.51 | 2.22 | 1.01 | 0.17 | 0.18 | 0.26 | 8.09 |
| 2000 | 1.35 | 0.70 | 1.26 | 0.32 | 0.43 | 0.34 | 0.20 | 0.60 | 0.60 | 1.18 | 0.35 | 0.18 | 7.51 |
| 2001 | 0.43 | 0.67 | 0.98 | 0.58 | 0.57 | T | 1.20 | 1.46 | 0.15 | 0.99 | 1.06 | 0.31 | 8.40 |
| 2002 | 0.25 | 0.09 | 0.64 | 0.18 | 0.51 | 0.08 | 0.13 | 1.15 | 2.54 | 1.27 | 0.81 | 0.22 | 7.87 |
| 2003 | 0.13 | 1.03 | 0.74 | 0.22 | 1.88 | 0.10 | 0.02 | 0.40 | 1.05 | 0.11 | 1.06 | 0.57 | 7.31 |
| 2004 | 0.82 | 0.73 | 0.02 | 2.30 | 0.19 | 0.08 | 0.10 | 0.20 | 2.00 | 1.09 | 1.86 | 0.21 | 9.60 |
| 2005 | 1.66 | 0.78 | 0.65 | 0.66 | 0.11 | 1.59 | 0.60 | 0.85 | 2.52 | 1.42 | 0.23 | 0.75 | 11.82 |
| 2006 | 0.41 | 0.02 | 1.23 | 0.34 | 0.11 | 0.31 | 0.91 | 1.46 | 1.41 | 2.75 | 0.55 | 0.37 | 9.87 |
| 2007 | 0.49 | 0.66 | 0.46 | 0.99 | 0.27 | 0.50 | 0.96 | 0.84 | 1.99 | 0.46 | 0.05 | 2.05 | 9.72 |
| 2008 | 0.63 | 0.61 | 0.41 | 0.86 | 0.89 | 0.50 | 0.02 | 1.19 | 0.23 | 0.11 | 0.95 | 0.86 | 7.26 |
| 2009 | 0.31 | 0.35 | 0.48 | 1.31 | 1.60 | 1.12 | 0.12 | 0.30 | 0.43 | 0.62 | 0.05 | 1.10 | 7.79 |
| POR=110 YRS | 0.61 | 0.58 | 0.81 | 0.79 | 0.79 | 0.45 | 0.61 | 0.99 | 0.96 | 0.91 | 0.63 | 0.59 | 8.72 |

WBAN : 23066

## AVERAGE TEMPERATURE (°F) 2009  GRAND JUNCTION (KGJT)

| YEAR | JAN | FEB | MAR | APR | MAY | JUN | JUL | AUG | SEP | OCT | NOV | DEC | ANNUAL |
|------|-----|-----|-----|-----|-----|-----|-----|-----|-----|-----|-----|-----|--------|
| 1980 | 32.6 | 39.2 | 40.9 | 51.4 | 59.1 | 74.0 | 78.6 | 75.3 | 67.9 | 53.8 | 42.3 | 40.1 | 54.6 |
| 1981 | 36.8 | 37.9 | 44.0 | 56.8 | 60.9 | 76.4 | 79.4 | 76.8 | 69.2 | 50.7 | 41.6 | 31.3 | 55.2 |
| 1982 | 26.0 | 34.7 | 46.0 | 51.3 | 61.4 | 72.2 | 79.0 | 78.2 | 67.5 | 51.9 | 41.2 | 33.0 | 53.5 |
| 1983 | 34.3 | 40.9 | 45.9 | 48.7 | 58.7 | 69.8 | 78.3 | 80.4 | 71.4 | 58.2 | 42.1 | 30.5 | 54.9 |
| 1984 | 20.7 | 31.7 | 44.4 | 49.0 | 66.2 | 70.1 | 78.0 | 76.6 | 68.3 | 50.2 | 40.8 | 32.6 | 52.4 |
| 1985 | 31.1 | 32.0 | 43.9 | 54.5 | 63.6 | 73.0 | 77.6 | 77.1 | 62.4 | 52.8 | 38.8 | 31.9 | 53.2 |
| 1986 | 34.2 | 40.4 | 49.0 | 52.5 | 60.6 | 74.3 | 76.0 | 75.6 | 63.2 | 51.5 | 40.8 | 32.4 | 54.2 |
| 1987 | 27.4 | 36.8 | 40.1 | 54.5 | 61.2 | 73.5 | 75.2 | 72.9 | 66.2 | 56.8 | 39.6 | 27.9 | 52.7 |
| 1988 | 17.4 | 29.2 | 41.0 | 53.1 | 60.9 | 76.5 | 80.6 | 76.2 | 64.5 | 58.9 | 40.7 | 30.0 | 52.4 |
| 1989 | 20.2 | 27.7 | 47.5 | 57.0 | 63.8 | 71.0 | 80.5 | 74.4 | 68.3 | 54.9 | 40.6 | 29.2 | 52.9 |
| 1990 | 28.5 | 35.4 | 46.8 | 55.7 | 61.3 | 75.2 | 78.0 | 76.6 | 69.7 | 53.2 | 39.5 | 20.6 | 53.4 |
| 1991 | 17.6 | 32.0 | 42.0 | 48.8 | 62.0 | 72.5 | 77.6 | 76.2 | 66.5 | 55.0 | 37.7 | 26.2 | 51.2 |
| 1992 | 19.9 | 37.6 | 47.2 | 59.0 | 64.5 | 71.2 | 75.1 | 75.1 | 67.7 | 57.8 | 35.9 | 24.5 | 53.0 |
| 1993 | 31.9 | 36.2 | 45.6 | 50.0 | 61.9 | 70.1 | 76.7 | 73.2 | 65.9 | 51.7 | 35.6 | 29.2 | 52.3 |
| 1994 | 31.7 | 34.2 | 47.4 | 53.2 | 65.3 | 77.8 | 81.3 | 79.4 | 68.3 | 52.9 | 37.1 | 33.1 | 55.1 |
| 1995 | 33.7 | 43.4 | 46.2 | 50.6 | 56.9 | 67.8 | 76.1 | 79.0 | 68.5 | 53.3 | 43.6 | 35.3 | 54.5 |
| 1996 | 30.1 | 40.2 | 44.9 | 51.0 | 64.2 | 73.3 | 79.2 | 77.6 | 63.0 | 51.3 | 39.9 | 30.7 | 53.8 |
| 1997 | 30.6 | 34.2 | 45.0 | 46.5 | 62.2 | 72.7 | 76.2 | 74.0 | 66.4 | 51.7 | 38.2 | 28.2 | 52.2 |
| 1998 | 33.5 | 35.8 | 41.8 | 48.8 | 62.1 | 67.8 | 79.1 | 76.8 | 70.0 | 53.3 | 40.8 | 25.6 | 53.0 |
| 1999 | 32.9 | 37.4 | 48.5 | 47.8 | 58.7 | 70.6 | 77.6 | 73.6 | 63.5 | 54.2 | 43.0 | 30.3 | 53.2 |
| 2000 | 32.8 | 39.5 | 43.4 | 56.4 | 65.8 | 73.1 | 80.1 | 78.4 | 67.2 | 54.4 | 32.8 | 30.8 | 54.6 |
| 2001 | 28.0 | 36.2 | 45.3 | 53.8 | 64.3 | 74.1 | 78.9 | 75.7 | 69.7 | 55.0 | 43.7 | 28.3 | 54.4 |
| 2002 | 27.9 | 30.6 | 40.0 | 56.2 | 62.8 | 76.0 | 81.8 | 75.3 | 65.6 | 51.0 | 37.7 | 32.3 | 53.1 |
| 2003 | 35.3 | 34.9 | 44.4 | 53.0 | 63.1 | 72.2 | 84.1 | 78.2 | 65.3 | 57.5 | 36.6 | 31.2 | 54.7 |
| 2004 | 21.7 | 29.9 | 49.9 | 53.2 | 63.8 | 72.8 | 78.6 | 74.7 | 64.6 | 53.4 | 38.5 | 28.2 | 52.4 |
| 2005 | 36.9 | 38.1 | 42.4 | 51.5 | 62.0 | 68.6 | 80.0 | 74.1 | 66.9 | 54.8 | 41.0 | 29.6 | 53.8 |
| 2006 | 32.2 | 34.7 | 42.0 | 54.6 | 65.1 | 75.8 | 79.9 | 74.8 | 61.4 | 50.3 | 40.0 | 28.1 | 53.2 |
| 2007 | 22.6 | 36.6 | 46.5 | 53.2 | 62.6 | 74.3 | 81.7 | 78.3 | 67.5 | 52.9 | 41.5 | 25.1 | 53.6 |
| 2008 | 21.5 | 32.0 | 41.4 | 48.5 | 58.5 | 71.6 | 80.3 | 76.7 | 66.5 | 53.7 | 41.3 | 26.0 | 51.5 |
| 2009 | 24.7 | 38.5 | 44.6 | 50.6 | 64.5 | 69.8 | 79.7 | 75.5 | 67.7 | 49.4 | 41.2 | 17.6 | 52.0 |
| POR=110 YRS | 26.3 | 32.7 | 43.1 | 51.5 | 61.9 | 71.0 | 78.5 | 75.7 | 65.9 | 54.0 | 39.1 | 28.7 | 52.4 |

## HEATING DEGREE DAYS (base 65°F) 2009 GRAND JUNCTION (KGJT)

| YEAR | JUL | AUG | SEP | OCT | NOV | DEC | JAN | FEB | MAR | APR | MAY | JUN | TOTAL |
|------|-----|-----|-----|-----|-----|-----|-----|-----|-----|-----|-----|-----|-------|
| 1980-81 | 0 | 2 | 21 | 359 | 674 | 765 | 864 | 754 | 645 | 247 | 153 | 15 | 4499 |
| 1981-82 | 0 | 0 | 12 | 439 | 696 | 1039 | 1203 | 841 | 581 | 405 | 136 | 6 | 5358 |
| 1982-83 | 2 | 0 | 61 | 397 | 704 | 983 | 946 | 668 | 586 | 482 | 238 | 22 | 5089 |
| 1983-84 | 0 | 0 | 27 | 208 | 678 | 1064 | 1366 | 959 | 631 | 474 | 89 | 44 | 5540 |
| 1984-85 | 0 | 0 | 54 | 452 | 719 | 996 | 1044 | 919 | 646 | 310 | 81 | 12 | 5233 |
| 1985-86 | 0 | 0 | 139 | 371 | 779 | 1018 | 949 | 685 | 489 | 366 | 168 | 3 | 4967 |
| 1986-87 | 0 | 0 | 130 | 414 | 718 | 1001 | 1159 | 785 | 765 | 314 | 143 | 0 | 5429 |
| 1987-88 | 0 | 6 | 34 | 248 | 754 | 1147 | 1469 | 1031 | 741 | 350 | 172 | 8 | 5960 |
| 1988-89 | 0 | 0 | 106 | 183 | 724 | 1078 | 1379 | 1038 | 534 | 258 | 113 | 8 | 5421 |
| 1989-90 | 0 | 0 | 40 | 316 | 729 | 1103 | 1124 | 820 | 557 | 271 | 139 | 20 | 5119 |
| 1990-91 | 0 | 0 | 28 | 360 | 759 | 1371 | 1464 | 919 | 706 | 478 | 136 | 18 | 6239 |
| 1991-92 | 0 | 2 | 37 | 304 | 815 | 1193 | 1390 | 788 | 540 | 195 | 53 | 8 | 5325 |
| 1992-93 | 0 | 6 | 25 | 222 | 868 | 1245 | 1018 | 799 | 597 | 446 | 144 | 33 | 5403 |
| 1993-94 | 4 | 0 | 59 | 410 | 875 | 1102 | 1025 | 853 | 540 | 360 | 64 | 0 | 5292 |
| 1994-95 | 0 | 0 | 24 | 368 | 832 | 984 | 962 | 596 | 578 | 425 | 256 | 47 | 5072 |
| 1995-96 | 8 | 0 | 73 | 357 | 634 | 914 | 1073 | 712 | 614 | 415 | 88 | 2 | 4890 |
| 1996-97 | 0 | 0 | 135 | 421 | 748 | 1055 | 1056 | 857 | 613 | 547 | 122 | 4 | 5558 |
| 1997-98 | 0 | 0 | 42 | 412 | 799 | 1138 | 970 | 813 | 709 | 478 | 137 | 55 | 5553 |
| 1998-99 | 0 | 0 | 9 | 355 | 715 | 1217 | 987 | 767 | 501 | 508 | 219 | 33 | 5311 |
| 1999-00 | 0 | 0 | 88 | 331 | 651 | 1067 | 992 | 732 | 663 | 256 | 100 | 1 | 4881 |
| 2000-01 | 0 | 0 | 58 | 339 | 959 | 1053 | 1140 | 802 | 605 | 333 | 105 | 31 | 5425 |
| 2001-02 | 0 | 0 | 23 | 316 | 633 | 1130 | 1144 | 955 | 767 | 259 | 149 | 6 | 5382 |
| 2002-03 | 0 | 1 | 55 | 426 | 817 | 1010 | 915 | 835 | 631 | 358 | 163 | 1 | 5212 |
| 2003-04 | 0 | 0 | 66 | 235 | 847 | 1040 | 1336 | 1011 | 464 | 350 | 104 | 8 | 5461 |
| 2004-05 | 0 | 0 | 109 | 353 | 789 | 1137 | 865 | 746 | 694 | 397 | 157 | 58 | 5305 |
| 2005-06 | 0 | 0 | 30 | 318 | 713 | 1093 | 980 | 841 | 704 | 308 | 88 | 0 | 5075 |
| 2006-07 | 0 | 2 | 164 | 450 | 743 | 1137 | 1307 | 788 | 569 | 354 | 142 | 19 | 5675 |
| 2007-08 | 0 | 0 | 77 | 372 | 698 | 1231 | 1340 | 950 | 724 | 485 | 216 | 44 | 6137 |
| 2008-09 | 0 | 0 | 25 | 353 | 706 | 1201 | 1241 | 738 | 625 | 423 | 66 | 4 | 5382 |
| 2009- | 0 | 0 | 57 | 481 | 708 | 1460 | | | | | | | |

WBAN : 23066

## COOLING DEGREE DAYS (base 65°F) 2009 GRAND JUNCTION (KGJT)

| YEAR | JAN | FEB | MAR | APR | MAY | JUN | JUL | AUG | SEP | OCT | NOV | DEC | TOTAL |
|------|-----|-----|-----|-----|-----|-----|-----|-----|-----|-----|-----|-----|-------|
| 1980 | 0 | 0 | 0 | 1 | 16 | 280 | 427 | 325 | 115 | 19 | 0 | 0 | 1183 |
| 1981 | 0 | 0 | 0 | 9 | 31 | 367 | 456 | 375 | 143 | 0 | 0 | 0 | 1381 |
| 1982 | 0 | 0 | 0 | 0 | 33 | 229 | 443 | 415 | 144 | 0 | 0 | 0 | 1264 |
| 1983 | 0 | 0 | 0 | 0 | 49 | 171 | 421 | 483 | 226 | 3 | 0 | 0 | 1353 |
| 1984 | 0 | 0 | 0 | 0 | 134 | 200 | 408 | 368 | 159 | 0 | 0 | 0 | 1269 |
| 1985 | 0 | 0 | 0 | 4 | 45 | 261 | 396 | 382 | 67 | 0 | 0 | 0 | 1155 |
| 1986 | 0 | 0 | 0 | 0 | 39 | 289 | 348 | 334 | 82 | 0 | 0 | 0 | 1092 |
| 1987 | 0 | 0 | 0 | 5 | 30 | 262 | 324 | 256 | 76 | 2 | 0 | 0 | 955 |
| 1988 | 0 | 0 | 0 | 0 | 51 | 360 | 489 | 357 | 98 | 4 | 0 | 0 | 1359 |
| 1989 | 0 | 0 | 0 | 26 | 85 | 195 | 489 | 300 | 145 | 11 | 0 | 0 | 1251 |
| 1990 | 0 | 0 | 0 | 1 | 34 | 331 | 412 | 368 | 174 | 3 | 0 | 0 | 1323 |
| 1991 | 0 | 0 | 0 | 0 | 50 | 247 | 398 | 356 | 88 | 0 | 0 | 0 | 1139 |
| 1992 | 0 | 0 | 0 | 21 | 43 | 203 | 319 | 328 | 114 | 7 | 0 | 0 | 1035 |
| 1993 | 0 | 0 | 0 | 0 | 56 | 193 | 371 | 260 | 92 | 6 | 0 | 0 | 978 |
| 1994 | 0 | 0 | 0 | 13 | 82 | 388 | 514 | 454 | 126 | 0 | 0 | 0 | 1577 |
| 1995 | 0 | 0 | 0 | 0 | 9 | 138 | 362 | 444 | 185 | 3 | 0 | 0 | 1141 |
| 1996 | 0 | 0 | 0 | 1 | 67 | 256 | 448 | 396 | 82 | 4 | 0 | 0 | 1254 |
| 1997 | 0 | 0 | 0 | 0 | 44 | 240 | 353 | 285 | 93 | 8 | 0 | 0 | 1023 |
| 1998 | 0 | 0 | 0 | 0 | 51 | 148 | 445 | 375 | 163 | 0 | 0 | 0 | 1182 |
| 1999 | 0 | 0 | 0 | 0 | 27 | 204 | 401 | 274 | 45 | 4 | 0 | 0 | 955 |
| 2000 | 0 | 0 | 0 | 5 | 131 | 249 | 476 | 421 | 129 | 15 | 0 | 0 | 1426 |
| 2001 | 0 | 0 | 0 | 3 | 89 | 311 | 437 | 341 | 170 | 13 | 0 | 0 | 1364 |
| 2002 | 0 | 0 | 0 | 0 | 85 | 341 | 526 | 327 | 80 | 0 | 0 | 0 | 1359 |
| 2003 | 0 | 0 | 0 | 2 | 111 | 226 | 596 | 414 | 83 | 7 | 0 | 0 | 1439 |
| 2004 | 0 | 0 | 0 | 2 | 73 | 250 | 426 | 308 | 105 | 0 | 0 | 0 | 1164 |
| 2005 | 0 | 0 | 0 | 0 | 69 | 177 | 469 | 288 | 95 | 11 | 0 | 0 | 1109 |
| 2006 | 0 | 0 | 0 | 1 | 96 | 331 | 468 | 311 | 64 | 3 | 0 | 0 | 1274 |
| 2007 | 0 | 0 | 0 | 5 | 73 | 305 | 523 | 420 | 163 | 3 | 0 | 0 | 1492 |
| 2008 | 0 | 0 | 0 | 0 | 18 | 247 | 485 | 373 | 77 | 9 | 0 | 0 | 1209 |
| 2009 | 0 | 0 | 0 | 0 | 55 | 156 | 462 | 331 | 145 | 3 | 0 | 0 | 1152 |

## SNOWFALL (inches) 2009 GRAND JUNCTION (KGJT)

| YEAR | JUL | AUG | SEP | OCT | NOV | DEC | JAN | FEB | MAR | APR | MAY | JUN | TOTAL |
|------|-----|-----|-----|-----|-----|-----|-----|-----|-----|-----|-----|-----|-------|
| 1980-81 | 0.0 | 0.0 | 0.0 | 0.0 | T | 0.0 | 3.9 | 0.8 | 1.2 | T | 0.0 | 0.0 | 5.9 |
| 1981-82 | 0.0 | 0.0 | 0.0 | 0.5 | 3.3 | 3.4 | 3.4 | 4.0 | 0.8 | T | 0.0 | 0.0 | 15.4 |
| 1982-83 | 0.0 | 0.0 | 0.0 | T | T | 1.9 | 6.1 | 3.1 | 1.5 | 2.2 | T | 0.0 | 14.8 |
| 1983-84 | 0.0 | 0.0 | 0.0 | 0.0 | 4.2 | 19.0 | 3.7 | 0.6 | 6.1 | 2.9 | 0.0 | 0.0 | 36.5 |
| 1984-85 | 0.0 | 0.0 | 0.0 | 0.7 | 2.0 | 2.7 | 5.0 | 2.7 | 5.6 | 0.1 | 0.0 | 0.0 | 18.8 |
| 1985-86 | 0.0 | 0.0 | 0.0 | 0.0 | 4.6 | 4.4 | 1.8 | 0.7 | T | 0.2 | T | 0.0 | 11.7 |
| 1986-87 | 0.0 | 0.0 | 0.0 | 2.2 | 1.2 | 1.0 | 3.0 | 5.5 | 9.4 | 0.6 | T | 0.0 | 22.9 |
| 1987-88 | 0.0 | 0.0 | 0.0 | 0.0 | 1.1 | 7.1 | 12.2 | 2.2 | 4.3 | 0.0 | 0.0 | 0.0 | 26.9 |
| 1988-89 | 0.0 | 0.0 | 0.0 | 0.0 | 0.9 | 3.1 | 10.2 | 16.0 | 1.1 | T | T | 0.0 | 31.3 |
| 1989-90 | 0.0 | 0.0 | 0.0 | 0.0 | 0.0 | 1.1 | 6.2 | 8.6 | 1.8 | 0.8 | 0.0 | 0.0 | 18.5 |
| 1990-91 | 0.0 | 0.0 | 0.0 | 0.0 | 1.5 | 5.1 | 12.7 | 0.3 | 3.7 | 4.7 | 0.0 | 0.0 | 28.0 |
| 1991-92 | 0.0 | 0.0 | T | 2.5 | 1.9 | 7.6 | 2.7 | 1.9 | T | 0.0 | T | 0.0 | 16.6 |
| 1992-93 | 0.0 | 0.0 | 0.0 | 0.0 | 2.0 | 4.4 | 6.0 | 8.4 | 0.0 | 0.4 | T | 0.0 | 21.2 |
| 1993-94 | 0.0 | T | 0.0 | 0.0 | 0.6 | 5.9 | 1.1 | 4.6 | 0.3 | T | 0.0 | 0.0 | 12.5 |
| 1994-95 | 0.0 | 0.0 | 0.0 | T | 5.4 | 2.2 | 4.0 | 1.1 | 3.7 | T | T | 0.0 | 16.4 |
| 1995-96 | 0.0 | 0.0 | T | 1.2 | T | 3.7 | 2.1 | 0.5 | 3.0 | T | 0.0 | 0.0 | 10.5 |
| 1996-97 | 0.0 | 0.0 | 0.0 | 1.1 | 4.2 | 3.5 | 4.9 | 3.2 | 1.2 | 6.5 | T | T | 24.6 |
| 1997-98 | 0.0 | T | 0.0 | 0.5 | T | 1.7 | 3.8 | 0.5 | 3.8 | 0.1 | 0.0 | T | 10.4 |
| 1998-99 | 0.0 | 0.0 | T | 0.0 | 2.5 | 7.0 | 0.5 | 1.4 | 0.0 | 3.8 | T | 0.0 | 15.2 |
| 1999-00 | 0.0 | T | 0.0 | 0.0 | T | 3.3 | 3.1 | 0.6 | 6.2 | 0.0 | 0.0 | 0.0 | 13.2 |
| 2000-01 | 0.0 | 0.0 | 0.0 | 0.0 | 3.6 | 1.5 | 4.3 | 3.3 | 3.7 | 0.0 | 0.0 | 0.0 | 16.4 |
| 2001-02 | T | 0.0 | T | 0.0 | 2.8 | 3.3 | 4.4 | 0.8 | 6.2 | 0.0 | 0.0 | 0.0 | 17.5 |
| 2002-03 | 0.0 | 0.0 | T | 0.0 | 0.3 | 1.9 | T | 4.1 | 3.9 | 0.6 | 0.0 | 0.0 | 10.8 |
| 2003-04 | 0.0 | 0.0 | 0.0 | 0.0 | 3.7 | 3.7 | 8.1 | 7.9 | T | 0.0 | 2.7 | 0.0 | 26.1 |
| 2004-05 | 0.0 | 0.0 | 0.0 | 0.5 | 10.5 | 1.2 | 6.6 | 1.8 | 0.9 | 0.0 | T | 0.0 | 21.5 |
| 2005-06 | 0.0 | 0.0 | T | 0.0 | T | 1.9 | 5.1 | 0.2 | 5.5 | 0.0 | 0.0 | 0.0 | 12.7 |
| 2006-07 | 0.0 | 0.0 | 0.0 | 0.2 | 3.9 | 2.7 | 7.1 | 2.3 | 0.1 | 0.5 | 0.0 | 0.0 | 16.8 |
| 2007-08 | 0.0 | 0.0 | 0.0 | 0.0 | T | 10.1 | 7.2 | 4.6 | 0.9 | T | 0.4 | 0.0 | 23.2 |
| 2008-09 | 0.0 | 0.0 | 0.0 | 0.0 | 0.0 | 13.9 | 1.5 | 2.3 | 2.6 | 1.9 | T | T | 22.2 |
| 2009- | 0.0 | 0.0 | 0.0 | 1.9 | 0.4 | 14.4 | | | | | | | |
| POR= 110 YRS | T | T | T | 0.4 | 2.4 | 5.1 | 6.0 | 3.8 | 3.1 | 0.9 | 0.1 | T | 21.8 |

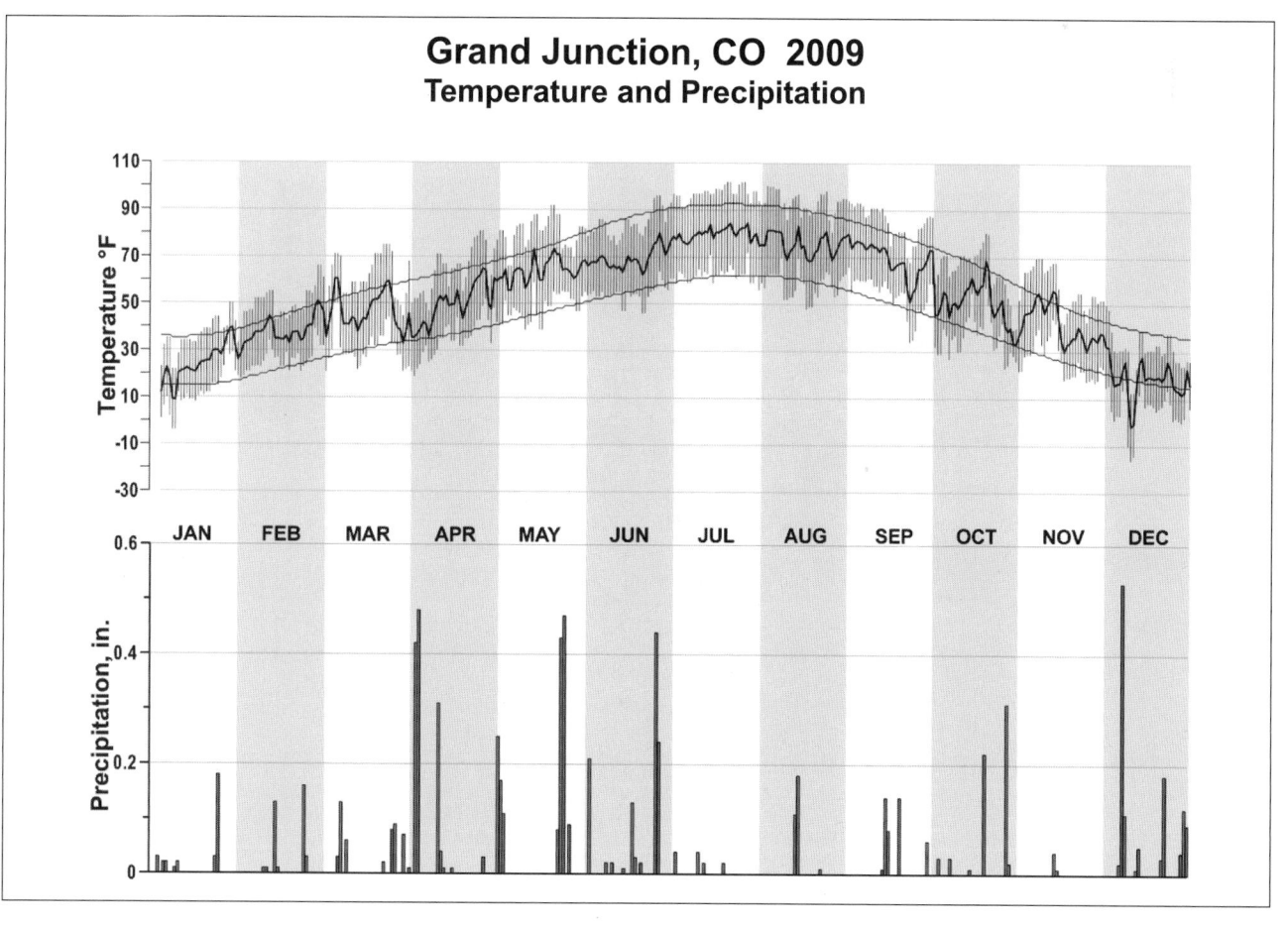

# Grand Junction, CO 2009
## Temperature and Precipitation

# 2009
# WINDSOR LOCKS
# CONNECTICUT (KBDL)

Bradley International Airport is located about 3 miles west of the Connecticut River on a slight rise of ground in a broad portion of the Connecticut River Valley between north-south mountain ranges whose heights do not exceed 1,200 feet.

The station is in the northern temperate climate zone. The prevailing west to east movement of air brings the majority of weather systems into Connecticut from the west. The average wintertime position of the Polar Front boundary between cold, dry polar air and warm, moist tropical air is just south of New England, which helps to explain the extensive winter storm activity and day to day variability of local weather. In summer, the Polar Front has an average position along the New England-Canada border with this station in a warm and pleasant atmosphere.

The location of Hartford, relative to continent and ocean, is also significant. Rapid weather changes result when storms move northward along the mid-Atlantic coast, frequently producing strong and persistent northeast winds associated with storms known locally as coastals or northeasters. Seasonally, weather characteristics vary from the cold and dry continental-polar air of winter to the warm and humid maritime air of summer.

Summer thunderstorms develop in the Berkshire Mountains to the west and northwest, move over the Connecticut Valley, and when accompanied by wind and hail, sometimes cause considerable damage to crops, particularly tobacco. During the winter, rain often falls through cold air trapped in the valley, creating extremely hazardous ice conditions. On clear nights in the late summer or early autumn, cool air drainage into the valley, and moisture from the Connecticut River, produce steam and/or ground fog which becomes quite dense throughout the valley, hampering ground and air transportation.

The mean date of the last springtime temperature of 32 degrees or lower is April 22, and the mean date of the first autumn temperature of 32 degrees is October 15.

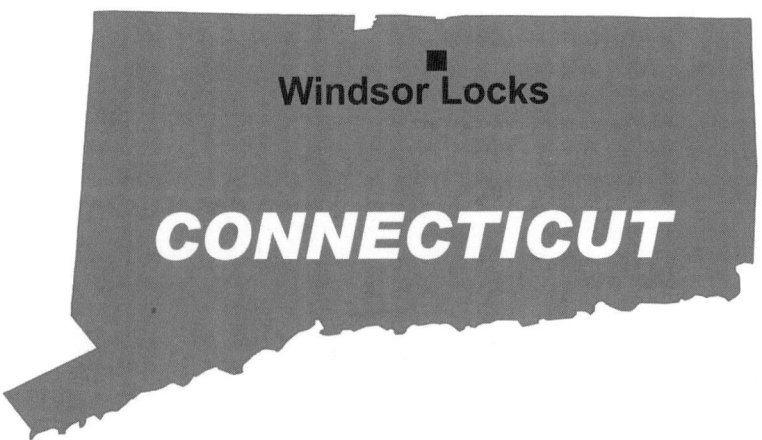

# NORMALS, MEANS, AND EXTREMES
## WINDSOR LOCKS (KBDL)

LATITUDE: 41 ° 56'N  LONGITUDE: -72 ° 40'W  ELEVATION (FT): GRND: 170  BARO: 165  TIME ZONE: EASTERN (UTC -5)  WBAN: 14740

| | ELEMENT | POR | JAN | FEB | MAR | APR | MAY | JUN | JUL | AUG | SEP | OCT | NOV | DEC | YEAR |
|---|---|---|---|---|---|---|---|---|---|---|---|---|---|---|---|
| TEMPERATURE °F | NORMAL DAILY MAXIMUM | 30 | 34.1 | 37.7 | 47.7 | 59.9 | 71.7 | 80.0 | 84.9 | 82.5 | 74.3 | 63.1 | 50.9 | 39.0 | 60.5 |
| | MEAN DAILY MAXIMUM | 61 | 34.2 | 37.4 | 46.4 | 60.1 | 71.1 | 79.7 | 84.6 | 82.5 | 74.6 | 63.6 | 51.1 | 38.6 | 60.3 |
| | HIGHEST DAILY MAXIMUM | 55 | 72 | 73 | 89 | 96 | 99 | 100 | 102 | 102 | 99 | 91 | 81 | 76 | 102 |
| | YEAR OF OCCURRENCE | | 2007 | 1985 | 1998 | 1976 | 1996 | 1964 | 1966 | 2001 | 1983 | 1963 | 1974 | 1998 | AUG 2001 |
| | MEAN OF EXTREME MAXS. | 61 | 54.1 | 55.2 | 68.1 | 81.8 | 88.8 | 93.6 | 95.3 | 93.8 | 89.2 | 80.8 | 70.0 | 58.7 | 77.5 |
| | NORMAL DAILY MINIMUM | 30 | 17.2 | 19.9 | 28.3 | 37.9 | 48.1 | 57.0 | 62.4 | 60.7 | 52.1 | 40.6 | 32.6 | 22.6 | 40.0 |
| | MEAN DAILY MINIMUM | 61 | 17.3 | 19.7 | 27.6 | 37.9 | 47.4 | 56.8 | 62.2 | 60.4 | 52.1 | 41.1 | 32.8 | 22.1 | 39.8 |
| | LOWEST DAILY MINIMUM | 55 | -26 | -21 | -6 | 9 | 28 | 35 | 44 | 36 | 30 | 17 | 1 | -14 | -26 |
| | YEAR OF OCCURRENCE | | 1961 | 1961 | 2003 | 1970 | 2008 | 2002 | 1962 | 1965 | 2000 | 1978 | 1989 | 1980 | JAN 1961 |
| | MEAN OF EXTREME MINS. | 61 | -2.5 | 0.3 | 10.7 | 25.0 | 33.8 | 43.7 | 50.8 | 47.3 | 36.2 | 26.1 | 17.5 | 3.6 | 24.4 |
| | NORMAL DRY BULB | 30 | 25.7 | 28.8 | 38.0 | 48.9 | 59.9 | 68.5 | 73.7 | 71.6 | 63.2 | 51.9 | 41.8 | 30.8 | 50.2 |
| | MEAN DRY BULB | 61 | 25.8 | 29.0 | 37.0 | 49.0 | 59.3 | 68.5 | 73.4 | 71.6 | 63.3 | 52.3 | 41.9 | 30.4 | 50.1 |
| | MEAN WET BULB | 26 | 22.9 | 24.7 | 31.1 | 40.9 | 51.2 | 60.9 | 65.5 | 64.5 | 57.7 | 46.6 | 37.2 | 27.6 | 44.2 |
| | MEAN DEW POINT | 26 | 18.0 | 19.0 | 25.4 | 34.8 | 46.7 | 57.3 | 62.5 | 61.8 | 54.9 | 42.9 | 32.7 | 22.7 | 39.9 |
| | NORMAL NO. DAYS WITH: | | | | | | | | | | | | | | |
| | MAXIMUM >= 90 | 30 | 0.0 | 0.0 | 0.0 | 0.3 | 1.2 | 3.4 | 7.4 | 4.4 | 1.0 | 0.0 | 0.0 | 0.0 | 17.7 |
| | MAXIMUM <= 32 | 30 | 13.4 | 8.7 | 1.7 | 0.1 | 0.0 | 0.0 | 0.0 | 0.0 | 0.0 | 0.0 | 0.5 | 7.5 | 31.9 |
| | MINIMUM <= 32 | 30 | 28.1 | 24.9 | 20.9 | 7.6 | 0.5 | 0.0 | 0.0 | 0.0 | 0.3 | 6.1 | 16.2 | 26.4 | 131.0 |
| | MINIMUM <= 0 | 30 | 2.8 | 1.4 | 0.0 | 0.0 | 0.0 | 0.0 | 0.0 | 0.0 | 0.0 | 0.0 | 0.0 | 0.6 | 4.8 |
| H/C | NORMAL HEATING DEG. DAYS | 30 | 1218 | 1024 | 844 | 486 | 195 | 38 | 3 | 12 | 120 | 413 | 697 | 1054 | 6104 |
| | NORMAL COOLING DEG. DAYS | 30 | 0 | 0 | 1 | 5 | 38 | 144 | 277 | 220 | 68 | 5 | 1 | 0 | 759 |
| RH | NORMAL (PERCENT) | 30 | 65 | 63 | 61 | 59 | 65 | 68 | 69 | 71 | 74 | 70 | 68 | 68 | 67 |
| | HOUR 01 LST | 30 | 70 | 68 | 69 | 69 | 77 | 81 | 82 | 85 | 86 | 82 | 75 | 72 | 76 |
| | HOUR 07 LST | 30 | 72 | 72 | 72 | 69 | 74 | 77 | 78 | 83 | 86 | 83 | 78 | 75 | 77 |
| | HOUR 13 LST | 30 | 56 | 52 | 49 | 45 | 48 | 51 | 51 | 54 | 55 | 52 | 55 | 57 | 52 |
| | HOUR 19 LST | 30 | 63 | 59 | 56 | 53 | 56 | 61 | 62 | 67 | 71 | 68 | 66 | 66 | 62 |
| S | PERCENT POSSIBLE SUNSHINE | 42 | 53 | 56 | 57 | 55 | 57 | 60 | 62 | 62 | 59 | 57 | 45 | 47 | 56 |
| W/O | MEAN NO. DAYS WITH: | | | | | | | | | | | | | | |
| | HEAVY FOG(VISBY <= 1/4 MI) | 46 | 2.7 | 2.1 | 2.4 | 1.2 | 1.3 | 1.8 | 1.6 | 1.9 | 2.5 | 2.9 | 2.1 | 2.8 | 25.3 |
| | THUNDERSTORMS | 60 | 0.1 | 0.2 | 0.7 | 1.2 | 2.6 | 4.2 | 4.7 | 3.8 | 1.9 | 0.9 | 0.4 | 0.2 | 20.9 |
| CLOUDNESS | MEAN: | | | | | | | | | | | | | | |
| | SUNRISE-SUNSET (OKTAS) | 42 | 5.2 | 5.2 | 5.4 | 5.4 | 5.4 | 5.3 | 5.1 | 4.9 | 4.8 | 4.6 | 5.4 | 5.2 | 5.2 |
| | MIDNIGHT-MIDNIGHT (OKTAS) | 32 | 5.0 | 4.9 | 5.1 | 5.0 | 5.1 | 5.0 | 4.8 | 4.7 | 4.6 | 4.3 | 5.0 | 5.0 | 4.9 |
| | MEAN NO. DAYS WITH: | | | | | | | | | | | | | | |
| | CLEAR | 42 | 7.7 | 6.6 | 6.7 | 6.3 | 5.6 | 5.7 | 5.7 | 6.9 | 8.4 | 9.4 | 5.7 | 6.9 | 81.6 |
| | PARTLY CLOUDY | 42 | 7.8 | 7.7 | 8.4 | 8.5 | 9.7 | 10.3 | 12.1 | 10.8 | 8.9 | 8.5 | 8.2 | 7.5 | 108.4 |
| | CLOUDY | 42 | 15.5 | 13.9 | 16.0 | 15.2 | 15.7 | 14.0 | 13.3 | 13.3 | 12.7 | 13.0 | 16.1 | 16.6 | 175.2 |
| PR | MEAN STATION PRESSURE(IN) | 26 | 29.84 | 29.83 | 29.82 | 29.77 | 29.78 | 29.76 | 29.77 | 29.82 | 29.86 | 29.87 | 29.86 | 29.85 | 29.82 |
| | MEAN SEA-LEVEL PRES. (IN) | 26 | 30.04 | 30.03 | 30.02 | 29.96 | 29.97 | 29.95 | 29.96 | 30.01 | 30.06 | 30.06 | 30.06 | 30.05 | 30.01 |
| WINDS | MEAN SPEED (MPH) | 26 | 8.4 | 8.9 | 9.4 | 9.0 | 8.2 | 7.3 | 6.7 | 6.4 | 6.8 | 7.5 | 8.0 | 8.2 | 7.9 |
| | PREVAIL.DIR(TENS OF DEGS) | 35 | 33 | 32 | 33 | 19 | 19 | 19 | 19 | 19 | 19 | 19 | 36 | 36 | 19 |
| | MAXIMUM 2-MINUTE: | | | | | | | | | | | | | | |
| | SPEED (MPH) | 13 | 46 | 40 | 43 | 41 | 43 | 45 | 39 | 39 | 38 | 41 | 39 | 41 | 46 |
| | DIR. (TENS OF DEGS) | | 24 | 29 | 30 | 20 | 30 | 36 | 31 | 27 | 02 | 29 | 29 | 26 | 24 |
| | YEAR OF OCCURRENCE | | 1999 | 2006 | 1997 | 2000 | 2008 | 2000 | 2005 | 2005 | 1999 | 2007 | 2003 | 2000 | JAN 1999 |
| | MAXIMUM 3-SECOND | | | | | | | | | | | | | | |
| | SPEED (MPH) | 13 | 56 | 58 | 54 | 53 | 69 | 52 | 60 | 44 | 46 | 52 | 54 | 55 | 69 |
| | DIR. (TENS OF DEGS) | | 24 | 32 | 29 | 20 | 29 | 34 | 27 | 27 | 32 | 28 | 24 | 25 | 29 |
| | YEAR OF OCCURRENCE | | 1999 | 2003 | 1997 | 2000 | 2008 | 2000 | 1999 | 2005 | 2002 | 2007 | 1998 | 2000 | MAY 2008 |
| PRECIPITATION | NORMAL (IN) | 30 | 3.84 | 2.96 | 3.88 | 3.86 | 4.39 | 3.85 | 3.67 | 3.98 | 4.13 | 3.94 | 4.06 | 3.60 | 46.16 |
| | MAXIMUM MONTHLY (IN) | 55 | 9.61 | 8.90 | 6.86 | 9.90 | 12.00 | 13.60 | 11.17 | 21.87 | 11.22 | 16.32 | 8.53 | 8.36 | 21.87 |
| | YEAR OF OCCURRENCE | | 1978 | 2008 | 1983 | 1983 | 1989 | 1982 | 2009 | 1955 | 1999 | 2005 | 1972 | 1969 | AUG 1955 |
| | MINIMUM MONTHLY (IN) | 55 | 0.38 | 0.45 | 0.27 | 1.10 | 0.73 | 0.67 | 0.97 | 0.54 | 0.84 | 0.35 | 0.51 | 0.78 | 0.27 |
| | YEAR OF OCCURRENCE | | 1981 | 1987 | 1981 | 1999 | 1959 | 1988 | 2001 | 1981 | 1986 | 1963 | 1976 | 1955 | MAR 1981 |
| | MAXIMUM IN 24 HOURS (IN) | 55 | 2.56 | 2.78 | 2.62 | 3.74 | 4.90 | 6.14 | 3.48 | 12.12 | 5.72 | 5.81 | 2.90 | 3.29 | 12.12 |
| | YEAR OF OCCURRENCE | | 1979 | 2008 | 1987 | 2007 | 1989 | 1982 | 1960 | 1955 | 1999 | 2005 | 1988 | 2008 | AUG 1955 |
| | NORMAL NO. DAYS WITH: | | | | | | | | | | | | | | |
| | PRECIPITATION >= 0.01 | 30 | 11.4 | 10.1 | 11.9 | 11.3 | 12.6 | 11.3 | 9.9 | 10.0 | 9.7 | 9.0 | 10.3 | 11.6 | 129.1 |
| | PRECIPITATION >= 1.00 | 30 | 0.9 | 0.7 | 1.0 | 1.0 | 1.1 | 1.0 | 0.9 | 1.2 | 1.1 | 1.1 | 1.2 | 1.0 | 12.2 |
| SNOWFALL | NORMAL (IN) | 30 | 14.4 | 11.1 | 7.9 | 1.5 | 0.1 | 0.0 | 0.0 | 0.0 | 0.0 | 0.1 | 2.5 | 8.4 | 46.0 |
| | MAXIMUM MONTHLY (IN) | 50 | 42.8 | 32.2 | 43.3 | 14.3 | 1.3 | 0.0 | 0.0 | T | 0.0 | 1.7 | 8.7 | 35.4 | 43.3 |
| | YEAR OF OCCURRENCE | | 1996 | 1969 | 1956 | 1982 | 1977 | 1993 | | 2008 | | 1979 | 1986 | 1969 | MAR 1956 |
| | MAXIMUM IN 24 HOURS (IN) | 50 | 14.9 | 21.0 | 14.8 | 14.1 | 1.3 | T | 0.0 | T | 0.0 | 1.7 | 8.6 | 13.9 | 21.0 |
| | YEAR OF OCCURRENCE' | | 1996 | 1983 | 1993 | 1982 | 1977 | 1993 | | 2008 | | 1979 | 1980 | 1969 | FEB 1983 |
| | MAXIMUM SNOW DEPTH (IN) | 49 | 25 | 29 | 20 | 14 | 0 | 0 | 0 | 0 | 0 | 1 | 8 | 16 | 29 |
| | YEAR OF OCCURRENCE | | 1961 | 1961 | 1956 | 1982 | | | | | | 1979 | 1971 | 1969 | FEB 1961 |
| | NORMAL NO. DAYS WITH: | | | | | | | | | | | | | | |
| | SNOWFALL >= 1.0 | 30 | 3.3 | 2.5 | 2.1 | 0.4 | 0.0 | 0.0 | 0.0 | 0.0 | 0.0 | 0.0 | 0.7 | 2.6 | 11.6 |

## PRECIPITATION (inches) 2009 WINDSOR LOCKS (KBDL)

| YEAR | JAN | FEB | MAR | APR | MAY | JUN | JUL | AUG | SEP | OCT | NOV | DEC | ANNUAL |
|------|-----|-----|-----|-----|-----|-----|-----|-----|-----|-----|-----|-----|--------|
| 1980 | 0.72 | 0.98 | 5.87 | 5.39 | 1.65 | 3.81 | 2.65 | 1.60 | 1.40 | 2.58 | 4.22 | 0.82 | 31.69 |
| 1981 | 0.38 | 7.27 | 0.27 | 2.92 | 2.17 | 1.37 | 4.21 | 0.54 | 4.49 | 5.19 | 2.34 | 4.00 | 35.15 |
| 1982 | 4.76 | 2.83 | 2.23 | 4.12 | 3.30 | 13.60 | 2.60 | 4.41 | 2.41 | 3.31 | 3.12 | 1.32 | 48.01 |
| 1983 | 4.68 | 3.83 | 6.86 | 9.90 | 4.82 | 2.61 | 1.07 | 2.55 | 2.10 | 5.52 | 6.09 | 5.97 | 56.00 |
| 1984 | 1.80 | 4.72 | 3.93 | 4.24 | 11.55 | 2.16 | 4.22 | 1.32 | 1.20 | 2.76 | 2.49 | 2.46 | 42.85 |
| 1985 | 0.73 | 1.72 | 2.16 | 1.54 | 2.77 | 3.55 | 4.55 | 6.44 | 3.83 | 2.27 | 6.04 | 1.28 | 36.88 |
| 1986 | 5.34 | 3.02 | 2.72 | 1.55 | 2.28 | 6.79 | 4.44 | 3.44 | 0.84 | 2.18 | 5.57 | 6.15 | 44.32 |
| 1987 | 6.20 | 0.45 | 4.44 | 5.23 | 2.18 | 3.66 | 2.27 | 4.25 | 7.19 | 3.67 | 3.66 | 1.57 | 44.77 |
| 1988 | 3.36 | 3.99 | 2.06 | 2.35 | 3.46 | 0.67 | 8.43 | 2.12 | 1.88 | 2.29 | 7.84 | 1.35 | 39.80 |
| 1989 | 0.88 | 1.85 | 3.02 | 3.33 | 12.00 | 6.65 | 3.40 | 6.81 | 4.67 | 7.62 | 2.89 | 1.49 | 54.61 |
| 1990 | 4.03 | 3.37 | 2.46 | 4.55 | 6.38 | 3.59 | 2.09 | 8.32 | 2.13 | 7.63 | 3.76 | 4.86 | 53.17 |
| 1991 | 2.45 | 1.78 | 4.52 | 3.54 | 5.18 | 2.37 | 2.90 | 8.69 | 5.67 | 3.17 | 4.03 | 2.96 | 47.26 |
| 1992 | 2.73 | 2.23 | 3.79 | 3.13 | 3.21 | 5.77 | 4.62 | 3.60 | 2.43 | 1.95 | 4.19 | 4.33 | 41.98 |
| 1993 | 2.63 | 2.90 | 6.67 | 4.71 | 1.92 | 2.63 | 4.90 | 1.80 | 5.35 | 4.15 | 3.27 | 4.16 | 45.09 |
| 1994 | 5.83 | 3.38 | 5.70 | 2.51 | 4.12 | 3.84 | 5.32 | 5.33 | 5.47 | 1.53 | 4.57 | 5.38 | 52.98 |
| 1995 | 3.84 | 3.24 | 1.89 | 2.60 | 2.63 | 1.02 | 2.58 | 3.81 | 3.15 | 9.46 | 4.38 | 2.32 | 40.92 |
| 1996 | 6.99 | 2.86 | 2.45 | 6.29 | 2.98 | 2.39 | 6.97 | 1.67 | 7.53 | 5.25 | 4.14 | 5.69 | 55.21 |
| 1997 | 3.15 | 1.38 | 3.60 | 2.43 | 3.37 | 1.90 | 3.92 | 7.33 | 0.97 | 1.65 | 5.87 | 2.18 | 37.75 |
| 1998 | 3.37 | 3.12 | 4.87 | 3.35 | 7.84 | 7.18 | 2.23 | 1.98 | 2.33 | 5.67 | 2.34 | 0.83 | 45.11 |
| 1999 | 5.26 | 3.50 | 4.28 | 1.10 | 3.23 | 0.72 | 2.59 | 2.66 | 11.22 | 3.54 | 3.54 | 2.47 | 44.11 |
| 2000 | 2.83 | 2.24 | 3.69 | 4.21 | 4.45 | 6.74 | 5.48 | 3.33 | 3.88 | 1.07 | 0.95 | 3.30 | 42.17 |
| 2001 | 1.35 | 2.90 | 6.13 | 1.22 | 4.71 | 5.12 | 0.97 | 3.71 | 3.10 | 0.76 | 0.86 | 2.20 | 33.03 |
| 2002 | 1.25 | 1.45 | 3.75 | 3.32 | 5.33 | 5.12 | 2.21 | 2.81 | 3.22 | 4.35 | 4.99 | 3.74 | 41.54 |
| 2003 | 2.21 | 3.39 | 3.67 | 2.62 | 4.95 | 6.27 | 2.65 | 5.76 | 11.13 | 5.16 | 3.14 | 4.96 | 55.91 |
| 2004 | 1.47 | 1.76 | 2.88 | 5.45 | 3.16 | 2.21 | 4.27 | 4.26 | 8.25 | 1.65 | 2.68 | 4.23 | 42.27 |
| 2005 | 4.47 | 2.83 | 3.69 | 5.69 | 2.07 | 2.84 | 7.38 | 2.34 | 1.47 | 16.32 | 4.35 | 3.67 | 57.12 |
| 2006 | 5.43 | 3.06 | 0.78 | 3.89 | 7.23 | 9.16 | 2.14 | 4.36 | 2.25 | 6.69 | 4.99 | 1.83 | 51.81 |
| 2007 | 2.81 | 1.54 | 3.71 | 7.54 | 3.73 | 3.62 | 4.54 | 0.98 | 1.17 | 3.39 | 3.03 | 4.33 | 40.39 |
| 2008 | 2.24 | 8.90 | 5.24 | 3.72 | 2.63 | 5.87 | 7.88 | 6.74 | 9.33 | 2.38 | 3.77 | 6.65 | 65.35 |
| 2009 | 2.90 | 1.30 | 2.59 | 3.37 | 3.43 | 6.27 | 11.17 | 2.85 | 1.78 | 4.86 | 2.27 | 5.50 | 48.29 |
| POR= 61 YRS | 3.40 | 3.07 | 3.82 | 3.87 | 3.84 | 3.80 | 3.58 | 3.93 | 4.00 | 3.84 | 3.90 | 3.82 | 44.87 |

WBAN : 14740

## AVERAGE TEMPERATURE (°F) 2009 WINDSOR LOCKS (KBDL)

| YEAR | JAN | FEB | MAR | APR | MAY | JUN | JUL | AUG | SEP | OCT | NOV | DEC | ANNUAL |
|------|-----|-----|-----|-----|-----|-----|-----|-----|-----|-----|-----|-----|--------|
| 1980 | 27.6 | 24.3 | 35.2 | 49.2 | 61.0 | 66.4 | 74.2 | 73.2 | 64.9 | 50.3 | 37.9 | 24.6 | 49.1 |
| 1981 | 17.8 | 35.3 | 38.1 | 52.0 | 61.6 | 69.6 | 74.8 | 70.6 | 62.5 | 49.3 | 43.7 | 31.0 | 50.5 |
| 1982 | 18.8 | 29.2 | 36.7 | 45.8 | 61.4 | 65.0 | 74.4 | 69.5 | 63.0 | 51.5 | 45.8 | 36.0 | 49.8 |
| 1983 | 27.1 | 29.1 | 39.2 | 48.9 | 56.8 | 69.9 | 74.9 | 72.7 | 66.5 | 52.5 | 42.7 | 28.1 | 50.7 |
| 1984 | 21.8 | 34.3 | 31.4 | 48.0 | 56.0 | 69.8 | 71.8 | 73.2 | 59.8 | 55.2 | 41.5 | 35.7 | 49.9 |
| 1985 | 21.5 | 29.9 | 39.7 | 50.7 | 60.6 | 63.7 | 72.4 | 70.2 | 63.4 | 51.9 | 43.2 | 27.5 | 49.6 |
| 1986 | 27.4 | 26.2 | 38.7 | 51.0 | 61.7 | 66.0 | 72.3 | 69.5 | 61.8 | 51.4 | 38.3 | 33.1 | 49.8 |
| 1987 | 25.0 | 26.7 | 39.8 | 49.7 | 60.8 | 68.8 | 74.2 | 69.0 | 62.9 | 49.2 | 41.4 | 33.2 | 50.1 |
| 1988 | 23.1 | 28.3 | 38.5 | 47.4 | 59.7 | 66.7 | 75.2 | 74.5 | 62.2 | 47.6 | 42.3 | 29.3 | 49.6 |
| 1989 | 30.8 | 28.6 | 37.4 | 46.5 | 60.4 | 68.3 | 72.6 | 71.4 | 63.9 | 53.4 | 40.9 | 18.1 | 49.4 |
| 1990 | 34.7 | 33.0 | 40.2 | 49.2 | 56.7 | 69.0 | 74.4 | 73.3 | 64.0 | 57.4 | 44.5 | 36.7 | 52.8 |
| 1991 | 27.0 | 33.9 | 40.5 | 53.3 | 65.8 | 70.5 | 73.7 | 73.1 | 62.1 | 55.1 | 42.7 | 32.8 | 52.5 |
| 1992 | 28.6 | 30.3 | 34.6 | 46.4 | 58.5 | 66.4 | 69.9 | 69.1 | 62.6 | 49.2 | 40.6 | 31.2 | 49.0 |
| 1993 | 28.8 | 54.3 | 34.5 | 49.6 | 61.2 | 68.6 | 74.3 | 73.4 | 63.0 | 49.8 | 40.8 | 30.9 | 52.4 |
| 1994 | 18.8 | 23.2 | 36.1 | 50.9 | 58.3 | 71.1 | 77.1 | 70.0 | 63.7 | 52.6 | 46.1 | 35.0 | 50.2 |
| 1995 | 32.4 | 25.8 | 41.1 | 46.8 | 58.0 | 69.3 | 76.5 | 72.1 | 62.2 | 55.8 | 38.1 | 27.1 | 50.4 |
| 1996 | 25.1 | 27.7 | 34.3 | 49.6 | 58.3 | 69.3 | 71.2 | 71.9 | 62.9 | 51.1 | 38.1 | 35.8 | 49.6 |
| 1997 | 27.1 | 34.7 | 36.2 | 47.2 | 56.1 | 68.7 | 72.3 | 70.1 | 63.2 | 50.6 | 38.8 | 31.4 | 49.7 |
| 1998 | 32.6 | 36.2 | 40.3 | 49.9 | 62.9 | 67.0 | 72.9 | 73.6 | 65.7 | 52.6 | 41.6 | 36.8 | 52.7 |
| 1999 | 25.9 | 31.8 | 38.5 | 49.2 | 59.9 | 71.0 | 76.5 | 71.3 | 66.0 | 51.1 | 46.3 | 34.6 | 51.8 |
| 2000 | 24.0 | 31.0 | 43.7 | 47.8 | 59.6 | 67.7 | 69.6 | 69.6 | 62.1 | 51.8 | 40.9 | 25.4 | 49.4 |
| 2001 | 25.4 | 28.4 | 34.3 | 49.7 | 59.5 | 69.6 | 69.7 | 75.2 | 63.4 | 53.5 | 45.9 | 36.9 | 51.0 |
| 2002 | 34.0 | 34.6 | 39.1 | 52.0 | 57.1 | 67.3 | 75.2 | 73.9 | 66.7 | 50.2 | 40.4 | 29.9 | 51.7 |
| 2003 | 20.9 | 23.7 | 37.2 | 46.3 | 56.9 | 67.0 | 73.2 | 74.3 | 64.5 | 50.0 | 44.4 | 33.0 | 49.3 |
| 2004 | 18.7 | 29.5 | 38.9 | 49.6 | 61.5 | 66.8 | 72.1 | 71.6 | 64.9 | 51.4 | 42.5 | 30.4 | 49.8 |
| 2005 | 23.5 | 30.1 | 33.1 | 51.3 | 54.6 | 72.3 | 74.5 | 75.1 | 67.5 | 53.4 | 42.9 | 28.2 | 50.5 |
| 2006 | 33.1 | 30.0 | 37.8 | 51.2 | 58.8 | 69.0 | 76.6 | 72.0 | 63.1 | 52.0 | 46.8 | 38.0 | 52.4 |
| 2007 | 31.5 | 23.9 | 36.2 | 46.7 | 62.1 | 69.0 | 73.5 | 72.7 | 67.0 | 59.7 | 40.7 | 30.1 | 51.1 |
| 2008 | 29.5 | 29.6 | 37.4 | 51.8 | 57.6 | 70.7 | 75.6 | 70.0 | 65.0 | 51.1 | 41.3 | 32.2 | 51.0 |
| 2009 | 21.3 | 30.4 | 37.7 | 51.0 | 59.1 | 66.3 | 70.2 | 73.1 | 62.1 | 50.5 | 46.5 | 30.3 | 49.9 |
| POR= 61 YRS | 25.8 | 29.0 | 37.0 | 49.0 | 59.3 | 68.5 | 73.4 | 71.6 | 63.3 | 52.3 | 41.9 | 30.4 | 50.1 |

## HEATING DEGREE DAYS (base 65°F) 2009  WINDSOR LOCKS (KBDL)

| YEAR | JUL | AUG | SEP | OCT | NOV | DEC | JAN | FEB | MAR | APR | MAY | JUN | TOTAL |
|------|-----|-----|-----|-----|-----|-----|-----|-----|-----|-----|-----|-----|-------|
| 1980-81 | 0 | 0 | 99 | 449 | 808 | 1246 | 1456 | 824 | 828 | 380 | 149 | 10 | 6249 |
| 1981-82 | 0 | 9 | 115 | 481 | 635 | 1048 | 1427 | 996 | 871 | 569 | 128 | 64 | 6343 |
| 1982-83 | 1 | 30 | 96 | 416 | 575 | 894 | 1170 | 1002 | 793 | 483 | 261 | 24 | 5745 |
| 1983-84 | 0 | 7 | 106 | 404 | 662 | 1135 | 1332 | 884 | 1035 | 503 | 286 | 32 | 6386 |
| 1984-85 | 3 | 3 | 186 | 298 | 698 | 896 | 1341 | 975 | 776 | 428 | 167 | 76 | 5847 |
| 1985-86 | 0 | 14 | 119 | 401 | 648 | 1157 | 1159 | 1081 | 809 | 413 | 174 | 63 | 6038 |
| 1986-87 | 14 | 32 | 135 | 422 | 793 | 981 | 1230 | 1065 | 773 | 452 | 191 | 29 | 6117 |
| 1987-88 | 1 | 31 | 100 | 481 | 700 | 981 | 1292 | 1057 | 817 | 523 | 186 | 75 | 6244 |
| 1988-89 | 9 | 23 | 112 | 539 | 672 | 1101 | 1054 | 1012 | 847 | 553 | 175 | 31 | 6128 |
| 1989-90 | 0 | 22 | 103 | 354 | 715 | 1444 | 935 | 890 | 763 | 478 | 251 | 21 | 5976 |
| 1990-91 | 5 | 0 | 112 | 276 | 608 | 873 | 1170 | 863 | 755 | 373 | 107 | 16 | 5158 |
| 1991-92 | 1 | 0 | 156 | 311 | 663 | 990 | 1122 | 1002 | 936 | 553 | 218 | 37 | 5989 |
| 1992-93 | 9 | 16 | 138 | 486 | 722 | 1042 | 1114 | 1148 | 935 | 454 | 139 | 43 | 6246 |
| 1993-94 | 3 | 4 | 142 | 464 | 722 | 1049 | 1424 | 1163 | 888 | 417 | 226 | 15 | 6517 |
| 1994-95 | 0 | 8 | 77 | 379 | 561 | 923 | 1005 | 1088 | 737 | 539 | 224 | 13 | 5554 |
| 1995-96 | 0 | 4 | 130 | 283 | 802 | 1169 | 1227 | 1078 | 944 | 465 | 241 | 19 | 6362 |
| 1996-97 | 1 | 2 | 120 | 424 | 802 | 900 | 1167 | 841 | 883 | 526 | 274 | 49 | 5989 |
| 1997-98 | 3 | 6 | 104 | 445 | 777 | 1036 | 998 | 801 | 769 | 447 | 101 | 59 | 5546 |
| 1998-99 | 1 | 3 | 61 | 378 | 694 | 873 | 1207 | 923 | 815 | 466 | 168 | 9 | 5598 |
| 1999-00 | 0 | 11 | 67 | 422 | 555 | 934 | 1264 | 978 | 657 | 512 | 210 | 65 | 5675 |
| 2000-01 | 3 | 17 | 154 | 398 | 716 | 1221 | 1217 | 1023 | 946 | 466 | 202 | 27 | 6390 |
| 2001-02 | 12 | 0 | 101 | 361 | 568 | 863 | 953 | 845 | 797 | 427 | 262 | 55 | 5244 |
| 2002-03 | 2 | 11 | 46 | 468 | 730 | 1081 | 1358 | 1151 | 853 | 555 | 248 | 51 | 6554 |
| 2003-04 | 0 | 3 | 57 | 459 | 612 | 986 | 1431 | 1021 | 801 | 459 | 140 | 47 | 6016 |
| 2004-05 | 2 | 5 | 64 | 411 | 667 | 1066 | 1284 | 972 | 980 | 409 | 317 | 14 | 6191 |
| 2005-06 | 4 | 0 | 47 | 369 | 658 | 1131 | 986 | 975 | 837 | 410 | 232 | 35 | 5684 |
| 2006-07 | 0 | 9 | 100 | 400 | 541 | 826 | 1035 | 1145 | 886 | 545 | 154 | 32 | 5673 |
| 2007-08 | 4 | 14 | 52 | 218 | 722 | 1077 | 1093 | 1019 | 847 | 391 | 237 | 11 | 5685 |
| 2008-09 | 0 | 4 | 85 | 424 | 704 | 1008 | 1350 | 964 | 842 | 447 | 198 | 42 | 6068 |
| 2009- | 3 | 12 | 118 | 443 | 548 | 1067 | | | | | | | |

WBAN : 14740

## COOLING DEGREE DAYS (base 65°F) 2009  WINDSOR LOCKS (KBDL)

| YEAR | JAN | FEB | MAR | APR | MAY | JUN | JUL | AUG | SEP | OCT | NOV | DEC | TOTAL |
|------|-----|-----|-----|-----|-----|-----|-----|-----|-----|-----|-----|-----|-------|
| 1980 | 0 | 0 | 0 | 0 | 31 | 117 | 296 | 263 | 107 | 1 | 0 | 0 | 815 |
| 1981 | 0 | 0 | 0 | 0 | 53 | 152 | 311 | 190 | 48 | 0 | 0 | 0 | 754 |
| 1982 | 0 | 0 | 0 | 0 | 22 | 70 | 298 | 176 | 45 | 2 | 3 | 0 | 616 |
| 1983 | 0 | 0 | 0 | 5 | 16 | 177 | 313 | 253 | 158 | 23 | 0 | 0 | 945 |
| 1984 | 0 | 0 | 0 | 0 | 11 | 182 | 218 | 265 | 38 | 4 | 0 | 0 | 718 |
| 1985 | 0 | 0 | 0 | 3 | 37 | 44 | 234 | 182 | 78 | 3 | 0 | 0 | 581 |
| 1986 | 0 | 0 | 0 | 0 | 79 | 103 | 249 | 179 | 48 | 7 | 0 | 0 | 665 |
| 1987 | 0 | 0 | 0 | 3 | 70 | 150 | 292 | 161 | 42 | 0 | 0 | 0 | 718 |
| 1988 | 0 | 0 | 0 | 0 | 27 | 134 | 331 | 326 | 37 | 6 | 0 | 0 | 861 |
| 1989 | 0 | 0 | 0 | 0 | 37 | 136 | 240 | 224 | 77 | 0 | 0 | 0 | 714 |
| 1990 | 0 | 0 | 0 | 13 | 1 | 146 | 305 | 263 | 89 | 48 | 1 | 0 | 866 |
| 1991 | 0 | 0 | 0 | 29 | 139 | 191 | 278 | 257 | 76 | 12 | 0 | 0 | 982 |
| 1992 | 0 | 0 | 0 | 0 | 21 | 85 | 170 | 151 | 74 | 2 | 0 | 0 | 503 |
| 1993 | 0 | 0 | 0 | 0 | 28 | 155 | 297 | 271 | 89 | 0 | 0 | 0 | 840 |
| 1994 | 0 | 0 | 0 | 1 | 25 | 205 | 381 | 173 | 43 | 0 | 0 | 0 | 828 |
| 1995 | 0 | 0 | 0 | 0 | 11 | 151 | 363 | 230 | 57 | 1 | 0 | 0 | 813 |
| 1996 | 0 | 0 | 0 | 9 | 42 | 155 | 200 | 221 | 63 | 0 | 0 | 0 | 690 |
| 1997 | 0 | 0 | 0 | 0 | 5 | 168 | 237 | 172 | 58 | 6 | 0 | 0 | 646 |
| 1998 | 0 | 0 | 13 | 1 | 42 | 127 | 254 | 273 | 89 | 0 | 0 | 0 | 799 |
| 1999 | 0 | 0 | 0 | 0 | 17 | 194 | 368 | 210 | 101 | 0 | 0 | 0 | 890 |
| 2000 | 0 | 0 | 0 | 0 | 46 | 154 | 152 | 166 | 77 | 1 | 0 | 0 | 596 |
| 2001 | 0 | 0 | 0 | 11 | 40 | 170 | 166 | 322 | 60 | 12 | 0 | 0 | 781 |
| 2002 | 0 | 0 | 0 | 43 | 25 | 132 | 322 | 293 | 104 | 17 | 0 | 0 | 936 |
| 2003 | 0 | 0 | 0 | 0 | 3 | 119 | 261 | 299 | 46 | 0 | 0 | 0 | 728 |
| 2004 | 0 | 0 | 0 | 2 | 40 | 108 | 228 | 216 | 68 | 0 | 0 | 0 | 662 |
| 2005 | 0 | 0 | 0 | 3 | 0 | 238 | 306 | 321 | 129 | 16 | 0 | 0 | 1013 |
| 2006 | 0 | 0 | 0 | 1 | 44 | 164 | 367 | 234 | 49 | 1 | 0 | 0 | 860 |
| 2007 | 0 | 0 | 0 | 3 | 69 | 158 | 272 | 259 | 120 | 59 | 0 | 0 | 940 |
| 2008 | 0 | 0 | 0 | 1 | 12 | 190 | 336 | 165 | 91 | 2 | 0 | 0 | 797 |
| 2009 | 0 | 0 | 0 | 33 | 21 | 89 | 172 | 271 | 38 | 0 | 0 | 0 | 624 |

### SNOWFALL (inches) 2009 WINDSOR LOCKS (KBDL)

| YEAR | JUL | AUG | SEP | OCT | NOV | DEC | JAN | FEB | MAR | APR | MAY | JUN | TOTAL |
|------|-----|-----|-----|-----|-----|-----|-----|-----|-----|-----|-----|-----|-------|
| 1980-81 | 0.0 | 0.0 | 0.0 | 0.0 | 8.6 | 3.9 | 4.1 | 0.9 | 0.2 | 0.0 | 0.0 | 0.0 | 17.7 |
| 1981-82 | 0.0 | 0.0 | 0.0 | T | T | 13.1 | 16.7 | 5.8 | 6.5 | 14.3 | 0.0 | 0.0 | 56.4 |
| 1982-83 | 0.0 | 0.0 | 0.0 | 0.0 | T | 5.7 | 10.2 | 29.4 | 0.2 | 0.9 | 0.0 | 0.0 | 46.4 |
| 1983-84 | 0.0 | 0.0 | 0.0 | 0.0 | T | 7.9 | 14.7 | 1.3 | 19.3 | T | 0.0 | 0.0 | 43.2 |
| 1984-85 | 0.0 | 0.0 | 0.0 | 0.0 | 0.1 | 3.8 | 6.9 | 9.4 | 2.1 | 1.4 | 0.0 | 0.0 | 23.7 |
| 1985-86 | 0.0 | 0.0 | 0.0 | 0.0 | 2.0 | 5.4 | 5.1 | 11.8 | 0.2 | 0.8 | 0.0 | 0.0 | 25.3 |
| 1986-87 | 0.0 | 0.0 | 0.0 | 0.0 | 8.7 | 4.9 | 34.0 | 1.6 | 1.7 | 0.4 | 0.0 | 0.0 | 51.3 |
| 1987-88 | 0.0 | 0.0 | 0.0 | T | 8.6 | 5.8 | 22.6 | 17.6 | 4.9 | T | T | 0.0 | 59.5 |
| 1988-89 | 0.0 | 0.0 | 0.0 | 0.0 | 0.0 | 6.3 | 0.6 | 4.6 | 3.4 | T | T | 0.0 | 14.9 |
| 1989-90 | 0.0 | 0.0 | 0.0 | 0.0 | 5.3 | 12.4 | 10.5 | 9.0 | 4.3 | 1.5 | 0.0 | 0.0 | 43.0 |
| 1990-91 | 0.0 | 0.0 | 0.0 | 0.0 | T | 8.1 | 10.2 | 5.8 | 5.7 | 0.0 | 0.0 | 0.0 | 29.8 |
| 1991-92 | 0.0 | 0.0 | 0.0 | 0.0 | 0.7 | 6.0 | 1.7 | 5.3 | 7.3 | 2.6 | 0.0 | 0.0 | 23.6 |
| 1992-93 | 0.0 | 0.0 | 0.0 | T | T | 6.7 | 6.7 | 13.8 | 31.1 | T | 0.0 | T | 62.1 |
| 1993-94 | 0.0 | 0.0 | 0.0 | 0.0 | T | 6.7 | 31.3 | 29.4 | 17.5 | 0.0 | T | 0.0 | 84.9 |
| 1994-95 | 0.0 | 0.0 | 0.0 | 0.0 | 3.9 | 1.1 | 5.7 | 10.1 | 0.0 | 1.5 | 0.0 | 0.0 | 22.3 |
| 1995-96 | 0.0 | 0.0 | 0.0 | 0.0 | 5.6 | 20.3 | 42.8 | 20.6 | 17.8 | | | | |
| 1996-97 | | | | | | | | | | | | | |
| 1997-98 | | | | | | | | | | | | | |
| 1998-99 | | | | | | | | | | | | | |
| 1999-00 | | | | | | | | | | | | | |
| 2000-01 | | | | | | 8.0 | 10.4 | 21.4 | 13.5 | 0.0 | | | |
| 2001-02 | | | | | | 3.0 | | | | | | | |
| 2002-03 | | | | | | 13.6 | 15.6 | 21.4 | 11.1 | 2.3 | 0.0 | 0.0 | |
| 2003-04 | 0.0 | 0.0 | 0.0 | T | T | 23.4 | 11.1 | 4.0 | 7.0 | 0.0 | 0.0 | 0.0 | 45.5 |
| 2004-05 | 0.0 | 0.0 | 0.0 | 0.0 | 1.5 | 9.6 | 27.7 | 16.3 | 19.0 | 0.0 | 0.0 | 0.0 | 74.1 |
| 2005-06 | 0.0 | 0.0 | 0.0 | 0.0 | 3.3 | 16.6 | 21.8 | 22.2 | 4.8 | 1.2 | 0.0 | 0.0 | 69.9 |
| 2006-07 | 0.0 | 0.0 | 0.0 | 0.0 | 0.0 | T | 1.2 | 11.2 | 11.6 | T | 0.0 | 0.0 | 24.0 |
| 2007-08 | 0.0 | 0.0 | 0.0 | 0.0 | 1.0 | 18.8 | 10.6 | 12.9 | 4.2 | 0.0 | T | 0.0 | 47.5 |
| 2008-09 | 0.0 | T | 0.0 | 0.0 | T | 20.7 | 12.9 | 3.9 | 8.0 | 0.0 | 0.0 | 0.0 | 45.5 |
| 2009- | 0.0 | 0.0 | 0.0 | T | 0.0 | 12.8 | | | | | | | |
| POR= 53 YRS | 0.0 | T | 0.0 | T | 1.9 | 10.2 | 13.1 | 12.1 | 9.3 | 1.2 | T | T | 47.8 |

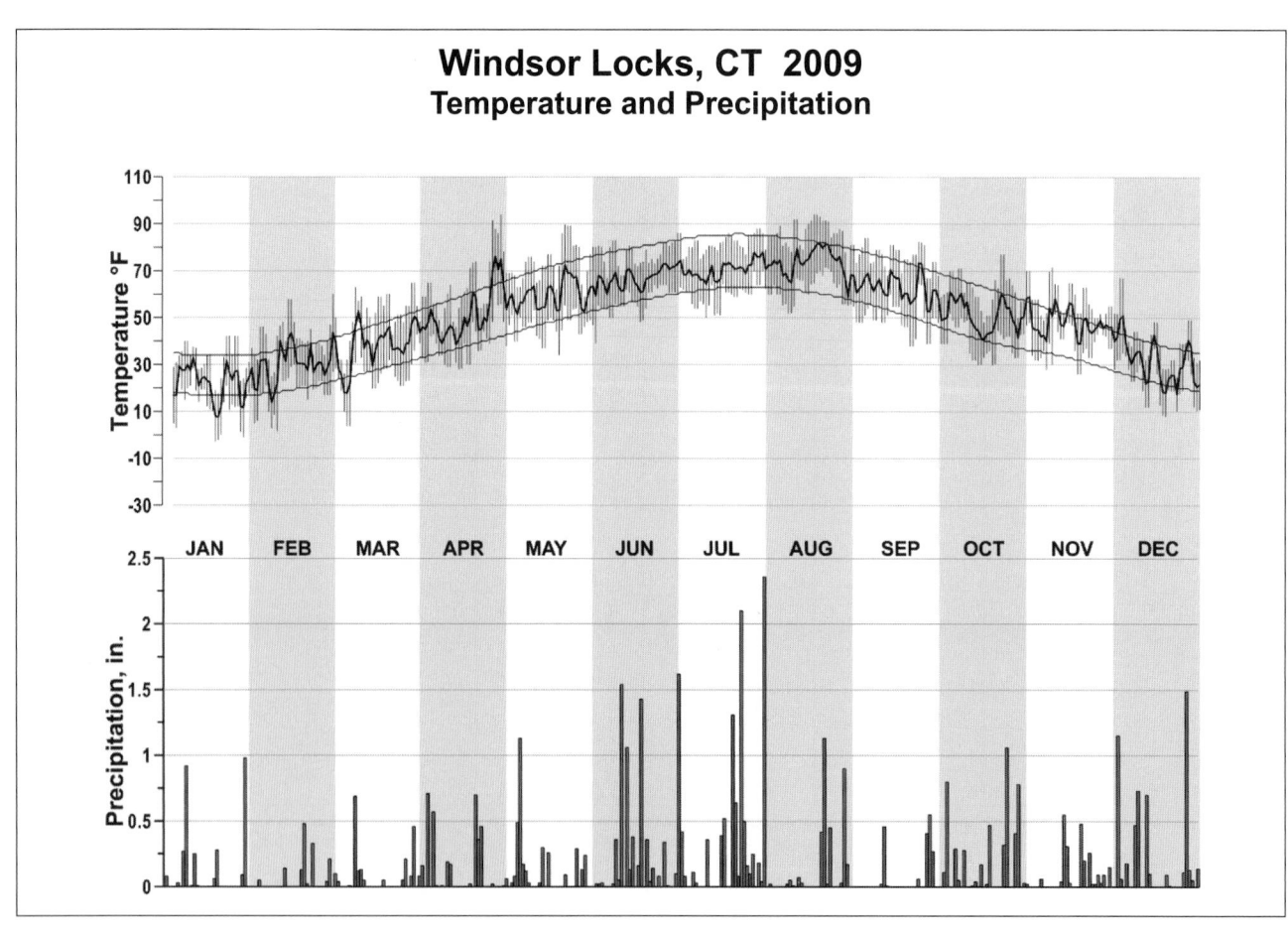

# Windsor Locks, CT 2009
## Temperature and Precipitation

# 2009
# WASHINGTON
# D.C. (KDCA)

Washington lies at the western edge of the mid Atlantic Coastal Plain, about 50 miles east of the Blue Ridge Mountains and 35 miles west of Chesapeake Bay, adjacent to the Potomac and Anacostia Rivers. Elevations range from a few feet above sea level to about 400 feet in parts of the northwest section of the city.

Observations have been kept continuously since November 1870. Since June 1941 the official observations have been taken at Washington National Airport.

National Airport is located at the center of the urban heat island. As a result, low temperatures are the highest for the area. Differences between the airport and suburban locations are often 10 to 15 degrees. There is less variation in the high temperatures.

Summers are warm and humid and winters are cold, but not severe. Periods of pleasant weather often occur in the spring and fall. The summertime temperature is in the upper 80s and the winter is in the upper 20s. Precipitation is rather uniformly distributed throughout the year.

Thunderstorms can occur at any time but are most frequent during the late spring and summer. The storms are most often accompanied by downpours and gusty winds, but are not usually severe.

Tornadoes, which infrequently occur, have resulted in significant damage. Severe hailstorms have occurred in the spring.

Tropical storms can bring heavy rain, high winds and flooding, but extensive damage from wind and tidal flooding is rare. Wind gusts of nearly 100 mph and rainfall over 7 inches have occurred during the passage of tropical storms and hurricanes.

Major flooding of the Potomac River can result from heavy rains over the basin, occasionally augmented by snowmelt, and above normal tides associated with hurricanes or severe storms along the coast. Flooding may also occur after a cold winter when the Potomac may be blocked with ice.

Although a snowfall of 10 inches or more in 24 hours is unusual, several notable falls of more than 25 inches have occurred. Normal snowfall during the winter season is 18 inches.

The average date of the last freezing temperature in the spring is April 1 and the average date for the first freezing temperature in the fall is November 10.

# NORMALS, MEANS, AND EXTREMES
## WASHINGTON (KDCA)

| LATITUDE: 38° 51'N | LONGITUDE: -77° 2 'W | | ELEVATION (FT): GRND: 10   BARO: 3 | | | | TIME ZONE: EASTERN   (UTC -5) | | | WBAN: 13743 | | |

| ELEMENT | POR | JAN | FEB | MAR | APR | MAY | JUN | JUL | AUG | SEP | OCT | NOV | DEC | YEAR |
|---|---|---|---|---|---|---|---|---|---|---|---|---|---|---|
| **TEMPERATURE °F** | | | | | | | | | | | | | | |
| NORMAL DAILY MAXIMUM | 30 | 42.5 | 46.5 | 55.7 | 66.3 | 75.4 | 83.9 | 88.3 | 86.3 | 79.3 | 68.0 | 57.3 | 47.0 | 66.4 |
| MEAN DAILY MAXIMUM | 64 | 43.5 | 46.6 | 55.4 | 66.7 | 75.6 | 83.8 | 87.9 | 86.2 | 79.6 | 68.8 | 57.7 | 46.7 | 66.5 |
| HIGHEST DAILY MAXIMUM | 68 | 79 | 82 | 89 | 95 | 99 | 101 | 104 | 105 | 101 | 94 | 86 | 79 | 105 |
| YEAR OF OCCURRENCE | | 1950 | 1948 | 1990 | 2002 | 1991 | 1994 | 1988 | 1997 | 1980 | 2007 | 1974 | 1998 | AUG 1997 |
| MEAN OF EXTREME MAXS. | 64 | 64.9 | 67.3 | 77.5 | 86.1 | 90.1 | 95.0 | 96.8 | 95.5 | 92.0 | 83.8 | 75.4 | 66.1 | 82.5 |
| NORMAL DAILY MINIMUM | 30 | 27.3 | 29.7 | 37.3 | 45.9 | 55.8 | 65.0 | 70.1 | 68.6 | 61.8 | 49.6 | 40.0 | 32.0 | 48.6 |
| MEAN DAILY MINIMUM | 64 | 28.4 | 30.0 | 37.1 | 46.5 | 56.2 | 65.3 | 70.3 | 69.1 | 62.1 | 50.3 | 40.6 | 31.9 | 49.0 |
| LOWEST DAILY MINIMUM | 68 | -5 | 4 | 11 | 24 | 34 | 47 | 54 | 49 | 39 | 29 | 16 | 1 | -5 |
| YEAR OF OCCURRENCE | | 1982 | 1961 | 1943 | 1982 | 1947 | 1972 | 1988 | 1986 | 1963 | 1969 | 1955 | 1942 | JAN 1982 |
| MEAN OF EXTREME MINS. | 64 | 12.7 | 15.4 | 22.8 | 33.4 | 43.8 | 53.8 | 61.2 | 59.0 | 48.7 | 37.0 | 27.1 | 17.6 | 36.0 |
| NORMAL DRY BULB | 30 | 34.9 | 38.1 | 46.5 | 56.1 | 65.6 | 74.5 | 79.2 | 77.4 | 70.5 | 58.8 | 48.7 | 39.5 | 57.5 |
| MEAN DRY BULB | 64 | 36.0 | 38.3 | 46.2 | 56.6 | 65.9 | 74.7 | 79.1 | 77.7 | 70.9 | 59.6 | 49.1 | 39.3 | 57.8 |
| MEAN WET BULB | 26 | 30.8 | 32.5 | 38.9 | 48.3 | 57.7 | 66.3 | 70.1 | 69.3 | 63.3 | 52.8 | 43.2 | 34.4 | 50.6 |
| MEAN DEW POINT | 26 | 25.7 | 27.2 | 33.2 | 43.2 | 53.9 | 63.1 | 67.2 | 66.7 | 60.4 | 49.3 | 38.8 | 29.4 | 46.5 |
| NORMAL NO. DAYS WITH: | | | | | | | | | | | | | | |
| MAXIMUM >= 90 | 30 | 0.0 | 0.0 | 0.0 | 0.5 | 1.5 | 6.9 | 14.5 | 9.9 | 3.3 | 0.1 | 0.0 | 0.0 | 36.7 |
| MAXIMUM <= 32 | 30 | 4.6 | 2.3 | 0.3 | 0.0 | 0.0 | 0.0 | 0.0 | 0.0 | 0.0 | 0.0 | * | 2.1 | 9.3 |
| MINIMUM <= 32 | 30 | 20.7 | 16.9 | 7.7 | 0.6 | 0.0 | 0.0 | 0.0 | 0.0 | 0.0 | 0.2 | 4.4 | 13.9 | 64.4 |
| MINIMUM <= 0 | 30 | 0.1 | 0.0 | 0.0 | 0.0 | 0.0 | 0.0 | 0.0 | 0.0 | 0.0 | 0.0 | 0.0 | 0.0 | 0.1 |
| **H/C** | | | | | | | | | | | | | | |
| NORMAL HEATING DEG. DAYS | 30 | 917 | 742 | 563 | 272 | 73 | 5 | 0 | 7 | 19 | 205 | 477 | 775 | 4055 |
| NORMAL COOLING DEG. DAYS | 30 | 0 | 0 | 4 | 21 | 107 | 304 | 456 | 407 | 200 | 28 | 4 | 0 | 1531 |
| **RH** | | | | | | | | | | | | | | |
| NORMAL (PERCENT) | 30 | 64 | 63 | 60 | 60 | 66 | 68 | 68 | 70 | 71 | 70 | 66 | 66 | 66 |
| HOUR 01 LST | 30 | 68 | 68 | 66 | 68 | 76 | 78 | 78 | 81 | 81 | 79 | 73 | 70 | 74 |
| HOUR 07 LST | 30 | 71 | 72 | 71 | 70 | 76 | 76 | 77 | 81 | 82 | 81 | 77 | 73 | 76 |
| HOUR 13 LST | 30 | 57 | 53 | 50 | 49 | 54 | 54 | 54 | 56 | 57 | 55 | 55 | 57 | 54 |
| HOUR 19 LST | 30 | 62 | 58 | 54 | 52 | 60 | 62 | 62 | 65 | 67 | 66 | 62 | 62 | 61 |
| **S** PERCENT POSSIBLE SUNSHINE | 50 | 46 | 50 | 55 | 57 | 58 | 64 | 62 | 62 | 61 | 59 | 51 | 46 | 56 |
| **W/O** MEAN NO. DAYS WITH: | | | | | | | | | | | | | | |
| HEAVY FOG(VISBY <= 1/4 MI) | 46 | 1.7 | 1.3 | 0.7 | 0.5 | 0.4 | 0.3 | 0.2 | 0.0 | 0.2 | 0.9 | 1.0 | 1.3 | 8.5 |
| THUNDERSTORMS | 64 | 0.3 | 0.3 | 1.1 | 2.7 | 4.9 | 6.3 | 6.9 | 5.3 | 2.4 | 1.0 | 0.6 | 0.1 | 31.9 |
| **CLOUDNESS** MEAN: | | | | | | | | | | | | | | |
| SUNRISE-SUNSET (OKTAS) | 50 | 5.4 | 5.2 | 5.2 | 5.1 | 5.1 | 4.8 | 4.8 | 4.6 | 4.6 | 4.3 | 5.0 | 5.2 | 4.9 |
| MIDNIGHT-MIDNIGHT (OKTAS) | 34 | 5.1 | 5.0 | 5.0 | 4.8 | 5.0 | 4.7 | 4.8 | 4.5 | 4.5 | 4.2 | 4.7 | 5.1 | 4.8 |
| MEAN NO. DAYS WITH: | | | | | | | | | | | | | | |
| CLEAR | 50 | 7.2 | 7.0 | 7.5 | 7.1 | 6.9 | 7.3 | 7.4 | 8.6 | 9.4 | 10.7 | 7.7 | 7.9 | 94.7 |
| PARTLY CLOUDY | 50 | 7.4 | 6.7 | 8.3 | 9.1 | 10.1 | 11.4 | 11.8 | 10.2 | 8.4 | 7.9 | 7.8 | 6.6 | 105.7 |
| CLOUDY | 50 | 16.4 | 14.5 | 15.2 | 13.9 | 14.0 | 11.4 | 11.4 | 11.7 | 11.6 | 11.9 | 13.8 | 15.9 | 161.7 |
| **PR** | | | | | | | | | | | | | | |
| MEAN STATION PRESSURE(IN) | 26 | 30.03 | 30.03 | 29.99 | 29.92 | 29.93 | 29.91 | 29.92 | 29.96 | 30.00 | 30.03 | 30.04 | 30.06 | 29.99 |
| MEAN SEA-LEVEL PRES. (IN) | 26 | 30.10 | 30.10 | 30.06 | 29.99 | 29.99 | 29.98 | 29.99 | 30.03 | 30.07 | 30.10 | 30.11 | 30.12 | 30.05 |
| **WINDS** | | | | | | | | | | | | | | |
| MEAN SPEED (MPH) | 26 | 9.6 | 9.8 | 10.6 | 10.0 | 9.2 | 8.8 | 8.2 | 8.0 | 8.5 | 8.4 | 9.1 | 9.2 | 9.1 |
| PREVAIL.DIR(TENS OF DEGS) | 40 | 34 | 33 | 33 | 19 | 19 | 19 | 19 | 19 | 19 | 19 | 34 | 34 | 19 |
| MAXIMUM 2-MINUTE: | | | | | | | | | | | | | | |
| SPEED (MPH) | 11 | 43 | 39 | 46 | 39 | 39 | 49 | 47 | 41 | 45 | 37 | 43 | 40 | 49 |
| DIR. (TENS OF DEGS) | | 32 | 33 | 25 | 31 | 36 | 31 | 05 | 35 | 11 | 29 | 33 | 29 | 31 |
| YEAR OF OCCURRENCE | | 2006 | 2002 | 2008 | 2000 | 2008 | 1999 | 2000 | 2003 | 2003 | 2006 | 2005 | 2004 | JUN 1999 |
| MAXIMUM 3-SECOND | | | | | | | | | | | | | | |
| SPEED (MPH) | 11 | 55 | 55 | 74 | 52 | 55 | 60 | 53 | 47 | 58 | 51 | 54 | 55 | 74 |
| DIR. (TENS OF DEGS) | | 31 | 32 | 22 | 33 | 25 | 31 | 27 | 34 | 13 | 26 | 35 | 32 | 22 |
| YEAR OF OCCURRENCE | | 2000 | 2007 | 2008 | 2007 | 2000 | 1999 | 2006 | 2003 | 2003 | 2004 | 2003 | 2008 | MAR 2008 |
| **PRECIPITATION** | | | | | | | | | | | | | | |
| NORMAL (IN) | 30 | 3.21 | 2.63 | 3.60 | 2.77 | 3.82 | 3.13 | 3.66 | 3.44 | 3.79 | 3.22 | 3.03 | 3.05 | 39.35 |
| MAXIMUM MONTHLY (IN) | 68 | 7.11 | 5.71 | 8.45 | 6.88 | 10.69 | 14.02 | 11.06 | 14.31 | 12.36 | 9.41 | 6.70 | 6.79 | 14.31 |
| YEAR OF OCCURRENCE | | 1978 | 1961 | 1994 | 1983 | 1953 | 2006 | 1945 | 1955 | 1975 | 2005 | 1963 | 2009 | AUG 1955 |
| MINIMUM MONTHLY (IN) | 68 | 0.31 | 0.35 | 0.05 | 0.03 | 0.75 | 0.95 | 0.93 | 0.55 | .11 | T | 0.29 | 0.22 | T |
| YEAR OF OCCURRENCE | | 1955 | 2009 | 2006 | 1985 | 1986 | 1988 | 1966 | 1962 | 2005 | 1963 | 1981 | 1955 | OCT 1963 |
| MAXIMUM IN 24 HOURS (IN) | 68 | 2.13 | 2.56 | 3.43 | 3.08 | 4.32 | 7.94 | 4.69 | 6.39 | 5.31 | 5.91 | 4.03 | 2.86 | 7.94 |
| YEAR OF OCCURRENCE | | 1976 | 2003 | 1958 | 1970 | 1953 | 2006 | 1970 | 1955 | 1975 | 2005 | 1993 | 1977 | JUN 2006 |
| NORMAL NO. DAYS WITH: | | | | | | | | | | | | | | |
| PRECIPITATION >= 0.01 | 30 | 10.5 | 9.3 | 10.6 | 9.6 | 11.2 | 10.2 | 10.4 | 8.6 | 8.1 | 7.8 | 8.5 | 9.5 | 114.3 |
| PRECIPITATION >= 1.00 | 30 | 0.7 | 0.4 | 0.9 | 0.5 | 0.8 | 0.7 | 1.0 | 1.1 | 1.1 | 1.0 | 0.8 | 0.6 | 9.6 |
| **SNOWFALL** | | | | | | | | | | | | | | |
| NORMAL (IN) | 30 | 6.2 | 5.2 | 1.6 | 0.* | 0.0 | 0.0 | 0.0 | 0.0 | 0.0 | 0.* | 0.7 | 1.5 | 15.2 |
| MAXIMUM MONTHLY (IN) | 66 | 23.8 | 30.6 | 17.1 | 0.6 | T | T | T | T | T | 0.3 | 11.5 | 16.6 | 30.6 |
| YEAR OF OCCURRENCE | | 1996 | 1979 | 1960 | 1972 | 1993 | 2009 | 2006 | 2008 | | 1979 | 1987 | 2009 | FEB 1979 |
| MAXIMUM IN 24 HOURS (IN) | 66 | 13.8 | 18.7 | 8.4 | 0.6 | T | T | T | T | 0.0 | 0.3 | 11.5 | 15.0 | 18.7 |
| YEAR OF OCCURRENCE' | | 1966 | 1979 | 1999 | 1972 | 1993 | 2009 | 1990 | 1992 | | 1979 | 1987 | 2009 | FEB 1979 |
| MAXIMUM SNOW DEPTH (IN) | 63 | 18 | 22 | 8 | 1 | 0 | 0 | 0 | 0 | 0 | 0 | 12 | 16 | 22 |
| YEAR OF OCCURRENCE | | 1987 | 1979 | 1960 | 1972 | | | | | | | 1987 | 2009 | FEB 1979 |
| NORMAL NO. DAYS WITH: | | | | | | | | | | | | | | |
| SNOWFALL >= 1.0 | 30 | 1.6 | 1.2 | 0.4 | 0.0 | 0.0 | 0.0 | 0.0 | 0.0 | 0.0 | 0.0 | 0.2 | 0.4 | 3.8 |

## PRECIPITATION (inches) 2009 WASHINGTON (KDCA)

| YEAR | JAN | FEB | MAR | APR | MAY | JUN | JUL | AUG | SEP | OCT | NOV | DEC | ANNUAL |
|------|------|------|------|------|------|------|------|------|------|------|------|------|--------|
| 1980 | 2.85 | 1.16 | 5.04 | 3.28 | 2.64 | 1.68 | 3.86 | 1.11 | 1.90 | 2.59 | 2.56 | 0.65 | 29.32 |
| 1981 | 0.38 | 2.82 | 1.49 | 2.63 | 3.42 | 2.55 | 5.69 | 3.02 | 1.94 | 3.64 | 0.29 | 2.80 | 30.67 |
| 1982 | 2.27 | 3.33 | 2.64 | 3.19 | 5.11 | 5.41 | 2.98 | 2.68 | 1.71 | 1.75 | 2.96 | 1.74 | 35.77 |
| 1983 | 1.69 | 3.09 | 4.84 | 6.88 | 4.62 | 7.09 | 1.78 | 3.11 | 2.90 | 4.87 | 5.09 | 5.91 | 51.87 |
| 1984 | 1.71 | 3.43 | 6.14 | 3.71 | 3.80 | 2.01 | 4.09 | 2.30 | 2.51 | 3.18 | 3.66 | 1.19 | 37.73 |
| 1985 | 2.11 | 3.07 | 1.88 | 0.03 | 5.79 | 2.05 | 2.91 | 2.35 | 6.67 | 3.85 | 4.47 | 0.68 | 35.86 |
| 1986 | 2.38 | 3.49 | 0.74 | 1.98 | 0.75 | 1.29 | 3.79 | 5.33 | 0.60 | 2.01 | 5.23 | 4.98 | 32.57 |
| 1987 | 4.90 | 2.11 | 1.54 | 2.28 | 2.54 | 3.90 | 2.59 | 2.07 | 5.11 | 2.53 | 4.49 | 2.57 | 36.63 |
| 1988 | 3.14 | 2.52 | 2.27 | 2.00 | 4.50 | 0.95 | 3.74 | 2.39 | 1.85 | 1.75 | 5.33 | 1.30 | 31.74 |
| 1989 | 2.49 | 2.80 | 4.30 | 3.50 | 7.77 | 6.02 | 5.66 | 1.15 | 6.68 | 5.48 | 2.37 | 2.10 | 50.32 |
| 1990 | 2.95 | 1.30 | 2.57 | 4.09 | 5.20 | 3.14 | 3.78 | 6.74 | 0.87 | 3.30 | 2.17 | 4.73 | 40.84 |
| 1991 | 2.90 | 0.83 | 4.42 | 1.39 | 1.57 | 1.27 | 3.76 | 2.03 | 3.50 | 2.03 | 0.85 | 5.07 | 29.62 |
| 1992 | 2.78 | 2.23 | 3.48 | 2.55 | 3.41 | 2.35 | 5.34 | 2.48 | 3.49 | 2.03 | 3.38 | 2.86 | 36.38 |
| 1993 | 2.90 | 2.27 | 6.82 | 3.62 | 3.40 | 1.73 | 1.36 | 3.87 | 3.68 | 2.62 | 4.76 | 4.38 | 41.41 |
| 1994 | 4.28 | 4.20 | 8.45 | 1.58 | 1.56 | 1.59 | 3.61 | 4.35 | 2.84 | 1.19 | 1.57 | 2.35 | 37.57 |
| 1995 | 3.22 | 1.71 | 2.14 | 1.89 | 4.19 | 2.42 | 4.03 | 0.88 | 3.73 | 8.65 | 4.77 | 2.17 | 39.80 |
| 1996 | 5.01 | 1.99 | 3.60 | 3.17 | 4.96 | 3.14 | 5.60 | 2.63 | 7.79 | 4.04 | 3.58 | 5.51 | 51.02 |
| 1997 | 2.55 | 2.43 | 4.15 | 2.41 | 3.04 | 2.94 | 1.14 | 3.51 | 1.59 | 3.72 | 4.60 | 1.74 | 33.82 |
| 1998 | 5.43 | 5.23 | 5.40 | 3.96 | 4.05 | 4.42 | 1.79 | 0.59 | 1.83 | 0.59 | 0.91 | 1.74 | 35.94 |
| 1999 | 5.42 | 2.54 | 3.87 | 2.09 | 1.28 | 2.26 | 1.01 | 5.02 | 10.27 | 2.16 | 1.82 | 2.49 | 40.23 |
| 2000 | 3.66 | 2.06 | 3.98 | 5.13 | 3.08 | 4.93 | 5.51 | 3.77 | 4.91 | 0.02 | 1.60 | 2.01 | 40.66 |
| 2001 | 2.22 | 1.83 | 3.88 | 1.68 | 3.71 | 4.69 | 4.78 | 2.98 | 1.41 | 0.69 | 0.55 | 1.53 | 29.95 |
| 2002 | 1.32 | 0.47 | 3.37 | 3.47 | 2.17 | 3.81 | 2.20 | 1.63 | 2.10 | 5.00 | 4.34 | 4.45 | 34.33 |
| 2003 | 2.41 | 5.68 | 4.20 | 2.55 | 7.06 | 7.87 | 5.76 | 4.65 | 6.87 | 3.93 | 4.23 | 4.32 | 59.53 |
| 2004 | 1.35 | 2.28 | 2.09 | 3.84 | 2.98 | 4.60 | 6.98 | 5.09 | 3.99 | 1.74 | 4.50 | 3.05 | 42.49 |
| 2005 | 3.31 | 1.63 | 4.46 | 4.33 | 4.61 | 2.87 | 6.06 | 2.33 | 0.11 | 9.41 | 1.92 | 3.34 | 44.38 |
| 2006 | 3.25 | 2.46 | 0.05 | 3.10 | 2.21 | 14.02 | 3.56 | 1.03 | 6.31 | 5.06 | 5.16 | 1.56 | 47.77 |
| 2007 | 2.46 | 2.22 | 3.19 | 4.17 | 1.75 | 1.38 | 2.40 | 3.47 | 0.60 | 6.55 | 1.46 | 3.28 | 32.93 |
| 2008 | 1.37 | 4.17 | 2.80 | 4.92 | 10.66 | 4.80 | 3.60 | 1.23 | 6.41 | 1.13 | 2.43 | 2.97 | 46.49 |
| 2009 | 2.68 | 0.35 | 1.97 | 4.12 | 8.05 | 5.86 | 1.07 | 2.46 | 3.31 | 5.71 | 4.43 | 6.79 | 46.80 |
| POR=<br>64 YRS | 2.86 | 2.56 | 3.41 | 2.96 | 3.97 | 3.61 | 3.94 | 3.81 | 3.49 | 3.09 | 3.03 | 3.12 | 39.85 |

WBAN : 13743

## AVERAGE TEMPERATURE (°F) 2009 WASHINGTON (KDCA)

| YEAR | JAN | FEB | MAR | APR | MAY | JUN | JUL | AUG | SEP | OCT | NOV | DEC | ANNUAL |
|------|------|------|------|------|------|------|------|------|------|------|------|------|--------|
| 1980 | 37.2 | 36.1 | 46.2 | 60.1 | 69.5 | 74.8 | 82.3 | 82.8 | 77.1 | 59.9 | 48.6 | 39.8 | 59.5 |
| 1981 | 33.0 | 43.7 | 47.6 | 62.1 | 66.2 | 78.7 | 80.2 | 77.0 | 71.0 | 58.3 | 51.4 | 38.5 | 59.0 |
| 1982 | 28.1 | 38.3 | 45.7 | 54.0 | 69.0 | 72.8 | 80.3 | 75.4 | 70.6 | 60.2 | 51.8 | 45.5 | 57.6 |
| 1983 | 38.1 | 38.7 | 48.8 | 53.3 | 64.9 | 75.0 | 81.2 | 81.0 | 72.6 | 60.5 | 50.3 | 36.0 | 58.4 |
| 1984 | 32.2 | 43.8 | 41.8 | 54.9 | 64.9 | 76.9 | 76.5 | 77.8 | 68.3 | 65.2 | 46.0 | 45.6 | 57.8 |
| 1985 | 30.8 | 37.8 | 47.7 | 61.6 | 68.1 | 72.3 | 79.0 | 76.7 | 71.9 | 61.2 | 54.3 | 36.4 | 58.2 |
| 1986 | 35.4 | 35.3 | 47.4 | 56.2 | 68.1 | 76.6 | 81.1 | 74.6 | 70.9 | 61.1 | 46.5 | 39.8 | 57.8 |
| 1987 | 34.7 | 37.0 | 47.7 | 54.8 | 67.2 | 76.4 | 82.6 | 78.7 | 72.1 | 54.4 | 49.9 | 41.5 | 58.1 |
| 1988 | 31.0 | 37.3 | 47.2 | 54.4 | 65.8 | 74.4 | 81.9 | 80.7 | 68.9 | 54.4 | 49.9 | 38.7 | 57.1 |
| 1989 | 39.9 | 37.8 | 46.1 | 55.4 | 64.1 | 76.8 | 78.3 | 77.1 | 71.4 | 60.5 | 48.0 | 27.9 | 56.9 |
| 1990 | 43.6 | 45.2 | 50.2 | 56.8 | 64.3 | 75.0 | 79.4 | 76.5 | 69.6 | 62.8 | 52.0 | 44.5 | 60.0 |
| 1991 | 38.6 | 43.0 | 48.8 | 58.2 | 73.0 | 76.8 | 81.4 | 80.0 | 71.0 | 60.4 | 48.8 | 42.3 | 60.2 |
| 1992 | 38.2 | 41.2 | 45.0 | 55.4 | 62.3 | 71.7 | 79.5 | 74.0 | 69.2 | 56.0 | 48.8 | 39.6 | 56.7 |
| 1993 | 39.7 | 34.3 | 42.2 | 54.8 | 67.4 | 75.3 | 83.1 | 79.6 | 71.0 | 58.1 | 48.8 | 38.1 | 57.7 |
| 1994 | 28.8 | 36.3 | 45.4 | 62.0 | 63.1 | 79.4 | 81.8 | 75.5 | 70.2 | 59.1 | 53.4 | 44.2 | 58.3 |
| 1995 | 39.6 | 34.4 | 49.2 | 56.3 | 65.8 | 74.6 | 81.5 | 81.3 | 71.0 | 62.3 | 43.1 | 35.6 | 57.9 |
| 1996 | 32.9 | 37.3 | 42.9 | 56.8 | 63.9 | 77.2 | 77.6 | 76.5 | 70.6 | 59.4 | 44.1 | 43.0 | 56.9 |
| 1997 | 37.0 | 44.7 | 48.7 | 54.0 | 62.9 | 73.1 | 80.4 | 77.6 | 70.5 | 59.6 | 46.2 | 41.0 | 58.0 |
| 1998 | 43.0 | 43.4 | 46.9 | 57.4 | 67.5 | 73.0 | 78.9 | 79.4 | 75.4 | 60.1 | 50.3 | 44.4 | 60.0 |
| 1999 | 38.2 | 41.0 | 44.5 | 56.5 | 67.3 | 74.7 | 83.0 | 79.7 | 70.0 | 57.2 | 53.1 | 42.0 | 58.9 |
| 2000 | 35.9 | 42.5 | 51.7 | 55.6 | 67.8 | 74.7 | 74.7 | 75.1 | 67.6 | 60.2 | 46.7 | 31.8 | 57.0 |
| 2001 | 35.5 | 40.9 | 43.8 | 57.6 | 65.9 | 75.2 | 75.3 | 78.9 | 68.9 | 59.7 | 54.8 | 45.5 | 58.5 |
| 2002 | 41.6 | 42.6 | 47.7 | 60.0 | 65.2 | 76.1 | 80.9 | 81.1 | 73.0 | 58.7 | 47.1 | 37.2 | 59.3 |
| 2003 | 31.1 | 33.7 | 47.1 | 55.1 | 61.7 | 71.4 | 77.8 | 78.8 | 70.5 | 57.5 | 53.1 | 39.2 | 56.4 |
| 2004 | 30.6 | 38.2 | 48.8 | 57.4 | 71.8 | 73.4 | 78.6 | 75.9 | 71.6 | 58.3 | 51.0 | 40.1 | 58.0 |
| 2005 | 36.1 | 39.6 | 43.1 | 57.3 | 61.9 | 75.7 | 80.5 | 80.1 | 75.0 | 60.5 | 50.2 | 36.4 | 58.0 |
| 2006 | 43.1 | 38.6 | 47.9 | 59.5 | 65.1 | 74.5 | 80.4 | 80.6 | 67.8 | 57.3 | 50.7 | 44.2 | 59.1 |
| 2007 | 40.7 | 30.9 | 47.7 | 53.6 | 67.8 | 76.1 | 79.3 | 79.7 | 72.9 | 67.1 | 49.8 | 41.8 | 59.0 |
| 2008 | 40.0 | 41.0 | 49.0 | 58.9 | 64.7 | 77.9 | 80.8 | 77.9 | 74.0 | 58.9 | 46.6 | 40.3 | 59.2 |
| 2009 | 31.7 | 39.8 | 45.2 | 57.0 | 65.5 | 73.7 | 76.9 | 79.9 | 70.2 | 58.5 | 52.2 | 37.9 | 57.4 |
| POR=<br>64 YRS | 36.0 | 38.3 | 46.2 | 56.6 | 65.9 | 74.7 | 79.1 | 77.7 | 70.9 | 59.6 | 49.1 | 39.3 | 57.8 |

## HEATING DEGREE DAYS (base 65°F) 2009 WASHINGTON (KDCA)

| YEAR | JUL | AUG | SEP | OCT | NOV | DEC | JAN | FEB | MAR | APR | MAY | JUN | TOTAL |
|---|---|---|---|---|---|---|---|---|---|---|---|---|---|
| 1980-81 | 0 | 0 | 4 | 189 | 487 | 774 | 984 | 592 | 536 | 133 | 75 | 0 | 3774 |
| 1981-82 | 0 | 0 | 19 | 219 | 399 | 818 | 1135 | 743 | 592 | 328 | 19 | 3 | 4275 |
| 1982-83 | 0 | 2 | 9 | 193 | 402 | 597 | 827 | 730 | 497 | 365 | 77 | 0 | 3699 |
| 1983-84 | 0 | 0 | 32 | 177 | 433 | 890 | 1009 | 610 | 710 | 302 | 95 | 4 | 4262 |
| 1984-85 | 0 | 0 | 54 | 59 | 561 | 594 | 1053 | 757 | 533 | 166 | 30 | 6 | 3813 |
| 1985-86 | 0 | 0 | 14 | 147 | 320 | 879 | 913 | 824 | 542 | 267 | 61 | 3 | 3970 |
| 1986-87 | 0 | 13 | 18 | 180 | 548 | 775 | 931 | 777 | 527 | 304 | 68 | 0 | 4141 |
| 1987-88 | 0 | 0 | 4 | 325 | 448 | 719 | 1047 | 796 | 544 | 317 | 69 | 25 | 4294 |
| 1988-89 | 0 | 0 | 18 | 330 | 442 | 807 | 771 | 755 | 596 | 297 | 112 | 0 | 4128 |
| 1989-90 | 0 | 0 | 35 | 167 | 507 | 1144 | 656 | 550 | 481 | 285 | 64 | 4 | 3893 |
| 1990-91 | 0 | 0 | 38 | 153 | 381 | 630 | 810 | 608 | 502 | 237 | 22 | 0 | 3381 |
| 1991-92 | 0 | 0 | 27 | 175 | 486 | 696 | 824 | 686 | 614 | 295 | 127 | 5 | 3935 |
| 1992-93 | 0 | 0 | 43 | 282 | 477 | 781 | 779 | 855 | 700 | 307 | 32 | 3 | 4259 |
| 1993-94 | 0 | 0 | 33 | 217 | 487 | 825 | 1115 | 796 | 599 | 135 | 132 | 0 | 4339 |
| 1994-95 | 0 | 0 | 5 | 190 | 348 | 639 | 782 | 853 | 485 | 274 | 59 | 0 | 3635 |
| 1995-96 | 0 | 0 | 30 | 144 | 651 | 901 | 993 | 798 | 675 | 268 | 131 | 0 | 4591 |
| 1996-97 | 0 | 0 | 9 | 182 | 617 | 677 | 861 | 564 | 499 | 322 | 110 | 30 | 3871 |
| 1997-98 | 0 | 0 | 14 | 232 | 557 | 737 | 675 | 596 | 581 | 229 | 50 | 11 | 3682 |
| 1998-99 | 0 | 0 | 4 | 153 | 433 | 638 | 824 | 667 | 630 | 251 | 34 | 3 | 3637 |
| 1999-00 | 0 | 0 | 22 | 240 | 348 | 707 | 896 | 649 | 408 | 280 | 46 | 2 | 3598 |
| 2000-01 | | 0 | 60 | 171 | 544 | 1022 | 908 | 670 | 654 | 246 | 54 | 4 | |
| 2001-02 | 0 | 0 | 44 | 207 | 302 | 598 | 717 | 621 | 532 | 225 | 101 | 0 | 3347 |
| 2002-03 | 0 | 0 | 1 | 257 | 531 | 855 | 1043 | 871 | 549 | 303 | 128 | 15 | 4553 |
| 2003-04 | 0 | 0 | 13 | 231 | 366 | 791 | 1062 | 771 | 494 | 267 | 34 | 4 | 4033 |
| 2004-05 | 0 | 0 | 7 | 209 | 414 | 766 | 887 | 702 | 675 | 240 | 121 | 4 | 4025 |
| 2005-06 | 0 | 0 | 5 | 179 | 439 | 881 | 672 | 733 | 529 | 174 | 81 | 0 | 3693 |
| 2006-07 | 0 | 0 | 22 | 250 | 419 | 639 | 746 | 950 | 535 | 353 | 53 | 0 | 3967 |
| 2007-08 | 0 | 0 | 10 | 74 | 451 | 713 | 769 | 688 | 487 | 204 | 67 | 0 | 3463 |
| 2008-09 | 0 | 0 | 2 | 225 | 543 | 759 | 1028 | 698 | 607 | 270 | 73 | 5 | 4210 |
| 2009- | 0 | 0 | 8 | 208 | 376 | 835 | | | | | | | |

WBAN : 13743

## COOLING DEGREE DAYS (base 65°F) 2009 WASHINGTON (KDCA)

| YEAR | JAN | FEB | MAR | APR | MAY | JUN | JUL | AUG | SEP | OCT | NOV | DEC | TOTAL |
|---|---|---|---|---|---|---|---|---|---|---|---|---|---|
| 1980 | 0 | 0 | 0 | 9 | 174 | 301 | 546 | 563 | 374 | 38 | 1 | 0 | 2006 |
| 1981 | 0 | 0 | 6 | 49 | 118 | 417 | 478 | 380 | 204 | 18 | 0 | 0 | 1670 |
| 1982 | 0 | 0 | 0 | 6 | 155 | 244 | 479 | 330 | 185 | 51 | 13 | 1 | 1464 |
| 1983 | 0 | 0 | 0 | 21 | 81 | 310 | 510 | 504 | 269 | 42 | 0 | 0 | 1737 |
| 1984 | 0 | 0 | 0 | 4 | 99 | 368 | 365 | 404 | 157 | 73 | 0 | 0 | 1470 |
| 1985 | 0 | 0 | 6 | 70 | 135 | 232 | 444 | 373 | 228 | 37 | 6 | 0 | 1531 |
| 1986 | 0 | 0 | 5 | 10 | 162 | 358 | 503 | 318 | 202 | 70 | 1 | 0 | 1629 |
| 1987 | 0 | 0 | 0 | 8 | 146 | 347 | 554 | 431 | 222 | 0 | 0 | 0 | 1708 |
| 1988 | 0 | 0 | 1 | 4 | 101 | 313 | 534 | 490 | 144 | 11 | 0 | 0 | 1598 |
| 1989 | 0 | 0 | 16 | 14 | 91 | 362 | 417 | 381 | 233 | 33 | 1 | 0 | 1548 |
| 1990 | 0 | 0 | 30 | 46 | 50 | 309 | 451 | 364 | 183 | 88 | 0 | 0 | 1521 |
| 1991 | 0 | 0 | 5 | 38 | 278 | 362 | 517 | 472 | 214 | 41 | 5 | 0 | 1932 |
| 1992 | 0 | 0 | 0 | 16 | 53 | 214 | 457 | 285 | 175 | 8 | 0 | 0 | 1208 |
| 1993 | 0 | 0 | 0 | 5 | 114 | 319 | 569 | 460 | 218 | 9 | 8 | 0 | 1702 |
| 1994 | 0 | 0 | 1 | 53 | 82 | 439 | 530 | 333 | 166 | 11 | 6 | 0 | 1621 |
| 1995 | 0 | 0 | 0 | 20 | 91 | 297 | 515 | 511 | 216 | 67 | 1 | 0 | 1718 |
| 1996 | 0 | 0 | 0 | 29 | 105 | 373 | 397 | 364 | 181 | 14 | 0 | 0 | 1463 |
| 1997 | 0 | 0 | 1 | 0 | 52 | 282 | 484 | 399 | 184 | 72 | 0 | 0 | 1474 |
| 1998 | 0 | 0 | 27 | 11 | 132 | 257 | 438 | 456 | 322 | 9 | 0 | 5 | 1657 |
| 1999 | 0 | 0 | 0 | 4 | 111 | 301 | 565 | 462 | 180 | 6 | 0 | 0 | 1629 |
| 2000 | 0 | 0 | 4 | 5 | 143 | 300 | 307 | 319 | 146 | 29 | 0 | 0 | 1253 |
| 2001 | 0 | 0 | 0 | 31 | 88 | 318 | 323 | 441 | 171 | 47 | 5 | 0 | 1424 |
| 2002 | 0 | 0 | 1 | 82 | 117 | 340 | 500 | 507 | 248 | 66 | 0 | 0 | 1861 |
| 2003 | 0 | 0 | 0 | 12 | 32 | 211 | 402 | 434 | 183 | 9 | 15 | 0 | 1298 |
| 2004 | 0 | 0 | 0 | 45 | 250 | 262 | 426 | 345 | 213 | 8 | 0 | 0 | 1549 |
| 2005 | 0 | 0 | 0 | 16 | 32 | 331 | 490 | 474 | 315 | 46 | 1 | 0 | 1705 |
| 2006 | 0 | 0 | 6 | 16 | 90 | 291 | 487 | 489 | 115 | 20 | 0 | 0 | 1514 |
| 2007 | 0 | 0 | 5 | 19 | 150 | 341 | 451 | 462 | 254 | 146 | 0 | 0 | 1828 |
| 2008 | 0 | 0 | 0 | 29 | 65 | 394 | 495 | 406 | 280 | 44 | 0 | 0 | 1713 |
| 2009 | 0 | 0 | 0 | 37 | 93 | 272 | 378 | 467 | 172 | 14 | 0 | 0 | 1433 |

## SNOWFALL (inches) 2009 WASHINGTON (KDCA)

| YEAR | JUL | AUG | SEP | OCT | NOV | DEC | JAN | FEB | MAR | APR | MAY | JUN | TOTAL |
|---|---|---|---|---|---|---|---|---|---|---|---|---|---|
| 1980-81 | 0.0 | 0.0 | 0.0 | 0.0 | T | 0.3 | 4.2 | T | T | 0.0 | 0.0 | 0.0 | 4.5 |
| 1981-82 | 0.0 | 0.0 | 0.0 | 0.0 | T | 1.7 | 15.3 | 5.3 | 0.2 | T | 0.0 | 0.0 | 22.5 |
| 1982-83 | 0.0 | 0.0 | 0.0 | 0.0 | 0.0 | 6.6 | T | 21.0 | 0.0 | T | 0.0 | 0.0 | 27.6 |
| 1983-84 | 0.0 | 0.0 | 0.0 | 0.0 | 0.3 | T | 6.5 | T | 1.8 | T | 0.0 | 0.0 | 8.6 |
| 1984-85 | 0.0 | 0.0 | 0.0 | 0.0 | T | 0.3 | 10.0 | T | T | T | 0.0 | 0.0 | 10.3 |
| 1985-86 | 0.0 | 0.0 | 0.0 | 0.0 | 0.0 | 0.7 | 1.8 | 12.9 | T | T | 0.0 | 0.0 | 15.4 |
| 1986-87 | 0.0 | 0.0 | 0.0 | 0.0 | 0.0 | T | 20.8 | 10.3 | T | T | 0.0 | 0.0 | 31.1 |
| 1987-88 | 0.0 | 0.0 | 0.0 | 0.0 | 11.5 | T | 13.1 | T | 0.4 | T | 0.0 | 0.0 | 25.0 |
| 1988-89 | 0.0 | 0.0 | 0.0 | 0.0 | 0.0 | 1.2 | 2.9 | 1.2 | 0.4 | 0.0 | 0.0 | 0.0 | 5.7 |
| 1989-90 | 0.0 | 0.0 | 0.0 | 0.0 | 3.5 | 9.0 | 0.2 | T | 2.4 | 0.2 | 0.0 | 0.0 | 15.3 |
| 1990-91 | T | 0.0 | 0.0 | 0.0 | 0.0 | 3.0 | 4.8 | 0.3 | T | 0.0 | T | 0.0 | 8.1 |
| 1991-92 | 0.0 | 0.0 | 0.0 | 0.0 | T | 0.0 | 4.0 | 2.6 | T | 0.0 | 0.0 | 0.0 | 6.6 |
| 1992-93 | 0.0 | T | 0.0 | 0.0 | T | 1.0 | T | 4.1 | 6.6 | 0.0 | T | 0.0 | 11.7 |
| 1993-94 | 0.0 | 0.0 | 0.0 | 0.0 | T | 2.6 | 3.5 | 3.1 | 4.0 | 0.0 | 0.0 | 0.0 | 13.2 |
| 1994-95 | 0.0 | 0.0 | 0.0 | 0.0 | T | 0.0 | 3.9 | 5.8 | 0.4 | 0.0 | 0.0 | 0.0 | 10.1 |
| 1995-96 | 0.0 | 0.0 | 0.0 | 0.0 | 0.5 | 1.3 | 23.8 | 15.2 | 5.2 | T | 0.0 | 0.0 | 46.0 |
| 1996-97 | 0.0 | T | 0.0 | 0.0 | .2 | .2 | 2.3 | 4.0 | T | 0.0 | 0.0 | 0.0 | 6.7 |
| 1997-98 | 0.0 | 0.0 | 0.0 | 0.0 | T | 0.1 | T | T | T | 0.0 | T | T | 0.1 |
| 1998-99 | 0.0 | 0.0 | 0.0 | 0.0 | 0.0 | 0.5 | 2.2 | 0.2 | 8.7 | 0.0 | 0.0 | 0.0 | 11.6 |
| 1999-00 | 0.0 | 0.0 | 0.0 | 0.0 | 0.0 | T | 14.5 | 0.9 | 0.0 | T | 0.0 | 0.0 | 15.4 |
| 2000-01 | 0.0 | T | 0.0 | 0.0 | 0.0 | 2.0 | 2.4 | 2.8 | 0.2 | T | 0.0 | 0.0 | 7.4 |
| 2001-02 | 0.0 | 0.0 | 0.0 | 0.0 | 0.0 | 0.0 | 2.7 | 0.5 | T | 0.0 | 0.0 | T | 3.2 |
| 2002-03 | 0.0 | 0.0 | 0.0 | 0.0 | 0.0 | 7.1 | 4.5 | 28.7 | 0.1 | 0.0 | 0.0 | 0.0 | 40.4 |
| 2003-04 | T | 0.0 | 0.0 | 0.0 | 0.0 | 6.2 | 6.2 | T | T | 0.0 | T | 0.0 | 12.4 |
| 2004-05 | T | T | 0.0 | 0.0 | 0.0 | 0.1 | 6.4 | 5.2 | 0.8 | 0.0 | T | 0.0 | 12.5 |
| 2005-06 | 0.0 | 0.0 | 0.0 | 0.0 | T | 4.8 | T | 8.8 | T | T | 0.0 | 0.0 | 13.6 |
| 2006-07 | T | 0.0 | 0.0 | 0.0 | 0.0 | T | 1.3 | 5.9 | 1.9 | 0.4 | 0.0 | 0.0 | 9.5 |
| 2007-08 | 0.0 | 0.0 | 0.0 | 0.0 | 0.0 | 2.6 | 1.3 | 1.0 | 0.0 | 0.0 | 0.0 | T | 4.9 |
| 2008-09 | 0.0 | T | 0.0 | 0.0 | T | T | 1.9 | 0.1 | 5.5 | 0.0 | 0.0 | T | 7.5 |
| 2009- | 0.0 | 0.0 | 0.0 | 0.0 | 0.0 | 16.6 | | | | | | | |
| POR= 65 YRS | T | T | 0.0 | 0.0 | 0.7 | 3.0 | 5.3 | 5.3 | 2.0 | T | T | T | 16.3 |

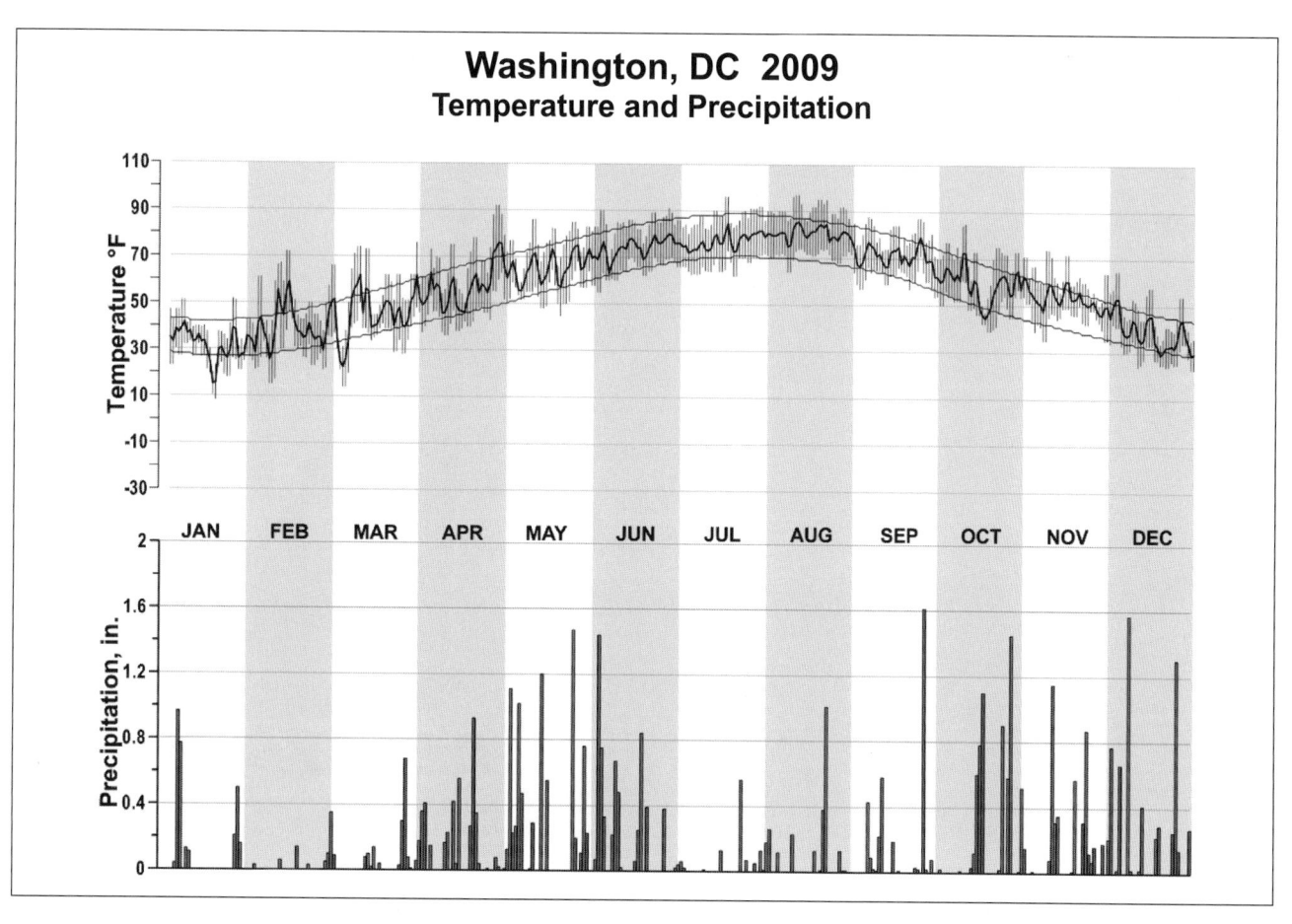

## Washington, DC 2009
### Temperature and Precipitation

# 2009
# WILMINGTON
# DELAWARE (KILG)

Delaware is part of the Atlantic Coastal Plain consisting mainly of flat low land with many marshes. Small streams and tidal estuaries comprise the drainage of the State. Wilmington, at the northern end of the State, marks the beginning of low rolling hills extending northward and northwestward into Pennsylvania. The Delaware River, the Delaware Bay, and the Atlantic Ocean are along the eastern boundary of the State. The broad Chesapeake Bay lies 35 miles, or less, to the west of the western boundary of nearly the entire State. These large water areas considerably influence the climate of the Wilmington, Delaware region.

Summers are warm and humid, winters are usually mild. During the summer maximum temperatures are usually in the 80s. The temperature reaches 100 degrees on the average once in six years. During January, the coldest month of the year, the daily average temperature is 32 degrees. Temperatures of zero may be expected once in four years. Most of the winter precipitation falls as rain. Seasonal snowfall has been as little as 1 inch, and as much as 50 inches. Snow is frequently mixed with rain and sleet, and seldom remains on the ground more than a few days.

The proximity of large water areas and the inflow of southerly winds cause the relative humidity to be quite high all year. During the summer months the relative humidity is approximately 75 percent. Fog is relatively frequent and may occur in any month. Light southeast winds blowing up the Delaware Bay favor the formation of fog. Light north-northeast winds bring in smoke from Philadelphia and from the heavy industry area located along the Delaware River north of Wilmington.

Rainfall distribution throughout the year is fairly uniform, however, the greatest amounts normally come during the summer months. Mostly, the summer rainfall comes in the form of thunderstorms. Moisture deficiencies for crops occur occasionally, but severe droughts are rare. During the fall, winter, and spring seasons, much of the rainfall comes from storms forming over the southern states or the South Atlantic and moving northward along the coast. During the late summer and early fall, hurricanes occasionally cause heavy rainfall, but winds seldom reach hurricane force in Wilmington. Heavy rains occasionally cause minor flooding, but the streams and rivers of northern Delaware are not subject to major flooding. Strong easterly and southeasterly winds sometimes cause high tides in the Delaware Bay and the Delaware River, resulting in the flooding of lowlands and damage to bay front and river front properties.

Based on the 1951-1980 period, the average first occurrence of 32 degrees Fahrenheit in the fall is October 29 and the average last occurrence in the spring is April 13.

**Wilmington**

**DELAWARE**

# NORMALS, MEANS, AND EXTREMES
## WILMINGTON (KILG)

LATITUDE: 39 ° 40'N　　LONGITUDE: -75 ° 36'W　　ELEVATION (FT): GRND: 75　BARO: 77　　TIME ZONE: EASTERN (UTC -5)　　WBAN: 13781

| ELEMENT | POR | JAN | FEB | MAR | APR | MAY | JUN | JUL | AUG | SEP | OCT | NOV | DEC | YEAR |
|---|---|---|---|---|---|---|---|---|---|---|---|---|---|---|
| **TEMPERATURE °F** | | | | | | | | | | | | | | |
| NORMAL DAILY MAXIMUM | 30 | 39.3 | 42.5 | 51.9 | 62.6 | 72.5 | 81.1 | 86.0 | 84.1 | 77.2 | 65.9 | 55.0 | 44.4 | 63.5 |
| MEAN DAILY MAXIMUM | 62 | 40.0 | 42.8 | 51.6 | 63.1 | 72.7 | 81.3 | 85.8 | 84.1 | 77.5 | 66.5 | 55.3 | 44.2 | 63.7 |
| HIGHEST DAILY MAXIMUM | 62 | 75 | 78 | 86 | 94 | 96 | 100 | 102 | 101 | 100 | 91 | 85 | 75 | 102 |
| YEAR OF OCCURRENCE | | 1950 | 1985 | 1998 | 1985 | 1996 | 1994 | 1966 | 1955 | 1983 | 1951 | 1950 | 1998 | JUL 1966 |
| MEAN OF EXTREME MAXS. | 62 | 60.1 | 62.4 | 73.2 | 82.5 | 88.0 | 93.1 | 95.3 | 93.3 | 89.7 | 81.8 | 72.7 | 63.7 | 79.7 |
| NORMAL DAILY MINIMUM | 30 | 23.7 | 25.8 | 33.4 | 42.1 | 52.4 | 61.8 | 67.3 | 65.8 | 58.1 | 45.6 | 36.9 | 28.4 | 45.1 |
| MEAN DAILY MINIMUM | 62 | 24.0 | 25.6 | 32.9 | 42.2 | 52.0 | 61.5 | 66.8 | 65.4 | 57.9 | 46.0 | 36.8 | 27.9 | 44.9 |
| LOWEST DAILY MINIMUM | 62 | -14 | -6 | 2 | 18 | 30 | 41 | 48 | 43 | 36 | 24 | 14 | -7 | -14 |
| YEAR OF OCCURRENCE | | 1985 | 1979 | 1984 | 1982 | 1978 | 1972 | 1988 | 1982 | 1974 | 1976 | 1955 | 1983 | JAN 1985 |
| MEAN OF EXTREME MINS. | 62 | 7.6 | 9.7 | 17.9 | 29.1 | 38.5 | 49.1 | 56.0 | 53.6 | 43.0 | 31.9 | 22.4 | 12.7 | 31.0 |
| NORMAL DRY BULB | 30 | 31.5 | 34.2 | 42.7 | 52.4 | 62.5 | 71.5 | 76.6 | 75.0 | 67.7 | 55.8 | 45.9 | 36.4 | 54.4 |
| MEAN DRY BULB | 62 | 32.0 | 34.2 | 42.3 | 52.7 | 62.4 | 71.6 | 76.3 | 74.9 | 67.7 | 56.3 | 46.1 | 36.1 | 54.4 |
| MEAN WET BULB | 26 | 28.5 | 29.9 | 36.3 | 45.8 | 55.5 | 64.6 | 68.8 | 67.9 | 61.6 | 50.8 | 41.7 | 32.6 | 48.7 |
| MEAN DEW POINT | 26 | 24.2 | 24.9 | 31.2 | 40.9 | 51.9 | 61.3 | 66.2 | 65.4 | 59.0 | 47.5 | 37.3 | 27.9 | 44.8 |
| NORMAL NO. DAYS WITH: | | | | | | | | | | | | | | |
| MAXIMUM >= 90 | 30 | 0.0 | 0.0 | 0.0 | 0.2 | 1.0 | 3.5 | 9.0 | 5.4 | 1.5 | 0.0 | 0.0 | 0.0 | 20.6 |
| MAXIMUM <= 32 | 30 | 7.7 | 5.0 | 0.7 | 0.0 | 0.0 | 0.0 | 0.0 | 0.0 | 0.0 | 0.0 | 0.1 | 3.5 | 17.0 |
| MINIMUM <= 32 | 30 | 24.9 | 21.0 | 13.7 | 2.8 | * | 0.0 | 0.0 | 0.0 | 0.0 | 1.3 | 9.8 | 21.2 | 94.7 |
| MINIMUM <= 0 | 30 | 0.5 | 0.3 | 0.0 | 0.0 | 0.0 | 0.0 | 0.0 | 0.0 | 0.0 | 0.0 | 0.0 | 0.1 | 0.9 |
| **H/C** | | | | | | | | | | | | | | |
| NORMAL HEATING DEG. DAYS | 30 | 1029 | 864 | 687 | 376 | 132 | 15 | 1 | 2 | 49 | 297 | 564 | 872 | 4888 |
| NORMAL COOLING DEG. DAYS | 30 | 0 | 0 | 2 | 9 | 62 | 215 | 368 | 317 | 135 | 16 | 1 | 0 | 1125 |
| **RH** | | | | | | | | | | | | | | |
| NORMAL (PERCENT) | 30 | 68 | 65 | 63 | 63 | 68 | 69 | 70 | 72 | 73 | 72 | 69 | 69 | 68 |
| HOUR 01 LST | 30 | 73 | 71 | 70 | 71 | 79 | 81 | 81 | 83 | 84 | 82 | 76 | 74 | 77 |
| HOUR 07 LST | 30 | 76 | 74 | 73 | 72 | 76 | 77 | 79 | 83 | 85 | 85 | 80 | 76 | 78 |
| HOUR 13 LST | 30 | 60 | 55 | 51 | 50 | 54 | 55 | 55 | 57 | 57 | 55 | 56 | 59 | 55 |
| HOUR 19 LST | 30 | 67 | 62 | 59 | 57 | 63 | 64 | 64 | 68 | 71 | 69 | 67 | 67 | 65 |
| **S** PERCENT POSSIBLE SUNSHINE | | | | | | | | | | | | | | |
| **W/O** MEAN NO. DAYS WITH: | | | | | | | | | | | | | | |
| HEAVY FOG(VISBY <= 1/4 MI) | 46 | 3.4 | 2.6 | 2.3 | 1.7 | 1.7 | 1.3 | 1.0 | 1.2 | 1.6 | 3.0 | 3.0 | 2.9 | 25.7 |
| THUNDERSTORMS | 62 | 0.2 | 0.3 | 1.0 | 2.1 | 3.8 | 5.4 | 5.9 | 5.1 | 2.3 | 0.8 | 0.6 | 0.2 | 27.7 |
| **CLOUDNESS** MEAN: | | | | | | | | | | | | | | |
| SUNRISE-SUNSET (OKTAS) | | | | | | | | | | | | | | |
| MIDNIGHT-MIDNIGHT (OKTAS) | | | | | | | | | | | | | | |
| MEAN NO. DAYS WITH: | | | | | | | | | | | | | | |
| CLEAR | 1 | 2.0 | 2.0 | 6.0 | | 8.0 | 9.0 | 3.0 | 7.0 | 5.0 | 9.0 | | 5.0 | |
| PARTLY CLOUDY | 1 | 1.0 | 1.0 | 6.0 | | 4.0 | 5.0 | 1.0 | 4.0 | 3.0 | 2.0 | | 2.0 | |
| CLOUDY | 1 | 4.0 | 5.0 | 11.0 | | 6.0 | 8.0 | | 2.0 | 7.0 | 3.0 | | 8.0 | |
| **PR** MEAN STATION PRESSURE(IN) | 26 | 30.01 | 30.00 | 29.97 | 29.90 | 29.91 | 29.89 | 29.90 | 29.94 | 29.98 | 30.01 | 30.01 | 30.02 | 29.96 |
| MEAN SEA-LEVEL PRES. (IN) | 26 | 30.10 | 30.08 | 30.05 | 29.98 | 29.99 | 29.98 | 29.98 | 30.03 | 30.07 | 30.10 | 30.10 | 30.11 | 30.05 |
| **WINDS** MEAN SPEED (MPH) | 26 | 9.2 | 9.7 | 10.3 | 9.9 | 8.6 | 7.9 | 7.5 | 7.0 | 7.6 | 7.6 | 8.6 | 8.9 | 8.6 |
| PREVAIL.DIR(TENS OF DEGS) | 34 | 31 | 31 | 31 | 31 | 31 | 17 | 31 | 19 | 32 | 32 | 31 | 31 | 31 |
| MAXIMUM 2-MINUTE: | | | | | | | | | | | | | | |
| SPEED (MPH) | 15 | 51 | 43 | 47 | 46 | 48 | 52 | 45 | 40 | 43 | 38 | 47 | 45 | 52 |
| DIR. (TENS OF DEGS) | | 15 | 31 | 24 | 33 | 24 | 31 | 32 | 14 | 13 | 29 | 29 | 31 | 31 |
| YEAR OF OCCURRENCE | | 1999 | 2006 | 2008 | 1995 | 1999 | 2009 | 1995 | 1997 | 2003 | 2009 | 2003 | 2008 | JUN 2009 |
| MAXIMUM 3-SECOND | | | | | | | | | | | | | | |
| SPEED (MPH) | 15 | 61 | 60 | 56 | 60 | 61 | 76 | 56 | 53 | 53 | 52 | 61 | 59 | 76 |
| DIR. (TENS OF DEGS) | | 23 | 26 | 24 | 29 | 23 | 31 | 26 | 15 | 12 | 29 | 29 | 32 | 31 |
| YEAR OF OCCURRENCE | | 1999 | 2009 | 2008 | 2007 | 1999 | 2009 | 2006 | 1997 | 2003 | 2009 | 2005 | 2008 | JUN 2009 |
| **PRECIPITATION** NORMAL (IN) | 30 | 3.43 | 2.81 | 3.97 | 3.39 | 4.15 | 3.59 | 4.28 | 3.51 | 4.01 | 3.08 | 3.19 | 3.40 | 42.81 |
| MAXIMUM MONTHLY (IN) | 62 | 8.41 | 7.02 | 9.17 | 8.55 | 7.38 | 9.90 | 12.63 | 12.09 | 12.68 | 8.01 | 7.84 | 8.58 | 12.68 |
| YEAR OF OCCURRENCE | | 1978 | 1979 | 2000 | 2007 | 1983 | 2003 | 1989 | 1955 | 1999 | 1995 | 1972 | 2009 | SEP 1999 |
| MINIMUM MONTHLY (IN) | 62 | 0.52 | 0.30 | 0.29 | 0.35 | 0.22 | 0.21 | 0.16 | 0.25 | .44 | 0.08 | 0.49 | 0.19 | 0.08 |
| YEAR OF OCCURRENCE | | 1981 | 2009 | 2006 | 1985 | 1964 | 1988 | 1955 | 1972 | 2005 | 2000 | 1976 | 1955 | OCT 2000 |
| MAXIMUM IN 24 HOURS (IN) | 62 | 2.53 | 2.35 | 4.87 | 4.39 | 2.72 | 4.35 | 6.83 | 4.11 | 8.43 | 3.88 | 3.83 | 2.38 | 8.43 |
| YEAR OF OCCURRENCE | | 1998 | 2003 | 2000 | 2007 | 1990 | 1972 | 1989 | 1971 | 1999 | 1966 | 1956 | 2008 | SEP 1999 |
| NORMAL NO. DAYS WITH: | | | | | | | | | | | | | | |
| PRECIPITATION >= 0.01 | 30 | 10.9 | 9.5 | 10.5 | 10.7 | 11.5 | 10.4 | 9.3 | 8.5 | 9.0 | 8.0 | 9.2 | 10.3 | 117.8 |
| PRECIPITATION >= 1.00 | 30 | 1.0 | 0.6 | 1.1 | 0.8 | 0.9 | 0.8 | 1.2 | 1.1 | 1.2 | 0.8 | 0.8 | 1.0 | 11.3 |
| **SNOWFALL** NORMAL (IN) | 30 | 7.5 | 6.3 | 2.2 | 0.3 | 0.0 | 0.0 | 0.0 | 0.0 | 0.0 | 0.1 | 0.6 | 1.9 | 18.9 |
| MAXIMUM MONTHLY (IN) | 58 | 26.2 | 31.6 | 20.3 | 2.6 | T | T | T | 0.0 | 0.0 | 2.5 | 11.9 | 21.5 | 31.6 |
| YEAR OF OCCURRENCE | | 1996 | 2003 | 1958 | 1982 | 1991 | 1992 | 2007 | | | 1979 | 1953 | 1966 | FEB 2003 |
| MAXIMUM IN 24 HOURS (IN) | 58 | 22.0 | 17.0 | 15.6 | 2.4 | T | T | T | 0.0 | 0.0 | 2.5 | 11.9 | 17.0 | 22.0 |
| YEAR OF OCCURRENCE' | | 1996 | 2003 | 1958 | 1987 | 1991 | 1992 | 1990 | | | 1979 | 1953 | 2009 | JAN 1996 |
| MAXIMUM SNOW DEPTH (IN) | 53 | 13 | 25 | 8 | 2 | 0 | 0 | 0 | 0 | 0 | T | 9 | 13 | 25 |
| YEAR OF OCCURRENCE | | 1987 | 2003 | 1956 | 1987 | | | | | | 1962 | 1953 | 2009 | FEB 2003 |
| NORMAL NO. DAYS WITH: | | | | | | | | | | | | | | |
| SNOWFALL >= 1.0 | 30 | 2.2 | 1.4 | 0.5 | 0.2 | 0.0 | 0.0 | 0.0 | 0.0 | 0.0 | 0.0 | 0.2 | 0.7 | 5.2 |

## PRECIPITATION (inches) 2009 WILMINGTON (KILG)

| YEAR | JAN | FEB | MAR | APR | MAY | JUN | JUL | AUG | SEP | OCT | NOV | DEC | ANNUAL |
|------|-----|-----|-----|-----|-----|-----|-----|-----|-----|-----|-----|-----|--------|
| 1980 | 2.44 | 0.83 | 6.22 | 4.55 | 2.40 | 4.23 | 3.49 | 1.09 | 1.44 | 3.99 | 2.41 | 0.83 | 33.92 |
| 1981 | 0.52 | 3.23 | 1.26 | 3.54 | 5.05 | 4.50 | 2.52 | 3.38 | 3.82 | 2.84 | 0.67 | 3.95 | 35.28 |
| 1982 | 3.75 | 2.71 | 2.87 | 5.41 | 3.72 | 4.70 | 2.70 | 4.68 | 2.30 | 1.97 | 3.87 | 2.39 | 41.07 |
| 1983 | 2.98 | 3.55 | 6.84 | 6.80 | 7.38 | 3.94 | 2.33 | 1.29 | 3.44 | 3.87 | 5.48 | 6.80 | 54.70 |
| 1984 | 1.25 | 4.27 | 5.40 | 4.24 | 5.03 | 4.54 | 6.53 | 1.56 | 2.02 | 3.31 | 1.63 | 1.94 | 41.72 |
| 1985 | 1.56 | 2.05 | 2.03 | 0.35 | 5.52 | 1.37 | 6.91 | 2.28 | 4.56 | 1.84 | 4.46 | 0.80 | 33.73 |
| 1986 | 4.21 | 2.77 | 1.19 | 2.77 | 1.69 | 4.05 | 3.99 | 2.88 | 2.75 | 4.04 | 6.42 | 6.11 | 42.87 |
| 1987 | 4.35 | 1.52 | 1.16 | 2.63 | 3.15 | 2.31 | 4.09 | 4.21 | 4.85 | 2.31 | 3.50 | 1.90 | 35.98 |
| 1988 | 2.46 | 4.14 | 1.82 | 2.59 | 4.95 | 0.21 | 8.29 | 3.03 | 2.18 | 1.94 | 5.29 | 0.90 | 37.80 |
| 1989 | 2.48 | 2.75 | 3.69 | 2.76 | 6.57 | 5.43 | 12.63 | 1.97 | 4.31 | 3.92 | 1.99 | 1.27 | 49.77 |
| 1990 | 3.56 | 1.35 | 2.15 | 3.42 | 7.03 | 3.94 | 4.27 | 6.15 | 2.64 | 2.85 | 1.61 | 5.16 | 44.13 |
| 1991 | 4.30 | 0.97 | 4.64 | 3.28 | 1.98 | 3.41 | 3.71 | 5.38 | 5.36 | 1.27 | 1.26 | 4.26 | 39.82 |
| 1992 | 1.05 | 1.81 | 4.36 | 1.76 | 4.48 | 3.14 | 4.34 | 2.21 | 4.30 | 1.11 | 4.27 | 4.21 | 37.04 |
| 1993 | 2.64 | 3.11 | 7.50 | 5.87 | 3.95 | 1.60 | 4.04 | 2.65 | 6.26 | 2.77 | 2.85 | 3.51 | 46.75 |
| 1994 | 5.00 | 3.55 | 7.36 | 2.85 | 3.69 | 2.11 | 7.01 | 5.68 | 2.10 | 0.85 | 2.96 | 2.24 | 45.40 |
| 1995 | 3.08 | 2.28 | 2.47 | 2.10 | 3.50 | 1.26 | 2.89 | 2.03 | 5.17 | 8.01 | 4.31 | 2.17 | 39.27 |
| 1996 | 4.58 | 1.19 | 3.63 | 4.98 | 3.27 | 5.00 | 6.25 | 3.04 | 4.05 | 4.70 | 3.25 | 7.96 | 51.90 |
| 1997 | 1.83 | 1.83 | 3.49 | 1.49 | 0.82 | 1.75 | 3.08 | 3.66 | 1.93 | 2.33 | 3.24 | 2.57 | 28.02 |
| 1998 | 4.80 | 2.95 | 4.86 | 2.91 | 4.13 | 4.66 | 2.18 | 3.14 | 1.76 | 2.80 | 1.26 | 1.01 | 36.46 |
| 1999 | 5.41 | 3.51 | 3.96 | 3.36 | 3.56 | 1.62 | 0.89 | 4.24 | 12.68 | 3.42 | 2.09 | 2.94 | 47.68 |
| 2000 | 3.83 | 2.00 | 9.17 | 3.43 | 2.94 | 4.83 | 4.64 | 2.47 | 7.30 | 0.08 | 2.54 | 2.80 | 46.03 |
| 2001 | 3.13 | 2.81 | 5.62 | 1.43 | 5.33 | 4.28 | 2.35 | 2.65 | 2.57 | 0.74 | 0.99 | 1.95 | 33.85 |
| 2002 | 2.72 | 0.43 | 4.05 | 2.26 | 3.40 | 4.96 | 1.40 | 2.03 | 3.42 | 6.16 | 4.47 | 4.50 | 39.80 |
| 2003 | 1.79 | 5.21 | 4.75 | 2.62 | 3.92 | 9.90 | 2.85 | 4.21 | 7.39 | 4.39 | 3.37 | 4.93 | 55.33 |
| 2004 | 1.66 | 2.33 | 2.96 | 5.61 | 4.41 | 6.98 | 8.24 | 5.33 | 9.29 | 2.43 | 4.65 | 2.84 | 56.73 |
| 2005 | 3.84 | 2.65 | 4.10 | 5.12 | 2.26 | 2.25 | 4.82 | 1.35 | 0.44 | 7.79 | 2.41 | 3.22 | 40.25 |
| 2006 | 4.14 | 2.38 | 0.29 | 4.36 | 2.22 | 9.40 | 6.05 | 2.59 | 6.18 | 5.56 | 4.31 | 1.93 | 49.41 |
| 2007 | 3.52 | 1.94 | 4.61 | 8.55 | 1.02 | 2.72 | 3.15 | 3.38 | 0.49 | 5.92 | 1.69 | 4.82 | 41.81 |
| 2008 | 1.57 | 4.32 | 4.00 | 1.97 | 5.12 | 2.71 | 4.69 | 1.16 | 5.19 | 1.81 | 3.51 | 4.39 | 40.44 |
| 2009 | 2.90 | 0.30 | 1.89 | 4.03 | 3.89 | 6.67 | 3.94 | 6.73 | 4.91 | 5.90 | 2.32 | 8.58 | 52.06 |
| POR= 62 YRS | 3.13 | 2.80 | 3.87 | 3.41 | 3.64 | 3.68 | 4.18 | 3.67 | 3.92 | 3.08 | 3.24 | 3.52 | 42.14 |

WBAN : 13781

## AVERAGE TEMPERATURE (°F) 2009 WILMINGTON (KILG)

| YEAR | JAN | FEB | MAR | APR | MAY | JUN | JUL | AUG | SEP | OCT | NOV | DEC | ANNUAL |
|------|-----|-----|-----|-----|-----|-----|-----|-----|-----|-----|-----|-----|--------|
| 1980 | 32.4 | 29.9 | 40.0 | 54.6 | 64.9 | 68.9 | 77.7 | 78.2 | 71.3 | 54.8 | 43.3 | 33.1 | 54.1 |
| 1981 | 25.4 | 37.9 | 40.2 | 55.0 | 62.5 | 72.3 | 77.1 | 73.5 | 66.9 | 53.0 | 45.3 | 34.2 | 53.6 |
| 1982 | 24.2 | 34.2 | 41.8 | 50.6 | 65.0 | 69.9 | 77.3 | 72.0 | 67.4 | 56.0 | 47.5 | 41.3 | 53.9 |
| 1983 | 35.2 | 35.3 | 45.9 | 53.1 | 61.0 | 71.8 | 77.6 | 77.0 | 69.3 | 56.9 | 46.7 | 32.1 | 55.2 |
| 1984 | 24.8 | 38.6 | 35.6 | 50.7 | 61.2 | 73.8 | 75.2 | 75.2 | 63.7 | 61.2 | 43.3 | 42.1 | 53.8 |
| 1985 | 27.5 | 37.4 | 47.1 | 58.0 | 65.9 | 70.6 | 76.6 | 74.4 | 69.1 | 58.3 | 51.0 | 32.9 | 55.7 |
| 1986 | 32.2 | 31.6 | 43.6 | 52.4 | 65.7 | 72.1 | 77.1 | 72.5 | 67.5 | 57.2 | 44.1 | 37.3 | 54.4 |
| 1987 | 31.4 | 31.9 | 44.6 | 52.3 | 63.1 | 73.5 | 79.1 | 74.3 | 68.3 | 51.7 | 47.4 | 38.6 | 54.7 |
| 1988 | 27.4 | 34.8 | 44.2 | 50.8 | 62.9 | 71.6 | 79.4 | 77.3 | 65.8 | 51.0 | 46.7 | 35.1 | 53.9 |
| 1989 | 36.0 | 34.3 | 42.1 | 51.6 | 62.1 | 74.3 | 75.9 | 74.4 | 68.4 | 57.6 | 44.6 | 25.0 | 53.9 |
| 1990 | 40.5 | 41.1 | 46.0 | 53.7 | 61.5 | 72.1 | 77.4 | 74.6 | 66.7 | 60.0 | 48.4 | 41.0 | 56.9 |
| 1991 | 34.3 | 39.7 | 44.9 | 54.7 | 73.6 | 73.6 | 77.3 | 76.7 | 67.2 | 57.6 | 46.8 | 39.5 | 56.8 |
| 1992 | 35.2 | 37.2 | 41.5 | 52.0 | 60.7 | 69.6 | 76.3 | 72.2 | 67.3 | 53.2 | 47.2 | 38.3 | 54.2 |
| 1993 | 37.6 | 31.0 | 39.0 | 52.8 | 65.3 | 72.8 | 79.4 | 77.8 | 68.7 | 55.7 | 47.4 | 36.4 | 55.3 |
| 1994 | 26.3 | 31.6 | 41.8 | 58.4 | 60.0 | 75.8 | 79.8 | 73.2 | 67.2 | 53.9 | 49.7 | 40.0 | 54.8 |
| 1995 | 36.6 | 30.3 | 45.3 | 51.5 | 62.2 | 72.0 | 78.7 | 77.1 | 68.2 | 59.3 | 41.3 | 31.3 | 54.5 |
| 1996 | 29.9 | 33.3 | 38.4 | 52.6 | 60.1 | 73.0 | 74.1 | 74.0 | 67.7 | 55.1 | 40.1 | 39.2 | 53.1 |
| 1997 | 31.8 | 38.9 | 43.2 | 50.3 | 58.6 | 69.6 | 75.9 | 72.8 | 65.5 | 55.6 | 43.6 | 37.6 | 53.6 |
| 1998 | 39.8 | 41.0 | 44.7 | 54.2 | 65.3 | 70.5 | 75.1 | 75.5 | 70.9 | 57.0 | 46.9 | 41.0 | 56.8 |
| 1999 | 34.8 | 36.9 | 41.7 | 52.7 | 62.9 | 71.6 | 80.0 | 75.8 | 68.1 | 53.8 | 49.4 | 38.8 | 55.5 |
| 2000 | 31.5 | 36.6 | 47.3 | 51.9 | 63.8 | 72.0 | 72.9 | 72.7 | 65.0 | 56.4 | 44.2 | 30.1 | 53.7 |
| 2001 | 31.6 | 36.2 | 39.8 | 53.3 | 62.7 | 73.0 | 72.7 | 77.3 | 65.6 | 56.7 | 50.8 | 42.5 | 55.2 |
| 2002 | 37.8 | 39.4 | 44.3 | 55.7 | 61.4 | 71.8 | 77.5 | 77.8 | 69.3 | 55.5 | 43.6 | 34.3 | 55.7 |
| 2003 | 28.7 | 28.7 | 42.8 | 51.0 | 58.1 | 68.8 | 75.7 | 76.5 | 68.5 | 54.2 | 49.3 | 35.8 | 53.2 |
| 2004 | 25.4 | 34.0 | 44.6 | 53.7 | 68.2 | 70.1 | 74.7 | 73.1 | 68.8 | 54.6 | 47.0 | 36.7 | 54.2 |
| 2005 | 31.8 | 35.1 | 38.6 | 53.8 | 57.5 | 72.9 | 77.0 | 77.9 | 71.8 | 57.7 | 47.9 | 33.7 | 54.6 |
| 2006 | 39.7 | 34.9 | 44.1 | 55.6 | 62.7 | 71.3 | 78.1 | 76.8 | 65.5 | 54.9 | 49.6 | 41.6 | 56.2 |
| 2007 | 38.3 | 28.1 | 43.6 | 50.6 | 64.8 | 73.1 | 76.6 | 76.4 | 70.5 | 63.5 | 45.4 | 37.6 | 55.7 |
| 2008 | 35.7 | 36.9 | 44.5 | 54.7 | 60.2 | 75.1 | 78.1 | 73.2 | 69.7 | 55.6 | 45.3 | 38.5 | 55.6 |
| 2009 | 28.5 | 36.5 | 42.3 | 54.5 | 63.4 | 70.5 | 73.5 | 76.7 | 66.6 | 55.3 | 49.7 | 35.4 | 54.4 |
| POR= 62 YRS | 32.0 | 34.2 | 42.3 | 52.7 | 62.4 | 71.6 | 76.3 | 74.9 | 67.7 | 56.3 | 46.1 | 36.1 | 54.4 |

## HEATING DEGREE DAYS (base 65°F) 2009 WILMINGTON (KILG)

| YEAR | JUL | AUG | SEP | OCT | NOV | DEC | JAN | FEB | MAR | APR | MAY | JUN | TOTAL |
|------|-----|-----|-----|-----|-----|-----|-----|-----|-----|-----|-----|-----|-------|
| 1980-81 | 0 | 0 | 20 | 322 | 645 | 985 | 1222 | 752 | 763 | 299 | 135 | 4 | 5147 |
| 1981-82 | 0 | 0 | 57 | 370 | 585 | 947 | 1259 | 855 | 715 | 426 | 69 | 12 | 5295 |
| 1982-83 | 0 | 14 | 29 | 305 | 519 | 724 | 919 | 822 | 587 | 368 | 163 | 7 | 4457 |
| 1983-84 | 0 | 0 | 74 | 275 | 542 | 1013 | 1240 | 758 | 904 | 422 | 162 | 5 | 5395 |
| 1984-85 | 0 | 2 | 113 | 149 | 641 | 701 | 1154 | 766 | 550 | 248 | 73 | 7 | 4404 |
| 1985-86 | 0 | 0 | 45 | 213 | 411 | 986 | 1011 | 930 | 653 | 373 | 99 | 11 | 4732 |
| 1986-87 | 0 | 27 | 36 | 276 | 619 | 848 | 1032 | 923 | 628 | 374 | 143 | 2 | 4908 |
| 1987-88 | 0 | 2 | 22 | 406 | 521 | 811 | 1159 | 869 | 637 | 419 | 121 | 38 | 5005 |
| 1988-89 | 3 | 0 | 52 | 434 | 541 | 923 | 893 | 854 | 710 | 395 | 142 | 0 | 4947 |
| 1989-90 | 0 | 2 | 54 | 236 | 605 | 1231 | 749 | 661 | 593 | 368 | 127 | 6 | 4632 |
| 1990-91 | 2 | 1 | 69 | 214 | 494 | 734 | 943 | 700 | 617 | 320 | 61 | 5 | 4160 |
| 1991-92 | 0 | 0 | 64 | 244 | 541 | 785 | 914 | 799 | 723 | 386 | 169 | 12 | 4637 |
| 1992-93 | 0 | 1 | 57 | 363 | 527 | 817 | 843 | 945 | 799 | 360 | 64 | 12 | 4788 |
| 1993-94 | 0 | 0 | 55 | 286 | 526 | 879 | 1193 | 929 | 715 | 223 | 189 | 2 | 4997 |
| 1994-95 | 0 | 0 | 29 | 343 | 454 | 770 | 875 | 967 | 598 | 401 | 122 | 0 | 4559 |
| 1995-96 | 0 | 0 | 41 | 208 | 703 | 1033 | 1083 | 912 | 816 | 378 | 206 | 10 | 5390 |
| 1996-97 | 0 | 0 | 43 | 304 | 741 | 794 | 1025 | 729 | 672 | 436 | 201 | 56 | 5001 |
| 1997-98 | 2 | 0 | 71 | 324 | 635 | 844 | 775 | 667 | 645 | 317 | 89 | 27 | 4396 |
| 1998-99 | 0 | 0 | 22 | 241 | 536 | 738 | 926 | 782 | 715 | 362 | 95 | 7 | 4424 |
| 1999-00 | 0 | 1 | 36 | 340 | 462 | 807 | 1032 | 816 | 544 | 390 | 117 | 15 | 4560 |
| 2000-01 | 0 | 2 | 99 | 274 | 619 | 1074 | 1029 | 801 | 776 | 361 | 110 | 10 | 5155 |
| 2001-02 | 1 | 0 | 73 | 271 | 418 | 691 | 836 | 712 | 633 | 324 | 159 | 7 | 4125 |
| 2002-03 | 0 | 1 | 11 | 325 | 636 | 945 | 1118 | 1011 | 683 | 419 | 215 | 49 | 5413 |
| 2003-04 | 0 | 0 | 21 | 330 | 466 | 898 | 1221 | 894 | 623 | 338 | 60 | 15 | 4866 |
| 2004-05 | 0 | 2 | 14 | 314 | 533 | 871 | 1024 | 831 | 810 | 334 | 225 | 13 | 4971 |
| 2005-06 | 0 | 0 | 15 | 243 | 507 | 964 | 777 | 834 | 640 | 277 | 122 | 9 | 4388 |
| 2006-07 | 0 | 0 | 53 | 321 | 454 | 718 | 820 | 1025 | 658 | 436 | 96 | 5 | 4586 |
| 2007-08 | 0 | 3 | 20 | 136 | 582 | 840 | 901 | 809 | 629 | 301 | 170 | 0 | 4391 |
| 2008-09 | 0 | 0 | 18 | 296 | 584 | 811 | 1126 | 789 | 697 | 347 | 107 | 19 | 4794 |
| 2009- | 0 | 0 | 33 | 302 | 453 | 910 | | | | | | | |

WBAN : 13781

## COOLING DEGREE DAYS (base 65°F) 2009 WILMINGTON (KILG)

| YEAR | JAN | FEB | MAR | APR | MAY | JUN | JUL | AUG | SEP | OCT | NOV | DEC | TOTAL |
|------|-----|-----|-----|-----|-----|-----|-----|-----|-----|-----|-----|-----|-------|
| 1980 | 0 | 0 | 0 | 0 | 83 | 159 | 400 | 417 | 214 | 10 | 0 | 0 | 1283 |
| 1981 | 0 | 0 | 0 | 9 | 62 | 228 | 381 | 270 | 120 | 3 | 0 | 0 | 1073 |
| 1982 | 0 | 0 | 0 | 2 | 75 | 163 | 391 | 238 | 107 | 32 | 1 | 0 | 1009 |
| 1983 | 0 | 0 | 0 | 17 | 47 | 218 | 398 | 378 | 209 | 29 | 0 | 0 | 1296 |
| 1984 | 0 | 0 | 0 | 0 | 50 | 276 | 321 | 327 | 80 | 34 | 0 | 0 | 1088 |
| 1985 | 0 | 0 | 4 | 47 | 106 | 181 | 366 | 300 | 174 | 13 | 0 | 0 | 1191 |
| 1986 | 0 | 0 | 0 | 0 | 129 | 227 | 379 | 267 | 116 | 40 | 0 | 0 | 1158 |
| 1987 | 0 | 0 | 0 | 3 | 91 | 264 | 446 | 295 | 129 | 0 | 0 | 0 | 1228 |
| 1988 | 0 | 0 | 0 | 0 | 62 | 242 | 455 | 389 | 80 | 5 | 0 | 0 | 1233 |
| 1989 | 0 | 0 | 6 | 0 | 61 | 287 | 345 | 299 | 162 | 17 | 0 | 0 | 1177 |
| 1990 | 0 | 0 | 10 | 36 | 23 | 227 | 395 | 304 | 127 | 66 | 0 | 0 | 1188 |
| 1991 | 0 | 0 | 1 | 18 | 197 | 271 | 390 | 371 | 135 | 23 | 2 | 0 | 1408 |
| 1992 | 0 | 0 | 0 | 3 | 39 | 156 | 358 | 230 | 134 | 3 | 0 | 0 | 923 |
| 1993 | 0 | 0 | 0 | 1 | 79 | 252 | 452 | 405 | 176 | 3 | 4 | 0 | 1372 |
| 1994 | 0 | 0 | 0 | 32 | 41 | 334 | 466 | 260 | 99 | 4 | 0 | 0 | 1236 |
| 1995 | 0 | 0 | 0 | 4 | 42 | 218 | 434 | 381 | 148 | 39 | 1 | 0 | 1267 |
| 1996 | 0 | 0 | 0 | 13 | 64 | 254 | 291 | 285 | 129 | 2 | 0 | 0 | 1038 |
| 1997 | 0 | 0 | 0 | 0 | 9 | 204 | 348 | 248 | 92 | 40 | 0 | 0 | 941 |
| 1998 | 0 | 0 | 22 | 0 | 106 | 195 | 320 | 337 | 204 | 2 | 0 | 0 | 1186 |
| 1999 | 0 | 0 | 0 | 0 | 40 | 211 | 476 | 342 | 139 | 1 | 0 | 0 | 1209 |
| 2000 | 0 | 0 | 0 | 3 | 89 | 230 | 253 | 247 | 106 | 11 | 0 | 0 | 939 |
| 2001 | 0 | 0 | 0 | 16 | 46 | 257 | 247 | 388 | 100 | 21 | 0 | 0 | 1075 |
| 2002 | 0 | 0 | 0 | 51 | 53 | 219 | 397 | 405 | 147 | 39 | 0 | 0 | 1311 |
| 2003 | 0 | 0 | 0 | 8 | 10 | 169 | 337 | 362 | 133 | 3 | 0 | 0 | 1022 |
| 2004 | 0 | 0 | 0 | 6 | 168 | 173 | 308 | 261 | 132 | 1 | 0 | 0 | 1049 |
| 2005 | 0 | 0 | 0 | 6 | 3 | 258 | 377 | 408 | 228 | 26 | 0 | 0 | 1306 |
| 2006 | 0 | 0 | 0 | 4 | 58 | 204 | 412 | 370 | 76 | 12 | 0 | 0 | 1136 |
| 2007 | 0 | 0 | 1 | 8 | 96 | 254 | 368 | 362 | 192 | 96 | 0 | 0 | 1377 |
| 2008 | 0 | 0 | 0 | 1 | 28 | 311 | 413 | 259 | 166 | 12 | 0 | 0 | 1190 |
| 2009 | 0 | 0 | 0 | 39 | 66 | 190 | 271 | 369 | 89 | 7 | 0 | 0 | 1031 |

## SNOWFALL (inches) 2009 WILMINGTON (KILG)

| YEAR | JUL | AUG | SEP | OCT | NOV | DEC | JAN | FEB | MAR | APR | MAY | JUN | TOTAL |
|------|-----|-----|-----|-----|-----|-----|-----|-----|-----|-----|-----|-----|-------|
| 1980-81 | 0.0 | 0.0 | 0.0 | 0.0 | 0.5 | 1.4 | 6.5 | T | 3.7 | 0.0 | 0.0 | 0.0 | 12.1 |
| 1981-82 | 0.0 | 0.0 | 0.0 | 0.0 | T | 2.8 | 14.6 | 4.5 | 0.4 | 2.6 | 0.0 | 0.0 | 24.9 |
| 1982-83 | 0.0 | 0.0 | 0.0 | 0.0 | T | 5.8 | T | 18.5 | 0.3 | 0.5 | 0.0 | 0.0 | 25.1 |
| 1983-84 | 0.0 | 0.0 | 0.0 | 0.0 | T | T | 9.7 | T | 5.2 | T | 0.0 | 0.0 | 14.9 |
| 1984-85 | 0.0 | 0.0 | 0.0 | 0.0 | T | 0.3 | 14.2 | 0.7 | T | 0.4 | 0.0 | 0.0 | 15.6 |
| 1985-86 | 0.0 | 0.0 | 0.0 | 0.0 | 0.0 | 1.4 | 3.1 | 9.7 | T | T | 0.0 | 0.0 | 14.2 |
| 1986-87 | 0.0 | 0.0 | 0.0 | 0.0 | T | 0.3 | 21.4 | 15.7 | 0.2 | 2.4 | 0.0 | 0.0 | 40.0 |
| 1987-88 | 0.0 | 0.0 | 0.0 | 0.0 | 0.7 | 2.1 | 10.8 | 1.1 | T | T | 0.0 | 0.0 | 14.7 |
| 1988-89 | 0.0 | 0.0 | 0.0 | 0.0 | T | 0.2 | 6.7 | 2.9 | 1.2 | 0.0 | 0.0 | 0.0 | 11.0 |
| 1989-90 | 0.0 | 0.0 | 0.0 | 0.0 | 5.6 | 8.9 | 1.5 | 1.0 | 1.3 | 1.6 | 0.0 | 0.0 | 19.9 |
| 1990-91 | T | 0.0 | 0.0 | 0.0 | 0.0 | 6.4 | 5.2 | 0.6 | 0.9 | T | T | 0.0 | 13.1 |
| 1991-92 | 0.0 | 0.0 | 0.0 | 0.0 | T | 0.2 | 1.5 | 1.3 | 0.5 | T | 0.0 | T | 3.5 |
| 1992-93 | 0.0 | 0.0 | 0.0 | 0.0 | T | 0.1 | 1.4 | 10.0 | 13.9 | 0.0 | 0.0 | 0.0 | 25.4 |
| 1993-94 | 0.0 | 0.0 | 0.0 | 0.0 | T | 2.4 | 2.7 | 9.2 | 3.4 | 0.0 | 0.0 | 0.0 | 17.7 |
| 1994-95 | 0.0 | 0.0 | 0.0 | 0.0 | T | 0.0 | T | 8.3 | T | 0.0 | 0.0 | 0.0 | 8.3 |
| 1995-96 | 0.0 | 0.0 | 0.0 | 0.0 | 3.2 | 7.1 | 26.2 | 7.5 | 6.0 | 5.9 | T | 0.0 | 55.9 |
| 1996-97 | 0.0 | 0.0 | 0.0 | 0.0 | 0.0 | T | 0.8 | 5.6 | 6.5 | 2.8 | 0.0 | 0.0 | 15.7 |
| 1997-98 | 0.0 | 0.0 | 0.0 | 0.0 | 0.0 | T | T | 0.0 | T | 0.0 | 0.0 | 0.0 | T |
| 1998-99 | 0.0 | 0.0 | 0.0 | 0.0 | 0.0 | 2.0 | 4.5 | 0.0 | 2.0 | 0.0 | 0.0 | 0.0 | 8.5 |
| 1999-00 | 0.0 | 0.0 | 0.0 | 0.0 | 0.0 | 0.0 | 14.2 | 4.0 | 0.0 | 2.1 | 0.0 | 0.0 | 20.3 |
| 2000-01 | 0.0 | 0.0 | 0.0 | 0.0 | 0.0 | 2.1 | 4.8 | 5.0 | 0.7 | T | 0.0 | 0.0 | 12.6 |
| 2001-02 | 0.0 | 0.0 | 0.0 | 0.0 | 0.0 | 0.0 | 2.4 | T | T | T | 0.0 | 0.0 | 2.4 |
| 2002-03 | 0.0 | 0.0 | 0.0 | T | T | 9.0 | 5.4 | 31.6 | T | T | 0.0 | 0.0 | 46.0 |
| 2003-04 | 0.0 | 0.0 | 0.0 | 0.0 | 0.0 | 6.5 | 9.8 | 0.1 | 2.4 | T | 0.0 | 0.0 | 18.8 |
| 2004-05 | 0.0 | 0.0 | 0.0 | 0.0 | 0.0 | 0.3 | 12.1 | 13.5 | 1.0 | 0.0 | 0.0 | 0.0 | 26.9 |
| 2005-06 | 0.0 | 0.0 | 0.0 | 0.0 | 0.2 | 7.1 | T | 14.6 | 0.0 | 0.2 | 0.0 | 0.0 | 22.1 |
| 2006-07 | 0.0 | 0.0 | 0.0 | 0.0 | 0.0 | T | 1.5 | 7.3 | 3.8 | T | 0.0 | 0.0 | 12.6 |
| 2007-08 | T | 0.0 | 0.0 | 0.0 | 0.0 | 4.1 | 2.8 | 2.3 | T | 0.0 | 0.0 | 0.0 | 9.2 |
| 2008-09 | 0.0 | 0.0 | 0.0 | 0.0 | 0.2 | 1.0 | 2.9 | 1.5 | 10.5 | T | 0.0 | 0.0 | 16.1 |
| 2009- | 0.0 | 0.0 | 0.0 | 0.0 | 0.0 | 19.7 | | | | | | | |
| POR= 62 YRS | T | 0.0 | 0.0 | T | 0.7 | 3.4 | 6.1 | 6.1 | 3.1 | 0.3 | T | T | 19.7 |

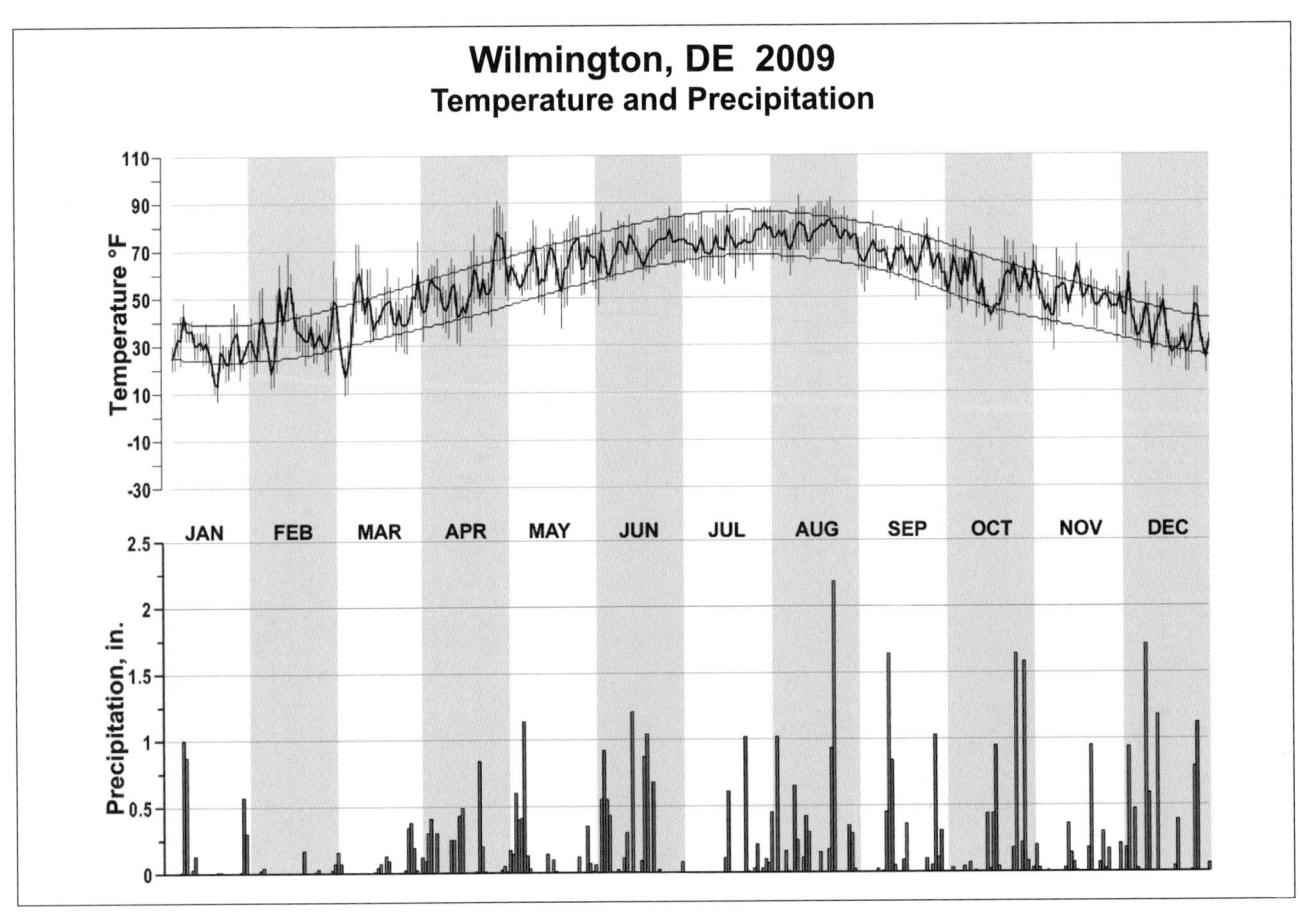

# Wilmington, DE 2009
## Temperature and Precipitation

# 2009
# DAYTONA BEACH
# FLORIDA (KDAB)

Daytona Beach is located on the Atlantic Ocean. The Halifax River, part of the Florida Inland Waterway, runs through the city. The terrain in the area is flat and the soil is mostly sandy. Elevations in the area range from 3 to 15 feet above mean sea level near the ocean to about 31 feet at the airport and on a ridge running along the western city limits.

Nearness to the ocean results in a climate tempered by the effect of land and sea breezes. In the summer, while maximum temperatures reach 90 degrees or above during the late morning or early afternoon, the number of hours of 90 degrees or above is relatively small due to the beginning of the sea breeze near midday and the occurrence of local afternoon convective thunderstorms which lower the temperature to the comfortable 80s. Winters, although subject to invasions of cold air, are relatively mild due to the nearness of the ocean and latitudinal location.

The rainy season from June through mid-October produces 60 percent of the annual rainfall. The major portion of the summer rainfall occurs in the form of local convective thunderstorms which are occasionally heavy and produce as much as 2 or 3 inches of rain. The more severe thunderstorms may be attended by strong gusty winds. Almost all rainfall during the winter months is associated with frontal passages.

Long periods of cloudiness and rain are infrequent, usually not lasting over 2 or 3 days. These periods are usually associated with a stationary front, a so-called northeaster, or a tropical disturbance.

Tropical disturbances or hurricanes are not considered a great threat to this area of the state. Generally hurricanes in this latitude tend to pass well offshore or lose much of their intensity while crossing the state before reaching this area. Only in gusts have hurricane-force winds been recorded at this station.

Heavy fog occurs mostly during the winter and early spring. These fogs usually form by radiational cooling at night and dissipate soon after sunrise. On rare occasions sea fog moves in from the ocean and persists for two or three days. There is no significant source in the area for air pollution.

# NORMALS, MEANS, AND EXTREMES
## DAYTONA BEACH (KDAB)

LATITUDE: 29 ° 10'N  LONGITUDE: -81 ° 3 'W  ELEVATION (FT): GRND: 30  BARO: 34  TIME ZONE: EASTERN  (UTC -5)  WBAN: 12834

| ELEMENT | POR | JAN | FEB | MAR | APR | MAY | JUN | JUL | AUG | SEP | OCT | NOV | DEC | YEAR |
|---|---|---|---|---|---|---|---|---|---|---|---|---|---|---|
| **TEMPERATURE °F** | | | | | | | | | | | | | | |
| NORMAL DAILY MAXIMUM | 30 | 69.7 | 71.1 | 75.6 | 79.8 | 85.0 | 88.8 | 91.0 | 90.1 | 87.9 | 82.6 | 76.9 | 71.4 | 80.8 |
| MEAN DAILY MAXIMUM | 62 | 68.8 | 70.5 | 74.9 | 79.7 | 85.0 | 88.3 | 90.0 | 89.4 | 87.0 | 81.7 | 75.5 | 70.4 | 80.1 |
| HIGHEST DAILY MAXIMUM | 66 | 87 | 89 | 92 | 96 | 100 | 102 | 102 | 100 | 99 | 95 | 89 | 88 | 102 |
| YEAR OF OCCURRENCE | | 1991 | 1985 | 1994 | 1968 | 1953 | 1944 | 1981 | 1999 | 1944 | 1959 | 1948 | 1990 | JUL 1981 |
| MEAN OF EXTREME MAXS. | 62 | 81.7 | 83.4 | 87.1 | 89.8 | 93.2 | 94.9 | 95.5 | 94.6 | 92.2 | 89.3 | 84.7 | 82.1 | 89.0 |
| NORMAL DAILY MINIMUM | 30 | 47.1 | 48.8 | 53.7 | 58.0 | 64.5 | 70.6 | 72.4 | 72.8 | 71.9 | 65.3 | 57.0 | 50.1 | 61.0 |
| MEAN DAILY MINIMUM | 62 | 47.5 | 49.4 | 53.9 | 58.7 | 65.1 | 70.7 | 72.6 | 72.9 | 72.1 | 65.6 | 56.3 | 50.1 | 61.2 |
| LOWEST DAILY MINIMUM | 66 | 15 | 24 | 26 | 35 | 44 | 52 | 60 | 65 | 52 | 41 | 27 | 19 | 15 |
| YEAR OF OCCURRENCE | | 1985 | 1958 | 1980 | 1950 | 1971 | 1984 | 1981 | 1984 | 1956 | 1993 | 1950 | 1983 | JAN 1985 |
| MEAN OF EXTREME MINS. | 62 | 30.3 | 33.3 | 37.7 | 45.0 | 54.6 | 64.0 | 68.3 | 69.1 | 65.5 | 50.9 | 40.4 | 32.9 | 49.3 |
| NORMAL DRY BULB | 30 | 58.4 | 60.0 | 64.7 | 68.9 | 74.8 | 79.7 | 81.7 | 81.5 | 79.9 | 74.0 | 67.0 | 60.8 | 71.0 |
| MEAN DRY BULB | 62 | 58.2 | 60.0 | 64.4 | 69.2 | 75.1 | 79.6 | 81.3 | 81.2 | 79.6 | 73.7 | 65.9 | 60.3 | 70.7 |
| MEAN WET BULB | 26 | 53.2 | 54.7 | 58.3 | 61.6 | 67.7 | 72.9 | 74.7 | 75.0 | 73.7 | 68.0 | 61.4 | 55.9 | 64.8 |
| MEAN DEW POINT | 26 | 50.7 | 52.1 | 55.5 | 58.6 | 65.3 | 71.4 | 73.2 | 73.7 | 72.3 | 66.1 | 59.5 | 53.4 | 62.7 |
| NORMAL NO. DAYS WITH: | | | | | | | | | | | | | | |
| MAXIMUM >= 90 | 30 | 0.0 | 0.0 | 0.2 | 1.5 | 5.3 | 11.2 | 18.8 | 14.2 | 6.4 | 0.7 | 0.0 | 0.0 | 58.3 |
| MAXIMUM <= 32 | 30 | 0.0 | * | 0.0 | 0.0 | 0.0 | 0.0 | 0.0 | 0.0 | 0.0 | 0.0 | 0.0 | 0.0 | 0.0 |
| MINIMUM <= 32 | 30 | 2.2 | 1.2 | 0.3 | 0.0 | 0.0 | 0.0 | 0.0 | 0.0 | 0.0 | 0.0 | 0.0 | 1.4 | 5.1 |
| MINIMUM <= 0 | 30 | 0.0 | 0.0 | 0.0 | 0.0 | 0.0 | 0.0 | 0.0 | 0.0 | 0.0 | 0.0 | 0.0 | 0.0 | 0.0 |
| **H/C** | | | | | | | | | | | | | | |
| NORMAL HEATING DEG. DAYS | 30 | 245 | 183 | 99 | 29 | 1 | 0 | 0 | 0 | 0 | 6 | 67 | 185 | 815 |
| NORMAL COOLING DEG. DAYS | 30 | 36 | 40 | 86 | 150 | 306 | 441 | 513 | 502 | 436 | 277 | 122 | 52 | 2961 |
| **RH** | | | | | | | | | | | | | | |
| NORMAL (PERCENT) | 30 | 77 | 74 | 73 | 71 | 73 | 78 | 79 | 81 | 81 | 77 | 78 | 78 | 77 |
| HOUR 01 LST | 30 | 85 | 84 | 84 | 83 | 85 | 88 | 89 | 90 | 88 | 84 | 86 | 85 | 86 |
| HOUR 07 LST | 30 | 87 | 86 | 87 | 85 | 85 | 87 | 90 | 92 | 90 | 87 | 88 | 87 | 88 |
| HOUR 13 LST | 30 | 59 | 56 | 55 | 53 | 57 | 63 | 63 | 66 | 67 | 62 | 61 | 61 | 60 |
| HOUR 19 LST | 30 | 77 | 72 | 69 | 67 | 70 | 75 | 76 | 80 | 80 | 77 | 80 | 80 | 75 |
| **S** PERCENT POSSIBLE SUNSHINE | | | | | | | | | | | | | | |
| **W/O** MEAN NO. DAYS WITH: | | | | | | | | | | | | | | |
| HEAVY FOG(VISBY <= 1/4 MI) | 46 | 4.1 | 3.3 | 2.5 | 1.5 | 1.3 | 0.8 | 0.9 | 0.9 | 0.5 | 1.3 | 2.2 | 3.8 | 23.1 |
| THUNDERSTORMS | 62 | 1.1 | 1.9 | 3.5 | 3.4 | 7.1 | 13.8 | 17.0 | 15.2 | 8.8 | 3.0 | 1.2 | 1.0 | 77.0 |
| **CLOUDNESS** MEAN: | | | | | | | | | | | | | | |
| SUNRISE-SUNSET (OKTAS) | | | | | | | | | | | | | | |
| MIDNIGHT-MIDNIGHT (OKTAS) | | | | | | | | | | | | | | |
| MEAN NO. DAYS WITH: | | | | | | | | | | | | | | |
| CLEAR | 1 | 1.0 | 4.0 | 7.0 | | 13.0 | 4.0 | | | | | | | |
| PARTLY CLOUDY | 1 | 2.0 | 4.0 | 4.0 | | 5.0 | 11.0 | | | | | | | |
| CLOUDY | 1 | 2.0 | 2.0 | 8.0 | | 3.0 | 6.0 | | | | | | | |
| **PR** MEAN STATION PRESSURE(IN) | 26 | 30.10 | 30.08 | 30.03 | 30.00 | 29.99 | 29.98 | 30.02 | 29.99 | 29.95 | 29.99 | 30.06 | 30.10 | 30.02 |
| MEAN SEA-LEVEL PRES. (IN) | 26 | 30.15 | 30.12 | 30.08 | 30.04 | 30.03 | 30.02 | 30.06 | 30.03 | 30.00 | 30.03 | 30.09 | 30.14 | 30.07 |
| **WINDS** MEAN SPEED (MPH) | 26 | 7.5 | 7.8 | 8.4 | 8.1 | 7.8 | 6.5 | 5.9 | 5.8 | 6.9 | 7.7 | 7.2 | 6.9 | 7.2 |
| PREVAIL.DIR(TENS OF DEGS) | 30 | 32 | 36 | 25 | 25 | 09 | 08 | 23 | 08 | 07 | 07 | 32 | 32 | 07 |
| MAXIMUM 2-MINUTE: | 14 | | | | | | | | | | | | | |
| SPEED (MPH) | | 36 | 44 | 47 | 33 | 48 | 38 | 43 | 69 | 52 | 46 | 39 | 30 | 69 |
| DIR. (TENS OF DEGS) | | 23 | 23 | 30 | 26 | 22 | 07 | 34 | 01 | 06 | 03 | 05 | 01 | 01 |
| YEAR OF OCCURRENCE | | 1998 | 1998 | 2005 | 2007 | 2005 | 2004 | 2004 | 2004 | 2004 | 1999 | 2001 | 2006 | AUG 2004 |
| MAXIMUM 3-SECOND | 14 | | | | | | | | | | | | | |
| SPEED (MPH) | | 45 | 56 | 63 | 46 | 55 | 49 | 54 | 83 | 75 | 56 | 49 | 39 | 83 |
| DIR. (TENS OF DEGS) | | 25 | 22 | 30 | 26 | 28 | 07 | 36 | 01 | 06 | 01 | 04 | 27 | 01 |
| YEAR OF OCCURRENCE | | 2002 | 1998 | 2005 | 2009 | 1996 | 2004 | 1998 | 2004 | 2004 | 1999 | 2001 | 2006 | AUG 2004 |
| **PRECIPITATION** NORMAL (IN) | 30 | 3.13 | 2.74 | 3.84 | 2.54 | 3.26 | 5.69 | 5.17 | 6.09 | 6.61 | 4.48 | 3.03 | 2.71 | 49.29 |
| MAXIMUM MONTHLY (IN) | 66 | 7.16 | 9.13 | 12.15 | 7.12 | 22.33 | 15.19 | 14.58 | 19.89 | 16.46 | 13.51 | 12.91 | 11.98 | 22.33 |
| YEAR OF OCCURRENCE | | 1986 | 1960 | 1996 | 1949 | 2009 | 1966 | 1944 | 1953 | 2004 | 2005 | 1994 | 1983 | MAY 2009 |
| MINIMUM MONTHLY (IN) | 66 | 0.15 | 0.29 | 0.08 | T | 0.08 | 0.83 | 0.16 | 2.01 | 0.42 | 0.19 | T | 0.06 | T |
| YEAR OF OCCURRENCE | | 1950 | 1944 | 2006 | 1967 | 1965 | 1998 | 1992 | 1963 | 1972 | 1967 | 1967 | 1956 | NOV 1967 |
| MAXIMUM IN 24 HOURS (IN) | 66 | 5.73 | 4.39 | 7.45 | 4.03 | 8.09 | 6.28 | 4.21 | 7.98 | 9.02 | 9.29 | 10.15 | 5.22 | 10.15 |
| YEAR OF OCCURRENCE | | 1989 | 1971 | 2000 | 1982 | 2009 | 1966 | 1986 | 2004 | 2004 | 1953 | 1994 | 1983 | NOV 1994 |
| NORMAL NO. DAYS WITH: | | | | | | | | | | | | | | |
| PRECIPITATION >= 0.01 | 30 | 7.9 | 7.5 | 7.7 | 5.6 | 8.2 | 12.9 | 12.4 | 13.8 | 13.2 | 10.5 | 7.8 | 7.8 | 115.3 |
| PRECIPITATION >= 1.00 | 30 | 0.8 | 0.8 | 1.4 | 0.8 | 0.8 | 1.5 | 1.4 | 1.8 | 2.5 | 1.4 | 0.6 | 0.8 | 14.6 |
| **SNOWFALL** NORMAL (IN) | 30 | 0.* | 0.0 | 0.0 | 0.0 | 0.0 | 0.0 | 0.0 | 0.0 | 0.0 | 0.0 | 0.0 | 0.* | 0.0 |
| MAXIMUM MONTHLY (IN) | 53 | T | T | T | 0.0 | 0.0 | T | 0.0 | T | 0.0 | 0.0 | 0.0 | T | T |
| YEAR OF OCCURRENCE | | 1977 | 1951 | 1993 | | | 1989 | | 1994 | | | | 1989 | AUG 1994 |
| MAXIMUM IN 24 HOURS (IN) | 53 | T | T | T | 0.0 | 0.0 | T | 0.0 | T | 0.0 | 0.0 | 0.0 | T | T |
| YEAR OF OCCURRENCE' | | 1977 | 1951 | 1993 | | | 1989 | | 1994 | | | | 1989 | AUG 1994 |
| MAXIMUM SNOW DEPTH (IN) | 48 | 0 | 0 | 0 | 0 | 0 | 0 | 0 | 0 | 0 | 0 | 0 | 0 | 0 |
| YEAR OF OCCURRENCE | | | | | | | | | | | | | | |
| NORMAL NO. DAYS WITH: | | | | | | | | | | | | | | |
| SNOWFALL >= 1.0 | 30 | 0.0 | 0.0 | 0.0 | 0.0 | 0.0 | 0.0 | 0.0 | 0.0 | 0.0 | 0.0 | 0.0 | 0.0 | 0.0 |

## PRECIPITATION (inches) 2009 DAYTONA BEACH (KDAB)

| YEAR | JAN | FEB | MAR | APR | MAY | JUN | JUL | AUG | SEP | OCT | NOV | DEC | ANNUAL |
|------|-----|-----|-----|-----|-----|-----|-----|-----|-----|-----|-----|-----|--------|
| 1980 | 3.75 | 0.76 | 2.41 | 2.54 | 3.62 | 5.57 | 5.82 | 4.13 | 1.83 | 2.42 | 3.12 | 1.39 | 37.36 |
| 1981 | 0.32 | 5.54 | 3.00 | 0.29 | 1.74 | 1.03 | 4.69 | 7.19 | 7.59 | 1.08 | 2.57 | 4.64 | 39.68 |
| 1982 | 2.46 | 2.08 | 5.81 | 6.04 | 4.68 | 8.29 | 5.31 | 3.21 | 4.96 | 3.23 | 1.58 | 2.53 | 50.18 |
| 1983 | 2.51 | 5.96 | 7.71 | 6.17 | 3.86 | 6.37 | 1.92 | 6.82 | 8.57 | 10.11 | 2.01 | 11.98 | 73.99 |
| 1984 | 1.46 | 3.44 | 1.31 | 5.29 | 6.04 | 2.84 | 6.77 | 4.02 | 10.73 | 1.09 | 3.52 | 0.20 | 46.71 |
| 1985 | 0.79 | 0.58 | 1.49 | 3.14 | 3.42 | 6.81 | 2.16 | 9.83 | 10.62 | 4.08 | 0.41 | 2.05 | 45.38 |
| 1986 | 7.16 | 1.28 | 1.85 | 0.44 | 0.99 | 3.50 | 14.43 | 3.47 | 3.58 | 3.47 | 5.08 | 2.76 | 48.01 |
| 1987 | 2.21 | 6.64 | 7.94 | 0.28 | 2.65 | 3.81 | 2.78 | 4.89 | 5.63 | 2.77 | 5.87 | 0.25 | 45.72 |
| 1988 | 5.36 | 1.72 | 4.57 | 1.68 | 1.78 | 2.39 | 2.94 | 4.79 | 6.81 | 1.24 | 6.70 | 0.93 | 40.91 |
| 1989 | 6.82 | 0.64 | 2.01 | 2.92 | 2.02 | 1.84 | 2.44 | 4.47 | 5.04 | 11.64 | 0.88 | 3.93 | 44.65 |
| 1990 | 1.42 | 5.61 | 1.94 | 1.48 | 1.45 | 2.71 | 5.85 | 7.00 | 1.61 | 5.88 | 0.83 | 0.34 | 36.12 |
| 1991 | 2.25 | 1.65 | 8.11 | 5.57 | 6.79 | 12.67 | 11.97 | 7.60 | 5.52 | 2.94 | 0.61 | 1.51 | 67.19 |
| 1992 | 2.42 | 1.71 | 2.28 | 2.81 | 3.13 | 10.64 | 0.16 | 8.86 | 6.57 | 5.21 | 2.15 | 0.47 | 46.41 |
| 1993 | 4.29 | 3.02 | 5.56 | 0.33 | 0.65 | 2.19 | 5.05 | 2.66 | 2.74 | 5.53 | 1.83 | 1.86 | 35.71 |
| 1994 | 5.60 | 2.66 | 3.44 | 5.05 | 3.09 | 6.54 | 6.91 | 7.08 | 5.93 | 4.72 | 12.91 | 2.71 | 66.64 |
| 1995 | 1.53 | 1.39 | 2.01 | 1.34 | 1.26 | 6.60 | 6.59 | 10.71 | 14.13 | 3.99 | 1.44 | 3.44 | 54.43 |
| 1996 | 5.53 | 1.32 | 12.15 | 2.22 | 2.28 | 11.35 | 1.90 | 5.70 | 3.92 | 11.15 | .96 | 2.01 | 60.49 |
| 1997 | 2.03 | 0.46 | 2.30 | 3.30 | 3.77 | 6.38 | 7.69 | 7.91 | 4.78 | 5.29 | 3.02 | 7.76 | 54.69 |
| 1998 | 4.33 | 7.25 | 3.97 | 0.14 | 0.16 | 0.83 | 5.63 | 7.56 | 5.79 | 1.84 | 1.66 | 1.35 | 40.51 |
| 1999 | 4.88 | 1.81 | 1.01 | 1.48 | 1.47 | 8.54 | 4.03 | 3.58 | 7.05 | 7.84 | 3.12 | 1.56 | 46.37 |
| 2000 | 1.80 | 0.65 | 8.48 | 1.15 | 0.32 | 3.08 | 5.09 | 3.17 | 13.55 | 0.93 | 1.14 | 0.80 | 40.16 |
| 2001 | 0.88 | 0.38 | 9.98 | 0.28 | 1.77 | 5.26 | 9.55 | 3.57 | 16.11 | 3.22 | 6.92 | 0.35 | 58.27 |
| 2002 | 2.01 | 2.76 | 1.51 | 2.53 | 1.66 | 12.30 | 7.35 | 11.56 | 3.86 | 2.94 | 1.85 | 9.61 | 59.94 |
| 2003 | 0.51 | 5.17 | 10.57 | 0.81 | 0.96 | 7.05 | 7.30 | 6.55 | 4.15 | 7.95 | 4.75 | 1.53 | 57.30 |
| 2004 | 1.25 | 4.47 | 1.10 | 1.19 | 0.49 | 5.20 | 10.34 | 17.96 | 16.46 | 1.34 | 0.93 | 2.24 | 62.97 |
| 2005 | 2.60 | 1.25 | 5.51 | 3.17 | 7.97 | 13.67 | 2.73 | 4.29 | 7.35 | 13.51 | 1.87 | 1.85 | 65.77 |
| 2006 | 0.24 | 4.33 | 0.08 | 1.11 | 0.78 | 5.72 | 4.48 | 4.81 | 2.97 | 2.53 | 1.10 | 3.21 | 31.36 |
| 2007 | 1.53 | 2.64 | 0.70 | 1.34 | 0.91 | 5.78 | 10.23 | 2.88 | 11.36 | 3.49 | 2.32 | 1.84 | 45.02 |
| 2008 | 1.30 | 2.12 | 3.20 | 1.34 | 0.63 | 3.64 | 9.48 | 10.33 | 4.29 | 4.45 | 0.96 | 0.93 | 42.67 |
| 2009 | 0.82 | 0.80 | 1.39 | 1.47 | 22.33 | 5.03 | 5.19 | 3.77 | 3.65 | 1.44 | 0.60 | 3.81 | 50.30 |
| POR= 62 YRS | 2.46 | 2.84 | 3.57 | 2.32 | 3.14 | 6.09 | 5.78 | 6.44 | 6.90 | 4.81 | 2.60 | 2.46 | 49.41 |

WBAN : 12834

## AVERAGE TEMPERATURE (°F) 2009 DAYTONA BEACH (KDAB)

| YEAR | JAN | FEB | MAR | APR | MAY | JUN | JUL | AUG | SEP | OCT | NOV | DEC | ANNUAL |
|------|-----|-----|-----|-----|-----|-----|-----|-----|-----|-----|-----|-----|--------|
| 1980 | 57.7 | 55.3 | 66.3 | 68.8 | 74.9 | 79.2 | 82.8 | 82.1 | 80.3 | 72.7 | 65.3 | 57.0 | 70.2 |
| 1981 | 48.8 | 59.2 | 60.4 | 70.5 | 73.5 | 82.0 | 82.8 | 81.5 | 77.7 | 73.8 | 62.7 | 57.1 | 69.2 |
| 1982 | 56.6 | 64.4 | 66.8 | 69.4 | 72.6 | 79.5 | 80.0 | 79.9 | 77.9 | 71.5 | 68.8 | 64.0 | 71.0 |
| 1983 | 53.9 | 57.2 | 60.2 | 64.3 | 72.4 | 77.0 | 81.7 | 81.1 | 77.8 | 73.5 | 62.6 | 58.1 | 68.3 |
| 1984 | 55.1 | 58.0 | 61.8 | 66.8 | 72.4 | 76.3 | 79.0 | 81.4 | 79.5 | 75.7 | 66.5 | 65.2 | 69.8 |
| 1985 | 53.7 | 61.0 | 66.6 | 69.4 | 76.1 | 81.7 | 80.6 | 81.4 | 78.3 | 76.8 | 71.2 | 56.1 | 71.1 |
| 1986 | 56.7 | 62.4 | 63.1 | 66.3 | 73.8 | 79.9 | 81.4 | 81.4 | 79.6 | 75.3 | 72.5 | 64.7 | 71.4 |
| 1987 | 55.8 | 59.7 | 63.3 | 65.1 | 74.3 | 79.7 | 81.9 | 82.3 | 79.6 | 70.0 | 66.5 | 61.9 | 70.0 |
| 1988 | 55.1 | 56.8 | 62.8 | 69.1 | 72.6 | 79.0 | 81.2 | 81.5 | 80.6 | 70.7 | 67.5 | 59.8 | 69.7 |
| 1989 | 64.8 | 61.9 | 67.8 | 69.5 | 75.4 | 80.3 | 82.7 | 81.8 | 80.5 | 73.4 | 65.9 | 53.3 | 71.4 |
| 1990 | 62.7 | 67.5 | 66.4 | 69.6 | 77.3 | 80.7 | 81.9 | 81.9 | 80.5 | 76.0 | 67.4 | 65.1 | 73.1 |
| 1991 | 63.4 | 61.9 | 65.6 | 73.7 | 78.5 | 80.3 | 82.5 | 82.4 | 80.8 | 73.8 | 64.4 | 63.5 | 72.6 |
| 1992 | 56.5 | 61.7 | 63.3 | 67.2 | 72.5 | 80.5 | 83.4 | 80.7 | 79.8 | 72.3 | 69.1 | 61.4 | 70.7 |
| 1993 | 64.5 | 57.5 | 62.9 | 65.5 | 72.7 | 80.0 | 82.5 | 81.9 | 79.6 | 73.7 | 66.9 | 56.4 | 70.3 |
| 1994 | 58.1 | 63.8 | 65.8 | 72.3 | 75.0 | 80.2 | 80.5 | 79.9 | 78.2 | 74.9 | 70.1 | 62.9 | 71.8 |
| 1995 | 56.5 | 58.7 | 66.0 | 70.3 | 79.0 | 79.0 | 81.2 | 81.3 | 79.5 | 76.4 | 62.8 | 58.1 | 70.7 |
| 1996 | 56.8 | 59.2 | 60.7 | 66.7 | 75.6 | 78.2 | 81.2 | 79.5 | 78.9 | 72.9 | 66.2 | 60.8 | 69.7 |
| 1997 | 59.0 | 64.7 | 69.9 | 67.8 | 74.0 | 78.4 | 81.6 | 81.5 | 79.3 | 73.2 | 64.3 | 59.4 | 71.1 |
| 1998 | 60.6 | 60.2 | 61.2 | 69.9 | 76.6 | 84.5 | 83.5 | 82.4 | 80.7 | 76.5 | 70.8 | 66.0 | 72.7 |
| 1999 | 62.0 | 62.1 | 62.2 | 72.4 | 74.5 | 79.6 | 82.4 | 82.9 | 79.0 | 74.4 | 67.3 | 60.0 | 71.6 |
| 2000 | 59.0 | 59.7 | 67.4 | 68.2 | 76.3 | 79.5 | 81.4 | 81.0 | 80.5 | 71.6 | 63.2 | 56.6 | 70.4 |
| 2001 | 53.9 | 64.1 | 64.5 | 69.1 | 74.6 | 79.9 | 81.2 | 81.3 | 77.6 | 72.9 | 68.9 | 65.4 | 71.1 |
| 2002 | 60.1 | 59.9 | 66.8 | 73.5 | 76.4 | 78.9 | 81.6 | 81.1 | 81.4 | 76.7 | 64.1 | 57.7 | 71.5 |
| 2003 | 52.1 | 60.9 | 69.3 | 69.4 | 78.6 | 79.5 | 81.0 | 80.5 | 78.8 | 74.8 | 70.3 | 58.3 | 71.1 |
| 2004 | 58.0 | 59.9 | 65.3 | 66.8 | 75.5 | 81.4 | 81.7 | 82.1 | 81.6 | 74.7 | 68.8 | 58.8 | 71.2 |
| 2005 | 60.4 | 61.0 | 62.4 | 66.3 | 72.8 | 79.3 | 82.9 | 83.6 | 80.8 | 74.7 | 68.1 | 58.0 | 70.9 |
| 2006 | 61.4 | 58.6 | 64.4 | 72.8 | 76.3 | 79.9 | 81.1 | 82.3 | 79.6 | 72.2 | 64.6 | 65.9 | 71.6 |
| 2007 | 62.3 | 58.6 | 65.4 | 68.6 | 74.0 | 80.3 | 82.8 | 84.0 | 80.7 | 78.1 | 66.0 | 65.7 | 72.2 |
| 2008 | 60.1 | 63.7 | 65.5 | 69.2 | 76.7 | 80.7 | 81.0 | 81.3 | 81.5 | 73.9 | 63.1 | 64.6 | 71.8 |
| 2009 | 57.9 | 57.9 | 65.7 | 69.9 | 76.8 | 81.5 | 81.3 | 82.5 | 80.5 | 76.6 | 68.0 | 62.5 | 71.8 |
| POR= 62 YRS | 58.2 | 60.0 | 64.4 | 69.2 | 75.1 | 79.6 | 81.3 | 81.2 | 79.6 | 73.7 | 65.9 | 60.3 | 70.7 |

## HEATING DEGREE DAYS (base 65°F) 2009  DAYTONA BEACH (KDAB)

| YEAR | JUL | AUG | SEP | OCT | NOV | DEC | JAN | FEB | MAR | APR | MAY | JUN | TOTAL |
|---|---|---|---|---|---|---|---|---|---|---|---|---|---|
| 1980-81 | 0 | 0 | 0 | 11 | 93 | 247 | 497 | 184 | 171 | 0 | 1 | 0 | 1204 |
| 1981-82 | 0 | 0 | 0 | 0 | 127 | 284 | 273 | 72 | 63 | 26 | 0 | 0 | 845 |
| 1982-83 | 0 | 0 | 0 | 24 | 21 | 125 | 345 | 220 | 167 | 74 | 2 | 0 | 978 |
| 1983-84 | 0 | 0 | 0 | 2 | 126 | 255 | 323 | 215 | 148 | 37 | 3 | 0 | 1109 |
| 1984-85 | 0 | 0 | 0 | 0 | 63 | 77 | 372 | 173 | 44 | 21 | 0 | 0 | 750 |
| 1985-86 | 0 | 0 | 0 | 0 | 24 | 303 | 261 | 119 | 141 | 30 | 0 | 0 | 878 |
| 1986-87 | 0 | 0 | 0 | 0 | 11 | 84 | 301 | 160 | 99 | 81 | 0 | 0 | 736 |
| 1987-88 | 0 | 0 | 0 | 10 | 74 | 146 | 316 | 259 | 120 | 23 | 0 | 0 | 948 |
| 1988-89 | 0 | 0 | 0 | 1 | 39 | 187 | 70 | 154 | 68 | 20 | 1 | 0 | 540 |
| 1989-90 | 0 | 0 | 0 | 31 | 59 | 369 | 120 | 47 | 37 | 14 | 0 | 0 | 677 |
| 1990-91 | 0 | 0 | 0 | 9 | 35 | 96 | 114 | 126 | 83 | 9 | 0 | 0 | 472 |
| 1991-92 | 0 | 0 | 0 | 0 | 110 | 114 | 264 | 139 | 106 | 43 | 15 | 0 | 791 |
| 1992-93 | 0 | 0 | 0 | 0 | 70 | 136 | 104 | 210 | 107 | 49 | 0 | 0 | 676 |
| 1993-94 | 0 | 0 | 0 | 12 | 62 | 281 | 230 | 100 | 70 | 10 | 0 | 0 | 765 |
| 1994-95 | 0 | 0 | 0 | 0 | 10 | 117 | 269 | 200 | 41 | 16 | 0 | 0 | 653 |
| 1995-96 | 0 | 0 | 0 | 3 | 135 | 239 | 272 | 212 | 197 | 55 | 0 | 0 | 1113 |
| 1996-97 | 0 | 0 | 0 | 11 | 62 | 155 | 204 | 80 | 12 | 23 | 1 | 0 | 548 |
| 1997-98 | 0 | 0 | 0 | 8 | 69 | 215 | 171 | 159 | 174 | 22 | 0 | 0 | 818 |
| 1998-99 | 0 | 0 | 0 | 0 | 10 | 88 | 141 | 126 | 105 | 17 | 6 | 0 | 493 |
| 1999-00 | 0 | 0 | 0 | 6 | 30 | 174 | 201 | 164 | 16 | 32 | 0 | 0 | 623 |
| 2000-01 | 0 | 0 | 0 | 6 | 118 | 293 | 355 | 87 | 88 | 26 | 0 | 0 | 973 |
| 2001-02 | 0 | 0 | 0 | 14 | 4 | 108 | 213 | 160 | 64 | 2 | 0 | 0 | 565 |
| 2002-03 | 0 | 0 | 0 | 0 | 110 | 235 | 389 | 148 | 25 | 30 | 0 | 0 | 937 |
| 2003-04 | 0 | 0 | 0 | 0 | 31 | 215 | 227 | 166 | 54 | 44 | 1 | 0 | 738 |
| 2004-05 | 0 | 0 | 0 | 0 | 15 | 230 | 169 | 138 | 143 | 37 | 4 | 0 | 736 |
| 2005-06 | 0 | 0 | 0 | 19 | 27 | 218 | 156 | 200 | 91 | 0 | 0 | 0 | 711 |
| 2006-07 | 0 | 0 | 0 | 13 | 96 | 76 | 158 | 197 | 64 | 31 | 0 | 0 | 635 |
| 2007-08 | 0 | 0 | 0 | 0 | 50 | 63 | 180 | 110 | 67 | 29 | 0 | 0 | 499 |
| 2008-09 | 0 | 0 | 0 | 27 | 123 | 93 | 237 | 214 | 70 | 20 | 0 | 0 | 784 |
| 2009- | 0 | 0 | 0 | 14 | 40 | 158 | | | | | | | |

WBAN : 12834

## COOLING DEGREE DAYS (base 65°F) 2009  DAYTONA BEACH (KDAB)

| YEAR | JAN | FEB | MAR | APR | MAY | JUN | JUL | AUG | SEP | OCT | NOV | DEC | TOTAL |
|---|---|---|---|---|---|---|---|---|---|---|---|---|---|
| 1980 | 12 | 21 | 131 | 135 | 315 | 435 | 559 | 538 | 467 | 258 | 109 | 5 | 2985 |
| 1981 | 0 | 25 | 37 | 172 | 269 | 516 | 559 | 521 | 385 | 282 | 65 | 47 | 2878 |
| 1982 | 19 | 61 | 127 | 166 | 240 | 440 | 472 | 470 | 392 | 234 | 141 | 97 | 2859 |
| 1983 | 6 | 6 | 28 | 57 | 238 | 238 | 369 | 521 | 504 | 391 | 270 | 62 | 46 | 2498 |
| 1984 | 22 | 20 | 55 | 96 | 238 | 345 | 442 | 515 | 441 | 338 | 114 | 91 | 2717 |
| 1985 | 29 | 67 | 101 | 160 | 348 | 506 | 490 | 511 | 405 | 373 | 217 | 35 | 3242 |
| 1986 | 13 | 50 | 89 | 79 | 280 | 452 | 516 | 515 | 444 | 324 | 246 | 82 | 3090 |
| 1987 | 20 | 17 | 52 | 92 | 297 | 449 | 530 | 543 | 442 | 171 | 125 | 58 | 2796 |
| 1988 | 17 | 27 | 62 | 155 | 242 | 425 | 509 | 518 | 474 | 185 | 121 | 32 | 2767 |
| 1989 | 71 | 74 | 162 | 162 | 331 | 468 | 553 | 530 | 474 | 299 | 94 | 11 | 3229 |
| 1990 | 55 | 124 | 85 | 161 | 385 | 478 | 531 | 528 | 470 | 355 | 114 | 107 | 3393 |
| 1991 | 71 | 45 | 113 | 278 | 426 | 465 | 548 | 548 | 480 | 278 | 98 | 74 | 3424 |
| 1992 | 8 | 51 | 60 | 118 | 256 | 472 | 577 | 492 | 448 | 231 | 196 | 35 | 2944 |
| 1993 | 95 | 7 | 48 | 71 | 245 | 456 | 551 | 534 | 446 | 289 | 126 | 23 | 2891 |
| 1994 | 23 | 73 | 101 | 237 | 318 | 466 | 487 | 468 | 402 | 315 | 170 | 57 | 3117 |
| 1995 | 12 | 29 | 81 | 182 | 442 | 426 | 507 | 511 | 441 | 365 | 74 | 32 | 3102 |
| 1996 | 24 | 50 | 72 | 114 | 338 | 402 | 505 | 459 | 427 | 262 | 106 | 29 | 2788 |
| 1997 | 26 | 77 | 170 | 113 | 283 | 405 | 522 | 519 | 436 | 269 | 55 | 47 | 2922 |
| 1998 | 42 | 29 | 64 | 176 | 369 | 592 | 581 | 546 | 479 | 363 | 191 | 124 | 3556 |
| 1999 | 54 | 47 | 26 | 249 | 307 | 442 | 548 | 561 | 429 | 304 | 106 | 29 | 3102 |
| 2000 | 21 | 18 | 98 | 134 | 355 | 440 | 514 | 503 | 471 | 216 | 71 | 40 | 2881 |
| 2001 | 19 | 69 | 82 | 156 | 305 | 455 | 511 | 512 | 383 | 263 | 129 | 126 | 3010 |
| 2002 | 66 | 23 | 126 | 263 | 361 | 423 | 523 | 506 | 499 | 370 | 89 | 15 | 3264 |
| 2003 | 0 | 39 | 168 | 168 | 427 | 438 | 503 | 489 | 420 | 311 | 197 | 15 | 3175 |
| 2004 | 17 | 24 | 73 | 106 | 330 | 497 | 522 | 535 | 505 | 308 | 134 | 45 | 3096 |
| 2005 | 34 | 32 | 69 | 84 | 251 | 436 | 562 | 585 | 482 | 328 | 125 | 7 | 2995 |
| 2006 | 51 | 28 | 78 | 240 | 360 | 454 | 506 | 545 | 445 | 244 | 91 | 113 | 3155 |
| 2007 | 80 | 26 | 84 | 144 | 284 | 466 | 559 | 597 | 476 | 411 | 86 | 92 | 3305 |
| 2008 | 34 | 78 | 89 | 161 | 368 | 478 | 504 | 512 | 503 | 310 | 71 | 89 | 3197 |
| 2009 | 23 | 20 | 99 | 171 | 371 | 502 | 512 | 549 | 470 | 381 | 136 | 90 | 3324 |

## SNOWFALL (inches)  2009  DAYTONA BEACH (KDAB)

| YEAR | JUL | AUG | SEP | OCT | NOV | DEC | JAN | FEB | MAR | APR | MAY | JUN | TOTAL |
|------|-----|-----|-----|-----|-----|-----|-----|-----|-----|-----|-----|-----|-------|
| 1979-80 | 0.0 | 0.0 | 0.0 | 0.0 | 0.0 | 0.0 | 0.0 | 0.0 | 0.0 | 0.0 | 0.0 | 0.0 | 0.0 |
| 1980-81 | 0.0 | 0.0 | 0.0 | 0.0 | 0.0 | 0.0 | 0.0 | 0.0 | 0.0 | 0.0 | 0.0 | 0.0 | 0.0 |
| 1981-82 | 0.0 | 0.0 | 0.0 | 0.0 | 0.0 | 0.0 | 0.0 | 0.0 | 0.0 | 0.0 | 0.0 | 0.0 | 0.0 |
| 1982-83 | 0.0 | 0.0 | 0.0 | 0.0 | 0.0 | 0.0 | 0.0 | 0.0 | 0.0 | 0.0 | 0.0 | 0.0 | 0.0 |
| 1983-84 | 0.0 | 0.0 | 0.0 | 0.0 | 0.0 | 0.0 | 0.0 | 0.0 | 0.0 | 0.0 | 0.0 | 0.0 | 0.0 |
| 1984-85 | 0.0 | 0.0 | 0.0 | 0.0 | 0.0 | 0.0 | 0.0 | 0.0 | 0.0 | 0.0 | 0.0 | 0.0 | 0.0 |
| 1985-86 | 0.0 | 0.0 | 0.0 | 0.0 | 0.0 | 0.0 | 0.0 | 0.0 | 0.0 | 0.0 | 0.0 | 0.0 | 0.0 |
| 1986-87 | 0.0 | 0.0 | 0.0 | 0.0 | 0.0 | 0.0 | 0.0 | 0.0 | 0.0 | 0.0 | 0.0 | 0.0 | 0.0 |
| 1987-88 | 0.0 | 0.0 | 0.0 | 0.0 | 0.0 | 0.0 | 0.0 | 0.0 | 0.0 | 0.0 | 0.0 | 0.0 | 0.0 |
| 1988-89 | 0.0 | 0.0 | 0.0 | 0.0 | 0.0 | 0.0 | 0.0 | 0.0 | 0.0 | 0.0 | 0.0 | T | T |
| 1989-90 | 0.0 | 0.0 | 0.0 | 0.0 | 0.0 | T | 0.0 | 0.0 | 0.0 | 0.0 | 0.0 | 0.0 | T |
| 1990-91 | 0.0 | 0.0 | 0.0 | 0.0 | 0.0 | 0.0 | 0.0 | 0.0 | 0.0 | 0.0 | 0.0 | 0.0 | 0.0 |
| 1991-92 | 0.0 | 0.0 | 0.0 | 0.0 | 0.0 | 0.0 | 0.0 | 0.0 | 0.0 | 0.0 | 0.0 | 0.0 | 0.0 |
| 1992-93 | 0.0 | 0.0 | 0.0 | 0.0 | 0.0 | 0.0 | 0.0 | 0.0 | T | 0.0 | 0.0 | 0.0 | T |
| 1993-94 | 0.0 | 0.0 | 0.0 | 0.0 | 0.0 | 0.0 | 0.0 | 0.0 | 0.0 | 0.0 | 0.0 | 0.0 | 0.0 |
| 1994-95 | 0.0 | T | 0.0 | 0.0 | 0.0 | 0.0 | 0.0 | 0.0 | 0.0 | 0.0 | 0.0 | 0.0 | T |
| 1995-96 | 0.0 | 0.0 | 0.0 | 0.0 | 0.0 | 0.0 | 0.0 | | | | | | |
| 1996-97 | | | | | | | | | | | | | |
| 1997-98 | | | | | | | | | | | | | |
| 1998-99 | | | | | | | | | | | | | |
| 1999-00 | | | | | | | | | | | | | |
| 2000-01 | | | | | | | | | | | | | |
| 2001-02 | | | | | | | | | | | | | |
| 2002-03 | | | | | | | | | | | | | |
| 2003-04 | | | | | | | | | | | | | |
| 2004-05 | | | | | | | | | | | | | |
| 2005-06 | | | | | | | | | | | | | |
| 2006-07 | | | | | | | | | | | | | |
| 2007-08 | | | | | | | | | | | | | |
| 2008-09 | | | | | | | | | | | | | |
| POR=<br>48 YRS | 0.0 | T | 0.0 | 0.0 | 0.0 | T | 0.0 | 0.0 | T | 0.0 | 0.0 | T | T |

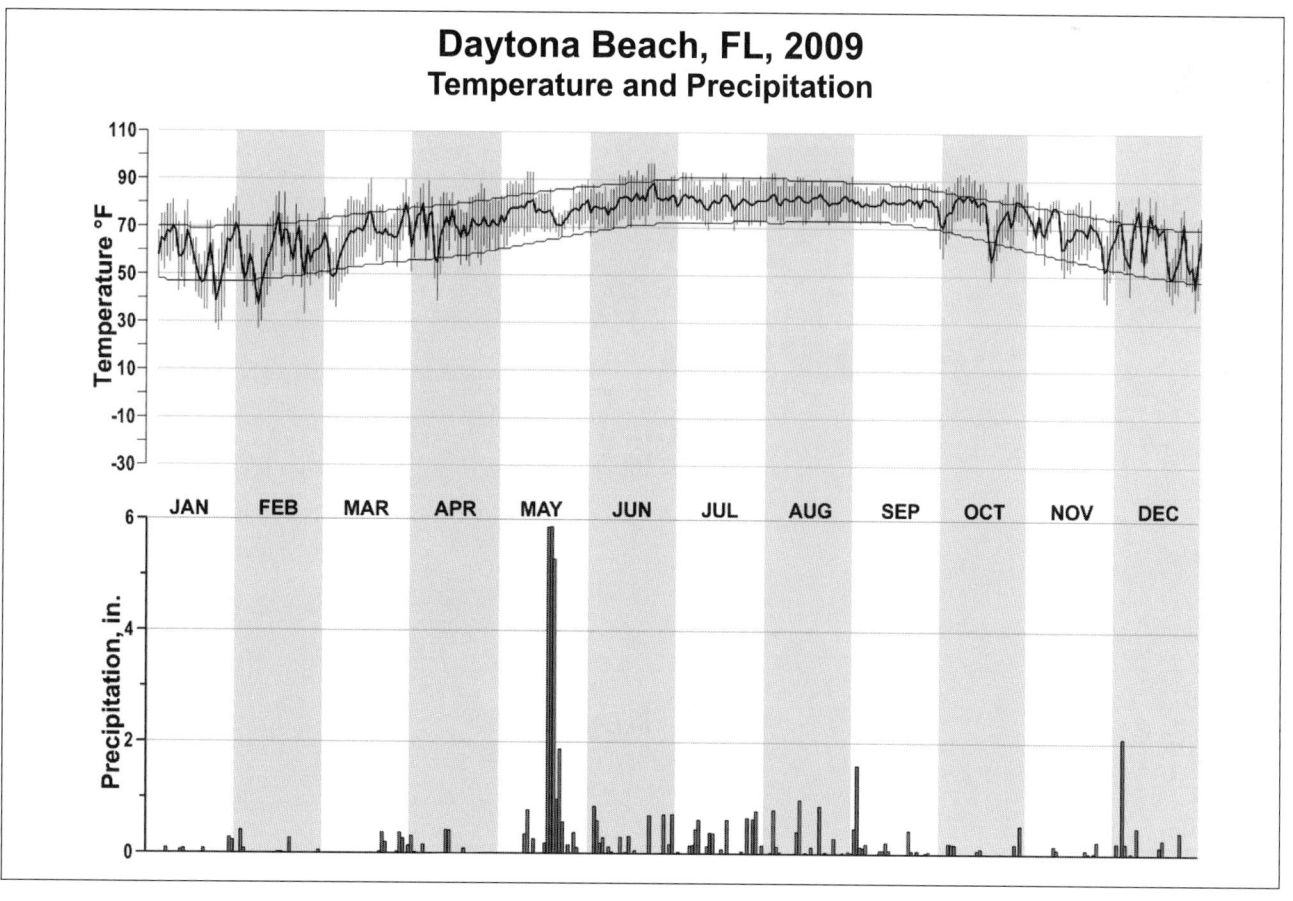

## Daytona Beach, FL, 2009
### Temperature and Precipitation

# 2009
# JACKSONVILLE
# FLORIDA (KJAX)

Jacksonville, a very large metropolitan area covering 840 square miles, extends from the Atlantic Ocean to about 40 miles inland. Downtown Jacksonville is located some 16 miles inland on the St. Johns River. The surrounding terrain is level. Easterly winds blowing about 40 percent of the time produce a maritime influence that modifies to some extent the heat of summer and the cold of winter. Summers are long, warm and relatively humid. Winters, although punctuated with periodic invasions of cool to occasionally cold air from the north, are mild because of the southern latitude and the proximity to the warm Atlantic Ocean waters. Because of the nearness to the ocean, climatic features across the city vary. For example, during the summer months temperatures at Jacksonville International Airport, located 17 miles inland, usually reach into the low and mid-90s before being tempered by sea breezes. Temperatures along the beaches rarely exceed 90 degrees. Summer thunderstorms usually occur before the noon hour along the beaches, while afternoon thunderstorms are the rule inland.

The annual temperature for Jacksonville is between 68 and 69 degrees. June, July, and August are the hottest months, with temperatures averaging near 80 degrees. December, January, and February are the coolest months, with temperatures near the middle 50s. Temperatures exceed 95 degrees only about ten times a year. Night temperatures in summer are usually comfortable, rarely failing to drop below 80 degrees.

The greatest rainfall, mostly in the form of local thundershowers, occurs during the summer months when a measurable amount can be expected one day in two. Rainfall of 1 inch or more in 24 hours normally occurs about fourteen times a year, and very infrequently heavy rains, associated with tropical storms, reach amounts of several inches with durations of more than 24 hours.

The atmosphere is moist, with an average relative humidity of about 75 percent, ranging from about 90 percent in early morning hours to about 55 percent during the afternoon.

Prevailing winds are northeasterly in the fall and winter months, and southwesterly in spring and summer. Wind movement, which averages slightly less than 9 mph, is 2 to 3 mph higher in the early afternoon than the early morning hours, and slightly higher in spring than in other seasons of the year. Although this area is in the Hurricane Belt, this section of the coast has been very fortunate in escaping hurricane-force winds. Most hurricanes reaching this latitude have tended to move parallel to the coastline, keeping well out to sea. Others have lost much of their force moving over land before reaching this area.

# NORMALS, MEANS, AND EXTREMES
## JACKSONVILLE (KJAX)

LATITUDE: 30 ° 29'N  LONGITUDE: -81 ° 41'W  ELEVATION (FT): GRND: 26  BARO: 34  TIME ZONE: EASTERN  (UTC -5)  WBAN: 13889

| | ELEMENT | POR | JAN | FEB | MAR | APR | MAY | JUN | JUL | AUG | SEP | OCT | NOV | DEC | YEAR |
|---|---|---|---|---|---|---|---|---|---|---|---|---|---|---|---|
| **TEMPERATURE °F** | NORMAL DAILY MAXIMUM | 30 | 64.2 | 67.3 | 73.4 | 78.6 | 84.3 | 88.7 | 90.8 | 89.4 | 86.1 | 79.1 | 72.5 | 65.8 | 78.4 |
| | MEAN DAILY MAXIMUM | 62 | 65.1 | 67.9 | 73.7 | 79.7 | 85.9 | 89.7 | 91.9 | 90.8 | 87.1 | 80.2 | 73.0 | 66.6 | 79.3 |
| | HIGHEST DAILY MAXIMUM | 68 | 85 | 88 | 91 | 95 | 100 | 103 | 105 | 102 | 100 | 96 | 88 | 84 | 105 |
| | YEAR OF OCCURRENCE | | 1947 | 1962 | 1974 | 1968 | 1967 | 1998 | 1942 | 1999 | 1944 | 1951 | 1986 | 2009 | JUL 1942 |
| | MEAN OF EXTREME MAXS. | 62 | 79.6 | 82.3 | 86.3 | 89.9 | 94.0 | 96.9 | 97.6 | 96.6 | 93.9 | 89.3 | 83.9 | 80.4 | 89.2 |
| | NORMAL DAILY MINIMUM | 30 | 41.9 | 44.3 | 49.8 | 54.6 | 62.5 | 69.4 | 72.4 | 72.2 | 69.4 | 59.7 | 50.8 | 44.1 | 57.6 |
| | MEAN DAILY MINIMUM | 62 | 42.6 | 45.1 | 50.4 | 56.1 | 63.7 | 70.0 | 72.8 | 72.7 | 70.1 | 60.7 | 50.8 | 44.5 | 58.3 |
| | LOWEST DAILY MINIMUM | 68 | 7 | 19 | 23 | 31 | 45 | 47 | 61 | 59 | 48 | 33 | 21 | 11 | 7 |
| | YEAR OF OCCURRENCE | | 1985 | 1996 | 1980 | 2007 | 1992 | 1984 | 1972 | 2000 | 1981 | 2008 | 1970 | 1983 | JAN 1985 |
| | MEAN OF EXTREME MINS. | 62 | 25.2 | 28.4 | 33.2 | 41.4 | 51.7 | 61.5 | 67.8 | 67.7 | 61.1 | 44.4 | 33.0 | 27.5 | 45.2 |
| | NORMAL DRY BULB | 30 | 53.1 | 55.8 | 61.6 | 66.6 | 73.4 | 79.1 | 81.6 | 80.8 | 77.8 | 69.4 | 61.7 | 55.0 | 68.0 |
| | MEAN DRY BULB | 62 | 53.9 | 56.5 | 62.1 | 67.9 | 74.8 | 80.0 | 82.3 | 81.8 | 78.6 | 70.5 | 61.9 | 55.6 | 68.8 |
| | MEAN WET BULB | 26 | 48.4 | 50.6 | 55.1 | 59.4 | 66.5 | 72.5 | 74.7 | 74.8 | 72.3 | 64.9 | 57.1 | 50.7 | 62.3 |
| | MEAN DEW POINT | 26 | 45.4 | 47.6 | 51.9 | 56.1 | 63.8 | 70.8 | 73.0 | 73.4 | 70.9 | 63.2 | 55.1 | 48.1 | 59.9 |
| | NORMAL NO. DAYS WITH: | | | | | | | | | | | | | | |
| | MAXIMUM >= 90 | 30 | 0.0 | 0.0 | * | 1.2 | 6.2 | 16.3 | 23.4 | 20.1 | 10.0 | 1.2 | 0.0 | 0.0 | 78.4 |
| | MAXIMUM <= 32 | 30 | * | 0.1 | 0.0 | 0.0 | 0.0 | 0.0 | 0.0 | 0.0 | 0.0 | 0.0 | 0.0 | * | 0.1 |
| | MINIMUM <= 32 | 30 | 7.0 | 4.3 | 1.0 | 0.0 | 0.0 | 0.0 | 0.0 | 0.0 | 0.0 | 0.0 | 1.0 | 5.0 | 18.3 |
| | MINIMUM <= 0 | 30 | 0.0 | 0.0 | 0.0 | 0.0 | 0.0 | 0.0 | 0.0 | 0.0 | 0.0 | 0.0 | 0.0 | 0.0 | 0.0 |
| **H/C** | NORMAL HEATING DEG. DAYS | 30 | 374 | 272 | 155 | 55 | 5 | 0 | 0 | 0 | 0 | 30 | 148 | 315 | 1354 |
| | NORMAL COOLING DEG. DAYS | 30 | 15 | 21 | 58 | 116 | 277 | 437 | 530 | 506 | 400 | 182 | 64 | 21 | 2627 |
| **RH** | NORMAL (PERCENT) | 30 | 76 | 74 | 73 | 71 | 74 | 78 | 78 | 81 | 82 | 81 | 80 | 79 | 77 |
| | HOUR 01 LST | 30 | 86 | 85 | 86 | 86 | 88 | 90 | 91 | 92 | 93 | 92 | 91 | 88 | 89 |
| | HOUR 07 LST | 30 | 88 | 88 | 89 | 89 | 90 | 90 | 91 | 93 | 94 | 93 | 92 | 90 | 91 |
| | HOUR 13 LST | 30 | 59 | 55 | 52 | 49 | 53 | 59 | 60 | 62 | 65 | 61 | 59 | 61 | 58 |
| | HOUR 19 LST | 30 | 76 | 71 | 68 | 65 | 68 | 75 | 76 | 80 | 83 | 84 | 84 | 82 | 76 |
| **S** | PERCENT POSSIBLE SUNSHINE | 49 | 59 | 62 | 68 | 73 | 71 | 65 | 65 | 64 | 58 | 60 | 60 | 54 | 63 |
| **W/O** | MEAN NO. DAYS WITH: | | | | | | | | | | | | | | |
| | HEAVY FOG(VISBY <= 1/4 MI) | 46 | 5.3 | 3.5 | 3.2 | 2.5 | 3.0 | 1.6 | 1.3 | 2.1 | 2.2 | 3.4 | 5.2 | 6.2 | 39.5 |
| | THUNDERSTORMS | 62 | 1.1 | 1.8 | 3.1 | 3.4 | 5.8 | 11.8 | 15.9 | 13.5 | 6.9 | 2.2 | 0.9 | 1.0 | 67.4 |
| **CLOUDNESS** | MEAN: | | | | | | | | | | | | | | |
| | SUNRISE-SUNSET (OKTAS) | 47 | 4.9 | 4.7 | 4.6 | 4.2 | 4.4 | 5.0 | 4.9 | 4.9 | 5.1 | 4.5 | 4.3 | 4.8 | 4.7 |
| | MIDNIGHT-MIDNIGHT (OKTAS) | 32 | 4.6 | 4.5 | 4.1 | 3.6 | 4.0 | 4.5 | 4.5 | 4.5 | 4.3 | 4.0 | 4.0 | 4.3 | 4.2 |
| | MEAN NO. DAYS WITH: | | | | | | | | | | | | | | |
| | CLEAR | 48 | 9.0 | 8.6 | 9.0 | 10.3 | 9.0 | 5.1 | 4.7 | 4.9 | 5.3 | 9.5 | 10.0 | 8.7 | 94.1 |
| | PARTLY CLOUDY | 48 | 8.0 | 7.2 | 9.3 | 10.2 | 12.1 | 13.1 | 14.5 | 15.0 | 11.7 | 9.3 | 8.7 | 8.2 | 127.3 |
| | CLOUDY | 48 | 14.1 | 12.4 | 12.7 | 9.5 | 9.9 | 11.7 | 11.8 | 11.1 | 13.0 | 12.2 | 11.4 | 14.1 | 143.9 |
| **PR** | MEAN STATION PRESSURE(IN) | 26 | 30.12 | 30.09 | 30.04 | 30.00 | 29.99 | 29.98 | 30.01 | 29.95 | 29.97 | 30.02 | 30.07 | 30.12 | 30.03 |
| | MEAN SEA-LEVEL PRES. (IN) | 26 | 30.15 | 30.12 | 30.08 | 30.03 | 30.02 | 30.01 | 30.04 | 30.02 | 30.00 | 30.05 | 30.10 | 30.15 | 30.06 |
| **WINDS** | MEAN SPEED (MPH) | 26 | 7.1 | 7.4 | 8.0 | 7.5 | 7.1 | 6.4 | 5.9 | 5.7 | 6.5 | 6.5 | 6.4 | 6.3 | 6.7 |
| | PREVAIL.DIR(TENS OF DEGS) | 42 | 31 | 31 | 25 | 26 | 25 | 25 | 25 | 24 | 06 | 05 | 36 | 36 | 05 |
| | MAXIMUM 2-MINUTE: | | | | | | | | | | | | | | |
| | SPEED (MPH) | 13 | 38 | 34 | 38 | 44 | 44 | 47 | 57 | 46 | 46 | 31 | 35 | 37 | 57 |
| | DIR. (TENS OF DEGS) | | 28 | 14 | 27 | 28 | 08 | 01 | 26 | 05 | 08 | 05 | 29 | 23 | 26 |
| | YEAR OF OCCURRENCE | | 2009 | 1998 | 2009 | 2003 | 2009 | 2008 | 1998 | 2008 | 2004 | 2007 | 2005 | 2004 | JUL 1998 |
| | MAXIMUM 3-SECOND | | | | | | | | | | | | | | |
| | SPEED (MPH) | 13 | 58 | 47 | 49 | 52 | 55 | 60 | 77 | 66 | 62 | 40 | 43 | 62 | 77 |
| | DIR. (TENS OF DEGS) | | 28 | 28 | 27 | 28 | 07 | 01 | 26 | 36 | 10 | 03 | 28 | 33 | 26 |
| | YEAR OF OCCURRENCE | | 2009 | 1998 | 2009 | 2003 | 2009 | 2008 | 1998 | 2008 | 2004 | 2007 | 2005 | 1997 | JUL 1998 |
| **PRECIPITATION** | NORMAL (IN) | 30 | 3.69 | 3.15 | 3.93 | 3.14 | 3.48 | 5.37 | 5.97 | 6.87 | 7.90 | 3.86 | 2.34 | 2.64 | 52.34 |
| | MAXIMUM MONTHLY (IN) | 68 | 10.20 | 11.12 | 10.71 | 11.61 | 13.51 | 17.15 | 16.21 | 16.83 | 19.36 | 13.44 | 7.85 | 9.77 | 19.36 |
| | YEAR OF OCCURRENCE | | 1991 | 1998 | 2003 | 1973 | 2009 | 2004 | 1960 | 2008 | 1949 | 1956 | 1947 | 1997 | SEP 1949 |
| | MINIMUM MONTHLY (IN) | 68 | 0.06 | 0.52 | 0.18 | 0.14 | 0.18 | 1.59 | 1.97 | 1.83 | 1.02 | 0.16 | 0.11 | 0.04 | 0.04 |
| | YEAR OF OCCURRENCE | | 1950 | 1962 | 1945 | 1987 | 1990 | 1990 | 1977 | 2003 | 1961 | 1942 | 1960 | 1956 | DEC 1956 |
| | MAXIMUM IN 24 HOURS (IN) | 68 | 3.02 | 6.22 | 7.12 | 8.25 | 6.98 | 6.15 | 10.09 | 9.14 | 10.17 | 7.90 | 5.44 | 4.79 | 10.17 |
| | YEAR OF OCCURRENCE | | 1963 | 1970 | 1970 | 1973 | 2009 | 2004 | 1966 | 2008 | 1950 | 1992 | 1969 | 2005 | SEP 1950 |
| | NORMAL NO. DAYS WITH: | | | | | | | | | | | | | | |
| | PRECIPITATION >= 0.01 | 30 | 8.7 | 7.7 | 8.2 | 6.4 | 8.1 | 13.1 | 13.6 | 14.5 | 12.5 | 8.2 | 6.9 | 8.0 | 115.9 |
| | PRECIPITATION >= 1.00 | 30 | 1.2 | 0.9 | 1.3 | 1.0 | 0.8 | 1.3 | 2.0 | 2.0 | 2.5 | 1.0 | 0.8 | 0.7 | 15.5 |
| **SNOWFALL** | NORMAL (IN) | 30 | 0.* | 0.* | 0.* | 0.0 | 0.0 | 0.0 | 0.0 | 0.0 | 0.0 | 0.0 | 0.0 | 0.* | 0.0 |
| | MAXIMUM MONTHLY (IN) | 60 | T | 1.5 | 0.5 | T | 0.0 | T | T | 0.0 | 0.0 | 0.0 | 0.0 | 0.8 | 1.5 |
| | YEAR OF OCCURRENCE | | 1985 | 1958 | 1986 | 1997 | | 1996 | 1990 | | | | | 1989 | FEB 1958 |
| | MAXIMUM IN 24 HOURS (IN) | 60 | T | 1.5 | 0.5 | T | 0.0 | T | T | 0.0 | 0.0 | 0.0 | 0.0 | 0.8 | 1.5 |
| | YEAR OF OCCURRENCE' | | 1985 | 1958 | 1986 | 1997 | | 1996 | 1990 | | | | | 1989 | FEB 1958 |
| | MAXIMUM SNOW DEPTH (IN) | 53 | 0 | 0 | 0 | 0 | 0 | 0 | 0 | 0 | 0 | 0 | 0 | 1 | 1 |
| | YEAR OF OCCURRENCE | | | | | | | | | | | | | 1989 | DEC 1989 |
| | NORMAL NO. DAYS WITH: | | | | | | | | | | | | | | |
| | SNOWFALL >= 1.0 | 30 | 0.0 | 0.0 | 0.0 | 0.0 | 0.0 | 0.0 | 0.0 | 0.0 | 0.0 | 0.0 | 0.0 | 0.0 | 0.0 |

## PRECIPITATION (inches) 2009 JACKSONVILLE (KJAX)

| YEAR | JAN | FEB | MAR | APR | MAY | JUN | JUL | AUG | SEP | OCT | NOV | DEC | ANNUAL |
|------|-----|-----|-----|-----|-----|-----|-----|-----|-----|-----|-----|-----|--------|
| 1980 | 2.61 | 1.06 | 6.83 | 3.91 | 3.02 | 4.59 | 5.29 | 3.97 | 3.03 | 2.69 | 2.32 | 0.21 | 39.53 |
| 1981 | 0.92 | 4.53 | 5.41 | 0.32 | 1.48 | 3.31 | 2.46 | 6.47 | 1.22 | 1.35 | 4.92 | 3.38 | 35.77 |
| 1982 | 3.00 | 1.67 | 4.26 | 3.60 | 3.55 | 8.06 | 3.81 | 6.93 | 9.32 | 3.37 | 1.93 | 2.02 | 51.52 |
| 1983 | 7.19 | 4.27 | 8.46 | 4.65 | 1.38 | 6.86 | 6.11 | 4.63 | 4.61 | 4.29 | 3.32 | 6.42 | 62.19 |
| 1984 | 2.13 | 4.67 | 5.77 | 3.14 | 1.46 | 4.76 | 6.01 | 3.78 | 12.28 | 1.53 | 3.30 | 0.13 | 48.96 |
| 1985 | 1.05 | 1.45 | 1.26 | 2.76 | 2.08 | 3.71 | 6.33 | 8.93 | 16.82 | 8.34 | 2.07 | 3.59 | 58.39 |
| 1986 | 4.19 | 4.72 | 5.44 | 0.93 | 2.13 | 2.53 | 3.27 | 9.60 | 1.99 | 1.80 | 2.85 | 4.65 | 44.10 |
| 1987 | 4.09 | 6.47 | 6.27 | 0.14 | 0.75 | 4.18 | 4.40 | 4.48 | 7.13 | 0.30 | 5.02 | 0.16 | 43.39 |
| 1988 | 6.36 | 6.08 | 2.65 | 3.44 | 1.35 | 3.71 | 4.50 | 8.48 | 16.36 | 2.35 | 4.27 | 1.13 | 60.68 |
| 1989 | 1.73 | 1.77 | 2.14 | 2.79 | 1.55 | 3.66 | 8.98 | 9.16 | 14.37 | 1.39 | 0.51 | 3.40 | 51.45 |
| 1990 | 1.84 | 4.07 | 1.59 | 1.34 | 0.18 | 1.59 | 6.53 | 3.81 | 2.60 | 4.54 | 1.17 | 1.94 | 31.20 |
| 1991 | 10.20 | 1.52 | 7.33 | 6.31 | 9.35 | 11.70 | 15.90 | 3.48 | 6.20 | 6.36 | 0.71 | 0.57 | 79.63 |
| 1992 | 5.79 | 2.64 | 4.09 | 5.33 | 5.97 | 7.04 | 3.32 | 10.76 | 7.33 | 8.34 | 1.92 | 0.65 | 63.18 |
| 1993 | 3.86 | 2.89 | 5.98 | 0.85 | 1.60 | 2.52 | 7.54 | 2.96 | 7.60 | 8.84 | 3.58 | 1.90 | 50.12 |
| 1994 | 6.58 | 0.92 | 2.14 | 1.51 | 3.15 | 13.96 | 8.26 | 3.29 | 9.79 | 10.23 | 3.49 | 3.94 | 67.26 |
| 1995 | 1.91 | 2.07 | 3.67 | 1.77 | 1.77 | 5.35 | 9.45 | 9.93 | 5.41 | 3.53 | 3.20 | 2.19 | 50.25 |
| 1996 | 1.11 | 1.11 | 6.83 | 2.85 | 0.72 | 11.41 | 4.20 | 7.83 | 8.49 | 11.46 | 1.39 | 3.23 | 60.63 |
| 1997 | 2.91 | 1.28 | 1.84 | 4.56 | 3.43 | 6.33 | 7.69 | 8.24 | 3.97 | 4.84 | 2.41 | 9.77 | 57.27 |
| 1998 | 3.49 | 11.12 | 2.64 | 4.71 | 0.96 | 2.95 | 7.29 | 10.09 | 7.65 | 3.01 | 2.39 | 0.42 | 56.72 |
| 1999 | 4.49 | 1.70 | 0.40 | 1.92 | 1.02 | 7.75 | 3.56 | 3.51 | 13.00 | 3.24 | 0.83 | 0.88 | 42.30 |
| 2000 | 2.77 | 1.17 | 1.79 | 2.60 | 1.15 | 2.43 | 5.69 | 7.38 | 11.64 | 0.23 | 1.55 | 1.37 | 39.77 |
| 2001 | 0.91 | 0.68 | 5.48 | 0.62 | 2.56 | 5.59 | 8.31 | 3.58 | 16.03 | 0.81 | 1.44 | 3.13 | 49.14 |
| 2002 | 4.48 | 0.82 | 4.38 | 2.41 | 0.47 | 6.24 | 7.80 | 8.14 | 9.31 | 2.58 | 2.68 | 5.41 | 54.72 |
| 2003 | 0.07 | 4.66 | 10.71 | 2.63 | 2.54 | 6.75 | 7.33 | 1.83 | 3.04 | 2.98 | 0.74 | 1.19 | 44.47 |
| 2004 | 1.64 | 4.47 | 1.36 | 2.02 | 1.24 | 17.15 | 8.60 | 9.85 | 16.31 | 1.32 | 2.85 | 2.66 | 69.47 |
| 2005 | 1.90 | 3.56 | 3.67 | 4.54 | 3.51 | 14.79 | 7.37 | 4.43 | 5.76 | 6.49 | 1.05 | 7.38 | 64.45 |
| 2006 | 2.30 | 3.91 | 0.68 | 1.22 | 2.01 | 7.25 | 3.97 | 7.08 | 4.55 | 1.81 | 0.39 | 2.90 | 38.07 |
| 2007 | 2.32 | 2.40 | 2.22 | 1.02 | 1.12 | 6.68 | 9.48 | 3.57 | 5.44 | 8.85 | 0.17 | 2.74 | 46.01 |
| 2008 | 2.63 | 5.22 | 3.50 | 2.34 | 0.66 | 8.21 | 8.73 | 16.83 | 5.84 | 1.62 | 1.01 | 0.59 | 57.18 |
| 2009 | 3.01 | 0.98 | 4.79 | 5.90 | 13.51 | 3.24 | 5.49 | 7.05 | 6.48 | 2.02 | 0.85 | 5.88 | 59.20 |
| POR= 62 YRS | 3.08 | 3.41 | 3.80 | 2.97 | 3.40 | 6.04 | 6.57 | 7.24 | 7.88 | 4.01 | 1.89 | 2.70 | 52.99 |

WBAN : 13889

## AVERAGE TEMPERATURE (°F) 2009 JACKSONVILLE (KJAX)

| YEAR | JAN | FEB | MAR | APR | MAY | JUN | JUL | AUG | SEP | OCT | NOV | DEC | ANNUAL |
|------|-----|-----|-----|-----|-----|-----|-----|-----|-----|-----|-----|-----|--------|
| 1980 | 53.3 | 51.2 | 62.4 | 68.1 | 75.2 | 80.2 | 83.7 | 83.1 | 80.9 | 68.5 | 61.3 | 52.7 | 68.4 |
| 1981 | 46.5 | 55.4 | 59.1 | 70.4 | 72.5 | 83.3 | 84.4 | 80.8 | 76.0 | 69.7 | 59.9 | 53.9 | 67.7 |
| 1982 | 53.9 | 61.4 | 65.4 | 68.2 | 73.1 | 81.6 | 82.6 | 82.1 | 77.6 | 69.0 | 64.5 | 59.2 | 69.9 |
| 1983 | 49.0 | 52.8 | 57.9 | 62.7 | 72.3 | 76.5 | 82.5 | 82.2 | 76.4 | 71.5 | 60.1 | 52.4 | 66.4 |
| 1984 | 50.4 | 56.0 | 60.8 | 66.6 | 73.2 | 77.5 | 79.6 | 80.3 | 75.4 | 72.5 | 58.2 | 61.6 | 67.7 |
| 1985 | 48.2 | 56.4 | 64.0 | 67.3 | 75.5 | 80.7 | 81.1 | 81.2 | 77.7 | 75.3 | 69.5 | 51.9 | 69.1 |
| 1986 | 51.5 | 59.3 | 60.7 | 66.0 | 73.3 | 81.3 | 83.8 | 81.5 | 79.7 | 72.0 | 68.7 | 57.3 | 69.6 |
| 1987 | 51.9 | 54.4 | 60.0 | 64.0 | 73.2 | 80.4 | 82.6 | 83.9 | 79.2 | 63.9 | 62.3 | 57.5 | 67.8 |
| 1988 | 48.7 | 52.3 | 59.8 | 67.8 | 71.4 | 78.4 | 81.6 | 82.8 | 79.4 | 66.8 | 64.7 | 54.4 | 67.3 |
| 1989 | 60.6 | 58.6 | 65.4 | 67.3 | 73.7 | 80.8 | 82.5 | 82.2 | 79.5 | 70.6 | 62.4 | 47.7 | 69.3 |
| 1990 | 58.3 | 63.1 | 65.1 | 66.8 | 75.3 | 80.9 | 83.5 | 82.8 | 79.5 | 72.7 | 63.4 | 60.8 | 71.0 |
| 1991 | 56.3 | 58.6 | 64.8 | 72.3 | 79.3 | 80.0 | 83.8 | 83.1 | 79.3 | 70.9 | 58.9 | 57.6 | 70.4 |
| 1992 | 52.6 | 58.5 | 60.7 | 65.0 | 72.1 | 81.1 | 84.3 | 82.3 | 78.9 | 68.6 | 64.0 | 56.4 | 68.7 |
| 1993 | 59.1 | 53.3 | 60.2 | 63.5 | 72.8 | 80.6 | 84.1 | 81.9 | 79.6 | 69.8 | 61.7 | 51.4 | 68.2 |
| 1994 | 51.5 | 60.1 | 63.3 | 69.5 | 73.6 | 80.8 | 82.2 | 80.8 | 77.0 | 70.8 | 66.4 | 58.4 | 69.5 |
| 1995 | 52.9 | 55.1 | 63.1 | 68.1 | 76.4 | 79.2 | 82.8 | 81.5 | 77.6 | 72.9 | 57.8 | 53.0 | 68.4 |
| 1996 | 52.5 | 56.5 | 57.4 | 65.5 | 75.7 | 78.6 | 82.5 | 79.2 | 77.4 | 69.9 | 59.7 | 55.4 | 67.5 |
| 1997 | 54.5 | 60.1 | 67.6 | 65.5 | 72.3 | 76.9 | 81.9 | 80.4 | 78.6 | 69.1 | 59.4 | 54.7 | 68.4 |
| 1998 | 56.7 | 57.6 | 58.8 | 67.7 | 76.6 | 84.0 | 83.3 | 81.5 | 78.8 | 72.1 | 66.7 | 59.8 | 70.3 |
| 1999 | 56.5 | 58.4 | 58.4 | 70.8 | 72.1 | 77.9 | 82.6 | 82.7 | 76.6 | 70.6 | 62.2 | 54.3 | 68.6 |
| 2000 | 53.8 | 56.3 | 63.7 | 64.5 | 75.5 | 79.0 | 81.4 | 81.2 | 77.7 | 67.5 | 58.9 | 49.7 | 67.4 |
| 2001 | 50.0 | 60.3 | 60.2 | 67.0 | 73.7 | 79.3 | 81.4 | 81.1 | 75.3 | 67.7 | 64.9 | 59.7 | 68.4 |
| 2002 | 55.6 | 55.1 | 63.9 | 71.5 | 73.8 | 79.2 | 81.6 | 80.0 | 80.2 | 73.9 | 59.3 | 52.7 | 68.9 |
| 2003 | 47.6 | 56.0 | 65.3 | 66.7 | 76.5 | 79.0 | 80.6 | 80.4 | 76.4 | 70.4 | 65.3 | 51.3 | 68.0 |
| 2004 | 52.3 | 53.8 | 62.8 | 65.5 | 76.1 | 80.9 | 81.6 | 81.1 | 79.1 | 72.5 | 64.2 | 53.0 | 68.6 |
| 2005 | 56.7 | 56.8 | 59.7 | 63.5 | 71.5 | 79.5 | 83.2 | 83.3 | 79.6 | 70.8 | 63.7 | 52.2 | 68.4 |
| 2006 | 57.2 | 54.3 | 60.7 | 69.9 | 74.6 | 79.3 | 81.3 | 82.9 | 78.7 | 68.7 | 60.0 | 60.4 | 69.0 |
| 2007 | 56.6 | 54.4 | 62.8 | 65.5 | 71.8 | 78.5 | 81.8 | 83.0 | 78.9 | 73.7 | 59.9 | 59.5 | 68.9 |
| 2008 | 52.5 | 57.6 | 60.1 | 66.2 | 73.7 | 79.8 | 80.5 | 81.0 | 77.9 | 68.0 | 57.7 | 59.0 | 67.8 |
| 2009 | 52.9 | 53.0 | 62.0 | 66.6 | 75.3 | 81.7 | 81.3 | 81.3 | 78.2 | 72.9 | 62.4 | 56.3 | 68.7 |
| POR= 62 YRS | 53.9 | 56.5 | 62.1 | 67.9 | 74.8 | 80.0 | 82.3 | 81.8 | 78.6 | 70.5 | 61.9 | 55.6 | 68.8 |

## HEATING DEGREE DAYS (base 65°F) 2009  JACKSONVILLE (KJAX)

| YEAR | JUL | AUG | SEP | OCT | NOV | DEC | JAN | FEB | MAR | APR | MAY | JUN | TOTAL |
|------|-----|-----|-----|-----|-----|-----|-----|-----|-----|-----|-----|-----|-------|
| 1980-81 | 0 | 0 | 0 | 26 | 146 | 379 | 570 | 273 | 202 | 9 | 6 | 0 | 1611 |
| 1981-82 | 0 | 0 | 1 | 13 | 180 | 362 | 360 | 126 | 99 | 42 | 0 | 0 | 1183 |
| 1982-83 | 0 | 0 | 0 | 49 | 95 | 218 | 490 | 336 | 233 | 109 | 1 | 0 | 1531 |
| 1983-84 | 0 | 0 | 0 | 10 | 181 | 412 | 447 | 262 | 172 | 57 | 9 | 1 | 1551 |
| 1984-85 | 0 | 0 | 0 | 5 | 235 | 122 | 530 | 269 | 95 | 47 | 0 | 0 | 1303 |
| 1985-86 | 0 | 0 | 0 | 2 | 36 | 420 | 411 | 185 | 184 | 53 | 7 | 0 | 1298 |
| 1986-87 | 0 | 0 | 0 | 24 | 41 | 249 | 401 | 293 | 189 | 101 | 0 | 0 | 1298 |
| 1987-88 | 0 | 0 | 0 | 80 | 147 | 256 | 507 | 374 | 176 | 50 | 7 | 0 | 1597 |
| 1988-89 | 0 | 0 | 0 | 34 | 79 | 332 | 157 | 231 | 111 | 63 | 5 | 0 | 1012 |
| 1989-90 | 0 | 0 | 0 | 47 | 133 | 531 | 221 | 117 | 70 | 49 | 1 | 0 | 1169 |
| 1990-91 | 0 | 0 | 0 | 29 | 96 | 191 | 285 | 199 | 100 | 10 | 0 | 0 | 910 |
| 1991-92 | 0 | 0 | 0 | 9 | 221 | 261 | 384 | 207 | 165 | 86 | 16 | 0 | 1349 |
| 1992-93 | 0 | 0 | 0 | 19 | 147 | 277 | 209 | 323 | 185 | 95 | 1 | 0 | 1256 |
| 1993-94 | 0 | 0 | 0 | 14 | 150 | 414 | 417 | 174 | 120 | 21 | 3 | 0 | 1313 |
| 1994-95 | 0 | 0 | 0 | 12 | 57 | 232 | 369 | 289 | 103 | 40 | 0 | 0 | 1102 |
| 1995-96 | 0 | 0 | 0 | 22 | 249 | 371 | 393 | 271 | 267 | 79 | 6 | 0 | 1658 |
| 1996-97 | 0 | 0 | 0 | 19 | 186 | 302 | 333 | 176 | 43 | 69 | 7 | 0 | 1135 |
| 1997-98 | 0 | 0 | 0 | 33 | 189 | 324 | 275 | 217 | 228 | 39 | 0 | 0 | 1305 |
| 1998-99 | 0 | 0 | 0 | 11 | 44 | 197 | 272 | 199 | 204 | 40 | 11 | 0 | 978 |
| 1999-00 | 0 | 0 | 0 | 34 | 112 | 331 | 346 | 255 | 81 | 69 | 1 | 0 | 1229 |
| 2000-01 | 0 | 0 | 0 | 29 | 221 | 476 | 466 | 170 | 189 | 52 | 0 | 0 | 1603 |
| 2001-02 | 0 | 0 | 1 | 64 | 63 | 208 | 311 | 283 | 126 | 17 | 9 | 0 | 1082 |
| 2002-03 | 0 | 0 | 0 | 5 | 206 | 376 | 533 | 258 | 74 | 49 | 0 | 0 | 1501 |
| 2003-04 | 0 | 0 | 0 | 1 | 97 | 417 | 402 | 329 | 120 | 77 | 6 | 0 | 1449 |
| 2004-05 | 0 | 0 | 0 | 7 | 93 | 393 | 275 | 239 | 205 | 88 | 2 | 0 | 1302 |
| 2005-06 | 0 | 0 | 0 | 56 | 88 | 389 | 253 | 297 | 176 | 16 | 2 | 0 | 1277 |
| 2006-07 | 0 | 0 | 0 | 53 | 189 | 184 | 287 | 298 | 106 | 83 | 0 | 0 | 1200 |
| 2007-08 | 0 | 0 | 0 | 1 | 164 | 188 | 384 | 226 | 172 | 57 | 5 | 0 | 1197 |
| 2008-09 | 0 | 0 | 0 | 65 | 247 | 213 | 377 | 334 | 144 | 41 | 4 | 0 | 1425 |
| 2009- | 0 | 0 | 0 | 32 | 108 | 305 | | | | | | | |

WBAN : 13889

## COOLING DEGREE DAYS (base 65°F) 2009  JACKSONVILLE (KJAX)

| YEAR | JAN | FEB | MAR | APR | MAY | JUN | JUL | AUG | SEP | OCT | NOV | DEC | TOTAL |
|------|-----|-----|-----|-----|-----|-----|-----|-----|-----|-----|-----|-----|-------|
| 1980 | 1 | 15 | 63 | 122 | 322 | 466 | 585 | 568 | 483 | 143 | 41 | 3 | 2812 |
| 1981 | 0 | 6 | 23 | 177 | 245 | 557 | 608 | 497 | 336 | 164 | 33 | 21 | 2667 |
| 1982 | 22 | 32 | 117 | 144 | 257 | 505 | 552 | 535 | 385 | 179 | 84 | 46 | 2858 |
| 1983 | 0 | 0 | 18 | 46 | 236 | 352 | 549 | 541 | 349 | 220 | 38 | 26 | 2375 |
| 1984 | 0 | 8 | 51 | 111 | 271 | 383 | 459 | 481 | 318 | 245 | 39 | 22 | 2388 |
| 1985 | 19 | 35 | 70 | 122 | 332 | 479 | 503 | 506 | 388 | 330 | 177 | 21 | 2982 |
| 1986 | 0 | 31 | 57 | 89 | 272 | 497 | 587 | 520 | 448 | 245 | 158 | 20 | 2924 |
| 1987 | 1 | 0 | 38 | 75 | 261 | 469 | 551 | 589 | 430 | 54 | 73 | 33 | 2574 |
| 1988 | 7 | 10 | 26 | 142 | 210 | 411 | 523 | 557 | 442 | 95 | 77 | 13 | 2513 |
| 1989 | 27 | 58 | 127 | 141 | 281 | 482 | 549 | 541 | 441 | 229 | 59 | 1 | 2936 |
| 1990 | 21 | 66 | 81 | 109 | 327 | 484 | 581 | 557 | 443 | 276 | 56 | 67 | 3068 |
| 1991 | 22 | 25 | 99 | 236 | 453 | 456 | 590 | 568 | 433 | 201 | 43 | 40 | 3166 |
| 1992 | 5 | 27 | 40 | 93 | 243 | 491 | 605 | 542 | 425 | 137 | 124 | 18 | 2750 |
| 1993 | 37 | 0 | 43 | 57 | 252 | 477 | 600 | 534 | 445 | 169 | 55 | 1 | 2670 |
| 1994 | 3 | 44 | 73 | 158 | 279 | 484 | 539 | 501 | 367 | 198 | 105 | 34 | 2785 |
| 1995 | 2 | 19 | 51 | 138 | 356 | 434 | 557 | 521 | 386 | 274 | 39 | 6 | 2783 |
| 1996 | 12 | 32 | 38 | 101 | 344 | 416 | 549 | 445 | 381 | 179 | 32 | 11 | 2540 |
| 1997 | 15 | 45 | 131 | 90 | 236 | 365 | 531 | 485 | 415 | 164 | 30 | 12 | 2519 |
| 1998 | 25 | 15 | 42 | 125 | 368 | 579 | 572 | 515 | 423 | 238 | 102 | 45 | 3049 |
| 1999 | 16 | 23 | 6 | 221 | 236 | 392 | 552 | 556 | 356 | 213 | 36 | 4 | 2611 |
| 2000 | 4 | 7 | 48 | 60 | 335 | 426 | 515 | 508 | 389 | 110 | 46 | 8 | 2456 |
| 2001 | 5 | 45 | 48 | 119 | 274 | 435 | 514 | 509 | 319 | 154 | 65 | 50 | 2537 |
| 2002 | 29 | 9 | 97 | 217 | 289 | 433 | 523 | 475 | 464 | 289 | 42 | 0 | 2867 |
| 2003 | 0 | 12 | 89 | 108 | 365 | 429 | 491 | 487 | 350 | 176 | 109 | 0 | 2616 |
| 2004 | 15 | 11 | 60 | 100 | 359 | 484 | 521 | 505 | 428 | 246 | 76 | 29 | 2834 |
| 2005 | 25 | 17 | 47 | 50 | 211 | 442 | 571 | 576 | 447 | 242 | 55 | 0 | 2683 |
| 2006 | 19 | 5 | 50 | 171 | 307 | 435 | 511 | 562 | 419 | 172 | 46 | 45 | 2742 |
| 2007 | 33 | 8 | 45 | 106 | 219 | 412 | 526 | 565 | 424 | 281 | 18 | 24 | 2661 |
| 2008 | 3 | 20 | 27 | 100 | 281 | 448 | 487 | 504 | 395 | 165 | 34 | 35 | 2499 |
| 2009 | 9 | 5 | 58 | 94 | 332 | 511 | 510 | 511 | 402 | 287 | 36 | 41 | 2796 |

## SNOWFALL (inches)  2009  JACKSONVILLE (KJAX)

| YEAR | JUL | AUG | SEP | OCT | NOV | DEC | JAN | FEB | MAR | APR | MAY | JUN | TOTAL |
|------|-----|-----|-----|-----|-----|-----|-----|-----|-----|-----|-----|-----|-------|
| 1976-77 | 0.0 | 0.0 | 0.0 | 0.0 | 0.0 | 0.0 | T | T | 0.0 | 0.0 | 0.0 | 0.0 | T |
| 1977-78 | 0.0 | 0.0 | 0.0 | 0.0 | 0.0 | 0.0 | T | T | 0.0 | 0.0 | 0.0 | 0.0 | T |
| 1978-79 | 0.0 | 0.0 | 0.0 | 0.0 | 0.0 | 0.0 | 0.0 | 0.0 | 0.0 | 0.0 | 0.0 | 0.0 | 0.0 |
| 1979-80 | 0.0 | 0.0 | 0.0 | 0.0 | 0.0 | 0.0 | 0.0 | 0.0 | T | 0.0 | 0.0 | 0.0 | T |
| 1980-81 | 0.0 | 0.0 | 0.0 | 0.0 | 0.0 | 0.0 | 0.0 | 0.0 | 0.0 | 0.0 | 0.0 | 0.0 | 0.0 |
| 1981-82 | 0.0 | 0.0 | 0.0 | 0.0 | 0.0 | 0.0 | T | 0.0 | 0.0 | 0.0 | 0.0 | 0.0 | T |
| 1982-83 | 0.0 | 0.0 | 0.0 | 0.0 | 0.0 | 0.0 | 0.0 | 0.0 | 0.0 | 0.0 | 0.0 | 0.0 | 0.0 |
| 1983-84 | 0.0 | 0.0 | 0.0 | 0.0 | 0.0 | 0.0 | 0.0 | 0.0 | 0.0 | 0.0 | 0.0 | 0.0 | 0.0 |
| 1984-85 | 0.0 | 0.0 | 0.0 | 0.0 | 0.0 | 0.0 | T | 0.0 | 0.0 | 0.0 | 0.0 | 0.0 | T |
| 1985-86 | 0.0 | 0.0 | 0.0 | 0.0 | 0.0 | 0.0 | 0.0 | 0.0 | 0.5 | 0.0 | 0.0 | 0.0 | 0.5 |
| 1986-87 | 0.0 | 0.0 | 0.0 | 0.0 | 0.0 | 0.0 | 0.0 | 0.0 | 0.0 | 0.0 | 0.0 | 0.0 | 0.0 |
| 1987-88 | 0.0 | 0.0 | 0.0 | 0.0 | 0.0 | 0.0 | 0.0 | T | 0.0 | 0.0 | 0.0 | 0.0 | T |
| 1988-89 | 0.0 | 0.0 | 0.0 | 0.0 | 0.0 | 0.0 | 0.0 | T | 0.0 | 0.0 | 0.0 | 0.0 | T |
| 1989-90 | 0.0 | 0.0 | 0.0 | 0.0 | 0.0 | 0.8 | 0.0 | 0.0 | 0.0 | 0.0 | 0.0 | 0.0 | 0.8 |
| 1990-91 | T | 0.0 | 0.0 | 0.0 | 0.0 | 0.0 | 0.0 | 0.0 | T | 0.0 | 0.0 | 0.0 | T |
| 1991-92 | 0.0 | 0.0 | 0.0 | 0.0 | 0.0 | 0.0 | 0.0 | 0.0 | T | 0.0 | 0.0 | 0.0 | T |
| 1992-93 | 0.0 | 0.0 | 0.0 | 0.0 | 0.0 | 0.0 | 0.0 | 0.0 | T | 0.0 | 0.0 | 0.0 | T |
| 1993-94 | 0.0 | 0.0 | 0.0 | 0.0 | 0.0 | T | 0.0 | T | 0.0 | 0.0 | 0.0 | 0.0 | T |
| 1994-95 | 0.0 | 0.0 | 0.0 | 0.0 | 0.0 | 0.0 | 0.0 | 0.0 | 0.0 | 0.0 | 0.0 | 0.0 | 0.0 |
| 1995-96 | 0.0 | 0.0 | 0.0 | 0.0 | 0.0 | T | 0.0 | 0.0 | 0.0 | 0.0 | 0.0 | T | T |
| 1996-97 | 0.0 | 0.0 | 0.0 | 0.0 | 0.0 | 0.0 | 0.0 | 0.0 | 0.0 | T | 0.0 | 0.0 | T |
| 1997-98 | 0.0 | 0.0 | 0.0 | 0.0 | 0.0 | 0.0 | 0.0 | 0.0 | 0.0 | 0.0 | 0.0 | 0.0 | 0.0 |
| 1998-99 | 0.0 | 0.0 | 0.0 | 0.0 | 0.0 | 0.0 | 0.0 | 0.0 | 0.0 | 0.0 | 0.0 | 0.0 | 0.0 |
| 1999-00 | 0.0 | 0.0 | 0.0 | 0.0 | 0.0 | 0.0 | 0.0 | 0.0 | 0.0 | 0.0 | 0.0 | 0.0 | 0.0 |
| 2000-01 | 0.0 | 0.0 | 0.0 | 0.0 | 0.0 | 0.0 | 0.0 | 0.0 | 0.0 | 0.0 | 0.0 | 0.0 | 0.0 |
| 2001-02 | 0.0 | 0.0 | 0.0 | 0.0 | 0.0 | 0.0 | | | | | | | |
| 2002-03 | | | | | | | | | | | | | |
| 2003-04 | | | | | | | | | | | | | |
| 2004-05 | | | | | | | | | | | | | |
| 2005- | | | | | | | | | | | | | |
| POR= 54 YRS | T | 0.0 | 0.0 | 0.0 | 0.0 | T | T | T | T | T | 0.0 | T | T |

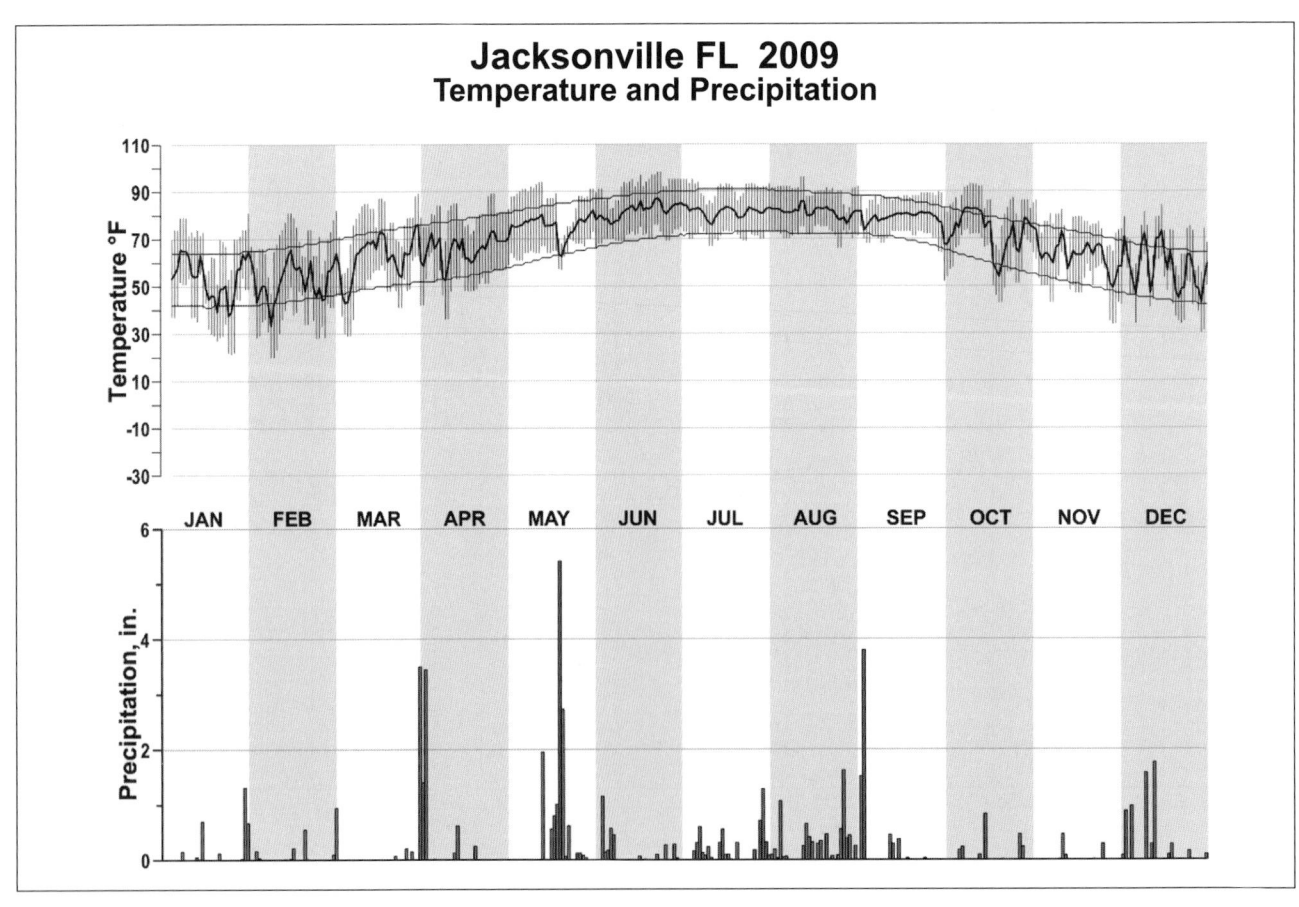

# Jacksonville FL  2009
## Temperature and Precipitation

# 2009
# KEY WEST
# FLORIDA (KEYW)

Key West is located at the end of the Overseas Highway and near the western end of the Florida Keys, which are a chain of islands swinging in a southwesterly arc from the southeast coast of the Florida peninsula. The nearest point of the mainland is about 60 statute miles to the northeast, while Cuba at its closest point is 98 miles south. The city occupies the island of the same name which is 3 1/2 miles long and 1 mile wide. Its mean elevation is around 8 feet. The maximum elevation of 18 feet covers only about one acre in the western portion. Soil is a thin layer of sand, or marlfill, overlying a stratum of Oolitic limestone. Vegetation on the eastern end of the island is scanty, chiefly of low growth. The western end, where settlement and landscaping are older, has a little heavier growth. The airport and Weather Service Office are located on the southeast shore on partially filled mangrove swamp.

The waters surrounding the key are quite shallow up to the mainland on the northeast and for 6 miles to the reef on the south. There is little wave action because the reef disrupts any established wave pattern.

Because of the nearness of the Gulf Stream in the Straits of Florida, about 12 miles south and southeast, and the tempering effects of the Gulf of Mexico to the west and north, Key West has a notably mild, tropical-maritime climate in which the average temperatures during the winter are about 14 degrees lower than in summer. Cold fronts are strongly modified by the warm water as they move in from northerly quadrants in winter. There is no known record of frost, ice, sleet, or snow in Key West. Prevailing easterly tradewinds and sea breezes suppress the usual summertime heating. Diurnal variations throughout the year average only about 10 degrees.

Precipitation is characterized by dry and wet seasons. The period of December through April receives abundant sunshine and slightly less than 25 percent of the annual rainfall. This rainfall usually occurs in advance of cold fronts in a few heavy showers, or occasionally five to eight light showers per month. June through October is normally the wet season, receiving approximately 53 percent of the yearly total in numerous showers and thunderstorms. Early morning is the favored time for diurnal showers. Easterly waves during this season occasionally bring excessive rainfall, while infrequent hurricanes may be accompanied by unusually heavy amounts. Humidity remains relatively high during the entire year.

# NORMALS, MEANS, AND EXTREMES
## KEY WEST (KEYW)

| LATITUDE: | LONGITUDE: | ELEVATION (FT): | TIME ZONE: | WBAN: 12836 |
|---|---|---|---|---|
| 24 ° 33'N | -81 ° 45'W | GRND: 5   BARO: 21 | EASTERN   (UTC -5) | |

| | ELEMENT | POR | JAN | FEB | MAR | APR | MAY | JUN | JUL | AUG | SEP | OCT | NOV | DEC | YEAR |
|---|---|---|---|---|---|---|---|---|---|---|---|---|---|---|---|
| TEMPERATURE °F | NORMAL DAILY MAXIMUM | 30 | 75.3 | 75.9 | 78.8 | 81.9 | 85.4 | 88.1 | 89.4 | 89.5 | 88.2 | 84.7 | 80.6 | 76.7 | 82.9 |
| | MEAN DAILY MAXIMUM | 58 | 74.9 | 75.9 | 78.9 | 81.7 | 85.2 | 87.9 | 89.4 | 89.6 | 88.2 | 84.7 | 80.0 | 76.3 | 82.7 |
| | HIGHEST DAILY MAXIMUM | 57 | 86 | 85 | 88 | 90 | 91 | 94 | 95 | 95 | 95 | 93 | 89 | 86 | 95 |
| | YEAR OF OCCURRENCE | | 1991 | 1991 | 1994 | 1991 | 2008 | 1952 | 1951 | 1957 | 2007 | 1962 | 1988 | 1994 | SEP 2007 |
| | MEAN OF EXTREME MAXS. | 58 | 81.7 | 82.2 | 84.3 | 86.1 | 88.7 | 90.7 | 91.9 | 92.1 | 91.0 | 88.9 | 85.4 | 82.5 | 87.1 |
| | NORMAL DAILY MINIMUM | 30 | 65.2 | 65.7 | 68.8 | 72.1 | 75.9 | 78.7 | 79.6 | 79.2 | 78.5 | 75.7 | 71.9 | 67.3 | 73.2 |
| | MEAN DAILY MINIMUM | 58 | 64.8 | 65.7 | 68.9 | 72.0 | 75.7 | 78.3 | 79.6 | 79.4 | 78.3 | 75.7 | 71.3 | 66.8 | 73.0 |
| | LOWEST DAILY MINIMUM | 57 | 41 | 45 | 47 | 48 | 64 | 68 | 69 | 68 | 69 | 60 | 49 | 44 | 41 |
| | YEAR OF OCCURRENCE | | 1981 | 1996 | 1986 | 1987 | 1992 | 1961 | 1952 | 1952 | 1985 | 1957 | 1959 | 1989 | JAN 1981 |
| | MEAN OF EXTREME MINS. | 58 | 52.4 | 54.3 | 58.4 | 63.5 | 69.7 | 72.6 | 73.8 | 73.3 | 73.3 | 69.3 | 61.6 | 55.3 | 64.8 |
| | NORMAL DRY BULB | 30 | 70.3 | 70.8 | 73.8 | 77.0 | 80.7 | 83.4 | 84.5 | 84.4 | 83.4 | 80.2 | 76.3 | 72.0 | 78.1 |
| | MEAN DRY BULB | 58 | 69.8 | 70.8 | 73.9 | 76.9 | 80.4 | 83.2 | 84.5 | 84.5 | 83.3 | 80.2 | 75.6 | 71.5 | 77.9 |
| | MEAN WET BULB | 26 | 64.4 | 65.5 | 67.1 | 69.1 | 73.1 | 76.6 | 77.2 | 77.5 | 77.0 | 74.0 | 70.1 | 66.5 | 71.5 |
| | MEAN DEW POINT | 26 | 62.2 | 63.2 | 64.8 | 66.6 | 71.0 | 74.7 | 75.4 | 75.6 | 75.4 | 72.2 | 67.9 | 64.4 | 69.5 |
| | NORMAL NO. DAYS WITH: | | | | | | | | | | | | | | |
| | MAXIMUM >= 90 | 30 | 0.0 | 0.0 | 0.0 | 0.1 | 0.8 | 7.4 | 15.2 | 16.0 | 7.6 | 0.5 | 0.0 | 0.0 | 47.6 |
| | MAXIMUM <= 32 | 30 | 0.0 | 0.0 | 0.0 | 0.0 | 0.0 | 0.0 | 0.0 | 0.0 | 0.0 | 0.0 | 0.0 | 0.0 | 0.0 |
| | MINIMUM <= 32 | 30 | 0.0 | 0.0 | 0.0 | 0.0 | 0.0 | 0.0 | 0.0 | 0.0 | 0.0 | 0.0 | 0.0 | 0.0 | 0.0 |
| | MINIMUM <= 0 | 30 | 0.0 | 0.0 | 0.0 | 0.0 | 0.0 | 0.0 | 0.0 | 0.0 | 0.0 | 0.0 | 0.0 | 0.0 | 0.0 |
| H/C | NORMAL HEATING DEG. DAYS | 30 | 24 | 18 | 6 | 0 | 0 | 0 | 0 | 0 | 0 | 0 | 0 | 14 | 62 |
| | NORMAL COOLING DEG. DAYS | 30 | 172 | 183 | 277 | 361 | 485 | 552 | 604 | 600 | 550 | 473 | 340 | 233 | 4830 |
| RH | NORMAL (PERCENT) | 30 | 77 | 75 | 74 | 71 | 73 | 74 | 73 | 74 | 76 | 76 | 77 | 77 | 75 |
| | HOUR 01 LST | 30 | 81 | 79 | 78 | 76 | 78 | 78 | 77 | 78 | 80 | 80 | 81 | 81 | 79 |
| | HOUR 07 LST | 30 | 82 | 81 | 80 | 77 | 77 | 78 | 77 | 79 | 81 | 82 | 83 | 82 | 80 |
| | HOUR 13 LST | 30 | 69 | 67 | 66 | 64 | 66 | 68 | 67 | 68 | 68 | 70 | 70 | 70 | 68 |
| | HOUR 19 LST | 30 | 77 | 75 | 73 | 71 | 72 | 73 | 72 | 73 | 76 | 75 | 77 | 77 | 74 |
| S | PERCENT POSSIBLE SUNSHINE | 37 | 74 | 77 | 82 | 84 | 82 | 76 | 77 | 76 | 72 | 71 | 71 | 71 | 76 |
| W/O | MEAN NO. DAYS WITH: | | | | | | | | | | | | | | |
| | HEAVY FOG(VISBY <= 1/4 MI) | 46 | 0.4 | 0.3 | 0.0 | 0.1 | 0.0 | 0.2 | 0.1 | 0.2 | 0.3 | 0.2 | 0.1 | 0.4 | 2.3 |
| | THUNDERSTORMS | 58 | 1.0 | 1.2 | 1.8 | 1.8 | 3.9 | 9.4 | 12.4 | 13.8 | 11.2 | 4.1 | 1.2 | 0.9 | 62.7 |
| CLOUDNESS | MEAN: | | | | | | | | | | | | | | |
| | SUNRISE-SUNSET (OKTAS) | 44 | 4.1 | 3.8 | 3.7 | 3.6 | 4.1 | 5.0 | 5.0 | 5.0 | 5.2 | 4.4 | 4.0 | 4.0 | 4.3 |
| | MIDNIGHT-MIDNIGHT (OKTAS) | 32 | 3.7 | 3.6 | 3.3 | 3.2 | 3.8 | 4.6 | 4.7 | 4.7 | 5.0 | 4.0 | 3.6 | 3.6 | 4.0 |
| | MEAN NO. DAYS WITH: | | | | | | | | | | | | | | |
| | CLEAR | 44 | 11.2 | 12.0 | 13.1 | 13.2 | 10.0 | 5.0 | 3.4 | 3.3 | 2.7 | 8.7 | 10.5 | 10.9 | 104.0 |
| | PARTLY CLOUDY | 44 | 11.1 | 9.8 | 11.5 | 11.2 | 13.3 | 14.0 | 16.2 | 16.6 | 15.2 | 12.9 | 11.1 | 11.3 | 154.2 |
| | CLOUDY | 44 | 8.7 | 6.5 | 6.5 | 5.6 | 7.7 | 11.0 | 11.4 | 11.1 | 12.1 | 9.4 | 8.4 | 8.8 | 107.2 |
| PR | MEAN STATION PRESSURE(IN) | 26 | 30.09 | 30.07 | 30.03 | 29.99 | 29.97 | 29.98 | 30.02 | 29.98 | 29.93 | 29.95 | 30.01 | 30.07 | 30.01 |
| | MEAN SEA-LEVEL PRES. (IN) | 26 | 30.11 | 30.09 | 30.05 | 30.02 | 29.99 | 30.00 | 30.04 | 30.00 | 29.95 | 29.97 | 30.03 | 30.09 | 30.03 |
| WINDS | MEAN SPEED (MPH) | 26 | 11.1 | 10.9 | 11.3 | 11.1 | 10.0 | 9.1 | 8.7 | 8.5 | 9.1 | 10.5 | 11.6 | 10.9 | 10.2 |
| | PREVAIL.DIR(TENS OF DEGS) | 30 | 05 | 05 | 12 | 12 | 12 | 12 | 12 | 12 | 12 | 06 | 05 | 05 | 12 |
| | MAXIMUM 2-MINUTE: | | | | | | | | | | | | | | |
| | SPEED (MPH) | 13 | 35 | 57 | 35 | 44 | 37 | 40 | 61 | 61 | 62 | 71 | 46 | 40 | 71 |
| | DIR. (TENS OF DEGS) | | 31 | 12 | 33 | 35 | 28 | 18 | 12 | 18 | 12 | 15 | 18 | 23 | 15 |
| | YEAR OF OCCURRENCE | | 2006 | 1998 | 2003 | 2003 | 2005 | 2006 | 2005 | 2005 | 2005 | 2005 | 1998 | 2009 | OCT 2005 |
| | MAXIMUM 3-SECOND | | | | | | | | | | | | | | |
| | SPEED (MPH) | 13 | 47 | 68 | 41 | 55 | 48 | 48 | 74 | 74 | 76 | 83 | 55 | 48 | 83 |
| | DIR. (TENS OF DEGS) | | 30 | 12 | 33 | 34 | 28 | 18 | 12 | 17 | 12 | 15 | 18 | 23 | 15 |
| | YEAR OF OCCURRENCE | | 2006 | 1998 | 2003 | 2003 | 2005 | 2006 | 2005 | 2005 | 2005 | 2005 | 1998 | 2009 | OCT 2005 |
| PRECIPITATION | NORMAL (IN) | 30 | 2.22 | 1.51 | 1.86 | 2.06 | 3.48 | 4.57 | 3.27 | 5.40 | 5.45 | 4.34 | 2.64 | 2.14 | 38.94 |
| | MAXIMUM MONTHLY (IN) | 61 | 17.64 | 4.87 | 9.69 | 12.83 | 12.90 | 14.43 | 11.69 | 13.49 | 18.45 | 21.57 | 27.67 | 11.18 | 27.67 |
| | YEAR OF OCCURRENCE | | 1983 | 1998 | 1987 | 1948 | 1960 | 1972 | 1970 | 2005 | 1963 | 1969 | 1980 | 1986 | NOV 1980 |
| | MINIMUM MONTHLY (IN) | 61 | T | 0.02 | T | 0.00 | 0.12 | 0.33 | 0.44 | 1.20 | 1.70 | 0.74 | T | .05 | 0.00 |
| | YEAR OF OCCURRENCE | | 1990 | 1948 | 1971 | 1959 | 1945 | 1994 | 1993 | 2007 | 1951 | 1972 | 1995 | 2005 | APR 1959 |
| | MAXIMUM IN 24 HOURS (IN) | 61 | 10.32 | 2.85 | 5.31 | 6.55 | 8.89 | 6.17 | 4.25 | 9.67 | 6.65 | 8.47 | 23.28 | 6.99 | 23.28 |
| | YEAR OF OCCURRENCE | | 1983 | 1998 | 1987 | 1985 | 1960 | 1982 | 2005 | 2005 | 1963 | 1971 | 1980 | 1986 | NOV 1980 |
| | NORMAL NO. DAYS WITH: | | | | | | | | | | | | | | |
| | PRECIPITATION >= 0.01 | 30 | 6.5 | 5.6 | 5.0 | 4.7 | 7.6 | 11.0 | 11.7 | 14.5 | 15.6 | 10.4 | 6.5 | 6.5 | 105.6 |
| | PRECIPITATION >= 1.00 | 30 | 0.5 | 0.4 | 0.4 | 0.6 | 1.1 | 1.3 | 0.8 | 1.6 | 1.3 | 1.0 | 0.6 | 0.6 | 10.2 |
| SNOWFALL | NORMAL (IN) | 30 | 0.0 | 0.0 | 0.0 | 0.0 | 0.0 | 0.0 | 0.0 | 0.0 | 0.0 | 0.0 | 0.0 | 0.0 | 0.0 |
| | MAXIMUM MONTHLY (IN) | 0 | 0.0 | 0.0 | 0.0 | 0.0 | 0.0 | 0.0 | 0.0 | 0.0 | 0.0 | 0.0 | 0.0 | 0.0 | 0.0 |
| | YEAR OF OCCURRENCE | | | | | | | | | | | | | | |
| | MAXIMUM IN 24 HOURS (IN) | 48 | 0.0 | 0.0 | 0.0 | 0.0 | 0.0 | 0.0 | 0.0 | 0.0 | 0.0 | 0.0 | 0.0 | 0.0 | 0.0 |
| | YEAR OF OCCURRENCE' | | | | | | | | | | | | | | |
| | MAXIMUM SNOW DEPTH (IN) | 44 | 0 | 0 | 0 | 0 | 0 | 0 | 0 | 0 | 0 | 0 | 0 | 0 | 0 |
| | YEAR OF OCCURRENCE | | | | | | | | | | | | | | |
| | NORMAL NO. DAYS WITH: | | | | | | | | | | | | | | |
| | SNOWFALL >= 1.0 | 30 | 0.0 | 0.0 | 0.0 | 0.0 | 0.0 | 0.0 | 0.0 | 0.0 | 0.0 | 0.0 | 0.0 | 0.0 | 0.0 |

## PRECIPITATION (inches) 2009  KEY WEST (KEYW)

| YEAR | JAN | FEB | MAR | APR | MAY | JUN | JUL | AUG | SEP | OCT | NOV | DEC | ANNUAL |
|------|-----|-----|-----|-----|-----|-----|-----|-----|-----|-----|-----|-----|--------|
| 1980 | 0.98 | 0.92 | 0.83 | 2.84 | 2.14 | 2.58 | 6.65 | 6.38 | 6.01 | 2.95 | 27.67 | 0.47 | 60.42 |
| 1981 | 0.46 | 2.27 | 0.84 | 0.39 | 0.55 | 0.90 | 1.15 | 8.88 | 6.68 | 1.39 | 3.50 | 0.07 | 27.08 |
| 1982 | 0.38 | 1.68 | 3.44 | 1.74 | 5.78 | 8.95 | 1.94 | 2.61 | 6.36 | 2.27 | 1.20 | 0.30 | 36.65 |
| 1983 | 17.64 | 3.48 | 6.57 | 1.88 | 2.24 | 3.82 | 4.64 | 2.79 | 2.25 | 1.20 | 0.91 | 4.97 | 52.39 |
| 1984 | 0.10 | 4.07 | 2.18 | 4.16 | 5.79 | 7.09 | 3.29 | 2.59 | 8.11 | 1.32 | 1.15 | 0.10 | 39.95 |
| 1985 | 0.32 | 0.29 | 2.15 | 10.60 | 4.39 | 0.57 | 3.64 | 5.92 | 6.42 | 3.04 | 2.62 | 2.28 | 42.24 |
| 1986 | 1.26 | 2.02 | 0.99 | 0.93 | 0.91 | 5.37 | 4.48 | 5.66 | 4.58 | 2.52 | 0.92 | 11.18 | 40.82 |
| 1987 | 0.89 | 0.57 | 9.69 | 0.20 | 2.39 | 4.35 | 3.07 | 3.87 | 6.88 | 8.39 | 5.84 | 2.76 | 48.90 |
| 1988 | 4.78 | 0.46 | 1.68 | 0.64 | 8.12 | 3.60 | 5.78 | 6.29 | 3.48 | 0.76 | 0.78 | 0.19 | 36.56 |
| 1989 | 0.44 | 0.33 | 1.05 | 1.37 | 2.91 | 1.80 | 5.90 | 3.30 | 3.15 | 2.70 | 6.54 | 1.67 | 31.16 |
| 1990 | T | 1.05 | 0.75 | 1.37 | 5.54 | 0.94 | 4.09 | 8.01 | 5.29 | 5.14 | 2.93 | 1.32 | 36.43 |
| 1991 | 2.24 | 0.27 | 2.90 | 1.45 | 8.07 | 8.42 | 2.21 | 2.23 | 5.39 | 7.12 | 0.13 | 0.72 | 41.15 |
| 1992 | 0.28 | 1.22 | 2.81 | 1.42 | 1.60 | 11.77 | 1.07 | 3.82 | 7.15 | 2.04 | 3.83 | 0.08 | 37.09 |
| 1993 | 7.95 | 0.33 | 1.43 | 1.78 | 4.41 | 1.71 | 0.44 | 3.81 | 5.75 | 6.01 | 0.22 | 2.22 | 36.06 |
| 1994 | 2.32 | 2.28 | 0.34 | 2.44 | 3.13 | 0.33 | 1.91 | 8.97 | 13.00 | 6.29 | 1.30 | 4.52 | 46.83 |
| 1995 | 4.23 | 0.11 | 1.85 | 2.16 | 0.66 | 7.89 | 2.87 | 2.87 | 4.42 | 8.73 | T | 2.93 | 38.72 |
| 1996 | 0.88 | 0.24 | 1.85 | 2.81 | 2.60 | 3.21 | 3.18 | 6.48 | 6.55 | 11.87 | .12 | 1.44 | 41.23 |
| 1997 | 3.75 | 0.63 | 1.87 | 2.62 | 1.00 | 4.84 | 6.90 | 3.63 | 6.50 | 1.15 | 2.15 | 4.41 | 39.45 |
| 1998 | 2.42 | 4.87 | 3.13 | 1.01 | 0.86 | 0.86 | 1.12 | 3.37 |  | 4.78 | 2.29 | 1.60 |  |
| 1999 | 2.20 | 1.88 | 0.71 | 1.24 | 2.99 | 5.97 | 1.99 | 9.55 | 7.72 | 12.55 | 0.42 | 0.65 | 47.87 |
| 2000 | 0.50 | 0.66 | 1.48 | 2.22 | 1.27 | 3.47 | 3.87 | 8.84 | 4.02 | 6.04 | 0.57 | 1.96 | 34.90 |
| 2001 | 0.31 | 0.11 | 1.96 | 3.18 | 2.40 | 2.39 | 6.82 | 6.12 | 10.01 | 7.27 | 2.82 | 3.56 | 46.95 |
| 2002 | 0.09 | 2.06 | 1.19 | 0.11 | 5.16 | 4.94 | 4.96 | 6.71 | 8.30 | 2.92 | 1.00 | 4.06 | 41.50 |
| 2003 | 0.40 | 1.29 | 3.10 | 5.20 | 2.68 | 6.42 | 1.57 | 5.07 | 5.69 | 2.05 | 3.50 | 1.03 | 38.00 |
| 2004 | 2.50 | 2.71 | 0.47 | 2.69 | 0.83 | 2.33 | 3.50 | 4.25 | 6.09 | 1.72 | 2.27 | 0.75 | 30.11 |
| 2005 | 1.55 | 0.20 | 3.76 | 2.99 | 1.25 | 6.35 | 6.36 | 13.49 | 7.35 | 4.64 | 5.74 | 0.05 | 53.73 |
| 2006 | 0.23 | 0.71 | 0.05 | 0.01 | 4.30 | 6.04 | 6.46 | 4.04 | 6.86 | 2.92 | 3.21 | 4.82 | 39.65 |
| 2007 | 0.65 | 1.39 | 0.63 | 2.16 | 2.82 | 4.21 | 3.22 | 1.20 | 9.31 | 11.25 | 0.70 | 0.80 | 38.34 |
| 2008 | 0.51 | 2.23 | 1.56 | 1.49 | 0.86 | 0.63 | 1.85 | 6.52 | 8.12 | 12.12 | 2.60 | 0.89 | 39.38 |
| 2009 | 0.81 | 0.65 | 0.72 | 0.70 | 3.87 | 2.10 | 2.16 | 4.91 | 5.31 | 1.11 | 6.63 | 4.49 | 33.46 |
| POR=<br>58 YRS | 1.97 | 1.52 | 1.67 | 1.96 | 3.07 | 4.59 | 3.65 | 5.31 | 6.51 | 5.00 | 2.48 | 2.07 | 39.80 |

WBAN : 12836

## AVERAGE TEMPERATURE (°F) 2009  KEY WEST (KEYW)

| YEAR | JAN | FEB | MAR | APR | MAY | JUN | JUL | AUG | SEP | OCT | NOV | DEC | ANNUAL |
|------|-----|-----|-----|-----|-----|-----|-----|-----|-----|-----|-----|-----|--------|
| 1980 | 70.9 | 66.5 | 75.2 | 77.7 | 81.3 | 83.8 | 84.7 | 85.0 | 84.4 | 81.6 | 75.7 | 68.5 | 77.9 |
| 1981 | 61.3 | 69.1 | 71.0 | 78.5 | 81.0 | 85.1 | 85.8 | 84.0 | 83.0 | 81.5 | 73.9 | 70.5 | 77.1 |
| 1982 | 71.3 | 75.9 | 77.1 | 80.8 | 80.1 | 83.1 | 85.5 | 83.9 | 82.3 | 78.9 | 75.4 | 73.4 | 79.0 |
| 1983 | 67.6 | 68.3 | 69.7 | 74.0 | 78.8 | 82.1 | 83.4 | 83.9 | 82.7 | 80.5 | 75.7 | 72.7 | 76.6 |
| 1984 | 68.6 | 70.8 | 72.7 | 76.4 | 80.8 | 81.2 | 82.9 | 83.5 | 81.5 | 80.1 | 74.7 | 74.1 | 77.3 |
| 1985 | 67.0 | 71.8 | 75.3 | 76.3 | 81.4 | 85.1 | 83.3 | 84.0 | 82.2 | 82.1 | 78.2 | 69.0 | 78.0 |
| 1986 | 67.8 | 72.0 | 71.5 | 74.3 | 80.1 | 83.5 | 85.2 | 83.7 | 83.5 | 81.5 | 80.6 | 75.0 | 78.2 |
| 1987 | 68.9 | 71.8 | 73.0 | 71.2 | 80.2 | 84.5 | 85.1 | 85.3 | 84.9 | 77.1 | 75.9 | 71.5 | 77.5 |
| 1988 | 69.1 | 68.4 | 71.9 | 76.6 | 77.9 | 82.5 | 84.2 | 84.2 | 83.6 | 79.8 | 78.5 | 71.4 | 77.3 |
| 1989 | 74.1 | 72.1 | 74.3 | 78.2 | 81.7 | 83.9 | 83.9 | 85.4 | 85.1 | 80.4 | 77.0 | 67.3 | 78.6 |
| 1990 | 73.2 | 75.5 | 75.0 | 76.6 | 81.7 | 84.4 | 84.7 | 85.6 | 83.8 | 80.9 | 75.6 | 74.6 | 79.3 |
| 1991 | 75.5 | 72.9 | 75.7 | 80.5 | 82.2 | 83.8 | 85.7 | 85.5 | 83.8 | 80.2 | 74.4 | 72.9 | 79.4 |
| 1992 | 69.6 | 72.5 | 73.4 | 75.6 | 77.8 | 83.5 | 85.0 | 85.1 | 83.1 | 79.6 | 77.9 | 72.8 | 78.0 |
| 1993 | 74.5 | 70.7 | 71.9 | 74.3 | 79.2 | 83.8 | 85.9 | 85.3 | 83.9 | 81.2 | 77.1 | 70.9 | 78.2 |
| 1994 | 70.3 | 74.6 | 74.6 | 79.2 | 81.6 | 85.3 | 84.4 | 83.5 | 82.6 | 80.5 | 78.0 | 73.8 | 79.0 |
| 1995 | 69.1 | 69.4 | 73.9 | 78.4 | 83.5 | 83.2 | 84.8 | 85.3 | 84.2 | 81.8 | 75.4 | 69.2 | 78.2 |
| 1996 | 68.6 | 67.4 | 69.9 | 75.8 | 80.7 | 82.6 | 84.8 | 83.7 | 83.5 | 79.0 | 75.6 | 72.5 | 77.0 |
| 1997 | 70.9 | 75.1 | 77.5 | 77.5 | 82.0 | 83.2 | 84.0 | 84.4 | 81.9 | 79.0 | 75.6 | 70.6 | 78.5 |
| 1998 | 70.7 | 70.7 | 70.6 | 76.5 | 79.8 | 85.1 | 85.8 | 84.6 |  | 81.1 | 77.3 | 75.0 |  |
| 1999 | 72.2 | 71.4 | 71.9 | 78.4 | 80.4 | 82.8 | 84.1 | 84.0 | 83.4 | 79.8 | 74.8 | 72.1 | 77.9 |
| 2000 | 70.2 | 70.8 | 75.7 | 75.9 | 81.0 | 83.4 | 85.1 | 83.2 | 83.9 | 78.2 | 74.4 | 69.9 | 77.6 |
| 2001 | 64.7 | 75.0 | 74.1 | 77.1 | 78.2 | 83.8 | 83.4 | 84.4 | 81.8 | 79.3 | 74.6 | 73.9 | 77.5 |
| 2002 | 70.9 | 71.0 | 75.4 | 79.1 | 81.5 | 82.4 | 84.3 | 84.4 | 84.0 | 82.4 | 74.7 | 71.3 | 78.5 |
| 2003 | 63.8 | 72.0 | 77.8 | 75.6 | 82.0 | 83.1 | 84.9 | 86.1 | 85.0 | 82.8 | 78.5 | 68.5 | 78.3 |
| 2004 | 68.6 | 71.1 | 73.5 | 74.9 | 79.1 | 84.1 | 84.5 | 84.6 | 82.6 | 81.0 | 77.5 | 70.8 | 77.7 |
| 2005 | 69.1 | 70.3 | 72.0 | 74.5 | 79.5 | 82.8 | 85.0 | 85.9 | 83.3 | 79.9 | 76.0 | 70.3 | 77.4 |
| 2006 | 69.8 | 69.4 | 73.2 | 76.6 | 78.8 | 81.9 | 83.9 | 84.6 | 83.4 | 80.8 | 73.3 | 74.3 | 77.5 |
| 2007 | 74.1 | 70.2 | 74.7 | 76.8 | 80.0 | 83.1 | 86.8 | 87.5 | 85.0 | 83.5 | 78.1 | 74.8 | 79.6 |
| 2008 | 70.5 | 75.1 | 75.1 | 76.6 | 81.4 | 84.5 | 84.5 | 84.8 | 82.4 | 78.9 | 72.3 | 71.3 | 78.1 |
| 2009 | 68.5 | 67.8 | 73.0 | 77.0 | 80.7 | 83.1 | 86.4 | 86.3 | 84.7 | 82.7 | 75.9 | 73.4 | 78.3 |
| POR=<br>58 YRS | 69.8 | 70.8 | 73.9 | 76.9 | 80.4 | 83.2 | 84.5 | 84.5 | 83.3 | 80.2 | 75.6 | 71.5 | 77.9 |

## HEATING DEGREE DAYS (base 65°F) 2009  KEY WEST (KEYW)

| YEAR | JUL | AUG | SEP | OCT | NOV | DEC | JAN | FEB | MAR | APR | MAY | JUN | TOTAL |
|------|-----|-----|-----|-----|-----|-----|-----|-----|-----|-----|-----|-----|-------|
| 1980-81 | 0 | 0 | 0 | 0 | 0 | 26 | 128 | 18 | 1 | 0 | 0 | 0 | 173 |
| 1981-82 | 0 | 0 | 0 | 0 | 0 | 25 | 24 | 0 | 0 | 0 | 0 | 0 | 49 |
| 1982-83 | 0 | 0 | 0 | 0 | 0 | 10 | 24 | 8 | 2 | 0 | 0 | 0 | 44 |
| 1983-84 | 0 | 0 | 0 | 0 | 0 | 36 | 25 | 8 | 12 | 0 | 0 | 0 | 81 |
| 1984-85 | 0 | 0 | 0 | 0 | 0 | 2 | 39 | 16 | 0 | 0 | 0 | 0 | 57 |
| 1985-86 | 0 | 0 | 0 | 0 | 0 | 33 | 31 | 9 | 33 | 0 | 0 | 0 | 106 |
| 1986-87 | 0 | 0 | 0 | 0 | 0 | 0 | 24 | 7 | 2 | 10 | 0 | 0 | 43 |
| 1987-88 | 0 | 0 | 0 | 0 | 0 | 5 | 22 | 20 | 17 | 0 | 0 | 0 | 64 |
| 1988-89 | 0 | 0 | 0 | 0 | 0 | 10 | 0 | 27 | 3 | 0 | 0 | 0 | 40 |
| 1989-90 | 0 | 0 | 0 | 0 | 0 | 61 | 2 | 0 | 0 | 0 | 0 | 0 | 63 |
| 1990-91 | 0 | 0 | 0 | 0 | 0 | 3 | 0 | 8 | 2 | 0 | 0 | 0 | 13 |
| 1991-92 | 0 | 0 | 0 | 0 | 0 | 0 | 12 | 2 | 0 | 0 | 0 | 0 | 14 |
| 1992-93 | 0 | 0 | 0 | 0 | 0 | 3 | 0 | 5 | 12 | 0 | 0 | 0 | 20 |
| 1993-94 | 0 | 0 | 0 | 0 | 0 | 9 | 9 | 5 | 0 | 0 | 0 | 0 | 23 |
| 1994-95 | 0 | 0 | 0 | 0 | 0 | 0 | 15 | 24 | 0 | 0 | 0 | 0 | 39 |
| 1995-96 | 0 | 0 | 0 | 0 | 0 | 37 | 37 | 61 | 29 | 0 | 0 | 0 | 164 |
| 1996-97 | 0 | 0 | 0 | 0 | 0 | 12 | 27 | 2 | 0 | 0 | 0 | 0 | 41 |
| 1997-98 | 0 | 0 | 0 | 0 | 0 | 15 | 12 | 7 | 19 | 0 | 0 | 0 | 53 |
| 1998-99 | 0 | 0 | | 0 | 0 | 0 | 16 | 8 | 2 | 0 | 0 | 0 | |
| 1999-00 | 0 | 0 | 0 | 0 | 0 | 4 | 10 | 11 | 0 | 0 | 0 | 0 | 25 |
| 2000-01 | 0 | 0 | 0 | 0 | 0 | 24 | 86 | 0 | 2 | 0 | 0 | 0 | 112 |
| 2001-02 | 0 | 0 | 0 | 0 | 0 | 1 | 35 | 8 | 3 | 0 | 0 | 0 | 47 |
| 2002-03 | 0 | 0 | 0 | 0 | 8 | 8 | 88 | 6 | 3 | 0 | 0 | 0 | 113 |
| 2003-04 | 0 | 0 | 0 | 0 | 0 | 27 | 22 | 15 | 0 | 0 | 0 | 0 | 64 |
| 2004-05 | 0 | 0 | 0 | 0 | 0 | 12 | 31 | 15 | 4 | 0 | 0 | 0 | 62 |
| 2005-06 | 0 | 0 | 0 | 0 | 0 | 17 | 24 | 24 | 0 | 0 | 0 | 0 | 65 |
| 2006-07 | 0 | 0 | 0 | 0 | 6 | 0 | 4 | 17 | 0 | 0 | 0 | 0 | 27 |
| 2007-08 | 0 | 0 | 0 | 0 | 0 | 0 | 23 | 3 | 1 | 0 | 0 | 0 | 27 |
| 2008-09 | 0 | 0 | 0 | 0 | 2 | 1 | 31 | 26 | 6 | 0 | 0 | 0 | 66 |
| 2009- | 0 | 0 | 0 | 0 | 1 | 8 | | | | | | | |

WBAN : 12836

## COOLING DEGREE DAYS (base 65°F) 2009  KEY WEST (KEYW)

| YEAR | JAN | FEB | MAR | APR | MAY | JUN | JUL | AUG | SEP | OCT | NOV | DEC | TOTAL |
|------|-----|-----|-----|-----|-----|-----|-----|-----|-----|-----|-----|-----|-------|
| 1980 | 200 | 96 | 340 | 387 | 514 | 575 | 620 | 625 | 589 | 524 | 328 | 140 | 4938 |
| 1981 | 21 | 141 | 195 | 411 | 505 | 613 | 651 | 598 | 548 | 521 | 274 | 201 | 4679 |
| 1982 | 224 | 309 | 384 | 484 | 473 | 549 | 641 | 593 | 525 | 437 | 320 | 277 | 5216 |
| 1983 | 110 | 106 | 156 | 278 | 436 | 519 | 576 | 595 | 537 | 487 | 329 | 282 | 4411 |
| 1984 | 147 | 182 | 258 | 350 | 498 | 492 | 558 | 582 | 502 | 474 | 298 | 291 | 4632 |
| 1985 | 110 | 213 | 326 | 349 | 516 | 610 | 576 | 596 | 521 | 534 | 403 | 163 | 4917 |
| 1986 | 123 | 209 | 239 | 286 | 477 | 563 | 634 | 585 | 560 | 521 | 475 | 315 | 4987 |
| 1987 | 151 | 203 | 258 | 202 | 477 | 594 | 629 | 637 | 603 | 380 | 335 | 211 | 4680 |
| 1988 | 155 | 125 | 239 | 356 | 408 | 531 | 604 | 602 | 562 | 465 | 411 | 216 | 4674 |
| 1989 | 289 | 235 | 298 | 401 | 524 | 572 | 593 | 640 | 611 | 482 | 369 | 138 | 5152 |
| 1990 | 263 | 299 | 315 | 356 | 522 | 588 | 620 | 644 | 575 | 499 | 326 | 309 | 5316 |
| 1991 | 334 | 234 | 344 | 472 | 540 | 573 | 650 | 643 | 569 | 482 | 287 | 253 | 5381 |
| 1992 | 164 | 223 | 266 | 323 | 405 | 561 | 625 | 627 | 548 | 462 | 395 | 252 | 4851 |
| 1993 | 298 | 169 | 231 | 282 | 445 | 569 | 656 | 638 | 574 | 508 | 370 | 197 | 4937 |
| 1994 | 181 | 279 | 302 | 431 | 520 | 612 | 607 | 579 | 534 | 489 | 397 | 277 | 5208 |
| 1995 | 147 | 152 | 283 | 409 | 583 | 551 | 618 | 638 | 580 | 526 | 318 | 170 | 4975 |
| 1996 | 158 | 137 | 188 | 333 | 490 | 533 | 621 | 586 | 562 | 439 | 323 | 249 | 4619 |
| 1997 | 217 | 293 | 391 | 385 | 534 | 556 | 597 | 607 | 515 | 441 | 325 | 195 | 5056 |
| 1998 | 195 | 175 | 201 | 351 | 466 | 611 | 652 | 616 | | 508 | 376 | 318 | |
| 1999 | 244 | 195 | 223 | 410 | 484 | 541 | 599 | 595 | 556 | 466 | 300 | 232 | 4845 |
| 2000 | 177 | 186 | 337 | 333 | 503 | 559 | 628 | 571 | 574 | 418 | 289 | 183 | 4758 |
| 2001 | 82 | 285 | 290 | 372 | 419 | 568 | 578 | 610 | 513 | 447 | 292 | 280 | 4736 |
| 2002 | 227 | 182 | 334 | 430 | 516 | 528 | 605 | 612 | 577 | 547 | 307 | 207 | 5072 |
| 2003 | 58 | 211 | 407 | 328 | 535 | 549 | 626 | 660 | 608 | 559 | 410 | 142 | 5093 |
| 2004 | 139 | 200 | 274 | 301 | 440 | 578 | 612 | 617 | 533 | 505 | 379 | 201 | 4779 |
| 2005 | 167 | 169 | 226 | 294 | 459 | 543 | 628 | 654 | 555 | 469 | 335 | 190 | 4689 |
| 2006 | 179 | 152 | 245 | 357 | 438 | 514 | 591 | 612 | 559 | 496 | 264 | 296 | 4703 |
| 2007 | 296 | 167 | 310 | 361 | 472 | 551 | 685 | 703 | 604 | 581 | 400 | 314 | 5444 |
| 2008 | 201 | 300 | 321 | 355 | 516 | 590 | 611 | 622 | 528 | 437 | 229 | 204 | 4914 |
| 2009 | 144 | 111 | 261 | 367 | 492 | 548 | 669 | 668 | 596 | 557 | 335 | 275 | 5023 |

## SNOWFALL (inches) 2009 KEY WEST (KEYW)

| YEAR | JUL | AUG | SEP | OCT | NOV | DEC | JAN | FEB | MAR | APR | MAY | JUN | TOTAL |
|------|-----|-----|-----|-----|-----|-----|-----|-----|-----|-----|-----|-----|-------|
| 1979-80 | 0.0 | 0.0 | 0.0 | 0.0 | 0.0 | 0.0 | 0.0 | 0.0 | 0.0 | 0.0 | 0.0 | 0.0 | 0.0 |
| 1980-81 | 0.0 | 0.0 | 0.0 | 0.0 | 0.0 | 0.0 | 0.0 | 0.0 | 0.0 | 0.0 | 0.0 | 0.0 | 0.0 |
| 1981-82 | 0.0 | 0.0 | 0.0 | 0.0 | 0.0 | 0.0 | 0.0 | 0.0 | 0.0 | 0.0 | 0.0 | 0.0 | 0.0 |
| 1982-83 | 0.0 | 0.0 | 0.0 | 0.0 | 0.0 | 0.0 | 0.0 | 0.0 | 0.0 | 0.0 | 0.0 | 0.0 | 0.0 |
| 1983-84 | 0.0 | 0.0 | 0.0 | 0.0 | 0.0 | 0.0 | 0.0 | 0.0 | 0.0 | 0.0 | 0.0 | 0.0 | 0.0 |
| 1984-85 | 0.0 | 0.0 | 0.0 | 0.0 | 0.0 | 0.0 | 0.0 | 0.0 | 0.0 | 0.0 | 0.0 | 0.0 | 0.0 |
| 1985-86 | 0.0 | 0.0 | 0.0 | 0.0 | 0.0 | 0.0 | 0.0 | 0.0 | 0.0 | 0.0 | 0.0 | 0.0 | 0.0 |
| 1986-87 | 0.0 | 0.0 | 0.0 | 0.0 | 0.0 | 0.0 | 0.0 | 0.0 | 0.0 | 0.0 | 0.0 | 0.0 | 0.0 |
| 1987-88 | 0.0 | 0.0 | 0.0 | 0.0 | 0.0 | 0.0 | 0.0 | 0.0 | 0.0 | 0.0 | 0.0 | 0.0 | 0.0 |
| 1988-89 | 0.0 | 0.0 | 0.0 | 0.0 | 0.0 | 0.0 | 0.0 | 0.0 | 0.0 | 0.0 | 0.0 | 0.0 | 0.0 |
| 1989-90 | 0.0 | 0.0 | 0.0 | 0.0 | 0.0 | 0.0 | 0.0 | 0.0 | 0.0 | 0.0 | 0.0 | 0.0 | 0.0 |
| 1990-91 | 0.0 | 0.0 | 0.0 | 0.0 | 0.0 | 0.0 | 0.0 | 0.0 | 0.0 | 0.0 | 0.0 | 0.0 | 0.0 |
| 1991-92 | 0.0 | 0.0 | 0.0 | 0.0 | 0.0 | 0.0 | 0.0 | 0.0 | 0.0 | 0.0 | 0.0 | 0.0 | 0.0 |
| 1992-93 | 0.0 | 0.0 | 0.0 | 0.0 | 0.0 | 0.0 | 0.0 | 0.0 | 0.0 | 0.0 | 0.0 | 0.0 | 0.0 |
| 1993-94 | 0.0 | 0.0 | 0.0 | 0.0 | 0.0 | 0.0 | 0.0 | 0.0 | 0.0 | 0.0 | 0.0 | 0.0 | 0.0 |
| 1994-95 | 0.0 | 0.0 | 0.0 | 0.0 | 0.0 | 0.0 | 0.0 | 0.0 | 0.0 | 0.0 | 0.0 | 0.0 | 0.0 |
| 1995-96 | 0.0 | 0.0 | 0.0 | 0.0 | 0.0 | 0.0 | 0.0 | 0.0 | 0.0 | 0.0 | 0.0 | 0.0 | 0.0 |
| 1996-97 | | | | | | | | | | | | | |
| 1997-98 | | | | | | | | | | | | | |
| 1998-99 | | | | | | | | | | | | | |
| 1999-00 | | | | | | | | | | | | | |
| 2000-01 | | | | | | | | | | | | | |
| 2001-02 | | | | | | | | | | | | | |
| 2002-03 | | | | | | | | | | | | | |
| 2003-04 | | | | | | | | | | | | | |
| 2004-05 | | | | | | | | | | | | | |
| 2005-06 | | | | | | | | | | | | | |
| 2006-07 | | | | | | | | | | | | | |
| 2007-08 | | | | | | | | | | | | | |
| 2008-09 | | | | | | | | | | | | | |
| POR= 44 YRS | 0.0 | 0.0 | 0.0 | 0.0 | 0.0 | 0.0 | 0.0 | 0.0 | 0.0 | 0.0 | 0.0 | 0.0 | 0.0 |

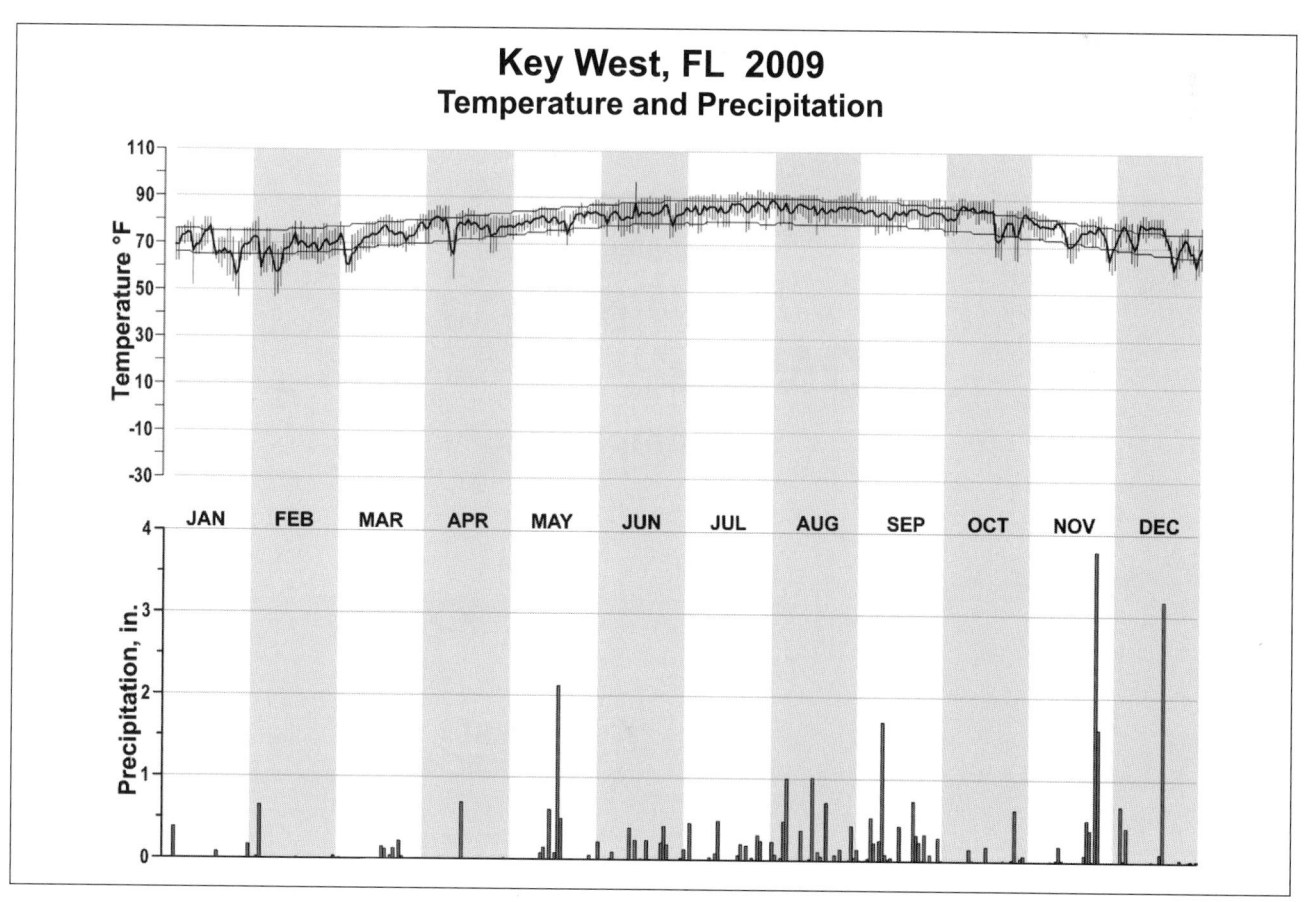

**Key West, FL 2009**
**Temperature and Precipitation**

# 2009
# MIAMI
# FLORIDA (KMIA)

Miami is located on the lower east coast of Florida. To the east of the city lies Biscayne Bay, an arm of the ocean, about 15 miles long and 3 miles wide. East of the bay is the island of Miami Beach, a mile or less wide and about 10 miles long, and beyond Miami Beach is the Atlantic Ocean. The surrounding countryside is level and sparsely wooded.

The climate of Miami is essentially subtropical marine, featured by a long and warm summer, with abundant rainfall, followed by a mild, dry winter. The marine influence is evidenced by the low daily range of temperature and the rapid warming of cold air masses which pass to the east of the state. The Miami area is subject to winds from the east or southeast about half the time, and in several specific respects has a climate whose features differ from those farther inland.

One of these features is the annual precipitation for the area. During the early morning hours more rainfall occurs at Miami Beach than at the airport, while during the afternoon the reverse is true. The airport office is about 9 miles inland.

An even more striking difference appears in the annual number of days with temperatures reaching 90 degrees or higher, with inland stations having about four times more than the beach. Minimum temperature contrasts also are particularly marked under proper conditions, with the difference between inland locations and the Miami Beach station frequently reaching to 15 degrees or more, especially in winter.

Freezing temperatures occur occasionally in the suburbs and farming districts southwest, west, and northwest of the city, but rarely near the ocean.

Hurricanes occasionally affect the area. The months of greatest frequency are September and October. Destructive tornadoes are very rare. Funnel clouds are occasionally sighted and a few touch the ground briefly but significant damage is seldom reported. Waterspouts are often visible from the beaches during the summer months, however, significant damage is seldom reported. June, July, and August have the highest frequency of dangerous lightning events.

# NORMALS, MEANS, AND EXTREMES
## MIAMI (KMIA)

| LATITUDE: 25 ° 49'N | LONGITUDE: -80 ° 17'W | | ELEVATION (FT): GRND: 6   BARO: 29 | | | | | TIME ZONE: EASTERN   (UTC -5) | | | WBAN: 12839 | |

| ELEMENT | POR | JAN | FEB | MAR | APR | MAY | JUN | JUL | AUG | SEP | OCT | NOV | DEC | YEAR |
|---|---|---|---|---|---|---|---|---|---|---|---|---|---|---|
| **TEMPERATURE °F** | | | | | | | | | | | | | | |
| NORMAL DAILY MAXIMUM | 30 | 76.5 | 77.7 | 80.7 | 83.8 | 87.2 | 89.5 | 90.9 | 90.6 | 89.0 | 85.4 | 81.2 | 77.5 | 84.2 |
| MEAN DAILY MAXIMUM | 62 | 75.7 | 77.1 | 79.8 | 82.7 | 85.9 | 88.2 | 89.6 | 89.9 | 88.4 | 85.0 | 80.6 | 77.0 | 83.3 |
| HIGHEST DAILY MAXIMUM | 67 | 88 | 89 | 93 | 96 | 96 | 98 | 98 | 98 | 98 | 97 | 95 | 91 | 89 | 98 |
| YEAR OF OCCURRENCE | | 1987 | 2008 | 2003 | 1971 | 2008 | 2009 | 1998 | 1990 | 1987 | 1980 | 2002 | 2009 | JUN 2009 |
| MEAN OF EXTREME MAXS. | 62 | 83.5 | 85.3 | 87.7 | 90.0 | 91.4 | 92.9 | 93.4 | 93.8 | 92.3 | 89.9 | 86.1 | 83.8 | 89.2 |
| NORMAL DAILY MINIMUM | 30 | 59.6 | 60.5 | 64.0 | 67.6 | 72.0 | 75.2 | 76.5 | 76.5 | 75.7 | 72.2 | 67.5 | 62.2 | 69.1 |
| MEAN DAILY MINIMUM | 62 | 59.7 | 61.0 | 64.5 | 68.0 | 72.1 | 75.0 | 76.5 | 76.7 | 75.9 | 72.4 | 66.7 | 62.0 | 69.2 |
| LOWEST DAILY MINIMUM | 67 | 30 | 32 | 32 | 46 | 53 | 60 | 69 | 68 | 68 | 51 | 39 | 30 | 30 |
| YEAR OF OCCURRENCE | | 1985 | 1947 | 1980 | 1971 | 1945 | 1984 | 2002 | 1950 | 1983 | 1943 | 1950 | 1989 | DEC 1989 |
| MEAN OF EXTREME MINS. | 62 | 42.4 | 45.6 | 49.2 | 56.7 | 64.2 | 70.0 | 72.1 | 72.3 | 71.9 | 63.0 | 52.5 | 45.5 | 58.8 |
| NORMAL DRY BULB | 30 | 68.1 | 69.1 | 72.4 | 75.7 | 79.6 | 82.4 | 83.7 | 83.6 | 82.4 | 78.8 | 74.4 | 69.9 | 76.7 |
| MEAN DRY BULB | 62 | 67.7 | 69.1 | 72.2 | 75.4 | 79.0 | 81.6 | 83.1 | 83.3 | 82.2 | 78.7 | 73.6 | 69.5 | 76.3 |
| MEAN WET BULB | 26 | 61.7 | 63.0 | 64.6 | 66.6 | 71.0 | 74.9 | 76.0 | 76.3 | 75.7 | 72.3 | 67.7 | 64.1 | 69.5 |
| MEAN DEW POINT | 26 | 59.1 | 60.3 | 61.8 | 63.6 | 68.5 | 73.1 | 74.0 | 74.5 | 74.1 | 70.3 | 65.6 | 61.5 | 67.2 |
| NORMAL NO. DAYS WITH: | | | | | | | | | | | | | | |
| MAXIMUM >= 90 | 30 | 0.0 | 0.0 | 0.3 | 1.7 | 4.8 | 10.8 | 18.0 | 16.9 | 10.8 | 2.6 | 0.0 | 0.0 | 65.9 |
| MAXIMUM <= 32 | 30 | 0.0 | 0.0 | 0.0 | 0.0 | 0.0 | 0.0 | 0.0 | 0.0 | 0.0 | 0.0 | 0.0 | 0.0 | 0.0 |
| MINIMUM <= 32 | 30 | 0.1 | 0.0 | * | 0.0 | 0.0 | 0.0 | 0.0 | 0.0 | 0.0 | 0.0 | 0.0 | 0.1 | 0.2 |
| MINIMUM <= 0 | 30 | 0.0 | 0.0 | 0.0 | 0.0 | 0.0 | 0.0 | 0.0 | 0.0 | 0.0 | 0.0 | 0.0 | 0.0 | 0.0 |
| **H/C** | | | | | | | | | | | | | | |
| NORMAL HEATING DEG. DAYS | 30 | 52 | 39 | 15 | 1 | 0 | 0 | 0 | 0 | 0 | 0 | 4 | 38 | 149 |
| NORMAL COOLING DEG. DAYS | 30 | 133 | 154 | 236 | 315 | 442 | 510 | 568 | 568 | 517 | 433 | 291 | 194 | 4361 |
| **RH** | | | | | | | | | | | | | | |
| NORMAL (PERCENT) | 30 | 73 | 71 | 70 | 67 | 71 | 76 | 74 | 76 | 78 | 75 | 74 | 73 | 73 |
| HOUR 01 LST | 30 | 81 | 80 | 78 | 76 | 79 | 83 | 82 | 83 | 84 | 82 | 81 | 80 | 81 |
| HOUR 07 LST | 30 | 85 | 84 | 82 | 79 | 80 | 83 | 83 | 85 | 87 | 85 | 84 | 84 | 83 |
| HOUR 13 LST | 30 | 59 | 57 | 56 | 54 | 58 | 65 | 63 | 65 | 66 | 63 | 63 | 60 | 61 |
| HOUR 19 LST | 30 | 70 | 68 | 66 | 64 | 69 | 74 | 72 | 75 | 77 | 73 | 72 | 71 | 71 |
| **S** | | | | | | | | | | | | | | |
| PERCENT POSSIBLE SUNSHINE | 20 | 66 | 68 | 74 | 76 | 72 | 68 | 72 | 71 | 70 | 70 | 67 | 63 | 70 |
| **W/O** MEAN NO. DAYS WITH: | | | | | | | | | | | | | | |
| HEAVY FOG(VISBY <= 1/4 MI) | 46 | 0.9 | 0.6 | 0.5 | 0.3 | 0.2 | 0.2 | 0.1 | 0.2 | 0.2 | 0.3 | 0.5 | 0.7 | 4.7 |
| THUNDERSTORMS | 62 | 0.8 | 1.1 | 1.8 | 2.8 | 6.4 | 12.5 | 14.6 | 15.4 | 11.4 | 4.2 | 1.0 | 0.6 | 72.6 |
| **CLOUDINESS** MEAN: | | | | | | | | | | | | | | |
| SUNRISE-SUNSET (OKTAS) | 48 | 4.3 | 4.2 | 4.3 | 4.2 | 4.6 | 5.4 | 5.1 | 5.1 | 5.3 | 4.6 | 4.3 | 4.2 | 4.6 |
| MIDNIGHT-MIDNIGHT (OKTAS) | 32 | 3.8 | 3.8 | 3.8 | 3.5 | 4.1 | 4.9 | 4.4 | 4.4 | 4.7 | 4.0 | 3.8 | 3.6 | 4.1 |
| MEAN NO. DAYS WITH: | | | | | | | | | | | | | | |
| CLEAR | 47 | 9.2 | 8.6 | 8.5 | 8.4 | 6.3 | 3.1 | 2.6 | 2.5 | 2.1 | 6.6 | 7.5 | 8.9 | 74.3 |
| PARTLY CLOUDY | 47 | 13.1 | 12.1 | 14.1 | 14.9 | 15.3 | 14.3 | 17.4 | 17.8 | 15.5 | 14.3 | 14.0 | 12.9 | 175.7 |
| CLOUDY | 47 | 8.7 | 7.6 | 8.3 | 6.7 | 9.3 | 12.6 | 11.0 | 10.7 | 12.4 | 10.1 | 8.5 | 9.1 | 115.0 |
| **PR** | | | | | | | | | | | | | | |
| MEAN STATION PRESSURE(IN) | 26 | 30.10 | 30.08 | 30.04 | 30.00 | 29.98 | 29.99 | 30.03 | 29.99 | 29.94 | 29.96 | 30.02 | 30.08 | 30.02 |
| MEAN SEA-LEVEL PRES. (IN) | 26 | 30.12 | 30.10 | 30.06 | 30.03 | 30.00 | 30.01 | 30.06 | 30.01 | 29.96 | 29.98 | 30.04 | 30.10 | 30.04 |
| **WINDS** | | | | | | | | | | | | | | |
| MEAN SPEED (MPH) | 26 | 8.9 | 9.2 | 10.1 | 9.9 | 9.1 | 7.8 | 7.5 | 7.4 | 7.7 | 8.8 | 9.2 | 8.6 | 8.7 |
| PREVAIL.DIR(TENS OF DEGS) | 41 | 35 | 12 | 09 | 11 | 10 | 12 | 12 | 11 | 11 | 06 | 10 | 35 | 12 |
| MAXIMUM 2-MINUTE: | | | | | | | | | | | | | | |
| SPEED (MPH) | 13 | 32 | 55 | 43 | 37 | 43 | 41 | 41 | 60 | 43 | 69 | 36 | 29 | 69 |
| DIR. (TENS OF DEGS) | | 33 | 19 | 26 | 16 | 10 | 14 | 10 | 13 | 10 | 15 | 18 | 22 | 15 |
| YEAR OF OCCURRENCE | | 2009 | 1998 | 2003 | 2008 | 1999 | 2007 | 2005 | 2005 | 1998 | 2005 | 1998 | 1997 | OCT 2005 |
| MAXIMUM 3-SECOND | | | | | | | | | | | | | | |
| SPEED (MPH) | 13 | 40 | 104 | 51 | 52 | 63 | 53 | 56 | 78 | 51 | 92 | 44 | 40 | 104 |
| DIR. (TENS OF DEGS) | | 26 | 19 | 26 | 15 | 33 | 14 | 07 | 12 | 28 | 15 | 31 | 23 | 19 |
| YEAR OF OCCURRENCE | | 2004 | 1998 | 2003 | 2008 | 1998 | 2007 | 2009 | 2005 | 2004 | 2005 | 1998 | 1997 | FEB 1998 |
| **PRECIPITATION** | | | | | | | | | | | | | | |
| NORMAL (IN) | 30 | 1.88 | 2.07 | 2.56 | 3.36 | 5.52 | 8.54 | 5.79 | 8.63 | 8.38 | 6.19 | 3.43 | 2.18 | 58.53 |
| MAXIMUM MONTHLY (IN) | 67 | 6.66 | 8.07 | 10.57 | 17.29 | 18.54 | 22.36 | 13.51 | 16.88 | 24.40 | 21.64 | 13.84 | 6.39 | 24.40 |
| YEAR OF OCCURRENCE | | 1969 | 1983 | 1986 | 1979 | 1968 | 1968 | 1947 | 1943 | 1960 | 1991 | 1992 | 1958 | SEP 1960 |
| MINIMUM MONTHLY (IN) | 67 | 0.04 | 0.01 | 0.02 | 0.05 | 0.44 | 1.81 | 1.77 | 1.65 | 2.63 | 0.72 | 0.09 | 0.12 | 0.01 |
| YEAR OF OCCURRENCE | | 1951 | 1944 | 1956 | 1981 | 1965 | 1945 | 1963 | 1954 | 1951 | 2002 | 1970 | 1988 | FEB 1944 |
| MAXIMUM IN 24 HOURS (IN) | 67 | 2.68 | 5.73 | 7.07 | 16.21 | 11.59 | 8.20 | 4.67 | 6.92 | 7.58 | 12.66 | 8.01 | 5.26 | 16.21 |
| YEAR OF OCCURRENCE | | 1973 | 1966 | 1949 | 1979 | 1977 | 1977 | 2003 | 1964 | 1960 | 2000 | 1992 | 2000 | APR 1979 |
| NORMAL NO. DAYS WITH: | | | | | | | | | | | | | | |
| PRECIPITATION >= 0.01 | 30 | 7.5 | 6.8 | 6.2 | 6.1 | 10.3 | 15.6 | 16.0 | 18.9 | 17.4 | 13.4 | 9.0 | 7.3 | 134.5 |
| PRECIPITATION >= 1.00 | 30 | 0.4 | 0.5 | 0.8 | 0.9 | 1.4 | 2.7 | 1.6 | 2.5 | 2.7 | 1.7 | 0.9 | 0.5 | 16.6 |
| **SNOWFALL** | | | | | | | | | | | | | | |
| NORMAL (IN) | 30 | 0.0 | 0.0 | 0.0 | 0.0 | 0.0 | 0.0 | 0.0 | 0.0 | 0.0 | 0.0 | 0.0 | 0.0 | 0.0 |
| MAXIMUM MONTHLY (IN) | 5 | 0.0 | 0.0 | 0.0 | 0.0 | T | 0.0 | 0.0 | 0.0 | 0.0 | 0.0 | 0.0 | 0.0 | T |
| YEAR OF OCCURRENCE | | | | | | 1998 | | | | | | | | MAY 1998 |
| MAXIMUM IN 24 HOURS (IN) | 59 | 0.0 | 0.0 | 0.0 | 0.0 | T | 0.0 | 0.0 | 0.0 | 0.0 | 0.0 | 0.0 | 0.0 | T |
| YEAR OF OCCURRENCE' | | | | | | 1998 | | | | | | | | MAY 1998 |
| MAXIMUM SNOW DEPTH (IN) | 53 | 0 | 0 | 0 | 0 | 0 | 0 | 0 | 0 | 0 | 0 | 0 | 0 | 0 |
| YEAR OF OCCURRENCE | | | | | | | | | | | | | | |
| NORMAL NO. DAYS WITH: | | | | | | | | | | | | | | |
| SNOWFALL >= 1.0 | 30 | 0.0 | 0.0 | 0.0 | 0.0 | 0.0 | 0.0 | 0.0 | 0.0 | 0.0 | 0.0 | 0.0 | 0.0 | 0.0 |

## PRECIPITATION (inches) 2009 MIAMI (KMIA)

| YEAR | JAN | FEB | MAR | APR | MAY | JUN | JUL | AUG | SEP | OCT | NOV | DEC | ANNUAL |
|------|-----|-----|-----|-----|-----|-----|-----|-----|-----|-----|-----|-----|--------|
| 1980 | 1.89 | 0.88 | 3.17 | 10.20 | 2.14 | 3.02 | 9.40 | 11.32 | 5.60 | 6.05 | 3.47 | 0.20 | 57.34 |
| 1981 | 0.61 | 4.66 | 1.32 | 0.05 | 4.94 | 5.49 | 2.78 | 12.25 | 14.79 | 1.62 | 2.14 | 0.14 | 50.79 |
| 1982 | 0.44 | 1.22 | 4.22 | 9.27 | 8.80 | 10.82 | 3.84 | 5.79 | 7.62 | 7.12 | 7.09 | 1.18 | 67.41 |
| 1983 | 5.36 | 8.07 | 2.82 | 1.79 | 1.44 | 8.66 | 6.20 | 5.88 | 7.48 | 3.52 | 2.01 | 4.19 | 57.42 |
| 1984 | 0.18 | 0.70 | 6.12 | 4.51 | 10.91 | 7.24 | 7.38 | 5.44 | 10.45 | 2.35 | 4.04 | 0.70 | 60.02 |
| 1985 | 0.35 | 0.06 | 1.35 | 3.27 | 3.19 | 6.33 | 11.23 | 11.88 | 8.59 | 5.17 | 1.37 | 3.47 | 56.26 |
| 1986 | 5.04 | 1.72 | 10.57 | 0.71 | 8.24 | 9.06 | 7.81 | 7.67 | 4.38 | 3.96 | 4.75 | 2.21 | 66.12 |
| 1987 | 0.87 | 2.62 | 3.82 | 0.38 | 4.99 | 5.48 | 5.17 | 3.24 | 10.17 | 4.33 | 4.92 | 4.28 | 50.27 |
| 1988 | 1.88 | 0.61 | 0.39 | 1.82 | 5.28 | 10.36 | 10.90 | 7.89 | 3.09 | 1.49 | 0.76 | 0.12 | 44.59 |
| 1989 | 0.67 | 0.71 | 0.89 | 2.14 | 0.99 | 10.83 | 3.53 | 12.78 | 5.83 | 2.65 | 0.99 | 0.62 | 42.63 |
| 1990 | 0.24 | 1.19 | 2.28 | 6.96 | 7.79 | 6.84 | 4.31 | 11.06 | 3.52 | 4.82 | 1.67 | 1.03 | 51.71 |
| 1991 | 1.59 | 2.04 | 2.32 | 5.16 | 2.50 | 7.51 | 7.29 | 8.84 | 11.17 | 21.64 | 1.18 | 0.18 | 71.42 |
| 1992 | 1.80 | 1.49 | 2.67 | 2.43 | 0.55 | 13.17 | 4.21 | 7.22 | 6.48 | 2.02 | 13.84 | 1.94 | 57.82 |
| 1993 | 5.04 | 2.14 | 5.98 | 3.08 | 4.13 | 3.64 | 7.28 | 5.13 | 12.59 | 7.23 | 6.06 | 0.49 | 62.79 |
| 1994 | 3.59 | 5.66 | 1.94 | 2.14 | 4.72 | 4.97 | 3.03 | 16.64 | 13.50 | 9.50 | 8.92 | 4.95 | 79.56 |
| 1995 | 3.13 | 1.41 | 4.60 | 3.73 | 2.94 | 20.33 | 6.36 | 13.13 | 10.37 | 9.91 | 2.53 | 0.86 | 79.30 |
| 1996 | 2.33 | 0.80 | 1.40 | 3.37 | 8.30 | 11.67 | 5.25 | 5.55 | 7.21 | 10.10 | .69 | 1.04 | 57.71 |
| 1997 | 1.71 | 1.57 | 2.06 | 5.16 | 9.80 | 13.18 | 7.62 | 6.28 | 12.47 | 2.60 | 2.89 | 5.27 | 70.61 |
| 1998 | 1.04 | 6.62 | 5.97 | 0.66 | 3.45 | 6.67 | 5.41 | 11.66 | 14.41 | 5.70 | 6.66 | 1.98 | 70.23 |
| 1999 | 2.98 | 0.27 | 0.25 | 1.46 | 4.89 | 11.08 | 3.60 | 13.87 | 7.01 | 14.55 | 1.45 | 2.68 | 64.09 |
| 2000 | 0.52 | 1.24 | 0.35 | 3.36 | 1.80 | 5.19 | 5.29 | 7.42 | 10.58 | 18.65 | 0.50 | 6.15 | 61.05 |
| 2001 | 0.60 | 0.05 | 4.76 | 1.79 | 6.10 | 8.94 | 6.92 | 7.27 | 17.99 | 13.16 | 1.42 | 3.03 | 72.03 |
| 2002 | 0.22 | 3.58 | 0.89 | 1.32 | 8.23 | 15.41 | 12.76 | 6.55 | 6.48 | 0.72 | 3.74 | 3.39 | 63.29 |
| 2003 | 0.43 | 0.83 | 3.89 | 2.87 | 11.05 | 11.94 | 6.32 | 9.65 | 13.18 | 4.09 | 6.64 | 1.24 | 72.13 |
| 2004 | 2.52 | 3.08 | 1.50 | 4.00 | 2.45 | 6.79 | 6.74 | 10.09 | 10.88 | 5.54 | 0.34 | 0.51 | 54.44 |
| 2005 | 1.92 | 0.62 | 3.97 | 3.27 | 7.47 | 17.60 | 5.00 | 9.27 | 9.91 | 5.48 | 2.70 | 1.00 | 68.21 |
| 2006 | 0.32 | 3.47 | 1.10 | 0.23 | 8.62 | 7.05 | 7.32 | 12.95 | 16.73 | 1.64 | 1.63 | 3.11 | 64.17 |
| 2007 | 0.54 | 2.13 | 2.70 | 5.33 | 5.28 | 15.22 | 9.03 | 4.44 | 8.22 | 9.63 | 0.66 | 0.79 | 63.97 |
| 2008 | 1.25 | 4.11 | 5.24 | 3.78 | 1.71 | 9.63 | 8.93 | 9.99 | 7.87 | 6.51 | 0.97 | 0.28 | 60.27 |
| 2009 | 0.34 | 0.12 | 1.78 | 1.17 | 7.53 | 11.64 | 6.17 | 7.91 | 6.83 | 2.62 | 2.97 | 3.01 | 52.09 |
| POR= 62 YRS | 1.85 | 2.06 | 2.45 | 3.19 | 5.95 | 9.30 | 6.22 | 7.98 | 9.00 | 7.03 | 2.93 | 1.96 | 59.92 |

WBAN : 12839

## AVERAGE TEMPERATURE (°F) 2009 MIAMI (KMIA)

| YEAR | JAN | FEB | MAR | APR | MAY | JUN | JUL | AUG | SEP | OCT | NOV | DEC | ANNUAL |
|------|-----|-----|-----|-----|-----|-----|-----|-----|-----|-----|-----|-----|--------|
| 1980 | 67.5 | 64.0 | 73.2 | 75.4 | 79.0 | 81.4 | 82.6 | 82.8 | 82.1 | 80.1 | 74.3 | 67.3 | 75.8 |
| 1981 | 59.7 | 69.5 | 70.1 | 77.8 | 79.6 | 83.7 | 85.0 | 83.2 | 81.2 | 79.7 | 71.4 | 67.8 | 75.7 |
| 1982 | 67.8 | 74.4 | 74.7 | 77.9 | 77.2 | 82.0 | 84.3 | 84.0 | 82.7 | 77.9 | 75.0 | 72.6 | 77.5 |
| 1983 | 67.2 | 67.5 | 67.6 | 71.9 | 78.2 | 81.8 | 85.0 | 83.3 | 81.6 | 78.3 | 72.5 | 69.8 | 75.4 |
| 1984 | 67.0 | 68.6 | 70.4 | 73.2 | 77.1 | 79.8 | 81.9 | 82.6 | 80.1 | 78.2 | 71.5 | 71.1 | 75.1 |
| 1985 | 62.1 | 68.4 | 72.5 | 74.2 | 79.1 | 82.4 | 81.0 | 82.4 | 80.6 | 80.5 | 75.6 | 66.0 | 75.4 |
| 1986 | 65.2 | 69.4 | 68.6 | 71.7 | 77.5 | 81.3 | 83.1 | 83.5 | 83.3 | 80.3 | 79.3 | 73.6 | 76.4 |
| 1987 | 66.1 | 70.8 | 71.9 | 70.6 | 78.7 | 84.2 | 84.2 | 85.4 | 83.6 | 77.6 | 75.3 | 69.8 | 76.5 |
| 1988 | 67.9 | 67.7 | 70.7 | 76.1 | 77.9 | 82.0 | 83.1 | 83.6 | 84.0 | 79.1 | 76.9 | 70.5 | 76.6 |
| 1989 | 72.7 | 70.8 | 73.6 | 77.1 | 81.0 | 82.7 | 83.3 | 84.3 | 84.0 | 79.0 | 76.2 | 65.0 | 77.5 |
| 1990 | 73.6 | 74.0 | 73.7 | 75.2 | 80.3 | 83.0 | 83.5 | 83.7 | 83.1 | 80.4 | 74.4 | 72.9 | 78.2 |
| 1991 | 72.9 | 69.7 | 73.9 | 78.4 | 81.5 | 82.9 | 83.5 | 84.6 | 82.4 | 78.9 | 73.1 | 72.2 | 77.8 |
| 1992 | 67.4 | 70.5 | 71.9 | 74.0 | 77.8 | 81.5 | 84.9 | 84.4 | 83.2 | 79.5 | 76.8 | 71.6 | 77.0 |
| 1993 | 73.2 | 68.9 | 71.5 | 74.0 | 79.2 | 83.3 | 84.6 | 84.8 | 83.0 | 80.8 | 75.9 | 68.9 | 77.3 |
| 1994 | 69.4 | 73.3 | 74.0 | 78.2 | 81.0 | 83.6 | 83.7 | 82.8 | 81.9 | 80.3 | 77.2 | 72.0 | 78.1 |
| 1995 | 67.3 | 67.9 | 73.5 | 77.5 | 82.1 | 81.8 | 84.5 | 84.2 | 83.6 | 81.5 | 73.8 | 68.2 | 77.2 |
| 1996 | 68.1 | 66.7 | 69.7 | 76.0 | 81.2 | 82.5 | 84.5 | 83.1 | 83.2 | 78.4 | 74.3 | 70.2 | 76.5 |
| 1997 | 68.3 | 74.3 | 76.3 | 75.7 | 80.6 | 82.2 | 84.1 | 84.3 | 81.5 | 78.6 | 74.1 | 68.9 | 77.4 |
| 1998 | 70.1 | 69.2 | 69.5 | 76.0 | 80.7 | 85.4 | 84.8 | 84.9 | 83.2 | 80.8 | 76.3 | 73.4 | 77.9 |
| 1999 | 70.0 | 69.6 | 70.5 | 77.8 | 78.6 | 80.9 | 84.0 | 83.6 | 81.9 | 79.2 | 74.3 | 70.0 | 76.7 |
| 2000 | 68.6 | 69.5 | 75.0 | 75.1 | 80.2 | 82.1 | 83.8 | 83.4 | 83.0 | 78.1 | 73.5 | 68.8 | 76.8 |
| 2001 | 63.2 | 74.2 | 73.5 | 76.0 | 77.7 | 82.6 | 82.6 | 84.4 | 83.3 | 79.1 | 74.2 | 73.3 | 76.9 |
| 2002 | 69.8 | 70.0 | 75.3 | 78.5 | 80.8 | 80.3 | 83.0 | 83.9 | 83.3 | 82.1 | 73.8 | 70.6 | 77.6 |
| 2003 | 63.2 | 73.0 | 78.7 | 75.8 | 81.4 | 82.3 | 84.5 | 83.1 | 82.3 | 80.7 | 76.0 | 67.0 | 77.3 |
| 2004 | 66.8 | 70.8 | 72.8 | 73.9 | 79.6 | 84.0 | 84.4 | 83.8 | 82.6 | 79.2 | 75.1 | 68.7 | 76.8 |
| 2005 | 68.0 | 68.6 | 71.3 | 73.6 | 79.2 | 81.3 | 85.2 | 85.2 | 83.2 | 79.4 | 75.5 | 69.4 | 76.7 |
| 2006 | 69.3 | 67.6 | 72.1 | 77.3 | 79.5 | 82.7 | 83.0 | 83.8 | 82.8 | 80.2 | 73.0 | 74.5 | 77.2 |
| 2007 | 72.5 | 69.3 | 73.9 | 75.1 | 78.6 | 81.4 | 83.9 | 85.0 | 83.2 | 81.5 | 74.9 | 74.9 | 77.9 |
| 2008 | 70.3 | 74.3 | 74.5 | 76.8 | 81.4 | 83.3 | 82.9 | 83.9 | 83.5 | 78.3 | 71.8 | 71.7 | 77.7 |
| 2009 | 67.6 | 68.1 | 72.7 | 77.0 | 80.6 | 83.5 | 85.0 | 85.4 | 83.9 | 82.4 | 75.8 | 73.0 | 77.9 |
| POR= 62 YRS | 67.7 | 69.1 | 72.2 | 75.4 | 79.0 | 81.6 | 83.1 | 83.3 | 82.2 | 78.7 | 73.6 | 69.5 | 76.3 |

## HEATING DEGREE DAYS (base 65°F) 2009  MIAMI (KMIA)

| YEAR | JUL | AUG | SEP | OCT | NOV | DEC | JAN | FEB | MAR | APR | MAY | JUN | TOTAL |
|------|-----|-----|-----|-----|-----|-----|-----|-----|-----|-----|-----|-----|-------|
| 1980-81 | 0 | 0 | 0 | 0 | 7 | 59 | 168 | 25 | 12 | 0 | 0 | 0 | 271 |
| 1981-82 | 0 | 0 | 0 | 0 | 1 | 80 | 65 | 1 | 3 | 0 | 0 | 0 | 150 |
| 1982-83 | 0 | 0 | 0 | 0 | 0 | 22 | 50 | 25 | 38 | 2 | 0 | 0 | 137 |
| 1983-84 | 0 | 0 | 0 | 0 | 4 | 69 | 54 | 37 | 17 | 0 | 0 | 0 | 181 |
| 1984-85 | 0 | 0 | 0 | 0 | 9 | 18 | 135 | 61 | 4 | 1 | 0 | 0 | 228 |
| 1985-86 | 0 | 0 | 0 | 0 | 2 | 78 | 76 | 22 | 54 | 0 | 0 | 0 | 232 |
| 1986-87 | 0 | 0 | 0 | 0 | 0 | 0 | 83 | 15 | 6 | 27 | 0 | 0 | 131 |
| 1987-88 | 0 | 0 | 0 | 0 | 3 | 29 | 49 | 38 | 26 | 0 | 0 | 0 | 145 |
| 1988-89 | 0 | 0 | 0 | 0 | 0 | 36 | 1 | 49 | 18 | 0 | 0 | 0 | 104 |
| 1989-90 | 0 | 0 | 0 | 1 | 0 | 110 | 7 | 4 | 0 | 0 | 0 | 0 | 122 |
| 1990-91 | 0 | 0 | 0 | 0 | 0 | 4 | 2 | 31 | 5 | 0 | 0 | 0 | 42 |
| 1991-92 | 0 | 0 | 0 | 0 | 7 | 0 | 38 | 7 | 6 | 0 | 0 | 0 | 58 |
| 1992-93 | 0 | 0 | 0 | 0 | 2 | 10 | 5 | 7 | 21 | 0 | 0 | 0 | 45 |
| 1993-94 | 0 | 0 | 0 | 0 | 4 | 31 | 26 | 15 | 1 | 0 | 0 | 0 | 77 |
| 1994-95 | 0 | 0 | 0 | 0 | 0 | 14 | 39 | 51 | 1 | 0 | 0 | 0 | 105 |
| 1995-96 | 0 | 0 | 0 | 0 | 3 | 77 | 65 | 77 | 41 | 0 | 0 | 0 | 263 |
| 1996-97 | 0 | 0 | 0 | 0 | 0 | 26 | 58 | 2 | 0 | 0 | 0 | 0 | 86 |
| 1997-98 | 0 | 0 | 0 | 0 | 2 | 49 | 20 | 29 | 25 | 0 | 0 | 0 | 125 |
| 1998-99 | 0 | 0 | 0 | 0 | 0 | 8 | 35 | 19 | 5 | 0 | 0 | 0 | 67 |
| 1999-00 | 0 | 0 | 0 | 0 | 0 | 26 | 37 | 19 | 0 | 0 | 0 | 0 | 82 |
| 2000-01 | 0 | 0 | 0 | 0 | 4 | 51 | 121 | 0 | 6 | 0 | 0 | 0 | 182 |
| 2001-02 | 0 | 0 | 0 | 0 | 0 | 11 | 53 | 14 | 4 | 0 | 0 | 0 | 82 |
| 2002-03 | 0 | 0 | 0 | 0 | 5 | 21 | 101 | 2 | 5 | 2 | 0 | 0 | 136 |
| 2003-04 | 0 | 0 | 0 | 0 | 0 | 54 | 34 | 24 | 0 | 0 | 0 | 0 | 112 |
| 2004-05 | 0 | 0 | 0 | 0 | 0 | 41 | 44 | 24 | 16 | 0 | 0 | 0 | 125 |
| 2005-06 | 0 | 0 | 0 | 0 | 2 | 18 | 33 | 45 | 7 | 0 | 0 | 0 | 105 |
| 2006-07 | 0 | 0 | 0 | 0 | 22 | 3 | 12 | 38 | 0 | 1 | 0 | 0 | 76 |
| 2007-08 | 0 | 0 | 0 | 0 | 0 | 4 | 27 | 5 | 3 | 1 | 0 | 0 | 40 |
| 2008-09 | 0 | 0 | 0 | 1 | 3 | 6 | 40 | 36 | 14 | 1 | 0 | 0 | 101 |
| 2009- | 0 | 0 | 0 | 0 | 2 | 16 | | | | | | | |

WBAN : 12839

## COOLING DEGREE DAYS (base 65°F) 2009  MIAMI (KMIA)

| YEAR | JAN | FEB | MAR | APR | MAY | JUN | JUL | AUG | SEP | OCT | NOV | DEC | TOTAL |
|------|-----|-----|-----|-----|-----|-----|-----|-----|-----|-----|-----|-----|-------|
| 1980 | 138 | 75 | 296 | 321 | 441 | 501 | 555 | 563 | 519 | 476 | 292 | 135 | 4312 |
| 1981 | 10 | 154 | 177 | 389 | 460 | 568 | 625 | 570 | 492 | 460 | 198 | 173 | 4276 |
| 1982 | 161 | 270 | 311 | 394 | 385 | 518 | 606 | 596 | 537 | 406 | 304 | 264 | 4752 |
| 1983 | 125 | 101 | 124 | 213 | 417 | 514 | 628 | 576 | 503 | 419 | 236 | 221 | 4077 |
| 1984 | 124 | 144 | 194 | 252 | 380 | 452 | 532 | 554 | 460 | 416 | 213 | 214 | 3935 |
| 1985 | 55 | 164 | 244 | 285 | 445 | 529 | 505 | 546 | 476 | 488 | 329 | 114 | 4180 |
| 1986 | 86 | 150 | 175 | 207 | 395 | 495 | 569 | 582 | 556 | 483 | 432 | 272 | 4402 |
| 1987 | 122 | 186 | 227 | 202 | 430 | 580 | 603 | 639 | 565 | 401 | 314 | 182 | 4451 |
| 1988 | 145 | 123 | 209 | 339 | 408 | 516 | 571 | 584 | 578 | 445 | 364 | 216 | 4498 |
| 1989 | 247 | 219 | 292 | 367 | 502 | 540 | 576 | 603 | 578 | 442 | 346 | 114 | 4826 |
| 1990 | 279 | 262 | 276 | 314 | 479 | 547 | 578 | 587 | 552 | 486 | 287 | 254 | 4901 |
| 1991 | 254 | 167 | 288 | 408 | 515 | 547 | 583 | 614 | 531 | 437 | 255 | 231 | 4830 |
| 1992 | 121 | 173 | 226 | 277 | 404 | 503 | 624 | 609 | 553 | 454 | 366 | 222 | 4532 |
| 1993 | 269 | 123 | 227 | 277 | 449 | 557 | 613 | 622 | 550 | 497 | 338 | 159 | 4681 |
| 1994 | 167 | 252 | 288 | 403 | 503 | 566 | 589 | 557 | 512 | 479 | 374 | 239 | 4929 |
| 1995 | 119 | 138 | 274 | 377 | 537 | 509 | 612 | 602 | 565 | 519 | 272 | 178 | 4702 |
| 1996 | 168 | 134 | 193 | 335 | 512 | 532 | 609 | 566 | 555 | 421 | 284 | 195 | 4504 |
| 1997 | 164 | 270 | 360 | 326 | 489 | 523 | 600 | 607 | 502 | 427 | 282 | 179 | 4729 |
| 1998 | 186 | 152 | 174 | 337 | 493 | 616 | 623 | 621 | 555 | 498 | 348 | 277 | 4880 |
| 1999 | 195 | 152 | 184 | 390 | 429 | 482 | 594 | 583 | 513 | 447 | 285 | 190 | 4444 |
| 2000 | 152 | 154 | 317 | 312 | 480 | 518 | 592 | 579 | 545 | 411 | 267 | 178 | 4505 |
| 2001 | 71 | 266 | 277 | 335 | 400 | 532 | 552 | 605 | 505 | 444 | 283 | 276 | 4546 |
| 2002 | 211 | 160 | 330 | 409 | 497 | 465 | 565 | 591 | 555 | 537 | 275 | 199 | 4794 |
| 2003 | 53 | 233 | 436 | 333 | 516 | 523 | 611 | 569 | 520 | 494 | 338 | 123 | 4749 |
| 2004 | 98 | 197 | 247 | 274 | 464 | 575 | 608 | 589 | 531 | 449 | 311 | 165 | 4508 |
| 2005 | 144 | 131 | 216 | 266 | 446 | 495 | 632 | 635 | 552 | 451 | 323 | 158 | 4449 |
| 2006 | 174 | 125 | 234 | 378 | 457 | 541 | 567 | 591 | 540 | 477 | 269 | 302 | 4655 |
| 2007 | 252 | 162 | 285 | 308 | 430 | 501 | 594 | 626 | 550 | 518 | 302 | 318 | 4846 |
| 2008 | 200 | 283 | 303 | 362 | 514 | 555 | 561 | 591 | 562 | 418 | 213 | 221 | 4783 |
| 2009 | 128 | 128 | 259 | 368 | 489 | 563 | 626 | 638 | 574 | 543 | 331 | 270 | 4917 |

## SNOWFALL (inches) 2009 MIAMI (KMIA)

| YEAR | JUL | AUG | SEP | OCT | NOV | DEC | JAN | FEB | MAR | APR | MAY | JUN | TOTAL |
|---|---|---|---|---|---|---|---|---|---|---|---|---|---|
| 1976-77 | 0.0 | 0.0 | 0.0 | 0.0 | 0.0 | 0.0 | 0.0 | 0.0 | 0.0 | 0.0 | 0.0 | 0.0 | 0.0 |
| 1977-78 | 0.0 | 0.0 | 0.0 | 0.0 | 0.0 | 0.0 | 0.0 | 0.0 | 0.0 | 0.0 | 0.0 | 0.0 | 0.0 |
| 1978-79 | 0.0 | 0.0 | 0.0 | 0.0 | 0.0 | 0.0 | 0.0 | 0.0 | 0.0 | 0.0 | 0.0 | 0.0 | 0.0 |
| 1979-80 | 0.0 | 0.0 | 0.0 | 0.0 | 0.0 | 0.0 | 0.0 | 0.0 | 0.0 | 0.0 | 0.0 | 0.0 | 0.0 |
| 1980-81 | 0.0 | 0.0 | 0.0 | 0.0 | 0.0 | 0.0 | 0.0 | 0.0 | 0.0 | 0.0 | 0.0 | 0.0 | 0.0 |
| 1981-82 | 0.0 | 0.0 | 0.0 | 0.0 | 0.0 | 0.0 | 0.0 | 0.0 | 0.0 | 0.0 | 0.0 | 0.0 | 0.0 |
| 1982-83 | 0.0 | 0.0 | 0.0 | 0.0 | 0.0 | 0.0 | 0.0 | 0.0 | 0.0 | 0.0 | 0.0 | 0.0 | 0.0 |
| 1983-84 | 0.0 | 0.0 | 0.0 | 0.0 | 0.0 | 0.0 | 0.0 | 0.0 | 0.0 | 0.0 | 0.0 | 0.0 | 0.0 |
| 1984-85 | 0.0 | 0.0 | 0.0 | 0.0 | 0.0 | 0.0 | 0.0 | 0.0 | 0.0 | 0.0 | 0.0 | 0.0 | 0.0 |
| 1985-86 | 0.0 | 0.0 | 0.0 | 0.0 | 0.0 | 0.0 | 0.0 | 0.0 | 0.0 | 0.0 | 0.0 | 0.0 | 0.0 |
| 1986-87 | 0.0 | 0.0 | 0.0 | 0.0 | 0.0 | 0.0 | 0.0 | 0.0 | 0.0 | 0.0 | 0.0 | 0.0 | 0.0 |
| 1987-88 | 0.0 | 0.0 | 0.0 | 0.0 | 0.0 | 0.0 | 0.0 | 0.0 | 0.0 | 0.0 | 0.0 | 0.0 | 0.0 |
| 1988-89 | 0.0 | 0.0 | 0.0 | 0.0 | 0.0 | 0.0 | 0.0 | 0.0 | 0.0 | 0.0 | 0.0 | 0.0 | 0.0 |
| 1989-90 | 0.0 | 0.0 | 0.0 | 0.0 | 0.0 | 0.0 | 0.0 | 0.0 | 0.0 | 0.0 | 0.0 | 0.0 | 0.0 |
| 1990-91 | 0.0 | 0.0 | 0.0 | 0.0 | 0.0 | 0.0 | 0.0 | 0.0 | 0.0 | 0.0 | 0.0 | 0.0 | 0.0 |
| 1991-92 | 0.0 | 0.0 | 0.0 | 0.0 | 0.0 | 0.0 | 0.0 | 0.0 | 0.0 | 0.0 | 0.0 | 0.0 | 0.0 |
| 1992-93 | 0.0 | 0.0 | 0.0 | 0.0 | 0.0 | 0.0 | 0.0 | 0.0 | 0.0 | 0.0 | 0.0 | 0.0 | 0.0 |
| 1993-94 | 0.0 | 0.0 | 0.0 | 0.0 | 0.0 | 0.0 | 0.0 | 0.0 | 0.0 | 0.0 | 0.0 | 0.0 | 0.0 |
| 1994-95 | 0.0 | 0.0 | 0.0 | 0.0 | 0.0 | 0.0 | 0.0 | 0.0 | 0.0 | 0.0 | 0.0 | 0.0 | 0.0 |
| 1995-96 | 0.0 | 0.0 | 0.0 | 0.0 | 0.0 | 0.0 | 0.0 | 0.0 | 0.0 | 0.0 | 0.0 | 0.0 | 0.0 |
| 1996-97 | 0.0 | 0.0 | 0.0 | 0.0 | 0.0 | 0.0 | 0.0 | 0.0 | 0.0 | 0.0 | 0.0 | 0.0 | 0.0 |
| 1997-98 | 0.0 | 0.0 | 0.0 | 0.0 | 0.0 | 0.0 | 0.0 | 0.0 | 0.0 | 0.0 | T | 0.0 | T |
| 1998-99 | 0.0 | 0.0 | 0.0 | 0.0 | 0.0 | 0.0 | 0.0 | 0.0 | 0.0 | 0.0 | T | 0.0 | T |
| 1999-00 | 0.0 | 0.0 | 0.0 | 0.0 | 0.0 | 0.0 | 0.0 | 0.0 | 0.0 | 0.0 | T | 0.0 | T |
| 2000-01 | 0.0 | 0.0 | 0.0 | 0.0 | 0.0 | 0.0 | 0.0 | 0.0 | 0.0 | | | | |
| 2001-02 | | | | | | | | | | | | | |
| 2002-03 | | | | | | | | | | | | | |
| 2003-04 | | | | | | | | | | | | | |
| 2004-05 | | | | | | | | | | | | | |
| 2005- | | | | | | | | | | | | | |
| POR= 53 YRS | 0.0 | 0.0 | 0.0 | 0.0 | 0.0 | 0.0 | 0.0 | 0.0 | 0.0 | 0.0 | T | 0.0 | T |

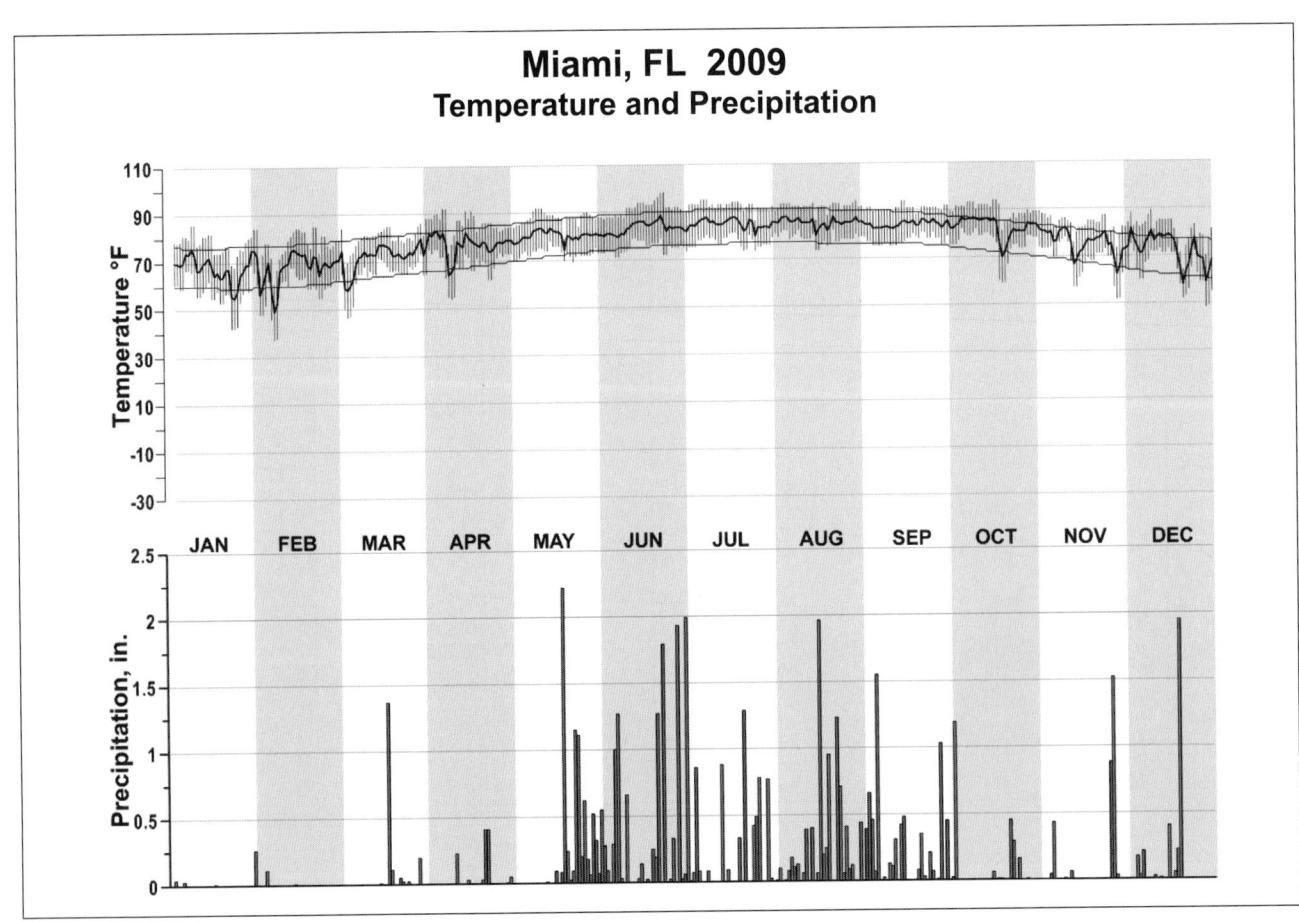

# Miami, FL 2009
## Temperature and Precipitation

# 2009
# ORLANDO
# FLORIDA (KMCO)

Orlando is located in the central section of the Florida peninsula, surrounded by many lakes. Relative humidities remain high the year-round, with values near 90 percent at night and 40 to 50 percent in the afternoon. On some winter days, the humidity may drop to 20 percent.

The rainy season extends from June through September, sometimes through October when tropical storms are near. During this period, scattered afternoon thunderstorms are an almost daily occurrence, and these bring a drop in temperature to make the climate bearable. Summer temperatures above 95 degrees are rather rare. There is usually a breeze which contributes to the general comfort.

During the winter months rainfall is light. While temperatures, on infrequent occasion, may drop at night to near freezing, they rise rapidly during the day and, in brilliant sunshine, afternoons are pleasant.

Frozen precipitation in the form of snowflakes, snow pellets, or sleet is rare. However, hail is occasionally reported during thunderstorms.

Hurricanes are usually not considered a great threat to Orlando, since, to reach this area, they must pass over a substantial stretch of land and, in so doing, lose much of their punch. Sustained hurricane winds of 75 mph or higher rarely occur. Orlando, being inland, is relatively safe from high water, although heavy rains sometimes briefly flood sections of the city.

# NORMALS, MEANS, AND EXTREMES
## ORLANDO (KMCO)

LATITUDE: 28 ° 26'N  LONGITUDE: -81 ° 19'W  ELEVATION (FT): GRND: 90  BARO: 96  TIME ZONE: EASTERN  (UTC -5)  WBAN: 12815

| ELEMENT | POR | JAN | FEB | MAR | APR | MAY | JUN | JUL | AUG | SEP | OCT | NOV | DEC | YEAR |
|---|---|---|---|---|---|---|---|---|---|---|---|---|---|---|
| **TEMPERATURE °F** | | | | | | | | | | | | | | |
| NORMAL DAILY MAXIMUM | 30 | 71.8 | 73.9 | 78.8 | 83.0 | 88.2 | 91.0 | 92.2 | 92.0 | 90.3 | 85.0 | 78.9 | 73.3 | 83.2 |
| MEAN DAILY MAXIMUM | 53 | 70.5 | 73.0 | 77.5 | 82.5 | 87.6 | 90.0 | 91.1 | 90.9 | 89.0 | 83.7 | 77.6 | 72.3 | 82.1 |
| HIGHEST DAILY MAXIMUM | 67 | 87 | 90 | 92 | 96 | 102 | 100 | 100 | 101 | 100 | 98 | 95 | 89 | 90 | 102 |
| YEAR OF OCCURRENCE | | 1991 | 1962 | 1994 | 1968 | 1945 | 1998 | 1998 | 1980 | 1988 | 2009 | 1992 | 1978 | MAY 1945 |
| MEAN OF EXTREME MAXS. | 57 | 82.7 | 84.4 | 87.6 | 90.5 | 94.0 | 95.2 | 95.6 | 95.1 | 93.5 | 90.7 | 85.9 | 83.0 | 89.9 |
| NORMAL DAILY MINIMUM | 30 | 49.9 | 51.3 | 55.9 | 59.9 | 65.9 | 71.3 | 72.6 | 73.0 | 71.9 | 65.5 | 57.9 | 52.6 | 62.4 |
| MEAN DAILY MINIMUM | 53 | 49.0 | 51.4 | 55.8 | 60.4 | 66.4 | 71.5 | 73.3 | 73.7 | 72.5 | 66.0 | 57.9 | 51.8 | 62.5 |
| LOWEST DAILY MINIMUM | 67 | 19 | 26 | 25 | 38 | 48 | 53 | 64 | 64 | 56 | 43 | 29 | 20 | 19 |
| YEAR OF OCCURRENCE | | 1985 | 1996 | 1980 | 1987 | 1992 | 1984 | 1981 | 1957 | 1956 | 2008 | 1950 | 1983 | JAN 1985 |
| MEAN OF EXTREME MINS. | 57 | 31.9 | 35.7 | 40.5 | 48.4 | 57.9 | 66.4 | 69.7 | 70.1 | 67.1 | 52.7 | 42.9 | 35.1 | 51.5 |
| NORMAL DRY BULB | 30 | 60.9 | 62.6 | 67.4 | 71.5 | 77.1 | 81.2 | 82.4 | 82.5 | 81.1 | 75.3 | 68.8 | 63.0 | 72.8 |
| MEAN DRY BULB | 57 | 60.3 | 62.2 | 66.8 | 71.6 | 77.2 | 81.0 | 82.4 | 82.4 | 80.9 | 75.0 | 67.9 | 62.3 | 72.5 |
| MEAN WET BULB | 26 | 54.7 | 56.2 | 59.4 | 62.4 | 68.2 | 73.2 | 74.9 | 75.2 | 73.9 | 68.3 | 62.2 | 57.2 | 65.5 |
| MEAN DEW POINT | 26 | 51.9 | 53.2 | 56.2 | 59.2 | 65.6 | 71.5 | 73.3 | 73.7 | 72.4 | 66.5 | 60.1 | 54.5 | 63.2 |
| NORMAL NO. DAYS WITH: | | | | | | | | | | | | | | |
| MAXIMUM >= 90 | 30 | 0.0 | 0.0 | 0.5 | 3.2 | 11.9 | 19.8 | 25.3 | 25.4 | 19.0 | 3.6 | 0.0 | * | 108.7 |
| MAXIMUM <= 32 | 30 | 0.0 | 0.0 | 0.0 | 0.0 | 0.0 | 0.0 | 0.0 | 0.0 | 0.0 | 0.0 | 0.0 | 0.0 | 0.0 |
| MINIMUM <= 32 | 30 | 1.5 | 0.5 | 0.1 | 0.0 | 0.0 | 0.0 | 0.0 | 0.0 | 0.0 | 0.0 | 0.0 | 0.6 | 2.7 |
| MINIMUM <= 0 | 30 | 0.0 | 0.0 | 0.0 | 0.0 | 0.0 | 0.0 | 0.0 | 0.0 | 0.0 | 0.0 | 0.0 | 0.0 | 0.0 |
| **H/C** | | | | | | | | | | | | | | |
| NORMAL HEATING DEG. DAYS | 30 | 202 | 128 | 57 | 9 | 0 | 0 | 0 | 0 | 0 | 2 | 40 | 142 | 580 |
| NORMAL COOLING DEG. DAYS | 30 | 62 | 60 | 129 | 201 | 373 | 485 | 539 | 543 | 484 | 319 | 154 | 79 | 3428 |
| **RH** | | | | | | | | | | | | | | |
| NORMAL (PERCENT) | 30 | 75 | 72 | 71 | 69 | 72 | 77 | 79 | 80 | 80 | 77 | 77 | 77 | 76 |
| HOUR 01 LST | 30 | 86 | 85 | 85 | 85 | 87 | 90 | 91 | 92 | 92 | 89 | 89 | 88 | 88 |
| HOUR 07 LST | 30 | 89 | 89 | 90 | 88 | 89 | 91 | 92 | 93 | 93 | 91 | 91 | 90 | 91 |
| HOUR 13 LST | 30 | 57 | 53 | 51 | 48 | 50 | 58 | 59 | 60 | 61 | 57 | 57 | 59 | 56 |
| HOUR 19 LST | 30 | 70 | 64 | 62 | 60 | 64 | 73 | 76 | 78 | 79 | 76 | 75 | 74 | 71 |
| **S** PERCENT POSSIBLE SUNSHINE | | | | | | | | | | | | | | |
| **W/O** MEAN NO. DAYS WITH: | | | | | | | | | | | | | | |
| HEAVY FOG(VISBY <= 1/4 MI) | 42 | 3.2 | 2.6 | 1.5 | 1.0 | 1.0 | 0.7 | 0.4 | 0.6 | 0.8 | 1.0 | 1.8 | 3.2 | 17.8 |
| THUNDERSTORMS | 53 | 1.0 | 1.7 | 2.9 | 3.1 | 7.0 | 14.8 | 18.2 | 17.6 | 9.7 | 2.4 | 1.0 | 0.9 | 80.3 |
| **CLOUDNESS** MEAN: | | | | | | | | | | | | | | |
| SUNRISE-SUNSET (OKTAS) | 48 | 4.6 | 4.6 | 4.5 | 4.2 | 4.3 | 5.2 | 5.1 | 5.1 | 5.1 | 4.4 | 4.2 | 4.5 | 4.7 |
| MIDNIGHT-MIDNIGHT (OKTAS) | 22 | 4.4 | 4.3 | 4.2 | 3.6 | 4.1 | 4.9 | 4.6 | 4.6 | 4.5 | 3.9 | 4.1 | 4.3 | 4.3 |
| MEAN NO. DAYS WITH: | | | | | | | | | | | | | | |
| CLEAR | 48 | 8.9 | 8.6 | 9.1 | 10.0 | 8.7 | 4.1 | 3.3 | 3.2 | 3.7 | 9.4 | 10.1 | 9.6 | 88.7 |
| PARTLY CLOUDY | 48 | 10.2 | 8.5 | 10.5 | 11.4 | 13.6 | 14.1 | 16.7 | 16.7 | 14.6 | 11.1 | 10.2 | 9.2 | 146.8 |
| CLOUDY | 48 | 11.9 | 11.1 | 11.5 | 8.6 | 8.7 | 11.8 | 11.1 | 11.1 | 11.7 | 10.6 | 9.7 | 12.2 | 130.0 |
| **PR** MEAN STATION PRESSURE(IN) | 26 | 30.03 | 30.00 | 29.96 | 29.92 | 29.90 | 29.91 | 29.95 | 29.91 | 29.88 | 29.91 | 29.97 | 30.02 | 29.95 |
| MEAN SEA-LEVEL PRES. (IN) | 26 | 30.14 | 30.12 | 30.07 | 30.04 | 30.01 | 30.02 | 30.06 | 30.02 | 29.99 | 30.02 | 30.09 | 30.13 | 30.06 |
| **WINDS** MEAN SPEED (MPH) | 26 | 8.3 | 8.6 | 9.3 | 8.8 | 8.2 | 7.1 | 6.5 | 6.4 | 7.2 | 7.9 | 8.1 | 7.8 | 7.9 |
| PREVAIL.DIR(TENS OF DEGS) | 44 | 36 | 36 | 10 | 11 | 09 | 09 | 19 | 10 | 06 | 36 | 36 | 36 | 36 |
| MAXIMUM 2-MINUTE: | 13 | | | | | | | | | | | | | |
| SPEED (MPH) | | 36 | 38 | 36 | 40 | 51 | 49 | 49 | 79 | 61 | 41 | 37 | 35 | 79 |
| DIR. (TENS OF DEGS) | | 27 | 18 | 23 | 28 | 35 | 35 | 28 | 12 | 06 | 01 | 23 | 20 | 12 |
| YEAR OF OCCURRENCE | | 2008 | 1998 | 2000 | 2006 | 1997 | 2009 | 2007 | 2004 | 2004 | 1999 | 2008 | 2002 | AUG 2004 |
| MAXIMUM 3-SECOND | 13 | | | | | | | | | | | | | |
| SPEED (MPH) | | 49 | 47 | 54 | 49 | 70 | 66 | 61 | 105 | 78 | 48 | 48 | 48 | 105 |
| DIR. (TENS OF DEGS) | | 25 | 20 | 23 | 20 | 34 | 20 | 27 | 12 | 04 | 36 | 23 | 20 | 12 |
| YEAR OF OCCURRENCE | | 2008 | 1998 | 1998 | 2000 | 1997 | 2005 | 2007 | 2004 | 2004 | 2005 | 2008 | 2002 | AUG 2004 |
| **PRECIPITATION** NORMAL (IN) | 30 | 2.43 | 2.35 | 3.54 | 2.42 | 3.74 | 7.35 | 7.15 | 6.25 | 5.76 | 2.73 | 2.32 | 2.31 | 48.35 |
| MAXIMUM MONTHLY (IN) | 67 | 7.23 | 8.74 | 11.38 | 9.10 | 14.56 | 18.28 | 19.57 | 16.11 | 15.87 | 14.51 | 10.29 | 12.63 | 19.57 |
| YEAR OF OCCURRENCE | | 1986 | 1998 | 1987 | 1992 | 2009 | 1968 | 1960 | 1972 | 1945 | 1950 | 1987 | 1997 | JUL 1960 |
| MINIMUM MONTHLY (IN) | 67 | 0.15 | 0.10 | 0.02 | 0.14 | 0.43 | 1.58 | 2.60 | 2.83 | 0.43 | 0.35 | 0.03 | T | T |
| YEAR OF OCCURRENCE | | 1950 | 1944 | 2006 | 1977 | 1961 | 1998 | 1992 | 2001 | 1972 | 1967 | 1967 | 1944 | DEC 1944 |
| MAXIMUM IN 24 HOURS (IN) | 67 | 4.19 | 4.38 | 5.03 | 5.65 | 3.95 | 8.40 | 8.19 | 5.29 | 9.67 | 7.74 | 5.87 | 3.61 | 9.67 |
| YEAR OF OCCURRENCE | | 1986 | 1970 | 1960 | 1992 | 2009 | 1945 | 1960 | 1949 | 1945 | 1950 | 1988 | 1969 | SEP 1945 |
| NORMAL NO. DAYS WITH: | | | | | | | | | | | | | | |
| PRECIPITATION >= 0.01 | 30 | 7.2 | 7.0 | 7.0 | 5.4 | 8.7 | 14.8 | 16.4 | 16.2 | 13.2 | 8.0 | 6.5 | 6.6 | 117.0 |
| PRECIPITATION >= 1.00 | 30 | 0.6 | 0.5 | 1.1 | 0.7 | 1.3 | 2.4 | 2.2 | 2.0 | 1.8 | 0.8 | 0.8 | 0.6 | 14.8 |
| **SNOWFALL** NORMAL (IN) | 30 | 0.* | 0.0 | 0.0 | 0.0 | 0.0 | 0.0 | 0.0 | 0.0 | 0.0 | 0.0 | 0.0 | 0.0 | 0.0 |
| MAXIMUM MONTHLY (IN) | 37 | T | 0.0 | T | T | T | T | T | T | 0.0 | 0.0 | 0.0 | 0.0 | T |
| YEAR OF OCCURRENCE | | 1977 | | 1992 | 2009 | 1997 | 2008 | 2007 | 1989 | | | | | APR 2009 |
| MAXIMUM IN 24 HOURS (IN) | 37 | T | 0.0 | T | T | T | T | T | T | 0.0 | 0.0 | 0.0 | 0.0 | T |
| YEAR OF OCCURRENCE' | | 1977 | | 1992 | 2009 | 1997 | 2008 | 1993 | 1989 | | | | | APR 2009 |
| MAXIMUM SNOW DEPTH (IN) | 52 | 0 | 0 | 0 | 0 | 0 | 0 | 0 | 0 | 0 | 0 | 0 | 0 | |
| YEAR OF OCCURRENCE | | | | | | | | | | | | | | |
| NORMAL NO. DAYS WITH: | | | | | | | | | | | | | | |
| SNOWFALL >= 1.0 | 30 | 0.0 | 0.0 | 0.0 | 0.0 | 0.0 | 0.0 | 0.0 | 0.0 | 0.0 | 0.0 | 0.0 | 0.0 | 0.0 |

## PRECIPITATION (inches) 2009 ORLANDO (KMCO)

| YEAR | JAN | FEB | MAR | APR | MAY | JUN | JUL | AUG | SEP | OCT | NOV | DEC | ANNUAL |
|------|-----|-----|-----|-----|-----|-----|-----|-----|-----|-----|-----|-----|--------|
| 1980 | 2.45 | 1.64 | 1.51 | 4.07 | 6.96 | 5.25 | 5.14 | 2.92 | 3.70 | 0.55 | 6.55 | 0.47 | 41.21 |
| 1981 | 0.21 | 4.36 | 1.85 | 0.18 | 2.02 | 12.49 | 3.53 | 5.60 | 8.26 | 3.13 | 2.50 | 2.97 | 47.10 |
| 1982 | 1.72 | 1.34 | 4.85 | 6.27 | 5.29 | 6.06 | 11.81 | 5.03 | 6.96 | 0.74 | 0.53 | 1.01 | 51.61 |
| 1983 | 2.08 | 8.32 | 5.37 | 3.21 | 1.77 | 7.82 | 6.49 | 4.83 | 5.16 | 3.78 | 1.36 | 5.33 | 55.52 |
| 1984 | 2.01 | 2.73 | 1.85 | 6.21 | 3.20 | 5.32 | 6.19 | 7.89 | 6.19 | 0.56 | 2.10 | 0.19 | 44.44 |
| 1985 | 0.91 | 1.27 | 4.59 | 1.69 | 3.00 | 4.54 | 7.28 | 11.63 | 5.45 | 2.55 | 0.82 | 3.46 | 47.19 |
| 1986 | 7.23 | 1.84 | 2.63 | 0.49 | 0.88 | 9.50 | 5.85 | 5.99 | 4.50 | 5.63 | 1.69 | 3.60 | 49.83 |
| 1987 | 1.27 | 1.74 | 11.38 | 0.59 | 1.40 | 3.54 | 7.95 | 6.07 | 8.64 | 3.41 | 10.29 | 0.51 | 56.79 |
| 1988 | 3.12 | 1.38 | 6.07 | 2.02 | 2.82 | 4.17 | 9.44 | 7.94 | 5.67 | 1.42 | 7.44 | 1.00 | 52.49 |
| 1989 | 3.80 | 0.15 | 1.35 | 2.28 | 2.38 | 6.79 | 4.74 | 6.20 | 10.29 | 1.75 | 1.44 | 4.49 | 45.66 |
| 1990 | 0.23 | 4.13 | 1.92 | 1.73 | 0.55 | 6.22 | 6.68 | 3.78 | 2.46 | 2.10 | 1.05 | 0.83 | 31.68 |
| 1991 | 2.37 | 0.98 | 6.66 | 7.72 | 9.48 | 5.98 | 10.78 | 7.13 | 4.53 | 4.76 | 0.27 | 0.24 | 60.90 |
| 1992 | 1.35 | 2.42 | 3.67 | 9.10 | 1.19 | 8.68 | 2.60 | 8.03 | 7.13 | 5.17 | 2.74 | 0.88 | 52.96 |
| 1993 | 4.89 | 1.48 | 6.26 | 1.78 | 2.32 | 4.47 | 6.49 | 5.95 | 5.35 | 4.61 | 0.17 | 0.76 | 44.53 |
| 1994 | 4.00 | 3.58 | 1.21 | 3.03 | 2.87 | 10.28 | 13.27 | 6.23 | 7.84 | 5.18 | 7.32 | 3.04 | 67.85 |
| 1995 | 1.50 | 1.13 | 2.12 | 0.81 | 4.24 | 8.23 | 5.10 | 9.48 | 3.59 | 4.35 | 1.74 | 0.76 | 43.05 |
| 1996 | 5.39 | 1.52 | 9.87 | 0.68 | 5.12 | 6.51 | 4.06 | 11.33 | 6.04 | 3.28 | .72 | 2.14 | 56.66 |
| 1997 | 1.13 | 2.44 | 3.46 | 4.02 | 3.17 | 8.20 | 11.51 | 7.99 | 2.59 | 4.22 | 3.15 | 12.63 | 64.51 |
| 1998 | 1.99 | 8.74 | 5.26 | 0.52 | 3.17 | 1.58 | 8.61 | 5.59 | 5.36 | 0.64 | 1.67 | 0.62 | 43.75 |
| 1999 | 2.99 | 0.36 | 0.56 | 2.40 | 5.43 | 13.84 | 5.14 | 4.50 | 6.40 | 8.40 | 2.13 | 2.65 | 54.80 |
| 2000 | 1.23 | 0.36 | 0.45 | 2.22 | 1.00 | 6.19 | 4.07 | 4.48 | 6.37 | 1.33 | 1.10 | 1.58 | 30.38 |
| 2001 | 0.66 | 0.22 | 3.72 | 2.09 | 5.84 | 7.65 | 18.27 | 2.83 | 10.47 | 1.01 | 1.68 | 0.48 | 54.92 |
| 2002 | 1.10 | 3.48 | 0.52 | 1.20 | 2.47 | 13.01 | 10.21 | 10.97 | 4.74 | 4.98 | 2.32 | 11.39 | 66.39 |
| 2003 | 0.80 | 1.58 | 5.28 | 4.32 | 2.43 | 6.56 | 8.26 | 11.90 | 5.75 | 1.82 | 2.42 | 1.56 | 52.68 |
| 2004 | 3.27 | 4.53 | 0.72 | 2.41 | 1.91 | 8.76 | 4.56 | 14.88 | 13.02 | 1.24 | 2.18 | 1.76 | 59.24 |
| 2005 | 3.33 | 1.29 | 6.38 | 1.43 | 4.57 | 17.39 | 3.95 | 7.66 | 1.67 | 10.20 | 0.60 | 2.04 | 60.51 |
| 2006 | 0.43 | 2.36 | 0.02 | 1.05 | 3.36 | 6.61 | 7.01 | 4.33 | 4.09 | 1.95 | 1.54 | 3.60 | 36.35 |
| 2007 | 1.73 | 0.91 | 0.52 | 2.05 | 0.54 | 5.91 | 6.52 | 4.47 | 8.96 | 5.41 | 0.42 | 1.05 | 38.49 |
| 2008 | 4.10 | 1.65 | 5.15 | 3.21 | 3.48 | 9.73 | 7.35 | 10.71 | 4.02 | 2.65 | 1.09 | 0.66 | 53.80 |
| 2009 | 2.08 | 0.62 | 0.48 | 1.06 | 14.56 | 8.05 | 6.05 | 4.74 | 4.58 | 2.85 | 1.02 | 5.39 | 51.48 |
| POR= 57 YRS | 2.25 | 2.64 | 3.42 | 2.28 | 3.51 | 7.53 | 7.34 | 6.67 | 5.87 | 3.18 | 1.95 | 2.29 | 48.93 |

WBAN : 12815

## AVERAGE TEMPERATURE (°F) 2009 ORLANDO (KMCO)

| YEAR | JAN | FEB | MAR | APR | MAY | JUN | JUL | AUG | SEP | OCT | NOV | DEC | ANNUAL |
|------|-----|-----|-----|-----|-----|-----|-----|-----|-----|-----|-----|-----|--------|
| 1980 | 60.5 | 57.2 | 68.2 | 70.4 | 76.4 | 80.1 | 83.6 | 83.6 | 81.7 | 75.4 | 67.1 | 59.0 | 71.9 |
| 1981 | 51.3 | 61.7 | 64.0 | 73.1 | 76.7 | 83.2 | 84.1 | 82.9 | 80.0 | 76.4 | 65.3 | 60.5 | 71.6 |
| 1982 | 60.0 | 68.4 | 70.4 | 72.6 | 75.3 | 82.0 | 82.6 | 82.2 | 80.2 | 74.1 | 70.8 | 66.7 | 73.8 |
| 1983 | 58.0 | 59.9 | 63.5 | 68.6 | 76.4 | 80.5 | 83.2 | 83.5 | 80.6 | 76.5 | 65.8 | 61.2 | 71.5 |
| 1984 | 57.8 | 61.2 | 64.7 | 69.2 | 75.6 | 78.4 | 80.7 | 81.5 | 78.9 | 75.4 | 65.8 | 66.0 | 71.3 |
| 1985 | 54.7 | 62.2 | 68.4 | 70.7 | 77.2 | 82.4 | 82.1 | 82.3 | 79.8 | 79.4 | 73.0 | 58.8 | 72.6 |
| 1986 | 59.8 | 64.3 | 65.4 | 69.3 | 76.7 | 81.7 | 82.3 | 83.3 | 81.7 | 77.5 | 75.8 | 67.3 | 73.8 |
| 1987 | 58.8 | 62.7 | 65.9 | 66.8 | 76.8 | 83.1 | 83.5 | 85.0 | 82.7 | 72.2 | 69.0 | 64.2 | 72.6 |
| 1988 | 58.5 | 60.4 | 65.5 | 72.0 | 75.5 | 80.3 | 80.7 | 82.8 | 83.9 | 73.7 | 70.5 | 62.4 | 72.2 |
| 1989 | 66.9 | 64.5 | 69.7 | 71.9 | 77.9 | 81.9 | 83.2 | 83.3 | 82.2 | 75.3 | 69.0 | 55.5 | 73.4 |
| 1990 | 65.8 | 69.1 | 69.3 | 71.5 | 79.4 | 81.9 | 82.8 | 83.5 | 82.0 | 77.1 | 69.3 | 66.3 | 74.8 |
| 1991 | 66.3 | 64.2 | 67.7 | 75.3 | 79.5 | 81.1 | 82.6 | 83.0 | 81.7 | 75.3 | 65.8 | 65.5 | 74.0 |
| 1992 | 59.7 | 64.8 | 66.4 | 69.9 | 74.9 | 81.2 | 84.5 | 82.3 | 81.6 | 73.4 | 70.4 | 63.5 | 72.7 |
| 1993 | 66.3 | 60.6 | 64.8 | 68.3 | 75.5 | 81.9 | 83.9 | 83.2 | 81.2 | 75.1 | 68.3 | 58.9 | 72.3 |
| 1994 | 60.9 | 66.4 | 68.2 | 74.1 | 77.4 | 81.3 | 81.5 | 81.0 | 79.2 | 76.1 | 71.9 | 64.9 | 73.6 |
| 1995 | 58.8 | 61.1 | 68.2 | 72.6 | 80.2 | 79.9 | 82.4 | 83.0 | 82.1 | 77.3 | 65.6 | 61.0 | 72.7 |
| 1996 | 59.1 | 61.3 | 62.9 | 71.6 | 80.2 | 81.8 | 82.9 | 81.6 | 80.6 | 73.8 | 67.5 | 62.3 | 72.1 |
| 1997 | 60.9 | 67.7 | 72.3 | 69.8 | 76.1 | 79.9 | 82.4 | 82.6 | 80.5 | 73.9 | 66.1 | 61.4 | 72.8 |
| 1998 | 62.4 | 61.7 | 63.8 | 71.3 | 77.7 | 85.0 | 84.0 | 83.5 | 80.9 | 77.4 | 71.2 | 67.3 | 73.9 |
| 1999 | 63.9 | 64.0 | 64.6 | 74.2 | 75.5 | 80.1 | 82.8 | 83.4 | 80.5 | 75.1 | 68.4 | 61.5 | 72.8 |
| 2000 | 60.6 | 62.6 | 69.9 | 70.4 | 77.9 | 81.3 | 82.6 | 82.3 | 81.4 | 72.9 | 65.7 | 59.8 | 72.3 |
| 2001 | 56.2 | 67.2 | 66.8 | 71.7 | 76.2 | 80.6 | 81.4 | 82.3 | 78.7 | 73.8 | 69.4 | 66.4 | 72.6 |
| 2002 | 61.4 | 61.6 | 69.5 | 74.5 | 78.1 | 79.6 | 81.7 | 81.3 | 82.2 | 78.2 | 65.7 | 59.8 | 72.8 |
| 2003 | 54.0 | 63.9 | 71.8 | 71.2 | 79.8 | 80.6 | 82.4 | 81.5 | 80.3 | 75.8 | 70.8 | 59.8 | 72.7 |
| 2004 | 59.5 | 62.6 | 67.5 | 69.2 | 77.2 | 82.7 | 82.8 | 82.7 | 81.4 | 75.9 | 70.1 | 61.0 | 72.7 |
| 2005 | 62.1 | 63.7 | 64.9 | 68.8 | 76.1 | 80.3 | 84.7 | 85.1 | 82.1 | 75.1 | 69.2 | 60.2 | 72.7 |
| 2006 | 63.2 | 60.9 | 67.4 | 74.5 | 77.0 | 81.2 | 82.7 | 83.2 | 81.1 | 74.2 | 66.0 | 67.6 | 73.3 |
| 2007 | 64.1 | 60.4 | 68.3 | 70.4 | 76.1 | 80.7 | 82.9 | 84.4 | 81.9 | 78.5 | 67.7 | 67.5 | 73.6 |
| 2008 | 61.6 | 66.2 | 67.2 | 71.0 | 78.1 | 81.1 | 81.4 | 81.6 | 80.9 | 73.8 | 63.8 | 64.5 | 72.6 |
| 2009 | 59.2 | 60.5 | 67.6 | 71.6 | 77.8 | 82.1 | 82.0 | 83.3 | 81.5 | 77.5 | 69.4 | 63.8 | 73.0 |
| POR= 57 YRS | 60.3 | 62.2 | 66.8 | 71.6 | 77.2 | 81.0 | 82.4 | 82.4 | 80.9 | 75.0 | 67.9 | 62.3 | 72.5 |

## HEATING DEGREE DAYS (base 65°F) 2009 ORLANDO (KMCO)

| YEAR | JUL | AUG | SEP | OCT | NOV | DEC | JAN | FEB | MAR | APR | MAY | JUN | TOTAL |
|---|---|---|---|---|---|---|---|---|---|---|---|---|---|
| 1980-81 | 0 | 0 | 0 | 1 | 67 | 190 | 416 | 119 | 76 | 1 | 0 | 0 | 870 |
| 1981-82 | 0 | 0 | 0 | 0 | 75 | 205 | 204 | 21 | 33 | 7 | 0 | 0 | 545 |
| 1982-83 | 0 | 0 | 0 | 14 | 16 | 94 | 233 | 148 | 105 | 13 | 0 | 0 | 623 |
| 1983-84 | 0 | 0 | 0 | 0 | 63 | 188 | 252 | 137 | 86 | 18 | 0 | 0 | 744 |
| 1984-85 | 0 | 0 | 0 | 0 | 68 | 71 | 340 | 146 | 22 | 12 | 0 | 0 | 659 |
| 1985-86 | 0 | 0 | 0 | 0 | 14 | 228 | 180 | 82 | 105 | 4 | 0 | 0 | 613 |
| 1986-87 | 0 | 0 | 0 | 0 | 0 | 42 | 216 | 97 | 48 | 66 | 0 | 0 | 469 |
| 1987-88 | 0 | 0 | 0 | 0 | 39 | 97 | 221 | 169 | 71 | 7 | 0 | 0 | 604 |
| 1988-89 | 0 | 0 | 0 | 0 | 11 | 135 | 32 | 119 | 59 | 4 | 0 | 0 | 360 |
| 1989-90 | 0 | 0 | 0 | 21 | 27 | 308 | 71 | 34 | 11 | 5 | 0 | 0 | 477 |
| 1990-91 | 0 | 0 | 0 | 6 | 14 | 69 | 75 | 88 | 52 | 0 | 0 | 0 | 304 |
| 1991-92 | 0 | 0 | 0 | 0 | 85 | 76 | 187 | 79 | 51 | 19 | 8 | 0 | 505 |
| 1992-93 | 0 | 0 | 0 | 0 | 47 | 102 | 73 | 131 | 80 | 12 | 0 | 0 | 445 |
| 1993-94 | 0 | 0 | 0 | 10 | 45 | 201 | 158 | 61 | 42 | 3 | 0 | 0 | 520 |
| 1994-95 | 0 | 0 | 0 | 0 | 6 | 88 | 205 | 153 | 24 | 5 | 0 | 0 | 481 |
| 1995-96 | 0 | 0 | 0 | 0 | 76 | 191 | 208 | 173 | 149 | 12 | 0 | 0 | 809 |
| 1996-97 | 0 | 0 | 0 | 6 | 38 | 122 | 164 | 43 | 0 | 7 | 0 | 0 | 380 |
| 1997-98 | 0 | 0 | 0 | 6 | 45 | 170 | 130 | 124 | 127 | 8 | 0 | 0 | 610 |
| 1998-99 | 0 | 0 | 0 | 0 | 7 | 64 | 109 | 92 | 61 | 12 | 5 | 0 | 350 |
| 1999-00 | 0 | 0 | 0 | 3 | 20 | 140 | 164 | 105 | 2 | 18 | 0 | 0 | 452 |
| 2000-01 | 0 | 0 | 0 | 4 | 70 | 217 | 300 | 49 | 55 | 11 | 0 | 0 | 706 |
| 2001-02 | 0 | 0 | 0 | 11 | 0 | 87 | 196 | 119 | 44 | 0 | 0 | 0 | 457 |
| 2002-03 | 0 | 0 | 0 | 0 | 77 | 178 | 337 | 89 | 14 | 19 | 0 | 0 | 714 |
| 2003-04 | 0 | 0 | 0 | 0 | 24 | 171 | 193 | 111 | 35 | 20 | 0 | 0 | 554 |
| 2004-05 | 0 | 0 | 0 | 0 | 6 | 166 | 136 | 78 | 101 | 14 | 0 | 0 | 501 |
| 2005-06 | 0 | 0 | 0 | 18 | 22 | 155 | 115 | 153 | 36 | 0 | 0 | 0 | 499 |
| 2006-07 | 0 | 0 | 0 | 9 | 72 | 48 | 108 | 160 | 37 | 19 | 0 | 0 | 453 |
| 2007-08 | 0 | 0 | 0 | 0 | 24 | 45 | 150 | 64 | 36 | 19 | 0 | 0 | 338 |
| 2008-09 | 0 | 0 | 0 | 23 | 108 | 85 | 200 | 155 | 50 | 16 | 0 | 0 | 637 |
| 2009- | 0 | 0 | 0 | 11 | 32 | 127 | | | | | | | |

WBAN : 12815

## COOLING DEGREE DAYS (base 65°F) 2009 ORLANDO (KMCO)

| YEAR | JAN | FEB | MAR | APR | MAY | JUN | JUL | AUG | SEP | OCT | NOV | DEC | TOTAL |
|---|---|---|---|---|---|---|---|---|---|---|---|---|---|
| 1980 | 27 | 25 | 169 | 172 | 362 | 459 | 586 | 582 | 508 | 331 | 138 | 12 | 3371 |
| 1981 | 0 | 34 | 52 | 253 | 372 | 552 | 602 | 559 | 458 | 359 | 89 | 73 | 3403 |
| 1982 | 56 | 123 | 211 | 241 | 325 | 518 | 550 | 542 | 465 | 303 | 196 | 152 | 3682 |
| 1983 | 22 | 11 | 68 | 129 | 361 | 473 | 573 | 582 | 476 | 362 | 95 | 77 | 3229 |
| 1984 | 37 | 35 | 84 | 151 | 332 | 411 | 490 | 520 | 426 | 331 | 99 | 107 | 3023 |
| 1985 | 27 | 74 | 137 | 191 | 386 | 531 | 539 | 548 | 451 | 454 | 262 | 45 | 3645 |
| 1986 | 25 | 69 | 124 | 139 | 372 | 506 | 543 | 573 | 507 | 392 | 333 | 121 | 3704 |
| 1987 | 32 | 38 | 82 | 127 | 376 | 549 | 582 | 627 | 540 | 230 | 163 | 78 | 3424 |
| 1988 | 26 | 43 | 95 | 223 | 336 | 466 | 496 | 559 | 573 | 275 | 182 | 61 | 3335 |
| 1989 | 101 | 111 | 213 | 216 | 408 | 509 | 573 | 579 | 523 | 346 | 153 | 19 | 3751 |
| 1990 | 102 | 156 | 156 | 206 | 453 | 514 | 559 | 581 | 518 | 388 | 149 | 116 | 3898 |
| 1991 | 121 | 71 | 143 | 315 | 455 | 490 | 553 | 565 | 508 | 326 | 118 | 98 | 3763 |
| 1992 | 28 | 79 | 101 | 175 | 325 | 496 | 612 | 540 | 507 | 265 | 217 | 62 | 3407 |
| 1993 | 120 | 14 | 80 | 118 | 334 | 514 | 591 | 573 | 492 | 331 | 146 | 18 | 3331 |
| 1994 | 39 | 107 | 149 | 284 | 393 | 495 | 517 | 501 | 436 | 352 | 218 | 89 | 3580 |
| 1995 | 19 | 52 | 132 | 241 | 476 | 455 | 546 | 567 | 518 | 391 | 103 | 74 | 3574 |
| 1996 | 33 | 72 | 93 | 217 | 477 | 512 | 561 | 524 | 478 | 283 | 124 | 46 | 3420 |
| 1997 | 46 | 125 | 234 | 155 | 353 | 452 | 546 | 552 | 475 | 290 | 82 | 69 | 3379 |
| 1998 | 55 | 38 | 96 | 206 | 402 | 605 | 599 | 579 | 485 | 390 | 201 | 139 | 3795 |
| 1999 | 82 | 69 | 54 | 297 | 336 | 460 | 560 | 578 | 471 | 321 | 130 | 38 | 3396 |
| 2000 | 35 | 42 | 159 | 188 | 410 | 496 | 553 | 542 | 499 | 257 | 97 | 63 | 3341 |
| 2001 | 32 | 117 | 117 | 217 | 352 | 474 | 515 | 542 | 419 | 289 | 140 | 134 | 3348 |
| 2002 | 93 | 30 | 192 | 291 | 414 | 445 | 525 | 515 | 523 | 414 | 105 | 25 | 3572 |
| 2003 | 1 | 63 | 232 | 214 | 469 | 475 | 546 | 519 | 466 | 342 | 206 | 17 | 3550 |
| 2004 | 27 | 48 | 115 | 151 | 388 | 538 | 558 | 555 | 502 | 346 | 169 | 50 | 3447 |
| 2005 | 55 | 47 | 105 | 133 | 351 | 463 | 619 | 629 | 523 | 336 | 153 | 10 | 3424 |
| 2006 | 66 | 45 | 118 | 291 | 381 | 490 | 554 | 568 | 488 | 301 | 107 | 136 | 3545 |
| 2007 | 86 | 37 | 149 | 188 | 353 | 480 | 565 | 609 | 511 | 426 | 109 | 128 | 3641 |
| 2008 | 54 | 105 | 111 | 209 | 411 | 492 | 514 | 522 | 483 | 306 | 80 | 75 | 3362 |
| 2009 | 26 | 33 | 135 | 220 | 405 | 519 | 533 | 573 | 504 | 405 | 167 | 96 | 3616 |

## SNOWFALL (inches) 2009 ORLANDO (KMCO)

| YEAR | JUL | AUG | SEP | OCT | NOV | DEC | JAN | FEB | MAR | APR | MAY | JUN | TOTAL |
|------|-----|-----|-----|-----|-----|-----|-----|-----|-----|-----|-----|-----|-------|
| 1980-81 | 0.0 | 0.0 | 0.0 | 0.0 | 0.0 | 0.0 | 0.0 | 0.0 | 0.0 | 0.0 | 0.0 | 0.0 | 0.0 |
| 1981-82 | 0.0 | 0.0 | 0.0 | 0.0 | 0.0 | 0.0 | 0.0 | 0.0 | 0.0 | 0.0 | 0.0 | 0.0 | 0.0 |
| 1982-83 | 0.0 | 0.0 | 0.0 | 0.0 | 0.0 | 0.0 | 0.0 | 0.0 | 0.0 | 0.0 | 0.0 | 0.0 | 0.0 |
| 1983-84 | 0.0 | 0.0 | 0.0 | 0.0 | 0.0 | 0.0 | 0.0 | 0.0 | 0.0 | 0.0 | 0.0 | 0.0 | 0.0 |
| 1984-85 | 0.0 | 0.0 | 0.0 | 0.0 | 0.0 | 0.0 | 0.0 | 0.0 | 0.0 | 0.0 | 0.0 | 0.0 | 0.0 |
| 1985-86 | 0.0 | 0.0 | 0.0 | 0.0 | 0.0 | 0.0 | 0.0 | 0.0 | 0.0 | 0.0 | 0.0 | 0.0 | 0.0 |
| 1986-87 | 0.0 | 0.0 | 0.0 | 0.0 | 0.0 | 0.0 | 0.0 | 0.0 | 0.0 | 0.0 | 0.0 | 0.0 | 0.0 |
| 1987-88 | 0.0 | 0.0 | 0.0 | 0.0 | 0.0 | 0.0 | 0.0 | 0.0 | 0.0 | 0.0 | 0.0 | 0.0 | 0.0 |
| 1988-89 | 0.0 | 0.0 | 0.0 | 0.0 | 0.0 | 0.0 | 0.0 | 0.0 | 0.0 | 0.0 | 0.0 | 0.0 | 0.0 |
| 1989-90 | 0.0 | T | 0.0 | 0.0 | 0.0 | 0.0 | 0.0 | 0.0 | 0.0 | 0.0 | 0.0 | 0.0 | T |
| 1990-91 | 0.0 | 0.0 | 0.0 | 0.0 | 0.0 | 0.0 | 0.0 | 0.0 | 0.0 | 0.0 | 0.0 | 0.0 | 0.0 |
| 1991-92 | T | 0.0 | 0.0 | 0.0 | 0.0 | 0.0 | 0.0 | 0.0 | T | T | 0.0 | 0.0 | T |
| 1992-93 | 0.0 | 0.0 | 0.0 | 0.0 | 0.0 | 0.0 | 0.0 | 0.0 | 0.0 | 0.0 | 0.0 | 0.0 | 0.0 |
| 1993-94 | T | 0.0 | 0.0 | 0.0 | 0.0 | 0.0 | 0.0 | 0.0 | 0.0 | 0.0 | 0.0 | 0.0 | T |
| 1994-95 | 0.0 | 0.0 | 0.0 | 0.0 | 0.0 | 0.0 | 0.0 | 0.0 | 0.0 | 0.0 | 0.0 | 0.0 | 0.0 |
| 1995-96 | 0.0 | 0.0 | 0.0 | 0.0 | 0.0 | 0.0 | 0.0 | 0.0 | 0.0 | 0.0 | 0.0 | 0.0 | 0.0 |
| 1996-97 | 0.0 | 0.0 | 0.0 | 0.0 | 0.0 | 0.0 | 0.0 | 0.0 | 0.0 | T | T | 0.0 | T |
| 1997-98 | 0.0 | 0.0 | 0.0 | 0.0 | 0.0 | 0.0 | 0.0 | 0.0 | 0.0 | 0.0 | 0.0 | 0.0 | 0.0 |
| 1998-99 | 0.0 | 0.0 | 0.0 | 0.0 | 0.0 | 0.0 | 0.0 | 0.0 | 0.0 | 0.0 | 0.0 | 0.0 | 0.0 |
| 1999-00 | 0.0 | 0.0 | 0.0 | 0.0 | 0.0 | 0.0 | 0.0 | 0.0 | 0.0 | 0.0 | 0.0 | 0.0 | 0.0 |
| 2000-01 | T | 0.0 | 0.0 | 0.0 | 0.0 | 0.0 | 0.0 | 0.0 | 0.0 | 0.0 | 0.0 | 0.0 | T |
| 2001-02 | 0.0 | 0.0 | 0.0 | 0.0 | 0.0 | 0.0 | 0.0 | 0.0 | 0.0 | 0.0 | 0.0 | 0.0 | 0.0 |
| 2002-03 | 0.0 | 0.0 | 0.0 | 0.0 | 0.0 | 0.0 | 0.0 | 0.0 | 0.0 | 0.0 | 0.0 | 0.0 | 0.0 |
| 2003-04 | 0.0 | 0.0 | 0.0 | 0.0 | 0.0 | 0.0 | 0.0 | 0.0 | 0.0 | 0.0 | 0.0 | 0.0 | 0.0 |
| 2004-05 | 0.0 | 0.0 | 0.0 | 0.0 | 0.0 | 0.0 | 0.0 | 0.0 | 0.0 | T | 0.0 | 0.0 | T |
| 2005-06 | 0.0 | 0.0 | 0.0 | 0.0 | 0.0 | 0.0 | 0.0 | 0.0 | 0.0 | 0.0 | 0.0 | 0.0 | 0.0 |
| 2006-07 | 0.0 | 0.0 | 0.0 | 0.0 | 0.0 | 0.0 | 0.0 | 0.0 | 0.0 | 0.0 | 0.0 | 0.0 | 0.0 |
| 2007-08 | T | 0.0 | 0.0 | 0.0 | 0.0 | 0.0 | 0.0 | 0.0 | 0.0 | 0.0 | 0.0 | 0.0 | T |
| 2008-09 | 0.0 | 0.0 | 0.0 | 0.0 | 0.0 | 0.0 | 0.0 | 0.0 | 0.0 | T | 0.0 | T | T |
| 2009- | 0.0 | 0.0 | 0.0 | 0.0 | 0.0 | 0.0 | | | | | | | |
| POR= 55 YRS | T | T | 0.0 | 0.0 | 0.0 | 0.0 | 0.0 | 0.0 | T | T | T | T | T |

## Orlando, FL 2009
### Temperature and Precipitation

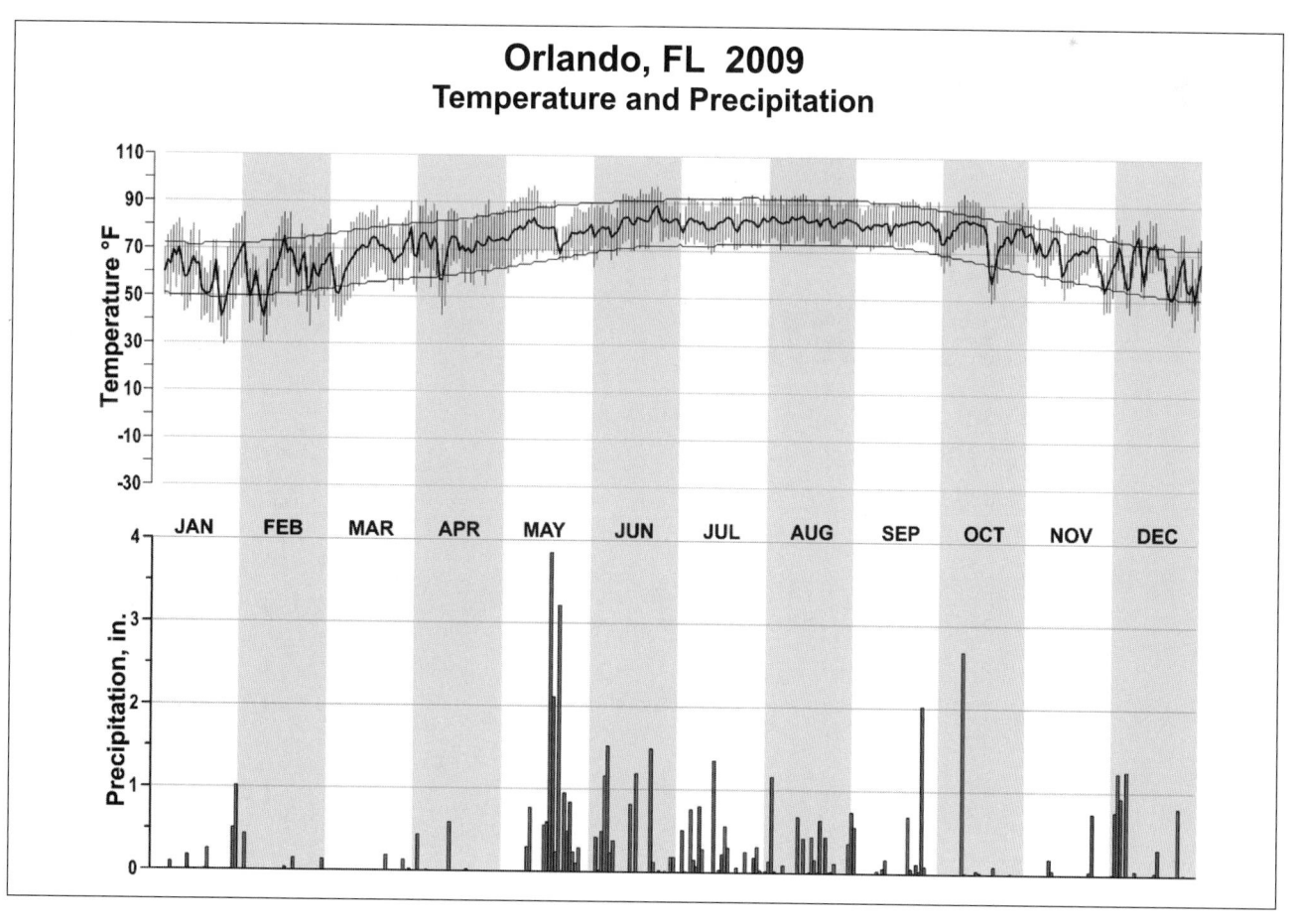

# 2009
# TAMPA
# FLORIDA (KTPA)

Tampa is on west central coast of the Florida Peninsula. Very near the Gulf of Mexico at the upper end of Tampa Bay, land and sea breezes modify the subtropical climate. Major rivers flowing into the area are the Hillsborough, the Alafia, and the Little Manatee.

Winters are mild. Summers are long, rather warm, and humid. Low temperatures are about 50 degrees in the winter and 70 degrees during the summer. Afternoon highs range from the low 70s in the winter to around 90 degrees from June through September. Invasions of cold northern air produce an occasional cool winter morning. Freezing temperatures occur on one or two mornings per year during December, January, and February. In some years no freezing temperatures occur. Temperatures rarely fail to recover to the 60s on the cooler winter days. Temperatures above the low 90s are uncommon because of the afternoon sea breezes and thunderstorms. An outstanding feature of the Tampa climate is the summer thunderstorm season. Most of the thunderstorms occur in the late afternoon hours from June through September. The resulting sudden drop in temperature from about 90 degrees to around 70 degrees makes for a pleasant change. Between a dry spring and a dry fall, some 30 inches of rain, about 60 percent of the annual total, falls during the summer months. Snowfall is very rare. Measurable snows under 1/2 inch have occurred only a few times in the last one hundred years.

A large part of the generally flat sandy land near the coast has an elevation of under 15 feet above sea level. This does make the area vulnerable to tidal surges. Tropical storms threaten the area on a few occasions most years. The greatest risk of hurricanes has been during the months of June and October. Many hurricanes, by replenishing the soil moisture and raising the water table, do far more good than harm. The heaviest rains in a 24-hour period, around 12 inches, have been associated with hurricanes.

Fittingly named the Suncoast, the sun shines more than 65 percent of the possible, with the sunniest months being April and May. Afternoon humidities are usually 60 percent or higher in the summer months, but range from 50 to 60 percent the remainder of the year.

Night ground fogs occur frequently during the cooler winter months. Prevailing winds are easterly, but westerly afternoon and early evening sea breezes occur most months of the year. Winds in excess of 25 mph are not common and usually occur only with thunderstorms or tropical disturbances.

Based on the 1951-1980 period, the average first occurrence of 32 degrees Fahrenheit in the fall is December 26 and the average last occurrence in the spring is February 3.

# NORMALS, MEANS, AND EXTREMES
## TAMPA (KTPA)

| LATITUDE: 27° 57'N | LONGITUDE: -82° 32'W | ELEVATION (FT): GRND: 8   BARO: 40 | TIME ZONE: EASTERN (UTC -5) | WBAN: 12842 |
|---|---|---|---|---|

| | ELEMENT | POR | JAN | FEB | MAR | APR | MAY | JUN | JUL | AUG | SEP | OCT | NOV | DEC | YEAR |
|---|---|---|---|---|---|---|---|---|---|---|---|---|---|---|---|
| **TEMPERATURE °F** | NORMAL DAILY MAXIMUM | 30 | 70.1 | 71.6 | 76.3 | 80.6 | 86.3 | 88.9 | 89.7 | 90.0 | 89.0 | 84.1 | 78.0 | 72.0 | 81.4 |
| | MEAN DAILY MAXIMUM | 77 | 70.6 | 70.9 | 75.6 | 81.1 | 87.2 | 89.0 | 90.1 | 90.3 | 88.4 | 84.1 | 76.9 | 72.1 | 81.4 |
| | HIGHEST DAILY MAXIMUM | 63 | 86 | 88 | 91 | 93 | 98 | 99 | 97 | 98 | 96 | 94 | 90 | 86 | 99 |
| | YEAR OF OCCURRENCE | | 2002 | 1971 | 1949 | 1975 | 1975 | 1985 | 1995 | 1975 | 1991 | 1990 | 2006 | 1994 | JUN 1985 |
| | MEAN OF EXTREME MAXS. | 77 | 80.9 | 82.0 | 85.3 | 88.7 | 93.0 | 94.6 | 94.5 | 94.4 | 93.4 | 90.5 | 85.5 | 82.0 | 88.7 |
| | NORMAL DAILY MINIMUM | 30 | 52.4 | 53.8 | 58.5 | 62.4 | 68.9 | 74.0 | 75.3 | 75.4 | 74.3 | 67.6 | 60.7 | 54.7 | 64.8 |
| | MEAN DAILY MINIMUM | 77 | 51.0 | 51.5 | 56.5 | 61.5 | 68.1 | 72.5 | 74.6 | 74.7 | 72.8 | 66.4 | 57.7 | 52.7 | 63.3 |
| | LOWEST DAILY MINIMUM | 63 | 21 | 24 | 29 | 40 | 49 | 53 | 63 | 67 | 57 | 40 | 23 | 18 | 18 |
| | YEAR OF OCCURRENCE | | 1985 | 1958 | 1980 | 1987 | 1992 | 1984 | 1970 | 1973 | 1981 | 1964 | 1970 | 1962 | DEC 1962 |
| | MEAN OF EXTREME MINS. | 77 | 32.7 | 36.1 | 41.2 | 48.6 | 58.3 | 66.9 | 70.5 | 70.9 | 67.5 | 52.6 | 41.9 | 35.2 | 51.9 |
| | NORMAL DRY BULB | 30 | 61.3 | 62.7 | 67.4 | 71.5 | 77.6 | 81.5 | 82.5 | 82.7 | 81.6 | 75.8 | 69.3 | 63.3 | 73.1 |
| | MEAN DRY BULB | 77 | 60.8 | 61.2 | 66.1 | 71.3 | 77.7 | 80.8 | 82.4 | 82.6 | 80.6 | 75.2 | 67.3 | 62.4 | 72.4 |
| | MEAN WET BULB | 26 | 55.3 | 56.9 | 60.2 | 63.6 | 69.4 | 74.2 | 75.6 | 75.9 | 74.3 | 68.6 | 62.3 | 57.5 | 66.2 |
| | MEAN DEW POINT | 26 | 52.8 | 54.2 | 57.4 | 60.6 | 66.8 | 72.3 | 74.0 | 74.3 | 72.7 | 66.5 | 60.0 | 54.9 | 63.9 |
| | NORMAL NO. DAYS WITH: | | | | | | | | | | | | | | |
| | MAXIMUM >= 90 | 30 | 0.0 | 0.0 | 0.0 | 0.7 | 8.5 | 17.4 | 21.8 | 22.3 | 16.4 | 2.9 | * | 0.0 | 90.0 |
| | MAXIMUM <= 32 | 30 | 0.0 | * | 0.0 | 0.0 | 0.0 | 0.0 | 0.0 | 0.0 | 0.0 | 0.0 | 0.0 | 0.0 | 0.0 |
| | MINIMUM <= 32 | 30 | 1.3 | 0.6 | 0.1 | 0.0 | 0.0 | 0.0 | 0.0 | 0.0 | 0.0 | 0.0 | 0.0 | 0.7 | 2.7 |
| | MINIMUM <= 0 | 30 | 0.0 | 0.0 | 0.0 | 0.0 | 0.0 | 0.0 | 0.0 | 0.0 | 0.0 | 0.0 | 0.0 | 0.0 | 0.0 |
| **H/C** | NORMAL HEATING DEG. DAYS | 30 | 187 | 136 | 63 | 13 | 0 | 0 | 0 | 0 | 0 | 4 | 44 | 144 | 591 |
| | NORMAL COOLING DEG. DAYS | 30 | 57 | 59 | 124 | 204 | 393 | 501 | 550 | 549 | 489 | 323 | 157 | 76 | 3482 |
| **RH** | NORMAL (PERCENT) | 30 | 76 | 74 | 73 | 70 | 71 | 75 | 77 | 79 | 79 | 76 | 76 | 77 | 75 |
| | HOUR 01 LST | 30 | 85 | 84 | 83 | 82 | 82 | 85 | 86 | 88 | 88 | 87 | 86 | 86 | 85 |
| | HOUR 07 LST | 30 | 87 | 87 | 87 | 86 | 85 | 86 | 88 | 90 | 91 | 90 | 89 | 88 | 88 |
| | HOUR 13 LST | 30 | 60 | 57 | 55 | 52 | 54 | 60 | 64 | 65 | 63 | 58 | 59 | 61 | 59 |
| | HOUR 19 LST | 30 | 74 | 70 | 68 | 64 | 64 | 70 | 74 | 76 | 76 | 73 | 74 | 75 | 72 |
| **S** | PERCENT POSSIBLE SUNSHINE | 56 | 66 | 66 | 71 | 75 | 75 | 67 | 62 | 62 | 62 | 66 | 64 | 61 | 66 |
| **W/O** | MEAN NO. DAYS WITH: | | | | | | | | | | | | | | |
| | HEAVY FOG(VISBY <= 1/4 MI) | 46 | 3.8 | 2.5 | 1.8 | 0.7 | 0.2 | 0.3 | 0.2 | 0.2 | 0.2 | 0.6 | 1.8 | 2.7 | 15.0 |
| | THUNDERSTORMS | 62 | 1.0 | 1.6 | 2.4 | 2.4 | 4.8 | 13.4 | 19.9 | 19.4 | 10.5 | 2.7 | 1.0 | 1.1 | 80.2 |
| **CLOUDINESS** | MEAN: | | | | | | | | | | | | | | |
| | SUNRISE-SUNSET (OKTAS) | | | | | | | | | | | | | | |
| | MIDNIGHT-MIDNIGHT (OKTAS) | | | | | | | | | | | | | | |
| | MEAN NO. DAYS WITH: | | | | | | | | | | | | | | |
| | CLEAR | 1 | 1.0 | 2.0 | 9.0 | | 9.0 | 9.0 | | | | | | | |
| | PARTLY CLOUDY | 1 | 1.0 | 2.0 | 3.0 | | 10.0 | 9.0 | | | | | | | |
| | CLOUDY | 1 | 3.0 | 1.0 | 8.0 | | 2.0 | 5.0 | | | | | | | |
| **PR** | MEAN STATION PRESSURE(IN) | 26 | 30.13 | 30.10 | 30.06 | 30.02 | 30.00 | 30.00 | 30.04 | 30.00 | 29.96 | 30.00 | 30.07 | 30.12 | 30.04 |
| | MEAN SEA-LEVEL PRES. (IN) | 26 | 30.14 | 30.11 | 30.07 | 30.03 | 30.01 | 30.01 | 30.05 | 30.01 | 29.97 | 30.01 | 30.08 | 30.13 | 30.05 |
| **WINDS** | MEAN SPEED (MPH) | 26 | 7.3 | 7.6 | 8.1 | 7.9 | 7.4 | 6.7 | 5.8 | 5.6 | 6.4 | 6.9 | 7.2 | 6.9 | 7.0 |
| | PREVAIL.DIR(TENS OF DEGS) | 41 | 01 | 06 | 09 | 09 | 28 | 28 | 28 | 09 | 06 | 06 | 03 | 05 | 06 |
| | MAXIMUM 2-MINUTE: | | | | | | | | | | | | | | |
| | SPEED (MPH) | 14 | 44 | 36 | 31 | 44 | 36 | 33 | 37 | 30 | 45 | 40 | 35 | 43 | 45 |
| | DIR. (TENS OF DEGS) | | 32 | 28 | 24 | 28 | 25 | 20 | 26 | 31 | 35 | 21 | 19 | 32 | 35 |
| | YEAR OF OCCURRENCE | | 1999 | 1998 | 2005 | 1997 | 1999 | 2006 | 2004 | 2005 | 2004 | 1996 | 2000 | 2004 | SEP 2004 |
| | MAXIMUM 3-SECOND | | | | | | | | | | | | | | |
| | SPEED (MPH) | 14 | 51 | 44 | 40 | 49 | 52 | 47 | 47 | 43 | 54 | 53 | 41 | 52 | 54 |
| | DIR. (TENS OF DEGS) | | 32 | 09 | 24 | 28 | 09 | 13 | 20 | 12 | 35 | 21 | 17 | 31 | 35 |
| | YEAR OF OCCURRENCE | | 1999 | 1998 | 2005 | 1997 | 2009 | 1996 | 2001 | 1996 | 2004 | 1996 | 2000 | 2004 | SEP 2004 |
| **PRECIPITATION** | NORMAL (IN) | 30 | 2.27 | 2.67 | 2.84 | 1.80 | 2.85 | 5.50 | 6.49 | 7.60 | 6.54 | 2.29 | 1.62 | 2.30 | 44.77 |
| | MAXIMUM MONTHLY (IN) | 63 | 8.02 | 10.82 | 12.64 | 10.71 | 17.64 | 13.75 | 20.59 | 18.59 | 13.98 | 7.36 | 6.12 | 15.57 | 20.59 |
| | YEAR OF OCCURRENCE | | 1948 | 1998 | 1959 | 1997 | 1979 | 1974 | 1960 | 1949 | 1979 | 1952 | 1963 | 1997 | JUL 1960 |
| | MINIMUM MONTHLY (IN) | 63 | T | 0.21 | T | T | 0.02 | 1.46 | 1.65 | 2.35 | .79 | 0.06 | T | 0.07 | T |
| | YEAR OF OCCURRENCE | | 1950 | 1950 | 2006 | 1981 | 2001 | 1997 | 1981 | 1952 | 2005 | 2000 | 1960 | 1984 | MAR 2006 |
| | MAXIMUM IN 24 HOURS (IN) | 63 | 3.81 | 8.54 | 5.20 | 5.44 | 11.84 | 5.53 | 12.11 | 5.37 | 8.45 | 2.93 | 4.48 | 4.76 | 12.11 |
| | YEAR OF OCCURRENCE | | 1996 | 2006 | 1960 | 1997 | 1979 | 1974 | 1960 | 1949 | 1997 | 1985 | 1988 | 1997 | JUL 1960 |
| | NORMAL NO. DAYS WITH: | | | | | | | | | | | | | | |
| | PRECIPITATION >= 0.01 | 30 | 7.1 | 6.4 | 6.6 | 4.7 | 6.2 | 11.7 | 14.9 | 16.0 | 12.4 | 6.5 | 5.5 | 6.3 | 104.3 |
| | PRECIPITATION >= 1.00 | 30 | 0.5 | 0.7 | 0.8 | 0.4 | 0.6 | 1.6 | 1.8 | 2.5 | 2.3 | 0.8 | 0.3 | 0.6 | 12.9 |
| **SNOWFALL** | NORMAL (IN) | 30 | 0.* | 0.0 | 0.0 | 0.0 | 0.0 | 0.0 | 0.0 | 0.0 | 0.0 | 0.0 | 0.0 | 0.* | 0.0 |
| | MAXIMUM MONTHLY (IN) | 63 | 0.2 | T | 0.0 | T | 0.0 | 0.0 | T | T | 0.0 | 0.0 | 0.0 | T | 0.2 |
| | YEAR OF OCCURRENCE | | 1977 | 1951 | 2006 | 1997 | | | 1998 | 2006 | | | | 1989 | JAN 1977 |
| | MAXIMUM IN 24 HOURS (IN) | 63 | 0.2 | T | T | T | 0.0 | 0.0 | T | T | 0.0 | 0.0 | 0.0 | T | 0.2 |
| | YEAR OF OCCURRENCE' | | 1977 | 1951 | 1980 | 1997 | | | 1998 | 2006 | | | | 1989 | JAN 1977 |
| | MAXIMUM SNOW DEPTH (IN) | 61 | 0 | 0 | 0 | 0 | 0 | 0 | 0 | 0 | 0 | 0 | 0 | 0 | 0 |
| | YEAR OF OCCURRENCE | | | | | | | | | | | | | | |
| | NORMAL NO. DAYS WITH: | | | | | | | | | | | | | | |
| | SNOWFALL >= 1.0 | 30 | 0.0 | 0.0 | 0.0 | 0.0 | 0.0 | 0.0 | 0.0 | 0.0 | 0.0 | 0.0 | 0.0 | 0.0 | 0.0 |

## PRECIPITATION (inches) 2009 TAMPA (KTPA)

| YEAR | JAN | FEB | MAR | APR | MAY | JUN | JUL | AUG | SEP | OCT | NOV | DEC | ANNUAL |
|------|-----|-----|-----|-----|-----|-----|-----|-----|-----|-----|-----|-----|--------|
| 1980 | 1.72 | 2.01 | 3.09 | 4.38 | 3.94 | 3.81 | 5.66 | 7.62 | 4.05 | 1.27 | 2.68 | 0.37 | 40.60 |
| 1981 | 0.44 | 5.34 | 1.70 | T | 1.68 | 9.37 | 1.65 | 7.71 | 5.87 | 0.87 | 0.43 | 3.58 | 38.64 |
| 1982 | 1.86 | 2.09 | 2.99 | 1.87 | 5.90 | 8.34 | 10.49 | 7.20 | 10.76 | 2.17 | 0.85 | 1.29 | 55.81 |
| 1983 | 1.25 | 7.35 | 7.59 | 2.76 | 4.10 | 7.17 | 6.37 | 8.89 | 6.61 | 1.74 | 2.33 | 4.71 | 60.87 |
| 1984 | 1.62 | 3.32 | 1.31 | 1.51 | 3.19 | 3.24 | 7.15 | 5.68 | 4.21 | 0.29 | 0.72 | 0.07 | 32.31 |
| 1985 | 2.06 | 2.07 | 1.80 | 0.96 | 0.22 | 6.43 | 6.48 | 8.65 | 9.04 | 4.77 | 0.99 | 1.13 | 44.60 |
| 1986 | 2.37 | 1.49 | 4.27 | 0.95 | 2.46 | 5.00 | 6.24 | 5.46 | 3.87 | 6.21 | 1.33 | 1.95 | 41.60 |
| 1987 | 3.29 | 1.50 | 12.01 | 0.39 | 2.86 | 3.39 | 6.06 | 8.50 | 4.76 | 1.46 | 4.36 | 0.50 | 49.08 |
| 1988 | 2.76 | 1.44 | 4.09 | 1.83 | 1.27 | 5.19 | 3.40 | 11.09 | 13.56 | 0.09 | 5.97 | 1.64 | 52.33 |
| 1989 | 1.54 | 0.41 | 1.79 | 0.71 | 0.24 | 7.41 | 8.86 | 7.90 | 6.11 | 1.89 | 2.05 | 4.72 | 43.63 |
| 1990 | 0.53 | 4.58 | 1.71 | 1.47 | 1.76 | 5.16 | 10.01 | 3.27 | 2.42 | 2.63 | 0.66 | 0.19 | 34.39 |
| 1991 | 2.41 | 0.41 | 4.73 | 1.54 | 6.88 | 3.78 | 9.92 | 7.35 | 3.43 | 0.78 | 1.26 | 0.67 | 43.16 |
| 1992 | 1.47 | 3.67 | 0.95 | 2.17 | 0.10 | 7.03 | 2.80 | 8.22 | 2.95 | 2.20 | 2.43 | 0.99 | 34.98 |
| 1993 | 3.60 | 2.32 | 3.93 | 2.45 | 1.74 | 3.18 | 2.92 | 5.06 | 6.60 | 4.23 | 0.22 | 1.28 | 37.53 |
| 1994 | 3.68 | 0.43 | 0.66 | 3.43 | 0.07 | 5.98 | 11.31 | 8.37 | 8.20 | 3.29 | 0.24 | 1.57 | 47.23 |
| 1995 | 3.51 | 2.02 | 2.02 | 1.48 | 1.67 | 9.79 | 10.12 | 13.75 | 2.80 | 4.71 | 1.24 | 1.02 | 54.13 |
| 1996 | 5.42 | 3.04 | 4.65 | 4.20 | 1.45 | 8.96 | 2.72 | 7.39 | 5.44 | 3.12 | .91 | 2.11 | 49.41 |
| 1997 | 0.95 | 0.66 | 1.28 | 10.71 | 1.70 | 1.46 | 6.73 | 8.20 | 12.84 | 4.20 | 3.41 | 15.57 | 67.71 |
| 1998 | 4.64 | 10.82 | 5.16 | 0.41 | 1.96 | 2.65 | 12.95 | 6.55 | 8.42 | 0.47 | 0.40 | 0.92 | 55.35 |
| 1999 | 3.04 | 0.29 | 0.72 | 0.40 | 1.52 | 4.65 | 3.65 | 8.35 | 6.05 | 2.85 | 1.78 | 1.02 | 34.32 |
| 2000 | 1.95 | 0.30 | 0.41 | 0.43 | 0.02 | 4.53 | 8.14 | 5.44 | 5.14 | 0.06 | 2.04 | 1.39 | 29.85 |
| 2001 | 1.03 | 1.18 | 6.73 | 0.02 | T | 6.81 | 6.01 | 2.83 | 11.76 | 2.39 | 0.10 | 0.89 | 39.75 |
| 2002 | 2.49 | 2.84 | 0.63 | 1.84 | 1.07 | 11.57 | 7.33 | 8.82 | 7.51 | 2.11 | 1.76 | 14.10 | 62.07 |
| 2003 | 0.11 | 2.90 | 3.94 | 4.19 | 2.50 | 13.19 | 3.63 | 14.90 | 4.01 | 0.46 | 0.86 | 1.30 | 51.99 |
| 2004 | 3.73 | 4.02 | 1.11 | 2.04 | 1.44 | 9.01 | 10.19 | 14.03 | 9.77 | 1.70 | 0.73 | 1.54 | 59.31 |
| 2005 | 0.57 | 1.80 | 3.32 | 2.76 | 3.61 | 12.26 | 3.38 | 4.09 | 0.79 | 4.20 | 0.90 | 1.27 | 38.95 |
| 2006 | 0.74 | 9.09 | T | 1.03 | 1.43 | 8.93 | 9.46 | 6.78 | 12.40 | 0.87 | 2.76 | 3.17 | 56.66 |
| 2007 | 1.43 | 1.77 | 0.92 | 1.92 | 0.35 | 8.70 | 7.78 | 10.73 | 4.87 | 2.11 | 0.11 | 1.30 | 41.99 |
| 2008 | 2.54 | 4.28 | 3.67 | 2.64 | 0.73 | 7.54 | 9.84 | 4.86 | 2.27 | 3.52 | 0.65 | 1.23 | 43.77 |
| 2009 | 2.38 | 0.71 | 0.98 | 1.22 | 9.11 | 5.10 | 10.23 | 3.52 | 5.24 | 2.24 | 2.82 | 2.32 | 45.87 |
| POR= 77 YRS | 2.15 | 2.96 | 3.06 | 2.03 | 2.69 | 6.61 | 7.61 | 7.83 | 6.39 | 2.43 | 1.56 | 2.25 | 47.57 |

WBAN : 12842

## AVERAGE TEMPERATURE (°F) 2009 TAMPA (KTPA)

| YEAR | JAN | FEB | MAR | APR | MAY | JUN | JUL | AUG | SEP | OCT | NOV | DEC | ANNUAL |
|------|-----|-----|-----|-----|-----|-----|-----|-----|-----|-----|-----|-----|--------|
| 1980 | 62.0 | 56.6 | 68.1 | 70.1 | 77.2 | 81.6 | 84.0 | 83.0 | 81.3 | 74.0 | 66.4 | 57.5 | 71.8 |
| 1981 | 50.4 | 61.4 | 62.8 | 72.4 | 75.4 | 81.5 | 82.5 | 81.7 | 78.6 | 74.5 | 64.4 | 59.2 | 70.4 |
| 1982 | 59.8 | 67.9 | 68.1 | 71.4 | 74.4 | 81.5 | 82.1 | 82.1 | 80.2 | 74.3 | 70.8 | 67.6 | 73.4 |
| 1983 | 58.9 | 60.3 | 63.3 | 68.6 | 76.8 | 80.9 | 82.2 | 82.2 | 79.4 | 75.8 | 65.9 | 59.9 | 71.2 |
| 1984 | 58.0 | 62.6 | 66.0 | 71.0 | 78.0 | 80.4 | 81.5 | 82.4 | 79.9 | 75.7 | 64.9 | 67.3 | 72.3 |
| 1985 | 55.9 | 63.6 | 69.4 | 72.5 | 79.8 | 83.7 | 82.4 | 83.1 | 80.5 | 79.2 | 73.6 | 59.0 | 73.6 |
| 1986 | 59.3 | 65.0 | 65.4 | 69.1 | 77.4 | 81.8 | 83.0 | 82.6 | 82.3 | 76.8 | 76.3 | 66.5 | 73.8 |
| 1987 | 59.2 | 63.2 | 66.4 | 66.4 | 77.8 | 82.7 | 83.1 | 83.7 | 81.4 | 71.3 | 68.9 | 64.3 | 72.4 |
| 1988 | 58.6 | 59.1 | 65.6 | 70.6 | 75.3 | 81.0 | 82.7 | 82.9 | 82.0 | 73.5 | 70.8 | 63.0 | 72.1 |
| 1989 | 67.1 | 64.9 | 69.8 | 72.0 | 78.4 | 82.4 | 83.3 | 83.0 | 82.4 | 75.4 | 68.9 | 56.2 | 73.7 |
| 1990 | 66.1 | 69.2 | 69.7 | 72.1 | 80.5 | 82.7 | 82.5 | 83.9 | 82.8 | 77.6 | 70.2 | 66.9 | 75.4 |
| 1991 | 66.7 | 64.2 | 68.4 | 76.8 | 81.2 | 81.3 | 82.3 | 83.2 | 81.9 | 75.3 | 65.8 | 64.6 | 74.3 |
| 1992 | 59.8 | 63.6 | 64.8 | 69.4 | 74.3 | 82.1 | 83.8 | 82.4 | 82.0 | 72.7 | 70.0 | 64.3 | 72.4 |
| 1993 | 67.0 | 60.2 | 64.3 | 67.2 | 76.1 | 81.8 | 83.8 | 83.7 | 81.9 | 75.8 | 69.1 | 59.5 | 72.5 |
| 1994 | 60.6 | 66.9 | 68.0 | 75.4 | 78.2 | 82.6 | 81.7 | 81.7 | 80.2 | 76.5 | 72.4 | 65.1 | 74.1 |
| 1995 | 58.8 | 61.4 | 68.7 | 73.5 | 81.8 | 80.2 | 83.0 | 83.3 | 82.0 | 77.8 | 65.2 | 61.0 | 73.1 |
| 1996 | 59.2 | 60.0 | 62.4 | 70.4 | 79.3 | 80.9 | 83.7 | 83.2 | 82.0 | 75.8 | 68.1 | 63.5 | 72.4 |
| 1997 | 62.6 | 68.6 | 73.9 | 71.6 | 77.7 | 81.8 | 82.7 | 82.8 | 81.7 | 74.7 | 66.8 | 61.3 | 73.9 |
| 1998 | 63.6 | 62.5 | 64.6 | 72.3 | 79.2 | 85.6 | 83.5 | 83.6 | 81.7 | 77.5 | 72.5 | 68.0 | 74.6 |
| 1999 | 63.8 | 64.2 | 65.1 | 74.3 | 77.8 | 81.2 | 83.6 | 83.7 | 80.9 | 76.3 | 68.9 | 63.2 | 73.6 |
| 2000 | 61.3 | 63.7 | 71.0 | 71.6 | 80.3 | 82.4 | 82.1 | 82.8 | 82.3 | 74.3 | 66.7 | 60.3 | 73.2 |
| 2001 | 55.2 | 68.2 | 65.7 | 72.7 | 77.4 | 82.2 | 82.1 | 83.2 | 79.3 | 75.1 | 71.3 | 68.8 | 73.4 |
| 2002 | 63.1 | 62.4 | 69.7 | 76.3 | 79.9 | 81.4 | 82.3 | 82.4 | 82.8 | 78.9 | 65.7 | 60.1 | 73.8 |
| 2003 | 54.8 | 63.3 | 71.7 | 72.1 | 80.0 | 81.7 | 82.7 | 81.9 | 81.3 | 76.9 | 71.4 | 59.8 | 73.1 |
| 2004 | 59.6 | 63.0 | 68.0 | 70.2 | 78.9 | 83.3 | 82.7 | 82.3 | 81.2 | 76.8 | 70.2 | 61.3 | 73.1 |
| 2005 | 62.6 | 64.0 | 65.1 | 69.0 | 76.6 | 80.7 | 84.2 | 84.8 | 82.9 | 76.0 | 70.4 | 59.3 | 73.0 |
| 2006 | 62.6 | 60.6 | 67.3 | 74.2 | 77.3 | 81.7 | 82.7 | 83.3 | 82.1 | 76.1 | 67.9 | 69.5 | 73.8 |
| 2007 | 64.9 | 60.8 | 69.5 | 71.0 | 78.0 | 81.5 | 83.7 | 84.9 | 82.6 | 79.8 | 68.7 | 68.4 | 74.5 |
| 2008 | 62.1 | 66.0 | 67.7 | 72.5 | 79.8 | 83.3 | 81.9 | 82.2 | 81.9 | 75.1 | 65.3 | 66.1 | 73.7 |
| 2009 | 61.2 | 61.8 | 69.3 | 73.8 | 78.7 | 83.0 | 83.2 | 83.9 | 82.6 | 78.0 | 68.9 | 63.2 | 74.0 |
| POR= 77 YRS | 60.8 | 61.2 | 66.1 | 71.3 | 77.7 | 80.8 | 82.4 | 82.6 | 80.6 | 75.2 | 67.3 | 62.4 | 72.4 |

## HEATING DEGREE DAYS (base 65°F) 2009  TAMPA (KTPA)

| YEAR | JUL | AUG | SEP | OCT | NOV | DEC | JAN | FEB | MAR | APR | MAY | JUN | TOTAL |
|---|---|---|---|---|---|---|---|---|---|---|---|---|---|
| 1980-81 | 0 | 0 | 0 | 1 | 65 | 233 | 447 | 127 | 103 | 2 | 0 | 0 | 978 |
| 1981-82 | 0 | 0 | 0 | 0 | 83 | 223 | 209 | 24 | 53 | 8 | 0 | 0 | 600 |
| 1982-83 | 0 | 0 | 0 | 12 | 18 | 95 | 218 | 148 | 103 | 20 | 0 | 0 | 614 |
| 1983-84 | 0 | 0 | 0 | 0 | 57 | 214 | 252 | 115 | 68 | 5 | 0 | 0 | 711 |
| 1984-85 | 0 | 0 | 0 | 0 | 87 | 61 | 306 | 119 | 17 | 5 | 0 | 0 | 595 |
| 1985-86 | 0 | 0 | 0 | 0 | 9 | 238 | 185 | 78 | 105 | 7 | 0 | 0 | 622 |
| 1986-87 | 0 | 0 | 0 | 0 | 0 | 53 | 202 | 88 | 42 | 64 | 0 | 0 | 449 |
| 1987-88 | 0 | 0 | 0 | 4 | 46 | 107 | 221 | 195 | 85 | 14 | 0 | 0 | 672 |
| 1988-89 | 0 | 0 | 0 | 0 | 9 | 127 | 41 | 116 | 45 | 7 | 0 | 0 | 345 |
| 1989-90 | 0 | 0 | 0 | 17 | 27 | 285 | 70 | 32 | 13 | 5 | 0 | 0 | 449 |
| 1990-91 | 0 | 0 | 0 | 7 | 11 | 70 | 72 | 84 | 46 | 0 | 0 | 0 | 290 |
| 1991-92 | 0 | 0 | 0 | 0 | 93 | 94 | 179 | 90 | 69 | 32 | 5 | 0 | 562 |
| 1992-93 | 0 | 0 | 0 | 0 | 57 | 83 | 58 | 137 | 84 | 24 | 0 | 0 | 443 |
| 1993-94 | 0 | 0 | 0 | 6 | 44 | 185 | 158 | 62 | 50 | 5 | 0 | 0 | 510 |
| 1994-95 | 0 | 0 | 0 | 0 | 7 | 81 | 200 | 151 | 23 | 1 | 0 | 0 | 463 |
| 1995-96 | 0 | 0 | 0 | 0 | 83 | 180 | 198 | 188 | 152 | 16 | 0 | 0 | 817 |
| 1996-97 | 0 | 0 | 0 | 3 | 36 | 101 | 132 | 39 | 0 | 7 | 0 | 0 | 318 |
| 1997-98 | 0 | 0 | 0 | 7 | 36 | 163 | 108 | 103 | 113 | 4 | 0 | 0 | 534 |
| 1998-99 | 0 | 0 | 0 | 0 | 4 | 56 | 118 | 97 | 44 | 6 | 5 | 0 | 330 |
| 1999-00 | 0 | 0 | 0 | 5 | 20 | 110 | 154 | 99 | 3 | 10 | 0 | 0 | 401 |
| 2000-01 | 0 | 0 | 0 | 3 | 61 | 212 | 318 | 48 | 60 | 8 | 0 | 0 | 710 |
| 2001-02 | 0 | 0 | 0 | 9 | 0 | 65 | 165 | 100 | 43 | 0 | 0 | 0 | 382 |
| 2002-03 | 0 | 0 | 0 | 0 | 84 | 174 | 311 | 87 | 11 | 16 | 0 | 0 | 683 |
| 2003-04 | 0 | 0 | 0 | 0 | 21 | 175 | 195 | 110 | 29 | 15 | 0 | 0 | 545 |
| 2004-05 | 0 | 0 | 0 | 0 | 7 | 160 | 131 | 71 | 93 | 13 | 0 | 0 | 475 |
| 2005-06 | 0 | 0 | 0 | 13 | 18 | 176 | 130 | 145 | 41 | 0 | 0 | 0 | 523 |
| 2006-07 | 0 | 0 | 0 | 4 | 55 | 40 | 95 | 147 | 34 | 27 | 0 | 0 | 402 |
| 2007-08 | 0 | 0 | 0 | 0 | 24 | 42 | 141 | 63 | 34 | 13 | 0 | 0 | 317 |
| 2008-09 | 0 | 0 | 0 | 22 | 84 | 67 | 162 | 126 | 38 | 9 | 0 | 0 | 508 |
| 2009- | 0 | 0 | 0 | 11 | 29 | 125 | | | | | | | |

WBAN : 12842

## COOLING DEGREE DAYS (base 65°F) 2009  TAMPA (KTPA)

| YEAR | JAN | FEB | MAR | APR | MAY | JUN | JUL | AUG | SEP | OCT | NOV | DEC | TOTAL |
|---|---|---|---|---|---|---|---|---|---|---|---|---|---|
| 1980 | 45 | 22 | 164 | 165 | 386 | 506 | 598 | 564 | 493 | 284 | 115 | 7 | 3349 |
| 1981 | 0 | 32 | 43 | 230 | 331 | 501 | 552 | 525 | 414 | 303 | 71 | 49 | 3051 |
| 1982 | 56 | 114 | 156 | 208 | 299 | 499 | 537 | 537 | 467 | 311 | 197 | 182 | 3563 |
| 1983 | 36 | 24 | 57 | 137 | 369 | 487 | 540 | 541 | 439 | 342 | 91 | 64 | 3127 |
| 1984 | 42 | 52 | 104 | 190 | 410 | 468 | 517 | 546 | 454 | 337 | 92 | 135 | 3347 |
| 1985 | 30 | 88 | 163 | 237 | 464 | 569 | 547 | 566 | 475 | 445 | 275 | 58 | 3917 |
| 1986 | 15 | 85 | 123 | 139 | 391 | 510 | 565 | 551 | 526 | 374 | 348 | 107 | 3734 |
| 1987 | 27 | 43 | 91 | 114 | 405 | 538 | 567 | 583 | 497 | 207 | 169 | 91 | 3332 |
| 1988 | 30 | 32 | 110 | 188 | 326 | 489 | 554 | 562 | 517 | 271 | 191 | 74 | 3344 |
| 1989 | 112 | 120 | 202 | 224 | 425 | 528 | 575 | 564 | 529 | 344 | 151 | 18 | 3792 |
| 1990 | 107 | 154 | 164 | 225 | 487 | 537 | 549 | 592 | 541 | 406 | 176 | 139 | 4077 |
| 1991 | 131 | 68 | 158 | 361 | 509 | 498 | 543 | 572 | 515 | 326 | 126 | 88 | 3895 |
| 1992 | 25 | 55 | 70 | 170 | 301 | 519 | 589 | 544 | 518 | 248 | 212 | 72 | 3323 |
| 1993 | 126 | 12 | 72 | 95 | 352 | 511 | 593 | 587 | 513 | 347 | 174 | 21 | 3403 |
| 1994 | 32 | 118 | 148 | 321 | 415 | 535 | 524 | 524 | 461 | 364 | 233 | 91 | 3766 |
| 1995 | 15 | 59 | 145 | 264 | 526 | 464 | 561 | 574 | 518 | 405 | 95 | 63 | 3689 |
| 1996 | 27 | 50 | 76 | 182 | 450 | 482 | 589 | 570 | 516 | 348 | 135 | 63 | 3488 |
| 1997 | 62 | 145 | 283 | 213 | 397 | 510 | 556 | 559 | 508 | 314 | 97 | 55 | 3699 |
| 1998 | 75 | 41 | 107 | 229 | 446 | 623 | 582 | 583 | 506 | 394 | 236 | 160 | 3982 |
| 1999 | 85 | 84 | 52 | 294 | 409 | 492 | 583 | 589 | 481 | 363 | 144 | 62 | 3638 |
| 2000 | 48 | 68 | 197 | 213 | 482 | 529 | 537 | 558 | 524 | 300 | 121 | 71 | 3648 |
| 2001 | 22 | 141 | 92 | 245 | 393 | 522 | 536 | 571 | 434 | 326 | 196 | 187 | 3665 |
| 2002 | 114 | 34 | 192 | 347 | 466 | 499 | 543 | 544 | 541 | 437 | 114 | 28 | 3859 |
| 2003 | 2 | 48 | 226 | 236 | 473 | 505 | 553 | 532 | 495 | 375 | 219 | 19 | 3683 |
| 2004 | 33 | 58 | 130 | 179 | 440 | 553 | 558 | 542 | 493 | 371 | 169 | 53 | 3579 |
| 2005 | 61 | 47 | 103 | 143 | 366 | 479 | 601 | 619 | 547 | 362 | 186 | 7 | 3521 |
| 2006 | 63 | 30 | 120 | 280 | 388 | 508 | 556 | 575 | 519 | 357 | 150 | 188 | 3734 |
| 2007 | 99 | 35 | 179 | 214 | 409 | 503 | 588 | 624 | 539 | 466 | 142 | 155 | 3953 |
| 2008 | 60 | 99 | 125 | 243 | 463 | 555 | 531 | 540 | 512 | 342 | 101 | 107 | 3678 |
| 2009 | 48 | 40 | 175 | 279 | 432 | 548 | 571 | 593 | 533 | 425 | 155 | 77 | 3876 |

## SNOWFALL (inches) 2009 TAMPA (KTPA)

| YEAR | JUL | AUG | SEP | OCT | NOV | DEC | JAN | FEB | MAR | APR | MAY | JUN | TOTAL |
|------|-----|-----|-----|-----|-----|-----|-----|-----|-----|-----|-----|-----|-------|
| 1980-81 | 0.0 | 0.0 | 0.0 | 0.0 | 0.0 | 0.0 | 0.0 | 0.0 | 0.0 | 0.0 | 0.0 | 0.0 | 0.0 |
| 1981-82 | 0.0 | 0.0 | 0.0 | 0.0 | 0.0 | 0.0 | 0.0 | 0.0 | 0.0 | 0.0 | 0.0 | 0.0 | 0.0 |
| 1982-83 | 0.0 | 0.0 | 0.0 | 0.0 | 0.0 | 0.0 | 0.0 | 0.0 | 0.0 | 0.0 | 0.0 | 0.0 | 0.0 |
| 1983-84 | 0.0 | 0.0 | 0.0 | 0.0 | 0.0 | 0.0 | 0.0 | 0.0 | 0.0 | 0.0 | 0.0 | 0.0 | 0.0 |
| 1984-85 | 0.0 | 0.0 | 0.0 | 0.0 | 0.0 | 0.0 | 0.0 | 0.0 | 0.0 | 0.0 | 0.0 | 0.0 | 0.0 |
| 1985-86 | 0.0 | 0.0 | 0.0 | 0.0 | 0.0 | 0.0 | 0.0 | 0.0 | 0.0 | 0.0 | 0.0 | 0.0 | 0.0 |
| 1986-87 | 0.0 | 0.0 | 0.0 | 0.0 | 0.0 | 0.0 | 0.0 | 0.0 | 0.0 | 0.0 | 0.0 | 0.0 | 0.0 |
| 1987-88 | 0.0 | 0.0 | 0.0 | 0.0 | 0.0 | 0.0 | 0.0 | 0.0 | 0.0 | 0.0 | 0.0 | 0.0 | 0.0 |
| 1988-89 | 0.0 | 0.0 | 0.0 | 0.0 | 0.0 | 0.0 | 0.0 | 0.0 | 0.0 | 0.0 | 0.0 | 0.0 | 0.0 |
| 1989-90 | 0.0 | 0.0 | 0.0 | 0.0 | 0.0 | T | 0.0 | 0.0 | 0.0 | 0.0 | 0.0 | 0.0 | T |
| 1990-91 | 0.0 | 0.0 | 0.0 | 0.0 | 0.0 | 0.0 | 0.0 | 0.0 | 0.0 | 0.0 | 0.0 | 0.0 | 0.0 |
| 1991-92 | 0.0 | 0.0 | 0.0 | 0.0 | 0.0 | 0.0 | 0.0 | 0.0 | 0.0 | 0.0 | 0.0 | 0.0 | 0.0 |
| 1992-93 | 0.0 | 0.0 | 0.0 | 0.0 | 0.0 | 0.0 | 0.0 | 0.0 | 0.0 | 0.0 | 0.0 | 0.0 | 0.0 |
| 1993-94 | 0.0 | 0.0 | 0.0 | 0.0 | 0.0 | 0.0 | 0.0 | 0.0 | 0.0 | 0.0 | 0.0 | 0.0 | 0.0 |
| 1994-95 | 0.0 | 0.0 | 0.0 | 0.0 | 0.0 | 0.0 | 0.0 | 0.0 | 0.0 | 0.0 | 0.0 | 0.0 | 0.0 |
| 1995-96 | 0.0 | 0.0 | 0.0 | 0.0 | 0.0 | 0.0 | 0.0 | 0.0 | 0.0 | 0.0 | 0.0 | 0.0 | 0.0 |
| 1996-97 | 0.0 | 0.0 | 0.0 | 0.0 | 0.0 | 0.0 | T | 0.0 | 0.0 | T | 0.0 | 0.0 | T |
| 1997-98 | 0.0 | 0.0 | 0.0 | 0.0 | 0.0 | 0.0 | 0.0 | 0.0 | 0.0 | 0.0 | 0.0 | 0.0 | 0.0 |
| 1998-99 | T | 0.0 | 0.0 | 0.0 | 0.0 | 0.0 | 0.0 | 0.0 | 0.0 | 0.0 | 0.0 | 0.0 | T |
| 1999-00 | 0.0 | 0.0 | 0.0 | 0.0 | 0.0 | 0.0 | 0.0 | 0.0 | 0.0 | 0.0 | 0.0 | 0.0 | 0.0 |
| 2000-01 | 0.0 | 0.0 | 0.0 | 0.0 | 0.0 | 0.0 | 0.0 | 0.0 | 0.0 | 0.0 | 0.0 | 0.0 | 0.0 |
| 2001-02 | 0.0 | 0.0 | 0.0 | 0.0 | 0.0 | 0.0 | 0.0 | 0.0 | 0.0 | 0.0 | 0.0 | 0.0 | 0.0 |
| 2002-03 | 0.0 | 0.0 | 0.0 | 0.0 | 0.0 | 0.0 | 0.0 | 0.0 | 0.0 | 0.0 | 0.0 | 0.0 | 0.0 |
| 2003-04 | 0.0 | 0.0 | 0.0 | 0.0 | 0.0 | 0.0 | 0.0 | 0.0 | 0.0 | 0.0 | 0.0 | 0.0 | 0.0 |
| 2004-05 | 0.0 | 0.0 | 0.0 | 0.0 | 0.0 | 0.0 | 0.0 | 0.0 | 0.0 | 0.0 | 0.0 | 0.0 | 0.0 |
| 2005-06 | 0.0 | 0.0 | 0.0 | 0.0 | 0.0 | 0.0 | 0.0 | 0.0 | 0.0 | 0.0 | 0.0 | 0.0 | 0.0 |
| 2006-07 | 0.0 | T | 0.0 | 0.0 | 0.0 | 0.0 | 0.0 | 0.0 | 0.0 | 0.0 | 0.0 | 0.0 | T |
| 2007-08 | 0.0 | 0.0 | 0.0 | 0.0 | 0.0 | 0.0 | 0.0 | 0.0 | 0.0 | 0.0 | 0.0 | 0.0 | 0.0 |
| 2008-09 | 0.0 | 0.0 | 0.0 | 0.0 | 0.0 | 0.0 | 0.0 | 0.0 | 0.0 | 0.0 | 0.0 | 0.0 | 0.0 |
| 2009- | 0.0 | 0.0 | 0.0 | 0.0 | 0.0 | 0.0 | | | | | | | |
| POR= 77 YRS | T | T | 0.0 | 0.0 | 0.0 | T | T | 0.0 | 0.0 | T | 0.0 | 0.0 | T |

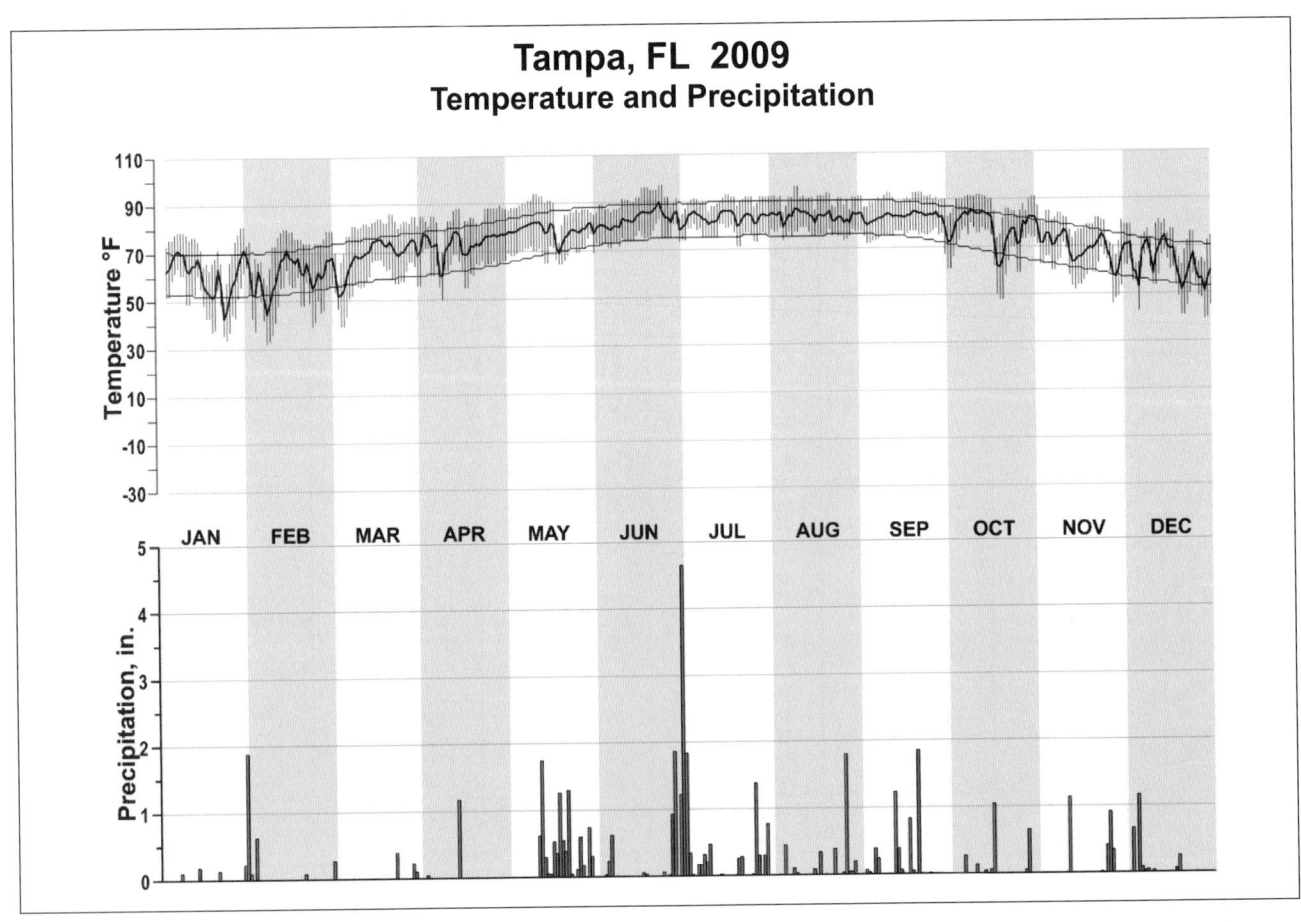

# Tampa, FL 2009
## Temperature and Precipitation

# 2009
# ATLANTA
# GEORGIA (KATL)

Atlanta is located in the foothills of the southern Appalachians in north-central Georgia. The terrain is rolling to hilly and slopes downward toward the east, west, and south so that drainage of the major river systems is generally into the Gulf of Mexico from the western and southern sections of the city and to the Atlantic from the eastern portions of the city.

The Gulf of Mexico and the Atlantic Ocean are approximately 250 miles south and southeast of the city, respectively. Both the Appalachian chain of mountains and the two nearby maritime bodies exert an important influence on the Atlanta climate. Temperatures are moderated throughout the year while abundant precipitation fosters natural vegetation and growth of crops. Summer temperatures in Atlanta are moderated somewhat by elevation but are still rather warm. However, prolonged periods of hot weather are unusual and 100 degree heat is rarely experienced.

With the mountains to the north tending to retard the southward movement of Polar air masses, Atlanta winters are rather mild. Cold spells are not unusual but they are rather short-lived and seldom disrupt outdoor activities for an extended period of time. Late March is the average date of the last temperature of 32 degrees in the spring and mid-November is the average date of the first temperature of 32 degrees in the fall, which gives an average growing season of about 234 days.

Minimum dry precipitation periods occur mainly during the late summer and early autumn. Maximum thunderstorm activity occurs during July, but severe local thunderstorms occur most frequently in March, April, and May, some spawning highly damaging tornadoes.

The average annual snowfall varies widely from year to year. A fall of 4 inches or more occurs about once every five years. Most snows melt in a short period of time due to the rapid warming which often follows the storm. Ice storms, freezing rain or glaze, occur about two out of every three years, causing hazardous travel and disruption of utilities. Severe ice storms occur about once in ten years, causing major disruption of utilities and significant property damage.

The Bermuda High pressure area has a dominant effect on Atlanta weather, particularly in the summer months. East or northeast winds produce the most unpleasant weather although southerly winds are quite humid during the summer. The generally light wind conditions contribute to the formation of an occasional early morning fog.

# NORMALS, MEANS, AND EXTREMES
## ATLANTA (KATL)

| LATITUDE: 33°38'N | LONGITUDE: -84°25'W | ELEVATION (FT): GRND: 998 BARO: 974 | TIME ZONE: EASTERN (UTC -5) | WBAN: 13874 |
|---|---|---|---|---|

| | ELEMENT | POR | JAN | FEB | MAR | APR | MAY | JUN | JUL | AUG | SEP | OCT | NOV | DEC | YEAR |
|---|---|---|---|---|---|---|---|---|---|---|---|---|---|---|---|
| **TEMPERATURE °F** | NORMAL DAILY MAXIMUM | 30 | 51.9 | 56.8 | 65.0 | 72.9 | 80.0 | 86.5 | 89.4 | 87.9 | 82.3 | 72.9 | 63.3 | 54.6 | 72.0 |
| | MEAN DAILY MAXIMUM | 80 | 52.3 | 55.2 | 63.8 | 72.1 | 80.3 | 86.1 | 88.7 | 87.9 | 81.9 | 73.0 | 62.3 | 54.0 | 71.5 |
| | HIGHEST DAILY MAXIMUM | 61 | 79 | 80 | 89 | 93 | 95 | 101 | 105 | 104 | 98 | 95 | 84 | 79 | 105 |
| | YEAR OF OCCURRENCE | | 1949 | 1996 | 1995 | 1986 | 1996 | 1952 | 1980 | 2007 | 1954 | 1954 | 1961 | 1991 | JUL 1980 |
| | MEAN OF EXTREME MAXS. | 80 | 69.6 | 72.7 | 80.0 | 84.9 | 89.3 | 94.3 | 95.5 | 94.6 | 91.3 | 84.2 | 77.2 | 70.8 | 83.7 |
| | NORMAL DAILY MINIMUM | 30 | 33.5 | 36.5 | 43.6 | 50.4 | 59.5 | 67.1 | 70.6 | 69.9 | 64.3 | 52.8 | 43.5 | 36.2 | 52.3 |
| | MEAN DAILY MINIMUM | 80 | 34.2 | 35.8 | 42.8 | 50.6 | 59.6 | 66.5 | 70.1 | 69.6 | 63.8 | 52.9 | 42.5 | 36.0 | 52.0 |
| | LOWEST DAILY MINIMUM | 61 | -8 | 5 | 10 | 26 | 37 | 46 | 53 | 55 | 36 | 28 | 3 | 0 | -8 |
| | YEAR OF OCCURRENCE | | 1985 | 1958 | 1960 | 1973 | 1971 | 1956 | 1967 | 1992 | 1967 | 1976 | 1950 | 1983 | JAN 1985 |
| | MEAN OF EXTREME MINS. | 80 | 15.1 | 19.6 | 26.0 | 35.0 | 46.7 | 57.0 | 64.0 | 62.8 | 51.3 | 37.8 | 26.9 | 19.1 | 38.4 |
| | NORMAL DRY BULB | 30 | 42.7 | 46.7 | 54.3 | 61.6 | 69.8 | 76.8 | 80.0 | 78.9 | 73.3 | 62.8 | 53.4 | 45.4 | 62.1 |
| | MEAN DRY BULB | 80 | 43.3 | 45.5 | 53.3 | 61.4 | 69.9 | 76.4 | 79.4 | 78.7 | 72.9 | 63.0 | 52.4 | 45.0 | 61.8 |
| | MEAN WET BULB | 26 | 37.9 | 40.6 | 46.4 | 53.0 | 61.4 | 67.9 | 71.1 | 70.8 | 65.4 | 56.1 | 47.3 | 40.2 | 54.8 |
| | MEAN DEW POINT | 26 | 33.2 | 35.9 | 41.0 | 47.9 | 57.9 | 65.3 | 68.8 | 68.4 | 62.6 | 52.5 | 43.3 | 35.8 | 51.1 |
| | NORMAL NO. DAYS WITH: | | | | | | | | | | | | | | |
| | MAXIMUM >= 90 | 30 | 0.0 | 0.0 | 0.0 | 0.1 | 1.3 | 9.5 | 15.5 | 11.3 | 3.7 | 0.0 | 0.0 | 0.0 | 41.4 |
| | MAXIMUM <= 32 | 30 | 1.2 | 0.5 | 0.1 | 0.0 | 0.0 | 0.0 | 0.0 | 0.0 | 0.0 | 0.0 | 0.0 | 0.5 | 2.3 |
| | MINIMUM <= 32 | 30 | 14.3 | 10.4 | 4.0 | 0.5 | 0.0 | 0.0 | 0.0 | 0.0 | 0.0 | * | 4.1 | 11.3 | 44.6 |
| | MINIMUM <= 0 | 30 | 0.2 | 0.0 | 0.0 | 0.0 | 0.0 | 0.0 | 0.0 | 0.0 | 0.0 | 0.0 | 0.0 | * | 0.2 |
| **H/C** | NORMAL HEATING DEG. DAYS | 30 | 692 | 523 | 346 | 150 | 26 | 1 | 0 | 0 | 11 | 126 | 352 | 600 | 2827 |
| | NORMAL COOLING DEG. DAYS | 30 | 0 | 1 | 11 | 52 | 170 | 354 | 463 | 430 | 262 | 58 | 8 | 1 | 1810 |
| **RH** | NORMAL (PERCENT) | 30 | 69 | 64 | 62 | 60 | 67 | 69 | 73 | 73 | 73 | 69 | 69 | 69 | 68 |
| | HOUR 01 LST | 30 | 74 | 69 | 68 | 69 | 77 | 79 | 83 | 84 | 82 | 79 | 76 | 75 | 76 |
| | HOUR 07 LST | 30 | 79 | 77 | 77 | 78 | 82 | 83 | 87 | 89 | 87 | 85 | 82 | 80 | 82 |
| | HOUR 13 LST | 30 | 60 | 54 | 52 | 48 | 53 | 55 | 58 | 59 | 59 | 54 | 56 | 59 | 56 |
| | HOUR 19 LST | 30 | 64 | 56 | 53 | 49 | 57 | 59 | 64 | 65 | 66 | 63 | 64 | 64 | 60 |
| **S** | PERCENT POSSIBLE SUNSHINE | 63 | 49 | 54 | 58 | 66 | 68 | 67 | 62 | 64 | 62 | 66 | 58 | 50 | 60 |
| **W/O** | MEAN NO. DAYS WITH: | | | | | | | | | | | | | | |
| | HEAVY FOG(VISBY <= 1/4 MI) | 46 | 4.4 | 2.9 | 2.2 | 1.2 | 1.2 | 0.9 | 1.0 | 1.3 | 2.0 | 2.0 | 3.0 | 3.8 | 25.9 |
| | THUNDERSTORMS | 64 | 1.3 | 1.8 | 3.5 | 4.0 | 6.1 | 8.0 | 10.2 | 7.6 | 3.0 | 1.0 | 1.1 | 0.9 | 48.5 |
| **CLOUDNESS** | MEAN: | | | | | | | | | | | | | | |
| | SUNRISE-SUNSET (OKTAS) | | | | 7.2 | | 2.4 | 3.2 | | | | | | | |
| | MIDNIGHT-MIDNIGHT (OKTAS) | | | | 6.4 | | | 3.2 | | | | | | | |
| | MEAN NO. DAYS WITH: | | | | | | | | | | | | | | |
| | CLEAR | 1 | 2.0 | 1.0 | 10.0 | | 12.0 | 9.0 | | | | | | | |
| | PARTLY CLOUDY | | | | 1.0 | | 6.0 | 9.0 | | | | | | | |
| | CLOUDY | 1 | 3.0 | 2.0 | 11.0 | | 2.0 | 3.0 | | | | | | | |
| **PR** | MEAN STATION PRESSURE(IN) | 26 | 29.03 | 29.00 | 28.96 | 28.93 | 28.93 | 28.93 | 28.96 | 28.96 | 28.93 | 29.00 | 29.02 | 29.05 | 28.98 |
| | MEAN SEA-LEVEL PRES. (IN) | 26 | 30.14 | 30.11 | 30.06 | 30.01 | 30.01 | 30.00 | 30.00 | 30.02 | 30.02 | 30.04 | 30.09 | 30.13 | 30.15 | 30.07 |
| **WINDS** | MEAN SPEED (MPH) | 26 | 9.4 | 9.7 | 9.7 | 9.1 | 8.3 | 7.6 | 7.4 | 7.0 | 7.9 | 8.0 | 8.4 | 9.0 | 8.5 |
| | PREVAIL.DIR(TENS OF DEGS) | 45 | 32 | 33 | 33 | 31 | 31 | 28 | 28 | 08 | 08 | 08 | 33 | 32 | 32 |
| | MAXIMUM 2-MINUTE: | | | | | | | | | | | | | | |
| | SPEED (MPH) | 14 | 39 | 48 | 43 | 44 | 46 | 38 | 46 | 48 | 41 | 31 | 39 | 37 | 48 |
| | DIR. (TENS OF DEGS) | | 32 | 26 | 32 | 03 | 34 | 34 | 31 | 36 | 09 | 34 | 29 | 25 | 36 |
| | YEAR OF OCCURRENCE | | 2005 | 2001 | 2003 | 1996 | 2003 | 2006 | 2001 | 2008 | 2004 | 2008 | 2004 | 2004 | AUG 2008 |
| | MAXIMUM 3-SECOND | | | | | | | | | | | | | | |
| | SPEED (MPH) | 14 | 48 | 70 | 53 | 52 | 58 | 51 | 61 | 71 | 55 | 40 | 46 | 46 | 71 |
| | DIR. (TENS OF DEGS) | | 32 | 26 | 33 | 15 | 35 | 28 | 31 | 01 | 07 | 27 | 32 | 26 | 01 |
| | YEAR OF OCCURRENCE | | 2005 | 2001 | 2008 | 2009 | 2003 | 1998 | 2001 | 2008 | 2004 | 2006 | 2005 | 2004 | AUG 2008 |
| **PRECIPITATION** | NORMAL (IN) | 30 | 5.03 | 4.68 | 5.38 | 3.62 | 3.95 | 3.63 | 5.12 | 3.67 | 4.09 | 3.11 | 4.10 | 3.82 | 50.20 |
| | MAXIMUM MONTHLY (IN) | 75 | 10.82 | 12.77 | 11.66 | 11.86 | 9.94 | 9.99 | 17.71 | 8.69 | 13.65 | 11.04 | 15.72 | 9.92 | 17.71 |
| | YEAR OF OCCURRENCE | | 1936 | 1961 | 1980 | 1979 | 2003 | 1991 | 1994 | 1967 | 2004 | 1995 | 1948 | 1961 | JUL 1994 |
| | MINIMUM MONTHLY (IN) | 75 | 0.84 | 0.77 | 1.04 | 0.49 | 0.32 | 0.16 | 0.57 | 0.50 | 0.04 | T | 0.41 | 0.69 | T |
| | YEAR OF OCCURRENCE | | 1981 | 1978 | 2004 | 1986 | 1936 | 1988 | 1995 | 1976 | 1984 | 1963 | 1939 | 1979 | OCT 1963 |
| | MAXIMUM IN 24 HOURS (IN) | 75 | 3.91 | 5.67 | 5.74 | 5.58 | 5.13 | 4.22 | 6.73 | 5.05 | 5.87 | 7.27 | 4.46 | 3.85 | 7.27 |
| | YEAR OF OCCURRENCE | | 1973 | 1961 | 1990 | 1979 | 1948 | 1991 | 2005 | 1940 | 1992 | 1995 | 2009 | 1961 | OCT 1995 |
| | NORMAL NO. DAYS WITH: | | | | | | | | | | | | | | |
| | PRECIPITATION >= 0.01 | 30 | 12.1 | 9.8 | 10.9 | 8.4 | 9.6 | 9.8 | 11.5 | 9.5 | 8.3 | 6.4 | 9.4 | 10.4 | 116.1 |
| | PRECIPITATION >= 1.00 | 30 | 1.4 | 1.5 | 1.7 | 1.0 | 1.0 | 1.0 | 1.6 | 1.0 | 1.3 | 1.0 | 1.4 | 1.0 | 14.9 |
| **SNOWFALL** | NORMAL (IN) | 30 | 1.0 | 0.5 | 0.6 | 0.* | 0.0 | 0.0 | 0.0 | 0.0 | 0.0 | 0.* | 0.* | 0.3 | 2.4 |
| | MAXIMUM MONTHLY (IN) | 70 | 8.3 | 4.4 | 7.9 | T | 0.0 | 0.0 | T | 0.0 | 0.0 | T | 1.0 | 3.0 | 8.3 |
| | YEAR OF OCCURRENCE | | 1940 | 1979 | 1983 | 2009 | | | 2001 | | | 1993 | 1968 | 2000 | JAN 1940 |
| | MAXIMUM IN 24 HOURS (IN) | 70 | 8.3 | 4.2 | 7.9 | T | 0.0 | 0.0 | T | 0.0 | 0.0 | T | 1.0 | 2.8 | 8.3 |
| | YEAR OF OCCURRENCE' | | 1940 | 1979 | 1983 | 2009 | | | 2001 | | | 1993 | 1968 | 1993 | JAN 1940 |
| | MAXIMUM SNOW DEPTH (IN) | 59 | 9 | 4 | 4 | 0 | 0 | 0 | 0 | 0 | 0 | 0 | 1 | 2 | 9 |
| | YEAR OF OCCURRENCE | | 1948 | 1979 | 1993 | | | | | | | | 1975 | 2000 | JAN 1948 |
| | NORMAL NO. DAYS WITH: | | | | | | | | | | | | | | |
| | SNOWFALL >= 1.0 | 30 | 0.4 | 0.2 | 0.1 | 0.0 | 0.0 | 0.0 | 0.0 | 0.0 | 0.0 | 0.0 | 0.0 | 0.2 | 0.9 |

## PRECIPITATION (inches) 2009 ATLANTA (KATL)

| YEAR | JAN | FEB | MAR | APR | MAY | JUN | JUL | AUG | SEP | OCT | NOV | DEC | ANNUAL |
|------|-----|-----|-----|-----|-----|-----|-----|-----|-----|-----|-----|-----|--------|
| 1980 | 5.69 | 2.69 | 11.66 | 1.88 | 8.37 | 4.49 | 0.76 | 1.59 | 4.77 | 1.61 | 2.14 | 1.29 | 46.94 |
| 1981 | 0.84 | 6.62 | 3.93 | 2.06 | 3.89 | 2.69 | 2.74 | 2.76 | 5.27 | 3.01 | 1.85 | 6.25 | 41.91 |
| 1982 | 4.75 | 6.99 | 3.79 | 6.02 | 2.60 | 6.09 | 6.31 | 1.45 | 3.00 | 5.83 | 4.15 | 5.23 | 56.21 |
| 1983 | 3.09 | 4.99 | 6.68 | 4.79 | 1.42 | 1.52 | 1.85 | 1.06 | 7.52 | 1.97 | 7.46 | 9.27 | 51.62 |
| 1984 | 4.66 | 5.97 | 5.83 | 6.62 | 6.57 | 0.74 | 11.21 | 6.46 | 0.04 | 1.54 | 2.10 | 3.65 | 55.39 |
| 1985 | 4.11 | 4.98 | 1.86 | 2.75 | 4.69 | 2.04 | 9.92 | 4.57 | 2.63 | 5.74 | 4.23 | 2.28 | 49.80 |
| 1986 | 0.88 | 2.46 | 4.13 | 0.49 | 2.95 | 2.18 | 3.27 | 6.08 | 3.68 | 5.15 | 6.20 | 3.03 | 40.50 |
| 1987 | 5.63 | 6.13 | 5.44 | 1.16 | 2.74 | 6.36 | 7.35 | 1.22 | 3.02 | 0.70 | 2.36 | 4.13 | 46.24 |
| 1988 | 4.64 | 3.32 | 2.57 | 6.06 | 1.71 | 0.16 | 5.04 | 4.92 | 6.35 | 5.00 | 4.87 | 1.21 | 45.85 |
| 1989 | 2.57 | 4.30 | 3.85 | 5.24 | 6.42 | 9.34 | 7.65 | 2.13 | 11.64 | 1.71 | 3.97 | 4.49 | 63.31 |
| 1990 | 8.47 | 9.75 | 8.36 | 2.76 | 5.26 | 1.39 | 3.49 | 4.64 | 3.01 | 6.12 | 1.27 | 3.04 | 57.56 |
| 1991 | 4.66 | 3.10 | 6.98 | 5.28 | 7.35 | 9.99 | 5.82 | 4.37 | 2.03 | 0.39 | 3.19 | 2.69 | 55.85 |
| 1992 | 3.58 | 3.94 | 3.81 | 1.03 | 1.73 | 4.14 | 9.03 | 5.04 | 8.55 | 2.84 | 10.04 | 6.38 | 60.11 |
| 1993 | 3.94 | 4.43 | 5.73 | 2.77 | 4.87 | 6.01 | 3.05 | 2.96 | 3.91 | 3.83 | 4.01 | 2.54 | 48.05 |
| 1994 | 5.11 | 3.76 | 5.77 | 3.68 | 2.16 | 2.44 | 17.71 | 4.16 | 5.86 | 4.51 | 3.27 | 1.59 | 60.02 |
| 1995 | 3.36 | 6.74 | 2.66 | 3.00 | 2.12 | 3.97 | 0.57 | 5.82 | 2.52 | 11.04 | 7.40 | 3.57 | 52.77 |
| 1996 | 8.26 | 3.82 | 6.42 | 2.91 | 2.12 | 1.70 | 2.14 | 4.66 | 4.32 | 0.89 | 3.22 | 4.14 | 44.60 |
| 1997 | 5.65 | 7.93 | 2.18 | 4.28 | 3.36 | 3.91 | 4.71 | 1.32 | 4.83 | 5.12 | 3.34 | 5.05 | 51.68 |
| 1998 | 5.83 | 7.10 | 6.25 | 5.12 | 1.23 | 3.58 | 2.93 | 5.54 | 4.45 | 0.26 | 1.97 | 1.90 | 46.16 |
| 1999 | 5.33 | 1.97 | 3.32 | 1.14 | 4.42 | 5.83 | 3.43 | 1.26 | 4.19 | 2.41 | 3.34 | 2.21 | 38.85 |
| 2000 | 4.89 | 1.26 | 3.63 | 2.63 | 1.86 | 1.11 | 2.70 | 4.03 | 4.93 | 0.88 | 5.02 | 2.62 | 35.56 |
| 2001 | 2.77 | 3.61 | 9.08 | 3.29 | 3.31 | 6.69 | 2.54 | 1.03 | 2.19 | 0.79 | 0.87 | 2.22 | 38.39 |
| 2002 | 5.35 | 2.54 | 5.49 | 1.83 | 3.52 | 2.81 | 2.59 | 0.77 | 6.39 | 5.94 | 5.36 | 5.23 | 47.82 |
| 2003 | 2.00 | 3.51 | 7.08 | 3.44 | 9.94 | 7.34 | 5.35 | 3.48 | 2.42 | 1.49 | 4.17 | 2.69 | 52.91 |
| 2004 | 2.84 | 4.60 | 1.04 | 2.80 | 2.58 | 5.99 | 2.20 | 3.63 | 13.65 | 2.19 | 7.26 | 4.82 | 53.60 |
| 2005 | 2.57 | 5.58 | 7.49 | 4.36 | 1.98 | 2.91 | 14.63 | 8.28 | 0.07 | 1.98 | 2.91 | 3.67 | 56.43 |
| 2006 | 5.10 | 5.50 | 2.93 | 2.48 | 2.86 | 5.80 | 1.31 | 8.66 | 3.31 | 3.04 | 4.39 | 3.08 | 48.46 |
| 2007 | 3.95 | 2.63 | 1.31 | 1.79 | 2.05 | 3.66 | 1.85 | 3.48 | 2.92 | 2.47 | 0.96 | 4.78 | 31.85 |
| 2008 | 2.85 | 4.61 | 5.17 | 3.22 | 2.80 | 0.58 | 7.17 | 3.77 | 0.75 | 3.48 | 2.64 | 4.39 | 41.43 |
| 2009 | 2.88 | 3.70 | 7.13 | 5.18 | 4.54 | 2.34 | 5.02 | 6.14 | 8.94 | 8.71 | 5.75 | 9.10 | 69.43 |
| POR=<br>80 YRS | 4.52 | 4.50 | 5.40 | 4.04 | 3.68 | 3.81 | 4.85 | 3.72 | 3.63 | 2.86 | 3.62 | 4.20 | 48.83 |

WBAN : 13874

## AVERAGE TEMPERATURE (°F) 2009 ATLANTA (KATL)

| YEAR | JAN | FEB | MAR | APR | MAY | JUN | JUL | AUG | SEP | OCT | NOV | DEC | ANNUAL |
|------|-----|-----|-----|-----|-----|-----|-----|-----|-----|-----|-----|-----|--------|
| 1980 | 44.9 | 41.9 | 52.1 | 62.6 | 72.0 | 79.1 | 85.1 | 83.8 | 78.9 | 61.5 | 51.9 | 44.9 | 63.2 |
| 1981 | 39.5 | 46.8 | 51.8 | 67.7 | 67.6 | 81.3 | 82.2 | 77.7 | 72.4 | 60.2 | 54.5 | 39.1 | 61.7 |
| 1982 | 38.5 | 47.4 | 56.5 | 58.4 | 72.5 | 76.3 | 79.1 | 77.5 | 70.5 | 62.7 | 53.7 | 49.9 | 61.9 |
| 1983 | 40.4 | 44.4 | 51.3 | 56.4 | 67.8 | 74.0 | 81.4 | 81.4 | 70.8 | 62.1 | 51.5 | 39.9 | 60.1 |
| 1984 | 39.6 | 47.5 | 51.7 | 58.1 | 67.5 | 78.3 | 76.8 | 77.5 | 71.3 | 69.8 | 50.6 | 53.7 | 61.9 |
| 1985 | 36.3 | 44.2 | 56.8 | 64.0 | 69.9 | 77.5 | 78.4 | 77.6 | 72.5 | 66.4 | 62.0 | 41.4 | 62.3 |
| 1986 | 43.4 | 49.8 | 54.4 | 62.9 | 71.0 | 80.0 | 84.1 | 77.4 | 74.6 | 64.0 | 57.9 | 45.1 | 63.7 |
| 1987 | 41.9 | 45.7 | 53.2 | 60.3 | 73.2 | 77.8 | 81.0 | 82.0 | 74.1 | 59.7 | 55.7 | 48.8 | 62.8 |
| 1988 | 39.2 | 45.5 | 54.9 | 63.0 | 70.0 | 78.6 | 80.5 | 81.0 | 73.4 | 59.3 | 55.0 | 46.6 | 62.3 |
| 1989 | 49.7 | 47.5 | 56.8 | 62.9 | 68.8 | 76.9 | 79.8 | 79.4 | 72.9 | 64.2 | 54.3 | 39.1 | 62.7 |
| 1990 | 49.8 | 54.4 | 57.7 | 61.9 | 70.4 | 78.6 | 80.6 | 80.6 | 75.7 | 64.4 | 56.5 | 49.1 | 65.0 |
| 1991 | 44.3 | 49.2 | 56.2 | 65.9 | 72.8 | 76.7 | 81.0 | 79.2 | 74.8 | 64.6 | 51.0 | 49.1 | 63.7 |
| 1992 | 45.1 | 51.8 | 54.0 | 61.9 | 68.1 | 74.5 | 80.2 | 76.1 | 73.2 | 61.9 | 51.5 | 44.5 | 61.9 |
| 1993 | 47.0 | 45.2 | 51.7 | 59.4 | 71.1 | 79.3 | 85.4 | 82.1 | 76.7 | 63.3 | 53.9 | 45.3 | 63.4 |
| 1994 | 40.5 | 50.1 | 57.4 | 67.6 | 69.4 | 80.5 | 79.1 | 79.1 | 73.9 | 64.1 | 58.3 | 50.4 | 64.2 |
| 1995 | 46.3 | 46.4 | 58.9 | 65.8 | 74.4 | 77.0 | 84.3 | 80.7 | 71.0 | 62.0 | 47.7 | 42.8 | 63.1 |
| 1996 | 41.1 | 47.1 | 50.6 | 61.2 | 74.9 | 79.1 | 81.8 | 79.5 | 73.2 | 63.4 | 52.0 | 48.8 | 62.7 |
| 1997 | 46.7 | 51.1 | 60.6 | 57.7 | 64.8 | 71.5 | 78.9 | 76.6 | 73.3 | 61.8 | 47.8 | 42.9 | 61.1 |
| 1998 | 46.0 | 47.0 | 50.3 | 59.0 | 72.7 | 78.9 | 80.7 | 77.8 | 75.6 | 65.9 | 56.7 | 50.0 | 63.4 |
| 1999 | 48.0 | 49.8 | 50.5 | 65.1 | 68.9 | 74.9 | 79.2 | 81.9 | 73.1 | 62.5 | 56.7 | 47.1 | 63.1 |
| 2000 | 43.1 | 50.9 | 57.3 | 58.6 | 72.9 | 77.9 | 81.4 | 79.7 | 70.8 | 64.1 | 51.0 | 37.2 | 62.1 |
| 2001 | 41.7 | 50.8 | 50.4 | 63.5 | 70.0 | 74.4 | 78.6 | 78.8 | 71.3 | 60.7 | 59.8 | 50.1 | 62.5 |
| 2002 | 46.9 | 45.4 | 54.5 | 64.8 | 68.3 | 76.4 | 80.4 | 80.3 | 76.1 | 65.4 | 50.9 | 43.9 | 62.8 |
| 2003 | 40.1 | 46.4 | 55.7 | 61.7 | 69.0 | 74.4 | 77.9 | 79.1 | 72.1 | 63.5 | 57.6 | 43.3 | 61.7 |
| 2004 | 42.9 | 43.7 | 58.4 | 62.0 | 72.9 | 75.9 | 79.6 | 76.8 | 72.2 | 67.0 | 56.4 | 44.6 | 62.7 |
| 2005 | 46.9 | 48.8 | 52.2 | 60.4 | 66.8 | 75.3 | 79.4 | 79.6 | 76.6 | 64.5 | 55.2 | 41.9 | 62.3 |
| 2006 | 49.4 | 44.2 | 54.2 | 65.9 | 69.9 | 77.2 | 81.1 | 81.0 | 71.9 | 61.8 | 53.8 | 50.1 | 63.4 |
| 2007 | 46.2 | 45.3 | 60.2 | 60.3 | 71.6 | 79.6 | 78.5 | 85.6 | 75.9 | 66.0 | 53.2 | 50.8 | 64.4 |
| 2008 | 42.3 | 48.3 | 54.1 | 61.3 | 69.5 | 79.9 | 79.9 | 78.2 | 74.6 | 62.0 | 50.6 | 48.6 | 62.4 |
| 2009 | 43.9 | 47.7 | 55.0 | 60.7 | 70.2 | 79.8 | 78.1 | 78.9 | 73.5 | 61.0 | 53.8 | 42.3 | 62.1 |
| POR=<br>80 YRS | 43.3 | 45.5 | 53.3 | 61.4 | 69.9 | 76.4 | 79.4 | 78.7 | 72.9 | 63.0 | 52.4 | 45.0 | 61.8 |

## HEATING DEGREE DAYS (base 65°F) 2009 ATLANTA (KATL)

| YEAR | JUL | AUG | SEP | OCT | NOV | DEC | JAN | FEB | MAR | APR | MAY | JUN | TOTAL |
|---|---|---|---|---|---|---|---|---|---|---|---|---|---|
| 1980-81 | 0 | 0 | 18 | 154 | 391 | 618 | 786 | 502 | 410 | 36 | 43 | 0 | 2958 |
| 1981-82 | 0 | 0 | 17 | 179 | 314 | 795 | 819 | 486 | 282 | 204 | 2 | 0 | 3098 |
| 1982-83 | 0 | 0 | 16 | 139 | 341 | 466 | 755 | 571 | 423 | 261 | 24 | 0 | 2996 |
| 1983-84 | 0 | 0 | 32 | 123 | 400 | 770 | 780 | 503 | 409 | 221 | 50 | 0 | 3288 |
| 1984-85 | 0 | 0 | 13 | 22 | 426 | 346 | 882 | 576 | 265 | 111 | 14 | 1 | 2656 |
| 1985-86 | 0 | 0 | 15 | 71 | 131 | 725 | 663 | 422 | 331 | 133 | 14 | 0 | 2505 |
| 1986-87 | 0 | 11 | 2 | 107 | 243 | 609 | 709 | 534 | 359 | 191 | 6 | 0 | 2771 |
| 1987-88 | 0 | 0 | 0 | 172 | 279 | 494 | 791 | 559 | 310 | 104 | 6 | 0 | 2715 |
| 1988-89 | 0 | 0 | 0 | 188 | 291 | 566 | 468 | 490 | 284 | 160 | 44 | 0 | 2491 |
| 1989-90 | 0 | 0 | 29 | 103 | 318 | 797 | 462 | 297 | 250 | 150 | 20 | 0 | 2426 |
| 1990-91 | 0 | 0 | 12 | 109 | 252 | 488 | 636 | 437 | 281 | 54 | 8 | 0 | 2277 |
| 1991-92 | 0 | 0 | 8 | 76 | 419 | 499 | 611 | 377 | 345 | 161 | 50 | 2 | 2548 |
| 1992-93 | 0 | 0 | 12 | 110 | 398 | 627 | 548 | 549 | 417 | 184 | 13 | 0 | 2858 |
| 1993-94 | 0 | 0 | 6 | 129 | 346 | 604 | 753 | 412 | 245 | 55 | 20 | 0 | 2570 |
| 1994-95 | 0 | 0 | 0 | 79 | 207 | 446 | 573 | 515 | 207 | 59 | 11 | 0 | 2097 |
| 1995-96 | 0 | 0 | 25 | 135 | 514 | 680 | 735 | 517 | 437 | 159 | 8 | 0 | 3210 |
| 1996-97 | 0 | 0 | 4 | 104 | 389 | 497 | 561 | 385 | 156 | 226 | 63 | 18 | 2403 |
| 1997-98 | 0 | 0 | 5 | 159 | 508 | 679 | 580 | 499 | 465 | 175 | 12 | 3 | 3085 |
| 1998-99 | 0 | 0 | 0 | 53 | 242 | 463 | 519 | 421 | 443 | 92 | 8 | 0 | 2241 |
| 1999-00 | 0 | 0 | 7 | 110 | 246 | 548 | 669 | 401 | 232 | 186 | 0 | 0 | 2399 |
| 2000-01 | 0 | 0 | 16 | 86 | 428 | 856 | 715 | 391 | 446 | 108 | 5 | 0 | 3051 |
| 2001-02 | 0 | 0 | 26 | 172 | 163 | 452 | 557 | 543 | 340 | 85 | 51 | 0 | 2389 |
| 2002-03 | 0 | 0 | 0 | 90 | 422 | 647 | 765 | 516 | 281 | 119 | 9 | 0 | 2849 |
| 2003-04 | 0 | 0 | 14 | 80 | 250 | 664 | 679 | 610 | 216 | 133 | 24 | 0 | 2670 |
| 2004-05 | 0 | 0 | 1 | 30 | 272 | 626 | 557 | 446 | 395 | 147 | 51 | 3 | 2528 |
| 2005-06 | 0 | 0 | 0 | 125 | 305 | 710 | 474 | 577 | 342 | 65 | 34 | 0 | 2632 |
| 2006-07 | 0 | 0 | 8 | 167 | 336 | 452 | 577 | 545 | 190 | 172 | 5 | 0 | 2452 |
| 2007-08 | 0 | 0 | 0 | 85 | 351 | 441 | 700 | 478 | 332 | 135 | 9 | 0 | 2531 |
| 2008-09 | 0 | 0 | 0 | 159 | 428 | 505 | 645 | 477 | 312 | 165 | 16 | 0 | 2707 |
| 2009- | 0 | 0 | 6 | 152 | 329 | 701 | | | | | | | |

WBAN : 13874

## COOLING DEGREE DAYS (base 65°F) 2009 ATLANTA (KATL)

| YEAR | JAN | FEB | MAR | APR | MAY | JUN | JUL | AUG | SEP | OCT | NOV | DEC | TOTAL |
|---|---|---|---|---|---|---|---|---|---|---|---|---|---|
| 1980 | 0 | 4 | 4 | 49 | 227 | 428 | 632 | 589 | 440 | 51 | 0 | 0 | 2424 |
| 1981 | 0 | 0 | 9 | 124 | 131 | 494 | 540 | 398 | 246 | 36 | 4 | 0 | 1982 |
| 1982 | 2 | 0 | 25 | 13 | 243 | 346 | 446 | 394 | 192 | 73 | 8 | 6 | 1748 |
| 1983 | 0 | 0 | 3 | 10 | 118 | 278 | 515 | 512 | 212 | 40 | 0 | 0 | 1688 |
| 1984 | 0 | 0 | 2 | 21 | 132 | 405 | 372 | 397 | 210 | 178 | 1 | 2 | 1720 |
| 1985 | 0 | 0 | 18 | 88 | 172 | 381 | 423 | 401 | 248 | 119 | 49 | 0 | 1899 |
| 1986 | 0 | 0 | 11 | 74 | 208 | 455 | 599 | 401 | 300 | 83 | 34 | 0 | 2165 |
| 1987 | 0 | 0 | 2 | 60 | 266 | 391 | 502 | 531 | 281 | 12 | 6 | 2 | 2053 |
| 1988 | 0 | 0 | 5 | 49 | 169 | 416 | 490 | 502 | 258 | 18 | 0 | 0 | 1907 |
| 1989 | 0 | 7 | 36 | 101 | 170 | 364 | 467 | 452 | 273 | 85 | 6 | 0 | 1961 |
| 1990 | 0 | 5 | 26 | 66 | 194 | 415 | 490 | 488 | 341 | 98 | 2 | 0 | 2125 |
| 1991 | 0 | 0 | 20 | 89 | 258 | 358 | 502 | 446 | 305 | 70 | 4 | 13 | 2065 |
| 1992 | 0 | 1 | 13 | 73 | 155 | 292 | 478 | 349 | 265 | 20 | 0 | 0 | 1646 |
| 1993 | 0 | 0 | 11 | 22 | 208 | 435 | 639 | 536 | 364 | 80 | 19 | 0 | 2314 |
| 1994 | 0 | 2 | 19 | 141 | 167 | 470 | 445 | 447 | 274 | 60 | 13 | 0 | 2038 |
| 1995 | 0 | 0 | 23 | 91 | 311 | 368 | 608 | 494 | 211 | 47 | 2 | 0 | 2155 |
| 1996 | 0 | 8 | 2 | 54 | 322 | 429 | 527 | 456 | 255 | 64 | 7 | 2 | 2126 |
| 1997 | 3 | 3 | 28 | 13 | 62 | 221 | 438 | 366 | 264 | 68 | 0 | 0 | 1466 |
| 1998 | 0 | 0 | 16 | 4 | 257 | 426 | 494 | 407 | 323 | 88 | 3 | 5 | 2023 |
| 1999 | 0 | 0 | 0 | 103 | 136 | 305 | 446 | 531 | 256 | 39 | 1 | 0 | 1817 |
| 2000 | 0 | 0 | 3 | 4 | 253 | 396 | 515 | 464 | 196 | 64 | 16 | 0 | 1911 |
| 2001 | 0 | 0 | 0 | 69 | 166 | 288 | 429 | 435 | 224 | 45 | 13 | 0 | 1669 |
| 2002 | 3 | 0 | 21 | 89 | 161 | 350 | 486 | 477 | 338 | 112 | 6 | 0 | 2043 |
| 2003 | 0 | 0 | 3 | 26 | 140 | 290 | 408 | 444 | 235 | 40 | 35 | 0 | 1621 |
| 2004 | 0 | 0 | 18 | 48 | 278 | 336 | 457 | 374 | 226 | 99 | 20 | 0 | 1856 |
| 2005 | 0 | 0 | 3 | 14 | 111 | 318 | 455 | 461 | 354 | 114 | 21 | 0 | 1851 |
| 2006 | 0 | 0 | 15 | 98 | 191 | 373 | 504 | 505 | 225 | 73 | 4 | 0 | 1988 |
| 2007 | 0 | 0 | 48 | 38 | 218 | 444 | 426 | 646 | 332 | 124 | 2 | 5 | 2283 |
| 2008 | 0 | 0 | 0 | 31 | 158 | 454 | 466 | 416 | 293 | 75 | 0 | 4 | 1897 |
| 2009 | 0 | 0 | 12 | 46 | 185 | 450 | 413 | 442 | 267 | 34 | 0 | 0 | 1849 |

## SNOWFALL (inches) 2009 ATLANTA (KATL)

| YEAR | JUL | AUG | SEP | OCT | NOV | DEC | JAN | FEB | MAR | APR | MAY | JUN | TOTAL |
|------|-----|-----|-----|-----|-----|-----|-----|-----|-----|-----|-----|-----|-------|
| 1980-81 | 0.0 | 0.0 | 0.0 | 0.0 | 0.0 | 0.0 | T | T | 0.0 | 0.0 | 0.0 | 0.0 | T |
| 1981-82 | 0.0 | 0.0 | 0.0 | 0.0 | 0.0 | T | 7.0 | 0.7 | 0.0 | 0.0 | 0.0 | 0.0 | 7.7 |
| 1982-83 | 0.0 | 0.0 | 0.0 | 0.0 | 0.0 | 0.0 | 1.9 | 0.5 | 7.9 | 0.0 | 0.0 | 0.0 | 10.3 |
| 1983-84 | 0.0 | 0.0 | 0.0 | 0.0 | 0.0 | T | T | 1.3 | T | 0.0 | 0.0 | 0.0 | 1.3 |
| 1984-85 | 0.0 | 0.0 | 0.0 | 0.0 | 0.0 | T | 0.4 | 1.5 | 0.0 | 0.0 | 0.0 | 0.0 | 1.9 |
| 1985-86 | 0.0 | 0.0 | 0.0 | 0.0 | 0.0 | T | 0.4 | T | 0.0 | 0.0 | 0.0 | 0.0 | 0.4 |
| 1986-87 | 0.0 | 0.0 | 0.0 | 0.0 | 0.0 | 0.0 | 3.6 | T | 1.2 | T | 0.0 | 0.0 | 4.8 |
| 1987-88 | 0.0 | 0.0 | 0.0 | 0.0 | 0.0 | 0.0 | 4.2 | T | 0.0 | 0.0 | 0.0 | 0.0 | 4.2 |
| 1988-89 | 0.0 | 0.0 | 0.0 | 0.0 | 0.0 | T | 0.0 | 0.7 | 0.0 | T | 0.0 | 0.0 | 0.7 |
| 1989-90 | 0.0 | 0.0 | 0.0 | 0.0 | 0.0 | 1.3 | 0.0 | 0.0 | 0.0 | T | 0.0 | 0.0 | 1.3 |
| 1990-91 | 0.0 | 0.0 | 0.0 | 0.0 | 0.0 | 0.0 | 2.1 | T | T | 0.0 | 0.0 | 0.0 | 2.1 |
| 1991-92 | 0.0 | 0.0 | 0.0 | 0.0 | 0.0 | 0.0 | 5.0 | 0.0 | T | 0.0 | 0.0 | 0.0 | 5.0 |
| 1992-93 | 0.0 | 0.0 | 0.0 | 0.0 | 0.0 | T | 0.0 | T | 4.2 | 0.0 | 0.0 | 0.0 | 4.2 |
| 1993-94 | 0.0 | 0.0 | 0.0 | T | 0.0 | 2.8 | T | 0.0 | 0.0 | 0.0 | 0.0 | 0.0 | 2.8 |
| 1994-95 | 0.0 | 0.0 | 0.0 | 0.0 | 0.0 | 0.0 | T | 0.4 | 0.0 | 0.0 | 0.0 | 0.0 | 0.4 |
| 1995-96 | 0.0 | 0.0 | 0.0 | 0.0 | 0.0 | 0.0 | 1.4 | | | | | | |
| 1996-97 | | | | | | | | | | | | | |
| 1997-98 | | | | | | | | | | | | | |
| 1998-99 | | | | | | | | | | | | | |
| 1999-00 | | | | | | | T | | | | | | |
| 2000-01 | 0.0 | 0.0 | 0.0 | 0.0 | T | 3.0 | 0.1 | 0.0 | 0.0 | 0.0 | 0.0 | 0.0 | 3.1 |
| 2001-02 | T | 0.0 | 0.0 | 0.0 | 0.0 | 0.0 | 4.6 | | 0.0 | | | | |
| 2002-03 | | | | | | | | | | | | | |
| 2003-04 | | | | | | | | | | | | | |
| 2004-05 | | | | | | | 0.5 | 0.0 | T | 0.0 | 0.0 | 0.0 | |
| 2005-06 | 0.0 | 0.0 | 0.0 | 0.0 | T | 0.0 | T | T | 0.0 | 0.0 | 0.0 | 0.0 | T |
| 2006-07 | 0.0 | 0.0 | 0.0 | 0.0 | 0.0 | 0.0 | T | 0.1 | 0.0 | 0.0 | 0.0 | 0.0 | 0.1 |
| 2007-08 | 0.0 | 0.0 | 0.0 | 0.0 | 0.0 | T | 1.4 | T | T | 0.0 | 0.0 | 0.0 | 1.4 |
| 2008-09 | 0.0 | 0.0 | 0.0 | 0.0 | 0.0 | 0.0 | T | T | 4.2 | T | 0.0 | 0.0 | 4.2 |
| 2009- | 0.0 | 0.0 | 0.0 | 0.0 | 0.0 | T | | | | | | | |
| POR= 79 YRS | T | 0.0 | 0.0 | T | T | 0.2 | 1.0 | 0.4 | 0.4 | T | 0.0 | 0.0 | 2.0 |

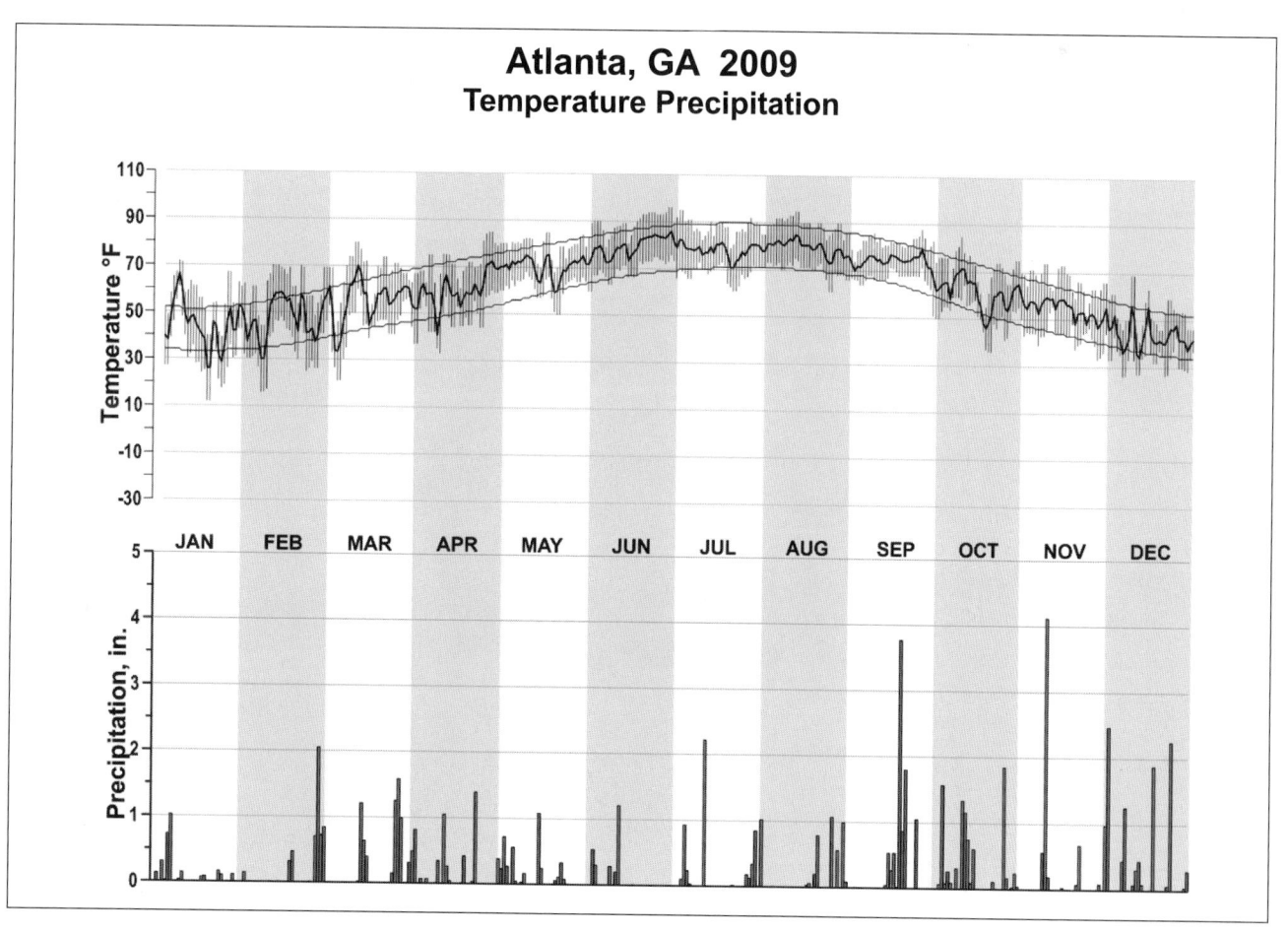

# Atlanta, GA 2009
## Temperature Precipitation

# 2009
# SAVANNAH
# GEORGIA (KSAV)

Savannah is surrounded by flat terrain, low and marshy to the north and east, and rising to several feet above sea level to the west and south. About half the land to the west and south is cleared and the other half is wooded and swampy.

The area has a temperate climate, with a seasonal low temperature of 51 degrees in winter, 66 degrees in spring, 80 degrees in summer, and 66 degrees in autumn. The lowest temperatures are below 10 degrees and the highest temperatures are about 100 degrees.

The normal annual rainfall is about 49 inches. About half falls in the thunderstorm season of June 15 through September 15. The remainder, produced principally by squall-line and frontal showers, is spread over the other nine months with a minor peak in March. Considerable periods of fair, mild weather are experienced in October, November, April, and to a less extent, in May. Snow is a rarity and even a trace does not occur on an average of once a year. The heaviest snowfalls are under 5 inches. Severe tropical storms affect this area about once in ten years. Rainfall from these storms constitute the heaviest sustained precipitation. Accumulations exceeding 22 inches have occurred.

The present exposure of the thermometers gives readings more nearly commensurate with those of suburban street levels of Savannah than was the case of previous locations atop various buildings. During that time, especially on still, clear nights, temperatures near the ground and in lower inland areas were as much as 15 degrees lower than the official low temperature. Present differences on comparable nights range from 3 - 8 degrees.

Sunshine is adequate at all seasons and seldom are there two or more days in succession without it. Sea- and land-breeze effect is usually not felt in Savannah, though it is a daily feature on the nearby islands. Dry, continental air masses reach this area in summer mostly by sliding down the Atlantic coast and giving cooler northeast winds. Such masses reaching this area from the northwest or west in summer bring mostly clear skies and high temperatures.

Based on the 1951-1980 period, the average first occurrence of 32 degrees Fahrenheit in the fall is November 15 and the average last occurrence in the spring is March 10.

# NORMALS, MEANS, AND EXTREMES
## SAVANNAH (KSAV)

| LATITUDE: 32 ° 7 'N | LONGITUDE: -81 ° 12'W | | ELEVATION (FT): GRND: 25   BARO: 143 | | TIME ZONE: EASTERN  (UTC -5) | | WBAN: 03822 | |

| ELEMENT | POR | JAN | FEB | MAR | APR | MAY | JUN | JUL | AUG | SEP | OCT | NOV | DEC | YEAR |
|---|---|---|---|---|---|---|---|---|---|---|---|---|---|---|
| NORMAL DAILY MAXIMUM | 30 | 60.4 | 64.1 | 71.0 | 77.7 | 84.3 | 89.5 | 92.3 | 90.3 | 86.0 | 78.1 | 70.5 | 62.6 | 77.2 |
| MEAN DAILY MAXIMUM | 59 | 60.4 | 63.8 | 70.5 | 77.7 | 84.4 | 89.1 | 91.6 | 90.3 | 85.7 | 78.0 | 70.0 | 62.5 | 77.0 |
| HIGHEST DAILY MAXIMUM | 59 | 84 | 86 | 91 | 95 | 100 | 104 | 105 | 104 | 98 | 97 | 89 | 83 | 105 |
| YEAR OF OCCURRENCE | | 1957 | 1989 | 1974 | 1986 | 1953 | 1985 | 1986 | 1954 | 1986 | 1986 | 1961 | 1971 | JUL 1986 |
| MEAN OF EXTREME MAXS. | 59 | 76.4 | 79.6 | 84.6 | 89.3 | 93.6 | 97.3 | 98.2 | 97.1 | 93.6 | 88.4 | 82.5 | 77.7 | 88.2 |
| NORMAL DAILY MINIMUM | 30 | 38.0 | 40.9 | 47.5 | 52.9 | 61.3 | 68.1 | 71.8 | 71.3 | 67.3 | 56.1 | 46.9 | 40.1 | 55.2 |
| MEAN DAILY MINIMUM | 59 | 38.5 | 41.1 | 47.4 | 54.0 | 62.3 | 69.1 | 72.3 | 72.0 | 67.8 | 56.9 | 46.8 | 40.3 | 55.7 |
| LOWEST DAILY MINIMUM | 59 | 3 | 14 | 20 | 28 | 39 | 51 | 61 | 57 | 43 | 28 | 15 | 9 | 3 |
| YEAR OF OCCURRENCE | | 1985 | 1958 | 1980 | 2007 | 1963 | 1984 | 1972 | 1986 | 1967 | 1952 | 1970 | 1983 | JAN 1985 |
| MEAN OF EXTREME MINS. | 59 | 21.3 | 24.8 | 30.3 | 38.5 | 49.0 | 60.5 | 66.7 | 66.0 | 56.4 | 39.9 | 29.8 | 23.4 | 42.2 |
| NORMAL DRY BULB | 30 | 49.2 | 52.5 | 59.3 | 65.3 | 72.8 | 78.8 | 82.1 | 80.8 | 76.7 | 67.1 | 58.7 | 51.4 | 66.2 |
| MEAN DRY BULB | 59 | 49.5 | 52.5 | 59.0 | 65.9 | 73.3 | 79.2 | 81.9 | 81.2 | 76.8 | 67.5 | 58.4 | 51.4 | 66.4 |
| MEAN WET BULB | 26 | 44.3 | 47.1 | 52.1 | 57.3 | 65.2 | 71.9 | 74.5 | 74.5 | 70.4 | 61.7 | 53.6 | 46.8 | 60.0 |
| MEAN DEW POINT | 26 | 40.2 | 42.6 | 48.0 | 53.3 | 62.2 | 70.0 | 72.6 | 72.8 | 68.6 | 59.2 | 50.8 | 43.1 | 57.0 |
| NORMAL NO. DAYS WITH: | | | | | | | | | | | | | | |
| MAXIMUM >= 90 | 30 | 0.0 | 0.0 | * | 1.3 | 5.5 | 15.2 | 23.0 | 18.6 | 8.1 | 0.6 | 0.0 | 0.0 | 72.3 |
| MAXIMUM <= 32 | 30 | 0.1 | 0.1 | * | 0.0 | 0.0 | 0.0 | 0.0 | 0.0 | 0.0 | 0.0 | 0.0 | 0.1 | 0.3 |
| MINIMUM <= 32 | 30 | 8.7 | 5.5 | 1.2 | * | 0.0 | 0.0 | 0.0 | 0.0 | 0.0 | * | 1.5 | 6.7 | 23.6 |
| MINIMUM <= 0 | 30 | 0.0 | 0.0 | 0.0 | 0.0 | 0.0 | 0.0 | 0.0 | 0.0 | 0.0 | 0.0 | 0.0 | 0.0 | 0.0 |
| NORMAL HEATING DEG. DAYS | 30 | 480 | 350 | 202 | 74 | 7 | 0 | 0 | 0 | 2 | 57 | 211 | 416 | 1799 |
| NORMAL COOLING DEG. DAYS | 30 | 6 | 11 | 39 | 99 | 265 | 430 | 546 | 506 | 366 | 138 | 38 | 10 | 2454 |
| NORMAL (PERCENT) | 30 | 70 | 68 | 67 | 66 | 70 | 74 | 76 | 78 | 78 | 75 | 74 | 72 | 72 |
| HOUR 01 LST | 30 | 78 | 77 | 79 | 80 | 85 | 87 | 88 | 90 | 89 | 86 | 85 | 81 | 84 |
| HOUR 07 LST | 30 | 82 | 82 | 83 | 84 | 86 | 88 | 89 | 91 | 91 | 89 | 87 | 84 | 86 |
| HOUR 13 LST | 30 | 56 | 51 | 49 | 46 | 51 | 55 | 57 | 60 | 60 | 54 | 54 | 56 | 54 |
| HOUR 19 LST | 30 | 67 | 62 | 61 | 59 | 64 | 69 | 72 | 75 | 76 | 74 | 74 | 71 | 69 |
| PERCENT POSSIBLE SUNSHINE | 46 | 54 | 57 | 62 | 71 | 68 | 65 | 63 | 62 | 57 | 63 | 61 | 55 | 62 |
| MEAN NO. DAYS WITH: | | | | | | | | | | | | | | |
| HEAVY FOG(VISBY <= 1/4 MI) | 46 | 4.3 | 3.0 | 2.9 | 2.4 | 3.0 | 2.2 | 1.2 | 1.8 | 3.0 | 3.2 | 4.2 | 4.5 | 35.7 |
| THUNDERSTORMS | 59 | 1.1 | 1.4 | 2.8 | 3.5 | 6.4 | 10.4 | 14.4 | 11.9 | 5.3 | 1.6 | 0.6 | 0.6 | 60.0 |
| MEAN: | | | | | | | | | | | | | | |
| SUNRISE-SUNSET (OKTAS) | 46 | 5.0 | 4.9 | 4.8 | 4.3 | 4.6 | 4.9 | 4.9 | 4.9 | 4.9 | 4.1 | 4.2 | 4.8 | 4.7 |
| MIDNIGHT-MIDNIGHT (OKTAS) | 32 | 4.6 | 4.5 | 4.3 | 3.7 | 4.1 | 4.4 | 4.5 | 4.4 | 4.3 | 3.7 | 3.8 | 4.3 | 4.2 |
| MEAN NO. DAYS WITH: | | | | | | | | | | | | | | |
| CLEAR | 46 | 9.3 | 8.5 | 9.0 | 10.6 | 9.2 | 6.8 | 5.6 | 5.8 | 7.0 | 11.8 | 11.0 | 9.2 | 103.8 |
| PARTLY CLOUDY | 46 | 6.0 | 6.2 | 8.5 | 8.6 | 10.0 | 11.2 | 13.6 | 13.6 | 10.3 | 7.9 | 7.0 | 7.1 | 110.0 |
| CLOUDY | 46 | 15.7 | 13.7 | 13.5 | 10.7 | 11.8 | 12.0 | 11.9 | 11.6 | 12.6 | 11.4 | 12.0 | 14.7 | 151.6 |
| MEAN STATION PRESSURE(IN) | 26 | 30.09 | 30.06 | 30.02 | 29.97 | 29.96 | 29.95 | 29.98 | 29.97 | 29.97 | 30.02 | 30.07 | 30.10 | 30.01 |
| MEAN SEA-LEVEL PRES. (IN) | 26 | 30.15 | 30.12 | 30.07 | 30.02 | 30.02 | 30.00 | 30.03 | 30.02 | 30.02 | 30.07 | 30.12 | 30.15 | 30.07 |
| MEAN SPEED (MPH) | 26 | 7.3 | 7.7 | 8.2 | 7.6 | 7.0 | 6.5 | 6.1 | 5.9 | 6.7 | 6.5 | 6.4 | 6.7 | 6.9 |
| PREVAIL.DIR(TENS OF DEGS) | 33 | 28 | 28 | 19 | 19 | 19 | 19 | 21 | 19 | 03 | 03 | 04 | 28 | 03 |
| MAXIMUM 2-MINUTE: | | | | | | | | | | | | | | |
| SPEED (MPH) | 13 | 36 | 37 | 43 | 35 | 38 | 43 | 45 | 38 | 40 | 32 | 33 | 32 | 45 |
| DIR. (TENS OF DEGS) | | 32 | 27 | 16 | 26 | 32 | 05 | 04 | 36 | 01 | 36 | 16 | 24 | 04 |
| YEAR OF OCCURRENCE | | 2005 | 2009 | 2001 | 2009 | 2005 | 1998 | 1997 | 2009 | 1999 | 1999 | 2006 | 2008 | JUL 1997 |
| MAXIMUM 3-SECOND | | | | | | | | | | | | | | |
| SPEED (MPH) | 13 | 44 | 44 | 58 | 49 | 52 | 56 | 51 | 54 | 54 | 45 | 44 | 43 | 58 |
| DIR. (TENS OF DEGS) | | 24 | 27 | 16 | 29 | 24 | 06 | 04 | 06 | 17 | 11 | 16 | 19 | 16 |
| YEAR OF OCCURRENCE | | 2009 | 2009 | 2001 | 2005 | 2004 | 1998 | 1997 | 1997 | 1998 | 2008 | 2006 | 2002 | MAR 2001 |
| NORMAL (IN) | 30 | 3.95 | 2.92 | 3.64 | 3.32 | 3.61 | 5.49 | 6.04 | 7.20 | 5.08 | 3.12 | 2.40 | 2.81 | 49.58 |
| MAXIMUM MONTHLY (IN) | 59 | 8.98 | 7.92 | 9.57 | 10.57 | 10.08 | 14.39 | 20.10 | 17.03 | 13.47 | 19.84 | 5.69 | 10.71 | 20.10 |
| YEAR OF OCCURRENCE | | 1991 | 1964 | 1959 | 1991 | 1957 | 1963 | 1964 | 1995 | 1953 | 1994 | 2008 | 2009 | JUL 1964 |
| MINIMUM MONTHLY (IN) | 59 | 0.45 | 0.26 | 0.09 | 0.33 | 0.36 | 0.84 | 1.35 | 1.02 | 0.35 | 0.02 | 0.05 | 0.12 | 0.02 |
| YEAR OF OCCURRENCE | | 1989 | 1991 | 2004 | 2007 | 2001 | 1954 | 1972 | 1980 | 1991 | 2000 | 2007 | 1984 | OCT 2000 |
| MAXIMUM IN 24 HOURS (IN) | 59 | 3.74 | 3.46 | 4.65 | 5.62 | 5.67 | 6.77 | 6.36 | 8.71 | 6.80 | 8.86 | 5.02 | 7.31 | 8.86 |
| YEAR OF OCCURRENCE | | 1998 | 1964 | 1959 | 1976 | 1976 | 1999 | 1957 | 1995 | 1979 | 1994 | 1969 | 2007 | OCT 1994 |
| NORMAL NO. DAYS WITH: | | | | | | | | | | | | | | |
| PRECIPITATION >= 0.01 | 30 | 10.0 | 8.0 | 8.6 | 6.6 | 8.3 | 11.2 | 12.7 | 13.2 | 10.5 | 6.3 | 6.8 | 8.6 | 110.8 |
| PRECIPITATION >= 1.00 | 30 | 1.0 | 0.7 | 1.0 | 0.9 | 1.0 | 1.4 | 1.8 | 2.2 | 1.4 | 0.6 | 0.7 | 0.8 | 13.5 |
| NORMAL (IN) | 30 | 0.1 | 0.2 | 0.1 | 0.0 | 0.0 | 0.0 | 0.0 | 0.0 | 0.0 | 0.0 | 0.0 | 0.2 | 0.6 |
| MAXIMUM MONTHLY (IN) | 54 | 2.0 | 3.6 | 1.1 | 0.0 | 0.0 | T | T | 0.0 | 0.0 | 0.0 | T | 3.6 | 3.6 |
| YEAR OF OCCURRENCE | | 1977 | 1968 | 1986 | 2006 | | 2008 | 2008 | | | | 2006 | 1989 | DEC 1989 |
| MAXIMUM IN 24 HOURS (IN) | 54 | 1.3 | 3.6 | 1.1 | T | 0.0 | T | T | 0.0 | 0.0 | 0.0 | T | 3.4 | 3.6 |
| YEAR OF OCCURRENCE' | | 1977 | 1968 | 1986 | 2006 | | 1989 | 2008 | | | | 2006 | 1989 | FEB 1968 |
| MAXIMUM SNOW DEPTH (IN) | 52 | 0 | 4 | 0 | 0 | 0 | 0 | 0 | 0 | 0 | 0 | 0 | 4 | 4 |
| YEAR OF OCCURRENCE | | | 1968 | | | | | | | | | | 1989 | DEC 1989 |
| NORMAL NO. DAYS WITH: | | | | | | | | | | | | | | |
| SNOWFALL >= 1.0 | 30 | 0.0 | 0.1 | 0.0 | 0.0 | 0.0 | 0.0 | 0.0 | 0.0 | 0.0 | 0.0 | 0.0 | 0.0 | 0.1 |

## PRECIPITATION (inches) 2009 SAVANNAH (KSAV)

| YEAR | JAN | FEB | MAR | APR | MAY | JUN | JUL | AUG | SEP | OCT | NOV | DEC | ANNUAL |
|------|-----|-----|-----|-----|-----|-----|-----|-----|-----|-----|-----|-----|--------|
| 1980 | 2.95 | 1.29 | 7.75 | 3.68 | 4.50 | 3.47 | 2.38 | 1.02 | 5.81 | 1.62 | 2.04 | 1.33 | 37.84 |
| 1981 | 1.03 | 2.94 | 3.91 | 1.75 | 2.10 | 3.01 | 5.42 | 10.91 | 2.88 | 1.29 | 1.65 | 3.17 | 40.06 |
| 1982 | 3.47 | 2.94 | 1.64 | 6.25 | 4.18 | 9.15 | 6.70 | 9.18 | 2.98 | 1.74 | 0.40 | 3.63 | 52.26 |
| 1983 | 5.90 | 5.23 | 9.01 | 5.15 | 1.07 | 5.81 | 5.30 | 3.67 | 3.39 | 1.03 | 4.18 | 4.77 | 54.51 |
| 1984 | 8.87 | 3.21 | 5.13 | 3.41 | 5.29 | 1.48 | 7.88 | 3.46 | 7.43 | 1.23 | 3.15 | 0.12 | 50.66 |
| 1985 | 0.51 | 1.37 | 1.65 | 1.37 | 2.18 | 6.72 | 5.00 | 9.42 | 0.76 | 3.37 | 4.28 | 2.01 | 38.64 |
| 1986 | 2.03 | 5.28 | 2.85 | 0.38 | 2.06 | 2.98 | 5.49 | 12.31 | 0.49 | 1.99 | 4.40 | 5.07 | 45.33 |
| 1987 | 8.62 | 4.39 | 5.33 | 0.50 | 3.82 | 8.03 | 4.37 | 9.46 | 8.16 | 0.33 | 2.06 | 1.41 | 56.48 |
| 1988 | 3.44 | 4.09 | 2.11 | 5.05 | 3.52 | 2.63 | 1.80 | 10.68 | 9.62 | 2.81 | 1.43 | 0.99 | 48.17 |
| 1989 | 0.45 | 0.67 | 1.41 | 3.59 | 3.10 | 7.30 | 4.91 | 6.29 | 7.98 | 4.71 | 1.26 | 5.20 | 46.87 |
| 1990 | 3.91 | 3.08 | 3.79 | 1.75 | 2.07 | 0.97 | 1.92 | 7.25 | 1.26 | 12.50 | 2.48 | 2.10 | 43.08 |
| 1991 | 8.98 | 0.26 | 5.48 | 10.57 | 7.13 | 5.12 | 15.41 | 10.51 | 0.35 | 1.60 | 1.26 | 1.75 | 68.42 |
| 1992 | 6.60 | 2.24 | 3.97 | 2.08 | 2.04 | 13.01 | 2.03 | 7.69 | 8.42 | 3.60 | 5.15 | 1.53 | 58.36 |
| 1993 | 5.52 | 3.35 | 7.96 | 3.22 | 1.31 | 2.48 | 4.34 | 3.04 | 6.84 | 2.58 | 5.26 | 2.15 | 48.05 |
| 1994 | 4.70 | 0.78 | 3.75 | 2.13 | 6.17 | 7.80 | 6.90 | 2.64 | 6.60 | 19.84 | 3.67 | 4.46 | 69.44 |
| 1995 | 2.44 | 4.47 | 0.72 | 0.69 | 3.52 | 5.89 | 6.65 | 17.03 | 4.10 | 3.42 | 1.29 | 0.89 | 51.11 |
| 1996 | 2.01 | 1.35 | 4.29 | 2.43 | 1.44 | 2.44 | 5.23 | 7.12 | 2.84 | 3.45 | .88 | 2.72 | 36.20 |
| 1997 | 3.12 | 2.34 | 1.54 | 3.62 | 2.99 | 5.90 | 11.64 | 4.56 | 5.32 | 5.65 | 5.04 | 3.98 | 55.70 |
| 1998 | 7.51 | 6.88 | 3.98 | 5.68 | 1.97 | 1.93 | 8.27 | 3.79 | 5.56 | 1.28 | 0.30 | 2.32 | 49.47 |
| 1999 | 4.73 | 1.95 | 1.25 | 1.68 | 2.54 | 14.25 | 7.15 | 4.26 | 6.50 | 2.04 | 0.49 | 1.94 | 48.78 |
| 2000 | 2.71 | 1.56 | 4.30 | 2.84 | 0.96 | 5.47 | 3.38 | 4.18 | 7.45 | T | 1.77 | 2.82 | 37.44 |
| 2001 | 1.71 | 0.77 | 6.65 | 0.71 | 0.36 | 6.41 | 4.85 | 4.65 | 4.72 | 0.16 | 0.16 | 0.49 | 31.64 |
| 2002 | 2.37 | 1.54 | 5.52 | 0.39 | 0.99 | 10.49 | 3.46 | 4.31 | 5.54 | 4.33 | 4.60 | 3.88 | 47.42 |
| 2003 | 0.65 | 3.23 | 7.84 | 6.21 | 5.99 | 4.82 | 6.38 | 2.82 | 3.61 | 3.02 | 1.67 | 1.47 | 47.71 |
| 2004 | 2.03 | 2.76 | 0.09 | 4.57 | 1.64 | 8.51 | 3.41 | 3.70 | 4.60 | 3.03 | 1.08 | 1.76 | 37.18 |
| 2005 | 1.41 | 1.96 | 7.06 | 2.96 | 3.89 | 6.60 | 3.87 | 5.32 | 0.48 | 6.93 | 2.83 | 2.72 | 46.03 |
| 2006 | 2.94 | 3.91 | 0.32 | 1.91 | 0.65 | 6.31 | 2.98 | 4.37 | 5.22 | 1.02 | 2.03 | 2.79 | 34.45 |
| 2007 | 2.72 | 2.00 | 1.88 | 0.33 | 1.52 | 8.95 | 7.01 | 4.33 | 7.42 | 4.29 | 0.05 | 9.44 | 49.94 |
| 2008 | 2.93 | 4.55 | 1.35 | 2.56 | 1.32 | 3.69 | 7.32 | 6.61 | 1.42 | 9.29 | 5.69 | 0.56 | 47.29 |
| 2009 | 1.02 | 1.33 | 4.42 | 6.97 | 9.69 | 4.39 | 6.57 | 7.86 | 2.43 | 3.41 | 2.31 | 10.71 | 61.11 |
| POR= 59 YRS | 3.34 | 2.97 | 3.80 | 3.15 | 3.80 | 5.82 | 6.56 | 6.62 | 4.91 | 3.05 | 2.14 | 2.87 | 49.03 |

WBAN : 03822

## AVERAGE TEMPERATURE (°F) 2009 SAVANNAH (KSAV)

| YEAR | JAN | FEB | MAR | APR | MAY | JUN | JUL | AUG | SEP | OCT | NOV | DEC | ANNUAL |
|------|-----|-----|-----|-----|-----|-----|-----|-----|-----|-----|-----|-----|--------|
| 1980 | 50.7 | 48.5 | 57.1 | 66.2 | 73.3 | 79.7 | 84.4 | 83.4 | 80.4 | 65.4 | 56.7 | 48.1 | 66.2 |
| 1981 | 43.5 | 52.6 | 56.8 | 69.0 | 71.6 | 84.5 | 84.4 | 79.2 | 75.2 | 65.4 | 57.7 | 48.3 | 65.7 |
| 1982 | 48.7 | 55.9 | 62.1 | 64.8 | 74.3 | 80.2 | 81.3 | 81.0 | 75.5 | 67.6 | 61.8 | 57.4 | 67.6 |
| 1983 | 46.1 | 50.8 | 58.0 | 62.8 | 73.0 | 78.2 | 84.1 | 82.9 | 75.7 | 69.7 | 57.7 | 48.5 | 65.6 |
| 1984 | 47.7 | 53.7 | 59.3 | 65.7 | 72.9 | 79.2 | 80.6 | 81.5 | 75.0 | 73.2 | 55.7 | 59.5 | 67.0 |
| 1985 | 45.3 | 53.0 | 61.5 | 66.9 | 74.6 | 81.3 | 82.7 | 80.7 | 76.6 | 72.2 | 67.5 | 48.9 | 67.6 |
| 1986 | 47.6 | 56.9 | 59.1 | 66.8 | 75.0 | 82.5 | 85.7 | 81.6 | 79.7 | 70.0 | 65.0 | 53.9 | 68.7 |
| 1987 | 49.0 | 50.4 | 58.2 | 63.9 | 74.1 | 80.6 | 83.6 | 84.5 | 77.8 | 61.5 | 60.4 | 54.7 | 66.6 |
| 1988 | 45.0 | 50.2 | 58.2 | 66.1 | 72.4 | 77.7 | 82.7 | 82.5 | 77.4 | 64.2 | 61.3 | 51.1 | 65.7 |
| 1989 | 56.6 | 56.1 | 60.6 | 65.6 | 72.3 | 81.0 | 82.9 | 80.7 | 76.8 | 68.6 | 59.5 | 43.7 | 67.0 |
| 1990 | 55.7 | 60.0 | 62.6 | 65.0 | 74.3 | 81.5 | 84.4 | 82.4 | 78.9 | 70.7 | 60.8 | 56.5 | 69.4 |
| 1991 | 52.2 | 56.2 | 62.4 | 70.0 | 77.5 | 79.4 | 83.2 | 82.2 | 77.9 | 68.4 | 56.6 | 55.6 | 68.5 |
| 1992 | 49.9 | 55.7 | 59.2 | 64.5 | 71.7 | 78.6 | 84.3 | 81.0 | 77.1 | 66.6 | 60.6 | 52.0 | 66.8 |
| 1993 | 54.8 | 50.3 | 57.0 | 62.4 | 73.6 | 81.3 | 86.7 | 83.1 | 79.6 | 68.1 | 59.9 | 49.7 | 67.2 |
| 1994 | 47.7 | 56.0 | 62.9 | 69.6 | 72.4 | 80.7 | 82.2 | 81.2 | 76.6 | 67.8 | 63.7 | 54.9 | 68.0 |
| 1995 | 50.7 | 52.4 | 62.5 | 69.1 | 76.5 | 79.2 | 83.8 | 82.4 | 76.1 | 71.5 | 55.6 | 50.1 | 67.5 |
| 1996 | 50.1 | 54.6 | 56.0 | 64.0 | 75.2 | 78.8 | 82.4 | 79.3 | 76.1 | 66.5 | 55.9 | 52.9 | 66.0 |
| 1997 | 50.9 | 55.1 | 64.7 | 63.1 | 68.9 | 75.4 | 81.2 | 79.3 | 76.9 | 66.6 | 55.1 | 50.5 | 65.6 |
| 1998 | 52.9 | 54.1 | 56.4 | 65.3 | 75.6 | 83.0 | 83.4 | 81.2 | 77.5 | 68.9 | 62.6 | 55.5 | 68.0 |
| 1999 | 52.4 | 54.3 | 55.8 | 69.0 | 70.8 | 77.1 | 82.7 | 83.1 | 75.7 | 67.8 | 59.7 | 50.8 | 66.6 |
| 2000 | 48.1 | 53.2 | 61.7 | 62.8 | 75.3 | 79.2 | 82.0 | 81.0 | 75.5 | 64.8 | 55.6 | 43.7 | 65.2 |
| 2001 | 46.8 | 56.0 | 57.4 | 65.1 | 73.0 | 78.8 | 80.8 | 81.3 | 74.2 | 64.9 | 63.0 | 55.9 | 66.4 |
| 2002 | 52.2 | 51.8 | 60.7 | 70.5 | 72.7 | 79.2 | 82.4 | 80.6 | 79.5 | 71.3 | 56.3 | 49.0 | 67.2 |
| 2003 | 45.4 | 51.6 | 61.6 | 65.4 | 74.3 | 78.8 | 81.8 | 81.8 | 76.1 | 69.0 | 63.0 | 48.1 | 66.4 |
| 2004 | 48.5 | 49.5 | 60.0 | 64.3 | 74.9 | 80.4 | 81.7 | 79.6 | 76.6 | 69.2 | 60.6 | 49.7 | 66.3 |
| 2005 | 51.9 | 52.4 | 55.8 | 61.7 | 70.1 | 78.5 | 83.1 | 82.3 | 79.2 | 67.7 | 60.0 | 48.5 | 65.9 |
| 2006 | 54.0 | 50.2 | 57.2 | 68.0 | 72.2 | 78.1 | 81.5 | 82.8 | 76.3 | 65.4 | 57.0 | 56.1 | 66.6 |
| 2007 | 52.5 | 50.3 | 60.9 | 64.4 | 72.3 | 78.7 | 81.2 | 83.6 | 77.7 | 71.6 | 57.7 | 56.9 | 67.3 |
| 2008 | 49.1 | 56.0 | 59.7 | 65.9 | 73.2 | 82.5 | 82.3 | 82.1 | 77.8 | 66.3 | 55.4 | 56.8 | 67.3 |
| 2009 | 50.6 | 50.4 | 59.5 | 65.1 | 74.2 | 81.9 | 80.9 | 81.2 | 77.1 | 68.7 | 59.5 | 52.1 | 66.8 |
| POR= 59 YRS | 49.5 | 52.5 | 59.0 | 65.9 | 73.3 | 79.2 | 81.9 | 81.2 | 76.8 | 67.5 | 58.4 | 51.4 | 66.4 |

## HEATING DEGREE DAYS (base 65°F) 2009  SAVANNAH (KSAV)

| YEAR | JUL | AUG | SEP | OCT | NOV | DEC | JAN | FEB | MAR | APR | MAY | JUN | TOTAL |
|------|-----|-----|-----|-----|-----|-----|-----|-----|-----|-----|-----|-----|-------|
| 1980-81 | 0 | 0 | 0 | 72 | 252 | 518 | 659 | 342 | 263 | 25 | 8 | 0 | 2139 |
| 1981-82 | 0 | 0 | 3 | 59 | 231 | 513 | 501 | 258 | 149 | 76 | 0 | 0 | 1790 |
| 1982-83 | 0 | 0 | 0 | 73 | 139 | 266 | 579 | 392 | 228 | 115 | 0 | 0 | 1792 |
| 1983-84 | 0 | 0 | 2 | 19 | 232 | 513 | 531 | 320 | 200 | 68 | 7 | 0 | 1892 |
| 1984-85 | 0 | 0 | 1 | 8 | 299 | 185 | 615 | 360 | 157 | 60 | 2 | 0 | 1687 |
| 1985-86 | 0 | 0 | 1 | 16 | 51 | 504 | 531 | 240 | 215 | 59 | 4 | 0 | 1621 |
| 1986-87 | 0 | 5 | 0 | 48 | 101 | 349 | 491 | 401 | 231 | 110 | 5 | 0 | 1741 |
| 1987-88 | 0 | 0 | 0 | 122 | 185 | 332 | 612 | 426 | 218 | 52 | 1 | 0 | 1948 |
| 1988-89 | 0 | 0 | 0 | 84 | 141 | 423 | 268 | 289 | 193 | 110 | 13 | 0 | 1521 |
| 1989-90 | 0 | 0 | 1 | 59 | 191 | 653 | 286 | 175 | 135 | 81 | 0 | 0 | 1581 |
| 1990-91 | 0 | 0 | 0 | 55 | 143 | 279 | 395 | 260 | 139 | 14 | 0 | 0 | 1285 |
| 1991-92 | 0 | 0 | 0 | 34 | 270 | 315 | 461 | 274 | 204 | 101 | 22 | 0 | 1681 |
| 1992-93 | 0 | 0 | 1 | 45 | 194 | 401 | 320 | 406 | 254 | 113 | 0 | 0 | 1734 |
| 1993-94 | 0 | 0 | 1 | 45 | 193 | 470 | 530 | 260 | 123 | 20 | 7 | 0 | 1649 |
| 1994-95 | 0 | 0 | 0 | 32 | 95 | 317 | 433 | 358 | 114 | 30 | 1 | 0 | 1380 |
| 1995-96 | 0 | 0 | 6 | 25 | 313 | 457 | 457 | 313 | 288 | 109 | 7 | 0 | 1975 |
| 1996-97 | 0 | 0 | 0 | 51 | 287 | 374 | 435 | 290 | 88 | 107 | 24 | 5 | 1661 |
| 1997-98 | 0 | 0 | 0 | 80 | 300 | 442 | 373 | 303 | 283 | 63 | 0 | 0 | 1844 |
| 1998-99 | 0 | 0 | 0 | 44 | 107 | 305 | 390 | 294 | 283 | 60 | 17 | 0 | 1500 |
| 1999-00 | 0 | 0 | 0 | 56 | 179 | 436 | 518 | 340 | 122 | 97 | 1 | 0 | 1749 |
| 2000-01 | 0 | 0 | 1 | 72 | 304 | 655 | 560 | 262 | 248 | 88 | 0 | 0 | 2190 |
| 2001-02 | 0 | 0 | 4 | 107 | 103 | 294 | 405 | 373 | 200 | 25 | 20 | 0 | 1531 |
| 2002-03 | 0 | 0 | 0 | 14 | 277 | 486 | 599 | 368 | 131 | 65 | 0 | 0 | 1940 |
| 2003-04 | 0 | 0 | 3 | 6 | 134 | 518 | 517 | 446 | 176 | 90 | 6 | 0 | 1896 |
| 2004-05 | 0 | 0 | 0 | 22 | 184 | 482 | 405 | 353 | 287 | 110 | 14 | 0 | 1857 |
| 2005-06 | 0 | 0 | 0 | 85 | 173 | 502 | 335 | 411 | 259 | 43 | 13 | 0 | 1821 |
| 2006-07 | 0 | 0 | 0 | 86 | 247 | 281 | 394 | 407 | 153 | 111 | 2 | 0 | 1681 |
| 2007-08 | 0 | 0 | 0 | 14 | 227 | 261 | 490 | 269 | 185 | 65 | 2 | 0 | 1513 |
| 2008-09 | 0 | 0 | 0 | 88 | 298 | 280 | 442 | 403 | 194 | 61 | 12 | 0 | 1778 |
| 2009- | 0 | 0 | 0 | 50 | 176 | 406 | | | | | | | |

WBAN : 03822

## COOLING DEGREE DAYS (base 65°F) 2009  SAVANNAH (KSAV)

| YEAR | JAN | FEB | MAR | APR | MAY | JUN | JUL | AUG | SEP | OCT | NOV | DEC | TOTAL |
|------|-----|-----|-----|-----|-----|-----|-----|-----|-----|-----|-----|-----|-------|
| 1980 | 0 | 16 | 19 | 87 | 270 | 449 | 607 | 579 | 468 | 90 | 8 | 2 | 2595 |
| 1981 | 0 | 1 | 14 | 150 | 219 | 589 | 609 | 444 | 319 | 80 | 17 | 4 | 2446 |
| 1982 | 6 | 10 | 67 | 79 | 292 | 463 | 514 | 503 | 323 | 158 | 48 | 42 | 2505 |
| 1983 | 0 | 0 | 16 | 55 | 253 | 400 | 598 | 562 | 332 | 171 | 18 | 6 | 2411 |
| 1984 | 0 | 0 | 31 | 94 | 261 | 431 | 486 | 517 | 309 | 270 | 24 | 19 | 2442 |
| 1985 | 11 | 31 | 58 | 123 | 307 | 496 | 557 | 496 | 355 | 248 | 133 | 7 | 2822 |
| 1986 | 0 | 18 | 40 | 121 | 321 | 532 | 651 | 525 | 449 | 212 | 107 | 13 | 2989 |
| 1987 | 0 | 0 | 26 | 85 | 292 | 474 | 583 | 611 | 392 | 22 | 50 | 21 | 2556 |
| 1988 | 0 | 3 | 11 | 92 | 238 | 386 | 555 | 547 | 378 | 66 | 35 | 1 | 2312 |
| 1989 | 13 | 45 | 64 | 135 | 248 | 488 | 563 | 493 | 362 | 177 | 33 | 0 | 2621 |
| 1990 | 4 | 41 | 69 | 90 | 296 | 503 | 608 | 549 | 422 | 237 | 25 | 24 | 2868 |
| 1991 | 7 | 18 | 64 | 169 | 397 | 441 | 572 | 540 | 393 | 143 | 25 | 29 | 2798 |
| 1992 | 0 | 11 | 29 | 94 | 234 | 414 | 603 | 504 | 370 | 101 | 69 | 7 | 2436 |
| 1993 | 12 | 0 | 10 | 43 | 271 | 497 | 678 | 567 | 447 | 149 | 46 | 1 | 2721 |
| 1994 | 0 | 14 | 62 | 165 | 243 | 476 | 542 | 509 | 355 | 125 | 63 | 14 | 2568 |
| 1995 | 1 | 9 | 46 | 160 | 366 | 433 | 591 | 547 | 348 | 235 | 36 | 4 | 2776 |
| 1996 | 3 | 20 | 19 | 84 | 331 | 419 | 549 | 451 | 339 | 105 | 20 | 2 | 2342 |
| 1997 | 4 | 20 | 84 | 57 | 155 | 322 | 510 | 452 | 363 | 137 | 11 | 0 | 2115 |
| 1998 | 10 | 5 | 24 | 77 | 336 | 546 | 575 | 508 | 383 | 173 | 44 | 18 | 2699 |
| 1999 | 7 | 1 | 4 | 187 | 205 | 368 | 556 | 568 | 328 | 153 | 26 | 1 | 2404 |
| 2000 | 3 | 5 | 27 | 37 | 325 | 435 | 538 | 503 | 325 | 73 | 28 | 0 | 2299 |
| 2001 | 2 | 17 | 20 | 99 | 253 | 422 | 499 | 515 | 287 | 106 | 49 | 19 | 2288 |
| 2002 | 16 | 9 | 73 | 196 | 265 | 433 | 547 | 487 | 443 | 218 | 20 | 0 | 2707 |
| 2003 | 0 | 0 | 32 | 83 | 292 | 420 | 528 | 529 | 346 | 136 | 81 | 0 | 2447 |
| 2004 | 12 | 2 | 30 | 77 | 324 | 465 | 524 | 458 | 353 | 162 | 57 | 13 | 2477 |
| 2005 | 6 | 6 | 12 | 20 | 182 | 415 | 567 | 543 | 433 | 175 | 28 | 0 | 2387 |
| 2006 | 0 | 0 | 20 | 141 | 245 | 400 | 518 | 560 | 347 | 105 | 15 | 12 | 2363 |
| 2007 | 13 | 0 | 31 | 101 | 233 | 420 | 510 | 585 | 388 | 223 | 14 | 15 | 2533 |
| 2008 | 4 | 14 | 29 | 99 | 260 | 529 | 545 | 536 | 392 | 135 | 14 | 29 | 2586 |
| 2009 | 4 | 0 | 33 | 72 | 304 | 515 | 500 | 509 | 369 | 171 | 15 | 13 | 2505 |

## SNOWFALL (inches)  2009  SAVANNAH (KSAV)

| YEAR | JUL | AUG | SEP | OCT | NOV | DEC | JAN | FEB | MAR | APR | MAY | JUN | TOTAL |
|------|-----|-----|-----|-----|-----|-----|-----|-----|-----|-----|-----|-----|-------|
| 1980-81 | 0.0 | 0.0 | 0.0 | 0.0 | 0.0 | T | 0.0 | 0.0 | 0.0 | 0.0 | 0.0 | 0.0 | T |
| 1981-82 | 0.0 | 0.0 | 0.0 | 0.0 | 0.0 | 0.0 | 0.0 | 0.0 | 0.0 | 0.0 | 0.0 | 0.0 | 0.0 |
| 1982-83 | 0.0 | 0.0 | 0.0 | 0.0 | 0.0 | 0.0 | T | 0.0 | T | 0.0 | 0.0 | 0.0 | T |
| 1983-84 | 0.0 | 0.0 | 0.0 | 0.0 | 0.0 | 0.0 | T | 0.0 | 0.0 | 0.0 | 0.0 | 0.0 | T |
| 1984-85 | 0.0 | 0.0 | 0.0 | 0.0 | 0.0 | 0.0 | 0.0 | 0.0 | 0.0 | 0.0 | 0.0 | 0.0 | 0.0 |
| 1985-86 | 0.0 | 0.0 | 0.0 | 0.0 | 0.0 | 0.0 | 0.3 | 0.0 | 1.1 | 0.0 | 0.0 | 0.0 | 1.4 |
| 1986-87 | 0.0 | 0.0 | 0.0 | 0.0 | 0.0 | 0.0 | 0.0 | T | 0.0 | 0.0 | 0.0 | 0.0 | T |
| 1987-88 | 0.0 | 0.0 | 0.0 | 0.0 | 0.0 | 0.0 | T | T | 0.0 | 0.0 | 0.0 | 0.0 | T |
| 1988-89 | 0.0 | 0.0 | 0.0 | 0.0 | 0.0 | T | 0.0 | 1.0 | 0.0 | 0.0 | 0.0 | T | 1.0 |
| 1989-90 | 0.0 | 0.0 | 0.0 | 0.0 | 0.0 | 3.6 | 0.0 | 0.0 | 0.0 | 0.0 | 0.0 | 0.0 | 3.6 |
| 1990-91 | 0.0 | 0.0 | 0.0 | 0.0 | 0.0 | 0.0 | 0.0 | T | 0.0 | 0.0 | 0.0 | 0.0 | T |
| 1991-92 | 0.0 | 0.0 | 0.0 | 0.0 | 0.0 | 0.0 | 0.0 | 0.0 | 0.0 | 0.0 | 0.0 | 0.0 | 0.0 |
| 1992-93 | 0.0 | 0.0 | 0.0 | 0.0 | 0.0 | 0.0 | 0.0 | 0.0 | 0.2 | 0.0 | 0.0 | 0.0 | 0.2 |
| 1993-94 | 0.0 | 0.0 | 0.0 | 0.0 | 0.0 | T | 0.0 | T | 0.0 | 0.0 | 0.0 | 0.0 | T |
| 1994-95 | 0.0 | 0.0 | 0.0 | 0.0 | 0.0 | 0.0 | 0.0 | T | 0.0 | 0.0 | 0.0 | 0.0 | T |
| 1995-96 | 0.0 | 0.0 | 0.0 | 0.0 | 0.0 | 0.0 | T | 0.2 | 0.0 | | | | |
| 1996-97 | | | | | | | | | | | | | |
| 1997-98 | | | | | | | | | | | | | |
| 1998-99 | | | | | | | | | | | | | |
| 1999-00 | | | | | | | | | | | | | |
| 2000-01 | | | | | | | T | | | | | | |
| 2001-02 | | | | | | | | | | | | | |
| 2002-03 | | | | | | | T | 0.0 | 0.0 | 0.0 | 0.0 | 0.0 | |
| 2003-04 | 0.0 | 0.0 | 0.0 | 0.0 | 0.0 | 0.0 | 0.0 | 0.0 | 0.0 | 0.0 | 0.0 | 0.0 | 0.0 |
| 2004-05 | 0.0 | 0.0 | 0.0 | 0.0 | 0.0 | T | 0.0 | 0.0 | 0.0 | 0.0 | 0.0 | 0.0 | T |
| 2005-06 | 0.0 | 0.0 | 0.0 | 0.0 | 0.0 | 0.0 | 0.0 | 0.0 | 0.0 | T | 0.0 | 0.0 | T |
| 2006-07 | 0.0 | 0.0 | 0.0 | 0.0 | T | 0.0 | 0.0 | 0.0 | 0.0 | 0.0 | 0.0 | 0.0 | T |
| 2007-08 | 0.0 | 0.0 | 0.0 | 0.0 | 0.0 | 0.0 | 0.0 | 0.0 | 0.0 | 0.0 | 0.0 | T | T |
| 2008-09 | T | 0.0 | 0.0 | 0.0 | 0.0 | 0.0 | 0.0 | 0.0 | 0.0 | 0.0 | 0.0 | 0.0 | T |
| 2009- | 0.0 | 0.0 | 0.0 | 0.0 | 0.0 | 0.0 | | | | | | | |
| POR= 55 YRS | T | 0.0 | 0.0 | 0.0 | T | 0.1 | T | 0.2 | T | T | 0.0 | T | 0.3 |

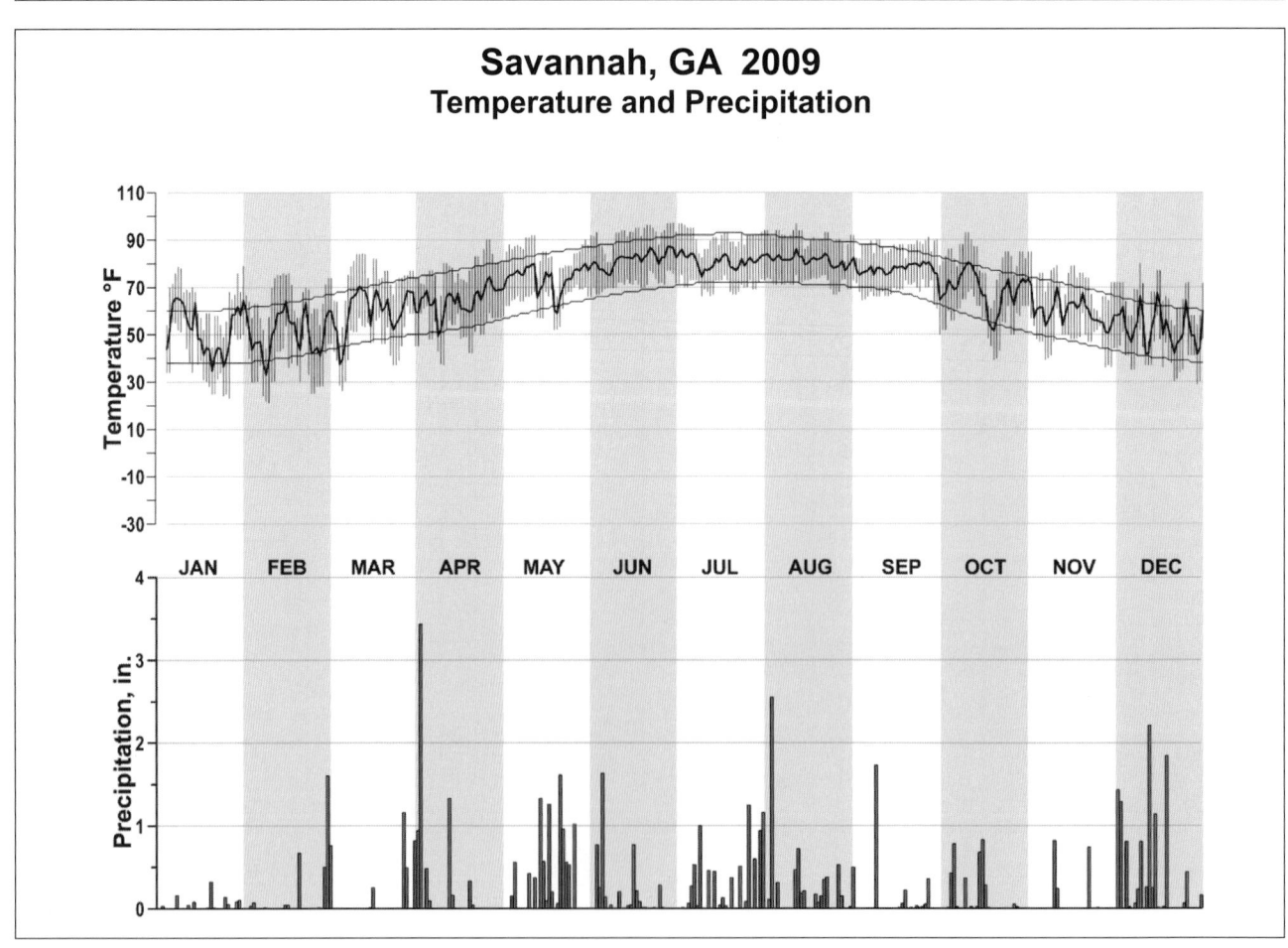

# Savannah, GA  2009
## Temperature and Precipitation

# 2009
# HILO
# HAWAII (PHTO)

The city of Hilo is located near the midpoint of the eastern shore of the Island of Hawaii. This island is by far the largest of the Hawaiian group, with an area of 4,038 square miles, more than twice that of all the other islands combined. Its topography is dominated by the great volcanic masses of Mauna Loa (13,653 feet), Mauna Kea (13,796 feet), and of Haulalai, the Kohala Mountains, and Kilauea. In fact, the island consists entirely of the slopes of these mountains and of the broad saddles between them. Mauna Loa and Kilauea, which occupy the southern half of the island, are still active volcanoes.

Hawaii lies well within the belt of northeasterly trade winds generated by the semi-permanent Pacific high pressure cell to the north and east. The climate provides equable temperatures from day to day and season to season. In Hilo, July and August are the warmest months, with average daily highs and lows of 83 and 68 degrees. January and February, the coolest months, have highs of 80 degrees and lows of 63 degrees. Greater variations occur in localities with less rain and cloud, but temperatures in the mid-90s and low 50s are uncommon anywhere on the island near sea level.

Over the windward slopes of Hawaii, rainfall occurs principally as orographic showers within the ascending moist trade winds. Mean annual rainfall, except for the semi-sheltered Hamakua district, increases from 100 inches or more along the coasts to a maximum of over 300 inches at elevations of 2,000 to 3,000 feet, and then declines to about 15 inches at the summits of Mauna Kea and Mauna Loa. Leeward areas are topographically sheltered from the trades and are therefore drier, although sea breezes created by daytime heating of the land move onshore and upslope, causing afternoon and evening cloudiness and showers. The driest locality on the island, and in the State, with an annual rainfall of less than 10 inches, is the coastal strip just leeward of the southern portion of the Kohala Mountains and of the saddle between the Kohalas and Mauna Kea.

Within the city of Hilo, average rainfall varies from about 130 inches a year near the shore to as much as 200 upslope. The wettest part of the island, with a mean annual rainfall exceeding 300 inches, lies about 6 miles upslope from the city limits. Relative humidity at Hilo is in the moderate range, however, due to the natural ventilation provided by the prevailing winds, the weather is seldom oppressive.

The trade winds prevail throughout the year and profoundly influence the climate. The islands entire western coast is sheltered from the trades by high mountains, except that unusually strong trade winds may sweep through the saddle between the Kohala Mountains and Mauna Kea and reach the areas to the lee. But even places exposed to the trades may be affected by local mountain circulations. Except for heavy rain, really severe weather seldom occurs. During the winter, cold fronts or the cyclonic storms of subtropical origin may bring blizzards to the upper slopes of Mauna Loa and Mauna Kea, with snow extending at times to 9,000 feet or below and icing nearer the summit.

Storms crossing the Pacific a thousand miles to the north, low pressure or tropical storms, may generate seas that cause heavy swell and surf.

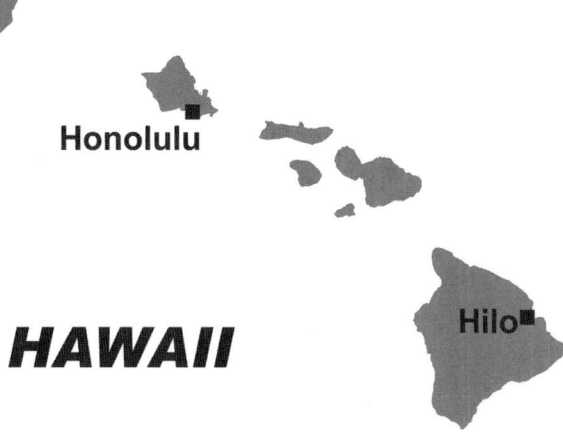

Honolulu

HAWAII

Hilo

# NORMALS, MEANS, AND EXTREMES
## HILO (PHTO)

LATITUDE: 19°43'N  LONGITUDE: -155°3'W  ELEVATION (FT): GRND: 30  BARO: 47  TIME ZONE: HAWAII (UTC -10)  WBAN: 21504

| | ELEMENT | POR | JAN | FEB | MAR | APR | MAY | JUN | JUL | AUG | SEP | OCT | NOV | DEC | YEAR |
|---|---|---|---|---|---|---|---|---|---|---|---|---|---|---|---|
| **TEMPERATURE °F** | NORMAL DAILY MAXIMUM | 30 | 79.2 | 79.4 | 79.2 | 79.3 | 80.6 | 82.2 | 82.5 | 83.2 | 83.4 | 82.7 | 80.7 | 79.5 | 81.0 |
| | MEAN DAILY MAXIMUM | 60 | 79.5 | 79.2 | 79.2 | 79.6 | 81.0 | 82.4 | 82.9 | 83.4 | 83.6 | 83.0 | 81.1 | 79.6 | 81.2 |
| | HIGHEST DAILY MAXIMUM | 63 | 92 | 92 | 93 | 89 | 94 | 90 | 91 | 92 | 92 | 91 | 92 | 93 | 94 |
| | YEAR OF OCCURRENCE | | 1997 | 1968 | 1972 | 1978 | 1966 | 1969 | 2004 | 1950 | 1951 | 1979 | 1996 | 1980 | MAY 1966 |
| | MEAN OF EXTREME MAXS. | 60 | 85.5 | 84.9 | 84.5 | 84.0 | 84.8 | 85.6 | 86.1 | 86.9 | 86.8 | 87.2 | 85.5 | 84.5 | 85.5 |
| | NORMAL DAILY MINIMUM | 30 | 63.6 | 63.5 | 64.7 | 65.6 | 66.7 | 68.0 | 69.2 | 69.4 | 69.0 | 68.5 | 67.2 | 64.9 | 66.7 |
| | MEAN DAILY MINIMUM | 60 | 63.6 | 63.5 | 64.3 | 65.3 | 66.4 | 67.6 | 68.6 | 69.0 | 68.6 | 67.9 | 66.8 | 64.8 | 66.4 |
| | LOWEST DAILY MINIMUM | 63 | 54 | 53 | 54 | 56 | 58 | 60 | 62 | 63 | 61 | 62 | 58 | 55 | 53 |
| | YEAR OF OCCURRENCE | | 1995 | 1962 | 1983 | 1949 | 1947 | 1946 | 1970 | 1955 | 1970 | 1999 | 1985 | 1977 | FEB 1962 |
| | MEAN OF EXTREME MINS. | 60 | 58.9 | 58.3 | 59.7 | 61.5 | 62.5 | 64.4 | 65.1 | 65.4 | 65.0 | 64.2 | 62.6 | 59.9 | 62.3 |
| | NORMAL DRY BULB | 30 | 71.4 | 71.5 | 72.0 | 72.5 | 73.7 | 75.1 | 75.9 | 76.3 | 76.2 | 75.6 | 74.0 | 72.2 | 73.9 |
| | MEAN DRY BULB | 60 | 71.6 | 71.3 | 71.8 | 72.5 | 73.7 | 75.1 | 75.9 | 76.3 | 76.1 | 75.5 | 74.0 | 72.2 | 73.8 |
| | MEAN WET BULB | 26 | 65.3 | 64.9 | 66.0 | 66.8 | 68.0 | 69.2 | 70.4 | 70.9 | 70.5 | 70.0 | 68.9 | 66.6 | 68.1 |
| | MEAN DEW POINT | 26 | 63.3 | 62.8 | 63.9 | 65.0 | 66.2 | 67.5 | 68.9 | 69.4 | 68.9 | 68.4 | 67.5 | 64.8 | 66.4 |
| | NORMAL NO. DAYS WITH: | | | | | | | | | | | | | | |
| | MAXIMUM >= 90 | 30 | 0.2 | 0.1 | * | 0.0 | 0.0 | 0.0 | 0.0 | 0.1 | 0.3 | 0.3 | 0.1 | 0.1 | 1.2 |
| | MAXIMUM <= 32 | 30 | 0.0 | 0.0 | 0.0 | 0.0 | 0.0 | 0.0 | 0.0 | 0.0 | 0.0 | 0.0 | 0.0 | 0.0 | 0.0 |
| | MINIMUM <= 32 | 30 | 0.0 | 0.0 | 0.0 | 0.0 | 0.0 | 0.0 | 0.0 | 0.0 | 0.0 | 0.0 | 0.0 | 0.0 | 0.0 |
| | MINIMUM <= 0 | 30 | 0.0 | 0.0 | 0.0 | 0.0 | 0.0 | 0.0 | 0.0 | 0.0 | 0.0 | 0.0 | 0.0 | 0.0 | 0.0 |
| **H/C** | NORMAL HEATING DEG. DAYS | 30 | 0 | 0 | 0 | 0 | 0 | 0 | 0 | 0 | 0 | 0 | 0 | 0 | 0 |
| | NORMAL COOLING DEG. DAYS | 30 | 198 | 180 | 215 | 223 | 268 | 303 | 336 | 350 | 336 | 328 | 268 | 223 | 3228 |
| **RH** | NORMAL (PERCENT) | 30 | 77 | 76 | 78 | 79 | 79 | 79 | 81 | 80 | 80 | 81 | 82 | 79 | 79 |
| | HOUR 02 LST | 30 | 83 | 83 | 85 | 88 | 87 | 87 | 88 | 88 | 87 | 88 | 87 | 85 | 86 |
| | HOUR 08 LST | 30 | 78 | 77 | 80 | 81 | 79 | 79 | 81 | 81 | 80 | 80 | 82 | 80 | 80 |
| | HOUR 14 LST | 30 | 67 | 65 | 67 | 69 | 68 | 66 | 69 | 70 | 69 | 70 | 72 | 69 | 68 |
| | HOUR 20 LST | 30 | 83 | 81 | 82 | 83 | 82 | 82 | 83 | 83 | 85 | 86 | 86 | 85 | 83 |
| **S** | PERCENT POSSIBLE SUNSHINE | 54 | 46 | 46 | 42 | 37 | 38 | 44 | 41 | 41 | 43 | 38 | 33 | 37 | 41 |
| **W/O** | MEAN NO. DAYS WITH: | | | | | | | | | | | | | | |
| | HEAVY FOG(VISBY <= 1/4 MI) | 46 | 0.1 | 0.0 | 0.0 | 0.0 | 0.0 | 0.0 | 0.0 | 0.0 | 0.1 | 0.1 | 0.1 | 0.0 | 0.4 |
| | THUNDERSTORMS | 60 | 1.0 | 1.4 | 1.4 | 0.9 | 0.6 | 0.1 | 0.3 | 0.3 | 0.5 | 1.3 | 1.3 | 1.1 | 10.2 |
| **CLOUDINESS** | MEAN: | | | | | | | | | | | | | | |
| | SUNRISE-SUNSET (OKTAS) | 51 | 5.0 | 5.3 | 6.0 | 6.4 | 6.2 | 5.9 | 6.1 | 5.9 | 5.6 | 5.7 | 5.9 | 5.5 | 5.8 |
| | MIDNIGHT-MIDNIGHT (OKTAS) | 33 | 5.0 | 5.2 | 5.9 | 6.4 | 6.2 | 6.1 | 6.3 | 5.9 | 5.6 | 5.8 | 6.0 | 5.5 | 5.8 |
| | MEAN NO. DAYS WITH: | | | | | | | | | | | | | | |
| | CLEAR | 51 | 6.5 | 5.3 | 2.7 | 1.2 | 1.2 | 1.7 | 1.3 | 1.8 | 2.9 | 2.7 | 3.2 | 5.0 | 35.5 |
| | PARTLY CLOUDY | 51 | 11.4 | 10.3 | 10.2 | 9.2 | 10.6 | 11.3 | 11.5 | 12.2 | 12.0 | 11.8 | 10.0 | 10.8 | 131.3 |
| | CLOUDY | 51 | 13.1 | 12.7 | 18.0 | 19.7 | 19.1 | 17.1 | 17.7 | 16.5 | 14.5 | 16.1 | 16.2 | 14.6 | 195.3 |
| **PR** | MEAN STATION PRESSURE(IN) | 26 | 29.97 | 29.99 | 30.02 | 30.03 | 30.01 | 30.01 | 29.99 | 29.97 | 29.95 | 29.95 | 29.97 | 29.98 | 29.99 |
| | MEAN SEA-LEVEL PRES. (IN) | 26 | 30.01 | 30.03 | 30.06 | 30.07 | 30.05 | 30.04 | 30.03 | 30.01 | 29.98 | 29.99 | 30.00 | 30.02 | 30.02 |
| **WINDS** | MEAN SPEED (MPH) | 26 | 7.0 | 7.3 | 7.4 | 7.2 | 7.0 | 6.8 | 6.7 | 6.6 | 6.6 | 6.5 | 6.5 | 6.7 | 6.9 |
| | PREVAIL.DIR(TENS OF DEGS) | 30 | 23 | 23 | 23 | 23 | 23 | 23 | 23 | 23 | 23 | 23 | 23 | 23 | 23 |
| | MAXIMUM 2-MINUTE: | | | | | | | | | | | | | | |
| | SPEED (MPH) | 12 | 31 | 37 | 29 | 26 | 23 | 24 | 24 | 24 | 24 | 24 | 33 | 29 | 37 |
| | DIR. (TENS OF DEGS) | | 01 | 36 | 01 | 36 | 08 | 06 | 07 | 10 | 07 | 02 | 05 | 18 | 36 |
| | YEAR OF OCCURRENCE | | 2003 | 2006 | 2002 | 2004 | 1999 | 2001 | 2006 | 1998 | 2003 | 2001 | 2009 | 2007 | FEB 2006 |
| | MAXIMUM 3-SECOND | | | | | | | | | | | | | | |
| | SPEED (MPH) | 12 | 38 | 44 | 36 | 32 | 29 | 28 | 30 | 32 | 32 | 35 | 46 | 39 | 46 |
| | DIR. (TENS OF DEGS) | | 01 | 36 | 01 | 10 | 06 | 11 | 33 | 10 | 04 | 32 | 03 | 19 | 03 |
| | YEAR OF OCCURRENCE | | 2004 | 2006 | 2002 | 2008 | 2004 | 2007 | 2004 | 1998 | 2008 | 2007 | 2009 | 2007 | NOV 2009 |
| **PRECIPITATION** | NORMAL (IN) | 30 | 9.74 | 8.86 | 14.35 | 12.54 | 8.07 | 7.36 | 10.71 | 9.78 | 9.14 | 9.64 | 15.58 | 10.50 | 126.27 |
| | MAXIMUM MONTHLY (IN) | 67 | 32.24 | 45.55 | 49.93 | 43.24 | 25.01 | 22.70 | 28.59 | 26.92 | 21.82 | 26.10 | 45.90 | 50.82 | 50.82 |
| | YEAR OF OCCURRENCE | | 1979 | 1979 | 1980 | 1986 | 1964 | 1997 | 1982 | 1991 | 1994 | 1951 | 2000 | 1954 | DEC 1954 |
| | MINIMUM MONTHLY (IN) | 67 | 0.13 | 0.52 | 0.88 | 2.93 | 1.18 | 1.80 | 3.54 | 2.66 | 1.59 | 2.40 | 1.01 | 0.28 | 0.13 |
| | YEAR OF OCCURRENCE | | 1998 | 2000 | 1972 | 1962 | 1945 | 1985 | 1999 | 1971 | 1974 | 1962 | 1989 | 1980 | JAN 1998 |
| | MAXIMUM IN 24 HOURS (IN) | 67 | 12.56 | 22.30 | 17.05 | 11.07 | 10.26 | 4.72 | 7.11 | 11.57 | 9.49 | 8.88 | 27.36 | 11.45 | 27.36 |
| | YEAR OF OCCURRENCE | | 2002 | 1979 | 1980 | 1971 | 1965 | 1997 | 1982 | 1991 | 1994 | 1951 | 2000 | 1987 | NOV 2000 |
| | NORMAL NO. DAYS WITH: | | | | | | | | | | | | | | |
| | PRECIPITATION >= 0.01 | 30 | 16.3 | 15.8 | 22.6 | 25.4 | 25.4 | 25.4 | 27.4 | 25.9 | 23.3 | 24.0 | 23.5 | 19.8 | 274.8 |
| | PRECIPITATION >= 1.00 | 30 | 2.8 | 2.3 | 3.8 | 2.8 | 1.6 | 1.3 | 2.2 | 1.8 | 2.2 | 2.0 | 4.2 | 2.5 | 29.5 |
| **SNOWFALL** | NORMAL (IN) | 30 | 0.0 | 0.0 | 0.0 | 0.0 | 0.0 | 0.0 | 0.0 | 0.0 | 0.0 | 0.0 | 0.0 | 0.0 | 0.0 |
| | MAXIMUM MONTHLY (IN) | 1 | 0.0 | 0.0 | 0.0 | 0.0 | 0.0 | 0.0 | 0.0 | 0.0 | 0.0 | 0.0 | 0.0 | 0.0 | 0.0 |
| | YEAR OF OCCURRENCE | | | | | | | | | | | | | | |
| | MAXIMUM IN 24 HOURS (IN) | 55 | 0.0 | 0.0 | 0.0 | 0.0 | 0.0 | 0.0 | 0.0 | 0.0 | 0.0 | 0.0 | 0.0 | 0.0 | 0.0 |
| | YEAR OF OCCURRENCE' | | | | | | | | | | | | | | |
| | MAXIMUM SNOW DEPTH (IN) | 47 | 0 | 0 | 0 | 0 | 0 | 0 | 0 | 0 | 0 | 0 | 0 | 0 | 0 |
| | YEAR OF OCCURRENCE | | | | | | | | | | | | | | |
| | NORMAL NO. DAYS WITH: | | | | | | | | | | | | | | |
| | SNOWFALL >= 1.0 | 30 | 0.0 | 0.0 | 0.0 | 0.0 | 0.0 | 0.0 | 0.0 | 0.0 | 0.0 | 0.0 | 0.0 | 0.0 | 0.0 |

## PRECIPITATION (inches) 2009 HILO (PHTO)

| YEAR | JAN | FEB | MAR | APR | MAY | JUN | JUL | AUG | SEP | OCT | NOV | DEC | ANNUAL |
|------|-----|-----|-----|-----|-----|-----|-----|-----|-----|-----|-----|-----|--------|
| 1980 | 0.91 | 4.14 | 49.93 | 11.01 | 5.88 | 9.66 | 9.17 | 8.24 | 13.70 | 7.69 | 7.13 | 0.28 | 127.74 |
| 1981 | 1.51 | 4.95 | 5.66 | 4.63 | 4.16 | 2.43 | 4.32 | 8.97 | 12.79 | 10.23 | 11.73 | 18.53 | 89.91 |
| 1982 | 13.58 | 1.35 | 48.50 | 12.00 | 6.89 | 6.03 | 28.59 | 25.45 | 9.92 | 6.53 | 4.74 | 6.78 | 170.36 |
| 1983 | 0.90 | 0.83 | 1.98 | 10.31 | 9.60 | 3.94 | 7.21 | 7.48 | 12.08 | 8.06 | 2.33 | 3.37 | 68.09 |
| 1984 | 10.76 | 10.06 | 3.37 | 12.08 | 6.59 | 4.28 | 6.63 | 9.36 | 4.05 | 2.52 | 18.38 | 12.00 | 100.08 |
| 1985 | 2.25 | 16.14 | 21.28 | 10.61 | 17.04 | 1.80 | 9.86 | 6.71 | 11.78 | 8.19 | 4.71 | 2.59 | 112.96 |
| 1986 | 4.95 | 0.58 | 15.37 | 43.24 | 8.61 | 9.11 | 11.17 | 10.64 | 14.36 | 11.53 | 35.72 | 5.75 | 171.03 |
| 1987 | 9.02 | 5.06 | 4.79 | 9.24 | 15.65 | 12.91 | 18.26 | 3.69 | 11.56 | 14.21 | 15.83 | 22.19 | 142.41 |
| 1988 | 10.31 | 9.95 | 13.09 | 12.90 | 7.77 | 5.11 | 5.50 | 16.56 | 11.30 | 8.50 | 25.74 | 13.46 | 140.19 |
| 1989 | 27.46 | 6.54 | 7.33 | 37.19 | 19.80 | 7.03 | 22.93 | 8.82 | 9.73 | 13.16 | 1.01 | 5.71 | 166.71 |
| 1990 | 29.13 | 15.24 | 10.80 | 4.02 | 8.13 | 10.04 | 10.78 | 7.80 | 18.47 | 20.96 | 45.75 | 30.10 | 211.22 |
| 1991 | 3.81 | 9.32 | 37.88 | 11.02 | 8.08 | 9.85 | 9.82 | 26.92 | 9.41 | 5.15 | 6.74 | 15.04 | 153.04 |
| 1992 | 1.33 | 1.29 | 3.90 | 6.62 | 2.99 | 9.36 | 17.63 | 13.62 | 17.59 | 3.38 | 25.16 | 17.02 | 119.89 |
| 1993 | 2.17 | 2.67 | 11.96 | 9.04 | 7.54 | 6.63 | 18.43 | 11.38 | 4.99 | 12.83 | 10.74 | 16.11 | 114.49 |
| 1994 | 10.39 | 25.52 | 18.48 | 8.59 | 7.18 | 13.29 | 11.71 | 14.58 | 21.82 | 8.73 | 35.91 | 6.61 | 182.81 |
| 1995 | 4.52 | 1.56 | 4.17 | 8.14 | 8.68 | 5.35 | 15.13 | 13.93 | 4.20 | 7.62 | 8.52 | 4.10 | 85.92 |
| 1996 | 14.29 | 11.81 | 16.66 | 6.27 | 3.65 | 10.33 | 13.22 | 4.77 | 7.03 | 11.07 | 14.22 | 6.89 | 120.21 |
| 1997 | 2.33 | 7.84 | 19.25 | 6.03 | 10.75 | 22.70 | 19.38 | 4.75 | 8.98 | 12.64 | 8.86 | 8.10 | 131.61 |
| 1998 | 0.13 | 2.40 | 3.67 | 8.86 | 15.65 | 11.27 | 6.09 | 8.48 | 10.76 | 16.01 | 15.57 | 9.89 | 108.78 |
| 1999 | 16.68 | 19.34 | 12.13 | 16.04 | 2.84 | 4.66 | 3.54 | 10.14 | 5.65 | 3.61 | 7.74 | 14.41 | 116.78 |
| 2000 | 17.87 | 0.52 | 5.81 | 7.25 | 3.36 | 8.19 | 13.16 | 10.54 | 9.20 | 17.65 | 45.90 | 4.59 | 144.04 |
| 2001 | 2.28 | 12.47 | 8.35 | 12.56 | 2.94 | 3.64 | 6.54 | 7.90 | 9.01 | 13.16 | 19.89 | 13.77 | 112.51 |
| 2002 | 26.14 | 19.00 | 10.76 | 7.41 | 14.95 | 7.16 | 6.98 | 13.65 | 8.14 | 6.53 | 2.86 | 10.45 | 134.03 |
| 2003 | 1.24 | 5.44 | 1.50 | 14.13 | 4.71 | 5.84 | 10.26 | 8.26 | 7.76 | 3.88 | 18.32 | 10.04 | 91.38 |
| 2004 | 13.14 | 8.29 | 27.25 | 20.51 | 8.91 | 6.28 | 4.44 | 6.83 | 5.69 | 14.13 | 11.02 | 11.00 | 137.49 |
| 2005 | 3.94 | 15.19 | 15.07 | 7.10 | 3.29 | 10.27 | 9.24 | 7.64 | 19.73 | 13.86 | 12.75 | 5.24 | 123.32 |
| 2006 | 11.43 | 8.46 | 26.42 | 8.69 | 22.51 | 4.19 | 7.82 | 5.69 | 9.52 | 7.43 | 3.21 | 6.66 | 122.03 |
| 2007 | 12.23 | 14.23 | 4.25 | 7.39 | 2.32 | 6.38 | 7.26 | 7.77 | 8.74 | 8.24 | 10.38 | 17.56 | 106.75 |
| 2008 | 14.20 | 39.08 | 5.21 | 5.91 | 4.12 | 2.18 | 6.17 | 3.88 | 4.27 | 5.40 | 6.73 | 30.38 | 127.53 |
| 2009 | 8.72 | 10.36 | 29.28 | 11.38 | 2.13 | 5.37 | 8.14 | 4.92 | 6.94 | 9.48 | 23.60 | 11.49 | 131.81 |
| POR= 60 YRS | 9.48 | 12.03 | 13.60 | 12.83 | 8.98 | 6.74 | 9.85 | 9.95 | 8.50 | 9.96 | 15.17 | 12.38 | 129.47 |

WBAN : 21504

## AVERAGE TEMPERATURE (°F) 2009 HILO (PHTO)

| YEAR | JAN | FEB | MAR | APR | MAY | JUN | JUL | AUG | SEP | OCT | NOV | DEC | ANNUAL |
|------|-----|-----|-----|-----|-----|-----|-----|-----|-----|-----|-----|-----|--------|
| 1980 | 71.6 | 72.6 | 72.3 | 74.5 | 77.3 | 77.6 | 77.8 | 75.0 | 75.7 | 74.8 | 73.8 | 74.2 | 74.8 |
| 1981 | 73.5 | 72.7 | 71.6 | 72.8 | 74.2 | 76.0 | 76.1 | 76.1 | 76.2 | 74.6 | 73.9 | 72.0 | 74.1 |
| 1982 | 71.9 | 71.8 | 70.3 | 71.2 | 72.9 | 76.3 | 76.7 | 76.9 | 76.1 | 74.9 | 74.6 | 71.8 | 73.8 |
| 1983 | 71.4 | 71.9 | 72.5 | 71.9 | 72.6 | 74.3 | 74.8 | 75.2 | 74.9 | 74.1 | 73.8 | 72.9 | 73.4 |
| 1984 | 72.4 | 71.5 | 73.8 | 73.0 | 74.0 | 74.7 | 75.2 | 75.3 | 75.4 | 76.5 | 73.6 | 71.1 | 73.9 |
| 1985 | 69.8 | 70.5 | 69.4 | 69.8 | 71.4 | 74.4 | 75.4 | 75.7 | 75.7 | 74.3 | 73.0 | 71.6 | 72.6 |
| 1986 | 71.1 | 73.6 | 74.7 | 73.6 | 75.4 | 76.6 | 77.8 | 78.5 | 77.9 | 76.4 | 75.1 | 72.8 | 75.3 |
| 1987 | 71.8 | 70.7 | 71.6 | 72.2 | 72.5 | 75.4 | 76.7 | 77.9 | 77.8 | 76.6 | 74.7 | 73.1 | 74.3 |
| 1988 | 71.9 | 72.3 | 72.2 | 72.6 | 74.2 | 74.7 | 75.7 | 76.0 | 76.6 | 77.9 | 76.3 | 74.9 | 74.6 |
| 1989 | 72.2 | 71.4 | 72.4 | 71.1 | 72.7 | 74.7 | 75.2 | 75.0 | 74.6 | 75.6 | 73.6 | 71.3 | 73.3 |
| 1990 | 72.1 | 70.4 | 71.2 | 73.5 | 74.1 | 75.0 | 76.0 | 77.0 | 77.2 | 76.2 | 75.4 | 72.5 | 74.2 |
| 1991 | 72.0 | 72.8 | 70.8 | 72.6 | 74.2 | 74.8 | 76.0 | 76.9 | 76.9 | 76.2 | 75.8 | 72.9 | 74.3 |
| 1992 | 71.2 | 71.4 | 72.3 | 72.4 | 74.8 | 76.2 | 76.2 | 77.2 | 77.8 | 77.7 | 75.2 | 73.6 | 74.7 |
| 1993 | 71.1 | 70.1 | 71.6 | 73.5 | 73.3 | 75.4 | 75.8 | 77.0 | 77.1 | 76.0 | 73.4 | 71.7 | 73.8 |
| 1994 | 70.0 | 71.3 | 71.7 | 73.4 | 74.9 | 76.0 | 78.1 | 78.6 | 78.1 | 77.4 | 74.9 | 73.0 | 74.8 |
| 1995 | 72.6 | 72.9 | 74.8 | 74.1 | 75.5 | 76.9 | 77.9 | 77.7 | 78.2 | 76.2 | 75.6 | 74.6 | 75.6 |
| 1996 | 73.4 | 70.9 | 71.5 | 74.4 | 76.1 | 76.9 | 77.5 | 77.5 | 77.5 | 77.3 | 75.8 | 73.2 | 75.2 |
| 1997 | 73.1 | 72.7 | 73.3 | 74.0 | 75.0 | 76.7 | 77.4 | 78.0 | 77.7 | 76.8 | 74.3 | 72.4 | 75.1 |
| 1998 | 71.8 | 71.7 | 72.9 | 71.8 | 72.2 | 73.8 | 74.8 | 76.1 | 74.5 | 74.2 | 72.4 | 69.9 | 73.0 |
| 1999 | 69.7 | 69.0 | 70.5 | 71.1 | 72.9 | 73.4 | 74.3 | 74.3 | 74.0 | 74.0 | 72.4 | 71.3 | 72.2 |
| 2000 | 69.1 | 71.3 | 71.8 | 71.3 | 73.9 | 75.5 | 75.5 | 76.1 | 75.4 | 75.3 | 73.0 | 71.7 | 73.3 |
| 2001 | 71.6 | 71.0 | 71.2 | 71.5 | 72.3 | 73.9 | 75.4 | 76.1 | 76.2 | 75.1 | 74.0 | 72.7 | 73.4 |
| 2002 | 71.8 | 70.1 | 71.2 | 73.5 | 74.3 | 75.4 | 76.6 | 77.0 | 76.0 | 76.1 | 74.6 | 73.0 | 74.1 |
| 2003 | 72.4 | 71.6 | 73.5 | 72.7 | 74.1 | 76.2 | 77.3 | 77.6 | 77.5 | 77.3 | 73.9 | 73.0 | 74.8 |
| 2004 | 72.6 | 73.5 | 73.1 | 73.6 | 75.9 | 76.3 | 77.5 | 78.2 | 77.9 | 76.7 | 75.3 | 72.8 | 75.3 |
| 2005 | 72.7 | 72.0 | 73.2 | 74.0 | 76.1 | 75.8 | 77.2 | 76.9 | 77.2 | 74.6 | 74.3 | 73.1 | 74.8 |
| 2006 | 71.8 | 70.5 | 72.2 | 72.0 | 72.2 | 75.2 | 76.3 | 76.5 | 75.7 | 76.2 | 75.7 | 72.8 | 73.9 |
| 2007 | 71.2 | 70.5 | 72.7 | 72.9 | 75.0 | 75.9 | 76.9 | 77.0 | 76.6 | 75.0 | 74.4 | 72.6 | 74.2 |
| 2008 | 70.3 | 71.2 | 72.8 | 72.8 | 73.9 | 75.4 | 76.7 | 76.1 | 75.5 | 74.3 | 73.1 | 71.9 | 73.7 |
| 2009 | 70.1 | 69.5 | 68.8 | 69.1 | 74.5 | 75.3 | 75.8 | 76.1 | 75.0 | 76.4 | 73.3 | 72.5 | 73.0 |
| POR= 60 YRS | 71.6 | 71.3 | 71.8 | 72.5 | 73.7 | 75.1 | 75.9 | 76.3 | 76.1 | 75.5 | 74.0 | 72.2 | 73.8 |

## HEATING DEGREE DAYS (base 65°F) 2009  HILO (PHTO)

| YEAR | JUL | AUG | SEP | OCT | NOV | DEC | JAN | FEB | MAR | APR | MAY | JUN | TOTAL |
|------|-----|-----|-----|-----|-----|-----|-----|-----|-----|-----|-----|-----|-------|
| 1983-84 | 0 | 0 | 0 | 0 | 0 | 0 | 0 | 0 | 0 | 0 | 0 | 0 | 0 |
| 1984-85 | 0 | 0 | 0 | 0 | 0 | 0 | 0 | 0 | 0 | 0 | 0 | 0 | 0 |
| 1985-86 | 0 | 0 | 0 | 0 | 0 | 0 | 0 | 0 | 0 | 0 | 0 | 0 | 0 |
| 1986-87 | 0 | 0 | 0 | 0 | 0 | 0 | 0 | 0 | 0 | 0 | 0 | 0 | 0 |
| 1987-88 | 0 | 0 | 0 | 0 | 0 | 0 | 0 | 0 | 0 | 0 | 0 | 0 | 0 |
| 1988-89 | 0 | 0 | 0 | 0 | 0 | 0 | 0 | 0 | 0 | 0 | 0 | 0 | 0 |
| 1989-90 | 0 | 0 | 0 | 0 | 0 | 0 | 0 | 0 | 0 | 0 | 0 | 0 | 0 |
| 1990-91 | 0 | 0 | 0 | 0 | 0 | 0 | 0 | 0 | 0 | 0 | 0 | 0 | 0 |
| 1991-92 | 0 | 0 | 0 | 0 | 0 | 0 | 0 | 0 | 0 | 0 | 0 | 0 | 0 |
| 1992-93 | 0 | 0 | 0 | 0 | 0 | 0 | 0 | 0 | 0 | 0 | 0 | 0 | 0 |
| 1993-94 | 0 | 0 | 0 | 0 | 0 | 0 | 0 | 0 | 0 | 0 | 0 | 0 | 0 |
| 1994-95 | 0 | 0 | 0 | 0 | 0 | 0 | 0 | 0 | 0 | 0 | 0 | 0 | 0 |
| 1995-96 | 0 | 0 | 0 | 0 | 0 | 0 | 0 | 0 | 0 | 0 | 0 | 0 | 0 |
| 1996-97 | 0 | 0 | 0 | 0 | 0 | 0 | 0 | 0 | 0 | 0 | 0 | 0 | 0 |
| 1997-98 | 0 | 0 | 0 | 0 | 0 | 0 | 0 | 0 | 0 | 0 | 0 | 0 | 0 |
| 1998-99 | 0 | 0 | 0 | 0 | 0 | 0 | 0 | 0 | 0 | 0 | 0 | 0 | 0 |
| 1999-00 | 0 | 0 | 0 | 0 | 0 | 0 | 0 | 0 | 0 | 0 | 0 | 0 | 0 |
| 2000-01 | 0 | 0 | 0 | 0 | 0 | 0 | 0 | 0 | 0 | 0 | 0 | 0 | 0 |
| 2001-02 | 0 | 0 | 0 | 0 | 0 | 0 | 0 | 0 | 0 | 0 | 0 | 0 | 0 |
| 2002-03 | 0 | 0 | 0 | 0 | 0 | 0 | 0 | 0 | 0 | 0 | 0 | 0 | 0 |
| 2003-04 | 0 | 0 | 0 | 0 | 0 | 0 | 0 | 0 | 0 | 0 | 0 | 0 | 0 |
| 2004-05 | 0 | 0 | 0 | 0 | 0 | 0 | 0 | 0 | 0 | 0 | 0 | 0 | 0 |
| 2005-06 | 0 | 0 | 0 | 0 | 0 | 0 | 0 | 0 | 0 | 0 | 0 | 0 | 0 |
| 2006-07 | 0 | 0 | 0 | 0 | 0 | 0 | 0 | 0 | 0 | 0 | 0 | 0 | 0 |
| 2007-08 | 0 | 0 | 0 | 0 | 0 | 0 | 0 | 0 | 0 | 0 | 0 | 0 | 0 |
| 2008-09 | 0 | 0 | 0 | 0 | 0 | 0 | 0 | 0 | 1 | 0 | 0 | 0 | 1 |
| 2009- | 0 | 0 | 0 | 0 | 0 | 0 | | | | | | | |

WBAN : 21504

## COOLING DEGREE DAYS (base 65°F) 2009  HILO (PHTO)

| YEAR | JAN | FEB | MAR | APR | MAY | JUN | JUL | AUG | SEP | OCT | NOV | DEC | TOTAL |
|------|-----|-----|-----|-----|-----|-----|-----|-----|-----|-----|-----|-----|-------|
| 1980 | 213 | 227 | 234 | 293 | 390 | 385 | 405 | 316 | 328 | 313 | 269 | 295 | 3668 |
| 1981 | 271 | 220 | 210 | 242 | 293 | 338 | 350 | 348 | 345 | 302 | 274 | 225 | 3418 |
| 1982 | 220 | 196 | 170 | 194 | 252 | 348 | 369 | 379 | 340 | 317 | 293 | 219 | 3297 |
| 1983 | 207 | 200 | 239 | 214 | 240 | 287 | 313 | 324 | 303 | 288 | 272 | 250 | 3137 |
| 1984 | 236 | 194 | 282 | 247 | 284 | 298 | 324 | 326 | 320 | 363 | 261 | 195 | 3330 |
| 1985 | 154 | 161 | 142 | 152 | 204 | 290 | 329 | 339 | 329 | 294 | 248 | 211 | 2853 |
| 1986 | 196 | 246 | 308 | 264 | 329 | 356 | 404 | 423 | 396 | 363 | 309 | 250 | 3844 |
| 1987 | 218 | 163 | 212 | 226 | 241 | 319 | 369 | 407 | 389 | 365 | 299 | 259 | 3467 |
| 1988 | 221 | 216 | 233 | 238 | 293 | 298 | 338 | 349 | 353 | 405 | 345 | 315 | 3604 |
| 1989 | 227 | 188 | 238 | 189 | 248 | 297 | 327 | 315 | 294 | 335 | 264 | 202 | 3124 |
| 1990 | 227 | 157 | 200 | 260 | 290 | 308 | 349 | 379 | 376 | 353 | 317 | 237 | 3453 |
| 1991 | 223 | 222 | 188 | 234 | 296 | 301 | 348 | 378 | 365 | 351 | 333 | 251 | 3490 |
| 1992 | 197 | 192 | 235 | 229 | 312 | 343 | 355 | 384 | 387 | 402 | 315 | 275 | 3626 |
| 1993 | 193 | 148 | 213 | 263 | 263 | 318 | 343 | 380 | 370 | 350 | 260 | 217 | 3318 |
| 1994 | 161 | 183 | 214 | 261 | 312 | 338 | 412 | 427 | 401 | 389 | 305 | 254 | 3657 |
| 1995 | 241 | 228 | 310 | 281 | 335 | 364 | 410 | 402 | 401 | 355 | 325 | 303 | 3955 |
| 1996 | 266 | 178 | 209 | 291 | 349 | 360 | 396 | 394 | 379 | 387 | 331 | 265 | 3805 |
| 1997 | 261 | 221 | 261 | 277 | 317 | 358 | 388 | 407 | 389 | 373 | 287 | 239 | 3778 |
| 1998 | 216 | 194 | 253 | 211 | 230 | 269 | 311 | 351 | 293 | 289 | 228 | 162 | 3007 |
| 1999 | 149 | 117 | 179 | 189 | 252 | 257 | 292 | 296 | 278 | 284 | 228 | 200 | 2721 |
| 2000 | 133 | 191 | 221 | 194 | 280 | 322 | 332 | 350 | 317 | 328 | 247 | 213 | 3128 |
| 2001 | 209 | 174 | 201 | 204 | 234 | 276 | 330 | 350 | 341 | 319 | 278 | 246 | 3162 |
| 2002 | 219 | 149 | 202 | 261 | 298 | 320 | 365 | 378 | 339 | 352 | 295 | 253 | 3431 |
| 2003 | 236 | 189 | 270 | 238 | 287 | 343 | 392 | 398 | 385 | 391 | 274 | 256 | 3659 |
| 2004 | 243 | 251 | 257 | 265 | 345 | 346 | 395 | 418 | 393 | 368 | 315 | 252 | 3848 |
| 2005 | 242 | 203 | 262 | 277 | 349 | 332 | 384 | 375 | 372 | 303 | 285 | 257 | 3641 |
| 2006 | 219 | 161 | 227 | 216 | 230 | 315 | 357 | 362 | 329 | 354 | 329 | 248 | 3347 |
| 2007 | 201 | 158 | 247 | 242 | 314 | 334 | 376 | 379 | 352 | 317 | 287 | 242 | 3449 |
| 2008 | 172 | 185 | 250 | 240 | 282 | 322 | 370 | 353 | 323 | 298 | 248 | 221 | 3264 |
| 2009 | 166 | 134 | 128 | 129 | 302 | 318 | 344 | 349 | 306 | 359 | 255 | 243 | 3033 |

**SNOWFALL (inches)  2009  HILO (PHTO)**

| YEAR | JUL | AUG | SEP | OCT | NOV | DEC | JAN | FEB | MAR | APR | MAY | JUN | TOTAL |
|------|-----|-----|-----|-----|-----|-----|-----|-----|-----|-----|-----|-----|-------|
| 1977-78 | 0.0 | 0.0 | 0.0 | 0.0 | 0.0 | 0.0 | 0.0 | 0.0 | 0.0 | 0.0 | 0.0 | 0.0 | 0.0 |
| 1978-79 | 0.0 | 0.0 | 0.0 | 0.0 | 0.0 | 0.0 | 0.0 | 0.0 | 0.0 | 0.0 | 0.0 | 0.0 | 0.0 |
| 1979-80 | 0.0 | 0.0 | 0.0 | 0.0 | 0.0 | 0.0 | 0.0 | 0.0 | 0.0 | 0.0 | 0.0 | 0.0 | 0.0 |
| 1980-81 | 0.0 | 0.0 | 0.0 | 0.0 | 0.0 | 0.0 | 0.0 | 0.0 | 0.0 | 0.0 | 0.0 | 0.0 | 0.0 |
| 1981-82 | 0.0 | 0.0 | 0.0 | 0.0 | 0.0 | 0.0 | 0.0 | 0.0 | 0.0 | 0.0 | 0.0 | 0.0 | 0.0 |
| 1982-83 | 0.0 | 0.0 | 0.0 | 0.0 | 0.0 | 0.0 | 0.0 | 0.0 | 0.0 | 0.0 | 0.0 | 0.0 | 0.0 |
| 1983-84 | 0.0 | 0.0 | 0.0 | 0.0 | 0.0 | 0.0 | 0.0 | 0.0 | 0.0 | 0.0 | 0.0 | 0.0 | 0.0 |
| 1984-85 | 0.0 | 0.0 | 0.0 | 0.0 | 0.0 | 0.0 | 0.0 | 0.0 | 0.0 | 0.0 | 0.0 | 0.0 | 0.0 |
| 1985-86 | 0.0 | 0.0 | 0.0 | 0.0 | 0.0 | 0.0 | 0.0 | 0.0 | 0.0 | 0.0 | 0.0 | 0.0 | 0.0 |
| 1986-87 | 0.0 | 0.0 | 0.0 | 0.0 | 0.0 | 0.0 | 0.0 | 0.0 | 0.0 | 0.0 | 0.0 | 0.0 | 0.0 |
| 1987-88 | 0.0 | 0.0 | 0.0 | 0.0 | 0.0 | 0.0 | 0.0 | 0.0 | 0.0 | 0.0 | 0.0 | 0.0 | 0.0 |
| 1988-89 | 0.0 | 0.0 | 0.0 | 0.0 | 0.0 | 0.0 | 0.0 | 0.0 | 0.0 | 0.0 | 0.0 | 0.0 | 0.0 |
| 1989-90 | 0.0 | 0.0 | 0.0 | 0.0 | 0.0 | 0.0 | 0.0 | 0.0 | 0.0 | 0.0 | 0.0 | 0.0 | 0.0 |
| 1990-91 | 0.0 | 0.0 | 0.0 | 0.0 | 0.0 | 0.0 | 0.0 | 0.0 | 0.0 | 0.0 | 0.0 | 0.0 | 0.0 |
| 1991-92 | 0.0 | 0.0 | 0.0 | 0.0 | 0.0 | 0.0 | 0.0 | 0.0 | 0.0 | 0.0 | 0.0 | 0.0 | 0.0 |
| 1992-93 | 0.0 | 0.0 | 0.0 | 0.0 | 0.0 | 0.0 | 0.0 | 0.0 | 0.0 | 0.0 | 0.0 | 0.0 | 0.0 |
| 1993-94 | 0.0 | 0.0 | 0.0 | 0.0 | 0.0 | 0.0 | 0.0 | 0.0 | 0.0 | 0.0 | 0.0 | 0.0 | 0.0 |
| 1994-95 | 0.0 | 0.0 | 0.0 | 0.0 | 0.0 | 0.0 | 0.0 | 0.0 | 0.0 | 0.0 | 0.0 | 0.0 | 0.0 |
| 1995-96 | 0.0 | 0.0 | 0.0 | 0.0 | 0.0 | 0.0 | 0.0 | 0.0 | 0.0 | 0.0 | 0.0 | 0.0 | 0.0 |
| 1996-97 | 0.0 | 0.0 | 0.0 | 0.0 | 0.0 | 0.0 | 0.0 | 0.0 | 0.0 | 0.0 | 0.0 | 0.0 | 0.0 |
| 1997-98 | 0.0 | 0.0 | 0.0 | 0.0 | 0.0 | 0.0 |     |     |     |     |     |     |       |
| 1998-99 |     |     |     |     |     |     |     |     |     |     |     |     |       |
| 1999-00 |     |     |     |     |     |     |     |     |     |     |     |     |       |
| 2000-01 |     |     |     |     |     |     |     |     |     |     |     |     |       |
| 2001-02 |     |     |     |     |     |     |     |     |     |     |     |     |       |
| 2002-03 |     |     |     |     |     |     |     |     |     |     |     |     |       |
| 2003-04 |     |     |     |     |     |     |     |     |     |     |     |     |       |
| 2004-05 |     |     |     |     |     |     |     |     |     |     |     |     |       |
| 2005- |     |     |     |     |     |     |     |     |     |     |     |     |       |
| 2008-09 | 0.0 |     |     |     |     |     |     |     |     |     |     |     |       |
| POR= 48 YRS | 0.0 | 0.0 | 0.0 | 0.0 | 0.0 | 0.0 | 0.0 | 0.0 | 0.0 | 0.0 | 0.0 | 0.0 | 0.0 |

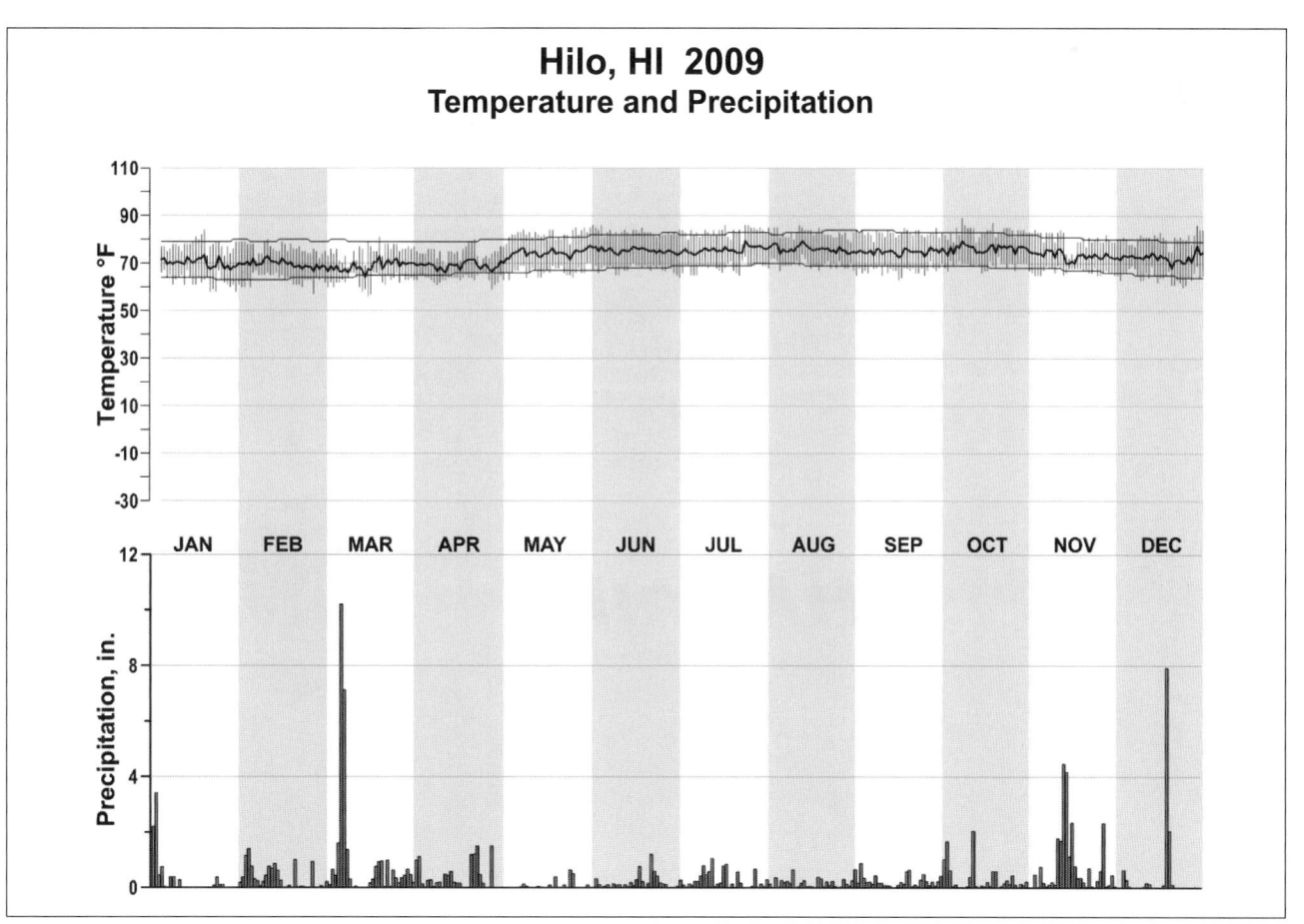

# Hilo, HI  2009
## Temperature and Precipitation

# 2009
# HONOLULU
# HAWAII (PHNL)

Oahu, on which Honolulu is located, is the third largest of the Hawaiian Islands. The Koolau Range, at an average elevation of 2,000 feet parallels the northeastern coast. The Waianae Mountains, somewhat higher in elevation, parallel the west coast. Honolulu Airport, the business and Waikiki districts, and a number of the residential areas of Honolulu lie along the southern coastal plain.

The climate of Hawaii is unusually pleasant for the tropics. Its outstanding features are the persistence of the trade winds, the remarkable variability in rainfall over short distances, the sunniness of the leeward lowlands in contrast to the persistent cloudiness over nearby mountain crests, the equable temperature, and the general infrequency of severe storms.

The prevailing wind throughout the year is the northeasterly trade wind, although its average frequency varies from more than 90 percent during the summer to only 50 percent in January.

Heavy mountain rainfall sustains extensive irrigation of cane fields and the water supply for Honolulu. Oahu is driest along the coast west of the Waianaes where rainfall drops to about 20 inches a year. Daytime showers, usually light, often occur while the sun continues to shine, a phenomenon referred to locally as liquid sunshine.

The moderate temperature range is associated with the small seasonal variation in the energy received from the sun and the tempering effect of the surrounding ocean. Honolulu Airport has recorded as high as the lower 90s and as low as the lower 50s.

Because of the trade winds, even the warmest months are usually comfortable. But when the trades diminish or give way to southerly winds, a situation known locally as kona weather, or kona storms when stormy, the humidity may become oppressively high.

Intense rains of the October to April winter season sometimes cause serious, flash flooding. Thunderstorms are infrequent and usually mild and hail seldom occurs. Infrequently, a small tornado or a waterspout may do some damage. Only a few tropical cyclones have struck Hawaii, although others have come near enough for their outlying winds, waves, clouds, and rain to affect the Islands.

Honolulu

HAWAII

Hilo

# NORMALS, MEANS, AND EXTREMES
## HONOLULU (PHNL)

| LATITUDE: 21 ° 19'N | LONGITUDE: -157° 56'W | | ELEVATION (FT): GRND: 8   BARO: 18 | | | | TIME ZONE: HAWAII   (UTC -10) | | | WBAN: 22521 | | |

| | ELEMENT | POR | JAN | FEB | MAR | APR | MAY | JUN | JUL | AUG | SEP | OCT | NOV | DEC | YEAR |
|---|---|---|---|---|---|---|---|---|---|---|---|---|---|---|---|
| **TEMPERATURE °F** | NORMAL DAILY MAXIMUM | 30 | 80.4 | 80.7 | 81.7 | 83.1 | 84.9 | 86.9 | 87.8 | 88.9 | 88.9 | 87.2 | 84.3 | 81.7 | 84.7 |
| | MEAN DAILY MAXIMUM | 60 | 80.0 | 80.2 | 81.0 | 82.4 | 84.1 | 86.1 | 86.9 | 87.9 | 87.9 | 86.4 | 83.8 | 81.0 | 84.0 |
| | HIGHEST DAILY MAXIMUM | 40 | 88 | 88 | 88 | 91 | 93 | 92 | 94 | 93 | 95 | 94 | 93 | 89 | 95 |
| | YEAR OF OCCURRENCE | | 1996 | 1984 | 1998 | 1996 | 1988 | 1996 | 1995 | 1997 | 1994 | 1984 | 1986 | 1995 | SEP 1994 |
| | MEAN OF EXTREME MAXS. | 60 | 83.8 | 83.9 | 84.7 | 85.9 | 87.4 | 88.4 | 89.3 | 90.3 | 90.5 | 89.6 | 87.2 | 84.9 | 87.2 |
| | NORMAL DAILY MINIMUM | 30 | 65.7 | 65.4 | 66.9 | 68.2 | 69.6 | 72.1 | 73.8 | 74.7 | 74.2 | 73.2 | 71.1 | 67.8 | 70.2 |
| | MEAN DAILY MINIMUM | 60 | 66.0 | 66.0 | 67.3 | 68.9 | 70.4 | 72.4 | 73.6 | 74.4 | 73.8 | 72.6 | 70.7 | 67.8 | 70.3 |
| | LOWEST DAILY MINIMUM | 40 | 53 | 53 | 55 | 57 | 60 | 65 | 66 | 65 | 66 | 61 | 57 | 54 | 53 |
| | YEAR OF OCCURRENCE | | 1998 | 1983 | 1976 | 1985 | 2006 | 1982 | 1990 | 2009 | 1985 | 1993 | 1990 | 1962 | JAN 1998 |
| | MEAN OF EXTREME MINS. | 60 | 59.0 | 59.2 | 60.7 | 63.4 | 65.3 | 68.3 | 69.7 | 70.3 | 69.5 | 67.4 | 64.8 | 60.7 | 64.9 |
| | NORMAL DRY BULB | 30 | 73.0 | 73.0 | 74.3 | 75.6 | 77.2 | 79.5 | 80.8 | 81.8 | 81.5 | 80.2 | 77.7 | 74.8 | 77.5 |
| | MEAN DRY BULB | 60 | 73.0 | 73.1 | 74.2 | 75.7 | 77.4 | 79.4 | 80.5 | 81.3 | 80.9 | 79.5 | 77.2 | 74.4 | 77.2 |
| | MEAN WET BULB | 26 | 66.2 | 65.5 | 66.5 | 67.0 | 68.5 | 69.9 | 70.9 | 71.4 | 71.7 | 71.2 | 69.7 | 67.5 | 68.8 |
| | MEAN DEW POINT | 26 | 63.8 | 62.9 | 63.5 | 64.1 | 65.2 | 66.5 | 67.9 | 68.4 | 69.1 | 68.7 | 67.3 | 65.0 | 66.0 |
| | NORMAL NO. DAYS WITH: | | | | | | | | | | | | | | |
| | MAXIMUM >= 90 | 30 | 0.0 | 0.0 | 0.0 | 0.1 | 0.4 | 2.0 | 5.7 | 11.7 | 10.5 | 4.7 | 0.4 | 0.0 | 35.5 |
| | MAXIMUM <= 32 | 30 | 0.0 | 0.0 | 0.0 | 0.0 | 0.0 | 0.0 | 0.0 | 0.0 | 0.0 | 0.0 | 0.0 | 0.0 | 0.0 |
| | MINIMUM <= 32 | 30 | 0.0 | 0.0 | 0.0 | 0.0 | 0.0 | 0.0 | 0.0 | 0.0 | 0.0 | 0.0 | 0.0 | 0.0 | 0.0 |
| | MINIMUM <= 0 | 30 | 0.0 | 0.0 | 0.0 | 0.0 | 0.0 | 0.0 | 0.0 | 0.0 | 0.0 | 0.0 | 0.0 | 0.0 | 0.0 |
| **H/C** | NORMAL HEATING DEG. DAYS | 30 | 0 | 0 | 0 | 0 | 0 | 0 | 0 | 0 | 0 | 0 | 0 | 0 | 0 |
| | NORMAL COOLING DEG. DAYS | 30 | 249 | 225 | 288 | 319 | 379 | 436 | 490 | 521 | 497 | 472 | 382 | 303 | 4561 |
| **RH** | NORMAL (PERCENT) | 30 | 73 | 71 | 69 | 67 | 66 | 65 | 66 | 66 | 68 | 69 | 71 | 73 | 69 |
| | HOUR 02 LST | 30 | 82 | 80 | 77 | 76 | 75 | 74 | 74 | 74 | 76 | 77 | 78 | 80 | 77 |
| | HOUR 08 LST | 30 | 81 | 79 | 73 | 70 | 67 | 66 | 68 | 68 | 70 | 71 | 75 | 79 | 72 |
| | HOUR 14 LST | 30 | 61 | 59 | 57 | 55 | 54 | 52 | 52 | 52 | 53 | 56 | 59 | 60 | 56 |
| | HOUR 20 LST | 30 | 74 | 72 | 71 | 70 | 69 | 69 | 69 | 69 | 70 | 71 | 73 | 74 | 71 |
| **S** | PERCENT POSSIBLE SUNSHINE | 46 | 65 | 68 | 72 | 70 | 72 | 74 | 76 | 77 | 77 | 71 | 64 | 63 | 71 |
| **W/O** | MEAN NO. DAYS WITH: | | | | | | | | | | | | | | |
| | HEAVY FOG(VISBY <= 1/4 MI) | 46 | 0.0 | 0.0 | 0.0 | 0.0 | 0.0 | 0.0 | 0.0 | 0.0 | 0.0 | 0.0 | 0.0 | 0.0 | 0.0 |
| | THUNDERSTORMS | 60 | 0.8 | 1.0 | 0.8 | 0.5 | 0.3 | 0.1 | 0.2 | 0.1 | 0.4 | 0.8 | 0.9 | 0.8 | 6.7 |
| **CLOUDNESS** | MEAN: | | | | | | | | | | | | | | |
| | SUNRISE-SUNSET (OKTAS) | 52 | 4.3 | 4.4 | 4.6 | 4.9 | 4.6 | 4.4 | 4.2 | 4.1 | 4.2 | 4.5 | 4.6 | 4.4 | 4.4 |
| | MIDNIGHT-MIDNIGHT (OKTAS) | 34 | 4.1 | 4.1 | 4.3 | 4.5 | 4.3 | 4.2 | 4.2 | 3.9 | 3.9 | 4.1 | 4.5 | 4.3 | 4.2 |
| | MEAN NO. DAYS WITH: | | | | | | | | | | | | | | |
| | CLEAR | 49 | 9.5 | 8.1 | 7.4 | 5.9 | 6.7 | 6.5 | 7.4 | 8.0 | 7.9 | 7.5 | 7.2 | 7.9 | 90.0 |
| | PARTLY CLOUDY | 49 | 12.9 | 12.5 | 14.1 | 14.6 | 15.6 | 17.4 | 18.0 | 16.9 | 15.9 | 14.8 | 13.4 | 13.7 | 179.8 |
| | CLOUDY | 49 | 8.5 | 7.6 | 9.3 | 9.6 | 8.7 | 6.2 | 5.1 | 5.7 | 5.7 | 8.1 | 8.8 | 8.7 | 92.0 |
| **PR** | MEAN STATION PRESSURE(IN) | 26 | 29.99 | 30.01 | 30.03 | 30.05 | 30.03 | 30.02 | 30.00 | 29.98 | 29.96 | 29.96 | 29.98 | 29.99 | 30.00 |
| | MEAN SEA-LEVEL PRES. (IN) | 26 | 30.00 | 30.02 | 30.05 | 30.06 | 30.04 | 30.04 | 30.02 | 30.00 | 29.98 | 29.98 | 29.99 | 30.01 | 30.02 |
| **WINDS** | MEAN SPEED (MPH) | 26 | 8.8 | 9.3 | 10.3 | 11.2 | 10.6 | 12.0 | 12.1 | 11.9 | 10.3 | 9.7 | 9.7 | 9.4 | 10.4 |
| | PREVAIL.DIR(TENS OF DEGS) | 13 | 06 | 06 | 06 | 06 | 06 | 06 | 06 | 06 | 06 | 06 | 06 | 06 | 06 |
| | MAXIMUM 2-MINUTE: | | | | | | | | | | | | | | |
| | SPEED (MPH) | 11 | 40 | 37 | 32 | 35 | 29 | 30 | 30 | 31 | 30 | 31 | 35 | 39 | 40 |
| | DIR. (TENS OF DEGS) | | 25 | 24 | 31 | 05 | 07 | 07 | 06 | 07 | 08 | 08 | 07 | 25 | 25 |
| | YEAR OF OCCURRENCE | | 2004 | 2004 | 2005 | 2000 | 2004 | 1999 | 1999 | 2001 | 2003 | 2007 | 2009 | 2007 | JAN 2004 |
| | MAXIMUM 3-SECOND | | | | | | | | | | | | | | |
| | SPEED (MPH) | 11 | 47 | 47 | 41 | 44 | 36 | 37 | 40 | 37 | 37 | 37 | 44 | 48 | 48 |
| | DIR. (TENS OF DEGS) | | 25 | 25 | 25 | 07 | 04 | 03 | 06 | 07 | 05 | 07 | 08 | 25 | 25 |
| | YEAR OF OCCURRENCE | | 2004 | 2004 | 2007 | 2000 | 2004 | 2005 | 2009 | 2009 | 2002 | 2001 | 2009 | 2007 | DEC 2007 |
| **PRECIPITATION** | NORMAL (IN) | 30 | 2.73 | 2.35 | 1.89 | 1.11 | 0.78 | 0.43 | 0.50 | 0.46 | 0.74 | 2.18 | 2.27 | 2.85 | 18.29 |
| | MAXIMUM MONTHLY (IN) | 63 | 14.74 | 13.68 | 20.79 | 8.92 | 7.23 | 2.46 | 2.33 | 3.74 | 2.74 | 11.15 | 18.79 | 17.29 | 20.79 |
| | YEAR OF OCCURRENCE | | 1949 | 1955 | 1951 | 1963 | 1965 | 1971 | 1989 | 2004 | 1947 | 1978 | 1996 | 1987 | MAR 1951 |
| | MINIMUM MONTHLY (IN) | 63 | 0.18 | 0.06 | 0.01 | 0.01 | 0.03 | T | 0.03 | T | 0.05 | 0.07 | 0.03 | 0.04 | T |
| | YEAR OF OCCURRENCE | | 1986 | 1983 | 1957 | 1960 | 2000 | 1959 | 1950 | 1974 | 1977 | 1996 | 1962 | 2002 | AUG 1974 |
| | MAXIMUM IN 24 HOURS (IN) | 63 | 6.72 | 6.88 | 17.07 | 4.21 | 3.44 | 2.28 | 2.20 | 3.03 | 1.40 | 7.57 | 9.15 | 8.25 | 17.07 |
| | YEAR OF OCCURRENCE | | 1963 | 1955 | 1958 | 1972 | 1965 | 1967 | 1989 | 2004 | 1963 | 1978 | 1954 | 1987 | MAR 1958 |
| | NORMAL NO. DAYS WITH: | | | | | | | | | | | | | | |
| | PRECIPITATION >= 0.01 | 30 | 8.8 | 7.9 | 9.0 | 8.6 | 7.3 | 5.8 | 7.2 | 5.4 | 6.9 | 7.3 | 9.1 | 9.7 | 93.0 |
| | PRECIPITATION >= 1.00 | 30 | 0.7 | 0.7 | 0.6 | 0.1 | 0.1 | 0.1 | 0.1 | 0.1 | 0.1 | 0.5 | 0.6 | 0.7 | 4.4 |
| **SNOWFALL** | NORMAL (IN) | 30 | 0.0 | 0.0 | 0.0 | 0.0 | 0.0 | 0.0 | 0.0 | 0.0 | 0.0 | 0.0 | 0.0 | 0.0 | 0.0 |
| | MAXIMUM MONTHLY (IN) | 2 | 0.0 | 0.0 | 0.0 | 0.0 | 0.0 | 0.0 | 0.0 | 0.0 | 0.0 | 0.0 | 0.0 | 0.0 | 0.0 |
| | YEAR OF OCCURRENCE | | | | | | | | | | | | | | |
| | MAXIMUM IN 24 HOURS (IN) | 52 | 0.0 | 0.0 | 0.0 | 0.0 | 0.0 | 0.0 | 0.0 | 0.0 | 0.0 | 0.0 | 0.0 | 0.0 | 0.0 |
| | YEAR OF OCCURRENCE' | | | | | | | | | | | | | | |
| | MAXIMUM SNOW DEPTH (IN) | 48 | 0 | 0 | 0 | 0 | 0 | 0 | 0 | 0 | 0 | 0 | 0 | 0 | 0 |
| | YEAR OF OCCURRENCE | | | | | | | | | | | | | | |
| | NORMAL NO. DAYS WITH: | | | | | | | | | | | | | | |
| | SNOWFALL >= 1.0 | 30 | 0.0 | 0.0 | 0.0 | 0.0 | 0.0 | 0.0 | 0.0 | 0.0 | 0.0 | 0.0 | 0.0 | 0.0 | 0.0 |

## PRECIPITATION (inches) 2009 HONOLULU (PHNL)

| YEAR | JAN | FEB | MAR | APR | MAY | JUN | JUL | AUG | SEP | OCT | NOV | DEC | ANNUAL |
|------|------|------|------|------|------|------|------|------|------|------|------|------|--------|
| 1980 | 8.91 | 2.26 | 3.04 | 1.13 | 0.78 | 1.76 | 0.37 | 0.36 | 0.41 | 0.30 | 0.21 | 7.37 | 26.90 |
| 1981 | 0.81 | 0.97 | 0.71 | 1.01 | 0.94 | 0.14 | 0.42 | 0.70 | 0.39 | 1.84 | 1.01 | 4.47 | 13.41 |
| 1982 | 12.82 | 2.16 | 3.73 | 1.28 | 0.13 | 0.35 | 0.20 | 1.98 | 0.52 | 7.24 | 1.32 | 3.19 | 34.92 |
| 1983 | 0.32 | 0.06 | 0.53 | 0.42 | 0.35 | 0.26 | 0.22 | 0.29 | 1.16 | 0.23 | 0.13 | 1.06 | 5.03 |
| 1984 | 0.21 | 0.60 | 1.08 | 2.41 | 0.16 | 0.08 | 0.23 | 0.04 | 1.36 | 1.89 | 3.58 | 5.44 | 17.08 |
| 1985 | 1.46 | 3.87 | 1.26 | 0.20 | 1.11 | 0.13 | 0.53 | 0.16 | 1.28 | 5.08 | 2.11 | 0.19 | 17.38 |
| 1986 | 0.18 | 1.38 | 0.17 | 0.35 | 0.81 | 0.36 | 1.54 | 0.90 | 2.00 | 1.23 | 4.23 | 0.78 | 13.93 |
| 1987 | 0.42 | 0.86 | 0.31 | 0.65 | 0.73 | 0.46 | 0.33 | 0.22 | 1.13 | 0.20 | 0.93 | 17.29 | 23.53 |
| 1988 | 3.05 | 1.31 | 0.67 | 0.50 | 1.25 | 0.04 | 0.12 | 0.34 | 0.86 | 0.23 | 1.39 | 6.71 | 16.47 |
| 1989 | 2.07 | 6.48 | 2.58 | 1.23 | 0.29 | 0.11 | 2.33 | 0.08 | 0.15 | 10.37 | 0.51 | 1.32 | 27.52 |
| 1990 | 4.32 | 4.15 | 0.86 | 0.30 | 0.30 | 0.08 | 0.49 | 0.01 | 0.98 | 0.47 | 2.96 | 4.92 | 19.84 |
| 1991 | 0.80 | 2.09 | 6.24 | 1.00 | 0.48 | 0.26 | 0.16 | 0.16 | 0.56 | 3.43 | 1.52 | 1.24 | 17.94 |
| 1992 | 0.43 | 1.35 | 0.72 | 0.11 | 1.13 | 0.10 | 2.01 | 0.97 | 2.08 | 2.57 | 1.04 | 6.49 | 19.00 |
| 1993 | 0.70 | 0.41 | 0.02 | 0.23 | 0.19 | 0.11 | 0.69 | 1.03 | 0.10 | 1.63 | 0.46 | 0.27 | 5.84 |
| 1994 | 0.78 | 7.04 | 3.77 | 0.46 | 0.09 | 0.23 | 0.48 | 0.14 | 0.39 | 1.56 | 0.20 | 0.45 | 15.59 |
| 1995 | 0.54 | 6.53 | 1.33 | 1.15 | 0.45 | 0.05 | 0.19 | 0.44 | 0.33 | 0.36 | 0.92 | 1.31 | 13.60 |
| 1996 | 3.52 | 1.00 | 2.68 | 0.34 | 0.13 | 1.01 | .74 | 2.17 | .78 | .07 | 18.79 | 1.89 | 33.12 |
| 1997 | 6.92 | 0.95 | 4.90 | 1.02 | 0.63 | 0.40 | 0.52 | 0.13 | 0.67 | 2.57 | 0.84 | 0.44 | 19.99 |
| 1998 | 0.77 | 0.21 | 0.03 | 0.54 | 0.16 | 0.29 | 0.19 | 0.15 | 0.05 | 0.13 | 0.85 | 1.15 | 4.52 |
| 1999 | 2.12 | 0.73 | 0.46 | 0.68 | 2.13 | 0.11 | 0.57 | 0.12 | 0.20 | 2.01 | 0.21 | 2.65 | 11.99 |
| 2000 | 1.26 | 0.07 | 0.38 | 0.46 | 0.03 | 0.03 | 0.41 | 1.17 | 0.78 | 0.25 | 2.09 | 0.17 | 7.10 |
| 2001 | 0.18 | 0.57 | 0.62 | 0.29 | 0.14 | 1.16 | 0.13 | 0.05 | 0.28 | 1.05 | 3.91 | 0.76 | 9.14 |
| 2002 | 3.91 | 0.69 | 2.51 | 0.08 | 1.96 | 0.07 | 0.08 | 0.27 | 0.10 | 2.11 | 0.36 | 0.04 | 12.18 |
| 2003 | 1.21 | 1.09 | 1.95 | 0.81 | 0.09 | 0.20 | 0.52 | 0.02 | 0.28 | 1.14 | 0.57 | 4.81 | 12.69 |
| 2004 | 6.88 | 9.47 | 0.56 | 0.58 | 1.29 | 0.31 | 0.09 | 3.74 | 1.01 | 1.25 | 7.87 | 5.96 | 39.01 |
| 2005 | 6.23 | 1.28 | 1.88 | 0.64 | 0.27 | 0.28 | 0.29 | 0.10 | 0.90 | 1.60 | 1.77 | 0.36 | 15.60 |
| 2006 | 1.53 | 2.62 | 16.92 | 0.75 | 1.11 | 0.09 | 0.08 | 0.11 | 0.66 | 2.50 | 2.50 | 0.58 | 29.45 |
| 2007 | 1.10 | 0.40 | 0.68 | 0.20 | 0.12 | 0.16 | 0.05 | 0.09 | 0.50 | 0.15 | 5.46 | 3.08 | 11.99 |
| 2008 | 0.29 | 0.42 | 0.08 | 0.20 | 0.53 | 0.55 | 0.80 | 0.38 | 0.43 | 0.57 | 2.93 | 7.58 | 14.76 |
| 2009 | 3.69 | 0.25 | 2.25 | 0.55 | 0.15 | 0.04 | 0.34 | 0.55 | 0.15 | 1.27 | 1.56 | 0.75 | 11.55 |
| POR= 60 YRS | 3.28 | 2.40 | 2.79 | 1.18 | 0.91 | 0.37 | 0.52 | 0.59 | 0.66 | 1.84 | 2.81 | 3.23 | 20.58 |

WBAN : 22521

## AVERAGE TEMPERATURE (°F) 2009 HONOLULU (PHNL)

| YEAR | JAN | FEB | MAR | APR | MAY | JUN | JUL | AUG | SEP | OCT | NOV | DEC | ANNUAL |
|------|------|------|------|------|------|------|------|------|------|------|------|------|--------|
| 1980 | 71.9 | 72.4 | 75.0 | 76.1 | 78.3 | 79.5 | 80.9 | 81.0 | 81.6 | 80.1 | 78.0 | 74.4 | 77.4 |
| 1981 | 73.2 | 73.6 | 74.7 | 75.9 | 77.3 | 80.6 | 79.7 | 80.1 | 80.7 | 78.3 | 76.7 | 74.0 | 77.1 |
| 1982 | 73.2 | 71.7 | 74.0 | 75.4 | 78.3 | 79.6 | 80.6 | 81.4 | 81.4 | 79.4 | 75.7 | 72.0 | 76.9 |
| 1983 | 71.9 | 71.3 | 73.5 | 74.6 | 75.7 | 78.9 | 79.7 | 82.4 | 82.3 | 81.1 | 80.1 | 75.1 | 77.2 |
| 1984 | 74.6 | 74.6 | 75.8 | 77.0 | 78.7 | 79.3 | 81.0 | 81.7 | 81.3 | 80.2 | 79.0 | 74.1 | 78.1 |
| 1985 | 71.4 | 73.9 | 74.5 | 74.5 | 76.5 | 79.2 | 81.6 | 81.9 | 81.1 | 79.8 | 75.1 | 73.3 | 76.9 |
| 1986 | 72.8 | 72.6 | 76.5 | 77.5 | 78.3 | 80.0 | 81.6 | 82.9 | 82.1 | 80.6 | 79.2 | 75.1 | 78.3 |
| 1987 | 73.4 | 71.2 | 74.0 | 76.0 | 75.7 | 80.4 | 82.1 | 82.7 | 82.9 | 81.4 | 78.8 | 75.8 | 77.9 |
| 1988 | 73.1 | 74.7 | 76.0 | 77.3 | 78.9 | 80.8 | 81.8 | 82.1 | 82.1 | 80.1 | 79.9 | 75.6 | 78.5 |
| 1989 | 74.5 | 73.6 | 75.3 | 74.5 | 78.4 | 80.9 | 81.6 | 81.4 | 81.9 | 78.6 | 76.7 | 72.9 | 77.5 |
| 1990 | 74.7 | 71.5 | 73.1 | 76.6 | 78.1 | 80.0 | 80.8 | 82.3 | 82.3 | 80.9 | 77.3 | 74.1 | 77.6 |
| 1991 | 72.4 | 73.4 | 72.9 | 75.9 | 77.8 | 79.4 | 81.2 | 82.4 | 81.5 | 80.0 | 79.5 | 76.2 | 77.7 |
| 1992 | 72.9 | 73.2 | 74.9 | 75.6 | 77.8 | 81.3 | 81.5 | 82.2 | 81.3 | 79.4 | 77.0 | 76.7 | 77.8 |
| 1993 | 70.9 | 71.1 | 74.0 | 77.4 | 77.2 | 80.2 | 80.6 | 81.3 | 81.1 | 79.7 | 76.3 | 75.0 | 77.1 |
| 1994 | 72.0 | 73.6 | 73.2 | 76.0 | 79.3 | 80.9 | 82.9 | 84.3 | 84.0 | 82.5 | 80.8 | 76.6 | 78.8 |
| 1995 | 74.2 | 73.4 | 75.6 | 76.4 | 78.6 | 81.3 | 83.3 | 83.4 | 83.2 | 82.7 | 80.2 | 79.1 | 79.3 |
| 1996 | 76.2 | 74.0 | 74.3 | 79.9 | 79.0 | 82.1 | 82.1 | 82.8 | 81.4 | 81.7 | 76.9 | 73.1 | 78.6 |
| 1997 | 72.3 | 74.7 | 75.3 | 76.3 | 76.2 | 81.1 | 81.5 | 82.6 | 82.7 | 80.6 | 76.4 | 74.0 | 77.8 |
| 1998 | 72.5 | 72.8 | 75.1 | 75.1 | 76.6 | 78.3 | 79.7 | 81.1 | 81.0 | 79.8 | 77.8 | 74.8 | 77.1 |
| 1999 | 73.3 | 73.7 | 74.6 | 75.4 | 77.1 | 78.8 | 79.3 | 80.8 | 80.2 | 78.3 | 76.9 | 74.1 | 76.9 |
| 2000 | 72.5 | 73.6 | 75.4 | 75.3 | 78.3 | 80.5 | 81.1 | 81.4 | 80.5 | 80.4 | 77.6 | 74.7 | 77.6 |
| 2001 | 75.4 | 74.1 | 75.0 | 76.6 | 78.2 | 79.7 | 81.5 | 82.2 | 82.1 | 79.9 | 77.4 | 76.4 | 78.2 |
| 2002 | 74.2 | 73.1 | 74.1 | 76.7 | 78.2 | 80.8 | 81.0 | 82.2 | 81.2 | 80.2 | 77.5 | 75.4 | 77.9 |
| 2003 | 72.5 | 73.8 | 75.7 | 77.3 | 79.0 | 80.5 | 82.4 | 83.2 | 82.1 | 81.1 | 78.8 | 75.7 | 78.5 |
| 2004 | 73.6 | 76.0 | 75.0 | 77.1 | 79.4 | 81.3 | 82.7 | 82.7 | 82.8 | 81.4 | 77.5 | 75.2 | 78.7 |
| 2005 | 72.7 | 73.9 | 73.8 | 78.7 | 81.3 | 82.7 | 83.2 | 83.6 | 81.1 | 78.6 | 77.1 | 73.5 | 78.4 |
| 2006 | 74.3 | 72.1 | 73.6 | 74.8 | 75.3 | 80.1 | 80.7 | 81.2 | 80.2 | 79.1 | 78.2 | 76.1 | 77.1 |
| 2007 | 74.9 | 73.4 | 74.2 | 76.7 | 78.4 | 80.7 | 81.6 | 82.1 | 81.6 | 80.2 | 76.6 | 75.6 | 78.0 |
| 2008 | 73.6 | 74.8 | 77.3 | 77.2 | 79.2 | 80.5 | 82.2 | 81.6 | 80.5 | 79.7 | 77.2 | 75.2 | 78.3 |
| 2009 | 72.5 | 73.4 | 73.7 | 74.7 | | | 81.9 | 81.6 | 82.5 | 81.2 | 78.9 | 74.1 | |
| POR= 60 YRS | 73.0 | 73.1 | 74.2 | 75.7 | 77.4 | 79.4 | 80.5 | 81.3 | 80.9 | 79.5 | 77.2 | 74.4 | 77.2 |

## HEATING DEGREE DAYS (base 65°F) 2009  HONOLULU (PHNL)

| YEAR | JUL | AUG | SEP | OCT | NOV | DEC | JAN | FEB | MAR | APR | MAY | JUN | TOTAL |
|---|---|---|---|---|---|---|---|---|---|---|---|---|---|
| 1983-84 | 0 | 0 | 0 | 0 | 0 | 0 | 0 | 0 | 0 | 0 | 0 | 0 | 0 |
| 1984-85 | 0 | 0 | 0 | 0 | 0 | 0 | 0 | 0 | 0 | 0 | 0 | 0 | 0 |
| 1985-86 | 0 | 0 | 0 | 0 | 0 | 0 | 0 | 0 | 0 | 0 | 0 | 0 | 0 |
| 1986-87 | 0 | 0 | 0 | 0 | 0 | 0 | 0 | 0 | 0 | 0 | 0 | 0 | 0 |
| 1987-88 | 0 | 0 | 0 | 0 | 0 | 0 | 0 | 0 | 0 | 0 | 0 | 0 | 0 |
| 1988-89 | 0 | 0 | 0 | 0 | 0 | 0 | 0 | 0 | 0 | 0 | 0 | 0 | 0 |
| 1989-90 | 0 | 0 | 0 | 0 | 0 | 0 | 0 | 0 | 0 | 0 | 0 | 0 | 0 |
| 1990-91 | 0 | 0 | 0 | 0 | 0 | 0 | 0 | 0 | 0 | 0 | 0 | 0 | 0 |
| 1991-92 | 0 | 0 | 0 | 0 | 0 | 0 | 0 | 0 | 0 | 0 | 0 | 0 | 0 |
| 1992-93 | 0 | 0 | 0 | 0 | 0 | 0 | 0 | 0 | 0 | 0 | 0 | 0 | 0 |
| 1993-94 | 0 | 0 | 0 | 0 | 0 | 0 | 0 | 0 | 0 | 0 | 0 | 0 | 0 |
| 1994-95 | 0 | 0 | 0 | 0 | 0 | 0 | 0 | 0 | 0 | 0 | 0 | 0 | 0 |
| 1995-96 | 0 | 0 | 0 | 0 | 0 | 0 | 0 | 0 | 0 | 0 | 0 | 0 | 0 |
| 1996-97 | 0 | 0 | 0 | 0 | 0 | 0 | 0 | 0 | 0 | 0 | 0 | 0 | 0 |
| 1997-98 | 0 | 0 | 0 | 0 | 0 | 0 | 0 | 0 | 0 | 0 | 0 | 0 | 0 |
| 1998-99 | 0 | 0 | 0 | 0 | 0 | 0 |  | 0 | 0 | 0 | 0 | 0 | |
| 1999-00 | 0 | 0 | 0 | 0 | 0 | 0 | 0 | 0 | 0 | 0 | 0 | 0 | 0 |
| 2000-01 | 0 | 0 | 0 | 0 | 0 | 0 | 0 | 0 | 0 | 0 | 0 | 0 | 0 |
| 2001-02 | 0 | 0 | 0 | 0 | 0 | 0 | 0 | 0 | 0 | 0 | 0 | 0 | 0 |
| 2002-03 | 0 | 0 | | 0 | 0 | 0 | 0 | 0 | 0 | 0 | 0 | 0 | 0 |
| 2003-04 | 0 | 0 | 0 | 0 | 0 | 0 | 0 | 0 | 0 | 0 | 0 | | |
| 2004-05 | 0 | 0 | 0 | 0 | 0 | 0 | 0 | 0 | 0 | 0 | 0 | 0 | 0 |
| 2005-06 | 0 | 0 | 0 | 0 | 0 | 0 | 0 | 0 | 0 | 0 | 0 | 0 | 0 |
| 2006-07 | 0 | 0 | 0 | 0 | 0 | 0 | 0 | 0 | 0 | 0 | 0 | 0 | 0 |
| 2007-08 | 0 | 0 | 0 | 0 | 0 | 0 | 0 | 0 | 0 | 0 | 0 | 0 | 0 |
| 2008-09 | 0 | 0 | 0 | 0 | 0 | 0 | 0 | 0 | 0 | 0 | | | 0 |
| 2009- | 0 | 0 | 0 | 0 | 0 | 0 | | | | | | | |

WBAN : 22521

## COOLING DEGREE DAYS (base 65°F) 2009  HONOLULU (PHNL)

| YEAR | JAN | FEB | MAR | APR | MAY | JUN | JUL | AUG | SEP | OCT | NOV | DEC | TOTAL |
|---|---|---|---|---|---|---|---|---|---|---|---|---|---|
| 1980 | 222 | 220 | 317 | 340 | 418 | 442 | 501 | 504 | 503 | 476 | 395 | 295 | 4633 |
| 1981 | 263 | 249 | 311 | 335 | 385 | 474 | 463 | 477 | 477 | 419 | 355 | 284 | 4492 |
| 1982 | 261 | 195 | 288 | 318 | 421 | 442 | 493 | 514 | 499 | 452 | 326 | 225 | 4434 |
| 1983 | 223 | 182 | 270 | 295 | 335 | 425 | 461 | 544 | 525 | 508 | 461 | 318 | 4547 |
| 1984 | 304 | 285 | 340 | 366 | 432 | 438 | 501 | 527 | 494 | 475 | 425 | 291 | 4878 |
| 1985 | 205 | 256 | 300 | 293 | 364 | 437 | 521 | 532 | 491 | 464 | 310 | 264 | 4437 |
| 1986 | 251 | 217 | 366 | 384 | 421 | 457 | 519 | 561 | 521 | 491 | 433 | 318 | 4939 |
| 1987 | 267 | 178 | 285 | 337 | 337 | 465 | 537 | 556 | 544 | 516 | 418 | 342 | 4782 |
| 1988 | 260 | 289 | 346 | 373 | 437 | 482 | 527 | 537 | 520 | 478 | 455 | 336 | 5040 |
| 1989 | 301 | 244 | 325 | 291 | 425 | 482 | 521 | 517 | 512 | 431 | 358 | 252 | 4659 |
| 1990 | 306 | 189 | 258 | 354 | 412 | 456 | 498 | 543 | 525 | 501 | 377 | 289 | 4708 |
| 1991 | 236 | 243 | 250 | 333 | 404 | 442 | 508 | 547 | 501 | 474 | 443 | 353 | 4734 |
| 1992 | 251 | 244 | 316 | 327 | 404 | 495 | 518 | 540 | 495 | 452 | 369 | 368 | 4779 |
| 1993 | 190 | 180 | 285 | 381 | 387 | 463 | 491 | 515 | 491 | 465 | 345 | 314 | 4507 |
| 1994 | 227 | 248 | 262 | 337 | 451 | 486 | 564 | 606 | 575 | 547 | 481 | 367 | 5151 |
| 1995 | 294 | 239 | 334 | 351 | 428 | 498 | 572 | 578 | 555 | 554 | 464 | 442 | 5309 |
| 1996 | 354 | 266 | 297 | 453 | 441 | 494 | 538 | 560 | 502 | 525 | 365 | 257 | 5052 |
| 1997 | 231 | 279 | 326 | 346 | 355 | 489 | 521 | 553 | 537 | 492 | 350 | 288 | 4767 |
| 1998 | 238 | 227 | 322 | 309 | 368 | 407 | 463 | 506 | 489 | 464 | 390 | 310 | 4493 |
| 1999 | 267 | 248 | 305 | 317 | 383 | 417 | 452 | 500 | 460 | 421 | 361 | 289 | 4420 |
| 2000 | 241 | 256 | 331 | 318 | 419 | 470 | 508 | 517 | 476 | 482 | 382 | 307 | 4707 |
| 2001 | 332 | 262 | 319 | 356 | 415 | 451 | 515 | 541 | 522 | 467 | 375 | 357 | 4912 |
| 2002 | 292 | 234 | 290 | 360 | 414 | 481 | 503 | 539 | 495 | 479 | 383 | 329 | 4799 |
| 2003 | 240 | 250 | 338 | 374 | 439 | 471 | 545 | 573 | 518 | 504 | 420 | 336 | 5008 |
| 2004 | 272 | 323 | 317 | 371 | 453 | 496 | 556 | 558 | 537 | 517 | 383 | 324 | 5107 |
| 2005 | 245 | 255 | 280 | 419 | 514 | 537 | 570 | 588 | 491 | 430 | 371 | 270 | 4970 |
| 2006 | 292 | 205 | 274 | 299 | 324 | 458 | 496 | 509 | 464 | 446 | 400 | 352 | 4519 |
| 2007 | 311 | 240 | 294 | 356 | 421 | 478 | 524 | 538 | 504 | 477 | 357 | 335 | 4835 |
| 2008 | 273 | 291 | 390 | 374 | 448 | 473 | 539 | 520 | 471 | 463 | 376 | 325 | 4943 |
| 2009 | 241 | 242 | 276 | 298 | | | 529 | 525 | 534 | 511 | 425 | 288 | |

## SNOWFALL (inches)  2009  HONOLULU (PHNL)

| YEAR | JUL | AUG | SEP | OCT | NOV | DEC | JAN | FEB | MAR | APR | MAY | JUN | TOTAL |
|---|---|---|---|---|---|---|---|---|---|---|---|---|---|
| 1976-77 | 0.0 | 0.0 | 0.0 | 0.0 | 0.0 | 0.0 | 0.0 | 0.0 | 0.0 | 0.0 | 0.0 | 0.0 | 0.0 |
| 1977-78 | 0.0 | 0.0 | 0.0 | 0.0 | 0.0 | 0.0 | 0.0 | 0.0 | 0.0 | 0.0 | 0.0 | 0.0 | 0.0 |
| 1978-79 | 0.0 | 0.0 | 0.0 | 0.0 | 0.0 | 0.0 | 0.0 | 0.0 | 0.0 | 0.0 | 0.0 | 0.0 | 0.0 |
| 1979-80 | 0.0 | 0.0 | 0.0 | 0.0 | 0.0 | 0.0 | 0.0 | 0.0 | 0.0 | 0.0 | 0.0 | 0.0 | 0.0 |
| 1980-81 | 0.0 | 0.0 | 0.0 | 0.0 | 0.0 | 0.0 | 0.0 | 0.0 | 0.0 | 0.0 | 0.0 | 0.0 | 0.0 |
| 1981-82 | 0.0 | 0.0 | 0.0 | 0.0 | 0.0 | 0.0 | 0.0 | 0.0 | 0.0 | 0.0 | 0.0 | 0.0 | 0.0 |
| 1982-83 | 0.0 | 0.0 | 0.0 | 0.0 | 0.0 | 0.0 | 0.0 | 0.0 | 0.0 | 0.0 | 0.0 | 0.0 | 0.0 |
| 1983-84 | 0.0 | 0.0 | 0.0 | 0.0 | 0.0 | 0.0 | 0.0 | 0.0 | 0.0 | 0.0 | 0.0 | 0.0 | 0.0 |
| 1984-85 | 0.0 | 0.0 | 0.0 | 0.0 | 0.0 | 0.0 | 0.0 | 0.0 | 0.0 | 0.0 | 0.0 | 0.0 | 0.0 |
| 1985-86 | 0.0 | 0.0 | 0.0 | 0.0 | 0.0 | 0.0 | 0.0 | 0.0 | 0.0 | 0.0 | 0.0 | 0.0 | 0.0 |
| 1986-87 | 0.0 | 0.0 | 0.0 | 0.0 | 0.0 | 0.0 | 0.0 | 0.0 | 0.0 | 0.0 | 0.0 | 0.0 | 0.0 |
| 1987-88 | 0.0 | 0.0 | 0.0 | 0.0 | 0.0 | 0.0 | 0.0 | 0.0 | 0.0 | 0.0 | 0.0 | 0.0 | 0.0 |
| 1988-89 | 0.0 | 0.0 | 0.0 | 0.0 | 0.0 | 0.0 | 0.0 | 0.0 | 0.0 | 0.0 | 0.0 | 0.0 | 0.0 |
| 1989-90 | 0.0 | 0.0 | 0.0 | 0.0 | 0.0 | 0.0 | 0.0 | 0.0 | 0.0 | 0.0 | 0.0 | 0.0 | 0.0 |
| 1990-91 | 0.0 | 0.0 | 0.0 | 0.0 | 0.0 | 0.0 | 0.0 | 0.0 | 0.0 | 0.0 | 0.0 | 0.0 | 0.0 |
| 1991-92 | 0.0 | 0.0 | 0.0 | 0.0 | 0.0 | 0.0 | 0.0 | 0.0 | 0.0 | 0.0 | 0.0 | 0.0 | 0.0 |
| 1992-93 | 0.0 | 0.0 | 0.0 | 0.0 | 0.0 | 0.0 | 0.0 | 0.0 | 0.0 | 0.0 | 0.0 | 0.0 | 0.0 |
| 1993-94 | 0.0 | 0.0 | 0.0 | 0.0 | 0.0 | 0.0 | 0.0 | 0.0 | 0.0 | 0.0 | 0.0 | 0.0 | 0.0 |
| 1994-95 | 0.0 | 0.0 | 0.0 | 0.0 | 0.0 | 0.0 | 0.0 | 0.0 | 0.0 | 0.0 | 0.0 | 0.0 | 0.0 |
| 1995-96 | 0.0 | 0.0 | 0.0 | 0.0 | 0.0 | 0.0 | 0.0 | 0.0 | 0.0 | 0.0 | 0.0 | 0.0 | 0.0 |
| 1996-97 | 0.0 | 0.0 | 0.0 | 0.0 | 0.0 | 0.0 | 0.0 | 0.0 | 0.0 | 0.0 | 0.0 | 0.0 | 0.0 |
| 1997-98 | 0.0 | 0.0 | 0.0 | 0.0 | 0.0 | 0.0 | 0.0 |  |  |  |  |  |  |
| 1998-99 |  |  |  |  |  |  |  |  |  |  |  |  |  |
| 1999-00 |  |  |  |  |  |  |  |  |  |  |  |  |  |
| 2000-01 |  |  |  |  |  |  |  |  |  |  |  |  |  |
| 2001-02 |  |  |  |  |  |  |  |  |  |  |  |  |  |
| 2002-03 |  |  |  |  |  |  |  |  |  |  |  |  |  |
| 2003-04 |  |  |  |  |  |  |  |  |  |  |  |  |  |
| 2004-05 |  |  |  |  |  |  |  |  |  |  |  |  |  |
| 2005- |  |  |  |  |  |  |  |  |  |  |  |  |  |
| POR=<br>48 YRS | 0.0 | 0.0 | 0.0 | 0.0 | 0.0 | 0.0 | 0.0 | 0.0 | 0.0 | 0.0 | 0.0 | 0.0 | 0.0 |

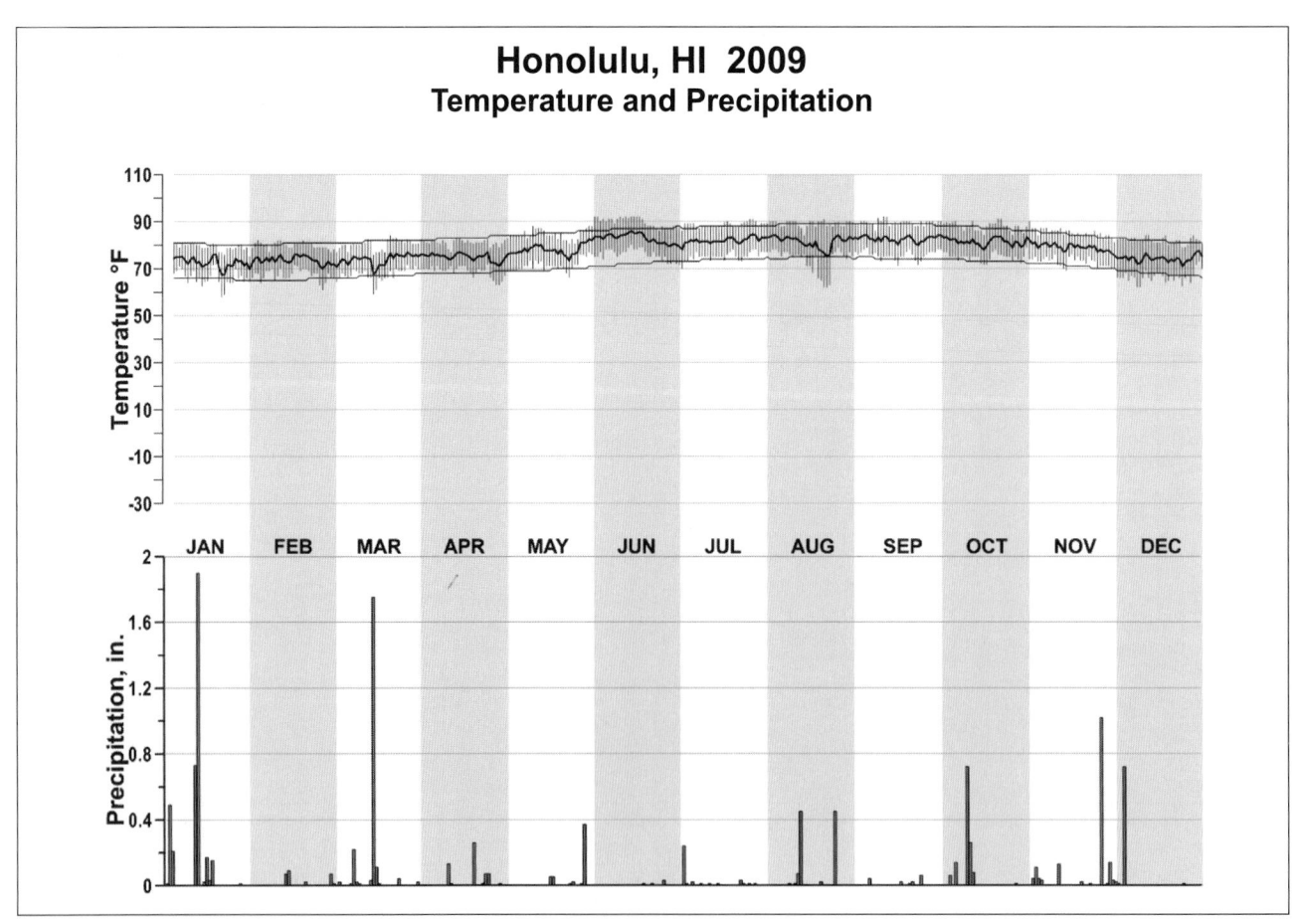

# Honolulu, HI  2009
## Temperature and Precipitation

# 2009
# BOISE
# IDAHO (KBOI)

Boise is situated in the Boise River Valley about 8 miles below the mouth of a mountain canyon where the valley proper begins. Sheltered by large shade trees and averaging 2,710 feet in elevation, the denser part of the city covers a gentle alluvial slope about 2 miles wide, stretching southwest from the foothills of the Boise Mountains to the river. The Boise Mountains immediately north of the city rise 5,000 to 6,000 feet above sea level in about 8 miles, the slopes partly mantled with sagebrush and then chaparral giving way near the summit to ridges of fir, spruce, and pine. Across the river, the land rises in two irregular steps, or benches, for several miles, finally reaching the low divide between the Boise and Snake Rivers. Downstream the valley widens, merging with the valley of the Snake about 40 miles to the northwest. Once semi-arid, the entire area is now irrigated from the upstream reservoirs.

Although air masses from the Pacific are considerably modified by the time they reach Boise, their influence, particularly in winter, alternates with that of atmospheric developments from other directions. The result is almost a typical upland continental type of climate in summer, while winters are usually tempered by periods of cloudy or stormy and mild weather. Autumns have prolonged periods of near ideal weather, while springtime is noted by changeable weather and varied temperatures. The Boise climate in general may be described as dry and temperate, with sufficient variation to be stimulating.

Summer hot periods rarely last longer than a few days. Temperatures of 100 degrees or higher occur nearly every year.

Winter cold spells with temperatures of 10 degrees or lower generally last longer than the summer hot spells. During cold weather, however, there is ordinarily little wind to add to the discomfort.

The normal precipitation pattern in the Boise area shows a winter high and a very pronounced summer low. Total amounts and intensity are generally greatest near the foothills, dwindling to westward and southward.

Tornadoes are very rare as are destructive force winds. Northwesterly winds, drying and rather raw in character, although of moderate velocity, are common from March through May. Diurnal southeasterly winds, descending from nearby foothills at night, frequently have a moderating effect on winter temperatures. There is an occasional, but moderate, duststorm during the warmer months, usually occurring at times of cold frontal passage.

Relative humidity is low but widespread irrigation maintains humidity several percent above the general dryness of western arid conditions in summer. Thunderstorms occur primarily during spring and summer, with less frequency during fall and occasionally during winter. December and January are the months of heavy fog or low stratus cloud conditions. Only a moderate amount of sunshine is received in the average winter, but protracted periods of clear, sunny weather are the rule in summer. Ice storms are practically unknown.

Based on the 1951-1980 period, the average first occurrence of 32 degrees Fahrenheit in the fall is October 9 and the average last occurrence in the spring is May 8.

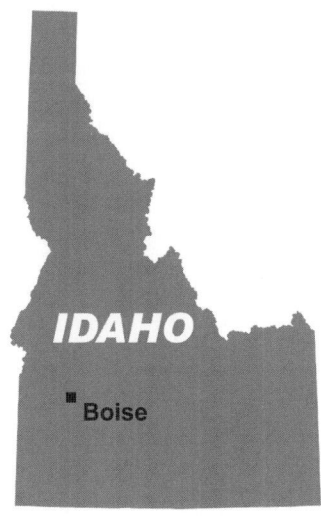

# NORMALS, MEANS, AND EXTREMES
## BOISE (KBOI)

LATITUDE: 43°33'N  LONGITUDE: -116°13'W  ELEVATION (FT): GRND: 2814  BARO: 2861  TIME ZONE: MOUNTAIN (UTC -7)  WBAN: 24131

| | ELEMENT | POR | JAN | FEB | MAR | APR | MAY | JUN | JUL | AUG | SEP | OCT | NOV | DEC | YEAR |
|---|---|---|---|---|---|---|---|---|---|---|---|---|---|---|---|
| TEMPERATURE °F | NORMAL DAILY MAXIMUM | 30 | 36.7 | 44.5 | 53.6 | 61.7 | 70.7 | 80.3 | 89.2 | 88.0 | 77.2 | 64.3 | 47.5 | 37.2 | 62.6 |
| | MEAN DAILY MAXIMUM | 70 | 36.9 | 43.9 | 53.0 | 61.5 | 71.2 | 79.6 | 90.8 | 88.5 | 77.6 | 64.7 | 48.3 | 38.5 | 62.9 |
| | HIGHEST DAILY MAXIMUM | 70 | 63 | 71 | 81 | 92 | 99 | 109 | 111 | 111 | 110 | 102 | 94 | 78 | 65 | 111 |
| | YEAR OF OCCURRENCE | | 1953 | 1992 | 1978 | 1987 | 2003 | 1940 | 1960 | 1961 | 1945 | 1997 | 1999 | 1964 | JUL 1960 |
| | MEAN OF EXTREME MAXS. | 70 | 51.4 | 58.5 | 68.5 | 78.9 | 89.7 | 97.8 | 102.9 | 100.9 | 94.0 | 82.2 | 64.9 | 53.4 | 78.6 |
| | NORMAL DAILY MINIMUM | 30 | 23.6 | 28.8 | 34.0 | 39.4 | 46.6 | 54.2 | 60.3 | 59.8 | 51.2 | 41.3 | 32.4 | 24.1 | 41.3 |
| | MEAN DAILY MINIMUM | 70 | 22.4 | 27.1 | 32.0 | 37.2 | 44.7 | 51.6 | 58.8 | 57.4 | 49.0 | 39.5 | 30.6 | 24.1 | 39.5 |
| | LOWEST DAILY MINIMUM | 70 | -17 | -15 | 6 | 19 | 22 | 31 | 35 | 34 | 23 | 11 | -3 | -25 | -25 |
| | YEAR OF OCCURRENCE | | 1950 | 1989 | 1971 | 1968 | 1982 | 1995 | 1986 | 1992 | 1970 | 1971 | 1985 | 1990 | DEC 1990 |
| | MEAN OF EXTREME MINS. | 70 | 4.4 | 12.0 | 20.0 | 25.0 | 30.7 | 39.1 | 46.8 | 45.2 | 35.2 | 24.8 | 16.8 | 7.7 | 25.6 |
| | NORMAL DRY BULB | 30 | 30.2 | 36.7 | 43.8 | 50.6 | 58.6 | 67.2 | 74.7 | 73.9 | 64.2 | 52.8 | 39.9 | 30.6 | 51.9 |
| | MEAN DRY BULB | 70 | 29.7 | 35.5 | 42.5 | 49.4 | 58.0 | 65.7 | 74.8 | 73.0 | 63.3 | 52.1 | 39.5 | 31.3 | 51.2 |
| | MEAN WET BULB | 26 | 27.0 | 30.6 | 36.1 | 40.6 | 46.9 | 52.2 | 55.3 | 53.8 | 48.2 | 41.6 | 33.9 | 26.8 | 41.1 |
| | MEAN DEW POINT | 26 | 24.6 | 26.9 | 30.4 | 34.1 | 39.3 | 43.0 | 45.9 | 44.0 | 39.7 | 33.2 | 30.0 | 24.0 | 34.6 |
| | NORMAL NO. DAYS WITH: | | | | | | | | | | | | | | |
| | MAXIMUM >= 90 | 30 | 0.0 | 0.0 | 0.0 | 0.1 | 1.4 | 6.2 | 17.9 | 16.7 | 3.6 | 0.1 | 0.0 | 0.0 | 46.0 |
| | MAXIMUM <= 32 | 30 | 9.3 | 3.1 | 0.2 | 0.0 | 0.0 | 0.0 | 0.0 | 0.0 | 0.0 | 0.0 | 1.7 | 7.8 | 22.1 |
| | MINIMUM <= 32 | 30 | 25.7 | 19.4 | 15.1 | 7.3 | 1.6 | 0.1 | 0.0 | 0.0 | 0.4 | 5.7 | 17.3 | 25.6 | 118.2 |
| | MINIMUM <= 0 | 30 | 1.6 | 0.5 | 0.0 | 0.0 | 0.0 | 0.0 | 0.0 | 0.0 | 0.0 | 0.0 | 0.1 | 1.9 | 4.1 |
| H/C | NORMAL HEATING DEG. DAYS | 30 | 1095 | 807 | 672 | 451 | 245 | 71 | 12 | 15 | 110 | 398 | 768 | 1083 | 5727 |
| | NORMAL COOLING DEG. DAYS | 30 | 0 | 0 | 0 | 3 | 31 | 122 | 297 | 275 | 74 | 5 | 0 | 0 | 807 |
| RH | NORMAL (PERCENT) | 30 | 77 | 70 | 61 | 54 | 50 | 44 | 37 | 37 | 44 | 52 | 69 | 76 | 56 |
| | HOUR 05 LST | 30 | 81 | 78 | 74 | 70 | 69 | 65 | 56 | 53 | 59 | 65 | 76 | 80 | 69 |
| | HOUR 11 LST | 30 | 74 | 66 | 55 | 47 | 43 | 38 | 33 | 32 | 38 | 45 | 64 | 73 | 51 |
| | HOUR 17 LST | 30 | 69 | 58 | 44 | 36 | 33 | 28 | 22 | 22 | 27 | 35 | 57 | 68 | 42 |
| | HOUR 23 LST | 30 | 80 | 76 | 67 | 61 | 57 | 50 | 41 | 40 | 49 | 58 | 73 | 80 | 61 |
| S | PERCENT POSSIBLE SUNSHINE | 60 | 40 | 50 | 62 | 68 | 72 | 76 | 87 | 85 | 82 | 69 | 43 | 38 | 64 |
| W/O | MEAN NO. DAYS WITH: | | | | | | | | | | | | | | |
| | HEAVY FOG(VISBY <= 1/4 MI) | 46 | 6.0 | 2.7 | 0.8 | 0.3 | 0.1 | 0.1 | 0.0 | 0.0 | 0.1 | 0.5 | 2.6 | 5.1 | 18.3 |
| | THUNDERSTORMS | 62 | 0.0 | 0.2 | 0.6 | 0.9 | 2.4 | 2.5 | 2.4 | 2.2 | 1.4 | 0.4 | 0.2 | 0.1 | 13.3 |
| CLOUDNESS | MEAN: | | | | | | | | | | | | | | |
| | SUNRISE-SUNSET (OKTAS) | | | | | | | | | | | | | | |
| | MIDNIGHT-MIDNIGHT (OKTAS) | | | | | | | | | | | | | | |
| | MEAN NO. DAYS WITH: | | | | | | | | | | | | | | |
| | CLEAR | | | | | | | | | | | | | | |
| | PARTLY CLOUDY | | | | | | | | | | | | | | |
| | CLOUDY | | | | | | | | | | | | | | |
| PR | MEAN STATION PRESSURE(IN) | 26 | 27.17 | 27.10 | 27.04 | 27.00 | 26.98 | 26.97 | 26.99 | 27.00 | 27.03 | 27.08 | 27.13 | 27.16 | 27.05 |
| | MEAN SEA-LEVEL PRES. (IN) | 26 | 30.22 | 30.13 | 30.03 | 29.97 | 29.91 | 29.89 | 29.89 | 29.90 | 29.96 | 30.06 | 30.15 | 30.22 | 30.03 |
| WINDS | MEAN SPEED (MPH) | 26 | 6.8 | 7.8 | 8.8 | 8.9 | 8.4 | 8.3 | 7.7 | 7.5 | 7.4 | 7.2 | 7.2 | 7.1 | 7.8 |
| | PREVAIL.DIR(TENS OF DEGS) | 39 | 14 | 14 | 14 | 14 | 32 | 32 | 32 | 14 | 14 | 14 | 14 | 14 | 14 |
| | MAXIMUM 2-MINUTE: | 14 | | | | | | | | | | | | | |
| | SPEED (MPH) | | 39 | 39 | 45 | 46 | 41 | 46 | 44 | 45 | 39 | 39 | 45 | 45 | 46 |
| | DIR. (TENS OF DEGS) | | 12 | 31 | 30 | 31 | 31 | 29 | 04 | 30 | 29 | 31 | 30 | 31 | 29 |
| | YEAR OF OCCURRENCE | | 2008 | 2003 | 2009 | 1998 | 2008 | 2008 | 2003 | 2006 | 2007 | 2009 | 2008 | 2000 | JUN 2008 |
| | MAXIMUM 3-SECOND | | | | | | | | | | | | | | |
| | SPEED (MPH) | 14 | 49 | 51 | 53 | 52 | 47 | 58 | 51 | 67 | 56 | 58 | 53 | 63 | 67 |
| | DIR. (TENS OF DEGS) | | 12 | 20 | 30 | 30 | 28 | 29 | 04 | 30 | 29 | 26 | 30 | 30 | 30 |
| | YEAR OF OCCURRENCE | | 2008 | 2009 | 2009 | 2006 | 2008 | 2008 | 2003 | 2006 | 2007 | 2009 | 2008 | 2000 | AUG 2006 |
| PRECIPITATION | NORMAL (IN) | 30 | 1.39 | 1.14 | 1.41 | 1.27 | 1.27 | 0.74 | 0.39 | 0.30 | 0.76 | 0.76 | 1.38 | 1.38 | 12.19 |
| | MAXIMUM MONTHLY (IN) | 70 | 3.87 | 3.70 | 3.46 | 3.04 | 4.40 | 3.41 | 1.62 | 2.37 | 2.93 | 2.59 | 3.36 | 4.23 | 4.40 |
| | YEAR OF OCCURRENCE | | 1970 | 1986 | 1989 | 1955 | 1998 | 1941 | 1982 | 1968 | 1986 | 2000 | 1988 | 1983 | MAY 1998 |
| | MINIMUM MONTHLY (IN) | 70 | 0.12 | 0.18 | 0.17 | 0.09 | T | 0.01 | 0.00 | 0.00 | 0.00 | 0.00 | 0.14 | 0.09 | 0.00 |
| | YEAR OF OCCURRENCE | | 1949 | 1997 | 1994 | 1949 | 1992 | 1966 | 1947 | 1998 | 1987 | 1988 | 1976 | 1976 | AUG 1998 |
| | MAXIMUM IN 24 HOURS (IN) | 70 | 1.48 | 1.00 | 1.65 | 1.27 | 2.05 | 2.24 | 0.94 | 1.61 | 1.74 | 1.06 | 0.88 | 1.16 | 2.24 |
| | YEAR OF OCCURRENCE | | 1953 | 1951 | 1981 | 1969 | 1990 | 1958 | 1960 | 1979 | 1976 | 2000 | 1971 | 1955 | JUN 1958 |
| | NORMAL NO. DAYS WITH: | | | | | | | | | | | | | | |
| | PRECIPITATION >= 0.01 | 30 | 11.3 | 10.1 | 10.3 | 8.6 | 7.6 | 5.1 | 3.0 | 2.7 | 3.9 | 5.2 | 10.5 | 11.4 | 89.7 |
| | PRECIPITATION >= 1.00 | 30 | 0.0 | 0.0 | * | * | * | 0.1 | 0.0 | * | * | 0.0 | 0.0 | 0.0 | 0.1 |
| SNOWFALL | NORMAL (IN) | 30 | 4.9 | 3.3 | 1.5 | 0.3 | 0.1 | 0.0 | 0.0 | 0.0 | 0.0 | 0.1 | 2.9 | 6.4 | 19.5 |
| | MAXIMUM MONTHLY (IN) | 70 | 21.4 | 25.2 | 11.9 | 8.0 | 4.0 | T | T | 0.0 | T | 2.7 | 18.6 | 26.2 | 26.2 |
| | YEAR OF OCCURRENCE | | 1964 | 1949 | 1951 | 1967 | 1964 | 2008 | 1995 | | 2008 | 1998 | 1971 | 1985 | 1983 | DEC 1983 |
| | MAXIMUM IN 24 HOURS (IN) | 70 | 8.5 | 13.0 | 6.4 | 7.2 | 4.0 | T | T | T | T | 1.7 | 6.5 | 9.8 | 13.0 |
| | YEAR OF OCCURRENCE' | | 1950 | 1949 | 1952 | 1969 | 1964 | 1995 | 1995 | 1989 | 1998 | 1971 | 1964 | 1996 | FEB 1949 |
| | MAXIMUM SNOW DEPTH (IN) | 61 | 12 | 9 | 6 | 1 | 0 | 0 | 0 | 0 | 0 | 0 | 11 | 13 | 13 |
| | YEAR OF OCCURRENCE | | 1982 | 1949 | 1952 | 1975 | | | | | | | 1985 | 1985 | DEC 1985 |
| | NORMAL NO. DAYS WITH: | | | | | | | | | | | | | | |
| | SNOWFALL >= 1.0 | 30 | 1.7 | 1.1 | 0.4 | 0.2 | 0.0 | 0.0 | 0.0 | 0.0 | 0.0 | 0.1 | 1.2 | 2.3 | 7.0 |

## PRECIPITATION (inches) 2009 BOISE (KBOI)

| YEAR | JAN | FEB | MAR | APR | MAY | JUN | JUL | AUG | SEP | OCT | NOV | DEC | ANNUAL |
|---|---|---|---|---|---|---|---|---|---|---|---|---|---|
| 1980 | 1.56 | 1.29 | 2.14 | 1.20 | 3.77 | 0.58 | 0.03 | T | 1.59 | 0.30 | 1.26 | 1.49 | 15.21 |
| 1981 | 1.20 | 1.02 | 2.76 | 1.93 | 0.95 | 0.77 | 0.23 | 0.13 | 0.36 | 0.97 | 2.24 | 2.72 | 15.28 |
| 1982 | 1.42 | 1.54 | 1.39 | 0.79 | 0.39 | 0.35 | 1.62 | 0.19 | 1.38 | 1.74 | 1.10 | 1.92 | 13.83 |
| 1983 | 1.67 | 1.26 | 2.70 | 2.29 | 1.93 | 0.17 | 1.16 | 0.28 | 0.65 | 0.56 | 1.87 | 4.23 | 18.77 |
| 1984 | 0.80 | 0.86 | 1.43 | 1.62 | 1.06 | 1.47 | 0.23 | 1.24 | 0.69 | 0.85 | 2.36 | 0.63 | 13.24 |
| 1985 | 0.20 | 0.55 | 0.97 | 0.90 | 1.52 | 0.37 | 0.85 | 0.04 | 1.81 | 0.84 | 1.85 | 1.24 | 11.14 |
| 1986 | 0.98 | 3.70 | 2.01 | 1.55 | 1.10 | 0.35 | 0.17 | 0.07 | 2.93 | 0.33 | 1.00 | 0.12 | 14.31 |
| 1987 | 0.73 | 1.24 | 2.01 | 0.38 | 0.69 | 0.58 | 0.70 | 0.11 | 0.00 | T | 1.00 | 1.05 | 8.49 |
| 1988 | 1.30 | 0.43 | 1.45 | 1.80 | 1.33 | 0.47 | 0.02 | 0.09 | 0.24 | 0.00 | 3.36 | 0.81 | 11.30 |
| 1989 | 1.14 | 1.15 | 3.46 | 0.46 | 0.21 | 0.08 | 0.03 | 0.78 | 1.20 | 1.24 | 0.59 | 0.10 | 10.44 |
| 1990 | 0.84 | 0.79 | 0.77 | 2.14 | 4.07 | 0.11 | 0.42 | 0.39 | 0.50 | 0.45 | 0.61 | 0.98 | 12.07 |
| 1991 | 0.96 | 0.46 | 0.55 | 1.65 | 1.57 | 0.64 | 0.37 | 0.04 | 0.21 | 0.91 | 1.76 | 0.35 | 9.47 |
| 1992 | 0.36 | 0.92 | 0.17 | 0.66 | T | 2.07 | 0.03 | T | 0.30 | 0.90 | 1.37 | 0.89 | 7.67 |
| 1993 | 1.65 | 0.96 | 2.45 | 2.09 | 0.92 | 2.10 | 0.52 | 0.24 | T | 0.47 | 0.38 | 0.98 | 12.76 |
| 1994 | 1.28 | 0.90 | 0.17 | 1.25 | 1.01 | 0.24 | 0.10 | T | 0.10 | 0.78 | 1.78 | 1.79 | 9.40 |
| 1995 | 2.10 | 0.51 | 1.51 | 1.03 | 2.36 | 0.88 | 0.59 | 0.06 | 0.22 | 0.42 | 2.20 | 2.14 | 14.02 |
| 1996 | 1.33 | 1.07 | 1.97 | 1.47 | 1.67 | 0.21 | .12 | T | .46 | .68 | 1.71 | 3.43 | 14.12 |
| 1997 | 2.74 | 0.18 | 0.52 | 1.89 | 1.14 | 1.35 | 0.45 | 0.27 | 0.67 | 0.55 | 0.68 | 0.65 | 11.09 |
| 1998 | 2.73 | 1.39 | 0.99 | 0.81 | 4.40 | 1.21 | 0.49 | 0.00 | 1.96 | 0.11 | 0.97 | 1.65 | 16.71 |
| 1999 | 1.40 | 1.96 | 0.75 | 0.61 | 1.10 | 0.47 | T | 0.29 | 0.00 | 0.11 | 1.00 | 0.90 | 8.59 |
| 2000 | 1.51 | 2.06 | 1.69 | 1.01 | 0.83 | 0.14 | 0.03 | 0.10 | 0.60 | 2.59 | 0.68 | 0.80 | 12.04 |
| 2001 | 1.07 | 0.49 | 1.07 | 1.20 | 0.27 | 0.32 | 0.15 | T | 0.44 | 0.86 | 1.52 | 1.15 | 8.54 |
| 2002 | 0.94 | 0.19 | 1.06 | 0.83 | 0.01 | 0.19 | 0.09 | 0.05 | 0.39 | 0.31 | 0.87 | 2.03 | 6.96 |
| 2003 | 1.56 | 0.87 | 1.48 | 1.39 | 1.35 | 0.18 | 0.27 | 0.33 | 0.04 | T | 1.02 | 1.62 | 10.11 |
| 2004 | 1.85 | 1.46 | 0.51 | 0.36 | 2.37 | 0.26 | 0.59 | 0.44 | 0.25 | 1.53 | 0.70 | 1.24 | 11.56 |
| 2005 | 0.23 | 0.34 | 1.20 | 1.16 | 3.96 | 0.91 | 0.02 | T | 0.32 | 0.44 | 1.71 | 3.37 | 13.66 |
| 2006 | 1.77 | 0.41 | 2.19 | 1.82 | 1.19 | 0.82 | 0.24 | 0.02 | 0.16 | 0.40 | 1.43 | 1.69 | 12.14 |
| 2007 | 0.19 | 1.27 | 0.32 | 1.06 | 0.23 | 0.94 | 0.02 | 0.04 | 0.65 | 1.03 | 1.06 | 1.27 | 8.08 |
| 2008 | 0.94 | 0.55 | 1.21 | 0.26 | 0.65 | 0.53 | 0.28 | T | 0.94 | 0.75 | 1.39 | 1.75 | 9.25 |
| 2009 | 0.87 | 0.20 | 1.26 | 0.72 | 0.98 | 1.54 | 0.04 | 1.79 | 0.01 | 1.39 | 0.71 | 1.75 | 11.26 |
| POR= 70 YRS | 1.40 | 1.11 | 1.22 | 1.18 | 1.27 | 0.86 | 0.26 | 0.30 | 0.57 | 0.79 | 1.33 | 1.41 | 11.70 |

WBAN : 24131

## AVERAGE TEMPERATURE (°F) 2009 BOISE (KBOI)

| YEAR | JAN | FEB | MAR | APR | MAY | JUN | JUL | AUG | SEP | OCT | NOV | DEC | ANNUAL |
|---|---|---|---|---|---|---|---|---|---|---|---|---|---|
| 1980 | 30.3 | 39.8 | 41.1 | 52.6 | 57.2 | 62.6 | 72.9 | 67.2 | 62.6 | 51.6 | 39.8 | 33.1 | 50.9 |
| 1981 | 33.9 | 36.8 | 44.5 | 50.3 | 54.7 | 63.2 | 71.0 | 74.2 | 63.7 | 48.7 | 44.0 | 35.3 | 51.7 |
| 1982 | 24.8 | 30.0 | 41.3 | 45.2 | 54.9 | 65.8 | 70.1 | 72.7 | 60.4 | 50.9 | 36.0 | 31.5 | 48.6 |
| 1983 | 35.9 | 41.5 | 44.6 | 47.1 | 56.5 | 63.8 | 69.4 | 74.9 | 60.8 | 53.8 | 42.0 | 23.2 | 51.1 |
| 1984 | 21.2 | 30.2 | 41.9 | 45.9 | 54.8 | 61.7 | 74.2 | 75.5 | 60.1 | 46.9 | 39.1 | 22.9 | 47.9 |
| 1985 | 19.1 | 25.8 | 36.0 | 51.6 | 58.5 | 67.2 | 77.7 | 69.2 | 56.2 | 48.4 | 27.7 | 12.6 | 45.8 |
| 1986 | 29.4 | 41.0 | 48.0 | 48.3 | 59.1 | 72.0 | 69.6 | 75.9 | 57.4 | 52.6 | 40.4 | 28.0 | 51.8 |
| 1987 | 27.8 | 37.6 | 44.2 | 56.0 | 62.2 | 70.2 | 71.2 | 70.0 | 65.6 | 54.9 | 40.7 | 32.9 | 52.8 |
| 1988 | 26.1 | 37.8 | 42.6 | 52.7 | 57.9 | 70.7 | 74.6 | 71.6 | 61.7 | 59.9 | 40.7 | 27.0 | 51.9 |
| 1989 | 24.7 | 23.2 | 43.7 | 53.1 | 56.3 | 68.5 | 77.0 | 70.0 | 63.3 | 51.2 | 39.4 | 30.4 | 50.1 |
| 1990 | 34.2 | 34.1 | 44.3 | 54.7 | 55.7 | 66.9 | 76.2 | 74.0 | 69.9 | 51.0 | 41.2 | 18.1 | 51.7 |
| 1991 | 24.4 | 41.6 | 43.0 | 48.3 | 54.9 | 62.6 | 75.7 | 76.6 | 65.8 | 51.8 | 37.4 | 31.7 | 51.2 |
| 1992 | 33.1 | 42.1 | 48.9 | 54.9 | 64.3 | 69.8 | 71.4 | 73.9 | 63.4 | 54.4 | 34.5 | 28.5 | 53.3 |
| 1993 | 24.8 | 28.9 | 41.7 | 48.1 | 63.1 | 61.9 | 65.0 | 68.6 | 63.0 | 53.2 | 32.8 | 33.3 | 48.7 |
| 1994 | 32.7 | 34.7 | 46.1 | 52.6 | 61.0 | 67.9 | 77.0 | 76.1 | 66.8 | 50.3 | 32.6 | 30.4 | 52.4 |
| 1995 | 36.2 | 41.5 | 43.6 | 48.2 | 56.9 | 63.9 | 73.4 | 71.2 | 65.7 | 49.2 | 44.4 | 33.1 | 52.3 |
| 1996 | 32.5 | 35.1 | 44.2 | 50.0 | 55.4 | 66.7 | 76.7 | 74.8 | 61.8 | 52.4 | 41.3 | 36.3 | 52.3 |
| 1997 | 32.1 | 36.8 | 45.4 | 48.9 | 62.9 | 66.0 | 72.8 | 75.6 | 67.7 | 52.0 | 43.0 | 31.2 | 52.9 |
| 1998 | 38.9 | 40.2 | 44.3 | 49.9 | 56.3 | 63.4 | 79.2 | 76.7 | 69.8 | 51.7 | 43.7 | 30.7 | 53.7 |
| 1999 | 35.3 | 37.0 | 43.8 | 47.5 | 55.8 | 66.3 | 73.7 | 75.1 | 64.0 | 54.3 | 46.9 | 31.9 | 52.6 |
| 2000 | 34.2 | 41.7 | 43.4 | 54.3 | 59.7 | 68.6 | 76.1 | 75.9 | 63.0 | 52.3 | 32.7 | 31.1 | 52.8 |
| 2001 | 27.2 | 33.9 | 45.9 | 48.1 | 61.4 | 67.3 | 74.3 | 78.7 | 68.4 | 53.5 | 44.0 | 30.7 | 52.8 |
| 2002 | 31.6 | 34.0 | 40.7 | 50.5 | 59.2 | 69.6 | 79.6 | 70.9 | 65.2 | 49.5 | 40.6 | 37.4 | 52.4 |
| 2003 | 38.7 | 36.9 | 46.5 | 49.4 | 58.8 | 69.7 | 80.6 | 77.4 | 66.7 | 58.1 | 37.7 | 37.2 | 54.8 |
| 2004 | 28.4 | 35.0 | 48.2 | 53.8 | 58.7 | 69.3 | 78.0 | 74.9 | 63.7 | 53.5 | 38.2 | 35.4 | 53.1 |
| 2005 | 32.8 | 36.9 | 45.7 | 50.5 | 59.4 | 64.4 | 78.3 | 76.7 | 63.3 | 53.7 | 37.5 | 29.6 | 52.4 |
| 2006 | 36.4 | 34.8 | 42.3 | 51.7 | 61.0 | 71.0 | 81.5 | 74.0 | 65.4 | 51.2 | 43.4 | 31.7 | 53.7 |
| 2007 | 29.2 | 38.9 | 48.3 | 51.7 | 62.5 | 70.6 | 83.1 | 75.5 | 64.7 | 52.0 | 40.9 | 32.4 | 54.2 |
| 2008 | 29.0 | 36.0 | 40.8 | 46.6 | 60.5 | 67.1 | 77.7 | 75.9 | 66.1 | 53.5 | 43.6 | 31.5 | 52.4 |
| 2009 | 31.3 | 36.9 | 41.7 | 50.6 | 61.6 | 67.8 | 78.6 | 75.2 | 70.6 | 48.5 | 41.4 | 26.8 | 52.6 |
| POR= 70 YRS | 29.7 | 35.5 | 42.5 | 49.4 | 58.0 | 65.7 | 74.8 | 73.0 | 63.3 | 52.1 | 39.5 | 31.3 | 51.2 |

## HEATING DEGREE DAYS (base 65°F) 2009 BOISE (KBOI)

| YEAR | JUL | AUG | SEP | OCT | NOV | DEC | JAN | FEB | MAR | APR | MAY | JUN | TOTAL |
|---|---|---|---|---|---|---|---|---|---|---|---|---|---|
| 1980-81 | 0 | 41 | 104 | 409 | 750 | 983 | 957 | 783 | 631 | 432 | 315 | 97 | 5502 |
| 1981-82 | 15 | 5 | 137 | 497 | 624 | 915 | 1240 | 974 | 729 | 586 | 312 | 86 | 6120 |
| 1982-83 | 27 | 2 | 182 | 432 | 863 | 1030 | 897 | 653 | 622 | 530 | 309 | 82 | 5629 |
| 1983-84 | 38 | 0 | 145 | 338 | 682 | 1290 | 1353 | 1004 | 710 | 566 | 328 | 162 | 6616 |
| 1984-85 | 0 | 8 | 204 | 557 | 771 | 1299 | 1412 | 1093 | 895 | 398 | 226 | 53 | 6916 |
| 1985-86 | 0 | 26 | 259 | 509 | 1113 | 1619 | 1097 | 668 | 522 | 499 | 280 | 15 | 6607 |
| 1986-87 | 35 | 2 | 259 | 376 | 733 | 1141 | 1149 | 761 | 639 | 287 | 140 | 41 | 5563 |
| 1987-88 | 23 | 18 | 86 | 306 | 722 | 990 | 1198 | 780 | 686 | 359 | 261 | 59 | 5488 |
| 1988-89 | 4 | 5 | 157 | 178 | 724 | 1169 | 1242 | 1166 | 656 | 356 | 276 | 30 | 5963 |
| 1989-90 | 0 | 29 | 97 | 421 | 759 | 1064 | 951 | 858 | 633 | 303 | 286 | 82 | 5483 |
| 1990-91 | 6 | 10 | 26 | 430 | 710 | 1449 | 1252 | 651 | 676 | 493 | 306 | 100 | 6109 |
| 1991-92 | 0 | 0 | 55 | 409 | 822 | 1026 | 982 | 657 | 492 | 308 | 92 | 54 | 4897 |
| 1992-93 | 6 | 40 | 118 | 340 | 907 | 1124 | 1239 | 1004 | 715 | 501 | 140 | 155 | 6289 |
| 1993-94 | 53 | 45 | 124 | 367 | 960 | 975 | 993 | 839 | 579 | 378 | 148 | 64 | 5525 |
| 1994-95 | 9 | 1 | 34 | 449 | 967 | 1066 | 887 | 652 | 657 | 499 | 255 | 111 | 5587 |
| 1995-96 | 5 | 23 | 86 | 485 | 610 | 984 | 1000 | 859 | 636 | 447 | 292 | 55 | 5482 |
| 1996-97 | 1 | 6 | 162 | 405 | 703 | 883 | 1013 | 783 | 601 | 473 | 128 | 41 | 5199 |
| 1997-98 | 18 | 0 | 61 | 410 | 651 | 1040 | 803 | 684 | 635 | 452 | 270 | 85 | 5109 |
| 1998-99 | 0 | 0 | 51 | 408 | 635 | 1058 | 914 | 778 | 652 | 520 | 311 | 95 | 5422 |
| 1999-00 | 8 | 13 | 101 | 324 | 535 | 1017 | 949 | 673 | 663 | 315 | 179 | 41 | 4818 |
| 2000-01 | 4 | 0 | 128 | 389 | 960 | 1044 | 1165 | 867 | 585 | 504 | 191 | 69 | 5906 |
| 2001-02 | 6 | 0 | 23 | 355 | 626 | 1056 | 1034 | 863 | 747 | 429 | 234 | 65 | 5438 |
| 2002-03 | 0 | 2 | 95 | 474 | 725 | 849 | 809 | 780 | 567 | 462 | 262 | 28 | 5053 |
| 2003-04 | 0 | 0 | 71 | 257 | 814 | 856 | 1128 | 866 | 515 | 328 | 206 | 45 | 5086 |
| 2004-05 | 0 | 10 | 95 | 351 | 796 | 909 | 990 | 779 | 591 | 430 | 183 | 102 | 5236 |
| 2005-06 | 0 | 3 | 104 | 344 | 819 | 1092 | 878 | 840 | 695 | 390 | 189 | 9 | 5363 |
| 2006-07 | 0 | 10 | 104 | 419 | 641 | 1024 | 1101 | 725 | 510 | 395 | 142 | 36 | 5107 |
| 2007-08 | 0 | 0 | 100 | 400 | 716 | 1003 | 1106 | 833 | 741 | 546 | 177 | 98 | 5720 |
| 2008-09 | 0 | 7 | 49 | 363 | 634 | 1032 | 1037 | 779 | 716 | 430 | 187 | 24 | 5258 |
| 2009- | 0 | 14 | 40 | 505 | 701 | 1179 | | | | | | | |

WBAN : 24131

## COOLING DEGREE DAYS (base 65°F) 2009 BOISE (KBOI)

| YEAR | JAN | FEB | MAR | APR | MAY | JUN | JUL | AUG | SEP | OCT | NOV | DEC | TOTAL |
|---|---|---|---|---|---|---|---|---|---|---|---|---|---|
| 1980 | 0 | 0 | 0 | 3 | 25 | 68 | 251 | 117 | 38 | 2 | 0 | 0 | 504 |
| 1981 | 0 | 0 | 0 | 1 | 3 | 52 | 205 | 296 | 101 | 0 | 0 | 0 | 658 |
| 1982 | 0 | 0 | 0 | 0 | 2 | 117 | 194 | 248 | 50 | 0 | 0 | 0 | 611 |
| 1983 | 0 | 0 | 0 | 0 | 55 | 50 | 180 | 313 | 26 | 0 | 0 | 0 | 624 |
| 1984 | 0 | 0 | 0 | 0 | 19 | 70 | 291 | 340 | 64 | 2 | 0 | 0 | 786 |
| 1985 | 0 | 0 | 0 | 2 | 28 | 125 | 402 | 165 | 4 | 0 | 0 | 0 | 726 |
| 1986 | 0 | 0 | 0 | 1 | 103 | 235 | 184 | 348 | 37 | 0 | 0 | 0 | 908 |
| 1987 | 0 | 0 | 0 | 23 | 61 | 202 | 223 | 180 | 111 | 0 | 0 | 0 | 800 |
| 1988 | 0 | 0 | 0 | 0 | 46 | 237 | 308 | 215 | 66 | 24 | 0 | 0 | 896 |
| 1989 | 0 | 0 | 0 | 6 | 14 | 140 | 376 | 191 | 56 | 0 | 0 | 0 | 783 |
| 1990 | 0 | 0 | 0 | 3 | 5 | 145 | 357 | 293 | 180 | 6 | 0 | 0 | 989 |
| 1991 | 0 | 0 | 0 | 0 | 0 | 36 | 337 | 368 | 85 | 6 | 0 | 0 | 832 |
| 1992 | 0 | 0 | 0 | 11 | 76 | 202 | 208 | 321 | 77 | 20 | 0 | 0 | 915 |
| 1993 | 0 | 0 | 0 | 0 | 86 | 68 | 61 | 162 | 71 | 8 | 0 | 0 | 456 |
| 1994 | 0 | 0 | 0 | 15 | 30 | 156 | 386 | 354 | 96 | 0 | 0 | 0 | 1037 |
| 1995 | 0 | 0 | 0 | 0 | 12 | 83 | 272 | 225 | 114 | 0 | 0 | 0 | 706 |
| 1996 | 0 | 0 | 0 | 0 | 4 | 111 | 372 | 314 | 73 | 22 | 0 | 0 | 896 |
| 1997 | 0 | 0 | 0 | 0 | 71 | 80 | 266 | 337 | 151 | 13 | 0 | 0 | 918 |
| 1998 | 0 | 0 | 0 | 4 | 9 | 42 | 449 | 371 | 201 | 3 | 0 | 0 | 1079 |
| 1999 | 0 | 0 | 0 | 0 | 32 | 142 | 282 | 334 | 80 | 1 | 0 | 0 | 871 |
| 2000 | 0 | 0 | 0 | 3 | 22 | 155 | 354 | 347 | 74 | 2 | 0 | 0 | 957 |
| 2001 | 0 | 0 | 0 | 4 | 85 | 146 | 303 | 433 | 131 | 7 | 0 | 0 | 1109 |
| 2002 | 0 | 0 | 0 | 0 | 61 | 208 | 460 | 193 | 105 | 0 | 0 | 0 | 1027 |
| 2003 | 0 | 0 | 0 | 0 | 76 | 173 | 488 | 391 | 131 | 49 | 0 | 0 | 1308 |
| 2004 | 0 | 0 | 0 | 0 | 19 | 179 | 411 | 323 | 61 | 5 | 0 | 0 | 998 |
| 2005 | 0 | 0 | 0 | 0 | 19 | 89 | 420 | 372 | 63 | 2 | 0 | 0 | 965 |
| 2006 | 0 | 0 | 0 | 0 | 73 | 196 | 518 | 297 | 125 | 0 | 0 | 0 | 1209 |
| 2007 | 0 | 0 | 0 | 4 | 70 | 211 | 569 | 333 | 99 | 1 | 0 | 0 | 1287 |
| 2008 | 0 | 0 | 0 | 0 | 45 | 168 | 401 | 355 | 85 | 14 | 0 | 0 | 1068 |
| 2009 | 0 | 0 | 0 | 7 | 87 | 114 | 428 | 338 | 214 | 0 | 0 | 0 | 1188 |

## SNOWFALL (inches) 2009 BOISE (KBOI)

| YEAR | JUL | AUG | SEP | OCT | NOV | DEC | JAN | FEB | MAR | APR | MAY | JUN | TOTAL |
|------|-----|-----|-----|-----|-----|-----|-----|-----|-----|-----|-----|-----|-------|
| 1980-81 | 0.0 | 0.0 | 0.0 | 0.0 | 3.2 | 1.7 | 3.6 | 0.7 | T | 0.5 | 0.0 | 0.0 | 9.7 |
| 1981-82 | 0.0 | 0.0 | 0.0 | 0.0 | 2.8 | 11.1 | 12.2 | 1.4 | 3.6 | 1.4 | 0.0 | 0.0 | 32.5 |
| 1982-83 | 0.0 | 0.0 | 0.0 | 0.0 | 2.1 | 6.4 | 1.6 | 0.9 | T | T | 0.8 | 0.0 | 11.8 |
| 1983-84 | 0.0 | 0.0 | 0.0 | 0.0 | 2.2 | 26.2 | 4.3 | 4.4 | T | 0.3 | T | 0.0 | 37.4 |
| 1984-85 | 0.0 | 0.0 | 0.0 | T | 0.2 | 7.7 | 2.6 | 5.3 | 1.5 | 1.0 | 0.0 | 0.0 | 18.3 |
| 1985-86 | 0.0 | 0.0 | 0.0 | T | 18.6 | 12.6 | 3.9 | 4.4 | 0.0 | T | 0.0 | 0.0 | 39.5 |
| 1986-87 | 0.0 | 0.0 | 0.0 | 0.0 | 5.9 | 0.9 | 0.5 | 0.6 | T | 0.0 | 0.0 | 0.0 | 7.9 |
| 1987-88 | 0.0 | 0.0 | 0.0 | 0.0 | 0.5 | 3.0 | 3.9 | 0.3 | 2.9 | 1.2 | 0.0 | 0.0 | 11.8 |
| 1988-89 | 0.0 | 0.0 | 0.0 | 0.0 | 2.5 | 10.8 | 8.0 | 0.7 | T | T | 0.0 | 0.0 | 22.0 |
| 1989-90 | 0.0 | T | 0.0 | T | 0.4 | T | 5.2 | 6.5 | 0.4 | T | T | 0.0 | 12.5 |
| 1990-91 | 0.0 | 0.0 | 0.0 | 0.0 | 0.1 | 15.7 | 1.5 | T | 1.2 | T | 0.0 | 0.0 | 18.5 |
| 1991-92 | 0.0 | 0.0 | 0.0 | 1.1 | 2.8 | T | 0.3 | T | 0.0 | T | 0.0 | 0.0 | 4.2 |
| 1992-93 | 0.0 | 0.0 | 0.0 | T | 3.6 | 5.7 | 14.6 | 10.4 | 0.2 | T | T | 0.0 | 34.5 |
| 1993-94 | 0.0 | 0.0 | 0.0 | 0.0 | 1.9 | 0.2 | 2.8 | 3.8 | T | T | 0.0 | 0.0 | 8.7 |
| 1994-95 | 0.0 | 0.0 | 0.0 | 0.0 | 8.8 | 8.5 | 1.5 | 5.7 | 1.2 | T | T | T | 25.7 |
| 1995-96 | T | 0.0 | 0.0 | 0.0 | 0.2 | 5.9 | 6.7 | 6.9 | 1.8 | T | T | 0.0 | 21.5 |
| 1996-97 | 0.0 | 0.0 | 0.0 | T | T | 13.0 | 3.4 | T | 0.3 | T | 0.0 | 0.0 | 16.7 |
| 1997-98 | 0.0 | 0.0 | 0.0 | 0.0 | 0.0 | 0.6 | 4.2 | 1.4 | 1.1 | T | 0.0 | 0.0 | 7.3 |
| 1998-99 | T | 0.0 | T | 0.0 | T | 8.8 | 1.7 | 5.7 | 6.7 | 0.4 | 0.0 | 0.0 | 23.3 |
| 1999-00 | 0.0 | 0.0 | 0.0 | 0.0 | 0.9 | 2.7 | 6.2 | 2.1 | 2.3 | 0.0 | T | 0.0 | 14.2 |
| 2000-01 | 0.0 | 0.0 | 0.0 | 0.0 | 4.0 | 1.6 | 8.9 | 3.1 | 1.3 | T | 0.0 | 0.0 | 18.9 |
| 2001-02 | 0.0 | 0.0 | 0.0 | 0.0 | 3.5 | 9.9 | 14.4 | T | 4.3 | 0.0 | 0.0 | 0.0 | 32.1 |
| 2002-03 | 0.0 | 0.0 | 0.0 | T | 0.0 | 1.7 | T | 0.4 | 0.8 | 2.1 | 0.0 | 0.0 | 5.0 |
| 2003-04 | 0.0 | 0.0 | 0.0 | T | 0.3 | 4.2 | 10.0 | 6.5 | 0.7 | 0.0 | 0.0 | 0.0 | 21.7 |
| 2004-05 | 0.0 | 0.0 | 0.0 | 0.0 | 1.0 | T | 1.4 | 0.4 | T | T | 0.0 | 0.0 | 2.8 |
| 2005-06 | 0.0 | 0.0 | 0.0 | 0.0 | 2.0 | 5.2 | 1.9 | T | 1.6 | T | 0.0 | 0.0 | 10.7 |
| 2006-07 | 0.0 | 0.0 | 0.0 | T | 1.2 | 1.9 | 1.9 | 3.3 | 0.5 | 0.0 | 0.0 | 0.0 | 8.8 |
| 2007-08 | 0.0 | 0.0 | 0.0 | 0.0 | 2.0 | 7.2 | 13.6 | 8.8 | 0.2 | T | 0.0 | T | 31.8 |
| 2008-09 | 0.0 | 0.0 | 0.0 | 1.7 | 0.3 | 20.2 | 7.6 | 1.0 | 2.8 | 0.1 | 0.0 | 0.0 | 33.7 |
| 2009- | 0.0 | 0.0 | 0.0 | T | 1.7 | 10.7 | | | | | | | |
| POR=<br>70 YRS | T | T | T | 0.1 | 2.1 | 5.9 | 6.5 | 3.5 | 1.7 | 0.5 | 0.1 | T | 20.4 |

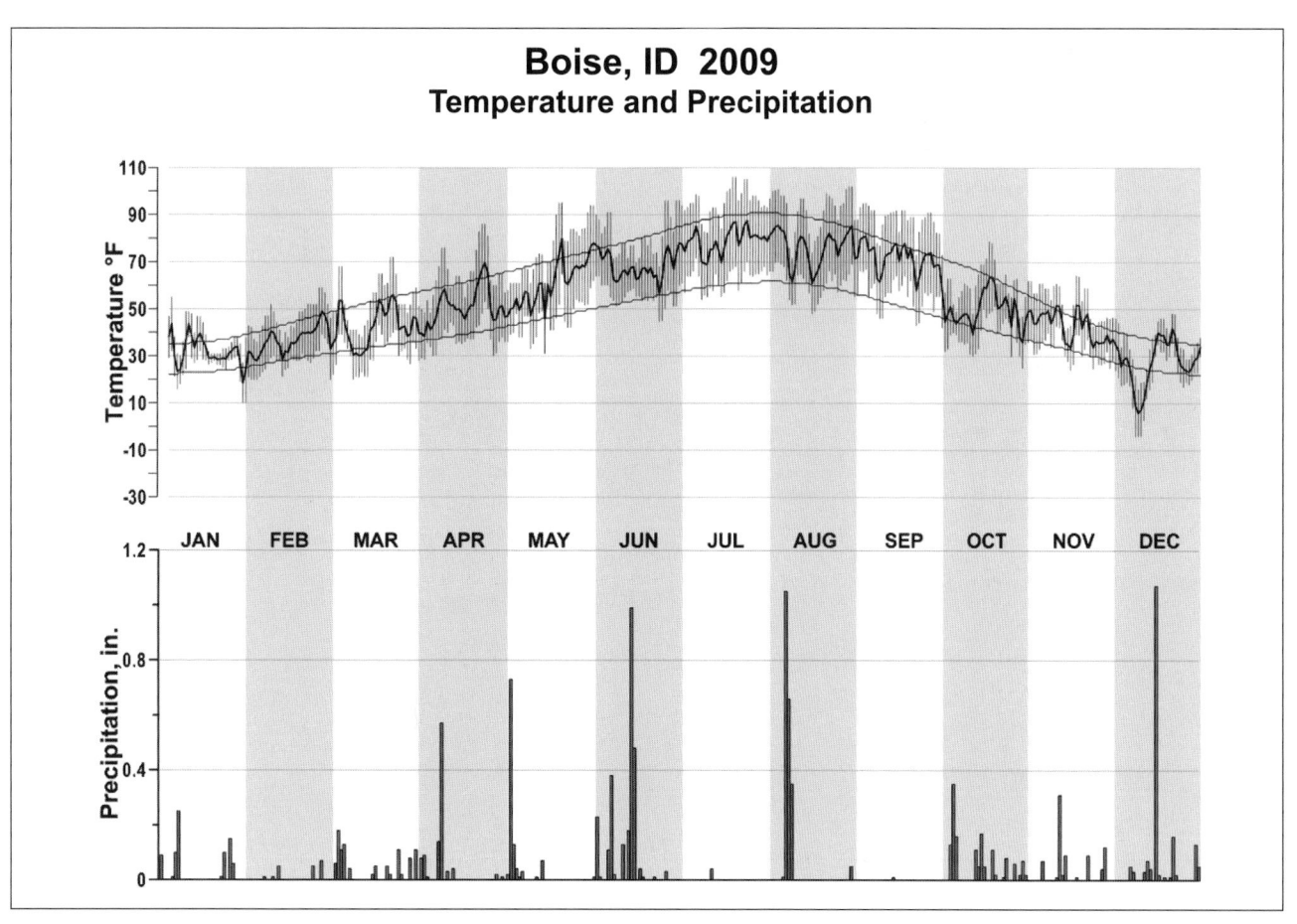

# Boise, ID 2009
## Temperature and Precipitation

# 2009
# CHICAGO
# ILLINOIS (KORD)

Chicago is located along the southwest shore of Lake Michigan and occupies a plain which, for the most part, is only some tens of feet above the lake. Lake Michigan averages 579 feet above sea level. Natural water drainage over most of the city would be into Lake Michigan, and from areas west of the city is into the Mississippi River System. But actual drainage over most of the city is artificially channeled also into the Mississippi system. Topography does not significantly affect air flow in or near the city except that lesser frictional drag over Lake Michigan causes winds to be frequently stronger along the lakeshore, and often permits air masses moving from the north to reach shore areas an hour or more before affecting western parts of the city.

Chicago is in a region of frequently changeable weather. The climate is predominately continental, ranging from relatively warm in summer to relatively cold in winter. However, the continentality is partially modified by Lake Michigan, and to a lesser extent by other Great Lakes. In late autumn and winter, air masses that are initially very cold often reach the city only after being tempered by passage over one or more of the lakes. Similarly, in late spring and summer, air masses reaching the city from the north, northeast, or east are cooler because of movement over the Great Lakes. Very low winter temperatures most often occur in air that flows southward to the west of Lake Superior before reaching the Chicago area. In summer the higher temperatures are with south or southwest flow and are therefore not influenced by the lakes, the only modifying effect being a local lake breeze. Strong south or southwest flow may overcome the lake breeze and cause high temperatures to extend over the entire city.

During the warm season, when the lake is cold relative to land, there is frequently a lake breeze that reduces daytime temperature near the shore, sometimes by 10 degrees or more below temperatures farther inland. When the breeze off the lake is light this effect usually reaches inland only a mile or two, but with stronger on-shore winds the whole city is cooled. On the other hand, temperatures at night are warmer near the lake so that 24-hour averages on the whole are only slightly different in various parts of the city and suburbs.

At the O'Hare International Airport temperatures of 96 degrees or higher occur in about half the summers, while about half the winters have a minimum as low as -15 degrees. The average occurrence of the first temperature as low as 32 degrees in the fall is mid-October and the average occurrence of the last temperature as low as 32 degrees in the spring is late April.

Precipitation falls mostly from air that has passed over the Gulf of Mexico. But in winter there is sometimes snowfall, light inland but locally heavy near the lakeshore, with Lake Michigan as the principal moisture source. The heavy lakeshore snow occurs when initially colder air moves from the north with a long trajectory over Lake Michigan and impinges on the Chicago lakeshore. In this situation the air mass is warmed and its moisture content increased up to a height of several thousand feet. Snowfall is produced by upward currents that become stronger, because of frictional effects, when the air moves from the lake onto land. This type of snowfall therefore tends to be heavier and to extend farther inland in south-shore areas of the city and in Indiana suburbs, where the angle between wind-flow and shoreline is greatest. The effect of Lake Michigan, both on winter temperatures and lake-produced snowfall, is enhanced by non-freezing of much of the lake during the winter, even though areas and harbors are often ice-choked.

Summer thunderstorms are often locally heavy and variable, parts of the city may receive substantial rainfall and other parts none. Longer periods of continuous precipitation are mostly in autumn, winter, and spring. About one-half the precipitation in winter, and about 10 percent of the yearly total precipitation, falls as snow. Snowfall from month to month and year to year is greatly variable. There is a 50 percent likelihood that the first and last 1-inch snowfall of a season will occur by December 5 and March 20, respectively.

Channeling of winds between tall buildings often causes locally stronger gusts in the central business area. However, the nickname, windy city, is a misnomer as the average wind speed is not greater than in many other parts of the U.S.

# NORMALS, MEANS, AND EXTREMES
## CHICAGO (KORD)

| LATITUDE: 41° 59'N | LONGITUDE: -87° 54'W | ELEVATION (FT): GRND: 662  BARO: 658 | TIME ZONE: CENTRAL  (UTC -6) | WBAN: 94846 |
|---|---|---|---|---|

| | ELEMENT | POR | JAN | FEB | MAR | APR | MAY | JUN | JUL | AUG | SEP | OCT | NOV | DEC | YEAR |
|---|---|---|---|---|---|---|---|---|---|---|---|---|---|---|---|
| **TEMPERATURE °F** | NORMAL DAILY MAXIMUM | 30 | 29.6 | 34.7 | 46.1 | 58.0 | 69.9 | 79.2 | 83.5 | 81.2 | 73.9 | 62.1 | 47.1 | 34.4 | 58.3 |
| | MEAN DAILY MAXIMUM | 51 | 29.9 | 34.3 | 45.6 | 58.7 | 70.0 | 79.4 | 83.6 | 82.0 | 75.0 | 62.8 | 48.2 | 34.7 | 58.7 |
| | HIGHEST DAILY MAXIMUM | 51 | 65 | 72 | 88 | 91 | 93 | 104 | 104 | 101 | 99 | 91 | 78 | 71 | 104 |
| | YEAR OF OCCURRENCE | | 2008 | 2000 | 1986 | 1980 | 1977 | 1988 | 1995 | 1991 | 1985 | 1963 | 1978 | 1982 | JUL 1995 |
| | MEAN OF EXTREME MAXS. | 51 | 50.3 | 54.7 | 71.9 | 81.1 | 87.2 | 92.9 | 94.8 | 93.0 | 89.7 | 81.5 | 68.1 | 55.8 | 76.8 |
| | NORMAL DAILY MINIMUM | 30 | 14.3 | 19.2 | 28.5 | 37.6 | 47.5 | 57.2 | 63.2 | 62.2 | 53.7 | 42.1 | 31.6 | 20.4 | 39.8 |
| | MEAN DAILY MINIMUM | 51 | 14.8 | 18.7 | 28.3 | 38.6 | 48.1 | 57.6 | 63.1 | 62.2 | 54.2 | 42.7 | 32.0 | 20.3 | 40.1 |
| | LOWEST DAILY MINIMUM | 51 | -27 | -19 | -8 | 7 | 24 | 36 | 40 | 41 | 28 | 17 | 1 | -25 | -27 |
| | YEAR OF OCCURRENCE | | 1985 | 1996 | 1962 | 1982 | 1966 | 1972 | 1965 | 1965 | 1974 | 1981 | 1976 | 1983 | JAN 1985 |
| | MEAN OF EXTREME MINS. | 51 | -7.2 | -2.5 | 10.2 | 23.5 | 34.4 | 43.7 | 50.8 | 51.1 | 39.0 | 27.5 | 15.8 | -1.2 | 23.8 |
| | NORMAL DRY BULB | 30 | 22.0 | 27.0 | 37.3 | 47.8 | 58.7 | 68.2 | 73.3 | 71.7 | 63.8 | 52.1 | 39.3 | 27.4 | 49.1 |
| | MEAN DRY BULB | 51 | 22.4 | 26.5 | 37.0 | 48.7 | 59.1 | 68.6 | 73.4 | 72.1 | 64.6 | 52.7 | 40.1 | 27.5 | 49.4 |
| | MEAN WET BULB | 26 | 21.9 | 24.7 | 33.0 | 41.9 | 51.5 | 61.0 | 65.7 | 65.1 | 58.0 | 46.8 | 36.1 | 25.7 | 44.3 |
| | MEAN DEW POINT | 26 | 18.5 | 21.3 | 28.8 | 37.1 | 47.5 | 57.6 | 62.7 | 62.5 | 54.7 | 43.1 | 32.6 | 22.4 | 40.7 |
| | NORMAL NO. DAYS WITH: | | | | | | | | | | | | | | |
| | MAXIMUM >= 90 | 30 | 0.0 | 0.0 | 0.0 | * | 0.8 | 3.8 | 7.2 | 4.3 | 1.7 | * | 0.0 | 0.0 | 17.8 |
| | MAXIMUM <= 32 | 30 | 17.2 | 11.9 | 3.7 | 0.1 | 0.0 | 0.0 | 0.0 | 0.0 | 0.0 | 0.0 | 2.1 | 10.9 | 45.9 |
| | MINIMUM <= 32 | 30 | 28.5 | 24.1 | 20.4 | 7.5 | 0.6 | 0.0 | 0.0 | 0.0 | 0.2 | 4.5 | 16.1 | 26.6 | 128.5 |
| | MINIMUM <= 0 | 30 | 5.8 | 2.6 | 0.1 | 0.0 | 0.0 | 0.0 | 0.0 | 0.0 | 0.0 | 0.0 | 0.0 | 2.4 | 10.9 |
| **H/C** | NORMAL HEATING DEG. DAYS | 30 | 1333 | 1075 | 858 | 513 | 232 | 49 | 6 | 9 | 112 | 401 | 759 | 1151 | 6498 |
| | NORMAL COOLING DEG. DAYS | 30 | 0 | 0 | 1 | 9 | 48 | 159 | 279 | 233 | 91 | 10 | 0 | 0 | 830 |
| **RH** | NORMAL (PERCENT) | 30 | 75 | 74 | 70 | 66 | 66 | 67 | 69 | 72 | 71 | 70 | 73 | 76 | 71 |
| | HOUR 00 LST | 30 | 77 | 78 | 76 | 73 | 75 | 77 | 79 | 82 | 81 | 77 | 77 | 79 | 78 |
| | HOUR 06 LST | 30 | 79 | 80 | 80 | 78 | 78 | 79 | 82 | 86 | 86 | 82 | 81 | 81 | 81 |
| | HOUR 12 LST | 30 | 69 | 66 | 61 | 56 | 54 | 56 | 56 | 58 | 57 | 57 | 65 | 70 | 60 |
| | HOUR 18 LST | 30 | 74 | 71 | 65 | 58 | 55 | 56 | 58 | 63 | 63 | 64 | 70 | 75 | 64 |
| **S** | PERCENT POSSIBLE SUNSHINE | 16 | 44 | 49 | 51 | 50 | 58 | 67 | 66 | 62 | 59 | 55 | 38 | 43 | 54 |
| **W/O** | MEAN NO. DAYS WITH: | | | | | | | | | | | | | | |
| | HEAVY FOG(VISBY <= 1/4 MI) | 46 | 1.8 | 1.7 | 1.9 | 0.6 | 1.0 | 0.5 | 0.3 | 0.5 | 0.5 | 0.7 | 1.0 | 1.8 | 12.3 |
| | THUNDERSTORMS | 51 | 0.4 | 0.5 | 2.1 | 3.9 | 5.3 | 6.5 | 6.0 | 6.0 | 4.1 | 1.7 | 1.0 | 0.5 | 38.0 |
| **CLOUDNESS** | MEAN: | | | | | | | | | | | | | | |
| | SUNRISE-SUNSET (OKTAS) | 38 | 5.5 | 5.4 | 5.8 | 5.5 | 5.0 | 4.7 | 4.4 | 4.4 | 4.5 | 4.7 | 5.7 | 5.6 | 5.1 |
| | MIDNIGHT-MIDNIGHT (OKTAS) | 32 | 5.3 | 5.0 | 5.1 | 5.0 | 4.5 | 4.3 | 4.0 | 4.0 | 4.2 | 4.3 | 5.4 | 5.5 | 4.7 |
| | MEAN NO. DAYS WITH: | | | | | | | | | | | | | | |
| | CLEAR | 38 | 6.8 | 6.0 | 4.9 | 6.0 | 7.2 | 7.3 | 8.0 | 8.6 | 8.5 | 8.6 | 5.2 | 5.7 | 82.8 |
| | PARTLY CLOUDY | 38 | 6.2 | 6.5 | 8.5 | 7.6 | 9.9 | 11.5 | 12.1 | 11.2 | 9.5 | 8.5 | 6.2 | 5.9 | 103.6 |
| | CLOUDY | 38 | 18.0 | 15.7 | 17.6 | 16.4 | 13.9 | 11.2 | 10.1 | 10.4 | 11.2 | 13.3 | 17.7 | 18.5 | 174.0 |
| **PR** | MEAN STATION PRESSURE(IN) | 26 | 29.34 | 29.35 | 29.31 | 29.24 | 29.25 | 29.25 | 29.27 | 29.31 | 29.33 | 29.33 | 29.32 | 29.34 | 29.30 |
| | MEAN SEA-LEVEL PRES. (IN) | 26 | 30.09 | 30.09 | 30.05 | 29.97 | 29.97 | 29.96 | 29.98 | 30.02 | 30.04 | 30.05 | 30.05 | 30.09 | 30.03 |
| **WINDS** | MEAN SPEED (MPH) | 26 | 11.3 | 11.1 | 11.3 | 11.2 | 10.1 | 8.8 | 8.3 | 8.0 | 8.4 | 9.7 | 10.7 | 10.6 | 10.0 |
| | PREVAIL.DIR(TENS OF DEGS) | 41 | 28 | 30 | 05 | 03 | 03 | 22 | 24 | 24 | 19 | 22 | 22 | 27 | 28 |
| | MAXIMUM 2-MINUTE: | | | | | | | | | | | | | | |
| | SPEED (MPH) | 13 | 37 | 38 | 41 | 43 | 43 | 39 | 48 | 46 | 37 | 43 | 41 | 48 | 48 |
| | DIR. (TENS OF DEGS) | | 09 | 29 | 02 | 26 | 33 | 30 | 17 | 28 | 28 | 07 | 23 | 19 | 19 |
| | YEAR OF OCCURRENCE | | 1999 | 2003 | 1998 | 1997 | 2004 | 1998 | 2005 | 2007 | 1997 | 1997 | 1998 | 2007 | DEC 2007 |
| | MAXIMUM 3-SECOND | | | | | | | | | | | | | | |
| | SPEED (MPH) | 13 | 49 | 56 | 54 | 56 | 58 | 54 | 61 | 59 | 48 | 52 | 57 | 63 | 63 |
| | DIR. (TENS OF DEGS) | | 18 | 24 | 23 | 25 | 34 | 24 | 20 | 25 | 28 | 07 | 24 | 20 | 20 |
| | YEAR OF OCCURRENCE | | 2008 | 1999 | 2004 | 1997 | 2004 | 2008 | 1999 | 2000 | 2006 | 1997 | 1998 | 2007 | DEC 2007 |
| **PRECIPITATION** | NORMAL (IN) | 30 | 1.75 | 1.63 | 2.65 | 3.68 | 3.38 | 3.63 | 3.51 | 4.62 | 3.27 | 2.71 | 3.01 | 2.43 | 36.27 |
| | MAXIMUM MONTHLY (IN) | 51 | 4.47 | 5.56 | 5.91 | 7.69 | 7.22 | 9.96 | 8.33 | 17.10 | 13.63 | 8.54 | 8.22 | 8.56 | 17.10 |
| | YEAR OF OCCURRENCE | | 1999 | 1997 | 1976 | 1983 | 2004 | 1993 | 1982 | 1987 | 2008 | 2001 | 1985 | 1982 | AUG 1987 |
| | MINIMUM MONTHLY (IN) | 51 | 0.10 | 0.12 | 0.63 | 0.74 | 0.30 | .76 | 1.18 | 0.51 | 0.02 | 0.16 | 0.44 | 0.23 | 0.02 |
| | YEAR OF OCCURRENCE | | 1981 | 1969 | 1981 | 2004 | 1992 | 2005 | 1977 | 1969 | 1979 | 1964 | 1999 | 1962 | SEP 1979 |
| | MAXIMUM IN 24 HOURS (IN) | 51 | 2.00 | 3.78 | 2.39 | 2.78 | 3.45 | 3.97 | 2.90 | 9.35 | 6.83 | 4.62 | 2.99 | 4.53 | 9.35 |
| | YEAR OF OCCURRENCE | | 1960 | 1997 | 1985 | 1983 | 1981 | 2009 | 1993 | 1987 | 2008 | 1969 | 1990 | 1982 | AUG 1987 |
| | NORMAL NO. DAYS WITH: | | | | | | | | | | | | | | |
| | PRECIPITATION >= 0.01 | 30 | 11.3 | 9.4 | 11.7 | 12.6 | 11.4 | 10.1 | 9.5 | 10.1 | 8.8 | 9.5 | 11.4 | 11.2 | 127.0 |
| | PRECIPITATION >= 1.00 | 30 | 0.2 | 0.2 | 0.4 | 0.9 | 0.6 | 1.0 | 1.0 | 1.3 | 0.9 | 0.5 | 0.5 | 0.2 | 8.1 |
| **SNOWFALL** | NORMAL (IN) | 30 | 11.3 | 8.3 | 6.0 | 1.6 | 0.* | 0.0 | 0.0 | 0.0 | 0.0 | 0.3 | 1.8 | 8.7 | 38.0 |
| | MAXIMUM MONTHLY (IN) | 50 | 34.3 | 26.2 | 24.7 | 11.1 | 1.6 | T | T | T | T | 6.6 | 10.4 | 35.3 | 35.3 |
| | YEAR OF OCCURRENCE | | 1979 | 1994 | 1965 | 1975 | 1966 | 2006 | 1995 | 2008 | 2006 | 1967 | 1959 | 1978 | DEC 1978 |
| | MAXIMUM IN 24 HOURS (IN) | 50 | 18.6 | 11.1 | 10.6 | 10.9 | 1.6 | T | T | T | T | 6.6 | 5.8 | 11.0 | 18.6 |
| | YEAR OF OCCURRENCE' | | 1999 | 2000 | 1970 | 1975 | 1966 | 1992 | 1995 | 1989 | 1967 | 1967 | 1975 | 1969 | JAN 1999 |
| | MAXIMUM SNOW DEPTH (IN) | 49 | 28 | 27 | 20 | 11 | 1 | 0 | 0 | 0 | 0 | 3 | 6 | 17 | 28 |
| | YEAR OF OCCURRENCE | | 1979 | 1967 | 1965 | 1975 | 1966 | | | | | 1989 | 1975 | 2000 | JAN 1979 |
| | NORMAL NO. DAYS WITH: | | | | | | | | | | | | | | |
| | SNOWFALL >= 1.0 | 30 | 3.5 | 2.7 | 2.0 | 0.4 | 0.0 | 0.0 | 0.0 | 0.0 | 0.0 | 0.1 | 0.6 | 2.6 | 11.9 |

## PRECIPITATION (inches) 2009 CHICAGO (KORD)

| YEAR | JAN | FEB | MAR | APR | MAY | JUN | JUL | AUG | SEP | OCT | NOV | DEC | ANNUAL |
|------|-----|-----|-----|-----|-----|-----|-----|-----|-----|-----|-----|-----|--------|
| 1980 | 1.04 | 1.24 | 1.96 | 3.41 | 3.22 | 3.42 | 3.56 | 8.54 | 5.65 | 2.09 | 1.10 | 3.43 | 38.66 |
| 1981 | 0.10 | 2.35 | 0.63 | 6.14 | 5.85 | 4.46 | 4.50 | 6.60 | 3.25 | 1.80 | 2.46 | 1.05 | 39.19 |
| 1982 | 2.90 | 0.41 | 4.15 | 2.78 | 2.08 | 1.56 | 8.33 | 3.93 | 1.15 | 1.88 | 6.95 | 8.56 | 44.68 |
| 1983 | 0.66 | 2.06 | 3.56 | 7.69 | 6.26 | 4.11 | 4.25 | 2.08 | 5.41 | 4.41 | 5.87 | 2.99 | 49.35 |
| 1984 | 1.15 | 1.39 | 3.00 | 4.11 | 4.49 | 2.02 | 3.19 | 2.10 | 3.84 | 3.15 | 2.64 | 2.92 | 34.00 |
| 1985 | 1.48 | 3.46 | 4.73 | 1.48 | 2.79 | 1.97 | 3.75 | 3.90 | 1.82 | 4.98 | 8.22 | 1.49 | 40.07 |
| 1986 | 0.39 | 2.58 | 1.49 | 1.85 | 3.11 | 3.49 | 4.30 | 1.15 | 7.12 | 3.75 | 1.41 | 1.09 | 31.73 |
| 1987 | 1.67 | 0.99 | 1.59 | 2.34 | 2.21 | 2.19 | 4.19 | 17.10 | 0.94 | 1.59 | 2.77 | 3.77 | 41.35 |
| 1988 | 1.88 | 1.29 | 2.15 | 2.08 | 1.19 | 1.05 | 2.74 | 3.29 | 3.79 | 5.05 | 6.45 | 2.40 | 33.36 |
| 1989 | 0.82 | 0.77 | 1.67 | 1.37 | 1.59 | 2.01 | 5.89 | 7.31 | 3.91 | 1.49 | 2.16 | 0.46 | 29.45 |
| 1990 | 1.97 | 2.25 | 3.09 | 1.79 | 6.85 | 4.50 | 2.25 | 7.75 | 1.03 | 4.10 | 5.60 | 1.94 | 43.12 |
| 1991 | 1.41 | 0.62 | 3.54 | 4.00 | 5.20 | 0.95 | 1.32 | 2.81 | 2.51 | 7.36 | 3.59 | 1.71 | 35.02 |
| 1992 | 0.87 | 1.39 | 2.67 | 2.21 | 0.30 | 1.35 | 3.77 | 3.56 | 4.31 | 1.79 | 5.41 | 2.49 | 30.12 |
| 1993 | 3.83 | 0.82 | 4.52 | 4.57 | 1.83 | 9.96 | 4.45 | 5.74 | 4.47 | 2.19 | 1.52 | 1.00 | 44.90 |
| 1994 | 1.77 | 2.56 | 1.09 | 2.20 | 0.58 | 6.09 | 1.62 | 4.05 | 1.04 | 3.23 | 3.75 | 1.61 | 29.59 |
| 1995 | 3.21 | 0.41 | 1.43 | 5.79 | 4.47 | 1.40 | 3.17 | 3.49 | 1.04 | 4.20 | 3.68 | 0.59 | 32.88 |
| 1996 | 1.58 | 0.71 | 0.95 | 2.59 | 6.95 | 4.80 | 3.95 | 1.45 | 2.73 | 2.32 | 1.48 | 1.21 | 30.72 |
| 1997 | 1.38 | 5.56 | 1.57 | 1.76 | 2.69 | 3.81 | 3.04 | 4.50 | 1.69 | 2.75 | 1.46 | 1.50 | 31.71 |
| 1998 | 2.67 | 1.70 | 4.29 | 3.56 | 3.02 | 2.90 | 1.75 | 6.88 | 2.34 | 5.22 | 2.00 | 1.20 | 37.53 |
| 1999 | 4.47 | 1.64 | 1.73 | 7.51 | 4.46 | 4.95 | 3.73 | 2.30 | 3.27 | 1.07 | 0.44 | 2.68 | 38.25 |
| 2000 | 1.35 | 1.97 | 1.18 | 5.15 | 4.02 | 4.32 | 3.58 | 2.26 | 3.59 | 1.12 | 2.71 | 2.11 | 33.36 |
| 2001 | 1.12 | 2.57 | 1.30 | 2.82 | 3.34 | 2.61 | 2.96 | 12.25 | 6.05 | 1.22 | 0.99 | 45.77 |
| 2002 | 1.20 | 0.96 | 2.73 | 3.00 | 4.39 | 4.61 | 2.68 | 8.06 | 1.72 | 1.60 | 1.04 | 1.93 | 33.92 |
| 2003 | 0.36 | 0.19 | 1.82 | 4.33 | 5.29 | 1.46 | 4.50 | 4.19 | 1.72 | 1.88 | 4.46 | 1.82 | 32.02 |
| 2004 | 0.91 | 0.71 | 2.68 | 0.74 | 7.22 | 2.82 | 2.66 | 5.30 | 0.26 | 2.85 | 4.28 | 1.15 | 31.58 |
| 2005 | 4.00 | 2.19 | 1.48 | 1.53 | 1.99 | 0.76 | 1.95 | 2.47 | 2.66 | 1.39 | 2.31 | 1.36 | 24.09 |
| 2006 | 2.76 | 1.80 | 2.73 | 3.60 | 3.65 | 4.05 | 3.70 | 2.95 | 5.85 | 4.04 | 3.65 | 3.18 | 41.96 |
| 2007 | 1.72 | 1.61 | 3.66 | 3.49 | 1.80 | 2.29 | 3.86 | 9.70 | 1.23 | 1.69 | 1.26 | 3.49 | 35.80 |
| 2008 | 1.93 | 3.53 | 2.63 | 2.72 | 4.10 | 4.18 | 4.76 | 3.73 | 13.63 | 2.07 | 1.81 | 5.77 | 50.86 |
| 2009 | 1.16 | 3.39 | 5.20 | 5.19 | 3.63 | 7.18 | 1.53 | 4.26 | 1.03 | 6.04 | 1.23 | 2.73 | 42.57 |
| POR= 51 YRS | 1.73 | 1.57 | 2.59 | 3.56 | 3.42 | 3.72 | 3.50 | 4.35 | 3.51 | 2.73 | 2.69 | 2.22 | 35.59 |

WBAN : 94846

## AVERAGE TEMPERATURE (°F) 2009 CHICAGO (KORD)

| YEAR | JAN | FEB | MAR | APR | MAY | JUN | JUL | AUG | SEP | OCT | NOV | DEC | ANNUAL |
|------|-----|-----|-----|-----|-----|-----|-----|-----|-----|-----|-----|-----|--------|
| 1980 | 23.4 | 21.5 | 32.6 | 46.5 | 59.7 | 65.3 | 75.7 | 75.7 | 66.0 | 48.4 | 39.9 | 28.0 | 48.6 |
| 1981 | 22.6 | 28.0 | 37.6 | 51.8 | 55.3 | 69.8 | 72.5 | 71.2 | 61.7 | 49.1 | 40.8 | 24.9 | 48.8 |
| 1982 | 12.2 | 21.5 | 35.1 | 44.5 | 64.3 | 62.1 | 74.1 | 68.8 | 62.1 | 53.2 | 39.1 | 36.0 | 47.8 |
| 1983 | 26.3 | 30.5 | 37.4 | 43.4 | 53.2 | 69.7 | 76.7 | 77.3 | 64.6 | 52.8 | 41.1 | 14.3 | 48.9 |
| 1984 | 17.1 | 33.9 | 29.5 | 45.8 | 55.5 | 70.3 | 70.3 | 72.8 | 61.1 | 54.7 | 37.9 | 31.0 | 48.3 |
| 1985 | 14.4 | 20.4 | 39.4 | 52.6 | 60.2 | 63.6 | 71.4 | 69.2 | 65.4 | 52.5 | 37.8 | 17.0 | 47.0 |
| 1986 | 22.8 | 24.0 | 40.4 | 51.5 | 59.5 | 66.3 | 74.9 | 68.5 | 66.8 | 53.7 | 36.0 | 30.6 | 49.6 |
| 1987 | 25.9 | 33.9 | 40.8 | 50.6 | 63.4 | 72.4 | 76.7 | 71.9 | 65.1 | 47.3 | 43.9 | 32.2 | 52.0 |
| 1988 | 19.8 | 22.7 | 38.1 | 48.2 | 61.0 | 71.7 | 76.8 | 76.8 | 65.9 | 46.1 | 41.7 | 27.7 | 49.7 |
| 1989 | 32.4 | 19.6 | 36.6 | 46.8 | 57.8 | 67.5 | 73.9 | 71.4 | 62.0 | 54.0 | 37.7 | 17.4 | 48.1 |
| 1990 | 33.9 | 31.3 | 41.3 | 49.9 | 56.2 | 69.6 | 71.7 | 71.9 | 65.9 | 51.6 | 44.7 | 28.6 | 51.4 |
| 1991 | 20.8 | 31.0 | 40.4 | 52.0 | 65.6 | 71.9 | 75.5 | 73.6 | 63.7 | 53.2 | 35.2 | 30.3 | 51.1 |
| 1992 | 28.1 | 33.3 | 37.5 | 46.1 | 56.9 | 64.9 | 69.3 | 67.0 | 62.7 | 50.4 | 38.3 | 28.6 | 48.6 |
| 1993 | 26.2 | 24.4 | 34.2 | 45.0 | 59.7 | 66.4 | 74.3 | 73.3 | 59.2 | 49.5 | 38.7 | 29.8 | 48.4 |
| 1994 | 15.9 | 22.1 | 38.5 | 51.1 | 58.2 | 70.2 | 73.4 | 68.7 | 66.8 | 54.7 | 44.4 | 34.8 | 49.9 |
| 1995 | 24.0 | 26.5 | 40.2 | 46.0 | 58.8 | 72.3 | 77.6 | 79.0 | 62.5 | 53.7 | 32.8 | 26.3 | 50.0 |
| 1996 | 23.4 | 26.0 | 30.8 | 45.2 | 55.0 | 68.0 | 69.9 | 72.3 | 63.5 | 51.9 | 33.4 | 27.7 | 47.3 |
| 1997 | 19.3 | 29.0 | 37.9 | 45.2 | 53.8 | 68.3 | 73.2 | 69.5 | 64.2 | 53.2 | 36.4 | 31.5 | 48.5 |
| 1998 | 29.6 | 38.7 | 39.0 | 49.8 | 64.8 | 69.3 | 74.5 | 73.5 | 67.7 | 55.5 | 44.8 | 34.7 | 53.5 |
| 1999 | 22.6 | 34.0 | 35.6 | 49.6 | 61.7 | 70.4 | 78.4 | 70.3 | 63.4 | 52.9 | 45.1 | 29.9 | 51.2 |
| 2000 | 25.3 | 34.1 | 44.2 | 47.2 | 62.0 | 67.3 | 71.1 | 72.4 | 64.7 | 56.1 | 37.0 | 16.0 | 49.8 |
| 2001 | 24.6 | 26.1 | 34.2 | 52.5 | 60.0 | 67.4 | 74.6 | 73.2 | 61.9 | 52.1 | 48.2 | 33.4 | 50.7 |
| 2002 | 31.9 | 32.2 | 34.6 | 49.9 | 55.2 | 71.0 | 77.1 | 73.2 | 67.4 | 49.8 | 37.6 | 30.2 | 50.8 |
| 2003 | 21.3 | 23.6 | 36.7 | 48.3 | 56.3 | 65.5 | 72.4 | 73.6 | 63.4 | 51.8 | 41.9 | 31.6 | 48.9 |
| 2004 | 20.3 | 27.4 | 41.2 | 50.4 | 60.0 | 67.1 | 71.2 | 67.5 | 66.5 | 53.9 | 43.7 | 28.9 | 49.8 |
| 2005 | 24.5 | 32.5 | 35.1 | 51.6 | 57.1 | 74.2 | 75.6 | 74.3 | 69.4 | 55.2 | 42.0 | 23.4 | 51.2 |
| 2006 | 35.8 | 28.2 | 38.3 | 53.1 | 59.6 | 68.3 | 76.5 | 74.3 | 62.4 | 49.0 | 42.9 | 33.8 | 51.9 |
| 2007 | 27.9 | 18.0 | 42.5 | 46.8 | 63.8 | 71.4 | 73.7 | 74.8 | 68.1 | 59.0 | 39.4 | 27.9 | 51.1 |
| 2008 | 23.5 | 23.1 | 34.9 | 49.5 | 55.5 | 70.8 | 74.0 | 72.7 | 66.2 | 52.7 | 39.3 | 22.9 | 48.8 |
| 2009 | 15.9 | 28.3 | 39.6 | 47.3 | 59.9 | 67.6 | 69.4 | 70.5 | 65.4 | 48.9 | 45.4 | 26.5 | 48.7 |
| POR= 51 YRS | 22.4 | 26.5 | 37.0 | 48.7 | 59.1 | 68.6 | 73.4 | 72.1 | 64.6 | 52.7 | 40.1 | 27.5 | 49.4 |

## HEATING DEGREE DAYS (base 65°F) 2009  CHICAGO (KORD)

| YEAR | JUL | AUG | SEP | OCT | NOV | DEC | JAN | FEB | MAR | APR | MAY | JUN | TOTAL |
|---|---|---|---|---|---|---|---|---|---|---|---|---|---|
| 1980-81 | 0 | 3 | 71 | 511 | 746 | 1140 | 1308 | 1031 | 846 | 397 | 313 | 6 | 6372 |
| 1981-82 | 8 | 6 | 135 | 489 | 719 | 1236 | 1632 | 1213 | 922 | 608 | 93 | 118 | 7179 |
| 1982-83 | 7 | 37 | 152 | 372 | 772 | 891 | 1194 | 961 | 847 | 643 | 364 | 38 | 6278 |
| 1983-84 | 16 | 0 | 125 | 383 | 714 | 1568 | 1479 | 894 | 1095 | 575 | 300 | 18 | 7167 |
| 1984-85 | 19 | 1 | 189 | 320 | 807 | 1046 | 1563 | 1245 | 787 | 418 | 183 | 103 | 6681 |
| 1985-86 | 0 | 6 | 141 | 380 | 813 | 1480 | 1302 | 1142 | 765 | 417 | 202 | 74 | 6722 |
| 1986-87 | 3 | 29 | 64 | 343 | 863 | 1060 | 1205 | 866 | 742 | 432 | 162 | 14 | 5783 |
| 1987-88 | 4 | 19 | 74 | 541 | 629 | 1011 | 1396 | 1221 | 828 | 503 | 176 | 40 | 6442 |
| 1988-89 | 0 | 9 | 63 | 583 | 693 | 1149 | 1003 | 1265 | 882 | 540 | 261 | 43 | 6491 |
| 1989-90 | 0 | 5 | 131 | 344 | 813 | 1471 | 956 | 938 | 733 | 491 | 271 | 33 | 6186 |
| 1990-91 | 10 | 5 | 103 | 425 | 605 | 1120 | 1365 | 945 | 756 | 393 | 142 | 13 | 5882 |
| 1991-92 | 0 | 0 | 163 | 367 | 887 | 1066 | 1137 | 913 | 847 | 560 | 284 | 77 | 6301 |
| 1992-93 | 9 | 37 | 136 | 449 | 795 | 1122 | 1196 | 1133 | 948 | 595 | 184 | 69 | 6673 |
| 1993-94 | 0 | 3 | 185 | 479 | 784 | 1084 | 1516 | 1197 | 817 | 433 | 253 | 51 | 6802 |
| 1994-95 | 1 | 23 | 63 | 322 | 611 | 932 | 1262 | 1074 | 760 | 561 | 199 | 25 | 5833 |
| 1995-96 | 1 | 0 | 150 | 349 | 958 | 1193 | 1284 | 1124 | 1054 | 589 | 343 | 58 | 7103 |
| 1996-97 | 9 | 0 | 119 | 399 | 940 | 1148 | 1410 | 1003 | 832 | 587 | 344 | 53 | 6844 |
| 1997-98 | 9 | 4 | 77 | 406 | 852 | 1030 | 1091 | 732 | 813 | 449 | 87 | 69 | 5619 |
| 1998-99 | 0 | 0 | 35 | 289 | 598 | 933 | 1309 | 860 | 903 | 456 | 149 | 34 | 5566 |
| 1999-00 | 1 | 4 | 110 | 368 | 591 | 1081 | 1224 | 892 | 640 | 528 | 148 | 57 | 5644 |
| 2000-01 | 6 | 3 | 112 | 286 | 833 | 1512 | 1248 | 1085 | 948 | 374 | 205 | 88 | 6700 |
| 2001-02 | 5 | 0 | 128 | 394 | 496 | 973 | 1018 | 913 | 934 | 490 | 331 | 33 | 5715 |
| 2002-03 | 0 | 0 | 58 | 473 | 811 | 1072 | 1346 | 1152 | 868 | 507 | 267 | 71 | 6625 |
| 2003-04 | 0 | 0 | 117 | 404 | 686 | 1025 | 1379 | 1086 | 730 | 445 | 200 | 47 | 6119 |
| 2004-05 | 3 | 39 | 47 | 338 | 632 | 1113 | 1248 | 906 | 926 | 401 | 256 | 8 | 5917 |
| 2005-06 | 0 | 0 | 37 | 340 | 681 | 1281 | 896 | 1024 | 821 | 358 | 231 | 28 | 5697 |
| 2006-07 | 0 | 0 | 115 | 497 | 659 | 960 | 1141 | 1310 | 697 | 543 | 116 | 18 | 6056 |
| 2007-08 | 0 | 1 | 56 | 234 | 763 | 1146 | 1280 | 1211 | 926 | 461 | 290 | 4 | 6372 |
| 2008-09 | 3 | 0 | 50 | 387 | 766 | 1302 | 1516 | 1023 | 781 | 526 | 171 | 71 | 6596 |
| 2009- | 9 | 18 | 50 | 493 | 583 | 1188 | | | | | | | |

WBAN : 94846

## COOLING DEGREE DAYS (base 65°F) 2009  CHICAGO (KORD)

| YEAR | JAN | FEB | MAR | APR | MAY | JUN | JUL | AUG | SEP | OCT | NOV | DEC | TOTAL |
|---|---|---|---|---|---|---|---|---|---|---|---|---|---|
| 1980 | 0 | 0 | 0 | 10 | 43 | 101 | 338 | 342 | 107 | 2 | 0 | 0 | 943 |
| 1981 | 0 | 0 | 0 | 9 | 20 | 157 | 248 | 204 | 44 | 0 | 0 | 0 | 682 |
| 1982 | 0 | 0 | 0 | 0 | 79 | 38 | 295 | 161 | 69 | 14 | 0 | 0 | 656 |
| 1983 | 0 | 0 | 1 | 0 | 4 | 189 | 385 | 388 | 122 | 10 | 0 | 0 | 1099 |
| 1984 | 0 | 0 | 0 | 5 | 11 | 184 | 190 | 254 | 77 | 8 | 0 | 0 | 729 |
| 1985 | 0 | 0 | 0 | 53 | 42 | 71 | 204 | 142 | 158 | 0 | 0 | 0 | 670 |
| 1986 | 0 | 0 | 7 | 17 | 37 | 118 | 318 | 145 | 123 | 3 | 0 | 0 | 768 |
| 1987 | 0 | 0 | 0 | 6 | 116 | 241 | 377 | 238 | 83 | 0 | 1 | 0 | 1062 |
| 1988 | 0 | 0 | 0 | 5 | 59 | 247 | 373 | 383 | 96 | 1 | 0 | 0 | 1164 |
| 1989 | 0 | 0 | 2 | 0 | 44 | 121 | 282 | 207 | 48 | 11 | 0 | 0 | 715 |
| 1990 | 0 | 0 | 7 | 43 | 8 | 179 | 226 | 224 | 137 | 11 | 1 | 0 | 836 |
| 1991 | 0 | 0 | 0 | 11 | 167 | 226 | 334 | 273 | 132 | 10 | 0 | 0 | 1153 |
| 1992 | 0 | 0 | 0 | 1 | 40 | 79 | 152 | 106 | 75 | 4 | 0 | 0 | 457 |
| 1993 | 0 | 0 | 0 | 0 | 28 | 118 | 294 | 266 | 19 | 5 | 0 | 0 | 730 |
| 1994 | 0 | 0 | 0 | 23 | 47 | 212 | 268 | 143 | 126 | 10 | 0 | 0 | 829 |
| 1995 | 0 | 0 | 0 | 0 | 13 | 254 | 398 | 445 | 81 | 8 | 0 | 0 | 1199 |
| 1996 | 0 | 0 | 0 | 0 | 41 | 154 | 166 | 235 | 79 | 2 | 0 | 0 | 677 |
| 1997 | 0 | 0 | 0 | 0 | 4 | 158 | 265 | 154 | 59 | 44 | 0 | 0 | 684 |
| 1998 | 0 | 0 | 13 | 0 | 88 | 205 | 301 | 267 | 123 | 5 | 0 | 0 | 1002 |
| 1999 | 0 | 0 | 0 | 0 | 52 | 201 | 422 | 176 | 70 | 2 | 2 | 0 | 925 |
| 2000 | 0 | 0 | 3 | 0 | 63 | 131 | 199 | 240 | 112 | 18 | 0 | 0 | 766 |
| 2001 | 0 | 0 | 0 | 7 | 58 | 168 | 309 | 262 | 44 | 1 | 0 | 0 | 849 |
| 2002 | 0 | 0 | 0 | 44 | 37 | 220 | 380 | 259 | 134 | 11 | 0 | 0 | 1085 |
| 2003 | 0 | 0 | 0 | 13 | 5 | 93 | 235 | 273 | 74 | 2 | 0 | 0 | 695 |
| 2004 | 0 | 0 | 0 | 13 | 55 | 118 | 203 | 121 | 100 | 5 | 0 | 0 | 615 |
| 2005 | 0 | 0 | 1 | 6 | 17 | 291 | 334 | 297 | 177 | 43 | 0 | 0 | 1166 |
| 2006 | 0 | 0 | 0 | 10 | 72 | 134 | 365 | 298 | 46 | 9 | 0 | 0 | 934 |
| 2007 | 0 | 0 | 6 | 3 | 85 | 215 | 279 | 313 | 154 | 56 | 0 | 0 | 1111 |
| 2008 | 0 | 0 | 0 | 0 | 5 | 187 | 290 | 246 | 90 | 10 | 0 | 0 | 828 |
| 2009 | 0 | 0 | 0 | 2 | 18 | 157 | 150 | 194 | 68 | 0 | 0 | 0 | 589 |

## SNOWFALL (inches)  2009  CHICAGO (KORD)

| YEAR | JUL | AUG | SEP | OCT | NOV | DEC | JAN | FEB | MAR | APR | MAY | JUN | TOTAL |
|---|---|---|---|---|---|---|---|---|---|---|---|---|---|
| 1980-81 | 0.0 | 0.0 | 0.0 | T | 5.1 | 9.7 | 2.0 | 15.9 | 2.3 | 0.0 | 0.0 | 0.0 | 35.0 |
| 1981-82 | 0.0 | 0.0 | 0.0 | T | 3.6 | 4.9 | 21.1 | 4.8 | 14.3 | 10.6 | 0.0 | 0.0 | 59.3 |
| 1982-83 | 0.0 | 0.0 | 0.0 | 0.0 | 0.4 | 2.1 | 5.0 | 8.9 | 9.0 | 1.2 | 0.0 | 0.0 | 26.6 |
| 1983-84 | 0.0 | 0.0 | 0.0 | 0.0 | 1.0 | 16.5 | 17.2 | 1.9 | 9.7 | 2.7 | 0.0 | 0.0 | 49.0 |
| 1984-85 | 0.0 | 0.0 | 0.0 | 0.0 | T | 6.6 | 18.9 | 13.3 | 0.3 | T | 0.0 | 0.0 | 39.1 |
| 1985-86 | 0.0 | 0.0 | 0.0 | 0.0 | 1.1 | 5.2 | 6.9 | 10.9 | 4.1 | 0.8 | 0.0 | 0.0 | 29.0 |
| 1986-87 | 0.0 | 0.0 | 0.0 | T | 3.8 | 0.4 | 17.3 | T | 4.7 | T | 0.0 | 0.0 | 26.2 |
| 1987-88 | 0.0 | 0.0 | 0.0 | 0.1 | 1.0 | 18.7 | 5.4 | 15.5 | 1.9 | T | 0.0 | 0.0 | 42.6 |
| 1988-89 | 0.0 | 0.0 | 0.0 | T | 0.9 | 5.0 | 0.4 | 15.1 | 2.0 | 0.6 | 0.5 | 0.0 | 24.5 |
| 1989-90 | 0.0 | T | 0.0 | 6.3 | 3.9 | 5.4 | 3.2 | 13.6 | 1.3 | 0.1 | T | 0.0 | 33.8 |
| 1990-91 | 0.0 | 0.0 | 0.0 | T | T | 3.2 | 11.1 | 3.3 | 5.9 | T | 0.0 | 0.0 | 23.5 |
| 1991-92 | 0.0 | 0.0 | 0.0 | T | 1.2 | 7.6 | 5.6 | 1.3 | 11.6 | 1.1 | 0.0 | T | 28.4 |
| 1992-93 | T | 0.0 | 0.0 | 0.3 | 0.2 | 5.7 | 15.2 | 8.0 | 13.8 | 3.7 | 0.0 | 0.0 | 46.9 |
| 1993-94 | 0.0 | 0.0 | 0.0 | T | 0.2 | 1.2 | 14.2 | 26.2 | T | T | 0.0 | 0.0 | 41.8 |
| 1994-95 | 0.0 | 0.0 | 0.0 | 0.0 | T | 7.0 | 13.1 | 0.4 | 3.5 | 0.1 | 0.0 | 0.0 | 24.1 |
| 1995-96 | T | 0.0 | 0.0 | T | 3.9 | 9.9 | 5.9 | 0.3 | 3.9 | T | T | 0.0 | 23.9 |
| 1996-97 | 0.0 | 0.0 | | | | | | | | | | | |
| 1997-98 | | | | | | | 11.0 | T | 8.2 | 0.0 | 0.0 | T | |
| 1998-99 | 0.0 | 0.0 | 0.0 | 0.0 | 0.2 | 1.0 | 29.6 | 1.9 | 18.2 | 0.0 | 0.0 | 0.0 | 50.9 |
| 1999-00 | 0.0 | 0.0 | 0.0 | 0.0 | 0.0 | 3.5 | 13.6 | 11.6 | T | 1.6 | 0.0 | T | 30.3 |
| 2000-01 | 0.0 | 0.0 | 0.0 | T | 0.1 | 30.9 | 1.5 | 2.2 | 4.2 | 0.3 | T | T | 39.2 |
| 2001-02 | T | 0.0 | 0.0 | T | 0.0 | 1.6 | 15.5 | 1.8 | 11.2 | 1.0 | 0.0 | T | 31.1 |
| 2002-03 | 0.0 | 0.0 | 0.0 | 0.0 | 4.7 | 8.0 | 4.3 | 1.5 | 7.1 | 3.0 | 0.0 | 0.0 | 28.6 |
| 2003-04 | T | T | 0.0 | 0.0 | T | 1.5 | 14.6 | 6.5 | 2.2 | 0.0 | 0.0 | 0.0 | 24.8 |
| 2004-05 | 0.0 | 0.0 | 0.0 | 0.0 | 5.1 | 0.6 | 27.8 | 2.7 | 3.2 | T | T | 0.0 | 39.4 |
| 2005-06 | 0.0 | 0.0 | 0.0 | T | 1.9 | 10.4 | 5.5 | 2.5 | 5.6 | T | 0.0 | T | 25.9 |
| 2006-07 | 0.0 | 0.0 | T | 0.3 | 0.4 | 5.8 | 3.5 | 20.3 | 2.3 | 3.0 | 0.0 | 0.0 | 35.6 |
| 2007-08 | 0.0 | 0.0 | 0.0 | 0.0 | 0.3 | 17.6 | 12.7 | 21.8 | 7.9 | T | 0.0 | 0.0 | 60.3 |
| 2008-09 | 0.0 | T | 0.0 | T | 0.6 | 21.9 | 21.5 | 4.5 | 2.1 | 2.1 | 0.0 | 0.0 | 52.7 |
| 2009- | 0.0 | 0.0 | 0.0 | T | T | 20.8 | | | | | | | |
| POR=<br>50 YRS | T | T | T | 0.3 | 1.7 | 8.8 | 11.3 | 7.8 | 6.5 | 1.4 | T | T | 37.8 |

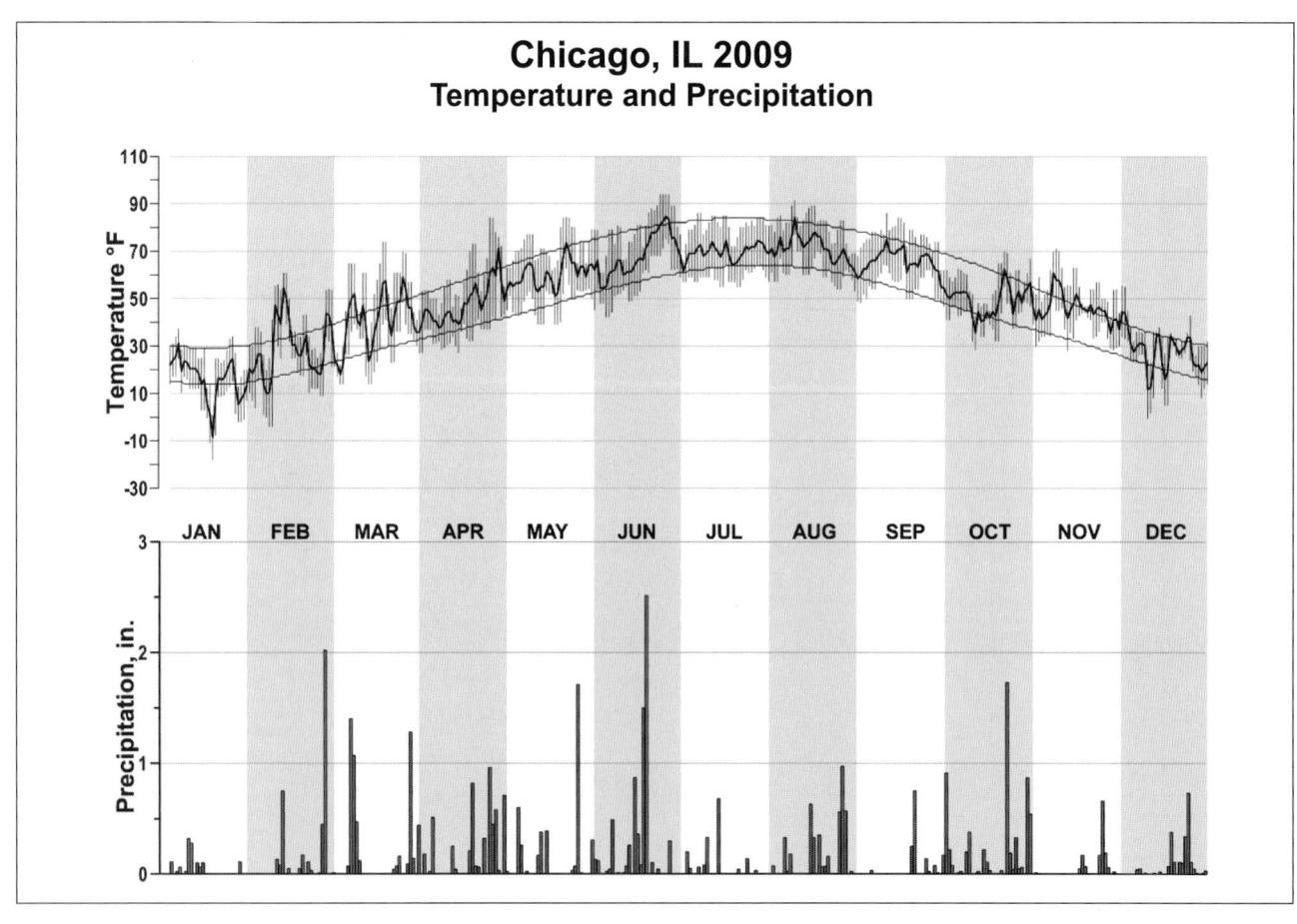

# Chicago, IL 2009
## Temperature and Precipitation

# 2009
# PEORIA
# ILLINOIS (KPIA)

The airport station is situated on a rather level tableland surrounded by well-drained and gently rolling terrain. It is set back a mile from the rim of the Illinois River Valley and is almost 200 feet above the river bed. Exposures of all instruments are good. The climate of this area is typically continental as shown by its changeable weather and the wide range of temperature extremes.

June and September are usually the most pleasant months of the year. Then during October or the first of November, Indian Summer is often experienced with an extended period of warm, dry weather.

Precipitation is normally heaviest during the growing season and lowest during midwinter.

The earliest snowfalls have occurred in September and the latest in the spring have occurred as late as May. Heavy snowfalls have rarely exceeded 20 inches.

Based on the 1951-1980 period, the average first occurrence of 32 degrees Fahrenheit in the fall is October 20 and the average last occurrence in the spring is April 24.

# NORMALS, MEANS, AND EXTREMES
## PEORIA (KPIA)

LATITUDE: 40 ° 40'N  LONGITUDE: -89 ° 41'W  ELEVATION (FT): GRND: 652  BARO: 716  TIME ZONE: CENTRAL (UTC -6)  WBAN: 14842

| | ELEMENT | POR | JAN | FEB | MAR | APR | MAY | JUN | JUL | AUG | SEP | OCT | NOV | DEC | YEAR |
|---|---|---|---|---|---|---|---|---|---|---|---|---|---|---|---|
| **TEMPERATURE °F** | NORMAL DAILY MAXIMUM | 30 | 30.7 | 36.6 | 49.4 | 62.0 | 73.0 | 82.2 | 85.7 | 83.6 | 76.7 | 64.4 | 48.8 | 35.5 | 60.7 |
| | MEAN DAILY MAXIMUM | 117 | 32.3 | 35.8 | 48.5 | 61.4 | 73.0 | 82.1 | 86.6 | 84.5 | 77.1 | 65.2 | 49.2 | 36.2 | 61.0 |
| | HIGHEST DAILY MAXIMUM | 71 | 70 | 72 | 86 | 92 | 94 | 105 | 104 | 103 | 100 | 93 | 81 | 71 | 105 |
| | YEAR OF OCCURRENCE | | 1989 | 1976 | 1986 | 1986 | 2006 | 1988 | 2005 | 1988 | 1953 | 2006 | 1950 | 1982 | JUN 1988 |
| | MEAN OF EXTREME MAXS. | 117 | 53.3 | 57.9 | 72.9 | 82.0 | 88.1 | 94.1 | 96.3 | 94.8 | 91.2 | 83.1 | 70.5 | 57.4 | 78.5 |
| | NORMAL DAILY MINIMUM | 30 | 14.3 | 19.7 | 30.2 | 40.3 | 50.8 | 60.1 | 64.6 | 62.6 | 54.0 | 42.3 | 31.4 | 20.1 | 40.9 |
| | MEAN DAILY MINIMUM | 117 | 16.5 | 19.5 | 30.1 | 40.7 | 51.2 | 60.6 | 65.0 | 63.2 | 55.1 | 43.7 | 31.9 | 21.2 | 41.6 |
| | LOWEST DAILY MINIMUM | 71 | -25 | -19 | -10 | 14 | 25 | 39 | 47 | 41 | 26 | 19 | -2 | -23 | -25 |
| | YEAR OF OCCURRENCE | | 1977 | 1996 | 1960 | 1982 | 1966 | 1993 | 1972 | 1986 | 1942 | 1972 | 1977 | 1989 | JAN 1977 |
| | MEAN OF EXTREME MINS. | 117 | -6.5 | -1.3 | 11.0 | 25.6 | 35.7 | 46.8 | 53.3 | 50.8 | 38.6 | 27.2 | 14.3 | -1.0 | 24.5 |
| | NORMAL DRY BULB | 30 | 22.5 | 28.2 | 39.8 | 51.2 | 61.9 | 71.1 | 75.1 | 73.1 | 65.4 | 53.4 | 40.1 | 27.8 | 50.8 |
| | MEAN DRY BULB | 117 | 24.4 | 27.6 | 39.3 | 51.1 | 62.1 | 71.4 | 75.8 | 73.8 | 66.2 | 54.5 | 40.5 | 28.7 | 51.3 |
| | MEAN WET BULB | 26 | 22.9 | 26.2 | 35.3 | 45.2 | 55.1 | 64.0 | 68.1 | 66.8 | 58.9 | 47.6 | 36.8 | 26.4 | 46.1 |
| | MEAN DEW POINT | 26 | 19.9 | 23.1 | 31.1 | 40.4 | 51.5 | 61.1 | 65.8 | 64.6 | 55.9 | 44.1 | 33.7 | 23.6 | 42.9 |
| | NORMAL NO. DAYS WITH: | | | | | | | | | | | | | | |
| | MAXIMUM >= 90 | 30 | 0.0 | 0.0 | 0.0 | 0.1 | 0.5 | 4.4 | 8.6 | 5.8 | 1.9 | 0.0 | 0.0 | 0.0 | 21.3 |
| | MAXIMUM <= 32 | 30 | 16.4 | 10.6 | 2.7 | 0.1 | 0.0 | 0.0 | 0.0 | 0.0 | 0.0 | 0.0 | 2.1 | 11.0 | 42.9 |
| | MINIMUM <= 32 | 30 | 29.2 | 24.3 | 18.1 | 5.4 | 0.3 | 0.0 | 0.0 | 0.0 | 0.0 | 0.1 | 4.4 | 16.8 | 125.3 |
| | MINIMUM <= 0 | 30 | 5.8 | 3.1 | 0.1 | 0.0 | 0.0 | 0.0 | 0.0 | 0.0 | 0.0 | 0.0 | 0.1 | 2.6 | 11.7 |
| **H/C** | NORMAL HEATING DEG. DAYS | 30 | 1316 | 1045 | 788 | 423 | 159 | 19 | 2 | 7 | 94 | 368 | 738 | 1138 | 6097 |
| | NORMAL COOLING DEG. DAYS | 30 | 0 | 0 | 1 | 11 | 64 | 210 | 325 | 263 | 112 | 12 | 0 | 0 | 998 |
| **RH** | NORMAL (PERCENT) | 30 | 76 | 75 | 70 | 65 | 67 | 69 | 73 | 75 | 73 | 71 | 75 | 79 | 72 |
| | HOUR 00 LST | 30 | 80 | 80 | 75 | 72 | 76 | 78 | 83 | 85 | 83 | 79 | 80 | 82 | 79 |
| | HOUR 06 LST | 30 | 81 | 82 | 81 | 79 | 82 | 83 | 88 | 90 | 89 | 85 | 84 | 84 | 84 |
| | HOUR 12 LST | 30 | 70 | 68 | 60 | 55 | 57 | 57 | 59 | 62 | 58 | 57 | 66 | 72 | 62 |
| | HOUR 18 LST | 30 | 74 | 72 | 62 | 55 | 57 | 57 | 61 | 65 | 63 | 64 | 71 | 77 | 65 |
| **S** | PERCENT POSSIBLE SUNSHINE | 52 | 47 | 50 | 51 | 55 | 60 | 67 | 69 | 67 | 64 | 61 | 43 | 42 | 56 |
| **W/O** | MEAN NO. DAYS WITH: | | | | | | | | | | | | | | |
| | HEAVY FOG(VISBY <= 1/4 MI) | 46 | 3.0 | 2.5 | 2.0 | 0.8 | 1.0 | 0.6 | 1.0 | 1.4 | 1.5 | 1.7 | 1.9 | 3.7 | 21.1 |
| | THUNDERSTORMS | 104 | 0.5 | 0.6 | 2.6 | 4.6 | 6.8 | 8.3 | 7.6 | 6.8 | 4.7 | 2.3 | 1.3 | 0.5 | 46.6 |
| **CLOUDNESS** | MEAN: | | | | | | | | | | | | | | |
| | SUNRISE-SUNSET (OKTAS) | | | | | | | | | | | | | | |
| | MIDNIGHT-MIDNIGHT (OKTAS) | | | | | | | | | | | | | | |
| | MEAN NO. DAYS WITH: | | | | | | | | | | | | | | |
| | CLEAR | | | | | | | | | | | | | | |
| | PARTLY CLOUDY | | | | | | | | | | | | | | |
| | CLOUDY | | | | | | | | | | | | | | |
| **PR** | MEAN STATION PRESSURE(IN) | 26 | 29.38 | 29.38 | 29.33 | 29.25 | 29.25 | 29.25 | 29.29 | 29.32 | 29.34 | 29.35 | 29.35 | 29.39 | 29.32 |
| | MEAN SEA-LEVEL PRES. (IN) | 26 | 30.13 | 30.12 | 30.05 | 29.96 | 29.96 | 29.95 | 29.98 | 30.02 | 30.04 | 30.06 | 30.07 | 30.12 | 30.04 |
| **WINDS** | MEAN SPEED (MPH) | 26 | 9.8 | 9.6 | 10.4 | 10.3 | 8.6 | 7.5 | 6.7 | 6.2 | 6.7 | 8.2 | 9.4 | 9.3 | 8.6 |
| | PREVAIL.DIR(TENS OF DEGS) | 41 | 30 | 31 | 19 | 19 | 19 | 19 | 19 | 19 | 19 | 19 | 19 | 19 | 19 |
| | MAXIMUM 2-MINUTE: | | | | | | | | | | | | | | |
| | SPEED (MPH) | 15 | 38 | 39 | 54 | 45 | 45 | 44 | 37 | 44 | 35 | 44 | 44 | 33 | 54 |
| | DIR. (TENS OF DEGS) | | 23 | 25 | 23 | 24 | 21 | 27 | 27 | 27 | 26 | 27 | 23 | 26 | 23 |
| | YEAR OF OCCURRENCE | | 1996 | 2001 | 1998 | 1997 | 2003 | 2009 | 2003 | 2008 | 2001 | 1996 | 1998 | 2009 | MAR 1998 |
| | MAXIMUM 3-SECOND | | | | | | | | | | | | | | |
| | SPEED (MPH) | 15 | 48 | 52 | 64 | 56 | 64 | 59 | 51 | 59 | 60 | 60 | 58 | 47 | 64 |
| | DIR. (TENS OF DEGS) | | 28 | 28 | 20 | 24 | 23 | 22 | 32 | 23 | 27 | 23 | 24 | 26 | 23 |
| | YEAR OF OCCURRENCE | | 2008 | 2001 | 1998 | 1997 | 2003 | 2001 | 2003 | 2009 | 2007 | 1996 | 1998 | 2009 | MAY 2003 |
| **PRECIPITATION** | NORMAL (IN) | 30 | 1.50 | 1.67 | 2.83 | 3.56 | 4.17 | 3.84 | 4.02 | 3.16 | 3.12 | 2.77 | 2.99 | 2.40 | 36.03 |
| | MAXIMUM MONTHLY (IN) | 71 | 8.11 | 5.37 | 7.49 | 8.66 | 10.19 | 11.69 | 10.15 | 8.61 | 13.09 | 10.80 | 7.62 | 6.34 | 13.09 |
| | YEAR OF OCCURRENCE | | 1965 | 1997 | 2009 | 1947 | 1995 | 1974 | 1993 | 1965 | 1961 | 1941 | 1985 | 1949 | SEP 1961 |
| | MINIMUM MONTHLY (IN) | 71 | 0.22 | 0.33 | 0.39 | 0.71 | .69 | 0.60 | 0.33 | 0.25 | 0.03 | 0.03 | 0.22 | 0.32 | 0.03 |
| | YEAR OF OCCURRENCE | | 1986 | 1947 | 1958 | 1971 | 2005 | 1988 | 1988 | 1992 | 1979 | 1964 | 1999 | 1995 | SEP 1979 |
| | MAXIMUM IN 24 HOURS (IN) | 71 | 4.45 | 3.34 | 3.39 | 5.06 | 3.62 | 4.44 | 4.22 | 4.32 | 4.72 | 3.70 | 4.32 | 3.38 | 5.06 |
| | YEAR OF OCCURRENCE | | 1965 | 1997 | 1944 | 1950 | 1956 | 1974 | 1993 | 1955 | 2008 | 1969 | 1990 | 1949 | APR 1950 |
| | NORMAL NO. DAYS WITH: | | | | | | | | | | | | | | |
| | PRECIPITATION >= 0.01 | 30 | 9.9 | 8.4 | 10.9 | 11.7 | 11.6 | 9.6 | 9.2 | 9.3 | 8.4 | 9.2 | 10.2 | 10.1 | 118.5 |
| | PRECIPITATION >= 1.00 | 30 | 0.2 | 0.3 | 0.4 | 0.8 | 0.8 | 1.0 | 0.8 | 1.3 | 0.8 | 0.9 | 0.6 | 0.6 | 8.3 |
| **SNOWFALL** | NORMAL (IN) | 30 | 8.2 | 5.6 | 3.2 | 1.0 | 0.* | 0.0 | 0.0 | 0.0 | 0.0 | 0.* | 2.1 | 7.1 | 27.2 |
| | MAXIMUM MONTHLY (IN) | 66 | 24.7 | 17.0 | 16.9 | 13.4 | 0.1 | T | T | 0.0 | T | 1.8 | 9.1 | 21.7 | 24.7 |
| | YEAR OF OCCURRENCE | | 1979 | 2008 | 1960 | 1982 | 1966 | 2000 | 1990 | | 1992 | 1967 | 1974 | 1977 | JAN 1979 |
| | MAXIMUM IN 24 HOURS (IN) | 66 | 12.2 | 7.6 | 9.0 | 6.1 | 0.1 | T | T | 0.0 | T | 1.8 | 7.2 | 10.2 | 12.2 |
| | YEAR OF OCCURRENCE' | | 1979 | 1944 | 1946 | 1982 | 1966 | 2000 | 1990 | | 1992 | 1967 | 1951 | 1973 | JAN 1979 |
| | MAXIMUM SNOW DEPTH (IN) | 103 | 20 | 17 | 10 | 10 | 0 | 0 | 0 | 0 | 1 | 7 | 13 | 20 |
| | YEAR OF OCCURRENCE | | 1979 | 1979 | 1960 | 1982 | | | | | 1942 | 1929 | 1975 | 1973 | JAN 1979 |
| | NORMAL NO. DAYS WITH: | | | | | | | | | | | | | | |
| | SNOWFALL >= 1.0 | 30 | 2.6 | 1.8 | 1.1 | 0.2 | 0.0 | 0.0 | 0.0 | 0.0 | 0.0 | 0.0 | 0.6 | 1.9 | 8.2 |

## PRECIPITATION (inches) 2009  PEORIA (KPIA)

| YEAR | JAN | FEB | MAR | APR | MAY | JUN | JUL | AUG | SEP | OCT | NOV | DEC | ANNUAL |
|------|-----|-----|-----|-----|-----|-----|-----|-----|-----|-----|-----|-----|--------|
| 1980 | 0.59 | 1.06 | 2.79 | 2.78 | 2.05 | 8.94 | 1.43 | 6.16 | 4.09 | 2.44 | 0.67 | 2.25 | 35.25 |
| 1981 | 0.48 | 2.41 | 0.92 | 5.71 | 5.77 | 6.22 | 7.08 | 5.61 | 1.31 | 1.37 | 1.64 | 1.24 | 39.76 |
| 1982 | 2.88 | 1.13 | 4.80 | 5.40 | 3.15 | 3.15 | 7.53 | 3.97 | 1.24 | 1.47 | 4.95 | 5.45 | 45.12 |
| 1983 | 0.53 | 1.01 | 2.84 | 7.06 | 6.66 | 4.48 | 1.99 | 1.09 | 5.08 | 3.01 | 5.58 | 2.65 | 41.98 |
| 1984 | 0.59 | 2.28 | 3.95 | 5.18 | 4.84 | 2.90 | 5.02 | 0.78 | 2.38 | 5.07 | 3.95 | 3.82 | 40.76 |
| 1985 | 0.99 | 2.62 | 5.77 | 1.14 | 3.14 | 5.11 | 3.43 | 3.70 | 3.43 | 4.61 | 7.62 | 2.24 | 43.80 |
| 1986 | 0.22 | 1.79 | 0.87 | 1.39 | 2.95 | 6.53 | 7.00 | 1.74 | 6.39 | 4.64 | 1.32 | 2.60 | 37.44 |
| 1987 | 1.49 | 0.84 | 1.98 | 1.84 | 1.69 | 3.27 | 2.90 | 4.02 | 1.62 | 0.73 | 2.88 | 4.15 | 27.41 |
| 1988 | 1.99 | 0.71 | 2.83 | 1.59 | 1.68 | 0.60 | 0.33 | 2.11 | 2.82 | 1.08 | 4.19 | 2.23 | 22.16 |
| 1989 | 1.00 | 1.17 | 1.14 | 4.39 | 2.23 | 1.28 | 2.22 | 2.86 | 2.87 | 1.57 | 0.93 | 0.87 | 22.53 |
| 1990 | 1.73 | 3.59 | 3.95 | 2.32 | 6.19 | 7.99 | 9.18 | 5.31 | 1.03 | 3.17 | 7.19 | 3.70 | 55.35 |
| 1991 | 1.19 | 0.57 | 3.67 | 2.97 | 5.94 | 1.50 | 0.35 | 3.41 | 3.59 | 7.31 | 3.57 | 2.06 | 36.13 |
| 1992 | 1.06 | 1.55 | 2.58 | 1.61 | 0.82 | 0.80 | 8.19 | 0.25 | 5.81 | 1.33 | 5.58 | 2.99 | 32.57 |
| 1993 | 3.55 | 1.68 | 4.08 | 4.89 | 3.25 | 5.70 | 10.15 | 7.38 | 7.56 | 2.42 | 2.22 | 1.19 | 54.07 |
| 1994 | 0.92 | 1.98 | 0.66 | 3.86 | 1.52 | 1.81 | 1.37 | 2.92 | 1.27 | 3.19 | 3.49 | 2.21 | 25.20 |
| 1995 | 2.83 | 0.54 | 1.56 | 4.77 | 10.19 | 1.57 | 2.83 | 1.95 | 1.47 | 3.23 | 2.48 | 0.32 | 33.74 |
| 1996 | 1.44 | 0.82 | 2.04 | 2.55 | 7.60 | 2.08 | 4.57 | 1.03 | 1.62 | 2.51 | 2.16 | 1.06 | 29.48 |
| 1997 | 1.05 | 5.37 | 1.68 | 2.76 | 2.65 | 1.38 | 0.89 | 6.10 | 3.13 | 2.29 | 2.85 | 1.69 | 31.84 |
| 1998 | 2.55 | 2.64 | 4.67 | 4.96 | 5.50 | 5.19 | 1.64 | 5.26 | 2.30 | 3.21 | 2.63 | 1.75 | 42.30 |
| 1999 | 3.07 | 1.16 | 0.94 | 4.31 | 4.92 | 3.21 | 4.22 | 2.78 | 1.54 | 1.35 | 0.22 | 2.55 | 30.27 |
| 2000 | 0.80 | 1.82 | 1.63 | 2.53 | 4.04 | 3.76 | 1.95 | 0.91 | 2.64 | 1.95 | 3.22 | 0.96 | 26.21 |
| 2001 | 3.29 | 2.82 | 1.09 | 4.21 | 5.33 | 2.79 | 1.46 | 4.24 | 4.15 | 5.08 | 1.90 | 1.35 | 37.71 |
| 2002 | 2.81 | 1.15 | 1.82 | 4.61 | 6.80 | 4.24 | 2.18 | 4.57 | 0.78 | 1.46 | 0.68 | 2.21 | 33.31 |
| 2003 | 0.47 | 0.77 | 1.97 | 2.63 | 3.94 | 4.45 | 5.73 | 2.44 | 1.53 | 1.78 | 3.38 | 1.92 | 31.01 |
| 2004 | 0.67 | 0.52 | 3.07 | 2.20 | 5.44 | 2.99 | 3.94 | 4.92 | 0.99 | 4.25 | 4.06 | 1.40 | 34.45 |
| 2005 | 4.18 | 1.73 | 1.65 | 1.82 | 0.69 | 0.87 | 2.39 | 1.87 | 3.91 | 1.48 | 3.52 | 1.31 | 25.42 |
| 2006 | 3.40 | 0.56 | 3.10 | 5.18 | 1.29 | 1.99 | 2.04 | 3.53 | 1.67 | 2.17 | 3.86 | 3.14 | 31.93 |
| 2007 | 2.40 | 2.56 | 5.79 | 4.35 | 2.62 | 3.73 | 2.74 | 4.12 | 1.53 | 1.94 | 1.84 | 3.40 | 37.02 |
| 2008 | 3.22 | 3.94 | 2.05 | 2.70 | 5.21 | 5.02 | 3.68 | 1.04 | 12.34 | 2.12 | 1.17 | 4.05 | 46.54 |
| 2009 | 0.72 | 2.02 | 7.49 | 6.68 | 5.74 | 4.76 | 3.91 | 4.65 | 3.53 | 7.95 | 2.89 | 4.17 | 54.51 |
| POR= 117 YRS | 1.83 | 1.65 | 2.87 | 3.66 | 4.02 | 3.74 | 3.73 | 3.14 | 3.66 | 2.52 | 2.44 | 2.00 | 35.26 |

WBAN : 14842

## AVERAGE TEMPERATURE (°F) 2009  PEORIA (KPIA)

| YEAR | JAN | FEB | MAR | APR | MAY | JUN | JUL | AUG | SEP | OCT | NOV | DEC | ANNUAL |
|------|-----|-----|-----|-----|-----|-----|-----|-----|-----|-----|-----|-----|--------|
| 1980 | 23.6 | 19.9 | 35.6 | 49.2 | 63.0 | 69.3 | 78.5 | 76.9 | 67.9 | 49.6 | 40.7 | 28.9 | 50.3 |
| 1981 | 23.7 | 27.8 | 40.7 | 55.4 | 58.7 | 73.2 | 75.4 | 72.8 | 65.9 | 53.4 | 45.0 | 27.3 | 51.6 |
| 1982 | 15.8 | 24.8 | 37.7 | 46.5 | 68.4 | 67.2 | 75.7 | 71.3 | 64.9 | 55.0 | 41.5 | 37.4 | 50.5 |
| 1983 | 28.2 | 33.5 | 40.3 | 46.8 | 58.6 | 72.6 | 80.2 | 80.8 | 67.9 | 55.3 | 44.9 | 15.2 | 52.0 |
| 1984 | 20.6 | 35.5 | 31.3 | 49.6 | 58.8 | 74.3 | 73.0 | 74.9 | 64.0 | 57.6 | 40.3 | 33.9 | 51.2 |
| 1985 | 16.8 | 23.2 | 43.8 | 57.0 | 64.3 | 68.7 | 73.7 | 70.2 | 66.6 | 55.5 | 39.0 | 18.6 | 49.8 |
| 1986 | 26.6 | 23.5 | 43.2 | 55.4 | 64.0 | 72.3 | 77.4 | 68.9 | 68.9 | 55.2 | 36.2 | 30.9 | 51.9 |
| 1987 | 25.2 | 35.7 | 44.1 | 54.1 | 68.4 | 74.0 | 79.0 | 73.7 | 65.7 | 48.0 | 44.6 | 32.4 | 53.7 |
| 1988 | 22.7 | 23.5 | 39.9 | 51.1 | 65.4 | 73.1 | 78.6 | 78.5 | 68.2 | 48.2 | 41.5 | 29.7 | 51.7 |
| 1989 | 33.9 | 18.7 | 39.2 | 50.7 | 59.1 | 69.8 | 75.3 | 72.4 | 62.2 | 54.9 | 39.8 | 16.2 | 49.4 |
| 1990 | 34.8 | 33.7 | 43.4 | 50.0 | 57.8 | 71.0 | 72.6 | 72.1 | 66.1 | 51.9 | 45.2 | 27.7 | 52.2 |
| 1991 | 21.5 | 32.4 | 42.8 | 55.8 | 68.0 | 74.7 | 75.4 | 74.3 | 65.4 | 53.9 | 36.0 | 32.0 | 52.7 |
| 1992 | 30.1 | 35.6 | 42.3 | 50.0 | 61.0 | 69.7 | 72.4 | 68.6 | 64.2 | 53.1 | 39.8 | 29.5 | 51.4 |
| 1993 | 27.4 | 25.1 | 36.4 | 49.8 | 63.6 | 70.4 | 76.1 | 75.0 | 61.4 | 51.7 | 39.4 | 31.4 | 50.6 |
| 1994 | 18.2 | 26.2 | 41.0 | 53.2 | 61.7 | 74.4 | 74.6 | 71.6 | 66.7 | 56.2 | 44.8 | 34.7 | 51.9 |
| 1995 | 24.1 | 28.3 | 42.6 | 49.3 | 58.3 | 72.1 | 76.2 | 79.4 | 62.7 | 54.2 | 34.4 | 27.5 | 50.8 |
| 1996 | 21.9 | 28.9 | 34.0 | 48.1 | 60.0 | 71.1 | 72.0 | 73.4 | 64.4 | 54.0 | 33.7 | 28.7 | 49.2 |
| 1997 | 19.3 | 31.9 | 41.4 | 47.1 | 57.1 | 70.7 | 75.4 | 71.5 | 66.1 | 54.9 | 38.1 | 30.9 | 50.4 |
| 1998 | 29.7 | 39.6 | 39.5 | 52.1 | 66.8 | 70.7 | 75.7 | 75.2 | 70.4 | 55.5 | 45.2 | 33.3 | 54.5 |
| 1999 | 22.9 | 36.1 | 38.2 | 53.7 | 63.9 | 72.0 | 78.1 | 71.2 | 64.1 | 54.4 | 47.2 | 31.5 | 52.8 |
| 2000 | 25.7 | 36.7 | 46.5 | 51.6 | 64.5 | 69.2 | 73.3 | 75.9 | 66.5 | 57.4 | 37.6 | 16.1 | 51.8 |
| 2001 | 24.2 | 28.6 | 36.4 | 57.8 | 63.8 | 69.7 | 76.8 | 74.6 | 64.0 | 53.0 | 50.0 | 34.5 | 52.8 |
| 2002 | 33.1 | 34.2 | 37.2 | 52.4 | 58.9 | 74.2 | 79.1 | 75.3 | 68.7 | 50.9 | 38.5 | 32.1 | 52.9 |
| 2003 | 22.1 | 24.4 | 40.0 | 53.3 | 60.6 | 68.1 | 74.8 | 75.6 | 63.6 | 54.5 | 43.0 | 32.9 | 51.1 |
| 2004 | 22.9 | 30.0 | 44.2 | 54.1 | 65.3 | 69.7 | 72.9 | 68.2 | 67.9 | 54.6 | 44.0 | 31.0 | 52.1 |
| 2005 | 26.9 | 35.1 | 39.0 | 55.5 | 60.8 | 76.1 | 77.8 | 76.1 | 71.6 | 55.8 | 43.4 | 24.1 | 53.5 |
| 2006 | 37.6 | 29.9 | 41.4 | 56.4 | 62.6 | 71.3 | 78.6 | 74.7 | 64.3 | 50.8 | 44.3 | 34.1 | 53.8 |
| 2007 | 28.1 | 19.7 | 47.3 | 49.9 | 67.5 | 72.9 | 73.9 | 78.3 | 70.9 | 59.9 | 41.7 | 28.7 | 53.2 |
| 2008 | 25.4 | 24.5 | 39.0 | 51.7 | 59.3 | 73.2 | 74.9 | 73.0 | 67.6 | 54.6 | 40.4 | 24.3 | 50.7 |
| 2009 | 18.5 | 31.3 | 42.5 | 50.9 | 62.2 | 72.4 | 70.7 | 70.5 | 66.3 | 49.3 | 46.6 | 27.8 | 50.8 |
| POR= 117 YRS | 24.4 | 27.6 | 39.3 | 51.1 | 62.1 | 71.4 | 75.8 | 73.8 | 66.2 | 54.5 | 40.5 | 28.7 | 51.3 |

## HEATING DEGREE DAYS (base 65°F) 2009  PEORIA (KPIA)

| YEAR | JUL | AUG | SEP | OCT | NOV | DEC | JAN | FEB | MAR | APR | MAY | JUN | TOTAL |
|---|---|---|---|---|---|---|---|---|---|---|---|---|---|
| 1980-81 | 0 | 0 | 65 | 470 | 722 | 1112 | 1273 | 1037 | 748 | 295 | 221 | 1 | 5944 |
| 1981-82 | 1 | 0 | 60 | 360 | 594 | 1163 | 1520 | 1119 | 839 | 548 | 29 | 28 | 6261 |
| 1982-83 | 0 | 13 | 94 | 325 | 697 | 849 | 1133 | 875 | 758 | 537 | 206 | 17 | 5504 |
| 1983-84 | 2 | 0 | 92 | 311 | 595 | 1541 | 1371 | 849 | 1038 | 467 | 206 | 1 | 6473 |
| 1984-85 | 1 | 1 | 153 | 246 | 734 | 956 | 1489 | 1164 | 656 | 284 | 71 | 38 | 5793 |
| 1985-86 | 0 | 6 | 111 | 287 | 774 | 1432 | 1184 | 1156 | 683 | 314 | 92 | 7 | 6046 |
| 1986-87 | 0 | 26 | 37 | 305 | 858 | 1048 | 1228 | 814 | 640 | 340 | 53 | 2 | 5351 |
| 1987-88 | 0 | 16 | 68 | 520 | 609 | 1001 | 1306 | 1198 | 772 | 409 | 64 | 12 | 5975 |
| 1988-89 | 0 | 4 | 38 | 517 | 698 | 1090 | 958 | 1290 | 796 | 442 | 231 | 25 | 6089 |
| 1989-90 | 0 | 3 | 134 | 317 | 749 | 1509 | 929 | 871 | 672 | 475 | 226 | 16 | 5901 |
| 1990-91 | 8 | 3 | 99 | 409 | 589 | 1148 | 1341 | 909 | 683 | 285 | 86 | 0 | 5560 |
| 1991-92 | 0 | 0 | 143 | 347 | 863 | 1017 | 1078 | 847 | 698 | 444 | 180 | 19 | 5636 |
| 1992-93 | 0 | 23 | 109 | 367 | 748 | 1094 | 1159 | 1110 | 878 | 450 | 96 | 48 | 6082 |
| 1993-94 | 0 | 0 | 135 | 408 | 761 | 1035 | 1448 | 1078 | 734 | 371 | 157 | 12 | 6139 |
| 1994-95 | 0 | 8 | 67 | 282 | 602 | 932 | 1258 | 1022 | 687 | 466 | 207 | 7 | 5538 |
| 1995-96 | 2 | 0 | 133 | 332 | 911 | 1154 | 1328 | 1041 | 954 | 499 | 210 | 28 | 6592 |
| 1996-97 | 2 | 0 | 97 | 338 | 930 | 1120 | 1412 | 922 | 723 | 530 | 252 | 18 | 6344 |
| 1997-98 | 5 | 3 | 54 | 364 | 802 | 1051 | 1088 | 707 | 793 | 381 | 50 | 44 | 5342 |
| 1998-99 | 0 | 0 | 19 | 297 | 587 | 975 | 1300 | 806 | 823 | 333 | 91 | 18 | 5249 |
| 1999-00 | 0 | 0 | 102 | 330 | 529 | 1031 | 1214 | 814 | 566 | 394 | 103 | 22 | 5105 |
| 2000-01 | 0 | 0 | 92 | 252 | 818 | 1507 | 1256 | 1013 | 881 | 248 | 120 | 36 | 6223 |
| 2001-02 | 1 | 0 | 102 | 371 | 444 | 938 | 981 | 856 | 853 | 419 | 225 | 5 | 5195 |
| 2002-03 | 0 | 0 | 38 | 459 | 788 | 1013 | 1322 | 1133 | 770 | 358 | 149 | 45 | 6075 |
| 2003-04 | 0 | 2 | 107 | 328 | 655 | 986 | 1295 | 1005 | 641 | 344 | 100 | 11 | 5474 |
| 2004-05 | 0 | 29 | 30 | 326 | 625 | 1048 | 1173 | 833 | 800 | 292 | 174 | 0 | 5330 |
| 2005-06 | 0 | 0 | 30 | 317 | 642 | 1264 | 843 | 978 | 721 | 266 | 170 | 18 | 5249 |
| 2006-07 | 0 | 0 | 88 | 470 | 614 | 951 | 1135 | 1261 | 546 | 463 | 46 | 0 | 5574 |
| 2007-08 | 0 | 0 | 32 | 219 | 694 | 1117 | 1222 | 1170 | 799 | 402 | 202 | 0 | 5857 |
| 2008-09 | 0 | 0 | 37 | 334 | 734 | 1256 | 1436 | 937 | 690 | 427 | 122 | 13 | 5986 |
| 2009- | 3 | 14 | 36 | 477 | 544 | 1149 | | | | | | | |

WBAN : 14842

## COOLING DEGREE DAYS (base 65°F) 2009  PEORIA (KPIA)

| YEAR | JAN | FEB | MAR | APR | MAY | JUN | JUL | AUG | SEP | OCT | NOV | DEC | TOTAL |
|---|---|---|---|---|---|---|---|---|---|---|---|---|---|
| 1980 | 0 | 0 | 0 | 6 | 70 | 160 | 425 | 378 | 156 | 1 | 0 | 0 | 1196 |
| 1981 | 0 | 0 | 0 | 13 | 33 | 250 | 331 | 250 | 93 | 6 | 1 | 0 | 977 |
| 1982 | 0 | 0 | 0 | 0 | 141 | 101 | 338 | 215 | 96 | 26 | 0 | 0 | 917 |
| 1983 | 0 | 0 | 0 | 0 | 14 | 250 | 479 | 494 | 188 | 17 | 0 | 0 | 1442 |
| 1984 | 0 | 0 | 0 | 12 | 24 | 285 | 256 | 315 | 129 | 22 | 0 | 0 | 1043 |
| 1985 | 0 | 0 | 3 | 48 | 54 | 155 | 279 | 173 | 164 | 0 | 0 | 0 | 876 |
| 1986 | 0 | 0 | 15 | 30 | 68 | 234 | 392 | 157 | 161 | 6 | 0 | 0 | 1063 |
| 1987 | 0 | 0 | 0 | 19 | 166 | 278 | 440 | 293 | 92 | 0 | 3 | 0 | 1291 |
| 1988 | 0 | 0 | 0 | 2 | 84 | 266 | 431 | 428 | 140 | 5 | 0 | 0 | 1356 |
| 1989 | 0 | 0 | 2 | 20 | 56 | 177 | 324 | 240 | 57 | 11 | 0 | 0 | 887 |
| 1990 | 0 | 0 | 9 | 34 | 8 | 204 | 251 | 230 | 138 | 9 | 0 | 0 | 883 |
| 1991 | 0 | 0 | 4 | 16 | 189 | 296 | 331 | 293 | 159 | 11 | 0 | 0 | 1299 |
| 1992 | 0 | 0 | 0 | 1 | 62 | 171 | 237 | 140 | 90 | 6 | 0 | 0 | 707 |
| 1993 | 0 | 0 | 0 | 2 | 60 | 218 | 350 | 321 | 33 | 6 | 0 | 0 | 990 |
| 1994 | 0 | 0 | 0 | 25 | 61 | 300 | 307 | 218 | 123 | 16 | 0 | 0 | 1050 |
| 1995 | 0 | 0 | 0 | 0 | 6 | 225 | 355 | 453 | 73 | 5 | 0 | 0 | 1117 |
| 1996 | 0 | 0 | 0 | 0 | 63 | 217 | 224 | 267 | 88 | 5 | 0 | 0 | 864 |
| 1997 | 0 | 0 | 0 | 0 | 12 | 199 | 334 | 210 | 94 | 59 | 0 | 0 | 908 |
| 1998 | 0 | 0 | 9 | 0 | 114 | 222 | 337 | 318 | 187 | 10 | 0 | 0 | 1197 |
| 1999 | 0 | 0 | 0 | 2 | 68 | 233 | 414 | 197 | 82 | 8 | 0 | 0 | 1004 |
| 2000 | 0 | 0 | 2 | 0 | 92 | 155 | 262 | 345 | 146 | 25 | 4 | 0 | 1031 |
| 2001 | 0 | 0 | 0 | 38 | 89 | 184 | 372 | 307 | 78 | 4 | 0 | 0 | 1072 |
| 2002 | 0 | 0 | 0 | 46 | 45 | 287 | 445 | 322 | 157 | 26 | 0 | 0 | 1328 |
| 2003 | 0 | 0 | 0 | 13 | 22 | 147 | 311 | 338 | 70 | 8 | 0 | 0 | 909 |
| 2004 | 0 | 0 | 0 | 23 | 120 | 160 | 254 | 133 | 124 | 7 | 0 | 0 | 821 |
| 2005 | 0 | 0 | 0 | 14 | 48 | 338 | 406 | 350 | 236 | 40 | 0 | 0 | 1432 |
| 2006 | 0 | 0 | 0 | 13 | 101 | 215 | 427 | 310 | 71 | 36 | 0 | 0 | 1173 |
| 2007 | 0 | 0 | 7 | 14 | 128 | 244 | 283 | 417 | 217 | 70 | 0 | 0 | 1380 |
| 2008 | 0 | 0 | 0 | 10 | 31 | 250 | 313 | 256 | 119 | 19 | 0 | 0 | 998 |
| 2009 | 0 | 0 | 0 | 12 | 40 | 241 | 186 | 191 | 83 | 0 | 0 | 0 | 753 |

## SNOWFALL (inches)  2009  PEORIA (KPIA)

| YEAR | JUL | AUG | SEP | OCT | NOV | DEC | JAN | FEB | MAR | APR | MAY | JUN | TOTAL |
|------|-----|-----|-----|-----|-----|-----|-----|-----|-----|-----|-----|-----|-------|
| 1980-81 | 0.0 | 0.0 | 0.0 | T | 4.1 | 3.3 | 5.9 | 10.5 | T | 0.0 | 0.0 | 0.0 | 23.8 |
| 1981-82 | 0.0 | 0.0 | 0.0 | 0.0 | 0.1 | 9.8 | 11.0 | 6.1 | 6.5 | 13.4 | 0.0 | 0.0 | 46.9 |
| 1982-83 | 0.0 | 0.0 | 0.0 | 0.0 | 0.9 | 2.0 | 5.6 | 4.8 | 5.7 | 0.1 | 0.0 | 0.0 | 19.1 |
| 1983-84 | 0.0 | 0.0 | 0.0 | 0.0 | 3.3 | 15.9 | 6.9 | 4.1 | 6.0 | 0.0 | 0.0 | 0.0 | 36.2 |
| 1984-85 | 0.0 | 0.0 | 0.0 | 0.0 | T | 0.9 | 9.8 | 6.2 | T | T | 0.0 | 0.0 | 16.9 |
| 1985-86 | 0.0 | 0.0 | 0.0 | 0.0 | 1.0 | 6.2 | 1.3 | 13.9 | 0.4 | 0.1 | 0.0 | 0.0 | 22.9 |
| 1986-87 | 0.0 | 0.0 | 0.0 | 0.0 | 1.0 | T | 18.0 | 0.1 | T | 0.0 | 0.0 | 0.0 | 19.1 |
| 1987-88 | 0.0 | 0.0 | 0.0 | T | 0.3 | 9.8 | 1.9 | 9.7 | 1.9 | 0.0 | 0.0 | 0.0 | 23.6 |
| 1988-89 | 0.0 | 0.0 | 0.0 | 0.0 | 0.7 | 4.7 | 0.3 | 15.2 | T | 0.9 | T | 0.0 | 21.8 |
| 1989-90 | 0.0 | 0.0 | 0.0 | 0.6 | T | 10.5 | 4.8 | 6.2 | T | T | 0.0 | 0.0 | 22.1 |
| 1990-91 | T | 0.0 | 0.0 | T | T | 3.8 | 9.8 | 1.0 | 6.9 | 0.0 | 0.0 | 0.0 | 21.5 |
| 1991-92 | 0.0 | 0.0 | 0.0 | T | 3.6 | 5.1 | 6.4 | 0.4 | 4.1 | T | 0.0 | 0.0 | 19.6 |
| 1992-93 | 0.0 | 0.0 | T | T | 0.7 | 6.1 | 6.8 | 12.4 | 3.5 | 0.7 | 0.0 | 0.0 | 30.2 |
| 1993-94 | 0.0 | 0.0 | 0.0 | T | 3.6 | 2.4 | 9.4 | 12.6 | T | 0.1 | 0.0 | 0.0 | 28.1 |
| 1994-95 | 0.0 | 0.0 | 0.0 | 0.0 | T | 0.5 | 3.8 | 2.8 | 0.7 | T | T | 0.0 | 7.8 |
| 1995-96 | 0.0 | 0.0 | 0.0 | 0.0 | 0.2 | 3.6 | 5.9 | 1.9 | 1.1 | 0.5 | 0.0 | 0.0 | 13.2 |
| 1996-97 | 0.0 | 0.0 | 0.0 | | | | | | | | 0.0 | | |
| 1997-98 | | 0.0 | | T | 1.9 | 13.0 | 7.1 | T | 7.3 | T | 0.0 | 0.0 | |
| 1998-99 | 0.0 | 0.0 | 0.0 | 0.0 | 0.5 | 5.2 | 18.1 | 1.0 | 6.9 | T | T | 0.0 | 31.7 |
| 1999-00 | 0.0 | 0.0 | 0.0 | 0.0 | 0.0 | 4.7 | 8.8 | 0.6 | T | T | T | T | 14.1 |
| 2000-01 | 0.0 | 0.0 | 0.0 | 0.0 | T | 21.2 | 5.1 | 1.8 | 2.1 | T | 0.0 | 0.0 | 30.2 |
| 2001-02 | 0.0 | 0.0 | 0.0 | 0.0 | T | 0.4 | 2.4 | 1.4 | 7.5 | T | 0.0 | T | 11.7 |
| 2002-03 | 0.0 | 0.0 | 0.0 | 0.0 | T | 3.3 | 6.3 | 7.9 | 1.3 | 0.4 | T | 0.0 | 19.2 |
| 2003-04 | 0.0 | 0.0 | 0.0 | 0.0 | T | 7.3 | 8.4 | 4.1 | 3.4 | 0.0 | 0.0 | 0.0 | 23.2 |
| 2004-05 | 0.0 | 0.0 | 0.0 | 0.0 | 7.9 | T | 6.2 | 2.1 | 1.6 | 0.0 | 0.0 | 0.0 | 17.8 |
| 2005-06 | 0.0 | 0.0 | 0.0 | 0.0 | T | 11.3 | 1.0 | T | 6.0 | 0.0 | 0.0 | 0.0 | 18.3 |
| 2006-07 | 0.0 | 0.0 | 0.0 | 0.0 | 1.3 | 7.5 | 2.2 | 13.3 | T | T | 0.0 | 0.0 | 24.3 |
| 2007-08 | 0.0 | 0.0 | 0.0 | T | T | 10.5 | 6.2 | 17.0 | T | T | 0.0 | 0.0 | 33.7 |
| 2008-09 | 0.0 | 0.0 | 0.0 | T | 4.6 | 5.3 | 12.9 | 3.5 | 4.9 | 1.8 | 0.0 | 0.0 | 33.0 |
| 2009- | 0.0 | 0.0 | 0.0 | 0.0 | T | 14.6 | | | | | | | |
| POR= 105 YRS | T | 0.0 | T | 0.1 | 1.7 | 5.4 | 6.5 | 5.2 | 3.8 | 0.8 | T | T | 23.5 |

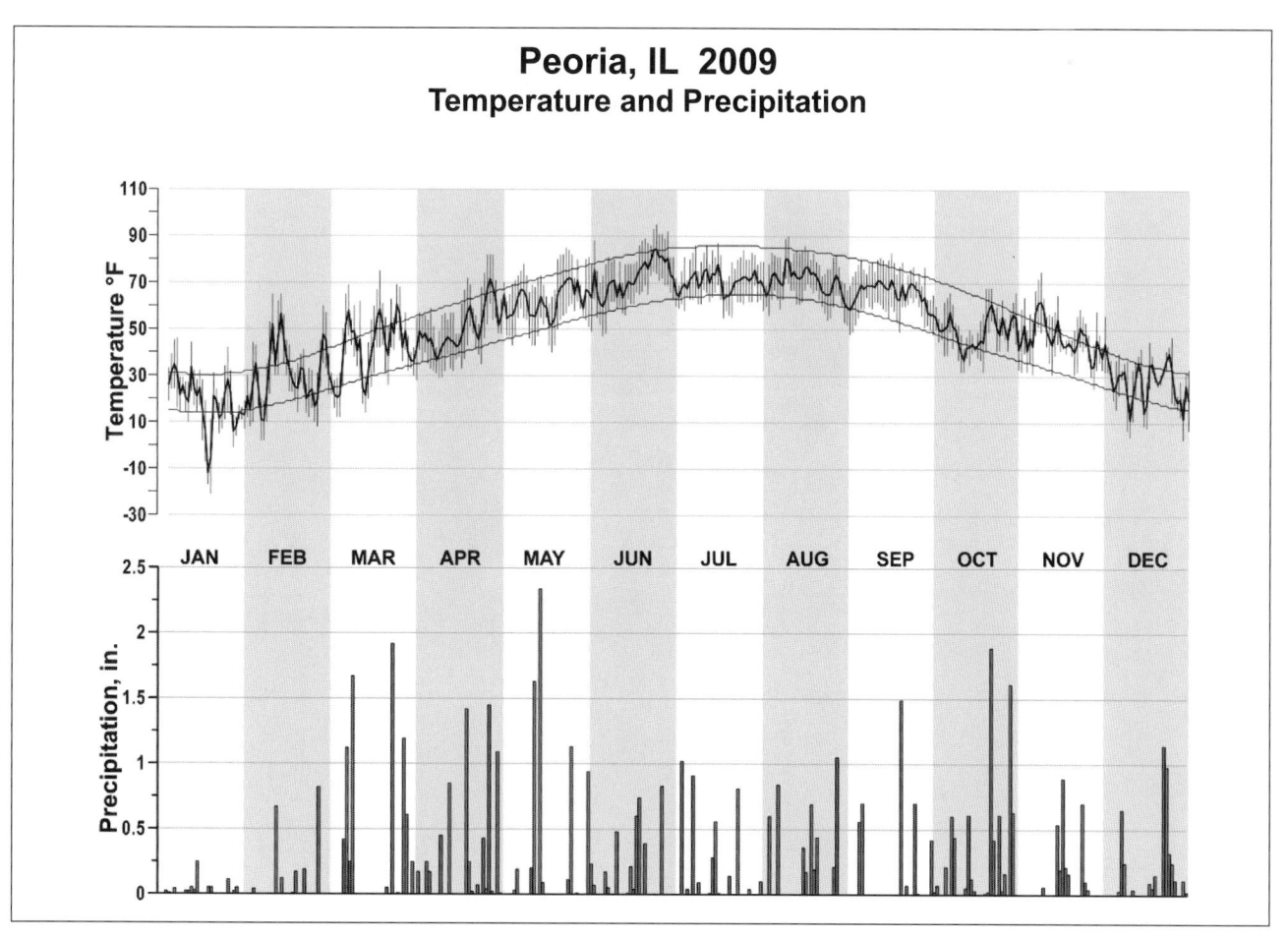

# Peoria, IL  2009
## Temperature and Precipitation

# 2009
# SPRINGFIELD
# ILLINOIS (KSPI)

The location of Springfield near the center of North America gives it a typical continental climate with warm summers and fairly cold winters. The surrounding country is nearly level. There are no large hills in the vicinity, but rolling terrain is found near the Sangamon River and Spring Creek.

Monthly temperatures range from the upper 20s for January to the upper 70s for July. Considerable variation may take place within the seasons. Temperatures of 70 degrees or higher may occur in winter and temperatures near 50 degrees are sometimes recorded during the summer months.

There are no wet and dry seasons. Monthly precipitation ranges from a little over 4 inches in May and June to about 2 inches in January. There is some variation in rainfall totals from year to year. Thunderstorms are common during hot weather, and these are sometimes locally severe with brief but heavy showers. The average year has about fifty thunderstorms of which two-thirds occur during the months of May through August. Damaging hail accompanies only a few of the thunderstorms and the areas affected are usually small.

Sunshine is particularly abundant during the summer months when days are long and not very cloudy. January is the cloudiest month, with only about a third as much sunshine as July or August. March is the windiest month, and August the month with the least wind. Velocities of more than 40 mph are not unusual for brief periods in most months of the year. The prevailing wind direction is southerly during most of the year with northwesterly winds during the late fall and early spring months.

An overall description of the climate of Springfield would be one indicating pleasant conditions with sharp seasonal changes, but no extended periods of severely cold weather. Summer weather is often uncomfortably warm and humid.

Based on the 1951-1980 period, the average first occurrence of 32 degrees Fahrenheit in the fall is October 19 and the average last occurrence in the spring is April 17.

# NORMALS, MEANS, AND EXTREMES
## SPRINGFIELD (KSPI)

| LATITUDE: 39 ° 50'N | LONGITUDE: -89 ° 41'W | ELEVATION (FT): GRND: 590  BARO: 594 | TIME ZONE: CENTRAL  (UTC -6) | WBAN: 93822 |
|---|---|---|---|---|

| | ELEMENT | POR | JAN | FEB | MAR | APR | MAY | JUN | JUL | AUG | SEP | OCT | NOV | DEC | YEAR |
|---|---|---|---|---|---|---|---|---|---|---|---|---|---|---|---|
| **TEMPERATURE °F** | NORMAL DAILY MAXIMUM | 30 | 33.1 | 38.9 | 51.1 | 63.4 | 74.4 | 83.3 | 86.5 | 84.5 | 78.5 | 66.6 | 50.9 | 38.0 | 62.4 |
| | MEAN DAILY MAXIMUM | 109 | 34.7 | 38.7 | 50.6 | 63.5 | 74.3 | 83.4 | 87.5 | 85.2 | 78.8 | 66.9 | 51.4 | 38.4 | 62.8 |
| | HIGHEST DAILY MAXIMUM | 62 | 71 | 74 | 87 | 90 | 95 | 103 | 112 | 103 | 101 | 93 | 83 | 74 | 112 |
| | YEAR OF OCCURRENCE | | 1950 | 1996 | 1981 | 1986 | 1967 | 1954 | 1954 | 1964 | 1984 | 2006 | 1950 | 1984 | JUL 1954 |
| | MEAN OF EXTREME MAXS. | 109 | 56.9 | 61.2 | 74.8 | 83.3 | 89.2 | 95.0 | 97.1 | 95.5 | 92.1 | 84.5 | 72.3 | 60.4 | 80.2 |
| | NORMAL DAILY MINIMUM | 30 | 17.1 | 22.2 | 32.4 | 42.2 | 52.7 | 61.9 | 66.0 | 63.9 | 55.4 | 44.4 | 33.7 | 22.6 | 42.9 |
| | MEAN DAILY MINIMUM | 109 | 19.1 | 22.4 | 32.0 | 42.7 | 53.1 | 62.4 | 66.6 | 64.7 | 56.6 | 45.6 | 34.0 | 23.6 | 43.6 |
| | LOWEST DAILY MINIMUM | 62 | -21 | -22 | -12 | 19 | 28 | 39 | 48 | 43 | 32 | 17 | -3 | -21 | -22 |
| | YEAR OF OCCURRENCE | | 1999 | 1963 | 1960 | 1997 | 1966 | 2003 | 1975 | 1986 | 2003 | 1952 | 1964 | 1989 | FEB 1963 |
| | MEAN OF EXTREME MINS. | 109 | -3.5 | 1.1 | 13.3 | 27.7 | 38.0 | 48.8 | 55.0 | 52.4 | 40.4 | 29.1 | 16.5 | 2.1 | 26.7 |
| | NORMAL DRY BULB | 30 | 25.1 | 30.6 | 41.8 | 52.8 | 63.6 | 72.6 | 76.3 | 74.2 | 67.0 | 55.5 | 42.3 | 30.3 | 52.7 |
| | MEAN DRY BULB | 109 | 26.9 | 30.6 | 41.3 | 53.1 | 63.7 | 73.0 | 77.0 | 74.9 | 67.7 | 56.3 | 42.7 | 31.0 | 53.2 |
| | MEAN WET BULB | 26 | 24.6 | 28.0 | 36.9 | 46.5 | 56.2 | 65.0 | 68.9 | 67.5 | 59.4 | 48.7 | 38.4 | 28.3 | 47.4 |
| | MEAN DEW POINT | 26 | 21.5 | 24.9 | 32.7 | 41.9 | 52.4 | 62.1 | 66.6 | 65.3 | 56.2 | 44.8 | 35.0 | 25.2 | 44.1 |
| | NORMAL NO. DAYS WITH: | | | | | | | | | | | | | | |
| | MAXIMUM >= 90 | 3 0 | 0.0 | 0.0 | 0.0 | * | 1.4 | 6.7 | 10.7 | 6.9 | 3.2 | 0.1 | 0.0 | 0.0 | 29.0 |
| | MAXIMUM <= 32 | 3 0 | 14.2 | 8.6 | 2.1 | * | 0.0 | 0.0 | 0.0 | 0.0 | 0.0 | 0.0 | 1.3 | 8.4 | 34.6 |
| | MINIMUM <= 32 | 30 | 27.6 | 22.0 | 15.9 | 4.3 | * | 0.0 | 0.0 | 0.0 | 0.1 | 3.7 | 14.4 | 24.8 | 112.8 |
| | MINIMUM <= 0 | 30 | 4.1 | 2.5 | 0.1 | 0.0 | 0.0 | 0.0 | 0.0 | 0.0 | 0.0 | 0.0 | 0.0 | 1.8 | 8.5 |
| **H/C** | NORMAL HEATING DEG. DAYS | 30 | 1239 | 979 | 726 | 376 | 126 | 14 | 1 | 4 | 77 | 319 | 674 | 1061 | 5596 |
| | NORMAL COOLING DEG. DAYS | 30 | 0 | 0 | 2 | 17 | 84 | 249 | 358 | 291 | 140 | 23 | 1 | 0 | 1165 |
| **RH** | NORMAL (PERCENT) | 30 | 77 | 75 | 71 | 66 | 67 | 70 | 73 | 75 | 72 | 69 | 74 | 79 | 72 |
| | HOUR 00 LST | 30 | 79 | 79 | 76 | 73 | 77 | 79 | 83 | 86 | 83 | 78 | 79 | 82 | 80 |
| | HOUR 06 LST | 30 | 81 | 82 | 82 | 81 | 81 | 84 | 87 | 91 | 89 | 84 | 83 | 84 | 84 |
| | HOUR 12 LST | 30 | 70 | 68 | 62 | 56 | 55 | 56 | 59 | 61 | 56 | 55 | 65 | 72 | 61 |
| | HOUR 18 LST | 30 | 74 | 72 | 63 | 57 | 55 | 56 | 61 | 65 | 62 | 62 | 71 | 77 | 65 |
| **S** | PERCENT POSSIBLE SUNSHINE | 48 | 48 | 52 | 51 | 56 | 63 | 68 | 71 | 70 | 68 | 63 | 48 | 44 | 59 |
| **W/O** | MEAN NO. DAYS WITH: | | | | | | | | | | | | | | |
| | HEAVY FOG(VISBY <= 1/4 MI) | 46 | 2.8 | 2.3 | 1.7 | 0.9 | 0.9 | 0.7 | 1.0 | 1.5 | 1.2 | 1.4 | 1.4 | 2.4 | 18.2 |
| | THUNDERSTORMS | 109 | 0.5 | 0.8 | 2.9 | 4.8 | 7.0 | 8.4 | 7.9 | 6.8 | 4.4 | 2.3 | 1.5 | 0.5 | 47.8 |
| **CLOUDINESS** | MEAN: | | | | | | | | | | | | | | |
| | SUNRISE-SUNSET (OKTAS) | | | | | | | | | | | | | | |
| | MIDNIGHT-MIDNIGHT (OKTAS) | | | | | | | | | | | | | | |
| | MEAN NO. DAYS WITH: | | | | | | | | | | | | | | |
| | CLEAR | | | | | | | | | | | | | | |
| | PARTLY CLOUDY | | | | | | | | | | | | | | |
| | CLOUDY | | | | | | | | | | | | | | |
| **PR** | MEAN STATION PRESSURE(IN) | 26 | 29.48 | 29.43 | 29.38 | 29.30 | 29.31 | 29.30 | 29.34 | 29.37 | 29.39 | 29.41 | 29.41 | 29.45 | 29.38 |
| | MEAN SEA-LEVEL PRES. (IN) | 26 | 30.15 | 30.11 | 30.05 | 29.96 | 29.96 | 29.95 | 29.98 | 30.01 | 30.04 | 30.06 | 30.08 | 30.13 | 30.04 |
| **WINDS** | MEAN SPEED (MPH) | 26 | 11.1 | 10.9 | 11.6 | 11.5 | 9.7 | 7.9 | 7.0 | 6.7 | 7.3 | 9.0 | 10.8 | 10.7 | 9.5 |
| | PREVAIL.DIR(TENS OF DEGS) | 35 | 31 | 19 | 19 | 19 | 19 | 19 | 20 | 19 | 19 | 19 | 19 | 19 | 19 |
| | MAXIMUM 2-MINUTE: | | | | | | | | | | | | | | |
| | SPEED (MPH) | 14 | 39 | 51 | 52 | 61 | 46 | 46 | 52 | 44 | 35 | 40 | 40 | 49 | 61 |
| | DIR. (TENS OF DEGS) | | 25 | 29 | 31 | 30 | 28 | 29 | 04 | 33 | 31 | 25 | 27 | 26 | 30 |
| | YEAR OF OCCURRENCE | | 1996 | 2001 | 2006 | 2002 | 2001 | 2008 | 2001 | 1999 | 2005 | 2004 | 2005 | 2007 | APR 2002 |
| | MAXIMUM 3-SECOND | | | | | | | | | | | | | | |
| | SPEED (MPH) | 14 | 52 | 62 | 58 | 74 | 68 | 64 | 62 | 60 | 47 | 53 | 53 | 59 | 74 |
| | DIR. (TENS OF DEGS) | | 29 | 27 | 25 | 30 | 28 | 29 | 06 | 29 | 31 | 25 | 17 | 28 | 30 |
| | YEAR OF OCCURRENCE | | 2006 | 2001 | 2009 | 2002 | 2008 | 2008 | 2001 | 2009 | 2005 | 1996 | 1998 | 2008 | APR 2002 |
| **PRECIPITATION** | NORMAL (IN) | 30 | 1.62 | 1.80 | 3.15 | 3.36 | 4.06 | 3.77 | 3.53 | 3.41 | 2.83 | 2.62 | 2.87 | 2.54 | 35.56 |
| | MAXIMUM MONTHLY (IN) | 62 | 5.67 | 4.89 | 7.89 | 9.91 | 10.72 | 9.22 | 10.76 | 8.37 | 8.57 | 11.32 | 6.94 | 8.94 | 11.32 |
| | YEAR OF OCCURRENCE | | 1949 | 1990 | 1973 | 1964 | 1996 | 1990 | 1981 | 1981 | 1986 | 2009 | 1985 | 1982 | OCT 2009 |
| | MINIMUM MONTHLY (IN) | 62 | 0.04 | 0.47 | 0.63 | 0.73 | 0.52 | 0.23 | 0.89 | 0.46 | T | 0.16 | 0.25 | 0.15 | T |
| | YEAR OF OCCURRENCE | | 1986 | 2004 | 1956 | 1971 | 1992 | 1959 | 1997 | 1992 | 1979 | 1964 | 1999 | 1955 | SEP 1979 |
| | MAXIMUM IN 24 HOURS (IN) | 62 | 2.78 | 2.54 | 2.84 | 4.45 | 3.95 | 4.99 | 4.43 | 4.79 | 5.12 | 3.51 | 2.74 | 6.12 | 6.12 |
| | YEAR OF OCCURRENCE | | 1975 | 1990 | 1972 | 1979 | 1990 | 2008 | 1981 | 1956 | 1959 | 1973 | 2003 | 1982 | DEC 1982 |
| | NORMAL NO. DAYS WITH: | | | | | | | | | | | | | | |
| | PRECIPITATION >= 0.01 | 30 | 9.1 | 8.5 | 11.4 | 11.4 | 11.0 | 9.6 | 8.6 | 8.3 | 8.0 | 8.3 | 9.8 | 9.9 | 113.9 |
| | PRECIPITATION >= 1.00 | 30 | 0.2 | 0.3 | 0.8 | 0.7 | 1.0 | 1.0 | 0.9 | 1.0 | 0.7 | 0.6 | 0.5 | 0.6 | 8.2 |
| **SNOWFALL** | NORMAL (IN) | 30 | 7.4 | 5.8 | 3.4 | 0.7 | 0.0 | 0.0 | 0.0 | 0.0 | 0.0 | 0.* | 1.6 | 6.0 | 24.9 |
| | MAXIMUM MONTHLY (IN) | 62 | 21.1 | 16.0 | 20.3 | 7.3 | T | 0.0 | T | 0.0 | 0.0 | 0.3 | 9.2 | 22.7 | 22.7 |
| | YEAR OF OCCURRENCE | | 1977 | 1993 | 1960 | 1980 | 1996 | | 1994 | | | 1989 | 1951 | 1973 | DEC 1973 |
| | MAXIMUM IN 24 HOURS (IN) | 62 | 8.8 | 10.3 | 8.2 | 6.1 | T | 0.0 | T | 0.0 | 0.0 | 0.3 | 8.0 | 10.9 | 10.9 |
| | YEAR OF OCCURRENCE' | | 1964 | 1965 | 1978 | 1980 | 1996 | | 1994 | | | 1989 | 1951 | 1973 | DEC 1973 |
| | MAXIMUM SNOW DEPTH (IN) | 108 | 16 | 12 | 16 | 5 | 0 | 0 | 0 | 0 | 0 | 2 | 7 | 15 | 16 |
| | YEAR OF OCCURRENCE | | 1918 | 1965 | 1978 | 1920 | | | | | | 1929 | 1951 | 1973 | MAR 1978 |
| | NORMAL NO. DAYS WITH: | | | | | | | | | | | | | | |
| | SNOWFALL >= 1.0 | 30 | 2.2 | 1.6 | 0.9 | 0.2 | 0.0 | 0.0 | 0.0 | 0.0 | 0.0 | 0.0 | 0.5 | 1.7 | 7.1 |

## PRECIPITATION (inches) 2009 SPRINGFIELD (KSPI)

| YEAR | JAN | FEB | MAR | APR | MAY | JUN | JUL | AUG | SEP | OCT | NOV | DEC | ANNUAL |
|------|-----|-----|-----|-----|-----|-----|-----|-----|-----|-----|-----|-----|--------|
| 1980 | 0.72 | 1.42 | 4.29 | 2.22 | 2.22 | 3.23 | 2.08 | 3.91 | 4.95 | 1.47 | 0.57 | 1.99 | 29.07 |
| 1981 | 0.43 | 2.12 | 2.27 | 4.57 | 6.17 | 5.80 | 10.76 | 8.37 | 1.13 | 1.94 | 2.21 | 2.35 | 48.12 |
| 1982 | 4.48 | 1.81 | 3.04 | 3.40 | 4.12 | 2.54 | 2.53 | 3.68 | 2.75 | 2.69 | 4.50 | 8.94 | 44.48 |
| 1983 | 0.46 | 0.96 | 3.44 | 5.02 | 4.53 | 2.62 | 1.60 | 0.84 | 1.36 | 3.63 | 4.71 | 3.50 | 32.67 |
| 1984 | 0.70 | 1.97 | 4.00 | 5.45 | 6.32 | 2.26 | 3.46 | 0.63 | 4.80 | 4.74 | 4.36 | 3.91 | 42.60 |
| 1985 | 0.65 | 2.96 | 4.19 | 1.46 | 1.75 | 5.82 | 2.95 | 6.03 | 0.64 | 3.08 | 6.94 | 2.43 | 38.90 |
| 1986 | 0.04 | 1.80 | 1.45 | 1.57 | 2.56 | 6.23 | 5.39 | 1.13 | 8.57 | 3.63 | 1.95 | 1.40 | 35.72 |
| 1987 | 1.46 | 0.73 | 2.08 | 2.59 | 0.56 | 4.08 | 4.12 | 3.23 | 0.99 | 1.26 | 3.25 | 5.00 | 29.35 |
| 1988 | 2.17 | 1.39 | 2.69 | 1.27 | 1.76 | 0.62 | 1.74 | 1.56 | 2.84 | 1.68 | 4.37 | 3.22 | 25.31 |
| 1989 | 0.88 | 1.27 | 1.68 | 5.50 | 4.18 | 0.89 | 3.13 | 2.57 | 5.49 | 1.02 | 0.84 | 0.58 | 28.03 |
| 1990 | 1.49 | 4.89 | 3.41 | 1.28 | 8.84 | 9.22 | 5.48 | 2.68 | 1.91 | 5.03 | 3.47 | 4.97 | 52.67 |
| 1991 | 1.29 | 0.71 | 2.64 | 3.57 | 5.74 | 1.26 | 3.22 | 4.03 | 4.35 | 6.41 | 3.31 | 1.38 | 37.91 |
| 1992 | 1.03 | 1.59 | 2.39 | 2.82 | 0.52 | 2.31 | 6.86 | 0.46 | 1.54 | 1.47 | 6.14 | 2.36 | 29.49 |
| 1993 | 3.97 | 1.48 | 2.48 | 4.59 | 2.05 | 7.26 | 9.46 | 3.21 | 6.05 | 2.92 | 3.04 | 1.07 | 47.58 |
| 1994 | 1.15 | 1.01 | 1.27 | 8.16 | 3.99 | 6.05 | 1.93 | 3.19 | 0.93 | 2.73 | 3.09 | 1.99 | 35.49 |
| 1995 | 4.01 | 0.51 | 3.41 | 2.71 | 7.54 | 1.29 | 2.70 | 3.67 | 0.65 | 2.24 | 1.38 | 1.34 | 31.45 |
| 1996 | 1.54 | 1.04 | 1.93 | 3.93 | 10.72 | 1.95 | 3.32 | 1.81 | 1.12 | 1.59 | 2.96 | .72 | 32.63 |
| 1997 | 1.58 | 2.74 | 2.50 | 1.48 | 3.10 | 1.54 | 0.89 | 4.64 | 3.53 | 1.79 | 4.50 | 1.75 | 30.04 |
| 1998 | 2.43 | 2.71 | 4.63 | 4.05 | 5.65 | 8.81 | 3.32 | 5.30 | 1.27 | 3.30 | 2.81 | 0.64 | 44.92 |
| 1999 | 1.94 | 2.15 | 0.97 | 4.61 | 2.90 | 2.95 | 2.08 | 4.64 | 2.42 | 1.78 | 0.25 | 2.20 | 28.89 |
| 2000 | 0.54 | 1.27 | 2.80 | 1.94 | 1.35 | 7.46 | 3.16 | 3.33 | 2.92 | 2.55 | 2.99 | 0.91 | 31.22 |
| 2001 | 2.06 | 3.01 | 1.11 | 1.99 | 3.50 | 4.42 | 3.41 | 3.34 | 2.50 | 4.96 | 2.61 | 2.09 | 35.00 |
| 2002 | 2.45 | 1.38 | 2.08 | 6.48 | 7.86 | 5.29 | 2.62 | 5.41 | 1.22 | 3.12 | 0.51 | 1.70 | 40.12 |
| 2003 | 0.76 | 1.15 | 1.79 | 2.83 | 3.17 | 6.77 | 3.91 | 3.82 | 1.58 | 2.99 | 4.58 | 1.42 | 34.77 |
| 2004 | 1.40 | 0.47 | 3.31 | 2.48 | 3.60 | 3.61 | 5.16 | 2.66 | 0.24 | 5.25 | 4.91 | 1.23 | 34.32 |
| 2005 | 5.55 | 1.71 | 1.11 | 1.97 | 2.21 | 1.36 | 2.68 | 2.88 | 3.96 | 2.79 | 3.94 | 1.47 | 31.63 |
| 2006 | 2.12 | 0.53 | 4.78 | 4.40 | 1.64 | 2.46 | 2.28 | 2.45 | 2.02 | 2.80 | 3.93 | 3.22 | 32.63 |
| 2007 | 2.46 | 2.75 | 2.21 | 2.76 | 1.51 | 5.57 | 1.52 | 2.50 | 1.60 | 2.91 | 2.28 | 3.65 | 31.72 |
| 2008 | 3.66 | 4.78 | 2.80 | 3.64 | 5.04 | 7.49 | 9.45 | 1.61 | 8.53 | 1.83 | 0.98 | 3.92 | 53.73 |
| 2009 | 0.65 | 1.24 | 4.00 | 5.73 | 5.63 | 7.64 | 3.26 | 3.04 | 2.18 | 11.32 | 3.48 | 4.45 | 52.62 |
| POR= 109 YRS | 1.86 | 1.75 | 3.00 | 3.59 | 3.87 | 4.02 | 3.34 | 3.17 | 3.21 | 2.75 | 2.46 | 2.12 | 35.14 |

WBAN : 93822

## AVERAGE TEMPERATURE (°F) 2009 SPRINGFIELD (KSPI)

| YEAR | JAN | FEB | MAR | APR | MAY | JUN | JUL | AUG | SEP | OCT | NOV | DEC | ANNUAL |
|------|-----|-----|-----|-----|-----|-----|-----|-----|-----|-----|-----|-----|--------|
| 1980 | 27.8 | 21.8 | 37.1 | 51.1 | 64.6 | 72.0 | 81.4 | 79.3 | 68.7 | 52.8 | 42.3 | 31.7 | 52.6 |
| 1981 | 26.4 | 32.4 | 44.1 | 60.4 | 60.3 | 74.9 | 77.1 | 73.8 | 67.2 | 54.6 | 45.3 | 26.8 | 53.6 |
| 1982 | 17.0 | 24.2 | 40.3 | 48.1 | 70.0 | 68.3 | 77.1 | 72.5 | 66.1 | 55.4 | 43.0 | 38.8 | 51.7 |
| 1983 | 28.7 | 34.9 | 40.6 | 47.8 | 59.9 | 73.8 | 80.4 | 80.0 | 69.1 | 57.2 | 45.7 | 16.1 | 52.9 |
| 1984 | 22.3 | 35.9 | 31.8 | 50.4 | 60.3 | 75.2 | 74.5 | 76.9 | 65.7 | 59.9 | 41.7 | 35.9 | 52.5 |
| 1985 | 18.5 | 24.9 | 45.7 | 57.8 | 65.2 | 69.7 | 74.3 | 70.8 | 67.9 | 58.1 | 42.8 | 21.5 | 51.4 |
| 1986 | 29.1 | 26.9 | 45.3 | 57.4 | 65.9 | 74.7 | 78.8 | 70.0 | 70.4 | 56.1 | 37.9 | 32.6 | 53.8 |
| 1987 | 26.1 | 36.6 | 45.4 | 54.7 | 70.5 | 75.4 | 78.8 | 75.3 | 67.6 | 50.6 | 46.2 | 34.6 | 55.2 |
| 1988 | 25.6 | 25.6 | 41.5 | 52.8 | 65.8 | 73.8 | 78.6 | 78.7 | 68.7 | 49.7 | 43.0 | 31.6 | 53.0 |
| 1989 | 36.0 | 20.3 | 40.6 | 52.3 | 59.3 | 70.5 | 75.4 | 73.3 | 63.8 | 56.9 | 42.0 | 18.7 | 50.8 |
| 1990 | 37.1 | 36.1 | 45.5 | 50.8 | 59.9 | 73.1 | 75.2 | 74.0 | 68.4 | 53.5 | 47.8 | 30.0 | 54.3 |
| 1991 | 24.0 | 35.3 | 45.1 | 57.3 | 69.2 | 75.6 | 75.3 | 74.5 | 66.8 | 55.8 | 38.5 | 34.7 | 54.3 |
| 1992 | 32.1 | 38.1 | 43.7 | 52.2 | 62.3 | 69.6 | 74.3 | 70.9 | 67.0 | 56.1 | 42.4 | 32.8 | 53.5 |
| 1993 | 29.8 | 27.6 | 38.6 | 50.9 | 64.3 | 72.0 | 77.2 | 76.0 | 62.4 | 52.9 | 40.5 | 33.1 | 52.1 |
| 1994 | 21.4 | 29.4 | 42.9 | 53.9 | 62.3 | 75.2 | 75.4 | 72.8 | 66.7 | 57.8 | 47.8 | 37.3 | 53.6 |
| 1995 | 26.4 | 30.3 | 45.1 | 51.6 | 61.2 | 73.7 | 78.3 | 80.2 | 64.7 | 56.0 | 37.5 | 28.0 | 52.8 |
| 1996 | 23.2 | 31.4 | 36.0 | 49.3 | 63.3 | 72.5 | 72.1 | 73.6 | 64.2 | 55.6 | 35.3 | 31.5 | 50.7 |
| 1997 | 20.4 | 34.4 | 43.4 | 48.2 | 58.8 | 70.9 | 75.6 | 71.8 | 66.7 | 56.4 | 39.2 | 32.4 | 51.5 |
| 1998 | 32.7 | 40.4 | 40.4 | 53.0 | 68.3 | 71.7 | 75.4 | 74.9 | 71.1 | 56.7 | 46.0 | 34.0 | 55.4 |
| 1999 | 25.1 | 37.3 | 38.7 | 55.0 | 64.7 | 72.3 | 77.9 | 71.5 | 64.5 | 54.9 | 48.4 | 33.9 | 53.7 |
| 2000 | 27.5 | 39.0 | 46.2 | 52.2 | 65.9 | 69.8 | 73.5 | 75.4 | 67.1 | 58.1 | 38.1 | 17.4 | 52.5 |
| 2001 | 25.8 | 31.4 | 37.4 | 59.4 | 65.2 | 70.4 | 76.6 | 74.1 | 64.4 | 54.1 | 49.9 | 35.5 | 53.7 |
| 2002 | 34.0 | 35.1 | 38.3 | 54.5 | 60.0 | 74.8 | 78.0 | 75.0 | 68.9 | 51.5 | 39.7 | 33.2 | 53.6 |
| 2003 | 22.1 | 26.2 | 43.0 | 54.2 | 61.3 | 68.5 | 75.0 | 75.3 | 62.9 | 54.9 | 44.8 | 34.7 | 51.9 |
| 2004 | 24.8 | 31.2 | 45.4 | 55.3 | 67.1 | 69.4 | 73.1 | 68.8 | 68.2 | 55.5 | 45.7 | 33.2 | 53.1 |
| 2005 | 29.1 | 36.8 | 40.5 | 56.5 | 62.6 | 75.4 | 76.7 | 76.4 | 71.3 | 56.4 | 45.0 | 26.6 | 54.4 |
| 2006 | 39.0 | 31.2 | 42.9 | 58.2 | 64.0 | 71.9 | 78.1 | 74.9 | 64.5 | 51.3 | 45.6 | 35.5 | 54.8 |
| 2007 | 30.6 | 22.5 | 49.8 | 51.1 | 68.7 | 72.9 | 73.0 | 79.3 | 71.1 | 60.2 | 43.2 | 31.2 | 54.5 |
| 2008 | 28.5 | 26.6 | 40.3 | 51.1 | 60.2 | 74.2 | 74.2 | 71.6 | 66.8 | 55.1 | 41.7 | 28.5 | 51.6 |
| 2009 | 23.1 | 34.7 | 45.5 | 53.4 | 64.3 | 74.3 | 71.3 | 71.7 | 66.5 | 51.4 | 48.9 | 30.2 | 52.9 |
| POR= 109 YRS | 26.9 | 30.6 | 41.3 | 53.1 | 63.7 | 73.0 | 77.0 | 74.9 | 67.7 | 56.3 | 42.7 | 31.0 | 53.2 |

## HEATING DEGREE DAYS (base 65°F) 2009  SPRINGFIELD (KSPI)

| YEAR | JUL | AUG | SEP | OCT | NOV | DEC | JAN | FEB | MAR | APR | MAY | JUN | TOTAL |
|---|---|---|---|---|---|---|---|---|---|---|---|---|---|
| 1980-81 | 0 | 0 | 49 | 395 | 675 | 1028 | 1193 | 906 | 648 | 181 | 184 | 0 | 5259 |
| 1981-82 | 0 | 0 | 51 | 332 | 581 | 1175 | 1483 | 1139 | 760 | 502 | 9 | 20 | 6052 |
| 1982-83 | 0 | 5 | 86 | 325 | 656 | 806 | 1117 | 836 | 749 | 510 | 169 | 14 | 5273 |
| 1983-84 | 0 | 0 | 75 | 269 | 574 | 1512 | 1320 | 835 | 1023 | 446 | 170 | 1 | 6225 |
| 1984-85 | 0 | 0 | 127 | 194 | 691 | 899 | 1437 | 1116 | 601 | 262 | 67 | 32 | 5426 |
| 1985-86 | 0 | 11 | 99 | 223 | 658 | 1340 | 1105 | 1060 | 618 | 261 | 74 | 1 | 5450 |
| 1986-87 | 0 | 19 | 25 | 284 | 807 | 999 | 1199 | 788 | 599 | 325 | 34 | 0 | 5079 |
| 1987-88 | 0 | 3 | 44 | 440 | 565 | 934 | 1216 | 1139 | 720 | 360 | 59 | 11 | 5491 |
| 1988-89 | 0 | 5 | 33 | 475 | 654 | 1027 | 890 | 1242 | 749 | 405 | 224 | 19 | 5723 |
| 1989-90 | 0 | 4 | 106 | 269 | 683 | 1431 | 856 | 803 | 613 | 459 | 168 | 10 | 5402 |
| 1990-91 | 7 | 2 | 72 | 360 | 512 | 1080 | 1262 | 827 | 612 | 243 | 69 | 0 | 5046 |
| 1991-92 | 0 | 0 | 120 | 295 | 787 | 932 | 1007 | 771 | 651 | 382 | 154 | 20 | 5119 |
| 1992-93 | 0 | 8 | 82 | 296 | 671 | 991 | 1083 | 1041 | 813 | 419 | 85 | 35 | 5524 |
| 1993-94 | 0 | 0 | 117 | 382 | 730 | 979 | 1347 | 992 | 679 | 351 | 142 | 12 | 5731 |
| 1994-95 | 0 | 4 | 64 | 242 | 512 | 851 | 1187 | 964 | 610 | 395 | 141 | 7 | 4977 |
| 1995-96 | 0 | 0 | 109 | 289 | 820 | 1140 | 1289 | 973 | 892 | 462 | 144 | 19 | 6137 |
| 1996-97 | 1 | 0 | 97 | 305 | 884 | 1031 | 1375 | 852 | 664 | 500 | 212 | 20 | 5941 |
| 1997-98 | 2 | 5 | 57 | 342 | 768 | 1003 | 993 | 681 | 768 | 354 | 39 | 37 | 5049 |
| 1998-99 | 0 | 0 | 17 | 266 | 567 | 953 | 1231 | 771 | 807 | 305 | 77 | 21 | 5015 |
| 1999-00 | 0 | 0 | 100 | 319 | 496 | 955 | 1155 | 748 | 581 | 376 | 89 | 24 | 4843 |
| 2000-01 | 0 | 1 | 78 | 244 | 804 | 1470 | 1208 | 933 | 852 | 220 | 109 | 34 | 5953 |
| 2001-02 | 0 | 0 | 103 | 339 | 443 | 907 | 954 | 829 | 820 | 357 | 197 | 1 | 4950 |
| 2002-03 | 0 | 0 | 36 | 443 | 751 | 980 | 1321 | 1082 | 675 | 336 | 134 | 44 | 5802 |
| 2003-04 | 0 | 0 | 113 | 315 | 599 | 932 | 1241 | 973 | 604 | 313 | 93 | 20 | 5203 |
| 2004-05 | 0 | 33 | 29 | 296 | 572 | 979 | 1106 | 785 | 751 | 268 | 140 | 0 | 4959 |
| 2005-06 | 0 | 0 | 35 | 310 | 592 | 1185 | 797 | 938 | 675 | 223 | 137 | 2 | 4894 |
| 2006-07 | 0 | 0 | 90 | 459 | 577 | 908 | 1062 | 1183 | 484 | 427 | 35 | 0 | 5225 |
| 2007-08 | 0 | 0 | 30 | 220 | 646 | 1042 | 1124 | 1105 | 759 | 416 | 182 | 0 | 5524 |
| 2008-09 | 0 | 0 | 39 | 316 | 694 | 1122 | 1293 | 842 | 599 | 378 | 93 | 9 | 5385 |
| 2009- | 0 | 12 | 38 | 413 | 476 | 1073 | | | | | | | |

WBAN : 93822

## COOLING DEGREE DAYS (base 65°F) 2009  SPRINGFIELD (KSPI)

| YEAR | JAN | FEB | MAR | APR | MAY | JUN | JUL | AUG | SEP | OCT | NOV | DEC | TOTAL |
|---|---|---|---|---|---|---|---|---|---|---|---|---|---|
| 1980 | 0 | 0 | 0 | 10 | 88 | 228 | 515 | 452 | 169 | 20 | 0 | 0 | 1482 |
| 1981 | 0 | 0 | 8 | 50 | 43 | 303 | 380 | 280 | 121 | 13 | 0 | 0 | 1198 |
| 1982 | 0 | 0 | 0 | 1 | 173 | 131 | 379 | 244 | 126 | 32 | 1 | 1 | 1088 |
| 1983 | 0 | 0 | 1 | 1 | 17 | 289 | 483 | 471 | 205 | 34 | 0 | 0 | 1501 |
| 1984 | 0 | 0 | 0 | 16 | 30 | 312 | 303 | 375 | 155 | 44 | 0 | 1 | 1236 |
| 1985 | 0 | 0 | 7 | 52 | 80 | 178 | 293 | 202 | 191 | 17 | 0 | 0 | 1020 |
| 1986 | 0 | 0 | 14 | 38 | 109 | 299 | 434 | 182 | 193 | 14 | 0 | 0 | 1283 |
| 1987 | 0 | 0 | 0 | 21 | 212 | 318 | 436 | 329 | 130 | 0 | 8 | 0 | 1454 |
| 1988 | 0 | 0 | 0 | 3 | 93 | 281 | 430 | 437 | 150 | 6 | 0 | 0 | 1400 |
| 1989 | 0 | 0 | 2 | 30 | 57 | 190 | 332 | 268 | 77 | 24 | 0 | 0 | 980 |
| 1990 | 0 | 0 | 14 | 39 | 18 | 261 | 328 | 289 | 182 | 12 | 1 | 0 | 1144 |
| 1991 | 0 | 0 | 5 | 20 | 207 | 327 | 326 | 302 | 184 | 18 | 0 | 0 | 1389 |
| 1992 | 0 | 0 | 0 | 4 | 75 | 166 | 294 | 199 | 150 | 28 | 0 | 0 | 916 |
| 1993 | 0 | 0 | 0 | 1 | 71 | 253 | 386 | 352 | 46 | 13 | 0 | 0 | 1122 |
| 1994 | 0 | 0 | 0 | 26 | 67 | 327 | 329 | 257 | 123 | 27 | 2 | 0 | 1158 |
| 1995 | 0 | 0 | 0 | 0 | 33 | 276 | 418 | 479 | 105 | 16 | 0 | 0 | 1327 |
| 1996 | 0 | 0 | 0 | 0 | 99 | 250 | 231 | 274 | 80 | 20 | 0 | 0 | 954 |
| 1997 | 0 | 0 | 0 | 0 | 28 | 204 | 341 | 223 | 115 | 81 | 0 | 0 | 992 |
| 1998 | 0 | 0 | 11 | 0 | 145 | 246 | 327 | 312 | 207 | 15 | 0 | 1 | 1264 |
| 1999 | 0 | 0 | 0 | 12 | 74 | 249 | 409 | 210 | 94 | 14 | 1 | 0 | 1063 |
| 2000 | 0 | 0 | 4 | 0 | 123 | 174 | 273 | 332 | 148 | 37 | 3 | 0 | 1094 |
| 2001 | 0 | 0 | 0 | 60 | 122 | 203 | 365 | 288 | 88 | 8 | 0 | 0 | 1134 |
| 2002 | 0 | 0 | 0 | 50 | 301 | 413 | 321 | 162 | 34 | 0 | 0 | 0 | 1332 |
| 2003 | 0 | 0 | 0 | 20 | 29 | 156 | 318 | 327 | 57 | 8 | 1 | 0 | 916 |
| 2004 | 0 | 0 | 0 | 28 | 165 | 161 | 257 | 157 | 133 | 9 | 0 | 0 | 910 |
| 2005 | 0 | 0 | 0 | 21 | 74 | 320 | 371 | 360 | 231 | 48 | 1 | 0 | 1426 |
| 2006 | 0 | 0 | 0 | 29 | 114 | 214 | 412 | 313 | 82 | 40 | 0 | 0 | 1204 |
| 2007 | 0 | 0 | 19 | 14 | 157 | 242 | 256 | 451 | 221 | 78 | 0 | 0 | 1438 |
| 2008 | 0 | 0 | 0 | 7 | 41 | 280 | 295 | 211 | 99 | 17 | 0 | 0 | 950 |
| 2009 | 0 | 0 | 0 | 36 | 80 | 294 | 199 | 226 | 93 | 0 | 0 | 0 | 928 |

## SNOWFALL (inches) 2009 SPRINGFIELD (KSPI)

| YEAR | JUL | AUG | SEP | OCT | NOV | DEC | JAN | FEB | MAR | APR | MAY | JUN | TOTAL |
|---|---|---|---|---|---|---|---|---|---|---|---|---|---|
| 1980-81 | 0.0 | 0.0 | 0.0 | T | 3.5 | 2.1 | 2.7 | 8.6 | 0.6 | 0.0 | 0.0 | 0.0 | 17.5 |
| 1981-82 | 0.0 | 0.0 | 0.0 | 0.0 | 0.1 | 21.6 | 12.0 | 11.4 | 0.7 | 4.6 | 0.0 | 0.0 | 50.4 |
| 1982-83 | 0.0 | 0.0 | 0.0 | 0.0 | 0.2 | 0.9 | 2.4 | 1.3 | 5.4 | 0.2 | 0.0 | 0.0 | 10.4 |
| 1983-84 | 0.0 | 0.0 | 0.0 | T | T | 16.2 | 3.9 | 9.8 | 5.7 | 0.0 | 0.0 | 0.0 | 35.6 |
| 1984-85 | 0.0 | 0.0 | 0.0 | 0.0 | 1.8 | 0.7 | 9.3 | 2.5 | T | T | 0.0 | 0.0 | 14.3 |
| 1985-86 | 0.0 | 0.0 | 0.0 | 0.0 | T | 4.3 | 0.2 | 15.1 | 0.6 | 0.1 | 0.0 | 0.0 | 20.3 |
| 1986-87 | 0.0 | 0.0 | 0.0 | 0.0 | 0.5 | T | 20.3 | T | 0.0 | 0.0 | 0.0 | 0.0 | 20.8 |
| 1987-88 | 0.0 | 0.0 | 0.0 | 0.0 | 0.1 | 5.7 | 0.4 | 10.2 | 5.0 | 0.0 | 0.0 | 0.0 | 21.4 |
| 1988-89 | 0.0 | 0.0 | 0.0 | 0.0 | T | 5.5 | 0.3 | 13.7 | 5.2 | T | T | 0.0 | 24.7 |
| 1989-90 | 0.0 | 0.0 | 0.0 | 0.3 | T | 6.7 | 0.6 | 0.5 | 1.2 | T | 0.0 | 0.0 | 9.3 |
| 1990-91 | 0.0 | 0.0 | 0.0 | 0.0 | T | 9.4 | 6.8 | 0.8 | 3.1 | 0.0 | 0.0 | 0.0 | 20.1 |
| 1991-92 | 0.0 | 0.0 | 0.0 | T | 3.2 | 0.5 | 4.6 | 0.9 | 2.2 | T | 0.0 | 0.0 | 11.4 |
| 1992-93 | 0.0 | 0.0 | 0.0 | 0.0 | 1.2 | 2.5 | 7.0 | 16.0 | 2.9 | 0.4 | 0.0 | 0.0 | 30.0 |
| 1993-94 | 0.0 | 0.0 | 0.0 | 0.2 | 0.8 | 1.7 | 5.8 | 4.4 | 0.9 | 2.7 | T | 0.0 | 16.5 |
| 1994-95 | T | 0.0 | 0.0 | 0.0 | T | 0.2 | 3.7 | 4.4 | 0.2 | 0.0 | 0.0 | 0.0 | 8.5 |
| 1995-96 | 0.0 | 0.0 | 0.0 | 0.0 | 1.2 | 5.9 | 11.2 | 1.1 | 1.6 | 0.0 | T | 0.0 | 21.0 |
| 1996-97 | 0.0 | 0.0 | 0.0 | 0.0 | 0.2 | 2.4 | 19.3 | T | T | T | 0.0 | 0.0 | 21.9 |
| 1997-98 | 0.0 | 0.0 | 0.0 | T | 1.2 | 7.6 | 3.5 | 0.0 | 6.9 | 0.0 | 0.0 | 0.0 | 19.2 |
| 1998-99 | 0.0 | 0.0 | 0.0 | 0.0 | T | 2.5 | 16.1 | 3.2 | 6.4 | T | 0.0 | 0.0 | 28.2 |
| 1999-00 | 0.0 | 0.0 | 0.0 | 0.0 | 0.0 | 2.4 | 6.8 | 0.3 | 5.5 | T | 0.0 | 0.0 | 15.0 |
| 2000-01 | 0.0 | 0.0 | 0.0 | 0.0 | T | 10.0 | 2.1 | 0.9 | 0.4 | T | 0.0 | 0.0 | 13.4 |
| 2001-02 | 0.0 | 0.0 | 0.0 | 0.0 | T | 3.3 | 1.1 | 4.0 | 2.4 | 0.0 | 0.0 | 0.0 | 10.8 |
| 2002-03 | 0.0 | 0.0 | 0.0 | 0.0 | T | 5.3 | 9.1 | 10.5 | 0.2 | T | 0.0 | 0.0 | 25.1 |
| 2003-04 | 0.0 | 0.0 | 0.0 | 0.0 | T | 3.7 | 6.1 | 2.7 | 1.8 | 0.0 | T | 0.0 | 14.3 |
| 2004-05 | 0.0 | 0.0 | 0.0 | 0.0 | 6.1 | T | 8.6 | 2.1 | 0.3 | 0.0 | 0.0 | 0.0 | 17.1 |
| 2005-06 | 0.0 | 0.0 | 0.0 | 0.0 | T | 14.5 | 0.9 | 4.4 | 6.0 | 0.0 | 0.0 | 0.0 | 25.8 |
| 2006-07 | 0.0 | 0.0 | 0.0 | T | 0.7 | 5.9 | 4.6 | 13.3 | 0.2 | T | 0.0 | 0.0 | 24.7 |
| 2007-08 | 0.0 | 0.0 | 0.0 | 0.0 | 0.2 | 11.0 | 7.0 | 13.1 | 4.2 | T | 0.0 | 0.0 | 35.5 |
| 2008-09 | 0.0 | 0.0 | 0.0 | T | 1.2 | 2.6 | 6.9 | 0.3 | 5.8 | 0.2 | 0.0 | 0.0 | 17.0 |
| 2009- | 0.0 | 0.0 | 0.0 | 0.0 | 0.0 | 6.9 | | | | | | | |
| POR=<br>109 YRS | T | 0.0 | 0.0 | 0.1 | 1.2 | 4.7 | 5.7 | 5.3 | 3.8 | 0.5 | T | 0.0 | 21.3 |

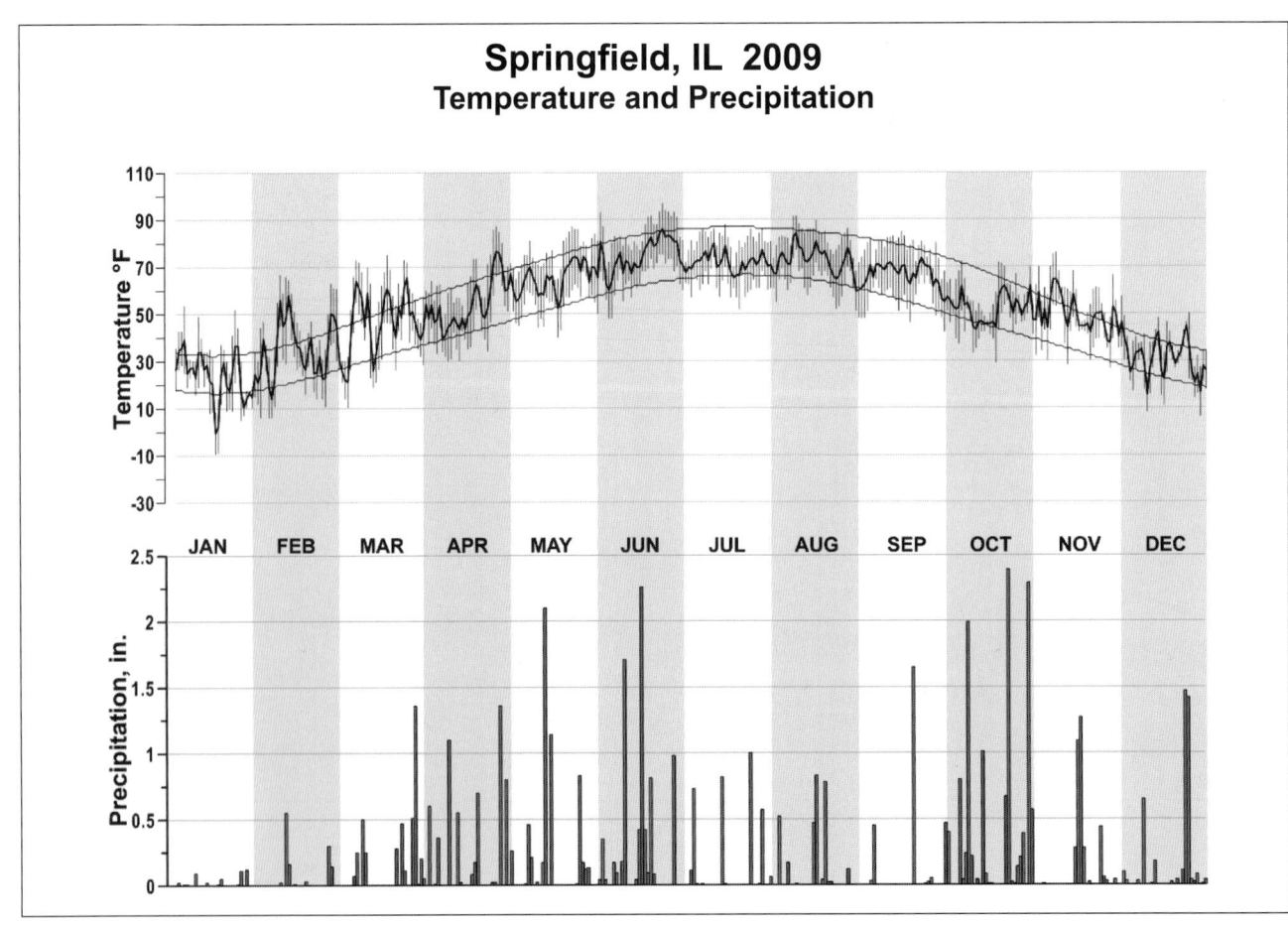

# Springfield, IL 2009
## Temperature and Precipitation

# 2009
# EVANSVILLE
# INDIANA (KEVV)

Evansville, Indiana, is located on the Ohio River. The country around Evansville ranges from level to areas of rolling terrain near the river. Dress Regional Airport, where the observations have been taken since August 31, 1940, is located in a shallow valley with low hills to the east and west which parallel the valley, but slope down to the south. There are hills 5 miles to the north which are about 100 feet higher than the field. The open end of the valley slopes down and south toward the city of Evansville and the Ohio River.

Records of precipitation, temperature, and wind are available from the city office locations prior to August 1940. Both precipitation and temperature records were from roof-top exposures in the city and from ground exposures at the airport. The airport exposure is not subject to the effect of an early morning smoke blanket that was prevalent over the city during the downtown exposure.

Prevailing wind direction is from the south-southwest. The strongest winds occur during a deep winter storm passage through the Lower Ohio Valley. Strong and cold north to northwest winds occur from late autumn to early spring, most often, in January and February, as large domes of arctic high pressure moves into the midwest.

Geographically, Evansville lies in the path of moisture-bearing low pressure formations that move from the western Gulf region, northeastward over the Mississippi and Ohio Valleys to the Great Lakes and northern Atlantic Coast. Much of the precipitation results from these storm systems, especially in the cooler part of the year.

Both temperature and precipitation are closely related to the movement of the polar front and the storms which move along the front. This is especially true in the winter and spring months.

In summer and early autumn changes are less severe and periods of polar air invasions are less prolonged. There is considerable variation in seasonal and monthly temperature and precipitation from year to year as these factors depend greatly on the frequency of storm and frontal passages. A comparatively few miles difference in the distance of the paths of these storms, often spells the difference between whether the precipitation is snow, rain, or freezing rain during winter months.

Convective thunderstorms, developing in the maritime tropical air from the Gulf of Mexico and squall line activity, seem to be the factors which combine to supply the summer rainfall. The greatest precipitation intensities for short periods of time come in the months of greatest thunderstorm frequency. The greatest intensities for 24 hours or more are confined to the winter months when storm centers to the south produce a sustained flow of overrunning Gulf air.

Severe storms are rather infrequent but thunderstorms cause some wind damage each year. Hail often occurs with the stronger thunderstorms. Evansville is in tornado alley with the most frequent occurrence in early spring and late fall. The tornado frequency would probably be less than one every ten years for Evansville.

Snowfall varies greatly from season to season, as do rainfall and temperature. Of note is the fact that snowfalls of 2 or more inches are very infrequent, and these amounts are usually melted within a day or two.

The growing season averages 199 days, but has been as long as 250 days and as short as 169 days.

# NORMALS, MEANS, AND EXTREMES
## EVANSVILLE (KEVV)

LATITUDE: 38 ° 2 'N  LONGITUDE: -87 ° 32'W  ELEVATION (FT): GRND: 400  BARO: 421  TIME ZONE: CENTRAL (UTC -6)  WBAN: 93817

| | ELEMENT | POR | JAN | FEB | MAR | APR | MAY | JUN | JUL | AUG | SEP | OCT | NOV | DEC | YEAR |
|---|---|---|---|---|---|---|---|---|---|---|---|---|---|---|---|
| **TEMPERATURE °F** | NORMAL DAILY MAXIMUM | 30 | 39.5 | 45.4 | 56.4 | 67.2 | 77.1 | 86.1 | 89.4 | 87.8 | 81.3 | 70.0 | 55.7 | 44.1 | 66.7 |
| | MEAN DAILY MAXIMUM | 113 | 41.2 | 42.8 | 55.3 | 65.6 | 76.3 | 84.1 | 88.7 | 87.4 | 80.1 | 70.0 | 54.7 | 43.9 | 65.8 |
| | HIGHEST DAILY MAXIMUM | 69 | 76 | 79 | 84 | 91 | 95 | 104 | 105 | 104 | 103 | 94 | 83 | 77 | 105 |
| | YEAR OF OCCURRENCE | | 1943 | 1962 | 1986 | 1989 | 1975 | 1954 | 1954 | 2007 | 1954 | 1953 | 1961 | 1982 | JUL 1954 |
| | MEAN OF EXTREME MAXS. | 113 | 63.5 | 67.6 | 76.9 | 84.0 | 89.2 | 95.1 | 96.3 | 95.7 | 92.6 | 85.5 | 75.1 | 65.0 | 82.2 |
| | NORMAL DAILY MINIMUM | 30 | 22.6 | 26.2 | 35.2 | 43.8 | 54.0 | 63.5 | 67.8 | 65.1 | 57.0 | 44.6 | 36.0 | 27.0 | 45.2 |
| | MEAN DAILY MINIMUM | 113 | 25.1 | 26.2 | 36.2 | 45.3 | 55.6 | 63.7 | 68.4 | 66.6 | 58.2 | 47.3 | 36.5 | 28.3 | 46.5 |
| | LOWEST DAILY MINIMUM | 69 | -21 | -23 | -9 | 23 | 28 | 41 | 47 | 43 | 31 | 21 | -3 | -15 | -23 |
| | YEAR OF OCCURRENCE | | 1977 | 1951 | 1960 | 1990 | 1963 | 1966 | 1947 | 1986 | 1942 | 1952 | 1950 | 1989 | FEB 1951 |
| | MEAN OF EXTREME MINS. | 113 | 2.0 | 7.0 | 18.0 | 29.1 | 39.6 | 50.2 | 56.4 | 54.2 | 41.9 | 29.8 | 19.9 | 8.0 | 29.7 |
| | NORMAL DRY BULB | 30 | 31.0 | 35.8 | 45.8 | 55.5 | 65.6 | 74.8 | 78.6 | 76.5 | 69.1 | 57.3 | 45.9 | 35.6 | 56.0 |
| | MEAN DRY BULB | 113 | 33.2 | 34.5 | 45.8 | 55.4 | 66.0 | 73.9 | 78.6 | 77.0 | 69.2 | 58.6 | 45.6 | 36.1 | 56.2 |
| | MEAN WET BULB | 26 | 29.8 | 32.8 | 40.6 | 49.6 | 59.2 | 67.1 | 70.4 | 69.1 | 61.9 | 51.2 | 41.7 | 32.9 | 50.5 |
| | MEAN DEW POINT | 26 | 26.6 | 29.3 | 36.3 | 45.4 | 55.9 | 64.4 | 68.2 | 66.9 | 59.0 | 47.7 | 38.2 | 29.6 | 47.3 |
| | NORMAL NO. DAYS WITH: | | | | | | | | | | | | | | |
| | MAXIMUM >= 90 | 30 | 0.0 | 0.0 | 0.0 | 0.1 | 1.6 | 9.6 | 15.3 | 10.9 | 4.4 | 0.1 | 0.0 | 0.0 | 42.0 |
| | MAXIMUM <= 32 | 30 | 9.0 | 5.4 | 0.7 | 0.0 | 0.0 | 0.0 | 0.0 | 0.0 | 0.0 | 0.0 | 0.4 | 5.0 | 20.5 |
| | MINIMUM <= 32 | 30 | 24.4 | 19.0 | 12.6 | 3.0 | * | 0.0 | 0.0 | 0.0 | 0.0 | 2.4 | 11.5 | 20.9 | 93.8 |
| | MINIMUM <= 0 | 30 | 1.6 | 0.9 | 0.1 | 0.0 | 0.0 | 0.0 | 0.0 | 0.0 | 0.0 | 0.0 | 0.0 | 0.6 | 3.2 |
| **H/C** | NORMAL HEATING DEG. DAYS | 30 | 1047 | 825 | 591 | 295 | 85 | 5 | 0 | 1 | 45 | 262 | 565 | 896 | 4617 |
| | NORMAL COOLING DEG. DAYS | 30 | 0 | 0 | 4 | 23 | 108 | 304 | 425 | 356 | 173 | 27 | 2 | 0 | 1422 |
| **RH** | NORMAL (PERCENT) | 30 | 74 | 72 | 68 | 66 | 69 | 69 | 71 | 73 | 73 | 70 | 72 | 76 | 71 |
| | HOUR 00 LST | 30 | 77 | 77 | 75 | 75 | 81 | 81 | 84 | 85 | 85 | 81 | 78 | 79 | 80 |
| | HOUR 06 LST | 30 | 80 | 80 | 79 | 80 | 83 | 83 | 85 | 88 | 89 | 85 | 81 | 81 | 83 |
| | HOUR 12 LST | 30 | 68 | 64 | 59 | 54 | 56 | 55 | 57 | 58 | 55 | 53 | 62 | 69 | 59 |
| | HOUR 18 LST | 30 | 72 | 67 | 61 | 55 | 58 | 56 | 59 | 62 | 65 | 64 | 68 | 74 | 63 |
| **S** | PERCENT POSSIBLE SUNSHINE | 56 | 42 | 48 | 55 | 60 | 64 | 71 | 73 | 73 | 69 | 65 | 48 | 42 | 59 |
| **W/O** | MEAN NO. DAYS WITH: | | | | | | | | | | | | | | |
| | HEAVY FOG(VISBY <= 1/4 MI) | 46 | 1.7 | 1.6 | 1.0 | 0.7 | 0.9 | 0.6 | 1.0 | 1.2 | 1.8 | 1.7 | 1.4 | 1.5 | 15.1 |
| | THUNDERSTORMS | 62 | 1.0 | 1.2 | 3.4 | 4.8 | 6.3 | 6.9 | 7.1 | 5.0 | 3.1 | 1.9 | 1.5 | 0.6 | 42.8 |
| **CLOUDNESS** | MEAN: | | | | | | | | | | | | | | |
| | SUNRISE-SUNSET (OKTAS) | 56 | 5.7 | 5.4 | 5.4 | 5.2 | 4.9 | 4.6 | 4.2 | 4.0 | 4.0 | 3.9 | 5.1 | 5.5 | 4.8 |
| | MIDNIGHT-MIDNIGHT (OKTAS) | 32 | 5.4 | 4.9 | 5.0 | 4.7 | 4.3 | 4.0 | 3.8 | 3.6 | 3.8 | 3.8 | 4.9 | 5.2 | 4.5 |
| | MEAN NO. DAYS WITH: | | | | | | | | | | | | | | |
| | CLEAR | 56 | 6.6 | 6.5 | 6.5 | 6.8 | 8.5 | 8.1 | 9.2 | 11.1 | 11.3 | 12.3 | 7.7 | 6.5 | 101.1 |
| | PARTLY CLOUDY | 56 | 5.3 | 6.3 | 8.1 | 8.3 | 8.6 | 11.4 | 12.1 | 11.0 | 8.3 | 7.4 | 6.6 | 6.0 | 99.4 |
| | CLOUDY | 56 | 19.1 | 15.4 | 16.4 | 14.9 | 13.9 | 10.5 | 9.1 | 8.4 | 9.9 | 10.8 | 15.2 | 18.0 | 161.6 |
| **PR** | MEAN STATION PRESSURE(IN) | 26 | 29.72 | 29.70 | 29.64 | 29.57 | 29.57 | 29.56 | 29.59 | 29.61 | 29.64 | 29.67 | 29.69 | 29.73 | 29.64 |
| | MEAN SEA-LEVEL PRES. (IN) | 26 | 30.15 | 30.13 | 30.06 | 29.98 | 29.98 | 29.97 | 29.99 | 30.02 | 30.05 | 30.09 | 30.11 | 30.15 | 30.06 |
| **WINDS** | MEAN SPEED (MPH) | 26 | 8.5 | 8.4 | 8.6 | 8.4 | 7.1 | 6.2 | 5.6 | 5.2 | 5.5 | 6.2 | 7.6 | 7.8 | 7.1 |
| | PREVAIL.DIR(TENS OF DEGS) | 41 | 35 | 33 | 33 | 23 | 23 | 23 | 25 | 24 | 18 | 19 | 22 | 31 | 25 |
| | MAXIMUM 2-MINUTE: | 13 | | | | | | | | | | | | | |
| | SPEED (MPH) | | 52 | 44 | 40 | 44 | 52 | 46 | 45 | 46 | 47 | 33 | 38 | 45 | 52 |
| | DIR. (TENS OF DEGS) | | 27 | 23 | 26 | 26 | 28 | 28 | 28 | 02 | 19 | 32 | 24 | 24 | 27 |
| | YEAR OF OCCURRENCE | | 2008 | 2009 | 2009 | 2005 | 2004 | 2000 | 2003 | 2006 | 2008 | 2001 | 1998 | 2008 | JAN 2008 |
| | MAXIMUM 3-SECOND | 13 | | | | | | | | | | | | | |
| | SPEED (MPH) | | 71 | 59 | 48 | 64 | 68 | 59 | 64 | 56 | 64 | 46 | 48 | 61 | 71 |
| | DIR. (TENS OF DEGS) | | 26 | 20 | 26 | 29 | 23 | 19 | 28 | 02 | 20 | 19 | 25 | 25 | 26 |
| | YEAR OF OCCURRENCE | | 2008 | 2009 | 2009 | 2006 | 1999 | 2007 | 2003 | 2006 | 2008 | 2007 | 1998 | 2008 | JAN 2008 |
| **PRECIPITATION** | NORMAL (IN) | 30 | 2.91 | 3.10 | 4.29 | 4.48 | 5.01 | 4.10 | 3.75 | 3.14 | 2.99 | 2.78 | 4.18 | 3.54 | 44.27 |
| | MAXIMUM MONTHLY (IN) | 69 | 13.50 | 7.26 | 12.84 | 11.83 | 13.51 | 9.30 | 9.69 | 8.51 | 9.89 | 8.33 | 8.49 | 8.23 | 13.51 |
| | YEAR OF OCCURRENCE | | 1950 | 2000 | 1964 | 1996 | 1995 | 1943 | 1958 | 2005 | 1945 | 1941 | 1957 | 1982 | MAY 1995 |
| | MINIMUM MONTHLY (IN) | 69 | 0.51 | 0.27 | 0.89 | 1.10 | 0.91 | 0.65 | 0.18 | 0.13 | 0.09 | 0.01 | 0.51 | 0.56 | 0.01 |
| | YEAR OF OCCURRENCE | | 1981 | 1947 | 1941 | 1959 | 1965 | 1991 | 1974 | 1943 | 2004 | 1964 | 1999 | 1976 | OCT 1964 |
| | MAXIMUM IN 24 HOURS (IN) | 69 | 3.74 | 3.84 | 6.85 | 7.26 | 6.05 | 3.67 | 4.09 | 3.70 | 4.37 | 3.00 | 3.65 | 3.43 | 7.26 |
| | YEAR OF OCCURRENCE | | 2000 | 2000 | 2008 | 1996 | 1961 | 1996 | 1978 | 1977 | 2006 | 1976 | 2005 | 2001 | APR 1996 |
| | NORMAL NO. DAYS WITH: | | | | | | | | | | | | | | |
| | PRECIPITATION >= 0.01 | 30 | 10.5 | 9.1 | 11.8 | 12.0 | 11.8 | 10.0 | 8.3 | 7.3 | 7.5 | 7.8 | 9.9 | 10.7 | 116.7 |
| | PRECIPITATION >= 1.00 | 30 | 0.5 | 0.9 | 0.9 | 1.2 | 1.4 | 1.0 | 1.1 | 1.1 | 0.8 | 0.6 | 1.3 | 0.9 | 11.7 |
| **SNOWFALL** | NORMAL (IN) | 30 | 4.6 | 3.8 | 2.0 | 0.4 | 0.0 | 0.0 | 0.0 | 0.0 | 0.0 | T | 0.2 | 2.8 | 14.2 |
| | MAXIMUM MONTHLY (IN) | 65 | 21.3 | 18.4 | 20.2 | 8.6 | T | T | 0.0 | 0.0 | 0.0 | 4.6 | 6.9 | 11.5 | 21.3 |
| | YEAR OF OCCURRENCE | | 1977 | 1993 | 1960 | 1971 | 1993 | 1994 | | | 1990 | 1993 | 1958 | 2000 | JAN 1977 |
| | MAXIMUM IN 24 HOURS (IN) | 65 | 8.7 | 10.9 | 10.6 | 8.6 | T | T | 0.0 | 0.0 | T | 4.1 | 6.9 | 7.0 | 10.9 |
| | YEAR OF OCCURRENCE' | | 1978 | 1993 | 1960 | 1971 | 1993 | 1994 | | | 1990 | 1993 | 1958 | 1963 | FEB 1993 |
| | MAXIMUM SNOW DEPTH (IN) | 57 | 14 | 12 | 13 | 4 | 0 | 0 | 0 | 0 | 0 | 2 | 7 | 7 | 14 |
| | YEAR OF OCCURRENCE | | 1978 | 1998 | 1960 | 1971 | | | | | | 1993 | 1958 | 1984 | JAN 1978 |
| | NORMAL NO. DAYS WITH: | | | | | | | | | | | | | | |
| | SNOWFALL >= 1.0 | 30 | 1.3 | 1.0 | 0.7 | 0.1 | 0.0 | 0.0 | 0.0 | 0.0 | 0.0 | 0.0 | 0.2 | 0.8 | 4.1 |

## PRECIPITATION (inches) 2009 EVANSVILLE (KEVV)

| YEAR | JAN | FEB | MAR | APR | MAY | JUN | JUL | AUG | SEP | OCT | NOV | DEC | ANNUAL |
|------|-----|-----|-----|-----|-----|-----|-----|-----|-----|-----|-----|-----|--------|
| 1980 | 1.77 | 1.25 | 4.38 | 2.73 | 4.10 | 6.01 | 4.50 | 2.15 | 2.51 | 3.13 | 2.34 | 0.89 | 35.76 |
| 1981 | 0.51 | 2.89 | 1.70 | 2.50 | 12.89 | 1.78 | 5.08 | 6.04 | 2.00 | 2.36 | 3.40 | 2.20 | 43.35 |
| 1982 | 9.15 | 1.65 | 5.07 | 3.24 | 4.29 | 2.95 | 2.62 | 3.41 | 6.07 | 1.75 | 4.25 | 8.23 | 52.68 |
| 1983 | 1.79 | 0.74 | 4.33 | 10.26 | 8.87 | 4.59 | 1.51 | 0.94 | 0.73 | 5.62 | 5.55 | 3.55 | 48.48 |
| 1984 | 0.85 | 2.55 | 7.02 | 5.75 | 2.89 | 3.35 | 1.50 | 2.70 | 6.97 | 5.13 | 5.05 | 5.99 | 49.75 |
| 1985 | 1.76 | 4.24 | 6.10 | 3.80 | 2.97 | 4.68 | 1.18 | 3.76 | 3.59 | 4.46 | 7.61 | 1.74 | 45.89 |
| 1986 | 1.15 | 5.77 | 2.64 | 2.29 | 2.93 | 3.77 | 5.39 | 2.07 | 3.84 | 3.30 | 2.35 | 2.18 | 37.68 |
| 1987 | 0.77 | 3.51 | 2.11 | 2.31 | 3.90 | 5.97 | 3.19 | 0.47 | 1.98 | 1.23 | 3.36 | 5.71 | 34.51 |
| 1988 | 3.28 | 3.94 | 2.89 | 1.77 | 1.33 | 1.11 | 6.63 | 2.72 | 1.19 | 2.86 | 7.96 | 2.75 | 38.43 |
| 1989 | 3.35 | 7.00 | 6.40 | 4.19 | 3.72 | 4.00 | 7.83 | 3.46 | 2.21 | 2.16 | 1.64 | 1.38 | 47.34 |
| 1990 | 4.26 | 5.60 | 2.15 | 3.75 | 11.34 | 3.22 | 1.01 | 3.47 | 2.54 | 4.81 | 2.92 | 7.45 | 52.52 |
| 1991 | 3.02 | 2.99 | 4.27 | 2.56 | 3.11 | 0.65 | 2.58 | 0.46 | 2.60 | 3.05 | 3.67 | 3.72 | 32.68 |
| 1992 | 0.85 | 1.51 | 4.50 | 1.19 | 3.44 | 1.44 | 8.40 | 4.39 | 2.89 | 1.17 | 4.34 | 1.69 | 35.81 |
| 1993 | 3.57 | 2.61 | 3.23 | 4.38 | 4.20 | 4.65 | 2.37 | 2.17 | 5.59 | 3.76 | 6.62 | 2.68 | 45.83 |
| 1994 | 3.18 | 2.32 | 1.88 | 5.77 | 0.94 | 3.45 | 2.30 | 2.52 | 2.61 | 2.67 | 6.52 | 2.59 | 36.75 |
| 1995 | 2.82 | 2.98 | 2.53 | 5.59 | 13.51 | 4.56 | 2.88 | 3.60 | 0.47 | 2.01 | 2.32 | 3.19 | 46.46 |
| 1996 | 3.51 | 1.50 | 5.19 | 11.83 | 7.32 | 7.78 | 4.56 | 1.20 | 8.45 | 2.53 | 6.66 | 3.50 | 64.03 |
| 1997 | 4.20 | 3.35 | 6.90 | 4.16 | 7.57 | 6.12 | 1.71 | 4.02 | 1.31 | 1.73 | 4.17 | 2.34 | 47.58 |
| 1998 | 2.24 | 2.71 | 3.07 | 8.50 | 5.91 | 5.31 | 3.89 | 3.91 | 0.49 | 3.38 | 2.78 | 3.48 | 45.67 |
| 1999 | 6.00 | 1.94 | 4.30 | 6.15 | 3.21 | 6.27 | 2.00 | 0.64 | 0.39 | 2.80 | 0.51 | 5.13 | 39.34 |
| 2000 | 4.36 | 7.26 | 3.21 | 2.35 | 2.60 | 5.86 | 4.14 | 5.60 | 5.03 | 0.59 | 3.43 | 4.12 | 48.55 |
| 2001 | 1.29 | 3.26 | 2.23 | 1.60 | 3.82 | 3.82 | 5.54 | 6.09 | 2.40 | 7.27 | 5.40 | 7.16 | 49.88 |
| 2002 | 3.72 | 0.74 | 6.20 | 8.58 | 5.70 | 2.86 | 4.32 | 0.63 | 5.22 | 3.75 | 2.97 | 5.65 | 50.34 |
| 2003 | 0.90 | 4.92 | 2.60 | 3.91 | 6.48 | 4.50 | 4.38 | 1.88 | 3.17 | 1.61 | 4.36 | 1.20 | 39.91 |
| 2004 | 2.95 | 0.59 | 2.17 | 1.91 | 9.31 | 1.66 | 7.56 | 3.08 | 0.09 | 5.62 | 6.23 | 2.31 | 43.48 |
| 2005 | 4.59 | 2.77 | 2.85 | 2.13 | 2.33 | 4.88 | 2.69 | 8.51 | 2.00 | 0.73 | 5.93 | 1.76 | 41.17 |
| 2006 | 4.09 | 2.17 | 9.36 | 3.44 | 5.77 | 3.73 | 6.46 | 7.41 | 8.75 | 5.46 | 4.95 | 4.59 | 66.18 |
| 2007 | 5.47 | 3.41 | 2.66 | 2.88 | 2.73 | 2.71 | 1.97 | 0.99 | 2.22 | 4.64 | 1.77 | 6.34 | 37.79 |
| 2008 | 3.97 | 5.97 | 12.34 | 5.07 | 8.07 | 3.09 | 3.90 | 0.52 | 1.16 | 1.61 | 3.42 | 4.76 | 53.88 |
| 2009 | 2.85 | 2.76 | 3.32 | 6.01 | 6.47 | 2.20 | 6.46 | 1.91 | 5.17 | 8.21 | 1.22 | 3.62 | 50.20 |
| POR= 113 YRS | 3.50 | 2.95 | 4.35 | 4.01 | 4.35 | 3.73 | 3.65 | 3.06 | 3.07 | 2.92 | 3.38 | 3.45 | 42.42 |

WBAN : 93817

## AVERAGE TEMPERATURE (°F) 2009 EVANSVILLE (KEVV)

| YEAR | JAN | FEB | MAR | APR | MAY | JUN | JUL | AUG | SEP | OCT | NOV | DEC | ANNUAL |
|------|-----|-----|-----|-----|-----|-----|-----|-----|-----|-----|-----|-----|--------|
| 1980 | 33.1 | 26.8 | 40.3 | 52.8 | 65.4 | 73.2 | 82.0 | 81.6 | 72.4 | 55.4 | 45.1 | 37.3 | 55.5 |
| 1981 | 29.6 | 37.2 | 44.7 | 61.7 | 61.4 | 76.6 | 78.5 | 75.9 | 67.2 | 57.0 | 48.3 | 34.5 | 56.1 |
| 1982 | 27.4 | 32.1 | 47.8 | 52.1 | 70.7 | 70.7 | 79.6 | 74.1 | 67.5 | 59.2 | 48.9 | 45.2 | 56.3 |
| 1983 | 35.2 | 39.3 | 46.5 | 51.5 | 62.7 | 74.7 | 81.4 | 81.9 | 70.6 | 59.3 | 47.7 | 26.2 | 56.4 |
| 1984 | 27.1 | 39.0 | 40.0 | 54.4 | 62.5 | 78.6 | 76.0 | 76.0 | 66.3 | 62.9 | 43.5 | 43.7 | 55.8 |
| 1985 | 23.7 | 29.6 | 51.9 | 59.1 | 66.5 | 73.7 | 79.2 | 75.1 | 68.5 | 60.7 | 50.1 | 28.2 | 55.5 |
| 1986 | 32.9 | 37.5 | 47.5 | 58.1 | 67.6 | 76.7 | 80.5 | 72.9 | 72.4 | 58.0 | 43.7 | 35.7 | 57.0 |
| 1987 | 32.2 | 38.5 | 47.9 | 54.2 | 71.7 | 76.5 | 78.3 | 77.9 | 71.0 | 51.2 | 49.5 | 39.7 | 57.4 |
| 1988 | 29.0 | 33.1 | 45.4 | 55.6 | 67.0 | 75.7 | 79.0 | 78.8 | 69.5 | 51.5 | 46.4 | 36.5 | 55.6 |
| 1989 | 40.1 | 32.5 | 46.9 | 56.6 | 63.2 | 73.8 | 77.7 | 76.8 | 68.3 | 58.3 | 45.6 | 23.0 | 55.2 |
| 1990 | 41.9 | 43.2 | 49.8 | 53.8 | 63.0 | 74.9 | 77.2 | 75.7 | 71.0 | 56.1 | 50.5 | 38.1 | 57.9 |
| 1991 | 31.3 | 39.7 | 47.9 | 58.9 | 71.1 | 79.1 | 79.1 | 77.0 | 70.2 | 59.2 | 43.2 | 39.2 | 57.8 |
| 1992 | 35.4 | 41.6 | 47.0 | 57.7 | 63.9 | 71.7 | 79.0 | 73.4 | 68.6 | 57.7 | 46.9 | 36.4 | 56.6 |
| 1993 | 36.5 | 32.9 | 43.1 | 54.1 | 67.0 | 75.8 | 83.1 | 79.5 | 67.0 | 55.5 | 44.7 | 36.4 | 56.3 |
| 1994 | 27.3 | 37.3 | 45.4 | 59.2 | 64.0 | 78.8 | 79.2 | 75.4 | 68.1 | 59.7 | 51.4 | 42.2 | 57.3 |
| 1995 | 35.0 | 36.1 | 49.9 | 58.3 | 66.4 | 76.0 | 81.1 | 83.4 | 68.7 | 60.3 | 41.2 | 35.6 | 57.7 |
| 1996 | 30.9 | 36.2 | 38.3 | 51.5 | 67.3 | 74.0 | 75.2 | 75.7 | 66.0 | 56.8 | 40.4 | 38.0 | 54.2 |
| 1997 | 29.0 | 39.7 | 47.3 | 50.5 | 60.3 | 72.0 | 77.2 | 74.1 | 67.8 | 57.3 | 41.8 | 35.0 | 54.3 |
| 1998 | 39.5 | 42.1 | 46.2 | 54.8 | 69.8 | 75.1 | 77.7 | 76.7 | 73.9 | 60.0 | 48.3 | 38.8 | 58.6 |
| 1999 | 34.7 | 40.9 | 41.5 | 57.7 | 65.5 | 74.5 | 79.4 | 74.4 | 68.2 | 56.6 | 51.3 | 37.3 | 56.8 |
| 2000 | 33.2 | 42.7 | 48.0 | 53.9 | 67.4 | 73.7 | 75.3 | 76.3 | 67.3 | 60.5 | 43.0 | 23.6 | 55.4 |
| 2001 | 30.8 | 39.2 | 41.5 | 61.3 | 67.6 | 72.7 | 77.6 | 76.7 | 67.2 | 56.6 | 50.8 | 40.5 | 56.9 |
| 2002 | 38.2 | 38.4 | 43.7 | 58.8 | 63.9 | 76.3 | 80.3 | 78.9 | 73.5 | 56.6 | 43.2 | 36.7 | 57.4 |
| 2003 | 27.4 | 31.5 | 46.5 | 57.4 | 64.4 | 69.5 | 77.1 | 77.3 | 66.8 | 57.7 | 50.3 | 38.4 | 55.4 |
| 2004 | 30.9 | 36.1 | 49.3 | 57.7 | 70.1 | 74.0 | 75.4 | 72.0 | 69.9 | 59.9 | 50.2 | 33.8 | 56.6 |
| 2005 | 38.1 | 40.5 | 42.7 | 56.4 | 63.2 | 75.4 | 77.6 | 79.3 | 72.3 | 59.0 | 47.8 | 32.8 | 57.1 |
| 2006 | 42.8 | 35.4 | 46.5 | 60.3 | 64.1 | 74.1 | 78.4 | 78.1 | 65.2 | 54.4 | 46.8 | 41.0 | 57.3 |
| 2007 | 37.1 | 29.0 | 53.7 | 53.7 | 69.6 | 75.6 | 76.7 | 83.4 | 73.5 | 63.3 | 45.9 | 38.8 | 58.4 |
| 2008 | 32.1 | 35.3 | 45.0 | 55.1 | 63.1 | 76.4 | 77.6 | 76.0 | 71.7 | 58.1 | 43.6 | 35.5 | 55.8 |
| 2009 | 28.8 | 38.5 | 49.4 | 56.7 | 66.2 | 77.0 | 73.4 | 74.8 | 70.1 | 53.7 | 49.9 | 35.0 | 56.1 |
| POR= 113 YRS | 33.2 | 34.5 | 45.8 | 55.4 | 66.0 | 73.9 | 78.6 | 77.0 | 69.2 | 58.6 | 45.6 | 36.1 | 56.2 |

## HEATING DEGREE DAYS (base 65°F) 2009 EVANSVILLE (KEVV)

| YEAR | JUL | AUG | SEP | OCT | NOV | DEC | JAN | FEB | MAR | APR | MAY | JUN | TOTAL |
|---|---|---|---|---|---|---|---|---|---|---|---|---|---|
| 1980-81 | 0 | 0 | 24 | 329 | 591 | 852 | 1090 | 771 | 624 | 161 | 155 | 0 | 4597 |
| 1981-82 | 0 | 0 | 53 | 256 | 498 | 940 | 1160 | 914 | 534 | 386 | 16 | 0 | 4757 |
| 1982-83 | 0 | 0 | 52 | 233 | 486 | 618 | 918 | 711 | 567 | 406 | 106 | 4 | 4101 |
| 1983-84 | 0 | 0 | 61 | 186 | 514 | 1195 | 1169 | 747 | 769 | 329 | 131 | 0 | 5101 |
| 1984-85 | 0 | 0 | 79 | 108 | 638 | 653 | 1276 | 985 | 411 | 208 | 55 | 9 | 4422 |
| 1985-86 | 0 | 0 | 75 | 185 | 446 | 1135 | 989 | 762 | 538 | 226 | 70 | 0 | 4426 |
| 1986-87 | 0 | 15 | 14 | 240 | 632 | 900 | 1007 | 735 | 528 | 330 | 19 | 0 | 4420 |
| 1987-88 | 0 | 0 | 15 | 423 | 456 | 777 | 1108 | 917 | 602 | 284 | 46 | 4 | 4632 |
| 1988-89 | 0 | 0 | 18 | 418 | 548 | 877 | 765 | 902 | 558 | 308 | 142 | 1 | 4537 |
| 1989-90 | 0 | 1 | 54 | 225 | 577 | 1297 | 707 | 603 | 487 | 358 | 97 | 15 | 4421 |
| 1990-91 | 2 | 1 | 35 | 291 | 432 | 828 | 1037 | 702 | 528 | 191 | 42 | 0 | 4089 |
| 1991-92 | 0 | 0 | 88 | 227 | 647 | 791 | 913 | 673 | 549 | 259 | 118 | 10 | 4275 |
| 1992-93 | 0 | 0 | 46 | 236 | 538 | 875 | 879 | 892 | 671 | 322 | 46 | 12 | 4517 |
| 1993-94 | 0 | 0 | 55 | 296 | 600 | 879 | 1164 | 770 | 603 | 213 | 96 | 5 | 4681 |
| 1994-95 | 0 | 0 | 44 | 180 | 403 | 702 | 922 | 804 | 465 | 229 | 62 | 0 | 3811 |
| 1995-96 | 0 | 0 | 50 | 168 | 710 | 904 | 1048 | 829 | 819 | 407 | 75 | 5 | 5015 |
| 1996-97 | 0 | 0 | 70 | 253 | 732 | 830 | 1108 | 702 | 542 | 428 | 172 | 11 | 4848 |
| 1997-98 | 0 | 0 | 24 | 300 | 692 | 922 | 786 | 635 | 604 | 301 | 30 | 20 | 4314 |
| 1998-99 | 0 | 0 | 4 | 181 | 492 | 810 | 931 | 667 | 720 | 220 | 39 | 4 | 4068 |
| 1999-00 | 0 | 0 | 53 | 257 | 404 | 854 | 976 | 641 | 518 | 327 | 45 | 5 | 4080 |
| 2000-01 | 0 | 0 | 71 | 191 | 659 | 1277 | 1053 | 716 | 723 | 183 | 37 | 12 | 4922 |
| 2001-02 | 0 | 0 | 67 | 267 | 421 | 754 | 827 | 738 | 654 | 240 | 126 | 0 | 4094 |
| 2002-03 | 0 | 0 | 10 | 297 | 647 | 871 | 1160 | 932 | 566 | 234 | 72 | 28 | 4817 |
| 2003-04 | 0 | 0 | 53 | 233 | 436 | 815 | 1051 | 834 | 487 | 245 | 48 | 0 | 4202 |
| 2004-05 | 1 | 9 | 10 | 169 | 440 | 959 | 829 | 679 | 684 | 259 | 122 | 0 | 4161 |
| 2005-06 | 0 | 0 | 16 | 226 | 514 | 991 | 683 | 824 | 567 | 175 | 111 | 0 | 4107 |
| 2006-07 | 0 | 0 | 59 | 346 | 542 | 737 | 858 | 1002 | 371 | 357 | 26 | 0 | 4298 |
| 2007-08 | 0 | 0 | 9 | 163 | 566 | 806 | 1014 | 856 | 615 | 306 | 107 | 0 | 4442 |
| 2008-09 | 0 | 0 | 4 | 248 | 634 | 906 | 1112 | 734 | 478 | 290 | 62 | 4 | 4472 |
| 2009- | 0 | 0 | 18 | 345 | 447 | 923 | | | | | | | |

WBAN : 93817

## COOLING DEGREE DAYS (base 65°F) 2009 EVANSVILLE (KEVV)

| YEAR | JAN | FEB | MAR | APR | MAY | JUN | JUL | AUG | SEP | OCT | NOV | DEC | TOTAL |
|---|---|---|---|---|---|---|---|---|---|---|---|---|---|
| 1980 | 0 | 0 | 0 | 5 | 102 | 264 | 535 | 521 | 257 | 39 | 3 | 0 | 1726 |
| 1981 | 0 | 0 | 1 | 69 | 50 | 355 | 425 | 343 | 128 | 15 | 3 | 0 | 1389 |
| 1982 | 0 | 0 | 10 | 3 | 198 | 179 | 458 | 290 | 134 | 59 | 9 | 11 | 1351 |
| 1983 | 0 | 0 | 3 | 8 | 42 | 303 | 514 | 532 | 236 | 17 | 0 | 0 | 1655 |
| 1984 | 0 | 0 | 0 | 16 | 60 | 416 | 348 | 349 | 127 | 49 | 0 | 0 | 1365 |
| 1985 | 0 | 3 | 13 | 36 | 108 | 276 | 447 | 319 | 190 | 58 | 5 | 0 | 1455 |
| 1986 | 0 | 0 | 2 | 27 | 156 | 360 | 487 | 265 | 246 | 32 | 0 | 0 | 1575 |
| 1987 | 0 | 0 | 0 | 8 | 235 | 350 | 420 | 408 | 201 | 0 | 1 | 0 | 1623 |
| 1988 | 0 | 0 | 0 | 11 | 113 | 329 | 441 | 436 | 162 | 8 | 0 | 0 | 1500 |
| 1989 | 0 | 0 | 3 | 64 | 96 | 272 | 403 | 369 | 161 | 28 | 0 | 0 | 1396 |
| 1990 | 0 | 0 | 21 | 29 | 43 | 318 | 387 | 336 | 220 | 23 | 3 | 0 | 1380 |
| 1991 | 0 | 0 | 4 | 15 | 241 | 369 | 445 | 379 | 249 | 55 | 0 | 0 | 1757 |
| 1992 | 0 | 0 | 0 | 47 | 90 | 219 | 440 | 268 | 162 | 14 | 0 | 0 | 1240 |
| 1993 | 0 | 0 | 0 | 3 | 115 | 342 | 566 | 456 | 122 | 9 | 0 | 0 | 1613 |
| 1994 | 0 | 1 | 0 | 42 | 74 | 423 | 449 | 330 | 144 | 24 | 2 | 0 | 1489 |
| 1995 | 0 | 0 | 3 | 33 | 116 | 336 | 510 | 577 | 169 | 29 | 0 | 0 | 1773 |
| 1996 | 0 | 0 | 0 | 8 | 154 | 282 | 321 | 338 | 105 | 6 | 0 | 0 | 1214 |
| 1997 | 0 | 0 | 0 | 0 | 32 | 227 | 386 | 288 | 116 | 70 | 0 | 0 | 1119 |
| 1998 | 0 | 0 | 28 | 3 | 188 | 330 | 398 | 370 | 276 | 35 | 0 | 1 | 1629 |
| 1999 | 0 | 0 | 0 | 10 | 59 | 297 | 454 | 298 | 158 | 8 | 0 | 0 | 1284 |
| 2000 | 0 | 0 | 0 | 0 | 126 | 271 | 327 | 356 | 146 | 57 | 6 | 0 | 1289 |
| 2001 | 0 | 0 | 0 | 79 | 126 | 251 | 400 | 369 | 141 | 11 | 0 | 0 | 1377 |
| 2002 | 0 | 0 | 0 | 60 | 97 | 347 | 481 | 436 | 273 | 43 | 0 | 0 | 1737 |
| 2003 | 0 | 0 | 0 | 15 | 61 | 170 | 382 | 387 | 113 | 13 | 2 | 0 | 1143 |
| 2004 | 0 | 0 | 6 | 33 | 211 | 276 | 331 | 231 | 161 | 18 | 2 | 0 | 1269 |
| 2005 | 1 | 0 | 0 | 9 | 75 | 319 | 398 | 450 | 241 | 46 | 5 | 0 | 1544 |
| 2006 | 0 | 0 | 2 | 42 | 91 | 278 | 420 | 415 | 70 | 24 | 0 | 0 | 1342 |
| 2007 | 0 | 0 | 26 | 26 | 175 | 325 | 368 | 574 | 272 | 120 | 0 | 0 | 1886 |
| 2008 | 0 | 0 | 0 | 17 | 56 | 351 | 399 | 344 | 212 | 42 | 0 | 0 | 1421 |
| 2009 | 0 | 0 | 3 | 50 | 103 | 368 | 270 | 309 | 178 | 0 | 0 | 0 | 1281 |

## SNOWFALL (inches)  2009  EVANSVILLE (KEVV)

| YEAR | JUL | AUG | SEP | OCT | NOV | DEC | JAN | FEB | MAR | APR | MAY | JUN | TOTAL |
|---|---|---|---|---|---|---|---|---|---|---|---|---|---|
| 1980-81 | 0.0 | 0.0 | 0.0 | 0.0 | 0.5 | 0.2 | 2.2 | 0.5 | T | 0.0 | 0.0 | 0.0 | 3.4 |
| 1981-82 | 0.0 | 0.0 | 0.0 | 0.0 | 0.1 | 0.7 | 4.1 | 6.8 | 0.3 | 3.0 | 0.0 | 0.0 | 15.0 |
| 1982-83 | 0.0 | 0.0 | 0.0 | 0.0 | T | T | 1.3 | 1.2 | 1.0 | 0.6 | 0.0 | 0.0 | 4.1 |
| 1983-84 | 0.0 | 0.0 | 0.0 | 0.0 | T | 1.6 | 5.5 | 9.7 | 2.8 | 0.0 | 0.0 | 0.0 | 19.6 |
| 1984-85 | 0.0 | 0.0 | 0.0 | 0.0 | T | 6.7 | 10.3 | 9.4 | 0.0 | T | 0.0 | 0.0 | 26.4 |
| 1985-86 | 0.0 | 0.0 | 0.0 | 0.0 | 0.0 | 2.8 | 1.1 | 6.8 | T | 0.0 | 0.0 | 0.0 | 10.7 |
| 1986-87 | 0.0 | 0.0 | 0.0 | 0.0 | T | 0.1 | 2.7 | 3.2 | 1.7 | 0.0 | 0.0 | 0.0 | 7.7 |
| 1987-88 | 0.0 | 0.0 | 0.0 | 0.0 | T | 1.3 | 4.0 | 1.4 | 0.7 | 0.0 | 0.0 | 0.0 | 7.4 |
| 1988-89 | 0.0 | 0.0 | 0.0 | 0.0 | T | 3.0 | T | 0.5 | 0.1 | 0.0 | 0.0 | 0.0 | 3.6 |
| 1989-90 | 0.0 | 0.0 | 0.0 | 0.9 | T | 6.0 | 1.6 | 0.2 | 4.6 | 0.0 | 0.0 | 0.0 | 13.3 |
| 1990-91 | 0.0 | 0.0 | T | 0.0 | 0.0 | 7.2 | 0.8 | 1.2 | 0.2 | 0.0 | 0.0 | 0.0 | 9.4 |
| 1991-92 | 0.0 | 0.0 | 0.0 | 0.0 | 1.0 | T | 1.0 | T | 1.3 | T | 0.0 | 0.0 | 3.3 |
| 1992-93 | 0.0 | 0.0 | 0.0 | T | 0.1 | 0.7 | T | 18.4 | 1.5 | T | T | 0.0 | 20.7 |
| 1993-94 | 0.0 | 0.0 | 0.0 | 4.6 | 0.3 | 3.4 | 7.4 | 0.7 | 8.0 | 0.0 | 0.0 | T | 24.4 |
| 1994-95 | 0.0 | 0.0 | 0.0 | 0.0 | 0.0 | T | 0.3 | 1.7 | 0.1 | 0.0 | 0.0 | 0.0 | 2.1 |
| 1995-96 | 0.0 | 0.0 | 0.0 | 0.0 | 0.1 | 5.4 | 13.5 | 1.1 | 9.6 | T | | | |
| 1996-97 | 0.0 | | 0.0 | | | T | 1.2 | 4.0 | 1.8 | T | T | | 0.0 |
| 1997-98 | | 0.0 | | | T | 6.6 | 0.7 | 12.2 | 1.1 | 0.0 | 0.0 | 0.0 | |
| 1998-99 | 0.0 | 0.0 | 0.0 | 0.0 | 0.0 | 1.5 | 1.1 | 3.3 | 2.8 | 0.0 | 0.0 | 0.0 | 8.7 |
| 1999-00 | 0.0 | 0.0 | 0.0 | 0.0 | 0.0 | 0.3 | 4.1 | T | T | 0.0 | 0.0 | 0.0 | 4.4 |
| 2000-01 | 0.0 | 0.0 | 0.0 | 0.0 | T | 11.5 | 4.7 | 1.0 | T | 0.1 | 0.0 | 0.0 | 17.3 |
| 2001-02 | 0.0 | 0.0 | 0.0 | 0.0 | 0.0 | 0.9 | 4.3 | 0.9 | T | 0.0 | 0.0 | 0.0 | 6.1 |
| 2002-03 | 0.0 | 0.0 | 0.0 | 0.0 | T | 7.8 | 1.8 | 11.4 | 0.0 | T | 0.0 | 0.0 | 21.0 |
| 2003-04 | 0.0 | 0.0 | 0.0 | 0.0 | 0.3 | 3.0 | 2.2 | 1.5 | 0.0 | 0.0 | 0.0 | 0.0 | 7.0 |
| 2004-05 | 0.0 | 0.0 | 0.0 | 0.0 | 0.0 | 22.3 | T | 0.3 | T | 0.0 | 0.0 | 0.0 | 22.6 |
| 2005-06 | 0.0 | 0.0 | 0.0 | 0.0 | 0.0 | 2.3 | 1.0 | 2.5 | T | 0.0 | 0.0 | 0.0 | 5.8 |
| 2006-07 | 0.0 | 0.0 | 0.0 | 0.0 | 0.0 | T | T | 3.6 | T | T | 0.0 | 0.0 | 3.6 |
| 2007-08 | 0.0 | 0.0 | 0.0 | 0.0 | T | 0.1 | 3.3 | 5.2 | 1.9 | 0.0 | 0.0 | 0.0 | 10.5 |
| 2008-09 | 0.0 | 0.0 | 0.0 | T | T | 0.8 | 6.5 | T | 0.0 | T | 0.0 | 0.0 | 7.3 |
| 2009- | 0.0 | 0.0 | 0.0 | 0.0 | 0.0 | T | | | | | | | |
| POR=<br>61 YRS | 0.0 | 0.0 | T | 0.1 | 0.5 | 2.7 | 4.1 | 3.6 | 2.3 | 0.3 | T | T | 13.6 |

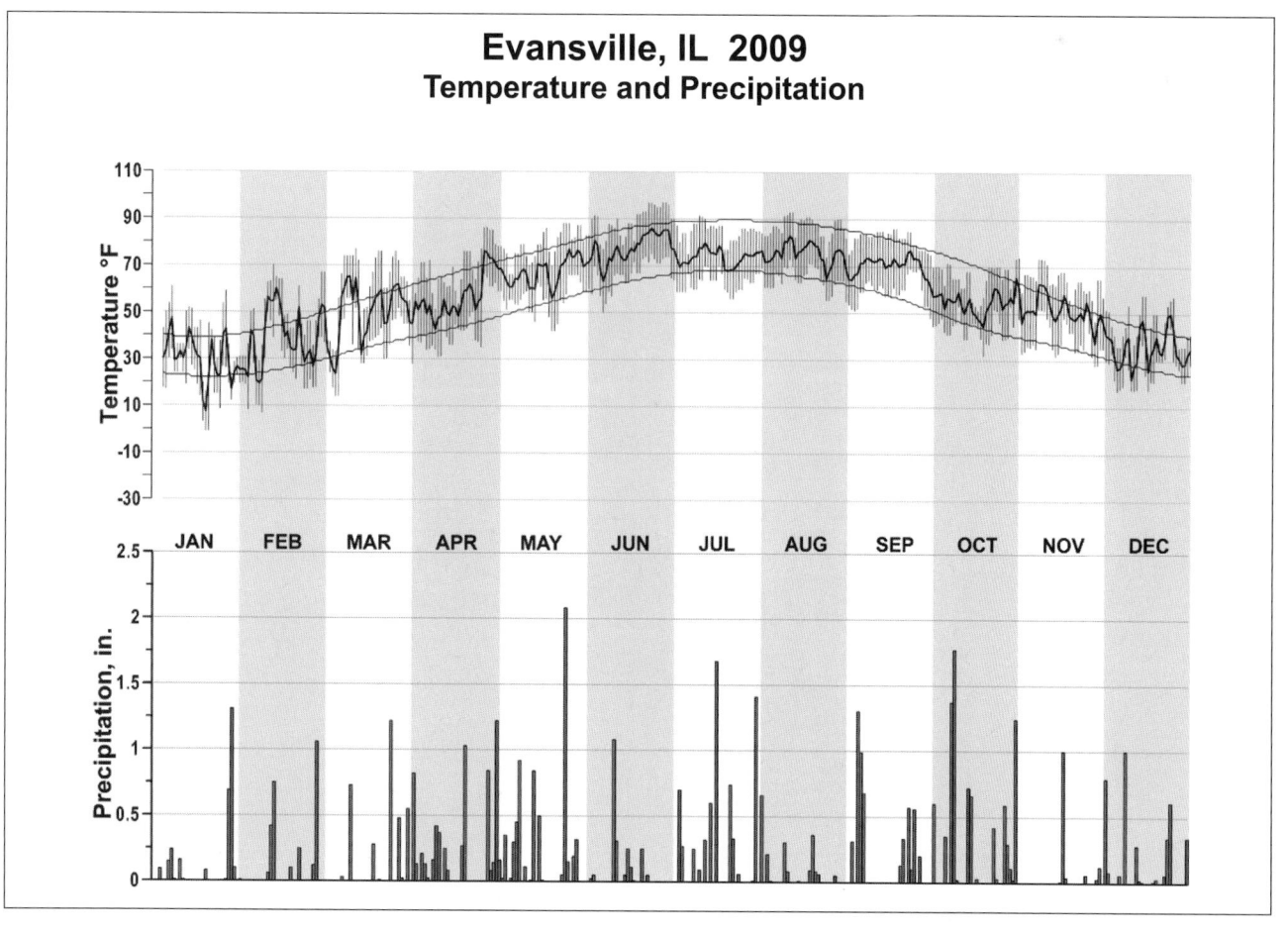

## Evansville, IL  2009
### Temperature and Precipitation

# 2009
# INDIANAPOLIS
# INDIANA (KIND)

Indianapolis is located in the central part of the state and is situated on level or slightly rolling terrain. The greater part of the city lies east of the White River which flows in a general north to south direction.

The National Weather Service Forecast Office is located approximately 7 miles southwest of the central part of the city at the Indianapolis International Airport. From a field elevation of 797 feet above sea level at the Indianapolis International Airport the terrain slopes gradually downward to a little below 645 feet at the White River, then upward to just over 910 feet in the northwest corner and eastern sections of the county. The street elevation at the former city office located in the Old Federal Building is 718 feet.

Indianapolis has a temperate climate, with very warm summers and without a dry season. Very cold temperatures may be produced by the invasion of continental polar air in the winter from northern latitudes. The polar air can be quite frigid with very low humidity. The arrival of maritime tropical air from the Gulf in the summer brings warm temperatures and moderate humidity. One of the longest and most severe heat waves brought temperatures of 100 degrees or more for nine consecutive days.

Precipitation is distributed fairly evenly throughout the year, and therefore there is no pronounced wet or dry season. Rainfall in the spring and summer is produced mostly by showers and thunderstorms. A rainfall of about 2 1/2 inches in a 24-hour period can be expected about once a year. Snowfalls of 3 inches or more occur on an average of two or three times in the winter.

Local levees and/or channel improvements now protect some formerly flood-prone areas.

Based on the 1951-1980 period, the average first occurrence of 32 degrees Fahrenheit in the fall is October 20 and the average last occurrence in the spring is April 22.

# NORMALS, MEANS, AND EXTREMES
## INDIANAPOLIS (KIND)

| | LATITUDE: 39 ° 42'N | LONGITUDE: -86 ° 16'W | | ELEVATION (FT): GRND: 790　BARO: 797 | | | | TIME ZONE: EASTERN　(UTC -5) | | | | WBAN: 93819 | |
|---|---|---|---|---|---|---|---|---|---|---|---|---|---|

| | ELEMENT | POR | JAN | FEB | MAR | APR | MAY | JUN | JUL | AUG | SEP | OCT | NOV | DEC | YEAR |
|---|---|---|---|---|---|---|---|---|---|---|---|---|---|---|---|
| **TEMPERATURE °F** | NORMAL DAILY MAXIMUM | 30 | 34.5 | 39.9 | 51.4 | 62.9 | 73.5 | 82.1 | 85.6 | 83.7 | 77.4 | 65.6 | 51.6 | 39.2 | 62.3 |
| | MEAN DAILY MAXIMUM | 67 | 35.5 | 39.5 | 50.5 | 62.9 | 73.1 | 81.9 | 85.2 | 83.9 | 77.4 | 66.0 | 51.2 | 39.2 | 62.2 |
| | HIGHEST DAILY MAXIMUM | 71 | 71 | 76 | 85 | 89 | 93 | 93 | 104 | 102 | 100 | 91 | 81 | 71 | 104 |
| | YEAR OF OCCURRENCE | | 1950 | 2000 | 1981 | 1970 | 1988 | 2009 | 1954 | 1988 | 1954 | 2007 | 1950 | 1950 | JUL 1954 |
| | MEAN OF EXTREME MAXS. | 78 | 57.8 | 62.1 | 74.0 | 81.6 | 86.8 | 92.2 | 93.7 | 92.4 | 89.9 | 82.1 | 71.1 | 61.1 | 78.7 |
| | NORMAL DAILY MINIMUM | 30 | 18.5 | 22.5 | 32.0 | 41.2 | 51.8 | 61.3 | 65.2 | 63.3 | 55.2 | 43.6 | 34.1 | 24.0 | 42.7 |
| | MEAN DAILY MINIMUM | 67 | 19.4 | 22.4 | 31.5 | 41.7 | 51.8 | 61.1 | 65.0 | 63.3 | 55.4 | 44.1 | 33.7 | 23.9 | 42.8 |
| | LOWEST DAILY MINIMUM | 71 | -27 | -21 | -7 | 16 | 28 | -27 | 44 | 41 | 28 | 17 | -2 | -27 | -27 |
| | YEAR OF OCCURRENCE | | 1994 | 1982 | 1980 | 1940 | 1966 | 1994 | 1942 | 1965 | 1942 | 1942 | 1958 | 1994 | DEC 1994 |
| | MEAN OF EXTREME MINS. | 78 | -3.2 | 1.8 | 12.8 | 26.1 | 36.7 | 47.5 | 53.9 | 52.3 | 39.8 | 28.5 | 17.7 | 3.0 | 26.4 |
| | NORMAL DRY BULB | 30 | 26.5 | 31.2 | 41.7 | 52.0 | 62.6 | 71.7 | 75.4 | 73.5 | 66.3 | 54.6 | 42.9 | 31.6 | 52.5 |
| | MEAN DRY BULB | 67 | 27.5 | 30.9 | 41.0 | 52.3 | 62.5 | 71.6 | 75.1 | 73.6 | 66.4 | 55.1 | 42.5 | 31.6 | 52.5 |
| | MEAN WET BULB | 25 | 26.3 | 28.9 | 36.4 | 46.1 | 55.6 | 64.2 | 67.7 | 66.8 | 59.7 | 48.8 | 39.1 | 29.2 | 47.4 |
| | MEAN DEW POINT | 25 | 22.9 | 25.0 | 32.2 | 41.8 | 52.2 | 61.4 | 65.4 | 64.7 | 56.9 | 45.5 | 36.0 | 26.4 | 44.2 |
| | NORMAL NO. DAYS WITH: | | | | | | | | | | | | | | |
| | MAXIMUM >= 90 | 30 | 0.0 | 0.0 | 0.0 | 0.0 | 0.4 | 3.3 | 7.3 | 4.7 | 1.8 | 0.0 | 0.0 | 0.0 | 17.5 |
| | MAXIMUM <= 32 | 30 | 13.1 | 8.3 | 1.7 | * | 0.0 | 0.0 | 0.0 | 0.0 | 0.0 | 0.0 | 0.9 | 7.8 | 31.8 |
| | MINIMUM <= 32 | 30 | 27.1 | 22.0 | 16.4 | 5.2 | 0.2 | 0.0 | 0.0 | 0.0 | * | 4.0 | 13.9 | 24.3 | 113.1 |
| | MINIMUM <= 0 | 30 | 3.5 | 2.0 | 0.1 | 0.0 | 0.0 | 0.0 | 0.0 | 0.0 | 0.0 | 0.0 | 0.0 | 1.4 | 7.0 |
| **H/C** | NORMAL HEATING DEG. DAYS | 30 | 1192 | 957 | 724 | 394 | 141 | 16 | 2 | 4 | 77 | 335 | 659 | 1020 | 5521 |
| | NORMAL COOLING DEG. DAYS | 30 | 0 | 0 | 2 | 10 | 69 | 221 | 331 | 272 | 122 | 14 | 1 | 0 | 1042 |
| **RH** | NORMAL (PERCENT) | 30 | 77 | 75 | 70 | 67 | 68 | 70 | 73 | 76 | 74 | 72 | 75 | 79 | 73 |
| | HOUR 01 LST | 30 | 80 | 78 | 75 | 74 | 77 | 81 | 84 | 87 | 85 | 81 | 80 | 81 | 80 |
| | HOUR 07 LST | 30 | 82 | 82 | 80 | 79 | 82 | 83 | 87 | 91 | 91 | 87 | 84 | 84 | 84 |
| | HOUR 13 LST | 30 | 72 | 68 | 61 | 56 | 56 | 57 | 59 | 61 | 58 | 57 | 66 | 72 | 62 |
| | HOUR 19 LST | 30 | 75 | 71 | 65 | 59 | 58 | 59 | 63 | 67 | 66 | 65 | 72 | 77 | 66 |
| **S** | PERCENT POSSIBLE SUNSHINE | 52 | 40 | 49 | 50 | 54 | 60 | 40 | 66 | 68 | 65 | 61 | 41 | 40 | 53 |
| **W/O** | MEAN NO. DAYS WITH: | | | | | | | | | | | | | | |
| | HEAVY FOG(VISBY <= 1/4 MI) | 46 | 2.8 | 2.2 | 1.3 | 0.5 | 0.7 | 0.7 | 0.8 | 1.3 | 1.4 | 1.2 | 1.3 | 2.3 | 16.5 |
| | THUNDERSTORMS | 62 | 0.8 | 0.7 | 2.7 | 4.3 | 6.4 | 7.4 | 7.6 | 5.9 | 3.5 | 1.8 | 1.1 | 0.4 | 42.6 |
| **CLOUDNESS** | MEAN: | | | | | | | | | | | | | | |
| | SUNRISE-SUNSET (OKTAS) | | | | | | | | | | | | | | |
| | MIDNIGHT-MIDNIGHT (OKTAS) | | | | | | | | | | | | | | |
| | MEAN NO. DAYS WITH: | | | | | | | | | | | | | | |
| | CLEAR | | | | | | | | | | | | | | |
| | PARTLY CLOUDY | | | | | | | | | | | | | | |
| | CLOUDY | | | | | | | | | | | | | | |
| **PR** | MEAN STATION PRESSURE(IN) | 26 | 29.23 | 29.22 | 29.18 | 29.11 | 29.12 | 29.12 | 29.15 | 29.18 | 29.20 | 29.21 | 29.21 | 29.23 | 29.18 |
| | MEAN SEA-LEVEL PRES. (IN) | 26 | 30.12 | 30.11 | 30.06 | 29.97 | 29.98 | 29.97 | 30.00 | 30.03 | 30.06 | 30.08 | 30.09 | 30.12 | 30.05 |
| **WINDS** | MEAN SPEED (MPH) | 26 | 11.0 | 10.8 | 11.2 | 11.2 | 9.6 | 8.4 | 7.7 | 7.3 | 7.9 | 8.9 | 10.3 | 10.5 | 9.6 |
| | PREVAIL.DIR(TENS OF DEGS) | 39 | 25 | 25 | 25 | 24 | 24 | 23 | 24 | 23 | 19 | 23 | 24 | 25 | 24 |
| | MAXIMUM 2-MINUTE: | 14 | | | | | | | | | | | | | |
| | SPEED (MPH) | | 44 | 48 | 45 | 52 | 60 | 47 | 49 | 53 | 49 | 44 | 47 | 45 | 60 |
| | DIR. (TENS OF DEGS) | | 26 | 22 | 27 | 29 | 19 | 27 | 28 | 27 | 29 | 25 | 28 | 25 | 19 |
| | YEAR OF OCCURRENCE | | 2008 | 2009 | 2002 | 2006 | 2009 | 2009 | 1998 | 2009 | 2008 | 2004 | 2003 | 2009 | MAY 2009 |
| | MAXIMUM 3-SECOND | | | | | | | | | | | | | | |
| | SPEED (MPH) | 14 | 58 | 60 | 62 | 85 | 70 | 58 | 61 | 67 | 63 | 51 | 58 | 58 | 85 |
| | DIR. (TENS OF DEGS) | | 25 | 23 | 25 | 28 | 19 | 25 | 31 | 27 | 28 | 26 | 25 | 25 | 28 |
| | YEAR OF OCCURRENCE | | 2008 | 2009 | 2002 | 2006 | 2009 | 2008 | 2008 | 2009 | 2008 | 2004 | 2005 | 2009 | APR 2006 |
| **PRECIPITATION** | NORMAL (IN) | 30 | 2.48 | 2.41 | 3.44 | 3.61 | 4.36 | 4.13 | 4.42 | 3.82 | 2.88 | 2.76 | 3.61 | 3.03 | 40.95 |
| | MAXIMUM MONTHLY (IN) | 71 | 12.69 | 5.35 | 10.74 | 8.09 | 10.10 | 12.69 | 11.79 | 8.34 | 10.37 | 8.36 | 8.50 | 12.69 | 12.69 |
| | YEAR OF OCCURRENCE | | 1950 | 1971 | 1963 | 1964 | 1943 | 1950 | 1992 | 1980 | 2003 | 1941 | 1985 | 1950 | JAN 1950 |
| | MINIMUM MONTHLY (IN) | 71 | 0.21 | 0.36 | 0.64 | 0.98 | 1.06 | 0.21 | 0.55 | 0.68 | 0.24 | 0.17 | 0.69 | 0.21 | 0.17 |
| | YEAR OF OCCURRENCE | | 1944 | 1978 | 2001 | 1976 | 1988 | 1944 | 1997 | 1964 | 1963 | 1963 | 1999 | 1944 | OCT 1963 |
| | MAXIMUM IN 24 HOURS (IN) | 71 | 3.47 | 2.50 | 3.05 | 2.56 | 3.80 | 3.47 | 5.32 | 4.72 | 7.31 | 3.90 | 4.15 | 3.47 | 7.31 |
| | YEAR OF OCCURRENCE | | 1950 | 1977 | 1963 | 1961 | 2004 | 1950 | 1987 | 1976 | 2003 | 1959 | 1993 | 1950 | SEP 2003 |
| | NORMAL NO. DAYS WITH: | | | | | | | | | | | | | | |
| | PRECIPITATION >= 0.01 | 30 | 13.1 | 10.7 | 13.1 | 12.6 | 12.0 | 10.4 | 9.8 | 9.4 | 8.1 | 8.8 | 10.9 | 12.6 | 131.5 |
| | PRECIPITATION >= 1.00 | 30 | 0.3 | 0.4 | 0.5 | 0.7 | 1.2 | 1.1 | 1.3 | 1.2 | 0.7 | 0.6 | 1.0 | 0.5 | 9.5 |
| **SNOWFALL** | NORMAL (IN) | 30 | 9.3 | 6.1 | 3.1 | 0.4 | 0.* | 0.0 | 0.0 | 0.0 | 0.0 | 0.4 | 1.3 | 6.4 | 27.0 |
| | MAXIMUM MONTHLY (IN) | 79 | 30.6 | 21.7 | 12.5 | 4.0 | 0.2 | 30.6 | 0.0 | T | T | 9.3 | 8.3 | 30.6 | 30.6 |
| | YEAR OF OCCURRENCE | | 1978 | 2003 | 1996 | 1940 | 1989 | 1978 | | 1989 | 2006 | 1989 | 1966 | 1978 | DEC 1978 |
| | MAXIMUM IN 24 HOURS (IN) | 68 | 12.2 | 12.5 | 11.3 | 3.1 | 0.2 | 12.2 | 0.0 | T | T | 7.5 | 8.2 | 12.2 | 12.5 |
| | YEAR OF OCCURRENCE' | | 1978 | 1965 | 1996 | 1953 | 1989 | 1978 | | 1989 | 2006 | 1989 | 1966 | 1978 | FEB 1965 |
| | MAXIMUM SNOW DEPTH (IN) | 62 | 20 | 15 | 9 | 3 | 0 | 20 | 0 | 0 | 0 | 2 | 8 | 20 | 20 |
| | YEAR OF OCCURRENCE | | 1978 | 1978 | 1960 | 1953 | | 1978 | | | | 1989 | 1966 | 1978 | DEC 1978 |
| | NORMAL NO. DAYS WITH: | | | | | | | | | | | | | | |
| | SNOWFALL >= 1.0 | 30 | 2.7 | 1.9 | 0.9 | 0.1 | 0.0 | 0.0 | 0.0 | 0.0 | 0.0 | 0.1 | 0.5 | 1.9 | 8.1 |

## PRECIPITATION (inches) 2009 INDIANAPOLIS (KIND)

| YEAR | JAN | FEB | MAR | APR | MAY | JUN | JUL | AUG | SEP | OCT | NOV | DEC | ANNUAL |
|---|---|---|---|---|---|---|---|---|---|---|---|---|---|
| 1980 | 1.67 | 1.84 | 4.26 | 2.10 | 2.26 | 4.15 | 2.87 | 8.34 | 3.31 | 1.87 | 1.41 | 0.78 | 34.86 |
| 1981 | 0.36 | 2.88 | 1.22 | 5.81 | 9.23 | 1.64 | 5.75 | 1.69 | 2.04 | 2.35 | 1.12 | 3.40 | 37.49 |
| 1982 | 5.64 | 1.62 | 4.73 | 2.40 | 5.94 | 5.16 | 3.44 | 1.00 | 1.20 | 0.91 | 4.16 | 5.78 | 41.98 |
| 1983 | 1.05 | 1.03 | 2.94 | 4.47 | 4.68 | 4.53 | 1.58 | 2.79 | 1.28 | 3.87 | 4.55 | 3.43 | 36.20 |
| 1984 | 0.97 | 3.16 | 3.14 | 3.90 | 4.35 | 1.51 | 4.83 | 3.27 | 4.69 | 2.60 | 5.38 | 4.33 | 42.13 |
| 1985 | 1.37 | 3.73 | 5.94 | 2.60 | 4.60 | 3.06 | 4.06 | 5.29 | 2.71 | 1.82 | 8.50 | 3.30 | 46.98 |
| 1986 | 0.73 | 2.84 | 3.93 | 4.34 | 7.37 | 3.58 | 4.88 | 1.18 | 5.68 | 7.84 | 2.32 | 1.71 | 46.40 |
| 1987 | 1.55 | 1.28 | 1.84 | 2.68 | 1.77 | 4.11 | 9.22 | 0.86 | 1.41 | 1.36 | 2.60 | 4.77 | 33.45 |
| 1988 | 2.35 | 3.04 | 3.22 | 4.02 | 1.06 | 0.36 | 4.71 | 1.46 | 1.14 | 3.07 | 4.39 | 2.50 | 31.32 |
| 1989 | 1.75 | 1.32 | 3.72 | 4.32 | 5.79 | 3.80 | 6.15 | 8.05 | 8.06 | 2.92 | 2.79 | 1.90 | 50.57 |
| 1990 | 1.79 | 5.17 | 3.93 | 2.44 | 7.59 | 3.11 | 3.68 | 4.46 | 2.68 | 4.64 | 3.23 | 7.72 | 50.44 |
| 1991 | 1.59 | 1.94 | 6.41 | 4.34 | 4.64 | 0.91 | 2.17 | 3.54 | 1.12 | 5.47 | 3.95 | 1.45 | 37.53 |
| 1992 | 1.40 | 1.15 | 2.61 | 4.17 | 1.56 | 4.07 | 11.79 | 1.42 | 3.40 | 2.84 | 7.88 | 1.89 | 44.18 |
| 1993 | 3.29 | 2.31 | 3.72 | 3.75 | 2.76 | 5.15 | 4.94 | 6.42 | 5.69 | 2.25 | 8.30 | 2.18 | 50.76 |
| 1994 | 2.68 | 1.39 | 0.92 | 5.62 | 2.03 | 7.00 | 1.27 | 2.59 | 2.19 | 0.86 | 3.40 | 1.66 | 31.61 |
| 1995 | 2.44 | 0.90 | 2.99 | 2.79 | 7.37 | 3.49 | 2.61 | 4.16 | 1.27 | 2.08 | 2.64 | 2.72 | 35.46 |
| 1996 | 4.08 | 1.16 | 3.51 | 7.02 | 8.89 | 5.16 | 6.26 | 2.27 | 7.61 | 1.88 | 6.22 | 2.75 | 56.81 |
| 1997 | 4.47 | 3.92 | 4.59 | 1.97 | 4.93 | 3.17 | 0.55 | 2.96 | 1.51 | 1.55 | 1.94 | 1.33 | 32.89 |
| 1998 | 2.51 | 1.23 | 5.49 | 4.84 | 6.13 | 10.26 | 3.97 | 3.66 | 0.48 | 4.89 | 2.52 | 1.00 | 46.98 |
| 1999 | 6.35 | 3.57 | 1.71 | 4.09 | 3.75 | 2.57 | 2.96 | 1.50 | 0.75 | 1.82 | 0.69 | 2.61 | 32.37 |
| 2000 | 2.07 | 2.86 | 1.64 | 3.80 | 5.00 | 4.55 | 2.95 | 4.26 | 4.82 | 3.08 | 2.67 | 2.76 | 40.46 |
| 2001 | 0.74 | 1.95 | 0.64 | 1.82 | 4.10 | 4.46 | 8.34 | 2.47 | 4.66 | 7.01 | 2.69 | 3.01 | 41.89 |
| 2002 | 2.48 | 1.67 | 4.22 | 5.47 | 7.46 | 3.11 | 1.65 | 1.42 | 3.70 | 2.64 | 2.89 | 3.02 | 39.73 |
| 2003 | 1.27 | 3.45 | 3.72 | 2.55 | 6.39 | 2.36 | 8.01 | 4.64 | 10.37 | 2.68 | 3.64 | 3.48 | 52.56 |
| 2004 | 4.65 | 0.77 | 4.33 | 1.54 | 8.55 | 4.44 | 7.72 | 4.73 | 0.75 | 4.31 | 5.14 | 1.95 | 48.88 |
| 2005 | 9.39 | 2.59 | 1.14 | 4.69 | 2.51 | 3.59 | 2.98 | 4.20 | 4.67 | 1.64 | 3.64 | 2.69 | 43.73 |
| 2006 | 3.30 | 1.89 | 6.79 | 3.63 | 4.34 | 5.63 | 3.98 | 3.01 | 3.53 | 5.45 | 4.25 | 5.24 | 51.04 |
| 2007 | 4.41 | 2.77 | 4.91 | 3.30 | 1.97 | 2.22 | 1.90 | 3.43 | 1.59 | 2.80 | 1.85 | 5.55 | 36.70 |
| 2008 | 2.14 | 4.34 | 7.49 | 1.54 | 5.54 | 8.00 | 6.58 | 1.83 | 2.04 | 1.97 | 1.97 | 5.58 | 49.02 |
| 2009 | 1.72 | 2.69 | 2.28 | 7.23 | 6.69 | 6.60 | 4.60 | 4.94 | 1.86 | 5.59 | 1.16 | 3.30 | 48.66 |
| POR= 78 YRS | 2.84 | 2.37 | 3.63 | 3.77 | 4.25 | 4.21 | 4.12 | 3.35 | 3.11 | 2.76 | 3.26 | 2.97 | 40.64 |

WBAN : 93819

## AVERAGE TEMPERATURE (°F) 2009 INDIANAPOLIS (KIND)

| YEAR | JAN | FEB | MAR | APR | MAY | JUN | JUL | AUG | SEP | OCT | NOV | DEC | ANNUAL |
|---|---|---|---|---|---|---|---|---|---|---|---|---|---|
| 1980 | 28.5 | 22.5 | 36.0 | 49.1 | 64.0 | 69.1 | 78.5 | 76.6 | 67.9 | 50.9 | 40.3 | 31.8 | 51.3 |
| 1981 | 23.5 | 33.0 | 39.9 | 57.4 | 59.9 | 73.3 | 75.4 | 72.8 | 64.6 | 53.1 | 44.2 | 27.8 | 52.1 |
| 1982 | 20.1 | 26.4 | 42.5 | 48.6 | 68.6 | 67.8 | 76.2 | 71.7 | 64.4 | 55.7 | 44.2 | 40.2 | 52.2 |
| 1983 | 30.6 | 35.5 | 42.9 | 48.2 | 58.2 | 71.8 | 79.7 | 80.1 | 69.4 | 57.5 | 44.1 | 20.2 | 53.2 |
| 1984 | 22.8 | 37.3 | 33.0 | 50.0 | 58.8 | 75.3 | 72.4 | 74.3 | 64.2 | 61.3 | 43.0 | 38.9 | 52.6 |
| 1985 | 20.4 | 26.1 | 44.6 | 57.1 | 64.9 | 70.6 | 74.3 | 71.2 | 66.2 | 57.6 | 46.6 | 22.5 | 51.8 |
| 1986 | 28.5 | 31.4 | 43.9 | 53.8 | 63.5 | 72.8 | 77.5 | 70.1 | 69.8 | 55.6 | 39.8 | 32.3 | 53.3 |
| 1987 | 27.6 | 35.5 | 44.5 | 52.5 | 68.1 | 73.7 | 75.8 | 73.6 | 68.1 | 48.6 | 46.3 | 35.8 | 54.2 |
| 1988 | 25.9 | 27.2 | 41.4 | 51.9 | 64.4 | 73.4 | 78.3 | 77.5 | 67.3 | 48.2 | 44.2 | 31.7 | 52.6 |
| 1989 | 36.3 | 27.2 | 42.5 | 51.4 | 59.4 | 71.4 | 75.7 | 71.8 | 64.4 | 55.4 | 40.9 | 18.8 | 51.3 |
| 1990 | 37.3 | 37.6 | 46.2 | 51.4 | 60.1 | 71.3 | 73.9 | 72.5 | 66.9 | 53.9 | 47.5 | 34.6 | 54.4 |
| 1991 | 26.9 | 35.3 | 43.7 | 55.3 | 68.3 | 74.6 | 76.7 | 75.2 | 67.9 | 57.5 | 40.2 | 35.7 | 54.8 |
| 1992 | 31.5 | 38.4 | 43.2 | 52.1 | 60.5 | 68.1 | 73.5 | 69.5 | 65.1 | 53.4 | 43.2 | 33.3 | 52.7 |
| 1993 | 31.6 | 27.4 | 39.1 | 50.7 | 63.5 | 71.0 | 77.2 | 75.3 | 62.2 | 51.6 | 42.2 | 32.4 | 52.0 |
| 1994 | 21.8 | 29.7 | 41.0 | 54.1 | 60.4 | 75.0 | 75.5 | 71.8 | 66.5 | 57.0 | 48.6 | 38.9 | 53.4 |
| 1995 | 28.6 | 30.0 | 45.0 | 51.9 | 61.6 | 72.8 | 76.6 | 79.6 | 64.7 | 55.7 | 37.0 | 28.6 | 52.7 |
| 1996 | 25.1 | 31.2 | 35.6 | 49.3 | 61.9 | 72.1 | 73.1 | 74.9 | 65.5 | 55.3 | 37.3 | 34.4 | 51.3 |
| 1997 | 24.3 | 36.0 | 43.0 | 48.3 | 56.7 | 69.6 | 75.4 | 71.6 | 66.1 | 54.6 | 39.0 | 33.4 | 51.5 |
| 1998 | 36.6 | 40.8 | 42.0 | 53.3 | 67.1 | 71.5 | 74.9 | 75.3 | 71.4 | 57.0 | 45.5 | 36.5 | 56.0 |
| 1999 | 28.5 | 37.5 | 38.4 | 54.9 | 64.3 | 73.7 | 79.2 | 72.4 | 67.1 | 55.2 | 48.8 | 34.6 | 54.6 |
| 2000 | 27.9 | 38.7 | 47.0 | 51.4 | 64.9 | 71.0 | 73.0 | 73.6 | 65.2 | 57.8 | 40.7 | 19.8 | 52.6 |
| 2001 | 28.0 | 34.6 | 38.4 | 58.5 | 64.4 | 70.2 | 74.4 | 75.4 | 65.1 | 54.3 | 49.5 | 38.2 | 54.3 |
| 2002 | 35.1 | 35.6 | 40.1 | 54.8 | 59.7 | 73.8 | 78.3 | 77.4 | 71.3 | 53.3 | 40.3 | 33.0 | 54.4 |
| 2003 | 21.7 | 25.4 | 43.3 | 54.6 | 61.1 | 68.2 | 74.1 | 74.9 | 64.6 | 54.5 | 47.3 | 34.4 | 52.0 |
| 2004 | 25.2 | 32.1 | 45.4 | 55.0 | 66.9 | 71.0 | 73.1 | 69.8 | 69.1 | 55.9 | 46.8 | 31.8 | 53.5 |
| 2005 | 30.7 | 36.3 | 38.3 | 55.0 | 60.2 | 74.8 | 76.4 | 76.6 | 70.5 | 56.3 | 44.5 | 27.1 | 53.9 |
| 2006 | 39.8 | 32.3 | 41.7 | 57.1 | 61.3 | 70.5 | 76.7 | 75.0 | 64.1 | 52.0 | 45.5 | 38.8 | 54.6 |
| 2007 | 32.5 | 19.7 | 49.0 | 50.8 | 68.1 | 74.2 | 74.3 | 80.0 | 71.7 | 62.1 | 43.4 | 34.1 | 55.0 |
| 2008 | 28.7 | 29.8 | 40.4 | 53.8 | 59.8 | 73.2 | 75.0 | 73.9 | 70.1 | 56.0 | 41.7 | 30.7 | 52.8 |
| 2009 | 23.0 | 33.9 | 47.3 | 53.9 | 63.6 | 73.4 | 70.9 | 73.0 | 68.5 | 51.6 | 47.4 | 31.1 | 53.1 |
| POR= 67 YRS | 27.5 | 30.9 | 41.0 | 52.3 | 62.5 | 71.6 | 75.1 | 73.6 | 66.4 | 55.1 | 42.5 | 31.6 | 52.5 |

## HEATING DEGREE DAYS (base 65°F) 2009  INDIANAPOLIS (KIND)

| YEAR | JUL | AUG | SEP | OCT | NOV | DEC | JAN | FEB | MAR | APR | MAY | JUN | TOTAL |
|------|-----|-----|-----|-----|-----|-----|-----|-----|-----|-----|-----|-----|-------|
| 1980-81 | 0 | 0 | 45 | 438 | 734 | 1022 | 1279 | 889 | 769 | 244 | 180 | 1 | 5601 |
| 1981-82 | 1 | 0 | 94 | 368 | 621 | 1146 | 1388 | 1075 | 690 | 486 | 26 | 18 | 5913 |
| 1982-83 | 0 | 2 | 110 | 325 | 621 | 764 | 1062 | 819 | 681 | 498 | 211 | 21 | 5114 |
| 1983-84 | 1 | 0 | 64 | 246 | 619 | 1386 | 1304 | 796 | 987 | 447 | 211 | 1 | 6062 |
| 1984-85 | 1 | 0 | 119 | 138 | 653 | 803 | 1375 | 1082 | 631 | 269 | 85 | 16 | 5172 |
| 1985-86 | 0 | 1 | 97 | 245 | 544 | 1311 | 1126 | 935 | 652 | 344 | 114 | 7 | 5376 |
| 1986-87 | 0 | 24 | 30 | 312 | 750 | 1009 | 1150 | 820 | 627 | 378 | 51 | 0 | 5151 |
| 1987-88 | 2 | 5 | 37 | 504 | 553 | 900 | 1205 | 1090 | 726 | 390 | 84 | 15 | 5511 |
| 1988-89 | 0 | 2 | 37 | 517 | 618 | 1024 | 882 | 1052 | 693 | 419 | 226 | 11 | 5481 |
| 1989-90 | 0 | 8 | 106 | 313 | 718 | 1426 | 851 | 760 | 588 | 432 | 153 | 19 | 5374 |
| 1990-91 | 4 | 1 | 86 | 348 | 518 | 932 | 1175 | 824 | 651 | 292 | 71 | 0 | 4902 |
| 1991-92 | 0 | 0 | 103 | 254 | 739 | 902 | 1035 | 766 | 670 | 390 | 185 | 34 | 5078 |
| 1992-93 | 2 | 7 | 88 | 353 | 645 | 973 | 1028 | 1046 | 796 | 424 | 103 | 35 | 5500 |
| 1993-94 | 0 | 1 | 129 | 409 | 680 | 1005 | 1332 | 984 | 738 | 336 | 192 | 3 | 5809 |
| 1994-95 | 0 | 4 | 61 | 256 | 483 | 805 | 1118 | 973 | 615 | 392 | 131 | 6 | 4844 |
| 1995-96 | 1 | 0 | 94 | 288 | 835 | 1125 | 1230 | 971 | 908 | 469 | 162 | 7 | 6090 |
| 1996-97 | 0 | 0 | 85 | 296 | 823 | 942 | 1254 | 807 | 676 | 494 | 261 | 30 | 5668 |
| 1997-98 | 3 | 5 | 47 | 366 | 772 | 973 | 876 | 670 | 720 | 344 | 48 | 44 | 4868 |
| 1998-99 | 0 | 0 | 14 | 257 | 579 | 878 | 1125 | 767 | 814 | 297 | 81 | 8 | 4820 |
| 1999-00 | 0 | 0 | 70 | 298 | 481 | 936 | 1143 | 755 | 553 | 401 | 83 | 18 | 4738 |
| 2000-01 | 0 | 0 | 101 | 237 | 724 | 1394 | 1141 | 845 | 816 | 246 | 100 | 31 | 5635 |
| 2001-02 | 5 | 0 | 91 | 331 | 457 | 826 | 919 | 819 | 764 | 348 | 211 | 5 | 4776 |
| 2002-03 | 0 | 0 | 22 | 389 | 731 | 983 | 1335 | 1101 | 666 | 320 | 144 | 45 | 5736 |
| 2003-04 | 0 | 0 | 89 | 328 | 528 | 942 | 1227 | 949 | 604 | 316 | 77 | 8 | 5068 |
| 2004-05 | 3 | 20 | 19 | 280 | 538 | 1024 | 1060 | 796 | 820 | 309 | 176 | 2 | 5047 |
| 2005-06 | 0 | 0 | 18 | 295 | 610 | 1165 | 774 | 910 | 716 | 248 | 187 | 10 | 4933 |
| 2006-07 | 0 | 0 | 84 | 415 | 579 | 807 | 997 | 1261 | 508 | 430 | 45 | 1 | 5127 |
| 2007-08 | 0 | 0 | 19 | 179 | 639 | 952 | 1118 | 1015 | 755 | 342 | 179 | 0 | 5198 |
| 2008-09 | 0 | 0 | 9 | 298 | 692 | 1055 | 1295 | 863 | 543 | 367 | 103 | 15 | 5240 |
| 2009- | 2 | 7 | 25 | 409 | 522 | 1043 | | | | | | | |

WBAN : 93819

## COOLING DEGREE DAYS (base 65°F) 2009  INDIANAPOLIS (KIND)

| YEAR | JAN | FEB | MAR | APR | MAY | JUN | JUL | AUG | SEP | OCT | NOV | DEC | TOTAL |
|------|-----|-----|-----|-----|-----|-----|-----|-----|-----|-----|-----|-----|-------|
| 1980 | 0 | 0 | 0 | 3 | 68 | 168 | 425 | 368 | 139 | 6 | 0 | 0 | 1177 |
| 1981 | 0 | 0 | 1 | 23 | 29 | 256 | 332 | 249 | 88 | 5 | 0 | 0 | 983 |
| 1982 | 0 | 0 | 0 | 0 | 146 | 109 | 356 | 214 | 98 | 40 | 3 | 1 | 967 |
| 1983 | 0 | 0 | 1 | 3 | 9 | 231 | 464 | 474 | 202 | 18 | 0 | 0 | 1402 |
| 1984 | 0 | 0 | 0 | 6 | 25 | 318 | 237 | 291 | 103 | 28 | 0 | 0 | 1008 |
| 1985 | 0 | 0 | 5 | 36 | 90 | 190 | 296 | 199 | 143 | 21 | 0 | 0 | 980 |
| 1986 | 0 | 0 | 6 | 12 | 74 | 249 | 395 | 189 | 181 | 24 | 0 | 0 | 1130 |
| 1987 | 0 | 0 | 0 | 6 | 156 | 265 | 343 | 279 | 137 | 0 | 0 | 0 | 1186 |
| 1988 | 0 | 0 | 1 | 3 | 72 | 274 | 422 | 395 | 114 | 2 | 0 | 0 | 1283 |
| 1989 | 0 | 0 | 1 | 20 | 57 | 215 | 338 | 227 | 94 | 21 | 0 | 0 | 973 |
| 1990 | 0 | 0 | 13 | 30 | 10 | 215 | 289 | 241 | 147 | 10 | 1 | 0 | 956 |
| 1991 | 0 | 0 | 0 | 7 | 178 | 294 | 367 | 323 | 196 | 30 | 0 | 0 | 1395 |
| 1992 | 0 | 0 | 0 | 10 | 53 | 137 | 274 | 152 | 99 | 1 | 0 | 0 | 726 |
| 1993 | 0 | 0 | 0 | 0 | 64 | 223 | 389 | 328 | 51 | 0 | 0 | 0 | 1055 |
| 1994 | 0 | 0 | 0 | 16 | 56 | 307 | 330 | 221 | 114 | 13 | 0 | 0 | 1057 |
| 1995 | 0 | 0 | 0 | 4 | 32 | 250 | 371 | 459 | 92 | 6 | 0 | 0 | 1214 |
| 1996 | 0 | 0 | 0 | 7 | 71 | 229 | 261 | 313 | 107 | 3 | 0 | 0 | 991 |
| 1997 | 0 | 0 | 0 | 0 | 12 | 178 | 327 | 217 | 87 | 49 | 0 | 0 | 870 |
| 1998 | 0 | 0 | 15 | 0 | 120 | 246 | 312 | 331 | 214 | 16 | 0 | 1 | 1255 |
| 1999 | 0 | 0 | 0 | 5 | 67 | 278 | 450 | 236 | 138 | 3 | 1 | 0 | 1178 |
| 2000 | 0 | 0 | 2 | 0 | 88 | 205 | 256 | 274 | 113 | 22 | 0 | 0 | 960 |
| 2001 | 0 | 0 | 0 | 60 | 90 | 194 | 302 | 328 | 101 | 6 | 0 | 0 | 1081 |
| 2002 | 0 | 0 | 0 | 48 | 53 | 274 | 419 | 392 | 218 | 33 | 0 | 0 | 1437 |
| 2003 | 0 | 0 | 0 | 17 | 27 | 149 | 287 | 314 | 83 | 9 | 3 | 0 | 889 |
| 2004 | 0 | 0 | 3 | 21 | 147 | 192 | 258 | 177 | 149 | 5 | 0 | 0 | 952 |
| 2005 | 0 | 0 | 0 | 15 | 34 | 304 | 361 | 367 | 189 | 36 | 0 | 0 | 1306 |
| 2006 | 0 | 0 | 0 | 20 | 78 | 178 | 367 | 319 | 65 | 18 | 0 | 0 | 1045 |
| 2007 | 0 | 0 | 18 | 11 | 150 | 285 | 297 | 470 | 223 | 97 | 0 | 0 | 1551 |
| 2008 | 0 | 0 | 0 | 14 | 24 | 252 | 317 | 283 | 168 | 28 | 0 | 0 | 1086 |
| 2009 | 0 | 0 | 0 | 38 | 67 | 274 | 192 | 261 | 136 | 0 | 0 | 0 | 968 |

## SNOWFALL (inches) 2009 INDIANAPOLIS (KIND)

| YEAR | JUL | AUG | SEP | OCT | NOV | DEC | JAN | FEB | MAR | APR | MAY | JUN | TOTAL |
|---|---|---|---|---|---|---|---|---|---|---|---|---|---|
| 1980-81 | 0.0 | 0.0 | 0.0 | T | 3.4 | 2.1 | 3.9 | 7.2 | 0.7 | 0.0 | 0.0 | 0.0 | 17.3 |
| 1981-82 | 0.0 | 0.0 | 0.0 | T | 0.4 | 15.6 | 21.8 | 13.6 | 3.5 | 3.3 | 0.0 | 0.0 | 58.2 |
| 1982-83 | 0.0 | 0.0 | 0.0 | 0.0 | 0.1 | 0.4 | 2.8 | 2.5 | 1.3 | T | 0.0 | 0.0 | 7.1 |
| 1983-84 | 0.0 | 0.0 | 0.0 | 0.0 | 0.1 | 8.3 | 7.2 | 17.1 | 9.2 | T | 0.0 | 0.0 | 41.9 |
| 1984-85 | 0.0 | 0.0 | 0.0 | 0.0 | 2.5 | 3.6 | 10.6 | 11.0 | T | 0.1 | 0.0 | 0.0 | 27.8 |
| 1985-86 | 0.0 | 0.0 | 0.0 | 0.0 | T | 8.1 | 1.7 | 9.5 | 1.1 | T | 0.0 | 0.0 | 20.4 |
| 1986-87 | 0.0 | 0.0 | 0.0 | 0.0 | T | 1.6 | 11.6 | 5.3 | 1.4 | T | 0.0 | 0.0 | 19.9 |
| 1987-88 | 0.0 | 0.0 | 0.0 | 0.0 | T | 1.1 | 2.9 | 4.8 | 2.5 | T | 0.0 | 0.0 | 11.3 |
| 1988-89 | 0.0 | 0.0 | 0.0 | T | 0.4 | 6.8 | 0.1 | 2.4 | 1.9 | 1.7 | 0.2 | 0.0 | 13.5 |
| 1989-90 | 0.0 | T | 0.0 | 9.3 | 0.6 | 8.2 | 4.1 | 2.5 | 1.3 | T | 0.0 | 0.0 | 26.0 |
| 1990-91 | 0.0 | 0.0 | 0.0 | 0.0 | 0.0 | 9.6 | 5.4 | 2.4 | 0.1 | 0.0 | T | 0.0 | 17.5 |
| 1991-92 | 0.0 | 0.0 | 0.0 | 0.0 | 3.0 | 0.2 | 7.1 | 0.2 | 2.9 | 1.3 | 0.0 | T | 14.7 |
| 1992-93 | 0.0 | 0.0 | 0.0 | 0.1 | 1.4 | 3.7 | 3.1 | 17.2 | 3.0 | T | 0.0 | 0.0 | 28.5 |
| 1993-94 | 0.0 | 0.0 | 0.0 | 2.4 | 1.1 | 5.6 | 14.0 | 3.9 | 3.7 | 0.8 | 0.0 | T | 31.5 |
| 1994-95 | 0.0 | 0.0 | 0.0 | 0.0 | 0.0 | 2.5 | 13.1 | 3.6 | 0.6 | 0.0 | 0.0 | T | 19.8 |
| 1995-96 | 0.0 | 0.0 | 0.0 | 0.0 | 1.4 | 11.0 | 25.2 | 1.6 | 12.5 | T | 0.0 | 0.0 | 51.7 |
| 1996-97 | 0.0 | 0.0 | 0.0 | 0.0 | .9 | 14.6 | 12.9 | 3.1 | T | T | 0.0 | 0.0 | 31.5 |
| 1997-98 | 0.0 | 0.0 | 0.0 | 0.0 | 5.8 | 2.5 | 0.8 | 0.6 | 0.7 | 0.0 | 0.0 | 0.0 | 10.4 |
| 1998-99 | 0.0 | 0.0 | 0.0 | 0.0 | 0.0 | 3.3 | 18.3 | 2.9 | 5.2 | T | 0.0 | 0.0 | 29.7 |
| 1999-00 | 0.0 | 0.0 | 0.0 | 0.0 | 0.0 | 3.1 | 11.3 | 1.6 | 8.1 | T | 0.0 | 0.0 | 24.1 |
| 2000-01 | 0.0 | T | 0.0 | 0.0 | 0.2 | 16.3 | 2.1 | 0.8 | T | 0.2 | T | 0.0 | 19.6 |
| 2001-02 | 0.0 | 0.0 | 0.0 | T | T | 0.7 | 4.6 | 4.8 | 0.8 | T | T | 0.0 | 10.9 |
| 2002-03 | 0.0 | 0.0 | 0.0 | 0.0 | 0.5 | 10.6 | 14.6 | 21.7 | 2.6 | T | 0.0 | 0.0 | 50.0 |
| 2003-04 | 0.0 | 0.0 | 0.0 | 0.0 | 0.4 | 4.8 | 10.0 | 1.8 | 3.9 | 0.0 | 0.0 | 0.0 | 20.9 |
| 2004-05 | 0.0 | 0.0 | 0.0 | 0.0 | 0.1 | 13.1 | 10.7 | 2.5 | 1.2 | T | 0.0 | 0.0 | 27.6 |
| 2005-06 | 0.0 | 0.0 | 0.0 | 0.0 | 0.8 | 14.1 | 2.7 | 3.9 | 5.5 | T | 0.0 | 0.0 | 27.0 |
| 2006-07 | 0.0 | 0.0 | T | T | T | 0.9 | 5.8 | 18.2 | 0.9 | T | 0.0 | 0.0 | 25.8 |
| 2007-08 | 0.0 | 0.0 | 0.0 | 0.0 | T | 10.8 | 4.8 | 5.4 | 2.3 | 0.0 | 0.0 | T | 23.3 |
| 2008-09 | 0.0 | 0.0 | 0.0 | T | 0.5 | 2.1 | 16.1 | 5.6 | T | T | T | 0.0 | 24.3 |
| 2009- | 0.0 | 0.0 | 0.0 | 0.0 | T | 7.7 | | | | | | | |
| POR= 62 YRS | 0.0 | T | T | 0.2 | 1.5 | 5.7 | 7.8 | 6.0 | 3.3 | 0.4 | T | T | 24.9 |

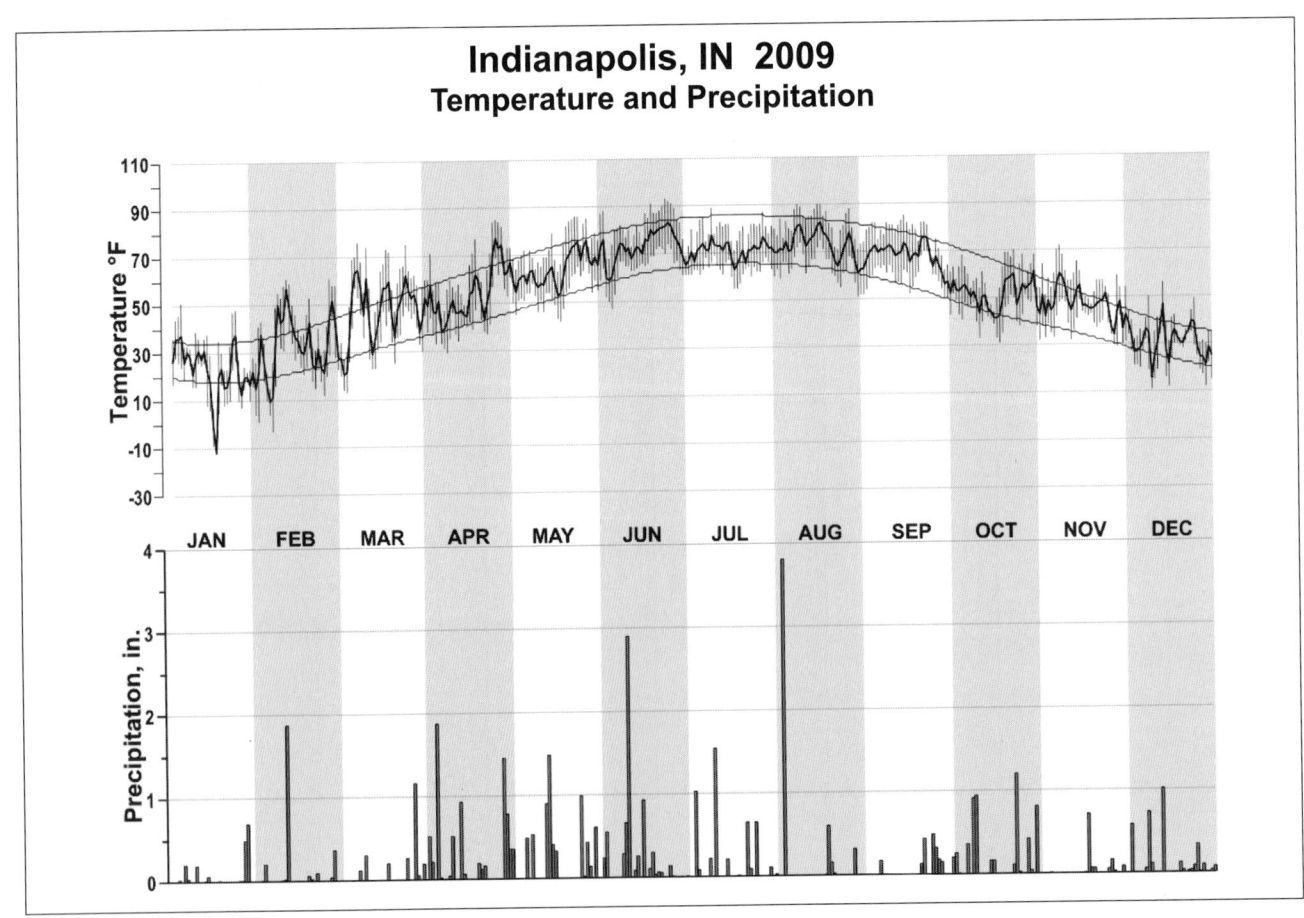

## Indianapolis, IN 2009
### Temperature and Precipitation

# 2009
# SOUTH BEND
# INDIANA (KSBN)

South Bend is located on the Saint Joseph River in the northern portion of Saint Joseph County, situated on mostly level to gently rolling terrain and some former marshland. Drainage for the area is through the Saint Joseph River and Kankakee River.

South Bend is under the climatic influence of Lake Michigan with its nearest shore 20 miles to the northwest. The lake has a moderating effect on the temperature. Temperatures of 100 degrees or higher are rare and cold waves are less severe than at many locations at the same latitude. This results in favorable conditions for orchard and vegetable growth.

Based on the 1951-1980 period, the average first occurrence of 32 degrees Fahrenheit in the fall is October 18 and the average last occurrence in the spring is May 1.

Precipitation is fairly evenly distributed throughout the year with the greatest amounts during the growing season. The predominant snow season is from November through March, although there are also generally lighter amounts in October and April.

Winter is marked by considerable cloudiness and rather high humidity along with frequent periods of snow. Heavy snowfalls, resulting from a cold northwest wind passing over Lake Michigan are not uncommon.

# NORMALS, MEANS, AND EXTREMES
## SOUTH BEND (KSBN)

LATITUDE: 41° 42'N  LONGITUDE: -86° 19'W  ELEVATION (FT): GRND: 776  BARO: 780  TIME ZONE: EASTERN  (UTC -5)  WBAN: 14848

| | ELEMENT | POR | JAN | FEB | MAR | APR | MAY | JUN | JUL | AUG | SEP | OCT | NOV | DEC | YEAR |
|---|---|---|---|---|---|---|---|---|---|---|---|---|---|---|---|
| **TEMPERATURE °F** | NORMAL DAILY MAXIMUM | 30 | 31.0 | 35.5 | 46.8 | 58.9 | 70.7 | 79.6 | 83.1 | 80.7 | 73.6 | 61.8 | 47.7 | 35.6 | 58.8 |
| | MEAN DAILY MAXIMUM | 113 | 31.7 | 32.6 | 45.3 | 57.9 | 70.1 | 78.6 | 83.9 | 82.0 | 73.9 | 63.0 | 47.0 | 35.3 | 58.4 |
| | HIGHEST DAILY MAXIMUM | 70 | 68 | 74 | 85 | 91 | 95 | 104 | 102 | 103 | 99 | 92 | 82 | 70 | 104 |
| | YEAR OF OCCURRENCE | | 1950 | 2000 | 1981 | 1942 | 1942 | 1988 | 1999 | 1988 | 1953 | 1963 | 1950 | 2001 | JUN 1988 |
| | MEAN OF EXTREME MAXS. | 114 | 51.4 | 54.0 | 70.7 | 79.9 | 86.2 | 92.5 | 93.6 | 91.9 | 88.8 | 80.6 | 68.4 | 55.7 | 76.1 |
| | NORMAL DAILY MINIMUM | 30 | 15.7 | 19.0 | 28.2 | 37.7 | 48.4 | 58.3 | 62.8 | 61.3 | 53.3 | 42.3 | 32.6 | 21.7 | 40.1 |
| | MEAN DAILY MINIMUM | 113 | 17.0 | 17.6 | 27.7 | 37.4 | 48.1 | 57.1 | 62.3 | 60.7 | 53.0 | 42.7 | 31.8 | 21.8 | 39.8 |
| | LOWEST DAILY MINIMUM | 70 | -22 | -17 | -13 | 11 | 24 | 35 | 42 | 40 | 29 | 20 | -7 | -16 | -22 |
| | YEAR OF OCCURRENCE | | 1943 | 1951 | 1943 | 1972 | 1968 | 1972 | 2001 | 1965 | 1942 | 1988 | 1950 | 1960 | JAN 1943 |
| | MEAN OF EXTREME MINS. | 114 | -4.9 | -0.9 | 9.8 | 23.0 | 33.3 | 43.8 | 50.8 | 49.3 | 38.5 | 29.1 | 17.3 | 1.5 | 24.2 |
| | NORMAL DRY BULB | 30 | 23.4 | 27.3 | 37.5 | 48.3 | 59.6 | 69.0 | 73.0 | 71.0 | 63.4 | 52.1 | 40.1 | 28.7 | 49.5 |
| | MEAN DRY BULB | 113 | 24.4 | 25.1 | 36.5 | 47.7 | 59.1 | 67.9 | 73.1 | 71.3 | 63.5 | 52.9 | 39.4 | 28.6 | 49.1 |
| | MEAN WET BULB | 26 | 22.9 | 25.1 | 32.6 | 42.3 | 52.1 | 61.3 | 65.3 | 64.7 | 57.6 | 46.6 | 36.7 | 26.7 | 44.5 |
| | MEAN DEW POINT | 26 | 20.1 | 21.8 | 28.4 | 37.5 | 48.1 | 57.8 | 62.6 | 62.3 | 54.7 | 43.2 | 33.6 | 24.1 | 41.2 |
| | NORMAL NO. DAYS WITH: | | | | | | | | | | | | | | |
| | MAXIMUM >= 90 | 30 | 0.0 | 0.0 | 0.0 | 0.0 | 0.6 | 3.2 | 5.7 | 2.9 | 0.9 | 0.0 | 0.0 | 0.0 | 13.3 |
| | MAXIMUM <= 32 | 30 | 16.6 | 11.9 | 3.4 | 0.1 | 0.0 | 0.0 | 0.0 | 0.0 | 0.0 | 0.0 | 1.8 | 9.8 | 43.6 |
| | MINIMUM <= 32 | 30 | 27.9 | 23.8 | 19.7 | 8.2 | 0.4 | 0.0 | 0.0 | 0.0 | * | 2.9 | 14.6 | 25.3 | 122.8 |
| | MINIMUM <= 0 | 30 | 3.7 | 2.4 | 0.1 | 0.0 | 0.0 | 0.0 | 0.0 | 0.0 | 0.0 | 0.0 | 0.0 | 1.5 | 7.7 |
| **H/C** | NORMAL HEATING DEG. DAYS | 30 | 1274 | 1055 | 844 | 498 | 213 | 41 | 6 | 13 | 117 | 393 | 731 | 1109 | 6294 |
| | NORMAL COOLING DEG. DAYS | 30 | 0 | 0 | 1 | 10 | 53 | 172 | 268 | 214 | 85 | 9 | 0 | 0 | 812 |
| **RH** | NORMAL (PERCENT) | 30 | 78 | 75 | 70 | 66 | 66 | 68 | 71 | 75 | 74 | 72 | 75 | 79 | 72 |
| | HOUR 01 LST | 30 | 80 | 79 | 76 | 74 | 76 | 79 | 83 | 86 | 84 | 80 | 79 | 82 | 80 |
| | HOUR 07 LST | 30 | 82 | 82 | 80 | 79 | 80 | 81 | 84 | 89 | 89 | 85 | 82 | 83 | 83 |
| | HOUR 13 LST | 30 | 73 | 68 | 60 | 56 | 55 | 55 | 57 | 60 | 59 | 59 | 68 | 75 | 62 |
| | HOUR 19 LST | 30 | 77 | 73 | 65 | 59 | 57 | 57 | 60 | 66 | 68 | 68 | 74 | 79 | 67 |
| **S** | PERCENT POSSIBLE SUNSHINE | | | | | | | | | | | | | | |
| **W/O** | MEAN NO. DAYS WITH: | | | | | | | | | | | | | | |
| | HEAVY FOG(VISBY <= 1/4 MI) | 46 | 2.4 | 2.4 | 2.4 | 1.2 | 1.2 | 0.9 | 1.3 | 2.0 | 1.6 | 1.7 | 1.9 | 3.2 | 22.2 |
| | THUNDERSTORMS | 62 | 0.4 | 0.5 | 2.3 | 4.3 | 4.9 | 7.4 | 7.2 | 6.2 | 4.0 | 1.9 | 1.0 | 0.5 | 40.6 |
| **CLOUDNESS** | MEAN: | | | | | | | | | | | | | | |
| | SUNRISE-SUNSET (OKTAS) | 51 | 6.4 | 6.2 | 6.0 | 5.5 | 5.0 | 4.9 | 4.5 | 4.5 | 4.6 | 4.9 | 6.1 | 6.4 | 5.4 |
| | MIDNIGHT-MIDNIGHT (OKTAS) | 32 | 6.3 | 5.9 | 5.7 | 5.3 | 4.8 | 4.6 | 4.2 | 4.2 | 4.4 | 4.7 | 6.0 | 6.4 | 5.2 |
| | MEAN NO. DAYS WITH: | | | | | | | | | | | | | | |
| | CLEAR | 57 | 3.4 | 3.7 | 4.7 | 5.8 | 6.9 | 7.2 | 8.2 | 8.6 | 8.8 | 8.7 | 3.6 | 3.0 | 72.6 |
| | PARTLY CLOUDY | 57 | 5.9 | 5.6 | 7.3 | 7.6 | 9.6 | 10.4 | 12.7 | 12.1 | 9.0 | 8.0 | 6.0 | 5.6 | 99.8 |
| | CLOUDY | 57 | 21.7 | 18.9 | 19.1 | 16.6 | 14.5 | 12.4 | 10.1 | 10.3 | 12.2 | 14.4 | 20.4 | 22.5 | 193.1 |
| **PR** | MEAN STATION PRESSURE(IN) | 26 | 29.22 | 29.23 | 29.21 | 29.14 | 29.15 | 29.14 | 29.18 | 29.21 | 29.23 | 29.23 | 29.22 | 29.23 | 29.20 |
| | MEAN SEA-LEVEL PRES. (IN) | 26 | 30.08 | 30.09 | 30.05 | 29.97 | 29.97 | 29.96 | 29.99 | 30.03 | 30.06 | 30.06 | 30.06 | 30.09 | 30.03 |
| **WINDS** | MEAN SPEED (MPH) | 26 | 11.0 | 10.5 | 10.7 | 10.6 | 9.3 | 8.0 | 7.4 | 6.9 | 7.4 | 8.6 | 10.0 | 10.4 | 9.2 |
| | PREVAIL.DIR(TENS OF DEGS) | 35 | 24 | 23 | 22 | 34 | 22 | 22 | 23 | 23 | 20 | 22 | 24 | 24 | 24 |
| | MAXIMUM 2-MINUTE: | | | | | | | | | | | | | | |
| | SPEED (MPH) | 13 | 38 | 45 | 43 | 44 | 44 | 41 | 45 | 39 | 36 | 56 | 41 | 43 | 56 |
| | DIR. (TENS OF DEGS) | | 27 | 20 | 24 | 23 | 31 | 28 | 31 | 31 | 25 | 25 | 23 | 21 | 25 |
| | YEAR OF OCCURRENCE | | 2008 | 1999 | 2009 | 2008 | 1997 | 2004 | 2004 | 1997 | 1997 | 2001 | 1998 | 2007 | OCT 2001 |
| | MAXIMUM 3-SECOND | | | | | | | | | | | | | | |
| | SPEED (MPH) | 13 | 48 | 56 | 56 | 61 | 55 | 52 | 55 | 51 | 45 | 90 | 52 | 62 | 90 |
| | DIR. (TENS OF DEGS) | | 27 | 20 | 23 | 24 | 32 | 29 | 27 | 32 | 31 | 26 | 24 | 21 | 26 |
| | YEAR OF OCCURRENCE | | 2008 | 1999 | 2004 | 2008 | 2007 | 2004 | 2003 | 1997 | 2009 | 2001 | 1998 | 2007 | OCT 2001 |
| **PRECIPITATION** | NORMAL (IN) | 30 | 2.27 | 1.98 | 2.89 | 3.62 | 3.50 | 4.19 | 3.73 | 3.98 | 3.79 | 3.27 | 3.39 | 3.09 | 39.70 |
| | MAXIMUM MONTHLY (IN) | 70 | 5.34 | 5.23 | 7.96 | 9.20 | 8.09 | 10.86 | 8.66 | 8.88 | 13.92 | 9.75 | 6.72 | 5.50 | 13.92 |
| | YEAR OF OCCURRENCE | | 2008 | 1976 | 1976 | 1947 | 1996 | 1993 | 2006 | 2007 | 2008 | 1954 | 1985 | 1965 | SEP 2008 |
| | MINIMUM MONTHLY (IN) | 70 | 0.44 | 0.54 | 0.54 | 0.50 | 0.80 | 0.48 | 0.02 | 0.32 | 0.01 | 0.42 | 1.24 | 0.60 | 0.01 |
| | YEAR OF OCCURRENCE | | 1945 | 1969 | 1958 | 1971 | 1994 | 1988 | 1946 | 1950 | 1979 | 1950 | 1998 | 1943 | SEP 1979 |
| | MAXIMUM IN 24 HOURS (IN) | 70 | 3.15 | 2.64 | 2.33 | 3.14 | 2.99 | 4.70 | 3.64 | 4.88 | 6.88 | 3.49 | 3.95 | 3.33 | 6.88 |
| | YEAR OF OCCURRENCE | | 2008 | 1954 | 1972 | 1947 | 1976 | 1968 | 1989 | 1995 | 2008 | 1988 | 1990 | 1965 | SEP 2008 |
| | NORMAL NO. DAYS WITH: | | | | | | | | | | | | | | |
| | PRECIPITATION >= 0.01 | 30 | 16.4 | 12.2 | 13.7 | 13.5 | 11.7 | 10.4 | 9.4 | 10.5 | 9.7 | 11.0 | 13.7 | 15.6 | 147.8 |
| | PRECIPITATION >= 1.00 | 30 | 0.3 | 0.2 | 0.3 | 0.7 | 0.5 | 1.1 | 0.9 | 1.1 | 0.9 | 0.8 | 0.6 | 0.5 | 7.9 |
| **SNOWFALL** | NORMAL (IN) | 30 | 23.2 | 15.5 | 8.7 | 1.7 | 0.* | 0.0 | 0.0 | 0.0 | 0.* | 0.5 | 7.7 | 19.2 | 76.5 |
| | MAXIMUM MONTHLY (IN) | 70 | 86.1 | 35.1 | 33.9 | 14.0 | 0.6 | T | T | T | 1.2 | 8.8 | 30.3 | 44.6 | 86.1 |
| | YEAR OF OCCURRENCE | | 1978 | 1958 | 1960 | 1982 | 1966 | 2008 | 2006 | 1996 | 1942 | 1989 | 1977 | 2000 | JAN 1978 |
| | MAXIMUM IN 24 HOURS (IN) | 70 | 16.7 | 14.4 | 14.8 | 8.7 | 0.6 | T | T | T | 1.0 | 8.8 | 17.5 | 13.7 | 17.5 |
| | YEAR OF OCCURRENCE' | | 1978 | 1993 | 1960 | 1982 | 1966 | 1992 | 1994 | 1989 | 1994 | 1989 | 1977 | 1981 | NOV 1977 |
| | MAXIMUM SNOW DEPTH (IN) | 61 | 41 | 35 | 21 | 6 | 0 | 0 | 0 | 0 | 0 | 6 | 20 | 23 | 41 |
| | YEAR OF OCCURRENCE | | 1978 | 1978 | 1960 | 1982 | | | | | | 1989 | 1977 | 1962 | JAN 1978 |
| | NORMAL NO. DAYS WITH: | | | | | | | | | | | | | | |
| | SNOWFALL >= 1.0 | 30 | 5.4 | 4.7 | 2.6 | 0.5 | 0.0 | 0.0 | 0.0 | 0.0 | 0.0 | 0.1 | 2.4 | 5.5 | 21.2 |

## PRECIPITATION (inches) 2009 SOUTH BEND (KSBN)

| YEAR | JAN | FEB | MAR | APR | MAY | JUN | JUL | AUG | SEP | OCT | NOV | DEC | ANNUAL |
|------|-----|-----|-----|-----|-----|-----|-----|-----|-----|-----|-----|-----|--------|
| 1980 | 1.52 | 1.51 | 3.74 | 3.44 | 1.65 | 5.97 | 3.29 | 7.84 | 5.64 | 3.35 | 1.47 | 3.91 | 43.33 |
| 1981 | 0.68 | 1.92 | 0.88 | 5.28 | 6.79 | 6.97 | 3.71 | 2.30 | 3.81 | 1.23 | 2.23 | 1.81 | 37.61 |
| 1982 | 2.95 | 1.17 | 4.54 | 1.46 | 5.51 | 3.12 | 7.47 | 2.84 | 2.51 | 0.91 | 4.52 | 3.40 | 40.40 |
| 1983 | 0.77 | 0.79 | 2.46 | 5.36 | 4.83 | 2.04 | 2.45 | 1.28 | 2.81 | 1.66 | 2.60 | 3.23 | 30.28 |
| 1984 | 0.86 | 1.45 | 2.10 | 4.22 | 4.02 | 3.43 | 1.76 | 1.47 | 4.02 | 4.38 | 2.73 | 4.42 | 34.86 |
| 1985 | 2.58 | 4.32 | 3.86 | 1.93 | 1.50 | 2.88 | 3.80 | 3.82 | 1.88 | 3.36 | 6.72 | 2.51 | 39.16 |
| 1986 | 1.24 | 2.46 | 2.09 | 1.87 | 3.42 | 5.06 | 6.15 | 1.90 | 4.27 | 3.81 | 2.90 | 1.67 | 36.84 |
| 1987 | 2.31 | 1.32 | 1.18 | 2.67 | 3.50 | 3.57 | 3.61 | 3.34 | 3.64 | 3.20 | 2.11 | 4.12 | 34.57 |
| 1988 | 2.21 | 1.98 | 3.03 | 2.91 | 1.40 | 0.48 | 1.28 | 5.63 | 4.42 | 6.68 | 5.72 | 2.91 | 38.65 |
| 1989 | 1.58 | 1.05 | 2.27 | 2.83 | 2.72 | 3.49 | 5.90 | 5.65 | 3.78 | 1.45 | 3.55 | 1.83 | 36.10 |
| 1990 | 2.36 | 3.66 | 2.79 | 2.91 | 6.86 | 4.40 | 5.45 | 4.60 | 3.76 | 7.09 | 6.69 | 5.04 | 55.61 |
| 1991 | 1.64 | 1.79 | 2.79 | 4.58 | 4.01 | 0.62 | 1.32 | 3.68 | 2.71 | 8.75 | 2.75 | 1.67 | 36.31 |
| 1992 | 1.64 | 1.73 | 2.93 | 2.19 | 1.17 | 1.74 | 5.24 | 2.07 | 8.84 | 1.60 | 5.54 | 3.99 | 38.68 |
| 1993 | 3.35 | 1.20 | 2.62 | 3.64 | 2.34 | 10.86 | 1.51 | 4.38 | 7.76 | 4.09 | 2.39 | 1.50 | 45.64 |
| 1994 | 2.46 | 1.45 | 0.80 | 2.80 | 0.80 | 5.10 | 4.97 | 4.19 | 4.68 | 3.49 | 4.36 | 2.50 | 37.60 |
| 1995 | 2.46 | 2.00 | 1.84 | 4.56 | 3.67 | 2.36 | 6.50 | 8.29 | 0.89 | 3.22 | 4.40 | 1.89 | 42.08 |
| 1996 | 1.66 | 2.09 | 1.24 | 3.60 | 8.09 | 7.20 | 6.69 | 1.75 | 3.30 | 3.27 | 3.70 | 2.90 | 45.49 |
| 1997 | 2.63 | 3.86 | 2.02 | 1.36 | 3.77 | 3.16 | 1.98 | 5.05 | 2.86 | 2.16 | 2.45 | 2.14 | 33.44 |
| 1998 | 3.76 | 1.69 | 4.03 | 3.78 | 2.49 | 3.98 | 2.27 | 5.84 | 1.54 | 2.50 | 1.24 | 2.04 | 35.16 |
| 1999 | 3.06 | 1.58 | 1.18 | 7.48 | 1.64 | 2.60 | 2.39 | 4.12 | 1.25 | 1.37 | 1.32 | 2.66 | 30.65 |
| 2000 | 2.57 | 1.63 | 1.81 | 3.69 | 4.60 | 7.75 | 2.88 | 1.49 | 3.22 | 2.28 | 3.07 | 2.29 | 37.28 |
| 2001 | 0.83 | 3.47 | 1.11 | 3.65 | 4.31 | 4.25 | 2.97 | 3.75 | 3.65 | 7.06 | 2.66 | 2.25 | 39.96 |
| 2002 | 2.48 | 2.10 | 2.77 | 3.40 | 5.75 | 1.26 | 2.47 | 2.31 | 1.16 | 1.43 | 1.91 | 1.80 | 28.84 |
| 2003 | 1.21 | 0.91 | 1.52 | 3.30 | 6.34 | 1.16 | 6.22 | 1.74 | 3.69 | 2.68 | 4.15 | 1.70 | 34.62 |
| 2004 | 1.56 | 0.76 | 3.37 | 0.83 | 5.67 | 5.09 | 4.12 | 5.62 | 0.92 | 1.97 | 4.56 | 2.21 | 36.68 |
| 2005 | 5.10 | 1.92 | 2.05 | 1.14 | 1.06 | 2.07 | 3.46 | 2.20 | 3.07 | 1.24 | 2.21 | 1.79 | 27.31 |
| 2006 | 2.87 | 0.95 | 3.08 | 2.41 | 5.45 | 2.00 | 8.66 | 4.66 | 3.53 | 4.51 | 3.33 | 3.55 | 45.00 |
| 2007 | 3.50 | 1.38 | 2.44 | 4.48 | 1.70 | 1.80 | 5.40 | 8.88 | 1.48 | 4.02 | 2.18 | 3.03 | 40.29 |
| 2008 | 5.34 | 3.39 | 2.50 | 2.51 | 2.64 | 2.84 | 2.38 | 1.90 | 13.92 | 3.13 | 2.02 | 3.79 | 46.36 |
| 2009 | 1.57 | 3.00 | 5.55 | 3.57 | 2.82 | 6.76 | 2.64 | 7.08 | 0.48 | 5.16 | 1.32 | 1.88 | 41.83 |
| POR= 114 YRS | 2.32 | 1.84 | 2.79 | 3.39 | 3.58 | 3.64 | 3.44 | 3.57 | 3.40 | 3.09 | 2.87 | 2.63 | 36.56 |

WBAN : 14848

## AVERAGE TEMPERATURE (°F) 2009 SOUTH BEND (KSBN)

| YEAR | JAN | FEB | MAR | APR | MAY | JUN | JUL | AUG | SEP | OCT | NOV | DEC | ANNUAL |
|------|-----|-----|-----|-----|-----|-----|-----|-----|-----|-----|-----|-----|--------|
| 1980 | 27.1 | 23.6 | 35.5 | 49.0 | 62.1 | 67.8 | 76.6 | 75.0 | 66.3 | 50.4 | 41.6 | 31.1 | 50.5 |
| 1981 | 23.4 | 32.1 | 40.7 | 52.5 | 56.6 | 69.2 | 71.4 | 70.6 | 62.8 | 50.0 | 41.5 | 27.6 | 49.9 |
| 1982 | 15.6 | 22.7 | 35.1 | 44.7 | 66.0 | 64.5 | 73.3 | 68.9 | 63.3 | 54.0 | 42.5 | 39.0 | 49.1 |
| 1983 | 29.3 | 33.2 | 40.2 | 45.5 | 55.3 | 72.0 | 78.7 | 78.3 | 66.5 | 53.8 | 43.8 | 18.4 | 51.3 |
| 1984 | 18.5 | 35.9 | 30.0 | 48.2 | 56.1 | 72.3 | 71.7 | 74.4 | 63.8 | 57.1 | 41.0 | 34.4 | 50.3 |
| 1985 | 19.6 | 23.7 | 41.3 | 55.2 | 63.1 | 66.8 | 73.1 | 69.8 | 65.6 | 54.6 | 41.6 | 20.1 | 49.5 |
| 1986 | 25.8 | 24.8 | 40.4 | 51.3 | 59.9 | 67.7 | 74.8 | 67.6 | 65.9 | 53.0 | 36.1 | 31.2 | 49.9 |
| 1987 | 25.3 | 30.8 | 40.0 | 50.4 | 64.7 | 72.8 | 75.8 | 71.7 | 64.6 | 46.8 | 43.7 | 32.8 | 51.6 |
| 1988 | 21.3 | 22.7 | 37.3 | 48.4 | 62.2 | 72.0 | 76.4 | 75.9 | 64.5 | 45.9 | 42.5 | 28.8 | 49.8 |
| 1989 | 33.4 | 22.2 | 37.0 | 47.3 | 57.2 | 68.0 | 73.9 | 70.6 | 61.8 | 52.9 | 38.8 | 17.7 | 48.4 |
| 1990 | 34.0 | 31.3 | 40.9 | 48.7 | 56.7 | 68.8 | 71.2 | 69.9 | 65.0 | 52.3 | 45.9 | 31.8 | 51.4 |
| 1991 | 24.2 | 32.3 | 41.5 | 52.6 | 67.5 | 74.0 | 75.7 | 73.2 | 63.2 | 54.9 | 36.6 | 32.1 | 52.3 |
| 1992 | 29.1 | 33.2 | 37.4 | 46.1 | 58.4 | 65.1 | 70.0 | 67.4 | 62.3 | 51.4 | 40.0 | 30.9 | 49.3 |
| 1993 | 28.3 | 24.4 | 35.0 | 46.9 | 60.7 | 67.7 | 74.5 | 73.6 | 58.9 | 50.5 | 39.4 | 29.7 | 49.1 |
| 1994 | 16.3 | 22.5 | 36.6 | 51.0 | 57.4 | 71.4 | 72.9 | 67.9 | 65.0 | 54.0 | 44.6 | 35.1 | 49.6 |
| 1995 | 25.3 | 25.3 | 40.3 | 46.5 | 57.9 | 71.3 | 74.9 | 77.2 | 62.2 | 54.3 | 34.8 | 26.0 | 49.7 |
| 1996 | 24.1 | 27.8 | 32.3 | 46.0 | 57.3 | 68.9 | 68.5 | 71.1 | 62.6 | 52.4 | 33.8 | 29.4 | 47.9 |
| 1997 | 20.6 | 29.6 | 37.6 | 43.7 | 52.5 | 68.3 | 72.2 | 67.9 | 62.4 | 51.6 | 36.2 | 30.7 | 47.8 |
| 1998 | 31.6 | 37.0 | 38.6 | 49.5 | 65.0 | 68.3 | 72.5 | 73.0 | 67.4 | 53.8 | 43.6 | 34.6 | 52.9 |
| 1999 | 22.7 | 33.0 | 33.5 | 51.0 | 62.7 | 71.3 | 78.0 | 70.1 | 64.9 | 52.6 | 45.9 | 31.9 | 51.5 |
| 2000 | 25.5 | 35.0 | 44.1 | 46.7 | 61.4 | 67.4 | 70.0 | 70.9 | 63.1 | 55.7 | 37.4 | 17.1 | 49.5 |
| 2001 | 24.7 | 28.1 | 34.1 | 53.7 | 61.2 | 67.3 | 72.4 | 72.9 | 61.8 | 52.7 | 48.6 | 35.1 | 51.1 |
| 2002 | 32.7 | 33.0 | 34.2 | 49.8 | 54.3 | 71.2 | 76.8 | 72.8 | 68.1 | 49.8 | 38.1 | 28.7 | 50.8 |
| 2003 | 20.3 | 23.1 | 37.9 | 50.4 | 56.0 | 66.1 | 71.1 | 72.9 | 62.4 | 51.1 | 43.3 | 32.1 | 48.9 |
| 2004 | 21.1 | 27.0 | 41.5 | 50.9 | 61.8 | 67.1 | 71.3 | 67.4 | 66.1 | 52.3 | 42.4 | 29.3 | 49.9 |
| 2005 | 23.9 | 30.9 | 33.0 | 52.2 | 57.2 | 73.9 | 74.5 | 73.3 | 67.7 | 53.4 | 42.5 | 24.1 | 50.6 |
| 2006 | 35.9 | 28.8 | 37.7 | 52.2 | 57.7 | 67.7 | 74.6 | 71.4 | 61.4 | 48.8 | 42.7 | 35.6 | 51.2 |
| 2007 | 29.0 | 17.8 | 42.4 | 47.0 | 64.6 | 71.3 | 72.2 | 73.0 | 67.5 | 59.0 | 39.8 | 29.5 | 51.1 |
| 2008 | 24.7 | 23.9 | 34.3 | 50.6 | 56.0 | 70.3 | 73.0 | 70.5 | 65.9 | 51.7 | 38.3 | 26.0 | 48.8 |
| 2009 | 15.6 | 28.7 | 40.1 | 48.4 | 59.2 | 69.1 | 68.3 | 69.3 | 64.0 | 48.8 | 44.2 | 27.6 | 48.6 |
| POR= 113 YRS | 24.4 | 25.1 | 36.5 | 47.7 | 59.1 | 67.9 | 73.1 | 71.3 | 63.5 | 52.9 | 39.4 | 28.6 | 49.1 |

## HEATING DEGREE DAYS (base 65°F) 2009 SOUTH BEND (KSBN)

| YEAR | JUL | AUG | SEP | OCT | NOV | DEC | JAN | FEB | MAR | APR | MAY | JUN | TOTAL |
|---|---|---|---|---|---|---|---|---|---|---|---|---|---|
| 1980-81 | 0 | 1 | 62 | 449 | 694 | 1047 | 1282 | 915 | 749 | 374 | 271 | 6 | 5850 |
| 1981-82 | 4 | 8 | 132 | 460 | 700 | 1154 | 1523 | 1178 | 922 | 604 | 64 | 72 | 6821 |
| 1982-83 | 2 | 30 | 114 | 353 | 668 | 798 | 1100 | 886 | 760 | 581 | 298 | 30 | 5620 |
| 1983-84 | 6 | 0 | 94 | 352 | 628 | 1440 | 1431 | 838 | 1080 | 503 | 290 | 4 | 6666 |
| 1984-85 | 7 | 0 | 128 | 244 | 714 | 940 | 1401 | 1153 | 727 | 339 | 116 | 47 | 5816 |
| 1985-86 | 0 | 4 | 121 | 319 | 694 | 1381 | 1208 | 1117 | 760 | 425 | 192 | 46 | 6267 |
| 1986-87 | 3 | 48 | 81 | 369 | 858 | 1038 | 1224 | 950 | 766 | 436 | 139 | 16 | 5928 |
| 1987-88 | 5 | 25 | 78 | 558 | 638 | 993 | 1347 | 1220 | 851 | 498 | 158 | 37 | 6408 |
| 1988-89 | 1 | 11 | 72 | 581 | 670 | 1116 | 972 | 1190 | 865 | 528 | 271 | 40 | 6317 |
| 1989-90 | 0 | 12 | 147 | 381 | 779 | 1462 | 954 | 936 | 751 | 521 | 257 | 46 | 6246 |
| 1990-91 | 6 | 14 | 110 | 408 | 565 | 1021 | 1260 | 910 | 722 | 378 | 106 | 2 | 5502 |
| 1991-92 | 0 | 1 | 167 | 318 | 844 | 1012 | 1104 | 915 | 850 | 561 | 247 | 75 | 6094 |
| 1992-93 | 6 | 25 | 139 | 417 | 742 | 1049 | 1130 | 1130 | 925 | 536 | 157 | 55 | 6311 |
| 1993-94 | 0 | 7 | 198 | 450 | 765 | 1087 | 1505 | 1186 | 875 | 441 | 273 | 28 | 6815 |
| 1994-95 | 0 | 30 | 92 | 336 | 606 | 921 | 1224 | 1105 | 759 | 548 | 221 | 20 | 5862 |
| 1995-96 | 7 | 0 | 135 | 337 | 896 | 1202 | 1258 | 1072 | 1005 | 566 | 276 | 23 | 6777 |
| 1996-97 | 22 | 1 | 140 | 384 | 928 | 1098 | 1368 | 983 | 844 | 631 | 383 | 39 | 6821 |
| 1997-98 | 12 | 20 | 120 | 437 | 858 | 1057 | 1026 | 775 | 824 | 458 | 74 | 80 | 5741 |
| 1998-99 | 0 | 0 | 52 | 346 | 632 | 936 | 1304 | 893 | 968 | 412 | 134 | 41 | 5718 |
| 1999-00 | 0 | 3 | 92 | 377 | 566 | 1020 | 1218 | 863 | 646 | 540 | 167 | 59 | 5551 |
| 2000-01 | 7 | 6 | 153 | 291 | 820 | 1476 | 1242 | 1026 | 951 | 342 | 168 | 87 | 6569 |
| 2001-02 | 22 | 0 | 139 | 381 | 489 | 923 | 993 | 891 | 948 | 487 | 347 | 25 | 5645 |
| 2002-03 | 0 | 2 | 41 | 487 | 800 | 1116 | 1376 | 1167 | 832 | 441 | 279 | 62 | 6603 |
| 2003-04 | 0 | 2 | 136 | 427 | 643 | 1012 | 1356 | 1096 | 719 | 441 | 177 | 44 | 6053 |
| 2004-05 | 1 | 43 | 56 | 387 | 670 | 1101 | 1267 | 948 | 984 | 384 | 250 | 17 | 6108 |
| 2005-06 | 2 | 3 | 50 | 385 | 666 | 1260 | 894 | 1007 | 840 | 377 | 282 | 26 | 5792 |
| 2006-07 | 1 | 1 | 127 | 498 | 663 | 906 | 1107 | 1318 | 702 | 537 | 97 | 22 | 5979 |
| 2007-08 | 0 | 7 | 63 | 235 | 751 | 1091 | 1240 | 1185 | 944 | 429 | 285 | 19 | 6249 |
| 2008-09 | 0 | 4 | 45 | 417 | 796 | 1201 | 1526 | 1010 | 763 | 500 | 205 | 33 | 6500 |
| 2009- | 12 | 33 | 74 | 495 | 614 | 1153 | | | | | | | |

WBAN : 14848

## COOLING DEGREE DAYS (base 65°F) 2009 SOUTH BEND (KSBN)

| YEAR | JAN | FEB | MAR | APR | MAY | JUN | JUL | AUG | SEP | OCT | NOV | DEC | TOTAL |
|---|---|---|---|---|---|---|---|---|---|---|---|---|---|
| 1980 | 0 | 0 | 0 | 9 | 65 | 145 | 367 | 319 | 107 | 6 | 0 | 0 | 1018 |
| 1981 | 0 | 0 | 2 | 6 | 15 | 137 | 211 | 191 | 72 | 0 | 0 | 0 | 634 |
| 1982 | 0 | 0 | 0 | 3 | 105 | 65 | 266 | 159 | 71 | 17 | 0 | 0 | 686 |
| 1983 | 0 | 0 | 0 | 1 | 3 | 247 | 440 | 417 | 146 | 13 | 0 | 0 | 1267 |
| 1984 | 0 | 0 | 0 | 5 | 19 | 228 | 226 | 298 | 98 | 7 | 0 | 0 | 881 |
| 1985 | 0 | 0 | 0 | 52 | 64 | 109 | 260 | 159 | 147 | 3 | 0 | 0 | 794 |
| 1986 | 0 | 0 | 4 | 22 | 41 | 135 | 312 | 136 | 113 | 2 | 0 | 0 | 765 |
| 1987 | 0 | 0 | 0 | 3 | 136 | 256 | 345 | 240 | 71 | 0 | 3 | 0 | 1054 |
| 1988 | 0 | 0 | 0 | 3 | 77 | 254 | 362 | 357 | 65 | 0 | 0 | 0 | 1118 |
| 1989 | 0 | 0 | 1 | 6 | 37 | 137 | 283 | 194 | 55 | 11 | 0 | 0 | 724 |
| 1990 | 0 | 0 | 9 | 37 | 8 | 167 | 203 | 171 | 117 | 21 | 0 | 0 | 733 |
| 1991 | 0 | 0 | 0 | 13 | 189 | 281 | 338 | 264 | 121 | 11 | 0 | 0 | 1217 |
| 1992 | 0 | 0 | 0 | 2 | 50 | 90 | 168 | 103 | 66 | 2 | 0 | 0 | 481 |
| 1993 | 0 | 0 | 0 | 0 | 29 | 144 | 302 | 284 | 18 | 9 | 0 | 0 | 786 |
| 1994 | 0 | 0 | 0 | 26 | 45 | 226 | 255 | 124 | 98 | 2 | 0 | 0 | 776 |
| 1995 | 0 | 0 | 0 | 0 | 9 | 216 | 320 | 382 | 57 | 10 | 0 | 0 | 994 |
| 1996 | 0 | 0 | 0 | 1 | 44 | 147 | 136 | 201 | 74 | 0 | 0 | 0 | 603 |
| 1997 | 0 | 0 | 0 | 0 | 0 | 146 | 241 | 117 | 48 | 29 | 0 | 0 | 581 |
| 1998 | 0 | 0 | 9 | 0 | 82 | 185 | 238 | 253 | 131 | 8 | 0 | 0 | 906 |
| 1999 | 0 | 0 | 0 | 0 | 67 | 236 | 410 | 166 | 96 | 3 | 0 | 0 | 978 |
| 2000 | 0 | 0 | 4 | 0 | 62 | 138 | 171 | 197 | 102 | 9 | 0 | 0 | 683 |
| 2001 | 0 | 0 | 0 | 11 | 56 | 164 | 259 | 253 | 53 | 6 | 0 | 0 | 802 |
| 2002 | 0 | 0 | 0 | 40 | 24 | 218 | 372 | 252 | 144 | 21 | 0 | 0 | 1071 |
| 2003 | 0 | 0 | 0 | 10 | 7 | 103 | 197 | 255 | 63 | 2 | 0 | 0 | 637 |
| 2004 | 0 | 0 | 0 | 27 | 85 | 117 | 204 | 123 | 96 | 3 | 0 | 0 | 655 |
| 2005 | 0 | 0 | 0 | 7 | 16 | 288 | 303 | 266 | 138 | 33 | 0 | 0 | 1051 |
| 2006 | 0 | 0 | 0 | 3 | 60 | 117 | 304 | 202 | 26 | 3 | 0 | 0 | 715 |
| 2007 | 0 | 0 | 9 | 3 | 90 | 217 | 232 | 261 | 145 | 57 | 0 | 0 | 1014 |
| 2008 | 0 | 0 | 0 | 6 | 13 | 184 | 256 | 179 | 81 | 10 | 0 | 0 | 729 |
| 2009 | 0 | 0 | 0 | 11 | 32 | 162 | 123 | 173 | 48 | 0 | 0 | 0 | 549 |

## SNOWFALL (inches)  2009  SOUTH BEND (KSBN)

| YEAR | JUL | AUG | SEP | OCT | NOV | DEC | JAN | FEB | MAR | APR | MAY | JUN | TOTAL |
|------|-----|-----|-----|-----|-----|-----|-----|-----|-----|-----|-----|-----|-------|
| 1980-81 | 0.0 | 0.0 | 0.0 | 1.1 | 8.8 | 24.3 | 23.8 | 20.7 | 6.3 | T | 0.0 | 0.0 | 85.0 |
| 1981-82 | 0.0 | 0.0 | 0.0 | 0.1 | 9.1 | 41.3 | 41.3 | 19.2 | 10.2 | 14.0 | 0.0 | 0.0 | 135.2 |
| 1982-83 | 0.0 | 0.0 | 0.0 | 0.0 | 2.1 | 2.5 | 8.0 | 9.6 | 12.0 | 1.1 | 0.0 | 0.0 | 35.3 |
| 1983-84 | 0.0 | 0.0 | 0.0 | 0.0 | 1.4 | 35.6 | 16.7 | 15.9 | 11.1 | 0.4 | 0.0 | 0.0 | 81.1 |
| 1984-85 | 0.0 | 0.0 | 0.0 | 0.0 | 0.6 | 14.1 | 40.0 | 28.9 | 1.6 | 3.1 | 0.0 | 0.0 | 88.3 |
| 1985-86 | 0.0 | 0.0 | 0.0 | 0.0 | 2.2 | 40.4 | 26.3 | 11.3 | 3.6 | 0.2 | 0.0 | 0.0 | 84.0 |
| 1986-87 | 0.0 | 0.0 | 0.0 | 0.0 | 9.7 | 4.8 | 31.4 | 5.9 | 2.0 | 1.5 | 0.0 | 0.0 | 55.3 |
| 1987-88 | 0.0 | 0.0 | T | T | 1.6 | 13.1 | 11.4 | 22.9 | 12.1 | T | 0.0 | 0.0 | 61.1 |
| 1988-89 | 0.0 | 0.0 | 0.0 | 0.3 | 7.8 | 14.8 | 3.1 | 16.3 | 2.5 | 1.7 | T | T | 46.5 |
| 1989-90 | T | T | 0.0 | 8.8 | 15.2 | 29.4 | 1.2 | 13.9 | 3.4 | 1.1 | 0.0 | 0.0 | 73.0 |
| 1990-91 | 0.0 | 0.0 | 0.0 | T | T | 17.5 | 15.7 | 23.4 | 2.4 | T | T | 0.0 | 59.0 |
| 1991-92 | 0.0 | 0.0 | 0.0 | T | 7.6 | 11.7 | 19.9 | 10.9 | 16.7 | 0.9 | 0.0 | T | 67.7 |
| 1992-93 | 0.0 | 0.0 | 0.0 | 1.2 | 7.7 | 12.6 | 8.1 | 32.2 | 19.0 | 1.4 | 0.0 | 0.0 | 82.2 |
| 1993-94 | 0.0 | 0.0 | 0.1 | T | 2.2 | 11.3 | 21.5 | 20.0 | 0.1 | 0.8 | T | 0.0 | 56.0 |
| 1994-95 | T | 0.0 | 1.0 | 0.0 | T | 2.4 | 17.3 | 25.8 | 2.8 | 0.6 | 0.0 | 0.0 | 49.9 |
| 1995-96 | 0.0 | 0.0 | 0.0 | T | 6.2 | 22.1 | 12.2 | 7.9 | 8.3 | 0.9 | T | 0.0 | 57.6 |
| 1996-97 | T | T | 0.0 | 0.0 | 15.3 | 10.5 | 36.4 | 10.2 | 3.3 | 0.4 | T | 0.0 | 76.1 |
| 1997-98 | 0.0 | 0.0 | 0.0 | 0.2 | 5.6 | 18.1 | 7.8 | T | 20.8 | 0.0 | 0.0 | T | 52.5 |
| 1998-99 | T | 0.0 | 0.0 | 0.0 | T | 17.1 | 37.0 | 10.4 | 14.2 | T | T | 0.0 | 78.7 |
| 1999-00 | 0.0 | 0.0 | 0.0 | 0.0 | 1.9 | 7.4 | 37.1 | 11.4 | T | 1.1 | 0.0 | 0.0 | 58.9 |
| 2000-01 | T | 0.0 | 0.0 | T | 10.4 | 44.6 | 2.9 | 5.4 | 10.6 | 2.7 | T | 0.0 | 76.6 |
| 2001-02 | T | 0.0 | 0.0 | T | 0.0 | 19.5 | 10.4 | 18.4 | 14.2 | 0.3 | 0.0 | 0.0 | 62.8 |
| 2002-03 | 0.0 | 0.0 | 0.0 | T | 4.6 | 9.0 | 27.2 | 14.5 | 3.8 | 1.0 | 0.0 | 0.0 | 60.1 |
| 2003-04 | T | 0.0 | 0.0 | 0.0 | 0.2 | 10.5 | 27.6 | 5.3 | 0.7 | T | 0.0 | 0.0 | 44.3 |
| 2004-05 | T | 0.0 | 0.0 | 0.0 | 5.0 | 20.1 | 28.8 | 8.0 | 15.4 | 1.1 | T | 0.0 | 78.4 |
| 2005-06 | 0.0 | 0.0 | 0.0 | 0.0 | 6.0 | 19.1 | 3.4 | 5.9 | 3.6 | 0.0 | 0.0 | T | 38.0 |
| 2006-07 | T | 0.0 | 0.0 | 1.8 | 0.2 | 4.0 | 25.3 | 17.6 | 9.5 | 6.1 | 0.0 | 0.0 | 64.5 |
| 2007-08 | 0.0 | 0.0 | 0.0 | T | 2.6 | 18.4 | 25.3 | 26.7 | 2.9 | T | 0.0 | T | 75.9 |
| 2008-09 | 0.0 | 0.0 | 0.0 | 0.7 | 16.2 | 16.8 | 29.4 | 12.0 | 0.1 | 2.0 | 0.0 | 0.0 | 77.2 |
| 2009- | 0.0 | 0.0 | 0.0 | T | 0.0 | 11.5 | | | | | | | |
| POR= 75 YRS | T | T | T | 1.2 | 6.6 | 14.7 | 16.5 | 12.1 | 8.2 | 2.1 | 0.2 | T | 61.6 |

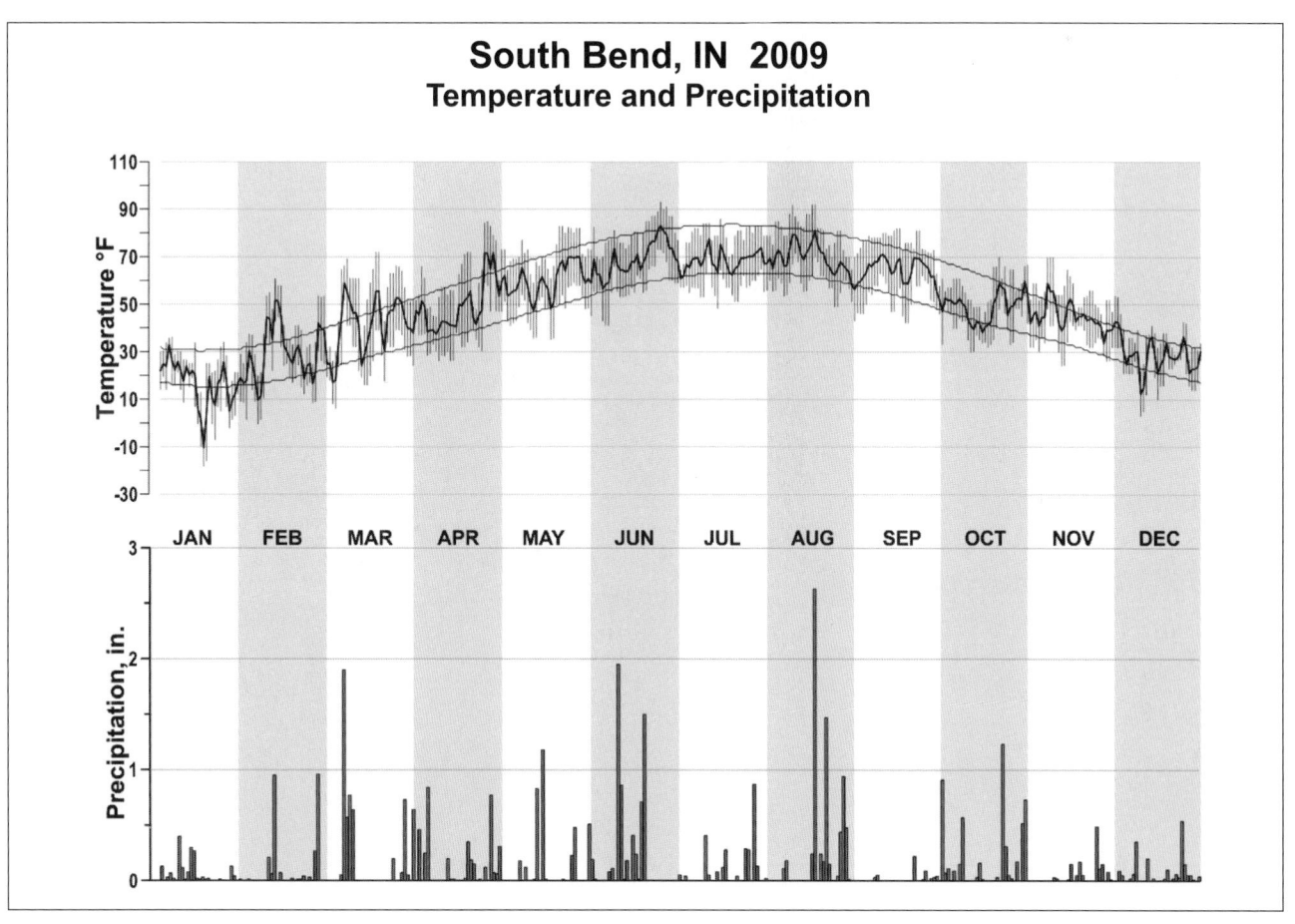

# South Bend, IN  2009
## Temperature and Precipitation

# 2009
# DES MOINES
# IOWA (KDSM)

Located in the heart of North America, Des Moines has a climate which is continental in character. This results in a marked seasonal contrast in both temperature and precipitation. There is a gently rolling terrain in and around the Des Moines metropolitan area. Drainage of the area is generally to the southeast to the Des Moines River and its tributaries.

Since agriculture and services for it are the mainstay of the area, it is convenient to separate the year into arbitrary seasons corresponding to the growing seasons of the principal crops of the section. The winter season, when most plant life is dormant, is from mid-November to late March. The summer season, when corn and soybeans can be grown, lasts from early May to early October. The spring growing season, including part of the growing season of oats and forage crops, and the fall harvest season, each runs about 6 weeks. There is a large variation in annual precipitation from a minimum of about 17 inches to a maximum of about 56 inches. The average annual snowfall is 32 inches. Annual variation of snowfall is also large, ranging from a minimum of about 8 inches to as much as 72 inches.

The winter is a season of cold dry air, interrupted by occasional storms of short duration. At the beginning and the end of the season, the precipitation may occur as rain, but during the major portion of the season it falls as snow. Drifting snow may be extensive and impede transportation. The average precipitation for this season is approximately 20 percent of the annual amount. Although occasional cold waves follow the storms, bitterly cold days on which the temperatures fail to rise above zero occur on an average of only 3 days in 4 years.

The average growing season with temperatures above 32 degrees normally spans 160 to 165 days between late April and mid-October. The growing season is characterized by prevailing southerly winds and precipitation falling primarily as showers and thunderstorms, occasionally with damaging wind, erosive downpours or hail. Some 60 percent of the annual precipitation falls during the crop season with the maximum rate normally in late May and June. The autumn is characteristically sunny with diminishing precipitation, a condition favorable for drying and harvesting crops.

# NORMALS, MEANS, AND EXTREMES
## DES MOINES (KDSM)

| LATITUDE: 41 ° 32'N | LONGITUDE: -93 ° 39'W | ELEVATION (FT): GRND: 910  BARO: 971 | TIME ZONE: CENTRAL  (UTC -6) | WBAN: 14933 |
|---|---|---|---|---|

| | ELEMENT | POR | JAN | FEB | MAR | APR | MAY | JUN | JUL | AUG | SEP | OCT | NOV | DEC | YEAR |
|---|---|---|---|---|---|---|---|---|---|---|---|---|---|---|---|
| **TEMPERATURE °F** | NORMAL DAILY MAXIMUM | 30 | 29.1 | 35.4 | 48.2 | 61.3 | 72.3 | 81.8 | 86.0 | 83.9 | 75.9 | 63.5 | 46.7 | 33.1 | 59.8 |
| | MEAN DAILY MAXIMUM | 64 | 29.2 | 34.6 | 46.4 | 61.4 | 72.3 | 81.5 | 85.9 | 83.9 | 75.9 | 64.3 | 47.6 | 33.9 | 59.7 |
| | HIGHEST DAILY MAXIMUM | 70 | 67 | 73 | 91 | 93 | 98 | 103 | 105 | 108 | 101 | 95 | 81 | 69 | 108 |
| | YEAR OF OCCURRENCE | | 2003 | 1972 | 1986 | 1980 | 1967 | 1988 | 1955 | 1983 | 1939 | 1963 | 1999 | 1984 | AUG 1983 |
| | MEAN OF EXTREME MAXS. | 72 | 51.9 | 56.8 | 73.0 | 83.5 | 87.8 | 93.2 | 96.8 | 95.7 | 90.8 | 83.3 | 69.1 | 56.9 | 78.2 |
| | NORMAL DAILY MINIMUM | 30 | 11.7 | 17.8 | 28.7 | 39.9 | 51.4 | 61.0 | 66.1 | 63.9 | 54.3 | 42.2 | 29.0 | 16.7 | 40.2 |
| | MEAN DAILY MINIMUM | 64 | 12.0 | 16.9 | 27.5 | 40.1 | 51.4 | 61.1 | 65.9 | 63.9 | 54.3 | 43.0 | 29.6 | 17.5 | 40.3 |
| | LOWEST DAILY MINIMUM | 70 | -24 | -26 | -22 | 9 | 30 | 38 | 47 | 40 | 0 | 14 | -4 | -22 | -26 |
| | YEAR OF OCCURRENCE | | 1970 | 1996 | 1962 | 1975 | 2005 | 1945 | 1971 | 1950 | 1996 | 1972 | 1991 | 1989 | FEB 1996 |
| | MEAN OF EXTREME MINS. | 72 | -10.6 | -5.0 | 7.2 | 23.7 | 36.5 | 48.5 | 55.2 | 52.4 | 37.9 | 26.4 | 11.4 | -3.5 | 23.3 |
| | NORMAL DRY BULB | 30 | 20.4 | 26.6 | 38.4 | 50.6 | 61.9 | 71.4 | 76.1 | 73.9 | 65.1 | 52.8 | 37.9 | 24.9 | 50.0 |
| | MEAN DRY BULB | 64 | 20.7 | 25.8 | 36.9 | 50.7 | 61.8 | 71.5 | 75.9 | 73.9 | 65.1 | 53.7 | 38.6 | 25.8 | 50.0 |
| | MEAN WET BULB | 26 | 19.9 | 23.6 | 33.4 | 43.7 | 54.1 | 63.4 | 67.9 | 66.5 | 58.0 | 46.2 | 33.9 | 23.2 | 44.5 |
| | MEAN DEW POINT | 26 | 16.5 | 20.0 | 28.6 | 38.5 | 50.2 | 60.3 | 65.5 | 64.1 | 54.7 | 42.1 | 30.4 | 20.2 | 40.9 |
| | NORMAL NO. DAYS WITH: | | | | | | | | | | | | | | |
| | MAXIMUM >= 90 | 30 | 0.0 | 0.0 | * | 0.1 | 0.3 | 4.5 | 9.6 | 6.9 | 2.1 | 0.1 | 0.0 | 0.0 | 23.6 |
| | MAXIMUM <= 32 | 30 | 16.8 | 12.1 | 3.9 | 0.3 | 0.0 | 0.0 | 0.0 | 0.0 | 0.0 | 0.0 | 3.7 | 13.4 | 50.2 |
| | MINIMUM <= 32 | 30 | 30.0 | 24.6 | 19.4 | 6.2 | 0.1 | 0.0 | 0.0 | 0.0 | 0.2 | 4.7 | 19.0 | 28.7 | 132.9 |
| | MINIMUM <= 0 | 30 | 7.3 | 3.3 | 0.3 | 0.0 | 0.0 | 0.0 | 0.0 | 0.0 | 0.0 | 0.0 | 0.2 | 4.0 | 15.1 |
| **H/C** | NORMAL HEATING DEG. DAYS | 30 | 1385 | 1090 | 826 | 439 | 153 | 16 | 1 | 6 | 103 | 386 | 804 | 1227 | 6436 |
| | NORMAL COOLING DEG. DAYS | 30 | 0 | 0 | 1 | 12 | 60 | 219 | 353 | 285 | 110 | 12 | 0 | 0 | 1052 |
| **RH** | NORMAL (PERCENT) | 30 | 72 | 71 | 67 | 63 | 64 | 66 | 69 | 72 | 70 | 67 | 72 | 76 | 69 |
| | HOUR 00 LST | 30 | 75 | 76 | 72 | 69 | 71 | 73 | 77 | 80 | 78 | 74 | 76 | 79 | 75 |
| | HOUR 06 LST | 30 | 77 | 78 | 78 | 77 | 78 | 80 | 83 | 86 | 85 | 81 | 80 | 81 | 80 |
| | HOUR 12 LST | 30 | 67 | 65 | 58 | 54 | 55 | 56 | 58 | 60 | 58 | 56 | 65 | 70 | 60 |
| | HOUR 18 LST | 30 | 70 | 67 | 58 | 52 | 53 | 54 | 57 | 60 | 59 | 58 | 67 | 73 | 61 |
| **S** | PERCENT POSSIBLE SUNSHINE | 55 | 51 | 54 | 57 | 56 | 61 | 68 | 72 | 70 | 66 | 62 | 49 | 45 | 59 |
| **W/O** | MEAN NO. DAYS WITH: | | | | | | | | | | | | | | |
| | HEAVY FOG(VISBY <= 1/4 MI) | 46 | 2.5 | 2.1 | 2.0 | 1.0 | 0.8 | 0.7 | 0.6 | 1.2 | 1.2 | 1.1 | 1.6 | 3.0 | 17.8 |
| | THUNDERSTORMS | 64 | 0.3 | 0.4 | 2.0 | 4.3 | 6.9 | 8.8 | 8.1 | 7.1 | 4.8 | 2.6 | 1.1 | 0.3 | 46.7 |
| **CLOUDNESS** | MEAN: | | | | | | | | | | | | | | |
| | SUNRISE-SUNSET (OKTAS) | | | | | | | 4.8 | | | | | | | |
| | MIDNIGHT-MIDNIGHT (OKTAS) | | | | | | | | | | | | | | |
| | MEAN NO. DAYS WITH: | | | | | | | | | | | | | | |
| | CLEAR | 1 | 1.0 | 1.0 | 6.0 | | 3.0 | 10.0 | | | | | | | |
| | PARTLY CLOUDY | 1 | 2.0 | | 3.0 | | 2.0 | 1.0 | | | | | | | |
| | CLOUDY | 1 | 3.0 | 3.0 | 7.0 | | 12.0 | 10.0 | | | | | | | |
| **PR** | MEAN STATION PRESSURE(IN) | 26 | 29.06 | 29.06 | 29.00 | 28.91 | 28.92 | 28.92 | 28.96 | 28.99 | 29.00 | 29.01 | 29.01 | 29.05 | 28.99 |
| | MEAN SEA-LEVEL PRES. (IN) | 26 | 30.13 | 30.12 | 30.04 | 29.94 | 29.94 | 29.92 | 29.96 | 30.00 | 30.02 | 30.04 | 30.06 | 30.12 | 30.02 |
| **WINDS** | MEAN SPEED (MPH) | 26 | 10.6 | 10.4 | 11.3 | 11.4 | 10.2 | 9.2 | 8.4 | 7.9 | 8.8 | 9.8 | 10.5 | 10.3 | 9.9 |
| | PREVAIL.DIR(TENS OF DEGS) | 38 | 33 | 33 | 33 | 33 | 15 | 19 | 19 | 16 | 17 | 19 | 33 | 33 | 33 |
| | MAXIMUM 2-MINUTE: | | | | | | | | | | | | | | |
| | SPEED (MPH) | 14 | 44 | 41 | 48 | 51 | 53 | 56 | 49 | 53 | 52 | 46 | 52 | 44 | 56 |
| | DIR. (TENS OF DEGS) | | 32 | 30 | 27 | 29 | 27 | 31 | 02 | 28 | 29 | 26 | 25 | 31 | 31 |
| | YEAR OF OCCURRENCE | | 2005 | 1999 | 2002 | 1996 | 2008 | 2007 | 2000 | 1998 | 2001 | 2004 | 1998 | 2004 | JUN 2007 |
| | MAXIMUM 3-SECOND | | | | | | | | | | | | | | |
| | SPEED (MPH) | 14 | 53 | 51 | 59 | 64 | 77 | 75 | 61 | 70 | 60 | 60 | 62 | 53 | 77 |
| | DIR. (TENS OF DEGS) | | 32 | 29 | 26 | 28 | 27 | 29 | 01 | 29 | 28 | 31 | 26 | 33 | 27 |
| | YEAR OF OCCURRENCE | | 2006 | 1996 | 2002 | 1996 | 2008 | 2007 | 2000 | 1998 | 2001 | 2008 | 1998 | 2009 | MAY 2008 |
| **PRECIPITATION** | NORMAL (IN) | 30 | 1.03 | 1.19 | 2.21 | 3.58 | 4.25 | 4.57 | 4.18 | 4.51 | 3.15 | 2.62 | 2.10 | 1.33 | 34.72 |
| | MAXIMUM MONTHLY (IN) | 70 | 4.38 | 2.99 | 5.82 | 7.76 | 12.13 | 14.19 | 10.51 | 13.68 | 10.19 | 7.29 | 6.52 | 3.43 | 14.19 |
| | YEAR OF OCCURRENCE | | 1960 | 1951 | 1990 | 1976 | 1996 | 1947 | 1958 | 1977 | 1961 | 1941 | 1983 | 1982 | JUN 1947 |
| | MINIMUM MONTHLY (IN) | 70 | 0.04 | 0.13 | 0.17 | 0.23 | 1.23 | 1.02 | 0.04 | 0.25 | 0.41 | 0.03 | 0.03 | 0.03 | 0.03 |
| | YEAR OF OCCURRENCE | | 1997 | 1968 | 1994 | 1985 | 1949 | 1992 | 1975 | 1984 | 1950 | 1952 | 1969 | 2002 | DEC 2002 |
| | MAXIMUM IN 24 HOURS (IN) | 70 | 2.97 | 1.77 | 2.42 | 3.80 | 3.23 | 5.50 | 5.14 | 6.18 | 4.47 | 2.81 | 3.55 | 1.69 | 6.18 |
| | YEAR OF OCCURRENCE | | 1960 | 1961 | 1945 | 1974 | 1996 | 1947 | 1958 | 1975 | 1961 | 1947 | 2003 | 1982 | AUG 1975 |
| | NORMAL NO. DAYS WITH: | | | | | | | | | | | | | | |
| | PRECIPITATION >= 0.01 | 30 | 7.6 | 7.7 | 9.5 | 11.0 | 12.0 | 11.0 | 9.9 | 9.4 | 8.5 | 8.1 | 8.6 | 8.5 | 111.8 |
| | PRECIPITATION >= 1.00 | 30 | 0.1 | 0.1 | 0.5 | 0.6 | 0.9 | 1.4 | 1.2 | 1.5 | 0.7 | 0.8 | 0.3 | 0.1 | 8.2 |
| **SNOWFALL** | NORMAL (IN) | 30 | 8.8 | 8.2 | 4.1 | 2.7 | 0.* | 0.0 | 0.0 | 0.0 | 0.* | 0.4 | 4.5 | 7.7 | 36.4 |
| | MAXIMUM MONTHLY (IN) | 66 | 22.3 | 22.7 | 18.8 | 15.6 | 0.2 | T | T | 0.0 | T | 7.4 | 14.7 | 28.2 | 28.2 |
| | YEAR OF OCCURRENCE | | 1996 | 2008 | 1948 | 1982 | 1944 | 1993 | 1992 | | 1992 | 1980 | 1991 | 2009 | DEC 2009 |
| | MAXIMUM IN 24 HOURS (IN) | 65 | 19.8 | 12.1 | 15.6 | 10.4 | 0.2 | T | T | 0.0 | T | 7.4 | 11.8 | 11.0 | 19.8 |
| | YEAR OF OCCURRENCE' | | 1942 | 1950 | 2004 | 1973 | 1944 | 1993 | 1992 | | 1992 | 1980 | 1968 | 1961 | JAN 1942 |
| | MAXIMUM SNOW DEPTH (IN) | 59 | 16 | 17 | 18 | 12 | 0 | 0 | 0 | 0 | 0 | 5 | 10 | 30 | 30 |
| | YEAR OF OCCURRENCE | | 2001 | 1979 | 1960 | 1973 | | | | | | 1980 | 1968 | 1962 | DEC 1962 |
| | NORMAL NO. DAYS WITH: | | | | | | | | | | | | | | |
| | SNOWFALL >= 1.0 | 30 | 2.5 | 2.5 | 1.2 | 0.6 | 0.0 | 0.0 | 0.0 | 0.0 | 0.0 | 0.1 | 1.3 | 2.3 | 10.5 |

## PRECIPITATION (inches) 2009 DES MOINES (KDSM)

| YEAR | JAN | FEB | MAR | APR | MAY | JUN | JUL | AUG | SEP | OCT | NOV | DEC | ANNUAL |
|------|-----|-----|-----|-----|-----|-----|-----|-----|-----|-----|-----|-----|--------|
| 1980 | 1.80 | 0.64 | 1.15 | 0.86 | 1.94 | 5.56 | 1.52 | 7.24 | 1.03 | 1.90 | 0.45 | 1.00 | 25.09 |
| 1981 | 0.25 | 0.97 | 0.39 | 2.00 | 2.46 | 5.02 | 5.76 | 6.32 | 2.30 | 2.06 | 2.63 | 1.14 | 31.30 |
| 1982 | 2.63 | 0.78 | 3.30 | 5.03 | 5.79 | 2.59 | 7.00 | 5.25 | 2.94 | 3.44 | 2.62 | 3.43 | 44.80 |
| 1983 | 1.17 | 1.95 | 3.72 | 3.80 | 3.93 | 3.65 | 2.44 | 3.01 | 3.87 | 5.54 | 6.52 | 1.57 | 41.17 |
| 1984 | 0.99 | 0.82 | 1.65 | 5.85 | 5.58 | 7.81 | 6.22 | 0.25 | 2.76 | 6.28 | 1.16 | 2.41 | 41.78 |
| 1985 | 0.64 | 1.98 | 3.37 | 0.23 | 1.56 | 3.72 | 2.04 | 2.83 | 5.42 | 3.75 | 1.65 | 1.31 | 28.50 |
| 1986 | 0.12 | 1.76 | 2.92 | 5.66 | 4.35 | 7.08 | 3.90 | 4.52 | 6.41 | 3.89 | 0.99 | 0.98 | 42.58 |
| 1987 | 0.42 | 1.38 | 2.99 | 2.92 | 3.75 | 2.10 | 5.08 | 10.04 | 1.40 | 1.03 | 3.27 | 2.59 | 36.97 |
| 1988 | 0.37 | 0.59 | 0.66 | 0.75 | 1.46 | 2.75 | 4.78 | 3.05 | 2.89 | 0.59 | 3.38 | 0.84 | 22.11 |
| 1989 | 1.30 | 1.05 | 0.37 | 1.95 | 3.62 | 2.22 | 3.65 | 6.53 | 5.41 | 2.28 | 0.19 | 0.57 | 29.14 |
| 1990 | 1.43 | 0.89 | 5.82 | 3.43 | 4.36 | 9.52 | 8.75 | 1.83 | 1.40 | 1.80 | 2.52 | 2.18 | 43.93 |
| 1991 | 0.95 | 0.17 | 3.90 | 7.54 | 7.88 | 2.87 | 1.14 | 3.65 | 0.90 | 4.96 | 3.61 | 2.20 | 39.77 |
| 1992 | 0.97 | 2.12 | 2.13 | 3.99 | 1.45 | 1.02 | 7.76 | 1.39 | 4.99 | 0.51 | 5.20 | 1.98 | 33.51 |
| 1993 | 1.59 | 1.52 | 3.22 | 2.96 | 7.51 | 7.68 | 9.75 | 12.24 | 5.79 | 1.70 | 1.06 | 0.86 | 55.88 |
| 1994 | 1.22 | 1.71 | 0.17 | 2.71 | 1.76 | 5.29 | 4.30 | 2.81 | 3.11 | 1.02 | 1.68 | 2.42 | 28.20 |
| 1995 | 0.98 | 0.68 | 2.75 | 6.04 | 6.29 | 2.88 | 4.36 | 1.04 | 3.12 | 1.24 | 1.57 | 0.08 | 31.03 |
| 1996 | 2.99 | 0.57 | 1.38 | 2.12 | 12.13 | 2.95 | 4.02 | 2.54 | 4.05 | 2.87 | 2.40 | .52 | 38.54 |
| 1997 | 0.04 | 0.77 | 0.85 | 4.11 | 3.95 | 4.90 | 2.89 | 2.36 | 2.14 | 4.62 | 1.13 | 0.77 | 28.53 |
| 1998 | 0.60 | 1.23 | 3.03 | 1.79 | 4.97 | 9.95 | 4.28 | 5.65 | 0.69 | 3.39 | 2.09 | 0.03 | 37.70 |
| 1999 | 0.26 | 0.63 | 0.90 | 4.43 | 5.82 | 3.46 | 2.78 | 4.56 | 2.32 | 0.36 | 1.28 | 0.35 | 27.15 |
| 2000 | 0.27 | 1.32 | 0.35 | 2.36 | 3.37 | 7.60 | 4.12 | 1.71 | 1.84 | 1.07 | 1.57 | 0.56 | 26.14 |
| 2001 | 1.31 | 2.08 | 1.08 | 2.20 | 7.46 | 2.61 | 1.04 | 2.22 | 4.77 | 2.12 | 0.86 | 0.70 | 28.45 |
| 2002 | 0.82 | 0.82 | 1.13 | 3.93 | 3.39 | 2.70 | 3.80 | 3.51 | 1.09 | 3.11 | 0.23 | T | 24.53 |
| 2003 | 0.49 | 1.69 | 0.96 | 6.09 | 5.31 | 4.21 | 4.06 | 0.31 | 2.15 | 1.16 | 5.30 | 1.18 | 32.91 |
| 2004 | 1.20 | 1.77 | 3.48 | 4.21 | 9.07 | 2.46 | 4.06 | 4.92 | 1.77 | 1.60 | 2.53 | 0.60 | 37.67 |
| 2005 | 0.84 | 1.79 | 1.61 | 4.73 | 5.17 | 4.68 | 3.28 | 1.33 | 1.41 | 0.92 | 1.34 | 0.95 | 28.05 |
| 2006 | 0.75 | 0.22 | 4.10 | 4.30 | 1.41 | 1.83 | 4.35 | 5.82 | 3.98 | 1.55 | 2.55 | 2.53 | 33.39 |
| 2007 | 0.74 | 2.42 | 3.16 | 4.59 | 5.94 | 3.02 | 2.56 | 7.08 | 3.53 | 5.49 | 0.28 | 2.90 | 41.71 |
| 2008 | 0.44 | 2.46 | 1.61 | 5.82 | 3.84 | 13.45 | 8.18 | 1.94 | 3.62 | 3.73 | 2.37 | 1.99 | 49.45 |
| 2009 | 0.80 | 0.58 | 4.73 | 5.65 | 3.79 | 4.69 | 2.36 | 4.29 | 1.22 | 6.56 | 1.30 | 2.83 | 38.80 |
| POR= 72 YRS | 1.05 | 1.19 | 2.26 | 3.42 | 4.29 | 4.54 | 3.66 | 3.90 | 2.95 | 2.39 | 1.75 | 1.24 | 32.64 |

WBAN : 14933

## AVERAGE TEMPERATURE (°F) 2009 DES MOINES (KDSM)

| YEAR | JAN | FEB | MAR | APR | MAY | JUN | JUL | AUG | SEP | OCT | NOV | DEC | ANNUAL |
|------|-----|-----|-----|-----|-----|-----|-----|-----|-----|-----|-----|-----|--------|
| 1980 | 23.4 | 21.3 | 34.7 | 52.0 | 63.5 | 71.2 | 79.9 | 76.2 | 66.8 | 49.6 | 41.6 | 26.6 | 50.6 |
| 1981 | 25.7 | 29.8 | 42.7 | 57.6 | 60.4 | 72.9 | 76.1 | 72.4 | 66.0 | 51.7 | 42.8 | 25.5 | 52.0 |
| 1982 | 9.6 | 22.9 | 35.4 | 46.9 | 64.7 | 67.1 | 76.9 | 73.2 | 64.9 | 54.6 | 38.4 | 31.6 | 48.9 |
| 1983 | 27.3 | 32.3 | 39.4 | 45.4 | 58.4 | 73.2 | 80.9 | 83.3 | 67.7 | 52.6 | 40.7 | 9.8 | 50.9 |
| 1984 | 19.7 | 35.5 | 31.1 | 48.8 | 58.7 | 73.2 | 75.9 | 77.5 | 62.4 | 52.8 | 39.5 | 27.6 | 50.2 |
| 1985 | 15.8 | 22.2 | 42.0 | 55.2 | 65.1 | 68.4 | 76.5 | 71.9 | 64.8 | 52.6 | 30.0 | 13.4 | 48.2 |
| 1986 | 26.9 | 21.7 | 42.5 | 53.9 | 62.6 | 73.3 | 77.2 | 69.3 | 67.4 | 52.6 | 33.3 | 28.8 | 50.8 |
| 1987 | 26.6 | 35.7 | 42.7 | 54.6 | 66.8 | 74.7 | 78.5 | 72.1 | 65.2 | 48.2 | 43.1 | 30.0 | 53.2 |
| 1988 | 19.6 | 21.6 | 40.2 | 51.4 | 67.4 | 75.6 | 78.5 | 78.8 | 66.9 | 47.9 | 39.8 | 28.8 | 51.4 |
| 1989 | 32.5 | 15.4 | 37.1 | 52.3 | 61.0 | 68.9 | 77.1 | 73.3 | 62.2 | 54.1 | 36.0 | 16.9 | 48.9 |
| 1990 | 31.7 | 31.1 | 42.0 | 50.0 | 57.9 | 71.6 | 73.8 | 74.4 | 68.1 | 52.4 | 43.7 | 22.9 | 51.6 |
| 1991 | 16.8 | 33.5 | 42.5 | 53.6 | 66.7 | 74.8 | 76.1 | 73.7 | 65.0 | 52.5 | 30.9 | 30.6 | 51.4 |
| 1992 | 30.7 | 34.9 | 42.4 | 48.9 | 62.0 | 70.3 | 71.2 | 68.3 | 63.1 | 53.4 | 34.9 | 27.1 | 50.6 |
| 1993 | 21.3 | 21.5 | 34.1 | 47.1 | 61.0 | 69.0 | 73.8 | 73.8 | 59.2 | 50.4 | 36.4 | 28.4 | 48.0 |
| 1994 | 13.0 | 21.3 | 40.5 | 50.9 | 62.5 | 72.4 | 73.0 | 71.3 | 66.4 | 55.5 | 42.1 | 30.3 | 49.9 |
| 1995 | 20.8 | 28.8 | 39.8 | 47.4 | 58.9 | 71.0 | 76.4 | 79.1 | 63.2 | 52.9 | 33.3 | 26.0 | 49.8 |
| 1996 | 17.3 | 25.0 | 32.5 | 47.8 | 57.7 | 70.9 | 72.7 | 72.3 | .0 | 53.4 | 32.1 | 22.2 | 42.0 |
| 1997 | 17.5 | 27.6 | 40.0 | 45.9 | 55.9 | 72.1 | 75.6 | 72.1 | 66.4 | 53.7 | 34.5 | 28.8 | 49.2 |
| 1998 | 25.9 | 36.3 | 33.6 | 50.8 | 66.5 | 68.1 | 75.4 | 75.2 | 70.4 | 54.0 | 42.4 | 30.3 | 52.4 |
| 1999 | 20.2 | 35.2 | 38.7 | 51.4 | 61.3 | 70.7 | 79.7 | 72.4 | 62.9 | 53.7 | 47.4 | 29.9 | 52.0 |
| 2000 | 25.4 | 36.5 | 45.1 | 52.0 | 64.8 | 69.3 | 73.8 | 75.5 | 67.2 | 57.7 | 33.9 | 11.6 | 51.1 |
| 2001 | 23.6 | 20.3 | 32.2 | 56.2 | 61.7 | 70.6 | 78.0 | 75.2 | 63.5 | 52.2 | 49.8 | 32.4 | 51.3 |
| 2002 | 31.0 | 32.0 | 34.4 | 50.6 | 59.3 | 74.5 | 78.2 | 72.8 | 68.1 | 47.4 | 36.2 | 31.4 | 51.3 |
| 2003 | 21.1 | 21.8 | 37.5 | 52.7 | 61.3 | 70.8 | 76.6 | 77.8 | 63.4 | 55.0 | 37.9 | 29.5 | 50.6 |
| 2004 | 19.9 | 25.3 | 42.5 | 53.9 | 63.7 | 68.8 | 72.3 | 68.8 | 68.4 | 53.5 | 42.3 | 29.5 | 50.7 |
| 2005 | 21.5 | 33.2 | 39.6 | 56.0 | 60.2 | 74.6 | 77.6 | 75.5 | 70.1 | 54.7 | 42.1 | 23.6 | 52.4 |
| 2006 | 35.6 | 27.7 | 39.4 | 56.6 | 63.5 | 73.5 | 79.1 | 74.6 | 62.8 | 50.2 | 42.0 | 34.7 | 53.3 |
| 2007 | 23.7 | 18.7 | 45.3 | 49.0 | 66.4 | 72.7 | 76.8 | 77.4 | 66.7 | 57.2 | 39.0 | 23.4 | 51.4 |
| 2008 | 19.4 | 19.8 | 35.6 | 47.9 | 60.4 | 71.4 | 75.4 | 72.7 | 65.5 | 53.9 | 39.0 | 21.5 | 48.5 |
| 2009 | 17.8 | 29.9 | 41.4 | 50.9 | 62.9 | 72.5 | 72.1 | 72.6 | 67.2 | 47.7 | 46.9 | 22.7 | 50.4 |
| POR= 64 YRS | 20.7 | 25.8 | 36.9 | 50.7 | 61.8 | 71.5 | 75.9 | 73.9 | 65.1 | 53.7 | 38.6 | 25.8 | 50.0 |

## HEATING DEGREE DAYS (base 65°F) 2009  DES MOINES (KDSM)

| YEAR | JUL | AUG | SEP | OCT | NOV | DEC | JAN | FEB | MAR | APR | MAY | JUN | TOTAL |
|------|-----|-----|-----|-----|-----|-----|-----|-----|-----|-----|-----|-----|-------|
| 1980-81 | 0 | 0 | 77 | 473 | 695 | 1182 | 1214 | 979 | 684 | 241 | 177 | 0 | 5722 |
| 1981-82 | 6 | 2 | 57 | 406 | 660 | 1218 | 1713 | 1175 | 911 | 536 | 73 | 30 | 6787 |
| 1982-83 | 0 | 6 | 113 | 326 | 791 | 1026 | 1162 | 908 | 787 | 587 | 219 | 17 | 5942 |
| 1983-84 | 0 | 0 | 96 | 394 | 720 | 1709 | 1401 | 851 | 1043 | 488 | 217 | 1 | 6920 |
| 1984-85 | 0 | 0 | 172 | 376 | 759 | 1154 | 1520 | 1192 | 707 | 335 | 59 | 27 | 6301 |
| 1985-86 | 0 | 0 | 172 | 378 | 1046 | 1596 | 1181 | 1208 | 702 | 344 | 116 | 2 | 6745 |
| 1986-87 | 0 | 25 | 51 | 378 | 941 | 1114 | 1184 | 813 | 687 | 336 | 58 | 6 | 5593 |
| 1987-88 | 0 | 24 | 54 | 513 | 648 | 1083 | 1399 | 1254 | 764 | 400 | 33 | 3 | 6175 |
| 1988-89 | 0 | 6 | 35 | 524 | 749 | 1114 | 1002 | 1384 | 866 | 417 | 170 | 31 | 6298 |
| 1989-90 | 0 | 6 | 140 | 345 | 865 | 1489 | 1025 | 943 | 703 | 469 | 221 | 21 | 6227 |
| 1990-91 | 3 | 0 | 85 | 394 | 633 | 1301 | 1491 | 875 | 699 | 355 | 112 | 0 | 5948 |
| 1991-92 | 0 | 1 | 144 | 395 | 1019 | 1058 | 1054 | 868 | 691 | 486 | 152 | 9 | 5877 |
| 1992-93 | 0 | 26 | 135 | 363 | 897 | 1169 | 1344 | 1210 | 953 | 528 | 151 | 42 | 6818 |
| 1993-94 | 0 | 2 | 185 | 460 | 851 | 1129 | 1606 | 1221 | 751 | 437 | 132 | 17 | 6791 |
| 1994-95 | 0 | 11 | 74 | 292 | 681 | 1071 | 1363 | 1007 | 774 | 521 | 191 | 20 | 6005 |
| 1995-96 | 3 | 0 | 136 | 381 | 945 | 1201 | 1471 | 1155 | 1002 | 514 | 264 | 31 | 7103 |
| 1996-97 | | 0 | | 364 | 982 | 1322 | 1468 | 1043 | 764 | 568 | 288 | 0 | |
| 1997-98 | 4 | 0 | 52 | 409 | 908 | 1118 | 1207 | 799 | 967 | 419 | 45 | 67 | 5995 |
| 1998-99 | 0 | 0 | 26 | 339 | 674 | 1067 | 1384 | 824 | 808 | 402 | 147 | 27 | 5698 |
| 1999-00 | 0 | 2 | 130 | 355 | 522 | 1080 | 1218 | 821 | 615 | 388 | 90 | 20 | 5241 |
| 2000-01 | 0 | 0 | 85 | 241 | 930 | 1647 | 1274 | 1244 | 1009 | 286 | 154 | 33 | 6903 |
| 2001-02 | 0 | 0 | 105 | 399 | 449 | 1005 | 1044 | 918 | 941 | 449 | 220 | 2 | 5532 |
| 2002-03 | 0 | 5 | 54 | 544 | 858 | 1033 | 1354 | 1203 | 843 | 389 | 139 | 17 | 6439 |
| 2003-04 | 0 | 2 | 135 | 320 | 808 | 1052 | 1395 | 1148 | 693 | 350 | 127 | 21 | 6051 |
| 2004-05 | 1 | 27 | 28 | 352 | 673 | 1093 | 1342 | 882 | 781 | 282 | 180 | 0 | 5641 |
| 2005-06 | 0 | 0 | 42 | 357 | 681 | 1275 | 908 | 1036 | 788 | 262 | 139 | 12 | 5500 |
| 2006-07 | 0 | 1 | 108 | 486 | 685 | 931 | 1274 | 1289 | 618 | 491 | 58 | 0 | 5941 |
| 2007-08 | 0 | 0 | 71 | 277 | 774 | 1281 | 1409 | 1306 | 901 | 506 | 173 | 0 | 6698 |
| 2008-09 | 0 | 0 | 73 | 355 | 775 | 1344 | 1453 | 976 | 723 | 423 | 113 | 7 | 6242 |
| 2009- | 2 | 12 | 31 | 529 | 537 | 1304 | | | | | | | |

WBAN : 14933

## COOLING DEGREE DAYS (base 65°F) 2009  DES MOINES (KDSM)

| YEAR | JAN | FEB | MAR | APR | MAY | JUN | JUL | AUG | SEP | OCT | NOV | DEC | TOTAL |
|------|-----|-----|-----|-----|-----|-----|-----|-----|-----|-----|-----|-----|-------|
| 1980 | 0 | 0 | 0 | 22 | 83 | 200 | 469 | 353 | 138 | 2 | 0 | 0 | 1267 |
| 1981 | 0 | 0 | 1 | 27 | 42 | 243 | 358 | 239 | 95 | 2 | 0 | 0 | 1007 |
| 1982 | 0 | 0 | 0 | 0 | 71 | 101 | 374 | 269 | 120 | 11 | 0 | 0 | 946 |
| 1983 | 0 | 0 | 0 | 4 | 20 | 272 | 502 | 574 | 183 | 19 | 0 | 0 | 1574 |
| 1984 | 0 | 0 | 0 | 8 | 27 | 254 | 345 | 397 | 101 | 2 | 0 | 0 | 1134 |
| 1985 | 0 | 0 | 0 | 46 | 69 | 134 | 361 | 221 | 174 | 0 | 0 | 0 | 1005 |
| 1986 | 0 | 0 | 11 | 17 | 46 | 258 | 386 | 162 | 130 | 0 | 0 | 0 | 1010 |
| 1987 | 0 | 0 | 0 | 31 | 121 | 304 | 426 | 250 | 65 | 0 | 0 | 0 | 1197 |
| 1988 | 0 | 0 | 0 | 1 | 112 | 329 | 425 | 444 | 101 | 3 | 0 | 0 | 1415 |
| 1989 | 0 | 0 | 6 | 44 | 53 | 156 | 379 | 269 | 61 | 15 | 0 | 0 | 983 |
| 1990 | 0 | 0 | 1 | 26 | 5 | 226 | 283 | 297 | 184 | 11 | 2 | 0 | 1035 |
| 1991 | 0 | 0 | 9 | 19 | 172 | 302 | 349 | 279 | 150 | 14 | 0 | 0 | 1294 |
| 1992 | 0 | 0 | 0 | 10 | 64 | 174 | 197 | 134 | 86 | 10 | 0 | 0 | 675 |
| 1993 | 0 | 0 | 0 | 0 | 31 | 167 | 281 | 280 | 19 | 15 | 0 | 0 | 793 |
| 1994 | 0 | 0 | 0 | 22 | 65 | 245 | 253 | 213 | 125 | 4 | 0 | 0 | 927 |
| 1995 | 0 | 0 | 0 | 0 | 9 | 206 | 361 | 440 | 90 | 11 | 0 | 0 | 1117 |
| 1996 | 0 | 0 | 0 | 4 | 44 | 212 | 245 | 231 | | 10 | 0 | 0 | |
| 1997 | 0 | 0 | 0 | 0 | 14 | 223 | 338 | 228 | 101 | 65 | 0 | 0 | 969 |
| 1998 | 0 | 0 | 2 | 1 | 99 | 165 | 330 | 323 | 190 | 3 | 0 | 0 | 1113 |
| 1999 | 0 | 0 | 0 | 0 | 40 | 204 | 463 | 236 | 72 | 11 | 1 | 0 | 1027 |
| 2000 | 0 | 0 | 3 | 4 | 90 | 156 | 278 | 334 | 157 | 25 | 0 | 0 | 1047 |
| 2001 | 0 | 0 | 0 | 28 | 61 | 207 | 409 | 322 | 67 | 6 | 0 | 0 | 1100 |
| 2002 | 0 | 0 | 0 | 25 | 52 | 295 | 416 | 253 | 151 | 6 | 0 | 0 | 1198 |
| 2003 | 0 | 0 | 0 | 23 | 28 | 199 | 367 | 404 | 91 | 18 | 0 | 0 | 1130 |
| 2004 | 0 | 0 | 1 | 23 | 92 | 142 | 236 | 151 | 139 | 3 | 0 | 0 | 787 |
| 2005 | 0 | 0 | 0 | 19 | 37 | 295 | 399 | 332 | 203 | 44 | 0 | 0 | 1329 |
| 2006 | 0 | 0 | 0 | 18 | 97 | 276 | 444 | 304 | 48 | 33 | 0 | 0 | 1220 |
| 2007 | 0 | 0 | 12 | 18 | 106 | 238 | 374 | 391 | 129 | 40 | 0 | 0 | 1308 |
| 2008 | 0 | 0 | 0 | 2 | 37 | 199 | 329 | 246 | 95 | 19 | 1 | 0 | 928 |
| 2009 | 0 | 0 | 0 | 9 | 54 | 242 | 228 | 253 | 108 | 0 | 0 | 0 | 894 |

## SNOWFALL (inches) 2009 DES MOINES (KDSM)

| YEAR | JUL | AUG | SEP | OCT | NOV | DEC | JAN | FEB | MAR | APR | MAY | JUN | TOTAL |
|------|-----|-----|-----|-----|-----|-----|-----|-----|-----|-----|-----|-----|-------|
| 1980-81 | 0.0 | 0.0 | 0.0 | 7.4 | T | 2.7 | 3.7 | 6.6 | T | 0.0 | 0.0 | 0.0 | 20.4 |
| 1981-82 | 0.0 | 0.0 | 0.0 | 0.4 | 3.2 | 10.7 | 18.5 | 2.6 | 11.9 | 15.6 | 0.0 | 0.0 | 62.9 |
| 1982-83 | 0.0 | 0.0 | 0.0 | 0.8 | 0.3 | 3.6 | 4.2 | 16.8 | 13.2 | 12.6 | 0.0 | 0.0 | 51.5 |
| 1983-84 | 0.0 | 0.0 | 0.0 | T | 9.8 | 19.6 | 12.5 | 1.3 | 13.7 | 0.1 | 0.0 | 0.0 | 57.0 |
| 1984-85 | 0.0 | 0.0 | 0.0 | T | 2.7 | 7.6 | 7.7 | 6.9 | 6.7 | T | 0.0 | 0.0 | 31.6 |
| 1985-86 | 0.0 | 0.0 | T | 0.0 | 8.1 | 15.9 | 1.3 | 6.7 | 0.2 | 0.1 | 0.0 | 0.0 | 32.3 |
| 1986-87 | 0.0 | 0.0 | 0.0 | T | 3.1 | 2.9 | 5.1 | 2.8 | 5.5 | 0.0 | 0.0 | 0.0 | 19.4 |
| 1987-88 | 0.0 | 0.0 | 0.0 | T | 0.1 | 13.0 | 1.2 | 7.6 | 0.8 | 0.0 | 0.0 | 0.0 | 22.7 |
| 1988-89 | 0.0 | 0.0 | 0.0 | 0.0 | 1.1 | 1.0 | 0.1 | 16.3 | 1.2 | 0.6 | T | 0.0 | 20.3 |
| 1989-90 | 0.0 | 0.0 | 0.0 | T | 1.7 | 6.6 | 11.3 | 7.4 | 0.1 | T | 0.0 | 0.0 | 27.1 |
| 1990-91 | 0.0 | 0.0 | 0.0 | T | 1.3 | 12.1 | 9.9 | 0.4 | 0.2 | T | T | 0.0 | 23.9 |
| 1991-92 | 0.0 | 0.0 | 0.0 | T | 14.7 | 1.2 | 1.9 | 4.5 | 0.2 | 0.1 | 0.0 | 0.0 | 22.6 |
| 1992-93 | T | 0.0 | T | T | 10.0 | 3.7 | 11.1 | 10.1 | 5.1 | 1.5 | T | T | 41.5 |
| 1993-94 | 0.0 | 0.0 | 0.0 | 0.2 | 0.9 | 2.0 | 8.8 | 15.2 | 0.5 | 0.7 | 0.0 | 0.0 | 28.3 |
| 1994-95 | 0.0 | 0.0 | 0.0 | T | 0.2 | 12.4 | 8.7 | 2.0 | 5.3 | 0.4 | T | 0.0 | 29.0 |
| 1995-96 | 0.0 | 0.0 | 0.0 | 1.8 | 6.5 | 6.7 | 22.3 | 3.0 | 1.6 | 0.2 | 0.0 | 0.0 | 42.1 |
| 1996-97 | 0.0 | 0.0 | 0.0 | 0.0 | 1.0 | 5.9 | 11.4 | 15.7 | T | 9.6 | 0.0 | 0.0 | 43.6 |
| 1997-98 | 0.0 | 0.0 | 0.0 | 6.6 | 3.6 | 11.7 | 8.3 | 4.3 | 13.3 | T | 0.0 | 0.0 | 47.8 |
| 1998-99 | 0.0 | 0.0 | 0.0 | 0.0 | 2.0 | 7.8 | 10.5 | 9.3 | 12.7 | 1.5 | 0.0 | 0.0 | 43.8 |
| 1999-00 | 0.0 | 0.0 | 0.0 | T | T | 6.4 | 5.7 | 8.9 | 1.1 | 0.7 | 0.0 | 0.0 | 22.8 |
| 2000-01 | 0.0 | 0.0 | 0.0 | 0.0 | 0.4 | 26.9 | 7.1 | 9.7 | 5.2 | T | 0.0 | 0.0 | 49.3 |
| 2001-02 | 0.0 | 0.0 | 0.0 | 0.0 | T | 1.0 | 9.1 | 0.9 | 6.8 | T | 0.0 | 0.0 | 17.8 |
| 2002-03 | 0.0 | 0.0 | T | 1.1 | 1.7 | 0.1 | 4.7 | 14.4 | 5.3 | 5.3 | 0.0 | 0.0 | 32.6 |
| 2003-04 | T | 0.0 | 0.0 | 0.0 | T | 11.4 | 14.5 | 15.0 | 17.5 | 0.0 | 0.0 | 0.0 | 58.4 |
| 2004-05 | 0.0 | 0.0 | 0.0 | 0.0 | 1.1 | 1.1 | 10.1 | 6.3 | 1.2 | 0.0 | T | 0.0 | 19.8 |
| 2005-06 | 0.0 | 0.0 | 0.0 | 0.0 | 4.7 | 11.0 | 1.3 | 3.3 | 5.0 | T | T | 0.0 | 25.3 |
| 2006-07 | 0.0 | 0.0 | 0.0 | T | 0.3 | 1.1 | 11.5 | 14.4 | 7.9 | 4.1 | 0.0 | 0.0 | 39.3 |
| 2007-08 | 0.0 | 0.0 | 0.0 | 0.0 | 4.8 | 14.1 | 11.0 | 22.7 | 4.8 | 1.1 | 0.0 | 0.0 | 58.5 |
| 2008-09 | 0.0 | 0.0 | 0.0 | 0.0 | 4.1 | 11.6 | 15.4 | 8.8 | 0.1 | 1.3 | 0.0 | 0.0 | 41.3 |
| 2009- | 0.0 | 0.0 | 0.0 | 1.1 | T | 28.2 | | | | | | | |
| POR=65 YRS | T | 0.0 | T | 0.4 | 2.8 | 7.2 | 8.4 | 8.0 | 6.0 | 1.9 | T | T | 34.7 |

# Des Moines, IA  2009
## Temperature and Precipitation

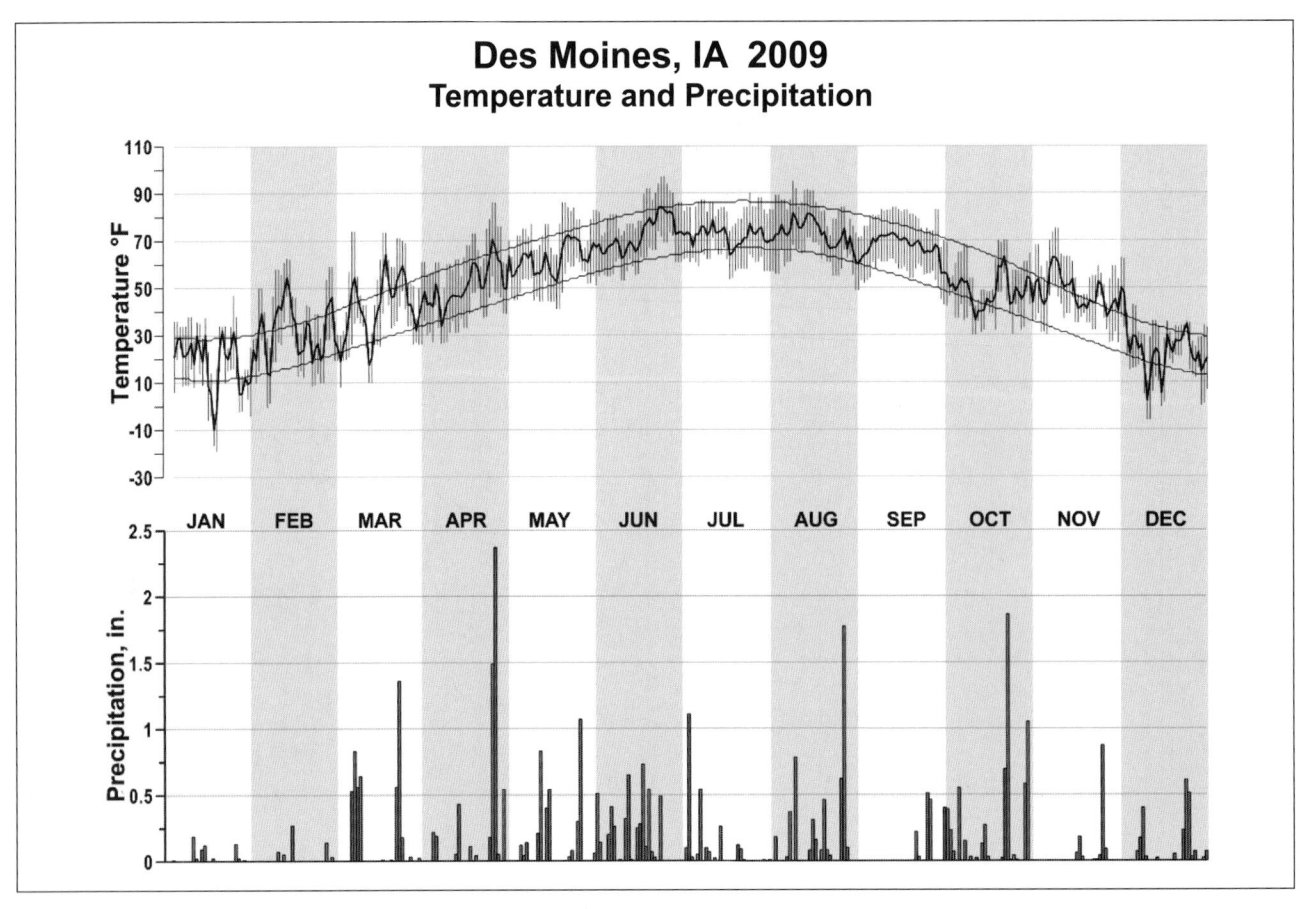

# 2009
# DODGE CITY
# KANSAS (KDDC)

The climate of Dodge City and southwestern Kansas is classified as semi-arid. Dodge City is nearly 300 miles east of the Rocky Mountains, but the weather reflects the influence of the mountains. The mountains form a barricade against all except high level moisture from the southwest, west, and northwest. Chinook winds occur occasionally but with less frequency and effect than at stations farther to the west. Relatively dry air predominating with an abundance of sunshine contribute to broad diurnal temperature ranges.

Thunderstorms during the growing season contribute most of the moisture. In general, the thunderstorms are widely scattered, occurring during the late afternoons and evenings. They are occasionally accompanied by hail and strong winds, but due to the local nature of the storms, damage to crops and buildings is spotty and variable. Winter is the dry season. However, the moisture accumulated during the winter months is important for the hard winter wheat. The duration of snow cover is generally brief due to mild temperatures and an abundance of sunshine. The exception results from the occasional blizzard that spreads across the flat treeless prairies of the high plains.

Afternoon temperatures in the 90s prevail during the summer months. Temperatures above 100 degrees are the exception. Due to low humidity and a continual breeze, these high temperatures are moderated. Temperatures normally drop sharply after sunset, allowing cool comfortable nights. During the winter months, large temperature changes are frequent, but the duration of extreme cold spells is brief.

The visibility at Dodge City is generally unrestricted as the terrain is favorable for unrestricted movement of air and air masses. Western Kansas is noted for clear skies and an abundance of sunshine.

Based on the 1951-1980 period, the average first occurrence of 32 degrees Fahrenheit in the fall is October 23 and the average last occurrence in the spring is April 21.

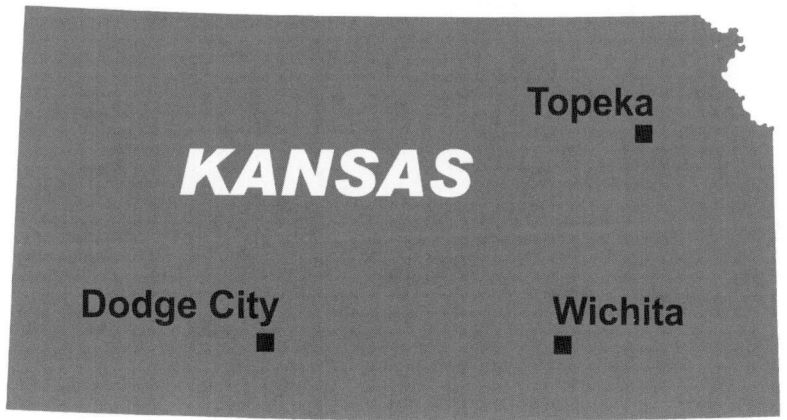

# NORMALS, MEANS, AND EXTREMES
## DODGE CITY (KDDC)

LATITUDE: 37°46'N  LONGITUDE: -99°58'W  ELEVATION (FT): GRND: 2576  BARO: 2590  TIME ZONE: CENTRAL (UTC -6)  WBAN: 13985

| | ELEMENT | POR | JAN | FEB | MAR | APR | MAY | JUN | JUL | AUG | SEP | OCT | NOV | DEC | YEAR |
|---|---|---|---|---|---|---|---|---|---|---|---|---|---|---|---|
| TEMPERATURE °F | NORMAL DAILY MAXIMUM | 30 | 41.4 | 48.3 | 57.3 | 67.1 | 75.9 | 86.9 | 92.8 | 90.8 | 82.0 | 70.4 | 54.5 | 44.4 | 67.7 |
| | MEAN DAILY MAXIMUM | 62 | 42.5 | 47.8 | 56.1 | 67.4 | 76.5 | 86.7 | 92.4 | 90.6 | 81.7 | 70.4 | 55.2 | 44.8 | 67.7 |
| | HIGHEST DAILY MAXIMUM | 66 | 80 | 85 | 93 | 100 | 105 | 110 | 109 | 108 | 106 | 98 | 91 | 86 | 110 |
| | YEAR OF OCCURRENCE | | 1989 | 2006 | 1989 | 1989 | 1996 | 1998 | 1986 | 2008 | 1947 | 2006 | 1980 | 1955 | JUN 1998 |
| | MEAN OF EXTREME MAXS. | 62 | 67.2 | 73.0 | 81.4 | 88.4 | 93.5 | 100.0 | 103.2 | 101.9 | 96.7 | 90.2 | 76.5 | 68.0 | 86.7 |
| | NORMAL DAILY MINIMUM | 30 | 18.7 | 23.6 | 31.2 | 40.7 | 51.7 | 61.6 | 66.8 | 65.6 | 56.5 | 43.8 | 30.2 | 21.7 | 42.7 |
| | MEAN DAILY MINIMUM | 62 | 19.0 | 23.2 | 30.0 | 40.9 | 51.8 | 61.4 | 66.9 | 65.5 | 56.3 | 44.0 | 30.5 | 21.8 | 42.6 |
| | LOWEST DAILY MINIMUM | 66 | -13 | -15 | -15 | 14 | 26 | 41 | 46 | 47 | 29 | 14 | 0 | -21 | -21 |
| | YEAR OF OCCURRENCE | | 1984 | 1951 | 1948 | 1997 | 1967 | 1954 | 1990 | 1950 | 1985 | 1993 | 1958 | 1989 | DEC 1989 |
| | MEAN OF EXTREME MINS. | 62 | -0.2 | 4.1 | 11.1 | 24.6 | 37.0 | 48.9 | 57.1 | 55.4 | 40.3 | 28.1 | 13.3 | 3.7 | 27.0 |
| | NORMAL DRY BULB | 30 | 30.1 | 36.0 | 44.3 | 53.9 | 63.8 | 74.3 | 79.8 | 78.2 | 69.3 | 57.1 | 42.4 | 33.1 | 55.2 |
| | MEAN DRY BULB | 62 | 30.7 | 35.5 | 43.1 | 54.2 | 64.2 | 74.2 | 79.7 | 78.1 | 69.1 | 57.2 | 42.9 | 33.4 | 55.2 |
| | MEAN WET BULB | 26 | 25.5 | 29.3 | 35.9 | 44.4 | 55.2 | 62.8 | 66.3 | 65.8 | 57.9 | 47.0 | 34.9 | 27.0 | 46.0 |
| | MEAN DEW POINT | 26 | 21.0 | 23.9 | 30.7 | 38.8 | 51.2 | 59.2 | 62.2 | 61.9 | 53.3 | 42.0 | 29.9 | 22.3 | 41.4 |
| | NORMAL NO. DAYS WITH: | | | | | | | | | | | | | | |
| | MAXIMUM >= 90 | 30 | 0.0 | 0.0 | 0.1 | 0.6 | 2.4 | 12.6 | 21.3 | 19.4 | 8.4 | 1.2 | * | 0.0 | 66.0 |
| | MAXIMUM <= 32 | 30 | 8.6 | 5.2 | 1.8 | 0.1 | 0.0 | 0.0 | 0.0 | 0.0 | 0.0 | 0.2 | 1.9 | 5.5 | 23.3 |
| | MINIMUM <= 32 | 30 | 29.4 | 22.7 | 17.4 | 5.3 | 0.2 | 0.0 | 0.0 | 0.0 | 0.2 | 3.2 | 17.9 | 28.3 | 124.6 |
| | MINIMUM <= 0 | 30 | 1.9 | 1.2 | 0.1 | 0.0 | 0.0 | 0.0 | 0.0 | 0.0 | 0.0 | 0.0 | 0.0 | 1.3 | 4.5 |
| H/C | NORMAL HEATING DEG. DAYS | 30 | 1087 | 826 | 647 | 351 | 121 | 12 | 1 | 2 | 65 | 273 | 674 | 978 | 5037 |
| | NORMAL COOLING DEG. DAYS | 30 | 0 | 0 | 2 | 18 | 79 | 291 | 462 | 407 | 193 | 28 | 1 | 0 | 1481 |
| RH | NORMAL (PERCENT) | 30 | 68 | 65 | 63 | 60 | 66 | 61 | 57 | 60 | 61 | 60 | 64 | 67 | 63 |
| | HOUR 00 LST | 30 | 74 | 72 | 70 | 70 | 76 | 71 | 67 | 69 | 69 | 68 | 72 | 73 | 71 |
| | HOUR 06 LST | 30 | 77 | 77 | 78 | 77 | 83 | 80 | 78 | 80 | 79 | 76 | 77 | 77 | 78 |
| | HOUR 12 LST | 30 | 59 | 55 | 52 | 47 | 53 | 48 | 45 | 47 | 47 | 46 | 52 | 57 | 51 |
| | HOUR 18 LST | 30 | 61 | 53 | 48 | 44 | 50 | 45 | 40 | 43 | 45 | 47 | 57 | 62 | 50 |
| S | PERCENT POSSIBLE SUNSHINE | 66 | 67 | 65 | 65 | 67 | 68 | 76 | 80 | 78 | 75 | 73 | 67 | 66 | 71 |
| W/O | MEAN NO. DAYS WITH: | | | | | | | | | | | | | | |
| | HEAVY FOG(VISBY <= 1/4 MI) | 46 | 2.9 | 3.2 | 2.7 | 1.8 | 1.8 | 1.1 | 0.6 | 1.2 | 1.5 | 2.7 | 2.6 | 3.1 | 25.2 |
| | THUNDERSTORMS | 62 | 0.1 | 0.4 | 1.4 | 3.5 | 7.5 | 9.4 | 9.7 | 8.0 | 4.2 | 2.1 | 0.5 | 0.2 | 47.0 |
| CLOUDNESS | MEAN: | | | | | | | | | | | | | | |
| | SUNRISE-SUNSET (OKTAS) | 1 | 4.8 | 4.0 | 4.8 | 5.1 | 3.6 | 2.0 | 2.4 | 2.4 | 4.0 | 3.2 | 4.0 | 3.2 | 3.6 |
| | MIDNIGHT-MIDNIGHT (OKTAS) | 1 | 4.0 | 4.0 | 5.3 | 5.3 | 3.6 | 1.6 | 2.4 | 2.4 | 3.2 | 4.0 | 4.0 | 3.2 | 3.6 |
| | MEAN NO. DAYS WITH: | | | | | | | | | | | | | | |
| | CLEAR | 3 | 6.3 | 5.0 | 9.7 | 9.0 | 12.3 | 14.0 | 16.0 | 16.5 | 9.0 | 9.5 | 8.0 | 13.0 | 128.3 |
| | PARTLY CLOUDY | 3 | 3.0 | 6.5 | 4.7 | 4.0 | 4.0 | 5.3 | 4.5 | 5.0 | 1.0 | 5.5 | 4.5 | 3.0 | 51.0 |
| | CLOUDY | 3 | 9.0 | 6.0 | 4.7 | 7.5 | 4.7 | 2.7 | 3.0 | 3.0 | 2.5 | 4.5 | 4.5 | 5.0 | 57.1 |
| PR | MEAN STATION PRESSURE(IN) | 26 | 27.36 | 27.34 | 27.27 | 27.24 | 27.24 | 27.26 | 27.31 | 27.33 | 27.33 | 27.33 | 27.33 | 27.35 | 27.31 |
| | MEAN SEA-LEVEL PRES. (IN) | 26 | 30.12 | 30.08 | 29.99 | 29.91 | 29.88 | 29.88 | 29.92 | 29.95 | 29.97 | 30.01 | 30.06 | 30.11 | 29.99 |
| WINDS | MEAN SPEED (MPH) | 26 | 12.7 | 13.1 | 14.5 | 15.0 | 14.0 | 13.5 | 12.7 | 12.0 | 13.1 | 13.0 | 13.0 | 12.6 | 13.3 |
| | PREVAIL.DIR(TENS OF DEGS) | 38 | 20 | 36 | 19 | 19 | 19 | 19 | 19 | 19 | 19 | 19 | 19 | 20 | 19 |
| | MAXIMUM 2-MINUTE: | | | | | | | | | | | | | | |
| | SPEED (MPH) | 17 | 56 | 47 | 49 | 51 | 55 | 60 | 56 | 63 | 51 | 49 | 49 | 46 | 63 |
| | DIR. (TENS OF DEGS) | | 34 | 01 | 28 | 31 | 30 | 29 | 36 | 32 | 14 | 34 | 34 | 32 | 32 |
| | YEAR OF OCCURRENCE | | 1996 | 2009 | 2000 | 2009 | 2006 | 1996 | 1996 | 2002 | 1996 | 2001 | 2006 | 2000 | AUG 2002 |
| | MAXIMUM 3-SECOND | | | | | | | | | | | | | | |
| | SPEED (MPH) | 17 | 66 | 64 | 60 | 67 | 67 | 70 | 79 | 75 | 59 | 72 | 62 | 56 | 79 |
| | DIR. (TENS OF DEGS) | | 34 | 34 | 15 | 35 | 20 | 28 | 25 | 32 | 14 | 34 | 34 | 32 | 25 |
| | YEAR OF OCCURRENCE | | 1996 | 2007 | 2007 | 2008 | 2007 | 1996 | 1996 | 2002 | 1996 | 2001 | 2006 | 2000 | JUL 1996 |
| PRECIPITATION | NORMAL (IN) | 30 | 0.62 | 0.66 | 1.84 | 2.25 | 3.00 | 3.15 | 3.17 | 2.73 | 1.70 | 1.45 | 1.01 | 0.77 | 22.35 |
| | MAXIMUM MONTHLY (IN) | 66 | 1.96 | 2.87 | 8.84 | 6.26 | 8.69 | 7.95 | 9.13 | 7.44 | 6.80 | 5.00 | 3.75 | 4.26 | 9.13 |
| | YEAR OF OCCURRENCE | | 1949 | 1993 | 1973 | 1976 | 1951 | 1951 | 1962 | 1977 | 1973 | 2008 | 1971 | 2006 | JUL 1962 |
| | MINIMUM MONTHLY (IN) | 66 | 0.00 | T | 0.02 | 0.07 | 0.25 | 0.12 | 0.17 | 0.68 | 0.01 | T | T | T | 0.00 |
| | YEAR OF OCCURRENCE | | 1986 | 2006 | 2008 | 1963 | 2004 | 1952 | 1946 | 1970 | 1980 | 1952 | 1989 | 1957 | JAN 1986 |
| | MAXIMUM IN 24 HOURS (IN) | 66 | 1.35 | 1.82 | 2.54 | 4.64 | 5.57 | 3.28 | 4.17 | 4.48 | 3.27 | 4.55 | 2.42 | 2.68 | 5.57 |
| | YEAR OF OCCURRENCE | | 1990 | 1993 | 1973 | 1978 | 1978 | 1944 | 1944 | 2003 | 1959 | 1968 | 1996 | 2006 | MAY 1978 |
| | NORMAL NO. DAYS WITH: | | | | | | | | | | | | | | |
| | PRECIPITATION >= 0.01 | 30 | 4.6 | 4.5 | 7.3 | 7.3 | 9.8 | 8.7 | 8.4 | 7.8 | 5.8 | 5.4 | 4.9 | 4.5 | 79.0 |
| | PRECIPITATION >= 1.00 | 30 | * | 0.1 | 0.3 | 0.5 | 0.6 | 0.9 | 0.7 | 0.5 | 0.4 | 0.2 | 0.3 | 0.1 | 4.6 |
| SNOWFALL | NORMAL (IN) | 30 | 4.9 | 4.4 | 5.1 | 1.0 | 0.* | 0.0 | 0.0 | 0.0 | 0.1 | 0.5 | 1.9 | 3.7 | 21.6 |
| | MAXIMUM MONTHLY (IN) | 66 | 15.7 | 19.6 | 24.0 | 10.0 | 0.9 | T | T | T | T | 4.5 | 16.7 | 14.9 | 24.0 |
| | YEAR OF OCCURRENCE | | 1990 | 1993 | 1970 | 2007 | 1978 | 2009 | 2006 | 1992 | 1985 | 1991 | 1992 | 1997 | MAR 1970 |
| | MAXIMUM IN 24 HOURS (IN) | 66 | 11.6 | 12.0 | 15.1 | 7.4 | 0.9 | T | T | T | T | 4.3 | 6.7 | 11.6 | 15.1 |
| | YEAR OF OCCURRENCE' | | 1990 | 1993 | 1999 | 1983 | 1978 | 2009 | 1992 | 1992 | 1985 | 1996 | 1948 | 1997 | MAR 1999 |
| | MAXIMUM SNOW DEPTH (IN) | 64 | 12 | 10 | 14 | 6 | T | 0 | 0 | 0 | 0 | 3 | 8 | 12 | 14 |
| | YEAR OF OCCURRENCE | | 1988 | 2003 | 1957 | 2007 | 1979 | | | | | 1976 | 1948 | 1997 | MAR 1957 |
| | NORMAL NO. DAYS WITH: | | | | | | | | | | | | | | |
| | SNOWFALL >= 1.0 | 30 | 1.5 | 1.2 | 1.4 | 0.3 | 0.0 | 0.0 | 0.0 | 0.0 | 0.0 | 0.1 | 0.6 | 1.1 | 6.2 |

## PRECIPITATION (inches) 2009  DODGE CITY (KDDC)

| YEAR | JAN | FEB | MAR | APR | MAY | JUN | JUL | AUG | SEP | OCT | NOV | DEC | ANNUAL |
|------|-----|-----|-----|-----|-----|-----|-----|-----|-----|-----|-----|-----|--------|
| 1980 | 1.06 | 1.46 | 2.87 | 1.89 | 3.60 | 3.85 | 2.00 | 2.06 | 0.01 | 0.23 | 0.01 | 0.76 | 19.80 |
| 1981 | 0.31 | 0.04 | 2.27 | 0.70 | 5.73 | 1.39 | 5.30 | 2.26 | 3.07 | 1.40 | 2.26 | 0.39 | 25.12 |
| 1982 | 0.18 | 1.30 | 0.81 | 0.66 | 2.74 | 3.65 | 5.54 | 1.01 | 1.31 | 0.84 | 0.44 | 1.04 | 19.52 |
| 1983 | 0.56 | 1.29 | 2.80 | 2.88 | 3.52 | 5.05 | 0.57 | 1.44 | 2.64 | 1.87 | 1.15 | 0.62 | 24.39 |
| 1984 | 0.68 | 0.20 | 2.73 | 4.38 | 1.38 | 1.89 | 0.63 | 1.35 | 0.35 | 2.87 | 0.10 | 2.41 | 18.97 |
| 1985 | 0.92 | 1.28 | 1.08 | 2.18 | 0.43 | 2.78 | 3.42 | 2.91 | 3.63 | 2.69 | 1.05 | 0.11 | 22.48 |
| 1986 | 0.00 | 0.44 | 0.18 | 2.19 | 1.31 | 4.70 | 2.12 | 5.60 | 1.22 | 0.93 | 0.63 | 0.96 | 20.28 |
| 1987 | 0.58 | 1.38 | 4.34 | 0.92 | 2.54 | 3.79 | 4.06 | 2.79 | 1.53 | 0.49 | 0.52 | 0.72 | 23.66 |
| 1988 | 0.89 | 0.13 | 0.53 | 3.08 | 3.56 | 0.15 | 1.95 | 1.11 | 2.80 | 0.56 | 0.10 | 0.10 | 14.96 |
| 1989 | 0.25 | 0.27 | 0.85 | 0.41 | 3.40 | 6.59 | 3.42 | 2.80 | 2.36 | 0.05 | T | 0.55 | 20.95 |
| 1990 | 1.60 | 1.17 | 1.64 | 3.64 | 4.91 | 1.01 | 3.07 | 1.19 | 1.45 | 0.18 | 0.42 | 0.60 | 20.88 |
| 1991 | 0.31 | T | 1.05 | 1.12 | 2.02 | 0.91 | 0.78 | 1.43 | 0.20 | 1.28 | 1.52 | 2.10 | 12.72 |
| 1992 | 0.39 | 0.35 | 1.43 | 0.21 | 3.06 | 6.98 | 2.67 | 2.90 | 0.48 | 0.71 | 1.58 | 0.90 | 21.66 |
| 1993 | 0.81 | 2.87 | 2.01 | 2.89 | 2.13 | 3.22 | 6.33 | 3.08 | 0.82 | 1.01 | 0.63 | 1.02 | 26.82 |
| 1994 | 0.41 | 0.45 | 0.06 | 1.90 | 0.89 | 1.88 | 4.74 | 3.48 | 0.90 | 1.72 | 1.73 | 0.94 | 19.10 |
| 1995 | 0.64 | 0.27 | 1.24 | 2.98 | 5.16 | 5.06 | 1.35 | 0.77 | 1.30 | 0.08 | 0.04 | 0.57 | 19.46 |
| 1996 | 0.41 | 0.07 | 1.46 | 0.54 | 3.70 | 3.31 | 6.09 | 7.34 | 5.08 | 1.40 | 3.03 | T | 32.43 |
| 1997 | 0.04 | 1.18 | T | 2.02 | 3.04 | 5.14 | 2.91 | 5.45 | 2.08 | 4.94 | 0.36 | 2.59 | 29.75 |
| 1998 | 0.75 | 0.33 | 2.73 | 1.11 | 2.66 | 1.56 | 4.81 | 1.30 | 0.01 | 4.20 | 2.02 | 0.29 | 21.77 |
| 1999 | 1.92 | 0.05 | 2.89 | 3.34 | 1.95 | 2.97 | 2.05 | 2.91 | 2.05 | 0.68 | 0.01 | 0.31 | 21.13 |
| 2000 | 0.50 | 0.16 | 4.97 | 2.03 | 1.82 | 2.83 | 4.19 | 1.91 | 0.19 | 2.74 | 0.13 | 0.46 | 21.93 |
| 2001 | 1.54 | 1.58 | 0.51 | 0.97 | 7.84 | 0.95 | 1.21 | 1.27 | 2.24 | 0.01 | 0.14 | 0.03 | 18.29 |
| 2002 | 0.65 | 0.38 | 0.29 | 1.05 | 1.35 | 1.42 | 0.57 | 5.14 | 0.19 | 2.85 | 0.02 | 0.61 | 14.52 |
| 2003 | 0.07 | 1.30 | 2.30 | 1.14 | 2.49 | 4.10 | 0.48 | 5.03 | 5.09 | 0.30 | 0.03 | 0.50 | 22.83 |
| 2004 | 0.05 | 1.03 | 1.96 | 2.53 | 0.25 | 4.36 | 5.20 | 2.12 | 2.82 | 1.64 | 2.34 | 0.16 | 24.46 |
| 2005 | 1.70 | 0.89 | 1.51 | 0.98 | 2.04 | 4.40 | 1.29 | 2.62 | 1.16 | 3.47 | 0.34 | 0.14 | 20.54 |
| 2006 | 0.11 | T | 1.01 | 0.79 | 3.94 | 2.47 | 2.04 | 4.13 | 0.64 | 1.67 | 0.07 | 4.26 | 21.13 |
| 2007 | 0.39 | 0.42 | 3.13 | 1.46 | 2.35 | 2.16 | 1.82 | 3.69 | 0.24 | 1.47 | 0.07 | 1.92 | 19.12 |
| 2008 | 0.17 | 0.61 | 0.02 | 1.79 | 3.88 | 1.56 | 1.14 | 1.89 | 1.89 | 5.00 | 0.19 | 0.15 | 18.29 |
| 2009 | 0.02 | 0.18 | 0.68 | 3.60 | 1.29 | 6.34 | 4.29 | 1.73 | 2.50 | 3.85 | 0.66 | 0.38 | 25.52 |
| POR= 62 YRS | 0.53 | 0.64 | 1.52 | 1.81 | 3.05 | 3.10 | 3.03 | 2.76 | 1.83 | 1.52 | 0.75 | 0.66 | 21.20 |

WBAN : 13985

## AVERAGE TEMPERATURE (°F) 2009  DODGE CITY (KDDC)

| YEAR | JAN | FEB | MAR | APR | MAY | JUN | JUL | AUG | SEP | OCT | NOV | DEC | ANNUAL |
|------|-----|-----|-----|-----|-----|-----|-----|-----|-----|-----|-----|-----|--------|
| 1980 | 26.9 | 30.4 | 39.6 | 53.2 | 62.6 | 78.4 | 87.2 | 82.1 | 73.5 | 58.3 | 46.7 | 38.3 | 56.4 |
| 1981 | 36.1 | 38.8 | 44.0 | 61.0 | 61.6 | 78.5 | 82.2 | 78.0 | 71.6 | 56.4 | 47.5 | 34.1 | 57.5 |
| 1982 | 28.2 | 29.9 | 44.4 | 52.4 | 64.7 | 69.8 | 80.7 | 81.1 | 72.1 | 58.4 | 41.4 | 35.5 | 54.9 |
| 1983 | 33.2 | 37.0 | 43.1 | 48.0 | 61.1 | 71.8 | 84.1 | 86.5 | 74.2 | 59.5 | 44.3 | 18.4 | 55.1 |
| 1984 | 28.6 | 40.7 | 38.7 | 49.0 | 64.4 | 76.7 | 82.4 | 82.5 | 70.0 | 56.1 | 45.7 | 35.1 | 55.8 |
| 1985 | 25.9 | 32.4 | 45.9 | 58.1 | 67.2 | 73.7 | 80.1 | 77.2 | 67.0 | 54.4 | 35.5 | 29.8 | 53.9 |
| 1986 | 40.1 | 37.4 | 51.0 | 56.8 | 64.3 | 75.5 | 81.3 | 74.6 | 70.1 | 55.6 | 40.7 | 35.5 | 56.9 |
| 1987 | 33.0 | 40.3 | 43.2 | 55.3 | 65.3 | 73.2 | 76.9 | 76.1 | 68.8 | 55.7 | 45.2 | 33.6 | 55.6 |
| 1988 | 27.0 | 35.1 | 43.8 | 52.1 | 65.4 | 77.2 | 79.7 | 80.3 | 68.2 | 55.4 | 45.2 | 37.9 | 55.6 |
| 1989 | 37.4 | 25.3 | 45.3 | 58.9 | 65.1 | 68.0 | 76.7 | 75.0 | 65.1 | 58.7 | 44.1 | 27.2 | 53.9 |
| 1990 | 35.3 | 36.3 | 45.5 | 52.2 | 60.1 | 77.5 | 77.4 | 78.6 | 72.9 | 57.3 | 48.4 | 27.9 | 55.8 |
| 1991 | 28.7 | 43.6 | 47.0 | 56.2 | 69.2 | 77.4 | 81.6 | 78.7 | 70.4 | 57.2 | 36.3 | 37.3 | 57.0 |
| 1992 | 38.0 | 42.1 | 48.4 | 56.1 | 62.7 | 68.1 | 75.5 | 71.9 | 68.4 | 56.6 | 35.7 | 26.4 | 54.2 |
| 1993 | 25.1 | 28.6 | 41.1 | 50.3 | 62.8 | 72.0 | 78.4 | 76.5 | 65.5 | 53.0 | 38.2 | 36.1 | 52.3 |
| 1994 | 29.5 | 30.7 | 47.1 | 52.2 | 65.6 | 78.3 | 76.9 | 77.9 | 68.7 | 56.9 | 43.3 | 36.7 | 55.3 |
| 1995 | 33.0 | 40.5 | 43.1 | 50.2 | 57.6 | 69.4 | 77.8 | 82.0 | 67.1 | 56.4 | 44.8 | 32.8 | 54.6 |
| 1996 | 28.4 | 37.7 | 39.1 | 54.4 | 67.5 | 74.9 | 77.8 | 73.8 | 65.1 | 56.1 | 39.8 | 34.6 | 54.1 |
| 1997 | 30.3 | 35.4 | 47.3 | 48.1 | 61.3 | 71.7 | 77.7 | 75.1 | 70.4 | 57.0 | 40.3 | 31.5 | 53.8 |
| 1998 | 32.5 | 39.9 | 38.1 | 52.1 | 68.2 | 75.5 | 80.4 | 78.4 | 75.4 | 59.4 | 47.4 | 34.4 | 56.8 |
| 1999 | 33.6 | 44.0 | 42.4 | 53.1 | 63.3 | 70.8 | 80.6 | 78.9 | 65.5 | 56.6 | 49.9 | 37.4 | 56.3 |
| 2000 | 33.5 | 41.7 | 45.1 | 54.1 | 67.2 | 72.6 | 79.0 | 81.4 | 71.2 | 58.3 | 36.8 | 26.3 | 55.6 |
| 2001 | 32.4 | 29.9 | 41.9 | 57.8 | 63.7 | 73.8 | 84.7 | 79.6 | 69.1 | 56.8 | 48.2 | 37.2 | 56.3 |
| 2002 | 33.8 | 35.8 | 39.0 | 57.7 | 64.1 | 77.9 | 81.4 | 77.6 | 69.9 | 49.5 | 42.2 | 34.8 | 55.3 |
| 2003 | 34.0 | 31.1 | 44.0 | 55.6 | 63.4 | 70.8 | 82.9 | 79.7 | 65.0 | 58.4 | 41.8 | 35.3 | 55.2 |
| 2004 | 32.1 | 33.3 | 49.9 | 53.5 | 68.5 | 72.3 | 76.5 | 72.7 | 71.1 | 57.0 | 42.4 | 37.1 | 55.5 |
| 2005 | 31.3 | 39.9 | 45.2 | 54.6 | 65.4 | 75.0 | 79.4 | 78.2 | 74.1 | 59.7 | 48.1 | 33.5 | 57.0 |
| 2006 | 42.9 | 35.7 | 46.5 | 60.5 | 66.5 | 77.2 | 82.1 | 77.8 | 64.6 | 55.7 | 43.9 | 36.4 | 57.5 |
| 2007 | 28.7 | 32.6 | 52.5 | 50.2 | 65.0 | 72.1 | 78.6 | 81.9 | 71.2 | 60.2 | 43.9 | 29.9 | 55.6 |
| 2008 | 32.8 | 35.4 | 43.1 | 51.5 | 63.3 | 75.0 | 80.4 | 77.7 | 67.6 | 57.2 | 45.1 | 32.7 | 55.2 |
| 2009 | 34.2 | 40.6 | 45.4 | 53.5 | 62.8 | 74.9 | 78.0 | 75.6 | 65.9 | 49.1 | 46.7 | 27.8 | 54.5 |
| POR= 62 YRS | 30.7 | 35.5 | 43.1 | 54.2 | 64.2 | 74.2 | 79.7 | 78.1 | 69.1 | 57.2 | 42.9 | 33.4 | 55.2 |

## HEATING DEGREE DAYS (base 65°F) 2009  DODGE CITY (KDDC)

| YEAR | JUL | AUG | SEP | OCT | NOV | DEC | JAN | FEB | MAR | APR | MAY | JUN | TOTAL |
|------|-----|-----|-----|-----|-----|-----|-----|-----|-----|-----|-----|-----|-------|
| 1980-81 | 0 | 0 | 26 | 257 | 548 | 822 | 889 | 728 | 644 | 180 | 164 | 0 | 4258 |
| 1981-82 | 0 | 0 | 24 | 284 | 517 | 954 | 1131 | 979 | 632 | 381 | 67 | 26 | 4995 |
| 1982-83 | 0 | 0 | 30 | 221 | 701 | 907 | 980 | 778 | 670 | 509 | 164 | 20 | 4980 |
| 1983-84 | 0 | 0 | 49 | 209 | 617 | 1442 | 1122 | 699 | 805 | 473 | 96 | 3 | 5515 |
| 1984-85 | 0 | 0 | 117 | 295 | 570 | 916 | 1206 | 907 | 585 | 217 | 48 | 10 | 4871 |
| 1985-86 | 0 | 0 | 154 | 331 | 878 | 1086 | 766 | 765 | 438 | 269 | 66 | 0 | 4753 |
| 1986-87 | 0 | 6 | 36 | 285 | 722 | 911 | 988 | 685 | 669 | 321 | 63 | 3 | 4689 |
| 1987-88 | 4 | 8 | 24 | 290 | 585 | 964 | 1171 | 861 | 653 | 382 | 88 | 0 | 5030 |
| 1988-89 | 0 | 1 | 48 | 301 | 589 | 835 | 847 | 1105 | 616 | 270 | 110 | 50 | 4772 |
| 1989-90 | 0 | 0 | 130 | 236 | 623 | 1165 | 914 | 795 | 596 | 384 | 190 | 1 | 5034 |
| 1990-91 | 4 | 0 | 35 | 258 | 499 | 1147 | 1116 | 591 | 555 | 275 | 58 | 0 | 4538 |
| 1991-92 | 0 | 0 | 55 | 311 | 853 | 851 | 828 | 654 | 505 | 282 | 135 | 30 | 4504 |
| 1992-93 | 0 | 8 | 40 | 289 | 875 | 1188 | 1228 | 1013 | 732 | 439 | 129 | 23 | 5964 |
| 1993-94 | 0 | 11 | 87 | 392 | 797 | 887 | 1095 | 955 | 551 | 402 | 83 | 0 | 5260 |
| 1994-95 | 0 | 1 | 51 | 253 | 644 | 868 | 983 | 679 | 670 | 442 | 239 | 27 | 4857 |
| 1995-96 | 0 | 0 | 115 | 276 | 597 | 990 | 1128 | 785 | 795 | 328 | 94 | 7 | 5115 |
| 1996-97 | 6 | 0 | 80 | 297 | 748 | 932 | 1068 | 825 | 541 | 503 | 148 | 5 | 5153 |
| 1997-98 | 0 | 6 | 40 | 313 | 734 | 1030 | 999 | 694 | 830 | 388 | 61 | 36 | 5131 |
| 1998-99 | 0 | 0 | 20 | 200 | 518 | 943 | 965 | 582 | 692 | 352 | 100 | 21 | 4393 |
| 1999-00 | 1 | 0 | 103 | 272 | 447 | 848 | 968 | 670 | 611 | 329 | 86 | 9 | 4344 |
| 2000-01 | 0 | 0 | 81 | 220 | 840 | 1193 | 1003 | 974 | 710 | 233 | 98 | 17 | 5369 |
| 2001-02 | 0 | 0 | 41 | 276 | 498 | 854 | 962 | 814 | 801 | 259 | 124 | 0 | 4629 |
| 2002-03 | 0 | 0 | 40 | 486 | 674 | 931 | 953 | 944 | 642 | 298 | 118 | 21 | 5107 |
| 2003-04 | 0 | 7 | 94 | 235 | 688 | 914 | 1011 | 914 | 472 | 349 | 74 | 11 | 4769 |
| 2004-05 | 1 | 8 | 17 | 256 | 669 | 859 | 1037 | 696 | 609 | 319 | 110 | 2 | 4583 |
| 2005-06 | 0 | 0 | 15 | 233 | 500 | 971 | 680 | 814 | 564 | 192 | 116 | 0 | 4085 |
| 2006-07 | 0 | 0 | 69 | 337 | 629 | 879 | 1118 | 900 | 386 | 444 | 60 | 9 | 4831 |
| 2007-08 | 0 | 0 | 25 | 200 | 627 | 1079 | 989 | 851 | 672 | 415 | 142 | 0 | 5000 |
| 2008-09 | 0 | 0 | 52 | 266 | 589 | 995 | 948 | 678 | 599 | 353 | 141 | 2 | 4623 |
| 2009- | 0 | 1 | 66 | 492 | 540 | 1147 | | | | | | | |

WBAN : 13985

## COOLING DEGREE DAYS (base 65°F) 2009  DODGE CITY (KDDC)

| YEAR | JAN | FEB | MAR | APR | MAY | JUN | JUL | AUG | SEP | OCT | NOV | DEC | TOTAL |
|------|-----|-----|-----|-----|-----|-----|-----|-----|-----|-----|-----|-----|-------|
| 1980 | 0 | 0 | 0 | 18 | 62 | 410 | 695 | 538 | 288 | 56 | 5 | 0 | 2072 |
| 1981 | 0 | 0 | 0 | 64 | 65 | 411 | 540 | 407 | 231 | 21 | 0 | 0 | 1739 |
| 1982 | 0 | 0 | 0 | 9 | 64 | 177 | 494 | 504 | 253 | 24 | 0 | 0 | 1525 |
| 1983 | 0 | 0 | 0 | 3 | 50 | 233 | 600 | 674 | 332 | 44 | 3 | 0 | 1939 |
| 1984 | 0 | 0 | 0 | 0 | 86 | 365 | 547 | 552 | 272 | 28 | 0 | 0 | 1850 |
| 1985 | 0 | 0 | 2 | 18 | 121 | 277 | 472 | 383 | 224 | 8 | 0 | 0 | 1505 |
| 1986 | 0 | 0 | 8 | 32 | 49 | 324 | 511 | 311 | 196 | 1 | 0 | 0 | 1432 |
| 1987 | 0 | 0 | 0 | 38 | 79 | 256 | 379 | 359 | 144 | 9 | 1 | 0 | 1265 |
| 1988 | 0 | 0 | 0 | 3 | 108 | 373 | 463 | 480 | 149 | 12 | 0 | 0 | 1588 |
| 1989 | 0 | 0 | 12 | 95 | 119 | 145 | 369 | 317 | 138 | 47 | 0 | 0 | 1242 |
| 1990 | 0 | 0 | 0 | 9 | 44 | 383 | 395 | 429 | 278 | 24 | 4 | 0 | 1566 |
| 1991 | 0 | 0 | 4 | 17 | 195 | 376 | 522 | 430 | 226 | 75 | 0 | 0 | 1845 |
| 1992 | 0 | 0 | 0 | 20 | 70 | 132 | 335 | 232 | 149 | 34 | 0 | 0 | 972 |
| 1993 | 0 | 0 | 0 | 3 | 67 | 239 | 420 | 372 | 106 | 26 | 0 | 0 | 1233 |
| 1994 | 0 | 0 | 2 | 23 | 106 | 406 | 376 | 409 | 168 | 9 | 0 | 0 | 1499 |
| 1995 | 0 | 0 | 0 | 8 | 19 | 168 | 405 | 534 | 186 | 16 | 0 | 0 | 1336 |
| 1996 | 0 | 0 | 0 | 16 | 176 | 310 | 406 | 280 | 89 | 28 | 0 | 0 | 1305 |
| 1997 | 0 | 0 | 0 | 3 | 41 | 213 | 403 | 329 | 208 | 71 | 0 | 0 | 1268 |
| 1998 | 0 | 0 | 5 | 5 | 168 | 358 | 484 | 420 | 338 | 36 | 0 | 0 | 1814 |
| 1999 | 0 | 0 | 1 | 3 | 57 | 203 | 491 | 439 | 127 | 19 | 0 | 0 | 1340 |
| 2000 | 0 | 0 | 0 | 10 | 161 | 242 | 441 | 516 | 274 | 22 | 0 | 0 | 1666 |
| 2001 | 0 | 0 | 0 | 26 | 69 | 287 | 620 | 460 | 169 | 30 | 0 | 0 | 1661 |
| 2002 | 0 | 0 | 0 | 47 | 103 | 395 | 515 | 395 | 193 | 12 | 0 | 0 | 1660 |
| 2003 | 0 | 0 | 0 | 24 | 72 | 202 | 564 | 468 | 97 | 39 | 0 | 0 | 1466 |
| 2004 | 0 | 0 | 10 | 12 | 190 | 238 | 363 | 254 | 208 | 13 | 0 | 0 | 1288 |
| 2005 | 0 | 0 | 0 | 10 | 131 | 310 | 452 | 414 | 292 | 73 | 0 | 0 | 1682 |
| 2006 | 0 | 0 | 0 | 64 | 167 | 375 | 536 | 405 | 62 | 56 | 3 | 0 | 1668 |
| 2007 | 0 | 0 | 5 | 8 | 66 | 231 | 429 | 531 | 218 | 58 | 0 | 0 | 1546 |
| 2008 | 0 | 0 | 0 | 16 | 96 | 310 | 485 | 400 | 136 | 28 | 0 | 0 | 1471 |
| 2009 | 0 | 0 | 0 | 16 | 79 | 302 | 410 | 336 | 100 | 5 | 0 | 0 | 1248 |

## SNOWFALL (inches) 2009 DODGE CITY (KDDC)

| YEAR | JUL | AUG | SEP | OCT | NOV | DEC | JAN | FEB | MAR | APR | MAY | JUN | TOTAL |
|------|-----|-----|-----|-----|-----|-----|-----|-----|-----|-----|-----|-----|-------|
| 1980-81 | 0.0 | 0.0 | 0.0 | 0.1 | 0.1 | 3.9 | 2.9 | 0.4 | 4.4 | 0.0 | 0.0 | 0.0 | 11.8 |
| 1981-82 | 0.0 | 0.0 | 0.0 | 0.0 | T | 3.6 | 0.6 | 13.6 | 1.4 | 0.0 | 0.0 | 0.0 | 19.2 |
| 1982-83 | 0.0 | 0.0 | 0.0 | 0.0 | 5.0 | 2.6 | 4.7 | 4.1 | 7.6 | 9.0 | 0.0 | 0.0 | 33.0 |
| 1983-84 | 0.0 | 0.0 | 0.0 | 0.0 | 2.8 | 6.2 | 6.8 | 0.1 | 11.0 | 0.3 | 0.0 | 0.0 | 27.2 |
| 1984-85 | 0.0 | 0.0 | T | T | 0.8 | 2.8 | 8.5 | 1.6 | 5.5 | 0.0 | 0.0 | 0.0 | 19.2 |
| 1985-86 | 0.0 | 0.0 | T | 0.0 | 9.8 | 1.1 | 0.0 | 2.8 | T | 0.0 | 0.0 | 0.0 | 13.7 |
| 1986-87 | 0.0 | 0.0 | 0.0 | T | T | 5.6 | 7.2 | 1.4 | 16.3 | T | 0.0 | 0.0 | 30.5 |
| 1987-88 | 0.0 | 0.0 | 0.0 | 0.0 | 2.5 | 7.2 | 11.8 | T | 3.5 | 0.0 | 0.0 | 0.0 | 25.0 |
| 1988-89 | 0.0 | 0.0 | 0.0 | 0.0 | 1.0 | T | 0.4 | 2.7 | T | 5.7 | 0.0 | 0.0 | 9.8 |
| 1989-90 | T | 0.0 | 0.0 | 0.0 | T | 5.0 | 15.7 | 7.7 | 0.5 | 0.4 | 0.0 | 0.0 | 29.3 |
| 1990-91 | 0.0 | 0.0 | 0.0 | 0.0 | 0.7 | 2.8 | 3.3 | T | 3.4 | 0.0 | T | 0.0 | 10.2 |
| 1991-92 | 0.0 | 0.0 | 0.0 | 4.5 | 1.0 | T | T | 0.6 | 0.0 | 0.0 | T | 0.0 | 6.1 |
| 1992-93 | T | T | T | T | 16.3 | 8.3 | 10.4 | 19.6 | 6.5 | T | 0.0 | 0.0 | 61.1 |
| 1993-94 | 0.0 | 0.0 | 0.0 | 0.2 | T | 2.8 | 1.8 | 1.0 | 1.0 | 0.1 | 0.0 | 0.0 | 6.9 |
| 1994-95 | T | 0.0 | T | 0.0 | T | 0.4 | 3.3 | 7.3 | 12.3 | 0.0 | 0.0 | T | 23.3 |
| 1995-96 | 0.0 | 0.0 | 1.4 | 0.0 | 0.1 | 6.9 | 1.7 | 1.0 | 3.4 | T | 0.0 | 0.0 | 14.5 |
| 1996-97 | T | 0.0 | 0.0 | 4.3 | T | 1.8 | 6.7 | 0.0 | 2.1 | T | 0.0 | 14.9 |
| 1997-98 | T | 0.0 | 0.0 | 2.0 | 1.4 | 14.9 | 0.7 | 0.3 | 18.5 | 0.1 | T | T | 37.9 |
| 1998-99 | T | 0.0 | 0.0 | 0.0 | T | 2.6 | 9.1 | 1.0 | 19.0 | T | 0.0 | T | 31.7 |
| 1999-00 | 0.0 | 0.0 | 0.0 | 0.0 | T | 2.5 | 4.8 | T | 3.5 | T | 0.0 | 0.0 | 10.8 |
| 2000-01 | T | T | 0.0 | T | 0.0 | 5.8 | 14.7 | 8.8 | 5.8 | T | T | T | 35.1 |
| 2001-02 | 0.0 | 0.0 | T | 0.0 | 0.0 | 0.3 | 6.2 | 5.3 | 0.6 | 0.0 | T | 0.0 | 12.4 |
| 2002-03 | 0.0 | T | 0.0 | 0.5 | T | 6.3 | 1.5 | 14.0 | 0.6 | 0.0 | 0.0 | T | 22.9 |
| 2003-04 | 0.0 | 0.0 | 0.0 | 0.0 | T | 5.9 | 0.4 | 5.7 | 0.8 | 2.9 | 0.0 | T | 15.7 |
| 2004-05 | T | 0.0 | 0.0 | 0.0 | 8.5 | 0.4 | 10.7 | 0.1 | T | T | 0.0 | 0.0 | 19.7 |
| 2005-06 | 0.0 | T | 0.0 | T | 1.0 | 3.9 | 0.4 | 0.2 | 7.9 | 0.0 | T | T | 13.4 |
| 2006-07 | T | 0.0 | 0.0 | 0.0 | 0.4 | 0.4 | 5.9 | 1.8 | 0.2 | 10.0 | 0.0 | 0.0 | 18.7 |
| 2007-08 | 0.0 | 0.0 | 0.0 | 0.0 | 1.5 | 14.7 | 3.5 | 1.0 | T | T | T | 0.0 | 20.7 |
| 2008-09 | 0.0 | 0.0 | 0.0 | T | T | 1.2 | 2.0 | T | 12.6 | 1.5 | 0.0 | T | 17.3 |
| 2009- | 0.0 | 0.0 | 0.0 | T | T | 2.6 | | | | | | | |
| POR= 62 YRS | T | T | T | 0.3 | 1.9 | 3.5 | 4.4 | 3.8 | 5.0 | 0.9 | T | T | 19.8 |

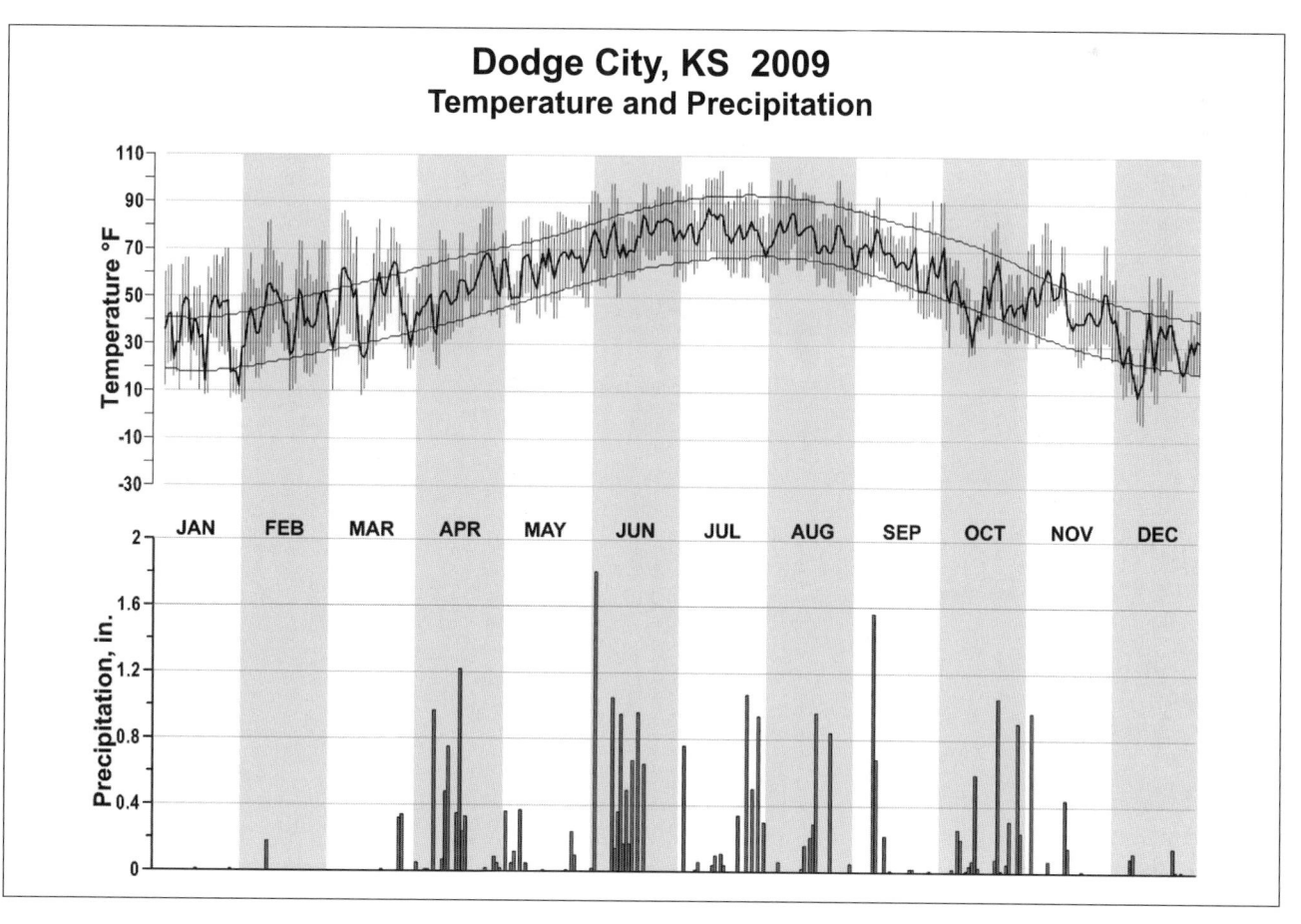

# Dodge City, KS  2009
## Temperature and Precipitation

# 2009
# WICHITA
# KANSAS (KICT)

Wichita is in the Central Great Plains where masses of warm, moist air from the Gulf of Mexico collide with cold, dry air from the Arctic region to create a wide range of weather the year around. Summers are usually warm and humid, and can be very hot and dry. The winters are usually mild, with brief periods of very cold weather.

The elevation is just over 1,300 feet above sea level. The terrain is basically flat with natural tree areas mainly along the Arkansas River and its tributaries.

The temperature extremes for the period of weather records at Wichita range from more than 110 degrees to less than -20 degrees. Temperatures above 90 degrees occur an average of 63 days per year, while very cold temperatures below zero occur about 2 days per year.

Precipitation averages about 30 inches per year, with 70 percent of that falling from April through September during the growing season. The wettest years have recorded over 50 inches. The driest years less than 15 inches.

Thunderstorms occur mainly during the spring and early summer. They can be severe and cause damage from heavy rain, large hail, strong winds and tornadoes.

The city of Wichita is protected against floods from the Arkansas River and its local tributaries by the Wichita-Vally Center Flood Control Project, which is designed to protect against floods up to the 75 to 100 year frequency class.

Snowfall normally is 15 inches per year, falling from December through March. Monthly snowfalls in excess of 20 inches and 24-hour snowfalls of more than 13 inches have occurred.

The prevailing wind direction is south with the windiest months March and April. July has the least wind. Strong north winds often occur with the passage of cold fronts from late fall through early spring. Extremely low wind chill factors are experienced with very cold outbreaks during the mid winter. On rare occasions during the summer, strong, hot, dry southwest winds can do considerable damage to crops.

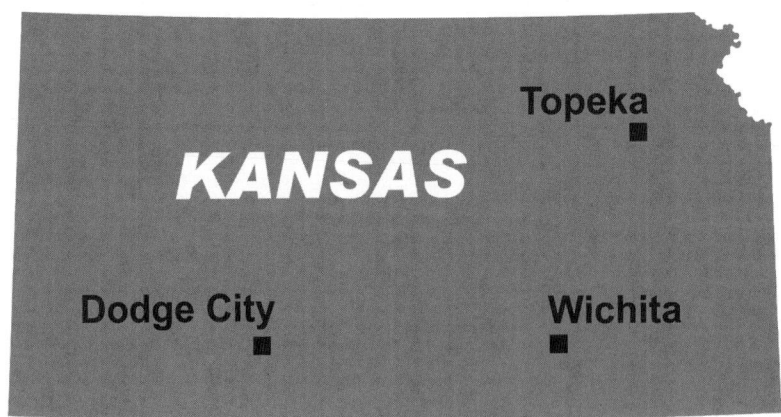

# NORMALS, MEANS, AND EXTREMES
## WICHITA (KICT)

| LATITUDE: 37°38'N | LONGITUDE: -97°25'W | ELEVATION (FT): GRND: 1320  BARO: 1341 | TIME ZONE: CENTRAL (UTC -6) | WBAN: 03928 |
|---|---|---|---|---|

| | ELEMENT | POR | JAN | FEB | MAR | APR | MAY | JUN | JUL | AUG | SEP | OCT | NOV | DEC | YEAR |
|---|---|---|---|---|---|---|---|---|---|---|---|---|---|---|---|
| **TEMPERATURE °F** | NORMAL DAILY MAXIMUM | 30 | 40.1 | 47.2 | 57.3 | 66.9 | 76.0 | 87.1 | 92.9 | 91.6 | 82.2 | 70.2 | 54.5 | 43.1 | 67.4 |
| | MEAN DAILY MAXIMUM | 62 | 41.0 | 47.1 | 56.6 | 67.8 | 76.9 | 86.8 | 92.3 | 91.0 | 82.0 | 70.5 | 55.6 | 44.1 | 67.6 |
| | HIGHEST DAILY MAXIMUM | 57 | 75 | 87 | 89 | 96 | 100 | 110 | 113 | 110 | 108 | 97 | 86 | 83 | 113 |
| | YEAR OF OCCURRENCE | | 2002 | 1996 | 1989 | 2006 | 1996 | 1980 | 1954 | 1984 | 2000 | 2006 | 2006 | 1955 | JUL 1954 |
| | MEAN OF EXTREME MAXS. | 62 | 63.2 | 69.8 | 79.2 | 85.6 | 91.5 | 99.1 | 103.4 | 102.7 | 96.5 | 88.0 | 74.4 | 64.7 | 84.8 |
| | NORMAL DAILY MINIMUM | 30 | 20.3 | 25.3 | 34.4 | 43.7 | 54.0 | 63.9 | 69.1 | 67.9 | 59.3 | 46.9 | 33.9 | 24.0 | 45.2 |
| | MEAN DAILY MINIMUM | 62 | 20.7 | 25.2 | 33.6 | 44.5 | 55.1 | 64.6 | 69.6 | 68.2 | 59.3 | 47.1 | 34.2 | 24.3 | 45.5 |
| | LOWEST DAILY MINIMUM | 57 | -12 | -21 | -2 | 15 | 31 | 43 | 51 | 48 | 31 | 18 | 1 | -16 | -21 |
| | YEAR OF OCCURRENCE | | 1962 | 1982 | 1960 | 1975 | 1976 | 1969 | 1975 | 1967 | 1984 | 1993 | 1975 | 1989 | FEB 1982 |
| | MEAN OF EXTREME MINS. | 62 | 1.9 | 6.3 | 14.7 | 27.9 | 39.9 | 51.9 | 59.5 | 57.6 | 43.2 | 30.7 | 17.9 | 6.7 | 29.9 |
| | NORMAL DRY BULB | 30 | 30.2 | 36.3 | 45.9 | 55.3 | 65.0 | 75.5 | 81.0 | 79.8 | 70.8 | 58.6 | 44.2 | 33.6 | 56.4 |
| | MEAN DRY BULB | 62 | 30.9 | 36.2 | 45.1 | 56.2 | 66.0 | 75.8 | 81.0 | 79.7 | 70.7 | 58.8 | 44.9 | 34.2 | 56.6 |
| | MEAN WET BULB | 26 | 27.5 | 31.6 | 39.6 | 48.6 | 59.1 | 66.6 | 69.9 | 68.9 | 61.2 | 50.6 | 39.1 | 29.9 | 49.4 |
| | MEAN DEW POINT | 26 | 23.8 | 27.5 | 34.9 | 44.3 | 56.0 | 63.5 | 66.4 | 65.4 | 57.4 | 46.4 | 35.1 | 26.3 | 45.6 |
| | NORMAL NO. DAYS WITH: | | | | | | | | | | | | | | |
| | MAXIMUM >= 90 | 30 | 0.0 | 0.0 | 0.0 | 0.2 | 1.4 | 12.2 | 21.4 | 19.6 | 8.6 | 0.8 | 0.0 | 0.0 | 64.2 |
| | MAXIMUM <= 32 | 30 | 9.1 | 5.4 | 0.9 | * | 0.0 | 0.0 | 0.0 | 0.0 | 0.0 | 0.0 | 0.9 | 5.3 | 21.6 |
| | MINIMUM <= 32 | 30 | 28.4 | 21.3 | 12.8 | 2.9 | 0.1 | 0.0 | 0.0 | 0.0 | 0.1 | 1.4 | 14.3 | 26.5 | 107.8 |
| | MINIMUM <= 0 | 30 | 1.7 | 1.1 | 0.0 | 0.0 | 0.0 | 0.0 | 0.0 | 0.0 | 0.0 | 0.0 | 0.0 | 0.8 | 3.6 |
| **H/C** | NORMAL HEATING DEG. DAYS | 30 | 1087 | 819 | 594 | 302 | 89 | 5 | 0 | 0 | 49 | 235 | 620 | 965 | 4765 |
| | NORMAL COOLING DEG. DAYS | 30 | 0 | 0 | 2 | 19 | 93 | 330 | 503 | 454 | 221 | 35 | 1 | 0 | 1658 |
| **RH** | NORMAL (PERCENT) | 30 | 71 | 68 | 66 | 65 | 70 | 66 | 61 | 62 | 65 | 65 | 70 | 73 | 67 |
| | HOUR 00 LST | 30 | 77 | 74 | 72 | 72 | 79 | 76 | 70 | 70 | 74 | 73 | 77 | 78 | 74 |
| | HOUR 06 LST | 30 | 80 | 79 | 79 | 79 | 85 | 83 | 80 | 81 | 83 | 80 | 81 | 81 | 81 |
| | HOUR 12 LST | 30 | 63 | 60 | 56 | 54 | 58 | 53 | 49 | 51 | 53 | 53 | 60 | 64 | 56 |
| | HOUR 18 LST | 30 | 66 | 59 | 54 | 52 | 57 | 50 | 46 | 47 | 51 | 55 | 65 | 69 | 56 |
| **S** | PERCENT POSSIBLE SUNSHINE | 53 | 59 | 60 | 63 | 66 | 64 | 69 | 76 | 77 | 70 | 64 | 59 | 57 | 65 |
| **W/O** | MEAN NO. DAYS WITH: | | | | | | | | | | | | | | |
| | HEAVY FOG(VISBY <= 1/4 MI) | 46 | 2.5 | 2.1 | 1.4 | 0.6 | 0.8 | 0.3 | 0.1 | 0.1 | 0.9 | 1.1 | 1.7 | 3.3 | 14.9 |
| | THUNDERSTORMS | 62 | 0.3 | 1.0 | 2.8 | 5.1 | 8.6 | 9.8 | 8.0 | 7.4 | 5.3 | 3.1 | 1.1 | 0.4 | 52.9 |
| **CLOUDINESS** | MEAN: | | | | | | | | | | | | | | |
| | SUNRISE-SUNSET (OKTAS) | 1 | 6.4 | 6.0 | 5.6 | 6.0 | 4.4 | 3.2 | 2.4 | 2.1 | 3.2 | 3.6 | 4.0 | 5.6 | 4.4 |
| | MIDNIGHT-MIDNIGHT (OKTAS) | 1 | 6.4 | 6.0 | 6.4 | 5.6 | 4.4 | 2.8 | 2.4 | 2.4 | 3.2 | 4.8 | 4.0 | 4.0 | 4.4 |
| | MEAN NO. DAYS WITH: | | | | | | | | | | | | | | |
| | CLEAR | 3 | 6.3 | 6.7 | 7.0 | 6.5 | 11.0 | 10.5 | 15.0 | 16.5 | 6.0 | 10.5 | 7.0 | 11.0 | 114.0 |
| | PARTLY CLOUDY | 2 | 5.0 | 6.5 | 3.5 | 2.5 | 8.0 | 7.0 | 4.5 | 5.0 | 3.0 | 4.0 | 2.5 | 3.5 | 55.0 |
| | CLOUDY | 3 | 8.3 | 8.0 | 8.0 | 9.0 | 5.5 | 3.0 | 3.0 | 2.0 | 2.5 | 5.0 | 8.0 | 7.5 | 69.8 |
| **PR** | MEAN STATION PRESSURE(IN) | 26 | 28.68 | 28.65 | 28.58 | 28.51 | 28.51 | 28.52 | 28.56 | 28.58 | 28.60 | 28.61 | 28.63 | 28.67 | 28.59 |
| | MEAN SEA-LEVEL PRES. (IN) | 26 | 30.14 | 30.10 | 30.01 | 29.93 | 29.91 | 29.90 | 29.94 | 29.96 | 29.99 | 30.03 | 30.07 | 30.13 | 30.01 |
| **WINDS** | MEAN SPEED (MPH) | 26 | 11.3 | 11.7 | 13.0 | 13.3 | 11.9 | 11.7 | 11.1 | 10.3 | 11.1 | 11.3 | 11.6 | 11.0 | 11.6 |
| | PREVAIL.DIR(TENS OF DEGS) | 37 | 36 | 36 | 19 | 19 | 19 | 19 | 19 | 19 | 19 | 19 | 19 | 19 | 19 |
| | MAXIMUM 2-MINUTE: | 17 | | | | | | | | | | | | | |
| | SPEED (MPH) | | 48 | 44 | 49 | 56 | 61 | 48 | 70 | 66 | 43 | 49 | 48 | 44 | 70 |
| | DIR. (TENS OF DEGS) | | 35 | 33 | 24 | 23 | 18 | 30 | 34 | 05 | 15 | 31 | 33 | 32 | 34 |
| | YEAR OF OCCURRENCE | | 1996 | 2007 | 2000 | 2001 | 1998 | 2006 | 1996 | 2003 | 2006 | 1993 | 1997 | 1997 | JUL 1996 |
| | MAXIMUM 3-SECOND | 17 | | | | | | | | | | | | | |
| | SPEED (MPH) | | 59 | 55 | 61 | 66 | 70 | 62 | 101 | 74 | 56 | 53 | 57 | 54 | 101 |
| | DIR. (TENS OF DEGS) | | 35 | 33 | 32 | 23 | 18 | 20 | 00 | 05 | 18 | 31 | 35 | 01 | 00 |
| | YEAR OF OCCURRENCE | | 1996 | 2007 | 2002 | 2001 | 1998 | 2008 | 1993 | 2003 | 2006 | 1993 | 1997 | 2008 | JUL 1993 |
| **PRECIPITATION** | NORMAL (IN) | 30 | 0.84 | 1.02 | 2.71 | 2.57 | 4.16 | 4.25 | 3.31 | 2.94 | 2.96 | 2.45 | 1.82 | 1.35 | 30.38 |
| | MAXIMUM MONTHLY (IN) | 56 | 3.17 | 3.89 | 9.17 | 9.94 | 13.14 | 10.46 | 9.22 | 11.96 | 12.96 | 9.42 | 5.88 | 4.71 | 13.14 |
| | YEAR OF OCCURRENCE | | 2005 | 2001 | 1973 | 2009 | 2008 | 1957 | 1962 | 2005 | 2008 | 1998 | 1964 | 1984 | MAY 2008 |
| | MINIMUM MONTHLY (IN) | 56 | T | T | 0.01 | 0.22 | 0.52 | 0.40 | 0.05 | 0.14 | 0.03 | T | T | 0.03 | T |
| | YEAR OF OCCURRENCE | | 1986 | 2006 | 1971 | 1963 | 1973 | 1998 | 1975 | 2000 | 1956 | 1958 | 1989 | 1955 | FEB 2006 |
| | MAXIMUM IN 24 HOURS (IN) | 56 | 1.89 | 2.05 | 2.65 | 6.04 | 4.70 | 4.98 | 3.86 | 4.50 | 10.57 | 5.84 | 4.33 | 2.60 | 10.57 |
| | YEAR OF OCCURRENCE | | 2005 | 2001 | 1961 | 2009 | 1963 | 1965 | 1983 | 1991 | 2008 | 1998 | 1964 | 1984 | SEP 2008 |
| | NORMAL NO. DAYS WITH: | | | | | | | | | | | | | | |
| | PRECIPITATION >= 0.01 | 30 | 5.4 | 5.4 | 8.1 | 8.5 | 11.2 | 9.7 | 7.2 | 7.6 | 7.2 | 6.4 | 5.8 | 5.7 | 88.2 |
| | PRECIPITATION >= 1.00 | 30 | 0.1 | 0.1 | 0.9 | 0.7 | 1.2 | 1.4 | 1.0 | 0.9 | 0.8 | 0.6 | 0.6 | 0.2 | 8.5 |
| **SNOWFALL** | NORMAL (IN) | 30 | 4.6 | 4.1 | 2.2 | 0.3 | 0.0 | 0.0 | 0.0 | 0.0 | 0.0 | 0.1 | 1.6 | 3.4 | 16.3 |
| | MAXIMUM MONTHLY (IN) | 56 | 19.7 | 16.7 | 16.5 | 4.6 | T | T | T | T | T | 1.5 | 7.1 | 13.8 | 19.7 |
| | YEAR OF OCCURRENCE | | 1987 | 1971 | 1970 | 1979 | 2009 | 1992 | 1993 | 2009 | 1992 | 1991 | 1972 | 1983 | JAN 1987 |
| | MAXIMUM IN 24 HOURS (IN) | 56 | 13.0 | 11.9 | 13.5 | 4.6 | T | T | T | T | T | 1.5 | 6.8 | 9.0 | 13.5 |
| | YEAR OF OCCURRENCE' | | 1962 | 1971 | 1970 | 1979 | 2009 | 1992 | 1993 | 2009 | 1992 | 1991 | 1984 | 1983 | MAR 1970 |
| | MAXIMUM SNOW DEPTH (IN) | 57 | 17 | 13 | 13 | 3 | 0 | 0 | 0 | 0 | 0 | T | 5 | 8 | 17 |
| | YEAR OF OCCURRENCE | | 1962 | 1971 | 1970 | 1979 | | | | | | 1993 | 2006 | 1987 | JAN 1962 |
| | NORMAL NO. DAYS WITH: | | | | | | | | | | | | | | |
| | SNOWFALL >= 1.0 | 30 | 1.4 | 1.1 | 0.4 | 0.1 | 0.0 | 0.0 | 0.0 | 0.0 | 0.0 | 0.0 | 0.5 | 1.0 | 4.5 |

## PRECIPITATION (inches) 2009 WICHITA (KICT)

| YEAR | JAN | FEB | MAR | APR | MAY | JUN | JUL | AUG | SEP | OCT | NOV | DEC | ANNUAL |
|------|-----|-----|-----|-----|-----|-----|-----|-----|-----|-----|-----|-----|--------|
| 1980 | 1.82 | 0.81 | 3.99 | 1.07 | 2.66 | 1.34 | 0.47 | 3.76 | 0.67 | 1.25 | 0.54 | 2.11 | 20.49 |
| 1981 | 0.25 | 0.22 | 2.15 | 0.38 | 6.33 | 4.25 | 1.27 | 2.65 | 2.25 | 4.69 | 2.93 | 0.29 | 27.66 |
| 1982 | 1.68 | 0.77 | 2.05 | 0.73 | 7.82 | 8.28 | 0.56 | 1.51 | 1.08 | 0.41 | 0.73 | 1.51 | 27.13 |
| 1983 | 1.66 | 1.23 | 4.26 | 3.80 | 4.08 | 7.38 | 3.86 | 1.39 | 2.53 | 2.97 | 2.39 | 1.13 | 36.68 |
| 1984 | 0.20 | 1.23 | 7.57 | 3.71 | 1.15 | 2.30 | 0.30 | 0.75 | 2.18 | 2.78 | 1.44 | 4.71 | 28.32 |
| 1985 | 0.26 | 2.07 | 1.64 | 2.28 | 2.01 | 4.79 | 3.97 | 2.86 | 5.97 | 5.58 | 1.60 | 0.61 | 33.64 |
| 1986 | T | 1.26 | 1.22 | 1.80 | 2.98 | 5.39 | 3.42 | 6.00 | 3.81 | 3.61 | 0.58 | 1.22 | 31.29 |
| 1987 | 1.40 | 3.33 | 4.13 | 0.61 | 8.01 | 4.50 | 2.14 | 7.69 | 2.10 | 0.90 | 1.50 | 2.25 | 38.56 |
| 1988 | 0.51 | 0.18 | 2.91 | 4.46 | 2.40 | 1.86 | 0.91 | 1.10 | 0.53 | 0.94 | 0.77 | 0.50 | 17.07 |
| 1989 | 0.79 | 0.39 | 2.38 | 0.23 | 4.96 | 7.96 | 4.07 | 5.72 | 7.38 | 0.37 | T | 0.44 | 34.69 |
| 1990 | 1.73 | 2.19 | 2.68 | 0.80 | 1.29 | 1.91 | 1.72 | 2.01 | 1.95 | 0.64 | 2.01 | 0.78 | 19.71 |
| 1991 | 0.58 | T | 0.71 | 2.27 | 4.09 | 1.34 | 2.65 | 7.57 | 2.08 | 1.00 | 2.38 | 2.07 | 26.74 |
| 1992 | 0.67 | 0.47 | 3.60 | 1.42 | 3.44 | 7.26 | 4.77 | 1.94 | 2.94 | 2.80 | 4.91 | 1.16 | 35.38 |
| 1993 | 1.12 | 2.25 | 1.75 | 2.11 | 9.62 | 4.44 | 6.21 | 1.31 | 1.77 | 1.68 | 0.55 | 0.30 | 33.11 |
| 1994 | 0.03 | 0.27 | 0.25 | 3.80 | 0.95 | 2.72 | 5.89 | 2.30 | 1.08 | 3.45 | 3.08 | 1.04 | 24.86 |
| 1995 | 0.44 | 0.41 | 2.45 | 3.40 | 5.99 | 8.86 | 4.20 | 5.51 | 2.00 | 0.28 | 0.08 | 0.68 | 34.30 |
| 1996 | 0.10 | 0.10 | 1.82 | 1.48 | 4.42 | 2.06 | 3.97 | 5.24 | 3.80 | 1.90 | 3.56 | 0.03 | 28.48 |
| 1997 | 0.36 | 2.18 | 0.46 | 3.81 | 4.78 | 4.35 | 5.89 | 4.36 | 3.50 | 3.39 | 1.30 | 2.65 | 37.03 |
| 1998 | 1.04 | 0.31 | 4.22 | 3.37 | 1.64 | 0.40 | 5.17 | 1.58 | 3.25 | 9.42 | 3.20 | 1.06 | 34.66 |
| 1999 | 1.35 | 0.39 | 1.89 | 6.02 | 7.17 | 7.55 | 3.48 | 1.25 | 10.69 | 0.16 | 1.45 | 4.06 | 45.46 |
| 2000 | 0.92 | 2.72 | 5.99 | 1.21 | 3.00 | 7.00 | 3.66 | 0.14 | 1.02 | 4.82 | 0.98 | 0.36 | 31.82 |
| 2001 | 1.47 | 3.89 | 2.25 | 1.42 | 3.12 | 4.31 | 1.24 | 1.90 | 3.10 | 1.06 | 0.65 | 0.08 | 24.49 |
| 2002 | 1.12 | 0.82 | 0.43 | 2.65 | 5.70 | 5.37 | 1.96 | 4.90 | 0.81 | 8.29 | 0.18 | 1.22 | 33.45 |
| 2003 | 0.12 | 1.94 | 3.53 | 4.06 | 2.99 | 4.08 | 0.47 | 6.82 | 4.39 | 3.19 | 0.11 | 0.89 | 32.59 |
| 2004 | 1.84 | 1.30 | 3.58 | 3.41 | 3.77 | 8.04 | 6.88 | 2.17 | 0.55 | 3.41 | 2.55 | 0.30 | 37.80 |
| 2005 | 3.17 | 1.83 | 1.82 | 1.03 | 1.86 | 7.30 | 4.35 | 11.96 | 0.62 | 2.16 | 0.02 | 0.59 | 36.71 |
| 2006 | 0.11 | T | 2.28 | 1.99 | 6.76 | 6.32 | 2.16 | 5.93 | 0.72 | 0.94 | 0.46 | 1.69 | 29.36 |
| 2007 | 1.21 | 0.47 | 5.60 | 2.70 | 4.11 | 8.53 | 4.05 | 3.61 | 0.72 | 4.21 | 0.14 | 2.62 | 37.97 |
| 2008 | 0.35 | 1.57 | 3.08 | 1.82 | 13.14 | 7.42 | 3.82 | 3.00 | 12.96 | 4.03 | 1.37 | 1.26 | 53.82 |
| 2009 | 0.08 | 0.60 | 2.04 | 9.94 | 2.94 | 4.51 | 3.55 | 3.94 | 5.16 | 3.82 | 0.56 | 0.39 | 37.53 |
| POR= 62 YRS | 0.86 | 1.04 | 2.34 | 2.52 | 4.13 | 4.63 | 3.76 | 3.35 | 3.22 | 2.59 | 1.44 | 1.09 | 30.97 |

WBAN : 03928

## AVERAGE TEMPERATURE (°F) 2009 WICHITA (KICT)

| YEAR | JAN | FEB | MAR | APR | MAY | JUN | JUL | AUG | SEP | OCT | NOV | DEC | ANNUAL |
|------|-----|-----|-----|-----|-----|-----|-----|-----|-----|-----|-----|-----|--------|
| 1980 | 31.4 | 28.2 | 41.5 | 54.3 | 63.5 | 79.9 | 90.5 | 85.3 | 75.2 | 58.8 | 46.9 | 36.9 | 57.7 |
| 1981 | 34.1 | 40.1 | 47.6 | 63.7 | 62.6 | 77.9 | 83.6 | 78.1 | 72.0 | 55.9 | 47.0 | 32.9 | 58.0 |
| 1982 | 25.5 | 28.0 | 46.0 | 53.5 | 65.3 | 70.4 | 81.5 | 82.0 | 71.9 | 58.0 | 43.0 | 36.1 | 55.1 |
| 1983 | 31.7 | 35.5 | 43.3 | 48.0 | 60.5 | 71.5 | 81.6 | 85.0 | 72.7 | 58.5 | 45.5 | 16.4 | 54.2 |
| 1984 | 26.7 | 41.4 | 40.7 | 51.8 | 63.6 | 77.7 | 81.6 | 82.8 | 70.3 | 58.3 | 45.5 | 37.2 | 56.5 |
| 1985 | 25.2 | 31.2 | 49.1 | 59.9 | 67.7 | 74.0 | 81.7 | 77.5 | 69.8 | 57.4 | 39.4 | 28.8 | 55.1 |
| 1986 | 38.1 | 37.8 | 51.9 | 58.7 | 66.7 | 78.7 | 83.0 | 75.9 | 73.6 | 57.6 | 39.9 | 35.8 | 58.1 |
| 1987 | 29.3 | 42.3 | 47.0 | 57.4 | 69.7 | 76.4 | 80.1 | 78.7 | 70.4 | 55.7 | 47.7 | 34.9 | 57.5 |
| 1988 | 27.0 | 34.1 | 44.2 | 53.9 | 68.3 | 78.6 | 80.9 | 83.0 | 72.0 | 56.7 | 47.5 | 38.5 | 57.1 |
| 1989 | 38.5 | 27.5 | 47.0 | 59.6 | 66.0 | 72.0 | 79.0 | 77.0 | 65.7 | 60.7 | 45.5 | 25.2 | 55.3 |
| 1990 | 39.5 | 38.8 | 46.9 | 54.3 | 63.7 | 81.7 | 81.7 | 80.8 | 74.3 | 58.4 | 50.4 | 30.2 | 58.4 |
| 1991 | 28.9 | 44.6 | 49.7 | 58.9 | 70.4 | 79.3 | 83.8 | 80.5 | 69.6 | 59.4 | 39.5 | 39.1 | 58.6 |
| 1992 | 39.0 | 44.9 | 50.0 | 56.7 | 63.3 | 71.2 | 78.9 | 73.8 | 71.0 | 59.1 | 41.3 | 33.3 | 56.9 |
| 1993 | 28.6 | 32.2 | 43.1 | 52.5 | 63.5 | 74.5 | 81.7 | 81.0 | 67.4 | 55.3 | 40.8 | 37.8 | 54.9 |
| 1994 | 31.0 | 33.9 | 50.3 | 54.7 | 66.4 | 79.3 | 77.6 | 79.0 | 70.5 | 60.5 | 46.7 | 37.1 | 57.3 |
| 1995 | 32.9 | 40.6 | 45.4 | 52.3 | 60.2 | 72.3 | 80.1 | 80.9 | 68.5 | 59.1 | 43.8 | 33.5 | 55.8 |
| 1996 | 28.3 | 38.0 | 40.6 | 55.2 | 69.6 | 77.2 | 79.1 | 77.4 | 67.7 | 58.4 | 40.8 | 34.0 | 55.5 |
| 1997 | 30.9 | 37.5 | 47.4 | 51.0 | 62.9 | 73.9 | 79.3 | 76.7 | 72.6 | 59.2 | 42.1 | 34.6 | 55.7 |
| 1998 | 34.5 | 41.2 | 40.6 | 54.3 | 69.9 | 78.4 | 81.8 | 81.6 | 78.6 | 60.8 | 49.5 | 36.0 | 58.9 |
| 1999 | 32.7 | 45.3 | 44.6 | 56.0 | 65.3 | 73.0 | 82.3 | 81.2 | 67.1 | 59.2 | 53.3 | 37.7 | 58.1 |
| 2000 | 33.8 | 42.0 | 48.5 | 55.1 | 68.9 | 73.1 | 80.3 | 86.6 | 74.4 | 61.6 | 40.2 | 23.9 | 57.4 |
| 2001 | 32.3 | 33.5 | 42.8 | 60.2 | 67.5 | 75.3 | 86.3 | 82.7 | 70.0 | 59.1 | 50.9 | 39.0 | 58.3 |
| 2002 | 36.0 | 37.6 | 42.0 | 58.2 | 63.7 | 76.5 | 81.6 | 80.3 | 73.7 | 52.3 | 44.2 | 35.6 | 56.8 |
| 2003 | 32.3 | 33.3 | 45.6 | 58.1 | 64.6 | 72.6 | 84.3 | 82.4 | 66.0 | 58.2 | 45.8 | 37.2 | 56.7 |
| 2004 | 32.1 | 33.7 | 50.3 | 56.8 | 69.5 | 73.6 | 77.3 | 75.4 | 74.4 | 60.9 | 47.1 | 37.8 | 57.4 |
| 2005 | 31.3 | 40.8 | 46.5 | 56.9 | 67.4 | 77.6 | 80.0 | 78.7 | 74.5 | 60.0 | 47.9 | 32.7 | 57.9 |
| 2006 | 43.2 | 36.3 | 49.0 | 63.1 | 67.9 | 76.6 | 83.6 | 81.7 | 67.9 | 57.9 | 48.2 | 39.1 | 59.5 |
| 2007 | 29.9 | 34.2 | 54.9 | 53.2 | 68.2 | 73.9 | 79.6 | 83.0 | 74.0 | 61.4 | 46.1 | 31.3 | 57.5 |
| 2008 | 32.6 | 34.8 | 45.5 | 53.9 | 66.1 | 76.8 | 80.2 | 77.9 | 67.9 | 57.7 | 46.2 | 33.1 | 56.1 |
| 2009 | 32.2 | 42.5 | 47.8 | 55.1 | 65.7 | 79.1 | 78.3 | 76.3 | 68.5 | 51.3 | 50.3 | 30.0 | 56.4 |
| POR= 62 YRS | 30.9 | 36.2 | 45.1 | 56.2 | 66.0 | 75.8 | 81.0 | 79.7 | 70.7 | 58.8 | 44.9 | 34.2 | 56.6 |

## HEATING DEGREE DAYS (base 65°F) 2009 WICHITA (KICT)

| YEAR | JUL | AUG | SEP | OCT | NOV | DEC | JAN | FEB | MAR | APR | MAY | JUN | TOTAL |
|------|-----|-----|-----|-----|-----|-----|-----|-----|-----|-----|-----|-----|-------|
| 1980-81 | 0 | 0 | 28 | 239 | 535 | 864 | 954 | 692 | 533 | 104 | 126 | 0 | 4075 |
| 1981-82 | 0 | 0 | 24 | 292 | 537 | 990 | 1214 | 1033 | 583 | 356 | 54 | 17 | 5100 |
| 1982-83 | 0 | 0 | 37 | 239 | 653 | 889 | 1022 | 818 | 664 | 507 | 168 | 20 | 5017 |
| 1983-84 | 0 | 0 | 47 | 221 | 582 | 1504 | 1180 | 680 | 747 | 394 | 95 | 0 | 5450 |
| 1984-85 | 0 | 0 | 103 | 237 | 576 | 856 | 1224 | 938 | 487 | 184 | 33 | 8 | 4646 |
| 1985-86 | 0 | 0 | 111 | 230 | 762 | 1116 | 826 | 755 | 416 | 220 | 41 | 0 | 4477 |
| 1986-87 | 0 | 3 | 11 | 233 | 747 | 899 | 1099 | 631 | 551 | 263 | 14 | 0 | 4451 |
| 1987-88 | 0 | 2 | 7 | 282 | 523 | 924 | 1170 | 891 | 637 | 330 | 33 | 0 | 4799 |
| 1988-89 | 0 | 0 | 16 | 265 | 519 | 813 | 817 | 1044 | 556 | 238 | 90 | 8 | 4366 |
| 1989-90 | 0 | 0 | 105 | 193 | 578 | 1228 | 783 | 728 | 555 | 332 | 112 | 0 | 4614 |
| 1990-91 | 0 | 0 | 18 | 238 | 445 | 1074 | 1114 | 567 | 475 | 201 | 53 | 0 | 4185 |
| 1991-92 | 0 | 0 | 71 | 242 | 759 | 793 | 802 | 574 | 461 | 259 | 113 | 10 | 4084 |
| 1992-93 | 0 | 2 | 25 | 209 | 703 | 975 | 1122 | 913 | 670 | 368 | 100 | 10 | 5097 |
| 1993-94 | 0 | 0 | 59 | 321 | 718 | 837 | 1046 | 862 | 458 | 321 | 66 | 0 | 4688 |
| 1994-95 | 0 | 0 | 33 | 186 | 542 | 856 | 989 | 675 | 603 | 375 | 164 | 2 | 4425 |
| 1995-96 | 0 | 0 | 90 | 205 | 627 | 969 | 1129 | 776 | 749 | 308 | 49 | 5 | 4907 |
| 1996-97 | 0 | 0 | 48 | 231 | 719 | 953 | 1050 | 763 | 541 | 415 | 106 | 0 | 4826 |
| 1997-98 | 0 | 0 | 9 | 263 | 680 | 934 | 941 | 660 | 754 | 322 | 30 | 21 | 4614 |
| 1998-99 | 0 | 0 | 4 | 157 | 459 | 892 | 994 | 545 | 625 | 271 | 64 | 10 | 4021 |
| 1999-00 | 0 | 0 | 69 | 198 | 345 | 838 | 960 | 659 | 506 | 289 | 46 | 5 | 3915 |
| 2000-01 | 0 | 0 | 55 | 154 | 740 | 1266 | 1009 | 878 | 679 | 167 | 46 | 0 | 4994 |
| 2001-02 | 0 | 0 | 23 | 199 | 419 | 796 | 888 | 759 | 704 | 245 | 112 | 2 | 4147 |
| 2002-03 | 0 | 0 | 8 | 404 | 617 | 905 | 1008 | 881 | 597 | 232 | 68 | 13 | 4733 |
| 2003-04 | 0 | 0 | 71 | 217 | 568 | 854 | 1011 | 901 | 457 | 257 | 73 | 0 | 4409 |
| 2004-05 | 0 | 0 | 0 | 153 | 530 | 835 | 1039 | 673 | 569 | 257 | 80 | 0 | 4136 |
| 2005-06 | 0 | 0 | 14 | 214 | 509 | 993 | 665 | 798 | 490 | 143 | 80 | 0 | 3906 |
| 2006-07 | 0 | 0 | 30 | 281 | 499 | 795 | 1080 | 855 | 320 | 362 | 14 | 0 | 4236 |
| 2007-08 | 0 | 0 | 5 | 181 | 566 | 1040 | 995 | 867 | 596 | 340 | 71 | 0 | 4661 |
| 2008-09 | 0 | 0 | 30 | 251 | 563 | 982 | 1010 | 626 | 526 | 314 | 88 | 0 | 4390 |
| 2009- | 0 | 1 | 37 | 424 | 434 | 1082 | | | | | | | |

WBAN : 03928

## COOLING DEGREE DAYS (base 65°F) 2009 WICHITA (KICT)

| YEAR | JAN | FEB | MAR | APR | MAY | JUN | JUL | AUG | SEP | OCT | NOV | DEC | TOTAL |
|------|-----|-----|-----|-----|-----|-----|-----|-----|-----|-----|-----|-----|-------|
| 1980 | 0 | 0 | 0 | 3 | 75 | 456 | 796 | 635 | 340 | 52 | 1 | 0 | 2358 |
| 1981 | 0 | 0 | 0 | 72 | 56 | 393 | 582 | 412 | 240 | 17 | 0 | 0 | 1772 |
| 1982 | 0 | 0 | 1 | 16 | 70 | 186 | 516 | 534 | 253 | 29 | 0 | 0 | 1605 |
| 1983 | 0 | 0 | 0 | 2 | 34 | 220 | 521 | 628 | 286 | 28 | 5 | 0 | 1724 |
| 1984 | 0 | 0 | 0 | 6 | 61 | 388 | 520 | 558 | 272 | 35 | 0 | 0 | 1840 |
| 1985 | 0 | 0 | 0 | 36 | 122 | 285 | 523 | 394 | 262 | 1 | 0 | 0 | 1623 |
| 1986 | 0 | 0 | 17 | 40 | 102 | 419 | 563 | 349 | 275 | 10 | 0 | 0 | 1775 |
| 1987 | 0 | 0 | 0 | 42 | 166 | 350 | 473 | 434 | 177 | 3 | 10 | 0 | 1655 |
| 1988 | 0 | 0 | 0 | 3 | 140 | 415 | 497 | 566 | 235 | 15 | 0 | 0 | 1871 |
| 1989 | 0 | 0 | 5 | 81 | 129 | 225 | 442 | 379 | 132 | 70 | 0 | 0 | 1463 |
| 1990 | 0 | 0 | 0 | 17 | 79 | 508 | 525 | 497 | 306 | 41 | 11 | 0 | 1984 |
| 1991 | 0 | 0 | 9 | 25 | 225 | 433 | 592 | 484 | 214 | 77 | 0 | 0 | 2059 |
| 1992 | 0 | 0 | 0 | 17 | 66 | 203 | 441 | 281 | 213 | 37 | 0 | 0 | 1258 |
| 1993 | 0 | 0 | 0 | 0 | 59 | 304 | 526 | 502 | 134 | 27 | 0 | 0 | 1552 |
| 1994 | 0 | 0 | 7 | 22 | 117 | 434 | 399 | 441 | 202 | 55 | 0 | 0 | 1677 |
| 1995 | 0 | 0 | 6 | 4 | 24 | 231 | 478 | 502 | 204 | 32 | 0 | 0 | 1481 |
| 1996 | 0 | 0 | 0 | 19 | 197 | 377 | 444 | 393 | 136 | 35 | 0 | 0 | 1601 |
| 1997 | 0 | 0 | 0 | 0 | 47 | 274 | 448 | 369 | 244 | 89 | 0 | 0 | 1471 |
| 1998 | 0 | 0 | 5 | 7 | 189 | 427 | 528 | 522 | 416 | 37 | 0 | 0 | 2131 |
| 1999 | 0 | 0 | 0 | 7 | 78 | 259 | 543 | 511 | 136 | 28 | 1 | 0 | 1563 |
| 2000 | 0 | 0 | 0 | 2 | 174 | 256 | 483 | 675 | 341 | 57 | 0 | 0 | 1988 |
| 2001 | 0 | 0 | 0 | 31 | 129 | 317 | 670 | 557 | 178 | 25 | 2 | 0 | 1909 |
| 2002 | 0 | 0 | 0 | 50 | 80 | 354 | 522 | 483 | 278 | 22 | 0 | 0 | 1789 |
| 2003 | 0 | 0 | 1 | 33 | 63 | 247 | 604 | 544 | 108 | 12 | 0 | 0 | 1612 |
| 2004 | 0 | 0 | 7 | 20 | 216 | 265 | 390 | 327 | 291 | 32 | 0 | 0 | 1548 |
| 2005 | 0 | 0 | 0 | 25 | 164 | 384 | 471 | 430 | 305 | 64 | 2 | 0 | 1845 |
| 2006 | 0 | 0 | 1 | 93 | 175 | 354 | 583 | 525 | 124 | 68 | 1 | 0 | 1924 |
| 2007 | 0 | 0 | 10 | 15 | 122 | 271 | 459 | 564 | 279 | 76 | 5 | 0 | 1801 |
| 2008 | 0 | 0 | 0 | 15 | 113 | 359 | 477 | 403 | 126 | 32 | 5 | 0 | 1530 |
| 2009 | 0 | 0 | 2 | 21 | 118 | 428 | 417 | 357 | 148 | 4 | 0 | 0 | 1495 |

## SNOWFALL (inches) 2009 WICHITA (KICT)

| YEAR | JUL | AUG | SEP | OCT | NOV | DEC | JAN | FEB | MAR | APR | MAY | JUN | TOTAL |
|------|-----|-----|-----|-----|-----|-----|-----|-----|-----|-----|-----|-----|-------|
| 1980-81 | 0.0 | 0.0 | 0.0 | T | T | 0.4 | T | 2.5 | 0.2 | 0.0 | 0.0 | 0.0 | 3.1 |
| 1981-82 | 0.0 | 0.0 | 0.0 | 0.0 | T | 1.2 | 0.0 | 12.7 | T | 0.0 | 0.0 | 0.0 | 13.9 |
| 1982-83 | 0.0 | 0.0 | 0.0 | 0.0 | T | 1.4 | 13.0 | 8.9 | 1.5 | 0.7 | 0.0 | 0.0 | 25.5 |
| 1983-84 | 0.0 | 0.0 | 0.0 | 0.0 | 4.1 | 13.8 | 4.3 | T | 6.9 | 0.0 | 0.0 | 0.0 | 29.1 |
| 1984-85 | 0.0 | 0.0 | 0.0 | 0.0 | 6.8 | 7.6 | 3.5 | 3.6 | 0.2 | 0.0 | 0.0 | 0.0 | 21.7 |
| 1985-86 | 0.0 | 0.0 | 0.0 | 0.0 | 1.0 | 3.0 | 0.0 | 7.5 | 0.0 | 0.0 | 0.0 | 0.0 | 11.5 |
| 1986-87 | 0.0 | 0.0 | 0.0 | 0.0 | 0.0 | 0.4 | 19.7 | 6.0 | T | 0.0 | 0.0 | 0.0 | 26.1 |
| 1987-88 | 0.0 | 0.0 | 0.0 | 0.0 | 6.2 | 12.6 | 8.5 | 1.1 | 11.0 | 0.0 | 0.0 | 0.0 | 39.4 |
| 1988-89 | 0.0 | 0.0 | 0.0 | 0.0 | 3.0 | 0.7 | 0.3 | 0.8 | T | T | T | T | 4.8 |
| 1989-90 | 0.0 | 0.0 | 0.0 | 0.0 | T | 4.4 | 0.3 | 7.0 | 0.7 | 0.1 | 0.0 | 0.0 | 12.5 |
| 1990-91 | 0.0 | 0.0 | 0.0 | 0.0 | 0.0 | 4.8 | 2.1 | 0.0 | T | T | T | T | 6.9 |
| 1991-92 | 0.0 | 0.0 | 0.0 | 1.5 | 0.3 | 0.0 | T | T | T | T | 0.0 | T | 1.8 |
| 1992-93 | 0.0 | 0.0 | T | T | 3.8 | 3.3 | 7.7 | 5.1 | 2.2 | T | 0.0 | 0.0 | 22.1 |
| 1993-94 | T | 0.0 | 0.0 | T | T | 0.1 | T | 1.2 | T | T | 0.0 | 0.0 | 1.3 |
| 1994-95 | 0.0 | T | 0.0 | 0.0 | 0.0 | 0.0 | 1.9 | 2.6 | 5.6 | 0.0 | 0.0 | 0.0 | 10.1 |
| 1995-96 | 0.0 | 0.0 | 0.0 | 0.0 | 0.3 | 5.7 | 1.9 | 0.6 | T | T | 0.0 | 0.0 | 8.5 |
| 1996-97 | 0.0 | 0.0 | 0.0 | 0.2 | 4.6 | 0.4 | 3.8 | 5.5 | T | 2.0 | 0.0 | 0.0 | 16.5 |
| 1997-98 | T | 0.0 | 0.0 | 0.1 | 0.8 | 5.4 | 0.6 | T | 13.6 | 0.0 | 0.0 | 0.0 | 20.5 |
| 1998-99 | 0.0 | 0.0 | 0.0 | 0.0 | 0.0 | 0.9 | 0.1 | T | 8.3 | T | T | 0.0 | 9.3 |
| 1999-00 | 0.0 | 0.0 | T | 0.0 | 0.0 | 7.5 | 10.2 | T | 4.2 | 0.0 | 0.0 | 0.0 | 21.9 |
| 2000-01 | 0.0 | 0.0 | 0.0 | 0.0 | T | 5.5 | 2.5 | 1.1 | T | T | T | T | 9.1 |
| 2001-02 | 0.0 | 0.0 | T | 0.0 | 0.0 | 0.0 | 1.1 | 0.7 | 3.0 | T | T | 0.0 | 4.8 |
| 2002-03 | 0.0 | 0.0 | 0.0 | 0.0 | T | 11.3 | 1.3 | 13.1 | T | T | T | 0.0 | 25.7 |
| 2003-04 | 0.0 | 0.0 | 0.0 | 0.0 | 0.0 | 7.4 | 0.8 | 7.2 | T | T | T | 0.0 | 15.4 |
| 2004-05 | T | 0.0 | 0.0 | 0.0 | 1.5 | T | 5.7 | T | 0.1 | 0.0 | T | T | 7.3 |
| 2005-06 | T | 0.0 | 0.0 | 0.0 | T | 7.6 | 0.1 | T | 1.5 | T | T | 0.0 | 9.2 |
| 2006-07 | 0.0 | 0.0 | 0.0 | 0.0 | 5.5 | 1.3 | 9.5 | 0.2 | 0.0 | 1.8 | 0.0 | 0.0 | 18.3 |
| 2007-08 | 0.0 | 0.0 | 0.0 | 0.0 | 2.5 | 10.6 | 5.0 | 4.0 | 0.0 | T | T | 0.0 | 22.1 |
| 2008-09 | 0.0 | 0.0 | 0.0 | 0.0 | T | 4.6 | 1.6 | T | 6.8 | 0.6 | T | 0.0 | 13.6 |
| 2009- | 0.0 | T | 0.0 | 0.0 | T | 4.0 | | | | | | | |
| POR= 62 YRS | T | T | T | T | 1.3 | 3.2 | 4.1 | 3.4 | 2.1 | 0.3 | T | T | 14.4 |

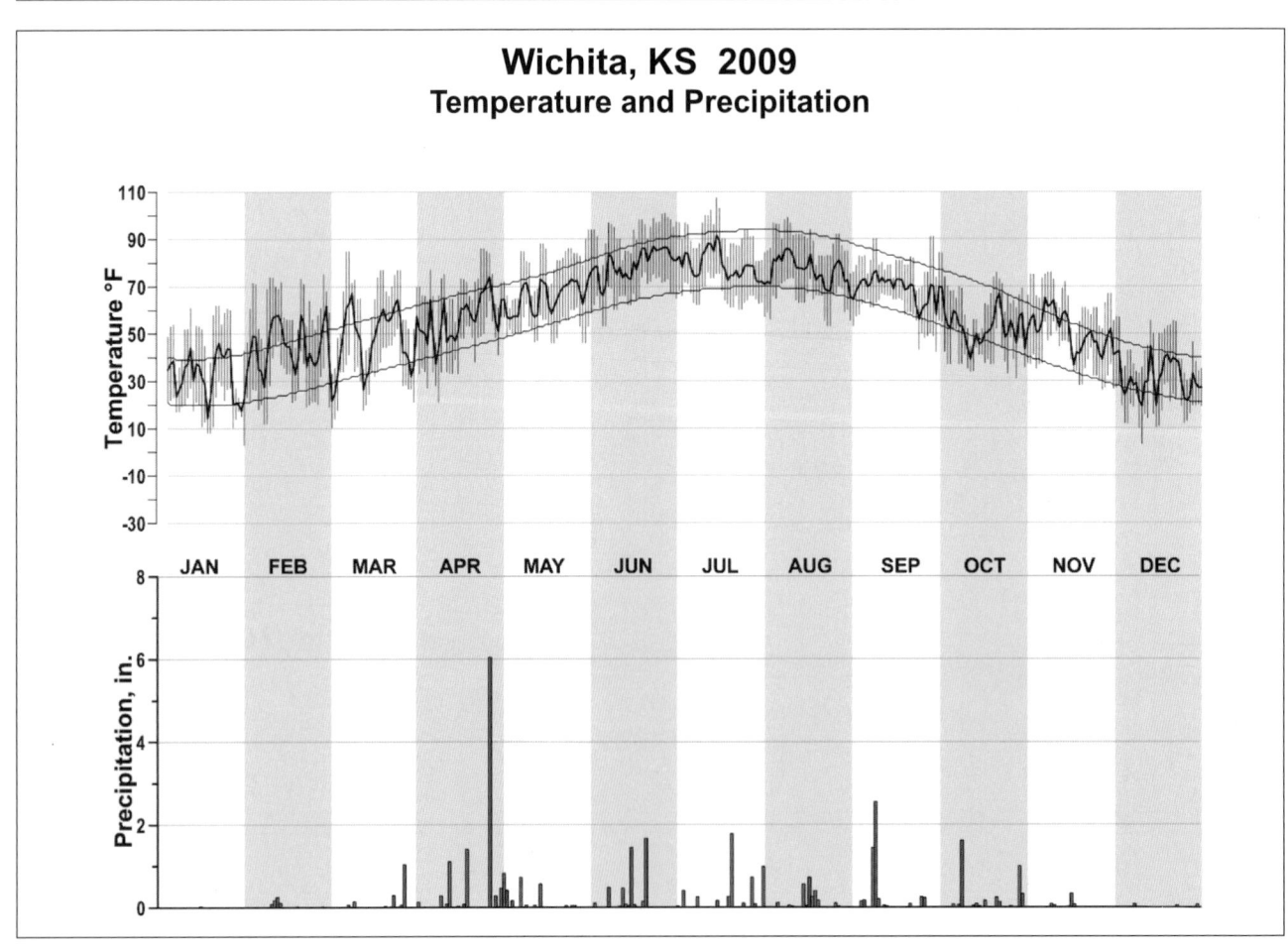

## Wichita, KS 2009
### Temperature and Precipitation

# 2009
# TOPEKA
# KANSAS (KTOP)

Topeka, is located near the geographical center of the United States, and the middle of the temperate zone. The city straddles the Kansas River about 60 miles above its junction with the Missouri River. The Kansas River flows in an easterly direction through northeastern Kansas. Near Topeka, the river valley ranges from 2 to 4 miles wide, and is bordered on both sides by rolling prairie uplands of some 200 to 300 feet. The city is built on both banks of the Kansas River and along two tributaries, Soldier Creek in north Topeka and Shunganunga Creek in the south and east part of town. Flooding is always a threat following periods of heavy rains but protective construction has reduced the problem.

Seventy percent of the annual precipitation normally falls during the six crop-growing months, April through September. The rains of this period are usually of short duration, predominantly of the thunderstorm type. They occur more frequently during the nighttime and early morning hours than at other times of the day. Excessive precipitation rates may occur with warm-season thunderstorms. Rainfall accumulations over 8 inches in 24 hours have occurred in Topeka. Tornadoes have occurred in the area on several occasions and caused severe damage and numerous injuries.

Individual summers show wide departures from average conditions. Hottest summers may produce temperatures of 100 degrees or higher on more than 50 days. On the other hand, 25 percent of the summers pass with two or fewer 100 degree days. Similarly, precipitation has shown a wide range for June, July, and August, varying from under 3 inches to more than 27 inches during the 3 months. Summers are hot with low relative humidity and persistent southerly winds. Oppressively warm periods with high relative humidity are usually of short duration.

Winter temperatures average about 45 degrees cooler than summer. Cold spells are seldom prolonged. Only on rare occasions do daytime temperatures fail to rise above freezing. Winter precipitation is often in the form of snow, sleet, or glaze, but storms of such severity to prevent normal movement of traffic or to interfere with scheduled activity are not common.

In the transitional spring and fall seasons, the numerous days of fair weather are interspersed with short intervals of stormy weather. Strong, blustery winds are quite common in late winter and spring. Autumn is characteristically a season of warm days, cool nights, and infrequent precipitation, with cold air invasions gradually increasing in intensity as the season progresses.

Nearly all crops of the temperate zone can be produced in the vicinity of Topeka. Wheat and other small grains, clover, soybeans, fruit, and berries do well, and the area supports an extensive dairy industry.

Based on the 1951-1980 period, the average first occurrence of 32 degrees Fahrenheit in the fall is October 14 and the average last occurrence in the spring is April 21.

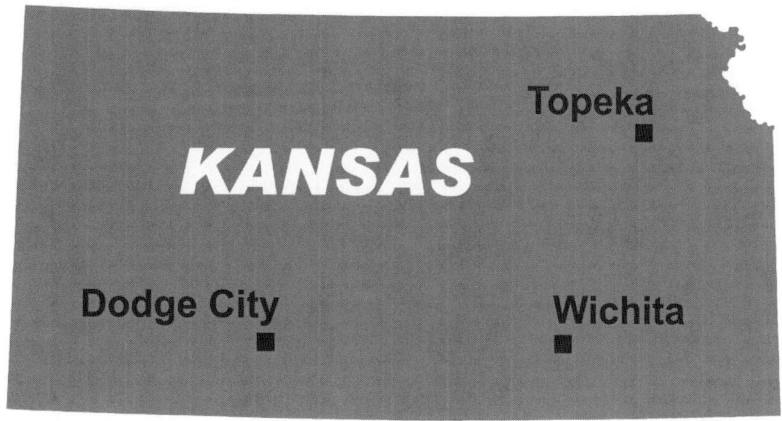

# NORMALS, MEANS, AND EXTREMES
## TOPEKA (KTOP)

LATITUDE: 39 ° 4 'N  LONGITUDE: -95 ° 37'W  ELEVATION (FT): GRND: 884  BARO: 883  TIME ZONE: CENTRAL  (UTC -6)  WBAN: 13996

| | ELEMENT | POR | JAN | FEB | MAR | APR | MAY | JUN | JUL | AUG | SEP | OCT | NOV | DEC | YEAR |
|---|---|---|---|---|---|---|---|---|---|---|---|---|---|---|---|
| **TEMPERATURE °F** | NORMAL DAILY MAXIMUM | 30 | 37.2 | 43.8 | 55.5 | 66.1 | 75.3 | 84.5 | 89.1 | 87.9 | 80.3 | 68.9 | 53.1 | 40.9 | 65.2 |
| | MEAN DAILY MAXIMUM | 62 | 38.1 | 44.0 | 54.6 | 66.6 | 76.0 | 84.5 | 89.4 | 88.4 | 80.5 | 69.3 | 54.2 | 41.8 | 65.6 |
| | HIGHEST DAILY MAXIMUM | 62 | 74 | 84 | 89 | 95 | 97 | 107 | 110 | 110 | 109 | 96 | 85 | 73 | 110 |
| | YEAR OF OCCURRENCE | | 2003 | 1972 | 1986 | 2006 | 1998 | 1953 | 1980 | 1984 | 2000 | 2006 | 2006 | 2001 | AUG 1984 |
| | MEAN OF EXTREME MAXS. | 62 | 62.0 | 67.9 | 78.3 | 86.3 | 90.1 | 95.4 | 99.6 | 99.7 | 94.3 | 87.1 | 74.6 | 64.1 | 83.3 |
| | NORMAL DAILY MINIMUM | 30 | 17.2 | 23.0 | 32.9 | 42.9 | 53.4 | 63.2 | 67.7 | 65.4 | 55.9 | 44.3 | 32.1 | 21.8 | 43.3 |
| | MEAN DAILY MINIMUM | 62 | 17.6 | 22.8 | 31.6 | 42.9 | 53.7 | 63.2 | 67.8 | 65.8 | 56.1 | 44.4 | 32.2 | 22.1 | 43.4 |
| | LOWEST DAILY MINIMUM | 62 | -20 | -23 | -7 | 10 | 26 | 43 | 43 | 41 | 29 | 19 | 2 | -26 | -26 |
| | YEAR OF OCCURRENCE | | 1974 | 1979 | 1978 | 1975 | 1963 | 1993 | 1972 | 1988 | 1984 | 1993 | 1976 | 1989 | DEC 1989 |
| | MEAN OF EXTREME MINS. | 62 | -2.7 | 2.4 | 12.7 | 26.3 | 37.5 | 49.8 | 55.4 | 53.0 | 38.7 | 27.3 | 15.6 | 2.5 | 26.5 |
| | NORMAL DRY BULB | 30 | 27.2 | 33.4 | 44.2 | 54.5 | 64.4 | 73.9 | 78.4 | 76.7 | 68.1 | 56.6 | 42.6 | 31.4 | 54.3 |
| | MEAN DRY BULB | 62 | 27.9 | 33.5 | 43.1 | 54.8 | 64.9 | 74.0 | 78.6 | 77.1 | 68.3 | 56.8 | 43.2 | 32.0 | 54.5 |
| | MEAN WET BULB | 26 | 25.7 | 29.3 | 38.1 | 47.5 | 57.9 | 66.4 | 70.3 | 68.9 | 60.6 | 49.6 | 38.0 | 28.4 | 48.4 |
| | MEAN DEW POINT | 26 | 21.8 | 25.3 | 33.1 | 42.9 | 54.6 | 63.8 | 67.7 | 66.3 | 57.5 | 45.8 | 34.2 | 24.7 | 44.8 |
| | NORMAL NO. DAYS WITH: | | | | | | | | | | | | | | |
| | MAXIMUM >= 90 | 30 | 0.0 | 0.0 | 0.0 | 0.4 | 1.1 | 7.9 | 15.7 | 14.0 | 5.5 | 0.5 | 0.0 | 0.0 | 45.1 |
| | MAXIMUM <= 32 | 30 | 10.6 | 6.7 | 1.3 | * | 0.0 | 0.0 | 0.0 | 0.0 | 0.0 | 0.0 | 1.4 | 6.7 | 26.7 |
| | MINIMUM <= 32 | 30 | 28.9 | 22.3 | 15.2 | 4.0 | 0.2 | 0.0 | 0.0 | 0.0 | 0.2 | 3.8 | 16.1 | 26.6 | 117.3 |
| | MINIMUM <= 0 | 30 | 3.3 | 2.1 | 0.1 | 0.0 | 0.0 | 0.0 | 0.0 | 0.0 | 0.0 | 0.0 | 0.0 | 1.5 | 7.0 |
| **H/C** | NORMAL HEATING DEG. DAYS | 30 | 1174 | 898 | 647 | 336 | 106 | 7 | 1 | 1 | 73 | 287 | 665 | 1030 | 5225 |
| | NORMAL COOLING DEG. DAYS | 30 | 0 | 0 | 3 | 22 | 85 | 278 | 419 | 357 | 166 | 26 | 1 | 0 | 1357 |
| **RH** | NORMAL (PERCENT) | 30 | 71 | 70 | 66 | 65 | 70 | 71 | 70 | 72 | 71 | 68 | 71 | 73 | 70 |
| | HOUR 00 LST | 30 | 77 | 76 | 73 | 73 | 80 | 81 | 80 | 81 | 81 | 77 | 78 | 79 | 78 |
| | HOUR 06 LST | 30 | 79 | 79 | 79 | 81 | 86 | 87 | 87 | 88 | 88 | 83 | 81 | 81 | 83 |
| | HOUR 12 LST | 30 | 64 | 62 | 57 | 55 | 59 | 58 | 59 | 58 | 56 | 54 | 61 | 66 | 59 |
| | HOUR 18 LST | 30 | 66 | 61 | 54 | 52 | 57 | 57 | 56 | 58 | 58 | 57 | 65 | 69 | 59 |
| **S** | PERCENT POSSIBLE SUNSHINE | 51 | 57 | 55 | 57 | 58 | 61 | 67 | 71 | 71 | 66 | 64 | 56 | 51 | 61 |
| **W/O** | MEAN NO. DAYS WITH: | | | | | | | | | | | | | | |
| | HEAVY FOG(VISBY <= 1/4 MI) | 46 | 1.9 | 1.7 | 1.0 | 1.1 | 1.2 | 0.8 | 0.8 | 1.5 | 1.7 | 2.0 | 1.5 | 2.2 | 17.4 |
| | THUNDERSTORMS | 62 | 0.4 | 0.7 | 2.5 | 5.2 | 8.7 | 9.7 | 8.4 | 7.8 | 5.9 | 3.4 | 1.2 | 0.4 | 54.3 |
| **CLOUDNESS** | MEAN: | | | | | | | | | | | | | | |
| | SUNRISE-SUNSET (OKTAS) | 1 | 7.2 | 5.6 | 4.0 | 6.4 | 4.8 | 4.0 | 5.2 | 3.2 | 2.8 | 4.4 | 4.4 | 3.6 | 4.6 |
| | MIDNIGHT-MIDNIGHT (OKTAS) | 1 | 7.2 | 5.6 | 4.0 | 6.4 | 4.8 | 4.0 | 5.2 | 3.2 | 2.8 | 4.8 | 4.4 | 2.4 | 4.6 |
| | MEAN NO. DAYS WITH: | | | | | | | | | | | | | | |
| | CLEAR | 3 | 4.3 | 8.7 | 7.0 | 8.0 | 11.0 | 11.0 | 6.5 | 12.0 | 7.0 | 10.0 | 6.5 | 10.5 | 102.5 |
| | PARTLY CLOUDY | 3 | 4.0 | 3.3 | 5.0 | 4.0 | 4.3 | 6.0 | 9.0 | 7.0 | 1.0 | 2.0 | 2.5 | 3.0 | 51.1 |
| | CLOUDY | 3 | 9.7 | 7.3 | 6.0 | 9.5 | 6.3 | 5.3 | 5.5 | 4.5 | 4.5 | 7.5 | 8.5 | 8.0 | 82.6 |
| **PR** | MEAN STATION PRESSURE(IN) | 26 | 29.17 | 29.15 | 29.08 | 29.00 | 28.99 | 29.00 | 29.03 | 29.06 | 29.08 | 29.10 | 29.12 | 29.17 | 29.08 |
| | MEAN SEA-LEVEL PRES. (IN) | 26 | 30.15 | 30.12 | 30.03 | 29.94 | 29.92 | 29.92 | 29.95 | 29.98 | 30.01 | 30.04 | 30.07 | 30.14 | 30.02 |
| **WINDS** | MEAN SPEED (MPH) | 26 | 8.4 | 8.5 | 9.9 | 10.0 | 8.5 | 7.9 | 7.0 | 6.4 | 7.1 | 7.8 | 8.5 | 8.0 | 8.2 |
| | PREVAIL.DIR(TENS OF DEGS) | 37 | 36 | 36 | 36 | 19 | 19 | 19 | 19 | 19 | 19 | 19 | 19 | 31 | 19 |
| | MAXIMUM 2-MINUTE: | | | | | | | | | | | | | | |
| | SPEED (MPH) | 16 | 39 | 37 | 44 | 43 | 47 | 54 | 44 | 44 | 43 | 37 | 45 | 37 | 54 |
| | DIR. (TENS OF DEGS) | | 31 | 18 | 17 | 21 | 34 | 36 | 34 | 24 | 32 | 14 | 31 | 33 | 36 |
| | YEAR OF OCCURRENCE | | 1996 | 2009 | 2009 | 2001 | 1999 | 2004 | 1996 | 2005 | 2000 | 2001 | 1997 | 2000 | JUN 2004 |
| | MAXIMUM 3-SECOND | | | | | | | | | | | | | | |
| | SPEED (MPH) | 16 | 49 | 51 | 58 | 61 | 59 | 74 | 62 | 67 | 52 | 52 | 54 | 49 | 74 |
| | DIR. (TENS OF DEGS) | | 31 | 18 | 16 | 23 | 34 | 36 | 33 | 24 | 32 | 23 | 30 | 33 | 36 |
| | YEAR OF OCCURRENCE | | 1996 | 2009 | 2009 | 2008 | 1999 | 2004 | 1994 | 2005 | 2000 | 1996 | 1997 | 2003 | JUN 2004 |
| **PRECIPITATION** | NORMAL (IN) | 30 | 0.95 | 1.18 | 2.56 | 3.14 | 4.86 | 4.88 | 3.83 | 3.81 | 3.71 | 2.99 | 2.31 | 1.42 | 35.64 |
| | MAXIMUM MONTHLY (IN) | 62 | 5.24 | 3.49 | 8.44 | 8.69 | 10.25 | 15.20 | 12.02 | 11.18 | 12.71 | 7.24 | 6.27 | 4.30 | 15.20 |
| | YEAR OF OCCURRENCE | | 1949 | 1971 | 1973 | 1999 | 2007 | 1967 | 1950 | 1977 | 1973 | 1980 | 1964 | 1973 | JUN 1967 |
| | MINIMUM MONTHLY (IN) | 62 | T | 0.02 | 0.10 | 0.62 | 0.41 | 0.56 | 0.59 | 0.26 | 0.66 | 0.04 | T | 0.04 | T |
| | YEAR OF OCCURRENCE | | 1986 | 2006 | 1966 | 1989 | 1966 | 1980 | 1983 | 1971 | 1952 | 1952 | 1989 | 1996 | NOV 1989 |
| | MAXIMUM IN 24 HOURS (IN) | 62 | 1.55 | 2.33 | 3.76 | 3.59 | 7.47 | 5.52 | 4.19 | 4.48 | 5.61 | 4.10 | 4.66 | 2.65 | 7.47 |
| | YEAR OF OCCURRENCE | | 1988 | 1971 | 1987 | 1967 | 2007 | 1967 | 1951 | 1962 | 2005 | 1985 | 1964 | 1980 | MAY 2007 |
| | NORMAL NO. DAYS WITH: | | | | | | | | | | | | | | |
| | PRECIPITATION >= 0.01 | 30 | 6.2 | 6.1 | 9.2 | 10.1 | 11.8 | 10.5 | 8.6 | 8.7 | 7.9 | 7.2 | 7.3 | 6.4 | 100.0 |
| | PRECIPITATION >= 1.00 | 30 | 0.1 | 0.1 | 0.4 | 0.7 | 1.5 | 1.4 | 1.3 | 1.4 | 1.1 | 1.0 | 0.5 | 0.2 | 9.7 |
| **SNOWFALL** | NORMAL (IN) | 30 | 6.1 | 4.9 | 2.2 | 0.5 | 0.0 | 0.0 | 0.0 | 0.0 | 0.0 | 0.0 | 1.5 | 4.8 | 20.3 |
| | MAXIMUM MONTHLY (IN) | 62 | 23.0 | 22.4 | 22.1 | 6.8 | T | T | T | T | T | 8.0 | 9.4 | 19.2 | 23.0 |
| | YEAR OF OCCURRENCE | | 1993 | 1971 | 1960 | 1970 | 2006 | 2009 | 2008 | 1994 | 2008 | 1996 | 1972 | 2009 | JAN 1993 |
| | MAXIMUM IN 24 HOURS (IN) | 62 | 15.2 | 15.2 | 8.4 | 7.6 | T | T | T | T | T | 8.0 | 7.4 | 9.0 | 15.2 |
| | YEAR OF OCCURRENCE' | | 1993 | 1971 | 1960 | 1970 | 1991 | 2009 | 1992 | 1994 | 2002 | 1996 | 1975 | 1973 | JAN 1993 |
| | MAXIMUM SNOW DEPTH (IN) | 64 | 12 | 12 | 18 | 4 | 0 | T | 0 | T | 0 | 6 | 8 | 9 | 18 |
| | YEAR OF OCCURRENCE | | 1979 | 1971 | 1960 | 1970 | | 1993 | | 1949 | | 1996 | 1975 | 1983 | MAR 1960 |
| | NORMAL NO. DAYS WITH: | | | | | | | | | | | | | | |
| | SNOWFALL >= 1.0 | 30 | 1.9 | 1.5 | 0.7 | 0.2 | 0.0 | 0.0 | 0.0 | 0.0 | 0.0 | 0.0 | 0.6 | 1.6 | 6.5 |

## PRECIPITATION (inches) 2009  TOPEKA (KTOP)

| YEAR | JAN | FEB | MAR | APR | MAY | JUN | JUL | AUG | SEP | OCT | NOV | DEC | ANNUAL |
|------|-----|-----|-----|-----|-----|-----|-----|-----|-----|-----|-----|-----|--------|
| 1980 | 1.34 | 0.91 | 4.15 | 1.03 | 4.85 | 0.56 | 0.87 | 5.86 | 1.19 | 7.24 | 0.25 | 3.86 | 32.11 |
| 1981 | 0.32 | 0.21 | 1.61 | 1.98 | 5.93 | 9.40 | 7.63 | 3.92 | 2.03 | 3.72 | 3.63 | 0.22 | 40.60 |
| 1982 | 1.67 | 0.59 | 1.14 | 1.58 | 9.39 | 5.99 | 5.08 | 4.53 | 1.17 | 1.25 | 2.26 | 3.61 | 38.26 |
| 1983 | 0.69 | 0.63 | 4.39 | 6.29 | 4.93 | 6.08 | 0.59 | 0.62 | 2.25 | 5.19 | 3.61 | 1.34 | 36.61 |
| 1984 | 0.11 | 1.35 | 4.57 | 4.26 | 3.45 | 10.17 | 1.66 | 1.04 | 4.24 | 4.10 | 0.72 | 2.36 | 38.03 |
| 1985 | 0.70 | 2.02 | 2.38 | 3.60 | 3.79 | 5.15 | 2.90 | 7.97 | 8.16 | 5.20 | 2.02 | 0.71 | 44.60 |
| 1986 | T | 1.55 | 1.35 | 3.15 | 7.53 | 2.51 | 4.21 | 5.50 | 6.21 | 3.30 | 0.87 | 1.20 | 37.38 |
| 1987 | 1.09 | 2.71 | 5.92 | 2.33 | 3.89 | 4.86 | 2.78 | 5.90 | 1.81 | 1.86 | 1.94 | 1.87 | 36.96 |
| 1988 | 2.04 | 0.48 | 0.73 | 2.93 | 3.08 | 3.13 | 1.74 | 1.34 | 1.94 | 0.26 | 0.86 | 0.86 | 19.39 |
| 1989 | 1.24 | 0.86 | 3.11 | 0.62 | 4.05 | 4.76 | 5.21 | 6.22 | 8.65 | 3.44 | T | 0.61 | 38.77 |
| 1990 | 1.22 | 2.31 | 3.75 | 1.01 | 4.45 | 5.57 | 3.01 | 5.69 | 0.83 | 2.71 | 2.91 | 0.97 | 34.43 |
| 1991 | 0.76 | 0.02 | 2.98 | 3.63 | 7.09 | 1.49 | 1.47 | 1.76 | 2.15 | 3.20 | 2.20 | 2.44 | 29.19 |
| 1992 | 0.89 | 1.18 | 5.29 | 3.25 | 1.75 | 3.35 | 6.37 | 1.24 | 3.92 | 1.41 | 5.27 | 2.01 | 35.93 |
| 1993 | 1.11 | 1.61 | 2.56 | 5.43 | 6.95 | 2.18 | 10.98 | 5.32 | 7.03 | 1.37 | 1.12 | 0.90 | 46.56 |
| 1994 | 0.42 | 0.82 | 0.19 | 4.31 | 0.95 | 4.63 | 3.16 | 7.87 | 1.46 | 1.30 | 2.87 | 1.52 | 29.50 |
| 1995 | 1.50 | 0.71 | 2.11 | 3.32 | 11.82 | 3.43 | 5.10 | 4.29 | 2.90 | 0.21 | 0.66 | 0.57 | 36.62 |
| 1996 | 0.76 | 0.19 | 1.48 | 1.57 | 7.72 | 7.97 | 2.65 | 6.09 | 3.60 | 2.79 | 2.66 | 0.04 | 37.52 |
| 1997 | 0.24 | 2.67 | 0.26 | 4.99 | 3.54 | 1.36 | 2.59 | 4.65 | 2.15 | 3.58 | 2.14 | 2.41 | 30.58 |
| 1998 | 0.79 | 0.77 | 2.88 | 2.16 | 2.08 | 7.22 | 9.32 | 0.88 | 4.19 | 5.01 | 5.64 | 1.22 | 42.16 |
| 1999 | 1.17 | 0.94 | 0.99 | 8.69 | 6.38 | 6.20 | 0.59 | 1.09 | 4.43 | 0.87 | 1.60 | 1.76 | 34.71 |
| 2000 | 0.19 | 2.00 | 2.62 | 1.07 | 2.08 | 7.25 | 2.77 | 0.61 | 2.97 | 3.52 | 1.91 | 0.35 | 27.34 |
| 2001 | 1.22 | 2.90 | 3.56 | 4.27 | 3.85 | 6.39 | 2.31 | 5.95 | 7.46 | 3.51 | 1.13 | 0.13 | 42.68 |
| 2002 | 1.51 | 0.75 | 0.72 | 4.64 | 4.87 | 4.12 | 0.81 | 3.05 | 1.63 | 5.42 | 0.26 | 0.05 | 27.83 |
| 2003 | 0.50 | 1.37 | 0.86 | 5.91 | 3.70 | 3.70 | 0.70 | 6.25 | 2.91 | 0.69 | 0.45 | 2.36 | 29.40 |
| 2004 | 0.84 | 1.68 | 3.83 | 1.96 | 4.46 | 6.39 | 7.27 | 4.91 | 1.68 | 3.98 | 2.44 | 0.62 | 40.06 |
| 2005 | 2.50 | 2.26 | 0.74 | 1.34 | 4.58 | 9.59 | 2.08 | 10.91 | 7.71 | 5.00 | 0.91 | 1.00 | 48.62 |
| 2006 | 0.48 | 0.02 | 2.15 | 5.13 | 3.20 | 1.18 | 3.41 | 9.04 | 2.47 | 3.04 | 0.90 | 1.70 | 32.72 |
| 2007 | 0.67 | 1.48 | 3.65 | 3.20 | 10.25 | 4.39 | 1.99 | 2.79 | 1.35 | 6.61 | 0.10 | 4.13 | 40.61 |
| 2008 | 0.65 | 3.32 | 2.58 | 2.95 | 3.55 | 7.50 | 3.67 | 1.48 | 6.17 | 3.98 | 0.88 | 1.48 | 38.21 |
| 2009 | 0.12 | 0.45 | 4.79 | 7.09 | 1.44 | 6.54 | 7.80 | 4.53 | 1.68 | 3.13 | 2.23 | 1.94 | 41.74 |
| POR=<br>62 YRS | 0.94 | 1.17 | 2.36 | 3.29 | 4.46 | 5.22 | 4.08 | 4.06 | 3.45 | 2.93 | 1.76 | 1.33 | 35.05 |

WBAN : 13996

## AVERAGE TEMPERATURE (°F) 2009  TOPEKA (KTOP)

| YEAR | JAN | FEB | MAR | APR | MAY | JUN | JUL | AUG | SEP | OCT | NOV | DEC | ANNUAL |
|------|-----|-----|-----|-----|-----|-----|-----|-----|-----|-----|-----|-----|--------|
| 1980 | 28.6 | 26.0 | 40.8 | 53.7 | 62.8 | 76.5 | 86.4 | 80.7 | 70.0 | 53.9 | 45.0 | 32.6 | 54.8 |
| 1981 | 31.4 | 35.5 | 46.1 | 60.6 | 60.9 | 75.5 | 79.5 | 73.1 | 68.0 | 56.1 | 47.2 | 30.1 | 55.3 |
| 1982 | 21.9 | 28.5 | 43.2 | 50.2 | 63.7 | 69.0 | 78.7 | 75.5 | 66.5 | 55.9 | 42.0 | 35.8 | 52.6 |
| 1983 | 32.5 | 36.1 | 44.9 | 49.4 | 62.5 | 73.5 | 81.1 | 83.0 | 72.2 | 58.7 | 45.8 | 14.4 | 54.5 |
| 1984 | 26.0 | 40.2 | 38.1 | 51.7 | 62.4 | 73.9 | 77.0 | 78.0 | 66.5 | 56.6 | 45.5 | 36.8 | 54.4 |
| 1985 | 19.9 | 25.6 | 48.6 | 58.7 | 66.5 | 72.0 | 79.7 | 72.8 | 66.8 | 56.6 | 36.7 | 25.1 | 52.4 |
| 1986 | 35.8 | 32.5 | 49.8 | 57.7 | 65.9 | 77.0 | 80.4 | 72.3 | 71.6 | 56.6 | 38.3 | 34.6 | 56.0 |
| 1987 | 29.7 | 40.3 | 46.7 | 57.1 | 70.4 | 76.2 | 78.1 | 75.5 | 68.2 | 52.6 | 47.4 | 35.9 | 56.5 |
| 1988 | 28.1 | 30.8 | 43.4 | 53.9 | 68.8 | 75.1 | 76.7 | 79.5 | 70.3 | 52.8 | 45.2 | 35.3 | 55.0 |
| 1989 | 38.0 | 22.9 | 44.4 | 57.9 | 64.2 | 71.4 | 77.6 | 74.8 | 62.3 | 57.1 | 42.3 | 21.0 | 52.8 |
| 1990 | 37.3 | 36.2 | 45.5 | 51.9 | 60.3 | 77.2 | 77.7 | 76.5 | 71.6 | 57.0 | 49.1 | 29.6 | 55.8 |
| 1991 | 25.2 |      | 48.2 | 57.7 | 69.4 | 77.1 | 80.2 | 77.3 | 69.3 | 58.6 | 37.9 | 37.4 |       |
| 1992 | 37.2 | 41.5 | 47.8 | 54.7 | 62.5 | 69.1 | 75.9 | 71.7 | 66.5 | 56.5 | 39.1 | 32.6 | 54.7 |
| 1993 | 26.4 | 29.9 | 40.8 | 50.2 | 63.1 | 72.9 | 78.2 | 77.8 | 63.4 | 54.0 | 39.3 | 34.8 | 52.6 |
| 1994 | 26.1 | 29.9 | 47.0 | 54.1 | 64.5 | 76.4 | 76.1 | 75.9 | 67.4 | 58.3 | 45.8 | 36.0 | 54.8 |
| 1995 | 29.3 | 37.0 | 45.1 | 52.0 | 59.3 | 72.5 | 80.2 | 80.9 | 65.8 | 57.2 | 40.1 | 30.6 | 54.2 |
| 1996 | 24.5 | 35.0 | 38.5 | 54.0 | 65.7 | 75.3 | 76.4 | 74.6 | 64.6 | 56.2 | 37.6 | 30.1 | 52.7 |
| 1997 | 26.2 | 35.0 | 46.2 | 49.9 | 60.5 | 73.5 | 78.5 | 75.5 | 69.7 | 58.1 | 40.3 | 32.8 | 53.9 |
| 1998 | 32.4 | 39.8 | 38.3 | 53.0 | 70.3 | 74.0 | 79.2 | 78.9 | 74.0 | 59.1 | 48.7 | 36.0 | 57.0 |
| 1999 | 28.6 | 42.9 | 43.3 | 55.1 | 64.8 | 73.4 | 82.4 | 77.6 | 65.6 | 56.9 | 51.3 | 36.2 | 56.5 |
| 2000 | 32.3 | 41.9 | 47.5 | 55.2 | 68.1 | 72.1 | 79.2 | 85.4 | 72.3 | 60.5 | 37.3 | 20.7 | 56.0 |
| 2001 | 30.3 | 30.5 | 40.8 | 60.6 | 67.6 | 73.4 | 82.9 | 79.0 | 66.5 | 56.7 | 51.1 | 37.8 | 56.4 |
| 2002 | 34.6 | 37.5 | 40.7 | 57.3 | 62.6 | 76.3 | 81.6 | 79.0 | 72.9 | 50.6 | 42.1 | 36.1 | 55.9 |
| 2003 | 28.8 | 31.3 | 44.0 | 57.9 | 64.2 | 72.1 | 82.2 | 81.0 | 64.9 | 57.9 | 43.7 | 36.2 | 55.4 |
| 2004 | 27.7 | 30.4 | 48.1 | 56.4 | 67.7 | 71.4 | 75.6 | 72.8 | 70.2 | 58.3 | 45.8 | 35.3 | 55.0 |
| 2005 | 29.2 | 38.3 | 44.7 | 57.0 | 65.0 | 76.2 | 78.6 | 77.8 | 71.9 | 57.3 | 46.4 | 29.3 | 56.0 |
| 2006 | 41.0 | 34.4 | 46.1 | 60.6 | 66.1 | 75.7 | 81.8 | 80.0 | 65.9 | 55.0 | 46.1 | 39.2 | 57.7 |
| 2007 | 29.4 | 29.9 | 53.8 | 52.6 | 68.7 | 74.2 | 79.0 | 83.5 | 71.4 | 60.0 | 44.4 | 30.7 | 56.5 |
| 2008 | 28.7 | 30.0 | 42.2 | 51.9 | 64.6 | 75.2 | 78.9 | 76.2 | 67.2 | 56.5 | 44.1 | 30.1 | 53.8 |
| 2009 | 29.5 | 38.6 | 45.9 | 54.3 | 65.7 | 76.3 | 74.7 | 74.4 | 65.8 | 50.6 | 48.8 | 28.1 | 54.4 |
| POR=<br>62 YRS | 27.9 | 33.5 | 43.1 | 54.8 | 64.9 | 74.0 | 78.6 | 77.1 | 68.3 | 56.8 | 43.2 | 32.0 | 54.5 |

## HEATING DEGREE DAYS (base 65°F) 2009 TOPEKA (KTOP)

| YEAR | JUL | AUG | SEP | OCT | NOV | DEC | JAN | FEB | MAR | APR | MAY | JUN | TOTAL |
|---|---|---|---|---|---|---|---|---|---|---|---|---|---|
| 1980-81 | 0 | 0 | 65 | 344 | 591 | 1001 | 1035 | 822 | 579 | 175 | 176 | 0 | 4788 |
| 1981-82 | 0 | 2 | 46 | 283 | 529 | 1076 | 1329 | 1014 | 664 | 449 | 76 | 32 | 5500 |
| 1982-83 | 0 | 0 | 93 | 303 | 683 | 896 | 1002 | 804 | 615 | 466 | 120 | 13 | 4995 |
| 1983-84 | 0 | 0 | 56 | 223 | 570 | 1565 | 1204 | 713 | 830 | 405 | 137 | 0 | 5703 |
| 1984-85 | 0 | 0 | 145 | 276 | 578 | 871 | 1389 | 1098 | 501 | 228 | 35 | 8 | 5129 |
| 1985-86 | 0 | 0 | 127 | 259 | 844 | 1228 | 899 | 906 | 491 | 252 | 49 | 0 | 5055 |
| 1986-87 | 0 | 9 | 27 | 263 | 792 | 934 | 1084 | 688 | 560 | 292 | 16 | 0 | 4665 |
| 1987-88 | 0 | 3 | 24 | 376 | 531 | 893 | 1136 | 988 | 662 | 331 | 16 | 5 | 4965 |
| 1988-89 | 2 | 4 | 24 | 383 | 587 | 912 | 832 | 1174 | 641 | 296 | 125 | 5 | 4985 |
| 1989-90 | 0 | 2 | 155 | 276 | 672 | 1360 | 851 | 801 | 600 | 413 | 176 | 4 | 5310 |
| 1990-91 | 1 | 1 | 39 | 276 | 477 | 1093 | 1227 |  | 523 | 233 | 48 | 0 |  |
| 1991-92 | 0 | 0 | 95 | 262 | 808 | 849 | 855 | 673 | 528 | 326 | 132 | 7 | 4535 |
| 1992-93 | 0 | 2 | 68 | 278 | 770 | 995 | 1189 | 979 | 744 | 440 | 101 | 22 | 5588 |
| 1993-94 | 0 | 1 | 108 | 356 | 763 | 930 | 1202 | 974 | 553 | 347 | 94 | 0 | 5328 |
| 1994-95 | 0 | 1 | 64 | 237 | 568 | 892 | 1097 | 774 | 613 | 382 | 193 | 1 | 4822 |
| 1995-96 | 0 | 0 | 107 | 246 | 740 | 1059 | 1250 | 867 | 814 | 347 | 93 | 6 | 5529 |
| 1996-97 | 0 | 0 | 98 | 294 | 813 | 1076 | 1193 | 833 | 577 | 450 | 173 | 0 | 5507 |
| 1997-98 | 1 | 0 | 21 | 286 | 737 | 994 | 1000 | 699 | 830 | 361 | 27 | 27 | 4983 |
| 1998-99 | 0 | 0 | 9 | 203 | 485 | 893 | 1122 | 613 | 667 | 295 | 58 | 12 | 4357 |
| 1999-00 | 0 | 0 | 92 | 261 | 408 | 886 | 1006 | 666 | 536 | 295 | 51 | 10 | 4211 |
| 2000-01 | 0 | 0 | 61 | 174 | 824 | 1368 | 1069 | 958 | 742 | 182 | 53 | 9 | 5440 |
| 2001-02 | 0 | 0 | 58 | 262 | 415 | 837 | 934 | 766 | 745 | 274 | 141 | 0 | 4432 |
| 2002-03 | 0 | 0 | 15 | 455 | 677 | 888 | 1115 | 937 | 643 | 256 | 78 | 21 | 5085 |
| 2003-04 | 0 | 0 | 90 | 229 | 629 | 885 | 1149 | 993 | 526 | 283 | 87 | 2 | 4873 |
| 2004-05 | 0 | 9 | 12 | 222 | 571 | 911 | 1105 | 740 | 626 | 269 | 107 | 0 | 4572 |
| 2005-06 | 0 | 0 | 31 | 285 | 550 | 1103 | 736 | 852 | 578 | 175 | 105 | 0 | 4415 |
| 2006-07 | 0 | 0 | 50 | 358 | 561 | 794 | 1097 | 977 | 364 | 388 | 22 | 0 | 4611 |
| 2007-08 | 0 | 0 | 28 | 213 | 618 | 1057 | 1118 | 1011 | 700 | 395 | 99 | 0 | 5239 |
| 2008-09 | 0 | 0 | 45 | 286 | 626 | 1077 | 1093 | 732 | 590 | 338 | 83 | 2 | 4872 |
| 2009- | 0 | 9 | 52 | 442 | 480 | 1136 |  |  |  |  |  |  |  |

WBAN : 13996

## COOLING DEGREE DAYS (base 65°F) 2009 TOPEKA (KTOP)

| YEAR | JAN | FEB | MAR | APR | MAY | JUN | JUL | AUG | SEP | OCT | NOV | DEC | TOTAL |
|---|---|---|---|---|---|---|---|---|---|---|---|---|---|
| 1980 | 0 | 0 | 0 | 9 | 69 | 356 | 670 | 496 | 220 | 9 | 0 | 0 | 1829 |
| 1981 | 0 | 0 | 0 | 53 | 58 | 321 | 457 | 260 | 143 | 17 | 0 | 0 | 1309 |
| 1982 | 0 | 0 | 0 | 11 | 43 | 157 | 432 | 334 | 147 | 28 | 0 | 0 | 1152 |
| 1983 | 0 | 0 | 0 | 7 | 50 | 274 | 509 | 564 | 278 | 33 | 2 | 0 | 1717 |
| 1984 | 0 | 0 | 0 | 14 | 67 | 274 | 379 | 407 | 196 | 20 | 0 | 3 | 1360 |
| 1985 | 0 | 0 | 0 | 46 | 88 | 225 | 461 | 249 | 188 | 6 | 0 | 0 | 1263 |
| 1986 | 0 | 0 | 26 | 42 | 85 | 363 | 488 | 243 | 233 | 9 | 0 | 0 | 1489 |
| 1987 | 0 | 0 | 0 | 61 | 192 | 344 | 410 | 335 | 126 | 0 | 9 | 0 | 1477 |
| 1988 | 0 | 0 | 0 | 4 | 140 | 314 | 375 | 458 | 191 | 11 | 0 | 0 | 1493 |
| 1989 | 0 | 0 | 11 | 90 | 107 | 206 | 399 | 311 | 81 | 41 | 0 | 0 | 1246 |
| 1990 | 0 | 0 | 1 | 26 | 39 | 377 | 403 | 366 | 241 | 37 | 7 | 0 | 1497 |
| 1991 | 0 | 0 | 11 | 22 | 192 | 371 | 478 | 387 | 229 | 69 | 0 | 0 | 1759 |
| 1992 | 0 | 0 | 0 | 25 | 61 | 134 | 344 | 217 | 162 | 20 | 0 | 0 | 963 |
| 1993 | 0 | 0 | 0 | 0 | 48 | 269 | 417 | 405 | 64 | 22 | 0 | 0 | 1225 |
| 1994 | 0 | 0 | 0 | 26 | 86 | 348 | 351 | 345 | 140 | 36 | 0 | 0 | 1332 |
| 1995 | 0 | 0 | 4 | 1 | 22 | 237 | 481 | 502 | 140 | 16 | 0 | 0 | 1403 |
| 1996 | 0 | 0 | 0 | 25 | 125 | 321 | 358 | 302 | 92 | 28 | 0 | 0 | 1251 |
| 1997 | 0 | 0 | 0 | 1 | 38 | 262 | 424 | 330 | 166 | 81 | 0 | 0 | 1302 |
| 1998 | 0 | 0 | 10 | 9 | 196 | 304 | 445 | 440 | 287 | 26 | 1 | 0 | 1718 |
| 1999 | 0 | 0 | 0 | 5 | 57 | 269 | 545 | 396 | 120 | 20 | 1 | 0 | 1413 |
| 2000 | 0 | 0 | 2 | 6 | 153 | 231 | 446 | 640 | 284 | 44 | 0 | 0 | 1806 |
| 2001 | 0 | 0 | 0 | 59 | 140 | 267 | 563 | 441 | 109 | 12 | 5 | 0 | 1596 |
| 2002 | 0 | 0 | 0 | 50 | 75 | 347 | 519 | 437 | 261 | 17 | 0 | 0 | 1706 |
| 2003 | 0 | 0 | 0 | 52 | 60 | 240 | 541 | 502 | 92 | 12 | 0 | 0 | 1499 |
| 2004 | 0 | 0 | 6 | 29 | 175 | 198 | 337 | 259 | 176 | 24 | 0 | 0 | 1204 |
| 2005 | 0 | 0 | 1 | 34 | 114 | 344 | 432 | 405 | 243 | 54 | 1 | 0 | 1628 |
| 2006 | 0 | 0 | 0 | 50 | 144 | 327 | 526 | 471 | 77 | 54 | 0 | 0 | 1649 |
| 2007 | 0 | 0 | 24 | 22 | 142 | 284 | 442 | 581 | 228 | 63 | 6 | 0 | 1792 |
| 2008 | 0 | 0 | 0 | 7 | 93 | 312 | 438 | 355 | 119 | 25 | 5 | 0 | 1354 |
| 2009 | 0 | 0 | 6 | 22 | 112 | 349 | 305 | 308 | 83 | 2 | 2 | 0 | 1189 |

## SNOWFALL (inches)  2009  TOPEKA (KTOP)

| YEAR | JUL | AUG | SEP | OCT | NOV | DEC | JAN | FEB | MAR | APR | MAY | JUN | TOTAL |
|------|-----|-----|-----|-----|-----|-----|-----|-----|-----|-----|-----|-----|-------|
| 1980-81 | 0.0 | 0.0 | 0.0 | T | 0.0 | 3.8 | 2.6 | 2.5 | 0.0 | 0.0 | 0.0 | 0.0 | 8.9 |
| 1981-82 | 0.0 | 0.0 | 0.0 | 0.0 | T | 1.4 | 3.2 | 8.0 | 0.3 | 0.5 | 0.0 | 0.0 | 13.4 |
| 1982-83 | 0.0 | 0.0 | 0.0 | 0.0 | 1.1 | 5.0 | 6.1 | 10.1 | 0.6 | 4.5 | 0.0 | 0.0 | 27.4 |
| 1983-84 | 0.0 | 0.0 | 0.0 | 0.0 | 4.1 | 18.8 | 2.6 | T | 4.2 | 0.0 | 0.0 | 0.0 | 29.7 |
| 1984-85 | 0.0 | 0.0 | 0.0 | 0.0 | T | 9.8 | 18.2 | 7.9 | 0.5 | 0.0 | 0.0 | 0.0 | 36.4 |
| 1985-86 | 0.0 | 0.0 | 0.0 | 0.0 | 3.3 | 5.8 | T | 1.5 | T | 0.0 | 0.0 | 0.0 | 10.6 |
| 1986-87 | 0.0 | 0.0 | 0.0 | T | 0.7 | 1.7 | 15.1 | 2.3 | 0.5 | 0.0 | 0.0 | 0.0 | 20.3 |
| 1987-88 | 0.0 | 0.0 | 0.0 | 0.0 | 0.9 | 9.6 | 0.6 | 6.0 | 4.7 | 0.0 | 0.0 | 0.0 | 21.8 |
| 1988-89 | 0.0 | 0.0 | 0.0 | 0.0 | 0.7 | 0.8 | T | 9.0 | 1.6 | 0.0 | T | T | 12.1 |
| 1989-90 | 0.0 | 0.0 | 0.0 | 0.0 | T | 9.5 | 1.0 | 0.1 | 7.6 | 0.0 | 0.0 | 0.0 | 18.2 |
| 1990-91 | 0.0 | 0.0 | 0.0 | 0.0 | 0.0 | 2.9 | 9.6 | | T | T | T | 0.0 | |
| 1991-92 | 0.0 | 0.0 | 0.0 | T | 6.2 | 0.1 | T | 0.9 | T | 0.0 | T | 7.2 |
| 1992-93 | T | 0.0 | 0.0 | T | 4.5 | 0.9 | 23.0 | 14.2 | 0.6 | T | | T | |
| 1993-94 | 0.0 | 0.0 | 0.0 | T | T | 3.3 | 2.0 | 6.0 | 1.9 | 1.4 | 0.0 | 0.0 | 14.6 |
| 1994-95 | 0.0 | T | 0.0 | 0.0 | T | 0.8 | 2.0 | 0.1 | 5.0 | | | 0.0 | |
| 1995-96 | 0.0 | 0.0 | 0.0 | 0.0 | 0.7 | 5.5 | 8.3 | T | T | T | 0.0 | T | 14.5 |
| 1996-97 | 0.0 | 0.0 | 0.0 | 8.0 | 1.1 | 0.8 | 3.9 | 5.8 | T | 1.6 | T | 0.0 | 21.2 |
| 1997-98 | 0.0 | 0.0 | 0.0 | T | T | 8.2 | 2.0 | T | 4.4 | T | 0.0 | T | 14.6 |
| 1998-99 | 0.0 | 0.0 | 0.0 | 0.0 | 0.0 | 0.5 | 4.0 | 2.0 | 1.8 | T | T | 0.0 | 8.3 |
| 1999-00 | 0.0 | 0.0 | 0.0 | 0.0 | 0.0 | 7.6 | 3.7 | 0.2 | T | 0.0 | 0.0 | 0.0 | 11.5 |
| 2000-01 | 0.0 | 0.0 | 0.0 | 0.0 | T | 8.3 | 1.2 | 8.3 | 1.3 | T | T | 0.0 | 19.1 |
| 2001-02 | 0.0 | 0.0 | 0.0 | 0.0 | 0.0 | T | 6.2 | 0.4 | 2.8 | 0.0 | T | 0.0 | 9.4 |
| 2002-03 | 0.0 | 0.0 | T | T | T | 0.4 | 5.4 | 6.4 | 0.5 | T | 0.0 | 0.0 | 12.7 |
| 2003-04 | 0.0 | 0.0 | 0.0 | 0.0 | 0.0 | 3.8 | 2.0 | 13.3 | 0.0 | 0.0 | 0.0 | 0.0 | 19.1 |
| 2004-05 | 0.0 | 0.0 | 0.0 | 0.0 | 4.5 | 0.0 | 3.5 | 3.2 | T | 0.0 | 0.0 | T | 11.2 |
| 2005-06 | 0.0 | 0.0 | 0.0 | 0.0 | 0.5 | 10.8 | 0.5 | 0.1 | 0.6 | T | T | 0.0 | 12.5 |
| 2006-07 | 0.0 | 0.0 | 0.0 | 0.0 | 0.4 | 0.1 | 4.9 | 4.2 | T | T | 0.0 | 0.0 | 9.6 |
| 2007-08 | 0.0 | 0.0 | 0.0 | 0.0 | 0.8 | 16.5 | 4.7 | 10.5 | 0.1 | T | 0.0 | 0.0 | 32.6 |
| 2008-09 | T | 0.0 | T | T | 0.5 | 5.1 | 1.6 | 2.5 | 0.9 | T | 0.0 | T | 10.6 |
| 2009- | 0.0 | 0.0 | 0.0 | 0.0 | T | 19.2 | | | | | | | |
| POR=<br>62 YRS | T | T | T | 0.1 | 1.1 | 5.0 | 5.4 | 4.5 | 3.0 | 0.5 | T | T | 19.6 |

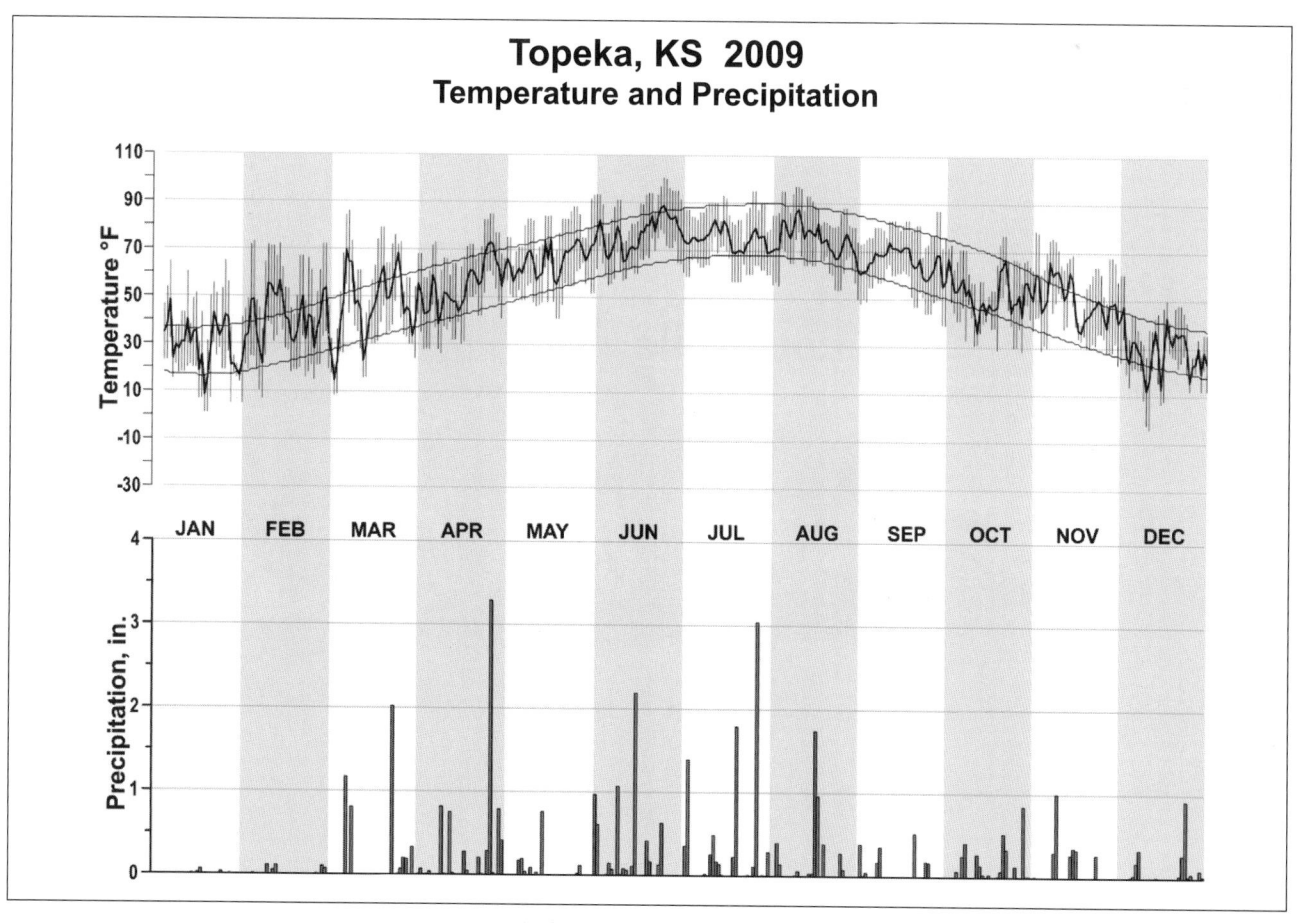

## Topeka, KS  2009
### Temperature and Precipitation

# 2009
# JACKSON
# KENTUCKY (KJKL)

Jackson, County Seat of Breathitt County, is located on the leading edge of the Eastern Kentucky Coal Fields. The topography of Breathitt County is mountainous, with 80 to 90 percent of the county area on a greater than 20 percent slope. The county contains a minimal portion of level land. Almost 80 percent of the land area is composed of slopes greater than 50 percent. The highest elevation is 1,547 feet above sea level. The terrain slopes gently westward into the beautiful Kentucky Bluegrass Region. To the east the mountains rise swiftly to heights of 4,000 to 5,000 feet above sea level.

The major industry in Breathitt County is coal. In addition, the county is rich in such natural resources as timber, petroleum, natural gas, sand, and clay. The roughness of the terrain effectively limits farming to small yields of tobacco and garden vegetables.

The climate of Jackson and Eastern Kentucky is temperate and well suited to a variety of plant and animal life. There are no large bodies of water close enough to have any significant effect on the climate. The North Fork of the Kentucky River flows through Breathitt County westward into the Kentucky River and eventually into the Ohio River. There are numerous small creeks and streams in the county. The steep slopes and narrow valleys make these creeks and streams especially prone to flash flooding during periods of heavy rainfall.

Jackson is subject to sudden and large changes in temperature. Extremes of cold and heat are rare and usually of short duration. Temperatures above 100 degrees or below zero are extremely rare. Average daily temperatures range from about 32 degrees in the winter to the low 70s in the summer, and in the low 50s during the spring and fall months. January is the coldest month with an average temperature of 31 degrees. The warmest month is July, with an average temperature of 73 degrees.

Total annual precipitation for the Jackson area averages nearly 44 inches and is fairly evenly distributed throughout the year. The spring and summer seasons average nearly 12 inches each, while winter averages 11 inches and fall slightly over 8 inches. July is the wettest month with an average of nearly 5 inches of precipitation. Snowfall amounts are variable and snow cover is normally limited to only a few days at a time.

# NORMALS, MEANS, AND EXTREMES
## JACKSON (KJKL)

LATITUDE: 37° 35'N   LONGITUDE: -83° 18'W   ELEVATION (FT): GRND: 1330   BARO: 1356   TIME ZONE: EASTERN   (UTC -5)   WBAN: 03889

| | ELEMENT | POR | JAN | FEB | MAR | APR | MAY | JUN | JUL | AUG | SEP | OCT | NOV | DEC | YEAR |
|---|---|---|---|---|---|---|---|---|---|---|---|---|---|---|---|
| TEMPERATURE °F | NORMAL DAILY MAXIMUM | 30 | 42.0 | 46.8 | 56.8 | 66.8 | 73.8 | 80.8 | 84.2 | 83.3 | 77.4 | 67.5 | 56.4 | 46.3 | 65.2 |
| | MEAN DAILY MAXIMUM | 29 | 43.1 | 48.1 | 57.5 | 67.9 | 74.6 | 81.5 | 84.6 | 84.0 | 77.9 | 68.0 | 57.3 | 46.4 | 65.9 |
| | HIGHEST DAILY MAXIMUM | 29 | 78 | 79 | 87 | 92 | 91 | 99 | 101 | 101 | 95 | 89 | 82 | 79 | 101 |
| | YEAR OF OCCURRENCE | | 2002 | 1996 | 1989 | 1986 | 2006 | 1988 | 1988 | 1988 | 2007 | 2007 | 2004 | 1982 | AUG 1988 |
| | MEAN OF EXTREME MAXS. | 29 | 66.0 | 70.7 | 79.2 | 85.7 | 86.3 | 90.0 | 92.2 | 92.3 | 88.7 | 82.3 | 76.3 | 66.8 | 81.4 |
| | NORMAL DAILY MINIMUM | 30 | 25.7 | 28.9 | 37.4 | 45.8 | 54.3 | 61.9 | 65.7 | 64.3 | 58.4 | 47.4 | 38.9 | 30.2 | 46.6 |
| | MEAN DAILY MINIMUM | 29 | 26.9 | 30.2 | 37.4 | 46.6 | 54.6 | 62.5 | 66.1 | 65.0 | 58.3 | 48.1 | 39.5 | 30.5 | 47.1 |
| | LOWEST DAILY MINIMUM | 29 | -18 | -8 | 7 | 20 | 32 | 44 | 52 | 45 | 34 | 26 | 13 | -13 | -18 |
| | YEAR OF OCCURRENCE | | 1994 | 1996 | 1996 | 1982 | 2004 | 1998 | 1983 | 1989 | 1989 | 1987 | 1986 | 1989 | JAN 1994 |
| | MEAN OF EXTREME MINS. | 29 | 4.7 | 10.1 | 18.6 | 28.6 | 40.4 | 50.9 | 57.1 | 55.0 | 43.0 | 32.3 | 22.9 | 10.7 | 31.2 |
| | NORMAL DRY BULB | 30 | 33.9 | 37.9 | 47.1 | 56.3 | 64.1 | 71.4 | 75.0 | 73.8 | 67.9 | 57.5 | 47.7 | 38.3 | 55.9 |
| | MEAN DRY BULB | 29 | 35.2 | 39.2 | 47.5 | 57.2 | 64.6 | 72.0 | 75.4 | 74.5 | 68.1 | 58.1 | 48.4 | 38.5 | 56.6 |
| | MEAN WET BULB | 26 | 30.5 | 32.8 | 39.1 | 47.3 | 56.7 | 64.9 | 68.3 | 67.2 | 60.9 | 50.5 | 41.1 | 33.4 | 49.4 |
| | MEAN DEW POINT | 26 | 26.3 | 28.0 | 33.4 | 41.3 | 53.5 | 62.7 | 66.6 | 65.4 | 58.4 | 46.6 | 36.3 | 29.3 | 45.7 |
| | NORMAL NO. DAYS WITH: | | | | | | | | | | | | | | |
| | MAXIMUM >= 90 | 30 | 0.0 | 0.0 | 0.0 | 0.2 | 0.1 | 1.9 | 6.8 | 4.6 | 1.2 | 0.0 | 0.0 | 0.0 | 14.8 |
| | MAXIMUM <= 32 | 30 | 6.5 | 3.6 | 0.7 | 0.1 | 0.0 | 0.0 | 0.0 | 0.0 | 0.0 | 0.0 | 0.4 | 5.2 | 16.5 |
| | MINIMUM <= 32 | 30 | 22.5 | 16.3 | 12.0 | 2.5 | 0.1 | 0.0 | 0.0 | 0.0 | 0.0 | 1.1 | 9.0 | 17.8 | 81.3 |
| | MINIMUM <= 0 | 30 | 0.7 | 0.2 | 0.0 | 0.0 | 0.0 | 0.0 | 0.0 | 0.0 | 0.0 | 0.0 | 0.0 | 0.5 | 1.4 |
| H/C | NORMAL HEATING DEG. DAYS | 30 | 966 | 761 | 557 | 273 | 128 | 10 | 0 | 4 | 44 | 263 | 522 | 830 | 4358 |
| | NORMAL COOLING DEG. DAYS | 30 | 0 | 0 | 0 | 11 | 100 | 201 | 310 | 277 | 130 | 29 | 1 | 0 | 1059 |
| RH | NORMAL (PERCENT) | 30 | 70 | 67 | 61 | 57 | 69 | 75 | 77 | 77 | 74 | 68 | 66 | 71 | 69 |
| | HOUR 01 LST | 30 | 73 | 70 | 64 | 62 | 75 | 82 | 85 | 85 | 82 | 74 | 69 | 74 | 75 |
| | HOUR 07 LST | 30 | 77 | 76 | 72 | 70 | 81 | 86 | 89 | 91 | 89 | 83 | 75 | 78 | 81 |
| | HOUR 13 LST | 30 | 64 | 61 | 53 | 48 | 57 | 62 | 63 | 62 | 60 | 55 | 58 | 65 | 59 |
| | HOUR 19 LST | 30 | 66 | 62 | 54 | 50 | 62 | 69 | 72 | 71 | 69 | 61 | 62 | 67 | 64 |
| S | PERCENT POSSIBLE SUNSHINE | 3 | 43 | 42 | 51 | 53 | 50 | 58 | 58 | 59 | 53 | 56 | 35 | 26 | 49 |
| W/O | MEAN NO. DAYS WITH: | | | | | | | | | | | | | | |
| | HEAVY FOG(VISBY <= 1/4 MI) | 29 | 4.2 | 4.1 | 4.0 | 2.3 | 6.1 | 7.5 | 8.5 | 8.0 | 6.2 | 4.4 | 3.5 | 4.5 | 63.3 |
| | THUNDERSTORMS | 29 | 0.7 | 1.0 | 2.8 | 4.2 | 7.9 | 9.4 | 10.3 | 7.4 | 3.0 | 1.4 | 1.1 | 0.3 | 49.5 |
| CLOUDNESS | MEAN: | | | | | | | | | | | | | | |
| | SUNRISE-SUNSET (OKTAS) | 14 | 6.2 | 6.2 | 5.8 | 5.4 | 5.5 | 5.2 | 4.5 | 4.5 | 4.4 | 4.5 | 5.3 | 5.7 | 5.3 |
| | MIDNIGHT-MIDNIGHT (OKTAS) | 15 | 5.6 | 5.4 | 5.0 | 4.7 | 4.9 | 4.5 | 4.3 | 4.1 | 3.9 | 4.1 | 4.9 | 5.5 | 4.7 |
| | MEAN NO. DAYS WITH: | | | | | | | | | | | | | | |
| | CLEAR | 1 | 2.0 | | 7.0 | | 7.0 | 10.0 | | | | | | | |
| | PARTLY CLOUDY | 1 | 1.0 | 1.0 | 2.0 | | 4.0 | 7.0 | | | | | | | |
| | CLOUDY | 1 | 1.0 | 7.0 | 9.0 | | 10.0 | 4.0 | | | | | | | |
| PR | MEAN STATION PRESSURE(IN) | 26 | 28.66 | 28.64 | 28.60 | 28.56 | 28.58 | 28.59 | 28.62 | 28.64 | 28.65 | 28.67 | 28.67 | 28.67 | 28.63 |
| | MEAN SEA-LEVEL PRES. (IN) | 26 | 30.12 | 30.10 | 30.05 | 29.99 | 30.00 | 30.00 | 30.02 | 30.04 | 30.07 | 30.10 | 30.12 | 30.14 | 30.06 |
| WINDS | MEAN SPEED (MPH) | 26 | 6.7 | 6.8 | 7.0 | 6.9 | 5.3 | 4.5 | 4.1 | 3.9 | 4.4 | 5.0 | 6.1 | 6.3 | 5.6 |
| | PREVAIL.DIR(TENS OF DEGS) | 29 | 20 | 20 | 20 | 20 | 20 | 19 | 21 | 19 | 19 | 19 | 20 | 20 | 20 |
| | MAXIMUM 2-MINUTE: | | | | | | | | | | | | | | |
| | SPEED (MPH) | 14 | 37 | 39 | 33 | 43 | 39 | 31 | 25 | 37 | 26 | 28 | 29 | 31 | 43 |
| | DIR. (TENS OF DEGS) | | 25 | 19 | 29 | 27 | 28 | 29 | 31 | 28 | 29 | 25 | 19 | 24 | 27 |
| | YEAR OF OCCURRENCE | | 1999 | 1998 | 2002 | 1996 | 1998 | 2001 | 2005 | 1999 | 2000 | 2001 | 1998 | 2009 | APR 1996 |
| | MAXIMUM 3-SECOND | | | | | | | | | | | | | | |
| | SPEED (MPH) | 14 | 56 | 58 | 56 | 57 | 47 | 45 | 40 | 51 | 40 | 44 | 45 | 58 | 58 |
| | DIR. (TENS OF DEGS) | | 25 | 18 | 29 | 30 | 28 | 34 | 32 | 32 | 33 | 25 | 23 | 23 | 23 |
| | YEAR OF OCCURRENCE | | 1999 | 1998 | 1997 | 1996 | 1998 | 2004 | 2005 | 2007 | 2005 | 2001 | 1998 | 2006 | DEC 2006 |
| PRECIPITATION | NORMAL (IN) | 30 | 3.56 | 3.68 | 4.38 | 3.79 | 5.16 | 4.67 | 4.59 | 4.13 | 3.77 | 3.18 | 4.20 | 4.27 | 49.38 |
| | MAXIMUM MONTHLY (IN) | 29 | 7.28 | 7.88 | 11.78 | 10.00 | 10.78 | 9.15 | 9.74 | 7.70 | 7.82 | 7.36 | 9.32 | 12.97 | 12.97 |
| | YEAR OF OCCURRENCE | | 1994 | 2003 | 1994 | 1998 | 2004 | 1997 | 1985 | 1993 | 1988 | 1989 | 1986 | 1990 | DEC 1990 |
| | MINIMUM MONTHLY (IN) | 29 | 0.80 | 1.20 | 1.47 | 0.78 | 1.82 | 1.37 | 1.77 | 1.16 | .51 | 0.51 | 0.80 | 1.74 | 0.51 |
| | YEAR OF OCCURRENCE | | 1981 | 2007 | 2003 | 1985 | 2007 | 1988 | 1995 | 2008 | 2005 | 1987 | 2009 | 1985 | SEP 2005 |
| | MAXIMUM IN 24 HOURS (IN) | 29 | 5.21 | 2.50 | 3.44 | 3.11 | 4.09 | 3.04 | 2.99 | 4.12 | 3.74 | 4.24 | 4.30 | 3.43 | 5.21 |
| | YEAR OF OCCURRENCE | | 2009 | 2004 | 1989 | 2005 | 2004 | 1997 | 2004 | 1999 | 2004 | 1989 | 1986 | 1990 | JAN 2009 |
| | NORMAL NO. DAYS WITH: | | | | | | | | | | | | | | |
| | PRECIPITATION >= 0.01 | 30 | 14.3 | 13.2 | 13.8 | 12.3 | 13.6 | 12.2 | 12.1 | 9.5 | 8.7 | 8.6 | 11.6 | 14.2 | 144.1 |
| | PRECIPITATION >= 1.00 | 30 | 0.7 | 0.8 | 1.0 | 0.7 | 1.4 | 1.2 | 0.9 | 1.3 | 1.1 | 0.9 | 1.1 | 1.2 | 12.3 |
| SNOWFALL | NORMAL (IN) | 30 | 7.5 | 7.0 | 3.5 | 1.2 | 0.1 | 0.0 | 0.0 | 0.0 | 0.0 | 0.1 | 0.7 | 4.4 | 24.5 |
| | MAXIMUM MONTHLY (IN) | 29 | 26.3 | 21.0 | 21.9 | 17.8 | 1.0 | T | T | T | 0.0 | 2.3 | 5.1 | 17.6 | 26.3 |
| | YEAR OF OCCURRENCE | | 1994 | 1985 | 1993 | 1987 | 1989 | 2008 | 2007 | 1991 | | 1993 | 1995 | 2009 | JAN 1994 |
| | MAXIMUM IN 24 HOURS (IN) | 29 | 15.2 | 12.7 | 19.8 | 8.5 | 1.0 | T | T | T | 0.0 | 2.3 | 4.3 | 8.1 | 19.8 |
| | YEAR OF OCCURRENCE' | | 1994 | 1985 | 1993 | 1987 | 1989 | 1995 | 2000 | 1991 | | 1993 | 1995 | 2009 | MAR 1993 |
| | MAXIMUM SNOW DEPTH (IN) | 28 | 17 | 14 | 20 | 11 | 0 | 0 | 0 | 0 | 0 | 0 | 4 | 10 | 20 |
| | YEAR OF OCCURRENCE | | 1994 | 1985 | 1993 | 1987 | | | | | | | 1995 | 2009 | MAR 1993 |
| | NORMAL NO. DAYS WITH: | | | | | | | | | | | | | | |
| | SNOWFALL >= 1.0 | 30 | 2.0 | 2.2 | 0.7 | 0.3 | 0.1 | 0.0 | 0.0 | 0.0 | 0.0 | 0.1 | 0.2 | 1.7 | 7.3 |

## PRECIPITATION (inches) 2009 JACKSON (KJKL)

| YEAR | JAN | FEB | MAR | APR | MAY | JUN | JUL | AUG | SEP | OCT | NOV | DEC | ANNUAL |
|------|-----|-----|-----|-----|-----|-----|-----|-----|-----|-----|-----|-----|--------|
| 1981 | 0.80 | 5.96 | 2.78 | 4.62 | 4.23 | 4.88 | 4.53 | 1.55 | 2.53 | 2.83 | 1.45 | 3.53 | 39.69 |
| 1982 | 5.28 | 4.34 | 5.24 | 2.19 | 5.43 | 6.13 | 4.12 | 6.14 | 1.53 | 1.62 | 3.55 | 3.70 | 49.27 |
| 1983 | 1.68 | 2.09 | 1.90 | 3.25 | 7.34 | 2.01 | 3.81 | 3.33 | 3.12 | 3.83 | 3.27 | 2.76 | 38.39 |
| 1984 | 1.31 | 2.88 | 3.43 | 5.72 | 7.36 | 2.09 | 6.71 | 2.58 | 2.43 | 4.09 | 5.54 | 4.08 | 48.22 |
| 1985 | 3.50 | 2.02 | 3.27 | 0.78 | 5.50 | 4.19 | 9.74 | 5.24 | 1.40 | 4.97 | 6.89 | 1.74 | 49.24 |
| 1986 | 1.84 | 5.44 | 1.56 | 0.95 | 2.42 | 2.15 | 2.73 | 2.49 | 3.27 | 2.38 | 9.32 | 3.06 | 37.61 |
| 1987 | 2.70 | 3.46 | 1.90 | 3.70 | 2.25 | 3.22 | 6.37 | 2.64 | 2.92 | 0.51 | 3.15 | 5.98 | 38.80 |
| 1988 | 2.59 | 2.00 | 3.09 | 2.97 | 4.50 | 1.37 | 4.56 | 4.15 | 7.82 | 1.85 | 6.12 | 4.03 | 45.05 |
| 1989 | 3.48 | 7.61 | 6.74 | 3.23 | 6.43 | 6.96 | 2.21 | 5.22 | 7.37 | 7.36 | 4.28 | 2.40 | 63.29 |
| 1990 | 2.56 | 6.27 | 3.16 | 2.95 | 5.08 | 4.02 | 4.18 | 4.21 | 1.86 | 4.73 | 2.91 | 12.97 | 54.90 |
| 1991 | 3.16 | 4.46 | 6.08 | 2.67 | 4.99 | 7.01 | 4.46 | 2.96 | 2.62 | 2.13 | 5.22 | 9.35 | 55.11 |
| 1992 | 1.87 | 3.12 | 5.80 | 1.66 | 4.61 | 4.08 | 6.67 | 3.20 | 3.66 | 1.60 | 3.23 | 4.92 | 44.42 |
| 1993 | 2.05 | 3.54 | 5.28 | 3.26 | 3.74 | 4.82 | 4.70 | 7.70 | 6.58 | 4.63 | 5.08 | 4.28 | 55.66 |
| 1994 | 7.28 | 7.42 | 11.78 | 5.52 | 3.37 | 4.85 | 3.74 | 6.11 | 2.45 | 2.53 | 2.73 | 2.98 | 60.76 |
| 1995 | 7.16 | 3.71 | 3.51 | 4.90 | 9.91 | 4.22 | 1.77 | 2.07 | 4.01 | 5.03 | 4.18 | 2.36 | 52.83 |
| 1996 | 5.63 | 3.11 | 5.46 | 5.95 | 5.86 | 4.35 | 4.96 | 3.01 | 6.47 | 4.00 | 7.28 | 2.72 | 58.80 |
| 1997 | 3.53 | 2.97 | 9.76 | 1.51 | 5.01 | 9.15 | 2.40 | 4.38 | 2.03 | 2.29 | 4.04 | 2.21 | 49.28 |
| 1998 | 3.76 | 4.45 | 2.86 | 10.00 | 6.28 | 8.29 | 2.46 | 2.47 | 2.09 | 2.59 | 2.98 | 5.16 | 53.39 |
| 1999 | 6.55 | 3.04 | 3.17 | 3.44 | 2.47 | 2.66 | 2.75 | 6.58 | 1.13 | 3.08 | 2.65 | 2.56 | 40.08 |
| 2000 | 2.63 | 3.53 | 1.94 | 4.97 | 4.33 | 6.80 | 5.69 | 4.38 | 4.92 | 1.07 | 1.47 | 4.35 | 46.08 |
| 2001 | 2.50 | 3.72 | 2.17 | 1.69 | 4.39 | 4.19 | 6.43 | 2.41 | 1.09 | 1.41 | 1.82 | 2.55 | 34.37 |
| 2002 | 4.09 | 1.24 | 7.96 | 4.11 | 5.23 | 4.98 | 5.50 | 1.72 | 3.48 | 6.39 | 3.61 | 4.28 | 52.59 |
| 2003 | 2.10 | 7.89 | 1.47 | 5.14 | 5.98 | 7.54 | 3.95 | 5.12 | 4.33 | 2.20 | 5.49 | 3.78 | 54.99 |
| 2004 | 4.23 | 3.77 | 3.87 | 4.01 | 10.78 | 6.18 | 7.02 | 2.39 | 7.55 | 4.96 | 4.37 | 3.27 | 62.40 |
| 2005 | 5.12 | 3.03 | 3.52 | 7.47 | 2.50 | 2.78 | 4.08 | 3.92 | 0.51 | 1.57 | 2.66 | 3.18 | 40.34 |
| 2006 | 5.57 | 1.85 | 2.89 | 4.57 | 3.61 | 3.24 | 3.87 | 3.69 | 6.39 | 5.49 | 2.43 | 2.03 | 45.63 |
| 2007 | 2.83 | 1.20 | 2.71 | 3.22 | 1.82 | 2.15 | 4.05 | 2.64 | 2.49 | 3.80 | 3.37 | 5.18 | 35.46 |
| 2008 | 2.46 | 3.41 | 4.14 | 4.00 | 3.24 | 3.94 | 6.13 | 1.16 | 0.67 | 1.46 | 3.03 | 6.86 | 40.50 |
| 2009 | 5.80 | 1.73 | 3.52 | 3.64 | 9.22 | 7.03 | 6.40 | 3.55 | 4.88 | 3.54 | 0.80 | 5.96 | 56.07 |
| POR= 29 YRS | 3.59 | 3.80 | 4.17 | 3.87 | 5.10 | 4.66 | 4.69 | 3.69 | 3.50 | 3.24 | 3.89 | 4.28 | 48.48 |

WBAN : 03889

## AVERAGE TEMPERATURE (°F) 2009 JACKSON (KJKL)

| YEAR | JAN | FEB | MAR | APR | MAY | JUN | JUL | AUG | SEP | OCT | NOV | DEC | ANNUAL |
|------|-----|-----|-----|-----|-----|-----|-----|-----|-----|-----|-----|-----|--------|
| 1981 | 29.3 | 38.4 | 42.7 | 61.4 | 60.4 | 74.2 | 75.0 | 72.8 | 66.6 | 56.3 | 48.3 | 35.3 | 55.1 |
| 1982 | 30.2 | 37.9 | 49.8 | 53.5 | 70.9 | 69.5 | 76.7 | 72.8 | 66.5 | 58.8 | 49.2 | 45.3 | 56.8 |
| 1983 | 34.4 | 38.0 | 47.2 | 51.8 | 60.8 | 71.1 | 75.9 | 77.5 | 68.2 | 58.3 | 47.7 | 30.6 | 55.1 |
| 1984 | 30.9 | 42.8 | 42.0 | 54.0 | 61.2 | 75.0 | 72.8 | 74.0 | 65.5 | 63.7 | 43.9 | 47.2 | 56.1 |
| 1985 | 24.5 | 33.0 | 49.3 | 60.8 | 65.3 | 69.9 | 74.0 | 71.0 | 67.4 | 62.2 | 54.4 | 32.3 | 55.3 |
| 1986 | 34.4 | 40.6 | 48.6 | 60.0 | 66.8 | 73.8 | 78.7 | 73.6 | 70.5 | 59.0 | 46.6 | 37.4 | 57.5 |
| 1987 | 33.5 | 38.9 | 48.7 | 54.3 | 69.9 | 73.6 | 75.5 | 75.6 | 68.1 | 52.9 | 51.6 | 40.2 | 56.9 |
| 1988 | 32.4 | 36.4 | 47.6 | 56.3 | 65.0 | 72.6 | 77.8 | 78.0 | 68.3 | 51.3 | 49.3 | 39.2 | 56.2 |
| 1989 | 42.7 | 35.4 | 50.0 | 56.4 | 60.4 | 71.2 | 75.4 | 73.2 | 67.0 | 57.8 | 47.3 | 26.7 | 55.3 |
| 1990 | 42.8 | 46.3 | 52.0 | 56.9 | 63.0 | 72.1 | 75.6 | 73.1 | 68.8 | 58.1 | 52.0 | 42.8 | 58.6 |
| 1991 | 36.2 | 41.2 | 48.8 | 61.3 | 70.7 | 73.3 | 76.7 | 74.8 | 69.3 | 61.4 | 45.9 | 41.8 | 58.5 |
| 1992 | 37.5 | 43.4 | 47.4 | 58.2 | 62.7 | 69.6 | 75.6 | 70.7 | 67.5 | 55.9 | 47.1 | 38.1 | 56.1 |
| 1993 | 39.7 | 35.4 | 42.7 | 54.5 | 65.5 | 71.8 | 79.1 | 75.6 | 66.9 | 55.6 | 47.2 | 36.5 | 55.9 |
| 1994 | 27.7 | 40.6 | 45.5 | 60.3 | 60.0 | 74.5 | 75.1 | 72.5 | 65.5 | 58.0 | 52.8 | 42.8 | 56.3 |
| 1995 | 35.3 | 35.6 | 50.4 | 57.6 | 64.0 | 71.7 | 77.8 | 80.0 | 67.8 | 58.6 | 42.5 | 34.3 | 56.3 |
| 1996 | 33.2 | 36.9 | 41.2 | 53.7 | 66.1 | 72.0 | 72.7 | 73.3 | 65.7 | 57.7 | 41.4 | 42.0 | 54.7 |
| 1997 | 34.5 | 43.6 | 49.3 | 51.7 | 59.4 | 69.4 | 75.5 | 71.9 | 66.3 | 57.3 | 42.8 | 36.7 | 54.9 |
| 1998 | 42.2 | 43.0 | 47.4 | 55.3 | 67.2 | 71.6 | 74.3 | 74.9 | 73.5 | 59.1 | 50.1 | 41.9 | 58.4 |
| 1999 | 40.2 | 41.6 | 41.9 | 59.3 | 65.6 | 72.5 | 78.4 | 73.9 | 68.4 | 57.6 | 52.2 | 40.9 | 57.7 |
| 2000 | 33.3 | 44.4 | 52.1 | 55.2 | 67.4 | 72.5 | 72.9 | 73.0 | 66.2 | 60.2 | 45.2 | 28.2 | 55.9 |
| 2001 | 33.6 | 42.8 | 42.0 | 61.0 | 65.6 | 70.2 | 73.4 | 74.9 | 66.1 | 57.7 | 54.9 | 43.6 | 57.2 |
| 2002 | 40.3 | 39.3 | 47.1 | 60.1 | 62.0 | 73.2 | 76.6 | 75.9 | 71.6 | 57.3 | 43.9 | 37.4 | 57.1 |
| 2003 | 28.3 | 33.9 | 51.1 | 59.3 | 63.2 | 68.3 | 73.8 | 74.6 | 66.1 | 58.0 | 51.8 | 37.1 | 55.5 |
| 2004 | 33.3 | 38.6 | 49.6 | 57.0 | 68.3 | 70.4 | 73.2 | 71.1 | 68.5 | 61.4 | 52.7 | 39.7 | 57.0 |
| 2005 | 40.0 | 42.1 | 44.2 | 58.8 | 63.0 | 74.4 | 77.3 | 78.2 | 73.5 | 59.9 | 51.2 | 35.8 | 58.2 |
| 2006 | 45.1 | 38.2 | 48.3 | 62.4 | 63.4 | 71.2 | 77.2 | 77.7 | 63.5 | 54.7 | 48.9 | 44.3 | 57.9 |
| 2007 | 38.1 | 30.5 | 54.2 | 54.6 | 68.2 | 73.7 | 74.2 | 79.9 | 72.6 | 63.4 | 47.2 | 43.0 | 58.3 |
| 2008 | 34.7 | 38.7 | 46.4 | 57.1 | 62.1 | 73.2 | 73.9 | 73.6 | 71.6 | 57.2 | 44.3 | 38.3 | 55.9 |
| 2009 | 31.0 | 39.8 | 49.0 | 57.8 | 65.0 | 72.3 | 70.6 | 72.9 | 68.3 | 54.8 | 50.9 | 35.8 | 55.7 |
| POR= 29 YRS | 35.2 | 39.2 | 47.5 | 57.2 | 64.6 | 72.0 | 75.4 | 74.5 | 68.1 | 58.1 | 48.4 | 38.5 | 56.6 |

## HEATING DEGREE DAYS (base 65°F) 2009  JACKSON (KJKL)

| YEAR | JUL | AUG | SEP | OCT | NOV | DEC | JAN | FEB | MAR | APR | MAY | JUN | TOTAL |
|------|-----|-----|-----|-----|-----|-----|-----|-----|-----|-----|-----|-----|-------|
| 1980-81 | | | | | | | 1100 | 739 | 689 | 153 | 174 | 0 | |
| 1981-82 | 0 | 0 | 68 | 266 | 495 | 911 | 1075 | 755 | 467 | 348 | 6 | 2 | 4393 |
| 1982-83 | 0 | 0 | 61 | 248 | 479 | 621 | 942 | 748 | 557 | 400 | 145 | 19 | 4220 |
| 1983-84 | 1 | 0 | 71 | 216 | 512 | 1056 | 1049 | 638 | 706 | 353 | 168 | 5 | 4775 |
| 1984-85 | 0 | 0 | 113 | 82 | 630 | 548 | 1250 | 890 | 483 | 195 | 67 | 27 | 4285 |
| 1985-86 | 0 | 1 | 75 | 138 | 321 | 1007 | 945 | 677 | 511 | 217 | 73 | 2 | 3967 |
| 1986-87 | 0 | 15 | 9 | 221 | 545 | 849 | 970 | 724 | 499 | 343 | 35 | 1 | 4211 |
| 1987-88 | 0 | 0 | 29 | 373 | 401 | 758 | 1005 | 824 | 540 | 276 | 78 | 31 | 4315 |
| 1988-89 | 0 | 0 | 24 | 421 | 467 | 792 | 683 | 822 | 482 | 302 | 199 | 4 | 4196 |
| 1989-90 | 0 | 11 | 68 | 240 | 526 | 1179 | 683 | 520 | 431 | 290 | 106 | 8 | 4062 |
| 1990-91 | 0 | 2 | 52 | 243 | 390 | 682 | 884 | 658 | 504 | 161 | 30 | 1 | 3607 |
| 1991-92 | 0 | 0 | 75 | 182 | 570 | 712 | 844 | 621 | 548 | 261 | 150 | 14 | 3977 |
| 1992-93 | 0 | 3 | 51 | 282 | 529 | 826 | 778 | 820 | 683 | 314 | 61 | 25 | 4372 |
| 1993-94 | 0 | 0 | 57 | 292 | 533 | 876 | 1152 | 678 | 598 | 184 | 191 | 0 | 4561 |
| 1994-95 | 0 | 1 | 42 | 223 | 366 | 683 | 913 | 817 | 448 | 255 | 102 | 6 | 3856 |
| 1995-96 | 0 | 0 | 49 | 218 | 670 | 946 | 976 | 812 | 731 | 358 | 91 | 4 | 4855 |
| 1996-97 | 0 | 0 | 64 | 231 | 702 | 707 | 938 | 595 | 483 | 402 | 195 | 29 | 4346 |
| 1997-98 | 0 | 5 | 40 | 284 | 659 | 868 | 700 | 610 | 584 | 284 | 60 | 28 | 4122 |
| 1998-99 | 0 | 0 | 11 | 212 | 439 | 713 | 770 | 648 | 711 | 200 | 41 | 7 | 3752 |
| 1999-00 | 0 | 0 | 43 | 228 | 378 | 739 | 973 | 598 | 403 | 297 | 33 | 11 | 3703 |
| 2000-01 | 0 | 0 | 82 | 183 | 591 | 1133 | 964 | 618 | 706 | 209 | 72 | 17 | 4575 |
| 2001-02 | 0 | 0 | 82 | 247 | 299 | 656 | 762 | 712 | 551 | 219 | 162 | 2 | 3692 |
| 2002-03 | 0 | 0 | 17 | 279 | 629 | 845 | 1133 | 864 | 426 | 200 | 92 | 33 | 4518 |
| 2003-04 | 0 | 0 | 47 | 232 | 404 | 859 | 979 | 760 | 477 | 265 | 52 | 11 | 4086 |
| 2004-05 | 0 | 13 | 23 | 122 | 376 | 778 | 769 | 636 | 641 | 220 | 120 | 0 | 3698 |
| 2005-06 | 0 | 0 | 5 | 216 | 418 | 898 | 606 | 746 | 516 | 129 | 123 | 1 | 3658 |
| 2006-07 | 0 | 0 | 94 | 330 | 474 | 636 | 827 | 958 | 374 | 349 | 52 | 0 | 4094 |
| 2007-08 | 0 | 0 | 14 | 141 | 530 | 675 | 934 | 757 | 571 | 258 | 125 | 0 | 4005 |
| 2008-09 | 0 | 0 | 0 | 272 | 609 | 820 | 1047 | 699 | 494 | 275 | 84 | 6 | 4306 |
| 2009- | 9 | 2 | 27 | 320 | 414 | 900 | | | | | | | |

WBAN : 03889

## COOLING DEGREE DAYS (base 65°F) 2009  JACKSON (KJKL)

| YEAR | JAN | FEB | MAR | APR | MAY | JUN | JUL | AUG | SEP | OCT | NOV | DEC | TOTAL |
|------|-----|-----|-----|-----|-----|-----|-----|-----|-----|-----|-----|-----|-------|
| 1981 | 0 | 0 | 5 | 50 | 40 | 281 | 316 | 249 | 121 | 4 | 2 | 0 | 1068 |
| 1982 | 0 | 0 | 2 | 8 | 195 | 143 | 368 | 249 | 115 | 63 | 10 | 18 | 1171 |
| 1983 | 0 | 0 | 11 | 11 | 22 | 208 | 350 | 393 | 175 | 17 | 0 | 0 | 1187 |
| 1984 | 0 | 0 | 0 | 31 | 58 | 313 | 247 | 285 | 133 | 48 | 3 | 2 | 1120 |
| 1985 | 0 | 0 | 4 | 76 | 81 | 180 | 284 | 194 | 155 | 60 | 9 | 0 | 1043 |
| 1986 | 0 | 0 | 12 | 71 | 136 | 271 | 428 | 289 | 178 | 42 | 0 | 0 | 1427 |
| 1987 | 0 | 0 | 2 | 29 | 194 | 268 | 332 | 335 | 127 | 2 | 7 | 0 | 1296 |
| 1988 | 0 | 0 | 6 | 19 | 83 | 263 | 404 | 410 | 128 | 6 | 1 | 0 | 1320 |
| 1989 | 0 | 0 | 22 | 48 | 66 | 195 | 326 | 271 | 135 | 24 | 1 | 0 | 1088 |
| 1990 | 0 | 0 | 37 | 52 | 50 | 225 | 335 | 262 | 172 | 36 | 6 | 0 | 1175 |
| 1991 | 0 | 0 | 11 | 58 | 213 | 258 | 372 | 312 | 213 | 78 | 3 | 0 | 1518 |
| 1992 | 0 | 0 | 11 | 66 | 89 | 160 | 337 | 183 | 133 | 8 | 0 | 0 | 987 |
| 1993 | 0 | 0 | 0 | 7 | 83 | 233 | 440 | 337 | 122 | 7 | 6 | 0 | 1235 |
| 1994 | 0 | 0 | 1 | 50 | 45 | 291 | 321 | 239 | 62 | 14 | 8 | 0 | 1031 |
| 1995 | 0 | 0 | 0 | 42 | 76 | 213 | 405 | 473 | 140 | 24 | 3 | 0 | 1376 |
| 1996 | 0 | 3 | 0 | 25 | 130 | 222 | 246 | 269 | 89 | 12 | 0 | 0 | 996 |
| 1997 | 0 | 0 | 2 | 11 | 29 | 166 | 333 | 223 | 88 | 51 | 0 | 0 | 903 |
| 1998 | 0 | 0 | 46 | 1 | 135 | 230 | 293 | 313 | 274 | 33 | 0 | 3 | 1328 |
| 1999 | 3 | 0 | 0 | 36 | 65 | 237 | 424 | 283 | 150 | 6 | 2 | 0 | 1206 |
| 2000 | 0 | 4 | 11 | 10 | 116 | 243 | 251 | 255 | 125 | 42 | 3 | 0 | 1060 |
| 2001 | 0 | 0 | 0 | 95 | 95 | 179 | 268 | 314 | 122 | 29 | 4 | 0 | 1106 |
| 2002 | 4 | 0 | 4 | 77 | 78 | 254 | 369 | 344 | 222 | 46 | 3 | 0 | 1401 |
| 2003 | 0 | 0 | 3 | 34 | 42 | 137 | 280 | 304 | 90 | 19 | 15 | 0 | 924 |
| 2004 | 0 | 0 | 8 | 30 | 158 | 182 | 266 | 207 | 133 | 19 | 13 | 0 | 1016 |
| 2005 | 0 | 0 | 2 | 40 | 66 | 290 | 386 | 416 | 266 | 66 | 11 | 0 | 1543 |
| 2006 | 0 | 0 | 4 | 56 | 82 | 193 | 385 | 399 | 56 | 16 | 0 | 0 | 1191 |
| 2007 | 0 | 0 | 45 | 44 | 159 | 270 | 289 | 470 | 246 | 98 | 0 | 0 | 1621 |
| 2008 | 0 | 0 | 0 | 28 | 39 | 252 | 282 | 273 | 203 | 38 | 0 | 0 | 1115 |
| 2009 | 0 | 0 | 4 | 65 | 91 | 234 | 192 | 254 | 132 | 15 | 0 | 0 | 987 |

## SNOWFALL (inches) 2009 JACKSON (KJKL)

| YEAR | JUL | AUG | SEP | OCT | NOV | DEC | JAN | FEB | MAR | APR | MAY | JUN | TOTAL |
|---|---|---|---|---|---|---|---|---|---|---|---|---|---|
| 1980-81 | | | | | | | 5.1 | 6.8 | 1.8 | 0.0 | 0.0 | 0.0 | |
| 1981-82 | 0.0 | 0.0 | 0.0 | 0.0 | T | 3.8 | 6.0 | 7.6 | 5.0 | 0.4 | 0.0 | 0.0 | 22.8 |
| 1982-83 | 0.0 | 0.0 | 0.0 | 0.0 | T | 4.1 | 2.0 | 6.9 | 2.8 | 0.1 | 0.0 | 0.0 | 15.9 |
| 1983-84 | 0.0 | 0.0 | 0.0 | 0.0 | T | 3.3 | 8.8 | 2.6 | 1.7 | T | 0.0 | 0.0 | 16.4 |
| 1984-85 | 0.0 | 0.0 | 0.0 | 0.0 | T | 1.9 | 12.1 | 21.0 | T | T | 0.0 | 0.0 | 35.0 |
| 1985-86 | 0.0 | 0.0 | 0.0 | 0.0 | 0.0 | 4.6 | 7.7 | 11.6 | 1.0 | T | 0.0 | 0.0 | 24.9 |
| 1986-87 | 0.0 | 0.0 | 0.0 | 0.0 | 0.8 | T | 11.6 | 6.0 | 0.7 | 17.8 | 0.0 | 0.0 | 36.9 |
| 1987-88 | 0.0 | 0.0 | 0.0 | 0.0 | T | 3.7 | 5.4 | 2.0 | 1.9 | 0.0 | 0.0 | 0.0 | 13.0 |
| 1988-89 | 0.0 | 0.0 | 0.0 | 0.0 | 0.2 | 2.5 | 0.4 | 5.6 | 0.1 | T | 1.0 | 0.0 | 9.8 |
| 1989-90 | 0.0 | 0.0 | 0.0 | 0.5 | 1.8 | 10.1 | 2.5 | 2.3 | 1.4 | T | 0.0 | 0.0 | 18.6 |
| 1990-91 | 0.0 | 0.0 | 0.0 | 0.0 | 0.0 | 0.3 | 0.9 | 4.8 | 3.0 | 0.0 | 0.0 | 0.0 | 9.0 |
| 1991-92 | 0.0 | T | 0.0 | 0.0 | T | 0.4 | 2.1 | 0.7 | 1.5 | 1.7 | 0.0 | 0.0 | 6.4 |
| 1992-93 | 0.0 | 0.0 | 0.0 | 0.0 | 1.1 | 4.5 | 0.1 | 8.3 | 21.9 | 0.2 | T | 0.0 | 36.1 |
| 1993-94 | 0.0 | 0.0 | 0.0 | 2.3 | 0.1 | 11.5 | 26.3 | 4.2 | 3.9 | T | 0.0 | 0.0 | 48.3 |
| 1994-95 | 0.0 | 0.0 | 0.0 | 0.0 | 0.0 | 0.2 | 7.3 | 6.9 | 6.7 | 0.0 | 0.0 | T | 21.1 |
| 1995-96 | 0.0 | 0.0 | 0.0 | 0.0 | 5.1 | 8.0 | 23.3 | 12.3 | 11.2 | 2.8 | | | |
| 1996-97 | 0.0 | 0.0 | 0.0 | 0.0 | 2.5 | 1.6 | 2.3 | 2.9 | T | T | 0.0 | 0.0 | 9.3 |
| 1997-98 | 0.0 | T | 0.0 | 0.0 | 1.1 | 11.3 | 0.5 | 17.6 | 3.2 | 0.0 | T | 0.0 | 33.7 |
| 1998-99 | 0.0 | 0.0 | 0.0 | 0.0 | T | 3.1 | 5.1 | 6.8 | 2.2 | 0.0 | 0.0 | 0.0 | 17.2 |
| 1999-00 | 0.0 | 0.0 | 0.0 | 0.0 | T | 2.4 | 8.3 | 3.2 | 0.1 | T | 0.0 | 0.0 | 14.0 |
| 2000-01 | T | 0.0 | 0.0 | 0.0 | 0.9 | 10.0 | 5.1 | 0.9 | 2.3 | 1.6 | T | 0.0 | 20.8 |
| 2001-02 | 0.0 | 0.0 | 0.0 | T | 0.0 | 1.2 | 15.0 | 3.2 | 0.8 | T | 0.0 | 0.0 | 20.2 |
| 2002-03 | 0.0 | 0.0 | 0.0 | 0.0 | 0.1 | 3.7 | 9.7 | 10.5 | 0.2 | 0.0 | 0.0 | 0.0 | 24.2 |
| 2003-04 | 0.0 | 0.0 | 0.0 | 0.0 | 0.3 | 9.1 | 3.9 | 2.6 | T | 0.0 | 0.0 | 0.0 | 15.9 |
| 2004-05 | 0.0 | 0.0 | 0.0 | 0.0 | 0.0 | 1.8 | 4.4 | 2.2 | 5.0 | 1.2 | 0.0 | 0.0 | 14.6 |
| 2005-06 | 0.0 | 0.0 | 0.0 | 0.0 | T | 1.6 | 2.6 | 9.4 | 0.4 | 0.0 | T | 0.0 | 14.0 |
| 2006-07 | 0.0 | 0.0 | 0.0 | T | T | 0.2 | 3.3 | 6.6 | 0.8 | 1.4 | 0.0 | T | 12.3 |
| 2007-08 | T | 0.0 | 0.0 | 0.0 | T | 0.5 | 3.5 | 3.9 | 2.9 | T | 0.0 | T | 10.8 |
| 2008-09 | 0.0 | 0.0 | 0.0 | T | 2.1 | 5.7 | 9.0 | 11.5 | 4.6 | 2.1 | 0.0 | 0.0 | 35.0 |
| 2009- | 0.0 | 0.0 | 0.0 | 0.0 | T | 17.6 | | | | | | | |
| POR= 29 YRS | T | T | 0.0 | 0.1 | 0.6 | 4.4 | 6.7 | 6.6 | 3.0 | 1.0 | T | T | 22.4 |

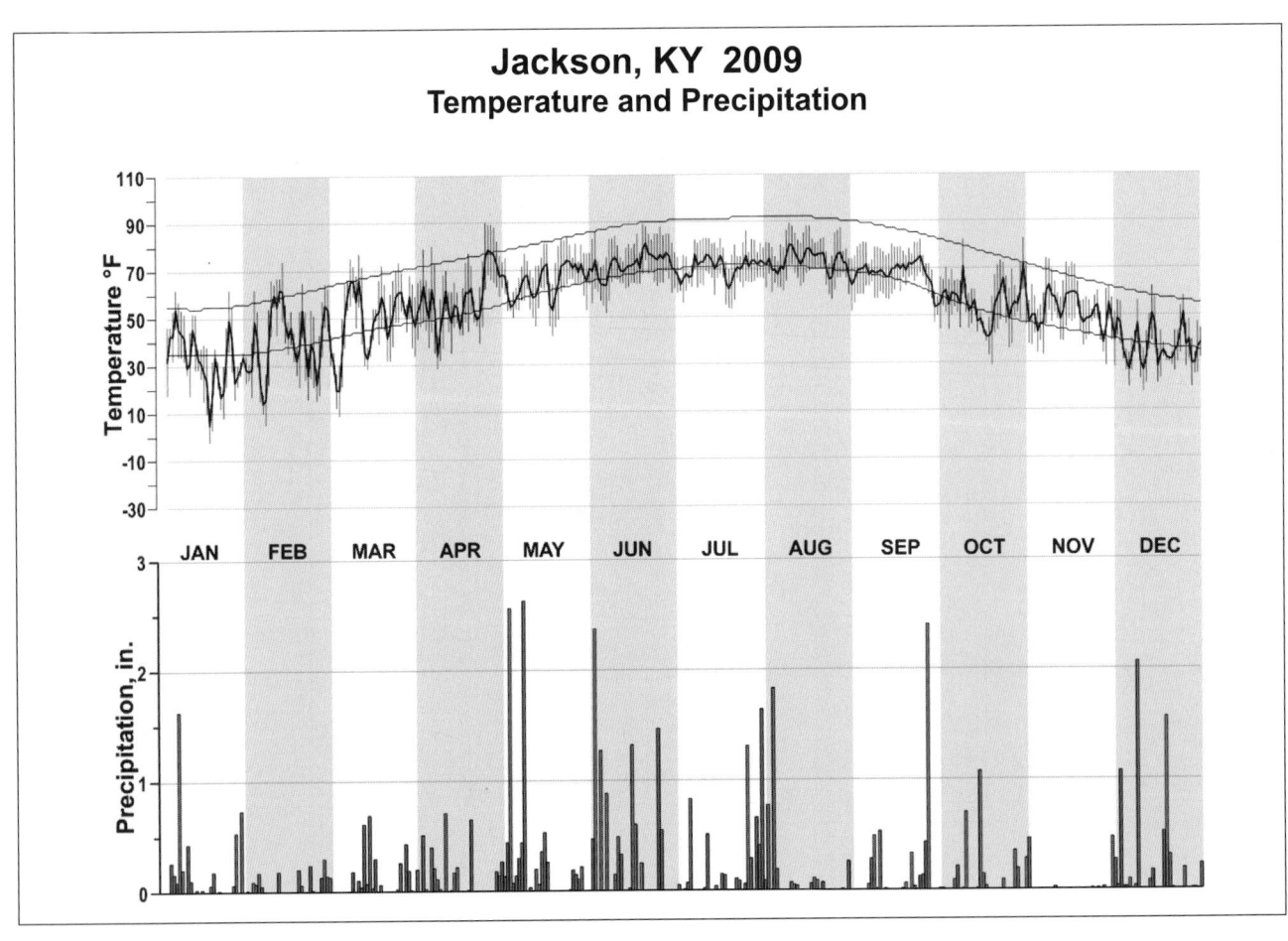

## Jackson, KY 2009
### Temperature and Precipitation

# 2009
# LEXINGTON
# KENTUCKY (KLEX)

Lexington, County Seat of Fayette County, is located in the heart of the famed Kentucky Blue Grass Region. Fayette County is a gently rolling plateau with the elevation varying between 900 and 1,050 feet above sea level. It is noted for its beauty, the fertility of its soil, excellent grass, stock farms, and burley tobacco. The soil has a high phosphorus content and this is very valuable in growing pasture grasses for the grazing of cattle and horses. Lexington has a decided continental climate with a rather large diurnal temperature range. The climate is temperate and well suited to a varied plant and animal life. There are no bodies of water close enough to have any effect on the climate. The closest river is the Kentucky which makes an arc about 15 to 20 miles to the southeast, south, and southwest on its course to the Ohio River. There are numerous small creeks that rise in the county and flow into the river. The reservoirs of the Lexington Water Company are about 5 miles southeast of the city and are the largest bodies of water in the area.

Lexington is subject to rather sudden and large changes in temperature with the spells generally of rather short duration. Temperatures above 100 degrees and below zero degrees are relatively rare. The average temperature for the winter is 35 degrees, spring 62 degrees, fall 50 degrees, and summer 74 degrees.

Precipitation is evenly distributed throughout the winter, spring, and summer, with about 12 inches recorded on the average for each of these seasons. The fall season averages nearly 8 1/2 inches. Snowfall amounts are variable and the ground does not retain snow cover more than a few days at a time.

The months of September and October are the most pleasant of the year. They have the least amount of precipitation, the greatest number of clear days, and generally comfortable temperatures are the rule during these months.

Based on the 1951-1980 period, the average first occurrence of 32 degrees Fahrenheit in the fall is October 25 and the average last occurrence in the spring is April 17.

# NORMALS, MEANS, AND EXTREMES
## LEXINGTON (KLEX)

LATITUDE: 38 ° 2 'N  LONGITUDE: -84 ° 36'W  ELEVATION (FT): GRND: 965  BARO: 984  TIME ZONE: EASTERN (UTC -5)  WBAN: 93820

| | ELEMENT | POR | JAN | FEB | MAR | APR | MAY | JUN | JUL | AUG | SEP | OCT | NOV | DEC | YEAR |
|---|---|---|---|---|---|---|---|---|---|---|---|---|---|---|---|
| **TEMPERATURE °F** | NORMAL DAILY MAXIMUM | 30 | 39.9 | 45.2 | 55.3 | 65.1 | 74.0 | 82.3 | 85.9 | 84.6 | 78.1 | 66.9 | 54.5 | 44.3 | 64.7 |
| | MEAN DAILY MAXIMUM | 113 | 41.2 | 42.3 | 54.2 | 63.8 | 74.4 | 81.4 | 86.1 | 85.0 | 78.1 | 67.9 | 53.5 | 43.7 | 64.3 |
| | HIGHEST DAILY MAXIMUM | 65 | 76 | 80 | 83 | 88 | 92 | 101 | 103 | 103 | 103 | 91 | 83 | 75 | 103 |
| | YEAR OF OCCURRENCE | | 1950 | 1996 | 1945 | 1962 | 1987 | 1988 | 1999 | 1983 | 1954 | 2007 | 1987 | 1982 | JUL 1999 |
| | MEAN OF EXTREME MAXS. | 113 | 63.3 | 66.3 | 75.4 | 82.2 | 86.7 | 91.8 | 93.7 | 93.4 | 90.2 | 82.6 | 73.9 | 65.0 | 80.4 |
| | NORMAL DAILY MINIMUM | 30 | 24.1 | 27.7 | 35.9 | 44.1 | 53.6 | 62.2 | 66.4 | 64.9 | 57.9 | 46.4 | 37.3 | 28.4 | 45.7 |
| | MEAN DAILY MINIMUM | 113 | 25.0 | 25.5 | 34.9 | 43.6 | 53.9 | 61.6 | 66.3 | 65.1 | 57.6 | 47.1 | 35.9 | 28.0 | 45.4 |
| | LOWEST DAILY MINIMUM | 65 | -21 | -15 | -2 | 18 | 26 | 39 | 47 | 42 | 34 | 20 | -3 | -19 | -21 |
| | YEAR OF OCCURRENCE | | 1963 | 1951 | 1960 | 1982 | 1966 | 1966 | 1972 | 1965 | 1993 | 1976 | 1950 | 1989 | JAN 1963 |
| | MEAN OF EXTREME MINS. | 113 | 1.5 | 5.3 | 16.6 | 27.7 | 37.9 | 48.8 | 55.5 | 53.7 | 42.3 | 30.0 | 19.0 | 8.0 | 28.9 |
| | NORMAL DRY BULB | 30 | 32.0 | 36.4 | 45.6 | 54.6 | 63.8 | 72.2 | 76.1 | 74.8 | 68.0 | 56.6 | 45.9 | 36.3 | 55.2 |
| | MEAN DRY BULB | 113 | 33.1 | 33.9 | 44.5 | 53.7 | 64.1 | 71.5 | 76.2 | 75.0 | 67.9 | 57.5 | 44.7 | 35.9 | 54.8 |
| | MEAN WET BULB | 26 | 29.6 | 32.3 | 39.1 | 48.1 | 57.5 | 65.4 | 68.4 | 67.5 | 61.1 | 50.8 | 41.0 | 32.7 | 49.5 |
| | MEAN DEW POINT | 26 | 26.3 | 28.5 | 34.6 | 43.7 | 54.5 | 62.9 | 66.2 | 65.2 | 58.3 | 47.1 | 37.5 | 29.6 | 46.2 |
| | NORMAL NO. DAYS WITH: | | | | | | | | | | | | | | |
| | MAXIMUM >= 90 | 3 0 | 0.0 | 0.0 | 0.0 | 0.0 | 0.2 | 3.0 | 8.1 | 6.4 | 2.0 | 0.0 | 0.0 | 0.0 | 19.7 |
| | MAXIMUM <= 32 | 3 0 | 8.9 | 5.7 | 0.9 | 0.0 | 0.0 | 0.0 | 0.0 | 0.0 | 0.0 | 0.0 | 0.3 | 5.1 | 20.9 |
| | MINIMUM <= 32 | 30 | 23.9 | 18.9 | 13.0 | 3.6 | 0.1 | 0.0 | 0.0 | 0.0 | 0.0 | 2.1 | 10.9 | 19.9 | 92.4 |
| | MINIMUM <= 0 | 30 | 1.3 | 0.5 | 0.1 | 0.0 | 0.0 | 0.0 | 0.0 | 0.0 | 0.0 | 0.0 | 0.0 | 0.4 | 2.3 |
| **H/C** | NORMAL HEATING DEG. DAYS | 30 | 1026 | 816 | 616 | 332 | 119 | 13 | 1 | 2 | 53 | 284 | 574 | 877 | 4713 |
| | NORMAL COOLING DEG. DAYS | 30 | 0 | 0 | 3 | 16 | 80 | 228 | 350 | 307 | 147 | 21 | 2 | 0 | 1154 |
| **RH** | NORMAL (PERCENT) | 30 | 75 | 71 | 67 | 64 | 70 | 72 | 73 | 74 | 74 | 71 | 72 | 76 | 72 |
| | HOUR 01 LST | 30 | 78 | 76 | 72 | 71 | 79 | 83 | 84 | 85 | 84 | 79 | 76 | 78 | 79 |
| | HOUR 07 LST | 30 | 81 | 80 | 78 | 76 | 82 | 84 | 86 | 89 | 89 | 86 | 81 | 82 | 83 |
| | HOUR 13 LST | 30 | 69 | 64 | 58 | 55 | 59 | 59 | 59 | 60 | 59 | 57 | 63 | 69 | 61 |
| | HOUR 19 LST | 30 | 73 | 67 | 60 | 57 | 62 | 64 | 64 | 65 | 67 | 66 | 69 | 73 | 66 |
| **S** | PERCENT POSSIBLE SUNSHINE | | | | | | | | | | | | | | |
| **W/O** | MEAN NO. DAYS WITH: | | | | | | | | | | | | | | |
| | HEAVY FOG(VISBY <= 1/4 MI) | 46 | 1.9 | 2.0 | 1.6 | 0.6 | 1.2 | 1.3 | 1.7 | 2.0 | 2.1 | 1.7 | 1.2 | 1.7 | 19.0 |
| | THUNDERSTORMS | 62 | 0.7 | 0.8 | 2.6 | 3.7 | 6.0 | 7.3 | 8.2 | 6.0 | 2.8 | 1.3 | 0.9 | 0.3 | 40.6 |
| **CLOUDNESS** | MEAN: | | | | | | | | | | | | | | |
| | SUNRISE-SUNSET (OKTAS) | 52 | 5.9 | 5.7 | 5.6 | 5.3 | 5.0 | 4.8 | 4.6 | 4.3 | 4.2 | 4.2 | 5.3 | 5.7 | 5.1 |
| | MIDNIGHT-MIDNIGHT (OKTAS) | 32 | 5.8 | 5.5 | 5.3 | 4.9 | 4.7 | 4.6 | 4.3 | 4.0 | 4.1 | 4.1 | 5.1 | 5.6 | 4.8 |
| | MEAN NO. DAYS WITH: | | | | | | | | | | | | | | |
| | CLEAR | 52 | 5.5 | 5.7 | 5.8 | 6.3 | 7.0 | 6.9 | 8.0 | 9.5 | 10.5 | 11.6 | 6.7 | 5.8 | 89.3 |
| | PARTLY CLOUDY | 52 | 5.8 | 5.6 | 7.3 | 8.5 | 10.0 | 11.8 | 12.3 | 11.9 | 8.4 | 7.2 | 6.7 | 5.7 | 101.2 |
| | CLOUDY | 52 | 19.7 | 17.0 | 17.9 | 15.2 | 14.0 | 11.3 | 10.7 | 9.6 | 11.1 | 12.1 | 16.6 | 19.5 | 174.7 |
| **PR** | MEAN STATION PRESSURE(IN) | 26 | 29.05 | 29.03 | 28.99 | 28.94 | 28.95 | 28.96 | 28.98 | 29.00 | 29.02 | 29.04 | 29.05 | 29.06 | 29.01 |
| | MEAN SEA-LEVEL PRES. (IN) | 26 | 30.13 | 30.10 | 30.05 | 29.99 | 29.99 | 29.99 | 30.01 | 30.03 | 30.06 | 30.09 | 30.11 | 30.14 | 30.06 |
| **WINDS** | MEAN SPEED (MPH) | 26 | 9.5 | 9.3 | 9.2 | 9.1 | 7.7 | 6.8 | 6.3 | 6.1 | 6.4 | 7.2 | 8.5 | 8.9 | 7.9 |
| | PREVAIL.DIR(TENS OF DEGS) | 38 | 19 | 19 | 19 | 19 | 19 | 19 | 23 | 19 | 19 | 19 | 19 | 19 | 19 |
| | MAXIMUM 2-MINUTE: | | | | | | | | | | | | | | |
| | SPEED (MPH) | 13 | 47 | 47 | 38 | 43 | 39 | 44 | 37 | 38 | 44 | 40 | 45 | 44 | 47 |
| | DIR. (TENS OF DEGS) | | 18 | 18 | 20 | 33 | 23 | 30 | 31 | 14 | 22 | 17 | 27 | 22 | 18 |
| | YEAR OF OCCURRENCE | | 1999 | 2008 | 2009 | 2007 | 2008 | 1998 | 1999 | 2005 | 2008 | 2002 | 2002 | 2008 | FEB 2008 |
| | MAXIMUM 3-SECOND | | | | | | | | | | | | | | |
| | SPEED (MPH) | 13 | 56 | 60 | 53 | 56 | 52 | 61 | 51 | 47 | 60 | 48 | 56 | 60 | 61 |
| | DIR. (TENS OF DEGS) | | 19 | 28 | 24 | 33 | 23 | 34 | 32 | 32 | 21 | 16 | 24 | 21 | 34 |
| | YEAR OF OCCURRENCE | | 1999 | 2008 | 2007 | 2007 | 2008 | 2007 | 1999 | 2007 | 2008 | 2002 | 2002 | 2008 | JUN 2007 |
| **PRECIPITATION** | NORMAL (IN) | 30 | 3.34 | 3.27 | 4.41 | 3.67 | 4.78 | 4.58 | 4.81 | 3.77 | 3.11 | 2.70 | 3.44 | 4.03 | 45.91 |
| | MAXIMUM MONTHLY (IN) | 65 | 16.65 | 10.12 | 13.82 | 9.30 | 10.91 | 11.69 | 10.64 | 11.18 | 10.25 | 6.97 | 7.32 | 10.17 | 16.65 |
| | YEAR OF OCCURRENCE | | 1950 | 1989 | 1997 | 1970 | 2004 | 1960 | 1958 | 1974 | 2006 | 2004 | 1990 | JAN 1950 |
| | MINIMUM MONTHLY (IN) | 65 | 0.37 | 0.67 | 0.99 | 0.79 | 1.20 | 0.61 | 1.26 | 0.29 | 0.24 | 0.33 | 0.45 | 0.61 | 0.24 |
| | YEAR OF OCCURRENCE | | 1981 | 1978 | 1966 | 1946 | 1965 | 1988 | 1995 | 1998 | 1959 | 1963 | 1976 | 1965 | SEP 1959 |
| | MAXIMUM IN 24 HOURS (IN) | 65 | 2.98 | 3.79 | 5.56 | 4.65 | 3.52 | 5.88 | 4.73 | 3.56 | 6.16 | 4.33 | 2.71 | 3.77 | 6.16 |
| | YEAR OF OCCURRENCE | | 1951 | 1989 | 1997 | 2008 | 2004 | 1960 | 1978 | 1968 | 2006 | 2007 | 1988 | 1978 | SEP 2006 |
| | NORMAL NO. DAYS WITH: | | | | | | | | | | | | | | |
| | PRECIPITATION >= 0.01 | 30 | 12.3 | 11.3 | 12.8 | 12.1 | 12.2 | 10.5 | 10.6 | 8.9 | 8.8 | 8.3 | 10.9 | 12.1 | 130.8 |
| | PRECIPITATION >= 1.00 | 30 | 0.8 | 0.7 | 0.9 | 0.7 | 1.0 | 0.9 | 1.6 | 1.1 | 0.7 | 0.6 | 0.9 | 0.8 | 10.7 |
| **SNOWFALL** | NORMAL (IN) | 30 | 5.9 | 4.5 | 2.3 | 0.4 | 0.* | 0.0 | 0.0 | 0.0 | 0.0 | 0.* | 0.6 | 2.0 | 15.7 |
| | MAXIMUM MONTHLY (IN) | 59 | 21.9 | 16.4 | 17.7 | 5.9 | T | T | T | T | 0.0 | 0.2 | 9.7 | 10.7 | 21.9 |
| | YEAR OF OCCURRENCE | | 1978 | 1960 | 1960 | 1987 | 1995 | 2009 | 1989 | 1989 | | 1972 | 1950 | 1967 | JAN 1978 |
| | MAXIMUM IN 24 HOURS (IN) | 59 | 10.2 | 7.3 | 9.5 | 4.9 | T | T | T | T | 0.0 | 0.2 | 7.5 | 7.8 | 10.2 |
| | YEAR OF OCCURRENCE' | | 1994 | 1971 | 1947 | 1987 | 1995 | 2009 | 1989 | 1989 | | 1972 | 1966 | 1967 | JAN 1994 |
| | MAXIMUM SNOW DEPTH (IN) | 55 | 14 | 9 | 12 | 2 | 0 | 0 | 0 | 0 | 0 | 0 | 8 | 5 | 14 |
| | YEAR OF OCCURRENCE | | 1978 | 1985 | 1960 | 1961 | | | | | | | 1950 | 1984 | JAN 1978 |
| | NORMAL NO. DAYS WITH: | | | | | | | | | | | | | | |
| | SNOWFALL >= 1.0 | 30 | 1.5 | 1.4 | 0.6 | 0.0 | 0.0 | 0.0 | 0.0 | 0.0 | 0.0 | 0.0 | 0.2 | 0.5 | 4.2 |

## PRECIPITATION (inches) 2009 LEXINGTON (KLEX)

| YEAR | JAN | FEB | MAR | APR | MAY | JUN | JUL | AUG | SEP | OCT | NOV | DEC | ANNUAL |
|------|-----|-----|-----|-----|-----|-----|-----|-----|-----|-----|-----|-----|--------|
| 1980 | 1.63 | 1.17 | 6.04 | 2.82 | 2.27 | 1.88 | 5.55 | 5.10 | 2.47 | 2.07 | 2.02 | 1.67 | 34.69 |
| 1981 | 0.37 | 4.76 | 1.76 | 4.88 | 5.10 | 2.29 | 5.27 | 2.72 | 1.97 | 2.44 | 1.99 | 3.10 | 36.65 |
| 1982 | 5.48 | 2.16 | 3.89 | 2.19 | 2.51 | 3.95 | 3.82 | 4.01 | 1.21 | 1.56 | 3.45 | 4.53 | 38.76 |
| 1983 | 1.29 | 1.61 | 1.48 | 5.18 | 10.84 | 2.18 | 2.41 | 1.26 | 1.33 | 6.13 | 3.59 | 3.46 | 40.76 |
| 1984 | 1.64 | 3.31 | 4.09 | 5.02 | 5.34 | 2.20 | 4.80 | 0.56 | 1.36 | 3.87 | 5.19 | 4.89 | 42.27 |
| 1985 | 1.91 | 1.11 | 3.69 | 2.34 | 4.34 | 4.98 | 3.37 | 3.76 | 1.93 | 4.23 | 4.96 | 1.13 | 37.75 |
| 1986 | 0.53 | 2.48 | 2.43 | 1.65 | 3.24 | 1.29 | 5.64 | 2.67 | 3.08 | 2.06 | 6.49 | 3.30 | 34.86 |
| 1987 | 1.30 | 3.62 | 3.13 | 2.23 | 1.80 | 6.59 | 3.48 | 4.18 | 0.91 | 0.55 | 2.72 | 6.17 | 36.68 |
| 1988 | 2.94 | 3.06 | 2.34 | 2.93 | 3.02 | 0.61 | 3.51 | 4.18 | 5.96 | 1.34 | 5.39 | 3.62 | 38.90 |
| 1989 | 3.99 | 10.12 | 6.08 | 2.60 | 5.39 | 4.26 | 4.20 | 3.98 | 4.98 | 3.38 | 2.38 | 1.80 | 53.16 |
| 1990 | 4.17 | 3.43 | 1.89 | 2.37 | 5.41 | 4.59 | 6.45 | 4.36 | 2.12 | 4.49 | 2.69 | 10.17 | 52.14 |
| 1991 | 2.57 | 3.91 | 5.80 | 2.70 | 3.95 | 2.91 | 3.60 | 3.08 | 2.09 | 2.70 | 1.27 | 7.22 | 41.80 |
| 1992 | 3.63 | 1.84 | 4.70 | 2.11 | 4.68 | 7.74 | 10.27 | 4.73 | 3.44 | 0.65 | 3.50 | 1.80 | 49.09 |
| 1993 | 2.42 | 4.15 | 3.77 | 3.53 | 2.43 | 5.46 | 3.38 | 4.52 | 3.00 | 4.19 | 5.42 | 3.31 | 45.58 |
| 1994 | 4.50 | 4.42 | 6.83 | 5.18 | 4.86 | 3.84 | 2.29 | 3.72 | 1.19 | 2.11 | 2.89 | 3.87 | 45.70 |
| 1995 | 5.01 | 2.26 | 3.32 | 3.90 | 8.97 | 8.17 | 1.26 | 4.89 | 2.76 | 3.64 | 3.19 | 2.71 | 50.08 |
| 1996 | 4.51 | 1.86 | 4.62 | 4.84 | 8.98 | 5.10 | 5.30 | 2.30 | 4.15 | 2.10 | 4.79 | 5.26 | 53.81 |
| 1997 | 3.70 | 3.97 | 13.82 | 1.89 | 8.85 | 9.54 | 3.29 | 2.58 | 2.38 | 2.37 | 4.06 | 2.68 | 59.13 |
| 1998 | 3.99 | 2.58 | 3.40 | 6.20 | 6.14 | 10.81 | 7.98 | 0.29 | 0.61 | 2.41 | 1.96 | 3.23 | 49.60 |
| 1999 | 5.77 | 2.38 | 3.80 | 2.23 | 1.31 | 5.38 | 2.47 | 0.99 | 1.39 | 1.63 | 1.82 | 2.70 | 31.87 |
| 2000 | 3.40 | 4.81 | 3.89 | 4.52 | 2.99 | 3.82 | 3.36 | 3.50 | 5.32 | 0.74 | 2.00 | 3.75 | 42.10 |
| 2001 | 1.35 | 3.56 | 3.27 | 1.14 | 6.00 | 2.58 | 5.78 | 2.93 | 2.46 | 3.71 | 3.30 | 2.89 | 38.97 |
| 2002 | 2.39 | 1.37 | 7.58 | 5.28 | 4.29 | 2.69 | 1.75 | 2.92 | 5.44 | 6.53 | 4.99 | 4.08 | 49.31 |
| 2003 | 0.95 | 4.85 | 2.44 | 4.10 | 8.35 | 6.41 | 5.08 | 4.53 | 5.07 | 1.75 | 5.93 | 3.93 | 53.39 |
| 2004 | 3.32 | 1.49 | 4.31 | 3.73 | 10.91 | 5.05 | 8.68 | 4.06 | 3.22 | 6.97 | 7.32 | 3.38 | 62.44 |
| 2005 | 4.27 | 2.23 | 3.49 | 3.47 | 2.64 | 2.28 | 3.05 | 6.10 | 0.89 | 0.93 | 1.77 | 2.40 | 33.52 |
| 2006 | 5.37 | 2.12 | 4.17 | 4.55 | 3.72 | 2.33 | 5.48 | 3.49 | 10.25 | 6.29 | 1.97 | 3.05 | 52.79 |
| 2007 | 3.37 | 2.47 | 2.38 | 4.12 | 1.21 | 2.68 | 6.39 | 4.00 | 0.88 | 6.53 | 2.75 | 6.93 | 43.71 |
| 2008 | 4.42 | 5.76 | 6.30 | 5.89 | 4.40 | 3.59 | 3.41 | 2.18 | 1.42 | 1.53 | 2.53 | 6.03 | 47.46 |
| 2009 | 4.32 | 2.54 | 2.39 | 4.78 | 6.04 | 5.19 | 7.57 | 4.53 | 5.90 | 5.77 | 0.96 | 4.02 | 54.01 |
| POR= 113 YRS | 3.91 | 3.17 | 4.48 | 3.71 | 4.21 | 4.19 | 4.46 | 3.51 | 2.99 | 2.58 | 3.23 | 3.78 | 44.22 |

WBAN : 93820

## AVERAGE TEMPERATURE (°F) 2009 LEXINGTON (KLEX)

| YEAR | JAN | FEB | MAR | APR | MAY | JUN | JUL | AUG | SEP | OCT | NOV | DEC | ANNUAL |
|------|-----|-----|-----|-----|-----|-----|-----|-----|-----|-----|-----|-----|--------|
| 1980 | 32.4 | 28.3 | 41.5 | 52.5 | 64.8 | 71.2 | 78.8 | 78.2 | 70.6 | 53.7 | 43.6 | 36.0 | 54.3 |
| 1981 | 27.5 | 37.0 | 42.6 | 59.6 | 60.4 | 73.8 | 75.8 | 73.4 | 65.8 | 55.6 | 45.9 | 33.0 | 54.2 |
| 1982 | 28.2 | 34.9 | 47.1 | 50.6 | 69.8 | 68.5 | 77.1 | 72.9 | 65.6 | 58.3 | 48.4 | 44.2 | 55.5 |
| 1983 | 33.8 | 37.2 | 46.3 | 50.8 | 60.7 | 72.8 | 79.8 | 80.5 | 70.2 | 59.0 | 46.5 | 28.4 | 55.5 |
| 1984 | 27.6 | 41.2 | 39.7 | 53.1 | 60.6 | 76.0 | 72.9 | 74.9 | 66.5 | 63.5 | 41.8 | 45.4 | 55.3 |
| 1985 | 23.8 | 30.5 | 48.5 | 58.5 | 64.7 | 70.3 | 75.1 | 72.8 | 67.6 | 60.5 | 53.0 | 29.6 | 54.6 |
| 1986 | 33.2 | 38.6 | 46.8 | 57.3 | 65.5 | 74.2 | 78.6 | 72.9 | 71.0 | 58.0 | 44.9 | 35.7 | 56.4 |
| 1987 | 31.9 | 38.0 | 46.8 | 53.7 | 70.3 | 75.0 | 77.1 | 77.5 | 70.0 | 52.0 | 50.0 | 38.9 | 56.8 |
| 1988 | 29.8 | 33.7 | 44.8 | 54.2 | 64.5 | 74.3 | 79.1 | 77.9 | 67.8 | 49.6 | 46.1 | 36.5 | 54.9 |
| 1989 | 40.5 | 33.1 | 47.3 | 54.2 | 60.6 | 71.6 | 76.5 | 74.1 | 67.6 | 57.0 | 45.0 | 23.0 | 54.2 |
| 1990 | 41.6 | 43.1 | 48.9 | 53.2 | 61.6 | 72.2 | 75.3 | 73.8 | 68.5 | 56.4 | 49.9 | 40.4 | 57.1 |
| 1991 | 33.9 | 39.1 | 47.4 | 58.1 | 70.5 | 74.3 | 77.6 | 75.8 | 68.8 | 58.7 | 43.4 | 40.1 | 57.3 |
| 1992 | 35.2 | 41.3 | 45.4 | 56.1 | 62.0 | 69.4 | 75.6 | 70.9 | 67.0 | 55.8 | 45.9 | 37.0 | 55.1 |
| 1993 | 37.4 | 33.1 | 42.0 | 52.7 | 65.0 | 72.1 | 80.1 | 76.5 | 66.0 | 55.1 | 44.6 | 35.0 | 55.0 |
| 1994 | 25.2 | 36.8 | 43.6 | 58.1 | 60.2 | 75.1 | 76.8 | 73.7 | 66.2 | 57.8 | 50.6 | 41.2 | 55.4 |
| 1995 | 33.9 | 33.5 | 47.2 | 55.5 | 63.3 | 72.8 | 77.3 | 79.5 | 66.5 | 56.9 | 39.7 | 33.0 | 54.9 |
| 1996 | 31.0 | 35.7 | 39.1 | 50.9 | 65.6 | 72.1 | 73.1 | 73.6 | 66.3 | 56.4 | 39.8 | 39.5 | 53.6 |
| 1997 | 31.6 | 40.8 | 46.2 | 49.1 | 58.0 | 69.3 | 76.1 | 72.7 | 66.6 | 56.1 | 41.6 | 35.6 | 53.6 |
| 1998 | 40.7 | 41.1 | 45.8 | 53.7 | 67.5 | 72.7 | 74.5 | 76.2 | 74.1 | 57.9 | 47.4 | 39.7 | 57.6 |
| 1999 | 36.2 | 40.2 | 40.3 | 56.5 | 65.3 | 73.5 | 79.6 | 75.7 | 68.8 | 56.7 | 50.3 | 37.3 | 56.7 |
| 2000 | 31.9 | 42.7 | 48.3 | 53.3 | 66.8 | 73.5 | 73.8 | 73.7 | 66.3 | 59.4 | 43.5 | 25.1 | 54.9 |
| 2001 | 31.1 | 39.9 | 40.6 | 59.7 | 66.5 | 71.3 | 75.3 | 76.1 | 66.1 | 56.7 | 52.0 | 40.9 | 56.4 |
| 2002 | 37.8 | 37.7 | 44.6 | 57.9 | 61.2 | 74.7 | 79.1 | 78.1 | 72.1 | 56.0 | 42.6 | 35.6 | 56.5 |
| 2003 | 26.1 | 32.0 | 48.2 | 56.9 | 63.6 | 69.0 | 75.1 | 75.7 | 65.1 | 56.2 | 49.5 | 35.5 | 54.4 |
| 2004 | 30.4 | 36.1 | 47.8 | 55.1 | 68.8 | 71.4 | 73.3 | 70.7 | 68.3 | 58.9 | 49.2 | 36.1 | 55.5 |
| 2005 | 37.5 | 39.6 | 40.6 | 56.3 | 61.5 | 75.0 | 77.9 | 78.5 | 71.7 | 58.1 | 47.1 | 32.4 | 56.4 |
| 2006 | 42.2 | 35.5 | 44.3 | 59.0 | 62.5 | 70.7 | 76.7 | 77.3 | 64.2 | 53.5 | 47.6 | 41.8 | 56.3 |
| 2007 | 37.0 | 27.5 | 52.3 | 52.9 | 67.9 | 74.3 | 74.8 | 80.7 | 72.3 | 62.7 | 45.5 | 40.2 | 57.3 |
| 2008 | 32.3 | 35.5 | 44.2 | 54.6 | 61.3 | 73.8 | 75.6 | 74.6 | 71.5 | 57.2 | 42.9 | 35.8 | 54.9 |
| 2009 | 28.5 | 37.8 | 48.0 | 55.5 | 64.5 | 73.6 | 72.0 | 73.3 | 68.5 | 53.2 | 48.0 | 35.1 | 54.8 |
| POR= 113 YRS | 33.1 | 33.9 | 44.5 | 53.7 | 64.1 | 71.5 | 76.2 | 75.0 | 67.9 | 57.5 | 44.7 | 35.9 | 54.8 |

## HEATING DEGREE DAYS (base 65°F) 2009 LEXINGTON (KLEX)

| YEAR | JUL | AUG | SEP | OCT | NOV | DEC | JAN | FEB | MAR | APR | MAY | JUN | TOTAL |
|------|-----|-----|-----|-----|-----|-----|-----|-----|-----|-----|-----|-----|-------|
| 1980-81 | 0 | 0 | 23 | 358 | 633 | 892 | 1156 | 777 | 687 | 182 | 180 | 0 | 4888 |
| 1981-82 | 0 | 0 | 77 | 286 | 568 | 985 | 1134 | 840 | 549 | 429 | 14 | 9 | 4891 |
| 1982-83 | 0 | 1 | 75 | 259 | 500 | 646 | 961 | 772 | 580 | 422 | 151 | 7 | 4374 |
| 1983-84 | 0 | 0 | 59 | 201 | 550 | 1128 | 1152 | 685 | 778 | 370 | 178 | 3 | 5104 |
| 1984-85 | 2 | 0 | 89 | 84 | 689 | 601 | 1275 | 959 | 510 | 228 | 66 | 23 | 4526 |
| 1985-86 | 0 | 0 | 72 | 179 | 360 | 1092 | 978 | 735 | 561 | 259 | 94 | 2 | 4332 |
| 1986-87 | 0 | 15 | 14 | 250 | 595 | 903 | 1016 | 749 | 559 | 342 | 39 | 0 | 4482 |
| 1987-88 | 0 | 0 | 17 | 399 | 447 | 804 | 1085 | 901 | 620 | 328 | 90 | 18 | 4709 |
| 1988-89 | 0 | 3 | 30 | 474 | 560 | 877 | 750 | 887 | 548 | 351 | 196 | 8 | 4684 |
| 1989-90 | 0 | 6 | 61 | 267 | 592 | 1297 | 720 | 608 | 505 | 378 | 128 | 17 | 4579 |
| 1990-91 | 0 | 3 | 57 | 288 | 453 | 757 | 955 | 719 | 544 | 215 | 34 | 0 | 4025 |
| 1991-92 | 0 | 0 | 77 | 230 | 642 | 765 | 915 | 682 | 600 | 293 | 159 | 17 | 4380 |
| 1992-93 | 0 | 5 | 64 | 288 | 566 | 863 | 847 | 884 | 705 | 363 | 64 | 27 | 4676 |
| 1993-94 | 0 | 0 | 67 | 313 | 608 | 922 | 1231 | 783 | 658 | 229 | 185 | 3 | 4999 |
| 1994-95 | 0 | 3 | 37 | 232 | 425 | 730 | 960 | 877 | 545 | 298 | 108 | 2 | 4217 |
| 1995-96 | 0 | 0 | 65 | 254 | 755 | 983 | 1048 | 844 | 799 | 426 | 96 | 8 | 5278 |
| 1996-97 | 0 | 0 | 64 | 268 | 750 | 784 | 1025 | 670 | 576 | 471 | 233 | 31 | 4872 |
| 1997-98 | 1 | 3 | 39 | 321 | 698 | 904 | 745 | 666 | 614 | 331 | 46 | 24 | 4392 |
| 1998-99 | 0 | 0 | 8 | 237 | 522 | 781 | 887 | 690 | 758 | 254 | 51 | 1 | 4189 |
| 1999-00 | 0 | 0 | 43 | 254 | 433 | 855 | 1019 | 640 | 513 | 346 | 53 | 6 | 4162 |
| 2000-01 | 0 | 0 | 84 | 215 | 642 | 1227 | 1043 | 692 | 752 | 226 | 56 | 20 | 4957 |
| 2001-02 | 0 | 0 | 84 | 275 | 387 | 741 | 835 | 756 | 628 | 251 | 178 | 1 | 4136 |
| 2002-03 | 0 | 0 | 18 | 314 | 668 | 905 | 1199 | 917 | 515 | 255 | 86 | 30 | 4907 |
| 2003-04 | 0 | 0 | 61 | 271 | 460 | 908 | 1069 | 833 | 529 | 301 | 53 | 3 | 4488 |
| 2004-05 | 0 | 13 | 28 | 193 | 471 | 888 | 844 | 703 | 746 | 264 | 143 | 0 | 4293 |
| 2005-06 | 0 | 0 | 10 | 258 | 531 | 1004 | 702 | 821 | 634 | 203 | 147 | 2 | 4312 |
| 2006-07 | 0 | 0 | 77 | 363 | 514 | 711 | 863 | 1045 | 409 | 377 | 59 | 0 | 4418 |
| 2007-08 | 0 | 0 | 13 | 164 | 577 | 765 | 1007 | 849 | 638 | 319 | 139 | 0 | 4471 |
| 2008-09 | 0 | 0 | 2 | 274 | 658 | 899 | 1126 | 757 | 521 | 316 | 89 | 9 | 4651 |
| 2009- | 3 | 3 | 27 | 364 | 502 | 919 | | | | | | | |

WBAN : 93820

## COOLING DEGREE DAYS (base 65°F) 2009 LEXINGTON (KLEX)

| YEAR | JAN | FEB | MAR | APR | MAY | JUN | JUL | AUG | SEP | OCT | NOV | DEC | TOTAL |
|------|-----|-----|-----|-----|-----|-----|-----|-----|-----|-----|-----|-----|-------|
| 1980 | 0 | 0 | 0 | 4 | 87 | 210 | 438 | 415 | 199 | 17 | 0 | 0 | 1370 |
| 1981 | 0 | 0 | 1 | 29 | 43 | 270 | 341 | 267 | 109 | 2 | 0 | 0 | 1062 |
| 1982 | 0 | 0 | 0 | 4 | 171 | 121 | 383 | 252 | 101 | 62 | 9 | 7 | 1110 |
| 1983 | 0 | 0 | 4 | 3 | 27 | 248 | 465 | 487 | 219 | 21 | 0 | 0 | 1474 |
| 1984 | 0 | 0 | 0 | 17 | 50 | 340 | 254 | 312 | 141 | 44 | 1 | 0 | 1159 |
| 1985 | 0 | 0 | 5 | 40 | 67 | 189 | 317 | 245 | 155 | 49 | 4 | 0 | 1071 |
| 1986 | 0 | 0 | 4 | 34 | 115 | 285 | 427 | 269 | 197 | 42 | 0 | 0 | 1373 |
| 1987 | 0 | 0 | 0 | 10 | 212 | 304 | 383 | 395 | 173 | 2 | 5 | 0 | 1484 |
| 1988 | 0 | 0 | 1 | 8 | 81 | 306 | 442 | 407 | 120 | 5 | 0 | 0 | 1370 |
| 1989 | 0 | 0 | 8 | 34 | 66 | 214 | 362 | 296 | 146 | 27 | 0 | 0 | 1153 |
| 1990 | 0 | 0 | 13 | 29 | 32 | 239 | 326 | 285 | 168 | 26 | 5 | 0 | 1123 |
| 1991 | 0 | 0 | 3 | 15 | 210 | 285 | 398 | 341 | 198 | 41 | 0 | 0 | 1491 |
| 1992 | 0 | 0 | 0 | 35 | 76 | 155 | 340 | 192 | 132 | 8 | 0 | 0 | 938 |
| 1993 | 0 | 0 | 0 | 2 | 74 | 247 | 474 | 365 | 103 | 13 | 3 | 0 | 1281 |
| 1994 | 0 | 0 | 0 | 31 | 43 | 312 | 372 | 280 | 80 | 16 | 0 | 0 | 1134 |
| 1995 | 0 | 0 | 0 | 22 | 65 | 246 | 391 | 457 | 117 | 10 | 0 | 0 | 1308 |
| 1996 | 0 | 0 | 0 | 7 | 122 | 228 | 258 | 274 | 109 | 10 | 0 | 0 | 1008 |
| 1997 | 0 | 0 | 0 | 2 | 24 | 166 | 354 | 249 | 94 | 51 | 0 | 0 | 940 |
| 1998 | 0 | 0 | 25 | 0 | 132 | 264 | 303 | 356 | 291 | 24 | 0 | 1 | 1396 |
| 1999 | 0 | 0 | 0 | 5 | 67 | 261 | 459 | 339 | 161 | 6 | 1 | 0 | 1299 |
| 2000 | 0 | 1 | 0 | 2 | 113 | 268 | 279 | 278 | 126 | 45 | 3 | 0 | 1115 |
| 2001 | 0 | 0 | 0 | 75 | 111 | 212 | 327 | 351 | 124 | 24 | 2 | 0 | 1226 |
| 2002 | 0 | 0 | 0 | 48 | 66 | 299 | 444 | 412 | 234 | 43 | 1 | 0 | 1547 |
| 2003 | 0 | 0 | 0 | 21 | 50 | 155 | 322 | 337 | 72 | 7 | 3 | 0 | 967 |
| 2004 | 0 | 0 | 4 | 10 | 177 | 204 | 264 | 195 | 132 | 9 | 0 | 0 | 995 |
| 2005 | 0 | 0 | 0 | 10 | 41 | 307 | 403 | 426 | 217 | 52 | 0 | 0 | 1456 |
| 2006 | 0 | 0 | 0 | 29 | 77 | 180 | 372 | 386 | 58 | 12 | 0 | 0 | 1114 |
| 2007 | 0 | 0 | 26 | 21 | 155 | 284 | 309 | 496 | 238 | 100 | 0 | 0 | 1629 |
| 2008 | 0 | 0 | 0 | 14 | 32 | 273 | 334 | 305 | 203 | 40 | 0 | 0 | 1201 |
| 2009 | 0 | 0 | 0 | 40 | 80 | 276 | 227 | 266 | 136 | 6 | 0 | 0 | 1031 |

## SNOWFALL (inches) 2009 LEXINGTON (KLEX)

| YEAR | JUL | AUG | SEP | OCT | NOV | DEC | JAN | FEB | MAR | APR | MAY | JUN | TOTAL |
|------|-----|-----|-----|-----|-----|-----|-----|-----|-----|-----|-----|-----|-------|
| 1980-81 | 0.0 | 0.0 | 0.0 | 0.0 | 0.1 | 0.5 | 2.2 | 0.4 | 0.5 | 0.0 | 0.0 | 0.0 | 3.7 |
| 1981-82 | 0.0 | 0.0 | 0.0 | 0.0 | 0.4 | 1.7 | 5.6 | 3.9 | 0.3 | 0.7 | 0.0 | 0.0 | 12.6 |
| 1982-83 | 0.0 | 0.0 | 0.0 | 0.0 | 0.0 | T | 0.2 | 7.5 | 0.3 | T | 0.0 | 0.0 | 8.0 |
| 1983-84 | 0.0 | 0.0 | 0.0 | 0.0 | T | 1.7 | 8.4 | 4.6 | 0.3 | 0.0 | 0.0 | 0.0 | 15.0 |
| 1984-85 | 0.0 | 0.0 | 0.0 | 0.0 | T | 4.9 | 10.2 | 10.7 | T | 0.5 | 0.0 | 0.0 | 26.3 |
| 1985-86 | 0.0 | 0.0 | 0.0 | 0.0 | 0.0 | 3.5 | 1.2 | 8.9 | 0.7 | T | 0.0 | 0.0 | 14.3 |
| 1986-87 | 0.0 | 0.0 | 0.0 | 0.0 | 0.2 | T | 3.6 | 3.5 | 2.1 | 5.9 | 0.0 | 0.0 | 15.3 |
| 1987-88 | 0.0 | 0.0 | 0.0 | 0.0 | 1.0 | 1.8 | 3.3 | 3.4 | 0.7 | 0.0 | 0.0 | 0.0 | 10.2 |
| 1988-89 | 0.0 | 0.0 | 0.0 | 0.0 | T | 0.7 | T | 1.5 | T | T | T | 0.0 | 2.2 |
| 1989-90 | T | T | 0.0 | T | 1.1 | 9.3 | 0.2 | T | 3.7 | T | 0.0 | 0.0 | 14.3 |
| 1990-91 | 0.0 | 0.0 | 0.0 | 0.0 | 0.0 | 0.8 | 0.1 | 3.2 | 1.7 | 0.0 | 0.0 | 0.0 | 5.8 |
| 1991-92 | 0.0 | 0.0 | 0.0 | 0.0 | T | 0.2 | 0.7 | 0.6 | 1.3 | 0.4 | 0.0 | 0.0 | 3.2 |
| 1992-93 | 0.0 | 0.0 | 0.0 | T | 1.8 | 2.5 | 0.4 | 11.5 | 7.1 | T | 0.0 | T | 23.3 |
| 1993-94 | 0.0 | 0.0 | 0.0 | T | 0.1 | 7.4 | 16.4 | 2.8 | 5.0 | 0.0 | 0.0 | 0.0 | 31.7 |
| 1994-95 | 0.0 | 0.0 | 0.0 | 0.0 | 0.0 | T | 2.0 | 3.7 | 3.4 | 0.0 | T | 0.0 | 9.1 |
| 1995-96 | 0.0 | 0.0 | 0.0 | T | 1.1 | 2.2 | 16.0 | 3.8 | 6.8 | 1.5 | 0.0 | 0.0 | 31.4 |
| 1996-97 | 0.0 | 0.0 | 0.0 | 0.0 | T | T | 1.1 | 2.9 | T | T | 0.0 | 0.0 | 4.0 |
| 1997-98 | 0.0 | 0.0 | 0.0 | 0.0 | T | 2.1 | 0.1 | 17.4 | T | 0.0 | 0.0 | 0.0 | 19.6 |
| 1998-99 | 0.0 | 0.0 | 0.0 | 0.0 | 0.0 | 0.8 | 1.4 | 1.8 | 0.1 | 0.0 | 0.0 | 0.0 | 4.1 |
| 1999-00 | 0.0 | 0.0 | 0.0 | 0.0 | 0.0 | 1.9 | 2.8 | 0.3 | 0.2 | 0.0 | 0.0 | 0.0 | 5.2 |
| 2000-01 | 0.0 | 0.0 | 0.0 | 0.0 | T | 5.8 | 4.2 | T | T | 0.3 | 0.0 | 0.0 | 10.3 |
| 2001-02 | 0.0 | 0.0 | 0.0 | 0.0 | 0.0 | 0.3 | 7.8 | 0.6 | T | 0.0 | 0.0 | 0.0 | 8.7 |
| 2002-03 | 0.0 | 0.0 | 0.0 | 0.0 | 1.0 | 4.3 | 6.3 | 8.3 | T | 0.0 | 0.0 | 0.0 | 19.9 |
| 2003-04 | 0.0 | 0.0 | 0.0 | 0.0 | T | 3.9 | 4.7 | T | T | T | 0.0 | 0.0 | 8.6 |
| 2004-05 | 0.0 | 0.0 | 0.0 | T | T | 0.3 | 2.3 | 0.4 | 2.8 | T | 0.0 | 0.0 | 5.8 |
| 2005-06 | 0.0 | 0.0 | 0.0 | 0.0 | T | 0.5 | 2.6 | 8.1 | T | 0.0 | 0.0 | 0.0 | 11.2 |
| 2006-07 | 0.0 | 0.0 | 0.0 | 0.0 | 0.4 | 0.3 | 0.4 | 5.9 | T | 0.8 | 0.0 | T | 7.8 |
| 2007-08 | 0.0 | 0.0 | 0.0 | 0.0 | T | T | 2.0 | 4.3 | 4.5 | 0.0 | 0.0 | 0.0 | 10.8 |
| 2008-09 | 0.0 | 0.0 | 0.0 | T | 0.1 | 2.7 | 5.0 | 4.5 | 0.6 | T | 0.0 | T | 12.9 |
| 2009- | 0.0 | 0.0 | 0.0 | 0.0 | 0.0 | 2.2 |  |  |  |  |  |  |  |
| POR= 62 YRS | T | T | 0.0 | T | 0.9 | 2.3 | 5.3 | 4.5 | 2.3 | 0.3 | T | T | 15.6 |

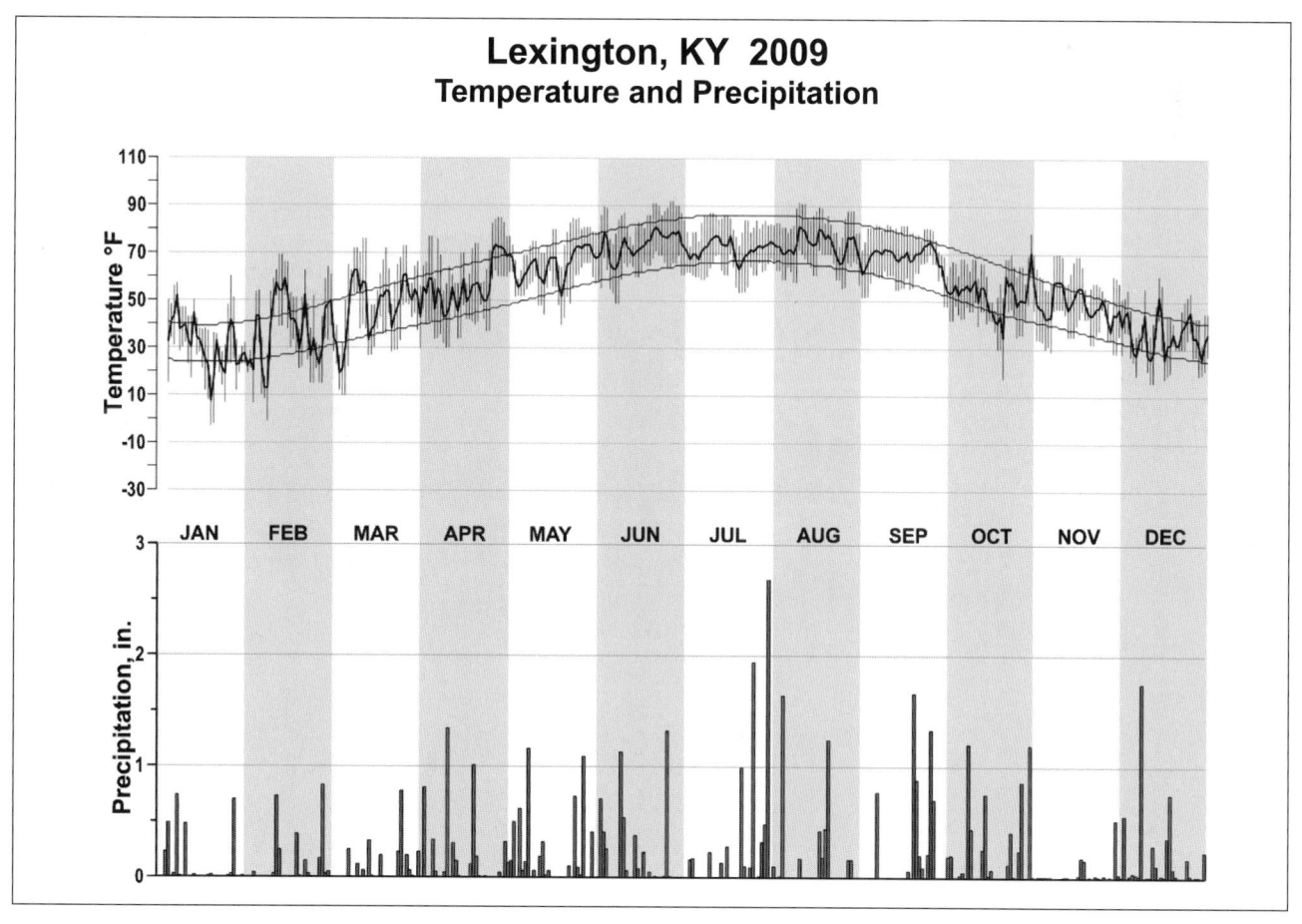

# Lexington, KY  2009
## Temperature and Precipitation

# 2009
# LOUISVILLE
# KENTUCKY (KSDF)

Louisville is located on the south bank of the Ohio River, 604 miles below Pittsburgh, Pennsylvania, and 377 miles above the mouth of the river at Cairo, Illinois. The city is divided by Beargrass Creek and its south fork into two portions with entirely different types of topography. The eastern portion is rolling, containing several creeks, and consists of plateaus and rolling hillsides. The highest elevation in this area is 565 feet. The western portion is mostly flat with an average elevation about 100 feet lower than the eastern area. Much of the western section lies in the flood plain of the Ohio River. Nearly all of the industries in the city are located in the western portion, while the eastern portion is almost entirely residential. A range of low hills about five miles northwest of Louisville, on the Indiana side of the Ohio River, present a partial barrier to arctic blasts in the winter months. During colder months, snow is frequently observed on the summits of these hills when there is no snow in the city of Louisville or in riverside communities on the Indiana side of the Ohio River.

The climate of Louisville, while continental in type, is of a variable nature because of its position with respect to the paths of high and low pressure systems and the occasional influx of warm moist air from the Gulf of Mexico. In winter and summer there are occasional cold and hot spells of short duration. As a whole, winters are moderately cold and summers are quite warm. Temperatures of 100 degrees or more in summer and zero degrees or less in winter are rare.

Thunderstorms with high rainfall intensities are common during the spring and summer months. The precipitation in Louisville is nonseasonal and varies from year to year. The fall months are usually the driest. Generally, March has the most rainfall and October the least. Snowfall usually occurs from November through March. As with rainfall, amounts vary from year to year and month to month. Some snow has also been recorded in the months of October and April. Mean total amounts for the months of January, February, and March are about the same with January showing a slight edge in total amount. Relative humidity remains rather high throughout the summer months. Cloud cover is about equally distributed throughout the year with the winter months showing somewhat of an increase in amount. The percentage of possible sunshine at Louisville varies from month to month with the greatest amount during the summer months as a result of the decreasing sky cover during that season. Heavy fog is unusual and there is only an average of 10 days during the year with heavy fog and these occur generally in the months of September through March.

The average date for the last occurrence in the spring of temperatures as low as 32 degrees is mid-April, and the first occurrence in the fall is generally in late October.

The prevailing direction of the wind has a southerly component and the velocity averages under 10 mph. The strongest winds are usually associated with thunderstorms.

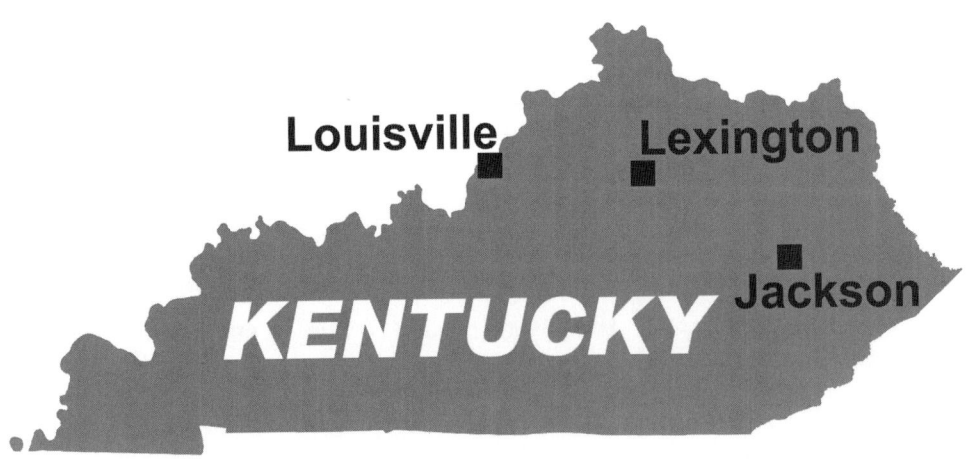

# NORMALS, MEANS, AND EXTREMES
## LOUISVILLE (KSDF)

| LATITUDE: 38 ° 10'N | LONGITUDE: -85 ° 43'W | | ELEVATION (FT): GRND: 480 BARO: 484 | | | | TIME ZONE: EASTERN (UTC -5) | | | WBAN: 93821 | | | |

| | ELEMENT | POR | JAN | FEB | MAR | APR | MAY | JUN | JUL | AUG | SEP | OCT | NOV | DEC | YEAR |
|---|---|---|---|---|---|---|---|---|---|---|---|---|---|---|---|
| **TEMPERATURE °F** | NORMAL DAILY MAXIMUM | 30 | 41.0 | 46.6 | 56.8 | 66.8 | 75.4 | 83.3 | 87.0 | 85.8 | 79.4 | 68.4 | 55.9 | 45.4 | 66.0 |
| | MEAN DAILY MAXIMUM | 62 | 41.9 | 46.3 | 56.1 | 67.8 | 76.6 | 84.5 | 88.0 | 87.2 | 80.6 | 69.3 | 56.2 | 45.5 | 66.7 |
| | HIGHEST DAILY MAXIMUM | 62 | 77 | 77 | 86 | 91 | 95 | 102 | 106 | 105 | 104 | 93 | 84 | 76 | 106 |
| | YEAR OF OCCURRENCE | | 1950 | 2000 | 1981 | 1960 | 1959 | 1952 | 1999 | 2007 | 1954 | 1958 | 1982 | JUL 1999 |
| | MEAN OF EXTREME MAXS. | 62 | 64.6 | 68.1 | 77.6 | 84.6 | 88.5 | 93.6 | 95.7 | 95.6 | 92.3 | 84.4 | 75.2 | 66.2 | 82.2 |
| | NORMAL DAILY MINIMUM | 30 | 24.9 | 28.5 | 37.1 | 46.0 | 56.1 | 65.1 | 69.8 | 68.2 | 60.9 | 48.5 | 39.3 | 29.9 | 47.9 |
| | MEAN DAILY MINIMUM | 62 | 25.5 | 28.3 | 36.3 | 46.2 | 55.5 | 64.2 | 68.4 | 66.9 | 59.4 | 47.3 | 37.7 | 29.3 | 47.1 |
| | LOWEST DAILY MINIMUM | 62 | -22 | -19 | -1 | 22 | 31 | 42 | 0 | 0 | 0 | 23 | -1 | -15 | -22 |
| | YEAR OF OCCURRENCE | | 1994 | 1951 | 1960 | 1982 | 1966 | 1966 | 1996 | 1996 | 1996 | 1952 | 1950 | 1989 | JAN 1994 |
| | MEAN OF EXTREME MINS. | 62 | 4.5 | 9.0 | 19.3 | 30.5 | 40.5 | 51.8 | 57.9 | 56.4 | 44.5 | 31.9 | 21.3 | 10.3 | 31.5 |
| | NORMAL DRY BULB | 30 | 33.0 | 37.6 | 46.9 | 56.4 | 65.8 | 74.2 | 78.4 | 77.0 | 70.1 | 58.5 | 47.6 | 37.6 | 56.9 |
| | MEAN DRY BULB | 62 | 33.7 | 37.3 | 46.2 | 57.0 | 66.0 | 74.4 | 78.2 | 77.1 | 70.0 | 58.3 | 47.0 | 37.4 | 56.9 |
| | MEAN WET BULB | 26 | 30.5 | 33.3 | 40.4 | 49.2 | 58.8 | 66.7 | 69.9 | 69.0 | 62.4 | 51.8 | 42.2 | 33.6 | 50.7 |
| | MEAN DEW POINT | 26 | 26.7 | 29.1 | 35.5 | 44.2 | 55.4 | 63.9 | 67.3 | 66.5 | 59.4 | 48.3 | 38.5 | 30.2 | 47.1 |
| | NORMAL NO. DAYS WITH: | | | | | | | | | | | | | | |
| | MAXIMUM >= 90 | 30 | 0.0 | 0.0 | 0.0 | * | 0.6 | 6.5 | 12.3 | 10.4 | 3.4 | 0.0 | 0.0 | 0.0 | 33.2 |
| | MAXIMUM <= 32 | 30 | 8.2 | 4.5 | 0.6 | 0.0 | 0.0 | 0.0 | 0.0 | 0.0 | 0.0 | 0.0 | 0.2 | 4.2 | 17.7 |
| | MINIMUM <= 32 | 30 | 23.1 | 17.8 | 11.1 | 2.1 | * | 0.0 | 0.0 | 0.0 | 0.0 | 1.0 | 8.7 | 18.8 | 82.6 |
| | MINIMUM <= 0 | 30 | 0.9 | 0.2 | 0.0 | 0.0 | 0.0 | 0.0 | 0.0 | 0.0 | 0.0 | 0.0 | 0.0 | 0.3 | 1.4 |
| **H/C** | NORMAL HEATING DEG. DAYS | 30 | 992 | 779 | 569 | 280 | 84 | 6 | 0 | 1 | 36 | 240 | 527 | 838 | 4352 |
| | NORMAL COOLING DEG. DAYS | 30 | 0 | 0 | 6 | 24 | 109 | 287 | 421 | 374 | 189 | 29 | 3 | 1 | 1443 |
| **RH** | NORMAL (PERCENT) | 30 | 71 | 68 | 64 | 62 | 68 | 69 | 70 | 72 | 72 | 70 | 70 | 73 | 69 |
| | HOUR 01 LST | 30 | 74 | 73 | 69 | 69 | 77 | 79 | 80 | 82 | 81 | 79 | 75 | 75 | 76 |
| | HOUR 07 LST | 30 | 78 | 77 | 76 | 76 | 82 | 83 | 85 | 87 | 88 | 85 | 79 | 79 | 81 |
| | HOUR 13 LST | 30 | 65 | 61 | 56 | 52 | 55 | 57 | 58 | 58 | 57 | 56 | 60 | 66 | 58 |
| | HOUR 19 LST | 30 | 68 | 63 | 56 | 53 | 57 | 59 | 60 | 61 | 62 | 62 | 65 | 69 | 61 |
| **S** | PERCENT POSSIBLE SUNSHINE | 48 | 41 | 49 | 51 | 56 | 61 | 67 | 68 | 67 | 65 | 61 | 46 | 40 | 56 |
| **W/O** | MEAN NO. DAYS WITH: | | | | | | | | | | | | | | |
| | HEAVY FOG(VISBY <= 1/4 MI) | 46 | 0.8 | 0.8 | 0.4 | 0.2 | 0.4 | 0.4 | 0.3 | 0.6 | 0.9 | 1.0 | 0.5 | 0.5 | 6.8 |
| | THUNDERSTORMS | 62 | 1.0 | 1.2 | 3.1 | 4.5 | 7.0 | 7.8 | 8.3 | 6.9 | 3.3 | 1.6 | 1.5 | 0.5 | 46.7 |
| **CLOUDNESS** | MEAN: | | | | | | | | | | | | | | |
| | SUNRISE-SUNSET (OKTAS) | | | | | | | | | | | | | | |
| | MIDNIGHT-MIDNIGHT (OKTAS) | | | | | | | | | | | | | | |
| | MEAN NO. DAYS WITH: | | | | | | | | | | | | | | |
| | CLEAR | | | | | | | | | | | | | | |
| | PARTLY CLOUDY | | | | | | | | | | | | | | |
| | CLOUDY | 1 | 1.0 | 4.0 | 6.0 | | 3.0 | 1.0 | | | 3.0 | 4.0 | | 2.0 | |
| **PR** | MEAN STATION PRESSURE(IN) | 26 | 29.60 | 29.58 | 29.52 | 29.46 | 29.46 | 29.46 | 29.48 | 29.50 | 29.53 | 29.56 | 29.58 | 29.61 | 29.53 |
| | MEAN SEA-LEVEL PRES. (IN) | 26 | 30.14 | 30.11 | 30.06 | 29.98 | 29.98 | 29.97 | 29.99 | 30.01 | 30.05 | 30.08 | 30.11 | 30.14 | 30.05 |
| **WINDS** | MEAN SPEED (MPH) | 26 | 9.2 | 9.1 | 9.1 | 9.1 | 7.8 | 7.1 | 6.7 | 6.3 | 6.4 | 7.0 | 8.3 | 8.4 | 7.9 |
| | PREVAIL.DIR(TENS OF DEGS) | 40 | 30 | 30 | 30 | 20 | 19 | 19 | 22 | 01 | 01 | 15 | 19 | 19 | 19 |
| | MAXIMUM 2-MINUTE: | | | | | | | | | | | | | | |
| | SPEED (MPH) | 15 | 45 | 46 | 47 | 56 | 54 | 54 | 55 | 47 | 53 | 40 | 44 | 40 | 56 |
| | DIR. (TENS OF DEGS) | | 26 | 28 | 21 | 22 | 30 | 04 | 36 | 21 | 22 | 29 | 23 | 27 | 22 |
| | YEAR OF OCCURRENCE | | 2008 | 2008 | 2006 | 1999 | 2006 | 1998 | 2004 | 1999 | 2008 | 2001 | 1998 | 2008 | APR 1999 |
| | MAXIMUM 3-SECOND | | | | | | | | | | | | | | |
| | SPEED (MPH) | 15 | 58 | 58 | 56 | 67 | 64 | 60 | 64 | 56 | 75 | 47 | 49 | 54 | 75 |
| | DIR. (TENS OF DEGS) | | 27 | 25 | 22 | 28 | 30 | 04 | 36 | 22 | 22 | 17 | 22 | 28 | 22 |
| | YEAR OF OCCURRENCE | | 2008 | 2009 | 2006 | 2003 | 2006 | 1998 | 2004 | 1999 | 2008 | 2009 | 1998 | 2008 | SEP 2008 |
| **PRECIPITATION** | NORMAL (IN) | 30 | 3.28 | 3.25 | 4.41 | 3.91 | 4.88 | 3.76 | 4.30 | 3.41 | 3.05 | 2.79 | 3.81 | 3.69 | 44.54 |
| | MAXIMUM MONTHLY (IN) | 62 | 11.38 | 9.02 | 14.91 | 11.10 | 11.57 | 10.11 | 10.05 | 8.79 | 10.49 | 8.86 | 9.12 | 8.86 | 14.91 |
| | YEAR OF OCCURRENCE | | 1950 | 1989 | 1964 | 1970 | 1990 | 1960 | 1979 | 1974 | 1979 | 1957 | 1990 | MAR 1964 |
| | MINIMUM MONTHLY (IN) | 62 | 0.45 | 0.76 | 1.02 | 0.76 | 1.37 | 0.49 | 0.34 | 0.23 | 0.01 | 0.39 | 0.72 | 0.65 | 0.01 |
| | YEAR OF OCCURRENCE | | 1981 | 1978 | 1966 | 1976 | 1977 | 1984 | 1999 | 1953 | 1995 | 1987 | 1976 | 1976 | SEP 1995 |
| | MAXIMUM IN 24 HOURS (IN) | 62 | 3.99 | 3.66 | 7.22 | 4.85 | 4.60 | 5.14 | 5.46 | 4.53 | 5.28 | 4.55 | 3.58 | 2.79 | 7.22 |
| | YEAR OF OCCURRENCE | | 2000 | 1990 | 1997 | 1970 | 1961 | 1960 | 1979 | 2009 | 2002 | 2007 | 1948 | 1978 | MAR 1997 |
| | NORMAL NO. DAYS WITH: | | | | | | | | | | | | | | |
| | PRECIPITATION >= 0.01 | 30 | 11.0 | 10.5 | 12.9 | 11.7 | 12.2 | 10.4 | 9.9 | 8.3 | 8.8 | 7.5 | 10.6 | 11.8 | 125.6 |
| | PRECIPITATION >= 1.00 | 30 | 0.7 | 0.7 | 0.8 | 0.8 | 1.3 | 0.8 | 1.2 | 0.8 | 0.7 | 0.7 | 1.0 | 1.0 | 10.5 |
| **SNOWFALL** | NORMAL (IN) | 30 | 5.1 | 4.5 | 2.1 | 0.2 | 0.* | 0.0 | 0.0 | 0.0 | 0.0 | 0.1 | 0.6 | 2.0 | 14.6 |
| | MAXIMUM MONTHLY (IN) | 62 | 28.4 | 19.3 | 22.9 | 1.6 | T | T | T | T | 0.0 | 2.4 | 13.2 | 10.0 | 28.4 |
| | YEAR OF OCCURRENCE | | 1978 | 1998 | 1960 | 1973 | 1989 | 1993 | 1994 | 2009 | | 1993 | 1966 | 2004 | JAN 1978 |
| | MAXIMUM IN 24 HOURS (IN) | 62 | 15.9 | 11.0 | 12.1 | 1.6 | T | T | T | T | 0.0 | 2.4 | 13.0 | 7.5 | 15.9 |
| | YEAR OF OCCURRENCE' | | 1994 | 1966 | 1968 | 1973 | 1989 | 1993 | 1994 | 2009 | | 1993 | 1966 | 2004 | JAN 1994 |
| | MAXIMUM SNOW DEPTH (IN) | 58 | 19 | 11 | 11 | 2 | 0 | 0 | 0 | 0 | 0 | T | 8 | 9 | 19 |
| | YEAR OF OCCURRENCE | | 1978 | 1966 | 1968 | 1987 | | | | | | 1989 | 1966 | 2004 | JAN 1978 |
| | NORMAL NO. DAYS WITH: | | | | | | | | | | | | | | |
| | SNOWFALL >= 1.0 | 30 | 1.5 | 1.1 | 0.6 | 0.1 | 0.0 | 0.0 | 0.0 | 0.0 | 0.0 | 0.1 | 0.2 | 0.7 | 4.3 |

## PRECIPITATION (inches) 2009  LOUISVILLE (KSDF)

| YEAR | JAN | FEB | MAR | APR | MAY | JUN | JUL | AUG | SEP | OCT | NOV | DEC | ANNUAL |
|------|-----|-----|-----|-----|-----|-----|-----|-----|-----|-----|-----|-----|--------|
| 1980 | 1.71 | 1.09 | 4.80 | 2.63 | 4.58 | 3.70 | 5.41 | 3.76 | 3.17 | 3.37 | 2.42 | 1.25 | 37.89 |
| 1981 | 0.45 | 3.23 | 1.54 | 4.44 | 4.63 | 3.23 | 3.98 | 3.21 | 3.22 | 1.60 | 2.40 | 2.02 | 33.95 |
| 1982 | 5.28 | 1.55 | 5.89 | 3.05 | 2.96 | 3.86 | 3.72 | 3.74 | 3.46 | 1.26 | 5.50 | 5.11 | 45.38 |
| 1983 | 1.63 | 1.52 | 2.16 | 7.10 | 10.58 | 4.42 | 0.99 | 2.39 | 1.13 | 6.47 | 5.03 | 3.96 | 47.38 |
| 1984 | 0.92 | 1.68 | 4.41 | 5.53 | 6.78 | 0.49 | 6.94 | 5.08 | 3.70 | 2.12 | 5.87 | 5.86 | 49.38 |
| 1985 | 2.20 | 2.08 | 4.43 | 1.69 | 3.93 | 4.37 | 3.45 | 4.49 | 1.48 | 4.24 | 4.43 | 0.96 | 37.75 |
| 1986 | 0.91 | 3.90 | 2.69 | 1.04 | 4.28 | 2.32 | 7.04 | 2.19 | 2.75 | 3.08 | 4.62 | 2.69 | 37.51 |
| 1987 | 0.81 | 4.42 | 3.05 | 2.35 | 1.61 | 3.58 | 5.31 | 2.66 | 1.15 | 0.39 | 2.62 | 4.70 | 32.65 |
| 1988 | 4.00 | 3.58 | 2.97 | 3.52 | 2.68 | 0.87 | 4.68 | 3.00 | 1.48 | 1.54 | 5.76 | 3.45 | 37.53 |
| 1989 | 3.68 | 9.02 | 5.50 | 4.93 | 4.39 | 5.26 | 6.90 | 2.20 | 2.42 | 2.65 | 2.57 | 1.45 | 50.97 |
| 1990 | 3.90 | 6.72 | 2.78 | 3.46 | 11.57 | 6.13 | 1.96 | 3.21 | 2.57 | 3.97 | 2.34 | 8.86 | 57.47 |
| 1991 | 3.29 | 3.72 | 4.79 | 2.61 | 4.02 | 1.23 | 2.99 | 3.35 | 2.74 | 2.31 | 1.87 | 5.23 | 38.15 |
| 1992 | 1.97 | 1.74 | 5.88 | 2.66 | 3.51 | 3.04 | 6.51 | 4.71 | 3.50 | 0.96 | 4.71 | 1.60 | 40.79 |
| 1993 | 3.50 | 4.20 | 5.20 | 3.57 | 2.80 | 4.05 | 4.58 | 5.74 | 3.90 | 4.03 | 3.26 | 2.56 | 47.39 |
| 1994 | 4.08 | 2.96 | 3.90 | 5.32 | 2.12 | 1.85 | 2.50 | 1.58 | 2.90 | 1.96 | 3.57 | 3.24 | 35.98 |
| 1995 | 3.20 | 2.00 | 2.17 | 2.64 | 9.48 | 2.84 | 3.39 | 4.07 | 0.01 | 5.42 | 2.39 | 3.28 | 40.89 |
| 1996 | 4.44 | 2.03 | 4.99 | 5.65 | 9.18 | 3.84 | | 1.31 | 5.66 | 2.59 | 3.35 | 4.56 | |
| 1997 | 3.35 | 3.39 | 12.58 | 2.01 | 6.01 | 8.11 | 1.74 | 3.70 | 1.28 | 1.41 | 3.63 | 2.50 | 49.71 |
| 1998 | 2.88 | 2.88 | 4.07 | 6.69 | 4.53 | 5.73 | 6.89 | 2.92 | 1.00 | 2.76 | 2.74 | 3.24 | 46.33 |
| 1999 | 7.23 | 2.20 | 3.47 | 3.04 | 3.12 | 6.36 | 0.34 | 0.97 | 1.74 | 2.46 | 1.61 | 4.81 | 37.35 |
| 2000 | 6.22 | 5.80 | 3.56 | 2.95 | 2.91 | 3.88 | 3.50 | 2.87 | 5.36 | 0.89 | 2.97 | 4.31 | 45.22 |
| 2001 | 1.46 | 3.42 | 2.27 | 1.04 | 5.19 | 2.61 | 4.47 | 3.42 | 4.08 | 6.39 | 5.16 | 4.48 | 43.99 |
| 2002 | 4.26 | 1.47 | 7.02 | 6.02 | 6.74 | 4.10 | 1.21 | 0.68 | 7.81 | 4.65 | 2.40 | 6.60 | 52.96 |
| 2003 | 1.13 | 4.12 | 2.04 | 5.99 | 6.42 | 3.22 | 3.44 | 6.72 | 6.44 | 1.90 | 4.51 | 3.13 | 49.06 |
| 2004 | 3.81 | 1.69 | 3.99 | 4.33 | 9.50 | 1.44 | 6.38 | 3.28 | 0.09 | 7.33 | 6.66 | 3.78 | 52.28 |
| 2005 | 5.07 | 2.35 | 3.85 | 3.56 | 4.67 | 2.46 | 3.02 | 7.17 | 1.32 | 0.82 | 3.53 | 2.04 | 39.86 |
| 2006 | 4.53 | 1.82 | 5.21 | 5.92 | 3.44 | 6.11 | 4.53 | 5.14 | 9.79 | 4.31 | 2.91 | 3.14 | 56.85 |
| 2007 | 3.63 | 2.90 | 2.99 | 4.55 | 2.37 | 1.58 | 4.13 | 1.61 | 1.95 | 8.86 | 2.44 | 7.52 | 44.53 |
| 2008 | 2.92 | 4.87 | 8.97 | 6.13 | 5.69 | 3.16 | 3.83 | 0.63 | 1.31 | 2.26 | 1.84 | 5.18 | 46.79 |
| 2009 | 3.63 | 2.20 | 1.36 | 4.43 | 4.59 | 9.22 | 6.02 | 5.88 | 5.70 | 7.00 | 1.05 | 2.85 | 53.93 |
| POR= 62 YRS | 3.49 | 3.30 | 4.54 | 4.07 | 4.66 | 3.76 | 4.09 | 3.31 | 3.16 | 2.92 | 3.57 | 3.68 | 44.55 |

WBAN : 93821

## AVERAGE TEMPERATURE (°F) 2009  LOUISVILLE (KSDF)

| YEAR | JAN | FEB | MAR | APR | MAY | JUN | JUL | AUG | SEP | OCT | NOV | DEC | ANNUAL |
|------|-----|-----|-----|-----|-----|-----|-----|-----|-----|-----|-----|-----|--------|
| 1980 | 33.5 | 29.6 | 41.8 | 53.6 | 66.8 | 73.4 | 81.5 | 81.0 | 73.5 | 55.8 | 46.3 | 38.3 | 56.3 |
| 1981 | 30.4 | 38.8 | 45.7 | 62.4 | 62.9 | 76.2 | 78.8 | 76.1 | 67.7 | 56.5 | 47.4 | 33.8 | 56.4 |
| 1982 | 28.6 | 34.9 | 47.1 | 51.3 | 70.3 | 69.3 | 78.0 | 73.5 | 66.8 | 59.0 | 48.7 | 44.9 | 56.0 |
| 1983 | 34.7 | 37.5 | 46.7 | 51.7 | 62.1 | 73.4 | 81.1 | 81.7 | 71.0 | 59.1 | 47.8 | 28.4 | 56.3 |
| 1984 | 28.9 | 41.5 | 40.4 | 55.0 | 62.6 | 77.7 | 75.5 | 76.0 | 67.2 | 63.9 | 44.0 | 45.9 | 56.6 |
| 1985 | 25.4 | 32.8 | 50.2 | 60.3 | 66.5 | 72.1 | 77.2 | 74.8 | 69.2 | 61.4 | 53.7 | 30.4 | 56.2 |
| 1986 | 34.5 | 39.9 | 48.3 | 58.5 | 67.0 | 75.7 | 80.3 | 74.3 | 73.1 | 59.5 | 45.9 | 36.7 | 57.8 |
| 1987 | 33.7 | 39.5 | 47.9 | 55.4 | 71.5 | 76.2 | 78.9 | 78.2 | 71.2 | 52.6 | 50.8 | 40.2 | 58.0 |
| 1988 | 31.0 | 34.7 | 46.1 | 57.0 | 67.1 | 75.6 | 80.3 | 80.0 | 70.1 | 52.3 | 47.8 | 38.0 | 56.7 |
| 1989 | 41.6 | 34.0 | 48.4 | 56.7 | 62.6 | 73.5 | 78.1 | 76.6 | 69.4 | 58.4 | 46.7 | 25.3 | 55.9 |
| 1990 | 43.1 | 44.3 | 51.2 | 55.5 | 64.2 | 75.1 | 78.5 | 77.5 | 71.8 | 58.7 | 52.0 | 40.8 | 59.4 |
| 1991 | 34.1 | 40.5 | 49.4 | 60.3 | 73.1 | 78.3 | 81.3 | 79.2 | 71.7 | 61.5 | 45.0 | 41.4 | 59.7 |
| 1992 | 37.1 | 43.7 | 47.9 | 58.2 | 63.9 | 72.1 | 78.5 | 73.2 | 69.1 | 58.1 | 47.9 | 38.6 | 57.4 |
| 1993 | 38.5 | 34.0 | 44.0 | 54.9 | 66.8 | 74.5 | 82.0 | 79.0 | 68.0 | 55.9 | 45.8 | 36.5 | 56.7 |
| 1994 | 26.8 | 38.0 | 45.4 | 59.9 | 63.0 | 77.5 | 79.0 | 76.1 | 68.7 | 59.4 | 52.2 | 42.4 | 57.4 |
| 1995 | 35.6 | 36.2 | 49.5 | 57.7 | 65.6 | 74.9 | 79.7 | 82.2 | 68.7 | 59.2 | 41.7 | 35.2 | 57.2 |
| 1996 | 32.4 | 37.7 | 41.0 | 53.6 | 68.5 | 74.6 | | | | 58.9 | 41.8 | 40.9 | |
| 1997 | 32.3 | 42.2 | 49.5 | 52.6 | 61.4 | 72.2 | 78.7 | 75.4 | 70.3 | 58.8 | 43.9 | 37.2 | 56.2 |
| 1998 | 42.4 | 43.6 | 48.1 | 56.2 | 70.3 | 75.0 | 78.3 | 78.8 | 76.2 | 62.6 | 51.3 | 42.0 | 60.4 |
| 1999 | 36.8 | 42.6 | 43.0 | 59.3 | 67.6 | 76.3 | 83.2 | 78.5 | 72.1 | 58.9 | 53.5 | 39.1 | 59.2 |
| 2000 | 34.4 | 45.1 | 51.4 | 55.8 | 69.2 | 75.0 | 76.8 | 77.1 | 68.0 | 61.4 | 44.6 | 26.2 | 57.1 |
| 2001 | 33.0 | 41.1 | 43.0 | 62.3 | 68.6 | 73.8 | 78.8 | 79.1 | 68.8 | 58.9 | 53.5 | 42.6 | 58.6 |
| 2002 | 40.8 | 40.6 | 46.1 | 60.6 | 64.7 | 77.9 | 81.4 | 80.9 | 75.2 | 58.3 | 45.3 | 38.2 | 59.2 |
| 2003 | 28.9 | 32.6 | 49.5 | 59.5 | 66.0 | 71.1 | 77.5 | 78.0 | 68.0 | 58.5 | 52.7 | 39.4 | 56.8 |
| 2004 | 33.1 | 38.7 | 51.3 | 59.1 | 71.7 | 76.1 | 76.6 | 74.3 | 71.9 | 61.4 | 51.5 | 36.6 | 58.6 |
| 2005 | 38.4 | 41.6 | 43.5 | 58.8 | 64.5 | 77.0 | 79.6 | 80.8 | 74.2 | 60.5 | 49.4 | 34.0 | 58.5 |
| 2006 | 44.2 | 37.9 | 47.0 | 61.5 | 65.0 | 73.7 | 79.3 | 79.3 | 66.6 | 55.7 | 49.1 | 43.1 | 58.5 |
| 2007 | 38.9 | 30.0 | 55.3 | 55.6 | 70.5 | 77.3 | 77.6 | 85.1 | 76.4 | 65.9 | 48.9 | 42.0 | 60.3 |
| 2008 | 34.7 | 37.7 | 46.5 | 57.9 | 64.8 | 78.6 | 79.8 | 78.8 | 73.9 | 59.7 | 45.5 | 37.5 | 58.0 |
| 2009 | 29.9 | 40.2 | 50.4 | 58.3 | 66.9 | 75.5 | 73.5 | 75.7 | 71.0 | 55.1 | 51.2 | 36.6 | 57.0 |
| POR= 62 YRS | 33.7 | 37.3 | 46.2 | 57.0 | 66.0 | 74.4 | 78.2 | 77.1 | 70.0 | 58.3 | 47.0 | 37.4 | 56.9 |

## HEATING DEGREE DAYS (base 65°F) 2009  LOUISVILLE (KSDF)

| YEAR | JUL | AUG | SEP | OCT | NOV | DEC | JAN | FEB | MAR | APR | MAY | JUN | TOTAL |
|---|---|---|---|---|---|---|---|---|---|---|---|---|---|
| 1980-81 | 0 | 0 | 12 | 309 | 555 | 821 | 1065 | 728 | 595 | 142 | 122 | 0 | 4349 |
| 1981-82 | 0 | 0 | 61 | 268 | 523 | 960 | 1124 | 837 | 549 | 408 | 13 | 3 | 4746 |
| 1982-83 | 0 | 1 | 56 | 246 | 495 | 624 | 933 | 763 | 571 | 399 | 121 | 5 | 4214 |
| 1983-84 | 0 | 0 | 54 | 196 | 509 | 1128 | 1115 | 673 | 757 | 315 | 141 | 0 | 4888 |
| 1984-85 | 0 | 0 | 73 | 84 | 623 | 584 | 1222 | 896 | 458 | 180 | 52 | 16 | 4188 |
| 1985-86 | 0 | 0 | 53 | 160 | 347 | 1067 | 941 | 696 | 516 | 224 | 69 | 0 | 4073 |
| 1986-87 | 0 | 12 | 5 | 210 | 570 | 869 | 962 | 706 | 526 | 294 | 21 | 0 | 4175 |
| 1987-88 | 0 | 0 | 9 | 377 | 423 | 762 | 1048 | 872 | 580 | 244 | 38 | 7 | 4360 |
| 1988-89 | 0 | 0 | 13 | 398 | 510 | 833 | 720 | 860 | 513 | 291 | 156 | 4 | 4298 |
| 1989-90 | 0 | 0 | 49 | 230 | 539 | 1222 | 672 | 574 | 445 | 320 | 82 | 13 | 4146 |
| 1990-91 | 0 | 0 | 34 | 229 | 387 | 745 | 949 | 677 | 482 | 167 | 27 | 0 | 3697 |
| 1991-92 | 0 | 0 | 52 | 168 | 590 | 725 | 855 | 610 | 523 | 244 | 124 | 8 | 3899 |
| 1992-93 | 0 | 0 | 40 | 219 | 505 | 813 | 819 | 859 | 644 | 299 | 44 | 18 | 4260 |
| 1993-94 | 0 | 0 | 48 | 289 | 572 | 875 | 1180 | 752 | 602 | 189 | 122 | 3 | 4632 |
| 1994-95 | 0 | 0 | 20 | 186 | 384 | 696 | 904 | 800 | 471 | 236 | 72 | 0 | 3769 |
| 1995-96 | 0 | 0 | 48 | 192 | 693 | 915 | 1002 | 782 | 738 | 353 | 66 | 2 | 4791 |
| 1996-97 |  |  |  | 202 | 689 | 741 | 1005 | 634 | 472 | 366 | 140 | 12 |  |
| 1997-98 | 0 | 0 | 9 | 263 | 621 | 854 | 696 | 594 | 561 | 261 | 27 | 15 | 3901 |
| 1998-99 | 0 | 0 | 0 | 119 | 405 | 711 | 866 | 620 | 676 | 183 | 16 | 0 | 3596 |
| 1999-00 | 0 | 0 | 20 | 197 | 346 | 796 | 942 | 575 | 416 | 270 | 26 | 2 | 3590 |
| 2000-01 | 0 | 0 | 60 | 174 | 608 | 1198 | 980 | 664 | 674 | 175 | 31 | 8 | 4572 |
| 2001-02 | 0 | 0 | 52 | 224 | 339 | 689 | 743 | 677 | 579 | 204 | 114 | 0 | 3621 |
| 2002-03 | 0 | 0 | 1 | 252 | 585 | 821 | 1114 | 898 | 474 | 205 | 48 | 14 | 4412 |
| 2003-04 | 0 | 0 | 37 | 214 | 380 | 788 | 982 | 758 | 435 | 209 | 40 | 0 | 3843 |
| 2004-05 | 0 | 1 | 8 | 128 | 402 | 873 | 820 | 648 | 656 | 207 | 95 | 0 | 3838 |
| 2005-06 | 0 | 0 | 9 | 206 | 469 | 954 | 638 | 756 | 548 | 152 | 98 | 0 | 3830 |
| 2006-07 | 0 | 0 | 44 | 311 | 471 | 670 | 805 | 971 | 346 | 329 | 27 | 0 | 3974 |
| 2007-08 | 0 | 0 | 3 | 114 | 480 | 712 | 935 | 787 | 569 | 237 | 65 | 0 | 3902 |
| 2008-09 | 0 | 0 | 0 | 212 | 578 | 849 | 1080 | 687 | 449 | 259 | 48 | 7 | 4169 |
| 2009- | 0 | 0 | 18 | 304 | 407 | 873 |  |  |  |  |  |  |  |

WBAN : 93821

## COOLING DEGREE DAYS (base 65°F) 2009  LOUISVILLE (KSDF)

| YEAR | JAN | FEB | MAR | APR | MAY | JUN | JUL | AUG | SEP | OCT | NOV | DEC | TOTAL |
|---|---|---|---|---|---|---|---|---|---|---|---|---|---|
| 1980 | 0 | 0 | 0 | 8 | 134 | 266 | 519 | 504 | 276 | 31 | 1 | 0 | 1739 |
| 1981 | 0 | 0 | 5 | 68 | 63 | 343 | 435 | 348 | 150 | 10 | 0 | 0 | 1422 |
| 1982 | 0 | 0 | 1 | 2 | 183 | 139 | 408 | 274 | 118 | 68 | 13 | 8 | 1214 |
| 1983 | 0 | 0 | 7 | 8 | 39 | 264 | 504 | 524 | 240 | 19 | 0 | 0 | 1605 |
| 1984 | 0 | 0 | 0 | 20 | 69 | 386 | 333 | 349 | 145 | 56 | 0 | 1 | 1359 |
| 1985 | 0 | 2 | 8 | 48 | 106 | 233 | 387 | 311 | 185 | 55 | 14 | 0 | 1349 |
| 1986 | 0 | 0 | 5 | 37 | 138 | 330 | 481 | 306 | 255 | 46 | 0 | 0 | 1598 |
| 1987 | 0 | 0 | 0 | 14 | 232 | 342 | 439 | 416 | 203 | 1 | 4 | 0 | 1651 |
| 1988 | 0 | 0 | 4 | 10 | 111 | 333 | 481 | 472 | 173 | 10 | 0 | 0 | 1594 |
| 1989 | 0 | 0 | 6 | 48 | 88 | 264 | 412 | 364 | 188 | 30 | 0 | 0 | 1400 |
| 1990 | 0 | 0 | 22 | 44 | 65 | 323 | 427 | 392 | 244 | 42 | 7 | 0 | 1566 |
| 1991 | 0 | 0 | 8 | 31 | 286 | 406 | 514 | 445 | 262 | 68 | 0 | 0 | 2020 |
| 1992 | 0 | 0 | 2 | 48 | 100 | 229 | 424 | 262 | 169 | 14 | 0 | 0 | 1248 |
| 1993 | 0 | 0 | 0 | 4 | 106 | 310 | 534 | 442 | 146 | 12 | 3 | 0 | 1557 |
| 1994 | 0 | 0 | 0 | 42 | 63 | 384 | 443 | 349 | 138 | 21 | 4 | 0 | 1444 |
| 1995 | 0 | 0 | 0 | 29 | 100 | 304 | 466 | 544 | 165 | 21 | 0 | 0 | 1629 |
| 1996 | 0 | 0 | 0 | 18 | 183 | 298 |  |  |  | 19 | 0 | 0 |  |
| 1997 | 0 | 0 | 0 | 2 | 35 | 237 | 435 | 330 | 174 | 79 | 0 | 0 | 1292 |
| 1998 | 0 | 0 | 44 | 4 | 200 | 321 | 416 | 436 | 345 | 52 | 0 | 7 | 1825 |
| 1999 | 0 | 0 | 0 | 18 | 103 | 345 | 573 | 427 | 238 | 12 | 6 | 0 | 1722 |
| 2000 | 0 | 5 | 2 | 0 | 162 | 308 | 375 | 385 | 156 | 68 | 2 | 0 | 1463 |
| 2001 | 0 | 0 | 0 | 103 | 150 | 280 | 433 | 445 | 174 | 40 | 1 | 0 | 1626 |
| 2002 | 0 | 0 | 0 | 77 | 110 | 391 | 514 | 496 | 314 | 51 | 3 | 0 | 1956 |
| 2003 | 0 | 0 | 0 | 46 | 87 | 203 | 398 | 411 | 136 | 17 | 17 | 0 | 1315 |
| 2004 | 0 | 0 | 18 | 41 | 255 | 339 | 398 | 296 | 221 | 24 | 2 | 0 | 1594 |
| 2005 | 1 | 0 | 0 | 29 | 89 | 365 | 459 | 494 | 292 | 72 | 5 | 0 | 1806 |
| 2006 | 0 | 0 | 0 | 53 | 106 | 269 | 452 | 452 | 102 | 30 | 2 | 0 | 1466 |
| 2007 | 0 | 0 | 51 | 51 | 202 | 376 | 396 | 629 | 350 | 149 | 2 | 0 | 2206 |
| 2008 | 2 | 0 | 0 | 31 | 68 | 413 | 466 | 436 | 273 | 57 | 0 | 1 | 1747 |
| 2009 | 0 | 0 | 1 | 64 | 113 | 329 | 268 | 339 | 206 | 4 | 0 | 0 | 1324 |

**SNOWFALL (inches) 2009 LOUISVILLE (KSDF)**

| YEAR | JUL | AUG | SEP | OCT | NOV | DEC | JAN | FEB | MAR | APR | MAY | JUN | TOTAL |
|---|---|---|---|---|---|---|---|---|---|---|---|---|---|
| 1980-81 | 0.0 | 0.0 | 0.0 | T | T | T | 2.5 | 0.3 | 0.1 | 0.0 | 0.0 | 0.0 | 2.9 |
| 1981-82 | 0.0 | 0.0 | 0.0 | 0.0 | 0.1 | 3.6 | 2.7 | 2.9 | 0.3 | 1.4 | 0.0 | 0.0 | 11.0 |
| 1982-83 | 0.0 | 0.0 | 0.0 | 0.0 | T | 0.0 | 0.6 | 4.5 | 0.1 | T | 0.0 | 0.0 | 5.2 |
| 1983-84 | 0.0 | 0.0 | 0.0 | 0.0 | 0.0 | 0.6 | 3.1 | 8.8 | 1.0 | T | 0.0 | 0.0 | 13.5 |
| 1984-85 | 0.0 | 0.0 | 0.0 | 0.0 | T | 4.8 | 7.4 | 6.7 | T | T | 0.0 | 0.0 | 18.9 |
| 1985-86 | 0.0 | 0.0 | 0.0 | 0.0 | 0.0 | 1.6 | 1.1 | 8.8 | 0.1 | 0.0 | 0.0 | 0.0 | 11.6 |
| 1986-87 | 0.0 | 0.0 | 0.0 | 0.0 | T | T | 2.2 | 6.7 | 9.3 | T | 0.0 | 0.0 | 18.2 |
| 1987-88 | 0.0 | 0.0 | 0.0 | 0.0 | T | T | 3.0 | 5.0 | 0.5 | 0.0 | 0.0 | 0.0 | 8.5 |
| 1988-89 | 0.0 | 0.0 | 0.0 | 0.0 | T | 0.3 | T | 0.6 | T | 0.0 | T | 0.0 | 0.9 |
| 1989-90 | 0.0 | 0.0 | 0.0 | 1.4 | T | 6.5 | 1.9 | 0.8 | 4.1 | T | 0.0 | 0.0 | 14.7 |
| 1990-91 | 0.0 | 0.0 | 0.0 | 0.0 | 0.0 | 4.1 | 0.3 | 1.5 | 0.2 | 0.0 | 0.0 | 0.0 | 6.1 |
| 1991-92 | 0.0 | 0.0 | 0.0 | 0.0 | 0.5 | 0.1 | 0.9 | 0.1 | 0.9 | 0.7 | 0.0 | 0.0 | 3.2 |
| 1992-93 | 0.0 | 0.0 | 0.0 | 0.0 | 0.9 | 1.0 | T | 15.9 | 1.1 | T | 0.0 | T | 18.9 |
| 1993-94 | 0.0 | 0.0 | 0.0 | 2.4 | T | 3.6 | 17.7 | 1.5 | 4.7 | T | 0.0 | 0.0 | 29.9 |
| 1994-95 | T | 0.0 | 0.0 | 0.0 | 0.0 | T | 0.1 | 2.8 | 1.1 | 0.0 | 0.0 | 0.0 | 4.0 |
| 1995-96 | 0.0 | 0.0 | 0.0 | 0.0 | T | 1.1 | 13.8 | 1.3 | 8.0 | T | | |  |
| 1996-97 | | | | 0.0 | T | 0.4 | 3.0 | 1.9 | T | T | 0.0 | 0.0 |  |
| 1997-98 | 0.0 | 0.0 | 0.0 | 0.0 | 0.4 | 1.9 | T | 19.3 | 1.2 | T | T | 0.0 | 22.8 |
| 1998-99 | 0.0 | 0.0 | 0.0 | 0.0 | 0.0 | 2.8 | 5.5 | 3.7 | 1.3 | 0.0 | 0.0 | T | 13.3 |
| 1999-00 | 0.0 | T | 0.0 | 0.0 | 0.0 | 4.9 | 6.7 | 0.1 | 0.3 | T | 0.0 | 0.0 | 12.0 |
| 2000-01 | 0.0 | 0.0 | 0.0 | 0.0 | T | 9.7 | 4.8 | 0.3 | T | T | 0.0 | 0.0 | 14.8 |
| 2001-02 | 0.0 | 0.0 | 0.0 | T | 0.0 | 0.6 | 8.1 | 0.6 | 0.4 | T | 0.0 | 0.0 | 9.7 |
| 2002-03 | 0.0 | 0.0 | 0.0 | 0.0 | 0.4 | 6.1 | 3.5 | 6.3 | T | T | 0.0 | 0.0 | 16.3 |
| 2003-04 | 0.0 | T | 0.0 | 0.0 | T | 3.6 | 3.6 | T | T | T | T | 0.0 | 7.2 |
| 2004-05 | 0.0 | 0.0 | 0.0 | 0.0 | 0.0 | 10.0 | 1.7 | 0.4 | 2.0 | T | 0.0 | 0.0 | 14.1 |
| 2005-06 | 0.0 | 0.0 | 0.0 | 0.0 | T | 0.8 | 1.7 | 3.1 | T | T | 0.0 | 0.0 | 5.6 |
| 2006-07 | 0.0 | 0.0 | 0.0 | 0.0 | T | 0.2 | 0.1 | 3.1 | T | T | 0.0 | 0.0 | 3.4 |
| 2007-08 | 0.0 | 0.0 | 0.0 | 0.0 | T | 1.4 | 2.1 | 5.8 | 10.6 | 0.0 | 0.0 | 0.0 | 19.9 |
| 2008-09 | 0.0 | 0.0 | 0.0 | 0.0 | T | 1.3 | 9.5 | 1.3 | T | T | 0.0 | 0.0 | 12.1 |
| 2009- | 0.0 | T | 0.0 | 0.0 | 0.0 | 0.3 | | | | | | |  |
| POR=<br>61 YRS | T | T | 0.0 | 0.1 | 0.8 | 2.3 | 5.1 | 4.2 | 2.8 | 0.1 | T | T | 15.4 |

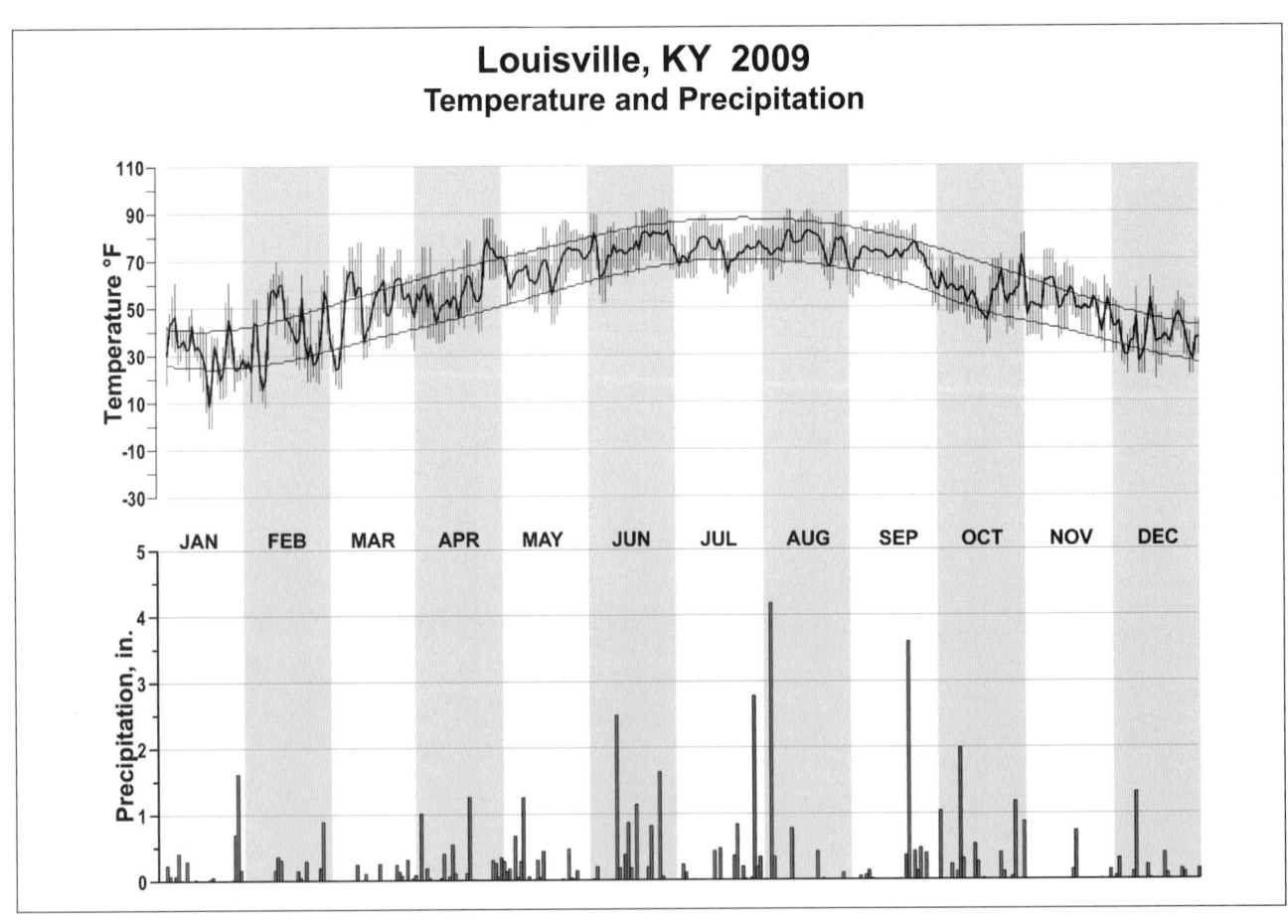

# Louisville, KY 2009
## Temperature and Precipitation

# 2009
# NEW ORLEANS
# LOUISIANA (KMSY)

The New Orleans metropolitan area is virtually surrounded by water. Lake Pontchartrain, some 610 square miles in area, borders the city on the north and is connected to the Gulf of Mexico through Lake Borgne on the east. In other directions there are bayous, lakes, and marshy delta land. The proximity of the Gulf of Mexico also has a great influence on the climate. Elevations in the city vary from a few feet below to a few feet above mean sea level. A massive levee system surrounding the city and along the Mississippi River offers protection against flooding from the river and tidal surges. The New Orleans International Airport is located 12 miles west of downtown New Orleans, between the Mississippi River and Lake Pontchartrain.

The climate of the city can best be described as humid with the surrounding water modifying the temperature and decreasing the range between the extremes. Almost daily sporadic afternoon thunderstorms from mid-June through September keep the temperature from rising much above 90 degrees. From about mid-November to mid-March, the area is subjected alternately to the southerly flow of warm tropical air and to the northerly flow of cold continental air in periods of varying lengths. The usual track of winter storms is to the north of New Orleans, but occasionally one moves this far south, bringing large and rather sudden drops in temperature. However, the cold spells seldom last over three or four days. The lowest temperatures observed are below 10 degrees. In about two-thirds of the years, the lowest temperature is about 24 degrees or warmer. The lowest temperatures in some years are entirely above freezing.

During the winter and spring, the cold Mississippi River water enhances the formation of river fogs, particularly when light southerly winds bring warm, moist air into the area from the Gulf of Mexico. The nearby lakes and marshes also contribute to fog formation. Even so, the fog usually does not seriously affect automobile traffic except for brief periods. However, air travel will be suspended for several hours and river traffic, at times, will be unable to move between New Orleans and the Gulf for several days.

Rather frequent and sometimes very heavy rains are typical for this area. There are an average of 120 days of measurable rain per year and an annual average accumulation of over 60 inches. A fairly definite rainy period occurs from mid-December to mid-March. Precipitation during this period is most likely to be steady rain for two to three day periods. April, May, October, and November are generally dry, but there have been some extremely heavy showers in those months. The greatest 24-hour amounts have exceeded 14 inches. Snowfall is rather infrequent and light. However, on rare occasions, snowstorms have produced accumulations over 8 inches.

While thunder occurs with most of the showers in the area, thunderstorms with damaging winds are infrequent. Hail of a damaging nature seldom occurs, and tornadoes are extremely rare. However, waterspouts are observed quite often on nearby lakes. Hurricanes have effected the area.

The lower Mississippi River floods result from runoff upstream. If the water level in the river becomes dangerously high, the spillways upriver can be opened to divert the floodwaters. Rainfall in the New Orleans area is pumped into the surrounding lakes and bayous. Local street and minor urban flooding of short duration result from occasional downpours.

Air pollution is not a serious problem. The area is not highly industrialized, and long periods of air stagnation are rare.

Based on the 1951-1980 period, the average first occurrence of 32 degrees Fahrenheit in the fall is December 5 and the average last occurrence in the spring is February 20.

# NORMALS, MEANS, AND EXTREMES
## NEW ORLEANS (KMSY)

| LATITUDE: 29°59'N | LONGITUDE: -90°15'W | | ELEVATION (FT): GRND: 0  BARO: 7 | | | | TIME ZONE: CENTRAL (UTC -6) | | | WBAN: 12916 | | | |

| | ELEMENT | POR | JAN | FEB | MAR | APR | MAY | JUN | JUL | AUG | SEP | OCT | NOV | DEC | YEAR |
|---|---|---|---|---|---|---|---|---|---|---|---|---|---|---|---|
| TEMPERATURE °F | NORMAL DAILY MAXIMUM | 30 | 61.8 | 65.3 | 72.1 | 78.0 | 84.8 | 89.4 | 91.1 | 91.0 | 87.1 | 79.7 | 71.0 | 64.5 | 78.0 |
| | MEAN DAILY MAXIMUM | 71 | 62.2 | 64.5 | 71.4 | 77.8 | 84.8 | 89.2 | 90.8 | 90.6 | 86.5 | 79.7 | 70.3 | 64.4 | 77.7 |
| | HIGHEST DAILY MAXIMUM | 63 | 83 | 85 | 89 | 92 | 96 | 101 | 101 | 102 | 101 | 94 | 87 | 84 | 102 |
| | YEAR OF OCCURRENCE | | 1982 | 1972 | 1982 | 1987 | 1953 | 2009 | 1981 | 1980 | 1980 | 1998 | 1997 | 1995 | AUG 1980 |
| | MEAN OF EXTREME MAXS. | 71 | 77.5 | 79.3 | 82.5 | 86.6 | 91.4 | 94.6 | 95.7 | 95.5 | 93.2 | 88.7 | 83.1 | 79.7 | 87.3 |
| | NORMAL DAILY MINIMUM | 30 | 43.4 | 46.1 | 52.7 | 58.4 | 66.4 | 72.0 | 74.2 | 73.9 | 70.6 | 60.2 | 51.8 | 45.6 | 59.6 |
| | MEAN DAILY MINIMUM | 71 | 44.1 | 46.1 | 52.5 | 58.9 | 66.4 | 71.6 | 74.0 | 73.9 | 70.4 | 60.8 | 51.1 | 45.9 | 59.6 |
| | LOWEST DAILY MINIMUM | 63 | 14 | 16 | 25 | 32 | 41 | 50 | 60 | 60 | 42 | 35 | 24 | 11 | 11 |
| | YEAR OF OCCURRENCE | | 1985 | 1996 | 1980 | 1971 | 1960 | 1984 | 1967 | 1968 | 1967 | 1993 | 1970 | 1989 | DEC 1989 |
| | MEAN OF EXTREME MINS. | 71 | 26.8 | 30.4 | 35.8 | 44.0 | 54.6 | 64.1 | 69.5 | 68.7 | 60.6 | 44.8 | 35.4 | 29.0 | 47.0 |
| | NORMAL DRY BULB | 30 | 52.6 | 55.7 | 62.4 | 68.2 | 75.6 | 80.7 | 82.7 | 82.5 | 78.9 | 70.0 | 61.4 | 55.1 | 68.8 |
| | MEAN DRY BULB | 71 | 53.1 | 55.3 | 62.0 | 68.4 | 75.6 | 80.4 | 82.4 | 82.3 | 78.5 | 70.3 | 60.7 | 55.1 | 68.7 |
| | MEAN WET BULB | 26 | 48.3 | 51.2 | 56.3 | 61.9 | 69.0 | 73.9 | 75.7 | 75.7 | 72.4 | 64.3 | 56.5 | 50.5 | 63.0 |
| | MEAN DEW POINT | 26 | 45.3 | 48.3 | 53.3 | 59.0 | 66.7 | 72.3 | 74.1 | 74.0 | 70.2 | 61.6 | 53.8 | 47.4 | 60.5 |
| | NORMAL NO. DAYS WITH: | | | | | | | | | | | | | | |
| | MAXIMUM >= 90 | 30 | 0.0 | 0.0 | 0.0 | 0.3 | 3.9 | 15.8 | 21.5 | 22.3 | 10.2 | 1.1 | 0.0 | 0.0 | 75.1 |
| | MAXIMUM <= 32 | 30 | 0.1 | 0.0 | 0.0 | 0.0 | 0.0 | 0.0 | 0.0 | 0.0 | 0.0 | 0.0 | 0.0 | 0.1 | 0.2 |
| | MINIMUM <= 32 | 30 | 4.6 | 2.4 | 0.4 | * | 0.0 | 0.0 | 0.0 | 0.0 | 0.0 | 0.0 | 0.7 | 3.5 | 11.6 |
| | MINIMUM <= 0 | 30 | 0.0 | 0.0 | 0.0 | 0.0 | 0.0 | 0.0 | 0.0 | 0.0 | 0.0 | 0.0 | 0.0 | 0.0 | 0.0 |
| H/C | NORMAL HEATING DEG. DAYS | 30 | 403 | 288 | 150 | 44 | 1 | 0 | 0 | 0 | 0 | 30 | 169 | 332 | 1417 |
| | NORMAL COOLING DEG. DAYS | 30 | 12 | 19 | 62 | 136 | 320 | 466 | 538 | 534 | 413 | 182 | 62 | 29 | 2773 |
| RH | NORMAL (PERCENT) | 30 | 75 | 74 | 73 | 73 | 74 | 78 | 79 | 79 | 77 | 75 | 76 | 77 | 76 |
| | HOUR 00 LST | 30 | 81 | 80 | 81 | 83 | 85 | 87 | 88 | 88 | 86 | 85 | 85 | 83 | 84 |
| | HOUR 06 LST | 30 | 84 | 84 | 85 | 87 | 89 | 90 | 92 | 92 | 89 | 88 | 87 | 85 | 88 |
| | HOUR 12 LST | 30 | 66 | 63 | 61 | 59 | 60 | 64 | 66 | 65 | 65 | 60 | 63 | 65 | 63 |
| | HOUR 18 LST | 30 | 71 | 66 | 65 | 64 | 65 | 68 | 71 | 72 | 72 | 71 | 75 | 74 | 70 |
| S | PERCENT POSSIBLE SUNSHINE | 22 | 46 | 50 | 56 | 62 | 62 | 63 | 59 | 61 | 61 | 64 | 54 | 48 | 57 |
| W/O | MEAN NO. DAYS WITH: | | | | | | | | | | | | | | |
| | HEAVY FOG(VISBY <= 1/4 MI) | 46 | 4.8 | 3.4 | 3.0 | 1.1 | 0.5 | 0.2 | 0.2 | 0.3 | 0.3 | 1.4 | 2.7 | 4.0 | 21.9 |
| | THUNDERSTORMS | 62 | 2.1 | 2.9 | 4.0 | 4.2 | 5.8 | 10.3 | 15.5 | 13.0 | 6.5 | 2.2 | 2.3 | 2.2 | 71.0 |
| CLOUDNESS | MEAN: | | | | | | | | | | | | | | |
| | SUNRISE-SUNSET (OKTAS) | 48 | 5.4 | 5.0 | 5.0 | 4.6 | 4.9 | 4.9 | 5.1 | 4.6 | 4.3 | 3.6 | 4.3 | 5.1 | 4.7 |
| | MIDNIGHT-MIDNIGHT (OKTAS) | 32 | 5.2 | 4.8 | 4.9 | 4.4 | 4.2 | 4.0 | 4.6 | 4.3 | 4.0 | 3.3 | 4.0 | 4.8 | 4.4 |
| | MEAN NO. DAYS WITH: | | | | | | | | | | | | | | |
| | CLEAR | 48 | 6.9 | 7.5 | 7.8 | 7.9 | 8.9 | 8.3 | 4.6 | 7.2 | 9.6 | 14.3 | 10.2 | 7.7 | 100.9 |
| | PARTLY CLOUDY | 48 | 7.1 | 6.4 | 8.0 | 10.4 | 11.2 | 12.5 | 14.6 | 13.8 | 10.6 | 7.9 | 8.2 | 7.4 | 118.1 |
| | CLOUDY | 48 | 16.9 | 14.3 | 15.2 | 11.7 | 10.9 | 9.2 | 11.8 | 10.0 | 9.8 | 8.9 | 11.5 | 15.9 | 146.1 |
| PR | MEAN STATION PRESSURE(IN) | 26 | 30.13 | 30.09 | 30.03 | 29.98 | 29.96 | 29.95 | 30.00 | 29.97 | 29.95 | 30.02 | 30.08 | 30.12 | 30.02 |
| | MEAN SEA-LEVEL PRES. (IN) | 26 | 30.16 | 30.12 | 30.06 | 30.02 | 29.99 | 29.98 | 29.98 | 30.03 | 30.00 | 30.05 | 30.11 | 30.15 | 30.05 |
| WINDS | MEAN SPEED (MPH) | 26 | 9.1 | 9.5 | 9.3 | 9.3 | 8.2 | 6.8 | 5.9 | 6.0 | 7.5 | 8.0 | 8.4 | 8.9 | 8.1 |
| | PREVAIL.DIR(TENS OF DEGS) | 31 | 01 | 36 | 17 | 17 | 19 | 19 | 25 | 05 | 05 | 05 | 01 | 36 | 19 |
| | MAXIMUM 2-MINUTE: | | | | | | | | | | | | | | |
| | SPEED (MPH) | 13 | 48 | 43 | 39 | 45 | 44 | 40 | 46 | 40 | 46 | 39 | 39 | 39 | 48 |
| | DIR. (TENS OF DEGS) | | 27 | 21 | 25 | 30 | 32 | 25 | 04 | 02 | 01 | 17 | 28 | 24 | 27 |
| | YEAR OF OCCURRENCE | | 1998 | 1998 | 2009 | 2004 | 2004 | 2004 | 2005 | 1997 | 2002 | 2002 | 2004 | 2000 | JAN 1998 |
| | MAXIMUM 3-SECOND | | | | | | | | | | | | | | |
| | SPEED (MPH) | 13 | 63 | 51 | 51 | 55 | 64 | 64 | 55 | 48 | 63 | 51 | 46 | 51 | 64 |
| | DIR. (TENS OF DEGS) | | 33 | 21 | 19 | 29 | 02 | 25 | 03 | 04 | 04 | 13 | 21 | 25 | 25 |
| | YEAR OF OCCURRENCE | | 1998 | 1998 | 2006 | 2004 | 2004 | 2004 | 2005 | 1997 | 2008 | 2002 | 2002 | 2005 | JUN 2004 |
| PRECIPITATION | NORMAL (IN) | 30 | 5.87 | 5.47 | 5.24 | 5.02 | 4.62 | 6.83 | 6.20 | 6.15 | 5.55 | 3.05 | 5.09 | 5.07 | 64.16 |
| | MAXIMUM MONTHLY (IN) | 63 | 19.28 | 12.59 | 19.09 | 16.12 | 21.18 | 17.62 | 13.15 | 16.12 | 18.98 | 13.20 | 19.81 | 25.92 | 25.92 |
| | YEAR OF OCCURRENCE | | 1998 | 1983 | 1948 | 1980 | 1995 | 2001 | 1991 | 1977 | 1998 | 1985 | 1989 | 2009 | DEC 2009 |
| | MINIMUM MONTHLY (IN) | 63 | 0.19 | 0.15 | 0.24 | 0.28 | 0.07 | 0.23 | 1.38 | 1.68 | 0.24 | 0.00 | 0.21 | 1.46 | 0.00 |
| | YEAR OF OCCURRENCE | | 2003 | 1989 | 1955 | 1976 | 2000 | 1979 | 2000 | 1980 | 1953 | 1978 | 1949 | 1958 | OCT 1978 |
| | MAXIMUM IN 24 HOURS (IN) | 63 | 6.08 | 5.60 | 7.87 | 8.08 | 12.40 | 7.40 | 4.43 | 4.96 | 9.55 | 6.08 | 12.66 | 8.76 | 12.66 |
| | YEAR OF OCCURRENCE | | 1978 | 1961 | 1948 | 1988 | 1995 | 1988 | 1996 | 1992 | 2002 | 2007 | 1989 | 2009 | NOV 1989 |
| | NORMAL NO. DAYS WITH: | | | | | | | | | | | | | | |
| | PRECIPITATION >= 0.01 | 30 | 10.5 | 8.4 | 8.6 | 7.2 | 8.0 | 11.8 | 13.9 | 13.2 | 10.2 | 5.9 | 8.6 | 9.4 | 115.7 |
| | PRECIPITATION >= 1.00 | 30 | 1.9 | 1.8 | 2.1 | 1.5 | 1.6 | 2.5 | 1.7 | 1.7 | 1.7 | 1.0 | 1.8 | 1.5 | 20.8 |
| SNOWFALL | NORMAL (IN) | 30 | 0.* | 0.* | 0.* | 0.0 | 0.0 | 0.0 | 0.0 | 0.0 | 0.0 | 0.0 | 0.0 | 0.* | 0.0 |
| | MAXIMUM MONTHLY (IN) | 51 | 0.4 | 2.0 | T | T | T | 0.0 | 0.0 | 0.0 | 0.0 | 0.0 | T | 2.7 | 2.7 |
| | YEAR OF OCCURRENCE | | 1985 | 1958 | 1993 | 1996 | 1989 | | | | | | 1950 | 1963 | DEC 1963 |
| | MAXIMUM IN 24 HOURS (IN) | 50 | 0.4 | 2.0 | T | T | T | 0.0 | 0.0 | 0.0 | 0.0 | 0.0 | T | 2.7 | 2.7 |
| | YEAR OF OCCURRENCE' | | 1985 | 1958 | 1993 | 1996 | 1989 | | | | | | 1950 | 1963 | DEC 1963 |
| | MAXIMUM SNOW DEPTH (IN) | 48 | 2 | 2 | 0 | 0 | 0 | 0 | 0 | 0 | 0 | 0 | 0 | 1 | 2 |
| | YEAR OF OCCURRENCE | | 1964 | 1958 | | | | | | | | | | 1989 | JAN 1964 |
| | NORMAL NO. DAYS WITH: | | | | | | | | | | | | | | |
| | SNOWFALL >= 1.0 | 30 | 0.0 | 0.0 | 0.0 | 0.0 | 0.0 | 0.0 | 0.0 | 0.0 | 0.0 | 0.0 | 0.0 | 0.0 | 0.0 |

## PRECIPITATION (inches) 2009 NEW ORLEANS (KMSY)

| YEAR | JAN | FEB | MAR | APR | MAY | JUN | JUL | AUG | SEP | OCT | NOV | DEC | ANNUAL |
|------|-----|-----|-----|-----|-----|-----|-----|-----|-----|-----|-----|-----|--------|
| 1980 | 6.37 | 3.09 | 10.08 | 16.12 | 9.65 | 3.69 | 4.84 | 1.68 | 6.31 | 5.87 | 3.85 | 1.54 | 73.09 |
| 1981 | 0.94 | 8.34 | 2.70 | 2.28 | 5.35 | 8.47 | 1.92 | 11.10 | 4.78 | 2.03 | 1.10 | 5.50 | 54.51 |
| 1982 | 2.76 | 7.88 | 2.56 | 5.86 | 1.19 | 5.43 | 13.07 | 1.92 | 5.40 | 3.84 | 5.45 | 10.26 | 65.62 |
| 1983 | 3.31 | 12.59 | 4.88 | 14.86 | 3.71 | 10.64 | 2.95 | 6.29 | 5.72 | 4.88 | 6.32 | 9.15 | 85.30 |
| 1984 | 4.10 | 5.27 | 4.90 | 1.72 | 3.54 | 7.21 | 3.86 | 9.51 | 3.79 | 2.84 | 2.80 | 2.53 | 52.07 |
| 1985 | 4.83 | 9.28 | 7.07 | 2.11 | 1.16 | 4.56 | 6.92 | 6.37 | 5.74 | 13.20 | 0.96 | 4.78 | 66.98 |
| 1986 | 3.49 | 2.93 | 1.88 | 1.50 | 1.61 | 8.87 | 3.60 | 6.74 | 1.42 | 2.87 | 7.90 | 5.05 | 47.86 |
| 1987 | 8.88 | 7.38 | 4.39 | 2.27 | 3.46 | 15.01 | 6.38 | 5.05 | 1.29 | 0.72 | 2.92 | 2.88 | 60.63 |
| 1988 | 3.74 | 11.31 | 8.90 | 9.25 | 1.68 | 11.28 | 6.78 | 7.53 | 5.86 | 2.87 | 1.26 | 3.94 | 74.40 |
| 1989 | 2.47 | 0.15 | 7.14 | 3.20 | 3.50 | 8.22 | 8.34 | 3.31 | 4.53 | 0.51 | 19.81 | 6.28 | 67.46 |
| 1990 | 7.59 | 11.45 | 5.98 | 4.59 | 5.87 | 1.01 | 2.30 | 2.45 | 4.55 | 2.38 | 3.21 | 9.67 | 61.05 |
| 1991 | 19.25 | 5.42 | 6.27 | 15.29 | 14.28 | 10.71 | 13.15 | 7.86 | 3.44 | 1.88 | 2.19 | 2.63 | 102.37 |
| 1992 | 9.94 | 8.73 | 6.69 | 2.52 | 0.95 | 9.52 | 5.75 | 9.64 | 6.63 | 0.55 | 15.27 | 5.68 | 81.87 |
| 1993 | 6.21 | 2.34 | 5.65 | 6.82 | 7.23 | 4.96 | 5.77 | 2.26 | 2.47 | 3.67 | 2.43 | 2.90 | 52.71 |
| 1994 | 3.25 | 0.54 | 4.82 | 2.83 | 3.67 | 9.35 | 8.95 | 4.59 | 5.61 | 2.30 | 1.39 | 4.61 | 51.91 |
| 1995 | 3.66 | 4.94 | 7.89 | 3.81 | 21.18 | 2.84 | 6.44 | 3.26 | 0.69 | 1.31 | 4.24 | 5.07 | 65.33 |
| 1996 | 4.66 | 1.56 | 2.97 | 3.87 | 1.37 | 8.60 | 10.32 | 8.76 | 3.96 | 2.59 | 3.10 | 5.55 | 57.31 |
| 1997 | 6.32 | 6.88 | 2.57 | 4.91 | 5.03 | 6.97 | 3.94 | 2.25 | 0.81 | 1.36 | 8.09 | 2.55 | 51.68 |
| 1998 | 19.28 | 4.28 | 5.97 | 4.39 | 0.43 | 3.38 | 6.56 | 8.30 | 18.98 | 1.82 | 3.40 | 2.25 | 79.04 |
| 1999 | 3.20 | 0.92 | 4.60 | 0.30 | 3.37 | 12.20 | 4.05 | 5.21 | 2.87 | 5.46 | 0.28 | 3.85 | 46.31 |
| 2000 | 2.25 | 1.81 | 2.41 | 1.13 | 0.07 | 5.46 | 1.38 | 2.35 | 6.50 | 1.10 | 11.72 | 2.70 | 38.88 |
| 2001 | 3.05 | 1.59 | 8.07 | 1.08 | 6.85 | 17.62 | 6.97 | 7.41 | 6.30 | 5.13 | 2.54 | 2.90 | 69.51 |
| 2002 | 3.29 | 2.76 | 3.58 | 2.14 | 3.04 | 4.83 | 4.54 | 4.09 | 14.23 | 10.09 | 5.10 | 4.82 | 62.51 |
| 2003 | 0.19 | 4.56 | 4.46 | 6.19 | 3.04 | 17.37 | 6.41 | 6.19 | 5.68 | 4.12 | 6.13 | 1.94 | 66.28 |
| 2004 | 3.00 | 8.41 | 1.11 | 14.81 | 10.04 | 13.58 | 3.89 | 3.75 | 1.29 | 8.53 | 7.43 | 3.44 | 79.28 |
| 2005 | 4.41 | 8.24 | 4.68 | 3.31 | 2.59 | 2.52 | 10.65 | 8.27 | 4.22 | 0.04 | 0.75 | 3.32 | 53.00 |
| 2006 | 2.95 | 3.16 | 0.63 | 3.78 | 0.28 | 2.78 | 4.86 | 6.70 | 4.72 | 3.24 | 2.75 | 10.03 | 45.88 |
| 2007 | 4.86 | 2.30 | 2.09 | 1.65 | 8.57 | 8.64 | 4.26 | 4.08 | 2.12 | 9.44 | 0.58 | 4.71 | 53.30 |
| 2008 | 3.34 | 2.88 | 2.15 | 6.70 | 7.19 | 4.40 | 3.24 | 7.21 | 11.17 | 1.78 | 1.77 | 2.21 | 54.04 |
| 2009 | 6.40 | 4.39 | 5.07 | 1.33 | 2.63 | 2.13 | 5.73 | 9.50 | 9.22 | 5.97 | 1.02 | 25.92 | 79.31 |
| POR= 71 YRS | 4.84 | 4.99 | 4.83 | 4.51 | 4.59 | 6.00 | 6.32 | 5.77 | 5.53 | 3.18 | 4.19 | 5.17 | 59.92 |

WBAN : 12916

## AVERAGE TEMPERATURE (°F) 2009 NEW ORLEANS (KMSY)

| YEAR | JAN | FEB | MAR | APR | MAY | JUN | JUL | AUG | SEP | OCT | NOV | DEC | ANNUAL |
|------|-----|-----|-----|-----|-----|-----|-----|-----|-----|-----|-----|-----|--------|
| 1980 | 56.1 | 52.0 | 62.0 | 66.2 | 77.9 | 83.3 | 85.8 | 85.5 | 83.5 | 68.8 | 60.0 | 53.6 | 69.6 |
| 1981 | 48.5 | 55.3 | 61.9 | 71.4 | 74.8 | 83.8 | 85.0 | 83.1 | 77.9 | 71.1 | 64.9 | 54.5 | 69.4 |
| 1982 | 54.5 | 55.0 | 65.9 | 69.8 | 76.5 | 81.5 | 81.4 | 82.2 | 76.8 | 70.4 | 62.5 | 59.4 | 69.7 |
| 1983 | 50.2 | 53.8 | 58.3 | 64.3 | 74.0 | 77.6 | 81.5 | 82.4 | 75.3 | 69.1 | 60.0 | 49.5 | 66.3 |
| 1984 | 46.6 | 53.9 | 59.3 | 67.5 | 73.7 | 77.4 | 78.8 | 79.2 | 76.3 | 73.5 | 58.8 | 62.4 | 67.3 |
| 1985 | 45.2 | 52.3 | 65.3 | 69.0 | 74.7 | 79.3 | 80.2 | 81.6 | 77.0 | 72.6 | 67.3 | 51.0 | 68.0 |
| 1986 | 51.2 | 59.2 | 60.6 | 67.2 | 76.7 | 81.0 | 83.2 | 81.6 | 81.0 | 70.4 | 66.3 | 53.2 | 69.3 |
| 1987 | 50.0 | 56.3 | 60.3 | 66.2 | 76.8 | 79.9 | 82.9 | 83.5 | 78.2 | 64.4 | 61.6 | 59.0 | 68.3 |
| 1988 | 49.4 | 53.2 | 60.9 | 68.4 | 73.3 | 78.5 | 81.6 | 81.4 | 79.8 | 68.0 | 65.6 | 56.0 | 68.0 |
| 1989 | 60.2 | 55.9 | 62.8 | 67.0 | 76.4 | 79.4 | 81.4 | 81.7 | 76.9 | 67.5 | 62.4 | 46.9 | 68.2 |
| 1990 | 57.2 | 61.3 | 63.3 | 67.6 | 76.2 | 82.6 | 82.3 | 83.0 | 79.6 | 68.1 | 62.3 | 59.0 | 70.2 |
| 1991 | 52.8 | 58.2 | 64.2 | 71.2 | 77.5 | 81.3 | 83.5 | 81.7 | 78.2 | 71.6 | 55.9 | 57.6 | 69.5 |
| 1992 | 51.1 | 58.3 | 62.0 | 66.4 | 72.8 | 80.6 | 83.1 | 79.6 | 78.5 | 69.5 | 57.4 | 58.2 | 68.1 |
| 1993 | 57.2 | 54.7 | 58.8 | 64.0 | 71.9 | 80.4 | 83.3 | 83.6 | 79.7 | 69.9 | 58.0 | 52.2 | 67.8 |
| 1994 | 50.0 | 56.9 | 60.8 | 69.6 | 75.1 | 81.4 | 80.8 | 81.3 | 77.8 | 70.8 | 65.3 | 56.5 | 68.9 |
| 1995 | 53.0 | 56.3 | 63.0 | 68.8 | 77.6 | 79.3 | 83.9 | 84.7 | 79.5 | 71.3 | 59.8 | 55.2 | 69.4 |
| 1996 | 52.9 | 55.9 | 58.4 | 67.1 | 78.1 | 80.6 | 82.8 | 81.1 | 79.0 | 70.4 | 63.3 | 57.3 | 68.9 |
| 1997 | 53.8 | 57.1 | 66.4 | 65.4 | 74.5 | 79.6 | 83.4 | 83.6 | 81.1 | 71.3 | 58.6 | 52.8 | 69.0 |
| 1998 | 56.1 | 55.9 | 59.8 | 67.5 | 78.8 | 83.7 | 85.3 | 84.6 | 81.1 | 73.3 | 65.4 | 59.1 | 70.9 |
| 1999 | 57.7 | 61.1 | 61.6 | 73.1 | 76.7 | 81.4 | 82.2 | 85.5 | 78.2 | 70.7 | 62.0 | 55.2 | 70.5 |
| 2000 | 56.4 | 60.6 | 66.0 | 68.9 | 80.1 | 81.7 | 84.6 | 84.4 | 79.3 | 69.8 | 59.9 | 49.6 | 70.1 |
| 2001 | 50.2 | 60.3 | 59.4 | 72.2 | 76.1 | 80.3 | 82.9 | 82.5 | 78.5 | 68.3 | 65.0 | 57.6 | 69.4 |
| 2002 | 54.7 | 51.9 | 62.1 | 72.5 | 76.6 | 80.8 | 83.4 | 82.8 | 80.9 | 74.0 | 59.9 | 54.2 | 69.5 |
| 2003 | 49.1 | 55.4 | 62.6 | 69.6 | 79.7 | 81.2 | 82.6 | 84.0 | 80.6 | 72.4 | 66.3 | 53.7 | 69.8 |
| 2004 | 53.7 | 53.0 | 65.4 | 67.6 | 75.5 | 81.0 | 82.6 | 81.6 | 80.7 | 76.6 | 65.4 | 52.4 | 69.6 |
| 2005 | 57.9 | 59.4 | 60.8 | 68.1 | 75.6 | 82.0 | 83.7 | 84.8 | 83.2 | 71.8 | 64.0 | 53.9 | 70.4 |
| 2006 | 59.3 | 56.7 | 65.8 | 73.5 | 76.9 | 83.8 | 83.1 | 83.5 | 79.2 | 71.6 | 59.9 | 55.1 | 70.7 |
| 2007 | 53.6 | 53.9 | 65.1 | 67.5 | 75.5 | 81.3 | 82.0 | 85.1 | 81.7 | 72.1 | 62.9 | 60.3 | 70.1 |
| 2008 | 53.0 | 59.1 | 62.4 | 69.2 | 76.7 | 82.6 | 83.8 | 82.5 | 78.9 | 70.2 | 61.0 | 59.3 | 69.9 |
| 2009 | 55.4 | 58.7 | 65.4 | 68.9 | 78.2 | 83.3 | 84.1 | 82.8 | 81.0 | 71.8 | 61.0 | 53.4 | 70.3 |
| POR= 71 YRS | 53.1 | 55.3 | 62.0 | 68.4 | 75.6 | 80.4 | 82.4 | 82.3 | 78.5 | 70.3 | 60.7 | 55.1 | 68.7 |

## HEATING DEGREE DAYS (base 65°F) 2009  NEW ORLEANS (KMSY)

| YEAR | JUL | AUG | SEP | OCT | NOV | DEC | JAN | FEB | MAR | APR | MAY | JUN | TOTAL |
|------|-----|-----|-----|-----|-----|-----|-----|-----|-----|-----|-----|-----|-------|
| 1980-81 | 0 | 0 | 0 | 35 | 195 | 363 | 504 | 275 | 123 | 12 | 0 | 0 | 1507 |
| 1981-82 | 0 | 0 | 0 | 36 | 100 | 333 | 365 | 278 | 127 | 29 | 0 | 0 | 1268 |
| 1982-83 | 0 | 0 | 0 | 31 | 146 | 234 | 453 | 309 | 217 | 81 | 1 | 0 | 1472 |
| 1983-84 | 0 | 0 | 1 | 37 | 183 | 483 | 564 | 321 | 197 | 48 | 2 | 0 | 1836 |
| 1984-85 | 0 | 0 | 2 | 14 | 214 | 146 | 605 | 359 | 62 | 28 | 0 | 0 | 1430 |
| 1985-86 | 0 | 0 | 0 | 12 | 49 | 443 | 421 | 195 | 160 | 28 | 0 | 0 | 1308 |
| 1986-87 | 0 | 0 | 0 | 28 | 85 | 370 | 464 | 242 | 168 | 75 | 0 | 0 | 1432 |
| 1987-88 | 0 | 0 | 0 | 58 | 149 | 222 | 490 | 351 | 166 | 23 | 0 | 0 | 1459 |
| 1988-89 | 0 | 0 | 0 | 12 | 92 | 301 | 186 | 292 | 155 | 60 | 0 | 0 | 1098 |
| 1989-90 | 0 | 0 | 0 | 53 | 142 | 559 | 253 | 136 | 101 | 41 | 0 | 0 | 1285 |
| 1990-91 | 0 | 0 | 0 | 62 | 122 | 244 | 371 | 196 | 105 | 8 | 0 | 0 | 1108 |
| 1991-92 | 0 | 0 | 0 | 22 | 312 | 262 | 426 | 203 | 128 | 54 | 5 | 0 | 1412 |
| 1992-93 | 0 | 0 | 0 | 2 | 240 | 218 | 248 | 285 | 209 | 82 | 0 | 0 | 1284 |
| 1993-94 | 0 | 0 | 0 | 42 | 259 | 399 | 464 | 263 | 177 | 49 | 0 | 0 | 1653 |
| 1994-95 | 0 | 0 | 0 | 16 | 72 | 268 | 375 | 257 | 123 | 33 | 0 | 0 | 1144 |
| 1995-96 | 0 | 0 | 0 | 16 | 186 | 358 | 380 | 307 | 248 | 54 | 1 | 0 | 1550 |
| 1996-97 | 0 | 0 | 0 | 17 | 116 | 253 | 373 | 248 | 58 | 44 | 0 | 0 | 1109 |
| 1997-98 | 0 | 0 | 0 | 38 | 202 | 383 | 273 | 251 | 210 | 25 | 0 | 0 | 1382 |
| 1998-99 | 0 | 0 | 0 | 2 | 48 | 243 | 263 | 150 | 118 | 21 | 0 | 0 | 845 |
| 1999-00 | 0 | 0 | 0 | 23 | 112 | 318 | 295 | 181 | 63 | 33 | 0 | 0 | 1025 |
| 2000-01 | 0 | 0 | 4 | 34 | 227 | 469 | 453 | 181 | 184 | 16 | 0 | 0 | 1568 |
| 2001-02 | 0 | 0 | 0 | 44 | 60 | 249 | 356 | 365 | 184 | 13 | 0 | 0 | 1271 |
| 2002-03 | 0 | 0 | 0 | 1 | 186 | 336 | 487 | 265 | 97 | 39 | 0 | 0 | 1411 |
| 2003-04 | 0 | 0 | 0 | 3 | 89 | 345 | 360 | 344 | 61 | 39 | 3 | 0 | 1244 |
| 2004-05 | 0 | 0 | 0 | 2 | 71 | 397 | 249 | 174 | 164 | 11 | 0 | 0 | 1068 |
| 2005-06 | 0 | 0 | 0 | 47 | 116 | 350 | 190 | 245 | 74 | 0 | 0 | 0 | 1022 |
| 2006-07 | 0 | 0 | 0 | 22 | 175 | 310 | 358 | 311 | 91 | 63 | 0 | 0 | 1330 |
| 2007-08 | 0 | 0 | 0 | 29 | 122 | 207 | 384 | 193 | 143 | 32 | 0 | 0 | 1110 |
| 2008-09 | 0 | 0 | 0 | 36 | 144 | 242 | 322 | 198 | 78 | 36 | 0 | 0 | 1056 |
| 2009- | 0 | 0 | 0 | 39 | 134 | 357 | | | | | | | |

WBAN : 12916

## COOLING DEGREE DAYS (base 65°F) 2009  NEW ORLEANS (KMSY)

| YEAR | JAN | FEB | MAR | APR | MAY | JUN | JUL | AUG | SEP | OCT | NOV | DEC | TOTAL |
|------|-----|-----|-----|-----|-----|-----|-----|-----|-----|-----|-----|-----|-------|
| 1980 | 10 | 13 | 70 | 85 | 409 | 554 | 653 | 640 | 561 | 160 | 51 | 17 | 3223 |
| 1981 | 0 | 12 | 35 | 210 | 311 | 570 | 627 | 565 | 396 | 231 | 102 | 12 | 3071 |
| 1982 | 49 | 6 | 160 | 182 | 366 | 504 | 517 | 541 | 363 | 208 | 78 | 66 | 3040 |
| 1983 | 0 | 0 | 16 | 67 | 286 | 385 | 518 | 545 | 317 | 171 | 42 | 10 | 2357 |
| 1984 | 0 | 6 | 31 | 130 | 281 | 379 | 436 | 448 | 351 | 286 | 33 | 71 | 2452 |
| 1985 | 0 | 10 | 78 | 154 | 308 | 437 | 480 | 521 | 366 | 251 | 124 | 13 | 2742 |
| 1986 | 0 | 40 | 32 | 99 | 370 | 487 | 573 | 524 | 488 | 203 | 127 | 9 | 2952 |
| 1987 | 3 | 4 | 30 | 120 | 373 | 456 | 562 | 580 | 402 | 48 | 53 | 42 | 2673 |
| 1988 | 14 | 15 | 49 | 131 | 263 | 411 | 523 | 513 | 448 | 113 | 118 | 30 | 2628 |
| 1989 | 46 | 43 | 95 | 124 | 363 | 439 | 515 | 525 | 365 | 137 | 70 | 6 | 2728 |
| 1990 | 17 | 40 | 56 | 127 | 353 | 538 | 545 | 567 | 448 | 166 | 50 | 62 | 2969 |
| 1991 | 2 | 12 | 88 | 202 | 396 | 496 | 580 | 524 | 402 | 233 | 47 | 40 | 3022 |
| 1992 | 0 | 18 | 41 | 103 | 258 | 472 | 568 | 459 | 408 | 149 | 20 | 14 | 2510 |
| 1993 | 13 | 4 | 26 | 61 | 222 | 468 | 572 | 585 | 447 | 200 | 57 | 8 | 2663 |
| 1994 | 4 | 40 | 54 | 191 | 320 | 501 | 495 | 513 | 390 | 204 | 94 | 12 | 2818 |
| 1995 | 10 | 20 | 67 | 155 | 399 | 434 | 594 | 615 | 444 | 216 | 39 | 59 | 3052 |
| 1996 | 13 | 50 | 47 | 124 | 412 | 473 | 558 | 507 | 427 | 189 | 70 | 21 | 2891 |
| 1997 | 34 | 31 | 107 | 67 | 301 | 445 | 576 | 583 | 489 | 238 | 19 | 9 | 2899 |
| 1998 | 4 | 1 | 59 | 108 | 435 | 568 | 635 | 613 | 489 | 267 | 68 | 70 | 3317 |
| 1999 | 41 | 48 | 20 | 273 | 369 | 497 | 543 | 643 | 403 | 209 | 29 | 23 | 3098 |
| 2000 | 35 | 58 | 103 | 159 | 477 | 506 | 616 | 610 | 439 | 192 | 79 | 0 | 3274 |
| 2001 | 0 | 56 | 19 | 239 | 353 | 469 | 564 | 550 | 410 | 153 | 69 | 29 | 2911 |
| 2002 | 45 | 4 | 100 | 248 | 368 | 482 | 577 | 559 | 483 | 289 | 40 | 7 | 3202 |
| 2003 | 2 | 6 | 29 | 184 | 463 | 493 | 551 | 594 | 473 | 237 | 135 | 2 | 3169 |
| 2004 | 19 | 3 | 80 | 127 | 337 | 490 | 553 | 524 | 479 | 368 | 91 | 15 | 3086 |
| 2005 | 33 | 21 | 43 | 108 | 334 | 517 | 586 | 620 | 551 | 263 | 94 | 14 | 3184 |
| 2006 | 21 | 20 | 105 | 262 | 375 | 568 | 566 | 583 | 431 | 235 | 32 | 7 | 3205 |
| 2007 | 12 | 5 | 96 | 145 | 337 | 496 | 534 | 630 | 510 | 258 | 65 | 68 | 3156 |
| 2008 | 19 | 29 | 67 | 164 | 367 | 537 | 591 | 551 | 422 | 204 | 32 | 72 | 3055 |
| 2009 | 30 | 29 | 98 | 159 | 415 | 557 | 601 | 555 | 487 | 258 | 21 | 7 | 3217 |

## SNOWFALL (inches)  2009  NEW ORLEANS (KMSY)

| YEAR | JUL | AUG | SEP | OCT | NOV | DEC | JAN | FEB | MAR | APR | MAY | JUN | TOTAL |
|---|---|---|---|---|---|---|---|---|---|---|---|---|---|
| 1976-77 | 0.0 | 0.0 | 0.0 | 0.0 | 0.0 | 0.0 | T | 0.0 | 0.0 | 0.0 | 0.0 | 0.0 | T |
| 1977-78 | 0.0 | 0.0 | 0.0 | 0.0 | 0.0 | 0.0 | T | T | 0.0 | 0.0 | 0.0 | 0.0 | T |
| 1978-79 | 0.0 | 0.0 | 0.0 | 0.0 | 0.0 | 0.0 | T | 0.0 | 0.0 | 0.0 | 0.0 | 0.0 | T |
| 1979-80 | 0.0 | 0.0 | 0.0 | 0.0 | 0.0 | 0.0 | 0.0 | 0.0 | T | 0.0 | 0.0 | 0.0 | T |
| 1980-81 | 0.0 | 0.0 | 0.0 | 0.0 | 0.0 | 0.0 | 0.0 | 0.0 | 0.0 | 0.0 | 0.0 | 0.0 | 0.0 |
| 1981-82 | 0.0 | 0.0 | 0.0 | 0.0 | 0.0 | 0.0 | T | 0.0 | 0.0 | 0.0 | 0.0 | 0.0 | T |
| 1982-83 | 0.0 | 0.0 | 0.0 | 0.0 | 0.0 | 0.0 | 0.0 | 0.0 | 0.0 | 0.0 | 0.0 | 0.0 | 0.0 |
| 1983-84 | 0.0 | 0.0 | 0.0 | 0.0 | 0.0 | 0.0 | 0.0 | 0.0 | 0.0 | 0.0 | 0.0 | 0.0 | 0.0 |
| 1984-85 | 0.0 | 0.0 | 0.0 | 0.0 | 0.0 | 0.0 | 0.4 | 0.0 | 0.0 | 0.0 | 0.0 | 0.0 | 0.4 |
| 1985-86 | 0.0 | 0.0 | 0.0 | 0.0 | 0.0 | 0.0 | 0.0 | 0.0 | 0.0 | 0.0 | 0.0 | 0.0 | 0.0 |
| 1986-87 | 0.0 | 0.0 | 0.0 | 0.0 | 0.0 | 0.0 | 0.0 | 0.0 | 0.0 | 0.0 | 0.0 | 0.0 | 0.0 |
| 1987-88 | 0.0 | 0.0 | 0.0 | 0.0 | 0.0 | 0.0 | 0.0 | T | 0.0 | 0.0 | 0.0 | 0.0 | T |
| 1988-89 | 0.0 | 0.0 | 0.0 | 0.0 | 0.0 | 0.0 | 0.0 | 0.0 | T | 0.0 | T | 0.0 | T |
| 1989-90 | 0.0 | 0.0 | 0.0 | 0.0 | 0.0 | 0.5 | 0.0 | 0.0 | 0.0 | T | 0.0 | 0.0 | 0.5 |
| 1990-91 | 0.0 | 0.0 | 0.0 | 0.0 | 0.0 | 0.0 | 0.0 | 0.0 | T | T | 0.0 | 0.0 | T |
| 1991-92 | 0.0 | 0.0 | 0.0 | 0.0 | 0.0 | 0.0 | 0.0 | 0.0 | T | 0.0 | 0.0 | 0.0 | T |
| 1992-93 | 0.0 | 0.0 | 0.0 | 0.0 | 0.0 | 0.0 | 0.0 | 0.0 | T | 0.0 | 0.0 | 0.0 | T |
| 1993-94 | 0.0 | 0.0 | 0.0 | 0.0 | 0.0 | 0.0 | 0.0 | T | 0.0 | 0.0 | 0.0 | 0.0 | T |
| 1994-95 | 0.0 | 0.0 | 0.0 | 0.0 | 0.0 | 0.0 | 0.0 | T | 0.0 | 0.0 | 0.0 | 0.0 | T |
| 1995-96 | 0.0 | 0.0 | 0.0 | 0.0 | 0.0 | T | 0.0 | 0.0 | 0.0 | T | | | |
| 1996-97 | | | | | | | | | | | | | |
| 1997-98 | | | | | | | | | | | | | |
| 1998-99 | | | | | | | | | | | | | |
| 1999-00 | | | | | | | | | | | | | |
| 2000-01 | | | | | | | T | | | | | | |
| 2001-02 | | | | | | | | | | | | | |
| 2002-03 | | | | | | | | | | | | | |
| 2003-04 | | | | | | | | | | | | | |
| 2004-05 | | | | | | | | | | | | | |
| 2005- | | | | | | | | | | | | | |
| POR=<br>49 YRS | 0.0 | 0.0 | 0.0 | 0.0 | 0.0 | 0.1 | T | 0.1 | T | T | T | 0.0 | 0.2 |

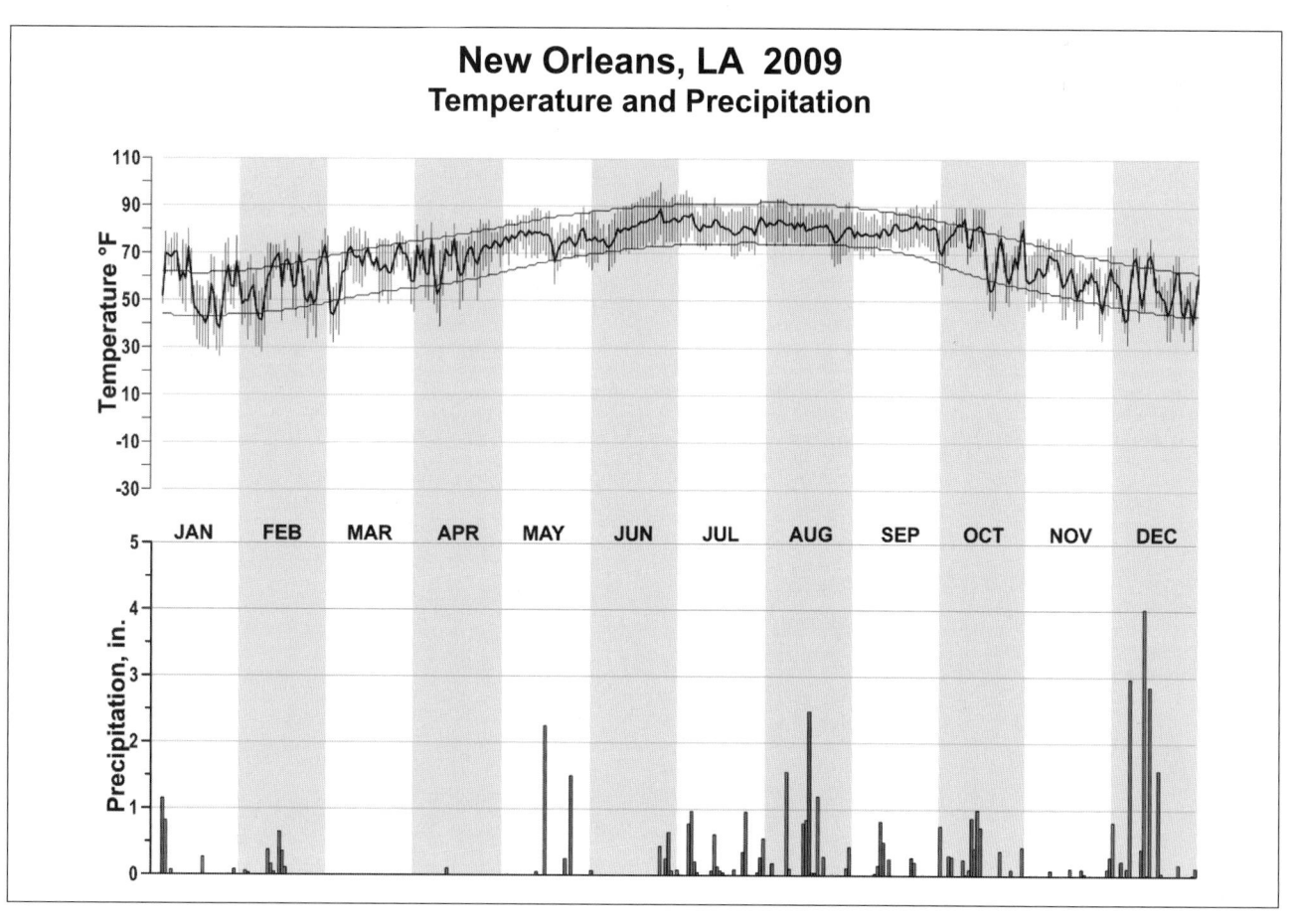

## New Orleans, LA  2009
### Temperature and Precipitation

# 2009
# SHREVEPORT
# LOUISIANA (KSHV)

Shreveport is located on the west side of the Red River, opposite Bossier City, in the northwestern section of Louisiana, some 30 miles south of Arkansas and 15 miles east of Texas. A portion of the city is situated in the Red River bottom lands and the remainder in gently rolling hills that begin about 1 mile west of the river. The NOAA National Weather Service Office is at the Shreveport Regional Airport, about 8 miles southwest of the downtown area. Elevations in the Shreveport area range from about 170 to 280 feet above sea level.

The climate of Shreveport is transitional between the subtropical humid type prevalent to the south and the continental climates of the Great Plains and Middle West to the north. During winter, masses of moderate to severely cold air move periodically through the area. The spring and fall seasons are usually mild, while the summer months are consistently quite warm and humid with high pressure and a moist southerly flow being the dominant feature. Rainfall is abundant with the normal annual just over 46 inches, with monthly averages ranging less than 3 inches in August to more than 5 inches in May. The average growing season for northwest Louisiana ranges between 230 and 240 days in length.

The majority of rainfall is of convective and air mass types-showery and brief-except during winter when nearly continuous frontal rains may persist for a few days. Extremes of precipitation occur in all seasons. While torrential rainfall is the exception in the Shreveport area, some heavy rainfall events of notes are 12.44 inches in a 24-hour peroid on July 24-25, 1933, and 19.08 inches over a three-day peroid on July 23-25, 1933. The July 1933 total of 25.44 inches was the greatest monthly total. The greatest annual rainfall of record was in 1991 with 81.99 inches, and the driest year of record was 1899 with 23.10 inches. The months with the fewest days of rain are August and October, with August having the least average precipitation.

The winter months are normally mild with cold spells generally of short duration. The typical pattern is turning cold one day, reaching the lowest temperature on the second day, and a warming trend on the third day. The coldest reading on record is -5 degrees F on February 12, 1899. Temperatures of freezing or below occur each winter with an average of 39 days during the year. Temperatures drop below 15 degrees F only about one out of every two winters. The average date of the first 32 degrees F in the fall is November 15 and the average date of the last freeze in the spring is March 10. Freezing temperatures have been recorded as early as October 19 and as late as April 11. Temperatures recorded at the NWS Office on clear, calm nights are normally 2 to 5 degrees warmer than those in the low-lying river bottom lands of the area.

Measurable snowfall amounts occur on an average of only once every other year; many consecuitive years may pass with no measurable snowfall. The heaviest snowstorm of record in the Shreveport area is 11.0 inches in december of 1929. This fell on the 21st and 22nd, and one-half inch remained on the ground December 25th, making this the only Christmas Day of record with snow on the ground. In 1948, 12.4 inches of snow was measured for the month of January for the greatest monthly amount on record. Occasional ice and sleet storms do considerable damage to trees, power and telephone lines, as well as make travel very difficult.

The summer months are consistently quite warm, with maximum temperatures exceeding 100 degrees about 6 days per year, exceeding 95 degrees about 32 days per year, and exceeding 90 degrees about 87 days per year. The highest temperature on record is 110 degrees F on August 18, 1909. Showers and thunderstorms at any one location in the area give about eight days in a month of measurable rainfall. The resulting point rainfall totals are usually less than one-half inch except on two or three days per month when heavier amounts are recorded.

Thunderstorms occur each month, but are most frequent in spring and summer months. The showers and thunderstorms during the spring and autumn months are most often produced by squall lines and fronts, and are generally heavier than the air mass showers which occur in the summer months. Severe local storms, including hailstorms, tornadoes, and local windstorms have occurred over small areas in all seasons, but are most frequent during the spring months, with a secondary peak from November to early January. large hail of a damaging nature is infrequent, although hail as large as grapefruit fell in March 1961, and baseball size hail fell in May 1974 and April 1995.

The average relative humidity is rather high in all seasons. These high humidity values may be experienced at any hour but occur mainly during the early morning hours, with two-thirds of the hours shortly before sunrise having relative humidity of 90 percent or higher. In contrast, more than half of the mid-afternoon hours have had relative humidity values of less than 50 percent.

Tropical cyclones are in the dissipating stages by the time they reach this portion of the state and winds from them are usually not a destructive factor. rainfall accompanying these systems can be heavy and can contribute to local flooding.

# NORMALS, MEANS, AND EXTREMES
## SHREVEPORT (KSHV)

| LATITUDE: 32 ° 26'N | LONGITUDE: -93 ° 49'W | ELEVATION (FT): GRND: 229　BARO: 274 | TIME ZONE: CENTRAL　(UTC -6) | WBAN: 13957 |
|---|---|---|---|---|

| ELEMENT | POR | JAN | FEB | MAR | APR | MAY | JUN | JUL | AUG | SEP | OCT | NOV | DEC | YEAR |
|---|---|---|---|---|---|---|---|---|---|---|---|---|---|---|
| **TEMPERATURE °F** | | | | | | | | | | | | | | |
| NORMAL DAILY MAXIMUM | 30 | 56.2 | 62.0 | 69.7 | 76.6 | 83.2 | 89.8 | 93.3 | 93.4 | 87.6 | 78.3 | 66.8 | 58.5 | 76.3 |
| MEAN DAILY MAXIMUM | 80 | 56.6 | 59.9 | 68.6 | 76.2 | 83.5 | 89.7 | 93.5 | 93.6 | 87.3 | 78.8 | 66.6 | 58.8 | 76.1 |
| HIGHEST DAILY MAXIMUM | 57 | 84 | 89 | 92 | 94 | 102 | 102 | 107 | 109 | 109 | 97 | 88 | 84 | 109 |
| YEAR OF OCCURRENCE | | 1972 | 1986 | 1974 | 1987 | 1998 | 2009 | 1998 | 2000 | 2000 | 2002 | 1984 | 2005 | SEP 2000 |
| MEAN OF EXTREME MAXS. | 80 | 76.8 | 79.3 | 84.2 | 87.3 | 91.8 | 95.8 | 99.1 | 100.1 | 96.5 | 90.9 | 82.7 | 76.9 | 88.5 |
| NORMAL DAILY MINIMUM | 30 | 36.5 | 40.3 | 47.2 | 53.8 | 62.7 | 69.9 | 73.4 | 72.3 | 66.4 | 55.0 | 45.3 | 38.3 | 55.1 |
| MEAN DAILY MINIMUM | 80 | 37.1 | 39.4 | 46.7 | 54.4 | 63.0 | 69.5 | 73.1 | 72.4 | 66.1 | 55.4 | 44.9 | 38.8 | 55.1 |
| LOWEST DAILY MINIMUM | 57 | 3 | 12 | 17 | 31 | 42 | 52 | 58 | 53 | 42 | 28 | 16 | 5 | 3 |
| YEAR OF OCCURRENCE | | 1962 | 1978 | 2002 | 1989 | 1960 | 1977 | 1972 | 1992 | 1984 | 1993 | 1976 | 1989 | JAN 1962 |
| MEAN OF EXTREME MINS. | 80 | 19.2 | 23.7 | 29.1 | 38.1 | 49.6 | 60.0 | 66.6 | 64.4 | 52.5 | 38.8 | 28.4 | 21.6 | 41.0 |
| NORMAL DRY BULB | 30 | 46.4 | 51.2 | 58.5 | 65.2 | 73.0 | 79.9 | 83.4 | 82.9 | 77.0 | 66.7 | 56.1 | 48.4 | 65.7 |
| MEAN DRY BULB | 80 | 46.9 | 49.7 | 57.6 | 65.3 | 73.3 | 79.7 | 83.3 | 83.0 | 76.7 | 67.1 | 55.8 | 48.8 | 65.6 |
| MEAN WET BULB | 26 | 41.5 | 45.0 | 50.9 | 57.8 | 66.2 | 71.8 | 74.1 | 73.5 | 68.2 | 59.5 | 50.8 | 43.4 | 58.6 |
| MEAN DEW POINT | 26 | 37.8 | 41.2 | 46.8 | 54.3 | 63.8 | 69.7 | 72.0 | 70.9 | 65.2 | 56.6 | 47.7 | 39.9 | 55.5 |
| NORMAL NO. DAYS WITH: | | | | | | | | | | | | | | |
| MAXIMUM >= 90 | 30 | 0.0 | 0.0 | 0.1 | 0.2 | 3.9 | 17.7 | 25.7 | 25.5 | 14.4 | 1.8 | 0.0 | 0.0 | 89.3 |
| MAXIMUM <= 32 | 30 | 0.9 | 0.5 | * | 0.0 | 0.0 | 0.0 | 0.0 | 0.0 | 0.0 | 0.0 | 0.0 | 0.5 | 1.9 |
| MINIMUM <= 32 | 30 | 12.7 | 7.2 | 2.4 | 0.2 | 0.0 | 0.0 | 0.0 | 0.0 | 0.0 | 0.2 | 3.0 | 10.8 | 36.5 |
| MINIMUM <= 0 | 30 | 0.0 | 0.0 | 0.0 | 0.0 | 0.0 | 0.0 | 0.0 | 0.0 | 0.0 | 0.0 | 0.0 | 0.0 | 0.0 |
| **H/C** | | | | | | | | | | | | | | |
| NORMAL HEATING DEG. DAYS | 30 | 597 | 408 | 247 | 89 | 8 | 0 | 0 | 0 | 6 | 78 | 296 | 522 | 2251 |
| NORMAL COOLING DEG. DAYS | 30 | 6 | 7 | 31 | 87 | 242 | 436 | 554 | 539 | 353 | 119 | 24 | 7 | 2405 |
| **RH** | | | | | | | | | | | | | | |
| NORMAL (PERCENT) | 30 | 73 | 70 | 68 | 70 | 75 | 75 | 73 | 72 | 72 | 74 | 76 | 75 | 73 |
| HOUR 00 LST | 30 | 79 | 77 | 76 | 80 | 85 | 86 | 85 | 84 | 84 | 84 | 83 | 81 | 82 |
| HOUR 06 LST | 30 | 85 | 84 | 85 | 88 | 92 | 92 | 92 | 92 | 91 | 91 | 88 | 86 | 89 |
| HOUR 12 LST | 30 | 63 | 59 | 56 | 56 | 61 | 60 | 58 | 56 | 57 | 56 | 61 | 63 | 59 |
| HOUR 18 LST | 30 | 66 | 58 | 55 | 55 | 62 | 61 | 59 | 58 | 61 | 65 | 70 | 69 | 62 |
| **S** PERCENT POSSIBLE SUNSHINE | 50 | 52 | 56 | 58 | 60 | 64 | 71 | 75 | 75 | 70 | 68 | 60 | 54 | 64 |
| **W/O** MEAN NO. DAYS WITH: | | | | | | | | | | | | | | |
| HEAVY FOG(VISBY <= 1/4 MI) | 46 | 2.9 | 1.8 | 1.2 | 1.3 | 0.9 | 0.4 | 0.3 | 0.3 | 0.9 | 2.0 | 2.8 | 2.8 | 17.6 |
| THUNDERSTORMS | 62 | 2.3 | 3.0 | 5.0 | 5.5 | 7.0 | 7.6 | 8.3 | 6.5 | 4.0 | 2.7 | 3.0 | 2.5 | 57.4 |
| **CLOUDNESS** MEAN: | | | | | | | | | | | | | | |
| SUNRISE-SUNSET (OKTAS) | | | | | | | | | | | | | | |
| MIDNIGHT-MIDNIGHT (OKTAS) | | | | | | | | | | | | | | |
| MEAN NO. DAYS WITH: | | | | | | | | | | | | | | |
| CLEAR | 1 | 6.0 | 5.0 | 10.0 | | 6.0 | 7.0 | | | | | | | |
| PARTLY CLOUDY | | | 2.0 | 3.0 | | 7.0 | 8.0 | | | | | | | |
| CLOUDY | 1 | 3.0 | 2.0 | 7.0 | | 3.0 | 4.0 | | | | | | | |
| **PR** MEAN STATION PRESSURE(IN) | 26 | 29.87 | 29.83 | 29.75 | 29.70 | 29.68 | 29.68 | 29.73 | 29.71 | 29.72 | 29.78 | 29.82 | 29.86 | 29.76 |
| MEAN SEA-LEVEL PRES. (IN) | 26 | 30.15 | 30.10 | 30.03 | 29.97 | 29.95 | 29.95 | 29.99 | 29.98 | 29.99 | 30.05 | 30.10 | 30.14 | 30.03 |
| **WINDS** MEAN SPEED (MPH) | 26 | 8.3 | 8.6 | 8.9 | 8.7 | 7.7 | 6.6 | 6.1 | 5.5 | 6.3 | 6.5 | 7.4 | 7.8 | 7.4 |
| PREVAIL.DIR(TENS OF DEGS) | 35 | 19 | 19 | 17 | 19 | 19 | 19 | 19 | 19 | 01 | 15 | 17 | 14 | 19 |
| MAXIMUM 2-MINUTE: | | | | | | | | | | | | | | |
| SPEED (MPH) | 14 | 38 | 43 | 54 | 45 | 63 | 48 | 41 | 40 | 40 | 38 | 41 | 43 | 63 |
| DIR. (TENS OF DEGS) | | 28 | 34 | 29 | 26 | 32 | 28 | 15 | 11 | 03 | 31 | 18 | 32 | 32 |
| YEAR OF OCCURRENCE | | 1999 | 2001 | 1999 | 1999 | 2000 | 2004 | 1998 | 1998 | 2005 | 2009 | 2004 | 2002 | MAY 2000 |
| MAXIMUM 3-SECOND | | | | | | | | | | | | | | |
| SPEED (MPH) | 14 | 47 | 56 | 68 | 55 | 81 | 59 | 67 | 59 | 56 | 51 | 52 | 51 | 81 |
| DIR. (TENS OF DEGS) | | 29 | 25 | 31 | 26 | 32 | 30 | 18 | 34 | 17 | 27 | 19 | 32 | 32 |
| YEAR OF OCCURRENCE | | 2008 | 2009 | 1999 | 1999 | 2000 | 2004 | 1998 | 1997 | 2008 | 2009 | 2004 | 2002 | MAY 2000 |
| **PRECIPITATION** NORMAL (IN) | 30 | 4.60 | 4.21 | 4.18 | 4.42 | 5.25 | 5.05 | 3.99 | 2.71 | 3.21 | 4.45 | 4.68 | 4.55 | 51.30 |
| MAXIMUM MONTHLY (IN) | 57 | 12.96 | 8.57 | 8.72 | 21.84 | 11.78 | 17.11 | 10.64 | 9.23 | 9.59 | 20.35 | 10.81 | 10.00 | 21.84 |
| YEAR OF OCCURRENCE | | 1999 | 1983 | 1997 | 1991 | 1967 | 1989 | 2007 | 1991 | 1968 | 2009 | 1987 | 1982 | APR 1991 |
| MINIMUM MONTHLY (IN) | 57 | 0.27 | 0.42 | 0.56 | 0.43 | 0.15 | 0.13 | 0.15 | 0.35 | 0.08 | 0.00 | 0.52 | 0.59 | 0.00 |
| YEAR OF OCCURRENCE | | 1971 | 1999 | 1966 | 1987 | 1998 | 1988 | 1964 | 2000 | 1994 | 1963 | 1999 | 1981 | OCT 1963 |
| MAXIMUM IN 24 HOURS (IN) | 57 | 7.00 | 3.53 | 3.63 | 10.44 | 10.76 | 7.28 | 4.96 | 4.64 | 5.52 | 6.42 | 6.51 | 3.94 | 10.76 |
| YEAR OF OCCURRENCE | | 1999 | 1965 | 1979 | 1991 | 2008 | 1993 | 2007 | 1955 | 2005 | 2009 | 1987 | 2001 | MAY 2008 |
| NORMAL NO. DAYS WITH: | | | | | | | | | | | | | | |
| PRECIPITATION >= 0.01 | 30 | 9.7 | 8.1 | 9.7 | 8.2 | 9.6 | 8.6 | 8.1 | 6.5 | 6.9 | 7.4 | 8.9 | 9.7 | 101.4 |
| PRECIPITATION >= 1.00 | 30 | 1.5 | 1.5 | 1.3 | 1.2 | 1.6 | 1.4 | 1.3 | 1.1 | 1.1 | 1.6 | 1.6 | 1.5 | 16.7 |
| **SNOWFALL** NORMAL (IN) | 30 | 0.8 | 0.4 | 0.1 | 0.* | 0.0 | 0.0 | 0.0 | 0.0 | 0.0 | 0.0 | 0.1 | 0.3 | 1.7 |
| MAXIMUM MONTHLY (IN) | 57 | 5.9 | 4.4 | 4.0 | 0.3 | 0.0 | 0.0 | 0.0 | T | 0.0 | T | 1.3 | 5.4 | 5.9 |
| YEAR OF OCCURRENCE | | 1978 | 1985 | 1965 | 1987 | 1994 | | | 1997 | | 1992 | 1980 | 1983 | JAN 1978 |
| MAXIMUM IN 24 HOURS (IN) | 57 | 5.6 | 4.4 | 4.0 | 0.3 | T | 0.0 | 0.0 | T | 0.0 | T | 1.3 | 5.4 | 5.6 |
| YEAR OF OCCURRENCE' | | 1982 | 1985 | 1965 | 1987 | 1994 | | | 1997 | | 1992 | 1980 | 1983 | JAN 1982 |
| MAXIMUM SNOW DEPTH (IN) | 61 | 9 | 4 | 2 | 0 | 0 | 0 | 0 | 0 | 0 | 0 | 1 | 3 | 9 |
| YEAR OF OCCURRENCE | | 1948 | 1951 | 1965 | | | | | | | | 1980 | 1963 | JAN 1948 |
| NORMAL NO. DAYS WITH: | | | | | | | | | | | | | | |
| SNOWFALL >= 1.0 | 30 | 0.3 | 0.1 | 0.0 | 0.0 | 0.0 | 0.0 | 0.0 | 0.0 | 0.0 | 0.0 | 0.0 | 0.1 | 0.5 |

## PRECIPITATION (inches) 2009 SHREVEPORT (KSHV)

| YEAR | JAN | FEB | MAR | APR | MAY | JUN | JUL | AUG | SEP | OCT | NOV | DEC | ANNUAL |
|------|-----|-----|-----|-----|-----|-----|-----|-----|-----|-----|-----|-----|--------|
| 1980 | 4.67 | 3.10 | 3.75 | 5.34 | 4.42 | 2.60 | 1.83 | 0.42 | 1.63 | 2.48 | 3.59 | 0.74 | 34.57 |
| 1981 | 1.43 | 3.83 | 3.33 | 1.97 | 9.96 | 6.45 | 2.36 | 0.94 | 3.32 | 5.63 | 1.49 | 0.59 | 41.30 |
| 1982 | 3.59 | 3.19 | 2.59 | 2.72 | 2.32 | 1.84 | 4.25 | 2.20 | 1.11 | 5.19 | 5.72 | 10.00 | 44.72 |
| 1983 | 2.45 | 8.57 | 3.68 | 1.47 | 8.22 | 6.60 | 1.18 | 1.67 | 3.12 | 0.79 | 4.90 | 7.18 | 49.83 |
| 1984 | 2.10 | 5.66 | 3.58 | 2.52 | 5.86 | 3.56 | 2.20 | 0.87 | 2.61 | 12.05 | 4.46 | 2.88 | 48.35 |
| 1985 | 2.38 | 4.42 | 4.28 | 3.05 | 1.96 | 4.57 | 8.40 | 0.35 | 4.40 | 9.87 | 4.25 | 3.37 | 51.30 |
| 1986 | 0.49 | 3.48 | 0.75 | 3.50 | 6.60 | 14.67 | 2.92 | 1.68 | 3.51 | 6.63 | 9.19 | 4.69 | 58.11 |
| 1987 | 2.26 | 7.80 | 1.48 | 0.43 | 6.67 | 5.43 | 1.21 | 3.50 | 0.94 | 5.49 | 10.81 | 8.12 | 54.14 |
| 1988 | 2.06 | 3.59 | 3.89 | 3.45 | 0.42 | 0.13 | 3.12 | 3.52 | 1.61 | 4.44 | 5.44 | 4.71 | 36.38 |
| 1989 | 7.20 | 4.06 | 3.41 | 2.41 | 10.07 | 17.11 | 4.46 | 3.94 | 1.08 | 1.50 | 2.32 | 3.34 | 60.90 |
| 1990 | 10.02 | 6.92 | 4.90 | 4.29 | 10.48 | 2.56 | 3.53 | 2.88 | 2.93 | 4.33 | 8.81 | 3.99 | 65.64 |
| 1991 | 7.70 | 5.13 | 2.89 | 21.84 | 10.71 | 2.53 | 3.47 | 9.23 | 3.45 | 3.59 | 3.94 | 7.51 | 81.99 |
| 1992 | 4.63 | 6.41 | 5.94 | 3.26 | 2.81 | 3.95 | 3.36 | 1.24 | 5.15 | 4.13 | 4.69 | 5.84 | 51.41 |
| 1993 | 4.63 | 4.80 | 5.94 | 4.19 | 3.30 | 15.73 | 0.27 | 4.09 | 3.51 | 4.43 | 4.85 | 1.44 | 57.18 |
| 1994 | 3.63 | 5.02 | 3.67 | 3.67 | 5.85 | 2.81 | 6.43 | 3.80 | 0.08 | 9.14 | 2.50 | 8.00 | 54.60 |
| 1995 | 5.44 | 3.75 | 4.05 | 7.80 | 3.26 | 1.09 | 5.68 | 0.83 | 3.36 | 1.65 | 1.94 | 5.11 | 43.96 |
| 1996 | 2.12 | 0.64 | 2.33 | 3.86 | 0.93 | 6.50 | 5.70 | 5.78 | 7.17 | 1.66 | 5.87 | 2.24 | 44.80 |
| 1997 | 4.47 | 8.09 | 8.72 | 11.93 | 3.19 | 6.14 | 1.73 | 5.48 | 2.41 | 7.50 | 3.44 | 6.10 | 69.20 |
| 1998 | 5.84 | 7.19 | 4.28 | 0.79 | 0.15 | 1.35 | 2.84 | 3.83 | 7.79 | 5.72 | 4.58 | 6.24 | 50.60 |
| 1999 | 12.96 | 0.42 | 5.10 | 7.88 | 3.96 | 7.98 | 2.80 | 1.47 | 4.90 | 3.21 | 0.52 | 3.82 | 55.02 |
| 2000 | 2.60 | 2.31 | 7.90 | 5.67 | 10.76 | 7.32 | 1.05 | T | 1.13 | 1.65 | 9.93 | 7.56 | 57.88 |
| 2001 | 5.76 | 6.52 | 6.47 | 0.86 | 4.31 | 7.33 | 1.75 | 4.10 | 6.84 | 5.17 | 4.16 | 6.10 | 59.37 |
| 2002 | 2.40 | 3.03 | 5.47 | 2.66 | 2.47 | 2.31 | 3.38 | 1.50 | 1.37 | 6.56 | 3.53 | 8.36 | 43.04 |
| 2003 | 0.44 | 7.66 | 2.19 | 2.12 | 2.04 | 4.61 | 3.07 | 3.19 | 2.93 | 1.92 | 2.81 | 3.61 | 36.59 |
| 2004 | 4.39 | 7.91 | 5.29 | 5.17 | 4.56 | 12.42 | 0.72 | 2.98 | 3.61 | 5.94 | 7.17 | 2.78 | 62.94 |
| 2005 | 4.37 | 3.76 | 1.91 | 4.59 | 0.73 | 0.38 | 4.60 | 3.27 | 5.66 | 1.41 | 1.06 | 1.24 | 32.98 |
| 2006 | 5.36 | 4.91 | 5.07 | 2.24 | 1.21 | 2.64 | 4.74 | 0.62 | 2.97 | 3.99 | 3.21 | 5.36 | 42.32 |
| 2007 | 7.64 | 3.32 | 2.09 | 1.64 | 4.26 | 6.00 | 10.64 | 0.61 | 1.32 | 2.36 | 3.06 | 4.58 | 47.52 |
| 2008 | 2.65 | 4.96 | 3.25 | 2.62 | 11.56 | 3.85 | 1.08 | 5.73 | 3.84 | 1.41 | 4.98 | 3.14 | 49.07 |
| 2009 | 2.13 | 1.63 | 6.48 | 3.97 | 7.44 | 1.22 | 6.49 | 1.69 | 2.58 | 20.35 | 1.42 | 4.64 | 60.04 |
| POR= 80 YRS | 4.44 | 4.10 | 4.14 | 4.47 | 4.87 | 3.97 | 3.59 | 2.70 | 2.95 | 3.76 | 4.24 | 4.55 | 47.78 |

WBAN : 13957

## AVERAGE TEMPERATURE (°F) 2009 SHREVEPORT (KSHV)

| YEAR | JAN | FEB | MAR | APR | MAY | JUN | JUL | AUG | SEP | OCT | NOV | DEC | ANNUAL |
|------|-----|-----|-----|-----|-----|-----|-----|-----|-----|-----|-----|-----|--------|
| 1980 | 48.3 | 47.9 | 54.9 | 63.0 | 74.3 | 83.4 | 86.9 | 85.5 | 82.1 | 63.5 | 54.2 | 49.1 | 66.1 |
| 1981 | 44.7 | 49.8 | 56.0 | 70.0 | 69.2 | 80.2 | 82.8 | 81.3 | 73.9 | 65.0 | 56.8 | 46.9 | 64.7 |
| 1982 | 46.1 | 45.6 | 61.5 | 63.3 | 74.4 | 78.6 | 83.3 | 83.2 | 75.6 | 64.6 | 55.4 | 51.2 | 65.2 |
| 1983 | 44.6 | 48.8 | 55.0 | 59.7 | 70.0 | 77.4 | 82.3 | 84.0 | 75.7 | 66.7 | 55.9 | 37.5 | 63.1 |
| 1984 | 40.6 | 50.4 | 58.0 | 64.9 | 72.1 | 79.3 | 81.1 | 82.1 | 75.0 | 70.6 | 56.4 | 60.0 | 65.9 |
| 1985 | 40.0 | 46.2 | 61.4 | 67.0 | 72.8 | 79.3 | 83.1 | 84.8 | 76.3 | 68.5 | 61.5 | 44.2 | 65.4 |
| 1986 | 49.0 | 54.4 | 59.7 | 66.5 | 72.3 | 79.9 | 83.7 | 80.8 | 79.4 | 65.2 | 55.6 | 46.1 | 66.1 |
| 1987 | 44.8 | 51.7 | 55.7 | 64.2 | 75.3 | 79.1 | 82.3 | 85.3 | 76.7 | 64.1 | 56.1 | 50.2 | 65.5 |
| 1988 | 42.2 | 49.3 | 56.3 | 65.2 | 71.8 | 79.8 | 83.3 | 83.7 | 77.8 | 64.1 | 58.6 | 49.2 | 65.1 |
| 1989 | 51.5 | 45.8 | 56.7 | 65.6 | 73.8 | 76.7 | 81.2 | 81.0 | 73.7 | 66.5 | 58.6 | 40.8 | 64.3 |
| 1990 | 52.5 | 56.4 | 59.4 | 65.6 | 72.6 | 82.7 | 82.2 | 83.3 | 79.7 | 65.0 | 58.9 | 48.5 | 67.2 |
| 1991 | 44.2 | 52.0 | 60.0 | 68.6 | 75.3 | 80.6 | 82.6 | 81.2 | 75.7 | 68.7 | 52.1 | 51.2 | 66.0 |
| 1992 | 47.2 | 54.7 | 59.5 | 65.1 | 71.4 | 78.5 | 82.9 | 78.6 | 76.0 | 67.0 | 52.0 | 49.5 | 65.2 |
| 1993 | 46.4 | 49.3 | 55.2 | 61.3 | 70.7 | 80.2 | 84.6 | 84.8 | 77.3 | 64.4 | 52.0 | 49.1 | 64.6 |
| 1994 | 46.4 | 51.2 | 58.0 | 66.7 | 70.9 | 81.4 | 82.1 | 81.1 | 76.3 | 67.0 | 60.0 | 51.1 | 66.0 |
| 1995 | 48.1 | 53.1 | 59.4 | 64.6 | 73.9 | 79.0 | 84.2 | 86.6 | 77.3 | 66.2 | 55.0 | 49.1 | 66.4 |
| 1996 | 46.2 | 52.4 | 53.4 | 63.8 | 77.0 | 78.3 | 82.3 | 80.0 | 74.1 | 66.0 | 55.7 | 51.5 | 65.1 |
| 1997 | 45.6 | 51.2 | 61.4 | 60.3 | 70.3 | 78.1 | 83.6 | 81.0 | 78.2 | 66.8 | 51.9 | 46.0 | 64.5 |
| 1998 | 51.7 | 51.6 | 56.1 | 63.5 | 77.3 | 84.9 | 88.5 | 84.6 | 81.7 | 68.5 | 58.0 | 49.3 | 68.0 |
| 1999 | 51.6 | 57.0 | 56.4 | 69.4 | 72.0 | 79.8 | 82.9 | 85.9 | 75.3 | 65.5 | 58.8 | 49.6 | 67.0 |
| 2000 | 50.4 | 57.2 | 61.7 | 63.6 | 75.8 | 79.1 | 83.9 | 86.9 | 78.1 | 67.8 | 52.6 | 39.8 | 66.4 |
| 2001 | 43.4 | 53.2 | 53.4 | 69.2 | 74.6 | 79.0 | 84.5 | 82.3 | 74.6 | 63.3 | 59.6 | 51.1 | 65.7 |
| 2002 | 48.9 | 47.0 | 55.4 | 69.2 | 72.9 | 79.7 | 83.1 | 83.6 | 79.7 | 66.6 | 53.0 | 48.2 | 65.6 |
| 2003 | 45.4 | 47.5 | 56.4 | 66.4 | 75.6 | 78.9 | 82.5 | 83.6 | 75.5 | 67.9 | 59.7 | 48.1 | 65.6 |
| 2004 | 48.4 | 46.7 | 62.5 | 65.4 | 74.5 | 79.0 | 82.2 | 80.2 | 78.4 | 73.3 | 58.6 | 48.9 | 66.5 |
| 2005 | 51.0 | 53.7 | 58.1 | 64.9 | 72.9 | 82.3 | 84.2 | 86.1 | 83.3 | 67.6 | 60.7 | 48.8 | 67.8 |
| 2006 | 55.0 | 48.7 | 61.1 | 70.9 | 75.4 | 81.0 | 85.4 | 86.4 | 77.1 | 67.1 | 57.7 | 51.3 | 68.1 |
| 2007 | 46.8 | 50.6 | 64.2 | 63.4 | 75.1 | 81.3 | 81.4 | 86.3 | 80.4 | 69.7 | 58.5 | 51.8 | 67.5 |
| 2008 | 45.5 | 52.9 | 59.8 | 64.8 | 73.2 | 81.2 | 84.2 | 82.6 | 74.6 | 65.0 | 58.1 | 48.3 | 65.6 |
| 2009 | 47.8 | 54.4 | 58.0 | 64.2 | 73.1 | 81.4 | 84.0 | 81.5 | 76.0 | 63.1 | 58.1 | 44.9 | 65.5 |
| POR= 80 YRS | 46.9 | 49.7 | 57.6 | 65.3 | 73.3 | 79.7 | 83.3 | 83.0 | 76.7 | 67.1 | 55.8 | 48.8 | 65.6 |

## HEATING DEGREE DAYS (base 65°F) 2009  SHREVEPORT (KSHV)

| YEAR | JUL | AUG | SEP | OCT | NOV | DEC | JAN | FEB | MAR | APR | MAY | JUN | TOTAL |
|---|---|---|---|---|---|---|---|---|---|---|---|---|---|
| 1980-81 | 0 | 0 | 3 | 128 | 340 | 488 | 620 | 425 | 279 | 14 | 20 | 0 | 2317 |
| 1981-82 | 0 | 0 | 8 | 129 | 246 | 554 | 588 | 537 | 202 | 125 | 4 | 0 | 2393 |
| 1982-83 | 0 | 0 | 9 | 120 | 309 | 457 | 624 | 449 | 308 | 186 | 12 | 0 | 2474 |
| 1983-84 | 0 | 0 | 15 | 69 | 305 | 848 | 747 | 421 | 247 | 81 | 11 | 0 | 2744 |
| 1984-85 | 0 | 0 | 19 | 36 | 286 | 208 | 770 | 528 | 151 | 42 | 1 | 0 | 2041 |
| 1985-86 | 0 | 0 | 11 | 49 | 174 | 638 | 490 | 331 | 176 | 44 | 1 | 0 | 1914 |
| 1986-87 | 0 | 0 | 0 | 86 | 299 | 579 | 618 | 366 | 286 | 117 | 0 | 0 | 2351 |
| 1987-88 | 0 | 0 | 0 | 79 | 279 | 456 | 701 | 453 | 278 | 54 | 1 | 0 | 2301 |
| 1988-89 | 0 | 0 | 0 | 76 | 218 | 482 | 418 | 535 | 295 | 93 | 2 | 0 | 2119 |
| 1989-90 | 0 | 0 | 17 | 85 | 244 | 743 | 382 | 243 | 216 | 92 | 3 | 0 | 2025 |
| 1990-91 | 0 | 0 | 6 | 126 | 208 | 509 | 634 | 357 | 197 | 26 | 6 | 0 | 2069 |
| 1991-92 | 0 | 0 | 9 | 36 | 401 | 422 | 544 | 297 | 185 | 84 | 23 | 0 | 2001 |
| 1992-93 | 0 | 0 | 2 | 33 | 381 | 480 | 569 | 435 | 307 | 154 | 3 | 0 | 2364 |
| 1993-94 | 0 | 0 | 5 | 132 | 404 | 490 | 577 | 392 | 239 | 81 | 20 | 0 | 2340 |
| 1994-95 | 0 | 0 | 5 | 74 | 179 | 440 | 528 | 334 | 224 | 80 | 10 | 0 | 1874 |
| 1995-96 | 0 | 0 | 10 | 56 | 311 | 507 | 576 | 387 | 372 | 115 | 2 | 0 | 2336 |
| 1996-97 | 0 | 0 | 10 | 59 | 277 | 427 | 604 | 383 | 133 | 157 | 5 | 0 | 2055 |
| 1997-98 | 0 | 0 | 0 | 95 | 392 | 580 | 406 | 368 | 324 | 88 | 0 | 0 | 2253 |
| 1998-99 | 0 | 0 | 0 | 37 | 215 | 501 | 422 | 250 | 261 | 41 | 1 | 0 | 1728 |
| 1999-00 | 0 | 0 | 2 | 79 | 189 | 477 | 462 | 247 | 149 | 103 | 0 | 0 | 1708 |
| 2000-01 | 0 | 0 | 10 | 70 | 388 | 774 | 662 | 332 | 352 | 35 | 0 | 0 | 2623 |
| 2001-02 | 0 | 0 | 5 | 118 | 175 | 441 | 515 | 497 | 317 | 48 | 9 | 0 | 2125 |
| 2002-03 | 0 | 0 | 0 | 48 | 363 | 525 | 598 | 483 | 262 | 49 | 0 | 0 | 2328 |
| 2003-04 | 0 | 0 | 1 | 32 | 206 | 516 | 518 | 524 | 113 | 75 | 7 | 0 | 1992 |
| 2004-05 | 0 | 0 | 0 | 15 | 196 | 496 | 446 | 328 | 231 | 50 | 19 | 0 | 1781 |
| 2005-06 | 0 | 0 | 0 | 90 | 204 | 501 | 309 | 453 | 179 | 16 | 0 | 0 | 1752 |
| 2006-07 | 0 | 0 | 0 | 73 | 238 | 424 | 565 | 399 | 111 | 118 | 0 | 0 | 1928 |
| 2007-08 | 0 | 0 | 0 | 59 | 233 | 424 | 610 | 353 | 206 | 88 | 10 | 0 | 1983 |
| 2008-09 | 0 | 0 | 0 | 89 | 298 | 516 | 536 | 311 | 258 | 103 | 2 | 0 | 2113 |
| 2009- | 0 | 0 | 0 | 128 | 210 | 617 | | | | | | | |

WBAN : 13957

## COOLING DEGREE DAYS (base 65°F) 2009  SHREVEPORT (KSHV)

| YEAR | JAN | FEB | MAR | APR | MAY | JUN | JUL | AUG | SEP | OCT | NOV | DEC | TOTAL |
|---|---|---|---|---|---|---|---|---|---|---|---|---|---|
| 1980 | 1 | 6 | 6 | 43 | 298 | 560 | 686 | 643 | 522 | 86 | 22 | 1 | 2874 |
| 1981 | 0 | 5 | 10 | 171 | 157 | 463 | 558 | 511 | 284 | 135 | 6 | 0 | 2300 |
| 1982 | 14 | 0 | 99 | 81 | 300 | 413 | 573 | 573 | 333 | 115 | 24 | 32 | 2557 |
| 1983 | 0 | 0 | 7 | 34 | 176 | 381 | 540 | 595 | 343 | 126 | 39 | 0 | 2241 |
| 1984 | 0 | 5 | 38 | 83 | 235 | 436 | 511 | 540 | 329 | 219 | 35 | 61 | 2492 |
| 1985 | 0 | 8 | 49 | 109 | 252 | 436 | 568 | 620 | 356 | 163 | 78 | 0 | 2639 |
| 1986 | 2 | 41 | 17 | 95 | 236 | 454 | 586 | 494 | 438 | 101 | 24 | 0 | 2488 |
| 1987 | 1 | 0 | 7 | 99 | 327 | 431 | 544 | 634 | 357 | 57 | 19 | 5 | 2481 |
| 1988 | 3 | 3 | 14 | 67 | 220 | 449 | 575 | 587 | 390 | 53 | 37 | 1 | 2399 |
| 1989 | 8 | 7 | 43 | 121 | 283 | 358 | 508 | 503 | 286 | 140 | 57 | 0 | 2314 |
| 1990 | 2 | 9 | 50 | 115 | 244 | 538 | 536 | 572 | 453 | 132 | 30 | 3 | 2684 |
| 1991 | 0 | 1 | 47 | 141 | 334 | 475 | 550 | 509 | 336 | 158 | 20 | 5 | 2576 |
| 1992 | 0 | 4 | 21 | 93 | 227 | 412 | 561 | 428 | 339 | 99 | 1 | 4 | 2189 |
| 1993 | 0 | 0 | 10 | 50 | 189 | 461 | 614 | 620 | 381 | 121 | 20 | 4 | 2470 |
| 1994 | 3 | 12 | 29 | 141 | 209 | 500 | 538 | 506 | 355 | 145 | 36 | 13 | 2487 |
| 1995 | 11 | 6 | 58 | 72 | 290 | 427 | 600 | 676 | 385 | 100 | 16 | 19 | 2660 |
| 1996 | 3 | 31 | 23 | 85 | 384 | 403 | 543 | 471 | 291 | 96 | 6 | 16 | 2352 |
| 1997 | 9 | 4 | 29 | 25 | 180 | 398 | 583 | 504 | 399 | 157 | 6 | 0 | 2294 |
| 1998 | 1 | 0 | 56 | 50 | 392 | 602 | 737 | 613 | 508 | 152 | 15 | 22 | 3148 |
| 1999 | 12 | 29 | 0 | 180 | 224 | 451 | 562 | 655 | 317 | 101 | 9 | 6 | 2546 |
| 2000 | 15 | 28 | 55 | 67 | 341 | 429 | 593 | 689 | 409 | 166 | 22 | 0 | 2814 |
| 2001 | 0 | 12 | 2 | 166 | 308 | 426 | 611 | 542 | 301 | 69 | 23 | 16 | 2476 |
| 2002 | 22 | 0 | 26 | 179 | 264 | 451 | 570 | 584 | 446 | 106 | 11 | 9 | 2668 |
| 2003 | 0 | 1 | 5 | 98 | 335 | 423 | 552 | 582 | 320 | 128 | 54 | 0 | 2498 |
| 2004 | 11 | 0 | 44 | 93 | 310 | 423 | 540 | 478 | 406 | 280 | 10 | 2 | 2597 |
| 2005 | 19 | 20 | 24 | 58 | 270 | 525 | 600 | 662 | 558 | 175 | 81 | 7 | 2999 |
| 2006 | 4 | 1 | 67 | 199 | 331 | 487 | 639 | 669 | 371 | 145 | 28 | 6 | 2947 |
| 2007 | 9 | 4 | 93 | 79 | 320 | 497 | 514 | 669 | 467 | 212 | 44 | 19 | 2927 |
| 2008 | 10 | 9 | 50 | 88 | 271 | 495 | 601 | 555 | 293 | 95 | 4 | 0 | 2475 |
| 2009 | 10 | 19 | 47 | 85 | 261 | 499 | 592 | 518 | 337 | 76 | 7 | 0 | 2451 |

## SNOWFALL (inches) 2009 SHREVEPORT (KSHV)

| YEAR | JUL | AUG | SEP | OCT | NOV | DEC | JAN | FEB | MAR | APR | MAY | JUN | TOTAL |
|------|-----|-----|-----|-----|-----|-----|-----|-----|-----|-----|-----|-----|-------|
| 1980-81 | 0.0 | 0.0 | 0.0 | 0.0 | 1.3 | 0.0 | 0.7 | 0.1 | 0.0 | 0.0 | 0.0 | 0.0 | 2.1 |
| 1981-82 | 0.0 | 0.0 | 0.0 | 0.0 | 0.0 | T | 5.6 | T | T | 0.0 | 0.0 | 0.0 | 5.6 |
| 1982-83 | 0.0 | 0.0 | 0.0 | 0.0 | 0.0 | T | T | T | 0.0 | 0.0 | 0.0 | 0.0 | T |
| 1983-84 | 0.0 | 0.0 | 0.0 | 0.0 | 0.0 | 5.4 | T | T | 0.0 | 0.0 | 0.0 | 0.0 | 5.4 |
| 1984-85 | 0.0 | 0.0 | 0.0 | 0.0 | 0.0 | 0.0 | 0.4 | 4.4 | 0.0 | 0.0 | 0.0 | 0.0 | 4.8 |
| 1985-86 | 0.0 | 0.0 | 0.0 | 0.0 | 0.0 | 0.0 | T | T | T | 0.0 | 0.0 | 0.0 | T |
| 1986-87 | 0.0 | 0.0 | 0.0 | 0.0 | 0.0 | T | 0.0 | T | T | 0.3 | 0.0 | 0.0 | 0.3 |
| 1987-88 | 0.0 | 0.0 | 0.0 | 0.0 | 0.0 | 0.0 | 1.2 | 0.8 | 0.0 | 0.0 | 0.0 | 0.0 | 2.0 |
| 1988-89 | 0.0 | 0.0 | 0.0 | 0.0 | 0.0 | 0.0 | T | T | T | 0.0 | T | 0.0 | T |
| 1989-90 | 0.0 | 0.0 | 0.0 | 0.0 | 0.0 | T | 0.0 | T | 0.0 | 0.0 | 0.0 | 0.0 | T |
| 1990-91 | 0.0 | 0.0 | 0.0 | 0.0 | 0.0 | T | 0.0 | T | 0.0 | T | T | 0.0 | T |
| 1991-92 | 0.0 | 0.0 | 0.0 | 0.0 | 0.0 | 0.0 | T | T | T | 0.0 | 0.0 | 0.0 | T |
| 1992-93 | 0.0 | 0.0 | 0.0 | T | T | 0.0 | T | T | 1.0 | T | 0.0 | 0.0 | 1.0 |
| 1993-94 | 0.0 | 0.0 | 0.0 | 0.0 | 0.0 | T | 0.0 | 0.4 | T | T | T | 0.0 | 0.4 |
| 1994-95 | 0.0 | 0.0 | 0.0 | 0.0 | T | 0.0 | T | T | 0.0 | T | 0.0 | 0.0 | T |
| 1995-96 | 0.0 | 0.0 | 0.0 | 0.0 | 0.0 | 0.0 | T | 0.4 |  |  | 0.0 | 0.0 | 0.4 |
| 1996-97 | 0.0 | 0.0 | 0.0 | 0.0 | 0.0 | 0.3 | 0.1 | T | 0.0 | T | 0.0 | 0.0 | 0.4 |
| 1997-98 | 0.0 | T | 0.0 | 0.0 | T | 0.0 | T | T | T | 0.0 | 0.0 | 0.0 | T |
| 1998-99 | 0.0 | 0.0 | 0.0 | 0.0 | 0.0 | 0.1 | 0.0 | 0.0 | T | 0.0 | 0.0 | 0.0 | 0.1 |
| 1999-00 | 0.0 | 0.0 | 0.0 | 0.0 | 0.0 | T | 1.1 | T | T | T | T | 0.0 | 1.1 |
| 2000-01 | 0.0 | 0.0 | 0.0 | 0.0 | T | 2.2 | T | 0.0 | 0.0 | 0.0 | T | 0.0 | 2.2 |
| 2001-02 | 0.0 | 0.0 | 0.0 | 0.0 | T | T | T | T | T | 0.0 | 0.0 | 0.0 | T |
| 2002-03 | 0.0 | 0.0 | 0.0 | 0.0 | 0.0 | 0.0 | T | T | 0.0 | 0.0 | 0.0 | 0.0 | T |
| 2003-04 | 0.0 | 0.0 | 0.0 | 0.0 | 0.0 | 0.0 | 0.0 | T | 0.0 | T | 0.0 | 0.0 | T |
| 2004-05 | 0.0 | 0.0 | 0.0 | 0.0 | 0.0 | T | 0.0 | 0.0 | T | 0.0 | 0.0 | 0.0 | T |
| 2005-06 | 0.0 | 0.0 | 0.0 | 0.0 | 0.0 | 0.0 | 0.0 | 0.0 | 0.0 | 0.0 | 0.0 | 0.0 | 0.0 |
| 2006-07 | 0.0 | 0.0 | 0.0 | 0.0 | 0.0 | 0.0 | 0.1 | T | T | 0.0 | 0.0 | 0.0 | 0.1 |
| 2007-08 | 0.0 | 0.0 | 0.0 | 0.0 | 0.0 | 0.0 | T | 0.0 | T | 0.0 | 0.0 | 0.0 | T |
| 2008-09 | 0.0 | 0.0 | 0.0 | 0.0 | 0.0 | T | 0.0 | 0.0 | 0.0 | 0.0 | 0.0 | 0.0 | T |
| 2009- | 0.0 | 0.0 | 0.0 | 0.0 | 0.0 | T |  |  |  |  |  |  |  |
| POR= 80 YRS | 0.0 | T | 0.0 | T | T | 0.2 | 0.9 | 0.3 | 0.1 | T | T | 0.0 | 1.5 |

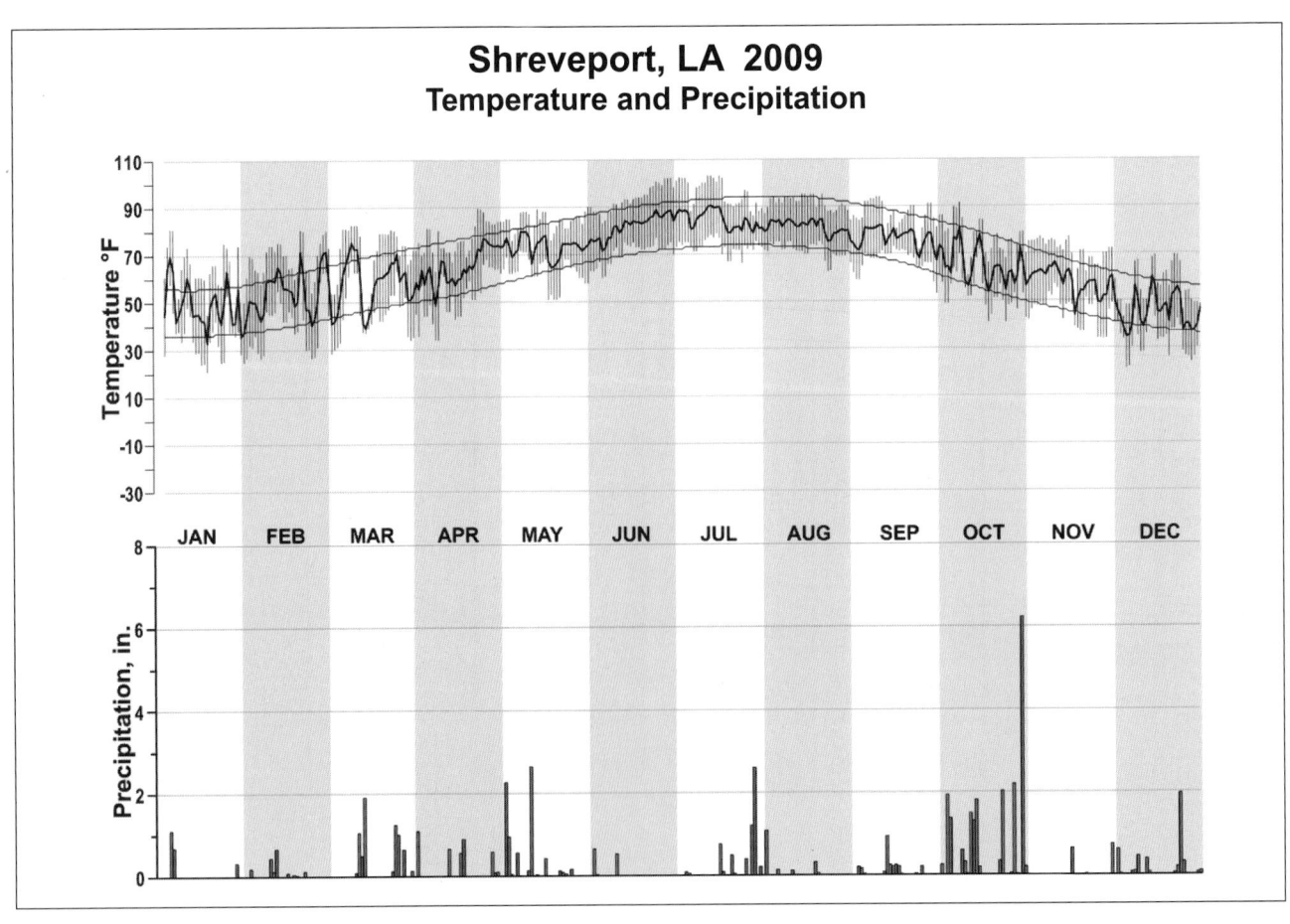

## Shreveport, LA 2009
### Temperature and Precipitation

# 2009
# CARIBOU
# MAINE (KCAR)

The Caribou Municipal Airport is located in Aroostook County, the largest and northernmost county in the state. The airport lies on top of high land which is about on the same level as most of the surrounding gently rolling hills. The Aroostook River, which runs about 1 mile to the east and southeast of the station, has little effect on the local weather. Even though Caribou is located only 150 miles from the Atlantic coast, its climate can be justly classed as a severe typical continental type. Winters are particularly long and windy, and seasonal snowfalls averaging over 100 inches are not unusual. While the extreme low temperatures may be less severe than one might expect, temperatures of zero or lower normally occur over 40 times per year. A study of heating degree day data will show the outstanding part that cold weather plays here.

Summers are cool and generally favored with abundant rainfall, which is one of the most important factors in the high yield of the potato and grain crops throughout the county. Our location high up in the St. Lawrence Valley allows Aroostook County to come under the influence of the Summer Polar Front, resulting in practically no dry periods of more than 3 or 4 days in the growing season. The growing season at Caribou averages more than 120 days, with the average last freeze in the spring in mid-May and the average first freeze in autumn in late September.

Autumn climate is nearly ideal, with mostly sunny warm days and crisp cool nights predominating. Aroostook County, even with its relatively short growing season, provides profitable farming. The principal crops are potatoes, peas, a variety of grains, and some hardy vegetables.

Probably unknown to many victims of hay fever and similar afflictions, the immediate Caribou area offers sparkling visibility and relatively pollen-free air in the late summer months. This latter condition is principally due to the extremely high degree of cultivation of all available land.

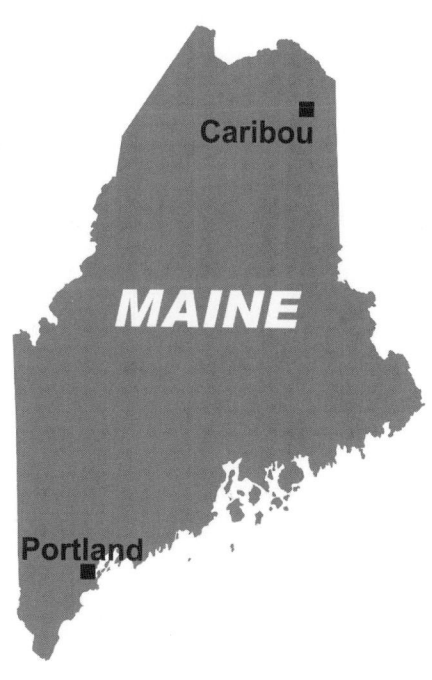

# NORMALS, MEANS, AND EXTREMES
## CARIBOU (KCAR)

| LATITUDE: 46 ° 52'N | LONGITUDE: -68 ° 1 'W | | ELEVATION (FT): GRND: 621   BARO: 626 | | | | TIME ZONE: EASTERN   (UTC -5) | | | WBAN: 14607 | | |

| | ELEMENT | POR | JAN | FEB | MAR | APR | MAY | JUN | JUL | AUG | SEP | OCT | NOV | DEC | YEAR |
|---|---|---|---|---|---|---|---|---|---|---|---|---|---|---|---|
| TEMPERATURE °F | NORMAL DAILY MAXIMUM | 30 | 19.3 | 23.2 | 34.1 | 47.0 | 62.6 | 71.8 | 76.3 | 74.2 | 64.1 | 51.4 | 37.4 | 24.8 | 48.9 |
| | MEAN DAILY MAXIMUM | 71 | 19.2 | 22.8 | 33.3 | 46.1 | 61.5 | 70.6 | 75.9 | 73.8 | 64.4 | 52.0 | 37.8 | 24.6 | 48.5 |
| | HIGHEST DAILY MAXIMUM | 70 | 53 | 59 | 73 | 86 | 96 | 96 | 95 | 95 | 91 | 79 | 68 | 58 | 96 |
| | YEAR OF OCCURRENCE | | 1995 | 1994 | 1962 | 1990 | 1977 | 1944 | 1991 | 1975 | 1945 | 2005 | 1956 | 1950 | MAY 1977 |
| | MEAN OF EXTREME MAXS. | 71 | 41.3 | 40.9 | 51.1 | 66.8 | 81.4 | 86.9 | 88.2 | 86.4 | 80.9 | 70.6 | 57.7 | 44.8 | 66.4 |
| | NORMAL DAILY MINIMUM | 30 | -.3 | 2.9 | 15.2 | 29.2 | 40.7 | 49.9 | 54.8 | 52.6 | 43.6 | 34.1 | 23.7 | 8.0 | 29.5 |
| | MEAN DAILY MINIMUM | 71 | 1.1 | 3.4 | 14.7 | 28.5 | 39.9 | 49.1 | 54.7 | 52.2 | 43.5 | 34.5 | 24.2 | 8.7 | 29.5 |
| | LOWEST DAILY MINIMUM | 70 | -37 | -41 | -28 | -2 | 18 | 30 | 36 | 34 | 23 | 14 | -8 | -31 | -41 |
| | YEAR OF OCCURRENCE | | 2009 | 1955 | 2001 | 1964 | 1974 | 1958 | 1969 | 2003 | 1980 | 1972 | 1995 | 1989 | FEB 1955 |
| | MEAN OF EXTREME MINS. | 71 | -21.1 | -18.0 | -7.9 | 14.4 | 28.0 | 36.4 | 43.7 | 39.4 | 29.9 | 21.3 | 6.5 | -12.6 | 13.3 |
| | NORMAL DRY BULB | 30 | 9.5 | 13.0 | 24.6 | 38.1 | 51.6 | 60.8 | 65.6 | 63.4 | 53.8 | 42.8 | 30.6 | 16.4 | 39.2 |
| | MEAN DRY BULB | 71 | 10.2 | 13.1 | 24.1 | 37.3 | 50.7 | 60.0 | 65.3 | 63.0 | 54.0 | 43.2 | 31.0 | 16.7 | 39.1 |
| | MEAN WET BULB | 26 | 9.4 | 12.3 | 21.6 | 33.3 | 45.0 | 55.0 | 60.1 | 58.8 | 50.9 | 39.9 | 29.3 | 16.6 | 36.0 |
| | MEAN DEW POINT | 26 | 5.7 | 7.7 | 16.4 | 27.8 | 39.4 | 50.8 | 57.1 | 55.9 | 48.0 | 36.6 | 26.1 | 13.4 | 32.1 |
| | NORMAL NO. DAYS WITH: | | | | | | | | | | | | | | |
| | MAXIMUM >= 90 | 30 | 0.0 | 0.0 | 0.0 | 0.0 | 0.2 | 0.4 | 0.8 | 0.4 | * | 0.0 | 0.0 | 0.0 | 1.8 |
| | MAXIMUM <= 32 | 30 | 25.7 | 21.9 | 12.5 | 1.7 | 0.0 | 0.0 | 0.0 | 0.0 | 0.0 | 0.3 | 9.8 | 22.5 | 94.4 |
| | MINIMUM <= 32 | 30 | 30.5 | 27.6 | 28.6 | 20.6 | 4.1 | * | 0.0 | 0.0 | 2.3 | 14.3 | 25.0 | 30.4 | 183.4 |
| | MINIMUM <= 0 | 30 | 16.9 | 12.8 | 4.5 | 0.0 | 0.0 | 0.0 | 0.0 | 0.0 | 0.0 | 0.0 | 0.2 | 9.9 | 44.3 |
| H/C | NORMAL HEATING DEG. DAYS | 30 | 1719 | 1466 | 1254 | 805 | 417 | 159 | 58 | 103 | 344 | 691 | 1039 | 1505 | 9560 |
| | NORMAL COOLING DEG. DAYS | 30 | 0 | 0 | 0 | 0 | 7 | 39 | 80 | 56 | 9 | 0 | 0 | 0 | 191 |
| RH | NORMAL (PERCENT) | 30 | | | | | | | | | | | | | |
| | HOUR 01 LST | 30 | | | | | | | | | | | | | |
| | HOUR 07 LST | 30 | 76 | 76 | 76 | 75 | 74 | 79 | 83 | 86 | 87 | 85 | 84 | 81 | 80 |
| | HOUR 13 LST | 30 | 66 | 61 | 59 | 55 | 52 | 56 | 58 | 59 | 61 | 62 | 69 | 70 | 61 |
| | HOUR 19 LST | 30 | 72 | 68 | 64 | 63 | 59 | 63 | 69 | 72 | 74 | 73 | 76 | 77 | 69 |
| S | PERCENT POSSIBLE SUNSHINE | | | | | | | | | | | | | | |
| W/O | MEAN NO. DAYS WITH: | | | | | | | | | | | | | | |
| | HEAVY FOG(VISBY <= 1/4 MI) | 28 | 2.8 | 2.4 | 3.0 | 2.0 | 1.2 | 1.4 | 2.4 | 2.4 | 2.2 | 2.0 | 2.3 | 2.6 | 26.7 |
| | THUNDERSTORMS | 44 | 0.0 | 0.0 | 0.1 | 0.3 | 1.7 | 4.6 | 6.8 | 4.2 | 1.1 | 0.7 | 0.0 | 0.0 | 19.5 |
| CLOUDNESS | MEAN: | | | | | | | | | | | | | | |
| | SUNRISE-SUNSET (OKTAS) | 51 | 5.4 | 5.4 | 5.4 | 5.8 | 5.8 | 5.8 | 5.5 | 5.3 | 5.3 | 5.6 | 6.3 | 5.7 | 5.6 |
| | MIDNIGHT-MIDNIGHT (OKTAS) | | | | | | | | | | | | | | |
| | MEAN NO. DAYS WITH: | | | | | | | | | | | | | | |
| | CLEAR | 55 | 7.0 | 6.2 | 7.0 | 5.1 | 4.1 | 3.3 | 3.0 | 4.5 | 5.3 | 4.8 | 2.8 | 5.5 | 58.6 |
| | PARTLY CLOUDY | 55 | 6.9 | 6.3 | 6.8 | 6.8 | 9.0 | 9.8 | 12.7 | 11.2 | 8.9 | 7.7 | 6.3 | 6.5 | 98.9 |
| | CLOUDY | 55 | 17.1 | 15.8 | 17.3 | 18.0 | 17.9 | 16.9 | 14.7 | 14.6 | 15.1 | 18.1 | 20.4 | 18.5 | 204.4 |
| PR | MEAN STATION PRESSURE(IN) | 26 | 29.23 | 29.24 | 29.26 | 29.25 | 29.25 | 29.22 | 29.20 | 29.28 | 29.32 | 29.31 | 29.29 | 29.25 | 29.26 |
| | MEAN SEA-LEVEL PRES. (IN) | 26 | 29.95 | 29.96 | 29.97 | 29.95 | 29.94 | 29.90 | 29.87 | 29.96 | 30.00 | 30.01 | 29.99 | 29.96 | 29.96 |
| WINDS | MEAN SPEED (MPH) | 26 | 8.6 | 8.7 | 9.3 | 9.0 | 8.6 | 7.8 | 6.9 | 6.7 | 7.4 | 8.2 | 8.5 | 8.5 | 8.2 |
| | PREVAIL.DIR(TENS OF DEGS) | 14 | 33 | 32 | 33 | 33 | 33 | 19 | 19 | 25 | 19 | 33 | 32 | 32 | 33 |
| | MAXIMUM 2-MINUTE: | | | | | | | | | | | | | | |
| | SPEED (MPH) | 15 | 43 | 41 | 38 | 39 | 37 | 36 | 30 | 39 | 32 | 40 | 41 | 39 | 43 |
| | DIR. (TENS OF DEGS) | | 25 | 25 | 34 | 33 | 34 | 35 | 31 | 22 | 30 | 30 | 32 | 31 | 25 |
| | YEAR OF OCCURRENCE | | 2008 | 2001 | 2007 | 2006 | 2005 | 2008 | 2002 | 2007 | 2006 | 1996 | 2001 | 1994 | JAN 2008 |
| | MAXIMUM 3-SECOND | | | | | | | | | | | | | | |
| | SPEED (MPH) | 13 | 54 | 56 | 48 | 49 | 47 | 55 | 46 | 55 | 44 | 52 | 52 | 55 | 56 |
| | DIR. (TENS OF DEGS) | | 25 | 26 | 35 | 33 | 19 | 23 | 22 | 20 | 31 | 30 | 31 | 30 | 26 |
| | YEAR OF OCCURRENCE | | 2008 | 2001 | 2007 | 2006 | 2009 | 2003 | 2004 | 2007 | 2006 | 1996 | 2001 | 2008 | FEB 2001 |
| PRECIPITATION | NORMAL (IN) | 30 | 2.97 | 2.06 | 2.57 | 2.64 | 3.28 | 3.31 | 3.89 | 4.15 | 3.27 | 2.99 | 3.12 | 3.19 | 37.44 |
| | MAXIMUM MONTHLY (IN) | 70 | 5.60 | 4.81 | 5.27 | 6.19 | 6.27 | 7.11 | 6.83 | 12.09 | 8.81 | 8.73 | 8.15 | 7.97 | 12.09 |
| | YEAR OF OCCURRENCE | | 1995 | 2008 | 2008 | 2005 | 1947 | 1940 | 1957 | 1981 | 1999 | 1990 | 1983 | 1973 | AUG 1981 |
| | MINIMUM MONTHLY (IN) | 70 | 0.12 | 0.26 | 0.66 | 0.54 | 0.47 | 0.88 | 0.96 | 0.55 | 0.86 | 0.63 | 0.45 | 0.74 | 0.12 |
| | YEAR OF OCCURRENCE | | 1944 | 1978 | 1965 | 1967 | 1982 | 1983 | 1991 | 2002 | 1968 | 1955 | 1939 | 1963 | JAN 1944 |
| | MAXIMUM IN 24 HOURS (IN) | 70 | 1.81 | 2.40 | 2.29 | 2.11 | 2.25 | 2.37 | 2.92 | 6.89 | 6.23 | 4.07 | 2.42 | 2.80 | 6.89 |
| | YEAR OF OCCURRENCE | | 1995 | 2003 | 2007 | 1958 | 1948 | 1957 | 1957 | 1981 | 1954 | 1970 | 2007 | 1973 | AUG 1981 |
| | NORMAL NO. DAYS WITH: | | | | | | | | | | | | | | |
| | PRECIPITATION >= 0.01 | 30 | 15.4 | 11.5 | 13.3 | 13.3 | 13.8 | 13.4 | 13.6 | 13.3 | 12.4 | 13.0 | 14.2 | 14.9 | 162.1 |
| | PRECIPITATION >= 1.00 | 30 | 0.4 | 0.1 | 0.2 | 0.1 | 0.4 | 0.4 | 0.8 | 0.8 | 0.5 | 0.4 | 0.4 | 0.3 | 4.8 |
| SNOWFALL | NORMAL (IN) | 30 | 27.0 | 20.7 | 19.7 | 9.9 | 0.6 | 0.* | 0.0 | 0.0 | 0.1 | 1.1 | 11.7 | 25.2 | 116.0 |
| | MAXIMUM MONTHLY (IN) | 69 | 44.5 | 47.7 | 47.1 | 36.4 | 10.9 | T | T | T | 2.5 | 12.1 | 34.9 | 59.9 | 59.9 |
| | YEAR OF OCCURRENCE | | 1994 | 2008 | 1955 | 1982 | 1967 | 1991 | 1992 | 1995 | 1991 | 1963 | 1974 | 1972 | DEC 1972 |
| | MAXIMUM IN 24 HOURS (IN) | 69 | 19.0 | 21.2 | 28.6 | 21.1 | 5.8 | T | T | T | 2.5 | 9.4 | 21.0 | 27.1 | 28.6 |
| | YEAR OF OCCURRENCE' | | 1994 | 1995 | 1984 | 1982 | 1967 | 1991 | 1992 | 1995 | 1991 | 1963 | 1986 | 2003 | MAR 1984 |
| | MAXIMUM SNOW DEPTH (IN) | 61 | 57 | 62 | 51 | 42 | 4 | 0 | 0 | 0 | 1 | 9 | 28 | 39 | 62 |
| | YEAR OF OCCURRENCE | | 1977 | 1977 | 1977 | 1982 | 1966 | | | | 1991 | 1963 | 1974 | 1978 | FEB 1977 |
| | NORMAL NO. DAYS WITH: | | | | | | | | | | | | | | |
| | SNOWFALL >= 1.0 | 30 | 6.0 | 5.1 | 5.1 | 2.8 | 0.1 | 0.0 | 0.0 | 0.0 | 0.0 | 0.3 | 2.8 | 6.7 | 28.9 |

## PRECIPITATION (inches) 2009 CARIBOU (KCAR)

| YEAR | JAN | FEB | MAR | APR | MAY | JUN | JUL | AUG | SEP | OCT | NOV | DEC | ANNUAL |
|------|-----|-----|-----|-----|-----|-----|-----|-----|-----|-----|-----|-----|--------|
| 1980 | 1.55 | 0.82 | 3.15 | 2.57 | 2.05 | 2.19 | 5.42 | 2.28 | 4.06 | 2.51 | 3.21 | 2.74 | 32.55 |
| 1981 | 1.68 | 2.39 | 3.43 | 2.17 | 3.18 | 4.15 | 2.62 | 12.09 | 2.38 | 6.28 | 2.51 | 3.81 | 46.69 |
| 1982 | 2.46 | 2.17 | 2.72 | 4.01 | 0.47 | 3.08 | 4.25 | 4.78 | 3.70 | 1.61 | 5.50 | 2.51 | 37.26 |
| 1983 | 2.95 | 1.77 | 3.84 | 4.20 | 5.28 | 0.88 | 5.92 | 3.86 | 2.70 | 1.81 | 8.15 | 5.01 | 46.37 |
| 1984 | 2.10 | 3.06 | 2.55 | 1.74 | 5.72 | 5.90 | 4.52 | 1.65 | 1.54 | 1.81 | 2.01 | 3.00 | 35.60 |
| 1985 | 0.99 | 2.77 | 1.87 | 1.80 | 2.64 | 2.89 | 5.05 | 1.74 | 2.30 | 1.42 | 3.50 | 2.24 | 29.21 |
| 1986 | 4.86 | 1.13 | 2.32 | 2.29 | 2.13 | 1.96 | 4.21 | 4.97 | 3.58 | 1.47 | 3.96 | 1.66 | 34.54 |
| 1987 | 2.29 | 0.33 | 1.24 | 1.75 | 2.46 | 3.59 | 3.16 | 1.82 | 4.37 | 2.18 | 2.33 | 2.56 | 28.08 |
| 1988 | 2.79 | 2.65 | 1.23 | 1.99 | 1.84 | 2.37 | 2.28 | 5.65 | 1.82 | 3.09 | 4.10 | 1.00 | 30.81 |
| 1989 | 1.88 | 1.43 | 1.40 | 2.24 | 4.13 | 2.29 | 2.63 | 5.41 | 3.52 | 1.62 | 3.88 | 2.35 | 32.78 |
| 1990 | 3.36 | 1.84 | 1.16 | 2.28 | 3.53 | 4.56 | 4.15 | 3.23 | 3.78 | 8.73 | 4.22 | 5.60 | 46.44 |
| 1991 | 2.06 | 1.13 | 4.71 | 2.62 | 3.51 | 1.98 | 0.96 | 6.72 | 3.45 | 4.46 | 1.61 | 2.10 | 35.31 |
| 1992 | 3.76 | 2.68 | 1.87 | 2.57 | 1.83 | 4.58 | 4.49 | 6.28 | 1.51 | 3.52 | 1.60 | 1.58 | 36.27 |
| 1993 | 2.18 | 2.10 | 1.72 | 3.48 | 3.58 | 4.54 | 2.48 | 3.57 | 5.11 | 4.31 | 2.94 | 3.99 | 40.00 |
| 1994 | 3.68 | 1.39 | 2.44 | 3.24 | 4.75 | 4.64 | 4.64 | 2.82 | 3.16 | 0.88 | 3.75 | 3.01 | 38.40 |
| 1995 | 5.60 | 2.70 | 2.23 | 2.12 | 2.46 | 1.18 | 1.48 | 2.94 | 1.90 | 5.13 | 4.88 | 1.79 | 34.41 |
| 1996 | 4.05 | 2.69 | 1.74 | 3.59 | 3.52 | 3.42 | 6.32 | 2.66 | 3.81 | 3.41 | 1.49 | 3.72 | 40.42 |
| 1997 | 3.60 | 2.52 | 2.47 | 1.68 | 5.10 | 4.37 | 2.64 | 4.12 | 2.67 | 1.31 | 2.08 | 2.81 | 35.37 |
| 1998 | 4.08 | 2.62 | 3.51 | 2.23 | 3.61 | 3.22 | 5.35 | 2.29 | 3.22 | 2.22 | 2.10 | 1.57 | 36.02 |
| 1999 | 3.59 | 1.38 | 2.30 | 1.43 | 2.40 | 3.20 | 2.94 | 3.69 | 8.81 | 3.48 | 2.70 | 2.76 | 38.68 |
| 2000 | 2.94 | 2.96 | 1.86 | 4.58 | 4.52 | 2.88 | 4.39 | 4.23 | 1.59 | 2.59 | 2.09 | 3.73 | 38.36 |
| 2001 | 0.81 | 2.66 | 3.15 | 0.96 | 2.01 | 2.77 | 5.82 | 1.51 | 3.65 | 2.69 | 2.08 | 0.99 | 29.10 |
| 2002 | 1.73 | 2.58 | 2.70 | 3.21 | 3.97 | 3.20 | 5.86 | 0.55 | 4.15 | 2.77 | 3.43 | 2.93 | 37.08 |
| 2003 | 0.69 | 3.78 | 2.51 | 1.43 | 3.18 | 3.99 | 6.10 | 3.53 | 1.95 | 6.58 | 3.83 | 5.07 | 42.64 |
| 2004 | 1.44 | 0.66 | 1.85 | 2.89 | 2.55 | 3.89 | 4.28 | 5.02 | 2.04 | 2.16 | 3.39 | 4.01 | 34.18 |
| 2005 | 1.99 | 2.43 | 4.90 | 6.19 | 3.20 | 2.31 | 4.10 | 4.11 | 5.04 | 7.38 | 6.06 | 6.60 | 54.31 |
| 2006 | 3.56 | 2.30 | 0.80 | 1.84 | 5.00 | 4.18 | 4.32 | 2.90 | 2.88 | 5.60 | 4.87 | 2.29 | 40.54 |
| 2007 | 2.31 | 1.94 | 4.19 | 3.33 | 2.90 | 2.44 | 3.51 | 3.48 | 2.07 | 3.66 | 7.39 | 4.80 | 42.02 |
| 2008 | 3.10 | 4.81 | 5.27 | 2.57 | 2.66 | 5.87 | 4.46 | 4.05 | 4.96 | 4.76 | 4.55 | 5.61 | 52.67 |
| 2009 | 3.07 | 1.95 | 2.06 | 2.95 | 5.29 | 2.69 | 5.08 | 1.60 | 1.95 | 4.74 | 3.76 | 3.79 | 38.93 |
| POR= 71 YRS | 2.41 | 2.14 | 2.48 | 2.60 | 3.13 | 3.41 | 4.05 | 3.79 | 3.37 | 3.32 | 3.43 | 3.05 | 37.18 |

WBAN : 14607

## AVERAGE TEMPERATURE (°F) 2009 CARIBOU (KCAR)

| YEAR | JAN | FEB | MAR | APR | MAY | JUN | JUL | AUG | SEP | OCT | NOV | DEC | ANNUAL |
|------|-----|-----|-----|-----|-----|-----|-----|-----|-----|-----|-----|-----|--------|
| 1980 | 13.3 | 11.9 | 24.0 | 42.6 | 51.2 | 59.4 | 64.8 | 66.0 | 50.8 | 40.9 | 30.1 | 9.0 | 38.7 |
| 1981 | 5.6 | 27.6 | 28.0 | 39.5 | 54.2 | 61.2 | 66.3 | 64.5 | 53.1 | 40.4 | 31.9 | 23.1 | 41.3 |
| 1982 | 4.0 | 11.3 | 24.5 | 35.1 | 53.6 | 60.3 | 66.4 | 58.7 | 54.9 | 44.4 | 32.7 | 22.1 | 39.0 |
| 1983 | 14.9 | 15.3 | 27.0 | 41.1 | 48.9 | 61.9 | 64.5 | 64.3 | 57.2 | 43.6 | 32.2 | 14.5 | 40.5 |
| 1984 | 6.7 | 22.4 | 19.3 | 39.8 | 49.2 | 59.2 | 65.8 | 65.5 | 51.7 | 44.5 | 33.4 | 18.4 | 39.7 |
| 1985 | 6.0 | 16.2 | 23.5 | 35.3 | 49.5 | 58.2 | 66.0 | 62.2 | 56.1 | 43.8 | 28.3 | 12.1 | 38.1 |
| 1986 | 11.8 | 12.2 | 24.4 | 42.8 | 52.2 | 56.9 | 62.6 | 60.7 | 50.5 | 41.2 | 25.9 | 16.2 | 38.1 |
| 1987 | 9.9 | 12.5 | 28.0 | 43.9 | 50.6 | 60.3 | 65.8 | 61.1 | 54.5 | 44.2 | 29.0 | 18.7 | 39.9 |
| 1988 | 12.1 | 13.6 | 23.1 | 39.4 | 54.6 | 58.5 | 67.4 | 64.4 | 52.4 | 40.5 | 33.5 | 13.8 | 39.4 |
| 1989 | 12.3 | 10.4 | 19.5 | 36.7 | 56.7 | 60.1 | 64.7 | 63.5 | 55.4 | 45.0 | 27.6 | 3.5 | 38.0 |
| 1990 | 17.3 | 11.7 | 24.7 | 39.6 | 49.0 | 62.9 | 66.2 | 66.6 | 53.5 | 44.7 | 31.0 | 20.5 | 40.6 |
| 1991 | 7.1 | 15.1 | 27.0 | 39.6 | 52.5 | 61.4 | 65.8 | 65.3 | 52.1 | 44.8 | 32.8 | 13.5 | 39.8 |
| 1992 | 10.4 | 10.9 | 20.2 | 36.4 | 53.2 | 60.2 | 60.3 | 63.7 | 56.2 | 41.1 | 29.3 | 19.6 | 38.5 |
| 1993 | 8.6 | 4.2 | 23.3 | 39.9 | 51.2 | 59.9 | 64.8 | 65.8 | 54.1 | 39.2 | 29.7 | 19.4 | 38.3 |
| 1994 | -0.7 | 7.7 | 25.3 | 37.0 | 48.7 | 63.6 | 68.1 | 61.6 | 53.5 | 46.1 | 33.9 | 20.1 | 38.7 |
| 1995 | 15.5 | 9.1 | 25.5 | 34.3 | 50.8 | 63.0 | 69.1 | 65.1 | 51.8 | 47.8 | 28.2 | 14.7 | 39.6 |
| 1996 | 11.2 | 13.5 | 23.7 | 38.8 | 48.7 | 62.1 | 64.6 | 63.7 | 55.0 | 42.2 | 29.1 | 26.2 | 39.9 |
| 1997 | 10.6 | 12.6 | 18.8 | 35.5 | 47.3 | 60.3 | 65.3 | 61.5 | 54.2 | 41.4 | 29.2 | 15.9 | 37.7 |
| 1998 | 13.8 | 19.7 | 27.4 | 40.1 | 56.2 | 61.2 | 66.3 | 64.1 | 55.5 | 43.4 | 30.0 | 21.4 | 41.6 |
| 1999 | 10.5 | 17.5 | 30.2 | 39.2 | 57.2 | 63.9 | 67.2 | 62.9 | 61.7 | 40.6 | 34.7 | 22.2 | 42.3 |
| 2000 | 11.1 | 14.5 | 29.9 | 37.2 | 49.4 | 58.7 | 63.4 | 62.4 | 53.6 | 42.7 | 35.0 | 15.5 | 39.5 |
| 2001 | 10.8 | 12.2 | 23.6 | 37.0 | 56.4 | 61.9 | 64.4 | 65.8 | 57.1 | 46.7 | 35.6 | 26.5 | 41.5 |
| 2002 | 15.4 | 15.8 | 24.2 | 38.2 | 48.6 | 57.5 | 64.1 | 65.1 | 58.0 | 40.3 | 28.4 | 17.3 | 39.4 |
| 2003 | 5.4 | 7.4 | 21.1 | 32.3 | 49.8 | 61.8 | 65.1 | 64.3 | 58.4 | 43.2 | 32.8 | 21.7 | 38.6 |
| 2004 | 3.2 | 14.9 | 26.1 | 39.2 | 50.6 | 56.3 | 65.2 | 63.6 | 54.8 | 44.9 | 29.9 | 14.6 | 38.6 |
| 2005 | 6.8 | 16.9 | 24.0 | 39.6 | 48.2 | 63.2 | 66.3 | 64.9 | 56.4 | 46.6 | 32.7 | 18.4 | 40.3 |
| 2006 | 18.5 | 14.9 | 28.3 | 41.7 | 54.5 | 64.0 | 68.3 | 60.4 | 54.9 | 43.3 | 36.7 | 23.6 | 42.4 |
| 2007 | 12.3 | 9.3 | 23.2 | 36.0 | 50.3 | 61.0 | 65.2 | 62.1 | 55.5 | 47.7 | 29.2 | 13.6 | 38.8 |
| 2008 | 13.2 | 13.0 | 18.3 | 39.9 | 49.9 | 60.7 | 67.6 | 63.2 | 55.3 | 43.4 | 33.5 | 16.0 | 39.5 |
| 2009 | 2.5 | 17.2 | 23.0 | 40.5 | 50.3 | 60.0 | 63.7 | 65.2 | 54.5 | 39.1 | 36.9 | 20.3 | 39.4 |
| POR= 71 YRS | 10.2 | 13.1 | 24.1 | 37.3 | 50.7 | 60.0 | 65.3 | 63.0 | 54.0 | 43.2 | 31.0 | 16.7 | 39.0 |

## HEATING DEGREE DAYS (base 65°F) 2009 CARIBOU (KCAR)

| YEAR | JUL | AUG | SEP | OCT | NOV | DEC | JAN | FEB | MAR | APR | MAY | JUN | TOTAL |
|------|-----|-----|-----|-----|-----|-----|-----|-----|-----|-----|-----|-----|-------|
| 1980-81 | 71 | 41 | 425 | 740 | 1042 | 1733 | 1839 | 1040 | 1141 | 757 | 333 | 125 | 9287 |
| 1981-82 | 37 | 77 | 355 | 757 | 984 | 1292 | 1891 | 1499 | 1251 | 890 | 355 | 145 | 9533 |
| 1982-83 | 60 | 199 | 310 | 632 | 961 | 1325 | 1544 | 1387 | 1171 | 712 | 493 | 148 | 8942 |
| 1983-84 | 75 | 78 | 257 | 656 | 978 | 1562 | 1804 | 1229 | 1412 | 748 | 493 | 199 | 9491 |
| 1984-85 | 38 | 57 | 396 | 630 | 940 | 1441 | 1822 | 1363 | 1281 | 884 | 472 | 198 | 9522 |
| 1985-86 | 43 | 118 | 272 | 650 | 1094 | 1636 | 1646 | 1473 | 1255 | 660 | 394 | 246 | 9487 |
| 1986-87 | 105 | 147 | 427 | 733 | 1167 | 1505 | 1704 | 1465 | 1140 | 629 | 442 | 142 | 9606 |
| 1987-88 | 79 | 152 | 314 | 641 | 1071 | 1430 | 1636 | 1485 | 1289 | 760 | 321 | 232 | 9410 |
| 1988-89 | 47 | 114 | 373 | 752 | 939 | 1583 | 1627 | 1524 | 1402 | 841 | 257 | 181 | 9640 |
| 1989-90 | 66 | 101 | 303 | 613 | 1116 | 1905 | 1471 | 1490 | 1241 | 759 | 490 | 107 | 9662 |
| 1990-91 | 57 | 47 | 337 | 623 | 1014 | 1370 | 1791 | 1392 | 1172 | 755 | 383 | 158 | 9099 |
| 1991-92 | 55 | 66 | 383 | 619 | 959 | 1595 | 1691 | 1564 | 1382 | 852 | 381 | 163 | 9710 |
| 1992-93 | 145 | 73 | 273 | 732 | 1064 | 1402 | 1747 | 1699 | 1284 | 748 | 423 | 175 | 9765 |
| 1993-94 | 73 | 53 | 327 | 791 | 1051 | 1407 | 2040 | 1601 | 1222 | 832 | 500 | 107 | 10004 |
| 1994-95 | 24 | 124 | 337 | 578 | 926 | 1388 | 1529 | 1565 | 1218 | 912 | 433 | 131 | 9165 |
| 1995-96 | 12 | 74 | 388 | 522 | 1098 | 1551 | 1666 | 1490 | 1271 | 782 | 496 | 116 | 9466 |
| 1996-97 | 48 | 83 | 304 | 702 | 1070 | 1196 | 1678 | 1462 | 1427 | 879 | 544 | 161 | 9554 |
| 1997-98 | 62 | 123 | 318 | 723 | 1068 | 1516 | 1579 | 1263 | 1159 | 740 | 269 | 151 | 8971 |
| 1998-99 | 42 | 69 | 275 | 660 | 1045 | 1346 | 1681 | 1323 | 1073 | 767 | 248 | 103 | 8632 |
| 1999-00 | 40 | 105 | 163 | 750 | 903 | 1320 | 1662 | 1456 | 1079 | 828 | 478 | 216 | 9000 |
| 2000-01 | 75 | 93 | 337 | 681 | 893 | 1525 | 1672 | 1474 | 1277 | 833 | 265 | 147 | 9272 |
| 2001-02 | 65 | 64 | 260 | 563 | 874 | 1186 | 1532 | 1372 | 1260 | 797 | 503 | 238 | 8714 |
| 2002-03 | 79 | 94 | 244 | 757 | 1093 | 1472 | 1842 | 1608 | 1356 | 975 | 468 | 161 | 10149 |
| 2003-04 | 36 | 104 | 199 | 668 | 959 | 1334 | 1907 | 1446 | 1198 | 768 | 443 | 254 | 9316 |
| 2004-05 | 50 | 79 | 302 | 618 | 1046 | 1556 | 1795 | 1342 | 1263 | 754 | 515 | 140 | 9460 |
| 2005-06 | 36 | 55 | 255 | 566 | 961 | 1441 | 1434 | 1394 | 1130 | 691 | 318 | 82 | 8363 |
| 2006-07 | 15 | 161 | 304 | 666 | 841 | 1277 | 1628 | 1551 | 1288 | 864 | 460 | 133 | 9188 |
| 2007-08 | 66 | 118 | 290 | 529 | 1069 | 1586 | 1593 | 1500 | 1440 | 746 | 460 | 139 | 9536 |
| 2008-09 | 12 | 74 | 288 | 662 | 939 | 1513 | 1931 | 1332 | 1292 | 728 | 449 | 170 | 9390 |
| 2009- | 70 | 80 | 311 | 795 | 838 | 1376 | | | | | | | |

WBAN : 14607

## COOLING DEGREE DAYS (base 65°F) 2009 CARIBOU (KCAR)

| YEAR | JAN | FEB | MAR | APR | MAY | JUN | JUL | AUG | SEP | OCT | NOV | DEC | TOTAL |
|------|-----|-----|-----|-----|-----|-----|-----|-----|-----|-----|-----|-----|-------|
| 1980 | 0 | 0 | 0 | 0 | 0 | 37 | 71 | 78 | 8 | 0 | 0 | 0 | 194 |
| 1981 | 0 | 0 | 0 | 0 | 3 | 17 | 83 | 68 | 4 | 0 | 0 | 0 | 175 |
| 1982 | 0 | 0 | 0 | 0 | 11 | 10 | 110 | 9 | 12 | 0 | 0 | 0 | 152 |
| 1983 | 0 | 0 | 0 | 0 | 0 | 62 | 65 | 66 | 32 | 1 | 0 | 0 | 226 |
| 1984 | 0 | 0 | 0 | 0 | 9 | 33 | 72 | 79 | 0 | 0 | 0 | 0 | 193 |
| 1985 | 0 | 0 | 0 | 0 | 0 | 0 | 82 | 40 | 10 | 0 | 0 | 0 | 132 |
| 1986 | 0 | 0 | 0 | 0 | 5 | 10 | 39 | 20 | 1 | 0 | 0 | 0 | 75 |
| 1987 | 0 | 0 | 0 | 0 | 2 | 10 | 111 | 38 | 8 | 0 | 0 | 0 | 169 |
| 1988 | 0 | 0 | 0 | 0 | 6 | 45 | 131 | 103 | 1 | 0 | 0 | 0 | 286 |
| 1989 | 0 | 0 | 0 | 0 | 9 | 37 | 64 | 63 | 19 | 0 | 0 | 0 | 192 |
| 1990 | 0 | 0 | 0 | 2 | 0 | 51 | 102 | 106 | 0 | 0 | 0 | 0 | 261 |
| 1991 | 0 | 0 | 0 | 0 | 2 | 55 | 89 | 85 | 4 | 0 | 0 | 0 | 235 |
| 1992 | 0 | 0 | 0 | 0 | 23 | 26 | 5 | 40 | 12 | 0 | 0 | 0 | 106 |
| 1993 | 0 | 0 | 0 | 0 | 0 | 31 | 72 | 83 | 6 | 0 | 0 | 0 | 192 |
| 1994 | 0 | 0 | 0 | 0 | 4 | 71 | 128 | 26 | 0 | 0 | 0 | 0 | 229 |
| 1995 | 0 | 0 | 0 | 0 | 0 | 76 | 147 | 82 | 0 | 0 | 0 | 0 | 305 |
| 1996 | 0 | 0 | 0 | 0 | 0 | 36 | 40 | 51 | 11 | 0 | 0 | 0 | 138 |
| 1997 | 0 | 0 | 0 | 0 | 0 | 30 | 77 | 21 | 1 | 0 | 0 | 0 | 129 |
| 1998 | 0 | 0 | 0 | 0 | 3 | 43 | 91 | 44 | 0 | 0 | 0 | 0 | 181 |
| 1999 | 0 | 0 | 0 | 0 | 12 | 76 | 115 | 48 | 70 | 0 | 0 | 0 | 321 |
| 2000 | 0 | 0 | 0 | 0 | 0 | 33 | 33 | 20 | 1 | 0 | 0 | 0 | 87 |
| 2001 | 0 | 0 | 0 | 0 | 5 | 59 | 56 | 94 | 31 | 0 | 0 | 0 | 245 |
| 2002 | 0 | 0 | 0 | 0 | 1 | 19 | 59 | 105 | 41 | 0 | 0 | 0 | 225 |
| 2003 | 0 | 0 | 0 | 0 | 1 | 70 | 47 | 90 | 10 | 0 | 0 | 0 | 218 |
| 2004 | 0 | 0 | 0 | 0 | 4 | 2 | 61 | 44 | 0 | 0 | 0 | 0 | 111 |
| 2005 | 0 | 0 | 0 | 0 | 0 | 92 | 86 | 59 | 4 | 2 | 0 | 0 | 243 |
| 2006 | 0 | 0 | 0 | 0 | 0 | 59 | 126 | 23 | 6 | 0 | 0 | 0 | 214 |
| 2007 | 0 | 0 | 0 | 0 | 13 | 22 | 78 | 36 | 11 | 0 | 0 | 0 | 160 |
| 2008 | 0 | 0 | 0 | 0 | 0 | 17 | 102 | 23 | 4 | 0 | 0 | 0 | 146 |
| 2009 | 0 | 0 | 0 | 0 | 1 | 25 | 36 | 97 | 2 | 0 | 0 | 0 | 161 |

## SNOWFALL (inches) 2009 CARIBOU (KCAR)

| YEAR | JUL | AUG | SEP | OCT | NOV | DEC | JAN | FEB | MAR | APR | MAY | JUN | TOTAL |
|---|---|---|---|---|---|---|---|---|---|---|---|---|---|
| 1980-81 | 0.0 | 0.0 | T | 0.2 | 12.0 | 28.9 | 35.2 | 8.6 | 36.4 | 1.4 | 0.2 | 0.0 | 122.9 |
| 1981-82 | 0.0 | 0.0 | T | 2.6 | 4.8 | 34.3 | 30.9 | 25.1 | 23.7 | 36.4 | 1.0 | 0.0 | 158.8 |
| 1982-83 | 0.0 | 0.0 | 0.0 | T | 14.6 | 4.0 | 22.8 | 22.8 | 13.9 | 4.8 | T | 0.0 | 82.9 |
| 1983-84 | 0.0 | 0.0 | 0.0 | T | 21.0 | 26.6 | 27.3 | 20.1 | 35.4 | 3.9 | 0.2 | 0.0 | 134.5 |
| 1984-85 | 0.0 | 0.0 | 0.0 | 0.8 | 3.8 | 30.7 | 10.5 | 26.8 | 11.9 | 5.0 | 1.3 | 0.0 | 90.8 |
| 1985-86 | 0.0 | 0.0 | 0.0 | T | 14.8 | 18.3 | 30.0 | 11.9 | 18.6 | 11.3 | T | 0.0 | 104.9 |
| 1986-87 | 0.0 | 0.0 | T | 0.4 | 28.1 | 8.5 | 26.4 | 4.1 | 12.3 | 5.2 | T | 0.0 | 85.0 |
| 1987-88 | 0.0 | 0.0 | T | T | 5.2 | 22.6 | 32.1 | 28.2 | 3.9 | 6.4 | T | 0.0 | 98.4 |
| 1988-89 | 0.0 | 0.0 | 0.0 | 1.2 | 13.8 | 11.5 | 18.4 | 16.6 | 11.9 | 9.4 | 0.0 | 0.0 | 82.8 |
| 1989-90 | 0.0 | 0.0 | T | 0.1 | 14.3 | 38.0 | 28.5 | 18.3 | 7.6 | 10.7 | 0.6 | T | 118.1 |
| 1990-91 | 0.0 | 0.0 | 0.0 | T | 12.0 | 22.6 | 21.5 | 12.2 | 24.0 | 2.5 | 0.0 | T | 94.8 |
| 1991-92 | 0.0 | 0.0 | 2.5 | T | 7.1 | 19.4 | 14.8 | 38.6 | 4.2 | 7.4 | T | 0.0 | 94.0 |
| 1992-93 | T | 0.0 | T | 0.4 | 5.3 | 15.9 | 5.6 | 22.8 | 23.4 | 7.1 | T | 0.0 | 80.5 |
| 1993-94 | 0.0 | 0.0 | 0.0 | 0.2 | 10.5 | 13.9 | 44.5 | 20.3 | 28.8 | 9.5 | T | 0.0 | 127.7 |
| 1994-95 | 0.0 | 0.0 | T | T | 9.8 | 25.9 |  | 40.9 | 15.7 | 7.9 | 1.1 | 0.0 |  |
| 1995-96 | 0.0 | T | 0.0 | T | 8.9 | 25.1 | 18.4 | 22.7 | 16.5 | 13.1 | 5.7 | 0.0 | 110.4 |
| 1996-97 | 0.0 | 0.0 | 0.0 | 0.8 | 8.5 | 14.4 | 35.5 | 30.2 | 28.9 | 11.7 | 0.9 | 0.0 | 130.9 |
| 1997-98 | 0.0 | 0.0 | 0.0 | 8.9 | 9.6 | 37.3 | 39.9 | 8.8 | 11.0 | 8.5 | T | 0.0 | 124.0 |
| 1998-99 | 0.0 | 0.0 | 0.0 | T | 12.6 | 21.9 | 36.7 | 18.5 | 30.3 | 8.8 | T | 0.0 | 128.8 |
| 1999-00 | 0.0 | 0.0 | 0.0 | 0.2 | 4.1 | 6.7 | 43.7 | 32.5 | 15.4 | 9.8 | T | T | 112.4 |
| 2000-01 | T | 0.0 | 0.0 | 10.3 | 5.2 | 32.7 | 14.2 | 30.5 | 30.6 | 8.6 | 0.0 | 0.0 | 132.1 |
| 2001-02 | 0.0 | 0.0 | 0.0 | 0.0 | 9.0 | 4.2 | 26.0 | 23.1 | 21.6 | 4.2 | 3.1 | 0.0 | 91.2 |
| 2002-03 | 0.0 | 0.0 | 0.0 | 1.3 | 19.7 | 15.0 | 15.2 | 39.1 | 23.3 | 3.7 | T | 0.0 | 117.3 |
| 2003-04 | 0.0 | 0.0 | 0.0 | 7.9 | 3.8 | 43.0 | 23.8 | 11.9 | 8.7 | 2.8 | T | 0.0 | 101.9 |
| 2004-05 | 0.0 | 0.0 | 0.0 | 0.0 | 11.3 | 24.2 | 24.5 | 25.3 | 37.6 | 5.5 | 0.0 | 0.0 | 128.4 |
| 2005-06 | 0.0 | 0.0 | 0.0 | 4.2 | 17.6 | 43.4 | 19.7 | 7.8 | 1.4 | 1.9 | 0.0 | 0.0 | 96.0 |
| 2006-07 | 0.0 | 0.0 | 0.0 | T | 3.2 | 8.5 | 17.6 | 21.3 | 26.2 | 29.0 | 0.8 | 0.0 | 106.6 |
| 2007-08 | 0.0 | 0.0 | 0.0 | 0.8 | 16.0 | 54.5 | 26.5 | 47.7 | 45.2 | 7.1 | T | 0.0 | 197.8 |
| 2008-09 | 0.0 | 0.0 | 0.0 | 1.0 | 9.6 | 39.7 | 29.5 | 18.4 | 16.6 | 3.3 | 0.0 | 0.0 | 118.1 |
| 2009- | 0.0 | 0.0 | 0.0 | 6.4 | 3.3 | 27.0 |  |  |  |  |  |  |  |
| POR= 71 YRS | T | T | T | 1.9 | 11.5 | 23.7 | 24.0 | 22.7 | 19.8 | 8.3 | 0.7 | T | 112.6 |

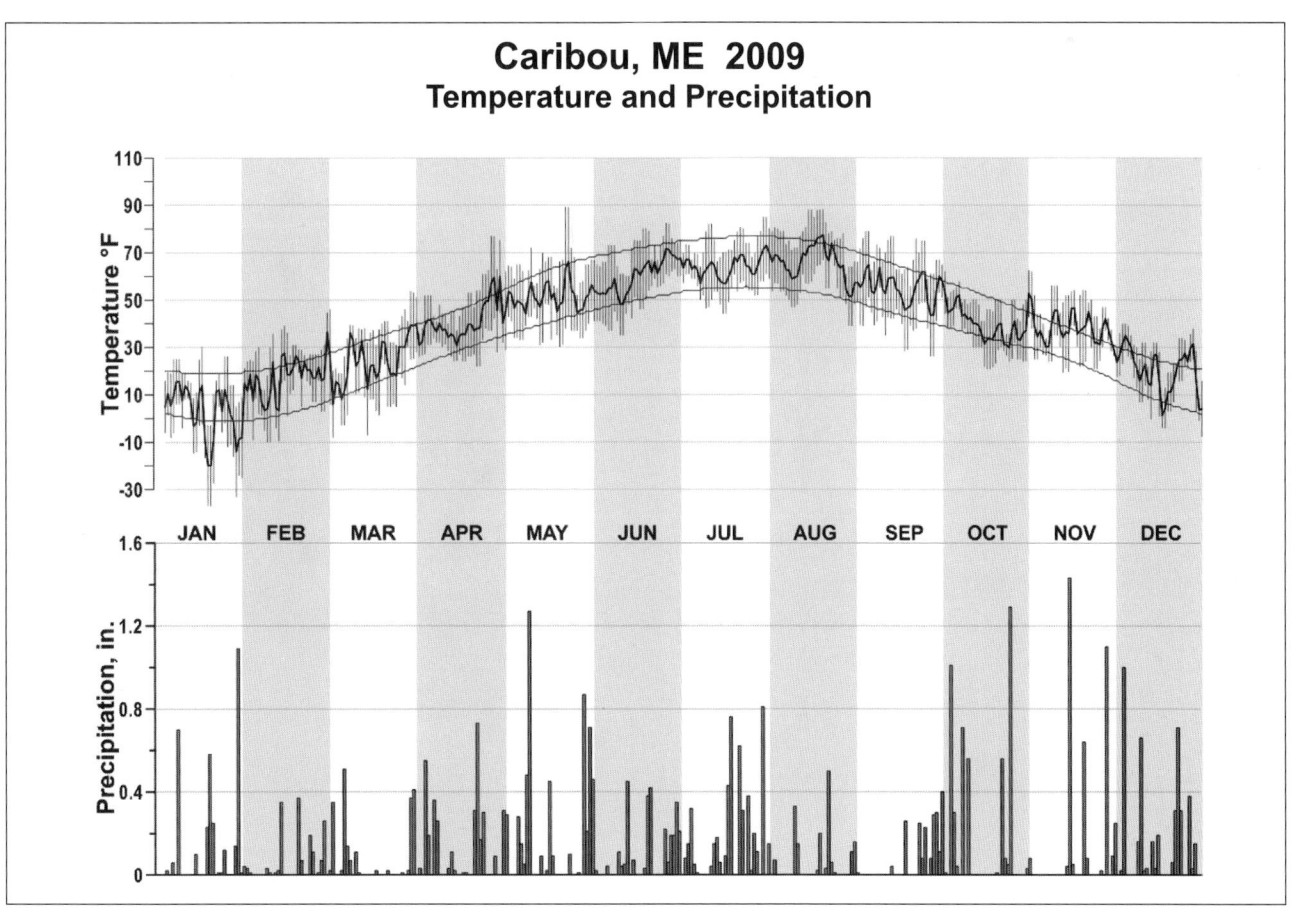

## Caribou, ME  2009
### Temperature and Precipitation

# 2009
# PORTLAND
# MAINE (KPWM)

The Portland City Airport is located 2 3/4 miles west of the site of the former city office. The surrounding country is mostly open, rolling and sloping generally toward the Fore River, a body of brackish water about 1,000 feet wide at a distance of about 1/2 mile from the station and forming one boundary (north through east) of the field. The airport is about 5 1/2 miles west-northwest of the open ocean. A slight rise reaching an elevation of 100 feet, lying northwest of the field, cuts down the wind slightly from that direction. The older portion of the city is situated on a hill rising abruptly from sea level to 170 feet, 1 1/2 miles east of the airport and on the opposite side of the Fore River. A line of low hills southeast of the airport, near the ocean, which reach a maximum height of 160 feet, shuts off sight of the ocean from the airport. Sebago Lake with an area of 44 square miles is situated about 15 miles to the northwest and 45 miles farther are the White Mountains, averaging 3,000 to 5,000 feet in height.

As a rule, Portland has very pleasant summers and falls, cold winters with frequent thaws, and disagreeable springs. Very few summer nights are too warm and humid for comfortable sleeping. Autumn has the greatest number of sunny days and the least cloudiness. Winters are quite severe, but begin late and then extend deeply into the normal springtime.

Heavy seasonal snowfalls, over 100 inches, normally occur about each 10 years. True blizzards are very rare. The White Mountains, to the northwest, keep considerable snow from reaching the Portland area and also moderate the temperature. Normal monthly precipitation is remarkably uniform throughout the year.

Winds are generally quite light with the highest velocities being confined mostly to March and November. Even in these months the occasional northeasterly gales have usually lost much of their severity before reaching the coast of Maine.

Temperatures well below zero are recorded frequently each winter. Cold waves sometimes come in on strong winds, but extremely low temperatures are generally accompanied by light winds.

The average freeze-free season at the airport station is 139 days. Mid-May is the average occurrence of the last freeze in spring, and the average occurrence of the first freeze in fall is late September. The freeze-free period is longer in the city proper, but may be even shorter at susceptible places further inland.

Daily maximum temperatures at the present airport site agree closely with those near the former intown office, but minimum temperatures on clear, quiet mornings range as much as 15 degrees lower at the airport.

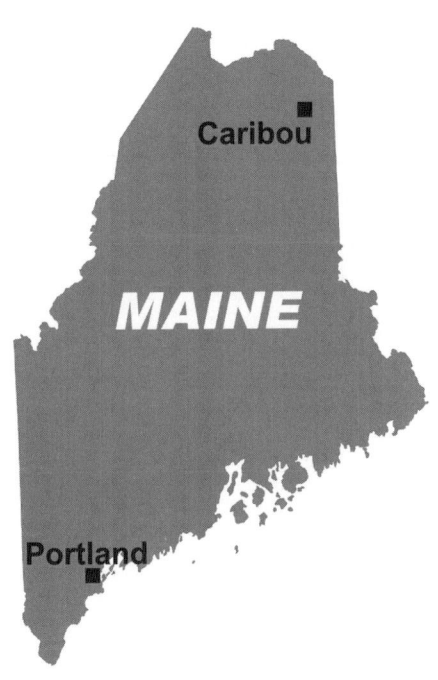

# NORMALS, MEANS, AND EXTREMES
## PORTLAND (KPWM)

| LATITUDE: 43 ° 38'N | LONGITUDE: -70 ° 18'W | ELEVATION (FT): GRND: 48  BARO: 72 | TIME ZONE: EASTERN  (UTC -5) | WBAN: 14764 |
|---|---|---|---|---|

| | ELEMENT | POR | JAN | FEB | MAR | APR | MAY | JUN | JUL | AUG | SEP | OCT | NOV | DEC | YEAR |
|---|---|---|---|---|---|---|---|---|---|---|---|---|---|---|---|
| **TEMPERATURE °F** | NORMAL DAILY MAXIMUM | 30 | 30.9 | 34.1 | 42.2 | 52.8 | 63.3 | 72.8 | 78.8 | 77.3 | 68.9 | 57.9 | 47.1 | 36.4 | 55.2 |
| | MEAN DAILY MAXIMUM | 90 | 31.1 | 32.3 | 41.1 | 51.7 | 62.9 | 71.7 | 78.2 | 77.0 | 68.7 | 58.8 | 46.9 | 35.7 | 54.7 |
| | HIGHEST DAILY MAXIMUM | 69 | 67 | 64 | 88 | 92 | 94 | 98 | 99 | 103 | 95 | 88 | 74 | 71 | 103 |
| | YEAR OF OCCURRENCE | | 2007 | 1957 | 1998 | 2009 | 1987 | 1991 | 1977 | 1975 | 1983 | 1963 | 2003 | 2001 | AUG 1975 |
| | MEAN OF EXTREME MAXS. | 90 | 49.3 | 50.3 | 59.3 | 72.9 | 83.1 | 89.1 | 91.6 | 90.2 | 85.9 | 75.5 | 64.7 | 54.5 | 72.2 |
| | NORMAL DAILY MINIMUM | 30 | 12.5 | 15.6 | 25.2 | 34.7 | 44.2 | 52.9 | 58.6 | 57.2 | 48.5 | 37.4 | 29.5 | 18.7 | 36.3 |
| | MEAN DAILY MINIMUM | 90 | 13.1 | 14.3 | 24.3 | 33.7 | 43.6 | 52.1 | 58.6 | 57.4 | 49.1 | 39.3 | 30.4 | 18.8 | 36.2 |
| | LOWEST DAILY MINIMUM | 69 | -26 | -39 | -21 | 8 | 23 | 33 | 40 | 33 | 23 | 15 | 3 | -21 | -39 |
| | YEAR OF OCCURRENCE | | 1971 | 1943 | 1950 | 1954 | 1956 | 1944 | 1965 | 1965 | 1941 | 1976 | 1989 | 1963 | FEB 1943 |
| | MEAN OF EXTREME MINS. | 90 | -8.4 | -6.3 | 4.1 | 22.0 | 30.9 | 40.7 | 48.0 | 44.9 | 33.9 | 23.9 | 15.1 | -1.6 | 20.6 |
| | NORMAL DRY BULB | 30 | 21.7 | 24.8 | 33.7 | 43.7 | 53.8 | 62.9 | 68.7 | 67.2 | 58.7 | 47.7 | 38.3 | 27.6 | 45.7 |
| | MEAN DRY BULB | 90 | 22.2 | 23.3 | 32.7 | 42.7 | 53.3 | 62.0 | 68.4 | 67.2 | 58.9 | 49.1 | 38.6 | 27.3 | 45.5 |
| | MEAN WET BULB | 26 | 19.6 | 21.5 | 28.3 | 38.2 | 47.7 | 57.5 | 62.9 | 62.3 | 55.2 | 44.6 | 35.3 | 25.2 | 41.5 |
| | MEAN DEW POINT | 26 | 14.4 | 15.8 | 23.1 | 32.8 | 44.0 | 54.2 | 60.2 | 59.7 | 52.5 | 41.2 | 31.1 | 20.2 | 37.4 |
| | NORMAL NO. DAYS WITH: | | | | | | | | | | | | | | |
| | MAXIMUM >= 90 | 30 | 0.0 | 0.0 | 0.0 | 0.0 | 0.2 | 1.1 | 1.9 | 1.2 | 0.3 | 0.0 | 0.0 | 0.0 | 4.7 |
| | MAXIMUM <= 32 | 30 | 16.7 | 12.4 | 4.2 | 0.2 | 0.0 | 0.0 | 0.0 | 0.0 | 0.0 | 0.0 | 1.2 | 10.4 | 45.1 |
| | MINIMUM <= 32 | 30 | 29.6 | 25.9 | 24.3 | 12.3 | 1.3 | 0.0 | 0.0 | 0.0 | 0.3 | 8.6 | 19.0 | 28.2 | 149.5 |
| | MINIMUM <= 0 | 30 | 5.4 | 3.0 | 0.3 | 0.0 | 0.0 | 0.0 | 0.0 | 0.0 | 0.0 | 0.0 | 0.0 | 1.7 | 10.4 |
| **H/C** | NORMAL HEATING DEG. DAYS | 30 | 1346 | 1141 | 985 | 649 | 361 | 116 | 19 | 37 | 199 | 523 | 790 | 1152 | 7318 |
| | NORMAL COOLING DEG. DAYS | 30 | 0 | 0 | 0 | 0 | 7 | 51 | 144 | 120 | 24 | 1 | 0 | 0 | 347 |
| **RH** | NORMAL (PERCENT) | 30 | 68 | 65 | 66 | 67 | 72 | 73 | 74 | 76 | 76 | 74 | 71 | 69 | 71 |
| | HOUR 01 LST | 30 | 72 | 70 | 73 | 76 | 83 | 86 | 87 | 88 | 87 | 83 | 77 | 75 | 80 |
| | HOUR 07 LST | 30 | 75 | 73 | 74 | 72 | 76 | 78 | 80 | 83 | 86 | 84 | 80 | 77 | 78 |
| | HOUR 13 LST | 30 | 59 | 55 | 56 | 55 | 59 | 61 | 60 | 60 | 61 | 58 | 59 | 59 | 59 |
| | HOUR 19 LST | 30 | 66 | 64 | 66 | 66 | 70 | 72 | 73 | 76 | 78 | 75 | 71 | 69 | 71 |
| **S** | PERCENT POSSIBLE SUNSHINE | 55 | 56 | 60 | 56 | 55 | 54 | 60 | 63 | 64 | 63 | 57 | 48 | 53 | 57 |
| **W/O** | MEAN NO. DAYS WITH: | | | | | | | | | | | | | | |
| | HEAVY FOG(VISBY <= 1/4 MI) | 46 | 1.9 | 2.4 | 3.5 | 2.6 | 3.9 | 4.9 | 5.7 | 5.2 | 4.3 | 3.7 | 2.8 | 2.2 | 43.1 |
| | THUNDERSTORMS | 62 | 0.1 | 0.0 | 0.3 | 0.5 | 1.9 | 3.7 | 4.2 | 3.3 | 1.3 | 0.5 | 0.3 | 0.1 | 16.2 |
| **CLOUDNESS** | MEAN: | | | | | | | | | | | | | | |
| | SUNRISE-SUNSET (OKTAS) | | | | | | 5.6 | 6.4 | | | | | | | |
| | MIDNIGHT-MIDNIGHT (OKTAS) | | | | | | | | | | | | | | |
| | MEAN NO. DAYS WITH: | | | | | | | | | | | | | | |
| | CLEAR | 1 | 4.0 | 2.0 | 7.0 | | 7.0 | 6.0 | 1.0 | 9.0 | 3.0 | 8.0 | | 5.0 | |
| | PARTLY CLOUDY | | | 1.0 | 3.0 | | 4.0 | 5.0 | 1.0 | 1.0 | 2.0 | | 4.0 | | |
| | CLOUDY | 1 | 4.0 | 5.0 | 7.0 | | 10.0 | 9.0 | 2.0 | 3.0 | 6.0 | 3.0 | | 8.0 | |
| **PR** | MEAN STATION PRESSURE(IN) | 26 | 29.93 | 29.93 | 29.93 | 29.89 | 29.90 | 29.86 | 29.88 | 29.93 | 29.98 | 29.98 | 29.97 | 29.94 | 29.93 |
| | MEAN SEA-LEVEL PRES. (IN) | 26 | 30.00 | 30.00 | 30.00 | 29.96 | 29.97 | 29.93 | 29.94 | 30.00 | 30.05 | 30.04 | 30.03 | 30.01 | 29.99 |
| **WINDS** | MEAN SPEED (MPH) | 26 | 8.2 | 8.6 | 9.3 | 9.4 | 8.6 | 7.7 | 7.0 | 6.9 | 7.2 | 7.9 | 8.1 | 8.2 | 8.1 |
| | PREVAIL.DIR(TENS OF DEGS) | 32 | 36 | 33 | 33 | 19 | 19 | 19 | 19 | 19 | 19 | 28 | 36 | 28 | 19 |
| | MAXIMUM 2-MINUTE: | | | | | | | | | | | | | | |
| | SPEED (MPH) | 15 | 38 | 45 | 41 | 43 | 39 | 38 | 37 | 57 | 33 | 37 | 41 | 44 | 57 |
| | DIR. (TENS OF DEGS) | | 28 | 00 | 11 | 07 | 07 | 33 | 29 | 28 | 29 | 24 | 10 | 03 | 28 |
| | YEAR OF OCCURRENCE | | 2004 | 1995 | 1999 | 2007 | 2005 | 2006 | 2002 | 1998 | 1995 | 2006 | 1997 | 2002 | AUG 1998 |
| | MAXIMUM 3-SECOND: | | | | | | | | | | | | | | |
| | SPEED (MPH) | 15 | 53 | 56 | 54 | 59 | 51 | 46 | 47 | 61 | 41 | 55 | 72 | 54 | 72 |
| | DIR. (TENS OF DEGS) | | 16 | 28 | 13 | 07 | 08 | 32 | 29 | 28 | 29 | 24 | 15 | 29 | 15 |
| | YEAR OF OCCURRENCE | | 1996 | 2006 | 2000 | 2007 | 2005 | 2006 | 2008 | 1998 | 2008 | 2006 | 1995 | 2006 | NOV 1995 |
| **PRECIPITATION** | NORMAL (IN) | 30 | 4.09 | 3.14 | 4.14 | 4.26 | 3.82 | 3.28 | 3.32 | 3.05 | 3.37 | 4.40 | 4.72 | 4.24 | 45.83 |
| | MAXIMUM MONTHLY (IN) | 69 | 11.92 | 8.06 | 9.97 | 9.90 | 12.34 | 9.18 | 8.60 | 15.22 | 10.84 | 16.83 | 13.50 | 9.69 | 16.83 |
| | YEAR OF OCCURRENCE | | 1979 | 2008 | 1953 | 1973 | 2006 | 2006 | 2009 | 1991 | 2008 | 1996 | 1983 | 1969 | OCT 1996 |
| | MINIMUM MONTHLY (IN) | 69 | 0.28 | 0.04 | 0.81 | 0.28 | 0.49 | 0.70 | 0.61 | 0.27 | 0.30 | 0.26 | 0.90 | 0.98 | 0.04 |
| | YEAR OF OCCURRENCE | | 2004 | 1987 | 1965 | 1999 | 1965 | 1941 | 1965 | 1947 | 1948 | 1947 | 1976 | 1955 | FEB 1987 |
| | MAXIMUM IN 24 HOURS (IN) | 69 | 3.56 | 3.41 | 3.47 | 5.26 | 4.66 | 5.58 | 3.01 | 7.83 | 7.49 | 13.32 | 5.25 | 3.82 | 13.32 |
| | YEAR OF OCCURRENCE | | 1977 | 1981 | 1951 | 1973 | 1989 | 1967 | 2009 | 1991 | 1954 | 1996 | 2009 | 1969 | OCT 1996 |
| | NORMAL NO. DAYS WITH: | | | | | | | | | | | | | | |
| | PRECIPITATION >= 0.01 | 30 | 11.7 | 9.6 | 12.1 | 11.7 | 12.4 | 11.5 | 10.6 | 9.3 | 9.8 | 10.0 | 11.3 | 12.0 | 132.0 |
| | PRECIPITATION >= 1.00 | 30 | 0.9 | 0.9 | 1.0 | 0.9 | 0.8 | 0.7 | 0.7 | 0.6 | 0.9 | 1.2 | 1.4 | 1.2 | 11.2 |
| **SNOWFALL** | NORMAL (IN) | 30 | 20.5 | 12.8 | 13.0 | 3.2 | 0.* | 0.0 | 0.0 | 0.0 | 0.* | 0.1 | 3.2 | 13.6 | 66.4 |
| | MAXIMUM MONTHLY (IN) | 68 | 62.4 | 61.2 | 49.0 | 15.9 | 7.0 | 0.0 | T | 0.0 | T | 3.8 | 20.5 | 54.8 | 62.4 |
| | YEAR OF OCCURRENCE | | 1979 | 1969 | 1993 | 1982 | 1945 | | 2002 | | 1992 | 1969 | 1997 | 1970 | JAN 1979 |
| | MAXIMUM IN 24 HOURS (IN) | 68 | 27.1 | 21.5 | 18.6 | 15.9 | 7.0 | 0.0 | T | 0.0 | T | 3.6 | 11.1 | 22.8 | 27.1 |
| | YEAR OF OCCURRENCE' | | 1979 | 1969 | 1993 | 1982 | 1945 | | 2002 | | 1992 | 1969 | 1972 | 1970 | JAN 1979 |
| | MAXIMUM SNOW DEPTH (IN) | 57 | 31 | 31 | 33 | 17 | 1 | 0 | 0 | 0 | 0 | 2 | 10 | 40 | 40 |
| | YEAR OF OCCURRENCE | | 1979 | 1967 | 1967 | 1956 | 1966 | | | | | 1969 | 1972 | 1970 | DEC 1970 |
| | NORMAL NO. DAYS WITH: | | | | | | | | | | | | | | |
| | SNOWFALL >= 1.0 | 30 | 4.7 | 2.8 | 3.2 | 0.8 | 0.0 | 0.0 | 0.0 | 0.0 | 0.0 | 0.0 | 1.0 | 3.8 | 16.3 |

## PRECIPITATION (inches) 2009 PORTLAND (KPWM)

| YEAR | JAN | FEB | MAR | APR | MAY | JUN | JUL | AUG | SEP | OCT | NOV | DEC | ANNUAL |
|------|-----|-----|-----|-----|-----|-----|-----|-----|-----|-----|-----|-----|--------|
| 1980 | 0.98 | 1.36 | 4.54 | 5.78 | 1.83 | 3.34 | 1.99 | 2.14 | 3.00 | 2.99 | 4.75 | 1.18 | 33.88 |
| 1981 | 0.93 | 7.10 | 1.44 | 3.46 | 2.27 | 4.59 | 5.44 | 2.31 | 6.14 | 4.71 | 2.80 | 4.51 | 45.70 |
| 1982 | 5.17 | 2.53 | 3.20 | 4.54 | 2.91 | 6.75 | 2.61 | 3.35 | 1.90 | 1.93 | 3.61 | 1.18 | 39.68 |
| 1983 | 4.59 | 3.94 | 9.75 | 6.82 | 5.98 | 1.35 | 4.31 | 2.58 | 1.35 | 3.38 | 13.50 | 8.78 | 66.33 |
| 1984 | 2.56 | 4.99 | 5.12 | 4.81 | 9.64 | 3.87 | 3.86 | 2.09 | 0.84 | 3.26 | 3.69 | 3.44 | 48.17 |
| 1985 | 1.03 | 1.54 | 3.15 | 1.25 | 2.03 | 2.74 | 3.30 | 3.18 | 2.97 | 4.07 | 6.36 | 2.39 | 34.01 |
| 1986 | 6.58 | 2.61 | 4.21 | 3.49 | 2.51 | 3.91 | 3.44 | 1.87 | 2.64 | 2.09 | 5.18 | 5.91 | 44.44 |
| 1987 | 5.21 | 0.04 | 4.29 | 6.33 | 2.62 | 5.01 | 1.79 | 2.48 | 4.64 | 2.54 | 3.82 | 2.01 | 40.78 |
| 1988 | 1.97 | 3.34 | 1.85 | 3.68 | 4.27 | 2.36 | 5.89 | 5.24 | 1.50 | 3.47 | 8.84 | 1.21 | 43.62 |
| 1989 | 1.15 | 2.37 | 2.14 | 2.94 | 8.74 | 4.49 | 2.50 | 1.73 | 4.48 | 4.81 | 3.97 | 2.23 | 41.55 |
| 1990 | 3.19 | 2.49 | 1.42 | 5.16 | 5.23 | 4.12 | 3.21 | 1.89 | 3.12 | 7.46 | 7.50 | 7.90 | 52.69 |
| 1991 | 2.91 | 1.85 | 6.19 | 6.71 | 3.77 | 1.47 | 2.35 | 15.22 | 5.44 | 2.83 | 4.34 | 4.06 | 57.14 |
| 1992 | 4.80 | 3.42 | 3.93 | 2.47 | 1.15 | 3.99 | 4.06 | 2.59 | 3.11 | 2.48 | 4.39 | 2.12 | 38.51 |
| 1993 | 2.80 | 3.84 | 6.26 | 5.69 | 1.14 | 2.89 | 2.73 | 1.21 | 4.10 | 3.74 | 4.05 | 5.40 | 43.85 |
| 1994 | 5.34 | 1.29 | 6.71 | 2.94 | 4.87 | 1.25 | 1.73 | 2.82 | 6.60 | 0.63 | 3.67 | 6.20 | 44.05 |
| 1995 | 4.92 | 3.55 | 2.14 | 2.36 | 3.31 | 2.60 | 3.14 | 0.47 | 2.40 | 4.78 | 7.27 | 4.35 | 41.29 |
| 1996 | 5.30 | 3.03 | 2.68 | 6.44 | 3.72 | 1.80 | 6.19 | 0.50 | 3.55 | 16.86 | 1.79 | 6.53 | 58.39 |
| 1997 | 4.59 | 2.59 | 4.05 | 5.21 | 2.41 | 0.76 | 2.01 | 4.15 | 2.68 | 1.37 | 5.22 | 2.57 | 37.61 |
| 1998 | 4.83 | 5.72 | 4.23 | 3.36 | 3.83 | 9.01 | 2.92 | 4.03 | 3.02 | 10.45 | 1.80 | 1.56 | 54.76 |
| 1999 | 5.97 | 3.87 | 4.52 | 0.28 | 4.98 | 0.95 | 1.62 | 1.53 | 8.80 | 3.89 | 2.25 | 2.00 | 40.66 |
| 2000 | 3.41 | 2.90 | 3.66 | 5.44 | 3.07 | 2.06 | 4.03 | 1.81 | 2.46 | 3.17 | 4.19 | 4.49 | 40.69 |
| 2001 | 1.33 | 2.58 | 8.01 | 1.26 | 1.14 | 5.37 | 2.05 | 1.28 | 4.14 | 1.72 | 2.20 | 2.03 | 33.11 |
| 2002 | 2.98 | 2.48 | 4.18 | 3.85 | 4.42 | 4.38 | 3.33 | 1.28 | 3.58 | 4.17 | 5.08 | 4.51 | 44.24 |
| 2003 | 1.14 | 3.51 | 3.65 | 3.20 | 3.12 | 2.15 | 1.50 | 2.10 | 4.68 | 5.60 | 3.06 | 4.54 | 38.25 |
| 2004 | 0.28 | 1.94 | 1.94 | 6.46 | 4.69 | 2.19 | 3.83 | 6.14 | 3.24 | 2.19 | 4.04 | 4.28 | 41.22 |
| 2005 | 3.33 | 3.34 | 4.81 | 8.72 | 6.40 | 5.12 | 2.83 | 2.37 | 2.13 | 14.38 | 7.67 | 5.30 | 66.40 |
| 2006 | 3.71 | 2.74 | 1.02 | 3.02 | 12.34 | 9.18 | 5.65 | 2.37 | 3.57 | 8.38 | 5.50 | 3.36 | 60.84 |
| 2007 | 2.86 | 0.96 | 3.01 | 8.97 | 2.32 | 3.33 | 4.62 | 3.23 | 3.22 | 6.37 | 4.21 | 3.79 | 46.89 |
| 2008 | 3.21 | 8.06 | 5.54 | 4.47 | 1.09 | 3.88 | 4.71 | 6.28 | 10.84 | 3.19 | 5.35 | 4.62 | 61.24 |
| 2009 | 2.35 | 2.79 | 2.66 | 4.63 | 4.52 | 8.56 | 8.60 | 5.15 | 1.38 | 4.99 | 7.74 | 5.24 | 58.61 |
| POR= 90 YRS | 3.88 | 3.60 | 4.03 | 4.05 | 3.55 | 3.39 | 3.08 | 2.87 | 3.41 | 3.82 | 4.49 | 4.14 | 44.31 |

WBAN : 14764

## AVERAGE TEMPERATURE (°F) 2009 PORTLAND (KPWM)

| YEAR | JAN | FEB | MAR | APR | MAY | JUN | JUL | AUG | SEP | OCT | NOV | DEC | ANNUAL |
|------|-----|-----|-----|-----|-----|-----|-----|-----|-----|-----|-----|-----|--------|
| 1980 | 22.7 | 20.5 | 32.0 | 44.3 | 53.7 | 61.0 | 69.6 | 71.2 | 61.5 | 45.9 | 36.2 | 21.3 | 45.0 |
| 1981 | 13.8 | 32.3 | 35.7 | 45.2 | 55.3 | 63.9 | 68.8 | 66.0 | 58.6 | 46.6 | 38.9 | 29.2 | 46.2 |
| 1982 | 15.0 | 23.3 | 32.1 | 42.0 | 54.6 | 58.3 | 69.2 | 64.4 | 59.2 | 48.1 | 41.3 | 32.0 | 45.0 |
| 1983 | 25.2 | 26.2 | 35.8 | 44.4 | 52.0 | 63.8 | 69.7 | 67.6 | 63.0 | 47.9 | 40.2 | 26.1 | 46.8 |
| 1984 | 19.8 | 32.0 | 28.2 | 43.3 | 53.0 | 63.9 | 69.6 | 69.1 | 57.8 | 49.6 | 39.2 | 31.2 | 46.4 |
| 1985 | 16.2 | 26.4 | 34.8 | 44.1 | 53.7 | 61.7 | 69.8 | 66.6 | 60.7 | 50.8 | 39.9 | 24.6 | 45.8 |
| 1986 | 25.0 | 23.4 | 34.6 | 46.5 | 53.6 | 60.9 | 66.3 | 66.2 | 57.4 | 47.9 | 36.3 | 29.6 | 45.6 |
| 1987 | 21.6 | 23.0 | 34.0 | 44.9 | 54.6 | 63.8 | 68.0 | 66.2 | 59.6 | 47.1 | 38.1 | 30.2 | 45.9 |
| 1988 | 21.6 | 25.7 | 34.2 | 43.5 | 54.7 | 63.8 | 71.1 | 71.1 | 59.4 | 46.5 | 41.0 | 26.0 | 46.6 |
| 1989 | 26.8 | 24.0 | 31.6 | 41.1 | 55.6 | 63.9 | 69.4 | 68.0 | 60.5 | 49.8 | 37.3 | 14.1 | 45.2 |
| 1990 | 30.2 | 25.7 | 34.7 | 44.6 | 51.8 | 62.4 | 70.3 | 69.8 | 59.8 | 52.4 | 41.8 | 33.7 | 48.1 |
| 1991 | 23.4 | 29.8 | 37.0 | 45.9 | 58.1 | 65.5 | 70.1 | 69.9 | 58.2 | 50.8 | 40.4 | 26.4 | 48.0 |
| 1992 | 23.9 | 25.9 | 30.9 | 41.4 | 51.9 | 63.4 | 65.8 | 66.8 | 58.9 | 47.3 | 37.6 | 28.9 | 45.2 |
| 1993 | 23.4 | 16.8 | 29.8 | 43.3 | 54.8 | 64.5 | 69.8 | 69.5 | 59.6 | 46.2 | 39.0 | 29.9 | 45.6 |
| 1994 | 14.0 | 19.6 | 33.5 | 44.5 | 52.3 | 66.0 | 72.4 | 66.4 | 58.4 | 49.5 | 41.5 | 31.5 | 45.8 |
| 1995 | 27.1 | 21.2 | 34.8 | 41.2 | 52.1 | 64.1 | 70.0 | 67.4 | 56.4 | 52.4 | 35.7 | 24.5 | 45.6 |
| 1996 | 21.4 | 23.9 | 29.8 | 42.9 | 52.5 | 63.0 | 66.4 | 67.4 | 59.8 | 46.8 | 35.1 | 34.5 | 45.3 |
| 1997 | 23.4 | 28.7 | 30.4 | 41.8 | 51.1 | 63.3 | 69.0 | 67.3 | 58.7 | 47.0 | 36.2 | 29.5 | 45.5 |
| 1998 | 27.1 | 31.3 | 36.7 | 45.8 | 56.8 | 60.7 | 69.5 | 69.5 | 61.2 | 49.2 | 39.4 | 32.5 | 48.3 |
| 1999 | 22.3 | 29.0 | 35.6 | 44.9 | 54.5 | 66.3 | 71.3 | 67.7 | 63.3 | 46.6 | 42.0 | 31.9 | 48.0 |
| 2000 | 21.5 | 26.4 | 37.8 | 43.3 | 53.2 | 63.7 | 67.1 | 67.3 | 59.9 | 48.9 | 40.4 | 24.6 | 46.2 |
| 2001 | 20.6 | 24.3 | 30.6 | 43.3 | 54.9 | 67.2 | 66.5 | 69.4 | 61.0 | 50.2 | 41.4 | 34.8 | 47.0 |
| 2002 | 30.4 | 28.8 | 34.9 | 44.4 | 52.5 | 61.4 | 69.5 | 70.1 | 63.1 | 47.0 | 37.0 | 27.4 | 47.2 |
| 2003 | 16.3 | 20.3 | 31.6 | 40.9 | 52.2 | 62.9 | 69.3 | 68.8 | 61.4 | 48.0 | 40.8 | 29.6 | 45.2 |
| 2004 | 15.0 | 25.9 | 34.9 | 44.5 | 54.2 | 60.5 | 66.9 | 67.8 | 60.9 | 49.7 | 38.9 | 28.0 | 45.6 |
| 2005 | 19.4 | 25.5 | 30.1 | 45.0 | 49.3 | 64.7 | 69.4 | 70.2 | 63.6 | 50.1 | 40.7 | 27.8 | 46.3 |
| 2006 | 30.2 | 25.8 | 34.5 | 45.7 | 54.6 | 65.0 | 72.0 | 67.0 | 60.3 | 49.0 | 43.8 | 34.5 | 48.5 |
| 2007 | 25.9 | 19.2 | 32.2 | 41.8 | 56.0 | 63.6 | 68.0 | 67.9 | 61.6 | 53.7 | 37.6 | 25.4 | 46.1 |
| 2008 | 26.6 | 25.9 | 31.9 | 44.8 | 54.2 | 64.5 | 70.7 | 66.2 | 60.4 | 48.8 | 39.3 | 28.6 | 46.8 |
| 2009 | 17.5 | 25.9 | 33.0 | 46.7 | 55.1 | 61.0 | 66.2 | 69.7 | 59.1 | 46.9 | 43.0 | 29.3 | 46.1 |
| POR= 90 YRS | 22.2 | 23.3 | 32.7 | 42.7 | 53.3 | 62.0 | 68.4 | 67.2 | 58.9 | 49.1 | 38.6 | 27.3 | 45.5 |

## HEATING DEGREE DAYS (base 65°F) 2009  PORTLAND (KPWM)

| YEAR | JUL | AUG | SEP | OCT | NOV | DEC | JAN | FEB | MAR | APR | MAY | JUN | TOTAL |
|------|-----|-----|-----|-----|-----|-----|-----|-----|-----|-----|-----|-----|-------|
| 1980-81 | 16 | 6 | 163 | 584 | 855 | 1349 | 1578 | 910 | 901 | 588 | 312 | 54 | 7316 |
| 1981-82 | 16 | 45 | 189 | 566 | 778 | 1102 | 1543 | 1161 | 1014 | 684 | 320 | 198 | 7616 |
| 1982-83 | 20 | 78 | 185 | 519 | 704 | 1015 | 1225 | 1080 | 895 | 612 | 393 | 101 | 6827 |
| 1983-84 | 8 | 38 | 139 | 527 | 738 | 1198 | 1397 | 949 | 1132 | 642 | 368 | 110 | 7246 |
| 1984-85 | 11 | 13 | 223 | 469 | 767 | 1043 | 1506 | 1076 | 930 | 620 | 347 | 115 | 7120 |
| 1985-86 | 4 | 32 | 157 | 433 | 747 | 1245 | 1236 | 1161 | 935 | 548 | 354 | 138 | 6990 |
| 1986-87 | 47 | 52 | 242 | 523 | 855 | 1092 | 1336 | 1172 | 955 | 597 | 343 | 77 | 7291 |
| 1987-88 | 20 | 58 | 171 | 548 | 798 | 1070 | 1339 | 1130 | 950 | 641 | 323 | 112 | 7160 |
| 1988-89 | 13 | 32 | 180 | 569 | 713 | 1201 | 1174 | 1141 | 1028 | 708 | 286 | 91 | 7136 |
| 1989-90 | 6 | 25 | 167 | 464 | 824 | 1573 | 1071 | 1093 | 935 | 607 | 402 | 107 | 7274 |
| 1990-91 | 12 | 24 | 170 | 388 | 690 | 964 | 1283 | 979 | 861 | 568 | 236 | 76 | 6251 |
| 1991-92 | 11 | 16 | 228 | 433 | 730 | 1191 | 1267 | 1127 | 1051 | 700 | 414 | 84 | 7252 |
| 1992-93 | 47 | 30 | 208 | 543 | 814 | 1112 | 1280 | 1344 | 1084 | 645 | 311 | 81 | 7499 |
| 1993-94 | 15 | 5 | 192 | 575 | 774 | 1080 | 1577 | 1266 | 968 | 609 | 389 | 61 | 7511 |
| 1994-95 | 1 | 35 | 201 | 472 | 697 | 1028 | 1167 | 1216 | 929 | 706 | 391 | 85 | 6928 |
| 1995-96 | 9 | 35 | 265 | 380 | 873 | 1248 | 1343 | 1181 | 1083 | 659 | 384 | 90 | 7550 |
| 1996-97 | 23 | 17 | 179 | 556 | 889 | 940 | 1284 | 1011 | 1065 | 689 | 424 | 135 | 7212 |
| 1997-98 | 11 | 21 | 192 | 554 | 857 | 1093 | 1164 | 938 | 872 | 569 | 264 | 143 | 6678 |
| 1998-99 | 3 | 9 | 124 | 482 | 761 | 1002 | 1314 | 1002 | 903 | 599 | 322 | 61 | 6582 |
| 1999-00 | 6 | 23 | 105 | 563 | 686 | 1020 | 1341 | 1113 | 838 | 645 | 361 | 112 | 6813 |
| 2000-01 | 11 | 18 | 184 | 491 | 729 | 1247 | 1371 | 1134 | 1060 | 643 | 330 | 57 | 7275 |
| 2001-02 | 32 | 15 | 140 | 456 | 702 | 928 | 1066 | 1006 | 927 | 609 | 380 | 149 | 6410 |
| 2002-03 | 23 | 21 | 103 | 563 | 830 | 1158 | 1501 | 1246 | 1028 | 717 | 391 | 114 | 7695 |
| 2003-04 | 9 | 22 | 110 | 521 | 719 | 1088 | 1541 | 1128 | 930 | 609 | 329 | 159 | 7165 |
| 2004-05 | 15 | 20 | 131 | 470 | 776 | 1140 | 1410 | 1099 | 1074 | 596 | 477 | 100 | 7308 |
| 2005-06 | 17 | 8 | 85 | 454 | 723 | 1146 | 1069 | 1089 | 937 | 575 | 324 | 64 | 6491 |
| 2006-07 | 0 | 43 | 149 | 491 | 630 | 940 | 1206 | 1276 | 1012 | 689 | 295 | 101 | 6832 |
| 2007-08 | 17 | 38 | 147 | 348 | 814 | 1219 | 1183 | 1126 | 1015 | 598 | 331 | 60 | 6896 |
| 2008-09 | 0 | 23 | 162 | 495 | 764 | 1121 | 1464 | 1087 | 986 | 544 | 315 | 120 | 7081 |
| 2009- | 41 | 27 | 183 | 557 | 651 | 1101 | | | | | | | |

WBAN : 14764

## COOLING DEGREE DAYS (base 65°F) 2009  PORTLAND (KPWM)

| YEAR | JAN | FEB | MAR | APR | MAY | JUN | JUL | AUG | SEP | OCT | NOV | DEC | TOTAL |
|------|-----|-----|-----|-----|-----|-----|-----|-----|-----|-----|-----|-----|-------|
| 1980 | 0 | 0 | 0 | 0 | 1 | 50 | 163 | 205 | 67 | 0 | 0 | 0 | 486 |
| 1981 | 0 | 0 | 0 | 0 | 20 | 30 | 138 | 84 | 5 | 0 | 0 | 0 | 277 |
| 1982 | 0 | 0 | 0 | 0 | 6 | 4 | 158 | 66 | 17 | 0 | 0 | 0 | 251 |
| 1983 | 0 | 0 | 0 | 0 | 0 | 73 | 161 | 125 | 84 | 3 | 0 | 0 | 446 |
| 1984 | 0 | 0 | 0 | 0 | 0 | 84 | 162 | 147 | 12 | 0 | 0 | 0 | 405 |
| 1985 | 0 | 0 | 0 | 0 | 5 | 25 | 161 | 92 | 35 | 1 | 0 | 0 | 319 |
| 1986 | 0 | 0 | 0 | 0 | 8 | 23 | 93 | 96 | 19 | 0 | 0 | 0 | 239 |
| 1987 | 0 | 0 | 0 | 0 | 28 | 47 | 121 | 103 | 17 | 0 | 0 | 0 | 316 |
| 1988 | 0 | 0 | 0 | 0 | 11 | 85 | 209 | 227 | 17 | 2 | 0 | 0 | 551 |
| 1989 | 0 | 0 | 0 | 0 | 2 | 67 | 151 | 126 | 38 | 0 | 0 | 0 | 384 |
| 1990 | 0 | 0 | 0 | 0 | 0 | 34 | 181 | 179 | 19 | 7 | 0 | 0 | 420 |
| 1991 | 0 | 0 | 0 | 0 | 29 | 96 | 178 | 176 | 31 | 0 | 0 | 0 | 510 |
| 1992 | 0 | 0 | 0 | 0 | 13 | 43 | 76 | 93 | 30 | 0 | 0 | 0 | 255 |
| 1993 | 0 | 0 | 0 | 0 | 1 | 72 | 170 | 150 | 37 | 0 | 0 | 0 | 430 |
| 1994 | 0 | 0 | 0 | 0 | 1 | 101 | 237 | 89 | 10 | 0 | 0 | 0 | 438 |
| 1995 | 0 | 0 | 0 | 0 | 3 | 67 | 174 | 118 | 14 | 0 | 0 | 0 | 376 |
| 1996 | 0 | 0 | 0 | 0 | 5 | 38 | 75 | 99 | 29 | 0 | 0 | 0 | 246 |
| 1997 | 0 | 0 | 0 | 0 | 0 | 90 | 145 | 100 | 8 | 2 | 0 | 0 | 345 |
| 1998 | 0 | 0 | 2 | 0 | 15 | 21 | 152 | 154 | 17 | 0 | 0 | 0 | 361 |
| 1999 | 0 | 0 | 0 | 0 | 4 | 109 | 208 | 116 | 59 | 0 | 0 | 0 | 496 |
| 2000 | 0 | 0 | 0 | 0 | 1 | 76 | 81 | 97 | 37 | 1 | 0 | 0 | 293 |
| 2001 | 0 | 0 | 0 | 0 | 23 | 127 | 85 | 161 | 25 | 3 | 0 | 0 | 424 |
| 2002 | 0 | 0 | 0 | 0 | 0 | 49 | 169 | 183 | 55 | 14 | 0 | 0 | 470 |
| 2003 | 0 | 0 | 0 | 0 | 0 | 56 | 149 | 148 | 9 | 0 | 0 | 0 | 362 |
| 2004 | 0 | 0 | 0 | 0 | 1 | 31 | 80 | 114 | 16 | 0 | 0 | 0 | 242 |
| 2005 | 0 | 0 | 0 | 0 | 0 | 97 | 159 | 174 | 52 | 2 | 0 | 0 | 484 |
| 2006 | 0 | 0 | 0 | 0 | 9 | 73 | 224 | 110 | 16 | 0 | 0 | 0 | 432 |
| 2007 | 0 | 0 | 0 | 0 | 22 | 66 | 119 | 133 | 53 | 4 | 0 | 0 | 397 |
| 2008 | 0 | 0 | 0 | 0 | 4 | 48 | 183 | 67 | 29 | 0 | 0 | 0 | 331 |
| 2009 | 0 | 0 | 0 | 4 | 15 | 7 | 86 | 178 | 14 | 0 | 0 | 0 | 304 |

## SNOWFALL (inches) 2009 PORTLAND (KPWM)

| YEAR | JUL | AUG | SEP | OCT | NOV | DEC | JAN | FEB | MAR | APR | MAY | JUN | TOTAL |
|------|-----|-----|-----|-----|-----|-----|-----|-----|-----|-----|-----|-----|-------|
| 1980-81 | 0.0 | 0.0 | 0.0 | 0.0 | 8.9 | 13.0 | 9.2 | 4.6 | 3.1 | T | 0.0 | 0.0 | 38.8 |
| 1981-82 | 0.0 | 0.0 | 0.0 | 0.0 | T | 24.0 | 25.9 | 11.0 | 8.5 | 15.9 | 0.0 | 0.0 | 85.3 |
| 1982-83 | 0.0 | 0.0 | 0.0 | 0.0 | 0.6 | 5.7 | 12.4 | 24.5 | 2.1 | T | 0.0 | 0.0 | 45.3 |
| 1983-84 | 0.0 | 0.0 | 0.0 | 0.0 | T | 12.6 | 28.3 | 3.3 | 26.4 | T | 0.0 | 0.0 | 70.6 |
| 1984-85 | 0.0 | 0.0 | 0.0 | 0.0 | T | 17.0 | 12.1 | 7.2 | 13.1 | 2.4 | 0.0 | 0.0 | 51.8 |
| 1985-86 | 0.0 | 0.0 | 0.0 | 0.0 | 3.1 | 11.2 | 18.6 | 12.0 | 6.4 | T | 0.0 | 0.0 | 51.3 |
| 1986-87 | 0.0 | 0.0 | 0.0 | 0.0 | 5.2 | 4.0 | 50.7 | 0.8 | 14.3 | 3.4 | 0.0 | 0.0 | 78.4 |
| 1987-88 | 0.0 | 0.0 | T | 0.0 | 5.4 | 9.1 | 19.8 | 20.8 | 3.0 | 4.2 | 0.0 | 0.0 | 62.3 |
| 1988-89 | 0.0 | 0.0 | 0.0 | T | T | 3.5 | 4.0 | 13.8 | 8.9 | 0.7 | 0.0 | 0.0 | 30.9 |
| 1989-90 | 0.0 | 0.0 | 0.0 | 0.0 | 5.0 | 15.6 | 20.4 | 25.6 | 3.2 | T | 0.0 | 0.0 | 69.8 |
| 1990-91 | 0.0 | 0.0 | T | 0.0 | 0.2 | 6.8 | 13.4 | 6.3 | 5.7 | T | 0.0 | 0.0 | 32.4 |
| 1991-92 | 0.0 | 0.0 | 0.0 | 0.0 | T | 22.5 | 2.4 | 10.3 | 13.4 | 10.0 | 0.0 | 0.0 | 58.6 |
| 1992-93 | 0.0 | 0.0 | T | T | 2.8 | 2.1 | 17.1 | 33.5 | 49.0 | 11.1 | 0.0 | 0.0 | 115.6 |
| 1993-94 | 0.0 | 0.0 | 0.0 | 0.2 | T | 12.3 | 39.3 | 12.2 | 12.2 | T | 0.0 | 0.0 | 76.2 |
| 1994-95 | 0.0 | 0.0 | 0.0 | 0.0 | 2.5 | 2.0 | 15.7 | 17.3 | 1.1 | T | 0.0 | 0.0 | 38.6 |
| 1995-96 | 0.0 | 0.0 | 0.0 | 0.0 | 2.0 | 37.3 | 37.1 | 13.1 | 25.0 | 8.5 | 0.0 | 0.0 | 123.0 |
| 1996-97 | 0.0 | 0.0 | 0.0 | 0.0 | 2.9 | 1.4 | 15.9 | 6.2 | 15.5 | 1.4 | 0.0 | 0.0 | 43.3 |
| 1997-98 | 0.0 | 0.0 | 0.0 | 0.0 | 20.5 | 6.3 | 17.1 | 0.9 | 9.7 | 0.0 | 0.0 | 0.0 | 54.5 |
| 1998-99 | 0.0 | 0.0 | 0.0 | 0.0 | T | 11.7 | 19.2 | 5.1 | 17.5 | 0.0 | 0.0 | 0.0 | 53.5 |
| 1999-00 | 0.0 | 0.0 | 0.0 | 0.0 | T | T | 14.9 | 14.6 | 11.6 | T | 0.0 | 0.0 | 41.1 |
| 2000-01 | 0.0 | 0.0 | 0.0 | T | T | 18.8 | 15.5 | 24.2 | 40.5 | 0.3 | 0.0 | 0.0 | 99.3 |
| 2001-02 | 0.0 | 0.0 | 0.0 | T | T | 5.0 | 15.8 | 1.1 | 10.4 | 0.3 | 0.0 | 0.0 | 32.6 |
| 2002-03 | T | 0.0 | 0.0 | 0.0 | 4.1 | 17.4 | 18.7 | 16.8 | 2.4 | 4.5 | 0.0 | 0.0 | 63.9 |
| 2003-04 | 0.0 | 0.0 | 0.0 | T | T | 24.0 | 2.8 | 12.3 | 6.1 | 1.5 | 0.0 | 0.0 | 46.7 |
| 2004-05 | 0.0 | 0.0 | 0.0 | 0.0 | T | 8.8 | 37.3 | 23.5 | 32.5 | T | 0.0 | 0.0 | 102.1 |
| 2005-06 | 0.0 | 0.0 | 0.0 | 0.0 | 2.3 | 20.4 | 13.5 | 9.5 | 1.0 | 0.5 | 0.0 | 0.0 | 47.2 |
| 2006-07 | 0.0 | 0.0 | 0.0 | 0.0 | 0.0 | 4.6 | 6.8 | 14.3 | 14.1 | 15.1 | 0.0 | 0.0 | 54.9 |
| 2007-08 | 0.0 | 0.0 | 0.0 | 0.0 | T | 38.8 | 19.1 | 32.8 | 12.3 | T | 0.0 | 0.0 | 103.0 |
| 2008-09 | 0.0 | 0.0 | 0.0 | T | T | 27.2 | 28.0 | 15.3 | 10.6 | 0.0 | 0.0 | 0.0 | 81.1 |
| 2009- | 0.0 | 0.0 | 0.0 | T | T | 13.8 | | | | | | | |
| POR=<br>72 YRS | T | 0.0 | T | 0.2 | 3.2 | 14.3 | 19.8 | 17.3 | 12.5 | 3.2 | 0.1 | 0.0 | 70.6 |

# Portland, ME 2009
## Temperature and Precipitation

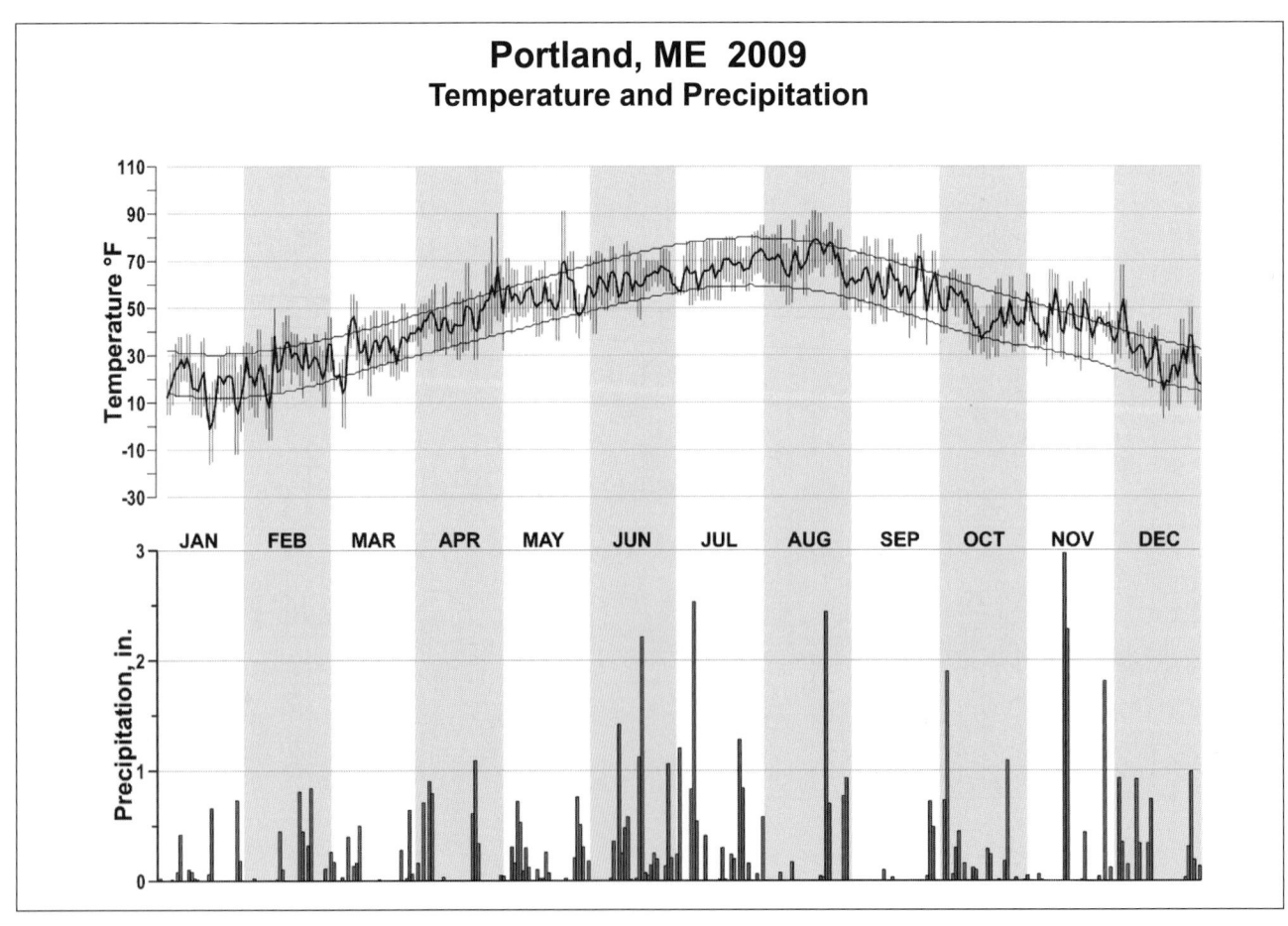

# 2009
# BALTIMORE
# MARYLAND (KBWI)

Baltimore-Washington International Airport lies in a region about midway between the rigorous climates of the North and the mild climates of the South, and adjacent to the modifying influences of the Chesapeake Bay and Atlantic Ocean to the east and the Appalachian Mountains to the west.  Since this region is near the average path of the low pressure systems which move across the country, changes in wind direction are frequent and contribute to the changeable character of the weather.  The net effect of the mountains to the west and the bay and ocean to the east is to produce a more equable climate compared with other continental locations farther inland at the same latitude.

Rainfall distribution throughout the year is rather uniform, however, the greatest intensities are confined to the summer and early fall months, the season for hurricanes and severe thunderstorms. Moisture deficiencies for crops occur occasionally during the growing season, but severe droughts are rare.  Rainfall during the growing season occurs principally in the form of thunderstorms, and rainfall totals during these months vary appreciably.

The average date for the last occurrence in spring of temperatures as low as 32 degrees is mid-April. The average date for the first occurrence in fall of temperatures as low as 32 degrees is late October. The freeze-free period is approximately 194 days.

In summer, the area is under the influence of the large semi-permanent high pressure system commonly known as the Bermuda High and centered over the Atlantic Ocean near 30 degrees N Latitude. This pressure system brings warm humid air to the area. The proximity of large water areas and the inflow of southerly winds contribute to high relative humidities during much of the year.

January is the coldest month, and July, the warmest.  Snowfall occurs on about eleven days per year on the average, however, an average of only about six days annually produces snowfalls of 1 inch or greater.  Snow is frequently mixed with rain and sleet, and snow seldom remains on the ground more than a few days.

Glaze or freezing rain which is hazardous to highway traffic occurs on an average of two to three times per year, generally in January or February.  Some years pass without the occurrence of freezing rain, while in others it occurs on as many as eight to ten days.  Sleet is observed on about five days annually with the greatest frequency of occurrence in January.

The annual prevailing wind direction is from the west.  Winter and spring months have the highest average wind speed.  Destructive velocities are rare and occur mostly during summer thunderstorms. Only rarely have hurricanes in the vicinity caused widespread damage, then primarily through flooding.

# NORMALS, MEANS, AND EXTREMES
## BALTIMORE (KBWI)

| LATITUDE: 39° 10'N | LONGITUDE: -76° 41'W | ELEVATION (FT): GRND: 143  BARO: 196 | TIME ZONE: EASTERN  (UTC -5) | WBAN: 93721 |
|---|---|---|---|---|

| | ELEMENT | POR | JAN | FEB | MAR | APR | MAY | JUN | JUL | AUG | SEP | OCT | NOV | DEC | YEAR |
|---|---|---|---|---|---|---|---|---|---|---|---|---|---|---|---|
| **TEMPERATURE °F** | NORMAL DAILY MAXIMUM | 30 | 41.2 | 44.8 | 53.9 | 64.5 | 73.9 | 82.7 | 87.2 | 85.1 | 78.2 | 67.0 | 56.3 | 46.0 | 65.1 |
| | MEAN DAILY MAXIMUM | 59 | 41.5 | 44.7 | 53.5 | 65.0 | 74.3 | 83.0 | 87.3 | 85.5 | 78.6 | 67.5 | 56.4 | 45.4 | 65.2 |
| | HIGHEST DAILY MAXIMUM | 59 | 75 | 79 | 89 | 94 | 98 | 101 | 104 | 105 | 100 | 94 | 83 | 77 | 105 |
| | YEAR OF OCCURRENCE | | 1975 | 2000 | 1998 | 1960 | 1991 | 1994 | 1988 | 1983 | 1983 | 2007 | 1974 | 1998 | AUG 1983 |
| | MEAN OF EXTREME MAXS. | 59 | 62.7 | 65.8 | 76.2 | 85.2 | 89.9 | 94.9 | 96.9 | 95.5 | 91.7 | 83.2 | 74.7 | 65.4 | 81.8 |
| | NORMAL DAILY MINIMUM | 30 | 23.5 | 26.1 | 33.6 | 42.0 | 51.8 | 60.8 | 65.8 | 63.9 | 56.6 | 43.7 | 34.7 | 27.3 | 44.2 |
| | MEAN DAILY MINIMUM | 59 | 24.7 | 26.5 | 33.7 | 43.1 | 52.5 | 61.7 | 66.9 | 65.7 | 58.4 | 46.1 | 36.8 | 28.3 | 45.4 |
| | LOWEST DAILY MINIMUM | 59 | -7 | -3 | 6 | 20 | 32 | 40 | 50 | 45 | 35 | 25 | 13 | 0 | -7 |
| | YEAR OF OCCURRENCE | | 1984 | 1979 | 1960 | 1965 | 1966 | 1972 | 2001 | 1986 | 1963 | 1969 | 1955 | 1983 | JAN 1984 |
| | MEAN OF EXTREME MINS. | 59 | 7.3 | 10.8 | 18.8 | 28.9 | 38.4 | 49.0 | 56.2 | 53.9 | 43.1 | 31.8 | 21.9 | 12.7 | 31.1 |
| | NORMAL DRY BULB | 30 | 32.3 | 35.5 | 43.7 | 53.2 | 62.9 | 71.8 | 76.5 | 74.5 | 67.4 | 55.4 | 45.5 | 36.7 | 54.6 |
| | MEAN DRY BULB | 59 | 33.1 | 35.6 | 43.6 | 54.1 | 63.4 | 72.5 | 77.1 | 75.6 | 68.5 | 56.8 | 46.6 | 36.9 | 55.3 |
| | MEAN WET BULB | 26 | 29.2 | 30.8 | 37.1 | 46.5 | 56.0 | 65.0 | 68.9 | 68.0 | 61.8 | 51.1 | 41.7 | 32.8 | 49.1 |
| | MEAN DEW POINT | 26 | 24.4 | 25.5 | 31.7 | 41.4 | 52.3 | 62.0 | 66.3 | 65.5 | 59.0 | 47.9 | 37.4 | 28.0 | 45.1 |
| | NORMAL NO. DAYS WITH: | | | | | | | | | | | | | | |
| | MAXIMUM >= 90 | 30 | 0.0 | 0.0 | 0.0 | 0.3 | 1.5 | 5.7 | 12.0 | 7.3 | 2.6 | 0.0 | 0.0 | 0.0 | 29.4 |
| | MAXIMUM <= 32 | 30 | 6.4 | 3.9 | 0.5 | 0.0 | 0.0 | 0.0 | 0.0 | 0.0 | 0.0 | 0.0 | 0.1 | 2.9 | 13.8 |
| | MINIMUM <= 32 | 30 | 24.4 | 20.5 | 13.3 | 3.0 | 0.0 | 0.0 | 0.0 | 0.0 | 0.0 | 1.4 | 10.4 | 20.6 | 93.6 |
| | MINIMUM <= 0 | 30 | 0.4 | 0.1 | 0.0 | 0.0 | 0.0 | 0.0 | 0.0 | 0.0 | 0.0 | 0.0 | 0.0 | * | 0.5 |
| **H/C** | NORMAL HEATING DEG. DAYS | 30 | 1000 | 816 | 648 | 349 | 120 | 16 | 0 | 1 | 42 | 296 | 570 | 862 | 4720 |
| | NORMAL COOLING DEG. DAYS | 30 | 0 | 0 | 4 | 11 | 71 | 236 | 372 | 311 | 129 | 13 | 0 | 0 | 1147 |
| **RH** | NORMAL (PERCENT) | 30 | 66 | 63 | 61 | 61 | 67 | 69 | 70 | 72 | 73 | 72 | 68 | 67 | 67 |
| | HOUR 01 LST | 30 | 70 | 68 | 67 | 69 | 78 | 82 | 82 | 85 | 84 | 82 | 75 | 72 | 76 |
| | HOUR 07 LST | 30 | 73 | 72 | 72 | 72 | 77 | 79 | 80 | 84 | 86 | 85 | 79 | 75 | 78 |
| | HOUR 13 LST | 30 | 57 | 53 | 50 | 49 | 53 | 54 | 54 | 56 | 56 | 55 | 55 | 57 | 54 |
| | HOUR 19 LST | 30 | 64 | 59 | 55 | 54 | 61 | 63 | 64 | 68 | 70 | 71 | 66 | 65 | 63 |
| **S** | PERCENT POSSIBLE SUNSHINE | 40 | 51 | 55 | 56 | 56 | 56 | 62 | 64 | 62 | 60 | 58 | 51 | 49 | 57 |
| **W/O** | MEAN NO. DAYS WITH: | | | | | | | | | | | | | | |
| | HEAVY FOG(VISBY <= 1/4 MI) | 46 | 2.7 | 2.6 | 2.1 | 1.3 | 1.3 | 0.8 | 0.5 | 0.8 | 0.9 | 2.3 | 2.2 | 2.8 | 20.3 |
| | THUNDERSTORMS | 59 | 0.3 | 0.3 | 0.9 | 2.4 | 4.0 | 5.6 | 6.1 | 4.9 | 2.0 | 0.8 | 0.5 | 0.1 | 27.9 |
| **CLOUDNESS** | MEAN: | | | | | | | | | | | | | | |
| | SUNRISE-SUNSET (OKTAS) | 46 | 5.1 | 5.0 | 5.0 | 5.0 | 5.0 | 4.6 | 4.4 | 4.4 | 4.2 | 4.1 | 4.8 | 5.0 | 4.7 |
| | MIDNIGHT-MIDNIGHT (OKTAS) | 32 | 4.9 | 4.7 | 4.8 | 4.6 | 4.7 | 4.4 | 4.3 | 4.2 | 4.2 | 3.9 | 4.5 | 4.8 | 4.5 |
| | MEAN NO. DAYS WITH: | | | | | | | | | | | | | | |
| | CLEAR | 46 | 8.0 | 7.7 | 7.9 | 7.7 | 7.7 | 8.4 | 8.9 | 9.2 | 10.3 | 11.6 | 8.1 | 8.1 | 103.6 |
| | PARTLY CLOUDY | 46 | 7.5 | 6.8 | 8.7 | 9.0 | 10.3 | 11.4 | 11.5 | 10.5 | 8.3 | 7.9 | 8.4 | 6.9 | 106.8 |
| | CLOUDY | 46 | 15.5 | 13.8 | 14.4 | 13.4 | 13.0 | 10.2 | 9.9 | 10.6 | 10.7 | 10.9 | 13.2 | 15.3 | 150.9 |
| **PR** | MEAN STATION PRESSURE(IN) | 26 | 29.92 | 29.91 | 29.88 | 29.81 | 29.82 | 29.80 | 29.82 | 29.82 | 29.91 | 29.92 | 29.93 | 29.90 | 29.87 |
| | MEAN SEA-LEVEL PRES. (IN) | 26 | 30.09 | 30.09 | 30.05 | 29.98 | 29.99 | 29.97 | 29.98 | 30.02 | 30.06 | 30.09 | 30.10 | 30.11 | 30.04 |
| **WINDS** | MEAN SPEED (MPH) | 26 | 8.2 | 8.5 | 9.1 | 8.8 | 7.6 | 7.0 | 6.6 | 6.2 | 6.6 | 6.7 | 7.5 | 7.8 | 7.6 |
| | PREVAIL.DIR(TENS OF DEGS) | 13 | 30 | 30 | 30 | 30 | 29 | 27 | 26 | 28 | 28 | 28 | 29 | 29 | 30 |
| | MAXIMUM 2-MINUTE: | | | | | | | | | | | | | | |
| | SPEED (MPH) | 13 | 40 | 40 | 41 | 43 | 39 | 44 | 40 | 39 | 44 | 37 | 45 | 45 | 45 |
| | DIR. (TENS OF DEGS) | | 30 | 27 | 28 | 28 | 29 | 27 | 28 | 32 | 09 | 28 | 28 | 27 | 27 |
| | YEAR OF OCCURRENCE | | 2000 | 2009 | 1997 | 2009 | 1997 | 2008 | 2002 | 2003 | 2003 | 2003 | 2003 | 2004 | DEC 2004 |
| | MAXIMUM 3-SECOND | | | | | | | | | | | | | | |
| | SPEED (MPH) | 13 | 53 | 54 | 53 | 58 | 51 | 60 | 48 | 53 | 55 | 49 | 59 | 61 | 61 |
| | DIR. (TENS OF DEGS) | | 28 | 28 | 28 | 28 | 31 | 31 | 28 | 27 | 05 | 28 | 27 | 29 | 29 |
| | YEAR OF OCCURRENCE | | 1999 | 2009 | 1997 | 2009 | 2002 | 2007 | 2002 | 1997 | 2003 | 2009 | 2003 | 2007 | DEC 2007 |
| **PRECIPITATION** | NORMAL (IN) | 30 | 3.47 | 3.02 | 3.93 | 3.00 | 3.89 | 3.43 | 3.85 | 3.74 | 3.98 | 3.16 | 3.12 | 3.35 | 41.94 |
| | MAXIMUM MONTHLY (IN) | 59 | 7.84 | 7.16 | 8.64 | 8.15 | 8.71 | 9.95 | 8.77 | 18.35 | 11.50 | 9.23 | 7.68 | 8.06 | 18.35 |
| | YEAR OF OCCURRENCE | | 1979 | 1979 | 1994 | 1952 | 1989 | 1972 | 2005 | 1955 | 1999 | 2005 | 1952 | 2009 | AUG 1955 |
| | MINIMUM MONTHLY (IN) | 59 | 0.29 | 0.26 | 0.18 | 0.39 | 0.37 | 0.15 | 0.30 | 0.77 | 0.21 | T | 0.31 | 0.20 | T |
| | YEAR OF OCCURRENCE | | 1955 | 2009 | 2006 | 1985 | 1986 | 1954 | 1955 | 1951 | 1967 | 1963 | 1981 | 1955 | OCT 1963 |
| | MAXIMUM IN 24 HOURS (IN) | 59 | 3.11 | 3.26 | 3.18 | 2.80 | 3.65 | 5.23 | 5.86 | 8.35 | 6.04 | 5.98 | 3.43 | 3.39 | 8.35 |
| | YEAR OF OCCURRENCE | | 1976 | 1983 | 1958 | 1952 | 2008 | 1972 | 1952 | 1955 | 1985 | 2005 | 1952 | 1977 | AUG 1955 |
| | NORMAL NO. DAYS WITH: | | | | | | | | | | | | | | |
| | PRECIPITATION >= 0.01 | 30 | 10.8 | 9.3 | 10.4 | 10.2 | 11.5 | 10.0 | 10.0 | 9.1 | 8.4 | 8.2 | 8.9 | 9.7 | 116.5 |
| | PRECIPITATION >= 1.00 | 30 | 0.8 | 0.7 | 0.9 | 0.6 | 0.7 | 1.0 | 1.1 | 1.0 | 1.0 | 1.0 | 0.7 | 0.9 | 10.4 |
| **SNOWFALL** | NORMAL (IN) | 30 | 7.0 | 6.4 | 2.4 | 0.1 | 0.0 | 0.0 | 0.0 | 0.0 | 0.0 | 0.* | 0.6 | 1.7 | 18.2 |
| | MAXIMUM MONTHLY (IN) | 59 | 32.6 | 40.5 | 21.6 | 0.7 | T | T | T | 0.0 | 0.0 | 0.3 | 8.4 | 20.4 | 40.5 |
| | YEAR OF OCCURRENCE | | 1996 | 2003 | 1960 | 1985 | 1963 | 2002 | 1992 | | | 1979 | 1967 | 1966 | FEB 2003 |
| | MAXIMUM IN 24 HOURS (IN) | 59 | 16.8 | 22.8 | 13.0 | 0.7 | T | T | T | 0.0 | 0.0 | 0.3 | 8.4 | 17.0 | 22.8 |
| | YEAR OF OCCURRENCE' | | 1996 | 1983 | 1962 | 1985 | 1963 | 2002 | 1992 | | | 1979 | 1967 | 2009 | FEB 1983 |
| | MAXIMUM SNOW DEPTH (IN) | 58 | 30 | 23 | 70 | 0 | 0 | 0 | 0 | 0 | 0 | 0 | 6 | 18 | 70 |
| | YEAR OF OCCURRENCE | | 1957 | 1983 | 1960 | | | | | | | | 1987 | 2009 | MAR 1960 |
| | NORMAL NO. DAYS WITH: | | | | | | | | | | | | | | |
| | SNOWFALL >= 1.0 | 30 | 1.8 | 1.3 | 0.7 | 0.0 | 0.0 | 0.0 | 0.0 | 0.0 | 0.0 | 0.0 | 0.1 | 0.5 | 4.4 |

## PRECIPITATION (inches) 2009 BALTIMORE (KBWI)

| YEAR | JAN | FEB | MAR | APR | MAY | JUN | JUL | AUG | SEP | OCT | NOV | DEC | ANNUAL |
|------|-----|-----|-----|-----|-----|-----|-----|-----|-----|-----|-----|-----|--------|
| 1980 | 2.58 | 1.06 | 5.46 | 4.24 | 3.58 | 3.04 | 3.25 | 4.00 | 1.00 | 3.08 | 2.72 | 0.70 | 34.71 |
| 1981 | 0.49 | 2.93 | 1.14 | 2.04 | 3.63 | 5.40 | 4.59 | 1.93 | 2.89 | 2.57 | 0.31 | 3.30 | 31.22 |
| 1982 | 3.37 | 4.04 | 3.03 | 3.61 | 1.85 | 5.70 | 2.16 | 0.95 | 3.63 | 2.31 | 3.13 | 2.39 | 36.17 |
| 1983 | 2.21 | 4.81 | 6.80 | 6.55 | 5.47 | 5.23 | 1.31 | 1.57 | 1.76 | 3.58 | 5.02 | 6.72 | 51.03 |
| 1984 | 1.96 | 3.90 | 5.79 | 2.95 | 4.29 | 1.65 | 3.27 | 4.11 | 2.38 | 1.94 | 3.01 | 1.71 | 36.96 |
| 1985 | 2.03 | 3.03 | 2.37 | 0.39 | 6.01 | 2.44 | 2.53 | 3.72 | 6.22 | 2.48 | 4.71 | 0.84 | 36.77 |
| 1986 | 2.16 | 3.78 | 0.96 | 2.64 | 0.37 | 1.46 | 4.12 | 4.26 | 0.58 | 1.86 | 5.96 | 5.52 | 33.67 |
| 1987 | 5.85 | 2.22 | 0.99 | 1.86 | 4.16 | 2.63 | 5.05 | 1.61 | 7.34 | 2.25 | 5.05 | 2.07 | 41.08 |
| 1988 | 3.24 | 3.25 | 2.35 | 2.44 | 4.37 | 0.84 | 3.78 | 2.64 | 2.05 | 1.59 | 4.78 | 0.97 | 32.30 |
| 1989 | 3.07 | 3.36 | 4.24 | 3.16 | 8.71 | 5.98 | 7.35 | 3.38 | 3.64 | 4.90 | 1.97 | 2.12 | 51.88 |
| 1990 | 3.71 | 1.48 | 2.54 | 4.23 | 4.92 | 2.55 | 5.68 | 6.17 | 1.07 | 2.57 | 2.10 | 4.86 | 41.88 |
| 1991 | 3.54 | 0.73 | 5.65 | 1.68 | 1.16 | 1.08 | 1.76 | 2.54 | 3.05 | 3.20 | 1.69 | 4.08 | 30.16 |
| 1992 | 1.27 | 2.49 | 4.58 | 1.76 | 2.92 | 1.89 | 5.07 | 2.19 | 5.96 | 2.73 | 3.44 | 4.63 | 38.93 |
| 1993 | 2.73 | 2.84 | 8.12 | 3.68 | 3.66 | 2.56 | 1.71 | 2.55 | 4.09 | 3.02 | 3.09 | 4.45 | 42.50 |
| 1994 | 4.59 | 4.07 | 8.64 | 2.53 | 3.02 | 2.84 | 4.54 | 3.44 | 3.93 | 1.82 | 1.95 | 1.95 | 43.32 |
| 1995 | 2.87 | 1.88 | 2.12 | 1.92 | 3.40 | 1.80 | 3.65 | 2.98 | 3.29 | 6.24 | 4.12 | 2.66 | 36.93 |
| 1996 | 6.80 | 2.36 | 3.57 | 3.76 | 5.68 | 4.08 | 7.38 | 4.17 | 5.65 | 4.32 | 3.77 | 6.77 | 58.31 |
| 1997 | 2.83 | 2.23 | 5.67 | 2.40 | 3.03 | 3.74 | 1.49 | 4.21 | 1.47 | 3.43 | 5.79 | 2.05 | 38.34 |
| 1998 | 5.65 | 6.40 | 5.56 | 3.02 | 3.46 | 3.22 | 1.42 | 0.91 | 1.27 | 1.06 | 1.13 | 1.27 | 34.37 |
| 1999 | 4.70 | 2.65 | 3.46 | 2.27 | 1.73 | 2.04 | 2.06 | 6.14 | 11.50 | 2.48 | 1.95 | 2.96 | 43.94 |
| 2000 | 3.64 | 2.01 | 4.35 | 5.06 | 2.82 | 5.54 | 5.64 | 3.18 | 5.55 | 0.08 | 1.73 | 2.31 | 41.91 |
| 2001 | 2.68 | 2.35 | 4.76 | 1.32 | 5.34 | 3.58 | 3.85 | 5.74 | 1.43 | 0.78 | 1.01 | 1.73 | 34.57 |
| 2002 | 2.19 | 0.36 | 3.75 | 4.08 | 2.99 | 2.39 | 2.26 | 3.66 | 3.17 | 6.01 | 3.78 | 4.96 | 39.60 |
| 2003 | 2.59 | 6.70 | 4.17 | 2.40 | 6.81 | 6.96 | 5.56 | 4.61 | 7.47 | 5.82 | 4.86 | 4.71 | 62.66 |
| 2004 | 1.26 | 2.40 | 2.73 | 5.33 | 5.05 | 4.17 | 8.69 | 2.71 | 3.94 | 1.44 | 5.02 | 2.93 | 45.67 |
| 2005 | 3.75 | 1.66 | 5.13 | 3.81 | 2.64 | 3.74 | 8.77 | 3.71 | 0.67 | 9.23 | 2.12 | 3.90 | 49.13 |
| 2006 | 3.48 | 2.64 | 0.18 | 3.27 | 1.60 | 7.32 | 1.86 | 1.45 | 7.56 | 5.75 | 6.25 | 1.88 | 43.24 |
| 2007 | 2.48 | 2.04 | 4.17 | 5.00 | 0.94 | 2.20 | 3.31 | 3.08 | 0.35 | 5.85 | 1.52 | 4.03 | 34.97 |
| 2008 | 1.47 | 3.80 | 2.37 | 4.62 | 7.77 | 3.70 | 5.47 | 1.48 | 7.22 | 1.27 | 2.61 | 3.19 | 44.97 |
| 2009 | 2.73 | 0.26 | 2.07 | 5.80 | 8.42 | 5.52 | 3.29 | 4.76 | 3.48 | 6.24 | 4.94 | 8.06 | 55.57 |
| POR= 59 YRS | 3.04 | 2.92 | 3.78 | 3.29 | 3.72 | 3.64 | 3.96 | 3.92 | 3.71 | 3.21 | 3.20 | 3.41 | 41.80 |

WBAN : 93721

## AVERAGE TEMPERATURE (°F) 2009 BALTIMORE (KBWI)

| YEAR | JAN | FEB | MAR | APR | MAY | JUN | JUL | AUG | SEP | OCT | NOV | DEC | ANNUAL |
|------|-----|-----|-----|-----|-----|-----|-----|-----|-----|-----|-----|-----|--------|
| 1980 | 33.8 | 31.5 | 41.5 | 55.7 | 65.5 | 71.3 | 78.2 | 78.7 | 72.2 | 55.3 | 44.2 | 35.5 | 55.3 |
| 1981 | 27.9 | 38.8 | 41.9 | 57.0 | 62.2 | 74.3 | 77.3 | 74.4 | 67.7 | 53.2 | 46.2 | 34.5 | 54.6 |
| 1982 | 25.5 | 35.8 | 42.9 | 50.7 | 66.1 | 69.4 | 77.1 | 73.0 | 67.3 | 56.3 | 48.4 | 42.0 | 54.5 |
| 1983 | 34.6 | 34.7 | 45.4 | 51.8 | 61.5 | 72.1 | 78.7 | 78.0 | 69.5 | 57.3 | 47.1 | 33.2 | 55.3 |
| 1984 | 28.5 | 41.7 | 38.2 | 51.5 | 61.3 | 73.4 | 73.9 | 75.0 | 64.8 | 62.2 | 43.9 | 44.1 | 54.9 |
| 1985 | 29.3 | 38.7 | 46.0 | 57.9 | 65.1 | 70.4 | 76.4 | 74.5 | 69.4 | 58.8 | 52.4 | 33.8 | 56.1 |
| 1986 | 33.2 | 32.9 | 45.0 | 53.5 | 66.7 | 74.4 | 79.4 | 73.1 | 68.9 | 58.9 | 44.8 | 38.2 | 55.8 |
| 1987 | 32.5 | 34.3 | 46.2 | 53.1 | 65.0 | 74.5 | 80.0 | 76.1 | 69.3 | 51.5 | 47.8 | 39.8 | 55.8 |
| 1988 | 28.7 | 35.9 | 45.1 | 52.0 | 64.0 | 73.0 | 80.3 | 78.5 | 66.8 | 51.3 | 48.1 | 36.3 | 55.0 |
| 1989 | 37.9 | 36.5 | 43.8 | 52.5 | 62.0 | 73.9 | 76.0 | 74.4 | 69.0 | 58.3 | 44.8 | 25.4 | 54.5 |
| 1990 | 42.0 | 42.3 | 47.6 | 54.8 | 62.3 | 73.3 | 78.4 | 74.6 | 67.3 | 60.7 | 49.6 | 42.2 | 57.9 |
| 1991 | 35.5 | 40.7 | 46.7 | 55.9 | 70.6 | 74.6 | 79.5 | 77.8 | 69.0 | 57.8 | 45.8 | 38.7 | 57.7 |
| 1992 | 34.6 | 37.1 | 41.3 | 52.0 | 60.8 | 70.1 | 77.4 | 72.3 | 67.7 | 54.3 | 47.2 | 38.9 | 54.5 |
| 1993 | 37.9 | 31.4 | 39.4 | 52.5 | 65.0 | 72.2 | 80.2 | 76.7 | 68.8 | 55.5 | 46.5 | 36.2 | 55.2 |
| 1994 | 27.1 | 34.0 | 43.0 | 59.6 | 60.6 | 77.2 | 80.1 | 74.1 | 68.1 | 56.8 | 51.9 | 42.6 | 56.3 |
| 1995 | 39.0 | 33.2 | 47.8 | 55.2 | 64.5 | 74.5 | 81.5 | 80.1 | 70.4 | 61.1 | 42.6 | 33.9 | 57.0 |
| 1996 | 31.7 | 35.7 | 39.9 | 54.0 | 60.6 | 73.3 | 74.3 | 73.2 | 67.8 | 55.6 | 40.2 | 39.6 | 53.8 |
| 1997 | 32.8 | 41.0 | 45.5 | 51.6 | 59.5 | 70.1 | 77.3 | 74.0 | 67.3 | 56.5 | 43.7 | 38.4 | 54.8 |
| 1998 | 40.9 | 41.7 | 45.9 | 55.2 | 66.5 | 71.7 | 76.6 | 75.7 | 71.8 | 56.3 | 46.1 | 41.1 | 57.5 |
| 1999 | 35.1 | 37.6 | 41.8 | 53.2 | 64.2 | 71.5 | 80.0 | 75.7 | 68.2 | 53.9 | 49.9 | 39.1 | 55.9 |
| 2000 | 32.5 | 38.1 | 48.5 | 52.9 | 64.7 | 72.8 | 72.7 | 73.4 | 65.3 | 57.1 | 44.2 | 30.0 | 54.4 |
| 2001 | 33.1 | 38.5 | 41.8 | 55.4 | 63.4 | 74.1 | 72.8 | 77.0 | 65.2 | 56.0 | 50.7 | 42.1 | 55.8 |
| 2002 | 39.1 | 39.3 | 45.0 | 56.7 | 62.2 | 73.8 | 78.6 | 78.4 | 69.5 | 56.0 | 44.4 | 34.3 | 56.4 |
| 2003 | 28.3 | 30.2 | 43.9 | 52.7 | 59.3 | 69.8 | 75.6 | 76.3 | 68.0 | 55.1 | 50.6 | 36.4 | 53.9 |
| 2004 | 27.8 | 34.8 | 45.6 | 54.7 | 69.8 | 70.9 | 76.2 | 74.2 | 69.4 | 55.4 | 48.5 | 37.5 | 55.4 |
| 2005 | 34.1 | 36.7 | 40.6 | 55.2 | 59.2 | 73.6 | 78.0 | 77.6 | 72.0 | 57.8 | 48.1 | 34.0 | 55.6 |
| 2006 | 41.6 | 36.1 | 45.6 | 57.5 | 63.4 | 73.1 | 79.9 | 78.4 | 65.5 | 55.2 | 49.6 | 42.4 | 57.4 |
| 2007 | 38.7 | 29.1 | 45.2 | 51.5 | 65.5 | 73.8 | 76.9 | 77.5 | 70.6 | 63.4 | 46.2 | 37.8 | 56.4 |
| 2008 | 35.4 | 37.1 | 45.0 | 55.8 | 60.5 | 75.3 | 77.5 | 73.6 | 69.4 | 55.6 | 45.4 | 38.5 | 55.8 |
| 2009 | 29.3 | 37.5 | 43.1 | 54.9 | 63.7 | 71.4 | 74.7 | 76.6 | 66.7 | 55.3 | 49.7 | 34.8 | 54.8 |
| POR= 59 YRS | 33.1 | 35.6 | 43.6 | 54.1 | 63.4 | 72.5 | 77.1 | 75.6 | 68.5 | 56.8 | 46.6 | 36.9 | 55.3 |

## HEATING DEGREE DAYS (base 65°F) 2009 BALTIMORE (KBWI)

| YEAR | JUL | AUG | SEP | OCT | NOV | DEC | JAN | FEB | MAR | APR | MAY | JUN | TOTAL |
|---|---|---|---|---|---|---|---|---|---|---|---|---|---|
| 1980-81 | 0 | 0 | 20 | 311 | 620 | 908 | 1145 | 727 | 706 | 252 | 148 | 1 | 4838 |
| 1981-82 | 0 | 0 | 51 | 363 | 557 | 940 | 1218 | 808 | 677 | 422 | 58 | 20 | 5114 |
| 1982-83 | 0 | 5 | 42 | 289 | 495 | 707 | 936 | 842 | 602 | 410 | 152 | 6 | 4486 |
| 1983-84 | 0 | 0 | 70 | 257 | 530 | 979 | 1123 | 671 | 825 | 397 | 169 | 9 | 5030 |
| 1984-85 | 0 | 1 | 96 | 123 | 625 | 643 | 1101 | 731 | 589 | 252 | 79 | 10 | 4250 |
| 1985-86 | 0 | 0 | 41 | 201 | 378 | 962 | 980 | 892 | 613 | 342 | 86 | 6 | 4501 |
| 1986-87 | 0 | 23 | 34 | 236 | 598 | 822 | 1002 | 853 | 576 | 357 | 106 | 1 | 4608 |
| 1987-88 | 0 | 1 | 15 | 412 | 511 | 774 | 1120 | 838 | 613 | 389 | 96 | 27 | 4796 |
| 1988-89 | 2 | 0 | 39 | 424 | 504 | 882 | 834 | 792 | 663 | 374 | 145 | 0 | 4659 |
| 1989-90 | 0 | 0 | 51 | 229 | 600 | 1221 | 707 | 631 | 552 | 341 | 102 | 5 | 4439 |
| 1990-91 | 1 | 0 | 63 | 195 | 454 | 701 | 907 | 674 | 562 | 289 | 55 | 4 | 3905 |
| 1991-92 | 0 | 0 | 49 | 246 | 570 | 809 | 936 | 802 | 730 | 387 | 161 | 8 | 4698 |
| 1992-93 | 0 | 1 | 51 | 328 | 529 | 801 | 834 | 934 | 787 | 369 | 61 | 11 | 4706 |
| 1993-94 | 0 | 0 | 52 | 292 | 553 | 886 | 1169 | 861 | 677 | 190 | 180 | 1 | 4861 |
| 1994-95 | 0 | 0 | 13 | 256 | 391 | 684 | 798 | 885 | 525 | 307 | 77 | 0 | 3936 |
| 1995-96 | 0 | 0 | 30 | 176 | 669 | 958 | 1024 | 840 | 772 | 345 | 199 | 12 | 5025 |
| 1996-97 | 0 | 0 | 42 | 283 | 736 | 778 | 994 | 667 | 597 | 394 | 182 | 53 | 4726 |
| 1997-98 | 0 | 0 | 49 | 307 | 633 | 815 | 737 | 647 | 625 | 295 | 59 | 22 | 4189 |
| 1998-99 | 0 | 1 | 25 | 263 | 560 | 734 | 919 | 762 | 714 | 349 | 71 | 9 | 4407 |
| 1999-00 | 0 | 0 | 37 | 336 | 445 | 794 | 999 | 774 | 508 | 362 | 102 | 8 | 4365 |
| 2000-01 | 0 | 1 | 97 | 254 | 616 | 1079 | 984 | 736 | 715 | 309 | 99 | 12 | 4902 |
| 2001-02 | 2 | 0 | 76 | 289 | 424 | 706 | 795 | 715 | 612 | 302 | 154 | 3 | 4078 |
| 2002-03 | 0 | 1 | 10 | 316 | 611 | 945 | 1131 | 967 | 649 | 370 | 191 | 31 | 5222 |
| 2003-04 | 0 | 0 | 24 | 303 | 434 | 881 | 1148 | 866 | 593 | 323 | 49 | 13 | 4634 |
| 2004-05 | 0 | 2 | 14 | 291 | 487 | 845 | 952 | 786 | 750 | 298 | 183 | 9 | 4617 |
| 2005-06 | 0 | 0 | 12 | 239 | 502 | 955 | 720 | 802 | 598 | 228 | 117 | 4 | 4177 |
| 2006-07 | 0 | 0 | 53 | 314 | 456 | 693 | 805 | 1000 | 608 | 406 | 92 | 1 | 4428 |
| 2007-08 | 0 | 0 | 26 | 143 | 559 | 835 | 909 | 799 | 614 | 274 | 165 | 0 | 4324 |
| 2008-09 | 0 | 0 | 18 | 304 | 584 | 814 | 1103 | 768 | 673 | 339 | 106 | 13 | 4722 |
| 2009- | 0 | 0 | 38 | 299 | 451 | 929 | | | | | | | |

WBAN : 93721

## COOLING DEGREE DAYS (base 65°F) 2009 BALTIMORE (KBWI)

| YEAR | JAN | FEB | MAR | APR | MAY | JUN | JUL | AUG | SEP | OCT | NOV | DEC | TOTAL |
|---|---|---|---|---|---|---|---|---|---|---|---|---|---|
| 1980 | 0 | 0 | 0 | 0 | 97 | 203 | 415 | 431 | 245 | 17 | 0 | 0 | 1408 |
| 1981 | 0 | 0 | 0 | 19 | 69 | 287 | 389 | 296 | 141 | 5 | 0 | 0 | 1206 |
| 1982 | 0 | 0 | 0 | 4 | 99 | 160 | 381 | 259 | 119 | 26 | 4 | 1 | 1053 |
| 1983 | 0 | 0 | 0 | 18 | 51 | 228 | 430 | 410 | 214 | 24 | 0 | 0 | 1375 |
| 1984 | 0 | 0 | 0 | 0 | 59 | 268 | 281 | 316 | 98 | 41 | 0 | 2 | 1065 |
| 1985 | 0 | 2 | 7 | 43 | 89 | 179 | 363 | 298 | 178 | 17 | 5 | 0 | 1181 |
| 1986 | 0 | 0 | 0 | 1 | 143 | 295 | 452 | 281 | 158 | 54 | 0 | 0 | 1384 |
| 1987 | 0 | 0 | 0 | 7 | 115 | 292 | 473 | 352 | 152 | 0 | 0 | 0 | 1391 |
| 1988 | 0 | 0 | 2 | 4 | 71 | 274 | 485 | 427 | 100 | 8 | 0 | 0 | 1371 |
| 1989 | 0 | 0 | 14 | 5 | 58 | 276 | 351 | 298 | 178 | 25 | 1 | 0 | 1206 |
| 1990 | 0 | 0 | 19 | 38 | 26 | 261 | 422 | 303 | 137 | 68 | 0 | 0 | 1274 |
| 1991 | 0 | 0 | 2 | 24 | 233 | 303 | 462 | 402 | 177 | 29 | 2 | 0 | 1634 |
| 1992 | 0 | 0 | 0 | 6 | 39 | 168 | 392 | 232 | 139 | 4 | 0 | 0 | 980 |
| 1993 | 0 | 0 | 0 | 0 | 70 | 235 | 476 | 371 | 175 | 3 | 5 | 0 | 1335 |
| 1994 | 0 | 0 | 0 | 38 | 49 | 374 | 476 | 292 | 112 | 6 | 3 | 0 | 1350 |
| 1995 | 0 | 0 | 0 | 20 | 72 | 289 | 520 | 475 | 199 | 60 | 3 | 0 | 1638 |
| 1996 | 0 | 0 | 0 | 19 | 70 | 265 | 295 | 259 | 135 | 1 | 0 | 0 | 1044 |
| 1997 | 0 | 0 | 0 | 0 | 20 | 211 | 385 | 287 | 124 | 51 | 0 | 0 | 1078 |
| 1998 | 0 | 0 | 39 | 9 | 115 | 228 | 367 | 341 | 235 | 0 | 0 | 0 | 1334 |
| 1999 | 0 | 0 | 0 | 0 | 54 | 210 | 471 | 340 | 138 | 0 | 0 | 0 | 1213 |
| 2000 | 0 | 0 | 3 | 6 | 102 | 248 | 245 | 269 | 115 | 17 | 0 | 0 | 1005 |
| 2001 | 0 | 0 | 27 | 54 | 75 | 290 | 249 | 381 | 90 | 12 | 2 | 0 | 1105 |
| 2002 | 0 | 0 | 0 | 59 | 75 | 275 | 430 | 425 | 154 | 43 | 0 | 0 | 1461 |
| 2003 | 0 | 0 | 0 | 6 | 21 | 181 | 336 | 358 | 120 | 2 | 8 | 0 | 1032 |
| 2004 | 0 | 0 | 0 | 23 | 203 | 197 | 355 | 295 | 152 | 2 | 0 | 0 | 1227 |
| 2005 | 0 | 0 | 0 | 13 | 10 | 273 | 408 | 399 | 227 | 20 | 0 | 0 | 1350 |
| 2006 | 0 | 0 | 3 | 9 | 74 | 252 | 468 | 422 | 72 | 15 | 0 | 0 | 1315 |
| 2007 | 0 | 0 | 2 | 9 | 114 | 268 | 376 | 394 | 203 | 99 | 0 | 0 | 1465 |
| 2008 | 0 | 0 | 0 | 6 | 33 | 317 | 395 | 274 | 158 | 19 | 0 | 0 | 1202 |
| 2009 | 0 | 0 | 0 | 39 | 72 | 213 | 306 | 368 | 95 | 7 | 0 | 0 | 1100 |

## SNOWFALL (inches)  2009  BALTIMORE (KBWI)

| YEAR | JUL | AUG | SEP | OCT | NOV | DEC | JAN | FEB | MAR | APR | MAY | JUN | TOTAL |
|---|---|---|---|---|---|---|---|---|---|---|---|---|---|
| 1980-81 | 0.0 | 0.0 | 0.0 | 0.0 | T | 0.2 | 4.1 | T | 0.3 | 0.0 | 0.0 | 0.0 | 4.6 |
| 1981-82 | 0.0 | 0.0 | 0.0 | 0.0 | T | 2.4 | 14.8 | 7.6 | 0.7 | T | 0.0 | 0.0 | 25.5 |
| 1982-83 | 0.0 | 0.0 | 0.0 | 0.0 | 0.0 | 7.2 | 1.2 | 27.2 | T | T | 0.0 | 0.0 | 35.6 |
| 1983-84 | 0.0 | 0.0 | 0.0 | 0.0 | T | T | 8.4 | T | 6.1 | T | 0.0 | 0.0 | 14.5 |
| 1984-85 | 0.0 | 0.0 | 0.0 | 0.0 | T | 0.1 | 9.1 | 0.4 | T | 0.7 | 0.0 | 0.0 | 10.3 |
| 1985-86 | 0.0 | 0.0 | 0.0 | 0.0 | 0.0 | 0.7 | 1.9 | 13.0 | T | T | 0.0 | 0.0 | 15.6 |
| 1986-87 | 0.0 | 0.0 | 0.0 | 0.0 | 0.0 | T | 25.1 | 10.1 | T | T | 0.0 | 0.0 | 35.2 |
| 1987-88 | 0.0 | 0.0 | 0.0 | 0.0 | 0.0 | 6.0 | 0.5 | 13.7 | 0.2 | T | 0.0 | 0.0 | 20.4 |
| 1988-89 | 0.0 | 0.0 | 0.0 | 0.0 | 0.0 | 0.9 | 6.0 | 1.1 | 0.3 | 0.0 | 0.0 | 0.0 | 8.3 |
| 1989-90 | 0.0 | 0.0 | 0.0 | 0.0 | 3.8 | 10.2 | 0.5 | T | 2.7 | 0.1 | 0.0 | 0.0 | 17.3 |
| 1990-91 | 0.0 | 0.0 | 0.0 | 0.0 | 0.0 | 4.8 | 4.2 | 0.1 | 0.3 | 0.0 | 0.0 | 0.0 | 9.4 |
| 1991-92 | T | 0.0 | 0.0 | 0.0 | T | T | 2.2 | 1.9 | T | T | 0.0 | 0.0 | 4.1 |
| 1992-93 | T | 0.0 | 0.0 | 0.0 | T | 1.5 | 1.4 | 8.8 | 12.7 | T | 0.0 | 0.0 | 24.4 |
| 1993-94 | 0.0 | 0.0 | 0.0 | 0.0 | T | 2.9 | 4.9 | 5.3 | 4.2 | T | 0.0 | 0.0 | 17.3 |
| 1994-95 | 0.0 | 0.0 | 0.0 | 0.0 | 0.2 | 0.0 | 0.3 | 7.5 | 0.2 | 0.0 | 0.0 | 0.0 | 8.2 |
| 1995-96 | 0.0 | 0.0 | 0.0 | 0.0 | 1.0 | 2.3 | 32.6 | 19.0 | 7.6 | | | | |
| 1996-97 | | | 0.0 | | .3 | 0.2 | 5.0 | 7.1 | 2.7 | T | T | 0.0 | |
| 1997-98 | 0.0 | 0.0 | 0.0 | 0.0 | 0.0 | 0.4 | 0.7 | T | 2.1 | 0.0 | T | 0.0 | 3.2 |
| 1998-99 | 0.0 | 0.0 | 0.0 | 0.0 | 0.0 | 3.0 | 4.0 | 0.6 | 7.6 | 0.0 | 0.0 | 0.0 | 15.2 |
| 1999-00 | 0.0 | 0.0 | 0.0 | 0.0 | 0.0 | 0.2 | 23.1 | 2.6 | 0.0 | 0.2 | 0.0 | 0.0 | 26.1 |
| 2000-01 | 0.0 | 0.0 | 0.0 | 0.0 | T | 1.3 | 3.7 | 3.7 | T | T | 0.0 | 0.0 | 8.7 |
| 2001-02 | 0.0 | 0.0 | 0.0 | 0.0 | 0.0 | 0.0 | 2.3 | T | T | T | 0.0 | T | 2.3 |
| 2002-03 | 0.0 | 0.0 | 0.0 | 0.0 | 0.0 | 9.7 | 5.3 | 40.5 | 2.6 | T | 0.0 | 0.0 | 58.1 |
| 2003-04 | 0.0 | 0.0 | 0.0 | T | 0.0 | 9.6 | 8.4 | 0.1 | 0.2 | 0.0 | 0.0 | 0.0 | 18.3 |
| 2004-05 | 0.0 | 0.0 | 0.0 | 0.0 | 0.0 | T | 7.6 | 10.0 | 0.4 | 0.0 | 0.0 | 0.0 | 18.0 |
| 2005-06 | 0.0 | 0.0 | 0.0 | 0.0 | 0.5 | 6.0 | T | 13.1 | T | T | 0.0 | 0.0 | 19.6 |
| 2006-07 | 0.0 | 0.0 | 0.0 | 0.0 | 0.0 | T | 0.9 | 8.5 | 1.4 | 0.2 | 0.0 | 0.0 | 11.0 |
| 2007-08 | 0.0 | 0.0 | 0.0 | 0.0 | T | 4.8 | 2.4 | 1.3 | 0.0 | 0.0 | 0.0 | 0.0 | 8.5 |
| 2008-09 | 0.0 | 0.0 | 0.0 | 0.0 | T | 0.6 | 2.1 | 0.6 | 5.8 | T | 0.0 | 0.0 | 9.1 |
| 2009- | 0.0 | 0.0 | 0.0 | 0.0 | 0.0 | 20.1 | | | | | | | |
| POR=  59 YRS | T | 0.0 | 0.0 | T | 0.8 | 3.5 | 6.0 | 6.8 | 3.3 | T | T | T | 20.4 |

# Baltimore, MD  2009
## Temperature and Precipitation

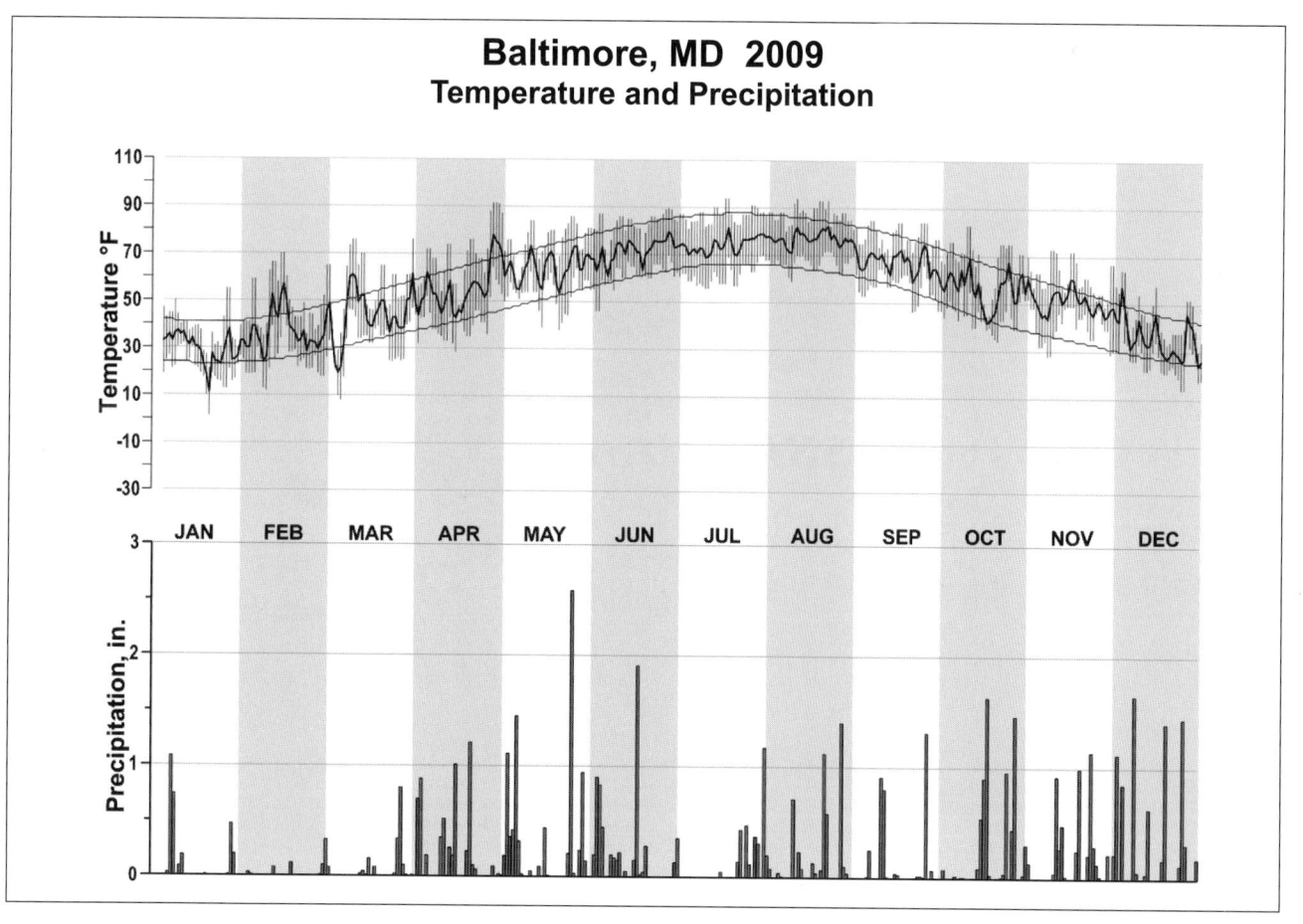

# 2009
# BOSTON
# MASSACHUSETTS (KBOS)

Climate is the composite of numerous weather elements. Three important influences are responsible for the main features of the Boston climate. First, the latitude places the city in the zone of prevailing west to east atmospheric flow. Both polar and tropical air masses influence the region. Secondly, Boston is situated on or near several tracks frequently followed by low pressure storm systems. The weather fluctuates regularly from fair to cloudy to stormy conditions and assures an adequate amount of precipitation. The third factor is the east-coast location of Boston. The ocean has a moderating influence on temperature extremes of winter and summer.

Hot summer afternoons are frequently relieved by the locally celebrated sea breeze, as air flows inland from the cool water surface to displace the warm air over the land. This refreshing east wind is more commonly experienced along the shore than in the interior of the city or the western suburbs. In winter, under appropriate conditions, the severity of cold waves is reduced by the nearness of the relatively warm ocean. The average last occurrence of freezing temperature in spring is early April and the first occurrence of freezing temperature in autumn is early November. In suburban areas, especially away from the coast, these dates are later in spring and earlier in autumn by up to one month in the more susceptible localities.

Boston has no dry season. Most growing seasons have several shorter dry spells during which irrigation for high-value crops may be useful. Much of the rainfall from June to September comes from showers and thunderstorms. During the rest of the year, low pressure systems pass more or less regularly and produce precipitation on an average of roughly one day in three. Coastal storms, or northeasters, are prolific producers of rain and snow. The main snow season extends from December through March. Periods when the ground is bare or nearly bare of snow may occur at any time in the winter.

Relative humidity has been known to fall as low as 5 percent but such desert dryness is very rare. Heavy fog occurs on an average of about two days per month with its prevalence increasing eastward from the interior of Boston Bay to the open waters beyond.

Although winds of 30 mph or higher may be expected on at least one day in every month of the year, gales are both more common and more severe in winter.

# NORMALS, MEANS, AND EXTREMES
## BOSTON (KBOS)

LATITUDE: 42 ° 21'N  LONGITUDE: -71 ° 0 'W  ELEVATION (FT): GRND: 19  BARO: 180  TIME ZONE: EASTERN (UTC -5)  WBAN: 14739

| ELEMENT | POR | JAN | FEB | MAR | APR | MAY | JUN | JUL | AUG | SEP | OCT | NOV | DEC | YEAR |
|---|---|---|---|---|---|---|---|---|---|---|---|---|---|---|
| **TEMPERATURE °F** | | | | | | | | | | | | | | |
| NORMAL DAILY MAXIMUM | 30 | 36.5 | 38.7 | 46.3 | 56.1 | 66.7 | 76.6 | 82.2 | 80.1 | 72.5 | 61.8 | 51.8 | 41.7 | 59.3 |
| MEAN DAILY MAXIMUM | 90 | 36.3 | 37.0 | 45.5 | 55.3 | 66.6 | 75.5 | 81.5 | 79.7 | 71.9 | 62.4 | 51.3 | 40.6 | 58.6 |
| HIGHEST DAILY MAXIMUM | 58 | 69 | 70 | 89 | 94 | 95 | 100 | 102 | 102 | 100 | 90 | 79 | 76 | 102 |
| YEAR OF OCCURRENCE | | 2007 | 1985 | 1998 | 1976 | 1979 | 1952 | 1977 | 1975 | 1953 | 1963 | 1994 | 1998 | JUL 1977 |
| MEAN OF EXTREME MAXS. | 90 | 56.2 | 56.2 | 66.6 | 78.9 | 87.0 | 92.7 | 94.9 | 93.2 | 88.6 | 79.8 | 69.9 | 60.5 | 77.0 |
| NORMAL DAILY MINIMUM | 30 | 22.1 | 24.2 | 31.5 | 40.5 | 50.2 | 59.4 | 65.5 | 64.5 | 56.8 | 46.4 | 37.9 | 27.8 | 43.9 |
| MEAN DAILY MINIMUM | 90 | 22.0 | 22.7 | 30.8 | 39.7 | 49.7 | 58.3 | 64.9 | 63.8 | 56.3 | 46.7 | 37.5 | 26.9 | 43.3 |
| LOWEST DAILY MINIMUM | 58 | -12 | -4 | 5 | 16 | 34 | 45 | 50 | 47 | 38 | 28 | 15 | -7 | -12 |
| YEAR OF OCCURRENCE | | 1957 | 1961 | 2007 | 1982 | 1956 | 1986 | 1988 | 1986 | 2000 | 1976 | 1989 | 1980 | JAN 1957 |
| MEAN OF EXTREME MINS. | 90 | 3.6 | 6.5 | 15.5 | 29.9 | 40.3 | 49.3 | 57.2 | 54.8 | 44.6 | 34.6 | 24.4 | 9.9 | 30.9 |
| NORMAL DRY BULB | 30 | 29.3 | 31.5 | 38.9 | 48.3 | 58.5 | 68.0 | 73.9 | 72.3 | 64.7 | 54.1 | 44.9 | 34.8 | 51.6 |
| MEAN DRY BULB | 90 | 29.2 | 29.8 | 38.2 | 47.5 | 58.2 | 67.0 | 73.2 | 71.8 | 64.1 | 54.6 | 44.4 | 33.8 | 51.0 |
| MEAN WET BULB | 26 | 25.3 | 26.6 | 32.0 | 40.8 | 50.3 | 60.0 | 65.3 | 64.8 | 58.3 | 48.1 | 39.4 | 30.0 | 45.1 |
| MEAN DEW POINT | 26 | 19.7 | 20.8 | 27.0 | 35.9 | 46.5 | 56.6 | 62.2 | 62.2 | 55.4 | 44.2 | 34.6 | 24.6 | 40.8 |
| NORMAL NO. DAYS WITH: | | | | | | | | | | | | | | |
| MAXIMUM >= 90 | 30 | 0.0 | 0.0 | 0.0 | 0.1 | 0.5 | 2.9 | 6.0 | 3.4 | 0.7 | 0.0 | 0.0 | 0.0 | 13.6 |
| MAXIMUM <= 32 | 30 | 10.8 | 7.6 | 2.1 | 0.1 | 0.0 | 0.0 | 0.0 | 0.0 | 0.0 | 0.0 | 0.4 | 5.1 | 26.1 |
| MINIMUM <= 32 | 30 | 25.6 | 22.4 | 16.0 | 2.4 | 0.0 | 0.0 | 0.0 | 0.0 | 0.0 | 0.4 | 7.3 | 21.2 | 95.3 |
| MINIMUM <= 0 | 30 | 0.5 | 0.3 | 0.0 | 0.0 | 0.0 | 0.0 | 0.0 | 0.0 | 0.0 | 0.0 | 0.0 | 0.1 | 0.9 |
| **H/C** | | | | | | | | | | | | | | |
| NORMAL HEATING DEG. DAYS | 30 | 1104 | 951 | 815 | 503 | 233 | 48 | 4 | 8 | 84 | 344 | 604 | 932 | 5630 |
| NORMAL COOLING DEG. DAYS | 30 | 0 | 0 | 1 | 4 | 32 | 139 | 282 | 235 | 76 | 7 | 1 | 0 | 777 |
| **RH** | | | | | | | | | | | | | | |
| NORMAL (PERCENT) | 30 | 64 | 62 | 64 | 64 | 68 | 69 | 69 | 72 | 73 | 69 | 66 | 65 | 67 |
| HOUR 01 LST | 30 | 66 | 66 | 68 | 69 | 75 | 77 | 78 | 80 | 80 | 76 | 71 | 68 | 73 |
| HOUR 07 LST | 30 | 69 | 68 | 70 | 69 | 73 | 73 | 74 | 78 | 80 | 77 | 74 | 70 | 73 |
| HOUR 13 LST | 30 | 59 | 56 | 57 | 56 | 61 | 58 | 57 | 60 | 60 | 58 | 58 | 58 | 58 |
| HOUR 19 LST | 30 | 62 | 60 | 62 | 62 | 66 | 67 | 66 | 70 | 72 | 68 | 65 | 63 | 65 |
| **S** PERCENT POSSIBLE SUNSHINE | 61 | 53 | 56 | 57 | 56 | 58 | 63 | 65 | 65 | 63 | 60 | 50 | 52 | 58 |
| **W/O** MEAN NO. DAYS WITH: | | | | | | | | | | | | | | |
| HEAVY FOG(VISBY <= 1/4 MI) | 46 | 1.7 | 1.6 | 2.2 | 1.5 | 2.3 | 2.0 | 1.8 | 1.5 | 1.6 | 1.9 | 1.7 | 1.4 | 21.2 |
| THUNDERSTORMS | 64 | 0.2 | 0.2 | 0.5 | 0.9 | 2.2 | 3.3 | 4.1 | 3.2 | 1.4 | 0.7 | 0.4 | 0.2 | 17.3 |
| **CLOUDNESS** MEAN: | | | | | | | | | | | | | | |
| SUNRISE-SUNSET (OKTAS) | 61 | 5.0 | 5.0 | 5.1 | 5.3 | 5.3 | 5.0 | 4.9 | 4.5 | 4.4 | 4.4 | 5.0 | 5.0 | 4.9 |
| MIDNIGHT-MIDNIGHT (OKTAS) | 32 | 4.8 | 4.7 | 4.9 | 4.9 | 5.0 | 4.7 | 4.7 | 4.3 | 4.3 | 4.1 | 4.8 | 4.7 | 4.7 |
| MEAN NO. DAYS WITH: | | | | | | | | | | | | | | |
| CLEAR | 61 | 9.0 | 8.2 | 7.7 | 7.0 | 6.3 | 6.6 | 6.6 | 9.1 | 10.0 | 10.7 | 7.7 | 8.5 | 97.4 |
| PARTLY CLOUDY | 61 | 6.8 | 6.6 | 8.0 | 8.1 | 9.9 | 10.4 | 12.0 | 10.5 | 7.9 | 7.7 | 7.2 | 7.3 | 102.4 |
| CLOUDY | 61 | 15.2 | 13.4 | 15.3 | 14.9 | 14.8 | 12.9 | 11.9 | 10.9 | 11.6 | 12.1 | 14.7 | 14.9 | 162.6 |
| **PR** MEAN STATION PRESSURE(IN) | 26 | 29.94 | 29.97 | 29.97 | 29.93 | 29.94 | 29.91 | 29.92 | 29.97 | 30.02 | 30.02 | 30.01 | 29.99 | 29.97 |
| MEAN SEA-LEVEL PRES. (IN) | 26 | 30.02 | 30.02 | 30.01 | 29.97 | 29.98 | 29.95 | 29.96 | 30.01 | 30.05 | 30.06 | 30.05 | 30.03 | 30.01 |
| **WINDS** MEAN SPEED (MPH) | 26 | 12.8 | 12.9 | 13.0 | 12.3 | 11.6 | 10.9 | 10.5 | 10.2 | 10.7 | 11.5 | 11.9 | 12.8 | 11.8 |
| PREVAIL.DIR(TENS OF DEGS) | 42 | 30 | 30 | 30 | 30 | 21 | 24 | 24 | 24 | 24 | 30 | 30 | 30 | 30 |
| MAXIMUM 2-MINUTE: | | | | | | | | | | | | | | |
| SPEED (MPH) | 13 | 49 | 48 | 46 | 43 | 45 | 45 | 45 | 39 | 47 | 44 | 48 | 46 | 49 |
| DIR. (TENS OF DEGS) | | 18 | 29 | 03 | 11 | 28 | 30 | 30 | 26 | 23 | 05 | 11 | 18 | 18 |
| YEAR OF OCCURRENCE | | 2006 | 2006 | 1997 | 2007 | 2006 | 2000 | 1999 | 2005 | 1998 | 2005 | 1997 | 2000 | JAN 2006 |
| MAXIMUM 3-SECOND | | | | | | | | | | | | | | |
| SPEED (MPH) | 13 | 61 | 58 | 56 | 55 | 55 | 62 | 74 | 46 | 63 | 56 | 57 | 55 | 74 |
| DIR. (TENS OF DEGS) | | 18 | 29 | 06 | 12 | 28 | 29 | 25 | 26 | 23 | 25 | 11 | 18 | 25 |
| YEAR OF OCCURRENCE | | 2006 | 2006 | 2001 | 2007 | 2006 | 2000 | 1999 | 2005 | 1998 | 2006 | 1997 | 2000 | JUL 1999 |
| **PRECIPITATION** NORMAL (IN) | 30 | 3.92 | 3.30 | 3.85 | 3.60 | 3.24 | 3.22 | 3.06 | 3.37 | 3.47 | 3.79 | 3.98 | 3.73 | 42.53 |
| MAXIMUM MONTHLY (IN) | 58 | 10.55 | 7.94 | 11.00 | 9.57 | 13.38 | 13.20 | 8.12 | 17.09 | 9.86 | 10.66 | 8.89 | 9.74 | 17.09 |
| YEAR OF OCCURRENCE | | 1979 | 2008 | 1953 | 2004 | 1954 | 1982 | 1959 | 1955 | 1999 | 1996 | 1983 | 1969 | AUG 1955 |
| MINIMUM MONTHLY (IN) | 58 | 0.61 | 0.72 | 0.56 | 0.83 | 0.53 | 0.48 | 0.52 | 0.66 | 0.35 | 0.41 | 0.64 | 0.81 | 0.35 |
| YEAR OF OCCURRENCE | | 1989 | 1987 | 2006 | 1999 | 1964 | 1999 | 1952 | 2007 | 1957 | 1994 | 1976 | 1989 | SEP 1957 |
| MAXIMUM IN 24 HOURS (IN) | 58 | 2.72 | 2.81 | 4.13 | 4.29 | 5.74 | 5.69 | 3.36 | 8.40 | 5.64 | 6.63 | 3.76 | 5.14 | 8.40 |
| YEAR OF OCCURRENCE | | 1979 | 2008 | 1968 | 2004 | 1954 | 1998 | 1996 | 1955 | 1954 | 1996 | 1992 | 1992 | AUG 1955 |
| NORMAL NO. DAYS WITH: | | | | | | | | | | | | | | |
| PRECIPITATION >= 0.01 | 30 | 11.7 | 10.0 | 12.0 | 11.0 | 11.8 | 10.4 | 9.3 | 9.8 | 9.0 | 8.9 | 10.2 | 12.0 | 126.1 |
| PRECIPITATION >= 1.00 | 30 | 1.1 | 0.9 | 1.0 | 0.9 | 0.9 | 0.5 | 0.6 | 0.7 | 0.8 | 0.8 | 0.9 | 1.1 | 0.8 | 10.0 |
| **SNOWFALL** NORMAL (IN) | 30 | 13.5 | 11.2 | 8.1 | 1.1 | 0.* | 0.0 | 0.0 | 0.0 | 0.0 | 0.* | 1.3 | 6.6 | 41.8 |
| MAXIMUM MONTHLY (IN) | 72 | 43.3 | 41.6 | 38.9 | 13.3 | 0.5 | T | T | T | 0.0 | 1.1 | 10.0 | 27.9 | 43.3 |
| YEAR OF OCCURRENCE | | 1996 | 2003 | 1993 | 1982 | 1977 | 2002 | 2004 | 1994 | | 1979 | 1938 | 1970 | JAN 1996 |
| MAXIMUM IN 24 HOURS (IN) | 72 | 21.0 | 23.6 | 17.7 | 13.2 | 0.5 | T | T | T | 0.0 | 1.1 | 8.0 | 13.0 | 23.6 |
| YEAR OF OCCURRENCE' | | 1978 | 1978 | 1960 | 1982 | 1977 | 2002 | 2004 | 1994 | | 2005 | 1987 | 1960 | FEB 1978 |
| MAXIMUM SNOW DEPTH (IN) | 54 | 26 | 29 | 20 | 12 | 0 | 0 | 0 | 0 | 0 | 0 | 6 | 14 | 29 |
| YEAR OF OCCURRENCE | | 1978 | 1978 | 1978 | 1982 | | | | | | | 1987 | 1975 | FEB 1978 |
| NORMAL NO. DAYS WITH: | | | | | | | | | | | | | | |
| SNOWFALL >= 1.0 | 30 | 3.5 | 2.7 | 2.2 | 0.3 | 0.0 | 0.0 | 0.0 | 0.0 | 0.0 | 0.0 | 0.5 | 1.9 | 11.1 |

## PRECIPITATION (inches) 2009 BOSTON (KBOS)

| YEAR | JAN | FEB | MAR | APR | MAY | JUN | JUL | AUG | SEP | OCT | NOV | DEC | ANNUAL |
|------|-----|-----|-----|-----|-----|-----|-----|-----|-----|-----|-----|-----|--------|
| 1980 | 0.74 | 0.88 | 5.37 | 4.36 | 2.30 | 3.05 | 2.20 | 1.55 | 0.82 | 4.14 | 3.01 | 0.97 | 29.39 |
| 1981 | 0.95 | 6.65 | 0.62 | 3.14 | 1.17 | 1.65 | 3.47 | 1.04 | 2.54 | 3.43 | 4.78 | 6.27 | 35.71 |
| 1982 | 4.69 | 2.66 | 2.17 | 3.42 | 2.58 | 13.20 | 4.22 | 2.22 | 1.57 | 3.19 | 3.42 | 1.27 | 44.61 |
| 1983 | 5.03 | 5.00 | 9.72 | 6.86 | 2.94 | 1.07 | 1.07 | 3.28 | 1.06 | 3.74 | 8.89 | 4.94 | 53.60 |
| 1984 | 2.31 | 7.81 | 6.82 | 4.43 | 8.77 | 3.06 | 4.43 | 1.60 | 1.22 | 5.18 | 1.68 | 2.93 | 50.24 |
| 1985 | 1.12 | 1.83 | 2.29 | 1.62 | 3.36 | 3.94 | 3.51 | 6.67 | 3.00 | 1.65 | 6.39 | 1.21 | 36.59 |
| 1986 | 3.42 | 2.83 | 3.42 | 1.59 | 1.31 | 7.74 | 3.96 | 3.32 | 1.08 | 3.27 | 6.01 | 6.38 | 44.33 |
| 1987 | 7.28 | 0.72 | 4.27 | 9.46 | 1.75 | 2.62 | 0.82 | 2.93 | 7.29 | 2.73 | 3.49 | 2.12 | 45.48 |
| 1988 | 2.50 | 3.93 | 3.52 | 1.47 | 2.86 | 1.29 | 7.62 | 1.11 | 1.29 | 1.60 | 6.57 | 1.02 | 34.78 |
| 1989 | 0.61 | 2.51 | 3.07 | 3.58 | 3.54 | 2.84 | 5.09 | 5.92 | 4.61 | 5.71 | 4.13 | 0.81 | 42.42 |
| 1990 | 3.78 | 3.60 | 1.71 | 5.94 | 6.53 | 0.69 | 4.08 | 6.57 | 1.67 | 7.36 | 1.39 | 3.18 | 46.50 |
| 1991 | 3.24 | 1.58 | 4.33 | 4.84 | 0.92 | 2.89 | 1.95 | 5.27 | 6.32 | 4.27 | 4.06 | 2.58 | 42.25 |
| 1992 | 3.11 | 2.28 | 3.59 | 2.34 | 1.40 | 4.61 | 2.66 | 4.25 | 3.46 | 1.62 | 6.14 | 8.26 | 43.72 |
| 1993 | 2.17 | 4.94 | 7.67 | 4.86 | 1.04 | 1.75 | 1.75 | 1.32 | 4.64 | 3.61 | 2.86 | 6.60 | 43.21 |
| 1994 | 5.22 | 2.95 | 7.49 | 2.25 | 5.35 | 0.86 | 1.80 | 7.03 | 4.58 | 0.41 | 4.31 | 5.37 | 47.62 |
| 1995 | 4.33 | 2.57 | 2.20 | 1.40 | 1.82 | 1.55 | 2.06 | 0.82 | 3.60 | 6.42 | 5.13 | 3.20 | 35.10 |
| 1996 | 7.44 | 3.17 | 2.36 | 4.38 | 2.73 | 1.03 | 5.23 | 1.54 | 6.09 | 10.66 | 2.29 | 5.76 | 52.68 |
| 1997 | 2.34 | 1.28 | 4.68 | 3.46 | 2.63 | 1.41 | 0.63 | 3.01 | 1.02 | 1.78 | 5.86 | 2.29 | 30.39 |
| 1998 | 4.76 | 5.54 | 4.15 | 3.58 | 6.84 | 11.58 | 2.47 | 3.37 | 3.03 | 5.38 | 1.38 | 1.59 | 53.67 |
| 1999 | 5.69 | 3.51 | 2.52 | 0.83 | 2.70 | T | 3.51 | 1.33 | 9.86 | 4.30 | 2.14 | 1.52 | 37.91 |
| 2000 | 2.62 | 2.55 | 3.59 | 5.02 | 2.88 | 6.61 | 5.20 | 2.22 | 2.87 | 2.86 | 4.51 | 4.67 | 45.60 |
| 2001 | 1.58 | 1.37 | 7.57 | 0.88 | 1.23 | 4.99 | 2.13 | 4.14 | 2.29 | 0.98 | 0.73 | 2.83 | 30.72 |
| 2002 | 3.14 | 1.81 | 3.52 | 2.61 | 4.48 | 4.77 | 1.42 | 2.13 | 3.39 | 3.47 | 5.03 | 5.30 | 41.07 |
| 2003 | 1.81 | 4.21 | 4.00 | 4.00 | 4.12 | 4.69 | 2.11 | 2.89 | 2.65 | 6.20 | 2.63 | 5.06 | 44.37 |
| 2004 | 1.01 | 1.45 | 3.38 | 9.57 | 3.07 | 1.95 | 3.87 | 4.38 | 7.44 | 1.88 | 2.91 | 3.66 | 44.57 |
| 2005 | 4.45 | 2.70 | 3.89 | 3.17 | 3.98 | 1.46 | 3.37 | 2.88 | 1.78 | 9.41 | 3.71 | 2.87 | 43.67 |
| 2006 | 4.55 | 2.64 | 0.56 | 1.83 | 12.48 | 10.09 | 3.58 | 3.20 | 1.72 | 4.50 | 5.80 | 1.89 | 52.84 |
| 2007 | 2.57 | 2.20 | 4.31 | 6.71 | 3.70 | 2.12 | 5.26 | 0.66 | 1.81 | 2.08 | 2.80 | 5.25 | 39.47 |
| 2008 | 2.69 | 7.94 | 4.66 | 2.98 | 2.73 | 3.46 | 6.00 | 4.47 | 6.45 | 1.41 | 4.57 | 7.10 | 54.46 |
| 2009 | 3.35 | 1.94 | 2.51 | 4.13 | 2.69 | 3.22 | 6.90 | 3.24 | 3.09 | 5.17 | 3.34 | 3.91 | 43.49 |
| POR=<br>90 YRS | 3.59 | 3.32 | 3.87 | 3.68 | 3.34 | 3.39 | 3.10 | 3.36 | 3.30 | 3.33 | 3.92 | 3.88 | 42.08 |

WBAN : 14739

## AVERAGE TEMPERATURE (°F) 2009 BOSTON (KBOS)

| YEAR | JAN | FEB | MAR | APR | MAY | JUN | JUL | AUG | SEP | OCT | NOV | DEC | ANNUAL |
|------|-----|-----|-----|-----|-----|-----|-----|-----|-----|-----|-----|-----|--------|
| 1980 | 29.4 | 27.9 | 36.9 | 48.7 | 59.4 | 66.3 | 75.8 | 74.2 | 67.0 | 52.4 | 41.2 | 28.6 | 50.7 |
| 1981 | 21.4 | 36.4 | 39.1 | 51.7 | 60.4 | 70.7 | 74.6 | 72.1 | 63.7 | 51.2 | 43.9 | 33.2 | 51.5 |
| 1982 | 22.9 | 30.8 | 38.7 | 48.2 | 57.8 | 63.3 | 74.9 | 70.3 | 64.1 | 54.2 | 47.6 | 39.6 | 51.0 |
| 1983 | 31.2 | 32.8 | 40.6 | 49.1 | 58.2 | 70.7 | 78.0 | 73.6 | 70.6 | 55.2 | 46.1 | 32.1 | 53.2 |
| 1984 | 26.7 | 37.6 | 31.9 | 46.1 | 58.0 | 70.5 | 74.7 | 74.6 | 62.1 | 53.3 | 44.6 | 39.5 | 51.6 |
| 1985 | 24.4 | 32.8 | 40.4 | 49.3 | 59.3 | 64.8 | 73.5 | 70.4 | 65.4 | 55.4 | 45.4 | 31.3 | 51.0 |
| 1986 | 31.4 | 28.9 | 40.7 | 48.4 | 58.4 | 66.1 | 71.0 | 70.5 | 63.2 | 54.0 | 42.3 | 35.5 | 50.9 |
| 1987 | 28.9 | 29.1 | 38.5 | 45.1 | 57.2 | 65.1 | 71.7 | 70.3 | 65.4 | 54.3 | 43.9 | 36.1 | 50.5 |
| 1988 | 27.8 | 32.2 | 39.2 | 46.8 | 57.6 | 68.5 | 73.7 | 75.5 | 64.6 | 50.8 | 46.7 | 32.8 | 51.4 |
| 1989 | 34.5 | 30.5 | 37.3 | 45.9 | 59.4 | 67.8 | 72.8 | 71.6 | 64.7 | 55.3 | 42.8 | 21.7 | 50.4 |
| 1990 | 36.4 | 34.1 | 40.1 | 47.6 | 54.9 | 66.6 | 73.1 | 73.3 | 64.6 | 58.3 | 48.5 | 40.7 | 53.2 |
| 1991 | 29.4 | 36.1 | 41.6 | 51.3 | 63.3 | 70.0 | 74.6 | 73.8 | 63.7 | 56.4 | 45.2 | 36.0 | 53.5 |
| 1992 | 31.0 | 32.4 | 35.4 | 46.4 | 55.6 | 67.8 | 69.5 | 70.4 | 63.9 | 52.5 | 42.9 | 34.8 | 50.2 |
| 1993 | 32.4 | 27.1 | 36.4 | 48.3 | 60.3 | 69.5 | 74.7 | 73.6 | 64.8 | 52.3 | 45.6 | 34.2 | 51.6 |
| 1994 | 22.2 | 26.9 | 38.2 | 51.4 | 58.4 | 71.9 | 77.5 | 72.4 | 64.2 | 55.5 | 49.0 | 38.5 | 52.2 |
| 1995 | 34.6 | 28.5 | 38.8 | 46.1 | 57.2 | 68.6 | 75.9 | 72.8 | 63.1 | 58.4 | 41.9 | 31.7 | 51.5 |
| 1996 | 30.1 | 30.9 | 36.5 | 47.9 | 57.4 | 68.1 | 71.8 | 70.9 | 64.2 | 53.2 | 40.3 | 39.3 | 50.9 |
| 1997 | 29.2 | 36.0 | 36.7 | 46.3 | 56.1 | 68.2 | 73.7 | 71.2 | 64.2 | 52.8 | 41.7 | 35.2 | 50.9 |
| 1998 | 33.9 | 35.3 | 41.5 | 49.4 | 60.3 | 64.7 | 74.4 | 72.5 | 66.3 | 54.5 | 44.6 | 39.1 | 53.0 |
| 1999 | 29.5 | 33.6 | 39.4 | 49.2 | 58.2 | 71.0 | 75.7 | 71.3 | 67.1 | 53.0 | 48.0 | 37.3 | 52.8 |
| 2000 | 27.5 | 34.2 | 43.3 | 47.3 | 57.2 | 67.3 | 70.0 | 70.3 | 63.5 | 53.9 | 43.8 | 29.2 | 50.6 |
| 2001 | 30.0 | 31.8 | 35.3 | 48.7 | 59.6 | 71.1 | 69.9 | 73.9 | 65.3 | 56.2 | 48.3 | 40.5 | 52.6 |
| 2002 | 36.7 | 36.3 | 40.2 | 49.7 | 57.7 | 66.5 | 75.2 | 75.3 | 68.4 | 52.4 | 42.9 | 33.2 | 52.9 |
| 2003 | 24.1 | 26.4 | 37.7 | 44.5 | 55.1 | 65.4 | 74.3 | 74.3 | 65.7 | 53.0 | 45.9 | 36.1 | 50.2 |
| 2004 | 20.7 | 33.0 | 39.4 | 49.5 | 58.9 | 66.5 | 71.0 | 72.1 | 65.2 | 53.8 | 44.4 | 35.0 | 50.8 |
| 2005 | 26.8 | 31.2 | 35.2 | 49.7 | 52.2 | 68.4 | 73.3 | 74.4 | 67.9 | 54.7 | 45.9 | 32.5 | 51.0 |
| 2006 | 36.5 | 31.0 | 38.7 | 49.8 | 57.1 | 68.1 | 76.0 | 71.4 | 64.6 | 54.3 | 49.1 | 41.1 | 53.1 |
| 2007 | 32.8 | 26.2 | 37.5 | 45.2 | 61.4 | 68.5 | 72.9 | 72.7 | 67.7 | 59.2 | 43.1 | 32.4 | 51.6 |
| 2008 | 33.4 | 32.5 | 38.3 | 49.4 | 57.7 | 70.3 | 75.0 | 70.1 | 65.3 | 53.4 | 43.4 | 35.7 | 52.0 |
| 2009 | 24.9 | 32.9 | 37.5 | 50.5 | 59.5 | 63.4 | 70.5 | 73.6 | 63.2 | 51.9 | 48.9 | 33.2 | 50.8 |
| POR=<br>90 YRS | 29.2 | 29.8 | 38.2 | 47.5 | 58.2 | 67.0 | 73.2 | 71.8 | 64.1 | 54.6 | 44.4 | 33.8 | 51.0 |

## HEATING DEGREE DAYS (base 65°F) 2009  BOSTON (KBOS)

| YEAR | JUL | AUG | SEP | OCT | NOV | DEC | JAN | FEB | MAR | APR | MAY | JUN | TOTAL |
|------|-----|-----|-----|-----|-----|-----|-----|-----|-----|-----|-----|-----|-------|
| 1980-81 | 2 | 5 | 72 | 387 | 706 | 1120 | 1344 | 794 | 796 | 393 | 200 | 7 | 5826 |
| 1981-82 | 2 | 6 | 91 | 419 | 628 | 979 | 1300 | 948 | 811 | 496 | 231 | 113 | 6024 |
| 1982-83 | 2 | 19 | 71 | 338 | 515 | 783 | 1040 | 896 | 749 | 478 | 223 | 22 | 5136 |
| 1983-84 | 0 | 8 | 42 | 327 | 561 | 1012 | 1182 | 790 | 1020 | 563 | 239 | 36 | 5780 |
| 1984-85 | 3 | 0 | 142 | 359 | 605 | 781 | 1255 | 897 | 758 | 471 | 204 | 71 | 5546 |
| 1985-86 | 3 | 11 | 65 | 298 | 580 | 1035 | 1035 | 1008 | 746 | 490 | 258 | 66 | 5595 |
| 1986-87 | 21 | 16 | 98 | 344 | 674 | 904 | 1112 | 997 | 814 | 588 | 285 | 76 | 5929 |
| 1987-88 | 8 | 18 | 57 | 326 | 626 | 888 | 1145 | 945 | 792 | 541 | 253 | 61 | 5660 |
| 1988-89 | 9 | 10 | 64 | 443 | 541 | 992 | 938 | 959 | 853 | 565 | 196 | 51 | 5621 |
| 1989-90 | 2 | 4 | 88 | 294 | 660 | 1336 | 880 | 857 | 762 | 524 | 307 | 60 | 5774 |
| 1990-91 | 4 | 5 | 84 | 236 | 496 | 744 | 1096 | 803 | 721 | 407 | 126 | 35 | 4757 |
| 1991-92 | 1 | 8 | 111 | 273 | 586 | 894 | 1049 | 937 | 913 | 552 | 317 | 37 | 5678 |
| 1992-93 | 21 | 14 | 109 | 386 | 656 | 930 | 1002 | 1056 | 880 | 493 | 167 | 31 | 5745 |
| 1993-94 | 3 | 1 | 89 | 387 | 579 | 946 | 1320 | 1059 | 827 | 404 | 226 | 7 | 5848 |
| 1994-95 | 1 | 3 | 61 | 288 | 479 | 813 | 932 | 1016 | 804 | 561 | 258 | 30 | 5246 |
| 1995-96 | 1 | 2 | 113 | 214 | 685 | 1023 | 1074 | 981 | 875 | 510 | 260 | 20 | 5758 |
| 1996-97 | 1 | 6 | 80 | 358 | 739 | 792 | 1104 | 806 | 868 | 551 | 269 | 87 | 5661 |
| 1997-98 | 5 | 3 | 83 | 383 | 693 | 917 | 955 | 826 | 736 | 462 | 192 | 66 | 5321 |
| 1998-99 | 0 | 2 | 36 | 321 | 606 | 798 | 1092 | 872 | 790 | 468 | 215 | 29 | 5229 |
| 1999-00 | 2 | 7 | 39 | 363 | 505 | 851 | 1152 | 887 | 665 | 523 | 250 | 74 | 5318 |
| 2000-01 | 6 | 2 | 108 | 341 | 628 | 1103 | 1079 | 925 | 914 | 488 | 217 | 18 | 5829 |
| 2001-02 | 2 | 1 | 67 | 286 | 492 | 753 | 869 | 798 | 763 | 459 | 235 | 81 | 4806 |
| 2002-03 | 1 | 6 | 30 | 407 | 653 | 980 | 1264 | 1077 | 840 | 614 | 301 | 77 | 6250 |
| 2003-04 | 2 | 5 | 27 | 365 | 566 | 891 | 1369 | 924 | 784 | 462 | 210 | 52 | 5657 |
| 2004-05 | 2 | 3 | 59 | 339 | 617 | 926 | 1178 | 941 | 914 | 456 | 388 | 66 | 5889 |
| 2005-06 | 10 | 0 | 32 | 325 | 565 | 1000 | 878 | 947 | 810 | 449 | 258 | 55 | 5329 |
| 2006-07 | 0 | 11 | 64 | 331 | 470 | 735 | 991 | 1079 | 846 | 593 | 174 | 50 | 5344 |
| 2007-08 | 0 | 12 | 50 | 203 | 652 | 1004 | 972 | 934 | 822 | 462 | 232 | 27 | 5370 |
| 2008-09 | 0 | 0 | 78 | 356 | 641 | 902 | 1236 | 894 | 847 | 447 | 191 | 78 | 5670 |
| 2009- | 13 | 5 | 87 | 398 | 479 | 981 | | | | | | | |

WBAN : 14739

## COOLING DEGREE DAYS (base 65°F) 2009  BOSTON (KBOS)

| YEAR | JAN | FEB | MAR | APR | MAY | JUN | JUL | AUG | SEP | OCT | NOV | DEC | TOTAL |
|------|-----|-----|-----|-----|-----|-----|-----|-----|-----|-----|-----|-----|-------|
| 1980 | 0 | 0 | 0 | 0 | 18 | 114 | 347 | 299 | 137 | 1 | 0 | 0 | 916 |
| 1981 | 0 | 0 | 0 | 0 | 67 | 185 | 306 | 232 | 60 | 0 | 0 | 0 | 850 |
| 1982 | 0 | 0 | 0 | 0 | 15 | 67 | 314 | 192 | 49 | 10 | 2 | 0 | 649 |
| 1983 | 0 | 0 | 0 | 7 | 18 | 200 | 410 | 283 | 217 | 27 | 0 | 0 | 1162 |
| 1984 | 0 | 0 | 0 | 3 | 31 | 207 | 312 | 306 | 62 | 3 | 0 | 0 | 924 |
| 1985 | 0 | 0 | 0 | 5 | 30 | 72 | 271 | 183 | 83 | 8 | 0 | 0 | 652 |
| 1986 | 0 | 0 | 0 | 0 | 60 | 105 | 211 | 190 | 55 | 10 | 0 | 0 | 631 |
| 1987 | 0 | 0 | 0 | 0 | 48 | 87 | 221 | 189 | 76 | 0 | 2 | 0 | 623 |
| 1988 | 0 | 0 | 0 | 0 | 31 | 173 | 287 | 342 | 59 | 11 | 0 | 0 | 903 |
| 1989 | 0 | 0 | 1 | 0 | 29 | 142 | 248 | 214 | 89 | 0 | 0 | 0 | 723 |
| 1990 | 0 | 0 | 0 | 10 | 2 | 116 | 261 | 268 | 77 | 34 | 8 | 0 | 776 |
| 1991 | 0 | 0 | 0 | 3 | 79 | 189 | 304 | 287 | 79 | 15 | 0 | 0 | 956 |
| 1992 | 0 | 0 | 0 | 0 | 30 | 126 | 165 | 189 | 83 | 5 | 0 | 0 | 598 |
| 1993 | 0 | 0 | 0 | 0 | 28 | 173 | 310 | 273 | 89 | 0 | 2 | 0 | 875 |
| 1994 | 0 | 0 | 0 | 1 | 29 | 221 | 395 | 241 | 44 | 1 | 4 | 0 | 936 |
| 1995 | 0 | 0 | 0 | 0 | 24 | 147 | 344 | 252 | 64 | 15 | 0 | 0 | 846 |
| 1996 | 0 | 0 | 0 | 3 | 30 | 120 | 222 | 196 | 63 | 0 | 1 | 0 | 635 |
| 1997 | 0 | 0 | 0 | 0 | 0 | 189 | 280 | 202 | 66 | 12 | 0 | 0 | 749 |
| 1998 | 0 | 0 | 16 | 2 | 52 | 59 | 298 | 240 | 83 | 1 | 0 | 0 | 751 |
| 1999 | 0 | 0 | 0 | 0 | 11 | 215 | 344 | 213 | 110 | 0 | 0 | 0 | 893 |
| 2000 | 0 | 0 | 0 | 0 | 15 | 151 | 167 | 174 | 71 | 5 | 0 | 0 | 583 |
| 2001 | 0 | 0 | 0 | 9 | 55 | 210 | 160 | 283 | 83 | 20 | 0 | 0 | 820 |
| 2002 | 0 | 0 | 0 | 8 | 18 | 133 | 324 | 330 | 138 | 21 | 0 | 0 | 972 |
| 2003 | 0 | 0 | 0 | 3 | 3 | 94 | 298 | 300 | 57 | 0 | 0 | 0 | 755 |
| 2004 | 0 | 0 | 0 | 3 | 24 | 104 | 198 | 231 | 72 | 0 | 0 | 0 | 632 |
| 2005 | 0 | 0 | 0 | 5 | 1 | 176 | 274 | 299 | 126 | 13 | 0 | 0 | 894 |
| 2006 | 0 | 0 | 0 | 0 | 20 | 158 | 341 | 218 | 60 | 4 | 0 | 0 | 801 |
| 2007 | 0 | 0 | 0 | 4 | 68 | 159 | 253 | 257 | 135 | 31 | 0 | 0 | 907 |
| 2008 | 0 | 0 | 0 | 2 | 13 | 195 | 318 | 165 | 95 | 1 | 0 | 0 | 789 |
| 2009 | 0 | 0 | 0 | 18 | 28 | 33 | 191 | 281 | 40 | 0 | 0 | 0 | 591 |

## SNOWFALL (inches) 2009 BOSTON (KBOS)

| YEAR | JUL | AUG | SEP | OCT | NOV | DEC | JAN | FEB | MAR | APR | MAY | JUN | TOTAL |
|------|-----|-----|-----|-----|-----|-----|-----|-----|-----|-----|-----|-----|-------|
| 1980-81 | 0.0 | 0.0 | 0.0 | 0.0 | 2.4 | 5.6 | 11.9 | 1.9 | 0.5 | 0.0 | 0.0 | 0.0 | 22.3 |
| 1981-82 | 0.0 | 0.0 | 0.0 | 0.0 | T | 17.6 | 18.0 | 7.6 | 5.3 | 13.3 | 0.0 | 0.0 | 61.8 |
| 1982-83 | 0.0 | 0.0 | 0.0 | 0.0 | T | 5.5 | 4.7 | 22.3 | 0.2 | T | 0.0 | 0.0 | 32.7 |
| 1983-84 | 0.0 | 0.0 | 0.0 | 0.0 | T | 2.6 | 21.1 | 0.3 | 19.0 | T | 0.0 | 0.0 | 43.0 |
| 1984-85 | 0.0 | 0.0 | 0.0 | 0.0 | T | 3.7 | 7.0 | 10.2 | 3.7 | 2.0 | 0.0 | 0.0 | 26.6 |
| 1985-86 | 0.0 | 0.0 | 0.0 | 0.0 | 3.0 | 1.3 | 0.8 | 10.4 | 2.6 | T | 0.0 | 0.0 | 18.1 |
| 1986-87 | 0.0 | 0.0 | 0.0 | 0.0 | 3.5 | 3.4 | 24.3 | 3.7 | 3.5 | 4.1 | 0.0 | 0.0 | 42.5 |
| 1987-88 | 0.0 | 0.0 | 0.0 | 0.0 | 9.0 | 7.5 | 17.0 | 14.1 | 5.0 | T | 0.0 | 0.0 | 52.6 |
| 1988-89 | 0.0 | 0.0 | 0.0 | T | 0.0 | 3.7 | 1.5 | 6.7 | 3.2 | 0.4 | 0.0 | 0.0 | 15.5 |
| 1989-90 | 0.0 | 0.0 | 0.0 | 0.0 | 4.5 | 6.2 | 7.0 | 16.9 | 4.1 | 0.5 | 0.0 | 0.0 | 39.2 |
| 1990-91 | 0.0 | 0.0 | 0.0 | 0.0 | T | 1.2 | 11.7 | 2.8 | 3.4 | 0.0 | 0.0 | 0.0 | 19.1 |
| 1991-92 | 0.0 | 0.0 | 0.0 | 0.0 | T | 5.8 | 0.4 | 4.0 | 10.8 | 1.0 | 0.0 | 0.0 | 22.0 |
| 1992-93 | 0.0 | 0.0 | 0.0 | 0.0 | 0.6 | 9.7 | 12.9 | 19.6 | 38.9 | 2.2 | 0.0 | 0.0 | 83.9 |
| 1993-94 | 0.0 | 0.0 | 0.0 | 0.0 | T | 11.6 | 33.7 | 36.2 | 14.8 | 0.0 | 0.0 | 0.0 | 96.3 |
| 1994-95 | 0.0 | T | 0.0 | 0.0 | 0.1 | 1.5 | 4.4 | 8.5 | 0.4 | T | 0.0 | 0.0 | 14.9 |
| 1995-96 | 0.0 | 0.0 | 0.0 | 0.0 | 4.1 | 24.1 | 39.8 | 15.5 | 16.8 | 7.3 | 0.0 | 0.0 | 107.6 |
| 1996-97 | 0.0 | 0.0 | 0.0 | 0.0 | 1.8 | 5.0 | 9.7 | 4.8 | 8.2 | 22.4 | 0.0 | 0.0 | 51.9 |
| 1997-98 | 0.0 | 0.0 | 0.0 | 0.0 | 3.9 | 7.8 | 10.0 | 0.3 | 3.6 | T | 0.0 | 0.0 | 25.6 |
| 1998-99 | 0.0 | 0.0 | 0.0 | 0.0 | 0.0 | 0.8 | 16.8 | 7.4 | 11.4 | T | 0.0 | 0.0 | 36.4 |
| 1999-00 | 0.0 | 0.0 | 0.0 | 0.0 | 0.0 | 0.0 | 13.7 | 9.2 | 2.0 | T | 0.0 | 0.0 | 24.9 |
| 2000-01 | 0.0 | 0.0 | 0.0 | T | T | 4.5 | 12.4 | 9.8 | 19.2 | T | 0.0 | 0.0 | 45.9 |
| 2001-02 | 0.0 | 0.0 | 0.0 | T | 0.0 | 5.0 | 7.9 | 0.5 | 1.4 | 0.3 | 0.0 | T | 15.1 |
| 2002-03 | 0.0 | 0.0 | 0.0 | T | 3.6 | 11.1 | 4.2 | 41.6 | 8.1 | 2.3 | 0.0 | 0.0 | 70.9 |
| 2003-04 | 0.0 | 0.0 | 0.0 | T | 0.0 | 21.5 | 4.9 | 2.4 | 10.6 | T | 0.0 | 0.0 | 39.4 |
| 2004-05 | T | 0.0 | 0.0 | 0.0 | 3.9 | 7.2 | 43.3 | 17.7 | 14.5 | T | 0.0 | 0.0 | 86.6 |
| 2005-06 | 0.0 | 0.0 | 0.0 | 1.1 | T | 10.7 | 7.6 | 20.0 | T | T | 0.0 | 0.0 | 39.4 |
| 2006-07 | 0.0 | 0.0 | 0.0 | 0.0 | 0.0 | 0.8 | 1.0 | 4.6 | 10.2 | 0.5 | 0.0 | 0.0 | 17.1 |
| 2007-08 | 0.0 | 0.0 | 0.0 | 0.0 | T | 26.9 | 8.3 | 15.0 | 1.0 | 0.0 | 0.0 | 0.0 | 51.2 |
| 2008-09 | 0.0 | 0.0 | 0.0 | 0.0 | T | 25.3 | 23.7 | 6.2 | 10.7 | 0.0 | 0.0 | 0.0 | 65.9 |
| 2009- | 0.0 | 0.0 | 0.0 | 0.1 | 0.0 | 15.2 | | | | | | | |
| POR=<br>87 YRS | T | T | 0.0 | T | 1.2 | 7.8 | 12.5 | 12.4 | 7.5 | 1.1 | 0.0 | T | 42.5 |

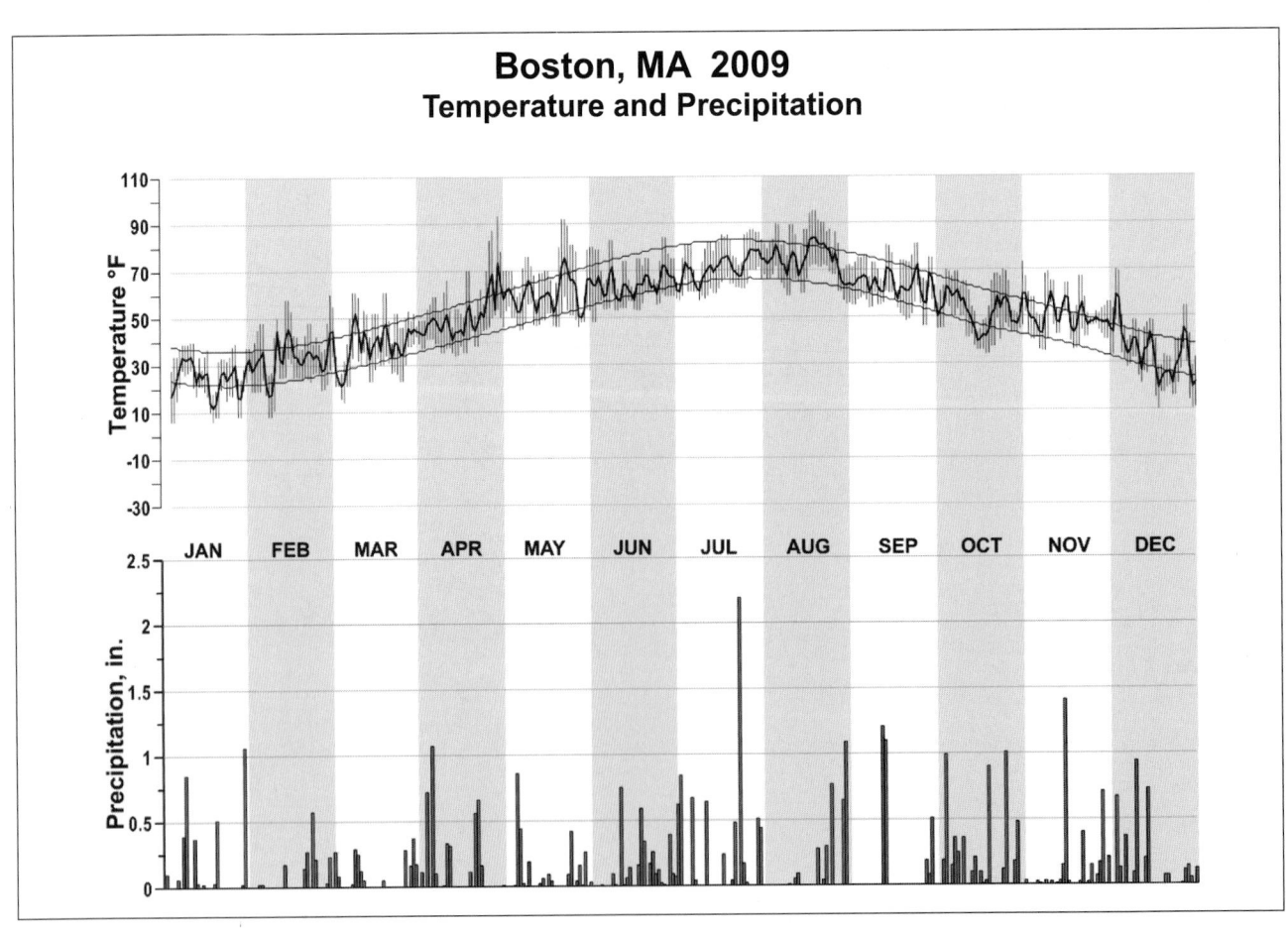

**Boston, MA 2009**
**Temperature and Precipitation**

# 2009
# WORCESTER
# MASSACHUSETTS (KORH)

Worcester Municipal Airport is located on the crest of a hill, 1,000 feet above sea level. It is about 500 feet above and 3 1/2 miles northwest of the city proper. The airport is surrounded by ridges and valleys with many of the valleys containing reservoirs. Only two of the ridges extend above the airport elevation. One is 400 feet higher and 2 1/2 miles to the northwest, and the other is 1,000 feet higher and 15 miles to the north.

The proximity to the Atlantic Ocean, Long Island Sound, and the Berkshire Hills plays an important part in determining the weather and, hence, the climate of Worcester. Rapid weather changes occur when storms move up the east coast after developing off the Carolina Coast. In the majority of these cases, they pass to the south and east, resulting in northeast and easterly winds with rain or snow and fog. Storms developing in the Texas-Oklahoma area normally travel up the St. Lawrence River Valley and, depending on the movement and intensity, usually deposit little precipitation over the area. However, they do bring an influx of warm air into the region. Wintertime cold snaps are quite frequent, but temperatures are usually modified by the passage of the air over land and mountains before reaching the county. Summertime thunderstorms develop over the hills to the west, with a majority moving toward the northeast. From the use of radar, we find many break up just before reaching Worcester, or pass either north or south of the city proper.

Airport site temperatures are moderate. The normal mean for the warmest month, July, is around 70 degrees. Though winters are reasonably cold, prolonged periods of severe cold weather are extremely rare. The three coldest months, December through February, have an average temperature of over 25 degrees. A review of Worcester Cooperative records since 1901 shows maximum temperatures above 100 degrees and minimum temperatures below -24 degrees.

Precipitation is usually plentiful and well distributed throughout the year. The annual snowfall for all Worcester sites since 1901, averages slightly less than 60 inches. The airport location averages slightly higher.

Based on the 1951-1980 period, the average first occurrence of 32 degrees Fahrenheit in the fall is October 17 and the average last occurrence in the spring is April 27.

# NORMALS, MEANS, AND EXTREMES
## WORCESTER (KORH)

| LATITUDE: 42° 16'N | LONGITUDE: -71° 52'W | ELEVATION (FT): GRND: 1003  BARO: 966 | TIME ZONE: EASTERN  (UTC -5) | WBAN: 94746 |
|---|---|---|---|---|

| | ELEMENT | POR | JAN | FEB | MAR | APR | MAY | JUN | JUL | AUG | SEP | OCT | NOV | DEC | YEAR |
|---|---|---|---|---|---|---|---|---|---|---|---|---|---|---|---|
| **TEMPERATURE °F** | NORMAL DAILY MAXIMUM | 30 | 31.4 | 34.1 | 43.0 | 54.4 | 66.3 | 74.4 | 79.3 | 77.1 | 69.0 | 58.4 | 47.1 | 36.2 | 55.9 |
| | MEAN DAILY MAXIMUM | 61 | 31.5 | 33.6 | 41.9 | 54.7 | 66.0 | 74.1 | 79.2 | 77.4 | 69.3 | 59.1 | 47.3 | 35.7 | 55.8 |
| | HIGHEST DAILY MAXIMUM | 53 | 66 | 67 | 84 | 91 | 92 | 95 | 96 | 96 | 91 | 85 | 78 | 72 | 96 |
| | YEAR OF OCCURRENCE | | 2007 | 1985 | 1998 | 1976 | 1962 | 2004 | 1988 | 1975 | 2007 | 1963 | 1982 | 1998 | JUL 1988 |
| | MEAN OF EXTREME MAXS. | 61 | 52.7 | 52.7 | 64.7 | 77.0 | 83.8 | 88.1 | 89.8 | 87.9 | 83.8 | 75.9 | 66.4 | 56.8 | 73.3 |
| | NORMAL DAILY MINIMUM | 30 | 15.8 | 17.8 | 25.6 | 35.5 | 46.2 | 55.0 | 60.8 | 59.5 | 51.3 | 40.7 | 32.0 | 21.6 | 38.5 |
| | MEAN DAILY MINIMUM | 61 | 16.4 | 17.8 | 25.5 | 35.9 | 45.9 | 55.0 | 61.0 | 59.7 | 51.7 | 41.6 | 32.4 | 21.4 | 38.7 |
| | LOWEST DAILY MINIMUM | 54 | -19 | -12 | -4 | 11 | 28 | 36 | 43 | 38 | 30 | 20 | 6 | -13 | -19 |
| | YEAR OF OCCURRENCE | | 1957 | 1967 | 1986 | 1982 | 1970 | 1986 | 1988 | 1965 | 1992 | 1969 | 1989 | 1962 | JAN 1957 |
| | MEAN OF EXTREME MINS. | 61 | -2.4 | -0.6 | 7.7 | 22.9 | 34.7 | 43.3 | 51.7 | 48.7 | 38.3 | 27.7 | 17.4 | 3.1 | 24.4 |
| | NORMAL DRY BULB | 30 | 23.6 | 26.0 | 34.3 | 45.0 | 56.3 | 64.7 | 70.1 | 68.3 | 60.2 | 49.6 | 39.6 | 28.9 | 47.2 |
| | MEAN DRY BULB | 61 | 23.9 | 25.7 | 33.7 | 45.3 | 55.9 | 64.7 | 70.1 | 68.5 | 60.5 | 50.3 | 39.9 | 28.6 | 47.3 |
| | MEAN WET BULB | 26 | 21.2 | 22.5 | 28.7 | 38.2 | 47.9 | 57.7 | 63.1 | 62.2 | 55.5 | 44.7 | 35.6 | 25.8 | 41.9 |
| | MEAN DEW POINT | 26 | 16.7 | 16.4 | 23.3 | 32.4 | 43.1 | 54.0 | 60.2 | 59.5 | 52.6 | 40.6 | 31.2 | 21.1 | 37.6 |
| | NORMAL NO. DAYS WITH: | | | | | | | | | | | | | | |
| | MAXIMUM >= 90 | 30 | 0.0 | 0.0 | 0.0 | 0.1 | 0.1 | 0.6 | 1.6 | 0.7 | * | 0.0 | 0.0 | 0.0 | 3.1 |
| | MAXIMUM <= 32 | 30 | 17.2 | 13.1 | 4.4 | 0.3 | 0.0 | 0.0 | 0.0 | 0.0 | 0.0 | 0.0 | 1.8 | 11.2 | 48.0 |
| | MINIMUM <= 32 | 30 | 29.1 | 25.9 | 24.2 | 9.9 | 0.4 | 0.0 | 0.0 | 0.0 | 0.1 | 4.5 | 16.8 | 27.3 | 138.2 |
| | MINIMUM <= 0 | 30 | 2.9 | 1.4 | 0.1 | 0.0 | 0.0 | 0.0 | 0.0 | 0.0 | 0.0 | 0.0 | 0.0 | 0.6 | 5.0 |
| **H/C** | NORMAL HEATING DEG. DAYS | 30 | 1284 | 1094 | 952 | 601 | 278 | 74 | 9 | 20 | 158 | 478 | 764 | 1119 | 6831 |
| | NORMAL COOLING DEG. DAYS | 30 | 0 | 0 | 0 | 0 | 7 | 64 | 166 | 122 | 12 | 0 | 0 | 0 | 371 |
| **RH** | NORMAL (PERCENT) | 30 | 66 | 64 | 63 | 61 | 65 | 70 | 71 | 73 | 74 | 70 | 69 | 69 | 68 |
| | HOUR 01 LST | 30 | 68 | 67 | 68 | 69 | 75 | 80 | 81 | 83 | 84 | 78 | 75 | 72 | 75 |
| | HOUR 07 LST | 30 | 72 | 72 | 72 | 69 | 72 | 76 | 78 | 81 | 83 | 79 | 77 | 74 | 75 |
| | HOUR 13 LST | 30 | 60 | 58 | 57 | 55 | 54 | 58 | 59 | 61 | 62 | 57 | 60 | 61 | 58 |
| | HOUR 19 LST | 30 | 66 | 63 | 60 | 58 | 61 | 67 | 70 | 74 | 76 | 70 | 69 | 68 | 67 |
| **S** | PERCENT POSSIBLE SUNSHINE | | | | | | | | | | | | | | |
| **W/O** | MEAN NO. DAYS WITH: | | | | | | | | | | | | | | |
| | HEAVY FOG(VISBY <= 1/4 MI) | 41 | 5.4 | 5.1 | 7.0 | 6.1 | 6.1 | 6.8 | 5.2 | 4.8 | 6.7 | 6.5 | 6.6 | 6.8 | 73.1 |
| | THUNDERSTORMS | 50 | 0.1 | 0.1 | 0.4 | 0.8 | 2.0 | 3.5 | 4.4 | 3.0 | 1.1 | 0.6 | 0.3 | 0.1 | 16.4 |
| **CLOUDNESS** | MEAN: | | | | | | | | | | | | | | |
| | SUNRISE-SUNSET (OKTAS) | | | | | | | | | | | | | | |
| | MIDNIGHT-MIDNIGHT (OKTAS) | | | | | | | | | | | | | | |
| | MEAN NO. DAYS WITH: | | | | | | | | | | | | | | |
| | CLEAR | | | | | | | | | | | | | | |
| | PARTLY CLOUDY | 1 | 1.0 | | 1.0 | | 2.0 | | 12.1 | | | | | | |
| | CLOUDY | 1 | 1.0 | 1.0 | 1.0 | | | | 13.0 | | | | | | |
| **PR** | MEAN STATION PRESSURE(IN) | 26 | 28.88 | 28.88 | 28.88 | 28.86 | 28.89 | 28.88 | 28.91 | 28.95 | 28.98 | 28.97 | 28.94 | 28.90 | 28.91 |
| | MEAN SEA-LEVEL PRES. (IN) | 26 | 30.00 | 30.00 | 30.00 | 29.96 | 29.98 | 29.96 | 29.98 | 30.03 | 30.07 | 30.07 | 30.05 | 30.03 | 30.01 |
| **WINDS** | MEAN SPEED (MPH) | 26 | 11.3 | 11.7 | 11.4 | 10.8 | 10.1 | 9.4 | 8.9 | 8.3 | 8.9 | 10.0 | 10.7 | 11.4 | 10.2 |
| | PREVAIL.DIR(TENS OF DEGS) | 24 | 29 | 29 | 30 | 30 | 28 | 26 | 27 | 26 | 27 | 26 | 29 | 29 | 29 |
| | MAXIMUM 2-MINUTE: | | | | | | | | | | | | | | |
| | SPEED (MPH) | 13 | 43 | 49 | 52 | 43 | 51 | 41 | 40 | 43 | 41 | 39 | 43 | 45 | 52 |
| | DIR. (TENS OF DEGS) | | 17 | 28 | 30 | 21 | 31 | 25 | 31 | 31 | 32 | 25 | 28 | 27 | 30 |
| | YEAR OF OCCURRENCE | | 2006 | 1996 | 2005 | 2000 | 1998 | 2008 | 2008 | 2002 | 1999 | 2006 | 2003 | 2000 | MAR 2005 |
| | MAXIMUM 3-SECOND | | | | | | | | | | | | | | |
| | SPEED (MPH) | 13 | 57 | 70 | 61 | 58 | 94 | 68 | 49 | 56 | 55 | 53 | 58 | 71 | 94 |
| | DIR. (TENS OF DEGS) | | 30 | 29 | 29 | 19 | 26 | 22 | 30 | | 31 | 28 | 26 | 07 | 26 |
| | YEAR OF OCCURRENCE | | 2000 | 1996 | 2008 | 2000 | 1998 | 2008 | 2008 | 2006 | 2002 | 2003 | 2003 | 2008 | MAY 1998 |
| **PRECIPITATION** | NORMAL (IN) | 30 | 4.07 | 3.10 | 4.23 | 3.92 | 4.35 | 4.02 | 4.19 | 4.09 | 4.27 | 4.67 | 4.34 | 3.80 | 49.05 |
| | MAXIMUM MONTHLY (IN) | 54 | 11.16 | 9.68 | 7.96 | 8.79 | 9.94 | 12.17 | 10.81 | 8.01 | 13.13 | 15.56 | 10.40 | 9.83 | 15.56 |
| | YEAR OF OCCURRENCE | | 1979 | 2008 | 1972 | 1987 | 1984 | 1982 | 2009 | 1991 | 1974 | 2005 | 1972 | 1973 | OCT 2005 |
| | MINIMUM MONTHLY (IN) | 54 | 0.89 | 0.25 | 0.50 | 0.75 | 0.86 | 0.32 | 0.74 | 1.01 | 0.69 | 0.70 | 0.67 | 0.74 | 0.25 |
| | YEAR OF OCCURRENCE | | 1970 | 1987 | 2006 | 2001 | 1959 | 1999 | 1987 | 2007 | 1986 | 2001 | 1976 | 1989 | FEB 1987 |
| | MAXIMUM IN 24 HOURS (IN) | 54 | 2.97 | 2.79 | 4.56 | 3.94 | 3.03 | 3.98 | 3.87 | 5.00 | 4.91 | 5.06 | 2.98 | 3.00 | 5.06 |
| | YEAR OF OCCURRENCE | | 1978 | 2008 | 1987 | 2007 | 1967 | 1986 | 3087 | 1991 | 2008 | 2005 | 1972 | 1986 | OCT 2005 |
| | NORMAL NO. DAYS WITH: | | | | | | | | | | | | | | |
| | PRECIPITATION >= 0.01 | 30 | 11.9 | 10.6 | 11.9 | 11.9 | 12.3 | 11.3 | 9.7 | 9.1 | 9.5 | 8.9 | 10.8 | 12.2 | 130.1 |
| | PRECIPITATION >= 1.00 | 30 | 0.9 | 0.9 | 1.0 | 0.9 | 1.1 | 0.9 | 1.2 | 1.2 | 1.1 | 1.5 | 1.2 | 1.0 | 12.9 |
| **SNOWFALL** | NORMAL (IN) | 30 | 15.9 | 13.6 | 11.3 | 3.3 | 0.5 | 0.0 | 0.0 | 0.0 | 0.* | 0.3 | 4.2 | 11.7 | 60.8 |
| | MAXIMUM MONTHLY (IN) | 47 | 50.9 | 45.2 | 44.1 | 21.0 | 12.7 | 0.1 | 0.0 | T | T | 7.5 | 20.7 | 37.0 | 50.9 |
| | YEAR OF OCCURRENCE | | 1987 | 1962 | 1993 | 1987 | 1977 | 1992 | | 1994 | 1994 | 1979 | 1971 | 1992 | JAN 1987 |
| | MAXIMUM IN 24 HOURS (IN) | 47 | 18.7 | 24.0 | 19.7 | 17.0 | 12.7 | 0.1 | 0.0 | T | T | 7.5 | 14.8 | 28.1 | 28.1 |
| | YEAR OF OCCURRENCE' | | 1961 | 1962 | 1993 | 1987 | 1977 | 1992 | | 1994 | 1994 | 1979 | 1971 | 1992 | DEC 1992 |
| | MAXIMUM SNOW DEPTH (IN) | 42 | 35 | 42 | 33 | 17 | 10 | 0 | 0 | 0 | 0 | 8 | 15 | 29 | 42 |
| | YEAR OF OCCURRENCE | | 1966 | 1961 | 1969 | 1987 | 1977 | | | | | 1979 | 1971 | 1992 | FEB 1961 |
| | NORMAL NO. DAYS WITH: | | | | | | | | | | | | | | |
| | SNOWFALL >= 1.0 | 30 | 4.1 | 3.2 | 2.8 | 0.8 | 0.1 | 0.0 | 0.0 | 0.0 | 0.0 | 0.0 | 1.2 | 3.2 | 15.4 |

## PRECIPITATION (inches) 2009 WORCESTER (KORH)

| YEAR | JAN | FEB | MAR | APR | MAY | JUN | JUL | AUG | SEP | OCT | NOV | DEC | ANNUAL |
|------|-----|-----|-----|-----|-----|-----|-----|-----|-----|-----|-----|-----|--------|
| 1980 | 0.95 | 0.73 | 6.86 | 4.77 | 2.23 | 4.55 | 3.59 | 1.95 | 1.82 | 6.16 | 4.58 | 1.06 | 39.25 |
| 1981 | 0.93 | 8.37 | 0.74 | 3.85 | 4.48 | 2.45 | 7.90 | 1.03 | 4.66 | 5.49 | 3.13 | 5.94 | 48.97 |
| 1982 | 5.00 | 3.22 | 3.67 | 4.30 | 2.96 | 12.17 | 3.61 | 3.36 | 2.69 | 2.67 | 4.32 | 1.70 | 49.67 |
| 1983 | 4.85 | 4.67 | 7.84 | 8.59 | 5.97 | 2.56 | 1.32 | 6.26 | 1.38 | 5.77 | 8.75 | 6.37 | 64.33 |
| 1984 | 2.44 | 5.78 | 5.47 | 4.23 | 9.94 | 2.85 | 5.69 | 1.17 | 1.68 | 3.99 | 2.71 | 2.84 | 48.79 |
| 1985 | 1.16 | 2.72 | 2.89 | 1.26 | 5.46 | 5.24 | 6.35 | 3.74 | 3.77 | 3.12 | 6.41 | 1.93 | 44.05 |
| 1986 | 5.56 | 3.14 | 2.93 | 1.59 | 3.14 | 7.21 | 4.83 | 3.20 | 0.69 | 2.72 | 5.63 | 7.25 | 47.89 |
| 1987 | 5.52 | 0.25 | 6.57 | 8.79 | 1.55 | 4.55 | 0.74 | 4.61 | 6.37 | 4.18 | 2.77 | 1.85 | 47.75 |
| 1988 | 2.71 | 2.78 | 3.46 | 3.45 | 4.47 | 1.25 | 6.27 | 2.19 | 2.70 | 3.66 | 7.91 | 1.42 | 42.27 |
| 1989 | 1.18 | 2.47 | 2.66 | 4.25 | 6.17 | 5.27 | 5.67 | 5.65 | 4.71 | 8.21 | 4.00 | 0.74 | 50.98 |
| 1990 | 3.75 | 3.88 | 1.52 | 4.78 | 7.65 | 1.74 | 2.44 | 6.84 | 1.73 | 10.19 | 2.41 | 5.46 | 52.39 |
| 1991 | 2.98 | 2.08 | 4.92 | 5.04 | 4.16 | 3.06 | 2.78 | 8.01 | 6.40 | 3.44 | 5.47 | 2.89 | 51.23 |
| 1992 | 3.01 | 2.51 | 4.15 | 2.59 | 2.54 | 4.68 | 5.25 | 4.83 | 3.58 | 2.36 | 4.94 | 4.61 | 45.05 |
| 1993 | 2.56 | 2.38 | 5.46 | 4.00 | 1.79 | 2.36 | 3.34 | 1.90 | 8.85 | 3.88 | 4.85 | 5.11 | 46.48 |
| 1994 | 4.78 | 1.86 | 5.38 | 2.73 | 5.87 | 2.48 | 3.09 | 7.64 | 4.84 | 1.24 | 4.54 | 4.81 | 49.26 |
| 1995 | 3.71 | 2.86 | 1.85 | 2.19 | 2.39 | 1.51 | 4.33 | 2.02 | 3.15 | 8.65 | 4.61 | 1.04 | 38.31 |
| 1996 | 6.30 | 2.66 | 2.20 | 6.70 | 3.34 | 3.01 | 6.50 | 3.99 | 6.07 | 5.81 | 2.93 | 6.91 | 56.42 |
| 1997 | 3.25 | 1.71 | 4.66 | 3.22 | 2.72 | 1.60 | 2.97 | 4.34 | 1.44 | 2.11 | 5.50 | 2.32 | 35.84 |
| 1998 | 4.59 | 3.17 | 5.82 | 3.30 | 5.89 | 9.68 | 1.76 | 2.38 | 1.69 | 4.93 | 2.28 | 1.46 | 46.95 |
| 1999 | 6.01 | 3.38 | 4.09 | 0.92 | 2.77 | 0.32 | 4.14 | 1.87 | 8.81 | 3.57 | 3.38 | 2.55 | 41.81 |
| 2000 | 3.11 | 2.59 | 3.82 | 6.85 | 3.51 | 5.84 | 4.04 | 2.09 | 3.01 | 2.05 | 3.61 | 3.62 | 44.14 |
| 2001 | 1.64 | 2.40 | 6.53 | 0.75 | 2.26 | 6.27 | 1.91 | 2.31 | 3.42 | 0.70 | 1.36 | 2.77 | 32.32 |
| 2002 | 2.36 | 1.43 | 4.21 | 3.67 | 5.55 | 4.83 | 2.65 | 2.94 | 3.97 | 4.39 | 3.83 | 4.24 | 44.07 |
| 2003 | 2.36 | 4.43 | 4.06 | 3.43 | 4.12 | 6.16 | 3.06 | 5.34 | 4.25 | 5.41 | 2.19 | 5.71 | 50.52 |
| 2004 | 1.43 | 1.45 | 3.35 | 6.57 | 3.23 | 1.45 | 4.84 | 5.07 | 7.52 | 2.22 | 3.93 | 4.78 | 45.84 |
| 2005 | 5.84 | 3.03 | 4.18 | 6.49 | 3.71 | 1.77 | 5.02 | 2.64 | 2.83 | 15.56 | 4.77 | 3.74 | 59.58 |
| 2006 | 5.23 | 3.51 | 0.50 | 2.35 | 6.64 | 6.91 | 6.91 | 4.07 | 2.37 | 6.91 | 7.09 | 2.49 | 51.29 |
| 2007 | 3.11 | 1.73 | 4.70 | 8.30 | 5.12 | 2.16 | 4.30 | 1.01 | 1.98 | 3.12 | 3.24 | 4.57 | 43.34 |
| 2008 | 2.45 | 9.68 | 5.62 | 4.24 | 2.45 | 5.55 | 7.97 | 3.53 | 9.22 | 2.62 | 4.25 | 5.63 | 63.21 |
| 2009 | 3.49 | 1.91 | 2.81 | 3.80 | 2.96 | 6.51 | 10.81 | 2.81 | 1.87 | 5.02 | 3.41 | 4.68 | 50.08 |
| POR= 61 YRS | 3.68 | 3.23 | 4.04 | 4.05 | 4.02 | 3.79 | 3.95 | 4.02 | 4.01 | 4.32 | 4.38 | 3.95 | 47.44 |

WBAN : 94746

## AVERAGE TEMPERATURE (°F) 2009 WORCESTER (KORH)

| YEAR | JAN | FEB | MAR | APR | MAY | JUN | JUL | AUG | SEP | OCT | NOV | DEC | ANNUAL |
|------|-----|-----|-----|-----|-----|-----|-----|-----|-----|-----|-----|-----|--------|
| 1980 | 25.0 | 22.2 | 33.1 | 46.0 | 57.0 | 61.5 | 71.3 | 69.9 | 61.1 | 47.3 | 36.4 | 24.2 | 46.3 |
| 1981 | 15.9 | 32.8 | 34.7 | 47.4 | 57.9 | 66.0 | 70.7 | 67.4 | 58.2 | 46.7 | 39.1 | 27.8 | 47.1 |
| 1982 | 17.1 | 25.7 | 33.8 | 43.6 | 58.3 | 61.5 | 71.5 | 65.9 | 61.3 | 50.8 | 43.8 | 34.8 | 47.3 |
| 1983 | 26.1 | 29.1 | 36.6 | 46.6 | 54.7 | 67.9 | 72.3 | 70.2 | 64.9 | 50.3 | 42.0 | 27.3 | 49.0 |
| 1984 | 22.7 | 34.0 | 29.2 | 45.6 | 55.4 | 68.1 | 70.0 | 71.4 | 59.6 | 53.9 | 41.1 | 34.9 | 48.8 |
| 1985 | 20.1 | 28.6 | 37.2 | 47.4 | 57.8 | 61.4 | 70.4 | 67.9 | 62.0 | 51.7 | 40.9 | 25.3 | 47.6 |
| 1986 | 25.9 | 23.1 | 36.4 | 48.1 | 57.9 | 62.6 | 68.2 | 66.8 | 59.4 | 49.6 | 37.1 | 30.3 | 47.1 |
| 1987 | 24.0 | 24.5 | 36.8 | 45.7 | 57.6 | 65.8 | 70.7 | 66.6 | 60.6 | 48.8 | 39.4 | 30.8 | 47.6 |
| 1988 | 23.0 | 26.6 | 35.0 | 43.8 | 57.3 | 64.6 | 72.6 | 72.3 | 60.2 | 45.5 | 41.7 | 27.2 | 47.5 |
| 1989 | 28.5 | 24.4 | 33.0 | 42.4 | 57.0 | 64.0 | 68.7 | 67.3 | 60.6 | 50.7 | 36.5 | 15.1 | 45.7 |
| 1990 | 31.3 | 28.8 | 36.2 | 43.9 | 51.5 | 64.6 | 69.2 | 68.8 | 59.3 | 53.0 | 42.1 | 34.0 | 48.6 |
| 1991 | 23.8 | 30.3 | 36.5 | 48.2 | 61.2 | 64.7 | 70.0 | 69.1 | 58.7 | 51.9 | 40.2 | 29.7 | 48.7 |
| 1992 | 25.3 | 27.1 | 30.7 | 42.3 | 55.5 | 63.5 | 65.9 | 65.5 | 59.2 | 46.8 | 37.8 | 28.7 | 45.7 |
| 1993 | 26.0 | 19.6 | 31.3 | 45.7 | 57.8 | 64.5 | 70.4 | 70.0 | 59.5 | 47.2 | 39.7 | 28.9 | 46.7 |
| 1994 | 15.9 | 21.1 | 33.5 | 47.7 | 54.3 | 67.2 | 72.9 | 66.3 | 59.0 | 50.9 | 43.5 | 33.3 | 47.1 |
| 1995 | 29.8 | 23.0 | 36.1 | 42.6 | 54.4 | 65.5 | | | 59.4 | 53.5 | 35.5 | 24.1 | |
| 1996 | 23.4 | 24.9 | 31.6 | 45.3 | 54.9 | 65.9 | | 68.9 | 61.1 | 49.7 | 35.9 | 33.2 | |
| 1997 | 23.6 | 0.0 | 31.9 | 43.5 | 52.2 | 65.8 | 69.1 | 67.6 | 60.3 | 48.8 | 36.2 | 30.0 | 44.1 |
| 1998 | | 32.9 | 37.0 | 47.2 | 59.7 | 62.9 | 70.0 | 70.3 | 63.1 | 50.3 | 40.2 | 35.7 | |
| 1999 | 26.0 | 31.3 | 36.3 | 46.7 | 58.1 | 68.2 | 72.4 | 68.5 | 64.6 | 49.1 | 44.5 | 32.9 | 49.9 |
| 2000 | 23.0 | 30.0 | 40.4 | 44.7 | 56.2 | 64.6 | 66.5 | 66.9 | 59.5 | 50.1 | 39.2 | 23.7 | 47.1 |
| 2001 | 24.7 | 26.1 | 31.1 | 46.8 | 58.1 | 67.0 | 66.6 | 72.2 | 62.2 | 52.5 | 44.5 | 35.4 | 48.9 |
| 2002 | 31.8 | 32.1 | 36.0 | 47.7 | 54.2 | 63.4 | 70.8 | 71.1 | 64.7 | 47.7 | 37.6 | 28.1 | 48.8 |
| 2003 | 17.7 | 21.6 | 35.0 | 42.5 | 54.0 | 63.8 | 70.6 | 70.7 | 62.5 | 48.7 | 43.4 | 31.3 | 46.8 |
| 2004 | 15.1 | 27.4 | 35.9 | 47.3 | 58.4 | 64.1 | 69.0 | 68.9 | 62.7 | 49.9 | 40.6 | 29.4 | 47.4 |
| 2005 | 22.2 | 28.6 | 30.9 | 49.3 | 51.4 | 68.6 | 71.5 | 72.5 | 65.4 | 51.5 | 42.1 | 27.2 | 48.4 |
| 2006 | 32.4 | 26.6 | 35.7 | 49.3 | 55.6 | 66.1 | 73.4 | 68.7 | 60.8 | 50.0 | 45.9 | 36.4 | 50.1 |
| 2007 | 28.4 | 20.7 | 33.5 | 43.4 | 59.4 | 65.4 | 71.0 | 70.1 | 64.4 | 56.5 | 38.2 | 27.5 | 48.2 |
| 2008 | 27.9 | 26.9 | 34.3 | 49.4 | 55.0 | 67.8 | 72.6 | 68.1 | 63.4 | 49.8 | 38.5 | 29.2 | 48.6 |
| 2009 | 18.9 | 28.2 | 35.3 | 48.4 | 57.1 | 61.9 | 66.8 | 70.4 | 60.0 | 47.9 | 44.6 | 27.9 | 47.3 |
| POR= 61 YRS | 23.9 | 25.7 | 33.7 | 45.3 | 55.9 | 64.7 | 70.1 | 68.5 | 60.6 | 50.3 | 39.9 | 28.6 | 47.3 |

## HEATING DEGREE DAYS (base 65°F) 2009  WORCESTER (KORH)

| YEAR | JUL | AUG | SEP | OCT | NOV | DEC | JAN | FEB | MAR | APR | MAY | JUN | TOTAL |
|---|---|---|---|---|---|---|---|---|---|---|---|---|---|
| 1980-81 | 2 | 19 | 165 | 540 | 853 | 1259 | 1516 | 894 | 934 | 521 | 241 | 42 | 6986 |
| 1981-82 | 1 | 29 | 204 | 562 | 772 | 1145 | 1478 | 1094 | 961 | 639 | 212 | 127 | 7224 |
| 1982-83 | 7 | 55 | 140 | 436 | 632 | 929 | 1199 | 996 | 871 | 548 | 318 | 44 | 6175 |
| 1983-84 | 5 | 20 | 115 | 459 | 682 | 1163 | 1303 | 892 | 1099 | 577 | 300 | 54 | 6669 |
| 1984-85 | 9 | 1 | 184 | 338 | 713 | 928 | 1382 | 1011 | 855 | 518 | 230 | 121 | 6290 |
| 1985-86 | 2 | 25 | 128 | 406 | 713 | 1223 | 1206 | 1166 | 879 | 501 | 262 | 119 | 6630 |
| 1986-87 | 48 | 49 | 182 | 471 | 830 | 1069 | 1266 | 1130 | 867 | 571 | 273 | 58 | 6814 |
| 1987-88 | 9 | 56 | 152 | 495 | 761 | 1052 | 1298 | 1107 | 922 | 629 | 253 | 115 | 6849 |
| 1988-89 | 19 | 32 | 155 | 597 | 693 | 1166 | 1123 | 1129 | 984 | 674 | 257 | 90 | 6919 |
| 1989-90 | 7 | 38 | 172 | 436 | 846 | 1540 | 1035 | 1006 | 884 | 632 | 413 | 58 | 7067 |
| 1990-91 | 22 | 30 | 192 | 379 | 683 | 957 | 1270 | 965 | 877 | 501 | 167 | 80 | 6123 |
| 1991-92 | 13 | 16 | 222 | 402 | 739 | 1090 | 1222 | 1092 | 1056 | 676 | 321 | 82 | 6931 |
| 1992-93 | 40 | 55 | 203 | 558 | 810 | 1117 | 1204 | 1259 | 1039 | 574 | 229 | 74 | 7162 |
| 1993-94 | 10 | 5 | 201 | 545 | 754 | 1115 | 1515 | 1226 | 968 | 515 | 337 | 49 | 7240 |
| 1994-95 | 5 | 48 | 182 | 431 | 636 | 975 | 1086 | 1170 | 887 | 667 | 329 | 58 | 6474 |
| 1995-96 |  |  | 191 | 352 | 877 | 1260 | 1279 | 1158 | 1031 | 587 | 331 | 36 |  |
| 1996-97 |  | 14 | 159 | 467 | 863 | 980 | 1277 |  | 1020 | 638 | 394 | 93 |  |
| 1997-98 | 15 | 18 | 158 | 498 | 858 | 1079 |  | 891 | 872 | 527 | 180 | 111 |  |
| 1998-99 | 3 | 8 | 98 | 450 | 736 | 902 | 1201 | 938 | 882 | 540 | 226 | 39 | 6023 |
| 1999-00 | 7 | 27 | 91 | 486 | 609 | 989 | 1295 | 1008 | 758 | 602 | 289 | 100 | 6261 |
| 2000-01 | 26 | 38 | 190 | 455 | 769 | 1273 | 1243 | 1081 | 1045 | 544 | 258 | 44 | 6966 |
| 2001-02 | 38 | 1 | 117 | 391 | 608 | 910 | 1022 | 913 | 892 | 527 | 337 | 125 | 5881 |
| 2002-03 | 13 | 28 | 73 | 537 | 816 | 1137 | 1456 | 1209 | 925 | 667 | 337 | 104 | 7302 |
| 2003-04 | 4 | 16 | 91 | 497 | 641 | 1040 | 1540 | 1083 | 897 | 526 | 220 | 88 | 6643 |
| 2004-05 | 11 | 22 | 94 | 459 | 728 | 1095 | 1320 | 1014 | 1047 | 466 | 411 | 61 | 6728 |
| 2005-06 | 14 | 0 | 59 | 420 | 682 | 1163 | 1005 | 1069 | 900 | 465 | 313 | 76 | 6166 |
| 2006-07 | 0 | 33 | 144 | 458 | 568 | 883 | 1129 | 1233 | 969 | 643 | 212 | 82 | 6354 |
| 2007-08 | 14 | 29 | 81 | 282 | 796 | 1155 | 1141 | 1098 | 948 | 464 | 303 | 39 | 6350 |
| 2008-09 | 1 | 10 | 101 | 468 | 790 | 1104 | 1421 | 1024 | 913 | 510 | 252 | 111 | 6705 |
| 2009- | 31 | 29 | 165 | 520 | 610 | 1148 |  |  |  |  |  |  |  |

WBAN : 94746

## COOLING DEGREE DAYS (base 65°F) 2009  WORCESTER (KORH)

| YEAR | JAN | FEB | MAR | APR | MAY | JUN | JUL | AUG | SEP | OCT | NOV | DEC | TOTAL |
|---|---|---|---|---|---|---|---|---|---|---|---|---|---|
| 1980 | 0 | 0 | 0 | 0 | 8 | 48 | 206 | 178 | 57 | 0 | 0 | 0 | 497 |
| 1981 | 0 | 0 | 0 | 0 | 29 | 78 | 184 | 110 | 10 | 0 | 0 | 0 | 411 |
| 1982 | 0 | 0 | 0 | 0 | 8 | 29 | 216 | 92 | 34 | 0 | 0 | 0 | 379 |
| 1983 | 0 | 0 | 0 | 4 | 4 | 138 | 238 | 188 | 118 | 11 | 0 | 0 | 701 |
| 1984 | 0 | 0 | 0 | 2 | 10 | 156 | 171 | 209 | 30 | 0 | 0 | 0 | 578 |
| 1985 | 0 | 0 | 0 | 0 | 15 | 22 | 177 | 120 | 47 | 0 | 0 | 0 | 381 |
| 1986 | 0 | 0 | 0 | 0 | 50 | 54 | 151 | 111 | 21 | 2 | 0 | 0 | 389 |
| 1987 | 0 | 0 | 0 | 0 | 49 | 90 | 193 | 110 | 25 | 0 | 0 | 0 | 467 |
| 1988 | 0 | 0 | 0 | 0 | 24 | 112 | 260 | 266 | 17 | 1 | 0 | 0 | 680 |
| 1989 | 0 | 0 | 0 | 0 | 16 | 70 | 131 | 116 | 47 | 0 | 0 | 0 | 380 |
| 1990 | 0 | 0 | 0 | 6 | 0 | 52 | 159 | 153 | 25 | 10 | 0 | 0 | 405 |
| 1991 | 0 | 0 | 0 | 4 | 57 | 79 | 173 | 149 | 40 | 2 | 0 | 0 | 504 |
| 1992 | 0 | 0 | 0 | 0 | 30 | 44 | 75 | 78 | 37 | 0 | 0 | 0 | 264 |
| 1993 | 0 | 0 | 0 | 0 | 12 | 67 | 181 | 163 | 44 | 0 | 0 | 0 | 467 |
| 1994 | 0 | 0 | 0 | 0 | 12 | 122 | 260 | 97 | 9 | 0 | 0 | 0 | 500 |
| 1995 | 0 | 0 | 0 | 0 | 6 | 77 |  |  | 29 | 3 | 0 | 0 |  |
| 1996 | 0 | 0 |  | 2 | 23 | 70 |  | 143 | 47 | 0 | 0 | 0 |  |
| 1997 | 0 |  | 0 | 0 | 0 | 123 | 148 | 106 | 25 | 2 | 0 | 0 |  |
| 1998 |  | 0 | 10 | 0 | 21 | 58 | 165 | 181 | 47 | 0 | 0 | 0 |  |
| 1999 | 0 | 0 | 0 | 0 | 19 | 142 | 245 | 140 | 86 | 0 | 0 | 0 | 632 |
| 2000 | 0 | 0 | 0 | 0 | 22 | 95 | 82 | 104 | 31 | 1 | 0 | 0 | 335 |
| 2001 | 0 | 0 | 0 | 4 | 51 | 109 | 95 | 232 | 41 | 7 | 0 | 0 | 539 |
| 2002 | 0 | 0 | 0 | 16 | 10 | 82 | 199 | 226 | 70 | 6 | 0 | 0 | 609 |
| 2003 | 0 | 0 | 0 | 0 | 4 | 76 | 185 | 200 | 23 | 0 | 0 | 0 | 488 |
| 2004 | 0 | 0 | 0 | 1 | 23 | 66 | 143 | 149 | 33 | 0 | 0 | 0 | 415 |
| 2005 | 0 | 0 | 0 | 2 | 0 | 177 | 219 | 242 | 80 | 8 | 0 | 0 | 728 |
| 2006 | 0 | 0 | 0 | 0 | 29 | 113 | 271 | 155 | 24 | 1 | 0 | 0 | 593 |
| 2007 | 0 | 0 | 0 | 4 | 45 | 103 | 206 | 195 | 70 | 27 | 0 | 0 | 650 |
| 2008 | 0 | 0 | 0 | 2 | 2 | 128 | 245 | 112 | 60 | 0 | 0 | 0 | 549 |
| 2009 | 0 | 0 | 0 | 18 | 15 | 25 | 94 | 202 | 20 | 0 | 0 | 0 | 374 |

## SNOWFALL (inches) 2009 WORCESTER (KORH)

| YEAR | JUL | AUG | SEP | OCT | NOV | DEC | JAN | FEB | MAR | APR | MAY | JUN | TOTAL |
|---|---|---|---|---|---|---|---|---|---|---|---|---|---|
| 1980-81 | 0.0 | 0.0 | 0.0 | 0.0 | 9.0 | 6.8 | 12.5 | 11.4 | 3.3 | T | T | 0.0 | 43.0 |
| 1981-82 | 0.0 | 0.0 | 0.0 | T | T | 24.6 | 16.7 | 6.5 | 11.0 | 15.1 | 0.0 | 0.0 | 73.9 |
| 1982-83 | 0.0 | 0.0 | 0.0 | 0.0 | 0.5 | 6.4 | 18.6 | 32.1 | 3.5 | 2.3 | T | 0.0 | 63.4 |
| 1983-84 | 0.0 | 0.0 | 0.0 | T | 1.1 | 17.2 | 24.1 | 3.3 | 30.9 | T | T | 0.0 | 76.6 |
| 1984-85 | 0.0 | 0.0 | 0.0 | 0.0 | T | 7.0 | 9.7 | 11.0 | 7.2 | 4.9 | 0.0 | 0.0 | 39.8 |
| 1985-86 | 0.0 | 0.0 | 0.0 | 0.0 | 6.9 | 9.1 | 5.8 | 14.5 | 2.3 | 0.1 | 0.3 | 0.0 | 39.0 |
| 1986-87 | 0.0 | 0.0 | 0.0 | 0.0 | 11.5 | 4.9 | 46.8 | 3.0 | 6.4 | 21.0 | 0.0 | 0.0 | 93.6 |
| 1987-88 | 0.0 | 0.0 | 0.0 | 0.0 | 10.2 | 12.9 | 25.2 | 15.8 | 6.4 | 0.6 | 0.0 | 0.0 | 71.1 |
| 1988-89 | 0.0 | 0.0 | 0.0 | 0.4 | T | 5.0 | 2.8 | 7.7 | 8.5 | 3.7 | 0.0 | 0.0 | 28.1 |
| 1989-90 | 0.0 | 0.0 | 0.0 | 0.0 | 7.9 | 10.2 | 11.3 | 15.2 | 6.4 | 2.1 | 0.0 | T | 53.1 |
| 1990-91 | 0.0 | 0.0 | 0.0 | 0.0 | 0.7 | 5.0 | 11.3 | 9.1 | 9.3 | 0.2 | 0.0 | 0.0 | 35.6 |
| 1991-92 | 0.0 | 0.0 | 0.0 | 0.0 | 5.8 | 14.5 | 2.7 | 8.4 | 11.8 | 2.7 | 0.0 | 0.1 | 46.0 |
| 1992-93 | 0.0 | 0.0 | T | 0.0 | 1.9 | 37.0 | 14.6 | 19.7 | 44.1 | 2.8 | 0.0 | T | 120.1 |
| 1993-94 | 0.0 | 0.0 | 0.0 | T | 0.2 | 12.9 | 34.1 | 25.9 | 27.1 | 0.0 | T | T | 100.2 |
| 1994-95 | 0.0 | T | T | 0.0 | 2.9 | 3.2 | 4.5 | 14.3 | T | T | 0.0 | 0.0 | 24.9 |
| 1995-96 | 0.0 | 0.0 | 0.0 | 0.0 | | | | | | | | | |
| 1996-97 | | | | | | | | | | | | | |
| 1997-98 | | | | | | | | | | | | | |
| 1998-99 | | | | | | | | | | | | | |
| 1999-00 | | | | | | | | | | | | | |
| 2000-01 | | | | | | | | | | | | | |
| 2001-02 | | | | | | | | | | | | | |
| 2002-03 | | | | | | | 22.0 | 38.9 | 8.3 | 5.1 | 0.0 | 0.0 | |
| 2003-04 | 0.0 | 0.0 | 0.0 | 1.5 | T | 21.5 | 13.0 | 5.9 | 7.6 | T | 0.0 | 0.0 | 49.5 |
| 2004-05 | 0.0 | 0.0 | 0.0 | 0.0 | 3.0 | 11.9 | 50.9 | 23.9 | 24.6 | 0.0 | 0.0 | T | 114.3 |
| 2005-06 | 0.0 | 0.0 | 0.0 | T | 3.2 | 18.0 | 24.1 | 18.9 | T | 1.5 | 0.0 | 0.0 | 65.7 |
| 2006-07 | 0.0 | 0.0 | 0.0 | 0.0 | 0.0 | 1.1 | 3.9 | 20.8 | 20.8 | 2.5 | 0.0 | 0.0 | 49.1 |
| 2007-08 | 0.0 | 0.0 | 0.0 | 0.0 | 0.7 | 27.1 | 12.3 | 22.6 | 7.4 | 0.0 | 0.0 | 0.0 | 70.1 |
| 2008-09 | 0.0 | 0.0 | 0.0 | T | T | 31.4 | 24.7 | 7.7 | 13.1 | T | 0.0 | 0.0 | 76.9 |
| 2009- | 0.0 | 0.0 | 0.0 | 2.4 | T | 20.7 | | | | | | | |
| POR= 61 YRS | 0.0 | T | T | 0.4 | 2.7 | 11.6 | 15.2 | 14.8 | 11.8 | 2.8 | 0.2 | T | 59.5 |

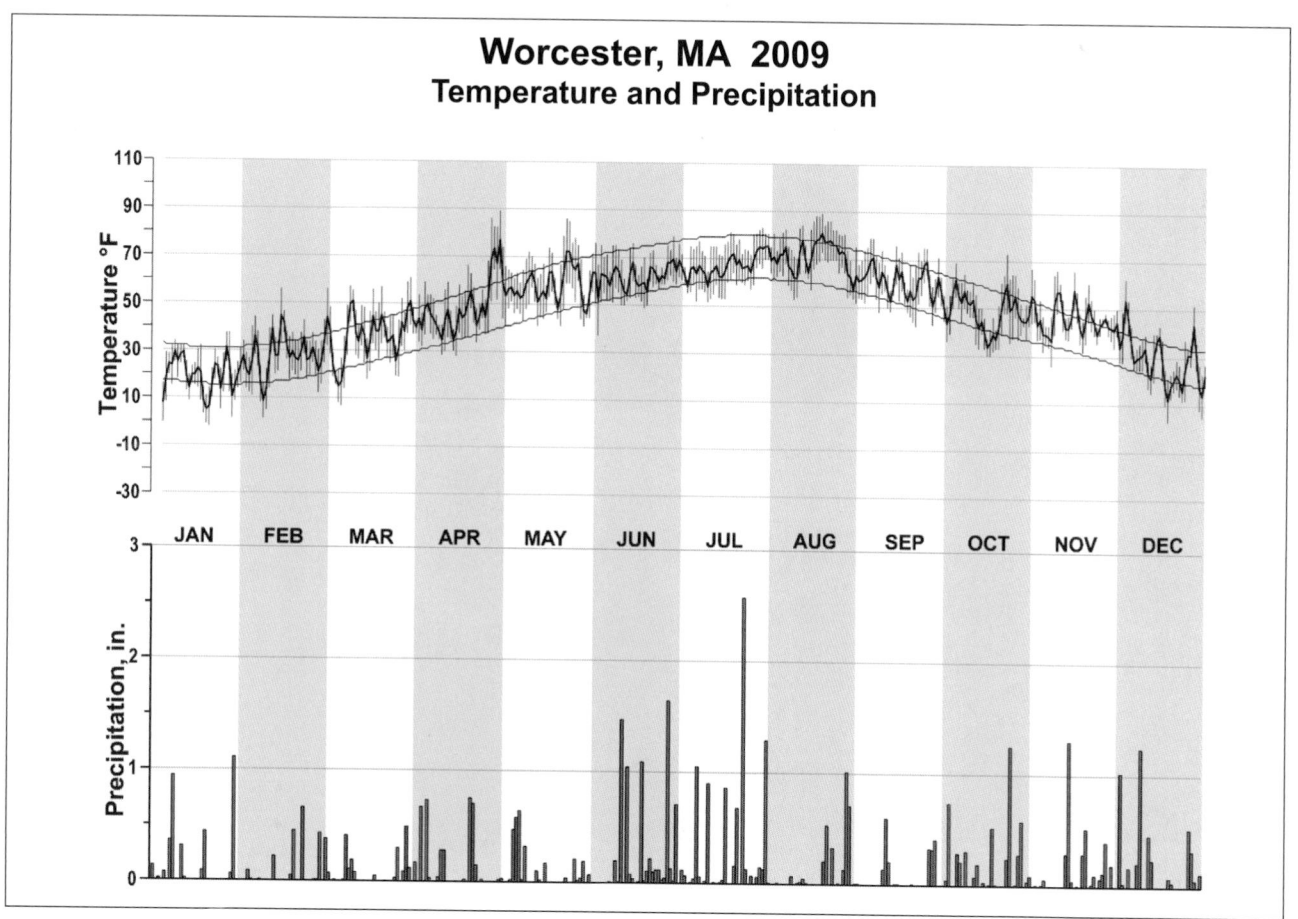

# Worcester, MA 2009
## Temperature and Precipitation

# 2009
# DETROIT
# MICHIGAN (KDTW)

Detroit and the immediate suburbs, including nearby urban areas in Canada, occupy an area approximately 25 miles in radius. The waterway, consisting of the Detroit and St. Clair Rivers, Lake St. Clair, and the west end of Lake Erie, lies at an elevation of 568 to 580 feet above sea level. Nearly flat land slopes up gently from the waters edge northwestward for about 10 miles and then gives way to increasingly rolling terrain. The Irish Hills, parallel to and about 40 miles northwest of the waterway, have tops 1,000 to 1,250 feet above sea level. On the Canadian side of the waterway the land is relatively level.

Northwest winds in winter bring snow flurry accumulations to all of Michigan except in the Detroit Metropolitan area while summer showers moving from the northwest weaken and sometimes dissipate as they approach Detroit. On the other hand, much of the heaviest precipitation in winter comes from southeast winds, especially to the northwest suburbs of the city.

The climate of Detroit is influenced by its location with respect to major storm tracks and the influence of the Great Lakes. The normal wintertime storm track is south of the city, which brings on the average, about 3 inch snowfalls. Winter storms can bring combinations of rain, snow, freezing rain, and sleet with heavy snowfall accumulations possible at times. In summer, most storms pass to the north allowing for intervals of warm, humid, sunny skies with occasional thunderstorms followed by days of mild, dry, and fair weather. Temperatures of 90 degrees or higher are reached during each summer.

The most pronounced lake effect occurs in the winter when arctic air moving across the lakes is warmed and moistened. This produces an excess of cloudiness but a moderation of cold wave temperatures.

Local climatic variations are due largely to the immediate effect of Lake St. Clair and the urban heat island. On warm days in late spring or early summer, lake breezes often lower temperatures by 10 to 15 degrees in the eastern part of the city and the northeastern suburbs. The urban heat island effect shows up mainly at night where minimum temperatures at the Metropolitan Airport average 4 degrees lower than downtown Detroit. On humid summer nights or on very cold winter nights, this difference can exceed 10 degrees.

The growing season averages 180 days and has ranged from 145 days to 205 days. On average, the last freezing temperature occurs in late April while the average first freezing temperature occurs in late October. A freeze has occurred as late as mid-May and as early as late September.

Air pollution comes primarily from heavy industry spread along both shores of the waterway from Port Huron to Toledo. However, wind dispersion is usually sufficient to keep it from becoming a major hazard.

# NORMALS, MEANS, AND EXTREMES
## DETROIT (KDTW)

LATITUDE: 42 ° 12'N    LONGITUDE: -83 ° 20'W    ELEVATION (FT): GRND: 631  BARO: 631    TIME ZONE: EASTERN  (UTC -5)    WBAN: 94847

| | ELEMENT | POR | JAN | FEB | MAR | APR | MAY | JUN | JUL | AUG | SEP | OCT | NOV | DEC | YEAR |
|---|---|---|---|---|---|---|---|---|---|---|---|---|---|---|---|
| **TEMPERATURE °F** | NORMAL DAILY MAXIMUM | 30 | 31.1 | 34.4 | 45.2 | 57.8 | 70.2 | 79.0 | 83.4 | 81.4 | 73.7 | 61.2 | 47.8 | 35.9 | 58.4 |
| | MEAN DAILY MAXIMUM | 51 | 31.1 | 34.1 | 44.6 | 58.3 | 69.7 | 79.0 | 83.1 | 81.4 | 74.2 | 61.6 | 48.2 | 35.6 | 58.4 |
| | HIGHEST DAILY MAXIMUM | 51 | 64 | 70 | 81 | 89 | 93 | 104 | 102 | 100 | 98 | 91 | 77 | 69 | 104 |
| | YEAR OF OCCURRENCE | | 2008 | 1999 | 2007 | 1977 | 1988 | 1988 | 1988 | 1988 | 1976 | 1963 | 1968 | 1998 | JUN 1988 |
| | MEAN OF EXTREME MAXS. | 51 | 50.3 | 52.8 | 69.0 | 79.7 | 85.9 | 91.9 | 93.5 | 91.8 | 88.6 | 79.9 | 67.4 | 54.9 | 75.5 |
| | NORMAL DAILY MINIMUM | 30 | 17.8 | 20.0 | 28.5 | 38.4 | 49.4 | 58.9 | 63.6 | 62.2 | 54.1 | 42.5 | 33.5 | 23.4 | 41.0 |
| | MEAN DAILY MINIMUM | 51 | 17.0 | 18.9 | 27.2 | 37.8 | 48.0 | 57.5 | 62.1 | 60.9 | 53.5 | 41.9 | 32.9 | 22.6 | 40.0 |
| | LOWEST DAILY MINIMUM | 51 | -21 | -15 | -4 | 10 | 25 | 36 | 41 | 38 | 29 | 17 | 9 | -10 | -21 |
| | YEAR OF OCCURRENCE | | 1984 | 1985 | 2003 | 1982 | 1966 | 1972 | 1965 | 1982 | 1974 | 1974 | 1969 | 1983 | JAN 1984 |
| | MEAN OF EXTREME MINS. | 51 | -2.4 | 0.5 | 9.7 | 23.7 | 34.4 | 44.4 | 50.5 | 49.3 | 38.2 | 27.4 | 18.3 | 3.4 | 24.8 |
| | NORMAL DRY BULB | 30 | 24.5 | 27.2 | 36.9 | 48.1 | 59.8 | 69.0 | 73.5 | 71.8 | 63.9 | 51.9 | 40.7 | 29.6 | 49.7 |
| | MEAN DRY BULB | 51 | 24.1 | 26.5 | 36.0 | 48.0 | 58.8 | 68.4 | 72.6 | 71.2 | 63.8 | 51.8 | 40.6 | 29.2 | 49.3 |
| | MEAN WET BULB | 26 | 23.1 | 24.6 | 31.6 | 41.6 | 51.7 | 61.0 | 65.0 | 64.4 | 57.7 | 46.5 | 36.9 | 27.2 | 44.3 |
| | MEAN DEW POINT | 26 | 19.7 | 20.9 | 27.2 | 36.7 | 47.5 | 57.5 | 62.0 | 61.8 | 54.7 | 43.1 | 33.4 | 24.2 | 40.7 |
| | NORMAL NO. DAYS WITH: | | | | | | | | | | | | | | |
| | MAXIMUM >= 90 | 30 | 0.0 | 0.0 | 0.0 | 0.0 | 0.5 | 2.8 | 5.0 | 2.9 | 0.8 | 0.0 | 0.0 | 0.0 | 12.0 |
| | MAXIMUM <= 32 | 30 | 16.7 | 12.9 | 4.1 | 0.2 | 0.0 | 0.0 | 0.0 | 0.0 | 0.0 | 0.0 | 1.4 | 10.3 | 45.6 |
| | MINIMUM <= 32 | 30 | 28.5 | 24.7 | 21.7 | 8.7 | 0.5 | 0.0 | 0.0 | 0.0 | 0.1 | 4.0 | 15.8 | 25.8 | 129.8 |
| | MINIMUM <= 0 | 30 | 3.1 | 2.0 | 0.1 | 0.0 | 0.0 | 0.0 | 0.0 | 0.0 | 0.0 | 0.0 | 0.0 | 1.2 | 6.4 |
| **H/C** | NORMAL HEATING DEG. DAYS | 30 | 1270 | 1074 | 886 | 527 | 219 | 41 | 5 | 12 | 121 | 426 | 742 | 1099 | 6422 |
| | NORMAL COOLING DEG. DAYS | 30 | 0 | 0 | 0 | 6 | 42 | 145 | 254 | 208 | 75 | 6 | 0 | 0 | 736 |
| **RH** | NORMAL (PERCENT) | 30 | 76 | 73 | 69 | 65 | 65 | 67 | 69 | 72 | 73 | 72 | 74 | 77 | 71 |
| | HOUR 01 LST | 30 | 79 | 78 | 75 | 73 | 75 | 79 | 81 | 84 | 84 | 80 | 79 | 80 | 79 |
| | HOUR 07 LST | 30 | 81 | 80 | 79 | 77 | 77 | 79 | 83 | 86 | 87 | 84 | 82 | 81 | 81 |
| | HOUR 13 LST | 30 | 70 | 65 | 60 | 53 | 53 | 55 | 55 | 57 | 57 | 58 | 65 | 70 | 60 |
| | HOUR 19 LST | 30 | 74 | 71 | 65 | 57 | 56 | 58 | 59 | 63 | 66 | 67 | 72 | 76 | 65 |
| **S** | PERCENT POSSIBLE SUNSHINE | 31 | 40 | 46 | 52 | 53 | 60 | 65 | 68 | 67 | 61 | 51 | 35 | 31 | 52 |
| **W/O** | MEAN NO. DAYS WITH: | | | | | | | | | | | | | | |
| | HEAVY FOG(VISBY <= 1/4 MI) | 46 | 2.3 | 2.3 | 2.1 | 1.0 | 0.8 | 0.5 | 0.5 | 1.0 | 1.4 | 1.5 | 1.5 | 2.9 | 17.8 |
| | THUNDERSTORMS | 51 | 0.2 | 0.4 | 1.5 | 2.9 | 3.9 | 6.1 | 6.1 | 5.3 | 3.7 | 1.1 | 0.7 | 0.3 | 32.2 |
| **CLOUDNESS** | MEAN: | | | | | | | | | | | | | | |
| | SUNRISE-SUNSET (OKTAS) | | | | | | | | | | | | | | |
| | MIDNIGHT-MIDNIGHT (OKTAS) | | | | | | | | | | | | | | |
| | MEAN NO. DAYS WITH: | | | | | | | | | | | | | | |
| | CLEAR | | | | | | | | | | | | | | |
| | PARTLY CLOUDY | | | | | | | | | | | | | | |
| | CLOUDY | | | | | | | | | | | | | | |
| **PR** | MEAN STATION PRESSURE(IN) | 26 | 29.33 | 29.37 | 29.32 | 29.26 | 29.26 | 29.26 | 29.28 | 29.32 | 29.34 | 29.35 | 29.33 | 29.34 | 29.31 |
| | MEAN SEA-LEVEL PRES. (IN) | 26 | 30.07 | 30.10 | 30.05 | 29.97 | 29.97 | 29.96 | 29.98 | 30.02 | 30.05 | 30.06 | 30.06 | 30.08 | 30.03 |
| **WINDS** | MEAN SPEED (MPH) | 26 | 11.4 | 10.8 | 10.8 | 10.8 | 9.6 | 8.7 | 8.3 | 7.6 | 8.0 | 9.4 | 10.7 | 10.9 | 9.8 |
| | PREVAIL.DIR(TENS OF DEGS) | 42 | 24 | 24 | 30 | 24 | 30 | 24 | 23 | 23 | 24 | 23 | 24 | 24 | 24 |
| | MAXIMUM 2-MINUTE: | | | | | | | | | | | | | | |
| | SPEED (MPH) | 14 | 46 | 51 | 46 | 47 | 61 | 45 | 53 | 44 | 35 | 47 | 47 | 49 | 61 |
| | DIR. (TENS OF DEGS) | | 24 | 22 | 23 | 22 | 22 | 30 | 28 | 24 | 27 | 22 | 27 | 29 | 22 |
| | YEAR OF OCCURRENCE | | 2008 | 1997 | 2004 | 2001 | 2004 | 2005 | 1998 | 2003 | 2001 | 2004 | 2003 | 1998 | MAY 2004 |
| | MAXIMUM 3-SECOND | | | | | | | | | | | | | | |
| | SPEED (MPH) | 14 | 56 | 60 | 59 | 57 | 78 | 60 | 67 | 53 | 45 | 56 | 58 | 60 | 78 |
| | DIR. (TENS OF DEGS) | | 27 | 24 | 24 | 24 | 22 | 23 | 28 | 23 | 28 | 24 | 25 | 31 | 22 |
| | YEAR OF OCCURRENCE | | 2008 | 2001 | 2004 | 1997 | 2004 | 2008 | 1998 | 2003 | 1997 | 2004 | 1998 | 1998 | MAY 2004 |
| **PRECIPITATION** | NORMAL (IN) | 30 | 1.91 | 1.88 | 2.52 | 3.05 | 3.05 | 3.55 | 3.16 | 3.10 | 3.27 | 2.23 | 2.66 | 2.51 | 32.89 |
| | MAXIMUM MONTHLY (IN) | 51 | 3.92 | 5.02 | 4.48 | 5.40 | 8.46 | 7.04 | 6.02 | 7.83 | 7.52 | 6.76 | 5.68 | 6.00 | 8.46 |
| | YEAR OF OCCURRENCE | | 1993 | 1990 | 1973 | 1961 | 2004 | 1987 | 1969 | 1975 | 1986 | 2001 | 1982 | 1965 | MAY 2004 |
| | MINIMUM MONTHLY (IN) | 51 | 0.27 | 0.15 | .74 | 0.69 | 0.87 | 0.97 | 0.59 | 0.27 | 0.43 | .13 | 0.62 | 0.46 | 0.13 |
| | YEAR OF OCCURRENCE | | 1961 | 1969 | 2005 | 2004 | 1988 | 1988 | 1974 | 2008 | 1960 | 2005 | 2009 | 1960 | OCT 2005 |
| | MAXIMUM IN 24 HOURS (IN) | 51 | 1.72 | 2.41 | 1.82 | 3.58 | 2.87 | 3.11 | 4.34 | 3.21 | 4.08 | 2.57 | 2.30 | 3.71 | 4.34 |
| | YEAR OF OCCURRENCE | | 1967 | 1998 | 1997 | 2000 | 1968 | 2009 | 1998 | 1964 | 2000 | 1985 | 2005 | 1965 | JUL 1998 |
| | NORMAL NO. DAYS WITH: | | | | | | | | | | | | | | |
| | PRECIPITATION >= 0.01 | 30 | 13.4 | 11.3 | 12.7 | 12.6 | 11.6 | 10.1 | 9.6 | 9.5 | 9.9 | 9.8 | 12.3 | 13.9 | 136.7 |
| | PRECIPITATION >= 1.00 | 30 | 0.1 | 0.2 | 0.3 | 0.4 | 0.6 | 0.9 | 0.8 | 0.7 | 0.6 | 0.3 | 0.3 | 0.2 | 5.3 |
| **SNOWFALL** | NORMAL (IN) | 30 | 11.9 | 9.3 | 7.0 | 1.7 | 0.* | 0.0 | 0.0 | 0.0 | 0.0 | 0.3 | 2.7 | 11.1 | 44.0 |
| | MAXIMUM MONTHLY (IN) | 50 | 29.6 | 24.2 | 21.0 | 9.0 | .1 | T | 0.0 | 0.0 | T | 2.9 | 11.8 | 34.9 | 34.9 |
| | YEAR OF OCCURRENCE | | 1978 | 2008 | 2008 | 1982 | 2005 | 2009 | | | 2009 | 1980 | 1966 | 1974 | DEC 1974 |
| | MAXIMUM IN 24 HOURS (IN) | 50 | 12.2 | 10.3 | 9.2 | 7.4 | .1 | T | 0.0 | 0.0 | T | 2.9 | 5.6 | 19.2 | 19.2 |
| | YEAR OF OCCURRENCE' | | 2005 | 1965 | 1973 | 1982 | 2005 | 2009 | | | 2009 | 1980 | 1977 | 1974 | DEC 1974 |
| | MAXIMUM SNOW DEPTH (IN) | 49 | 24 | 18 | 9 | 6 | 0 | 0 | 0 | 0 | 0 | 1 | 6 | 19 | 24 |
| | YEAR OF OCCURRENCE | | 1999 | 1982 | 1982 | 1982 | | | | | | 1980 | 1966 | 1974 | JAN 1999 |
| | NORMAL NO. DAYS WITH: | | | | | | | | | | | | | | |
| | SNOWFALL >= 1.0 | 30 | 3.6 | 2.9 | 2.1 | 0.5 | 0.0 | 0.0 | 0.0 | 0.0 | 0.0 | 0.1 | 0.9 | 3.5 | 13.6 |

## PRECIPITATION (inches) 2009 DETROIT (KDTW)

| YEAR | JAN | FEB | MAR | APR | MAY | JUN | JUL | AUG | SEP | OCT | NOV | DEC | ANNUAL |
|---|---|---|---|---|---|---|---|---|---|---|---|---|---|
| 1980 | 0.69 | 1.00 | 3.88 | 4.23 | 3.22 | 6.42 | 4.33 | 6.09 | 2.94 | 1.26 | 0.88 | 2.30 | 37.24 |
| 1981 | 0.57 | 3.13 | 0.82 | 3.44 | 2.60 | 3.33 | 4.29 | 2.32 | 5.47 | 3.92 | 1.26 | 2.38 | 33.53 |
| 1982 | 3.43 | 1.10 | 3.14 | 1.60 | 2.83 | 4.11 | 4.78 | 0.72 | 2.55 | 1.01 | 5.68 | 3.29 | 34.24 |
| 1983 | 0.84 | 0.89 | 1.87 | 4.20 | 5.47 | 4.88 | 4.53 | 1.57 | 2.49 | 2.85 | 4.28 | 3.78 | 37.65 |
| 1984 | 0.78 | 1.31 | 3.12 | 2.48 | 3.62 | 1.04 | 0.95 | 3.00 | 2.30 | 2.28 | 2.49 | 2.90 | 26.27 |
| 1985 | 2.63 | 3.83 | 4.42 | 2.11 | 3.11 | 1.62 | 3.96 | 4.88 | 2.59 | 3.91 | 5.51 | 1.51 | 40.08 |
| 1986 | 1.30 | 3.46 | 2.29 | 2.73 | 1.36 | 5.75 | 2.47 | 3.52 | 7.52 | 3.05 | 1.88 | 2.28 | 37.61 |
| 1987 | 2.35 | 0.53 | 2.19 | 2.14 | 2.50 | 7.04 | 2.20 | 6.87 | 2.69 | 2.00 | 3.17 | 4.60 | 38.28 |
| 1988 | 1.30 | 2.02 | 1.16 | 1.50 | 0.87 | 0.97 | 2.43 | 3.13 | 3.65 | 3.57 | 4.29 | 1.97 | 26.86 |
| 1989 | 1.28 | 0.77 | 2.16 | 2.22 | 4.16 | 3.79 | 4.21 | 2.14 | 3.03 | 1.73 | 2.53 | 1.24 | 29.26 |
| 1990 | 1.80 | 5.02 | 1.91 | 2.72 | 3.74 | 4.92 | 1.47 | 3.85 | 6.06 | 4.14 | 2.64 | 4.37 | 42.64 |
| 1991 | 1.44 | 0.94 | 1.41 | 2.66 | 6.20 | 1.89 | 1.23 | 4.31 | 0.90 | 4.14 | 2.61 | 1.91 | 29.64 |
| 1992 | 1.78 | 1.54 | 3.34 | 4.34 | 1.33 | 2.35 | 5.91 | 2.50 | 5.55 | 2.01 | 4.33 | 2.35 | 37.33 |
| 1993 | 3.92 | 1.27 | 2.12 | 3.32 | 1.24 | 6.05 | 2.17 | 1.60 | 4.26 | 2.21 | 1.69 | 0.78 | 30.63 |
| 1994 | 2.79 | 1.38 | 2.29 | 4.04 | 1.18 | 3.97 | 3.20 | 3.30 | 2.38 | 1.35 | 2.74 | 2.39 | 31.01 |
| 1995 | 2.47 | 0.89 | 1.73 | 3.44 | 3.55 | 1.55 | 3.40 | 3.71 | 0.62 | 3.53 | 3.08 | 0.85 | 28.82 |
| 1996 | 1.85 | 1.76 | 1.56 | 3.39 | 2.82 | 2.37 | 2.64 | 0.43 | 4.42 | 1.59 | 1.99 | 2.57 | 27.39 |
| 1997 | 1.57 | 3.90 | 3.22 | 1.56 | 5.23 | 3.17 | 2.68 | 3.22 | 3.41 | 1.91 | 0.94 | 1.61 | 32.42 |
| 1998 | 2.60 | 3.56 | 3.62 | 3.86 | 2.46 | 2.69 | 5.72 | 4.19 | 1.50 | 1.41 | 1.36 | 1.16 | 34.13 |
| 1999 | 3.00 | 1.98 | 1.12 | 5.13 | 2.20 | 5.46 | 3.62 | 1.31 | 3.11 | 1.56 | 1.49 | 2.22 | 32.20 |
| 2000 | 1.29 | 0.84 | 1.55 | 4.35 | 5.11 | 4.90 | 5.40 | 4.63 | 6.71 | 3.05 | 1.69 | 2.63 | 42.15 |
| 2001 | 0.69 | 2.88 | 0.93 | 3.20 | 3.70 | 3.40 | 1.16 | 2.87 | 4.28 | 6.76 | 2.35 | 2.23 | 34.45 |
| 2002 | 3.36 | 1.91 | 2.12 | 4.48 | 3.76 | 1.07 | 3.50 | 3.32 | 1.95 | 1.15 | 2.72 | 1.16 | 30.50 |
| 2003 | 0.42 | 0.66 | 1.46 | 2.07 | 4.73 | 2.50 | 2.59 | 4.36 | 4.27 | 2.74 | 2.97 | 2.62 | 31.39 |
| 2004 | 1.43 | 0.63 | 3.29 | 0.69 | 8.46 | 2.86 | 2.85 | 4.51 | 0.65 | 2.08 | 3.21 | 2.91 | 33.57 |
| 2005 | 3.40 | 3.02 | 0.74 | 1.66 | 1.85 | 1.95 | 5.38 | 1.32 | 1.63 | 0.13 | 4.70 | 2.52 | 28.30 |
| 2006 | 3.24 | 2.71 | 3.21 | 2.71 | 4.60 | 3.95 | 4.38 | 2.05 | 1.73 | 4.11 | 2.90 | 3.65 | 39.24 |
| 2007 | 3.02 | 0.82 | 3.09 | 2.68 | 2.56 | 3.10 | 2.10 | 6.61 | 1.44 | 2.00 | 1.77 | 3.48 | 32.67 |
| 2008 | 2.13 | 3.61 | 3.17 | 0.96 | 2.03 | 4.05 | 3.24 | 0.27 | 5.99 | 1.15 | 3.31 | 4.07 | 33.98 |
| 2009 | 1.10 | 2.12 | 4.17 | 5.03 | 2.89 | 5.27 | 2.56 | 2.76 | 1.46 | 3.23 | 0.62 | 2.90 | 34.11 |
| POR= 51 YRS | 1.96 | 1.82 | 2.42 | 3.03 | 3.12 | 3.52 | 3.20 | 3.28 | 2.88 | 2.26 | 2.56 | 2.54 | 32.59 |

WBAN : 94847

## AVERAGE TEMPERATURE (°F) 2009 DETROIT (KDTW)

| YEAR | JAN | FEB | MAR | APR | MAY | JUN | JUL | AUG | SEP | OCT | NOV | DEC | ANNUAL |
|---|---|---|---|---|---|---|---|---|---|---|---|---|---|
| 1980 | 24.5 | 22.2 | 31.3 | 45.9 | 59.8 | 63.7 | 72.7 | 72.7 | 63.8 | 46.3 | 37.4 | 26.0 | 47.2 |
| 1981 | 19.0 | 28.8 | 36.5 | 49.8 | 55.9 | 68.0 | 72.4 | 70.0 | 60.9 | 47.6 | 41.1 | 27.8 | 48.2 |
| 1982 | 17.1 | 20.7 | 33.0 | 43.2 | 64.3 | 64.2 | 72.4 | 67.7 | 61.8 | 52.6 | 41.6 | 37.3 | 48.0 |
| 1983 | 28.7 | 31.6 | 38.4 | 44.2 | 54.4 | 68.2 | 74.5 | 73.6 | 64.0 | 51.6 | 41.2 | 20.8 | 49.3 |
| 1984 | 18.0 | 33.3 | 28.9 | 47.8 | 54.5 | 70.8 | 70.8 | 72.7 | 61.2 | 54.9 | 38.6 | 34.0 | 48.8 |
| 1985 | 20.4 | 23.5 | 38.4 | 51.0 | 60.1 | 62.8 | 71.3 | 69.2 | 64.3 | 53.0 | 42.4 | 22.2 | 48.2 |
| 1986 | 23.9 | 24.6 | 37.6 | 50.6 | 61.3 | 67.3 | 75.0 | 68.9 | 65.9 | 52.6 | 37.3 | 31.7 | 49.7 |
| 1987 | 26.1 | 29.6 | 39.8 | 50.8 | 63.3 | 71.3 | 76.1 | 71.6 | 64.6 | 46.6 | 43.5 | 33.6 | 51.4 |
| 1988 | 23.8 | 23.4 | 36.9 | 48.5 | 62.0 | 70.4 | 77.1 | 75.1 | 63.3 | 46.0 | 42.2 | 28.7 | 49.8 |
| 1989 | 32.8 | 24.1 | 35.2 | 45.1 | 57.5 | 67.5 | 73.0 | 69.9 | 61.9 | 52.1 | 38.2 | 18.0 | 47.9 |
| 1990 | 33.6 | 30.7 | 39.5 | 49.0 | 56.6 | 68.5 | 72.2 | 71.2 | 64.5 | 52.8 | 44.2 | 32.8 | 51.3 |
| 1991 | 25.0 | 31.2 | 40.3 | 52.0 | 66.5 | 72.4 | 74.9 | 73.4 | 63.1 | 54.8 | 38.5 | 32.1 | 52.0 |
| 1992 | 28.3 | 30.8 | 35.5 | 46.3 | 58.3 | 65.5 | 68.8 | 66.7 | 61.4 | 49.7 | 40.5 | 33.2 | 48.8 |
| 1993 | 29.4 | 24.2 | 34.7 | 47.8 | 60.2 | 67.5 | 75.5 | 74.5 | 61.0 | 51.9 | 41.2 | 30.8 | 49.9 |
| 1994 | 17.3 | 23.5 | 37.1 | 51.2 | 58.7 | 72.3 | 74.2 | 69.6 | 66.2 | 53.8 | 45.5 | 35.4 | 50.4 |
| 1995 | 28.4 | 24.9 | 39.2 | 45.7 | 59.7 | 71.6 | 74.8 | 77.2 | 62.8 | 55.1 | 35.5 | 25.6 | 50.0 |
| 1996 | 24.3 | 26.0 | 30.8 | 45.2 | 56.7 | 70.7 | 70.6 | 72.9 | 64.1 | 52.0 | 34.3 | 31.5 | 48.3 |
| 1997 | 23.0 | 30.6 | 37.2 | 45.8 | 52.0 | 69.5 | 72.2 | 68.1 | 62.7 | 51.8 | 37.1 | 32.2 | 48.5 |
| 1998 | 32.8 | 36.7 | 39.5 | 50.4 | 65.6 | 69.1 | 73.4 | 73.2 | 68.0 | 53.8 | 43.8 | 35.3 | 53.5 |
| 1999 | 23.1 | 32.8 | 34.8 | 50.7 | 62.4 | 70.8 | 76.8 | 70.2 | 65.5 | 51.6 | 45.2 | 32.0 | 51.3 |
| 2000 | 24.6 | 31.9 | 44.0 | 48.0 | 61.8 | 69.4 | 70.3 | 70.8 | 62.5 | 55.1 | 40.2 | 19.3 | 49.8 |
| 2001 | 26.2 | 29.7 | 35.1 | 51.2 | 61.2 | 69.6 | 73.6 | 74.1 | 62.3 | 52.5 | 47.6 | 35.9 | 51.6 |
| 2002 | 32.7 | 32.9 | 34.8 | 49.8 | 54.5 | 70.9 | 76.5 | 73.1 | 68.9 | 50.0 | 39.2 | 28.7 | 51.0 |
| 2003 | 20.5 | 23.2 | 35.6 | 48.4 | 56.5 | 66.6 | 72.6 | 72.9 | 63.2 | 50.9 | 44.3 | 33.1 | 49.0 |
| 2004 | 20.0 | 28.5 | 40.4 | 50.9 | 60.9 | 67.1 | 71.8 | 68.5 | 67.5 | 53.3 | 43.0 | 29.7 | 50.1 |
| 2005 | 24.1 | 28.5 | 33.2 | 50.7 | 56.6 | 74.1 | 75.4 | 74.9 | 68.4 | 55.2 | 43.2 | 25.8 | 50.8 |
| 2006 | 35.3 | 29.8 | 38.1 | 52.3 | 60.7 | 69.4 | 76.1 | 72.9 | 62.0 | 49.7 | 42.6 | 37.4 | 52.2 |
| 2007 | 29.6 | 19.4 | 40.2 | 47.7 | 61.7 | 71.4 | 72.0 | 73.9 | 66.7 | 59.1 | 40.0 | 29.6 | 50.9 |
| 2008 | 29.0 | 25.2 | 33.4 | 51.8 | 57.4 | 70.6 | 73.2 | 72.1 | 66.3 | 50.6 | 39.0 | 27.5 | 49.7 |
| 2009 | 17.3 | 28.5 | 38.7 | 49.8 | 59.5 | 67.8 | 68.9 | 71.2 | 66.1 | 50.0 | 45.3 | 29.3 | 49.4 |
| POR= 51 YRS | 24.1 | 26.5 | 36.0 | 48.0 | 58.8 | 68.4 | 72.6 | 71.2 | 63.8 | 51.8 | 40.6 | 29.2 | 49.3 |

## HEATING DEGREE DAYS (base 65°F) 2009 DETROIT (KDTW)

| YEAR | JUL | AUG | SEP | OCT | NOV | DEC | JAN | FEB | MAR | APR | MAY | JUN | TOTAL |
|------|-----|-----|-----|-----|-----|-----|-----|-----|-----|-----|-----|-----|-------|
| 1980-81 | 0 | 0 | 110 | 578 | 822 | 1201 | 1418 | 1008 | 878 | 452 | 293 | 19 | 6779 |
| 1981-82 | 3 | 9 | 167 | 534 | 710 | 1144 | 1477 | 1237 | 985 | 647 | 75 | 70 | 7058 |
| 1982-83 | 2 | 39 | 145 | 383 | 696 | 852 | 1119 | 928 | 816 | 618 | 323 | 59 | 5980 |
| 1983-84 | 6 | 0 | 125 | 418 | 708 | 1367 | 1450 | 912 | 1112 | 507 | 334 | 9 | 6948 |
| 1984-85 | 11 | 4 | 164 | 310 | 785 | 955 | 1377 | 1154 | 818 | 435 | 177 | 93 | 6283 |
| 1985-86 | 2 | 8 | 129 | 366 | 672 | 1317 | 1271 | 1125 | 842 | 435 | 166 | 48 | 6381 |
| 1986-87 | 1 | 33 | 76 | 380 | 824 | 1028 | 1198 | 984 | 776 | 423 | 158 | 11 | 5892 |
| 1987-88 | 4 | 30 | 69 | 566 | 639 | 969 | 1273 | 1201 | 864 | 486 | 138 | 46 | 6285 |
| 1988-89 | 2 | 3 | 90 | 590 | 679 | 1118 | 991 | 1138 | 916 | 591 | 254 | 33 | 6405 |
| 1989-90 | 0 | 11 | 151 | 400 | 797 | 1451 | 966 | 955 | 785 | 506 | 258 | 27 | 6307 |
| 1990-91 | 1 | 1 | 112 | 380 | 618 | 994 | 1234 | 939 | 761 | 394 | 125 | 5 | 5564 |
| 1991-92 | 0 | 0 | 151 | 319 | 788 | 1013 | 1129 | 985 | 906 | 555 | 224 | 59 | 6129 |
| 1992-93 | 15 | 30 | 153 | 469 | 725 | 976 | 1097 | 1138 | 931 | 506 | 161 | 51 | 6252 |
| 1993-94 | 0 | 2 | 155 | 404 | 710 | 1050 | 1472 | 1156 | 858 | 433 | 238 | 24 | 6502 |
| 1994-95 | 0 | 9 | 55 | 345 | 579 | 910 | 1129 | 1115 | 793 | 573 | 170 | 17 | 5695 |
| 1995-96 | 2 | 0 | 125 | 306 | 877 | 1215 | 1253 | 1122 | 1055 | 589 | 289 | 8 | 6841 |
| 1996-97 | 3 | 0 | 102 | 397 | 915 | 1033 | 1297 | 959 | 855 | 566 | 394 | 25 | 6546 |
| 1997-98 | 3 | 13 | 103 | 435 | 830 | 1008 | 991 | 787 | 791 | 431 | 69 | 80 | 5541 |
| 1998-99 | 0 | 2 | 44 | 350 | 629 | 914 | 1293 | 898 | 927 | 424 | 115 | 33 | 5629 |
| 1999-00 | 0 | 3 | 81 | 413 | 585 | 1016 | 1246 | 958 | 645 | 504 | 159 | 31 | 5641 |
| 2000-01 | 1 | 13 | 148 | 307 | 737 | 1412 | 1194 | 983 | 920 | 415 | 146 | 45 | 6321 |
| 2001-02 | 9 | 0 | 131 | 382 | 514 | 898 | 992 | 891 | 929 | 480 | 342 | 21 | 5589 |
| 2002-03 | 0 | 0 | 38 | 481 | 766 | 1119 | 1372 | 1165 | 907 | 499 | 256 | 48 | 6651 |
| 2003-04 | 0 | 0 | 102 | 432 | 610 | 980 | 1388 | 1052 | 757 | 431 | 174 | 36 | 5962 |
| 2004-05 | 5 | 15 | 35 | 356 | 653 | 1087 | 1261 | 1017 | 978 | 421 | 264 | 8 | 6100 |
| 2005-06 | 0 | 1 | 30 | 329 | 647 | 1207 | 912 | 979 | 828 | 377 | 200 | 15 | 5525 |
| 2006-07 | 0 | 0 | 113 | 471 | 664 | 851 | 1094 | 1272 | 765 | 515 | 163 | 17 | 5925 |
| 2007-08 | 3 | 11 | 62 | 229 | 744 | 1089 | 1112 | 1148 | 970 | 390 | 252 | 11 | 6021 |
| 2008-09 | 1 | 3 | 37 | 438 | 770 | 1154 | 1469 | 1014 | 808 | 469 | 180 | 37 | 6380 |
| 2009- | 5 | 14 | 54 | 457 | 584 | 1099 | | | | | | | |

WBAN : 94847

## COOLING DEGREE DAYS (base 65°F) 2009 DETROIT (KDTW)

| YEAR | JAN | FEB | MAR | APR | MAY | JUN | JUL | AUG | SEP | OCT | NOV | DEC | TOTAL |
|------|-----|-----|-----|-----|-----|-----|-----|-----|-----|-----|-----|-----|-------|
| 1980 | 0 | 0 | 0 | 3 | 38 | 69 | 246 | 248 | 79 | 3 | 0 | 0 | 686 |
| 1981 | 0 | 0 | 0 | 1 | 17 | 118 | 241 | 168 | 51 | 0 | 0 | 0 | 596 |
| 1982 | 0 | 0 | 0 | 0 | 58 | 55 | 237 | 129 | 57 | 5 | 0 | 0 | 541 |
| 1983 | 0 | 0 | 0 | 2 | 0 | 160 | 306 | 272 | 104 | 6 | 0 | 0 | 850 |
| 1984 | 0 | 0 | 0 | 0 | 15 | 189 | 197 | 252 | 55 | 2 | 0 | 0 | 710 |
| 1985 | 0 | 0 | 0 | 25 | 32 | 32 | 201 | 146 | 116 | 0 | 0 | 0 | 552 |
| 1986 | 0 | 0 | 0 | 10 | 55 | 120 | 319 | 160 | 110 | 3 | 0 | 0 | 777 |
| 1987 | 0 | 0 | 0 | 4 | 111 | 207 | 355 | 245 | 64 | 0 | 1 | 0 | 987 |
| 1988 | 0 | 0 | 0 | 0 | 52 | 214 | 385 | 322 | 46 | 8 | 0 | 0 | 1027 |
| 1989 | 0 | 0 | 0 | 0 | 29 | 114 | 256 | 171 | 64 | 5 | 0 | 0 | 639 |
| 1990 | 0 | 0 | 1 | 32 | 8 | 139 | 234 | 200 | 101 | 11 | 0 | 0 | 726 |
| 1991 | 0 | 0 | 0 | 10 | 179 | 233 | 315 | 268 | 104 | 9 | 0 | 0 | 1118 |
| 1992 | 0 | 0 | 0 | 0 | 27 | 80 | 143 | 91 | 51 | 0 | 0 | 0 | 392 |
| 1993 | 0 | 0 | 0 | 0 | 21 | 135 | 334 | 302 | 41 | 5 | 0 | 0 | 838 |
| 1994 | 0 | 0 | 0 | 24 | 49 | 248 | 290 | 160 | 97 | 2 | 0 | 0 | 870 |
| 1995 | 0 | 0 | 0 | 0 | 11 | 222 | 312 | 383 | 67 | 7 | 0 | 0 | 1002 |
| 1996 | 0 | 0 | 0 | 2 | 37 | 185 | 184 | 251 | 82 | 0 | 0 | 0 | 741 |
| 1997 | 0 | 0 | 0 | 0 | 0 | 165 | 231 | 114 | 40 | 29 | 0 | 0 | 579 |
| 1998 | 0 | 0 | 7 | 0 | 95 | 209 | 268 | 261 | 136 | 9 | 0 | 0 | 985 |
| 1999 | 0 | 0 | 0 | 0 | 43 | 213 | 372 | 171 | 104 | 2 | 0 | 0 | 905 |
| 2000 | 0 | 0 | 3 | 0 | 66 | 172 | 170 | 199 | 78 | 4 | 0 | 0 | 692 |
| 2001 | 0 | 0 | 0 | 6 | 36 | 190 | 279 | 290 | 58 | 4 | 0 | 0 | 863 |
| 2002 | 0 | 0 | 0 | 30 | 23 | 203 | 364 | 257 | 161 | 23 | 0 | 0 | 1061 |
| 2003 | 0 | 0 | 0 | 8 | 2 | 104 | 242 | 252 | 54 | 0 | 0 | 0 | 662 |
| 2004 | 0 | 0 | 0 | 12 | 53 | 106 | 222 | 132 | 116 | 1 | 0 | 0 | 642 |
| 2005 | 0 | 0 | 0 | 2 | 9 | 290 | 328 | 311 | 136 | 33 | 0 | 0 | 1109 |
| 2006 | 0 | 0 | 0 | 3 | 70 | 155 | 352 | 254 | 31 | 3 | 0 | 0 | 868 |
| 2007 | 0 | 0 | 2 | 2 | 67 | 216 | 227 | 294 | 120 | 57 | 0 | 0 | 985 |
| 2008 | 0 | 0 | 0 | 2 | 22 | 188 | 262 | 231 | 82 | 0 | 0 | 0 | 787 |
| 2009 | 0 | 0 | 0 | 20 | 12 | 128 | 131 | 214 | 92 | 0 | 0 | 0 | 597 |

## SNOWFALL (inches) 2009 DETROIT (KDTW)

| YEAR | JUL | AUG | SEP | OCT | NOV | DEC | JAN | FEB | MAR | APR | MAY | JUN | TOTAL |
|---|---|---|---|---|---|---|---|---|---|---|---|---|---|
| 1980-81 | 0.0 | 0.0 | 0.0 | 2.9 | 3.4 | 10.5 | 7.6 | 13.4 | 0.6 | 0.0 | 0.0 | 0.0 | 38.4 |
| 1981-82 | 0.0 | 0.0 | 0.0 | 0.1 | 0.7 | 17.3 | 20.0 | 13.3 | 13.6 | 9.0 | 0.0 | 0.0 | 74.0 |
| 1982-83 | 0.0 | 0.0 | 0.0 | T | 1.8 | 1.4 | 1.5 | 4.3 | 7.6 | 3.4 | 0.0 | 0.0 | 20.0 |
| 1983-84 | 0.0 | 0.0 | 0.0 | 0.0 | 3.5 | 19.9 | 9.9 | 8.7 | 9.7 | 0.1 | 0.0 | 0.0 | 51.8 |
| 1984-85 | 0.0 | 0.0 | 0.0 | 0.0 | 4.1 | 6.2 | 20.9 | 16.9 | 6.1 | 0.9 | 0.0 | 0.0 | 55.1 |
| 1985-86 | 0.0 | 0.0 | 0.0 | 0.0 | 2.0 | 14.1 | 8.6 | 20.8 | 7.4 | 1.3 | 0.0 | 0.0 | 54.2 |
| 1986-87 | 0.0 | 0.0 | 0.0 | T | 3.3 | 6.0 | 24.0 | 2.0 | 13.3 | 1.1 | 0.0 | 0.0 | 49.7 |
| 1987-88 | 0.0 | 0.0 | 0.0 | T | 0.7 | 15.3 | 7.0 | 19.2 | 2.7 | 0.2 | 0.0 | 0.0 | 45.1 |
| 1988-89 | 0.0 | 0.0 | 0.0 | T | 1.0 | 6.3 | 5.3 | 9.6 | 2.4 | 0.5 | T | 0.0 | 25.1 |
| 1989-90 | 0.0 | 0.0 | 0.0 | 2.7 | 2.4 | 11.8 | 4.0 | 11.1 | 7.8 | 2.0 | 0.0 | 0.0 | 41.8 |
| 1990-91 | 0.0 | 0.0 | T | 0.0 | T | 13.2 | 8.8 | 9.2 | 0.2 | T | T | 0.0 | 31.4 |
| 1991-92 | 0.0 | 0.0 | 0.0 | T | 2.2 | 8.6 | 18.4 | 2.4 | 11.7 | 0.2 | 0.0 | 0.0 | 43.5 |
| 1992-93 | 0.0 | 0.0 | 0.0 | 0.4 | 0.9 | 5.0 | 11.0 | 15.2 | 15.7 | 4.0 | 0.0 | 0.0 | 52.2 |
| 1993-94 | 0.0 | 0.0 | 0.0 | 0.4 | 0.6 | 1.9 | 17.9 | 17.1 | 3.7 | 4.2 | T | 0.0 | 45.8 |
| 1994-95 | 0.0 | 0.0 | T | 0.0 | T | 9.6 | 13.1 | 5.7 | 3.5 | 1.6 | 0.0 | 0.0 | 33.5 |
| 1995-96 | 0.0 | 0.0 | 0.0 | 0.0 | 1.3 | 4.5 | 6.3 | | 11.8 | | | | |
| 1996-97 | | | | | | | | | | | | | |
| 1997-98 | | | | | 4.6 | 6.0 | 8.3 | T | 4.4 | 0.0 | T | 0.0 | |
| 1998-99 | 0.0 | 0.0 | T | 0.0 | 0.0 | 1.2 | 27.3 | 7.8 | 13.2 | 0.0 | 0.0 | 0.0 | 49.5 |
| 1999-00 | 0.0 | 0.0 | 0.0 | T | T | 4.0 | 9.6 | 8.1 | 1.1 | 0.6 | T | 0.0 | 23.4 |
| 2000-01 | 0.0 | 0.0 | 0.0 | T | 1.3 | 25.1 | 3.4 | 2.9 | 5.4 | 0.9 | 0.0 | 0.0 | 39.0 |
| 2001-02 | 0.0 | 0.0 | 0.0 | T | 0.0 | 4.9 | 15.0 | 6.7 | 7.1 | T | 0.0 | 0.0 | 33.7 |
| 2002-03 | 0.0 | 0.0 | 0.0 | 0.0 | 1.4 | 13.1 | 13.9 | 19.2 | 8.1 | 5.0 | T | 0.0 | 60.7 |
| 2003-04 | 0.0 | 0.0 | 0.0 | 0.0 | 0.4 | 3.4 | 14.0 | 0.9 | 5.1 | T | T | 0.0 | 23.8 |
| 2004-05 | 0.0 | 0.0 | 0.0 | 0.0 | 0.1 | 12.5 | 26.9 | 12.5 | 7.4 | 4.3 | 0.1 | 0.0 | 63.8 |
| 2005-06 | 0.0 | 0.0 | 0.0 | 0.0 | 4.3 | 19.8 | 5.0 | 3.8 | 3.4 | T | 0.0 | T | 36.3 |
| 2006-07 | 0.0 | 0.0 | 0.0 | 0.2 | 0.1 | 2.4 | 6.4 | 14.1 | 5.5 | 1.6 | 0.0 | 0.0 | 30.3 |
| 2007-08 | 0.0 | 0.0 | 0.0 | 0.0 | 0.5 | 12.2 | 13.8 | 24.2 | 21.0 | T | 0.0 | T | 71.7 |
| 2008-09 | 0.0 | 0.0 | 0.0 | 0.0 | 2.2 | 21.4 | 25.2 | 8.5 | 1.0 | 7.4 | T | T | 65.7 |
| 2009- | 0.0 | 0.0 | T | 0.0 | 0.0 | 7.8 | | | | | | | |
| POR= 50 YRS | 0.0 | 0.0 | T | 0.1 | 2.4 | 10.0 | 11.3 | 9.2 | 6.9 | 1.7 | T | T | 41.6 |

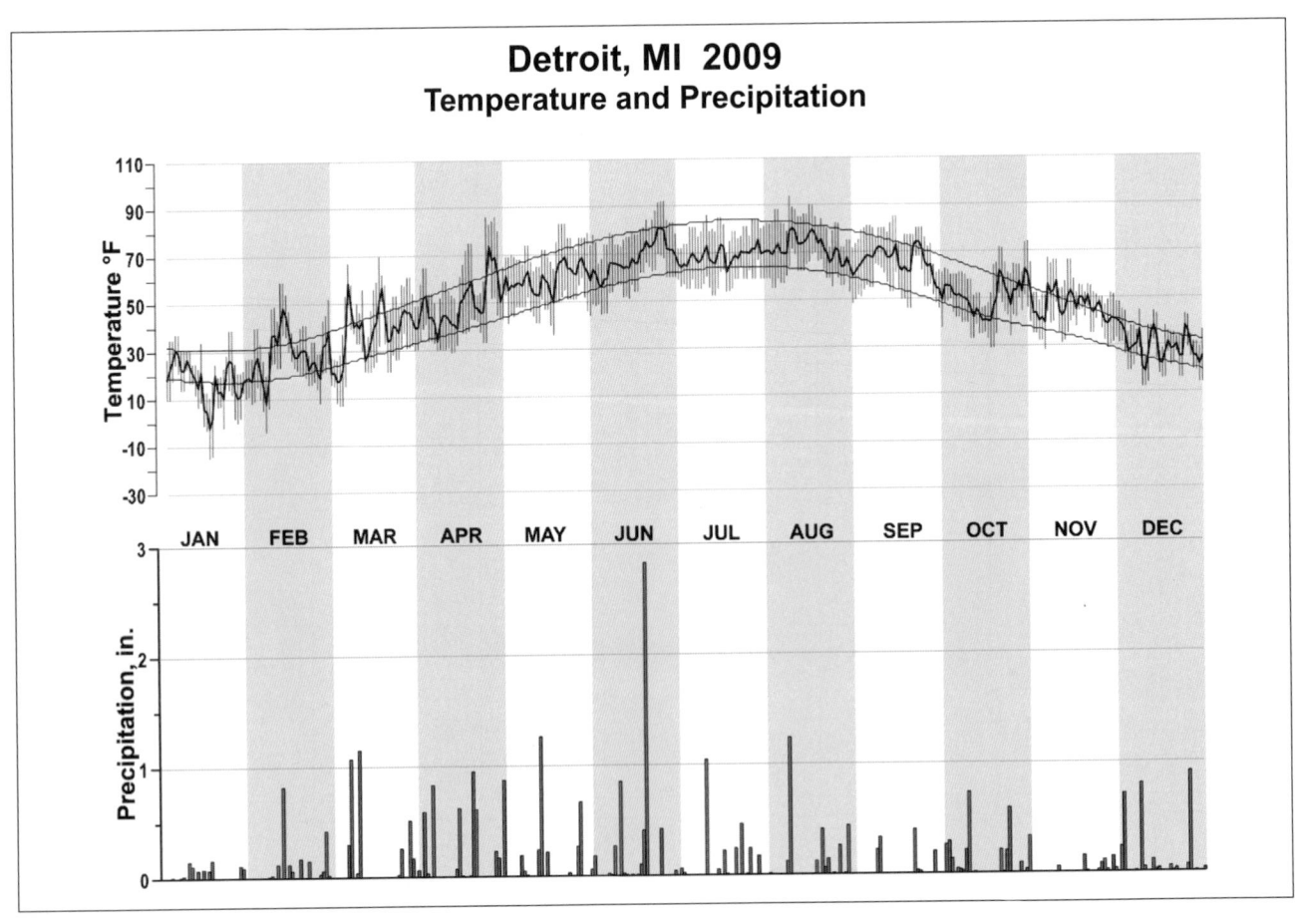

## Detroit, MI 2009
### Temperature and Precipitation

# 2009
# GRAND RAPIDS
# MICHIGAN (KGRR)

Grand Rapids, Michigan, is located in the west-central part of Kent County, in the picturesque Grand River valley about 30 air miles east of Lake Michigan. The Grand River, the longest stream in Michigan, flows through the city and bisects it into east and west sections. High hills rise on either side of the valley. Elevations range from 602 feet on the valley floor to 1,020 feet in the extreme southern part of Kent County, southwest of the airport.

Grand Rapids is under the natural climatic influence of Lake Michigan. In spring the cooling effect of Lake Michigan helps retard the growth of vegetation until the danger of frost has passed. The warming effect in the fall retards frost until most of the crops have matured. Fall is a colorful time of year in western Michigan, compensating for the late spring. During the winter, excessive cloudiness and numerous snow flurries occur with strong westerly winds. The tempering effect of Lake Michigan on cold waves coming in from the west and northwest is quite evident.

The tempering effect of the lake promotes the growth of a great variety of fruit trees and berries, especially apples, peaches, cherries, and blueberries. The intense cold of winter is modified, thus reducing winter kill of fruit trees. Summer days are pleasantly warm and most summer nights are quite comfortable, although there are about three weeks of hot, humid weather during most summers. Prolonged severe cold waves with below-zero temperatures are infrequent. The temperature usually rises to above zero during the daytime hours regardless of early morning readings.

July is the sunniest month and December is the month with the least sunshine. November through January is usually a period of excessive cloudiness and minimal sunshine.

Precipitation is usually ample for the growth and development of all vegetation. About one-half of the annual precipitation falls during the growing season, May through September. Droughts occur occasionally, but are seldom of protracted length. The snowfall season extends from mid-November to mid-March. Some winters have had continuous snow cover throughout this period, although there is usually a mid-winter thaw. The Grand River flows through the city and reaches critical heights a couple of times each year, generally once in January-February and again in March-April. Overflow is generally limited to the lowlands of the flood plain.

November is one of the windiest months and although violent windstorms are infrequent, gusts have on occasion exceeded 65 mph. Summer thunderstorms occasionally produce gusty winds over 60 mph.

# NORMALS, MEANS, AND EXTREMES
## GRAND RAPIDS (KGRR)

| LATITUDE: 42°52'N | LONGITUDE: -85°31'W | | ELEVATION (FT): GRND: 790 BARO: 788 | | TIME ZONE: EASTERN (UTC -5) | | WBAN: 94860 |

| | ELEMENT | POR | JAN | FEB | MAR | APR | MAY | JUN | JUL | AUG | SEP | OCT | NOV | DEC | YEAR |
|---|---|---|---|---|---|---|---|---|---|---|---|---|---|---|---|
| **TEMPERATURE °F** | NORMAL DAILY MAXIMUM | 30 | 29.3 | 32.6 | 43.3 | 56.6 | 69.6 | 78.4 | 82.3 | 79.7 | 71.7 | 59.6 | 45.5 | 33.7 | 56.9 |
| | MEAN DAILY MAXIMUM | 72 | 30.1 | 31.7 | 42.5 | 56.9 | 69.3 | 78.3 | 82.8 | 80.7 | 72.5 | 60.8 | 46.0 | 34.2 | 57.2 |
| | HIGHEST DAILY MAXIMUM | 46 | 63 | 69 | 78 | 88 | 84 | 96 | 100 | 100 | 93 | 88 | 77 | 69 | 100 |
| | YEAR OF OCCURRENCE | | 2008 | 1999 | 2000 | 1970 | 2009 | 2009 | 1988 | 1964 | 1973 | 2007 | 1975 | 2001 | JUL 1988 |
| | MEAN OF EXTREME MAXS. | 73 | 48.2 | 49.9 | 67.2 | 79.0 | 85.1 | 91.3 | 92.3 | 91.5 | 87.5 | 78.9 | 66.4 | 53.0 | 74.2 |
| | NORMAL DAILY MINIMUM | 30 | 15.6 | 17.4 | 25.9 | 36.1 | 46.6 | 55.8 | 60.5 | 59.0 | 51.0 | 40.2 | 31.2 | 21.4 | 38.4 |
| | MEAN DAILY MINIMUM | 72 | 16.3 | 16.6 | 25.0 | 36.1 | 46.4 | 55.8 | 60.7 | 59.0 | 50.9 | 41.0 | 31.2 | 21.7 | 38.4 |
| | LOWEST DAILY MINIMUM | 46 | -22 | -19 | -8 | 3 | -22 | -22 | 41 | 39 | 27 | 18 | 5 | -18 | -22 |
| | YEAR OF OCCURRENCE | | 1994 | 1973 | 1978 | 1982 | 1994 | 1994 | 1983 | 1976 | 1991 | 1988 | 1977 | 1983 | JUN 1994 |
| | MEAN OF EXTREME MINS. | 73 | -5.1 | -2.9 | 5.7 | 21.0 | 31.3 | 41.5 | 48.0 | 46.5 | 35.5 | 26.3 | 16.1 | 1.7 | 22.1 |
| | NORMAL DRY BULB | 30 | 22.4 | 25.0 | 34.6 | 46.3 | 58.1 | 67.1 | 71.4 | 69.4 | 61.3 | 49.9 | 38.4 | 27.6 | 47.6 |
| | MEAN DRY BULB | 72 | 23.2 | 24.2 | 33.8 | 46.5 | 57.9 | 67.2 | 71.8 | 69.9 | 61.7 | 50.9 | 38.6 | 27.9 | 47.8 |
| | MEAN WET BULB | 26 | 22.1 | 23.6 | 30.2 | 40.5 | 50.7 | 60.1 | 64.0 | 63.4 | 56.4 | 45.4 | 35.5 | 26.3 | 43.2 |
| | MEAN DEW POINT | 26 | 19.5 | 20.1 | 26.5 | 35.9 | 46.7 | 56.8 | 61.3 | 61.2 | 53.5 | 42.5 | 32.7 | 23.9 | 40.1 |
| | NORMAL NO. DAYS WITH: | | | | | | | | | | | | | | |
| | MAXIMUM >= 90 | 30 | 0.0 | 0.0 | 0.0 | 0.0 | 0.5 | 2.2 | 4.3 | 2.1 | 0.4 | 0.0 | 0.0 | 0.0 | 9.5 |
| | MAXIMUM <= 32 | 30 | 19.1 | 14.6 | 5.3 | 0.4 | 0.0 | 0.0 | 0.0 | 0.0 | 0.0 | 0.0 | 2.6 | 13.1 | 55.1 |
| | MINIMUM <= 32 | 30 | 29.0 | 25.6 | 23.2 | 11.4 | 1.7 | 0.0 | 0.0 | 0.0 | 0.4 | 5.7 | 17.6 | 27.8 | 142.4 |
| | MINIMUM <= 0 | 30 | 3.5 | 2.7 | 0.5 | 0.0 | 0.0 | 0.0 | 0.0 | 0.0 | 0.0 | 0.0 | 0.0 | 1.2 | 7.9 |
| **H/C** | NORMAL HEATING DEG. DAYS | 30 | 1317 | 1135 | 956 | 571 | 255 | 58 | 10 | 24 | 159 | 471 | 793 | 1147 | 6896 |
| | NORMAL COOLING DEG. DAYS | 30 | 0 | 0 | 2 | 6 | 38 | 124 | 217 | 165 | 56 | 5 | 0 | 0 | 613 |
| **RH** | NORMAL (PERCENT) | 30 | 79 | 76 | 71 | 66 | 66 | 69 | 71 | 75 | 76 | 75 | 77 | 80 | 73 |
| | HOUR 01 LST | 30 | 81 | 80 | 77 | 74 | 77 | 81 | 84 | 87 | 87 | 82 | 80 | 82 | 81 |
| | HOUR 07 LST | 30 | 82 | 82 | 81 | 79 | 79 | 82 | 85 | 90 | 90 | 86 | 83 | 83 | 84 |
| | HOUR 13 LST | 30 | 73 | 69 | 63 | 56 | 53 | 55 | 56 | 60 | 60 | 62 | 69 | 74 | 63 |
| | HOUR 19 LST | 30 | 78 | 73 | 66 | 58 | 55 | 57 | 59 | 64 | 69 | 72 | 75 | 79 | 67 |
| **S** | PERCENT POSSIBLE SUNSHINE | 36 | 28 | 39 | 46 | 51 | 28 | 28 | 64 | 61 | 54 | 44 | 27 | 23 | 41 |
| **W/O** | MEAN NO. DAYS WITH: | | | | | | | | | | | | | | |
| | HEAVY FOG(VISBY <= 1/4 MI) | 46 | 2.3 | 2.3 | 2.5 | 1.5 | 1.4 | 1.2 | 1.3 | 1.9 | 2.0 | 1.9 | 2.1 | 3.3 | 23.7 |
| | THUNDERSTORMS | 62 | 0.3 | 0.3 | 1.7 | 3.4 | 4.0 | 5.6 | 6.2 | 5.1 | 3.8 | 1.6 | 1.2 | 0.3 | 33.5 |
| **CLOUDNESS** | MEAN: | | | | | | | | | | | | | | |
| | SUNRISE-SUNSET (OKTAS) | | | | | | | | | | | | | | |
| | MIDNIGHT-MIDNIGHT (OKTAS) | | | | | | | | | | | | | | |
| | MEAN NO. DAYS WITH: | | | | | | | | | | | | | | |
| | CLEAR | | | | | | | | | | | | | | |
| | PARTLY CLOUDY | | | | | | | | | | | | | | |
| | CLOUDY | | | | | | | | | | | | | | |
| **PR** | MEAN STATION PRESSURE(IN) | 26 | 29.16 | 29.18 | 29.16 | 29.11 | 29.08 | 29.11 | 29.13 | 29.17 | 29.19 | 29.18 | 29.16 | 29.17 | 29.15 |
| | MEAN SEA-LEVEL PRES. (IN) | 26 | 30.06 | 30.07 | 30.05 | 29.99 | 29.97 | 29.96 | 29.98 | 30.02 | 30.05 | 30.05 | 30.04 | 30.06 | 30.03 |
| **WINDS** | MEAN SPEED (MPH) | 26 | 11.0 | 10.6 | 10.8 | 10.8 | 9.7 | 8.5 | 8.1 | 7.5 | 7.9 | 9.2 | 10.3 | 10.5 | 9.6 |
| | PREVAIL.DIR(TENS OF DEGS) | 35 | 26 | 25 | 26 | 08 | 26 | 25 | 25 | 25 | 19 | 19 | 19 | 21 | 25 |
| | MAXIMUM 2-MINUTE: | | | | | | | | | | | | | | |
| | SPEED (MPH) | 14 | 45 | 55 | 47 | 52 | 45 | 45 | 51 | 41 | 44 | 47 | 49 | 48 | 55 |
| | DIR. (TENS OF DEGS) | | 24 | 24 | 25 | 19 | 24 | 24 | 26 | 22 | 23 | 24 | 23 | 25 | 24 |
| | YEAR OF OCCURRENCE | | 1996 | 1999 | 2002 | 1999 | 1996 | 1996 | 1999 | 2007 | 2007 | 1996 | 1998 | 2008 | FEB 1999 |
| | MAXIMUM 3-SECOND | | | | | | | | | | | | | | |
| | SPEED (MPH) | 14 | 55 | 64 | 58 | 61 | 55 | 55 | 58 | 55 | 58 | 60 | 66 | 64 | 66 |
| | DIR. (TENS OF DEGS) | | 24 | 25 | 25 | 25 | 24 | 28 | 29 | 23 | 24 | 25 | 23 | 20 | 23 |
| | YEAR OF OCCURRENCE | | 1996 | 1999 | 2002 | 1997 | 1996 | 2008 | 1999 | 2007 | 2007 | 1996 | 1998 | 2007 | NOV 1998 |
| **PRECIPITATION** | NORMAL (IN) | 30 | 2.03 | 1.54 | 2.59 | 3.48 | 3.35 | 3.67 | 3.56 | 3.78 | 4.28 | 2.80 | 3.35 | 2.70 | 37.13 |
| | MAXIMUM MONTHLY (IN) | 46 | 4.67 | 4.80 | 5.12 | 6.69 | 4.67 | 6.17 | 8.83 | 8.46 | 11.85 | 7.64 | 7.90 | 6.63 | 11.85 |
| | YEAR OF OCCURRENCE | | 2005 | 1997 | 1974 | 1999 | 2005 | 2009 | 1992 | 1987 | 1986 | 2009 | 2003 | 1971 | SEP 1986 |
| | MINIMUM MONTHLY (IN) | 46 | 0.47 | 0.33 | 0.54 | .77 | 0.47 | 0.47 | 0.81 | 0.14 | T | 0.60 | 0.95 | T | T |
| | YEAR OF OCCURRENCE | | 1981 | 1969 | 2001 | 2005 | 1981 | 1981 | 1976 | 1969 | 1979 | 1964 | 1986 | 1952 | SEP 1979 |
| | MAXIMUM IN 24 HOURS (IN) | 46 | 2.16 | 3.05 | 1.78 | 2.66 | 2.16 | 3.15 | 3.68 | 3.68 | 4.55 | 3.19 | 3.00 | 2.79 | 4.55 |
| | YEAR OF OCCURRENCE | | 1993 | 1997 | 1985 | 1999 | 1993 | 2009 | 1994 | 1987 | 1986 | 1993 | 1990 | 1982 | SEP 1986 |
| | NORMAL NO. DAYS WITH: | | | | | | | | | | | | | | |
| | PRECIPITATION >= 0.01 | 30 | 16.7 | 11.9 | 12.2 | 13.2 | 10.7 | 9.9 | 9.4 | 9.7 | 10.5 | 11.3 | 13.3 | 15.5 | 144.3 |
| | PRECIPITATION >= 1.00 | 30 | 0.2 | 0.1 | 0.2 | 0.7 | 0.7 | 1.0 | 0.8 | 1.0 | 1.1 | 0.5 | 0.6 | 0.6 | 7.5 |
| **SNOWFALL** | NORMAL (IN) | 30 | 21.1 | 12.2 | 9.0 | 2.8 | 0.* | 0.0 | 0.0 | 0.0 | 0.0 | 0.6 | 7.7 | 18.8 | 72.2 |
| | MAXIMUM MONTHLY (IN) | 46 | 46.8 | 41.6 | 36.0 | 12.4 | 46.8 | 46.8 | T | T | T | 8.4 | 25.3 | 59.2 | 59.2 |
| | YEAR OF OCCURRENCE | | 1999 | 2008 | 1965 | 1982 | 1999 | 1999 | 1990 | 1993 | 1967 | 1967 | 1991 | 2000 | DEC 2000 |
| | MAXIMUM IN 24 HOURS (IN) | 46 | 16.1 | 9.1 | 13.6 | 9.8 | 16.1 | 16.1 | T | T | T | 8.4 | 10.4 | 15.1 | 16.1 |
| | YEAR OF OCCURRENCE' | | 1978 | 1985 | 2002 | 1975 | 1978 | 1978 | 1990 | 1993 | 1967 | 1967 | 1991 | 1970 | JUN 1978 |
| | MAXIMUM SNOW DEPTH (IN) | 61 | 27 | 21 | 15 | 10 | 0 | 0 | 0 | 0 | 0 | 5 | 11 | 22 | 27 |
| | YEAR OF OCCURRENCE | | 1978 | 1985 | 2007 | 1975 | | | | | | 1967 | 1951 | 1951 | JAN 1978 |
| | NORMAL NO. DAYS WITH: | | | | | | | | | | | | | | |
| | SNOWFALL >= 1.0 | 30 | 5.9 | 3.5 | 2.1 | 0.8 | 0.0 | 0.0 | 0.0 | 0.0 | 0.0 | 0.2 | 2.3 | 5.5 | 20.3 |

## PRECIPITATION (inches) 2009 GRAND RAPIDS (KGRR)

| YEAR | JAN | FEB | MAR | APR | MAY | JUN | JUL | AUG | SEP | OCT | NOV | DEC | ANNUAL |
|---|---|---|---|---|---|---|---|---|---|---|---|---|---|
| 1980 | 1.76 | 1.76 | 1.74 | 3.64 | 3.19 | 4.00 | 5.90 | 3.18 | 4.57 | 1.99 | 1.57 | 3.60 | 36.90 |
| 1981 | 0.47 | 2.03 | 1.29 | 6.11 | 8.29 | 4.22 | 3.74 | 2.95 | 9.52 | 2.54 | 2.58 | 1.10 | 44.84 |
| 1982 | 2.98 | 0.36 | 3.36 | 2.11 | 3.63 | 2.45 | 3.81 | 3.07 | 1.92 | 1.42 | 5.36 | 6.49 | 36.96 |
| 1983 | 1.33 | 1.12 | 3.30 | 5.06 | 4.64 | 2.09 | 4.76 | 1.49 | 4.87 | 2.66 | 3.00 | 2.79 | 37.11 |
| 1984 | 0.94 | 1.15 | 2.77 | 2.10 | 4.77 | 0.62 | 2.12 | 1.49 | 2.16 | 3.34 | 2.85 | 4.37 | 28.68 |
| 1985 | 1.94 | 3.26 | 4.20 | 2.54 | 1.36 | 1.68 | 3.09 | 6.48 | 4.26 | 4.64 | 5.45 | 2.00 | 40.90 |
| 1986 | 1.07 | 3.34 | 2.35 | 2.58 | 3.88 | 7.14 | 5.27 | 5.30 | 11.85 | 2.76 | 0.95 | 1.04 | 47.53 |
| 1987 | 0.67 | 0.37 | 1.15 | 2.40 | 0.94 | 3.56 | 2.93 | 8.46 | 4.47 | 2.33 | 2.49 | 3.29 | 33.06 |
| 1988 | 2.39 | 1.14 | 2.12 | 3.11 | 1.07 | 0.25 | 3.69 | 3.04 | 7.49 | 5.37 | 4.82 | 1.88 | 36.37 |
| 1989 | 0.95 | 1.01 | 2.47 | 1.79 | 4.33 | 5.02 | 1.29 | 4.78 | 4.90 | 1.53 | 4.86 | 0.97 | 33.90 |
| 1990 | 2.39 | 2.08 | 1.96 | 2.23 | 4.39 | 3.00 | 3.73 | 3.40 | 4.22 | 5.05 | 7.14 | 2.97 | 42.56 |
| 1991 | 1.32 | 0.64 | 3.58 | 5.58 | 4.44 | 1.76 | 6.24 | 3.79 | 2.93 | 5.61 | 6.41 | 2.63 | 44.93 |
| 1992 | 1.52 | 1.06 | 3.51 | 3.98 | 1.45 | 1.61 | 8.83 | 3.55 | 5.60 | 2.34 | 5.64 | 3.27 | 42.36 |
| 1993 | 4.21 | 1.13 | 2.31 | 4.93 | 2.17 | 6.05 | 1.83 | 7.73 | 8.20 | 4.32 | 2.12 | 1.47 | 46.47 |
| 1994 | 2.68 | 1.70 | 1.46 | 3.25 | 2.64 | 7.17 | 8.07 | 7.39 | 2.38 | 3.29 | 5.44 | 1.11 | 46.58 |
| 1995 | 2.73 | 0.94 | 1.49 | 3.85 | 2.85 | 3.97 | 6.11 | 3.10 | 1.54 | 2.78 | 4.37 | 1.52 | 35.25 |
| 1996 | 1.18 | 0.90 | 0.96 | 2.41 | 4.83 | 6.33 | 1.28 | .33 | 2.72 | 2.76 | 1.40 | 2.45 | 27.55 |
| 1997 | 2.63 | 4.80 | 1.40 | 1.70 | 3.05 | 2.76 | 1.95 | 1.99 | 4.85 | 2.02 | 1.49 | 0.98 | 29.62 |
| 1998 | 4.09 | 1.50 | 4.93 | 2.75 | 1.86 | 2.11 | 2.49 | 1.70 | 2.49 | 2.98 | 2.27 | 1.20 | 30.37 |
| 1999 | 3.54 | 1.49 | 0.96 | 6.69 | 2.46 | 3.81 | 2.88 | 3.22 | 3.21 | 1.00 | 0.95 | 2.31 | 32.52 |
| 2000 | 1.02 | 1.09 | 1.33 | 3.98 | 8.65 | 4.67 | 4.06 | 2.31 | 6.31 | 1.85 | 2.82 | 2.07 | 40.16 |
| 2001 | 0.71 | 2.58 | 0.54 | 2.05 | 10.01 | 3.35 | 2.00 | 3.48 | 7.38 | 2.26 | 2.37 | | 40.49 |
| 2002 | 1.12 | 1.48 | 2.65 | 3.72 | 4.32 | 2.06 | 2.09 | 4.67 | 0.97 | 2.04 | 2.40 | 1.97 | 29.49 |
| 2003 | 0.91 | 0.61 | 1.58 | 2.96 | 5.68 | 1.72 | 3.56 | 3.75 | 2.45 | 1.76 | 7.90 | 1.23 | 34.11 |
| 2004 | 2.01 | 1.03 | 4.79 | 2.04 | 9.29 | 3.09 | 2.56 | 3.41 | 0.67 | 3.98 | 4.03 | 2.38 | 39.28 |
| 2005 | 4.67 | 2.54 | 1.41 | 0.77 | 2.04 | 7.35 | 4.08 | 1.42 | 3.82 | 0.71 | 5.29 | 2.43 | 36.53 |
| 2006 | 4.30 | 2.39 | 3.31 | 2.22 | 4.92 | 1.77 | 6.90 | 1.62 | 5.38 | 5.02 | 2.80 | 3.76 | 44.39 |
| 2007 | 2.19 | 1.98 | 3.50 | 4.27 | 1.68 | 3.39 | 1.24 | 6.10 | 1.21 | 3.13 | 1.06 | 3.03 | 32.78 |
| 2008 | 3.76 | 4.16 | 2.47 | 3.56 | 2.13 | 5.66 | 5.42 | 1.05 | 9.54 | 2.71 | 2.07 | 6.27 | 48.80 |
| 2009 | 1.74 | 3.31 | 2.66 | 5.56 | 2.65 | 6.17 | 2.35 | 4.74 | 1.56 | 7.64 | 1.48 | 2.99 | 42.85 |
| POR= 73 YRS | 2.16 | 1.75 | 2.57 | 3.30 | 3.33 | 3.70 | 3.39 | 3.14 | 3.65 | 2.81 | 3.10 | 2.64 | 35.54 |

WBAN : 94860

## AVERAGE TEMPERATURE (°F) 2009 GRAND RAPIDS (KGRR)

| YEAR | JAN | FEB | MAR | APR | MAY | JUN | JUL | AUG | SEP | OCT | NOV | DEC | ANNUAL |
|---|---|---|---|---|---|---|---|---|---|---|---|---|---|
| 1980 | 25.1 | 22.7 | 32.5 | 47.0 | 59.9 | 64.4 | 72.7 | 72.6 | 62.7 | 45.8 | 37.5 | 25.1 | 47.3 |
| 1981 | 20.8 | 30.2 | 35.1 | 47.3 | 55.5 | 68.6 | 72.1 | 70.7 | 60.9 | 48.2 | 39.8 | 28.7 | 48.2 |
| 1982 | 17.2 | 22.1 | 32.5 | 41.8 | 65.0 | 62.8 | 73.1 | 68.6 | 61.9 | 53.6 | 40.8 | 36.2 | 48.0 |
| 1983 | 27.6 | 30.9 | 36.9 | 42.6 | 52.8 | 67.7 | 74.7 | 72.2 | 62.4 | 50.7 | 40.7 | 19.2 | 48.2 |
| 1984 | 17.1 | 34.0 | 28.9 | 47.3 | 53.5 | 70.1 | 70.1 | 73.0 | 60.5 | 54.2 | 40.3 | 32.2 | 48.4 |
| 1985 | 18.6 | 21.3 | 36.2 | 51.7 | 60.4 | 63.8 | 70.7 | 67.4 | 63.6 | 50.6 | 38.4 | 22.3 | 47.1 |
| 1986 | 22.7 | 22.5 | 36.7 | 49.7 | 58.8 | 64.4 | 72.5 | 66.0 | 62.8 | 50.2 | 35.0 | 29.8 | 47.6 |
| 1987 | 25.4 | 29.9 | 37.3 | 50.1 | 62.6 | 71.2 | 74.1 | 69.3 | 62.2 | 45.0 | 41.1 | 31.5 | 50.0 |
| 1988 | 20.8 | 20.7 | 34.1 | 47.1 | 60.9 | 68.5 | 74.7 | 73.4 | 61.3 | 44.2 | 40.7 | 27.3 | 47.8 |
| 1989 | 30.5 | 19.7 | 31.8 | 43.9 | 55.8 | 65.9 | 71.9 | 68.4 | 59.4 | 51.1 | 35.7 | 17.2 | 45.9 |
| 1990 | 32.1 | 28.1 | 37.1 | 47.7 | 55.0 | 67.1 | 70.3 | 69.0 | 62.8 | 49.9 | 43.3 | 30.0 | 49.4 |
| 1991 | 22.2 | 29.2 | 37.9 | 50.8 | 64.6 | 71.2 | 72.4 | 70.5 | 60.1 | 52.0 | 35.7 | 29.4 | 49.7 |
| 1992 | 27.5 | 30.3 | 33.7 | 44.0 | 57.4 | 64.3 | 67.2 | 65.0 | 59.7 | 48.2 | 38.2 | 30.5 | 47.2 |
| 1993 | 25.9 | 21.9 | 32.4 | 44.0 | 58.2 | 65.2 | 72.7 | 71.4 | 56.5 | 48.0 | 37.9 | 27.8 | 46.8 |
| 1994 | 14.5 | 19.2 | 34.1 | 47.9 | 56.1 | 68.6 | 71.1 | 66.6 | 63.8 | 51.7 | 42.3 | 33.2 | 47.4 |
| 1995 | 26.5 | 23.2 | 37.0 | 42.8 | 56.4 | 69.8 | 72.5 | 74.8 | 59.1 | 51.4 | 32.0 | 24.2 | 47.5 |
| 1996 | 21.7 | 24.4 | 28.4 | 43.0 | 55.6 | 67.7 | 68.2 | 72.2 | 62.3 | 50.2 | 33.2 | 28.7 | 46.3 |
| 1997 | 21.6 | 27.3 | 34.6 | 44.1 | 50.3 | 68.3 | 70.8 | 66.1 | 61.1 | 49.9 | 35.6 | 31.1 | 46.7 |
| 1998 | 29.6 | 34.1 | 36.1 | 48.6 | 64.3 | 67.4 | 72.2 | 72.2 | 66.0 | 52.7 | 42.1 | 33.3 | 51.6 |
| 1999 | 21.2 | 31.4 | 32.9 | 48.8 | 61.0 | 69.4 | 74.7 | 67.8 | 62.7 | 49.5 | 43.6 | 30.4 | 49.5 |
| 2000 | 23.4 | 31.4 | 42.2 | 45.9 | 59.6 | 66.1 | 68.6 | 69.1 | 60.4 | 53.6 | 37.6 | 19.0 | 48.1 |
| 2001 | 25.5 | 25.9 | 32.5 | 50.0 | 59.6 | 66.3 | 71.3 | 71.6 | 59.3 | 50.4 | 46.8 | 33.9 | 49.4 |
| 2002 | 30.2 | 30.7 | 31.0 | 47.0 | 52.7 | 69.4 | 74.9 | 70.4 | 65.6 | 47.0 | 36.2 | 28.2 | 48.6 |
| 2003 | 19.1 | 20.0 | 33.3 | 45.7 | 54.9 | 64.7 | 70.1 | 71.2 | 61.3 | 48.9 | 41.1 | 31.6 | 46.8 |
| 2004 | 19.5 | 25.7 | 39.3 | 49.9 | 59.7 | 66.3 | 70.5 | 66.1 | 65.5 | 50.5 | 40.4 | 27.9 | 48.4 |
| 2005 | 21.7 | 28.2 | 29.9 | 50.6 | 54.7 | 72.9 | 72.8 | 72.2 | 66.7 | 52.7 | 40.5 | 25.8 | 49.1 |
| 2006 | 33.2 | 26.4 | 36.5 | 50.6 | 58.2 | 67.7 | 74.7 | 71.2 | 60.1 | 47.2 | 41.3 | 35.2 | 50.2 |
| 2007 | 27.3 | 18.7 | 39.6 | 45.1 | 62.7 | 70.6 | 72.3 | 73.3 | 66.1 | 58.1 | 39.0 | 29.2 | 50.2 |
| 2008 | 26.6 | 22.3 | 32.8 | 50.5 | 55.5 | 68.9 | 72.2 | 70.7 | 64.7 | 49.9 | 39.0 | 26.2 | 48.3 |
| 2009 | 17.5 | 27.8 | 37.4 | 47.5 | 58.3 | 67.6 | 67.1 | 69.0 | 63.6 | 48.3 | 44.1 | 27.9 | 48.0 |
| POR= 72 YRS | 23.2 | 24.2 | 33.8 | 46.5 | 57.9 | 67.2 | 71.8 | 69.9 | 61.7 | 50.9 | 38.6 | 27.9 | 47.8 |

## HEATING DEGREE DAYS (base 65°F) 2009 GRAND RAPIDS (KGRR)

| YEAR | JUL | AUG | SEP | OCT | NOV | DEC | JAN | FEB | MAR | APR | MAY | JUN | TOTAL |
|---|---|---|---|---|---|---|---|---|---|---|---|---|---|
| 1980-81 | 0 | 4 | 115 | 588 | 821 | 1227 | 1363 | 969 | 919 | 525 | 298 | 8 | 6837 |
| 1981-82 | 7 | 5 | 173 | 513 | 749 | 1116 | 1475 | 1196 | 998 | 689 | 91 | 98 | 7110 |
| 1982-83 | 6 | 34 | 140 | 361 | 717 | 884 | 1151 | 949 | 866 | 663 | 373 | 61 | 6205 |
| 1983-84 | 16 | 2 | 149 | 440 | 721 | 1413 | 1480 | 892 | 1107 | 532 | 353 | 13 | 7118 |
| 1984-85 | 16 | 4 | 189 | 333 | 735 | 1014 | 1431 | 1216 | 886 | 428 | 173 | 81 | 6506 |
| 1985-86 | 9 | 17 | 157 | 436 | 790 | 1316 | 1302 | 1186 | 871 | 466 | 214 | 75 | 6839 |
| 1986-87 | 11 | 56 | 118 | 452 | 894 | 1084 | 1220 | 978 | 849 | 440 | 178 | 20 | 6300 |
| 1987-88 | 18 | 36 | 118 | 610 | 712 | 1032 | 1364 | 1277 | 950 | 531 | 169 | 60 | 6877 |
| 1988-89 | 3 | 21 | 135 | 639 | 722 | 1162 | 1062 | 1263 | 1023 | 625 | 297 | 51 | 7003 |
| 1989-90 | 2 | 23 | 203 | 424 | 874 | 1477 | 1014 | 1026 | 865 | 549 | 306 | 45 | 6808 |
| 1990-91 | 11 | 15 | 139 | 475 | 645 | 1077 | 1323 | 997 | 830 | 430 | 157 | 18 | 6117 |
| 1991-92 | 5 | 4 | 211 | 404 | 871 | 1098 | 1155 | 1001 | 964 | 626 | 262 | 75 | 6676 |
| 1992-93 | 25 | 66 | 197 | 512 | 794 | 1060 | 1204 | 1203 | 1004 | 624 | 227 | 84 | 7000 |
| 1993-94 | 1 | 16 | 259 | 524 | 810 | 1150 | 1561 | 1274 | 950 | 520 | 297 | 42 | 7404 |
| 1994-95 | 0 | 46 | 94 | 405 | 675 | 980 | 1186 | 1165 | 860 | 659 | 265 | 28 | 6363 |
| 1995-96 | 14 | 0 | 212 | 417 | 982 | 1260 | 1337 | 1166 | 1130 | 653 | 320 | 29 | 7520 |
| 1996-97 | 10 | 2 | 147 | 451 | 944 | 1119 | 1338 | 1048 | 935 | 616 | 448 | 24 | 7082 |
| 1997-98 | 8 | 43 | 136 | 473 | 876 | 1046 | 1091 | 858 | 899 | 486 | 76 | 98 | 6090 |
| 1998-99 | 0 | 2 | 58 | 380 | 682 | 975 | 1353 | 940 | 989 | 481 | 156 | 49 | 6065 |
| 1999-00 | 1 | 17 | 116 | 474 | 634 | 1066 | 1282 | 971 | 706 | 567 | 213 | 64 | 6111 |
| 2000-01 | 15 | 19 | 199 | 348 | 815 | 1420 | 1218 | 1089 | 1001 | 444 | 189 | 80 | 6837 |
| 2001-02 | 20 | 1 | 194 | 449 | 538 | 955 | 1072 | 955 | 1048 | 561 | 399 | 43 | 6235 |
| 2002-03 | 0 | 4 | 69 | 559 | 857 | 1132 | 1416 | 1255 | 976 | 579 | 309 | 72 | 7228 |
| 2003-04 | 3 | 5 | 152 | 495 | 709 | 1025 | 1407 | 1135 | 789 | 462 | 194 | 53 | 6429 |
| 2004-05 | 7 | 50 | 61 | 443 | 733 | 1149 | 1334 | 1023 | 1082 | 433 | 321 | 15 | 6651 |
| 2005-06 | 4 | 3 | 56 | 404 | 727 | 1206 | 979 | 1073 | 876 | 426 | 264 | 20 | 6038 |
| 2006-07 | 0 | 0 | 163 | 547 | 706 | 919 | 1159 | 1290 | 787 | 589 | 137 | 25 | 6322 |
| 2007-08 | 2 | 11 | 83 | 263 | 770 | 1102 | 1185 | 1231 | 990 | 431 | 298 | 20 | 6386 |
| 2008-09 | 1 | 3 | 62 | 468 | 776 | 1194 | 1466 | 1035 | 850 | 524 | 216 | 47 | 6642 |
| 2009- | 25 | 28 | 80 | 513 | 622 | 1142 | | | | | | | |

WBAN : 94860

## COOLING DEGREE DAYS (base 65°F) 2009 GRAND RAPIDS (KGRR)

| YEAR | JAN | FEB | MAR | APR | MAY | JUN | JUL | AUG | SEP | OCT | NOV | DEC | TOTAL |
|---|---|---|---|---|---|---|---|---|---|---|---|---|---|
| 1980 | 0 | 0 | 0 | 6 | 34 | 96 | 247 | 245 | 53 | 0 | 0 | 0 | 681 |
| 1981 | 0 | 0 | 0 | 0 | 13 | 124 | 236 | 190 | 55 | 0 | 0 | 0 | 618 |
| 1982 | 0 | 0 | 0 | 0 | 99 | 40 | 263 | 153 | 55 | 12 | 0 | 0 | 622 |
| 1983 | 0 | 0 | 0 | 0 | 2 | 146 | 325 | 234 | 76 | 5 | 0 | 0 | 788 |
| 1984 | 0 | 0 | 0 | 7 | 5 | 174 | 179 | 259 | 60 | 6 | 0 | 0 | 690 |
| 1985 | 0 | 0 | 0 | 36 | 39 | 52 | 193 | 99 | 119 | 0 | 0 | 0 | 538 |
| 1986 | 0 | 0 | 0 | 14 | 26 | 64 | 252 | 94 | 61 | 0 | 0 | 0 | 511 |
| 1987 | 0 | 0 | 0 | 2 | 110 | 213 | 306 | 174 | 38 | 0 | 0 | 0 | 843 |
| 1988 | 0 | 0 | 0 | 0 | 50 | 170 | 310 | 289 | 29 | 2 | 0 | 0 | 850 |
| 1989 | 0 | 0 | 0 | 0 | 19 | 84 | 223 | 135 | 40 | 0 | 0 | 0 | 501 |
| 1990 | 0 | 0 | 6 | 37 | 6 | 115 | 185 | 144 | 83 | 13 | 0 | 0 | 589 |
| 1991 | 0 | 0 | 0 | 12 | 155 | 211 | 240 | 182 | 71 | 7 | 0 | 0 | 878 |
| 1992 | 0 | 0 | 0 | 2 | 36 | 60 | 99 | 74 | 44 | 1 | 0 | 0 | 316 |
| 1993 | 0 | 0 | 0 | 0 | 21 | 94 | 244 | 222 | 10 | 5 | 0 | 0 | 596 |
| 1994 | 0 | 0 | 0 | 13 | 29 | 159 | 196 | 102 | 65 | 1 | 0 | 0 | 565 |
| 1995 | 0 | 0 | 0 | 0 | 5 | 179 | 252 | 315 | 40 | 6 | 0 | 0 | 797 |
| 1996 | 0 | 0 | 0 | 0 | 33 | 118 | 116 | 230 | 74 | 0 | 0 | 0 | 571 |
| 1997 | 0 | 0 | 0 | 0 | 0 | 131 | 194 | 81 | 26 | 13 | 0 | 0 | 445 |
| 1998 | 0 | 0 | 9 | 0 | 62 | 177 | 230 | 233 | 95 | 4 | 0 | 0 | 810 |
| 1999 | 0 | 0 | 0 | 3 | 39 | 186 | 308 | 110 | 55 | 0 | 0 | 0 | 701 |
| 2000 | 0 | 0 | 2 | 0 | 48 | 102 | 132 | 154 | 66 | 2 | 0 | 0 | 506 |
| 2001 | 0 | 0 | 0 | 0 | 27 | 128 | 223 | 212 | 30 | 3 | 0 | 0 | 623 |
| 2002 | 0 | 0 | 0 | 29 | 23 | 180 | 316 | 179 | 96 | 10 | 0 | 0 | 833 |
| 2003 | 0 | 0 | 0 | 8 | 3 | 69 | 165 | 205 | 46 | 1 | 0 | 0 | 497 |
| 2004 | 0 | 0 | 0 | 15 | 38 | 103 | 186 | 92 | 83 | 0 | 0 | 0 | 517 |
| 2005 | 0 | 0 | 0 | 4 | 8 | 259 | 255 | 234 | 112 | 30 | 0 | 0 | 902 |
| 2006 | 0 | 0 | 0 | 2 | 59 | 108 | 306 | 200 | 21 | 0 | 0 | 0 | 696 |
| 2007 | 0 | 0 | 5 | 1 | 73 | 199 | 234 | 279 | 124 | 56 | 0 | 0 | 971 |
| 2008 | 0 | 0 | 0 | 3 | 8 | 143 | 229 | 186 | 61 | 5 | 0 | 0 | 635 |
| 2009 | 0 | 0 | 0 | 6 | 16 | 129 | 97 | 160 | 47 | 0 | 0 | 0 | 455 |

## SNOWFALL (inches)  2009  GRAND RAPIDS (KGRR)

| YEAR | JUL | AUG | SEP | OCT | NOV | DEC | JAN | FEB | MAR | APR | MAY | JUN | TOTAL |
|------|-----|-----|-----|-----|-----|-----|-----|-----|-----|-----|-----|-----|-------|
| 1980-81 | 0.0 | 0.0 | 0.0 | 0.4 | 5.5 | 17.3 | 8.1 | 18.8 | 1.4 | T | 0.0 | 0.0 | 51.5 |
| 1981-82 | 0.0 | 0.0 | 0.0 | T | 4.4 | 8.9 | 30.3 | 6.7 | 11.8 | 12.4 | 0.0 | 0.0 | 74.5 |
| 1982-83 | 0.0 | 0.0 | 0.0 | 0.0 | 5.2 | 8.2 | 5.7 | 2.9 | 13.2 | 0.7 | 0.0 | 0.0 | 35.9 |
| 1983-84 | 0.0 | 0.0 | 0.0 | T | 4.7 | 34.8 | 19.6 | 1.6 | 10.6 | 0.1 | T | 0.0 | 71.4 |
| 1984-85 | 0.0 | 0.0 | 0.0 | 0.0 | T | 15.7 | 22.6 | 21.3 | 6.7 | 3.3 | 0.0 | 0.0 | 69.6 |
| 1985-86 | 0.0 | 0.0 | 0.0 | 0.0 | 3.5 | 30.7 | 18.4 | 20.2 | 6.1 | 0.2 | 0.0 | 0.0 | 79.1 |
| 1986-87 | 0.0 | 0.0 | 0.0 | 0.0 | 5.3 | 12.7 | 19.2 | 0.9 | 5.7 | 3.8 | 0.0 | 0.0 | 47.6 |
| 1987-88 | 0.0 | 0.0 | 0.0 | 1.6 | 0.7 | 18.2 | 21.9 | 18.1 | 3.4 | 0.3 | 0.0 | 0.0 | 64.2 |
| 1988-89 | 0.0 | 0.0 | 0.0 | 0.2 | 5.5 | 14.4 | 8.7 | 25.1 | 6.3 | 2.2 | T | 0.0 | 62.4 |
| 1989-90 | 0.0 | 0.0 | 0.0 | 5.8 | 19.4 | 25.2 | 10.6 | 23.8 | 2.7 | 2.1 | 0.2 | 0.0 | 89.8 |
| 1990-91 | T | 0.0 | 0.0 | T | 2.0 | 18.6 | 27.7 | 9.5 | 2.8 | T | 0.0 | 0.0 | 60.6 |
| 1991-92 | 0.0 | 0.0 | 0.0 | 0.3 | 25.3 | 27.9 | 13.4 | 3.5 | 15.1 | 2.3 | 0.0 | 0.0 | 87.8 |
| 1992-93 | 0.0 | 0.0 | 0.0 | 2.3 | 4.2 | 14.2 | 11.1 | 18.6 | 11.6 | 3.3 | 0.0 | 0.0 | 65.3 |
| 1993-94 | 0.0 | T | 0.0 | T | 1.9 | 17.7 | 25.3 | 29.6 | 1.9 | T | 0.1 | 0.0 | 76.5 |
| 1994-95 | 0.0 | 0.0 | 0.0 | 0.0 | 0.6 | 8.9 | 21.6 | 18.6 | 4.4 | 0.8 | 0.0 | 0.0 | 54.9 |
| 1995-96 | 0.0 | 0.0 | 0.0 | T | 20.8 | 17.4 | 13.5 | 6.9 | 19.3 | 1.8 | 0.0 | 0.0 | 79.7 |
| 1996-97 | 0.0 | 0.0 | 0.0 | 0.0 | 5.6 | 22.8 | 45.5 | 14.0 | 5.2 | 3.5 | T | 0.0 | 96.6 |
| 1997-98 | 0.0 | 0.0 | 0.0 | 2.4 | 10.1 | 11.5 | 20.3 | 0.5 | 13.8 | 0.0 | 0.0 | 0.0 | 58.6 |
| 1998-99 | 0.0 | 0.0 | 0.0 | 0.0 | 0.2 | 7.5 | 46.8 | 8.0 | 14.2 | T | T | 0.0 | 76.7 |
| 1999-00 | 0.0 | 0.0 | 0.0 | | 0.1 | 18.4 | 15.7 | 11.5 | 0.6 | 6.8 | T | 0.0 | |
| 2000-01 | 0.0 | 0.0 | 0.0 | 0.0 | 19.0 | 59.2 | 4.1 | 7.4 | 4.1 | 0.2 | T | 0.0 | 94.0 |
| 2001-02 | 0.0 | 0.0 | 0.0 | 0.5 | T | 53.9 | 17.5 | 7.5 | 22.6 | 2.0 | T | T | 104.0 |
| 2002-03 | 0.0 | 0.0 | 0.0 | T | 7.6 | 15.6 | 30.2 | 18.9 | 14.4 | 1.3 | T | T | 88.0 |
| 2003-04 | 0.0 | 0.0 | T | T | 4.3 | 8.4 | 44.2 | 13.0 | 3.9 | 0.2 | T | 0.0 | 74.0 |
| 2004-05 | 0.0 | 0.0 | 0.0 | T | 10.3 | 13.1 | 27.8 | 15.1 | 14.4 | 1.0 | T | T | 81.7 |
| 2005-06 | 0.0 | 0.0 | 0.0 | 0.0 | 17.3 | 28.9 | 9.6 | 10.1 | 3.3 | T | 0.0 | 0.0 | 69.2 |
| 2006-07 | 0.0 | 0.0 | 0.0 | 2.1 | 0.3 | 11.6 | 15.4 | 33.6 | 13.6 | 6.7 | 0.0 | 0.0 | 83.3 |
| 2007-08 | 0.0 | 0.0 | 0.0 | 0.0 | 2.9 | 25.0 | 28.3 | 41.6 | 9.2 | T | 0.0 | 0.0 | 107.0 |
| 2008-09 | 0.0 | 0.0 | 0.0 | T | 10.0 | 54.6 | 29.9 | 10.0 | 0.4 | T | 0.0 | 0.0 | 104.9 |
| 2009- | 0.0 | 0.0 | 0.0 | T | 0.7 | 35.4 | | | | | | | |
| POR= 65 YRS | T | T | T | 0.4 | 7.5 | 18.2 | 20.7 | 12.8 | 10.2 | 2.4 | 0.1 | T | 72.3 |

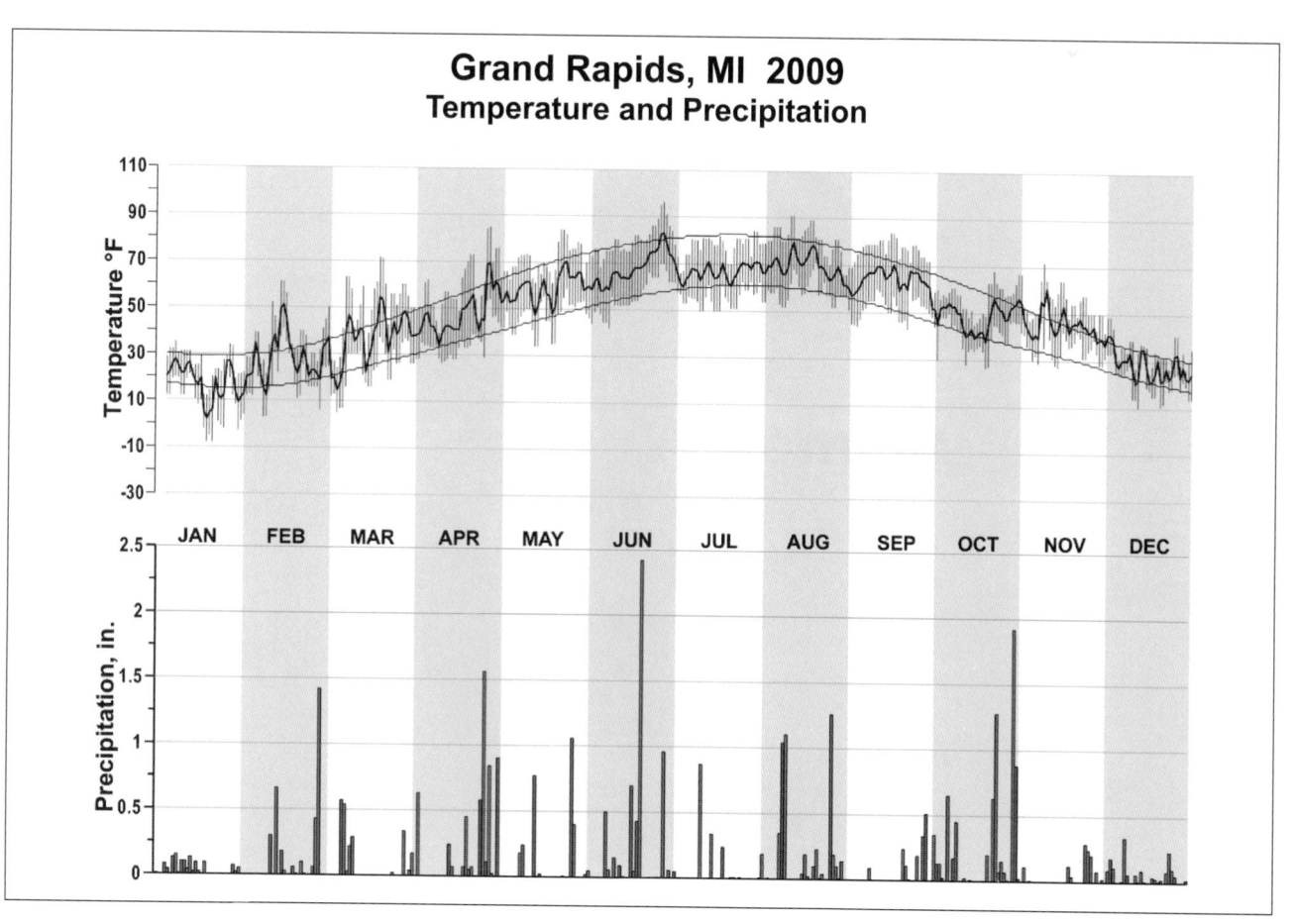

## Grand Rapids, MI  2009
### Temperature and Precipitation

# 2009
# HOUGHTON LAKE
# MICHIGAN (KHTL)

Houghton Lake is located in north-central lower Michigan. The present station is on the northeast shore of Houghton Lake, the largest inland lake in Michigan, with a circumference of about 32 miles. The Muskegon River source is Higgins Lake, 8 miles to the north. It flows through Houghton Lake, then southwestward to Lake Michigan. The station lies within an elongated bowl shaped 1,000-foot plateau, which extends roughly 50 miles north, 75 miles southwest, and about 20 miles southeast of Houghton Lake. In the immediate area, the land is level to rolling, but there are hills and ridges from 100 to 300 feet higher in elevation surrounding the station. Soils are generally sand, or sandy loam supporting little agricultural production, but the area is rich in natural resources of forests, lakes, and streams.

The interior location diminishes the influence of the larger Great Lakes, which lie 70 to 80 miles east and west of Houghton Lake. Hence, the daily temperature range is larger, especially in summer, and temperature extremes are greater than are found nearer the shores of either Lake Michigan or Lake Huron. Temperatures reach the 100 degree mark about one summer out of ten, and at the other extreme, fall below zero an average of twenty-two times during the winter season.

Precipitation is normally a little heavier during the summer season. About 60 percent of the annual total falls in the six-month period from April through September. The heaviest precipitation occurs with summertime thunderstorms.

Snowfall averages above 80 inches per year at Houghton Lake, with considerable variation from year to year. Much heavier snows, averaging over 100 inches a season, fall within a 30- to 60-mile radius to the north and west of Houghton Lake. Seasonal totals have ranged from 24 inches to over 124 inches. Measurable amounts of snow have occurred in nine of the twelve months, and the average number of months with measurable snowfall is six.

Cloudiness is greatest in the late fall and early winter, while sunshine percentage is highest in the spring and summer. Cloudiness is increased in the late fall due to the moisture and warmth picked up by the westerly and northwesterly winds while crossing Lake Michigan.

The growing season is normally quite short, averaging about 90 days between spring and fall freezes.

# NORMALS, MEANS, AND EXTREMES
## HOUGHTON LAKE (KHTL)

**LATITUDE:** 44 ° 22'N  **LONGITUDE:** -84 ° 41'W  **ELEVATION (FT):** GRND: 1151  BARO: 1151  **TIME ZONE:** EASTERN  (UTC -5)  **WBAN:** 94814

| | ELEMENT | POR | JAN | FEB | MAR | APR | MAY | JUN | JUL | AUG | SEP | OCT | NOV | DEC | YEAR |
|---|---|---|---|---|---|---|---|---|---|---|---|---|---|---|---|
| **TEMPERATURE °F** | NORMAL DAILY MAXIMUM | 30 | 25.9 | 29.3 | 39.4 | 53.0 | 67.2 | 75.5 | 80.0 | 77.1 | 68.3 | 56.0 | 41.9 | 30.5 | 53.7 |
| | MEAN DAILY MAXIMUM | 45 | 25.9 | 28.8 | 38.9 | 53.3 | 66.1 | 75.2 | 79.6 | 76.9 | 68.7 | 55.7 | 42.2 | 30.5 | 53.5 |
| | HIGHEST DAILY MAXIMUM | 45 | 54 | 59 | 76 | 86 | 90 | 103 | 98 | 96 | 92 | 85 | 70 | 64 | 103 |
| | YEAR OF OCCURRENCE | | 1996 | 2000 | 1990 | 2002 | 2006 | 1995 | 1995 | 2001 | 1985 | 2007 | 2008 | 2001 | JUN 1995 |
| | MEAN OF EXTREME MAXS. | 45 | 42.9 | 45.8 | 60.2 | 75.1 | 82.8 | 88.1 | 89.8 | 88.2 | 83.1 | 74.4 | 60.6 | 46.6 | 69.8 |
| | NORMAL DAILY MINIMUM | 30 | 9.7 | 10.5 | 19.2 | 30.6 | 40.7 | 48.9 | 53.4 | 52.2 | 45.3 | 36.2 | 27.6 | 16.8 | 32.6 |
| | MEAN DAILY MINIMUM | 45 | 10.1 | 10.0 | 18.7 | 31.6 | 42.1 | 50.9 | 55.2 | 53.6 | 46.5 | 37.0 | 28.4 | 17.4 | 33.5 |
| | LOWEST DAILY MINIMUM | 45 | -26 | -34 | -23 | 2 | 21 | 29 | 33 | 29 | 21 | 16 | -5 | -21 | -34 |
| | YEAR OF OCCURRENCE | | 1981 | 1979 | 1967 | 2003 | 1966 | 1998 | 1965 | 1982 | 1989 | 1969 | 1995 | 1976 | FEB 1979 |
| | MEAN OF EXTREME MINS. | 45 | -12.9 | -13.9 | -6.1 | 15.8 | 27.8 | 36.0 | 40.5 | 38.4 | 29.8 | 22.4 | 12.3 | -4.4 | 15.5 |
| | NORMAL DRY BULB | 30 | 17.8 | 19.9 | 29.3 | 41.8 | 53.9 | 62.2 | 66.7 | 64.6 | 56.8 | 46.1 | 34.8 | 23.7 | 43.1 |
| | MEAN DRY BULB | 45 | 18.0 | 19.4 | 28.9 | 42.4 | 54.1 | 63.2 | 67.4 | 65.3 | 57.6 | 46.4 | 35.3 | 24.0 | 43.5 |
| | MEAN WET BULB | 26 | 18.5 | 19.6 | 26.5 | 37.4 | 47.9 | 57.8 | 61.6 | 60.8 | 54.1 | 43.0 | 32.9 | 23.4 | 40.3 |
| | MEAN DEW POINT | 26 | 15.6 | 16.0 | 21.9 | 31.3 | 42.6 | 53.6 | 58.1 | 58.0 | 51.1 | 39.8 | 30.2 | 20.8 | 36.6 |
| | NORMAL NO. DAYS WITH: | | | | | | | | | | | | | | |
| | MAXIMUM >= 90 | 30 | 0.0 | 0.0 | 0.0 | 0.0 | 0.1 | 0.9 | 2.0 | 0.5 | * | 0.0 | 0.0 | 0.0 | 3.5 |
| | MAXIMUM <= 32 | 30 | 23.5 | 17.9 | 8.3 | 1.1 | 0.0 | 0.0 | 0.0 | 0.0 | 0.0 | * | 5.2 | 18.2 | 74.2 |
| | MINIMUM <= 32 | 30 | 30.6 | 27.5 | 27.4 | 16.9 | 4.3 | 0.2 | 0.0 | * | 1.6 | 9.2 | 21.9 | 29.6 | 169.2 |
| | MINIMUM <= 0 | 30 | 8.0 | 7.3 | 2.6 | 0.0 | 0.0 | 0.0 | 0.0 | 0.0 | 0.0 | 0.0 | 0.1 | 2.8 | 20.8 |
| **H/C** | NORMAL HEATING DEG. DAYS | 30 | 1468 | 1278 | 1115 | 685 | 348 | 131 | 56 | 82 | 254 | 577 | 901 | 1271 | 8166 |
| | NORMAL COOLING DEG. DAYS | 30 | 0 | 0 | 0 | 3 | 20 | 61 | 124 | 85 | 24 | 1 | 0 | 0 | 318 |
| **RH** | NORMAL (PERCENT) | 30 | | | | 64 | | 67 | | | | | | | |
| | HOUR 01 LST | 30 | | | | 77 | | 85 | | | | | | | |
| | HOUR 07 LST | 30 | 83 | 84 | 84 | 80 | 78 | 82 | 86 | 91 | 92 | 88 | 87 | 85 | 85 |
| | HOUR 13 LST | 30 | 73 | 68 | 62 | 53 | 50 | 54 | 55 | 60 | 62 | 64 | 73 | 76 | 63 |
| | HOUR 19 LST | 30 | 78 | 74 | 67 | 57 | 53 | 57 | 59 | 67 | 73 | 73 | 79 | 81 | 68 |
| **S** | PERCENT POSSIBLE SUNSHINE | | | | | | | | | | | | | | |
| **W/O** | MEAN NO. DAYS WITH: | | | | | | | | | | | | | | |
| | HEAVY FOG(VISBY <= 1/4 MI) | 45 | 2.5 | 2.2 | 3.2 | 1.7 | 1.6 | 2.5 | 3.1 | 4.6 | 4.8 | 3.2 | 2.9 | 2.9 | 35.2 |
| | THUNDERSTORMS | 45 | 0.2 | 0.2 | 0.9 | 2.0 | 4.2 | 5.7 | 5.9 | 5.5 | 3.5 | 1.3 | 0.7 | 0.2 | 30.3 |
| **CLOUDNESS** | MEAN: | | | | | | | | | | | | | | |
| | SUNRISE-SUNSET (OKTAS) | 31 | 6.4 | 5.9 | 5.6 | 5.5 | 5.1 | 4.8 | 4.5 | 4.6 | 5.0 | 5.6 | 6.5 | 6.4 | 5.5 |
| | MIDNIGHT-MIDNIGHT (OKTAS) | 16 | 5.9 | 5.2 | 5.1 | 4.7 | 4.4 | 4.4 | 3.9 | 3.9 | 4.7 | 5.0 | 6.1 | 6.2 | 5.0 |
| | MEAN NO. DAYS WITH: | | | | | | | | | | | | | | |
| | CLEAR | 31 | 2.9 | 4.2 | 5.9 | 6.2 | 7.0 | 7.0 | 7.3 | 7.8 | 6.3 | 5.3 | 2.4 | 2.4 | 64.7 |
| | PARTLY CLOUDY | 31 | 6.3 | 6.7 | 7.2 | 7.2 | 9.3 | 11.2 | 13.2 | 11.1 | 9.3 | 7.7 | 5.0 | 5.9 | 100.1 |
| | CLOUDY | 31 | 21.8 | 17.4 | 17.8 | 16.6 | 14.1 | 11.8 | 10.1 | 12.1 | 14.4 | 18.0 | 22.6 | 22.8 | 199.5 |
| **PR** | MEAN STATION PRESSURE(IN) | 26 | 28.73 | 28.76 | 28.76 | 28.72 | 28.73 | 28.73 | 28.75 | 28.80 | 28.81 | 28.78 | 28.75 | 28.74 | 28.76 |
| | MEAN SEA-LEVEL PRES. (IN) | 26 | 30.03 | 30.05 | 30.05 | 29.97 | 29.97 | 29.96 | 29.98 | 30.03 | 30.05 | 30.04 | 30.02 | 30.03 | 30.02 |
| **WINDS** | MEAN SPEED (MPH) | 26 | 8.8 | 8.6 | 8.7 | 9.4 | 8.8 | 7.7 | 7.1 | 6.8 | 7.3 | 8.4 | 9.2 | 8.9 | 8.3 |
| | PREVAIL.DIR(TENS OF DEGS) | 20 | 28 | 25 | 33 | 32 | 32 | 28 | 24 | 24 | 24 | 28 | 28 | 28 | 28 |
| | MAXIMUM 2-MINUTE: | | | | | | | | | | | | | | |
| | SPEED (MPH) | 13 | 37 | 37 | 37 | 37 | 40 | 51 | 37 | 35 | 33 | 38 | 39 | 39 | 51 |
| | DIR. (TENS OF DEGS) | | 28 | 24 | 22 | 26 | 26 | 19 | 13 | 26 | 27 | 24 | 22 | 23 | 19 |
| | YEAR OF OCCURRENCE | | 2008 | 2001 | 2006 | 2004 | 1998 | 1999 | 2000 | 2006 | 2007 | 2004 | 1998 | 2008 | JUN 1999 |
| | MAXIMUM 3-SECOND | | | | | | | | | | | | | | |
| | SPEED (MPH) | 13 | 51 | 48 | 46 | 52 | 54 | 61 | 49 | 48 | 43 | 52 | 53 | 53 | 61 |
| | DIR. (TENS OF DEGS) | | 27 | 23 | 27 | 22 | 23 | 19 | 29 | 26 | 26 | 24 | 23 | 23 | 19 |
| | YEAR OF OCCURRENCE | | 2008 | 1999 | 2009 | 1997 | 1998 | 1999 | 2007 | 2006 | 2007 | 2004 | 2007 | 2008 | JUN 1999 |
| **PRECIPITATION** | NORMAL (IN) | 30 | 1.61 | 1.25 | 2.05 | 2.29 | 2.57 | 2.93 | 2.75 | 3.72 | 3.11 | 2.26 | 2.14 | 1.75 | 28.43 |
| | MAXIMUM MONTHLY (IN) | 45 | 3.13 | 3.36 | 5.67 | 4.73 | 7.40 | 9.20 | 5.33 | 7.18 | 9.49 | 8.08 | 5.10 | 4.60 | 9.49 |
| | YEAR OF OCCURRENCE | | 1974 | 1971 | 1976 | 1991 | 2004 | 2008 | 1994 | 1975 | 1986 | 1991 | 1988 | 2008 | SEP 1986 |
| | MINIMUM MONTHLY (IN) | 45 | 0.20 | 0.15 | 0.19 | .74 | 0.40 | 0.85 | 0.55 | 0.85 | 0.01 | 0.47 | 0.45 | 0.34 | 0.01 |
| | YEAR OF OCCURRENCE | | 2003 | 2003 | 2001 | 2005 | 1966 | 1988 | 1989 | 1969 | 1979 | 1971 | 1986 | 1997 | SEP 1979 |
| | MAXIMUM IN 24 HOURS (IN) | 45 | 1.39 | 1.52 | 2.18 | 1.81 | 1.94 | 3.28 | 3.83 | 3.12 | 2.55 | 3.47 | 1.82 | 1.70 | 3.83 |
| | YEAR OF OCCURRENCE | | 1974 | 1997 | 1976 | 1991 | 1973 | 1996 | 1984 | 1981 | 1985 | 1998 | 1988 | 1971 | JUL 1984 |
| | NORMAL NO. DAYS WITH: | | | | | | | | | | | | | | |
| | PRECIPITATION >= 0.01 | 30 | 14.9 | 11.2 | 12.1 | 11.7 | 10.1 | 10.4 | 9.7 | 10.4 | 11.5 | 11.2 | 13.0 | 14.3 | 140.5 |
| | PRECIPITATION >= 1.00 | 30 | 0.0 | 0.1 | 0.1 | 0.3 | 0.3 | 0.7 | 0.6 | 0.9 | 0.8 | 0.3 | 0.1 | 0.2 | 4.4 |
| **SNOWFALL** | NORMAL (IN) | 30 | 18.3 | 12.3 | 10.1 | 4.3 | 0.3 | 0.0 | 0.0 | 0.0 | 0.* | 0.6 | 8.8 | 15.9 | 70.6 |
| | MAXIMUM MONTHLY (IN) | 37 | 38.0 | 23.6 | 28.7 | 17.3 | 2.3 | 0.0 | T | T | 0.1 | 4.4 | 41.9 | 45.8 | 45.8 |
| | YEAR OF OCCURRENCE | | 1982 | 1971 | 1971 | 2007 | 1979 | | 1996 | 1993 | 1967 | 1980 | 1995 | 2008 | DEC 2008 |
| | MAXIMUM IN 24 HOURS (IN) | 37 | 15.4 | 8.5 | 11.7 | 7.6 | 3.2 | 0.0 | T | T | 0.1 | 3.5 | 14.4 | 13.2 | 15.4 |
| | YEAR OF OCCURRENCE' | | 1978 | 1974 | 1970 | 1979 | 1994 | | 1970 | 1993 | 1967 | 1980 | 1981 | 1980 | JAN 1978 |
| | MAXIMUM SNOW DEPTH (IN) | 36 | 24 | 21 | 22 | 7 | 3 | 0 | 0 | 0 | 0 | 1 | 17 | 22 | 24 |
| | YEAR OF OCCURRENCE | | 1979 | 1979 | 1978 | 1973 | 1994 | | | | | 1992 | 1995 | 2008 | JAN 1979 |
| | NORMAL NO. DAYS WITH: | | | | | | | | | | | | | | |
| | SNOWFALL >= 1.0 | 30 | 5.6 | 4.6 | 3.3 | 1.5 | 0.1 | 0.0 | 0.0 | 0.0 | 0.0 | 0.2 | 2.8 | 5.0 | 23.1 |

## PRECIPITATION (inches) 2009 HOUGHTON LAKE (KHTL)

| YEAR | JAN | FEB | MAR | APR | MAY | JUN | JUL | AUG | SEP | OCT | NOV | DEC | ANNUAL |
|------|-----|-----|-----|-----|-----|-----|-----|-----|-----|-----|-----|-----|--------|
| 1980 | 1.61 | 0.69 | 1.10 | 3.30 | 1.55 | 3.44 | 2.29 | 2.01 | 3.75 | 1.91 | 1.55 | 1.95 | 25.15 |
| 1981 | 0.79 | 2.10 | 0.88 | 3.88 | 1.73 | 4.02 | 1.89 | 7.06 | 1.89 | 2.61 | 2.07 | 1.08 | 30.00 |
| 1982 | 2.43 | 0.29 | 2.41 | 2.46 | 2.98 | 3.21 | 3.31 | 3.25 | 3.58 | 1.86 | 2.52 | 3.47 | 31.77 |
| 1983 | 1.20 | 0.79 | 3.11 | 1.86 | 5.99 | 0.95 | 1.40 | 3.89 | 4.63 | 3.66 | 1.60 | 1.69 | 30.77 |
| 1984 | 1.06 | 0.88 | 2.28 | 1.96 | 2.60 | 3.01 | 4.30 | 2.95 | 2.74 | 2.17 | 1.99 | 2.93 | 28.87 |
| 1985 | 1.64 | 1.99 | 3.55 | 2.42 | 1.87 | 1.71 | 2.28 | 4.76 | 6.14 | 1.63 | 3.56 | 2.14 | 33.69 |
| 1986 | 1.06 | 1.73 | 2.20 | 1.73 | 3.20 | 5.43 | 4.38 | 1.76 | 9.49 | 1.75 | 0.45 | 0.85 | 34.03 |
| 1987 | 1.06 | 0.61 | 0.78 | 0.97 | 1.56 | 1.04 | 1.62 | 6.69 | 4.35 | 2.21 | 2.63 | 2.45 | 25.97 |
| 1988 | 2.09 | 0.75 | 2.39 | 2.37 | 0.56 | 0.85 | 2.49 | 4.50 | 3.63 | 3.38 | 5.10 | 1.86 | 29.97 |
| 1989 | 0.97 | 0.70 | 2.99 | 0.98 | 3.19 | 2.90 | 0.55 | 2.62 | 1.03 | 1.30 | 2.08 | 0.92 | 20.23 |
| 1990 | 2.43 | 1.09 | 1.47 | 1.77 | 4.15 | 2.57 | 3.82 | 3.28 | 3.00 | 2.94 | 2.90 | 1.55 | 30.97 |
| 1991 | 1.22 | 0.62 | 3.02 | 4.73 | 5.03 | 1.48 | 4.43 | 2.36 | 3.05 | 8.08 | 1.88 | 1.76 | 37.66 |
| 1992 | 1.29 | 1.35 | 1.84 | 3.13 | 0.49 | 1.91 | 2.86 | 3.45 | 3.63 | 2.75 | 4.65 | 1.98 | 29.33 |
| 1993 | 1.73 | 0.89 | 0.62 | 2.96 | 2.50 | 4.84 | 1.28 | 5.67 | 2.10 | 1.78 | 1.58 | 0.74 | 26.69 |
| 1994 | 1.88 | 1.44 | 1.57 | 2.72 | 1.45 | 1.72 | 5.33 | 5.52 | 1.56 | 1.04 | 2.86 | 0.53 | 27.62 |
| 1995 | 2.10 | 0.65 | 1.61 | 2.96 | 1.51 | 3.83 | 2.52 | 5.64 | 1.67 | 2.20 | 4.82 | 1.38 | 30.89 |
| 1996 | 1.71 | 1.46 | 0.96 | 2.99 | 1.59 | 5.84 | 2.76 | 3.56 | 4.41 | 2.84 | .72 | 2.29 | 31.13 |
| 1997 | 2.46 | 2.60 | 1.51 | 1.30 | 3.33 | 1.26 | 2.63 | 3.58 | 3.45 | 1.85 | 1.40 | 0.34 | 25.71 |
| 1998 | 2.12 | 0.89 | 4.40 | 0.86 | 2.31 | 2.24 | 0.65 | 1.30 | 2.10 | 4.00 | 2.53 | 1.04 | 24.44 |
| 1999 | 1.86 | 1.29 | 0.40 | 2.04 | 2.06 | 6.31 | 4.48 | 2.21 | 3.19 | 1.77 | 0.58 | 1.63 | 27.82 |
| 2000 | 1.38 | 1.45 | 1.17 | 1.39 | 5.15 | 2.79 | 4.98 | 2.15 | 2.00 | 0.64 | 2.32 | 0.60 | 26.02 |
| 2001 | 0.59 | 1.60 | 0.19 | 3.22 | 5.31 | 2.53 | 0.81 | 3.15 | 3.61 | 4.61 | 1.91 | 0.59 | 28.12 |
| 2002 | 0.34 | 1.90 | 1.99 | 2.87 | 2.30 | 1.93 | 1.60 | 1.97 | 1.91 | 1.83 | 0.97 | 0.41 | 20.02 |
| 2003 | 0.20 | 0.15 | 0.83 | 3.36 | 2.58 | 1.93 | 2.96 | 2.44 | 1.83 | 2.13 | 4.95 | 1.20 | 24.56 |
| 2004 | 1.26 | 0.63 | 2.36 | 3.26 | 7.40 | 1.51 | 1.49 | 1.79 | 0.77 | 4.42 | 1.58 | 1.78 | 28.25 |
| 2005 | 2.92 | 1.58 | 1.30 | 0.74 | 1.74 | 2.61 | 3.17 | 1.81 | 3.74 | 0.54 | 4.13 | 1.50 | 25.78 |
| 2006 | 2.72 | 1.31 | 2.27 | 2.15 | 3.99 | 3.87 | 1.48 | 3.46 | 3.99 | 3.89 | 2.05 | 2.61 | 33.79 |
| 2007 | 0.90 | 0.67 | 2.78 | 4.23 | 1.95 | 3.53 | 2.15 | 2.82 | 1.17 | 1.79 | 1.17 | 2.39 | 25.55 |
| 2008 | 2.23 | 2.03 | 1.66 | 2.57 | 1.10 | 9.20 | 2.31 | 3.18 | 2.81 | 1.54 | 1.80 | 4.60 | 35.03 |
| 2009 | 1.12 | 2.46 | 1.81 | 4.25 | 2.64 | 3.02 | 2.54 | 3.35 | 2.68 | 4.11 | 1.20 | 1.91 | 31.09 |
| POR= 45 YRS | 1.56 | 1.23 | 1.91 | 2.45 | 2.68 | 3.19 | 2.56 | 3.27 | 3.06 | 2.43 | 2.29 | 1.84 | 28.47 |

WBAN : 94814

## AVERAGE TEMPERATURE (°F) 2009 HOUGHTON LAKE (KHTL)

| YEAR | JAN | FEB | MAR | APR | MAY | JUN | JUL | AUG | SEP | OCT | NOV | DEC | ANNUAL |
|------|-----|-----|-----|-----|-----|-----|-----|-----|-----|-----|-----|-----|--------|
| 1980 | 19.1 | 15.7 | 25.0 | 41.9 | 56.0 | 59.2 | 67.4 | 68.0 | 56.7 | 41.6 | 33.9 | 18.9 | 42.0 |
| 1981 | 14.8 | 24.0 | 32.5 | 44.3 | 53.0 | 63.6 | 66.8 | 65.6 | 55.9 | 42.9 | 35.6 | 25.3 | 43.7 |
| 1982 | 11.4 | 17.9 | 25.8 | 37.3 | 60.9 | 57.4 | 68.6 | 61.9 | 57.2 | 48.9 | 36.2 | 31.2 | 42.9 |
| 1983 | 21.8 | 25.9 | 32.4 | 40.3 | 48.5 | 63.3 | 71.5 | 68.6 | 59.7 | 46.4 | 36.6 | 16.8 | 44.3 |
| 1984 | 12.4 | 28.2 | 24.1 | 44.8 | 50.1 | 65.2 | 66.4 | 68.8 | 56.1 | 49.2 | 35.4 | 27.0 | 44.0 |
| 1985 | 14.9 | 17.9 | 30.3 | 46.5 | 57.0 | 59.5 | 66.1 | 63.9 | 59.8 | 47.5 | 34.5 | 18.0 | 43.0 |
| 1986 | 17.3 | 18.4 | 30.9 | 47.3 | 56.9 | 60.5 | 69.7 | 62.8 | 58.1 | 46.8 | 32.3 | 26.8 | 44.0 |
| 1987 | 21.6 | 23.4 | 33.6 | 47.3 | 58.1 | 67.1 | 71.4 | 65.6 | 59.5 | 43.9 | 37.7 | 28.7 | 46.5 |
| 1988 | 17.3 | 16.3 | 28.0 | 43.7 | 58.6 | 64.8 | 71.4 | 68.8 | 57.9 | 42.4 | 36.6 | 23.7 | 44.1 |
| 1989 | 24.9 | 15.5 | 25.0 | 41.1 | 54.5 | 62.2 | 69.5 | 64.8 | 56.2 | 47.9 | 30.7 | 12.9 | 42.1 |
| 1990 | 26.2 | 21.3 | 32.1 | 44.9 | 51.9 | 63.4 | 66.8 | 65.0 | 57.9 | 45.3 | 38.9 | 25.8 | 45.0 |
| 1991 | 16.1 | 24.5 | 33.1 | 46.9 | 61.1 | 67.7 | 68.6 | 68.0 | 55.8 | 47.9 | 33.4 | 25.4 | 45.7 |
| 1992 | 22.8 | 24.1 | 28.6 | 39.4 | 54.3 | 60.4 | 63.0 | 62.0 | 56.8 | 44.8 | 33.6 | 26.1 | 43.0 |
| 1993 | 21.0 | 16.2 | 28.7 | 40.2 | 54.6 | 61.9 | 70.2 | 68.8 | 52.9 | 44.6 | 34.7 | 25.9 | 43.3 |
| 1994 | 9.3 | 12.9 | 30.1 | 43.2 | 55.0 | 65.3 | 68.1 | 64.4 | 61.9 | 50.9 | 39.3 | 30.7 | 44.3 |
| 1995 | 23.1 | 18.2 | 34.3 | 40.1 | 54.5 | 68.9 | 70.7 | 72.3 | 56.9 | 50.1 | 29.1 | 19.5 | 44.8 |
| 1996 | 15.9 | 19.8 | 25.7 | 36.8 | 50.8 | 64.0 | 64.0 | 66.3 | 58.6 | 47.0 | 30.6 | 24.8 | 42.0 |
| 1997 | 17.0 | 20.4 | 27.2 | 40.3 | 46.6 | 64.8 | 65.8 | 61.6 | 57.0 | 46.0 | 32.3 | 28.2 | 42.3 |
| 1998 | 24.1 | 30.2 | 31.2 | 45.2 | 60.3 | 63.0 | 66.7 | 66.7 | 60.4 | 48.8 | 37.9 | 28.7 | 46.9 |
| 1999 | 17.1 | 26.1 | 29.1 | 44.7 | 56.1 | 64.9 | 69.8 | 63.1 | 56.8 | 44.9 | 39.7 | 26.2 | 44.9 |
| 2000 | 17.8 | 25.0 | 38.1 | 41.4 | 55.9 | 62.7 | 64.7 | 64.7 | 57.6 | 49.6 | 35.8 | 16.0 | 44.1 |
| 2001 | 22.8 | 20.1 | 27.7 | 46.1 | 56.8 | 63.4 | 67.0 | 67.0 | 55.5 | 46.5 | 42.2 | 31.4 | 45.5 |
| 2002 | 26.5 | 25.7 | 25.7 | 41.6 | 49.6 | 64.5 | 71.3 | 65.1 | 61.3 | 41.4 | 32.3 | 24.8 | 44.2 |
| 2003 | 15.5 | 13.6 | 28.2 | 40.1 | 52.3 | 60.5 | 65.6 | 64.0 | 57.2 | 45.2 | 37.0 | 27.7 | 42.5 |
| 2004 | 12.1 | 19.3 | 32.1 | 44.1 | 52.8 | 62.2 | 64.9 | 61.9 | 61.3 | 47.7 | 36.4 | 23.1 | 43.2 |
| 2005 | 16.1 | 23.5 | 24.4 | 45.4 | 51.1 | 69.0 | 68.3 | 66.2 | 61.3 | 48.9 | 37.3 | 23.2 | 44.6 |
| 2006 | 28.0 | 19.9 | 31.3 | 46.2 | 56.1 | 62.8 | 70.1 | 64.6 | 55.4 | 43.0 | 38.1 | 31.3 | 45.6 |
| 2007 | 22.4 | 14.0 | 32.9 | 40.1 | 57.5 | 64.6 | 66.0 | 66.5 | 60.2 | 52.6 | 32.7 | 23.2 | 44.4 |
| 2008 | 22.3 | 16.9 | 25.1 | 46.2 | 50.6 | 64.6 | 67.6 | 64.7 | 58.3 | 44.7 | 33.9 | 20.2 | 42.9 |
| 2009 | 10.4 | 20.6 | 30.5 | 42.8 | 53.1 | 60.9 | 62.1 | 63.3 | 57.5 | 43.7 | 39.7 | 22.5 | 42.3 |
| POR= 45 YRS | 18.0 | 19.4 | 28.9 | 42.4 | 54.1 | 63.2 | 67.4 | 65.3 | 57.6 | 46.4 | 35.3 | 24.0 | 43.5 |

## HEATING DEGREE DAYS (base 65°F) 2009  HOUGHTON LAKE (KHTL)

| YEAR | JUL | AUG | SEP | OCT | NOV | DEC | JAN | FEB | MAR | APR | MAY | JUN | TOTAL |
|------|-----|-----|-----|-----|-----|-----|-----|-----|-----|-----|-----|-----|-------|
| 1980-81 | 18 | 27 | 258 | 716 | 928 | 1423 | 1553 | 1143 | 1001 | 613 | 368 | 80 | 8128 |
| 1981-82 | 50 | 44 | 280 | 676 | 872 | 1222 | 1658 | 1315 | 1209 | 824 | 148 | 224 | 8522 |
| 1982-83 | 12 | 139 | 255 | 494 | 855 | 1040 | 1331 | 1087 | 1003 | 733 | 505 | 116 | 7570 |
| 1983-84 | 25 | 29 | 209 | 569 | 846 | 1487 | 1628 | 1062 | 1262 | 599 | 458 | 53 | 8227 |
| 1984-85 | 42 | 28 | 270 | 481 | 881 | 1168 | 1547 | 1313 | 1071 | 571 | 248 | 172 | 7792 |
| 1985-86 | 46 | 88 | 209 | 535 | 908 | 1451 | 1474 | 1297 | 1048 | 532 | 274 | 148 | 8010 |
| 1986-87 | 23 | 109 | 212 | 557 | 975 | 1179 | 1335 | 1159 | 963 | 523 | 271 | 57 | 7363 |
| 1987-88 | 41 | 60 | 169 | 649 | 813 | 1117 | 1473 | 1406 | 1140 | 636 | 225 | 106 | 7835 |
| 1988-89 | 7 | 75 | 218 | 692 | 847 | 1276 | 1236 | 1381 | 1234 | 711 | 333 | 126 | 8136 |
| 1989-90 | 19 | 79 | 277 | 520 | 1025 | 1607 | 1196 | 1217 | 1009 | 627 | 398 | 97 | 8071 |
| 1990-91 | 32 | 51 | 248 | 604 | 773 | 1208 | 1509 | 1127 | 985 | 540 | 219 | 33 | 7329 |
| 1991-92 | 37 | 30 | 295 | 525 | 939 | 1219 | 1298 | 1178 | 1123 | 763 | 341 | 161 | 7909 |
| 1992-93 | 83 | 132 | 257 | 619 | 934 | 1196 | 1359 | 1358 | 1119 | 737 | 331 | 123 | 8248 |
| 1993-94 | 4 | 37 | 358 | 627 | 904 | 1205 | 1722 | 1457 | 1073 | 649 | 320 | 79 | 8435 |
| 1994-95 | 26 | 75 | 132 | 431 | 763 | 1059 | 1293 | 1307 | 947 | 741 | 320 | 35 | 7129 |
| 1995-96 | 22 | 2 | 260 | 461 | 1072 | 1403 | 1515 | 1307 | 1214 | 840 | 440 | 91 | 8627 |
| 1996-97 | 65 | 37 | 212 | 552 | 1021 | 1240 | 1481 | 1246 | 1162 | 737 | 561 | 64 | 8378 |
| 1997-98 | 79 | 134 | 239 | 582 | 972 | 1135 | 1260 | 968 | 1041 | 592 | 175 | 146 | 7323 |
| 1998-99 | 34 | 48 | 176 | 499 | 805 | 1118 | 1477 | 1083 | 1105 | 602 | 282 | 100 | 7329 |
| 1999-00 | 17 | 91 | 249 | 617 | 754 | 1193 | 1456 | 1154 | 826 | 701 | 299 | 114 | 7471 |
| 2000-01 | 83 | 69 | 256 | 473 | 867 | 1512 | 1303 | 1250 | 1149 | 558 | 262 | 127 | 7909 |
| 2001-02 | 62 | 62 | 296 | 567 | 677 | 1032 | 1188 | 1213 | 703 | 471 | 112 | 7477 |  |
| 2002-03 | 22 | 65 | 157 | 730 | 973 | 1238 | 1526 | 1433 | 1132 | 741 | 387 | 155 | 8559 |
| 2003-04 | 36 | 42 | 235 | 606 | 834 | 1150 | 1633 | 1320 | 1013 | 620 | 378 | 110 | 7977 |
| 2004-05 | 58 | 136 | 155 | 529 | 851 | 1293 | 1510 | 1155 | 1253 | 582 | 431 | 47 | 8000 |
| 2005-06 | 33 | 41 | 148 | 517 | 823 | 1290 | 1140 | 1258 | 1038 | 557 | 299 | 106 | 7250 |
| 2006-07 | 16 | 76 | 289 | 674 | 798 | 1038 | 1315 | 1423 | 986 | 737 | 258 | 85 | 7695 |
| 2007-08 | 60 | 62 | 177 | 397 | 959 | 1290 | 1316 | 1387 | 1233 | 560 | 437 | 71 | 7949 |
| 2008-09 | 19 | 68 | 201 | 623 | 927 | 1384 | 1684 | 1238 | 1061 | 656 | 364 | 176 | 8401 |
| 2009- | 115 | 108 | 219 | 656 | 754 | 1311 | | | | | | | |

WBAN : 94814

## COOLING DEGREE DAYS (base 65°F) 2009  HOUGHTON LAKE (KHTL)

| YEAR | JAN | FEB | MAR | APR | MAY | JUN | JUL | AUG | SEP | OCT | NOV | DEC | TOTAL |
|------|-----|-----|-----|-----|-----|-----|-----|-----|-----|-----|-----|-----|-------|
| 1980 | 0 | 0 | 0 | 2 | 17 | 43 | 98 | 124 | 16 | 0 | 0 | 0 | 300 |
| 1981 | 0 | 0 | 0 | 0 | 3 | 47 | 114 | 70 | 13 | 0 | 0 | 0 | 247 |
| 1982 | 0 | 0 | 0 | 0 | 27 | 3 | 132 | 52 | 29 | 3 | 0 | 0 | 246 |
| 1983 | 0 | 0 | 0 | 0 | 0 | 73 | 235 | 148 | 56 | 0 | 0 | 0 | 512 |
| 1984 | 0 | 0 | 0 | 1 | 2 | 65 | 90 | 151 | 11 | 0 | 0 | 0 | 320 |
| 1985 | 0 | 0 | 0 | 22 | 8 | 11 | 86 | 61 | 60 | 0 | 0 | 0 | 248 |
| 1986 | 0 | 0 | 0 | 8 | 29 | 22 | 175 | 46 | 15 | 0 | 0 | 0 | 295 |
| 1987 | 0 | 0 | 0 | 1 | 64 | 125 | 250 | 86 | 12 | 0 | 0 | 0 | 538 |
| 1988 | 0 | 0 | 0 | 0 | 35 | 108 | 213 | 199 | 11 | 0 | 0 | 0 | 566 |
| 1989 | 0 | 0 | 0 | 0 | 10 | 49 | 164 | 83 | 19 | 0 | 0 | 0 | 325 |
| 1990 | 0 | 0 | 0 | 29 | 0 | 55 | 94 | 56 | 41 | 0 | 0 | 0 | 275 |
| 1991 | 0 | 0 | 0 | 2 | 104 | 120 | 157 | 130 | 26 | 0 | 0 | 0 | 539 |
| 1992 | 0 | 0 | 0 | 0 | 15 | 31 | 28 | 46 | 19 | 0 | 0 | 0 | 139 |
| 1993 | 0 | 0 | 0 | 0 | 16 | 34 | 175 | 161 | 4 | 0 | 0 | 0 | 390 |
| 1994 | 0 | 0 | 0 | 0 | 19 | 97 | 130 | 64 | 45 | 0 | 0 | 0 | 355 |
| 1995 | 0 | 0 | 0 | 0 | 4 | 160 | 202 | 233 | 24 | 5 | 0 | 0 | 628 |
| 1996 | 0 | 0 | 0 | 0 | 10 | 67 | 41 | 85 | 25 | 0 | 0 | 0 | 228 |
| 1997 | 0 | 0 | 0 | 0 | 0 | 63 | 112 | 39 | 7 | 0 | 0 | 0 | 221 |
| 1998 | 0 | 0 | 0 | 0 | 34 | 93 | 98 | 110 | 45 | 0 | 0 | 0 | 380 |
| 1999 | 0 | 0 | 0 | 0 | 15 | 107 | 174 | 41 | 10 | 0 | 0 | 0 | 347 |
| 2000 | 0 | 0 | 0 | 0 | 24 | 51 | 79 | 65 | 42 | 1 | 0 | 0 | 262 |
| 2001 | 0 | 0 | 0 | 0 | 12 | 87 | 132 | 130 | 19 | 0 | 0 | 0 | 380 |
| 2002 | 0 | 0 | 0 | 10 | 4 | 105 | 223 | 73 | 55 | 6 | 0 | 0 | 476 |
| 2003 | 0 | 0 | 0 | 0 | 0 | 28 | 60 | 106 | 10 | 0 | 0 | 0 | 204 |
| 2004 | 0 | 0 | 0 | 0 | 8 | 35 | 60 | 48 | 50 | 0 | 0 | 0 | 201 |
| 2005 | 0 | 0 | 0 | 0 | 5 | 171 | 143 | 83 | 46 | 24 | 0 | 0 | 472 |
| 2006 | 0 | 0 | 0 | 0 | 30 | 47 | 179 | 68 | 6 | 0 | 0 | 0 | 330 |
| 2007 | 0 | 0 | 0 | 0 | 33 | 76 | 101 | 115 | 39 | 20 | 0 | 0 | 384 |
| 2008 | 0 | 0 | 0 | 0 | 0 | 65 | 104 | 66 | 9 | 0 | 0 | 0 | 244 |
| 2009 | 0 | 0 | 0 | 0 | 2 | 59 | 30 | 64 | 4 | 0 | 0 | 0 | 159 |

**SNOWFALL (inches) 2009 HOUGHTON LAKE (KHTL)**

| YEAR | JUL | AUG | SEP | OCT | NOV | DEC | JAN | FEB | MAR | APR | MAY | JUN | TOTAL |
|------|-----|-----|-----|-----|-----|-----|-----|-----|-----|-----|-----|-----|-------|
| 1980-81 | 0.0 | 0.0 | 0.0 | 4.4 | 3.2 | 25.6 | 19.3 | 16.3 | 5.4 | 0.2 | 0.0 | 0.0 | 74.4 |
| 1981-82 | 0.0 | 0.0 | 0.0 | 3.0 | 16.1 | 15.7 | 38.0 | 6.1 | 14.3 | 5.5 | 0.0 | 0.0 | 98.7 |
| 1982-83 | 0.0 | 0.0 | 0.0 | 0.2 | 7.7 | 4.8 | 15.4 | 6.4 | 13.3 | 2.7 | 1.0 | 0.0 | 51.5 |
| 1983-84 | 0.0 | 0.0 | 0.0 | 0.0 | 5.0 | 19.8 | 17.0 | 6.3 | 6.5 | 5.1 | 0.4 | 0.0 | 60.1 |
| 1984-85 | 0.0 | 0.0 | 0.0 | 0.0 | 0.6 | 11.7 | 25.3 | 18.7 | 14.1 | 10.0 | 0.0 | 0.0 | 80.4 |
| 1985-86 | 0.0 | 0.0 | 0.0 | 0.0 | 12.6 | 23.1 | 12.7 | 14.0 | 8.9 | 0.3 | 0.0 | 0.0 | 71.6 |
| 1986-87 | 0.0 | 0.0 | 0.0 | T | 3.4 | 10.3 | 14.1 | 6.4 | 2.3 | 2.0 | 0.0 | 0.0 | 38.5 |
| 1987-88 | 0.0 | 0.0 | 0.0 | 2.4 | 8.7 | 18.6 | 13.4 | 12.9 | 10.3 | 1.4 | 0.0 | 0.0 | 67.7 |
| 1988-89 | 0.0 | 0.0 | 0.0 | | 7.8 | 12.3 | 9.3 | 13.5 | 18.7 | 3.1 | T | 0.0 | |
| 1989-90 | 0.0 | 0.0 | T | 0.6 | 12.2 | 15.0 | 21.7 | 11.0 | 2.4 | 2.1 | 0.9 | 0.0 | 65.9 |
| 1990-91 | 0.0 | 0.0 | 0.0 | T | 9.1 | 15.3 | 22.6 | 8.5 | 4.5 | 1.8 | 0.0 | 0.0 | 61.8 |
| 1991-92 | 0.0 | 0.0 | T | T | 4.9 | 17.5 | 15.0 | 15.8 | 9.5 | 3.6 | 0.0 | 0.0 | 66.3 |
| 1992-93 | 0.0 | 0.0 | 0.0 | 3.2 | 11.8 | 14.3 | 14.7 | 13.1 | 8.1 | 2.9 | 0.0 | 0.0 | 68.1 |
| 1993-94 | 0.0 | T | 0.0 | 0.1 | 2.3 | 9.0 | 26.2 | 12.7 | 9.0 | 3.9 | 0.1 | 0.0 | 63.3 |
| 1994-95 | 0.0 | 0.0 | 0.0 | 0.0 | 6.6 | 7.6 | 12.2 | 10.2 | 7.4 | 1.9 | 0.0 | 0.0 | 45.9 |
| 1995-96 | 0.0 | 0.0 | T | T | 41.9 | 16.4 | 26.0 | 8.7 | 13.4 | 10.6 | | 0.0 | |
| 1996-97 | T | 0.0 | T | T | 8.9 | | | | | | | | |
| 1997-98 | | | | | | | | | | | | | |
| 1998-99 | | | | | | | | | | | | | |
| 1999-00 | | | | | 1.6 | 3.7 | 16.1 | 10.4 | T | | | | |
| 2000-01 | | | | | | | | | | | | | |
| 2001-02 | | | | | | | | | | | | | |
| 2002-03 | | | | | | | | | | | | | |
| 2003-04 | | | | | | | | | | | | | |
| 2004-05 | | | | | | | | | | | | | |
| 2005-06 | | 0.0 | 0.0 | 0.0 | 19.5 | 15.6 | 16.8 | 12.1 | 4.9 | 1.6 | T | 0.0 | |
| 2006-07 | 0.0 | 0.0 | 0.0 | 0.4 | 3.7 | 11.9 | 13.9 | 12.5 | 14.3 | 17.3 | 0.0 | 0.0 | 74.0 |
| 2007-08 | 0.0 | 0.0 | 0.0 | 0.0 | 6.0 | 25.3 | 15.2 | 22.9 | 6.4 | T | 0.0 | 0.0 | 75.8 |
| 2008-09 | 0.0 | 0.0 | 0.0 | T | 18.3 | 45.8 | 16.4 | 13.2 | 5.2 | 0.2 | 0.0 | 0.0 | 99.1 |
| 2009- | 0.0 | 0.0 | 0.0 | T | 0.7 | 20.4 | | | | | | | |
| POR=<br>46 YRS | T | T | T | 0.6 | 8.4 | 14.3 | 15.8 | 10.8 | 8.9 | 3.5 | 0.2 | 0.0 | 62.5 |

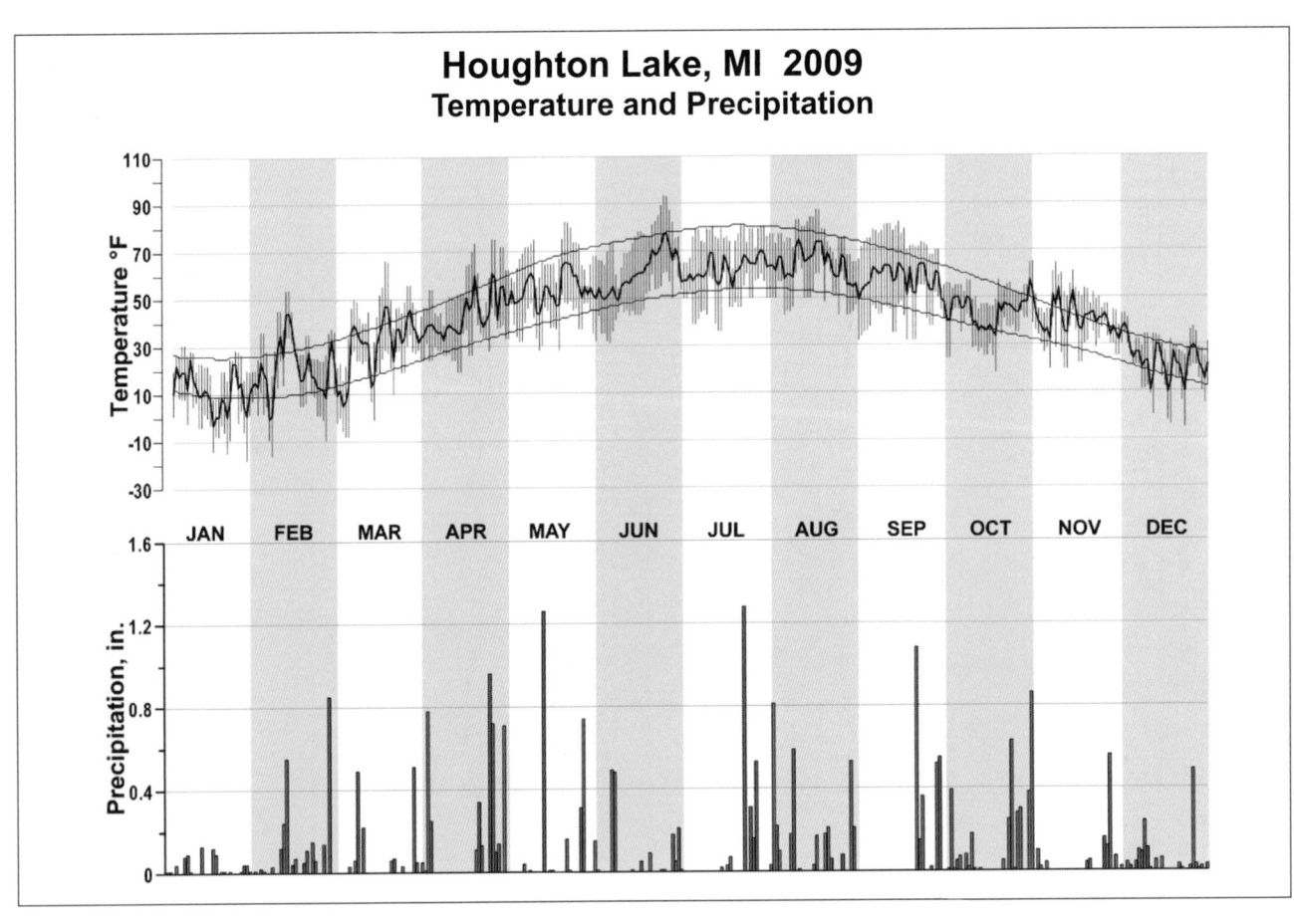

# Houghton Lake, MI 2009
## Temperature and Precipitation

# 2009
# MARQUETTE COUNTY AIRPORT
# MICHIGAN (KMQT)

The Marquette County Airport lies about 7.5 miles southwest of the nearest shoreline of Lake Superior and about 8 miles west of the city of Marquette. Lake Superior is the largest body of fresh water in the world and the deepest and coldest of the Great Lakes. An irregular northwest-southeast ridge line lies just to the east of the airport. There are several water storage basins in the vicinity of the station. One basin, about 20 miles long, is 3 miles northwest and another, about 8 miles in diameter, is 3 miles west.

The climate is influenced considerably by the proximity of Lake Superior. As a consequence of the cool expanse of water in the summer, there is rarely a long period of sweltering hot weather. Periods of drought are extremely rare. In the winter, cold outbreaks are tempered considerably by the waters of Lake Superior if the lake is unfrozen. However, winds blowing across these relatively warmer waters pick up moisture and cause cloudy weather throughout the winter, as well as frequent periods of light snow. Lake-formed snow showers and snow squalls are intensified near the station by upslope winds, especially from the northwest through northeast. With a northeast through east wind, especially in autumn, the upslope condition will cause light snow at the airport, while along the lakeshore, only drizzle or no precipitation may occur.

The growing season averages 117 days. Precipitation is rather evenly distributed throughout the year, with an average precipitation of 4 inches or more in June and September and less than 2 inch averages only in January and February. One hundred inches or more of snow occur in nine of ten winter seasons.

# NORMALS, MEANS, AND EXTREMES
## MARQUETTE (KMQT)

| LATITUDE: 46°31'N | LONGITUDE: -87°32'W | | ELEVATION (FT): GRND: 1415 BARO: 1415 | | | | TIME ZONE: EASTERN (UTC -5) | | | WBAN: 94850 | | |

| | ELEMENT | POR | JAN | FEB | MAR | APR | MAY | JUN | JUL | AUG | SEP | OCT | NOV | DEC | YEAR |
|---|---|---|---|---|---|---|---|---|---|---|---|---|---|---|---|
| **TEMPERATURE °F** | NORMAL DAILY MAXIMUM | 30 | 19.7 | 24.2 | 33.1 | 45.8 | 61.5 | 70.3 | 75.2 | 72.6 | 63.2 | 50.9 | 35.4 | 24.1 | 48.0 |
| | MEAN DAILY MAXIMUM | 48 | 21.0 | 24.5 | 33.9 | 47.5 | 61.8 | 71.4 | 76.2 | 73.7 | 64.7 | 52.0 | 36.7 | 25.3 | 49.1 |
| | HIGHEST DAILY MAXIMUM | 31 | 49 | 61 | 71 | 92 | 93 | 96 | 99 | 96 | 93 | 87 | 73 | 59 | 99 |
| | YEAR OF OCCURRENCE | | 2006 | 1981 | 2000 | 1980 | 2006 | 1995 | 1988 | 2001 | 2002 | 1992 | 1999 | 1998 | JUL 1988 |
| | MEAN OF EXTREME MAXS. | 48 | 38.0 | 43.7 | 55.1 | 72.9 | 84.0 | 87.9 | 89.6 | 88.1 | 82.4 | 73.7 | 56.4 | 41.0 | 67.7 |
| | NORMAL DAILY MINIMUM | 30 | 3.3 | 5.4 | 14.3 | 26.9 | 39.1 | 48.3 | 53.5 | 52.0 | 43.8 | 34.0 | 22.4 | 10.2 | 29.4 |
| | MEAN DAILY MINIMUM | 48 | 4.6 | 5.0 | 13.8 | 27.3 | 38.9 | 48.4 | 53.6 | 52.1 | 44.8 | 34.9 | 23.2 | 11.1 | 29.8 |
| | LOWEST DAILY MINIMUM | 31 | -27 | -34 | -30 | -9 | 17 | 28 | 36 | 34 | 24 | 14 | -8 | -28 | -34 |
| | YEAR OF OCCURRENCE | | 1996 | 1979 | 2003 | 2003 | 1983 | 1986 | 2000 | 1992 | 1993 | 1984 | 2005 | 1983 | FEB 1979 |
| | MEAN OF EXTREME MINS. | 48 | -17.0 | -17.7 | -11.6 | 8.2 | 24.6 | 34.0 | 40.3 | 38.1 | 29.7 | 21.7 | 4.8 | -10.8 | 12.0 |
| | NORMAL DRY BULB | 30 | 11.5 | 14.8 | 23.7 | 36.4 | 50.3 | 59.3 | 64.4 | 62.3 | 53.5 | 42.5 | 28.9 | 17.2 | 38.7 |
| | MEAN DRY BULB | 48 | 13.9 | 16.3 | 25.2 | 38.1 | 50.6 | 60.1 | 65.4 | 63.7 | 55.7 | 44.4 | 31.1 | 19.4 | 40.3 |
| | MEAN WET BULB | | | | | | | | | | | | | | |
| | MEAN DEW POINT | | | | | | | | | | | | | | |
| | NORMAL NO. DAYS WITH: | | | | | | | | | | | | | | |
| | MAXIMUM >= 90 | 30 | 0.0 | 0.0 | 0.0 | * | 0.1 | 0.7 | 1.4 | 0.6 | 0.1 | 0.0 | 0.0 | 0.0 | 2.9 |
| | MAXIMUM <= 32 | 30 | 27.2 | 21.2 | 13.6 | 3.3 | 0.1 | 0.0 | 0.0 | 0.0 | 0.0 | 0.6 | 11.1 | 23.9 | 101.0 |
| | MINIMUM <= 32 | 30 | 31.0 | 27.9 | 29.1 | 22.5 | 8.7 | 0.6 | 0.0 | * | 2.7 | 15.0 | 26.4 | 30.5 | 194.4 |
| | MINIMUM <= 0 | 30 | 12.2 | 10.5 | 4.6 | 0.2 | 0.0 | 0.0 | 0.0 | 0.0 | 0.0 | 0.0 | 0.6 | 7.1 | 35.2 |
| **H/C** | NORMAL HEATING DEG. DAYS | 30 | 1659 | 1405 | 1280 | 859 | 468 | 200 | 92 | 134 | 348 | 700 | 1083 | 1484 | 9712 |
| | NORMAL COOLING DEG. DAYS | 30 | 0 | 0 | 0 | 0 | 12 | 30 | 72 | 50 | 3 | 0 | 0 | 0 | 167 |
| **RH** | NORMAL (PERCENT) | 30 | | | | | | | | | | | | | |
| | HOUR 01 LST | 30 | | | | | | | | | | | | | |
| | HOUR 07 LST | 30 | | | | | | | | | | | | | |
| | HOUR 13 LST | 30 | | | | | | | | | | | | | |
| | HOUR 19 LST | 30 | | | | | | | | | | | | | |
| **S** | PERCENT POSSIBLE SUNSHINE | 24 | 37 | 44 | 50 | 53 | 62 | 66 | 67 | 63 | 57 | 46 | 36 | 34 | 51 |
| **W/O** | MEAN NO. DAYS WITH: | | | | | | | | | | | | | | |
| | HEAVY FOG(VISBY <= 1/4 MI) | 43 | 1.0 | 1.0 | 2.1 | 2.6 | 2.5 | 2.0 | 1.7 | 2.8 | 2.2 | 2.2 | 1.8 | 1.6 | 23.5 |
| | THUNDERSTORMS | 44 | 0.0 | 0.0 | 0.4 | 0.9 | 2.2 | 4.5 | 4.8 | 4.1 | 2.8 | 1.3 | 0.2 | 0.0 | 21.2 |
| **CLOUDNESS** | MEAN: | | | | | | | | | | | | | | |
| | SUNRISE-SUNSET (OKTAS) | | | | | | | | | | | | | | |
| | MIDNIGHT-MIDNIGHT (OKTAS) | | | | | | | | | | | | | | |
| | MEAN NO. DAYS WITH: | | | | | | | | | | | | | | |
| | CLEAR | | | | | | | | | | | | | | |
| | PARTLY CLOUDY | | | | | | | | | | | | | | |
| | CLOUDY | | | | | | | | | | | | | | |
| **PR** | MEAN STATION PRESSURE(IN) | | | | | | | | | | | | | | |
| | MEAN SEA-LEVEL PRES. (IN) | | | | | | | | | | | | | | |
| **WINDS** | MEAN SPEED (MPH) | | | | | | | | | | | | | | |
| | PREVAIL.DIR(TENS OF DEGS) | | | | | | | | | | | | | | |
| | MAXIMUM 2-MINUTE: | 6 | | | | | | | | | | | | | |
| | SPEED (MPH) | | 44 | 31 | 40 | 44 | 34 | 38 | 35 | 37 | 35 | 38 | 31 | 35 | 44 |
| | DIR. (TENS OF DEGS) | | 32 | 32 | 32 | 32 | 36 | 32 | 32 | 32 | 28 | 14 | 32 | 22 | 32 |
| | YEAR OF OCCURRENCE | | 1980 | 1985 | 1982 | 1982 | 1981 | 1984 | 1982 | 1984 | 1983 | 1984 | 1979 | 1982 | APR 1982 |
| | MAXIMUM 3-SECOND | | | | | | | | | | | | | | |
| | SPEED (MPH) | | | | | | | | | | | | | | |
| | DIR. (TENS OF DEGS) | | | | | | | | | | | | | | |
| | YEAR OF OCCURRENCE | | | | | | | | | | | | | | |
| **PRECIPITATION** | NORMAL (IN) | 30 | 2.60 | 1.85 | 3.13 | 2.79 | 3.07 | 3.21 | 3.01 | 3.55 | 3.74 | 3.66 | 3.27 | 2.43 | 36.31 |
| | MAXIMUM MONTHLY (IN) | 31 | 6.61 | 5.35 | 6.08 | 6.56 | 6.95 | 6.61 | 5.40 | 8.59 | 7.48 | 7.59 | 8.25 | 5.05 | 8.59 |
| | YEAR OF OCCURRENCE | | 1997 | 2002 | 1979 | 1985 | 2006 | 1981 | 1991 | 1988 | 2007 | 1979 | 1988 | 2004 | AUG 1988 |
| | MINIMUM MONTHLY (IN) | 31 | 0.70 | 0.48 | 0.56 | 0.90 | 0.06 | 0.61 | 0.57 | 0.44 | 1.21 | 1.65 | 1.00 | 0.37 | 0.06 |
| | YEAR OF OCCURRENCE | | 2003 | 1994 | 1980 | 1998 | 1986 | 1992 | 1981 | 2008 | 1989 | 1994 | 1990 | 1994 | MAY 1986 |
| | MAXIMUM IN 24 HOURS (IN) | 31 | 2.23 | 2.05 | 2.42 | 3.09 | 3.77 | 2.80 | 2.64 | 2.34 | 4.29 | 3.66 | 2.97 | 2.48 | 4.29 |
| | YEAR OF OCCURRENCE | | 1988 | 1983 | 2006 | 1985 | 2006 | 1989 | 1985 | 1988 | 2007 | 1985 | 1988 | 1985 | SEP 2007 |
| | NORMAL NO. DAYS WITH: | | | | | | | | | | | | | | |
| | PRECIPITATION >= 0.01 | 30 | 17.6 | 13.3 | 14.3 | 11.5 | 10.4 | 11.5 | 11.0 | 11.5 | 13.8 | 14.1 | 15.9 | 17.2 | 162.1 |
| | PRECIPITATION >= 1.00 | 30 | 0.3 | 0.1 | 0.7 | 0.5 | 0.7 | 0.6 | 0.5 | 0.8 | 0.7 | 0.8 | 0.6 | 0.2 | 6.5 |
| **SNOWFALL** | NORMAL (IN) | 30 | 42.1 | 29.7 | 31.8 | 12.5 | 1.5 | 0.* | 0.0 | 0.* | 0.1 | 5.9 | 22.6 | 38.3 | 184.5 |
| | MAXIMUM MONTHLY (IN) | 31 | 91.7 | 91.9 | 83.1 | 55.2 | 22.6 | T | T | T | 1.7 | 18.6 | 48.9 | 89.5 | 91.9 |
| | YEAR OF OCCURRENCE | | 1997 | 2002 | 2002 | 2007 | 1990 | 2009 | 2007 | 2009 | 1993 | 1979 | 1991 | 2000 | FEB 2002 |
| | MAXIMUM IN 24 HOURS (IN) | 30 | 23.3 | 20.6 | 28.0 | 20.0 | 17.2 | T | T | T | 1.7 | 12.7 | 24.4 | 25.8 | 28.0 |
| | YEAR OF OCCURRENCE' | | 1988 | 1983 | 1997 | 1985 | 1990 | 2009 | 2000 | 2009 | 1993 | 1989 | 2001 | 1985 | MAR 1997 |
| | MAXIMUM SNOW DEPTH (IN) | 46 | 54 | 56 | 63 | 41 | 12 | 0 | 0 | 0 | 1 | 10 | 27 | 47 | 63 |
| | YEAR OF OCCURRENCE | | 1969 | 1971 | 1997 | 1997 | 1990 | | | | 1974 | 1989 | 1975 | 1983 | MAR 1997 |
| | NORMAL NO. DAYS WITH: | | | | | | | | | | | | | | |
| | SNOWFALL >= 1.0 | 30 | 9.4 | 6.5 | 6.5 | 3.1 | 0.3 | 0.0 | 0.0 | 0.0 | 0.1 | 1.8 | 6.1 | 8.5 | 42.3 |

## PRECIPITATION (inches) 2009 MARQUETTE (KMQT)

| YEAR | JAN | FEB | MAR | APR | MAY | JUN | JUL | AUG | SEP | OCT | NOV | DEC | ANNUAL |
|------|-----|-----|-----|-----|-----|-----|-----|-----|-----|-----|-----|-----|--------|
| 1980 | 2.95 | 1.47 | 0.56 | 4.11 | 1.87 | 2.51 | 3.18 | 3.49 | 6.94 | 2.71 | 1.92 | 2.21 | 33.92 |
| 1981 | 1.96 | 2.18 | 2.30 | 3.10 | 2.43 | 6.61 | 0.57 | 2.28 | 2.11 | 4.63 | 2.00 | 4.33 | 34.50 |
| 1982 | 3.80 | 0.59 | 1.89 | 4.45 | 2.76 | 1.24 | 4.65 | 3.45 | 5.00 | 3.87 | 2.53 | 2.83 | 37.06 |
| 1983 | 2.67 | 3.14 | 4.85 | 3.17 | 6.49 | 1.62 | 1.45 | 3.21 | 4.98 | 5.75 | 5.68 | 3.35 | 46.36 |
| 1984 | 1.27 | 3.68 | 3.99 | 2.46 | 0.79 | 2.81 | 1.92 | 4.29 | 4.29 | 2.03 | 2.35 | 2.60 | 32.48 |
| 1985 | 2.96 | 2.91 | 4.63 | 6.56 | 3.59 | 1.49 | 3.76 | 4.78 | 6.32 | 4.88 | 5.74 | 3.97 | 51.59 |
| 1986 | 2.94 | 1.21 | 4.77 | 2.36 | 0.06 | 2.45 | 3.07 | 4.62 | 2.85 | 4.50 | 1.04 | 0.52 | 30.39 |
| 1987 | 1.42 | 1.53 | 2.07 | 2.30 | 3.93 | 2.01 | 4.88 | 3.58 | 2.66 | 4.31 | 4.05 | 3.56 | 36.30 |
| 1988 | 4.02 | 0.95 | 3.66 | 1.48 | 1.34 | 0.71 | 1.46 | 8.59 | 3.97 | 5.01 | 8.25 | 2.36 | 41.80 |
| 1989 | 1.82 | 1.60 | 2.89 | 1.48 | 2.15 | 5.36 | 0.91 | 2.61 | 1.21 | 3.49 | 3.43 | 2.58 | 29.53 |
| 1990 | 1.74 | 1.96 | 1.88 | 2.32 | 4.25 | 3.13 | 2.01 | 2.70 | 4.76 | 5.27 | 1.00 | 1.71 | 32.73 |
| 1991 | 0.92 | 1.78 | 3.75 | 3.50 | 2.64 | 3.06 | 5.40 | 0.81 | 3.10 | 4.73 | 5.87 | 1.51 | 37.07 |
| 1992 | 1.28 | 2.66 | 2.57 | 2.12 | 1.76 | 0.61 | 4.93 | 3.28 | 3.35 | 1.98 | 4.27 | 2.46 | 31.27 |
| 1993 | 2.39 | 0.68 | 0.99 | 4.11 | 4.61 | 2.42 | 1.95 | 3.98 | 4.41 | 3.32 | 2.84 | 1.31 | 33.01 |
| 1994 | 1.65 | 0.48 | 2.27 | 2.27 | 2.31 | 2.60 | 1.48 | 3.98 | 2.92 | 1.65 | 2.17 | 0.37 | 24.15 |
| 1995 | 1.45 | 2.89 | 3.31 | 2.31 | 3.85 | 1.49 | 3.93 | 4.31 | 3.60 | 6.37 | 2.97 | 2.48 | 38.96 |
| 1996 | 5.22 | 2.91 | 2.09 | 6.50 | 1.70 | 3.94 | 4.43 | 2.11 | 3.58 | 3.64 | 1.87 | 4.45 | 42.44 |
| 1997 | 6.61 | 0.97 | 3.65 | 1.03 | 2.96 | 3.44 | 1.60 | 3.01 | 1.36 | 2.38 | 3.48 | 2.40 | 32.89 |
| 1998 | 2.71 | 1.12 | 5.35 | 0.90 | 1.93 | 3.06 | 0.59 | 2.70 | 3.56 | 1.93 | 2.89 | 1.60 | 28.34 |
| 1999 | 4.93 | 3.15 | 1.04 | 2.21 | 5.74 | 3.05 | 4.78 | 2.35 | 3.09 | 3.47 | 2.27 | 1.54 | 37.62 |
| 2000 | 2.34 | 1.26 | 3.34 | 1.11 | 1.22 | 3.63 | 2.10 | 3.50 | 1.90 | 2.80 | 3.66 | 3.02 | 29.88 |
| 2001 | 2.80 | 2.00 | 1.72 | 3.18 | 2.37 | 4.21 | 3.03 | 1.26 | 3.93 | 3.19 | 4.89 | 1.66 | 34.24 |
| 2002 | 1.08 | 5.35 | 5.71 | 5.12 | 3.13 | 3.81 | 3.39 | 3.07 | 5.71 | 5.09 | 2.13 | 0.56 | 44.15 |
| 2003 | 0.70 | 1.93 | 3.35 | 3.49 | 6.17 | 1.62 | 2.86 | 1.36 | 4.57 | 3.36 | 3.13 | 2.77 | 35.31 |
| 2004 | 1.99 | 2.63 | 4.69 | 2.91 | 5.52 | 2.86 | 3.18 | 3.18 | 1.64 | 3.57 | 2.11 | 5.05 | 39.33 |
| 2005 | 1.67 | 1.91 | 2.54 | 2.12 | 2.80 | 3.41 | 2.07 | 1.95 | 3.97 | 5.18 | 4.30 | 1.72 | 33.64 |
| 2006 | 2.15 | 2.90 | 3.58 | 1.27 | 6.95 | 1.28 | 2.36 | 2.46 | 2.77 | 1.72 | 1.93 | 2.91 | 32.28 |
| 2007 | 2.27 | 1.62 | 4.46 | 5.07 | 2.06 | 1.93 | 2.36 | 0.85 | 7.48 | 5.14 | 2.19 | 3.18 | 38.61 |
| 2008 | 2.18 | 1.53 | 2.07 | 6.39 | 2.03 | 3.28 | 2.85 | 0.44 | 3.69 | 2.62 | 3.25 | 3.46 | 33.79 |
| 2009 | 1.59 | 3.14 | 0.96 | 3.43 | 2.51 | 1.79 | 1.93 | 4.82 | 1.84 | 6.95 | 1.16 | 1.59 | 31.71 |
| POR= 48 YRS | 2.33 | 1.94 | 2.88 | 2.85 | 3.15 | 3.22 | 2.89 | 3.18 | 3.85 | 3.73 | 2.57 | 2.52 | 35.11 |

WBAN : 94850

## AVERAGE TEMPERATURE (°F) 2009 MARQUETTE (KMQT)

| YEAR | JAN | FEB | MAR | APR | MAY | JUN | JUL | AUG | SEP | OCT | NOV | DEC | ANNUAL |
|------|-----|-----|-----|-----|-----|-----|-----|-----|-----|-----|-----|-----|--------|
| 1980 | 13.1 | 12.8 | 21.3 | 39.1 | 54.5 | 56.9 | 65.7 | 64.5 | 52.9 | 38.6 | 29.8 | 15.4 | 38.7 |
| 1981 | 13.9 | 18.4 | 27.3 | 39.0 | 48.1 | 59.6 | 65.7 | 63.8 | 52.0 | 40.1 | 34.5 | 18.7 | 40.1 |
| 1982 | 4.8 | 12.1 | 23.1 | 32.8 | 54.9 | 54.0 | 66.0 | 59.8 | 53.8 | 45.4 | 29.1 | 23.3 | 38.3 |
| 1983 | 18.0 | 21.9 | 26.0 | 33.8 | 44.1 | 60.5 | 70.1 | 67.9 | 57.5 | 43.2 | 31.9 | 8.7 | 40.3 |
| 1984 | 9.0 | 25.3 | 17.5 | 40.3 | 48.2 | 62.0 | 64.4 | 65.4 | 52.4 | 46.7 | 30.5 | 18.7 | 40.0 |
| 1985 | 11.3 | 11.8 | 26.7 | 40.9 | 52.1 | 56.8 | 62.6 | 61.4 | 54.7 | 44.0 | 26.2 | 9.7 | 38.2 |
| 1986 | 14.4 | 14.8 | 26.4 | 42.0 | 54.8 | 57.1 | 66.3 | 60.0 | 53.2 | 42.7 | 25.2 | 21.3 | 39.9 |
| 1987 | 18.0 | 22.9 | 30.1 | 44.6 | 53.7 | 63.8 | 67.8 | 62.7 | 56.9 | 39.0 | 32.6 | 23.5 | 43.0 |
| 1988 | 11.6 | 10.8 | 23.1 | 38.5 | 54.7 | 62.3 | 68.6 | 65.0 | 54.9 | 38.0 | 31.8 | 16.3 | 39.6 |
| 1989 | 19.1 | 8.4 | 19.4 | 34.7 | 50.8 | 57.5 | 66.7 | 63.2 | 55.7 | 45.3 | 24.8 | 9.2 | 37.9 |
| 1990 | 22.0 | 18.1 | 28.5 | 42.3 | 46.9 | 59.9 | 63.7 | 63.2 | 53.7 | 41.3 | 34.0 | 18.0 | 41.0 |
| 1991 | 10.0 | 20.1 | 26.7 | 42.2 | 55.7 | 62.9 | 64.9 | 65.2 | 52.4 | 41.0 | 27.0 | 19.8 | 40.7 |
| 1992 | 17.0 | 19.8 | 22.6 | 34.6 | 52.6 | 56.3 | 58.4 | 59.2 | 53.5 | 41.4 | 28.4 | 20.2 | 38.7 |
| 1993 | 16.9 | 13.9 | 25.6 | 34.4 | 49.8 | 57.8 | 66.1 | 65.4 | 49.9 | 40.3 | 27.0 | 20.8 | 39.0 |
| 1994 | 2.8 | 7.7 | 27.3 | 36.8 | 50.2 | 61.9 | 64.0 | 60.7 | 58.2 | 47.3 | 33.7 | 27.5 | 39.8 |
| 1995 | 17.4 | 15.0 | 28.2 | 32.1 | 50.4 | 66.3 | 65.6 | 68.1 | 53.9 | 44.3 | 22.8 | 16.1 | 40.0 |
| 1996 | 10.5 | 11.9 | 19.5 | 31.5 | 46.6 | 62.3 | 61.6 | 64.6 | 56.3 | 43.8 | 26.8 | 19.5 | 37.9 |
| 1997 | 12.9 | 16.9 | 21.2 | 35.8 | 44.8 | 64.6 | 63.6 | 59.6 | 56.0 | 44.0 | 27.3 | 24.7 | 39.3 |
| 1998 | 17.8 | 28.4 | 24.7 | 41.7 | 55.0 | 59.5 | 65.5 | 65.8 | 58.3 | 47.4 | 34.1 | 22.4 | 43.4 |
| 1999 | 11.7 | 21.5 | 26.8 | 39.3 | 54.0 | 62.3 | 68.8 | 62.2 | 56.7 | 41.8 | 37.2 | 20.7 | 41.9 |
| 2000 | 12.9 | 20.1 | 32.9 | 35.8 | 53.9 | 58.0 | 62.8 | 62.4 | 54.4 | 47.3 | 31.4 | 12.1 | 40.3 |
| 2001 | 20.8 | 12.6 | 23.7 | 41.8 | 53.3 | 61.4 | 65.2 | 66.2 | 55.4 | 43.2 | 39.9 | 25.9 | 42.5 |
| 2002 | 21.3 | 21.8 | 17.8 | 36.2 | 45.2 | 62.1 | 69.6 | 65.0 | 59.9 | 38.0 | 25.9 | 22.2 | 40.4 |
| 2003 | 11.8 | 8.1 | 20.4 | 35.2 | 49.9 | 60.0 | 64.8 | 65.9 | 56.9 | 43.7 | 30.4 | 23.2 | 39.2 |
| 2004 | 8.1 | 20.4 | 25.3 | 36.3 | 48.2 | 58.2 | 61.9 | 59.4 | 61.8 | 45.1 | 33.9 | 16.8 | 39.6 |
| 2005 | 12.2 | 20.0 | 21.0 | 42.6 | 49.1 | 65.1 | 67.3 | 65.6 | 60.9 | 47.8 | 29.4 | 18.5 | 41.6 |
| 2006 | 24.0 | 14.5 | 27.3 | 43.9 | 52.8 | 60.9 | 68.7 | 63.6 | 53.6 | 40.0 | 34.7 | 25.5 | 42.5 |
| 2007 | 5.8 | 10.5 | 28.5 | 35.8 | 54.8 | 63.8 | 66.2 | 64.7 | 57.9 | 49.2 | 28.6 | 17.8 | 40.3 |
| 2008 | 4.6 | 12.0 | 20.0 | 37.9 | 45.9 | 60.4 | 64.8 | 63.6 | 56.8 | 44.4 | 31.2 | 13.4 | 37.9 |
| 2009 | 7.2 | 15.9 | 25.4 | 37.2 | 47.9 | 57.7 | 59.8 | 61.9 | 59.5 | 39.1 | 37.3 | -4.3 | 37.1 |
| POR= 48 YRS | 13.9 | 16.3 | 25.2 | 38.1 | 50.6 | 60.1 | 65.4 | 63.7 | 55.7 | 44.4 | 31.1 | 19.4 | 40.3 |

## HEATING DEGREE DAYS (base 65°F) 2009 MARQUETTE (KMQT)

| YEAR | JUL | AUG | SEP | OCT | NOV | DEC | JAN | FEB | MAR | APR | MAY | JUN | TOTAL |
|---|---|---|---|---|---|---|---|---|---|---|---|---|---|
| 1980-81 | 53 | 68 | 361 | 813 | 1050 | 1532 | 1581 | 1303 | 1160 | 776 | 519 | 168 | 9384 |
| 1981-82 | 73 | 78 | 384 | 766 | 907 | 1425 | 1864 | 1480 | 1292 | 960 | 313 | 326 | 9868 |
| 1982-83 | 37 | 190 | 350 | 598 | 1069 | 1289 | 1446 | 1201 | 1205 | 929 | 638 | 207 | 9159 |
| 1983-84 | 35 | 38 | 264 | 672 | 989 | 1740 | 1737 | 1145 | 1469 | 733 | 517 | 124 | 9463 |
| 1984-85 | 71 | 72 | 369 | 560 | 1028 | 1431 | 1658 | 1486 | 1178 | 724 | 400 | 249 | 9226 |
| 1985-86 | 111 | 146 | 323 | 645 | 1156 | 1709 | 1564 | 1400 | 1191 | 684 | 334 | 250 | 9513 |
| 1986-87 | 71 | 169 | 349 | 684 | 1185 | 1348 | 1450 | 1171 | 1075 | 606 | 387 | 106 | 8601 |
| 1987-88 | 58 | 130 | 247 | 800 | 964 | 1278 | 1650 | 1569 | 1295 | 787 | 347 | 166 | 9291 |
| 1988-89 | 37 | 101 | 305 | 833 | 988 | 1504 | 1416 | 1584 | 1412 | 902 | 432 | 245 | 9759 |
| 1989-90 | 65 | 117 | 283 | 610 | 1199 | 1727 | 1324 | 1308 | 1123 | 702 | 556 | 176 | 9190 |
| 1990-91 | 104 | 111 | 347 | 726 | 921 | 1454 | 1703 | 1250 | 1182 | 678 | 334 | 115 | 8925 |
| 1991-92 | 74 | 112 | 385 | 739 | 1133 | 1394 | 1483 | 1308 | 1306 | 905 | 398 | 276 | 9513 |
| 1992-93 | 208 | 202 | 340 | 732 | 1092 | 1383 | 1485 | 1424 | 1215 | 910 | 474 | 233 | 9698 |
| 1993-94 | 38 | 78 | 446 | 760 | 1130 | 1366 | 1927 | 1601 | 1164 | 844 | 470 | 135 | 9959 |
| 1994-95 | 84 | 153 | 220 | 538 | 932 | 1155 | 1471 | 1395 | 1137 | 978 | 452 | 106 | 8621 |
| 1995-96 | 68 | 21 | 336 | 637 | 1259 | 1509 | 1688 | 1538 | 1407 | 997 | 566 | 141 | 10167 |
| 1996-97 | 126 | 73 | 294 | 649 | 1140 | 1402 | 1607 | 1343 | 1350 | 867 | 620 | 89 | 9560 |
| 1997-98 | 111 | 193 | 276 | 649 | 1127 | 1243 | 1455 | 1019 | 1243 | 690 | 326 | 203 | 8535 |
| 1998-99 | 65 | 51 | 221 | 538 | 920 | 1314 | 1646 | 1213 | 1178 | 764 | 346 | 153 | 8409 |
| 1999-00 | 40 | 111 | 289 | 711 | 827 | 1364 | 1609 | 1295 | 990 | 867 | 349 | 208 | 8660 |
| 2000-01 | 117 | 108 | 317 | 543 | 1000 | 1632 | 1365 | 1460 | 1268 | 693 | 357 | 189 | 9049 |
| 2001-02 | 94 | 63 | 302 | 672 | 745 | 1205 | 1345 | 1203 | 1456 | 861 | 614 | 150 | 8710 |
| 2002-03 | 34 | 62 | 209 | 830 | 1169 | 1321 | 1641 | 1588 | 1375 | 888 | 460 | 194 | 9771 |
| 2003-04 | 78 | 62 | 272 | 652 | 1031 | 1289 | 1759 | 1285 | 1223 | 852 | 519 | 221 | 9243 |
| 2004-05 | 137 | 195 | 143 | 610 | 926 | 1486 | 1630 | 1257 | 1358 | 662 | 486 | 75 | 8965 |
| 2005-06 | 77 | 72 | 177 | 542 | 1060 | 1435 | 1264 | 1411 | 1162 | 629 | 400 | 159 | 8388 |
| 2006-07 | 41 | 89 | 347 | 765 | 901 | 1218 | 1472 | 1520 | 1124 | 872 | 341 | 121 | 8811 |
| 2007-08 | 80 | 83 | 247 | 496 | 1088 | 1452 | 1511 | 1533 | 1389 | 806 | 587 | 157 | 9429 |
| 2008-09 | 54 | 79 | 268 | 631 | 1007 | 1594 | 1783 | 1366 | 1222 | 826 | 524 | 251 | 9605 |
| 2009- | 166 | 148 | 177 | 798 | 826 | 1783 | | | | | | | |

WBAN : 94850

## COOLING DEGREE DAYS (base 65°F) 2009 MARQUETTE (KMQT)

| YEAR | JAN | FEB | MAR | APR | MAY | JUN | JUL | AUG | SEP | OCT | NOV | DEC | TOTAL |
|---|---|---|---|---|---|---|---|---|---|---|---|---|---|
| 1980 | 0 | 0 | 0 | 2 | 21 | 36 | 83 | 61 | 5 | 0 | 0 | 0 | 208 |
| 1981 | 0 | 0 | 0 | 0 | 1 | 13 | 103 | 50 | 2 | 0 | 0 | 0 | 169 |
| 1982 | 0 | 0 | 0 | 0 | 7 | 2 | 75 | 35 | 23 | 0 | 0 | 0 | 142 |
| 1983 | 0 | 0 | 0 | 0 | 0 | 78 | 200 | 135 | 47 | 1 | 0 | 0 | 461 |
| 1984 | 0 | 0 | 0 | 0 | 3 | 42 | 59 | 89 | 0 | 0 | 0 | 0 | 193 |
| 1985 | 0 | 0 | 0 | 7 | 7 | 10 | 44 | 41 | 21 | 0 | 0 | 0 | 130 |
| 1986 | 0 | 0 | 0 | 1 | 1 | 17 | 119 | 21 | 0 | 0 | 0 | 0 | 179 |
| 1987 | 0 | 0 | 0 | 3 | 41 | 77 | 150 | 65 | 13 | 0 | 0 | 0 | 349 |
| 1988 | 0 | 0 | 0 | 0 | 32 | 89 | 157 | 106 | 8 | 0 | 0 | 0 | 392 |
| 1989 | 0 | 0 | 0 | 0 | 1 | 27 | 123 | 70 | 13 | 1 | 0 | 0 | 235 |
| 1990 | 0 | 0 | 0 | 24 | 0 | 29 | 70 | 63 | 14 | 0 | 0 | 0 | 200 |
| 1991 | 0 | 0 | 0 | 0 | 51 | 60 | 77 | 126 | 13 | 0 | 0 | 0 | 327 |
| 1992 | 0 | 0 | 0 | 0 | 21 | 22 | 8 | 30 | 3 | 5 | 0 | 0 | 89 |
| 1993 | 0 | 0 | 0 | 0 | 10 | 22 | 79 | 97 | 0 | 0 | 0 | 0 | 208 |
| 1994 | 0 | 0 | 0 | 0 | 18 | 49 | 62 | 29 | 24 | 0 | 0 | 0 | 182 |
| 1995 | 0 | 0 | 0 | 0 | 8 | 152 | 93 | 123 | 8 | 5 | 0 | 0 | 389 |
| 1996 | 0 | 0 | 0 | 0 | 3 | 70 | 29 | 67 | 40 | 0 | 0 | 0 | 209 |
| 1997 | 0 | 0 | 0 | 0 | 0 | 80 | 74 | 33 | 9 | 3 | 0 | 0 | 199 |
| 1998 | 0 | 0 | 0 | 0 | 25 | 47 | 86 | 83 | 26 | 0 | 0 | 0 | 267 |
| 1999 | 0 | 0 | 0 | 0 | 13 | 77 | 162 | 31 | 47 | 0 | 0 | 0 | 330 |
| 2000 | 0 | 0 | 0 | 0 | 14 | 7 | 57 | 35 | 6 | 1 | 0 | 0 | 120 |
| 2001 | 0 | 0 | 0 | 5 | 1 | 91 | 110 | 109 | 21 | 0 | 0 | 0 | 337 |
| 2002 | 0 | 0 | 0 | 7 | 8 | 73 | 185 | 71 | 63 | 0 | 0 | 0 | 407 |
| 2003 | 0 | 0 | 0 | 0 | 0 | 49 | 75 | 100 | 36 | 2 | 0 | 0 | 262 |
| 2004 | 0 | 0 | 0 | 0 | 4 | 23 | 48 | 29 | 52 | 0 | 0 | 0 | 156 |
| 2005 | 0 | 0 | 0 | 0 | 0 | 86 | 155 | 100 | 60 | 15 | 0 | 0 | 416 |
| 2006 | 0 | 0 | 0 | 0 | 29 | 43 | 163 | 52 | 10 | 0 | 0 | 0 | 297 |
| 2007 | 0 | 0 | 0 | 1 | 32 | 95 | 125 | 80 | 41 | 12 | 0 | 0 | 386 |
| 2008 | 0 | 0 | 0 | 0 | 0 | 25 | 55 | 44 | 28 | 0 | 0 | 0 | 152 |
| 2009 | 0 | 0 | 0 | 0 | 0 | 38 | 10 | 58 | 18 | 0 | 0 | 0 | 124 |

## SNOWFALL (inches) 2009 MARQUETTE (KMQT)

| YEAR | JUL | AUG | SEP | OCT | NOV | DEC | JAN | FEB | MAR | APR | MAY | JUN | TOTAL |
|---|---|---|---|---|---|---|---|---|---|---|---|---|---|
| 1980-81 | 0.0 | 0.0 | 0.1 | 11.9 | 13.1 | 41.5 | 41.9 | 29.5 | 34.0 | 4.1 | T | 0.0 | 176.1 |
| 1981-82 | 0.0 | 0.0 | 0.0 | 14.2 | 15.3 | 82.6 | 68.8 | 9.6 | 24.1 | 29.2 | 0.0 | 0.0 | 243.8 |
| 1982-83 | 0.0 | 0.0 | 0.0 | 7.2 | 15.1 | 17.4 | 42.4 | 42.9 | 54.1 | 20.2 | T | 0.0 | 199.3 |
| 1983-84 | 0.0 | 0.0 | T | 2.5 | 31.6 | 54.3 | 30.2 | 38.1 | 46.3 | 1.1 | T | 0.0 | 204.1 |
| 1984-85 | 0.0 | 0.0 | T | T | 13.6 | 30.6 | 56.2 | 51.4 | 59.1 | 18.1 | 0.0 | 0.0 | 229.0 |
| 1985-86 | 0.0 | 0.0 | 0.0 | T | 28.1 | 56.4 | 49.8 | 17.4 | 49.1 | 6.8 | 0.1 | 0.0 | 207.7 |
| 1986-87 | 0.0 | T | 0.0 | 2.3 | 15.7 | 10.5 | 21.9 | 27.7 | 19.3 | 11.3 | 0.1 | 0.0 | 108.8 |
| 1987-88 | 0.0 | 0.0 | T | 8.3 | 21.4 | 36.1 | 62.8 | 28.0 | 49.0 | 5.8 | 0.0 | 0.0 | 211.4 |
| 1988-89 | 0.0 | 0.0 | 0.0 | 10.9 | 42.0 | 41.7 | 43.1 | 33.7 | 44.3 | 7.8 | 0.2 | 0.0 | 223.7 |
| 1989-90 | 0.0 | T | 0.5 | 16.6 | 41.7 | 58.4 | 26.4 | 35.8 | 16.3 | 17.1 | 22.6 | T | 235.4 |
| 1990-91 | 0.0 | 0.0 | T | 9.6 | 2.5 | 27.8 | 25.7 | 35.2 | 21.1 | 9.0 | 0.6 | 0.0 | 131.5 |
| 1991-92 | 0.0 | 0.0 | T | 2.9 | 48.9 | 30.7 | 33.3 | 49.6 | 23.3 | 16.7 | T | 0.0 | 205.4 |
| 1992-93 | 0.0 | 0.0 | T | 11.5 | 30.0 | 25.7 | 40.9 | 19.9 | 18.1 | 29.1 | T | 0.0 | 175.2 |
| 1993-94 | 0.0 | 0.0 | 1.7 | 8.7 | 29.0 | 22.9 | 46.8 | 12.4 | 28.3 | 2.7 | 0.1 | T | 152.6 |
| 1994-95 | 0.0 | 0.0 | 0.0 | T | 14.6 | 7.4 | 23.8 | 63.6 | 45.9 | 10.2 | 0.0 | 0.0 | 165.5 |
| 1995-96 | 0.0 | 0.0 | T | 3.3 | 31.7 | 45.1 | 66.6 | 37.8 | 22.9 | 43.4 | 0.6 | 0.0 | 251.4 |
| 1996-97 | 0.0 | 0.0 | 0.0 | 1.5 | 15.6 | 79.8 | 91.7 | 18.7 | 50.4 | 4.8 | 9.1 | 0.0 | 271.6 |
| 1997-98 | 0.0 | 0.0 | 0.0 | 8.3 | 37.2 | 38.8 | 41.7 | 15.3 | 35.4 | 5.7 | 0.0 | T | 182.4 |
| 1998-99 | 0.0 | 0.0 | T | 0.3 | 9.3 | 28.0 | 67.7 | 34.4 | 20.9 | 1.8 | 0.0 | 0.0 | 162.4 |
| 1999-00 | 0.0 | 0.0 | 0.0 | 3.5 | 7.5 | 30.0 | 44.4 | 30.8 | 27.4 | 8.8 | 0.0 | T | 152.4 |
| 2000-01 | T | 0.0 | 0.0 | 12.1 | 38.0 | 89.5 | 44.5 | 41.5 | 40.2 | 2.2 | 0.0 | 0.0 | 268.0 |
| 2001-02 | 0.0 | 0.0 | 0.0 | 13.7 | 39.3 | 37.4 | 35.2 | 91.9 | 83.1 | 18.0 | 1.2 | 0.0 | 319.8 |
| 2002-03 | 0.0 | 0.0 | 0.0 | 17.0 | 38.4 | 11.4 | 23.2 | 38.5 | 32.1 | 22.5 | 0.0 | 0.0 | 183.1 |
| 2003-04 | 0.0 | 0.0 | 0.1 | 10.5 | 23.5 | 41.4 | 51.5 | 39.6 | 43.1 | 7.0 | T | 0.0 | 216.7 |
| 2004-05 | 0.0 | 0.0 | 0.0 | 2.8 | 8.0 | 81.1 | 38.2 | 40.7 | 44.1 | 3.5 | 1.6 | 0.0 | 220.0 |
| 2005-06 | 0.0 | 0.0 | T | 0.5 | 40.3 | 34.6 | 31.0 | 54.9 | 29.4 | 2.0 | 0.7 | T | 193.4 |
| 2006-07 | 0.0 | 0.0 | T | 7.3 | 8.6 | 36.1 | 52.7 | 25.5 | 45.1 | 55.2 | T | T | 230.5 |
| 2007-08 | T | 0.0 | 0.0 | 0.1 | 26.5 | 48.4 | 35.1 | 24.6 | 31.4 | 38.5 | T | T | 204.6 |
| 2008-09 | 0.0 | 0.0 | 0.0 | 2.2 | 43.6 | 68.9 | 43.1 | 58.9 | 8.1 | 21.1 | 0.1 | T | 246.0 |
| 2009- | 0.0 | T | 0.0 | 2.0 | 4.6 | 43.1 | | | | | | | |
| POR= 48 YRS | T | T | 0.2 | 5.4 | 20.5 | 36.4 | 37.5 | 30.1 | 29.7 | 12.0 | 1.3 | T | 173.1 |

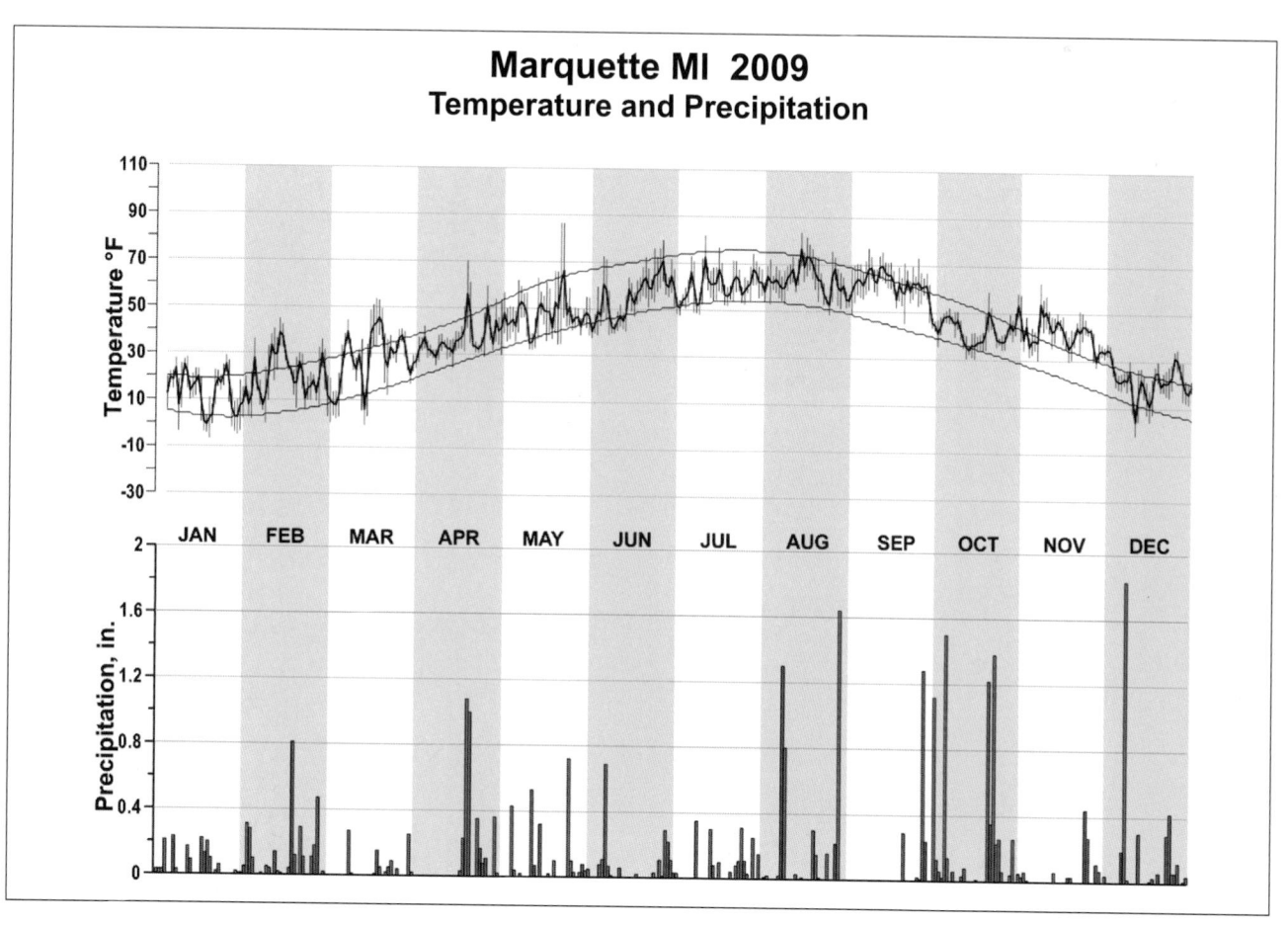

**Marquette MI 2009**
**Temperature and Precipitation**

# 2009
# DULUTH
# MINNESOTA (KDLH)

Duluth, Minnesota is located at the western tip of Lake Superior. The city, about 20 miles long, lies at the base of a range of hills that rise abruptly to 600 - 800 feet above the level of Lake Superior. The range runs in a northeast and southwest direction. Two or 3 miles from the lake the land becomes a slightly rolling plateau.

Duluth in the summer is known as the Air Conditioned City. Being situated below high terrain and along the lake, any easterly component winds automatically cool the city. However, with westerly flow in the summer, the wind generally abates at night, thus, allowing cool lake air to move back into the city area near the lake.

An important influence on the climate is the passage of a succession of high and low pressure systems west and east. The proximity of Lake Superior, which is the largest and coldest of the Great Lakes, modifies the local weather. Summer temperatures are cooler and winter temperatures are warmer. The lake effect at Duluth is most prevalent when low pressure systems pass to the south creating easterly winds. In the summer, warm, moist air flowing over the cold lake surface has a stabilizing effect that results in cool, cloudy weather over Duluth. However, during the winter cold air flowing over the warm open lake surface absorbs moisture that is later precipitated over Duluth as snow. The lake effect is further reflected from the low frequency of severe storms such as wind, hail, tornadoes, freezing rain (glaze), and blizzards when compared to other areas that are a further distance from the lake.

Easterly component winds at Duluth occur 40 to 50 percent of the time from March through August and 20 to 25 percent of the time from November through February. During the winter 60 to 70 percent of the winds are from a westerly component.

The climate of Duluth is predominantly continental with significant local Lake Superior effects. Duluth averages 143 days between the last occurrence of 32 degrees in mid-May and the first in early October. At the Duluth Airport about six miles away from the lake, the average first and last occurrences of 32 degrees are late May and late September, giving a freeze-free period of 123 days.

Fall colors throughout this area are outstanding. Reds, yellows, browns, and combinations of these are an experience to see. Recreation is superb from December through March for cross-country and down-hill skiing and snowmobiling. The snow is dry.

Ice in the harbor forms about mid-November and generally is gone by mid-April. The shipping season can vary from year to year depending on temperatures and the winds that move the ice around. In most years there is little or no shipping during February and March on Lake Superior.

# NORMALS, MEANS, AND EXTREMES
## DULUTH (KDLH)

LATITUDE: 46 ° 50'N  LONGITUDE: -92 ° 11'W  ELEVATION (FT): GRND: 1422  BARO: 1429  TIME ZONE: CENTRAL (UTC -6)  WBAN: 14913

| ELEMENT | POR | JAN | FEB | MAR | APR | MAY | JUN | JUL | AUG | SEP | OCT | NOV | DEC | YEAR |
|---|---|---|---|---|---|---|---|---|---|---|---|---|---|---|
| **TEMPERATURE °F** | | | | | | | | | | | | | | |
| NORMAL DAILY MAXIMUM | 30 | 17.9 | 24.4 | 34.2 | 49.0 | 63.4 | 71.2 | 76.3 | 73.9 | 64.5 | 52.5 | 35.2 | 22.3 | 48.7 |
| MEAN DAILY MAXIMUM | 82 | 17.7 | 22.1 | 32.9 | 47.4 | 61.1 | 69.5 | 76.3 | 74.1 | 64.0 | 52.6 | 35.1 | 22.4 | 47.9 |
| HIGHEST DAILY MAXIMUM | 68 | 52 | 55 | 78 | 88 | 90 | 94 | 97 | 97 | 95 | 86 | 71 | 55 | 97 |
| YEAR OF OCCURRENCE | | 1942 | 2000 | 1946 | 1952 | 1986 | 1995 | 2006 | 1947 | 1976 | 1953 | 1999 | 1962 | JUL 2006 |
| MEAN OF EXTREME MAXS. | 82 | 36.6 | 41.2 | 51.6 | 71.0 | 81.4 | 85.6 | 88.7 | 87.2 | 81.7 | 73.0 | 55.1 | 39.4 | 66.0 |
| NORMAL DAILY MINIMUM | 30 | -1.2 | 5.1 | 16.5 | 28.9 | 40.2 | 48.5 | 54.6 | 53.5 | 44.8 | 34.5 | 20.7 | 5.6 | 29.3 |
| MEAN DAILY MINIMUM | 82 | 0.4 | 4.1 | 15.6 | 29.1 | 39.8 | 48.2 | 55.1 | 54.2 | 45.3 | 35.5 | 21.4 | 7.3 | 29.7 |
| LOWEST DAILY MINIMUM | 68 | -39 | -39 | -29 | -5 | 17 | 27 | 35 | 32 | 22 | 8 | -23 | -34 | -39 |
| YEAR OF OCCURRENCE | | 1972 | 1996 | 1989 | 1975 | 1967 | 1972 | 1988 | 1986 | 1942 | 1976 | 1964 | 1983 | FEB 1996 |
| MEAN OF EXTREME MINS. | 82 | -24.4 | -19.4 | -9.6 | 12.6 | 26.9 | 36.0 | 43.0 | 40.9 | 30.0 | 20.4 | 0.6 | -15.9 | 11.8 |
| NORMAL DRY BULB | 30 | 8.4 | 14.8 | 25.4 | 39.0 | 51.8 | 59.9 | 65.5 | 63.7 | 54.7 | 43.5 | 28.0 | 14.0 | 39.1 |
| MEAN DRY BULB | 82 | 9.1 | 13.1 | 24.3 | 38.3 | 50.4 | 59.0 | 65.7 | 64.2 | 54.7 | 44.1 | 28.2 | 14.9 | 38.8 |
| MEAN WET BULB | 26 | 9.2 | 13.0 | 22.3 | 33.0 | 44.1 | 53.8 | 59.1 | 58.4 | 50.9 | 38.4 | 25.7 | 13.9 | 35.2 |
| MEAN DEW POINT | 26 | 6.8 | 9.2 | 17.8 | 27.6 | 39.0 | 50.7 | 56.6 | 56.1 | 48.1 | 34.8 | 22.7 | 11.3 | 31.7 |
| NORMAL NO. DAYS WITH: | | | | | | | | | | | | | | |
| MAXIMUM >= 90 | 30 | 0.0 | 0.0 | 0.0 | 0.0 | * | 0.2 | 1.1 | 0.6 | 0.1 | 0.0 | 0.0 | 0.0 | 2.0 |
| MAXIMUM <= 32 | 30 | 28.2 | 21.4 | 14.2 | 2.2 | 0.0 | 0.0 | 0.0 | 0.0 | 0.0 | 0.8 | 13.2 | 26.1 | 106.1 |
| MINIMUM <= 32 | 30 | 31.0 | 27.8 | 28.6 | 19.4 | 5.1 | 0.3 | 0.0 | * | 2.3 | 12.4 | 26.2 | 30.8 | 183.9 |
| MINIMUM <= 0 | 30 | 16.8 | 11.1 | 4.0 | 0.1 | 0.0 | 0.0 | 0.0 | 0.0 | 0.0 | 0.0 | 1.7 | 11.7 | 45.4 |
| **H/C** | | | | | | | | | | | | | | |
| NORMAL HEATING DEG. DAYS | 30 | 1771 | 1422 | 1244 | 787 | 421 | 180 | 69 | 106 | 331 | 682 | 1124 | 1587 | 9724 |
| NORMAL COOLING DEG. DAYS | 30 | 0 | 0 | 0 | 0 | 7 | 28 | 82 | 60 | 12 | 0 | 0 | 0 | 189 |
| **RH** | | | | | | | | | | | | | | |
| NORMAL (PERCENT) | 30 | 75 | 73 | 71 | 63 | 63 | 71 | 73 | 76 | 77 | 73 | 77 | 78 | 73 |
| HOUR 00 LST | 30 | 77 | 75 | 75 | 70 | 72 | 81 | 84 | 86 | 85 | 78 | 79 | 80 | 79 |
| HOUR 06 LST | 30 | 79 | 79 | 80 | 76 | 76 | 83 | 87 | 90 | 89 | 83 | 82 | 81 | 82 |
| HOUR 12 LST | 30 | 71 | 67 | 64 | 55 | 53 | 60 | 61 | 64 | 64 | 63 | 71 | 74 | 64 |
| HOUR 18 LST | 30 | 71 | 67 | 64 | 54 | 52 | 59 | 61 | 65 | 69 | 67 | 73 | 75 | 65 |
| **S** PERCENT POSSIBLE SUNSHINE | 47 | 48 | 53 | 55 | 56 | 57 | 58 | 65 | 61 | 52 | 46 | 35 | 39 | 52 |
| **W/O** MEAN NO. DAYS WITH: | | | | | | | | | | | | | | |
| HEAVY FOG(VISBY <= 1/4 MI) | 46 | 2.5 | 2.4 | 4.1 | 3.6 | 4.9 | 5.7 | 4.8 | 5.7 | 5.5 | 4.3 | 3.5 | 3.4 | 50.4 |
| THUNDERSTORMS | 62 | 0.1 | 0.0 | 0.5 | 1.6 | 3.5 | 6.4 | 7.9 | 6.5 | 3.9 | 1.4 | 0.3 | 0.1 | 32.2 |
| **CLOUDINESS** MEAN: | | | | | | | | | | | | | | |
| SUNRISE-SUNSET (OKTAS) | 48 | 5.4 | 5.2 | 5.4 | 5.4 | 5.3 | 5.3 | 4.7 | 4.7 | 5.2 | 5.3 | 6.0 | 5.6 | 5.3 |
| MIDNIGHT-MIDNIGHT (OKTAS) | 32 | 5.0 | 5.0 | 5.2 | 5.0 | 4.8 | 4.8 | 4.3 | 4.3 | 4.7 | 5.0 | 5.7 | 5.4 | 4.9 |
| MEAN NO. DAYS WITH: | | | | | | | | | | | | | | |
| CLEAR | 48 | 7.4 | 7.3 | 7.0 | 6.2 | 6.2 | 5.1 | 6.7 | 7.2 | 6.3 | 6.4 | 4.4 | 6.0 | 76.2 |
| PARTLY CLOUDY | 48 | 7.0 | 6.2 | 6.9 | 7.9 | 9.6 | 11.0 | 12.7 | 11.9 | 8.4 | 7.6 | 5.5 | 5.8 | 100.5 |
| CLOUDY | 48 | 16.7 | 14.7 | 17.0 | 15.9 | 15.2 | 13.9 | 10.9 | 11.3 | 14.6 | 16.4 | 19.5 | 18.6 | 184.7 |
| **PR** MEAN STATION PRESSURE(IN) | 26 | 28.45 | 28.48 | 28.49 | 28.45 | 28.43 | 28.43 | 28.46 | 28.50 | 28.48 | 28.46 | 28.44 | 28.45 | 28.46 |
| MEAN SEA-LEVEL PRES. (IN) | 26 | 30.07 | 30.08 | 30.06 | 29.99 | 29.96 | 29.94 | 29.96 | 30.01 | 30.00 | 30.00 | 30.01 | 30.05 | 30.01 |
| **WINDS** MEAN SPEED (MPH) | 26 | 10.7 | 10.5 | 10.9 | 11.1 | 10.6 | 9.5 | 8.7 | 8.7 | 9.6 | 10.5 | 10.7 | 10.5 | 10.2 |
| PREVAIL.DIR(TENS OF DEGS) | 35 | 31 | 31 | 09 | 09 | 09 | 11 | 11 | 12 | 31 | 31 | 32 | 33 | 09 |
| MAXIMUM 2-MINUTE: | | | | | | | | | | | | | | |
| SPEED (MPH) | 13 | 37 | 44 | 47 | 46 | 44 | 44 | 43 | 46 | 39 | 39 | 44 | 41 | 47 |
| DIR. (TENS OF DEGS) | | 28 | 29 | 08 | 08 | 28 | 25 | 29 | 29 | 13 | 31 | 08 | 28 | 08 |
| YEAR OF OCCURRENCE | | 2009 | 2002 | 1998 | 2003 | 2002 | 2007 | 2007 | 2001 | 2007 | 2006 | 1998 | 2001 | MAR 1998 |
| MAXIMUM 3-SECOND: | | | | | | | | | | | | | | |
| SPEED (MPH) | 13 | 52 | 54 | 57 | 62 | 52 | 53 | 60 | 62 | 49 | 49 | 55 | 56 | 62 |
| DIR. (TENS OF DEGS) | | 28 | 28 | 08 | 07 | 29 | 23 | 27 | 29 | 13 | 01 | 09 | 31 | 07 |
| YEAR OF OCCURRENCE | | 2009 | 2002 | 1998 | 2008 | 2002 | 2008 | 2007 | 2001 | 2007 | 1999 | 1998 | 1999 | APR 2008 |
| **PRECIPITATION** NORMAL (IN) | 30 | 1.12 | 0.83 | 1.69 | 2.09 | 2.95 | 4.25 | 4.20 | 4.22 | 4.13 | 2.46 | 2.12 | 0.94 | 31.00 |
| MAXIMUM MONTHLY (IN) | 68 | 4.70 | 2.72 | 5.12 | 8.18 | 7.67 | 8.04 | 8.74 | 10.31 | 9.38 | 7.53 | 5.08 | 3.70 | 10.31 |
| YEAR OF OCCURRENCE | | 1969 | 1998 | 1965 | 2001 | 1962 | 1986 | 1999 | 1972 | 1991 | 1949 | 2000 | 1968 | AUG 1972 |
| MINIMUM MONTHLY (IN) | 68 | 0.13 | 0.13 | 0.22 | 0.24 | 0.15 | 0.55 | .82 | 0.71 | 0.19 | 0.13 | 0.19 | 0.13 | 0.13 |
| YEAR OF OCCURRENCE | | 2008 | 1988 | 1959 | 1987 | 1976 | 1995 | 2005 | 1970 | 1952 | 1944 | 1976 | 1997 | JAN 2008 |
| MAXIMUM IN 24 HOURS (IN) | 68 | 1.74 | 1.38 | 2.38 | 2.60 | 3.25 | 4.05 | 3.68 | 5.79 | 3.77 | 2.90 | 2.64 | 2.12 | 5.79 |
| YEAR OF OCCURRENCE | | 1975 | 1965 | 1977 | 2001 | 1979 | 1958 | 1987 | 1978 | 1972 | 1973 | 1968 | 1950 | AUG 1978 |
| NORMAL NO. DAYS WITH: | | | | | | | | | | | | | | |
| PRECIPITATION >= 0.01 | 30 | 11.6 | 9.2 | 10.7 | 10.4 | 11.1 | 12.7 | 11.8 | 11.7 | 12.8 | 10.6 | 11.1 | 10.3 | 134.0 |
| PRECIPITATION >= 1.00 | 30 | * | * | 0.2 | 0.1 | 0.4 | 0.9 | 1.3 | 1.3 | 1.0 | 0.4 | 0.4 | * | 6.0 |
| **SNOWFALL** NORMAL (IN) | 30 | 19.6 | 11.4 | 13.7 | 6.3 | 0.3 | 0.* | 0.0 | 0.0 | 0.1 | 1.6 | 15.2 | 14.9 | 83.1 |
| MAXIMUM MONTHLY (IN) | 66 | 46.8 | 32.1 | 45.5 | 31.5 | 8.1 | 0.2 | T | T | 2.4 | 9.7 | 50.1 | 44.3 | 50.1 |
| YEAR OF OCCURRENCE | | 1969 | 2001 | 1965 | 1950 | 1954 | 1945 | 1992 | 2008 | 1991 | 1995 | 1991 | 1950 | NOV 1991 |
| MAXIMUM IN 24 HOURS (IN) | 66 | 23.2 | 17.0 | 19.4 | 11.6 | 4.3 | 0.2 | T | T | 2.4 | 7.9 | 24.1 | 25.4 | 25.4 |
| YEAR OF OCCURRENCE' | | 2004 | 1948 | 1965 | 1983 | 1954 | 1945 | 1992 | 1995 | 1991 | 1966 | 1991 | 1950 | DEC 1950 |
| MAXIMUM SNOW DEPTH (IN) | 61 | 42 | 38 | 48 | 41 | 9 | 0 | 0 | 0 | 0 | 6 | 30 | 32 | 48 |
| YEAR OF OCCURRENCE | | 1969 | 1969 | 1965 | 1965 | 1950 | | | | | 1966 | 1991 | 1983 | MAR 1965 |
| NORMAL NO. DAYS WITH: | | | | | | | | | | | | | | |
| SNOWFALL >= 1.0 | 30 | 4.5 | 3.6 | 3.9 | 1.7 | 0.1 | 0.0 | 0.0 | 0.0 | 0.0 | 0.6 | 3.7 | 4.1 | 22.2 |

## PRECIPITATION (inches) 2009 DULUTH (KDLH)

| YEAR | JAN | FEB | MAR | APR | MAY | JUN | JUL | AUG | SEP | OCT | NOV | DEC | ANNUAL |
|------|-----|-----|-----|-----|-----|-----|-----|-----|-----|-----|-----|-----|--------|
| 1980 | 1.55 | 0.56 | 1.02 | 0.41 | 0.82 | 2.35 | 3.94 | 5.34 | 6.61 | 1.64 | 0.70 | 0.63 | 25.57 |
| 1981 | 0.32 | 1.50 | 1.05 | 4.48 | 1.15 | 5.83 | 3.26 | 2.84 | 2.42 | 3.59 | 0.96 | 0.97 | 28.37 |
| 1982 | 2.02 | 0.48 | 2.06 | 2.06 | 4.30 | 1.97 | 6.21 | 1.60 | 4.19 | 5.07 | 3.08 | 1.19 | 34.23 |
| 1983 | 1.34 | 0.49 | 2.05 | 2.28 | 2.12 | 2.00 | 3.51 | 3.37 | 5.57 | 2.32 | 5.01 | 1.97 | 32.03 |
| 1984 | 0.78 | 0.61 | 0.54 | 2.34 | 1.83 | 5.70 | 1.33 | 1.96 | 3.82 | 5.19 | 0.82 | 1.91 | 26.83 |
| 1985 | 0.39 | 0.66 | 1.85 | 2.35 | 4.44 | 3.18 | 4.16 | 3.91 | 6.02 | 1.76 | 2.33 | 0.78 | 31.83 |
| 1986 | 0.66 | 0.74 | 0.88 | 4.11 | 2.59 | 8.04 | 4.58 | 5.29 | 6.26 | 0.66 | 2.01 | 0.45 | 36.27 |
| 1987 | 0.69 | 0.31 | 0.60 | 0.24 | 4.02 | 0.83 | 5.46 | 1.87 | 2.93 | 0.96 | 1.26 | 0.67 | 19.84 |
| 1988 | 0.78 | 0.13 | 2.55 | 0.44 | 3.96 | 4.56 | 1.14 | 6.82 | 6.18 | 1.05 | 3.44 | 1.12 | 32.17 |
| 1989 | 1.87 | 0.34 | 1.49 | 2.11 | 3.50 | 3.81 | 1.09 | 5.02 | 4.40 | 1.02 | 1.01 | 0.63 | 26.29 |
| 1990 | 0.51 | 0.51 | 3.35 | 3.76 | 1.48 | 4.83 | 2.42 | 5.39 | 6.49 | 3.51 | 0.65 | 0.49 | 33.39 |
| 1991 | 0.52 | 0.55 | 1.17 | 3.90 | 6.11 | 5.64 | 5.33 | 2.49 | 9.38 | 2.85 | 4.89 | 0.61 | 43.44 |
| 1992 | 0.60 | 0.58 | 0.84 | 2.87 | 2.87 | 5.04 | 3.64 | 4.13 | 3.90 | 1.13 | 1.84 | 1.23 | 28.67 |
| 1993 | 1.79 | 0.38 | 0.44 | 2.42 | 4.74 | 6.95 | 5.75 | 3.63 | 1.81 | 0.56 | 2.60 | 1.28 | 32.35 |
| 1994 | 1.85 | 0.62 | 1.11 | 4.11 | 1.91 | 4.54 | 2.20 | 3.45 | 5.61 | 2.27 | 2.53 | 0.33 | 30.53 |
| 1995 | 1.38 | 1.04 | 1.97 | 2.19 | 4.13 | 0.55 | 5.73 | 6.67 | 3.02 | 4.24 | 1.84 | 1.54 | 34.30 |
| 1996 | 1.35 | 1.10 | 0.50 | 1.47 | 1.61 | 5.10 | 8.37 | 2.92 | 5.34 | 3.08 | 3.95 | .86 | 35.65 |
| 1997 | 0.94 | 0.23 | 1.48 | 1.07 | 1.90 | 5.17 | 4.33 | 1.26 | 1.75 | 2.29 | 0.43 | 0.13 | 20.98 |
| 1998 | 0.36 | 2.72 | 1.84 | 1.42 | 2.21 | 6.13 | 1.72 | 2.59 | 3.33 | 4.21 | 3.42 | 1.57 | 31.52 |
| 1999 | 0.87 | 0.64 | 0.76 | 2.96 | 2.93 | 4.72 | 8.74 | 7.41 | 5.80 | 2.35 | 0.70 | 0.23 | 38.11 |
| 2000 | 0.77 | 1.08 | 2.64 | 1.38 | 3.07 | 4.18 | 3.50 | 4.42 | 2.02 | 2.27 | 5.08 | 0.93 | 31.34 |
| 2001 | 1.25 | 1.76 | 0.69 | 8.18 | 3.06 | 1.89 | 3.31 | 1.41 | 2.42 | 2.21 | 0.55 | 30.22 |
| 2002 | 0.40 | 0.64 | 1.87 | 2.72 | 2.04 | 5.75 | 5.40 | 4.22 | 3.93 | 3.05 | 0.34 | 0.81 | 31.17 |
| 2003 | 0.23 | 0.23 | 1.03 | 2.03 | 3.54 | 3.47 | 4.79 | 2.31 | 3.52 | 1.57 | 1.70 | 0.54 | 24.96 |
| 2004 | 1.89 | 1.48 | 2.11 | 1.38 | 3.56 | 1.87 | 3.72 | 3.72 | 4.31 | 3.17 | 0.49 | 2.13 | 29.83 |
| 2005 | 2.51 | 1.06 | 0.52 | 1.24 | 4.43 | 5.46 | 0.82 | 2.33 | 3.60 | 4.86 | 3.02 | 2.54 | 32.39 |
| 2006 | 0.48 | 0.94 | 1.56 | 1.61 | 4.72 | 3.53 | 3.47 | 1.16 | 2.86 | 1.78 | 1.22 | 1.22 | 24.55 |
| 2007 | 0.20 | 1.49 | 2.49 | 2.22 | 3.39 | 2.67 | 1.88 | 1.39 | 4.38 | 6.80 | 0.66 | 2.45 | 30.02 |
| 2008 | 0.13 | 0.37 | 0.81 | 3.77 | 2.73 | 5.21 | 4.58 | 2.90 | 4.17 | 2.95 | 1.44 | 1.94 | 31.00 |
| 2009 | 0.53 | 0.98 | 3.37 | 0.89 | 2.13 | 1.81 | 3.33 | 6.02 | 1.40 | 4.60 | 1.15 | 2.89 | 29.10 |
| POR= 68 YRS | 1.09 | 0.86 | 1.71 | 2.35 | 3.20 | 4.14 | 3.85 | 3.88 | 3.54 | 2.50 | 1.85 | 1.22 | 30.19 |

WBAN : 14913

## AVERAGE TEMPERATURE (°F) 2009 DULUTH (KDLH)

| YEAR | JAN | FEB | MAR | APR | MAY | JUN | JUL | AUG | SEP | OCT | NOV | DEC | ANNUAL |
|------|-----|-----|-----|-----|-----|-----|-----|-----|-----|-----|-----|-----|--------|
| 1980 | 8.7 | 10.7 | 19.7 | 41.4 | 54.9 | 59.5 | 67.6 | 63.6 | 52.9 | 38.9 | 30.1 | 11.6 | 38.3 |
| 1981 | 11.9 | 16.3 | 28.3 | 39.2 | 50.2 | 58.7 | 65.5 | 64.3 | 52.9 | 40.5 | 34.7 | 14.8 | 39.8 |
| 1982 | -3.2 | 10.6 | 20.8 | 35.7 | 52.8 | 54.8 | 64.5 | 60.8 | 54.3 | 45.1 | 24.4 | 19.6 | 36.7 |
| 1983 | 13.2 | 21.1 | 25.6 | 35.1 | 46.7 | 59.9 | 69.6 | 69.7 | 56.6 | 44.9 | 31.6 | 1.8 | 39.7 |
| 1984 | 7.1 | 23.0 | 18.7 | 42.4 | 50.5 | 61.4 | 66.6 | 67.7 | 51.9 | 45.7 | 28.4 | 13.2 | 39.7 |
| 1985 | 6.5 | 11.6 | 30.3 | 42.1 | 54.9 | 56.5 | 64.6 | 59.7 | 52.1 | 42.7 | 19.9 | 3.0 | 37.0 |
| 1986 | 11.7 | 11.5 | 28.5 | 42.1 | 51.8 | 58.9 | 64.7 | 60.8 | 52.7 | 43.4 | 24.0 | 18.8 | 39.1 |
| 1987 | 15.3 | 23.7 | 31.6 | 46.1 | 53.1 | 63.1 | 67.6 | 63.5 | 57.2 | 39.4 | 32.2 | 20.4 | 42.8 |
| 1988 | 4.8 | 6.2 | 24.4 | 40.1 | 56.0 | 62.8 | 70.0 | 64.5 | 55.4 | 38.6 | 28.9 | 13.5 | 38.8 |
| 1989 | 14.3 | 3.4 | 18.9 | 36.8 | 51.7 | 58.4 | 68.6 | 64.8 | 55.9 | 44.4 | 23.2 | 4.1 | 37.0 |
| 1990 | 18.7 | 14.9 | 27.1 | 40.2 | 48.3 | 61.5 | 64.6 | 64.1 | 56.2 | 42.1 | 31.4 | 12.6 | 40.1 |
| 1991 | 5.9 | 19.4 | 26.1 | 42.7 | 54.7 | 61.7 | 63.8 | 66.3 | 53.0 | 40.4 | 22.0 | 15.9 | 39.3 |
| 1992 | 16.7 | 21.2 | 25.9 | 36.2 | 53.8 | 56.3 | 59.4 | 60.5 | 53.8 | 41.7 | 27.3 | 15.9 | 39.1 |
| 1993 | 11.9 | 14.5 | 27.1 | 37.2 | 50.3 | 57.1 | 63.6 | 65.0 | 49.5 | 40.0 | 24.4 | 17.0 | 38.1 |
| 1994 | -2.1 | 6.5 | 28.6 | 39.6 | 53.8 | 62.5 | 64.5 | 62.5 | 58.5 | 47.2 | 33.6 | 21.3 | 39.7 |
| 1995 | 13.7 | 11.7 | 28.2 | 34.7 | 52.4 | 65.1 | 66.8 | 67.4 | 55.1 | 43.7 | 20.6 | 12.1 | 39.3 |
| 1996 | 2.2 | 11.9 | 19.4 | 33.7 | 48.2 | 60.4 | 62.1 | 65.4 | 56.2 | 43.6 | 21.6 | 11.2 | 36.3 |
| 1997 | 7.0 | 14.5 | 23.5 | 35.7 | 45.8 | 62.2 | 63.8 | 61.0 | 57.5 | 43.9 | 24.9 | 22.7 | 38.5 |
| 1998 | 16.4 | 29.0 | 25.4 | 44.1 | 55.9 | 58.8 | 65.7 | 67.1 | 59.3 | 45.9 | 32.1 | 18.7 | 43.2 |
| 1999 | 9.3 | 21.3 | 27.8 | 41.4 | 53.6 | 60.6 | 67.6 | 63.5 | 54.5 | 42.3 | 36.2 | 19.7 | 41.5 |
| 2000 | 10.3 | 21.4 | 33.4 | 38.1 | 52.9 | 56.0 | 64.6 | 64.7 | 53.9 | 46.9 | 28.6 | 4.4 | 39.6 |
| 2001 | 17.4 | 7.8 | 24.5 | 39.8 | 52.8 | 60.9 | 65.7 | 67.0 | 55.4 | 43.4 | 39.5 | 22.2 | 41.4 |
| 2002 | 17.8 | 21.5 | 18.3 | 37.2 | 46.6 | 61.8 | 69.1 | 63.6 | 58.0 | 35.6 | 25.6 | 20.2 | 39.6 |
| 2003 | 9.0 | 9.1 | 22.9 | 38.3 | 50.7 | 59.8 | 66.2 | 67.6 | 56.2 | 44.3 | 26.7 | 20.5 | 39.3 |
| 2004 | 4.0 | 18.9 | 28.3 | 39.4 | 46.7 | 57.3 | 63.4 | 58.8 | 59.2 | 45.3 | 33.4 | 14.0 | 39.1 |
| 2005 | 7.7 | 19.4 | 23.2 | 43.1 | 48.9 | 61.8 | 67.6 | 64.7 | 59.7 | 45.9 | 30.7 | 17.7 | 40.9 |
| 2006 | 23.8 | 11.8 | 28.7 | 45.7 | 53.1 | 62.9 | 71.9 | 66.0 | 54.4 | 39.8 | 31.8 | 23.8 | 42.8 |
| 2007 | 13.7 | 8.0 | 29.4 | 38.9 | 53.9 | 63.1 | 67.4 | 65.1 | 57.4 | 47.7 | 28.9 | 13.4 | 40.6 |
| 2008 | 10.0 | 10.2 | 22.3 | 37.9 | 48.5 | 60.0 | 64.5 | 64.4 | 56.2 | 44.6 | 28.9 | 7.0 | 37.9 |
| 2009 | 3.3 | 15.3 | 24.6 | 39.7 | 50.2 | 58.8 | 62.2 | 62.2 | 62.1 | 39.1 | 37.9 | 12.4 | 39.0 |
| POR= 82 YRS | 9.1 | 13.1 | 24.3 | 38.3 | 50.4 | 59.0 | 65.7 | 64.2 | 54.7 | 44.1 | 28.2 | 14.9 | 38.8 |

## HEATING DEGREE DAYS (base 65°F) 2009 DULUTH (KDLH)

| YEAR | JUL | AUG | SEP | OCT | NOV | DEC | JAN | FEB | MAR | APR | MAY | JUN | TOTAL |
|------|-----|-----|-----|-----|-----|-----|-----|-----|-----|-----|-----|-----|-------|
| 1980-81 | 39 | 76 | 357 | 800 | 1043 | 1650 | 1644 | 1358 | 1133 | 769 | 455 | 185 | 9509 |
| 1981-82 | 74 | 62 | 363 | 752 | 903 | 1549 | 2117 | 1523 | 1363 | 875 | 370 | 303 | 10254 |
| 1982-83 | 66 | 161 | 332 | 609 | 1212 | 1398 | 1598 | 1226 | 1214 | 887 | 562 | 185 | 9450 |
| 1983-84 | 30 | 9 | 285 | 615 | 996 | 1959 | 1792 | 1210 | 1429 | 671 | 446 | 116 | 9558 |
| 1984-85 | 40 | 40 | 391 | 588 | 1093 | 1601 | 1809 | 1494 | 1067 | 679 | 314 | 252 | 9368 |
| 1985-86 | 57 | 178 | 394 | 686 | 1349 | 1925 | 1646 | 1492 | 1127 | 680 | 416 | 196 | 10146 |
| 1986-87 | 76 | 151 | 361 | 663 | 1224 | 1426 | 1537 | 1153 | 1027 | 561 | 376 | 112 | 8667 |
| 1987-88 | 34 | 100 | 234 | 782 | 977 | 1377 | 1862 | 1704 | 1253 | 741 | 300 | 146 | 9510 |
| 1988-89 | 22 | 97 | 287 | 812 | 1081 | 1590 | 1567 | 1721 | 1424 | 839 | 405 | 206 | 10051 |
| 1989-90 | 27 | 78 | 272 | 633 | 1249 | 1887 | 1428 | 1398 | 1166 | 745 | 506 | 130 | 9519 |
| 1990-91 | 76 | 94 | 273 | 700 | 1003 | 1623 | 1833 | 1273 | 1201 | 661 | 332 | 129 | 9198 |
| 1991-92 | 97 | 67 | 363 | 754 | 1284 | 1515 | 1492 | 1266 | 1207 | 859 | 355 | 271 | 9530 |
| 1992-93 | 174 | 167 | 332 | 717 | 1121 | 1518 | 1638 | 1408 | 1171 | 825 | 451 | 239 | 9761 |
| 1993-94 | 73 | 71 | 459 | 769 | 1214 | 1485 | 2082 | 1635 | 1123 | 756 | 351 | 105 | 10123 |
| 1994-95 | 58 | 114 | 197 | 544 | 932 | 1347 | 1584 | 1487 | 1136 | 902 | 389 | 98 | 8788 |
| 1995-96 | 43 | 16 | 305 | 655 | 1327 | 1634 | 1946 | 1537 | 1410 | 932 | 514 | 172 | 10491 |
| 1996-97 | 105 | 38 | 277 | 655 | 1296 | 1660 | 1789 | 1407 | 1280 | 873 | 588 | 106 | 10074 |
| 1997-98 | 96 | 155 | 228 | 649 | 1193 | 1309 | 1498 | 1001 | 1218 | 620 | 282 | 193 | 8442 |
| 1998-99 | 50 | 21 | 196 | 581 | 981 | 1431 | 1722 | 1221 | 1147 | 699 | 357 | 158 | 8564 |
| 1999-00 | 38 | 76 | 329 | 698 | 857 | 1396 | 1691 | 1256 | 975 | 800 | 369 | 274 | 8759 |
| 2000-01 | 75 | 52 | 336 | 556 | 1084 | 1872 | 1469 | 1595 | 1250 | 748 | 369 | 164 | 9570 |
| 2001-02 | 66 | 48 | 300 | 661 | 760 | 1320 | 1456 | 1211 | 1441 | 826 | 574 | 157 | 8820 |
| 2002-03 | 27 | 68 | 242 | 902 | 1176 | 1383 | 1730 | 1554 | 1301 | 795 | 438 | 171 | 9787 |
| 2003-04 | 49 | 36 | 277 | 638 | 1142 | 1373 | 1886 | 1329 | 1132 | 762 | 561 | 236 | 9421 |
| 2004-05 | 90 | 200 | 194 | 604 | 938 | 1572 | 1769 | 1270 | 1290 | 650 | 491 | 137 | 9205 |
| 2005-06 | 56 | 74 | 186 | 582 | 1021 | 1456 | 1269 | 1486 | 1117 | 572 | 372 | 91 | 8282 |
| 2006-07 | 10 | 32 | 319 | 777 | 988 | 1267 | 1581 | 1592 | 1094 | 778 | 348 | 100 | 8886 |
| 2007-08 | 60 | 63 | 254 | 526 | 1075 | 1594 | 1698 | 1582 | 1318 | 807 | 506 | 167 | 9650 |
| 2008-09 | 59 | 63 | 274 | 625 | 1077 | 1789 | 1908 | 1387 | 1246 | 751 | 454 | 212 | 9845 |
| 2009- | 99 | 124 | 106 | 796 | 806 | 1624 | | | | | | | |

WBAN : 14913

## COOLING DEGREE DAYS (base 65°F) 2009 DULUTH (KDLH)

| YEAR | JAN | FEB | MAR | APR | MAY | JUN | JUL | AUG | SEP | OCT | NOV | DEC | TOTAL |
|------|-----|-----|-----|-----|-----|-----|-----|-----|-----|-----|-----|-----|-------|
| 1980 | 0 | 0 | 0 | 0 | 25 | 46 | 126 | 40 | 3 | 0 | 0 | 0 | 240 |
| 1981 | 0 | 0 | 0 | 0 | 2 | 2 | 97 | 48 | 6 | 0 | 0 | 0 | 155 |
| 1982 | 0 | 0 | 0 | 0 | 0 | 0 | 58 | 36 | 18 | 0 | 0 | 0 | 112 |
| 1983 | 0 | 0 | 0 | 0 | 0 | 42 | 179 | 165 | 42 | 0 | 0 | 0 | 428 |
| 1984 | 0 | 0 | 0 | 0 | 4 | 13 | 96 | 133 | 4 | 0 | 0 | 0 | 250 |
| 1985 | 0 | 0 | 0 | 0 | 9 | 4 | 54 | 20 | 15 | 0 | 0 | 0 | 102 |
| 1986 | 0 | 0 | 0 | 0 | 13 | 18 | 74 | 26 | 0 | 0 | 0 | 0 | 131 |
| 1987 | 0 | 0 | 0 | 0 | 13 | 13 | 62 | 121 | 60 | 7 | 0 | 0 | 263 |
| 1988 | 0 | 0 | 0 | 0 | 27 | 83 | 183 | 89 | 4 | 0 | 0 | 0 | 386 |
| 1989 | 0 | 0 | 0 | 0 | 0 | 17 | 147 | 80 | 7 | 0 | 0 | 0 | 251 |
| 1990 | 0 | 0 | 0 | 7 | 0 | 32 | 70 | 73 | 16 | 0 | 0 | 0 | 198 |
| 1991 | 0 | 0 | 0 | 0 | 20 | 35 | 69 | 117 | 8 | 0 | 0 | 0 | 249 |
| 1992 | 0 | 0 | 0 | 0 | 13 | 18 | 9 | 37 | 3 | 3 | 0 | 0 | 83 |
| 1993 | 0 | 0 | 0 | 0 | 0 | 8 | 36 | 76 | 0 | 0 | 0 | 0 | 120 |
| 1994 | 0 | 0 | 0 | 0 | 11 | 36 | 52 | 43 | 8 | 0 | 0 | 0 | 150 |
| 1995 | 0 | 0 | 0 | 0 | 7 | 107 | 105 | 99 | 14 | 4 | 0 | 0 | 336 |
| 1996 | 0 | 0 | 0 | 0 | 0 | 44 | 25 | 58 | 21 | 0 | 0 | 0 | 148 |
| 1997 | 0 | 0 | 0 | 0 | 0 | 27 | 67 | 37 | 8 | 0 | 0 | 0 | 139 |
| 1998 | 0 | 0 | 0 | 0 | 8 | 12 | 81 | 93 | 32 | 0 | 0 | 0 | 226 |
| 1999 | 0 | 0 | 0 | 0 | 12 | 33 | 127 | 35 | 20 | 0 | 0 | 0 | 227 |
| 2000 | 0 | 0 | 0 | 0 | 2 | 10 | 69 | 51 | 9 | 0 | 0 | 0 | 141 |
| 2001 | 0 | 0 | 0 | 0 | 0 | 49 | 96 | 117 | 17 | 0 | 0 | 0 | 279 |
| 2002 | 0 | 0 | 0 | 0 | 10 | 68 | 161 | 33 | 39 | 0 | 0 | 0 | 311 |
| 2003 | 0 | 0 | 0 | 0 | 0 | 21 | 94 | 122 | 20 | 3 | 0 | 0 | 260 |
| 2004 | 0 | 0 | 0 | 0 | 0 | 12 | 47 | 15 | 26 | 0 | 0 | 0 | 100 |
| 2005 | 0 | 0 | 0 | 0 | 0 | 46 | 142 | 74 | 35 | 0 | 0 | 0 | 297 |
| 2006 | 0 | 0 | 0 | 0 | 8 | 34 | 232 | 68 | 7 | 0 | 0 | 0 | 349 |
| 2007 | 0 | 0 | 0 | 0 | 8 | 48 | 142 | 73 | 35 | 0 | 0 | 0 | 306 |
| 2008 | 0 | 0 | 0 | 0 | 0 | 23 | 50 | 54 | 16 | 0 | 0 | 0 | 143 |
| 2009 | 0 | 0 | 0 | 0 | 0 | 33 | 17 | 44 | 24 | 0 | 0 | 0 | 118 |

## SNOWFALL (inches) 2009 DULUTH (KDLH)

| YEAR | JUL | AUG | SEP | OCT | NOV | DEC | JAN | FEB | MAR | APR | MAY | JUN | TOTAL |
|------|-----|-----|-----|-----|-----|-----|-----|-----|-----|-----|-----|-----|-------|
| 1980-81 | 0.0 | 0.0 | 0.0 | 1.3 | 1.6 | 7.1 | 4.7 | 13.3 | 4.2 | 4.3 | T | 0.0 | 36.5 |
| 1981-82 | 0.0 | 0.0 | T | 1.4 | 12.5 | 15.5 | 34.2 | 8.0 | 17.8 | 6.3 | 0.0 | 0.0 | 95.7 |
| 1982-83 | 0.0 | 0.0 | 0.0 | 0.8 | 16.0 | 21.4 | 15.7 | 9.0 | 9.9 | 23.7 | T | 0.0 | 96.5 |
| 1983-84 | 0.0 | 0.0 | T | T | 37.7 | 32.1 | 20.0 | 4.0 | 9.9 | 3.3 | 0.3 | 0.0 | 107.3 |
| 1984-85 | 0.0 | 0.0 | T | 3.5 | 2.4 | 8.2 | 15.3 | 12.0 | 26.7 | 0.1 | 0.0 | 0.0 | 68.2 |
| 1985-86 | 0.0 | 0.0 | 0.7 | 2.3 | 34.1 | 18.8 | 11.5 | 10.3 | 11.4 | 0.2 | T | 0.0 | 89.3 |
| 1986-87 | 0.0 | 0.0 | 0.0 | 1.8 | 8.2 | 7.3 | 11.2 | 5.1 | 6.7 | 0.3 | T | 0.0 | 40.6 |
| 1987-88 | 0.0 | 0.0 | 0.0 | 3.9 | 5.8 | 11.3 | 16.7 | 2.3 | 13.7 | 0.1 | 0.0 | 0.0 | 53.8 |
| 1988-89 | 0.0 | 0.0 | 0.0 | 0.3 | 24.3 | 20.7 | 31.2 | 5.3 | 17.6 | 19.0 | 0.7 | 0.0 | 119.1 |
| 1989-90 | 0.0 | T | 0.0 | 0.6 | 8.4 | 14.2 | 5.5 | 15.0 | 11.6 | 2.4 | 0.6 | T | 58.3 |
| 1990-91 | 0.0 | 0.0 | 0.0 | 3.2 | 5.1 | 13.5 | 11.4 | 9.7 | 9.5 | 8.5 | 2.9 | 0.0 | 63.8 |
| 1991-92 | 0.0 | 0.0 | 2.4 | 4.3 | 50.1 | 12.9 | 9.3 | 10.6 | 0.7 | 9.7 | 0.0 | 0.0 | 100.0 |
| 1992-93 | T | 0.0 | T | 5.1 | 29.4 | 15.5 | 24.9 | 10.0 | 5.8 | 3.5 | T | 0.0 | 94.2 |
| 1993-94 | 0.0 | 0.0 | T | 0.6 | 28.5 | 15.1 | 35.3 | 7.9 | 13.0 | 10.0 | 0.0 | T | 110.4 |
| 1994-95 | 0.0 | 0.0 | 0.0 | 0.2 | 15.5 | 6.2 | 12.8 | 17.6 | 20.9 | 16.2 | T | 0.0 | 89.4 |
| 1995-96 | 0.0 | T | 0.1 | 9.7 | 20.4 | 25.9 | 34.0 | 21.3 | 11.4 | 12.6 | T | | |
| 1996-97 | | | | | 15.9 | 41.7 | 35.5 | 8.9 | 23.6 | 1.7 | 0.1 | 0.0 | |
| 1997-98 | 0.0 | 0.0 | 0.0 | 1.2 | 16.9 | 6.0 | 29.9 | 9.6 | 14.2 | 2.3 | 0.0 | T | 80.1 |
| 1998-99 | 0.0 | T | 0.0 | T | 15.7 | 12.8 | 18.9 | 13.5 | 17.8 | 11.5 | 0.0 | 0.0 | 90.2 |
| 1999-00 | 0.0 | 0.0 | 0.0 | 0.0 | 6.2 | 4.4 | 16.5 | 13.3 | 9.5 | 5.6 | T | 0.0 | 55.5 |
| 2000-01 | 0.0 | 0.0 | 0.0 | T | 17.5 | 19.2 | 14.9 | 32.1 | 7.9 | 7.7 | 0.0 | 0.0 | 99.3 |
| 2001-02 | 0.0 | 0.0 | 0.0 | 2.0 | 11.4 | 7.3 | 11.3 | 11.2 | 29.7 | 11.6 | 1.5 | 0.0 | 86.0 |
| 2002-03 | 0.0 | 0.0 | T | 6.3 | 4.3 | 3.4 | 5.9 | 6.4 | 17.1 | 12.9 | 0.0 | 0.0 | 56.3 |
| 2003-04 | 0.0 | 0.0 | T | 4.5 | 19.4 | 8.4 | 42.3 | 23.4 | 11.6 | 0.3 | 0.0 | 0.0 | 109.9 |
| 2004-05 | 0.0 | 0.0 | 0.0 | T | 1.4 | 17.3 | 45.7 | 20.2 | 6.3 | 0.5 | 0.1 | 0.0 | 91.5 |
| 2005-06 | 0.0 | 0.0 | 0.0 | 0.1 | 11.4 | 34.5 | 7.3 | 23.4 | 12.5 | T | T | T | 89.2 |
| 2006-07 | 0.0 | 0.0 | 0.0 | 4.1 | 4.0 | 10.1 | 4.5 | 19.7 | 25.5 | 12.7 | 0.1 | 0.0 | 80.7 |
| 2007-08 | 0.0 | 0.0 | 0.0 | T | 6.0 | 35.4 | 2.8 | 7.1 | 13.6 | 14.6 | 0.5 | 0.0 | 80.0 |
| 2008-09 | 0.0 | T | 0.0 | 0.2 | 3.2 | 37.1 | 9.3 | 7.1 | 15.5 | 1.2 | T | 0.0 | 73.6 |
| 2009- | 0.0 | 0.0 | 0.0 | 3.7 | 1.1 | 33.4 | | | | | | | |
| POR= 61 YRS | T | T | 0.1 | 1.7 | 12.0 | 16.7 | 18.0 | 12.3 | 14.3 | 6.9 | 0.6 | T | 82.6 |

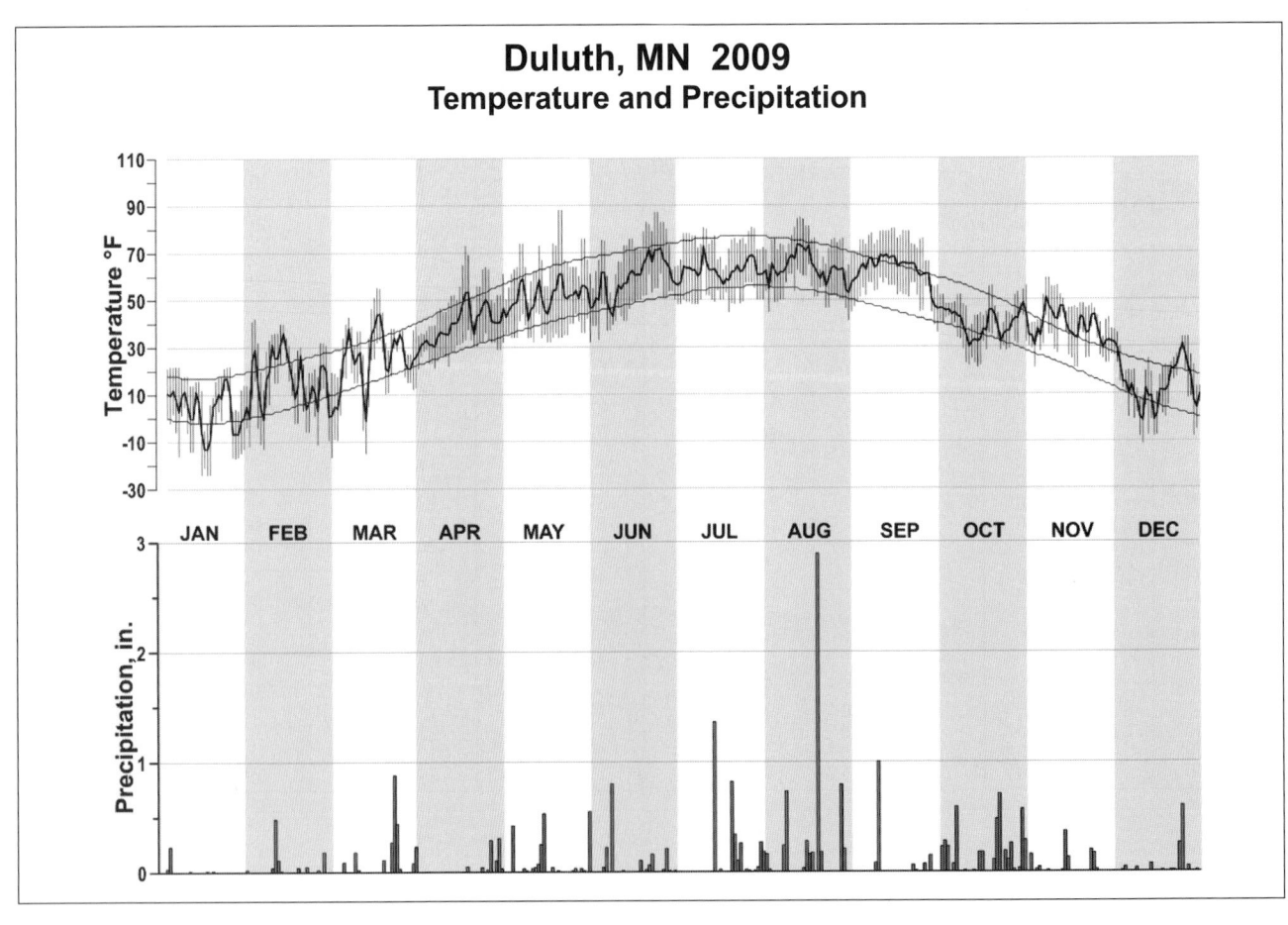

## Duluth, MN  2009
### Temperature and Precipitation

# 2009
# INTERNATIONAL FALLS
# MINNESOTA (KINL)

Situated on the Canadian border, International Falls is subjected to frequent outbreaks of continental polar air throughout most of the year. These are tempered to mildness during June, July, and August, when the land and lake areas to the north and northwest have been warmed by long days of sunshine. Periods of fine, mild weather occur, interspersed with showers and an occasional three or four day period of cloudy, rainy weather. The area of small lakes, covering up to 30 percent of the area to the north and northwest, supplies a good deal of the moisture for the late afternoon and evening showers and stores heat that tempers southward flow of cold air during September and October. This prolongs the fall season until early November. In November the water surfaces freeze and snow returns to International Falls. From December through February, temperatures fall below zero on most days and occasionally fail to rise above zero for a week or more.

In winter, frost penetrates into the ground to depths of 36 to 60 inches. If winter begins abruptly so that a heavy blanket of snow covers the ground before protracted freezing occurs, it may freeze to only a few inches deep. This is very important to loggers, who depend upon deep soil freezing for road foundations into otherwise inaccessible places. The wide expanse of deep snow and ice prolongs winter. The transition to summer is rapid after the spring thaw. Spring lasts only about a month.

By June 1st, the ground generally is warm enough for successful planting, but vigilance against freezing temperatures is required through most of June. Crops that do not mature by September 1st have little chance of providing a harvest. Heaviest precipitation coincides with the growing season.

Based on the 1951-1980 period, the average first occurrence of 32 degrees Fahrenheit in the fall is September 15 and the average last occurrence in the spring is May 26.

Heavy deposits of glaze occur only about once a year at International Falls. Occasional storms that intensify over the southern plateau or plains states and move rapidly northeastward, drawing up moist gulf air, bring the most violent weather changes. They often produce severe thunderstorms and windstorms in early fall and blizzards with heavy snowfall and drifting in winter. Quite often such a storm brings an abrupt end to fall weather. During winter, a variation of 100 miles in the paths of such storms as they approach the border is of tremendous importance to local transportation and road maintenance.

Surrounding terrain is generally level. Forests of varying density and swampland surround the station for many miles to the east, south, and west. Rainy Lake, approximately 300 square miles in area, lies to the north. The lake is 5 miles from the station at its closest point.

# NORMALS, MEANS, AND EXTREMES
## INTERNATIONAL FALLS (KINL)

| LATITUDE: 48 ° 33'N | LONGITUDE: -93 ° 24'W | | ELEVATION (FT): GRND: 1184  BARO: 1185 | | | TIME ZONE: CENTRAL  (UTC -6) | | WBAN: 14918 | | | | |

| ELEMENT | POR | JAN | FEB | MAR | APR | MAY | JUN | JUL | AUG | SEP | OCT | NOV | DEC | YEAR |
|---|---|---|---|---|---|---|---|---|---|---|---|---|---|---|
| **TEMPERATURE °F** | | | | | | | | | | | | | | |
| NORMAL DAILY MAXIMUM | 30 | 13.8 | 22.4 | 34.9 | 51.5 | 66.6 | 74.2 | 78.6 | 76.3 | 64.7 | 51.7 | 32.5 | 18.1 | 48.8 |
| MEAN DAILY MAXIMUM | 84 | 12.5 | 19.5 | 32.7 | 49.4 | 64.0 | 71.7 | 76.6 | 75.2 | 64.8 | 51.6 | 32.9 | 18.7 | 47.5 |
| HIGHEST DAILY MAXIMUM | 70 | 48 | 58 | 76 | 93 | 95 | 99 | 98 | 95 | 95 | 88 | 73 | 57 | 99 |
| YEAR OF OCCURRENCE | | 1973 | 2000 | 1946 | 1952 | 1964 | 1995 | 1988 | 2003 | 1976 | 1963 | 1975 | 1939 | JUN 1995 |
| MEAN OF EXTREME MAXS. | 92 | 35.3 | 40.8 | 52.5 | 73.3 | 84.1 | 87.5 | 90.1 | 88.9 | 83.2 | 74.0 | 54.4 | 38.1 | 66.9 |
| NORMAL DAILY MINIMUM | 30 | -8.4 | -.7 | 12.3 | 27.1 | 40.0 | 49.1 | 53.6 | 51.3 | 41.6 | 31.5 | 16.4 | -1.1 | 26.1 |
| MEAN DAILY MINIMUM | 84 | -8.0 | -2.3 | 10.5 | 27.0 | 38.8 | 48.2 | 53.5 | 50.9 | 41.7 | 32.5 | 17.3 | 0.8 | 25.9 |
| LOWEST DAILY MINIMUM | 70 | -46 | -45 | -38 | -14 | 11 | 23 | 34 | 30 | 20 | 2 | -32 | -41 | -46 |
| YEAR OF OCCURRENCE | | 1968 | 1996 | 1962 | 1954 | 1967 | 1964 | 2001 | 1982 | 1965 | 1988 | 1985 | 1955 | JAN 1968 |
| MEAN OF EXTREME MINS. | 92 | -32.5 | -28.2 | -16.4 | 9.0 | 24.4 | 34.2 | 41.2 | 37.9 | 26.9 | 17.3 | -5.3 | -24.5 | 7.0 |
| NORMAL DRY BULB | 30 | 2.7 | 10.9 | 23.6 | 39.3 | 53.3 | 61.6 | 66.1 | 63.8 | 53.2 | 41.6 | 24.4 | 8.5 | 37.4 |
| MEAN DRY BULB | 84 | 2.3 | 8.6 | 21.7 | 38.2 | 51.4 | 60.0 | 65.1 | 63.1 | 53.2 | 42.3 | 25.1 | 9.7 | 36.7 |
| MEAN WET BULB | 26 | 5.0 | 10.0 | 20.3 | 33.0 | 45.1 | 55.2 | 59.6 | 58.0 | 49.8 | 37.2 | 23.4 | 10.1 | 33.9 |
| MEAN DEW POINT | 26 | 1.9 | 5.7 | 15.3 | 26.9 | 39.7 | 51.7 | 56.9 | 55.5 | 47.1 | 33.8 | 20.6 | 7.5 | 30.2 |
| NORMAL NO. DAYS WITH: | | | | | | | | | | | | | | |
| MAXIMUM >= 90 | 30 | 0.0 | 0.0 | 0.0 | * | 0.3 | 0.9 | 1.6 | 1.0 | 0.1 | 0.0 | 0.0 | 0.0 | 3.9 |
| MAXIMUM <= 32 | 30 | 28.7 | 22.1 | 13.1 | 1.9 | * | 0.0 | 0.0 | 0.0 | 0.0 | 1.1 | 15.6 | 26.9 | 109.4 |
| MINIMUM <= 32 | 30 | 31.0 | 27.9 | 29.2 | 21.5 | 6.1 | 0.4 | 0.0 | 0.1 | 4.0 | 17.1 | 28.2 | 30.9 | 196.4 |
| MINIMUM <= 0 | 30 | 21.3 | 15.0 | 6.8 | 0.4 | 0.0 | 0.0 | 0.0 | 0.0 | 0.0 | 0.0 | 3.6 | 16.6 | 63.7 |
| **H/C** | | | | | | | | | | | | | | |
| NORMAL HEATING DEG. DAYS | 30 | 1946 | 1531 | 1298 | 775 | 378 | 140 | 55 | 102 | 360 | 723 | 1217 | 1744 | 10269 |
| NORMAL COOLING DEG. DAYS | 30 | 0 | 0 | 0 | 1 | 17 | 47 | 91 | 67 | 10 | 0 | 0 | 0 | 233 |
| **RH** | | | | | | | | | | | | | | |
| NORMAL (PERCENT) | 30 | 74 | 71 | 66 | 60 | 61 | 69 | 73 | 75 | 77 | 74 | 78 | 78 | 71 |
| HOUR 00 LST | 30 | 76 | 74 | 71 | 67 | 72 | 80 | 85 | 87 | 87 | 81 | 82 | 80 | 79 |
| HOUR 06 LST | 30 | 76 | 77 | 77 | 76 | 77 | 83 | 89 | 92 | 91 | 85 | 84 | 81 | 82 |
| HOUR 12 LST | 30 | 70 | 65 | 58 | 49 | 48 | 56 | 58 | 60 | 63 | 64 | 73 | 74 | 62 |
| HOUR 18 LST | 30 | 71 | 64 | 56 | 48 | 47 | 55 | 57 | 60 | 67 | 67 | 75 | 76 | 62 |
| **S** PERCENT POSSIBLE SUNSHINE | | | | | | | | | | | | | | |
| **W/O** MEAN NO. DAYS WITH: | | | | | | | | | | | | | | |
| HEAVY FOG(VISBY <= 1/4 MI) | 46 | 1.2 | 1.2 | 1.5 | 1.4 | 0.9 | 1.2 | 1.8 | 1.8 | 2.1 | 1.5 | 1.2 | 1.4 | 17.2 |
| THUNDERSTORMS | 62 | 0.0 | 0.0 | 0.3 | 1.1 | 3.9 | 7.4 | 9.0 | 7.0 | 4.0 | 1.0 | 0.1 | 0.0 | 33.8 |
| **CLOUDNESS** MEAN: | | | | | | | | | | | | | | |
| SUNRISE-SUNSET (OKTAS) | 47 | 5.4 | 5.0 | 5.3 | 5.2 | 5.3 | 5.4 | 5.0 | 4.9 | 5.5 | 5.8 | 6.3 | 5.6 | 5.4 |
| MIDNIGHT-MIDNIGHT (OKTAS) | 32 | 5.1 | 4.8 | 4.9 | 4.9 | 5.0 | 5.0 | 4.7 | 4.4 | 5.0 | 5.6 | 5.9 | 5.3 | 5.1 |
| MEAN NO. DAYS WITH: | | | | | | | | | | | | | | |
| CLEAR | 57 | 7.5 | 7.9 | 7.7 | 6.9 | 6.8 | 4.8 | 6.1 | 7.1 | 5.6 | 5.8 | 3.7 | 6.2 | 76.1 |
| PARTLY CLOUDY | 57 | 7.3 | 6.4 | 7.6 | 8.3 | 9.0 | 10.8 | 13.2 | 11.3 | 8.8 | 7.1 | 5.2 | 5.8 | 100.8 |
| CLOUDY | 57 | 16.2 | 14.1 | 15.6 | 14.8 | 15.2 | 14.4 | 11.2 | 12.1 | 15.3 | 18.1 | 21.1 | 18.9 | 187.0 |
| **PR** MEAN STATION PRESSURE(IN) | 26 | 28.72 | 28.75 | 28.75 | 28.70 | 28.67 | 28.65 | 28.67 | 28.72 | 28.70 | 28.70 | 28.69 | 28.71 | 28.70 |
| MEAN SEA-LEVEL PRES. (IN) | 26 | 30.06 | 30.08 | 30.05 | 29.98 | 29.93 | 29.90 | 29.71 | 29.75 | 29.77 | 29.98 | 30.00 | 30.04 | 29.94 |
| **WINDS** MEAN SPEED (MPH) | 26 | 8.0 | 8.2 | 8.8 | 8.9 | 8.7 | 7.8 | 6.9 | 7.0 | 7.7 | 8.4 | 8.4 | 7.9 | 8.1 |
| PREVAIL.DIR(TENS OF DEGS) | 36 | 30 | 30 | 30 | 06 | 19 | 19 | 31 | 19 | 19 | 30 | 30 | 30 | 30 |
| MAXIMUM 2-MINUTE: | | | | | | | | | | | | | | |
| SPEED (MPH) | 13 | 43 | 35 | 35 | 35 | 41 | 48 | 47 | 43 | 38 | 35 | 37 | 36 | 48 |
| DIR. (TENS OF DEGS) | | 27 | 30 | 31 | 33 | 34 | 32 | 26 | 30 | 34 | 30 | 33 | 31 | 32 |
| YEAR OF OCCURRENCE | | 2009 | 2002 | 2007 | 2002 | 2008 | 2005 | 2003 | 2001 | 2002 | 1999 | 2005 | 1999 | JUN 2005 |
| MAXIMUM 3-SECOND | | | | | | | | | | | | | | |
| SPEED (MPH) | 13 | 62 | 47 | 47 | 44 | 59 | 60 | 62 | 59 | 45 | 46 | 47 | 49 | 62 |
| DIR. (TENS OF DEGS) | | 27 | 28 | 17 | 33 | 35 | 32 | 33 | 32 | 18 | 31 | 24 | 30 | 27 |
| YEAR OF OCCURRENCE | | 2009 | 2009 | 2008 | 2002 | 2008 | 2005 | 2007 | 2001 | 2004 | 1999 | 1999 | 1999 | JAN 2009 |
| **PRECIPITATION** NORMAL (IN) | 30 | 0.84 | 0.64 | 0.96 | 1.38 | 2.55 | 3.98 | 3.37 | 3.14 | 3.03 | 1.98 | 1.36 | 0.70 | 23.93 |
| MAXIMUM MONTHLY (IN) | 70 | 3.03 | 1.81 | 3.80 | 3.54 | 6.67 | 8.29 | 9.52 | 11.26 | 7.36 | 4.84 | 3.49 | 2.00 | 11.26 |
| YEAR OF OCCURRENCE | | 1975 | 1955 | 2009 | 2001 | 1985 | 2002 | 1966 | 1942 | 1961 | 1971 | 1977 | 2004 | AUG 1942 |
| MINIMUM MONTHLY (IN) | 70 | 0.07 | 0.04 | 0.13 | 0.08 | 0.20 | 0.70 | 1.00 | 0.58 | 0.28 | 0.14 | 0.10 | 0.16 | 0.04 |
| YEAR OF OCCURRENCE | | 2003 | 2002 | 2001 | 1987 | 1976 | 1961 | 1941 | 1991 | 1952 | 1992 | 1999 | 1940 | FEB 2002 |
| MAXIMUM IN 24 HOURS (IN) | 70 | 1.50 | 1.14 | 1.80 | 1.65 | 2.76 | 4.32 | 4.87 | 4.82 | 3.37 | 2.62 | 1.56 | 1.25 | 4.87 |
| YEAR OF OCCURRENCE | | 1975 | 1946 | 1957 | 2005 | 1991 | 2002 | 1966 | 1942 | 1973 | 1979 | 1977 | 1960 | JUL 1966 |
| NORMAL NO. DAYS WITH: | | | | | | | | | | | | | | |
| PRECIPITATION >= 0.01 | 30 | 10.9 | 8.7 | 9.4 | 8.4 | 11.1 | 13.3 | 11.6 | 11.2 | 12.0 | 10.9 | 10.4 | 10.9 | 128.8 |
| PRECIPITATION >= 1.00 | 30 | * | 0.0 | 0.0 | 0.1 | 0.3 | 0.7 | 0.6 | 0.5 | 0.7 | 0.2 | * | 0.0 | 3.1 |
| **SNOWFALL** NORMAL (IN) | 30 | 15.2 | 10.5 | 9.0 | 5.4 | 0.3 | 0.* | 0.0 | 0.0 | 0.1 | 2.3 | 13.3 | 13.9 | 70.0 |
| MAXIMUM MONTHLY (IN) | 69 | 43.0 | 32.3 | 31.5 | 23.5 | 13.4 | 0.3 | T | 0.0 | T | 8.5 | 29.7 | 43.9 | 43.9 |
| YEAR OF OCCURRENCE | | 1975 | 2008 | 1951 | 2008 | 1954 | 1969 | 1992 | 1996 | 1942 | 1981 | 1965 | 1992 | DEC 1992 |
| MAXIMUM IN 24 HOURS (IN) | 69 | 17.7 | 12.1 | 17.0 | 13.9 | 7.7 | 0.3 | T | T | 1.5 | 5.9 | 14.7 | 14.4 | 17.7 |
| YEAR OF OCCURRENCE' | | 1975 | 1996 | 1966 | 1950 | 1954 | 1969 | 1992 | 1995 | 1951 | 1981 | 1991 | 1990 | JAN 1975 |
| MAXIMUM SNOW DEPTH (IN) | 60 | 38 | 35 | 38 | 25 | 12 | 0 | 0 | 0 | 0 | 5 | 26 | 25 | 38 |
| YEAR OF OCCURRENCE | | 1982 | 1969 | 1966 | 1975 | 1954 | | | | | 1988 | 1965 | 1985 | JAN 1982 |
| NORMAL NO. DAYS WITH: | | | | | | | | | | | | | | |
| SNOWFALL >= 1.0 | 30 | 4.3 | 3.3 | 2.7 | 1.5 | 0.1 | 0.0 | 0.0 | 0.0 | 0.1 | 0.8 | 3.8 | 3.9 | 20.5 |

## PRECIPITATION (inches) 2009 INTERNATIONAL FALLS (KINL)

| YEAR | JAN | FEB | MAR | APR | MAY | JUN | JUL | AUG | SEP | OCT | NOV | DEC | ANNUAL |
|------|-----|-----|-----|-----|-----|-----|-----|-----|-----|-----|-----|-----|--------|
| 1980 | 0.92 | 0.55 | 0.87 | 0.45 | 0.83 | 1.70 | 2.23 | 4.03 | 4.08 | 1.81 | 1.62 | 0.56 | 19.65 |
| 1981 | 0.26 | 0.22 | 1.18 | 1.49 | 2.47 | 3.71 | 2.33 | 2.03 | 4.12 | 2.86 | 0.67 | 0.76 | 22.10 |
| 1982 | 1.24 | 0.51 | 1.58 | 0.84 | 3.51 | 2.68 | 2.37 | 2.88 | 3.63 | 3.67 | 1.52 | 0.29 | 24.72 |
| 1983 | 0.36 | 0.98 | 0.72 | 0.62 | 1.21 | 5.02 | 2.98 | 3.66 | 4.23 | 2.58 | 1.95 | 0.66 | 24.97 |
| 1984 | 0.30 | 0.76 | 0.22 | 0.89 | 1.77 | 6.50 | 2.14 | 1.30 | 1.14 | 4.11 | 0.91 | 1.27 | 21.31 |
| 1985 | 0.38 | 0.70 | 0.72 | 3.17 | 6.67 | 6.15 | 1.22 | 4.27 | 2.97 | 1.97 | 1.57 | 0.51 | 30.30 |
| 1986 | 0.61 | 0.95 | 0.26 | 3.33 | 0.50 | 3.67 | 2.59 | 1.52 | 2.42 | 0.64 | 1.27 | 0.35 | 18.11 |
| 1987 | 0.37 | 0.48 | 1.05 | 0.08 | 3.13 | 1.11 | 7.86 | 2.50 | 1.38 | 0.66 | 0.72 | 0.20 | 19.54 |
| 1988 | 0.44 | 0.14 | 1.71 | 0.30 | 1.18 | 4.39 | 3.04 | 6.66 | 3.93 | 1.03 | 1.39 | 0.76 | 24.97 |
| 1989 | 1.42 | 0.28 | 0.81 | 0.54 | 1.93 | 6.59 | 1.51 | 4.82 | 1.92 | 1.21 | 0.99 | 0.45 | 22.47 |
| 1990 | 0.62 | 0.50 | 1.35 | 1.47 | 0.98 | 5.50 | 3.13 | 2.15 | 1.23 | 1.42 | 0.75 | 1.21 | 20.31 |
| 1991 | 0.79 | 0.91 | 1.10 | 2.52 | 4.09 | 4.58 | 4.10 | 0.58 | 4.94 | 1.47 | 1.83 | 0.94 | 27.85 |
| 1992 | 0.78 | 1.57 | 0.56 | 1.43 | 2.08 | 2.17 | 4.00 | 5.48 | 2.34 | 0.14 | 1.17 | 1.70 | 23.42 |
| 1993 | 0.68 | 0.10 | 0.32 | 1.67 | 2.82 | 2.01 | 6.03 | 3.84 | 2.04 | 0.73 | 0.81 | 0.44 | 21.49 |
| 1994 | 0.46 | 0.34 | 0.58 | 1.47 | 1.65 | 4.56 | 4.02 | 2.73 | 3.67 | 1.97 | 2.46 | 0.67 | 24.58 |
| 1995 | 1.03 | 0.65 | 0.70 | 1.21 | 2.36 | 2.66 | 3.29 | 3.13 | 3.56 | 3.12 | 1.32 | 1.26 | 24.29 |
| 1996 | 1.78 | 1.47 | 0.43 | 1.54 | 2.10 | 4.41 | 3.05 | 3.52 | 2.64 | 3.17 | 3.00 | 1.48 | 28.59 |
| 1997 | 0.72 | 0.19 | 1.18 | 1.31 | 1.44 | 2.57 | 2.65 | 1.41 | 3.90 | 2.37 | 0.99 | 0.23 | 18.96 |
| 1998 | 0.64 | 1.04 | 0.30 | 0.73 | 4.81 | 4.33 | 2.18 | 1.95 | 0.28 | 4.20 | 1.70 | 0.37 | 22.53 |
| 1999 | 0.09 | 0.51 | 2.03 | 1.82 | 5.70 | 3.31 | 6.21 | 2.67 | 5.89 | 0.65 | T | 0.18 | 29.06 |
| 2000 | 0.60 | 0.23 | 0.73 | 1.66 | 2.57 | 4.05 | 2.71 | 4.11 | 2.00 | 1.20 | 2.75 | 0.20 | 22.81 |
| 2001 | 0.18 | 0.25 | 0.13 | 3.54 | 4.50 | 3.07 | 8.29 | 1.65 | 2.11 | 2.59 | 0.90 | 0.27 | 27.48 |
| 2002 | 0.07 | 0.04 | 0.29 | 1.48 | 2.08 | 8.29 | 3.79 | 4.35 | 1.19 | 1.07 | 0.22 | 0.26 | 23.13 |
| 2003 | T | 0.09 | 0.51 | 0.65 | 1.44 | 3.99 | 3.30 | 2.36 | 2.17 | 1.03 | 0.93 | 0.38 | 16.85 |
| 2004 | 0.65 | 0.14 | 0.63 | 0.90 | 5.00 | 1.54 | 4.33 | 1.50 | 5.99 | 3.98 | 0.31 | 2.00 | 26.97 |
| 2005 | 0.92 | 0.28 | 0.36 | 2.32 | 4.60 | 3.86 | 3.45 | 3.79 | 1.33 | 3.20 | 3.29 | 0.61 | 28.01 |
| 2006 | 0.91 | 0.62 | 2.20 | 1.16 | 2.12 | 2.20 | 3.14 | 1.03 | 1.38 | 1.19 | 1.09 | 0.99 | 18.03 |
| 2007 | 0.32 | 0.53 | 1.52 | 1.64 | 3.59 | 4.70 | 1.62 | 1.18 | 4.98 | 3.21 | 0.56 | 1.11 | 24.96 |
| 2008 | 0.19 | 0.38 | 0.70 | 2.95 | 3.62 | 4.94 | 3.13 | 0.85 | 4.19 | 2.40 | 2.81 | 1.42 | 27.58 |
| 2009 | 0.79 | 1.21 | 3.80 | 1.11 | 2.68 | 1.90 | 3.63 | 2.54 | 1.87 | 3.24 | 1.20 | 1.53 | 25.50 |
| POR= 92 YRS | 0.96 | 0.69 | 1.01 | 1.56 | 2.60 | 3.79 | 3.62 | 3.11 | 3.00 | 1.93 | 1.27 | 0.84 | 24.38 |

WBAN : 14918

## AVERAGE TEMPERATURE (°F) 2009 INTERNATIONAL FALLS (KINL)

| YEAR | JAN | FEB | MAR | APR | MAY | JUN | JUL | AUG | SEP | OCT | NOV | DEC | ANNUAL |
|------|-----|-----|-----|-----|-----|-----|-----|-----|-----|-----|-----|-----|--------|
| 1980 | 2.1 | 8.1 | 16.8 | 44.3 | 58.8 | 62.3 | 68.8 | 64.9 | 52.8 | 38.9 | 26.1 | 4.6 | 37.4 |
| 1981 | 6.2 | 14.5 | 28.6 | 40.9 | 53.3 | 61.5 | 68.6 | 67.8 | 54.0 | 40.1 | 34.9 | 10.5 | 40.1 |
| 1982 | -10.5 | 6.5 | 18.7 | 35.4 | 55.8 | 56.2 | 67.3 | 61.1 | 53.7 | 45.0 | 21.6 | 16.4 | 35.6 |
| 1983 | 11.2 | 16.6 | 27.8 | 38.2 | 49.0 | 61.9 | 69.6 | 68.6 | 55.1 | 41.7 | 27.7 | -4.3 | 38.6 |
| 1984 | 0.7 | 20.7 | 17.5 | 44.0 | 48.9 | 61.5 | 65.1 | 67.7 | 49.0 | 44.6 | 25.6 | 3.4 | 37.4 |
| 1985 | -0.1 | 4.7 | 26.7 | 41.7 | 54.4 | 54.5 | 63.6 | 60.1 | 51.2 | 42.1 | 14.3 | 0.2 | 34.5 |
| 1986 | 7.7 | 9.4 | 27.6 | 43.2 | 56.0 | 60.4 | 66.7 | 61.2 | 52.4 | 41.4 | 19.9 | 15.9 | 38.5 |
| 1987 | 11.0 | 23.5 | 29.8 | 47.9 | 55.8 | 64.1 | 67.9 | 62.5 | 56.2 | 38.6 | 31.0 | 18.6 | 42.2 |
| 1988 | 0.2 | 1.9 | 21.7 | 39.5 | 58.5 | 67.6 | 68.4 | 65.2 | 53.8 | 37.8 | 26.1 | 8.8 | 37.5 |
| 1989 | 8.0 | -1.7 | 16.8 | 36.0 | 53.4 | 60.2 | 69.7 | 65.2 | 55.0 | 43.4 | 21.0 | -1.5 | 35.5 |
| 1990 | 13.4 | 9.6 | 27.0 | 38.7 | 49.7 | 63.2 | 65.8 | 66.0 | 55.6 | 40.6 | 28.1 | 5.9 | 38.6 |
| 1991 | -0.1 | 15.4 | 25.3 | 44.8 | 58.1 | 66.1 | 66.4 | 67.9 | 51.8 | 38.1 | 20.2 | 13.0 | 38.9 |
| 1992 | 12.3 | 16.5 | 23.9 | 35.6 | 55.0 | 57.7 | 59.4 | 60.3 | 53.0 | 40.4 | 25.4 | 8.6 | 37.3 |
| 1993 | 4.7 | 9.1 | 25.5 | 38.6 | 50.7 | 58.6 | 64.4 | 65.0 | 48.5 | 37.4 | 22.9 | 13.1 | 36.5 |
| 1994 | -7.8 | 5.5 | 29.2 | 38.7 | 54.2 | 64.1 | 64.8 | 61.8 | 57.8 | 47.5 | 30.9 | 19.1 | 38.8 |
| 1995 | 9.6 | 6.4 | 27.0 | 33.9 | 52.5 | 68.1 | 65.9 | 67.1 | 53.2 | 41.6 | 15.4 | 6.5 | 37.3 |
| 1996 | -5.4 | 7.3 | 15.2 | 32.0 | 49.9 | 63.8 | 65.0 | 66.1 | 55.5 | 42.5 | 17.5 | 6.1 | 34.6 |
| 1997 | 2.1 | 10.2 | 18.4 | 35.9 | 46.0 | 62.7 | 64.2 | 60.6 | 56.4 | 43.1 | 23.0 | 20.4 | 36.9 |
| 1998 | 12.0 | 27.1 | 25.9 | 44.0 | 56.0 | 60.0 | 64.3 | 66.2 | 56.5 | 44.2 | 27.2 | 10.7 | 41.2 |
| 1999 | 3.7 | 19.5 | 26.9 | 42.0 | 54.3 | 61.0 | 66.6 | 62.1 | 51.9 | 40.0 | 33.5 | 17.6 | 39.9 |
| 2000 | 4.7 | 18.9 | 32.2 | 39.2 | 53.7 | 57.5 | 66.2 | 63.8 | 52.1 | 44.3 | 28.8 | -1.3 | 38.3 |
| 2001 | 14.0 | 3.3 | 23.1 | 40.8 | 54.0 | 62.4 | 66.3 | 65.6 | 54.3 | 40.4 | 36.9 | 18.7 | 40.0 |
| 2002 | 11.9 | 18.8 | 15.4 | 37.4 | 46.0 | 63.4 | 68.5 | 64.1 | 56.9 | 33.2 | 23.0 | 17.3 | 38.0 |
| 2003 | 4.9 | 4.1 | 20.8 | 39.5 | 53.2 | 61.0 | 64.2 | 66.7 | 54.2 | 42.4 | 22.3 | 16.6 | 37.5 |
| 2004 | -4.1 | 14.2 | 24.1 | 37.3 | 45.1 | 56.1 | 62.9 | 56.0 | 58.5 | 43.6 | 30.7 | 9.1 | 36.1 |
| 2005 | .9 | 13.8 | 19.5 | 44.2 | 49.8 | 62.6 | 66.0 | 61.4 | 55.8 | 43.1 | 27.4 | 15.0 | 38.3 |
| 2006 | 18.8 | 4.3 | 25.9 | 46.2 | 54.6 | 62.0 | 68.6 | 63.1 | 53.0 | 37.8 | 28.8 | 18.3 | 40.1 |
| 2007 | 9.5 | 2.6 | 26.3 | 37.2 | 55.1 | 63.0 | 65.6 | 61.6 | 53.9 | 44.4 | 26.0 | 8.6 | 37.8 |
| 2008 | 5.1 | 4.0 | 18.8 | 35.9 | 45.5 | 57.9 | 62.4 | 62.2 | 53.5 | 42.7 | 26.7 | -0.2 | 34.5 |
| 2009 | -3.0 | 8.0 | 21.7 | 38.6 | 46.4 | 57.8 | 58.8 | 60.3 | 59.7 | 37.8 | 34.3 | 7.8 | 35.7 |
| POR= 84 YRS | 2.3 | 8.6 | 21.7 | 38.2 | 51.4 | 60.0 | 65.1 | 63.1 | 53.2 | 42.3 | 25.1 | 9.7 | 36.7 |

## HEATING DEGREE DAYS (base 65°F) 2009 INTERNATIONAL FALLS (KINL)

| YEAR | JUL | AUG | SEP | OCT | NOV | DEC | JAN | FEB | MAR | APR | MAY | JUN | TOTAL |
|---|---|---|---|---|---|---|---|---|---|---|---|---|---|
| 1980-81 | 19 | 56 | 365 | 805 | 1161 | 1873 | 1818 | 1411 | 1119 | 716 | 359 | 108 | 9810 |
| 1981-82 | 32 | 17 | 324 | 764 | 894 | 1685 | 2343 | 1638 | 1431 | 882 | 280 | 256 | 10546 |
| 1982-83 | 21 | 161 | 354 | 612 | 1298 | 1500 | 1664 | 1352 | 1145 | 797 | 490 | 153 | 9547 |
| 1983-84 | 27 | 23 | 320 | 715 | 1116 | 2149 | 1995 | 1276 | 1470 | 621 | 494 | 112 | 10318 |
| 1984-85 | 50 | 50 | 474 | 626 | 1176 | 1908 | 2015 | 1687 | 1182 | 688 | 324 | 307 | 10487 |
| 1985-86 | 74 | 163 | 411 | 704 | 1520 | 2011 | 1773 | 1555 | 1152 | 646 | 313 | 161 | 10483 |
| 1986-87 | 34 | 145 | 367 | 723 | 1346 | 1517 | 1671 | 1158 | 1084 | 513 | 299 | 103 | 8960 |
| 1987-88 | 31 | 127 | 257 | 809 | 1018 | 1433 | 2011 | 1828 | 1337 | 756 | 237 | 53 | 9897 |
| 1988-89 | 29 | 85 | 331 | 834 | 1162 | 1743 | 1765 | 1868 | 1490 | 863 | 363 | 170 | 10703 |
| 1989-90 | 11 | 73 | 308 | 670 | 1312 | 2062 | 1596 | 1548 | 1171 | 796 | 464 | 88 | 10099 |
| 1990-91 | 39 | 59 | 284 | 748 | 1101 | 1833 | 2017 | 1383 | 1219 | 601 | 261 | 35 | 9580 |
| 1991-92 | 44 | 43 | 393 | 826 | 1338 | 1610 | 1631 | 1404 | 1267 | 875 | 335 | 233 | 9999 |
| 1992-93 | 173 | 155 | 354 | 756 | 1178 | 1746 | 1868 | 1560 | 1221 | 787 | 434 | 211 | 10443 |
| 1993-94 | 58 | 73 | 492 | 845 | 1257 | 1606 | 2263 | 1666 | 1102 | 784 | 339 | 80 | 10565 |
| 1994-95 | 58 | 135 | 221 | 535 | 1019 | 1416 | 1714 | 1640 | 1174 | 924 | 390 | 80 | 9306 |
| 1995-96 | 50 | 36 | 358 | 719 | 1484 | 1809 | 2183 | 1672 | 1540 | 984 | 462 | 111 | 11408 |
| 1996-97 | 50 | 37 | 309 | 691 | 1419 | 1819 | 1943 | 1531 | 1436 | 865 | 579 | 88 | 10767 |
| 1997-98 | 105 | 164 | 262 | 677 | 1253 | 1377 | 1637 | 1053 | 1205 | 621 | 291 | 169 | 8814 |
| 1998-99 | 68 | 40 | 259 | 637 | 1129 | 1676 | 1895 | 1268 | 1174 | 682 | 339 | 153 | 9320 |
| 1999-00 | 57 | 110 | 387 | 768 | 936 | 1465 | 1865 | 1330 | 1011 | 767 | 351 | 223 | 9270 |
| 2000-01 | 49 | 70 | 384 | 634 | 1078 | 2051 | 1574 | 1723 | 1297 | 724 | 335 | 148 | 10067 |
| 2001-02 | 62 | 72 | 339 | 756 | 836 | 1429 | 1639 | 1288 | 1529 | 822 | 590 | 121 | 9483 |
| 2002-03 | 25 | 69 | 289 | 979 | 1254 | 1470 | 1856 | 1699 | 1363 | 758 | 358 | 150 | 10270 |
| 2003-04 | 65 | 52 | 332 | 699 | 1272 | 1494 | 2135 | 1466 | 1260 | 825 | 609 | 267 | 10476 |
| 2004-05 | 111 | 272 | 223 | 653 | 1022 | 1726 | 1979 | 1428 | 1404 | 620 | 464 | 115 | 10017 |
| 2005-06 | 71 | 151 | 296 | 676 | 1125 | 1542 | 1427 | 1692 | 1205 | 555 | 340 | 110 | 9190 |
| 2006-07 | 22 | 75 | 363 | 835 | 1083 | 1441 | 1716 | 1743 | 1194 | 820 | 315 | 122 | 9729 |
| 2007-08 | 75 | 135 | 343 | 632 | 1162 | 1744 | 1850 | 1761 | 1422 | 866 | 596 | 211 | 10797 |
| 2008-09 | 100 | 119 | 352 | 682 | 1143 | 2016 | 2102 | 1587 | 1334 | 785 | 571 | 228 | 11019 |
| 2009- | 186 | 177 | 165 | 836 | 912 | 1767 | | | | | | | |

WBAN : 14918

## COOLING DEGREE DAYS (base 65°F) 2009 INTERNATIONAL FALLS (KINL)

| YEAR | JAN | FEB | MAR | APR | MAY | JUN | JUL | AUG | SEP | OCT | NOV | DEC | TOTAL |
|---|---|---|---|---|---|---|---|---|---|---|---|---|---|
| 1980 | 0 | 0 | 0 | 2 | 69 | 51 | 142 | 62 | 5 | 0 | 0 | 0 | 331 |
| 1981 | 0 | 0 | 0 | 0 | 5 | 9 | 151 | 112 | 3 | 0 | 0 | 0 | 280 |
| 1982 | 0 | 0 | 0 | 2 | 1 | 0 | 101 | 46 | 22 | 0 | 0 | 0 | 172 |
| 1983 | 0 | 0 | 0 | 0 | 0 | 67 | 178 | 142 | 30 | 0 | 0 | 0 | 417 |
| 1984 | 0 | 0 | 0 | 0 | 4 | 16 | 61 | 139 | 0 | 1 | 0 | 0 | 221 |
| 1985 | 0 | 0 | 0 | 0 | 0 | 1 | 41 | 19 | 2 | 0 | 0 | 0 | 63 |
| 1986 | 0 | 0 | 0 | 0 | 41 | 32 | 96 | 34 | 0 | 0 | 0 | 0 | 203 |
| 1987 | 0 | 0 | 0 | 7 | 20 | 83 | 130 | 56 | 0 | 0 | 0 | 0 | 296 |
| 1988 | 0 | 0 | 0 | 0 | 40 | 136 | 141 | 96 | 3 | 0 | 0 | 0 | 416 |
| 1989 | 0 | 0 | 0 | 0 | 11 | 32 | 165 | 85 | 14 | 5 | 0 | 0 | 312 |
| 1990 | 0 | 0 | 0 | 11 | 0 | 40 | 72 | 96 | 10 | 0 | 0 | 0 | 229 |
| 1991 | 0 | 0 | 0 | 0 | 54 | 76 | 90 | 137 | 6 | 0 | 0 | 0 | 363 |
| 1992 | 0 | 0 | 0 | 0 | 33 | 22 | 5 | 16 | 1 | 0 | 0 | 0 | 77 |
| 1993 | 0 | 0 | 0 | 0 | 0 | 26 | 47 | 81 | 0 | 0 | 0 | 0 | 154 |
| 1994 | 0 | 0 | 0 | 0 | 14 | 60 | 56 | 42 | 13 | 0 | 0 | 0 | 185 |
| 1995 | 0 | 0 | 0 | 0 | 9 | 180 | 84 | 108 | 9 | 0 | 0 | 0 | 390 |
| 1996 | 0 | 0 | 0 | 0 | 0 | 81 | 60 | 78 | 32 | 0 | 0 | 0 | 251 |
| 1997 | 0 | 0 | 0 | 0 | 0 | 24 | 85 | 37 | 10 | 4 | 0 | 0 | 160 |
| 1998 | 0 | 0 | 0 | 0 | 18 | 25 | 53 | 84 | 13 | 0 | 0 | 0 | 193 |
| 1999 | 0 | 0 | 0 | 0 | 14 | 40 | 112 | 26 | 2 | 0 | 0 | 0 | 194 |
| 2000 | 0 | 0 | 0 | 0 | 10 | 3 | 93 | 42 | 5 | 0 | 0 | 0 | 153 |
| 2001 | 0 | 0 | 0 | 4 | 0 | 78 | 109 | 98 | 26 | 0 | 0 | 0 | 315 |
| 2002 | 0 | 0 | 0 | 0 | 8 | 79 | 140 | 47 | 50 | 0 | 0 | 0 | 324 |
| 2003 | 0 | 0 | 0 | 0 | 0 | 35 | 44 | 114 | 16 | 4 | 0 | 0 | 213 |
| 2004 | 0 | 0 | 0 | 0 | 0 | 6 | 51 | 2 | 35 | 0 | 0 | 0 | 94 |
| 2005 | 0 | 0 | 0 | 0 | 0 | 49 | 109 | 47 | 26 | 4 | 0 | 0 | 235 |
| 2006 | 0 | 0 | 0 | 0 | 25 | 26 | 143 | 24 | 11 | 0 | 0 | 0 | 229 |
| 2007 | 0 | 0 | 0 | 0 | 15 | 70 | 98 | 35 | 17 | 0 | 0 | 0 | 235 |
| 2008 | 0 | 0 | 0 | 0 | 0 | 5 | 24 | 39 | 13 | 0 | 0 | 0 | 81 |
| 2009 | 0 | 0 | 0 | 0 | 0 | 20 | 1 | 37 | 13 | 0 | 0 | 0 | 71 |

## SNOWFALL (inches) 2009 INTERNATIONAL FALLS (KINL)

| YEAR | JUL | AUG | SEP | OCT | NOV | DEC | JAN | FEB | MAR | APR | MAY | JUN | TOTAL |
|------|-----|-----|-----|-----|-----|-----|-----|-----|-----|-----|-----|-----|-------|
| 1980-81 | 0.0 | 0.0 | 0.0 | 1.7 | 19.1 | 9.3 | 6.8 | 3.9 | 1.7 | 3.3 | T | 0.0 | 45.8 |
| 1981-82 | 0.0 | 0.0 | 1.4 | 8.5 | 5.3 | 16.2 | 28.4 | 4.5 | 14.8 | 10.8 | 0.0 | 0.0 | 89.9 |
| 1982-83 | 0.0 | 0.0 | 0.0 | T | 13.9 | 3.9 | 5.7 | 16.4 | 2.3 | 3.7 | 0.1 | 0.0 | 46.0 |
| 1983-84 | 0.0 | 0.0 | T | 0.9 | 26.5 | 15.4 | 5.7 | 6.1 | 3.9 | T | 0.2 | 0.0 | 58.7 |
| 1984-85 | 0.0 | 0.0 | T | 2.1 | 10.9 | 12.4 | 9.8 | 15.3 | 13.9 | 2.0 | T | 0.0 | 66.4 |
| 1985-86 | 0.0 | 0.0 | T | 4.8 | 27.1 | 15.5 | 16.4 | 15.4 | 5.2 | 0.1 | 0.3 | 0.0 | 84.8 |
| 1986-87 | 0.0 | 0.0 | 0.0 | 0.7 | 11.5 | 8.4 | 14.1 | 12.6 | 6.9 | 0.5 | 0.0 | 0.0 | 54.7 |
| 1987-88 | 0.0 | 0.0 | 0.0 | 1.4 | 0.1 | 8.1 | 14.7 | 4.9 | 16.1 | 0.2 | 0.0 | 0.0 | 45.5 |
| 1988-89 | 0.0 | 0.0 | 0.0 | 5.4 | 20.8 | 21.4 | 28.9 | 6.7 | 12.5 | 7.1 | 1.9 | 0.0 | 104.7 |
| 1989-90 | 0.0 | 0.0 | T | T | 8.5 | 10.8 | 9.8 | 11.7 | 3.0 | 17.1 | T | T | 60.9 |
| 1990-91 | T | 0.0 | T | 2.5 | 7.8 | 31.1 | 18.5 | 18.3 | 12.1 | 5.1 | 1.8 | 0.0 | 97.2 |
| 1991-92 | 0.0 | 0.0 | T | 3.9 | 23.8 | 22.2 | 16.7 | 32.3 | 2.7 | 9.3 | 0.1 | T | 111.0 |
| 1992-93 | T | 0.0 | 0.1 | 1.8 | 24.9 | 43.9 | 17.6 | 1.6 | 6.8 | 2.0 | T | 0.0 | 98.7 |
| 1993-94 | 0.0 | 0.0 | 1.3 | 4.7 | 18.4 | 5.7 | 15.4 | 8.8 | 10.0 | 10.6 | T | T | 74.9 |
| 1994-95 | 0.0 | T | T | 1.1 | 10.6 | 10.1 | 17.5 | 13.8 | 4.0 | 12.8 | T | 0.0 | 69.9 |
| 1995-96 | 0.0 | T | T | 4.7 | 16.5 | 21.4 | 30.0 | 18.1 | 7.7 | 17.6 | T | 0.0 | 116.0 |
| 1996-97 | 0.0 | T | T | 5.4 | 23.9 | 24.7 | 11.3 | 5.4 | 13.0 | 1.4 | 0.2 | 0.0 | 85.3 |
| 1997-98 | 0.0 | 0.0 | 0.0 | 3.3 | 10.8 | 4.7 | 17.1 | 5.6 | 4.2 | 5.7 | T | 0.0 | 51.4 |
| 1998-99 | 0.0 | 0.0 | 0.0 | T | 22.4 | 12.9 | 4.0 | 9.2 | 11.4 | 5.9 | T | 0.0 | 65.8 |
| 1999-00 | 0.0 | 0.0 | 0.0 | T | T | 3.3 | 17.0 | 1.9 | 1.4 | 3.5 | 0.0 | 0.0 | 27.1 |
| 2000-01 | 0.0 | 0.0 | 0.0 | T | 3.4 | 15.4 | 7.7 | 10.4 | 2.9 | 21.6 | 0.0 | 0.0 | 61.4 |
| 2001-02 | 0.0 | 0.0 | 0.0 | 1.6 | 17.5 | | | | | | | | |
| 2002-03 | | | | | | | 7.0 | 9.4 | 11.8 | 13.0 | 0.0 | 0.0 | |
| 2003-04 | 0.0 | 0.0 | 1.2 | 3.9 | 19.9 | 7.9 | 22.9 | 5.7 | 8.5 | T | T | 0.0 | 70.0 |
| 2004-05 | 0.0 | 0.0 | 0.0 | T | 1.8 | 17.5 | 28.2 | 8.1 | 2.9 | 0.8 | 0.1 | 0.0 | 59.4 |
| 2005-06 | 0.0 | 0.0 | 0.0 | 0.1 | 9.7 | 11.7 | 11.6 | 15.1 | 3.8 | T | 0.2 | 0.0 | 52.2 |
| 2006-07 | 0.0 | 0.0 | 0.0 | 4.0 | 3.5 | 9.5 | 4.4 | 8.9 | 4.3 | 4.7 | T | 0.0 | 39.3 |
| 2007-08 | 0.0 | 0.0 | T | T | 13.9 | 23.8 | 9.0 | 13.5 | 11.8 | 23.5 | 0.3 | 0.0 | 95.8 |
| 2008-09 | 0.0 | 0.0 | 0.0 | 0.3 | 10.9 | 35.9 | 21.3 | 24.6 | 30.1 | 1.9 | 0.6 | 0.0 | 125.6 |
| 2009- | 0.0 | 0.0 | 0.0 | 2.8 | 2.0 | 26.2 | | | | | | | |
| POR= 63 YRS | T | T | 0.1 | 1.9 | 10.1 | 12.2 | 12.3 | 9.3 | 8.5 | 5.9 | 0.7 | T | 61.0 |

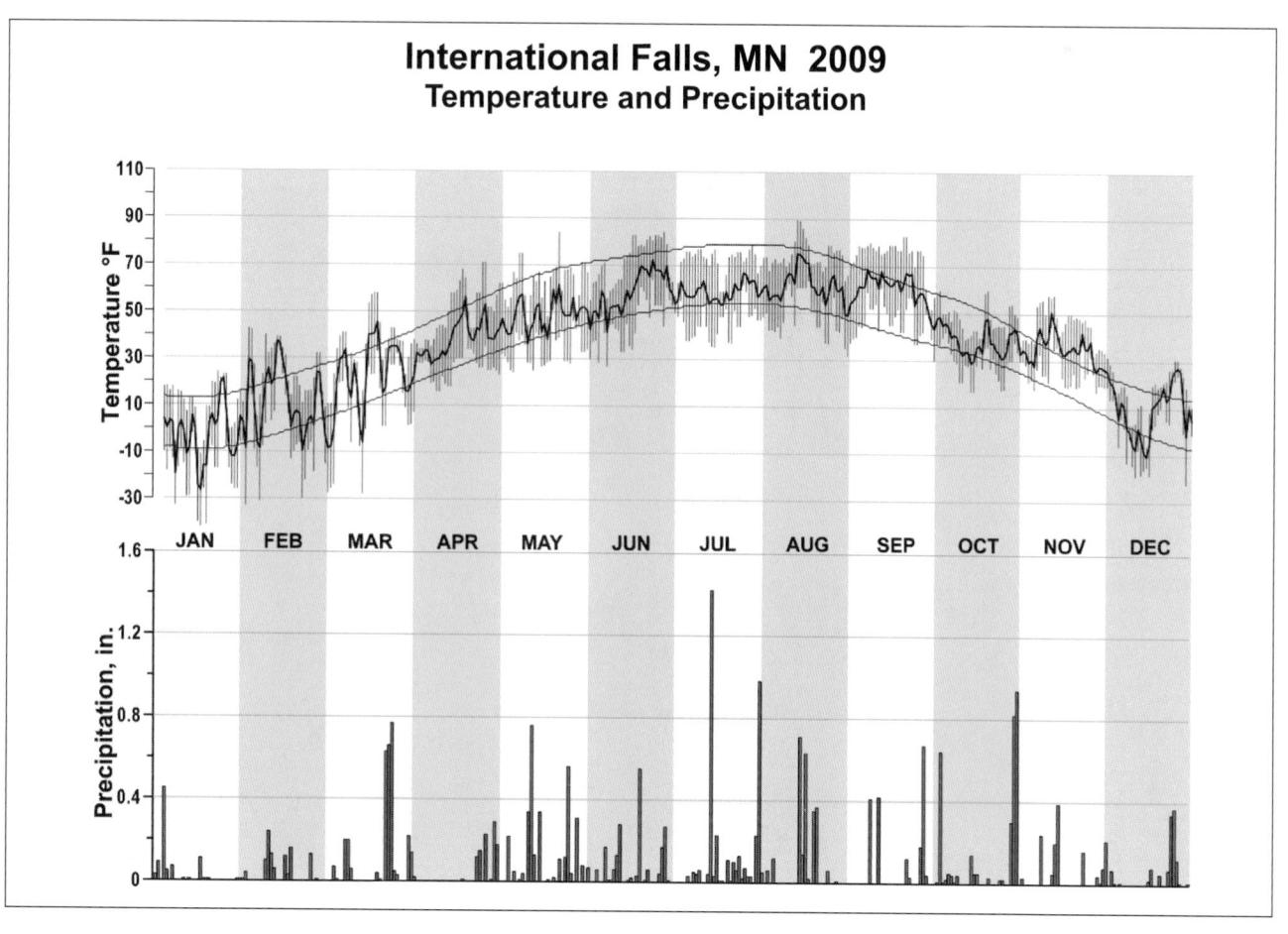

# International Falls, MN  2009
## Temperature and Precipitation

# 2009
# MINNEAPOLIS
# MINNESOTA (KMSP)

The Twin Cities of Minneapolis and St. Paul are located at the confluence of the Mississippi and Minnesota Rivers over the heart of an artesian water basin. Its flat or gently rolling terrain varies little in elevation from that of the official observation station at International Airport. Numerous lakes dot the surrounding area. Minneapolis alone boasts of 22 lakes within the city park system. The largest body of water, nearly 15,000 acres, is Lake Minnetonka, located about 15 miles west of the airport. Most bodies of water are relatively small and shallow and are ice covered during winter.

The climate of the Minneapolis-St. Paul area is predominantly continental. Seasonal temperature variations are quite large. Temperatures range from less than -30 degrees to over 100 degrees. The growing season is 166 days. Because of this favorable growing season, all crops generally mature before the autumn freeze occurs.

The Twin Cities lie near the northern edge of the influx of moisture from the Gulf of Mexico. Severe storms such as blizzards, freezing rain (glaze), tornadoes, wind and hail storms do occur. The total annual precipitation is important. Even more significant is its proper distribution during the growing season. During the five month growing season, May through September, the major crops produced are corn, soybeans, small grains, and hay. During this period, the normal rainfall is over 16 inches, approximately 65 percent of the annual precipitation. Winter snowfall is nearly 48 inches. Winter recreational weather is excellent because of the dry snow. These conditions exist from about Christmas into early March. Snow depths average 6 to 8 inches in the city and 8 to 10 inches in the suburbs during this period.

Floods occur along the Mississippi River due to spring snow melt, excessive rainfall, or both. Occasionally an ice jam forms and creates a local flood condition. The flood problem at St. Paul is complicated because the Minnesota River empties into the Mississippi River between the two cities. Consequently, high water or flooding on the Minnesota River creates a greater flood potential at St. Paul. Flood stage at St. Paul can be expected on the average once in every eight years.

# NORMALS, MEANS, AND EXTREMES
## MINNEAPOLIS (KMSP)

| LATITUDE: 44° 52'N | LONGITUDE: -93° 13'W | ELEVATION (FT): GRND: 815 BARO: 874 | TIME ZONE: CENTRAL (UTC -6) | WBAN: 14922 |
|---|---|---|---|---|

| | ELEMENT | POR | JAN | FEB | MAR | APR | MAY | JUN | JUL | AUG | SEP | OCT | NOV | DEC | YEAR |
|---|---|---|---|---|---|---|---|---|---|---|---|---|---|---|---|
| TEMPERATURE °F | NORMAL DAILY MAXIMUM | 30 | 21.9 | 28.4 | 40.6 | 57.0 | 70.1 | 79.0 | 83.3 | 80.4 | 71.1 | 58.4 | 40.1 | 26.4 | 54.7 |
| | MEAN DAILY MAXIMUM | 119 | 22.0 | 25.2 | 38.7 | 55.2 | 68.5 | 76.7 | 83.2 | 80.5 | 70.5 | 58.8 | 40.1 | 27.0 | 53.9 |
| | HIGHEST DAILY MAXIMUM | 71 | 58 | 61 | 83 | 95 | 97 | 102 | 105 | 102 | 98 | 90 | 77 | 68 | 105 |
| | YEAR OF OCCURRENCE | | 1944 | 2000 | 1986 | 1980 | 2009 | 1985 | 1988 | 1947 | 1976 | 1997 | 1999 | 1998 | JUL 1988 |
| | MEAN OF EXTREME MAXS. | 119 | 41.1 | 45.5 | 62.1 | 79.5 | 87.7 | 93.2 | 94.8 | 93.3 | 88.6 | 79.9 | 61.9 | 45.8 | 72.8 |
| | NORMAL DAILY MINIMUM | 30 | 4.3 | 11.8 | 23.5 | 36.2 | 48.5 | 57.8 | 63.0 | 60.8 | 50.8 | 38.9 | 24.8 | 10.9 | 35.9 |
| | MEAN DAILY MINIMUM | 119 | 5.2 | 8.9 | 22.0 | 35.9 | 48.0 | 57.3 | 63.3 | 60.9 | 51.1 | 40.2 | 25.2 | 12.2 | 35.9 |
| | LOWEST DAILY MINIMUM | 71 | -34 | -32 | -32 | 2 | 18 | 34 | 43 | 39 | 26 | 13 | -17 | -29 | -34 |
| | YEAR OF OCCURRENCE | | 1970 | 1996 | 1962 | 1962 | 1967 | 1945 | 1972 | 1967 | 1974 | 1997 | 1964 | 1983 | JAN 1970 |
| | MEAN OF EXTREME MINS. | 119 | -17.7 | -12.6 | 0.0 | 20.9 | 33.1 | 44.1 | 52.0 | 49.1 | 35.2 | 25.1 | 6.0 | -10.1 | 18.8 |
| | NORMAL DRY BULB | 30 | 13.1 | 20.1 | 32.1 | 46.6 | 59.3 | 68.4 | 73.2 | 70.6 | 61.0 | 48.7 | 32.5 | 18.7 | 45.4 |
| | MEAN DRY BULB | 119 | 13.6 | 17.2 | 30.4 | 45.5 | 58.3 | 67.1 | 73.3 | 70.7 | 60.8 | 49.5 | 32.6 | 19.6 | 44.9 |
| | MEAN WET BULB | 26 | 13.9 | 17.8 | 27.9 | 39.2 | 50.3 | 60.0 | 64.5 | 63.1 | 55.2 | 42.7 | 29.5 | 18.0 | 40.2 |
| | MEAN DEW POINT | 26 | 10.3 | 14.1 | 23.1 | 33.1 | 45.4 | 56.4 | 61.5 | 60.2 | 51.8 | 38.3 | 25.9 | 14.7 | 36.2 |
| | NORMAL NO. DAYS WITH: | | | | | | | | | | | | | | |
| | MAXIMUM >= 90 | 30 | 0.0 | 0.0 | 0.0 | 0.1 | 0.7 | 2.7 | 5.6 | 3.0 | 0.9 | * | 0.0 | 0.0 | 13.0 |
| | MAXIMUM <= 32 | 30 | 23.6 | 16.7 | 7.1 | 0.4 | 0.0 | 0.0 | 0.0 | 0.0 | 0.0 | 0.1 | 8.0 | 20.4 | 76.3 |
| | MINIMUM <= 32 | 30 | 30.8 | 26.8 | 24.1 | 10.1 | 0.7 | 0.0 | 0.0 | 0.0 | 0.0 | 0.5 | 7.3 | 30.2 | 154.2 |
| | MINIMUM <= 0 | 30 | 12.9 | 6.9 | 1.7 | 0.0 | 0.0 | 0.0 | 0.0 | 0.0 | 0.0 | 0.0 | 0.9 | 7.5 | 29.9 |
| H/C | NORMAL HEATING DEG. DAYS | 30 | 1616 | 1273 | 1034 | 560 | 222 | 44 | 7 | 20 | 178 | 516 | 978 | 1428 | 7876 |
| | NORMAL COOLING DEG. DAYS | 30 | 0 | 0 | 0 | 4 | 41 | 146 | 259 | 190 | 56 | 3 | 0 | 0 | 699 |
| RH | NORMAL (PERCENT) | 30 | 71 | 71 | 67 | 59 | 60 | 64 | 66 | 69 | 70 | 67 | 72 | 75 | 68 |
| | HOUR 00 LST | 30 | 74 | 75 | 72 | 65 | 66 | 72 | 75 | 78 | 77 | 73 | 76 | 77 | 73 |
| | HOUR 06 LST | 30 | 75 | 77 | 77 | 74 | 75 | 79 | 81 | 84 | 84 | 80 | 80 | 79 | 79 |
| | HOUR 12 LST | 30 | 68 | 65 | 60 | 50 | 51 | 54 | 55 | 58 | 58 | 58 | 66 | 70 | 59 |
| | HOUR 18 LST | 30 | 69 | 67 | 60 | 49 | 48 | 52 | 54 | 58 | 60 | 59 | 68 | 73 | 60 |
| S | PERCENT POSSIBLE SUNSHINE | 58 | 53 | 59 | 57 | 58 | 61 | 66 | 72 | 69 | 62 | 55 | 39 | 42 | 58 |
| W/O | MEAN NO. DAYS WITH: | | | | | | | | | | | | | | |
| | HEAVY FOG(VISBY <= 1/4 MI) | 46 | 1.0 | 1.3 | 1.5 | 0.4 | 0.3 | 0.4 | 0.2 | 0.5 | 0.8 | 0.6 | 0.9 | 1.3 | 9.2 |
| | THUNDERSTORMS | 64 | 0.0 | 0.2 | 1.0 | 2.6 | 4.9 | 7.6 | 7.3 | 6.4 | 4.4 | 1.8 | 0.5 | 0.2 | 36.9 |
| CLOUDNESS | MEAN: | | | | | | | | | | | | | | |
| | SUNRISE-SUNSET (OKTAS) | 58 | 5.1 | 5.0 | 5.4 | 5.3 | 5.1 | 4.9 | 4.2 | 4.2 | 4.4 | 4.6 | 5.7 | 5.5 | 5.0 |
| | MIDNIGHT-MIDNIGHT (OKTAS) | 32 | 4.9 | 4.8 | 5.1 | 5.0 | 4.8 | 4.5 | 4.0 | 4.0 | 4.2 | 4.6 | 5.3 | 5.2 | 4.7 |
| | MEAN NO. DAYS WITH: | | | | | | | | | | | | | | |
| | CLEAR | 58 | 8.3 | 7.7 | 7.1 | 6.9 | 7.0 | 7.2 | 9.8 | 10.0 | 9.8 | 9.8 | 5.5 | 6.4 | 95.5 |
| | PARTLY CLOUDY | 58 | 7.3 | 6.9 | 7.4 | 7.7 | 9.1 | 10.4 | 11.8 | 11.2 | 8.5 | 7.5 | 6.4 | 6.4 | 100.6 |
| | CLOUDY | 58 | 15.4 | 13.7 | 16.5 | 15.3 | 14.8 | 12.4 | 9.5 | 9.8 | 11.8 | 13.7 | 18.1 | 18.2 | 169.2 |
| PR | MEAN STATION PRESSURE(IN) | 26 | 29.15 | 29.17 | 29.13 | 29.05 | 29.04 | 29.02 | 29.07 | 29.11 | 29.10 | 29.11 | 29.11 | 29.14 | 29.10 |
| | MEAN SEA-LEVEL PRES. (IN) | 26 | 30.10 | 30.11 | 30.06 | 29.97 | 29.94 | 29.91 | 29.95 | 29.99 | 30.00 | 30.02 | 30.04 | 30.09 | 30.02 |
| WINDS | MEAN SPEED (MPH) | 26 | 9.9 | 9.8 | 10.5 | 11.4 | 10.5 | 9.6 | 9.0 | 8.7 | 9.5 | 10.2 | 10.0 | 9.7 | 9.9 |
| | PREVAIL.DIR(TENS OF DEGS) | 41 | 32 | 32 | 32 | 32 | 15 | 14 | 16 | 15 | 16 | 32 | 32 | 32 | 32 |
| | MAXIMUM 2-MINUTE: | | | | | | | | | | | | | | |
| | SPEED (MPH) | 13 | 35 | 37 | 37 | 45 | 49 | 49 | 40 | 47 | 41 | 38 | 39 | 39 | 49 |
| | DIR. (TENS OF DEGS) | | 29 | 31 | 29 | 28 | 22 | 31 | 27 | 26 | 29 | 31 | 28 | 31 | 31 |
| | YEAR OF OCCURRENCE | | 2004 | 2002 | 2004 | 2000 | 1998 | 2008 | 2007 | 2007 | 2009 | 2001 | 2005 | 2004 | JUN 2008 |
| | MAXIMUM 3-SECOND | | | | | | | | | | | | | | |
| | SPEED (MPH) | 13 | 47 | 47 | 49 | 59 | 64 | 62 | 54 | 71 | 54 | 53 | 49 | 52 | 71 |
| | DIR. (TENS OF DEGS) | | 32 | 33 | 32 | 28 | 26 | 32 | 28 | 26 | 28 | 28 | 28 | 31 | 26 |
| | YEAR OF OCCURRENCE | | 2004 | 2002 | 2000 | 2000 | 1998 | 2008 | 2007 | 2007 | 2009 | 2008 | 2005 | 2004 | AUG 2007 |
| PRECIPITATION | NORMAL (IN) | 30 | 1.04 | 0.79 | 1.86 | 2.31 | 3.24 | 4.34 | 4.04 | 4.05 | 2.69 | 2.11 | 1.94 | 1.00 | 29.41 |
| | MAXIMUM MONTHLY (IN) | 71 | 3.63 | 2.14 | 4.75 | 7.00 | 8.03 | 9.82 | 17.90 | 9.32 | 7.53 | 5.68 | 5.29 | 4.27 | 17.90 |
| | YEAR OF OCCURRENCE | | 1967 | 1981 | 1965 | 2001 | 1962 | 1990 | 1987 | 2007 | 1942 | 1971 | 1991 | 1982 | JUL 1987 |
| | MINIMUM MONTHLY (IN) | 71 | 0.10 | 0.06 | 0.32 | 0.16 | 0.53 | 0.22 | 0.58 | 0.43 | 0.41 | 0.01 | 0.02 | T | T |
| | YEAR OF OCCURRENCE | | 1990 | 1964 | 1994 | 1987 | 2009 | 1988 | 1975 | 1946 | 1940 | 1952 | 1939 | 1943 | DEC 1943 |
| | MAXIMUM IN 24 HOURS (IN) | 71 | 1.21 | 1.10 | 1.66 | 2.58 | 3.03 | 3.28 | 10.00 | 7.36 | 3.55 | 4.83 | 2.91 | 2.47 | 10.00 |
| | YEAR OF OCCURRENCE | | 1967 | 1966 | 1965 | 2006 | 1965 | 2003 | 1987 | 1977 | 1942 | 2005 | 1940 | 1982 | JUL 1987 |
| | NORMAL NO. DAYS WITH: | | | | | | | | | | | | | | |
| | PRECIPITATION >= 0.01 | 30 | 9.9 | 7.5 | 10.2 | 11.3 | 10.9 | 11.1 | 10.4 | 10.4 | 9.8 | 8.4 | 9.1 | 9.7 | 118.7 |
| | PRECIPITATION >= 1.00 | 30 | * | * | 0.1 | 0.4 | 0.5 | 1.2 | 1.0 | 1.0 | 0.6 | 0.4 | 0.3 | 0.1 | 5.5 |
| SNOWFALL | NORMAL (IN) | 30 | 13.5 | 8.2 | 10.4 | 3.1 | 0.1 | 0.0 | 0.0 | 0.0 | 0.* | 0.6 | 10.0 | 10.0 | 55.9 |
| | MAXIMUM MONTHLY (IN) | 66 | 46.4 | 26.5 | 40.0 | 21.8 | 3.0 | T | T | T | 1.7 | 8.2 | 46.9 | 33.2 | 46.9 |
| | YEAR OF OCCURRENCE | | 1982 | 1962 | 1951 | 1983 | 1946 | 1995 | 2008 | 2007 | 1942 | 1991 | 1991 | 1969 | NOV 1991 |
| | MAXIMUM IN 24 HOURS (IN) | 66 | 18.5 | 9.3 | 14.7 | 13.6 | 3.0 | T | T | T | 1.7 | 8.2 | 21.0 | 16.5 | 21.0 |
| | YEAR OF OCCURRENCE' | | 1982 | 1939 | 1985 | 1983 | 1946 | 1995 | 1994 | 1992 | 1942 | 1991 | 1991 | 1982 | NOV 1991 |
| | MAXIMUM SNOW DEPTH (IN) | 59 | 38 | 30 | 27 | 10 | 2 | 0 | 0 | 0 | 0 | 1 | 23 | 21 | 38 |
| | YEAR OF OCCURRENCE | | 1982 | 1967 | 1965 | 1985 | 1984 | | | | | 2009 | 1991 | 1991 | JAN 1982 |
| | NORMAL NO. DAYS WITH: | | | | | | | | | | | | | | |
| | SNOWFALL >= 1.0 | 30 | 4.1 | 3.0 | 3.1 | 0.9 | 0.0 | 0.0 | 0.0 | 0.0 | 0.0 | 0.1 | 2.8 | 2.9 | 16.9 |

## PRECIPITATION (inches) 2009 MINNEAPOLIS (KMSP)

| YEAR | JAN | FEB | MAR | APR | MAY | JUN | JUL | AUG | SEP | OCT | NOV | DEC | ANNUAL |
|------|-----|-----|-----|-----|-----|-----|-----|-----|-----|-----|-----|-----|--------|
| 1980 | 0.94 | 0.67 | 1.12 | 0.83 | 2.29 | 5.52 | 2.30 | 3.26 | 3.68 | 0.66 | 0.26 | 0.24 | 21.77 |
| 1981 | 0.30 | 2.14 | 0.71 | 2.17 | 2.18 | 4.42 | 4.09 | 4.73 | 1.46 | 2.69 | 2.16 | 0.92 | 27.97 |
| 1982 | 2.45 | 0.43 | 2.09 | 1.62 | 4.99 | 1.44 | 0.92 | 3.80 | 1.50 | 3.45 | 3.27 | 4.27 | 30.23 |
| 1983 | 0.67 | 1.19 | 3.22 | 3.97 | 6.20 | 5.22 | 3.07 | 3.12 | 3.34 | 2.61 | 4.93 | 1.53 | 39.07 |
| 1984 | 0.88 | 1.64 | 1.47 | 3.86 | 2.29 | 7.95 | 3.03 | 5.15 | 2.65 | 5.48 | 0.31 | 2.24 | 36.95 |
| 1985 | 0.87 | 0.50 | 4.48 | 1.81 | 3.65 | 2.18 | 2.20 | 5.02 | 4.37 | 3.66 | 1.72 | 1.20 | 31.66 |
| 1986 | 0.90 | 0.84 | 2.03 | 5.88 | 3.48 | 5.34 | 4.11 | 4.44 | 6.90 | 1.77 | 0.62 | 0.31 | 36.62 |
| 1987 | 0.63 | 0.13 | 0.64 | 0.16 | 1.88 | 1.95 | 17.90 | 3.67 | 1.28 | 0.60 | 2.07 | 1.25 | 32.16 |
| 1988 | 1.37 | 0.30 | 1.33 | 1.58 | 1.70 | 0.22 | 1.17 | 4.29 | 2.79 | 0.80 | 2.86 | 0.67 | 19.08 |
| 1989 | 0.52 | 1.04 | 2.19 | 2.66 | 3.38 | 3.50 | 3.50 | 2.92 | 1.28 | 0.53 | 1.38 | 0.42 | 23.32 |
| 1990 | 0.10 | 0.77 | 3.66 | 3.80 | 3.36 | 9.82 | 5.06 | 1.71 | 1.88 | 1.23 | 0.65 | 1.01 | 33.05 |
| 1991 | 0.49 | 1.03 | 2.29 | 3.58 | 6.35 | 2.57 | 2.95 | 3.14 | 5.43 | 2.52 | 5.29 | 1.05 | 36.69 |
| 1992 | 0.66 | 0.57 | 1.56 | 1.99 | 1.15 | 3.68 | 5.21 | 4.54 | 5.20 | 2.11 | 1.95 | 1.05 | 29.67 |
| 1993 | 1.25 | 0.39 | 1.25 | 1.99 | 4.02 | 6.28 | 5.58 | 6.50 | 2.04 | 0.79 | 1.57 | 0.55 | 32.21 |
| 1994 | 1.17 | 0.78 | 0.32 | 3.77 | 2.21 | 3.09 | 4.12 | 2.90 | 4.74 | 4.65 | 1.39 | 0.53 | 29.67 |
| 1995 | 0.36 | 0.25 | 2.11 | 1.90 | 2.43 | 3.38 | 2.72 | 4.59 | 2.21 | 3.68 | 0.88 | 1.15 | 25.66 |
| 1996 | 1.87 | 0.24 | 1.39 | 0.76 | 2.37 | 4.76 | 2.09 | 1.43 | 1.30 | 3.01 | 5.08 | 1.75 | 26.05 |
| 1997 | 1.71 | 0.30 | 1.18 | 1.01 | 1.70 | 3.70 | 12.60 | 6.01 | 3.19 | 2.03 | 0.69 | 0.31 | 34.43 |
| 1998 | 1.64 | 0.80 | 4.56 | 1.56 | 4.40 | 6.52 | 2.63 | 5.99 | 1.32 | 2.19 | 1.32 | 0.46 | 33.39 |
| 1999 | 2.67 | 0.40 | 1.86 | 3.43 | 6.56 | 3.68 | 4.55 | 2.64 | 2.73 | 0.92 | 0.77 | 0.33 | 30.54 |
| 2000 | 0.90 | 1.08 | 1.12 | 1.12 | 4.56 | 4.56 | 6.10 | 3.19 | 2.15 | 1.09 | 3.38 | 1.23 | 30.48 |
| 2001 | 1.21 | 1.33 | 1.09 | 7.00 | 4.53 | 6.35 | 2.12 | 2.31 | 3.50 | 1.28 | 2.77 | 0.74 | 34.23 |
| 2002 | 0.46 | 0.41 | 1.38 | 3.15 | 2.83 | 8.30 | 5.19 | 8.30 | 3.90 | 4.18 | 0.09 | 0.22 | 38.41 |
| 2003 | 0.22 | 0.54 | 1.44 | 2.40 | 6.14 | 4.66 | 2.05 | 1.12 | 2.20 | 0.62 | 0.71 | 0.62 | 22.72 |
| 2004 | 0.23 | 1.09 | 2.11 | 2.06 | 6.39 | 3.06 | 3.36 | 1.19 | 4.21 | 2.32 | 0.93 | 0.44 | 27.39 |
| 2005 | 1.21 | 0.96 | 1.37 | 2.30 | 2.78 | 4.24 | 2.94 | 5.22 | 4.44 | 5.45 | 1.53 | 0.97 | 33.41 |
| 2006 | 0.71 | 0.32 | 2.01 | 5.97 | 1.66 | 2.81 | 1.29 | 6.90 | 2.44 | 0.41 | 0.92 | 2.13 | 27.57 |
| 2007 | 0.31 | 1.37 | 3.64 | 1.11 | 1.99 | 2.05 | 3.29 | 9.32 | 6.04 | 3.63 | 0.09 | 1.48 | 34.32 |
| 2008 | 0.15 | 0.40 | 1.97 | 3.12 | 2.53 | 2.70 | 2.13 | 3.35 | 1.78 | 1.96 | 1.14 | 1.15 | 22.38 |
| 2009 | 0.57 | 0.93 | 1.50 | 1.57 | 0.53 | 2.86 | 2.17 | 6.43 | 0.46 | 5.57 | 0.38 | 1.83 | 24.80 |
| POR= 118 YRS | 0.85 | 0.82 | 1.64 | 2.23 | 3.43 | 4.20 | 3.56 | 3.55 | 2.88 | 2.09 | 1.45 | 0.94 | 27.64 |

WBAN : 14922

## AVERAGE TEMPERATURE (°F) 2009 MINNEAPOLIS (KMSP)

| YEAR | JAN | FEB | MAR | APR | MAY | JUN | JUL | AUG | SEP | OCT | NOV | DEC | ANNUAL |
|------|-----|-----|-----|-----|-----|-----|-----|-----|-----|-----|-----|-----|--------|
| 1980 | 15.3 | 15.3 | 27.3 | 49.2 | 61.5 | 67.6 | 75.2 | 70.7 | 59.5 | 45.1 | 36.6 | 19.8 | 45.3 |
| 1981 | 18.0 | 23.4 | 37.7 | 49.1 | 57.1 | 67.0 | 70.9 | 69.3 | 60.0 | 46.7 | 38.0 | 17.5 | 46.2 |
| 1982 | 2.3 | 15.8 | 29.0 | 43.8 | 62.5 | 63.7 | 75.6 | 71.8 | 60.9 | 50.3 | 31.5 | 25.7 | 44.4 |
| 1983 | 19.6 | 26.9 | 34.2 | 42.3 | 54.6 | 68.0 | 77.2 | 76.8 | 62.6 | 48.4 | 34.0 | 3.7 | 45.7 |
| 1984 | 12.0 | 27.5 | 24.8 | 47.1 | 56.0 | 69.7 | 72.2 | 73.5 | 57.2 | 50.7 | 33.3 | 17.9 | 45.2 |
| 1985 | 10.1 | 16.5 | 35.6 | 52.1 | 62.2 | 63.9 | 73.9 | 67.6 | 59.9 | 47.5 | 24.8 | 7.7 | 43.5 |
| 1986 | 17.5 | 15.7 | 33.9 | 49.6 | 59.4 | 68.6 | 73.9 | 67.1 | 59.8 | 49.2 | 28.2 | 24.7 | 45.6 |
| 1987 | 21.2 | 31.6 | 38.7 | 53.5 | 63.5 | 72.8 | 76.0 | 69.0 | 62.5 | 44.6 | 37.9 | 25.0 | 49.7 |
| 1988 | 10.4 | 13.9 | 33.8 | 47.4 | 65.4 | 74.4 | 78.1 | 73.9 | 62.4 | 44.0 | 32.7 | 20.5 | 46.4 |
| 1989 | 21.2 | 8.6 | 26.6 | 45.3 | 57.5 | 68.4 | 76.4 | 70.8 | 60.9 | 49.9 | 28.0 | 10.6 | 43.7 |
| 1990 | 26.3 | 23.7 | 35.7 | 46.8 | 56.3 | 69.5 | 71.3 | 70.6 | 64.4 | 48.1 | 37.4 | 16.9 | 47.3 |
| 1991 | 12.5 | 24.4 | 34.3 | 49.1 | 61.9 | 72.3 | 72.3 | 71.1 | 59.0 | 47.2 | 24.5 | 21.2 | 45.9 |
| 1992 | 21.9 | 28.0 | 33.1 | 43.6 | 60.5 | 65.6 | 65.8 | 65.9 | 59.6 | 47.4 | 31.4 | 21.2 | 45.3 |
| 1993 | 14.6 | 17.2 | 29.5 | 44.2 | 57.2 | 64.5 | 70.3 | 70.4 | 55.0 | 46.5 | 30.6 | 22.2 | 43.5 |
| 1994 | 4.4 | 13.2 | 34.7 | 45.9 | 60.7 | 69.9 | 70.1 | 67.4 | 64.3 | 52.2 | 38.0 | 24.5 | 45.4 |
| 1995 | 18.5 | 19.3 | 35.0 | 42.2 | 56.9 | 71.2 | 73.1 | 74.7 | 60.2 | 48.6 | 27.4 | 19.1 | 45.5 |
| 1996 | 10.2 | 18.0 | 25.3 | 41.4 | 55.6 | 67.4 | 70.0 | 70.5 | 62.2 | 48.8 | 25.4 | 13.7 | 42.4 |
| 1997 | 10.3 | 19.9 | 29.3 | 43.0 | 53.4 | 70.0 | 71.0 | 68.8 | 62.4 | 50.2 | 28.1 | 26.9 | 44.4 |
| 1998 | 19.1 | 31.9 | 31.9 | 50.7 | 63.4 | 64.9 | 72.6 | 71.6 | 66.6 | 51.2 | 37.2 | 24.6 | 48.8 |
| 1999 | 12.4 | 27.9 | 33.8 | 49.0 | 60.1 | 67.3 | 76.2 | 70.1 | 61.1 | 49.6 | 41.8 | 25.6 | 47.9 |
| 2000 | 15.9 | 27.9 | 41.1 | 46.7 | 60.9 | 66.1 | 72.4 | 72.2 | 61.6 | 53.3 | 31.2 | 7.6 | 46.4 |
| 2001 | 20.0 | 11.8 | 27.5 | 48.4 | 59.7 | 69.1 | 75.9 | 74.2 | 60.9 | 48.6 | 46.4 | 27.6 | 47.5 |
| 2002 | 24.6 | 28.3 | 24.9 | 45.7 | 54.6 | 71.1 | 77.0 | 70.9 | 65.5 | 41.8 | 33.0 | 26.2 | 47.0 |
| 2003 | 15.3 | 15.7 | 31.3 | 48.3 | 57.7 | 68.2 | 73.7 | 75.3 | 62.5 | 51.1 | 32.1 | 25.0 | 46.4 |
| 2004 | 11.2 | 21.6 | 36.0 | 50.0 | 56.6 | 65.5 | 72.2 | 66.3 | 67.4 | 50.1 | 37.7 | 22.6 | 46.4 |
| 2005 | 15.6 | 26.5 | 31.8 | 52.0 | 56.4 | 73.4 | 76.8 | 71.7 | 66.3 | 52.4 | 36.6 | 19.4 | 48.2 |
| 2006 | 28.6 | 20.0 | 33.6 | 53.6 | 61.9 | 71.0 | 79.7 | 72.1 | 59.7 | 45.9 | 36.8 | 29.1 | 49.3 |
| 2007 | 19.7 | 13.5 | 38.4 | 47.2 | 64.2 | 72.7 | 76.0 | 72.1 | 64.8 | 54.3 | 34.5 | 16.6 | 47.8 |
| 2008 | 13.2 | 15.2 | 28.3 | 44.0 | 56.3 | 68.7 | 75.6 | 72.5 | 63.6 | 50.4 | 34.7 | 13.5 | 44.7 |
| 2009 | 8.3 | 20.8 | 32.2 | 47.6 | 60.8 | 67.7 | 70.0 | 69.5 | 66.5 | 43.2 | 42.7 | 17.3 | 45.6 |
| POR= 119 YRS | 13.6 | 17.2 | 30.4 | 45.5 | 58.3 | 67.1 | 73.3 | 70.7 | 60.8 | 49.5 | 32.6 | 19.6 | 44.9 |

## HEATING DEGREE DAYS (base 65°F) 2009  MINNEAPOLIS (KMSP)

| YEAR | JUL | AUG | SEP | OCT | NOV | DEC | JAN | FEB | MAR | APR | MAY | JUN | TOTAL |
|---|---|---|---|---|---|---|---|---|---|---|---|---|---|
| 1980-81 | 0 | 12 | 194 | 611 | 845 | 1396 | 1453 | 1160 | 838 | 472 | 249 | 28 | 7258 |
| 1981-82 | 11 | 11 | 172 | 564 | 803 | 1466 | 1945 | 1374 | 1111 | 629 | 117 | 71 | 8274 |
| 1982-83 | 0 | 14 | 168 | 448 | 997 | 1212 | 1400 | 1061 | 947 | 673 | 313 | 49 | 7282 |
| 1983-84 | 2 | 0 | 161 | 514 | 923 | 1901 | 1641 | 1082 | 1240 | 531 | 284 | 7 | 8286 |
| 1984-85 | 5 | 12 | 251 | 435 | 943 | 1453 | 1694 | 1355 | 904 | 403 | 123 | 104 | 7682 |
| 1985-86 | 0 | 28 | 240 | 537 | 1201 | 1774 | 1466 | 1377 | 957 | 454 | 212 | 30 | 8276 |
| 1986-87 | 0 | 43 | 177 | 480 | 1096 | 1243 | 1352 | 929 | 809 | 347 | 134 | 13 | 6623 |
| 1987-88 | 2 | 29 | 106 | 623 | 804 | 1236 | 1688 | 1479 | 962 | 523 | 76 | 4 | 7532 |
| 1988-89 | 1 | 16 | 116 | 646 | 963 | 1373 | 1353 | 1576 | 1184 | 583 | 251 | 44 | 8106 |
| 1989-90 | 0 | 6 | 159 | 470 | 1105 | 1683 | 1194 | 1151 | 899 | 569 | 274 | 37 | 7547 |
| 1990-91 | 2 | 5 | 136 | 516 | 820 | 1484 | 1624 | 1130 | 945 | 481 | 197 | 3 | 7343 |
| 1991-92 | 7 | 8 | 228 | 548 | 1206 | 1354 | 1333 | 1067 | 981 | 636 | 190 | 72 | 7630 |
| 1992-93 | 32 | 52 | 182 | 542 | 1003 | 1351 | 1557 | 1335 | 1096 | 617 | 243 | 70 | 8080 |
| 1993-94 | 3 | 18 | 302 | 566 | 1025 | 1322 | 1879 | 1445 | 932 | 569 | 180 | 27 | 8268 |
| 1994-95 | 2 | 45 | 99 | 390 | 802 | 1250 | 1434 | 1274 | 924 | 678 | 247 | 47 | 7192 |
| 1995-96 | 6 | 0 | 201 | 511 | 1123 | 1416 | 1697 | 1360 | 1222 | 699 | 304 | 62 | 8601 |
| 1996-97 | 3 | 2 | 167 | 500 | 1182 | 1583 | 1688 | 1255 | 1100 | 653 | 351 | 6 | 8490 |
| 1997-98 | 27 | 26 | 113 | 483 | 1101 | 1173 | 1414 | 917 | 1019 | 423 | 104 | 107 | 6907 |
| 1998-99 | 0 | 0 | 74 | 422 | 829 | 1249 | 1625 | 1034 | 958 | 473 | 171 | 76 | 6911 |
| 1999-00 | 0 | 2 | 174 | 471 | 690 | 1214 | 1515 | 1070 | 734 | 542 | 176 | 72 | 6660 |
| 2000-01 | 12 | 1 | 146 | 364 | 1008 | 1771 | 1386 | 1483 | 1155 | 497 | 197 | 54 | 8074 |
| 2001-02 | 8 | 2 | 162 | 505 | 552 | 1152 | 1243 | 1021 | 1234 | 588 | 348 | 30 | 6845 |
| 2002-03 | 0 | 4 | 119 | 711 | 951 | 1197 | 1532 | 1372 | 1037 | 505 | 228 | 30 | 7686 |
| 2003-04 | 0 | 0 | 175 | 441 | 979 | 1232 | 1661 | 1251 | 892 | 456 | 260 | 60 | 7407 |
| 2004-05 | 8 | 50 | 59 | 457 | 810 | 1308 | 1525 | 1073 | 1022 | 394 | 268 | 0 | 6974 |
| 2005-06 | 0 | 3 | 61 | 416 | 845 | 1403 | 1123 | 1253 | 967 | 334 | 187 | 19 | 6611 |
| 2006-07 | 0 | 0 | 196 | 597 | 839 | 1103 | 1399 | 1434 | 817 | 539 | 102 | 9 | 7035 |
| 2007-08 | 0 | 10 | 106 | 352 | 908 | 1494 | 1598 | 1436 | 1130 | 621 | 271 | 11 | 7937 |
| 2008-09 | 0 | 0 | 101 | 449 | 905 | 1588 | 1747 | 1229 | 1008 | 515 | 162 | 69 | 7773 |
| 2009- | 8 | 18 | 55 | 669 | 662 | 1472 | | | | | | | |

WBAN : 14922

## COOLING DEGREE DAYS (base 65°F) 2009  MINNEAPOLIS (KMSP)

| YEAR | JAN | FEB | MAR | APR | MAY | JUN | JUL | AUG | SEP | OCT | NOV | DEC | TOTAL |
|---|---|---|---|---|---|---|---|---|---|---|---|---|---|
| 1980 | 0 | 0 | 0 | 16 | 82 | 121 | 322 | 194 | 38 | 1 | 0 | 0 | 774 |
| 1981 | 0 | 0 | 0 | 0 | 10 | 96 | 200 | 151 | 28 | 0 | 0 | 0 | 485 |
| 1982 | 0 | 0 | 0 | 0 | 46 | 40 | 338 | 232 | 53 | 0 | 0 | 0 | 709 |
| 1983 | 0 | 0 | 0 | 0 | 0 | 145 | 389 | 368 | 98 | 8 | 0 | 0 | 1008 |
| 1984 | 0 | 0 | 0 | 0 | 13 | 155 | 237 | 280 | 24 | 0 | 0 | 0 | 709 |
| 1985 | 0 | 0 | 0 | 22 | 43 | 77 | 284 | 118 | 93 | 0 | 0 | 0 | 637 |
| 1986 | 0 | 0 | 0 | 1 | 45 | 148 | 286 | 115 | 32 | 0 | 0 | 0 | 627 |
| 1987 | 0 | 0 | 0 | 11 | 95 | 253 | 348 | 159 | 37 | 0 | 0 | 0 | 903 |
| 1988 | 0 | 0 | 0 | 1 | 96 | 296 | 412 | 302 | 45 | 0 | 0 | 0 | 1152 |
| 1989 | 0 | 0 | 0 | 0 | 26 | 153 | 359 | 192 | 41 | 8 | 0 | 0 | 779 |
| 1990 | 0 | 0 | 0 | 28 | 11 | 178 | 206 | 191 | 125 | 1 | 0 | 0 | 740 |
| 1991 | 0 | 0 | 0 | 8 | 109 | 246 | 238 | 205 | 51 | 0 | 0 | 0 | 857 |
| 1992 | 0 | 0 | 0 | 3 | 56 | 96 | 64 | 88 | 28 | 2 | 0 | 0 | 337 |
| 1993 | 0 | 0 | 0 | 0 | 12 | 60 | 176 | 195 | 8 | 0 | 0 | 0 | 451 |
| 1994 | 0 | 0 | 0 | 3 | 52 | 183 | 167 | 126 | 86 | 0 | 0 | 0 | 617 |
| 1995 | 0 | 0 | 0 | 0 | 3 | 240 | 264 | 308 | 63 | 9 | 0 | 0 | 887 |
| 1996 | 0 | 0 | 0 | 0 | 20 | 142 | 168 | 181 | 87 | 4 | 0 | 0 | 602 |
| 1997 | 0 | 0 | 0 | 0 | 1 | 163 | 222 | 150 | 41 | 33 | 0 | 0 | 610 |
| 1998 | 0 | 0 | 0 | 0 | 62 | 111 | 243 | 212 | 130 | 0 | 0 | 0 | 758 |
| 1999 | 0 | 0 | 0 | 0 | 28 | 151 | 357 | 166 | 64 | 0 | 0 | 0 | 766 |
| 2000 | 0 | 0 | 0 | 0 | 55 | 111 | 249 | 228 | 53 | 8 | 0 | 0 | 704 |
| 2001 | 0 | 0 | 0 | 8 | 38 | 184 | 351 | 293 | 46 | 2 | 0 | 0 | 922 |
| 2002 | 0 | 0 | 0 | 18 | 33 | 221 | 379 | 195 | 141 | 0 | 0 | 0 | 987 |
| 2003 | 0 | 0 | 0 | 13 | 8 | 130 | 278 | 326 | 108 | 16 | 0 | 0 | 879 |
| 2004 | 0 | 0 | 0 | 10 | 8 | 81 | 239 | 98 | 140 | 2 | 0 | 0 | 578 |
| 2005 | 0 | 0 | 0 | 7 | 6 | 263 | 372 | 217 | 106 | 30 | 0 | 0 | 1001 |
| 2006 | 0 | 0 | 0 | 1 | 99 | 205 | 460 | 230 | 44 | 12 | 0 | 0 | 1051 |
| 2007 | 0 | 0 | 1 | 14 | 83 | 247 | 349 | 234 | 110 | 32 | 0 | 0 | 1070 |
| 2008 | 0 | 0 | 0 | 0 | 12 | 131 | 336 | 239 | 64 | 5 | 0 | 0 | 787 |
| 2009 | 0 | 0 | 0 | 2 | 40 | 158 | 171 | 166 | 110 | 0 | 0 | 0 | 647 |

## SNOWFALL (inches) 2009 MINNEAPOLIS (KMSP)

| YEAR | JUL | AUG | SEP | OCT | NOV | DEC | JAN | FEB | MAR | APR | MAY | JUN | TOTAL |
|------|-----|-----|-----|-----|-----|-----|-----|-----|-----|-----|-----|-----|-------|
| 1980-81 | 0.0 | 0.0 | 0.0 | T | 0.9 | 2.8 | 4.6 | 11.0 | 0.1 | 1.7 | 0.0 | 0.0 | 21.1 |
| 1981-82 | 0.0 | 0.0 | 0.0 | 0.9 | 14.0 | 10.6 | 46.4 | 7.4 | 10.9 | 4.8 | 0.0 | 0.0 | 95.0 |
| 1982-83 | 0.0 | 0.0 | 0.0 | 1.4 | 3.6 | 19.3 | 3.2 | 10.8 | 14.3 | 21.8 | 0.0 | 0.0 | 74.4 |
| 1983-84 | 0.0 | 0.0 | 0.0 | T | 30.4 | 21.0 | 10.6 | 9.3 | 17.3 | 9.8 | 0.0 | 0.0 | 98.4 |
| 1984-85 | 0.0 | 0.0 | 0.0 | 0.3 | 2.0 | 16.3 | 13.1 | 4.2 | 36.8 | T | 0.0 | 0.0 | 72.7 |
| 1985-86 | 0.0 | 0.0 | 0.4 | T | 23.9 | 13.5 | 10.3 | 12.3 | 8.7 | 0.4 | 0.0 | 0.0 | 69.5 |
| 1986-87 | 0.0 | 0.0 | 0.0 | T | 4.4 | 4.2 | 5.5 | 1.2 | 2.1 | T | 0.0 | 0.0 | 17.4 |
| 1987-88 | 0.0 | 0.0 | 0.0 | 0.3 | 4.5 | 7.5 | 19.5 | 4.5 | 3.7 | 2.4 | 0.0 | 0.0 | 42.4 |
| 1988-89 | 0.0 | 0.0 | 0.0 | 0.2 | 15.8 | 7.2 | 6.0 | 17.3 | 22.7 | 0.8 | 0.1 | T | 70.1 |
| 1989-90 | 0.0 | 0.0 | 0.0 | 0.0 | 11.3 | 7.0 | 1.1 | 10.7 | 3.2 | 2.2 | 0.0 | 0.0 | 35.5 |
| 1990-91 | 0.0 | 0.0 | 0.0 | T | 5.0 | 11.7 | 6.5 | 14.2 | 4.4 | 1.5 | 0.3 | 0.0 | 43.6 |
| 1991-92 | 0.0 | T | 0.0 | 8.2 | 46.9 | 6.7 | 5.0 | 5.9 | 10.8 | 0.6 | 0.0 | 0.0 | 84.1 |
| 1992-93 | 0.0 | T | T | 1.3 | 12.2 | 9.2 | 12.0 | 5.3 | 6.9 | 0.5 | 0.0 | 0.0 | 47.4 |
| 1993-94 | T | 0.0 | 0.0 | T | 7.7 | 4.5 | 24.3 | 12.0 | 1.7 | 5.5 | 0.0 | T | 55.7 |
| 1994-95 | T | 0.0 | 0.0 | T | 6.2 | 6.5 | 4.2 | 2.1 | 10.4 | 0.2 | 0.0 | T | 29.6 |
| 1995-96 | 0.0 | 0.0 | T | 0.7 | 6.6 | 16.1 | 14.5 | 1.2 | 14.1 | 2.3 | T | 0.0 | 55.5 |
| 1996-97 | 0.0 | 0.0 | 0.0 | T | 15.3 | 23.5 | 14.2 | 4.0 | 14.3 | 0.6 | T | T | 71.9 |
| 1997-98 | T | 0.0 | 0.0 | T | 8.6 | 3.3 | 20.4 | 1.1 | 11.6 | T | T | T | 45.0 |
| 1998-99 | 0.0 | 0.0 | T | 0.0 | 0.1 | 3.1 | 33.1 | 4.2 | 16.0 | T | 0.0 | 0.0 | 56.5 |
| 1999-00 | 0.0 | 0.0 | 0.0 | T | 0.7 | 7.3 | 18.2 | 7.7 | 1.0 | 1.3 | 0.0 | 0.0 | 36.2 |
| 2000-01 | 0.0 | 0.0 | 0.0 | 0.0 | | | | | | | | | |
| 2001-02 | | | | | | | | | | | | | |
| 2002-03 | | | | | | | | | | | | | |
| 2003-04 | | | | | | | | 8.6 | 8.0 | 6.6 | T | T | 0.0 | |
| 2004-05 | | | | | | | | | | | | | |
| 2005-06 | 0.0 | 0.0 | 0.0 | T | 5.1 | 14.5 | 2.3 | 2.1 | 20.4 | T | 0.0 | 0.0 | 44.4 |
| 2006-07 | 0.0 | 0.0 | 0.0 | T | 0.2 | 4.3 | 5.5 | 12.6 | 11.0 | 1.9 | 0.0 | 0.0 | 35.5 |
| 2007-08 | T | T | T | 0.0 | 0.4 | 18.1 | 2.0 | 4.8 | 18.0 | 1.6 | T | 0.0 | 44.9 |
| 2008-09 | T | 0.0 | 0.0 | T | 4.3 | 17.4 | 8.4 | 10.9 | 1.5 | 2.5 | 0.0 | 0.0 | 45.0 |
| 2009- | 0.0 | 0.0 | 0.0 | 2.8 | T | 20.9 | | | | | | | |
| POR= 65 YRS | T | T | T | 0.5 | 6.7 | 9.4 | 10.2 | 7.3 | 10.2 | 2.5 | 0.1 | T | 46.9 |

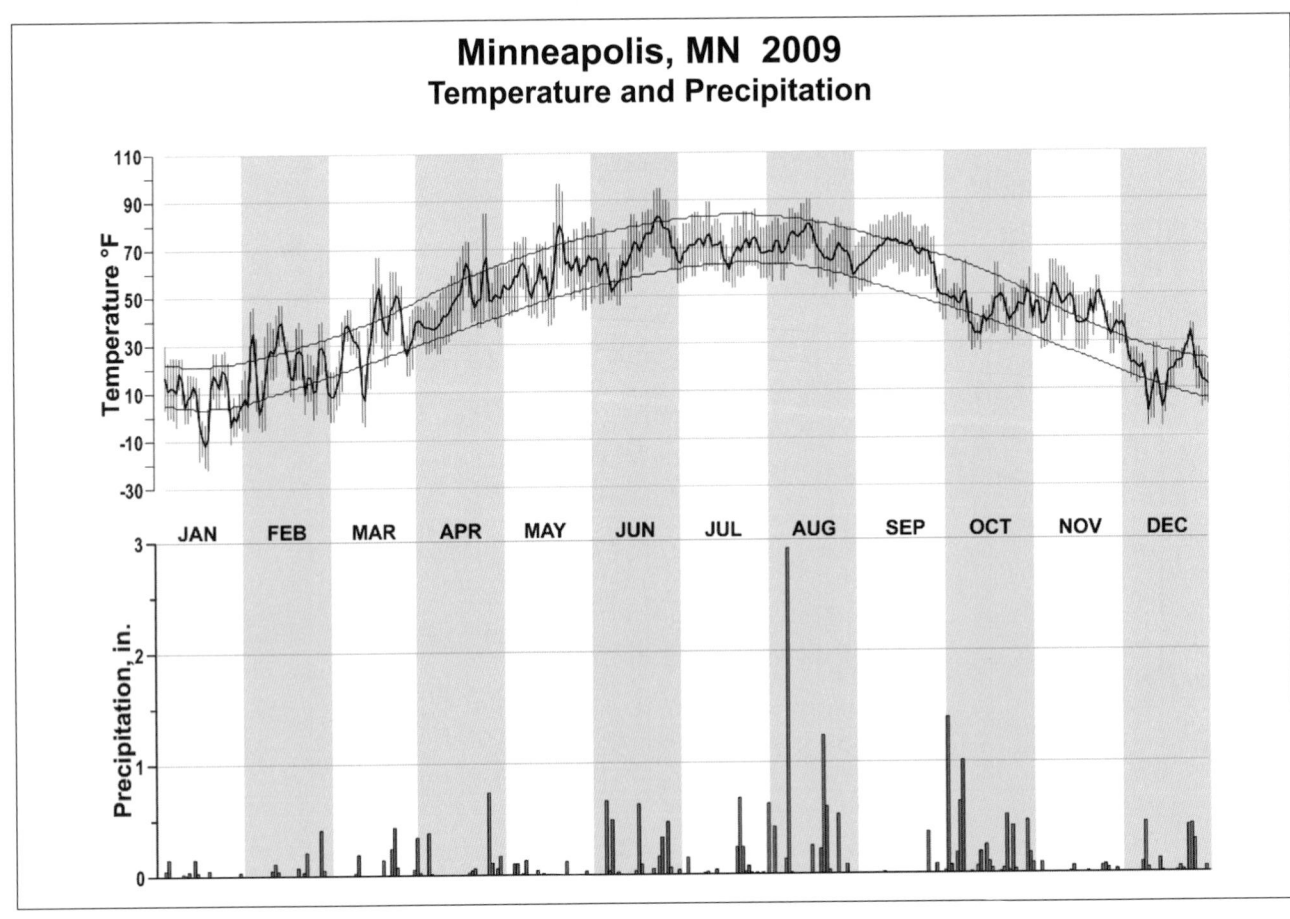

# Minneapolis, MN 2009
## Temperature and Precipitation

# 2009
# JACKSON
# MISSISSIPPI (KJAN)

Jackson is located on the west bank of the Pearl River, about 45 miles east of the Mississippi River and 150 miles north of the Gulf of Mexico. The nearby terrain is gently rolling with no topographic features that appreciably influence the weather. The National Weather Service Office is nearly 7 miles east-northeast of the Jackson Post Office and over 5 miles southwest of the Ross Barnett Reservoir. Alluvial plains up to 3 miles wide extend along the river near Jackson, where some levees have been built on both sides of the river.

The climate is significantly humid during most of the year, with relatively short mild winters and long warm summers. The Gulf of Mexico has a moderating effect on the climate. Cold spells are fairly frequent in winter, but are usually of short duration. Sub-zero temperatures rarely occur. Temperatures occasionally exceed 80 degrees in mid-winter. In summer, temperatures reach 90 degrees or higher on about two-thirds of the days, 100 degree readings are infrequent. Extended periods of very hot weather are rare. On unusual occasions, temperatures at night may drop into the 50s, even in July or August.

Snowfall averages less than two inches per season, with nearly two-thirds of the seasons having only a trace of snow or none at all. Ice storms occasionally cause major damage to trees and power lines during the winter or early spring season. Rainfall is abundant and fairly well-distributed throughout the year. The area does not have a true dry season. However, the six-month period, June through November, is relatively dry in comparison with the December through May period when 60 percent of the annual precipitation can be expected.

Excessive rainfall may occur in any season. In spite of the normally abundant rainfall, fairly serious droughts occasionally occur during the summer or fall season. Tropical disturbances, including hurricanes and their remnants, are infrequent. However, those that pass near or visit the Mississippi Coast in the summer or early fall may bring several days of heavy rain to the Jackson area.

Thunderstorms can be expected on an average of 65 days a year, usually occurring in each month. They are most frequent in summer when they occur on about one-third of the days. At other times of the year, thunderstorms are usually associated with passing weather systems and are likely to be attended by higher winds than in summer. Severe thunderstorms normally affect portions of the metropolitan area a few times each year.

# NORMALS, MEANS, AND EXTREMES
## JACKSON (KJAN)

LATITUDE: 32 ° 19'N  LONGITUDE: -90 ° 4 'W  ELEVATION (FT): GRND: 300  BARO: 296  TIME ZONE: CENTRAL  (UTC -6)  WBAN: 03940

| | ELEMENT | POR | JAN | FEB | MAR | APR | MAY | JUN | JUL | AUG | SEP | OCT | NOV | DEC | YEAR |
|---|---|---|---|---|---|---|---|---|---|---|---|---|---|---|---|
| **TEMPERATURE °F** | NORMAL DAILY MAXIMUM | 30 | 55.1 | 60.3 | 68.1 | 75.0 | 82.1 | 88.9 | 91.4 | 91.4 | 86.4 | 76.8 | 66.3 | 57.9 | 75.0 |
| | MEAN DAILY MAXIMUM | 46 | 56.4 | 60.7 | 69.0 | 76.9 | 83.6 | 90.1 | 92.1 | 91.8 | 87.0 | 77.8 | 67.8 | 59.2 | 76.0 |
| | HIGHEST DAILY MAXIMUM | 45 | 83 | 85 | 89 | 94 | 99 | 105 | 106 | 107 | 104 | 95 | 88 | 84 | 107 |
| | YEAR OF OCCURRENCE | | 2002 | 1989 | 1982 | 1987 | 1964 | 1988 | 1980 | 2000 | 2000 | 2006 | 1971 | 1978 | AUG 2000 |
| | MEAN OF EXTREME MAXS. | 46 | 74.7 | 78.1 | 83.7 | 87.2 | 91.8 | 96.1 | 98.0 | 97.6 | 94.6 | 89.0 | 82.4 | 76.7 | 87.5 |
| | NORMAL DAILY MINIMUM | 30 | 35.0 | 38.2 | 45.4 | 51.7 | 61.0 | 68.1 | 71.4 | 70.3 | 64.6 | 52.0 | 43.4 | 37.3 | 53.2 |
| | MEAN DAILY MINIMUM | 46 | 35.1 | 37.5 | 44.9 | 52.5 | 61.0 | 68.1 | 71.4 | 70.5 | 64.9 | 52.3 | 43.4 | 37.2 | 53.2 |
| | LOWEST DAILY MINIMUM | 45 | 2 | 10 | 15 | 27 | 38 | 47 | 51 | 54 | 35 | 26 | 17 | 4 | 2 |
| | YEAR OF OCCURRENCE | | 1985 | 1996 | 1980 | 1987 | 1971 | 1984 | 1967 | 2004 | 1967 | 1993 | 1976 | 1989 | JAN 1985 |
| | MEAN OF EXTREME MINS. | 46 | 17.1 | 20.4 | 26.9 | 35.5 | 46.4 | 56.3 | 63.9 | 62.6 | 50.1 | 35.3 | 26.5 | 20.1 | 38.4 |
| | NORMAL DRY BULB | 30 | 45.0 | 49.2 | 56.8 | 63.4 | 71.5 | 78.5 | 81.4 | 80.9 | 75.5 | 64.4 | 54.8 | 47.6 | 64.1 |
| | MEAN DRY BULB | 46 | 45.8 | 49.1 | 56.9 | 64.7 | 72.3 | 79.1 | 81.7 | 81.2 | 76.0 | 65.1 | 55.6 | 48.2 | 64.6 |
| | MEAN WET BULB | 26 | 41.4 | 44.5 | 50.3 | 57.0 | 65.3 | 70.9 | 73.6 | 73.1 | 68.1 | 58.7 | 50.2 | 43.6 | 58.1 |
| | MEAN DEW POINT | 26 | 38.2 | 41.1 | 46.5 | 53.6 | 62.6 | 69.0 | 71.9 | 71.2 | 65.6 | 55.9 | 47.3 | 40.5 | 55.3 |
| | NORMAL NO. DAYS WITH: | | | | | | | | | | | | | | |
| | MAXIMUM >= 90 | 30 | 0.0 | 0.0 | 0.0 | 0.2 | 4.4 | 17.9 | 24.3 | 23.5 | 12.7 | 1.0 | 0.0 | 0.0 | 84.0 |
| | MAXIMUM <= 32 | 30 | 0.9 | 0.4 | * | 0.0 | 0.0 | 0.0 | 0.0 | 0.0 | 0.0 | 0.0 | * | 0.3 | 1.6 |
| | MINIMUM <= 32 | 30 | 14.2 | 9.4 | 3.6 | 0.4 | 0.0 | 0.0 | 0.0 | 0.0 | 0.0 | 0.4 | 5.5 | 12.6 | 46.1 |
| | MINIMUM <= 0 | 30 | 0.0 | 0.0 | 0.0 | 0.0 | 0.0 | 0.0 | 0.0 | 0.0 | 0.0 | 0.0 | 0.0 | 0.0 | 0.0 |
| **H/C** | NORMAL HEATING DEG. DAYS | 30 | 611 | 440 | 272 | 115 | 11 | 0 | 0 | 0 | 7 | 102 | 315 | 528 | 2401 |
| | NORMAL COOLING DEG. DAYS | 30 | 0 | 8 | 32 | 82 | 228 | 419 | 524 | 505 | 338 | 99 | 25 | 4 | 2264 |
| **RH** | NORMAL (PERCENT) | 30 | 76 | 73 | 71 | 71 | 74 | 75 | 77 | 77 | 76 | 75 | 77 | 77 | 75 |
| | HOUR 00 LST | 30 | 84 | 82 | 81 | 85 | 87 | 88 | 91 | 90 | 89 | 88 | 87 | 85 | 86 |
| | HOUR 06 LST | 30 | 87 | 87 | 87 | 90 | 92 | 92 | 94 | 95 | 94 | 93 | 90 | 88 | 91 |
| | HOUR 12 LST | 30 | 65 | 60 | 56 | 53 | 56 | 57 | 59 | 57 | 57 | 54 | 58 | 63 | 58 |
| | HOUR 18 LST | 30 | 71 | 63 | 59 | 57 | 61 | 62 | 66 | 65 | 69 | 71 | 75 | 73 | 66 |
| **S** | PERCENT POSSIBLE SUNSHINE | 31 | 49 | 55 | 61 | 66 | 63 | 71 | 64 | 65 | 63 | 67 | 57 | 47 | 61 |
| **W/O** | MEAN NO. DAYS WITH: | | | | | | | | | | | | | | |
| | HEAVY FOG(VISBY <= 1/4 MI) | 46 | 2.9 | 1.9 | 1.7 | 1.5 | 1.0 | 0.8 | 1.3 | 1.2 | 1.1 | 2.0 | 2.2 | 3.1 | 20.7 |
| | THUNDERSTORMS | 46 | 2.3 | 2.7 | 5.5 | 5.6 | 7.1 | 8.6 | 12.7 | 10.2 | 4.8 | 2.4 | 2.7 | 2.3 | 66.9 |
| **CLOUDNESS** | MEAN: | | | | | | | | | | | | | | |
| | SUNRISE-SUNSET (OKTAS) | 1 | 6.4 | 5.6 | 4.8 | 4.0 | 4.0 | 3.6 | 5.6 | 3.2 | 2.4 | 5.2 | 5.6 | 6.4 | 4.7 |
| | MIDNIGHT-MIDNIGHT (OKTAS) | 1 | 6.4 | 5.6 | 4.8 | 4.0 | 4.0 | 3.2 | 3.6 | 2.8 | 2.4 | 5.6 | 5.6 | 6.4 | 4.5 |
| | MEAN NO. DAYS WITH: | | | | | | | | | | | | | | |
| | CLEAR | 2 | 8.5 | 12.5 | 10.0 | 12.0 | 16.0 | 14.5 | 9.0 | 9.0 | 5.0 | 6.0 | 5.0 | 3.0 | 110.5 |
| | PARTLY CLOUDY | 2 | 3.5 | 3.5 | 4.0 | 6.0 | 8.0 | 12.0 | 4.0 | 7.0 | | | 1.0 | 3.0 | |
| | CLOUDY | 2 | 15.0 | 13.0 | 10.0 | 5.0 | 11.5 | 9.0 | 14.0 | 5.0 | 1.0 | 12.0 | 7.0 | 5.0 | 107.5 |
| **PR** | MEAN STATION PRESSURE(IN) | 26 | 29.81 | 29.77 | 29.70 | 29.66 | 29.64 | 29.63 | 29.67 | 29.66 | 29.66 | 29.72 | 29.77 | 29.77 | 29.71 |
| | MEAN SEA-LEVEL PRES. (IN) | 26 | 30.16 | 30.12 | 30.05 | 30.00 | 29.98 | 29.98 | 30.01 | 30.00 | 30.00 | 30.07 | 30.12 | 30.16 | 30.05 |
| **WINDS** | MEAN SPEED (MPH) | 26 | 7.2 | 7.5 | 7.6 | 7.1 | 6.2 | 5.3 | 4.6 | 4.5 | 5.6 | 5.6 | 6.2 | 6.9 | 6.2 |
| | PREVAIL.DIR(TENS OF DEGS) | 39 | 16 | 17 | 17 | 17 | 17 | 16 | 17 | 15 | 15 | 15 | 17 | 16 | 17 |
| | MAXIMUM 2-MINUTE: | 16 | | | | | | | | | | | | | |
| | SPEED (MPH) | | 37 | 43 | 39 | 45 | 33 | 39 | 44 | 47 | 55 | 30 | 30 | 41 | 55 |
| | DIR. (TENS OF DEGS) | | 33 | 13 | 33 | 14 | 14 | 34 | 02 | 35 | 06 | 31 | 30 | 15 | 06 |
| | YEAR OF OCCURRENCE | | 2002 | 1998 | 2005 | 2006 | 2008 | 1994 | 1998 | 2005 | 1998 | 2002 | 2008 | 1994 | SEP 1998 |
| | MAXIMUM 3-SECOND | | | | | | | | | | | | | | |
| | SPEED (MPH) | 16 | 47 | 52 | 49 | 54 | 54 | 53 | 53 | 64 | 63 | 44 | 46 | 51 | 64 |
| | DIR. (TENS OF DEGS) | | 10 | 31 | 28 | 14 | 29 | 05 | 02 | 36 | 06 | 15 | 31 | 12 | 36 |
| | YEAR OF OCCURRENCE | | 2009 | 2008 | 2005 | 2006 | 2009 | 2007 | 1998 | 2005 | 1998 | 2002 | 2008 | 2007 | AUG 2005 |
| **PRECIPITATION** | NORMAL (IN) | 30 | 5.67 | 4.50 | 5.74 | 5.98 | 4.86 | 3.82 | 4.69 | 3.66 | 3.23 | 3.42 | 5.04 | 5.34 | 55.95 |
| | MAXIMUM MONTHLY (IN) | 45 | 14.10 | 10.28 | 15.13 | 15.95 | 10.82 | 8.45 | 13.25 | 11.51 | 9.61 | 9.80 | 9.98 | 17.70 | 17.70 |
| | YEAR OF OCCURRENCE | | 1979 | 1987 | 1976 | 1991 | 1967 | 1997 | 1979 | 2008 | 1965 | 2009 | 1977 | 1982 | DEC 1982 |
| | MINIMUM MONTHLY (IN) | 45 | 0.75 | 1.29 | 0.90 | 1.21 | 0.29 | 0.10 | 1.04 | 0.26 | 0.56 | 0.00 | 0.51 | 0.91 | 0.00 |
| | YEAR OF OCCURRENCE | | 1986 | 2000 | 2007 | 1987 | 1988 | 1988 | 1987 | 2000 | 1969 | 1963 | 1985 | 1980 | OCT 1963 |
| | MAXIMUM IN 24 HOURS (IN) | 45 | 5.63 | 4.78 | 4.69 | 8.50 | 3.58 | 6.49 | 5.49 | 4.79 | 5.86 | 6.99 | 4.34 | 6.71 | 8.50 |
| | YEAR OF OCCURRENCE | | 1979 | 2003 | 1991 | 2003 | 2004 | 1997 | 2001 | 1992 | 1965 | 1975 | 1983 | 1982 | APR 2003 |
| | NORMAL NO. DAYS WITH: | | | | | | | | | | | | | | |
| | PRECIPITATION >= 0.01 | 30 | 10.9 | 9.2 | 10.3 | 8.4 | 9.2 | 9.1 | 10.6 | 9.1 | 7.9 | 6.3 | 8.9 | 9.8 | 109.7 |
| | PRECIPITATION >= 1.00 | 30 | 1.9 | 1.5 | 2.0 | 2.1 | 1.7 | 1.2 | 1.3 | 1.1 | 0.9 | 1.0 | 1.7 | 1.7 | 18.1 |
| **SNOWFALL** | NORMAL (IN) | 30 | 0.5 | 0.1 | 0.1 | 0.* | 0.0 | 0.0 | 0.0 | 0.0 | 0.0 | 0.0 | 0.* | 0.2 | 0.9 |
| | MAXIMUM MONTHLY (IN) | 37 | 6.3 | 3.6 | 5.3 | 1.1 | 0.0 | 0.0 | 0.0 | T | 0.0 | 0.0 | 0.2 | 4.8 | 6.3 |
| | YEAR OF OCCURRENCE | | 1982 | 1968 | 1968 | 1987 | | | | 2001 | | | 1976 | 1997 | JAN 1982 |
| | MAXIMUM IN 24 HOURS (IN) | 37 | 6.0 | 3.6 | 5.3 | 1.1 | 0.0 | 0.0 | 0.0 | T | 0.0 | 0.0 | 0.2 | 4.8 | 6.0 |
| | YEAR OF OCCURRENCE' | | 1982 | 1968 | 1968 | 1987 | | | | 2001 | | | 1976 | 1997 | JAN 1982 |
| | MAXIMUM SNOW DEPTH (IN) | 36 | 6 | 1 | 2 | 1 | 0 | 0 | 0 | 0 | 0 | 0 | T | 5 | 6 |
| | YEAR OF OCCURRENCE | | 1982 | 1985 | 1968 | 1987 | | | | | | | 1976 | 1997 | JAN 1982 |
| | NORMAL NO. DAYS WITH: | | | | | | | | | | | | | | |
| | SNOWFALL >= 1.0 | 30 | 0.1 | 0.0 | 0.0 | 0.0 | 0.0 | 0.0 | 0.0 | 0.0 | 0.0 | 0.0 | 0.0 | 0.1 | 0.2 |

## PRECIPITATION (inches) 2009 JACKSON (KJAN)

| YEAR | JAN | FEB | MAR | APR | MAY | JUN | JUL | AUG | SEP | OCT | NOV | DEC | ANNUAL |
|------|-----|-----|-----|-----|-----|-----|-----|-----|-----|-----|-----|-----|--------|
| 1980 | 7.53 | 3.19 | 13.57 | 14.33 | 6.60 | 1.74 | 2.91 | 1.45 | 3.25 | 3.47 | 4.11 | 0.91 | 63.06 |
| 1981 | 1.41 | 2.63 | 6.19 | 1.26 | 6.64 | 3.66 | 6.51 | 2.81 | 3.51 | 5.12 | 1.97 | 4.90 | 46.61 |
| 1982 | 4.48 | 5.22 | 5.13 | 6.59 | 0.77 | 6.27 | 9.29 | 4.97 | 1.05 | 6.73 | 7.43 | 17.70 | 75.63 |
| 1983 | 8.17 | 6.55 | 6.00 | 15.53 | 9.41 | 2.93 | 1.70 | 3.70 | 2.70 | 1.52 | 8.11 | 6.95 | 73.27 |
| 1984 | 2.64 | 4.64 | 4.84 | 3.96 | 5.61 | 3.18 | 3.07 | 4.56 | 0.93 | 7.68 | 6.48 | 2.17 | 49.76 |
| 1985 | 4.05 | 7.55 | 3.13 | 3.31 | 0.86 | 1.74 | 4.43 | 7.06 | 3.94 | 7.17 | 0.51 | 3.61 | 47.36 |
| 1986 | 0.75 | 1.53 | 3.34 | 1.75 | 10.00 | 3.72 | 4.78 | 2.03 | 2.63 | 5.10 | 9.40 | 4.98 | 50.01 |
| 1987 | 4.66 | 10.28 | 5.47 | 1.21 | 4.98 | 6.17 | 1.04 | 4.03 | 1.50 | 0.27 | 4.20 | 3.50 | 47.31 |
| 1988 | 2.25 | 3.89 | 7.46 | 5.37 | 0.29 | 0.10 | 2.73 | 3.02 | 2.28 | 6.14 | 5.66 | 4.80 | 43.99 |
| 1989 | 4.38 | 2.52 | 4.53 | 2.13 | 7.92 | 8.17 | 4.47 | 1.74 | 5.40 | 0.23 | 6.86 | 4.20 | 52.55 |
| 1990 | 12.17 | 8.30 | 3.55 | 3.66 | 6.34 | 1.46 | 2.84 | 0.61 | 4.83 | 1.24 | 3.33 | 5.71 | 54.04 |
| 1991 | 4.98 | 5.30 | 7.50 | 15.95 | 7.11 | 3.35 | 3.32 | 1.49 | 3.45 | 2.95 | 2.39 | 5.27 | 63.06 |
| 1992 | 4.27 | 3.83 | 2.11 | 1.48 | 1.22 | 4.75 | 4.48 | 8.33 | 3.80 | 1.55 | 8.27 | 4.32 | 48.41 |
| 1993 | 4.72 | 3.40 | 3.84 | 4.09 | 4.31 | 3.79 | 7.00 | 2.47 | 2.04 | 4.08 | 5.39 | 2.89 | 48.02 |
| 1994 | 8.31 | 6.14 | 5.63 | 4.32 | 2.25 | 7.37 | 5.09 | 4.29 | 1.81 | 5.12 | 2.57 | 3.31 | 56.21 |
| 1995 | 3.68 | 2.01 | 6.63 | 8.59 | 4.30 | 3.60 | 5.29 | 5.26 | 1.52 | 6.51 | 6.20 | 5.44 | 59.03 |
| 1996 | 8.85 | 2.01 | 8.58 | 6.01 | 0.72 | 3.85 | 6.99 | 6.70 | 1.49 | 1.59 | 4.84 | 3.33 | 54.96 |
| 1997 | 4.46 | 6.75 | 2.26 | 7.77 | 6.83 | 8.45 | 1.15 | 5.02 | 2.69 | 4.26 | 3.53 | 5.77 | 58.94 |
| 1998 | 9.52 | 5.51 | 6.17 | 3.68 | 1.47 | 3.26 | 6.44 | 2.02 | 2.41 | 0.51 | 4.32 | 6.24 | 51.55 |
| 1999 | 8.47 | 2.04 | 4.65 | 2.09 | 2.66 | 4.34 | 4.41 | 1.71 | 2.48 | 5.74 | 1.88 | 2.76 | 43.23 |
| 2000 | 1.88 | 1.29 | 4.41 | 7.82 | 2.67 | 5.82 | 1.86 | 0.26 | 3.50 | 1.46 | 7.75 | 3.87 | 42.59 |
| 2001 | 5.41 | 3.84 | 9.41 | 1.73 | 4.71 | 6.50 | 9.45 | 4.71 | 4.52 | 3.58 | 6.29 | 4.07 | 64.22 |
| 2002 | 5.60 | 3.05 | 7.29 | 2.73 | 3.90 | 3.73 | 9.47 | 5.44 | 9.57 | 7.22 | 4.18 | 6.30 | 68.48 |
| 2003 | 1.26 | 9.16 | 4.07 | 11.89 | 4.57 | 6.46 | 2.30 | 4.47 | 2.61 | 3.13 | 5.49 | 3.33 | 58.74 |
| 2004 | 4.23 | 6.50 | 1.23 | 2.89 | 9.25 | 5.96 | 5.43 | 7.70 | 1.41 | 4.81 | 8.14 | 5.23 | 62.78 |
| 2005 | 3.72 | 4.90 | 7.68 | 7.33 | 3.22 | 1.66 | 4.59 | 7.89 | 3.14 | 0.00 | 3.16 | 4.88 | 52.17 |
| 2006 | 6.66 | 7.10 | 4.75 | 2.55 | 3.01 | 2.53 | 5.33 | 1.49 | 2.19 | 8.01 | 2.04 | 5.55 | 51.21 |
| 2007 | 5.54 | 2.52 | 0.90 | 2.42 | 2.02 | 1.39 | 7.34 | 1.31 | 4.23 | 1.84 | 1.74 | 3.57 | 34.82 |
| 2008 | 3.91 | 7.03 | 2.19 | 4.78 | 5.95 | 2.78 | 1.82 | 11.51 | 4.89 | 2.04 | 3.79 | 8.91 | 59.60 |
| 2009 | 3.50 | 3.08 | 8.71 | 4.13 | 4.77 | 0.50 | 8.20 | 2.30 | 4.33 | 9.80 | 0.89 | 6.58 | 56.79 |
| POR= 46 YRS | 5.12 | 4.61 | 5.51 | 5.59 | 4.80 | 3.50 | 4.79 | 4.07 | 3.48 | 3.57 | 4.56 | 5.40 | 55.00 |

WBAN : 03940

## AVERAGE TEMPERATURE (°F) 2009 JACKSON (KJAN)

| YEAR | JAN | FEB | MAR | APR | MAY | JUN | JUL | AUG | SEP | OCT | NOV | DEC | ANNUAL |
|------|-----|-----|-----|-----|-----|-----|-----|-----|-----|-----|-----|-----|--------|
| 1980 | 47.4 | 45.5 | 54.3 | 61.6 | 72.8 | 80.7 | 85.8 | 84.7 | 82.4 | 61.4 | 53.2 | 46.2 | 64.7 |
| 1981 | 41.8 | 48.7 | 55.1 | 71.0 | 69.9 | 81.8 | 83.6 | 82.6 | 73.2 | 65.4 | 58.6 | 46.6 | 64.9 |
| 1982 | 47.2 | 48.8 | 62.1 | 63.8 | 74.7 | 79.5 | 82.4 | 81.8 | 74.1 | 65.1 | 56.4 | 52.9 | 65.7 |
| 1983 | 43.7 | 47.4 | 53.9 | 60.1 | 70.7 | 76.6 | 82.9 | 82.7 | 74.1 | 65.8 | 55.1 | 41.8 | 62.9 |
| 1984 | 39.9 | 49.7 | 56.5 | 63.9 | 71.0 | 78.9 | 80.6 | 79.9 | 74.6 | 71.2 | 53.6 | 58.3 | 64.8 |
| 1985 | 38.0 | 44.8 | 61.3 | 65.2 | 72.2 | 78.8 | 80.5 | 80.3 | 74.0 | 68.0 | 62.5 | 42.7 | 64.0 |
| 1986 | 45.3 | 52.6 | 57.0 | 64.1 | 72.9 | 80.4 | 83.2 | 80.5 | 79.8 | 66.1 | 60.1 | 46.6 | 65.7 |
| 1987 | 44.4 | 51.3 | 55.5 | 62.2 | 75.9 | 79.0 | 82.0 | 82.9 | 75.2 | 59.9 | 56.6 | 52.1 | 64.8 |
| 1988 | 41.7 | 47.5 | 55.8 | 64.6 | 70.8 | 79.8 | 82.5 | 82.6 | 77.9 | 61.6 | 59.3 | 49.3 | 64.5 |
| 1989 | 52.5 | 48.2 | 58.8 | 63.2 | 72.0 | 78.4 | 80.6 | 80.8 | 74.1 | 63.9 | 57.1 | 40.4 | 64.2 |
| 1990 | 51.0 | 56.2 | 60.0 | 63.7 | 71.8 | 81.3 | 81.5 | 82.3 | 78.4 | 63.7 | 58.0 | 51.8 | 66.6 |
| 1991 | 46.6 | 51.7 | 59.3 | 68.0 | 75.6 | 79.9 | 82.9 | 81.4 | 76.0 | 67.2 | 52.4 | 52.0 | 66.1 |
| 1992 | 45.6 | 53.7 | 57.6 | 64.1 | 71.1 | 77.7 | 82.0 | 77.8 | 75.6 | 65.0 | 52.7 | 50.1 | 64.4 |
| 1993 | 49.9 | 49.2 | 54.4 | 61.0 | 71.0 | 80.9 | 82.5 | 82.9 | 75.9 | 62.3 | 52.1 | 45.8 | 64.0 |
| 1994 | 41.9 | 49.4 | 55.8 | 66.2 | 70.6 | 79.4 | 78.9 | 79.0 | 74.0 | 66.2 | 58.5 | 49.7 | 64.1 |
| 1995 | 46.9 | 49.5 | 58.5 | 64.7 | 73.5 | 76.0 | 82.4 | 83.5 | 75.2 | 63.8 | 52.1 | 47.6 | 64.5 |
| 1996 | 45.2 | 49.8 | 51.9 | 61.5 | 75.7 | 78.6 | 81.3 | 78.9 | 73.6 | 65.2 | 55.3 | 50.9 | 64.0 |
| 1997 | 46.0 | 51.5 | 61.5 | 59.6 | 69.8 | 77.1 | 82.3 | 79.5 | 76.7 | 65.1 | 50.7 | 44.9 | 63.7 |
| 1998 | 49.2 | 50.0 | 55.7 | 62.6 | 74.4 | 82.2 | 83.5 | 82.7 | 80.6 | 68.4 | 58.7 | 51.4 | 66.6 |
| 1999 | 51.9 | 53.9 | 54.4 | 69.3 | 71.5 | 79.5 | 82.0 | 84.0 | 75.2 | 65.0 | 55.8 | 48.4 | 65.9 |
| 2000 | 48.8 | 54.7 | 60.4 | 61.8 | 75.3 | 78.7 | 83.5 | 85.0 | 76.5 | 65.7 | 52.1 | 38.5 | 65.1 |
| 2001 | 41.6 | 53.5 | 52.1 | 67.8 | 72.3 | 77.2 | 81.6 | 80.5 | 74.0 | 62.1 | 59.8 | 52.0 | 64.5 |
| 2002 | 49.1 | 46.4 | 57.4 | 67.9 | 72.6 | 78.6 | 81.6 | 80.9 | 78.8 | 67.8 | 52.4 | 47.4 | 65.1 |
| 2003 | 41.1 | 47.8 | 56.7 | 64.9 | 74.2 | 77.7 | 80.7 | 81.6 | 74.2 | 65.3 | 58.9 | 45.5 | 64.1 |
| 2004 | 47.2 | 46.5 | 60.5 | 63.0 | 72.8 | 77.9 | 80.5 | 77.6 | 76.0 | 72.1 | 58.9 | 46.6 | 65.0 |
| 2005 | 51.0 | 52.1 | 55.1 | 63.0 | 70.5 | 78.8 | 82.6 | 82.9 | 78.3 | 64.9 | 57.4 | 45.1 | 65.1 |
| 2006 | 52.4 | 47.7 | 58.0 | 68.8 | 72.9 | 79.9 | 82.5 | 84.0 | 74.8 | 65.3 | 54.0 | 50.5 | 65.9 |
| 2007 | 47.4 | 48.1 | 61.7 | 62.1 | 73.3 | 81.2 | 79.9 | 85.3 | 78.1 | 67.7 | 57.1 | 53.7 | 66.3 |
| 2008 | 45.1 | 52.3 | 58.6 | 64.7 | 72.7 | 80.7 | 83.0 | 79.4 | 75.2 | 63.9 | 53.2 | 50.1 | 64.9 |
| 2009 | 47.4 | 52.4 | 58.7 | 64.1 | 73.3 | 81.6 | 81.0 | 80.1 | 77.2 | 63.4 | 55.2 | 45.6 | 65.0 |
| POR= 46 YRS | 45.8 | 49.1 | 56.9 | 64.7 | 72.3 | 79.1 | 81.7 | 81.2 | 76.0 | 65.1 | 55.6 | 48.2 | 64.7 |

## HEATING DEGREE DAYS (base 65°F) 2009 JACKSON (KJAN)

| YEAR | JUL | AUG | SEP | OCT | NOV | DEC | JAN | FEB | MAR | APR | MAY | JUN | TOTAL |
|---|---|---|---|---|---|---|---|---|---|---|---|---|---|
| 1980-81 | 0 | 0 | 0 | 158 | 363 | 575 | 711 | 454 | 312 | 20 | 18 | 0 | 2611 |
| 1981-82 | 0 | 0 | 16 | 104 | 217 | 568 | 569 | 451 | 199 | 123 | 4 | 0 | 2251 |
| 1982-83 | 0 | 0 | 19 | 120 | 286 | 421 | 653 | 487 | 347 | 176 | 12 | 0 | 2521 |
| 1983-84 | 0 | 0 | 21 | 83 | 315 | 716 | 770 | 440 | 289 | 113 | 29 | 0 | 2776 |
| 1984-85 | 0 | 0 | 12 | 27 | 350 | 249 | 832 | 562 | 158 | 92 | 2 | 0 | 2284 |
| 1985-86 | 0 | 0 | 14 | 54 | 148 | 685 | 606 | 360 | 261 | 90 | 8 | 0 | 2226 |
| 1986-87 | 0 | 0 | 0 | 79 | 183 | 566 | 632 | 378 | 297 | 150 | 0 | 0 | 2285 |
| 1987-88 | 0 | 0 | 0 | 161 | 271 | 408 | 716 | 502 | 292 | 74 | 4 | 0 | 2428 |
| 1988-89 | 0 | 0 | 0 | 134 | 208 | 484 | 392 | 495 | 239 | 135 | 20 | 0 | 2107 |
| 1989-90 | 0 | 0 | 11 | 109 | 265 | 754 | 429 | 261 | 209 | 125 | 9 | 0 | 2172 |
| 1990-91 | 0 | 0 | 5 | 145 | 227 | 427 | 566 | 368 | 229 | 35 | 2 | 0 | 2004 |
| 1991-92 | 0 | 0 | 4 | 67 | 404 | 403 | 593 | 323 | 233 | 118 | 18 | 0 | 2163 |
| 1992-93 | 0 | 0 | 1 | 45 | 364 | 452 | 461 | 437 | 327 | 163 | 5 | 0 | 2255 |
| 1993-94 | 0 | 0 | 10 | 161 | 401 | 588 | 709 | 437 | 286 | 88 | 13 | 0 | 2693 |
| 1994-95 | 0 | 0 | 7 | 67 | 211 | 470 | 558 | 430 | 229 | 80 | 14 | 0 | 2066 |
| 1995-96 | 0 | 0 | 4 | 106 | 389 | 533 | 603 | 460 | 410 | 161 | 4 | 0 | 2670 |
| 1996-97 | 0 | 0 | 11 | 80 | 298 | 435 | 592 | 381 | 143 | 170 | 16 | 0 | 2126 |
| 1997-98 | 0 | 0 | 0 | 129 | 420 | 617 | 481 | 414 | 336 | 124 | 0 | 1 | 2522 |
| 1998-99 | 0 | 0 | 0 | 46 | 197 | 442 | 412 | 320 | 324 | 53 | 3 | 0 | 1797 |
| 1999-00 | 0 | 0 | 8 | 87 | 268 | 509 | 496 | 308 | 165 | 119 | 0 | 0 | 1960 |
| 2000-01 | 0 | 0 | 7 | 73 | 404 | 813 | 717 | 330 | 392 | 65 | 1 | 0 | 2802 |
| 2001-02 | 0 | 0 | 16 | 149 | 168 | 408 | 510 | 517 | 293 | 57 | 15 | 0 | 2133 |
| 2002-03 | 0 | 0 | 0 | 48 | 383 | 537 | 734 | 475 | 251 | 76 | 2 | 0 | 2506 |
| 2003-04 | 0 | 0 | 8 | 61 | 219 | 600 | 558 | 529 | 168 | 119 | 20 | 0 | 2282 |
| 2004-05 | 0 | 0 | 0 | 18 | 203 | 570 | 441 | 359 | 315 | 98 | 33 | 0 | 2037 |
| 2005-06 | 0 | 0 | 0 | 118 | 274 | 608 | 388 | 477 | 251 | 28 | 10 | 0 | 2154 |
| 2006-07 | 0 | 0 | 3 | 106 | 334 | 445 | 551 | 470 | 154 | 134 | 5 | 0 | 2202 |
| 2007-08 | 0 | 0 | 0 | 85 | 248 | 382 | 611 | 381 | 242 | 97 | 9 | 0 | 2055 |
| 2008-09 | 0 | 0 | 0 | 116 | 348 | 467 | 536 | 359 | 220 | 97 | 4 | 0 | 2147 |
| 2009- | 0 | 0 | 1 | 131 | 290 | 596 | | | | | | | |

WBAN : 03940

## COOLING DEGREE DAYS (base 65°F) 2009 JACKSON (KJAN)

| YEAR | JAN | FEB | MAR | APR | MAY | JUN | JUL | AUG | SEP | OCT | NOV | DEC | TOTAL |
|---|---|---|---|---|---|---|---|---|---|---|---|---|---|
| 1980 | 0 | 11 | 11 | 28 | 253 | 476 | 652 | 619 | 527 | 52 | 14 | 1 | 2644 |
| 1981 | 0 | 3 | 12 | 207 | 174 | 514 | 582 | 555 | 267 | 127 | 30 | 2 | 2473 |
| 1982 | 25 | 3 | 116 | 93 | 310 | 443 | 544 | 530 | 297 | 130 | 35 | 51 | 2577 |
| 1983 | 0 | 0 | 13 | 35 | 194 | 354 | 564 | 555 | 300 | 115 | 24 | 6 | 2160 |
| 1984 | 0 | 5 | 35 | 87 | 221 | 422 | 489 | 470 | 310 | 227 | 19 | 49 | 2334 |
| 1985 | 0 | 3 | 54 | 105 | 233 | 424 | 486 | 480 | 290 | 153 | 79 | 0 | 2307 |
| 1986 | 0 | 21 | 18 | 70 | 259 | 469 | 572 | 487 | 446 | 122 | 44 | 2 | 2510 |
| 1987 | 0 | 0 | 9 | 72 | 344 | 429 | 533 | 560 | 311 | 14 | 22 | 15 | 2309 |
| 1988 | 0 | 1 | 14 | 67 | 193 | 452 | 552 | 552 | 395 | 34 | 45 | 3 | 2308 |
| 1989 | 8 | 30 | 54 | 87 | 245 | 406 | 489 | 498 | 293 | 84 | 36 | 0 | 2230 |
| 1990 | 3 | 25 | 60 | 92 | 226 | 494 | 515 | 544 | 412 | 113 | 24 | 24 | 2532 |
| 1991 | 0 | 3 | 59 | 129 | 337 | 456 | 562 | 515 | 341 | 142 | 37 | 7 | 2588 |
| 1992 | 0 | 1 | 10 | 97 | 213 | 383 | 532 | 403 | 324 | 51 | 2 | 0 | 2016 |
| 1993 | 0 | 1 | 6 | 48 | 197 | 485 | 547 | 562 | 340 | 82 | 19 | 0 | 2287 |
| 1994 | 0 | 6 | 9 | 132 | 196 | 437 | 437 | 439 | 283 | 110 | 24 | 4 | 2077 |
| 1995 | 5 | 2 | 38 | 79 | 282 | 339 | 545 | 581 | 318 | 76 | 9 | 4 | 2278 |
| 1996 | 0 | 28 | 12 | 64 | 343 | 418 | 512 | 439 | 275 | 92 | 13 | 5 | 2201 |
| 1997 | 13 | 9 | 40 | 15 | 175 | 368 | 540 | 459 | 357 | 139 | 0 | 0 | 2115 |
| 1998 | 2 | 0 | 55 | 60 | 298 | 522 | 582 | 554 | 474 | 156 | 12 | 29 | 2744 |
| 1999 | 12 | 14 | 0 | 188 | 213 | 441 | 533 | 596 | 319 | 96 | 1 | 1 | 2414 |
| 2000 | 2 | 17 | 28 | 29 | 328 | 417 | 581 | 628 | 357 | 102 | 23 | 0 | 2512 |
| 2001 | 0 | 16 | 0 | 156 | 238 | 375 | 522 | 487 | 290 | 67 | 17 | 10 | 2178 |
| 2002 | 24 | 0 | 65 | 151 | 256 | 415 | 520 | 500 | 420 | 142 | 11 | 0 | 2504 |
| 2003 | 0 | 1 | 3 | 79 | 290 | 388 | 492 | 524 | 289 | 80 | 41 | 0 | 2187 |
| 2004 | 15 | 0 | 31 | 65 | 271 | 395 | 488 | 394 | 338 | 247 | 27 | 6 | 2277 |
| 2005 | 13 | 5 | 13 | 45 | 208 | 421 | 554 | 564 | 408 | 122 | 49 | 1 | 2403 |
| 2006 | 5 | 0 | 41 | 151 | 261 | 452 | 549 | 596 | 307 | 121 | 10 | 2 | 2495 |
| 2007 | 9 | 1 | 58 | 55 | 268 | 494 | 469 | 637 | 401 | 178 | 16 | 39 | 2625 |
| 2008 | 3 | 18 | 50 | 93 | 257 | 476 | 563 | 453 | 311 | 91 | 1 | 10 | 2326 |
| 2009 | 0 | 14 | 30 | 74 | 269 | 504 | 503 | 476 | 371 | 87 | 4 | 0 | 2332 |

## SNOWFALL (inches)  2009  JACKSON (KJAN)

| YEAR | JUL | AUG | SEP | OCT | NOV | DEC | JAN | FEB | MAR | APR | MAY | JUN | TOTAL |
|------|-----|-----|-----|-----|-----|-----|-----|-----|-----|-----|-----|-----|-------|
| 1976-77 | 0.0 | 0.0 | 0.0 | 0.0 | 0.2 | 0.0 | 5.8 | 0.0 | 0.0 | 0.0 | 0.0 | 0.0 | 6.0 |
| 1977-78 | 0.0 | 0.0 | 0.0 | 0.0 | 0.0 | 0.0 | 1.1 | T | 0.1 | 0.0 | 0.0 | 0.0 | 1.2 |
| 1978-79 | 0.0 | 0.0 | 0.0 | 0.0 | 0.0 | T | T | T | 0.0 | 0.0 | 0.0 | 0.0 | T |
| 1979-80 | 0.0 | 0.0 | 0.0 | 0.0 | 0.0 | 0.0 | T | T | T | 0.0 | 0.0 | 0.0 | T |
| 1980-81 | 0.0 | 0.0 | 0.0 | 0.0 | 0.0 | T | T | T | 0.0 | 0.0 | 0.0 | 0.0 | T |
| 1981-82 | 0.0 | 0.0 | 0.0 | 0.0 | 0.0 | T | 6.3 | T | T | 0.0 | 0.0 | 0.0 | 6.3 |
| 1982-83 | 0.0 | 0.0 | 0.0 | 0.0 | 0.0 | 0.0 | T | T | 0.0 | T | 0.0 | 0.0 | T |
| 1983-84 | 0.0 | 0.0 | 0.0 | 0.0 | 0.0 | T | T | T | 0.0 | 0.0 | 0.0 | 0.0 | T |
| 1984-85 | 0.0 | 0.0 | 0.0 | 0.0 | 0.0 | T | 0.3 | 1.4 | 0.0 | 0.0 | 0.0 | 0.0 | 1.7 |
| 1985-86 | 0.0 | 0.0 | 0.0 | 0.0 | 0.0 | T | 0.0 | T | 0.0 | 0.0 | 0.0 | 0.0 | T |
| 1986-87 | 0.0 | 0.0 | 0.0 | 0.0 | 0.0 | 0.0 | T | 0.0 | T | 1.1 | 0.0 | 0.0 | 1.1 |
| 1987-88 | 0.0 | 0.0 | 0.0 | 0.0 | 0.0 | 0.0 | T | T | 0.0 | 0.0 | 0.0 | 0.0 | T |
| 1988-89 | 0.0 | 0.0 | 0.0 | 0.0 | 0.0 | T | 0.0 | T | T | T | 0.0 | 0.0 | T |
| 1989-90 | 0.0 | 0.0 | 0.0 | 0.0 | T | 0.1 | 0.0 | 0.0 | 0.0 | 0.0 | 0.0 | 0.0 | 0.1 |
| 1990-91 | 0.0 | 0.0 | 0.0 | 0.0 | 0.0 | T | 0.0 | 0.0 | 0.0 | T | 0.0 | 0.0 | T |
| 1991-92 | 0.0 | 0.0 | 0.0 | 0.0 | T | 0.0 | 0.2 | 0.0 | 0.0 | 0.0 | 0.0 | 0.0 | 0.2 |
| 1992-93 | 0.0 | 0.0 | 0.0 | 0.0 | T | 0.0 | T | 0.0 | 1.6 | 0.0 | 0.0 | 0.0 | 1.6 |
| 1993-94 | 0.0 | 0.0 | 0.0 | 0.0 | 0.0 | T | 0.0 | 0.0 | 0.0 | 0.0 | 0.0 | 0.0 | T |
| 1994-95 | 0.0 | 0.0 | 0.0 | 0.0 | 0.0 | 0.0 | 0.0 | 0.0 | 0.0 | 0.0 | 0.0 | 0.0 | 0.0 |
| 1995-96 | 0.0 | 0.0 | 0.0 | 0.0 | 0.0 | | T | T | T | 0.0 | 0.0 | 0.0 | |
| 1996-97 | 0.0 | 0.0 | 0.0 | 0.0 | 0.0 | 0.7 | T | T | 0.0 | 0.0 | 0.0 | 0.0 | 0.7 |
| 1997-98 | 0.0 | 0.0 | 0.0 | 0.0 | 0.0 | 4.8 | 0.0 | 0.0 | 0.0 | 0.0 | 0.0 | 0.0 | 4.8 |
| 1998-99 | 0.0 | 0.0 | 0.0 | 0.0 | 0.0 | T | T | 0.0 | T | 0.0 | 0.0 | 0.0 | T |
| 1999-00 | 0.0 | 0.0 | 0.0 | 0.0 | 0.0 | 0.0 | 0.4 | 0.0 | 0.0 | T | 0.0 | 0.0 | 0.4 |
| 2000-01 | 0.0 | 0.0 | 0.0 | 0.0 | T | 1.0 | T | 0.0 | 0.0 | 0.0 | 0.0 | 0.0 | 1.0 |
| 2001-02 | 0.0 | T | 0.0 | 0.0 | 0.0 | 0.0 | | | | | | | |
| 2002-03 | | | | | | | | | | | | | |
| 2003-04 | | | | | | | | | | | | | |
| 2004-05 | | | | | | | | | | | | | |
| 2005- | | | | | | | | | | | | | |
| POR=<br>40 YRS | 0.0 | T | 0.0 | 0.0 | T | 0.3 | 0.4 | 0.1 | 0.2 | T | 0.0 | 0.0 | 1.0 |

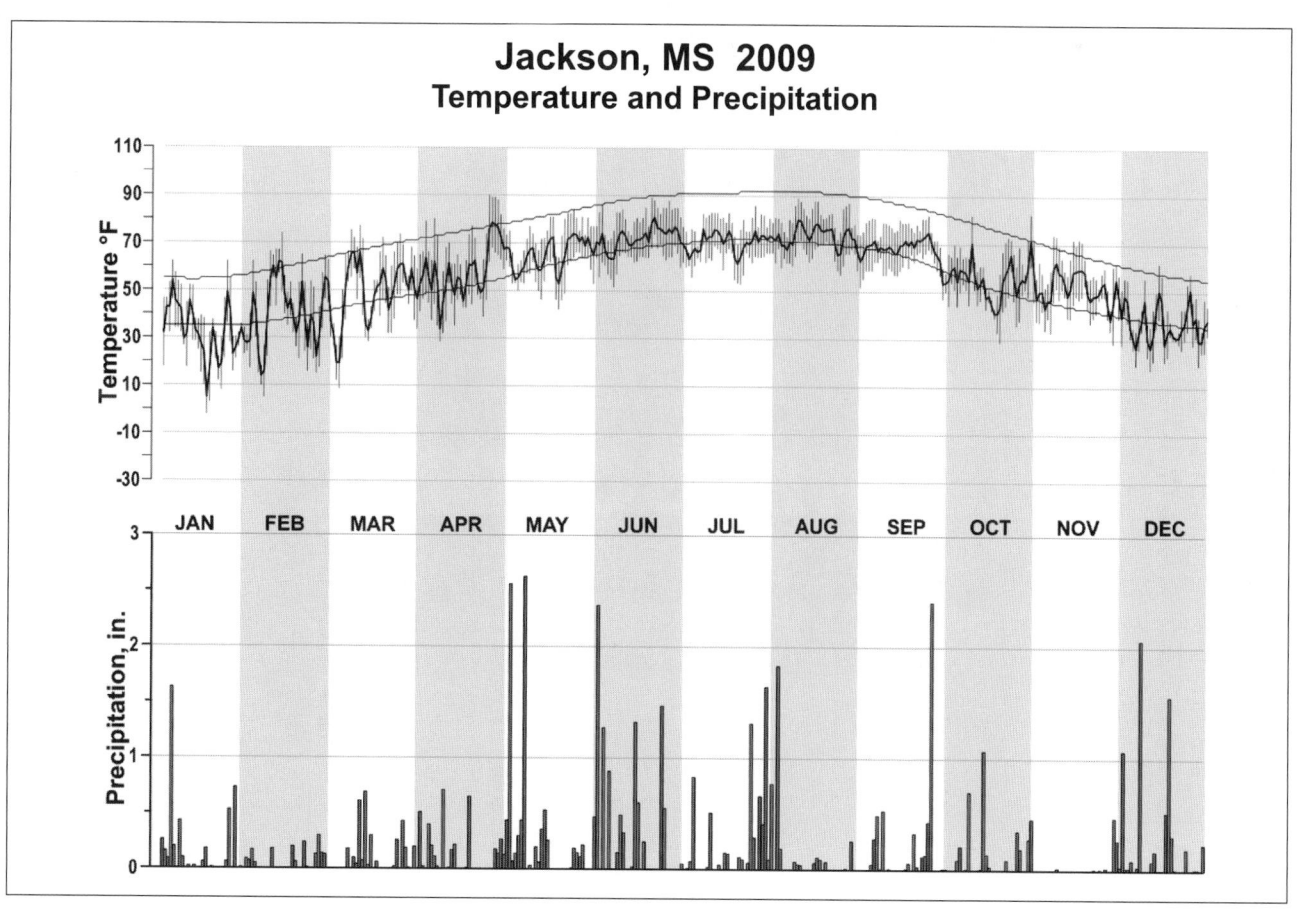

# Jackson, MS  2009
## Temperature and Precipitation

# 2009
# INTERNATIONAL AIRPORT
# KANSAS CITY, MISSOURI (KMCI)

The National Weather Service Office at Kansas City is very near the geographical center of the United States.  The surrounding terrain is gently rolling.  It has a modified continental climate. There are no natural topographic obstructions to prevent the free sweep of air from all directions. The influx of moist air from the Gulf of Mexico, or dry air from the semi-arid regions of the southwest, determine whether wet or dry conditions will prevail.  There is often conflict between the warm moist gulf air and the cold polar continental air from the north in this area.

Early spring brings a period of frequent and rapid fluctuations in weather, with the fluctuations generally less frequent as spring progresses.  The summer season is characterized by warm days and mild nights, with moderate humidities.   July is the warmest month.  The fall season is normally mild and usually includes a period near the middle of the season characterized by mild, sunny days, and cool nights. Winters are not severely cold. January is the coldest month.  Falls of snow to a depth of 10 inches or more are comparatively rare. The distribution of measurable snow normally extends from November to April.

Nearly 60 percent of the annual precipitation occurs during the six months from April through September. More than 75 percent of the annual moisture normally falls during the growing season. The frequency and distribution of precipitation over a normal day is also important. The maximum frequency of precipitation, from April through October, occurs during the six hours following midnight and the minimum frequency occurs during the six hours following noon.

# NORMALS, MEANS, AND EXTREMES
## KANSAS CITY (KMCI)

| LATITUDE:<br>39 ° 17'N | LONGITUDE:<br>-94 ° 43'W | ELEVATION (FT):<br>GRND: 976   BARO: 1008 | TIME ZONE:<br>CENTRAL   (UTC -6) | WBAN: 03947 |
|---|---|---|---|---|

| | ELEMENT | POR | JAN | FEB | MAR | APR | MAY | JUN | JUL | AUG | SEP | OCT | NOV | DEC | YEAR |
|---|---|---|---|---|---|---|---|---|---|---|---|---|---|---|---|
| **TEMPERATURE °F** | NORMAL DAILY MAXIMUM | 30 | 36.0 | 42.6 | 54.4 | 65.2 | 74.6 | 83.9 | 88.8 | 87.1 | 79.0 | 67.6 | 52.0 | 40.0 | 64.3 |
| | MEAN DAILY MAXIMUM | 37 | 36.6 | 42.5 | 54.3 | 65.2 | 74.6 | 83.2 | 88.7 | 87.2 | 78.7 | 66.7 | 52.5 | 40.2 | 64.2 |
| | HIGHEST DAILY MAXIMUM | 37 | 71 | 78 | 86 | 93 | 95 | 105 | 107 | 109 | 106 | 95 | 82 | 74 | 109 |
| | YEAR OF OCCURRENCE | | 2003 | 2006 | 1986 | 1987 | 2006 | 1980 | 1974 | 1984 | 2000 | 2006 | 2005 | 2001 | AUG 1984 |
| | MEAN OF EXTREME MAXS. | 43 | 59.0 | 66.2 | 77.7 | 84.2 | 87.9 | 93.6 | 98.5 | 98.8 | 92.4 | 85.2 | 72.6 | 62.2 | 81.5 |
| | NORMAL DAILY MINIMUM | 30 | 17.8 | 23.3 | 33.2 | 43.5 | 53.9 | 63.2 | 68.2 | 66.1 | 57.2 | 45.9 | 33.4 | 22.5 | 44.0 |
| | MEAN DAILY MINIMUM | 37 | 18.6 | 23.3 | 33.3 | 44.1 | 54.1 | 63.1 | 68.4 | 66.6 | 57.3 | 45.8 | 34.0 | 22.8 | 44.3 |
| | LOWEST DAILY MINIMUM | 37 | -17 | -19 | -10 | 12 | 30 | 42 | 51 | 43 | 31 | 17 | 1 | -23 | -23 |
| | YEAR OF OCCURRENCE | | 1982 | 1982 | 1978 | 1975 | 1976 | 1990 | 1997 | 1986 | 1995 | 1993 | 1991 | 1989 | DEC 1989 |
| | MEAN OF EXTREME MINS. | 43 | -1.8 | 2.2 | 12.8 | 26.8 | 39.8 | 50.1 | 57.7 | 55.5 | 40.8 | 28.7 | 16.7 | 1.6 | 27.6 |
| | NORMAL DRY BULB | 30 | 26.9 | 33.0 | 43.8 | 54.4 | 64.3 | 73.6 | 78.5 | 76.6 | 68.1 | 56.8 | 42.7 | 31.3 | 54.2 |
| | MEAN DRY BULB | 43 | 27.8 | 33.0 | 43.9 | 55.1 | 64.6 | 73.6 | 78.6 | 77.1 | 68.4 | 56.6 | 43.4 | 31.9 | 54.5 |
| | MEAN WET BULB | 26 | 25.3 | 28.8 | 37.7 | 47.1 | 57.5 | 66.0 | 70.0 | 68.7 | 60.7 | 49.6 | 38.1 | 28.3 | 48.2 |
| | MEAN DEW POINT | 26 | 21.7 | 25.1 | 32.8 | 42.5 | 54.2 | 63.4 | 67.6 | 66.2 | 57.6 | 45.9 | 34.3 | 24.8 | 44.7 |
| | NORMAL NO. DAYS WITH: | | | | | | | | | | | | | | |
| | MAXIMUM >= 90 | 30 | 0.0 | 0.0 | 0.0 | 0.3 | 0.4 | 5.6 | 14.5 | 11.7 | 3.7 | 0.1 | 0.0 | 0.0 | 36.3 |
| | MAXIMUM <= 32 | 30 | 12.1 | 7.3 | 1.8 | 0.1 | 0.0 | 0.0 | 0.0 | 0.0 | 0.0 | 0.0 | 1.9 | 8.0 | 31.2 |
| | MINIMUM <= 32 | 30 | 27.9 | 21.2 | 14.9 | 3.7 | * | 0.0 | 0.0 | 0.0 | * | 2.3 | 13.8 | 26.0 | 109.8 |
| | MINIMUM <= 0 | 30 | 3.6 | 1.9 | 0.2 | 0.0 | 0.0 | 0.0 | 0.0 | 0.0 | 0.0 | 0.0 | 0.0 | 1.9 | 7.6 |
| **H/C** | NORMAL HEATING DEG. DAYS | 30 | 1182 | 897 | 658 | 331 | 124 | 8 | 0 | 7 | 58 | 269 | 668 | 1047 | 5249 |
| | NORMAL COOLING DEG. DAYS | 30 | 0 | 0 | 0 | 12 | 101 | 264 | 418 | 367 | 151 | 12 | 0 | 0 | 1325 |
| **RH** | NORMAL (PERCENT) | 30 | 71 | 70 | 66 | 64 | 69 | 71 | 70 | 72 | 71 | 68 | 71 | 73 | 70 |
| | HOUR 00 LST | 30 | 74 | 74 | 71 | 70 | 77 | 79 | 78 | 79 | 79 | 74 | 75 | 76 | 76 |
| | HOUR 06 LST | 30 | 78 | 78 | 78 | 78 | 84 | 86 | 85 | 87 | 86 | 81 | 80 | 80 | 82 |
| | HOUR 12 LST | 30 | 65 | 63 | 58 | 56 | 59 | 60 | 59 | 60 | 59 | 56 | 62 | 66 | 60 |
| | HOUR 18 LST | 30 | 67 | 63 | 56 | 53 | 58 | 59 | 57 | 60 | 61 | 60 | 66 | 69 | 61 |
| **S** | PERCENT POSSIBLE SUNSHINE | 23 | 58 | 55 | 58 | 62 | 61 | 66 | 72 | 67 | 66 | 60 | 49 | 49 | 60 |
| **W/O** | MEAN NO. DAYS WITH: | | | | | | | | | | | | | | |
| | HEAVY FOG(VISBY <= 1/4 MI) | 37 | 2.5 | 2.4 | 1.8 | 0.9 | 1.1 | 0.8 | 0.5 | 1.1 | 1.1 | 1.6 | 1.8 | 2.9 | 18.5 |
| | THUNDERSTORMS | 37 | 0.3 | 0.9 | 2.8 | 5.1 | 7.9 | 8.9 | 7.8 | 7.2 | 5.4 | 3.2 | 1.3 | 0.4 | 51.2 |
| **CLOUDNESS** | MEAN: | | | | | | | | | | | | | | |
| | SUNRISE-SUNSET (OKTAS) | | | | 5.6 | | | 4.0 | | | | | | | |
| | MIDNIGHT-MIDNIGHT (OKTAS) | | | | 6.4 | | | | | | | | | | |
| | MEAN NO. DAYS WITH: | | | | | | | | | | | | | | |
| | CLEAR | 1 | 4.0 | 5.0 | 10.0 | | 6.0 | 7.0 | | | | | | | |
| | PARTLY CLOUDY | 1 | 1.0 | 1.0 | 5.0 | | | 9.0 | | | | | | | |
| | CLOUDY | 1 | 3.0 | | 9.0 | | 9.0 | 5.0 | | | | | | | |
| **PR** | MEAN STATION PRESSURE(IN) | 26 | 29.01 | 28.99 | 28.93 | 28.85 | 28.86 | 28.86 | 28.90 | 28.93 | 28.94 | 28.95 | 28.96 | 29.00 | 28.93 |
| | MEAN SEA-LEVEL PRES. (IN) | 26 | 30.14 | 30.12 | 30.03 | 29.94 | 29.93 | 29.93 | 29.96 | 29.99 | 30.02 | 30.04 | 30.07 | 30.13 | 30.03 |
| **WINDS** | MEAN SPEED (MPH) | 26 | 11.1 | 10.9 | 12.1 | 12.3 | 10.4 | 9.7 | 9.1 | 8.7 | 9.4 | 10.3 | 11.0 | 10.7 | 10.5 |
| | PREVAIL.DIR(TENS OF DEGS) | 37 | 20 | 20 | 19 | 20 | 19 | 19 | 20 | 20 | 19 | 19 | 19 | 20 | 20 |
| | MAXIMUM 2-MINUTE: | | | | | | | | | | | | | | |
| | SPEED (MPH) | 14 | 40 | 40 | 46 | 48 | 51 | 51 | 58 | 39 | 41 | 40 | 37 | 40 | 58 |
| | DIR. (TENS OF DEGS) | | 21 | 20 | 23 | 20 | 27 | 01 | 02 | 30 | 14 | 21 | 31 | 19 | 02 |
| | YEAR OF OCCURRENCE | | 2008 | 2000 | 2000 | 2001 | 2008 | 2000 | 2000 | 2006 | 1996 | 1996 | 2002 | 2004 | JUL 2000 |
| | MAXIMUM 3-SECOND | | | | | | | | | | | | | | |
| | SPEED (MPH) | 14 | 52 | 52 | 58 | 59 | 66 | 61 | 74 | 55 | 49 | 52 | 52 | 52 | 74 |
| | DIR. (TENS OF DEGS) | | 19 | 19 | 23 | 18 | 25 | 36 | 01 | 14 | 14 | 23 | 33 | 18 | 01 |
| | YEAR OF OCCURRENCE | | 2008 | 2000 | 2000 | 2001 | 2008 | 2000 | 2000 | 2001 | 1996 | 1996 | 1997 | 2008 | JUL 2000 |
| **PRECIPITATION** | NORMAL (IN) | 30 | 1.15 | 1.31 | 2.44 | 3.38 | 5.39 | 4.44 | 4.42 | 3.54 | 4.64 | 3.33 | 2.30 | 1.64 | 37.98 |
| | MAXIMUM MONTHLY (IN) | 37 | 2.66 | 3.25 | 9.08 | 8.43 | 12.75 | 11.86 | 15.47 | 9.58 | 11.34 | 8.15 | 5.12 | 5.42 | 15.47 |
| | YEAR OF OCCURRENCE | | 1982 | 2001 | 1973 | 1999 | 1995 | 2001 | 1992 | 1982 | 1977 | 1998 | 1992 | 1980 | JUL 1992 |
| | MINIMUM MONTHLY (IN) | 37 | 0.02 | 0.04 | 0.33 | 0.66 | 1.05 | 1.27 | 0.12 | 0.50 | 1.13 | 0.21 | 0.01 | 0.03 | 0.01 |
| | YEAR OF OCCURRENCE | | 1986 | 2006 | 1994 | 2000 | 1992 | 2006 | 2003 | 2000 | 1974 | 1988 | 1995 | 1996 | NOV 1995 |
| | MAXIMUM IN 24 HOURS (IN) | 37 | 1.83 | 2.21 | 3.07 | 4.69 | 4.26 | 4.48 | 5.08 | 6.19 | 8.82 | 4.92 | 2.15 | 3.67 | 8.82 |
| | YEAR OF OCCURRENCE | | 1982 | 2008 | 2001 | 1975 | 1974 | 2001 | 1986 | 1982 | 1977 | 1973 | 1998 | 1980 | SEP 1977 |
| | NORMAL NO. DAYS WITH: | | | | | | | | | | | | | | |
| | PRECIPITATION >= 0.01 | 30 | 7.3 | 7.1 | 10.0 | 11.0 | 11.5 | 10.5 | 8.6 | 8.5 | 8.4 | 7.4 | 7.9 | 7.5 | 105.7 |
| | PRECIPITATION >= 1.00 | 30 | 0.1 | 0.1 | 0.3 | 0.7 | 1.5 | 1.3 | 1.4 | 0.9 | 1.4 | 1.0 | 0.6 | 0.4 | 9.7 |
| **SNOWFALL** | NORMAL (IN) | 30 | 5.8 | 5.0 | 2.6 | 0.8 | 0.0 | 0.0 | 0.0 | 0.0 | 0.0 | 0.3 | 1.3 | 4.3 | 20.1 |
| | MAXIMUM MONTHLY (IN) | 37 | 14.2 | 15.7 | 11.4 | 7.2 | T | T | T | T | T | 6.5 | 7.1 | 15.1 | 15.7 |
| | YEAR OF OCCURRENCE | | 1977 | 1993 | 1978 | 1983 | 2009 | 2008 | 2007 | 2009 | 1992 | 1996 | 1975 | 2009 | FEB 1993 |
| | MAXIMUM IN 24 HOURS (IN) | 37 | 9.5 | 10.8 | 9.2 | 4.0 | T | T | T | T | T | 6.5 | 6.1 | 10.8 | 10.8 |
| | YEAR OF OCCURRENCE' | | 1993 | 1993 | 1990 | 1983 | 2009 | 1994 | 1992 | 2009 | 1992 | 1996 | 1975 | 1987 | FEB 1993 |
| | MAXIMUM SNOW DEPTH (IN) | 36 | 12 | 11 | 9 | 2 | 0 | 0 | 0 | 0 | 0 | 0 | 7 | 11 | 12 |
| | YEAR OF OCCURRENCE | | 1979 | 1979 | 1990 | 1994 | | | | | | | 1975 | 1987 | JAN 1979 |
| | NORMAL NO. DAYS WITH: | | | | | | | | | | | | | | |
| | SNOWFALL >= 1.0 | 30 | 2.0 | 1.7 | 0.8 | 0.4 | 0.0 | 0.0 | 0.0 | 0.0 | 0.0 | 0.1 | 0.5 | 1.3 | 6.8 |

## PRECIPITATION (inches) 2009 KANSAS CITY (KMCI)

| YEAR | JAN | FEB | MAR | APR | MAY | JUN | JUL | AUG | SEP | OCT | NOV | DEC | ANNUAL |
|------|-----|-----|-----|-----|-----|-----|-----|-----|-----|-----|-----|-----|--------|
| 1980 | 1.60 | 1.44 | 3.64 | 1.02 | 3.06 | 2.52 | 1.99 | 4.89 | 1.63 | 4.13 | 0.45 | 5.42 | 31.79 |
| 1981 | 0.49 | 0.31 | 1.43 | 1.94 | 9.46 | 7.44 | 8.43 | 2.43 | 2.71 | 4.14 | 2.84 | 0.45 | 42.07 |
| 1982 | 2.66 | 1.13 | 2.94 | 1.55 | 9.81 | 6.04 | 2.73 | 9.58 | 1.58 | 3.04 | 2.21 | 3.94 | 47.21 |
| 1983 | 0.58 | 0.57 | 2.93 | 5.52 | 6.03 | 5.03 | 0.26 | 0.86 | 1.89 | 3.85 | 3.94 | 1.42 | 32.88 |
| 1984 | 0.14 | 1.96 | 4.52 | 6.82 | 2.26 | 4.14 | 3.91 | 0.75 | 3.42 | 6.04 | 1.24 | 3.57 | 38.77 |
| 1985 | 0.94 | 2.69 | 2.05 | 1.75 | 7.00 | 3.56 | 5.82 | 6.98 | 9.23 | 7.51 | 3.95 | 1.24 | 52.72 |
| 1986 | 0.02 | 1.25 | 1.34 | 2.12 | 4.76 | 2.48 | 8.36 | 3.16 | 10.40 | 3.17 | 1.18 | 1.20 | 39.44 |
| 1987 | 0.77 | 2.26 | 2.85 | 2.24 | 4.74 | 4.58 | 3.00 | 4.64 | 3.66 | 1.32 | 1.88 | 2.05 | 33.99 |
| 1988 | 1.40 | 0.72 | 1.43 | 2.15 | 2.14 | 1.80 | 1.21 | 1.87 | 8.48 | 0.21 | 1.96 | 0.85 | 24.22 |
| 1989 | 0.98 | 0.59 | 2.13 | 1.50 | 4.56 | 3.44 | 4.76 | 7.38 | 8.87 | 2.88 | T | 0.55 | 37.64 |
| 1990 | 1.20 | 2.11 | 3.90 | 2.47 | 7.36 | 6.27 | 4.40 | 5.04 | 1.28 | 2.46 | 3.01 | 1.11 | 40.61 |
| 1991 | 1.37 | 0.20 | 2.36 | 4.99 | 3.69 | 3.06 | 1.72 | 1.35 | 2.12 | 3.71 | 2.05 | 2.08 | 28.70 |
| 1992 | 1.21 | 2.01 | 3.79 | 4.92 | 1.05 | 3.84 | 15.47 | 2.37 | 5.69 | 1.38 | 5.12 | 3.78 | 50.63 |
| 1993 | 1.96 | 1.28 | 2.21 | 5.59 | 7.30 | 5.67 | 10.90 | 3.98 | 7.63 | 1.75 | 2.07 | 1.12 | 51.46 |
| 1994 | 0.63 | 1.47 | 0.33 | 6.98 | 1.29 | 2.45 | 2.79 | 3.54 | 2.65 | 1.27 | 3.18 | 1.76 | 28.34 |
| 1995 | 1.42 | 1.35 | 1.12 | 2.12 | 12.75 | 3.36 | 4.64 | 4.00 | 1.85 | 0.50 | 1.18 | 0.40 | 34.69 |
| 1996 | 1.12 | 0.35 | 1.28 | 1.80 | 10.29 | 7.51 | 4.83 | 2.97 | 3.44 | 3.67 | 3.15 | .03 | 40.44 |
| 1997 | 0.69 | 2.94 | 1.16 | 4.13 | 4.63 | 2.90 | 3.53 | 2.49 | 3.34 | 2.98 | 1.95 | 2.33 | 33.07 |
| 1998 | 0.97 | 1.10 | 3.44 | 2.15 | 1.75 | 9.22 | 4.97 | 3.61 | 8.69 | 8.15 | 4.29 | 1.19 | 49.53 |
| 1999 | 2.30 | 1.71 | 1.49 | 8.43 | 5.62 | 8.67 | 0.51 | 1.56 | 5.32 | 0.67 | 1.63 | 2.18 | 40.09 |
| 2000 | 0.46 | 2.21 | 2.93 | 0.66 | 4.55 | 7.55 | 6.02 | 0.50 | 3.13 | 3.55 | 2.59 | 0.81 | 34.96 |
| 2001 | 2.08 | 3.25 | 3.88 | 4.03 | 4.81 | 11.86 | 6.26 | 5.48 | 7.98 | 2.56 | 0.56 | 0.75 | 53.50 |
| 2002 | 1.66 | 0.73 | 1.03 | 4.53 | 6.97 | 1.44 | 1.18 | 2.06 | 1.31 | 3.51 | 0.32 | 0.03 | 24.77 |
| 2003 | 0.47 | 0.74 | 1.27 | 4.80 | 2.64 | 6.02 | 0.12 | 4.72 | 2.61 | 0.84 | 1.61 | 2.11 | 27.95 |
| 2004 | 0.61 | 1.62 | 3.59 | 2.43 | 5.12 | 6.20 | 4.26 | 4.15 | 3.48 | 3.08 | 2.67 | 0.38 | 37.59 |
| 2005 | 2.51 | 2.39 | 0.87 | 2.32 | 5.66 | 10.22 | 1.24 | 8.34 | 3.61 | 4.10 | 1.15 | 1.73 | 44.14 |
| 2006 | 1.11 | 0.04 | 1.78 | 4.15 | 1.67 | 1.27 | 3.20 | 7.66 | 2.22 | 3.29 | 2.72 | 1.76 | 30.87 |
| 2007 | 0.84 | 1.40 | 2.85 | 3.09 | 5.94 | 4.16 | 0.99 | 1.71 | 2.47 | 6.46 | 0.20 | 2.91 | 33.02 |
| 2008 | 0.97 | 3.10 | 2.72 | 4.53 | 3.96 | 4.31 | 6.63 | 1.19 | 9.82 | 4.01 | 1.59 | 1.83 | 44.66 |
| 2009 | 0.05 | 0.81 | 4.62 | 7.12 | 2.84 | 6.87 | 4.51 | 8.51 | 2.02 | 3.66 | 2.25 | 1.69 | 44.95 |
| POR= 43 YRS | 1.13 | 1.28 | 2.38 | 3.68 | 5.04 | 4.99 | 4.11 | 3.85 | 4.58 | 3.54 | 2.00 | 1.61 | 38.19 |

WBAN : 03947

## AVERAGE TEMPERATURE (°F) 2009 KANSAS CITY (KMCI)

| YEAR | JAN | FEB | MAR | APR | MAY | JUN | JUL | AUG | SEP | OCT | NOV | DEC | ANNUAL |
|------|-----|-----|-----|-----|-----|-----|-----|-----|-----|-----|-----|-----|--------|
| 1980 | 28.7 | 25.2 | 38.7 | 54.6 | 63.9 | 75.3 | 85.2 | 80.3 | 69.6 | 54.1 | 44.5 | 32.1 | 54.4 |
| 1981 | 30.3 | 33.4 | 45.2 | 61.1 | 60.5 | 74.1 | 78.3 | 72.9 | 68.4 | 55.3 | 45.6 | 29.0 | 54.5 |
| 1982 | 18.6 | 27.8 | 42.5 | 51.1 | 65.5 | 68.8 | 79.4 | 75.0 | 67.5 | 55.6 | 41.9 | 35.5 | 52.4 |
| 1983 | 30.1 | 35.9 | 43.1 | 46.3 | 59.6 | 70.8 | 81.5 | 83.5 | 71.2 | 57.2 | 44.3 | 13.2 | 53.1 |
| 1984 | 25.0 | 38.9 | 36.0 | 50.3 | 60.4 | 74.3 | 76.1 | 79.0 | 66.1 | 56.8 | 43.9 | 35.5 | 53.5 |
| 1985 | 18.7 | 25.3 | 47.4 | 57.9 | 66.0 | 68.8 | 77.0 | 72.1 | 66.7 | 56.5 | 36.8 | 22.9 | 51.3 |
| 1986 | 34.5 | 30.5 | 48.5 | 57.1 | 65.2 | 76.5 | 79.7 | 72.0 | 71.8 | 56.9 | 37.9 | 34.5 | 55.4 |
| 1987 | 29.7 | 39.4 | 47.1 | 56.8 | 70.6 | 76.0 | 79.9 | 76.4 | 67.8 | 52.0 | 46.7 | 35.1 | 56.5 |
| 1988 | 26.7 | 27.9 | 43.2 | 54.5 | 69.1 | 78.1 | 79.6 | 81.3 | 70.5 | 52.2 | 44.8 | 35.2 | 55.3 |
| 1989 | 37.7 | 22.8 | 43.8 | 56.9 | 63.2 | 71.1 | 77.8 | 75.5 | 63.2 | 57.9 | 42.3 | 21.1 | 52.8 |
| 1990 | 37.9 | 36.2 | 45.7 | 52.7 | 60.4 | 75.5 | 77.3 | 77.1 | 72.1 | 57.1 | 50.1 | 29.3 | 56.0 |
| 1991 | 22.9 | 39.4 | 47.2 | 56.7 | 67.6 | 76.1 | 80.6 | 77.7 | 68.8 | 57.6 | 36.7 | 36.1 | 55.6 |
| 1992 | 35.9 | 39.8 | 46.9 | 53.1 | 62.6 | 69.7 | 74.4 | 70.7 | 66.1 | 56.6 | 39.0 | 32.9 | 54.0 |
| 1993 | 25.9 | 28.6 | 39.8 | 50.4 | 63.5 | 72.9 | 77.6 | 77.8 | 62.8 | 53.6 | 39.4 | 34.5 | 52.2 |
| 1994 | 25.2 | 30.0 | 46.0 | 53.8 | 64.6 | 75.8 | 76.0 | 75.3 | 67.7 | 57.5 | 46.3 | 36.5 | 54.6 |
| 1995 | 28.2 | 35.9 | 45.0 | 52.5 | 59.4 | 72.7 | 78.2 | 78.9 | 65.2 | 57.1 | 40.1 | 31.1 | 53.7 |
| 1996 | 23.1 | 34.2 | 37.7 | 52.8 | 64.5 | 73.8 | 75.2 | 74.9 | 64.8 | 56.3 | 37.0 | 29.7 | 52.0 |
| 1997 | 24.4 | 33.8 | 45.3 | 48.8 | 59.9 | 73.1 | 77.7 | 74.5 | 69.2 | 57.9 | 40.7 | 33.5 | 53.2 |
| 1998 | 33.8 | 41.3 | 39.1 | 54.3 | 71.1 | 73.5 | 77.8 | 77.0 | 72.4 | 58.6 | 48.0 | 34.8 | 56.8 |
| 1999 | 27.8 | 41.2 | 42.6 | 54.6 | 63.4 | 71.5 | 81.0 | 76.3 | 65.7 | 57.0 | 51.9 | 35.8 | 55.7 |
| 2000 | 31.8 | 40.8 | 47.1 | 55.0 | 67.3 | 71.2 | 76.8 | 81.8 | 71.0 | 59.9 | 36.6 | 19.1 | 54.9 |
| 2001 | 29.2 | 30.0 | 40.0 | 60.3 | 66.1 | 72.1 | 80.7 | 77.4 | 65.7 | 56.0 | 51.2 | 37.3 | 55.5 |
| 2002 | 34.1 | 37.0 | 40.1 | 56.9 | 61.9 | 76.1 | 81.3 | 78.6 | 72.9 | 50.7 | 41.5 | 36.6 | 55.6 |
| 2003 | 27.3 | 30.3 | 43.5 | 57.4 | 63.3 | 70.9 | 81.1 | 81.5 | 64.6 | 57.6 | 42.9 | 35.4 | 54.7 |
| 2004 | 27.4 | 31.5 | 47.6 | 56.9 | 67.3 | 70.6 | 75.0 | 72.5 | 70.0 | 57.8 | 46.8 | 34.5 | 54.8 |
| 2005 | 28.3 | 38.3 | 43.7 | 56.8 | 64.6 | 75.5 | 78.9 | 78.1 | 72.6 | 58.5 | 47.5 | 30.0 | 56.1 |
| 2006 | 42.7 | 35.3 | 46.6 | 62.2 | 67.1 | 76.1 | 81.6 | 80.2 | 66.4 | 56.3 | 46.4 | 38.3 | 58.3 |
| 2007 | 28.8 | 28.1 | 52.6 | 51.9 | 68.2 | 73.2 | 77.4 | 82.9 | 71.2 | 59.6 | 43.9 | 29.9 | 55.6 |
| 2008 | 27.5 | 27.7 | 41.6 | 51.6 | 63.7 | 73.9 | 77.2 | 75.4 | 66.6 | 55.6 | 43.2 | 28.8 | 52.7 |
| 2009 | 27.5 | 36.8 | 45.2 | 53.0 | 64.8 | 74.8 | 73.6 | 74.0 | 67.0 | 50.6 | 49.9 | 28.5 | 53.8 |
| POR= 43 YRS | 27.8 | 33.0 | 43.9 | 55.1 | 64.6 | 73.6 | 78.6 | 77.1 | 68.4 | 56.6 | 43.4 | 31.9 | 54.5 |

## HEATING DEGREE DAYS (base 65°F) 2009  KANSAS CITY (KMCI)

| YEAR | JUL | AUG | SEP | OCT | NOV | DEC | JAN | FEB | MAR | APR | MAY | JUN | TOTAL |
|---|---|---|---|---|---|---|---|---|---|---|---|---|---|
| 1980-81 | 0 | 0 | 63 | 347 | 609 | 1011 | 1069 | 880 | 607 | 169 | 179 | 2 | 4936 |
| 1981-82 | 0 | 2 | 40 | 309 | 573 | 1112 | 1432 | 1037 | 690 | 416 | 51 | 32 | 5694 |
| 1982-83 | 0 | 2 | 78 | 307 | 688 | 911 | 1074 | 810 | 675 | 557 | 180 | 29 | 5311 |
| 1983-84 | 0 | 0 | 57 | 271 | 617 | 1602 | 1234 | 750 | 891 | 443 | 175 | 1 | 6041 |
| 1984-85 | 0 | 0 | 143 | 269 | 624 | 907 | 1431 | 1102 | 538 | 256 | 41 | 19 | 5330 |
| 1985-86 | 0 | 3 | 131 | 260 | 841 | 1297 | 940 | 960 | 528 | 267 | 60 | 0 | 5287 |
| 1986-87 | 0 | 12 | 23 | 251 | 805 | 938 | 1088 | 712 | 549 | 298 | 15 | 0 | 4691 |
| 1987-88 | 0 | 3 | 30 | 398 | 552 | 922 | 1180 | 1069 | 668 | 311 | 19 | 0 | 5152 |
| 1988-89 | 2 | 1 | 18 | 394 | 599 | 915 | 836 | 1176 | 658 | 319 | 135 | 7 | 5060 |
| 1989-90 | 0 | 1 | 138 | 267 | 675 | 1360 | 836 | 800 | 601 | 398 | 167 | 10 | 5253 |
| 1990-91 | 1 | 0 | 44 | 278 | 452 | 1104 | 1296 | 709 | 553 | 258 | 62 | 0 | 4757 |
| 1991-92 | 0 | 0 | 96 | 277 | 841 | 888 | 898 | 724 | 554 | 367 | 137 | 6 | 4788 |
| 1992-93 | 0 | 7 | 80 | 271 | 773 | 987 | 1203 | 1011 | 775 | 431 | 88 | 26 | 5652 |
| 1993-94 | 0 | 2 | 118 | 365 | 760 | 938 | 1227 | 974 | 581 | 353 | 91 | 1 | 5410 |
| 1994-95 | 0 | 4 | 69 | 252 | 552 | 878 | 1135 | 807 | 613 | 370 | 196 | 1 | 4877 |
| 1995-96 | 0 | 0 | 111 | 255 | 743 | 1047 | 1295 | 891 | 840 | 381 | 110 | 10 | 5683 |
| 1996-97 | 0 | 0 | 95 | 291 | 835 | 1089 | 1252 | 865 | 603 | 476 | 184 | 1 | 5691 |
| 1997-98 | 3 | 0 | 25 | 290 | 720 | 971 | 960 | 658 | 806 | 323 | 18 | 27 | 4801 |
| 1998-99 | 0 | 0 | 9 | 213 | 505 | 930 | 1145 | 656 | 690 | 313 | 84 | 12 | 4557 |
| 1999-00 | 0 | 0 | 94 | 261 | 392 | 898 | 1024 | 694 | 554 | 303 | 58 | 9 | 4287 |
| 2000-01 | 0 | 0 | 52 | 195 | 845 | 1416 | 1102 | 972 | 768 | 190 | 67 | 16 | 5623 |
| 2001-02 | 0 | 0 | 71 | 280 | 407 | 854 | 949 | 777 | 765 | 283 | 150 | 0 | 4536 |
| 2002-03 | 0 | 0 | 14 | 457 | 698 | 876 | 1162 | 963 | 659 | 270 | 91 | 27 | 5217 |
| 2003-04 | 0 | 0 | 93 | 235 | 657 | 912 | 1158 | 967 | 540 | 267 | 85 | 5 | 4919 |
| 2004-05 | 0 | 4 | 10 | 237 | 539 | 938 | 1128 | 742 | 654 | 266 | 104 | 0 | 4622 |
| 2005-06 | 0 | 0 | 27 | 258 | 524 | 1077 | 682 | 826 | 563 | 145 | 94 | 0 | 4196 |
| 2006-07 | 0 | 0 | 46 | 323 | 551 | 822 | 1114 | 1027 | 397 | 406 | 26 | 0 | 4712 |
| 2007-08 | 0 | 0 | 28 | 217 | 628 | 1082 | 1155 | 1075 | 719 | 400 | 109 | 0 | 5413 |
| 2008-09 | 0 | 0 | 55 | 306 | 651 | 1120 | 1154 | 785 | 609 | 377 | 86 | 4 | 5147 |
| 2009- | 0 | 6 | 36 | 441 | 447 | 1124 | | | | | | | |

WBAN : 03947

## COOLING DEGREE DAYS (base 65°F) 2009  KANSAS CITY (KMCI)

| YEAR | JAN | FEB | MAR | APR | MAY | JUN | JUL | AUG | SEP | OCT | NOV | DEC | TOTAL |
|---|---|---|---|---|---|---|---|---|---|---|---|---|---|
| 1980 | 0 | 0 | 0 | 21 | 69 | 316 | 632 | 483 | 210 | 15 | 0 | 0 | 1746 |
| 1981 | 0 | 0 | 0 | 58 | 44 | 279 | 418 | 253 | 149 | 14 | 0 | 0 | 1215 |
| 1982 | 0 | 0 | 0 | 8 | 75 | 154 | 452 | 320 | 158 | 26 | 0 | 0 | 1193 |
| 1983 | 0 | 0 | 3 | 5 | 19 | 210 | 517 | 582 | 251 | 31 | 1 | 0 | 1619 |
| 1984 | 0 | 0 | 0 | 9 | 41 | 287 | 353 | 445 | 184 | 23 | 0 | 0 | 1342 |
| 1985 | 0 | 0 | 0 | 49 | 77 | 137 | 379 | 230 | 191 | 5 | 0 | 0 | 1068 |
| 1986 | 0 | 0 | 24 | 35 | 73 | 352 | 466 | 237 | 232 | 7 | 0 | 0 | 1426 |
| 1987 | 0 | 0 | 0 | 58 | 196 | 336 | 474 | 364 | 118 | 1 | 9 | 0 | 1556 |
| 1988 | 0 | 0 | 2 | 5 | 151 | 400 | 459 | 514 | 188 | 5 | 0 | 0 | 1724 |
| 1989 | 0 | 0 | 9 | 85 | 88 | 196 | 402 | 334 | 87 | 56 | 0 | 0 | 1257 |
| 1990 | 0 | 0 | 8 | 33 | 33 | 331 | 394 | 384 | 263 | 40 | 11 | 0 | 1497 |
| 1991 | 0 | 0 | 9 | 16 | 154 | 339 | 490 | 399 | 219 | 54 | 0 | 0 | 1680 |
| 1992 | 0 | 0 | 1 | 20 | 71 | 154 | 298 | 191 | 121 | 19 | 0 | 0 | 875 |
| 1993 | 0 | 0 | 0 | 0 | 49 | 271 | 398 | 406 | 55 | 19 | 0 | 0 | 1198 |
| 1994 | 0 | 0 | 0 | 23 | 86 | 331 | 345 | 333 | 154 | 28 | 0 | 0 | 1300 |
| 1995 | 0 | 0 | 0 | 1 | 27 | 239 | 417 | 438 | 123 | 18 | 0 | 0 | 1263 |
| 1996 | 0 | 0 | 0 | 19 | 105 | 282 | 325 | 313 | 94 | 29 | 0 | 0 | 1167 |
| 1997 | 0 | 0 | 0 | 0 | 33 | 248 | 403 | 300 | 159 | 75 | 0 | 0 | 1218 |
| 1998 | 0 | 0 | 10 | 10 | 213 | 291 | 407 | 378 | 237 | 20 | 0 | 0 | 1566 |
| 1999 | 0 | 0 | 0 | 7 | 42 | 215 | 501 | 358 | 123 | 24 | 4 | 0 | 1274 |
| 2000 | 0 | 0 | 4 | 10 | 134 | 202 | 369 | 528 | 239 | 41 | 0 | 0 | 1527 |
| 2001 | 0 | 0 | 0 | 58 | 109 | 239 | 495 | 392 | 100 | 8 | 1 | 0 | 1402 |
| 2002 | 0 | 0 | 0 | 47 | 60 | 341 | 512 | 426 | 255 | 18 | 0 | 0 | 1659 |
| 2003 | 0 | 0 | 0 | 47 | 41 | 209 | 505 | 517 | 85 | 15 | 0 | 0 | 1419 |
| 2004 | 0 | 0 | 6 | 32 | 164 | 178 | 316 | 244 | 166 | 18 | 0 | 0 | 1124 |
| 2005 | 0 | 0 | 2 | 28 | 101 | 319 | 438 | 415 | 263 | 62 | 7 | 0 | 1635 |
| 2006 | 0 | 0 | 0 | 65 | 166 | 340 | 519 | 478 | 96 | 60 | 0 | 0 | 1724 |
| 2007 | 0 | 0 | 16 | 20 | 133 | 253 | 393 | 560 | 222 | 56 | 1 | 0 | 1654 |
| 2008 | 0 | 0 | 0 | 7 | 76 | 272 | 383 | 329 | 108 | 21 | 3 | 0 | 1199 |
| 2009 | 0 | 0 | 5 | 23 | 87 | 305 | 273 | 295 | 102 | 0 | 0 | 0 | 1090 |

**SNOWFALL (inches) 2009 KANSAS CITY (KMCI)**

| YEAR | JUL | AUG | SEP | OCT | NOV | DEC | JAN | FEB | MAR | APR | MAY | JUN | TOTAL |
|------|-----|-----|-----|-----|-----|-----|-----|-----|-----|-----|-----|-----|-------|
| 1980-81 | 0.0 | 0.0 | 0.0 | T | T | 3.2 | 4.0 | 2.9 | 0.1 | 0.0 | 0.0 | 0.0 | 10.2 |
| 1981-82 | 0.0 | 0.0 | 0.0 | 0.0 | 0.1 | 5.3 | 6.0 | 12.7 | 4.0 | 1.3 | 0.0 | 0.0 | 29.4 |
| 1982-83 | 0.0 | 0.0 | 0.0 | 0.0 | 0.5 | 0.7 | 6.3 | 7.4 | 1.3 | 7.2 | 0.0 | 0.0 | 23.4 |
| 1983-84 | 0.0 | 0.0 | 0.0 | 0.0 | 0.7 | 13.2 | 1.3 | 0.5 | 8.7 | 0.0 | 0.0 | 0.0 | 24.4 |
| 1984-85 | 0.0 | 0.0 | 0.0 | 0.0 | 0.4 | 7.0 | 11.8 | 6.9 | 0.3 | 0.0 | 0.0 | 0.0 | 26.4 |
| 1985-86 | 0.0 | 0.0 | 0.0 | 0.0 | 3.5 | 5.4 | T | 4.5 | T | T | 0.0 | 0.0 | 13.4 |
| 1986-87 | 0.0 | 0.0 | 0.0 | T | 0.6 | 1.2 | 10.5 | 5.0 | T | 0.0 | 0.0 | 0.0 | 17.3 |
| 1987-88 | 0.0 | 0.0 | 0.0 | T | 2.0 | 11.9 | 0.9 | 9.3 | 2.2 | 0.0 | 0.0 | 0.0 | 26.3 |
| 1988-89 | 0.0 | 0.0 | 0.0 | 0.0 | 0.1 | 0.1 | 0.2 | 6.5 | T | 0.0 | T | T | 6.9 |
| 1989-90 | 0.0 | 0.0 | 0.0 | 0.0 | T | 6.8 | 1.0 | 2.1 | 9.6 | 0.0 | T | 0.0 | 19.5 |
| 1990-91 | 0.0 | 0.0 | 0.0 | 0.0 | 1.7 | 1.6 | 12.1 | T | 1.2 | T | 0.0 | 0.0 | 16.6 |
| 1991-92 | 0.0 | 0.0 | 0.0 | 0.0 | 4.6 | 0.2 | T | 2.0 | 0.5 | 2.8 | 0.0 | 0.0 | 10.1 |
| 1992-93 | T | 0.0 | T | T | 4.1 | 0.8 | 12.0 | 15.7 | 1.1 | 0.6 | 0.0 | T | 34.3 |
| 1993-94 | 0.0 | 0.0 | 0.0 | T | 0.5 | 2.7 | 1.4 | 10.3 | 0.5 | 2.6 | 0.0 | T | 18.0 |
| 1994-95 | 0.0 | 0.0 | 0.0 | 0.0 | 0.0 | 2.5 | 1.3 | 0.7 | 2.4 | T | 0.0 | 0.0 | 6.9 |
| 1995-96 | 0.0 | 0.0 | 0.0 | 0.0 | 0.7 | 5.3 | 11.4 | T | 0.7 | 1.0 | 0.0 | | |
| 1996-97 | | | | 6.5 | 4.8 | .5 | 9.8 | 6.3 | 0.0 | 1.3 | T | 0.0 | |
| 1997-98 | 0.0 | 0.0 | 0.0 | 1.0 | 0.5 | 10.9 | 0.6 | 1.0 | 5.6 | 0.0 | T | T | 19.6 |
| 1998-99 | 0.0 | 0.0 | 0.0 | 0.0 | 0.0 | 1.4 | 4.3 | 4.7 | 2.5 | T | 0.0 | 0.0 | 12.9 |
| 1999-00 | 0.0 | 0.0 | 0.0 | 0.0 | 0.0 | 3.9 | 4.9 | 2.1 | 2.0 | 0.0 | T | 0.0 | 12.9 |
| 2000-01 | 0.0 | 0.0 | 0.0 | 0.0 | T | 11.8 | 2.0 | 7.7 | 1.3 | 0.0 | T | T | 22.8 |
| 2001-02 | 0.0 | T | T | T | 0.0 | T | 5.1 | T | 3.5 | 0.0 | 0.0 | 0.0 | 8.6 |
| 2002-03 | 0.0 | 0.0 | 0.0 | T | T | 0.3 | 4.8 | 2.7 | 1.6 | T | 0.0 | 0.0 | 9.4 |
| 2003-04 | 0.0 | 0.0 | T | 0.0 | T | 7.1 | 1.8 | 11.3 | 0.0 | 0.0 | 0.0 | T | 20.2 |
| 2004-05 | 0.0 | 0.0 | 0.0 | 0.0 | 6.8 | 0.0 | 4.3 | 2.2 | T | 0.0 | T | T | 13.3 |
| 2005-06 | 0.0 | 0.0 | 0.0 | 0.0 | T | 11.1 | 0.8 | 0.5 | 1.0 | T | 0.0 | 0.0 | 13.4 |
| 2006-07 | 0.0 | 0.0 | 0.0 | 0.0 | 0.4 | T | 6.0 | 3.6 | 0.2 | T | 0.0 | 0.0 | 10.2 |
| 2007-08 | T | 0.0 | 0.0 | 0.0 | 0.4 | 9.4 | 4.8 | 8.9 | 0.6 | 0.0 | 0.0 | T | 24.1 |
| 2008-09 | 0.0 | 0.0 | 0.0 | 0.0 | 0.9 | 6.2 | 0.9 | 5.3 | 1.3 | T | T | 0.0 | 14.6 |
| 2009- | 0.0 | T | 0.0 | 0.0 | 1.2 | 15.1 | | | | | | | |
| POR= 36 YRS | T | T | T | 0.2 | 1.2 | 4.4 | 5.0 | 4.7 | 2.3 | 0.7 | T | T | 18.5 |

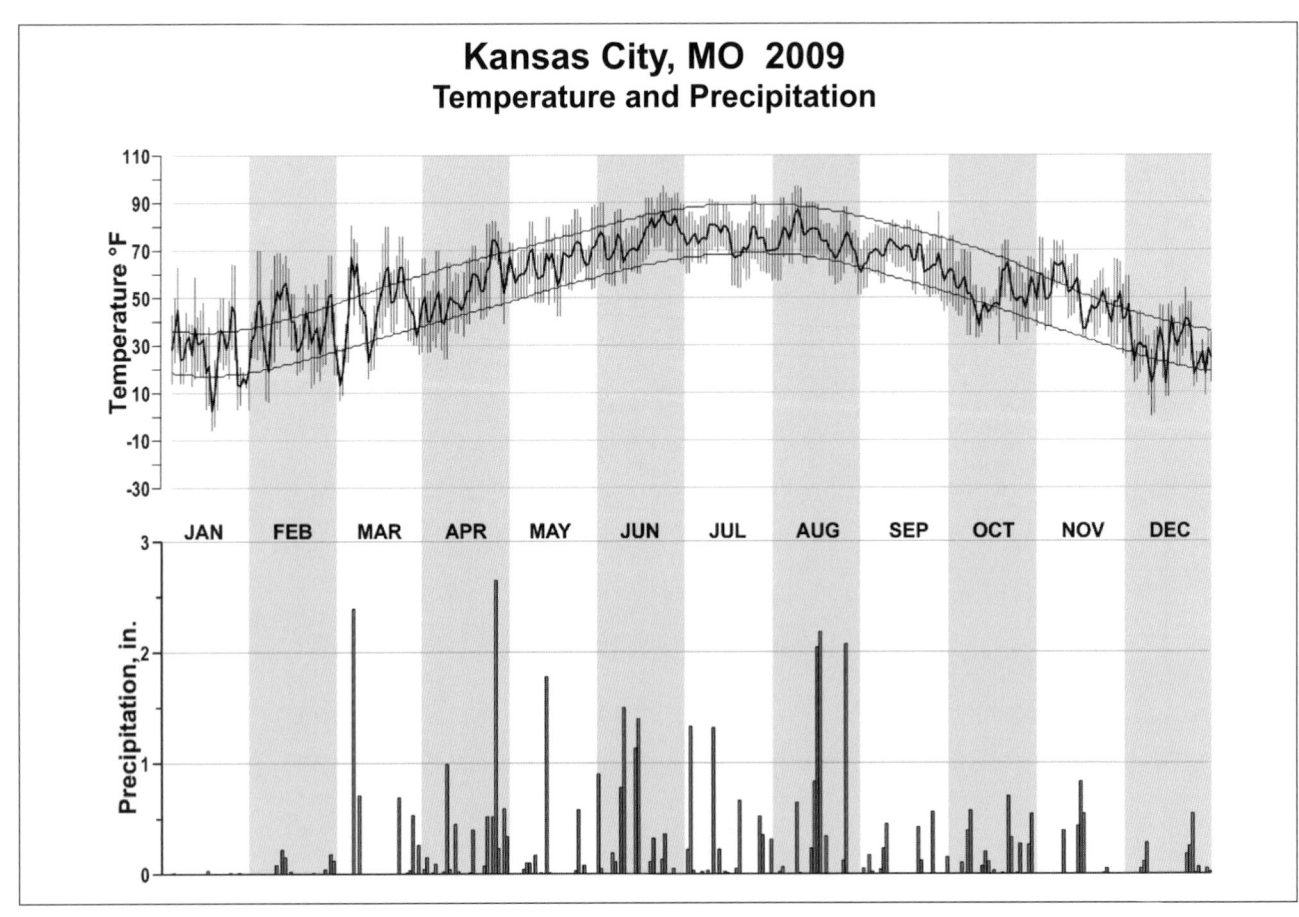

# Kansas City, MO 2009
## Temperature and Precipitation

# 2009
# ST. LOUIS
# MISSOURI (KSTL)

St Louis is located at the confluence of the Missouri and Mississippi Rivers and near the geographical center of the United States. Its position in the middle latitudes allows the area to be affected by warm moist air that originates in the Gulf of Mexico, as well as cold air masses that originate in Canada. The alternate invasion of these airmasses produces a wide variety of weather conditions, and allows the region to enjoy a true four-season climate.

During the summer months, air originating from the Gulf of Mexico tends to dominate the area, producing warm and humid conditions. Since 1870, records indicate that temperatures of 90 degrees or higher occur on about 35-40 days per year. Extrmely hot days (100 degrees or more) are expected on no more than five days per year.

Winters are brisk and stimulating, but prolonged periods of extremely cold weather are rare. Records show that temperatures drop to zero or below an average of 2 or 3 days per year, and temperatures as cold as 32 degrees or lower occur less than 25 days in most years. Snowfall has averaged a little over 18 inches per winter season, and snowfall of an inch or less is received on 5 to 10 days in most years.

Normal annual precipitation for St. Louis is a little less than 34 inches. The three winter months are the driest, with an average total of about 6 inches of precipitation. The spring months of March through May are normally the wettest with normal total rainfall of just under 10.5 inches. It is not unusual to have extended dry periods of one to two weeks during growing season.

Thunderstorms normally occur on between 40 and 50 days per year. During any year, there are usually a few of these thunderstorms that are severe, and produce large hail and damaging winds. Tornadoes have produced extensive damage and loss of life in the St. Louis area.

# NORMALS, MEANS, AND EXTREMES
## ST LOUIS (KSTL)

| LATITUDE: 38° 45'N | LONGITUDE: -90° 22'W | | ELEVATION (FT): GRND: 654 BARO: 710 | | TIME ZONE: CENTRAL (UTC -6) | | WBAN: 13994 |
|---|---|---|---|---|---|---|---|

| | ELEMENT | POR | JAN | FEB | MAR | APR | MAY | JUN | JUL | AUG | SEP | OCT | NOV | DEC | YEAR |
|---|---|---|---|---|---|---|---|---|---|---|---|---|---|---|---|
| **TEMPERATURE °F** | NORMAL DAILY MAXIMUM | 30 | 37.9 | 44.3 | 55.4 | 66.7 | 76.5 | 85.3 | 89.8 | 87.9 | 80.1 | 68.3 | 53.8 | 42.0 | 65.7 |
| | MEAN DAILY MAXIMUM | 64 | 39.0 | 44.1 | 54.4 | 66.9 | 76.2 | 85.0 | 88.9 | 87.6 | 80.2 | 68.9 | 54.5 | 42.7 | 65.7 |
| | HIGHEST DAILY MAXIMUM | 52 | 76 | 85 | 89 | 93 | 94 | 102 | 107 | 107 | 104 | 94 | 85 | 76 | 107 |
| | YEAR OF OCCURRENCE | | 1970 | 1972 | 1985 | 2002 | 1996 | 1988 | 1980 | 1984 | 1984 | 2006 | 1989 | 1970 | AUG 1984 |
| | MEAN OF EXTREME MAXS. | 64 | 64.1 | 68.7 | 79.1 | 86.8 | 89.5 | 95.3 | 98.6 | 98.4 | 93.2 | 86.2 | 75.3 | 66.4 | 83.5 |
| | NORMAL DAILY MINIMUM | 30 | 21.2 | 26.5 | 36.2 | 46.5 | 56.6 | 65.9 | 70.6 | 68.6 | 60.3 | 48.2 | 36.7 | 25.8 | 46.9 |
| | MEAN DAILY MINIMUM | 64 | 22.1 | 26.1 | 34.8 | 46.0 | 55.9 | 65.3 | 69.6 | 67.9 | 59.5 | 48.0 | 36.6 | 26.6 | 46.5 |
| | LOWEST DAILY MINIMUM | 52 | -18 | -12 | -5 | 22 | 31 | 43 | 51 | 47 | 36 | 23 | 1 | -16 | -18 |
| | YEAR OF OCCURRENCE | | 1985 | 1996 | 1960 | 1975 | 1976 | 1969 | 1972 | 1986 | 1974 | 1976 | 1964 | 1989 | JAN 1985 |
| | MEAN OF EXTREME MINS. | 64 | 1.2 | 6.7 | 16.2 | 30.3 | 40.8 | 52.4 | 58.6 | 56.7 | 44.1 | 32.7 | 19.6 | 7.0 | 30.5 |
| | NORMAL DRY BULB | 30 | 29.6 | 35.4 | 45.8 | 56.6 | 66.5 | 75.6 | 80.2 | 78.2 | 70.2 | 58.3 | 45.3 | 33.9 | 56.3 |
| | MEAN DRY BULB | 64 | 30.6 | 35.1 | 44.6 | 56.5 | 66.0 | 75.3 | 79.3 | 77.7 | 69.9 | 58.5 | 45.5 | 34.7 | 56.1 |
| | MEAN WET BULB | 26 | 28.0 | 31.4 | 39.6 | 49.0 | 58.6 | 66.8 | 70.3 | 69.4 | 61.9 | 51.3 | 40.7 | 31.2 | 49.9 |
| | MEAN DEW POINT | 26 | 24.3 | 27.4 | 34.5 | 44.2 | 54.7 | 63.5 | 67.5 | 66.7 | 58.4 | 47.1 | 36.4 | 27.3 | 46.0 |
| | NORMAL NO. DAYS WITH: | | | | | | | | | | | | | | |
| | MAXIMUM >= 90 | 30 | 0.0 | 0.0 | 0.0 | 0.3 | 1.3 | 8.6 | 15.8 | 12.2 | 4.5 | 0.1 | 0.0 | 0.0 | 42.8 |
| | MAXIMUM <= 32 | 30 | 11.0 | 6.6 | 1.1 | 0.0 | 0.0 | 0.0 | 0.0 | 0.0 | 0.0 | 0.0 | 0.7 | 6.3 | 25.7 |
| | MINIMUM <= 32 | 30 | 25.6 | 19.2 | 12.2 | 2.5 | * | 0.0 | 0.0 | 0.0 | 0.0 | 1.3 | 10.2 | 22.1 | 93.1 |
| | MINIMUM <= 0 | 30 | 1.8 | 0.6 | * | 0.0 | 0.0 | 0.0 | 0.0 | 0.0 | 0.0 | 0.0 | 0.0 | 0.8 | 3.2 |
| **H/C** | NORMAL HEATING DEG. DAYS | 30 | 1097 | 844 | 613 | 294 | 79 | 6 | 0 | 1 | 46 | 246 | 583 | 949 | 4758 |
| | NORMAL COOLING DEG. DAYS | 30 | 0 | 0 | 7 | 32 | 114 | 316 | 461 | 396 | 196 | 36 | 3 | 0 | 1561 |
| **RH** | NORMAL (PERCENT) | 30 | 75 | 73 | 68 | 64 | 67 | 67 | 68 | 70 | 70 | 69 | 72 | 76 | 70 |
| | HOUR 00 LST | 30 | 79 | 77 | 74 | 70 | 76 | 77 | 77 | 80 | 79 | 76 | 76 | 79 | 77 |
| | HOUR 06 LST | 30 | 82 | 83 | 81 | 79 | 83 | 83 | 84 | 87 | 87 | 84 | 82 | 83 | 83 |
| | HOUR 12 LST | 30 | 69 | 65 | 59 | 55 | 56 | 56 | 56 | 57 | 57 | 56 | 63 | 69 | 60 |
| | HOUR 18 LST | 30 | 71 | 66 | 59 | 54 | 56 | 56 | 56 | 58 | 59 | 60 | 66 | 73 | 61 |
| **S** | PERCENT POSSIBLE SUNSHINE | 37 | 50 | 52 | 54 | 56 | 59 | 66 | 69 | 64 | 63 | 60 | 46 | 43 | 57 |
| **W/O** | MEAN NO. DAYS WITH: | | | | | | | | | | | | | | |
| | HEAVY FOG(VISBY <= 1/4 MI) | 46 | 2.1 | 1.6 | 1.2 | 0.6 | 0.6 | 0.3 | 0.3 | 0.4 | 0.5 | 0.7 | 0.9 | 1.8 | 11.0 |
| | THUNDERSTORMS | 64 | 0.7 | 0.9 | 3.0 | 5.5 | 6.8 | 7.7 | 6.9 | 6.2 | 3.6 | 2.4 | 1.6 | 0.8 | 46.1 |
| **CLOUDNESS** | MEAN: | | | | | | | | | | | | | | |
| | SUNRISE-SUNSET (OKTAS) | 48 | 5.4 | 5.3 | 5.4 | 5.2 | 5.1 | 4.8 | 4.4 | 4.2 | 4.1 | 4.0 | 4.9 | 5.3 | 4.8 |
| | MIDNIGHT-MIDNIGHT (OKTAS) | 32 | 5.0 | 5.1 | 5.1 | 4.9 | 4.7 | 4.3 | 3.8 | 3.8 | 3.9 | 3.9 | 4.8 | 5.1 | 4.5 |
| | MEAN NO. DAYS WITH: | | | | | | | | | | | | | | |
| | CLEAR | 48 | 7.3 | 6.6 | 6.5 | 6.9 | 7.1 | 7.1 | 9.3 | 10.1 | 11.4 | 12.3 | 8.6 | 7.1 | 100.3 |
| | PARTLY CLOUDY | 48 | 6.6 | 6.5 | 8.0 | 8.2 | 9.5 | 11.0 | 11.3 | 11.1 | 8.2 | 7.5 | 6.6 | 6.6 | 101.1 |
| | CLOUDY | 48 | 17.1 | 15.1 | 16.5 | 14.9 | 14.4 | 11.9 | 10.4 | 9.8 | 10.4 | 11.2 | 14.9 | 17.3 | 163.9 |
| **PR** | MEAN STATION PRESSURE(IN) | 26 | 29.46 | 29.46 | 29.38 | 29.30 | 29.31 | 29.31 | 29.34 | 29.37 | 29.39 | 29.41 | 29.42 | 29.46 | 29.38 |
| | MEAN SEA-LEVEL PRES. (IN) | 26 | 30.14 | 30.12 | 30.04 | 29.96 | 29.95 | 29.94 | 29.97 | 30.00 | 30.03 | 30.06 | 30.08 | 30.14 | 30.04 |
| **WINDS** | MEAN SPEED (MPH) | 26 | 10.3 | 10.3 | 10.8 | 10.7 | 9.3 | 8.3 | 7.8 | 7.4 | 7.7 | 8.7 | 9.8 | 9.9 | 9.3 |
| | PREVAIL.DIR(TENS OF DEGS) | 40 | 31 | 31 | 31 | 31 | 19 | 19 | 25 | 19 | 19 | 17 | 31 | 31 | 31 |
| | MAXIMUM 2-MINUTE: | | | | | | | | | | | | | | |
| | SPEED (MPH) | 13 | 43 | 40 | 46 | 53 | 41 | 49 | 43 | 45 | 41 | 43 | 41 | 46 | 53 |
| | DIR. (TENS OF DEGS) | | 27 | 12 | 26 | 26 | 27 | 20 | 27 | 31 | 25 | 31 | 24 | 26 | 26 |
| | YEAR OF OCCURRENCE | | 2008 | 2007 | 2009 | 2006 | 2008 | 2009 | 2004 | 2005 | 2005 | 2001 | 1998 | 2009 | APR 2006 |
| | MAXIMUM 3-SECOND | | | | | | | | | | | | | | |
| | SPEED (MPH) | 13 | 59 | 54 | 60 | 70 | 60 | 64 | 55 | 58 | 51 | 52 | 53 | 56 | 70 |
| | DIR. (TENS OF DEGS) | | 28 | 23 | 26 | 27 | 04 | 17 | 06 | 26 | 26 | 26 | 24 | 27 | 27 |
| | YEAR OF OCCURRENCE | | 2008 | 1999 | 2009 | 2006 | 2004 | 2002 | 2006 | 2008 | 2005 | 2007 | 1998 | 2009 | APR 2006 |
| **PRECIPITATION** | NORMAL (IN) | 30 | 2.14 | 2.28 | 3.60 | 3.69 | 4.11 | 3.76 | 3.90 | 2.98 | 2.96 | 2.76 | 3.71 | 2.86 | 38.75 |
| | MAXIMUM MONTHLY (IN) | 52 | 9.01 | 4.68 | 8.39 | 10.32 | 12.92 | 12.35 | 10.71 | 6.44 | 9.77 | 12.38 | 9.95 | 7.82 | 12.92 |
| | YEAR OF OCCURRENCE | | 2005 | 1986 | 2008 | 1994 | 1995 | 2003 | 1981 | 1970 | 2008 | 2009 | 1985 | 1982 | MAY 1995 |
| | MINIMUM MONTHLY (IN) | 52 | 0.10 | 0.25 | 1.09 | 0.99 | .78 | 0.44 | 0.60 | 0.08 | T | 0.21 | 0.44 | 0.32 | T |
| | YEAR OF OCCURRENCE | | 1986 | 1963 | 1966 | 1977 | 2005 | 1991 | 1970 | 1971 | 1979 | 1975 | 1969 | 1958 | SEP 1979 |
| | MAXIMUM IN 24 HOURS (IN) | 52 | 3.07 | 2.60 | 2.95 | 4.91 | 6.55 | 3.78 | 3.47 | 3.17 | 4.58 | 2.88 | 3.71 | 4.03 | 6.55 |
| | YEAR OF OCCURRENCE | | 2005 | 1999 | 1977 | 1979 | 1995 | 2003 | 1982 | 1993 | 2008 | 2009 | 1985 | 1982 | MAY 1995 |
| | NORMAL NO. DAYS WITH: | | | | | | | | | | | | | | |
| | PRECIPITATION >= 0.01 | 30 | 9.4 | 8.2 | 11.1 | 11.4 | 11.3 | 9.6 | 8.3 | 8.1 | 7.5 | 8.5 | 10.1 | 9.4 | 112.9 |
| | PRECIPITATION >= 1.00 | 30 | 0.3 | 0.5 | 0.5 | 0.6 | 1.0 | 1.1 | 1.2 | 0.9 | 0.9 | 0.6 | 1.0 | 0.7 | 9.3 |
| **SNOWFALL** | NORMAL (IN) | 30 | 7.4 | 4.8 | 3.3 | 0.6 | 0.0 | 0.0 | 0.0 | 0.0 | 0.0 | 0.* | 1.5 | 4.9 | 22.5 |
| | MAXIMUM MONTHLY (IN) | 73 | 23.9 | 20.8 | 22.3 | 6.5 | 0.6 | T | T | 0.0 | 0.0 | T | 11.3 | 26.3 | 26.3 |
| | YEAR OF OCCURRENCE | | 1977 | 1993 | 1960 | 1971 | 1993 | 1993 | 1997 | | | 1993 | 1951 | 1973 | DEC 1973 |
| | MAXIMUM IN 24 HOURS (IN) | 73 | 13.9 | 11.7 | 10.7 | 6.1 | 0.6 | T | T | 0.0 | 0.0 | T | 10.3 | 12.0 | 13.9 |
| | YEAR OF OCCURRENCE' | | 1982 | 1993 | 1989 | 1971 | 1993 | 1993 | 1997 | | | 1993 | 1951 | 1973 | JAN 1982 |
| | MAXIMUM SNOW DEPTH (IN) | 61 | 13 | 20 | 17 | 6 | 0 | 0 | 0 | 0 | 0 | 0 | 10 | 12 | 20 |
| | YEAR OF OCCURRENCE | | 1978 | 1982 | 1978 | 1971 | | | | | | | 1951 | 1973 | FEB 1982 |
| | NORMAL NO. DAYS WITH: | | | | | | | | | | | | | | |
| | SNOWFALL >= 1.0 | 30 | 2.0 | 1.2 | 0.7 | 0.3 | 0.0 | 0.0 | 0.0 | 0.0 | 0.0 | 0.0 | 0.4 | 1.5 | 6.1 |

## PRECIPITATION (inches) 2009 ST LOUIS (KSTL)

| YEAR | JAN | FEB | MAR | APR | MAY | JUN | JUL | AUG | SEP | OCT | NOV | DEC | ANNUAL |
|------|-----|-----|-----|-----|-----|-----|-----|-----|-----|-----|-----|-----|--------|
| 1980 | 0.63 | 1.54 | 3.98 | 1.54 | 3.40 | 2.19 | 3.56 | 2.72 | 3.12 | 2.89 | 1.25 | 0.66 | 27.48 |
| 1981 | 0.64 | 2.18 | 2.97 | 3.40 | 6.79 | 5.82 | 10.71 | 3.31 | 1.17 | 3.81 | 2.71 | 2.01 | 45.52 |
| 1982 | 4.90 | 1.37 | 2.88 | 2.55 | 4.85 | 5.96 | 7.91 | 5.27 | 5.27 | 2.30 | 3.89 | 7.82 | 54.97 |
| 1983 | 0.72 | 0.95 | 3.54 | 7.30 | 6.32 | 4.32 | 1.23 | 2.24 | 1.24 | 5.40 | 7.79 | 3.75 | 44.80 |
| 1984 | 0.84 | 3.43 | 5.37 | 6.29 | 5.19 | 2.74 | 0.76 | 0.64 | 8.88 | 7.12 | 5.50 | 4.89 | 51.65 |
| 1985 | 0.53 | 3.77 | 5.18 | 3.60 | 3.30 | 9.43 | 5.23 | 3.66 | 0.43 | 1.96 | 9.95 | 3.69 | 50.73 |
| 1986 | 0.10 | 4.68 | 1.22 | 1.23 | 2.42 | 4.43 | 2.61 | 2.22 | 7.99 | 5.34 | 1.58 | 1.06 | 34.88 |
| 1987 | 1.98 | 1.40 | 2.16 | 1.74 | 2.00 | 3.59 | 5.04 | 5.56 | 1.62 | 1.74 | 4.09 | 7.46 | 38.38 |
| 1988 | 3.30 | 2.27 | 4.73 | 1.15 | 1.44 | 1.97 | 3.02 | 2.31 | 1.99 | 1.86 | 6.65 | 3.24 | 33.93 |
| 1989 | 2.58 | 1.43 | 4.53 | 2.10 | 4.11 | 2.34 | 4.59 | 3.00 | 1.69 | 0.95 | 0.59 | 0.69 | 28.60 |
| 1990 | 1.42 | 3.53 | 2.66 | 3.07 | 9.59 | 3.02 | 3.34 | 2.84 | 0.78 | 4.96 | 3.36 | 6.52 | 45.09 |
| 1991 | 1.52 | 0.98 | 3.20 | 3.27 | 3.87 | 0.44 | 5.18 | 0.98 | 2.98 | 5.70 | 3.26 | 2.10 | 33.48 |
| 1992 | 1.12 | 1.89 | 3.45 | 2.46 | 1.45 | 1.19 | 4.31 | 3.45 | 2.98 | 1.21 | 6.32 | 3.66 | 33.49 |
| 1993 | 3.54 | 2.75 | 3.31 | 6.16 | 3.94 | 7.12 | 5.06 | 4.78 | 9.16 | 2.61 | 4.85 | 1.48 | 54.76 |
| 1994 | 2.09 | 1.51 | 1.27 | 10.32 | 1.72 | 2.16 | 1.42 | 3.76 | 1.18 | 2.85 | 4.90 | 1.52 | 34.70 |
| 1995 | 4.39 | 1.33 | 3.19 | 3.33 | 12.92 | 2.96 | 2.16 | 4.52 | 0.74 | 2.01 | 1.28 | 2.85 | 41.68 |
| 1996 | 3.27 | 0.52 | 3.06 | 7.97 | 4.34 | 3.72 | 6.33 | 1.57 | 2.86 | 2.67 | 6.50 | .86 | 43.67 |
| 1997 | 2.74 | 4.14 | 2.85 | 2.66 | 3.05 | 2.00 | 1.44 | 3.36 | 2.73 | 2.05 | 2.36 | 1.85 | 31.23 |
| 1998 | 2.88 | 2.93 | 6.00 | 4.63 | 3.62 | 6.90 | 6.39 | 2.35 | 1.86 | 2.51 | 2.72 | 0.83 | 43.62 |
| 1999 | 5.10 | 3.52 | 2.40 | 3.72 | 2.20 | 5.26 | 4.22 | 1.95 | 1.09 | 2.04 | 0.72 | 1.84 | 34.06 |
| 2000 | 1.23 | 3.11 | 1.88 | 1.84 | 5.84 | 8.22 | 2.25 | 3.64 | 2.62 | 2.60 | 2.79 | 1.35 | 37.37 |
| 2001 | 1.12 | 2.48 | 1.45 | 3.01 | 2.81 | 3.60 | 4.00 | 1.99 | 2.81 | 5.50 | 3.06 | 3.46 | 35.29 |
| 2002 | 3.16 | 0.83 | 3.67 | 4.25 | 7.81 | 5.26 | 1.47 | 4.12 | 2.44 | 4.78 | 1.14 | 2.02 | 40.95 |
| 2003 | 0.96 | 2.00 | 2.80 | 4.29 | 3.97 | 12.35 | 2.51 | 2.54 | 4.15 | 2.81 | 5.34 | 2.34 | 46.06 |
| 2004 | 3.97 | 0.85 | 4.36 | 1.94 | 9.75 | 0.83 | 5.52 | 4.10 | 0.23 | 3.21 | 5.74 | 1.77 | 42.27 |
| 2005 | 9.01 | 1.84 | 1.47 | 2.17 | 0.78 | 5.10 | 2.22 | 3.87 | 5.30 | 1.52 | 3.35 | 1.22 | 37.85 |
| 2006 | 1.63 | 0.46 | 3.26 | 2.10 | 2.88 | 2.37 | 2.73 | 2.27 | 1.28 | 3.66 | 5.25 | 2.04 | 29.93 |
| 2007 | 3.11 | 1.98 | 2.80 | 3.18 | 4.26 | 2.88 | 3.11 | 1.57 | 1.71 | 1.97 | 1.25 | 2.75 | 30.57 |
| 2008 | 1.98 | 4.60 | 8.39 | 3.76 | 10.84 | 1.89 | 7.50 | 1.59 | 9.77 | 1.23 | 1.86 | 4.55 | 57.96 |
| 2009 | 0.77 | 2.33 | 3.04 | 4.06 | 4.72 | 6.42 | 4.20 | 2.48 | 3.16 | 12.38 | 3.11 | 4.25 | 50.92 |
| POR= 64 YRS | 2.17 | 2.17 | 3.34 | 3.58 | 4.10 | 4.05 | 3.82 | 2.92 | 2.98 | 2.92 | 3.16 | 2.51 | 37.72 |

WBAN : 13994

## AVERAGE TEMPERATURE (°F) 2009 ST LOUIS (KSTL)

| YEAR | JAN | FEB | MAR | APR | MAY | JUN | JUL | AUG | SEP | OCT | NOV | DEC | ANNUAL |
|------|-----|-----|-----|-----|-----|-----|-----|-----|-----|-----|-----|-----|--------|
| 1980 | 31.4 | 27.9 | 41.0 | 54.5 | 66.9 | 75.5 | 85.0 | 83.5 | 72.5 | 55.9 | 46.4 | 36.6 | 56.4 |
| 1981 | 31.2 | 36.8 | 46.7 | 63.1 | 60.7 | 75.7 | 78.7 | 76.1 | 69.2 | 55.7 | 48.7 | 31.1 | 56.1 |
| 1982 | 22.5 | 28.6 | 45.3 | 51.5 | 70.7 | 70.6 | 79.3 | 75.2 | 68.0 | 58.3 | 46.3 | 41.6 | 54.8 |
| 1983 | 32.3 | 38.1 | 44.5 | 50.3 | 62.3 | 75.3 | 83.5 | 84.2 | 72.0 | 59.7 | 48.2 | 20.5 | 55.9 |
| 1984 | 28.3 | 40.4 | 37.1 | 54.1 | 63.2 | 79.5 | 78.3 | 80.7 | 68.2 | 61.8 | 44.3 | 40.7 | 56.4 |
| 1985 | 22.6 | 30.5 | 49.5 | 60.4 | 67.6 | 71.6 | 79.3 | 74.7 | 70.9 | 61.4 | 46.5 | 27.3 | 55.2 |
| 1986 | 34.9 | 34.5 | 49.2 | 60.8 | 68.2 | 78.3 | 82.8 | 74.0 | 73.3 | 58.3 | 41.5 | 35.4 | 57.6 |
| 1987 | 30.6 | 40.1 | 48.6 | 56.9 | 72.6 | 77.9 | 81.0 | 78.9 | 70.5 | 53.8 | 49.1 | 38.0 | 58.2 |
| 1988 | 29.2 | 30.5 | 45.2 | 57.1 | 69.0 | 77.7 | 81.6 | 82.7 | 72.5 | 53.9 | 47.2 | 37.2 | 57.0 |
| 1989 | 41.2 | 28.2 | 45.0 | 57.7 | 64.3 | 74.8 | 79.3 | 77.8 | 67.4 | 61.3 | 47.1 | 24.1 | 55.7 |
| 1990 | 42.9 | 41.3 | 49.8 | 55.7 | 63.6 | 77.2 | 80.2 | 77.9 | 74.1 | 58.1 | 52.7 | 34.7 | 59.0 |
| 1991 | 29.3 | 41.7 | 50.1 | 61.5 | 73.0 | 79.9 | 80.9 | 79.7 | 72.4 | 60.5 | 42.4 | 39.2 | 59.2 |
| 1992 | 37.0 | 42.7 | 48.5 | 57.8 | 65.1 | 73.8 | 79.0 | 73.4 | 69.2 | 59.4 | 44.3 | 35.7 | 57.2 |
| 1993 | 32.1 | 31.7 | 41.5 | 54.2 | 66.4 | 75.1 | 81.9 | 80.4 | 66.4 | 56.1 | 43.9 | 37.2 | 55.6 |
| 1994 | 26.8 | 35.0 | 47.7 | 57.7 | 65.0 | 79.0 | 79.5 | 76.6 | 70.2 | 61.5 | 52.1 | 41.5 | 57.7 |
| 1995 | 31.1 | 36.4 | 49.0 | 56.8 | 64.7 | 75.6 | 81.3 | 83.9 | 68.0 | 60.9 | 42.7 | 33.8 | 57.0 |
| 1996 | 29.4 | 37.4 | 41.0 | 54.6 | 68.6 | 74.6 | 75.8 | 77.6 | 67.2 | 58.1 | 39.0 | 35.5 | 54.9 |
| 1997 | 26.7 | 38.1 | 47.6 | 50.9 | 61.7 | 73.6 | 80.2 | 76.0 | 70.2 | 59.1 | 42.0 | 35.1 | 55.1 |
| 1998 | 36.4 | 42.6 | 43.2 | 55.3 | 71.2 | 75.5 | 78.9 | 79.1 | 75.0 | 60.2 | 49.5 | 37.2 | 58.7 |
| 1999 | 31.3 | 42.5 | 42.8 | 58.6 | 66.8 | 74.9 | 82.9 | 76.5 | 69.7 | 58.8 | 52.6 | 38.3 | 58.0 |
| 2000 | 33.0 | 42.9 | 49.1 | 55.2 | 69.0 | 72.9 | 77.6 | 80.4 | 69.6 | 61.3 | 41.3 | 21.6 | 56.2 |
| 2001 | 30.5 | 35.1 | 41.5 | 63.1 | 68.7 | 74.1 | 80.7 | 79.6 | 68.8 | 57.7 | 53.1 | 39.9 | 57.7 |
| 2002 | 38.1 | 39.3 | 42.2 | 59.5 | 63.6 | 78.0 | 82.8 | 80.0 | 73.4 | 55.3 | 44.1 | 37.5 | 57.9 |
| 2003 | 28.1 | 31.4 | 47.5 | 58.1 | 65.1 | 71.3 | 79.9 | 80.8 | 67.7 | 60.0 | 49.3 | 38.7 | 56.5 |
| 2004 | 30.9 | 35.9 | 49.9 | 59.3 | 70.8 | 74.4 | 77.9 | 73.6 | 71.5 | 60.2 | 49.3 | 36.9 | 57.6 |
| 2005 | 33.9 | 40.4 | 43.9 | 59.0 | 66.5 | 78.8 | 80.0 | 80.0 | 73.9 | 58.8 | 48.6 | 32.0 | 58.0 |
| 2006 | 42.3 | 35.3 | 46.6 | 62.1 | 66.2 | 76.2 | 82.6 | 80.1 | 67.5 | 55.3 | 46.9 | 40.7 | 58.5 |
| 2007 | 34.1 | 29.3 | 53.6 | 53.9 | 70.9 | 76.4 | 78.4 | 84.4 | 74.0 | 63.1 | 46.5 | 35.5 | 58.3 |
| 2008 | 33.0 | 32.4 | 43.9 | 54.2 | 63.2 | 76.8 | 79.3 | 76.7 | 70.3 | 58.3 | 44.8 | 32.5 | 55.5 |
| 2009 | 28.9 | 39.0 | 49.3 | 56.0 | 67.0 | 77.7 | 75.6 | 76.4 | 69.8 | 53.7 | 51.8 | 33.7 | 56.6 |
| POR= 64 YRS | 30.6 | 35.1 | 44.6 | 56.5 | 66.0 | 75.3 | 79.3 | 77.7 | 69.9 | 58.5 | 45.5 | 34.7 | 56.1 |

## HEATING DEGREE DAYS (base 65°F) 2009  ST LOUIS (KSTL)

| YEAR | JUL | AUG | SEP | OCT | NOV | DEC | JAN | FEB | MAR | APR | MAY | JUN | TOTAL |
|---|---|---|---|---|---|---|---|---|---|---|---|---|---|
| 1980-81 | 0 | 0 | 30 | 305 | 553 | 877 | 1039 | 784 | 569 | 127 | 168 | 0 | 4452 |
| 1981-82 | 0 | 0 | 35 | 298 | 483 | 1048 | 1308 | 1015 | 603 | 407 | 9 | 11 | 5217 |
| 1982-83 | 0 | 0 | 49 | 261 | 569 | 721 | 1008 | 745 | 632 | 437 | 117 | 7 | 4546 |
| 1983-84 | 0 | 0 | 58 | 192 | 498 | 1376 | 1133 | 705 | 860 | 342 | 115 | 0 | 5279 |
| 1984-85 | 0 | 0 | 103 | 151 | 616 | 746 | 1308 | 960 | 487 | 200 | 42 | 10 | 4623 |
| 1985-86 | 0 | 0 | 64 | 145 | 550 | 1159 | 929 | 850 | 506 | 194 | 44 | 0 | 4441 |
| 1986-87 | 0 | 11 | 12 | 221 | 699 | 910 | 1062 | 691 | 501 | 267 | 10 | 0 | 4384 |
| 1987-88 | 0 | 0 | 12 | 346 | 490 | 830 | 1102 | 995 | 610 | 241 | 17 | 3 | 4646 |
| 1988-89 | 0 | 0 | 5 | 354 | 528 | 854 | 730 | 1029 | 625 | 293 | 128 | 4 | 4550 |
| 1989-90 | 0 | 0 | 73 | 183 | 536 | 1261 | 679 | 657 | 496 | 327 | 85 | 9 | 4306 |
| 1990-91 | 3 | 0 | 24 | 250 | 375 | 934 | 1101 | 648 | 474 | 154 | 21 | 0 | 3984 |
| 1991-92 | 0 | 0 | 67 | 191 | 674 | 796 | 859 | 642 | 509 | 264 | 95 | 5 | 4102 |
| 1992-93 | 0 | 3 | 46 | 204 | 615 | 902 | 1014 | 927 | 726 | 327 | 52 | 13 | 4829 |
| 1993-94 | 0 | 0 | 58 | 292 | 628 | 852 | 1179 | 833 | 531 | 261 | 84 | 1 | 4719 |
| 1994-95 | 0 | 0 | 33 | 155 | 388 | 722 | 1045 | 794 | 488 | 260 | 88 | 0 | 3973 |
| 1995-96 | 0 | 0 | 65 | 167 | 664 | 961 | 1096 | 797 | 735 | 322 | 67 | 14 | 4888 |
| 1996-97 | 0 | 0 | 57 | 235 | 776 | 908 | 1180 | 746 | 530 | 418 | 142 | 5 | 4997 |
| 1997-98 | 0 | 0 | 13 | 271 | 684 | 922 | 880 | 621 | 689 | 293 | 19 | 24 | 4416 |
| 1998-99 | 0 | 0 | 1 | 175 | 460 | 859 | 1038 | 626 | 679 | 214 | 41 | 7 | 4100 |
| 1999-00 | 0 | 0 | 41 | 226 | 370 | 826 | 985 | 638 | 490 | 297 | 36 | 8 | 3917 |
| 2000-01 | 0 | 0 | 57 | 176 | 709 | 1338 | 1063 | 831 | 721 | 154 | 47 | 13 | 5109 |
| 2001-02 | 0 | 0 | 50 | 246 | 352 | 772 | 826 | 714 | 683 | 234 | 120 | 0 | 3997 |
| 2002-03 | 0 | 0 | 12 | 338 | 620 | 847 | 1135 | 937 | 536 | 249 | 64 | 27 | 4765 |
| 2003-04 | 0 | 0 | 50 | 176 | 470 | 812 | 1048 | 836 | 466 | 225 | 55 | 0 | 4138 |
| 2004-05 | 1 | 4 | 9 | 175 | 460 | 865 | 961 | 681 | 648 | 216 | 82 | 0 | 4102 |
| 2005-06 | 0 | 0 | 16 | 243 | 495 | 1017 | 696 | 826 | 567 | 135 | 95 | 0 | 4090 |
| 2006-07 | 0 | 0 | 40 | 344 | 535 | 747 | 953 | 992 | 385 | 360 | 18 | 0 | 4374 |
| 2007-08 | 0 | 0 | 15 | 167 | 550 | 908 | 982 | 940 | 644 | 332 | 107 | 0 | 4645 |
| 2008-09 | 0 | 0 | 15 | 229 | 600 | 1001 | 1111 | 723 | 494 | 315 | 61 | 3 | 4552 |
| 2009- | 0 | 4 | 18 | 342 | 389 | 964 | | | | | | | |

WBAN : 13994

## COOLING DEGREE DAYS (base 65°F) 2009  ST LOUIS (KSTL)

| YEAR | JAN | FEB | MAR | APR | MAY | JUN | JUL | AUG | SEP | OCT | NOV | DEC | TOTAL |
|---|---|---|---|---|---|---|---|---|---|---|---|---|---|
| 1980 | 0 | 0 | 0 | 23 | 120 | 320 | 626 | 580 | 262 | 31 | 2 | 0 | 1964 |
| 1981 | 0 | 0 | 7 | 77 | 42 | 327 | 431 | 353 | 166 | 13 | 1 | 0 | 1417 |
| 1982 | 0 | 0 | 0 | 7 | 191 | 186 | 453 | 322 | 146 | 63 | 15 | 4 | 1387 |
| 1983 | 0 | 0 | 3 | 3 | 41 | 322 | 578 | 603 | 274 | 36 | 2 | 0 | 1862 |
| 1984 | 0 | 0 | 0 | 24 | 67 | 442 | 423 | 493 | 202 | 57 | 2 | 1 | 1711 |
| 1985 | 0 | 0 | 14 | 70 | 128 | 214 | 451 | 310 | 245 | 43 | 2 | 0 | 1477 |
| 1986 | 0 | 0 | 25 | 75 | 150 | 407 | 561 | 298 | 267 | 21 | 0 | 0 | 1804 |
| 1987 | 0 | 0 | 0 | 32 | 251 | 393 | 501 | 439 | 183 | 5 | 20 | 0 | 1824 |
| 1988 | 0 | 0 | 4 | 10 | 144 | 389 | 521 | 556 | 238 | 16 | 0 | 0 | 1878 |
| 1989 | 0 | 0 | 11 | 80 | 111 | 305 | 450 | 403 | 151 | 75 | 6 | 0 | 1592 |
| 1990 | 0 | 0 | 30 | 55 | 47 | 382 | 480 | 408 | 304 | 41 | 12 | 0 | 1759 |
| 1991 | 0 | 0 | 19 | 56 | 277 | 452 | 498 | 463 | 295 | 58 | 3 | 0 | 2121 |
| 1992 | 0 | 0 | 7 | 54 | 104 | 277 | 440 | 268 | 182 | 36 | 0 | 0 | 1368 |
| 1993 | 0 | 0 | 1 | 9 | 102 | 320 | 529 | 483 | 107 | 23 | 0 | 0 | 1574 |
| 1994 | 0 | 0 | 4 | 48 | 90 | 428 | 453 | 365 | 194 | 53 | 7 | 0 | 1642 |
| 1995 | 0 | 0 | 0 | 22 | 85 | 322 | 510 | 595 | 164 | 47 | 0 | 0 | 1745 |
| 1996 | 0 | 5 | 0 | 16 | 188 | 309 | 342 | 402 | 132 | 25 | 0 | 0 | 1419 |
| 1997 | 0 | 0 | 1 | 1 | 44 | 268 | 477 | 349 | 173 | 99 | 0 | 0 | 1412 |
| 1998 | 0 | 0 | 21 | 7 | 217 | 347 | 439 | 446 | 305 | 33 | 2 | 3 | 1820 |
| 1999 | 0 | 0 | 0 | 28 | 104 | 309 | 563 | 364 | 190 | 37 | 5 | 0 | 1600 |
| 2000 | 0 | 5 | 6 | 8 | 167 | 251 | 398 | 483 | 202 | 69 | 5 | 0 | 1594 |
| 2001 | 0 | 0 | 0 | 106 | 166 | 293 | 492 | 457 | 167 | 29 | 2 | 1 | 1713 |
| 2002 | 0 | 0 | 1 | 75 | 81 | 396 | 560 | 471 | 270 | 43 | 0 | 0 | 1897 |
| 2003 | 0 | 0 | 1 | 49 | 71 | 221 | 470 | 498 | 140 | 29 | 8 | 0 | 1487 |
| 2004 | 0 | 0 | 4 | 60 | 242 | 291 | 408 | 279 | 209 | 31 | 0 | 0 | 1524 |
| 2005 | 0 | 0 | 2 | 41 | 135 | 420 | 473 | 469 | 289 | 55 | 9 | 0 | 1893 |
| 2006 | 0 | 0 | 5 | 54 | 138 | 326 | 556 | 474 | 123 | 50 | 0 | 0 | 1726 |
| 2007 | 0 | 0 | 38 | 32 | 206 | 350 | 422 | 611 | 293 | 113 | 0 | 0 | 2065 |
| 2008 | 0 | 0 | 0 | 17 | 56 | 361 | 451 | 369 | 181 | 30 | 1 | 0 | 1466 |
| 2009 | 0 | 0 | 13 | 52 | 127 | 389 | 339 | 364 | 167 | 0 | 4 | 0 | 1455 |

## SNOWFALL (inches) 2009 ST LOUIS (KSTL)

| YEAR | JUL | AUG | SEP | OCT | NOV | DEC | JAN | FEB | MAR | APR | MAY | JUN | TOTAL |
|---|---|---|---|---|---|---|---|---|---|---|---|---|---|
| 1980-81 | 0.0 | 0.0 | 0.0 | 0.0 | 8.0 | 1.1 | 0.6 | 8.2 | 0.2 | 0.0 | 0.0 | 0.0 | 18.1 |
| 1981-82 | 0.0 | 0.0 | 0.0 | 0.0 | T | 7.9 | 16.6 | 9.0 | 0.4 | 2.7 | 0.0 | 0.0 | 36.6 |
| 1982-83 | 0.0 | 0.0 | 0.0 | 0.0 | T | T | 3.3 | 0.3 | 1.4 | 2.4 | 0.0 | 0.0 | 7.4 |
| 1983-84 | 0.0 | 0.0 | 0.0 | 0.0 | T | 6.5 | 2.3 | 9.9 | 5.2 | 0.0 | 0.0 | 0.0 | 23.9 |
| 1984-85 | 0.0 | 0.0 | 0.0 | 0.0 | 1.7 | 1.8 | 5.1 | 1.3 | T | 0.0 | 0.0 | 0.0 | 9.9 |
| 1985-86 | 0.0 | 0.0 | 0.0 | 0.0 | T | 5.7 | 1.0 | 6.1 | 0.2 | T | 0.0 | 0.0 | 13.0 |
| 1986-87 | 0.0 | 0.0 | 0.0 | 0.0 | T | T | 23.6 | 0.6 | T | 0.0 | 0.0 | 0.0 | 24.2 |
| 1987-88 | 0.0 | 0.0 | 0.0 | 0.0 | T | 7.3 | 1.4 | 6.7 | 2.8 | 0.0 | 0.0 | 0.0 | 18.2 |
| 1988-89 | 0.0 | 0.0 | 0.0 | 0.0 | 2.9 | 5.9 | 0.1 | 3.9 | 11.0 | 0.0 | 0.0 | 0.0 | 23.8 |
| 1989-90 | 0.0 | 0.0 | 0.0 | T | T | 9.1 | 0.2 | 6.9 | 8.4 | T | T | T | 24.6 |
| 1990-91 | 0.0 | 0.0 | 0.0 | 0.0 | T | 13.2 | 1.9 | T | 1.7 | 0.0 | 0.0 | 0.0 | 16.8 |
| 1991-92 | 0.0 | 0.0 | 0.0 | T | 3.9 | 0.4 | 3.8 | 2.3 | 3.1 | 0.0 | 0.0 | 0.0 | 13.5 |
| 1992-93 | 0.0 | 0.0 | 0.0 | 0.0 | T | 0.7 | 5.5 | 20.8 | 3.0 | 0.3 | 0.6 | T | 30.9 |
| 1993-94 | 0.0 | 0.0 | 0.0 | T | 1.6 | 1.1 | 5.0 | 1.0 | 0.2 | 1.4 | 0.0 | 0.0 | 10.3 |
| 1994-95 | 0.0 | 0.0 | 0.0 | 0.0 | 0.0 | 0.1 | 3.5 | 1.6 | 0.5 | T | 0.0 | 0.0 | 5.7 |
| 1995-96 | 0.0 | 0.0 | 0.0 | 0.0 | 0.9 | 7.9 | 13.4 | 2.8 | 1.0 | T | 0.0 | 0.0 | 26.0 |
| 1996-97 | T | 0.0 | 0.0 | T | | 1.5 | 13.1 | 0.8 | T | 4.2 | 0.0 | 0.0 | |
| 1997-98 | T | 0.0 | 0.0 | 0.0 | 2.6 | 4.1 | 5.7 | 0.5 | 5.9 | T | 0.2 | 0.0 | 19.0 |
| 1998-99 | 0.0 | 0.0 | 0.0 | 0.0 | 0.0 | 1.8 | 12.0 | 1.5 | 0.5 | T | 0.0 | 0.0 | 15.8 |
| 1999-00 | 0.0 | 0.0 | 0.0 | 0.0 | 0.0 | 0.3 | 5.0 | T | 4.8 | T | 0.0 | 0.0 | 10.1 |
| 2000-01 | 0.0 | 0.0 | 0.0 | 0.0 | 0.0 | 15.9 | 0.7 | 1.6 | T | 0.4 | 0.0 | 0.0 | 18.6 |
| 2001-02 | 0.0 | 0.0 | 0.0 | T | 0.2 | T | 6.2 | 2.6 | 5.0 | 0.0 | 0.0 | 0.0 | 14.0 |
| 2002-03 | 0.0 | 0.0 | 0.0 | 0.0 | 0.2 | 9.4 | 7.4 | 12.7 | 0.1 | T | 0.0 | 0.0 | 29.8 |
| 2003-04 | 0.0 | 0.0 | 0.0 | 0.0 | T | 3.3 | 2.2 | 1.9 | T | 0.0 | T | 0.0 | 7.4 |
| 2004-05 | T | 0.0 | 0.0 | 0.0 | 1.8 | T | 7.7 | 3.2 | 0.1 | 0.0 | 0.0 | T | 12.8 |
| 2005-06 | 0.0 | 0.0 | 0.0 | 0.0 | T | 3.8 | 1.0 | 3.1 | 2.6 | T | 0.0 | 0.0 | 10.5 |
| 2006-07 | 0.0 | 0.0 | 0.0 | 0.0 | 2.3 | 2.1 | 3.7 | 4.7 | 0.4 | T | 0.0 | 0.0 | 13.2 |
| 2007-08 | 0.0 | 0.0 | 0.0 | 0.0 | T | 7.8 | 4.2 | 8.2 | 10.0 | T | T | 0.0 | 30.2 |
| 2008-09 | 0.0 | 0.0 | 0.0 | 0.0 | 2.4 | 1.5 | 7.0 | 0.1 | 1.8 | 0.1 | 0.0 | 0.0 | 12.9 |
| 2009- | 0.0 | 0.0 | 0.0 | 0.0 | 0.0 | 3.3 | | | | | | | |
| POR= 65 YRS | T | 0.0 | 0.0 | T | 1.3 | 3.7 | 5.6 | 4.2 | 3.7 | 0.4 | T | T | 18.9 |

# St. Louis, MO 2009
## Temperature and Precipitation

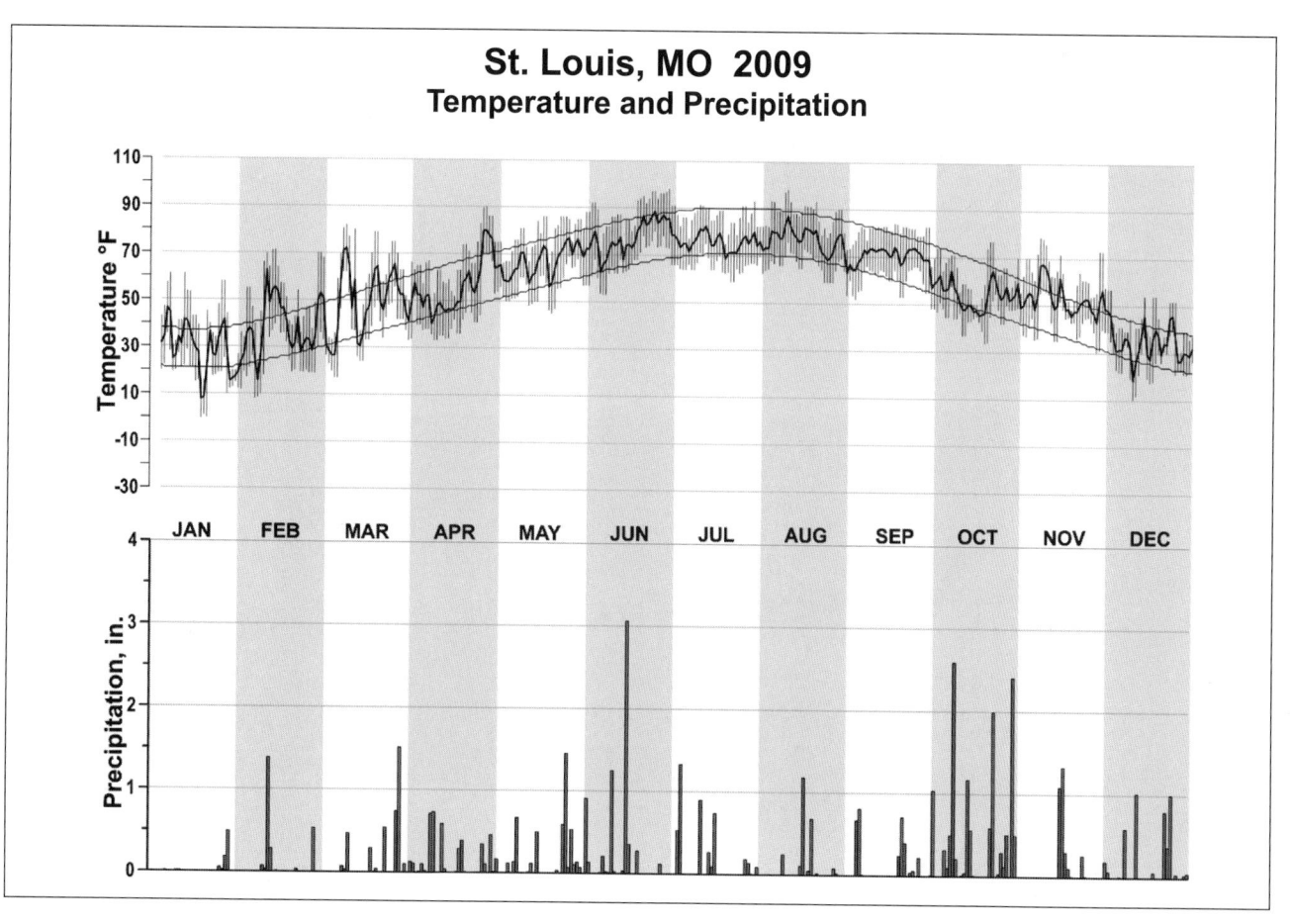

# 2009
# BILLINGS
# MONTANA (KBIL)

Billings, Montana, at an elevation of 3,100 to 3,600 feet above sea level, is situated in the borderline area between the Great Plains and the Rocky Mountains, and has a climate which takes on some of the characteristics of both regions. Its climate may be classified as semi-arid, but with irrigation and the favorable distribution of the precipitation, it is possible to raise a variety of crops in the area.

About a third of the annual precipitation falls during May and June, with June being the wettest month. The period of least precipitation is from November through February. These four months normally produce less than 20 percent of the annual precipitation. The heaviest snows occur during the spring and fall months when the temperature and moisture conditions are most favorable. Heavy snows of 6 inches or more also occur during November and December. The occurrence of thawing periods normally prevents the snow from accumulating to great depths on the ground. Thunderstorms are most frequent during the summer months. These storms are frequently accompanied by strong, gusty winds and occasionally by hail. Destructive hailstorms, however, are rather infrequent.

Winter is usually cold, though not extremely so, and generally affords several mild periods of a week to several weeks in length. The winter cold periods are ushered in by moderately strong north to northeast winds and snow. The coldest temperatures occur after the snow ends and the sky clears. True blizzard conditions are not observed very often in town, but in the surrounding rural areas, blizzard conditions may develop several times during the winter. Cold weather improves with the onset of moderate to strong southwest winds. This wind is sometimes a foehn condition (chinook), but is more often a drainage wind moving down the Yellowstone Valley which transports warmer air of Pacific origin to the area. Occasionally an open winter occurs when cold Arctic outbreaks pass far to the east and temperatures stay above zero degrees.

Spring brings a period of frequent and rapid fluctuations in the weather. It is usually cloudy and cool with frequent periods of rain and/or snow. As the season progresses, snows become less frequent until late May and June when rain is the rule. The last freezing temperatures in spring usually occur before mid-May though they have occurred as late as late June.

The summer season is characterized by warm days with abundant sunshine and low humidities. The nights are cool because of the altitude and the cool air drainage into the valley from the higher terrain. Seldom is there a protracted rainy spell during this season. Frequent thunderstorms bring threatening afternoon cloudiness but usually only small amounts of rain.

The first freezing temperatures of the fall season occur in late September, but they have been noted as early as late August. Over the years, the fall months have been about evenly distributed between cold, wet ones, and mild, dry, pleasant ones. The change to severe winter weather usually arrives after the middle of November. There have been years when the more severe type of winter weather have been delayed until late in December.

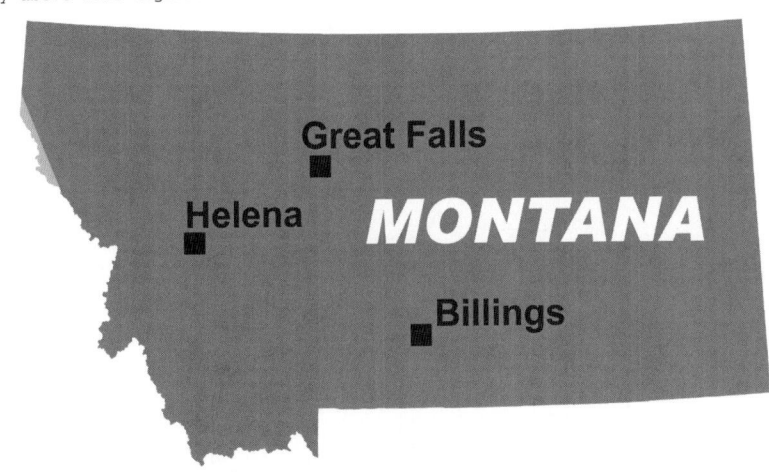

# NORMALS, MEANS, AND EXTREMES
## BILLINGS (KBIL)

| LATITUDE: 45° 48'N | LONGITUDE: -108° 32'W | ELEVATION (FT): GRND: 3581  BARO: 3582 | TIME ZONE: MOUNTAIN  (UTC -7) | WBAN: 24033 |
|---|---|---|---|---|

| | ELEMENT | POR | JAN | FEB | MAR | APR | MAY | JUN | JUL | AUG | SEP | OCT | NOV | DEC | YEAR |
|---|---|---|---|---|---|---|---|---|---|---|---|---|---|---|---|
| **TEMPERATURE °F** | NORMAL DAILY MAXIMUM | 30 | 32.8 | 39.5 | 47.6 | 57.5 | 67.4 | 78.0 | 85.8 | 84.5 | 71.8 | 58.9 | 42.7 | 34.5 | 58.4 |
| | MEAN DAILY MAXIMUM | 62 | 32.9 | 39.0 | 45.9 | 56.9 | 67.2 | 76.7 | 86.6 | 85.1 | 72.8 | 60.1 | 45.2 | 35.8 | 58.7 |
| | HIGHEST DAILY MAXIMUM | 75 | 68 | 72 | 80 | 92 | 96 | 105 | 108 | 105 | 103 | 90 | 77 | 69 | 108 |
| | YEAR OF OCCURRENCE | | 1953 | 1961 | 2004 | 1939 | 1936 | 1984 | 2002 | 1961 | 1983 | 1992 | 1999 | 1980 | JUL 2002 |
| | MEAN OF EXTREME MAXS. | 62 | 55.1 | 59.9 | 68.2 | 78.3 | 86.4 | 94.3 | 99.6 | 98.0 | 91.9 | 81.3 | 66.5 | 56.6 | 78.0 |
| | NORMAL DAILY MINIMUM | 30 | 15.1 | 20.1 | 26.4 | 34.7 | 44.0 | 52.5 | 58.3 | 57.3 | 47.1 | 37.2 | 25.6 | 17.7 | 36.3 |
| | MEAN DAILY MINIMUM | 62 | 14.5 | 19.6 | 24.9 | 34.0 | 43.4 | 51.6 | 58.4 | 56.8 | 47.1 | 37.3 | 26.2 | 18.3 | 36.0 |
| | LOWEST DAILY MINIMUM | 76 | -30 | -38 | -19 | -5 | 14 | 32 | 41 | 35 | 22 | -7 | -22 | -32 | -38 |
| | YEAR OF OCCURRENCE | | 1997 | 1936 | 1989 | 1936 | 1954 | 1969 | 1972 | 1992 | 1984 | 1991 | 1959 | 1983 | FEB 1936 |
| | MEAN OF EXTREME MINS. | 62 | -10.1 | -4.0 | 3.2 | 19.8 | 30.7 | 41.0 | 49.0 | 46.7 | 33.5 | 20.8 | 5.0 | -6.3 | 19.1 |
| | NORMAL DRY BULB | 30 | 24.0 | 29.8 | 37.0 | 46.1 | 55.7 | 65.2 | 72.0 | 70.9 | 59.5 | 48.1 | 34.1 | 26.1 | 47.4 |
| | MEAN DRY BULB | 62 | 23.7 | 29.3 | 35.4 | 45.5 | 55.3 | 64.2 | 72.5 | 71.0 | 59.9 | 48.7 | 35.7 | 27.1 | 47.4 |
| | MEAN WET BULB | 26 | 20.8 | 23.3 | 29.4 | 36.9 | 45.2 | 52.3 | 56.3 | 54.3 | 47.0 | 38.3 | 28.2 | 21.4 | 37.8 |
| | MEAN DEW POINT | 26 | 15.6 | 17.1 | 23.3 | 30.1 | 38.8 | 46.3 | 49.2 | 46.8 | 40.0 | 31.2 | 22.0 | 15.2 | 31.3 |
| | NORMAL NO. DAYS WITH: | | | | | | | | | | | | | | |
| | MAXIMUM >= 90 | 30 | 0.0 | 0.0 | 0.0 | * | 0.4 | 4.2 | 11.7 | 10.7 | 1.9 | * | 0.0 | 0.0 | 28.9 |
| | MAXIMUM <= 32 | 30 | 12.8 | 7.7 | 4.2 | 0.9 | 0.0 | 0.0 | 0.0 | 0.0 | 0.1 | 0.8 | 5.9 | 10.8 | 43.2 |
| | MINIMUM <= 32 | 30 | 27.4 | 23.6 | 22.8 | 11.4 | 1.6 | 0.0 | 0.0 | 0.0 | 1.2 | 8.8 | 21.9 | 27.5 | 146.2 |
| | MINIMUM <= 0 | 30 | 7.2 | 3.6 | 0.9 | 0.0 | 0.0 | 0.0 | 0.0 | 0.0 | 0.0 | 0.0 | 1.3 | 4.5 | 17.5 |
| **H/C** | NORMAL HEATING DEG. DAYS | 30 | 1280 | 1001 | 876 | 575 | 312 | 90 | 20 | 25 | 205 | 516 | 911 | 1195 | 7006 |
| | NORMAL COOLING DEG. DAYS | 30 | 0 | 0 | 0 | 2 | 13 | 90 | 227 | 204 | 44 | 3 | 0 | 0 | 583 |
| **RH** | NORMAL (PERCENT) | 30 | 62 | 59 | 60 | 56 | 56 | 54 | 48 | 45 | 51 | 55 | 60 | 61 | 56 |
| | HOUR 05 LST | 30 | 65 | 67 | 69 | 69 | 71 | 71 | 65 | 63 | 65 | 66 | 66 | 65 | 67 |
| | HOUR 11 LST | 30 | 61 | 57 | 55 | 48 | 47 | 44 | 39 | 40 | 46 | 51 | 57 | 60 | 50 |
| | HOUR 17 LST | 30 | 57 | 51 | 46 | 41 | 42 | 38 | 32 | 30 | 36 | 43 | 54 | 57 | 44 |
| | HOUR 23 LST | 30 | 64 | 63 | 63 | 59 | 61 | 59 | 51 | 48 | 53 | 58 | 62 | 63 | 59 |
| **S** | PERCENT POSSIBLE SUNSHINE | 56 | 47 | 53 | 61 | 60 | 61 | 64 | 76 | 75 | 68 | 61 | 46 | 45 | 60 |
| **W/O** | MEAN NO. DAYS WITH: | | | | | | | | | | | | | | |
| | HEAVY FOG(VISBY <= 1/4 MI) | 46 | 1.4 | 2.1 | 2.3 | 2.6 | 1.3 | 0.5 | 0.3 | 0.3 | 0.8 | 2.0 | 2.1 | 1.8 | 17.5 |
| | THUNDERSTORMS | 62 | 0.0 | 0.0 | 0.1 | 1.3 | 4.0 | 7.4 | 7.4 | 5.6 | 2.0 | 0.2 | 0.0 | 0.0 | 28.0 |
| **CLOUDNESS** | MEAN: | | | | | | | | | | | | | | |
| | SUNRISE-SUNSET (OKTAS) | | | | | | | | | | | | | | |
| | MIDNIGHT-MIDNIGHT (OKTAS) | | | | | | | | | | | | | | |
| | MEAN NO. DAYS WITH: | | | | | | | | | | | | | | |
| | CLEAR | | | | | | | | | | | | | | |
| | PARTLY CLOUDY | | | | | | | | | | | | | | |
| | CLOUDY | | | | | | | | | | | | | | |
| **PR** | MEAN STATION PRESSURE(IN) | 26 | 26.31 | 26.30 | 26.25 | 26.27 | 26.26 | 26.28 | 26.34 | 26.34 | 26.34 | 26.33 | 26.30 | 26.31 | 26.30 |
| | MEAN SEA-LEVEL PRES. (IN) | 26 | 30.08 | 30.07 | 30.00 | 29.95 | 29.90 | 29.89 | 29.92 | 29.93 | 29.97 | 30.02 | 30.04 | 30.08 | 29.99 |
| **WINDS** | MEAN SPEED (MPH) | 26 | 13.3 | 11.9 | 10.8 | 10.5 | 9.9 | 9.5 | 8.9 | 8.9 | 9.4 | 10.5 | 11.8 | 12.8 | 10.7 |
| | PREVAIL.DIR(TENS OF DEGS) | 30 | 24 | 24 | 24 | 23 | 23 | 23 | 23 | 23 | 23 | 24 | 24 | 24 | 24 |
| | MAXIMUM 2-MINUTE: | | | | | | | | | | | | | | |
| | SPEED (MPH) | 14 | 49 | 48 | 53 | 60 | 45 | 47 | 61 | 59 | 51 | 56 | 46 | 49 | 61 |
| | DIR. (TENS OF DEGS) | | 32 | 26 | 30 | 32 | 17 | 28 | 23 | 32 | 27 | 31 | 30 | 31 | 23 |
| | YEAR OF OCCURRENCE | | 2009 | 1999 | 1999 | 2001 | 2002 | 1998 | 2002 | 2002 | 1997 | 1999 | 2008 | 2008 | JUL 2002 |
| | MAXIMUM 3-SECOND | | | | | | | | | | | | | | |
| | SPEED (MPH) | 14 | 66 | 61 | 63 | 69 | 62 | 62 | 85 | 70 | 62 | 70 | 58 | 61 | 85 |
| | DIR. (TENS OF DEGS) | | 32 | 26 | 31 | 31 | 17 | 33 | 25 | 32 | 19 | 32 | 29 | 31 | 25 |
| | YEAR OF OCCURRENCE | | 2009 | 1999 | 1999 | 2001 | 2002 | 2001 | 2007 | 2002 | 2005 | 1999 | 2008 | 2008 | JUL 2007 |
| **PRECIPITATION** | NORMAL (IN) | 30 | 0.81 | 0.58 | 1.12 | 1.74 | 2.48 | 1.89 | 1.28 | 0.85 | 1.34 | 1.26 | 0.75 | 0.67 | 14.77 |
| | MAXIMUM MONTHLY (IN) | 75 | 2.35 | 1.77 | 2.70 | 4.42 | 7.71 | 7.64 | 5.08 | 3.50 | 4.99 | 3.80 | 2.34 | 2.00 | 7.71 |
| | YEAR OF OCCURRENCE | | 1972 | 1978 | 1954 | 1955 | 1981 | 1944 | 1993 | 1965 | 1941 | 1971 | 1978 | 1973 | MAY 1981 |
| | MINIMUM MONTHLY (IN) | 75 | 0.04 | 0.02 | 0.11 | 0.06 | 0.22 | 0.24 | 0.04 | 0.01 | 0.06 | 0.01 | T | 0.05 | T |
| | YEAR OF OCCURRENCE | | 1941 | 1997 | 2004 | 1962 | 2001 | 1961 | 2003 | 2001 | 1964 | 1987 | 1954 | 1957 | NOV 1954 |
| | MAXIMUM IN 24 HOURS (IN) | 75 | 1.41 | 0.65 | 1.88 | 3.19 | 2.83 | 2.78 | 2.32 | 2.47 | 2.19 | 1.98 | 1.37 | 0.96 | 3.19 |
| | YEAR OF OCCURRENCE | | 1972 | 1986 | 2006 | 1978 | 1952 | 1997 | 1993 | 1965 | 1966 | 1974 | 1959 | 1978 | APR 1978 |
| | NORMAL NO. DAYS WITH: | | | | | | | | | | | | | | |
| | PRECIPITATION >= 0.01 | 30 | 8.0 | 6.9 | 9.3 | 10.7 | 12.0 | 11.0 | 8.1 | 6.6 | 7.2 | 6.8 | 6.3 | 7.2 | 100.1 |
| | PRECIPITATION >= 1.00 | 30 | 0.1 | 0.0 | 0.0 | * | 0.3 | 0.2 | 0.2 | 0.1 | 0.2 | 0.2 | 0.0 | 0.0 | 1.3 |
| **SNOWFALL** | NORMAL (IN) | 30 | 10.9 | 6.5 | 10.3 | 7.6 | 1.8 | 0.* | 0.0 | 0.* | 1.3 | 4.2 | 7.5 | 8.9 | 59.0 |
| | MAXIMUM MONTHLY (IN) | 75 | 27.7 | 22.4 | 27.6 | 42.3 | 15.6 | 2.0 | 0.4 | T | 9.3 | 23.1 | 25.2 | 28.8 | 42.3 |
| | YEAR OF OCCURRENCE | | 1963 | 1978 | 1935 | 1955 | 1981 | 1950 | 1993 | 1992 | 1984 | 1949 | 1978 | 1955 | APR 1955 |
| | MAXIMUM IN 24 HOURS (IN) | 71 | 16.6 | 9.0 | 10.5 | 23.7 | 15.3 | 2.0 | 0.4 | T | 7.5 | 11.2 | 15.3 | 13.7 | 23.7 |
| | YEAR OF OCCURRENCE' | | 1972 | 1944 | 1964 | 1955 | 1981 | 1950 | 1993 | 1992 | 1983 | 1980 | 1959 | 1978 | APR 1955 |
| | MAXIMUM SNOW DEPTH (IN) | 61 | 18 | 22 | 22 | 33 | 10 | 0 | 0 | 0 | 7 | 14 | 17 | 24 | 33 |
| | YEAR OF OCCURRENCE | | 1972 | 1978 | 1978 | 1955 | 1983 | | | | 1984 | 1949 | 1978 | 1978 | APR 1955 |
| | NORMAL NO. DAYS WITH: | | | | | | | | | | | | | | |
| | SNOWFALL >= 1.0 | 30 | 3.1 | 2.1 | 3.2 | 2.2 | 0.4 | 0.0 | 0.0 | 0.0 | 0.4 | 1.2 | 2.4 | 2.6 | 17.6 |

## PRECIPITATION (inches) 2009 BILLINGS (KBIL)

| YEAR | JAN | FEB | MAR | APR | MAY | JUN | JUL | AUG | SEP | OCT | NOV | DEC | ANNUAL |
|------|-----|-----|-----|-----|-----|-----|-----|-----|-----|-----|-----|-----|--------|
| 1980 | 1.11 | 0.78 | 1.53 | 0.46 | 4.47 | 1.64 | 0.39 | 1.17 | 0.77 | 2.45 | 0.42 | 0.33 | 15.52 |
| 1981 | 0.21 | 0.24 | 1.75 | 0.35 | 7.71 | 1.58 | 1.65 | 0.55 | 0.14 | 1.33 | 0.41 | 0.53 | 16.45 |
| 1982 | 0.71 | 0.34 | 1.81 | 1.53 | 2.63 | 5.03 | 1.91 | 0.45 | 1.22 | 1.15 | 0.42 | 1.07 | 18.27 |
| 1983 | 0.11 | 0.31 | 0.73 | 0.56 | 2.23 | 0.88 | 1.52 | 1.12 | 2.26 | 1.32 | 0.90 | 0.92 | 12.86 |
| 1984 | 0.65 | 0.93 | 0.84 | 1.38 | 1.12 | 1.65 | 0.29 | 0.58 | 1.32 | 0.37 | 0.95 | 0.84 | 10.92 |
| 1985 | 0.31 | 0.39 | 2.05 | 0.31 | 1.27 | 1.07 | 1.40 | 1.66 | 1.89 | 0.69 | 1.43 | 0.20 | 12.67 |
| 1986 | 0.37 | 1.72 | 1.04 | 2.72 | 1.92 | 2.15 | 1.01 | 0.43 | 1.24 | 0.33 | 1.21 | 0.12 | 14.26 |
| 1987 | 0.07 | 0.49 | 1.36 | 0.42 | 3.84 | 1.03 | 2.23 | 1.73 | 0.68 | 0.01 | 0.29 | 0.31 | 12.46 |
| 1988 | 0.45 | 0.71 | 0.66 | 1.82 | 1.84 | 0.43 | 0.04 | 0.12 | 2.12 | 1.01 | 0.60 | 0.56 | 10.36 |
| 1989 | 1.27 | 0.56 | 2.04 | 2.36 | 2.06 | 1.18 | 0.55 | 0.76 | 0.70 | 2.05 | 0.52 | 1.36 | 15.41 |
| 1990 | 0.29 | 0.50 | 1.70 | 2.06 | 2.81 | 0.66 | 0.37 | 0.93 | 0.08 | 1.05 | 0.33 | 0.49 | 11.27 |
| 1991 | 0.82 | 0.49 | 0.62 | 3.87 | 2.25 | 5.62 | 1.04 | 0.35 | 3.11 | 1.29 | 0.96 | 0.31 | 20.73 |
| 1992 | 0.09 | 0.12 | 0.65 | 2.35 | 1.70 | 2.69 | 1.67 | 0.34 | 0.62 | 0.42 | 0.30 | 0.51 | 11.46 |
| 1993 | 0.47 | 0.32 | 0.50 | 1.86 | 0.40 | 2.05 | 5.08 | 0.69 | 1.76 | 2.11 | 0.26 | 0.20 | 15.70 |
| 1994 | 0.34 | 0.36 | 0.62 | 1.89 | 1.53 | 1.97 | 2.02 | 0.11 | 1.33 | 2.06 | 1.17 | 0.25 | 13.65 |
| 1995 | 0.53 | 0.28 | 1.87 | 1.84 | 3.69 | 3.10 | 1.62 | 1.00 | 1.01 | 0.94 | 0.51 | 0.34 | 16.73 |
| 1996 | 0.82 | 0.62 | 1.02 | 1.06 | 3.85 | 0.85 | .57 | .07 | 1.80 | .58 | .86 | .23 | 12.33 |
| 1997 | 0.95 | 0.02 | 0.80 | 1.13 | 1.49 | 4.14 | 2.76 | 0.94 | 0.28 | 1.16 | 0.49 | 0.41 | 14.57 |
| 1998 | 1.03 | 0.23 | 1.32 | 1.29 | 1.26 | 3.63 | 2.29 | 1.94 | 1.50 | 1.36 | 0.76 | 0.41 | 17.02 |
| 1999 | 0.48 | 0.26 | 0.54 | 2.41 | 1.76 | 2.17 | 0.36 | 1.61 | 1.49 | 0.12 | 0.25 | 0.20 | 11.65 |
| 2000 | 0.55 | 1.30 | 0.78 | 1.32 | 1.64 | 1.30 | 0.51 | 0.06 | 1.85 | 0.54 | 0.49 | 0.34 | 10.68 |
| 2001 | 0.30 | 0.60 | 0.79 | 1.51 | 0.22 | 4.11 | 1.05 | 0.01 | 1.06 | 0.76 | 0.37 | 0.17 | 10.95 |
| 2002 | 0.37 | 0.23 | 0.25 | 2.09 | 1.09 | 1.41 | 0.55 | 0.67 | 1.23 | 1.12 | 0.04 | 0.25 | 9.30 |
| 2003 | 0.40 | 0.81 | 0.83 | 1.40 | 1.89 | 1.79 | T | 0.03 | 0.15 | 1.38 | 0.30 | 0.76 | 9.74 |
| 2004 | 0.25 | 0.78 | 0.11 | 1.51 | 0.81 | 1.95 | 2.27 | 0.23 | 1.19 | 1.67 | 0.06 | 0.25 | 11.08 |
| 2005 | 0.21 | 0.25 | 0.67 | 3.31 | 1.78 | 2.35 | 1.77 | 0.30 | 0.83 | 1.97 | 1.39 | 0.44 | 15.27 |
| 2006 | 0.05 | 0.11 | 2.67 | 1.50 | 1.14 | 0.49 | 0.40 | 0.42 | 2.73 | 2.22 | 0.86 | 0.38 | 12.97 |
| 2007 | 0.34 | 0.56 | 1.37 | 2.51 | 3.93 | 1.12 | 1.63 | 0.07 | 1.73 | 2.48 | 0.43 | 0.28 | 16.45 |
| 2008 | 0.35 | 0.07 | 0.42 | 0.20 | 4.83 | 0.31 | 0.77 | 1.18 | 2.44 | 1.82 | 0.27 | 1.23 | 13.89 |
| 2009 | 0.43 | 0.37 | 1.36 | 1.83 | 0.64 | 1.55 | 0.61 | 1.20 | 0.65 | 1.45 | 0.17 | 0.65 | 10.91 |
| POR= 62 YRS | 0.72 | 0.59 | 1.06 | 1.76 | 2.26 | 2.03 | 1.10 | 0.85 | 1.30 | 1.18 | 0.70 | 0.65 | 14.20 |

WBAN : 24033

## AVERAGE TEMPERATURE (°F) 2009 BILLINGS (KBIL)

| YEAR | JAN | FEB | MAR | APR | MAY | JUN | JUL | AUG | SEP | OCT | NOV | DEC | ANNUAL |
|------|-----|-----|-----|-----|-----|-----|-----|-----|-----|-----|-----|-----|--------|
| 1980 | 17.0 | 29.2 | 34.4 | 54.6 | 61.0 | 66.5 | 75.7 | 67.4 | 62.0 | 50.7 | 40.6 | 30.3 | 49.1 |
| 1981 | 36.0 | 32.5 | 41.7 | 50.5 | 55.6 | 64.6 | 73.8 | 73.2 | 63.3 | 46.0 | 40.5 | 28.5 | 50.5 |
| 1982 | 13.1 | 28.0 | 32.9 | 42.6 | 52.4 | 61.4 | 70.8 | 75.8 | 60.4 | 50.2 | 33.8 | 28.7 | 45.8 |
| 1983 | 35.3 | 38.4 | 38.6 | 42.9 | 53.2 | 63.9 | 72.0 | 77.5 | 59.3 | 53.5 | 37.8 | 8.7 | 48.4 |
| 1984 | 29.3 | 38.3 | 38.1 | 45.0 | 55.5 | 64.5 | 74.5 | 74.7 | 54.0 | 42.1 | 37.6 | 19.6 | 47.8 |
| 1985 | 20.3 | 24.1 | 34.1 | 50.5 | 60.3 | 64.0 | 74.8 | 65.6 | 53.0 | 49.6 | 15.2 | 25.9 | 44.8 |
| 1986 | 37.2 | 24.2 | 46.2 | 44.5 | 54.4 | 69.7 | 69.3 | 70.5 | 53.7 | 50.0 | 32.0 | 32.3 | 48.7 |
| 1987 | 30.3 | 35.0 | 37.6 | 54.1 | 60.2 | 67.9 | 70.4 | 66.2 | 61.2 | 49.6 | 40.3 | 29.8 | 50.2 |
| 1988 | 23.6 | 28.7 | 39.8 | 48.3 | 59.7 | 75.9 | 76.2 | 72.2 | 58.7 | 52.0 | 37.0 | 29.3 | 50.1 |
| 1989 | 27.2 | 13.4 | 29.9 | 45.0 | 55.0 | 63.6 | 75.3 | 69.3 | 60.5 | 47.2 | 39.7 | 25.0 | 45.9 |
| 1990 | 31.2 | 29.0 | 38.1 | 46.1 | 53.6 | 65.1 | 72.1 | 72.6 | 66.5 | 48.7 | 40.6 | 19.2 | 48.6 |
| 1991 | 20.8 | 41.0 | 38.5 | 43.7 | 54.8 | 64.3 | 72.6 | 75.5 | 60.6 | 45.6 | 32.7 | 34.1 | 48.7 |
| 1992 | 35.2 | 39.4 | 44.0 | 48.9 | 58.4 | 65.9 | 65.8 | 67.2 | 60.7 | 49.8 | 36.9 | 19.7 | 49.3 |
| 1993 | 17.9 | 21.7 | 40.6 | 47.0 | 59.8 | 62.2 | 62.7 | 65.6 | 57.0 | 47.5 | 31.4 | 33.7 | 45.6 |
| 1994 | 28.2 | 22.5 | 42.0 | 47.1 | 59.9 | 66.4 | 72.1 | 73.4 | 64.5 | 48.4 | 34.6 | 31.0 | 49.2 |
| 1995 | 31.9 | 34.9 | 34.6 | 42.7 | 51.2 | 61.9 | 69.3 | 70.7 | 58.5 | 47.2 | 38.4 | 28.3 | 47.5 |
| 1996 | 16.5 | 29.3 | 28.1 | 47.3 | 50.4 | 67.0 | 72.5 | 73.7 | 57.9 | 46.2 | 24.6 | 19.4 | 44.4 |
| 1997 | 18.8 | 32.3 | 37.6 | 38.8 | 56.0 | 65.5 | 68.8 | 69.4 | 64.1 | 49.3 | 35.4 | 30.0 | 47.2 |
| 1998 | 24.5 | 35.9 | 32.8 | 47.8 | 57.4 | 58.2 | 75.3 | 72.7 | 66.6 | 48.8 | 38.1 | 26.2 | 48.7 |
| 1999 | 30.1 | 38.3 | 39.6 | 43.4 | 53.3 | 62.9 | 71.3 | 72.4 | 56.3 | 50.1 | 45.3 | 35.3 | 49.9 |
| 2000 | 27.6 | 31.3 | 40.9 | 47.2 | 56.3 | 64.4 | 75.7 | 73.6 | 59.8 | 47.4 | 26.8 | 20.1 | 47.6 |
| 2001 | 30.3 | 20.6 | 38.6 | 46.3 | 58.9 | 63.5 | 74.2 | 75.2 | 63.6 | 47.8 | 41.2 | 27.8 | 49.0 |
| 2002 | 28.0 | 32.8 | 24.7 | 40.8 | 52.2 | 65.4 | 76.8 | 66.7 | 61.4 | 41.2 | 39.3 | 31.6 | 46.7 |
| 2003 | 31.1 | 25.2 | 34.0 | 49.5 | 55.2 | 63.7 | 78.4 | 77.1 | 60.3 | 53.5 | 30.6 | 31.6 | 49.2 |
| 2004 | 23.6 | 32.3 | 44.4 | 49.4 | 53.5 | 61.6 | 72.2 | 68.9 | 59.8 | 48.8 | 39.4 | 32.9 | 48.9 |
| 2005 | 22.5 | 34.4 | 40.6 | 46.3 | 52.8 | 63.3 | 74.2 | 69.7 | 61.8 | 49.7 | 39.3 | 26.3 | 48.4 |
| 2006 | 38.0 | 30.0 | 34.7 | 49.8 | 58.2 | 68.6 | 78.0 | 71.4 | 59.6 | 44.5 | 35.7 | 31.6 | 50.0 |
| 2007 | 25.4 | 26.1 | 44.6 | 44.5 | 56.5 | 66.0 | 79.1 | 72.0 | 61.3 | 50.2 | 36.5 | 27.9 | 49.2 |
| 2008 | 25.0 | 32.3 | 37.7 | 43.8 | 54.8 | 63.6 | 73.9 | 72.6 | 58.5 | 48.4 | 42.3 | 19.2 | 47.7 |
| 2009 | 29.7 | 33.7 | 33.8 | 45.4 | 57.3 | 61.7 | 71.2 | 70.2 | 66.8 | 41.3 | 41.8 | 16.4 | 47.4 |
| POR= 62 YRS | 23.7 | 29.3 | 35.4 | 45.5 | 55.3 | 64.2 | 72.5 | 71.0 | 59.9 | 48.7 | 35.7 | 27.1 | 47.4 |

## HEATING DEGREE DAYS (base 65°F) 2009  BILLINGS (KBIL)

| YEAR | JUL | AUG | SEP | OCT | NOV | DEC | JAN | FEB | MAR | APR | MAY | JUN | TOTAL |
|------|-----|-----|-----|-----|-----|-----|-----|-----|-----|-----|-----|-----|-------|
| 1980-81 | 0 | 25 | 127 | 462 | 724 | 1073 | 891 | 905 | 717 | 427 | 292 | 79 | 5722 |
| 1981-82 | 12 | 6 | 124 | 583 | 729 | 1124 | 1603 | 1028 | 987 | 666 | 386 | 142 | 7390 |
| 1982-83 | 12 | 0 | 215 | 453 | 926 | 1118 | 911 | 741 | 810 | 656 | 381 | 82 | 6305 |
| 1983-84 | 29 | 5 | 234 | 359 | 811 | 1741 | 1101 | 769 | 828 | 592 | 316 | 97 | 6882 |
| 1984-85 | 12 | 3 | 351 | 701 | 812 | 1404 | 1381 | 1140 | 950 | 428 | 184 | 103 | 7469 |
| 1985-86 | 13 | 65 | 358 | 471 | 1492 | 1207 | 853 | 1136 | 579 | 610 | 347 | 18 | 7149 |
| 1986-87 | 8 | 2 | 331 | 457 | 982 | 1005 | 1070 | 829 | 841 | 337 | 183 | 44 | 6089 |
| 1987-88 | 39 | 56 | 134 | 473 | 734 | 1083 | 1276 | 1047 | 775 | 492 | 200 | 14 | 6323 |
| 1988-89 | 0 | 6 | 221 | 395 | 833 | 1099 | 1168 | 1441 | 1084 | 595 | 308 | 97 | 7247 |
| 1989-90 | 0 | 25 | 172 | 546 | 752 | 1235 | 1042 | 1002 | 829 | 560 | 346 | 108 | 6617 |
| 1990-91 | 9 | 0 | 73 | 500 | 725 | 1413 | 1365 | 665 | 814 | 630 | 311 | 50 | 6555 |
| 1991-92 | 3 | 0 | 171 | 612 | 963 | 951 | 918 | 737 | 641 | 481 | 232 | 73 | 5782 |
| 1992-93 | 53 | 95 | 166 | 475 | 837 | 1398 | 1456 | 1210 | 751 | 531 | 177 | 134 | 7283 |
| 1993-94 | 98 | 60 | 250 | 534 | 1004 | 963 | 1135 | 1186 | 707 | 534 | 161 | 77 | 6709 |
| 1994-95 | 13 | 20 | 67 | 505 | 908 | 1045 | 1020 | 838 | 936 | 661 | 425 | 141 | 6579 |
| 1995-96 | 9 | 12 | 227 | 551 | 790 | 1131 | 1500 | 1032 | 1139 | 525 | 446 | 44 | 7406 |
| 1996-97 | 0 | 3 | 224 | 578 | 1205 | 1406 | 1425 | 910 | 840 | 780 | 285 | 38 | 7694 |
| 1997-98 | 44 | 39 | 90 | 493 | 882 | 1077 | 1249 | 806 | 992 | 508 | 235 | 211 | 6626 |
| 1998-99 | 0 | 0 | 98 | 493 | 803 | 1195 | 1077 | 741 | 780 | 642 | 369 | 100 | 6298 |
| 1999-00 | 26 | 4 | 271 | 454 | 585 | 916 | 1152 | 968 | 742 | 524 | 272 | 86 | 6000 |
| 2000-01 | 0 | 14 | 212 | 540 | 1141 | 1386 | 1072 | 1235 | 812 | 555 | 209 | 126 | 7302 |
| 2001-02 | 0 | 0 | 119 | 531 | 705 | 1143 | 1140 | 895 | 1237 | 718 | 400 | 99 | 6987 |
| 2002-03 | 0 | 24 | 169 | 732 | 764 | 1030 | 1046 | 1107 | 954 | 458 | 336 | 108 | 6728 |
| 2003-04 | 0 | 10 | 191 | 372 | 1026 | 1029 | 1276 | 940 | 627 | 464 | 351 | 135 | 6421 |
| 2004-05 | 7 | 15 | 175 | 501 | 763 | 989 | 1313 | 850 | 750 | 557 | 377 | 122 | 6419 |
| 2005-06 | 10 | 45 | 150 | 470 | 764 | 1193 | 828 | 970 | 931 | 450 | 245 | 11 | 6067 |
| 2006-07 | 0 | 11 | 192 | 632 | 873 | 1029 | 1223 | 1082 | 622 | 611 | 264 | 60 | 6599 |
| 2007-08 | 0 | 9 | 174 | 452 | 847 | 1145 | 1236 | 943 | 838 | 630 | 328 | 114 | 6716 |
| 2008-09 | 1 | 9 | 208 | 513 | 677 | 1411 | 1086 | 867 | 957 | 586 | 265 | 158 | 6738 |
| 2009- | 4 | 10 | 72 | 731 | 693 | 1499 | | | | | | | |

WBAN : 24033

## COOLING DEGREE DAYS (base 65°F) 2009  BILLINGS (KBIL)

| YEAR | JAN | FEB | MAR | APR | MAY | JUN | JUL | AUG | SEP | OCT | NOV | DEC | TOTAL |
|------|-----|-----|-----|-----|-----|-----|-----|-----|-----|-----|-----|-----|-------|
| 1980 | 0 | 0 | 0 | 20 | 40 | 99 | 339 | 105 | 46 | 26 | 0 | 0 | 675 |
| 1981 | 0 | 0 | 0 | 0 | 6 | 74 | 291 | 268 | 76 | 3 | 0 | 0 | 718 |
| 1982 | 0 | 0 | 0 | 0 | 0 | 41 | 198 | 342 | 82 | 0 | 0 | 0 | 663 |
| 1983 | 0 | 0 | 0 | 0 | 24 | 54 | 256 | 400 | 69 | 7 | 2 | 0 | 812 |
| 1984 | 0 | 0 | 0 | 0 | 30 | 91 | 315 | 310 | 29 | 0 | 0 | 0 | 775 |
| 1985 | 0 | 0 | 0 | 0 | 42 | 83 | 325 | 92 | 6 | 0 | 0 | 0 | 548 |
| 1986 | 0 | 0 | 1 | 2 | 25 | 163 | 152 | 177 | 1 | 0 | 0 | 0 | 521 |
| 1987 | 0 | 0 | 0 | 17 | 41 | 134 | 215 | 100 | 30 | 4 | 0 | 0 | 541 |
| 1988 | 0 | 0 | 0 | 0 | 41 | 351 | 355 | 234 | 37 | 0 | 0 | 0 | 1018 |
| 1989 | 0 | 0 | 0 | 2 | 2 | 64 | 327 | 164 | 45 | 0 | 0 | 0 | 604 |
| 1990 | 0 | 0 | 0 | 0 | 0 | 117 | 239 | 245 | 123 | 0 | 0 | 0 | 724 |
| 1991 | 0 | 0 | 0 | 1 | 3 | 35 | 244 | 332 | 45 | 16 | 0 | 0 | 676 |
| 1992 | 0 | 0 | 0 | 4 | 36 | 106 | 84 | 166 | 42 | 10 | 0 | 0 | 448 |
| 1993 | 0 | 0 | 0 | 0 | 18 | 57 | 34 | 85 | 19 | 2 | 0 | 0 | 215 |
| 1994 | 0 | 0 | 0 | 3 | 9 | 126 | 241 | 289 | 58 | 0 | 0 | 0 | 726 |
| 1995 | 0 | 0 | 0 | 0 | 3 | 55 | 151 | 195 | 39 | 5 | 0 | 0 | 448 |
| 1996 | 0 | 0 | 0 | 0 | 0 | 108 | 241 | 280 | 15 | 1 | 0 | 0 | 645 |
| 1997 | 0 | 0 | 0 | 0 | 14 | 61 | 170 | 182 | 71 | 13 | 0 | 0 | 511 |
| 1998 | 0 | 0 | 0 | 0 | 6 | 12 | 326 | 243 | 152 | 0 | 0 | 0 | 739 |
| 1999 | 0 | 0 | 0 | 0 | 10 | 44 | 228 | 241 | 17 | 0 | 1 | 0 | 541 |
| 2000 | 0 | 0 | 0 | 0 | 11 | 73 | 337 | 287 | 63 | 0 | 0 | 0 | 771 |
| 2001 | 0 | 0 | 0 | 0 | 32 | 87 | 293 | 326 | 84 | 4 | 0 | 0 | 826 |
| 2002 | 0 | 0 | 0 | 0 | 12 | 120 | 373 | 84 | 69 | 0 | 0 | 0 | 658 |
| 2003 | 0 | 0 | 0 | 0 | 40 | 76 | 424 | 393 | 55 | 26 | 0 | 0 | 1014 |
| 2004 | 0 | 0 | 0 | 2 | 0 | 39 | 238 | 144 | 26 | 4 | 0 | 0 | 453 |
| 2005 | 0 | 0 | 0 | 0 | 3 | 76 | 302 | 198 | 61 | 6 | 0 | 0 | 646 |
| 2006 | 0 | 0 | 0 | 0 | 38 | 125 | 408 | 217 | 39 | 1 | 0 | 0 | 828 |
| 2007 | 0 | 0 | 0 | 2 | 9 | 98 | 445 | 231 | 69 | 0 | 0 | 0 | 854 |
| 2008 | 0 | 0 | 0 | 1 | 20 | 79 | 284 | 255 | 19 | 2 | 0 | 0 | 660 |
| 2009 | 0 | 0 | 0 | 4 | 35 | 68 | 202 | 182 | 132 | 0 | 0 | 0 | 623 |

## SNOWFALL (inches) 2009 BILLINGS (KBIL)

| YEAR | JUL | AUG | SEP | OCT | NOV | DEC | JAN | FEB | MAR | APR | MAY | JUN | TOTAL |
|------|-----|-----|-----|-----|-----|-----|-----|-----|-----|-----|-----|-----|-------|
| 1980-81 | 0.0 | 0.0 | T | 17.8 | 4.5 | 5.5 | 2.1 | 3.4 | 16.3 | 0.7 | 15.6 | 0.0 | 65.9 |
| 1981-82 | 0.0 | 0.0 | 0.0 | 5.7 | 3.2 | 5.4 | 9.8 | 5.3 | 18.2 | 13.5 | 2.0 | 0.0 | 63.1 |
| 1982-83 | 0.0 | 0.0 | 5.7 | 1.5 | 5.6 | 11.2 | 0.1 | 1.0 | 6.4 | 5.8 | 11.9 | 0.0 | 49.2 |
| 1983-84 | 0.0 | 0.0 | 7.5 | T | 5.5 | 10.9 | 5.3 | 6.8 | 4.5 | 9.0 | T | 0.0 | 49.5 |
| 1984-85 | 0.0 | 0.0 | 9.3 | 6.5 | 9.9 | 16.1 | 4.8 | 3.8 | 21.7 | 1.9 | 0.0 | 0.0 | 74.0 |
| 1985-86 | 0.0 | 0.0 | 3.6 | 6.0 | 17.1 | 2.0 | 3.3 | 13.8 | 6.4 | 12.9 | 8.3 | 0.0 | 73.4 |
| 1986-87 | 0.0 | 0.0 | 0.0 | 0.0 | 12.3 | 1.9 | 0.6 | 6.0 | 13.3 | 0.3 | 0.4 | 0.0 | 34.8 |
| 1987-88 | 0.0 | 0.0 | 0.0 | 0.3 | 2.6 | 3.6 | 7.4 | 8.8 | 1.9 | 10.7 | 2.0 | 0.0 | 37.3 |
| 1988-89 | 0.0 | 0.0 | T | 2.0 | 5.6 | 6.2 | 18.5 | 6.8 | 25.1 | 11.8 | T | 0.0 | 76.0 |
| 1989-90 | 0.0 | 0.0 | T | 7.2 | 5.8 | 17.1 | 3.3 | 8.9 | 13.0 | 11.2 | T | 0.0 | 66.5 |
| 1990-91 | T | T | 0.0 | 3.5 | 1.5 | 6.2 | 11.4 | 1.0 | 3.7 | 30.0 | 3.6 | T | 60.9 |
| 1991-92 | T | 0.0 | 0.0 | 15.6 | 7.6 | 3.5 | 0.9 | 1.1 | 3.1 | 3.4 | 0.0 | 0.0 | 35.2 |
| 1992-93 | 0.0 | T | 0.0 | 4.0 | 1.7 | 10.6 | 11.4 | 5.8 | 2.7 | 6.9 | 0.0 | T | 43.1 |
| 1993-94 | 0.4 | 0.0 | T | 7.8 | 5.6 | 3.0 | 8.7 | 6.8 | 7.9 | 10.1 | T | T | 50.3 |
| 1994-95 | T | 0.0 | 0.0 | T | 13.9 | 4.6 | 2.4 | 6.3 | 6.8 | 8.8 | 3.9 | T | 46.7 |
| 1995-96 |  | 0.0 | 1.6 | 4.9 | 3.3 | 3.2 | 13.4 | 10.2 | 18.8 | 7.1 | 0.9 | T |  |
| 1996-97 | T | 0.0 | T | 9.7 | 15.7 | 20.6 | 18.5 | 0.8 | 10.3 | 23.1 | T | T | 98.7 |
| 1997-98 | T | 0.0 | 0.0 | 0.9 | 5.3 | 6.4 | 18.6 | 1.7 | 20.5 | T | T | 0.0 | 53.4 |
| 1998-99 | T | 0.0 | 0.0 | 0.0 | 5.1 | 6.5 | 18.2 | 2.4 | 8.5 | 8.7 | T | 0.0 | 49.4 |
| 1999-00 | 0.0 | 0.0 | T | 0.9 | 0.0 | 3.7 | 10.6 | 13.7 | 3.1 | 3.1 | 1.6 | T | 36.7 |
| 2000-01 | 0.0 | 0.0 | 5.5 | T | 9.6 | 9.1 | 5.4 | 14.8 | 8.1 | 7.2 | T | T | 59.7 |
| 2001-02 | T | 0.0 | 0.0 | 1.0 | 5.5 | 3.5 | 9.4 | 6.9 | 12.5 | 17.3 | 3.2 | 0.0 | 59.3 |
| 2002-03 | T | 0.0 | T | 5.9 | 0.1 | 4.9 | 13.5 | 11.4 | 15.8 | 0.2 | T | T | 51.8 |
| 2003-04 | 0.0 | 0.0 | 0.0 | 3.6 | 5.6 | 11.2 | 7.5 | 4.1 | 1.1 | 3.6 | 1.1 | 0.0 | 37.8 |
| 2004-05 | T | 0.0 | 0.0 | T | 2.1 | 3.6 | 9.0 | 6.0 | 10.7 | 20.9 | 3.1 | T | 55.4 |
| 2005-06 | 0.0 | 0.0 | 0.0 | 10.8 | 4.6 | 7.7 | 0.1 | 1.4 | 8.5 | 2.0 | T | T | 35.1 |
| 2006-07 | 0.0 | 0.0 | 0.0 | 6.1 | 8.8 | 5.2 | 8.2 | 15.0 | 14.0 | 7.9 | T | 0.0 | 65.2 |
| 2007-08 | 0.0 | 0.0 | T | T | 6.8 | 4.3 | 9.7 | 2.4 | 6.3 | 1.8 | 1.8 | T | 33.1 |
| 2008-09 | T | 0.0 | T | 12.9 | 0.1 | 21.0 | 9.1 | 10.1 | 16.9 | 5.5 | T | T | 75.6 |
| 2009- | 0.0 | 0.0 | 0.0 | 6.6 | 1.4 | 15.3 |  |  |  |  |  |  |  |
| POR= 61 YRS | T | T | 1.1 | 4.2 | 6.5 | 8.8 | 10.0 | 7.3 | 10.4 | 8.6 | 1.6 | T | 58.5 |

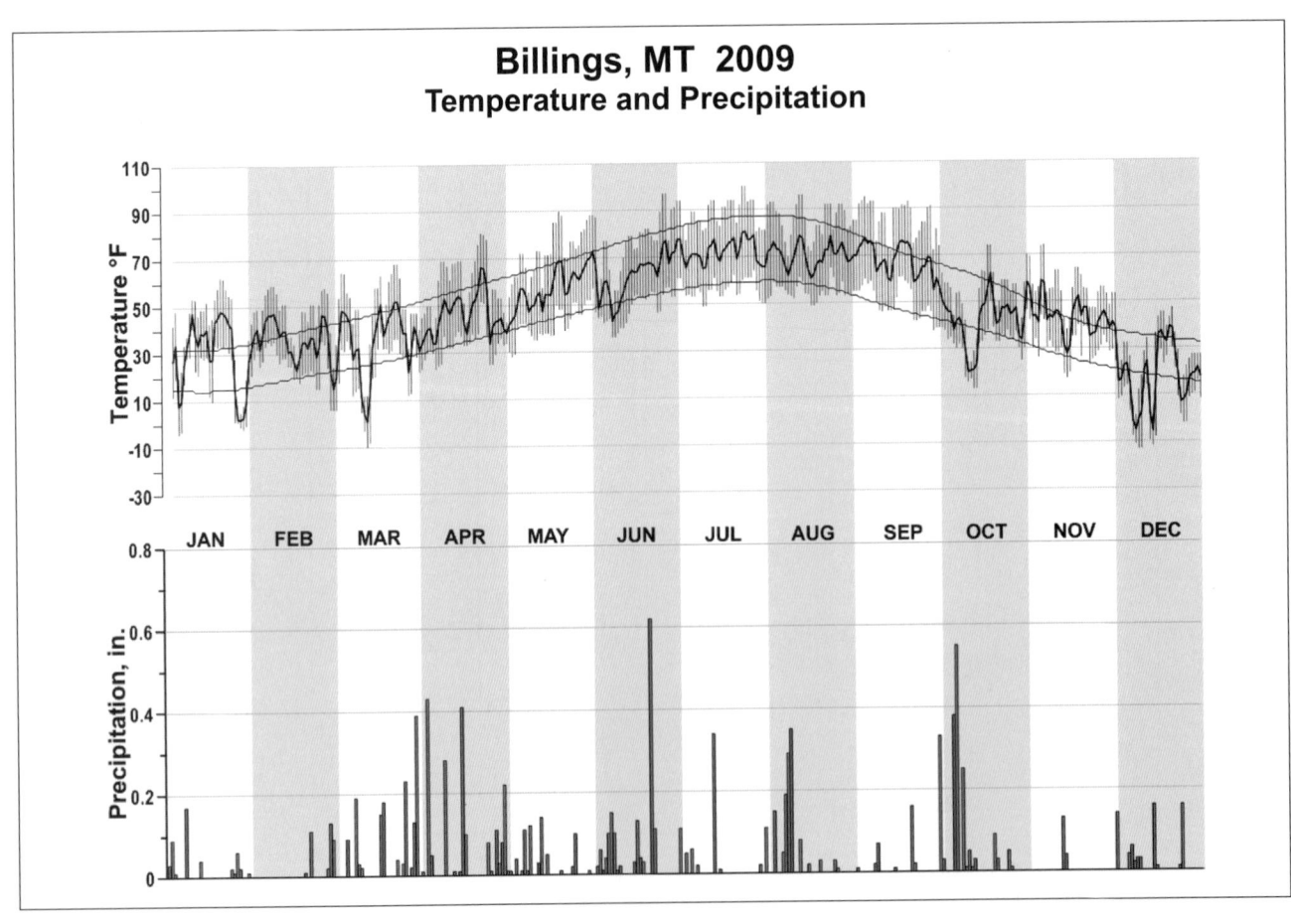

# Billings, MT 2009
## Temperature and Precipitation

# 2009
# GREAT FALLS
# MONTANA (KGTF)

The city of Great Falls is located along the main stem of the Missouri River at its confluence with the Sun River. The Weather Service Office is located at the Municipal Airport on a plateau between the Sun and Missouri Rivers. This plateau is about 200 feet higher than most of the immediate valley area, and the airport is about two miles southwest of the Sun and Missouri River Junction. Except to the north and northeast, the valley is encircled by mountain ranges, which lie about 30 miles away from east to south, 40 miles to the southwest, and 60 to 100 miles distant from west to northwest. Topography plays an important part in the climate of Great Falls. The Continental Divide to the west, and Big and Little Belt Ranges to the south, are primary factors in producing the frequent wintertime chinook winds observed in this part of Montana. The combination of valleys and plateaus in the immediate area, contributes to marked temperature differences between the airport and the city proper, either on calm, clear mornings, or when chinook winds reach the airport before they are felt at the lower elevations in town.

Summertime in the area generally is quite pleasant, with cool nights, moderately warm and sunny days, and very little hot, humid weather. Most of the summer rainfall occurs in showers or thunderstorms, and steady rains may occur during late spring or early summer. At the airport, freezing temperatures do not occur in July or August and very rarely in June. Frost occurs frequently in April and October, but more often in the valleys than on the surrounding hills or plateaus. However, frost may occur on rare occasions in nearby low lying areas at any time of the year.

Winters are not as cold as is usually expected of a continental location at this latitude, largely as a result of the chinook winds for which this area is noted. While sub-zero weather is experienced normally several times during a winter, the coldest weather seldom lasts more than a few days at a time, and is usually terminated by southwest chinook winds which can produce sharp temperature rises of 40 degrees or more in 24 hours.

As a result of recurring chinooks throughout the winter season, snow seldom lies on the ground for more than a few days. In fact, the ground usually is bare, or nearly bare, of snow most of the winter, except in the surrounding mountains and higher foothills. On the other hand, invasions of cold air from the polar regions occur a few times each winter, and sharp temperature falls from above freezing to below zero within 24 hours are observed occasionally.

Precipitation generally falls as snow during late fall, winter, and early spring, although rain can occur in any month. Late spring, summer, and early fall precipitation is almost always rain, but some hail is observed occasionally during summer thunderstorms.

Although average annual precipitation at Great Falls would normally classify the area as semi-arid, it is important to note that about 70 percent of the annual total falls normally during the April to September growing season. The combination of ideal temperatures during the peak of the growing season, long hours of summer sunshine, and adequate precipitation during the six critical months, makes the climate very favorable for dryland farming. Heavy fog occurs about one day per month, but each case lasts only a small part of the day. Although the average windspeed is relatively high, strong winds over 70 mph are seldom observed. Visibility normally is excellent.

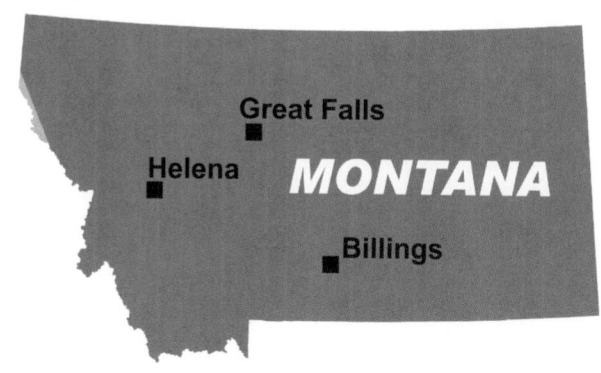

# NORMALS, MEANS, AND EXTREMES
## GREAT FALLS (KGTF)

LATITUDE: 47° 28'N  LONGITUDE: -111° 22'W  ELEVATION (FT): GRND: 3664  BARO: 3673  TIME ZONE: MOUNTAIN (UTC -7)  WBAN: 24143

| ELEMENT | POR | JAN | FEB | MAR | APR | MAY | JUN | JUL | AUG | SEP | OCT | NOV | DEC | YEAR |
|---|---|---|---|---|---|---|---|---|---|---|---|---|---|---|
| **TEMPERATURE °F** | | | | | | | | | | | | | | |
| NORMAL DAILY MAXIMUM | 30 | 32.1 | 37.7 | 45.3 | 55.6 | 64.7 | 73.9 | 82.0 | 81.2 | 69.6 | 58.0 | 42.1 | 34.2 | 56.4 |
| MEAN DAILY MAXIMUM | 62 | 31.5 | 37.4 | 43.3 | 54.8 | 65.0 | 73.4 | 83.5 | 81.8 | 70.2 | 58.3 | 43.7 | 35.0 | 56.5 |
| HIGHEST DAILY MAXIMUM | 72 | 67 | 70 | 78 | 89 | 93 | 101 | 105 | 106 | 98 | 91 | 76 | 69 | 106 |
| YEAR OF OCCURRENCE | | 1992 | 1992 | 2004 | 1980 | 1980 | 1990 | 1973 | 1969 | 1980 | 1992 | 1999 | 1939 | AUG 1969 |
| MEAN OF EXTREME MAXS. | 62 | 55.1 | 58.0 | 65.3 | 75.5 | 83.9 | 90.8 | 97.2 | 96.4 | 89.4 | 79.2 | 64.5 | 55.1 | 75.9 |
| NORMAL DAILY MINIMUM | 30 | 11.3 | 15.1 | 21.5 | 29.7 | 38.3 | 46.0 | 50.4 | 49.9 | 41.2 | 33.0 | 22.5 | 14.4 | 31.1 |
| MEAN DAILY MINIMUM | 62 | 12.5 | 17.4 | 22.1 | 31.6 | 40.6 | 48.2 | 53.7 | 52.4 | 43.9 | 35.2 | 25.1 | 16.8 | 33.3 |
| LOWEST DAILY MINIMUM | 72 | -37 | -35 | -29 | -8 | 12 | 31 | 36 | 30 | 16 | -11 | -25 | -43 | -43 |
| YEAR OF OCCURRENCE | | 1969 | 1996 | 1951 | 2008 | 2009 | 2009 | 1999 | 1992 | 2000 | 1991 | 1985 | 1968 | DEC 1968 |
| MEAN OF EXTREME MINS. | 62 | -16.0 | -8.4 | -3.3 | 15.8 | 28.1 | 37.6 | 44.0 | 41.8 | 30.6 | 16.7 | -0.1 | -12.2 | 14.6 |
| NORMAL DRY BULB | 30 | 21.7 | 26.4 | 33.4 | 42.6 | 51.5 | 60.0 | 66.2 | 65.6 | 55.4 | 45.5 | 32.3 | 24.3 | 43.7 |
| MEAN DRY BULB | 62 | 22.0 | 27.4 | 32.7 | 43.2 | 52.8 | 60.9 | 68.6 | 67.2 | 57.1 | 46.7 | 34.4 | 25.9 | 44.9 |
| MEAN WET BULB | 26 | 20.5 | 22.3 | 27.5 | 35.8 | 43.0 | 49.9 | 53.3 | 51.7 | 45.0 | 36.6 | 27.2 | 20.9 | 36.1 |
| MEAN DEW POINT | 26 | 15.7 | 15.8 | 21.9 | 28.5 | 36.7 | 44.4 | 46.3 | 44.5 | 38.8 | 30.1 | 21.8 | 15.1 | 30.0 |
| NORMAL NO. DAYS WITH: | | | | | | | | | | | | | | |
| MAXIMUM >= 90 | 30 | 0.0 | 0.0 | 0.0 | 0.0 | 0.1 | 2.2 | 7.1 | 7.3 | 1.1 | * | 0.0 | 0.0 | 17.8 |
| MAXIMUM <= 32 | 30 | 12.3 | 8.4 | 5.0 | 1.3 | 0.0 | 0.0 | 0.0 | 0.0 | * | 1.1 | 6.0 | 10.6 | 44.7 |
| MINIMUM <= 32 | 30 | 27.0 | 24.0 | 25.4 | 16.3 | 4.1 | 0.1 | 0.0 | 0.1 | 2.8 | 12.7 | 22.6 | 26.6 | 161.7 |
| MINIMUM <= 0 | 30 | 8.6 | 5.3 | 1.8 | 0.1 | 0.0 | 0.0 | 0.0 | 0.0 | 0.0 | 0.2 | 2.2 | 6.1 | 24.3 |
| **H/C** | | | | | | | | | | | | | | |
| NORMAL HEATING DEG. DAYS | 30 | 1327 | 1065 | 964 | 657 | 410 | 183 | 52 | 73 | 291 | 592 | 967 | 1247 | 7828 |
| NORMAL COOLING DEG. DAYS | 30 | 0 | 0 | 0 | 1 | 7 | 47 | 105 | 107 | 19 | 2 | 0 | 0 | 288 |
| **RH** | | | | | | | | | | | | | | |
| NORMAL (PERCENT) | 30 | 65 | 63 | 62 | 55 | 56 | 56 | 49 | 48 | 54 | 56 | 62 | 62 | 57 |
| HOUR 05 LST | 30 | 68 | 68 | 70 | 69 | 72 | 73 | 69 | 68 | 69 | 67 | 67 | 66 | 69 |
| HOUR 11 LST | 30 | 62 | 58 | 55 | 47 | 47 | 45 | 39 | 40 | 45 | 48 | 56 | 59 | 50 |
| HOUR 17 LST | 30 | 61 | 54 | 49 | 41 | 41 | 41 | 32 | 31 | 37 | 45 | 57 | 60 | 46 |
| HOUR 23 LST | 30 | 66 | 67 | 67 | 61 | 63 | 63 | 54 | 54 | 59 | 62 | 65 | 65 | 62 |
| **S** | | | | | | | | | | | | | | |
| PERCENT POSSIBLE SUNSHINE | 46 | 49 | 56 | 66 | 62 | 62 | 65 | 79 | 76 | 67 | 61 | 46 | 44 | 61 |
| **W/O** MEAN NO. DAYS WITH: | | | | | | | | | | | | | | |
| HEAVY FOG(VISBY <= 1/4 MI) | 46 | 1.4 | 1.6 | 2.4 | 1.9 | 0.6 | 0.5 | 0.1 | 0.3 | 0.6 | 1.4 | 2.0 | 1.2 | 14.0 |
| THUNDERSTORMS | 62 | 0.0 | 0.1 | 0.1 | 0.6 | 2.9 | 6.3 | 6.3 | 5.5 | 1.4 | 0.2 | 0.0 | 0.1 | 23.5 |
| **CLOUDNESS** MEAN: | | | | | | | | | | | | | | |
| SUNRISE-SUNSET (OKTAS) | | | | 9.6 | | | 8.0 | 4.0 | | | | | | |
| MIDNIGHT-MIDNIGHT (OKTAS) | | | | | | | 8.0 | 4.0 | | | | | | |
| MEAN NO. DAYS WITH: | | | | | | | | | | | | | | |
| CLEAR | | | 1.0 | | 2.0 | | 1.0 | 8.0 | | 8.0 | 6.0 | 2.0 | 5.0 | |
| PARTLY CLOUDY | 1 | 1.0 | 2.0 | 1.0 | | | 6.0 | 6.0 | | | 3.0 | 8.0 | 1.0 | |
| CLOUDY | 1 | 2.5 | 5.0 | 12.0 | | | 20.0 | 5.0 | | 1.0 | 4.0 | 7.0 | | |
| **PR** | | | | | | | | | | | | | | |
| MEAN STATION PRESSURE(IN) | 26 | 26.20 | 26.20 | 26.18 | 26.18 | 26.18 | 26.20 | 26.25 | 26.26 | 26.25 | 26.23 | 26.19 | 26.20 | 26.21 |
| MEAN SEA-LEVEL PRES. (IN) | 26 | 30.09 | 30.09 | 30.03 | 29.98 | 29.93 | 29.91 | 29.93 | 29.95 | 29.99 | 30.03 | 30.04 | 30.09 | 30.01 |
| **WINDS** | | | | | | | | | | | | | | |
| MEAN SPEED (MPH) | 26 | 14.2 | 12.4 | 11.8 | 11.4 | 10.8 | 10.0 | 9.2 | 9.2 | 10.1 | 12.0 | 13.8 | 14.2 | 11.6 |
| PREVAIL.DIR(TENS OF DEGS) | 33 | 23 | 23 | 24 | 24 | 24 | 24 | 24 | 24 | 24 | 23 | 23 | 23 | 23 |
| MAXIMUM 2-MINUTE: | | | | | | | | | | | | | | |
| SPEED (MPH) | 15 | 51 | 54 | 49 | 52 | 48 | 52 | 46 | 51 | 49 | 54 | 58 | 55 | 58 |
| DIR. (TENS OF DEGS) | | 23 | 24 | 23 | 26 | 24 | 25 | 25 | 22 | 24 | 25 | 23 | 25 | 23 |
| YEAR OF OCCURRENCE | | 2002 | 2002 | 2004 | 2002 | 1998 | 2006 | 2009 | 2002 | 1998 | 1999 | 2003 | 1999 | NOV 2003 |
| MAXIMUM 3-SECOND | | | | | | | | | | | | | | |
| SPEED (MPH) | 15 | 64 | 66 | 63 | 58 | 83 | 64 | 59 | 83 | 57 | 62 | 72 | 67 | 83 |
| DIR. (TENS OF DEGS) | | 29 | 24 | 26 | 27 | 33 | 25 | 26 | 17 | 23 | 27 | 23 | 25 | 33 |
| YEAR OF OCCURRENCE | | 2000 | 2002 | 2004 | 2002 | 2008 | 2006 | 2002 | 2003 | 1998 | 1999 | 2003 | 1999 | MAY 2008 |
| **PRECIPITATION** | | | | | | | | | | | | | | |
| NORMAL (IN) | 30 | 0.68 | 0.51 | 1.01 | 1.40 | 2.53 | 2.24 | 1.45 | 1.65 | 1.23 | 0.93 | 0.59 | 0.67 | 14.89 |
| MAXIMUM MONTHLY (IN) | 72 | 2.05 | 2.16 | 2.18 | 4.63 | 8.13 | 6.85 | 4.68 | 4.90 | 3.56 | 3.43 | 2.27 | 1.92 | 8.13 |
| YEAR OF OCCURRENCE | | 1969 | 1958 | 1967 | 1975 | 1953 | 2005 | 1993 | 1985 | 1941 | 1975 | 1955 | 1977 | MAY 1953 |
| MINIMUM MONTHLY (IN) | 72 | T | 0.01 | 0.10 | 0.05 | 0.51 | 0.52 | 0.04 | 0.03 | 0.09 | T | 0.02 | T | OCT 1965 |
| YEAR OF OCCURRENCE | | 1944 | 1950 | 1986 | 1981 | 2001 | 1960 | 1959 | 1969 | 1990 | 1965 | 2009 | 1954 | 1965 |
| MAXIMUM IN 24 HOURS (IN) | 72 | 0.74 | 11.87 | 1.14 | 2.43 | 3.42 | 2.90 | 2.40 | 2.40 | 2.74 | 1.82 | 1.15 | 0.97 | 11.87 |
| YEAR OF OCCURRENCE | | 1966 | 2008 | 1977 | 1951 | 1980 | 2002 | 1983 | 1989 | 1982 | 1954 | 1946 | 1972 | FEB 2008 |
| NORMAL NO. DAYS WITH: | | | | | | | | | | | | | | |
| PRECIPITATION >= 0.01 | 30 | 8.7 | 7.2 | 9.7 | 9.3 | 11.9 | 10.8 | 7.7 | 8.8 | 7.1 | 6.4 | 7.0 | 8.1 | 102.7 |
| PRECIPITATION >= 1.00 | 30 | 0.0 | 0.0 | * | 0.1 | 0.3 | 0.2 | 0.2 | 0.3 | 0.1 | * | 0.0 | 0.0 | 1.2 |
| **SNOWFALL** | | | | | | | | | | | | | | |
| NORMAL (IN) | 30 | 9.6 | 7.6 | 11.5 | 8.0 | 2.3 | 0.* | 0.0 | 0.3 | 1.4 | 4.4 | 7.4 | 8.4 | 60.9 |
| MAXIMUM MONTHLY (IN) | 71 | 22.6 | 26.1 | 27.8 | 35.4 | 11.6 | 11.1 | T | 8.3 | 10.4 | 16.6 | 24.9 | 30.5 | 35.4 |
| YEAR OF OCCURRENCE | | 1969 | 1958 | 1989 | 2009 | 1989 | 1950 | 1993 | 1992 | 1984 | 1975 | 1955 | 2008 | APR 2009 |
| MAXIMUM IN 24 HOURS (IN) | 71 | 10.2 | 11.0 | 11.5 | 16.8 | 11.6 | 11.0 | T | 8.3 | 8.4 | 8.3 | 10.8 | 9.8 | 16.8 |
| YEAR OF OCCURRENCE' | | 1984 | 1951 | 1987 | 1973 | 1989 | 1950 | 1993 | 1992 | 1988 | 1957 | 1946 | 1945 | APR 1973 |
| MAXIMUM SNOW DEPTH (IN) | 57 | 17 | 21 | 15 | 24 | 12 | 6 | 0 | 0 | 5 | 6 | 12 | 14 | 24 |
| YEAR OF OCCURRENCE | | 1978 | 1978 | 1977 | 1975 | 2009 | 2008 | | | 1988 | 1985 | 2005 | 1958 | APR 1975 |
| NORMAL NO. DAYS WITH: | | | | | | | | | | | | | | |
| SNOWFALL >= 1.0 | 30 | 3.2 | 2.6 | 3.3 | 2.2 | 0.7 | 0.0 | 0.0 | 0.1 | 0.3 | 1.5 | 2.4 | 2.9 | 19.2 |

## PRECIPITATION (inches) 2009 GREAT FALLS (KGTF)

| YEAR | JAN | FEB | MAR | APR | MAY | JUN | JUL | AUG | SEP | OCT | NOV | DEC | ANNUAL |
|---|---|---|---|---|---|---|---|---|---|---|---|---|---|
| 1980 | 0.67 | 1.03 | 0.74 | 0.62 | 5.12 | 3.91 | 0.27 | 0.67 | 0.98 | 1.75 | 0.19 | 0.27 | 16.22 |
| 1981 | 0.34 | 0.44 | 2.09 | 0.05 | 5.20 | 1.32 | 1.04 | 1.21 | 0.39 | 1.06 | 0.29 | 0.43 | 13.86 |
| 1982 | 1.09 | 0.99 | 1.97 | 1.04 | 3.63 | 3.09 | 0.66 | 0.41 | 2.43 | 0.75 | 0.63 | 0.99 | 17.68 |
| 1983 | 0.10 | 0.33 | 1.61 | 0.26 | 1.34 | 3.03 | 3.78 | 1.10 | 1.89 | 0.77 | 1.28 | 0.70 | 16.19 |
| 1984 | 0.72 | 0.69 | 1.31 | 0.94 | 1.34 | 2.10 | 0.05 | 1.01 | 0.71 | 1.20 | 0.49 | 1.25 | 11.81 |
| 1985 | 0.35 | 0.22 | 1.02 | 0.41 | 3.28 | 0.58 | 0.47 | 4.90 | 3.23 | 1.10 | 1.16 | 0.47 | 17.19 |
| 1986 | 0.57 | 0.75 | 0.10 | 2.83 | 1.74 | 1.72 | 1.67 | 0.81 | 1.52 | 0.90 | 0.45 | 0.27 | 13.33 |
| 1987 | 0.05 | 0.24 | 1.81 | 0.64 | 2.63 | 1.33 | 3.05 | 2.43 | 1.30 | 0.02 | 0.30 | 0.24 | 14.04 |
| 1988 | 0.76 | 0.47 | 0.44 | 0.77 | 1.60 | 1.42 | 1.82 | 0.26 | 2.33 | 0.66 | 0.30 | 0.97 | 11.80 |
| 1989 | 0.96 | 1.19 | 1.38 | 2.41 | 2.41 | 1.70 | 3.03 | 4.88 | 1.87 | 0.41 | 0.81 | 1.32 | 22.37 |
| 1990 | 0.29 | 0.17 | 1.69 | 0.84 | 3.97 | 1.23 | 1.03 | 3.19 | 0.09 | 0.13 | 0.70 | 0.73 | 14.06 |
| 1991 | 0.63 | 0.21 | 1.21 | 1.54 | 1.54 | 4.15 | 0.75 | 1.35 | 1.00 | 0.81 | 0.77 | 0.08 | 14.04 |
| 1992 | 0.48 | 0.23 | 0.43 | 1.32 | 2.14 | 3.22 | 1.81 | 1.37 | 0.25 | 2.61 | 0.29 | 0.31 | 14.46 |
| 1993 | 1.17 | 0.70 | 0.86 | 3.16 | 2.74 | 2.58 | 4.68 | 3.04 | 1.71 | 1.10 | 0.97 | 0.30 | 23.01 |
| 1994 | 0.47 | 0.53 | 0.20 | 1.90 | 1.81 | 1.56 | 0.72 | 0.61 | 0.35 | 1.77 | 0.42 | 0.24 | 10.58 |
| 1995 | 0.05 | 0.15 | 0.82 | 2.17 | 3.11 | 2.92 | 3.36 | 0.54 | 1.20 | 0.78 | 0.35 | 0.14 | 15.59 |
| 1996 | 0.49 | 0.26 | 0.83 | 1.40 | 2.57 | 1.14 | 0.17 | 0.67 | 1.47 | 0.54 | 0.35 | 1.25 | 11.14 |
| 1997 | 0.27 | 0.32 | 0.62 | 1.27 | 2.89 | 3.49 | 1.88 | 1.61 | 0.27 | 0.91 | 0.18 | 0.33 | 14.04 |
| 1998 | 0.72 | 0.42 | 1.10 | 0.42 | 3.08 | 5.18 | 1.73 | 1.72 | 1.00 | 0.80 | 0.96 | 0.22 | 17.35 |
| 1999 | 0.33 | 0.36 | 0.53 | 1.43 | 2.29 | 1.67 | 0.81 | 2.18 | 1.72 | 0.67 | 0.45 | 0.03 | 12.47 |
| 2000 | 0.34 | 0.69 | 0.74 | 0.33 | 2.10 | 1.55 | 1.04 | 0.12 | 1.32 | 1.34 | 0.49 | 0.19 | 10.25 |
| 2001 | 0.65 | 0.39 | 0.51 | 1.10 | 0.51 | 1.79 | 2.74 | 0.26 | 1.50 | 0.33 | 0.14 | 0.39 | 10.31 |
| 2002 | 0.30 | 0.26 | 0.70 | 0.42 | 1.61 | 5.03 | 1.50 | 2.49 | 1.65 | 0.47 | 0.27 | 0.29 | 14.99 |
| 2003 | 0.12 | 0.61 | 0.49 | 1.78 | 2.11 | 1.85 | 0.18 | 1.31 | 1.13 | 0.27 | 0.18 | 0.11 | 10.14 |
| 2004 | 0.23 | 0.06 | 0.29 | 1.06 | 2.91 | 2.82 | 0.42 | 2.55 | 1.99 | 1.05 | 0.16 | 0.43 | 13.97 |
| 2005 | 0.16 | 0.01 | 0.94 | 1.20 | 1.07 | 6.85 | 0.10 | 0.97 | 1.68 | 0.69 | 1.69 | 0.29 | 15.65 |
| 2006 | 0.71 | 0.44 | 1.70 | 2.88 | 2.64 | 4.24 | 0.27 | 1.33 | 1.82 | 1.48 | 0.43 | 0.59 | 18.53 |
| 2007 | 0.28 | 1.53 | 0.25 | 2.35 | 2.81 | 1.00 | 0.13 | 0.24 | 1.71 | 0.68 | 0.76 | 0.12 | 11.86 |
| 2008 | 0.84 | 0.43 | 0.35 | 1.51 | 3.82 | 3.08 | 1.25 | 1.31 | 1.87 | 0.53 | 0.74 | 1.43 | 17.16 |
| 2009 | 0.45 | 0.44 | 0.98 | 2.35 | 0.95 | 1.49 | 3.60 | 1.01 | 0.91 | 1.49 | 0.02 | 0.76 | 14.45 |
| POR= 62 YRS | 0.77 | 0.60 | 0.96 | 1.40 | 2.46 | 2.66 | 1.35 | 1.42 | 1.17 | 0.82 | 0.64 | 0.65 | 14.90 |

WBAN : 24143

## AVERAGE TEMPERATURE (°F) 2009 GREAT FALLS (KGTF)

| YEAR | JAN | FEB | MAR | APR | MAY | JUN | JUL | AUG | SEP | OCT | NOV | DEC | ANNUAL |
|---|---|---|---|---|---|---|---|---|---|---|---|---|---|
| 1980 | 15.2 | 28.1 | 32.4 | 52.9 | 57.1 | 60.9 | 69.4 | 62.4 | 58.0 | 49.0 | 39.4 | 23.8 | 45.7 |
| 1981 | 33.8 | 30.8 | 37.2 | 46.5 | 52.8 | 58.1 | 66.5 | 69.8 | 59.8 | 45.3 | 40.8 | 24.6 | 47.2 |
| 1982 | 6.3 | 19.5 | 27.9 | 37.9 | 48.4 | 60.2 | 66.8 | 65.0 | 53.9 | 46.6 | 32.1 | 26.8 | 41.0 |
| 1983 | 32.2 | 36.7 | 35.8 | 41.7 | 50.7 | 60.1 | 65.8 | 72.4 | 53.6 | 48.9 | 35.1 | 4.0 | 44.8 |
| 1984 | 29.5 | 36.9 | 35.3 | 44.4 | 51.7 | 59.8 | 69.8 | 71.6 | 51.8 | 40.2 | 35.5 | 13.0 | 45.0 |
| 1985 | 19.2 | 21.6 | 33.4 | 48.5 | 57.3 | 62.2 | 73.0 | 61.8 | 48.2 | 44.5 | 12.3 | 24.6 | 42.2 |
| 1986 | 36.6 | 18.6 | 43.9 | 42.4 | 53.2 | 66.0 | 64.9 | 69.0 | 51.5 | 49.6 | 32.0 | 33.2 | 46.7 |
| 1987 | 32.4 | 36.0 | 36.2 | 52.8 | 57.8 | 65.4 | 66.8 | 61.6 | 58.9 | 47.5 | 40.4 | 29.6 | 48.8 |
| 1988 | 23.6 | 28.9 | 37.3 | 46.8 | 56.4 | 69.5 | 69.1 | 67.1 | 56.0 | 49.7 | 35.7 | 29.5 | 47.5 |
| 1989 | 28.0 | 10.3 | 29.0 | 43.4 | 50.9 | 60.8 | 70.2 | 64.0 | 56.1 | 46.2 | 36.6 | 27.5 | 43.6 |
| 1990 | 30.0 | 28.0 | 35.7 | 44.3 | 49.8 | 59.8 | 67.5 | 68.5 | 62.4 | 46.0 | 37.7 | 17.8 | 45.6 |
| 1991 | 19.0 | 39.2 | 35.0 | 43.1 | 51.2 | 57.9 | 68.0 | 71.7 | 57.7 | 42.4 | 31.7 | 35.0 | 46.0 |
| 1992 | 34.7 | 36.6 | 40.9 | 47.1 | 55.3 | 62.8 | 61.6 | 62.9 | 56.8 | 47.6 | 35.2 | 18.4 | 46.7 |
| 1993 | 14.8 | 19.3 | 37.6 | 43.7 | 55.5 | 56.5 | 58.1 | 60.5 | 52.5 | 46.2 | 28.1 | 33.6 | 42.2 |
| 1994 | 27.3 | 17.2 | 39.6 | 44.5 | 54.7 | 60.5 | 68.7 | 67.2 | 60.1 | 44.1 | 30.4 | 28.6 | 45.2 |
| 1995 | 28.2 | 28.0 | 30.2 | 39.8 | 49.3 | 56.9 | 63.9 | 64.0 | 56.0 | 43.2 | 35.1 | 26.6 | 43.4 |
| 1996 | 12.1 | 26.3 | 24.5 | 44.4 | 46.8 | 60.7 | 67.0 | 67.3 | 54.1 | 43.3 | 20.7 | 15.3 | 40.2 |
| 1997 | 17.4 | 31.3 | 33.6 | 36.6 | 51.8 | 59.3 | 65.3 | 65.3 | 60.5 | 45.7 | 34.3 | 30.7 | 44.3 |
| 1998 | 21.6 | 33.6 | 31.4 | 45.1 | 54.1 | 55.4 | 69.2 | 69.1 | 63.1 | 46.2 | 35.5 | 24.7 | 45.8 |
| 1999 | 27.4 | 34.5 | 37.8 | 39.9 | 49.7 | 57.8 | 63.6 | 68.0 | 52.5 | 46.7 | 43.3 | 35.2 | 46.4 |
| 2000 | 25.8 | 28.0 | 36.8 | 45.5 | 53.1 | 59.0 | 69.8 | 68.4 | 55.8 | 45.4 | 25.1 | 19.2 | 44.3 |
| 2001 | 30.5 | 17.5 | 36.1 | 42.2 | 55.3 | 60.1 | 68.6 | 70.8 | 60.6 | 46.1 | 41.8 | 25.4 | 46.3 |
| 2002 | 26.2 | 30.3 | 17.4 | 38.2 | 49.8 | 59.4 | 71.2 | 60.8 | 57.0 | 37.6 | 38.6 | 30.8 | 43.1 |
| 2003 | 29.3 | 24.4 | 31.2 | 46.0 | 51.1 | 60.8 | 72.4 | 71.0 | 56.4 | 48.9 | 27.2 | 30.4 | 45.8 |
| 2004 | 20.5 | 31.1 | 40.2 | 45.0 | 48.0 | 56.3 | 68.1 | 64.5 | 54.9 | 44.1 | 37.1 | 31.3 | 45.1 |
| 2005 | 20.3 | 31.7 | 34.6 | 43.9 | 50.3 | 58.3 | 69.1 | 65.8 | 56.5 | 48.2 | 36.8 | 25.0 | 45.0 |
| 2006 | 37.1 | 27.0 | 31.9 | 46.5 | 54.3 | 62.8 | 73.5 | 67.2 | 58.3 | 43.2 | 31.9 | 32.0 | 47.1 |
| 2007 | 27.5 | 23.6 | 41.9 | 42.0 | 53.8 | 62.4 | 76.8 | 67.6 | 56.4 | 48.5 | 33.5 | 26.7 | 46.7 |
| 2008 | 22.5 | 29.6 | 34.2 | 38.6 | 52.0 | 58.4 | 67.9 | 68.1 | 55.0 | 46.6 | 40.6 | 16.5 | 44.2 |
| 2009 | 26.3 | 29.3 | 30.9 | 40.4 | 51.3 | 57.9 | 66.9 | 65.3 | 63.6 | 38.9 | 39.7 | 14.3 | 43.7 |
| POR= 62 YRS | 22.0 | 27.4 | 32.7 | 43.2 | 52.8 | 60.9 | 68.6 | 67.2 | 57.1 | 46.7 | 34.4 | 25.9 | 44.9 |

## HEATING DEGREE DAYS (base 65°F) 2009  GREAT FALLS (KGTF)

| YEAR | JUL | AUG | SEP | OCT | NOV | DEC | JAN | FEB | MAR | APR | MAY | JUN | TOTAL |
|---|---|---|---|---|---|---|---|---|---|---|---|---|---|
| 1980-81 | 16 | 110 | 225 | 504 | 763 | 1275 | 960 | 953 | 855 | 548 | 373 | 218 | 6800 |
| 1981-82 | 34 | 14 | 201 | 603 | 718 | 1244 | 1819 | 1271 | 1142 | 806 | 511 | 161 | 8524 |
| 1982-83 | 44 | 66 | 342 | 565 | 978 | 1181 | 1007 | 786 | 899 | 692 | 437 | 154 | 7151 |
| 1983-84 | 59 | 2 | 356 | 490 | 891 | 1888 | 1094 | 810 | 915 | 620 | 419 | 183 | 7727 |
| 1984-85 | 12 | 18 | 415 | 760 | 879 | 1611 | 1415 | 1212 | 971 | 489 | 249 | 134 | 8165 |
| 1985-86 | 4 | 147 | 498 | 629 | 1581 | 1246 | 872 | 1297 | 648 | 672 | 390 | 48 | 8032 |
| 1986-87 | 50 | 22 | 400 | 471 | 987 | 979 | 1004 | 803 | 888 | 372 | 238 | 70 | 6284 |
| 1987-88 | 66 | 136 | 189 | 540 | 729 | 1090 | 1278 | 1039 | 852 | 538 | 281 | 65 | 6803 |
| 1988-89 | 24 | 39 | 294 | 468 | 876 | 1090 | 1140 | 1529 | 1109 | 642 | 430 | 150 | 7791 |
| 1989-90 | 3 | 96 | 269 | 575 | 845 | 1155 | 1079 | 1031 | 902 | 613 | 462 | 204 | 7234 |
| 1990-91 | 34 | 37 | 118 | 583 | 813 | 1460 | 1425 | 714 | 922 | 649 | 417 | 209 | 7381 |
| 1991-92 | 19 | 5 | 233 | 705 | 992 | 925 | 935 | 819 | 740 | 532 | 307 | 128 | 6340 |
| 1992-93 | 130 | 171 | 260 | 538 | 886 | 1441 | 1556 | 1274 | 841 | 634 | 291 | 266 | 8288 |
| 1993-94 | 221 | 165 | 372 | 578 | 1101 | 966 | 1161 | 1335 | 780 | 608 | 316 | 162 | 7765 |
| 1994-95 | 35 | 48 | 158 | 640 | 1031 | 1122 | 1129 | 1031 | 1072 | 749 | 478 | 235 | 7728 |
| 1995-96 | 70 | 91 | 285 | 671 | 889 | 1181 | 1639 | 1114 | 1252 | 609 | 557 | 150 | 8508 |
| 1996-97 | 41 | 50 | 323 | 669 | 1326 | 1535 | 1467 | 938 | 967 | 846 | 408 | 177 | 8747 |
| 1997-98 | 56 | 66 | 155 | 594 | 915 | 1055 | 1336 | 877 | 1032 | 589 | 335 | 282 | 7292 |
| 1998-99 | 8 | 14 | 146 | 573 | 880 | 1241 | 1157 | 846 | 839 | 746 | 473 | 222 | 7145 |
| 1999-00 | 115 | 30 | 372 | 561 | 642 | 919 | 1210 | 1069 | 866 | 579 | 361 | 191 | 6915 |
| 2000-01 | 21 | 34 | 289 | 600 | 1188 | 1413 | 1059 | 1328 | 893 | 680 | 305 | 185 | 7995 |
| 2001-02 | 23 | 4 | 166 | 575 | 690 | 1222 | 1198 | 963 | 1466 | 799 | 468 | 199 | 7773 |
| 2002-03 | 21 | 138 | 261 | 842 | 785 | 1053 | 1099 | 1132 | 1045 | 562 | 435 | 163 | 7536 |
| 2003-04 | 7 | 25 | 301 | 496 | 1128 | 1066 | 1375 | 976 | 759 | 596 | 521 | 263 | 7513 |
| 2004-05 | 41 | 88 | 306 | 639 | 834 | 1036 | 1377 | 924 | 934 | 625 | 448 | 216 | 7468 |
| 2005-06 | 33 | 77 | 265 | 513 | 839 | 1234 | 858 | 1058 | 1017 | 548 | 337 | 103 | 6882 |
| 2006-07 | 3 | 50 | 220 | 668 | 987 | 1015 | 1156 | 1153 | 709 | 684 | 346 | 117 | 7108 |
| 2007-08 | 2 | 42 | 280 | 507 | 935 | 1179 | 1310 | 1020 | 947 | 786 | 403 | 214 | 7625 |
| 2008-09 | 19 | 45 | 295 | 567 | 725 | 1497 | 1194 | 998 | 1052 | 728 | 427 | 224 | 7771 |
| 2009- | 34 | 59 | 125 | 803 | 753 | 1562 | | | | | | | |

WBAN : 24143

## COOLING DEGREE DAYS (base 65°F) 2009  GREAT FALLS (KGTF)

| YEAR | JAN | FEB | MAR | APR | MAY | JUN | JUL | AUG | SEP | OCT | NOV | DEC | TOTAL |
|---|---|---|---|---|---|---|---|---|---|---|---|---|---|
| 1980 | 0 | 0 | 0 | 12 | 30 | 31 | 156 | 37 | 21 | 18 | 0 | 0 | 305 |
| 1981 | 0 | 0 | 0 | 0 | 3 | 17 | 85 | 168 | 49 | 0 | 0 | 0 | 322 |
| 1982 | 0 | 0 | 0 | 0 | 0 | 24 | 104 | 73 | 15 | 0 | 0 | 0 | 216 |
| 1983 | 0 | 0 | 0 | 0 | 4 | 14 | 90 | 241 | 19 | 0 | 0 | 0 | 368 |
| 1984 | 0 | 0 | 0 | 5 | 15 | 33 | 169 | 229 | 26 | 0 | 0 | 0 | 477 |
| 1985 | 0 | 0 | 0 | 0 | 20 | 58 | 260 | 51 | 0 | 0 | 0 | 0 | 389 |
| 1986 | 0 | 0 | 0 | 0 | 0 | 32 | 85 | 56 | 153 | 0 | 0 | 0 | 326 |
| 1987 | 0 | 0 | 0 | 13 | 23 | 90 | 128 | 36 | 13 | 4 | 0 | 0 | 307 |
| 1988 | 0 | 0 | 0 | 0 | 20 | 206 | 160 | 112 | 30 | 0 | 0 | 0 | 528 |
| 1989 | 0 | 0 | 0 | 0 | 0 | 31 | 170 | 72 | 8 | 0 | 0 | 0 | 281 |
| 1990 | 0 | 0 | 0 | 0 | 0 | 54 | 119 | 157 | 47 | 2 | 0 | 0 | 379 |
| 1991 | 0 | 0 | 0 | 0 | 0 | 1 | 119 | 217 | 19 | 13 | 0 | 0 | 369 |
| 1992 | 0 | 0 | 0 | 2 | 10 | 70 | 31 | 113 | 21 | 5 | 0 | 0 | 252 |
| 1993 | 0 | 0 | 0 | 0 | 4 | 19 | 13 | 30 | 3 | 0 | 0 | 0 | 69 |
| 1994 | 0 | 0 | 0 | 0 | 4 | 33 | 156 | 122 | 17 | 0 | 0 | 0 | 332 |
| 1995 | 0 | 0 | 0 | 0 | 0 | 1 | 44 | 68 | 24 | 0 | 0 | 0 | 137 |
| 1996 | 0 | 0 | 0 | 0 | 0 | 27 | 110 | 129 | 4 | 0 | 0 | 0 | 270 |
| 1997 | 0 | 0 | 0 | 0 | 5 | 15 | 71 | 85 | 26 | 2 | 0 | 0 | 204 |
| 1998 | 0 | 0 | 0 | 0 | 4 | 0 | 146 | 146 | 95 | 0 | 0 | 0 | 391 |
| 1999 | 0 | 0 | 0 | 0 | 4 | 9 | 80 | 131 | 3 | 0 | 0 | 0 | 227 |
| 2000 | 0 | 0 | 0 | 0 | 1 | 19 | 175 | 147 | 20 | 0 | 0 | 0 | 362 |
| 2001 | 0 | 0 | 0 | 0 | 12 | 42 | 144 | 190 | 43 | 0 | 0 | 0 | 431 |
| 2002 | 0 | 0 | 0 | 0 | 6 | 41 | 222 | 15 | 28 | 0 | 0 | 0 | 312 |
| 2003 | 0 | 0 | 0 | 0 | 13 | 46 | 245 | 216 | 49 | 5 | 0 | 0 | 574 |
| 2004 | 0 | 0 | 0 | 0 | 0 | 9 | 142 | 75 | 9 | 0 | 0 | 0 | 235 |
| 2005 | 0 | 0 | 0 | 0 | 0 | 19 | 167 | 108 | 15 | 0 | 0 | 0 | 309 |
| 2006 | 0 | 0 | 0 | 0 | 15 | 46 | 272 | 128 | 25 | 0 | 0 | 0 | 486 |
| 2007 | 0 | 0 | 0 | 0 | 3 | 47 | 376 | 131 | 29 | 0 | 0 | 0 | 586 |
| 2008 | 0 | 0 | 0 | 0 | 6 | 24 | 118 | 149 | 1 | 0 | 0 | 0 | 298 |
| 2009 | 0 | 0 | 0 | 0 | 8 | 16 | 100 | 73 | 90 | 0 | 0 | 0 | 287 |

## SNOWFALL (inches) 2009 GREAT FALLS (KGTF)

| YEAR | JUL | AUG | SEP | OCT | NOV | DEC | JAN | FEB | MAR | APR | MAY | JUN | TOTAL |
|---|---|---|---|---|---|---|---|---|---|---|---|---|---|
| 1980-81 | 0.0 | 0.0 | 0.0 | 7.7 | 3.3 | 5.4 | 4.1 | 7.1 | 11.5 | 0.1 | T | 0.0 | 39.2 |
| 1981-82 | 0.0 | 0.0 | 0.0 | 7.9 | 1.0 | 5.9 | 19.7 | 16.3 | 23.4 | 18.5 | 7.6 | T | 100.3 |
| 1982-83 | 0.0 | 0.0 | 0.7 | 1.5 | 8.8 | 13.0 | 0.9 | 4.1 | 6.6 | 1.4 | 8.6 | 0.0 | 45.6 |
| 1983-84 | 0.0 | 0.0 | 7.8 | T | 14.4 | 11.9 | 16.2 | 7.7 | 19.5 | 5.2 | 1.0 | 0.0 | 83.7 |
| 1984-85 | 0.0 | 0.0 | 10.4 | 10.9 | 5.5 | 16.6 | 5.4 | 3.8 | 11.9 | 2.4 | 0.0 | 0.0 | 66.9 |
| 1985-86 | 0.0 | T | 2.5 | 8.5 | 18.1 | 7.9 | 4.4 | 15.4 | 0.5 | 14.1 | 2.4 | 0.0 | 73.8 |
| 1986-87 | 0.0 | 0.0 | 0.1 | 1.2 | 7.9 | 4.5 | 1.0 | 1.8 | 16.5 | 0.6 | 5.3 | 0.0 | 38.9 |
| 1987-88 | 0.0 | 0.0 | 0.0 | 0.1 | 2.9 | 4.7 | 12.6 | 9.2 | 3.9 | 7.4 | 0.0 | 0.0 | 40.8 |
| 1988-89 | 0.0 | 0.0 | 9.1 | 5.3 | 6.0 | 10.9 | 16.0 | 18.7 | 24.2 | 15.7 | 11.6 | 0.0 | 117.5 |
| 1989-90 | T | 0.0 | 1.7 | 1.3 | 7.4 | 19.9 | 5.1 | 3.0 | 16.2 | 5.4 | T | T | 60.0 |
| 1990-91 | T | 0.0 | 0.0 | 0.4 | 6.7 | 8.5 | 9.2 | 3.1 | 23.9 | 6.0 | 4.0 | T | 61.8 |
| 1991-92 | 0.0 | T | 0.0 | 11.2 | 9.1 | 0.9 | 6.1 | 2.2 | 3.5 | 3.2 | 0.9 | 0.0 | 37.1 |
| 1992-93 | 0.0 | 8.3 | 0.0 | 8.3 | 3.3 | 6.4 | 14.7 | 10.1 | 7.0 | 4.4 | T | 0.0 | 62.5 |
| 1993-94 | T | T | T | 3.8 | 13.0 | 4.2 | 6.5 | 10.1 | 2.1 | 11.8 | T | T | 51.5 |
| 1994-95 | 0.0 | 0.0 | T | 3.8 | 10.2 | 2.7 | 0.2 | 4.1 | 12.9 | 0.0 | 0.0 | | |
| 1995-96 | | | T | 4.5 | 6.3 | 1.9 | 8.8 | 11.5 | 21.7 | 8.6 | 2.8 | T | |
| 1996-97 | | 0.0 | T | 8.3 | 8.9 | | | | | | | | |
| 1997-98 | 0.0 | 0.0 | 0.0 | 0.4 | 3.3 | 4.0 | 11.0 | 3.4 | 14.7 | 0.9 | T | 0.3 | 38.0 |
| 1998-99 | 0.0 | 0.0 | 0.0 | 1.0 | 13.0 | 5.0 | 9.6 | 7.0 | 9.9 | 8.4 | 2.8 | 0.1 | 56.8 |
| 1999-00 | 0.0 | T | T | 5.4 | 1.7 | 0.2 | 5.7 | 13.2 | 7.2 | 2.6 | 8.3 | T | 44.3 |
| 2000-01 | T | 0.0 | 1.6 | 0.9 | 9.4 | 6.6 | 8.6 | 12.1 | 9.3 | 7.6 | 1.6 | 0.0 | 57.7 |
| 2001-02 | T | 0.0 | 0.0 | 3.7 | 2.6 | 5.6 | 7.3 | 7.6 | 19.3 | 5.2 | 6.7 | T | 58.0 |
| 2002-03 | T | 0.0 | T | 5.7 | T | 7.4 | 2.9 | 12.9 | 12.9 | 2.8 | 4.3 | 0.0 | 48.9 |
| 2003-04 | 0.0 | 0.0 | 0.0 | 5.6 | 8.5 | 7.8 | 8.5 | 2.6 | 5.3 | 5.3 | 5.0 | 0.0 | 48.6 |
| 2004-05 | 0.0 | 0.0 | 0.0 | 2.2 | 0.9 | 11.3 | 8.3 | 0.7 | 27.8 | 4.4 | T | 0.0 | 55.6 |
| 2005-06 | 0.0 | 0.0 | 0.0 | 0.4 | 24.9 | 5.1 | 3.9 | 5.5 | 14.2 | 3.0 | 1.8 | 0.0 | 58.8 |
| 2006-07 | 0.0 | 0.0 | 1.3 | 7.4 | 9.8 | 9.1 | 5.7 | 22.4 | 2.1 | 13.1 | 0.0 | 0.0 | 70.9 |
| 2007-08 | 0.0 | 0.0 | 0.0 | T | 12.7 | 1.9 | 14.1 | 6.9 | 5.9 | 21.7 | 0.2 | 6.8 | 70.2 |
| 2008-09 | 0.0 | 0.0 | 0.0 | 6.7 | 2.7 | 30.5 | 12.3 | 9.1 | 12.0 | 35.4 | 1.5 | 0.5 | 110.7 |
| 2009- | 0.0 | 0.0 | 0.0 | 6.9 | 0.2 | 12.9 | | | | | | | |
| POR= 62 YRS | T | 0.1 | 1.2 | 3.9 | 7.3 | 8.4 | 9.8 | 8.1 | 11.0 | 8.5 | 2.1 | 0.4 | 60.8 |

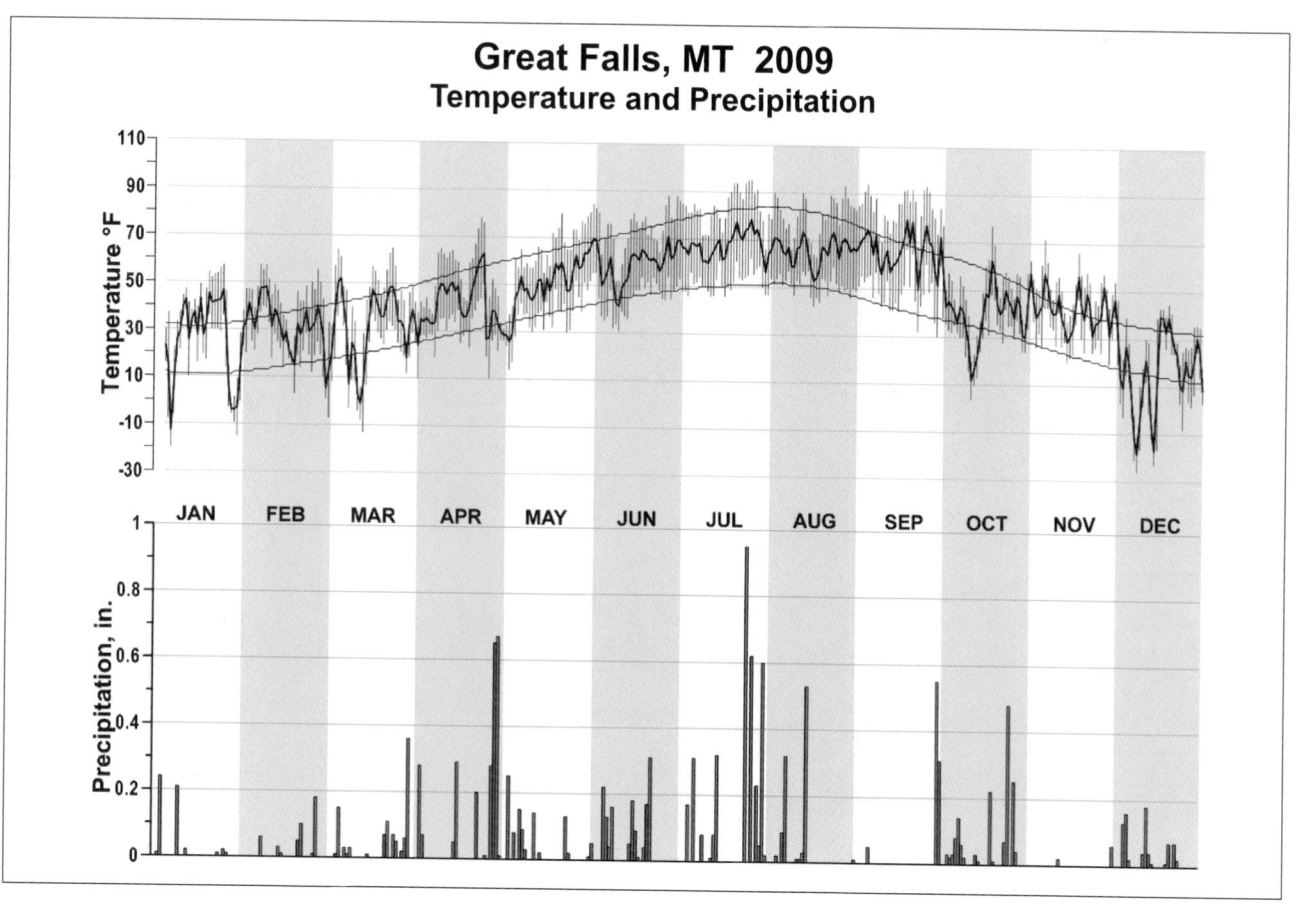

**Great Falls, MT 2009**
**Temperature and Precipitation**

# 2009
# HELENA
# MONTANA (KHLN)

Helena is located on the south side of an intermountain valley bounded on the west and south by the main chain of the Continental Divide. The valley is approximately 25 miles in width from north to south and 35 miles long from east to west. The average height of the mountains above the valley floor is about 3,000 feet.

The climate of Helena may be described as modified continental. Several factors enter into modifying the continental climate characteristics. Some of these are invasion by Pacific Ocean air masses, drainage of cool air into the valley from the surrounding mountains, and the protecting mountain shield in all directions.

The mountains to the north and east sometimes deflect shallow masses of invading cold Arctic air to the east. Following periods of extreme cold, when the return circulation of maritime air has brought warming to most of the eastern part of the state, cold air may remain trapped in the valley for several days before being replaced by warmer air. During these periods of transition from cold-to-warm temperatures, inversions are often quite pronounced.

As may be expected in a northern latitude, cold waves may occur from November through February, with temperatures occasionally dropping to zero or lower.

Summertime temperatures are moderate, with maximum readings generally under 90 degrees and very seldom reaching 100 degrees. Like all mountain stations, there is usually a marked change in temperature from day to night. During the summer this tends to produce an agreeable combination of fairly warm days and cool nights.

Most of the precipitation falls from April through July from frequent showers or thunderstorms, but usually with some steady rains in June, the wettest month of the year. Like summer, fall and winter months are relatively dry. During the April to September growing season, precipitation varies considerably.

Thunderstorms are rather frequent from May through August. Snow can be expected from September through May, but amounts during the spring and fall are usually light, and snow on the ground ordinarily lasts only a day or two. During the winter months snow may remain on the ground for several weeks at a time. There is little drifting of snow in the valley, and blizzard conditions are very infrequent.

Severe ice, sleet, and hailstorms are very seldom observed. Since 1880, only a few hailstorms have caused extensive damage in the city of Helena.

In winter, hours of sunshine are more than would be expected at a mountain location.

Due to the sheltering influence of the mountains, Foehn (Chinook) winds are not as pronounced as might be expected for a location on the eastern slopes of the Rocky Mountains. Strong winds can occur at any time throughout the year, but generally do not last more than a few hours at a time.

Based on the 1951-1980 period, the average first occurrence of 32 degrees Fahrenheit in the fall is September 18 and the average last occurrence in the spring is May 18.

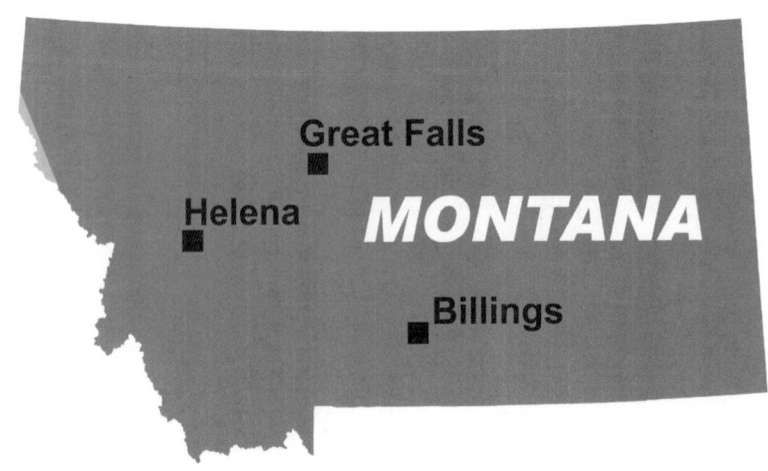

# NORMALS, MEANS, AND EXTREMES
## HELENA (KHLN)

LATITUDE: 46° 36'N  LONGITUDE: -111° 57'W  ELEVATION (FT): GRND: 3828  BARO: 3867  TIME ZONE: MOUNTAIN (UTC -7)  WBAN: 24144

| | ELEMENT | POR | JAN | FEB | MAR | APR | MAY | JUN | JUL | AUG | SEP | OCT | NOV | DEC | YEAR |
|---|---|---|---|---|---|---|---|---|---|---|---|---|---|---|---|
| TEMPERATURE °F | NORMAL DAILY MAXIMUM | 30 | 30.5 | 37.3 | 46.8 | 56.9 | 65.9 | 75.0 | 83.4 | 82.5 | 71.0 | 58.4 | 41.5 | 31.5 | 56.7 |
| | MEAN DAILY MAXIMUM | 117 | 29.9 | 33.6 | 43.4 | 54.4 | 64.5 | 71.6 | 83.0 | 81.2 | 68.4 | 57.3 | 41.5 | 32.6 | 55.1 |
| | HIGHEST DAILY MAXIMUM | 69 | 72 | 69 | 78 | 86 | 93 | 100 | 105 | 105 | 99 | 87 | 75 | 64 | 105 |
| | YEAR OF OCCURRENCE | | 2000 | 1995 | 2004 | 1992 | 2001 | 1988 | 2002 | 1969 | 1967 | 2001 | 1999 | 1980 | JUL 2002 |
| | MEAN OF EXTREME MAXS. | 117 | 52.3 | 55.5 | 64.7 | 75.5 | 84.0 | 91.1 | 96.5 | 95.6 | 89.2 | 77.5 | 62.4 | 53.0 | 74.8 |
| | NORMAL DAILY MINIMUM | 30 | 9.9 | 15.6 | 23.5 | 31.2 | 39.8 | 47.5 | 52.3 | 50.8 | 41.2 | 31.2 | 20.3 | 11.3 | 31.2 |
| | MEAN DAILY MINIMUM | 117 | 11.4 | 14.9 | 22.5 | 31.3 | 40.4 | 47.0 | 53.6 | 51.8 | 41.9 | 33.4 | 22.2 | 14.8 | 32.1 |
| | LOWEST DAILY MINIMUM | 69 | -42 | -42 | -30 | 1 | 17 | 30 | 36 | 28 | 18 | -8 | -39 | -38 | -42 |
| | YEAR OF OCCURRENCE | | 1957 | 1996 | 1955 | 1954 | 1954 | 1999 | 1971 | 1992 | 1970 | 1991 | 1959 | 1964 | FEB 1996 |
| | MEAN OF EXTREME MINS. | 117 | -16.8 | -8.0 | 0.4 | 16.8 | 27.9 | 36.3 | 43.3 | 40.8 | 28.5 | 15.5 | -0.0 | -12.2 | 14.4 |
| | NORMAL DRY BULB | 30 | 20.2 | 26.4 | 35.1 | 44.1 | 52.9 | 61.2 | 67.8 | 66.7 | 56.1 | 44.8 | 30.9 | 21.4 | 44.0 |
| | MEAN DRY BULB | 117 | 20.7 | 24.3 | 33.0 | 42.9 | 52.5 | 59.3 | 68.3 | 66.6 | 55.2 | 45.4 | 31.9 | 23.7 | 43.7 |
| | MEAN WET BULB | 26 | 18.4 | 21.8 | 28.3 | 35.5 | 43.1 | 49.8 | 53.6 | 52.2 | 45.2 | 36.6 | 26.4 | 18.1 | 35.8 |
| | MEAN DEW POINT | 26 | 15.2 | 16.9 | 22.8 | 29.0 | 36.8 | 43.8 | 46.4 | 45.2 | 38.9 | 30.5 | 22.0 | 14.6 | 30.2 |
| | NORMAL NO. DAYS WITH: | | | | | | | | | | | | | | |
| | MAXIMUM >= 90 | 30 | 0.0 | 0.0 | 0.0 | 0.0 | 0.2 | 2.4 | 7.7 | 7.2 | 1.2 | 0.0 | 0.0 | 0.0 | 18.7 |
| | MAXIMUM <= 32 | 30 | 14.2 | 8.1 | 3.3 | 0.5 | 0.0 | 0.0 | 0.0 | 0.0 | 0.0 | 0.6 | 6.1 | 14.5 | 47.3 |
| | MINIMUM <= 32 | 30 | 29.5 | 26.7 | 27.4 | 16.9 | 3.6 | 0.2 | 0.0 | * | 3.4 | 16.9 | 26.8 | 29.6 | 181.0 |
| | MINIMUM <= 0 | 30 | 8.6 | 4.5 | 0.9 | 0.0 | 0.0 | 0.0 | 0.0 | 0.0 | 0.0 | 0.0 | 2.0 | 6.3 | 22.5 |
| H/C | NORMAL HEATING DEG. DAYS | 30 | 1397 | 1093 | 932 | 634 | 384 | 157 | 42 | 56 | 283 | 631 | 1018 | 1348 | 7975 |
| | NORMAL COOLING DEG. DAYS | 30 | 0 | 0 | 0 | 0 | 3 | 39 | 122 | 100 | 13 | 0 | 0 | 0 | 277 |
| RH | NORMAL (PERCENT) | 30 | 71 | 67 | 62 | 55 | 55 | 54 | 48 | 49 | 54 | 60 | 67 | 71 | 59 |
| | HOUR 05 LST | 30 | 73 | 74 | 74 | 72 | 73 | 73 | 68 | 70 | 73 | 74 | 74 | 74 | 73 |
| | HOUR 11 LST | 30 | 68 | 63 | 54 | 46 | 45 | 44 | 40 | 42 | 47 | 53 | 63 | 68 | 53 |
| | HOUR 17 LST | 30 | 63 | 54 | 45 | 38 | 39 | 38 | 31 | 31 | 35 | 42 | 58 | 66 | 45 |
| | HOUR 23 LST | 30 | 72 | 70 | 67 | 60 | 61 | 59 | 53 | 54 | 59 | 64 | 70 | 73 | 64 |
| S | PERCENT POSSIBLE SUNSHINE | 55 | 47 | 56 | 61 | 59 | 61 | 64 | 78 | 75 | 68 | 60 | 44 | 42 | 60 |
| W/O | MEAN NO. DAYS WITH: | | | | | | | | | | | | | | |
| | HEAVY FOG(VISBY <= 1/4 MI) | 46 | 2.2 | 1.2 | 1.1 | 0.5 | 0.3 | 0.1 | 0.1 | 0.2 | 0.3 | 0.5 | 1.1 | 2.4 | 10.0 |
| | THUNDERSTORMS | 62 | 0.0 | 0.0 | 0.1 | 0.9 | 3.6 | 7.0 | 7.7 | 7.2 | 1.6 | 0.3 | 0.1 | 0.0 | 28.5 |
| CLOUDNESS | MEAN: | | | | | | | | | | | | | | |
| | SUNRISE-SUNSET (OKTAS) | | | | | | | | | | | | | | |
| | MIDNIGHT-MIDNIGHT (OKTAS) | | | | | | | | | | | | | | |
| | MEAN NO. DAYS WITH: | | | | | | | | | | | | | | |
| | CLEAR | 1 | 1.5 | 2.0 | 2.0 | | | 8.0 | 2.0 | 11.0 | 7.0 | 1.0 | | 6.0 | |
| | PARTLY CLOUDY | 1 | 1.0 | 5.0 | 1.0 | | 6.0 | 5.0 | 1.0 | 3.0 | 5.0 | 7.0 | | 2.0 | |
| | CLOUDY | 1 | 3.0 | 3.0 | 12.0 | | 23.0 | 6.0 | | | 4.0 | 7.0 | | 3.0 | |
| PR | MEAN STATION PRESSURE(IN) | 26 | 26.03 | 26.01 | 25.98 | 25.98 | 25.98 | 26.00 | 26.05 | 26.05 | 26.06 | 26.05 | 26.03 | 26.04 | 26.02 |
| | MEAN SEA-LEVEL PRES. (IN) | 26 | 30.19 | 30.14 | 30.06 | 30.00 | 29.95 | 29.94 | 29.97 | 29.99 | 30.03 | 30.09 | 30.13 | 30.20 | 30.06 |
| WINDS | MEAN SPEED (MPH) | 26 | 6.0 | 6.5 | 7.5 | 8.3 | 8.2 | 8.0 | 7.3 | 6.7 | 6.6 | 6.8 | 6.5 | 5.8 | 7.0 |
| | PREVAIL.DIR(TENS OF DEGS) | 29 | 28 | 28 | 28 | 28 | 28 | 28 | 29 | 28 | 28 | 28 | 28 | 28 | 28 |
| | MAXIMUM 2-MINUTE: | | | | | | | | | | | | | | |
| | SPEED (MPH) | 15 | 46 | 51 | 47 | 41 | 44 | 52 | 52 | 46 | 52 | 51 | 43 | 44 | 52 |
| | DIR. (TENS OF DEGS) | | 29 | 26 | 28 | 28 | 29 | 18 | 24 | 26 | 26 | 28 | 25 | 28 | 24 |
| | YEAR OF OCCURRENCE | | 1997 | 1999 | 2001 | 1997 | 1996 | 1999 | 2002 | 1999 | 1999 | 1999 | 2005 | 1995 | JUL 2002 |
| | MAXIMUM 3-SECOND | | | | | | | | | | | | | | |
| | SPEED (MPH) | 15 | 59 | 64 | 58 | 49 | 61 | 60 | 58 | 59 | 64 | 61 | 56 | 62 | 64 |
| | DIR. (TENS OF DEGS) | | 28 | 26 | 26 | 27 | 18 | 19 | 25 | 26 | 26 | 26 | 27 | 26 | 26 |
| | YEAR OF OCCURRENCE | | 2004 | 1999 | 2004 | 1997 | 2008 | 1999 | 2002 | 1999 | 1999 | 1999 | 2003 | 2008 | SEP 1999 |
| PRECIPITATION | NORMAL (IN) | 30 | 0.52 | 0.38 | 0.63 | 0.91 | 1.78 | 1.82 | 1.34 | 1.29 | 1.05 | 0.66 | 0.48 | 0.46 | 11.32 |
| | MAXIMUM MONTHLY (IN) | 69 | 2.78 | 1.20 | 1.62 | 3.00 | 6.09 | 4.74 | 4.70 | 4.23 | 3.37 | 2.68 | 1.50 | 1.48 | 6.09 |
| | YEAR OF OCCURRENCE | | 1969 | 1986 | 1982 | 1975 | 1981 | 1944 | 1993 | 1974 | 1965 | 1975 | 1950 | 1977 | MAY 1981 |
| | MINIMUM MONTHLY (IN) | 69 | T | 0.02 | 0.02 | 0.10 | 0.29 | 0.08 | .07 | 0.02 | 0.08 | 0.02 | 0.04 | 0.01 | T |
| | YEAR OF OCCURRENCE | | 1987 | 1991 | 1959 | 1977 | 1979 | 1985 | 2005 | 1988 | 1972 | 1978 | 1969 | 2007 | JAN 1987 |
| | MAXIMUM IN 24 HOURS (IN) | 69 | 0.77 | 0.58 | 1.01 | 1.57 | 2.31 | 1.78 | 2.26 | 1.86 | 1.61 | 1.01 | 0.82 | 0.51 | 2.31 |
| | YEAR OF OCCURRENCE | | 1969 | 1953 | 1957 | 2006 | 1981 | 1979 | 1983 | 1974 | 1980 | 2000 | 1959 | 1982 | MAY 1981 |
| | NORMAL NO. DAYS WITH: | | | | | | | | | | | | | | |
| | PRECIPITATION >= 0.01 | 30 | 7.5 | 6.3 | 8.1 | 8.6 | 11.4 | 10.4 | 8.1 | 7.7 | 6.3 | 5.7 | 6.7 | 7.3 | 94.1 |
| | PRECIPITATION >= 1.00 | 30 | 0.0 | 0.0 | 0.0 | 0.0 | 0.1 | 0.1 | 0.1 | 0.1 | 0.1 | 0.0 | 0.0 | 0.0 | 0.5 |
| SNOWFALL | NORMAL (IN) | 30 | 8.2 | 5.7 | 6.8 | 4.6 | 0.9 | 0.* | 0.* | 0.3 | 1.2 | 2.8 | 5.4 | 7.4 | 43.3 |
| | MAXIMUM MONTHLY (IN) | 62 | 35.6 | 19.7 | 21.6 | 20.6 | 12.7 | 2.7 | T | 6.2 | 13.7 | 11.0 | 32.9 | 22.8 | 35.6 |
| | YEAR OF OCCURRENCE | | 1969 | 1959 | 1955 | 1967 | 1967 | 1969 | 1993 | 1992 | 1965 | 1969 | 1959 | 1967 | JAN 1969 |
| | MAXIMUM IN 24 HOURS (IN) | 62 | 11.5 | 11.7 | 8.7 | 12.9 | 12.5 | 2.7 | T | 6.2 | 13.3 | 7.4 | 21.5 | 10.7 | 21.5 |
| | YEAR OF OCCURRENCE' | | 1969 | 1993 | 1955 | 1960 | 1967 | 1969 | 1993 | 1992 | 1957 | 1969 | 1959 | 1991 | NOV 1959 |
| | MAXIMUM SNOW DEPTH (IN) | 51 | 23 | 15 | 15 | 8 | 8 | T | 0 | 0 | 10 | 6 | 19 | 14 | 23 |
| | YEAR OF OCCURRENCE | | 1969 | 1975 | 1969 | 1986 | 1967 | 1969 | | | 1965 | 1969 | 1978 | 1985 | JAN 1969 |
| | NORMAL NO. DAYS WITH: | | | | | | | | | | | | | | |
| | SNOWFALL >= 1.0 | 30 | 2.7 | 1.7 | 2.1 | 1.4 | 0.3 | 0.0 | 0.0 | 0.1 | 0.4 | 1.0 | 1.6 | 2.0 | 13.3 |

## PRECIPITATION (inches) 2009  HELENA (KHLN)

| YEAR | JAN | FEB | MAR | APR | MAY | JUN | JUL | AUG | SEP | OCT | NOV | DEC | ANNUAL |
|------|-----|-----|-----|-----|-----|-----|-----|-----|-----|-----|-----|-----|--------|
| 1980 | 0.62 | 0.74 | 0.88 | 0.63 | 4.32 | 3.16 | 1.92 | 0.28 | 2.57 | 1.21 | 0.32 | 0.40 | 17.05 |
| 1981 | 0.15 | 0.10 | 1.10 | 0.75 | 6.09 | 1.15 | 1.78 | 0.10 | 0.90 | 0.82 | 0.54 | 0.33 | 13.81 |
| 1982 | 0.80 | 0.58 | 1.62 | 0.54 | 1.77 | 2.99 | 0.49 | 0.74 | 2.74 | 0.35 | 0.31 | 1.05 | 13.98 |
| 1983 | 0.24 | 0.07 | 0.36 | 0.29 | 1.79 | 2.20 | 3.48 | 2.67 | 1.56 | 0.35 | 0.26 | 0.76 | 14.03 |
| 1984 | 0.17 | 0.15 | 0.49 | 1.45 | 1.03 | 2.14 | 0.11 | 1.11 | 0.73 | 0.74 | 0.47 | 0.41 | 9.00 |
| 1985 | 0.16 | 0.38 | 0.32 | 0.46 | 0.75 | 0.08 | 0.10 | 2.64 | 2.11 | 0.76 | 0.84 | 0.35 | 8.95 |
| 1986 | 0.32 | 1.20 | 0.49 | 1.08 | 0.83 | 1.56 | 1.37 | 1.84 | 2.45 | 0.03 | 0.54 | 0.38 | 12.09 |
| 1987 | T | 0.03 | 1.19 | 0.76 | 1.90 | 1.50 | 2.88 | 0.38 | 0.80 | 0.05 | 0.12 | 0.42 | 10.03 |
| 1988 | 0.27 | 0.50 | 0.45 | 1.32 | 1.82 | 1.50 | 0.36 | 0.02 | 2.09 | 0.69 | 0.69 | 0.32 | 10.03 |
| 1989 | 1.42 | 0.82 | 1.35 | 0.72 | 1.00 | 1.43 | 1.55 | 1.61 | 1.31 | 0.54 | 0.26 | 0.48 | 12.49 |
| 1990 | 0.47 | 0.14 | 0.91 | 0.43 | 1.54 | 0.92 | 0.40 | 2.57 | 0.11 | 0.11 | 0.36 | 0.47 | 8.43 |
| 1991 | 0.27 | 0.02 | 0.90 | 0.75 | 1.71 | 3.27 | 0.72 | 0.70 | 1.26 | 0.65 | 0.88 | 0.79 | 11.92 |
| 1992 | 0.29 | 0.10 | 0.60 | 0.55 | 0.64 | 2.36 | 1.06 | 1.01 | 0.09 | 1.87 | 0.19 | 0.57 | 9.33 |
| 1993 | 0.80 | 1.03 | 0.56 | 1.63 | 1.71 | 3.14 | 4.70 | 2.79 | 1.25 | 0.71 | 0.36 | 0.13 | 18.81 |
| 1994 | 0.20 | 0.40 | 0.32 | 1.45 | 1.23 | 0.84 | 0.71 | 0.47 | 0.09 | 1.14 | 0.55 | 0.07 | 7.47 |
| 1995 | 0.20 | 0.08 | 0.49 | 1.15 | 3.09 | 2.93 | 1.51 | 0.33 | 1.59 | 0.10 | 0.62 | 0.28 | 12.37 |
| 1996 | 0.55 | 0.11 | 0.58 | 0.70 | 2.42 | 1.20 | 1.27 | 0.89 | 0.51 | 0.04 | 0.84 | 0.61 | 9.72 |
| 1997 | 0.28 | 0.10 | 0.10 | 0.20 | 2.35 | 2.43 | 1.25 | 1.79 | 0.31 | 1.62 | 0.13 | 0.01 | 10.57 |
| 1998 | 0.49 | 0.12 | 0.39 | 0.64 | 2.27 | 3.03 | 2.96 | 0.50 | 0.82 | 0.14 | 1.07 | 0.14 | 12.57 |
| 1999 | 0.38 | 0.26 | 0.02 | 1.05 | 2.19 | 2.15 | 0.41 | 1.92 | 0.54 | 0.39 | 0.13 | 0.10 | 9.54 |
| 2000 | 0.26 | 0.32 | 0.26 | 0.73 | 0.98 | 1.42 | 0.73 | 0.43 | 0.54 | 2.12 | 0.36 | 0.23 | 8.38 |
| 2001 | 0.27 | 0.17 | 0.44 | 1.39 | 1.23 | 2.11 | 1.94 | 0.43 | 1.38 | 0.54 | 0.13 | 0.28 | 10.31 |
| 2002 | 0.04 | 0.29 | 0.52 | 0.61 | 1.86 | 4.36 | 1.61 | 1.32 | 1.22 | 0.16 | 0.50 | 0.05 | 12.54 |
| 2003 | 0.41 | 0.29 | 0.74 | 2.27 | 1.25 | 1.49 | 0.23 | 1.03 | 0.74 | 0.34 | 0.20 | 0.35 | 9.34 |
| 2004 | 0.26 | 0.17 | 0.37 | 1.82 | 2.21 | 1.07 | 0.68 | 2.84 | 1.76 | 0.41 | 0.10 | 0.36 | 12.05 |
| 2005 | 0.26 | 0.06 | 0.86 | 0.90 | 2.11 | 4.55 | 0.07 | 0.29 | 0.72 | 0.94 | 0.77 | 0.63 | 12.16 |
| 2006 | 0.22 | 0.24 | 0.60 | 2.95 | 1.77 | 2.69 | 0.39 | 0.25 | 1.17 | 1.32 | 0.55 | 0.38 | 12.53 |
| 2007 | 0.09 | 0.63 | 0.14 | 0.82 | 3.25 | 1.44 | 0.31 | 0.39 | 1.69 | 0.96 | 0.63 | 0.01 | 10.36 |
| 2008 | 0.49 | 0.31 | 0.12 | 0.49 | 2.62 | 1.58 | 0.47 | 0.45 | 0.70 | 0.38 | 0.86 | 0.77 | 9.24 |
| 2009 | 0.40 | 0.22 | 1.17 | 0.60 | 0.43 | 1.45 | 1.82 | 1.86 | 0.97 | 0.89 | 0.13 | 0.31 | 10.25 |
| POR= 117 YRS | 0.60 | 0.46 | 0.72 | 0.99 | 1.91 | 2.11 | 1.10 | 0.99 | 1.11 | 0.73 | 0.63 | 0.59 | 11.94 |

WBAN : 24144

## AVERAGE TEMPERATURE (°F) 2009  HELENA (KHLN)

| YEAR | JAN | FEB | MAR | APR | MAY | JUN | JUL | AUG | SEP | OCT | NOV | DEC | ANNUAL |
|------|-----|-----|-----|-----|-----|-----|-----|-----|-----|-----|-----|-----|--------|
| 1980 | 14.3 | 25.3 | 31.3 | 49.0 | 55.4 | 59.8 | 67.3 | 62.9 | 56.8 | 45.4 | 34.9 | 26.9 | 44.1 |
| 1981 | 28.4 | 29.7 | 38.0 | 46.4 | 52.8 | 58.9 | 67.7 | 69.5 | 58.8 | 43.8 | 36.4 | 25.3 | 46.3 |
| 1982 | 16.5 | 23.7 | 33.9 | 40.5 | 50.9 | 61.2 | 68.6 | 69.1 | 55.5 | 44.9 | 27.4 | 22.7 | 42.9 |
| 1983 | 30.6 | 35.3 | 38.3 | 43.1 | 51.5 | 60.2 | 66.0 | 70.8 | 53.5 | 46.0 | 34.7 | 5.5 | 44.6 |
| 1984 | 27.3 | 32.4 | 37.2 | 43.5 | 52.8 | 60.0 | 70.0 | 69.8 | 52.3 | 41.7 | 33.1 | 11.6 | 44.3 |
| 1985 | 12.3 | 18.8 | 33.4 | 46.9 | 56.2 | 63.3 | 75.0 | 63.1 | 49.6 | 42.3 | 12.6 | 15.0 | 40.7 |
| 1986 | 25.5 | 21.8 | 42.9 | 43.3 | 53.6 | 66.4 | 64.2 | 68.0 | 51.2 | 45.3 | 29.0 | 18.6 | 44.2 |
| 1987 | 23.3 | 31.9 | 37.0 | 50.2 | 55.9 | 64.4 | 66.2 | 62.8 | 59.9 | 46.8 | 34.7 | 24.7 | 46.5 |
| 1988 | 18.8 | 29.1 | 36.1 | 47.1 | 55.5 | 68.4 | 71.3 | 68.6 | 56.4 | 50.3 | 33.8 | 23.1 | 46.5 |
| 1989 | 24.5 | 6.1 | 27.6 | 44.5 | 51.7 | 62.2 | 72.0 | 64.3 | 55.6 | 45.1 | 36.8 | 23.8 | 42.9 |
| 1990 | 28.8 | 26.9 | 34.9 | 45.8 | 51.0 | 61.5 | 69.4 | 68.4 | 63.6 | 45.3 | 37.4 | 14.0 | 45.6 |
| 1991 | 19.0 | 37.5 | 35.3 | 43.8 | 52.8 | 59.5 | 70.8 | 72.8 | 58.4 | 43.3 | 29.8 | 22.9 | 45.5 |
| 1992 | 23.7 | 34.2 | 42.7 | 47.7 | 57.5 | 64.9 | 64.3 | 64.7 | 56.7 | 46.7 | 32.8 | 14.7 | 45.9 |
| 1993 | 12.3 | 15.0 | 36.4 | 44.3 | 57.3 | 58.5 | 59.5 | 61.5 | 54.0 | 43.8 | 26.8 | 30.0 | 41.6 |
| 1994 | 29.8 | 20.7 | 39.4 | 45.6 | 56.2 | 62.7 | 69.0 | 69.9 | 61.6 | 44.3 | 28.3 | 23.6 | 45.9 |
| 1995 | 22.7 | 30.5 | 30.7 | 42.0 | 49.7 | 57.4 | 65.8 | 65.2 | 55.7 | 42.8 | 34.3 | 22.7 | 43.3 |
| 1996 | 9.9 | 21.1 | 25.7 |  | 48.2 | 61.3 | 68.3 | 66.6 | 55.5 | 43.8 | 23.1 | 16.6 | |
| 1997 | 13.3 | 28.2 | 36.9 | 38.4 | 53.2 | 60.1 | 65.8 | 65.4 | 58.8 | 44.1 | 31.2 | 24.4 | 43.3 |
| 1998 | 21.2 | 31.0 | 32.8 | 44.8 | 53.9 | 55.1 | 69.8 | 68.6 | 61.9 | 43.8 | 34.5 | 24.5 | 45.2 |
| 1999 | 26.8 | 32.4 | 37.1 | 40.8 | 50.2 | 58.8 | 65.3 | 68.4 | 52.9 | 45.8 | 39.2 | 28.1 | 45.5 |
| 2000 | 25.0 | 28.8 | 38.6 | 47.0 | 54.9 | 62.4 | 72.1 | 69.6 | 56.4 | 44.6 | 22.0 | 16.2 | 44.8 |
| 2001 | 20.1 | 17.6 | 35.4 | 44.0 | 59.2 | 64.1 | 72.0 | 74.7 | 64.6 | 48.4 | 39.2 | 24.5 | 47.0 |
| 2002 | 29.6 | 30.3 | 25.2 | 43.0 | 53.4 | 62.8 | 72.4 | 64.0 | 59.1 | 40.4 | 34.6 | 28.1 | 45.2 |
| 2003 | 29.4 | 25.3 | 34.2 | 46.0 | 53.3 | 63.2 | 76.4 | 73.5 | 59.3 | 50.6 | 28.8 | 28.3 | 47.4 |
| 2004 | 15.0 | 29.0 | 43.0 | 47.6 | 51.8 | 60.8 | 70.7 | 66.4 | 56.3 | 46.0 | 35.3 | 29.5 | 46.0 |
| 2005 | 16.5 | 31.3 | 37.1 | 44.9 | 52.4 | 59.8 | 72.5 | 68.6 | 58.0 | 49.1 | 34.6 | 19.9 | 45.4 |
| 2006 | 36.0 | 29.0 | 35.5 | 47.8 | 56.7 | 65.0 | 75.4 | 69.3 | 59.2 | 44.0 | 32.8 | 27.0 | 48.1 |
| 2007 | 25.6 | 27.4 | 43.1 | 46.0 | 56.2 | 65.6 | 78.8 | 69.7 | 58.3 | 47.6 | 34.4 | 27.8 | 48.4 |
| 2008 | 20.3 | 29.9 | 36.2 | 40.7 | 54.0 | 61.8 | 71.9 | 69.7 | 56.8 | 45.7 | 40.4 | 20.0 | 45.6 |
| 2009 | 27.4 | 32.3 | 31.0 | 42.3 | 55.4 | 60.3 | 69.4 | 68.5 | 64.1 | 40.2 | 36.7 | 10.8 | 44.9 |
| POR= 117 YRS | 20.7 | 24.3 | 33.0 | 42.9 | 52.5 | 59.3 | 68.3 | 66.6 | 55.2 | 45.4 | 31.9 | 23.7 | 43.6 |

## HEATING DEGREE DAYS (base 65°F) 2009 HELENA (KHLN)

| YEAR | JUL | AUG | SEP | OCT | NOV | DEC | JAN | FEB | MAR | APR | MAY | JUN | TOTAL |
|------|-----|-----|-----|-----|-----|-----|-----|-----|-----|-----|-----|-----|-------|
| 1980-81 | 25 | 81 | 242 | 602 | 899 | 1175 | 1127 | 986 | 832 | 552 | 371 | 191 | 7083 |
| 1981-82 | 21 | 16 | 195 | 650 | 853 | 1227 | 1497 | 1153 | 959 | 726 | 428 | 136 | 7861 |
| 1982-83 | 30 | 16 | 304 | 618 | 1120 | 1306 | 1059 | 828 | 823 | 649 | 417 | 152 | 7322 |
| 1983-84 | 76 | 0 | 351 | 584 | 901 | 1842 | 1164 | 941 | 856 | 640 | 380 | 174 | 7909 |
| 1984-85 | 2 | 7 | 377 | 716 | 954 | 1654 | 1625 | 1291 | 973 | 538 | 266 | 97 | 8500 |
| 1985-86 | 3 | 105 | 455 | 696 | 1571 | 1545 | 1218 | 1202 | 677 | 645 | 380 | 42 | 8539 |
| 1986-87 | 66 | 23 | 409 | 602 | 1077 | 1432 | 1288 | 923 | 862 | 437 | 276 | 77 | 7472 |
| 1987-88 | 75 | 104 | 163 | 556 | 901 | 1241 | 1426 | 1034 | 889 | 529 | 297 | 63 | 7278 |
| 1988-89 | 10 | 13 | 282 | 449 | 934 | 1292 | 1251 | 1650 | 1156 | 610 | 407 | 107 | 8161 |
| 1989-90 | 0 | 92 | 274 | 611 | 839 | 1268 | 1116 | 1058 | 925 | 573 | 426 | 177 | 7359 |
| 1990-91 | 15 | 31 | 78 | 604 | 823 | 1579 | 1420 | 767 | 914 | 630 | 373 | 159 | 7393 |
| 1991-92 | 2 | 0 | 220 | 666 | 1053 | 1297 | 1273 | 884 | 687 | 513 | 237 | 94 | 6926 |
| 1992-93 | 68 | 141 | 246 | 564 | 960 | 1552 | 1631 | 1397 | 879 | 617 | 228 | 214 | 8497 |
| 1993-94 | 179 | 122 | 323 | 652 | 1139 | 1080 | 1080 | 1236 | 783 | 576 | 268 | 128 | 7566 |
| 1994-95 | 18 | 17 | 110 | 633 | 1095 | 1275 | 1307 | 963 | 1059 | 681 | 465 | 222 | 7845 |
| 1995-96 | 37 | 61 | 293 | 683 | 912 | 1303 | 1703 | 1268 | 1210 |  | 510 | 127 |  |
| 1996-97 | 15 | 53 | 286 | 651 | 1251 | 1494 | 1596 | 1026 | 865 | 793 | 361 | 152 | 8543 |
| 1997-98 | 49 | 53 | 189 | 642 | 1006 | 1252 | 1352 | 942 | 990 | 600 | 336 | 293 | 7704 |
| 1998-99 | 2 | 8 | 158 | 650 | 909 | 1248 | 1176 | 905 | 859 | 718 | 451 | 188 | 7272 |
| 1999-00 | 79 | 20 | 354 | 590 | 767 | 1136 | 1232 | 1043 | 814 | 535 | 308 | 119 | 6997 |
| 2000-01 | 16 | 13 | 279 | 625 | 1283 | 1504 | 1385 | 1319 | 912 | 625 | 212 | 113 | 8286 |
| 2001-02 | 11 | 0 | 66 | 511 | 766 | 1249 | 1093 | 968 | 1229 | 652 | 366 | 126 | 7037 |
| 2002-03 | 1 | 52 | 201 | 755 | 904 | 1138 | 1096 | 1104 | 951 | 562 | 389 | 113 | 7266 |
| 2003-04 | 0 | 8 | 208 | 442 | 1081 | 1128 | 1542 | 1038 | 675 | 513 | 401 | 151 | 7187 |
| 2004-05 | 8 | 53 | 264 | 581 | 883 | 1093 | 1499 | 939 | 860 | 600 | 382 | 177 | 7339 |
| 2005-06 | 7 | 46 | 223 | 487 | 906 | 1395 | 892 | 1004 | 907 | 507 | 282 | 61 | 6717 |
| 2006-07 | 0 | 20 | 202 | 644 | 959 | 1171 | 1215 | 1045 | 673 | 566 | 274 | 69 | 6838 |
| 2007-08 | 0 | 15 | 236 | 529 | 911 | 1145 | 1378 | 1010 | 885 | 726 | 351 | 158 | 7344 |
| 2008-09 | 0 | 32 | 239 | 592 | 732 | 1387 | 1164 | 909 | 1047 | 675 | 310 | 167 | 7254 |
| 2009- | 17 | 34 | 101 | 763 | 843 | 1673 |  |  |  |  |  |  |  |

WBAN : 24144

## COOLING DEGREE DAYS (base 65°F) 2009 HELENA (KHLN)

| YEAR | JAN | FEB | MAR | APR | MAY | JUN | JUL | AUG | SEP | OCT | NOV | DEC | TOTAL |
|------|-----|-----|-----|-----|-----|-----|-----|-----|-----|-----|-----|-----|-------|
| 1980 | 0 | 0 | 0 | 0 | 14 | 14 | 104 | 25 | 4 | 0 | 0 | 0 | 161 |
| 1981 | 0 | 0 | 0 | 0 | 0 | 15 | 109 | 165 | 16 | 0 | 0 | 0 | 305 |
| 1982 | 0 | 0 | 0 | 0 | 0 | 30 | 147 | 151 | 25 | 0 | 0 | 0 | 353 |
| 1983 | 0 | 0 | 0 | 0 | 4 | 16 | 115 | 186 | 12 | 0 | 0 | 0 | 333 |
| 1984 | 0 | 0 | 0 | 0 | 10 | 31 | 165 | 163 | 4 | 0 | 0 | 0 | 373 |
| 1985 | 0 | 0 | 0 | 0 | 2 | 55 | 318 | 54 | 0 | 0 | 0 | 0 | 429 |
| 1986 | 0 | 0 | 0 | 0 | 35 | 91 | 45 | 123 | 1 | 0 | 0 | 0 | 295 |
| 1987 | 0 | 0 | 0 | 0 | 3 | 66 | 122 | 41 | 15 | 0 | 0 | 0 | 247 |
| 1988 | 0 | 0 | 0 | 0 | 8 | 170 | 211 | 132 | 30 | 0 | 0 | 0 | 551 |
| 1989 | 0 | 0 | 0 | 0 | 0 | 30 | 222 | 82 | 0 | 0 | 0 | 0 | 334 |
| 1990 | 0 | 0 | 0 | 0 | 0 | 77 | 159 | 142 | 42 | 0 | 0 | 0 | 420 |
| 1991 | 0 | 0 | 0 | 0 | 0 | 2 | 188 | 250 | 29 | 0 | 0 | 0 | 469 |
| 1992 | 0 | 0 | 0 | 0 | 10 | 96 | 55 | 140 | 3 | 0 | 0 | 0 | 304 |
| 1993 | 0 | 0 | 0 | 0 | 2 | 24 | 15 | 23 | 0 | 0 | 0 | 0 | 64 |
| 1994 | 0 | 0 | 0 | 0 | 1 | 67 | 146 | 176 | 16 | 0 | 0 | 0 | 406 |
| 1995 | 0 | 0 | 0 | 0 | 0 | 5 | 71 | 73 | 24 | 0 | 0 | 0 | 173 |
| 1996 | 0 | 0 | 0 |  | 0 | 25 | 125 | 109 | 9 | 0 | 0 | 0 |  |
| 1997 | 0 | 0 | 0 | 0 | 1 | 13 | 81 | 71 | 12 | 0 | 0 | 0 | 178 |
| 1998 | 0 | 0 | 0 | 0 | 0 | 0 | 161 | 124 | 72 | 0 | 0 | 0 | 357 |
| 1999 | 0 | 0 | 0 | 0 | 0 | 10 | 93 | 132 | 0 | 0 | 0 | 0 | 235 |
| 2000 | 0 | 0 | 0 | 0 | 2 | 50 | 245 | 165 | 27 | 0 | 0 | 0 | 489 |
| 2001 | 0 | 0 | 0 | 0 | 41 | 94 | 237 | 308 | 61 | 0 | 0 | 0 | 741 |
| 2002 | 0 | 0 | 0 | 0 | 15 | 71 | 238 | 29 | 30 | 0 | 0 | 0 | 383 |
| 2003 | 0 | 0 | 0 | 0 | 31 | 66 | 359 | 276 | 43 | 2 | 0 | 0 | 777 |
| 2004 | 0 | 0 | 0 | 0 | 0 | 31 | 191 | 102 | 10 | 0 | 0 | 0 | 334 |
| 2005 | 0 | 0 | 0 | 0 | 0 | 28 | 247 | 165 | 21 | 0 | 0 | 0 | 461 |
| 2006 | 0 | 0 | 0 | 0 | 31 | 69 | 332 | 163 | 33 | 0 | 0 | 0 | 628 |
| 2007 | 0 | 0 | 0 | 0 | 8 | 95 | 434 | 167 | 42 | 0 | 0 | 0 | 746 |
| 2008 | 0 | 0 | 0 | 0 | 18 | 69 | 220 | 183 | 0 | 0 | 0 | 0 | 490 |
| 2009 | 0 | 0 | 0 | 0 | 22 | 33 | 157 | 147 | 82 | 0 | 0 | 0 | 441 |

### SNOWFALL (inches) 2009 HELENA (KHLN)

| YEAR | JUL | AUG | SEP | OCT | NOV | DEC | JAN | FEB | MAR | APR | MAY | JUN | TOTAL |
|------|-----|-----|-----|-----|-----|-----|-----|-----|-----|-----|-----|-----|-------|
| 1980-81 | 0.0 | 0.0 | 0.0 | 3.7 | 1.2 | 3.8 | 2.7 | 2.1 | 3.3 | 0.1 | T | 0.0 | 16.9 |
| 1981-82 | 0.0 | 0.0 | 0.0 | 5.2 | 3.4 | 6.1 | 18.4 | 4.8 | 13.9 | 4.1 | 0.8 | 0.0 | 56.7 |
| 1982-83 | 0.0 | 0.0 | 6.5 | 0.5 | 4.1 | 11.3 | 3.2 | 0.2 | 2.2 | 1.1 | 9.9 | 0.0 | 39.0 |
| 1983-84 | 0.0 | 0.0 | 6.4 | 0.0 | 1.3 | 13.0 | 1.3 | 1.5 | 5.7 | 2.3 | 0.8 | 0.0 | 32.3 |
| 1984-85 | 0.0 | 0.0 | 6.3 | 9.0 | 5.7 | 7.5 | 3.9 | 6.2 | 4.0 | 0.8 | 0.0 | 0.0 | 43.4 |
| 1985-86 | 0.0 | 0.0 | 2.9 | 8.8 | 10.4 | 8.5 | 4.2 | 15.6 | 1.2 | 11.6 | 0.2 | 0.0 | 63.4 |
| 1986-87 | 0.0 | 0.0 | 0.0 | T | 7.6 | 5.0 | 0.2 | 0.2 | 9.1 | 4.3 | 3.8 | 0.0 | 30.2 |
| 1987-88 | 0.0 | 0.0 | 0.0 | 0.3 | 0.9 | 1.2 | 4.4 | 8.0 | 5.2 | 2.9 | 0.0 | 0.0 | 22.9 |
| 1988-89 | 0.0 | 0.0 | 5.9 | 1.5 | 6.0 | 5.0 | 23.0 | 13.0 | 20.7 | 7.9 | 3.5 | T | 86.5 |
| 1989-90 | 0.0 | T | T | 2.6 | 1.8 | 9.4 | 4.3 | 1.8 | 14.0 | 0.8 | 0.1 | 0.0 | 34.8 |
| 1990-91 | 0.0 | 0.5 | 0.0 | 0.5 | 8.6 | 11.6 | 5.9 | 0.8 | 12.2 | 5.6 | 0.4 | 0.5 | 46.6 |
| 1991-92 | 0.0 | T | 0.0 | 7.3 | 8.8 | 15.7 | 4.7 | 1.3 | T | 2.2 | T | 0.0 | 40.0 |
| 1992-93 | T | 6.2 | T | 8.6 | 1.5 | 11.8 | 10.2 | 17.9 | 2.3 | 2.5 | T | 0.2 | 61.2 |
| 1993-94 | T | 0.0 | T | 4.6 | 7.3 | 1.1 | 4.4 | 5.0 | 4.4 | 8.0 | T | 0.0 | 34.8 |
| 1994-95 | 0.0 | T | 0.0 | T | 8.6 | 2.0 | 0.3 | 1.6 | 8.1 | 0.0 | 0.0 | | |
| 1995-96 | T | 0.0 | T | | 7.4 | 2.7 | 14.2 | | | | | | |
| 1996-97 | | | | | | | | | | | | | |
| 1997-98 | | | | | | | | | | | | | |
| 1998-99 | | | | | | | | | | | | | |
| 1999-00 | | | | T | | | 3.9 | 5.7 | 2.3 | | | | |
| 2000-01 | | | | | | | | | | | | | |
| 2001-02 | | | | | | | | | | | | | |
| 2002-03 | | | | | | | | | | | | | |
| 2003-04 | | | | | | | | | | 4.8 | T | 0.0 | |
| 2004-05 | 0.0 | 0.0 | T | T | 1.3 | 6.7 | 11.2 | 0.4 | 10.9 | T | 0.0 | 0.0 | 30.5 |
| 2005-06 | 0.0 | 0.0 | 0.0 | 0.0 | 3.9 | 6.5 | 1.2 | 0.9 | 6.6 | 0.2 | 0.0 | 0.0 | 19.3 |
| 2006-07 | 0.0 | 0.0 | 0.0 | 1.2 | 6.1 | 0.4 | 1.7 | 8.8 | 0.3 | 0.3 | 0.0 | 0.0 | 18.8 |
| 2007-08 | 0.0 | 0.0 | 0.0 | 0.0 | 3.8 | T | 12.7 | 6.8 | 1.0 | 7.0 | T | T | 31.3 |
| 2008-09 | 0.0 | 0.0 | 0.0 | 1.2 | 0.2 | 15.9 | 5.7 | 4.7 | 8.3 | 5.7 | 0.0 | 0.2 | 41.9 |
| 2009- | 0.0 | 0.0 | 0.0 | 3.7 | 1.2 | 10.4 | | | | | | | |
| POR= 115 YRS | T | 0.1 | 1.1 | 2.9 | 6.6 | 8.2 | 9.8 | 7.5 | 8.4 | 4.9 | 1.5 | 0.1 | 51.1 |

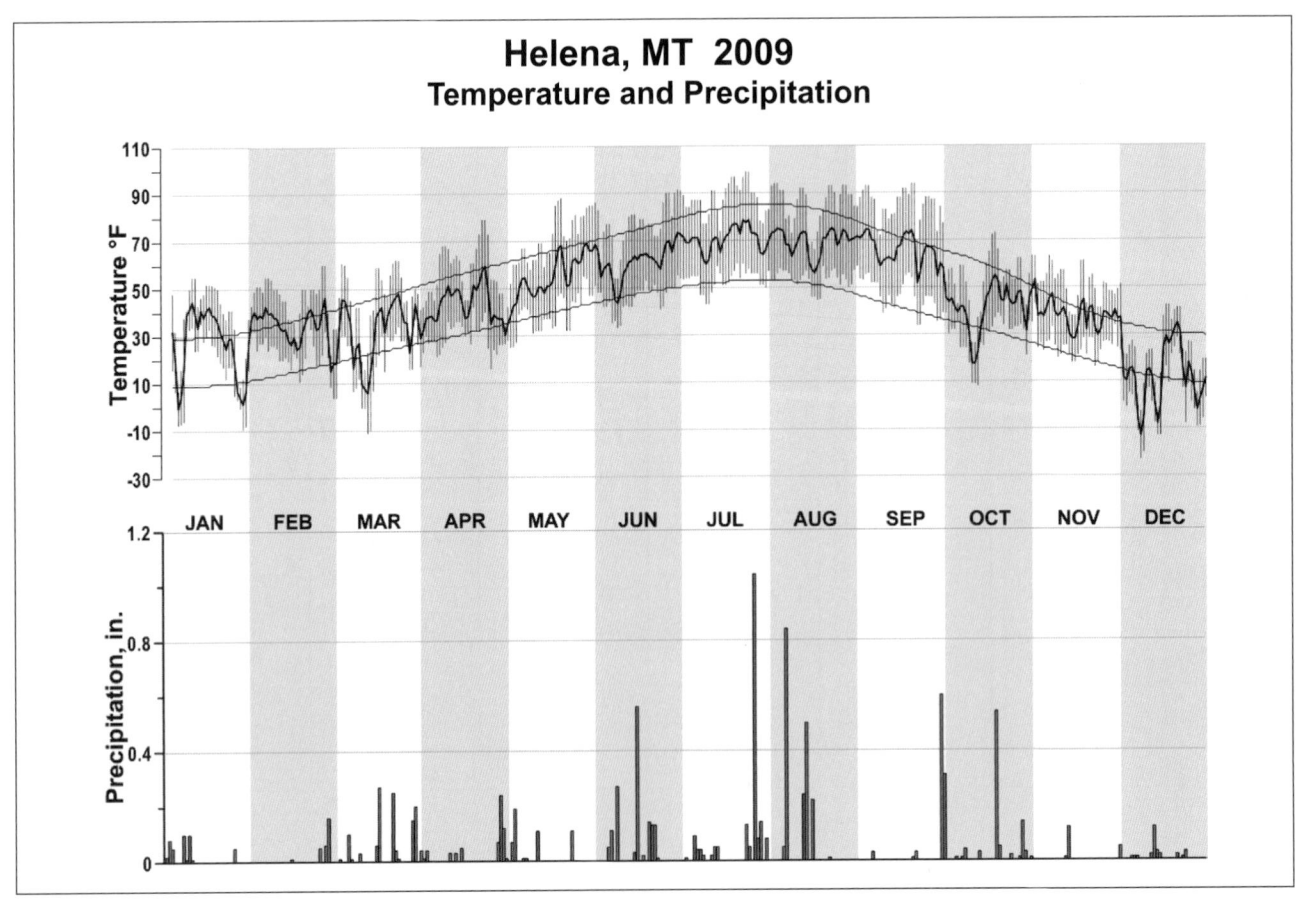

## Helena, MT  2009
### Temperature and Precipitation

# 2009
# SCOTTSBLUFF
# NEBRASKA (KBFF)

Scottsbluff is located in the North Platte river valley that extends from central Wyoming southeast across western Nebraska. The valley is approximately 20 miles wide in the vicinity of Scottsbluff with a range of hills both to the north and south, parallel to the river. To the south the hills average 600 to 700 feet above the river with some projections upward to 1,000 feet. To the north, rolling hills range from 300 to 400 feet higher than the river.

Due to the protection of the higher hills to the south, southerly winds in the valley are rare. Prevailing winds are west to northwest during the winter months and east to southeast during the summer months. West to northwest winds are intensified by the funneling action of the valley and velocities of 30 to 50 mph are common during the winter and early spring. Quite often these winds are warmed by the downslope (chinook) effect from the higher elevations to the west and bring rapid warming and melting of the snow. Outbreaks of Arctic air bring cold wave conditions about five times each season. Snow with strong winds causing blowing and drifting snow occur several times each winter with a severe blizzard of extended duration occurring about once every thirty years. Easterly winds during the winter and early spring cause upslope conditions with low cloudiness and precipitation.

The average temperature is in the upper 40s. Summertime highs generally range from the 80s to the 90s with lows around 60. Summer temperatures of 100 degrees are reached or exceeded at least once each summer. In winter, highs average about 40 degrees with lows in the teens. Temperatures of zero or below occur about 15 times each winter.

Most of the precipitation occurs as thunderstorms during the spring and summer months. Severe thunderstorms with destructive hail are quite common during the late spring and summer. Tornadoes are infrequent and usually of short duration.

The Platte River in the vicinity of Scottsbluff is a wide shallow stream and has very little effect on the climate. Water stored in numerous upstream reservoirs is used for extensive irrigation in the valley. Lowland flooding occurs when heavy rains fall upstream and a greater than normal amount of water is being released from the upstream reservoirs.

Based on the 1951-1980 period, the average first occurrence of 32 degrees Fahrenheit in the fall is September 29 and the average last occurrence in the spring is May 7.

# NORMALS, MEANS, AND EXTREMES
## SCOTTSBLUFF (KBFF)

LATITUDE: 41°52'N  LONGITUDE: -103°35'W  
ELEVATION (FT): GRND: 3948  BARO: 3949  
TIME ZONE: MOUNTAIN (UTC -7)  
WBAN: 24028

| | ELEMENT | POR | JAN | FEB | MAR | APR | MAY | JUN | JUL | AUG | SEP | OCT | NOV | DEC | YEAR |
|---|---|---|---|---|---|---|---|---|---|---|---|---|---|---|---|
| **TEMPERATURE °F** | NORMAL DAILY MAXIMUM | 30 | 38.0 | 44.3 | 51.7 | 61.0 | 71.1 | 82.2 | 88.7 | 86.8 | 77.3 | 64.4 | 48.2 | 39.8 | 62.8 |
| | MEAN DAILY MAXIMUM | 116 | 38.9 | 41.1 | 50.7 | 60.5 | 71.0 | 80.4 | 89.2 | 86.8 | 76.2 | 64.5 | 49.8 | 40.5 | 62.5 |
| | HIGHEST DAILY MAXIMUM | 67 | 74 | 77 | 87 | 93 | 103 | 106 | 109 | 104 | 102 | 92 | 80 | 77 | 109 |
| | YEAR OF OCCURRENCE | | 1982 | 1962 | 1943 | 1992 | 2003 | 1990 | 1989 | 2007 | 1998 | 1967 | 1989 | 1980 | JUL 1989 |
| | MEAN OF EXTREME MAXS. | 117 | 60.5 | 64.8 | 73.6 | 82.7 | 90.2 | 97.3 | 101.4 | 98.8 | 94.1 | 85.1 | 72.1 | 62.3 | 81.9 |
| | NORMAL DAILY MINIMUM | 30 | 11.0 | 15.8 | 23.0 | 31.4 | 42.4 | 52.1 | 57.4 | 54.9 | 43.7 | 32.7 | 21.3 | 14.1 | 33.3 |
| | MEAN DAILY MINIMUM | 116 | 12.2 | 14.8 | 22.4 | 31.5 | 42.3 | 51.1 | 57.8 | 54.9 | 43.9 | 32.7 | 21.3 | 14.1 | 33.3 |
| | LOWEST DAILY MINIMUM | 67 | -32 | -28 | -27 | -8 | 15 | 30 | 40 | 39 | 19 | -6 | -13 | -42 | -42 |
| | YEAR OF OCCURRENCE | | 1963 | 1962 | 1948 | 1975 | 1983 | 1969 | 1959 | 2004 | 1985 | 1991 | 2004 | 1989 | DEC 1989 |
| | MEAN OF EXTREME MINS. | 117 | -11.1 | -5.0 | 2.9 | 16.4 | 29.5 | 40.1 | 49.5 | 46.7 | 31.3 | 18.7 | 2.8 | -8.3 | 17.8 |
| | NORMAL DRY BULB | 30 | 24.5 | 30.0 | 37.3 | 46.2 | 56.8 | 67.2 | 73.0 | 70.9 | 60.5 | 47.8 | 34.0 | 25.7 | 47.8 |
| | MEAN DRY BULB | 116 | 25.6 | 28.0 | 36.5 | 46.0 | 56.7 | 65.9 | 73.5 | 70.9 | 60.1 | 48.6 | 35.6 | 27.4 | 47.9 |
| | MEAN WET BULB | 26 | 21.6 | 24.2 | 30.2 | 37.6 | 47.5 | 55.4 | 60.6 | 59.4 | 50.3 | 38.8 | 28.3 | 21.2 | 39.6 |
| | MEAN DEW POINT | 26 | 16.8 | 17.9 | 24.5 | 31.4 | 42.0 | 50.3 | 55.7 | 54.8 | 45.0 | 33.0 | 23.4 | 16.2 | 34.3 |
| | NORMAL NO. DAYS WITH: | | | | | | | | | | | | | | |
| | MAXIMUM >= 90 | 30 | 0.0 | 0.0 | 0.0 | 0.1 | 1.0 | 7.8 | 15.9 | 13.6 | 4.9 | 0.1 | 0.0 | 0.0 | 43.4 |
| | MAXIMUM <= 32 | 30 | 9.2 | 5.9 | 2.7 | 0.6 | 0.0 | 0.0 | 0.0 | 0.0 | * | 0.4 | 3.9 | 8.2 | 30.9 |
| | MINIMUM <= 32 | 30 | 30.0 | 27.1 | 26.6 | 14.7 | 1.9 | 0.0 | 0.0 | 0.0 | 1.6 | 12.8 | 26.8 | 30.2 | 171.7 |
| | MINIMUM <= 0 | 30 | 5.6 | 3.3 | 0.7 | 0.1 | 0.0 | 0.0 | 0.0 | 0.0 | 0.0 | 0.0 | 1.1 | 4.4 | 15.3 |
| **H/C** | NORMAL HEATING DEG. DAYS | 30 | 1241 | 969 | 843 | 549 | 259 | 53 | 9 | 12 | 170 | 517 | 917 | 1203 | 6742 |
| | NORMAL COOLING DEG. DAYS | 30 | 0 | 0 | 0 | 1 | 19 | 135 | 273 | 211 | 51 | 0 | 0 | 0 | 690 |
| **RH** | NORMAL (PERCENT) | 30 | 66 | 62 | 61 | 58 | 59 | 58 | 57 | 60 | 57 | 58 | 64 | 66 | 61 |
| | HOUR 05 LST | 30 | 74 | 75 | 77 | 77 | 80 | 80 | 81 | 84 | 80 | 76 | 76 | 75 | 78 |
| | HOUR 11 LST | 30 | 58 | 52 | 50 | 45 | 46 | 43 | 43 | 44 | 42 | 43 | 52 | 56 | 48 |
| | HOUR 17 LST | 30 | 58 | 48 | 44 | 40 | 42 | 38 | 37 | 39 | 37 | 40 | 54 | 58 | 45 |
| | HOUR 23 LST | 30 | 72 | 70 | 69 | 67 | 69 | 68 | 67 | 71 | 68 | 67 | 71 | 72 | 69 |
| **S** | PERCENT POSSIBLE SUNSHINE | | | | | | | | | | | | | | |
| **W/O** | MEAN NO. DAYS WITH: | | | | | | | | | | | | | | |
| | HEAVY FOG(VISBY <= 1/4 MI) | 46 | 1.3 | 1.0 | 1.8 | 1.0 | 0.5 | 0.5 | 0.6 | 1.0 | 1.0 | 1.0 | 1.8 | 1.0 | 12.5 |
| | THUNDERSTORMS | 62 | 0.0 | 0.0 | 0.4 | 2.2 | 8.1 | 11.6 | 11.2 | 8.9 | 4.5 | 1.0 | 0.1 | 0.1 | 48.1 |
| **CLOUDNESS** | MEAN: | | | | | | | | | | | | | | |
| | SUNRISE-SUNSET (OKTAS) | | | 4.0 | | | 7.2 | 3.2 | | | | | | | |
| | MIDNIGHT-MIDNIGHT (OKTAS) | | | | | | 8.0 | 3.2 | | | | | | | |
| | MEAN NO. DAYS WITH: | | | | | | | | | | | | | | |
| | CLEAR | 1 | 1.0 | 6.0 | 3.0 | | 2.0 | 16.0 | | | | | | | |
| | PARTLY CLOUDY | 1 | 2.0 | 8.0 | 3.0 | | 13.0 | 5.0 | | | | | | | |
| | CLOUDY | 1 | 7.0 | 3.0 | 9.0 | | 13.0 | 4.0 | | | | | | | |
| **PR** | MEAN STATION PRESSURE(IN) | 26 | 25.96 | 25.95 | 25.91 | 25.90 | 25.91 | 25.94 | 25.99 | 26.01 | 26.00 | 25.99 | 25.96 | 25.96 | 25.96 |
| | MEAN SEA-LEVEL PRES. (IN) | 26 | 30.10 | 30.07 | 29.99 | 29.93 | 29.88 | 29.87 | 29.91 | 29.94 | 29.97 | 30.02 | 30.06 | 30.11 | 29.99 |
| **WINDS** | MEAN SPEED (MPH) | 26 | 10.2 | 10.6 | 11.3 | 11.7 | 11.2 | 10.2 | 8.8 | 8.5 | 8.6 | 9.5 | 9.8 | 9.7 | 10.0 |
| | PREVAIL.DIR(TENS OF DEGS) | 35 | 31 | 31 | 31 | 31 | 31 | 12 | 11 | 11 | 31 | 31 | 31 | 31 | 31 |
| | MAXIMUM 2-MINUTE: | | | | | | | | | | | | | | |
| | SPEED (MPH) | 14 | 45 | 52 | 53 | 55 | 51 | 46 | 49 | 45 | 56 | 48 | 48 | 47 | 56 |
| | DIR. (TENS OF DEGS) | | 28 | 32 | 30 | 31 | 33 | 27 | 03 | 01 | 30 | 31 | 32 | 29 | 30 |
| | YEAR OF OCCURRENCE | | 1996 | 2007 | 1997 | 1997 | 1999 | 1997 | 2008 | 2007 | 2008 | 1996 | 2005 | 1997 | SEP 2008 |
| | MAXIMUM 3-SECOND | | | | | | | | | | | | | | |
| | SPEED (MPH) | 14 | 55 | 64 | 70 | 68 | 67 | 55 | 79 | 69 | 81 | 57 | 61 | 59 | 81 |
| | DIR. (TENS OF DEGS) | | 32 | 33 | 29 | 31 | 23 | 29 | 05 | 23 | 30 | 29 | 30 | 28 | 30 |
| | YEAR OF OCCURRENCE | | 2009 | 2007 | 2004 | 1997 | 2009 | 2009 | 2008 | 2009 | 2008 | 1996 | 2008 | 2004 | SEP 2008 |
| **PRECIPITATION** | NORMAL (IN) | 30 | 0.54 | 0.58 | 1.16 | 1.79 | 2.70 | 2.65 | 2.13 | 1.19 | 1.22 | 1.01 | 0.80 | 0.56 | 16.33 |
| | MAXIMUM MONTHLY (IN) | 67 | 1.26 | 1.93 | 2.64 | 3.89 | 7.25 | 8.33 | 4.82 | 3.48 | 4.22 | 3.16 | 2.15 | 1.54 | 8.33 |
| | YEAR OF OCCURRENCE | | 1978 | 1986 | 1990 | 1984 | 1987 | 1947 | 1978 | 2002 | 1973 | 2009 | 1993 | 1978 | JUN 1947 |
| | MINIMUM MONTHLY (IN) | 67 | T | T | 0.14 | 0.29 | 0.27 | 0.25 | 0.04 | 0.04 | T | 0.04 | T | 0.02 | T |
| | YEAR OF OCCURRENCE | | 1989 | 1996 | 2004 | 1962 | 1966 | 2007 | 2006 | 2001 | 1953 | 1956 | 1943 | 2002 | FEB 1996 |
| | MAXIMUM IN 24 HOURS (IN) | 67 | 0.92 | 0.88 | 1.68 | 2.11 | 2.62 | 3.74 | 2.53 | 2.01 | 3.28 | 1.64 | 1.43 | 0.90 | 3.74 |
| | YEAR OF OCCURRENCE | | 1976 | 1987 | 1974 | 1997 | 1988 | 1953 | 1948 | 1987 | 1951 | 2005 | 1993 | 1975 | JUN 1953 |
| | NORMAL NO. DAYS WITH: | | | | | | | | | | | | | | |
| | PRECIPITATION >= 0.01 | 30 | 6.2 | 5.4 | 7.4 | 9.3 | 11.2 | 10.2 | 9.1 | 7.3 | 7.1 | 5.6 | 5.3 | 5.7 | 89.8 |
| | PRECIPITATION >= 1.00 | 30 | 0.0 | 0.0 | 0.1 | 0.2 | 0.5 | 0.5 | 0.4 | 0.1 | 0.1 | 0.1 | * | 0.0 | 2.0 |
| **SNOWFALL** | NORMAL (IN) | 30 | 6.9 | 6.2 | 8.9 | 5.0 | 0.6 | 0.0 | 0.0 | 0.0 | 0.6 | 2.3 | 6.6 | 7.1 | 44.2 |
| | MAXIMUM MONTHLY (IN) | 67 | 23.7 | 23.4 | 23.5 | 22.2 | 7.5 | 0.1 | T | T | 5.7 | 31.3 | 18.5 | 18.2 | 31.3 |
| | YEAR OF OCCURRENCE | | 1949 | 1987 | 1980 | 1997 | 1967 | 1951 | 1993 | 1998 | 2000 | 2009 | 1983 | 2007 | OCT 2009 |
| | MAXIMUM IN 24 HOURS (IN) | 67 | 11.2 | 9.8 | 15.0 | 11.3 | 3.7 | 0.1 | T | T | 4.8 | 13.8 | 8.6 | 10.9 | 15.0 |
| | YEAR OF OCCURRENCE' | | 1976 | 1987 | 1974 | 1988 | 1979 | 1951 | 1993 | 1996 | 1985 | 2009 | 2004 | 1975 | MAR 1974 |
| | MAXIMUM SNOW DEPTH (IN) | 61 | 19 | 18 | 14 | 10 | 2 | 0 | 0 | 0 | 4 | 15 | 12 | 12 | 19 |
| | YEAR OF OCCURRENCE | | 1949 | 1949 | 1987 | 1975 | 1983 | | | | 1985 | 2009 | 1979 | 1987 | JAN 1949 |
| | NORMAL NO. DAYS WITH: | | | | | | | | | | | | | | |
| | SNOWFALL >= 1.0 | 30 | 2.0 | 2.1 | 2.9 | 1.6 | 0.2 | 0.0 | 0.0 | 0.0 | 0.2 | 0.7 | 2.1 | 2.2 | 14.0 |

## PRECIPITATION (inches) 2009 SCOTTSBLUFF (KBFF)

| YEAR | JAN | FEB | MAR | APR | MAY | JUN | JUL | AUG | SEP | OCT | NOV | DEC | ANNUAL |
|------|-----|-----|-----|-----|-----|-----|-----|-----|-----|-----|-----|-----|--------|
| 1980 | 1.21 | 0.99 | 2.16 | 0.57 | 2.82 | 0.79 | 1.07 | 0.47 | 0.47 | 0.76 | 0.57 | 0.15 | 12.03 |
| 1981 | 0.69 | 0.14 | 0.59 | 1.47 | 2.75 | 2.54 | 3.54 | 1.10 | 0.39 | 0.34 | 0.26 | 0.19 | 14.00 |
| 1982 | 0.32 | 0.20 | 0.46 | 0.50 | 2.93 | 6.63 | 4.78 | 1.66 | 1.78 | 1.22 | 0.80 | 0.57 | 21.85 |
| 1983 | 0.29 | 0.04 | 1.94 | 2.33 | 4.20 | 1.81 | 0.69 | 1.23 | 0.13 | 0.68 | 1.75 | 0.60 | 15.69 |
| 1984 | 0.44 | 0.50 | 1.47 | 3.89 | 1.23 | 1.23 | 1.80 | 0.57 | 0.45 | 0.88 | 0.28 | 0.50 | 13.24 |
| 1985 | 0.64 | 0.20 | 0.37 | 1.23 | 0.86 | 1.76 | 0.80 | 0.18 | 2.71 | 1.01 | 1.28 | 1.17 | 12.21 |
| 1986 | 0.07 | 1.93 | 0.83 | 2.49 | 1.51 | 5.55 | 4.00 | 1.01 | 1.86 | 1.42 | 0.81 | 0.26 | 21.74 |
| 1987 | 0.34 | 1.88 | 1.70 | 0.44 | 7.25 | 4.13 | 1.14 | 3.42 | 0.90 | 0.08 | 0.95 | 1.01 | 23.24 |
| 1988 | 0.80 | 0.11 | 1.11 | 2.27 | 5.19 | 2.29 | 0.85 | 0.80 | 0.97 | 0.11 | 0.46 | 0.40 | 15.36 |
| 1989 | T | 1.03 | 0.77 | 0.65 | 1.89 | 1.15 | 0.32 | 1.13 | 1.63 | 0.70 | 0.07 | 0.65 | 9.99 |
| 1990 | 0.59 | 0.72 | 2.64 | 1.75 | 2.94 | 1.14 | 3.10 | 1.23 | 0.97 | 0.99 | 1.25 | 0.36 | 17.68 |
| 1991 | 0.46 | 0.39 | 0.50 | 1.16 | 4.35 | 4.00 | 0.56 | 0.11 | 0.90 | 1.17 | 0.72 | 0.02 | 14.34 |
| 1992 | 0.81 | 0.86 | 1.22 | 0.34 | 2.03 | 3.00 | 2.96 | 1.65 | 0.17 | 1.15 | 0.98 | 0.66 | 15.83 |
| 1993 | 0.45 | 1.64 | 1.36 | 1.95 | 0.98 | 5.55 | 3.10 | 2.53 | 2.17 | 2.35 | 2.15 | 0.59 | 24.82 |
| 1994 | 0.59 | 0.77 | 0.73 | 1.96 | 1.10 | 2.80 | 2.56 | 0.45 | 0.66 | 2.76 | 0.64 | 0.95 | 15.97 |
| 1995 | 1.07 | 0.60 | 0.37 | 2.41 | 4.59 | 3.46 | 0.87 | 0.08 | 1.36 | 0.84 | 0.50 | 0.55 | 16.70 |
| 1996 | 0.83 | T | 1.03 | 0.91 | 4.48 | 1.02 | 2.06 | 2.24 | 2.44 | 0.42 | 0.89 | 0.22 | 16.54 |
| 1997 | 0.26 | 0.36 | 0.18 | 3.89 | 5.34 | 3.40 | 2.28 | 1.46 | 0.93 | 1.83 | 0.11 | 0.31 | 20.35 |
| 1998 | 0.20 | 0.64 | 1.30 | 1.53 | 1.46 | 2.32 | 3.38 | 1.19 | 0.41 | 2.76 | 1.20 | 0.86 | 17.25 |
| 1999 | 0.07 | 0.22 | 1.03 | 3.47 | 1.45 | 3.70 | 1.71 | 2.34 | 2.40 | 0.06 | 0.24 | 0.13 | 16.82 |
| 2000 | 0.48 | 0.89 | 1.04 | 2.80 | 1.48 | 0.68 | 1.70 | 0.33 | 2.31 | 2.47 | 0.37 | 0.24 | 14.79 |
| 2001 | 0.28 | 0.29 | 0.42 | 3.03 | 2.22 | 1.70 | 2.79 | 0.04 | 1.01 | 0.94 | 0.30 | T | 13.02 |
| 2002 | 0.05 | 0.03 | 0.66 | 0.44 | 0.73 | 0.59 | 0.08 | 3.48 | 0.69 | 0.87 | 0.15 | T | 7.77 |
| 2003 | 0.12 | 0.77 | 1.79 | 1.42 | 1.27 | 1.63 | 0.47 | 0.59 | 0.94 | 0.31 | 0.71 | 0.44 | 10.46 |
| 2004 | 0.13 | 0.73 | 0.14 | 0.90 | 0.57 | 1.70 | 2.24 | 0.21 | 2.81 | 1.20 | 1.35 | 0.06 | 12.04 |
| 2005 | 0.66 | 0.25 | 1.22 | 2.62 | 2.39 | 5.58 | 1.67 | 1.91 | 0.76 | 2.18 | 0.26 | 0.14 | 19.64 |
| 2006 | 0.49 | 0.84 | 1.36 | 0.84 | 1.12 | 3.59 | 0.04 | 1.34 | 0.63 | 0.53 | 0.06 | 1.19 | 12.03 |
| 2007 | 0.08 | 0.38 | 1.66 | 1.34 | 1.09 | 0.25 | 0.69 | 1.40 | 0.41 | 0.71 | 0.05 | 1.30 | 9.36 |
| 2008 | 0.01 | 0.33 | 0.84 | 1.26 | 2.24 | 2.17 | 1.37 | 3.10 | 1.69 | 0.86 | 0.20 | 0.20 | 14.27 |
| 2009 | 0.92 | 0.25 | 0.80 | 2.98 | 1.40 | 5.96 | 1.91 | 0.95 | 0.70 | 3.16 | 0.30 | 0.72 | 20.05 |
| POR= 116 YRS | 0.40 | 0.49 | 0.91 | 1.78 | 2.65 | 2.71 | 1.84 | 1.24 | 1.25 | 0.93 | 0.60 | 0.50 | 15.30 |

WBAN : 24028

## AVERAGE TEMPERATURE (°F) 2009 SCOTTSBLUFF (KBFF)

| YEAR | JAN | FEB | MAR | APR | MAY | JUN | JUL | AUG | SEP | OCT | NOV | DEC | ANNUAL |
|------|-----|-----|-----|-----|-----|-----|-----|-----|-----|-----|-----|-----|--------|
| 1980 | 23.1 | 29.7 | 36.1 | 49.4 | 58.6 | 72.3 | 78.2 | 73.1 | 66.3 | 51.6 | 40.8 | 37.0 | 51.4 |
| 1981 | 32.5 | 32.8 | 42.7 | 56.4 | 57.4 | 71.3 | 76.2 | 72.8 | 66.9 | 50.9 | 42.7 | 31.2 | 52.8 |
| 1982 | 22.5 | 27.7 | 37.6 | 43.9 | 55.6 | 62.8 | 72.8 | 74.2 | 62.1 | 48.5 | 34.2 | 27.9 | 47.5 |
| 1983 | 32.6 | 36.5 | 36.8 | 40.7 | 51.1 | 63.9 | 74.2 | 76.7 | 63.9 | 51.5 | 34.2 | 12.4 | 47.9 |
| 1984 | 27.3 | 35.5 | 37.4 | 42.2 | 58.8 | 66.8 | 74.8 | 76.4 | 59.9 | 46.2 | 37.6 | 24.8 | 49.0 |
| 1985 | 20.8 | 25.6 | 39.5 | 50.7 | 61.7 | 66.3 | 75.2 | 71.6 | 57.4 | 47.5 | 22.0 | 20.2 | 46.5 |
| 1986 | 32.9 | 27.9 | 44.3 | 47.3 | 55.3 | 69.5 | 72.6 | 70.9 | 59.6 | 48.5 | 35.2 | 28.7 | 49.4 |
| 1987 | 29.0 | 33.8 | 33.3 | 51.0 | 60.4 | 68.3 | 74.7 | 68.9 | 60.5 | 47.6 | 38.5 | 25.2 | 49.3 |
| 1988 | 18.1 | 30.4 | 35.9 | 46.8 | 58.2 | 72.7 | 74.6 | 72.6 | 60.4 | 50.4 | 37.2 | 28.1 | 48.8 |
| 1989 | 30.4 | 18.4 | 35.8 | 48.2 | 58.4 | 66.4 | 77.1 | 72.5 | 61.2 | 48.8 | 39.3 | 20.5 | 48.1 |
| 1990 | 32.3 | 29.8 | 37.4 | 46.8 | 54.3 | 70.4 | 71.7 | 71.1 | 65.9 | 48.8 | 39.1 | 20.6 | 49.0 |
| 1991 | 22.7 | 37.3 | 38.9 | 45.9 | 59.2 | 68.2 | 74.0 | 73.9 | 62.7 | 47.6 | 32.7 | 31.9 | 49.6 |
| 1992 | 31.2 | 37.8 | 41.2 | 50.4 | 59.7 | 65.9 | 68.6 | 67.1 | 62.4 | 49.8 | 30.8 | 18.6 | 48.6 |
| 1993 | 20.9 | 17.6 | 37.7 | 45.9 | 58.8 | 63.1 | 69.9 | 68.6 | 57.2 | 46.4 | 29.4 | 30.0 | 45.5 |
| 1994 | 27.8 | 26.7 | 40.6 | 47.3 | 62.6 | 71.1 | 71.3 | 73.4 | 64.9 | 49.2 | 35.9 | 29.4 | 50.0 |
| 1995 | 24.5 | 34.4 | 37.9 | 42.9 | 50.9 | 64.0 | 71.9 | 73.9 | 59.9 | 46.6 | 38.4 | 27.1 | 47.7 |
| 1996 | 21.4 | 31.0 | 32.9 | 46.9 | 56.4 | 69.1 | 72.3 | 71.0 | 60.2 | 47.8 | 31.9 | 27.6 | 47.4 |
| 1997 | 24.8 | 30.0 | 39.3 | 39.6 | 56.1 | 67.8 | 72.4 | 69.7 | 63.5 | 49.0 | 34.0 | 29.6 | 48.0 |
| 1998 | 28.8 | 34.8 | 34.6 | 45.9 | 58.8 | 62.2 | 74.9 | 73.2 | 68.1 | 48.3 | 39.8 | 26.2 | 49.6 |
| 1999 | 31.6 | 37.8 | 40.4 | 43.6 | 55.9 | 66.7 | 75.2 | 72.4 | 57.6 | 49.9 | 43.1 | 33.0 | 50.6 |
| 2000 | 29.9 | 35.8 | 40.1 | 47.2 | 59.5 | 67.1 | 76.3 | 75.8 | 62.6 | 49.9 | 26.8 | 23.8 | 49.6 |
| 2001 | 28.9 | 26.0 | 38.4 | 47.6 | 56.8 | 67.7 | 76.6 | 73.8 | 64.4 | 49.0 | 38.7 | 29.9 | 49.8 |
| 2002 | 29.3 | 31.3 | 30.1 | 48.5 | 55.6 | 72.8 | 77.5 | 70.8 | 61.6 | 42.4 | 36.5 | 31.5 | 49.0 |
| 2003 | 32.0 | 25.8 | 39.1 | 48.7 | 57.6 | 64.5 | 78.0 | 74.6 | 59.2 | 52.2 | 33.7 | 29.0 | 49.5 |
| 2004 | 28.3 | 30.6 | 42.9 | 47.9 | 58.2 | 63.9 | 71.6 | 67.7 | 62.0 | 50.5 | 34.5 | 31.0 | 49.1 |
| 2005 | 28.4 | 34.5 | 39.0 | 45.7 | 55.3 | 66.3 | 75.3 | 69.3 | 64.0 | 49.5 | 40.2 | 26.8 | 49.5 |
| 2006 | 34.8 | 26.9 | 35.2 | 50.6 | 58.9 | 71.5 | 77.1 | 72.9 | 56.7 | 46.4 | 36.2 | 28.7 | 49.7 |
| 2007 | 22.7 | 27.9 | 43.7 | 45.1 | 59.2 | 69.5 | 77.7 | 75.3 | 63.8 | 51.4 | 39.0 | 20.0 | 49.6 |
| 2008 | 19.1 | 29.2 | 37.9 | 44.9 | 54.6 | 65.3 | 75.3 | 70.9 | 59.7 | 48.3 | 39.4 | 23.4 | 47.3 |
| 2009 | 28.9 | 34.1 | 39.1 | 45.2 | 58.7 | 65.5 | 71.0 | 69.7 | 63.4 | 40.4 | 40.4 | 18.3 | 47.9 |
| POR= 116 YRS | 25.6 | 28.0 | 36.5 | 46.0 | 56.7 | 65.9 | 73.5 | 70.9 | 60.1 | 48.6 | 35.6 | 27.4 | 47.9 |

### HEATING DEGREE DAYS (base 65°F) 2009  SCOTTSBLUFF (KBFF)

| YEAR | JUL | AUG | SEP | OCT | NOV | DEC | JAN | FEB | MAR | APR | MAY | JUN | TOTAL |
|---|---|---|---|---|---|---|---|---|---|---|---|---|---|
| 1980-81 | 0 | 1 | 61 | 411 | 721 | 859 | 998 | 897 | 684 | 261 | 253 | 16 | 5162 |
| 1981-82 | 6 | 0 | 42 | 432 | 662 | 1044 | 1307 | 1039 | 842 | 629 | 289 | 101 | 6393 |
| 1982-83 | 4 | 1 | 155 | 503 | 921 | 1142 | 997 | 792 | 864 | 722 | 429 | 112 | 6642 |
| 1983-84 | 7 | 0 | 126 | 412 | 919 | 1627 | 1165 | 851 | 850 | 677 | 219 | 60 | 6913 |
| 1984-85 | 0 | 0 | 223 | 574 | 812 | 1238 | 1367 | 1096 | 780 | 422 | 139 | 75 | 6726 |
| 1985-86 | 0 | 16 | 286 | 534 | 1284 | 1378 | 986 | 1036 | 636 | 524 | 297 | 27 | 7004 |
| 1986-87 | 0 | 0 | 162 | 504 | 891 | 1118 | 1110 | 868 | 978 | 420 | 156 | 28 | 6235 |
| 1987-88 | 10 | 37 | 153 | 532 | 788 | 1228 | 1446 | 999 | 894 | 539 | 241 | 17 | 6884 |
| 1988-89 | 1 | 4 | 154 | 446 | 825 | 1138 | 1065 | 1303 | 897 | 511 | 214 | 71 | 6629 |
| 1989-90 | 0 | 0 | 169 | 497 | 764 | 1372 | 1009 | 977 | 850 | 541 | 327 | 30 | 6536 |
| 1990-91 | 18 | 4 | 79 | 497 | 771 | 1372 | 1307 | 771 | 802 | 570 | 220 | 19 | 6430 |
| 1991-92 | 5 | 0 | 137 | 536 | 965 | 1018 | 1042 | 783 | 731 | 438 | 194 | 45 | 5894 |
| 1992-93 | 22 | 57 | 107 | 465 | 1019 | 1430 | 1362 | 1323 | 841 | 567 | 204 | 121 | 7518 |
| 1993-94 | 3 | 23 | 237 | 568 | 1061 | 1078 | 1147 | 1066 | 751 | 521 | 121 | 13 | 6589 |
| 1994-95 | 10 | 14 | 84 | 483 | 868 | 1095 | 1251 | 851 | 832 | 658 | 429 | 101 | 6676 |
| 1995-96 | 10 | 1 | 211 | 561 | 791 | 1166 | 1346 | 981 | 991 | 537 | 286 | 30 | 6911 |
| 1996-97 | 2 | 2 | 190 | 526 | 985 | 1152 | 1242 | 972 | 790 | 755 | 279 | 24 | 6919 |
| 1997-98 | 13 | 14 | 121 | 496 | 926 | 1088 | 1115 | 840 | 936 | 565 | 203 | 144 | 6461 |
| 1998-99 | 4 | 0 | 50 | 511 | 748 | 1192 | 1030 | 757 | 757 | 635 | 279 | 47 | 6010 |
| 1999-00 | 0 | 7 | 231 | 461 | 652 | 983 | 1082 | 840 | 768 | 533 | 191 | 58 | 5806 |
| 2000-01 | 0 | 3 | 172 | 459 | 1142 | 1265 | 1112 | 1086 | 817 | 518 | 259 | 80 | 6913 |
| 2001-02 | 0 | 3 | 85 | 490 | 784 | 1082 | 1102 | 940 | 1076 | 492 | 310 | 18 | 6382 |
| 2002-03 | 0 | 14 | 163 | 691 | 849 | 1031 | 1018 | 1089 | 795 | 483 | 263 | 85 | 6481 |
| 2003-04 | 0 | 12 | 201 | 390 | 931 | 1112 | 1130 | 991 | 681 | 505 | 229 | 93 | 6275 |
| 2004-05 | 22 | 45 | 133 | 444 | 911 | 1049 | 1128 | 848 | 798 | 573 | 316 | 85 | 6352 |
| 2005-06 | 6 | 22 | 112 | 474 | 736 | 1176 | 930 | 1060 | 917 | 425 | 228 | 5 | 6091 |
| 2006-07 | 0 | 6 | 256 | 572 | 858 | 1120 | 1304 | 1033 | 653 | 588 | 213 | 52 | 6655 |
| 2007-08 | 0 | 8 | 119 | 417 | 774 | 1389 | 1418 | 1030 | 836 | 597 | 315 | 67 | 6970 |
| 2008-09 | 1 | 16 | 166 | 510 | 762 | 1283 | 1111 | 859 | 797 | 588 | 221 | 69 | 6383 |
| 2009- | 12 | 11 | 116 | 757 | 733 | 1441 | | | | | | | |

WBAN : 24028

### COOLING DEGREE DAYS (base 65°F) 2009  SCOTTSBLUFF (KBFF)

| YEAR | JAN | FEB | MAR | APR | MAY | JUN | JUL | AUG | SEP | OCT | NOV | DEC | TOTAL |
|---|---|---|---|---|---|---|---|---|---|---|---|---|---|
| 1980 | 0 | 0 | 0 | 1 | 22 | 234 | 417 | 263 | 104 | 3 | 0 | 0 | 1044 |
| 1981 | 0 | 0 | 0 | 10 | 23 | 213 | 361 | 250 | 107 | 3 | 0 | 0 | 967 |
| 1982 | 0 | 0 | 0 | 0 | 3 | 40 | 251 | 291 | 74 | 0 | 0 | 0 | 659 |
| 1983 | 0 | 0 | 0 | 0 | 6 | 82 | 297 | 369 | 99 | 0 | 0 | 0 | 853 |
| 1984 | 0 | 0 | 0 | 0 | 35 | 124 | 311 | 359 | 78 | 0 | 0 | 0 | 907 |
| 1985 | 0 | 0 | 0 | 3 | 42 | 121 | 324 | 225 | 66 | 0 | 0 | 0 | 781 |
| 1986 | 0 | 0 | 0 | 0 | 1 | 167 | 243 | 192 | 10 | 0 | 0 | 0 | 613 |
| 1987 | 0 | 0 | 0 | 5 | 20 | 134 | 320 | 164 | 26 | 0 | 0 | 0 | 669 |
| 1988 | 0 | 0 | 0 | 0 | 37 | 256 | 302 | 248 | 22 | 0 | 0 | 0 | 865 |
| 1989 | 0 | 0 | 0 | 10 | 18 | 118 | 382 | 242 | 61 | 0 | 0 | 0 | 831 |
| 1990 | 0 | 0 | 0 | 0 | 1 | 199 | 232 | 202 | 113 | 3 | 0 | 0 | 750 |
| 1991 | 0 | 0 | 0 | 2 | 46 | 125 | 289 | 284 | 73 | 2 | 0 | 0 | 821 |
| 1992 | 0 | 0 | 0 | 7 | 36 | 79 | 140 | 131 | 39 | 0 | 0 | 0 | 432 |
| 1993 | 0 | 0 | 0 | 0 | 18 | 71 | 159 | 142 | 10 | 0 | 0 | 0 | 400 |
| 1994 | 0 | 0 | 0 | 0 | 51 | 203 | 213 | 279 | 86 | 0 | 0 | 0 | 832 |
| 1995 | 0 | 0 | 0 | 0 | 0 | 78 | 232 | 281 | 64 | 0 | 0 | 0 | 655 |
| 1996 | 0 | 0 | 0 | 0 | 26 | 161 | 234 | 195 | 52 | 0 | 0 | 0 | 668 |
| 1997 | 0 | 0 | 0 | 0 | 8 | 115 | 247 | 168 | 85 | 5 | 0 | 0 | 628 |
| 1998 | 0 | 0 | 0 | 0 | 18 | 68 | 319 | 262 | 150 | 0 | 0 | 0 | 817 |
| 1999 | 0 | 0 | 0 | 0 | 5 | 104 | 326 | 243 | 14 | 0 | 0 | 0 | 692 |
| 2000 | 0 | 0 | 0 | 0 | 31 | 130 | 356 | 344 | 107 | 0 | 0 | 0 | 968 |
| 2001 | 0 | 0 | 0 | 0 | 13 | 169 | 369 | 278 | 74 | 0 | 0 | 0 | 903 |
| 2002 | 0 | 0 | 0 | 4 | 23 | 256 | 393 | 199 | 67 | 0 | 0 | 0 | 942 |
| 2003 | 0 | 0 | 0 | 0 | 40 | 77 | 407 | 315 | 32 | 0 | 0 | 0 | 871 |
| 2004 | 0 | 0 | 0 | 0 | 25 | 67 | 231 | 135 | 51 | 0 | 0 | 0 | 509 |
| 2005 | 0 | 0 | 0 | 0 | 22 | 131 | 330 | 161 | 86 | 2 | 0 | 0 | 732 |
| 2006 | 0 | 0 | 0 | 0 | 45 | 206 | 382 | 257 | 14 | 1 | 0 | 0 | 905 |
| 2007 | 0 | 0 | 0 | 0 | 38 | 194 | 401 | 337 | 92 | 4 | 0 | 0 | 1066 |
| 2008 | 0 | 0 | 0 | 0 | 3 | 82 | 329 | 206 | 15 | 0 | 0 | 0 | 635 |
| 2009 | 0 | 0 | 0 | 0 | 34 | 91 | 205 | 163 | 76 | 0 | 0 | 0 | 569 |

## SNOWFALL (inches) 2009 SCOTTSBLUFF (KBFF)

| YEAR | JUL | AUG | SEP | OCT | NOV | DEC | JAN | FEB | MAR | APR | MAY | JUN | TOTAL |
|---|---|---|---|---|---|---|---|---|---|---|---|---|---|
| 1980-81 | 0.0 | 0.0 | 0.0 | 1.2 | 3.7 | 1.7 | 8.3 | 2.1 | 3.2 | 1.3 | 0.0 | 0.0 | 21.5 |
| 1981-82 | 0.0 | 0.0 | 0.0 | 0.6 | 0.2 | 1.8 | 4.4 | 3.8 | 2.8 | 2.1 | T | 0.0 | 15.7 |
| 1982-83 | 0.0 | 0.0 | 0.0 | 5.3 | 4.3 | 4.9 | 2.8 | T | 16.4 | 8.9 | 2.6 | 0.0 | 45.2 |
| 1983-84 | 0.0 | 0.0 | T | 0.0 | 18.5 | 7.9 | 4.8 | 2.7 | 13.5 | 13.3 | 1.2 | 0.0 | 61.9 |
| 1984-85 | 0.0 | 0.0 | 1.6 | 2.2 | 2.5 | 7.1 | 7.2 | 2.3 | 3.2 | 1.7 | 0.0 | 0.0 | 27.8 |
| 1985-86 | 0.0 | 0.0 | 5.1 | 0.3 | 17.5 | 18.1 | 0.5 | 16.0 | 6.0 | 12.5 | 1.0 | 0.0 | 77.0 |
| 1986-87 | 0.0 | 0.0 | 0.0 | 1.5 | 5.1 | 2.8 | 3.7 | 23.4 | 16.6 | 0.4 | 0.0 | 0.0 | 53.5 |
| 1987-88 | 0.0 | 0.0 | 0.0 | 0.3 | 5.8 | 17.0 | 14.6 | 1.9 | 10.6 | 11.6 | T | 0.0 | 61.8 |
| 1988-89 | 0.0 | 0.0 | 0.0 | T | 3.5 | 4.1 | T | 13.0 | 9.2 | 2.7 | T | 0.0 | 32.5 |
| 1989-90 | 0.0 | 0.0 | T | 2.0 | 0.6 | 11.6 | 7.0 | 8.6 | 18.6 | 6.0 | 1.5 | 0.0 | 55.9 |
| 1990-91 | 0.0 | 0.0 | 0.0 | 7.6 | 10.7 | 4.4 | 5.7 | 2.2 | 4.2 | 2.6 | 1.8 | T | 39.2 |
| 1991-92 | 0.0 | 0.0 | 0.0 | 5.9 | 4.7 | 0.1 | 7.8 | 1.0 | 6.6 | T | 0.0 | T | 26.1 |
| 1992-93 | T | 0.0 | T | 4.1 | 10.3 | 10.4 | 7.9 | 19.0 | 11.5 | 2.0 | T | T | 65.2 |
| 1993-94 | T | 0.0 | 2.5 | 3.3 | 11.1 | 6.2 | 7.1 | 8.9 | 0.1 | 8.7 | 0.0 | T | 47.9 |
| 1994-95 | 0.0 | 0.0 | T | T | 4.2 | 11.0 | 11.5 | 9.9 | 3.4 | 7.6 | T | T | 47.6 |
| 1995-96 | 0.0 | 0.0 | 0.5 | 1.8 | 3.5 | 6.2 | 11.0 | T | 7.8 | 2.7 | T | 0.0 | 33.5 |
| 1996-97 | 0.0 | T | .0 | .7 | 8.3 | 3.7 | 2.7 |  |  | 22.2 | 0.0 | T |  |
| 1997-98 | 0.0 | 0.0 | 0.0 | 8.0 | 1.1 | 3.1 | 1.6 | 3.0 | 7.0 | 2.3 | T | 0.0 | 26.1 |
| 1998-99 | T | T | 0.0 | T | 11.9 | 8.0 | 0.6 | 1.6 | 7.2 | 4.1 | T | 0.0 | 33.4 |
| 1999-00 | 0.0 | 0.0 | 0.0 | 0.0 | 1.8 | 1.0 | 7.0 | 10.0 | 6.0 | 0.4 | 0.0 | 0.0 | 26.2 |
| 2000-01 | 0.0 | 0.0 | 5.7 | T | 2.5 | 5.0 | 4.5 | 8.8 | 4.8 | 7.0 | T | 0.0 | 38.3 |
| 2001-02 | 0.0 | 0.0 | 0.0 | 0.0 | 4.8 | T | T | 0.4 | 1.6 | 1.0 | 0.0 | 0.0 | 7.8 |
| 2002-03 | 0.0 | 0.0 | 0.0 | 9.0 | 2.5 | T | 4.0 | 11.0 | 7.2 | 7.0 | 0.0 | 0.0 | 40.7 |
| 2003-04 | 0.0 | 0.0 | 0.0 | 0.6 | 10.8 | 7.1 | 6.5 | 11.0 | 0.9 | 1.1 | 0.0 | 0.0 | 38.0 |
| 2004-05 | 0.0 | 0.0 | 0.0 | T | 11.7 | 1.2 | 7.9 | 2.6 | 7.4 | 7.1 | 2.0 | 0.0 | 39.9 |
| 2005-06 | 0.0 | 0.0 | 0.0 | 0.0 | 0.7 | 3.4 | 5.9 | 14.8 | 10.7 | 5.0 | T | T | 40.5 |
| 2006-07 | 0.0 | 0.0 | 0.0 | 3.0 | 0.4 | 15.9 | 1.2 | 2.9 | 5.0 | 2.8 | 0.0 | 0.0 | 31.2 |
| 2007-08 | 0.0 | 0.0 | 0.0 | 0.0 | 2.2 | 18.2 | 0.7 | 4.6 | 9.9 | 5.3 | 0.4 | 0.0 | 41.3 |
| 2008-09 | 0.0 | 0.0 | 0.0 | T | 1.2 | 4.6 | 14.3 | 2.4 | 7.7 | 9.3 | T | 0.0 | 39.5 |
| 2009- | 0.0 | 0.0 | 0.0 | 31.3 | 3.6 | 16.9 |  |  |  |  |  |  |  |
| POR= 84 YRS | T | T | 0.3 | 2.5 | 4.7 | 6.0 | 5.2 | 6.1 | 7.7 | 5.3 | 0.8 | T | 38.6 |

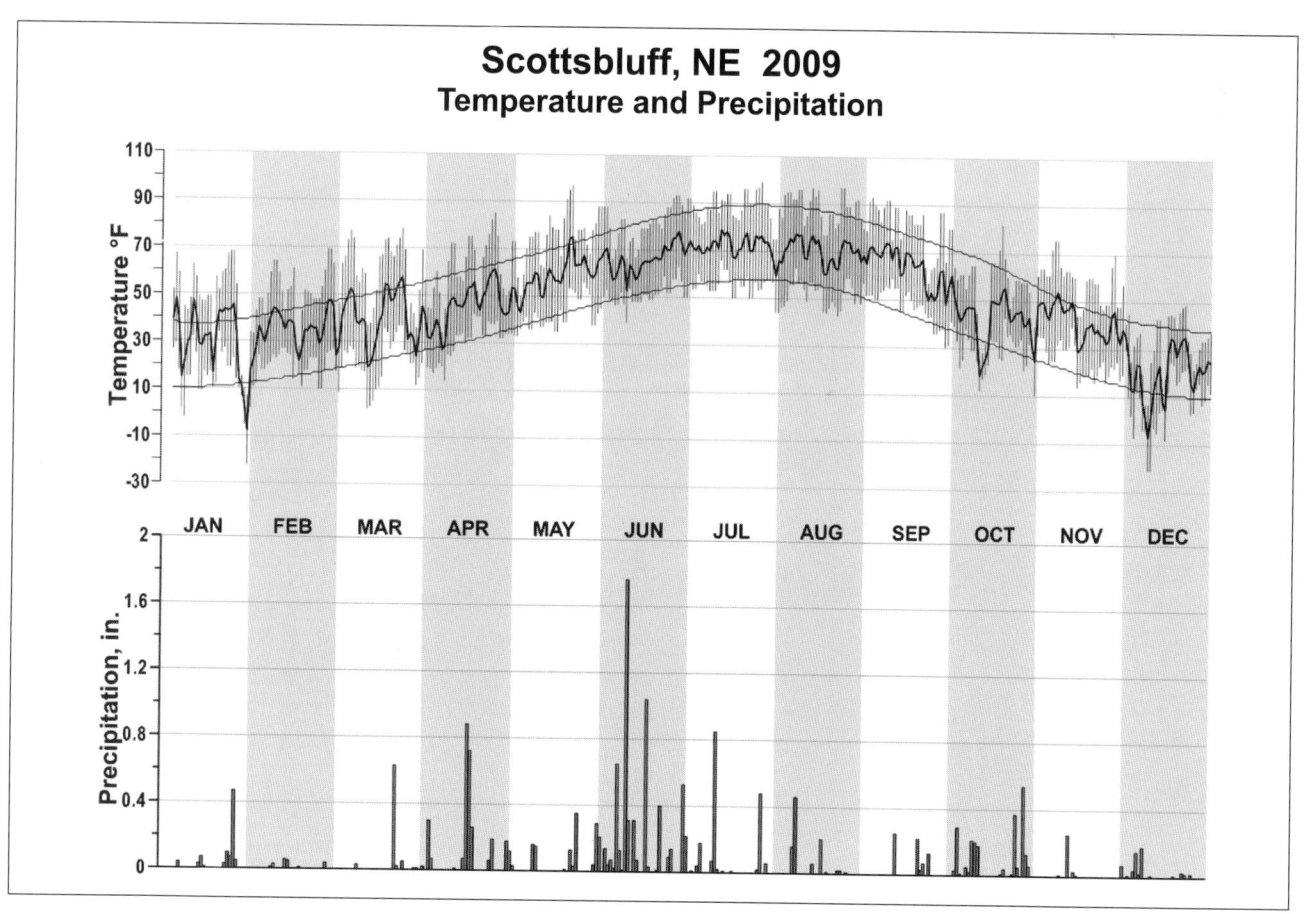

## Scottsbluff, NE 2009
### Temperature and Precipitation

# 2009
# NORTH PLATTE
# NEBRASKA (KLBF)

The climate of North Platte is characterized throughout the year by frequent rapid changes in the weather. During the winter, most North Pacific lows cross the country north of North Platte. The passage usually brings little or no snowfall, and only a moderate drop in temperature. Only when there is a major outbreak of cold air from Canada does the temperature fall to zero or below. The duration of below-zero temperature is hardly more than two mornings, and by the third or fourth day the temperature is ordinarily rising to the 40s or higher. Snowfall at the onset of a cold outbreak is usually less than 2 inches.

Only when a low moves from the middle Rockies through Nebraska, allowing easterly winds to draw moist air into the low circulation, does snowfall of appreciable amounts occur. Few of these storms move slowly enough, or are intense enough, to deposit much precipitation in the North Platte area. However, during some winters the cold outbreak and intense low from the mid-Rockies combine to produce severe cold and snow several inches in depth, with blizzard conditions following. During and after these snowfalls and blizzards, rail and highway traffic may be stalled until the snow is cleared. Widespread loss of unsheltered livestock and wild life results from such conditions.

The sudden and frequent weather changes of the winter continue through spring with decreasing intensity of temperature changes but increasing precipitation. The summer and fall months bring frequent changes from hot to cool weather. Most summer and fall precipitation is associated with thunderstorms, so the amounts are extremely variable. The surrounding area is occasionally damaged by locally severe winds and hailstorms.

Temperatures may reach into the upper 90s and lower 100s frequently during the summer months, but the elevation and clear skies bring rapid cooling after sunset to lows in the 60s or below by daybreak. Since the humidity is generally low, the extremely hot days of summer are not uncomfortable.

Based on the 1951-1980 period, the average first occurrence of 32 degrees Fahrenheit in the fall is September 24 and the average last occurrence in the spring is May 11.

# NORMALS, MEANS, AND EXTREMES
## NORTH PLATTE (KLBF)

LATITUDE: 41° 7 'N  LONGITUDE: -100° 40'W  ELEVATION (FT): GRND: 2763  BARO: 2781  TIME ZONE: CENTRAL (UTC -6)  WBAN: 24023

| | ELEMENT | POR | JAN | FEB | MAR | APR | MAY | JUN | JUL | AUG | SEP | OCT | NOV | DEC | YEAR |
|---|---|---|---|---|---|---|---|---|---|---|---|---|---|---|---|
| **TEMPERATURE °F** | NORMAL DAILY MAXIMUM | 30 | 36.5 | 43.3 | 52.1 | 62.7 | 72.0 | 82.6 | 88.4 | 86.8 | 78.0 | 65.6 | 48.5 | 39.2 | 63.0 |
| | MEAN DAILY MAXIMUM | 62 | 36.3 | 41.8 | 49.9 | 61.9 | 71.7 | 81.4 | 87.9 | 86.2 | 77.5 | 65.6 | 49.9 | 39.3 | 62.5 |
| | HIGHEST DAILY MAXIMUM | 58 | 73 | 79 | 86 | 98 | 99 | 107 | 112 | 108 | 102 | 94 | 83 | 75 | 112 |
| | YEAR OF OCCURRENCE | | 1990 | 1962 | 2004 | 1992 | 2006 | 1952 | 1954 | 2008 | 1990 | 1990 | 2006 | 1980 | JUL 1954 |
| | MEAN OF EXTREME MAXS. | 62 | 59.5 | 65.5 | 75.8 | 84.9 | 89.7 | 96.0 | 100.3 | 98.7 | 94.2 | 86.1 | 73.2 | 62.5 | 82.2 |
| | NORMAL DAILY MINIMUM | 30 | 9.9 | 15.4 | 23.8 | 33.4 | 44.5 | 54.2 | 60.2 | 58.4 | 46.7 | 33.7 | 20.7 | 12.1 | 34.4 |
| | MEAN DAILY MINIMUM | 62 | 9.7 | 14.9 | 22.6 | 33.6 | 44.9 | 54.7 | 60.7 | 58.5 | 47.1 | 34.0 | 21.2 | 12.8 | 34.6 |
| | LOWEST DAILY MINIMUM | 58 | -23 | -22 | -22 | 7 | 18 | 29 | 39 | 35 | 17 | 10 | -13 | -34 | -34 |
| | YEAR OF OCCURRENCE | | 2004 | 1981 | 2002 | 1975 | 2005 | 1969 | 1997 | 1976 | 1984 | 1993 | 1976 | 1989 | DEC 1989 |
| | MEAN OF EXTREME MINS. | 62 | -10.5 | -5.2 | 2.6 | 17.2 | 29.1 | 41.0 | 48.7 | 45.9 | 29.9 | 17.5 | 4.0 | -6.9 | 17.8 |
| | NORMAL DRY BULB | 30 | 23.2 | 29.4 | 38.0 | 48.1 | 58.3 | 68.4 | 74.3 | 72.6 | 62.4 | 49.7 | 34.6 | 25.7 | 48.7 |
| | MEAN DRY BULB | 62 | 23.0 | 28.4 | 36.3 | 47.7 | 58.3 | 68.3 | 74.3 | 72.4 | 62.3 | 49.8 | 35.6 | 26.0 | 48.5 |
| | MEAN WET BULB | 26 | 20.7 | 24.3 | 31.4 | 40.2 | 50.9 | 59.6 | 64.6 | 63.3 | 53.6 | 41.6 | 29.5 | 21.4 | 41.8 |
| | MEAN DEW POINT | 26 | 16.8 | 19.6 | 26.5 | 34.6 | 46.9 | 55.9 | 61.0 | 60.0 | 49.3 | 36.7 | 25.3 | 17.2 | 37.5 |
| | NORMAL NO. DAYS WITH: | | | | | | | | | | | | | | |
| | MAXIMUM >= 90 | 30 | 0.0 | 0.0 | 0.0 | 0.3 | 0.6 | 5.8 | 13.3 | 11.1 | 4.5 | 0.3 | 0.0 | 0.0 | 35.9 |
| | MAXIMUM <= 32 | 30 | 11.4 | 7.8 | 3.3 | 0.4 | 0.0 | 0.0 | 0.0 | 0.0 | 0.0 | 0.2 | 4.1 | 9.6 | 36.8 |
| | MINIMUM <= 32 | 30 | 31.0 | 27.7 | 26.3 | 13.5 | 2.3 | * | 0.0 | 0.0 | 2.3 | 13.8 | 27.6 | 30.8 | 175.3 |
| | MINIMUM <= 0 | 30 | 7.1 | 3.5 | 0.7 | 0.0 | 0.0 | 0.0 | 0.0 | 0.0 | 0.0 | 0.0 | 1.0 | 4.2 | 16.5 |
| **H/C** | NORMAL HEATING DEG. DAYS | 30 | 1312 | 1013 | 853 | 519 | 240 | 46 | 6 | 14 | 158 | 481 | 902 | 1222 | 6766 |
| | NORMAL COOLING DEG. DAYS | 30 | 0 | 0 | 0 | 4 | 22 | 139 | 279 | 234 | 70 | 2 | 0 | 0 | 750 |
| **RH** | NORMAL (PERCENT) | 30 | 71 | 69 | 66 | 62 | 66 | 65 | 64 | 67 | 64 | 63 | 68 | 71 | 66 |
| | HOUR 00 LST | 30 | 77 | 78 | 75 | 72 | 76 | 76 | 75 | 78 | 74 | 74 | 77 | 78 | 76 |
| | HOUR 06 LST | 30 | 80 | 81 | 82 | 81 | 85 | 85 | 84 | 86 | 84 | 82 | 82 | 80 | 83 |
| | HOUR 12 LST | 30 | 62 | 58 | 53 | 48 | 52 | 52 | 51 | 53 | 48 | 47 | 55 | 60 | 53 |
| | HOUR 18 LST | 30 | 64 | 56 | 49 | 44 | 50 | 48 | 47 | 49 | 46 | 48 | 59 | 63 | 52 |
| **S** | PERCENT POSSIBLE SUNSHINE | 52 | 63 | 63 | 63 | 64 | 65 | 71 | 77 | 75 | 72 | 70 | 61 | 61 | 67 |
| **W/O** | MEAN NO. DAYS WITH: | | | | | | | | | | | | | | |
| | HEAVY FOG(VISBY <= 1/4 MI) | 46 | 1.5 | 2.3 | 2.0 | 1.0 | 1.0 | 0.9 | 1.2 | 2.3 | 2.1 | 1.9 | 2.0 | 1.7 | 19.9 |
| | THUNDERSTORMS | 62 | 0.0 | 0.2 | 0.9 | 2.6 | 6.6 | 9.9 | 10.2 | 8.3 | 3.9 | 1.5 | 0.2 | 0.0 | 44.3 |
| **CLOUDNESS** | MEAN: | | | | | | | | | | | | | | |
| | SUNRISE-SUNSET (OKTAS) | 44 | 5.0 | 5.0 | 5.2 | 5.1 | 5.1 | 4.2 | 3.7 | 3.8 | 3.7 | 3.9 | 4.7 | 4.5 | 4.5 |
| | MIDNIGHT-MIDNIGHT (OKTAS) | 32 | 4.4 | 4.3 | 4.7 | 4.6 | 4.7 | 4.2 | 3.8 | 3.7 | 3.5 | 3.6 | 4.3 | 4.1 | 4.2 |
| | MEAN NO. DAYS WITH: | | | | | | | | | | | | | | |
| | CLEAR | 44 | 8.4 | 7.2 | 7.5 | 7.0 | 6.8 | 10.1 | 12.1 | 11.7 | 12.8 | 12.5 | 8.8 | 9.8 | 114.7 |
| | PARTLY CLOUDY | 44 | 8.7 | 7.5 | 7.7 | 9.0 | 10.0 | 10.7 | 12.0 | 11.4 | 8.4 | 8.4 | 8.4 | 7.8 | 109.6 |
| | CLOUDY | 44 | 13.9 | 13.6 | 15.8 | 14.0 | 14.3 | 9.2 | 6.9 | 8.0 | 8.8 | 10.1 | 13.2 | 13.4 | 141.2 |
| **PR** | MEAN STATION PRESSURE(IN) | 26 | 27.13 | 27.12 | 27.07 | 27.04 | 27.04 | 27.06 | 27.11 | 27.13 | 27.13 | 27.12 | 27.12 | 27.13 | 27.10 |
| | MEAN SEA-LEVEL PRES. (IN) | 26 | 30.12 | 30.10 | 30.01 | 29.93 | 29.90 | 29.89 | 29.93 | 29.96 | 29.99 | 30.03 | 30.07 | 30.12 | 30.00 |
| **WINDS** | MEAN SPEED (MPH) | 26 | 8.7 | 9.2 | 10.7 | 11.7 | 10.6 | 9.7 | 8.8 | 8.4 | 9.2 | 9.2 | 9.0 | 8.5 | 9.5 |
| | PREVAIL.DIR(TENS OF DEGS) | 38 | 32 | 32 | 32 | 12 | 17 | 12 | 12 | 12 | 12 | 32 | 32 | 32 | 32 |
| | MAXIMUM 2-MINUTE: | | | | | | | | | | | | | | |
| | SPEED (MPH) | 13 | 45 | 52 | 47 | 49 | 64 | 53 | 55 | 56 | 46 | 45 | 48 | 45 | 64 |
| | DIR. (TENS OF DEGS) | | 35 | 36 | 31 | 28 | 28 | 27 | 26 | 31 | 29 | 30 | 33 | 36 | 28 |
| | YEAR OF OCCURRENCE | | 2009 | 1999 | 2000 | 2002 | 2004 | 2009 | 2003 | 1999 | 2004 | 2009 | 2005 | 2000 | MAY 2004 |
| | MAXIMUM 3-SECOND | | | | | | | | | | | | | | |
| | SPEED (MPH) | 13 | 58 | 62 | 56 | 60 | 77 | 69 | 70 | 75 | 59 | 61 | 62 | 55 | 77 |
| | DIR. (TENS OF DEGS) | | 35 | 36 | 31 | 29 | 28 | 05 | 24 | 30 | 24 | 31 | 33 | 31 | 28 |
| | YEAR OF OCCURRENCE | | 2009 | 1999 | 2000 | 2006 | 2004 | 2006 | 1999 | 1999 | 2007 | 1997 | 2005 | 2005 | MAY 2004 |
| **PRECIPITATION** | NORMAL (IN) | 30 | 0.39 | 0.51 | 1.24 | 1.97 | 3.34 | 3.17 | 3.17 | 2.15 | 1.32 | 1.24 | 0.76 | 0.40 | 19.66 |
| | MAXIMUM MONTHLY (IN) | 58 | 1.12 | 1.98 | 2.98 | 5.94 | 8.01 | 6.81 | 7.05 | 6.30 | 6.03 | 4.78 | 2.89 | 2.56 | 8.01 |
| | YEAR OF OCCURRENCE | | 1960 | 1978 | 1992 | 2001 | 1962 | 1965 | 1979 | 1992 | 1963 | 2008 | 1979 | 2006 | MAY 1962 |
| | MINIMUM MONTHLY (IN) | 57 | T | T | 0.05 | 0.10 | 0.77 | 0.33 | 0.42 | 0.06 | T | 0.05 | 0.02 | T | T |
| | YEAR OF OCCURRENCE | | 1964 | 1996 | 1994 | 1989 | 1966 | 1952 | 1955 | 1967 | 1953 | 1988 | 2007 | 1988 | FEB 1996 |
| | MAXIMUM IN 24 HOURS (IN) | 57 | 0.75 | 3.52 | 2.26 | 2.42 | 3.62 | 3.80 | 3.15 | 2.93 | 2.53 | 2.30 | 1.48 | 1.44 | 3.80 |
| | YEAR OF OCCURRENCE | | 1992 | 2008 | 1959 | 1971 | 2007 | 1965 | 1964 | 1957 | 1963 | 2008 | 1979 | 2006 | JUN 1965 |
| | NORMAL NO. DAYS WITH: | | | | | | | | | | | | | | |
| | PRECIPITATION >= 0.01 | 30 | 5.0 | 5.4 | 7.2 | 8.5 | 11.4 | 9.7 | 10.6 | 8.5 | 6.8 | 5.5 | 5.0 | 4.0 | 87.6 |
| | PRECIPITATION >= 1.00 | 30 | 0.0 | * | 0.2 | 0.3 | 0.9 | 0.6 | 0.9 | 0.4 | 0.3 | 0.3 | * | 0.0 | 3.9 |
| **SNOWFALL** | NORMAL (IN) | 30 | 5.0 | 4.7 | 4.8 | 2.8 | 0.* | 0.0 | 0.0 | 0.0 | 0.2 | 1.0 | 4.9 | 4.4 | 27.8 |
| | MAXIMUM MONTHLY (IN) | 58 | 17.1 | 20.6 | 21.9 | 14.5 | 3.6 | T | T | T | 3.1 | 30.3 | 17.5 | 16.3 | 30.3 |
| | YEAR OF OCCURRENCE | | 1976 | 1978 | 1980 | 1984 | 1967 | 2009 | 2009 | 2008 | 1985 | 2009 | 1979 | 2006 | OCT 2009 |
| | MAXIMUM IN 24 HOURS (IN) | 58 | 11.9 | 9.7 | 15.1 | 8.5 | 2.3 | T | T | T | 3.1 | 11.8 | 9.9 | 8.6 | 15.1 |
| | YEAR OF OCCURRENCE' | | 1976 | 1955 | 1980 | 1984 | 1967 | 2009 | 2009 | 1992 | 1985 | 2009 | 2004 | 1968 | MAR 1980 |
| | MAXIMUM SNOW DEPTH (IN) | 61 | 18 | 13 | 18 | 9 | 1 | 0 | 0 | 0 | 3 | 12 | 13 | 10 | 18 |
| | YEAR OF OCCURRENCE | | 1949 | 1993 | 1980 | 1949 | 1967 | | | | 1985 | 2009 | 1979 | 2006 | MAR 1980 |
| | NORMAL NO. DAYS WITH: | | | | | | | | | | | | | | |
| | SNOWFALL >= 1.0 | 30 | 1.5 | 1.6 | 1.5 | 0.8 | 0.0 | 0.0 | 0.0 | 0.0 | 0.1 | 0.4 | 1.4 | 1.1 | 8.4 |

## PRECIPITATION (inches) 2009 NORTH PLATTE (KLBF)

| YEAR | JAN | FEB | MAR | APR | MAY | JUN | JUL | AUG | SEP | OCT | NOV | DEC | ANNUAL |
|---|---|---|---|---|---|---|---|---|---|---|---|---|---|
| 1980 | 0.52 | 0.82 | 2.56 | 0.77 | 2.59 | 1.89 | 0.64 | 3.04 | 0.34 | 1.00 | 0.13 | 0.02 | 14.32 |
| 1981 | 0.07 | 0.05 | 2.72 | 2.47 | 5.37 | 2.32 | 5.09 | 2.48 | 0.25 | 0.60 | 1.94 | 0.43 | 23.79 |
| 1982 | 0.20 | 0.15 | 0.99 | 1.42 | 6.32 | 2.35 | 1.78 | 1.18 | 1.26 | 2.44 | 0.73 | 1.08 | 19.90 |
| 1983 | 0.33 | 0.25 | 1.54 | 2.12 | 3.20 | 3.32 | 3.74 | 1.98 | 0.14 | 0.56 | 1.56 | 0.46 | 19.20 |
| 1984 | 0.36 | 0.87 | 1.20 | 5.01 | 2.82 | 4.37 | 0.94 | 1.38 | 0.39 | 2.41 | 0.69 | 0.72 | 21.16 |
| 1985 | 0.55 | 0.14 | 0.44 | 1.84 | 4.01 | 0.87 | 3.98 | 1.16 | 3.23 | 1.24 | 1.09 | 0.79 | 19.34 |
| 1986 | 0.02 | 1.10 | 0.70 | 3.77 | 2.80 | 1.70 | 2.57 | 1.22 | 1.02 | 1.58 | 0.19 | 0.27 | 16.94 |
| 1987 | 0.16 | 1.55 | 1.65 | 1.01 | 3.19 | 3.95 | 2.81 | 1.19 | 1.16 | 1.67 | 1.26 | 0.81 | 20.41 |
| 1988 | 0.72 | 0.03 | 0.37 | 2.02 | 3.59 | 3.12 | 3.03 | 3.93 | 1.59 | 0.05 | 0.40 | T | 18.85 |
| 1989 | 0.55 | 0.73 | 0.38 | 0.10 | 3.02 | 3.51 | 1.86 | 2.37 | 1.11 | 0.08 | 0.02 | 0.28 | 14.01 |
| 1990 | 0.27 | 0.18 | 1.75 | 1.52 | 3.65 | 1.90 | 1.99 | 1.79 | 0.31 | 1.48 | 0.87 | 0.09 | 15.80 |
| 1991 | 0.35 | 0.21 | 1.00 | 3.00 | 5.39 | 2.78 | 1.81 | 0.53 | 1.75 | 2.14 | 0.92 | 0.67 | 20.55 |
| 1992 | 0.89 | 1.42 | 2.98 | 0.18 | 3.18 | 2.61 | 3.75 | 6.30 | 0.25 | 0.92 | 0.20 | 0.33 | 23.01 |
| 1993 | 0.74 | 1.37 | 0.61 | 1.80 | 2.47 | 6.12 | 5.47 | 3.78 | 0.71 | 1.59 | 1.32 | 0.22 | 26.20 |
| 1994 | 0.53 | 0.29 | 0.05 | 1.47 | 1.16 | 4.92 | 4.52 | 1.24 | 0.60 | 2.63 | 0.78 | 0.66 | 18.85 |
| 1995 | 0.12 | 0.09 | 1.21 | 3.09 | 4.51 | 2.59 | 2.01 | 0.73 | 1.98 | 0.87 | 0.08 | 0.02 | 17.30 |
| 1996 | 0.50 | T | 0.33 | 0.84 | 4.12 | 3.87 | 5.57 | 3.25 | 5.55 | .33 | .40 | .03 | 24.79 |
| 1997 | T | 0.73 | 0.07 | 1.00 | 1.87 | 2.58 | 3.19 | 3.40 | 1.76 | 2.77 | 0.04 | 0.20 | 17.61 |
| 1998 | 0.27 | 0.40 | 1.30 | 0.67 | 2.93 | 4.84 | 5.81 | 1.85 | 1.13 | 1.92 | 1.31 | 0.01 | 22.44 |
| 1999 | 0.34 | 0.26 | 0.64 | 2.83 | 1.92 | 5.32 | 0.93 | 5.49 | 1.20 | 0.22 | 0.14 | 0.05 | 19.34 |
| 2000 | 0.31 | 0.46 | 1.22 | 1.65 | 1.21 | 1.53 | 2.86 | 2.05 | 1.22 | 3.24 | 0.53 | 0.04 | 16.32 |
| 2001 | 0.48 | 0.40 | 0.60 | 5.94 | 2.19 | 1.71 | 2.52 | 5.26 | 2.76 | 0.80 | 0.96 | 0.07 | 23.69 |
| 2002 | 0.08 | 0.01 | 0.68 | 1.17 | 1.56 | 2.20 | 0.45 | 1.29 | 1.14 | 2.38 | 0.10 | T | 11.06 |
| 2003 | 0.35 | 0.49 | 1.28 | 3.84 | 2.02 | 5.45 | 1.90 | 0.59 | 1.08 | 0.49 | 0.73 | 0.03 | 18.25 |
| 2004 | 0.26 | 0.77 | 0.16 | 1.90 | 1.77 | 5.06 | 4.97 | 1.25 | 1.98 | 0.96 | 1.30 | 0.06 | 20.44 |
| 2005 | 0.41 | 0.17 | 1.76 | 2.33 | 3.04 | 5.07 | 1.26 | 2.78 | 0.17 | 0.78 | 0.46 | 0.22 | 18.45 |
| 2006 | 0.21 | 0.13 | 0.73 | 1.40 | 0.83 | 5.03 | 3.27 | 1.95 | 2.73 | 0.86 | 0.07 | 2.56 | 19.77 |
| 2007 | 0.61 | 0.81 | 1.56 | 4.12 | 6.49 | 2.52 | 2.71 | 0.76 | 1.94 | 1.17 | 0.02 | 0.84 | 23.55 |
| 2008 | 0.03 | 0.10 | 0.90 | 3.67 | 7.45 | 2.79 | 2.23 | 2.75 | 1.34 | 4.78 | 0.34 | 0.24 | 26.62 |
| 2009 | 0.33 | 0.96 | 0.32 | 2.84 | 2.80 | 3.06 | 5.01 | 2.34 | 1.17 | 4.29 | 0.08 | 0.67 | 23.87 |
| POR= 62 YRS | 0.40 | 0.51 | 1.11 | 1.99 | 3.35 | 3.49 | 3.05 | 2.17 | 1.54 | 1.20 | 0.61 | 0.42 | 19.84 |

WBAN : 24023

## AVERAGE TEMPERATURE (°F) 2009 NORTH PLATTE (KLBF)

| YEAR | JAN | FEB | MAR | APR | MAY | JUN | JUL | AUG | SEP | OCT | NOV | DEC | ANNUAL |
|---|---|---|---|---|---|---|---|---|---|---|---|---|---|
| 1980 | 25.1 | 26.8 | 35.5 | 50.5 | 60.2 | 72.5 | 78.0 | 74.0 | 63.7 | 48.9 | 38.5 | 32.0 | 50.5 |
| 1981 | 29.7 | 29.1 | 40.2 | 55.8 | 54.8 | 68.6 | 73.8 | 70.1 | 63.0 | 49.0 | 39.9 | 26.7 | 50.1 |
| 1982 | 17.1 | 28.1 | 36.3 | 44.8 | 57.3 | 63.4 | 75.0 | 73.1 | 61.7 | 49.2 | 33.3 | 28.1 | 47.3 |
| 1983 | 27.1 | 35.1 | 37.2 | 42.3 | 54.0 | 65.2 | 75.3 | 78.0 | 64.8 | 51.3 | 36.7 | 7.5 | 47.9 |
| 1984 | 20.4 | 33.2 | 34.1 | 43.3 | 58.1 | 68.0 | 73.7 | 75.5 | 58.4 | 48.0 | 37.2 | 22.5 | 47.7 |
| 1985 | 18.3 | 23.1 | 40.5 | 52.2 | 60.8 | 65.4 | 75.3 | 70.1 | 59.8 | 48.6 | 24.6 | 19.2 | 46.5 |
| 1986 | 31.6 | 27.0 | 44.3 | 48.9 | 58.0 | 71.4 | 75.6 | 71.3 | 63.0 | 50.4 | 35.5 | 30.2 | 50.6 |
| 1987 | 29.5 | 35.8 | 36.7 | 51.3 | 63.1 | 70.2 | 75.7 | 70.3 | 61.3 | 47.0 | 38.2 | 27.7 | 50.6 |
| 1988 | 16.4 | 26.5 | 37.8 | 48.5 | 60.7 | 74.6 | 74.5 | 73.4 | 62.0 | 48.7 | 38.0 | 30.3 | 49.3 |
| 1989 | 30.2 | 17.8 | 35.7 | 51.1 | 59.0 | 65.8 | 74.3 | 71.7 | 61.0 | 50.7 | 37.6 | 20.6 | 48.0 |
| 1990 | 30.6 | 30.9 | 40.0 | 48.5 | 56.7 | 71.1 | 73.7 | 74.0 | 67.6 | 50.2 | 38.1 | 21.9 | 50.3 |
| 1991 | 23.1 | 37.5 | 40.5 | 49.3 | 61.8 | 71.0 | 74.9 | 73.7 | 63.6 | 48.4 | 32.2 | 33.4 | 50.8 |
| 1992 | 32.1 | 37.3 | 41.8 | 50.6 | 58.7 | 65.7 | 68.5 | 66.5 | 63.0 | 49.9 | 32.4 | 22.5 | 49.1 |
| 1993 | 18.5 | 18.0 | 37.0 | 45.5 | 58.7 | 65.3 | 71.5 | 70.3 | 58.3 | 48.5 | 32.7 | 30.4 | 46.2 |
| 1994 | 24.0 | 23.7 | 41.5 | 47.8 | 62.6 | 71.7 | 70.5 | 72.6 | 65.4 | 52.0 | 37.4 | 29.1 | 49.9 |
| 1995 | 26.8 | 34.0 | 37.8 | 43.8 | 52.6 | 66.7 | 73.8 | 79.3 | 62.6 | 48.2 | 38.6 | 28.8 | 49.4 |
| 1996 | 20.9 | 30.1 | 31.3 | 47.3 | 56.6 | 68.5 | 71.2 | 70.0 | 59.1 | 48.6 | 30.7 | 25.0 | 46.6 |
| 1997 | 23.3 | 29.1 | 39.5 | 42.3 | 55.1 | 69.5 | 73.9 | 70.9 | 64.2 | 51.1 | 33.9 | 28.2 | 48.4 |
| 1998 | 26.0 | 34.7 | 32.6 | 46.3 | 59.6 | 64.1 | 75.3 | 72.4 | 68.5 | 49.7 | 39.3 | 27.7 | 49.7 |
| 1999 | 27.6 | 36.7 | 38.5 | 45.3 | 57.1 | 66.9 | 75.6 | 70.7 | 58.2 | 50.0 | 41.8 | 30.7 | 49.9 |
| 2000 | 27.3 | 33.9 | 40.9 | 48.0 | 60.7 | 67.9 | 76.5 | 77.5 | 63.6 | 51.1 | 24.6 | 21.6 | 49.5 |
| 2001 | 27.7 | 23.4 | 36.9 | 49.1 | 57.0 | 67.5 | 77.5 | 71.1 | 61.7 | 48.4 | 38.9 | 27.4 | 48.9 |
| 2002 | 26.9 | 30.5 | 29.2 | 48.0 | 56.0 | 74.8 | 79.1 | 73.5 | 63.4 | 42.8 | 37.0 | 30.2 | 49.3 |
| 2003 | 27.9 | 25.8 | 40.4 | 49.5 | 57.9 | 66.1 | 77.4 | 75.6 | 60.9 | 53.8 | 34.8 | 29.9 | 50.0 |
| 2004 | 23.9 | 27.1 | 42.1 | 48.8 | 61.7 | 64.9 | 71.9 | 67.8 | 66.2 | 51.9 | 37.7 | 30.5 | 49.5 |
| 2005 | 23.8 | 34.2 | 38.7 | 47.7 | 57.1 | 70.9 | 77.2 | 72.8 | 67.9 | 50.9 | 39.5 | 25.5 | 50.5 |
| 2006 | 37.0 | 28.1 | 36.7 | 52.8 | 61.3 | 72.4 | 77.8 | 72.3 | 57.7 | 46.4 | 35.6 | 27.9 | 50.5 |
| 2007 | 17.2 | 23.1 | 46.2 | 46.1 | 60.4 | 68.5 | 75.8 | 76.1 | 65.7 | 52.4 | 38.4 | 21.1 | 49.3 |
| 2008 | 23.2 | 29.5 | 37.0 | 45.0 | 55.5 | 65.9 | 75.9 | 71.5 | 62.2 | 49.1 | 38.0 | 21.8 | 47.9 |
| 2009 | 27.4 | 31.1 | 37.3 | 46.7 | 58.5 | 66.5 | 70.8 | 69.0 | 60.7 | 40.9 | 40.0 | 17.7 | 47.2 |
| POR= 62 YRS | 23.0 | 28.4 | 36.3 | 47.7 | 58.3 | 68.3 | 74.3 | 72.4 | 62.3 | 49.8 | 35.6 | 26.0 | 48.5 |

## HEATING DEGREE DAYS (base 65°F) 2009 NORTH PLATTE (KLBF)

| YEAR | JUL | AUG | SEP | OCT | NOV | DEC | JAN | FEB | MAR | APR | MAY | JUN | TOTAL |
|---|---|---|---|---|---|---|---|---|---|---|---|---|---|
| 1980-81 | 0 | 6 | 107 | 491 | 790 | 1019 | 1089 | 1000 | 762 | 283 | 318 | 26 | 5891 |
| 1981-82 | 9 | 4 | 101 | 492 | 749 | 1179 | 1479 | 1030 | 885 | 601 | 239 | 111 | 6879 |
| 1982-83 | 0 | 18 | 160 | 484 | 946 | 1138 | 1167 | 833 | 854 | 672 | 343 | 90 | 6705 |
| 1983-84 | 2 | 0 | 128 | 419 | 840 | 1780 | 1379 | 915 | 953 | 647 | 236 | 33 | 7332 |
| 1984-85 | 0 | 0 | 247 | 519 | 829 | 1312 | 1440 | 1168 | 752 | 393 | 156 | 83 | 6899 |
| 1985-86 | 0 | 23 | 252 | 502 | 1205 | 1416 | 1029 | 1060 | 634 | 479 | 219 | 2 | 6821 |
| 1986-87 | 0 | 14 | 98 | 446 | 878 | 1074 | 1093 | 810 | 868 | 420 | 102 | 15 | 5818 |
| 1987-88 | 13 | 36 | 139 | 551 | 796 | 1152 | 1501 | 1109 | 839 | 490 | 170 | 3 | 6799 |
| 1988-89 | 0 | 13 | 128 | 498 | 803 | 1067 | 1072 | 1316 | 902 | 430 | 211 | 67 | 6507 |
| 1989-90 | 2 | 7 | 180 | 437 | 815 | 1374 | 1061 | 948 | 771 | 502 | 259 | 15 | 6371 |
| 1990-91 | 15 | 1 | 84 | 457 | 797 | 1331 | 1290 | 762 | 754 | 466 | 149 | 5 | 6111 |
| 1991-92 | 3 | 1 | 148 | 508 | 977 | 971 | 1010 | 797 | 714 | 436 | 219 | 45 | 5829 |
| 1992-93 | 18 | 60 | 113 | 466 | 970 | 1310 | 1436 | 1311 | 862 | 578 | 204 | 73 | 7401 |
| 1993-94 | 2 | 24 | 218 | 513 | 965 | 1066 | 1263 | 1151 | 724 | 517 | 131 | 5 | 6579 |
| 1994-95 | 3 | 8 | 90 | 395 | 820 | 1106 | 1180 | 863 | 840 | 631 | 377 | 67 | 6380 |
| 1995-96 | 17 | 2 | 167 | 517 | 786 | 1116 | 1359 | 1004 | 1037 | 527 | 283 | 40 | 6855 |
| 1996-97 | 3 | 1 | 214 | 503 | 1021 | 1233 | 1284 | 1001 | 782 | 673 | 306 | 14 | 7035 |
| 1997-98 | 10 | 17 | 113 | 452 | 926 | 1133 | 1202 | 841 | 998 | 551 | 195 | 114 | 6552 |
| 1998-99 | 2 | 0 | 39 | 467 | 764 | 1149 | 1151 | 785 | 815 | 586 | 248 | 54 | 6060 |
| 1999-00 | 8 | 9 | 227 | 458 | 690 | 1057 | 1164 | 894 | 738 | 503 | 169 | 49 | 5966 |
| 2000-01 |  | 0 | 142 | 424 | 1204 | 1337 | 1152 | 1156 | 868 | 473 | 255 | 87 |  |
| 2001-02 | 0 | 5 | 146 | 511 | 773 | 1160 | 1173 | 957 | 1104 | 506 | 299 | 11 | 6645 |
| 2002-03 | 0 | 8 | 138 | 680 | 831 | 1074 | 1142 | 1091 | 758 | 462 | 236 | 56 | 6476 |
| 2003-04 | 0 | 7 | 168 | 352 | 897 | 1079 | 1267 | 1091 | 705 | 477 | 153 | 101 | 6297 |
| 2004-05 | 11 | 46 | 87 | 402 | 812 | 1062 | 1269 | 856 | 807 | 513 | 266 | 28 | 6159 |
| 2005-06 | 6 | 3 | 72 | 438 | 757 | 1218 | 858 | 1026 | 868 | 363 | 187 | 3 | 5799 |
| 2006-07 | 0 | 10 | 232 | 586 | 875 | 1143 | 1475 | 1166 | 575 | 564 | 165 | 32 | 6823 |
| 2007-08 | 0 | 2 | 102 | 407 | 792 | 1355 | 1287 | 1021 | 863 | 596 | 294 | 35 | 6754 |
| 2008-09 | 0 | 0 | 126 | 488 | 805 | 1329 | 1158 | 945 | 853 | 544 | 233 | 69 | 6550 |
| 2009- | 8 | 22 | 154 | 741 | 742 | 1458 |  |  |  |  |  |  |  |

WBAN : 24023

## COOLING DEGREE DAYS (base 65°F) 2009 NORTH PLATTE (KLBF)

| YEAR | JAN | FEB | MAR | APR | MAY | JUN | JUL | AUG | SEP | OCT | NOV | DEC | TOTAL |
|---|---|---|---|---|---|---|---|---|---|---|---|---|---|
| 1980 | 0 | 0 | 0 | 10 | 27 | 243 | 411 | 289 | 74 | 0 | 0 | 0 | 1054 |
| 1981 | 0 | 0 | 0 | 10 | 9 | 141 | 288 | 168 | 47 | 1 | 0 | 0 | 664 |
| 1982 | 0 | 0 | 0 | 0 | 8 | 70 | 314 | 276 | 68 | 0 | 0 | 0 | 736 |
| 1983 | 0 | 0 | 0 | 0 | 8 | 103 | 331 | 412 | 128 | 1 | 0 | 0 | 983 |
| 1984 | 0 | 0 | 0 | 0 | 27 | 129 | 281 | 331 | 55 | 2 | 0 | 0 | 825 |
| 1985 | 0 | 0 | 0 | 14 | 32 | 100 | 326 | 189 | 100 | 0 | 0 | 0 | 761 |
| 1986 | 0 | 0 | 0 | 3 | 11 | 201 | 334 | 217 | 43 | 0 | 0 | 0 | 809 |
| 1987 | 0 | 0 | 0 | 16 | 48 | 176 | 352 | 208 | 35 | 0 | 0 | 0 | 835 |
| 1988 | 0 | 0 | 0 | 1 | 41 | 293 | 301 | 282 | 46 | 0 | 0 | 0 | 964 |
| 1989 | 0 | 0 | 0 | 21 | 35 | 99 | 295 | 220 | 67 | 2 | 0 | 0 | 739 |
| 1990 | 0 | 0 | 0 | 15 | 10 | 205 | 291 | 289 | 165 | 5 | 0 | 0 | 980 |
| 1991 | 0 | 0 | 0 | 3 | 59 | 194 | 317 | 276 | 112 | 2 | 0 | 0 | 963 |
| 1992 | 0 | 0 | 0 | 13 | 31 | 68 | 135 | 116 | 60 | 1 | 0 | 0 | 424 |
| 1993 | 0 | 0 | 0 | 0 | 15 | 88 | 212 | 194 | 24 | 6 | 0 | 0 | 539 |
| 1994 | 0 | 0 | 0 | 10 | 63 | 213 | 183 | 251 | 111 | 0 | 0 | 0 | 831 |
| 1995 | 0 | 0 | 0 | 0 | 0 | 125 | 298 | 449 | 103 | 4 | 0 | 0 | 979 |
| 1996 | 0 | 0 | 0 | 0 | 33 | 149 | 204 | 165 | 44 | 3 | 0 | 0 | 598 |
| 1997 | 0 | 0 | 0 | 0 | 5 | 156 | 295 | 206 | 96 | 29 | 0 | 0 | 787 |
| 1998 | 0 | 0 | 0 | 0 | 37 | 94 | 326 | 237 | 150 | 0 | 0 | 0 | 844 |
| 1999 | 0 | 0 | 0 | 0 | 9 | 119 | 343 | 192 | 28 | 0 | 0 | 0 | 691 |
| 2000 | 0 | 0 | 0 | 0 | 42 | 140 | 362 | 393 | 107 | 0 | 0 | 0 | 1044 |
| 2001 | 0 | 0 | 0 | 4 | 13 | 168 | 394 | 201 | 52 | 3 | 0 | 0 | 835 |
| 2002 | 0 | 0 | 0 | 2 | 27 | 312 | 446 | 277 | 96 | 0 | 0 | 0 | 1160 |
| 2003 | 0 | 0 | 0 | 5 | 24 | 98 | 393 | 341 | 50 | 12 | 0 | 0 | 923 |
| 2004 | 0 | 0 | 0 | 0 | 57 | 104 | 231 | 141 | 129 | 1 | 0 | 0 | 663 |
| 2005 | 0 | 0 | 0 | 0 | 29 | 196 | 394 | 252 | 164 | 9 | 0 | 0 | 1044 |
| 2006 | 0 | 0 | 0 | 3 | 77 | 232 | 403 | 241 | 21 | 17 | 0 | 0 | 994 |
| 2007 | 0 | 0 | 0 | 6 | 30 | 142 | 344 | 351 | 128 | 23 | 0 | 0 | 1024 |
| 2008 | 0 | 0 | 0 | 0 | 7 | 72 | 346 | 207 | 49 | 2 | 0 | 0 | 683 |
| 2009 | 0 | 0 | 0 | 2 | 39 | 120 | 194 | 151 | 28 | 0 | 0 | 0 | 534 |

**SNOWFALL (inches)  2009  NORTH PLATTE (KLBF)**

| YEAR | JUL | AUG | SEP | OCT | NOV | DEC | JAN | FEB | MAR | APR | MAY | JUN | TOTAL |
|------|-----|-----|-----|-----|-----|-----|-----|-----|-----|-----|-----|-----|-------|
| 1980-81 | 0.0 | 0.0 | 0.0 | 0.6 | 1.2 | T | 0.7 | 0.5 | 0.5 | 0.4 | 0.0 | 0.0 | 3.9 |
| 1981-82 | 0.0 | 0.0 | 0.0 | T | 5.3 | 6.4 | 4.4 | 1.7 | 6.4 | 0.9 | 0.0 | 0.0 | 25.1 |
| 1982-83 | 0.0 | 0.0 | 0.0 | 1.0 | 2.0 | 9.7 | 1.6 | 0.1 | 5.9 | 5.4 | 0.0 | 0.0 | 25.7 |
| 1983-84 | 0.0 | 0.0 | T | 0.0 | 12.1 | 7.3 | 5.3 | 9.6 | 8.8 | 14.5 | 0.2 | 0.0 | 57.8 |
| 1984-85 | 0.0 | 0.0 | T | 0.3 | 0.9 | 8.7 | 8.5 | 0.8 | 2.1 | 0.0 | 0.0 | 0.0 | 21.3 |
| 1985-86 | 0.0 | 0.0 | 3.1 | T | 13.0 | 8.1 | 0.2 | 8.7 | 3.8 | T | 0.0 | 0.0 | 36.9 |
| 1986-87 | 0.0 | 0.0 | 0.0 | 2.0 | 1.0 | 2.6 | 1.6 | 9.2 | 7.2 | 0.1 | 0.0 | 0.0 | 23.7 |
| 1987-88 | 0.0 | 0.0 | 0.0 | 1.3 | 8.2 | 7.2 | 12.6 | 0.6 | 3.8 | 2.1 | 0.0 | 0.0 | 35.8 |
| 1988-89 | 0.0 | 0.0 | 0.0 | T | 2.5 | T | 6.1 | 10.6 | 4.3 | 0.3 | T | 0.0 | 23.8 |
| 1989-90 | T | 0.0 | T | T | T | 2.2 | 5.9 | 1.9 | 5.3 | 0.3 | T | T | 15.6 |
| 1990-91 | 0.0 | 0.0 | 0.0 | 2.0 | 9.7 | 0.8 | 4.1 | 1.1 | 5.6 | T | T | 0.0 | 23.3 |
| 1991-92 | 0.0 | 0.0 | 0.0 | 7.3 | 1.6 | 1.8 | 5.4 | 1.2 | 5.1 | 0.3 | T | 0.0 | 22.7 |
| 1992-93 | T | T | 0.0 | T | 2.2 | 3.9 | 10.1 | 17.2 | 2.7 | 2.0 | T | T | 38.1 |
| 1993-94 | T | 0.0 | T | T | 3.9 | 3.3 | 10.6 | 6.0 | 0.4 | 11.5 | T | T | 35.7 |
| 1994-95 | T | 0.0 | 0.0 | 0.0 | 4.6 | 9.0 | 0.1 | 2.7 | 7.4 | 9.6 | 0.0 | T | 33.4 |
| 1995-96 | T | 0.0 | 0.7 | 4.1 | 0.8 | 0.5 | 4.9 | T | 3.4 | 4.0 | 0.0 | 0.0 | 18.4 |
| 1996-97 | T |  | 0.0 | T | 5.2 | 1.0 | 0.9 | 12.1 | 0.5 | 4.4 | 0.0 | 0.0 |  |
| 1997-98 | 0.0 | T | 0.0 | 5.7 | 1.6 | 3.2 | 4.4 | 4.2 | 8.8 | 0.0 | T | T | 27.9 |
| 1998-99 | T | 0.0 | 0.0 | T | 5.7 | 0.7 | 4.5 | 5.2 | 2.7 | 1.2 | 0.0 | T | 20.0 |
| 1999-00 | T | T | 0.0 | 0.0 | 1.4 | 1.1 | 8.1 | 3.6 | 0.8 | T | T | T | 15.0 |
| 2000-01 | T | T | 2.8 | 0.0 | 10.6 | 1.0 | 7.4 | 5.3 | 4.2 | 6.3 | T | T | 37.6 |
| 2001-02 | T | T | 0.0 | 0.0 | 1.6 | 3.4 | 2.0 | 0.2 | 10.5 | T | T | T | 17.7 |
| 2002-03 | 0.0 | 0.0 | 0.0 | 8.2 | 1.0 | T | 4.1 | 8.9 | 1.0 | 7.6 | 0.0 | 0.0 | 30.8 |
| 2003-04 | 0.0 | 0.0 | 0.0 | 0.0 | 1.2 | 0.7 | 5.3 | 5.2 | 0.7 | 1.5 | 0.0 | 0.0 | 14.6 |
| 2004-05 | 0.0 | T | 0.0 | 0.0 | 9.9 | 0.6 | 5.7 | 1.8 | 5.1 | 0.4 | T | 0.0 | 23.5 |
| 2005-06 | T | 0.0 | 0.0 | 0.0 | 3.9 | 3.5 | T | 2.7 | 9.7 | 0.1 | 0.0 | T | 19.9 |
| 2006-07 | T | 0.0 | 0.0 | 0.8 | 2.0 | 16.3 | 10.2 | 7.8 | T | 4.9 | T | 0.0 | 42.0 |
| 2007-08 | 0.0 | 0.0 | 0.0 | 0.0 | 0.6 | 9.9 | 1.3 | 1.7 | 8.3 | 9.6 | 0.6 | T | 32.0 |
| 2008-09 | T | T | 0.0 | 1.1 | 2.0 | 3.8 | 5.3 | 9.1 | 3.2 | 7.5 | 0.0 | T | 32.0 |
| 2009- | T | 0.0 | 0.0 | 30.3 | T | 12.3 |  |  |  |  |  |  |  |
| POR= 62 YRS | T | T | 0.1 | 1.8 | 3.7 | 4.6 | 5.3 | 5.0 | 6.3 | 3.0 | 0.2 | T | 30.0 |

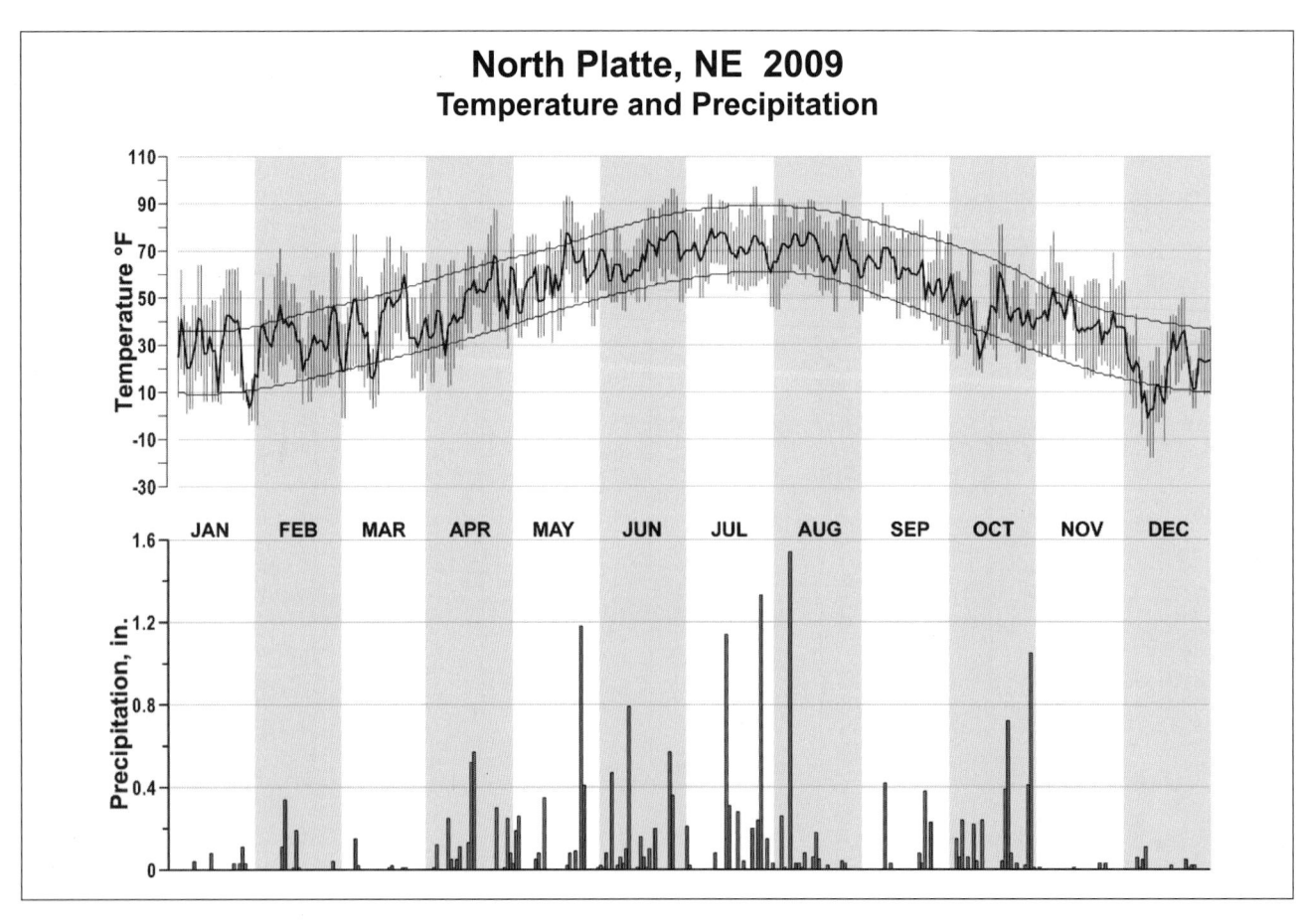

**North Platte, NE  2009**
**Temperature and Precipitation**

# 2009
# OMAHA (EPPLEY AIRFIELD)
# NEBRASKA (KOMA)

Omaha, Nebraska, is situated on the west bank of the Missouri River. The river level at Omaha is normally about 965 feet above sea level and the rolling hills in and around Omaha rise to about 1,300 feet above sea level. The climate is typically continental with relatively warm summers and cold, dry winters. It is situated midway between two distinctive climatic zones, the humid east and the dry west. Fluctuations between these two zones produce weather conditions for periods that are characteristic of either zone, or combinations of both. Omaha is also affected by most low pressure systems that cross the country. This causes periodic and rapid changes in weather, especially during the winter months.

Most of the precipitation in Omaha falls during sharp showers or thunderstorms, and these occur mostly during the growing season from April to September. Of the total precipitation, about 75 percent falls during this six-month period. The rain occurs mostly as evening or nighttime showers and thunderstorms. Although winters are relatively cold, precipitation is light, with only 10 percent of the total annual precipitation falling during the winter months.

Sunshine is fairly abundant, ranging around 50 percent of the possible in the winter to 75 percent of the possible in the summer.

# NORMALS, MEANS, AND EXTREMES
## OMAHA (KOMA)

| LATITUDE: 41°18'N | LONGITUDE: -95°53'W | ELEVATION (FT): GRND: 982  BARO: 1028 | TIME ZONE: CENTRAL  (UTC -6) | WBAN: 14942 |
|---|---|---|---|---|

| | ELEMENT | POR | JAN | FEB | MAR | APR | MAY | JUN | JUL | AUG | SEP | OCT | NOV | DEC | YEAR |
|---|---|---|---|---|---|---|---|---|---|---|---|---|---|---|---|
| **TEMPERATURE °F** | NORMAL DAILY MAXIMUM | 30 | 31.7 | 37.9 | 50.4 | 63.2 | 73.7 | 83.7 | 87.4 | 85.2 | 77.3 | 65.2 | 47.8 | 34.8 | 61.5 |
| | MEAN DAILY MAXIMUM | 61 | 31.7 | 37.3 | 48.8 | 63.4 | 74.1 | 83.4 | 87.7 | 85.4 | 77.2 | 65.7 | 49.4 | 36.0 | 61.7 |
| | HIGHEST DAILY MAXIMUM | 72 | 69 | 78 | 89 | 97 | 99 | 105 | 114 | 110 | 104 | 96 | 83 | 72 | 114 |
| | YEAR OF OCCURRENCE | | 1944 | 1972 | 1986 | 1989 | 1939 | 1953 | 1936 | 1936 | 1939 | 1938 | 1999 | 1939 | JUL 1936 |
| | MEAN OF EXTREME MAXS. | 61 | 54.1 | 60.3 | 75.0 | 86.3 | 90.0 | 96.4 | 99.1 | 97.3 | 92.4 | 84.8 | 71.1 | 58.4 | 80.4 |
| | NORMAL DAILY MINIMUM | 30 | 11.6 | 18.0 | 28.1 | 39.6 | 50.7 | 60.6 | 65.9 | 63.8 | 53.5 | 41.1 | 28.1 | 16.4 | 39.8 |
| | MEAN DAILY MINIMUM | 61 | 12.1 | 17.6 | 27.5 | 40.1 | 51.4 | 61.1 | 66.4 | 64.0 | 54.0 | 42.0 | 29.0 | 17.2 | 40.2 |
| | LOWEST DAILY MINIMUM | 72 | -23 | -21 | -16 | 5 | 27 | 38 | 44 | 43 | 25 | 13 | -9 | -23 | -23 |
| | YEAR OF OCCURRENCE | | 1982 | 1981 | 1948 | 1975 | 1980 | 1983 | 1972 | 1967 | 1984 | 1972 | 1964 | 1989 | DEC 1989 |
| | MEAN OF EXTREME MINS. | 61 | -9.6 | -3.9 | 7.0 | 23.7 | 36.3 | 47.3 | 54.9 | 51.8 | 37.2 | 25.3 | 11.3 | -2.6 | 23.2 |
| | NORMAL DRY BULB | 30 | 21.7 | 28.0 | 39.3 | 51.4 | 62.2 | 72.2 | 76.7 | 74.5 | 65.4 | 53.2 | 38.0 | 25.6 | 50.7 |
| | MEAN DRY BULB | 61 | 21.9 | 27.5 | 38.1 | 51.7 | 62.7 | 72.4 | 77.1 | 74.7 | 65.6 | 53.9 | 39.2 | 26.7 | 51.0 |
| | MEAN WET BULB | 25 | 21.1 | 24.3 | 34.1 | 43.9 | 54.6 | 63.9 | 68.6 | 67.2 | 58.2 | 46.3 | 33.7 | 23.4 | 44.9 |
| | MEAN DEW POINT | 25 | 17.5 | 20.8 | 29.3 | 38.6 | 50.8 | 60.9 | 66.3 | 64.8 | 54.9 | 42.2 | 30.0 | 20.4 | 41.4 |
| | NORMAL NO. DAYS WITH: | | | | | | | | | | | | | | |
| | MAXIMUM >= 90 | 30 | 0.0 | 0.0 | 0.0 | 0.5 | 1.0 | 7.1 | 11.9 | 9.0 | 3.6 | 0.2 | 0.0 | 0.0 | 33.3 |
| | MAXIMUM <= 32 | 30 | 15.2 | 10.6 | 3.1 | 0.1 | 0.0 | 0.0 | 0.0 | 0.0 | 0.0 | 0.1 | 3.1 | 11.8 | 44.0 |
| | MINIMUM <= 32 | 30 | 30.1 | 25.4 | 20.7 | 7.0 | 0.5 | 0.0 | 0.0 | 0.0 | 0.0 | 0.4 | 5.8 | 20.9 | 140.5 |
| | MINIMUM <= 0 | 30 | 6.9 | 3.5 | 0.6 | 0.0 | 0.0 | 0.0 | 0.0 | 0.0 | 0.0 | 0.0 | 0.2 | 3.8 | 15.0 |
| **H/C** | NORMAL HEATING DEG. DAYS | 30 | 1349 | 1052 | 805 | 424 | 151 | 17 | 1 | 6 | 105 | 384 | 806 | 1211 | 6311 |
| | NORMAL COOLING DEG. DAYS | 30 | 0 | 0 | 1 | 14 | 60 | 233 | 365 | 296 | 114 | 12 | 0 | 0 | 1095 |
| **RH** | NORMAL (PERCENT) | 30 | 73 | 73 | 68 | 63 | 66 | 68 | 71 | 74 | 71 | 68 | 72 | 76 | 70 |
| | HOUR 00 LST | 30 | 77 | 79 | 74 | 70 | 75 | 78 | 80 | 83 | 81 | 76 | 78 | 80 | 78 |
| | HOUR 06 LST | 30 | 80 | 81 | 80 | 78 | 82 | 84 | 86 | 89 | 87 | 83 | 82 | 82 | 83 |
| | HOUR 12 LST | 30 | 67 | 65 | 59 | 53 | 55 | 56 | 59 | 61 | 57 | 55 | 64 | 69 | 60 |
| | HOUR 18 LST | 30 | 68 | 66 | 56 | 50 | 53 | 53 | 57 | 60 | 58 | 57 | 66 | 73 | 60 |
| **S** | PERCENT POSSIBLE SUNSHINE | | | | | | | | | | | | | | |
| **W/O** | MEAN NO. DAYS WITH: | | | | | | | | | | | | | | |
| | HEAVY FOG(VISBY <= 1/4 MI) | 45 | 1.7 | 1.9 | 1.3 | 0.7 | 1.1 | 0.3 | 0.6 | 1.8 | 1.7 | 1.3 | 1.8 | 2.3 | 16.5 |
| | THUNDERSTORMS | 61 | 0.1 | 0.4 | 1.6 | 4.0 | 7.5 | 8.9 | 8.4 | 7.4 | 4.9 | 2.4 | 0.8 | 0.2 | 46.6 |
| **CLOUDNESS** | MEAN: | | | | | | | | | | | | | | |
| | SUNRISE-SUNSET (OKTAS) | 49 | 4.9 | 5.0 | 5.3 | 5.1 | 5.0 | 4.6 | 3.8 | 3.8 | 3.8 | 3.8 | 4.8 | 5.0 | 4.6 |
| | MIDNIGHT-MIDNIGHT (OKTAS) | 22 | 4.5 | 4.5 | 4.8 | 4.7 | 4.6 | 4.4 | 3.8 | 3.7 | 3.8 | 3.8 | 4.6 | 4.7 | 4.3 |
| | MEAN NO. DAYS WITH: | | | | | | | | | | | | | | |
| | CLEAR | 49 | 8.8 | 7.3 | 7.1 | 7.3 | 7.4 | 8.0 | 11.2 | 12.2 | 12.3 | 12.9 | 8.6 | 8.0 | 111.1 |
| | PARTLY CLOUDY | 49 | 7.9 | 7.4 | 8.0 | 8.5 | 9.6 | 11.4 | 12.1 | 10.1 | 7.6 | 8.1 | 7.4 | 7.5 | 105.6 |
| | CLOUDY | 49 | 14.3 | 13.5 | 15.9 | 14.2 | 14.1 | 10.6 | 7.7 | 8.7 | 10.2 | 10.1 | 13.9 | 15.4 | 148.6 |
| **PR** | MEAN STATION PRESSURE(IN) | 22 | 29.03 | 29.05 | 28.96 | 28.90 | 28.88 | 28.87 | 28.91 | 28.95 | 28.96 | 28.98 | 28.99 | 29.03 | 28.96 |
| | MEAN SEA-LEVEL PRES. (IN) | 22 | 30.13 | 30.14 | 30.03 | 29.94 | 29.92 | 29.90 | 29.94 | 29.98 | 30.00 | 30.04 | 30.06 | 30.12 | 30.02 |
| **WINDS** | MEAN SPEED (MPH) | 25 | 10.2 | 10.3 | 11.3 | 11.7 | 10.4 | 9.5 | 8.5 | 8.2 | 9.1 | 9.7 | 10.3 | 10.3 | 10.0 |
| | PREVAIL.DIR(TENS OF DEGS) | 39 | 34 | 35 | 35 | 17 | 17 | 17 | 17 | 17 | 17 | 17 | 17 | 17 | 17 |
| | MAXIMUM 2-MINUTE: | | | | | | | | | | | | | | |
| | SPEED (MPH) | 13 | 45 | 43 | 52 | 54 | 57 | 44 | 56 | 51 | 48 | 48 | 46 | 45 | 57 |
| | DIR. (TENS OF DEGS) | | 33 | 17 | 18 | 32 | 16 | 10 | 03 | 25 | 27 | 11 | 32 | 32 | 16 |
| | YEAR OF OCCURRENCE | | 1997 | 2009 | 2009 | 2004 | 1998 | 2004 | 2001 | 2007 | 2006 | 1998 | 2003 | 2008 | MAY 1998 |
| | MAXIMUM 3-SECOND | | | | | | | | | | | | | | |
| | SPEED (MPH) | 13 | 53 | 53 | 68 | 64 | 86 | 54 | 70 | 68 | 60 | 59 | 55 | 56 | 86 |
| | DIR. (TENS OF DEGS) | | 33 | 19 | 18 | 19 | 17 | 30 | 02 | 27 | 28 | 32 | 28 | 31 | 17 |
| | YEAR OF OCCURRENCE | | 1997 | 2009 | 2009 | 2001 | 1998 | 2008 | 2001 | 2007 | 2006 | 2008 | 1998 | 2008 | MAY 1998 |
| **PRECIPITATION** | NORMAL (IN) | 30 | 0.77 | 0.80 | 2.13 | 2.94 | 4.44 | 3.95 | 3.86 | 3.21 | 3.17 | 2.21 | 1.82 | 0.92 | 30.22 |
| | MAXIMUM MONTHLY (IN) | 72 | 3.70 | 2.97 | 5.96 | 8.48 | 10.63 | 10.81 | 10.34 | 12.26 | 13.75 | 6.23 | 4.70 | 5.42 | 13.75 |
| | YEAR OF OCCURRENCE | | 1949 | 1965 | 1973 | 1999 | 2007 | 1947 | 1993 | 1999 | 1965 | 2007 | 1983 | 1984 | SEP 1965 |
| | MINIMUM MONTHLY (IN) | 72 | T | 0.06 | 0.12 | 0.23 | 0.56 | 0.24 | 0.39 | 0.61 | 0.41 | T | 0.03 | T | T |
| | YEAR OF OCCURRENCE | | 1986 | 2006 | 1956 | 1936 | 1948 | 2007 | 1983 | 1984 | 1953 | 1952 | 2007 | 1943 | JAN 1986 |
| | MAXIMUM IN 24 HOURS (IN) | 72 | 1.52 | 2.24 | 2.44 | 2.82 | 5.63 | 4.27 | 3.37 | 10.48 | 6.47 | 3.13 | 2.53 | 3.03 | 10.48 |
| | YEAR OF OCCURRENCE | | 1967 | 1954 | 2004 | 1999 | 2007 | 1994 | 1958 | 1999 | 1965 | 1968 | 1948 | 1984 | AUG 1999 |
| | NORMAL NO. DAYS WITH: | | | | | | | | | | | | | | |
| | PRECIPITATION >= 0.01 | 30 | 5.9 | 6.1 | 8.3 | 10.4 | 12.0 | 10.0 | 10.1 | 8.5 | 8.3 | 6.6 | 7.0 | 6.7 | 99.9 |
| | PRECIPITATION >= 1.00 | 30 | 0.1 | 0.1 | 0.5 | 0.6 | 1.1 | 1.0 | 1.0 | 0.9 | 1.0 | 0.6 | 0.5 | 0.1 | 7.5 |
| **SNOWFALL** | NORMAL (IN) | 30 | 6.3 | 5.9 | 4.5 | 1.2 | 0.0 | 0.0 | 0.0 | 0.0 | 0.* | 0.3 | 3.4 | 5.5 | 27.1 |
| | MAXIMUM MONTHLY (IN) | 71 | 25.7 | 25.4 | 27.2 | 10.0 | 2.0 | T | T | T | T | 7.2 | 12.0 | 24.6 | 27.2 |
| | YEAR OF OCCURRENCE | | 1936 | 1965 | 1948 | 1992 | 1945 | 2009 | 2009 | 2002 | 1985 | 1941 | 1957 | 2009 | MAR 1948 |
| | MAXIMUM IN 24 HOURS (IN) | 64 | 13.1 | 18.3 | 13.0 | 9.9 | 2.0 | T | T | T | T | 7.2 | 8.7 | 10.2 | 18.3 |
| | YEAR OF OCCURRENCE' | | 1949 | 1965 | 1948 | 1992 | 1945 | 2009 | 2009 | 2002 | 1985 | 1941 | 1957 | 1969 | FEB 1965 |
| | MAXIMUM SNOW DEPTH (IN) | 60 | 17 | 26 | 27 | 8 | 0 | 0 | 0 | 0 | 0 | 5 | 9 | 17 | 27 |
| | YEAR OF OCCURRENCE | | 1984 | 2004 | 1960 | 1992 | | | | | | 1997 | 1957 | 1983 | MAR 1960 |
| | NORMAL NO. DAYS WITH: | | | | | | | | | | | | | | |
| | SNOWFALL >= 1.0 | 30 | 1.9 | 1.9 | 1.3 | 0.6 | 0.0 | 0.0 | 0.0 | 0.0 | 0.0 | 0.2 | 1.3 | 2.0 | 9.2 |

## PRECIPITATION (inches) 2009 OMAHA (KOMA)

| YEAR | JAN | FEB | MAR | APR | MAY | JUN | JUL | AUG | SEP | OCT | NOV | DEC | ANNUAL |
|------|-----|-----|-----|-----|-----|-----|-----|-----|-----|-----|-----|-----|--------|
| 1980 | 0.93 | 0.52 | 1.40 | 1.72 | 2.50 | 8.99 | 3.63 | 6.98 | 0.82 | 2.70 | 0.11 | 0.04 | 30.34 |
| 1981 | 0.20 | 0.09 | 0.88 | 1.33 | 4.13 | 2.14 | 1.87 | 4.80 | 1.51 | 1.92 | 2.60 | 0.86 | 22.33 |
| 1982 | 1.83 | 0.26 | 1.90 | 1.22 | 9.92 | 4.16 | 2.46 | 3.21 | 2.27 | 1.10 | 1.81 | 1.17 | 31.31 |
| 1983 | 0.86 | 0.68 | 3.65 | 1.00 | 2.81 | 6.52 | 0.39 | 1.24 | 2.45 | 2.16 | 4.70 | 0.63 | 27.09 |
| 1984 | 0.38 | 0.62 | 2.32 | 4.77 | 4.92 | 5.56 | 1.58 | 0.61 | 2.55 | 3.87 | 0.52 | 5.42 | 33.12 |
| 1985 | 0.56 | 1.88 | 1.36 | 3.16 | 2.46 | 1.73 | 3.27 | 1.50 | 2.71 | 1.36 | 0.85 | 0.37 | 21.21 |
| 1986 | T | 1.00 | 2.51 | 4.96 | 4.88 | 2.37 | 2.77 | 3.86 | 8.11 | 4.86 | 0.99 | 0.89 | 37.20 |
| 1987 | 0.08 | 0.55 | 4.14 | 2.24 | 8.64 | 3.29 | 6.72 | 10.16 | 1.56 | 1.33 | 1.60 | 1.01 | 41.32 |
| 1988 | 0.42 | 0.18 | 0.14 | 1.57 | 4.68 | 1.60 | 2.68 | 1.78 | 2.63 | 0.14 | 2.55 | 0.95 | 19.32 |
| 1989 | 1.10 | 0.86 | 0.40 | 1.80 | 0.83 | 5.05 | 3.06 | 1.80 | 6.46 | 1.55 | 0.15 | 0.74 | 23.80 |
| 1990 | 0.59 | 0.34 | 4.01 | 0.36 | 5.08 | 3.88 | 6.36 | 0.81 | 0.81 | 1.71 | 1.15 | 1.18 | 26.28 |
| 1991 | 1.08 | 0.26 | 2.85 | 4.46 | 4.07 | 7.79 | 2.96 | 3.67 | 1.37 | 3.76 | 3.51 | 1.75 | 37.53 |
| 1992 | 1.41 | 1.18 | 3.08 | 3.19 | 2.27 | 1.44 | 7.31 | 1.57 | 6.86 | 2.22 | 3.01 | 1.15 | 34.69 |
| 1993 | 1.42 | 0.93 | 2.67 | 2.26 | 4.90 | 8.03 | 10.34 | 7.53 | 2.29 | 1.18 | 0.66 | 0.51 | 42.72 |
| 1994 | 0.50 | 1.01 | 0.15 | 1.46 | 1.73 | 8.54 | 3.60 | 1.97 | 3.32 | 1.37 | 1.64 | 1.21 | 26.50 |
| 1995 | 0.80 | 0.47 | 2.50 | 4.26 | 7.07 | 1.28 | 3.14 | 2.52 | 2.75 |  |  |  |  |
| 1996 |  |  | 0.83 | 2.36 | 7.57 | 2.96 | 2.39 | 2.19 | 4.90 | 1.64 | 2.46 | .32 |  |
| 1997 | 0.29 | 0.69 | 1.08 | 3.66 | 1.54 | 4.51 | 4.69 | 1.33 | 4.30 | 5.25 | 2.30 | 0.57 | 30.21 |
| 1998 | 1.13 | 1.27 | 4.13 | 3.53 | 4.71 | 8.23 | 7.77 | 3.85 | 0.85 | 2.65 | 1.48 | 0.13 | 39.73 |
| 1999 | 0.59 | 1.40 | 1.31 | 8.48 | 4.50 | 3.75 | 3.07 | 12.26 | 1.64 | 0.03 | 1.11 | 0.57 | 38.71 |
| 2000 | 0.17 | 1.95 | 0.81 | 2.85 | 2.69 | 5.52 | 5.03 | 1.30 | 0.64 | 1.93 | 3.27 | 0.95 | 27.11 |
| 2001 | 1.61 | 1.53 | 1.38 | 2.37 | 8.78 | 2.29 | 2.06 | 1.95 | 2.39 | 2.10 | 1.55 | 0.67 | 28.68 |
| 2002 | 0.37 | 0.30 | 0.84 | 3.35 | 4.14 | 2.08 | 2.70 | 8.05 | 0.90 | 3.18 | 0.13 | T | 26.04 |
| 2003 | 0.34 | 1.32 | 0.50 | 3.66 | 4.37 | 3.25 | 2.49 | 0.74 | 1.41 | 1.43 | 2.91 | 0.84 | 23.26 |
| 2004 | 1.25 | 1.31 | 4.49 | 0.97 | 8.21 | 2.70 | 6.83 | 3.77 | 1.66 | 0.26 | 2.02 | 0.34 | 33.81 |
| 2005 | 0.52 | 1.90 | 0.98 | 2.63 | 4.60 | 2.69 | 1.93 | 4.66 | 0.92 | 0.75 | 1.04 | 0.81 | 23.43 |
| 2006 | 0.68 | 0.06 | 2.17 | 3.81 | 2.54 | 1.07 | 2.57 | 8.52 | 4.26 | 0.87 | 0.26 | 2.25 | 29.06 |
| 2007 | 0.59 | 1.12 | 4.15 | 3.71 | 10.63 | 0.24 | 1.66 | 6.62 | 2.31 | 6.23 | 0.03 | 1.79 | 39.08 |
| 2008 | 0.29 | 0.59 | 1.53 | 4.00 | 6.36 | 9.51 | 3.13 | 1.32 | 2.90 | 4.55 | 1.56 | 0.79 | 36.53 |
| 2009 | 0.24 | 0.75 | 1.05 | 2.21 | 1.38 | 4.58 | 3.65 | 6.24 | 1.72 | 3.46 | 0.36 | 2.28 | 27.92 |
| POR= 61 YRS | 0.77 | 0.93 | 1.92 | 2.90 | 4.51 | 4.02 | 3.74 | 3.96 | 3.10 | 2.19 | 1.48 | 0.90 | 30.42 |

WBAN : 14942

## AVERAGE TEMPERATURE (°F) 2009 OMAHA (KOMA)

| YEAR | JAN | FEB | MAR | APR | MAY | JUN | JUL | AUG | SEP | OCT | NOV | DEC | ANNUAL |
|------|-----|-----|-----|-----|-----|-----|-----|-----|-----|-----|-----|-----|--------|
| 1980 | 22.4 | 20.3 | 32.9 | 50.7 | 61.6 | 73.1 | 79.6 | 76.6 | 64.7 | 49.0 | 40.3 | 26.1 | 49.8 |
| 1981 | 24.1 | 28.4 | 40.8 | 57.4 | 58.6 | 72.6 | 76.6 | 71.0 | 64.8 | 50.2 | 40.6 | 22.9 | 50.7 |
| 1982 | 9.4 | 22.6 | 34.8 | 47.7 | 62.9 | 65.5 | 77.1 | 72.6 | 64.7 | 55.0 | 37.1 | 28.3 | 48.1 |
| 1983 | 24.9 | 30.1 | 37.3 | 43.5 | 56.5 | 69.6 | 79.4 | 81.5 | 67.0 | 52.2 | 38.5 | 7.3 | 49.0 |
| 1984 | 19.6 | 33.2 | 30.3 | 46.6 | 57.7 | 71.8 | 75.1 | 76.5 | 61.9 | 52.2 | 39.4 | 27.1 | 49.3 |
| 1985 | 19.1 | 23.6 | 43.4 | 54.9 | 63.4 | 67.2 | 74.1 | 69.4 | 62.1 | 52.7 | 28.5 | 16.4 | 47.9 |
| 1986 | 29.4 | 22.8 | 42.8 | 51.9 | 61.3 | 73.9 | 77.9 | 70.1 | 67.9 | 53.7 | 34.4 | 29.4 | 51.3 |
| 1987 | 28.6 | 36.8 | 42.8 | 55.3 | 67.3 | 74.2 | 77.8 | 70.9 | 64.8 | 48.3 | 43.4 | 30.9 | 53.4 |
| 1988 | 21.1 | 23.9 | 40.7 | 50.5 | 67.3 | 76.3 | 76.4 | 77.3 | 66.2 | 49.1 | 40.0 | 29.5 | 51.5 |
| 1989 | 32.4 | 16.0 | 37.9 | 54.6 | 62.4 | 69.4 | 77.4 | 74.4 | 63.0 | 53.9 | 36.2 | 17.7 | 49.6 |
| 1990 | 33.5 | 31.3 | 42.7 | 50.8 | 58.5 | 73.5 | 74.8 | 75.3 | 69.0 | 53.3 | 42.9 | 21.2 | 52.2 |
| 1991 | 16.3 | 34.2 | 42.5 | 54.2 | 67.3 | 74.6 | 75.8 | 74.3 | 66.0 | 52.1 | 30.6 | 31.8 | 51.6 |
| 1992 | 32.6 | 36.7 | 43.6 | 50.0 | 61.8 | 69.4 | 71.1 | 68.4 | 64.0 | 53.5 | 35.4 | 27.9 | 51.2 |
| 1993 | 19.5 | 22.2 | 35.6 | 47.7 | 61.4 | 70.0 | 75.1 | 75.2 | 59.7 | 50.9 | 35.0 | 29.4 | 48.5 |
| 1994 | 18.1 | 22.9 | 42.1 | 51.3 | 64.5 | 74.1 | 73.2 | 72.6 | 67.1 | 55.3 | 41.4 | 28.7 | 50.9 |
| 1995 | 23.0 | 31.5 | 39.3 | 47.9 | 58.1 | 71.8 | 78.7 | 79.6 |  |  |  |  |  |
| 1996 |  |  | 33.0 | 49.2 | 59.4 | 72.8 | 73.4 | 72.8 | 63.1 | 54.0 | 32.9 | 22.2 |  |
| 1997 | 19.0 | 29.5 | 41.6 | 45.7 | 58.0 | 73.4 | 77.2 | 73.4 | 66.5 | 54.0 | 35.3 | 29.9 | 50.3 |
| 1998 | 25.8 | 36.0 | 32.5 | 52.0 | 66.5 | 69.6 | 76.8 | 75.6 | 71.6 | 55.8 | 43.6 | 30.9 | 53.1 |
| 1999 | 22.6 | 35.5 | 39.5 | 51.6 | 62.3 | 70.8 | 80.4 | 72.8 | 63.8 | 53.3 | 47.0 | 31.0 | 52.6 |
| 2000 | 27.1 | 36.2 | 44.6 | 51.9 | 66.0 | 70.5 | 75.1 | 77.1 | 67.8 | 57.5 | 33.4 | 15.6 | 51.9 |
| 2001 | 26.6 | 21.4 | 35.0 | 55.5 | 63.5 | 71.6 | 79.0 | 75.7 | 65.0 | 53.5 | 49.4 | 32.0 | 52.4 |
| 2002 | 30.3 | 32.1 | 33.0 | 52.4 | 59.6 | 77.3 | 81.1 | 74.9 | 68.2 | 47.5 | 37.6 | 32.1 | 52.2 |
| 2003 | 22.8 | 23.7 | 39.6 | 53.4 | 59.6 | 69.1 | 78.4 | 78.0 | 63.4 | 55.4 | 37.9 | 30.2 | 51.0 |
| 2004 | 21.1 | 24.9 | 43.5 | 54.0 | 63.9 | 68.7 | 73.1 | 70.0 | 70.4 | 54.8 | 42.1 | 30.4 | 51.4 |
| 2005 | 20.9 | 32.9 | 40.7 | 55.5 | 61.8 | 75.6 | 79.7 | 75.0 | 71.1 | 55.0 | 42.6 | 24.4 | 52.9 |
| 2006 | 36.7 | 29.7 | 39.4 | 56.9 | 64.3 | 74.7 | 79.8 | 74.5 | 62.3 | 50.0 | 39.8 | 33.2 | 53.4 |
| 2007 | 22.3 | 21.3 | 46.3 | 49.9 | 66.5 | 72.9 | 78.5 | 78.2 | 67.0 | 56.9 | 38.7 | 22.8 | 51.8 |
| 2008 | 19.9 | 22.1 | 36.3 | 47.5 | 60.2 | 71.9 | 77.3 | 75.8 | 65.4 | 54.7 | 38.9 | 22.2 | 49.4 |
| 2009 | 20.8 | 29.8 | 39.4 | 50.0 | 63.7 | 71.7 | 71.9 | 71.7 | 65.5 | 46.8 | 45.7 | 20.4 | 49.8 |
| POR= 61 YRS | 21.9 | 27.5 | 38.1 | 51.7 | 62.7 | 72.4 | 77.1 | 74.7 | 65.6 | 53.9 | 39.2 | 26.7 | 51.0 |

## HEATING DEGREE DAYS (base 65°F) 2009 OMAHA (KOMA)

| YEAR | JUL | AUG | SEP | OCT | NOV | DEC | JAN | FEB | MAR | APR | MAY | JUN | TOTAL |
|---|---|---|---|---|---|---|---|---|---|---|---|---|---|
| 1980-81 | 0 | 3 | 108 | 491 | 735 | 1198 | 1259 | 1018 | 743 | 241 | 221 | 0 | 6017 |
| 1981-82 | 7 | 3 | 85 | 452 | 723 | 1299 | 1721 | 1183 | 930 | 518 | 102 | 56 | 7079 |
| 1982-83 | 0 | 13 | 115 | 315 | 829 | 1131 | 1240 | 971 | 854 | 638 | 278 | 37 | 6421 |
| 1983-84 | 0 | 0 | 102 | 405 | 789 | 1786 | 1401 | 916 | 1071 | 552 | 243 | 7 | 7272 |
| 1984-85 | 0 | 3 | 184 | 391 | 766 | 1166 | 1416 | 1153 | 666 | 325 | 88 | 45 | 6203 |
| 1985-86 | 0 | 13 | 217 | 378 | 1089 | 1501 | 1095 | 1176 | 689 | 389 | 134 | 1 | 6682 |
| 1986-87 | 0 | 15 | 40 | 338 | 913 | 1096 | 1122 | 784 | 685 | 322 | 67 | 7 | 5389 |
| 1987-88 | 1 | 32 | 67 | 512 | 639 | 1048 | 1353 | 1185 | 748 | 433 | 29 | 6 | 6053 |
| 1988-89 | 1 | 7 | 56 | 488 | 744 | 1095 | 1002 | 1368 | 844 | 380 | 143 | 23 | 6151 |
| 1989-90 | 0 | 7 | 140 | 356 | 855 | 1460 | 973 | 935 | 684 | 460 | 206 | 15 | 6091 |
| 1990-91 | 4 | 1 | 75 | 371 | 662 | 1350 | 1506 | 859 | 695 | 338 | 108 | 0 | 5969 |
| 1991-92 | 0 | 0 | 123 | 402 | 1027 | 1022 | 999 | 816 | 656 | 449 | 154 | 11 | 5659 |
| 1992-93 | 2 | 26 | 114 | 359 | 881 | 1141 | 1403 | 1192 | 904 | 514 | 140 | 34 | 6710 |
| 1993-94 | 0 | 1 | 171 | 448 | 895 | 1097 | 1448 | 1174 | 702 | 432 | 120 | 8 | 6496 |
| 1994-95 | 0 | 5 | 83 | 302 | 698 | 1116 | 1296 | 931 | 791 | 507 | 213 | 21 | 5963 |
| 1995-96 | 1 | 0 | | | | | | | 981 | 471 | 211 | 22 | |
| 1996-97 | 0 | 0 | 130 | 347 | 956 | 1321 | 1417 | 986 | 719 | 572 | 226 | 0 | 6674 |
| 1997-98 | 1 | 7 | 57 | 399 | 884 | 1082 | 1207 | 806 | 1000 | 390 | 52 | 58 | 5943 |
| 1998-99 | 0 | 0 | 22 | 287 | 634 | 1050 | 1307 | 818 | 781 | 396 | 122 | 27 | 5444 |
| 1999-00 | 0 | 2 | 112 | 366 | 535 | 1047 | 1168 | 828 | 625 | 391 | 72 | 16 | 5162 |
| 2000-01 | 0 | 1 | 87 | 242 | 940 | 1526 | 1185 | 1214 | 925 | 301 | 120 | 20 | 6561 |
| 2001-02 | 0 | 0 | 87 | 360 | 462 | 1014 | 1072 | 915 | 985 | 400 | 221 | 0 | 5516 |
| 2002-03 | 0 | 5 | 69 | 546 | 816 | 1013 | 1299 | 1149 | 781 | 369 | 183 | 35 | 6265 |
| 2003-04 | 0 | 0 | 128 | 315 | 807 | 1070 | 1357 | 1156 | 659 | 339 | 124 | 30 | 5985 |
| 2004-05 | 5 | 28 | 24 | 318 | 679 | 1065 | 1359 | 892 | 749 | 297 | 160 | 0 | 5576 |
| 2005-06 | 1 | 0 | 30 | 357 | 666 | 1249 | 871 | 983 | 788 | 270 | 126 | 0 | 5341 |
| 2006-07 | 0 | 0 | 123 | 490 | 746 | 978 | 1317 | 1219 | 581 | 466 | 48 | 5 | 5973 |
| 2007-08 | 0 | 0 | 74 | 279 | 782 | 1301 | 1389 | 1237 | 881 | 520 | 188 | 0 | 6651 |
| 2008-09 | 0 | 0 | 74 | 340 | 782 | 1320 | 1363 | 978 | 785 | 456 | 112 | 17 | 6227 |
| 2009- | 1 | 18 | 50 | 557 | 570 | 1377 | | | | | | | |

WBAN : 14942

## COOLING DEGREE DAYS (base 65°F) 2009 OMAHA (KOMA)

| YEAR | JAN | FEB | MAR | APR | MAY | JUN | JUL | AUG | SEP | OCT | NOV | DEC | TOTAL |
|---|---|---|---|---|---|---|---|---|---|---|---|---|---|
| 1980 | 0 | 0 | 0 | 15 | 61 | 254 | 459 | 368 | 107 | 0 | 0 | 0 | 1264 |
| 1981 | 0 | 0 | 0 | 24 | 29 | 235 | 372 | 196 | 85 | 0 | 0 | 0 | 941 |
| 1982 | 0 | 0 | 0 | 5 | 43 | 78 | 383 | 252 | 113 | 12 | 0 | 0 | 886 |
| 1983 | 0 | 0 | 0 | 0 | 20 | 183 | 453 | 519 | 167 | 17 | 0 | 0 | 1359 |
| 1984 | 0 | 0 | 0 | 6 | 22 | 220 | 320 | 366 | 96 | 4 | 0 | 0 | 1034 |
| 1985 | 0 | 0 | 0 | 30 | 44 | 116 | 290 | 156 | 137 | 1 | 0 | 0 | 774 |
| 1986 | 0 | 0 | 10 | 5 | 26 | 276 | 408 | 181 | 133 | 0 | 0 | 0 | 1039 |
| 1987 | 0 | 0 | 0 | 39 | 145 | 292 | 407 | 221 | 69 | 2 | 1 | 0 | 1176 |
| 1988 | 0 | 0 | 0 | 5 | 109 | 351 | 364 | 394 | 99 | 3 | 0 | 0 | 1325 |
| 1989 | 0 | 0 | 10 | 77 | 68 | 159 | 395 | 306 | 89 | 19 | 0 | 0 | 1123 |
| 1990 | 0 | 0 | 0 | 41 | 9 | 277 | 316 | 327 | 199 | 12 | 4 | 0 | 1185 |
| 1991 | 0 | 0 | 5 | 21 | 184 | 295 | 345 | 295 | 161 | 9 | 0 | 0 | 1315 |
| 1992 | 0 | 0 | 0 | 7 | 63 | 150 | 198 | 136 | 90 | 11 | 0 | 0 | 655 |
| 1993 | 0 | 0 | 0 | 0 | 35 | 188 | 322 | 324 | 21 | 19 | 0 | 0 | 909 |
| 1994 | 0 | 0 | 0 | 28 | 108 | 287 | 262 | 248 | 151 | 9 | 0 | 0 | 1093 |
| 1995 | 0 | 0 | 0 | 0 | 8 | 230 | 431 | 458 | | | 0 | 0 | |
| 1996 | | | 0 | 8 | 46 | 264 | 266 | 250 | 79 | 14 | 0 | 0 | |
| 1997 | 0 | 0 | 0 | 0 | 18 | 260 | 388 | 273 | 110 | 62 | 0 | 0 | 1111 |
| 1998 | 0 | 0 | 1 | 6 | 108 | 205 | 373 | 336 | 227 | 9 | 0 | 0 | 1265 |
| 1999 | 0 | 0 | 0 | 0 | 43 | 208 | 484 | 248 | 84 | 12 | 0 | 0 | 1079 |
| 2000 | 0 | 0 | 0 | 2 | 110 | 188 | 317 | 385 | 179 | 18 | 0 | 0 | 1199 |
| 2001 | 0 | 0 | 0 | 27 | 80 | 225 | 439 | 342 | 93 | 10 | 1 | 0 | 1217 |
| 2002 | 0 | 0 | 0 | 27 | 62 | 377 | 507 | 322 | 170 | 8 | 0 | 0 | 1473 |
| 2003 | 0 | 0 | 0 | 29 | 24 | 166 | 421 | 410 | 85 | 20 | 0 | 0 | 1155 |
| 2004 | 0 | 0 | 0 | 20 | 100 | 147 | 264 | 190 | 191 | 10 | 0 | 0 | 922 |
| 2005 | 0 | 0 | 1 | 20 | 67 | 324 | 466 | 319 | 220 | 52 | 0 | 0 | 1469 |
| 2006 | 0 | 0 | 0 | 36 | 109 | 296 | 448 | 302 | 48 | 29 | 0 | 0 | 1268 |
| 2007 | 0 | 0 | 7 | 20 | 100 | 246 | 427 | 415 | 139 | 33 | 0 | 0 | 1387 |
| 2008 | 0 | 0 | 0 | 1 | 45 | 211 | 386 | 342 | 94 | 27 | 5 | 0 | 1111 |
| 2009 | 0 | 0 | 0 | 13 | 77 | 225 | 222 | 235 | 70 | 0 | 0 | 0 | 842 |

## SNOWFALL (inches)  2009  OMAHA (KOMA)

| YEAR | JUL | AUG | SEP | OCT | NOV | DEC | JAN | FEB | MAR | APR | MAY | JUN | TOTAL |
|------|-----|-----|-----|-----|-----|-----|-----|-----|-----|-----|-----|-----|-------|
| 1980-81 | 0.0 | 0.0 | 0.0 | 2.0 | T | 0.5 | 3.6 | 3.0 | T | 0.0 | 0.0 | 0.0 | 9.1 |
| 1981-82 | 0.0 | 0.0 | 0.0 | T | 1.0 | 7.0 | 2.4 | 3.0 | 8.0 | 2.9 | 0.0 | 0.0 | 24.3 |
| 1982-83 | 0.0 | 0.0 | 0.0 | T | T | 4.0 | 6.5 | 9.0 | 9.0 | 3.0 | 0.0 | 0.0 | 31.5 |
| 1983-84 | 0.0 | 0.0 | 0.0 | 0.0 | 6.9 | 13.5 | 3.3 | 2.0 | 14.2 | T | 0.0 | 0.0 | 39.9 |
| 1984-85 | 0.0 | 0.0 | 0.0 | T | T | 5.0 | 4.0 | 4.0 | 6.0 | T | 0.0 | 0.0 | 19.0 |
| 1985-86 | 0.0 | 0.0 | T | 0.0 | 5.0 | 4.5 | T | 7.7 | T | 0.3 | 0.0 | 0.0 | 17.5 |
| 1986-87 | 0.0 | 0.0 | 0.0 | 0.0 | 1.8 | 5.5 | 1.2 | 1.6 | 10.5 | T | 0.0 | 0.0 | 20.6 |
| 1987-88 | 0.0 | 0.0 | 0.0 | T | 9.0 | 2.0 | 2.0 | 2.2 | 0.8 | 0.0 | 0.0 | 0.0 | 16.0 |
| 1988-89 | 0.0 | 0.0 | 0.0 | 0.0 | 4.3 | 2.7 | 1.4 | 11.8 | 3.3 | T | T | 0.0 | 23.5 |
| 1989-90 | 0.0 | 0.0 | 0.0 | T | 1.2 | 5.3 | 5.2 | 4.0 | 6.7 | 0.1 | 0.0 | T | 22.5 |
| 1990-91 | 0.0 | 0.0 | 0.0 | T | 1.1 | 10.1 | 14.9 | 0.3 | 4.1 | 1.1 | 0.0 | T | 31.6 |
| 1991-92 | 0.0 | 0.0 | 0.0 | 2.5 | 8.8 | T | 0.3 | 1.1 | 0.4 | 10.0 | 0.0 | 0.0 | 23.1 |
| 1992-93 | 0.0 | 0.0 | 0.0 | 0.0 | 5.8 | 3.9 | 13.0 | 8.5 | 4.1 | 1.5 | 0.0 | T | 36.8 |
| 1993-94 | 0.0 | 0.0 | 0.0 | T | 2.8 | 3.1 | 3.5 | 8.8 | 1.4 | 1.2 | 0.0 | T | 20.8 |
| 1994-95 | 0.0 | 0.0 | 0.0 | 0.0 | 2.0 | 12.1 | 5.5 | 3.0 | 4.8 | 0.4 | T | 0.0 | 27.8 |
| 1995-96 | T | 0.0 | 0.0 | | | | | | | | | | |
| 1996-97 | | | | | | 6.2 | | | T | | | | |
| 1997-98 | | | | | | | | | | | | | |
| 1998-99 | | | | | | 4.0 | 6.0 | 12.2 | 5.8 | 0.4 | 0.0 | 0.0 | |
| 1999-00 | | | | | T | 4.1 | 1.6 | 7.7 | T | T | 0.0 | T | |
| 2000-01 | 0.0 | 0.0 | 0.0 | 0.0 | 2.7 | 18.1 | 6.9 | 10.4 | 0.7 | T | T | T | 38.8 |
| 2001-02 | 0.0 | 0.0 | 0.0 | T | T | 1.8 | 6.6 | 2.2 | 11.1 | T | T | 0.0 | 21.7 |
| 2002-03 | T | T | 0.0 | 1.6 | 1.5 | T | 7.5 | 12.2 | 3.9 | 4.9 | T | 0.0 | 31.6 |
| 2003-04 | 0.0 | 0.0 | 0.0 | 0.0 | 0.3 | 6.8 | 19.8 | 17.7 | 3.2 | 0.0 | T | 0.0 | 47.8 |
| 2004-05 | T | T | 0.0 | 0.0 | 2.4 | T | 14.6 | 4.4 | T | 0.0 | T | T | 21.4 |
| 2005-06 | 0.0 | T | T | 0.0 | 4.1 | 5.4 | T | 1.2 | 9.4 | T | 0.0 | 0.0 | 20.1 |
| 2006-07 | 0.0 | 0.0 | 0.0 | T | T | 2.6 | 11.1 | 8.9 | 7.9 | T | T | 0.0 | 30.5 |
| 2007-08 | 0.0 | 0.0 | 0.0 | T | 0.2 | 9.2 | 6.5 | 6.4 | 0.9 | T | 0.0 | T | 23.2 |
| 2008-09 | T | 0.0 | 0.0 | 0.0 | 0.1 | 5.5 | 5.4 | 9.8 | T | 0.6 | T | T | 21.4 |
| 2009- | T | 0.0 | 0.0 | 3.7 | 0.0 | 24.6 | | | | | | | |
| POR=<br>58 YRS | T | T | T | 0.3 | 2.5 | 5.9 | 7.1 | 6.5 | 6.0 | 0.9 | T | T | 29.2 |

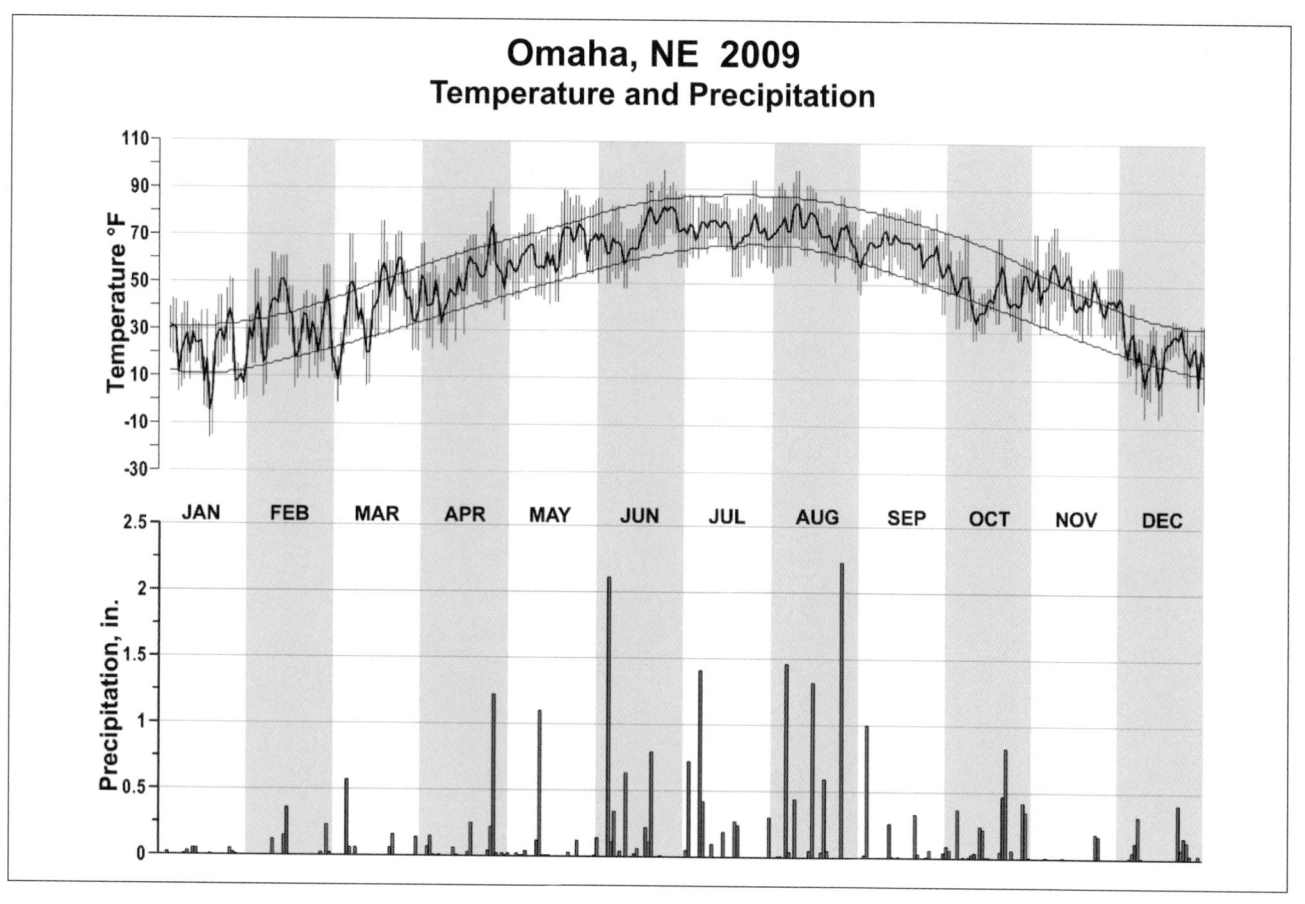

# 2009
# ELKO
# NEVADA (KEKO)

Elko is located in the Humbolt River Valley of northeastern Nevada. Weather observations are taken at the Flight Service Station which is located at the Municipal Airport on the west side of town. The elevation at the airport is just above 5,000 feet.

The Ruby mountain range, with many peaks near or exceeding 10,000 feet in height, dominates the landscape from about 40 miles northeast through 40 miles southeast of Elko. The immediate terrain consists of sagebrush-covered valleys and hills. The highest hills are approximately 2,500 feet above the valley floors. A few areas, mostly in the higher mountains, are covered with sparse stands of juniper, aspen, pinion pine, and spruce. The only heavily forested area in northeastern Nevada is in the Jarbidge Wilderness Area north of Elko near the Idaho border.

Because of the high elevation and proximity of the mountains, there is a wide range between the normal high and low temperatures. High radiative cooling at night makes cool nights the rule, even in mid summer.

Normal precipitation is light, especially during the summer months when the precipitation falls mostly as light showers which do not contribute much toward crop growth. The precipitation that falls between November and June (rain and snow) is critical to agriculture in the area. Not only is the precipitation that falls directly on the fields a benefit to farmers and ranchers, but the runoff from snowfall that accumulates in the mountains is used for irrigation.

The principal crop in northeast Nevada is hay. Cattle ranching is a major industry within the area. The ranges ordinarily furnish excellent summer pasture for cattle. Hay crops are needed for winter feeding.

Mining is another major industry. Many of the mines are located in the mountains at rather high elevations and are affected by daily weather. This is especially true during the winter when snow and rain may cause poor or impassable road conditions, thereby halting mining operations.

Transportation by air, rail, or road is seldom affected by the weather for more than short periods.

Based on the 1951-1980 period, the average first occurrence of 32 degrees Fahrenheit in the fall is September 8 and the average last occurrence in the spring is June 5.

# NORMALS, MEANS, AND EXTREMES
## ELKO (KEKO)

| LATITUDE: | LONGITUDE: | | ELEVATION (FT): | | | | TIME ZONE: | | | WBAN: 24121 | |
|---|---|---|---|---|---|---|---|---|---|---|---|
| 40 ° 49'N | -115° 47'W | | GRND: 5050   BARO: 5079 | | | | PACIFIC   (UTC -8) | | | | |

| | ELEMENT | POR | JAN | FEB | MAR | APR | MAY | JUN | JUL | AUG | SEP | OCT | NOV | DEC | YEAR |
|---|---|---|---|---|---|---|---|---|---|---|---|---|---|---|---|
| **TEMPERATURE °F** | NORMAL DAILY MAXIMUM | 30 | 37.1 | 42.9 | 51.2 | 59.3 | 68.6 | 79.9 | 89.6 | 88.1 | 78.2 | 65.0 | 48.1 | 38.2 | 62.2 |
| | MEAN DAILY MAXIMUM | 116 | 36.7 | 40.7 | 51.0 | 59.1 | 69.6 | 78.7 | 91.1 | 88.2 | 77.5 | 65.6 | 49.5 | 38.8 | 62.2 |
| | HIGHEST DAILY MAXIMUM | 79 | 65 | 70 | 78 | 86 | 97 | 104 | 107 | 107 | 99 | 88 | 78 | 65 | 107 |
| | YEAR OF OCCURRENCE | | 2003 | 1986 | 2004 | 1992 | 2003 | 1981 | 1981 | 1978 | 1950 | 1980 | 1980 | 1995 | JUL 1981 |
| | MEAN OF EXTREME MAXS. | 117 | 50.7 | 56.4 | 66.8 | 76.2 | 85.6 | 93.5 | 99.1 | 97.3 | 91.4 | 81.2 | 66.3 | 52.5 | 76.4 |
| | NORMAL DAILY MINIMUM | 30 | 14.1 | 19.7 | 25.9 | 29.9 | 36.8 | 43.5 | 48.6 | 47.0 | 38.1 | 28.3 | 20.9 | 13.8 | 30.6 |
| | MEAN DAILY MINIMUM | 116 | 10.9 | 16.6 | 23.5 | 28.7 | 35.7 | 41.7 | 48.1 | 45.4 | 36.0 | 27.7 | 20.0 | 12.9 | 28.9 |
| | LOWEST DAILY MINIMUM | 79 | -43 | -37 | -9 | -2 | 10 | 23 | 30 | 20 | 9 | 1 | -12 | -38 | -43 |
| | YEAR OF OCCURRENCE | | 1937 | 1933 | 1952 | 1936 | 1965 | 1976 | 1995 | 1992 | 1934 | 1996 | 1931 | 1932 | JAN 1937 |
| | MEAN OF EXTREME MINS. | 117 | -8.7 | -1.7 | 10.3 | 17.3 | 23.3 | 31.5 | 38.5 | 35.3 | 24.4 | 14.3 | 3.7 | -5.4 | 15.2 |
| | NORMAL DRY BULB | 30 | 25.6 | 31.3 | 38.6 | 44.6 | 52.7 | 61.7 | 69.1 | 67.6 | 58.2 | 46.7 | 34.5 | 26.0 | 46.4 |
| | MEAN DRY BULB | 116 | 23.8 | 28.7 | 37.3 | 43.9 | 52.7 | 60.2 | 69.6 | 66.8 | 56.8 | 46.7 | 34.8 | 25.9 | 45.6 |
| | MEAN WET BULB | 26 | 25.6 | 28.4 | 33.8 | 37.6 | 42.4 | 46.5 | 49.7 | 47.7 | 42.7 | 37.5 | 31.4 | 26.8 | 37.5 |
| | MEAN DEW POINT | 26 | 20.7 | 23.1 | 27.4 | 29.9 | 34.5 | 36.9 | 39.2 | 37.1 | 32.9 | 28.1 | 25.6 | 21.3 | 29.7 |
| | NORMAL NO. DAYS WITH: | | | | | | | | | | | | | | |
| | MAXIMUM >= 90 | 3 0 | 0.0 | 0.0 | 0.0 | 0.0 | 0.3 | 5.2 | 17.8 | 14.3 | 3.1 | 0.0 | 0.0 | 0.0 | 40.7 |
| | MAXIMUM <= 32 | 3 0 | 8.8 | 4.3 | 0.7 | 0.1 | 0.0 | 0.0 | 0.0 | 0.0 | 0.0 | 0.1 | 2.1 | 7.5 | 23.6 |
| | MINIMUM <= 32 | 30 | 29.3 | 26.4 | 25.4 | 19.5 | 7.6 | 1.3 | 0.1 | 0.4 | 7.5 | 21.7 | 26.3 | 29.0 | 194.5 |
| | MINIMUM <= 0 | 30 | 4.8 | 1.9 | 0.1 | 0.0 | 0.0 | 0.0 | 0.0 | 0.0 | 0.0 | 0.0 | 0.0 | 3.5 | 11.0 |
| **H/C** | NORMAL HEATING DEG. DAYS | 30 | 1222 | 943 | 820 | 612 | 383 | 161 | 53 | 57 | 237 | 569 | 916 | 1208 | 7181 |
| | NORMAL COOLING DEG. DAYS | 30 | 0 | 0 | 0 | 0 | 2 | 62 | 181 | 135 | 31 | 1 | 0 | 0 | 412 |
| **RH** | NORMAL (PERCENT) | 30 | 74 | 71 | 63 | 53 | 50 | 42 | 34 | 35 | 41 | 51 | 67 | 73 | 55 |
| | HOUR 04 LST | 30 | 79 | 79 | 78 | 73 | 72 | 66 | 56 | 55 | 61 | 68 | 77 | 79 | 70 |
| | HOUR 10 LST | 30 | 72 | 66 | 55 | 42 | 38 | 31 | 25 | 25 | 32 | 42 | 61 | 70 | 47 |
| | HOUR 16 LST | 30 | 59 | 52 | 43 | 33 | 30 | 24 | 19 | 19 | 22 | 29 | 48 | 57 | 36 |
| | HOUR 22 LST | 30 | 78 | 76 | 70 | 60 | 56 | 47 | 36 | 37 | 46 | 57 | 72 | 77 | 59 |
| **S** | PERCENT POSSIBLE SUNSHINE | | | | | | | | | | | | | | |
| **W/O** | MEAN NO. DAYS WITH: | | | | | | | | | | | | | | |
| | HEAVY FOG(VISBY <= 1/4 MI) | 43 | 2.0 | 1.1 | 0.8 | 0.7 | 0.3 | 0.1 | 0.0 | 0.1 | 0.1 | 0.4 | 0.6 | 1.8 | 8.0 |
| | THUNDERSTORMS | 59 | 0.1 | 0.3 | 0.3 | 0.9 | 2.7 | 3.2 | 4.2 | 3.8 | 1.7 | 0.5 | 0.2 | 0.1 | 18.0 |
| **CLOUDNESS** | MEAN: | | | | | | | | | | | | | | |
| | SUNRISE-SUNSET (OKTAS) | 43 | 5.4 | 5.3 | 5.4 | 5.3 | 4.8 | 3.5 | 2.7 | 2.7 | 2.6 | 3.4 | 4.8 | 5.0 | 4.2 |
| | MIDNIGHT-MIDNIGHT (OKTAS) | 5 | 4.3 | 4.3 | 3.4 | 3.8 | 3.3 | 2.8 | 2.0 | 2.6 | 2.1 | 2.0 | 4.2 | 4.3 | 3.3 |
| | MEAN NO. DAYS WITH: | | | | | | | | | | | | | | |
| | CLEAR | 56 | 6.7 | 6.3 | 6.3 | 6.3 | 8.1 | 12.9 | 17.0 | 17.5 | 18.0 | 13.9 | 8.3 | 7.7 | 129.0 |
| | PARTLY CLOUDY | 56 | 7.6 | 7.4 | 8.1 | 9.1 | 10.2 | 9.6 | 9.4 | 8.7 | 6.6 | 7.9 | 6.8 | 6.8 | 98.2 |
| | CLOUDY | 56 | 16.7 | 14.6 | 16.6 | 14.6 | 12.6 | 7.5 | 3.9 | 4.2 | 4.9 | 8.7 | 14.3 | 16.0 | 134.6 |
| **PR** | MEAN STATION PRESSURE(IN) | 26 | 25.01 | 24.97 | 24.92 | 24.90 | 24.90 | 24.92 | 24.97 | 24.97 | 24.97 | 24.98 | 24.99 | 24.99 | 24.96 |
| | MEAN SEA-LEVEL PRES. (IN) | 26 | 30.20 | 30.11 | 30.02 | 29.96 | 29.92 | 29.91 | 29.92 | 29.93 | 29.97 | 30.04 | 30.11 | 30.16 | 30.02 |
| **WINDS** | MEAN SPEED (MPH) | 26 | 4.9 | 5.4 | 6.1 | 6.7 | 6.4 | 6.3 | 6.0 | 5.7 | 5.4 | 5.0 | 5.1 | 5.0 | 5.7 |
| | PREVAIL.DIR(TENS OF DEGS) | 23 | 07 | 25 | 25 | 25 | 24 | 25 | 23 | 25 | 07 | 25 | 25 | 25 | 25 |
| | MAXIMUM 2-MINUTE: | | | | | | | | | | | | | | |
| | SPEED (MPH) | 8 | 40 | 43 | 40 | 41 | 43 | 43 | 45 | 45 | 49 | 36 | 38 | 45 | 49 |
| | DIR. (TENS OF DEGS) | | 17 | 28 | 25 | 26 | 26 | 20 | 23 | 24 | 26 | 27 | 29 | 27 | 26 |
| | YEAR OF OCCURRENCE | | 2008 | 2006 | 2006 | 2005 | 2004 | 2005 | 2002 | 2007 | 2006 | 2007 | 2009 | 2008 | SEP 2006 |
| | MAXIMUM 3-SECOND | | | | | | | | | | | | | | |
| | SPEED (MPH) | 8 | 54 | 48 | 49 | 61 | 52 | 63 | 60 | 54 | 55 | 47 | 45 | 59 | 63 |
| | DIR. (TENS OF DEGS) | | 16 | 28 | 25 | 20 | 17 | 29 | 22 | 24 | 27 | 27 | 29 | 27 | 29 |
| | YEAR OF OCCURRENCE | | 2008 | 2006 | 2006 | 2008 | 2009 | 2008 | 2002 | 2007 | 2006 | 2004 | 2009 | 2008 | JUN 2008 |
| **PRECIPITATION** | NORMAL (IN) | 30 | 1.14 | 0.88 | 0.98 | 0.81 | 1.08 | 0.67 | 0.30 | 0.36 | 0.68 | 0.71 | 1.05 | 0.93 | 9.59 |
| | MAXIMUM MONTHLY (IN) | 79 | 3.35 | 2.93 | 2.39 | 2.17 | 4.09 | 2.94 | 2.35 | 4.61 | 3.22 | 2.76 | 2.77 | 4.21 | 4.61 |
| | YEAR OF OCCURRENCE | | 1956 | 1932 | 1989 | 1963 | 1971 | 2009 | 1950 | 1970 | 1978 | 1938 | 1942 | 1983 | AUG 1970 |
| | MINIMUM MONTHLY (IN) | 79 | 0.04 | 0.06 | 0.04 | 0.02 | T | T | 0.00 | T | T | T | T | T | 0.00 |
| | YEAR OF OCCURRENCE | | 1961 | 1988 | 1988 | 1992 | 1974 | 1994 | 1963 | 2006 | 1951 | 1995 | 1959 | 1976 | JUL 1963 |
| | MAXIMUM IN 24 HOURS (IN) | 79 | 1.27 | 0.89 | 1.22 | 1.10 | 1.73 | 1.85 | 1.28 | 4.13 | 2.32 | 1.31 | 1.31 | 1.62 | 4.13 |
| | YEAR OF OCCURRENCE | | 1951 | 1936 | 2006 | 1943 | 1971 | 1968 | 2001 | 1970 | 1978 | 1939 | 1950 | 1950 | AUG 1970 |
| | NORMAL NO. DAYS WITH: | | | | | | | | | | | | | | |
| | PRECIPITATION >= 0.01 | 30 | 9.2 | 8.3 | 9.4 | 7.6 | 8.5 | 5.0 | 3.5 | 3.6 | 4.8 | 4.9 | 7.6 | 8.1 | 80.5 |
| | PRECIPITATION >= 1.00 | 30 | * | 0.0 | 0.0 | 0.0 | * | 0.0 | 0.0 | * | * | 0.0 | 0.0 | 0.0 | 0.0 |
| **SNOWFALL** | NORMAL (IN) | 30 | 9.4 | 6.2 | 4.4 | 2.6 | 1.2 | 0.* | 0.0 | 0.0 | 0.1 | 0.8 | 5.0 | 7.4 | 37.1 |
| | MAXIMUM MONTHLY (IN) | 78 | 45.7 | 26.1 | 23.2 | 16.6 | 11.3 | T | T | 0.0 | 2.0 | 5.6 | 16.8 | 33.2 | 45.7 |
| | YEAR OF OCCURRENCE | | 1996 | 1932 | 1967 | 1999 | 1971 | 2009 | 1995 | 2006 | 1982 | 1984 | 1985 | 1983 | JAN 1996 |
| | MAXIMUM IN 24 HOURS (IN) | 61 | 18.4 | 9.1 | 13.8 | 10.0 | 8.6 | T | T | T | 2.0 | 5.2 | 9.0 | 9.3 | 18.4 |
| | YEAR OF OCCURRENCE' | | 1996 | 1949 | 1967 | 1975 | 1971 | 2009 | 1995 | 1993 | 1982 | 1963 | 1965 | 1992 | JAN 1996 |
| | MAXIMUM SNOW DEPTH (IN) | 55 | 24 | 21 | 12 | 9 | 8 | 0 | 0 | 0 | 0 | 4 | 17 | 12 | 24 |
| | YEAR OF OCCURRENCE | | 1948 | 1949 | 1967 | 1975 | 1975 | | | | | 1984 | 1963 | 1968 | JAN 1948 |
| | NORMAL NO. DAYS WITH: | | | | | | | | | | | | | | |
| | SNOWFALL >= 1.0 | 30 | 3.2 | 2.0 | 1.8 | 0.9 | 0.3 | 0.0 | 0.0 | 0.0 | 0.0 | 0.3 | 2.2 | 2.7 | 13.4 |

## PRECIPITATION (inches) 2009 ELKO (KEKO)

| YEAR | JAN | FEB | MAR | APR | MAY | JUN | JUL | AUG | SEP | OCT | NOV | DEC | ANNUAL |
|------|------|------|------|------|------|------|------|------|------|------|------|------|--------|
| 1980 | 3.11 | 1.89 | 0.77 | 1.22 | 3.15 | 0.80 | 0.33 | 0.10 | 0.42 | 0.19 | 0.62 | 0.21 | 12.81 |
| 1981 | 0.64 | 0.33 | 1.20 | 0.75 | 0.80 | 0.24 | 0.02 | 0.19 | 0.13 | 0.69 | 0.60 | 3.19 | 8.78 |
| 1982 | 0.82 | 0.65 | 1.94 | 0.50 | 1.04 | 0.54 | 0.69 | 1.24 | 2.55 | 1.11 | 1.78 | 0.86 | 13.72 |
| 1983 | 1.73 | 1.34 | 1.91 | 1.28 | 0.60 | 0.47 | 0.01 | 1.25 | 1.57 | 1.21 | 2.76 | 4.21 | 18.34 |
| 1984 | 0.57 | 0.80 | 1.25 | 1.00 | 0.24 | 1.29 | 1.04 | 0.46 | 0.11 | 1.75 | 1.40 | 0.45 | 10.36 |
| 1985 | 0.54 | 0.15 | 1.09 | 0.23 | 0.60 | 0.17 | 0.25 | 0.02 | 1.17 | 0.16 | 2.14 | 0.78 | 7.30 |
| 1986 | 0.18 | 1.86 | 0.52 | 1.17 | 0.75 | 0.39 | 0.12 | 0.02 | 0.81 | 0.04 | 0.13 | 0.09 | 6.08 |
| 1987 | 0.54 | 0.68 | 1.13 | 0.26 | 1.80 | 0.69 | 0.14 | 0.01 | 0.09 | 0.55 | 1.97 | 0.76 | 8.62 |
| 1988 | 1.27 | 0.06 | 0.04 | 0.46 | 0.91 | 0.58 | 0.08 | 0.26 | 0.11 | T | 1.94 | 1.01 | 6.72 |
| 1989 | 0.46 | 0.93 | 2.39 | 0.28 | 0.36 | 0.50 | 0.18 | 0.52 | 0.69 | 0.27 | 0.79 | 0.51 | 7.88 |
| 1990 | 0.97 | 0.78 | 1.07 | 1.51 | 0.96 | 0.97 | 0.19 | 0.56 | 0.15 | 0.07 | 0.98 | 1.22 | 9.43 |
| 1991 | 0.49 | 0.46 | 0.62 | 0.86 | 1.71 | 0.06 | 0.20 | 0.25 | 0.58 | 1.29 | 1.29 | 0.04 | 7.85 |
| 1992 | 0.17 | 0.75 | 1.64 | 0.02 | 0.40 | 0.67 | 0.27 | 0.17 | 0.01 | 0.54 | 1.03 | 1.89 | 7.56 |
| 1993 | 1.98 | 0.93 | 0.68 | 0.24 | 0.44 | 1.43 | 0.36 | 0.09 | 0.41 | 0.76 | 0.07 | 0.22 | 7.61 |
| 1994 | 0.32 | 1.11 | 0.15 | 1.11 | 1.68 | T | 0.22 | 0.11 | 0.79 | 0.52 | 1.61 | 0.70 | 8.32 |
| 1995 | 1.56 | 0.33 | 2.04 | 1.15 | 2.35 | 1.66 | 0.24 | 0.02 | 0.31 | T | 0.39 | 1.41 | 11.46 |
| 1996 | 3.28 | 1.45 | 0.88 | 0.78 | 2.23 | 0.13 | .73 | T | .20 | 1.10 | 1.36 | 3.10 | 15.24 |
| 1997 | 2.44 | 0.21 | 0.21 | 0.93 | 0.22 | 1.69 | 1.08 | 1.37 | 0.63 | 0.77 | 1.23 | 0.21 | 10.99 |
| 1998 | 2.34 | 1.41 | 1.22 | 0.29 | 1.91 | 0.89 | 0.24 | T | 1.92 | 0.98 | 0.77 | 0.46 | 12.43 |
| 1999 | 1.56 | 0.74 | 0.28 | 1.75 | 0.83 | 1.18 | T | 0.19 | 0.02 | 0.52 | 0.41 | 0.07 | 7.55 |
| 2000 | 1.48 | 2.32 | 0.77 | 0.69 | 0.73 | 0.08 | 0.04 | 0.25 | 0.03 | 1.73 | 0.50 | 0.33 | 8.95 |
| 2001 | 0.53 | 0.80 | 1.00 | 1.10 | 0.03 | 0.03 | 1.46 | 0.02 | 0.26 | 0.03 | 1.62 | 1.61 | 8.49 |
| 2002 | 0.54 | 0.47 | 0.62 | 1.60 | 0.87 | 0.43 | 0.03 | 0.01 | 1.07 | 0.08 | 1.21 | 0.55 | 7.48 |
| 2003 | 0.95 | 0.55 | 0.48 | 1.66 | 1.68 | 0.01 | 0.92 | 1.69 | 0.20 | 0.05 | 0.71 | 1.96 | 10.86 |
| 2004 | 0.69 | 0.91 | 0.30 | 1.39 | 0.96 | 0.28 | 0.16 | 1.17 | 1.30 | 1.95 | 0.87 | 1.19 | 11.17 |
| 2005 | 2.12 | 0.84 | 1.37 | 1.56 | 1.87 | 0.74 | 0.66 | 0.07 | 0.46 | 1.61 | 1.27 | 2.81 | 15.38 |
| 2006 | 1.53 | 1.21 | 2.31 | 2.08 | 0.19 | 0.40 | 1.11 | T | 0.09 | 0.75 | 1.05 | 0.68 | 11.40 |
| 2007 | 0.27 | 1.05 | 0.48 | 0.61 | 0.17 | 0.37 | 0.08 | 0.14 | 0.17 | 1.05 | 0.32 | 1.02 | 5.73 |
| 2008 | 1.75 | 0.79 | 0.39 | 0.15 | 1.13 | 0.57 | 0.13 | 0.25 | 0.01 | 0.43 | 1.54 | 0.91 | 8.05 |
| 2009 | 1.28 | 0.59 | 0.78 | 1.83 | 0.47 | 2.94 | 0.20 | 0.68 | 0.27 | 0.49 | 0.01 | 1.68 | 11.22 |
| POR= 115 YRS | 1.22 | 0.93 | 0.92 | 0.84 | 1.09 | 0.77 | 0.40 | 0.41 | 0.45 | 0.73 | 1.01 | 1.10 | 9.87 |

WBAN : 24121

## AVERAGE TEMPERATURE (°F) 2009 ELKO (KEKO)

| YEAR | JAN | FEB | MAR | APR | MAY | JUN | JUL | AUG | SEP | OCT | NOV | DEC | ANNUAL |
|------|------|------|------|------|------|------|------|------|------|------|------|------|--------|
| 1980 | 32.2 | 39.0 | 37.0 | 47.3 | 51.5 | 60.5 | 71.6 | 67.2 | 61.6 | 49.6 | 40.0 | 34.2 | 49.3 |
| 1981 | 34.6 | 35.5 | 40.8 | 48.2 | 54.2 | 67.3 | 73.9 | 72.8 | 61.6 | 44.8 | 40.5 | 34.1 | 50.7 |
| 1982 | 23.2 | 29.6 | 37.8 | 42.3 | 52.5 | 61.6 | 69.1 | 70.6 | 58.6 | 45.4 | 34.1 | 28.0 | 46.1 |
| 1983 | 30.4 | 31.8 | 41.8 | 42.7 | 52.6 | 62.1 | 68.7 | 72.1 | 62.7 | 50.8 | 36.5 | 28.7 | 48.4 |
| 1984 | 17.1 | 23.8 | 36.0 | 41.9 | 54.7 | 59.7 | 71.9 | 70.7 | 60.7 | 43.3 | 37.7 | 20.9 | 44.9 |
| 1985 | 21.8 | 26.8 | 35.1 | 48.0 | 54.0 | 64.7 | 75.9 | 65.9 | 53.7 | 46.2 | 30.1 | 23.1 | 45.4 |
| 1986 | 32.3 | 37.6 | 43.8 | 45.0 | 52.3 | 65.7 | 70.5 | 70.5 | 53.0 | 46.0 | 34.0 | 24.9 | 47.7 |
| 1987 | 21.5 | 31.1 | 37.5 | 49.5 | 55.6 | 63.9 | 66.8 | 66.6 | 59.6 | 50.2 | 35.7 | 25.7 | 47.0 |
| 1988 | 20.3 | 29.3 | 37.2 | 46.5 | 51.5 | 65.2 | 72.0 | 67.7 | 56.8 | 53.0 | 33.5 | 21.7 | 46.2 |
| 1989 | 11.6 | 22.3 | 41.5 | 49.0 | 52.3 | 61.5 | 70.8 | 65.5 | 57.8 | 46.1 | 33.4 | 27.8 | 45.0 |
| 1990 | 27.8 | 26.7 | 40.9 | 50.0 | 50.1 | 61.8 | 70.5 | 67.0 | 63.9 | 45.8 | 33.6 | 14.5 | 46.1 |
| 1991 | 22.5 | 37.4 | 36.7 | 41.2 | 48.1 | 59.0 | 70.4 | 68.7 | 59.7 | 46.5 | 35.5 | 27.2 | 46.1 |
| 1992 | 24.4 | 35.6 | 41.1 | 48.1 | 56.5 | 60.7 | 65.1 | 66.7 | 56.9 | 47.6 | 28.2 | 17.2 | 45.7 |
| 1993 | 15.5 | 20.5 | 36.4 | 42.9 | 55.7 | 56.0 | 59.7 | 62.0 | 55.6 | 45.6 | 26.4 | 25.4 | 41.8 |
| 1994 | 28.5 | 26.5 | 40.1 | 44.7 | 54.1 | 60.6 | 68.5 | 67.8 | 57.6 | 43.2 | 25.4 | 28.1 | 45.4 |
| 1995 | 32.0 | 37.6 | 37.5 | 42.5 | 49.6 | 56.3 | 64.5 | 65.7 | 56.8 | 43.2 | 38.1 | 30.2 | 46.2 |
| 1996 | 28.4 | 25.2 | 38.7 | 45.3 | 51.9 | 62.6 | 69.2 | 64.3 | 53.4 | 43.3 | 34.8 | 28.5 | 45.5 |
| 1997 | 27.7 | 29.1 | 40.3 | 42.1 | 56.0 | 60.6 | 65.0 | 66.3 | 59.0 | 43.5 | 36.3 | 23.8 | 45.8 |
| 1998 | 31.6 | 31.7 | 37.6 | 42.8 | 49.5 | 56.7 | 69.4 | 67.0 | 59.6 | 43.1 | 36.8 | 24.3 | 45.8 |
| 1999 | 29.4 | 31.9 | 37.8 | 40.3 | 49.7 | 59.2 | 66.0 | 64.9 | 55.8 | 45.6 | 39.4 | 25.5 | 45.5 |
| 2000 | 30.0 | 36.0 | 37.3 | 48.4 | 53.6 | 61.7 | 66.1 | 67.3 | 55.4 | 43.7 | 27.5 | 26.5 | 46.1 |
| 2001 | 20.5 | 27.3 | 41.1 | 44.8 | 58.4 | 64.3 | 71.1 | 72.1 | 62.4 | 49.2 | 36.8 | 23.7 | 47.6 |
| 2002 | 23.2 | 24.3 | 34.5 | 46.3 | 51.9 | 63.8 | 73.6 | 66.1 | 59.5 | 43.3 | 35.7 | 31.6 | 46.2 |
| 2003 | 36.3 | 30.5 | 39.5 | 43.2 | 53.6 | 64.1 | 74.3 | 71.0 | 59.8 | 52.1 | 32.5 | 30.2 | 48.9 |
| 2004 | 16.6 | 24.4 | 43.5 | 46.9 | 53.2 | 64.0 | 71.3 | 67.2 | 57.3 | 46.9 | 34.2 | 29.3 | 46.2 |
| 2005 | 20.5 | 23.7 | 37.3 | 44.6 | 54.1 | 59.0 | 73.1 | 69.1 | 56.3 | 47.6 | 37.3 | 29.1 | 46.0 |
| 2006 | 31.5 | 31.0 | 34.5 | 47.3 | 57.1 | 66.6 | 75.4 | 67.8 | 58.3 | 45.7 | 34.9 | 25.1 | 47.9 |
| 2007 | 16.2 | 33.0 | 42.8 | 46.5 | 56.1 | 65.1 | 75.8 | 70.5 | 58.8 | 46.2 | 35.8 | 25.5 | 47.7 |
| 2008 | 17.4 | 26.2 | 36.5 | 40.9 | 52.8 | 61.7 | 71.5 | 70.3 | 60.4 | 47.6 | 41.6 | 24.3 | 45.9 |
| 2009 | 28.1 | 32.1 | 37.9 | 44.2 | 57.3 | 61.8 | 71.6 | 67.5 | 63.7 | 43.6 | 34.6 | 16.0 | 46.5 |
| POR= 116 YRS | 23.8 | 28.7 | 37.3 | 43.9 | 52.7 | 60.2 | 69.6 | 66.8 | 56.8 | 46.7 | 34.8 | 25.9 | 45.6 |

## HEATING DEGREE DAYS (base 65°F) 2009 ELKO (KEKO)

| YEAR | JUL | AUG | SEP | OCT | NOV | DEC | JAN | FEB | MAR | APR | MAY | JUN | TOTAL |
|------|-----|-----|-----|-----|-----|-----|-----|-----|-----|-----|-----|-----|-------|
| 1980-81 | 0 | 34 | 119 | 467 | 743 | 947 | 933 | 819 | 744 | 497 | 336 | 67 | 5706 |
| 1981-82 | 0 | 4 | 131 | 618 | 726 | 952 | 1289 | 987 | 836 | 676 | 381 | 130 | 6730 |
| 1982-83 | 32 | 0 | 225 | 598 | 925 | 1143 | 1066 | 924 | 713 | 661 | 390 | 114 | 6791 |
| 1983-84 | 29 | 0 | 105 | 434 | 847 | 1119 | 1480 | 1187 | 894 | 686 | 318 | 201 | 7300 |
| 1984-85 | 0 | 10 | 163 | 664 | 811 | 1360 | 1331 | 1060 | 921 | 505 | 335 | 69 | 7229 |
| 1985-86 | 0 | 42 | 338 | 573 | 1042 | 1294 | 1002 | 759 | 650 | 596 | 399 | 49 | 6744 |
| 1986-87 | 18 | 5 | 370 | 583 | 924 | 1235 | 1341 | 943 | 844 | 459 | 286 | | |
| 1987-88 | 50 | 32 | 175 | 451 | 872 | 1211 | 1381 | 1028 | 856 | 550 | 420 | 102 | 7128 |
| 1988-89 | 0 | 16 | 255 | 365 | 938 | 1338 | 1647 | 1194 | 723 | 475 | 387 | 111 | 7449 |
| 1989-90 | 4 | 59 | 214 | 578 | 940 | 1147 | 1147 | 1067 | 742 | 445 | 457 | 135 | 6935 |
| 1990-91 | 12 | 50 | 86 | 587 | 933 | 1559 | 1310 | 767 | 869 | 710 | 518 | 176 | 7577 |
| 1991-92 | 2 | 10 | 178 | 566 | 879 | 1164 | 1250 | 845 | 734 | 503 | 256 | 152 | 6539 |
| 1992-93 | 51 | 89 | 242 | 534 | 1100 | 1477 | 1528 | 1239 | 879 | 658 | 283 | 272 | 8352 |
| 1993-94 | 168 | 113 | 279 | 591 | 1153 | 1222 | 1125 | 1075 | 764 | 603 | 329 | 156 | 7578 |
| 1994-95 | 19 | 18 | 211 | 670 | 1180 | 1137 | 1017 | 760 | 845 | 668 | 471 | 257 | 7253 |
| 1995-96 | 55 | 47 | 257 | 666 | 803 | 1072 | 1125 | 1149 | 808 | 583 | 401 | 93 | 7059 |
| 1996-97 | 7 | 71 | 342 | 664 | 900 | 1125 | 1149 | 996 | 759 | 682 | 275 | 134 | 7104 |
| 1997-98 | 49 | 16 | 195 | 657 | 854 | 1268 | 1030 | 925 | 845 | 657 | 476 | 243 | 7215 |
| 1998-99 | 6 | 33 | 184 | 673 | 837 | 1253 | 1096 | 920 | 834 | 731 | 469 | 184 | 7220 |
| 1999-00 | 18 | 58 | 271 | 595 | 765 | 1217 | 1075 | 838 | 853 | 490 | 347 | 108 | 6635 |
| 2000-01 | 24 | 25 | 285 | 653 | 1116 | 1187 | 1372 | 1046 | 737 | 601 | 218 | 89 | 7353 |
| 2001-02 | 4 | 0 | 110 | 484 | 840 | 1272 | 1290 | 1129 | 936 | 552 | 409 | 108 | 7134 |
| 2002-03 | 0 | 43 | 173 | 663 | 874 | 1028 | 886 | 957 | 785 | 647 | 366 | 76 | 6498 |
| 2003-04 | 1 | 0 | 161 | 397 | 969 | 1074 | 1494 | 1171 | 659 | 537 | 362 | 72 | 6897 |
| 2004-05 | 0 | 37 | 231 | 553 | 919 | 1098 | 1373 | 1149 | 854 | 606 | 330 | 196 | 7346 |
| 2005-06 | 0 | 16 | 263 | 528 | 825 | 1106 | 1030 | 945 | 937 | 525 | 262 | 37 | 6474 |
| 2006-07 | 0 | 15 | 227 | 591 | 893 | 1230 | 1503 | 890 | 682 | 548 | 278 | 74 | 6931 |
| 2007-08 | 0 | 0 | 213 | 577 | 868 | 1218 | 1468 | 1115 | 877 | 717 | 380 | 148 | 7581 |
| 2008-09 | 0 | 0 | 141 | 534 | 694 | 1255 | 1136 | 915 | 831 | 615 | 239 | 130 | 6490 |
| 2009- | 7 | 47 | 82 | 655 | 905 | 1513 | | | | | | | |

WBAN : 24121

## COOLING DEGREE DAYS (base 65°F) 2009 ELKO (KEKO)

| YEAR | JAN | FEB | MAR | APR | MAY | JUN | JUL | AUG | SEP | OCT | NOV | DEC | TOTAL |
|------|-----|-----|-----|-----|-----|-----|-----|-----|-----|-----|-----|-----|-------|
| 1980 | 0 | 0 | 0 | 0 | 0 | 36 | 211 | 109 | 25 | 0 | 0 | 0 | 381 |
| 1981 | 0 | 0 | 0 | 2 | 9 | 145 | 282 | 254 | 36 | 0 | 0 | 0 | 728 |
| 1982 | 0 | 0 | 0 | 0 | 0 | 36 | 165 | 181 | 39 | 0 | 0 | 0 | 421 |
| 1983 | 0 | 0 | 0 | 0 | 11 | 33 | 151 | 228 | 40 | 0 | 0 | 0 | 463 |
| 1984 | 0 | 0 | 0 | 0 | 3 | 48 | 223 | 197 | 42 | 0 | 0 | 0 | 513 |
| 1985 | 0 | 0 | 0 | 0 | 0 | 66 | 342 | 77 | 7 | 0 | 0 | 0 | 492 |
| 1986 | 0 | 0 | 0 | 0 | 12 | 75 | 89 | 186 | 16 | 0 | 0 | 0 | 378 |
| 1987 | 0 | 0 | 0 | 0 | 3 | | 113 | 89 | 20 | 0 | 0 | 0 | |
| 1988 | 0 | 0 | 0 | 0 | 6 | 113 | 224 | 105 | 19 | 0 | 0 | 0 | 467 |
| 1989 | 0 | 0 | 0 | 0 | 0 | 13 | 188 | 78 | 4 | 0 | 0 | 0 | 283 |
| 1990 | 0 | 0 | 0 | 0 | 0 | 47 | 193 | 121 | 61 | 0 | 0 | 0 | 422 |
| 1991 | 0 | 0 | 0 | 0 | 0 | 2 | 176 | 131 | 26 | 0 | 0 | 0 | 335 |
| 1992 | 0 | 0 | 0 | 0 | 0 | 28 | 65 | 152 | 7 | 0 | 0 | 0 | 252 |
| 1993 | 0 | 0 | 0 | 0 | 1 | 9 | 10 | 26 | 5 | 0 | 0 | 0 | 51 |
| 1994 | 0 | 0 | 0 | 0 | 0 | 28 | 135 | 109 | 0 | 0 | 0 | 0 | 272 |
| 1995 | 0 | 0 | 0 | 0 | 0 | 3 | 45 | 77 | 17 | 0 | 0 | 0 | 142 |
| 1996 | 0 | 0 | 0 | 0 | 1 | 29 | 143 | 55 | 1 | 0 | 0 | 0 | 229 |
| 1997 | 0 | 0 | 0 | 0 | 3 | 7 | 57 | 65 | 22 | 0 | 0 | 0 | 154 |
| 1998 | 0 | 0 | 0 | 0 | 0 | 1 | 149 | 102 | 27 | 0 | 0 | 0 | 279 |
| 1999 | 0 | 0 | 0 | 0 | 0 | 17 | 57 | 61 | 1 | 0 | 0 | 0 | 136 |
| 2000 | 0 | 0 | 0 | 0 | 3 | 18 | 63 | 104 | 3 | 0 | 0 | 0 | 191 |
| 2001 | 0 | 0 | 0 | 0 | 21 | 76 | 201 | 227 | 38 | 1 | 0 | 0 | 564 |
| 2002 | 0 | 0 | 0 | 0 | 10 | 76 | 274 | 84 | 17 | 0 | 0 | 0 | 461 |
| 2003 | 0 | 0 | 0 | 0 | 23 | 54 | 295 | 197 | 13 | 3 | 0 | 0 | 585 |
| 2004 | 0 | 0 | 0 | 0 | 0 | 48 | 204 | 111 | 7 | 0 | 0 | 0 | 370 |
| 2005 | 0 | 0 | 0 | 0 | 0 | 23 | 259 | 149 | 10 | 0 | 0 | 0 | 441 |
| 2006 | 0 | 0 | 0 | 0 | 21 | 92 | 327 | 113 | 34 | 0 | 0 | 0 | 587 |
| 2007 | 0 | 0 | 0 | 0 | 5 | 86 | 341 | 178 | 34 | 0 | 0 | 0 | 644 |
| 2008 | 0 | 0 | 0 | 0 | 7 | 57 | 207 | 175 | 9 | 0 | 0 | 0 | 455 |
| 2009 | 0 | 0 | 0 | 0 | 6 | 39 | 217 | 134 | 51 | 0 | 0 | 0 | 447 |

## SNOWFALL (inches) 2009 ELKO (KEKO)

| YEAR | JUL | AUG | SEP | OCT | NOV | DEC | JAN | FEB | MAR | APR | MAY | JUN | TOTAL |
|------|-----|-----|-----|-----|-----|-----|-----|-----|-----|-----|-----|-----|-------|
| 1980-81 | 0.0 | 0.0 | 0.0 | 1.2 | 0.5 | 0.5 | 4.4 | 1.4 | 2.6 | 2.9 | 0.2 | T | 13.7 |
| 1981-82 | 0.0 | 0.0 | 0.0 | T | 0.3 | 9.8 | 11.2 | 0.3 | 13.8 | 3.5 | T | T | 38.9 |
| 1982-83 | 0.0 | 0.0 | 2.0 | T | 8.1 | 6.6 | 16.5 | 10.6 | 9.6 | 1.9 | 0.2 | 0.0 | 55.5 |
| 1983-84 | 0.0 | 0.0 | 0.0 | 0.0 | 13.4 | 33.2 | 6.6 | 5.8 | 5.1 | 5.9 | T | 0.0 | 70.0 |
| 1984-85 | 0.0 | 0.0 | 0.0 | 5.6 | 5.4 | 4.9 | 5.6 | 2.0 | 7.7 | 0.4 | 0.0 | 0.0 | 31.6 |
| 1985-86 | 0.0 | 0.0 | 0.0 | 0.7 | 16.8 | 5.1 | 0.8 | 2.0 | 1.9 | 1.4 | 0.1 | 0.0 | 28.8 |
| 1986-87 | 0.0 | 0.0 | T | 0.0 | 1.1 | 1.0 | 6.9 | 5.7 | 3.1 | T | 0.0 | 0.0 | 17.8 |
| 1987-88 | 0.0 | 0.0 | 0.0 | 0.0 | 0.3 | 6.1 | 14.5 | 0.2 | 1.0 | T | 3.7 | 0.0 | 25.8 |
| 1988-89 | 0.0 | 0.0 | 0.0 | 0.0 | 11.3 | 16.1 | 11.0 | 9.6 | 4.6 | T | 0.0 | 0.0 | 52.6 |
| 1989-90 | 0.0 | 0.0 | 0.0 | 0.1 | 4.3 | 0.0 | 6.9 | 9.4 | 3.5 | T | 2.1 | 0.0 | 26.3 |
| 1990-91 | T | T | 0.0 | 0.0 | 3.2 | 12.7 | 2.1 | 0.7 | 5.0 | 1.5 | 2.0 | 0.0 | 27.2 |
| 1991-92 | 0.0 | 0.0 | 0.0 | 2.9 | 5.9 | 0.7 | 2.4 | 3.2 | 1.8 | T | 0.0 | 0.0 | 16.9 |
| 1992-93 | 0.0 | 0.0 | 0.0 | 0.0 | 4.5 | 19.9 | 23.5 | 15.9 | 1.2 | 0.6 | T | 0.0 | 65.6 |
| 1993-94 | 0.0 | T | 0.0 | 0.0 | 1.4 | 0.9 | 5.0 | 13.2 | 0.1 | 1.4 | T | 0.0 | 22.0 |
| 1994-95 | 0.0 | 0.0 | 0.0 | T | 15.4 | 4.2 | 9.9 | 3.3 | 7.5 | 5.7 | 0.2 | T | 46.2 |
| 1995-96 | T | 0.0 | 0.0 | 0.0 | 0.8 | 2.9 | 45.7 | 11.6 | 6.5 | 0.9 | 0.3 | 0.0 | 68.7 |
| 1996-97 | 0.0 | 0.0 | 0.0 | 2.1 | 6.0 | 27.7 | 15.6 | 2.3 | 3.3 | 2.1 | 0.3 | 0.0 | 59.4 |
| 1997-98 | 0.0 | 0.0 | 0.0 | T | 1.1 | 3.2 | 5.5 | 11.0 | 10.5 | 0.6 | 0.2 | 0.0 | 32.1 |
| 1998-99 | 0.0 | 0.0 | 0.0 | 0.0 | 2.3 | 6.9 | 8.8 | 4.2 | 5.4 | 16.6 | 1.1 | 0.0 | 45.3 |
| 1999-00 | 0.0 | T | 0.0 | 0.0 | 0.7 | 1.2 | 5.1 | 12.3 | 8.2 | T | T | 0.0 | 27.5 |
| 2000-01 | 0.0 | 0.0 | 0.0 | 0.9 | 5.0 | 4.8 | 8.2 | 9.0 | 0.9 | 4.6 | 0.0 | T | 33.4 |
| 2001-02 | 0.0 | 0.0 | 0.0 | 0.0 | 9.4 | 20.9 | 9.1 | 1.2 | 2.5 | 3.8 | 1.5 | 0.0 | 48.4 |
| 2002-03 | 0.0 | 0.0 | 0.0 | T | 0.5 | 5.3 | 0.2 | 2.9 | 0.2 | 6.3 | 0.0 | 0.0 | 15.4 |
| 2003-04 | 0.0 | 0.0 | 0.0 | 0.5 | 2.2 | 20.8 | 13.4 | 11.7 | 0.5 | 3.3 | T | 0.0 | 52.4 |
| 2004-05 | 0.0 | 0.0 | T | 1.1 | 6.0 | 15.1 | 25.6 | 9.1 | 3.5 | 2.5 | 0.0 | 0.0 | 62.9 |
| 2005-06 | 0.0 | 0.0 | 0.0 | 0.0 | 7.8 | 5.8 | 5.6 | 3.5 | 20.0 | 4.7 | 0.0 | T | 47.4 |
| 2006-07 | 0.0 | 0.0 | 0.0 | T | 2.8 | 7.3 | 4.5 | 7.5 | 3.4 | 0.1 | T | 0.0 | 25.6 |
| 2007-08 | 0.0 | 0.0 | T | 1.1 | 0.1 | 8.2 | 28.0 | 11.0 | 2.8 | 1.2 | 0.0 | 0.0 | 52.4 |
| 2008-09 | 0.0 | 0.0 | 0.0 | T | 0.1 | 12.3 | 5.0 | 7.8 | 6.5 | 8.5 | 0.0 | T | 40.2 |
| 2009- | 0.0 | 0.0 | 0.0 | 1.7 | 0.1 | 20.4 | | | | | | | |
| POR=  94 YRS | T | T | T | 0.8 | 3.8 | 8.4 | 10.0 | 6.6 | 5.6 | 2.5 | 0.7 | T | 38.4 |

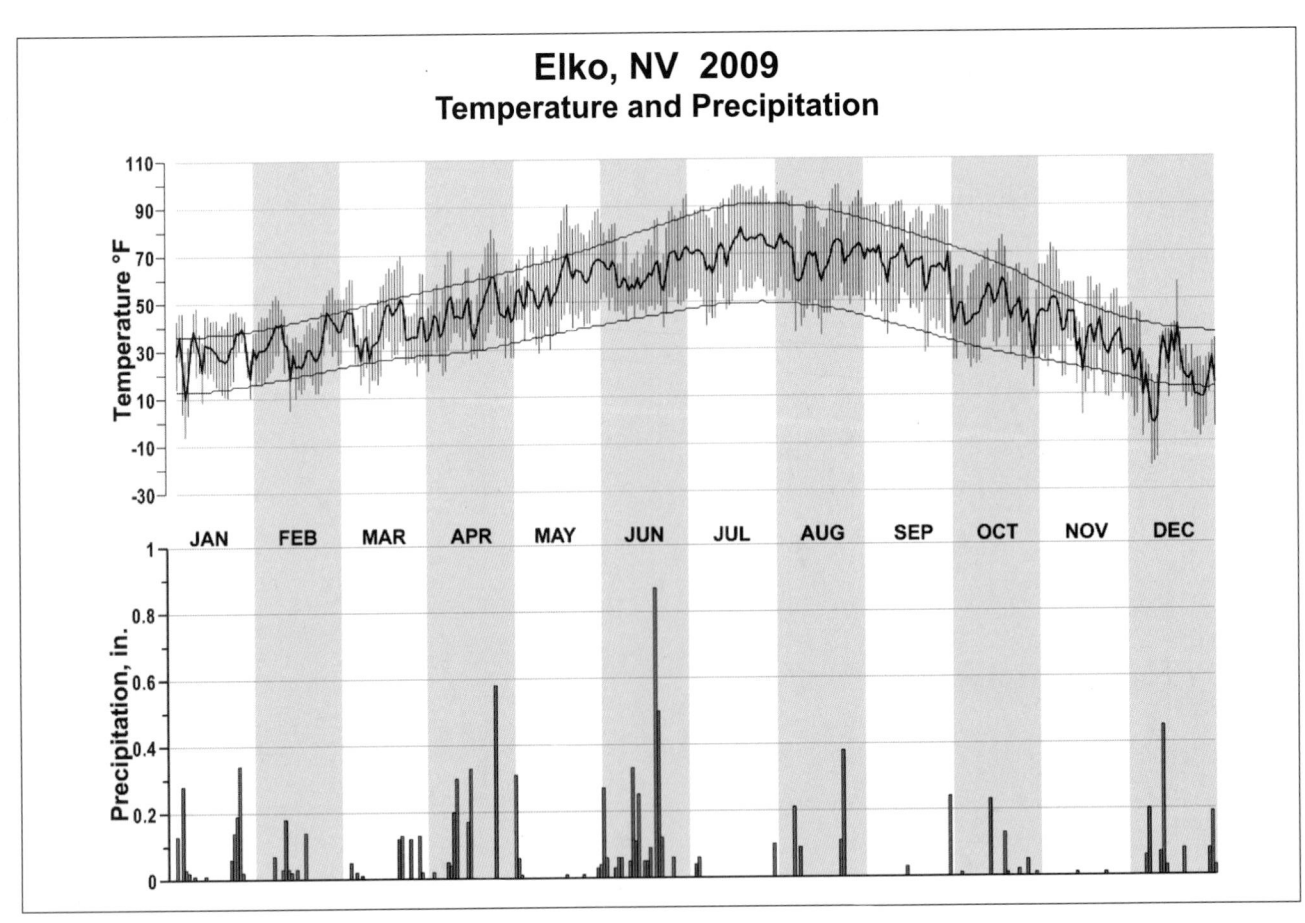

# Elko, NV  2009
## Temperature and Precipitation

# 2009
# LAS VEGAS
# NEVADA (KLAS)

Las Vegas is located in a broad desert valley in extreme southern Nevada and almost surrounded by mountains that are roughly 2,000 to 10,000 feet higher than the valley floor. The Las Vegas Valley itself is about 600 square miles and runs from the northwest to the southeast, sloping gradually upwards on each side towards the surrounding mountains. To the west of the Las Vegas Valley are the Spring Mountains, which includes Mount Charleston, the region's highest peak at 11,918 feet. The north side of the valley is bordered by the Sheep Mountain Range, while the southern end is marked by the Bird Spring Mountain Range, McCullough Mountain Range and Black Mountain. To the east, Sunrise and Frenchman Mountain separate the valley from Lake Mead. The Las Vegas Valley itself slopes downward from west to east. This affects the local climatology significantly in terms of driving variations in wind, temperature, precipitation and storm runoff.

The official climate station for the Las Vegas Valley is located at McCarran International Airport, which is located about 7 miles south of downtown Las Vegas near the southern end of the Las Vegas Strip. During the 1990s and most of the early 2000s, Las Vegas experienced a massive increase in population which resulted in explosive development of the Las Vegas Valley. This increase in urbanization has resulted in an urban heat island effect at the center of the valley, especially in areas near downtown and along The Strip, and most noted during the warmer summer months. As a result of this, McCarran International Airport frequently sees low temperatures some 5 to 15 degrees warmer than outlying areas of the valley, especially on nights with a clear sky and light winds. The lowest temperatures in the Las Vegas Valley are frequently recorded on the eastern side of the valley, which is lower and where colder air often likes to drain into at night, or on the higher elevations along the valley's west side.

Las Vegas is commonly noted for its abundant sunshine throughout the year and hot summer temperatures which reach into the triple digits. The coldest of winter nights will see temperatures drop into the 20s, with readings in the teens or lower experienced only in the most severe cold outbreaks.

The Spring Mountains immediately west of the valley as well as the Sierra Nevada Mountains in California frequently actas barriers to moisture moving in from the Pacific. It is primarily these features which limit the number of days each year that precipitation falls in Las Vegas and help make Las Vegas the driest major metropolitan area in the continental United States. During the cold season months, cold fronts and storm systems moving in from the Pacific occasionally bring precipitation and more often, gusty winds with them. While strong winds associated with cold season storms have been seen as early as late September and as late as early June, they are most common in the spring months and again in the fall when the majority of storms tend to pass through the area with no precipitation.  The strong winds that do occur usually reach this valley from the southwest or pass through from the northwest. Winds over 50 mph are infrequent, but when they do occur, are probably the most provoking of the elements experienced in the Las Vegas Valley because of the blowing dust and sand associated with them. However, outside of the wind, the springand fall months are usually considered the most ideal, though rather sharp temperature changes can occur during these months. Snow itself has fallen in about two-thirds of the winter seasons  at least once, however, it usually melts as it falls. Measurable snow at the official climate station typically occurs once every four or five years, however, higher elevations on the valley's west side such as the Summerlin area see snow about every three years or so.

In the warm season months, typically in July and August, a push of moisture associated with the monsoon moves into the Mojave Desert bringing higher than average humidity and triggering scattered thunderstorms. These storms typically  develop in the mountains surrounding the Las Vegas Valley and then move into the valley itself. While the gusty winds associated with them occasionally do cause damage, other times the main impact from these storms is the heavy rain they unleash that triggers flash flooding. The flash floods that do result from thunderstorms often sweep down normally dry washes or cause water to poor into low-lying areas.
By September, the monsoon typically wanes and the first break from the intense heat of summer is experienced.

# NORMALS, MEANS, AND EXTREMES
## LAS VEGAS (KLAS)

| | | |
|---|---|---|
| LATITUDE: 36 ° 4 'N | LONGITUDE: -115° 9 'W | ELEVATION (FT): GRND: 2127  BARO: 2091 |

TIME ZONE: PACIFIC (UTC -8)  WBAN: 23169

| | ELEMENT | POR | JAN | FEB | MAR | APR | MAY | JUN | JUL | AUG | SEP | OCT | NOV | DEC | YEAR |
|---|---|---|---|---|---|---|---|---|---|---|---|---|---|---|---|
| TEMPERATURE °F | NORMAL DAILY MAXIMUM | 30 | 57.1 | 63.0 | 69.5 | 78.1 | 87.8 | 98.9 | 104. | 101. | 93.8 | 80.8 | 66.0 | 57.3 | 79.8 |
| | MEAN DAILY MAXIMUM | 61 | 56.7 | 62.4 | 69.3 | 78.0 | 88.2 | 98.5 | 104.3 | 102.0 | 94.4 | 81.2 | 66.4 | 56.9 | 79.9 |
| | HIGHEST DAILY MAXIMUM | 61 | 77 | 87 | 92 | 99 | 109 | 115 | 117 | 116 | 113 | 103 | 87 | 77 | 117 |
| | YEAR OF OCCURRENCE | | 1975 | 1986 | 2004 | 2000 | 2003 | 1994 | 2005 | 1979 | 1950 | 1978 | 1988 | 1980 | JUL 2005 |
| | MEAN OF EXTREME MAXS. | 61 | 68.1 | 74.5 | 83.0 | 92.3 | 101.3 | 109.8 | 112.3 | 110.0 | 104.9 | 94.3 | 79.2 | 67.9 | 91.5 |
| | NORMAL DAILY MINIMUM | 30 | 36.8 | 41.4 | 47.0 | 53.9 | 62.9 | 72.3 | 78.2 | 76.7 | 68.8 | 56.5 | 44.0 | 36.6 | 56.3 |
| | MEAN DAILY MINIMUM | 61 | 34.9 | 39.3 | 44.8 | 52.0 | 61.4 | 70.4 | 77.3 | 75.5 | 67.2 | 54.9 | 42.7 | 35.0 | 54.6 |
| | LOWEST DAILY MINIMUM | 61 | 8 | 16 | 23 | 31 | 40 | 48 | 60 | 56 | 46 | 26 | 21 | 11 | 8 |
| | YEAR OF OCCURRENCE | | 1963 | 1989 | 1971 | 1975 | 1964 | 1993 | 1987 | 1968 | 1965 | 1971 | 1952 | 1990 | JAN 1963 |
| | MEAN OF EXTREME MINS. | 61 | 24.1 | 28.1 | 33.2 | 40.4 | 48.6 | 57.6 | 67.7 | 66.5 | 56.0 | 43.1 | 30.8 | 24.8 | 43.4 |
| | NORMAL DRY BULB | 30 | 47.0 | 52.2 | 58.3 | 66.0 | 75.4 | 85.6 | 91.2 | 89.3 | 81.3 | 68.7 | 55.0 | 47.0 | 68.1 |
| | MEAN DRY BULB | 61 | 45.8 | 50.9 | 57.1 | 65.0 | 74.8 | 84.6 | 90.8 | 88.7 | 80.8 | 68.0 | 54.6 | 46.0 | 67.3 |
| | MEAN WET BULB | 26 | 35.8 | 38.5 | 41.7 | 44.7 | 50.0 | 54.1 | 60.9 | 60.5 | 54.8 | 47.5 | 39.7 | 35.0 | 46.9 |
| | MEAN DEW POINT | 26 | 27.6 | 29.2 | 30.6 | 31.2 | 35.8 | 39.1 | 47.5 | 48.8 | 42.1 | 35.0 | 29.6 | 25.5 | 35.2 |
| | NORMAL NO. DAYS WITH: | | | | | | | | | | | | | | |
| | MAXIMUM >= 90 | 30 | 0.0 | 0.0 | 0.0 | 3.5 | 15.0 | 26.3 | 30.3 | 29.8 | 22.2 | 6.0 | 0.0 | 0.0 | 133.1 |
| | MAXIMUM <= 32 | 30 | 0.1 | 0.1 | 0.0 | 0.0 | 0.0 | 0.0 | 0.0 | 0.0 | 0.0 | 0.0 | 0.0 | * | 0.2 |
| | MINIMUM <= 32 | 30 | 9.1 | 3.2 | 0.6 | * | 0.0 | 0.0 | 0.0 | 0.0 | 0.0 | 0.1 | 2.0 | 9.0 | 24.0 |
| | MINIMUM <= 0 | 30 | 0.0 | 0.0 | 0.0 | 0.0 | 0.0 | 0.0 | 0.0 | 0.0 | 0.0 | 0.0 | 0.0 | 0.0 | 0.0 |
| H/C | NORMAL HEATING DEG. DAYS | 30 | 574 | 375 | 244 | 83 | 16 | 0 | 0 | 0 | 1 | 57 | 318 | 571 | 2239 |
| | NORMAL COOLING DEG. DAYS | 30 | 0 | 1 | 20 | 98 | 323 | 602 | 796 | 739 | 474 | 157 | 4 | 0 | 3214 |
| RH | NORMAL (PERCENT) | 30 | 48 | 43 | 37 | 26 | 23 | 17 | 21 | 25 | 26 | 29 | 36 | 44 | 31 |
| | HOUR 04 LST | 30 | 57 | 52 | 47 | 35 | 33 | 24 | 28 | 33 | 34 | 38 | 45 | 52 | 40 |
| | HOUR 10 LST | 30 | 43 | 38 | 32 | 23 | 20 | 15 | 20 | 23 | 23 | 26 | 32 | 38 | 28 |
| | HOUR 16 LST | 30 | 33 | 29 | 24 | 17 | 15 | 11 | 15 | 17 | 18 | 20 | 26 | 31 | 21 |
| | HOUR 22 LST | 30 | 51 | 45 | 39 | 27 | 24 | 17 | 21 | 25 | 27 | 30 | 39 | 46 | 33 |
| S | PERCENT POSSIBLE SUNSHINE | 47 | 77 | 81 | 83 | 87 | 88 | 93 | 88 | 88 | 91 | 87 | 81 | 78 | 85 |
| W/O | MEAN NO. DAYS WITH: | | | | | | | | | | | | | | |
| | HEAVY FOG(VISBY <= 1/4 MI) | 46 | 0.3 | 0.1 | 0.1 | 0.0 | 0.0 | 0.0 | 0.0 | 0.0 | 0.0 | 0.0 | 0.0 | 0.2 | 0.7 |
| | THUNDERSTORMS | 61 | 0.0 | 0.3 | 0.4 | 0.4 | 0.9 | 1.0 | 4.0 | 3.8 | 1.6 | 0.6 | 0.2 | 0.1 | 13.3 |
| CLOUDNESS | MEAN: | | | | | | | | | | | | | | |
| | SUNRISE-SUNSET (OKTAS) | | | | | | | | | | | | | | |
| | MIDNIGHT-MIDNIGHT (OKTAS) | | | | | | | | | | | | | | |
| | MEAN NO. DAYS WITH: | | | | | | | | | | | | | | |
| | CLEAR | 1 | 4.0 | 1.0 | 9.0 | | 27.0 | 16.0 | | | | | | | |
| | PARTLY CLOUDY | 1 | 2.0 | 3.0 | 6.0 | | 1.0 | | | | | | | | |
| | CLOUDY | 1 | 1.0 | 3.0 | | | 1.0 | 1.0 | | | | | | | |
| PR | MEAN STATION PRESSURE(IN) | 26 | 27.83 | 27.76 | 27.66 | 27.63 | 27.57 | 27.55 | 27.57 | 27.61 | 27.62 | 27.69 | 27.78 | 27.83 | 27.68 |
| | MEAN SEA-LEVEL PRES. (IN) | 26 | 30.12 | 30.04 | 29.94 | 29.86 | 29.78 | 29.74 | 29.78 | 29.80 | 29.83 | 29.92 | 30.05 | 30.12 | 29.92 |
| WINDS | MEAN SPEED (MPH) | 26 | 7.0 | 8.0 | 9.5 | 10.5 | 10.5 | 10.4 | 9.3 | 8.9 | 8.3 | 7.6 | 7.1 | 6.8 | 8.7 |
| | PREVAIL.DIR(TENS OF DEGS) | 30 | 26 | 26 | 21 | 22 | 21 | 20 | 20 | 20 | 21 | 25 | 25 | 26 | 20 |
| | MAXIMUM 2-MINUTE: | | | | | | | | | | | | | | |
| | SPEED (MPH) | 14 | 45 | 53 | 46 | 44 | 56 | 47 | 45 | 43 | 41 | 46 | 40 | 48 | 56 |
| | DIR. (TENS OF DEGS) | | 23 | 35 | 33 | 24 | 22 | 33 | 03 | 32 | 16 | 33 | 24 | 34 | 22 |
| | YEAR OF OCCURRENCE | | 1996 | 2008 | 2001 | 2002 | 2000 | 1998 | 1998 | 1998 | 1998 | 1996 | 2001 | 2000 | MAY 2000 |
| | MAXIMUM 3-SECOND | | | | | | | | | | | | | | |
| | SPEED (MPH) | 14 | 52 | 67 | 59 | 59 | 64 | 60 | 54 | 61 | 49 | 62 | 49 | 61 | 67 |
| | DIR. (TENS OF DEGS) | | 24 | 34 | 02 | 24 | 26 | 31 | 04 | 16 | 23 | 34 | 24 | 34 | 34 |
| | YEAR OF OCCURRENCE | | 1999 | 2008 | 2009 | 2002 | 2000 | 2008 | 1998 | 1998 | 1998 | 2007 | 2001 | 2009 | FEB 2008 |
| PRECIPITATION | NORMAL (IN) | 30 | 0.59 | 0.69 | 0.59 | 0.15 | 0.24 | 0.08 | 0.44 | 0.45 | 0.31 | 0.24 | 0.31 | 0.40 | 4.49 |
| | MAXIMUM MONTHLY (IN) | 61 | 3.00 | 2.89 | 4.80 | 2.44 | 0.96 | 0.97 | 2.48 | 2.59 | 2.06 | 1.45 | 2.22 | 2.10 | 4.80 |
| | YEAR OF OCCURRENCE | | 1995 | 1998 | 1992 | 1965 | 1969 | 1990 | 1984 | 1957 | 1997 | 2005 | 1965 | 2004 | MAR 1992 |
| | MINIMUM MONTHLY (IN) | 61 | T | 0.00 | 0.00 | 0.00 | 0.00 | 0.00 | 0.00 | 0.00 | 0.00 | 0.00 | 0.00 | 0.00 | 0.00 |
| | YEAR OF OCCURRENCE | | 1984 | 1977 | 1972 | 1962 | 1970 | 1982 | 1981 | 1980 | 1971 | 1979 | 1980 | 1981 | JUN 1982 |
| | MAXIMUM IN 24 HOURS (IN) | 61 | 1.09 | 1.30 | 1.27 | 0.97 | 0.83 | 0.97 | 1.36 | 2.59 | 1.07 | 1.37 | 1.78 | 2.10 | 2.59 |
| | YEAR OF OCCURRENCE | | 1990 | 1993 | 1992 | 1965 | 1987 | 1990 | 1984 | 1957 | 1963 | 2005 | 1960 | 2004 | AUG 1957 |
| | NORMAL NO. DAYS WITH: | | | | | | | | | | | | | | |
| | PRECIPITATION >= 0.01 | 30 | 3.4 | 3.5 | 3.6 | 1.8 | 1.6 | 0.7 | 2.6 | 3.0 | 1.9 | 1.8 | 1.8 | 2.9 | 28.6 |
| | PRECIPITATION >= 1.00 | 30 | 0.0 | 0.1 | * | 0.0 | 0.0 | 0.0 | 0.1 | 0.1 | 0.0 | * | * | 0.0 | 0.3 |
| SNOWFALL | NORMAL (IN) | 30 | 0.9 | 0.1 | 0.* | 0.0 | 0.0 | 0.0 | 0.0 | 0.0 | 0.0 | 0.0 | 0.* | 0.* | 1.0 |
| | MAXIMUM MONTHLY (IN) | 50 | 16.7 | 1.4 | 0.1 | T | 0.0 | T | 0.0 | T | T | 0.0 | 4.0 | 3.6 | 16.7 |
| | YEAR OF OCCURRENCE | | 1949 | 1949 | 1976 | 1970 | | 2008 | | 1989 | 2009 | 2009 | 1964 | 2008 | JAN 1949 |
| | MAXIMUM IN 24 HOURS (IN) | 50 | 9.0 | 6.9 | 0.1 | 0.0 | 0.0 | 0.0 | 0.0 | T | T | T | 4.0 | 3.6 | 9.0 |
| | YEAR OF OCCURRENCE' | | 1974 | 1979 | 1976 | | | | | 1989 | | 1956 | 1964 | 2008 | JAN 1974 |
| | MAXIMUM SNOW DEPTH (IN) | 49 | 8 | 6 | 0 | 0 | 0 | 0 | 0 | 0 | 0 | 0 | 3 | 3 | 8 |
| | YEAR OF OCCURRENCE | | 1974 | 1979 | | | | | | | | | 1964 | 2008 | JAN 1974 |
| | NORMAL NO. DAYS WITH: | | | | | | | | | | | | | | |
| | SNOWFALL >= 1.0 | 30 | 0.2 | 0.0 | 0.0 | 0.0 | 0.0 | 0.0 | 0.0 | 0.0 | 0.0 | 0.0 | 0.0 | 0.0 | 0.2 |

## PRECIPITATION (inches) 2009  LAS VEGAS (KLAS)

| YEAR | JAN | FEB | MAR | APR | MAY | JUN | JUL | AUG | SEP | OCT | NOV | DEC | ANNUAL |
|---|---|---|---|---|---|---|---|---|---|---|---|---|---|
| 1980 | 1.45 | 2.25 | 0.94 | 0.18 | 0.15 | T | 0.43 | 0.00 | 0.18 | 0.04 | 0.00 | 0.01 | 5.63 |
| 1981 | 0.09 | 0.20 | 1.44 | 0.02 | 0.50 | T | 0.00 | 0.20 | 0.25 | 0.15 | 0.29 | 0.00 | 3.14 |
| 1982 | 0.09 | 1.10 | 0.29 | 0.01 | 0.31 | 0.00 | 0.05 | 0.71 | 0.07 | 0.04 | 0.60 | 0.72 | 3.99 |
| 1983 | 0.43 | 0.32 | 0.90 | 0.45 | 0.16 | T | 0.06 | 1.25 | 0.50 | 0.26 | 0.10 | 0.43 | 4.86 |
| 1984 | T | 0.03 | T | 0.04 | 0.00 | 0.22 | 2.48 | 0.99 | 0.47 | T | 0.94 | 1.68 | 6.85 |
| 1985 | 0.19 | 0.02 | 0.06 | 0.31 | T | 0.02 | 0.13 | 0.00 | 0.08 | 0.07 | 0.37 | 0.02 | 1.27 |
| 1986 | 0.23 | 0.15 | 0.32 | 0.10 | 0.28 | T | 0.13 | 0.04 | 0.05 | 0.07 | 0.81 | 0.47 | 2.65 |
| 1987 | 1.13 | 0.45 | 0.49 | 0.17 | 0.90 | 0.13 | 0.13 | 0.01 | T | 0.49 | 1.80 | 0.89 | 6.59 |
| 1988 | 0.65 | 0.26 | 0.00 | 0.76 | T | 0.04 | 0.04 | 0.46 | T | 0.00 | T | 0.08 | 2.29 |
| 1989 | 0.51 | 0.06 | 0.05 | T | 0.64 | T | 0.05 | 0.80 | T | T | 0.00 | T | 2.11 |
| 1990 | 1.18 | 0.37 | T | 0.18 | T | 0.97 | 0.59 | T | 0.19 | 0.17 | 0.10 | T | 3.75 |
| 1991 | 0.21 | 0.54 | 1.01 | T | 0.05 | 0.19 | 0.54 | 0.78 | 0.06 | 0.06 | 0.38 | 0.24 | 4.06 |
| 1992 | 0.45 | 1.30 | 4.80 | 0.02 | 0.05 | 0.09 | 0.03 | 0.21 | 0.00 | 1.22 | 0.00 | 1.71 | 9.88 |
| 1993 | 1.63 | 2.52 | 0.14 | 0.01 | 0.01 | 0.08 | 0.00 | 0.26 | 0.00 | 0.02 | 0.17 | 0.21 | 5.05 |
| 1994 | 0.04 | 0.48 | 0.13 | T | 0.01 | 0.00 | 0.11 | 0.08 | 0.35 | T | 0.28 | 1.08 | 2.56 |
| 1995 | 3.00 | 0.03 | 0.39 | 0.03 | 0.16 | 0.02 | T | 0.05 | T | T | 0.00 | 0.01 | 3.69 |
| 1996 | 0.13 | 0.14 | 0.10 | 0.00 | 0.13 | T | 1.18 | T | .00 | .11 | .79 | .18 | 2.76 |
| 1997 | 0.30 | T | 0.00 | 0.04 | T | T | 0.60 | 0.33 | 2.06 | T | 0.23 | 0.07 | 3.63 |
| 1998 | 0.17 | 2.89 | 1.03 | 0.14 | 0.13 | 0.03 | 0.46 | 0.23 | 1.29 | 0.22 | 0.33 | 0.43 | 7.35 |
| 1999 | T | 0.08 | T | 0.73 | T | 0.14 | 2.18 | 0.25 | 0.35 | T | 0.00 | T | 3.73 |
| 2000 | T | 1.59 | 0.21 | T | T | T | T | 0.71 | 0.00 | 0.92 | T | 0.04 | 3.47 |
| 2001 | 0.87 | 2.21 | 0.16 | 0.04 | 0.02 | T | 0.39 | 0.05 | T | 0.00 | 0.09 | 0.11 | 3.94 |
| 2002 | T | T | 0.10 | 0.00 | 0.00 | 0.00 | 0.52 | 0.00 | 0.31 | 0.32 | 0.12 | 0.07 | 1.44 |
| 2003 | 0.02 | 2.13 | 0.32 | 0.38 | 0.01 | 0.00 | 1.08 | 0.83 | 0.52 | 0.00 | 0.61 | 0.96 | 6.86 |
| 2004 | 0.01 | 1.46 | 0.23 | 0.92 | 0.00 | T | 0.05 | 0.51 | 0.18 | 0.59 | 1.71 | 2.10 | 7.76 |
| 2005 | 2.07 | 2.45 | 0.47 | 0.06 | T | 0.07 | 0.52 | 0.26 | T | 1.45 | 0.00 | 0.02 | 7.37 |
| 2006 | 0.03 | 0.05 | 0.19 | T | T | T | 0.13 | 0.04 | T | 1.07 | 0.00 | 0.12 | 1.69 |
| 2007 | 0.06 | 0.16 | T | 0.08 | T | 0.00 | 0.29 | 0.76 | 0.67 | 0.00 | 0.64 | 0.07 | 2.73 |
| 2008 | 0.57 | 0.05 | 0.08 | 0.00 | 0.13 | T | 0.08 | 0.07 | 0.03 | 0.01 | 0.47 | 1.15 | 2.64 |
| 2009 | 0.04 | 0.78 | T | 0.05 | 0.00 | 0.10 | 0.29 | 0.02 | T | T | 0.02 | 0.29 | 1.59 |
| POR= 61 YRS | 0.52 | 0.59 | 0.42 | 0.18 | 0.16 | 0.08 | 0.45 | 0.44 | 0.28 | 0.24 | 0.39 | 0.37 | 4.12 |

WBAN : 23169

## AVERAGE TEMPERATURE (°F) 2009  LAS VEGAS (KLAS)

| YEAR | JAN | FEB | MAR | APR | MAY | JUN | JUL | AUG | SEP | OCT | NOV | DEC | ANNUAL |
|---|---|---|---|---|---|---|---|---|---|---|---|---|---|
| 1980 | 49.5 | 53.2 | 54.2 | 63.5 | 69.0 | 83.9 | 92.0 | 90.2 | 81.4 | 68.9 | 56.8 | 52.7 | 67.9 |
| 1981 | 51.1 | 52.5 | 56.4 | 70.6 | 74.3 | 88.8 | 92.7 | 90.0 | 82.5 | 64.4 | 58.0 | 48.8 | 69.2 |
| 1982 | 45.6 | 50.5 | 55.1 | 63.8 | 73.6 | 81.5 | 88.1 | 87.3 | 77.9 | 63.0 | 50.5 | 44.5 | 65.1 |
| 1983 | 46.6 | 51.7 | 56.4 | 58.5 | 72.8 | 82.8 | 88.5 | 83.8 | 82.5 | 67.8 | 55.3 | 47.9 | 66.2 |
| 1984 | 47.1 | 50.1 | 57.9 | 63.1 | 80.7 | 83.5 | 88.2 | 85.4 | 81.7 | 63.0 | 52.7 | 44.0 | 66.5 |
| 1985 | 44.4 | 47.4 | 54.9 | 68.2 | 76.9 | 87.4 | 92.0 | 89.9 | 75.4 | 67.3 | 51.7 | 48.3 | 67.0 |
| 1986 | 51.7 | 55.8 | 63.0 | 66.2 | 76.6 | 87.8 | 87.6 | 91.2 | 75.4 | 65.0 | 55.8 | 46.0 | 68.5 |
| 1987 | 44.7 | 51.4 | 54.6 | 68.4 | 74.5 | 86.3 | 86.9 | 88.2 | 81.2 | 71.0 | 53.4 | 42.5 | 66.9 |
| 1988 | 45.1 | 52.4 | 58.1 | 64.2 | 73.4 | 85.3 | 92.6 | 86.9 | 79.1 | 74.9 | 56.0 | 46.0 | 67.8 |
| 1989 | 43.9 | 50.0 | 63.4 | 72.7 | 75.7 | 85.3 | 93.4 | 86.9 | 80.0 | 67.2 | 57.3 | 48.0 | 68.7 |
| 1990 | 45.2 | 48.8 | 60.5 | 68.8 | 74.5 | 85.9 | 90.8 | 87.8 | 82.0 | 69.2 | 55.1 | 40.2 | 67.4 |
| 1991 | 45.5 | 55.9 | 52.7 | 64.2 | 69.9 | 82.1 | 90.2 | 87.8 | 81.9 | 72.2 | 55.2 | 47.0 | 67.1 |
| 1992 | 45.9 | 54.1 | 56.8 | 70.5 | 77.7 | 83.2 | 88.7 | 90.5 | 83.7 | 70.9 | 52.7 | 43.6 | 68.2 |
| 1993 | 45.7 | 50.1 | 60.9 | 67.5 | 77.0 | 82.5 | 89.4 | 88.5 | 81.3 | 69.1 | 51.5 | 46.3 | 67.5 |
| 1994 | 49.3 | 48.5 | 62.7 | 67.6 | 76.6 | 90.3 | 93.3 | 92.9 | 83.1 | 67.2 | 49.4 | 47.5 | 69.0 |
| 1995 | 47.5 | 58.7 | 57.9 | 64.8 | 71.0 | 80.9 | 92.4 | 93.1 | 83.7 | 69.4 | 59.8 | 48.9 | 69.0 |
| 1996 | 48.5 | 54.8 | 59.8 | 68.3 | 77.3 | 87.0 | 93.2 | 91.9 | 80.4 | 66.8 | 56.5 | 47.9 | 69.4 |
| 1997 | 48.3 | 51.7 | 62.7 | 65.4 | 81.6 | 84.3 | 88.2 | 90.7 | 81.3 | 67.3 | 56.2 | 45.9 | 68.6 |
| 1998 | 48.7 | 49.4 | 56.6 | 61.1 | 70.0 | 80.0 | 91.7 | 92.0 | 80.0 | 66.6 | 54.8 | 47.8 | 66.6 |
| 1999 | 50.5 | 52.7 | 60.6 | 60.9 | 75.3 | 85.2 | 88.2 | 88.0 | 81.6 | 71.6 | 58.8 | 48.7 | 68.5 |
| 2000 | 51.4 | 53.6 | 58.6 | 71.2 | 80.8 | 88.7 | 92.3 | 90.5 | 81.7 | 67.3 | 50.2 | 49.5 | 69.7 |
| 2001 | 46.4 | 49.8 | 60.6 | 65.0 | 82.2 | 87.9 | 90.3 | 91.9 | 85.1 | 72.1 | 58.6 | 45.4 | 69.6 |
| 2002 | 46.1 | 51.9 | 55.5 | 69.7 | 75.8 | 88.1 | 94.6 | 90.6 | 82.8 | 67.4 | 56.8 | 47.5 | 68.9 |
| 2003 | 54.2 | 51.6 | 59.6 | 62.9 | 77.9 | 87.9 | 94.8 | 90.3 | 84.4 | 75.4 | 52.6 | 47.9 | 70.0 |
| 2004 | 47.5 | 48.8 | 66.5 | 67.8 | 79.1 | 88.1 | 93.1 | 89.5 | 81.6 | 68.3 | 53.7 | 49.2 | 69.4 |
| 2005 | 51.4 | 53.2 | 59.6 | 65.7 | 79.0 | 85.0 | 95.3 | 89.6 | 82.0 | 70.4 | 59.3 | 49.7 | 70.0 |
| 2006 | 50.2 | 53.9 | 55.1 | 66.6 | 81.0 | 90.5 | 94.6 | 91.1 | 80.9 | 67.8 | 58.4 | 47.5 | 69.8 |
| 2007 | 46.0 | 54.7 | 64.4 | 70.5 | 80.5 | 89.5 | 95.4 | 92.5 | 82.7 | 69.8 | 60.9 | 45.2 | 71.0 |
| 2008 | 46.1 | 52.5 | 61.2 | 67.6 | 74.4 | 88.7 | 93.7 | 93.0 | 84.8 | 71.1 | 60.8 | 46.1 | 70.0 |
| 2009 | 51.0 | 52.0 | 59.8 | 66.0 | 83.6 | 83.5 | 94.7 | 90.6 | 86.2 | 67.0 | 59.2 | 45.5 | 69.9 |
| POR= 61 YRS | 45.8 | 50.9 | 57.1 | 65.0 | 74.8 | 84.6 | 90.8 | 88.7 | 80.8 | 68.0 | 54.6 | 46.0 | 67.3 |

## HEATING DEGREE DAYS (base 65°F) 2009  LAS VEGAS (KLAS)

| YEAR | JUL | AUG | SEP | OCT | NOV | DEC | JAN | FEB | MAR | APR | MAY | JUN | TOTAL |
|------|-----|-----|-----|-----|-----|-----|-----|-----|-----|-----|-----|-----|-------|
| 1981-82 | 0 | 0 | 0 | 74 | 214 | 497 | 594 | 398 | 301 | 98 | 9 | 0 | 2185 |
| 1982-83 | 0 | 0 | 10 | 84 | 429 | 631 | 564 | 364 | 263 | 198 | 21 | 0 | 2564 |
| 1983-84 | 0 | 0 | 0 | 3 | 297 | 524 | 548 | 424 | 216 | 111 | 0 | 0 | 2123 |
| 1984-85 | 0 | 0 | 0 | 127 | 363 | 641 | 629 | 487 | 308 | 41 | 0 | 0 | 2596 |
| 1985-86 | 0 | 0 | 1 | 31 | 393 | 512 | 404 | 270 | 125 | 57 | 11 | 0 | 1804 |
| 1986-87 | 0 | 0 | 14 | 53 | 268 | 586 | 622 | 375 | 316 | 40 | 1 | 0 | 2275 |
| 1987-88 | 0 | 0 | 0 | 18 | 342 | 689 | 612 | 357 | 225 | 83 | 33 | 0 | 2359 |
| 1988-89 | 0 | 0 | 0 | 0 | 291 | 581 | 647 | 425 | 118 | 23 | 16 | 0 | 2101 |
| 1989-90 | 0 | 0 | 0 | 70 | 224 | 519 | 606 | 449 | 172 | 12 | 0 | 0 | 2052 |
| 1990-91 | 0 | 0 | 0 | 23 | 290 | 761 | 597 | 247 | 376 | 57 | 25 | 2 | 2378 |
| 1991-92 | 0 | 0 | 0 | 77 | 297 | 552 | 584 | 308 | 248 | 7 | 0 | 0 | 2073 |
| 1992-93 | 0 | 0 | 0 | 16 | 364 | 655 | 591 | 410 | 143 | 32 | 3 | 8 | 2222 |
| 1993-94 | 0 | 0 | 0 | 33 | 398 | 573 | 480 | 455 | 93 | 60 | 1 | 0 | 2093 |
| 1994-95 | 0 | 0 | 0 | 35 | 465 | 537 | 537 | 170 | 230 | 90 | 18 | 6 | 2088 |
| 1995-96 | 0 | 0 | 0 | 22 | 151 | 490 | 504 | 287 | 169 | 22 | 11 | 0 | 1656 |
| 1996-97 | 0 | 0 | 0 | 138 | 249 | 524 | 512 | 368 | 115 | 117 | 0 | 0 | 2023 |
| 1997-98 | 0 | 0 | 0 | 57 | 259 | 583 | 500 | 426 | 270 | 166 | 23 | 0 | 2284 |
| 1998-99 | 0 | 0 | 0 | 39 | 296 | 525 | 444 | 337 | 143 | 188 | 5 | 6 | 1983 |
| 1999-00 | 0 | 0 | 0 | 11 | 184 | 499 | 412 | 325 | 195 | 16 | 3 | 0 | 1645 |
| 2000-01 | 0 | 0 | 0 | 80 | 436 | 474 | 570 | 421 | 175 | 114 | 2 | 0 | 2272 |
| 2001-02 | 0 | 0 | 0 | 0 | 204 | 602 | 578 | 363 | 304 | 34 | 7 | 0 | 2092 |
| 2002-03 | 0 | 0 | 0 | 39 | 244 | 536 | 331 | 368 | 187 | 109 | 10 | 0 | 1824 |
| 2003-04 | 0 | 0 | 0 | 9 | 365 | 524 | 536 | 462 | 71 | 29 | 0 | 0 | 1996 |
| 2004-05 | 0 | 0 | 0 | 80 | 332 | 483 | 414 | 326 | 174 | 46 | 1 | 0 | 1856 |
| 2005-06 | 0 | 0 | 0 | 7 | 182 | 465 | 453 | 305 | 298 | 41 | 0 | 0 | 1751 |
| 2006- | 0 | 0 | 0 | 27 | 204 | 537 | | | | | | | |
| 2006-07 | 0 | 0 | 0 | 27 | 204 | 537 | 553 | 286 | 102 | 35 | 1 | 0 | 1745 |
| 2007-08 | 0 | 0 | 0 | 14 | 149 | 607 | 578 | 358 | 135 | 21 | 17 | 0 | 1879 |
| 2008-09 | 0 | 0 | 0 | 29 | 131 | 577 | 427 | 361 | 180 | 77 | 0 | 0 | 1782 |
| 2009- | 0 | 0 | 0 | 53 | 192 | 599 | | | | | | | |

WBAN : 23169

## COOLING DEGREE DAYS (base 65°F) 2009  LAS VEGAS (KLAS)

| YEAR | JAN | FEB | MAR | APR | MAY | JUN | JUL | AUG | SEP | OCT | NOV | DEC | TOTAL |
|------|-----|-----|-----|-----|-----|-----|-----|-----|-----|-----|-----|-----|-------|
| 1980 | 0 | 0 | 0 | 68 | 160 | 575 | 842 | 788 | 498 | 211 | 15 | 0 | 3157 |
| 1981 | 0 | 0 | 5 | 205 | 296 | 721 | 866 | 781 | 531 | 64 | 12 | 0 | 3481 |
| 1982 | 0 | 0 | 2 | 70 | 281 | 501 | 721 | 699 | 404 | 30 | 0 | 0 | 2708 |
| 1983 | 0 | 0 | 2 | 9 | 269 | 541 | 735 | 589 | 534 | 94 | 10 | 0 | 2783 |
| 1984 | 0 | 0 | 3 | 61 | 496 | 563 | 724 | 641 | 508 | 74 | 1 | 0 | 3071 |
| 1985 | 0 | 0 | 0 | 143 | 377 | 678 | 844 | 778 | 319 | 110 | 2 | 0 | 3251 |
| 1986 | 0 | 20 | 69 | 98 | 379 | 693 | 707 | 821 | 332 | 59 | 0 | 0 | 3178 |
| 1987 | 0 | 0 | 0 | 148 | 302 | 645 | 685 | 729 | 495 | 211 | 0 | 0 | 3215 |
| 1988 | 0 | 0 | 16 | 64 | 300 | 615 | 864 | 685 | 434 | 312 | 31 | 0 | 3321 |
| 1989 | 0 | 11 | 74 | 259 | 351 | 614 | 887 | 687 | 456 | 143 | 0 | 0 | 3482 |
| 1990 | 0 | 0 | 42 | 134 | 302 | 634 | 810 | 713 | 516 | 163 | 0 | 0 | 3314 |
| 1991 | 0 | 0 | 0 | 42 | 187 | 524 | 788 | 714 | 515 | 307 | 12 | 0 | 3089 |
| 1992 | 0 | 0 | 0 | 180 | 402 | 552 | 742 | 798 | 571 | 206 | 3 | 0 | 3454 |
| 1993 | 0 | 0 | 21 | 114 | 381 | 537 | 765 | 737 | 494 | 166 | 0 | 0 | 3215 |
| 1994 | 0 | 0 | 31 | 145 | 369 | 768 | 883 | 870 | 551 | 108 | 0 | 0 | 3725 |
| 1995 | 0 | 0 | 13 | 91 | 211 | 490 | 856 | 880 | 570 | 168 | 2 | 0 | 3281 |
| 1996 | 0 | 3 | 16 | 129 | 402 | 665 | 883 | 841 | 471 | 201 | 0 | 0 | 3611 |
| 1997 | 0 | 0 | 48 | 136 | 522 | 584 | 727 | 805 | 494 | 137 | 4 | 0 | 3457 |
| 1998 | 0 | 0 | 18 | 55 | 186 | 457 | 834 | 842 | 456 | 95 | 0 | 0 | 2943 |
| 1999 | 0 | 0 | 13 | 74 | 333 | 620 | 726 | 719 | 508 | 221 | 3 | 0 | 3217 |
| 2000 | 0 | 0 | 3 | 208 | 498 | 719 | 851 | 799 | 509 | 158 | 0 | 0 | 3745 |
| 2001 | 0 | 0 | 47 | 122 | 541 | 692 | 792 | 843 | 609 | 231 | 17 | 0 | 3894 |
| 2002 | 0 | 0 | 13 | 182 | 349 | 700 | 923 | 802 | 539 | 123 | 5 | 0 | 3636 |
| 2003 | 0 | 1 | 26 | 52 | 416 | 695 | 930 | 792 | 591 | 335 | 0 | 0 | 3838 |
| 2004 | 0 | 0 | 124 | 122 | 445 | 702 | 874 | 767 | 506 | 189 | 1 | 0 | 3730 |
| 2005 | 0 | 0 | 13 | 73 | 438 | 607 | 947 | 769 | 518 | 182 | 18 | 0 | 3565 |
| 2006 | 0 | 0 | 0 | 105 | 502 | 772 | 925 | 815 | 482 | 118 | 12 | 0 | 3731 |
| 2007 | 0 | 0 | 90 | 207 | 491 | 742 | 947 | 860 | 538 | 168 | 32 | 0 | 4075 |
| 2008 | 0 | 0 | 25 | 106 | 316 | 720 | 895 | 873 | 602 | 224 | 14 | 0 | 3775 |
| 2009 | 0 | 0 | 24 | 111 | 583 | 559 | 927 | 799 | 641 | 122 | 24 | 0 | 3790 |

## SNOWFALL (inches) 2009 LAS VEGAS (KLAS)

| YEAR | JUL | AUG | SEP | OCT | NOV | DEC | JAN | FEB | MAR | APR | MAY | JUN | TOTAL |
|------|-----|-----|-----|-----|-----|-----|-----|-----|-----|-----|-----|-----|-------|
| 1980-81 | 0.0 | 0.0 | 0.0 | 0.0 | 0.0 | 0.0 | 0.0 | 0.0 | 0.0 | 0.0 | 0.0 | 0.0 | 0.0 |
| 1981-82 | 0.0 | 0.0 | 0.0 | 0.0 | 0.0 | 0.0 | 0.0 | 0.0 | 0.0 | 0.0 | 0.0 | 0.0 | 0.0 |
| 1982-83 | 0.0 | 0.0 | 0.0 | 0.0 | 0.0 | 0.0 | 0.0 | 0.0 | 0.0 | 0.0 | 0.0 | 0.0 | 0.0 |
| 1983-84 | 0.0 | 0.0 | 0.0 | 0.0 | 0.0 | 0.0 | 0.0 | 0.0 | 0.0 | 0.0 | 0.0 | 0.0 | 0.0 |
| 1984-85 | 0.0 | 0.0 | 0.0 | 0.0 | 0.0 | T | 0.0 | T | 0.0 | 0.0 | 0.0 | 0.0 | T |
| 1985-86 | 0.0 | 0.0 | 0.0 | 0.0 | 0.0 | T | 0.0 | 0.0 | 0.0 | 0.0 | 0.0 | 0.0 | T |
| 1986-87 | 0.0 | 0.0 | 0.0 | 0.0 | 0.0 | 0.0 | T | 0.6 | 0.0 | 0.0 | 0.0 | 0.0 | 0.6 |
| 1987-88 | 0.0 | 0.0 | 0.0 | 0.0 | 0.0 | 0.0 | 0.0 | T | 0.0 | 0.0 | 0.0 | 0.0 | T |
| 1988-89 | 0.0 | 0.0 | 0.0 | 0.0 | 0.0 | T | 0.0 | 0.3 | 0.0 | 0.0 | 0.0 | 0.0 | 0.3 |
| 1989-90 | 0.0 | T | 0.0 | 0.0 | 0.0 | 0.0 | T | 1.4 | 0.0 | 0.0 | 0.0 | 0.0 | 1.4 |
| 1990-91 | 0.0 | 0.0 | 0.0 | 0.0 | 0.0 | 0.0 | 0.0 | 0.0 | 0.0 | 0.0 | 0.0 | 0.0 | 0.0 |
| 1991-92 | 0.0 | 0.0 | 0.0 | 0.0 | 0.0 | 0.0 | 0.0 | 0.0 | T | 0.0 | 0.0 | 0.0 | T |
| 1992-93 | 0.0 | 0.0 | 0.0 | 0.0 | 0.0 | T | 0.0 | 0.0 | 0.0 | 0.0 | 0.0 | 0.0 | T |
| 1993-94 | 0.0 | 0.0 | 0.0 | 0.0 | 0.0 | 0.0 | 0.0 | 0.0 | 0.0 | 0.0 | 0.0 | 0.0 | 0.0 |
| 1994-95 | 0.0 | 0.0 | 0.0 | 0.0 | T | 0.0 | T | 0.0 | 0.0 | 0.0 | 0.0 | 0.0 | T |
| 1995-96 | 0.0 | 0.0 | 0.0 | 0.0 | 0.0 | 0.0 | 0.0 | T | 0.0 | 0.0 | 0.0 | 0.0 | T |
| 1996-97 | 0.0 | 0.0 | 0.0 | 0.0 | 0.0 | 0.0 | T | 0.0 | 0.0 | 0.0 | 0.0 | 0.0 | T |
| 1997-98 | 0.0 | 0.0 | 0.0 | 0.0 | 0.0 | 0.0 | 0.0 | 0.0 | 0.0 | 0.0 | 0.0 | 0.0 | 0.0 |
| 1998-99 | 0.0 | 0.0 | 0.0 | 0.0 | 0.0 | 1.0 | 0.0 | 0.0 | 0.0 | 0.0 | 0.0 | 0.0 | 1.0 |
| 1999-00 | 0.0 | 0.0 | 0.0 | 0.0 | 0.0 | 0.0 | 0.0 | 0.0 | 0.0 | 0.0 | 0.0 | 0.0 | 0.0 |
| 2000-01 | 0.0 | 0.0 | 0.0 | 0.0 | 0.0 | 0.0 | 0.0 | T | 0.0 | 0.0 | 0.0 | 0.0 | T |
| 2001-02 | 0.0 | 0.0 | 0.0 | 0.0 | 0.0 | 0.0 | T | 0.0 | 0.0 | 0.0 | 0.0 | 0.0 | T |
| 2002-03 | 0.0 | 0.0 | 0.0 | 0.0 | 0.0 | 0.0 | 0.0 | 0.0 | 0.0 | 0.0 | 0.0 | 0.0 | 0.0 |
| 2003-04 | 0.0 | 0.0 | 0.0 | 0.0 | 0.0 | 1.3 | 0.0 | T | 0.0 | 0.0 | 0.0 | 0.0 | 1.3 |
| 2004-05 | 0.0 | T | 0.0 | 0.0 | 0.0 | 0.0 | T | 0.0 | 0.0 | 0.0 | 0.0 | 0.0 | T |
| 2005-06 | 0.0 | 0.0 | 0.0 | 0.0 | 0.0 | 0.0 | 0.0 | 0.0 | 0.0 | 0.0 | 0.0 | 0.0 | 0.0 |
| 2006-07 | 0.0 | 0.0 | 0.0 | 0.0 | 0.0 | T | 0.0 | T | 0.0 | 0.0 | 0.0 | 0.0 | T |
| 2007-08 | 0.0 | 0.0 | 0.0 | 0.0 | 0.0 | 0.0 | 0.0 | 0.0 | 0.0 | 0.0 | 0.0 | 0.0 | 0.0 |
| 2008-09 | 0.0 | 0.0 | 0.0 | 0.0 | 0.0 | 3.6 | 0.0 | 0.0 | 0.0 | 0.0 | 0.0 | 0.0 | 3.6 |
| 2009- | 0.0 | 0.0 | 0.0 | 0.0 | 0.0 | 0.0 | | | | | | | |
| POR= 61 YRS | 0.0 | T | 0.0 | 0.0 | 0.1 | 0.1 | 0.7 | T | T | 0.0 | 0.0 | 0.0 | 0.9 |

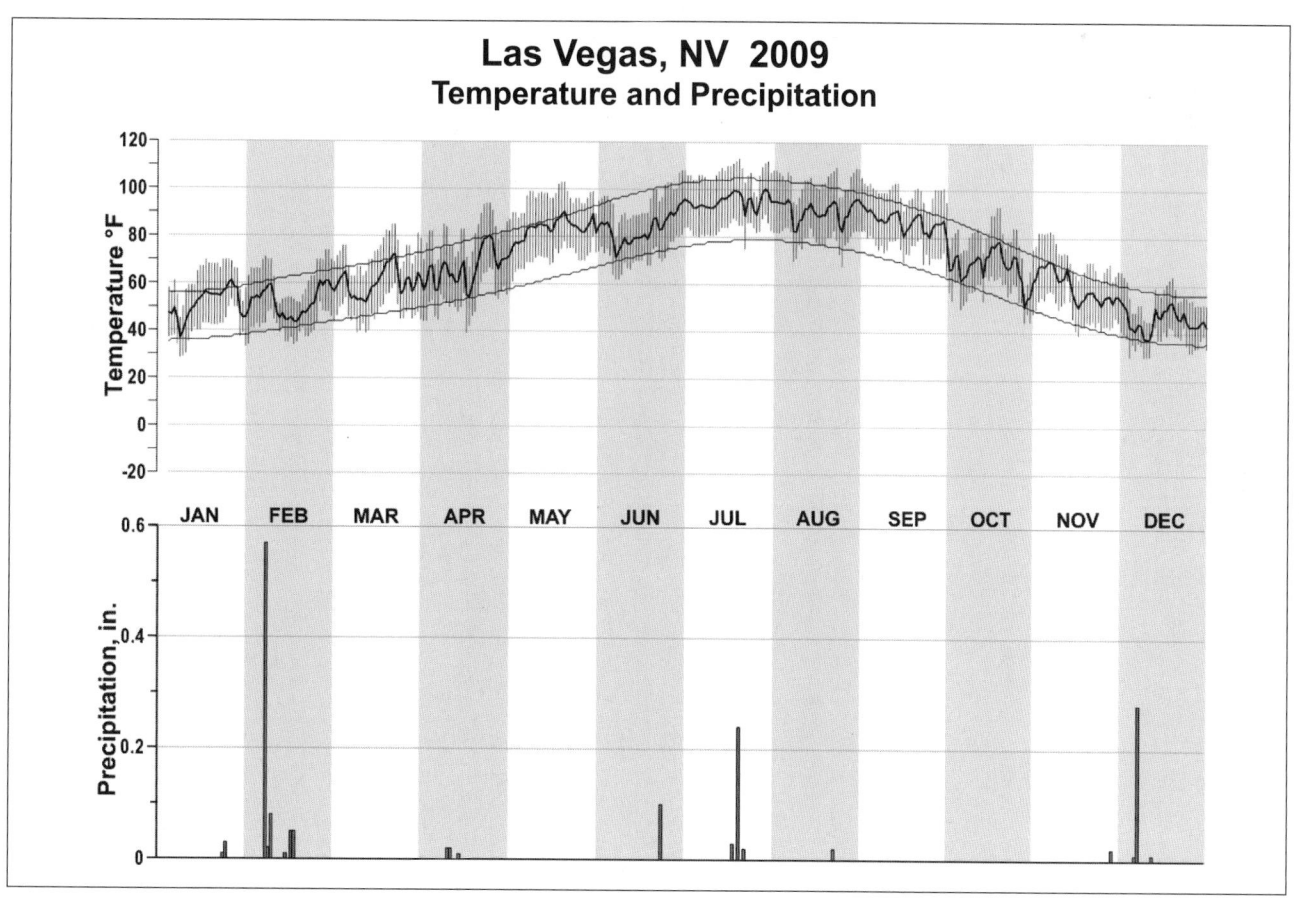

**Las Vegas, NV 2009**
**Temperature and Precipitation**

# 2009
# RENO
# NEVADA (KRNO)

At an elevation of 4,400 feet above mean sea level, Reno is located at the west edge of Truckee Meadows in a semi-arid plateau lying in the lee of the Sierra Nevada Mountain Range. To the west, the Sierras rise to elevations of 9,000 to 11,000 feet. Hills to the east reach 6,000 to 7,000 feet. The Truckee River, flowing from the Sierras eastward through Reno, drains into Pyramid Lake to the northeast of the city.

The daily temperatures on the whole are mild, but the difference between the high and low often exceeds 45 degrees. While the afternoon high may exceed 90 degrees, a light wrap is often needed shortly after sunset. Nights with low temperatures over 60 degrees are rare. Afternoon temperatures in winter are moderate.

Based on the 1951-1980 period, the average first occurrence of 32 degrees Fahrenheit in the fall is September 16 and the average last occurrence in the spring is June 1.

More than half of the precipitation in Reno occurs mainly as mixed rain and snow, and falls from December to March. Although there is an average of about 25 inches of snow a year, it seldom remains on the ground for more than three or four days at a time. Summer rain comes mainly as brief thunderstorms in the middle and late afternoons. While precipitation is scarce, considerable water is available from the high altitude reservoirs in the Sierra Nevada, where precipitation is heavy.

Humidity is very low during the summer months, and moderately low during the winter. Fogs are rare, and are usually confined to the early morning hours of midwinter. Sunshine is abundant throughout the year.

# NORMALS, MEANS, AND EXTREMES
## RENO (KRNO)

| LATITUDE: 39° 29'N | LONGITUDE: -119° 46'W | ELEVATION (FT): GRND: 4410  BARO: 4407 | TIME ZONE: PACIFIC (UTC -8) | WBAN: 23185 |
|---|---|---|---|---|

| | ELEMENT | POR | JAN | FEB | MAR | APR | MAY | JUN | JUL | AUG | SEP | OCT | NOV | DEC | YEAR |
|---|---|---|---|---|---|---|---|---|---|---|---|---|---|---|---|
| **TEMPERATURE °F** | NORMAL DAILY MAXIMUM | 30 | 45.5 | 51.7 | 57.2 | 64.1 | 72.6 | 82.8 | 91.2 | 89.9 | 81.7 | 69.9 | 55.3 | 46.4 | 67.4 |
| | MEAN DAILY MAXIMUM | 85 | 45.0 | 48.6 | 55.8 | 62.7 | 72.0 | 80.5 | 89.9 | 88.9 | 80.2 | 68.9 | 54.7 | 46.1 | 66.1 |
| | HIGHEST DAILY MAXIMUM | 68 | 71 | 75 | 83 | 89 | 97 | 103 | 108 | 105 | 101 | 91 | 77 | 70 | 108 |
| | YEAR OF OCCURRENCE | | 2003 | 1986 | 1966 | 1981 | 2003 | 1988 | 2007 | 1983 | 1950 | 1980 | 2005 | 1969 | JUL 2007 |
| | MEAN OF EXTREME MAXS. | 97 | 60.8 | 65.4 | 71.8 | 79.6 | 88.8 | 95.7 | 100.1 | 99.0 | 93.6 | 84.4 | 71.1 | 61.5 | 81.0 |
| | NORMAL DAILY MINIMUM | 30 | 21.8 | 25.4 | 29.3 | 33.2 | 40.2 | 46.5 | 51.4 | 49.9 | 43.1 | 34.0 | 26.4 | 20.7 | 35.2 |
| | MEAN DAILY MINIMUM | 85 | 20.6 | 23.6 | 27.7 | 31.7 | 39.5 | 45.3 | 51.0 | 48.8 | 41.8 | 33.4 | 25.9 | 21.0 | 34.2 |
| | LOWEST DAILY MINIMUM | 68 | -16 | -16 | -2 | 13 | 18 | 21 | 33 | 24 | 20 | 8 | 1 | -16 | -16 |
| | YEAR OF OCCURRENCE | | 1949 | 1989 | 1945 | 1956 | 1964 | 2005 | 1976 | 1962 | 1965 | 1971 | 1958 | 1972 | FEB 1989 |
| | MEAN OF EXTREME MINS. | 97 | 5.5 | 10.8 | 15.8 | 21.2 | 28.3 | 34.1 | 41.6 | 39.7 | 31.9 | 21.9 | 13.3 | 6.8 | 22.7 |
| | NORMAL DRY BULB | 30 | 33.6 | 38.5 | 43.3 | 48.6 | 56.4 | 64.7 | 71.3 | 69.9 | 62.4 | 52.0 | 40.9 | 33.6 | 51.3 |
| | MEAN DRY BULB | 85 | 32.8 | 36.1 | 41.8 | 47.2 | 55.8 | 63.0 | 70.5 | 68.9 | 61.0 | 51.1 | 40.3 | 33.6 | 50.2 |
| | MEAN WET BULB | 26 | 28.4 | 30.8 | 34.4 | 37.4 | 43.4 | 48.2 | 52.5 | 51.1 | 46.1 | 40.0 | 33.0 | 28.0 | 39.4 |
| | MEAN DEW POINT | 26 | 24.3 | 25.1 | 26.6 | 28.4 | 34.3 | 38.2 | 42.5 | 41.1 | 37.0 | 31.7 | 27.1 | 23.6 | 31.7 |
| | NORMAL NO. DAYS WITH: | | | | | | | | | | | | | | |
| | MAXIMUM >= 90 | 30 | 0.0 | 0.0 | 0.0 | 0.0 | 1.0 | 7.3 | 19.8 | 17.0 | 5.2 | 0.1 | 0.0 | 0.0 | 50.4 |
| | MAXIMUM <= 32 | 30 | 3.1 | 1.0 | * | 0.0 | 0.0 | 0.0 | 0.0 | 0.0 | 0.0 | 0.0 | 0.3 | 2.8 | 7.2 |
| | MINIMUM <= 32 | 30 | 27.6 | 23.9 | 21.2 | 14.5 | 4.2 | 0.4 | 0.0 | * | 2.1 | 13.1 | 24.1 | 28.3 | 159.4 |
| | MINIMUM <= 0 | 30 | 0.9 | 0.2 | * | 0.0 | 0.0 | 0.0 | 0.0 | 0.0 | 0.0 | 0.0 | 0.0 | 0.9 | 2.0 |
| **H/C** | NORMAL HEATING DEG. DAYS | 30 | 984 | 756 | 683 | 502 | 285 | 91 | 12 | 22 | 130 | 416 | 732 | 987 | 5600 |
| | NORMAL COOLING DEG. DAYS | 30 | 0 | 0 | 0 | 0 | 11 | 72 | 204 | 164 | 41 | 1 | 0 | 0 | 493 |
| **RH** | NORMAL (PERCENT) | 30 | 70 | 61 | 53 | 45 | 42 | 38 | 34 | 35 | 42 | 50 | 61 | 68 | 50 |
| | HOUR 04 LST | 30 | 81 | 76 | 71 | 66 | 66 | 62 | 60 | 61 | 66 | 71 | 75 | 79 | 70 |
| | HOUR 10 LST | 30 | 71 | 59 | 48 | 37 | 33 | 29 | 26 | 27 | 33 | 42 | 56 | 67 | 44 |
| | HOUR 16 LST | 30 | 52 | 42 | 34 | 28 | 26 | 22 | 18 | 19 | 22 | 27 | 40 | 48 | 32 |
| | HOUR 22 LST | 30 | 75 | 64 | 56 | 48 | 45 | 39 | 35 | 36 | 44 | 54 | 64 | 72 | 53 |
| **S** | PERCENT POSSIBLE SUNSHINE | 45 | 65 | 68 | 75 | 80 | 81 | 85 | 92 | 92 | 91 | 83 | 70 | 64 | 79 |
| **W/O** | MEAN NO. DAYS WITH: | | | | | | | | | | | | | | |
| | HEAVY FOG(VISBY <= 1/4 MI) | 46 | 1.8 | 0.9 | 0.3 | 0.2 | 0.0 | 0.0 | 0.0 | 0.0 | 0.0 | 0.1 | 0.3 | 1.8 | 5.4 |
| | THUNDERSTORMS | 64 | 0.0 | 0.0 | 0.1 | 0.4 | 1.8 | 2.6 | 3.3 | 2.9 | 1.2 | 0.4 | 0.0 | 0.0 | 12.7 |
| **CLOUDNESS** | MEAN: | | | | | | | | | | | | | | |
| | SUNRISE-SUNSET (OKTAS) | | | | | | | | | | | | | | |
| | MIDNIGHT-MIDNIGHT (OKTAS) | | | | | | | | | | | | | | |
| | MEAN NO. DAYS WITH: | | | | | | | | | | | | | | |
| | CLEAR | 1 | 2.0 | 3.0 | 4.0 | | 13.0 | 17.0 | | | | | | | |
| | PARTLY CLOUDY | 1 | 3.0 | 3.0 | 3.0 | | 8.0 | | | | | | | | |
| | CLOUDY | 1 | 3.0 | 5.0 | 5.0 | | 7.0 | 3.0 | | | | | | | |
| **PR** | MEAN STATION PRESSURE(IN) | 26 | 25.65 | 25.50 | 25.56 | 25.54 | 25.53 | 25.55 | 25.59 | 25.59 | 25.58 | 25.62 | 25.64 | 25.64 | 25.58 |
| | MEAN SEA-LEVEL PRES. (IN) | 26 | 30.17 | 30.10 | 30.00 | 29.96 | 29.91 | 29.89 | 29.91 | 29.92 | 29.95 | 30.03 | 30.12 | 30.16 | 30.01 |
| **WINDS** | MEAN SPEED (MPH) | 26 | 4.8 | 5.9 | 7.5 | 8.4 | 8.3 | 7.9 | 7.4 | 6.7 | 5.7 | 5.1 | 5.3 | 5.0 | 6.5 |
| | PREVAIL.DIR(TENS OF DEGS) | 30 | 19 | 19 | 28 | 30 | 29 | 29 | 30 | 29 | 30 | 29 | 19 | 19 | 30 |
| | MAXIMUM 2-MINUTE: | | | | | | | | | | | | | | |
| | SPEED (MPH) | 14 | 45 | 51 | 49 | 52 | 47 | 45 | 44 | 40 | 45 | 49 | 55 | 67 | 67 |
| | DIR. (TENS OF DEGS) | | 18 | 19 | 21 | 24 | 16 | 19 | 16 | 19 | 16 | 19 | 20 | 19 | 19 |
| | YEAR OF OCCURRENCE | | 2008 | 2004 | 1999 | 2008 | 2009 | 2005 | 2007 | 1999 | 2009 | 2007 | 2009 | 2002 | DEC 2002 |
| | MAXIMUM 3-SECOND | | | | | | | | | | | | | | |
| | SPEED (MPH) | 14 | 62 | 61 | 61 | 63 | 58 | 54 | 62 | 49 | 54 | 61 | 71 | 82 | 82 |
| | DIR. (TENS OF DEGS) | | 17 | 18 | 20 | 24 | 16 | 18 | 16 | 19 | 17 | 19 | 18 | 16 | 16 |
| | YEAR OF OCCURRENCE | | 2008 | 2004 | 1999 | 2008 | 2009 | 2005 | 2007 | 1999 | 2009 | 2007 | 2009 | 2002 | DEC 2002 |
| **PRECIPITATION** | NORMAL (IN) | 30 | 1.06 | 1.06 | 0.86 | 0.35 | 0.62 | 0.47 | 0.24 | 0.27 | 0.45 | 0.42 | 0.80 | 0.88 | 7.48 |
| | MAXIMUM MONTHLY (IN) | 68 | 4.13 | 4.84 | 2.87 | 2.04 | 2.89 | 1.53 | 1.06 | 1.65 | 2.31 | 2.14 | 3.08 | 5.25 | 5.25 |
| | YEAR OF OCCURRENCE | | 1969 | 1986 | 1995 | 1958 | 1963 | 1989 | 1971 | 1965 | 1982 | 1945 | 1983 | 1955 | DEC 1955 |
| | MINIMUM MONTHLY (IN) | 68 | T | T | T | T | T | 0.00 | 0.00 | 0.00 | 0.00 | 0.00 | 0.00 | T | 0.00 |
| | YEAR OF OCCURRENCE | | 1966 | 1967 | 1988 | 2008 | 1985 | 1959 | 1951 | 1957 | 1974 | 1995 | 1959 | 1989 | OCT 1995 |
| | MAXIMUM IN 24 HOURS (IN) | 68 | 2.37 | 1.80 | 1.25 | 1.64 | 1.76 | 0.79 | 0.80 | 0.97 | 0.91 | 1.55 | 1.65 | 2.16 | 2.37 |
| | YEAR OF OCCURRENCE | | 1943 | 1990 | 2004 | 1958 | 1987 | 1969 | 1949 | 1965 | 1982 | 1962 | 1988 | 1955 | JAN 1943 |
| | NORMAL NO. DAYS WITH: | | | | | | | | | | | | | | |
| | PRECIPITATION >= 0.01 | 30 | 6.9 | 7.0 | 6.8 | 3.9 | 4.3 | 3.6 | 2.0 | 2.1 | 3.3 | 3.3 | 5.0 | 5.4 | 53.6 |
| | PRECIPITATION >= 1.00 | 30 | 0.1 | 0.1 | * | 0.0 | * | 0.0 | 0.0 | 0.0 | 0.0 | * | 0.1 | 0.1 | 0.4 |
| **SNOWFALL** | NORMAL (IN) | 30 | 5.2 | 5.4 | 3.3 | 0.9 | 0.7 | 0.* | 0.0 | 0.0 | 0.1 | 0.5 | 3.1 | 4.3 | 23.5 |
| | MAXIMUM MONTHLY (IN) | 60 | 22.9 | 23.5 | 29.0 | 7.5 | 14.1 | 0.2 | T | 0.0 | 1.5 | 5.1 | 16.5 | 25.6 | 29.0 |
| | YEAR OF OCCURRENCE | | 1993 | 1969 | 1952 | 1958 | 1964 | 1995 | 2007 | 2008 | 1982 | 1971 | 1985 | 1971 | MAR 1952 |
| | MAXIMUM IN 24 HOURS (IN) | 60 | 12.0 | 18.0 | 16.9 | 7.3 | 9.0 | 0.2 | 0.0 | 0.0 | 1.5 | 3.7 | 15.4 | 16.4 | 18.0 |
| | YEAR OF OCCURRENCE' | | 1956 | 1990 | 1952 | 1958 | 1962 | 1995 | | | 1982 | 1971 | 1985 | 2004 | FEB 1990 |
| | MAXIMUM SNOW DEPTH (IN) | 53 | 20 | 13 | 10 | 4 | 6 | 0 | 0 | 0 | 1 | 3 | 10 | 16 | 20 |
| | YEAR OF OCCURRENCE | | 2005 | 1990 | 1952 | 1975 | 1971 | | | | 1982 | 1984 | 1985 | 2004 | JAN 2005 |
| | NORMAL NO. DAYS WITH: | | | | | | | | | | | | | | |
| | SNOWFALL >= 1.0 | 30 | 1.8 | 1.6 | 1.1 | 0.4 | 0.2 | 0.0 | 0.0 | 0.0 | 0.0 | 0.2 | 0.8 | 1.3 | 7.4 |

## PRECIPITATION (inches) 2009 RENO (KRNO)

| YEAR | JAN | FEB | MAR | APR | MAY | JUN | JUL | AUG | SEP | OCT | NOV | DEC | ANNUAL |
|------|------|------|------|------|------|------|------|------|------|------|------|------|--------|
| 1980 | 2.77 | 1.90 | 0.76 | 0.51 | 0.78 | 0.12 | 0.54 | 0.32 | 0.48 | 0.14 | 0.28 | 0.60 | 9.20 |
| 1981 | 0.85 | 0.21 | 0.58 | 0.21 | 0.57 | T | 0.01 | 0.36 | 0.07 | 0.64 | 2.13 | 1.05 | 6.68 |
| 1982 | 1.20 | 0.41 | 1.14 | 0.34 | 0.10 | 1.07 | 0.04 | 0.09 | 2.31 | 1.65 | 1.71 | 1.04 | 11.10 |
| 1983 | 1.72 | 1.58 | 1.31 | 1.35 | 0.21 | 0.53 | T | 0.78 | 0.84 | 0.36 | 3.08 | 1.47 | 13.23 |
| 1984 | 0.36 | 0.22 | 0.20 | 0.24 | 0.06 | 0.34 | 0.45 | 0.02 | 0.04 | 0.60 | 1.68 | 0.07 | 4.28 |
| 1985 | 0.24 | 0.68 | 1.07 | T | T | 0.12 | T | 0.01 | 0.63 | 0.46 | 1.23 | 0.55 | 4.99 |
| 1986 | 0.40 | 4.84 | 0.88 | 0.77 | 0.26 | 0.31 | 0.86 | 0.07 | 0.28 | 0.06 | 0.02 | 0.19 | 8.94 |
| 1987 | 0.49 | 0.78 | 0.80 | 0.49 | 2.29 | 1.12 | 0.01 | 0.01 | 0.01 | 0.54 | 0.37 | 0.59 | 7.50 |
| 1988 | 0.50 | 0.02 | T | 0.95 | 0.12 | 0.59 | 0.22 | 0.01 | 0.04 | 0.02 | 1.99 | 0.84 | 5.30 |
| 1989 | 0.20 | 0.80 | 0.46 | 0.03 | 1.33 | 1.53 | 0.00 | 0.82 | 1.19 | 0.43 | 0.55 | T | 7.34 |
| 1990 | 0.62 | 1.98 | 0.07 | 0.33 | 0.19 | 0.03 | 0.86 | 0.21 | 0.31 | 0.06 | 0.15 | 0.45 | 5.26 |
| 1991 | 0.01 | 0.21 | 1.42 | 0.47 | 0.50 | 0.39 | 0.04 | 0.24 | 0.60 | 0.23 | 0.89 | 0.15 | 5.15 |
| 1992 | 0.13 | 0.45 | 0.69 | 0.06 | 0.10 | 1.12 | 0.15 | 0.28 | T | 0.45 | 0.06 | 1.87 | 5.36 |
| 1993 | 2.42 | 1.27 | 0.55 | 0.01 | 0.27 | 0.35 | T | T | T | 1.42 | 0.13 | 0.16 | 6.58 |
| 1994 | 0.06 | 0.62 | 1.00 | 0.03 | 1.39 | 0.00 | 0.09 | 0.00 | 0.15 | 0.23 | 1.47 | 0.16 | 5.20 |
| 1995 | 3.31 | 0.20 | 2.87 | 0.40 | 1.81 | 1.29 | 0.22 | T | T | 0.00 | 0.19 | 2.27 | 12.56 |
| 1996 | 1.33 | 2.30 | 1.63 | 0.16 | 1.07 | 0.71 | .20 | .16 | .45 | .28 | .89 | 3.03 | 12.21 |
| 1997 | 3.32 | 0.71 | 0.01 | 0.22 | 0.13 | 1.17 | 0.04 | T | 0.55 | 0.16 | 0.86 | 0.58 | 7.75 |
| 1998 | 1.10 | 2.59 | 2.21 | 0.60 | 0.82 | 1.39 | T | T | 2.17 | 0.34 | 0.77 | 0.04 | 12.03 |
| 1999 | 0.76 | 1.25 | 0.11 | 0.55 | 0.20 | 0.06 | 0.10 | 0.82 | 0.07 | 0.42 | 0.01 | 0.07 | 4.42 |
| 2000 | 2.14 | 0.98 | 0.38 | 0.34 | 0.23 | 0.23 | 0.00 | 0.79 | 0.04 | 0.04 | 0.40 | 0.14 | 5.71 |
| 2001 | 0.31 | 0.18 | 0.15 | 0.66 | T | 0.09 | 0.07 | T | 0.09 | 0.14 | 0.83 | 1.83 | 4.35 |
| 2002 | 0.59 | 0.24 | 0.42 | 1.21 | 0.20 | 0.10 | 0.12 | 0.82 | T | 0.12 | 1.08 | 2.18 | 7.08 |
| 2003 | 0.17 | 0.23 | 0.31 | 0.83 | 0.04 | 0.38 | 0.23 | 1.01 | 0.01 | 0.03 | 0.12 | 1.22 | 4.58 |
| 2004 | 0.96 | 1.56 | 1.26 | T | 0.32 | 0.20 | T | 0.28 | 0.01 | 1.58 | 1.53 | 1.71 | 9.41 |
| 2005 | 1.78 | 0.84 | 0.42 | 0.61 | 0.59 | 0.37 | 0.59 | 0.10 | T | 0.03 | 0.18 | 3.88 | 9.39 |
| 2006 | 1.60 | 1.04 | 0.92 | 1.88 | 0.31 | T | 0.34 | T | 0.00 | 0.42 | 0.25 | 0.41 | 7.17 |
| 2007 | 0.13 | 1.01 | 0.03 | 0.18 | 0.16 | 0.12 | T | 0.16 | 0.44 | 0.19 | 0.25 | 1.06 | 3.73 |
| 2008 | 2.80 | 0.78 | 0.07 | T | 0.56 | T | 0.34 | T | 0.01 | 0.11 | 0.92 | 0.50 | 6.09 |
| 2009 | 0.51 | 0.21 | 1.61 | 0.35 | 0.50 | 1.52 | 0.01 | 0.01 | T | 1.50 | 0.24 | 1.79 | 8.25 |
| POR= 97 YRS | 1.26 | 1.03 | 0.77 | 0.45 | 0.59 | 0.40 | 0.27 | 0.27 | 0.31 | 0.40 | 0.64 | 1.00 | 7.39 |

WBAN : 23185

## AVERAGE TEMPERATURE (°F) 2009 RENO (KRNO)

| YEAR | JAN | FEB | MAR | APR | MAY | JUN | JUL | AUG | SEP | OCT | NOV | DEC | ANNUAL |
|------|------|------|------|------|------|------|------|------|------|------|------|------|--------|
| 1980 | 36.9 | 40.6 | 38.5 | 49.8 | 54.4 | 60.8 | 71.3 | 67.5 | 63.0 | 50.9 | 41.8 | 36.2 | 51.0 |
| 1981 | 36.1 | 38.8 | 41.7 | 50.7 | 57.5 | 68.3 | 67.9 | 69.4 | 64.7 | 46.9 | 42.5 | 39.0 | 52.0 |
| 1982 | 28.5 | 40.3 | 40.2 | 44.0 | 55.0 | 61.8 | 70.4 | 68.8 | 57.0 | 46.9 | 36.2 | 32.3 | 48.5 |
| 1983 | 34.2 | 38.6 | 40.5 | 43.8 | 53.3 | 62.5 | 67.2 | 69.9 | 62.9 | 54.1 | 41.2 | 38.8 | 50.6 |
| 1984 | 31.9 | 37.2 | 44.3 | 45.8 | 59.4 | 61.7 | 73.4 | 69.8 | 63.1 | 46.2 | 39.6 | 30.7 | 50.3 |
| 1985 | 30.6 | 37.0 | 38.7 | 52.7 | 56.3 | 68.6 | 73.4 | 68.5 | 57.4 | 50.4 | 34.8 | 31.2 | 50.0 |
| 1986 | 40.3 | 42.8 | 47.7 | 49.2 | 57.4 | 67.5 | 69.4 | 73.0 | 56.3 | 50.9 | 43.0 | 35.3 | 52.7 |
| 1987 | 31.6 | 38.4 | 43.4 | 54.8 | 59.7 | 67.9 | 68.2 | 71.3 | 65.0 | 56.3 | 41.8 | 31.9 | 52.5 |
| 1988 | 33.0 | 40.2 | 44.1 | 51.2 | 56.6 | 67.0 | 75.2 | 73.2 | 63.4 | 58.4 | 42.7 | 31.2 | 53.0 |
| 1989 | 30.9 | 31.1 | 46.4 | 54.0 | 57.0 | 66.3 | 72.6 | 67.7 | 61.9 | 51.4 | 41.8 | 35.9 | 51.4 |
| 1990 | 34.3 | 30.8 | 45.7 | 54.5 | 56.4 | 65.7 | 73.7 | 71.1 | 65.4 | 54.7 | 40.9 | 25.8 | 51.6 |
| 1991 | 31.8 | 43.7 | 39.9 | 46.9 | 51.8 | 62.5 | 74.3 | 71.5 | 65.8 | 56.6 | 43.4 | 33.7 | 51.8 |
| 1992 | 34.4 | 41.8 | 46.9 | 55.8 | 64.9 | 66.5 | 70.8 | 72.7 | 64.8 | 55.3 | 39.9 | 30.6 | 53.7 |
| 1993 | 25.7 | 34.3 | 48.5 | 48.7 | 58.4 | 61.9 | 69.0 | 70.0 | 65.3 | 54.9 | 40.4 | 36.5 | 51.1 |
| 1994 | 37.7 | 38.5 | 48.7 | 52.9 | 59.8 | 68.5 | 77.2 | 73.7 | 65.0 | 52.3 | 35.1 | 35.7 | 53.8 |
| 1995 | 38.2 | 46.0 | 43.3 | 47.5 | 55.0 | 62.1 | 72.1 | 72.7 | 63.0 | 52.4 | 46.4 | 38.1 | 53.1 |
| 1996 | 37.6 | 39.5 | 42.2 | 48.5 | 55.1 | 64.0 | 72.9 | 70.0 | 60.0 | 49.9 | 41.3 | 38.4 | 51.6 |
| 1997 | 34.2 | 37.8 | 46.8 | 47.4 | 60.7 | 62.6 | 69.4 | 69.8 | 62.6 | 49.5 | 42.0 | 31.0 | 51.2 |
| 1998 | 38.2 | 36.2 | 42.2 | 45.4 | 50.9 | 63.3 | 75.8 | 74.9 | 65.7 | 49.8 | 42.8 | 31.7 | 51.4 |
| 1999 | 37.7 | 39.7 | 43.9 | 46.7 | 58.8 | 67.3 | 73.7 | 70.0 | 65.8 | 56.0 | 46.7 | 34.9 | 53.4 |
| 2000 | 38.7 | 41.3 | 45.7 | 54.1 | 60.5 | 70.9 | 72.9 | 73.4 | 63.4 | 52.0 | 37.8 | 36.9 | 54.0 |
| 2001 | 33.0 | 35.8 | 48.7 | 47.8 | 66.5 | 69.9 | 74.2 | 76.1 | 68.0 | 58.2 | 44.6 | 36.5 | 54.9 |
| 2002 | 34.4 | 40.9 | 42.7 | 52.0 | 58.3 | 70.0 | 78.4 | 72.5 | 66.4 | 52.6 | 43.1 | 37.5 | 54.1 |
| 2003 | 43.1 | 37.6 | 46.5 | 45.5 | 60.4 | 71.6 | 79.2 | 74.1 | 68.1 | 59.4 | 40.0 | 38.1 | 55.3 |
| 2004 | 36.2 | 38.7 | 51.5 | 53.7 | 61.0 | 70.3 | 78.0 | 74.3 | 65.9 | 52.6 | 40.3 | 35.0 | 54.8 |
| 2005 | 28.9 | 38.6 | 46.3 | 49.2 | 60.0 | 64.3 | 80.0 | 75.7 | 62.8 | 55.4 | 45.4 | 39.0 | 53.8 |
| 2006 | 37.9 | 39.2 | 39.1 | 50.7 | 62.4 | 72.7 | 79.7 | 73.6 | 65.6 | 52.8 | 44.3 | 34.9 | 54.4 |
| 2007 | 31.2 | 40.8 | 49.4 | 53.1 | 63.4 | 72.4 | 80.0 | 76.3 | 63.6 | 52.5 | 44.4 | 34.1 | 55.1 |
| 2008 | 32.1 | 39.0 | 45.0 | 49.9 | 59.3 | 69.7 | 77.6 | 77.0 | 68.1 | 54.6 | 46.3 | 34.3 | 54.4 |
| 2009 | 37.6 | 40.2 | 43.8 | 50.3 | 65.0 | 66.2 | 77.7 | 73.9 | 69.6 | 52.0 | 43.8 | 26.9 | 53.9 |
| POR= 85 YRS | 32.8 | 36.1 | 41.8 | 47.2 | 55.8 | 63.0 | 70.5 | 68.9 | 61.0 | 51.1 | 40.3 | 33.6 | 50.2 |

## HEATING DEGREE DAYS (base 65°F) 2009 RENO (KRNO)

| YEAR | JUL | AUG | SEP | OCT | NOV | DEC | JAN | FEB | MAR | APR | MAY | JUN | TOTAL |
|------|-----|-----|-----|-----|-----|-----|-----|-----|-----|-----|-----|-----|-------|
| 1980-81 | 13 | 35 | 79 | 430 | 688 | 885 | 890 | 727 | 715 | 424 | 228 | 48 | 5162 |
| 1981-82 | 12 | 7 | 83 | 554 | 669 | 800 | 1123 | 687 | 760 | 623 | 307 | 133 | 5758 |
| 1982-83 | 15 | 11 | 278 | 556 | 855 | 1006 | 947 | 732 | 752 | 630 | 371 | 77 | 6230 |
| 1983-84 | 40 | 8 | 104 | 332 | 708 | 805 | 1019 | 801 | 637 | 570 | 183 | 133 | 5340 |
| 1984-85 | 0 | 8 | 111 | 575 | 753 | 1056 | 1060 | 781 | 810 | 359 | 266 | 45 | 5824 |
| 1985-86 | 5 | 12 | 230 | 446 | 896 | 1039 | 757 | 618 | 528 | 469 | 285 | 32 | 5317 |
| 1986-87 | 5 | 0 | 291 | 430 | 654 | 913 | 1028 | 737 | 661 | 299 | 182 | 34 | 5234 |
| 1987-88 | 38 | 5 | 45 | 265 | 690 | 1017 | 982 | 714 | 643 | 408 | 267 | 88 | 5162 |
| 1988-89 | 0 | 0 | 132 | 202 | 663 | 1042 | 1049 | 944 | 568 | 321 | 256 | 21 | 5198 |
| 1989-90 | 0 | 21 | 99 | 417 | 688 | 895 | 943 | 954 | 590 | 312 | 260 | 64 | 5243 |
| 1990-91 | 0 | 20 | 55 | 313 | 715 | 1209 | 1021 | 588 | 772 | 540 | 404 | 106 | 5743 |
| 1991-92 | 0 | 6 | 41 | 265 | 642 | 965 | 945 | 666 | 555 | 273 | 61 | 83 | 4502 |
| 1992-93 | 10 | 9 | 41 | 293 | 748 | 1056 | 1212 | 853 | 503 | 480 | 200 | 139 | 5544 |
| 1993-94 | 16 | 8 | 77 | 316 | 730 | 874 | 838 | 735 | 498 | 360 | 186 | 34 | 4672 |
| 1994-95 | 0 | 0 | 58 | 387 | 888 | 901 | 824 | 529 | 665 | 520 | 308 | 147 | 5227 |
| 1995-96 | 4 | 0 | 87 | 383 | 553 | 829 | 843 | 733 | 700 | 488 | 301 | 71 | 4992 |
| 1996-97 | 2 | 7 | 155 | 463 | 705 | 820 | 946 | 756 | 559 | 520 | 139 | 102 | 5174 |
| 1997-98 | 10 | 3 | 94 | 470 | 683 | 1047 | 824 | 801 | 697 | 582 | 429 | 87 | 5727 |
| 1998-99 | 0 | 0 | 81 | 464 | 658 | 1027 | 841 | 705 | 648 | 541 | 220 | 80 | 5265 |
| 1999-00 | 0 | 18 | 37 | 271 | 543 | 926 | 810 | 683 | 593 | 324 | 184 | 16 | 4405 |
| 2000-01 | 7 | 0 | 103 | 401 | 812 | 864 | 984 | 812 | 497 | 512 | 54 | 15 | 5061 |
| 2001-02 | 0 | 0 | 10 | 218 | 605 | 875 | 942 | 672 | 686 | 383 | 233 | 34 | 4658 |
| 2002-03 | 0 | 0 | 48 | 379 | 651 | 848 | 673 | 763 | 568 | 578 | 224 | 17 | 4749 |
| 2003-04 | 0 | 0 | 20 | 181 | 744 | 825 | 885 | 756 | 411 | 334 | 143 | 25 | 4324 |
| 2004-05 | 0 | 2 | 63 | 381 | 732 | 924 | 1113 | 734 | 576 | 468 | 179 | 91 | 5263 |
| 2005-06 | 0 | 1 | 103 | 288 | 580 | 797 | 833 | 717 | 794 | 423 | 123 | 11 | 4670 |
| 2006-07 | 0 | 0 | 88 | 371 | 614 | 926 | 1039 | 669 | 476 | 355 | 114 | 30 | 4682 |
| 2007-08 | 0 | 0 | 130 | 378 | 612 | 953 | 1012 | 751 | 613 | 445 | 220 | 35 | 5149 |
| 2008-09 | 0 | 0 | 21 | 324 | 557 | 945 | 841 | 690 | 653 | 435 | 87 | 67 | 4620 |
| 2009- | 0 | 9 | 24 | 395 | 630 | 1173 | | | | | | | |

WBAN : 23185

## COOLING DEGREE DAYS (base 65°F) 2009 RENO (KRNO)

| YEAR | JAN | FEB | MAR | APR | MAY | JUN | JUL | AUG | SEP | OCT | NOV | DEC | TOTAL |
|------|-----|-----|-----|-----|-----|-----|-----|-----|-----|-----|-----|-----|-------|
| 1980 | 0 | 0 | 0 | 0 | 2 | 32 | 218 | 119 | 25 | 1 | 0 | 0 | 397 |
| 1981 | 0 | 0 | 0 | 2 | 4 | 153 | 112 | 151 | 80 | 0 | 0 | 0 | 502 |
| 1982 | 0 | 0 | 0 | 0 | 2 | 45 | 188 | 135 | 47 | 0 | 0 | 0 | 417 |
| 1983 | 0 | 0 | 0 | 0 | 16 | 9 | 115 | 170 | 49 | 0 | 0 | 0 | 359 |
| 1984 | 0 | 0 | 0 | 0 | 16 | 42 | 264 | 162 | 61 | 0 | 0 | 0 | 545 |
| 1985 | 0 | 0 | 0 | 0 | 3 | 157 | 273 | 126 | 6 | 0 | 0 | 0 | 565 |
| 1986 | 0 | 0 | 0 | 0 | 53 | 112 | 148 | 253 | 39 | 0 | 0 | 0 | 605 |
| 1987 | 0 | 0 | 0 | 1 | 27 | 126 | 142 | 210 | 53 | 2 | 0 | 0 | 561 |
| 1988 | 0 | 0 | 0 | 0 | 11 | 152 | 323 | 264 | 92 | 2 | 0 | 0 | 844 |
| 1989 | 0 | 0 | 0 | 0 | 19 | 66 | 240 | 112 | 13 | 0 | 0 | 0 | 450 |
| 1990 | 0 | 0 | 0 | 1 | 0 | 95 | 278 | 216 | 76 | 0 | 0 | 0 | 666 |
| 1991 | 0 | 0 | 0 | 0 | 0 | 39 | 296 | 214 | 71 | 13 | 0 | 0 | 633 |
| 1992 | 0 | 0 | 0 | 5 | 61 | 135 | 197 | 257 | 45 | 1 | 0 | 0 | 701 |
| 1993 | 0 | 0 | 0 | 0 | 4 | 52 | 145 | 171 | 92 | 10 | 0 | 0 | 474 |
| 1994 | 0 | 0 | 0 | 3 | 31 | 148 | 381 | 274 | 65 | 0 | 0 | 0 | 902 |
| 1995 | 0 | 0 | 0 | 0 | 6 | 66 | 231 | 247 | 35 | 0 | 0 | 0 | 585 |
| 1996 | 0 | 0 | 0 | 0 | 0 | 50 | 254 | 169 | 14 | 2 | 0 | 0 | 489 |
| 1997 | 0 | 0 | 0 | 0 | 14 | 35 | 154 | 160 | 26 | 0 | 0 | 0 | 389 |
| 1998 | 0 | 0 | 0 | 0 | 0 | 43 | 344 | 316 | 108 | 0 | 0 | 0 | 811 |
| 1999 | 0 | 0 | 0 | 0 | 36 | 157 | 276 | 180 | 69 | 0 | 0 | 0 | 718 |
| 2000 | 0 | 0 | 0 | 0 | 54 | 199 | 256 | 271 | 62 | 6 | 0 | 0 | 848 |
| 2001 | 0 | 0 | 0 | 0 | 107 | 174 | 292 | 348 | 106 | 14 | 0 | 0 | 1041 |
| 2002 | 0 | 0 | 0 | 0 | 33 | 191 | 423 | 240 | 97 | 0 | 0 | 0 | 984 |
| 2003 | 0 | 0 | 0 | 0 | 90 | 223 | 449 | 292 | 119 | 13 | 0 | 0 | 1186 |
| 2004 | 0 | 0 | 0 | 1 | 24 | 195 | 409 | 297 | 97 | 1 | 0 | 0 | 1024 |
| 2005 | 0 | 0 | 0 | 0 | 31 | 77 | 470 | 341 | 44 | 0 | 0 | 0 | 963 |
| 2006 | 0 | 0 | 0 | 1 | 52 | 249 | 463 | 274 | 114 | 0 | 0 | 0 | 1153 |
| 2007 | 0 | 0 | 0 | 4 | 76 | 260 | 474 | 355 | 94 | 0 | 0 | 0 | 1263 |
| 2008 | 0 | 0 | 0 | 0 | 51 | 181 | 395 | 379 | 119 | 7 | 0 | 0 | 1132 |
| 2009 | 0 | 0 | 0 | 1 | 94 | 110 | 403 | 292 | 169 | 0 | 0 | 0 | 1069 |

## SNOWFALL (inches) 2009 RENO (KRNO)

| YEAR | JUL | AUG | SEP | OCT | NOV | DEC | JAN | FEB | MAR | APR | MAY | JUN | TOTAL |
|------|-----|-----|-----|-----|-----|-----|-----|-----|-----|-----|-----|-----|-------|
| 1980-81 | 0.0 | 0.0 | 0.0 | T | T | T | 3.9 | T | 2.2 | T | 0.0 | 0.0 | 6.1 |
| 1981-82 | 0.0 | 0.0 | 0.0 | 1.1 | 2.0 | 0.1 | 12.5 | 2.3 | 6.7 | 0.8 | 0.5 | 0.0 | 26.0 |
| 1982-83 | 0.0 | 0.0 | 1.5 | 0.0 | 8.6 | 1.8 | 1.5 | 1.0 | 3.0 | 2.9 | 3.5 | 0.0 | 23.8 |
| 1983-84 | 0.0 | 0.0 | 0.0 | 0.0 | 5.7 | 0.5 | 6.7 | 1.5 | 0.1 | T | 0.0 | 0.0 | 14.5 |
| 1984-85 | 0.0 | 0.0 | 0.0 | 3.4 | 3.0 | 1.3 | 4.3 | 0.8 | 7.0 | T | T | 0.0 | 19.8 |
| 1985-86 | 0.0 | 0.0 | T | 1.2 | 16.5 | 1.4 | 0.0 | T | 1.4 | 0.4 | T | 0.0 | 20.9 |
| 1986-87 | 0.0 | 0.0 | T | 0.0 | 0.2 | 0.6 | 1.8 | 8.0 | 2.5 | T | 0.0 | 0.0 | 13.1 |
| 1987-88 | 0.0 | 0.0 | 0.0 | 0.0 | 0.8 | 6.3 | 8.2 | 0.0 | T | T | T | 0.0 | 15.3 |
| 1988-89 | 0.0 | 0.0 | 0.0 | 0.0 | 4.1 | 11.7 | 3.3 | 13.3 | 2.2 | 0.8 | T | 0.0 | 35.4 |
| 1989-90 | 0.0 | 0.0 | 0.0 | T | T | T | 5.6 | 21.6 | 2.0 | 0.0 | 0.0 | 0.0 | 29.2 |
| 1990-91 | 0.0 | 0.0 | 0.0 | 0.0 | 0.4 | 2.7 | 0.1 | 0.0 | 4.5 | T | 1.6 | 0.0 | 9.3 |
| 1991-92 | 0.0 | 0.0 | 0.0 | T | 5.0 | 1.4 | 0.4 | 1.3 | T | 0.0 | 0.0 | 0.0 | 8.1 |
| 1992-93 | 0.0 | 0.0 | 0.0 | 0.0 | T | 14.3 | 22.9 | 13.0 | 0.0 | 0.0 | T | 0.0 | 50.2 |
| 1993-94 | 0.0 | 0.0 | T | T | T | 0.3 | 0.5 | 5.2 | T | T | T | 0.0 | 6.0 |
| 1994-95 | 0.0 | 0.0 | 0.0 | 0.0 | 15.3 | 0.5 | 8.2 | 1.9 | 1.7 | 1.1 | T | 0.2 | 28.9 |
| 1995-96 | 0.0 | 0.0 | 0.0 | 0.0 | 0.0 | | 1.2 | | | | | | |
| 1996-97 | | | | | | | | | | | | | |
| 1997-98 | | | | | | | | | | | | | |
| 1998-99 | | | | | | | | | | | | | |
| 1999-00 | | | | | | | | | | | | | |
| 2000-01 | | | | | | | | | | | | | |
| 2001-02 | | | | | | | | | | | | | |
| 2002-03 | | | | | | | | | | | | | |
| 2003-04 | | | | | | | | | | | | | |
| 2004-05 | | | | | 7.0 | 21.6 | 21.1 | T | T | T | 0.0 | 0.0 | |
| 2005-06 | 0.0 | 0.0 | 0.0 | 0.0 | T | 3.6 | 4.3 | 6.4 | 6.1 | 1.0 | T | 0.0 | 21.4 |
| 2006-07 | 0.0 | 0.0 | 0.0 | 0.0 | T | 2.4 | 1.7 | 4.1 | T | 0.2 | T | 0.0 | 8.4 |
| 2007-08 | 0.0 | 0.0 | 0.2 | 0.0 | T | 2.8 | 12.4 | 7.0 | 0.6 | T | 0.0 | 0.0 | 23.0 |
| 2008-09 | 0.0 | 0.0 | 0.0 | 1.0 | T | 5.9 | 0.5 | 0.9 | 2.2 | 3.0 | 0.0 | 0.0 | 13.5 |
| 2009- | 0.0 | 0.0 | 0.0 | 0.1 | 0.2 | 15.6 | | | | | | | |
| POR= 67 YRS | 0.0 | 0.0 | T | 0.4 | 1.9 | 4.5 | 5.6 | 5.1 | 3.6 | 1.1 | 0.7 | 0.0 | 22.9 |

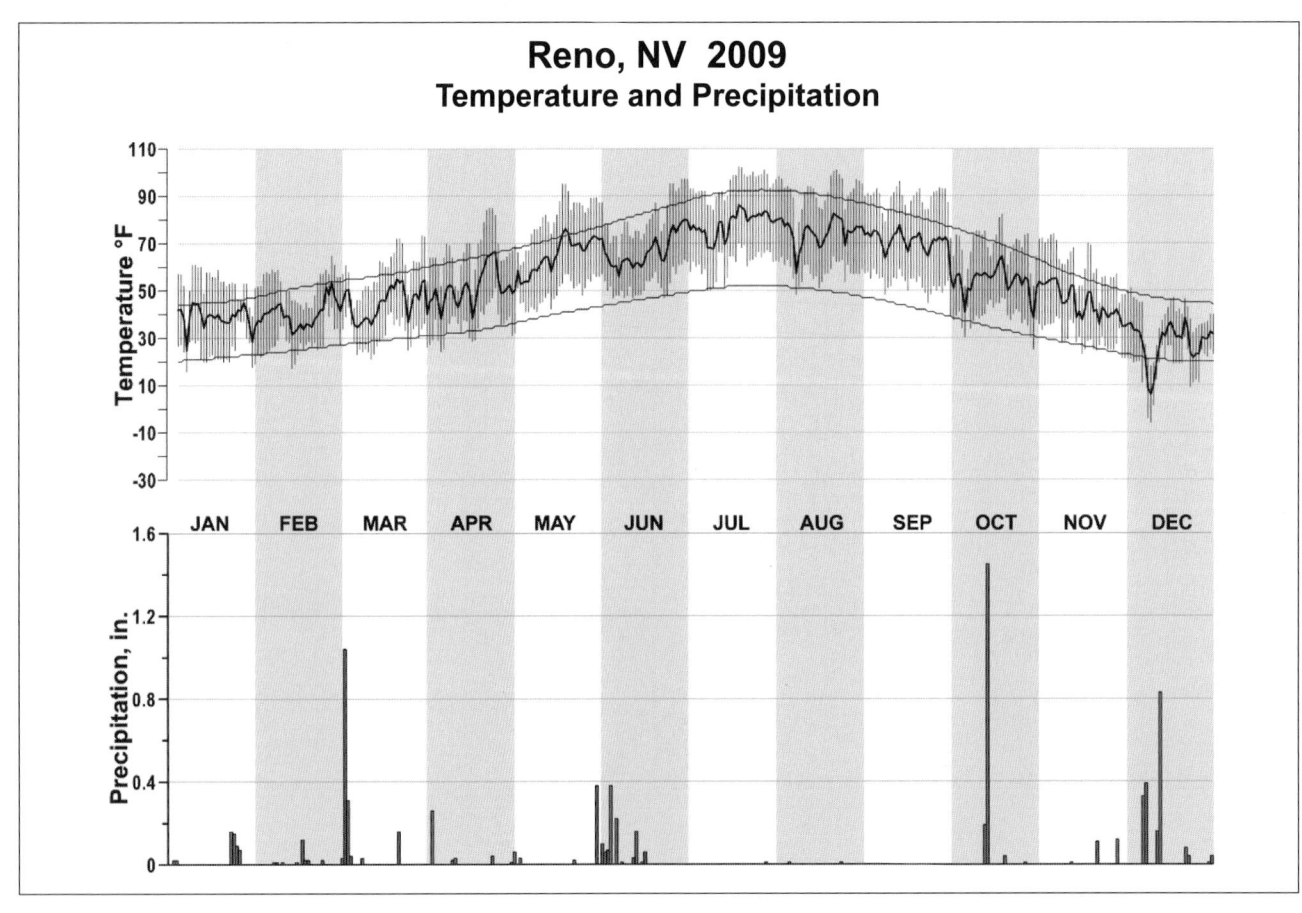

# 2009
# CONCORD
# NEW HAMPSHIRE (KCON)

Concord, the Capital of New Hampshire, is situated near the geographical center of New England at an altitude of approximately 300 feet above sea level on the Merrimack River. Its surroundings are hilly with many lakes and ponds. The countryside is generously wooded, mostly on land reclaimed from fields which were formerly cleared for farming. From the coast about 50 miles to the southeast, the terrain slopes gently upward to the city. West of the city, the land rises some 2,000 feet higher in only half that distance. Mount Washington, at an elevation of 6,288 feet is in the White Mountains 75 miles north of town.

Northwesterly winds are prevalent. They bring cold, dry air during the winter and pleasantly cool, dry air in the summer. Stronger southerly winds occur during July and August, and easterly winds usually accompany summer and winter storms. Winter breezes are somewhat lighter, and winds are frequently calm during the night and early morning hours. Low temperatures, as a rule, do not interrupt normal out-of-doors activity because winds are calm or light, producing a low wind chill factor.

Very hot summer weather is infrequent. During any month, temperatures considerably above the average maxima and much below the normal minima are observed.

The average amount of precipitation for the warmer half of the year differs little from that for the colder half. Precipitation occurrences average approximately one day of three for the year, with a somewhat higher frequency for the April-May period, offsetting the lower frequency of August-October. The more significant rains and heavier snowfalls are associated with easterly winds, especially northeasterly winds. The first snowfall of an inch or more is likely to come between the middle of November and the middle of December. The snow cover normally lasts from mid-December until the last week of March, but bare ground is not rare in the winter, nor is a snowscape rare earlier or later in the season. Rain, sleet, or freezing rain may also occur.

Agriculture is neither intensive nor large-scale in the vicinity of the station. Potatoes and other frost-resistant vegetables, hardy fruits such as apples, forage for the dairy industry, and maple sugar are the principal crops.

Based on the 1951-1980 period, the average first occurrence of 32 degrees Fahrenheit in the fall is September 22 and the average last occurrence in the spring is May 23. Freezing temperatures have occurred as late as June and as early as August.

# NORMALS, MEANS, AND EXTREMES
## CONCORD (KCON)

LATITUDE: 43 ° 11'N LONGITUDE: -71 ° 30'W  ELEVATION (FT): GRND: 340 BARO: 343  TIME ZONE: EASTERN (UTC -5)  WBAN: 14745

| ELEMENT | POR | JAN | FEB | MAR | APR | MAY | JUN | JUL | AUG | SEP | OCT | NOV | DEC | YEAR |
|---|---|---|---|---|---|---|---|---|---|---|---|---|---|---|
| **TEMPERATURE °F** | | | | | | | | | | | | | | |
| NORMAL DAILY MAXIMUM | 30 | 30.6 | 34.1 | 43.8 | 56.9 | 69.6 | 77.9 | 82.9 | 80.8 | 72.1 | 60.5 | 47.6 | 35.6 | 57.7 |
| MEAN DAILY MAXIMUM | 89 | 31.1 | 32.7 | 42.7 | 55.7 | 68.5 | 76.5 | 82.2 | 80.1 | 71.4 | 60.9 | 47.2 | 35.1 | 57.0 |
| HIGHEST DAILY MAXIMUM | 68 | 69 | 67 | 89 | 95 | 97 | 98 | 102 | 101 | 98 | 90 | 80 | 73 | 102 |
| YEAR OF OCCURRENCE | | 2007 | 1997 | 1998 | 1976 | 1962 | 2008 | 1966 | 1975 | 1953 | 1963 | 1950 | 1998 | JUL 1966 |
| MEAN OF EXTREME MAXS. | 89 | 50.6 | 52.2 | 64.7 | 79.9 | 88.1 | 92.5 | 93.5 | 92.1 | 88.1 | 79.4 | 68.0 | 55.4 | 75.4 |
| NORMAL DAILY MINIMUM | 30 | 9.7 | 12.6 | 22.7 | 32.2 | 42.4 | 51.8 | 57.1 | 55.6 | 46.6 | 35.1 | 27.6 | 16.2 | 34.1 |
| MEAN DAILY MINIMUM | 89 | 10.6 | 12.1 | 22.6 | 32.0 | 42.2 | 51.4 | 57.2 | 55.1 | 46.6 | 36.1 | 27.9 | 16.2 | 34.2 |
| LOWEST DAILY MINIMUM | 68 | -33 | -37 | -16 | 8 | 21 | 30 | 35 | 29 | 21 | 10 | -5 | -22 | -37 |
| YEAR OF OCCURRENCE | | 1984 | 1943 | 1967 | 2003 | 1966 | 1972 | 1965 | 1965 | 1947 | 1972 | 1989 | 1951 | FEB 1943 |
| MEAN OF EXTREME MINS. | 89 | -12.7 | -10.8 | 0.8 | 18.1 | 27.6 | 36.7 | 43.8 | 40.4 | 30.2 | 19.7 | 10.5 | -5.6 | 16.6 |
| NORMAL DRY BULB | 30 | 20.1 | 23.3 | 33.3 | 44.6 | 56.0 | 64.9 | 70.0 | 68.2 | 59.4 | 47.8 | 37.6 | 25.9 | 45.9 |
| MEAN DRY BULB | 89 | 20.8 | 22.4 | 32.7 | 43.8 | 55.4 | 64.1 | 69.7 | 67.7 | 59.0 | 48.5 | 37.5 | 25.7 | 45.6 |
| MEAN WET BULB | 26 | 18.6 | 20.5 | 27.7 | 37.9 | 48.3 | 58.4 | 63.2 | 62.2 | 54.7 | 43.4 | 34.1 | 23.8 | 41.1 |
| MEAN DEW POINT | 26 | 14.0 | 15.0 | 22.3 | 32.2 | 43.8 | 54.8 | 60.3 | 59.5 | 52.0 | 39.9 | 30.1 | 19.4 | 36.9 |
| NORMAL NO. DAYS WITH: | | | | | | | | | | | | | | |
| MAXIMUM >= 90 | 30 | 0.0 | 0.0 | 0.0 | 0.2 | 1.0 | 2.5 | 4.5 | 2.8 | 0.4 | 0.0 | 0.0 | 0.0 | 11.4 |
| MAXIMUM <= 32 | 30 | 17.7 | 12.8 | 4.5 | 0.2 | 0.0 | 0.0 | 0.0 | 0.0 | 0.0 | 0.0 | 1.8 | 12.0 | 49.0 |
| MINIMUM <= 32 | 30 | 30.1 | 27.1 | 26.1 | 16.5 | 4.7 | 0.2 | 0.0 | 0.1 | 2.3 | 14.3 | 21.5 | 29.1 | 172.0 |
| MINIMUM <= 0 | 30 | 8.6 | 5.9 | 1.0 | 0.0 | 0.0 | 0.0 | 0.0 | 0.0 | 0.0 | 0.0 | 0.1 | 4.2 | 19.8 |
| **H/C** | | | | | | | | | | | | | | |
| NORMAL HEATING DEG. DAYS | 30 | 1402 | 1183 | 997 | 623 | 302 | 90 | 22 | 44 | 212 | 548 | 835 | 1220 | 7478 |
| NORMAL COOLING DEG. DAYS | 30 | 0 | 0 | 0 | 2 | 18 | 82 | 173 | 133 | 33 | 1 | 0 | 0 | 442 |
| **RH** | | | | | | | | | | | | | | |
| NORMAL (PERCENT) | 30 | 69 | 66 | 64 | 62 | 66 | 70 | 72 | 75 | 76 | 73 | 72 | 72 | 70 |
| HOUR 01 LST | 30 | 75 | 73 | 73 | 76 | 83 | 87 | 89 | 91 | 91 | 86 | 80 | 78 | 82 |
| HOUR 07 LST | 30 | 77 | 76 | 77 | 75 | 77 | 80 | 83 | 88 | 90 | 87 | 83 | 80 | 81 |
| HOUR 13 LST | 30 | 59 | 54 | 52 | 46 | 47 | 51 | 51 | 53 | 54 | 52 | 58 | 60 | 53 |
| HOUR 19 LST | 30 | 67 | 62 | 59 | 55 | 58 | 62 | 64 | 70 | 76 | 72 | 71 | 71 | 66 |
| **S** PERCENT POSSIBLE SUNSHINE | 58 | 52 | 55 | 53 | 52 | 54 | 57 | 63 | 60 | 56 | 53 | 43 | 47 | 54 |
| **W/O** MEAN NO. DAYS WITH: | | | | | | | | | | | | | | |
| HEAVY FOG(VISBY <= 1/4 MI) | 46 | 2.9 | 2.5 | 3.6 | 1.9 | 2.9 | 3.5 | 4.2 | 5.4 | 8.2 | 6.1 | 3.7 | 3.5 | 48.4 |
| THUNDERSTORMS | 62 | 0.0 | 0.1 | 0.2 | 0.8 | 2.5 | 4.5 | 5.4 | 4.2 | 1.7 | 0.5 | 0.2 | 0.1 | 20.2 |
| **CLOUDNESS** MEAN: | | | | | | | | | | | | | | |
| SUNRISE-SUNSET (OKTAS) | 55 | 5.0 | 5.0 | 5.1 | 5.3 | 5.4 | 5.0 | 4.9 | 4.6 | 4.7 | 4.6 | 5.3 | 5.0 | 5.0 |
| MIDNIGHT-MIDNIGHT (OKTAS) | 32 | 4.7 | 4.6 | 4.7 | 4.9 | 5.0 | 4.7 | 4.6 | 4.4 | 4.6 | 4.4 | 5.0 | 4.8 | 4.7 |
| MEAN NO. DAYS WITH: | | | | | | | | | | | | | | |
| CLEAR | 55 | 9.0 | 7.5 | 7.7 | 7.0 | 6.1 | 6.1 | 6.5 | 8.1 | 8.4 | 9.0 | 6.2 | 7.7 | 89.3 |
| PARTLY CLOUDY | 55 | 7.1 | 7.6 | 8.0 | 8.1 | 9.9 | 11.6 | 12.2 | 11.1 | 8.9 | 8.7 | 7.5 | 7.7 | 108.4 |
| CLOUDY | 55 | 14.9 | 13.2 | 15.3 | 14.9 | 15.0 | 12.2 | 11.7 | 11.3 | 12.2 | 12.7 | 15.8 | 15.1 | 164.3 |
| **PR** | | | | | | | | | | | | | | |
| MEAN STATION PRESSURE(IN) | 26 | 29.63 | 29.62 | 29.62 | 29.58 | 29.59 | 29.57 | 29.59 | 29.64 | 29.68 | 29.68 | 29.66 | 29.64 | 29.63 |
| MEAN SEA-LEVEL PRES. (IN) | 26 | 30.02 | 30.02 | 30.01 | 29.96 | 29.97 | 29.94 | 29.95 | 30.01 | 30.06 | 30.06 | 30.05 | 30.03 | 30.01 |
| **WINDS** | | | | | | | | | | | | | | |
| MEAN SPEED (MPH) | 26 | 6.5 | 7.2 | 7.4 | 7.1 | 6.5 | 5.8 | 5.3 | 4.8 | 5.0 | 5.5 | 6.1 | 6.5 | 6.1 |
| PREVAIL.DIR(TENS OF DEGS) | 29 | 32 | 32 | 32 | 32 | 33 | 32 | 32 | 33 | 33 | 32 | 32 | 32 | 32 |
| MAXIMUM 2-MINUTE: | 13 | | | | | | | | | | | | | |
| SPEED (MPH) | | 38 | 40 | 40 | 39 | 38 | 43 | 43 | 30 | 36 | 37 | 36 | 39 | 43 |
| DIR. (TENS OF DEGS) | | 33 | 31 | 32 | 29 | 29 | 33 | 32 | 28 | 06 | 31 | 31 | 32 | 33 |
| YEAR OF OCCURRENCE | | 2000 | 2001 | 2005 | 2004 | 2002 | 2004 | 1999 | 2006 | 1999 | 1998 | 2009 | 1999 | JUN 2004 |
| MAXIMUM 3-SECOND | | | | | | | | | | | | | | |
| SPEED (MPH) | 13 | 48 | 55 | 52 | 53 | 47 | 66 | 52 | 47 | 47 | 49 | 52 | 52 | 66 |
| DIR. (TENS OF DEGS) | | 30 | 27 | 31 | 29 | 30 | 34 | 32 | 29 | 33 | 29 | 31 | 32 | 34 |
| YEAR OF OCCURRENCE | | 2007 | 2006 | 2005 | 2004 | 2002 | 2007 | 1999 | 2006 | 2002 | 1998 | 2009 | 1999 | JUN 2007 |
| **PRECIPITATION** | | | | | | | | | | | | | | |
| NORMAL (IN) | 30 | 2.97 | 2.36 | 3.04 | 3.07 | 3.33 | 3.10 | 3.37 | 3.21 | 3.16 | 3.46 | 3.57 | 2.96 | 37.60 |
| MAXIMUM MONTHLY (IN) | 68 | 8.09 | 8.96 | 7.81 | 7.22 | 11.09 | 10.10 | 7.55 | 7.26 | 9.44 | 14.57 | 7.36 | 7.52 | 14.57 |
| YEAR OF OCCURRENCE | | 1979 | 2008 | 1953 | 2007 | 2006 | 1944 | 2009 | 1991 | 1999 | 2005 | 1983 | 1973 | OCT 2005 |
| MINIMUM MONTHLY (IN) | 68 | 0.40 | 0.03 | 0.86 | 0.83 | 0.50 | 0.64 | 0.96 | 0.06 | 0.41 | 0.59 | 0.75 | 0.58 | 0.03 |
| YEAR OF OCCURRENCE | | 1970 | 1987 | 1981 | 1999 | 2008 | 1979 | 1955 | 1996 | 1948 | 1947 | 1976 | 1943 | FEB 1987 |
| MAXIMUM IN 24 HOURS (IN) | 68 | 2.12 | 2.98 | 2.48 | 2.88 | 5.60 | 4.47 | 2.76 | 3.97 | 5.20 | 5.24 | 2.89 | 3.31 | 5.60 |
| YEAR OF OCCURRENCE | | 1979 | 2008 | 2001 | 2007 | 2006 | 1944 | 2008 | 1991 | 2008 | 2005 | 1947 | 1969 | MAY 2006 |
| NORMAL NO. DAYS WITH: | | | | | | | | | | | | | | |
| PRECIPITATION >= 0.01 | 30 | 11.1 | 9.2 | 11.5 | 11.8 | 12.0 | 11.6 | 10.5 | 9.9 | 9.5 | 9.5 | 10.9 | 11.2 | 128.7 |
| PRECIPITATION >= 1.00 | 30 | 0.5 | 0.5 | 0.6 | 0.5 | 0.6 | 0.5 | 0.9 | 0.6 | 0.7 | 1.0 | 0.8 | 0.7 | 7.9 |
| **SNOWFALL** | | | | | | | | | | | | | | |
| NORMAL (IN) | 30 | 18.9 | 12.9 | 11.5 | 3.1 | 0.* | 0.0 | 0.0 | 0.0 | 0.* | 0.1 | 4.7 | 13.4 | 64.6 |
| MAXIMUM MONTHLY (IN) | 68 | 45.4 | 49.8 | 38.3 | 15.3 | 5.0 | T | 0.0 | 0.0 | | 2.1 | 18.4 | 44.5 | 49.8 |
| YEAR OF OCCURRENCE | | 1987 | 1969 | 1956 | 1982 | 1945 | 1993 | | | 1992 | 1969 | 1971 | 2007 | FEB 1969 |
| MAXIMUM IN 24 HOURS (IN) | 68 | 19.0 | 14.2 | 18.1 | 13.9 | 5.0 | T | 0.0 | 0.0 | T | 2.1 | 9.5 | 15.6 | 19.0 |
| YEAR OF OCCURRENCE' | | 1944 | 1972 | 2001 | 1982 | 1945 | 1993 | | | 1992 | 1969 | 1961 | 2005 | JAN 1944 |
| MAXIMUM SNOW DEPTH (IN) | 60 | 29 | 42 | 45 | 25 | 1 | 0 | 0 | 0 | 0 | 2 | 10 | 26 | 45 |
| YEAR OF OCCURRENCE | | 2008 | 2008 | 2008 | 2008 | 1966 | | | | | 1969 | 1971 | 1970 | MAR 2008 |
| NORMAL NO. DAYS WITH: | | | | | | | | | | | | | | |
| SNOWFALL >= 1.0 | 30 | 4.8 | 2.9 | 3.1 | 0.7 | 0.0 | 0.0 | 0.0 | 0.0 | 0.0 | 0.0 | 1.5 | 4.0 | 17.0 |

## PRECIPITATION (inches) 2009  CONCORD (KCON)

| YEAR | JAN | FEB | MAR | APR | MAY | JUN | JUL | AUG | SEP | OCT | NOV | DEC | ANNUAL |
|------|-----|-----|-----|-----|-----|-----|-----|-----|-----|-----|-----|-----|--------|
| 1980 | 0.43 | 0.78 | 3.37 | 3.72 | 0.86 | 2.83 | 2.35 | 3.99 | 2.19 | 2.63 | 3.12 | 0.79 | 27.06 |
| 1981 | 0.48 | 7.77 | 0.86 | 3.12 | 3.21 | 2.81 | 5.54 | 3.25 | 4.61 | 6.51 | 3.51 | 4.17 | 45.84 |
| 1982 | 3.98 | 2.88 | 2.47 | 3.08 | 1.91 | 7.84 | 2.83 | 2.54 | 1.85 | 1.52 | 2.93 | 0.91 | 34.74 |
| 1983 | 3.92 | 2.17 | 7.07 | 5.88 | 5.19 | 2.52 | 2.07 | 2.07 | 1.21 | 3.28 | 7.36 | 5.35 | 48.09 |
| 1984 | 1.89 | 5.06 | 2.92 | 3.74 | 9.52 | 2.83 | 4.44 | 0.97 | 1.08 | 4.42 | 2.67 | 2.70 | 42.24 |
| 1985 | 0.95 | 1.99 | 2.86 | 1.02 | 2.05 | 3.05 | 2.83 | 2.51 | 3.78 | 3.62 | 4.58 | 1.65 | 30.89 |
| 1986 | 4.78 | 2.23 | 3.58 | 1.85 | 1.44 | 4.95 | 4.77 | 3.72 | 2.27 | 1.71 | 4.48 | 4.50 | 40.28 |
| 1987 | 3.00 | 0.03 | 3.47 | 4.71 | 1.08 | 5.77 | 3.77 | 2.84 | 3.94 | 4.14 | 2.50 | 1.55 | 36.80 |
| 1988 | 1.97 | 2.24 | 1.32 | 2.75 | 3.35 | 0.80 | 6.53 | 5.44 | 1.56 | 1.23 | 5.06 | 1.05 | 33.30 |
| 1989 | 0.74 | 2.05 | 2.18 | 3.40 | 5.11 | 4.25 | 3.62 | 3.55 | 4.22 | 4.86 | 3.34 | 0.91 | 38.23 |
| 1990 | 2.82 | 2.63 | 1.64 | 3.00 | 5.09 | 2.51 | 1.79 | 7.19 | 2.31 | 4.93 | 3.25 | 4.12 | 41.28 |
| 1991 | 1.85 | 1.42 | 3.01 | 2.61 | 2.51 | 1.72 | 1.80 | 7.26 | 5.52 | 3.32 | 4.84 | 3.62 | 39.48 |
| 1992 | 1.97 | 1.30 | 2.77 | 1.78 | 2.39 | 3.28 | 3.50 | 2.13 | 2.48 | 2.41 | 3.54 | 2.23 | 29.78 |
| 1993 | 1.51 | 1.63 | 3.04 | 3.77 | 0.97 | 1.95 | 2.10 | 1.97 | 4.01 | 3.58 | 3.72 | 3.36 | 31.61 |
| 1994 | 3.48 | 0.80 | 5.03 | 2.38 | 4.33 | 1.94 | 4.02 | 2.85 | 3.55 | 0.91 | 2.67 | 4.19 | 36.15 |
| 1995 | 3.39 | 2.03 | 2.46 | 1.64 | 3.56 | 1.36 | 4.01 | 2.42 | 2.70 | 7.31 | 5.58 | 1.95 | 38.41 |
| 1996 | 4.87 | 2.58 | 1.55 | 6.11 | 4.81 | 3.09 | 5.17 | .06 | 3.05 | 8.11 | 2.06 | 5.84 | 47.30 |
| 1997 | 3.29 | 2.20 | 4.04 | 3.66 | 2.29 | 0.70 | 3.80 | 3.67 | 1.68 | 1.44 | 5.63 | 1.95 | 34.35 |
| 1998 | 3.62 | 2.85 | 3.70 | 2.21 | 3.97 | 7.95 | 1.90 | 1.41 | 1.63 | 3.58 | 2.20 | 0.92 | 35.94 |
| 1999 | 5.38 | 2.84 | 2.82 | 0.83 | 2.79 | 1.94 | 4.35 | 3.45 | 9.44 | 2.47 | 2.66 | 1.35 | 40.32 |
| 2000 | 2.26 | 2.83 | 3.34 | 4.82 | 2.89 | 3.14 | 4.33 | 1.89 | 3.54 | 2.29 | 3.42 | 2.99 | 37.74 |
| 2001 | 1.44 | 2.57 | 6.44 | 0.84 | 2.33 | 5.74 | 3.29 | 0.67 | 3.51 | 0.99 | 1.10 | 2.24 | 31.16 |
| 2002 | 2.01 | 1.98 | 3.48 | 2.97 | 4.38 | 5.33 | 2.24 | 1.96 | 3.63 | 3.07 | 5.31 | 3.57 | 39.93 |
| 2003 | 2.52 | 3.38 | 3.16 | 4.08 | 4.71 | 1.64 | 1.91 | 6.78 | 4.99 | 4.22 | 2.23 | 5.31 | 44.93 |
| 2004 | 0.67 | 1.10 | 2.81 | 6.61 | 5.13 | 2.30 | 3.89 | 4.62 | 6.11 | 1.66 | 3.20 | 4.05 | 42.15 |
| 2005 | 3.32 | 2.69 | 4.03 | 5.76 | 3.91 | 5.07 | 2.69 | 3.18 | 2.39 | 14.57 | 5.12 | 4.47 | 57.20 |
| 2006 | 3.60 | 2.60 | 1.31 | 2.64 | 11.09 | 8.85 | 4.16 | 3.45 | 2.27 | 6.94 | 4.80 | 3.53 | 55.24 |
| 2007 | 2.72 | 1.54 | 3.24 | 7.22 | 3.83 | 3.05 | 5.54 | 1.71 | 3.09 | 3.85 | 3.40 | 5.06 | 44.25 |
| 2008 | 2.66 | 8.96 | 5.87 | 3.74 | 0.50 | 4.89 | 6.48 | 4.76 | 8.56 | 3.05 | 3.87 | 4.63 | 57.97 |
| 2009 | 3.03 | 1.72 | 2.90 | 4.00 | 3.96 | 6.46 | 7.55 | 4.18 | 0.92 | 5.15 | 3.32 | 4.02 | 47.21 |
| POR= 89 YRS | 2.79 | 2.47 | 3.17 | 3.20 | 3.22 | 3.41 | 3.43 | 3.07 | 3.30 | 3.14 | 3.66 | 3.14 | 38.00 |

WBAN : 14745

## AVERAGE TEMPERATURE (°F) 2009  CONCORD (KCON)

| YEAR | JAN | FEB | MAR | APR | MAY | JUN | JUL | AUG | SEP | OCT | NOV | DEC | ANNUAL |
|------|-----|-----|-----|-----|-----|-----|-----|-----|-----|-----|-----|-----|--------|
| 1980 | 22.4 | 19.1 | 31.8 | 44.4 | 55.6 | 62.8 | 70.6 | 68.4 | 58.2 | 45.1 | 34.8 | 19.2 | 44.4 |
| 1981 | 12.5 | 30.8 | 34.2 | 47.1 | 57.5 | 66.1 | 69.9 | 66.6 | 58.7 | 45.2 | 37.8 | 25.4 | 46.0 |
| 1982 | 10.9 | 20.8 | 30.2 | 41.6 | 57.3 | 60.9 | 69.5 | 65.4 | 60.2 | 47.6 | 41.7 | 32.4 | 44.9 |
| 1983 | 23.1 | 26.1 | 35.9 | 45.2 | 53.1 | 65.1 | 70.1 | 69.3 | 62.3 | 48.2 | 39.4 | 23.4 | 46.8 |
| 1984 | 16.0 | 30.5 | 28.2 | 44.6 | 52.9 | 65.9 | 68.7 | 69.3 | 57.8 | 50.4 | 38.3 | 30.3 | 46.1 |
| 1985 | 15.9 | 25.8 | 35.7 | 45.2 | 56.1 | 62.2 | 70.3 | 66.7 | 60.5 | 49.2 | 38.5 | 22.0 | 45.7 |
| 1986 | 23.0 | 21.3 | 35.5 | 48.2 | 57.5 | 61.3 | 67.3 | 66.1 | 57.3 | 47.4 | 34.1 | 29.0 | 45.7 |
| 1987 | 20.0 | 22.0 | 34.5 | 46.8 | 56.7 | 64.4 | 70.6 | 65.1 | 58.7 | 45.8 | 36.9 | 28.1 | 45.8 |
| 1988 | 18.5 | 23.5 | 33.4 | 43.9 | 57.3 | 62.9 | 72.6 | 70.5 | 57.8 | 44.9 | 39.1 | 23.2 | 45.6 |
| 1989 | 25.7 | 22.6 | 31.8 | 41.4 | 58.3 | 65.0 | 69.5 | 67.7 | 60.8 | 49.1 | 35.8 | 11.9 | 45.0 |
| 1990 | 28.6 | 24.6 | 34.7 | 45.6 | 52.8 | 65.3 | 70.8 | 69.8 | 59.7 | 51.8 | 40.1 | 30.9 | 47.9 |
| 1991 | 20.2 | 28.3 | 36.0 | 47.6 | 60.2 | 65.6 | 69.2 | 69.9 | 57.7 | 50.8 | 39.7 | 25.5 | 47.6 |
| 1992 | 23.3 | 26.0 | 30.2 | 42.5 | 55.0 | 64.0 | 66.0 | 67.1 | 59.7 | 44.9 | 35.6 | 26.5 | 45.1 |
| 1993 | 23.5 | 15.7 | 30.5 | 45.7 | 56.9 | 64.9 | 71.4 | 70.0 | 58.8 | 45.3 | 36.4 | 26.8 | 45.5 |
| 1994 | 11.3 | 16.7 | 31.9 | 45.8 | 54.4 | 67.9 | 73.4 | 66.4 | 58.2 | 49.3 | 41.5 | 30.6 | 45.6 |
| 1995 | 28.1 | 20.7 | 36.0 | 41.5 | 55.1 | 66.1 | 73.0 | 69.4 | 56.9 | 52.0 | 34.2 | 22.9 | 46.3 |
| 1996 | 21.2 | 23.4 | 29.6 | 43.9 | 53.7 | 65.2 | 68.4 | 69.3 | 60.6 | 46.3 | 33.4 | 31.3 | 45.5 |
| 1997 | 21.8 | 28.3 | 29.8 | 42.3 | 51.6 | 65.9 | 69.1 | 68.1 | 59.8 | 46.8 | 36.2 | 28.0 | 45.6 |
| 1998 | 27.3 | 30.9 | 36.8 | 46.3 | 59.2 | 64.3 | 69.8 | 69.6 | 61.2 | 49.4 | 39.0 | 31.1 | 48.7 |
| 1999 | 20.7 | 27.8 | 34.5 | 44.9 | 57.0 | 68.4 | 71.7 | 67.2 | 63.4 | 46.3 | 41.9 | 29.7 | 47.8 |
| 2000 | 20.1 | 25.4 | 38.5 | 45.4 | 55.2 | 65.1 | 67.2 | 67.4 | 58.6 | 49.2 | 38.4 | 22.5 | 46.1 |
| 2001 | 19.5 | 23.5 | 29.4 | 44.0 | 56.4 | 66.7 | 66.6 | 72.1 | 60.3 | 49.2 | 40.8 | 32.1 | 46.7 |
| 2002 | 28.8 | 27.8 | 34.3 | 46.9 | 53.4 | 62.7 | 70.4 | 71.0 | 63.4 | 45.4 | 35.6 | 24.8 | 47.0 |
| 2003 | 12.6 | 18.5 | 31.2 | 41.1 | 53.9 | 64.8 | 71.5 | 71.0 | 61.6 | 47.9 | 40.6 | 27.9 | 45.2 |
| 2004 | 14.2 | 25.0 | 35.8 | 46.6 | 57.8 | 62.8 | 69.1 | 69.0 | 61.1 | 48.9 | 38.9 | 27.6 | 46.4 |
| 2005 | 18.6 | 23.7 | 28.6 | 46.9 | 51.2 | 68.4 | 70.6 | 71.0 | 63.2 | 51.1 | 37.9 | 24.0 | 46.3 |
| 2006 | 28.8 | 24.6 | 33.0 | 47.0 | 56.1 | 66.8 | 72.6 | 66.6 | 59.5 | 48.6 | 43.3 | 33.9 | 48.4 |
| 2007 | 25.5 | 18.8 | 32.0 | 42.5 | 57.9 | 65.2 | 69.5 | 68.6 | 61.7 | 54.2 | 35.9 | 23.7 | 46.3 |
| 2008 | 23.2 | 24.4 | 30.5 | 46.0 | 54.2 | 66.4 | 71.1 | 66.4 | 61.0 | 46.8 | 37.2 | 27.0 | 46.2 |
| 2009 | 14.3 | 24.4 | 33.0 | 47.2 | 55.7 | 63.0 | 66.5 | 69.2 | 57.7 | 45.5 | 41.7 | 25.2 | 45.3 |
| POR= 89 YRS | 20.8 | 22.4 | 32.7 | 43.8 | 55.4 | 64.1 | 69.7 | 67.7 | 59.0 | 48.5 | 37.5 | 25.7 | 45.6 |

## HEATING DEGREE DAYS (base 65°F) 2009  CONCORD (KCON)

| YEAR | JUL | AUG | SEP | OCT | NOV | DEC | JAN | FEB | MAR | APR | MAY | JUN | TOTAL |
|------|-----|-----|-----|-----|-----|-----|-----|-----|-----|-----|-----|-----|-------|
| 1980-81 | 13 | 33 | 245 | 611 | 899 | 1417 | 1626 | 953 | 951 | 530 | 267 | 40 | 7585 |
| 1981-82 | 12 | 43 | 192 | 608 | 810 | 1222 | 1674 | 1233 | 1072 | 695 | 246 | 136 | 7943 |
| 1982-83 | 25 | 66 | 169 | 535 | 692 | 1007 | 1291 | 1086 | 895 | 588 | 364 | 82 | 6800 |
| 1983-84 | 14 | 33 | 167 | 521 | 760 | 1283 | 1516 | 993 | 1135 | 607 | 382 | 85 | 7496 |
| 1984-85 | 19 | 27 | 238 | 446 | 792 | 1072 | 1514 | 1092 | 901 | 588 | 286 | 106 | 7081 |
| 1985-86 | 9 | 38 | 166 | 485 | 785 | 1326 | 1295 | 1216 | 907 | 499 | 267 | 140 | 7133 |
| 1986-87 | 41 | 67 | 251 | 538 | 919 | 1109 | 1390 | 1199 | 939 | 542 | 296 | 77 | 7368 |
| 1987-88 | 18 | 89 | 201 | 589 | 837 | 1138 | 1436 | 1194 | 971 | 626 | 254 | 137 | 7490 |
| 1988-89 | 19 | 60 | 219 | 622 | 769 | 1289 | 1211 | 1182 | 1022 | 703 | 219 | 80 | 7395 |
| 1989-90 | 6 | 53 | 169 | 484 | 865 | 1639 | 1121 | 1121 | 933 | 585 | 369 | 67 | 7412 |
| 1990-91 | 22 | 15 | 183 | 409 | 737 | 1049 | 1381 | 1024 | 893 | 517 | 186 | 58 | 6474 |
| 1991-92 | 19 | 7 | 238 | 438 | 754 | 1216 | 1286 | 1123 | 1070 | 669 | 319 | 90 | 7229 |
| 1992-93 | 52 | 33 | 203 | 617 | 875 | 1190 | 1277 | 1372 | 1062 | 574 | 248 | 85 | 7588 |
| 1993-94 | 6 | 15 | 229 | 600 | 851 | 1177 | 1661 | 1345 | 1021 | 568 | 338 | 50 | 7861 |
| 1994-95 | 1 | 54 | 210 | 479 | 699 | 1059 | 1137 | 1235 | 893 | 697 | 309 | 46 | 6819 |
| 1995-96 | 6 | 17 | 257 | 397 | 918 | 1298 | 1352 | 1198 | 1093 | 625 | 358 | 54 | 7573 |
| 1996-97 | 10 | 6 | 171 | 572 | 943 | 1038 | 1334 | 1023 | 1086 | 675 | 409 | 83 | 7350 |
| 1997-98 | 16 | 18 | 178 | 560 | 858 | 1138 | 1161 | 950 | 870 | 554 | 186 | 91 | 6580 |
| 1998-99 | 5 | 15 | 144 | 475 | 773 | 1044 | 1365 | 1037 | 940 | 598 | 249 | 51 | 6696 |
| 1999-00 | 10 | 38 | 116 | 571 | 687 | 1087 | 1387 | 1142 | 815 | 583 | 316 | 98 | 6850 |
| 2000-01 | 20 | 26 | 217 | 486 | 790 | 1310 | 1403 | 1157 | 1095 | 627 | 276 | 68 | 7475 |
| 2001-02 | 43 | 1 | 167 | 484 | 719 | 1011 | 1116 | 1038 | 943 | 551 | 370 | 123 | 6566 |
| 2002-03 | 26 | 20 | 111 | 608 | 876 | 1238 | 1619 | 1296 | 1038 | 710 | 335 | 96 | 7973 |
| 2003-04 | 6 | 28 | 112 | 521 | 723 | 1143 | 1568 | 1153 | 900 | 552 | 240 | 115 | 7061 |
| 2004-05 | 6 | 28 | 133 | 491 | 773 | 1154 | 1432 | 1151 | 1121 | 538 | 424 | 65 | 7316 |
| 2005-06 | 6 | 7 | 98 | 433 | 804 | 1263 | 1119 | 1123 | 987 | 531 | 294 | 58 | 6723 |
| 2006-07 | 0 | 48 | 176 | 502 | 643 | 957 | 1218 | 1285 | 1019 | 667 | 251 | 78 | 6844 |
| 2007-08 | 20 | 46 | 143 | 334 | 866 | 1274 | 1289 | 1170 | 1064 | 565 | 334 | 48 | 7153 |
| 2008-09 | 1 | 27 | 158 | 556 | 827 | 1171 | 1564 | 1130 | 985 | 536 | 288 | 89 | 7332 |
| 2009- | 36 | 36 | 223 | 597 | 691 | 1229 | | | | | | | |

WBAN : 14745

## COOLING DEGREE DAYS (base 65°F) 2009  CONCORD (KCON)

| YEAR | JAN | FEB | MAR | APR | MAY | JUN | JUL | AUG | SEP | OCT | NOV | DEC | TOTAL |
|------|-----|-----|-----|-----|-----|-----|-----|-----|-----|-----|-----|-----|-------|
| 1980 | 0 | 0 | 0 | 0 | 5 | 64 | 193 | 145 | 51 | 0 | 0 | 0 | 458 |
| 1981 | 0 | 0 | 0 | 0 | 38 | 80 | 172 | 101 | 11 | 0 | 0 | 0 | 402 |
| 1982 | 0 | 0 | 0 | 0 | 11 | 17 | 171 | 87 | 31 | 0 | 0 | 0 | 317 |
| 1983 | 0 | 0 | 0 | 0 | 2 | 93 | 179 | 172 | 93 | 5 | 0 | 0 | 544 |
| 1984 | 0 | 0 | 0 | 0 | 15 | 117 | 139 | 165 | 28 | 0 | 0 | 0 | 464 |
| 1985 | 0 | 0 | 0 | 0 | 16 | 27 | 184 | 93 | 36 | 1 | 0 | 0 | 357 |
| 1986 | 0 | 0 | 0 | 0 | 44 | 37 | 118 | 109 | 23 | 0 | 0 | 0 | 331 |
| 1987 | 0 | 0 | 0 | 1 | 45 | 64 | 198 | 96 | 21 | 0 | 0 | 0 | 425 |
| 1988 | 0 | 0 | 0 | 0 | 19 | 81 | 262 | 238 | 7 | 4 | 0 | 0 | 611 |
| 1989 | 0 | 0 | 0 | 0 | 17 | 87 | 153 | 141 | 50 | 0 | 0 | 0 | 448 |
| 1990 | 0 | 0 | 0 | 10 | 0 | 80 | 210 | 170 | 30 | 8 | 0 | 0 | 508 |
| 1991 | 0 | 0 | 0 | 2 | 44 | 84 | 157 | 166 | 29 | 4 | 0 | 0 | 486 |
| 1992 | 0 | 0 | 0 | 2 | 15 | 66 | 90 | 105 | 52 | 0 | 0 | 0 | 330 |
| 1993 | 0 | 0 | 0 | 0 | 7 | 90 | 212 | 176 | 49 | 0 | 0 | 0 | 534 |
| 1994 | 0 | 0 | 0 | 0 | 18 | 143 | 269 | 104 | 11 | 0 | 0 | 0 | 545 |
| 1995 | 0 | 0 | 0 | 0 | 9 | 84 | 264 | 159 | 22 | 0 | 0 | 0 | 538 |
| 1996 | 0 | 0 | 0 | 2 | 15 | 65 | 122 | 149 | 48 | 0 | 0 | 0 | 401 |
| 1997 | 0 | 0 | 0 | 0 | 2 | 113 | 149 | 119 | 28 | 3 | 0 | 0 | 414 |
| 1998 | 0 | 0 | 5 | 0 | 13 | 80 | 157 | 165 | 36 | 0 | 0 | 0 | 456 |
| 1999 | 0 | 0 | 0 | 0 | 9 | 157 | 228 | 112 | 76 | 0 | 0 | 0 | 582 |
| 2000 | 0 | 0 | 0 | 0 | 20 | 106 | 93 | 107 | 32 | 0 | 0 | 0 | 358 |
| 2001 | 0 | 0 | 0 | 5 | 18 | 123 | 101 | 226 | 32 | 0 | 0 | 0 | 505 |
| 2002 | 0 | 0 | 0 | 14 | 15 | 60 | 205 | 212 | 71 | 8 | 0 | 0 | 585 |
| 2003 | 0 | 0 | 0 | 0 | 0 | 97 | 217 | 220 | 15 | 0 | 0 | 0 | 549 |
| 2004 | 0 | 0 | 0 | 4 | 24 | 56 | 142 | 156 | 23 | 0 | 0 | 0 | 405 |
| 2005 | 0 | 0 | 0 | 2 | 0 | 172 | 189 | 200 | 49 | 7 | 0 | 0 | 619 |
| 2006 | 0 | 0 | 0 | 0 | 26 | 119 | 235 | 104 | 18 | 0 | 0 | 0 | 502 |
| 2007 | 0 | 0 | 0 | 0 | 35 | 90 | 167 | 163 | 50 | 5 | 0 | 0 | 510 |
| 2008 | 0 | 0 | 0 | 0 | 6 | 97 | 198 | 79 | 48 | 0 | 0 | 0 | 428 |
| 2009 | 0 | 0 | 0 | 10 | 8 | 37 | 91 | 173 | 10 | 0 | 0 | 0 | 329 |

## SNOWFALL (inches)  2009  CONCORD (KCON)

| YEAR | JUL | AUG | SEP | OCT | NOV | DEC | JAN | FEB | MAR | APR | MAY | JUN | TOTAL |
|------|-----|-----|-----|-----|-----|-----|-----|-----|-----|-----|-----|-----|-------|
| 1980-81 | 0.0 | 0.0 | 0.0 | T | 9.4 | 9.8 | 9.2 | 20.9 | 5.4 | T | 0.0 | 0.0 | 54.7 |
| 1981-82 | 0.0 | 0.0 | 0.0 | 0.0 | T | 33.0 | 26.2 | 9.0 | 6.5 | 15.3 | 0.0 | 0.0 | 90.0 |
| 1982-83 | 0.0 | 0.0 | 0.0 | 0.0 | 1.3 | 3.6 | 9.0 | 20.8 | 4.0 | T | T | 0.0 | 38.7 |
| 1983-84 | 0.0 | 0.0 | 0.0 | 0.0 | T | 17.5 | 20.4 | 12.7 | 25.0 | T | T | 0.0 | 75.6 |
| 1984-85 | 0.0 | 0.0 | 0.0 | 0.0 | T | 16.5 | 11.6 | 11.0 | 12.4 | 1.0 | 0.0 | 0.0 | 52.5 |
| 1985-86 | 0.0 | 0.0 | 0.0 | 0.0 | 8.3 | 11.2 | 15.1 | 11.5 | 4.4 | T | T | 0.0 | 50.5 |
| 1986-87 | 0.0 | 0.0 | 0.0 | 0.0 | 14.4 | 7.7 | 45.4 | 0.6 | 7.0 | 9.4 | 0.0 | 0.0 | 84.5 |
| 1987-88 | 0.0 | 0.0 | 0.0 | 0.0 | 5.8 | 12.0 | 19.3 | 23.7 | 4.6 | 0.1 | 0.0 | 0.0 | 65.5 |
| 1988-89 | 0.0 | 0.0 | 0.0 | T | 0.5 | 5.0 | 5.6 | 7.0 | 10.2 | 0.8 | T | 0.0 | 29.1 |
| 1989-90 | 0.0 | 0.0 | 0.0 | 0.0 | 4.7 | 12.0 | 23.1 | 22.0 | 1.3 | T | 0.0 | 0.0 | 63.1 |
| 1990-91 | 0.0 | 0.0 | 0.0 | 0.0 | 0.6 | 8.8 | 11.2 | 6.8 | 6.0 | 0.2 | 0.0 | 0.0 | 33.6 |
| 1991-92 | 0.0 | 0.0 | 0.0 | 0.0 | 1.3 | 13.5 | 3.8 | 5.3 | 6.8 | 4.5 | 0.0 | 0.0 | 35.2 |
| 1992-93 | 0.0 | 0.0 | T | T | 1.1 | 10.9 | 19.7 | 24.0 | 35.8 | 6.6 | 0.0 | T | 98.1 |
| 1993-94 | 0.0 | 0.0 | 0.0 | T | 0.2 | 10.0 | 36.8 | 14.4 | 23.6 | T | 0.0 | 0.0 | 85.0 |
| 1994-95 | 0.0 | 0.0 | 0.0 | 0.0 | 4.3 | 3.2 | 8.1 | 18.2 | 0.7 | T | 0.0 | 0.0 | 34.5 |
| 1995-96 | 0.0 | 0.0 | 0.0 | 0.0 | 7.0 | 28.0 | 27.3 | 15.6 |  | 14.3 | T | 0.0 |  |
| 1996-97 | 0.0 | 0.0 | 0.0 | 0.0 | 2.2 | 20.0 | 14.3 | 14.1 | 22.7 | 3.4 | 0.0 | 0.0 | 76.7 |
| 1997-98 | 0.0 | 0.0 | 0.0 | 0.0 | 12.4 | 17.3 | 16.3 | 3.3 | 9.6 | 0.0 | 0.0 | 0.0 | 58.9 |
| 1998-99 | 0.0 | 0.0 | 0.0 | 0.0 | 0.6 | 5.5 | 18.0 | 1.9 | 12.5 | T | 0.0 | 0.0 | 38.5 |
| 1999-00 | 0.0 | 0.0 | 0.0 | 0.0 | T | T | 16.6 | 15.1 |  |  |  |  |  |
| 2000-01 |  |  |  |  |  |  | 13.5 | 20.9 | 37.5 | T | 0.0 | 0.0 |  |
| 2001-02 | 0.0 | 0.0 | 0.0 | 0.0 | 0.0 | 6.0 | 14.4 | 0.8 | 12.6 | 1.0 | 0.0 | 0.0 | 34.8 |
| 2002-03 | 0.0 | 0.0 | 0.0 | 1.4 | 6.8 | 18.7 | 28.0 | 18.9 | 2.8 | 12.2 | 0.0 | 0.0 | 88.8 |
| 2003-04 | 0.0 | 0.0 | 0.0 | T | T | 33.9 | 6.2 | 7.0 | 8.2 | T | 0.0 | 0.0 | 55.3 |
| 2004-05 | 0.0 | 0.0 | 0.0 | 0.0 | 0.1 | 8.0 | 27.5 | 23.4 | 22.4 | T | 0.0 | 0.0 | 81.4 |
| 2005-06 | 0.0 | 0.0 | 0.0 | 0.0 | 4.0 | 20.8 | 14.9 | 9.7 | 0.3 | T | 0.0 | 0.0 | 49.7 |
| 2006-07 | 0.0 | 0.0 | 0.0 | 0.0 | 0.0 | 1.8 | 2.5 | 15.0 | 17.2 | 12.0 | 0.0 | 0.0 | 48.5 |
| 2007-08 | 0.0 | 0.0 | 0.0 | 0.0 | 2.1 | 44.5 | 22.9 | 33.1 | 14.1 | 1.3 | 0.0 | 0.0 | 118.0 |
| 2008-09 | 0.0 | 0.0 | 0.0 | 0.0 | 1.0 | 27.6 | 33.1 | 11.5 | 16.0 | 0.0 | 0.0 | 0.0 | 89.2 |
| 2009- | 0.0 | 0.0 | 0.0 | T | 0.0 | 16.4 |  |  |  |  |  |  |  |
| POR= 62 YRS | 0.0 | 0.0 | T | 0.1 | 3.6 | 13.3 | 17.8 | 14.6 | 11.4 | 2.9 | 0.1 | T | 63.8 |

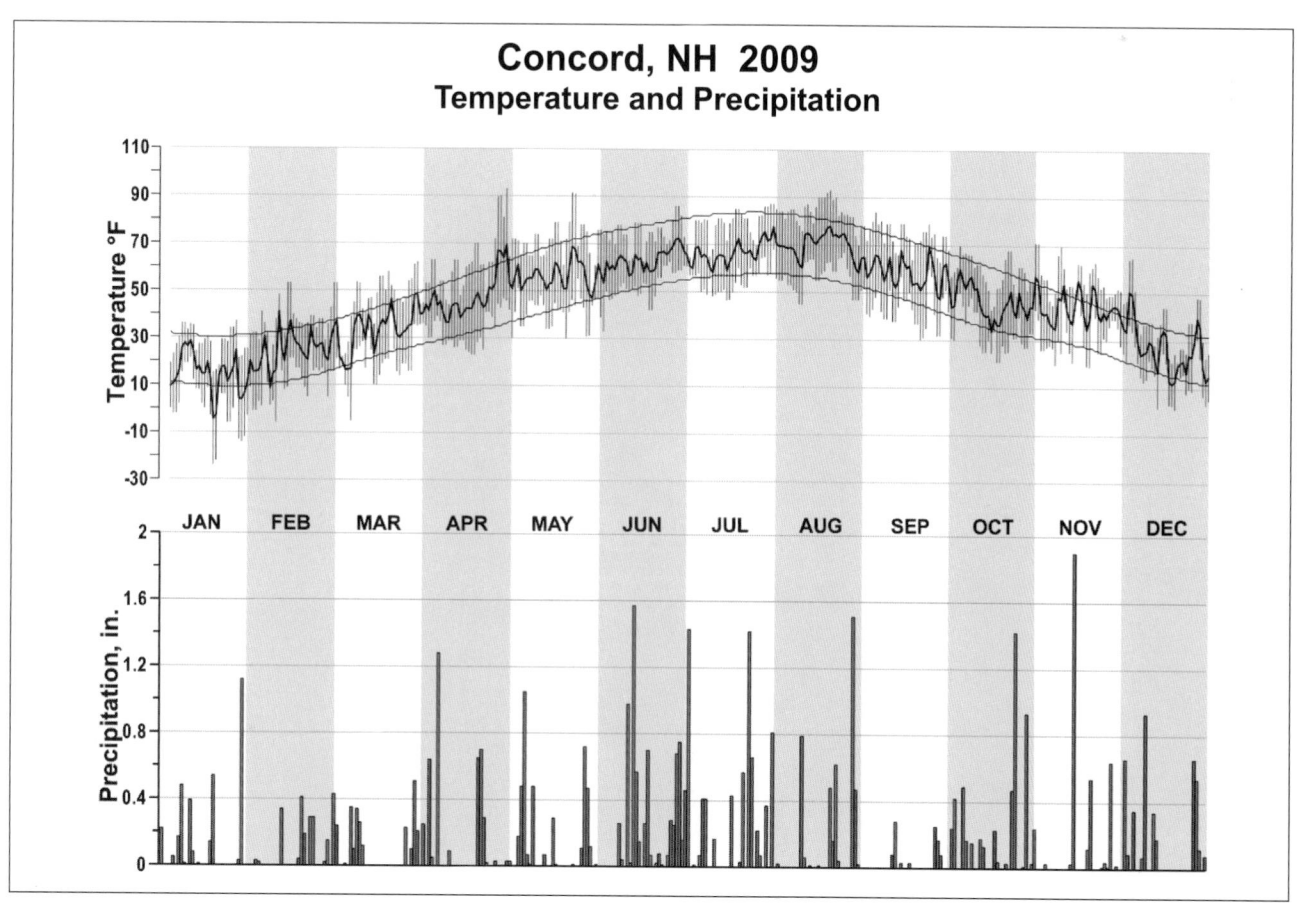

**Concord, NH  2009**
**Temperature and Precipitation**

# 2009
# ATLANTIC CITY
# N.A.F.E.C. (KACY)

The Atlantic City National Weather Service Office is located at the National Aviation Facilities Experimental Center, Pomona, which is about 10 miles west-northwest of Atlantic City and the Atlantic Ocean. The surrounding terrain is fairly flat at an elevation of 50 to 60 feet above sea level. Vegetation in the area consists of scrub pine and low underbrush, but clearing for the air facility has been quite extensive. Bays and salt marshes are as near as 6 miles east of the airport. Atlantic City is located on Abescon Island on the southeast coast of New Jersey. Surrounding terrain, composed of tidal marshes and beach sand, is flat and lies slightly above sea level. The climate is principally continental in character. However, the moderating influence of the Atlantic Ocean is apparent throughout the year, being more marked in the city than at the airport. As a result, summers are relatively cooler and winters milder than elsewhere at the same latitude.

Land and sea breezes, local circulations resulting from the differential heating and cooling of the land and sea, often prevail. These winds occur when moderate or intense storms are not present in the area, thus enabling the local circulation to overcome the general wind pattern. During the warm season sea breezes in the late morning and afternoon hours prevent excessive heating. Frequently, the temperature at Atlantic City during the afternoon hours in the summer averages several degrees lower than at the airport and the airport averages several degrees lower than localities farther inland. On occasions, sea breezes have lowered the temperature as much as 15 to 20 degrees within a half hour. However, the major effect of the sea breeze at the airport is preventing the temperature from rising above the 80s. Because the change in ocean temperature lags behind the air temperature from season to season, the weather tends to remain comparatively mild late into the fall, but on the other hand, warming is retarded in the spring. Normal ocean temperatures range from an average near 37 degrees in January to near 72 degrees in August.

Precipitation is moderate and well distributed throughout the year, with June the driest month and August the wettest. Tropical storms or hurricanes occasionally bring excessive rainfall to the area. The bulk of winter precipitation results from storms which move northeastward along or near the east coast of the United States. Snowfall is considerably less than elsewhere at the same latitude and does not remain long on the ground. Precipitation, often beginning as snow, will frequently become mixed with or change to rain while continuing as snow over more interior sections. In addition, ice storms and resultant glaze are relatively infrequent.

Newark

NEW JERSEY

Atlantic City

# NORMALS, MEANS, AND EXTREMES
## ATLANTIC CITY (KACY)

LATITUDE: 39°27'N　　LONGITUDE: -74°34'W　　ELEVATION (FT): GRND: 65　BARO: 117　　TIME ZONE: EASTERN (UTC -5)　　WBAN: 93730

| | ELEMENT | POR | JAN | FEB | MAR | APR | MAY | JUN | JUL | AUG | SEP | OCT | NOV | DEC | YEAR |
|---|---|---|---|---|---|---|---|---|---|---|---|---|---|---|---|
| **TEMPERATURE °F** | NORMAL DAILY MAXIMUM | 30 | 41.4 | 43.9 | 51.9 | 61.3 | 71.1 | 80.0 | 85.1 | 83.3 | 76.6 | 66.3 | 56.0 | 46.4 | 63.6 |
| | MEAN DAILY MAXIMUM | 51 | 41.3 | 43.4 | 51.4 | 61.7 | 71.3 | 80.0 | 84.8 | 83.5 | 77.0 | 66.4 | 56.2 | 45.7 | 63.6 |
| | HIGHEST DAILY MAXIMUM | 66 | 78 | 75 | 87 | 94 | 99 | 106 | 104 | 103 | 99 | 90 | 84 | 77 | 106 |
| | YEAR OF OCCURRENCE | | 1967 | 1985 | 1998 | 2002 | 1969 | 1969 | 1966 | 2001 | 1983 | 2007 | 1950 | 1998 | JUN 1969 |
| | MEAN OF EXTREME MAXS. | 51 | 61.8 | 63.1 | 72.9 | 82.5 | 88.3 | 93.3 | 95.4 | 93.9 | 89.2 | 81.6 | 72.4 | 63.5 | 79.8 |
| | NORMAL DAILY MINIMUM | 30 | 22.8 | 24.5 | 31.7 | 39.8 | 49.8 | 59.3 | 65.4 | 63.7 | 56.0 | 43.9 | 35.7 | 27.1 | 43.3 |
| | MEAN DAILY MINIMUM | 51 | 23.0 | 24.3 | 31.5 | 40.3 | 49.9 | 59.3 | 65.5 | 64.2 | 56.4 | 44.6 | 36.3 | 27.0 | 43.5 |
| | LOWEST DAILY MINIMUM | 66 | -10 | -11 | 3 | 12 | 25 | 37 | 42 | 40 | 32 | 20 | 10 | -7 | -11 |
| | YEAR OF OCCURRENCE | | 1977 | 1979 | 2009 | 1969 | 1966 | 1980 | 1988 | 1976 | 1969 | 1988 | 1989 | 1950 | FEB 1979 |
| | MEAN OF EXTREME MINS. | 51 | 4.2 | 7.1 | 15.2 | 25.7 | 35.1 | 45.2 | 53.3 | 51.0 | 40.5 | 29.5 | 20.0 | 9.9 | 28.1 |
| | NORMAL DRY BULB | 30 | 32.1 | 34.2 | 41.8 | 50.6 | 60.5 | 69.7 | 75.3 | 73.5 | 66.3 | 55.1 | 45.9 | 36.8 | 53.5 |
| | MEAN DRY BULB | 51 | 32.1 | 33.9 | 41.5 | 51.0 | 60.6 | 69.9 | 75.2 | 73.9 | 66.7 | 55.5 | 46.3 | 36.4 | 53.6 |
| | MEAN WET BULB | 26 | 29.5 | 30.5 | 36.1 | 44.9 | 54.2 | 63.8 | 68.7 | 68.0 | 61.7 | 51.1 | 42.3 | 33.4 | 48.7 |
| | MEAN DEW POINT | 26 | 25.6 | 26.1 | 31.6 | 40.6 | 50.9 | 61.0 | 66.5 | 66.1 | 59.5 | 48.3 | 38.6 | 29.4 | 45.4 |
| | NORMAL NO. DAYS WITH: | | | | | | | | | | | | | | |
| | MAXIMUM >= 90 | 30 | 0.0 | 0.0 | 0.0 | 0.1 | 1.0 | 3.3 | 7.8 | 4.4 | 1.1 | 0.0 | 0.0 | 0.0 | 17.7 |
| | MAXIMUM <= 32 | 30 | 6.4 | 4.7 | 0.6 | 0.0 | 0.0 | 0.0 | 0.0 | 0.0 | 0.0 | 0.0 | 0.1 | 3.2 | 15.0 |
| | MINIMUM <= 32 | 30 | 25.1 | 21.7 | 16.6 | 6.3 | 0.4 | 0.0 | 0.0 | 0.0 | 0.0 | 3.2 | 12.5 | 22.2 | 108.0 |
| | MINIMUM <= 0 | 30 | 0.6 | 0.4 | 0.0 | 0.0 | 0.0 | 0.0 | 0.0 | 0.0 | 0.0 | 0.0 | 0.0 | 0.2 | 1.2 |
| **H/C** | NORMAL HEATING DEG. DAYS | 30 | 1019 | 873 | 725 | 437 | 187 | 32 | 1 | 6 | 69 | 323 | 573 | 868 | 5113 |
| | NORMAL COOLING DEG. DAYS | 30 | 0 | 0 | 1 | 5 | 44 | 168 | 322 | 269 | 110 | 15 | 1 | 0 | 935 |
| **RH** | NORMAL (PERCENT) | 30 | 72 | 70 | 67 | 68 | 72 | 73 | 74 | 77 | 78 | 76 | 73 | 72 | 73 |
| | HOUR 01 LST | 30 | 76 | 76 | 76 | 78 | 84 | 87 | 88 | 89 | 89 | 87 | 81 | 77 | 82 |
| | HOUR 07 LST | 30 | 79 | 79 | 77 | 77 | 79 | 81 | 83 | 87 | 88 | 88 | 84 | 79 | 82 |
| | HOUR 13 LST | 30 | 59 | 56 | 53 | 52 | 56 | 57 | 57 | 59 | 59 | 56 | 57 | 58 | 57 |
| | HOUR 19 LST | 30 | 71 | 69 | 65 | 65 | 69 | 70 | 71 | 76 | 80 | 79 | 74 | 72 | 72 |
| **S** | PERCENT POSSIBLE SUNSHINE | 36 | 50 | 53 | 55 | 56 | 56 | 60 | 61 | 65 | 61 | 59 | 51 | 47 | 56 |
| **W/O** | MEAN NO. DAYS WITH: | | | | | | | | | | | | | | |
| | HEAVY FOG(VISBY <= 1/4 MI) | 46 | 2.8 | 2.6 | 2.8 | 3.2 | 3.9 | 3.5 | 3.4 | 3.0 | 3.0 | 3.7 | 3.1 | 2.0 | 37.0 |
| | THUNDERSTORMS | 51 | 0.1 | 0.3 | 0.8 | 1.8 | 2.8 | 3.8 | 5.3 | 4.4 | 1.6 | 0.7 | 0.4 | 0.2 | 22.2 |
| **CLOUDNESS** | MEAN: | | | | | | | | | | | | | | |
| | SUNRISE-SUNSET (OKTAS) | | | | | | | | | | | | | | |
| | MIDNIGHT-MIDNIGHT (OKTAS) | | | | | | | | | | | | | | |
| | MEAN NO. DAYS WITH: | | | | | | | | | | | | | | |
| | CLEAR | | | | | | | | | | | | | | |
| | PARTLY CLOUDY | | | | | | | | | | | | | | |
| | CLOUDY | | | | | | | | | | | | | | |
| **PR** | MEAN STATION PRESSURE(IN) | 26 | 30.00 | 29.99 | 29.96 | 29.90 | 29.91 | 29.89 | 29.91 | 29.95 | 29.99 | 30.01 | 30.01 | 30.01 | 29.96 |
| | MEAN SEA-LEVEL PRES. (IN) | 26 | 30.07 | 30.06 | 30.04 | 29.97 | 29.99 | 29.97 | 29.98 | 30.02 | 30.06 | 30.08 | 30.09 | 30.08 | 30.03 |
| **WINDS** | MEAN SPEED (MPH) | 26 | 9.6 | 10.1 | 10.7 | 10.5 | 9.0 | 8.0 | 7.5 | 7.0 | 7.4 | 7.8 | 8.9 | 9.3 | 8.8 |
| | PREVAIL.DIR(TENS OF DEGS) | 30 | 30 | 30 | 31 | 19 | 19 | 20 | 19 | 19 | 20 | 30 | 31 | 30 | 30 |
| | MAXIMUM 2-MINUTE: | | | | | | | | | | | | | | |
| | SPEED (MPH) | 14 | 47 | 41 | 49 | 41 | 43 | 36 | 41 | 31 | 39 | 38 | 43 | 41 | 49 |
| | DIR. (TENS OF DEGS) | | 30 | 28 | 05 | 13 | 04 | 30 | 26 | 28 | 05 | 28 | 27 | 10 | 05 |
| | YEAR OF OCCURRENCE | | 2000 | 2009 | 2007 | 2005 | 2008 | 1998 | 2006 | 2007 | 2009 | 2003 | 2003 | 2003 | MAR 2007 |
| | MAXIMUM 3-SECOND | | | | | | | | | | | | | | |
| | SPEED (MPH) | 14 | 56 | 54 | 60 | 54 | 59 | 46 | 59 | 41 | 51 | 54 | 54 | 58 | 60 |
| | DIR. (TENS OF DEGS) | | 30 | 27 | 05 | 23 | 04 | 31 | 25 | 28 | 05 | 28 | 28 | 09 | 05 |
| | YEAR OF OCCURRENCE | | 2000 | 2009 | 2007 | 1996 | 2008 | 1998 | 2006 | 2007 | 2009 | 2003 | 2003 | 2003 | MAR 2007 |
| **PRECIPITATION** | NORMAL (IN) | 30 | 3.60 | 2.85 | 4.06 | 3.45 | 3.38 | 2.66 | 3.86 | 4.32 | 3.14 | 2.86 | 3.26 | 3.15 | 40.59 |
| | MAXIMUM MONTHLY (IN) | 66 | 7.71 | 5.98 | 9.25 | 7.95 | 11.51 | 7.05 | 13.09 | 16.06 | 6.94 | 9.04 | 9.65 | 9.99 | 16.06 |
| | YEAR OF OCCURRENCE | | 1948 | 1958 | 1994 | 1952 | 1948 | 2009 | 1959 | 1997 | 2009 | 2005 | 1972 | 2009 | AUG 1997 |
| | MINIMUM MONTHLY (IN) | 66 | 0.26 | 0.68 | 0.37 | 0.84 | 0.40 | 0.10 | 0.51 | 0.34 | 0.41 | 0.06 | 0.68 | 0.62 | 0.06 |
| | YEAR OF OCCURRENCE | | 1955 | 2009 | 2006 | 1976 | 1957 | 1954 | 1983 | 1943 | 1970 | 2000 | 1976 | 1955 | OCT 2000 |
| | MAXIMUM IN 24 HOURS (IN) | 66 | 2.86 | 2.59 | 3.00 | 3.37 | 4.15 | 3.69 | 6.46 | 6.40 | 4.17 | 2.95 | 3.93 | 4.36 | 6.46 |
| | YEAR OF OCCURRENCE | | 1944 | 1966 | 2000 | 1952 | 1959 | 2007 | 1959 | 1997 | 2009 | 1958 | 1953 | 2008 | JUL 1959 |
| | NORMAL NO. DAYS WITH: | | | | | | | | | | | | | | |
| | PRECIPITATION >= 0.01 | 30 | 10.5 | 9.2 | 10.4 | 10.7 | 10.8 | 8.7 | 8.8 | 8.7 | 8.2 | 7.6 | 9.4 | 10.3 | 113.3 |
| | PRECIPITATION >= 1.00 | 30 | 0.8 | 0.6 | 1.2 | 0.8 | 0.7 | 0.7 | 1.0 | 1.3 | 0.8 | 0.8 | 0.8 | 0.8 | 10.3 |
| **SNOWFALL** | NORMAL (IN) | 30 | 4.6 | 5.5 | 1.3 | 0.3 | 0.* | 0.0 | 0.0 | 0.0 | 0.0 | 0.* | 0.3 | 1.5 | 13.5 |
| | MAXIMUM MONTHLY (IN) | 61 | 20.3 | 35.2 | 17.6 | 3.9 | T | T | T | 0.0 | 0.0 | T | 7.8 | 12.6 | 35.2 |
| | YEAR OF OCCURRENCE | | 1987 | 1967 | 1969 | 1990 | 1989 | 1994 | 1991 | | | 2008 | 1967 | 2009 | FEB 1967 |
| | MAXIMUM IN 24 HOURS (IN) | 61 | 16.3 | 17.1 | 11.5 | 3.9 | T | T | T | 0.0 | 0.0 | T | 7.8 | 11.4 | 17.1 |
| | YEAR OF OCCURRENCE' | | 1987 | 1979 | 1969 | 1990 | 1989 | 1994 | 1991 | | | 1990 | 1967 | 2009 | FEB 1979 |
| | MAXIMUM SNOW DEPTH (IN) | 41 | 19 | 23 | 11 | 3 | 0 | 0 | 0 | 0 | 0 | 0 | 6 | 11 | 23 |
| | YEAR OF OCCURRENCE | | 1987 | 1979 | 1969 | 1990 | | | | | | | 1989 | 2009 | FEB 1979 |
| | NORMAL NO. DAYS WITH: | | | | | | | | | | | | | | |
| | SNOWFALL >= 1.0 | 30 | 1.6 | 1.3 | 0.5 | 0.1 | 0.0 | 0.0 | 0.0 | 0.0 | 0.0 | 0.0 | 0.0 | 0.5 | 4.0 |

## PRECIPITATION (inches) 2009 ATLANTIC CITY (KACY)

| YEAR | JAN | FEB | MAR | APR | MAY | JUN | JUL | AUG | SEP | OCT | NOV | DEC | ANNUAL |
|------|-----|-----|-----|-----|-----|-----|-----|-----|-----|-----|-----|-----|--------|
| 1980 | 2.63 | 0.82 | 6.38 | 5.40 | 1.61 | 3.58 | 2.47 | 2.63 | 1.74 | 3.20 | 3.63 | 0.75 | 34.84 |
| 1981 | 0.56 | 3.72 | 1.41 | 6.20 | 3.18 | 4.91 | 1.28 | 3.25 | 1.96 | 2.96 | 1.12 | 3.94 | 34.49 |
| 1982 | 4.11 | 2.06 | 2.70 | 3.85 | 2.42 | 3.03 | 3.62 | 1.63 | 1.34 | 1.14 | 4.17 | 2.85 | 32.92 |
| 1983 | 2.46 | 3.32 | 5.85 | 7.45 | 5.21 | 3.01 | 0.51 | 2.90 | 2.22 | 3.48 | 6.70 | 5.06 | 48.17 |
| 1984 | 2.41 | 3.70 | 5.92 | 4.84 | 6.58 | 1.62 | 4.35 | 2.44 | 1.31 | 1.46 | 3.02 | 1.79 | 39.44 |
| 1985 | 2.07 | 1.71 | 2.38 | 1.02 | 5.04 | 1.55 | 4.36 | 3.94 | 2.26 | 0.90 | 3.81 | 0.93 | 29.97 |
| 1986 | 3.73 | 3.42 | 1.88 | 4.57 | 0.54 | 2.41 | 4.50 | 3.35 | 2.39 | 3.90 | 5.04 | 4.85 | 40.58 |
| 1987 | 6.23 | 1.54 | 3.36 | 5.83 | 2.96 | 2.10 | 6.12 | 2.64 | 3.51 | 2.56 | 2.27 | 2.19 | 41.31 |
| 1988 | 2.99 | 3.80 | 2.21 | 1.77 | 3.20 | 1.03 | 4.58 | 2.78 | 3.07 | 2.76 | 4.65 | 0.64 | 33.48 |
| 1989 | 2.41 | 3.47 | 4.67 | 4.55 | 4.64 | 3.74 | 6.40 | 5.68 | 5.92 | 4.45 | 2.87 | 1.61 | 50.41 |
| 1990 | 2.70 | 1.00 | 2.60 | 3.46 | 5.71 | 1.52 | 3.64 | 5.96 | 1.91 | 2.63 | 2.01 | 3.57 | 36.71 |
| 1991 | 5.12 | 1.21 | 5.52 | 3.07 | 0.85 | 1.43 | 5.28 | 3.44 | 2.62 | 1.36 | 0.76 | 3.75 | 34.41 |
| 1992 | 0.92 | 2.12 | 2.94 | 1.58 | 2.34 | 3.59 | 3.39 | 6.64 | 5.04 | 1.26 | 3.10 | 3.96 | 36.88 |
| 1993 | 2.62 | 2.38 | 8.80 | 2.23 | 2.24 | 1.36 | 1.76 | 5.40 | 2.67 | 3.82 | 1.20 | 4.22 | 38.70 |
| 1994 | 4.13 | 3.54 | 9.25 | 2.03 | 3.48 | 0.71 | 4.11 | 5.37 | 3.78 | 0.82 | 3.09 | 2.18 | 42.49 |
| 1995 | 2.85 | 1.84 | 0.93 | 2.17 | 4.82 | 3.11 | 2.07 | 3.24 | 3.90 | 5.16 | 4.12 | 1.49 | 35.70 |
| 1996 | 3.59 | 2.05 | 3.65 | 4.45 | 4.51 | 3.94 | 4.63 | 4.31 | 5.82 | 5.03 | 1.38 | 7.10 | 50.46 |
| 1997 | 2.76 | 3.09 | 5.33 | 4.31 | 2.69 | 2.20 | 5.89 | 16.06 | 1.55 | 2.67 | 4.54 | 4.25 | 55.34 |
| 1998 | 5.78 | 4.89 | 6.49 | 3.27 | 6.68 | 2.77 | 1.40 | 1.77 | 0.49 | 3.10 | 1.26 | 2.02 | 39.92 |
| 1999 | 5.78 | 2.88 | 4.56 | 3.22 | 1.45 | 2.84 | 1.68 | 6.28 | 4.91 | 3.45 | 1.46 | 2.75 | 41.26 |
| 2000 | 4.29 | 2.26 | 5.14 | 2.65 | 2.16 | 4.64 | 7.88 | 7.43 | 4.19 | 0.06 | 3.66 | 2.04 | 46.40 |
| 2001 | 3.35 | 3.42 | 5.71 | 1.55 | 1.24 | 3.29 | 3.29 | 2.90 | 1.28 | 1.00 | 1.06 | 1.83 | 29.92 |
| 2002 | 2.08 | 0.74 | 5.60 | 4.08 | 2.74 | 4.98 | 1.07 | 2.43 | 3.30 | 6.37 | 5.96 | 4.31 | 43.66 |
| 2003 | 3.01 | 5.39 | 3.96 | 2.79 | 2.58 | 6.99 | 4.06 | 1.67 | 2.94 | 4.82 | 4.37 | 5.74 | 48.32 |
| 2004 | 1.55 | 2.15 | 3.45 | 4.71 | 3.29 | 1.81 | 5.21 | 4.14 | 2.30 | 3.49 | 4.42 | 2.55 | 39.07 |
| 2005 | 4.01 | 3.23 | 3.68 | 3.40 | 3.53 | 3.90 | 4.43 | 1.02 | 0.53 | 9.04 | 2.80 | 4.38 | 43.95 |
| 2006 | 5.83 | 2.22 | 0.37 | 3.45 | 3.58 | 5.05 | 5.20 | 3.68 | 6.32 | 6.09 | 6.64 | 2.24 | 50.67 |
| 2007 | 3.41 | 2.36 | 3.52 | 5.47 | 1.39 | 5.18 | 1.77 | 3.51 | 1.37 | 4.76 | 1.40 | 7.21 | 41.35 |
| 2008 | 2.18 | 5.27 | 3.05 | 3.27 | 4.59 | 2.28 | 3.40 | 2.44 | 5.30 | 1.60 | 5.90 | 7.27 | 46.55 |
| 2009 | 2.76 | 0.68 | 2.53 | 6.23 | 3.43 | 7.05 | 3.86 | 6.99 | 6.94 | 7.97 | 3.12 | 9.99 | 61.55 |
| POR= 51 YRS | 3.40 | 2.95 | 3.94 | 3.44 | 3.22 | 2.91 | 4.14 | 4.41 | 3.12 | 3.28 | 3.36 | 3.59 | 41.76 |

WBAN : 93730

## AVERAGE TEMPERATURE (°F) 2009 ATLANTIC CITY (KACY)

| YEAR | JAN | FEB | MAR | APR | MAY | JUN | JUL | AUG | SEP | OCT | NOV | DEC | ANNUAL |
|------|-----|-----|-----|-----|-----|-----|-----|-----|-----|-----|-----|-----|--------|
| 1980 | 30.8 | 28.6 | 38.7 | 51.6 | 61.6 | 64.3 | 72.4 | 73.1 | 67.9 | 52.3 | 41.3 | 31.2 | 51.2 |
| 1981 | 22.8 | 34.2 | 36.7 | 50.4 | 57.8 | 70.2 | 76.5 | 73.8 | 67.6 | 53.4 | 44.8 | 34.4 | 51.9 |
| 1982 | 26.4 | 36.8 | 43.1 | 49.8 | 64.1 | 69.9 | 76.9 | 71.8 | 64.8 | 55.2 | 48.5 | 41.0 | 54.0 |
| 1983 | 33.4 | 35.5 | 44.9 | 51.7 | 60.3 | 70.7 | 78.7 | 75.9 | 67.9 | 56.5 | 46.5 | 33.9 | 54.7 |
| 1984 | 28.0 | 40.1 | 36.6 | 49.7 | 60.7 | 73.6 | 75.3 | 77.1 | 65.9 | 62.3 | 44.6 | 44.0 | 54.8 |
| 1985 | 26.8 | 35.4 | 44.8 | 55.1 | 64.5 | 70.0 | 76.9 | 73.7 | 68.4 | 58.1 | 53.4 | 33.1 | 55.0 |
| 1986 | 32.6 | 32.1 | 43.0 | 50.0 | 63.1 | 70.5 | 75.9 | 71.9 | 66.3 | 56.6 | 44.7 | 37.9 | 53.7 |
| 1987 | 31.7 | 31.0 | 42.0 | 49.4 | 61.3 | 71.9 | 76.6 | 72.0 | 66.8 | 50.7 | 47.4 | 38.4 | 53.3 |
| 1988 | 27.9 | 34.3 | 42.2 | 48.7 | 59.5 | 69.1 | 77.1 | 76.1 | 64.3 | 49.7 | 45.9 | 34.2 | 52.4 |
| 1989 | 36.3 | 33.8 | 40.6 | 49.4 | 59.9 | 72.2 | 74.0 | 73.3 | 67.3 | 57.2 | 44.2 | 24.7 | 52.7 |
| 1990 | 40.6 | 40.5 | 44.5 | 51.3 | 59.2 | 70.2 | 75.1 | 73.4 | 64.5 | 59.2 | 48.0 | 41.4 | 55.7 |
| 1991 | 33.7 | 38.4 | 43.9 | 53.3 | 66.0 | 71.1 | 76.9 | 75.6 | 65.4 | 56.4 | 46.5 | 38.9 | 55.5 |
| 1992 | 34.2 | 35.7 | 39.5 | 49.3 | 57.0 | 66.5 | 73.9 | 70.5 | 66.1 | 51.8 | 46.0 | 37.4 | 52.3 |
| 1993 | 37.3 | 30.5 | 38.7 | 50.5 | 61.2 | 69.1 | 76.5 | 73.5 | 67.0 | 53.3 | 45.3 | 35.2 | 53.2 |
| 1994 | 27.6 | 31.6 | 41.0 | 55.0 | 57.4 | 72.4 | 77.6 | 70.4 | 64.3 | 53.2 | 49.5 | 40.4 | 53.4 |
| 1995 | 37.8 | 31.6 | 43.1 | 49.6 | 59.7 | 68.3 | 77.0 | 73.4 | 66.5 | 58.8 | 41.5 | 31.9 | 53.3 |
| 1996 | 31.3 | 33.2 | 36.6 | 51.2 | 59.0 | 70.9 | 73.0 | 71.8 | 67.8 | 54.5 | 40.1 | 39.7 | 52.4 |
| 1997 | 32.4 | 39.3 | 42.1 | 48.8 | 57.5 | 67.5 | 74.9 | 71.6 | 65.2 | 54.2 | 43.1 | 37.0 | 52.8 |
| 1998 | 40.3 | 40.2 | 44.3 | 52.6 | 62.8 | 69.7 | 75.2 | 75.0 | 69.3 | 56.1 | 45.5 | 39.9 | 55.9 |
| 1999 | 35.8 | 36.2 | 40.9 | 50.2 | 60.6 | 70.0 | 78.6 | 74.3 | 68.1 | 54.6 | 49.9 | 39.6 | 54.9 |
| 2000 | 31.2 | 37.2 | 46.5 | 50.1 | 62.4 | 71.6 | 71.7 | 72.5 | 65.0 | 56.0 | 44.3 | 29.5 | 53.2 |
| 2001 | 31.7 | 35.8 | 39.3 | 52.2 | 61.9 | 73.1 | 71.6 | 76.8 | 65.5 | 56.3 | 49.3 | 43.0 | 54.7 |
| 2002 | 38.6 | 38.1 | 44.4 | 54.9 | 60.4 | 70.0 | 77.3 | 76.4 | 68.0 | 56.4 | 44.1 | 35.5 | 55.4 |
| 2003 | 27.9 | 29.6 | 42.2 | 49.2 | 56.2 | 68.5 | 75.8 | 76.4 | 68.4 | 54.1 | 50.0 | 37.2 | 53.0 |
| 2004 | 26.9 | 34.2 | 43.1 | 52.3 | 66.9 | 69.9 | 75.1 | 72.7 | 68.1 | 54.5 | 46.7 | 36.7 | 53.9 |
| 2005 | 32.3 | 35.0 | 37.8 | 51.1 | 55.8 | 72.4 | 77.2 | 78.0 | 71.3 | 58.0 | 48.4 | 34.7 | 54.3 |
| 2006 | 40.6 | 35.1 | 42.8 | 53.7 | 61.7 | 70.9 | 78.1 | 76.6 | 66.0 | 55.4 | 51.3 | 43.1 | 56.3 |
| 2007 | 39.2 | 29.4 | 42.6 | 50.1 | 63.0 | 71.2 | 75.5 | 75.8 | 69.0 | 64.1 | 45.8 | 38.5 | 55.4 |
| 2008 | 35.8 | 38.0 | 44.0 | 53.8 | 59.7 | 74.9 | 78.3 | 73.3 | 69.0 | 55.6 | 45.5 | 40.1 | 55.7 |
| 2009 | 30.0 | 37.7 | 41.9 | 54.6 | 62.7 | 68.7 | 73.9 | 77.1 | 66.4 | 56.1 | 50.7 | 36.9 | 54.7 |
| POR= 51 YRS | 32.1 | 33.9 | 41.5 | 51.0 | 60.6 | 69.9 | 75.2 | 73.9 | 66.7 | 55.5 | 46.3 | 36.4 | 53.6 |

## HEATING DEGREE DAYS (base 65°F) 2009  ATLANTIC CITY (KACY)

| YEAR | JUL | AUG | SEP | OCT | NOV | DEC | JAN | FEB | MAR | APR | MAY | JUN | TOTAL |
|------|-----|-----|-----|-----|-----|-----|-----|-----|-----|-----|-----|-----|-------|
| 1980-81 | 3 | 5 | 38 | 392 | 705 | 1042 | 1299 | 855 | 871 | 436 | 251 | 14 | 5911 |
| 1981-82 | 0 | 0 | 47 | 356 | 600 | 940 | 1187 | 787 | 670 | 448 | 79 | 13 | 5127 |
| 1982-83 | 0 | 14 | 64 | 318 | 498 | 740 | 973 | 820 | 617 | 405 | 166 | 16 | 4631 |
| 1983-84 | 0 | 6 | 94 | 288 | 547 | 958 | 1137 | 714 | 874 | 451 | 171 | 8 | 5248 |
| 1984-85 | 0 | 0 | 80 | 129 | 604 | 646 | 1179 | 821 | 621 | 319 | 106 | 14 | 4519 |
| 1985-86 | 0 | 0 | 53 | 225 | 346 | 981 | 997 | 915 | 674 | 444 | 157 | 25 | 4817 |
| 1986-87 | 0 | 30 | 56 | 289 | 600 | 834 | 1023 | 947 | 705 | 459 | 194 | 6 | 5143 |
| 1987-88 | 0 | 10 | 46 | 436 | 520 | 814 | 1145 | 885 | 698 | 481 | 200 | 68 | 5303 |
| 1988-89 | 5 | 9 | 79 | 476 | 565 | 949 | 883 | 867 | 751 | 464 | 184 | 3 | 5235 |
| 1989-90 | 0 | 1 | 67 | 260 | 617 | 1239 | 747 | 680 | 630 | 421 | 188 | 19 | 4869 |
| 1990-91 | 1 | 1 | 90 | 240 | 504 | 722 | 963 | 737 | 649 | 362 | 119 | 25 | 4413 |
| 1991-92 | 0 | 0 | 102 | 281 | 550 | 803 | 949 | 839 | 785 | 468 | 269 | 44 | 5090 |
| 1992-93 | 3 | 0 | 73 | 407 | 565 | 849 | 852 | 961 | 806 | 429 | 150 | 39 | 5134 |
| 1993-94 | 0 | 0 | 79 | 353 | 591 | 918 | 1150 | 932 | 735 | 307 | 250 | 14 | 5329 |
| 1994-95 | 0 | 15 | 73 | 359 | 458 | 755 | 835 | 929 | 674 | 456 | 185 | 14 | 4753 |
| 1995-96 | 0 | 1 | 50 | 223 | 702 | 1020 | 1038 | 917 | 872 | 412 | 238 | 19 | 5492 |
| 1996-97 | 0 | 1 | 46 | 318 | 738 | 776 | 1007 | 714 | 703 | 479 | 236 | 94 | 5112 |
| 1997-98 | 1 | 0 | 79 | 346 | 652 | 860 | 757 | 688 | 652 | 364 | 140 | 28 | 4567 |
| 1998-99 | 0 | 0 | 36 | 270 | 579 | 770 | 896 | 799 | 741 | 437 | 160 | 23 | 4711 |
| 1999-00 | 0 | 2 | 33 | 322 | 448 | 782 | 1039 | 799 | 569 | 442 | 155 | 24 | 4615 |
| 2000-01 | 0 | 4 | 93 | 285 | 616 | 1093 | 1027 | 812 | 790 | 395 | 133 | 10 | 5258 |
| 2001-02 | 3 | 0 | 77 | 288 | 464 | 678 | 812 | 746 | 631 | 355 | 178 | 21 | 4253 |
| 2002-03 | 0 | 0 | 15 | 304 | 620 | 904 | 1139 | 985 | 702 | 469 | 269 | 55 | 5462 |
| 2003-04 | 0 | 0 | 25 | 333 | 452 | 858 | 1177 | 886 | 669 | 378 | 66 | 25 | 4869 |
| 2004-05 | 0 | 4 | 22 | 317 | 538 | 869 | 1007 | 836 | 836 | 417 | 286 | 27 | 5159 |
| 2005-06 | 0 | 0 | 14 | 234 | 494 | 932 | 750 | 831 | 682 | 338 | 143 | 19 | 4437 |
| 2006-07 | 0 | 0 | 58 | 298 | 406 | 670 | 794 | 990 | 692 | 449 | 143 | 10 | 4510 |
| 2007-08 | 1 | 0 | 21 | 127 | 569 | 812 | 898 | 777 | 644 | 332 | 186 | 0 | 4367 |
| 2008-09 | 0 | 1 | 27 | 303 | 575 | 765 | 1078 | 758 | 708 | 350 | 121 | 27 | 4713 |
| 2009- | 0 | 0 | 39 | 284 | 422 | 866 | | | | | | | |

WBAN : 93730

## COOLING DEGREE DAYS (base 65°F) 2009  ATLANTIC CITY (KACY)

| YEAR | JAN | FEB | MAR | APR | MAY | JUN | JUL | AUG | SEP | OCT | NOV | DEC | TOTAL |
|------|-----|-----|-----|-----|-----|-----|-----|-----|-----|-----|-----|-----|-------|
| 1980 | 0 | 0 | 0 | 0 | 47 | 88 | 238 | 262 | 132 | 6 | 0 | 0 | 773 |
| 1981 | 0 | 0 | 0 | 6 | 31 | 177 | 364 | 279 | 132 | 1 | 0 | 0 | 990 |
| 1982 | 0 | 0 | 0 | 0 | 56 | 165 | 375 | 231 | 64 | 23 | 9 | 0 | 923 |
| 1983 | 0 | 0 | 0 | 15 | 30 | 196 | 431 | 350 | 190 | 33 | 0 | 0 | 1245 |
| 1984 | 0 | 0 | 0 | 0 | 43 | 273 | 326 | 382 | 115 | 53 | 0 | 1 | 1193 |
| 1985 | 0 | 0 | 3 | 30 | 99 | 170 | 376 | 278 | 163 | 18 | 3 | 0 | 1140 |
| 1986 | 0 | 0 | 0 | 0 | 104 | 198 | 343 | 250 | 99 | 38 | 0 | 0 | 1032 |
| 1987 | 0 | 0 | 0 | 0 | 87 | 217 | 366 | 231 | 106 | 0 | 0 | 0 | 1007 |
| 1988 | 0 | 0 | 0 | 0 | 38 | 197 | 388 | 360 | 63 | 5 | 0 | 0 | 1051 |
| 1989 | 0 | 0 | 3 | 1 | 35 | 225 | 289 | 267 | 145 | 23 | 2 | 0 | 990 |
| 1990 | 0 | 0 | 4 | 15 | 16 | 180 | 319 | 270 | 84 | 67 | 0 | 0 | 955 |
| 1991 | 0 | 0 | 2 | 15 | 154 | 216 | 376 | 334 | 119 | 19 | 2 | 0 | 1237 |
| 1992 | 0 | 0 | 0 | 3 | 27 | 98 | 287 | 178 | 110 | 6 | 0 | 0 | 709 |
| 1993 | 0 | 0 | 0 | 0 | 44 | 171 | 366 | 271 | 146 | 0 | 5 | 0 | 1003 |
| 1994 | 0 | 0 | 0 | 18 | 22 | 245 | 398 | 189 | 60 | 0 | 0 | 0 | 932 |
| 1995 | 0 | 0 | 0 | 1 | 28 | 122 | 379 | 271 | 100 | 41 | 4 | 0 | 946 |
| 1996 | 0 | 0 | 0 | 7 | 58 | 200 | 252 | 215 | 136 | 1 | 0 | 0 | 869 |
| 1997 | 0 | 0 | 0 | 0 | 12 | 178 | 315 | 211 | 91 | 19 | 0 | 0 | 826 |
| 1998 | 0 | 0 | 17 | 0 | 79 | 175 | 325 | 316 | 171 | 5 | 0 | 0 | 1088 |
| 1999 | 0 | 0 | 0 | 0 | 30 | 180 | 430 | 301 | 131 | 6 | 1 | 0 | 1079 |
| 2000 | 0 | 0 | 0 | 1 | 80 | 230 | 216 | 241 | 103 | 14 | 0 | 0 | 885 |
| 2001 | 0 | 0 | 0 | 17 | 43 | 258 | 374 | 98 | 23 | 1 | 0 | 1027 |
| 2002 | 0 | 0 | 0 | 57 | 41 | 195 | 388 | 362 | 113 | 43 | 0 | 0 | 1199 |
| 2003 | 0 | 0 | 0 | 5 | 3 | 166 | 345 | 363 | 131 | 3 | 10 | 0 | 1026 |
| 2004 | 0 | 0 | 0 | 3 | 135 | 179 | 318 | 248 | 123 | 0 | 0 | 0 | 1006 |
| 2005 | 0 | 0 | 0 | 6 | 7 | 254 | 384 | 407 | 210 | 22 | 0 | 0 | 1290 |
| 2006 | 0 | 0 | 0 | 4 | 49 | 203 | 412 | 368 | 91 | 11 | 0 | 0 | 1138 |
| 2007 | 0 | 0 | 2 | 7 | 89 | 201 | 334 | 342 | 149 | 105 | 0 | 0 | 1229 |
| 2008 | 0 | 0 | 0 | 2 | 30 | 306 | 418 | 267 | 155 | 17 | 0 | 0 | 1195 |
| 2009 | 0 | 0 | 0 | 47 | 55 | 145 | 283 | 380 | 86 | 12 | 0 | 0 | 1008 |

### SNOWFALL (inches) 2009 ATLANTIC CITY (KACY)

| YEAR | JUL | AUG | SEP | OCT | NOV | DEC | JAN | FEB | MAR | APR | MAY | JUN | TOTAL |
|---|---|---|---|---|---|---|---|---|---|---|---|---|---|
| 1980-81 | 0.0 | 0.0 | 0.0 | T | T | 0.1 | 3.2 | 0.0 | T | 0.0 | 0.0 | 0.0 | 3.3 |
| 1981-82 | 0.0 | 0.0 | 0.0 | 0.0 | 0.0 | 0.6 | 7.8 | 4.4 | T | 2.0 | 0.0 | 0.0 | 14.8 |
| 1982-83 | 0.0 | 0.0 | 0.0 | 0.0 | 0.0 | 6.7 | T | 14.9 | T | 0.7 | 0.0 | 0.0 | 22.3 |
| 1983-84 | 0.0 | 0.0 | 0.0 | 0.0 | 0.1 | 0.4 | 3.8 | T | 4.0 | T | 0.0 | 0.0 | 8.3 |
| 1984-85 | 0.0 | 0.0 | 0.0 | 0.0 | T | T | 15.1 | 0.0 | T | 1.3 | 0.0 | 0.0 | 16.4 |
| 1985-86 | 0.0 | 0.0 | 0.0 | 0.0 | 0.0 | 4.2 | 3.9 | 9.6 | T | T | 0.0 | 0.0 | 17.7 |
| 1986-87 | 0.0 | 0.0 | 0.0 | 0.0 | 0.0 | T | 20.3 | 10.7 | 1.6 | 0.7 | 0.0 | 0.0 | 33.3 |
| 1987-88 | 0.0 | 0.0 | 0.0 | 0.0 | T | T | 7.1 | 0.2 | T | T | 0.0 | 0.0 | 7.3 |
| 1988-89 | 0.0 | 0.0 | 0.0 | 0.0 | 0.0 | 0.4 | 0.9 | 12.8 | 3.4 | T | T | 0.0 | 17.5 |
| 1989-90 | 0.0 | 0.0 | 0.0 | 0.0 | 6.0 | 9.3 | T | T | 3.8 | 3.9 | 0.0 | 0.0 | 23.0 |
| 1990-91 | 0.0 | 0.0 | 0.0 | T | 0.0 | 3.1 | 4.7 | 1.7 | T | 0.0 | 0.0 | 0.0 | 9.5 |
| 1991-92 | T | 0.0 | 0.0 | 0.0 | T | T | 2.2 | 1.0 | T | T | 0.0 | 0.0 | 3.2 |
| 1992-93 | 0.0 | 0.0 | 0.0 | 0.0 | 0.0 | 0.2 | 0.8 | 6.0 | 3.7 | T | 0.0 | 0.0 | 10.7 |
| 1993-94 | 0.0 | 0.0 | 0.0 | 0.0 | 0.0 | 2.7 | 0.5 | 4.4 | 0.2 | 0.0 | 0.0 | T | 7.8 |
| 1994-95 | 0.0 | 0.0 | 0.0 | 0.0 | T | 0.0 | T | 0.8 | 0.0 | 0.0 | 0.0 | 0.0 | 0.8 |
| 1995-96 | 0.0 | 0.0 | 0.0 | 0.0 | 0.0 | 0.0 | 12.0 | 17.5 | 8.5 | 8.4 | 0.0 | 0.0 | 46.4 |
| 1996-97 | 0.0 | 0.0 | 0.0 | 0.0 | 0.0 | T | 0.8 | 3.9 | T | 0.0 | 0.0 | 0.0 | 4.7 |
| 1997-98 | 0.0 | 0.0 | 0.0 | 0.0 | 0.0 | 3.0 | 0.0 | 0.0 | 0.0 | 0.0 | 0.0 | 0.0 | 3.0 |
| 1998-99 | 0.0 | 0.0 | 0.0 | 0.0 | 0.0 | 2.7 | 3.0 | 0.0 | T | 0.0 | 0.0 | 0.0 | 5.7 |
| 1999-00 | 0.0 | 0.0 | 0.0 | 0.0 | 0.0 | 0.0 | 15.1 | 1.0 | 0.0 | 0.2 | 0.0 | 0.0 | 16.3 |
| 2000-01 | 0.0 | 0.0 | 0.0 | 0.0 | 0.0 | 10.5 | 1.4 | 8.1 | 0.1 | T | 0.0 | 0.0 | 20.1 |
| 2001-02 | 0.0 | 0.0 | 0.0 | 0.0 | 0.0 | 0.0 | 2.4 | 0.2 | T | T | 0.0 | 0.0 | 2.6 |
| 2002-03 | 0.0 | 0.0 | 0.0 | 0.0 | T | 5.5 | 3.6 | 33.2 | T | T | 0.0 | 0.0 | 42.3 |
| 2003-04 | 0.0 | 0.0 | 0.0 | 0.0 | T | 5.0 | 7.9 | 2.5 | 1.4 | 0.1 | 0.0 | 0.0 | 16.9 |
| 2004-05 | 0.0 | 0.0 | 0.0 | 0.0 | 0.0 | 0.6 | 4.8 | 8.0 | 0.5 | 0.0 | 0.0 | 0.0 | 13.9 |
| 2005-06 | 0.0 | 0.0 | 0.0 | 0.0 | T | 5.3 | 4.7 | 4.2 | T | T | 0.0 | 0.0 | 14.2 |
| 2006-07 | 0.0 | 0.0 | 0.0 | 0.0 | 0.0 | T | 2.4 | 1.9 | 2.2 | T | 0.0 | 0.0 | 6.5 |
| 2007-08 | 0.0 | 0.0 | 0.0 | 0.0 | 0.0 | 4.4 | 2.6 | 3.0 | 0.0 | 0.0 | 0.0 | 0.0 | 10.0 |
| 2008-09 | 0.0 | 0.0 | 0.0 | T | T | 0.3 | 3.8 | 3.2 | 7.0 | 0.0 | 0.0 | 0.0 | 14.3 |
| 2009- | 0.0 | 0.0 | 0.0 | 0.0 | 0.0 | 12.6 | | | | | | | |
| POR= 52 YRS | T | 0.0 | 0.0 | T | 0.3 | 2.6 | 5.2 | 6.1 | 1.9 | 0.4 | T | T | 16.5 |

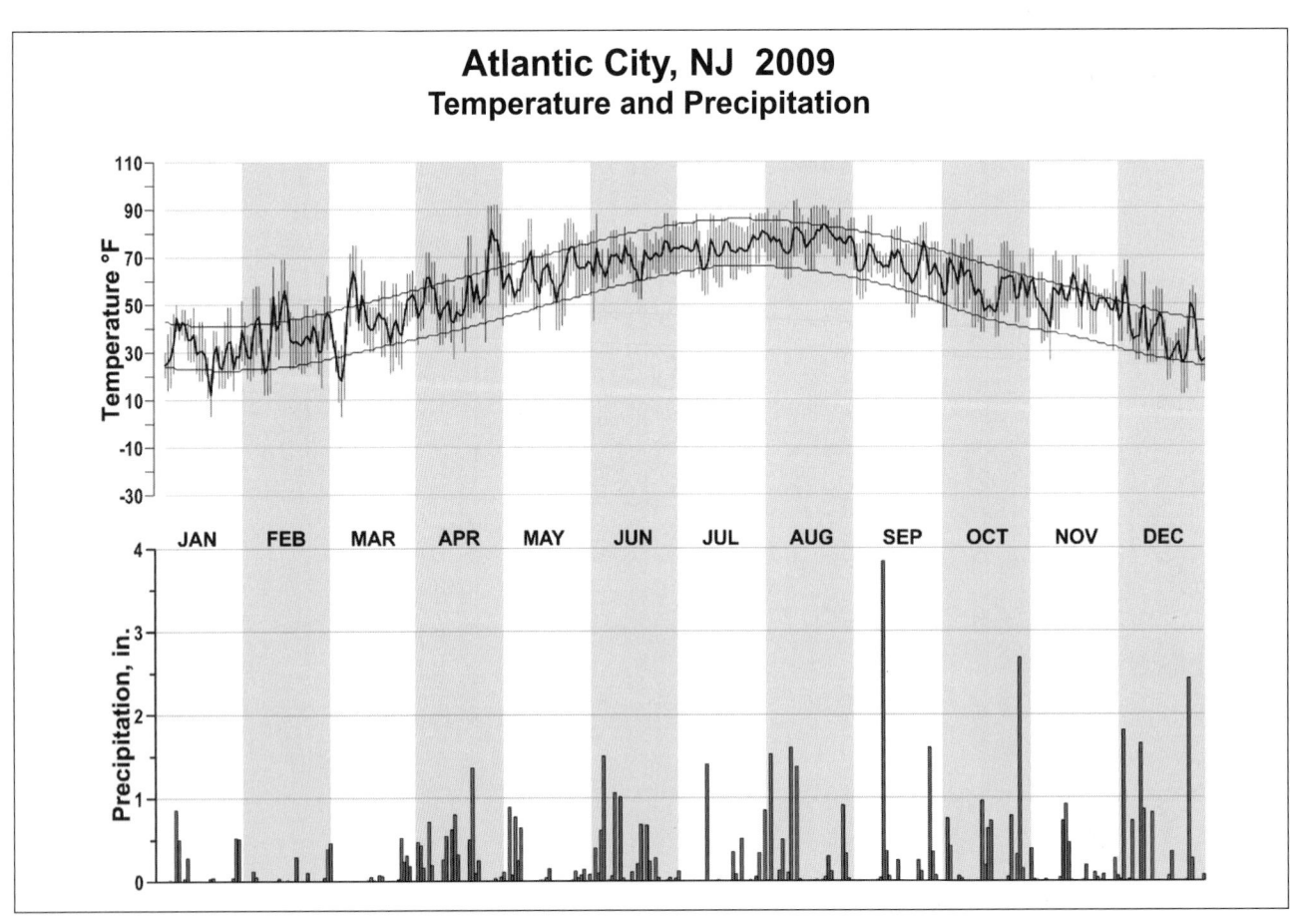

## Atlantic City, NJ  2009
### Temperature and Precipitation

# 2009
# NEWARK
# NEW JERSEY (KEWR)

Terrain in vicinity of the station is flat and rather marshy. To the northwest are ridges oriented roughly in a south-southwest to north-northeast direction. They rise to an elevation of about 200 feet at 4.5 to 5 miles and to 500 to 600 feet at 7 to 8 miles. All winds between west-northwest and north-northwest are downslope and therefore are subject to some adiabatic temperature increase. This effect is evident in the rapid improvement which normally occurs with shift of wind to westerly, following a coastal storm or frontal passage. The drying effect of the downslope winds accounts for the relatively few local thunderstorms occurring at the station, compared to areas to the west. Easterly winds, particularly southeasterly, moderate the temperature because of the influence of the Atlantic Ocean.

Temperature falls of 5 to 15 degrees, depending on the season, are not uncommon when the wind backs from southwesterly to southeasterly. Periods of very hot weather, lasting as long as a week, are associated with a west-southwest air flow which has a long trajectory over land. Extremes of cold are related to rapidly moving outbreaks of cold air traveling southeastward from the

Hudson Bay region. Temperatures of zero or below occur in one winter out of four, but are much more common several miles to the west of the station. Average dates of the last occurrence in spring and the first occurrence in autumn of temperatures as low as 32 degrees are in mid-April and the end of October or early November. Areas to the west of the station experience a growing season at least a month shorter than that at the airport.

A considerable amount of precipitation is realized from the Northeasters of the Atlantic coast. These storms, more typical of the fall and winter, generally last for a period of two days and commonly produce between 1 and 2 inches of precipitation. Storms producing 4 inches or more of snow occur from two to five times a winter. Snowstorms producing 8 inches or more have occurred in about one-half the winters. As many as three such storms have been experienced in one winter. The frequency and intensity of snow storms and the duration of snow cover increase dramatically within a few miles to the west of the station.

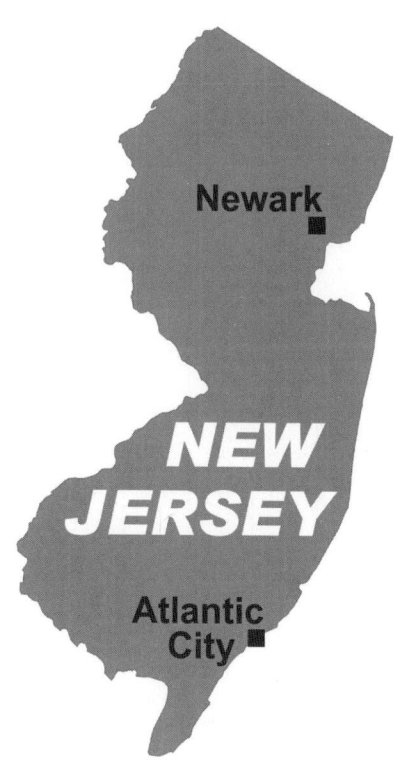

# NORMALS, MEANS, AND EXTREMES
## NEWARK (KEWR)

| LATITUDE: 40 ° 40'N | LONGITUDE: -74 ° 10'W | ELEVATION (FT): GRND: 7  BARO: 28 | TIME ZONE: EASTERN  (UTC -5) | WBAN: 14734 |
|---|---|---|---|---|

| ELEMENT | POR | JAN | FEB | MAR | APR | MAY | JUN | JUL | AUG | SEP | OCT | NOV | DEC | YEAR |
|---|---|---|---|---|---|---|---|---|---|---|---|---|---|---|
| **TEMPERATURE °F** | | | | | | | | | | | | | | |
| NORMAL DAILY MAXIMUM | 30 | 38.1 | 41.1 | 50.1 | 60.8 | 71.4 | 80.2 | 85.2 | 83.2 | 75.7 | 64.7 | 53.7 | 43.0 | 62.3 |
| MEAN DAILY MAXIMUM | 74 | 38.7 | 41.2 | 49.9 | 61.4 | 72.1 | 81.0 | 85.9 | 84.2 | 76.8 | 66.0 | 54.4 | 42.8 | 62.9 |
| HIGHEST DAILY MAXIMUM | 68 | 74 | 76 | 89 | 97 | 99 | 102 | 105 | 105 | 105 | 92 | 85 | 76 | 105 |
| YEAR OF OCCURRENCE | | 1950 | 1949 | 1945 | 2002 | 1996 | 1994 | 1993 | 2001 | 1953 | 1949 | 1950 | 1998 | AUG 2001 |
| MEAN OF EXTREME MAXS. | 74 | 58.7 | 60.1 | 71.7 | 83.3 | 89.8 | 95.2 | 97.4 | 95.1 | 90.7 | 82.2 | 72.1 | 61.9 | 79.9 |
| NORMAL DAILY MINIMUM | 30 | 24.4 | 26.6 | 34.2 | 43.7 | 54.1 | 63.5 | 69.1 | 67.7 | 59.9 | 48.2 | 39.1 | 29.8 | 46.7 |
| MEAN DAILY MINIMUM | 74 | 24.5 | 25.9 | 33.3 | 43.0 | 53.0 | 62.4 | 68.0 | 66.8 | 59.1 | 48.0 | 38.8 | 28.8 | 46.0 |
| LOWEST DAILY MINIMUM | 68 | -8 | -7 | 6 | 16 | 33 | 43 | 52 | 45 | 35 | 28 | 15 | -1 | -8 |
| YEAR OF OCCURRENCE | | 1985 | 1943 | 1943 | 1982 | 1947 | 1945 | 1952 | 1982 | 1947 | 1969 | 1955 | 1980 | JAN 1985 |
| MEAN OF EXTREME MINS. | 74 | 7.9 | 9.9 | 18.1 | 30.9 | 41.2 | 51.3 | 59.1 | 56.6 | 45.3 | 34.8 | 24.9 | 13.2 | 32.8 |
| NORMAL DRY BULB | 30 | 31.3 | 33.8 | 42.2 | 52.3 | 62.7 | 71.9 | 77.2 | 75.5 | 67.8 | 56.4 | 46.4 | 36.4 | 54.5 |
| MEAN DRY BULB | 74 | 31.6 | 33.6 | 41.6 | 52.2 | 62.5 | 71.8 | 76.9 | 75.5 | 68.0 | 57.0 | 46.6 | 35.9 | 54.4 |
| MEAN WET BULB | 26 | 27.6 | 28.9 | 35.0 | 44.2 | 54.0 | 63.4 | 67.7 | 67.1 | 60.8 | 50.3 | 41.2 | 31.8 | 47.7 |
| MEAN DEW POINT | 26 | 22.6 | 23.2 | 29.3 | 38.4 | 49.6 | 59.4 | 64.4 | 64.1 | 57.5 | 46.2 | 36.4 | 26.5 | 43.1 |
| NORMAL NO. DAYS WITH: | | | | | | | | | | | | | | |
| MAXIMUM >= 90 | 30 | 0.0 | 0.0 | 0.0 | 0.2 | 1.8 | 4.9 | 9.7 | 6.9 | 1.5 | 0.0 | 0.0 | 0.0 | 25.0 |
| MAXIMUM <= 32 | 30 | 8.8 | 5.4 | 0.7 | * | 0.0 | 0.0 | 0.0 | 0.0 | 0.0 | 0.0 | 0.1 | 3.9 | 18.9 |
| MINIMUM <= 32 | 30 | 23.4 | 20.1 | 11.9 | 1.4 | 0.0 | 0.0 | 0.0 | 0.0 | 0.0 | 0.4 | 6.2 | 18.3 | 81.7 |
| MINIMUM <= 0 | 30 | 0.5 | 0.2 | 0.0 | 0.0 | 0.0 | 0.0 | 0.0 | 0.0 | 0.0 | 0.0 | 0.0 | 0.1 | 0.8 |
| **H/C** | | | | | | | | | | | | | | |
| NORMAL HEATING DEG. DAYS | 30 | 1030 | 869 | 697 | 375 | 126 | 13 | 1 | 6 | 42 | 269 | 543 | 872 | 4843 |
| NORMAL COOLING DEG. DAYS | 30 | 0 | 0 | 2 | 10 | 70 | 236 | 394 | 347 | 142 | 18 | 1 | 0 | 1220 |
| **RH** | | | | | | | | | | | | | | |
| NORMAL (PERCENT) | 30 | 66 | 63 | 60 | 58 | 64 | 64 | 64 | 67 | 69 | 68 | 66 | 66 | 65 |
| HOUR 01 LST | 30 | 71 | 69 | 67 | 66 | 73 | 73 | 74 | 77 | 78 | 77 | 72 | 71 | 72 |
| HOUR 07 LST | 30 | 74 | 72 | 70 | 66 | 71 | 71 | 72 | 76 | 79 | 79 | 76 | 74 | 73 |
| HOUR 13 LST | 30 | 59 | 54 | 51 | 48 | 52 | 51 | 51 | 53 | 55 | 53 | 55 | 58 | 53 |
| HOUR 19 LST | 30 | 64 | 60 | 57 | 55 | 59 | 59 | 59 | 63 | 65 | 64 | 63 | 63 | 61 |
| **S** PERCENT POSSIBLE SUNSHINE | | | | | | | | | | | | | | |
| **W/O** MEAN NO. DAYS WITH: | | | | | | | | | | | | | | |
| HEAVY FOG(VISBY <= 1/4 MI) | 46 | 1.6 | 1.2 | 1.2 | 0.7 | 1.0 | 0.7 | 0.2 | 0.2 | 0.4 | 1.1 | 1.3 | 1.2 | 10.8 |
| THUNDERSTORMS | 74 | 0.3 | 0.3 | 1.0 | 1.7 | 3.7 | 5.1 | 6.0 | 4.6 | 2.1 | 1.0 | 0.5 | 0.3 | 26.6 |
| **CLOUDNESS** MEAN: | | | | | | | | | | | | | | |
| SUNRISE-SUNSET (OKTAS) | 50 | 5.2 | 5.0 | 5.0 | 5.1 | 5.2 | 5.0 | 4.9 | 4.7 | 4.5 | 4.3 | 5.0 | 5.1 | 4.9 |
| MIDNIGHT-MIDNIGHT (OKTAS) | 51 | 4.8 | 4.7 | 4.9 | 4.8 | 5.0 | 4.9 | 4.7 | 4.5 | 4.4 | 4.2 | 4.9 | 4.9 | 4.7 |
| MEAN NO. DAYS WITH: | | | | | | | | | | | | | | |
| CLEAR | 54 | 7.7 | 7.3 | 8.0 | 7.2 | 6.3 | 6.7 | 6.5 | 7.7 | 9.5 | 10.8 | 7.5 | 7.9 | 93.1 |
| PARTLY CLOUDY | 54 | 7.7 | 7.6 | 8.4 | 8.9 | 10.6 | 10.9 | 12.2 | 11.7 | 8.9 | 8.4 | 8.2 | 8.0 | 111.5 |
| CLOUDY | 54 | 15.7 | 13.4 | 14.6 | 14.0 | 14.1 | 12.4 | 12.3 | 11.6 | 11.6 | 11.8 | 14.3 | 15.1 | 160.9 |
| **PR** | | | | | | | | | | | | | | |
| MEAN STATION PRESSURE(IN) | 26 | 30.00 | 30.03 | 30.01 | 29.94 | 29.95 | 29.93 | 29.94 | 29.98 | 30.03 | 30.05 | 30.05 | 30.05 | 30.00 |
| MEAN SEA-LEVEL PRES. (IN) | 26 | 30.07 | 30.06 | 30.04 | 29.97 | 29.99 | 29.96 | 29.97 | 30.02 | 30.06 | 30.08 | 30.08 | 30.08 | 30.03 |
| **WINDS** | | | | | | | | | | | | | | |
| MEAN SPEED (MPH) | 26 | 10.8 | 11.1 | 11.3 | 10.7 | 9.8 | 9.5 | 8.8 | 8.6 | 9.0 | 9.4 | 10.2 | 10.6 | 10.0 |
| PREVAIL.DIR(TENS OF DEGS) | 39 | 26 | 32 | 33 | 33 | 24 | 25 | 25 | 24 | 03 | 03 | 24 | 26 | 26 |
| MAXIMUM 2-MINUTE: | | | | | | | | | | | | | | |
| SPEED (MPH) | 13 | 44 | 49 | 53 | 55 | 49 | 48 | 48 | 44 | 44 | 44 | 46 | 48 | 55 |
| DIR. (TENS OF DEGS) | | 28 | 27 | 26 | 29 | 28 | 35 | 32 | 35 | 03 | 27 | 28 | 27 | 29 |
| YEAR OF OCCURRENCE | | 1999 | 2009 | 2008 | 2002 | 2007 | 2005 | 2006 | 2002 | 2004 | 2006 | 2003 | 2000 | APR 2002 |
| MAXIMUM 3-SECOND | | | | | | | | | | | | | | |
| SPEED (MPH) | 13 | 54 | 60 | 64 | 76 | 56 | 62 | 64 | 53 | 55 | 55 | 58 | 62 | 76 |
| DIR. (TENS OF DEGS) | | 28 | 27 | 27 | 27 | 27 | 34 | 32 | 30 | 35 | 29 | 26 | 27 | 27 |
| YEAR OF OCCURRENCE | | 2000 | 2009 | 2008 | 2002 | 2007 | 2005 | 2006 | 1997 | 1998 | 2009 | 2003 | 2000 | APR 2002 |
| **PRECIPITATION** | | | | | | | | | | | | | | |
| NORMAL (IN) | 30 | 3.98 | 2.96 | 4.21 | 3.92 | 4.46 | 3.40 | 4.68 | 4.02 | 4.01 | 3.16 | 3.88 | 3.57 | 46.25 |
| MAXIMUM MONTHLY (IN) | 68 | 10.10 | 5.82 | 11.14 | 11.85 | 10.22 | 10.50 | 9.98 | 11.84 | 10.28 | 13.22 | 11.53 | 9.47 | 13.22 |
| YEAR OF OCCURRENCE | | 1979 | 2008 | 1983 | 1983 | 1984 | 2003 | 1988 | 1955 | 1944 | 2005 | 1977 | 1983 | OCT 2005 |
| MINIMUM MONTHLY (IN) | 68 | 0.45 | 0.52 | 0.79 | 0.90 | 0.52 | 0.07 | 0.89 | 0.36 | .45 | 0.21 | 0.51 | 0.27 | 0.07 |
| YEAR OF OCCURRENCE | | 1981 | 2002 | 2006 | 1963 | 1964 | 1949 | 1966 | 1995 | 2005 | 1963 | 1976 | 1955 | JUN 1949 |
| MAXIMUM IN 24 HOURS (IN) | 68 | 3.59 | 2.45 | 2.83 | 6.25 | 4.22 | 2.97 | 4.64 | 7.84 | 6.41 | 4.24 | 7.22 | 2.84 | 7.84 |
| YEAR OF OCCURRENCE | | 1979 | 1961 | 1991 | 2007 | 1979 | 1992 | 1997 | 1971 | 1999 | 2005 | 1977 | 2008 | AUG 1971 |
| NORMAL NO. DAYS WITH: | | | | | | | | | | | | | | |
| PRECIPITATION >= 0.01 | 30 | 10.5 | 9.9 | 10.9 | 10.8 | 11.7 | 10.7 | 10.0 | 9.6 | 9.0 | 8.3 | 9.5 | 10.7 | 121.6 |
| PRECIPITATION >= 1.00 | 30 | 1.0 | 0.6 | 0.9 | 1.0 | 1.1 | 1.0 | 1.2 | 1.1 | 1.0 | 0.9 | 1.0 | 1.0 | 11.8 |
| **SNOWFALL** | | | | | | | | | | | | | | |
| NORMAL (IN) | 30 | 8.9 | 8.4 | 4.3 | 0.8 | 0.* | 0.0 | 0.0 | 0.0 | 0.0 | 0.* | 0.6 | 2.9 | 25.9 |
| MAXIMUM MONTHLY (IN) | 68 | 31.6 | 33.4 | 26.0 | 13.8 | T | T | T | T | T | 0.3 | 5.7 | 29.1 | 33.4 |
| YEAR OF OCCURRENCE | | 1996 | 1994 | 1956 | 1982 | 1995 | 2008 | 2009 | 2008 | 2008 | 1952 | 1989 | 1947 | FEB 1994 |
| MAXIMUM IN 24 HOURS (IN) | 68 | 27.4 | 20.0 | 17.6 | 12.8 | T | T | T | T | T | 0.3 | 5.7 | 26.0 | 27.4 |
| YEAR OF OCCURRENCE' | | 1996 | 1961 | 1956 | 1982 | 1995 | 2001 | 2009 | 2002 | 1998 | 1952 | 1989 | 1947 | JAN 1996 |
| MAXIMUM SNOW DEPTH (IN) | 23 | 17 | 25 | 18 | 11 | 0 | 0 | 0 | 0 | 0 | 0 | 9 | 22 | 25 |
| YEAR OF OCCURRENCE | | 1978 | 1961 | 1956 | 1982 | | | | | | | 1938 | 1947 | FEB 1961 |
| NORMAL NO. DAYS WITH: | | | | | | | | | | | | | | |
| SNOWFALL >= 1.0 | 30 | 2.4 | 1.9 | 1.1 | 0.2 | 0.0 | 0.0 | 0.0 | 0.0 | 0.0 | 0.0 | 0.2 | 0.8 | 6.6 |

## PRECIPITATION (inches) 2009 NEWARK (KEWR)

| YEAR | JAN | FEB | MAR | APR | MAY | JUN | JUL | AUG | SEP | OCT | NOV | DEC | ANNUAL |
|------|-----|-----|-----|-----|-----|-----|-----|-----|-----|-----|-----|-----|--------|
| 1980 | 1.66 | 1.28 | 9.13 | 7.28 | 2.61 | 3.27 | 2.78 | 0.92 | 1.87 | 3.37 | 3.71 | 0.63 | 38.51 |
| 1981 | 0.45 | 4.81 | 1.10 | 3.15 | 3.88 | 2.61 | 4.51 | 0.57 | 3.42 | 3.47 | 1.75 | 5.32 | 35.04 |
| 1982 | 6.77 | 2.36 | 2.82 | 6.20 | 2.96 | 5.28 | 2.86 | 2.78 | 2.39 | 1.68 | 3.16 | 1.32 | 40.58 |
| 1983 | 4.37 | 3.03 | 11.14 | 11.14 | 4.22 | 2.81 | 1.59 | 3.46 | 2.93 | 5.80 | 5.54 | 9.47 | 65.50 |
| 1984 | 2.78 | 4.57 | 6.96 | 6.36 | 10.22 | 4.77 | 8.65 | 1.74 | 2.46 | 3.93 | 2.88 | 3.69 | 59.01 |
| 1985 | 1.22 | 2.58 | 1.59 | 1.17 | 4.23 | 4.29 | 4.52 | 2.58 | 4.19 | 1.29 | 8.32 | 1.31 | 37.29 |
| 1986 | 4.44 | 3.88 | 1.95 | 5.88 | 1.41 | 1.71 | 6.62 | 4.16 | 1.96 | 1.93 | 6.78 | 5.23 | 45.95 |
| 1987 | 6.21 | 1.30 | 3.81 | 5.06 | 2.55 | 4.13 | 4.66 | 5.26 | 3.87 | 3.37 | 2.94 | 2.37 | 45.53 |
| 1988 | 3.74 | 4.15 | 2.13 | 1.97 | 5.86 | 1.06 | 9.98 | 1.82 | 1.66 | 2.45 | 7.71 | 0.98 | 43.51 |
| 1989 | 1.98 | 2.70 | 4.42 | 3.25 | 8.80 | 5.41 | 5.23 | 7.03 | 6.45 | 5.40 | 2.57 | 0.75 | 53.99 |
| 1990 | 4.72 | 1.71 | 2.81 | 3.98 | 6.87 | 3.68 | 4.98 | 7.71 | 2.72 | 5.11 | 2.82 | 5.19 | 52.30 |
| 1991 | 3.72 | 1.81 | 5.49 | 3.91 | 4.80 | 2.95 | 5.21 | 5.63 | 3.24 | 1.29 | 2.04 | 3.67 | 43.76 |
| 1992 | 1.27 | 1.37 | 3.48 | 1.35 | 3.46 | 4.67 | 4.79 | 3.37 | 2.60 | 0.73 | 5.02 | 4.63 | 36.74 |
| 1993 | 2.75 | 2.87 | 7.22 | 4.59 | 1.77 | 1.21 | 2.15 | 2.84 | 6.29 | 3.98 | 1.95 | 4.89 | 42.51 |
| 1994 | 6.09 | 4.77 | 6.90 | 2.98 | 3.64 | 3.58 | 3.57 | 5.01 | 2.26 | 1.04 | 4.36 | 3.12 | 47.32 |
| 1995 | 3.29 | 3.36 | 1.30 | 2.24 | 3.27 | 1.64 | 5.98 | 0.36 | 3.64 | 4.77 | 5.79 | 2.03 | 37.67 |
| 1996 | 5.24 | 2.34 | 4.40 | 5.63 | 2.59 | 5.06 | 8.27 | 2.39 | 6.05 | 6.92 | 2.31 | 6.87 | 58.07 |
| 1997 | 3.50 | 2.18 | 5.19 | 3.08 | 3.12 | 2.42 | 7.05 | 2.89 | 2.20 | 2.02 | 4.54 | 4.16 | 42.35 |
| 1998 | 4.93 | 4.77 | 4.14 | 6.17 | 6.52 | 5.98 | 1.34 | 3.20 | 2.72 | 1.81 | 0.86 | 1.03 | 43.47 |
| 1999 | 6.87 | 3.10 | 3.63 | 1.90 | 4.19 | 0.41 | 1.01 | 5.51 | 9.38 | 2.90 | 2.90 | 2.95 | 44.75 |
| 2000 | 3.39 | 1.60 | 3.43 | 3.57 | 5.66 | 3.42 | 6.30 | 4.73 | 4.58 | 0.54 | 2.71 | 3.42 | 43.35 |
| 2001 | 2.57 | 1.79 | 6.69 | 1.71 | 2.88 | 3.97 | 2.29 | 1.97 | 4.29 | 0.46 | 0.81 | 2.01 | 31.44 |
| 2002 | 1.85 | 0.52 | 3.59 | 3.76 | 3.89 | 5.88 | 1.19 | 4.05 | 3.66 | 6.79 | 4.48 | 3.71 | 43.37 |
| 2003 | 2.94 | 3.90 | 3.98 | 2.42 | 3.45 | 10.50 | 2.59 | 8.21 | 5.57 | 3.72 | 3.94 | 5.11 | 56.33 |
| 2004 | 1.89 | 2.44 | 3.07 | 4.85 | 4.60 | 2.95 | 8.39 | 3.70 | 8.01 | 0.89 | 4.21 | 3.37 | 48.37 |
| 2005 | 3.93 | 2.81 | 4.16 | 3.42 | 1.21 | 2.99 | 4.05 | 0.51 | 0.45 | 13.22 | 3.74 | 3.65 | 44.14 |
| 2006 | 4.82 | 2.36 | 0.79 | 4.05 | 3.35 | 5.99 | 6.71 | 2.82 | 3.38 | 6.75 | 6.95 | 2.19 | 50.16 |
| 2007 | 3.50 | 1.43 | 3.93 | 11.85 | 1.87 | 5.24 | 6.71 | 7.32 | 1.81 | 3.70 | 2.35 | 4.78 | 54.49 |
| 2008 | 2.30 | 5.82 | 3.61 | 2.70 | 3.95 | 5.63 | 3.14 | 2.80 | 7.14 | 2.79 | 3.07 | 5.88 | 48.83 |
| 2009 | 2.86 | 0.58 | 1.61 | 4.61 | 4.08 | 7.96 | 6.60 | 4.14 | 1.73 | 5.43 | 1.20 | 7.13 | 47.93 |
| POR= 74 YRS | 3.45 | 2.90 | 3.97 | 3.80 | 3.85 | 3.59 | 4.31 | 4.09 | 3.76 | 3.26 | 3.63 | 3.53 | 44.14 |

WBAN : 14734

## AVERAGE TEMPERATURE (°F) 2009 NEWARK (KEWR)

| YEAR | JAN | FEB | MAR | APR | MAY | JUN | JUL | AUG | SEP | OCT | NOV | DEC | ANNUAL |
|------|-----|-----|-----|-----|-----|-----|-----|-----|-----|-----|-----|-----|--------|
| 1980 | 34.0 | 30.8 | 38.9 | 52.6 | 65.9 | 70.2 | 78.9 | 78.6 | 70.8 | 55.0 | 42.9 | 30.4 | 54.1 |
| 1981 | 24.1 | 37.6 | 40.2 | 55.3 | 64.0 | 74.6 | 79.3 | 75.1 | 67.2 | 53.1 | 46.0 | 34.6 | 54.3 |
| 1982 | 24.2 | 36.2 | 41.8 | 50.6 | 63.2 | 67.9 | 78.4 | 72.5 | 66.7 | 56.9 | 48.8 | 42.8 | 54.2 |
| 1983 | 35.0 | 35.9 | 44.7 | 52.2 | 60.8 | 73.5 | 79.6 | 77.6 | 70.6 | 57.8 | 47.8 | 34.2 | 55.8 |
| 1984 | 27.8 | 40.8 | 36.5 | 52.7 | 62.2 | 75.0 | 76.6 | 77.3 | 65.4 | 62.3 | 45.3 | 40.8 | 55.2 |
| 1985 | 24.9 | 33.5 | 44.5 | 57.0 | 67.1 | 69.4 | 76.3 | 75.6 | 70.2 | 58.5 | 49.5 | 33.3 | 55.0 |
| 1986 | 33.0 | 31.1 | 44.2 | 53.4 | 66.7 | 72.7 | 76.9 | 74.2 | 68.6 | 58.0 | 45.0 | 38.1 | 55.2 |
| 1987 | 31.5 | 33.0 | 45.0 | 53.9 | 63.9 | 74.5 | 79.4 | 75.3 | 68.7 | 53.7 | 47.6 | 38.4 | 55.4 |
| 1988 | 28.7 | 34.4 | 43.9 | 51.1 | 63.4 | 73.0 | 80.5 | 79.8 | 68.0 | 52.6 | 48.9 | 35.5 | 55.0 |
| 1989 | 37.0 | 34.2 | 42.4 | 52.5 | 63.2 | 74.3 | 77.2 | 76.3 | 69.9 | 59.1 | 45.0 | 25.6 | 54.7 |
| 1990 | 40.4 | 39.8 | 44.9 | 53.3 | 61.1 | 73.4 | 77.8 | 76.6 | 68.6 | 62.4 | 50.0 | 42.3 | 57.6 |
| 1991 | 33.6 | 38.6 | 44.4 | 54.8 | 68.9 | 74.2 | 77.9 | 77.7 | 68.0 | 58.3 | 47.6 | 38.8 | 56.9 |
| 1992 | 35.2 | 36.0 | 39.3 | 50.2 | 61.7 | 72.6 | 76.9 | 75.1 | 69.6 | 55.7 | 47.8 | 38.8 | 54.9 |
| 1993 | 37.6 | 31.0 | 40.2 | 54.3 | 67.0 | 75.8 | 82.6 | 79.2 | 69.2 | 56.4 | 47.8 | 37.2 | 56.5 |
| 1994 | 25.4 | 30.4 | 41.6 | 57.4 | 63.7 | 77.8 | 81.9 | 75.7 | 69.7 | 58.7 | 52.0 | 41.4 | 56.3 |
| 1995 | 37.5 | 30.8 | 45.5 | 52.6 | 62.7 | 73.1 | 79.6 | 78.5 | 68.6 | 61.0 | 42.9 | 31.7 | 55.4 |
| 1996 | 29.7 | 33.6 | 38.8 | 53.1 | 61.6 | 72.9 | 73.9 | 74.0 | 68.0 | 55.5 | 41.9 | 40.2 | 53.6 |
| 1997 | 31.1 | 39.4 | 41.8 | 50.9 | 59.2 | 70.9 | 76.8 | 73.6 | 66.9 | 56.5 | 43.7 | 37.6 | 54.0 |
| 1998 | 40.1 | 40.8 | 45.2 | 53.9 | 64.9 | 70.1 | 77.6 | 77.0 | 70.4 | 57.6 | 47.6 | 41.9 | 57.3 |
| 1999 | 33.5 | 37.8 | 42.9 | 53.4 | 63.3 | 74.2 | 80.9 | 76.2 | 69.3 | 55.5 | 50.1 | 39.4 | 56.4 |
| 2000 | 31.5 | 37.4 | 47.8 | 51.4 | 64.2 | 72.4 | 73.7 | 73.3 | 66.4 | 57.0 | 45.2 | 30.7 | 54.3 |
| 2001 | 32.2 | 35.7 | 40.0 | 53.4 | 64.0 | 73.9 | 74.1 | 79.1 | 67.4 | 57.8 | 51.9 | 43.6 | 56.1 |
| 2002 | 39.4 | 40.3 | 44.1 | 56.0 | 60.9 | 72.4 | 80.1 | 77.9 | 70.5 | 55.6 | 45.4 | 35.2 | 56.5 |
| 2003 | 27.5 | 29.4 | 43.1 | 49.9 | 59.0 | 69.3 | 77.3 | 77.6 | 68.6 | 55.1 | 49.7 | 36.6 | 53.6 |
| 2004 | 24.2 | 34.7 | 43.8 | 54.0 | 66.3 | 72.2 | 75.0 | 74.6 | 69.6 | 55.6 | 47.5 | 36.5 | 54.5 |
| 2005 | 30.0 | 35.6 | 38.5 | 54.4 | 59.1 | 74.7 | 78.3 | 80.4 | 73.5 | 57.8 | 48.7 | 33.8 | 55.4 |
| 2006 | 39.7 | 35.2 | 43.2 | 55.7 | 63.8 | 72.6 | 79.5 | 77.3 | 66.7 | 55.8 | 51.3 | 43.0 | 57.0 |
| 2007 | 37.1 | 27.9 | 42.0 | 50.1 | 65.1 | 72.8 | 75.9 | 75.3 | 70.1 | 63.5 | 44.9 | 36.5 | 55.1 |
| 2008 | 35.7 | 35.7 | 43.0 | 54.7 | 60.5 | 75.3 | 78.7 | 74.1 | 69.5 | 55.4 | 45.4 | 37.4 | 55.5 |
| 2009 | 27.9 | 36.7 | 42.2 | 54.5 | 63.3 | 68.9 | 74.3 | 77.3 | 66.9 | 55.6 | 50.8 | 35.6 | 54.5 |
| POR= 74 YRS | 31.6 | 33.6 | 41.6 | 52.2 | 62.5 | 71.8 | 76.9 | 75.5 | 68.0 | 57.0 | 46.6 | 35.9 | 54.4 |

## HEATING DEGREE DAYS (base 65°F) 2009 NEWARK (KEWR)

| YEAR | JUL | AUG | SEP | OCT | NOV | DEC | JAN | FEB | MAR | APR | MAY | JUN | TOTAL |
|---|---|---|---|---|---|---|---|---|---|---|---|---|---|
| 1980-81 | 0 | 0 | 28 | 314 | 654 | 1066 | 1261 | 762 | 764 | 290 | 96 | 0 | 5235 |
| 1981-82 | 0 | 0 | 52 | 360 | 563 | 934 | 1258 | 802 | 712 | 433 | 85 | 42 | 5241 |
| 1982-83 | 0 | 13 | 36 | 267 | 493 | 679 | 923 | 810 | 622 | 395 | 162 | 5 | 4405 |
| 1983-84 | 0 | 0 | 52 | 249 | 510 | 949 | 1144 | 696 | 874 | 366 | 128 | 9 | 4977 |
| 1984-85 | 0 | 0 | 83 | 114 | 584 | 745 | 1235 | 877 | 641 | 268 | 62 | 15 | 4624 |
| 1985-86 | 0 | 0 | 21 | 212 | 462 | 971 | 985 | 942 | 642 | 341 | 89 | 7 | 4672 |
| 1986-87 | 0 | 11 | 22 | 240 | 594 | 826 | 1030 | 893 | 616 | 331 | 140 | 3 | 4706 |
| 1987-88 | 0 | 1 | 25 | 342 | 518 | 818 | 1117 | 880 | 647 | 410 | 120 | 28 | 4906 |
| 1988-89 | 1 | 0 | 18 | 386 | 476 | 906 | 859 | 853 | 698 | 366 | 132 | 6 | 4701 |
| 1989-90 | 0 | 0 | 37 | 190 | 594 | 1215 | 756 | 699 | 622 | 369 | 122 | 2 | 4606 |
| 1990-91 | 1 | 1 | 50 | 163 | 446 | 697 | 967 | 734 | 630 | 330 | 63 | 4 | 4086 |
| 1991-92 | 0 | 0 | 55 | 227 | 513 | 804 | 917 | 834 | 790 | 441 | 148 | 4 | 4733 |
| 1992-93 | 0 | 0 | 38 | 295 | 510 | 807 | 842 | 946 | 765 | 318 | 42 | 4 | 4567 |
| 1993-94 | 0 | 0 | 48 | 263 | 513 | 853 | 1219 | 964 | 718 | 242 | 104 | 0 | 4924 |
| 1994-95 | 0 | 0 | 7 | 195 | 387 | 724 | 848 | 952 | 596 | 371 | 112 | 0 | 4192 |
| 1995-96 | 0 | 0 | 32 | 163 | 657 | 1026 | 1091 | 906 | 809 | 369 | 176 | 7 | 5236 |
| 1996-97 | 0 | 0 | 46 | 291 | 685 | 763 | 1043 | 714 | 713 | 418 | 179 | 43 | 4895 |
| 1997-98 | 1 | 0 | 51 | 294 | 635 | 842 | 765 | 672 | 633 | 327 | 94 | 22 | 4336 |
| 1998-99 | 0 | 0 | 19 | 227 | 516 | 711 | 970 | 754 | 674 | 345 | 92 | 2 | 4310 |
| 1999-00 | 0 | 2 | 22 | 290 | 439 | 786 | 1031 | 793 | 521 | 399 | 106 | 22 | 4411 |
| 2000-01 | 0 | 0 | 80 | 251 | 586 | 1056 | 1010 | 813 | 768 | 353 | 106 | 7 | 5030 |
| 2001-02 | 0 | 0 | 53 | 241 | 389 | 658 | 788 | 686 | 641 | 330 | 169 | 16 | 3971 |
| 2002-03 | 0 | 2 | 8 | 325 | 584 | 919 | 1155 | 989 | 672 | 456 | 188 | 37 | 5335 |
| 2003-04 | 0 | 0 | 14 | 306 | 456 | 875 | 1254 | 870 | 650 | 332 | 66 | 10 | 4833 |
| 2004-05 | 0 | 0 | 14 | 286 | 521 | 874 | 1078 | 817 | 813 | 322 | 194 | 13 | 4932 |
| 2005-06 | 0 | 0 | 9 | 245 | 481 | 959 | 779 | 828 | 667 | 281 | 98 | 11 | 4358 |
| 2006-07 | 0 | 0 | 38 | 293 | 403 | 676 | 858 | 1031 | 705 | 449 | 95 | 5 | 4553 |
| 2007-08 | 0 | 8 | 21 | 128 | 598 | 879 | 901 | 842 | 679 | 310 | 165 | 0 | 4531 |
| 2008-09 | 0 | 0 | 21 | 303 | 578 | 847 | 1142 | 785 | 700 | 352 | 101 | 20 | 4849 |
| 2009- | 0 | 0 | 37 | 292 | 421 | 905 | | | | | | | |

WBAN : 14734

## COOLING DEGREE DAYS (base 65°F) 2009 NEWARK (KEWR)

| YEAR | JAN | FEB | MAR | APR | MAY | JUN | JUL | AUG | SEP | OCT | NOV | DEC | TOTAL |
|---|---|---|---|---|---|---|---|---|---|---|---|---|---|
| 1980 | 0 | 0 | 0 | 0 | 97 | 187 | 435 | 427 | 209 | 10 | 0 | 0 | 1365 |
| 1981 | 0 | 0 | 0 | 6 | 75 | 293 | 446 | 319 | 124 | 0 | 0 | 0 | 1263 |
| 1982 | 0 | 0 | 0 | 6 | 39 | 136 | 421 | 249 | 95 | 24 | 12 | 0 | 982 |
| 1983 | 0 | 0 | 0 | 19 | 39 | 268 | 458 | 396 | 226 | 36 | 0 | 0 | 1442 |
| 1984 | 0 | 0 | 0 | 2 | 47 | 316 | 365 | 388 | 102 | 36 | 0 | 0 | 1256 |
| 1985 | 0 | 0 | 11 | 36 | 134 | 152 | 357 | 335 | 183 | 19 | 3 | 0 | 1230 |
| 1986 | 0 | 0 | 2 | 2 | 149 | 243 | 380 | 303 | 136 | 30 | 0 | 0 | 1245 |
| 1987 | 0 | 0 | 0 | 6 | 116 | 293 | 453 | 327 | 143 | 0 | 1 | 0 | 1339 |
| 1988 | 0 | 0 | 0 | 0 | 75 | 274 | 488 | 465 | 115 | 10 | 0 | 0 | 1427 |
| 1989 | 0 | 0 | 3 | 1 | 81 | 294 | 385 | 360 | 194 | 16 | 0 | 0 | 1334 |
| 1990 | 0 | 0 | 7 | 23 | 11 | 262 | 403 | 365 | 165 | 89 | 2 | 0 | 1327 |
| 1991 | 0 | 0 | 0 | 28 | 190 | 288 | 406 | 399 | 151 | 28 | 0 | 0 | 1490 |
| 1992 | 0 | 0 | 0 | 4 | 52 | 242 | 373 | 323 | 185 | 15 | 0 | 0 | 1194 |
| 1993 | 0 | 0 | 0 | 5 | 113 | 340 | 553 | 450 | 182 | 9 | 4 | 0 | 1656 |
| 1994 | 0 | 0 | 0 | 23 | 74 | 389 | 530 | 338 | 155 | 5 | 4 | 0 | 1518 |
| 1995 | 0 | 0 | 0 | 3 | 49 | 247 | 460 | 425 | 147 | 44 | 0 | 0 | 1375 |
| 1996 | 0 | 0 | 0 | 21 | 77 | 252 | 282 | 288 | 144 | 2 | 2 | 0 | 1068 |
| 1997 | 0 | 0 | 0 | 0 | 6 | 229 | 375 | 278 | 115 | 37 | 0 | 0 | 1040 |
| 1998 | 0 | 0 | 29 | 0 | 95 | 182 | 398 | 377 | 188 | 4 | 0 | 0 | 1273 |
| 1999 | 0 | 0 | 0 | 2 | 47 | 283 | 499 | 360 | 158 | 3 | 0 | 0 | 1352 |
| 2000 | 0 | 0 | 0 | 0 | 89 | 251 | 278 | 265 | 129 | 11 | 0 | 0 | 1023 |
| 2001 | 0 | 0 | 0 | 11 | 84 | 282 | 291 | 442 | 131 | 27 | 1 | 0 | 1269 |
| 2002 | 0 | 0 | 0 | 68 | 49 | 247 | 472 | 408 | 180 | 38 | 0 | 0 | 1462 |
| 2003 | 0 | 0 | 0 | 9 | 7 | 173 | 388 | 398 | 129 | 4 | 4 | 0 | 1112 |
| 2004 | 0 | 0 | 0 | 8 | 115 | 229 | 316 | 303 | 160 | 1 | 0 | 0 | 1132 |
| 2005 | 0 | 0 | 0 | 10 | 20 | 305 | 420 | 485 | 269 | 26 | 0 | 0 | 1535 |
| 2006 | 0 | 0 | 0 | 11 | 67 | 245 | 456 | 388 | 96 | 16 | 0 | 0 | 1279 |
| 2007 | 0 | 0 | 0 | 8 | 109 | 243 | 348 | 333 | 180 | 89 | 0 | 0 | 1310 |
| 2008 | 0 | 0 | 0 | 8 | 32 | 316 | 434 | 288 | 165 | 12 | 0 | 0 | 1255 |
| 2009 | 0 | 0 | 0 | 44 | 56 | 142 | 293 | 386 | 100 | 6 | 0 | 0 | 1027 |

## SNOWFALL (inches) 2009 NEWARK (KEWR)

| YEAR | JUL | AUG | SEP | OCT | NOV | DEC | JAN | FEB | MAR | APR | MAY | JUN | TOTAL |
|------|-----|-----|-----|-----|-----|-----|-----|-----|-----|-----|-----|-----|-------|
| 1980-81 | 0.0 | 0.0 | 0.0 | 0.0 | 0.4 | 3.1 | 6.9 | T | 9.1 | 0.0 | 0.0 | 0.0 | 19.5 |
| 1981-82 | 0.0 | 0.0 | 0.0 | 0.0 | T | 3.4 | 12.3 | 0.5 | 0.8 | 13.8 | 0.0 | 0.0 | 30.8 |
| 1982-83 | 0.0 | 0.0 | 0.0 | 0.0 | T | 2.9 | 2.3 | 21.5 | 0.2 | 4.1 | 0.0 | 0.0 | 31.0 |
| 1983-84 | 0.0 | 0.0 | 0.0 | 0.0 | 1.2 | 2.4 | 13.7 | 0.3 | 11.3 | T | 0.0 | 0.0 | 28.9 |
| 1984-85 | 0.0 | 0.0 | 0.0 | 0.0 | T | 6.8 | 8.9 | 7.4 | 0.1 | T | 0.0 | 0.0 | 23.2 |
| 1985-86 | 0.0 | 0.0 | 0.0 | 0.0 | 0.6 | 4.6 | 2.8 | 13.9 | T | 0.1 | 0.0 | 0.0 | 22.0 |
| 1986-87 | 0.0 | 0.0 | 0.0 | 0.0 | T | 2.3 | 21.4 | 6.5 | 2.4 | 0.0 | 0.0 | 0.0 | 32.6 |
| 1987-88 | 0.0 | 0.0 | 0.0 | 0.0 | 1.5 | 2.3 | 15.4 | 2.7 | 0.9 | T | 0.0 | | |
| 1988-89 | 0.0 | 0.0 | 0.0 | 0.0 | 0.0 | 0.1 | 4.1 | 0.6 | 2.7 | 0.0 | 0.0 | 0.0 | 7.5 |
| 1989-90 | 0.0 | 0.0 | 0.0 | 0.0 | 5.7 | 0.5 | 2.4 | 2.8 | 2.5 | 0.6 | 0.0 | 0.0 | 14.5 |
| 1990-91 | 0.0 | 0.0 | 0.0 | 0.0 | T | 7.6 | 8.5 | 5.2 | 0.2 | 0.0 | 0.0 | 0.0 | 21.5 |
| 1991-92 | 0.0 | 0.0 | 0.0 | 0.0 | T | 0.5 | 1.0 | 1.0 | 11.4 | T | 0.0 | 0.0 | 13.9 |
| 1992-93 | 0.0 | 0.0 | 0.0 | 0.0 | T | 0.5 | 10.7 | 16.8 | 0.0 | 0.0 | 0.0 | 0.0 | 28.8 |
| 1993-94 | 0.0 | 0.0 | 0.0 | 0.0 | T | 3.9 | 18.5 | 33.4 | 8.7 | 0.0 | T | 0.0 | 64.5 |
| 1994-95 | 0.0 | 0.0 | 0.0 | 0.0 | T | T | 0.1 | 10.2 | T | 0.0 | T | 0.0 | 10.3 |
| 1995-96 | 0.0 | 0.0 | 0.0 | 0.0 | 3.0 | 12.8 | 31.6 | 18.4 | 11.9 | 0.7 | 0.0 | 0.0 | 78.4 |
| 1996-97 | | | | | T | T | 3.4 | 4.4 | 7.1 | 1.4 | 0.0 | 0.0 | |
| 1997-98 | 0.0 | 0.0 | 0.0 | 0.0 | 0.2 | 1.4 | 2.2 | T | 3.1 | T | 0.0 | 0.0 | 6.9 |
| 1998-99 | 0.0 | 0.0 | T | 0.0 | 0.0 | 1.2 | 4.1 | 2.0 | 5.5 | 0.0 | 0.0 | 0.0 | 12.8 |
| 1999-00 | 0.0 | 0.0 | 0.0 | 0.0 | T | T | 12.2 | 5.3 | T | 0.9 | 0.0 | 0.0 | 18.4 |
| 2000-01 | 0.0 | 0.0 | 0.0 | 0.0 | 0.0 | 14.9 | 6.1 | 11.1 | 7.2 | 0.0 | T | T | 39.3 |
| 2001-02 | 0.0 | 0.0 | 0.0 | 0.0 | 0.0 | 0.0 | 3.6 | T | T | T | 0.0 | T | 3.6 |
| 2002-03 | 0.0 | T | 0.0 | T | 0.6 | 10.2 | 4.7 | 29.9 | 3.3 | 4.4 | 0.0 | 0.0 | 53.1 |
| 2003-04 | 0.0 | 0.0 | 0.0 | 0.0 | T | 21.0 | 16.9 | 0.4 | 9.5 | 0.0 | T | 0.0 | 47.8 |
| 2004-05 | 0.0 | 0.0 | 0.0 | 0.0 | T | 1.6 | 15.4 | 18.6 | 7.8 | 0.0 | 0.0 | 0.0 | 43.4 |
| 2005-06 | 0.0 | 0.0 | 0.0 | 0.0 | T | 11.0 | 2.9 | 21.5 | 1.2 | 1.3 | 0.0 | 0.0 | 37.9 |
| 2006-07 | T | 0.0 | 0.0 | 0.0 | 0.0 | T | 3.9 | 5.5 | 7.1 | T | 0.0 | T | 16.5 |
| 2007-08 | 0.0 | 0.0 | 0.0 | 0.0 | 0.4 | 3.9 | T | 10.3 | 0.0 | 0.0 | 0.0 | T | 14.6 |
| 2008-09 | T | T | T | T | T | 8.3 | 8.9 | 2.9 | 7.0 | T | 0.0 | 0.0 | 27.1 |
| 2009- | T | 0.0 | 0.0 | 0.0 | 0.0 | 13.3 | | | | | | | |
| POR= 74 YRS | T | T | T | T | 0.6 | 5.2 | 7.5 | 8.2 | 4.9 | 0.7 | T | T | 27.1 |

# Newark, NJ 2009
## Temperature and Precipitation

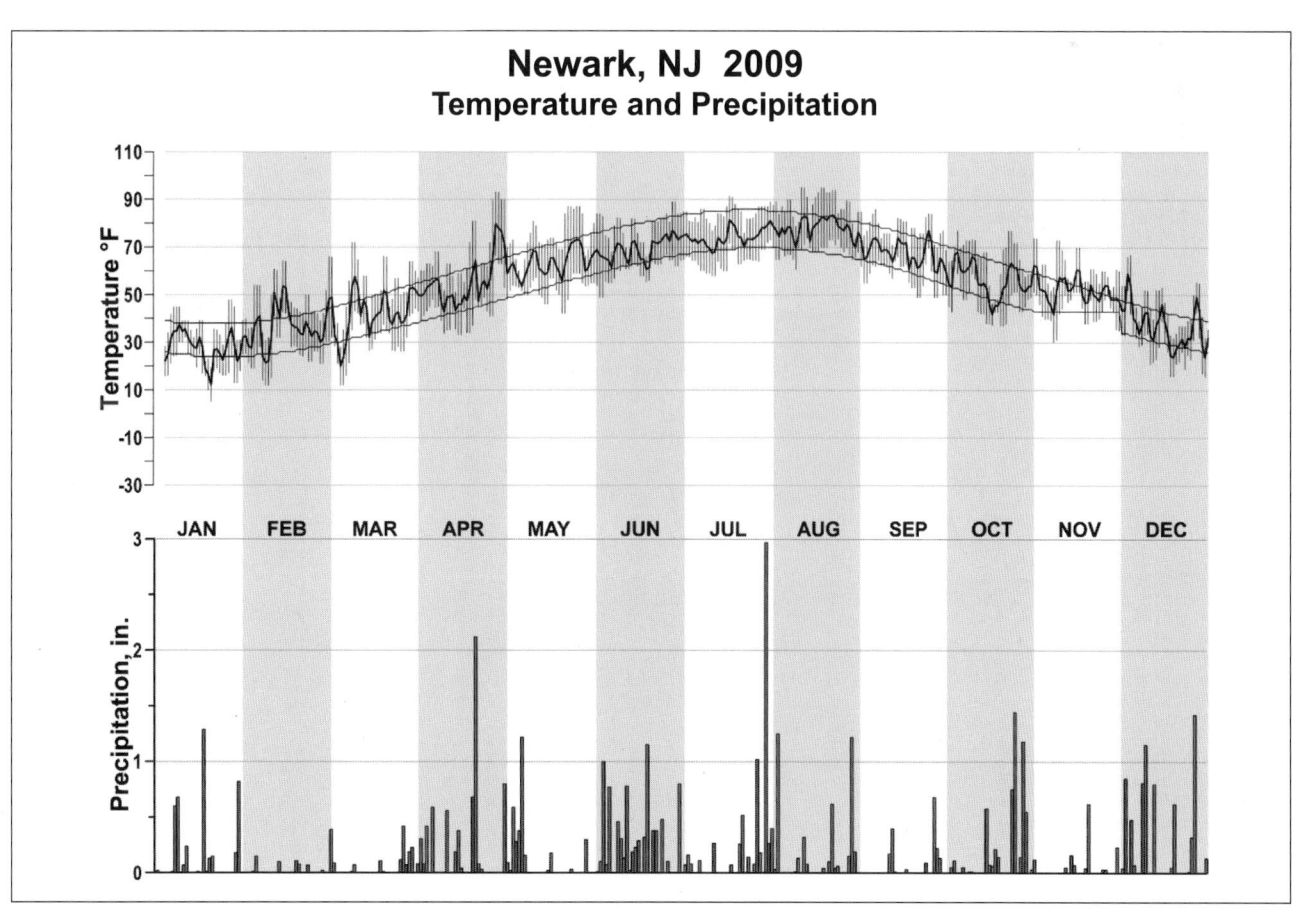

# 2009
# ALBUQUERQUE
# NEW MEXICO (KABQ)

The Albuquerque metropolitan area is largely situated in the Rio Grande Valley and on the mesas and piedmont slopes which rise either side of the valley floor. The Rio Grande flows from north to south through the area. The Sandia and Manzano Mountains rise abruptly at the eastern edge of the city with Tijeras Canyon separating the two ranges. West of the city the land gradually rises to the Continental Divide, some 90 miles away.

The climate of Albuquerque is best described as arid continental with abundant sunshine, low humidity, scant precipitation, and a wide yet tolerable seasonal range of temperatures. Sunny days and low humidity are renowned features of the climate. More than three-fourths of the daylight hours have sunshine, even in the winter months. The air is normally dry and muggy days are rare. The combination of dry air and plentiful solar radiation allows widespread use of energy-efficient devices such as evaporative coolers and solar collectors.

Precipitation within the valley area is adequate only for native desert vegetation and deep-rooted imports. However, irrigation supports successful farming and fruit growing in the Rio Grande Valley. On the east slopes of the Sandias and Manzanos, precipitation is sufficient for thick stands of timber and good grass cover.

Meager amounts of precipitation fall in the winter, much of it as snow. Snowfalls of an inch or more occur about four times a year in the Rio Grande Valley, while the mountains receive substantial snowfall on occasion. Snow seldom remains on the ground more than 24 hours in the city proper. However, snow cover on the east slopes of the Sandias is sufficient for skiing during most winters.

Nearly half of the annual precipitation in Albuquerque results from afternoon and evening thunderstorms during the summer. Thunderstorm frequency increases rapidly around July 1st, peaks during August, then tapers off by the end of September. Thunderstorms are usually brief, sometimes produce heavy rainfall, and often lower afternoon temperatures noticeably. Hailstorms are infrequent and tornadoes rare.

Temperatures in Albuquerque are those characteristic of a dry, high altitude, continental climate. The average daily range of temperature is relatively high, but extreme temperatures are rare. High temperatures during the winter are near 50 degrees with only a few days on which the temperature fails to rise above the freezing mark. In the summer, daytime maxima are about 90 degrees, but with the large daily range, the nights usually are comfortably cool.

The average number of days between the last freezing temperature in spring and the first freeze in fall varies widely across the Albuquerque metropolitan area. The growing season in Albuquerque and adjacent suburbs ranges from around 170 days in the Rio Grande Valley to about 200 days in parts of the northeast section of the city.

Sustained winds of 12 mph or less occur approximately 80 percent of the time at the Albuquerque International Airport, while sustained winds greater than 25 mph have a frequency less than 3 percent. Late winter and spring storms along with occasional east winds out of Tijeras Canyon are the main sources of strong wind conditions. Blowing dust, the least attractive feature of the climate, often accompanies the occasional strong winds of winter and spring.

# NORMALS, MEANS, AND EXTREMES
## ALBUQUERQUE (KABQ)

| LATITUDE:<br>35 ° 02'N | LONGITUDE:<br>-106° 36'W | | | ELEVATION (FT):<br>GRND: 5310　BARO: 5308 | | | | | TIME ZONE:<br>MOUNTAIN　(UTC -7) | | | WBAN: 23050 | |

| | ELEMENT | POR | JAN | FEB | MAR | APR | MAY | JUN | JUL | AUG | SEP | OCT | NOV | DEC | YEAR |
|---|---|---|---|---|---|---|---|---|---|---|---|---|---|---|---|
| **TEMPERATURE °F** | NORMAL DAILY MAXIMUM | 30 | 47.6 | 54.6 | 62.4 | 70.6 | 79.7 | 90.2 | 92.3 | 89.0 | 82.2 | 70.7 | 57.1 | 47.9 | 70.4 |
| | MEAN DAILY MAXIMUM | 102 | 47.0 | 51.5 | 60.4 | 68.4 | 78.2 | 87.3 | 91.5 | 88.6 | 79.9 | 68.5 | 55.8 | 46.5 | 68.6 |
| | HIGHEST DAILY MAXIMUM | 70 | 69 | 76 | 85 | 89 | 98 | 107 | 105 | 101 | 100 | 91 | 77 | 72 | 107 |
| | YEAR OF OCCURRENCE | | 1994 | 1986 | 1971 | 1989 | 1951 | 1994 | 1980 | 1979 | 1979 | 1979 | 1975 | 1958 | JUN 1994 |
| | MEAN OF EXTREME MAXS. | 113 | 60.8 | 67.5 | 75.5 | 83.1 | 91.2 | 99.0 | 99.7 | 96.2 | 92.1 | 83.2 | 70.6 | 60.8 | 81.6 |
| | NORMAL DAILY MINIMUM | 30 | 23.8 | 28.2 | 33.7 | 40.5 | 49.7 | 59.4 | 64.7 | 63.2 | 56.0 | 43.8 | 31.6 | 24.2 | 43.2 |
| | MEAN DAILY MINIMUM | 102 | 23.1 | 26.9 | 31.8 | 39.5 | 49.2 | 57.1 | 64.2 | 62.5 | 53.7 | 42.6 | 30.5 | 23.3 | 42.0 |
| | LOWEST DAILY MINIMUM | 70 | -17 | -5 | 8 | 19 | 28 | 40 | 52 | 50 | 37 | 21 | -7 | -7 | -17 |
| | YEAR OF OCCURRENCE | | 1971 | 1951 | 1948 | 1980 | 1975 | 1980 | 1985 | 1992 | 1971 | 1991 | 1976 | 1990 | JAN 1971 |
| | MEAN OF EXTREME MINS. | 113 | 10.3 | 14.5 | 20.2 | 28.6 | 37.8 | 48.8 | 59.0 | 57.0 | 45.7 | 31.8 | 19.3 | 11.5 | 32.0 |
| | NORMAL DRY BULB | 30 | 35.7 | 41.4 | 48.1 | 55.6 | 64.7 | 74.8 | 78.5 | 76.1 | 69.1 | 57.3 | 44.4 | 36.1 | 56.8 |
| | MEAN DRY BULB | 102 | 35.1 | 39.2 | 46.1 | 54.0 | 63.7 | 72.3 | 77.9 | 75.6 | 66.8 | 55.6 | 43.2 | 34.9 | 55.4 |
| | MEAN WET BULB | 26 | 28.0 | 30.6 | 34.2 | 38.3 | 45.9 | 52.0 | 58.9 | 59.1 | 52.8 | 43.5 | 33.8 | 28.2 | 42.1 |
| | MEAN DEW POINT | 26 | 21.2 | 22.7 | 23.7 | 26.5 | 33.1 | 40.5 | 50.3 | 52.7 | 44.9 | 35.4 | 26.0 | 20.9 | 33.2 |
| | NORMAL NO. DAYS WITH: | | | | | | | | | | | | | | |
| | MAXIMUM >= 90 | 3 0 | 0.0 | 0.0 | 0.0 | 0.0 | 2.4 | 17.3 | 22.3 | 15.9 | 4.1 | 0.1 | 0.0 | 0.0 | 62.1 |
| | MAXIMUM <= 32 | 3 0 | 1.9 | 0.7 | 0.0 | 0.0 | 0.0 | 0.0 | 0.0 | 0.0 | 0.0 | 0.0 | 0.2 | 1.6 | 4.4 |
| | MINIMUM <= 32 | 30 | 27.9 | 20.9 | 12.8 | 4.0 | 0.2 | 0.0 | 0.0 | 0.0 | 0.0 | 2.2 | 15.7 | 27.7 | 111.4 |
| | MINIMUM <= 0 | 30 | 0.2 | 0.0 | 0.0 | 0.0 | 0.0 | 0.0 | 0.0 | 0.0 | 0.0 | 0.0 | 0.1 | 0.1 | 0.4 |
| **H/C** | NORMAL HEATING DEG. DAYS | 30 | 914 | 670 | 525 | 294 | 85 | 4 | 0 | 0 | 29 | 248 | 614 | 898 | 4281 |
| | NORMAL COOLING DEG. DAYS | 30 | 0 | 0 | 0 | 6 | 70 | 297 | 417 | 343 | 148 | 9 | 0 | 0 | 1290 |
| **RH** | NORMAL (PERCENT) | 30 | 57 | 50 | 41 | 34 | 33 | 30 | 41 | 48 | 46 | 46 | 50 | 56 | 44 |
| | HOUR 05 LST | 30 | 71 | 65 | 57 | 50 | 49 | 46 | 59 | 67 | 63 | 63 | 65 | 70 | 60 |
| | HOUR 11 LST | 30 | 51 | 43 | 34 | 27 | 26 | 24 | 33 | 40 | 39 | 39 | 42 | 50 | 37 |
| | HOUR 17 LST | 30 | 42 | 32 | 25 | 20 | 20 | 18 | 27 | 31 | 30 | 31 | 36 | 43 | 30 |
| | HOUR 23 LST | 30 | 61 | 52 | 44 | 36 | 35 | 33 | 46 | 54 | 51 | 50 | 54 | 61 | 48 |
| **S** | PERCENT POSSIBLE SUNSHINE | 63 | 72 | 72 | 73 | 77 | 79 | 83 | 76 | 76 | 79 | 79 | 76 | 71 | 76 |
| **W/O** | MEAN NO. DAYS WITH: | | | | | | | | | | | | | | |
| | HEAVY FOG(VISBY <= 1/4 MI) | 46 | 1.4 | 0.9 | 0.6 | 0.3 | 0.1 | 0.0 | 0.1 | 0.0 | 0.2 | 0.4 | 0.5 | 1.6 | 6.1 |
| | THUNDERSTORMS | 62 | 0.1 | 0.3 | 0.9 | 1.4 | 3.8 | 5.3 | 10.7 | 10.4 | 4.7 | 2.2 | 0.6 | 0.1 | 40.5 |
| **CLOUDNESS** | MEAN: | | | | | | | | | | | | | | |
| | SUNRISE-SUNSET (OKTAS) | 57 | 3.9 | 4.1 | 4.1 | 3.7 | 3.4 | 2.7 | 3.5 | 3.5 | 2.8 | 2.8 | 3.1 | 3.6 | 3.4 |
| | MIDNIGHT-MIDNIGHT (OKTAS) | 32 | 3.5 | 3.8 | 3.6 | 3.2 | 3.1 | 2.7 | 3.7 | 3.7 | 2.9 | 2.6 | 2.9 | 3.3 | 3.3 |
| | MEAN NO. DAYS WITH: | | | | | | | | | | | | | | |
| | CLEAR | 58 | 12.8 | 10.8 | 11.3 | 12.5 | 13.9 | 17.4 | 11.9 | 13.1 | 16.3 | 17.0 | 14.8 | 13.7 | 165.5 |
| | PARTLY CLOUDY | 58 | 7.8 | 7.7 | 9.6 | 9.5 | 10.4 | 8.7 | 13.9 | 12.3 | 7.7 | 7.7 | 7.6 | 7.2 | 110.1 |
| | CLOUDY | 58 | 10.4 | 9.8 | 10.1 | 8.0 | 6.6 | 3.9 | 4.7 | 5.1 | 5.5 | 5.9 | 7.1 | 9.6 | 86.7 |
| **PR** | MEAN STATION PRESSURE(IN) | 26 | 24.79 | 24.74 | 24.69 | 24.68 | 24.70 | 24.74 | 24.81 | 24.83 | 24.80 | 24.79 | 24.78 | 24.78 | 24.76 |
| | MEAN SEA-LEVEL PRES. (IN) | 26 | 30.11 | 30.02 | 29.92 | 29.85 | 29.81 | 29.80 | 29.88 | 29.91 | 29.92 | 29.97 | 30.06 | 30.11 | 29.95 |
| **WINDS** | MEAN SPEED (MPH) | 26 | 7.5 | 8.3 | 9.1 | 10.1 | 9.8 | 9.2 | 8.2 | 7.7 | 7.6 | 7.6 | 7.5 | 7.1 | 8.3 |
| | PREVAIL.DIR(TENS OF DEGS) | 38 | 36 | 36 | 36 | 19 | 19 | 09 | 11 | 11 | 11 | 36 | 36 | 36 | 36 |
| | MAXIMUM 2-MINUTE: | | | | | | | | | | | | | | |
| | SPEED (MPH) | 13 | 49 | 47 | 49 | 51 | 48 | 53 | 46 | 51 | 43 | 46 | 51 | 51 | 53 |
| | DIR. (TENS OF DEGS) | | 09 | 27 | 09 | 28 | 25 | 09 | 09 | 08 | 34 | 18 | 29 | 06 | 09 |
| | YEAR OF OCCURRENCE | | 2003 | 2000 | 2000 | 2009 | 1999 | 2004 | 2000 | 2000 | 2000 | 2008 | 2003 | 1997 | JUN 2004 |
| | MAXIMUM 3-SECOND | | | | | | | | | | | | | | |
| | SPEED (MPH) | 13 | 58 | 59 | 58 | 66 | 64 | 63 | 56 | 61 | 53 | 58 | 61 | 57 | 66 |
| | DIR. (TENS OF DEGS) | | 09 | 29 | 25 | 27 | 28 | 08 | 06 | 09 | 33 | 17 | 29 | 07 | 27 |
| | YEAR OF OCCURRENCE | | 2003 | 2000 | 2009 | 2009 | 2001 | 2004 | 2000 | 2000 | 2000 | 2008 | 2003 | 1997 | APR 2009 |
| **PRECIPITATION** | NORMAL (IN) | 30 | 0.49 | 0.44 | 0.61 | 0.50 | 0.60 | 0.65 | 1.27 | 1.73 | 1.07 | 1.00 | 0.62 | 0.49 | 9.47 |
| | MAXIMUM MONTHLY (IN) | 70 | 1.38 | 1.82 | 2.34 | 3.00 | 3.07 | 2.86 | 3.55 | 3.74 | 2.83 | 3.08 | 1.93 | 1.85 | 3.74 |
| | YEAR OF OCCURRENCE | | 2005 | 1993 | 1998 | 2004 | 1941 | 1996 | 2006 | 2006 | 2005 | 1972 | 1991 | 1959 | AUG 2006 |
| | MINIMUM MONTHLY (IN) | 70 | T | T | T | T | T | T | 0.08 | T | T | 0.00 | 0.00 | 0.00 | 0.00 |
| | YEAR OF OCCURRENCE | | 2009 | 2009 | 2008 | 1996 | 2006 | 1975 | 1980 | 1962 | 1957 | 1952 | 1949 | 1981 | DEC 1981 |
| | MAXIMUM IN 24 HOURS (IN) | 70 | 0.87 | 1.04 | 1.45 | 2.29 | 1.14 | 1.64 | 1.77 | 2.13 | 1.92 | 1.80 | 1.67 | 1.35 | 2.29 |
| | YEAR OF OCCURRENCE | | 1962 | 2004 | 1998 | 2004 | 1969 | 1952 | 1961 | 1994 | 1955 | 1969 | 1991 | 1958 | APR 2004 |
| | NORMAL NO. DAYS WITH: | | | | | | | | | | | | | | |
| | PRECIPITATION >= 0.01 | 30 | 4.6 | 4.1 | 5.2 | 3.2 | 4.8 | 4.1 | 8.4 | 9.6 | 6.1 | 5.2 | 4.4 | 4.2 | 63.9 |
| | PRECIPITATION >= 1.00 | 30 | 0.0 | 0.0 | * | * | 0.0 | 0.1 | 0.1 | 0.2 | * | 0.1 | * | 0.0 | 0.5 |
| **SNOWFALL** | NORMAL (IN) | 30 | 3.1 | 2.2 | 1.8 | 0.9 | 0.* | 0.0 | 0.0 | 0.0 | 0.* | 0.3 | 1.1 | 2.6 | 12.0 |
| | MAXIMUM MONTHLY (IN) | 70 | 9.5 | 10.3 | 13.9 | 8.1 | 1.0 | T | T | T | T | 3.2 | 9.3 | 20.8 | 20.8 |
| | YEAR OF OCCURRENCE | | 1973 | 1986 | 1973 | 1973 | 1979 | 2007 | 2008 | 2008 | 2006 | 1986 | 1940 | 2006 | DEC 2006 |
| | MAXIMUM IN 24 HOURS (IN) | 70 | 5.1 | 6.0 | 10.7 | 10.9 | 1.0 | T | T | T | T | 3.2 | 5.5 | 14.2 | 14.2 |
| | YEAR OF OCCURRENCE' | | 1973 | 1986 | 1973 | 1988 | 1979 | 1996 | 1990 | 1993 | 1971 | 1986 | 1946 | 1958 | DEC 1958 |
| | MAXIMUM SNOW DEPTH (IN) | 61 | 47 | 16 | 8 | 11 | 0 | 0 | 0 | 0 | 0 | 3 | 12 | 25 | 47 |
| | YEAR OF OCCURRENCE | | 1977 | 1986 | 1973 | 1988 | | | | | | 1986 | 1992 | 1958 | JAN 1977 |
| | NORMAL NO. DAYS WITH: | | | | | | | | | | | | | | |
| | SNOWFALL >= 1.0 | 30 | 1.0 | 0.9 | 0.6 | 0.3 | 0.0 | 0.0 | 0.0 | 0.0 | 0.0 | 0.1 | 0.4 | 0.8 | 4.1 |

## PRECIPITATION (inches) 2009 ALBUQUERQUE (KABQ)

| YEAR | JAN | FEB | MAR | APR | MAY | JUN | JUL | AUG | SEP | OCT | NOV | DEC | ANNUAL |
|------|-----|-----|-----|-----|-----|-----|-----|-----|-----|-----|-----|-----|--------|
| 1980 | 0.87 | 0.58 | 0.60 | 0.60 | 0.56 | 0.01 | 0.08 | 2.61 | 1.83 | 0.09 | 0.30 | 0.74 | 8.87 |
| 1981 | 0.05 | 0.67 | 0.80 | 0.30 | 0.53 | 0.35 | 1.07 | 1.68 | 0.41 | 1.43 | 0.37 | 0.00 | 7.66 |
| 1982 | 0.32 | 0.20 | 0.84 | 0.05 | 0.52 | 0.09 | 1.32 | 1.09 | 1.34 | 0.26 | 0.60 | 0.78 | 7.41 |
| 1983 | 1.10 | 0.71 | 0.61 | 0.02 | 0.32 | 1.21 | 0.55 | 0.27 | 0.91 | 1.20 | 0.44 | 0.42 | 7.76 |
| 1984 | 0.33 | T | 0.62 | 0.50 | 0.16 | 0.48 | 1.13 | 2.70 | 1.13 | 3.04 | 0.63 | 1.36 | 12.08 |
| 1985 | 0.49 | 0.54 | 0.70 | 1.69 | 1.12 | 0.53 | 1.16 | 0.49 | 1.53 | 2.15 | 0.19 | 0.16 | 10.75 |
| 1986 | 0.22 | 1.01 | 0.17 | 0.33 | 1.11 | 2.57 | 1.51 | 2.26 | 0.53 | 1.54 | 1.29 | 0.44 | 12.98 |
| 1987 | 0.66 | 0.61 | 0.07 | 1.00 | 0.58 | 0.13 | 0.91 | 2.98 | 0.20 | 0.44 | 0.42 | 0.34 | 8.34 |
| 1988 | 0.15 | 0.07 | 0.85 | 1.42 | 0.62 | 1.25 | 2.26 | 3.29 | 2.63 | 0.32 | 0.22 | 0.03 | 13.11 |
| 1989 | 0.57 | 0.35 | 0.48 | T | 0.02 | 0.02 | 1.51 | 0.48 | 0.31 | 0.97 | T | 0.28 | 4.99 |
| 1990 | 0.21 | 0.49 | 0.41 | 1.71 | 0.45 | 0.27 | 2.36 | 1.79 | 0.96 | 0.15 | 0.86 | 0.59 | 10.25 |
| 1991 | 0.60 | 0.06 | 0.14 | T | 1.14 | 0.65 | 2.63 | 1.26 | 1.43 | 0.26 | 1.93 | 1.49 | 11.59 |
| 1992 | 0.60 | 0.20 | 0.63 | 0.22 | 1.81 | 0.67 | 2.01 | 2.17 | 0.79 | 0.70 | 1.12 | 1.16 | 12.08 |
| 1993 | 0.94 | 1.82 | 0.22 | T | 0.20 | 0.44 | 0.23 | 3.05 | 0.49 | 0.64 | 0.97 | 0.03 | 9.03 |
| 1994 | 0.02 | 0.26 | 0.59 | 0.07 | 1.87 | 0.28 | 0.61 | 2.70 | 1.21 | 1.54 | 1.38 | 0.62 | 11.15 |
| 1995 | 0.55 | 0.39 | 0.16 | 0.69 | 0.08 | 0.20 | 0.35 | 0.74 | 2.32 | T | 0.03 | 0.17 | 5.68 |
| 1996 | 0.17 | 0.19 | 0.02 | T | 0.02 | 2.86 | 1.03 | 1.54 | 1.45 | 1.52 | 0.95 | T | 9.75 |
| 1997 | 0.55 | 0.12 | 0.11 | 1.65 | 0.42 | 1.03 | 2.04 | 1.96 | 2.43 | 0.32 | 0.73 | 1.00 | 12.36 |
| 1998 | 0.14 | 0.66 | 2.34 | 0.64 | T | 0.17 | 2.37 | 0.88 | 0.15 | 1.80 | 0.46 | 0.22 | 9.83 |
| 1999 | 0.12 | T | 1.10 | 0.59 | 0.54 | 0.60 | 1.47 | 3.04 | 0.54 | 0.26 | T | 0.03 | 8.29 |
| 2000 | 0.30 | 0.30 | 1.27 | T | 0.07 | 0.72 | 0.83 | 0.57 | 0.37 | 2.66 | 0.91 | 0.24 | 8.24 |
| 2001 | 0.28 | 0.27 | 0.27 | 0.51 | 0.38 | 0.26 | 1.37 | 1.59 | 0.51 | 0.14 | 0.68 | 0.24 | 6.50 |
| 2002 | 0.34 | 0.07 | T | 0.39 | 0.02 | 0.18 | 0.88 | 1.59 | 1.53 | 0.54 | 0.49 | 0.36 | 6.39 |
| 2003 | T | 1.02 | 1.45 | T | 0.09 | 0.20 | 0.41 | 0.71 | 0.29 | 1.58 | 0.49 | 0.11 | 6.35 |
| 2004 | 0.10 | 1.17 | 0.67 | 3.00 | T | 0.61 | 2.25 | 0.23 | 0.97 | 1.13 | 1.37 | 0.30 | 11.80 |
| 2005 | 1.38 | 1.78 | 1.12 | 1.17 | 0.40 | 0.09 | 1.03 | 0.49 | 2.83 | 1.03 | T | 0.10 | 11.42 |
| 2006 | 0.04 | T | 0.14 | 0.13 | T | 1.14 | 3.55 | 3.74 | 1.10 | 1.70 | 0.02 | 1.50 | 13.06 |
| 2007 | 0.18 | 0.70 | 0.64 | 1.06 | 2.00 | 0.66 | 1.63 | 1.05 | 0.73 | 0.17 | 0.25 | 1.14 | 10.21 |
| 2008 | 0.39 | 0.41 | T | 0.11 | 0.18 | 0.50 | 3.38 | 1.04 | 0.08 | 1.38 | 0.23 | 0.65 | 8.35 |
| 2009 | T | T | 0.31 | 0.34 | 0.36 | 0.80 | 0.80 | 0.94 | 1.42 | 1.51 | 0.04 | 0.15 | 6.67 |
| POR= 113 YRS | 0.37 | 0.39 | 0.48 | 0.57 | 0.61 | 0.62 | 1.41 | 1.40 | 0.95 | 0.84 | 0.49 | 0.47 | 8.60 |

WBAN : 23050

## AVERAGE TEMPERATURE (°F) 2009 ALBUQUERQUE (KABQ)

| YEAR | JAN | FEB | MAR | APR | MAY | JUN | JUL | AUG | SEP | OCT | NOV | DEC | ANNUAL |
|------|-----|-----|-----|-----|-----|-----|-----|-----|-----|-----|-----|-----|--------|
| 1980 | 40.2 | 44.2 | 46.1 | 52.1 | 61.1 | 77.2 | 82.7 | 77.4 | 69.9 | 54.5 | 43.5 | 40.5 | 57.5 |
| 1981 | 38.0 | 42.9 | 46.2 | 59.0 | 64.5 | 77.0 | 79.8 | 76.4 | 69.7 | 55.7 | 47.0 | 40.5 | 58.1 |
| 1982 | 35.9 | 39.4 | 47.4 | 56.1 | 63.0 | 74.8 | 79.1 | 77.4 | 69.5 | 54.8 | 42.9 | 34.4 | 56.2 |
| 1983 | 35.0 | 39.7 | 46.9 | 50.2 | 63.0 | 73.4 | 80.4 | 79.4 | 73.4 | 58.3 | 45.1 | 36.7 | 56.8 |
| 1984 | 34.1 | 40.1 | 46.8 | 52.8 | 69.9 | 73.6 | 78.9 | 75.7 | 68.8 | 51.6 | 43.7 | 35.6 | 56.0 |
| 1985 | 33.8 | 38.3 | 47.5 | 57.4 | 64.0 | 74.1 | 77.1 | 76.6 | 65.9 | 57.5 | 45.4 | 37.6 | 56.3 |
| 1986 | 41.3 | 43.0 | 50.9 | 56.5 | 63.7 | 72.7 | 74.7 | 76.0 | 66.5 | 54.6 | 42.0 | 36.3 | 56.5 |
| 1987 | 32.3 | 39.2 | 43.7 | 54.8 | 62.8 | 73.0 | 77.8 | 74.7 | 68.8 | 61.3 | 45.2 | 35.3 | 55.7 |
| 1988 | 34.6 | 43.9 | 47.0 | 55.1 | 64.3 | 74.4 | 78.1 | 75.0 | 66.3 | 61.1 | 45.4 | 33.9 | 56.6 |
| 1989 | 35.5 | 41.9 | 52.8 | 61.4 | 68.8 | 75.6 | 78.6 | 74.3 | 69.4 | 56.7 | 46.4 | 35.1 | 58.0 |
| 1990 | 34.6 | 38.5 | 48.6 | 57.3 | 63.6 | 79.0 | 76.8 | 73.8 | 70.9 | 58.3 | 45.0 | 32.1 | 56.5 |
| 1991 | 35.7 | 44.6 | 46.1 | 56.0 | 65.5 | 73.4 | 76.9 | 75.5 | 68.1 | 59.6 | 43.4 | 37.3 | 56.8 |
| 1992 | 32.7 | 42.3 | 48.9 | 60.0 | 64.6 | 72.4 | 76.2 | 75.0 | 70.3 | 60.8 | 39.7 | 32.8 | 56.3 |
| 1993 | 39.7 | 42.5 | 48.8 | 57.1 | 65.7 | 75.1 | 79.9 | 75.6 | 69.1 | 56.2 | 43.3 | 37.3 | 57.5 |
| 1994 | 38.1 | 40.7 | 50.2 | 58.5 | 66.9 | 80.4 | 81.3 | 79.4 | 71.0 | 57.0 | 44.5 | 40.9 | 59.1 |
| 1995 | 39.2 | 49.3 | 50.7 | 54.2 | 64.5 | 74.8 | 80.0 | 79.8 | 69.5 | 59.5 | 50.8 | 40.9 | 59.4 |
| 1996 | 39.3 | 46.1 | 46.8 | 57.5 | 71.5 | 76.5 | 79.5 | 76.2 | 66.2 | 55.9 | 45.6 | 38.7 | 58.3 |
| 1997 | 33.4 | 40.7 | 51.8 | 52.6 | 65.9 | 77.6 | 77.6 | 77.1 | 71.5 | 56.5 | 43.6 | 32.7 | 56.3 |
| 1998 | 37.9 | 38.8 | 46.7 | 52.2 | 65.6 | 74.4 | 77.1 | 77.1 | 74.4 | 57.9 | 46.5 | 38.5 | 57.3 |
| 1999 | 40.7 | 44.0 | 50.7 | 53.7 | 63.7 | 72.8 | 76.7 | 74.7 | 68.2 | 58.6 | 49.8 | 35.6 | 57.4 |
| 2000 | 40.9 | 45.0 | 47.7 | 59.1 | 70.5 | 76.2 | 79.5 | 78.1 | 72.4 | 55.7 | 39.4 | 37.1 | 58.5 |
| 2001 | 33.8 | 42.7 | 48.4 | 57.8 | 68.9 | 76.9 | 79.4 | 75.9 | 72.4 | 60.8 | 47.9 | 36.3 | 58.4 |
| 2002 | 37.0 | 40.1 | 47.5 | 61.8 | 67.5 | 79.1 | 78.7 | 77.6 | 69.2 | 57.0 | 44.9 | 36.6 | 58.1 |
| 2003 | 43.5 | 40.7 | 47.5 | 56.7 | 67.6 | 74.9 | 83.9 | 78.7 | 70.6 | 62.2 | 45.8 | 37.2 | 59.1 |
| 2004 | 38.3 | 37.7 | 53.2 | 54.5 | 68.1 | 75.1 | 77.9 | 74.7 | 68.7 | 56.7 | 44.7 | 37.0 | 57.2 |
| 2005 | 41.7 | 43.2 | 45.9 | 56.1 | 66.5 | 75.1 | 81.3 | 76.6 | 71.6 | 57.8 | 47.8 | 38.8 | 58.5 |
| 2006 | 39.5 | 42.8 | 48.4 | 60.5 | 70.0 | 78.2 | 78.7 | 73.5 | 64.7 | 56.6 | 47.6 | 35.1 | 58.0 |
| 2007 | 32.0 | 41.5 | 50.9 | 55.7 | 64.4 | 75.8 | 79.3 | 79.6 | 71.5 | 59.8 | 48.5 | 36.5 | 58.0 |
| 2008 | 33.5 | 41.8 | 48.8 | 55.9 | 64.1 | 76.3 | 76.6 | 76.6 | 70.4 | 58.7 | 47.1 | 38.8 | 57.4 |
| 2009 | 40.5 | 44.6 | 50.2 | 55.7 | 68.8 | 73.2 | 80.4 | 77.5 | 68.4 | 56.1 | 47.2 | 33.9 | 58.0 |
| POR= 102 YRS | 35.1 | 39.2 | 46.1 | 54.0 | 63.7 | 72.3 | 77.9 | 75.6 | 66.8 | 55.6 | 43.2 | 34.9 | 55.4 |

## HEATING DEGREE DAYS (base 65°F) 2009  ALBUQUERQUE (KABQ)

| YEAR | JUL | AUG | SEP | OCT | NOV | DEC | JAN | FEB | MAR | APR | MAY | JUN | TOTAL |
|---|---|---|---|---|---|---|---|---|---|---|---|---|---|
| 1980-81 | 0 | 0 | 6 | 335 | 640 | 752 | 827 | 611 | 575 | 197 | 62 | 2 | 4007 |
| 1981-82 | 0 | 0 | 3 | 280 | 534 | 754 | 895 | 709 | 538 | 268 | 94 | 0 | 4075 |
| 1982-83 | 0 | 0 | 23 | 314 | 658 | 941 | 922 | 703 | 556 | 439 | 127 | 0 | 4683 |
| 1983-84 | 0 | 0 | 11 | 198 | 592 | 875 | 948 | 714 | 559 | 362 | 22 | 3 | 4284 |
| 1984-85 | 0 | 0 | 51 | 411 | 631 | 903 | 960 | 744 | 536 | 220 | 74 | 7 | 4537 |
| 1985-86 | 0 | 0 | 61 | 228 | 581 | 842 | 727 | 610 | 431 | 249 | 80 | 8 | 3817 |
| 1986-87 | 0 | 0 | 51 | 313 | 680 | 882 | 1004 | 717 | 653 | 300 | 81 | 2 | 4683 |
| 1987-88 | 0 | 0 | 2 | 133 | 589 | 914 | 937 | 605 | 551 | 290 | 103 | 2 | 4126 |
| 1988-89 | 0 | 5 | 39 | 118 | 579 | 959 | 909 | 640 | 373 | 133 | 31 | 0 | 3786 |
| 1989-90 | 0 | 0 | 10 | 260 | 551 | 918 | 934 | 735 | 501 | 233 | 103 | 0 | 4245 |
| 1990-91 | 0 | 0 | 14 | 202 | 595 | 1013 | 903 | 563 | 581 | 263 | 60 | 12 | 4206 |
| 1991-92 | 0 | 0 | 21 | 188 | 645 | 851 | 994 | 651 | 493 | 170 | 53 | 5 | 4071 |
| 1992-93 | 0 | 0 | 8 | 128 | 752 | 991 | 778 | 624 | 496 | 238 | 69 | 3 | 4087 |
| 1993-94 | 0 | 0 | 15 | 284 | 642 | 853 | 827 | 676 | 453 | 218 | 52 | 0 | 4020 |
| 1994-95 | 0 | 0 | 1 | 251 | 610 | 741 | 793 | 435 | 433 | 320 | 67 | 0 | 3651 |
| 1995-96 | 0 | 0 | 37 | 165 | 419 | 741 | 787 | 540 | 558 | 239 | 12 | 0 | 3498 |
| 1996-97 | 0 | 0 | 66 | 297 | 575 | 809 | 973 | 676 | 403 | 366 | 48 | 12 | 4225 |
| 1997-98 | 0 | 0 | 9 | 278 | 635 | 994 | 834 | 727 | 558 | 376 | 51 | 4 | 4466 |
| 1998-99 | 0 | 0 | 0 | 225 | 549 | 816 | 744 | 584 | 435 | 332 | 98 | 5 | 3788 |
| 1999-00 | 0 | 0 | 23 | 203 | 447 | 902 | 740 | 571 | 528 | 184 | 32 | 0 | 3630 |
| 2000-01 | 0 | 0 | 16 | 302 | 760 | 858 | 961 | 619 | 509 | 218 | 35 | 3 | 4281 |
| 2001-02 | 0 | 0 | 1 | 134 | 508 | 881 | 858 | 692 | 532 | 105 | 29 | 0 | 3740 |
| 2002-03 | 0 | 0 | 13 | 245 | 597 | 874 | 659 | 674 | 535 | 243 | 51 | 0 | 3891 |
| 2003-04 | 0 | 0 | 1 | 107 | 570 | 852 | 818 | 787 | 357 | 309 | 29 | 0 | 3830 |
| 2004-05 | 0 | 0 | 28 | 251 | 601 | 861 | 716 | 603 | 583 | 262 | 76 | 0 | 3981 |
| 2005-06 | 0 | 0 | 11 | 228 | 507 | 805 | 784 | 616 | 506 | 141 | 10 | 0 | 3608 |
| 2006-07 | 0 | 0 | 60 | 277 | 511 | 920 | 1017 | 650 | 432 | 279 | 72 | 1 | 4219 |
| 2007-08 | 0 | 0 | 1 | 172 | 488 | 875 | 971 | 667 | 498 | 269 | 111 | 5 | 4057 |
| 2008-09 | 0 | 0 | 0 | 200 | 530 | 804 | 749 | 565 | 451 | 278 | 14 | 0 | 3591 |
| 2009- | 0 | 0 | 38 | 276 | 527 | 958 | | | | | | | |

WBAN : 23050

## COOLING DEGREE DAYS (base 65°F) 2009  ALBUQUERQUE (KABQ)

| YEAR | JAN | FEB | MAR | APR | MAY | JUN | JUL | AUG | SEP | OCT | NOV | DEC | TOTAL |
|---|---|---|---|---|---|---|---|---|---|---|---|---|---|
| 1980 | 0 | 0 | 0 | 0 | 27 | 375 | 557 | 392 | 160 | 15 | 0 | 0 | 1526 |
| 1981 | 0 | 0 | 0 | 28 | 51 | 368 | 470 | 360 | 152 | 1 | 0 | 0 | 1430 |
| 1982 | 0 | 0 | 0 | 6 | 38 | 301 | 441 | 394 | 163 | 4 | 0 | 0 | 1347 |
| 1983 | 0 | 0 | 0 | 1 | 72 | 260 | 484 | 450 | 267 | 1 | 0 | 0 | 1535 |
| 1984 | 0 | 0 | 0 | 4 | 179 | 266 | 441 | 340 | 169 | 1 | 0 | 0 | 1400 |
| 1985 | 0 | 0 | 0 | 0 | 51 | 289 | 383 | 368 | 97 | 0 | 0 | 0 | 1188 |
| 1986 | 0 | 0 | 0 | 1 | 50 | 245 | 310 | 349 | 103 | 0 | 0 | 0 | 1058 |
| 1987 | 0 | 0 | 0 | 0 | 17 | 251 | 404 | 308 | 120 | 25 | 0 | 0 | 1125 |
| 1988 | 0 | 0 | 0 | 1 | 85 | 288 | 411 | 322 | 86 | 3 | 0 | 0 | 1196 |
| 1989 | 0 | 0 | 0 | 31 | 154 | 323 | 426 | 295 | 150 | 10 | 0 | 0 | 1389 |
| 1990 | 0 | 0 | 0 | 10 | 66 | 426 | 374 | 281 | 200 | 2 | 0 | 0 | 1359 |
| 1991 | 0 | 0 | 0 | 0 | 87 | 269 | 375 | 331 | 120 | 25 | 0 | 0 | 1207 |
| 1992 | 0 | 0 | 0 | 27 | 49 | 235 | 354 | 318 | 171 | 5 | 0 | 0 | 1159 |
| 1993 | 0 | 0 | 0 | 8 | 101 | 312 | 470 | 337 | 145 | 16 | 0 | 0 | 1389 |
| 1994 | 0 | 0 | 0 | 29 | 115 | 469 | 512 | 455 | 188 | 10 | 0 | 0 | 1778 |
| 1995 | 0 | 0 | 0 | 4 | 55 | 302 | 472 | 467 | 182 | 1 | 0 | 0 | 1483 |
| 1996 | 0 | 0 | 0 | 19 | 218 | 352 | 457 | 354 | 109 | 21 | 0 | 0 | 1530 |
| 1997 | 0 | 0 | 0 | 0 | 85 | 261 | 398 | 354 | 212 | 21 | 0 | 0 | 1331 |
| 1998 | 0 | 0 | 0 | 1 | 78 | 292 | 383 | 383 | 288 | 11 | 0 | 0 | 1436 |
| 1999 | 0 | 0 | 0 | 0 | 65 | 246 | 367 | 306 | 122 | 12 | 0 | 0 | 1118 |
| 2000 | 0 | 0 | 0 | 13 | 208 | 346 | 457 | 411 | 244 | 21 | 0 | 0 | 1700 |
| 2001 | 0 | 0 | 0 | 8 | 162 | 369 | 453 | 346 | 228 | 12 | 0 | 0 | 1578 |
| 2002 | 0 | 0 | 0 | 15 | 114 | 431 | 432 | 397 | 145 | 3 | 0 | 0 | 1537 |
| 2003 | 0 | 0 | 0 | 0 | 141 | 306 | 592 | 430 | 175 | 26 | 0 | 0 | 1670 |
| 2004 | 0 | 0 | 1 | 2 | 131 | 310 | 407 | 306 | 146 | 2 | 0 | 0 | 1305 |
| 2005 | 0 | 0 | 0 | 2 | 132 | 311 | 514 | 370 | 216 | 13 | 0 | 0 | 1558 |
| 2006 | 0 | 0 | 0 | 11 | 175 | 401 | 435 | 270 | 58 | 23 | 0 | 0 | 1373 |
| 2007 | 0 | 0 | 0 | 5 | 60 | 335 | 452 | 462 | 203 | 19 | 0 | 0 | 1536 |
| 2008 | 0 | 0 | 0 | 2 | 90 | 349 | 370 | 367 | 168 | 12 | 0 | 0 | 1358 |
| 2009 | 0 | 0 | 0 | 8 | 139 | 251 | 484 | 394 | 146 | 6 | 0 | 0 | 1428 |

## SNOWFALL (inches) 2009 ALBUQUERQUE (KABQ)

| YEAR | JUL | AUG | SEP | OCT | NOV | DEC | JAN | FEB | MAR | APR | MAY | JUN | TOTAL |
|---|---|---|---|---|---|---|---|---|---|---|---|---|---|
| 1980-81 | 0.0 | 0.0 | 0.0 | T | 2.8 | 7.4 | 0.5 | 2.6 | 0.9 | T | 0.0 | 0.0 | 14.2 |
| 1981-82 | 0.0 | 0.0 | 0.0 | 0.0 | 0.0 | 0.0 | 3.6 | 1.2 | 0.7 | T | 0.0 | 0.0 | 5.5 |
| 1982-83 | 0.0 | 0.0 | 0.0 | 0.0 | 0.9 | 3.3 | 7.3 | 4.2 | 1.0 | T | T | 0.0 | 16.7 |
| 1983-84 | 0.0 | 0.0 | 0.0 | 0.0 | 0.8 | 0.8 | 4.1 | T | 0.1 | 3.0 | 0.0 | 0.0 | 8.8 |
| 1984-85 | 0.0 | 0.0 | 0.0 | T | T | 3.4 | 2.0 | 2.9 | 0.6 | 0.0 | 0.0 | 0.0 | 8.9 |
| 1985-86 | 0.0 | 0.0 | 0.0 | 0.0 | 0.7 | 0.9 | 2.9 | 10.3 | 0.3 | 0.0 | T | 0.0 | 15.1 |
| 1986-87 | 0.0 | 0.0 | 0.0 | 3.2 | 0.6 | 0.2 | 4.9 | 4.9 | 0.2 | 2.2 | 0.0 | 0.0 | 16.2 |
| 1987-88 | 0.0 | 0.0 | 0.0 | 0.0 | 1.1 | 1.7 | 1.2 | T | 7.9 | 4.2 | 0.0 | 0.0 | 16.1 |
| 1988-89 | 0.0 | 0.0 | 0.0 | 0.0 | 1.7 | 0.3 | 3.4 | 3.2 | 3.1 | 0.0 | 0.0 | 0.0 | 11.7 |
| 1989-90 | 0.0 | 0.0 | 0.0 | 0.0 | T | 2.5 | 1.8 | 4.8 | T | 0.3 | T | T | 9.4 |
| 1990-91 | T | 0.0 | 0.0 | 0.0 | 2.2 | 6.3 | 0.9 | T | 0.8 | T | 0.0 | 0.0 | 10.2 |
| 1991-92 | 0.0 | 0.0 | 0.0 | 2.5 | 1.5 | 2.1 | 5.6 | T | 1.0 | 0.0 | T | T | 12.7 |
| 1992-93 | 0.0 | T | 0.0 | 0.0 | 5.9 | 7.6 | 0.8 | 2.0 | 0.2 | 0.0 | T | 0.0 | 16.5 |
| 1993-94 | 0.0 | T | 0.0 | T | 4.1 | 0.2 | T | T | 1.2 | 0.0 | T | 0.0 | 5.5 |
| 1994-95 | 0.0 | 0.0 | 0.0 | 0.0 | T | 0.5 | 5.3 | 0.0 | 0.4 | 3.2 | 0.0 | 0.0 | 9.4 |
| 1995-96 | 0.0 | 0.0 | 0.0 | 0.0 | 0.0 | 0.9 | 0.6 | 1.7 | 0.2 | 0.0 | 0.0 | T | 3.4 |
| 1996-97 | T | T | 0.0 | 1.1 | 2.5 | 0.0 | 4.7 | 0.9 | 0.3 | 2.9 | 0.0 | T | 12.4 |
| 1997-98 | 0.0 | 0.0 | 0.0 | T | 1.1 | 8.8 | 0.2 | 1.0 | 0.9 | 0.4 | 0.0 | 0.0 | 12.4 |
| 1998-99 | T | 0.0 | 0.0 | 0.0 | 1.3 | 2.7 | T | 0.0 | 3.3 | T | T | 0.0 | 7.3 |
| 1999-00 | 0.0 | 0.0 | 0.0 | T | 0.0 | 0.1 | 0.7 | 0.8 | 2.9 | T | 0.0 | 0.0 | 4.5 |
| 2000-01 | T | 0.0 | 0.0 | 0.0 | 0.1 | 6.3 | 2.7 | 0.8 | 0.1 | 0.0 | 0.0 | 0.0 | 10.0 |
| 2001-02 | 0.0 | 0.0 | 0.0 | 0.0 | T | 0.7 | 4.0 | T | T | | | | |
| 2002-03 | | | | | | T | T | T | 2.3 | 0.5 | T | 0.0 | 0.0 |
| 2003-04 | 0.0 | 0.0 | T | 0.0 | 0.0 | 2.0 | T | 2.7 | 0.7 | T | 0.0 | 0.0 | 5.4 |
| 2004-05 | 0.0 | 0.0 | 0.0 | 0.2 | 1.7 | 0.3 | T | T | 4.2 | 0.5 | 0.0 | 0.0 | 6.9 |
| 2005-06 | 0.0 | 0.0 | 0.0 | T | T | 0.9 | 0.3 | T | 1.6 | 0.0 | 0.0 | T | 2.8 |
| 2006-07 | T | T | T | 0.0 | 0.2 | 20.8 | 2.2 | 4.7 | 0.0 | T | T | T | 27.9 |
| 2007-08 | T | 0.0 | 0.0 | 0.0 | 1.0 | 0.7 | 0.6 | 2.8 | 0.1 | 0.0 | T | 0.0 | 5.2 |
| 2008-09 | T | T | 0.0 | 0.0 | 0.0 | 3.5 | T | T | T | T | 0.0 | 0.0 | 3.5 |
| 2009- | 0.0 | 0.0 | 0.0 | 0.8 | T | 0.7 | | | | | | | |
| POR= 79 YRS | T | T | T | 0.1 | 1.1 | 2.5 | 2.2 | 1.8 | 1.6 | 0.5 | T | T | 9.8 |

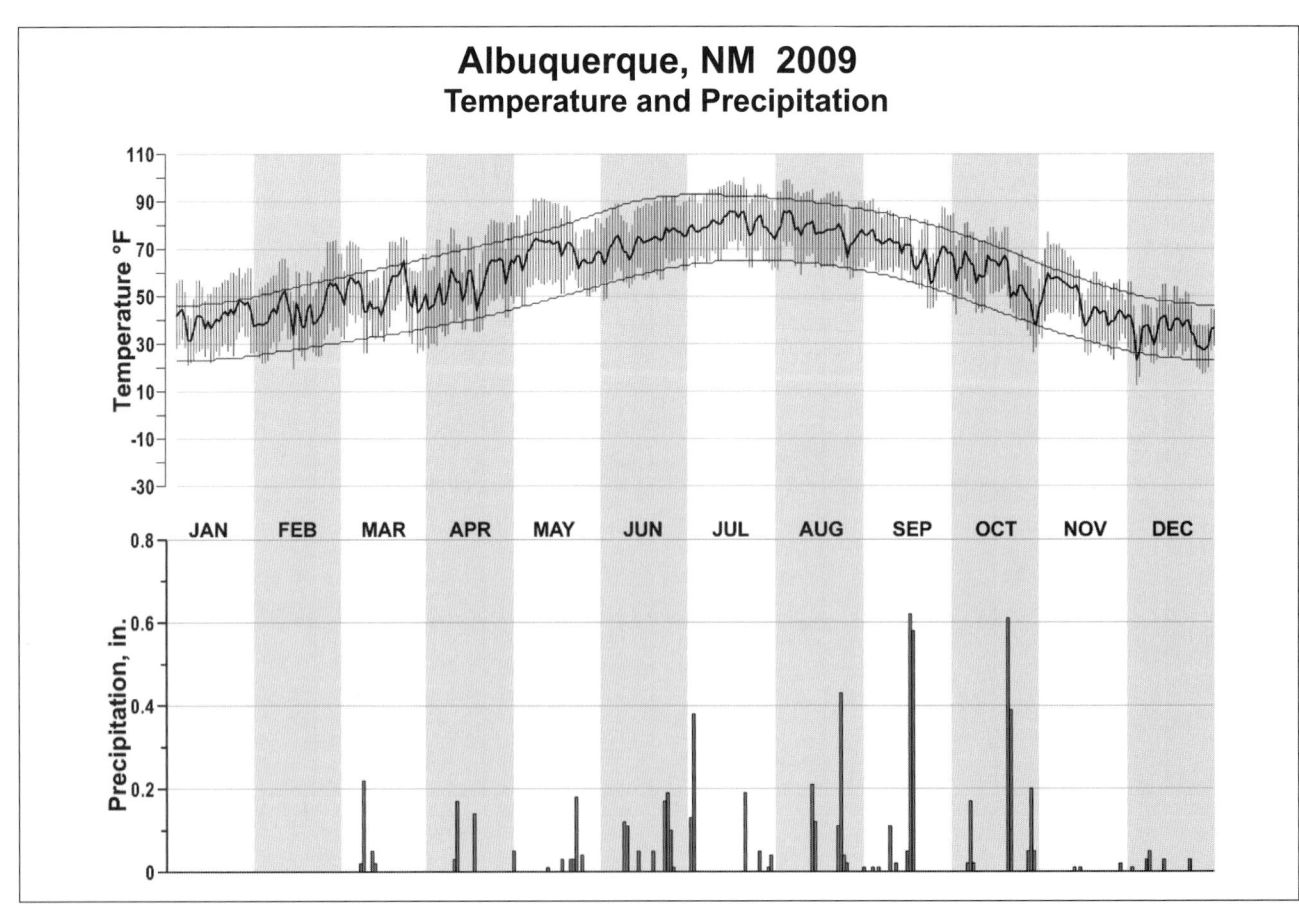

### Albuquerque, NM 2009
**Temperature and Precipitation**

# 2009
# ALBANY
# NEW YORK (KALB)

Albany is located on the west bank of the Hudson River some 150 miles north of New York City, and 8 miles south of the confluence of the Mohawk and Hudson Rivers. The river-front portion of the city is only a few feet above sea level, and there is a tidal effect upstream to Troy. Eleven miles west of Albany the Helderberg escarpment rises to 1,800 feet. Between it and the Hudson River the valley floor is gently rolling, ranging some 200 to 500 feet above sea level. East of the city there is more rugged terrain 5 or 6 miles wide with elevations of 300 to 600 feet. Farther to the east the terrain rises more sharply. It reaches a north-south range of hills 12 miles east of Albany with elevations ranging to 2,000 feet.

The climate at Albany is primarily continental in character, but is subjected to some modification by the Atlantic Ocean. The moderating effect on temperatures is more pronounced during the warmer months than in winter when outbursts of cold air sweep down from Canada. In the warmer seasons, temperatures rise rapidly in the daytime. However, temperatures also fall rapidly after sunset so that the nights are relatively cool. Occasionally there are extended periods of oppressive heat up to a week or more in duration.

Winters are usually cold and sometimes fairly severe. Maximum temperatures during the colder winters are often below freezing and nighttime lows are frequently below 10 degrees. Sub-zero readings occur about twelve times a year. Snowfall throughout the area is quite variable and snow flurries are quite frequent during the winter. Precipitation is sufficient to serve the economy of the region in most years, and only occasionally do periods of drought exist. Most of the rainfall in the summer is from thunderstorms. Tornadoes are quite rare and hail is not usually of any consequence.

Wind velocities are moderate. The north-south Hudson River Valley has a marked effect on the lighter winds and in the warm months, average wind direction is usually southerly. Destructive winds rarely occur.

The area enjoys one of the highest percentages of sunshine in the entire state. Seldom does the area experience long periods of cloudy days and long periods of smog are rare.

Based on the 1951-1980 period, the average first occurrence of 32 degrees Fahrenheit in the fall is September 29 and the average last occurrence in the spring is May 7.

# NORMALS, MEANS, AND EXTREMES
## ALBANY (KALB)

| LATITUDE: 42° 44'N | LONGITUDE: -73° 48'W | ELEVATION (FT): GRND: 280 BARO: 281 | TIME ZONE: EASTERN (UTC -5) | WBAN: 14735 |
|---|---|---|---|---|

| | ELEMENT | POR | JAN | FEB | MAR | APR | MAY | JUN | JUL | AUG | SEP | OCT | NOV | DEC | YEAR |
|---|---|---|---|---|---|---|---|---|---|---|---|---|---|---|---|
| **TEMPERATURE °F** | NORMAL DAILY MAXIMUM | 30 | 31.1 | 34.3 | 44.5 | 57.3 | 69.8 | 77.5 | 82.2 | 79.7 | 71.3 | 59.7 | 47.5 | 36.0 | 57.6 |
| | MEAN DAILY MAXIMUM | 71 | 30.5 | 33.2 | 43.2 | 57.4 | 69.4 | 77.8 | 82.8 | 80.8 | 72.3 | 61.0 | 47.8 | 35.1 | 57.6 |
| | HIGHEST DAILY MAXIMUM | 64 | 71 | 68 | 89 | 92 | 94 | 99 | 100 | 99 | 100 | 89 | 82 | 71 | 100 |
| | YEAR OF OCCURRENCE | | 2007 | 1997 | 1998 | 1990 | 1981 | 1952 | 1953 | 1955 | 1953 | 1963 | 1950 | 1984 | SEP 1953 |
| | MEAN OF EXTREME MAXS. | 71 | 51.3 | 51.9 | 66.5 | 80.0 | 86.5 | 91.8 | 93.1 | 91.6 | 87.0 | 78.6 | 67.5 | 54.8 | 75.1 |
| | NORMAL DAILY MINIMUM | 30 | 13.3 | 15.7 | 25.4 | 35.9 | 46.5 | 55.0 | 60.0 | 58.3 | 49.9 | 38.8 | 30.8 | 20.1 | 37.5 |
| | MEAN DAILY MINIMUM | 71 | 13.2 | 15.1 | 24.7 | 35.8 | 46.2 | 55.3 | 60.3 | 58.6 | 50.0 | 39.5 | 31.0 | 19.4 | 37.4 |
| | LOWEST DAILY MINIMUM | 64 | -28 | -21 | -21 | 10 | 26 | 36 | 40 | 34 | 24 | 16 | 5 | -22 | -28 |
| | YEAR OF OCCURRENCE | | 1971 | 1973 | 1948 | 1965 | 1968 | 2009 | 1978 | 1982 | 1947 | 1969 | 1972 | 1969 | JAN 1971 |
| | MEAN OF EXTREME MINS. | 71 | -8.4 | -6.5 | 5.6 | 21.9 | 31.9 | 40.8 | 48.1 | 45.0 | 34.4 | 24.4 | 15.2 | -1.9 | 20.9 |
| | NORMAL DRY BULB | 30 | 22.2 | 25.0 | 35.0 | 46.6 | 58.1 | 66.3 | 71.1 | 69.0 | 60.6 | 49.3 | 39.2 | 28.0 | 47.5 |
| | MEAN DRY BULB | 71 | 21.9 | 24.2 | 34.0 | 46.6 | 57.8 | 66.7 | 71.6 | 69.7 | 61.2 | 50.3 | 39.4 | 27.3 | 47.6 |
| | MEAN WET BULB | 26 | 20.7 | 22.3 | 29.5 | 40.2 | 50.9 | 60.4 | 64.7 | 63.8 | 56.8 | 45.5 | 36.1 | 25.9 | 43.1 |
| | MEAN DEW POINT | 26 | 16.7 | 17.7 | 24.7 | 34.9 | 46.7 | 57.0 | 61.9 | 61.4 | 54.4 | 42.5 | 32.7 | 22.2 | 39.4 |
| | NORMAL NO. DAYS WITH: | | | | | | | | | | | | | | |
| | MAXIMUM >= 90 | 30 | 0.0 | 0.0 | 0.0 | 0.2 | 0.4 | 1.5 | 4.1 | 1.9 | 0.3 | 0.0 | 0.0 | 0.0 | 8.4 |
| | MAXIMUM <= 32 | 30 | 16.8 | 12.5 | 4.2 | 0.2 | 0.0 | 0.0 | 0.0 | 0.0 | 0.0 | 0.0 | 1.2 | 10.1 | 45.0 |
| | MINIMUM <= 32 | 30 | 29.0 | 25.5 | 23.9 | 11.4 | 1.1 | 0.0 | 0.0 | 0.0 | 0.5 | 8.0 | 18.4 | 27.2 | 145.0 |
| | MINIMUM <= 0 | 30 | 5.7 | 3.8 | 0.4 | 0.0 | 0.0 | 0.0 | 0.0 | 0.0 | 0.0 | 0.0 | 0.0 | 1.6 | 11.5 |
| **H/C** | NORMAL HEATING DEG. DAYS | 30 | 1330 | 1135 | 938 | 553 | 240 | 62 | 10 | 26 | 168 | 484 | 772 | 1142 | 6860 |
| | NORMAL COOLING DEG. DAYS | 30 | 0 | 0 | 1 | 3 | 27 | 102 | 206 | 157 | 46 | 2 | 0 | 0 | 544 |
| **RH** | NORMAL (PERCENT) | 30 | 72 | 69 | 66 | 62 | 67 | 70 | 72 | 75 | 78 | 75 | 74 | 74 | 71 |
| | HOUR 01 LST | 30 | 76 | 74 | 73 | 72 | 78 | 83 | 85 | 88 | 89 | 84 | 79 | 78 | 80 |
| | HOUR 07 LST | 30 | 78 | 77 | 77 | 73 | 76 | 79 | 82 | 87 | 90 | 87 | 82 | 80 | 81 |
| | HOUR 13 LST | 30 | 64 | 59 | 54 | 50 | 53 | 56 | 56 | 59 | 60 | 58 | 63 | 65 | 58 |
| | HOUR 19 LST | 30 | 70 | 66 | 61 | 56 | 59 | 63 | 64 | 71 | 76 | 73 | 72 | 73 | 67 |
| **S** | PERCENT POSSIBLE SUNSHINE | 61 | 46 | 52 | 54 | 54 | 56 | 60 | 64 | 60 | 57 | 52 | 37 | 39 | 53 |
| **W/O** | MEAN NO. DAYS WITH: | | | | | | | | | | | | | | |
| | HEAVY FOG(VISBY <= 1/4 MI) | 46 | 1.6 | 1.3 | 1.6 | 0.6 | 1.0 | 1.1 | 1.3 | 2.2 | 3.6 | 3.7 | 1.5 | 1.8 | 21.3 |
| | THUNDERSTORMS | 64 | 0.1 | 0.1 | 0.5 | 1.1 | 3.1 | 4.9 | 5.6 | 4.6 | 2.0 | 0.7 | 0.4 | 0.1 | 23.2 |
| **CLOUDNESS** | MEAN: | | | | | | | | | | | | | | |
| | SUNRISE-SUNSET (OKTAS) | | | | 5.6 | | 6.4 | 6.4 | | | | | | | |
| | MIDNIGHT-MIDNIGHT (OKTAS) | | | | 6.4 | | | | | | | | | | |
| | MEAN NO. DAYS WITH: | | | | | | | | | | | | | | |
| | CLEAR | | | | | | | | | | | | | | |
| | PARTLY CLOUDY | | | | | | | | | | | | | | |
| | CLOUDY | | | | | | | | | | | | | | |
| **PR** | MEAN STATION PRESSURE(IN) | 26 | 29.74 | 29.75 | 29.72 | 29.66 | 29.67 | 29.64 | 29.66 | 29.71 | 29.76 | 29.76 | 29.76 | 29.75 | 29.72 |
| | MEAN SEA-LEVEL PRES. (IN) | 26 | 30.07 | 30.09 | 30.04 | 29.98 | 29.98 | 29.96 | 29.97 | 30.02 | 30.07 | 30.08 | 30.08 | 30.08 | 30.04 |
| **WINDS** | MEAN SPEED (MPH) | 26 | 8.9 | 9.2 | 9.8 | 9.3 | 8.1 | 7.5 | 6.9 | 6.2 | 6.6 | 7.3 | 8.5 | 8.6 | 8.1 |
| | PREVAIL.DIR(TENS OF DEGS) | 38 | 30 | 30 | 30 | 30 | 19 | 19 | 19 | 19 | 19 | 19 | 19 | 31 | 30 |
| | MAXIMUM 2-MINUTE: | | | | | | | | | | | | | | |
| | SPEED (MPH) | 14 | 44 | 46 | 46 | 40 | 55 | 43 | 41 | 47 | 35 | 39 | 40 | 43 | 55 |
| | DIR. (TENS OF DEGS) | | 31 | 29 | 30 | 29 | 32 | 23 | 29 | 35 | 32 | 28 | 28 | 29 | 32 |
| | YEAR OF OCCURRENCE | | 2006 | 2006 | 1997 | 2004 | 1998 | 2002 | 1999 | 1999 | 2002 | 1998 | 2003 | 2000 | MAY 1998 |
| | MAXIMUM 3-SECOND | | | | | | | | | | | | | | |
| | SPEED (MPH) | 14 | 56 | 60 | 56 | 52 | 82 | 54 | 52 | 68 | 49 | 58 | 53 | 54 | 82 |
| | DIR. (TENS OF DEGS) | | 25 | 31 | 27 | 32 | 32 | 23 | 28 | 18 | 01 | 33 | 30 | 19 | 32 |
| | YEAR OF OCCURRENCE | | 2008 | 2006 | 1997 | 2007 | 1998 | 2002 | 1999 | 2002 | 1999 | 2007 | 2009 | 2008 | MAY 1998 |
| **PRECIPITATION** | NORMAL (IN) | 30 | 2.71 | 2.27 | 3.17 | 3.25 | 3.67 | 3.74 | 3.50 | 3.68 | 3.31 | 3.23 | 3.31 | 2.76 | 38.60 |
| | MAXIMUM MONTHLY (IN) | 63 | 6.44 | 5.04 | 6.21 | 7.95 | 8.96 | 8.74 | 9.91 | 7.34 | 11.06 | 9.00 | 8.07 | 6.73 | 11.06 |
| | YEAR OF OCCURRENCE | | 1978 | 2008 | 2008 | 1983 | 1953 | 2006 | 2009 | 2004 | 1999 | 2005 | 1972 | 1973 | SEP 1999 |
| | MINIMUM MONTHLY (IN) | 63 | 0.42 | 0.24 | 0.26 | 0.60 | 1.05 | 0.65 | 0.49 | 0.73 | 0.40 | 0.20 | 0.91 | 0.64 | 0.20 |
| | YEAR OF OCCURRENCE | | 1980 | 1987 | 1981 | 1999 | 1980 | 1964 | 1968 | 1947 | 1964 | 1963 | 1978 | 1958 | OCT 1963 |
| | MAXIMUM IN 24 HOURS (IN) | 63 | 1.91 | 1.74 | 2.47 | 2.78 | 2.29 | 3.48 | 3.49 | 4.52 | 6.00 | 3.31 | 2.26 | 4.02 | 6.00 |
| | YEAR OF OCCURRENCE | | 1978 | 1990 | 2008 | 2007 | 2006 | 1952 | 1996 | 1971 | 1999 | 1987 | 1991 | 1948 | SEP 1999 |
| | NORMAL NO. DAYS WITH: | | | | | | | | | | | | | | |
| | PRECIPITATION >= 0.01 | 30 | 12.8 | 10.2 | 12.2 | 11.9 | 13.0 | 11.6 | 10.4 | 10.8 | 10.7 | 9.7 | 11.6 | 12.1 | 137.0 |
| | PRECIPITATION >= 1.00 | 30 | 0.4 | 0.2 | 0.6 | 0.6 | 0.6 | 0.7 | 0.7 | 0.9 | 0.7 | 0.7 | 0.6 | 0.3 | 7.0 |
| **SNOWFALL** | NORMAL (IN) | 30 | 18.0 | 12.7 | 10.9 | 2.9 | 0.1 | 0.0 | 0.0 | 0.0 | 0.0 | 0.2 | 5.1 | 12.8 | 62.7 |
| | MAXIMUM MONTHLY (IN) | 63 | 47.8 | 34.5 | 34.7 | 17.7 | 2.2 | T | T | 0.0 | T | 6.5 | 24.6 | 57.5 | 57.5 |
| | YEAR OF OCCURRENCE | | 1987 | 1962 | 1956 | 1982 | 2002 | 2008 | 2008 | | 1989 | 1987 | 1972 | 1969 | DEC 1969 |
| | MAXIMUM IN 24 HOURS (IN) | 63 | 21.2 | 17.9 | 26.6 | 17.5 | 2.2 | T | T | 0.0 | T | 6.5 | 21.9 | 19.2 | 26.6 |
| | YEAR OF OCCURRENCE' | | 1983 | 1958 | 1993 | 1982 | 2002 | 1991 | 1995 | | 1989 | 1987 | 1971 | 2002 | MAR 1993 |
| | MAXIMUM SNOW DEPTH (IN) | 61 | 36 | 22 | 28 | 13 | 0 | 0 | 0 | 0 | 0 | 2 | 18 | 36 | 36 |
| | YEAR OF OCCURRENCE | | 1970 | 1971 | 1993 | 1982 | | | | | | 1987 | 1971 | 1969 | JAN 1970 |
| | NORMAL NO. DAYS WITH: | | | | | | | | | | | | | | |
| | SNOWFALL >= 1.0 | 30 | 4.3 | 3.1 | 2.3 | 0.5 | 0.0 | 0.0 | 0.0 | 0.0 | 0.0 | 0.0 | 1.3 | 3.8 | 15.3 |

## PRECIPITATION (inches) 2009  ALBANY (KALB)

| YEAR | JAN | FEB | MAR | APR | MAY | JUN | JUL | AUG | SEP | OCT | NOV | DEC | ANNUAL |
|---|---|---|---|---|---|---|---|---|---|---|---|---|---|
| 1980 | 0.42 | 0.89 | 4.44 | 3.02 | 1.05 | 4.90 | 2.69 | 6.45 | 2.24 | 2.27 | 2.99 | 1.23 | 32.59 |
| 1981 | 0.59 | 5.02 | 0.26 | 1.99 | 2.44 | 2.78 | 3.50 | 1.76 | 3.45 | 3.55 | 1.56 | 3.54 | 30.44 |
| 1982 | 3.18 | 2.14 | 3.23 | 2.46 | 2.60 | 6.48 | 2.43 | 2.01 | 1.42 | 0.99 | 3.80 | 1.33 | 32.07 |
| 1983 | 3.73 | 2.03 | 5.33 | 7.95 | 6.26 | 1.95 | 1.34 | 3.41 | 2.28 | 2.18 | 4.73 | 5.10 | 46.29 |
| 1984 | 1.28 | 2.98 | 3.04 | 4.29 | 7.92 | 1.74 | 3.97 | 3.25 | 1.53 | 2.50 | 2.15 | 2.48 | 37.13 |
| 1985 | 0.81 | 1.18 | 3.67 | 1.44 | 2.71 | 4.12 | 1.86 | 2.23 | 3.07 | 1.81 | 5.00 | 2.05 | 29.95 |
| 1986 | 3.17 | 3.00 | 3.72 | 1.49 | 3.11 | 5.43 | 6.68 | 4.09 | 2.61 | 2.12 | 4.62 | 3.92 | 43.96 |
| 1987 | 4.23 | 0.24 | 1.99 | 4.25 | 1.57 | 3.54 | 2.50 | 3.67 | 6.98 | 6.90 | 1.78 | 1.64 | 39.29 |
| 1988 | 1.95 | 3.00 | 1.62 | 2.22 | 2.95 | 1.42 | 3.12 | 4.77 | 1.50 | 1.40 | 4.58 | 1.02 | 29.55 |
| 1989 | 0.46 | 1.60 | 2.69 | 2.68 | 5.92 | 6.52 | 5.91 | 2.90 | 2.81 | 5.53 | 1.90 | 0.75 | 39.67 |
| 1990 | 3.84 | 3.94 | 3.66 | 3.87 | 6.12 | 2.66 | 1.68 | 6.66 | 1.81 | 4.60 | 3.67 | 3.50 | 46.01 |
| 1991 | 2.15 | 1.67 | 2.53 | 4.14 | 2.74 | 1.69 | 1.65 | 4.32 | 3.33 | 3.82 | 4.76 | 2.92 | 35.72 |
| 1992 | 1.86 | 1.30 | 1.66 | 2.77 | 3.61 | 1.96 | 4.26 | 2.05 | 2.43 | 2.80 | 3.66 | 3.02 | 31.38 |
| 1993 | 2.14 | 2.86 | 5.12 | 5.39 | 1.37 | 2.87 | 6.55 | 1.54 | 3.22 | 3.31 | 3.80 | 3.08 | 41.25 |
| 1994 | 3.20 | 1.80 | 4.27 | 3.45 | 3.27 | 3.26 | 4.25 | 4.13 | 2.15 | 0.83 | 1.53 | 2.58 | 34.72 |
| 1995 | 2.11 | 1.95 | 2.20 | 1.94 | 1.35 | 2.27 | 2.23 | 3.66 | 2.28 | 8.03 | 3.76 | 2.30 | 34.08 |
| 1996 | 5.08 | 1.49 | 2.10 | 5.76 | 4.24 | 3.60 | 6.46 | 3.15 | 5.07 | 2.03 | 2.91 | 4.50 | 46.39 |
| 1997 | 1.67 | 2.00 | 4.41 | 2.30 | 2.60 | 0.74 | 2.34 | 4.64 | 4.10 | 1.91 | 5.91 | 2.10 | 34.72 |
| 1998 | 3.80 | 2.58 | 2.86 | 3.49 | 5.87 | 6.58 | 2.74 | 2.21 | 1.98 | 4.14 | 1.65 | 1.04 | 38.94 |
| 1999 | 4.78 | 1.59 | 4.15 | 0.60 | 2.77 | 2.08 | 2.24 | 3.45 | 11.06 | 2.42 | 2.07 | 1.42 | 38.63 |
| 2000 | 3.43 | 2.83 | 3.80 | 4.23 | 4.95 | 6.69 | 4.48 | 4.69 | 3.06 | 2.48 | 1.90 | 4.38 | 46.92 |
| 2001 | 1.00 | 1.85 | 5.50 | 1.33 | 3.21 | 3.78 | 3.59 | 2.10 | 1.64 | 1.26 | 1.38 | 1.95 | 28.59 |
| 2002 | 2.77 | 1.61 | 3.56 | 2.51 | 4.55 | 5.45 | 0.83 | 3.86 | 3.37 | 4.02 | 4.86 | 3.97 | 41.36 |
| 2003 | 3.45 | 2.15 | 2.26 | 2.89 | 5.08 | 2.84 | 4.52 | 4.41 | 4.91 | 4.67 | 3.66 | 5.48 | 46.32 |
| 2004 | 1.16 | 1.33 | 2.43 | 3.06 | 3.54 | 2.08 | 7.20 | 7.34 | 4.67 | 1.23 | 3.02 | 2.93 | 39.99 |
| 2005 | 4.27 | 1.38 | 3.99 | 2.36 | 1.44 | 3.87 | 7.54 | 3.01 | 2.20 | 9.00 | 5.71 | 2.95 | 47.72 |
| 2006 | 4.75 | 1.02 | 1.23 | 4.73 | 5.31 | 8.74 | 2.92 | 3.92 | 3.87 | 4.95 | 3.13 | 2.01 | 46.58 |
| 2007 | 2.14 | 1.54 | 3.29 | 5.96 | 3.51 | 3.36 | 7.03 | 2.34 | 2.74 | 5.53 | 3.04 | 4.74 | 45.22 |
| 2008 | 1.00 | 5.04 | 6.21 | 2.63 | 1.24 | 5.45 | 6.94 | 3.01 | 4.21 | 5.09 | 2.43 | 4.54 | 47.79 |
| 2009 | 2.24 | 0.64 | 2.63 | 1.47 | 4.08 | 5.02 | 9.91 | 3.58 | 1.73 | 4.16 | 2.17 | 3.59 | 41.22 |
| POR= 71 YRS | 2.45 | 2.17 | 2.98 | 2.99 | 3.44 | 3.50 | 3.61 | 3.29 | 3.47 | 3.07 | 3.07 | 2.96 | 37.00 |

WBAN : 14735

## AVERAGE TEMPERATURE (°F) 2009  ALBANY (KALB)

| YEAR | JAN | FEB | MAR | APR | MAY | JUN | JUL | AUG | SEP | OCT | NOV | DEC | ANNUAL |
|---|---|---|---|---|---|---|---|---|---|---|---|---|---|
| 1980 | 24.1 | 19.8 | 33.3 | 48.0 | 59.5 | 63.3 | 72.2 | 70.7 | 62.6 | 47.4 | 34.8 | 19.9 | 46.3 |
| 1981 | 14.0 | 33.1 | 34.7 | 48.1 | 58.9 | 66.7 | 69.3 | 68.5 | 58.8 | 44.8 | 37.7 | 25.7 | 46.7 |
| 1982 | 14.3 | 23.4 | 32.8 | 44.3 | 59.5 | 62.9 | 70.1 | 65.5 | 60.5 | 50.6 | 43.0 | 33.7 | 46.7 |
| 1983 | 24.3 | 26.8 | 37.6 | 46.7 | 54.9 | 67.2 | 72.2 | 69.8 | 62.6 | 49.6 | 39.2 | 24.0 | 47.9 |
| 1984 | 18.1 | 32.4 | 29.0 | 47.6 | 53.2 | 66.4 | 68.9 | 71.8 | 60.2 | 53.8 | 40.3 | 33.8 | 48.0 |
| 1985 | 19.9 | 26.8 | 37.3 | 49.7 | 60.0 | 62.2 | 70.7 | 68.7 | 63.3 | 50.2 | 40.1 | 24.5 | 47.8 |
| 1986 | 23.0 | 22.0 | 37.2 | 50.5 | 61.3 | 64.6 | 71.3 | 67.8 | 60.1 | 48.9 | 35.7 | 30.8 | 47.8 |
| 1987 | 21.7 | 21.7 | 37.7 | 50.4 | 60.0 | 68.3 | 73.5 | 67.2 | 60.6 | 46.6 | 40.1 | 30.7 | 48.2 |
| 1988 | 20.6 | 24.1 | 34.2 | 46.6 | 59.5 | 65.1 | 75.0 | 72.3 | 60.0 | 46.0 | 41.0 | 26.6 | 47.6 |
| 1989 | 27.8 | 24.2 | 33.5 | 44.6 | 59.5 | 68.0 | 71.6 | 69.8 | 62.5 | 51.5 | 39.3 | 13.7 | 47.2 |
| 1990 | 32.8 | 28.2 | 37.8 | 48.9 | 55.3 | 67.3 | 73.0 | 70.9 | 61.7 | 53.1 | 41.8 | 33.6 | 50.4 |
| 1991 | 23.2 | 30.0 | 37.4 | 51.2 | 63.2 | 69.0 | 71.6 | 71.2 | 59.9 | 53.2 | 40.1 | 28.9 | 49.9 |
| 1992 | 24.5 | 26.9 | 31.5 | 44.7 | 58.5 | 65.2 | 67.4 | 67.4 | 61.4 | 46.5 | 38.9 | 29.8 | 46.9 |
| 1993 | 26.6 | 18.3 | 31.4 | 48.4 | 59.5 | 66.3 | 73.1 | 71.7 | 60.5 | 48.6 | 38.4 | 27.4 | 47.5 |
| 1994 | 12.7 | 19.2 | 33.1 | 48.2 | 56.4 | 68.9 | 74.0 | 67.1 | 60.9 | 50.1 | 43.3 | 31.6 | 47.1 |
| 1995 | 31.3 | 22.8 | 40.0 | 43.9 | 57.0 | 66.9 | 74.0 | 70.9 | 59.1 | 53.4 | 35.7 | 23.9 | 48.2 |
| 1996 | 20.6 | 25.3 | 31.1 | 46.2 | 55.2 | 68.6 | 69.7 | 70.1 | 62.3 | 49.1 | 34.6 | 33.8 | 47.2 |
| 1997 | 22.7 | 30.4 | 33.2 | 44.2 | 53.6 | 67.9 | 70.6 | 68.6 | 60.7 | 48.0 | 35.8 | 29.8 | 47.1 |
| 1998 | 29.0 | 31.8 | 38.4 | 48.8 | 62.9 | 66.3 | 70.9 | 71.1 | 63.2 | 50.7 | 39.9 | 33.7 | 50.6 |
| 1999 | 21.8 | 28.2 | 34.4 | 46.6 | 59.4 | 69.7 | 74.2 | 69.2 | 64.8 | 48.8 | 44.1 | 31.0 | 49.4 |
| 2000 | 20.7 | 27.6 | 40.2 | 45.3 | 59.4 | 65.9 | 67.6 | 68.5 | 59.4 | 50.0 | 38.0 | 22.2 | 47.1 |
| 2001 | 24.6 | 26.9 | 30.9 | 47.4 | 58.9 | 68.4 | 68.9 | 73.7 | 62.4 | 51.9 | 44.8 | 34.2 | 49.4 |
| 2002 | 31.3 | 31.7 | 36.3 | 48.9 | 55.0 | 66.8 | 73.4 | 72.9 | 65.0 | 47.9 | 38.6 | 27.1 | 49.6 |
| 2003 | 15.6 | 21.1 | 34.4 | 44.5 | 56.8 | 66.3 | 72.2 | 72.7 | 63.0 | 48.2 | 42.3 | 29.3 | 47.2 |
| 2004 | 14.7 | 24.8 | 37.9 | 49.1 | 61.5 | 66.1 | 71.0 | 69.3 | 63.3 | 49.9 | 39.5 | 27.8 | 47.9 |
| 2005 | 19.6 | 27.0 | 31.2 | 50.2 | 54.6 | 72.9 | 73.8 | 73.8 | 66.2 | 52.0 | 42.4 | 26.7 | 49.2 |
| 2006 | 31.6 | 27.9 | 36.0 | 49.6 | 58.7 | 67.6 | 74.9 | 69.8 | 61.1 | 48.8 | 44.6 | 35.2 | 50.5 |
| 2007 | 27.4 | 19.5 | 31.8 | 44.3 | 60.3 | 68.8 | 70.6 | 70.8 | 64.3 | 56.8 | 37.6 | 27.8 | 48.3 |
| 2008 | 27.6 | 26.3 | 33.8 | 51.9 | 55.5 | 70.4 | 73.5 | 68.5 | 64.3 | 48.2 | 39.5 | 28.5 | 49.0 |
| 2009 | 18.3 | 27.5 | 36.3 | 49.6 | 58.0 | 66.1 | 68.3 | 71.0 | 60.4 | 48.0 | 43.4 | 27.4 | 47.9 |
| POR= 71 YRS | 21.9 | 24.2 | 34.0 | 46.6 | 57.8 | 66.7 | 71.6 | 69.7 | 61.2 | 50.3 | 39.4 | 27.3 | 47.5 |

## HEATING DEGREE DAYS (base 65°F) 2009 ALBANY (KALB)

| YEAR | JUL | AUG | SEP | OCT | NOV | DEC | JAN | FEB | MAR | APR | MAY | JUN | TOTAL |
|---|---|---|---|---|---|---|---|---|---|---|---|---|---|
| 1980-81 | 0 | 7 | 140 | 539 | 900 | 1393 | 1575 | 885 | 930 | 502 | 235 | 30 | 7136 |
| 1981-82 | 8 | 22 | 204 | 622 | 816 | 1209 | 1564 | 1160 | 992 | 617 | 182 | 87 | 7483 |
| 1982-83 | 20 | 65 | 156 | 436 | 657 | 969 | 1255 | 1062 | 843 | 539 | 312 | 58 | 6372 |
| 1983-84 | 5 | 24 | 150 | 479 | 766 | 1265 | 1448 | 939 | 1109 | 517 | 363 | 60 | 7125 |
| 1984-85 | 12 | 8 | 170 | 344 | 737 | 959 | 1389 | 1062 | 852 | 458 | 184 | 106 | 6281 |
| 1985-86 | 7 | 16 | 123 | 452 | 740 | 1246 | 1295 | 1177 | 859 | 432 | 154 | 75 | 6576 |
| 1986-87 | 17 | 46 | 173 | 495 | 872 | 1053 | 1332 | 1207 | 842 | 433 | 210 | 29 | 6709 |
| 1987-88 | 2 | 56 | 154 | 567 | 741 | 1056 | 1370 | 1181 | 946 | 546 | 198 | 99 | 6916 |
| 1988-89 | 8 | 30 | 160 | 584 | 714 | 1185 | 1146 | 1133 | 968 | 607 | 194 | 35 | 6764 |
| 1989-90 | 0 | 22 | 134 | 413 | 766 | 1584 | 990 | 1026 | 839 | 500 | 298 | 44 | 6616 |
| 1990-91 | 5 | 6 | 148 | 388 | 689 | 964 | 1290 | 973 | 850 | 417 | 141 | 22 | 5893 |
| 1991-92 | 6 | 0 | 197 | 372 | 740 | 1111 | 1248 | 1098 | 1034 | 605 | 210 | 56 | 6677 |
| 1992-93 | 17 | 27 | 167 | 565 | 773 | 1082 | 1183 | 1300 | 1034 | 492 | 185 | 67 | 6892 |
| 1993-94 | 0 | 11 | 185 | 500 | 791 | 1161 | 1619 | 1272 | 983 | 502 | 283 | 33 | 7340 |
| 1994-95 | 0 | 47 | 138 | 457 | 644 | 1027 | 1037 | 1177 | 766 | 627 | 252 | 41 | 6213 |
| 1995-96 | 2 | 12 | 196 | 355 | 872 | 1266 | 1369 | 1146 | 1046 | 559 | 316 | 18 | 7157 |
| 1996-97 | 1 | 2 | 133 | 488 | 903 | 961 | 1306 | 961 | 977 | 616 | 350 | 35 | 6733 |
| 1997-98 | 3 | 11 | 152 | 521 | 872 | 1083 | 1112 | 924 | 828 | 478 | 98 | 84 | 6166 |
| 1998-99 | 2 | 10 | 99 | 435 | 746 | 964 | 1334 | 1021 | 940 | 542 | 180 | 22 | 6295 |
| 1999-00 | 2 | 14 | 93 | 493 | 621 | 1048 | 1367 | 1077 | 762 | 585 | 209 | 74 | 6345 |
| 2000-01 | 14 | 25 | 207 | 456 | 801 | 1323 | 1246 | 1060 | 1048 | 522 | 202 | 46 | 6950 |
| 2001-02 | 21 | 0 | 116 | 405 | 604 | 949 | 1036 | 927 | 885 | 511 | 326 | 51 | 5831 |
| 2002-03 | 4 | 9 | 69 | 534 | 786 | 1171 | 1525 | 1221 | 942 | 611 | 246 | 46 | 7164 |
| 2003-04 | 0 | 9 | 88 | 511 | 671 | 1102 | 1554 | 1158 | 834 | 481 | 149 | 50 | 6607 |
| 2004-05 | 3 | 24 | 77 | 459 | 758 | 1148 | 1400 | 1060 | 1043 | 438 | 319 | 5 | 6734 |
| 2005-06 | 0 | 1 | 57 | 408 | 671 | 1177 | 1029 | 1033 | 892 | 458 | 228 | 43 | 5997 |
| 2006-07 | 0 | 17 | 131 | 494 | 605 | 914 | 1156 | 1268 | 1020 | 619 | 192 | 38 | 6454 |
| 2007-08 | 7 | 25 | 89 | 268 | 815 | 1145 | 1154 | 1116 | 959 | 387 | 296 | 16 | 6277 |
| 2008-09 | 0 | 10 | 97 | 517 | 757 | 1126 | 1442 | 1045 | 883 | 473 | 225 | 46 | 6621 |
| 2009- | 5 | 14 | 146 | 520 | 644 | 1158 | | | | | | | |

WBAN : 14735

## COOLING DEGREE DAYS (base 65°F) 2009 ALBANY (KALB)

| YEAR | JAN | FEB | MAR | APR | MAY | JUN | JUL | AUG | SEP | OCT | NOV | DEC | TOTAL |
|---|---|---|---|---|---|---|---|---|---|---|---|---|---|
| 1980 | 0 | 0 | 0 | 0 | 28 | 63 | 230 | 189 | 73 | 0 | 0 | 0 | 583 |
| 1981 | 0 | 0 | 0 | 2 | 53 | 87 | 149 | 137 | 25 | 0 | 0 | 0 | 453 |
| 1982 | 0 | 0 | 0 | 0 | 19 | 31 | 184 | 88 | 29 | 0 | 4 | 0 | 355 |
| 1983 | 0 | 0 | 0 | 0 | 8 | 134 | 236 | 179 | 86 | 6 | 0 | 0 | 649 |
| 1984 | 0 | 0 | 0 | 0 | 3 | 107 | 140 | 226 | 35 | 3 | 0 | 0 | 514 |
| 1985 | 0 | 0 | 0 | 5 | 37 | 27 | 191 | 140 | 80 | 2 | 0 | 0 | 482 |
| 1986 | 0 | 0 | 6 | 4 | 46 | 69 | 220 | 140 | 33 | 1 | 0 | 0 | 519 |
| 1987 | 0 | 0 | 0 | 4 | 62 | 136 | 271 | 133 | 29 | 0 | 0 | 0 | 635 |
| 1988 | 0 | 0 | 0 | 0 | 36 | 110 | 326 | 263 | 16 | 4 | 0 | 0 | 755 |
| 1989 | 0 | 0 | 1 | 0 | 31 | 132 | 213 | 178 | 63 | 0 | 0 | 0 | 618 |
| 1990 | 0 | 0 | 2 | 22 | 1 | 119 | 261 | 197 | 55 | 24 | 0 | 0 | 681 |
| 1991 | 0 | 0 | 0 | 9 | 92 | 147 | 221 | 198 | 50 | 14 | 0 | 0 | 731 |
| 1992 | 0 | 0 | 0 | 2 | 15 | 70 | 106 | 112 | 65 | 0 | 0 | 0 | 370 |
| 1993 | 0 | 0 | 0 | 0 | 17 | 116 | 259 | 224 | 55 | 2 | 0 | 0 | 673 |
| 1994 | 0 | 0 | 0 | 6 | 24 | 160 | 290 | 119 | 21 | 0 | 1 | 0 | 621 |
| 1995 | 0 | 0 | 0 | 0 | 11 | 102 | 289 | 200 | 27 | 0 | 0 | 0 | 629 |
| 1996 | 0 | 0 | 0 | 1 | 19 | 132 | 153 | 168 | 57 | 0 | 0 | 0 | 530 |
| 1997 | 0 | 0 | 0 | 0 | 3 | 128 | 186 | 129 | 28 | 1 | 0 | 0 | 475 |
| 1998 | 0 | 0 | 10 | 0 | 38 | 128 | 192 | 206 | 54 | 0 | 0 | 0 | 628 |
| 1999 | 0 | 0 | 0 | 0 | 15 | 170 | 291 | 151 | 94 | 0 | 0 | 0 | 721 |
| 2000 | 0 | 0 | 0 | 0 | 42 | 107 | 100 | 141 | 45 | 1 | 0 | 0 | 436 |
| 2001 | 0 | 0 | 0 | 3 | 16 | 152 | 150 | 279 | 45 | 8 | 0 | 0 | 653 |
| 2002 | 0 | 0 | 0 | 32 | 22 | 113 | 273 | 261 | 77 | 10 | 0 | 0 | 788 |
| 2003 | 0 | 0 | 0 | 4 | 0 | 94 | 229 | 258 | 37 | 0 | 0 | 0 | 622 |
| 2004 | 0 | 0 | 0 | 11 | 50 | 86 | 194 | 165 | 33 | 0 | 0 | 0 | 539 |
| 2005 | 0 | 0 | 0 | 0 | 4 | 248 | 279 | 281 | 99 | 14 | 0 | 0 | 925 |
| 2006 | 0 | 0 | 0 | 0 | 37 | 128 | 316 | 174 | 20 | 0 | 0 | 0 | 675 |
| 2007 | 0 | 0 | 0 | 5 | 54 | 158 | 188 | 212 | 75 | 22 | 0 | 0 | 714 |
| 2008 | 0 | 0 | 0 | 1 | 5 | 185 | 270 | 124 | 84 | 0 | 0 | 0 | 669 |
| 2009 | 0 | 0 | 0 | 16 | 13 | 82 | 115 | 208 | 16 | 0 | 0 | 0 | 450 |

### SNOWFALL (inches) 2009 ALBANY (KALB)

| YEAR | JUL | AUG | SEP | OCT | NOV | DEC | JAN | FEB | MAR | APR | MAY | JUN | TOTAL |
|---|---|---|---|---|---|---|---|---|---|---|---|---|---|
| 1980-81 | 0.0 | 0.0 | 0.0 | 0.0 | 11.8 | 12.8 | 11.9 | 6.9 | 1.5 | T | 0.0 | 0.0 | 44.9 |
| 1981-82 | 0.0 | 0.0 | 0.0 | 0.0 | 1.1 | 31.4 | 18.2 | 9.6 | 19.1 | 17.7 | 0.0 | 0.0 | 97.1 |
| 1982-83 | 0.0 | 0.0 | 0.0 | 0.0 | 0.6 | 5.5 | 27.5 | 17.4 | 9.2 | 14.7 | 0.1 | 0.0 | 75.0 |
| 1983-84 | 0.0 | 0.0 | 0.0 | 0.0 | 1.7 | 11.6 | 16.5 | 7.2 | 28.2 | T | 0.0 | 0.0 | 65.2 |
| 1984-85 | 0.0 | 0.0 | 0.0 | 0.0 | 2.2 | 11.7 | 8.4 | 10.1 | 8.7 | 0.2 | 0.0 | 0.0 | 41.3 |
| 1985-86 | 0.0 | 0.0 | 0.0 | 0.0 | 11.8 | 11.5 | 18.0 | 16.1 | 3.4 | 1.7 | T | 0.0 | 62.5 |
| 1986-87 | 0.0 | 0.0 | 0.0 | 0.0 | 8.3 | 20.3 | 47.8 | 2.8 | 0.8 | 0.6 | 0.0 | 0.0 | 80.6 |
| 1987-88 | 0.0 | 0.0 | 0.0 | 6.5 | 6.2 | 11.4 | 21.7 | 26.0 | 4.8 | 0.1 | 0.0 | 0.0 | 76.7 |
| 1988-89 | 0.0 | 0.0 | 0.0 | T | T | 7.8 | 1.3 | 5.1 | 4.7 | 0.1 | 0.0 | 0.0 | 19.0 |
| 1989-90 | T | 0.0 | T | 0.0 | 1.9 | 8.0 | 20.3 | 22.8 | 4.9 | T | 0.0 | 0.0 | 57.9 |
| 1990-91 | 0.0 | 0.0 | 0.0 | T | 0.4 | 8.5 | 11.2 | 5.3 | 3.3 | 0.0 | 0.0 | T | 28.7 |
| 1991-92 | 0.0 | 0.0 | 0.0 | 0.0 | 1.5 | 12.7 | 3.4 | 6.3 | 4.9 | 1.9 | 0.0 | 0.0 | 30.7 |
| 1992-93 | 0.0 | 0.0 | 0.0 | T | 2.8 | 12.6 | 14.3 | 28.6 | 34.3 | 1.6 | 0.0 | 0.0 | 94.2 |
| 1993-94 | 0.0 | 0.0 | 0.0 | T | 0.7 | 6.1 | 42.0 | 20.2 | 19.1 | T | T | 0.0 | 88.1 |
| 1994-95 | 0.0 | 0.0 | 0.0 | 0.0 | 4.1 | 2.9 | 3.9 | 15.4 | 4.6 | T | 0.0 | 0.0 | 30.9 |
| 1995-96 | T | 0.0 | 0.0 | 0.0 | 5.8 | 25.1 | 28.4 | 5.8 | 20.3 | 1.1 | 0.0 | 0.0 | 86.5 |
| 1996-97 | 0.0 | 0.0 | 0.0 | 0.0 | 4.0 | 11.1 | 16.7 | 8.2 | 23.6 | 3.0 | 0.0 | 0.0 | 66.6 |
| 1997-98 | 0.0 | 0.0 | 0.0 | T | 11.8 | 14.7 | 13.5 | 6.0 | 6.1 | 0.0 | 0.0 | 0.0 | 52.1 |
| 1998-99 | 0.0 | 0.0 | 0.0 | 0.0 | T | 3.2 | 20.4 | 5.8 | 14.7 | 0.0 | 0.0 | 0.0 | 44.1 |
| 1999-00 | 0.0 | 0.0 | 0.0 | 0.0 | 0.4 | 1.1 | 31.0 | 12.4 | 3.9 | 13.3 | 0.0 | 0.0 | 62.1 |
| 2000-01 | 0.0 | 0.0 | 0.0 | 0.4 | 2.5 | 20.0 | 7.6 | 16.0 | 30.6 | T | T | 0.0 | 77.1 |
| 2001-02 | T | 0.0 | 0.0 | T | T | 7.8 | 22.9 | 3.4 | 8.7 | 2.4 | 2.2 | T | 47.4 |
| 2002-03 | 0.0 | 0.0 | 0.0 | 0.5 | 12.0 | 33.2 | 32.2 | 16.7 | 5.2 | 5.6 | 0.0 | 0.0 | 105.4 |
| 2003-04 | 0.0 | 0.0 | 0.0 | T | 0.3 | 28.5 | 13.3 | 8.3 | 14.6 | 0.1 | 0.0 | 0.0 | 65.1 |
| 2004-05 | 0.0 | 0.0 | 0.0 | 0.0 | 0.5 | 10.7 | 31.8 | 7.0 | 25.9 | 0.0 | 0.0 | 0.0 | 75.9 |
| 2005-06 | 0.0 | 0.0 | 0.0 | 0.1 | 1.8 | 8.8 | 14.2 | 3.9 | 1.4 | T | 0.0 | 0.0 | 30.2 |
| 2006-07 | 0.0 | 0.0 | 0.0 | 0.0 | T | 0.3 | 3.9 | 23.2 | 14.9 | 3.6 | 0.0 | 0.0 | 45.9 |
| 2007-08 | 0.0 | 0.0 | 0.0 | 0.0 | 0.4 | 31.2 | 7.8 | 15.8 | 5.9 | T | 0.0 | T | 61.1 |
| 2008-09 | T | 0.0 | 0.0 | T | 0.5 | 27.5 | 19.4 | 2.5 | 2.7 | T | 0.0 | 0.0 | 52.6 |
| 2009- | 0.0 | 0.0 | 0.0 | 0.0 | T | 13.3 | | | | | | | |
| POR= 72 YRS | T | 0.0 | T | 0.1 | 3.9 | 14.0 | 16.6 | 13.1 | 11.4 | 2.5 | 0.2 | T | 61.8 |

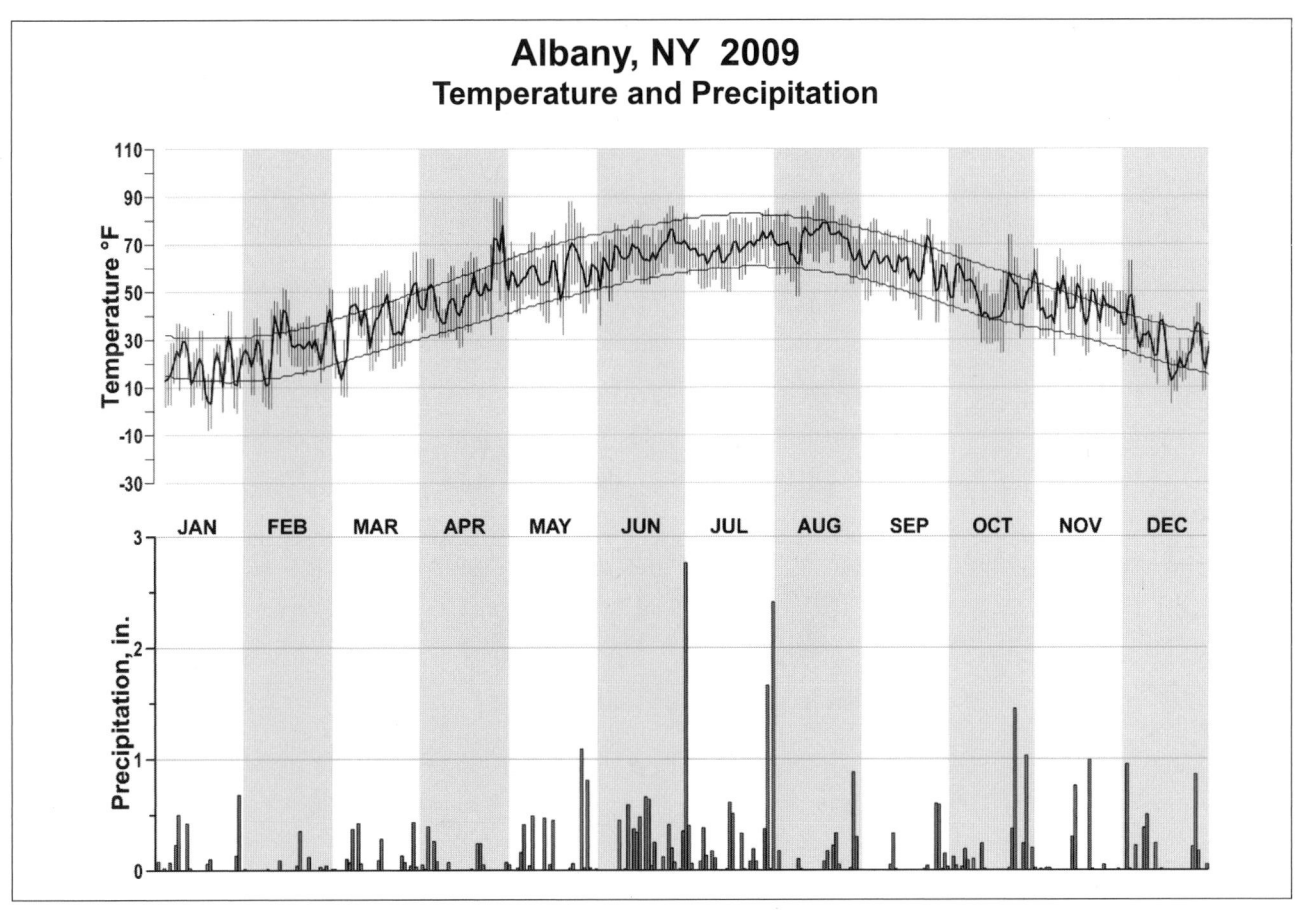

**Albany, NY 2009**
**Temperature and Precipitation**

# 2009
# BUFFALO
# NEW YORK (KBUF)

The Niagara Frontier experiences a fairly humid, continental type climate, but with a definite "maritime" flavor due to a strong modification from the Great Lakes (especially Lake Erie). Buffalo's weather repution is highly exaggerated, and due mainly to its propensity for localized heavy Lake-effect snowstorms in late fall and early winter. Summers, on the other hand, are among the most pleasant in the Northeast.

Winters in general are cloudy, cold and snowy...but are changeable and include frequent thaws and rain as well. Snow covers the ground more often than not from Christmas into early March...but periods of bare ground are not uncommon. Over half of the annual snowfall comes from "Lake-effect" process and is very localized. This feature develops when cold air crosses the warmer lake waters and becomes saturated.. creating clouds and precipitation downwind. The exact location of these snowbands are determined by the direction of the wind. Areas south of Buffalo derive much more snow from this process than the more densely populated northern suburbs. This snow machine can start as early as mid-November, peaks in December, then virtually shuts down after Lake Erie freezes in mid to late January. The Buffalo area is not subject to heavy general or "synoptic" snowstorms. Most of them pass by to the east. Total season snowfall ranges from about 60 inches in the far northern suburbs to 80-90 inches in the city and eastern suburbs to as much as 120 inches south of the city. The lakes do modify any extreme cold as the mercury falls below zero on only about four nights in an average winter...with anything below -10 extremely rare.

Spring comes slowly to the Niagara Frontier. The ice pack in lake Erie does not usually disappear until mid-April and the Lake remains chilly through most of May. As the prevailing flow is southwesterly, areas near the lake are often as much as 20 degrees colder than inland locations. Conversely, the cool Lake acts as a strong stabilizing influence so areas near the city and lakeshore experience fewer thunderstorms and more sunshine then inland areas in spring. The slow start to the growing season also diminishes the threat of damaging late season frosts. The average date of the last frost is around April 30 in the metro area...but mid-May well inland.

Summer is beautiful in the Buffalo area. Sunshine is plentiful, temperature are warm but seldom hot, and humidity levels moderate. Rainfall is adequate, but does show an overnight maximum and seldom is a problem for outdoor activities. The stabilizing effect of Lake Erie continues to inhibit thunderstorms and ehance sunshine in the immediate Buffalo area..at least through most of July. It also moderates most extreme heat approaching from the Ohio Valley. There usually are several periods of uncomfortably warm and muggy weather in an average summer...but 90-degree readings are relatively rare (only 3 per year). August usually turns a bit more humid and showery as the Lake is warmer and loses its stabilizing influence. In fact, a good nighttime thunderstorm or two is often a feature of late summmer in Buffalo. Overall though...Buffalo has the sunniest and driest summers of any major city in the Northeast.

Autumn is pleasant, but rather brief. September is usually very tame, and much of October as well. The first frosts can be expected in late September over interior sections, but not until mid-October in the metro area. The warm lake can extend the growing season into early November during some years close to the Lakeshore. The growing season is relatively long for the latitude...about 180 days...and is conducive to the many Fruit orchards and wineries, especially near Lake Ontario and along the Lake Erie shore. Cold air surges from Canada become more common starting in late October...with their passage over the warmer Great Lakes resulting in a drastic increase in cloud cover in late October and early November as the Lake-effect season begins. The first measurable snows can be expected in mid to late November, but ground cover is only sporadic until mid December. Many of Buffalo's greatest snowstorms however, have occurred in late November and early December, all due to the Lake effect phenomenon.

# NORMALS, MEANS, AND EXTREMES
## BUFFALO (KBUF)

| LATITUDE: 42°56'N | LONGITUDE: -78°44'W | ELEVATION (FT): GRND: 710  BARO: 717 | TIME ZONE: EASTERN (UTC -5) | WBAN: 14733 |
|---|---|---|---|---|

| | ELEMENT | POR | JAN | FEB | MAR | APR | MAY | JUN | JUL | AUG | SEP | OCT | NOV | DEC | YEAR |
|---|---|---|---|---|---|---|---|---|---|---|---|---|---|---|---|
| TEMPERATURE °F | NORMAL DAILY MAXIMUM | 30 | 31.1 | 33.2 | 42.5 | 54.1 | 66.4 | 74.8 | 79.6 | 77.8 | 70.1 | 58.9 | 46.7 | 36.0 | 55.9 |
| | MEAN DAILY MAXIMUM | 88 | 31.5 | 31.7 | 40.9 | 53.0 | 65.0 | 74.0 | 79.4 | 77.9 | 70.4 | 59.6 | 46.8 | 35.7 | 55.5 |
| | HIGHEST DAILY MAXIMUM | 66 | 72 | 71 | 81 | 94 | 91 | 96 | 97 | 99 | 98 | 87 | 80 | 74 | 99 |
| | YEAR OF OCCURRENCE | | 1950 | 2000 | 1945 | 1990 | 2006 | 1988 | 1995 | 1948 | 1953 | 1951 | 1961 | 1982 | AUG 1948 |
| | MEAN OF EXTREME MAXS. | 88 | 53.2 | 54.2 | 67.3 | 77.3 | 83.4 | 88.5 | 89.7 | 88.6 | 85.9 | 77.9 | 67.8 | 56.5 | 74.2 |
| | NORMAL DAILY MINIMUM | 30 | 17.8 | 18.6 | 26.1 | 36.4 | 47.7 | 56.9 | 62.1 | 60.5 | 52.9 | 42.6 | 33.7 | 23.6 | 39.9 |
| | MEAN DAILY MINIMUM | 88 | 18.4 | 18.0 | 25.7 | 35.6 | 46.5 | 56.3 | 62.1 | 60.7 | 53.2 | 43.5 | 33.8 | 23.6 | 39.8 |
| | LOWEST DAILY MINIMUM | 66 | -16 | -20 | -7 | 12 | 26 | 35 | 43 | 38 | 32 | 20 | 9 | -10 | -20 |
| | YEAR OF OCCURRENCE | | 1982 | 1961 | 1984 | 1982 | 1947 | 1945 | 1945 | 1982 | 1991 | 1965 | 1971 | 1980 | FEB 1961 |
| | MEAN OF EXTREME MINS. | 88 | -0.1 | 0.4 | 8.2 | 23.2 | 33.9 | 43.5 | 50.9 | 48.0 | 38.2 | 29.5 | 19.1 | 4.5 | 24.9 |
| | NORMAL DRY BULB | 30 | 24.5 | 25.9 | 34.3 | 45.3 | 57.0 | 65.8 | 70.8 | 69.1 | 61.5 | 50.7 | 40.2 | 29.8 | 47.9 |
| | MEAN DRY BULB | 88 | 25.0 | 24.9 | 33.3 | 44.3 | 55.8 | 65.3 | 70.8 | 69.3 | 61.8 | 51.5 | 40.3 | 29.7 | 47.7 |
| | MEAN WET BULB | 26 | 23.1 | 23.5 | 29.6 | 39.8 | 49.7 | 59.5 | 63.7 | 62.9 | 56.6 | 45.9 | 36.8 | 27.4 | 43.2 |
| | MEAN DEW POINT | 26 | 19.8 | 20.1 | 25.8 | 35.5 | 46.1 | 56.3 | 60.6 | 60.1 | 53.8 | 42.6 | 33.3 | 24.2 | 39.9 |
| | NORMAL NO. DAYS WITH: | | | | | | | | | | | | | | |
| | MAXIMUM >= 90 | 30 | 0.0 | 0.0 | 0.0 | * | 0.1 | 0.5 | 1.5 | 0.6 | * | 0.0 | 0.0 | 0.0 | 2.7 |
| | MAXIMUM <= 32 | 30 | 16.8 | 14.3 | 7.0 | 0.6 | 0.0 | 0.0 | 0.0 | 0.0 | 0.0 | 0.0 | 2.1 | 10.8 | 51.6 |
| | MINIMUM <= 32 | 30 | 27.9 | 25.4 | 23.0 | 9.8 | 0.3 | 0.0 | 0.0 | 0.0 | * | 3.0 | 14.0 | 25.2 | 128.6 |
| | MINIMUM <= 0 | 30 | 1.9 | 1.1 | 0.2 | 0.0 | 0.0 | 0.0 | 0.0 | 0.0 | 0.0 | 0.0 | 0.0 | 0.6 | 3.8 |
| H/C | NORMAL HEATING DEG. DAYS | 30 | 1256 | 1110 | 961 | 594 | 268 | 65 | 8 | 21 | 149 | 442 | 737 | 1081 | 6692 |
| | NORMAL COOLING DEG. DAYS | 30 | 0 | 0 | 0 | 4 | 28 | 101 | 203 | 158 | 50 | 4 | 0 | 0 | 548 |
| RH | NORMAL (PERCENT) | 30 | 77 | 76 | 72 | 68 | 67 | 69 | 68 | 72 | 74 | 72 | 75 | 77 | 72 |
| | HOUR 01 LST | 30 | 78 | 79 | 77 | 75 | 77 | 80 | 80 | 83 | 83 | 79 | 78 | 79 | 79 |
| | HOUR 07 LST | 30 | 80 | 80 | 79 | 76 | 75 | 78 | 78 | 83 | 85 | 82 | 79 | 81 | 80 |
| | HOUR 13 LST | 30 | 73 | 69 | 63 | 58 | 55 | 57 | 55 | 57 | 60 | 60 | 68 | 72 | 62 |
| | HOUR 19 LST | 30 | 77 | 76 | 71 | 64 | 61 | 62 | 61 | 66 | 71 | 72 | 75 | 77 | 69 |
| S | PERCENT POSSIBLE SUNSHINE | 65 | 31 | 38 | 46 | 50 | 58 | 64 | 67 | 64 | 57 | 50 | 29 | 27 | 48 |
| W/O | MEAN NO. DAYS WITH: | | | | | | | | | | | | | | |
| | HEAVY FOG(VISBY <= 1/4 MI) | 46 | 1.4 | 1.5 | 2.6 | 1.7 | 1.7 | 0.9 | 0.7 | 0.7 | 0.8 | 1.2 | 1.3 | 1.2 | 15.7 |
| | THUNDERSTORMS | 64 | 0.2 | 0.2 | 1.1 | 2.2 | 3.3 | 5.2 | 5.9 | 6.0 | 3.5 | 1.7 | 1.0 | 0.5 | 30.8 |
| CLOUDNESS | MEAN: | | | | | | | | | | | | | | |
| | SUNRISE-SUNSET (OKTAS) | | | | | | | | | | | | | | |
| | MIDNIGHT-MIDNIGHT (OKTAS) | | | | | | | | | | | | | | |
| | MEAN NO. DAYS WITH: | | | | | | | | | | | | | | |
| | CLEAR | | | | | | | | | | | | | | |
| | PARTLY CLOUDY | | | | | | | | | | | | | | |
| | CLOUDY | | | | | | | | | | | | | | |
| PR | MEAN STATION PRESSURE(IN) | 26 | 29.25 | 29.27 | 29.26 | 29.21 | 29.22 | 29.21 | 29.23 | 29.27 | 29.30 | 29.30 | 29.27 | 29.27 | 29.26 |
| | MEAN SEA-LEVEL PRES. (IN) | 26 | 30.04 | 30.06 | 30.05 | 29.98 | 29.98 | 29.96 | 29.97 | 30.03 | 30.05 | 30.06 | 30.05 | 30.05 | 30.02 |
| WINDS | MEAN SPEED (MPH) | 26 | 12.7 | 11.9 | 11.3 | 10.9 | 10.3 | 9.4 | 9.1 | 8.4 | 8.8 | 9.9 | 11.2 | 11.9 | 10.5 |
| | PREVAIL.DIR(TENS OF DEGS) | 42 | 25 | 25 | 24 | 24 | 24 | 24 | 24 | 24 | 25 | 24 | 27 | 27 | 24 |
| | MAXIMUM 2-MINUTE: | | | | | | | | | | | | | | |
| | SPEED (MPH) | 14 | 49 | 54 | 48 | 40 | 43 | 38 | 41 | 37 | 59 | 46 | 47 | 53 | 59 |
| | DIR. (TENS OF DEGS) | | 23 | 24 | 24 | 31 | 24 | 19 | 31 | 34 | 20 | 23 | 24 | 24 | 20 |
| | YEAR OF OCCURRENCE | | 2008 | 1997 | 2002 | 2004 | 1997 | 2007 | 1999 | 2009 | 2005 | 2001 | 2007 | 2008 | SEP 2005 |
| | MAXIMUM 3-SECOND | | | | | | | | | | | | | | |
| | SPEED (MPH) | 14 | 68 | 70 | 62 | 54 | 55 | 48 | 54 | 49 | 67 | 64 | 61 | 75 | 75 |
| | DIR. (TENS OF DEGS) | | 25 | 25 | 24 | 31 | 23 | 19 | 31 | 35 | 19 | 21 | 22 | 23 | 23 |
| | YEAR OF OCCURRENCE | | 2008 | 1997 | 2002 | 2004 | 1997 | 2008 | 1999 | 2009 | 2005 | 2002 | 1998 | 2008 | DEC 2008 |
| PRECIPITATION | NORMAL (IN) | 30 | 3.16 | 2.42 | 2.99 | 3.04 | 3.35 | 3.82 | 3.14 | 3.87 | 3.84 | 3.19 | 3.92 | 3.80 | 40.54 |
| | MAXIMUM MONTHLY (IN) | 66 | 6.88 | 5.90 | 5.97 | 5.90 | 7.22 | 8.36 | 8.93 | 10.67 | 8.99 | 9.13 | 9.75 | 8.71 | 10.67 |
| | YEAR OF OCCURRENCE | | 1982 | 1990 | 1991 | 1961 | 1989 | 1987 | 1992 | 1977 | 1977 | 1954 | 1985 | 1990 | AUG 1977 |
| | MINIMUM MONTHLY (IN) | 66 | 1.03 | 0.81 | 1.20 | 0.90 | 0.60 | 0.11 | 0.73 | 1.10 | 0.77 | 0.30 | 1.44 | 0.69 | 0.11 |
| | YEAR OF OCCURRENCE | | 1946 | 1968 | 1967 | 2003 | 2005 | 1955 | 2001 | 1948 | 1964 | 1963 | 1944 | 1943 | JUN 1955 |
| | MAXIMUM IN 24 HOURS (IN) | 66 | 3.68 | 2.31 | 2.14 | 2.09 | 3.52 | 5.01 | 3.38 | 3.88 | 4.94 | 3.49 | 2.51 | 2.33 | 5.01 |
| | YEAR OF OCCURRENCE | | 2007 | 1954 | 1954 | 1991 | 1986 | 1987 | 1963 | 1963 | 1979 | 1945 | 1949 | 1990 | JUN 1987 |
| | NORMAL NO. DAYS WITH: | | | | | | | | | | | | | | |
| | PRECIPITATION >= 0.01 | 30 | 19.8 | 17.2 | 15.7 | 13.6 | 12.6 | 11.9 | 10.5 | 10.5 | 11.6 | 12.8 | 15.8 | 19.4 | 171.4 |
| | PRECIPITATION >= 1.00 | 30 | 0.3 | 0.2 | 0.3 | 0.3 | 0.6 | 0.8 | 0.7 | 1.0 | 1.0 | 0.4 | 0.6 | 0.4 | 6.6 |
| SNOWFALL | NORMAL (IN) | 30 | 26.1 | 17.8 | 12.4 | 3.6 | 0.3 | 0.0 | 0.0 | 0.0 | 0.0 | 0.3 | 11.0 | 25.5 | 97.0 |
| | MAXIMUM MONTHLY (IN) | 66 | 68.3 | 54.2 | 32.8 | 15.0 | 7.9 | T | T | T | T | 22.6 | 45.6 | 82.7 | 82.7 |
| | YEAR OF OCCURRENCE | | 1977 | 1958 | 2001 | 1975 | 1989 | 1980 | 2009 | 2008 | 2009 | 2006 | 2000 | 2001 | DEC 2001 |
| | MAXIMUM IN 24 HOURS (IN) | 66 | 25.3 | 19.4 | 17.2 | 6.8 | 7.9 | T | T | T | T | 2.8 | 24.9 | 37.9 | 37.9 |
| | YEAR OF OCCURRENCE' | | 1982 | 1984 | 1993 | 1975 | 1989 | 1980 | 2009 | 1991 | 2009 | 1993 | 2000 | 1995 | DEC 1995 |
| | MAXIMUM SNOW DEPTH (IN) | 61 | 38 | 42 | 20 | 12 | 4 | 0 | 0 | 0 | 0 | 22 | 25 | 44 | 44 |
| | YEAR OF OCCURRENCE | | 1977 | 1977 | 1984 | 1975 | 1989 | | | | | 2006 | 2000 | 2001 | DEC 2001 |
| | NORMAL NO. DAYS WITH: | | | | | | | | | | | | | | |
| | SNOWFALL >= 1.0 | 30 | 7.5 | 5.6 | 3.5 | 1.0 | 0.0 | 0.0 | 0.0 | 0.0 | 0.0 | 0.1 | 2.7 | 6.5 | 26.9 |

## PRECIPITATION (inches) 2009 BUFFALO (KBUF)

| YEAR | JAN | FEB | MAR | APR | MAY | JUN | JUL | AUG | SEP | OCT | NOV | DEC | ANNUAL |
|------|-----|-----|-----|-----|-----|-----|-----|-----|-----|-----|-----|-----|--------|
| 1980 | 1.97 | 1.08 | 4.05 | 2.43 | 1.60 | 5.82 | 3.55 | 3.58 | 4.53 | 4.69 | 2.36 | 2.65 | 38.31 |
| 1981 | 1.11 | 3.50 | 1.70 | 3.09 | 2.56 | 3.68 | 5.05 | 3.13 | 4.24 | 3.31 | 2.22 | 2.87 | 36.46 |
| 1982 | 6.88 | 1.28 | 2.64 | 2.33 | 3.66 | 3.14 | 1.50 | 4.62 | 3.37 | 2.06 | 6.31 | 3.32 | 41.11 |
| 1983 | 1.44 | 1.30 | 3.20 | 2.55 | 3.28 | 2.99 | 2.01 | 3.51 | 2.11 | 4.62 | 5.19 | 7.30 | 39.50 |
| 1984 | 1.54 | 3.59 | 1.77 | 2.53 | 4.67 | 6.86 | 1.37 | 4.16 | 3.73 | 0.87 | 2.66 | 3.67 | 37.42 |
| 1985 | 4.27 | 3.34 | 4.42 | 1.33 | 3.46 | 3.21 | 1.81 | 4.63 | 1.20 | 3.73 | 9.75 | 4.85 | 46.00 |
| 1986 | 2.31 | 2.60 | 1.95 | 3.33 | 4.42 | 4.15 | 2.82 | 2.73 | 3.88 | 4.34 | 3.11 | 4.02 | 39.66 |
| 1987 | 2.90 | 0.85 | 3.66 | 3.40 | 1.35 | 8.36 | 3.09 | 3.38 | 5.32 | 2.62 | 4.44 | 2.78 | 42.15 |
| 1988 | 1.58 | 4.07 | 2.99 | 2.96 | 2.74 | 1.56 | 6.35 | 2.69 | 2.07 | 6.08 | 3.37 | 2.15 | 38.61 |
| 1989 | 1.77 | 2.54 | 3.15 | 1.88 | 7.22 | 7.83 | 0.93 | 1.84 | 3.85 | 2.98 | 4.83 | 2.34 | 41.16 |
| 1990 | 2.69 | 5.90 | 1.50 | 5.22 | 6.08 | 3.55 | 3.14 | 3.25 | 3.65 | 4.59 | 2.61 | 8.71 | 50.89 |
| 1991 | 2.07 | 2.06 | 5.97 | 5.83 | 3.10 | 0.86 | 3.34 | 2.84 | 3.19 | 3.11 | 4.02 | 3.81 | 40.20 |
| 1992 | 2.01 | 2.45 | 2.93 | 4.68 | 3.48 | 2.21 | 8.93 | 3.79 | 5.56 | 2.80 | 4.92 | 3.80 | 47.56 |
| 1993 | 4.35 | 1.92 | 3.02 | 2.55 | 1.79 | 4.99 | 1.78 | 3.86 | 5.53 | 3.69 | 3.58 | 3.60 | 40.66 |
| 1994 | 2.90 | 1.40 | 2.61 | 4.02 | 3.54 | 4.27 | 2.08 | 4.09 | 3.19 | 1.87 | 4.08 | 2.67 | 36.72 |
| 1995 | 4.89 | 2.62 | 1.33 | 1.41 | 2.40 | 1.33 | 3.53 | 2.07 | 1.32 | 6.07 | 4.14 | 2.88 | 33.99 |
| 1996 | 3.42 | 2.09 | 2.37 | 5.63 | 4.08 | 5.20 | 5.15 | 2.14 | 7.51 | 4.22 | 2.99 | 3.42 | 48.22 |
| 1997 | 4.25 | 2.97 | 4.47 | 1.65 | 3.61 | 3.06 | 1.85 | 4.67 | 5.06 | 2.29 | 4.32 | 2.88 | 41.08 |
| 1998 | 5.61 | 2.28 | 3.86 | 2.54 | 3.73 | 2.87 | 4.39 | 1.74 | 2.43 | 2.10 | 1.61 | 1.54 | 34.70 |
| 1999 | 5.78 | 1.10 | 2.43 | 2.21 | 2.82 | 1.93 | 1.00 | 4.38 | 3.95 | 2.95 | 3.33 | 2.20 | 34.08 |
| 2000 | 2.65 | 1.75 | 2.12 | 4.07 | 4.38 | 6.51 | 2.90 | 3.21 | 3.92 | 1.11 | 5.82 | 3.76 | 42.20 |
| 2001 | 2.15 | 2.33 | 3.31 | 1.27 | 4.28 | 1.36 | 0.73 | 2.13 | 3.45 | 4.34 | 3.35 | 6.48 | 35.18 |
| 2002 | 3.54 | 3.15 | 3.28 | 4.38 | 5.23 | 1.47 | 3.24 | 1.77 | 2.54 | 3.21 | 3.57 | 4.36 | 39.74 |
| 2003 | 2.30 | 2.69 | 2.81 | 0.90 | 5.43 | 1.79 | 3.69 | 2.47 | 3.91 | 3.43 | 4.10 | 3.64 | 37.16 |
| 2004 | 2.95 | 1.15 | 3.10 | 3.94 | 5.72 | 2.02 | 6.04 | 1.86 | 4.07 | 2.98 | 2.91 | 4.99 | 41.73 |
| 2005 | 3.57 | 2.42 | 1.38 | 4.50 | 0.60 | 3.27 | 1.82 | 5.92 | 4.89 | 2.64 | 5.70 | 2.36 | 39.07 |
| 2006 | 3.67 | 2.45 | 2.14 | 1.98 | 1.90 | 3.38 | 4.60 | 3.28 | 6.95 | 8.75 | 2.15 | 3.16 | 44.41 |
| 2007 | 4.77 | 1.71 | 2.61 | 2.96 | 0.87 | 1.82 | 3.31 | 1.13 | 3.55 | 2.73 | 5.38 | 4.28 | 35.12 |
| 2008 | 2.41 | 4.83 | 4.22 | 2.05 | 2.54 | 4.91 | 2.80 | 5.33 | 3.96 | 4.13 | 3.34 | 6.79 | 47.31 |
| 2009 | 2.27 | 2.65 | 3.25 | 3.15 | 1.89 | 2.92 | 4.37 | 5.32 | 5.65 | 4.77 | 2.94 | 5.13 | 44.31 |
| POR=<br>88 YRS | 3.05 | 2.50 | 2.86 | 2.87 | 2.95 | 2.99 | 2.89 | 3.35 | 3.44 | 2.98 | 3.57 | 3.44 | 36.89 |

WBAN : 14733

## AVERAGE TEMPERATURE (°F) 2009 BUFFALO (KBUF)

| YEAR | JAN | FEB | MAR | APR | MAY | JUN | JUL | AUG | SEP | OCT | NOV | DEC | ANNUAL |
|------|-----|-----|-----|-----|-----|-----|-----|-----|-----|-----|-----|-----|--------|
| 1980 | 25.8 | 21.2 | 31.8 | 46.1 | 58.1 | 61.9 | 71.7 | 72.6 | 62.4 | 48.7 | 39.4 | 25.3 | 47.1 |
| 1981 | 19.3 | 32.9 | 33.9 | 47.2 | 56.4 | 66.2 | 71.8 | 70.0 | 60.9 | 48.2 | 40.4 | 29.0 | 48.0 |
| 1982 | 17.2 | 23.2 | 32.5 | 41.6 | 61.0 | 62.2 | 71.8 | 65.0 | 61.6 | 52.6 | 43.0 | 37.5 | 47.4 |
| 1983 | 27.0 | 29.6 | 36.7 | 43.6 | 53.9 | 67.6 | 74.2 | 71.2 | 63.7 | 51.7 | 40.8 | 22.7 | 48.6 |
| 1984 | 20.4 | 33.8 | 27.1 | 47.7 | 52.9 | 67.8 | 70.3 | 70.3 | 58.5 | 53.2 | 39.0 | 35.6 | 48.1 |
| 1985 | 21.1 | 24.8 | 35.6 | 49.5 | 59.5 | 62.7 | 69.7 | 69.2 | 64.2 | 52.5 | 42.0 | 25.6 | 48.0 |
| 1986 | 25.5 | 24.5 | 36.2 | 47.8 | 59.7 | 64.1 | 71.1 | 67.9 | 61.8 | 50.9 | 37.7 | 32.4 | 48.3 |
| 1987 | 26.1 | 25.0 | 37.7 | 50.0 | 60.5 | 68.9 | 74.2 | 68.9 | 63.4 | 47.9 | 42.5 | 34.3 | 50.0 |
| 1988 | 26.6 | 24.3 | 35.2 | 46.1 | 59.7 | 64.0 | 74.8 | 72.4 | 62.1 | 46.9 | 43.0 | 30.0 | 48.8 |
| 1989 | 31.3 | 22.7 | 33.0 | 41.9 | 55.1 | 65.9 | 71.5 | 68.5 | 60.8 | 51.5 | 37.9 | 17.4 | 46.5 |
| 1990 | 33.4 | 29.3 | 36.9 | 48.5 | 54.9 | 66.7 | 71.4 | 70.4 | 61.7 | 52.5 | 43.4 | 34.4 | 50.3 |
| 1991 | 26.0 | 30.6 | 37.8 | 50.5 | 64.3 | 69.1 | 71.9 | 71.0 | 62.0 | 53.1 | 39.3 | 31.3 | 50.6 |
| 1992 | 27.1 | 27.7 | 31.6 | 43.8 | 57.3 | 63.4 | 66.8 | 66.3 | 61.6 | 47.9 | 40.2 | 31.9 | 47.1 |
| 1993 | 29.5 | 20.7 | 30.7 | 47.3 | 57.0 | 66.0 | 73.4 | 72.0 | 59.4 | 49.2 | 39.6 | 29.6 | 47.9 |
| 1994 | 17.2 | 22.8 | 33.4 | 48.2 | 54.7 | 69.0 | 73.3 | 68.0 | 61.9 | 52.2 | 45.1 | 34.0 | 48.3 |
| 1995 | 29.8 | 21.9 | 37.8 | 42.3 | 56.8 | 69.9 | 72.7 | 73.0 | 60.0 | 54.2 | 36.4 | 24.5 | 48.3 |
| 1996 | 22.5 | 24.2 | 29.0 | 42.2 | 54.5 | 67.8 | 68.5 | 70.5 | 62.7 | 51.7 | 35.4 | 33.5 | 46.9 |
| 1997 | 24.7 | 30.1 | 33.1 | 42.3 | 50.6 | 66.7 | 68.6 | 66.8 | 60.5 | 50.5 | 37.6 | 31.8 | 46.9 |
| 1998 | 31.1 | 34.1 | 36.5 | 46.8 | 62.8 | 65.3 | 69.6 | 71.2 | 63.7 | 52.6 | 42.0 | 35.3 | 50.9 |
| 1999 | 23.5 | 31.0 | 31.0 | 46.0 | 59.7 | 68.4 | 74.3 | 67.9 | 64.3 | 50.1 | 43.9 | 32.0 | 49.3 |
| 2000 | 23.6 | 29.9 | 40.0 | 44.2 | 57.5 | 64.9 | 67.6 | 68.0 | 61.2 | 52.3 | 38.8 | 22.1 | 47.5 |
| 2001 | 27.0 | 28.2 | 31.1 | 47.3 | 58.8 | 67.0 | 69.8 | 73.0 | 62.7 | 53.0 | 46.9 | 35.9 | 50.1 |
| 2002 | 31.6 | 31.2 | 34.2 | 46.2 | 51.8 | 67.0 | 73.4 | 71.5 | 66.9 | 49.3 | 39.4 | 28.4 | 49.2 |
| 2003 | 19.0 | 20.8 | 33.5 | 43.0 | 55.4 | 63.5 | 69.6 | 70.8 | 62.8 | 48.8 | 43.1 | 33.2 | 47.0 |
| 2004 | 17.4 | 25.5 | 37.1 | 46.0 | 58.2 | 63.6 | 69.1 | 67.2 | 65.2 | 51.6 | 42.4 | 29.7 | 47.8 |
| 2005 | 23.8 | 25.3 | 29.4 | 46.8 | 53.5 | 71.8 | 75.0 | 72.8 | 66.0 | 52.7 | 43.3 | 27.1 | 49.0 |
| 2006 | 34.9 | 27.9 | 35.2 | 48.0 | 60.0 | 68.3 | 73.7 | 69.7 | 60.5 | 49.1 | 44.6 | 37.2 | 50.8 |
| 2007 | 28.9 | 18.6 | 35.0 | 42.5 | 59.2 | 69.4 | 69.7 | 72.4 | 66.1 | 58.8 | 39.1 | 29.4 | 49.1 |
| 2008 | 29.7 | 25.1 | 31.6 | 50.9 | 53.4 | 67.9 | 71.4 | 68.5 | 64.2 | 49.7 | 39.9 | 29.4 | 48.5 |
| 2009 | 18.5 | 27.2 | 35.4 | 46.7 | 57.2 | 64.5 | 66.9 | 70.2 | 62.4 | 48.6 | 44.1 | 28.6 | 47.5 |
| POR=<br>88 YRS | 25.0 | 24.9 | 33.3 | 44.3 | 55.8 | 65.3 | 70.8 | 69.3 | 61.8 | 51.5 | 40.3 | 29.7 | 47.7 |

## HEATING DEGREE DAYS (base 65°F) 2009 BUFFALO (KBUF)

| YEAR | JUL | AUG | SEP | OCT | NOV | DEC | JAN | FEB | MAR | APR | MAY | JUN | TOTAL |
|------|-----|-----|-----|-----|-----|-----|-----|-----|-----|-----|-----|-----|-------|
| 1980-81 | 2 | 0 | 128 | 498 | 759 | 1224 | 1411 | 895 | 956 | 527 | 269 | 33 | 6702 |
| 1981-82 | 6 | 11 | 170 | 514 | 732 | 1108 | 1476 | 1163 | 1002 | 698 | 147 | 95 | 7122 |
| 1982-83 | 4 | 65 | 140 | 382 | 656 | 848 | 1172 | 987 | 868 | 636 | 342 | 71 | 6171 |
| 1983-84 | 5 | 10 | 125 | 418 | 722 | 1304 | 1378 | 899 | 1167 | 519 | 385 | 35 | 6967 |
| 1984-85 | 11 | 22 | 210 | 360 | 774 | 905 | 1354 | 1120 | 902 | 476 | 196 | 95 | 6425 |
| 1985-86 | 8 | 12 | 114 | 378 | 685 | 1215 | 1215 | 1128 | 885 | 519 | 197 | 80 | 6436 |
| 1986-87 | 4 | 42 | 137 | 430 | 811 | 1003 | 1199 | 1115 | 837 | 447 | 213 | 28 | 6266 |
| 1987-88 | 3 | 25 | 91 | 527 | 665 | 948 | 1184 | 1174 | 916 | 560 | 186 | 113 | 6392 |
| 1988-89 | 5 | 17 | 122 | 560 | 654 | 1078 | 1038 | 1177 | 985 | 687 | 321 | 60 | 6704 |
| 1989-90 | 1 | 28 | 170 | 411 | 806 | 1466 | 970 | 995 | 866 | 518 | 311 | 46 | 6588 |
| 1990-91 | 5 | 2 | 141 | 395 | 640 | 941 | 1203 | 956 | 836 | 431 | 141 | 22 | 5713 |
| 1991-92 | 1 | 1 | 166 | 376 | 762 | 1037 | 1169 | 1076 | 1027 | 633 | 254 | 93 | 6595 |
| 1992-93 | 28 | 41 | 148 | 525 | 738 | 1021 | 1095 | 1235 | 1053 | 526 | 257 | 60 | 6727 |
| 1993-94 | 0 | 8 | 212 | 486 | 752 | 1089 | 1476 | 1174 | 972 | 502 | 327 | 48 | 7046 |
| 1994-95 | 0 | 26 | 123 | 390 | 591 | 955 | 1085 | 1201 | 835 | 674 | 247 | 22 | 6149 |
| 1995-96 | 14 | 3 | 164 | 329 | 851 | 1250 | 1310 | 1178 | 1107 | 677 | 333 | 22 | 7238 |
| 1996-97 | 15 | 1 | 130 | 406 | 881 | 969 | 1241 | 970 | 983 | 673 | 438 | 40 | 6747 |
| 1997-98 | 17 | 25 | 150 | 457 | 814 | 1023 | 1045 | 862 | 878 | 538 | 96 | 104 | 6009 |
| 1998-99 | 0 | 9 | 88 | 378 | 682 | 912 | 1280 | 949 | 1048 | 566 | 193 | 58 | 6163 |
| 1999-00 | 0 | 17 | 97 | 454 | 628 | 1014 | 1276 | 1012 | 770 | 617 | 246 | 73 | 6204 |
| 2000-01 | 20 | 26 | 176 | 385 | 780 | 1323 | 1171 | 1023 | 1042 | 528 | 190 | 61 | 6725 |
| 2001-02 | 18 | 0 | 127 | 371 | 535 | 893 | 1029 | 940 | 946 | 561 | 416 | 62 | 5898 |
| 2002-03 | 1 | 3 | 51 | 498 | 758 | 1127 | 1420 | 1232 | 967 | 653 | 289 | 86 | 7085 |
| 2003-04 | 2 | 12 | 93 | 498 | 651 | 978 | 1466 | 1137 | 860 | 562 | 228 | 94 | 6581 |
| 2004-05 | 7 | 30 | 61 | 409 | 673 | 1090 | 1268 | 1104 | 1096 | 538 | 353 | 20 | 6649 |
| 2005-06 | 0 | 0 | 36 | 398 | 643 | 1169 | 927 | 1034 | 914 | 504 | 196 | 31 | 5852 |
| 2006-07 | 0 | 10 | 142 | 489 | 603 | 855 | 1110 | 1294 | 923 | 670 | 220 | 31 | 6347 |
| 2007-08 | 6 | 6 | 64 | 226 | 771 | 1097 | 1085 | 1149 | 1030 | 425 | 355 | 37 | 6251 |
| 2008-09 | 1 | 16 | 85 | 472 | 747 | 1094 | 1433 | 1049 | 909 | 549 | 245 | 60 | 6660 |
| 2009- | 19 | 16 | 108 | 503 | 622 | 1121 | | | | | | | |

WBAN : 14733

## COOLING DEGREE DAYS (base 65°F) 2009 BUFFALO (KBUF)

| YEAR | JAN | FEB | MAR | APR | MAY | JUN | JUL | AUG | SEP | OCT | NOV | DEC | TOTAL |
|------|-----|-----|-----|-----|-----|-----|-----|-----|-----|-----|-----|-----|-------|
| 1980 | 0 | 0 | 0 | 0 | 32 | 56 | 217 | 242 | 58 | 2 | 0 | 0 | 607 |
| 1981 | 0 | 0 | 0 | 2 | 13 | 78 | 225 | 173 | 55 | 0 | 0 | 0 | 546 |
| 1982 | 0 | 0 | 0 | 3 | 31 | 18 | 221 | 74 | 45 | 2 | 0 | 2 | 396 |
| 1983 | 0 | 0 | 0 | 0 | 5 | 157 | 300 | 214 | 90 | 15 | 0 | 0 | 781 |
| 1984 | 0 | 0 | 0 | 5 | 16 | 123 | 183 | 193 | 23 | 1 | 0 | 0 | 544 |
| 1985 | 0 | 0 | 0 | 18 | 32 | 32 | 161 | 151 | 96 | 0 | 1 | 0 | 491 |
| 1986 | 0 | 0 | 0 | 7 | 38 | 60 | 200 | 137 | 46 | 0 | 0 | 0 | 488 |
| 1987 | 0 | 0 | 0 | 4 | 79 | 151 | 298 | 152 | 49 | 0 | 0 | 0 | 733 |
| 1988 | 0 | 0 | 0 | 0 | 29 | 88 | 315 | 255 | 41 | 8 | 0 | 0 | 736 |
| 1989 | 0 | 0 | 0 | 0 | 21 | 97 | 207 | 143 | 50 | 0 | 0 | 0 | 518 |
| 1990 | 0 | 0 | 3 | 29 | 4 | 104 | 208 | 176 | 47 | 14 | 0 | 0 | 585 |
| 1991 | 0 | 0 | 0 | 3 | 125 | 153 | 221 | 193 | 83 | 13 | 0 | 0 | 791 |
| 1992 | 0 | 0 | 0 | 1 | 24 | 53 | 90 | 90 | 55 | 0 | 0 | 0 | 313 |
| 1993 | 0 | 0 | 0 | 0 | 14 | 97 | 267 | 231 | 51 | 3 | 0 | 0 | 663 |
| 1994 | 0 | 0 | 0 | 5 | 14 | 175 | 267 | 125 | 36 | 2 | 0 | 0 | 624 |
| 1995 | 0 | 0 | 0 | 0 | 2 | 176 | 262 | 256 | 21 | 1 | 0 | 0 | 718 |
| 1996 | 0 | 0 | 0 | 0 | 12 | 108 | 131 | 177 | 65 | 2 | 0 | 0 | 495 |
| 1997 | 0 | 0 | 0 | 0 | 0 | 99 | 135 | 84 | 22 | 12 | 0 | 0 | 352 |
| 1998 | 0 | 0 | 1 | 0 | 34 | 118 | 148 | 207 | 57 | 1 | 0 | 0 | 566 |
| 1999 | 0 | 0 | 0 | 0 | 33 | 165 | 297 | 112 | 81 | 0 | 0 | 0 | 688 |
| 2000 | 0 | 0 | 0 | 0 | 17 | 76 | 108 | 126 | 69 | 0 | 0 | 0 | 396 |
| 2001 | 0 | 0 | 0 | 3 | 8 | 129 | 174 | 255 | 64 | 5 | 0 | 0 | 638 |
| 2002 | 0 | 0 | 0 | 5 | 15 | 128 | 268 | 211 | 114 | 16 | 0 | 0 | 757 |
| 2003 | 0 | 0 | 0 | 1 | 0 | 49 | 151 | 199 | 32 | 4 | 0 | 0 | 436 |
| 2004 | 0 | 0 | 0 | 0 | 26 | 60 | 140 | 106 | 75 | 0 | 0 | 0 | 407 |
| 2005 | 0 | 0 | 0 | 0 | 3 | 232 | 315 | 247 | 72 | 20 | 0 | 0 | 889 |
| 2006 | 0 | 0 | 0 | 0 | 49 | 136 | 276 | 163 | 15 | 0 | 0 | 0 | 639 |
| 2007 | 0 | 0 | 0 | 0 | 49 | 173 | 160 | 242 | 103 | 43 | 0 | 0 | 770 |
| 2008 | 0 | 0 | 0 | 7 | 2 | 131 | 205 | 131 | 68 | 0 | 0 | 0 | 544 |
| 2009 | 0 | 0 | 0 | 5 | 11 | 50 | 88 | 182 | 33 | 0 | 0 | 0 | 369 |

**SNOWFALL (inches) 2009 BUFFALO (KBUF)**

| YEAR | JUL | AUG | SEP | OCT | NOV | DEC | JAN | FEB | MAR | APR | MAY | JUN | TOTAL |
|------|-----|-----|-----|-----|-----|-----|-----|-----|-----|-----|-----|-----|-------|
| 1980-81 | 0.0 | 0.0 | 0.0 | T | 6.7 | 21.6 | 14.4 | 5.0 | 13.2 | T | 0.0 | 0.0 | 60.9 |
| 1981-82 | 0.0 | 0.0 | 0.0 | T | 1.8 | 24.8 | 53.2 | 12.7 | 9.0 | 10.9 | 0.0 | 0.0 | 112.4 |
| 1982-83 | 0.0 | 0.0 | 0.0 | 0.0 | 15.8 | 12.9 | 9.0 | 5.5 | 6.9 | 2.3 | T | 0.0 | 52.4 |
| 1983-84 | 0.0 | 0.0 | 0.0 | T | 17.7 | 52.0 | 13.4 | 32.5 | 16.0 | 0.9 | T | 0.0 | 132.5 |
| 1984-85 | 0.0 | 0.0 | 0.0 | 0.0 | 1.4 | 11.2 | 65.9 | 20.9 | 6.3 | 1.5 | 0.0 | 0.0 | 107.2 |
| 1985-86 | 0.0 | 0.0 | 0.0 | 0.0 | 5.2 | 68.4 | 17.3 | 17.3 | 4.8 | 1.7 | T | 0.0 | 114.7 |
| 1986-87 | 0.0 | 0.0 | 0.0 | 0.0 | 13.7 | 4.8 | 28.5 | 7.7 | 10.8 | 2.0 | 0.0 | 0.0 | 67.5 |
| 1987-88 | 0.0 | 0.0 | 0.0 | T | 0.9 | 9.8 | 6.9 | 31.9 | 6.1 | 0.8 | 0.0 | 0.0 | 56.4 |
| 1988-89 | 0.0 | 0.0 | 0.0 | 0.5 | 0.6 | 5.4 | 29.6 | 10.1 | 2.5 | 7.9 | 0.0 | 67.4 |
| 1989-90 | 0.0 | 0.0 | 0.0 | T | 7.8 | 34.8 | 11.8 | 28.0 | 1.4 | 9.9 | T | 0.0 | 93.7 |
| 1990-91 | 0.0 | 0.0 | 0.0 | T | 0.7 | 15.4 | 16.6 | 16.1 | 8.5 | 0.2 | T | 0.0 | 57.5 |
| 1991-92 | 0.0 | T | 0.0 | 0.2 | 18.0 | 21.4 | 18.4 | 7.0 | 22.8 | 5.0 | 0.0 | 0.0 | 92.8 |
| 1992-93 | 0.0 | 0.0 | 0.0 | 0.6 | 13.7 | 16.5 | 13.1 | 19.5 | 29.3 | 0.5 | T | 0.0 | 93.2 |
| 1993-94 | T | 0.0 | T | 2.9 | 4.8 | 27.9 | 35.4 | 21.6 | 13.2 | 6.9 | 0.0 | 0.0 | 112.7 |
| 1994-95 | 0.0 | 0.0 | T | 0.0 | 0.9 | 7.8 | 23.1 | 34.6 | 4.3 | 3.9 | T | 0.0 | 74.6 |
| 1995-96 | 0.0 | 0.0 | 0.0 | 0.0 | 15.7 | 61.2 | 25.3 | 11.9 | 24.1 | 3.2 | T | | 97.6 |
| 1996-97 | 0.0 | 0.0 | 0.0 | 0.0 | 11.5 | 18.9 | 42.4 | 9.3 | 13.4 | 2.1 | 0.0 | 0.0 | 97.6 |
| 1997-98 | 0.0 | 0.0 | 0.0 | 0.2 | 16.5 | 16.8 | 13.6 | 1.8 | 25.3 | T | T | T | 74.2 |
| 1998-99 | 0.0 | 0.0 | 0.0 | 0.0 | 0.2 | 11.6 | 65.1 | 6.9 | 15.8 | 1.0 | 0.0 | 0.0 | 100.6 |
| 1999-00 | 0.0 | 0.0 | 0.0 | T | 0.9 | 12.7 | 19.4 | 16.2 | 10.7 | 3.7 | 0.0 | 0.0 | 63.6 |
| 2000-01 | 0.0 | 0.0 | 0.0 | T | 45.6 | 50.3 | 19.6 | 9.8 | 32.8 | 0.6 | 0.0 | 0.0 | 158.7 |
| 2001-02 | 0.0 | 0.0 | 0.0 | 0.4 | 0.0 | 82.7 | 13.7 | 17.2 | 15.9 | 2.5 | T | 0.0 | 132.4 |
| 2002-03 | 0.0 | 0.0 | 0.0 | T | 8.9 | 35.8 | 37.4 | 19.5 | 6.6 | 3.1 | 0.0 | 0.0 | 111.3 |
| 2003-04 | 0.0 | 0.0 | 0.0 | T | 4.2 | 21.6 | 45.2 | 5.9 | 20.7 | 3.3 | 0.0 | 0.0 | 100.9 |
| 2004-05 | 0.0 | 0.0 | 0.0 | T | 0.2 | 22.8 | 37.0 | 22.3 | 17.5 | 9.3 | 0.0 | 0.0 | 109.1 |
| 2005-06 | T | 0.0 | 0.0 | 0.0 | 17.9 | 20.3 | 7.1 | 26.3 | 6.5 | 0.1 | 0.0 | 0.0 | 78.2 |
| 2006-07 | 0.0 | 0.0 | 0.0 | 22.6 | 2.1 | 7.5 | 15.5 | 33.5 | 5.4 | 2.3 | 0.0 | 0.0 | 88.9 |
| 2007-08 | 0.0 | 0.0 | 0.0 | 0.0 | 3.4 | 31.3 | 17.5 | 22.5 | 29.1 | T | 0.0 | 0.0 | 103.8 |
| 2008-09 | 0.0 | T | 0.0 | 0.1 | 6.2 | 49.2 | 30.6 | 11.6 | 0.5 | 2.0 | 0.0 | 0.0 | 100.2 |
| 2009- | T | 0.0 | T | T | T | 25.1 | | | | | | | |
| POR= 88 YRS | T | T | T | 0.5 | 9.5 | 21.9 | 22.3 | 16.6 | 11.8 | 3.0 | 0.1 | T | 85.7 |

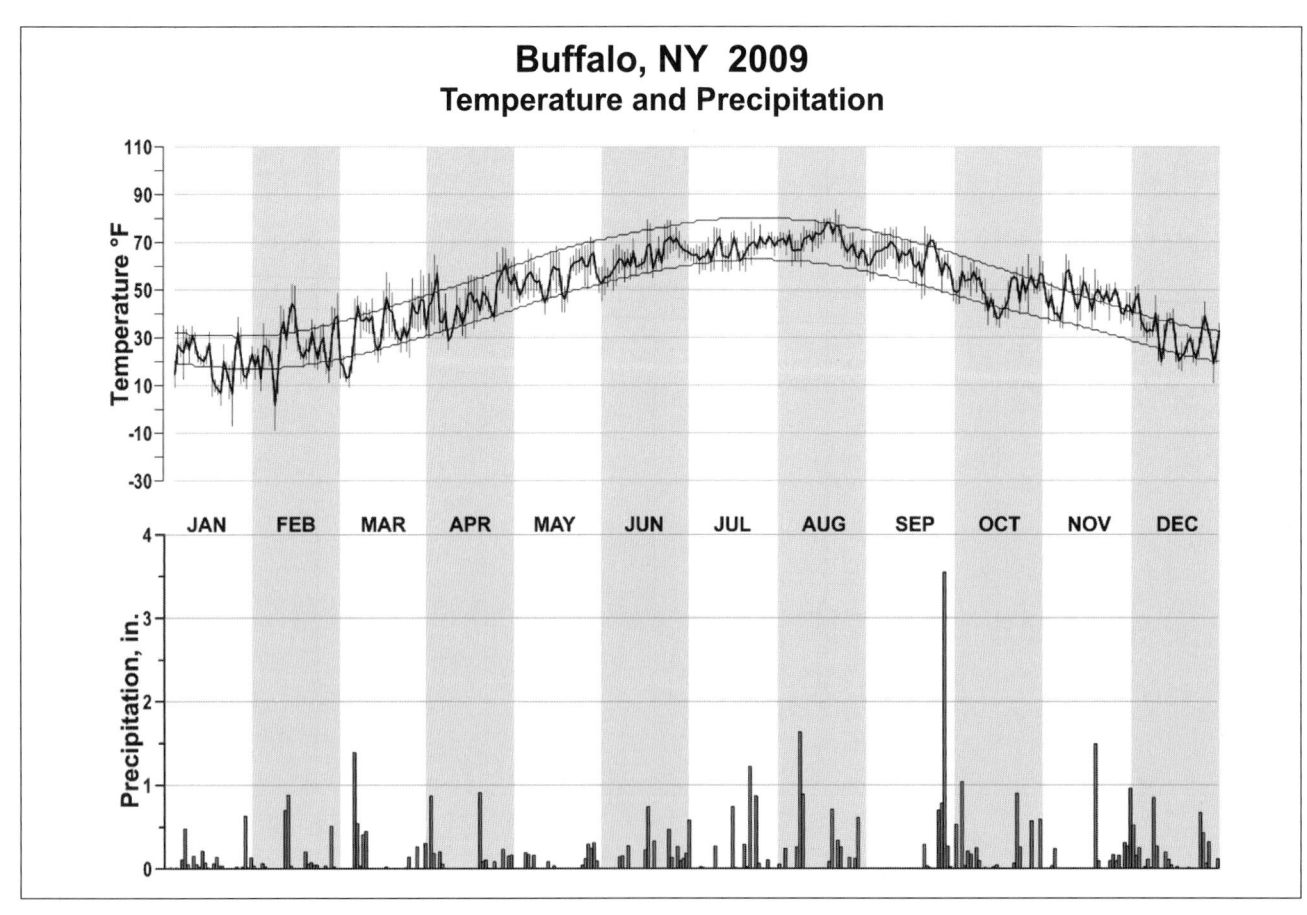

**Buffalo, NY 2009**
**Temperature and Precipitation**

# 2009
# NEW YORK
# NEW YORK (KNYC)

New York City, in area exceeding 300 square miles, is located on the Atlantic coastal plain at the mouth of the Hudson River. The terrain is laced with numerous waterways, all but one of the five boroughs in the city are situated on islands. Elevations range from less than 50 feet over most of Manhattan, Brooklyn, and Queens to almost 300 feet in northern Manhattan and the Bronx, and over 400 feet in Staten Island. Extensive suburban areas on Long Island, and in Connecticut, New York State and New Jersey border the city on the east, north, and west. About 30 miles to the west and northwest, hills rise to about 1,500 feet and to the north in upper Westchester County to 800 feet. To the southwest and to the east are the low-lying land areas of the New Jersey coastal plain and of Long Island, bordering on the Atlantic.

The New York Metropolitan area is close to the path of most storm and frontal systems which move across the North American continent. Therefore, weather conditions affecting the city most often approach from a westerly direction. New York City can thus experience higher temperatures in summer and lower ones in winter than would otherwise be expected in a coastal area. However, the frequent passage of weather systems often helps reduce the length of both warm and cold spells, and is also a major factor in keeping periods of prolonged air stagnation to a minimum.

Although continental influence predominates, oceanic influence is by no means absent. During the summer local sea breezes, winds blowing onshore from the cool water surface, often moderate the afternoon heat. The effect of the sea breeze diminishes inland. On winter mornings, ocean temperatures which are warm relative to the land reinforce the effect of the city heat island and low temperatures are often 10-20 degrees lower in the inland suburbs than in the central city. The relatively warm water temperatures also delay the advent of winter snows. Conversely, the lag in warming of water temperatures keeps spring temperatures relatively cool. One year-round measure of the ocean influence is the small average daily variation in temperature.

Precipitation is moderate and distributed fairly evenly throughout the year. Most of the rainfall from May through October comes from thunderstorms, usually of brief duration and sometimes intense. Heavy rains of long duration associated with tropical storms occur infrequently in late summer or fall. For the other months of the year precipitation is more likely to be associated with widespread storm areas, so that day-long rain, snow or a mixture of both is more common. Coastal storms, occurring most often in the fall and winter months, produce on occasion considerable amounts of precipitation and have been responsible for record rains, snows, and high winds.

The average annual precipitation is reasonably uniform within the city but is higher in the northern and western suburbs and less on eastern Long Island. Annual snowfall totals also show a consistent increase to the north and west of the city with lesser amounts along the south shores and the eastern end of Long Island, reflecting the influence of the ocean waters.

Local Climatological Data is published for three locations in New York City, Central Park, La Guardia Airport, and John F. Kennedy International Airport. Other nearby locations for which it is published are Newark, New Jersey, and Bridgeport, Connecticut.

Based on the 1951-1980 period, the average first occurrence of 32 degrees Fahrenheit in the fall is November 11 and the average last occurrence in the spring is April 1.

# NORMALS, MEANS, AND EXTREMES
## NEW YORK (KNYC)

LATITUDE: 40 ° 47'N  LONGITUDE: -73 ° 58'W  ELEVATION (FT): GRND: 156  BARO: 161  TIME ZONE: EASTERN (UTC -5)  WBAN: 94728

| | ELEMENT | POR | JAN | FEB | MAR | APR | MAY | JUN | JUL | AUG | SEP | OCT | NOV | DEC | YEAR |
|---|---|---|---|---|---|---|---|---|---|---|---|---|---|---|---|
| **TEMPERATURE °F** | NORMAL DAILY MAXIMUM | 30 | 38.0 | 41.0 | 49.8 | 60.7 | 70.9 | 79.0 | 84.2 | 82.4 | 74.7 | 63.5 | 53.1 | 42.9 | 61.7 |
| | MEAN DAILY MAXIMUM | 134 | 38.2 | 37.3 | 47.8 | 58.1 | 70.6 | 77.6 | 84.2 | 82.3 | 74.0 | 64.5 | 51.6 | 41.8 | 60.7 |
| | HIGHEST DAILY MAXIMUM | 141 | 72 | 75 | 86 | 96 | 99 | 101 | 106 | 104 | 102 | 94 | 84 | 75 | 106 |
| | YEAR OF OCCURRENCE | | 2007 | 1985 | 1998 | 2002 | 1962 | 1966 | 1936 | 1936 | 1918 | 1941 | 1950 | 1998 | JUL 1936 |
| | MEAN OF EXTREME MAXS. | 134 | 58.6 | 59.5 | 70.7 | 82.7 | 88.2 | 92.7 | 95.5 | 93.4 | 89.1 | 79.3 | 71.1 | 61.9 | 78.6 |
| | NORMAL DAILY MINIMUM | 30 | 26.2 | 28.1 | 35.1 | 44.2 | 54.2 | 63.3 | 68.8 | 67.7 | 60.3 | 49.6 | 41.0 | 31.6 | 47.5 |
| | MEAN DAILY MINIMUM | 134 | 25.6 | 24.6 | 33.1 | 42.1 | 53.3 | 61.2 | 68.1 | 66.7 | 58.9 | 49.6 | 39.2 | 29.9 | 46.0 |
| | LOWEST DAILY MINIMUM | 141 | -6 | -15 | 3 | 12 | 32 | 44 | 52 | 50 | 39 | 28 | 5 | -13 | -15 |
| | YEAR OF OCCURRENCE | | 1882 | 1934 | 1872 | 1923 | 1891 | 1945 | 1943 | 1986 | 1912 | 1936 | 1875 | 1917 | FEB 1934 |
| | MEAN OF EXTREME MINS. | 134 | 8.3 | 11.2 | 18.7 | 31.2 | 42.8 | 52.2 | 59.8 | 57.9 | 47.8 | 36.8 | 27.2 | 15.2 | 34.1 |
| | NORMAL DRY BULB | 30 | 32.1 | 34.6 | 42.5 | 52.5 | 62.6 | 71.2 | 76.5 | 75.1 | 67.5 | 56.6 | 47.1 | 37.3 | 54.6 |
| | MEAN DRY BULB | 134 | 31.9 | 30.9 | 40.4 | 50.1 | 62.0 | 69.5 | 76.2 | 74.6 | 66.5 | 57.1 | 45.4 | 35.9 | 53.4 |
| | MEAN WET BULB | 26 | 43.9 | 45.0 | 47.6 | 53.3 | 58.8 | 64.7 | 66.8 | 66.8 | 63.7 | 57.8 | 52.0 | 46.6 | 55.6 |
| | MEAN DEW POINT | 26 | 40.2 | 41.1 | 43.9 | 49.4 | 55.5 | 61.5 | 63.7 | 64.0 | 60.7 | 54.5 | 48.2 | 42.8 | 52.1 |
| | NORMAL NO. DAYS WITH: | | | | | | | | | | | | | | |
| | MAXIMUM >= 90 | 3 0 | 0.0 | 0.0 | 0.0 | 0.2 | 1.0 | 3.1 | 7.7 | 5.0 | 1.1 | 0.0 | 0.0 | 0.0 | 18.1 |
| | MAXIMUM <= 32 | 3 0 | 9.2 | 5.5 | 0.8 | * | 0.0 | 0.0 | 0.0 | 0.0 | 0.0 | 0.0 | 0.1 | 4.1 | 19.7 |
| | MINIMUM <= 32 | 30 | 22.0 | 18.4 | 10.5 | 1.2 | 0.0 | 0.0 | 0.0 | 0.0 | 0.0 | 0.3 | 4.0 | 15.1 | 71.5 |
| | MINIMUM <= 0 | 30 | 0.3 | * | 0.0 | 0.0 | 0.0 | 0.0 | 0.0 | 0.0 | 0.0 | 0.0 | 0.0 | * | 0.3 |
| **H/C** | NORMAL HEATING DEG. DAYS | 30 | 1009 | 853 | 695 | 372 | 127 | 16 | 7 | 2 | 43 | 261 | 525 | 844 | 4754 |
| | NORMAL COOLING DEG. DAYS | 30 | 0 | 0 | 2 | 10 | 63 | 214 | 379 | 331 | 134 | 16 | 2 | 0 | 1151 |
| **RH** | NORMAL (PERCENT) | 30 | 63 | 61 | 60 | 58 | 65 | 68 | 66 | 69 | 70 | 67 | 64 | 64 | 65 |
| | HOUR 01 LST | 30 | 64 | 63 | 63 | 62 | 70 | 73 | 71 | 75 | 76 | 72 | 68 | 66 | 69 |
| | HOUR 07 LST | 30 | 68 | 67 | 67 | 65 | 73 | 76 | 74 | 78 | 79 | 76 | 72 | 68 | 72 |
| | HOUR 13 LST | 30 | 58 | 55 | 53 | 48 | 54 | 58 | 55 | 56 | 58 | 56 | 56 | 59 | 56 |
| | HOUR 19 LST | 30 | 60 | 58 | 56 | 53 | 60 | 63 | 61 | 65 | 67 | 65 | 61 | 61 | 61 |
| **S** | PERCENT POSSIBLE SUNSHINE | 107 | 51 | 55 | 57 | 58 | 61 | 64 | 65 | 64 | 62 | 61 | 52 | 49 | 58 |
| **W/O** | MEAN NO. DAYS WITH: | | | | | | | | | | | | | | |
| | HEAVY FOG(VISBY <= 1/4 MI) | 43 | 0.6 | 0.6 | 0.4 | 0.2 | 0.2 | 0.6 | 0.5 | 0.5 | 0.3 | 0.1 | 0.2 | 0.4 | 4.6 |
| | THUNDERSTORMS | 44 | 0.1 | 0.2 | 0.6 | 0.9 | 1.6 | 2.7 | 3.0 | 2.5 | 0.9 | 0.5 | 0.3 | 0.1 | 13.4 |
| **CLOUDNESS** | MEAN: | | | | | | | | | | | | | | |
| | SUNRISE-SUNSET (OKTAS) | | | | | | | | | | | | | | |
| | MIDNIGHT-MIDNIGHT (OKTAS) | | | | | | | | | | | | | | |
| | MEAN NO. DAYS WITH: | | | | | | | | | | | | | | |
| | CLEAR | | | | | | | | | | | | | | |
| | PARTLY CLOUDY | | | | | | | | | | | | | | |
| | CLOUDY | | | | | | | | | | | | | | |
| **PR** | MEAN STATION PRESSURE(IN) | 26 | 29.92 | 29.91 | 29.90 | 29.87 | 29.87 | 29.86 | 29.87 | 29.89 | 29.90 | 29.92 | 29.92 | 29.91 | 29.90 |
| | MEAN SEA-LEVEL PRES. (IN) | 26 | 30.02 | 30.02 | 30.01 | 29.97 | 29.98 | 29.97 | 29.97 | 30.00 | 30.01 | 30.03 | 30.03 | 30.02 | 30.00 |
| **WINDS** | MEAN SPEED (MPH) | 26 | 7.6 | 7.6 | 7.7 | 7.3 | 6.7 | 6.2 | 6.1 | 5.8 | 6.1 | 6.4 | 6.8 | 7.3 | 6.8 |
| | PREVAIL.DIR(TENS OF DEGS) | 17 | 28 | 33 | 33 | 33 | 05 | 24 | 24 | 24 | 05 | 28 | 28 | 28 | 28 |
| | MAXIMUM 2-MINUTE: | | | | | | | | | | | | | | |
| | SPEED (MPH) | 14 | 40 | 34 | 37 | 35 | 30 | 29 | 24 | 33 | 29 | 29 | 32 | 34 | 40 |
| | DIR. (TENS OF DEGS) | | 00 | 08 | 00 | 08 | 05 | 28 | 17 | 30 | 09 | 28 | 28 | 07 | 00 |
| | YEAR OF OCCURRENCE | | 1996 | 1998 | 1996 | 2000 | 2008 | 2008 | 1996 | 1997 | 1999 | 2009 | 2003 | 1997 | JAN 1996 |
| | MAXIMUM 3-SECOND | | | | | | | | | | | | | | |
| | SPEED (MPH) | 14 | 53 | 52 | 52 | 51 | 45 | 45 | 41 | 57 | 46 | 46 | 47 | 51 | 57 |
| | DIR. (TENS OF DEGS) | | 00 | 07 | 00 | 07 | 33 | 27 | 16 | 31 | 30 | 26 | 28 | 06 | 31 |
| | YEAR OF OCCURRENCE | | 1996 | 1998 | 1996 | 1998 | 2008 | 2008 | 1996 | 1997 | 2002 | 2006 | 2003 | 1997 | AUG 1997 |
| **PRECIPITATION** | NORMAL (IN) | 30 | 4.13 | 3.15 | 4.37 | 4.28 | 4.69 | 3.84 | 4.62 | 4.22 | 4.23 | 3.85 | 4.36 | 3.95 | 49.69 |
| | MAXIMUM MONTHLY (IN) | 40 | 10.52 | 6.87 | 10.41 | 13.05 | 10.24 | 10.26 | 11.89 | 12.36 | 16.85 | 16.73 | 12.41 | 9.98 | 16.85 |
| | YEAR OF OCCURRENCE | | 1979 | 1869 | 1980 | 2007 | 1989 | 2003 | 1889 | 1990 | 1882 | 2005 | 1972 | 1973 | SEP 1882 |
| | MINIMUM MONTHLY (IN) | 140 | 0.58 | 0.46 | 0.80 | 0.95 | 0.30 | 0.02 | 0.44 | 0.18 | 0.21 | 0.14 | 0.34 | 0.25 | 0.02 |
| | YEAR OF OCCURRENCE | | 1981 | 1895 | 2006 | 1881 | 1903 | 1949 | 1999 | 1995 | 1884 | 1963 | 1976 | 1955 | JUN 1949 |
| | MAXIMUM IN 24 HOURS (IN) | 140 | 3.91 | 3.04 | 4.25 | 7.81 | 4.88 | 4.74 | 4.39 | 5.78 | 8.30 | 11.17 | 8.09 | 3.21 | 11.17 |
| | YEAR OF OCCURRENCE | | 1979 | 1973 | 1876 | 2007 | 1968 | 1884 | 1997 | 1971 | 1882 | 1903 | 1977 | 1909 | OCT 1903 |
| | NORMAL NO. DAYS WITH: | | | | | | | | | | | | | | |
| | PRECIPITATION >= 0.01 | 30 | 10.3 | 9.4 | 10.7 | 11.1 | 11.4 | 10.8 | 10.2 | 9.5 | 9.1 | 8.3 | 9.3 | 10.6 | 120.7 |
| | PRECIPITATION >= 1.00 | 30 | 1.0 | 0.9 | 1.0 | 1.2 | 1.1 | 0.9 | 1.3 | 1.1 | 1.2 | 1.0 | 1.2 | 1.2 | 13.1 |
| **SNOWFALL** | NORMAL (IN) | 30 | 8.1 | 7.6 | 3.2 | 0.5 | 0.* | 0.0 | 0.0 | 0.0 | 0.0 | 0.* | 0.4 | 2.6 | 22.4 |
| | MAXIMUM MONTHLY (IN) | 141 | 27.4 | 27.9 | 30.5 | 13.5 | T | 0.0 | T | 0.0 | 0.0 | 0.8 | 19.0 | 29.6 | 30.5 |
| | YEAR OF OCCURRENCE | | 1925 | 1934 | 1896 | 1875 | 1995 | | 1990 | | | 1925 | 1898 | 1947 | MAR 1896 |
| | MAXIMUM IN 24 HOURS (IN) | 41 | 19.2 | 17.6 | 18.1 | 10.2 | T | 0.0 | T | 0.0 | 0.0 | 0.8 | 10.0 | 26.4 | 26.4 |
| | YEAR OF OCCURRENCE' | | 1996 | 1983 | 1941 | 1915 | 1995 | | 1990 | | | 1925 | 1898 | 1947 | DEC 1947 |
| | MAXIMUM SNOW DEPTH (IN) | 133 | 15 | 22 | 9 | 9 | 0 | 0 | 0 | 0 | 0 | 0 | 5 | 10 | 22 |
| | YEAR OF OCCURRENCE | | 1978 | 1994 | 1967 | 1982 | | | | | | | 1989 | 2009 | FEB 1994 |
| | NORMAL NO. DAYS WITH: | | | | | | | | | | | | | | |
| | SNOWFALL >= 1.0 | 30 | 2.4 | 1.8 | 1.0 | 0.1 | 0.0 | 0.0 | 0.0 | 0.0 | 0.0 | 0.0 | 0.1 | 0.7 | 6.1 |

## PRECIPITATION (inches) 2009 NEW YORK (KNYC)

| YEAR | JAN | FEB | MAR | APR | MAY | JUN | JUL | AUG | SEP | OCT | NOV | DEC | ANNUAL |
|------|-----|-----|-----|-----|-----|-----|-----|-----|-----|-----|-----|-----|--------|
| 1980 | 1.72 | 1.04 | 10.41 | 8.26 | 2.33 | 3.84 | 5.26 | 1.16 | 1.98 | 3.86 | 4.11 | 0.58 | 44.55 |
| 1981 | 0.58 | 6.04 | 1.19 | 3.42 | 3.56 | 2.71 | 6.21 | 0.59 | 3.45 | 3.49 | 1.69 | 5.18 | 38.11 |
| 1982 | 6.46 | 2.37 | 2.56 | 5.67 | 2.43 | 5.12 | 3.14 | 4.66 | 1.77 | 2.31 | 3.44 | 1.47 | 41.40 |
| 1983 | | | | | | | | | | | | | |
| 1984 | 1.87 | 4.86 | 6.30 | 6.62 | 9.74 | 5.76 | 7.03 | 1.38 | 2.51 | 3.63 | 4.07 | 3.26 | 57.03 |
| 1985 | 1.00 | 2.41 | 1.91 | 1.41 | 5.72 | 4.41 | 4.41 | 2.58 | 4.75 | 1.30 | 8.09 | 0.83 | 38.82 |
| 1986 | 4.23 | 2.86 | 1.46 | 3.93 | 1.68 | 1.86 | 5.56 | 4.24 | 2.20 | 1.92 | 6.85 | 6.16 | 42.95 |
| 1987 | 5.81 | 1.01 | 4.93 | 5.90 | 1.45 | 3.94 | 4.12 | 4.89 | 5.25 | 3.89 | 3.08 | 2.17 | 46.44 |
| 1988 | 3.64 | 3.91 | 2.10 | 2.20 | 5.27 | 1.29 | 8.14 | 2.19 | 2.34 | 3.56 | 8.90 | 1.13 | 44.67 |
| 1989 | 2.29 | 3.03 | 4.93 | 4.26 | 10.24 | 8.79 | 5.13 | 8.44 | 6.90 | 7.48 | 2.79 | 0.83 | 65.11 |
| 1990 | 5.34 | 2.33 | 3.64 | 5.12 | 9.10 | 2.50 | 3.51 | 12.36 | 2.24 | 6.38 | 2.82 | 5.58 | 60.92 |
| 1991 | 3.38 | 1.93 | 5.16 | 3.68 | 3.11 | 4.16 | 4.57 | 7.13 | 3.71 | 2.13 | 1.96 | 4.26 | 45.18 |
| 1992 | 1.68 | 1.87 | 4.08 | 1.76 | 4.02 | 4.77 | 4.49 | 3.49 | 4.89 | 1.16 | 5.64 | 5.50 | 43.35 |
| 1993 | 3.44 | 2.81 | 6.64 | 4.28 | 1.56 | 1.49 | 1.70 | 5.41 | 5.25 | 4.55 | 2.20 | 4.95 | 44.28 |
| 1994 | 5.62 | 3.44 | 6.33 | 2.42 | 4.26 | 3.21 | 3.86 | 6.33 | 3.33 | 1.35 | 4.34 | 2.90 | 47.39 |
| 1995 | 3.75 | 3.13 | 1.26 | 2.29 | 2.84 | 2.09 | 6.13 | 0.18 | 3.03 | 7.82 | 5.78 | 2.12 | 40.42 |
| 1996 | 5.64 | 2.59 | 3.81 | 6.33 | 2.64 | 5.71 | 5.76 | 1.87 | 4.97 | 7.52 | 2.87 | 6.48 | 56.19 |
| 1997 | 3.65 | 2.54 | 5.18 | 2.86 | 3.05 | 1.93 | 8.36 | 3.21 | 2.10 | 2.10 | 4.68 | 4.27 | 43.93 |
| 1998 | 5.20 | 5.81 | 5.08 | 7.05 | 6.94 | 5.94 | 1.09 | 2.78 | 3.44 | 2.76 | 1.48 | 1.12 | 48.69 |
| 1999 | 7.02 | 3.49 | 4.01 | 1.93 | 4.04 | 0.59 | 0.44 | 2.89 | 8.81 | 2.73 | 2.33 | 3.23 | 41.51 |
| 2000 | 3.23 | 1.66 | 3.34 | 3.53 | 4.50 | 4.87 | 7.28 | 3.82 | 5.82 | 0.67 | 3.54 | 3.19 | 45.45 |
| 2001 | 3.16 | 1.95 | 7.48 | 1.58 | 2.03 | 5.29 | 2.04 | 2.56 | 5.30 | 0.66 | 1.36 | 2.24 | 35.65 |
| 2002 | 1.93 | 0.71 | 3.54 | 3.41 | 3.69 | 4.48 | 1.05 | 4.91 | 5.16 | 7.20 | 5.06 | 4.06 | 45.20 |
| 2003 | 2.30 | 4.55 | 4.57 | 3.20 | 3.40 | 10.26 | 3.76 | 5.85 | 6.03 | 4.90 | 4.18 | 5.42 | 58.42 |
| 2004 | 2.13 | 2.68 | 2.99 | 4.11 | 5.76 | 3.02 | 7.64 | 3.02 | 11.51 | 1.15 | 4.21 | 3.71 | 51.93 |
| 2005 | 4.67 | 3.04 | 4.96 | 4.81 | 1.48 | 3.21 | 3.56 | 3.96 | 0.48 | 16.73 | 4.47 | 4.60 | 55.97 |
| 2006 | 4.99 | 2.88 | 0.80 | 5.56 | 4.62 | 8.55 | 6.16 | 6.08 | 3.69 | 7.07 | 7.34 | 2.15 | 59.89 |
| 2007 | 3.63 | 1.99 | 5.35 | 13.05 | 1.88 | 6.55 | 6.89 | 7.18 | 1.81 | 4.65 | 3.47 | 5.22 | 61.67 |
| 2008 | 2.85 | 5.95 | 4.08 | 2.77 | 4.01 | 4.70 | 2.84 | 5.58 | 7.05 | 3.62 | 3.54 | 6.62 | 53.61 |
| 2009 | 2.98 | 0.93 | 1.75 | 4.69 | 5.17 | 10.05 | 7.11 | 4.22 | 2.26 | 5.58 | 1.61 | 7.27 | 53.62 |
| POR= 133 YRS | 3.48 | 3.32 | 3.93 | 3.62 | 3.71 | 3.67 | 4.30 | 4.26 | 3.87 | 3.64 | 3.56 | 3.59 | 44.95 |

WBAN : 94728

## AVERAGE TEMPERATURE (°F) 2009 NEW YORK (KNYC)

| YEAR | JAN | FEB | MAR | APR | MAY | JUN | JUL | AUG | SEP | OCT | NOV | DEC | ANNUAL |
|------|-----|-----|-----|-----|-----|-----|-----|-----|-----|-----|-----|-----|--------|
| 1980 | 33.7 | 31.4 | 41.2 | 54.5 | 65.6 | 70.3 | 79.3 | 80.3 | 70.8 | 55.2 | 44.6 | 32.5 | 55.0 |
| 1981 | 26.3 | 39.3 | 42.3 | 56.2 | 64.8 | 73.0 | 78.5 | 76.0 | 67.6 | 54.4 | 47.7 | 36.5 | 55.2 |
| 1982 | 26.1 | 35.3 | 42.0 | 51.2 | 64.1 | 68.6 | 77.9 | 73.2 | 68.3 | 58.5 | 50.4 | 42.8 | 54.9 |
| 1983 | 34.5 | 36.4 | 44.0 | 52.3 | 60.2 | 73.4 | 79.5 | 77.7 | 71.8 | 57.9 | 48.9 | 35.2 | 56.0 |
| 1984 | 29.9 | 40.6 | 36.7 | 51.9 | 61.6 | 74.5 | 74.7 | 76.7 | 65.9 | 61.8 | 47.3 | 43.8 | 55.5 |
| 1985 | 28.8 | 36.6 | 45.8 | 55.5 | 65.3 | 68.6 | 76.2 | 75.4 | 70.5 | 59.5 | 50.0 | 34.2 | 55.5 |
| 1986 | 34.1 | 32.0 | 45.1 | 54.5 | 66.0 | 71.6 | 76.0 | 73.1 | 67.9 | 58.0 | 45.7 | 39.0 | 55.3 |
| 1987 | 32.3 | 33.2 | 45.2 | 53.4 | 63.6 | 72.8 | 78.0 | 74.2 | 67.7 | 53.8 | 47.7 | 39.5 | 55.1 |
| 1988 | 29.5 | 35.0 | 43.6 | 51.2 | 62.7 | 71.8 | 79.3 | 78.8 | 67.4 | 52.8 | 49.4 | 35.9 | 54.8 |
| 1989 | 37.4 | 34.5 | 42.4 | 52.2 | 62.1 | 72.0 | 75.0 | 74.0 | 68.1 | 58.2 | 45.7 | 25.9 | 54.0 |
| 1990 | 41.4 | 39.8 | 45.1 | 53.5 | 60.2 | 72.1 | 76.8 | 75.3 | 67.5 | 61.9 | 50.4 | 42.6 | 57.2 |
| 1991 | 34.9 | 40.0 | 44.6 | 55.7 | 68.7 | 74.1 | 77.7 | 77.1 | 67.5 | 58.4 | 48.3 | 39.6 | 57.2 |
| 1992 | 35.7 | 36.4 | 40.0 | 50.5 | 61.0 | 70.3 | 74.2 | 73.0 | 67.2 | 54.5 | 46.5 | 37.9 | 53.9 |
| 1993 | 36.3 | 30.8 | 39.7 | 53.3 | 65.7 | 73.3 | 80.2 | 77.2 | 67.3 | 56.0 | 48.8 | 37.3 | 55.5 |
| 1994 | 25.6 | 30.6 | 40.7 | 55.6 | 61.8 | 75.2 | 79.4 | 74.0 | 67.6 | 58.0 | 52.0 | 42.2 | 55.2 |
| 1995 | 37.5 | 31.6 | 45.0 | 51.9 | 61.9 | 71.8 | 79.2 | 78.6 | 68.3 | 61.6 | 43.6 | 32.4 | 55.3 |
| 1996 | 30.5 | 33.8 | 38.9 | 52.2 | 61.1 | 71.4 | 73.3 | 74.5 | 68.0 | 56.4 | 43.0 | 41.3 | 53.7 |
| 1997 | 32.2 | 40.0 | 41.9 | 51.7 | 59.4 | 70.9 | 75.8 | 73.3 | 67.0 | 56.7 | 44.5 | 38.3 | 54.3 |
| 1998 | 40.0 | 40.6 | 45.4 | 54.0 | 64.3 | 69.2 | 76.5 | 76.7 | 70.2 | 57.6 | 48.1 | 43.2 | 57.2 |
| 1999 | 33.9 | 38.9 | 42.5 | 53.5 | 63.1 | 73.2 | 81.4 | 75.5 | 69.1 | 56.0 | 50.8 | 40.0 | 56.5 |
| 2000 | 31.3 | 37.3 | 47.2 | 51.0 | 63.5 | 71.3 | 72.3 | 72.5 | 66.0 | 57.0 | 45.3 | 31.1 | 53.8 |
| 2001 | 33.7 | 35.9 | 39.6 | 54.0 | 63.6 | 73.0 | 73.2 | 78.7 | 67.7 | 58.5 | 52.7 | 44.1 | 56.2 |
| 2002 | 40.0 | 40.6 | 44.2 | 56.1 | 60.7 | 71.5 | 78.8 | 77.8 | 70.3 | 55.2 | 46.0 | 36.0 | 56.4 |
| 2003 | 27.5 | 30.1 | 43.1 | 49.8 | 58.7 | 68.4 | 75.8 | 76.7 | 68.0 | 55.1 | 50.0 | 37.6 | 53.4 |
| 2004 | 24.8 | 35.0 | 43.6 | 53.6 | 65.2 | 71.3 | 74.5 | 74.3 | 69.4 | 56.0 | 48.2 | 38.4 | 54.5 |
| 2005 | 31.3 | 36.6 | 39.5 | 55.2 | 58.9 | 74.0 | 77.6 | 79.7 | 73.3 | 57.9 | 49.7 | 35.3 | 55.8 |
| 2006 | 40.9 | 35.8 | 43.1 | 55.7 | 63.1 | 71.0 | 78.0 | 75.8 | 66.6 | 56.3 | 51.9 | 43.6 | 56.8 |
| 2007 | 37.5 | 28.3 | 42.2 | 50.3 | 65.3 | 71.4 | 75.0 | 74.0 | 70.3 | 63.6 | 45.5 | 37.0 | 55.0 |
| 2008 | 36.5 | 35.8 | 42.7 | 55.0 | 60.1 | 74.0 | 78.4 | 73.8 | 68.9 | 55.2 | 45.9 | 38.1 | 55.4 |
| 2009 | 28.0 | 36.7 | 42.4 | 54.5 | 62.5 | 67.5 | 72.7 | 75.7 | 66.3 | 55.0 | 51.2 | 35.9 | 54.0 |
| POR= 134 YRS | 31.9 | 30.9 | 40.4 | 50.1 | 62.0 | 69.5 | 76.2 | 74.6 | 66.5 | 57.1 | 45.4 | 35.9 | 53.4 |

## HEATING DEGREE DAYS (base 65°F) 2009 NEW YORK (KNYC)

| YEAR | JUL | AUG | SEP | OCT | NOV | DEC | JAN | FEB | MAR | APR | MAY | JUN | TOTAL |
|------|-----|-----|-----|-----|-----|-----|-----|-----|-----|-----|-----|-----|-------|
| 1980-81 | 0 | 0 | 31 | 305 | 602 | 1000 | 1194 | 715 | 698 | 264 | 78 | 3 | 4890 |
| 1981-82 | 0 | 0 | 48 | 320 | 513 | 876 | 1198 | 825 | 707 | 413 | 74 | 36 | 5010 |
| 1982-83 | 0 | 5 | 24 | 229 | 446 | 679 | 936 | 793 | 644 | 393 | 161 | 3 | 4313 |
| 1983-84 | 0 | 0 | 34 | 249 | 480 | 914 | 1082 | 698 | 870 | 389 | 137 | 9 | 4862 |
| 1984-85 | 0 | 0 | 69 | 114 | 525 | 654 | 1113 | 789 | 596 | 305 | 79 | 24 | 4268 |
| 1985-86 | 0 | 0 | 17 | 188 | 448 | 947 | 950 | 917 | 615 | 312 | 89 | 11 | 4494 |
| 1986-87 | 0 | 10 | 27 | 236 | 572 | 797 | 1008 | 883 | 608 | 348 | 146 | 8 | 4643 |
| 1987-88 | 0 | 2 | 29 | 343 | 512 | 780 | 1093 | 867 | 656 | 409 | 133 | 31 | 4855 |
| 1988-89 | 3 | 0 | 23 | 385 | 459 | 896 | 844 | 849 | 696 | 376 | 143 | 14 | 4688 |
| 1989-90 | 0 | 1 | 54 | 217 | 572 | 1205 | 724 | 702 | 612 | 366 | 150 | 4 | 4607 |
| 1990-91 | 3 | 2 | 57 | 166 | 436 | 686 | 927 | 696 | 625 | 311 | 61 | 3 | 3973 |
| 1991-92 | 0 | 0 | 60 | 222 | 496 | 782 | 902 | 827 | 767 | 434 | 160 | 12 | 4662 |
| 1992-93 | 0 | 3 | 54 | 324 | 547 | 834 | 882 | 953 | 779 | 347 | 57 | 14 | 4794 |
| 1993-94 | 0 | 0 | 65 | 275 | 483 | 852 | 1215 | 958 | 749 | 282 | 142 | 0 | 5021 |
| 1994-95 | 0 | 0 | 18 | 212 | 388 | 700 | 846 | 931 | 614 | 386 | 130 | 2 | 4227 |
| 1995-96 | 0 | 0 | 31 | 146 | 637 | 1001 | 1065 | 894 | 801 | 389 | 183 | 8 | 5155 |
| 1996-97 | 0 | 0 | 46 | 263 | 656 | 726 | 1010 | 691 | 712 | 393 | 174 | 40 | 4711 |
| 1997-98 | 2 | 0 | 48 | 284 | 611 | 822 | 768 | 676 | 635 | 322 | 99 | 29 | 4296 |
| 1998-99 | 0 | 0 | 20 | 222 | 499 | 670 | 955 | 725 | 687 | 340 | 98 | 4 | 4220 |
| 1999-00 | 0 | 3 | 23 | 271 | 418 | 769 | 1038 | 795 | 544 | 411 | 118 | 31 | 4421 |
| 2000-01 | 0 | 0 | 81 | 256 | 586 | 1041 | 965 | 809 | 780 | 340 | 124 | 6 | 4988 |
| 2001-02 | 0 | 0 | 47 | 228 | 364 | 639 | 769 | 677 | 639 | 332 | 172 | 20 | 3887 |
| 2002-03 | 0 | 2 | 11 | 327 | 562 | 891 | 1156 | 972 | 671 | 462 | 195 | 47 | 5296 |
| 2003-04 | 0 | 0 | 18 | 299 | 450 | 843 | 1241 | 862 | 658 | 342 | 76 | 12 | 4801 |
| 2004-05 | 0 | 0 | 16 | 273 | 495 | 820 | 1035 | 789 | 782 | 300 | 194 | 9 | 4713 |
| 2005-06 | 1 | 0 | 6 | 249 | 453 | 915 | 739 | 813 | 674 | 279 | 113 | 19 | 4261 |
| 2006-07 | 0 | 0 | 33 | 279 | 383 | 655 | 845 | 1024 | 698 | 445 | 89 | 9 | 4460 |
| 2007-08 | 0 | 12 | 19 | 124 | 580 | 860 | 874 | 841 | 687 | 301 | 171 | 1 | 4470 |
| 2008-09 | 0 | 0 | 27 | 302 | 566 | 828 | 1139 | 785 | 690 | 350 | 119 | 29 | 4835 |
| 2009- | 0 | 0 | 37 | 310 | 407 | 894 | | | | | | | |

WBAN : 94728

## COOLING DEGREE DAYS (base 65°F) 2009 NEW YORK (KNYC)

| YEAR | JAN | FEB | MAR | APR | MAY | JUN | JUL | AUG | SEP | OCT | NOV | DEC | TOTAL |
|------|-----|-----|-----|-----|-----|-----|-----|-----|-----|-----|-----|-----|-------|
| 1980 | 0 | 0 | 0 | 1 | 94 | 188 | 448 | 480 | 213 | 11 | 0 | 0 | 1435 |
| 1981 | 0 | 0 | 0 | 4 | 78 | 252 | 425 | 347 | 129 | 0 | 0 | 0 | 1235 |
| 1982 | 0 | 0 | 0 | 7 | 55 | 152 | 405 | 266 | 129 | 36 | 16 | 0 | 1066 |
| 1983 | 0 | 0 | 0 | 19 | 16 | 259 | 460 | 404 | 244 | 35 | 0 | 0 | 1437 |
| 1984 | 0 | 0 | 0 | 3 | 39 | 301 | 306 | 367 | 106 | 26 | 0 | 0 | 1148 |
| 1985 | 0 | 0 | 8 | 28 | 95 | 139 | 353 | 329 | 189 | 21 | 5 | 0 | 1167 |
| 1986 | 0 | 0 | 5 | 4 | 127 | 214 | 348 | 269 | 120 | 27 | 0 | 0 | 1114 |
| 1987 | 0 | 0 | 0 | 5 | 110 | 251 | 406 | 295 | 118 | 0 | 2 | 0 | 1187 |
| 1988 | 0 | 0 | 0 | 0 | 66 | 243 | 455 | 435 | 104 | 12 | 0 | 0 | 1315 |
| 1989 | 0 | 0 | 4 | 0 | 61 | 231 | 313 | 287 | 151 | 10 | 0 | 0 | 1057 |
| 1990 | 0 | 0 | 4 | 25 | 8 | 225 | 375 | 328 | 140 | 77 | 4 | 0 | 1186 |
| 1991 | 0 | 0 | 0 | 38 | 182 | 280 | 403 | 382 | 142 | 24 | 1 | 0 | 1452 |
| 1992 | 0 | 0 | 0 | 5 | 46 | 174 | 292 | 256 | 127 | 8 | 0 | 0 | 908 |
| 1993 | 0 | 0 | 0 | 0 | 82 | 269 | 474 | 386 | 140 | 3 | 4 | 0 | 1358 |
| 1994 | 0 | 0 | 0 | 7 | 51 | 316 | 454 | 286 | 102 | 2 | 3 | 0 | 1221 |
| 1995 | 0 | 0 | 0 | 0 | 40 | 212 | 445 | 428 | 137 | 48 | 0 | 0 | 1310 |
| 1996 | 0 | 0 | 0 | 13 | 67 | 209 | 267 | 300 | 142 | 4 | 0 | 0 | 1002 |
| 1997 | 0 | 0 | 0 | 0 | 7 | 222 | 343 | 265 | 113 | 32 | 0 | 0 | 982 |
| 1998 | 0 | 0 | 36 | 0 | 89 | 162 | 366 | 368 | 184 | 1 | 0 | 2 | 1208 |
| 1999 | 0 | 0 | 0 | 3 | 46 | 258 | 517 | 336 | 152 | 3 | 0 | 0 | 1315 |
| 2000 | 0 | 0 | 0 | 0 | 81 | 227 | 234 | 240 | 117 | 12 | 0 | 0 | 911 |
| 2001 | 0 | 0 | 0 | 15 | 89 | 250 | 262 | 430 | 137 | 32 | 1 | 0 | 1216 |
| 2002 | 0 | 0 | 0 | 73 | 44 | 221 | 436 | 404 | 175 | 31 | 0 | 0 | 1384 |
| 2003 | 0 | 0 | 0 | 12 | 9 | 155 | 341 | 369 | 114 | 3 | 7 | 0 | 1010 |
| 2004 | 0 | 0 | 0 | 6 | 90 | 204 | 299 | 297 | 155 | 2 | 0 | 0 | 1053 |
| 2005 | 0 | 0 | 0 | 12 | 13 | 288 | 398 | 464 | 263 | 34 | 0 | 0 | 1472 |
| 2006 | 0 | 0 | 0 | 7 | 63 | 206 | 407 | 343 | 89 | 15 | 0 | 0 | 1130 |
| 2007 | 0 | 0 | 0 | 12 | 104 | 207 | 317 | 302 | 182 | 88 | 0 | 0 | 1212 |
| 2008 | 0 | 0 | 0 | 10 | 25 | 278 | 422 | 278 | 147 | 3 | 0 | 0 | 1163 |
| 2009 | 0 | 0 | 0 | 43 | 47 | 111 | 246 | 341 | 85 | 3 | 0 | 0 | 876 |

## SNOWFALL (inches)  2009  NEW YORK (KNYC)

| YEAR | JUL | AUG | SEP | OCT | NOV | DEC | JAN | FEB | MAR | APR | MAY | JUN | TOTAL |
|---|---|---|---|---|---|---|---|---|---|---|---|---|---|
| 1980-81 | 0.0 | 0.0 | 0.0 | 0.0 | T | 2.8 | 8.0 | T | 8.6 | 0.0 | 0.0 | 0.0 | 19.4 |
| 1981-82 | 0.0 | 0.0 | 0.0 | 0.0 | 0.0 | 2.1 | 11.8 | 0.4 | 0.7 | 9.6 | 0.0 | 0.0 | 24.6 |
| 1982-83 | 0.0 | 0.0 | 0.0 | 0.0 | 0.0 | 3.0 | 1.9 | 23.5 | T | 0.8 | 0.0 | 0.0 | 29.2 |
| 1983-84 | 0.0 | 0.0 | 0.0 | 0.0 | T | 1.6 | 11.7 | 0.2 | 11.9 | 0.0 | 0.0 | 0.0 | 25.4 |
| 1984-85 | 0.0 | 0.0 | 0.0 | 0.0 | T | 5.5 | 8.4 | 10.0 | 0.2 | T | 0.0 | 0.0 | 24.1 |
| 1985-86 | 0.0 | 0.0 | 0.0 | 0.0 | T | 0.9 | 2.2 | 9.9 | T | T | 0.0 | 0.0 | 13.0 |
| 1986-87 | 0.0 | 0.0 | 0.0 | 0.0 | T | 0.6 | 13.6 | 7.0 | 1.9 | 0.0 | 0.0 | 0.0 | 23.1 |
| 1987-88 | 0.0 | 0.0 | 0.0 | 0.0 | 1.1 | 2.6 | 13.9 | 1.5 | T | 0.0 | 0.0 | 0.0 | 19.1 |
| 1988-89 | 0.0 | 0.0 | 0.0 | 0.0 | 0.0 | 0.3 | 5.0 | 0.3 | 2.5 | 0.0 | 0.0 | 0.0 | 8.1 |
| 1989-90 | 0.0 | 0.0 | 0.0 | 0.0 | 4.7 | 1.4 | 1.8 | 1.8 | 3.1 | 0.6 | 0.0 | 0.0 | 13.4 |
| 1990-91 | T | 0.0 | 0.0 | 0.0 | 0.0 | 7.2 | 8.4 | 9.1 | 0.2 | 0.0 | 0.0 | 0.0 | 24.9 |
| 1991-92 | 0.0 | 0.0 | 0.0 | 0.0 | T | 0.7 | 1.5 | 1.0 | 9.4 | T | 0.0 | 0.0 | 12.6 |
| 1992-93 | 0.0 | 0.0 | 0.0 | 0.0 | 0.0 | 0.4 | 1.5 | 10.7 | 11.9 | 0.0 | 0.0 | 0.0 | 24.5 |
| 1993-94 | 0.0 | 0.0 | 0.0 | 0.0 | T | 6.9 | 12.0 | 26.4 | 8.1 | 0.0 | 0.0 | 0.0 | 53.4 |
| 1994-95 | 0.0 | 0.0 | 0.0 | 0.0 | T | T | 0.2 | 11.6 | T | T | T | 0.0 | 11.8 |
| 1995-96 | 0.0 | 0.0 | 0.0 | 0.0 | 2.9 | 11.5 | 26.1 | 21.2 | 13.2 | 0.7 | 0.0 | 0.0 | 75.6 |
| 1996-97 | 0.0 | 0.0 | 0.0 | 0.0 | .1 | T | 4.4 | 3.8 | 1.7 | T | 0.0 | 0.0 | 10.0 |
| 1997-98 | 0.0 | 0.0 | 0.0 | 0.0 | T | T | 0.5 | 0.0 | 5.0 | 0.0 | 0.0 | 0.0 | 5.5 |
| 1998-99 | 0.0 | 0.0 | 0.0 | 0.0 | 0.0 | 2.0 | 4.5 | 1.7 | 4.5 | 0.0 | 0.0 | 0.0 | 12.7 |
| 1999-00 | 0.0 | 0.0 | 0.0 | 0.0 | 0.0 | T | 9.5 | 5.2 | 0.4 | 1.2 | 0.0 | 0.0 | 16.3 |
| 2000-01 | 0.0 | 0.0 | 0.0 | T | 0.0 | 13.4 | 8.3 | 9.5 | 3.8 | 0.0 | 0.0 | 0.0 | 35.0 |
| 2001-02 | 0.0 | 0.0 | 0.0 | 0.0 | 0.0 | T | 3.5 | T | T | T | 0.0 | 0.0 | 3.5 |
| 2002-03 | 0.0 | 0.0 | 0.0 | T | T | 11.0 | 4.7 | 26.1 | 3.5 | 4.0 | 0.0 | 0.0 | 49.3 |
| 2003-04 | 0.0 | 0.0 | 0.0 | 0.0 | 0.0 | 19.8 | 17.3 | 0.7 | 4.8 | 0.0 | 0.0 | 0.0 | 42.6 |
| 2004-05 | 0.0 | 0.0 | 0.0 | 0.0 | T | 3.0 | 15.3 | 15.8 | 6.9 | 0.0 | 0.0 | 0.0 | 41.0 |
| 2005-06 | 0.0 | 0.0 | 0.0 | 0.0 | T | 9.7 | 2.0 | 26.9 | 1.3 | 0.1 | 0.0 | 0.0 | 40.0 |
| 2006-07 | 0.0 | 0.0 | 0.0 | 0.0 | 0.0 | 0.0 | 2.6 | 3.8 | 6.0 | T | 0.0 | 0.0 | 12.4 |
| 2007-08 | 0.0 | 0.0 | 0.0 | 0.0 | T | 2.9 | T | 9.0 | T | 0.0 | 0.0 | 0.0 | 11.9 |
| 2008-09 | 0.0 | 0.0 | 0.0 | 0.0 | T | 6.0 | 9.0 | 4.3 | 8.3 | T | 0.0 | 0.0 | 27.6 |
| 2009- | 0.0 | 0.0 | 0.0 | 0.0 | 0.0 | 12.4 | | | | | | | |
| POR=<br>98 YRS | 0.1 | 0.1 | 0.1 | 0.1 | 0.8 | 5.2 | 7.2 | 8.6 | 5.2 | 1.0 | 0.1 | 0.1 | 28.6 |

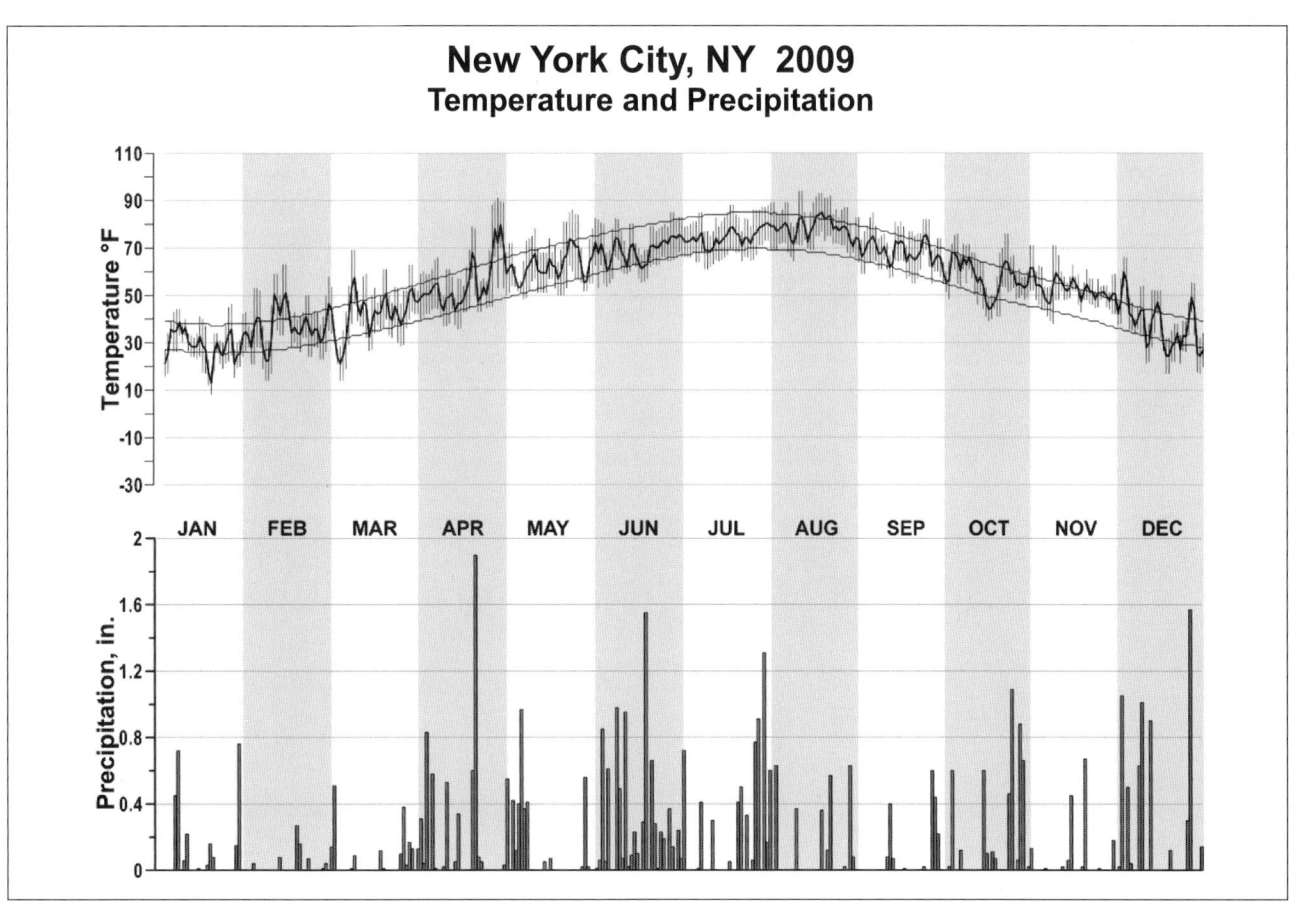

# New York City, NY  2009
## Temperature and Precipitation

# 2009
# ROCHESTER
# NEW YORK (KROC)

Rochester and the Genesee Valley experience a fairly humid, continental type climate, which is strongly modified by the proximity of the Great Lakes. Precipitation is rather evenly distributed throughout the year in quanity, but frequency is much higher during the cloudy winter months than in the sunny ones. Snowfall is heavy, but is highly variable over short distances.

Winters in general are cloudy, cold and snowy..but are changeable and include frequent thaws and rain as well. Snow covers the ground more often than not from christmas into early March..but periods of bare ground are not uncommon. About half of the annual snowfall comes from "lake-effect" process and is very localized. This feature develops when cold air crosses the warmer lake waters and becomes saturated.. creating clouds and precipitation downwind. The exact location of these snowbands are determined by the direction of the wind. Areas east of Rochester receive the most snow from this process..as northwest winds have a longer "fetch" off Lake Ontario..while areas south of the city get somewhat less. Lake Erie can even contribute some snow from this process if a west or southwest wind is storng enough. Since Lake Ontario does not freeze in most winters..this Lake effect machine can remain active throughout the winter. The Rochester area is also subject to occasional general or "synoptic" snowfalls..but the worst effects from these usually pass by to the east. Total season snowfall ranges from 70 inches south of the city to about 90 inches in Rochester to over 120 inches along the lakeshore east of the city. About 50 inches of this total results from general snows..the rest is due to the Lake effect machine. The lake does modify any extreme cold as the mercury falls below zero on only anout six nights in an average winter..with anything below -10 extremely rare.

Spring comes slowly to the region. The last frosts usually occur by April 30 near Lake Ontario..but as late as mid-May south of the Thruway. The spring months are actually the driest months statistically, due in part to the stabilizing effects of the Great Lakes, although soils are wet. Sunshine increases markedly in May.

Summers are warm and sunny across the region. The average temperature is in the 70 to 72 degree range. Rain can be expected on every third or fourth day.. almost always in the form of showers and thunderstorms. This activity is more common inland than near the Lake. Completely overcast days in summer are rare. Severe weather is not common..but a few cases of damaging winds and small tornadoes occur each year. The greatest risk of this type of activity is south of the Thruway. There usually are several periods of uncomfortably warm and muggy weather in an average summer..but only about nine days reach 90-degree mark in an average year. Still, the area usually experiences some of the most delightful summer weather in the East.

Autumn is pleasant, but rather brief. Mild and dry conditions predominate through September and much of October, but colder airmasses cross the Great Lakes with increasing frequency starting in late October, and result in a drastic increase in cloud cover across the region in late October and early November. Although the first frosts may not occur until late October near Lake Ontario, the firstlake effect snows of the season follow soon after...usually by mid November. These early snows melt off quickly, with a general snow cover seldom established before mid December. The growing season is relatively long for the latitude...average about 180 days. The long growing season...combined with ample spring moisture and abundant summer sunshine...is beneficial for the many fruit orchards and wineries...especially near the Lake Ontario shore and Finger Lakes.

# NORMALS, MEANS, AND EXTREMES
## ROCHESTER (KROC)

| LATITUDE: 43 ° 7 'N | LONGITUDE: -77 ° 40'W | ELEVATION (FT): GRND: 538  BARO: 588 | TIME ZONE: EASTERN  (UTC -5) | WBAN: 14768 |
|---|---|---|---|---|

| | ELEMENT | POR | JAN | FEB | MAR | APR | MAY | JUN | JUL | AUG | SEP | OCT | NOV | DEC | YEAR |
|---|---|---|---|---|---|---|---|---|---|---|---|---|---|---|---|
| **TEMPERATURE °F** | NORMAL DAILY MAXIMUM | 30 | 31.2 | 33.2 | 42.7 | 55.2 | 67.9 | 76.6 | 81.4 | 79.1 | 71.1 | 59.7 | 47.2 | 36.1 | 56.8 |
| | MEAN DAILY MAXIMUM | 84 | 31.8 | 32.3 | 41.7 | 54.8 | 67.5 | 76.5 | 81.7 | 79.7 | 71.8 | 60.9 | 47.5 | 35.9 | 56.8 |
| | HIGHEST DAILY MAXIMUM | 69 | 74 | 73 | 84 | 93 | 94 | 100 | 98 | 99 | 99 | 91 | 81 | 72 | 100 |
| | YEAR OF OCCURRENCE | | 1950 | 1997 | 1945 | 1990 | 1987 | 1953 | 1993 | 1948 | 1953 | 1951 | 1950 | 1982 | JUN 1953 |
| | MEAN OF EXTREME MAXS. | 84 | 53.8 | 53.7 | 68.4 | 79.2 | 85.8 | 90.9 | 92.5 | 91.0 | 87.8 | 80.0 | 68.9 | 56.9 | 75.7 |
| | NORMAL DAILY MINIMUM | 30 | 16.6 | 17.3 | 25.2 | 35.3 | 46.1 | 55.0 | 60.0 | 58.7 | 51.3 | 41.1 | 32.6 | 22.7 | 38.5 |
| | MEAN DAILY MINIMUM | 84 | 17.8 | 17.3 | 25.6 | 35.9 | 46.5 | 55.6 | 61.0 | 59.5 | 51.9 | 42.2 | 33.1 | 22.9 | 39.1 |
| | LOWEST DAILY MINIMUM | 69 | -17 | -19 | -7 | 13 | 26 | 35 | 42 | 36 | 28 | 20 | 5 | -16 | -19 |
| | YEAR OF OCCURRENCE | | 1994 | 1979 | 1999 | 1982 | 1979 | 1949 | 1963 | 1965 | 1947 | 1972 | 1971 | 1942 | FEB 1979 |
| | MEAN OF EXTREME MINS. | 84 | -1.7 | -1.2 | 6.9 | 22.5 | 32.6 | 42.0 | 48.9 | 46.7 | 37.5 | 28.4 | 18.2 | 3.6 | 23.7 |
| | NORMAL DRY BULB | 30 | 23.9 | 25.3 | 33.9 | 45.3 | 57.0 | 65.8 | 70.7 | 68.9 | 61.2 | 50.4 | 39.9 | 29.4 | 47.6 |
| | MEAN DRY BULB | 84 | 24.8 | 24.8 | 33.7 | 45.3 | 57.0 | 66.1 | 71.4 | 69.7 | 61.8 | 51.5 | 40.3 | 29.5 | 48.0 |
| | MEAN WET BULB | 26 | 22.7 | 23.2 | 29.4 | 39.8 | 49.9 | 59.6 | 63.8 | 62.9 | 56.6 | 45.9 | 36.7 | 27.2 | 43.1 |
| | MEAN DEW POINT | 26 | 19.5 | 19.7 | 25.7 | 35.4 | 46.3 | 56.6 | 61.1 | 60.5 | 54.1 | 42.9 | 33.4 | 24.1 | 39.9 |
| | NORMAL NO. DAYS WITH: | | | | | | | | | | | | | | |
| | MAXIMUM >= 90 | 30 | 0.0 | 0.0 | 0.0 | * | 0.3 | 1.3 | 3.8 | 1.9 | 0.4 | 0.0 | 0.0 | 0.0 | 7.7 |
| | MAXIMUM <= 32 | 30 | 16.4 | 13.9 | 6.3 | 0.5 | 0.0 | 0.0 | 0.0 | 0.0 | 0.0 | 0.0 | 1.7 | 10.0 | 48.8 |
| | MINIMUM <= 32 | 30 | 28.2 | 25.3 | 22.8 | 10.9 | 0.8 | 0.0 | 0.0 | 0.0 | 0.2 | 4.3 | 15.4 | 25.5 | 133.4 |
| | MINIMUM <= 0 | 30 | 2.6 | 1.8 | 0.2 | 0.0 | 0.0 | 0.0 | 0.0 | 0.0 | 0.0 | 0.0 | 0.0 | 0.6 | 5.2 |
| **H/C** | NORMAL HEATING DEG. DAYS | 30 | 1263 | 1117 | 958 | 582 | 266 | 69 | 17 | 25 | 154 | 447 | 741 | 1089 | 6728 |
| | NORMAL COOLING DEG. DAYS | 30 | 0 | 0 | 1 | 5 | 32 | 109 | 209 | 162 | 54 | 4 | 0 | 0 | 576 |
| **RH** | NORMAL (PERCENT) | 30 | 76 | 75 | 71 | 67 | 68 | 71 | 72 | 75 | 78 | 76 | 76 | 77 | 74 |
| | HOUR 01 LST | 30 | 78 | 79 | 77 | 76 | 80 | 84 | 85 | 88 | 88 | 84 | 80 | 80 | 82 |
| | HOUR 07 LST | 30 | 79 | 80 | 80 | 77 | 78 | 81 | 84 | 88 | 89 | 86 | 82 | 82 | 82 |
| | HOUR 13 LST | 30 | 70 | 67 | 62 | 56 | 55 | 57 | 55 | 59 | 62 | 61 | 67 | 72 | 62 |
| | HOUR 19 LST | 30 | 76 | 74 | 69 | 62 | 60 | 62 | 61 | 68 | 75 | 75 | 76 | 78 | 70 |
| **S** | PERCENT POSSIBLE SUNSHINE | 57 | 35 | 41 | 49 | 53 | 59 | 66 | 69 | 66 | 59 | 49 | 31 | 31 | 51 |
| **W/O** | MEAN NO. DAYS WITH: | | | | | | | | | | | | | | |
| | HEAVY FOG(VISBY <= 1/4 MI) | 46 | 0.9 | 0.8 | 1.5 | 0.8 | 1.0 | 0.7 | 0.6 | 0.8 | 1.2 | 1.5 | 0.8 | 1.2 | 11.8 |
| | THUNDERSTORMS | 62 | 0.1 | 0.1 | 0.9 | 1.9 | 3.4 | 5.3 | 6.1 | 5.6 | 2.7 | 0.9 | 0.4 | 0.2 | 27.6 |
| **CLOUDNESS** | MEAN: | | | | | | | | | | | | | | |
| | SUNRISE-SUNSET (OKTAS) | 56 | 6.6 | 6.3 | 5.8 | 5.4 | 5.3 | 4.9 | 4.6 | 4.7 | 4.9 | 5.3 | 6.4 | 6.6 | 5.6 |
| | MIDNIGHT-MIDNIGHT (OKTAS) | 32 | 6.5 | 6.2 | 5.7 | 5.3 | 5.0 | 4.7 | 4.3 | 4.5 | 4.7 | 5.2 | 6.3 | 6.5 | 5.4 |
| | MEAN NO. DAYS WITH: | | | | | | | | | | | | | | |
| | CLEAR | 56 | 2.0 | 2.4 | 4.6 | 6.1 | 6.0 | 7.2 | 7.8 | 7.7 | 7.0 | 6.4 | 2.1 | 2.0 | 61.3 |
| | PARTLY CLOUDY | 56 | 6.6 | 6.7 | 8.1 | 7.7 | 9.7 | 10.8 | 12.5 | 11.7 | 10.5 | 8.4 | 6.1 | 5.6 | 104.4 |
| | CLOUDY | 56 | 22.4 | 19.2 | 18.3 | 16.3 | 15.4 | 12.1 | 10.7 | 11.7 | 12.6 | 16.3 | 21.8 | 23.4 | 200.2 |
| **PR** | MEAN STATION PRESSURE(IN) | 26 | 29.43 | 29.44 | 29.44 | 29.37 | 29.38 | 29.37 | 29.38 | 29.43 | 29.46 | 29.47 | 29.44 | 29.44 | 29.42 |
| | MEAN SEA-LEVEL PRES. (IN) | 26 | 30.05 | 30.06 | 30.05 | 29.98 | 29.98 | 29.96 | 29.97 | 30.02 | 30.06 | 30.07 | 30.06 | 30.06 | 30.03 |
| **WINDS** | MEAN SPEED (MPH) | 26 | 10.7 | 10.3 | 10.2 | 10.1 | 9.0 | 8.1 | 7.5 | 7.1 | 7.5 | 8.4 | 9.5 | 10.1 | 9.0 |
| | PREVAIL.DIR(TENS OF DEGS) | 35 | 26 | 26 | 26 | 26 | 26 | 23 | 23 | 23 | 22 | 22 | 26 | 26 | 26 |
| | MAXIMUM 2-MINUTE: | | | | | | | | | | | | | | |
| | SPEED (MPH) | 13 | 60 | 59 | 55 | 45 | 45 | 44 | 52 | 36 | 68 | 40 | 51 | 48 | 68 |
| | DIR. (TENS OF DEGS) | | 23 | 26 | 25 | 24 | 18 | 25 | 20 | 30 | 27 | 27 | 26 | 24 | 27 |
| | YEAR OF OCCURRENCE | | 2008 | 2002 | 2002 | 2008 | 2002 | 2007 | 1999 | 1999 | 1998 | 2003 | 2003 | 2004 | SEP 1998 |
| | MAXIMUM 3-SECOND | | | | | | | | | | | | | | |
| | SPEED (MPH) | 13 | 75 | 77 | 70 | 62 | 63 | 51 | 64 | 53 | 89 | 54 | 66 | 63 | 89 |
| | DIR. (TENS OF DEGS) | | 25 | 25 | 24 | 25 | 28 | 25 | 01 | 31 | 29 | 27 | 28 | 27 | 29 |
| | YEAR OF OCCURRENCE | | 2008 | 2006 | 2002 | 1997 | 1998 | 2007 | 2005 | 1999 | 1998 | 2003 | 2003 | 2008 | SEP 1998 |
| **PRECIPITATION** | NORMAL (IN) | 30 | 2.34 | 2.04 | 2.58 | 2.75 | 2.82 | 3.36 | 2.93 | 3.54 | 3.45 | 2.60 | 2.84 | 2.73 | 33.98 |
| | MAXIMUM MONTHLY (IN) | 69 | 5.79 | 5.07 | 5.42 | 4.90 | 6.62 | 7.11 | 9.70 | 6.00 | 6.30 | 7.85 | 6.99 | 5.05 | 9.70 |
| | YEAR OF OCCURRENCE | | 1978 | 1950 | 1942 | 1944 | 1974 | 1998 | 1947 | 1984 | 1977 | 1955 | 1985 | 1944 | JUL 1947 |
| | MINIMUM MONTHLY (IN) | 69 | 0.72 | 0.66 | 0.47 | 1.18 | 0.24 | 0.22 | 0.61 | 0.76 | 0.28 | 0.23 | 0.44 | 0.62 | 0.22 |
| | YEAR OF OCCURRENCE | | 1988 | 1987 | 1958 | 1995 | 2007 | 1963 | 1994 | 1951 | 1960 | 1963 | 1976 | 1958 | JUN 1963 |
| | MAXIMUM IN 24 HOURS (IN) | 69 | 2.24 | 2.43 | 2.21 | 2.22 | 3.85 | 2.86 | 3.33 | 3.03 | 3.54 | 3.13 | 3.13 | 1.60 | 3.85 |
| | YEAR OF OCCURRENCE | | 1998 | 1950 | 1942 | 1991 | 1974 | 1950 | 2006 | 2005 | 1979 | 1995 | 1945 | 1978 | MAY 1974 |
| | NORMAL NO. DAYS WITH: | | | | | | | | | | | | | | |
| | PRECIPITATION >= 0.01 | 30 | 19.1 | 16.3 | 15.2 | 13.5 | 11.8 | 11.6 | 10.2 | 10.7 | 11.8 | 12.8 | 15.9 | 18.4 | 167.3 |
| | PRECIPITATION >= 1.00 | 30 | 0.2 | 0.1 | 0.2 | 0.2 | 0.3 | 0.7 | 0.5 | 0.8 | 0.8 | 0.2 | 0.4 | 0.2 | 4.6 |
| **SNOWFALL** | NORMAL (IN) | 30 | 25.8 | 22.2 | 16.6 | 5.1 | 0.5 | 0.0 | 0.0 | 0.0 | 0.0 | 0.1 | 8.1 | 21.9 | 100.3 |
| | MAXIMUM MONTHLY (IN) | 69 | 61.3 | 64.8 | 45.0 | 20.2 | 10.9 | T | T | T | T | 2.6 | 24.9 | 46.2 | 64.8 |
| | YEAR OF OCCURRENCE | | 2004 | 1958 | 1999 | 1979 | 1989 | 2006 | 2006 | 1965 | 2009 | 1993 | 1996 | 2008 | FEB 1958 |
| | MAXIMUM IN 24 HOURS (IN) | 69 | 23.0 | 22.8 | 23.3 | 10.4 | 10.8 | T | T | T | T | 2.6 | 14.1 | 19.1 | 23.3 |
| | YEAR OF OCCURRENCE' | | 1996 | 1978 | 1999 | 1990 | 1989 | 1998 | 1990 | 1965 | 2009 | 1993 | 1995 | 1978 | MAR 1999 |
| | MAXIMUM SNOW DEPTH (IN) | 61 | 32 | 34 | 34 | 10 | 4 | 0 | 0 | 0 | 0 | 1 | 8 | 40 | 40 |
| | YEAR OF OCCURRENCE | | 1978 | 1966 | 1999 | 1975 | 1989 | | | | | 1957 | 1972 | 1959 | DEC 1959 |
| | NORMAL NO. DAYS WITH: | | | | | | | | | | | | | | |
| | SNOWFALL >= 1.0 | 30 | 7.1 | 6.7 | 4.0 | 1.4 | 0.1 | 0.0 | 0.0 | 0.0 | 0.0 | 0.0 | 2.4 | 6.4 | 28.1 |

## PRECIPITATION (inches) 2009 ROCHESTER (KROC)

| YEAR | JAN | FEB | MAR | APR | MAY | JUN | JUL | AUG | SEP | OCT | NOV | DEC | ANNUAL |
|------|-----|-----|-----|-----|-----|-----|-----|-----|-----|-----|-----|-----|--------|
| 1980 | 1.11 | 1.16 | 3.83 | 2.35 | 1.49 | 6.77 | 1.90 | 3.44 | 3.57 | 3.73 | 2.52 | 2.45 | 34.32 |
| 1981 | 1.24 | 3.13 | 1.04 | 1.95 | 2.27 | 2.70 | 4.60 | 4.44 | 5.37 | 3.29 | 2.18 | 2.78 | 34.99 |
| 1982 | 4.16 | 1.01 | 1.73 | 1.63 | 1.77 | 3.92 | 3.13 | 3.00 | 3.57 | 1.79 | 3.95 | 2.17 | 31.83 |
| 1983 | 1.43 | 1.23 | 2.45 | 3.50 | 3.44 | 2.40 | 1.13 | 5.43 | 1.56 | 3.26 | 4.91 | 4.47 | 35.21 |
| 1984 | 1.62 | 2.97 | 2.08 | 3.05 | 5.47 | 1.67 | 1.90 | 6.00 | 3.34 | 0.76 | 1.47 | 3.31 | 33.64 |
| 1985 | 2.49 | 1.78 | 3.47 | 1.30 | 2.08 | 2.63 | 1.86 | 1.11 | 2.49 | 2.34 | 6.99 | 1.46 | 30.00 |
| 1986 | 1.63 | 2.46 | 1.90 | 3.80 | 1.64 | 4.27 | 3.13 | 3.29 | 5.11 | 3.56 | 1.93 | 3.56 | 36.28 |
| 1987 | 1.89 | 0.66 | 1.98 | 3.68 | 1.19 | 3.94 | 5.85 | 3.92 | 4.60 | 1.65 | 2.74 | 1.98 | 34.08 |
| 1988 | 0.72 | 2.18 | 1.62 | 2.32 | 1.73 | 1.10 | 4.30 | 3.81 | 1.69 | 2.34 | 1.68 | 1.11 | 24.60 |
| 1989 | 1.18 | 1.55 | 3.69 | 1.62 | 5.99 | 5.65 | 0.98 | 2.46 | 2.82 | 3.13 | 2.01 | 1.58 | 32.66 |
| 1990 | 1.61 | 3.93 | 1.56 | 3.58 | 5.76 | 2.88 | 3.05 | 3.59 | 3.36 | 4.37 | 2.27 | 4.18 | 40.14 |
| 1991 | 1.69 | 1.16 | 4.70 | 4.07 | 2.43 | 1.19 | 2.37 | 1.80 | 2.86 | 1.65 | 2.39 | 2.92 | 29.23 |
| 1992 | 1.46 | 1.87 | 3.53 | 3.43 | 2.83 | 1.98 | 6.03 | 4.45 | 3.02 | 1.78 | 2.90 | 2.98 | 36.26 |
| 1993 | 2.32 | 1.52 | 2.44 | 3.07 | 1.24 | 2.76 | 1.67 | 1.67 | 4.37 | 3.21 | 3.27 | 1.60 | 29.14 |
| 1994 | 2.68 | 1.63 | 1.70 | 4.08 | 2.56 | 2.43 | 0.61 | 4.27 | 2.68 | 1.34 | 3.24 | 2.32 | 29.54 |
| 1995 | 2.46 | 1.58 | 1.15 | 1.18 | 1.75 | 2.07 | 3.85 | 3.05 | 1.50 | 5.70 | 4.21 | 1.50 | 30.00 |
| 1996 | 3.18 | 1.72 | 2.07 | 4.84 | 3.51 | 6.65 | 2.18 | 3.33 | 5.09 | 5.40 | 4.12 | 2.97 | 45.06 |
| 1997 | 2.03 | 2.40 | 3.88 | 1.33 | 2.12 | 3.01 | 1.94 | 4.22 | 5.36 | 1.94 | 3.57 | 2.88 | 34.68 |
| 1998 | 5.63 | 2.34 | 3.50 | 1.81 | 2.63 | 7.11 | 6.09 | 5.39 | 3.00 | 1.45 | 1.41 | 1.60 | 41.96 |
| 1999 | 3.92 | 0.69 | 3.28 | 2.07 | 2.72 | 2.52 | 1.78 | 5.71 | 3.41 | 2.12 | 2.86 | 2.06 | 33.14 |
| 2000 | 2.98 | 1.97 | 2.04 | 4.35 | 4.70 | 4.47 | 3.66 | 4.11 | 3.53 | 1.36 | 2.19 | 2.47 | 37.83 |
| 2001 | 1.95 | 2.26 | 4.13 | 1.19 | 2.66 | 1.84 | 1.80 | 4.30 | 3.15 | 2.28 | 1.90 | 1.72 | 29.18 |
| 2002 | 2.97 | 1.61 | 2.09 | 3.44 | 5.87 | 4.29 | 1.59 | 0.84 | 2.61 | 2.09 | 3.11 | 3.85 | 34.36 |
| 2003 | 2.05 | 1.96 | 1.95 | 1.27 | 4.56 | 2.23 | 2.26 | 4.13 | 2.69 | 1.90 | 4.26 | 2.42 | 31.68 |
| 2004 | 2.81 | 0.72 | 2.04 | 3.48 | 4.53 | 3.11 | 6.35 | 3.68 | 4.30 | 1.49 | 2.31 | 2.99 | 37.81 |
| 2005 | 3.34 | 1.40 | 1.11 | 4.43 | 1.24 | 2.44 | 3.36 | 5.10 | 4.98 | 3.48 | 3.18 | 1.37 | 35.43 |
| 2006 | 2.42 | 2.13 | 1.80 | 2.18 | 1.77 | 3.72 | 8.02 | 2.75 | 5.39 | 4.96 | 2.89 | 3.03 | 41.06 |
| 2007 | 4.25 | 2.09 | 2.98 | 3.64 | 0.24 | 2.30 | 2.31 | 0.81 | 2.50 | 3.05 | 4.01 | 4.28 | 32.46 |
| 2008 | 1.60 | 4.27 | 3.75 | 1.94 | 1.41 | 2.59 | 3.91 | 2.97 | 1.66 | 3.38 | 2.11 | 3.61 | 33.20 |
| 2009 | 2.17 | 1.63 | 2.99 | 2.20 | 3.30 | 6.25 | 4.32 | 1.63 | 2.03 | 2.97 | 1.13 | 2.95 | 33.57 |
| POR= 84 YRS | 2.35 | 2.21 | 2.67 | 2.69 | 2.73 | 2.90 | 2.92 | 3.11 | 2.85 | 2.62 | 2.80 | 2.56 | 32.41 |

WBAN : 14768

## AVERAGE TEMPERATURE (°F) 2009 ROCHESTER (KROC)

| YEAR | JAN | FEB | MAR | APR | MAY | JUN | JUL | AUG | SEP | OCT | NOV | DEC | ANNUAL |
|------|-----|-----|-----|-----|-----|-----|-----|-----|-----|-----|-----|-----|--------|
| 1980 | 24.0 | 19.7 | 32.4 | 47.8 | 60.0 | 63.1 | 72.9 | 74.3 | 63.7 | 48.8 | 38.8 | 24.7 | 47.5 |
| 1981 | 15.7 | 32.3 | 34.5 | 48.0 | 57.2 | 67.3 | 71.9 | 69.4 | 59.8 | 47.3 | 39.9 | 28.7 | 47.7 |
| 1982 | 16.1 | 23.0 | 33.5 | 43.2 | 60.9 | 63.6 | 72.0 | 66.1 | 62.8 | 52.7 | 43.4 | 37.4 | 47.9 |
| 1983 | 27.4 | 29.1 | 37.2 | 43.9 | 53.8 | 66.7 | 73.8 | 70.7 | 63.9 | 52.9 | 40.7 | 25.1 | 48.8 |
| 1984 | 20.4 | 33.2 | 26.5 | 47.5 | 52.6 | 66.8 | 69.2 | 72.0 | 60.6 | 54.9 | 40.7 | 35.9 | 48.4 |
| 1985 | 21.9 | 25.6 | 36.7 | 49.6 | 58.6 | 61.7 | 68.8 | 68.7 | 63.8 | 51.0 | 41.4 | 25.0 | 47.7 |
| 1986 | 25.0 | 24.5 | 37.0 | 47.9 | 59.8 | 63.3 | 69.8 | 65.7 | 59.8 | 49.9 | 36.9 | 31.7 | 47.6 |
| 1987 | 25.3 | 23.6 | 37.1 | 49.7 | 59.9 | 67.9 | 72.7 | 67.3 | 61.6 | 47.1 | 40.6 | 32.6 | 48.8 |
| 1988 | 25.0 | 23.7 | 34.7 | 45.0 | 58.7 | 64.2 | 73.7 | 71.1 | 60.1 | 45.8 | 42.6 | 29.4 | 47.8 |
| 1989 | 30.3 | 22.5 | 32.3 | 42.1 | 56.3 | 67.4 | 72.8 | 68.5 | 61.7 | 52.6 | 38.1 | 17.1 | 46.8 |
| 1990 | 33.6 | 29.3 | 37.3 | 48.8 | 54.4 | 67.2 | 70.7 | 69.9 | 60.7 | 52.1 | 42.4 | 33.8 | 50.0 |
| 1991 | 25.1 | 30.5 | 37.0 | 50.0 | 62.8 | 68.3 | 72.3 | 70.3 | 60.5 | 52.1 | 39.0 | 30.7 | 49.9 |
| 1992 | 26.2 | 27.2 | 30.2 | 44.1 | 57.1 | 63.5 | 66.6 | 66.3 | 60.9 | 46.4 | 38.9 | 30.2 | 46.5 |
| 1993 | 27.5 | 18.7 | 30.0 | 46.9 | 56.6 | 65.5 | 72.4 | 71.4 | 59.0 | 48.0 | 39.0 | 28.4 | 47.0 |
| 1994 | 14.9 | 21.1 | 32.0 | 48.0 | 54.2 | 67.8 | 73.5 | 69.2 | 62.4 | 52.4 | 45.8 | 34.8 | 48.0 |
| 1995 | 32.1 | 23.8 | 38.5 | 41.6 | 57.2 | 69.5 | 73.0 | 73.3 | 60.4 | 55.2 | 35.2 | 25.3 | 48.8 |
| 1996 | 23.7 | 24.6 | 29.7 | 43.3 | 54.6 | 68.0 | 68.1 | 69.3 | 62.0 | 51.2 | 34.3 | 33.6 | 46.9 |
| 1997 | 24.3 | 30.5 | 32.9 | 43.8 | 50.4 | 67.1 | 67.8 | 66.3 | 59.3 | 49.1 | 37.4 | 31.2 | 46.7 |
| 1998 | 31.6 | 32.6 | 38.1 | 47.7 | 62.9 | 65.7 | 69.6 | 69.8 | 62.7 | 51.2 | 41.8 | 34.6 | 50.7 |
| 1999 | 22.9 | 30.6 | 30.8 | 45.3 | 59.6 | 68.3 | 74.3 | 67.1 | 64.0 | 50.9 | 44.9 | 32.2 | 49.2 |
| 2000 | 23.2 | 30.2 | 41.2 | 45.1 | 59.5 | 65.8 | 67.1 | 67.5 | 60.9 | 51.7 | 38.4 | 22.6 | 47.8 |
| 2001 | 26.6 | 28.6 | 30.3 | 47.7 | 59.5 | 66.7 | 69.0 | 72.2 | 61.2 | 52.4 | 47.1 | 35.9 | 49.8 |
| 2002 | 32.6 | 31.9 | 35.1 | 48.2 | 53.4 | 67.6 | 74.0 | 72.4 | 67.7 | 49.9 | 40.5 | 28.4 | 50.1 |
| 2003 | 18.2 | 20.9 | 34.1 | 43.3 | 55.2 | 64.8 | 70.6 | 70.7 | 62.5 | 48.2 | 42.2 | 32.3 | 46.9 |
| 2004 | 17.2 | 25.0 | 38.6 | 45.8 | 58.7 | 62.7 | 68.3 | 66.9 | 64.2 | 50.2 | 41.1 | 29.3 | 47.3 |
| 2005 | 21.7 | 25.8 | 29.7 | 46.0 | 52.2 | 70.5 | 73.0 | 72.8 | 65.5 | 53.4 | 43.7 | 28.1 | 48.5 |
| 2006 | 35.5 | 28.4 | 36.0 | 48.0 | 58.9 | 67.8 | 75.0 | 70.3 | 62.1 | 50.6 | 45.9 | 39.0 | 51.5 |
| 2007 | 30.0 | 20.3 | 35.3 | 44.6 | 59.6 | 69.6 | 69.7 | 71.7 | 64.7 | 58.0 | 38.6 | 29.3 | 49.3 |
| 2008 | 30.5 | 26.4 | 31.5 | 52.3 | 54.8 | 69.8 | 72.0 | 67.6 | 63.1 | 49.7 | 39.3 | 30.6 | 49.0 |
| 2009 | 19.1 | 29.0 | 35.8 | 46.6 | 56.8 | 63.5 | 66.2 | 69.3 | 60.7 | 48.2 | 41.9 | 27.9 | 47.1 |
| POR= 84 YRS | 24.8 | 24.8 | 33.7 | 45.3 | 57.0 | 66.1 | 71.4 | 69.7 | 61.8 | 51.5 | 40.3 | 29.5 | 48.0 |

## HEATING DEGREE DAYS (base 65°F) 2009 ROCHESTER (KROC)

| YEAR | JUL | AUG | SEP | OCT | NOV | DEC | JAN | FEB | MAR | APR | MAY | JUN | TOTAL |
|------|-----|-----|-----|-----|-----|-----|-----|-----|-----|-----|-----|-----|-------|
| 1980-81 | 1 | 0 | 108 | 498 | 782 | 1243 | 1522 | 908 | 938 | 507 | 260 | 26 | 6793 |
| 1981-82 | 6 | 12 | 201 | 546 | 748 | 1119 | 1510 | 1171 | 972 | 648 | 162 | 67 | 7162 |
| 1982-83 | 10 | 54 | 113 | 377 | 643 | 847 | 1161 | 998 | 854 | 627 | 347 | 78 | 6109 |
| 1983-84 | 9 | 8 | 121 | 387 | 723 | 1228 | 1376 | 917 | 1187 | 520 | 395 | 50 | 6921 |
| 1984-85 | 14 | 7 | 162 | 307 | 724 | 897 | 1330 | 1097 | 869 | 471 | 217 | 119 | 6214 |
| 1985-86 | 15 | 23 | 121 | 429 | 700 | 1231 | 1235 | 1129 | 864 | 506 | 206 | 100 | 6559 |
| 1986-87 | 16 | 62 | 175 | 462 | 840 | 1026 | 1223 | 1153 | 858 | 454 | 234 | 39 | 6542 |
| 1987-88 | 7 | 50 | 139 | 547 | 722 | 997 | 1232 | 1192 | 933 | 594 | 220 | 126 | 6759 |
| 1988-89 | 6 | 40 | 164 | 596 | 664 | 1095 | 1070 | 1184 | 1009 | 682 | 288 | 33 | 6831 |
| 1989-90 | 0 | 33 | 149 | 383 | 801 | 1478 | 967 | 993 | 853 | 520 | 327 | 46 | 6550 |
| 1990-91 | 7 | 6 | 171 | 406 | 669 | 959 | 1230 | 957 | 862 | 458 | 170 | 29 | 5924 |
| 1991-92 | 2 | 1 | 196 | 408 | 776 | 1057 | 1195 | 1088 | 1069 | 621 | 259 | 89 | 6761 |
| 1992-93 | 26 | 46 | 172 | 571 | 774 | 1071 | 1158 | 1289 | 1077 | 538 | 263 | 66 | 7051 |
| 1993-94 | 0 | 10 | 214 | 525 | 775 | 1127 | 1550 | 1221 | 1015 | 505 | 345 | 56 | 7343 |
| 1994-95 | 1 | 16 | 106 | 386 | 568 | 928 | 1015 | 1147 | 811 | 694 | 239 | 33 | 5944 |
| 1995-96 | 13 | 7 | 162 | 300 | 885 | 1224 | 1271 | 1163 | 1086 | 647 | 347 | 16 | 7121 |
| 1996-97 | 19 | 6 | 140 | 421 | 915 | 968 | 1255 | 960 | 989 | 628 | 448 | 45 | 6794 |
| 1997-98 | 32 | 29 | 181 | 492 | 822 | 1041 | 1029 | 901 | 837 | 513 | 109 | 99 | 6085 |
| 1998-99 | 4 | 14 | 115 | 424 | 690 | 935 | 1295 | 955 | 1054 | 583 | 194 | 57 | 6320 |
| 1999-00 | 2 | 25 | 100 | 431 | 595 | 1008 | 1289 | 1003 | 732 | 591 | 211 | 75 | 6062 |
| 2000-01 | 23 | 35 | 178 | 404 | 793 | 1308 | 1186 | 1014 | 1067 | 522 | 192 | 63 | 6785 |
| 2001-02 | 24 | 3 | 152 | 391 | 529 | 896 | 999 | 921 | 919 | 521 | 372 | 63 | 5790 |
| 2002-03 | 2 | 4 | 45 | 482 | 728 | 1125 | 1445 | 1229 | 951 | 646 | 298 | 76 | 7031 |
| 2003-04 | 0 | 15 | 101 | 510 | 678 | 1006 | 1475 | 1153 | 811 | 574 | 220 | 120 | 6663 |
| 2004-05 | 10 | 33 | 74 | 453 | 708 | 1098 | 1334 | 1091 | 1085 | 562 | 388 | 37 | 6873 |
| 2005-06 | 2 | 1 | 50 | 377 | 634 | 1138 | 908 | 1019 | 892 | 502 | 229 | 35 | 5787 |
| 2006-07 | 0 | 14 | 109 | 440 | 566 | 798 | 1077 | 1246 | 913 | 608 | 229 | 33 | 6033 |
| 2007-08 | 10 | 12 | 85 | 242 | 785 | 1102 | 1065 | 1113 | 1030 | 386 | 317 | 23 | 6170 |
| 2008-09 | 1 | 21 | 115 | 471 | 761 | 1060 | 1414 | 1001 | 896 | 556 | 259 | 81 | 6636 |
| 2009- | 32 | 22 | 139 | 514 | 686 | 1141 | | | | | | | |

WBAN : 14768

## COOLING DEGREE DAYS (base 65°F) 2009 ROCHESTER (KROC)

| YEAR | JAN | FEB | MAR | APR | MAY | JUN | JUL | AUG | SEP | OCT | NOV | DEC | TOTAL |
|------|-----|-----|-----|-----|-----|-----|-----|-----|-----|-----|-----|-----|-------|
| 1980 | 0 | 0 | 0 | 0 | 46 | 73 | 253 | 294 | 76 | 2 | 0 | 0 | 744 |
| 1981 | 0 | 0 | 0 | 5 | 23 | 102 | 228 | 156 | 50 | 0 | 0 | 0 | 564 |
| 1982 | 0 | 0 | 0 | 3 | 40 | 30 | 232 | 95 | 52 | 3 | 1 | 0 | 456 |
| 1983 | 0 | 0 | 0 | 0 | 7 | 136 | 289 | 192 | 96 | 20 | 0 | 0 | 740 |
| 1984 | 0 | 0 | 0 | 1 | 14 | 113 | 152 | 233 | 35 | 1 | 0 | 0 | 549 |
| 1985 | 0 | 0 | 0 | 15 | 23 | 27 | 139 | 145 | 90 | 0 | 0 | 0 | 439 |
| 1986 | 0 | 0 | 1 | 0 | 50 | 53 | 168 | 94 | 28 | 0 | 0 | 0 | 394 |
| 1987 | 0 | 0 | 0 | 1 | 82 | 131 | 254 | 127 | 42 | 0 | 0 | 0 | 637 |
| 1988 | 0 | 0 | 0 | 0 | 34 | 107 | 284 | 232 | 29 | 7 | 0 | 0 | 693 |
| 1989 | 0 | 0 | 0 | 0 | 26 | 111 | 248 | 153 | 60 | 3 | 0 | 0 | 601 |
| 1990 | 0 | 0 | 3 | 41 | 5 | 122 | 192 | 164 | 45 | 14 | 0 | 0 | 586 |
| 1991 | 0 | 0 | 0 | 14 | 108 | 135 | 234 | 175 | 68 | 14 | 0 | 0 | 748 |
| 1992 | 0 | 0 | 0 | 1 | 19 | 51 | 84 | 96 | 57 | 0 | 0 | 0 | 308 |
| 1993 | 0 | 0 | 0 | 0 | 9 | 86 | 239 | 214 | 42 | 5 | 0 | 0 | 595 |
| 1994 | 0 | 0 | 0 | 0 | 16 | 145 | 271 | 152 | 36 | 2 | 1 | 0 | 623 |
| 1995 | 0 | 0 | 0 | 0 | 3 | 175 | 266 | 269 | 33 | 5 | 0 | 0 | 751 |
| 1996 | 0 | 0 | 0 | 1 | 31 | 115 | 120 | 147 | 58 | 0 | 0 | 0 | 472 |
| 1997 | 0 | 0 | 0 | 1 | 0 | 114 | 126 | 75 | 15 | 8 | 0 | 0 | 339 |
| 1998 | 0 | 0 | 9 | 0 | 48 | 125 | 152 | 168 | 53 | 4 | 0 | 0 | 559 |
| 1999 | 0 | 0 | 0 | 0 | 38 | 161 | 296 | 93 | 77 | 0 | 0 | 0 | 665 |
| 2000 | 0 | 0 | 0 | 2 | 49 | 105 | 95 | 120 | 63 | 0 | 0 | 0 | 434 |
| 2001 | 0 | 0 | 0 | 9 | 28 | 119 | 154 | 233 | 44 | 6 | 0 | 0 | 593 |
| 2002 | 0 | 0 | 0 | 24 | 19 | 148 | 288 | 241 | 135 | 23 | 0 | 0 | 878 |
| 2003 | 0 | 0 | 0 | 3 | 0 | 73 | 181 | 197 | 32 | 1 | 0 | 0 | 487 |
| 2004 | 0 | 0 | 0 | 2 | 33 | 56 | 119 | 99 | 57 | 0 | 0 | 0 | 366 |
| 2005 | 0 | 0 | 0 | 0 | 0 | 209 | 256 | 249 | 71 | 23 | 0 | 0 | 808 |
| 2006 | 0 | 0 | 0 | 0 | 47 | 130 | 318 | 184 | 29 | 0 | 0 | 0 | 708 |
| 2007 | 0 | 0 | 2 | 0 | 67 | 174 | 163 | 227 | 84 | 33 | 0 | 0 | 750 |
| 2008 | 0 | 0 | 0 | 13 | 6 | 175 | 226 | 113 | 66 | 2 | 0 | 0 | 601 |
| 2009 | 0 | 0 | 0 | 10 | 12 | 44 | 78 | 162 | 19 | 0 | 0 | 0 | 325 |

## SNOWFALL (inches) 2009 ROCHESTER (KROC)

| YEAR | JUL | AUG | SEP | OCT | NOV | DEC | JAN | FEB | MAR | APR | MAY | JUN | TOTAL |
|---|---|---|---|---|---|---|---|---|---|---|---|---|---|
| 1980-81 | 0.0 | 0.0 | 0.0 | T | 8.4 | 31.8 | 31.5 | 9.3 | 12.0 | 1.4 | 0.0 | 0.0 | 94.4 |
| 1981-82 | 0.0 | 0.0 | 0.0 | 0.1 | 2.4 | 46.1 | 43.6 | 14.9 | 8.9 | 12.4 | 0.0 | 0.0 | 128.4 |
| 1982-83 | 0.0 | 0.0 | 0.0 | T | 3.0 | 11.6 | 10.2 | 13.6 | 9.3 | 12.2 | T | 0.0 | 59.9 |
| 1983-84 | 0.0 | 0.0 | 0.0 | 0.0 | 17.6 | 19.6 | 23.4 | 27.8 | 29.1 | 0.5 | T | 0.0 | 118.0 |
| 1984-85 | 0.0 | 0.0 | 0.0 | 0.0 | 1.6 | 11.6 | 36.8 | 26.1 | 8.4 | 2.6 | 0.0 | 0.0 | 87.1 |
| 1985-86 | 0.0 | 0.0 | 0.0 | 0.0 | 7.6 | 18.3 | 15.5 | 17.9 | 9.3 | 2.1 | T | 0.0 | 70.7 |
| 1986-87 | 0.0 | 0.0 | 0.0 | 0.0 | 7.4 | 9.3 | 29.6 | 13.0 | 5.3 | 2.5 | 0.0 | 0.0 | 67.1 |
| 1987-88 | 0.0 | 0.0 | 0.0 | T | 4.6 | 19.3 | 9.8 | 29.4 | 5.6 | 1.1 | 0.0 | 0.0 | 69.8 |
| 1988-89 | 0.0 | 0.0 | 0.0 | 0.1 | 0.2 | 10.3 | 15.0 | 30.6 | 15.6 | 3.9 | 10.9 | 0.0 | 86.6 |
| 1989-90 | 0.0 | 0.0 | 0.0 | T | 6.5 | 32.8 | 14.0 | 31.3 | 5.4 | 15.8 | T | 0.0 | 105.8 |
| 1990-91 | T | 0.0 | 0.0 | T | 4.4 | 18.2 | 26.5 | 16.1 | 2.0 | 1.1 | 0.0 | 0.0 | 68.3 |
| 1991-92 | 0.0 | 0.0 | 0.0 | 0.0 | 13.7 | 23.9 | 18.3 | 12.8 | 38.1 | 3.8 | 0.0 | 0.0 | 110.6 |
| 1992-93 | 0.0 | 0.0 | 0.0 | T | 9.5 | 29.3 | 22.4 | 31.2 | 37.1 | 2.0 | T | 0.0 | 131.5 |
| 1993-94 | 0.0 | 0.0 | T | 2.6 | 9.8 | 14.0 | 43.0 | 35.1 | 12.1 | 9.6 | 0.0 | 0.0 | 126.2 |
| 1994-95 | 0.0 | 0.0 | T | T | 2.8 | 7.6 | 12.8 | 23.6 | 5.3 | 4.1 | 0.0 | 0.0 | 56.2 |
| 1995-96 | 0.0 | 0.0 | 0.0 | T | 23.4 | 20.5 | 36.9 | 14.1 | 28.6 | 5.3 | 1.5 | 0.0 | 130.3 |
| 1996-97 | T | 0.0 | 0.0 | 0.0 | 24.9 | 14.0 | 24.8 | 13.3 | 26.4 | 1.3 | T | 0.0 | 104.7 |
| 1997-98 | 0.0 | 0.0 | 0.0 | 0.1 | 21.6 | 26.4 | 14.6 | 9.1 | 27.9 | 0.0 | 0.0 | T | 99.7 |
| 1998-99 | 0.0 | 0.0 | T | 0.0 | 0.1 | 10.1 | 48.8 | 4.7 | 45.0 | 2.9 | 0.0 | 0.0 | 111.6 |
| 1999-00 | 0.0 | 0.0 | 0.0 | 0.0 | 4.7 | 19.1 | 42.0 | 25.7 | 13.6 | 5.6 | 0.0 | 0.0 | 110.7 |
| 2000-01 | 0.0 | 0.0 | 0.0 | T | 7.3 | 39.3 | 21.6 | 23.3 | 41.4 | 0.1 | 0.0 | 0.0 | 133.0 |
| 2001-02 | 0.0 | 0.0 | 0.0 | T | 0.1 | 7.1 | 11.9 | 18.7 | 13.8 | 6.5 | T | 0.0 | 58.1 |
| 2002-03 | 0.0 | 0.0 | 0.0 | 0.0 | 16.9 | 41.1 | 43.4 | 21.9 | 5.6 | 6.3 | T | 0.0 | 135.2 |
| 2003-04 | 0.0 | 0.0 | 0.0 | T | 5.4 | 27.2 | 61.3 | 10.3 | 16.3 | 5.1 | 0.0 | 0.0 | 125.6 |
| 2004-05 | 0.0 | 0.0 | 0.0 | 0.0 | 1.9 | 16.8 | 49.7 | 27.8 | 17.4 | T | 0.0 | 0.0 | 113.6 |
| 2005-06 | 0.0 | 0.0 | 0.0 | T | 7.8 | 19.0 | 14.0 | 28.8 | 4.2 | 0.1 | T | T | 73.9 |
| 2006-07 | T | 0.0 | 0.0 | 0.4 | 0.5 | 4.3 | 29.8 | 46.5 | 19.5 | 6.2 | 0.0 | 0.0 | 107.2 |
| 2007-08 | 0.0 | 0.0 | 0.0 | 0.0 | 1.7 | 41.8 | 15.2 | 23.7 | 23.6 | T | 0.0 | 0.0 | 106.0 |
| 2008-09 | 0.0 | 0.0 | 0.0 | 0.1 | 11.0 | 46.2 | 29.3 | 13.9 | 1.4 | 1.8 | 0.0 | 0.0 | 103.7 |
| 2009- | 0.0 | 0.0 | T | T | T | 25.3 | | | | | | | |
| POR= 84 YRS | T | 0.0 | T | 0.2 | 6.6 | 18.8 | 23.0 | 20.9 | 15.3 | 3.3 | 0.2 | T | 88.3 |

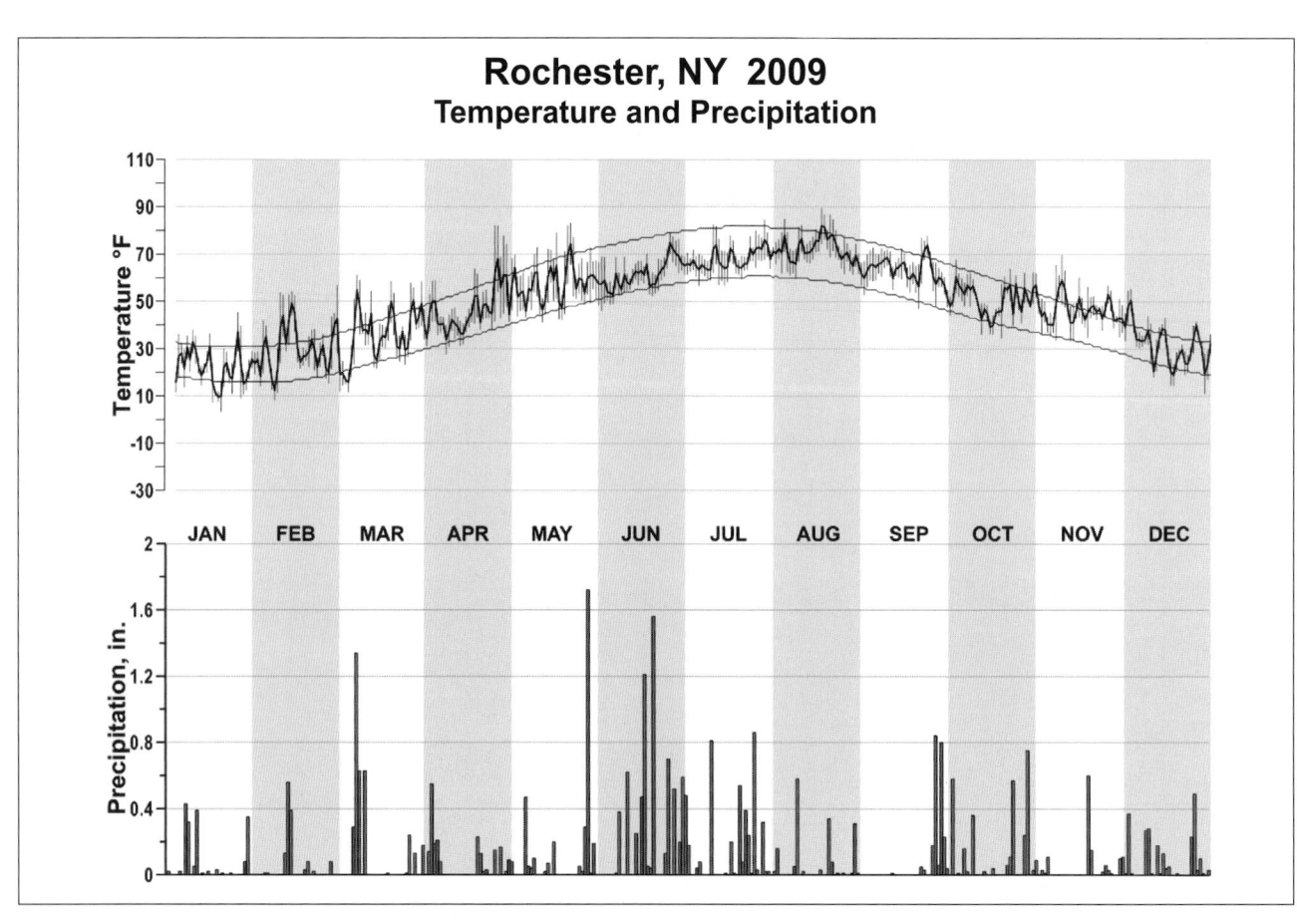

# Rochester, NY 2009
## Temperature and Precipitation

# 2009
# SYRACUSE
# NEW YORK (KSYR)

Syracuse is located approximately at the geographical center of the state. Gently rolling terrain stretches northward for about 30 miles to the eastern end of Lake Ontario. Oneida Lake is about 8 miles northeast of Syracuse. Approximately 5 miles south of the city, hills rise to 1,500 feet. Immediately to the west, the terrain is gently rolling with elevations 500 to 800 feet above sea level.

The climate of Syracuse is primarily continental in character and comparatively humid. Nearly all cyclonic systems moving from the interior of the country through the St. Lawrence Valley will affect the Syracuse area. Seasonal and diurnal changes are marked and produce an invigorating climate.

In the summer and in portions of the transitional seasons, temperatures usually rise rapidly during the daytime to moderate levels and as a rule fall rapidly after sunset. The nights are relatively cool and comfortable. There are only a few days in a year when atmospheric humidity causes great personal discomfort.

Winters are usually cold and are sometimes severe in part. Daytime temperatures average in the low 30s with nighttime lows in the teens. Low winter temperatures below -25 degrees have been recorded. The autumn, winter, and spring seasons display marked variability.

Based on the 1951-1980 period, the average first occurrence of 32 degrees Fahrenheit in the fall is October 16 and the average last occurrence in the spring is April 28.

Precipitation in the Syracuse area is derived principally from cyclonic storms which pass from the interior of the country through the St. Lawrence Valley. Lake Ontario provides the source of significant winter precipitation. The lake is quite deep and never freezes so cold air flowing over the lake is quickly saturated and produces the cloudiness and snow squalls which are a well-known feature of winter weather in the Syracuse area.

The area enjoys sufficient precipitation in most years to meet the needs of agriculture and water supplies. The precipitation is uncommonly well distributed, averaging about 3 inches per month throughout the year. Snowfall is moderately heavy with an average just over 100 inches. There are about 30 days per year with thunderstorms, mostly during the warmer months.

Wind velocities are moderate, but during the winter months there are numerous days with sufficient winds to cause blowing and drifting snow.

During December, January, and February there is much cloudiness. Syracuse receives only about one-third of possible sunshine during winter months. Approximately two-thirds of possible sunshine is received during the warm months.

# NORMALS, MEANS, AND EXTREMES
## SYRACUSE (KSYR)

| LATITUDE: | LONGITUDE: | ELEVATION (FT): | TIME ZONE: | WBAN: 14771 |
|---|---|---|---|---|
| 43 ° 6 'N | -76 ° 6 'W | GRND: 410  BARO: 417 | EASTERN  (UTC -5) | |

| | ELEMENT | POR | JAN | FEB | MAR | APR | MAY | JUN | JUL | AUG | SEP | OCT | NOV | DEC | YEAR |
|---|---|---|---|---|---|---|---|---|---|---|---|---|---|---|---|
| **TEMPERATURE °F** | NORMAL DAILY MAXIMUM | 30 | 31.4 | 33.5 | 43.1 | 55.7 | 68.5 | 77.0 | 81.7 | 79.6 | 71.4 | 59.8 | 47.4 | 36.3 | 57.1 |
| | MEAN DAILY MAXIMUM | 88 | 31.7 | 32.3 | 42.0 | 55.3 | 67.9 | 76.4 | 81.6 | 79.8 | 71.6 | 60.6 | 47.6 | 35.8 | 56.9 |
| | HIGHEST DAILY MAXIMUM | 59 | 70 | 69 | 87 | 92 | 96 | 98 | 98 | 101 | 97 | 87 | 81 | 72 | 101 |
| | YEAR OF OCCURRENCE | | 2008 | 1981 | 1986 | 1990 | 1977 | 1953 | 2002 | 2002 | 1953 | 1963 | 1950 | 2001 | AUG 2002 |
| | MEAN OF EXTREME MAXS. | 88 | 54.2 | 53.4 | 67.8 | 79.4 | 85.8 | 91.0 | 92.3 | 91.2 | 87.4 | 79.2 | 68.7 | 57.8 | 75.7 |
| | NORMAL DAILY MINIMUM | 30 | 14.0 | 15.5 | 24.2 | 34.9 | 45.8 | 54.6 | 60.1 | 58.8 | 51.1 | 40.4 | 32.0 | 20.9 | 37.7 |
| | MEAN DAILY MINIMUM | 88 | 15.9 | 16.3 | 25.2 | 35.8 | 46.4 | 55.4 | 60.9 | 59.4 | 51.6 | 41.7 | 32.9 | 21.8 | 38.6 |
| | LOWEST DAILY MINIMUM | 59 | -26 | -26 | -16 | 9 | 25 | 35 | 45 | 40 | 28 | 19 | 5 | -22 | -26 |
| | YEAR OF OCCURRENCE | | 1966 | 1979 | 1950 | 1972 | 1966 | 1966 | 1976 | 1965 | 1991 | 1976 | 1976 | 1980 | FEB 1979 |
| | MEAN OF EXTREME MINS. | 88 | -7.4 | -4.8 | 5.1 | 22.3 | 32.9 | 42.1 | 49.8 | 46.8 | 36.6 | 27.3 | 17.2 | -0.1 | 22.3 |
| | NORMAL DRY BULB | 30 | 22.7 | 24.5 | 33.6 | 45.3 | 57.1 | 65.8 | 70.9 | 69.2 | 61.3 | 50.1 | 39.7 | 28.6 | 47.4 |
| | MEAN DRY BULB | 88 | 23.8 | 24.3 | 33.7 | 45.6 | 57.2 | 66.1 | 71.2 | 69.7 | 61.7 | 51.2 | 40.3 | 28.9 | 47.8 |
| | MEAN WET BULB | 26 | 21.8 | 22.6 | 28.8 | 40.1 | 50.5 | 59.9 | 64.1 | 63.2 | 56.7 | 45.8 | 36.6 | 26.8 | 43.1 |
| | MEAN DEW POINT | 26 | 18.4 | 18.6 | 24.4 | 35.3 | 46.6 | 56.6 | 61.3 | 60.6 | 54.1 | 42.7 | 33.3 | 23.6 | 39.6 |
| | NORMAL NO. DAYS WITH: | | | | | | | | | | | | | | |
| | MAXIMUM >= 90 | 30 | 0.0 | 0.0 | 0.0 | * | 0.4 | 1.5 | 3.9 | 1.7 | 0.3 | 0.0 | 0.0 | 0.0 | 7.8 |
| | MAXIMUM <= 32 | 30 | 16.5 | 13.5 | 5.6 | 0.3 | 0.0 | 0.0 | 0.0 | 0.0 | 0.0 | 0.0 | 1.7 | 10.3 | 47.9 |
| | MINIMUM <= 32 | 30 | 28.2 | 25.1 | 23.4 | 11.2 | 0.5 | 0.0 | 0.0 | 0.0 | 0.2 | 5.0 | 15.3 | 26.0 | 134.9 |
| | MINIMUM <= 0 | 30 | 4.2 | 2.8 | 0.7 | 0.0 | 0.0 | 0.0 | 0.0 | 0.0 | 0.0 | 0.0 | 0.0 | 1.4 | 9.1 |
| **H/C** | NORMAL HEATING DEG. DAYS | 30 | 1296 | 1131 | 959 | 579 | 258 | 66 | 10 | 25 | 158 | 460 | 748 | 1113 | 6803 |
| | NORMAL COOLING DEG. DAYS | 30 | 0 | 0 | 1 | 4 | 29 | 105 | 203 | 158 | 48 | 3 | 0 | 0 | 551 |
| **RH** | NORMAL (PERCENT) | 30 | 75 | 72 | 69 | 66 | 68 | 71 | 72 | 75 | 77 | 75 | 76 | 77 | 73 |
| | HOUR 01 LST | 30 | 77 | 76 | 76 | 75 | 79 | 83 | 84 | 87 | 87 | 83 | 80 | 79 | 81 |
| | HOUR 07 LST | 30 | 78 | 79 | 79 | 76 | 77 | 79 | 81 | 87 | 88 | 85 | 82 | 81 | 81 |
| | HOUR 13 LST | 30 | 68 | 64 | 60 | 54 | 55 | 57 | 56 | 59 | 62 | 61 | 68 | 71 | 61 |
| | HOUR 19 LST | 30 | 75 | 72 | 67 | 60 | 60 | 63 | 63 | 70 | 76 | 75 | 76 | 78 | 70 |
| **S** | PERCENT POSSIBLE SUNSHINE | 57 | 35 | 39 | 46 | 49 | 54 | 59 | 63 | 58 | 53 | 44 | 27 | 26 | 46 |
| **W/O** | MEAN NO. DAYS WITH: | | | | | | | | | | | | | | |
| | HEAVY FOG(VISBY <= 1/4 MI) | 46 | 0.8 | 0.9 | 0.9 | 0.5 | 0.6 | 0.4 | 0.5 | 0.7 | 0.9 | 1.0 | 0.8 | 0.8 | 8.8 |
| | THUNDERSTORMS | 64 | 0.2 | 0.2 | 0.8 | 1.8 | 3.4 | 5.1 | 6.3 | 5.3 | 2.5 | 0.9 | 0.5 | 0.1 | 27.1 |
| **CLOUDNESS** | MEAN: | | | | | | | | | | | | | | |
| | SUNRISE-SUNSET (OKTAS) | | | | | | | | | | | | | | |
| | MIDNIGHT-MIDNIGHT (OKTAS) | | | | | | | | | | | | | | |
| | MEAN NO. DAYS WITH: | | | | | | | | | | | | | | |
| | CLEAR | | | | | | | | | | | | | | |
| | PARTLY CLOUDY | | | | | | | | | | | | | | |
| | CLOUDY | 1 | 3.0 | 5.0 | 11.0 | | 9.0 | 6.0 | | | | | | | |
| **PR** | MEAN STATION PRESSURE(IN) | 26 | 29.56 | 29.59 | 29.60 | 29.53 | 29.53 | 29.51 | 29.53 | 29.58 | 29.61 | 29.62 | 29.61 | 29.60 | 29.57 |
| | MEAN SEA-LEVEL PRES. (IN) | 26 | 30.04 | 30.07 | 30.05 | 29.97 | 29.97 | 29.95 | 29.96 | 30.01 | 30.05 | 30.06 | 30.05 | 30.06 | 30.02 |
| **WINDS** | MEAN SPEED (MPH) | 26 | 9.9 | 9.7 | 9.6 | 9.2 | 8.1 | 7.3 | 6.9 | 6.4 | 7.0 | 7.7 | 9.2 | 9.6 | 8.4 |
| | PREVAIL.DIR(TENS OF DEGS) | 33 | 26 | 27 | 28 | 30 | 29 | 28 | 27 | 28 | 10 | 26 | 26 | 26 | 26 |
| | MAXIMUM 2-MINUTE: | 16 | | | | | | | | | | | | | |
| | SPEED (MPH) | | 47 | 52 | 43 | 43 | 43 | 36 | 54 | 38 | 59 | 43 | 53 | 48 | 59 |
| | DIR. (TENS OF DEGS) | | 28 | 26 | 25 | 29 | 29 | 33 | 28 | 30 | 32 | 27 | 23 | 25 | 32 |
| | YEAR OF OCCURRENCE | | 2008 | 2006 | 1996 | 2008 | 1998 | 2007 | 1999 | 2007 | 1998 | 2003 | 2005 | 2000 | SEP 1998 |
| | MAXIMUM 3-SECOND | | | | | | | | | | | | | | |
| | SPEED (MPH) | 16 | 59 | 64 | 51 | 59 | 56 | 52 | 66 | 56 | 77 | 54 | 69 | 62 | 77 |
| | DIR. (TENS OF DEGS) | | 29 | 26 | 32 | 33 | 28 | 33 | 28 | 23 | 32 | 27 | 24 | 25 | 32 |
| | YEAR OF OCCURRENCE | | 2008 | 2006 | 2008 | 2007 | 1998 | 2007 | 1999 | 2001 | 1998 | 2003 | 2005 | 2000 | SEP 1998 |
| **PRECIPITATION** | NORMAL (IN) | 30 | 2.60 | 2.12 | 3.02 | 3.39 | 3.39 | 3.71 | 4.02 | 3.56 | 4.15 | 3.20 | 3.77 | 3.12 | 40.05 |
| | MAXIMUM MONTHLY (IN) | 59 | 5.77 | 5.38 | 6.84 | 8.12 | 7.82 | 12.30 | 10.12 | 8.41 | 8.81 | 8.29 | 6.79 | 5.50 | 12.30 |
| | YEAR OF OCCURRENCE | | 1978 | 1951 | 1955 | 1976 | 2004 | 1972 | 2006 | 1956 | 1975 | 1955 | 1972 | 1983 | JUN 1972 |
| | MINIMUM MONTHLY (IN) | 59 | 1.02 | 0.63 | 1.01 | 1.22 | .59 | 1.10 | 0.90 | 1.02 | 0.75 | 0.21 | 1.25 | 1.40 | 0.21 |
| | YEAR OF OCCURRENCE | | 1970 | 1987 | 1981 | 1985 | 2005 | 1962 | 1969 | 1999 | 1964 | 1963 | 1978 | 1999 | OCT 1963 |
| | MAXIMUM IN 24 HOURS (IN) | 59 | 2.49 | 1.99 | 1.77 | 2.85 | 3.13 | 3.88 | 4.29 | 4.27 | 4.14 | 3.60 | 2.09 | 2.18 | 4.29 |
| | YEAR OF OCCURRENCE | | 2008 | 1961 | 1993 | 1976 | 1969 | 1972 | 2006 | 1954 | 1975 | 1955 | 1996 | 1952 | JUL 2006 |
| | NORMAL NO. DAYS WITH: | | | | | | | | | | | | | | |
| | PRECIPITATION >= 0.01 | 30 | 19.7 | 15.5 | 16.5 | 14.0 | 12.7 | 12.2 | 11.3 | 11.1 | 12.6 | 13.2 | 16.8 | 18.3 | 173.9 |
| | PRECIPITATION >= 1.00 | 30 | 0.2 | 0.2 | 0.2 | 0.5 | 0.4 | 0.7 | 1.0 | 0.7 | 0.9 | 0.7 | 0.4 | 0.4 | 6.3 |
| **SNOWFALL** | NORMAL (IN) | 30 | 31.5 | 20.1 | 18.1 | 4.8 | 0.1 | 0.0 | 0.0 | 0.0 | 0.* | 0.5 | 10.7 | 26.1 | 111.9 |
| | MAXIMUM MONTHLY (IN) | 59 | 78.1 | 72.6 | 54.4 | 16.4 | 2.1 | T | T | T | T | 5.7 | 25.9 | 70.3 | 78.1 |
| | YEAR OF OCCURRENCE | | 2004 | 1958 | 1993 | 1983 | 1996 | 1992 | 2006 | 2008 | 1992 | 1988 | 1976 | 2000 | JAN 2004 |
| | MAXIMUM IN 24 HOURS (IN) | 59 | 24.5 | 21.4 | 35.6 | 7.1 | 2.1 | T | T | T | T | 2.9 | 12.1 | 19.6 | 35.6 |
| | YEAR OF OCCURRENCE' | | 1966 | 1961 | 1993 | 1975 | 1996 | 1992 | 1992 | 2008 | 1992 | 1988 | 1973 | 2003 | MAR 1993 |
| | MAXIMUM SNOW DEPTH (IN) | 59 | 39 | 48 | 25 | 8 | 1 | 0 | 0 | 0 | 0 | 2 | 14 | 23 | 48 |
| | YEAR OF OCCURRENCE | | 1966 | 1966 | 1971 | 1975 | 1996 | | | | | 1965 | 1973 | 1969 | FEB 1966 |
| | NORMAL NO. DAYS WITH: | | | | | | | | | | | | | | |
| | SNOWFALL >= 1.0 | 30 | 9.2 | 5.1 | 4.8 | 1.6 | 0.1 | 0.0 | 0.0 | 0.0 | 0.0 | 0.2 | 2.9 | 7.1 | 31.0 |

## PRECIPITATION (inches) 2009 SYRACUSE (KSYR)

| YEAR | JAN | FEB | MAR | APR | MAY | JUN | JUL | AUG | SEP | OCT | NOV | DEC | ANNUAL |
|------|-----|-----|-----|-----|-----|-----|-----|-----|-----|-----|-----|-----|--------|
| 1980 | 1.47 | 1.38 | 4.34 | 3.33 | 1.34 | 4.45 | 2.57 | 1.33 | 3.40 | 2.56 | 2.64 | 3.27 | 32.08 |
| 1981 | 1.34 | 2.72 | 1.01 | 2.04 | 2.61 | 1.89 | 2.68 | 2.63 | 5.58 | 6.66 | 3.09 | 2.96 | 35.21 |
| 1982 | 3.59 | 1.26 | 2.63 | 1.71 | 2.87 | 4.64 | 3.83 | 2.60 | 4.22 | 0.72 | 4.52 | 2.55 | 35.14 |
| 1983 | 1.92 | 1.07 | 2.30 | 6.34 | 3.33 | 1.50 | 2.31 | 2.80 | 2.98 | 1.98 | 4.30 | 5.50 | 36.33 |
| 1984 | 1.30 | 2.88 | 2.39 | 3.16 | 4.97 | 2.02 | 3.66 | 5.17 | 2.61 | 1.95 | 3.48 | 4.38 | 37.97 |
| 1985 | 2.49 | 1.55 | 2.61 | 1.22 | 3.39 | 2.80 | 2.75 | 1.44 | 3.88 | 3.39 | 5.18 | 1.80 | 32.50 |
| 1986 | 2.41 | 2.27 | 2.82 | 3.42 | 2.67 | 4.89 | 5.23 | 3.36 | 5.47 | 3.32 | 3.74 | 3.33 | 42.93 |
| 1987 | 3.03 | 0.63 | 1.86 | 3.31 | 1.41 | 5.04 | 2.16 | 2.12 | 5.99 | 3.13 | 3.02 | 1.99 | 33.69 |
| 1988 | 1.50 | 2.13 | 1.79 | 2.70 | 3.05 | 2.46 | 5.72 | 3.77 | 1.88 | 3.57 | 3.95 | 1.92 | 34.44 |
| 1989 | 1.06 | 1.71 | 3.13 | 1.52 | 4.27 | 5.41 | 2.20 | 2.68 | 5.96 | 4.08 | 2.78 | 2.13 | 36.93 |
| 1990 | 2.13 | 3.95 | 3.70 | 4.09 | 5.62 | 2.92 | 3.72 | 5.33 | 3.45 | 6.09 | 3.23 | 5.24 | 49.47 |
| 1991 | 2.44 | 1.54 | 4.07 | 3.90 | 3.90 | 1.67 | 2.86 | 4.03 | 4.20 | 2.62 | 2.72 | 3.10 | 37.05 |
| 1992 | 2.62 | 2.46 | 3.80 | 3.54 | 5.21 | 1.78 | 8.00 | 2.64 | 4.55 | 2.69 | 3.75 | 2.57 | 43.61 |
| 1993 | 3.08 | 2.45 | 3.75 | 6.55 | 2.25 | 2.93 | 4.76 | 4.71 | 3.83 | 2.91 | 3.19 | 3.20 | 43.61 |
| 1994 | 3.37 | 1.92 | 5.14 | 3.62 | 3.02 | 2.38 | 2.64 | 5.19 | 2.43 | 1.61 | 3.50 | 2.52 | 37.34 |
| 1995 | 1.80 | 2.19 | 1.31 | 1.88 | 1.70 | 1.00 | 1.98 | 3.50 | 2.53 | 6.57 | 4.83 | 2.05 | 31.34 |
| 1996 | 3.35 | 1.25 | 1.74 | 4.28 | 3.02 | 3.05 | 4.24 | 1.71 | 4.38 | 2.14 | 5.78 | 4.45 | 39.39 |
| 1997 | 1.46 | 2.25 | 3.57 | 1.77 | 2.43 | 1.64 | 2.78 | 4.06 | 2.75 | 1.50 | 4.28 | 4.13 | 32.62 |
| 1998 | 4.76 | 3.14 | 2.94 | 2.09 | 2.37 | 4.62 | 3.63 | 4.77 | 2.41 | 2.53 | 2.06 | 1.74 | 37.06 |
| 1999 | 5.33 | 1.43 | 3.53 | 1.75 | 0.81 | 1.78 | 2.55 | 1.02 | 5.35 | 2.77 | 3.16 | 1.40 | 30.88 |
| 2000 | 2.80 | 2.46 | 2.37 | 4.24 | 4.75 | 4.46 | 2.73 | 2.48 | 3.13 | 2.25 | 2.98 | 2.36 | 37.01 |
| 2001 | 1.57 | 1.77 | 5.38 | 1.53 | 2.24 | 3.58 | 2.08 | 4.84 | 4.05 | 2.15 | 2.92 | 2.19 | 34.30 |
| 2002 | 2.13 | 1.44 | 2.75 | 4.38 | 5.77 | 5.35 | 1.75 | 2.71 | 3.55 | 3.98 | 3.61 | 2.84 | 40.26 |
| 2003 | 1.44 | 2.58 | 2.89 | 2.61 | 5.27 | 2.83 | 3.30 | 3.03 | 3.14 | 4.27 | 3.14 | 3.10 | 37.60 |
| 2004 | 1.86 | 1.12 | 2.04 | 3.72 | 7.82 | 2.42 | 6.95 | 5.09 | 3.23 | 2.28 | 2.81 | 3.80 | 43.14 |
| 2005 | 2.96 | 1.57 | 1.39 | 5.71 | 0.59 | 1.95 | 4.61 | 5.95 | 1.75 | 6.40 | 4.66 | 2.56 | 40.10 |
| 2006 | 2.96 | 1.66 | 1.86 | 3.93 | 2.25 | 5.09 | 10.12 | 3.21 | 4.04 | 5.70 | 2.62 | 3.76 | 47.20 |
| 2007 | 4.20 | 2.49 | 4.16 | 4.42 | 0.86 | 3.67 | 3.61 | 1.76 | 3.20 | 4.02 | 4.17 | 5.04 | 41.60 |
| 2008 | 1.36 | 4.71 | 5.00 | 2.99 | 1.78 | 3.75 | 4.28 | 3.62 | 2.47 | 4.87 | 3.19 | 3.89 | 41.91 |
| 2009 | 1.86 | 1.30 | 3.80 | 2.31 | 3.53 | 5.24 | 2.04 | 4.57 | 2.27 | 4.08 | 2.16 | 2.20 | 35.36 |
| POR= 88 YRS | 2.70 | 2.44 | 3.11 | 3.25 | 3.10 | 3.49 | 3.58 | 3.45 | 3.27 | 3.21 | 3.34 | 3.05 | 37.99 |

WBAN : 14771

## AVERAGE TEMPERATURE (°F) 2009 SYRACUSE (KSYR)

| YEAR | JAN | FEB | MAR | APR | MAY | JUN | JUL | AUG | SEP | OCT | NOV | DEC | ANNUAL |
|------|-----|-----|-----|-----|-----|-----|-----|-----|-----|-----|-----|-----|--------|
| 1980 | 25.6 | 19.8 | 32.4 | 47.8 | 59.8 | 63.0 | 72.5 | 73.8 | 63.4 | 48.8 | 37.6 | 22.6 | 47.3 |
| 1981 | 15.0 | 33.7 | 36.4 | 50.0 | 59.2 | 68.0 | 73.3 | 70.4 | 61.6 | 47.9 | 39.0 | 29.0 | 48.6 |
| 1982 | 14.8 | 25.1 | 33.2 | 43.9 | 59.4 | 63.1 | 70.4 | 65.3 | 60.6 | 50.4 | 43.9 | 34.1 | 47.0 |
| 1983 | 23.4 | 26.4 | 35.7 | 44.3 | 53.7 | 66.7 | 72.0 | 69.0 | 62.5 | 50.3 | 39.0 | 22.5 | 47.1 |
| 1984 | 18.7 | 32.0 | 24.5 | 46.0 | 52.4 | 65.4 | 68.0 | 68.8 | 57.7 | 52.2 | 38.3 | 33.5 | 46.5 |
| 1985 | 22.0 | 27.3 | 36.3 | 47.8 | 59.5 | 62.0 | 69.8 | 68.9 | 63.5 | 51.4 | 41.2 | 26.0 | 48.0 |
| 1986 | 23.9 | 23.4 | 37.4 | 49.2 | 61.0 | 64.3 | 71.0 | 66.8 | 60.5 | 49.7 | 36.8 | 31.6 | 48.0 |
| 1987 | 23.8 | 21.4 | 38.0 | 51.9 | 60.3 | 68.3 | 73.6 | 68.5 | 61.1 | 47.7 | 40.9 | 32.3 | 49.0 |
| 1988 | 23.1 | 24.6 | 34.4 | 45.7 | 59.7 | 64.1 | 74.0 | 71.8 | 60.8 | 46.6 | 43.0 | 27.8 | 48.0 |
| 1989 | 28.6 | 22.7 | 32.9 | 43.5 | 58.2 | 67.3 | 71.1 | 68.2 | 61.8 | 51.7 | 38.8 | 14.7 | 46.6 |
| 1990 | 33.2 | 29.0 | 37.5 | 49.3 | 54.5 | 67.3 | 71.8 | 70.3 | 61.2 | 52.8 | 42.2 | 33.5 | 50.2 |
| 1991 | 24.3 | 29.8 | 37.7 | 51.0 | 62.8 | 68.4 | 72.4 | 71.8 | 60.5 | 53.1 | 40.0 | 30.7 | 50.2 |
| 1992 | 24.7 | 26.5 | 29.3 | 44.4 | 57.5 | 64.0 | 67.3 | 67.5 | 61.3 | 46.6 | 39.7 | 31.0 | 46.7 |
| 1993 | 27.5 | 17.0 | 30.1 | 46.9 | 58.0 | 65.2 | 72.5 | 70.7 | 60.0 | 48.2 | 38.6 | 26.9 | 46.8 |
| 1994 | 12.7 | 19.2 | 30.8 | 47.9 | 54.1 | 68.0 | 72.9 | 67.5 | 60.9 | 50.6 | 44.0 | 31.9 | 46.7 |
| 1995 | 30.4 | 20.7 | 37.4 | 42.4 | 56.9 | 69.4 | 73.4 | 71.8 | 59.1 | 54.7 | 35.3 | 24.2 | 48.0 |
| 1996 | 21.5 | 24.6 | 29.8 | 43.2 | 55.0 | 68.2 | 69.4 | 70.3 | 63.1 | 50.7 | 34.7 | 34.9 | 47.1 |
| 1997 | 23.9 | 30.4 | 33.6 | 44.2 | 52.2 | 67.9 | 69.8 | 68.4 | 60.1 | 49.1 | 37.2 | 30.4 | 47.3 |
| 1998 | 29.6 | 31.2 | 37.9 | 48.1 | 62.9 | 66.3 | 70.1 | 71.1 | 64.0 | 51.9 | 41.7 | 35.4 | 50.9 |
| 1999 | 22.5 | 29.6 | 31.5 | 46.5 | 60.7 | 69.8 | 75.0 | 68.9 | 64.8 | 49.5 | 44.3 | 30.9 | 49.5 |
| 2000 | 21.3 | 28.8 | 40.1 | 44.3 | 59.1 | 65.6 | 67.0 | 68.2 | 60.7 | 50.9 | 38.5 | 21.7 | 47.2 |
| 2001 | 25.6 | 27.6 | 29.9 | 47.8 | 59.3 | 67.2 | 69.4 | 73.7 | 62.3 | 53.3 | 47.3 | 36.8 | 50.0 |
| 2002 | 32.9 | 32.3 | 36.3 | 48.5 | 54.1 | 68.0 | 73.7 | 73.0 | 66.9 | 50.4 | 40.7 | 28.7 | 50.5 |
| 2003 | 18.9 | 21.6 | 34.2 | 43.8 | 56.1 | 64.6 | 71.2 | 71.5 | 63.3 | 48.6 | 42.2 | 30.0 | 47.2 |
| 2004 | 14.7 | 23.5 | 37.5 | 46.1 | 60.3 | 63.8 | 69.5 | 68.7 | 65.0 | 51.5 | 41.0 | 29.1 | 47.6 |
| 2005 | 21.0 | 25.7 | 30.7 | 48.3 | 54.2 | 72.7 | 74.7 | 73.8 | 65.4 | 52.1 | 44.0 | 26.2 | 49.1 |
| 2006 | 33.5 | 27.2 | 34.1 | 47.8 | 58.2 | 67.2 | 74.2 | 69.3 | 60.8 | 49.3 | 44.7 | 37.4 | 50.3 |
| 2007 | 27.3 | 18.5 | 31.6 | 43.8 | 58.6 | 68.4 | 69.8 | 70.9 | 65.4 | 58.0 | 37.9 | 27.9 | 48.2 |
| 2008 | 29.5 | 25.8 | 31.6 | 51.6 | 53.7 | 69.8 | 71.3 | 67.0 | 62.7 | 48.3 | 38.2 | 28.9 | 48.2 |
| 2009 | 18.3 | 26.4 | 35.3 | 48.2 | 57.5 | 64.8 | 68.0 | 70.9 | 61.3 | 49.0 | 43.5 | 28.1 | 47.6 |
| POR= 88 YRS | 23.8 | 24.3 | 33.7 | 45.6 | 57.2 | 66.1 | 71.2 | 69.7 | 61.7 | 51.2 | 40.3 | 28.9 | 47.8 |

## HEATING DEGREE DAYS (base 65°F) 2009 SYRACUSE (KSYR)

| YEAR | JUL | AUG | SEP | OCT | NOV | DEC | JAN | FEB | MAR | APR | MAY | JUN | TOTAL |
|------|-----|-----|-----|-----|-----|-----|-----|-----|-----|-----|-----|-----|-------|
| 1980-81 | 3 | 0 | 120 | 496 | 814 | 1307 | 1544 | 869 | 882 | 446 | 221 | 27 | 6729 |
| 1981-82 | 2 | 4 | 145 | 523 | 775 | 1110 | 1552 | 1114 | 978 | 626 | 183 | 79 | 7091 |
| 1982-83 | 13 | 57 | 152 | 449 | 628 | 951 | 1280 | 1073 | 902 | 615 | 351 | 67 | 6538 |
| 1983-84 | 11 | 25 | 140 | 457 | 769 | 1312 | 1432 | 949 | 1246 | 563 | 386 | 68 | 7358 |
| 1984-85 | 16 | 33 | 227 | 390 | 797 | 971 | 1329 | 1048 | 882 | 514 | 193 | 109 | 6509 |
| 1985-86 | 10 | 18 | 121 | 415 | 702 | 1200 | 1266 | 1156 | 856 | 471 | 172 | 76 | 6463 |
| 1986-87 | 12 | 50 | 155 | 468 | 838 | 1027 | 1270 | 1208 | 831 | 395 | 211 | 35 | 6500 |
| 1987-88 | 7 | 27 | 138 | 529 | 717 | 1007 | 1290 | 1167 | 942 | 571 | 187 | 131 | 6713 |
| 1988-89 | 9 | 33 | 150 | 574 | 653 | 1148 | 1120 | 1175 | 989 | 639 | 242 | 38 | 6770 |
| 1989-90 | 3 | 36 | 151 | 406 | 779 | 1554 | 976 | 1001 | 849 | 496 | 319 | 43 | 6613 |
| 1990-91 | 4 | 4 | 160 | 386 | 675 | 967 | 1253 | 980 | 839 | 428 | 153 | 24 | 5873 |
| 1991-92 | 1 | 0 | 189 | 378 | 743 | 1056 | 1240 | 1112 | 1099 | 617 | 245 | 79 | 6759 |
| 1992-93 | 15 | 33 | 164 | 562 | 753 | 1047 | 1156 | 1337 | 1074 | 537 | 230 | 68 | 6976 |
| 1993-94 | 2 | 10 | 190 | 515 | 785 | 1172 | 1618 | 1274 | 1054 | 507 | 345 | 53 | 7525 |
| 1994-95 | 0 | 32 | 146 | 439 | 621 | 1019 | 1065 | 1235 | 850 | 671 | 248 | 36 | 6362 |
| 1995-96 | 5 | 8 | 194 | 313 | 884 | 1256 | 1344 | 1160 | 1085 | 648 | 328 | 21 | 7246 |
| 1996-97 | 5 | 2 | 123 | 438 | 903 | 929 | 1267 | 966 | 964 | 613 | 389 | 31 | 6630 |
| 1997-98 | 12 | 7 | 156 | 491 | 828 | 1065 | 1089 | 941 | 844 | 500 | 102 | 96 | 6131 |
| 1998-99 | 3 | 13 | 89 | 403 | 692 | 911 | 1310 | 986 | 1032 | 545 | 161 | 41 | 6186 |
| 1999-00 | 0 | 14 | 96 | 473 | 613 | 1049 | 1349 | 1042 | 765 | 614 | 224 | 69 | 6308 |
| 2000-01 | 22 | 29 | 184 | 427 | 791 | 1337 | 1215 | 1038 | 1080 | 508 | 190 | 62 | 6883 |
| 2001-02 | 16 | 0 | 126 | 366 | 527 | 867 | 988 | 908 | 882 | 519 | 352 | 49 | 5600 |
| 2002-03 | 4 | 7 | 55 | 461 | 724 | 1117 | 1424 | 1208 | 946 | 634 | 277 | 71 | 6928 |
| 2003-04 | 0 | 17 | 81 | 501 | 675 | 1078 | 1553 | 1196 | 847 | 567 | 179 | 95 | 6789 |
| 2004-05 | 5 | 22 | 59 | 414 | 711 | 1104 | 1358 | 1093 | 1053 | 493 | 330 | 16 | 6658 |
| 2005-06 | 0 | 0 | 57 | 402 | 623 | 1196 | 968 | 1050 | 952 | 508 | 240 | 49 | 6045 |
| 2006-07 | 0 | 15 | 137 | 481 | 601 | 847 | 1161 | 1296 | 1029 | 631 | 237 | 35 | 6470 |
| 2007-08 | 12 | 22 | 72 | 239 | 804 | 1142 | 1092 | 1130 | 1026 | 399 | 350 | 18 | 6306 |
| 2008-09 | 2 | 25 | 126 | 509 | 797 | 1113 | 1442 | 1076 | 914 | 515 | 237 | 58 | 6814 |
| 2009- | 11 | 13 | 129 | 489 | 640 | 1138 | | | | | | | |

WBAN : 14771

## COOLING DEGREE DAYS (base 65°F) 2009 SYRACUSE (KSYR)

| YEAR | JAN | FEB | MAR | APR | MAY | JUN | JUL | AUG | SEP | OCT | NOV | DEC | TOTAL |
|------|-----|-----|-----|-----|-----|-----|-----|-----|-----|-----|-----|-----|-------|
| 1980 | 0 | 0 | 0 | 0 | 41 | 62 | 243 | 279 | 80 | 1 | 0 | 0 | 706 |
| 1981 | 0 | 0 | 3 | 4 | 47 | 125 | 264 | 180 | 49 | 0 | 0 | 0 | 672 |
| 1982 | 0 | 0 | 0 | 0 | 18 | 25 | 186 | 72 | 25 | 0 | 3 | 0 | 329 |
| 1983 | 0 | 0 | 0 | 0 | 2 | 125 | 236 | 155 | 70 | 7 | 0 | 0 | 595 |
| 1984 | 0 | 0 | 0 | 0 | 4 | 88 | 119 | 154 | 14 | 1 | 0 | 0 | 380 |
| 1985 | 0 | 0 | 0 | 7 | 30 | 26 | 165 | 144 | 87 | 0 | 0 | 0 | 459 |
| 1986 | 0 | 0 | 5 | 1 | 52 | 62 | 201 | 112 | 28 | 0 | 0 | 0 | 461 |
| 1987 | 0 | 0 | 0 | 7 | 73 | 142 | 280 | 143 | 29 | 0 | 0 | 0 | 674 |
| 1988 | 0 | 0 | 0 | 0 | 33 | 112 | 296 | 251 | 32 | 9 | 0 | 0 | 733 |
| 1989 | 0 | 0 | 0 | 0 | 37 | 112 | 198 | 144 | 59 | 0 | 0 | 0 | 550 |
| 1990 | 0 | 0 | 5 | 33 | 2 | 118 | 222 | 177 | 51 | 16 | 0 | 0 | 624 |
| 1991 | 0 | 0 | 0 | 16 | 89 | 136 | 237 | 218 | 61 | 16 | 0 | 0 | 773 |
| 1992 | 0 | 0 | 0 | 5 | 21 | 54 | 94 | 118 | 60 | 0 | 0 | 0 | 352 |
| 1993 | 0 | 0 | 0 | 0 | 17 | 81 | 241 | 195 | 48 | 0 | 0 | 0 | 582 |
| 1994 | 0 | 0 | 0 | 1 | 15 | 150 | 251 | 120 | 28 | 0 | 0 | 0 | 565 |
| 1995 | 0 | 0 | 0 | 0 | 8 | 178 | 275 | 226 | 26 | 2 | 0 | 0 | 715 |
| 1996 | 0 | 0 | 0 | 0 | 22 | 126 | 150 | 175 | 73 | 0 | 0 | 0 | 546 |
| 1997 | 0 | 0 | 0 | 0 | 0 | 123 | 167 | 122 | 15 | 3 | 0 | 0 | 430 |
| 1998 | 0 | 0 | 9 | 0 | 41 | 140 | 167 | 209 | 65 | 1 | 0 | 0 | 632 |
| 1999 | 0 | 0 | 0 | 0 | 34 | 190 | 315 | 141 | 99 | 0 | 0 | 0 | 779 |
| 2000 | 0 | 0 | 0 | 1 | 46 | 94 | 91 | 132 | 58 | 0 | 0 | 0 | 422 |
| 2001 | 0 | 0 | 0 | 2 | 22 | 137 | 158 | 276 | 51 | 10 | 1 | 0 | 657 |
| 2002 | 0 | 0 | 0 | 30 | 19 | 145 | 283 | 263 | 120 | 16 | 0 | 0 | 876 |
| 2003 | 0 | 0 | 0 | 6 | 5 | 67 | 198 | 227 | 34 | 0 | 0 | 0 | 537 |
| 2004 | 0 | 0 | 0 | 6 | 39 | 63 | 151 | 142 | 67 | 0 | 0 | 0 | 468 |
| 2005 | 0 | 0 | 0 | 0 | 4 | 254 | 307 | 279 | 75 | 11 | 0 | 0 | 930 |
| 2006 | 0 | 0 | 0 | 0 | 34 | 125 | 289 | 151 | 14 | 0 | 0 | 0 | 613 |
| 2007 | 0 | 0 | 0 | 0 | 45 | 142 | 167 | 214 | 88 | 32 | 0 | 0 | 688 |
| 2008 | 0 | 0 | 0 | 8 | 6 | 166 | 204 | 91 | 66 | 0 | 0 | 0 | 541 |
| 2009 | 0 | 0 | 0 | 17 | 12 | 57 | 113 | 206 | 27 | 0 | 0 | 0 | 432 |

## SNOWFALL (inches) 2009 SYRACUSE (KSYR)

| YEAR | JUL | AUG | SEP | OCT | NOV | DEC | JAN | FEB | MAR | APR | MAY | JUN | TOTAL |
|------|-----|-----|-----|-----|-----|-----|-----|-----|-----|-----|-----|-----|-------|
| 1980-81 | 0.0 | 0.0 | 0.0 | T | 7.3 | 28.8 | 23.4 | 8.5 | 10.6 | 0.4 | 0.0 | 0.0 | 79.0 |
| 1981-82 | 0.0 | 0.0 | 0.0 | 0.5 | 12.1 | 37.3 | 48.2 | 11.6 | 14.4 | 13.0 | 0.0 | 0.0 | 137.1 |
| 1982-83 | 0.0 | 0.0 | 0.0 | T | 1.9 | 10.9 | 20.3 | 8.2 | 8.3 | 16.4 | T | 0.0 | 66.0 |
| 1983-84 | 0.0 | 0.0 | 0.0 | 0.0 | 7.6 | 24.2 | 21.8 | 19.7 | 40.3 | T | 0.0 | 0.0 | 113.6 |
| 1984-85 | 0.0 | 0.0 | 0.0 | 0.0 | 5.0 | 23.4 | 57.3 | 21.6 | 7.1 | 2.0 | 0.0 | 0.0 | 116.4 |
| 1985-86 | 0.0 | 0.0 | 0.0 | 0.0 | 8.0 | 28.2 | 29.9 | 26.1 | 11.0 | 1.7 | T | 0.0 | 104.9 |
| 1986-87 | 0.0 | 0.0 | 0.0 | 0.0 | 16.1 | 8.8 | 49.2 | 15.1 | 3.0 | 1.3 | 0.0 | 0.0 | 93.5 |
| 1987-88 | 0.0 | 0.0 | 0.0 | T | 10.8 | 20.7 | 18.0 | 46.1 | 10.2 | 5.6 | 0.0 | 0.0 | 111.4 |
| 1988-89 | 0.0 | 0.0 | 0.0 | 5.7 | 0.2 | 34.4 | 19.4 | 21.7 | 9.9 | 6.5 | 0.0 | 0.0 | 97.8 |
| 1989-90 | 0.0 | 0.0 | T | T | 12.9 | 64.6 | 27.4 | 33.3 | 15.2 | 8.6 | 0.0 | 0.0 | 162.0 |
| 1990-91 | 0.0 | 0.0 | 0.0 | 0.2 | 7.8 | 24.5 | 30.9 | 27.7 | 2.8 | 3.0 | 0.0 | 0.0 | 96.9 |
| 1991-92 | 0.0 | 0.0 | 0.0 | 0.0 | 5.5 | 37.9 | 50.5 | 27.6 | 41.3 | 4.1 | 0.0 | T | 166.9 |
| 1992-93 | T | 0.0 | T | 1.4 | 10.1 | 19.8 | 42.9 | 51.3 | 54.4 | 12.2 | 0.0 | 0.0 | 192.1 |
| 1993-94 | 0.0 | 0.0 | 0.0 | 1.0 | 17.1 | 34.0 | 57.0 | 30.8 | 25.3 | 3.7 | 0.0 | 0.0 | 168.9 |
| 1994-95 | 0.0 | 0.0 | 0.0 | 0.0 | 3.5 | 5.9 | 13.4 | 32.3 | 7.1 | 0.0 | 0.0 | 0.0 | 62.2 |
| 1995-96 | 0.0 | 0.0 | 0.0 | 0.0 | 34.2 | 45.1 | 36.0 | 16.5 | 32.2 | 4.8 | 2.1 | 0.0 | 170.9 |
| 1996-97 | 0.0 | 0.0 | 0.0 | T | 25.9 | 21.2 | 38.7 | 19.1 | 23.4 | 2.8 | 0.0 | 0.0 | 131.1 |
| 1997-98 | 0.0 | 0.0 | 0.0 | 1.2 | 19.3 | 47.8 | 31.8 | 14.7 | 19.9 | 0.0 | 0.0 | 0.0 | 134.7 |
| 1998-99 | 0.0 | 0.0 | 0.0 | 0.0 | T | 13.5 | 50.7 | 5.7 | 28.4 | 0.0 | 0.0 | 0.0 | 98.3 |
| 1999-00 | 0.0 | 0.0 | 0.0 | 0.0 | 3.8 | 15.7 | 29.9 | 27.4 | 7.1 | 1.9 | T | 0.0 | 85.8 |
| 2000-01 | 0.0 | 0.0 | 0.0 | T | 20.2 | 70.3 | 28.4 | 27.8 | 45.0 | 0.2 | T | 0.0 | 191.9 |
| 2001-02 | 0.0 | 0.0 | 0.0 | 0.2 | 0.5 | 7.3 | 21.2 | 13.5 | 14.1 | 2.6 | T | 0.0 | 59.4 |
| 2002-03 | 0.0 | 0.0 | 0.0 | T | 17.2 | 40.0 | 44.2 | 37.1 | 11.7 | 3.0 | 0.0 | 0.0 | 153.2 |
| 2003-04 | 0.0 | 0.0 | 0.0 | T | 10.5 | 48.5 | 78.1 | 19.4 | 20.5 | 4.3 | 0.0 | T | 181.3 |
| 2004-05 | 0.0 | 0.0 | 0.0 | 0.0 | 2.6 | 19.0 | 44.5 | 42.0 | 28.1 | T | 0.0 | 0.0 | 136.2 |
| 2005-06 | T | 0.0 | 0.0 | T | 8.4 | 53.0 | 12.1 | 34.8 | 16.3 | T | T | 0.0 | 124.6 |
| 2006-07 | T | 0.0 | 0.0 | T | 0.1 | 12.1 | 37.9 | 59.5 | 19.9 | 10.7 | 0.0 | 0.0 | 140.2 |
| 2007-08 | 0.0 | 0.0 | 0.0 | T | 6.8 | 49.8 | 10.1 | 29.5 | 12.9 | T | 0.0 | 0.0 | 109.1 |
| 2008-09 | 0.0 | T | 0.0 | 0.6 | 16.1 | 57.3 | 49.8 | 24.2 | 0.9 | 0.7 | 0.0 | 0.0 | 149.6 |
| 2009- | 0.0 | 0.0 | 0.0 | 0.0 | 0.3 | 22.2 | | | | | | | |
| POR=<br>88 YRS | T | T | T | 0.5 | 8.5 | 24.3 | 26.7 | 22.9 | 16.4 | 3.7 | 0.1 | T | 103.1 |

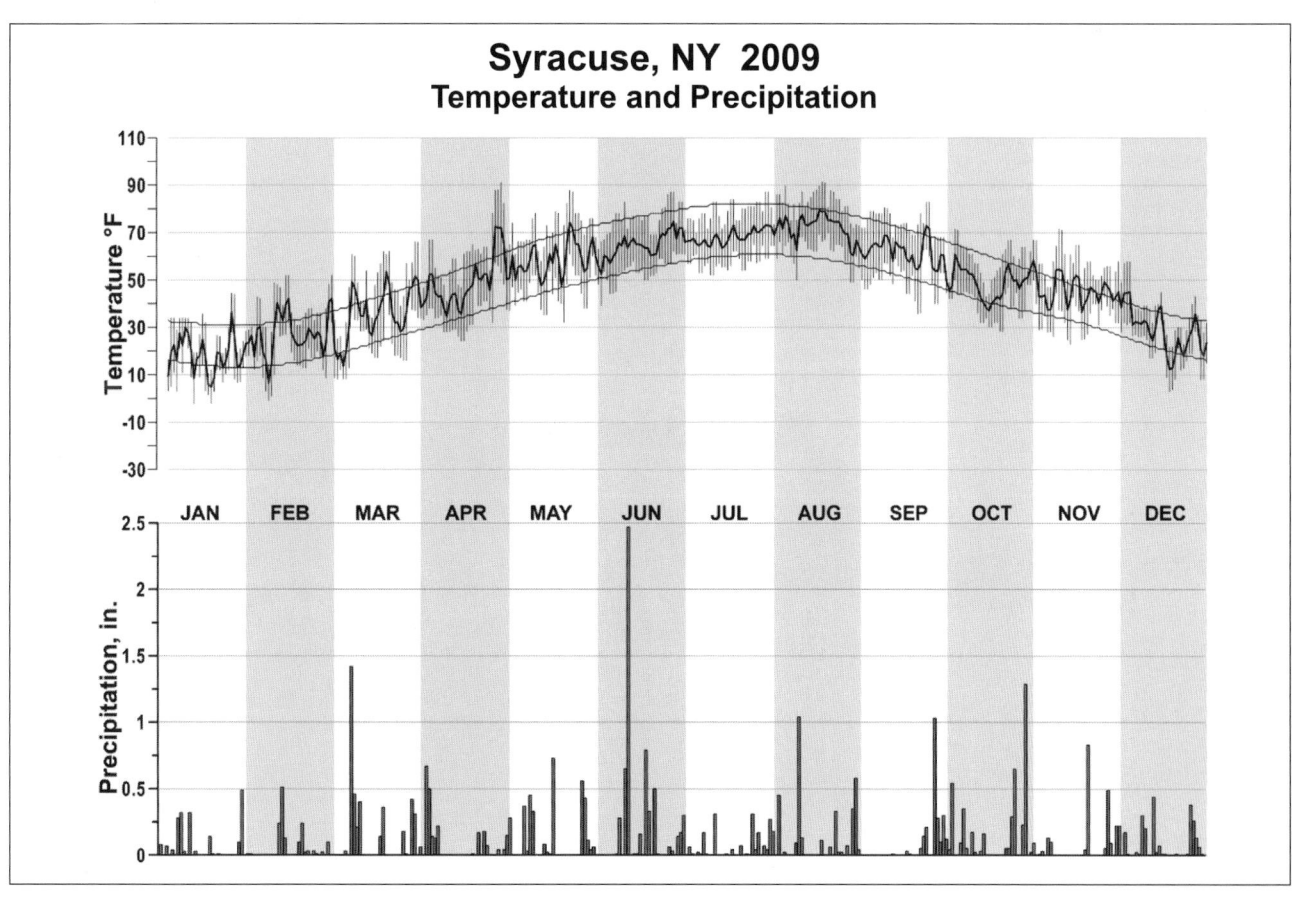

# 2009
# ASHEVILLE
# NORTH CAROLINA (KAVL)

The city of Asheville is located on both banks of the French Broad River, near the center of the French Broad Basin. Upstream from Asheville, the valley runs south for 18 miles and then curves toward the south-southwest. Downstream from the city, the valley is oriented toward the north-northwest. Two miles upstream from the principal section of Asheville, the Swannanoa River joins the French Broad from the east. The entire valley is known as the Asheville Plateau, having an average elevation near 2,200 feet above sea level, and is flanked by mountain ridges to the east and west, whose peaks range from 2,000 to 4,400 feet above the valley floor. At the Carolina-Tennessee border, about 25 miles north-northwest of Asheville, a relatively high ridge of mountains blocks the northern end of the valley. Thirty miles south, the Blue Ridge Mountains form an escarpment, having a general elevation of about 2,700 feet above sea level. The tallest peaks near Asheville are Mt. Mitchell, 6,684 feet above sea level, 20 miles northeast of the city, and Big Pisgah Mountain, 5,721 feet above sea level, 16 miles to the southwest.

Asheville has a temperate, but invigorating, climate. Considerable variation in temperature often occurs from day to day in summer, as well as during the other seasons.

While the office was located in the city, the combination of roof exposure conditions and a smoke blanket, caused by inversions in temperature in the valley on quiet nights, resulted in higher early morning temperatures at City Office sites than were experienced nearer ground level in nearby rural areas. The growing season in this area is of sufficient length for commercial crops, the average length of freeze-free period being about 195 days. The average last occurrence in spring of a temperature 32 degrees or lower is mid-April and the average first occurrence in fall of 32 degrees is late October.

The orientation of the French Broad Valley appears to have a pronounced influence on the wind direction. Prevailing winds are from the northwest during all months of the year. Also, the shielding effect of the nearby mountain barriers apparently has a direct bearing on the annual amount of precipitation received in this vicinity. In an area northwest of Asheville, the average annual precipitation is the lowest in North Carolina. Precipitation increases sharply in all other directions, especially to the south and southwest.

Destructive events caused directly by meteorological conditions are infrequent. The most frequent, occurring at approximately 12-year intervals, are floods on the French Broad River. These floods are usually associated with heavy rains caused by storms moving out of the Gulf of Mexico. Snowstorms which have seriously disrupted normal life in this community are infrequent. Hailstorms that cause property damage are extremely rare.

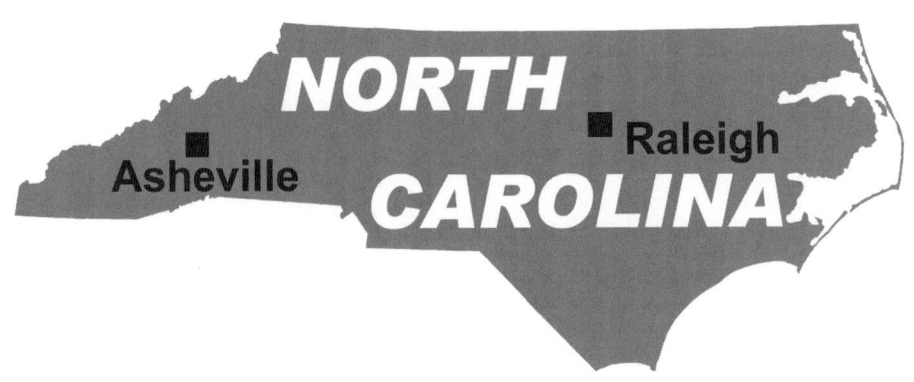

# NORMALS, MEANS, AND EXTREMES
## ASHEVILLE (KAVL)

| LATITUDE: 35 ° 25'N | LONGITUDE: -082° 32'W | | ELEVATION (FT): GRND: 2117   BARO: 2174 | | | | TIME ZONE: EASTERN   (UTC -5) | | | WBAN: 03812 | | |

| | ELEMENT | POR | JAN | FEB | MAR | APR | MAY | JUN | JUL | AUG | SEP | OCT | NOV | DEC | YEAR |
|---|---|---|---|---|---|---|---|---|---|---|---|---|---|---|---|
| TEMPERATURE °F | NORMAL DAILY MAXIMUM | 30 | 45.9 | 50.0 | 57.7 | 66.5 | 73.5 | 80.0 | 83.3 | 81.7 | 76.0 | 67.1 | 57.4 | 49.3 | 65.7 |
| | MEAN DAILY MAXIMUM | 58 | 47.9 | 50.5 | 58.5 | 68.0 | 75.2 | 81.1 | 84.0 | 83.2 | 77.0 | 68.2 | 58.5 | 50.4 | 66.9 |
| | HIGHEST DAILY MAXIMUM | 45 | 80 | 78 | 83 | 89 | 93 | 96 | 96 | 100 | 92 | 86 | 81 | 78 | 100 |
| | YEAR OF OCCURRENCE | | 1999 | 1996 | 1985 | 1972 | 1996 | 1969 | 1988 | 1983 | 1998 | 1986 | 2004 | 1971 | AUG 1983 |
| | MEAN OF EXTREME MAXS. | 58 | 66.1 | 69.1 | 76.6 | 82.9 | 85.9 | 90.0 | 91.6 | 90.5 | 86.8 | 80.9 | 73.9 | 67.3 | 80.1 |
| | NORMAL DAILY MINIMUM | 30 | 25.8 | 28.0 | 34.9 | 41.8 | 50.6 | 58.3 | 62.7 | 61.8 | 55.4 | 43.3 | 35.3 | 28.8 | 43.9 |
| | MEAN DAILY MINIMUM | 58 | 26.6 | 28.1 | 34.5 | 42.2 | 50.5 | 58.1 | 62.5 | 61.6 | 55.0 | 43.3 | 34.4 | 28.4 | 43.8 |
| | LOWEST DAILY MINIMUM | 45 | -16 | -2 | 2 | 20 | 28 | 35 | 44 | 42 | 30 | 21 | 8 | -7 | -16 |
| | YEAR OF OCCURRENCE | | 1985 | 1967 | 1993 | 2007 | 1989 | 1966 | 1988 | 1986 | 1967 | 1976 | 1970 | 1983 | JAN 1985 |
| | MEAN OF EXTREME MINS. | 58 | 8.4 | 12.2 | 19.4 | 27.5 | 35.6 | 45.9 | 54.4 | 52.5 | 41.2 | 27.9 | 19.5 | 12.2 | 29.7 |
| | NORMAL DRY BULB | 30 | 35.8 | 39.0 | 46.3 | 54.1 | 62.0 | 69.2 | 73.0 | 71.8 | 65.7 | 55.2 | 46.4 | 39.0 | 54.8 |
| | MEAN DRY BULB | 58 | 37.3 | 39.4 | 46.5 | 55.1 | 62.9 | 69.7 | 73.3 | 72.4 | 66.0 | 55.8 | 46.4 | 39.4 | 55.4 |
| | MEAN WET BULB | 26 | 32.1 | 34.2 | 40.2 | 47.2 | 56.1 | 63.5 | 66.9 | 66.4 | 60.5 | 50.3 | 41.4 | 34.3 | 49.4 |
| | MEAN DEW POINT | 26 | 28.1 | 29.9 | 35.4 | 42.5 | 53.3 | 61.6 | 65.2 | 64.9 | 58.7 | 47.4 | 37.7 | 30.3 | 46.3 |
| | NORMAL NO. DAYS WITH: | | | | | | | | | | | | | | |
| | MAXIMUM >= 90 | 30 | 0.0 | 0.0 | 0.0 | 0.0 | 0.1 | 1.5 | 5.1 | 2.5 | 0.4 | 0.0 | 0.0 | 0.0 | 9.6 |
| | MAXIMUM <= 32 | 30 | 2.7 | 1.6 | 0.3 | 0.0 | 0.0 | 0.0 | 0.0 | 0.0 | 0.0 | 0.0 | * | 1.3 | 5.9 |
| | MINIMUM <= 32 | 30 | 22.5 | 18.6 | 12.0 | 4.2 | 0.3 | 0.0 | 0.0 | 0.0 | 0.0 | 3.6 | 12.9 | 20.3 | 94.4 |
| | MINIMUM <= 0 | 30 | 0.4 | * | 0.0 | 0.0 | 0.0 | 0.0 | 0.0 | 0.0 | 0.0 | 0.0 | 0.0 | 0.1 | 0.5 |
| H/C | NORMAL HEATING DEG. DAYS | 30 | 890 | 714 | 566 | 317 | 122 | 17 | 7 | 2 | 63 | 296 | 542 | 790 | 4326 |
| | NORMAL COOLING DEG. DAYS | 30 | 0 | 0 | 0 | 6 | 45 | 159 | 271 | 229 | 100 | 8 | 0 | 0 | 818 |
| RH | NORMAL (PERCENT) | 30 | 73 | 70 | 68 | 65 | 75 | 79 | 81 | 83 | 82 | 77 | 74 | 73 | 75 |
| | HOUR 01 LST | 30 | 81 | 78 | 78 | 77 | 88 | 93 | 94 | 96 | 95 | 90 | 85 | 82 | 86 |
| | HOUR 07 LST | 30 | 85 | 84 | 84 | 84 | 91 | 93 | 95 | 97 | 96 | 93 | 87 | 85 | 90 |
| | HOUR 13 LST | 30 | 59 | 56 | 53 | 49 | 56 | 60 | 62 | 62 | 62 | 55 | 56 | 59 | 57 |
| | HOUR 19 LST | 30 | 69 | 62 | 59 | 55 | 65 | 70 | 73 | 77 | 79 | 73 | 69 | 69 | 68 |
| S | PERCENT POSSIBLE SUNSHINE | 32 | 55 | 59 | 61 | 66 | 61 | 62 | 60 | 54 | 56 | 62 | 58 | 55 | 59 |
| W/O | MEAN NO. DAYS WITH: | | | | | | | | | | | | | | |
| | HEAVY FOG(VISBY <= 1/4 MI) | 45 | 4.0 | 2.7 | 2.3 | 2.5 | 5.2 | 7.7 | 8.5 | 12.3 | 11.0 | 8.1 | 3.7 | 4.4 | 72.4 |
| | THUNDERSTORMS | 51 | 0.4 | 0.8 | 2.1 | 3.0 | 6.0 | 8.1 | 8.7 | 7.7 | 2.8 | 0.6 | 0.6 | 0.3 | 41.1 |
| CLOUDNESS | MEAN: | | | | | | | | | | | | | | |
| | SUNRISE-SUNSET (OKTAS) | 32 | 5.0 | 4.8 | 4.9 | 4.6 | 5.0 | 5.0 | 5.0 | 5.0 | 4.8 | 4.0 | 4.3 | 4.7 | 4.8 |
| | MIDNIGHT-MIDNIGHT (OKTAS) | 32 | 4.9 | 4.7 | 4.7 | 4.4 | 4.8 | 4.7 | 4.8 | 5.0 | 4.7 | 4.0 | 4.1 | 4.5 | 4.6 |
| | MEAN NO. DAYS WITH: | | | | | | | | | | | | | | |
| | CLEAR | 32 | 9.1 | 8.7 | 8.8 | 9.6 | 7.2 | 6.2 | 5.0 | 4.8 | 6.7 | 11.6 | 10.2 | 9.3 | 97.2 |
| | PARTLY CLOUDY | 32 | 7.3 | 6.4 | 8.1 | 8.5 | 10.4 | 12.0 | 13.4 | 13.1 | 10.2 | 7.8 | 6.9 | 7.0 | 111.1 |
| | CLOUDY | 32 | 14.6 | 13.2 | 14.1 | 11.8 | 13.3 | 11.8 | 11.7 | 12.2 | 12.2 | 10.7 | 11.9 | 13.7 | 151.2 |
| PR | MEAN STATION PRESSURE(IN) | 26 | 27.82 | 27.80 | 27.78 | 27.76 | 27.79 | 27.82 | 27.84 | 27.85 | 27.86 | 27.87 | 27.85 | 27.84 | 27.82 |
| | MEAN SEA-LEVEL PRES. (IN) | 26 | 30.12 | 30.09 | 30.06 | 30.01 | 30.03 | 30.03 | 30.05 | 30.07 | 30.09 | 30.12 | 30.13 | 30.14 | 30.08 |
| WINDS | MEAN SPEED (MPH) | 26 | 8.8 | 8.5 | 8.4 | 7.9 | 6.4 | 5.3 | 5.1 | 4.7 | 4.9 | 5.9 | 7.3 | 7.8 | 6.8 |
| | PREVAIL.DIR(TENS OF DEGS) | 38 | 35 | 35 | 35 | 35 | 35 | 35 | 35 | 35 | 35 | 35 | 35 | 35 | 35 |
| | MAXIMUM 2-MINUTE: | | | | | | | | | | | | | | |
| | SPEED (MPH) | 13 | 46 | 43 | 47 | 49 | 44 | 33 | 41 | 41 | 45 | 33 | 38 | 37 | 49 |
| | DIR. (TENS OF DEGS) | | 33 | 33 | 32 | 34 | 33 | 32 | 33 | 30 | 36 | 33 | 33 | 33 | 34 |
| | YEAR OF OCCURRENCE | | 2003 | 2001 | 2001 | 2007 | 2008 | 2002 | 2001 | 2000 | 1999 | 2008 | 2003 | 2006 | APR 2007 |
| | MAXIMUM 3-SECOND | | | | | | | | | | | | | | |
| | SPEED (MPH) | 13 | 53 | 49 | 56 | 66 | 52 | 46 | 55 | 53 | 54 | 43 | 46 | 46 | 66 |
| | DIR. (TENS OF DEGS) | | 33 | 32 | 31 | 33 | 33 | 10 | 16 | 30 | 36 | 33 | 33 | 34 | 33 |
| | YEAR OF OCCURRENCE | | 2003 | 2001 | 2001 | 2007 | 2008 | 2000 | 2007 | 2000 | 1999 | 2008 | 2003 | 2008 | APR 2007 |
| PRECIPITATION | NORMAL (IN) | 30 | 4.06 | 3.83 | 4.59 | 3.50 | 4.42 | 4.38 | 3.87 | 4.30 | 3.72 | 3.18 | 3.82 | 3.40 | 47.07 |
| | MAXIMUM MONTHLY (IN) | 45 | 9.96 | 8.07 | 9.86 | 8.70 | 9.18 | 10.73 | 10.88 | 11.28 | 13.71 | 8.82 | 7.76 | 9.16 | 13.71 |
| | YEAR OF OCCURRENCE | | 1998 | 1990 | 1975 | 1998 | 2009 | 1989 | 2003 | 1967 | 2004 | 1990 | 1979 | 2009 | SEP 2004 |
| | MINIMUM MONTHLY (IN) | 45 | 0.45 | 0.44 | 0.77 | 0.25 | 0.96 | 0.85 | 0.46 | 0.52 | 0.16 | 0.00 | 1.19 | 0.16 | 0.00 |
| | YEAR OF OCCURRENCE | | 1981 | 1978 | 1985 | 1976 | 2007 | 2008 | 1986 | 1981 | 1984 | 2000 | 1981 | 1965 | OCT 2000 |
| | MAXIMUM IN 24 HOURS (IN) | 45 | 4.67 | 3.47 | 5.13 | 3.06 | 4.95 | 4.36 | 4.02 | 5.10 | 4.23 | 4.22 | 4.03 | 2.66 | 5.13 |
| | YEAR OF OCCURRENCE | | 1998 | 1982 | 1968 | 1973 | 1973 | 1997 | 1969 | 1990 | 2004 | 1995 | 1977 | 1973 | MAR 1968 |
| | NORMAL NO. DAYS WITH: | | | | | | | | | | | | | | |
| | PRECIPITATION >= 0.01 | 30 | 11.0 | 9.5 | 11.8 | 9.5 | 11.4 | 11.8 | 11.8 | 12.2 | 9.7 | 7.4 | 9.7 | 9.9 | 125.7 |
| | PRECIPITATION >= 1.00 | 30 | 1.0 | 0.9 | 1.2 | 1.0 | 1.0 | 1.0 | 0.9 | 1.0 | 0.9 | 0.8 | 1.4 | 0.9 | 12.0 |
| SNOWFALL | NORMAL (IN) | 30 | 4.7 | 3.1 | 2.6 | 0.7 | 0.* | 0.0 | 0.0 | 0.0 | 0.0 | 0.* | 0.4 | 1.8 | 13.3 |
| | MAXIMUM MONTHLY (IN) | 45 | 17.6 | 25.5 | 18.2 | 11.5 | T | T | T | T | 0.0 | T | 9.6 | 16.3 | 25.5 |
| | YEAR OF OCCURRENCE | | 1966 | 1969 | 1993 | 1987 | 1993 | 1995 | 1994 | 1990 | | 2008 | 1968 | 1971 | FEB 1969 |
| | MAXIMUM IN 24 HOURS (IN) | 45 | 14.0 | 11.7 | 16.5 | 11.5 | T | T | T | T | 0.0 | T | 5.7 | 16.3 | 16.5 |
| | YEAR OF OCCURRENCE' | | 1988 | 1969 | 1993 | 1987 | 1993 | 1995 | 1994 | 1990 | | 1993 | 1968 | 1971 | MAR 1993 |
| | MAXIMUM SNOW DEPTH (IN) | 50 | 14 | 13 | 18 | 12 | 0 | 0 | 0 | 0 | 0 | 0 | 5 | 14 | 18 |
| | YEAR OF OCCURRENCE | | 1988 | 1969 | 1993 | 1987 | | | | | | | 1968 | 1971 | MAR 1993 |
| | NORMAL NO. DAYS WITH: | | | | | | | | | | | | | | |
| | SNOWFALL >= 1.0 | 30 | 1.3 | 0.9 | 0.7 | 0.2 | 0.0 | 0.0 | 0.0 | 0.0 | 0.0 | 0.0 | 0.1 | 0.4 | 3.6 |

## PRECIPITATION (inches) 2009 ASHEVILLE (KAVL)

| YEAR | JAN | FEB | MAR | APR | MAY | JUN | JUL | AUG | SEP | OCT | NOV | DEC | ANNUAL |
|------|------|------|------|------|------|------|------|------|------|------|------|------|--------|
| 1980 | 2.85 | 0.53 | 8.26 | 4.77 | 4.54 | 4.68 | 2.21 | 2.38 | 4.36 | 2.62 | 3.04 | 0.59 | 40.83 |
| 1981 | 0.45 | 4.80 | 3.24 | 2.07 | 7.50 | 4.41 | 2.06 | 0.52 | 1.36 | 2.19 | 1.19 | 4.79 | 34.58 |
| 1982 | 5.41 | 7.02 | 1.92 | 3.62 | 3.78 | 3.98 | 9.92 | 1.73 | 1.33 | 3.48 | 4.59 | 4.04 | 50.82 |
| 1983 | 3.39 | 5.63 | 6.27 | 5.27 | 3.48 | 3.71 | 1.06 | 0.95 | 5.66 | 4.43 | 4.77 | 8.30 | 52.92 |
| 1984 | 2.36 | 6.43 | 4.82 | 4.05 | 6.62 | 3.69 | 5.88 | 5.02 | 0.16 | 2.73 | 2.61 | 1.34 | 45.71 |
| 1985 | 2.95 | 4.74 | 0.77 | 2.74 | 1.59 | 1.47 | 4.37 | 7.04 | 1.25 | 3.41 | 4.91 | 0.70 | 35.94 |
| 1986 | 1.11 | 1.85 | 2.75 | 0.57 | 3.55 | 1.28 | 0.46 | 6.10 | 3.15 | 4.19 | 5.28 | 4.28 | 34.57 |
| 1987 | 3.49 | 6.17 | 2.85 | 3.67 | 1.87 | 8.94 | 1.86 | 1.79 | 6.79 | 0.36 | 3.09 | 2.33 | 43.21 |
| 1988 | 3.71 | 0.88 | 1.31 | 3.46 | 1.06 | 0.94 | 2.65 | 1.78 | 2.79 | 3.12 | 3.47 | 1.41 | 26.58 |
| 1989 | 1.65 | 4.61 | 2.91 | 3.17 | 5.54 | 10.73 | 8.33 | 4.98 | 8.17 | 2.98 | 4.27 | 3.29 | 60.63 |
| 1990 | 3.27 | 8.07 | 5.95 | 1.96 | 5.09 | 0.90 | 6.55 | 7.78 | 1.43 | 8.82 | 1.55 | 4.50 | 55.87 |
| 1991 | 3.25 | 1.66 | 6.13 | 5.38 | 2.41 | 5.27 | 6.07 | 3.83 | 1.27 | 0.19 | 3.34 | 4.86 | 43.66 |
| 1992 | 3.08 | 3.66 | 3.52 | 3.99 | 6.18 | 6.62 | 1.10 | 7.64 | 3.15 | 4.15 | 7.24 | 3.71 | 54.04 |
| 1993 | 3.82 | 2.03 | 6.16 | 3.21 | 4.59 | 1.12 | 2.07 | 5.29 | 1.56 | 1.21 | 3.32 | 3.59 | 37.97 |
| 1994 | 5.35 | 5.11 | 7.52 | 3.30 | 1.74 | 5.89 | 6.76 | 6.01 | 5.33 | 4.27 | 3.15 | 3.03 | 57.46 |
| 1995 | 7.03 | 2.93 | 2.42 | 0.98 | 6.04 | 8.89 | 3.61 | 9.22 | 1.95 | 7.23 | 3.66 | 1.43 | 55.39 |
| 1996 | 7.22 | 2.71 | 3.36 | 2.00 | 2.55 | 3.54 | 4.83 | 6.68 | 5.22 | 0.68 | 4.45 | 3.92 | 47.16 |
| 1997 | 4.44 | 5.29 | 5.48 | 5.26 | 2.91 | 8.29 | 2.97 | 1.37 | 4.89 | 3.90 | 1.60 | 2.98 | 49.38 |
| 1998 | 9.96 | 6.38 | 3.71 | 8.70 | 2.22 | 3.64 | 1.97 | 2.23 | 1.62 | 1.79 | 2.76 | 3.04 | 48.02 |
| 1999 | 6.38 | 3.29 | 2.82 | 2.44 | 2.53 | 4.39 | 3.85 | 3.37 | 2.20 | 3.29 | 3.31 | 1.98 | 39.85 |
| 2000 | 3.10 | 2.33 | 3.82 | 5.11 | 1.27 | 2.78 | 2.84 | 4.45 | 3.27 | 0.00 | 4.25 | 2.37 | 35.59 |
| 2001 | 2.63 | 2.73 | 5.00 | 1.32 | 2.47 | 2.91 | 5.50 | 3.20 | 4.37 | 0.60 | 1.42 | 2.34 | 34.49 |
| 2002 | 3.64 | 1.30 | 4.36 | 1.73 | 3.42 | 6.13 | 2.09 | 6.05 | 3.14 | 4.23 | 6.40 | 44.47 |
| 2003 | 1.19 | 4.47 | 4.34 | 5.25 | 8.36 | 6.16 | 10.88 | 6.80 | 3.01 | 2.33 | 3.89 | 2.78 | 59.46 |
| 2004 | 0.83 | 4.20 | 2.02 | 2.95 | 3.23 | 7.39 | 4.68 | 3.79 | 13.71 | 1.11 | 5.02 | 3.43 | 52.36 |
| 2005 | 2.00 | 2.57 | 3.33 | 2.86 | 1.65 | 10.09 | 10.26 | 5.71 | 0.34 | 1.20 | 3.74 | 3.51 | 47.26 |
| 2006 | 3.58 | 2.55 | 0.91 | 4.58 | 1.69 | 5.16 | 2.81 | 7.12 | 7.80 | 2.93 | 4.52 | 4.64 | 48.29 |
| 2007 | 3.35 | 1.45 | 4.29 | 1.77 | 0.96 | 2.91 | 4.85 | 2.84 | 3.40 | 3.02 | 1.49 | 4.06 | 34.39 |
| 2008 | 2.56 | 3.79 | 4.51 | 2.84 | 1.33 | 0.85 | 4.02 | 5.84 | 1.70 | 1.84 | 1.61 | 4.74 | 35.63 |
| 2009 | 2.40 | 1.87 | 4.07 | 3.54 | 9.18 | 6.41 | 2.88 | 3.69 | 8.17 | 5.50 | 5.26 | 9.16 | 62.13 |
| POR= 58 YRS | 3.56 | 3.56 | 4.41 | 3.48 | 3.96 | 4.31 | 4.49 | 4.47 | 3.75 | 3.16 | 3.54 | 3.60 | 46.29 |

WBAN : 03812

## AVERAGE TEMPERATURE (°F) 2009 ASHEVILLE (KAVL)

| YEAR | JAN | FEB | MAR | APR | MAY | JUN | JUL | AUG | SEP | OCT | NOV | DEC | ANNUAL |
|------|------|------|------|------|------|------|------|------|------|------|------|------|--------|
| 1980 | 40.5 | 35.1 | 46.2 | 56.5 | 64.8 | 71.7 | 77.5 | 74.8 | 70.2 | 54.7 | 47.1 | 39.6 | 56.6 |
| 1981 | 33.3 | 39.9 | 44.9 | 60.1 | 60.7 | 74.3 | 75.0 | 71.7 | 66.1 | 54.5 | 48.2 | 35.8 | 55.4 |
| 1982 | 32.3 | 41.2 | 50.0 | 53.6 | 67.3 | 71.5 | 74.6 | 71.7 | 64.5 | 56.3 | 47.1 | 44.9 | 56.3 |
| 1983 | 36.7 | 38.8 | 46.7 | 51.1 | 61.6 | 69.0 | 75.7 | 76.5 | 66.6 | 57.5 | 47.3 | 36.4 | 55.3 |
| 1984 | 34.0 | 40.5 | 44.8 | 51.7 | 59.9 | 70.0 | 70.6 | 71.6 | 62.8 | 62.7 | 43.1 | 46.3 | 54.8 |
| 1985 | 30.5 | 38.3 | 48.1 | 56.6 | 62.6 | 69.8 | 72.2 | 70.9 | 64.2 | 60.5 | 56.0 | 34.7 | 55.4 |
| 1986 | 35.0 | 42.2 | 46.0 | 56.0 | 63.3 | 71.7 | 76.1 | 70.9 | 68.0 | 57.4 | 50.7 | 39.8 | 56.4 |
| 1987 | 35.3 | 38.9 | 46.5 | 52.6 | 66.7 | 71.2 | 74.7 | 74.7 | 66.5 | 50.1 | 47.0 | 42.2 | 55.5 |
| 1988 | 32.1 | 37.1 | 47.1 | 54.6 | 61.1 | 69.3 | 73.8 | 74.6 | 66.3 | 50.2 | 46.7 | 38.7 | 54.3 |
| 1989 | 42.1 | 39.8 | 50.3 | 54.5 | 59.4 | 70.0 | 73.3 | 71.9 | 65.8 | 56.1 | 46.2 | 31.6 | 55.1 |
| 1990 | 42.8 | 45.6 | 50.4 | 54.2 | 63.3 | 70.9 | 73.8 | 73.9 | 67.6 | 57.8 | 49.9 | 45.5 | 58.0 |
| 1991 | 39.2 | 42.5 | 49.6 | 58.4 | 67.3 | 70.3 | 75.2 | 72.1 | 67.3 | 57.0 | 45.5 | 43.4 | 57.3 |
| 1992 | 40.7 | 44.2 | 46.3 | 55.9 | 59.9 | 68.1 | 74.9 | 70.4 | 67.8 | 54.4 | 46.2 | 39.2 | 55.7 |
| 1993 | 42.0 | 38.0 | 43.0 | 52.6 | 63.2 | 71.4 | 78.0 | 73.9 | 67.8 | 55.0 | 46.0 | 36.8 | 55.6 |
| 1994 | 32.4 | 40.6 | 47.3 | 58.5 | 60.2 | 72.6 | 73.1 | 71.4 | 64.6 | 55.9 | 50.1 | 43.6 | 55.9 |
| 1995 | 38.9 | 38.4 | 49.4 | 56.4 | 64.1 | 68.5 | 75.0 | 75.3 | 65.4 | 56.8 | 43.3 | 37.1 | 55.7 |
| 1996 | 35.3 | 39.5 | 41.4 | 53.6 | 66.1 | 69.6 | 72.0 | 71.5 | 64.4 | 55.8 | 42.3 | 40.5 | 54.3 |
| 1997 | 36.5 | 43.2 | 51.5 | 51.5 | 58.4 | 66.5 | 73.4 | 70.6 | 66.1 | 55.4 | 42.3 | 36.9 | 54.4 |
| 1998 | 41.3 | 42.2 | 43.9 | 54.1 | 66.5 | 71.7 | 74.9 | 73.4 | 70.3 | 58.3 | 49.2 | 43.9 | 57.5 |
| 1999 | 41.7 | 42.4 | 42.9 | 58.5 | 61.9 | 69.3 | 74.3 | 74.1 | 65.6 | 56.0 | 51.3 | 41.9 | 56.7 |
| 2000 | 36.0 | 43.3 | 49.9 | 53.0 | 66.2 | 70.6 | 72.8 | 71.7 | 65.2 | 57.2 | 44.6 | 32.5 | 55.3 |
| 2001 | 36.7 | 44.5 | 44.6 | 58.2 | 64.0 | 70.5 | 72.9 | 73.8 | 64.5 | 53.9 | 51.7 | 44.8 | 56.7 |
| 2002 | 41.1 | 40.3 | 47.5 | 59.1 | 62.2 | 70.8 | 74.6 | 73.3 | 68.8 | 59.7 | 44.8 | 38.6 | 56.7 |
| 2003 | 33.7 | 39.6 | 50.1 | 55.2 | 62.8 | 68.1 | 72.0 | 73.7 | 65.2 | 54.8 | 51.4 | 36.6 | 55.3 |
| 2004 | 35.9 | 38.3 | 49.5 | 55.2 | 66.3 | 70.5 | 72.9 | 70.8 | 66.5 | 61.0 | 50.3 | 38.5 | 56.3 |
| 2005 | 40.8 | 41.8 | 45.4 | 54.5 | 60.6 | 69.5 | 74.5 | 74.0 | 68.4 | 58.5 | 48.3 | 36.6 | 56.1 |
| 2006 | 43.2 | 39.0 | 47.5 | 58.6 | 61.5 | 69.5 | 73.6 | 74.3 | 64.3 | 54.2 | 47.0 | 42.6 | 56.3 |
| 2007 | 39.5 | 36.3 | 52.0 | 53.1 | 63.5 | 71.3 | 71.4 | 76.7 | 68.1 | 59.6 | 45.5 | 43.9 | 56.7 |
| 2008 | 36.4 | 42.5 | 46.1 | 55.3 | 62.9 | 72.5 | 72.9 | 72.8 | 67.1 | 54.1 | 43.0 | 42.0 | 55.6 |
| 2009 | 36.9 | 39.5 | 47.3 | 54.9 | 63.9 | 71.7 | 70.8 | 72.4 | 66.9 | 55.0 | 49.0 | 36.6 | 55.4 |
| POR= 58 YRS | 37.3 | 39.4 | 46.5 | 55.1 | 62.9 | 69.7 | 73.3 | 72.4 | 66.0 | 55.8 | 46.4 | 39.4 | 55.3 |

## HEATING DEGREE DAYS (base 65°F) 2009  ASHEVILLE (KAVL)

| YEAR | JUL | AUG | SEP | OCT | NOV | DEC | JAN | FEB | MAR | APR | MAY | JUN | TOTAL |
|---|---|---|---|---|---|---|---|---|---|---|---|---|---|
| 1980-81 | 0 | 0 | 37 | 315 | 533 | 778 | 978 | 696 | 615 | 152 | 152 | 0 | 4256 |
| 1981-82 | 0 | 1 | 57 | 326 | 499 | 897 | 1006 | 659 | 458 | 333 | 38 | 0 | 4274 |
| 1982-83 | 0 | 0 | 74 | 274 | 531 | 616 | 872 | 725 | 562 | 410 | 127 | 13 | 4204 |
| 1983-84 | 0 | 0 | 84 | 229 | 527 | 882 | 955 | 706 | 618 | 391 | 176 | 9 | 4577 |
| 1984-85 | 1 | 0 | 107 | 91 | 648 | 576 | 1064 | 737 | 520 | 249 | 109 | 19 | 4121 |
| 1985-86 | 0 | 6 | 111 | 156 | 266 | 932 | 923 | 633 | 581 | 273 | 91 | 2 | 3974 |
| 1986-87 | 0 | 32 | 16 | 268 | 419 | 774 | 913 | 725 | 567 | 369 | 40 | 1 | 4124 |
| 1987-88 | 0 | 0 | 47 | 452 | 532 | 702 | 1013 | 802 | 545 | 308 | 132 | 31 | 4564 |
| 1988-89 | 5 | 0 | 33 | 453 | 544 | 808 | 702 | 698 | 454 | 331 | 200 | 4 | 4232 |
| 1989-90 | 1 | 8 | 74 | 279 | 558 | 1028 | 679 | 535 | 446 | 321 | 91 | 3 | 4023 |
| 1990-91 | 0 | 0 | 55 | 229 | 445 | 601 | 793 | 627 | 472 | 204 | 53 | 16 | 3495 |
| 1991-92 | 0 | 1 | 64 | 242 | 578 | 663 | 748 | 596 | 573 | 274 | 169 | 25 | 3933 |
| 1992-93 | 0 | 0 | 34 | 324 | 558 | 794 | 708 | 751 | 677 | 365 | 82 | 9 | 4302 |
| 1993-94 | 0 | 0 | 44 | 310 | 563 | 866 | 1005 | 676 | 544 | 206 | 165 | 0 | 4379 |
| 1994-95 | 0 | 1 | 52 | 276 | 441 | 656 | 803 | 736 | 474 | 264 | 87 | 17 | 3807 |
| 1995-96 | 0 | 0 | 50 | 255 | 645 | 856 | 912 | 735 | 725 | 341 | 78 | 20 | 4617 |
| 1996-97 | 0 | 0 | 76 | 284 | 673 | 753 | 876 | 601 | 412 | 400 | 212 | 64 | 4351 |
| 1997-98 | 0 | 5 | 35 | 292 | 675 | 863 | 727 | 631 | 647 | 322 | 71 | 23 | 4291 |
| 1998-99 | 0 | 0 | 11 | 214 | 465 | 649 | 718 | 627 | 680 | 210 | 100 | 7 | 3681 |
| 1999-00 | 6 | 0 | 56 | 274 | 405 | 705 | 893 | 621 | 460 | 353 | 41 | 11 | 3825 |
| 2000-01 | 1 | 1 | 88 | 238 | 607 | 998 | 872 | 566 | 628 | 222 | 63 | 1 | 4285 |
| 2001-02 | 0 | 0 | 97 | 340 | 391 | 620 | 735 | 687 | 536 | 201 | 130 | 1 | 3738 |
| 2002-03 | 0 | 0 | 10 | 202 | 600 | 811 | 966 | 705 | 454 | 289 | 88 | 19 | 4144 |
| 2003-04 | 0 | 0 | 52 | 309 | 411 | 874 | 894 | 767 | 478 | 295 | 49 | 1 | 4130 |
| 2004-05 | 0 | 3 | 21 | 128 | 436 | 817 | 744 | 644 | 602 | 309 | 143 | 22 | 3869 |
| 2005-06 | 0 | 0 | 4 | 215 | 496 | 877 | 668 | 724 | 537 | 203 | 149 | 10 | 3883 |
| 2006-07 | 0 | 0 | 68 | 332 | 531 | 684 | 785 | 796 | 393 | 352 | 81 | 0 | 4022 |
| 2007-08 | 0 | 0 | 34 | 198 | 576 | 650 | 880 | 645 | 579 | 288 | 97 | 0 | 3947 |
| 2008-09 | 1 | 0 | 35 | 330 | 653 | 706 | 863 | 708 | 543 | 298 | 76 | 0 | 4213 |
| 2009- | 0 | 0 | 23 | 309 | 474 | 873 | | | | | | | |

WBAN : 03812

## COOLING DEGREE DAYS (base 65°F) 2009  ASHEVILLE (KAVL)

| YEAR | JAN | FEB | MAR | APR | MAY | JUN | JUL | AUG | SEP | OCT | NOV | DEC | TOTAL |
|---|---|---|---|---|---|---|---|---|---|---|---|---|---|
| 1980 | 0 | 0 | 0 | 8 | 64 | 210 | 396 | 311 | 198 | 4 | 0 | 0 | 1191 |
| 1981 | 0 | 0 | 0 | 10 | 25 | 286 | 316 | 213 | 98 | 7 | 0 | 0 | 955 |
| 1982 | 0 | 0 | 0 | 0 | 117 | 206 | 305 | 215 | 64 | 16 | 0 | 0 | 923 |
| 1983 | 0 | 0 | 0 | 0 | 25 | 139 | 335 | 362 | 141 | 5 | 0 | 0 | 1007 |
| 1984 | 0 | 0 | 0 | 0 | 25 | 165 | 180 | 211 | 49 | 27 | 0 | 0 | 657 |
| 1985 | 0 | 0 | 5 | 2 | 43 | 170 | 229 | 194 | 90 | 25 | 4 | 0 | 762 |
| 1986 | 0 | 0 | 0 | 8 | 43 | 209 | 353 | 222 | 112 | 38 | 0 | 0 | 985 |
| 1987 | 0 | 0 | 0 | 7 | 97 | 192 | 310 | 309 | 97 | 0 | 0 | 0 | 1012 |
| 1988 | 0 | 0 | 0 | 0 | 18 | 168 | 282 | 304 | 79 | 3 | 0 | 0 | 854 |
| 1989 | 0 | 0 | 5 | 23 | 34 | 159 | 264 | 229 | 107 | 11 | 0 | 0 | 832 |
| 1990 | 0 | 0 | 0 | 3 | 48 | 187 | 279 | 283 | 141 | 11 | 0 | 0 | 952 |
| 1991 | 0 | 0 | 3 | 13 | 132 | 181 | 324 | 227 | 139 | 3 | 0 | 0 | 1022 |
| 1992 | 0 | 0 | 0 | 7 | 20 | 125 | 313 | 174 | 125 | 0 | 0 | 0 | 764 |
| 1993 | 0 | 0 | 0 | 0 | 37 | 210 | 411 | 285 | 133 | 5 | 0 | 0 | 1081 |
| 1994 | 0 | 0 | 0 | 15 | 21 | 233 | 259 | 205 | 43 | 2 | 0 | 0 | 778 |
| 1995 | 0 | 0 | 1 | 12 | 65 | 127 | 315 | 325 | 68 | 8 | 0 | 0 | 921 |
| 1996 | 0 | 0 | 0 | 6 | 120 | 164 | 224 | 209 | 64 | 5 | 0 | 0 | 792 |
| 1997 | 0 | 0 | 0 | 0 | 13 | 116 | 265 | 184 | 76 | 1 | 0 | 0 | 655 |
| 1998 | 0 | 0 | 0 | 1 | 124 | 230 | 315 | 269 | 178 | 14 | 0 | 0 | 1131 |
| 1999 | 0 | 0 | 0 | 23 | 9 | 143 | 304 | 291 | 78 | 3 | 0 | 0 | 851 |
| 2000 | 0 | 0 | 0 | 0 | 83 | 187 | 251 | 215 | 103 | 4 | 1 | 0 | 844 |
| 2001 | 0 | 0 | 0 | 24 | 42 | 172 | 249 | 282 | 87 | 1 | 0 | 0 | 857 |
| 2002 | 0 | 0 | 0 | 32 | 50 | 181 | 305 | 267 | 130 | 47 | 0 | 0 | 1012 |
| 2003 | 0 | 0 | 0 | 0 | 29 | 117 | 223 | 281 | 65 | 0 | 9 | 0 | 724 |
| 2004 | 0 | 0 | 0 | 8 | 96 | 173 | 254 | 189 | 71 | 9 | 4 | 0 | 804 |
| 2005 | 0 | 0 | 0 | 0 | 14 | 167 | 302 | 285 | 112 | 20 | 0 | 0 | 900 |
| 2006 | 0 | 0 | 0 | 18 | 45 | 153 | 274 | 297 | 54 | 4 | 0 | 0 | 845 |
| 2007 | 0 | 0 | 0 | 1 | 42 | 192 | 208 | 370 | 133 | 38 | 0 | 0 | 984 |
| 2008 | 0 | 0 | 0 | 0 | 39 | 232 | 253 | 245 | 104 | 2 | 0 | 0 | 875 |
| 2009 | 0 | 0 | 0 | 4 | 50 | 209 | 186 | 237 | 84 | 4 | 0 | 0 | 774 |

## SNOWFALL (inches) 2009 ASHEVILLE (KAVL)

| YEAR | JUL | AUG | SEP | OCT | NOV | DEC | JAN | FEB | MAR | APR | MAY | JUN | TOTAL |
|------|-----|-----|-----|-----|-----|-----|-----|-----|-----|-----|-----|-----|-------|
| 1980-81 | 0.0 | 0.0 | 0.0 | 0.0 | T | T | 4.7 | T | 9.9 | 0.0 | 0.0 | 0.0 | 14.6 |
| 1981-82 | 0.0 | 0.0 | 0.0 | 0.0 | T | 2.0 | 8.6 | 8.1 | 0.1 | 3.0 | 0.0 | 0.0 | 21.8 |
| 1982-83 | 0.0 | 0.0 | 0.0 | 0.0 | 0.0 | 0.4 | 10.5 | 9.3 | 4.5 | 2.0 | 0.0 | 0.0 | 26.7 |
| 1983-84 | 0.0 | 0.0 | 0.0 | 0.0 | T | T | 0.2 | 2.9 | T | 0.0 | 0.0 | 0.0 | 3.1 |
| 1984-85 | 0.0 | 0.0 | 0.0 | 0.0 | 0.0 | T | 4.5 | 3.1 | 0.1 | 0.4 | 0.0 | 0.0 | 8.1 |
| 1985-86 | 0.0 | 0.0 | 0.0 | 0.0 | 0.0 | 0.4 | 0.8 | 3.7 | 0.1 | T | 0.0 | 0.0 | 5.0 |
| 1986-87 | 0.0 | 0.0 | 0.0 | 0.0 | T | T | 15.0 | 2.7 | 0.3 | 11.5 | 0.0 | 0.0 | 29.5 |
| 1987-88 | 0.0 | 0.0 | 0.0 | 0.0 | 0.3 | 0.5 | 14.2 | T | T | 1.2 | 0.0 | 0.0 | 16.2 |
| 1988-89 | 0.0 | 0.0 | 0.0 | 0.0 | 0.0 | T | 1.2 | 6.0 | T | 1.0 | 0.0 | 0.0 | 8.2 |
| 1989-90 | 0.0 | 0.0 | 0.0 | T | T | 3.0 | T | T | T | 0.0 | 0.0 | 0.0 | 3.0 |
| 1990-91 | 0.0 | T | 0.0 | 0.0 | 0.0 | T | T | 0.4 | 3.1 | 0.0 | 0.0 | T | 3.5 |
| 1991-92 | 0.0 | 0.0 | 0.0 | 0.0 | 1.0 | 0.0 | T | T | 0.5 | 0.0 | T | 0.0 | 1.5 |
| 1992-93 | 0.0 | 0.0 | 0.0 | 0.0 | T | T | T | 3.0 | 18.2 | T | T | 0.0 | 21.2 |
| 1993-94 | 0.0 | 0.0 | 0.0 | T | T | 8.0 | 2.6 | 0.3 | 0.1 | T | 0.0 | 0.0 | 11.0 |
| 1994-95 | T | 0.0 | 0.0 | 0.0 | 0.0 | T | 0.2 | 3.2 | T | 0.0 | 0.0 | T | 3.4 |
| 1995-96 | 0.0 | 0.0 | 0.0 | 0.0 | T | 0.1 | 15.7 | 4.4 | 1.5 | T | 0.0 | T | 21.7 |
| 1996-97 | 0.0 | 0.0 | 0.0 | 0.0 | 0.1 | 2.5 | 3.8 | 1.9 | T | T | 0.0 | 0.0 | 8.3 |
| 1997-98 | 0.0 | 0.0 | 0.0 | 0.0 | T | 4.5 | 12.7 | T | 0.4 | 0.0 | 0.0 | 0.0 | 17.6 |
| 1998-99 | 0.0 | 0.0 | 0.0 | 0.0 | 0.0 | 0.4 | 1.6 | T | 4.2 | 0.0 | 0.0 | 0.0 | 6.2 |
| 1999-00 | 0.0 | 0.0 | 0.0 | 0.0 | 0.1 | T | 6.2 | T | T | 0.1 | 0.0 | 0.0 | 6.4 |
| 2000-01 | 0.0 | 0.0 | 0.0 | 0.0 | 2.0 | 6.1 | T | 0.8 | 6.4 | 0.2 | 0.0 | 0.0 | 15.5 |
| 2001-02 | 0.0 | 0.0 | 0.0 | 0.0 | 0.0 | 0.0 | 1.3 | 0.3 | T | 0.0 | 0.0 | 0.0 | 1.6 |
| 2002-03 | 0.0 | 0.0 | 0.0 | 0.0 | T | 4.0 | 4.2 | 3.7 | 2.0 | 4.0 | T | 0.0 | 17.9 |
| 2003-04 | T | 0.0 | 0.0 | 0.0 | T | 3.5 | 4.0 | 6.8 | T | T | 0.0 | 0.0 | 14.3 |
| 2004-05 | T | 0.0 | 0.0 | 0.0 | 0.0 | 0.6 | 1.6 | 4.0 | 0.3 | 0.1 | 0.0 | 0.0 | 6.6 |
| 2005-06 | 0.0 | 0.0 | 0.0 | 0.0 | T | T | T | 1.4 | T | 0.0 | 0.0 | 0.0 | 1.4 |
| 2006-07 | 0.0 | 0.0 | 0.0 | 0.0 | T | T | 0.9 | 1.0 | T | 1.3 | 0.0 | 0.0 | 3.2 |
| 2007-08 | 0.0 | 0.0 | 0.0 | 0.0 | T | 0.1 | 3.7 | T | 0.0 | 0.0 | 0.0 | 0.0 | 3.8 |
| 2008-09 | 0.0 | 0.0 | 0.0 | T | T | T | 1.0 | T | 5.1 | T | 0.0 | 0.0 | 6.1 |
| 2009- | 0.0 | 0.0 | 0.0 | 0.0 | 0.0 | 10.4 | | | | | | | |
| POR= 58 YRS | T | T | 0.0 | T | 0.6 | 1.7 | 3.8 | 3.1 | 1.9 | 0.4 | T | T | 11.5 |

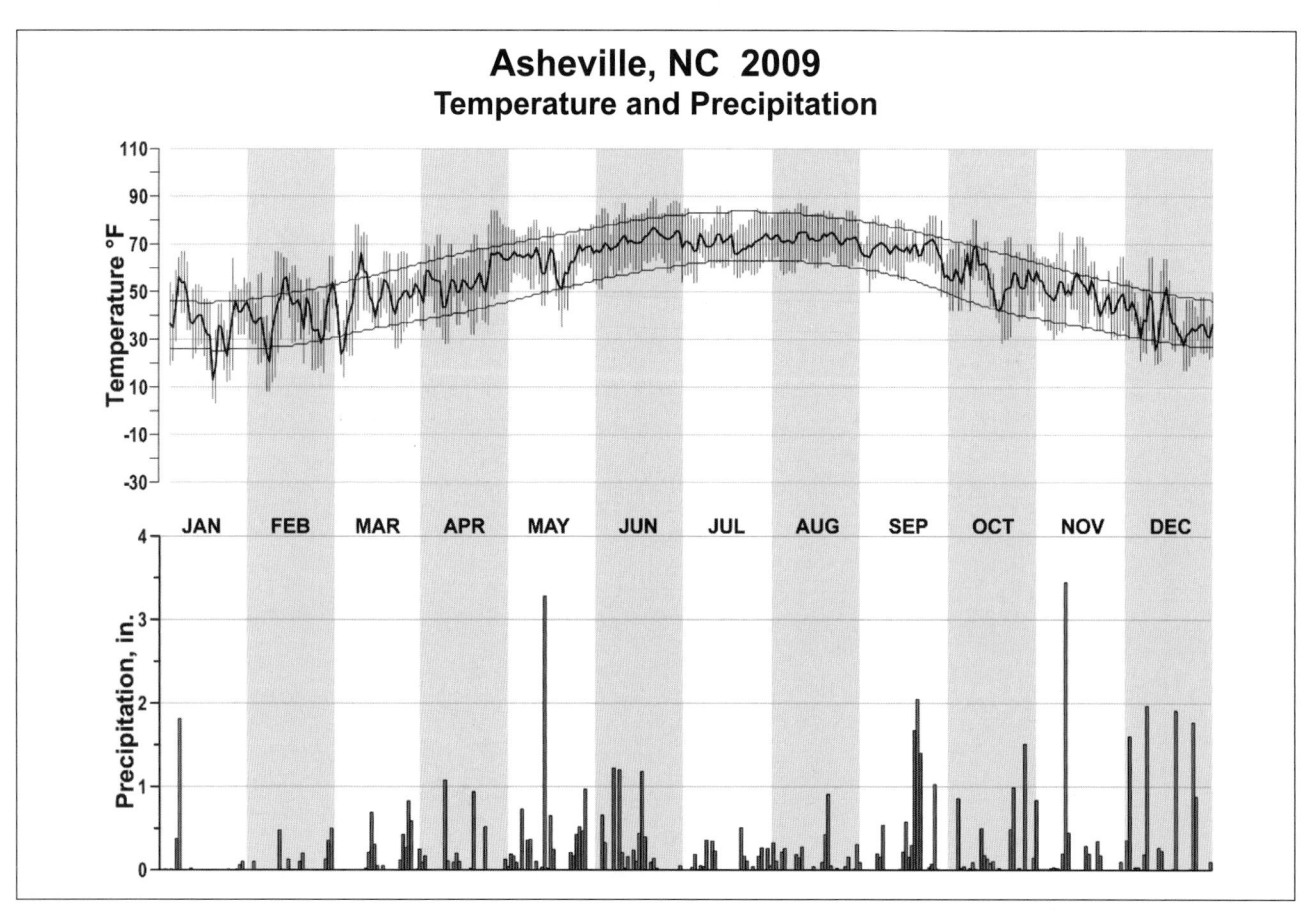

## Asheville, NC  2009
### Temperature and Precipitation

# 2009 RALEIGH/DURHAM
# NORTH CAROLINA (KRDU)

The Raleigh-Durham Airport is located in the zone of transition between the Coastal Plain and the Piedmont Plateau. The surrounding terrain is rolling, with an average elevation of around 400 feet, the range over a 10-mile radius is roughly between 200 and 550 feet. Being centrally located between the mountains on the west and the coast on the south and east, the Raleigh-Durham area enjoys a favorable climate. The mountains form a partial barrier to cold air masses moving eastward from the interior of the nation. As a result, there are few days in the heart of the winter season when the temperature falls below 20 degrees. Tropical air is present over the eastern and central sections of North Carolina during much of the summer season, bringing warm temperatures and rather high humidities to the Raleigh-Durham area. Afternoon temperatures reach 90 degrees or higher on about one-fourth of the days in the middle of summer, but reach 100 degrees less than once per year. Even in the hottest weather, early morning temperatures almost always drop into the lower 70s.

Rainfall is well distributed throughout the year as a whole. July and August have the greatest amount of rainfall, and October and November the least. There are times in spring and summer when soil moisture is scanty. This usually results from too many days between rains rather than from a shortage of total rainfall, but occasionally the accumulated total during the growing season falls short of plant needs. Most summer rain is produced by thunderstorms, which may occasionally be accompanied by strong winds, intense rains, and hail. The Raleigh-Durham area is far enough from the coast so that the bad weather effects of coastal storms are reduced. While snow and sleet usually occur each year, excessive accumulations of snow are rare.

From September 1887 to December 1950, the office was located in the downtown areas of Raleigh. The various buildings occupied were within an area of three blocks. All thermometers were exposed on the roof, and this, plus the smoke over the city, had an effect on the temperature record of that period. Lowest temperatures at the city office were frequently from 2 to 5 degrees higher than those recorded in surrounding rural areas. Maximum temperatures in the city were generally a degree or two lower. These observations are supported by a period of simultaneous record from the Municipal Airport and the city office location between 1937 and 1940.

From September 1946 to May 1954, simultaneous records were kept at a surface location on the North Carolina State College campus in Raleigh, and at the Raleigh-Durham Airport 10 1/2 air miles to the northwest.

Based on the 1951-1980 period, the average first occurrence of 32 degrees Fahrenheit in the fall is October 27 and the average last occurrence in the spring is April 11.

# NORMALS, MEANS, AND EXTREMES
## RALEIGH/DURHAM (KRDU)

| LATITUDE: 35°52'N | LONGITUDE: -78°47'W | ELEVATION (FT): GRND: 397  BARO: 430 | TIME ZONE: EASTERN  (UTC -5) | WBAN: 13722 |
|---|---|---|---|---|

| ELEMENT | POR | JAN | FEB | MAR | APR | MAY | JUN | JUL | AUG | SEP | OCT | NOV | DEC | YEAR |
|---|---|---|---|---|---|---|---|---|---|---|---|---|---|---|
| **TEMPERATURE °F** | | | | | | | | | | | | | | |
| NORMAL DAILY MAXIMUM | 30 | 49.8 | 54.0 | 62.5 | 71.8 | 78.7 | 85.5 | 89.1 | 87.2 | 81.3 | 71.8 | 62.4 | 53.3 | 70.6 |
| MEAN DAILY MAXIMUM | 62 | 50.6 | 54.0 | 62.0 | 72.1 | 79.1 | 85.8 | 88.9 | 87.4 | 81.4 | 71.9 | 62.5 | 53.1 | 70.7 |
| HIGHEST DAILY MAXIMUM | 65 | 80 | 84 | 92 | 95 | 97 | 104 | 105 | 105 | 104 | 98 | 88 | 81 | 105 |
| YEAR OF OCCURRENCE | | 2002 | 1977 | 1945 | 1980 | 1953 | 1954 | 1952 | 2007 | 1954 | 1954 | 1950 | 2007 | AUG 2007 |
| MEAN OF EXTREME MAXS. | 62 | 71.0 | 73.6 | 81.0 | 87.7 | 90.4 | 95.7 | 96.6 | 95.6 | 91.8 | 85.5 | 78.7 | 72.0 | 85.0 |
| NORMAL DAILY MINIMUM | 30 | 29.6 | 31.9 | 38.9 | 46.4 | 55.3 | 63.8 | 68.5 | 67.2 | 61.0 | 48.2 | 39.5 | 32.6 | 48.6 |
| MEAN DAILY MINIMUM | 62 | 30.0 | 31.9 | 38.5 | 46.9 | 55.5 | 63.6 | 68.2 | 67.2 | 60.6 | 48.5 | 39.1 | 32.1 | 48.5 |
| LOWEST DAILY MINIMUM | 65 | -9 | -1 | 11 | 23 | 31 | 38 | 48 | 46 | 37 | 19 | 11 | 4 | -9 |
| YEAR OF OCCURRENCE | | 1985 | 1996 | 1980 | 1985 | 1977 | 1977 | 1975 | 1965 | 1983 | 1962 | 1970 | 1983 | JAN 1985 |
| MEAN OF EXTREME MINS. | 62 | 12.0 | 15.6 | 22.5 | 30.5 | 40.6 | 51.0 | 58.4 | 57.0 | 46.0 | 31.8 | 22.8 | 15.3 | 33.6 |
| NORMAL DRY BULB | 30 | 39.7 | 43.0 | 50.7 | 59.1 | 67.0 | 74.7 | 78.8 | 77.2 | 71.2 | 60.0 | 51.0 | 43.0 | 59.6 |
| MEAN DRY BULB | 62 | 40.3 | 42.9 | 50.2 | 59.5 | 67.3 | 74.9 | 78.6 | 77.4 | 71.0 | 60.2 | 50.8 | 42.6 | 59.6 |
| MEAN WET BULB | 26 | 35.2 | 37.1 | 43.4 | 51.2 | 59.8 | 67.7 | 71.2 | 70.3 | 64.5 | 54.5 | 45.5 | 37.7 | 53.2 |
| MEAN DEW POINT | 26 | 30.1 | 32.2 | 38.2 | 46.5 | 56.6 | 65.2 | 69.3 | 68.4 | 62.4 | 51.7 | 41.5 | 32.9 | 49.6 |
| NORMAL NO. DAYS WITH: | | | | | | | | | | | | | | |
| MAXIMUM >= 90 | 30 | 0.0 | 0.0 | * | 0.4 | 1.7 | 8.4 | 14.4 | 11.0 | 3.2 | 0.2 | 0.0 | 0.0 | 39.3 |
| MAXIMUM <= 32 | 30 | 1.9 | 0.9 | 0.1 | 0.0 | 0.0 | 0.0 | 0.0 | 0.0 | 0.0 | 0.0 | * | 0.8 | 3.7 |
| MINIMUM <= 32 | 30 | 19.1 | 15.8 | 8.8 | 2.1 | * | 0.0 | 0.0 | 0.0 | 0.0 | 1.2 | 8.8 | 16.9 | 72.7 |
| MINIMUM <= 0 | 30 | 0.1 | * | 0.0 | 0.0 | 0.0 | 0.0 | 0.0 | 0.0 | 0.0 | 0.0 | 0.0 | 0.0 | 0.1 |
| **H/C** | | | | | | | | | | | | | | |
| NORMAL HEATING DEG. DAYS | 30 | 783 | 627 | 456 | 214 | 61 | 5 | 0 | 1 | 20 | 194 | 425 | 679 | 3465 |
| NORMAL COOLING DEG. DAYS | 30 | 0 | 1 | 9 | 38 | 119 | 293 | 429 | 379 | 206 | 39 | 6 | 2 | 1521 |
| **RH** | | | | | | | | | | | | | | |
| NORMAL (PERCENT) | 30 | 68 | 65 | 63 | 63 | 72 | 74 | 76 | 77 | 78 | 75 | 70 | 70 | 71 |
| HOUR 01 LST | 30 | 74 | 72 | 71 | 74 | 84 | 87 | 88 | 89 | 89 | 87 | 80 | 76 | 81 |
| HOUR 07 LST | 30 | 80 | 79 | 80 | 81 | 86 | 87 | 89 | 92 | 92 | 90 | 85 | 81 | 85 |
| HOUR 13 LST | 30 | 56 | 53 | 50 | 46 | 54 | 56 | 57 | 59 | 59 | 53 | 53 | 56 | 54 |
| HOUR 19 LST | 30 | 65 | 60 | 56 | 54 | 66 | 67 | 70 | 73 | 77 | 77 | 69 | 67 | 67 |
| **S** PERCENT POSSIBLE SUNSHINE | 42 | 52 | 56 | 60 | 63 | 59 | 60 | 60 | 58 | 58 | 60 | 57 | 53 | 58 |
| **W/O** MEAN NO. DAYS WITH: | | | | | | | | | | | | | | |
| HEAVY FOG(VISBY <= 1/4 MI) | 46 | 3.3 | 2.6 | 2.0 | 1.5 | 1.8 | 1.4 | 2.3 | 2.7 | 2.7 | 2.8 | 3.0 | 3.3 | 29.4 |
| THUNDERSTORMS | 62 | 0.5 | 0.8 | 2.2 | 3.5 | 5.9 | 7.3 | 10.4 | 7.8 | 2.9 | 1.2 | 0.8 | 0.4 | 43.7 |
| **CLOUDNESS** MEAN: | | | | | | | | | | | | | | |
| SUNRISE-SUNSET (OKTAS) | 47 | 5.0 | 4.7 | 4.7 | 4.4 | 4.7 | 4.6 | 4.8 | 4.7 | 4.7 | 3.9 | 4.2 | 4.6 | 4.6 |
| MIDNIGHT-MIDNIGHT (OKTAS) | 32 | 4.7 | 4.4 | 4.3 | 4.1 | 4.6 | 4.4 | 4.7 | 4.6 | 4.2 | 3.7 | 4.0 | 4.3 | 4.3 |
| MEAN NO. DAYS WITH: | | | | | | | | | | | | | | |
| CLEAR | 48 | 9.0 | 8.7 | 9.4 | 9.6 | 8.2 | 7.6 | 7.2 | 7.3 | 9.5 | 12.8 | 11.3 | 10.0 | 110.6 |
| PARTLY CLOUDY | 48 | 6.7 | 6.1 | 7.3 | 9.1 | 10.0 | 11.8 | 12.2 | 12.3 | 9.1 | 7.1 | 7.3 | 7.0 | 106.0 |
| CLOUDY | 48 | 15.2 | 13.5 | 14.3 | 11.3 | 12.8 | 10.7 | 11.6 | 11.3 | 11.4 | 11.1 | 11.4 | 14.0 | 148.6 |
| **PR** MEAN STATION PRESSURE(IN) | 26 | 29.62 | 29.63 | 29.59 | 29.53 | 29.55 | 29.54 | 29.56 | 29.58 | 29.60 | 29.64 | 29.66 | 29.64 | 29.60 |
| MEAN SEA-LEVEL PRES. (IN) | 26 | 30.13 | 30.11 | 30.07 | 30.00 | 30.01 | 30.00 | 30.02 | 30.04 | 30.07 | 30.11 | 30.13 | 30.15 | 30.07 |
| **WINDS** MEAN SPEED (MPH) | 26 | 7.3 | 7.5 | 8.2 | 8.1 | 7.3 | 6.5 | 6.2 | 5.8 | 6.2 | 5.7 | 6.1 | 6.5 | 6.8 |
| PREVAIL.DIR(TENS OF DEGS) | 40 | 23 | 23 | 23 | 24 | 24 | 23 | 24 | 23 | 04 | 03 | 23 | 24 | 24 |
| MAXIMUM 2-MINUTE: | | | | | | | | | | | | | | |
| SPEED (MPH) | 13 | 43 | 45 | 46 | 35 | 41 | 36 | 39 | 36 | 48 | 26 | 38 | 40 | 48 |
| DIR. (TENS OF DEGS) | | 24 | 23 | 24 | 23 | 23 | 03 | 23 | 24 | 20 | 24 | 22 | 22 | 20 |
| YEAR OF OCCURRENCE | | 2009 | 1997 | 1999 | 2009 | 2002 | 2007 | 2008 | 2001 | 2004 | 2009 | 2006 | 2000 | SEP 2004 |
| MAXIMUM 3-SECOND | | | | | | | | | | | | | | |
| SPEED (MPH) | 13 | 70 | 64 | 61 | 55 | 52 | 44 | 53 | 44 | 79 | 35 | 45 | 51 | 79 |
| DIR. (TENS OF DEGS) | | 24 | 23 | 23 | 26 | 22 | 03 | 22 | 18 | 18 | 23 | 22 | 24 | 18 |
| YEAR OF OCCURRENCE | | 1999 | 1997 | 1999 | 2006 | 2002 | 2007 | 2007 | 2005 | 2004 | 2009 | 2006 | 2009 | SEP 2004 |
| **PRECIPITATION** NORMAL (IN) | 30 | 4.02 | 3.47 | 4.03 | 2.80 | 3.79 | 3.42 | 4.29 | 3.78 | 4.26 | 3.18 | 2.97 | 3.04 | 43.05 |
| MAXIMUM MONTHLY (IN) | 65 | 7.52 | 6.42 | 7.78 | 6.10 | 7.67 | 10.45 | 10.27 | 12.18 | 21.79 | 9.35 | 9.03 | 6.65 | 21.79 |
| YEAR OF OCCURRENCE | | 1954 | 1989 | 1983 | 1978 | 1974 | 2006 | 1991 | 1986 | 1999 | 2002 | 2006 | 1983 | SEP 1999 |
| MINIMUM MONTHLY (IN) | 65 | 0.87 | 0.69 | 1.03 | 0.23 | 0.58 | 0.33 | 0.80 | 0.81 | 0.23 | 0.44 | 0.48 | 0.25 | 0.23 |
| YEAR OF OCCURRENCE | | 1981 | 1991 | 1985 | 1976 | 1999 | 1993 | 1953 | 1950 | 1985 | 2000 | 2007 | 1965 | SEP 1985 |
| MAXIMUM IN 24 HOURS (IN) | 65 | 3.11 | 3.22 | 3.70 | 4.04 | 4.40 | 5.65 | 4.27 | 5.20 | 5.41 | 5.78 | 4.70 | 3.18 | 5.78 |
| YEAR OF OCCURRENCE | | 1984 | 1973 | 1983 | 1978 | 1957 | 2006 | 1997 | 5020 | 1999 | 2002 | 1963 | 1958 | OCT 2002 |
| NORMAL NO. DAYS WITH: | | | | | | | | | | | | | | |
| PRECIPITATION >= 0.01 | 30 | 10.5 | 9.5 | 10.2 | 9.1 | 10.1 | 9.8 | 11.2 | 9.6 | 8.2 | 6.8 | 8.4 | 9.7 | 113.1 |
| PRECIPITATION >= 1.00 | 30 | 1.3 | 1.0 | 0.9 | 0.6 | 0.8 | 0.8 | 1.0 | 1.1 | 1.3 | 1.1 | 0.7 | 0.6 | 11.2 |
| **SNOWFALL** NORMAL (IN) | 30 | 2.3 | 3.0 | 1.0 | 0.1 | 0.0 | 0.0 | 0.0 | 0.0 | 0.0 | 0.0 | 0.2 | 0.5 | 7.1 |
| MAXIMUM MONTHLY (IN) | 65 | 25.8 | 17.2 | 14.0 | 1.8 | T | T | T | 0.0 | 0.0 | 0.0 | 2.6 | 10.6 | 25.8 |
| YEAR OF OCCURRENCE | | 2000 | 1979 | 1960 | 1983 | 2006 | 1996 | 1993 | | | | 1975 | 1958 | JAN 2000 |
| MAXIMUM IN 24 HOURS (IN) | 65 | 17.9 | 10.4 | 9.3 | 1.8 | T | T | T | 0.0 | 0.0 | 0.0 | 2.6 | 9.1 | 17.9 |
| YEAR OF OCCURRENCE' | | 2000 | 1979 | 1969 | 1983 | 1995 | 1996 | 1993 | | | | 1975 | 1958 | JAN 2000 |
| MAXIMUM SNOW DEPTH (IN) | 61 | 20 | 10 | 11 | 0 | 0 | 0 | 0 | 0 | 0 | 0 | 2 | 9 | 20 |
| YEAR OF OCCURRENCE | | 2000 | 1979 | 1980 | | | | | | | | 2000 | 1958 | JAN 2000 |
| NORMAL NO. DAYS WITH: | | | | | | | | | | | | | | |
| SNOWFALL >= 1.0 | 30 | 0.6 | 0.7 | 0.2 | 0.0 | 0.0 | 0.0 | 0.0 | 0.0 | 0.0 | 0.0 | 0.1 | 0.2 | 1.8 |

## PRECIPITATION (inches) 2009 RALEIGH/DURHAM (KRDU)

| YEAR | JAN | FEB | MAR | APR | MAY | JUN | JUL | AUG | SEP | OCT | NOV | DEC | ANNUAL |
|------|------|------|------|------|------|------|------|------|------|------|------|------|--------|
| 1980 | 4.39 | 1.91 | 5.87 | 1.97 | 2.33 | 4.89 | 2.11 | 1.87 | 3.76 | 2.25 | 2.87 | 1.42 | 35.64 |
| 1981 | 0.87 | 3.02 | 2.35 | 1.03 | 4.28 | 0.55 | 5.69 | 5.34 | 2.70 | 4.64 | 0.95 | 4.96 | 36.38 |
| 1982 | 3.43 | 4.97 | 3.02 | 3.33 | 4.20 | 8.39 | 3.34 | 1.83 | 1.55 | 3.93 | 2.34 | 4.02 | 44.35 |
| 1983 | 1.79 | 6.00 | 7.78 | 3.54 | 5.89 | 3.09 | 1.10 | 1.81 | 2.13 | 3.59 | 3.86 | 6.65 | 47.23 |
| 1984 | 4.93 | 5.65 | 5.40 | 4.45 | 5.43 | 3.08 | 9.20 | 1.13 | 2.31 | 0.73 | 1.64 | 2.32 | 46.27 |
| 1985 | 4.83 | 4.44 | 1.03 | 0.64 | 3.95 | 2.87 | 6.28 | 3.73 | 0.23 | 1.75 | 7.61 | 0.81 | 38.17 |
| 1986 | 1.88 | 1.65 | 3.06 | 1.01 | 2.98 | 1.92 | 4.32 | 12.18 | 0.95 | 1.28 | 2.77 | 2.95 | 36.95 |
| 1987 | 6.53 | 5.52 | 2.88 | 4.68 | 1.19 | 2.11 | 1.78 | 5.80 | 5.48 | 1.71 | 1.39 | 3.02 | 42.09 |
| 1988 | 3.15 | 2.42 | 1.76 | 3.56 | 2.85 | 2.88 | 2.69 | 3.40 | 4.90 | 5.67 | 3.34 | 1.04 | 37.66 |
| 1989 | 1.35 | 6.42 | 5.40 | 4.91 | 3.88 | 7.30 | 5.46 | 5.08 | 3.96 | 3.44 | 3.94 | 3.01 | 54.15 |
| 1990 | 3.07 | 3.82 | 5.02 | 2.19 | 6.97 | 1.03 | 2.22 | 2.65 | 0.30 | 5.69 | 1.51 | 3.08 | 37.55 |
| 1991 | 4.12 | 0.69 | 4.59 | 1.04 | 2.89 | 2.05 | 10.27 | 1.87 | 3.16 | 1.40 | 0.73 | 2.65 | 35.46 |
| 1992 | 3.80 | 2.23 | 2.95 | 1.93 | 2.60 | 5.12 | 3.45 | 7.63 | 2.22 | 3.79 | 5.02 | 2.44 | 43.18 |
| 1993 | 4.50 | 2.22 | 6.13 | 4.84 | 3.32 | 0.33 | 2.11 | 1.77 | 3.50 | 2.95 | 2.66 | 3.72 | 38.05 |
| 1994 | 3.55 | 2.97 | 5.91 | 0.86 | 2.85 | 2.20 | 4.67 | 4.20 | 1.99 | 4.62 | 1.32 | 1.27 | 36.41 |
| 1995 | 4.50 | 4.52 | 2.49 | 1.32 | 3.91 | 7.75 | 3.29 | 2.70 | 2.46 | 9.10 | 4.67 | 1.88 | 48.59 |
| 1996 | 4.24 | 2.94 | 3.39 | 3.98 | 3.26 | 3.28 | 6.98 | 3.72 | 16.65 | 3.48 | 4.33 | 2.89 | 59.14 |
| 1997 | 3.13 | 2.83 | 3.41 | 4.75 | 2.21 | 3.83 | 6.51 | 1.01 | 3.07 | 3.26 | 4.05 | 2.75 | 40.81 |
| 1998 | 7.49 | 5.79 | 7.36 | 3.12 | 3.79 | 3.45 | 4.84 | 4.66 | 3.55 | 2.79 | 2.40 | 3.44 | 52.68 |
| 1999 | 5.77 | 1.96 | 3.69 | 3.53 | 0.58 | 1.16 | 3.00 | 3.20 | 21.79 | 2.45 | 1.20 | 2.31 | 50.64 |
| 2000 | 6.27 | 2.20 | 1.76 | 4.66 | 1.23 | 2.50 | 6.19 | 6.64 | 3.82 | T | 2.56 | 1.51 | 39.34 |
| 2001 | 1.30 | 2.34 | 7.11 | 1.72 | 3.53 | 4.54 | 4.13 | 4.88 | 0.86 | 1.86 | 0.50 | 2.01 | 34.78 |
| 2002 | 5.97 | 1.27 | 4.18 | 1.13 | 1.13 | 2.75 | 4.76 | 4.71 | 3.49 | 9.35 | 3.56 | 5.04 | 47.34 |
| 2003 | 1.84 | 4.64 | 5.23 | 4.47 | 4.28 | 4.16 | 4.40 | 8.57 | 4.47 | 2.62 | 1.81 | 3.52 | 50.01 |
| 2004 | 1.23 | 3.32 | 3.31 | 1.72 | 3.44 | 4.22 | 8.16 | 9.26 | 4.49 | 2.51 | 3.89 | 1.48 | 47.03 |
| 2005 | 2.64 | 2.29 | 3.54 | 3.13 | 2.07 | 1.50 | 7.64 | 3.90 | 0.82 | 2.06 | 3.72 | 4.24 | 37.55 |
| 2006 | 2.07 | 1.65 | 1.25 | 4.93 | 1.58 | 10.45 | 3.31 | 4.42 | 8.53 | 3.47 | 9.03 | 3.00 | 53.69 |
| 2007 | 3.12 | 1.74 | 3.52 | 3.88 | 1.43 | 4.46 | 4.94 | 0.90 | 2.22 | 4.66 | 0.48 | 4.45 | 35.80 |
| 2008 | 1.26 | 3.16 | 5.53 | 3.92 | 3.12 | 4.08 | 5.96 | 5.92 | 9.24 | 1.17 | 4.05 | 3.06 | 50.47 |
| 2009 | 2.53 | 1.65 | 6.83 | 1.69 | 4.13 | 2.34 | 2.16 | 1.65 | 3.35 | 1.10 | 6.91 | 6.09 | 40.43 |
| POR= 62 YRS | 3.51 | 3.32 | 3.85 | 2.91 | 3.53 | 3.60 | 4.60 | 4.40 | 3.82 | 2.99 | 3.07 | 3.12 | 42.72 |

WBAN : 13722

## AVERAGE TEMPERATURE (°F) 2009 RALEIGH/DURHAM (KRDU)

| YEAR | JAN | FEB | MAR | APR | MAY | JUN | JUL | AUG | SEP | OCT | NOV | DEC | ANNUAL |
|------|------|------|------|------|------|------|------|------|------|------|------|------|--------|
| 1980 | 40.6 | 36.5 | 46.5 | 62.0 | 69.8 | 75.0 | 78.9 | 79.6 | 74.9 | 58.7 | 49.1 | 40.7 | 59.4 |
| 1981 | 33.4 | 43.9 | 46.2 | 61.7 | 64.0 | 78.9 | 80.8 | 74.6 | 68.3 | 57.2 | 50.7 | 39.7 | 58.3 |
| 1982 | 35.5 | 45.5 | 51.7 | 57.4 | 71.0 | 74.7 | 79.1 | 76.5 | 70.5 | 60.6 | 51.9 | 47.5 | 60.2 |
| 1983 | 38.1 | 40.7 | 50.7 | 55.1 | 65.4 | 72.5 | 79.1 | 79.1 | 70.7 | 60.4 | 50.9 | 39.5 | 58.5 |
| 1984 | 36.3 | 45.7 | 47.2 | 55.9 | 65.5 | 75.5 | 74.9 | 76.6 | 67.5 | 66.3 | 47.1 | 49.7 | 59.0 |
| 1985 | 34.0 | 41.9 | 52.7 | 62.0 | 67.3 | 73.9 | 76.8 | 75.2 | 69.7 | 63.7 | 58.4 | 39.4 | 59.6 |
| 1986 | 38.5 | 44.5 | 51.7 | 61.2 | 67.4 | 78.4 | 81.7 | 75.6 | 72.3 | 63.0 | 52.9 | 42.6 | 60.8 |
| 1987 | 38.3 | 40.4 | 49.2 | 56.9 | 69.3 | 76.3 | 81.2 | 79.2 | 73.0 | 54.6 | 52.8 | 44.5 | 59.6 |
| 1988 | 34.7 | 41.9 | 50.6 | 57.8 | 66.0 | 72.4 | 79.0 | 80.3 | 70.3 | 54.4 | 52.0 | 42.4 | 58.5 |
| 1989 | 44.8 | 42.9 | 50.7 | 58.0 | 65.0 | 77.0 | 78.1 | 76.2 | 71.7 | 61.3 | 51.4 | 34.6 | 59.3 |
| 1990 | 48.0 | 51.0 | 54.9 | 60.3 | 67.4 | 75.2 | 80.0 | 78.0 | 72.2 | 64.1 | 54.2 | 48.1 | 62.8 |
| 1991 | 41.9 | 46.8 | 54.3 | 62.3 | 72.3 | 75.8 | 80.6 | 77.8 | 71.6 | 61.1 | 50.0 | 47.0 | 61.8 |
| 1992 | 43.4 | 46.2 | 50.5 | 59.2 | 63.2 | 72.0 | 80.4 | 74.1 | 71.5 | 58.1 | 51.9 | 42.5 | 59.4 |
| 1993 | 43.3 | 40.9 | 47.9 | 56.9 | 69.0 | 76.1 | 82.5 | 78.2 | 73.7 | 59.6 | 51.8 | 40.3 | 60.0 |
| 1994 | 36.9 | 44.2 | 52.2 | 63.6 | 64.2 | 77.0 | 79.6 | 76.0 | 69.1 | 59.3 | 54.0 | 47.9 | 60.3 |
| 1995 | 42.9 | 41.0 | 52.7 | 61.0 | 68.5 | 73.8 | 80.4 | 80.6 | 70.5 | 63.4 | 47.1 | 39.5 | 60.1 |
| 1996 | 38.3 | 42.5 | 45.3 | 58.4 | 67.7 | 76.4 | 78.6 | 75.2 | 70.3 | 61.0 | 45.7 | 45.8 | 58.8 |
| 1997 | 41.3 | 46.7 | 54.6 | 55.2 | 63.9 | 76.1 | 79.5 | 76.6 | 70.9 | 59.6 | 47.3 | 40.9 | 59.0 |
| 1998 | 44.7 | 46.1 | 50.2 | 58.8 | 69.2 | 77.3 | 79.5 | 77.8 | 74.6 | 61.1 | 52.0 | 46.6 | 61.5 |
| 1999 | 45.7 | 45.7 | 47.5 | 61.5 | 66.8 | 74.5 | 81.3 | 80.0 | 69.1 | 58.6 | 54.4 | 43.9 | 60.8 |
| 2000 | 38.4 | 45.8 | 53.4 | 57.6 | 70.0 | 77.2 | 77.0 | 76.1 | 69.4 | 60.2 | 48.0 | 35.5 | 59.1 |
| 2001 | 41.6 | 47.2 | 49.1 | 60.8 | 68.0 | 76.9 | 76.2 | 79.6 | 69.0 | 59.4 | 55.8 | 47.6 | 60.9 |
| 2002 | 42.7 | 44.6 | 52.6 | 63.3 | 67.0 | 77.9 | 80.6 | 78.4 | 73.1 | 61.6 | 49.1 | 40.5 | 61.0 |
| 2003 | 36.7 | 41.1 | 53.3 | 58.2 | 66.3 | 73.7 | 78.3 | 78.9 | 69.9 | 59.6 | 56.5 | 41.0 | 59.5 |
| 2004 | 38.0 | 41.4 | 52.6 | 60.4 | 73.3 | 76.2 | 79.2 | 75.3 | 71.3 | 62.6 | 53.6 | 43.5 | 60.6 |
| 2005 | 43.7 | 44.6 | 47.8 | 59.2 | 64.8 | 76.3 | 82.3 | 80.2 | 76.0 | 62.6 | 52.9 | 40.7 | 60.9 |
| 2006 | 47.8 | 43.5 | 51.5 | 62.7 | 65.6 | 74.6 | 79.8 | 80.5 | 69.7 | 58.8 | 53.0 | 47.8 | 61.3 |
| 2007 | 45.2 | 40.8 | 55.9 | 60.0 | 68.2 | 76.5 | 78.6 | 84.1 | 75.0 | 67.4 | 50.7 | 48.5 | 62.6 |
| 2008 | 41.6 | 47.8 | 53.3 | 60.4 | 66.5 | 80.7 | 79.0 | 78.5 | 72.6 | 59.3 | 49.4 | 47.4 | 61.4 |
| 2009 | 40.1 | 46.1 | 52.2 | 62.4 | 71.3 | 78.4 | 79.6 | 80.4 | 70.9 | 60.3 | 53.2 | 40.5 | 61.3 |
| POR= 62 YRS | 40.3 | 42.9 | 50.2 | 59.5 | 67.3 | 74.9 | 78.6 | 77.4 | 71.0 | 60.2 | 50.8 | 42.6 | 59.6 |

## HEATING DEGREE DAYS (base 65°F) 2009 RALEIGH/DURHAM (KRDU)

| YEAR | JUL | AUG | SEP | OCT | NOV | DEC | JAN | FEB | MAR | APR | MAY | JUN | TOTAL |
|------|-----|-----|-----|-----|-----|-----|-----|-----|-----|-----|-----|-----|-------|
| 1980-81 | 0 | 0 | 16 | 225 | 477 | 747 | 973 | 583 | 579 | 149 | 99 | 0 | 3848 |
| 1981-82 | 0 | 4 | 31 | 253 | 425 | 776 | 907 | 538 | 411 | 244 | 15 | 0 | 3604 |
| 1982-83 | 0 | 0 | 14 | 182 | 392 | 542 | 828 | 675 | 438 | 305 | 79 | 7 | 3462 |
| 1983-84 | 0 | 0 | 59 | 180 | 417 | 784 | 882 | 553 | 545 | 283 | 83 | 5 | 3791 |
| 1984-85 | 0 | 0 | 63 | 42 | 530 | 468 | 954 | 644 | 395 | 146 | 42 | 4 | 3288 |
| 1985-86 | 0 | 0 | 36 | 96 | 207 | 789 | 812 | 569 | 415 | 157 | 59 | 0 | 3140 |
| 1986-87 | 0 | 11 | 12 | 149 | 370 | 687 | 820 | 681 | 484 | 248 | 29 | 0 | 3491 |
| 1987-88 | 0 | 0 | 1 | 319 | 362 | 631 | 932 | 665 | 444 | 228 | 62 | 22 | 3666 |
| 1988-89 | 0 | 0 | 8 | 336 | 386 | 695 | 619 | 623 | 459 | 257 | 102 | 0 | 3485 |
| 1989-90 | 0 | 3 | 30 | 167 | 404 | 934 | 518 | 390 | 357 | 186 | 37 | 0 | 3026 |
| 1990-91 | 0 | 0 | 18 | 124 | 323 | 520 | 709 | 501 | 354 | 153 | 18 | 0 | 2720 |
| 1991-92 | 0 | 0 | 24 | 156 | 451 | 562 | 659 | 537 | 446 | 226 | 114 | 3 | 3178 |
| 1992-93 | 0 | 0 | 29 | 224 | 390 | 691 | 666 | 670 | 524 | 244 | 15 | 0 | 3453 |
| 1993-94 | 0 | 0 | 19 | 198 | 405 | 758 | 863 | 576 | 394 | 113 | 101 | 0 | 3427 |
| 1994-95 | 0 | 0 | 8 | 190 | 326 | 527 | 679 | 665 | 380 | 165 | 36 | 0 | 2976 |
| 1995-96 | 0 | 0 | 18 | 123 | 532 | 787 | 819 | 646 | 601 | 227 | 72 | 2 | 3827 |
| 1996-97 | 0 | 0 | 9 | 142 | 569 | 590 | 729 | 510 | 326 | 300 | 97 | 45 | 3317 |
| 1997-98 | 0 | 0 | 14 | 223 | 524 | 742 | 624 | 525 | 475 | 207 | 21 | 7 | 3362 |
| 1998-99 | 0 | 0 | 7 | 146 | 383 | 573 | 593 | 533 | 536 | 160 | 36 | 2 | 2969 |
| 1999-00 | 2 | 0 | 29 | 216 | 314 | 646 | 818 | 548 | 365 | 235 | 24 | 0 | 3197 |
| 2000-01 | 0 | 0 | 40 | 181 | 511 | 910 | 718 | 492 | 490 | 200 | 36 | 0 | 3578 |
| 2001-02 | 0 | 0 | 50 | 211 | 287 | 532 | 687 | 566 | 393 | 151 | 91 | 0 | 2968 |
| 2002-03 | 0 | 0 | 0 | 186 | 479 | 754 | 871 | 663 | 365 | 216 | 55 | 5 | 3594 |
| 2003-04 | 0 | 0 | 16 | 178 | 274 | 739 | 830 | 677 | 387 | 190 | 26 | 0 | 3317 |
| 2004-05 | 0 | 0 | 7 | 120 | 354 | 660 | 662 | 564 | 527 | 193 | 77 | 3 | 3167 |
| 2005-06 | 0 | 0 | 0 | 154 | 368 | 748 | 524 | 595 | 432 | 135 | 76 | 0 | 3032 |
| 2006-07 | 0 | 0 | 19 | 217 | 357 | 524 | 603 | 670 | 312 | 207 | 59 | 0 | 2968 |
| 2007-08 | 0 | 0 | 3 | 73 | 419 | 511 | 718 | 500 | 359 | 169 | 37 | 0 | 2789 |
| 2008-09 | 0 | 0 | 8 | 205 | 459 | 539 | 765 | 524 | 408 | 130 | 29 | 0 | 3067 |
| 2009- | 0 | 0 | 8 | 176 | 349 | 752 | | | | | | | |

WBAN : 13722

## COOLING DEGREE DAYS (base 65°F) 2009 RALEIGH/DURHAM (KRDU)

| YEAR | JAN | FEB | MAR | APR | MAY | JUN | JUL | AUG | SEP | OCT | NOV | DEC | TOTAL |
|------|-----|-----|-----|-----|-----|-----|-----|-----|-----|-----|-----|-----|-------|
| 1980 | 0 | 0 | 0 | 45 | 190 | 306 | 441 | 460 | 321 | 38 | 6 | 0 | 1807 |
| 1981 | 0 | 0 | 2 | 56 | 75 | 425 | 497 | 309 | 139 | 19 | 0 | 0 | 1522 |
| 1982 | 0 | 0 | 3 | 24 | 208 | 299 | 443 | 363 | 183 | 53 | 5 | 7 | 1588 |
| 1983 | 0 | 0 | 0 | 16 | 97 | 238 | 441 | 447 | 239 | 42 | 0 | 0 | 1520 |
| 1984 | 0 | 0 | 0 | 16 | 108 | 324 | 311 | 366 | 143 | 90 | 0 | 0 | 1358 |
| 1985 | 0 | 3 | 20 | 65 | 121 | 277 | 373 | 323 | 181 | 64 | 14 | 0 | 1441 |
| 1986 | 0 | 0 | 7 | 51 | 142 | 408 | 526 | 349 | 237 | 96 | 15 | 0 | 1831 |
| 1987 | 0 | 0 | 0 | 11 | 170 | 347 | 508 | 447 | 250 | 0 | 2 | 0 | 1735 |
| 1988 | 0 | 3 | 5 | 17 | 98 | 249 | 438 | 482 | 172 | 14 | 3 | 0 | 1481 |
| 1989 | 0 | 11 | 23 | 54 | 110 | 367 | 412 | 359 | 237 | 59 | 4 | 0 | 1636 |
| 1990 | 0 | 3 | 49 | 51 | 117 | 312 | 472 | 410 | 239 | 102 | 4 | 5 | 1764 |
| 1991 | 0 | 0 | 28 | 78 | 253 | 333 | 493 | 403 | 230 | 40 | 9 | 13 | 1880 |
| 1992 | 0 | 0 | 3 | 58 | 65 | 218 | 487 | 285 | 228 | 15 | 3 | 0 | 1362 |
| 1993 | 0 | 0 | 0 | 12 | 147 | 338 | 550 | 417 | 286 | 34 | 17 | 0 | 1801 |
| 1994 | 0 | 0 | 6 | 78 | 86 | 367 | 462 | 350 | 139 | 21 | 3 | 1 | 1513 |
| 1995 | 3 | 0 | 3 | 53 | 154 | 270 | 486 | 490 | 191 | 81 | 3 | 0 | 1734 |
| 1996 | 0 | 0 | 0 | 35 | 164 | 352 | 427 | 324 | 176 | 26 | 0 | 0 | 1504 |
| 1997 | 1 | 3 | 14 | 11 | 68 | 249 | 451 | 368 | 199 | 64 | 0 | 0 | 1428 |
| 1998 | 3 | 0 | 23 | 29 | 158 | 382 | 457 | 401 | 302 | 32 | 0 | 12 | 1799 |
| 1999 | 3 | 0 | 0 | 64 | 100 | 292 | 513 | 474 | 161 | 24 | 0 | 0 | 1631 |
| 2000 | 0 | 0 | 9 | 19 | 185 | 374 | 380 | 351 | 176 | 38 | 8 | 0 | 1540 |
| 2001 | 0 | 0 | 3 | 83 | 137 | 364 | 356 | 461 | 176 | 46 | 17 | 2 | 1645 |
| 2002 | 4 | 0 | 15 | 104 | 160 | 391 | 489 | 422 | 253 | 86 | 6 | 0 | 1930 |
| 2003 | 0 | 0 | 7 | 19 | 101 | 274 | 419 | 436 | 171 | 16 | 24 | 0 | 1467 |
| 2004 | 0 | 0 | 10 | 58 | 294 | 342 | 445 | 329 | 200 | 52 | 17 | 0 | 1747 |
| 2005 | 6 | 0 | 0 | 23 | 80 | 351 | 544 | 481 | 339 | 85 | 12 | 0 | 1921 |
| 2006 | 0 | 0 | 18 | 75 | 101 | 294 | 456 | 485 | 168 | 29 | 5 | 0 | 1631 |
| 2007 | 1 | 0 | 34 | 65 | 163 | 352 | 430 | 602 | 311 | 153 | 1 | 7 | 2119 |
| 2008 | 0 | 4 | 6 | 36 | 90 | 475 | 441 | 426 | 244 | 36 | 0 | 2 | 1760 |
| 2009 | 0 | 0 | 18 | 60 | 231 | 406 | 458 | 483 | 189 | 36 | 1 | 0 | 1882 |

## SNOWFALL (inches) 2009 RALEIGH/DURHAM (KRDU)

| YEAR | JUL | AUG | SEP | OCT | NOV | DEC | JAN | FEB | MAR | APR | MAY | JUN | TOTAL |
|------|-----|-----|-----|-----|-----|-----|-----|-----|-----|-----|-----|-----|-------|
| 1980-81 | 0.0 | 0.0 | 0.0 | 0.0 | 0.0 | 3.1 | 2.6 | 0.0 | T | 0.0 | 0.0 | 0.0 | 5.7 |
| 1981-82 | 0.0 | 0.0 | 0.0 | 0.0 | 0.0 | T | 6.0 | 0.6 | 0.0 | 0.0 | 0.0 | 0.0 | 6.6 |
| 1982-83 | 0.0 | 0.0 | 0.0 | 0.0 | 0.0 | T | T | 2.7 | 7.3 | 1.8 | 0.0 | 0.0 | 11.8 |
| 1983-84 | 0.0 | 0.0 | 0.0 | 0.0 | 0.0 | 0.0 | T | 6.9 | T | 0.0 | 0.0 | 0.0 | 6.9 |
| 1984-85 | 0.0 | 0.0 | 0.0 | 0.0 | 0.0 | 0.0 | 4.1 | T | 0.0 | 0.0 | 0.0 | 0.0 | 4.1 |
| 1985-86 | 0.0 | 0.0 | 0.0 | 0.0 | 0.0 | T | T | 0.9 | T | 0.0 | 0.0 | 0.0 | 0.9 |
| 1986-87 | 0.0 | 0.0 | 0.0 | 0.0 | T | 0.0 | 0.6 | 10.2 | T | T | 0.0 | 0.0 | 10.8 |
| 1987-88 | 0.0 | 0.0 | 0.0 | 0.0 | 0.6 | 0.0 | 7.3 | T | 0.0 | 0.0 | 0.0 | 0.0 | 7.9 |
| 1988-89 | 0.0 | 0.0 | 0.0 | 0.0 | 0.0 | 0.1 | 0.0 | 11.1 | 0.5 | 0.3 | 0.0 | 0.0 | 12.0 |
| 1989-90 | 0.0 | 0.0 | 0.0 | 0.0 | 0.0 | 2.7 | 0.0 | T | T | 0.0 | 0.0 | 0.0 | 2.7 |
| 1990-91 | 0.0 | 0.0 | 0.0 | 0.0 | 0.0 | T | T | T | T | 0.0 | 0.0 | 0.0 | T |
| 1991-92 | 0.0 | 0.0 | 0.0 | 0.0 | T | 0.0 | T | T | 0.0 | 0.0 | 0.0 | 0.0 | T |
| 1992-93 | 0.0 | 0.0 | 0.0 | 0.0 | 0.0 | T | T | 1.6 | 0.9 | 0.0 | 0.0 | 0.0 | 2.5 |
| 1993-94 | T | 0.0 | 0.0 | 0.0 | 0.0 | 3.1 | T | 1.1 | 0.2 | 0.0 | 0.0 | 0.0 | 4.4 |
| 1994-95 | 0.0 | 0.0 | 0.0 | 0.0 | 0.0 | 0.0 | 0.8 | 1.4 | T | 0.0 | T | 0.0 | 2.2 |
| 1995-96 | 0.0 | 0.0 | 0.0 | 0.0 | T | T | 5.6 | 8.1 | 0.9 | 0.0 | T | T | 14.6 |
| 1996-97 |  |  |  |  |  |  | 0.4 | T | 0.0 | 0.0 | T | T |  |
| 1997-98 | 0.0 | 0.0 | 0.0 | 0.0 | 0.0 | 0.4 | 2.0 | 0.0 | T | 0.0 | 0.0 | T | 2.4 |
| 1998-99 | 0.0 | 0.0 | 0.0 | 0.0 | 0.0 | T | T | T | T | 0.0 | 0.0 | 0.0 | T |
| 1999-00 | 0.0 | 0.0 | 0.0 | 0.0 | 0.0 | T | 25.8 | 0.0 | 0.0 | 0.0 | 0.0 | T | 25.8 |
| 2000-01 | 0.0 | 0.0 | 0.0 | 0.0 | 2.2 | 0.1 | T | 0.3 | T | T | T | 0.0 | 2.6 |
| 2001-02 | 0.0 | 0.0 | 0.0 | 0.0 | 0.0 | 0.0 | 10.8 | T | 0.0 | 0.0 | 0.0 | 0.0 | 10.8 |
| 2002-03 | 0.0 | 0.0 | 0.0 | 0.0 | 0.0 | 2.3 | 3.5 | 1.6 | T | 0.0 | 0.0 | 0.0 | 7.4 |
| 2003-04 | 0.0 | 0.0 | 0.0 | 0.0 | 0.0 | T | 5.7 | 9.2 | T | 0.0 | 0.0 | 0.0 | 14.9 |
| 2004-05 | 0.0 | 0.0 | 0.0 | 0.0 | 0.0 | T | 0.7 | T | 0.2 | 0.0 | 0.0 | 0.0 | 0.9 |
| 2005-06 | T | 0.0 | 0.0 | 0.0 | 0.0 | T | T | T | T | T | T | 0.0 | T |
| 2006-07 | 0.0 | 0.0 | 0.0 | 0.0 | T | 0.0 | 1.0 | 0.6 | 0.0 | T | 0.0 | 0.0 | 1.6 |
| 2007-08 | 0.0 | 0.0 | 0.0 | 0.0 | 0.0 | T | 0.5 | T | 0.0 | 0.0 | 0.0 | 0.0 | 0.5 |
| 2008-09 | 0.0 | 0.0 | 0.0 | 0.0 | 0.4 | 0.0 | 3.5 | T | 3.2 | 0.0 | 0.0 | 0.0 | 7.1 |
| 2009- | 0.0 | 0.0 | 0.0 | 0.0 | 0.0 | 0.1 |  |  |  |  |  |  |  |
| POR= 61 YRS | T | 0.0 | 0.0 | 0.0 | 0.2 | 0.6 | 2.7 | 2.4 | 1.1 | T | T | T | 7.0 |

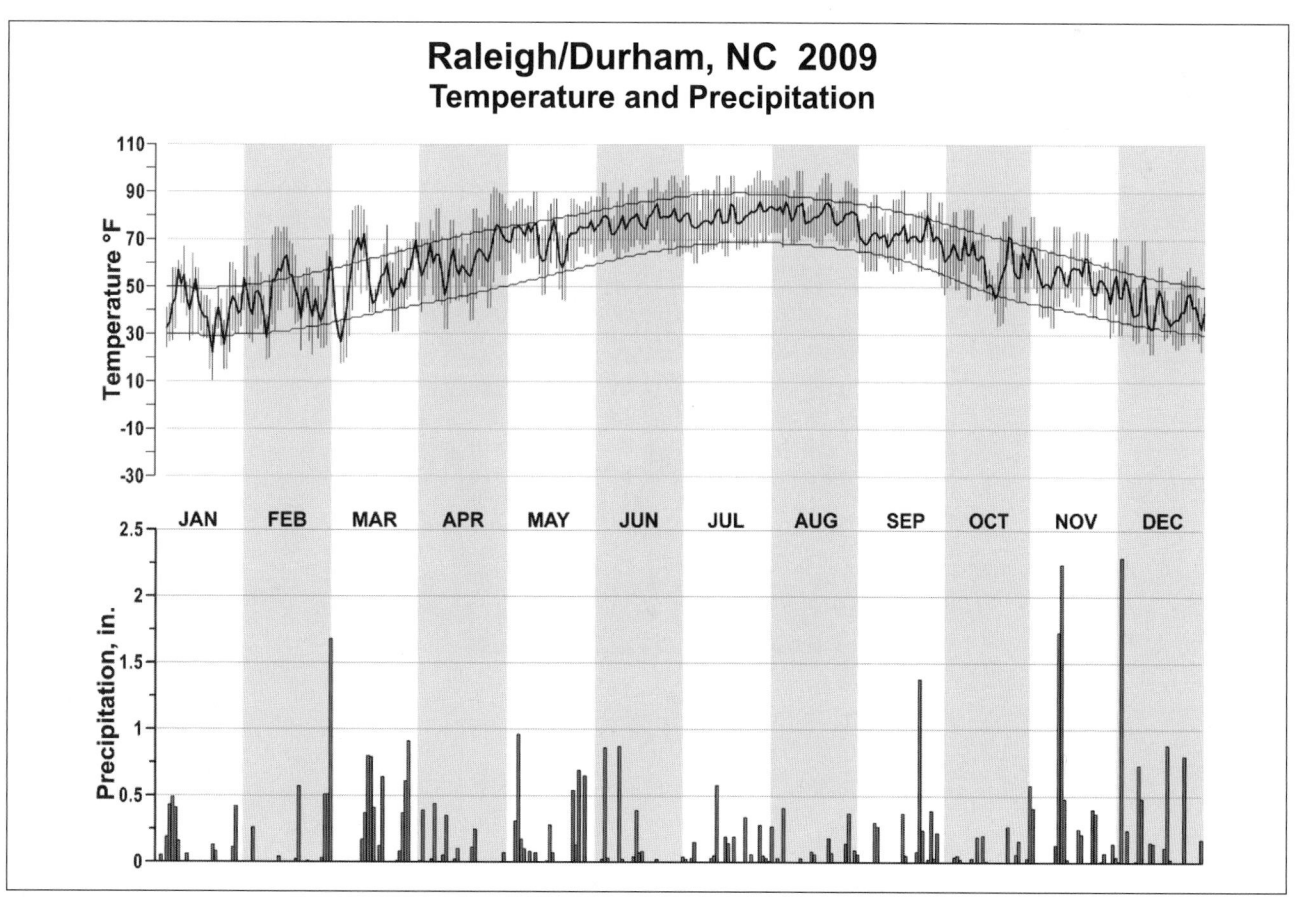

# Raleigh/Durham, NC  2009
## Temperature and Precipitation

# 2009
# FARGO
# NORTH DAKOTA (KFAR)

Moorhead, Minnesota, and Fargo are twin cities in the Red River Valley of the north. The Red River of the north flows northward between the two cities and is a part of the Hudson Bay drainage area. The Red River is approximately 2 miles east of the airport at its nearest point and has no significant effect on the weather. In recent years, spring floods due to melting snow have been common. Summer floods caused by heavy rains are infrequent.

The surrounding terrain is flat and open. Northerly winds blowing up the valley occasionally causing low cloudiness and fog. However, this upslope cloudiness is very infrequent. Aside from this, there are no pronounced climatic differences due to geographical features in the immediate area.

The summers are generally comfortable with very few days of hot and humid weather. Nights, with few exceptions, are comfortably cool. The winter months are cold and dry with temperatures rising above freezing only on an average of six days each month, and nighttime lows dropping below zero approximately half of the time.

Precipitation is the most important climatic factor in the area. The Red River Valley lies in an area where lighter amounts fall to the west and heavier amounts to the east. Seventy-five percent of the precipitation occurs during the growing season (April to September) and is often accompanied by electrical storms and heavy falls in a short time. Winter precipitation is light, indicating that heavy snowfall is the exception rather than the rule. The first light snow in the fall occasionally falls in September, but usually very little, if any, occurs until October or November. The latest fall is generally in April.

With the flat terrain, surface friction has little effect on the wind in the area and this fact has led to the legendary Dakota blizzards. Strong winds with even light snowfall cause much drifting and blowing snow, reducing visibility to near zero. Fortunately, these conditions occur only several times during the winter months.

# NORMALS, MEANS, AND EXTREMES
## FARGO (KFAR)

LATITUDE: 46 ° 55'N  LONGITUDE: -96 ° 48'W  ELEVATION (FT): GRND: 895  BARO: 911  TIME ZONE: CENTRAL  (UTC -6)  WBAN: 14914

| | ELEMENT | POR | JAN | FEB | MAR | APR | MAY | JUN | JUL | AUG | SEP | OCT | NOV | DEC | YEAR |
|---|---|---|---|---|---|---|---|---|---|---|---|---|---|---|---|
| TEMPERATURE °F | NORMAL DAILY MAXIMUM | 30 | 15.9 | 22.8 | 35.3 | 54.5 | 69.5 | 77.4 | 82.2 | 81.0 | 69.9 | 56.1 | 35.2 | 20.8 | 51.7 |
| | MEAN DAILY MAXIMUM | 68 | 16.1 | 21.5 | 34.5 | 53.6 | 68.0 | 76.4 | 82.5 | 81.1 | 70.0 | 56.9 | 36.7 | 21.7 | 51.6 |
| | HIGHEST DAILY MAXIMUM | 57 | 52 | 66 | 78 | 100 | 98 | 100 | 106 | 106 | 102 | 93 | 74 | 57 | 106 |
| | YEAR OF OCCURRENCE | | 1981 | 1958 | 1967 | 1980 | 1964 | 1995 | 1988 | 1976 | 1959 | 1963 | 1990 | 1962 | JUL 1988 |
| | MEAN OF EXTREME MAXS. | 68 | 37.6 | 41.3 | 54.7 | 78.6 | 87.9 | 91.9 | 94.9 | 94.9 | 89.9 | 79.2 | 59.4 | 41.6 | 71.0 |
| | NORMAL DAILY MINIMUM | 30 | -2.3 | 5.4 | 19.0 | 32.4 | 45.3 | 54.5 | 59.0 | 57.0 | 46.1 | 34.4 | 18.7 | 4.2 | 31.1 |
| | MEAN DAILY MINIMUM | 68 | -1.7 | 3.2 | 16.9 | 32.0 | 43.6 | 53.8 | 58.9 | 57.0 | 46.5 | 35.1 | 19.9 | 5.1 | 30.9 |
| | LOWEST DAILY MINIMUM | 57 | -36 | -39 | -23 | -7 | 20 | 30 | 36 | 33 | 19 | 7 | -24 | -32 | -39 |
| | YEAR OF OCCURRENCE | | 2004 | 1996 | 1980 | 1975 | 2005 | 1969 | 1967 | 1982 | 1965 | 1976 | 1985 | 1967 | FEB 1996 |
| | MEAN OF EXTREME MINS. | 68 | -23.5 | -20.0 | -8.4 | 15.1 | 27.4 | 39.9 | 46.2 | 43.1 | 30.4 | 19.2 | -1.0 | -16.9 | 12.6 |
| | NORMAL DRY BULB | 30 | 6.8 | 14.1 | 27.2 | 43.5 | 57.4 | 66.0 | 70.6 | 69.0 | 58.0 | 45.3 | 27.0 | 12.5 | 41.5 |
| | MEAN DRY BULB | 68 | 7.2 | 12.4 | 25.7 | 42.8 | 55.8 | 65.2 | 70.7 | 69.0 | 58.3 | 46.0 | 28.3 | 13.4 | 41.2 |
| | MEAN WET BULB | 26 | 8.2 | 13.0 | 24.6 | 37.3 | 48.4 | 58.3 | 62.9 | 61.1 | 52.4 | 39.6 | 25.2 | 13.0 | 37.0 |
| | MEAN DEW POINT | 26 | 5.2 | 10.0 | 21.3 | 31.6 | 42.8 | 54.7 | 59.9 | 57.7 | 48.6 | 35.1 | 22.2 | 10.3 | 33.3 |
| | NORMAL NO. DAYS WITH: | | | | | | | | | | | | | | |
| | MAXIMUM >= 90 | 30 | 0.0 | 0.0 | 0.0 | 0.2 | 0.8 | 2.3 | 4.5 | 4.5 | 1.2 | 0.1 | 0.0 | 0.0 | 13.6 |
| | MAXIMUM <= 32 | 30 | 27.1 | 20.2 | 11.3 | 1.0 | 0.0 | 0.0 | 0.0 | 0.0 | 0.0 | 0.6 | 12.8 | 24.4 | 97.4 |
| | MINIMUM <= 32 | 30 | 31.0 | 27.8 | 27.0 | 16.0 | 3.0 | 0.0 | 0.0 | 0.0 | 1.7 | 12.9 | 27.2 | 30.9 | 177.5 |
| | MINIMUM <= 0 | 30 | 17.5 | 11.1 | 3.9 | 0.1 | 0.0 | 0.0 | 0.0 | 0.0 | 0.0 | 0.0 | 2.2 | 12.8 | 47.6 |
| H/C | NORMAL HEATING DEG. DAYS | 30 | 1808 | 1441 | 1185 | 652 | 271 | 73 | 17 | 37 | 245 | 614 | 1137 | 1612 | 9092 |
| | NORMAL COOLING DEG. DAYS | 30 | 0 | 0 | 0 | 3 | 33 | 104 | 191 | 162 | 38 | 2 | 0 | 0 | 533 |
| RH | NORMAL (PERCENT) | 30 | 76 | 77 | 77 | 63 | 59 | 66 | 68 | 67 | 68 | 68 | 76 | 77 | 70 |
| | HOUR 00 LST | 30 | 77 | 79 | 81 | 71 | 67 | 75 | 79 | 77 | 77 | 74 | 79 | 79 | 76 |
| | HOUR 06 LST | 30 | 77 | 79 | 83 | 78 | 76 | 82 | 86 | 87 | 84 | 80 | 82 | 79 | 81 |
| | HOUR 12 LST | 30 | 74 | 74 | 71 | 54 | 49 | 55 | 56 | 55 | 57 | 59 | 71 | 74 | 62 |
| | HOUR 18 LST | 30 | 76 | 76 | 71 | 51 | 45 | 51 | 53 | 51 | 54 | 60 | 74 | 77 | 62 |
| S | PERCENT POSSIBLE SUNSHINE | 54 | 50 | 56 | 58 | 60 | 61 | 62 | 71 | 69 | 60 | 54 | 40 | 43 | 57 |
| W/O | MEAN NO. DAYS WITH: | | | | | | | | | | | | | | |
| | HEAVY FOG(VISBY <= 1/4 MI) | 46 | 1.3 | 1.6 | 2.5 | 1.0 | 0.5 | 0.7 | 1.0 | 1.2 | 1.0 | 1.0 | 1.6 | 2.0 | 15.4 |
| | THUNDERSTORMS | 62 | 0.0 | 0.1 | 0.2 | 1.1 | 3.4 | 6.4 | 7.3 | 6.1 | 2.8 | 0.9 | 0.1 | 0.0 | 28.4 |
| CLOUDNESS | MEAN: | | | | | | | | | | | | | | |
| | SUNRISE-SUNSET (OKTAS) | | | | | | | | | | | | | | |
| | MIDNIGHT-MIDNIGHT (OKTAS) | | | | | | | | | | | | | | |
| | MEAN NO. DAYS WITH: | | | | | | | | | | | | | | |
| | CLEAR | 1 | 2.0 | 2.0 | 7.0 | | 3.0 | 9.0 | | | | | | | |
| | PARTLY CLOUDY | | | 1.0 | 3.0 | | 3.0 | 5.0 | | | | | | | |
| | CLOUDY | 1 | 1.0 | 4.0 | 12.0 | | 7.0 | 5.0 | | | | | | | |
| PR | MEAN STATION PRESSURE(IN) | 26 | 29.09 | 29.11 | 29.07 | 29.00 | 28.95 | 28.93 | 28.98 | 29.01 | 29.00 | 29.02 | 29.03 | 29.08 | 29.02 |
| | MEAN SEA-LEVEL PRES. (IN) | 26 | 30.12 | 30.13 | 30.08 | 29.98 | 29.92 | 29.89 | 29.93 | 29.96 | 29.97 | 30.00 | 30.04 | 30.10 | 30.01 |
| WINDS | MEAN SPEED (MPH) | 26 | 11.4 | 11.7 | 12.2 | 12.4 | 12.2 | 10.5 | 9.4 | 9.8 | 10.6 | 11.5 | 11.5 | 11.4 | 11.2 |
| | PREVAIL.DIR(TENS OF DEGS) | 38 | 17 | 17 | 35 | 35 | 17 | 17 | 17 | 17 | 17 | 17 | 17 | 17 | 17 |
| | MAXIMUM 2-MINUTE: | | | | | | | | | | | | | | |
| | SPEED (MPH) | 14 | 49 | 51 | 49 | 49 | 51 | 48 | 74 | 51 | 40 | 49 | 47 | 43 | 74 |
| | DIR. (TENS OF DEGS) | | 34 | 33 | 34 | 32 | 18 | 29 | 33 | 34 | 30 | 33 | 31 | 35 | 33 |
| | YEAR OF OCCURRENCE | | 1996 | 1996 | 1999 | 2000 | 2004 | 2008 | 1999 | 2001 | 2005 | 1996 | 1999 | 2008 | JUL 1999 |
| | MAXIMUM 3-SECOND | | | | | | | | | | | | | | |
| | SPEED (MPH) | 14 | 56 | 58 | 55 | 59 | 63 | 63 | 91 | 58 | 48 | 61 | 56 | 54 | 91 |
| | DIR. (TENS OF DEGS) | | 34 | 28 | 34 | 32 | 18 | 26 | 34 | 34 | 18 | 33 | 30 | 35 | 34 |
| | YEAR OF OCCURRENCE | | 1997 | 2002 | 1999 | 2000 | 2004 | 2008 | 1999 | 2001 | 2008 | 1996 | 1999 | 2008 | JUL 1999 |
| PRECIPITATION | NORMAL (IN) | 30 | 0.76 | 0.59 | 1.17 | 1.37 | 2.61 | 3.51 | 2.88 | 2.52 | 2.18 | 1.97 | 1.06 | 0.57 | 21.19 |
| | MAXIMUM MONTHLY (IN) | 68 | 1.85 | 1.74 | 4.62 | 5.28 | 7.34 | 11.72 | 8.42 | 8.52 | 6.50 | 7.03 | 4.58 | 2.19 | 11.72 |
| | YEAR OF OCCURRENCE | | 1989 | 1979 | 2009 | 1986 | 1998 | 2000 | 1952 | 1944 | 1999 | 1982 | 1977 | 1951 | JUN 2000 |
| | MINIMUM MONTHLY (IN) | 68 | 0.09 | 0.03 | 0.03 | 0.01 | 0.46 | 0.58 | 0.42 | 0.18 | 0.13 | 0.05 | 0.02 | 0.04 | 0.01 |
| | YEAR OF OCCURRENCE | | 2008 | 1954 | 1958 | 1988 | 1976 | 1972 | 1950 | 1984 | 1974 | 1986 | 1999 | 1958 | APR 1988 |
| | MAXIMUM IN 24 HOURS (IN) | 68 | 1.19 | 1.22 | 1.16 | 1.91 | 4.10 | 4.64 | 5.10 | 4.72 | 3.97 | 3.22 | 1.99 | 0.87 | 5.10 |
| | YEAR OF OCCURRENCE | | 1996 | 1946 | 1950 | 1963 | 1977 | 2000 | 1993 | 1943 | 1957 | 1982 | 1977 | 1960 | JUL 1993 |
| | NORMAL NO. DAYS WITH: | | | | | | | | | | | | | | |
| | PRECIPITATION >= 0.01 | 30 | 9.3 | 7.3 | 8.2 | 8.0 | 9.8 | 11.0 | 9.9 | 9.2 | 7.8 | 7.3 | 6.7 | 8.0 | 102.5 |
| | PRECIPITATION >= 1.00 | 30 | 0.0 | 0.0 | * | 0.2 | 0.6 | 0.8 | 0.7 | 0.6 | 0.6 | 0.3 | 0.1 | 0.0 | 3.9 |
| SNOWFALL | NORMAL (IN) | 30 | 12.5 | 7.0 | 8.7 | 2.3 | 0.* | 0.0 | 0.0 | 0.0 | 0.* | 0.6 | 7.7 | 7.9 | 46.7 |
| | MAXIMUM MONTHLY (IN) | 68 | 31.5 | 19.5 | 28.1 | 16.9 | 1.0 | T | T | T | 0.6 | 8.1 | 26.4 | 33.5 | 33.5 |
| | YEAR OF OCCURRENCE | | 1989 | 1979 | 2009 | 2008 | 1950 | 1994 | 1999 | 1994 | 1942 | 1951 | 1996 | 2008 | DEC 2008 |
| | MAXIMUM IN 24 HOURS (IN) | 68 | 19.4 | 11.2 | 12.0 | 8.6 | 1.0 | T | T | T | 0.6 | 7.8 | 12.6 | 9.3 | 19.4 |
| | YEAR OF OCCURRENCE' | | 1989 | 1951 | 1997 | 1970 | 1950 | 1994 | 1999 | 1994 | 1942 | 1951 | 1977 | 1988 | JAN 1989 |
| | MAXIMUM SNOW DEPTH (IN) | 53 | 30 | 24 | 32 | 8 | 1 | 0 | 0 | 0 | 0 | 5 | 17 | 19 | 32 |
| | YEAR OF OCCURRENCE | | 1989 | 1994 | 1997 | 1975 | 1979 | | | | | 1951 | 1985 | 1985 | MAR 1997 |
| | NORMAL NO. DAYS WITH: | | | | | | | | | | | | | | |
| | SNOWFALL >= 1.0 | 30 | 3.0 | 2.1 | 2.4 | 0.7 | 0.0 | 0.0 | 0.0 | 0.0 | 0.0 | 0.3 | 2.1 | 2.4 | 13.0 |

## PRECIPITATION (inches) 2009  FARGO (KFAR)

| YEAR | JAN | FEB | MAR | APR | MAY | JUN | JUL | AUG | SEP | OCT | NOV | DEC | ANNUAL |
|------|-----|-----|-----|-----|-----|-----|-----|-----|-----|-----|-----|-----|--------|
| 1980 | 1.23 | 0.57 | 0.62 | 0.02 | 0.64 | 2.68 | 0.76 | 4.24 | 2.52 | 1.06 | 0.47 | 0.30 | 15.11 |
| 1981 | 0.11 | 0.49 | 0.67 | 0.61 | 3.46 | 2.56 | 3.21 | 1.76 | 1.11 | 2.36 | 0.40 | 0.85 | 17.59 |
| 1982 | 1.32 | 0.54 | 1.25 | 0.45 | 1.82 | 1.61 | 2.64 | 1.12 | 1.12 | 7.03 | 1.13 | 0.17 | 20.20 |
| 1983 | 0.46 | 0.21 | 2.27 | 0.42 | 2.00 | 2.34 | 4.16 | 2.56 | 1.63 | 1.62 | 1.04 | 0.96 | 19.67 |
| 1984 | 0.79 | 0.90 | 1.12 | 1.68 | 0.61 | 5.38 | 0.64 | 0.18 | 1.23 | 6.76 | 0.18 | 0.90 | 20.37 |
| 1985 | 0.20 | 0.18 | 1.35 | 0.60 | 5.03 | 1.44 | 3.91 | 2.30 | 1.39 | 1.12 | 1.06 | 0.59 | 19.17 |
| 1986 | 0.85 | 0.27 | 0.19 | 5.28 | 1.00 | 3.98 | 4.78 | 1.72 | 3.67 | 0.05 | 1.43 | 0.29 | 23.51 |
| 1987 | 0.27 | 0.86 | 0.49 | 0.12 | 3.46 | 0.66 | 2.86 | 3.23 | 1.70 | 0.18 | 0.48 | 0.69 | 15.00 |
| 1988 | 1.62 | 0.22 | 1.02 | 0.01 | 1.82 | 1.24 | 0.46 | 2.14 | 3.22 | 0.49 | 1.18 | 1.11 | 14.53 |
| 1989 | 1.85 | 0.21 | 1.49 | 1.03 | 2.60 | 1.51 | 0.62 | 6.07 | 2.10 | 0.31 | 1.18 | 0.24 | 19.21 |
| 1990 | 0.13 | 0.58 | 1.54 | 1.78 | 1.52 | 6.05 | 0.78 | 0.99 | 1.75 | 1.22 | 0.02 | 0.77 | 17.13 |
| 1991 | 0.29 | 1.27 | 0.97 | 3.15 | 2.38 | 6.26 | 1.86 | 1.87 | 1.28 | 0.71 | 0.46 | 0.37 | 20.87 |
| 1992 | 0.89 | 0.51 | 1.05 | 0.89 | 2.32 | 6.47 | 0.83 | 2.35 | 2.55 | 0.26 | 1.73 | 0.56 | 20.41 |
| 1993 | 0.79 | 0.19 | 0.83 | 0.74 | 2.67 | 4.28 | 7.71 | 1.13 | 0.49 | 0.19 | 1.88 | 1.00 | 21.90 |
| 1994 | 0.67 | 0.64 | 0.97 | 2.56 | 0.82 | 2.53 | 5.76 | 2.85 | 2.06 | 3.15 | 0.89 | 0.20 | 23.10 |
| 1995 | 0.76 | 0.62 | 2.62 | 0.69 | 2.07 | 1.41 | 5.27 | 1.75 | 2.58 | 2.04 | 0.99 | 0.73 | 21.53 |
| 1996 | 1.82 | 0.94 | 0.41 | 0.21 | 3.00 | 1.33 | 1.36 | 2.11 | 3.18 | 2.41 | .07 | .69 | 17.53 |
| 1997 | 1.79 | 0.59 | 1.89 | 3.12 | 2.54 | 4.86 | 2.73 | 2.60 | 2.31 | 2.89 | 0.45 | 0.07 | 25.84 |
| 1998 | 0.81 | 1.51 | 0.97 | 0.60 | 7.34 | 6.62 | 2.74 | 1.93 | 2.44 | 4.73 | 1.75 | 0.31 | 31.75 |
| 1999 | 1.15 | 0.20 | 1.83 | 1.04 | 3.50 | 2.83 | 2.34 | 4.43 | 6.50 | 1.04 | T | 0.45 | 25.31 |
| 2000 | 0.33 | 0.99 | 1.77 | 1.33 | 2.69 | 11.72 | 2.44 | 3.07 | 3.64 | 1.96 | 4.13 | 0.69 | 34.76 |
| 2001 | 0.20 | 0.74 | 0.26 | 2.70 | 2.88 | 2.73 | 3.14 | 2.19 | 1.45 | 2.74 | 1.00 | 0.22 | 20.25 |
| 2002 | 0.21 | 0.12 | 1.06 | 1.26 | 3.87 | 4.76 | 5.65 | 3.73 | 1.73 | 1.44 | 0.15 | 0.83 | 24.81 |
| 2003 | 0.26 | 0.18 | 0.63 | 1.32 | 4.24 | 4.56 | 1.72 | 1.06 | 1.40 | 1.34 | 0.53 | 1.18 | 18.42 |
| 2004 | 0.73 | 0.72 | 1.58 | 0.16 | 6.22 | 1.07 | 4.21 | 2.01 | 4.69 | 3.54 | 0.05 | 1.01 | 25.99 |
| 2005 | 1.12 | 0.61 | 0.13 | 0.87 | 2.42 | 8.47 | 1.06 | 7.52 | 1.69 | 2.39 | 2.84 | 1.32 | 30.44 |
| 2006 | 0.37 | 0.46 | 1.22 | 1.28 | 1.99 | 1.34 | 2.23 | 2.21 | 3.91 | 0.96 | 0.12 | 1.06 | 17.15 |
| 2007 | 0.10 | 0.73 | 2.18 | 3.16 | 3.87 | 5.78 | 1.20 | 2.39 | 3.39 | 1.76 | 0.09 | 1.59 | 26.24 |
| 2008 | 0.09 | 0.67 | 0.98 | 2.33 | 1.89 | 6.06 | 1.78 | 4.55 | 5.08 | 4.46 | 1.13 | 1.80 | 30.82 |
| 2009 | 0.55 | 1.29 | 4.62 | 0.81 | 1.62 | 2.93 | 1.18 | 2.13 | 2.06 | 5.44 | 0.41 | 1.85 | 24.89 |
| POR= 68 YRS | 0.62 | 0.53 | 1.07 | 1.69 | 2.57 | 3.43 | 2.97 | 2.74 | 2.10 | 1.67 | 0.85 | 0.68 | 20.92 |

WBAN : 14914

## AVERAGE TEMPERATURE (°F) 2009  FARGO (KFAR)

| YEAR | JAN | FEB | MAR | APR | MAY | JUN | JUL | AUG | SEP | OCT | NOV | DEC | ANNUAL |
|------|-----|-----|-----|-----|-----|-----|-----|-----|-----|-----|-----|-----|--------|
| 1980 | 6.6 | 8.3 | 20.7 | 49.0 | 61.4 | 65.7 | 71.9 | 67.5 | 56.9 | 42.4 | 33.1 | 12.7 | 41.4 |
| 1981 | 11.8 | 19.6 | 33.5 | 45.6 | 55.5 | 62.8 | 71.1 | 69.6 | 57.4 | 44.5 | 35.4 | 8.7 | 43.0 |
| 1982 | -7.0 | 8.9 | 22.9 | 40.7 | 58.1 | 59.1 | 70.9 | 68.5 | 57.5 | 45.7 | 24.1 | 20.9 | 39.2 |
| 1983 | 16.1 | 21.8 | 29.9 | 40.2 | 52.1 | 66.1 | 73.5 | 72.9 | 56.7 | 44.4 | 31.3 | -0.3 | 42.1 |
| 1984 | 9.7 | 24.9 | 23.4 | 45.6 | 54.2 | 65.8 | 70.6 | 73.3 | 54.4 | 47.4 | 29.7 | 9.6 | 42.4 |
| 1985 | 5.1 | 10.9 | 32.9 | 46.6 | 60.2 | 60.0 | 69.0 | 64.5 | 53.9 | 44.6 | 15.4 | 3.9 | 38.9 |
| 1986 | 13.8 | 10.5 | 31.6 | 43.9 | 57.5 | 67.5 | 71.5 | 65.6 | 56.1 | 45.3 | 23.1 | 20.8 | 42.3 |
| 1987 | 18.2 | 27.5 | 31.4 | 51.5 | 61.7 | 69.1 | 74.0 | 66.8 | 59.6 | 42.6 | 33.4 | 20.6 | 46.4 |
| 1988 | 5.9 | 9.3 | 29.5 | 44.5 | 63.9 | 73.8 | 75.8 | 72.2 | 58.5 | 42.9 | 27.5 | 15.2 | 43.3 |
| 1989 | 11.4 | 1.7 | 20.1 | 42.2 | 58.2 | 64.1 | 75.9 | 70.8 | 58.5 | 45.8 | 24.0 | 4.3 | 39.8 |
| 1990 | 21.8 | 17.6 | 31.4 | 43.6 | 55.0 | 67.0 | 70.0 | 71.1 | 62.3 | 45.6 | 32.1 | 12.2 | 44.1 |
| 1991 | 6.4 | 20.5 | 30.4 | 48.0 | 61.5 | 70.1 | 70.2 | 72.7 | 58.8 | 42.0 | 22.0 | 18.8 | 43.5 |
| 1992 | 17.3 | 23.5 | 32.6 | 41.4 | 58.7 | 62.0 | 64.3 | 64.8 | 56.8 | 45.1 | 27.3 | 10.5 | 42.0 |
| 1993 | 7.5 | 9.8 | 25.6 | 43.3 | 56.7 | 63.1 | 67.0 | 69.2 | 54.7 | 43.9 | 26.6 | 16.0 | 40.3 |
| 1994 | -3.9 | 6.3 | 30.5 | 43.7 | 59.9 | 68.2 | 67.6 | 66.5 | 61.9 | 50.4 | 34.0 | 20.7 | 42.2 |
| 1995 | 10.8 | 12.4 | 28.4 | 38.9 | 54.8 | 71.4 | 70.0 | 72.0 | 58.8 | 44.1 | 21.2 | 11.5 | 41.2 |
| 1996 | -1.8 | 11.2 | 17.4 | 37.8 | 53.6 | 67.0 | 67.9 | 71.1 | 59.1 | 45.3 | 17.7 | 5.9 | 37.7 |
| 1997 | 1.8 | 12.4 | 20.1 | 37.8 | 53.0 | 69.0 | 69.3 | 67.9 | 61.9 | 47.0 | 23.2 | 23.5 | 40.6 |
| 1998 | 11.3 | 28.0 | 26.6 | 49.2 | 60.9 | 63.4 | 71.7 | 72.7 | 63.9 | 47.6 | 29.3 | 17.3 | 45.2 |
| 1999 | 6.3 | 22.5 | 31.2 | 45.1 | 58.0 | 66.4 | 71.5 | 68.0 | 55.4 | 44.3 | 37.1 | 22.9 | 44.1 |
| 2000 | 9.8 | 21.6 | 35.2 | 42.4 | 57.3 | 62.7 | 70.6 | 69.6 | 58.0 | 48.3 | 26.0 | -.2 | 41.8 |
| 2001 | 14.3 | 3.9 | 23.0 | 44.4 | 58.5 | 65.9 | 72.6 | 70.5 | 59.4 | 44.4 | 39.7 | 20.1 | 43.1 |
| 2002 | 16.5 | 24.0 | 20.0 | 40.1 | 51.2 | 69.0 | 73.1 | 67.3 | 61.8 | 37.2 | 27.9 | 19.9 | 42.3 |
| 2003 | 9.3 | 8.2 | 24.7 | 45.3 | 55.7 | 65.5 | 70.4 | 72.6 | 58.5 | 48.9 | 24.9 | 19.9 | 42.0 |
| 2004 | 3.2 | 14.7 | 30.3 | 44.3 | 52.3 | 62.5 | 68.2 | 62.2 | 62.9 | 47.0 | 34.2 | 18.0 | 41.7 |
| 2005 | 5.9 | 17.4 | 28.1 | 49.1 | 54.4 | 68.2 | 71.3 | 68.3 | 63.2 | 48.3 | 31.3 | 17.2 | 43.6 |
| 2006 | 23.5 | 9.7 | 27.4 | 50.7 | 58.9 | 68.5 | 74.9 | 69.7 | 58.7 | 43.2 | 31.9 | 25.6 | 45.2 |
| 2007 | 13.1 | 6.5 | 31.6 | 42.9 | 60.4 | 69.8 | 74.0 | 67.2 | 60.6 | 50.0 | 31.2 | 10.3 | 43.1 |
| 2008 | 6.5 | 7.5 | 22.8 | 41.0 | 53.9 | 63.6 | 70.3 | 69.5 | 59.7 | 46.7 | 31.8 | 5.9 | 39.9 |
| 2009 | 1.7 | 11.6 | 24.1 | 42.0 | 53.8 | 63.6 | 66.5 | 66.0 | 65.1 | 40.4 | 38.5 | 9.7 | 40.3 |
| POR= 68 YRS | 7.2 | 12.4 | 25.7 | 42.8 | 55.8 | 65.2 | 70.7 | 69.0 | 58.3 | 46.0 | 28.3 | 13.4 | 41.3 |

## HEATING DEGREE DAYS (base 65°F) 2009  FARGO (KFAR)

| YEAR | JUL | AUG | SEP | OCT | NOV | DEC | JAN | FEB | MAR | APR | MAY | JUN | TOTAL |
|------|-----|-----|-----|-----|-----|-----|-----|-----|-----|-----|-----|-----|-------|
| 1980-81 | 3 | 35 | 267 | 696 | 951 | 1616 | 1645 | 1266 | 971 | 574 | 298 | 84 | 8406 |
| 1981-82 | 14 | 10 | 250 | 627 | 881 | 1742 | 2236 | 1570 | 1298 | 725 | 222 | 187 | 9762 |
| 1982-83 | 0 | 66 | 257 | 589 | 1219 | 1359 | 1513 | 1206 | 1082 | 738 | 390 | 74 | 8493 |
| 1983-84 | 16 | 2 | 301 | 631 | 1004 | 2023 | 1714 | 1154 | 1280 | 576 | 344 | 52 | 9097 |
| 1984-85 | 15 | 13 | 339 | 541 | 1053 | 1715 | 1853 | 1514 | 988 | 550 | 172 | 179 | 8932 |
| 1985-86 | 13 | 72 | 329 | 625 | 1487 | 1895 | 1585 | 1527 | 1027 | 627 | 266 | 45 | 9498 |
| 1986-87 | 0 | 69 | 268 | 602 | 1251 | 1360 | 1447 | 1047 | 1036 | 415 | 163 | 39 | 7697 |
| 1987-88 | 15 | 59 | 177 | 688 | 940 | 1369 | 1832 | 1614 | 1092 | 609 | 131 | 8 | 8534 |
| 1988-89 | 3 | 25 | 207 | 677 | 1118 | 1537 | 1658 | 1771 | 1386 | 677 | 224 | 96 | 9379 |
| 1989-90 | 0 | 17 | 224 | 599 | 1224 | 1881 | 1332 | 1324 | 1034 | 666 | 314 | 58 | 8673 |
| 1990-91 | 8 | 18 | 173 | 594 | 982 | 1637 | 1813 | 1242 | 1066 | 505 | 211 | 6 | 8255 |
| 1991-92 | 3 | 3 | 234 | 708 | 1284 | 1425 | 1473 | 1198 | 998 | 709 | 247 | 137 | 8419 |
| 1992-93 | 66 | 84 | 252 | 613 | 1125 | 1686 | 1780 | 1542 | 1214 | 643 | 269 | 111 | 9385 |
| 1993-94 | 35 | 19 | 310 | 652 | 1144 | 1514 | 2137 | 1640 | 1062 | 637 | 211 | 18 | 9379 |
| 1994-95 | 19 | 58 | 146 | 446 | 925 | 1366 | 1674 | 1469 | 1131 | 777 | 316 | 42 | 8369 |
| 1995-96 | 10 | 0 | 241 | 641 | 1308 | 1654 | 2071 | 1557 | 1472 | 812 | 355 | 65 | 10186 |
| 1996-97 | 12 | 6 | 234 | 604 | 1412 | 1826 | 1952 | 1466 | 1382 | 810 | 368 | 4 | 10076 |
| 1997-98 | 48 | 43 | 126 | 560 | 1249 | 1281 | 1659 | 1031 | 1185 | 467 | 151 | 111 | 7911 |
| 1998-99 | 5 | 0 | 124 | 530 | 1064 | 1471 | 1814 | 1186 | 1042 | 588 | 234 | 66 | 8124 |
| 1999-00 | 4 | 27 | 285 | 635 | 831 | 1297 | 1704 | 1251 | 918 | 672 | 252 | 119 | 7995 |
| 2000-01 | 26 | 23 | 226 | 511 | 1166 | 2014 | 1567 | 1705 | 1296 | 617 | 215 | 81 | 9447 |
| 2001-02 | 15 | 26 | 203 | 631 | 751 | 1385 | 1496 | 1140 | 1386 | 743 | 453 | 41 | 8270 |
| 2002-03 | 10 | 41 | 188 | 854 | 1103 | 1391 | 1718 | 1584 | 1242 | 590 | 289 | 61 | 9071 |
| 2003-04 | 9 | 9 | 246 | 517 | 1193 | 1389 | 1907 | 1453 | 1068 | 614 | 386 | 110 | 8901 |
| 2004-05 | 55 | 113 | 142 | 551 | 919 | 1449 | 1826 | 1328 | 1137 | 478 | 332 | 18 | 8348 |
| 2005-06 | 17 | 24 | 112 | 515 | 1005 | 1475 | 1279 | 1545 | 1160 | 427 | 235 | 24 | 7818 |
| 2006-07 | 3 | 7 | 228 | 678 | 985 | 1216 | 1597 | 1633 | 1030 | 658 | 179 | 27 | 8241 |
| 2007-08 | 2 | 41 | 198 | 463 | 1008 | 1689 | 1807 | 1663 | 1304 | 714 | 342 | 89 | 9320 |
| 2008-09 | 7 | 19 | 193 | 563 | 991 | 1827 | 1955 | 1492 | 1262 | 685 | 353 | 127 | 9474 |
| 2009- | 31 | 60 | 73 | 756 | 789 | 1706 | | | | | | | |

WBAN : 14914

## COOLING DEGREE DAYS (base 65°F) 2009  FARGO (KFAR)

| YEAR | JAN | FEB | MAR | APR | MAY | JUN | JUL | AUG | SEP | OCT | NOV | DEC | TOTAL |
|------|-----|-----|-----|-----|-----|-----|-----|-----|-----|-----|-----|-----|-------|
| 1980 | 0 | 0 | 0 | 18 | 102 | 89 | 222 | 119 | 31 | 1 | 0 | 0 | 582 |
| 1981 | 0 | 0 | 0 | 0 | 9 | 25 | 212 | 159 | 26 | 0 | 0 | 0 | 431 |
| 1982 | 0 | 0 | 0 | 2 | 11 | 20 | 189 | 179 | 39 | 0 | 0 | 0 | 440 |
| 1983 | 0 | 0 | 0 | 0 | 2 | 113 | 288 | 252 | 55 | 0 | 0 | 0 | 710 |
| 1984 | 0 | 0 | 0 | 0 | 18 | 81 | 196 | 279 | 24 | 4 | 0 | 0 | 602 |
| 1985 | 0 | 0 | 0 | 6 | 31 | 35 | 143 | 63 | 4 | 0 | 0 | 0 | 282 |
| 1986 | 0 | 0 | 0 | 0 | 41 | 126 | 208 | 92 | 10 | 0 | 0 | 0 | 477 |
| 1987 | 0 | 0 | 0 | 17 | 66 | 169 | 303 | 121 | 25 | 0 | 0 | 0 | 701 |
| 1988 | 0 | 0 | 0 | 0 | 102 | 280 | 346 | 252 | 22 | 0 | 0 | 0 | 1002 |
| 1989 | 0 | 0 | 0 | 0 | 19 | 76 | 345 | 201 | 34 | 11 | 0 | 0 | 686 |
| 1990 | 0 | 0 | 0 | 29 | 10 | 123 | 172 | 214 | 98 | 1 | 0 | 0 | 647 |
| 1991 | 0 | 0 | 0 | 2 | 107 | 166 | 171 | 250 | 54 | 0 | 0 | 0 | 750 |
| 1992 | 0 | 0 | 0 | 9 | 58 | 52 | 49 | 85 | 11 | 3 | 0 | 0 | 267 |
| 1993 | 0 | 0 | 0 | 0 | 19 | 61 | 107 | 155 | 9 | 4 | 0 | 0 | 355 |
| 1994 | 0 | 0 | 0 | 4 | 59 | 122 | 105 | 109 | 60 | 0 | 0 | 0 | 459 |
| 1995 | 0 | 0 | 0 | 0 | 5 | 243 | 175 | 227 | 62 | 3 | 0 | 0 | 715 |
| 1996 | 0 | 0 | 0 | 0 | 8 | 131 | 107 | 204 | 63 | 0 | 0 | 0 | 513 |
| 1997 | 0 | 0 | 0 | 0 | 3 | 131 | 186 | 136 | 38 | 9 | 0 | 0 | 503 |
| 1998 | 0 | 0 | 0 | 0 | 32 | 71 | 218 | 245 | 96 | 0 | 0 | 0 | 662 |
| 1999 | 0 | 0 | 0 | 0 | 23 | 117 | 214 | 127 | 4 | 0 | 0 | 0 | 485 |
| 2000 | 0 | 0 | 0 | 0 | 19 | 57 | 208 | 172 | 19 | 0 | 0 | 0 | 475 |
| 2001 | 0 | 0 | 0 | 9 | 20 | 112 | 257 | 202 | 44 | 0 | 0 | 0 | 644 |
| 2002 | 0 | 0 | 0 | 2 | 32 | 169 | 268 | 123 | 95 | 0 | 0 | 0 | 689 |
| 2003 | 0 | 0 | 0 | 7 | 7 | 81 | 183 | 252 | 58 | 25 | 0 | 0 | 613 |
| 2004 | 0 | 0 | 0 | 0 | 0 | 43 | 161 | 35 | 84 | 1 | 0 | 0 | 324 |
| 2005 | 0 | 0 | 0 | 8 | 11 | 122 | 219 | 133 | 64 | 6 | 0 | 0 | 563 |
| 2006 | 0 | 0 | 0 | 3 | 53 | 136 | 316 | 162 | 45 | 9 | 0 | 0 | 724 |
| 2007 | 0 | 0 | 0 | 1 | 42 | 181 | 290 | 116 | 76 | 7 | 0 | 0 | 713 |
| 2008 | 0 | 0 | 0 | 0 | 4 | 51 | 176 | 164 | 41 | 0 | 0 | 0 | 436 |
| 2009 | 0 | 0 | 0 | 0 | 8 | 90 | 85 | 95 | 83 | 0 | 0 | 0 | 361 |

## SNOWFALL (inches) 2009 FARGO (KFAR)

| YEAR | JUL | AUG | SEP | OCT | NOV | DEC | JAN | FEB | MAR | APR | MAY | JUN | TOTAL |
|------|-----|-----|-----|-----|-----|-----|-----|-----|-----|-----|-----|-----|-------|
| 1980-81 | 0.0 | 0.0 | 0.0 | 0.5 | 1.1 | 4.6 | 2.1 | 4.5 | 0.3 | T | 0.0 | 0.0 | 13.1 |
| 1981-82 | 0.0 | 0.0 | T | 2.3 | 2.2 | 9.9 | 30.0 | 10.9 | 14.0 | 0.2 | 0.0 | 0.0 | 69.5 |
| 1982-83 | 0.0 | 0.0 | 0.0 | 0.0 | 6.8 | 0.3 | 3.8 | 2.0 | 7.4 | 2.9 | T | 0.0 | 23.2 |
| 1983-84 | 0.0 | 0.0 | T | T | 5.3 | 11.8 | 11.5 | 3.1 | 7.7 | 0.5 | 0.0 | 0.0 | 39.9 |
| 1984-85 | 0.0 | 0.0 | T | T | 1.4 | 7.4 | 3.7 | 3.1 | 12.6 | T | 0.0 | 0.0 | 28.2 |
| 1985-86 | 0.0 | 0.0 | 0.0 | T | 24.3 | 10.4 | 11.2 | 6.7 | 0.7 | 3.7 | T | 0.0 | 57.0 |
| 1986-87 | 0.0 | 0.0 | 0.0 | T | 5.3 | 3.8 | 2.8 | 10.4 | 1.2 | T | 0.0 | 0.0 | 23.5 |
| 1987-88 | 0.0 | 0.0 | 0.0 | T | 3.0 | 6.6 | 24.3 | 4.4 | 6.2 | T | 0.0 | 0.0 | 44.5 |
| 1988-89 | 0.0 | 0.0 | 0.0 | T | 11.6 | 14.9 | 31.5 | 2.3 | 12.4 | 0.9 | T | 0.0 | 73.6 |
| 1989-90 | 0.0 | T | T | T | 16.3 | 2.6 | 0.8 | 7.9 | 11.5 | 7.2 | T | 0.0 | 46.3 |
| 1990-91 | 0.0 | 0.0 | 0.0 | 1.3 | 0.2 | 12.4 | 4.0 | 15.3 | 10.9 | 4.2 | T | T | 48.3 |
| 1991-92 | 0.0 | 0.0 | T | 0.3 | 5.2 | 5.9 | 10.5 | 2.1 | 0.2 | 3.3 | T | 0.0 | 27.5 |
| 1992-93 | 0.0 | 0.0 | T | 1.8 | 16.4 | 9.2 | 16.7 | 3.3 | 6.4 | T | 0.0 | 0.0 | 53.8 |
| 1993-94 | 0.0 | 0.0 | 0.0 | T | 21.5 | 13.8 | 18.0 | 12.8 | 12.1 | 10.9 | 0.0 | T | 89.1 |
| 1994-95 | 0.0 | T | 0.0 | 0.0 | 4.0 | 3.0 | 10.8 | 9.5 | 19.0 | 4.0 | 0.0 | 0.0 | 50.3 |
| 1995-96 | 0.0 | 0.0 | T | 1.0 | 9.6 | 12.7 | 27.2 | 8.9 | 15.0 | 0.2 | 0.0 | 0.0 | 74.6 |
| 1996-97 | 0.0 | 0.0 | 0.0 | T | 26.4 | 20.4 | 28.6 | 8.0 | 26.2 | 7.4 | T | 0.0 | 117.0 |
| 1997-98 | 0.0 | 0.0 | 0.0 | 0.3 | 11.1 | 7.4 | 12.6 | 3.6 | 5.4 | 0.7 | 0.0 | T | 41.1 |
| 1998-99 | 0.0 | 0.0 | 0.0 | 0.0 | 12.3 | 4.5 | 19.7 | 2.1 | 10.0 | T | 0.0 | 0.0 | 48.6 |
| 1999-00 | T | 0.0 | 0.0 | T | 0.0 | 4.3 | 6.5 | 11.4 | 5.6 | 6.2 | 0.0 | 0.0 | 34.0 |
| 2000-01 | 0.0 | 0.0 | 0.0 | T | 15.3 | 13.4 | 2.7 | 11.6 | 1.5 | 8.0 | 0.0 | 0.0 | 52.5 |
| 2001-02 | 0.0 | 0.0 | 0.0 | 5.4 | 11.2 | 5.0 | 3.8 | 1.9 | 15.7 | 6.3 | T | 0.0 | 49.3 |
| 2002-03 | 0.0 | 0.0 | T | 1.9 | 1.8 | 10.9 | 4.4 | 3.1 | 7.7 | 3.6 | 0.0 | 0.0 | 33.4 |
| 2003-04 | 0.0 | 0.0 | 0.0 | 0.5 | 4.9 | 14.4 | 17.0 | 5.9 | 10.6 | 0.5 | T | 0.0 | 53.8 |
| 2004-05 | 0.0 | 0.0 | 0.0 | T | 0.5 | 7.5 | 11.7 | 7.4 | 1.5 | T | 0.2 | 0.0 | 28.8 |
| 2005-06 | 0.0 | 0.0 | T | T | 9.9 | 13.9 | 5.2 | 8.1 | 5.2 | 0.0 | 0.0 | 0.0 | 42.3 |
| 2006-07 | 0.0 | 0.0 | 0.0 | 2.2 | 0.2 | 5.4 | 2.9 | 10.9 | 9.4 | 7.8 | 0.0 | 0.0 | 38.8 |
| 2007-08 | 0.0 | 0.0 | T | 0.0 | 0.8 | 19.3 | 2.8 | 8.8 | 11.2 | 16.9 | T | 0.0 | 59.8 |
| 2008-09 | 0.0 | 0.0 | 0.0 | 1.4 | 0.9 | 33.5 | 7.3 | 8.1 | 28.1 | 0.2 | 0.0 | 0.0 | 79.5 |
| 2009- | 0.0 | 0.0 | 0.0 | 1.2 | 0.2 | 24.4 | | | | | | | |
| POR=<br>68 YRS | T | T | T | 0.7 | 6.1 | 8.4 | 9.2 | 6.2 | 7.9 | 3.4 | 0.1 | T | 42.0 |

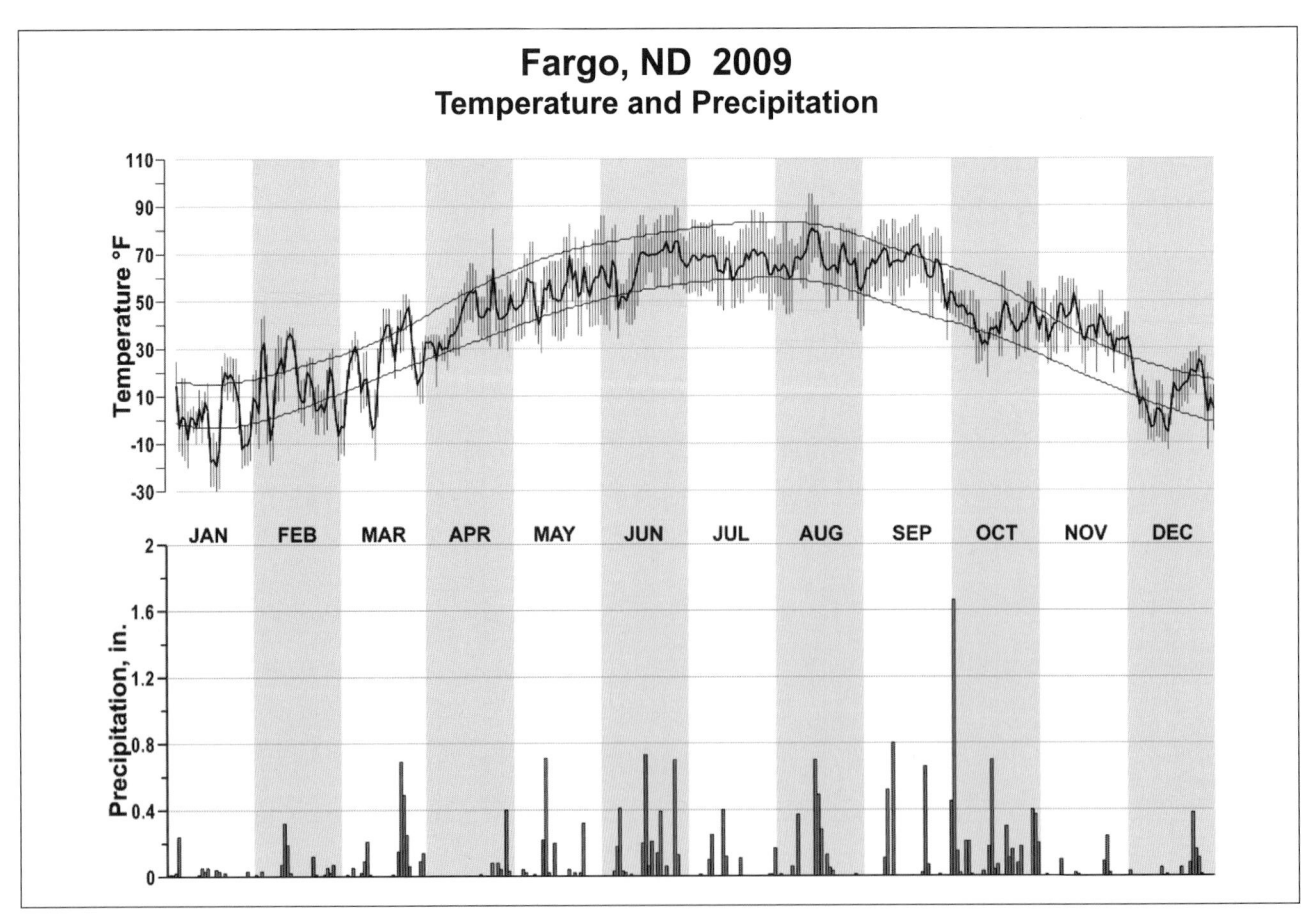

# Fargo, ND  2009
## Temperature and Precipitation

# 2009
# WILLISTON
# NORTH DAKOTA (KISN)

Williston lies in a flat valley at the junction of the Missouri River and Little Muddy Creek. The surrounding country is rolling. Hills to the east are highest, ranging from 250 to 300 feet in height at a distance of 5 to 7 miles. Across the Missouri River to the south, the bluffs are about 225 feet high at 4 miles distance.

Great extremes of temperatures are encountered, winters being cold, while summer days are usually warm. In winter, temperatures below zero are common and lows of -50 degrees have been recorded. When temperatures are lowest, however, the air is generally dry, with little or no wind and the weather is fine and invigorating. At the other extreme, temperatures above 100 degrees have been reached in all months from May to September. The low humidity that generally prevails on the hottest summer days keeps them from becoming oppressive.

The climate of Williston and vicinity is continental, semi-arid, characterized by marked season changes. Winter is the relatively dry season, with only about 1/2 inch of monthly precipitation occurring from November to February. There is considerably less than the average amount of snowfall for similar locations in the United States. Ice crystals, which rarely yield more than a trace of precipitation, are common in the cold months. Although snow has been observed every month except July and August, there is usually very little from April to November. Accumulated winter snow remains unmelted on the ground until about March. Summer precipitation is variable from year to year. The amount of rain occurring during the growing period is the most important element of climate for agricultural interests in the vicinity of Williston. Generally, considerably more precipitation occurs in the spring and summer months than in winter, but even so, the rainfall is just adequate for successful farming operations in normal years. A series of dry years, in addition to causing failure of crops, may result in erosion of the fertile topsoil by winds.

The growing season averages 131 days. It has ranged from 94 to 172 days during the period of record.

Clear and partly cloudy skies, nearly equally distributed, occur about 70 percent of the time. Heavy fog occurs on the average about ten times a year. Because of the northern latitude of Williston, it enjoys long hours of daylight in the spring and summer. Relatively little cloudiness occurs then, so that the duration of sunshine averages about two-thirds of the possible amount. These conditions are conducive to rapid growth of vegetation, making successful agricultural pursuits possible in spite of the relatively short growing season.

Summer storms are generally in the form of thunderstorms or rain showers, occasionally accompanied by hail and squally winds. Tornadoes are rare in this area. In the winter, cold waves and occasionally blizzard conditions occur. Cold waves result when extremely cold air advances southward from northwestern Canada. In blizzard conditions the advancing cold wave is accompanied by winds of gale force and the air is filled with fine, wind-driven snow. In extreme instances in the country, it becomes impossible for persons to ascertain their bearings or to remain alive many hours without shelter in such storms.

# NORMALS, MEANS, AND EXTREMES
## WILLISTON (KISN)

LATITUDE: 48° 11'N  LONGITUDE: -103° 38'W  ELEVATION (FT): GRND: 1902  BARO: 1902  TIME ZONE: CENTRAL (UTC -6)  WBAN: 94014

| ELEMENT | POR | JAN | FEB | MAR | APR | MAY | JUN | JUL | AUG | SEP | OCT | NOV | DEC | YEAR |
|---|---|---|---|---|---|---|---|---|---|---|---|---|---|---|
| **TEMPERATURE °F** | | | | | | | | | | | | | | |
| NORMAL DAILY MAXIMUM | 30 | 19.4 | 27.6 | 40.1 | 56.0 | 68.2 | 77.3 | 83.4 | 82.8 | 70.0 | 57.0 | 36.2 | 24.0 | 53.5 |
| MEAN DAILY MAXIMUM | 48 | 20.0 | 27.1 | 39.4 | 56.1 | 67.8 | 77.0 | 84.4 | 83.3 | 71.2 | 57.1 | 38.1 | 24.9 | 53.9 |
| HIGHEST DAILY MAXIMUM | 48 | 53 | 66 | 78 | 92 | 106 | 106 | 109 | 107 | 104 | 93 | 76 | 58 | 109 |
| YEAR OF OCCURRENCE | | 1981 | 1992 | 2007 | 1980 | 1980 | 1988 | 1980 | 1983 | 1983 | 1963 | 1999 | 1979 | JUL 1980 |
| MEAN OF EXTREME MAXS. | 48 | 42.8 | 48.0 | 64.4 | 79.7 | 88.4 | 94.0 | 98.9 | 98.5 | 92.3 | 80.0 | 61.0 | 45.9 | 74.5 |
| NORMAL DAILY MINIMUM | 30 | -3.3 | 5.9 | 17.2 | 29.1 | 40.9 | 50.1 | 55.2 | 53.8 | 42.2 | 30.2 | 14.9 | 2.1 | 28.2 |
| MEAN DAILY MINIMUM | 48 | -0.8 | 6.1 | 17.2 | 30.2 | 41.2 | 50.5 | 56.4 | 54.1 | 43.0 | 31.2 | 17.0 | 4.5 | 29.2 |
| LOWEST DAILY MINIMUM | 48 | -40 | -41 | -31 | -15 | 10 | 26 | 34 | 34 | 15 | -9 | -27 | -50 | -50 |
| YEAR OF OCCURRENCE | | 1966 | 1962 | 2003 | 1975 | 2005 | 1998 | 1967 | 1994 | 2003 | 2002 | 1996 | 1983 | DEC 1983 |
| MEAN OF EXTREME MINS. | 48 | -26.0 | -18.8 | -7.2 | 13.0 | 25.4 | 37.6 | 44.3 | 40.6 | 27.5 | 13.5 | -4.5 | -21.1 | 10.4 |
| NORMAL DRY BULB | 30 | 8.0 | 16.8 | 28.7 | 42.5 | 54.6 | 63.7 | 69.3 | 68.3 | 56.1 | 43.6 | 25.6 | 13.0 | 40.9 |
| MEAN DRY BULB | 48 | 9.6 | 16.6 | 28.3 | 43.2 | 54.6 | 63.9 | 70.4 | 68.7 | 57.1 | 44.2 | 27.6 | 14.7 | 41.6 |
| MEAN WET BULB | 26 | 10.5 | 15.3 | 25.1 | 35.8 | 45.9 | 54.7 | 59.2 | 57.2 | 47.9 | 36.4 | 23.5 | 13.4 | 35.4 |
| MEAN DEW POINT | 26 | 7.9 | 12.3 | 21.2 | 29.6 | 40.1 | 50.1 | 54.5 | 51.9 | 42.7 | 31.2 | 20.5 | 10.6 | 31.1 |
| NORMAL NO. DAYS WITH: | | | | | | | | | | | | | | |
| MAXIMUM >= 90 | 30 | 0.0 | 0.0 | 0.0 | 0.1 | 0.9 | 3.2 | 7.7 | 8.9 | 1.9 | 0.1 | 0.0 | 0.0 | 22.8 |
| MAXIMUM <= 32 | 30 | 22.7 | 15.4 | 8.1 | 1.1 | 0.0 | 0.0 | 0.0 | 0.0 | 0.0 | 0.9 | 10.9 | 19.9 | 79.0 |
| MINIMUM <= 32 | 30 | 30.9 | 27.7 | 28.5 | 17.4 | 4.2 | 0.2 | 0.0 | 0.0 | 2.9 | 16.4 | 28.5 | 30.9 | 187.6 |
| MINIMUM <= 0 | 30 | 16.3 | 9.5 | 3.4 | 0.1 | 0.0 | 0.0 | 0.0 | 0.0 | 0.0 | 0.1 | 3.5 | 12.0 | 44.9 |
| **H/C** | | | | | | | | | | | | | | |
| NORMAL HEATING DEG. DAYS | 30 | 1751 | 1336 | 1109 | 660 | 327 | 103 | 27 | 46 | 274 | 648 | 1167 | 1596 | 9044 |
| NORMAL COOLING DEG. DAYS | 30 | 0 | 0 | 0 | 1 | 20 | 79 | 176 | 164 | 23 | 0 | 0 | 0 | 463 |
| **RH** | | | | | | | | | | | | | | |
| NORMAL (PERCENT) | 30 | 77 | 77 | 73 | 61 | 60 | 61 | 59 | 57 | 62 | 65 | 76 | 78 | 67 |
| HOUR 00 LST | 30 | 79 | 81 | 79 | 70 | 69 | 72 | 70 | 66 | 70 | 72 | 80 | 81 | 74 |
| HOUR 06 LST | 30 | 80 | 81 | 83 | 80 | 80 | 81 | 81 | 79 | 81 | 80 | 82 | 81 | 81 |
| HOUR 12 LST | 30 | 74 | 71 | 64 | 49 | 48 | 48 | 46 | 45 | 50 | 55 | 70 | 74 | 58 |
| HOUR 18 LST | 30 | 76 | 71 | 60 | 45 | 43 | 44 | 39 | 37 | 43 | 52 | 72 | 77 | 55 |
| **S** PERCENT POSSIBLE SUNSHINE | 46 | 52 | 57 | 61 | 60 | 62 | 66 | 74 | 74 | 67 | 59 | 45 | 50 | 61 |
| **W/O** MEAN NO. DAYS WITH: | | | | | | | | | | | | | | |
| HEAVY FOG(VISBY <= 1/4 MI) | 44 | 1.8 | 1.4 | 1.2 | 1.0 | 0.5 | 0.3 | 0.4 | 0.4 | 0.6 | 1.2 | 1.7 | 1.3 | 11.8 |
| THUNDERSTORMS | 46 | 0.0 | 0.0 | 0.1 | 0.8 | 3.3 | 7.8 | 8.9 | 6.6 | 2.2 | 0.2 | 0.0 | 0.0 | 29.9 |
| **CLOUDNESS** MEAN: | | | | | | | | | | | | | | |
| SUNRISE-SUNSET (OKTAS) | 35 | 5.4 | 5.4 | 5.3 | 5.3 | 5.0 | 4.7 | 3.7 | 3.8 | 4.3 | 4.7 | 5.3 | 5.2 | 4.8 |
| MIDNIGHT-MIDNIGHT (OKTAS) | 29 | 5.0 | 4.9 | 4.9 | 4.6 | 4.6 | 4.3 | 3.4 | 3.4 | 3.8 | 4.3 | 4.9 | 4.8 | 4.4 |
| MEAN NO. DAYS WITH: | | | | | | | | | | | | | | |
| CLEAR | 35 | 6.2 | 5.7 | 6.9 | 6.0 | 6.4 | 7.7 | 11.6 | 11.4 | 9.8 | 8.8 | 6.3 | 6.7 | 93.5 |
| PARTLY CLOUDY | 35 | 8.3 | 7.5 | 8.2 | 9.1 | 11.3 | 11.0 | 12.4 | 11.9 | 8.9 | 8.1 | 7.4 | 7.4 | 111.5 |
| CLOUDY | 35 | 16.5 | 15.1 | 15.9 | 14.8 | 13.3 | 11.3 | 6.9 | 7.7 | 11.3 | 14.1 | 16.3 | 16.9 | 160.1 |
| **PR** MEAN STATION PRESSURE(IN) | 26 | 27.99 | 28.00 | 27.97 | 27.94 | 27.91 | 27.90 | 27.95 | 27.96 | 27.96 | 27.97 | 27.96 | 27.98 | 27.96 |
| MEAN SEA-LEVEL PRES. (IN) | 26 | 30.12 | 30.12 | 30.05 | 29.98 | 29.92 | 29.89 | 29.93 | 29.94 | 29.98 | 30.02 | 30.04 | 30.10 | 30.01 |
| **WINDS** MEAN SPEED (MPH) | 26 | 8.9 | 9.0 | 9.7 | 10.4 | 10.7 | 9.9 | 8.8 | 9.0 | 8.9 | 9.2 | 8.7 | 8.8 | 9.3 |
| PREVAIL.DIR(TENS OF DEGS) | 34 | 22 | 22 | 36 | 36 | 36 | 36 | 36 | 15 | 36 | 32 | 22 | 22 | 22 |
| MAXIMUM 2-MINUTE: | | | | | | | | | | | | | | |
| SPEED (MPH) | 13 | 41 | 46 | 48 | 55 | 51 | 62 | 61 | 44 | 40 | 55 | 52 | 46 | 62 |
| DIR. (TENS OF DEGS) | | 29 | 30 | 32 | 30 | 28 | 30 | 33 | 31 | 28 | 28 | 29 | 32 | 30 |
| YEAR OF OCCURRENCE | | 2009 | 2002 | 2004 | 2000 | 2005 | 2005 | 1999 | 2003 | 2002 | 1999 | 1999 | 2004 | JUN 2005 |
| MAXIMUM 3-SECOND | | | | | | | | | | | | | | |
| SPEED (MPH) | 13 | 51 | 58 | 62 | 66 | 62 | 69 | 79 | 55 | 51 | 67 | 63 | 55 | 79 |
| DIR. (TENS OF DEGS) | | 29 | 27 | 30 | 30 | 27 | 30 | 34 | 24 | 30 | 29 | 29 | 32 | 34 |
| YEAR OF OCCURRENCE | | 2009 | 2002 | 1999 | 2000 | 2005 | 2005 | 1999 | 2007 | 2009 | 1999 | 1999 | 2004 | JUL 1999 |
| **PRECIPITATION** NORMAL (IN) | 30 | 0.54 | 0.39 | 0.74 | 1.05 | 1.88 | 2.36 | 2.28 | 1.48 | 1.35 | 0.87 | 0.65 | 0.57 | 14.16 |
| MAXIMUM MONTHLY (IN) | 48 | 1.83 | 1.75 | 2.26 | 3.31 | 7.38 | 6.16 | 6.62 | 4.66 | 3.11 | 3.56 | 3.31 | 2.50 | 7.38 |
| YEAR OF OCCURRENCE | | 1999 | 1998 | 1975 | 1967 | 1965 | 1991 | 1997 | 1993 | 1986 | 1971 | 2000 | 2008 | MAY 1965 |
| MINIMUM MONTHLY (IN) | 48 | 0.03 | .01 | 0.01 | .01 | 0.15 | 0.46 | 0.22 | 0.03 | .08 | T | 0.01 | 0.02 | T |
| YEAR OF OCCURRENCE | | 1973 | 2005 | 1966 | 2005 | 1980 | 1997 | 2006 | 2001 | 2005 | 1965 | 1999 | 1997 | OCT 1965 |
| MAXIMUM IN 24 HOURS (IN) | 48 | 0.77 | 0.98 | 1.39 | 2.25 | 2.05 | 2.29 | 5.03 | 3.11 | 2.24 | 2.21 | 2.07 | 1.07 | 5.03 |
| YEAR OF OCCURRENCE | | 1995 | 2009 | 2003 | 2006 | 1965 | 1994 | 1963 | 1993 | 1971 | 1971 | 2000 | 2008 | JUL 1963 |
| NORMAL NO. DAYS WITH: | | | | | | | | | | | | | | |
| PRECIPITATION >= 0.01 | 30 | 8.0 | 6.1 | 7.8 | 8.1 | 10.0 | 10.6 | 9.0 | 8.0 | 7.4 | 5.6 | 6.7 | 8.7 | 96.0 |
| PRECIPITATION >= 1.00 | 30 | 0.0 | 0.0 | 0.0 | 0.1 | 0.3 | 0.3 | 0.3 | 0.2 | 0.2 | 0.1 | * | 0.0 | 1.5 |
| **SNOWFALL** NORMAL (IN) | 30 | 8.8 | 5.7 | 7.4 | 3.7 | 0.9 | 0.* | 0.0 | 0.0 | 0.3 | 1.7 | 6.7 | 8.2 | 43.4 |
| MAXIMUM MONTHLY (IN) | 48 | 28.3 | 21.7 | 30.9 | 22.2 | 15.5 | T | T | T | 4.0 | 14.2 | 21.2 | 32.0 | 32.0 |
| YEAR OF OCCURRENCE | | 1999 | 1994 | 1975 | 1970 | 1983 | 2009 | 2009 | 1993 | 1984 | 1985 | 2000 | 2008 | DEC 2008 |
| MAXIMUM IN 24 HOURS (IN) | 48 | 12.6 | 9.2 | 9.7 | 15.0 | 14.6 | T | T | T | 4.0 | 10.5 | 7.9 | 10.1 | 15.0 |
| YEAR OF OCCURRENCE' | | 1995 | 1994 | 1985 | 1986 | 1983 | 2009 | 2009 | 1993 | 1984 | 1985 | 1975 | 1978 | APR 1986 |
| MAXIMUM SNOW DEPTH (IN) | 47 | 24 | 24 | 22 | 17 | 12 | 0 | 0 | 0 | 2 | 5 | 13 | 17 | 24 |
| YEAR OF OCCURRENCE | | 2004 | 2004 | 1975 | 1975 | 1983 | | | | 1984 | 2008 | 2000 | 1978 | FEB 2004 |
| NORMAL NO. DAYS WITH: | | | | | | | | | | | | | | |
| SNOWFALL >= 1.0 | 30 | 2.3 | 1.8 | 2.3 | 1.0 | 0.2 | 0.0 | 0.0 | 0.0 | 0.1 | 0.5 | 2.0 | 2.7 | 12.9 |

## PRECIPITATION (inches) 2009 WILLISTON (KISN)

| YEAR | JAN | FEB | MAR | APR | MAY | JUN | JUL | AUG | SEP | OCT | NOV | DEC | ANNUAL |
|------|-----|-----|-----|-----|-----|-----|-----|-----|-----|-----|-----|-----|--------|
| 1980 | 0.58 | 0.28 | 0.16 | 0.40 | 0.15 | 1.80 | 0.54 | 1.83 | 2.24 | 1.62 | 0.48 | 0.72 | 10.80 |
| 1981 | 0.08 | 0.21 | 0.18 | 0.39 | 1.18 | 3.58 | 2.47 | 0.87 | 0.57 | 0.51 | 0.51 | 0.36 | 10.91 |
| 1982 | 1.28 | 0.45 | 1.34 | 1.02 | 1.79 | 2.77 | 1.17 | 1.96 | 1.86 | 1.91 | 0.07 | 1.43 | 17.05 |
| 1983 | 0.35 | 0.07 | 1.07 | 0.03 | 1.98 | 0.87 | 2.12 | 0.87 | 0.51 | 0.30 | 0.78 | 0.57 | 9.52 |
| 1984 | 0.81 | 0.12 | 0.90 | 1.48 | 0.47 | 1.87 | 0.93 | 0.65 | 2.24 | 0.49 | 0.33 | 0.27 | 10.56 |
| 1985 | 0.13 | 0.24 | 1.07 | 1.53 | 1.08 | 1.12 | 1.40 | 1.96 | 1.08 | 1.86 | 0.58 | 0.64 | 12.69 |
| 1986 | 0.35 | 0.62 | 0.83 | 2.57 | 3.39 | 2.19 | 5.70 | 1.06 | 3.11 | 0.78 | 1.15 | 0.09 | 21.84 |
| 1987 | 0.32 | 0.19 | 1.70 | 0.29 | 2.02 | 0.71 | 4.97 | 0.47 | 0.65 | 0.11 | 0.18 | 0.07 | 11.68 |
| 1988 | 0.75 | 0.28 | 0.63 | 0.15 | 1.39 | 3.46 | 0.52 | 0.39 | 1.49 | 0.28 | 0.51 | 0.79 | 10.64 |
| 1989 | 1.00 | 0.39 | 0.53 | 1.44 | 1.68 | 1.48 | 0.87 | 1.01 | 0.54 | 1.75 | 0.44 | 0.41 | 11.54 |
| 1990 | 0.36 | 0.04 | 0.44 | 0.61 | 2.08 | 1.77 | 1.32 | 1.73 | 0.31 | 0.10 | 0.14 | 0.40 | 9.30 |
| 1991 | 0.18 | 0.37 | 0.44 | 1.60 | 2.60 | 6.16 | 1.04 | 1.46 | 0.38 | 0.46 | 0.18 | 0.18 | 17.61 |
| 1992 | 0.48 | 0.10 | 0.56 | 2.28 | 1.44 | 1.92 | 1.04 | 2.18 | 0.76 | 0.52 | 0.77 | 0.44 | 12.49 |
| 1993 | 0.23 | 0.23 | 0.88 | 0.41 | 0.81 | 3.76 | 6.28 | 4.66 | 0.28 | 0.09 | 0.95 | 0.38 | 18.96 |
| 1994 | 0.96 | 0.89 | 0.22 | 1.37 | 2.70 | 3.94 | 0.31 | 1.78 | 0.69 | 1.73 | 0.25 | 1.04 | 15.88 |
| 1995 | 0.91 | 0.21 | 1.11 | 0.87 | 2.11 | 1.36 | 3.49 | 2.49 | 0.54 | 0.39 | 0.79 | 0.67 | 14.94 |
| 1996 | 0.79 | 0.18 | 0.75 | 0.54 | 1.08 | 2.01 | 3.37 | .52 | 2.07 | .82 | 1.27 | 1.29 | 14.69 |
| 1997 | 0.18 | 0.21 | 0.94 | 0.27 | 0.19 | 0.46 | 6.62 | 1.72 | 0.15 | 0.61 | 0.40 | 0.02 | 11.77 |
| 1998 | 0.36 | 1.75 | 0.28 | 0.10 | 1.03 | 3.01 | 1.74 | 3.41 | 0.89 | 3.24 | 1.27 | 0.74 | 17.82 |
| 1999 | 1.83 | 0.46 | 0.40 | 0.32 | 3.29 | 1.89 | 3.90 | 1.04 | 1.57 | 0.18 | 0.01 | 0.30 | 15.19 |
| 2000 | 0.38 | 0.46 | 0.60 | 1.65 | 2.54 | 2.84 | 3.66 | 0.93 | 1.60 | 0.76 | 3.31 | 0.52 | 19.25 |
| 2001 | 0.30 | 0.10 | 0.02 | 2.35 | 0.81 | 4.54 | 4.30 | 0.03 | 0.30 | 0.29 | 0.12 | 0.66 | 13.82 |
| 2002 | 0.70 | 0.55 | 0.85 | 0.88 | 1.21 | 3.83 | 2.65 | 1.95 | 0.49 | 1.15 | 0.26 | 0.62 | 15.14 |
| 2003 | 0.58 | 0.37 | 1.84 | 1.64 | 2.29 | 2.37 | 1.94 | 0.92 | 0.78 | 0.69 | 0.56 | 0.58 | 14.56 |
| 2004 | 1.25 | 0.24 | 0.29 | 0.42 | 3.20 | 1.83 | 2.37 | 0.88 | 1.02 | 0.99 | 0.12 | 0.50 | 13.11 |
| 2005 | 0.47 | 0.01 | 0.48 | 0.01 | 3.26 | 4.64 | 1.65 | 0.66 | 0.08 | 1.40 | 0.85 | 0.23 | 13.74 |
| 2006 | 0.27 | 0.17 | 1.27 | 3.18 | 1.53 | 1.05 | 0.22 | 1.41 | 1.32 | 0.84 | 0.23 | 0.32 | 11.81 |
| 2007 | 0.12 | 0.84 | 0.62 | 0.40 | 4.65 | 3.32 | 1.95 | 0.58 | 0.62 | 1.07 | 0.19 | 0.10 | 14.46 |
| 2008 | 0.19 | 0.27 | 0.45 | 0.27 | 1.06 | 1.93 | 1.05 | 1.27 | 1.64 | 1.89 | 1.57 | 2.50 | 14.09 |
| 2009 | 0.90 | 1.40 | 0.06 | 1.04 | 0.68 | 2.08 | 3.31 | 2.30 | 0.37 | 1.17 | 0.02 | 0.48 | 13.81 |
| POR= 48 YRS | 0.56 | 0.42 | 0.68 | 1.16 | 1.99 | 2.52 | 2.33 | 1.41 | 1.15 | 0.86 | 0.56 | 0.59 | 14.23 |

WBAN : 94014

## AVERAGE TEMPERATURE (°F) 2009 WILLISTON (KISN)

| YEAR | JAN | FEB | MAR | APR | MAY | JUN | JUL | AUG | SEP | OCT | NOV | DEC | ANNUAL |
|------|-----|-----|-----|-----|-----|-----|-----|-----|-----|-----|-----|-----|--------|
| 1980 | 7.7 | 14.9 | 25.6 | 49.7 | 60.6 | 67.1 | 73.2 | 65.7 | 58.1 | 46.6 | 33.2 | 16.2 | 43.2 |
| 1981 | 21.8 | 23.3 | 38.1 | 49.1 | 59.2 | 63.8 | 73.8 | 74.3 | 61.8 | 45.4 | 34.6 | 16.1 | 46.8 |
| 1982 | -5.8 | 11.8 | 26.3 | 41.8 | 54.9 | 64.3 | 71.7 | 70.3 | 56.4 | 44.4 | 24.0 | 16.0 | 39.7 |
| 1983 | 20.0 | 25.0 | 30.0 | 41.2 | 51.1 | 63.7 | 74.9 | 78.0 | 56.3 | 47.2 | 30.3 | -4.5 | 42.8 |
| 1984 | 15.2 | 30.9 | 28.3 | 47.4 | 54.5 | 63.4 | 73.0 | 73.7 | 50.2 | 41.0 | 27.2 | 5.5 | 42.5 |
| 1985 | 7.0 | 13.2 | 32.4 | 47.0 | 58.9 | 58.8 | 71.2 | 64.6 | 51.2 | 42.8 | 13.2 | 11.2 | 39.3 |
| 1986 | 22.2 | 13.3 | 39.8 | 41.9 | 55.8 | 66.8 | 67.3 | 66.8 | 52.0 | 45.0 | 22.6 | 22.1 | 43.0 |
| 1987 | 20.7 | 29.2 | 32.1 | 51.6 | 59.8 | 68.7 | 70.0 | 63.7 | 59.5 | 42.3 | 34.2 | 22.9 | 46.2 |
| 1988 | 7.6 | 15.1 | 33.2 | 45.3 | 61.6 | 77.3 | 72.9 | 69.3 | 55.7 | 44.2 | 27.7 | 17.1 | 43.9 |
| 1989 | 12.9 | 2.2 | 23.2 | 43.7 | 56.3 | 63.2 | 75.7 | 70.2 | 58.0 | 45.1 | 29.3 | 10.5 | 40.9 |
| 1990 | 22.9 | 20.5 | 34.9 | 43.2 | 54.5 | 65.5 | 69.9 | 71.4 | 61.9 | 44.6 | 30.3 | 9.1 | 44.1 |
| 1991 | 7.7 | 29.1 | 32.6 | 47.0 | 57.4 | 66.4 | 70.5 | 73.6 | 57.3 | 40.8 | 25.2 | 23.3 | 44.2 |
| 1992 | 22.7 | 28.2 | 36.2 | 43.4 | 55.6 | 63.4 | 64.0 | 63.9 | 55.4 | 44.0 | 27.6 | 9.9 | 42.8 |
| 1993 | 5.7 | 9.6 | 33.4 | 43.9 | 56.1 | 58.8 | 62.3 | 63.9 | 52.3 | 42.1 | 24.9 | 20.5 | 39.5 |
| 1994 | 2.2 | 3.2 | 34.2 | 43.4 | 56.7 | 62.4 | 67.7 | 68.2 | 61.1 | 46.6 | 29.0 | 15.7 | 40.9 |
| 1995 | 9.4 | 19.9 | 28.3 | 38.7 | 52.1 | 66.0 | 68.5 | 70.2 | 55.8 | 43.8 | 24.9 | 11.8 | 40.8 |
| 1996 | -.1 | 19.5 | 18.8 | 39.0 | 49.6 | 65.0 | 68.4 | 71.6 | 55.2 | 42.1 | 15.5 | 5.3 | 37.5 |
| 1997 | 1.4 | 17.3 | 24.0 | 38.7 | 52.3 | 69.5 | 68.7 | 68.7 | 61.2 | 44.8 | 26.0 | 25.3 | 41.5 |
| 1998 | 8.7 | 30.8 | 20.2 | 46.6 | 55.7 | 59.4 | 71.7 | 73.0 | 63.1 | 45.7 | 29.7 | 16.6 | 43.4 |
| 1999 | 7.4 | 22.4 | 32.5 | 43.6 | 52.9 | 62.3 | 68.1 | 69.5 | 53.1 | 43.5 | 37.4 | 26.1 | 43.2 |
| 2000 | 11.7 | 21.2 | 35.8 | 42.6 | 55.6 | 60.0 | 72.1 | 69.5 | 57.9 | 46.0 | 17.2 | 5.4 | 41.3 |
| 2001 | 17.5 | 5.7 | 30.3 | 42.9 | 54.2 | 61.1 | 71.1 | 70.8 | 59.0 | 42.3 | 35.3 | 14.4 | 42.1 |
| 2002 | 16.6 | 24.9 | 13.9 | 37.5 | 48.8 | 63.8 | 73.5 | 65.4 | 56.9 | 32.4 | 29.4 | 19.5 | 40.2 |
| 2003 | 9.7 | 7.4 | 21.6 | 46.1 | 52.5 | 61.2 | 72.3 | 74.7 | 56.1 | 47.8 | 15.0 | 17.2 | 40.1 |
| 2004 | 4.2 | 14.6 | 32.4 | 43.5 | 49.6 | 57.1 | 68.3 | 63.8 | 58.8 | 42.7 | 32.8 | 21.1 | 40.7 |
| 2005 | 5.2 | 23.2 | 33.2 | 47.5 | 51.9 | 64.8 | 71.1 | 67.4 | 59.2 | 44.5 | 32.5 | 17.7 | 43.2 |
| 2006 | 26.6 | 18.3 | 28.9 | 48.4 | 55.3 | 64.7 | 74.8 | 70.4 | 56.5 | 38.5 | 27.9 | 19.2 | 44.1 |
| 2007 | 15.7 | 10.2 | 35.8 | 41.2 | 55.1 | 64.6 | 75.9 | 68.1 | 57.8 | 45.8 | 29.3 | 13.8 | 42.8 |
| 2008 | 10.9 | 12.4 | 30.2 | 41.5 | 54.1 | 61.3 | 71.4 | 70.0 | 56.1 | 43.6 | 31.6 | 5.3 | 40.7 |
| 2009 | 7.8 | 10.0 | 20.6 | 41.4 | 51.9 | 60.8 | 65.9 | 65.3 | 64.3 | 38.0 | 35.6 | 5.8 | 39.0 |
| POR= 48 YRS | 9.6 | 16.6 | 28.3 | 43.2 | 54.6 | 63.9 | 70.4 | 68.7 | 57.1 | 44.2 | 27.6 | 14.7 | 41.6 |

## HEATING DEGREE DAYS (base 65°F) 2009 WILLISTON (KISN)

| YEAR | JUL | AUG | SEP | OCT | NOV | DEC | JAN | FEB | MAR | APR | MAY | JUN | TOTAL |
|------|-----|-----|-----|-----|-----|-----|-----|-----|-----|-----|-----|-----|-------|
| 1980-81 | 5 | 52 | 221 | 567 | 947 | 1511 | 1330 | 1162 | 825 | 473 | 205 | 70 | 7368 |
| 1981-82 | 4 | 9 | 149 | 600 | 903 | 1509 | 2198 | 1485 | 1191 | 692 | 322 | 79 | 9141 |
| 1982-83 | 2 | 46 | 278 | 633 | 1223 | 1516 | 1389 | 1115 | 1076 | 708 | 436 | 101 | 8523 |
| 1983-84 | 1 | 1 | 300 | 544 | 1031 | 2158 | 1538 | 983 | 1131 | 520 | 343 | 92 | 8642 |
| 1984-85 | 0 | 25 | 444 | 741 | 1128 | 1845 | 1793 | 1446 | 1001 | 535 | 210 | 201 | 9369 |
| 1985-86 | 21 | 73 | 415 | 683 | 1553 | 1664 | 1321 | 1446 | 775 | 685 | 303 | 35 | 8974 |
| 1986-87 | 14 | 48 | 386 | 612 | 1266 | 1324 | 1369 | 997 | 1012 | 402 | 190 | 33 | 7653 |
| 1987-88 | 31 | 88 | 183 | 698 | 920 | 1301 | 1781 | 1444 | 980 | 585 | 175 | 8 | 8194 |
| 1988-89 | 1 | 35 | 279 | 637 | 1114 | 1479 | 1613 | 1757 | 1292 | 640 | 275 | 115 | 9237 |
| 1989-90 | 0 | 35 | 225 | 609 | 1065 | 1690 | 1294 | 1241 | 927 | 649 | 333 | 88 | 8156 |
| 1990-91 | 19 | 7 | 166 | 626 | 1034 | 1728 | 1775 | 999 | 997 | 533 | 253 | 20 | 8157 |
| 1991-92 | 5 | 8 | 262 | 744 | 1188 | 1289 | 1306 | 1058 | 887 | 649 | 313 | 112 | 7821 |
| 1992-93 | 75 | 111 | 296 | 644 | 1116 | 1704 | 1838 | 1551 | 973 | 630 | 277 | 203 | 9418 |
| 1993-94 | 100 | 87 | 377 | 704 | 1197 | 1376 | 1945 | 1727 | 945 | 638 | 257 | 103 | 9456 |
| 1994-95 | 19 | 63 | 138 | 563 | 1072 | 1524 | 1719 | 1259 | 1136 | 784 | 396 | 101 | 8774 |
| 1995-96 | 24 | 16 | 290 | 648 | 1197 | 1644 | 2023 | 1315 | 1427 | 776 | 470 | 84 | 9914 |
| 1996-97 | 23 | 6 | 293 | 702 | 1480 | 1844 | 1961 | 1330 | 1266 | 781 | 396 | 20 | 10102 |
| 1997-98 | 34 | 37 | 130 | 630 | 1167 | 1223 | 1740 | 953 | 1381 | 546 | 290 | 180 | 8311 |
| 1998-99 | 9 | 0 | 161 | 591 | 1051 | 1490 | 1778 | 1188 | 1001 | 635 | 371 | 118 | 8393 |
| 1999-00 | 27 | 17 | 352 | 661 | 817 | 1198 | 1645 | 1264 | 898 | 668 | 289 | 166 | 8002 |
| 2000-01 | 15 | 34 | 231 | 580 | 1425 | 1841 | 1467 | 1652 | 1067 | 661 | 339 | 157 | 9469 |
| 2001-02 | 12 | 15 | 211 | 698 | 883 | 1561 | 1491 | 1115 | 1581 | 821 | 510 | 114 | 9012 |
| 2002-03 | 0 | 67 | 265 | 1003 | 1064 | 1404 | 1706 | 1607 | 1336 | 561 | 396 | 153 | 9562 |
| 2003-04 | 6 | 13 | 308 | 527 | 1495 | 1473 | 1878 | 1455 | 1002 | 638 | 469 | 245 | 9509 |
| 2004-05 | 50 | 86 | 214 | 685 | 960 | 1354 | 1850 | 1166 | 978 | 518 | 398 | 76 | 8335 |
| 2005-06 | 13 | 42 | 216 | 636 | 970 | 1458 | 1185 | 1302 | 1110 | 475 | 323 | 53 | 7783 |
| 2006-07 | 4 | 12 | 258 | 817 | 1105 | 1411 | 1519 | 1527 | 896 | 708 | 307 | 77 | 8641 |
| 2007-08 | 4 | 45 | 248 | 589 | 1063 | 1581 | 1668 | 1516 | 1071 | 701 | 335 | 132 | 8953 |
| 2008-09 | 4 | 32 | 266 | 658 | 994 | 1842 | 1769 | 1534 | 1369 | 704 | 402 | 171 | 9745 |
| 2009- | 45 | 66 | 91 | 832 | 876 | 1829 | | | | | | | |

WBAN : 94014

## COOLING DEGREE DAYS (base 65°F) 2009 WILLISTON (KISN)

| YEAR | JAN | FEB | MAR | APR | MAY | JUN | JUL | AUG | SEP | OCT | NOV | DEC | TOTAL |
|------|-----|-----|-----|-----|-----|-----|-----|-----|-----|-----|-----|-----|-------|
| 1980 | 0 | 0 | 0 | 7 | 107 | 114 | 265 | 84 | 22 | 4 | 0 | 0 | 603 |
| 1981 | 0 | 0 | 0 | 2 | 31 | 41 | 285 | 301 | 60 | 0 | 0 | 0 | 720 |
| 1982 | 0 | 0 | 0 | 0 | 13 | 64 | 217 | 215 | 29 | 0 | 0 | 0 | 538 |
| 1983 | 0 | 0 | 0 | 0 | 8 | 68 | 314 | 413 | 45 | 0 | 0 | 0 | 848 |
| 1984 | 0 | 0 | 0 | 0 | 21 | 54 | 251 | 304 | 8 | 0 | 0 | 0 | 638 |
| 1985 | 0 | 0 | 0 | 0 | 26 | 21 | 218 | 68 | 6 | 0 | 0 | 0 | 339 |
| 1986 | 0 | 0 | 0 | 0 | 0 | 27 | 96 | 94 | 110 | 0 | 0 | 0 | 327 |
| 1987 | 0 | 0 | 0 | 6 | 37 | 150 | 192 | 53 | 22 | 0 | 0 | 0 | 460 |
| 1988 | 0 | 0 | 0 | 0 | 75 | 383 | 253 | 175 | 4 | 0 | 0 | 0 | 890 |
| 1989 | 0 | 0 | 0 | 5 | 9 | 68 | 338 | 202 | 22 | 0 | 0 | 0 | 644 |
| 1990 | 0 | 0 | 0 | 4 | 17 | 110 | 175 | 211 | 75 | 0 | 0 | 0 | 592 |
| 1991 | 0 | 0 | 0 | 0 | 25 | 68 | 182 | 282 | 40 | 0 | 0 | 0 | 597 |
| 1992 | 0 | 0 | 0 | 3 | 29 | 71 | 50 | 87 | 13 | 2 | 0 | 0 | 255 |
| 1993 | 0 | 0 | 0 | 0 | 8 | 28 | 22 | 58 | 2 | 0 | 0 | 0 | 118 |
| 1994 | 0 | 0 | 0 | 0 | 5 | 31 | 110 | 169 | 27 | 0 | 0 | 0 | 342 |
| 1995 | 0 | 0 | 0 | 0 | 3 | 137 | 137 | 183 | 22 | 0 | 0 | 0 | 482 |
| 1996 | 0 | 0 | 0 | 0 | 0 | 89 | 137 | 215 | 7 | 0 | 0 | 0 | 448 |
| 1997 | 0 | 0 | 0 | 0 | 9 | 164 | 157 | 157 | 22 | 10 | 0 | 0 | 519 |
| 1998 | 0 | 0 | 0 | 0 | 6 | 20 | 223 | 257 | 108 | 0 | 0 | 0 | 614 |
| 1999 | 0 | 0 | 0 | 0 | 3 | 40 | 129 | 162 | 0 | 0 | 0 | 0 | 334 |
| 2000 | 0 | 0 | 0 | 0 | 3 | 24 | 240 | 181 | 28 | 0 | 0 | 0 | 476 |
| 2001 | 0 | 0 | 0 | 5 | 11 | 48 | 207 | 203 | 38 | 0 | 0 | 0 | 512 |
| 2002 | 0 | 0 | 0 | 0 | 15 | 84 | 269 | 83 | 29 | 0 | 0 | 0 | 480 |
| 2003 | 0 | 0 | 0 | 0 | 16 | 46 | 241 | 322 | 46 | 0 | 0 | 0 | 671 |
| 2004 | 0 | 0 | 0 | 0 | 0 | 16 | 164 | 57 | 34 | 0 | 0 | 0 | 271 |
| 2005 | 0 | 0 | 0 | 1 | 1 | 76 | 207 | 122 | 50 | 7 | 0 | 0 | 464 |
| 2006 | 0 | 0 | 0 | 0 | 26 | 52 | 313 | 185 | 9 | 0 | 0 | 0 | 585 |
| 2007 | 0 | 0 | 0 | 0 | 8 | 71 | 353 | 147 | 37 | 0 | 0 | 0 | 616 |
| 2008 | 0 | 0 | 0 | 0 | 3 | 30 | 211 | 192 | 6 | 1 | 0 | 0 | 443 |
| 2009 | 0 | 0 | 0 | 0 | 2 | 53 | 80 | 83 | 78 | 0 | 0 | 0 | 296 |

## SNOWFALL (inches) 2009 WILLISTON (KISN)

| YEAR | JUL | AUG | SEP | OCT | NOV | DEC | JAN | FEB | MAR | APR | MAY | JUN | TOTAL |
|------|-----|-----|-----|-----|-----|-----|-----|-----|-----|-----|-----|-----|-------|
| 1980-81 | 0.0 | 0.0 | 0.0 | 1.1 | 3.3 | 10.4 | 1.2 | 3.1 | T | 0.0 | T | 0.0 | 19.1 |
| 1981-82 | 0.0 | 0.0 | 0.0 | 0.3 | 4.6 | 5.7 | 24.3 | 8.2 | 17.0 | 9.8 | 0.5 | 0.0 | 70.4 |
| 1982-83 | 0.0 | 0.0 | T | T | 0.8 | 13.4 | 1.7 | 1.6 | 9.2 | 0.1 | 15.5 | 0.0 | 42.3 |
| 1983-84 | 0.0 | 0.0 | 1.0 | T | 6.0 | 9.2 | 7.2 | 0.5 | 7.9 | 13.5 | 0.3 | 0.0 | 45.6 |
| 1984-85 | 0.0 | 0.0 | 4.0 | 4.5 | 3.5 | 6.5 | 1.8 | 3.1 | 11.4 | 1.5 | 0.0 | 0.0 | 36.3 |
| 1985-86 | 0.0 | 0.0 | 0.1 | 14.2 | 11.3 | 9.2 | 3.6 | 8.7 | 4.4 | 17.6 | T | 0.0 | 69.1 |
| 1986-87 | 0.0 | 0.0 | 0.0 | T | 13.4 | 1.2 | 4.2 | 1.8 | 14.2 | 0.4 | 0.0 | 0.0 | 35.2 |
| 1987-88 | 0.0 | 0.0 | 0.0 | 0.2 | T | 1.2 | 8.9 | 3.4 | 3.7 | T | 0.0 | 0.0 | 17.4 |
| 1988-89 | 0.0 | 0.0 | 0.0 | T | 4.2 | 13.3 | 22.8 | 6.2 | 1.8 | 1.3 | 0.0 | T | 49.6 |
| 1989-90 | T | 0.0 | T | 2.2 | 4.6 | 5.5 | 6.0 | 0.8 | 4.3 | 2.4 | 2.1 | T | 27.9 |
| 1990-91 | T | T | 0.0 | T | 1.6 | 8.7 | 4.4 | 5.2 | 7.2 | 1.8 | 2.7 | T | 31.6 |
| 1991-92 | T | T | T | 5.3 | 3.3 | 2.7 | 6.3 | 2.2 | 0.9 | 3.5 | 0.0 | 0.0 | 24.2 |
| 1992-93 | 0.0 | T | 0.0 | 3.8 | 8.5 | 10.7 | 7.3 | 4.1 | 2.6 | T | 0.0 | 0.0 | 37.0 |
| 1993-94 | T | T | T | 0.4 | 16.4 | 5.2 | 21.6 | 21.7 | 2.4 | 4.5 | 0.0 | T | 72.2 |
| 1994-95 | 0.0 | 0.0 | 0.0 | T | 3.6 | 14.0 | 16.1 | 2.6 | 9.0 | 1.9 | 0.1 | 0.0 | 47.3 |
| 1995-96 | T | 0.0 | T | 0.2 | 8.2 | 15.6 | 18.0 | 3.7 | 16.5 | 1.3 | T | 0.0 | 63.5 |
| 1996-97 | 0.0 | 0.0 | 0.0 | 5.8 | 15.6 | 19.8 | 3.2 | 2.6 | 12.3 | 0.5 | T | 0.0 | 59.8 |
| 1997-98 | 0.0 | 0.0 | 0.0 | 0.3 | 4.7 | 0.1 | 7.1 | 20.3 | 4.8 | 0.2 | 0.1 | T | 37.6 |
| 1998-99 | 0.0 | T | 0.0 | T | 15.3 | 11.3 | 28.3 | 4.7 | 4.2 | 4.2 | 4.2 | 0.0 | 72.2 |
| 1999-00 | T | 0.0 | 0.0 | 0.1 | 0.1 | 3.6 | 8.1 | 5.7 | 6.2 | 11.3 | 0.2 | 0.0 | 35.3 |
| 2000-01 | 0.0 | 0.0 | T | 0.4 | 21.2 | 7.5 | 4.8 | 1.7 | 0.4 | 9.1 | 0.0 | 0.0 | 45.1 |
| 2001-02 | 0.0 | 0.0 | 0.0 | 1.7 | 0.6 | 6.8 | 4.8 | 6.6 | 15.1 | 2.8 | 11.2 | 0.0 | 49.6 |
| 2002-03 | 0.0 | 0.0 | 0.0 | 7.5 | 2.8 | 6.6 | 7.3 | 8.9 | 7.7 | 13.5 | T | 0.0 | 54.3 |
| 2003-04 | 0.0 | 0.0 | 0.0 | 7.0 | 9.7 | 12.3 | 25.4 | 3.3 | 2.7 | 2.8 | 1.0 | 0.0 | 64.2 |
| 2004-05 | 0.0 | 0.0 | 0.0 | 0.2 | 0.3 | 13.0 | 14.8 | 0.2 | 6.6 | T | 0.5 | 0.0 | 35.6 |
| 2005-06 | 0.0 | 0.0 | 0.0 | 5.3 | 4.9 | 3.2 | 3.3 | 2.9 | 4.0 | 2.3 | T | 0.0 | 25.9 |
| 2006-07 | 0.0 | 0.0 | 0.0 | 5.2 | 3.7 | 5.3 | 2.1 | 12.7 | 2.5 | 2.5 | T | T | 34.0 |
| 2007-08 | 0.0 | 0.0 | 0.0 | 0.0 | 1.0 | 2.1 | 4.0 | 4.2 | 3.6 | T | T | T | 14.9 |
| 2008-09 | 0.0 | 0.0 | 0.0 | 8.4 | 0.2 | 32.0 | 10.8 | 16.7 | 1.1 | 0.8 | T | T | 70.0 |
| 2009- | T | 0.0 | 0.0 | 1.5 | T | 8.7 | | | | | | | |
| POR= 48 YRS | T | T | 0.2 | 2.0 | 5.2 | 8.2 | 8.5 | 5.7 | 6.4 | 4.4 | 0.9 | T | 41.5 |

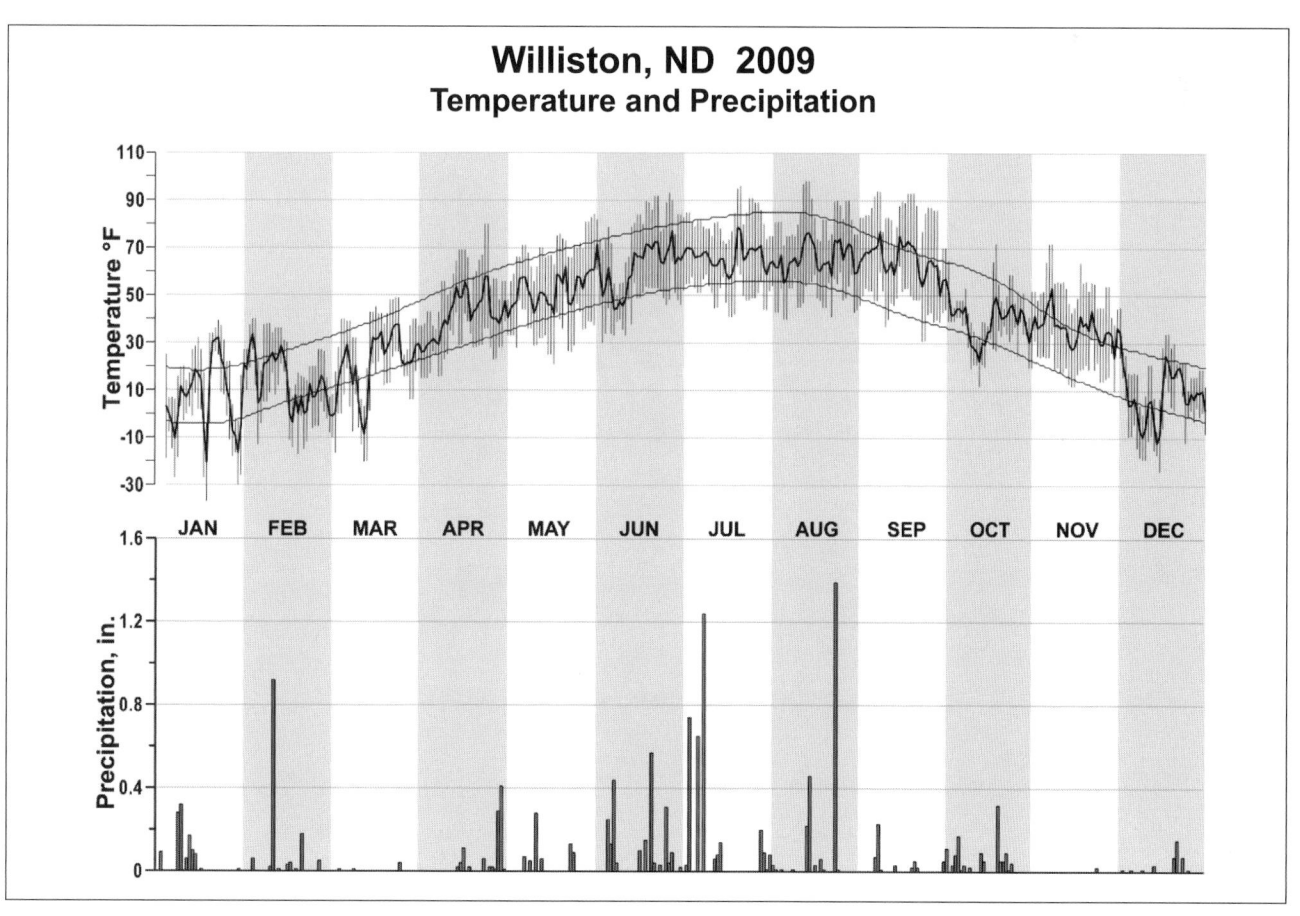

# 2009
# CLEVELAND
# OHIO (KCLE)

Cleveland is on the south shore of Lake Erie in northeast Ohio. The metropolitan area has a lake frontage of 31 miles. The surrounding terrain is generally level except for an abrupt ridge on the eastern edge of the city which rises some 500 feet above the shore terrain. The Cuyahoga River, which flows through a rather deep but narrow north-south valley, bisects the city.

Local climate is continental in character but with strong modifying influences by Lake Erie. West to northerly winds blowing off Lake Erie tend to lower daily high temperatures in summer and raise temperatures in winter. Temperatures at Hopkins Airport which is 5 miles south of the lakeshore average from 2-4 degrees higher than the lakeshore in summer, while overnight low temperatures average from 2-4 degrees lower than the lakefront during all seasons.

In this area, summers are moderately warm and humid with occasional days when temperatures exceed 90 degrees. Winters are relatively cold and cloudy with an average of 5 days with sub-zero temperatures. Weather changes occur every few days from the passing of cold fronts.

The daily range in temperature is usually greatest in late summer and least in winter. Annual extremes in temperature normally occur soon after late June and December. Maximum temperatures below freezing occur most often in December, January, and February. Temperatures of 100 degrees or higher are rare. On the average, freezing temperatures in fall are first recorded in October while the last freezing temperature in spring normally occurs in April.

As is characteristic of continental climates, precipitation varies widely from year to year. However, it is normally abundant and well distributed throughout the year with spring being the wettest season. Showers and thunderstorms account for most of the rainfall during the growing season. Thunderstorms are most frequent from April through August. Snowfall may fluctuate widely. Mean annual snowfall increases from west to east in Cuyahoga County ranging from about 45 inches in the west to more than 90 inches in the extreme east.

Damaging winds of 50 mph or greater are usually associated with thunderstorms. Tornadoes, one of the most destructive of all atmospheric storms, occasionally occur in Cuyahoga County.

# NORMALS, MEANS, AND EXTREMES
## CLEVELAND (KCLE)

| LATITUDE: | LONGITUDE: | ELEVATION (FT): | TIME ZONE: | WBAN: 14820 |
|---|---|---|---|---|
| 41 ° 24'N | -81 ° 51'W | GRND: 778  BARO: 805 | EASTERN  (UTC -5) | |

| | ELEMENT | POR | JAN | FEB | MAR | APR | MAY | JUN | JUL | AUG | SEP | OCT | NOV | DEC | YEAR |
|---|---|---|---|---|---|---|---|---|---|---|---|---|---|---|---|
| TEMPERATURE °F | NORMAL DAILY MAXIMUM | 30 | 32.6 | 35.8 | 46.1 | 57.3 | 68.6 | 77.4 | 81.4 | 79.2 | 72.3 | 60.8 | 48.7 | 37.4 | 58.1 |
| | MEAN DAILY MAXIMUM | 113 | 34.3 | 34.0 | 45.1 | 55.9 | 67.8 | 76.0 | 81.4 | 79.6 | 72.4 | 62.3 | 48.6 | 37.7 | 57.9 |
| | HIGHEST DAILY MAXIMUM | 68 | 73 | 74 | 83 | 88 | 92 | 104 | 103 | 102 | 101 | 90 | 82 | 77 | 104 |
| | YEAR OF OCCURRENCE | | 1950 | 2000 | 1945 | 1986 | 1959 | 1988 | 1941 | 1948 | 1953 | 1946 | 1950 | 1982 | JUN 1988 |
| | MEAN OF EXTREME MAXS. | 113 | 56.4 | 58.7 | 72.0 | 80.4 | 85.5 | 91.5 | 93.0 | 91.3 | 88.2 | 80.1 | 70.0 | 59.6 | 77.2 |
| | NORMAL DAILY MINIMUM | 30 | 18.8 | 21.0 | 28.9 | 37.9 | 48.3 | 57.7 | 62.3 | 61.2 | 54.3 | 43.7 | 34.9 | 24.9 | 41.2 |
| | MEAN DAILY MINIMUM | 113 | 20.4 | 19.9 | 28.8 | 38.1 | 49.2 | 57.8 | 63.6 | 62.3 | 54.9 | 45.2 | 34.8 | 25.1 | 41.7 |
| | LOWEST DAILY MINIMUM | 68 | -20 | -15 | -5 | 10 | 25 | 31 | 41 | 38 | 32 | 19 | 3 | -15 | -20 |
| | YEAR OF OCCURRENCE | | 1994 | 1963 | 1984 | 1964 | 1966 | 1972 | 1968 | 1982 | 1942 | 1988 | 1976 | 1989 | JAN 1994 |
| | MEAN OF EXTREME MINS. | 113 | -0.5 | 1.4 | 11.0 | 23.8 | 33.8 | 43.5 | 50.4 | 48.8 | 39.8 | 30.1 | 19.5 | 6.4 | 25.7 |
| | NORMAL DRY BULB | 30 | 25.7 | 28.4 | 37.5 | 47.6 | 58.5 | 67.5 | 71.9 | 70.2 | 63.3 | 52.2 | 41.8 | 31.1 | 49.6 |
| | MEAN DRY BULB | 113 | 27.4 | 26.9 | 37.0 | 47.1 | 58.5 | 67.0 | 72.5 | 70.9 | 63.7 | 53.8 | 41.7 | 31.4 | 49.8 |
| | MEAN WET BULB | 26 | 24.8 | 26.3 | 32.7 | 42.4 | 52.2 | 61.3 | 65.1 | 64.6 | 58.4 | 47.6 | 38.5 | 28.6 | 45.2 |
| | MEAN DEW POINT | 26 | 21.8 | 22.9 | 28.9 | 38.4 | 48.9 | 58.4 | 62.4 | 62.2 | 55.6 | 44.5 | 35.0 | 25.7 | 42.1 |
| | NORMAL NO. DAYS WITH: | | | | | | | | | | | | | | |
| | MAXIMUM >= 90 | 3 0 | 0.0 | 0.0 | 0.0 | 0.0 | 0.2 | 1.8 | 4.2 | 2.1 | 0.6 | 0.0 | 0.0 | 0.0 | 8.9 |
| | MAXIMUM <= 32 | 3 0 | 14.6 | 11.7 | 4.2 | 0.2 | 0.0 | 0.0 | 0.0 | 0.0 | 0.0 | 0.0 | 1.1 | 8.9 | 40.7 |
| | MINIMUM <= 32 | 30 | 27.5 | 23.3 | 20.7 | 8.7 | 0.6 | * | 0.0 | 0.0 | 0.0 | 2.4 | 13.0 | 24.1 | 120.3 |
| | MINIMUM <= 0 | 30 | 2.8 | 2.0 | 0.1 | 0.0 | 0.0 | 0.0 | 0.0 | 0.0 | 0.0 | 0.0 | 0.0 | 0.8 | 5.7 |
| H/C | NORMAL HEATING DEG. DAYS | 30 | 1205 | 1025 | 847 | 516 | 235 | 54 | 7 | 13 | 114 | 389 | 680 | 1036 | 6121 |
| | NORMAL COOLING DEG. DAYS | 30 | 0 | 0 | 2 | 7 | 40 | 140 | 236 | 190 | 79 | 8 | 0 | 0 | 702 |
| RH | NORMAL (PERCENT) | 30 | 75 | 74 | 70 | 67 | 69 | 70 | 71 | 74 | 74 | 72 | 73 | 76 | 72 |
| | HOUR 01 LST | 30 | 78 | 76 | 75 | 74 | 78 | 81 | 82 | 84 | 83 | 78 | 76 | 78 | 79 |
| | HOUR 07 LST | 30 | 80 | 79 | 79 | 77 | 79 | 80 | 82 | 86 | 85 | 82 | 78 | 79 | 81 |
| | HOUR 13 LST | 30 | 71 | 67 | 62 | 57 | 58 | 58 | 57 | 61 | 61 | 60 | 66 | 71 | 62 |
| | HOUR 19 LST | 30 | 74 | 73 | 68 | 62 | 61 | 62 | 62 | 68 | 70 | 71 | 72 | 75 | 68 |
| S | PERCENT POSSIBLE SUNSHINE | 66 | 31 | 37 | 45 | 53 | 58 | 65 | 67 | 63 | 60 | 52 | 33 | 27 | 49 |
| W/O | MEAN NO. DAYS WITH: | | | | | | | | | | | | | | |
| | HEAVY FOG(VISBY <= 1/4 MI) | 46 | 1.2 | 1.7 | 1.8 | 1.0 | 1.2 | 0.5 | 0.3 | 0.7 | 0.6 | 0.7 | 0.9 | 1.2 | 11.8 |
| | THUNDERSTORMS | 62 | 0.2 | 0.5 | 1.6 | 3.3 | 4.7 | 6.2 | 6.2 | 5.1 | 3.2 | 1.5 | 0.9 | 0.3 | 33.7 |
| CLOUDNESS | MEAN: | | | | | | | | | | | | | | |
| | SUNRISE-SUNSET (OKTAS) | | | | | | | | | | | | | | |
| | MIDNIGHT-MIDNIGHT (OKTAS) | | | | | | | | | | | | | | |
| | MEAN NO. DAYS WITH: | | | | | | | | | | | | | | |
| | CLEAR | | | | | | | | | | | | | | |
| | PARTLY CLOUDY | 1 | 2.0 | 1.0 | 4.0 | | | | | | | | | | |
| | CLOUDY | 1 | 9.0 | 6.0 | 11.0 | | 11.0 | 9.0 | | | | | | | |
| PR | MEAN STATION PRESSURE(IN) | 26 | 29.21 | 29.21 | 29.20 | 29.13 | 29.15 | 29.14 | 29.16 | 29.20 | 29.22 | 29.23 | 29.22 | 29.22 | 29.19 |
| | MEAN SEA-LEVEL PRES. (IN) | 26 | 30.10 | 30.10 | 30.08 | 30.00 | 30.00 | 29.99 | 30.01 | 30.05 | 30.08 | 30.09 | 30.09 | 30.11 | 30.06 |
| WINDS | MEAN SPEED (MPH) | 26 | 11.3 | 10.8 | 10.9 | 10.3 | 9.2 | 8.4 | 8.0 | 7.6 | 8.2 | 9.2 | 10.7 | 10.9 | 9.6 |
| | PREVAIL.DIR(TENS OF DEGS) | 42 | 25 | 24 | 21 | 02 | 21 | 21 | 22 | 22 | 21 | 21 | 21 | 24 | 24 |
| | MAXIMUM 2-MINUTE: | | | | | | | | | | | | | | |
| | SPEED (MPH) | 14 | 51 | 46 | 46 | 40 | 45 | 52 | 47 | 39 | 41 | 46 | 47 | 53 | 53 |
| | DIR. (TENS OF DEGS) | | 25 | 26 | 27 | 03 | 29 | 33 | 26 | 24 | 29 | 24 | 26 | 25 | 25 |
| | YEAR OF OCCURRENCE | | 2008 | 2009 | 2002 | 2005 | 2004 | 2009 | 2003 | 2000 | 2008 | 1996 | 2003 | 2004 | DEC 2004 |
| | MAXIMUM 3-SECOND | | | | | | | | | | | | | | |
| | SPEED (MPH) | 14 | 66 | 67 | 62 | 54 | 55 | 62 | 68 | 47 | 54 | 57 | 59 | 62 | 68 |
| | DIR. (TENS OF DEGS) | | 25 | 17 | 26 | 04 | 29 | 33 | 21 | 02 | 28 | 26 | 27 | 26 | 21 |
| | YEAR OF OCCURRENCE | | 2008 | 2009 | 2002 | 2005 | 2004 | 2009 | 1999 | 2007 | 2008 | 1996 | 2003 | 2004 | JUL 1999 |
| PRECIPITATION | NORMAL (IN) | 30 | 2.48 | 2.29 | 2.94 | 3.37 | 3.50 | 3.89 | 3.52 | 3.69 | 3.77 | 2.74 | 3.38 | 3.14 | 38.71 |
| | MAXIMUM MONTHLY (IN) | 68 | 7.01 | 5.54 | 6.07 | 6.61 | 9.14 | 9.06 | 9.12 | 9.03 | 11.05 | 9.50 | 8.80 | 8.59 | 11.05 |
| | YEAR OF OCCURRENCE | | 1950 | 2008 | 1954 | 1961 | 1989 | 1972 | 1992 | 2007 | 1996 | 1954 | 1985 | 1990 | SEP 1996 |
| | MINIMUM MONTHLY (IN) | 68 | 0.36 | 0.48 | 0.78 | 1.18 | 0.66 | 0.65 | 0.68 | 0.53 | 0.74 | 0.61 | 0.80 | 0.71 | 0.36 |
| | YEAR OF OCCURRENCE | | 1961 | 1978 | 1958 | 1946 | 2007 | 1988 | 2001 | 1969 | 1964 | 1952 | 1976 | 1958 | JAN 1961 |
| | MAXIMUM IN 24 HOURS (IN) | 68 | 2.53 | 2.53 | 2.76 | 2.24 | 3.73 | 4.00 | 2.87 | 3.65 | 5.24 | 3.44 | 2.73 | 2.81 | 5.24 |
| | YEAR OF OCCURRENCE | | 1995 | 2008 | 1948 | 1961 | 1955 | 1972 | 1969 | 1994 | 1996 | 1954 | 1985 | 1992 | SEP 1996 |
| | NORMAL NO. DAYS WITH: | | | | | | | | | | | | | | |
| | PRECIPITATION >= 0.01 | 30 | 16.9 | 13.7 | 14.7 | 14.5 | 12.6 | 11.2 | 10.5 | 10.4 | 10.3 | 11.7 | 14.0 | 16.3 | 156.8 |
| | PRECIPITATION >= 1.00 | 30 | 0.2 | 0.1 | 0.2 | 0.5 | 0.7 | 0.8 | 1.0 | 0.9 | 0.9 | 0.4 | 0.5 | 0.4 | 6.6 |
| SNOWFALL | NORMAL (IN) | 30 | 17.1 | 14.1 | 10.7 | 2.5 | 0.1 | 0.0 | 0.0 | 0.0 | 0.0 | 0.4 | 5.1 | 13.1 | 63.1 |
| | MAXIMUM MONTHLY (IN) | 68 | 42.8 | 39.1 | 30.4 | 19.0 | 2.1 | T | T | T | T | 8.0 | 23.4 | 35.1 | 42.8 |
| | YEAR OF OCCURRENCE | | 1978 | 1993 | 2008 | 1943 | 1974 | 1996 | 1993 | 2008 | 1993 | 1962 | 1996 | 2004 | JAN 1978 |
| | MAXIMUM IN 24 HOURS (IN) | 68 | 10.8 | 14.8 | 16.0 | 11.6 | 2.1 | T | T | T | T | 6.7 | 15.0 | 12.2 | 16.0 |
| | YEAR OF OCCURRENCE' | | 1996 | 1993 | 1987 | 1982 | 1974 | 1996 | 1993 | 2008 | 1993 | 1962 | 1950 | 1974 | MAR 1987 |
| | MAXIMUM SNOW DEPTH (IN) | 61 | 21 | 22 | 15 | 14 | 0 | 0 | 0 | 0 | 0 | 6 | 20 | 19 | 22 |
| | YEAR OF OCCURRENCE | | 1978 | 2009 | 2008 | 1987 | | | | | | 1962 | 1950 | 1962 | FEB 2009 |
| | NORMAL NO. DAYS WITH: | | | | | | | | | | | | | | |
| | SNOWFALL >= 1.0 | 30 | 5.3 | 4.0 | 2.9 | 0.7 | 0.0 | 0.0 | 0.0 | 0.0 | 0.0 | 0.2 | 1.8 | 3.8 | 18.7 |

## PRECIPITATION (inches) 2009 CLEVELAND (KCLE)

| YEAR | JAN | FEB | MAR | APR | MAY | JUN | JUL | AUG | SEP | OCT | NOV | DEC | ANNUAL |
|------|-----|-----|-----|-----|-----|-----|-----|-----|-----|-----|-----|-----|--------|
| 1980 | 1.18 | 1.27 | 3.66 | 2.65 | 3.13 | 2.69 | 4.77 | 4.38 | 3.11 | 2.38 | 1.29 | 2.10 | 32.61 |
| 1981 | 0.76 | 2.72 | 1.61 | 4.62 | 2.19 | 4.68 | 5.31 | 2.61 | 6.75 | 2.33 | 1.99 | 3.44 | 39.01 |
| 1982 | 4.00 | 1.41 | 3.77 | 1.62 | 2.65 | 5.01 | 1.21 | 2.66 | 4.82 | 0.93 | 5.17 | 3.68 | 36.93 |
| 1983 | 1.08 | 0.77 | 3.54 | 4.48 | 4.17 | 3.45 | 4.16 | 3.15 | 2.87 | 4.14 | 5.89 | 2.92 | 40.62 |
| 1984 | 1.25 | 3.82 | 3.80 | 2.29 | 5.95 | 3.40 | 3.35 | 5.51 | 2.43 | 2.20 | 3.95 | 3.38 | 41.33 |
| 1985 | 1.78 | 2.60 | 4.97 | 1.38 | 3.45 | 2.93 | 3.23 | 4.01 | 2.05 | 3.45 | 8.80 | 2.63 | 41.28 |
| 1986 | 2.23 | 3.08 | 2.44 | 3.90 | 4.34 | 2.97 | 3.10 | 3.58 | 6.41 | 2.83 | 3.01 | 2.82 | 40.71 |
| 1987 | 1.98 | 0.49 | 3.84 | 2.97 | 2.40 | 7.94 | 3.36 | 5.51 | 2.07 | 3.41 | 1.02 | 2.96 | 37.95 |
| 1988 | 1.03 | 2.84 | 2.20 | 3.47 | 1.33 | 0.65 | 3.42 | 3.35 | 1.77 | 2.51 | 4.63 | 2.49 | 29.69 |
| 1989 | 2.07 | 1.73 | 3.46 | 3.73 | 9.14 | 5.22 | 3.02 | 1.09 | 4.61 | 4.50 | 3.61 | 1.72 | 43.90 |
| 1990 | 2.35 | 4.70 | 0.86 | 4.57 | 6.10 | 1.72 | 5.62 | 4.79 | 7.33 | 4.92 | 2.28 | 8.59 | 53.83 |
| 1991 | 2.18 | 2.31 | 3.64 | 4.22 | 3.24 | 1.37 | 1.69 | 2.79 | 3.40 | 2.65 | 2.92 | 2.26 | 32.67 |
| 1992 | 3.32 | 2.65 | 3.05 | 3.77 | 3.01 | 2.66 | 9.12 | 4.58 | 3.25 | 2.27 | 6.54 | 4.31 | 48.53 |
| 1993 | 4.44 | 2.61 | 3.85 | 3.16 | 1.56 | 5.18 | 2.58 | 1.52 | 5.94 | 3.52 | 4.06 | 2.21 | 40.63 |
| 1994 | 2.66 | 0.83 | 1.30 | 3.70 | 1.67 | 3.35 | 2.46 | 5.35 | 1.73 | 1.05 | 2.52 | 2.94 | 29.56 |
| 1995 | 5.81 | 1.73 | 1.72 | 4.33 | 3.96 | 3.67 | 5.39 | 2.00 | 1.03 | 4.08 | 3.88 | 1.45 | 39.05 |
| 1996 | 2.69 | 1.63 | 2.81 | 5.61 | 2.08 | 3.89 | 3.18 | 0.79 | 11.05 | 4.65 | 5.03 | 3.03 | 46.44 |
| 1997 | 1.77 | 2.93 | 3.26 | 2.20 | 4.21 | 3.34 | 1.51 | 5.26 | 4.25 | 1.63 | 2.58 | 2.42 | 35.36 |
| 1998 | 3.92 | 1.89 | 3.25 | 6.07 | 1.92 | 2.97 | 2.72 | 3.02 | 1.20 | 2.36 | 1.59 | 1.92 | 32.83 |
| 1999 | 3.64 | 2.36 | 1.65 | 3.89 | 1.54 | 1.43 | 4.66 | 1.80 | 1.93 | 3.06 | 3.31 | 2.70 | 31.97 |
| 2000 | 2.63 | 2.05 | 1.57 | 3.72 | 5.46 | 5.72 | 2.57 | 4.72 | 3.29 | 3.56 | 2.55 | 2.75 | 40.59 |
| 2001 | 1.59 | 1.63 | 2.43 | 2.33 | 3.84 | 3.96 | 0.68 | 3.31 | 3.90 | 5.56 | 2.62 | 2.53 | 34.38 |
| 2002 | 2.21 | 2.43 | 4.13 | 3.67 | 5.77 | 0.92 | 2.87 | 2.00 | 3.50 | 1.52 | 3.65 | 3.71 | 36.38 |
| 2003 | 1.98 | 2.74 | 2.33 | 2.47 | 6.49 | 3.16 | 4.89 | 1.96 | 6.02 | 2.87 | 3.58 | 4.01 | 42.50 |
| 2004 | 2.69 | 0.76 | 4.82 | 3.74 | 5.90 | 2.87 | 2.88 | 2.46 | 3.22 | 2.34 | 3.24 | 4.47 | 39.39 |
| 2005 | 5.89 | 2.07 | 1.66 | 5.57 | 1.43 | 1.64 | 3.24 | 7.60 | 3.55 | 2.53 | 2.65 | 2.04 | 39.87 |
| 2006 | 1.92 | 2.80 | 1.54 | 2.45 | 4.54 | 4.84 | 4.46 | 2.21 | 3.14 | 5.83 | 3.40 | 3.51 | 40.64 |
| 2007 | 5.84 | 1.41 | 3.71 | 3.52 | 0.66 | 1.66 | 2.57 | 9.03 | 2.12 | 2.65 | 4.07 | 4.20 | 41.44 |
| 2008 | 3.31 | 5.54 | 5.47 | 2.21 | 4.17 | 5.21 | 3.02 | 1.43 | 3.79 | 2.60 | 3.97 | 3.77 | 44.49 |
| 2009 | 2.66 | 2.73 | 3.33 | 3.08 | 2.35 | 2.68 | 3.75 | 3.52 | 3.84 | 3.66 | 1.45 | 2.71 | 35.76 |
| POR=<br>113 YRS | 2.59 | 2.20 | 2.92 | 3.09 | 3.17 | 3.27 | 3.39 | 3.12 | 3.22 | 2.64 | 2.81 | 2.62 | 35.04 |

WBAN : 14820

## AVERAGE TEMPERATURE (°F) 2009 CLEVELAND (KCLE)

| YEAR | JAN | FEB | MAR | APR | MAY | JUN | JUL | AUG | SEP | OCT | NOV | DEC | ANNUAL |
|------|-----|-----|-----|-----|-----|-----|-----|-----|-----|-----|-----|-----|--------|
| 1980 | 25.5 | 21.9 | 33.6 | 46.1 | 58.5 | 64.0 | 72.3 | 73.2 | 64.7 | 47.9 | 39.4 | 28.5 | 48.0 |
| 1981 | 20.1 | 31.5 | 36.0 | 50.6 | 55.7 | 68.2 | 71.3 | 70.0 | 62.4 | 50.0 | 42.6 | 30.6 | 49.1 |
| 1982 | 19.8 | 25.2 | 37.1 | 44.6 | 64.9 | 64.1 | 73.6 | 67.9 | 62.7 | 55.3 | 45.4 | 40.5 | 50.1 |
| 1983 | 30.7 | 33.9 | 40.8 | 47.1 | 55.7 | 69.0 | 75.2 | 73.7 | 65.1 | 53.4 | 43.9 | 23.2 | 51.0 |
| 1984 | 20.7 | 34.5 | 28.4 | 46.8 | 54.0 | 69.5 | 68.7 | 70.6 | 61.1 | 56.3 | 40.9 | 36.5 | 49.0 |
| 1985 | 20.8 | 25.2 | 40.3 | 53.6 | 60.4 | 62.7 | 71.1 | 68.9 | 64.9 | 54.0 | 46.0 | 24.3 | 49.4 |
| 1986 | 26.7 | 28.8 | 39.5 | 49.8 | 60.8 | 67.2 | 73.1 | 69.0 | 67.0 | 54.3 | 40.3 | 32.6 | 50.8 |
| 1987 | 27.4 | 30.5 | 39.0 | 49.1 | 63.0 | 70.2 | 75.2 | 70.8 | 63.5 | 47.5 | 46.1 | 34.8 | 51.4 |
| 1988 | 25.6 | 25.8 | 37.5 | 47.9 | 59.7 | 68.9 | 75.9 | 74.2 | 64.0 | 47.1 | 43.8 | 31.3 | 50.1 |
| 1989 | 35.0 | 26.1 | 38.1 | 45.3 | 57.6 | 68.3 | 73.4 | 71.0 | 64.0 | 54.0 | 41.0 | 19.2 | 49.4 |
| 1990 | 35.8 | 34.1 | 42.0 | 49.4 | 56.3 | 67.6 | 71.2 | 69.8 | 63.4 | 53.7 | 45.3 | 35.6 | 52.0 |
| 1991 | 27.3 | 32.8 | 40.7 | 52.6 | 66.9 | 71.1 | 74.7 | 72.7 | 64.5 | 55.7 | 40.2 | 34.8 | 52.8 |
| 1992 | 30.2 | 32.7 | 36.6 | 47.9 | 57.9 | 64.1 | 70.7 | 67.6 | 63.3 | 49.9 | 42.0 | 34.1 | 49.8 |
| 1993 | 32.3 | 25.6 | 33.6 | 47.6 | 59.2 | 67.9 | 75.0 | 73.2 | 62.2 | 51.3 | 41.8 | 30.5 | 50.0 |
| 1994 | 19.3 | 26.5 | 36.3 | 50.5 | 54.7 | 69.7 | 73.3 | 69.3 | 63.9 | 54.5 | 47.8 | 36.9 | 50.2 |
| 1995 | 30.1 | 27.0 | 40.0 | 46.6 | 59.3 | 71.5 | 75.5 | 77.8 | 63.2 | 56.4 | 38.5 | 26.1 | 51.0 |
| 1996 | 25.9 | 27.6 | 30.9 | 46.3 | 56.7 | 69.3 | 69.6 | 70.9 | 64.0 | 54.0 | 36.3 | 34.7 | 48.9 |
| 1997 | 25.7 | 34.4 | 38.8 | 45.2 | 52.9 | 68.2 | 70.6 | 67.5 | 62.6 | 53.1 | 39.0 | 33.4 | 49.3 |
| 1998 | 35.2 | 34.7 | 41.2 | 49.3 | 64.4 | 68.5 | 71.4 | 72.1 | 67.0 | 53.5 | 44.9 | 37.0 | 53.5 |
| 1999 | 27.1 | 34.7 | 34.3 | 50.4 | 61.0 | 70.1 | 76.2 | 69.1 | 65.0 | 52.5 | 45.9 | 33.8 | 51.7 |
| 2000 | 26.6 | 34.3 | 43.2 | 47.0 | 61.5 | 68.6 | 68.0 | 68.9 | 63.1 | 54.7 | 39.8 | 22.3 | 49.8 |
| 2001 | 27.7 | 32.2 | 34.0 | 51.4 | 60.0 | 68.0 | 71.9 | 72.6 | 61.4 | 54.4 | 48.8 | 37.1 | 51.6 |
| 2002 | 34.6 | 34.0 | 37.4 | 50.5 | 55.5 | 70.4 | 75.5 | 73.6 | 68.7 | 51.8 | 41.3 | 30.7 | 52.0 |
| 2003 | 21.2 | 24.5 | 38.7 | 48.9 | 57.8 | 66.9 | 72.5 | 73.2 | 63.7 | 51.4 | 47.8 | 33.6 | 50.0 |
| 2004 | 21.6 | 29.7 | 40.4 | 49.0 | 62.2 | 66.7 | 71.4 | 68.1 | 65.8 | 53.6 | 44.5 | 31.0 | 50.3 |
| 2005 | 27.4 | 30.9 | 32.6 | 48.7 | 55.2 | 73.7 | 75.2 | 74.1 | 67.1 | 54.4 | 44.7 | 27.4 | 51.0 |
| 2006 | 38.6 | 30.5 | 37.7 | 51.5 | 58.8 | 66.1 | 74.2 | 72.5 | 62.2 | 50.1 | 44.9 | 38.4 | 52.1 |
| 2007 | 31.1 | 18.8 | 40.1 | 46.5 | 61.9 | 69.7 | 70.9 | 72.8 | 66.7 | 60.2 | 41.4 | 32.6 | 51.1 |
| 2008 | 30.2 | 27.2 | 33.5 | 52.3 | 56.3 | 70.9 | 73.5 | 71.1 | 66.2 | 51.5 | 40.2 | 31.0 | 50.3 |
| 2009 | 19.4 | 30.5 | 39.6 | 50.7 | 60.1 | 68.1 | 70.0 | 72.5 | 65.4 | 52.4 | 47.7 | 31.7 | 50.7 |
| POR=<br>113 YRS | 27.4 | 26.9 | 37.0 | 47.1 | 58.5 | 67.0 | 72.5 | 70.9 | 63.7 | 53.8 | 41.7 | 31.4 | 49.8 |

## HEATING DEGREE DAYS (base 65°F) 2009 CLEVELAND (KCLE)

| YEAR | JUL | AUG | SEP | OCT | NOV | DEC | JAN | FEB | MAR | APR | MAY | JUN | TOTAL |
|---|---|---|---|---|---|---|---|---|---|---|---|---|---|
| 1980-81 | 3 | 2 | 97 | 521 | 763 | 1125 | 1385 | 935 | 894 | 430 | 298 | 30 | 6483 |
| 1981-82 | 11 | 11 | 145 | 458 | 664 | 1059 | 1393 | 1109 | 860 | 608 | 78 | 75 | 6471 |
| 1982-83 | 5 | 42 | 136 | 310 | 586 | 760 | 1056 | 864 | 742 | 533 | 294 | 56 | 5384 |
| 1983-84 | 7 | 0 | 116 | 362 | 628 | 1291 | 1366 | 878 | 1126 | 544 | 347 | 19 | 6684 |
| 1984-85 | 16 | 17 | 174 | 270 | 716 | 877 | 1364 | 1110 | 757 | 370 | 187 | 99 | 5957 |
| 1985-86 | 2 | 7 | 118 | 338 | 565 | 1255 | 1180 | 1009 | 785 | 459 | 172 | 52 | 5942 |
| 1986-87 | 3 | 40 | 63 | 332 | 736 | 999 | 1158 | 958 | 795 | 473 | 170 | 23 | 5750 |
| 1987-88 | 3 | 22 | 90 | 535 | 562 | 929 | 1213 | 1129 | 848 | 506 | 208 | 60 | 6105 |
| 1988-89 | 8 | 5 | 83 | 557 | 629 | 1040 | 922 | 1084 | 831 | 585 | 272 | 33 | 6049 |
| 1989-90 | 0 | 6 | 108 | 350 | 716 | 1416 | 898 | 858 | 718 | 492 | 270 | 56 | 5888 |
| 1990-91 | 7 | 3 | 121 | 350 | 585 | 906 | 1163 | 897 | 748 | 379 | 111 | 11 | 5281 |
| 1991-92 | 0 | 0 | 123 | 310 | 738 | 930 | 1074 | 929 | 872 | 513 | 243 | 90 | 5822 |
| 1992-93 | 8 | 26 | 118 | 462 | 682 | 952 | 1009 | 1097 | 967 | 519 | 192 | 56 | 6088 |
| 1993-94 | 0 | 3 | 134 | 420 | 691 | 1063 | 1414 | 1073 | 880 | 443 | 330 | 57 | 6508 |
| 1994-95 | 4 | 10 | 83 | 322 | 507 | 865 | 1077 | 1061 | 769 | 546 | 190 | 13 | 5447 |
| 1995-96 | 3 | 0 | 103 | 271 | 787 | 1200 | 1203 | 1077 | 1052 | 556 | 297 | 19 | 6568 |
| 1996-97 | 8 | 0 | 98 | 333 | 850 | 932 | 1213 | 849 | 805 | 584 | 368 | 46 | 6086 |
| 1997-98 | 11 | 30 | 103 | 404 | 773 | 972 | 916 | 763 | 758 | 465 | 94 | 73 | 5362 |
| 1998-99 | 1 | 4 | 52 | 350 | 597 | 863 | 1171 | 840 | 942 | 433 | 155 | 56 | 5464 |
| 1999-00 | 3 | 3 | 83 | 384 | 569 | 961 | 1180 | 884 | 674 | 533 | 169 | 54 | 5497 |
| 2000-01 | 10 | 21 | 136 | 315 | 750 | 1317 | 1152 | 914 | 955 | 427 | 177 | 58 | 6232 |
| 2001-02 | 11 | 0 | 150 | 332 | 480 | 857 | 934 | 862 | 847 | 464 | 309 | 31 | 5277 |
| 2002-03 | 0 | 0 | 37 | 432 | 701 | 1053 | 1350 | 1128 | 809 | 490 | 224 | 52 | 6276 |
| 2003-04 | 0 | 1 | 88 | 426 | 511 | 966 | 1338 | 1016 | 754 | 482 | 153 | 52 | 5787 |
| 2004-05 | 0 | 24 | 55 | 350 | 610 | 1045 | 1159 | 947 | 999 | 484 | 299 | 12 | 5984 |
| 2005-06 | 0 | 1 | 22 | 350 | 604 | 1160 | 810 | 957 | 841 | 402 | 240 | 51 | 5438 |
| 2006-07 | 3 | 0 | 104 | 454 | 594 | 818 | 1041 | 1286 | 767 | 547 | 176 | 32 | 5822 |
| 2007-08 | 5 | 6 | 57 | 205 | 703 | 996 | 1074 | 1091 | 967 | 383 | 284 | 15 | 5786 |
| 2008-09 | 2 | 1 | 43 | 411 | 735 | 1049 | 1407 | 956 | 777 | 453 | 171 | 37 | 6042 |
| 2009- | 3 | 12 | 56 | 388 | 513 | 1027 | | | | | | | |

WBAN : 14820

## COOLING DEGREE DAYS (base 65°F) 2009 CLEVELAND (KCLE)

| YEAR | JAN | FEB | MAR | APR | MAY | JUN | JUL | AUG | SEP | OCT | NOV | DEC | TOTAL |
|---|---|---|---|---|---|---|---|---|---|---|---|---|---|
| 1980 | 0 | 0 | 0 | 0 | 27 | 83 | 235 | 263 | 97 | 0 | 0 | 0 | 705 |
| 1981 | 0 | 0 | 0 | 4 | 16 | 132 | 214 | 175 | 73 | 0 | 0 | 0 | 614 |
| 1982 | 0 | 0 | 0 | 3 | 84 | 54 | 278 | 140 | 73 | 17 | 6 | 6 | 661 |
| 1983 | 0 | 0 | 0 | 5 | 12 | 185 | 327 | 277 | 127 | 12 | 0 | 0 | 945 |
| 1984 | 0 | 0 | 0 | 3 | 13 | 159 | 139 | 197 | 60 | 5 | 0 | 0 | 576 |
| 1985 | 0 | 0 | 0 | 38 | 52 | 34 | 201 | 131 | 122 | 4 | 2 | 0 | 584 |
| 1986 | 0 | 0 | 1 | 9 | 48 | 128 | 259 | 168 | 131 | 8 | 0 | 0 | 752 |
| 1987 | 0 | 0 | 0 | 0 | 114 | 183 | 322 | 209 | 53 | 0 | 3 | 0 | 884 |
| 1988 | 0 | 0 | 0 | 0 | 47 | 185 | 348 | 297 | 58 | 9 | 0 | 0 | 944 |
| 1989 | 0 | 0 | 4 | 0 | 46 | 138 | 268 | 199 | 83 | 14 | 0 | 0 | 752 |
| 1990 | 0 | 0 | 10 | 31 | 8 | 141 | 208 | 158 | 80 | 8 | 1 | 0 | 645 |
| 1991 | 0 | 0 | 1 | 14 | 176 | 200 | 307 | 245 | 114 | 26 | 0 | 0 | 1083 |
| 1992 | 0 | 0 | 0 | 8 | 28 | 68 | 191 | 114 | 74 | 0 | 0 | 0 | 483 |
| 1993 | 0 | 0 | 0 | 0 | 17 | 147 | 316 | 262 | 62 | 2 | 0 | 0 | 806 |
| 1994 | 0 | 0 | 0 | 15 | 17 | 204 | 269 | 149 | 58 | 5 | 1 | 0 | 718 |
| 1995 | 0 | 0 | 0 | 0 | 21 | 216 | 336 | 404 | 58 | 11 | 0 | 0 | 1046 |
| 1996 | 0 | 0 | 0 | 2 | 48 | 155 | 160 | 193 | 76 | 1 | 0 | 0 | 635 |
| 1997 | 0 | 0 | 0 | 0 | 0 | 147 | 194 | 113 | 38 | 41 | 0 | 0 | 533 |
| 1998 | 0 | 0 | 25 | 0 | 82 | 183 | 207 | 231 | 118 | 4 | 0 | 0 | 850 |
| 1999 | 0 | 0 | 0 | 0 | 39 | 218 | 357 | 138 | 88 | 1 | 0 | 0 | 841 |
| 2000 | 0 | 0 | 4 | 1 | 69 | 169 | 114 | 145 | 86 | 7 | 0 | 0 | 595 |
| 2001 | 0 | 0 | 0 | 23 | 31 | 156 | 231 | 240 | 48 | 11 | 0 | 0 | 740 |
| 2002 | 0 | 0 | 0 | 37 | 23 | 199 | 330 | 275 | 153 | 31 | 0 | 0 | 1048 |
| 2003 | 0 | 0 | 0 | 10 | 9 | 116 | 238 | 261 | 56 | 8 | 3 | 0 | 701 |
| 2004 | 0 | 0 | 0 | 0 | 9 | 72 | 111 | 206 | 129 | 89 | 1 | 0 | 617 |
| 2005 | 0 | 0 | 0 | 4 | 3 | 279 | 323 | 286 | 93 | 30 | 0 | 0 | 1018 |
| 2006 | 0 | 0 | 0 | 4 | 54 | 92 | 296 | 238 | 26 | 0 | 0 | 0 | 710 |
| 2007 | 0 | 0 | 3 | 0 | 89 | 178 | 196 | 256 | 117 | 65 | 0 | 0 | 904 |
| 2008 | 0 | 0 | 0 | 8 | 21 | 198 | 274 | 196 | 84 | 1 | 0 | 0 | 782 |
| 2009 | 0 | 0 | 0 | 32 | 27 | 132 | 163 | 251 | 76 | 0 | 0 | 0 | 681 |

**SNOWFALL (inches) 2009 CLEVELAND (KCLE)**

| YEAR | JUL | AUG | SEP | OCT | NOV | DEC | JAN | FEB | MAR | APR | MAY | JUN | TOTAL |
|------|-----|-----|-----|-----|-----|-----|-----|-----|-----|-----|-----|-----|-------|
| 1980-81 | 0.0 | 0.0 | 0.0 | T | 5.4 | 13.5 | 15.0 | 9.7 | 16.9 | T | 0.0 | 0.0 | 60.5 |
| 1981-82 | 0.0 | 0.0 | 0.0 | 4.0 | 2.9 | 27.1 | 28.1 | 7.6 | 17.6 | 13.2 | 0.0 | 0.0 | 100.5 |
| 1982-83 | 0.0 | 0.0 | 0.0 | T | 2.2 | 6.3 | 6.5 | 8.3 | 11.3 | 3.4 | 0.0 | 0.0 | 38.0 |
| 1983-84 | 0.0 | 0.0 | 0.0 | 0.0 | 7.1 | 13.0 | 12.9 | 27.1 | 19.3 | T | 0.0 | 0.0 | 79.4 |
| 1984-85 | 0.0 | 0.0 | 0.0 | 0.0 | 4.0 | 8.9 | 25.5 | 18.2 | 1.2 | 5.9 | 0.0 | 0.0 | 63.7 |
| 1985-86 | 0.0 | 0.0 | 0.0 | 0.0 | T | 23.4 | 17.2 | 10.8 | 6.7 | 0.2 | 0.0 | 0.0 | 58.3 |
| 1986-87 | 0.0 | 0.0 | 0.0 | 0.0 | 3.1 | 1.1 | 16.4 | 5.0 | 26.2 | 4.0 | 0.0 | 0.0 | 55.8 |
| 1987-88 | 0.0 | 0.0 | 0.0 | T | 1.0 | 16.4 | 8.7 | 22.9 | 20.4 | 1.9 | 0.0 | 0.0 | 71.3 |
| 1988-89 | 0.0 | 0.0 | 0.0 | T | 1.7 | 17.9 | 6.6 | 13.8 | 9.9 | 4.9 | T | 0.0 | 54.8 |
| 1989-90 | 0.0 | 0.0 | 0.0 | T | 9.1 | 24.0 | 10.5 | 9.9 | 4.4 | 4.7 | 0.0 | 0.0 | 62.6 |
| 1990-91 | 0.0 | 0.0 | 0.0 | T | T | 7.4 | 16.6 | 18.9 | 4.2 | T | 0.0 | 0.0 | 47.1 |
| 1991-92 | 0.0 | 0.0 | 0.0 | 0.0 | 3.5 | 9.4 | 23.8 | 6.2 | 18.4 | 4.4 | 0.0 | 0.0 | 65.7 |
| 1992-93 | 0.0 | 0.0 | 0.0 | T | 7.1 | 7.1 | 8.7 | 39.1 | 25.4 | 1.1 | 0.0 | T | 88.5 |
| 1993-94 | T | 0.0 | T | 0.2 | 3.0 | 19.0 | 27.4 | 12.3 | 7.0 | 3.6 | 0.0 | 0.0 | 72.5 |
| 1994-95 | 0.0 | 0.0 | 0.0 | 0.0 | T | 1.0 | 23.4 | 14.7 | 4.3 | 0.2 | 0.0 | 0.0 | 43.6 |
| 1995-96 | 0.0 | 0.0 | 0.0 | 0.0 | 9.9 | 29.6 | 21.9 | 10.1 | 19.4 | 10.2 | 0.0 | T | 101.1 |
| 1996-97 | 0.0 | 0.0 | 0.0 | 0.0 | 23.4 | 5.0 | 13.0 | 8.4 | 5.3 | 0.8 | 0.0 | 0.0 | 55.9 |
| 1997-98 | 0.0 | 0.0 | 0.0 | T | 8.6 | 10.7 | 5.0 | 0.2 | 9.5 | T | 0.0 | 0.0 | 34.0 |
| 1998-99 | 0.0 | 0.0 | 0.0 | 0.0 | 0.1 | 6.9 | 29.6 | 14.2 | 11.6 | T | 0.0 | 0.0 | 62.4 |
| 1999-00 | 0.0 | 0.0 | 0.0 | T | 1.6 | 10.3 | 24.7 | 13.9 | 8.0 | 1.0 | 0.0 | 0.0 | 59.5 |
| 2000-01 | 0.0 | 0.0 | T | 0.1 | 11.2 | 21.9 | 14.9 | 3.2 | 26.5 | 0.3 | 0.0 | 0.0 | 78.1 |
| 2001-02 | 0.0 | 0.0 | 0.0 | 1.0 | T | 3.5 | 6.3 | 16.9 | 15.9 | 2.2 | T | 0.0 | 45.8 |
| 2002-03 | 0.0 | 0.0 | 0.0 | T | 6.1 | 22.4 | 30.3 | 30.1 | 6.7 | 0.1 | T | 0.0 | 95.7 |
| 2003-04 | 0.0 | 0.0 | 0.0 | 0.3 | 4.5 | 26.6 | 32.9 | 5.7 | 18.6 | 2.6 | T | T | 91.2 |
| 2004-05 | 0.0 | 0.0 | 0.0 | 0.0 | 0.9 | 35.1 | 32.8 | 10.3 | 19.8 | 19.0 | T | 0.0 | 117.9 |
| 2005-06 | 0.0 | 0.0 | 0.0 | 0.0 | 5.5 | 21.0 | 4.6 | 16.9 | 2.2 | 0.4 | 0.0 | 0.0 | 50.6 |
| 2006-07 | 0.0 | 0.0 | 0.0 | T | 1.4 | 9.5 | 21.1 | 23.8 | 7.3 | 13.4 | 0.0 | 0.0 | 76.5 |
| 2007-08 | 0.0 | 0.0 | 0.0 | 0.0 | 1.1 | 9.6 | 16.5 | 19.6 | 30.4 | T | T | 0.0 | 77.2 |
| 2008-09 | 0.0 | T | 0.0 | 0.3 | 10.1 | 8.4 | 40.5 | 17.1 | 1.0 | 2.6 | 0.0 | 0.0 | 80.0 |
| 2009- | 0.0 | 0.0 | 0.0 | 0.0 | 1.2 | 6.9 | | | | | | | |
| POR= 62 YRS | T | T | T | 0.5 | 5.3 | 12.5 | 15.2 | 12.6 | 11.1 | 2.6 | T | T | 59.8 |

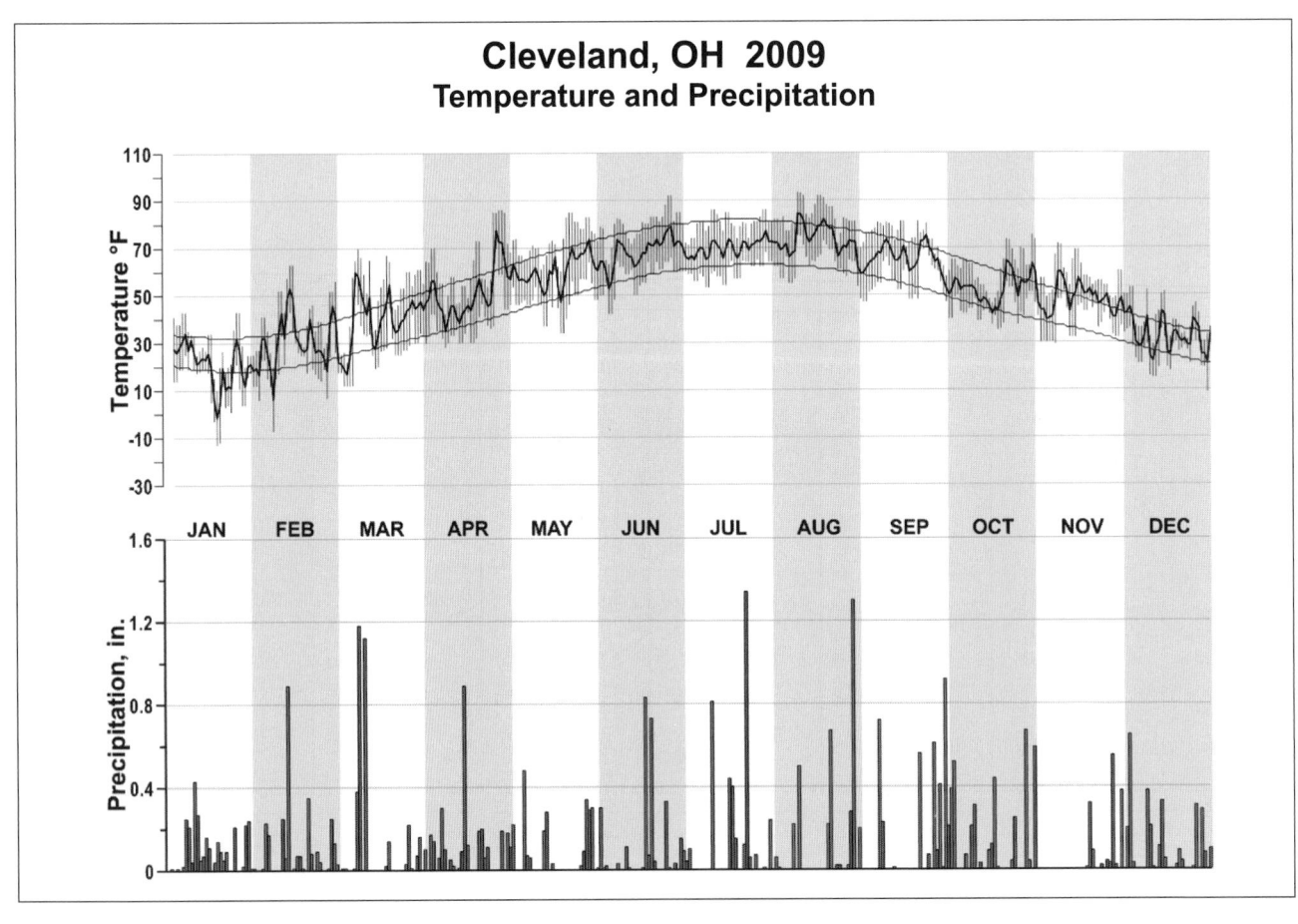

# Cleveland, OH 2009
## Temperature and Precipitation

# 2009
# COLUMBUS
# OHIO (KCMH)

Columbus is located in the center of the state and in the drainage area of the Ohio River. The airport is located at the eastern boundary of the city approximately 7 miles from the center of the business district.

Four nearly parallel streams run through or adjacent to the city. The Scioto River is the principal stream and flows from the northwest into the center of the city and then flows straight south toward the Ohio River. The Olentangy River runs almost due south and empties into the Scioto just west of the business district. Two minor streams run through portions of Columbus or skirt the eastern and southern fringes of the area. They are Alum Creek and Big Walnut Creek. Alum Creek empties into the Big Walnut southeast of the city and the Big Walnut empties into the Scioto a few miles downstream. The Scioto and Olentangy are gorge-like in character with very little flood plain and the two creeks have only a little more flood plain or bottomland.

The narrow valleys associated with the streams flowing through the city supply the only variation in the micro-climate of the area. The city proper shows the typical metropolitan effect with shrubs and flowers blossoming earlier than in the immediate surroundings and in retarding light frost on clear quiet nights. Many small areas to the southeast and to the north and northeast show marked effects of air drainage as evidenced by the frequent formation of shallow ground fog at daybreak during the summer and fall months and the higher frequency of frost in the spring and fall.

The average occurrence of the last freezing temperature in the spring within the city proper is mid-April, and the first freeze in the fall is very late October, but in the immediate surroundings there is much variation. For example, at Valley Crossing located at the southeastern outskirts of the city, the average occurrence of the last 32 degree temperature in the spring is very early May, while the first 32 degree temperature in the fall is mid-October.

The records show a high frequency of calm or very low wind speeds during the late evening and early morning hours, from June through September. The rolling landscape is conducive to air drainage and from the Weather Service location at the airport the air drainage is toward the northwest with the wind direction indicated as southeast. Air drainage takes place at speeds generally 4 mph or less and frequently provides the only perceptible breeze during the night.

Columbus is located in the area of changeable weather. Air masses from central and northwest Canada frequently invade this region. Air from the Gulf of Mexico often reachs central Ohio during the summer and to a much lesser extent in the fall and winter. There are also occasional weather changes brought about by cool outbreaks from the Hudson Bay region of Canada, especially during the spring months. At infrequent intervals the general circulation will bring showers or snow to Columbus from the Atlantic. Although Columbus does not have a wet or dry season as such, the month of October usually has the least amount of precipitation.

# NORMALS, MEANS, AND EXTREMES
## COLUMBUS (KCMH)

| LATITUDE: 39 ° 59'N | LONGITUDE: -82 ° 52'W | | ELEVATION (FT): GRND: 812   BARO: 812 | | | | | TIME ZONE: EASTERN   (UTC -5) | | | WBAN: 14821 | |

| ELEMENT | POR | JAN | FEB | MAR | APR | MAY | JUN | JUL | AUG | SEP | OCT | NOV | DEC | YEAR |
|---|---|---|---|---|---|---|---|---|---|---|---|---|---|---|
| **TEMPERATURE °F** | | | | | | | | | | | | | | |
| NORMAL DAILY MAXIMUM | 30 | 36.2 | 40.5 | 51.7 | 62.9 | 73.3 | 81.6 | 85.3 | 83.8 | 77.1 | 65.4 | 52.4 | 41.0 | 62.6 |
| MEAN DAILY MAXIMUM | 62 | 36.1 | 39.7 | 50.2 | 62.8 | 72.9 | 81.5 | 84.9 | 83.6 | 77.0 | 65.2 | 51.7 | 40.0 | 62.1 |
| HIGHEST DAILY MAXIMUM | 70 | 74 | 75 | 85 | 89 | 94 | 102 | 100 | 101 | 100 | 91 | 80 | 76 | 102 |
| YEAR OF OCCURRENCE | | 1950 | 2000 | 1945 | 1948 | 1941 | 1944 | 1999 | 1983 | 1951 | 2007 | 1987 | 1982 | JUN 1944 |
| MEAN OF EXTREME MAXS. | 62 | 58.5 | 62.1 | 74.0 | 82.0 | 87.4 | 92.7 | 94.0 | 93.0 | 89.7 | 82.0 | 71.3 | 61.7 | 79.0 |
| NORMAL DAILY MINIMUM | 30 | 20.3 | 23.5 | 32.2 | 41.2 | 51.8 | 60.7 | 64.9 | 63.2 | 55.9 | 44.0 | 34.9 | 25.9 | 43.2 |
| MEAN DAILY MINIMUM | 62 | 20.8 | 23.0 | 31.1 | 41.0 | 50.7 | 59.6 | 64.0 | 62.4 | 55.1 | 43.6 | 34.4 | 25.4 | 42.6 |
| LOWEST DAILY MINIMUM | 70 | -22 | -13 | -6 | 14 | 25 | 35 | 43 | 39 | 31 | 20 | 5 | -17 | -22 |
| YEAR OF OCCURRENCE | | 1994 | 1977 | 1984 | 1982 | 1966 | 1972 | 1972 | 1965 | 1963 | 1962 | 1976 | 1989 | JAN 1994 |
| MEAN OF EXTREME MINS. | 62 | -0.5 | 3.3 | 12.7 | 25.3 | 35.6 | 46.0 | 52.1 | 50.3 | 39.0 | 28.2 | 18.3 | 5.7 | 26.3 |
| NORMAL DRY BULB | 30 | 28.3 | 32.0 | 42.0 | 52.0 | 62.6 | 71.2 | 75.1 | 73.5 | 66.5 | 54.7 | 43.7 | 33.5 | 52.9 |
| MEAN DRY BULB | 62 | 28.4 | 31.3 | 40.7 | 51.9 | 61.8 | 70.7 | 74.4 | 73.0 | 66.1 | 54.4 | 43.1 | 32.7 | 52.4 |
| MEAN WET BULB | 26 | 26.4 | 28.4 | 35.4 | 45.1 | 54.4 | 63.0 | 66.6 | 65.7 | 59.2 | 48.2 | 39.0 | 29.8 | 46.8 |
| MEAN DEW POINT | 26 | 23.0 | 24.4 | 30.8 | 40.4 | 50.9 | 59.8 | 63.8 | 63.1 | 56.3 | 44.7 | 35.4 | 26.7 | 43.3 |
| NORMAL NO. DAYS WITH: | | | | | | | | | | | | | | |
| MAXIMUM >= 90 | 30 | 0.0 | 0.0 | 0.0 | 0.0 | 0.5 | 3.5 | 6.2 | 4.3 | 1.2 | 0.0 | 0.0 | 0.0 | 15.7 |
| MAXIMUM <= 32 | 30 | 12.3 | 8.7 | 2.2 | 0.1 | 0.0 | 0.0 | 0.0 | 0.0 | 0.0 | 0.0 | 0.8 | 7.5 | 31.6 |
| MINIMUM <= 32 | 30 | 26.6 | 22.3 | 17.9 | 6.4 | 0.4 | 0.0 | 0.0 | 0.0 | 0.0 | 3.1 | 13.3 | 22.7 | 112.7 |
| MINIMUM <= 0 | 30 | 2.4 | 1.4 | 0.1 | 0.0 | 0.0 | 0.0 | 0.0 | 0.0 | 0.0 | 0.0 | 0.0 | 0.5 | 4.4 |
| **H/C** | | | | | | | | | | | | | | |
| NORMAL HEATING DEG. DAYS | 30 | 1154 | 940 | 731 | 415 | 152 | 27 | 3 | 7 | 80 | 347 | 654 | 982 | 5492 |
| NORMAL COOLING DEG. DAYS | 30 | 0 | 0 | 2 | 9 | 61 | 198 | 305 | 254 | 109 | 12 | 1 | 0 | 951 |
| **RH** | | | | | | | | | | | | | | |
| NORMAL (PERCENT) | 30 | 74 | 70 | 65 | 63 | 67 | 69 | 70 | 72 | 72 | 70 | 71 | 75 | 70 |
| HOUR 01 LST | 30 | 76 | 74 | 71 | 71 | 77 | 81 | 82 | 85 | 83 | 80 | 76 | 77 | 78 |
| HOUR 07 LST | 30 | 78 | 77 | 76 | 76 | 79 | 81 | 84 | 87 | 87 | 83 | 80 | 80 | 81 |
| HOUR 13 LST | 30 | 68 | 64 | 56 | 52 | 54 | 55 | 56 | 57 | 57 | 55 | 63 | 69 | 59 |
| HOUR 19 LST | 30 | 71 | 67 | 59 | 54 | 56 | 58 | 60 | 62 | 64 | 64 | 68 | 73 | 63 |
| **S** PERCENT POSSIBLE SUNSHINE | 45 | 36 | 42 | 44 | 50 | 56 | 60 | 60 | 60 | 61 | 56 | 37 | 31 | 49 |
| **W/O** MEAN NO. DAYS WITH: | | | | | | | | | | | | | | |
| HEAVY FOG(VISBY <= 1/4 MI) | 46 | 1.4 | 1.5 | 1.0 | 0.5 | 0.7 | 0.7 | 0.8 | 1.1 | 1.5 | 1.0 | 0.8 | 1.4 | 12.4 |
| THUNDERSTORMS | 62 | 0.6 | 0.6 | 1.9 | 4.1 | 6.4 | 7.7 | 8.0 | 6.0 | 3.2 | 1.3 | 1.0 | 0.3 | 41.1 |
| **CLOUDNESS** MEAN: | | | | | | | | | | | | | | |
| SUNRISE-SUNSET (OKTAS) | 47 | 6.2 | 6.1 | 5.9 | 5.6 | 5.3 | 5.0 | 4.8 | 4.6 | 4.5 | 4.5 | 5.8 | 6.2 | 5.4 |
| MIDNIGHT-MIDNIGHT (OKTAS) | 32 | 6.0 | 5.5 | 5.5 | 5.1 | 4.8 | 4.7 | 4.4 | 4.3 | 4.3 | 4.3 | 5.6 | 6.0 | 5.0 |
| MEAN NO. DAYS WITH: | | | | | | | | | | | | | | |
| CLEAR | 1 | 2.0 | | | | | | | | | | | | |
| PARTLY CLOUDY | 1 | 6.0 | | | | | | | | | | | | |
| CLOUDY | 1 | 23.0 | | | | | | | | | | | | |
| **PR** MEAN STATION PRESSURE(IN) | 26 | 29.20 | 29.19 | 29.16 | 29.10 | 29.11 | 29.11 | 29.14 | 29.16 | 29.19 | 29.20 | 29.20 | 29.22 | 29.17 |
| MEAN SEA-LEVEL PRES. (IN) | 26 | 30.11 | 30.10 | 30.06 | 29.98 | 29.98 | 29.98 | 30.00 | 30.03 | 30.06 | 30.09 | 30.10 | 30.12 | 30.05 |
| **WINDS** MEAN SPEED (MPH) | 26 | 9.2 | 8.9 | 9.0 | 8.8 | 7.4 | 6.7 | 6.1 | 5.7 | 5.9 | 6.7 | 8.2 | 8.6 | 7.6 |
| PREVAIL.DIR(TENS OF DEGS) | 35 | 28 | 28 | 29 | 36 | 19 | 20 | 19 | 36 | 36 | 19 | 19 | 19 | 28 |
| MAXIMUM 2-MINUTE: | | | | | | | | | | | | | | |
| SPEED (MPH) | 14 | 44 | 49 | 47 | 47 | 47 | 41 | 47 | 45 | 54 | 45 | 45 | 47 | 54 |
| DIR. (TENS OF DEGS) | | 27 | 26 | 26 | 26 | 26 | 27 | 33 | 33 | 21 | 27 | 27 | 22 | 21 |
| YEAR OF OCCURRENCE | | 2008 | 2009 | 2002 | 1996 | 2001 | 2005 | 1997 | 2007 | 2008 | 2006 | 1996 | 2001 | SEP 2008 |
| MAXIMUM 3-SECOND | | | | | | | | | | | | | | |
| SPEED (MPH) | 14 | 55 | 66 | 55 | 60 | 60 | 55 | 60 | 58 | 75 | 55 | 56 | 59 | 75 |
| DIR. (TENS OF DEGS) | | 23 | 21 | 25 | 26 | 25 | 28 | 28 | 33 | 21 | 28 | 25 | 26 | 21 |
| YEAR OF OCCURRENCE | | 2008 | 2009 | 2002 | 2002 | 2006 | 2006 | 1997 | 2007 | 2008 | 2006 | 1998 | 2000 | SEP 2008 |
| **PRECIPITATION** NORMAL (IN) | 30 | 2.53 | 2.20 | 2.89 | 3.25 | 3.88 | 4.08 | 4.62 | 3.72 | 2.92 | 2.31 | 3.19 | 2.93 | 38.52 |
| MAXIMUM MONTHLY (IN) | 70 | 8.95 | 5.15 | 9.59 | 6.51 | 9.11 | 10.39 | 12.36 | 11.46 | 6.86 | 6.70 | 10.67 | 6.98 | 12.36 |
| YEAR OF OCCURRENCE | | 2005 | 1990 | 1964 | 1998 | 1968 | 2008 | 1992 | 2003 | 2003 | 2006 | 1985 | 1990 | JUL 1992 |
| MINIMUM MONTHLY (IN) | 70 | 0.53 | 0.29 | 0.61 | 0.67 | 0.95 | 0.65 | 0.48 | 0.58 | 0.51 | 0.11 | 0.42 | 0.46 | 0.11 |
| YEAR OF OCCURRENCE | | 1944 | 1978 | 1941 | 1971 | 1977 | 1999 | 1940 | 1951 | 1963 | 1963 | 2009 | 1955 | OCT 1963 |
| MAXIMUM IN 24 HOURS (IN) | 70 | 4.81 | 2.15 | 3.40 | 2.37 | 2.72 | 3.64 | 5.16 | 3.79 | 4.86 | 2.84 | 2.47 | 2.56 | 5.16 |
| YEAR OF OCCURRENCE | | 1959 | 1975 | 1964 | 1957 | 1968 | 2008 | 1992 | 1972 | 1979 | 2007 | 1985 | 1998 | JUL 1992 |
| NORMAL NO. DAYS WITH: | | | | | | | | | | | | | | |
| PRECIPITATION >= 0.01 | 30 | 13.8 | 11.4 | 13.1 | 13.2 | 12.6 | 10.9 | 10.7 | 10.5 | 8.5 | 9.4 | 11.6 | 13.2 | 138.9 |
| PRECIPITATION >= 1.00 | 30 | 0.3 | 0.3 | 0.2 | 0.5 | 0.6 | 1.1 | 1.1 | 1.0 | 0.7 | 0.4 | 0.6 | 0.4 | 7.2 |
| **SNOWFALL** NORMAL (IN) | 30 | 10.5 | 6.2 | 4.1 | 1.2 | 0.* | 0.0 | 0.0 | 0.0 | 0.0 | 0.2 | 1.6 | 5.0 | 28.8 |
| MAXIMUM MONTHLY (IN) | 61 | 34.4 | 26.3 | 21.8 | 12.6 | 0.8 | T | T | T | T | 4.6 | 15.2 | 17.3 | 34.4 |
| YEAR OF OCCURRENCE | | 1978 | 2003 | 2008 | 1987 | 1989 | 1995 | 1995 | 2007 | 1994 | 1993 | 1950 | 1960 | JAN 1978 |
| MAXIMUM IN 24 HOURS (IN) | 61 | 8.8 | 8.9 | 8.6 | 12.3 | 0.8 | T | T | T | T | 4.6 | 8.2 | 8.7 | 12.3 |
| YEAR OF OCCURRENCE' | | 1996 | 1971 | 1962 | 1987 | 1989 | 1995 | 1995 | 2007 | 1994 | 1993 | 1950 | 1960 | APR 1987 |
| MAXIMUM SNOW DEPTH (IN) | 60 | 17 | 13 | 18 | 10 | 0 | 0 | 0 | 0 | 0 | 0 | 13 | 10 | 18 |
| YEAR OF OCCURRENCE | | 1978 | 1979 | 2008 | 1987 | | | | | | | 1950 | 1960 | MAR 2008 |
| NORMAL NO. DAYS WITH: | | | | | | | | | | | | | | |
| SNOWFALL >= 1.0 | 30 | 3.1 | 2.2 | 1.2 | 0.2 | 0.0 | 0.0 | 0.0 | 0.0 | 0.0 | 0.1 | 0.6 | 1.8 | 9.2 |

## PRECIPITATION (inches) 2009 COLUMBUS (KCMH)

| YEAR | JAN | FEB | MAR | APR | MAY | JUN | JUL | AUG | SEP | OCT | NOV | DEC | ANNUAL |
|------|------|------|------|------|------|------|------|------|------|------|------|------|--------|
| 1980 | 1.69 | 1.38 | 3.77 | 1.59 | 4.56 | 5.17 | 4.58 | 6.26 | 1.86 | 2.53 | 2.07 | 1.96 | 37.42 |
| 1981 | 0.70 | 4.60 | 1.11 | 5.38 | 6.50 | 5.73 | 4.14 | 1.41 | 2.28 | 1.40 | 1.65 | 2.88 | 37.78 |
| 1982 | 4.77 | 1.49 | 3.99 | 1.90 | 4.68 | 3.37 | 3.90 | 1.02 | 4.25 | 0.92 | 5.19 | 3.84 | 39.32 |
| 1983 | 1.20 | 0.74 | 1.69 | 5.58 | 5.06 | 4.59 | 2.80 | 2.23 | 1.91 | 4.45 | 5.00 | 3.16 | 38.41 |
| 1984 | 1.07 | 1.97 | 3.89 | 3.10 | 4.93 | 0.71 | 3.15 | 2.96 | 1.48 | 2.91 | 4.41 | 2.84 | 33.42 |
| 1985 | 1.31 | 1.67 | 3.78 | 0.73 | 4.96 | 1.41 | 6.88 | 2.34 | 1.18 | 1.93 | 10.67 | 1.81 | 38.67 |
| 1986 | 1.54 | 2.96 | 2.61 | 1.31 | 2.47 | 5.53 | 3.60 | 1.61 | 3.44 | 4.16 | 3.00 | 2.81 | 35.04 |
| 1987 | 1.14 | 0.59 | 2.04 | 2.02 | 2.85 | 3.60 | 3.89 | 2.96 | 1.53 | 1.57 | 1.63 | 2.88 | 26.70 |
| 1988 | 2.14 | 4.26 | 2.54 | 2.24 | 2.27 | 1.34 | 7.80 | 2.68 | 3.52 | 1.70 | 3.59 | 2.49 | 36.57 |
| 1989 | 1.97 | 3.10 | 4.16 | 3.30 | 4.69 | 6.36 | 6.79 | 4.30 | 2.16 | 2.49 | 2.65 | 1.79 | 43.76 |
| 1990 | 2.43 | 5.15 | 1.32 | 2.82 | 7.01 | 5.25 | 8.00 | 1.86 | 5.26 | 5.05 | 2.03 | 6.98 | 53.16 |
| 1991 | 1.97 | 2.30 | 3.97 | 4.15 | 2.47 | 2.81 | 2.14 | 2.02 | 4.05 | 1.76 | 1.31 | 3.79 | 32.74 |
| 1992 | 1.79 | 0.85 | 3.40 | 2.83 | 3.40 | 2.33 | 12.36 | 3.75 | 2.14 | 1.40 | 4.03 | 1.32 | 39.60 |
| 1993 | 4.14 | 1.82 | 3.50 | 4.49 | 2.47 | 3.33 | 5.95 | 0.74 | 1.75 | 3.05 | 4.45 | 2.16 | 37.85 |
| 1994 | 3.79 | 1.56 | 1.94 | 3.64 | 1.69 | 1.93 | 6.02 | 3.29 | 1.68 | 0.92 | 2.94 | 2.22 | 31.62 |
| 1995 | 4.54 | 1.64 | 1.61 | 3.17 | 4.86 | 5.30 | 6.99 | 7.56 | 1.15 | 4.04 | 2.47 | 1.97 | 45.30 |
| 1996 | 3.73 | 2.14 | 3.40 | 6.39 | 5.81 | 3.82 | 5.09 | 1.58 | 5.50 | 1.44 | 3.20 | 3.46 | 45.56 |
| 1997 | 2.19 | 1.50 | 3.96 | 1.65 | 5.58 | 6.62 | 2.91 | 5.76 | 1.36 | 1.58 | 2.92 | 2.13 | 38.16 |
| 1998 | 2.32 | 2.48 | 1.88 | 6.51 | 3.09 | 6.99 | 2.75 | 1.99 | 1.27 | 3.05 | 1.99 | 3.25 | 37.57 |
| 1999 | 2.87 | 2.77 | 1.88 | 4.65 | 1.80 | 0.65 | 3.02 | 2.40 | 1.91 | 1.00 | 1.95 | 2.69 | 27.59 |
| 2000 | 3.53 | 2.79 | 2.70 | 4.15 | 5.42 | 3.50 | 4.10 | 4.10 | 4.18 | 2.70 | 2.13 | 3.59 | 42.89 |
| 2001 | 1.31 | 1.37 | 1.03 | 3.39 | 7.03 | 2.30 | 4.66 | 4.14 | 1.60 | 3.32 | 3.69 | 3.01 | 36.85 |
| 2002 | 1.92 | 1.72 | 3.45 | 4.02 | 6.60 | 3.46 | 4.13 | 2.12 | 4.35 | 2.68 | 3.00 | 2.76 | 40.21 |
| 2003 | 1.69 | 2.96 | 2.22 | 2.54 | 5.92 | 4.99 | 2.94 | 11.46 | 6.86 | 1.78 | 2.89 | 2.78 | 49.03 |
| 2004 | 5.08 | 2.02 | 3.27 | 3.96 | 5.93 | 5.34 | 6.46 | 3.42 | 2.98 | 3.33 | 4.12 | 3.36 | 49.27 |
| 2005 | 8.95 | 1.27 | 3.53 | 4.36 | 3.36 | 2.69 | 1.79 | 5.09 | 2.91 | 1.33 | 3.31 | 1.67 | 40.26 |
| 2006 | 2.67 | 1.35 | 3.48 | 2.52 | 3.25 | 4.30 | 5.77 | 2.94 | 5.35 | 6.70 | 2.12 | 3.18 | 43.63 |
| 2007 | 4.25 | 2.06 | 6.68 | 2.31 | 1.40 | 2.78 | 3.02 | 4.73 | 2.53 | 3.58 | 2.18 | 4.34 | 39.86 |
| 2008 | 1.64 | 3.88 | 7.58 | 2.16 | 3.14 | 10.39 | 1.65 | 3.73 | 2.50 | 1.44 | 2.49 | 4.84 | 45.44 |
| 2009 | 2.73 | 1.93 | 1.15 | 4.23 | 2.42 | 3.44 | 4.90 | 3.27 | 2.50 | 4.89 | 0.42 | 3.60 | 35.48 |
| POR=<br>62 YRS | 2.90 | 2.28 | 3.17 | 3.47 | 3.93 | 4.03 | 4.41 | 3.51 | 2.81 | 2.25 | 2.95 | 2.82 | 38.53 |

WBAN : 14821

## AVERAGE TEMPERATURE (°F) 2009 COLUMBUS (KCMH)

| YEAR | JAN | FEB | MAR | APR | MAY | JUN | JUL | AUG | SEP | OCT | NOV | DEC | ANNUAL |
|------|------|------|------|------|------|------|------|------|------|------|------|------|--------|
| 1980 | 29.3 | 25.2 | 37.2 | 49.5 | 62.4 | 67.4 | 75.9 | 75.9 | 68.3 | 50.8 | 40.8 | 32.5 | 51.3 |
| 1981 | 23.3 | 34.0 | 40.2 | 55.8 | 59.5 | 70.9 | 71.9 | 70.4 | 62.3 | 51.1 | 40.9 | 30.6 | 50.9 |
| 1982 | 21.2 | 29.2 | 40.4 | 46.4 | 66.8 | 65.8 | 74.4 | 69.2 | 63.5 | 56.2 | 45.4 | 40.4 | 51.6 |
| 1983 | 29.9 | 34.0 | 43.3 | 48.4 | 57.6 | 69.4 | 76.7 | 76.2 | 67.1 | 54.5 | 44.0 | 24.8 | 52.2 |
| 1984 | 23.3 | 37.4 | 32.3 | 50.0 | 57.6 | 73.1 | 71.2 | 72.9 | 63.1 | 59.4 | 40.6 | 39.5 | 51.7 |
| 1985 | 21.7 | 26.0 | 43.7 | 56.3 | 62.6 | 66.9 | 72.7 | 71.2 | 66.6 | 57.3 | 48.2 | 26.0 | 51.6 |
| 1986 | 30.1 | 32.7 | 42.5 | 54.5 | 64.3 | 70.6 | 75.7 | 71.0 | 69.2 | 56.3 | 41.3 | 33.3 | 53.5 |
| 1987 | 29.9 | 34.9 | 44.3 | 52.1 | 66.0 | 72.7 | 76.6 | 74.3 | 66.9 | 49.1 | 47.6 | 35.7 | 54.2 |
| 1988 | 26.5 | 29.3 | 40.2 | 50.3 | 62.6 | 69.6 | 77.5 | 75.3 | 65.2 | 47.4 | 43.9 | 31.6 | 51.6 |
| 1989 | 36.6 | 28.7 | 42.0 | 48.2 | 57.2 | 68.8 | 73.9 | 71.2 | 65.2 | 54.2 | 42.1 | 19.8 | 50.7 |
| 1990 | 37.7 | 37.5 | 45.3 | 50.7 | 59.1 | 70.3 | 73.6 | 72.5 | 66.4 | 55.1 | 46.2 | 37.2 | 54.3 |
| 1991 | 29.7 | 35.7 | 43.9 | 56.1 | 70.9 | 75.0 | 77.6 | 75.0 | 66.2 | 55.9 | 41.0 | 36.4 | 55.3 |
| 1992 | 32.2 | 36.8 | 40.7 | 51.8 | 59.9 | 67.3 | 73.5 | 69.4 | 64.7 | 51.9 | 44.8 | 34.7 | 52.3 |
| 1993 | 34.3 | 27.8 | 38.6 | 50.3 | 62.2 | 69.8 | 76.2 | 75.7 | 64.9 | 53.0 | 43.5 | 32.9 | 52.4 |
| 1994 | 21.3 | 30.0 | 39.5 | 53.9 | 58.4 | 73.9 | 75.2 | 71.7 | 65.4 | 55.5 | 48.2 | 38.8 | 52.7 |
| 1995 | 29.4 | 27.9 | 43.6 | 50.8 | 60.9 | 72.9 | 76.0 | 78.4 | 64.2 | 56.1 | 37.7 | 28.8 | 52.2 |
| 1996 | 27.8 | 30.5 | 35.6 | 50.2 | 60.9 | 72.3 | 72.8 | 74.0 | 65.8 | 55.0 | 37.5 | 37.1 | 51.6 |
| 1997 | 28.1 | 36.3 | 42.7 | 48.4 | 56.6 | 70.2 | 74.2 | 70.3 | 65.1 | 55.2 | 40.1 | 34.7 | 51.8 |
| 1998 | 37.6 | 40.5 | 43.6 | 53.1 | 67.3 | 74.1 | 74.8 | 76.3 | 71.6 | 55.7 | 45.9 | 38.1 | 56.4 |
| 1999 | 31.1 | 37.1 | 37.5 | 55.0 | 64.8 | 74.5 | 80.2 | 73.1 | 67.9 | 55.3 | 47.5 | 34.6 | 54.9 |
| 2000 | 27.0 | 37.5 | 45.9 | 51.4 | 64.9 | 71.6 | 72.5 | 71.3 | 64.8 | 57.2 | 41.0 | 23.3 | 52.4 |
| 2001 | 28.7 | 35.3 | 38.1 | 56.8 | 63.6 | 71.1 | 74.3 | 75.2 | 64.4 | 55.8 | 49.6 | 38.4 | 54.3 |
| 2002 | 35.6 | 35.9 | 41.3 | 54.6 | 58.8 | 73.5 | 77.9 | 76.2 | 70.7 | 53.5 | 41.4 | 32.4 | 54.3 |
| 2003 | 22.7 | 26.9 | 43.2 | 54.9 | 60.8 | 67.3 | 73.4 | 73.7 | 64.4 | 52.8 | 47.3 | 33.8 | 51.8 |
| 2004 | 24.2 | 32.0 | 43.5 | 52.6 | 66.8 | 70.1 | 73.7 | 70.6 | 68.0 | 55.1 | 46.2 | 32.7 | 53.0 |
| 2005 | 30.7 | 34.0 | 37.2 | 54.2 | 58.3 | 74.4 | 77.0 | 76.6 | 70.0 | 55.4 | 45.4 | 29.6 | 53.6 |
| 2006 | 40.8 | 33.6 | 40.2 | 56.8 | 61.1 | 69.3 | 76.6 | 75.8 | 63.7 | 52.3 | 45.9 | 40.1 | 54.7 |
| 2007 | 34.4 | 21.2 | 47.4 | 50.7 | 66.8 | 73.1 | 73.2 | 77.8 | 71.0 | 62.1 | 44.6 | 35.7 | 54.8 |
| 2008 | 31.7 | 31.0 | 39.4 | 55.6 | 60.4 | 73.2 | 75.4 | 73.9 | 70.1 | 54.9 | 41.7 | 33.2 | 53.4 |
| 2009 | 22.6 | 33.8 | 46.0 | 53.5 | 63.5 | 72.2 | 71.0 | 72.7 | 66.8 | 51.6 | 47.8 | 32.2 | 52.8 |
| POR=<br>62 YRS | 28.4 | 31.3 | 40.7 | 51.9 | 61.8 | 70.7 | 74.4 | 73.0 | 66.1 | 54.4 | 43.1 | 32.7 | 52.4 |

## HEATING DEGREE DAYS (base 65°F) 2009 COLUMBUS (KCMH)

| YEAR | JUL | AUG | SEP | OCT | NOV | DEC | JAN | FEB | MAR | APR | MAY | JUN | TOTAL |
|------|-----|-----|-----|-----|-----|-----|-----|-----|-----|-----|-----|-----|-------|
| 1980-81 | 0 | 0 | 46 | 435 | 717 | 1000 | 1286 | 864 | 761 | 287 | 195 | 14 | 5605 |
| 1981-82 | 8 | 5 | 141 | 429 | 713 | 1061 | 1351 | 997 | 758 | 556 | 45 | 33 | 6097 |
| 1982-83 | 3 | 19 | 107 | 304 | 585 | 759 | 1081 | 863 | 669 | 493 | 239 | 30 | 5152 |
| 1983-84 | 6 | 0 | 83 | 325 | 626 | 1236 | 1284 | 796 | 1007 | 447 | 254 | 3 | 6067 |
| 1984-85 | 6 | 3 | 143 | 182 | 727 | 782 | 1339 | 1086 | 654 | 286 | 134 | 35 | 5377 |
| 1985-86 | 0 | 2 | 96 | 249 | 500 | 1202 | 1076 | 901 | 694 | 328 | 113 | 19 | 5180 |
| 1986-87 | 0 | 26 | 41 | 287 | 702 | 974 | 1083 | 838 | 637 | 393 | 103 | 9 | 5093 |
| 1987-88 | 0 | 4 | 53 | 489 | 521 | 900 | 1187 | 1029 | 762 | 433 | 119 | 49 | 5546 |
| 1988-89 | 3 | 7 | 57 | 547 | 624 | 1032 | 873 | 1009 | 711 | 499 | 274 | 28 | 5664 |
| 1989-90 | 0 | 11 | 90 | 345 | 680 | 1394 | 840 | 766 | 613 | 444 | 190 | 26 | 5399 |
| 1990-91 | 0 | 3 | 83 | 310 | 558 | 857 | 1089 | 817 | 649 | 282 | 42 | 0 | 4690 |
| 1991-92 | 0 | 0 | 105 | 296 | 714 | 878 | 1011 | 814 | 747 | 402 | 190 | 35 | 5192 |
| 1992-93 | 0 | 8 | 101 | 403 | 600 | 932 | 942 | 1034 | 811 | 434 | 130 | 51 | 5446 |
| 1993-94 | 0 | 1 | 84 | 366 | 637 | 989 | 1351 | 973 | 787 | 340 | 233 | 12 | 5773 |
| 1994-95 | 0 | 8 | 51 | 295 | 497 | 805 | 1098 | 1031 | 657 | 427 | 150 | 6 | 5025 |
| 1995-96 | 0 | 0 | 78 | 274 | 810 | 1112 | 1147 | 992 | 905 | 448 | 191 | 11 | 5968 |
| 1996-97 | 2 | 0 | 74 | 307 | 821 | 859 | 1136 | 798 | 687 | 494 | 264 | 15 | 5457 |
| 1997-98 | 0 | 10 | 63 | 346 | 741 | 933 | 843 | 679 | 681 | 352 | 38 | 41 | 4727 |
| 1998-99 | 0 | 0 | 22 | 292 | 567 | 829 | 1045 | 776 | 848 | 299 | 68 | 9 | 4755 |
| 1999-00 | 0 | 0 | 46 | 295 | 517 | 936 | 1172 | 789 | 586 | 403 | 84 | 21 | 4849 |
| 2000-01 | 1 | 3 | 104 | 256 | 717 | 1285 | 1117 | 822 | 826 | 289 | 92 | 24 | 5536 |
| 2001-02 | 2 | 0 | 97 | 297 | 455 | 815 | 904 | 809 | 724 | 345 | 225 | 4 | 4677 |
| 2002-03 | 0 | 0 | 21 | 390 | 700 | 1002 | 1306 | 1059 | 668 | 306 | 144 | 54 | 5650 |
| 2003-04 | 0 | 0 | 76 | 374 | 526 | 960 | 1258 | 951 | 659 | 370 | 78 | 3 | 5255 |
| 2004-05 | 0 | 8 | 24 | 307 | 556 | 996 | 1056 | 861 | 857 | 323 | 215 | 1 | 5204 |
| 2005-06 | 0 | 0 | 15 | 324 | 580 | 1092 | 748 | 875 | 758 | 248 | 180 | 20 | 4840 |
| 2006-07 | 0 | 0 | 80 | 399 | 567 | 766 | 946 | 1221 | 548 | 431 | 72 | 2 | 5032 |
| 2007-08 | 0 | 0 | 25 | 171 | 607 | 899 | 1025 | 982 | 789 | 289 | 168 | 1 | 4956 |
| 2008-09 | 0 | 0 | 3 | 321 | 694 | 978 | 1305 | 869 | 584 | 369 | 106 | 9 | 5238 |
| 2009- | 0 | 11 | 33 | 404 | 507 | 1010 | | | | | | | |

WBAN : 14821

## COOLING DEGREE DAYS (base 65°F) 2009 COLUMBUS (KCMH)

| YEAR | JAN | FEB | MAR | APR | MAY | JUN | JUL | AUG | SEP | OCT | NOV | DEC | TOTAL |
|------|-----|-----|-----|-----|-----|-----|-----|-----|-----|-----|-----|-----|-------|
| 1980 | 0 | 0 | 0 | 0 | 61 | 132 | 343 | 344 | 151 | 3 | 0 | 0 | 1034 |
| 1981 | 0 | 0 | 0 | 16 | 32 | 198 | 231 | 181 | 64 | 4 | 0 | 0 | 726 |
| 1982 | 0 | 0 | 0 | 4 | 111 | 66 | 301 | 154 | 67 | 39 | 7 | 4 | 753 |
| 1983 | 0 | 0 | 1 | 2 | 17 | 167 | 377 | 355 | 152 | 9 | 0 | 0 | 1080 |
| 1984 | 0 | 0 | 0 | 8 | 30 | 253 | 205 | 254 | 94 | 14 | 0 | 0 | 858 |
| 1985 | 0 | 0 | 2 | 32 | 64 | 97 | 245 | 201 | 152 | 19 | 2 | 0 | 814 |
| 1986 | 0 | 0 | 2 | 19 | 95 | 194 | 339 | 221 | 171 | 25 | 0 | 0 | 1066 |
| 1987 | 0 | 0 | 0 | 11 | 142 | 246 | 366 | 299 | 116 | 0 | 5 | 0 | 1185 |
| 1988 | 0 | 0 | 0 | 0 | 54 | 194 | 396 | 333 | 70 | 5 | 0 | 0 | 1052 |
| 1989 | 0 | 0 | 5 | 2 | 40 | 149 | 282 | 211 | 106 | 12 | 0 | 0 | 807 |
| 1990 | 0 | 0 | 11 | 21 | 13 | 191 | 273 | 244 | 133 | 9 | 3 | 0 | 898 |
| 1991 | 0 | 0 | 0 | 21 | 232 | 307 | 402 | 317 | 147 | 23 | 0 | 0 | 1449 |
| 1992 | 0 | 0 | 0 | 13 | 37 | 115 | 272 | 152 | 99 | 2 | 0 | 0 | 690 |
| 1993 | 0 | 0 | 0 | 0 | 48 | 204 | 352 | 343 | 89 | 2 | 0 | 0 | 1038 |
| 1994 | 0 | 0 | 0 | 15 | 39 | 286 | 322 | 224 | 71 | 8 | 0 | 0 | 965 |
| 1995 | 0 | 0 | 0 | 6 | 32 | 251 | 347 | 424 | 61 | 4 | 0 | 0 | 1125 |
| 1996 | 0 | 0 | 0 | 11 | 72 | 238 | 251 | 285 | 102 | 2 | 0 | 0 | 961 |
| 1997 | 0 | 0 | 0 | 0 | 10 | 181 | 291 | 182 | 73 | 48 | 0 | 0 | 785 |
| 1998 | 0 | 0 | 27 | 0 | 118 | 248 | 313 | 357 | 227 | 10 | 0 | 3 | 1303 |
| 1999 | 0 | 0 | 0 | 4 | 69 | 301 | 476 | 258 | 139 | 0 | 0 | 0 | 1247 |
| 2000 | 0 | 0 | 0 | 2 | 87 | 225 | 239 | 209 | 105 | 18 | 0 | 0 | 885 |
| 2001 | 0 | 0 | 0 | 50 | 56 | 215 | 299 | 324 | 85 | 19 | 0 | 0 | 1048 |
| 2002 | 0 | 0 | 0 | 40 | 36 | 268 | 405 | 354 | 199 | 41 | 0 | 0 | 1343 |
| 2003 | 0 | 0 | 1 | 8 | 22 | 130 | 267 | 277 | 65 | 6 | 2 | 0 | 778 |
| 2004 | 0 | 0 | 0 | 8 | 141 | 162 | 274 | 189 | 120 | 4 | 0 | 0 | 898 |
| 2005 | 0 | 0 | 0 | 7 | 14 | 290 | 379 | 363 | 171 | 33 | 0 | 0 | 1257 |
| 2006 | 0 | 0 | 0 | 11 | 69 | 156 | 364 | 340 | 48 | 13 | 0 | 0 | 1001 |
| 2007 | 0 | 0 | 8 | 7 | 135 | 251 | 263 | 405 | 212 | 88 | 0 | 0 | 1369 |
| 2008 | 0 | 0 | 0 | 15 | 33 | 256 | 330 | 283 | 164 | 15 | 0 | 0 | 1096 |
| 2009 | 0 | 0 | 1 | 30 | 67 | 236 | 191 | 254 | 93 | 1 | 0 | 0 | 873 |

## SNOWFALL (inches)  2009  COLUMBUS (KCMH)

| YEAR | JUL | AUG | SEP | OCT | NOV | DEC | JAN | FEB | MAR | APR | MAY | JUN | TOTAL |
|------|-----|-----|-----|-----|-----|-----|-----|-----|-----|-----|-----|-----|-------|
| 1980-81 | 0.0 | 0.0 | 0.0 | T | 8.0 | 7.3 | 7.8 | 3.7 | 3.3 | 0.0 | 0.0 | 0.0 | 30.1 |
| 1981-82 | 0.0 | 0.0 | 0.0 | 0.0 | 1.9 | 9.8 | 11.8 | 3.7 | 3.2 | 4.7 | 0.0 | 0.0 | 35.1 |
| 1982-83 | 0.0 | 0.0 | 0.0 | 0.0 | T | 1.5 | 2.6 | 4.5 | 2.8 | 0.1 | 0.0 | 0.0 | 11.5 |
| 1983-84 | 0.0 | 0.0 | 0.0 | 0.0 | 0.5 | 5.7 | 9.0 | 10.8 | 9.8 | 0.3 | 0.0 | 0.0 | 36.1 |
| 1984-85 | 0.0 | 0.0 | 0.0 | 0.0 | 0.9 | 7.3 | 21.9 | 12.5 | T | 0.8 | 0.0 | 0.0 | 43.4 |
| 1985-86 | 0.0 | 0.0 | 0.0 | 0.0 | 0.0 | 8.6 | 4.8 | 9.8 | 1.8 | T | 0.0 | 0.0 | 25.0 |
| 1986-87 | 0.0 | 0.0 | 0.0 | 0.0 | 0.4 | 0.4 | 2.7 | 1.2 | 5.9 | 12.6 | 0.0 | 0.0 | 23.2 |
| 1987-88 | 0.0 | 0.0 | 0.0 | T | 0.6 | 4.6 | 8.4 | 6.5 | 3.8 | T | 0.0 | 0.0 | 23.9 |
| 1988-89 | 0.0 | 0.0 | 0.0 | T | 0.8 | 5.9 | 0.6 | 3.9 | 6.6 | 0.1 | 0.8 | 0.0 | 18.7 |
| 1989-90 | 0.0 | 0.0 | 0.0 | 0.4 | 0.3 | 9.4 | 3.3 | 6.0 | 1.4 | 0.4 | 0.0 | T | 21.2 |
| 1990-91 | 0.0 | 0.0 | 0.0 | 0.0 | 0.0 | 3.7 | 3.4 | 4.5 | 4.0 | T | 0.0 | 0.0 | 15.6 |
| 1991-92 | 0.0 | 0.0 | 0.0 | T | 0.6 | 1.6 | 12.2 | 1.8 | 1.6 | 1.1 | 0.0 | 0.0 | 18.9 |
| 1992-93 | 0.0 | 0.0 | 0.0 | T | 3.0 | 2.4 | 1.5 | 14.6 | 8.9 | 0.2 | 0.0 | 0.0 | 30.6 |
| 1993-94 | 0.0 | 0.0 | 0.0 | 4.6 | 0.8 | 4.2 | 19.5 | 2.9 | 4.6 | 1.1 | 0.0 | T | 37.7 |
| 1994-95 | T | 0.0 | T | 0.0 | 0.0 | 0.3 | 12.6 | 5.3 | 2.5 | T | 0.0 | T | 20.7 |
| 1995-96 | T | 0.0 | 0.0 | 0.0 | 2.7 | 11.8 | 24.5 | 4.3 | 7.6 | 3.2 | 0.0 | | |
| 1996-97 | | | | | 1.9 | | | | | | | | |
| 1997-98 | | | | | 0.9 | 2.9 | 1.2 | 2.2 | 2.8 | T | 0.0 | T | |
| 1998-99 | 0.0 | 0.0 | 0.0 | 0.0 | 0.0 | 2.8 | 20.6 | 7.3 | 9.8 | 0.0 | 0.0 | 0.0 | 40.5 |
| 1999-00 | 0.0 | 0.0 | 0.0 | 0.0 | 1.5 | 4.6 | 13.8 | 7.9 | 1.9 | 0.1 | 0.0 | 0.0 | 29.8 |
| 2000-01 | T | 0.0 | 0.0 | 0.0 | 1.3 | 13.4 | 6.3 | 3.4 | 1.1 | 0.8 | T | 0.0 | 26.3 |
| 2001-02 | 0.0 | 0.0 | 0.0 | T | 0.0 | 1.7 | 4.7 | 2.0 | 1.4 | 0.3 | 0.0 | 0.0 | 10.1 |
| 2002-03 | 0.0 | 0.0 | 0.0 | T | 3.1 | 5.1 | 14.4 | 26.3 | 1.7 | T | 0.0 | 0.0 | 50.6 |
| 2003-04 | 0.0 | 0.0 | 0.0 | 0.0 | 0.7 | 6.3 | 11.5 | 1.5 | 4.8 | 0.3 | 0.0 | 0.0 | 25.1 |
| 2004-05 | 0.0 | 0.0 | 0.0 | T | 0.1 | 9.6 | 10.6 | 3.7 | 8.6 | 4.7 | 0.0 | 0.0 | 37.3 |
| 2005-06 | 0.0 | 0.0 | 0.0 | 0.0 | 1.7 | 6.2 | 1.8 | 1.3 | 1.9 | 0.0 | 0.0 | 0.0 | 12.9 |
| 2006-07 | 0.0 | 0.0 | 0.0 | T | 0.1 | 0.1 | 6.8 | 13.1 | 1.9 | 0.3 | 0.0 | 0.0 | 22.3 |
| 2007-08 | 0.0 | T | 0.0 | 0.0 | T | 7.8 | 5.0 | 10.7 | 21.8 | T | 0.0 | 0.0 | 45.3 |
| 2008-09 | 0.0 | 0.0 | 0.0 | T | 0.8 | 2.1 | 20.0 | 0.2 | T | 0.1 | 0.0 | 0.0 | 23.2 |
| 2009- | 0.0 | 0.0 | 0.0 | 0.0 | T | 8.9 | | | | | | | |
| POR= 60 YRS | T | T | T | 0.1 | 1.9 | 5.5 | 9.0 | 6.2 | 4.6 | 0.9 | T | T | 28.2 |

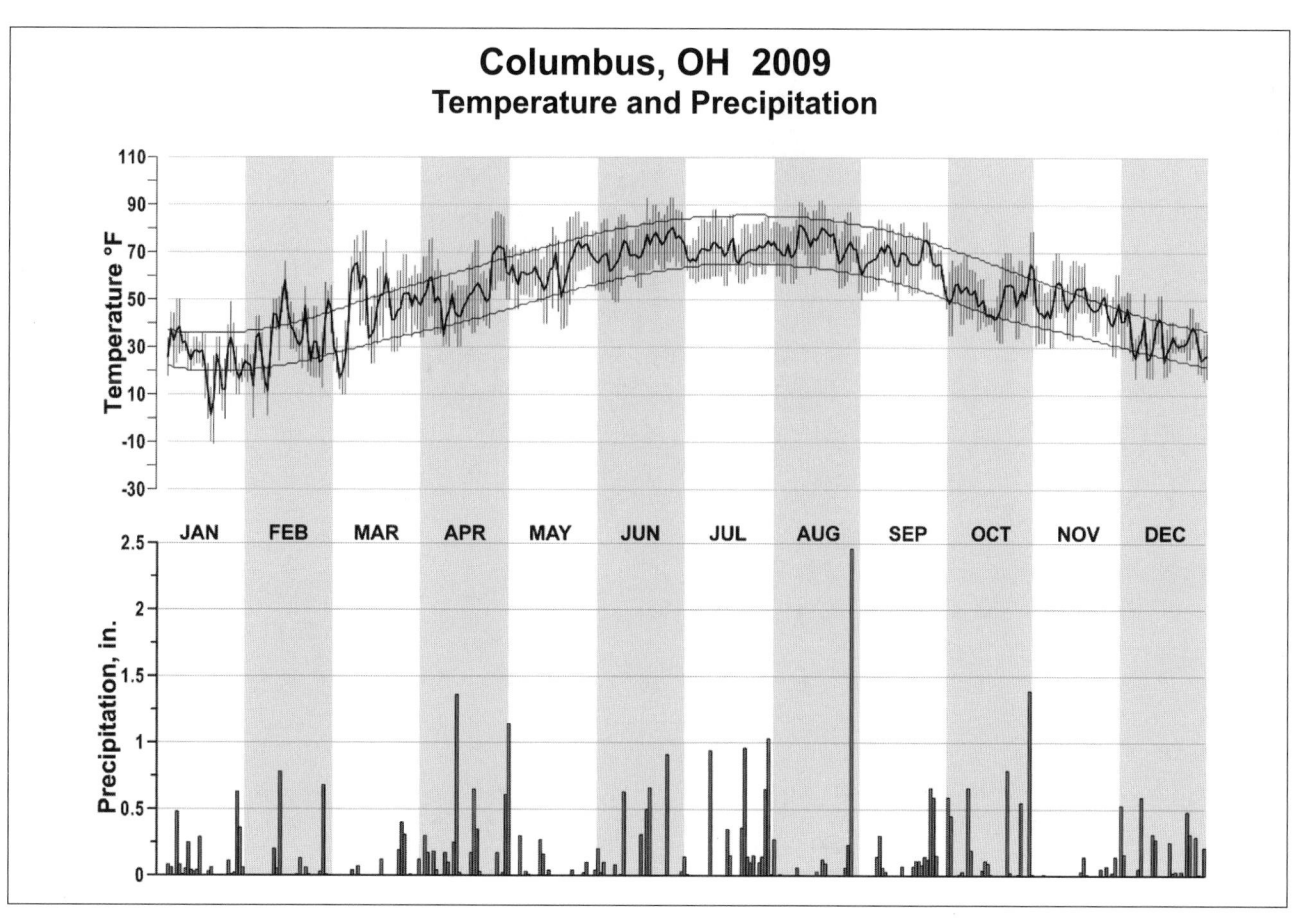

# Columbus, OH  2009
## Temperature and Precipitation

# 2009
# CINCINNATI, OH (KCVG)
## CINCINNATI-NORTHERN KENTUCKY INTERNATIONAL AIRPORT
## HEBRON, KY

Greater Cincinnati Airport is located on a gently rolling plateau about 12 miles southwest of downtown Cincinnati and 2 miles south of the Ohio River at its nearest point. The river valley is rather narrow and steep-sided varying from 1 to 3 miles in width and the river bed is 500 feet below the level of the airport.

The climate is continental with a rather wide range of temperatures from winter to summer. A precipitation maximum occurs during winter and spring with a late summer and fall minimum. On the average, the maximum snowfall occurs during January, although the heaviest 24-hour amounts have been recorded during late November and February.

The heaviest precipitation, as well as the precipitation of the longest duration, is normally associated with low pressure disturbances moving in a general southwest to northeast direction through the Ohio valley and south of the Cincinnati area.

Summers are warm and rather humid. The temperature will reach 100 degrees or more in 1 year out of 3. However, the temperature will reach 90 degrees or higher on about 19 days each year. Winters are moderately cold with frequent periods of extensive cloudiness.

The freeze free period lasts on the average 187 days from mid-April to the latter part of October.

# NORMALS, MEANS, AND EXTREMES
## COVINGTON (KCVG)

| LATITUDE: 39 ° 2 'N | LONGITUDE: -84 ° 40'W | | ELEVATION (FT): GRND: 859 BARO: 885 | | | | TIME ZONE: EASTERN (UTC -5) | | | WBAN: 93814 | | |

| | ELEMENT | POR | JAN | FEB | MAR | APR | MAY | JUN | JUL | AUG | SEP | OCT | NOV | DEC | YEAR |
|---|---|---|---|---|---|---|---|---|---|---|---|---|---|---|---|
| **TEMPERATURE °F** | NORMAL DAILY MAXIMUM | 30 | 38.0 | 43.1 | 53.9 | 64.7 | 74.4 | 82.4 | 86.4 | 84.8 | 78.0 | 66.4 | 53.6 | 42.7 | 64.0 |
| | MEAN DAILY MAXIMUM | 62 | 38.1 | 42.3 | 52.2 | 64.3 | 73.9 | 82.1 | 85.7 | 84.9 | 78.2 | 66.5 | 52.9 | 42.0 | 63.6 |
| | HIGHEST DAILY MAXIMUM | 48 | 69 | 75 | 84 | 89 | 93 | 102 | 103 | 102 | 98 | 91 | 81 | 75 | 103 |
| | YEAR OF OCCURRENCE | | 1967 | 2000 | 1986 | 1976 | 1962 | 1988 | 1988 | 1962 | 1964 | 2007 | 1987 | 1982 | JUL 1988 |
| | MEAN OF EXTREME MAXS. | 62 | 60.7 | 64.2 | 74.8 | 82.1 | 86.9 | 92.1 | 94.2 | 93.6 | 90.4 | 82.4 | 72.3 | 62.8 | 79.7 |
| | NORMAL DAILY MINIMUM | 30 | 21.3 | 25.0 | 33.8 | 42.7 | 52.9 | 61.6 | 66.1 | 64.2 | 56.8 | 44.9 | 35.7 | 26.4 | 44.3 |
| | MEAN DAILY MINIMUM | 62 | 21.9 | 24.6 | 32.8 | 42.9 | 52.2 | 61.0 | 65.2 | 63.8 | 56.4 | 45.0 | 35.1 | 26.2 | 43.9 |
| | LOWEST DAILY MINIMUM | 48 | -25 | -11 | -11 | 15 | 27 | 39 | 47 | 43 | 31 | 16 | 1 | -20 | -25 |
| | YEAR OF OCCURRENCE | | 1977 | 1996 | 1980 | 1997 | 1963 | 1972 | 2008 | 1986 | 1993 | 1962 | 1976 | 1989 | JAN 1977 |
| | MEAN OF EXTREME MINS. | 62 | -1.2 | 3.4 | 14.0 | 26.4 | 36.3 | 47.7 | 54.1 | 52.7 | 40.5 | 28.0 | 17.6 | 5.9 | 27.1 |
| | NORMAL DRY BULB | 30 | 29.7 | 34.1 | 43.9 | 53.7 | 63.7 | 72.0 | 76.3 | 74.5 | 67.4 | 55.7 | 44.7 | 34.6 | 54.2 |
| | MEAN DRY BULB | 62 | 30.0 | 33.5 | 42.5 | 53.6 | 63.1 | 71.6 | 75.4 | 74.4 | 67.3 | 55.8 | 44.0 | 34.1 | 53.8 |
| | MEAN WET BULB | 26 | 27.8 | 30.3 | 37.3 | 46.7 | 56.3 | 64.5 | 67.8 | 66.8 | 60.0 | 49.4 | 39.9 | 31.0 | 48.2 |
| | MEAN DEW POINT | 26 | 24.4 | 26.2 | 32.6 | 42.0 | 52.9 | 61.8 | 65.4 | 64.4 | 57.2 | 46.0 | 36.2 | 27.7 | 44.7 |
| | NORMAL NO. DAYS WITH: | | | | | | | | | | | | | | |
| | MAXIMUM >= 90 | 30 | 0.0 | 0.0 | 0.0 | 0.0 | 0.3 | 3.5 | 7.6 | 5.1 | 1.5 | 0.0 | 0.0 | 0.0 | 18.0 |
| | MAXIMUM <= 32 | 30 | 11.0 | 7.2 | 1.2 | 0.0 | 0.0 | 0.0 | 0.0 | 0.0 | 0.0 | 0.0 | 0.5 | 6.4 | 26.3 |
| | MINIMUM <= 32 | 30 | 26.1 | 21.1 | 15.6 | 4.7 | 0.3 | 0.0 | 0.0 | 0.0 | * | 3.3 | 12.8 | 21.7 | 105.6 |
| | MINIMUM <= 0 | 30 | 2.7 | 1.2 | 0.1 | 0.0 | 0.0 | 0.0 | 0.0 | 0.0 | 0.0 | 0.0 | 0.0 | 1.1 | 5.1 |
| **H/C** | NORMAL HEATING DEG. DAYS | 30 | 1110 | 881 | 670 | 368 | 130 | 19 | 1 | 3 | 68 | 319 | 626 | 953 | 5148 |
| | NORMAL COOLING DEG. DAYS | 30 | 0 | 0 | 3 | 13 | 73 | 215 | 335 | 282 | 126 | 16 | 1 | 0 | 1064 |
| **RH** | NORMAL (PERCENT) | 30 | 75 | 71 | 67 | 65 | 69 | 72 | 73 | 74 | 73 | 71 | 72 | 75 | 71 |
| | HOUR 01 LST | 30 | 78 | 75 | 73 | 71 | 79 | 82 | 84 | 85 | 83 | 79 | 76 | 78 | 79 |
| | HOUR 07 LST | 30 | 81 | 79 | 79 | 77 | 82 | 85 | 86 | 90 | 89 | 85 | 81 | 81 | 83 |
| | HOUR 13 LST | 30 | 70 | 65 | 59 | 54 | 57 | 58 | 58 | 59 | 58 | 57 | 63 | 69 | 61 |
| | HOUR 19 LST | 30 | 73 | 67 | 60 | 56 | 60 | 62 | 62 | 65 | 67 | 66 | 69 | 73 | 65 |
| **S** | PERCENT POSSIBLE SUNSHINE | 13 | 33 | 40 | 48 | 56 | 57 | 61 | 62 | 61 | 61 | 54 | 36 | 31 | 50 |
| **W/O** | MEAN NO. DAYS WITH: | | | | | | | | | | | | | | |
| | HEAVY FOG(VISBY <= 1/4 MI) | 46 | 2.4 | 2.2 | 1.5 | 0.8 | 1.2 | 1.2 | 1.4 | 2.3 | 2.5 | 2.1 | 1.4 | 2.3 | 21.3 |
| | THUNDERSTORMS | 62 | 0.7 | 0.8 | 2.5 | 4.2 | 6.0 | 7.4 | 7.5 | 6.0 | 3.1 | 1.5 | 1.1 | 0.4 | 41.2 |
| **CLOUDNESS** | MEAN: | | | | | | | | | | | | | | |
| | SUNRISE-SUNSET (OKTAS) | | | | | | | | | | | | | | |
| | MIDNIGHT-MIDNIGHT (OKTAS) | | | | | | | | | | | | | | |
| | MEAN NO. DAYS WITH: | | | | | | | | | | | | | | |
| | CLEAR | | | | 5.0 | | | | | | | | | | |
| | PARTLY CLOUDY | | | | 1.0 | | 1.0 | | | | | | | | |
| | CLOUDY | 1 | 1.0 | 4.0 | 7.0 | | 2.0 | 1.0 | | | | | | | |
| **PR** | MEAN STATION PRESSURE(IN) | 26 | 29.15 | 29.14 | 29.10 | 29.04 | 29.05 | 29.06 | 29.09 | 29.11 | 29.13 | 29.15 | 29.15 | 29.16 | 29.11 |
| | MEAN SEA-LEVEL PRES. (IN) | 26 | 30.12 | 30.10 | 30.06 | 29.98 | 29.98 | 29.98 | 30.00 | 30.03 | 30.06 | 30.09 | 30.10 | 30.12 | 30.05 |
| **WINDS** | MEAN SPEED (MPH) | 26 | 9.8 | 9.8 | 10.1 | 9.7 | 8.0 | 7.2 | 6.7 | 6.4 | 6.8 | 7.6 | 9.1 | 9.3 | 8.4 |
| | PREVAIL.DIR(TENS OF DEGS) | 39 | 22 | 22 | 21 | 22 | 21 | 21 | 22 | 22 | 21 | 21 | 21 | 22 | 21 |
| | MAXIMUM 2-MINUTE: | | | | | | | | | | | | | | |
| | SPEED (MPH) | 14 | 41 | 41 | 45 | 45 | 39 | 45 | 47 | 41 | 54 | 48 | 40 | 40 | 54 |
| | DIR. (TENS OF DEGS) | | 17 | 24 | 27 | 28 | 27 | 03 | 30 | 29 | 19 | 29 | 17 | 27 | 19 |
| | YEAR OF OCCURRENCE | | 1996 | 2009 | 2002 | 2002 | 2003 | 2009 | 2008 | 2000 | 2008 | 2001 | 2001 | 2000 | SEP 2008 |
| | MAXIMUM 3-SECOND | | | | | | | | | | | | | | |
| | SPEED (MPH) | 14 | 51 | 56 | 56 | 76 | 54 | 56 | 61 | 53 | 74 | 60 | 55 | 56 | 76 |
| | DIR. (TENS OF DEGS) | | 27 | 23 | 28 | 22 | 25 | 27 | 31 | 29 | 23 | 29 | 17 | 27 | 22 |
| | YEAR OF OCCURRENCE | | 2008 | 2009 | 2002 | 2006 | 1999 | 2001 | 2001 | 2000 | 2008 | 2001 | 2001 | 2006 | APR 2006 |
| **PRECIPITATION** | NORMAL (IN) | 30 | 2.92 | 2.75 | 3.90 | 3.96 | 4.59 | 4.42 | 3.75 | 3.79 | 2.82 | 2.96 | 3.46 | 3.28 | 42.60 |
| | MAXIMUM MONTHLY (IN) | 62 | 9.43 | 6.72 | 12.18 | 9.77 | 9.48 | 9.61 | 8.70 | 7.71 | 8.61 | 8.60 | 7.51 | 7.90 | 12.18 |
| | YEAR OF OCCURRENCE | | 1950 | 1955 | 1964 | 1998 | 1968 | 1998 | 2001 | 1982 | 1979 | 1983 | 1985 | 1990 | MAR 1964 |
| | MINIMUM MONTHLY (IN) | 62 | 0.57 | 0.25 | 1.14 | 1.04 | 0.91 | 0.95 | 0.63 | 0.31 | 0.18 | 0.25 | 0.43 | 0.51 | 0.18 |
| | YEAR OF OCCURRENCE | | 1981 | 1978 | 1960 | 1971 | 2007 | 1965 | 1997 | 1953 | 1963 | 1963 | 1949 | 1976 | SEP 1963 |
| | MAXIMUM IN 24 HOURS (IN) | 62 | 4.33 | 2.84 | 5.21 | 3.31 | 3.71 | 3.45 | 4.28 | 3.52 | 4.54 | 4.47 | 3.36 | 2.96 | 5.21 |
| | YEAR OF OCCURRENCE | | 1959 | 1990 | 1964 | 1996 | 1956 | 1974 | 1988 | 1995 | 1979 | 1985 | 1948 | 1948 | MAR 1964 |
| | NORMAL NO. DAYS WITH: | | | | | | | | | | | | | | |
| | PRECIPITATION >= 0.01 | 30 | 12.6 | 11.7 | 12.9 | 12.5 | 11.8 | 11.5 | 10.2 | 9.7 | 8.4 | 8.5 | 10.9 | 12.2 | 132.9 |
| | PRECIPITATION >= 1.00 | 30 | 0.5 | 0.6 | 0.7 | 0.8 | 1.1 | 1.2 | 1.0 | 1.0 | 0.7 | 0.6 | 0.6 | 0.6 | 9.4 |
| **SNOWFALL** | NORMAL (IN) | 30 | 7.8 | 6.0 | 3.8 | 0.6 | 0.* | 0.0 | 0.0 | 0.0 | 0.0 | 0.4 | 1.3 | 3.7 | 23.6 |
| | MAXIMUM MONTHLY (IN) | 62 | 31.5 | 19.9 | 13.0 | 3.7 | 0.2 | T | T | T | 0.0 | 6.2 | 12.1 | 12.5 | 31.5 |
| | YEAR OF OCCURRENCE | | 1978 | 1993 | 1968 | 1977 | 1989 | 2009 | 1994 | 2000 | | 1993 | 1966 | 1989 | JAN 1978 |
| | MAXIMUM IN 24 HOURS (IN) | 62 | 12.8 | 12.6 | 9.8 | 3.6 | 0.2 | T | T | T | 0.0 | 5.9 | 9.0 | 7.5 | 12.8 |
| | YEAR OF OCCURRENCE' | | 1996 | 1998 | 1968 | 1977 | 1989 | 2009 | 1994 | 2000 | | 1993 | 1966 | 1990 | JAN 1996 |
| | MAXIMUM SNOW DEPTH (IN) | 61 | 14 | 19 | 11 | 5 | 0 | 0 | 0 | 0 | 0 | 4 | 8 | 9 | 19 |
| | YEAR OF OCCURRENCE | | 1978 | 1998 | 2008 | 1987 | | | | | | 1993 | 1966 | 2004 | FEB 1998 |
| | NORMAL NO. DAYS WITH: | | | | | | | | | | | | | | |
| | SNOWFALL >= 1.0 | 30 | 2.4 | 1.9 | 1.1 | 0.3 | 0.0 | 0.0 | 0.0 | 0.0 | 0.0 | 0.1 | 0.4 | 1.1 | 7.3 |

## PRECIPITATION (inches) 2009 COVINGTON (KCVG)

| YEAR | JAN | FEB | MAR | APR | MAY | JUN | JUL | AUG | SEP | OCT | NOV | DEC | ANNUAL |
|---|---|---|---|---|---|---|---|---|---|---|---|---|---|
| 1980 | 2.26 | 1.04 | 4.50 | 1.96 | 4.59 | 4.13 | 5.51 | 4.19 | 1.83 | 3.28 | 2.58 | 1.26 | 37.13 |
| 1981 | 0.57 | 3.86 | 1.72 | 5.05 | 5.07 | 3.34 | 3.66 | 2.15 | 1.47 | 2.33 | 2.94 | 2.39 | 34.55 |
| 1982 | 7.17 | 1.17 | 4.67 | 2.18 | 4.60 | 3.61 | 2.44 | 7.71 | 1.27 | 0.99 | 5.08 | 4.25 | 45.14 |
| 1983 | 1.56 | 1.14 | 2.02 | 4.84 | 8.89 | 2.22 | 1.96 | 3.23 | 1.22 | 8.60 | 4.20 | 2.84 | 42.72 |
| 1984 | 0.75 | 2.40 | 3.61 | 4.88 | 4.82 | 2.11 | 2.57 | 3.30 | 3.50 | 3.85 | 6.00 | 4.21 | 42.00 |
| 1985 | 1.68 | 2.25 | 6.90 | 1.34 | 6.18 | 4.55 | 3.59 | 2.02 | 0.76 | 5.83 | 7.51 | 1.52 | 44.13 |
| 1986 | 1.01 | 2.85 | 3.07 | 1.57 | 3.59 | 1.46 | 3.33 | 3.78 | 3.53 | 3.08 | 3.79 | 2.58 | 33.64 |
| 1987 | 0.92 | 1.62 | 4.65 | 2.88 | 2.73 | 4.62 | 5.07 | 2.27 | 1.17 | 1.42 | 1.82 | 3.43 | 32.60 |
| 1988 | 2.75 | 4.94 | 3.42 | 3.92 | 1.99 | 1.19 | 6.85 | 2.44 | 3.05 | 1.86 | 4.78 | 2.78 | 39.97 |
| 1989 | 3.21 | 4.67 | 6.40 | 5.19 | 4.64 | 3.04 | 5.97 | 5.33 | 2.97 | 3.18 | 3.05 | 1.96 | 49.61 |
| 1990 | 2.59 | 5.82 | 2.75 | 3.22 | 9.41 | 5.01 | 3.68 | 5.67 | 4.13 | 5.09 | 2.31 | 7.90 | 57.58 |
| 1991 | 2.84 | 3.99 | 6.20 | 3.62 | 3.41 | 1.39 | 2.66 | 5.04 | 2.60 | 1.37 | 1.89 | 5.08 | 40.09 |
| 1992 | 2.99 | 0.93 | 4.19 | 2.71 | 2.84 | 3.65 | 7.00 | 3.17 | 3.23 | 1.11 | 4.31 | 1.36 | 37.49 |
| 1993 | 3.83 | 3.43 | 3.60 | 3.13 | 2.33 | 4.80 | 1.26 | 4.20 | 2.68 | 2.61 | 4.31 | 2.53 | 38.71 |
| 1994 | 3.22 | 1.68 | 2.22 | 6.46 | 2.06 | 4.08 | 5.64 | 5.14 | 0.55 | 1.49 | 2.87 | 2.88 | 38.29 |
| 1995 | 3.51 | 1.80 | 2.58 | 4.26 | 8.57 | 2.65 | 2.37 | 5.59 | 2.43 | 4.28 | 2.15 | 3.43 | 43.62 |
| 1996 | 4.36 | 1.98 | 5.58 | 8.20 | 9.20 | 5.83 | 2.62 | .76 | 5.41 | 1.74 | 3.40 | 4.33 | 53.41 |
| 1997 | 2.79 | 2.13 | 6.00 | 1.98 | 6.33 | 8.34 | 0.63 | 3.95 | 0.55 | 1.68 | 2.97 | 2.77 | 40.12 |
| 1998 | 3.27 | 3.04 | 3.52 | 9.77 | 5.12 | 9.61 | 4.75 | 2.67 | 0.67 | 2.82 | 2.33 | 3.82 | 51.39 |
| 1999 | 4.76 | 3.66 | 1.89 | 2.88 | 1.98 | 3.16 | 3.16 | 2.61 | 0.86 | 2.49 | 1.42 | 3.60 | 32.47 |
| 2000 | 4.45 | 5.71 | 3.34 | 4.27 | 5.21 | 4.74 | 3.53 | 2.90 | 4.79 | 1.37 | 2.33 | 3.18 | 45.82 |
| 2001 | 1.33 | 1.81 | 1.42 | 1.46 | 5.15 | 4.45 | 8.70 | 5.00 | 3.13 | 6.73 | 3.31 | 4.08 | 46.57 |
| 2002 | 2.33 | 1.81 | 4.60 | 5.97 | 8.03 | 3.56 | 1.38 | 1.50 | 4.87 | 4.51 | 2.29 | 4.90 | 45.75 |
| 2003 | 1.66 | 3.60 | 2.50 | 1.91 | 7.29 | 2.75 | 5.00 | 4.80 | 5.07 | 2.11 | 3.92 | 2.26 | 42.87 |
| 2004 | 4.55 | 1.25 | 2.97 | 4.50 | 6.85 | 2.93 | 6.14 | 3.51 | 1.53 | 6.13 | 5.15 | 2.80 | 48.31 |
| 2005 | 6.60 | 1.94 | 4.09 | 3.78 | 1.88 | 2.92 | 1.76 | 6.34 | 2.00 | 2.21 | 4.13 | 1.81 | 39.46 |
| 2006 | 4.21 | 1.34 | 6.92 | 5.06 | 3.13 | 3.67 | 4.03 | 1.90 | 6.21 | 4.48 | 2.08 | 3.46 | 46.49 |
| 2007 | 3.84 | 3.42 | 3.16 | 3.15 | 0.91 | 1.74 | 1.92 | 0.55 | 2.47 | 7.07 | 2.73 | 5.76 | 36.72 |
| 2008 | 2.33 | 5.21 | 9.67 | 2.75 | 6.32 | 5.21 | 3.39 | 1.78 | 1.22 | 1.63 | 1.73 | 4.42 | 45.66 |
| 2009 | 2.96 | 2.52 | 1.61 | 3.72 | 3.74 | 7.33 | 5.35 | 1.78 | 4.83 | 5.42 | 0.92 | 2.93 | 43.11 |
| POR= 62 YRS | 3.27 | 2.84 | 3.95 | 3.76 | 4.28 | 4.05 | 4.06 | 3.17 | 2.84 | 2.89 | 3.23 | 3.17 | 41.51 |

WBAN : 93814

## AVERAGE TEMPERATURE (°F) 2009 COVINGTON (KCVG)

| YEAR | JAN | FEB | MAR | APR | MAY | JUN | JUL | AUG | SEP | OCT | NOV | DEC | ANNUAL |
|---|---|---|---|---|---|---|---|---|---|---|---|---|---|
| 1980 | 29.9 | 24.0 | 38.5 | 50.3 | 64.9 | 70.1 | 76.6 | 76.5 | 68.6 | 50.6 | 41.5 | 32.9 | 52.0 |
| 1981 | 24.1 | 34.2 | 40.1 | 58.1 | 59.8 | 72.3 | 75.9 | 73.6 | 65.2 | 53.9 | 43.7 | 29.0 | 52.5 |
| 1982 | 23.9 | 30.6 | 44.3 | 49.6 | 68.1 | 67.4 | 77.0 | 71.3 | 66.9 | 59.3 | 48.4 | 42.9 | 54.1 |
| 1983 | 31.6 | 35.3 | 44.7 | 49.4 | 59.0 | 71.6 | 79.2 | 78.3 | 67.2 | 55.7 | 44.7 | 24.6 | 53.4 |
| 1984 | 23.7 | 38.2 | 34.4 | 51.1 | 58.9 | 74.1 | 72.2 | 74.3 | 65.7 | 61.5 | 42.0 | 42.4 | 53.2 |
| 1985 | 22.7 | 29.4 | 47.5 | 57.9 | 65.4 | 69.9 | 75.1 | 72.5 | 67.5 | 59.3 | 49.9 | 26.2 | 53.6 |
| 1986 | 30.9 | 35.2 | 45.2 | 55.3 | 64.5 | 72.9 | 77.6 | 72.0 | 70.1 | 56.4 | 42.7 | 34.0 | 54.7 |
| 1987 | 30.7 | 37.3 | 45.0 | 53.0 | 69.3 | 73.6 | 76.1 | 75.2 | 68.3 | 49.3 | 48.0 | 36.8 | 55.2 |
| 1988 | 27.5 | 30.5 | 42.2 | 52.4 | 64.4 | 72.4 | 78.5 | 77.5 | 67.2 | 48.5 | 45.0 | 34.1 | 53.4 |
| 1989 | 38.6 | 30.8 | 45.4 | 52.9 | 60.1 | 71.3 | 76.7 | 73.5 | 66.4 | 55.7 | 43.9 | 21.6 | 53.1 |
| 1990 | 40.0 | 40.8 | 48.2 | 52.8 | 61.6 | 71.8 | 74.7 | 73.7 | 67.7 | 55.9 | 48.9 | 38.3 | 56.2 |
| 1991 | 31.0 | 37.3 | 45.4 | 56.9 | 70.4 | 75.4 | 77.9 | 74.8 | 68.6 | 58.6 | 41.4 | 37.3 | 56.3 |
| 1992 | 32.9 | 39.7 | 43.6 | 54.1 | 61.2 | 68.0 | 73.5 | 69.6 | 64.5 | 53.3 | 44.2 | 35.6 | 53.4 |
| 1993 | 34.5 | 29.3 | 40.2 | 52.0 | 63.3 | 70.9 | 79.1 | 76.2 | 63.9 | 53.0 | 43.5 | 32.8 | 53.2 |
| 1994 | 23.3 | 33.2 | 41.4 | 55.5 | 59.3 | 75.0 | 75.2 | 71.9 | 65.7 | 57.1 | 49.9 | 40.0 | 54.0 |
| 1995 | 31.3 | 31.0 | 45.7 | 53.3 | 62.7 | 72.9 | 76.6 | 79.5 | 65.7 | 55.3 | 37.5 | 29.3 | 53.4 |
| 1996 | 27.9 | 32.4 | 36.7 | 49.2 | 63.2 | 71.0 | 72.1 | 74.3 | 65.4 | 55.1 | 37.8 | 37.6 | 51.9 |
| 1997 | 28.2 | 37.6 | 44.2 | 48.1 | 57.2 | 69.1 | 75.1 | 71.6 | 65.6 | 55.4 | 41.3 | 35.7 | 52.4 |
| 1998 | 38.6 | 40.6 | 43.6 | 53.0 | 66.6 | 71.3 | 74.3 | 75.1 | 71.6 | 56.2 | 45.5 | 38.3 | 56.2 |
| 1999 | 32.4 | 37.9 | 38.3 | 55.0 | 63.6 | 73.0 | 79.0 | 72.7 | 66.8 | 55.3 | 48.5 | 34.8 | 54.8 |
| 2000 | 28.6 | 40.4 | 46.8 | 52.3 | 65.4 | 71.4 | 72.6 | 72.3 | 64.9 | 57.8 | 41.7 | 23.2 | 53.1 |
| 2001 | 29.7 | 37.0 | 39.7 | 58.0 | 64.2 | 69.7 | 74.2 | 74.9 | 64.6 | 55.2 | 50.0 | 39.3 | 54.7 |
| 2002 | 36.9 | 36.8 | 41.9 | 55.7 | 59.6 | 72.9 | 78.2 | 77.5 | 71.6 | 53.9 | 41.3 | 34.2 | 55.0 |
| 2003 | 24.0 | 28.4 | 45.0 | 55.7 | 61.9 | 67.2 | 73.9 | 74.6 | 64.8 | 54.5 | 48.0 | 35.0 | 52.8 |
| 2004 | 27.7 | 33.9 | 45.4 | 53.5 | 67.0 | 71.0 | 73.3 | 70.6 | 68.5 | 56.6 | 47.7 | 33.4 | 54.1 |
| 2005 | 33.4 | 37.7 | 39.0 | 55.8 | 60.3 | 74.7 | 77.5 | 77.7 | 70.8 | 56.1 | 45.6 | 30.2 | 54.9 |
| 2006 | 41.6 | 34.8 | 42.4 | 57.9 | 61.8 | 70.1 | 76.9 | 77.4 | 64.1 | 53.0 | 46.2 | 40.4 | 55.6 |
| 2007 | 35.1 | 22.7 | 50.7 | 52.2 | 67.6 | 74.2 | 74.6 | 81.6 | 72.9 | 61.3 | 43.8 | 36.4 | 56.1 |
| 2008 | 30.0 | 32.3 | 41.1 | 54.7 | 60.5 | 72.8 | 74.4 | 74.5 | 71.0 | 56.5 | 42.4 | 33.9 | 53.7 |
| 2009 | 26.0 | 35.8 | 47.5 | 54.8 | 64.3 | 72.7 | 70.2 | 72.0 | 67.9 | 52.0 | 47.5 | 33.3 | 53.7 |
| POR= 62 YRS | 30.0 | 33.5 | 42.5 | 53.6 | 63.1 | 71.6 | 75.4 | 74.4 | 67.3 | 55.8 | 44.0 | 34.1 | 53.8 |

## HEATING DEGREE DAYS (base 65°F) 2009  COVINGTON (KCVG)

| YEAR | JUL | AUG | SEP | OCT | NOV | DEC | JAN | FEB | MAR | APR | MAY | JUN | TOTAL |
|---|---|---|---|---|---|---|---|---|---|---|---|---|---|
| 1980-81 | 0 | 0 | 48 | 446 | 697 | 988 | 1261 | 858 | 768 | 230 | 191 | 6 | 5493 |
| 1981-82 | 0 | 0 | 87 | 344 | 634 | 1107 | 1268 | 956 | 635 | 460 | 28 | 19 | 5538 |
| 1982-83 | 0 | 1 | 56 | 244 | 505 | 682 | 1029 | 825 | 627 | 466 | 199 | 21 | 4655 |
| 1983-84 | 1 | 0 | 89 | 288 | 600 | 1247 | 1274 | 773 | 939 | 425 | 219 | 4 | 5859 |
| 1984-85 | 0 | 0 | 101 | 128 | 684 | 692 | 1306 | 992 | 543 | 256 | 72 | 22 | 4796 |
| 1985-86 | 0 | 0 | 78 | 212 | 450 | 1195 | 1056 | 828 | 613 | 305 | 105 | 3 | 4845 |
| 1986-87 | 0 | 21 | 25 | 292 | 664 | 955 | 1058 | 766 | 612 | 365 | 52 | 2 | 4812 |
| 1987-88 | 0 | 1 | 39 | 477 | 505 | 868 | 1156 | 991 | 699 | 374 | 84 | 22 | 5216 |
| 1988-89 | 1 | 0 | 38 | 509 | 595 | 949 | 811 | 949 | 608 | 380 | 211 | 14 | 5065 |
| 1989-90 | 0 | 4 | 77 | 297 | 630 | 1335 | 770 | 671 | 531 | 390 | 127 | 21 | 4853 |
| 1990-91 | 0 | 1 | 66 | 296 | 477 | 821 | 1046 | 773 | 602 | 250 | 44 | 0 | 4376 |
| 1991-92 | 0 | 0 | 81 | 232 | 700 | 853 | 988 | 727 | 658 | 339 | 172 | 27 | 4777 |
| 1992-93 | 0 | 8 | 97 | 358 | 617 | 907 | 937 | 997 | 762 | 384 | 103 | 40 | 5210 |
| 1993-94 | 0 | 0 | 101 | 370 | 640 | 992 | 1291 | 885 | 724 | 298 | 211 | 5 | 5517 |
| 1994-95 | 0 | 6 | 56 | 250 | 447 | 766 | 1037 | 948 | 589 | 353 | 109 | 2 | 4563 |
| 1995-96 | 0 | 0 | 68 | 298 | 818 | 1099 | 1142 | 939 | 873 | 478 | 132 | 10 | 5857 |
| 1996-97 | 0 | 0 | 93 | 307 | 811 | 847 | 1136 | 765 | 640 | 498 | 252 | 36 | 5385 |
| 1997-98 | 1 | 7 | 53 | 339 | 703 | 900 | 808 | 679 | 679 | 355 | 51 | 47 | 4622 |
| 1998-99 | 0 | 0 | 17 | 284 | 576 | 826 | 1007 | 754 | 820 | 297 | 90 | 5 | 4676 |
| 1999-00 | 0 | 0 | 63 | 294 | 489 | 931 | 1119 | 709 | 560 | 375 | 73 | 13 | 4626 |
| 2000-01 | 0 | 0 | 107 | 248 | 693 | 1290 | 1084 | 778 | 777 | 268 | 96 | 28 | 5369 |
| 2001-02 | 1 | 0 | 96 | 316 | 442 | 786 | 865 | 786 | 711 | 309 | 217 | 6 | 4535 |
| 2002-03 | 0 | 0 | 20 | 375 | 702 | 947 | 1263 | 1019 | 613 | 287 | 122 | 47 | 5395 |
| 2003-04 | 0 | 0 | 74 | 330 | 502 | 923 | 1148 | 895 | 605 | 346 | 67 | 2 | 4892 |
| 2004-05 | 2 | 12 | 20 | 264 | 513 | 973 | 970 | 757 | 796 | 281 | 169 | 3 | 4760 |
| 2005-06 |  | 0 | 16 | 304 | 577 | 1072 | 717 | 840 | 691 | 230 | 163 | 9 |  |
| 2006-07 | 0 | 0 | 75 | 384 | 557 | 759 | 922 | 1177 | 458 | 395 | 58 | 0 | 4785 |
| 2007-08 | 0 | 0 | 15 | 188 | 629 | 881 | 1081 | 942 | 731 | 317 | 162 | 1 | 4947 |
| 2008-09 | 1 | 0 | 6 | 283 | 673 | 958 | 1205 | 813 | 534 | 340 | 83 | 10 | 4906 |
| 2009- | 4 | 7 | 30 | 401 | 518 | 974 |  |  |  |  |  |  |  |

WBAN : 93814

## COOLING DEGREE DAYS (base 65°F) 2009  COVINGTON (KCVG)

| YEAR | JAN | FEB | MAR | APR | MAY | JUN | JUL | AUG | SEP | OCT | NOV | DEC | TOTAL |
|---|---|---|---|---|---|---|---|---|---|---|---|---|---|
| 1980 | 0 | 0 | 0 | 0 | 98 | 187 | 364 | 363 | 166 | 5 | 0 | 0 | 1183 |
| 1981 | 0 | 0 | 1 | 31 | 34 | 234 | 343 | 275 | 99 | 9 | 0 | 0 | 1026 |
| 1982 | 0 | 0 | 0 | 5 | 129 | 99 | 381 | 203 | 120 | 73 | 13 | 8 | 1031 |
| 1983 | 0 | 0 | 4 | 4 | 18 | 225 | 448 | 417 | 161 | 8 | 0 | 0 | 1285 |
| 1984 | 0 | 0 | 0 | 13 | 38 | 289 | 233 | 295 | 130 | 29 | 0 | 0 | 1027 |
| 1985 | 0 | 0 | 6 | 47 | 93 | 174 | 318 | 241 | 162 | 41 | 5 | 0 | 1087 |
| 1986 | 0 | 0 | 4 | 22 | 97 | 247 | 399 | 243 | 183 | 30 | 0 | 0 | 1225 |
| 1987 | 0 | 0 | 0 | 12 | 193 | 266 | 353 | 325 | 147 | 0 | 4 | 0 | 1300 |
| 1988 | 0 | 0 | 2 | 3 | 70 | 251 | 425 | 392 | 111 | 6 | 0 | 0 | 1260 |
| 1989 | 0 | 0 | 7 | 26 | 67 | 210 | 369 | 275 | 125 | 17 | 0 | 0 | 1096 |
| 1990 | 0 | 0 | 17 | 32 | 27 | 230 | 309 | 276 | 155 | 21 | 3 | 0 | 1070 |
| 1991 | 0 | 0 | 0 | 14 | 218 | 317 | 408 | 310 | 195 | 42 | 0 | 0 | 1504 |
| 1992 | 0 | 0 | 0 | 19 | 59 | 126 | 273 | 158 | 88 | 2 | 0 | 0 | 725 |
| 1993 | 0 | 0 | 0 | 0 | 56 | 224 | 443 | 353 | 75 | 5 | 0 | 0 | 1156 |
| 1994 | 0 | 0 | 0 | 19 | 40 | 314 | 325 | 228 | 83 | 14 | 0 | 0 | 1023 |
| 1995 | 0 | 0 | 0 | 9 | 45 | 242 | 365 | 454 | 97 | 1 | 0 | 0 | 1213 |
| 1996 | 0 | 0 | 0 | 7 | 79 | 198 | 226 | 296 | 109 | 5 | 0 | 0 | 920 |
| 1997 | 0 | 0 | 0 | 0 | 16 | 163 | 320 | 218 | 77 | 48 | 0 | 0 | 842 |
| 1998 | 0 | 0 | 23 | 0 | 104 | 242 | 299 | 319 | 222 | 16 | 0 | 5 | 1230 |
| 1999 | 0 | 0 | 0 | 6 | 52 | 253 | 440 | 247 | 125 | 2 | 0 | 0 | 1125 |
| 2000 | 0 | 0 | 0 | 2 | 89 | 211 | 240 | 235 | 107 | 30 | 0 | 0 | 914 |
| 2001 | 0 | 0 | 0 | 64 | 78 | 174 | 296 | 315 | 90 | 16 | 0 | 0 | 1033 |
| 2002 | 0 | 0 | 0 | 36 | 56 | 250 | 415 | 395 | 228 | 37 | 0 | 0 | 1417 |
| 2003 | 0 | 0 | 0 | 15 | 31 | 122 | 283 | 309 | 78 | 10 | 1 | 0 | 849 |
| 2004 | 0 | 0 | 2 | 8 | 138 | 191 | 267 | 193 | 132 | 10 | 0 | 0 | 941 |
| 2005 | 0 | 0 | 0 | 9 | 28 | 299 | 395 | 400 | 194 | 36 | 0 | 0 | 1361 |
| 2006 | 0 | 0 | 0 | 26 | 70 | 167 | 373 | 394 | 56 | 19 | 0 | 0 | 1105 |
| 2007 | 0 | 0 | 24 | 19 | 145 | 286 | 306 | 523 | 259 | 83 | 0 | 0 | 1645 |
| 2008 | 0 | 0 | 0 | 15 | 28 | 242 | 298 | 300 | 196 | 27 | 0 | 0 | 1106 |
| 2009 | 0 | 0 | 0 | 39 | 66 | 247 | 172 | 230 | 122 | 5 | 0 | 0 | 881 |

## SNOWFALL (inches) 2009 COVINGTON (KCVG)

| YEAR | JUL | AUG | SEP | OCT | NOV | DEC | JAN | FEB | MAR | APR | MAY | JUN | TOTAL |
|------|-----|-----|-----|-----|-----|-----|-----|-----|-----|-----|-----|-----|-------|
| 1980-81 | 0.0 | 0.0 | 0.0 | T | 1.2 | 3.7 | 4.0 | 2.6 | 2.5 | 0.0 | 0.0 | 0.0 | 14.0 |
| 1981-82 | 0.0 | 0.0 | 0.0 | T | 0.3 | 10.9 | 7.1 | 3.9 | 0.5 | 1.5 | 0.0 | 0.0 | 24.2 |
| 1982-83 | 0.0 | 0.0 | 0.0 | 0.0 | T | T | 0.8 | 5.5 | 0.3 | T | 0.0 | 0.0 | 6.6 |
| 1983-84 | 0.0 | 0.0 | 0.0 | 0.0 | T | 1.7 | 4.1 | 6.7 | 4.1 | 0.0 | 0.0 | 0.0 | 16.6 |
| 1984-85 | 0.0 | 0.0 | 0.0 | 0.0 | 1.4 | 7.3 | 12.2 | 9.5 | 0.4 | 1.7 | 0.0 | 0.0 | 32.5 |
| 1985-86 | 0.0 | 0.0 | 0.0 | 0.0 | 0.0 | 5.0 | 2.8 | 11.3 | 0.8 | T | 0.0 | 0.0 | 19.9 |
| 1986-87 | 0.0 | 0.0 | 0.0 | 0.0 | T | 0.8 | 1.6 | 2.4 | 8.8 | 2.3 | 0.0 | 0.0 | 15.9 |
| 1987-88 | 0.0 | 0.0 | 0.0 | 0.0 | 0.1 | 0.2 | 4.3 | 4.7 | 2.3 | T | 0.0 | 0.0 | 11.6 |
| 1988-89 | 0.0 | 0.0 | 0.0 | 0.0 | 0.7 | 2.9 | T | 3.0 | 1.2 | 0.3 | 0.2 | 0.0 | 8.3 |
| 1989-90 | 0.0 | 0.0 | 0.0 | 5.9 | 0.2 | 12.5 | 1.3 | 3.6 | 5.6 | T | 0.0 | 0.0 | 29.1 |
| 1990-91 | 0.0 | 0.0 | 0.0 | 0.0 | 0.0 | 8.6 | 4.3 | 2.6 | T | 0.0 | 0.0 | 0.0 | 15.5 |
| 1991-92 | T | 0.0 | 0.0 | 0.0 | 1.9 | 0.5 | 3.6 | 1.2 | 3.6 | 2.9 | 0.1 | 0.0 | 13.8 |
| 1992-93 | 0.0 | 0.0 | 0.0 | T | 1.6 | 3.8 | 0.3 | 19.9 | 3.9 | T | 0.0 | T | 29.5 |
| 1993-94 | 0.0 | 0.0 | 0.0 | 6.2 | 0.8 | 5.4 | 13.3 | 0.5 | 6.6 | 0.2 | 0.0 | 0.0 | 33.0 |
| 1994-95 | T | 0.0 | 0.0 | 0.0 | 0.0 | 0.2 | 16.4 | 7.6 | 3.3 | 0.0 | 0.0 | 0.0 | 27.5 |
| 1995-96 | 0.0 | 0.0 | 0.0 | 0.0 | 0.9 | 4.1 | 27.0 | 1.7 | 8.4 | 2.5 | 0.0 | 0.0 | 44.6 |
| 1996-97 | 0.0 | 0.0 | 0.0 | 0.0 | 2.1 | 1.3 | 4.3 | 4.5 | T | T | 0.0 | 0.0 | 12.2 |
| 1997-98 | 0.0 | 0.0 | 0.0 | 0.0 | 0.4 | 4.2 | 1.2 | 18.5 | 7.1 | T | 0.0 | 0.0 | 31.4 |
| 1998-99 | 0.0 | 0.0 | 0.0 | 0.0 | 0.0 | 3.4 | 9.6 | 3.9 | 9.5 | T | 0.0 | 0.0 | 26.4 |
| 1999-00 | 0.0 | 0.0 | 0.0 | 0.0 | T | 2.3 | 7.6 | 0.6 | 0.1 | 0.2 | 0.0 | 0.0 | 10.8 |
| 2000-01 | 0.0 | T | 0.0 | 0.0 | 0.4 | 8.5 | 5.8 | 0.7 | 1.2 | 0.7 | 0.0 | T | 17.3 |
| 2001-02 | 0.0 | 0.0 | 0.0 | T | 0.0 | 0.7 | 5.5 | 1.5 | 0.4 | 1.0 | 0.0 | 0.0 | 9.1 |
| 2002-03 | T | 0.0 | 0.0 | 0.0 | 0.8 | 5.3 | 9.5 | 17.4 | 0.4 | T | 0.0 | 0.0 | 33.4 |
| 2003-04 | T | 0.0 | 0.0 | 0.0 | 0.2 | 6.1 | 12.6 | 1.5 | 1.8 | 1.3 | 0.0 | 0.0 | 23.5 |
| 2004-05 | 0.0 | 0.0 | 0.0 | 0.0 | T | 10.4 | 6.3 | 4.3 | 3.0 | 0.3 | 0.0 | 0.0 | 24.3 |
| 2005-06 | 0.0 | 0.0 | 0.0 | 0.0 | 0.3 | 7.3 | 1.1 | 4.6 | 4.2 | T | 0.0 | 0.0 | 17.5 |
| 2006-07 | 0.0 | 0.0 | 0.0 | T | T | 0.1 | 4.9 | 11.2 | 0.7 | 0.2 | 0.0 | 0.0 | 17.1 |
| 2007-08 | 0.0 | 0.0 | 0.0 | 0.0 | T | 4.2 | 3.4 | 7.5 | 11.1 | 0.0 | 0.0 | 0.0 | 26.2 |
| 2008-09 | 0.0 | 0.0 | 0.0 | 0.0 | T | 2.0 | 14.0 | 7.0 | 0.0 | T | 0.0 | T | 23.0 |
| 2009- | 0.0 | 0.0 | 0.0 | 0.0 | T | 5.7 | | | | | | | |
| POR= 62 YRS | T | T | 0.0 | 0.2 | 1.6 | 4.0 | 7.2 | 5.6 | 4.1 | 0.5 | 0.0 | T | 23.2 |

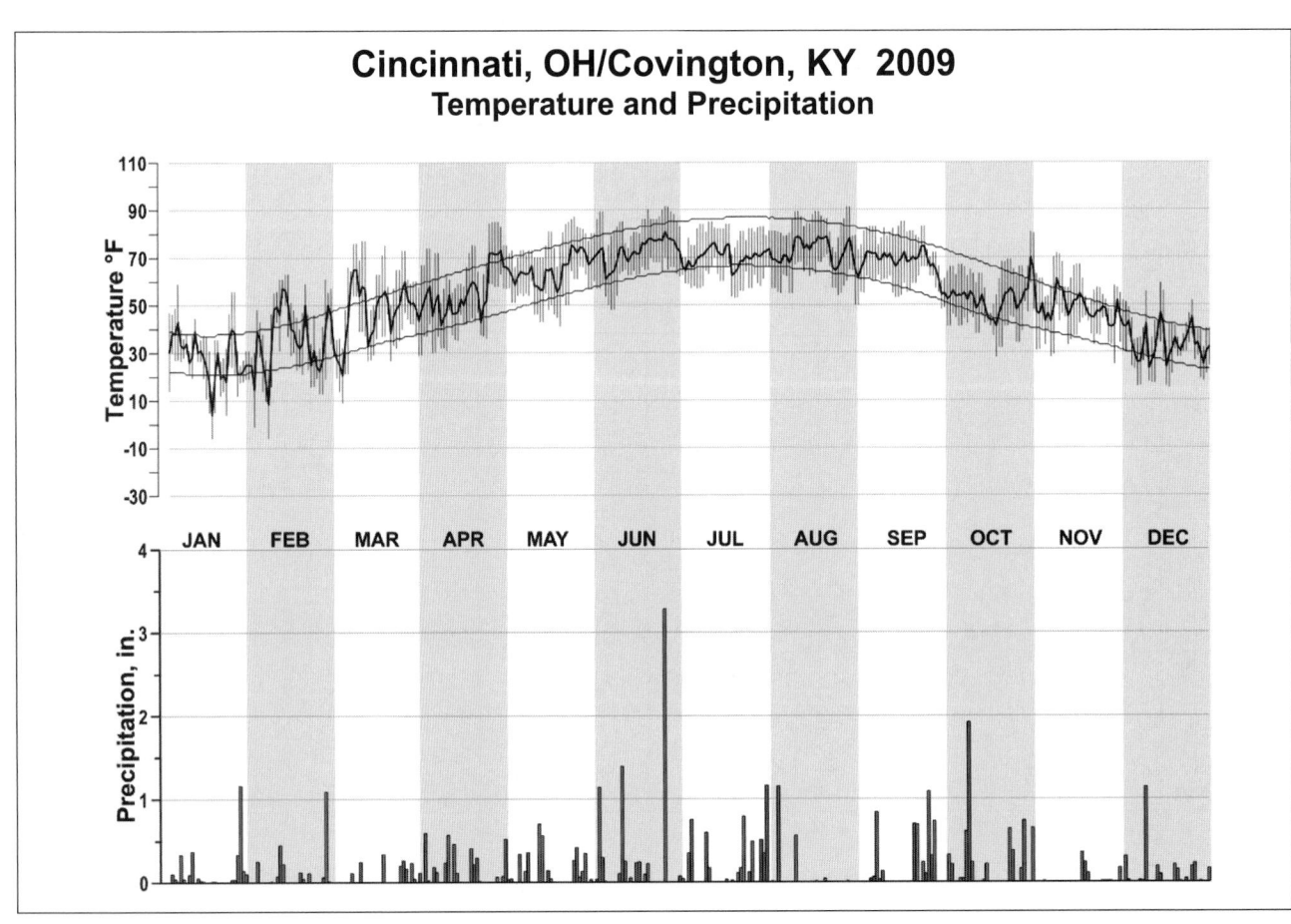

# Cincinnati, OH/Covington, KY 2009
## Temperature and Precipitation

# 2009
# TOLEDO
# OHIO (KTOL)

Toledo is located on the western end of Lake Erie at the mouth of the Maumee River. Except for a bank up from the river about 30 feet, the terrain is generally level with only a slight slope toward the river and Lake Erie. The city has quite a diversified industrial section and excellent harbor facilities, making it a large transportation center for rail, water, and motor freight. Generally rich agricultural land is found in the surrounding area, especially up the Maumee Valley toward the Indiana state line.

Rainfall is usually sufficient for general agriculture. The terrain is level and drainage rather poor, therefore, a little less than the normal precipitation during the growing season is better than excessive amounts. Snowfall is generally light in this area, distributed throughout the winter from November to March with frequent thaws.

The nearness of Lake Erie and the other Great Lakes has a moderating effect on the temperature, and extremes are seldom recorded. On average, only fifteen days a year experience temperatures of 90 degrees or higher, and only eight days when it drops to zero or lower. The growing season averages 160 days, but has ranged from over 220 to less than 125 days.

Humidity is rather high throughout the year in this area, and there is an excessive amount of cloudiness. In the winter months the sun shines during only about 30 percent of the daylight hours. December and January, the cloudiest months, sometimes have as little as 16 percent of the possible hours of sunshine.

Severe windstorms, causing more than minor damage, occur infrequently. There are on the average twenty-three days per year having a sustained wind velocity of 32 mph or more.

Flooding in the Toledo area is produced by several factors. Heavy rains of 1 inch or more will cause a sudden rise in creeks and drainage ditches to the point of overflow. The western shores of Lake Erie are subject to flooding when the lake level is high and prolonged periods of east to northeast winds prevail.

# NORMALS, MEANS, AND EXTREMES
## TOLEDO (KTOL)

| LATITUDE: 41° 35'N | LONGITUDE: -83° 48'W | ELEVATION (FT): GRND: 674  BARO: 693 | TIME ZONE: EASTERN  (UTC -5) | WBAN: 94830 |
|---|---|---|---|---|

| ELEMENT | POR | JAN | FEB | MAR | APR | MAY | JUN | JUL | AUG | SEP | OCT | NOV | DEC | YEAR |
|---|---|---|---|---|---|---|---|---|---|---|---|---|---|---|
| **TEMPERATURE °F** | | | | | | | | | | | | | | |
| NORMAL DAILY MAXIMUM | 30 | 31.4 | 35.1 | 46.5 | 58.9 | 70.7 | 79.5 | 83.4 | 81.0 | 74.0 | 62.1 | 48.3 | 36.0 | 58.9 |
| MEAN DAILY MAXIMUM | 55 | 31.2 | 34.6 | 45.6 | 59.4 | 70.7 | 79.9 | 83.8 | 81.9 | 75.1 | 62.8 | 48.7 | 35.9 | 59.1 |
| HIGHEST DAILY MAXIMUM | 54 | 66 | 71 | 81 | 88 | 95 | 104 | 104 | 99 | 98 | 91 | 80 | 70 | 104 |
| YEAR OF OCCURRENCE | | 2008 | 2000 | 1998 | 2002 | 1962 | 1988 | 1995 | 1993 | 1978 | 1963 | 2003 | 2001 | JUL 1995 |
| MEAN OF EXTREME MAXS. | 55 | 51.5 | 55.7 | 70.4 | 81.0 | 87.2 | 92.9 | 94.1 | 91.9 | 89.5 | 80.8 | 68.6 | 56.9 | 76.7 |
| NORMAL DAILY MINIMUM | 30 | 16.4 | 18.9 | 27.9 | 37.7 | 48.6 | 58.2 | 62.6 | 60.7 | 52.9 | 41.6 | 32.6 | 22.3 | 40.0 |
| MEAN DAILY MINIMUM | 55 | 16.4 | 18.7 | 27.1 | 37.6 | 47.5 | 56.8 | 61.3 | 59.7 | 52.0 | 40.9 | 32.0 | 21.8 | 39.3 |
| LOWEST DAILY MINIMUM | 54 | -20 | -14 | -6 | 8 | 25 | 32 | 40 | 34 | 26 | 15 | 2 | -19 | -20 |
| YEAR OF OCCURRENCE | | 1984 | 1982 | 1984 | 1982 | 2005 | 1972 | 1988 | 1982 | 1974 | 1976 | 1958 | 1989 | JAN 1984 |
| MEAN OF EXTREME MINS. | 55 | -4.3 | -0.9 | 8.8 | 21.8 | 32.6 | 42.9 | 49.0 | 47.0 | 35.7 | 25.2 | 16.1 | 1.3 | 22.9 |
| NORMAL DRY BULB | 30 | 23.9 | 27.0 | 37.2 | 48.3 | 59.6 | 68.8 | 73.0 | 70.8 | 63.5 | 51.8 | 40.5 | 29.2 | 49.5 |
| MEAN DRY BULB | 55 | 23.8 | 26.7 | 36.4 | 48.5 | 59.1 | 68.5 | 72.5 | 70.8 | 63.5 | 51.9 | 40.4 | 28.9 | 49.3 |
| MEAN WET BULB | 26 | 23.5 | 25.5 | 32.5 | 42.4 | 52.4 | 61.6 | 65.5 | 64.8 | 57.8 | 46.7 | 37.2 | 27.4 | 44.8 |
| MEAN DEW POINT | 26 | 20.5 | 21.9 | 28.3 | 37.7 | 48.5 | 58.3 | 62.7 | 62.5 | 55.0 | 43.5 | 34.0 | 24.8 | 41.5 |
| NORMAL NO. DAYS WITH: | | | | | | | | | | | | | | |
| MAXIMUM >= 90 | 30 | 0.0 | 0.0 | 0.0 | 0.0 | 0.9 | 3.4 | 5.9 | 3.2 | 1.2 | 0.0 | 0.0 | 0.0 | 14.6 |
| MAXIMUM <= 32 | 30 | 16.7 | 12.6 | 4.0 | 0.2 | 0.0 | 0.0 | 0.0 | 0.0 | 0.0 | 0.0 | 1.7 | 10.6 | 45.8 |
| MINIMUM <= 32 | 30 | 28.5 | 24.6 | 21.5 | 9.6 | 1.0 | * | 0.0 | 0.0 | 0.4 | 6.1 | 16.8 | 26.0 | 134.5 |
| MINIMUM <= 0 | 30 | 4.3 | 3.0 | 0.2 | 0.0 | 0.0 | 0.0 | 0.0 | 0.0 | 0.0 | 0.0 | 0.0 | 1.4 | 8.9 |
| **H/C** | | | | | | | | | | | | | | |
| NORMAL HEATING DEG. DAYS | 30 | 1281 | 1079 | 878 | 517 | 224 | 45 | 6 | 18 | 129 | 431 | 745 | 1107 | 6460 |
| NORMAL COOLING DEG. DAYS | 30 | 0 | 0 | 1 | 7 | 42 | 148 | 248 | 190 | 73 | 6 | 0 | 0 | 715 |
| **RH** | | | | | | | | | | | | | | |
| NORMAL (PERCENT) | 30 | 77 | 75 | 70 | 66 | 67 | 69 | 71 | 76 | 76 | 74 | 76 | 79 | 73 |
| HOUR 01 LST | 30 | 80 | 79 | 77 | 75 | 79 | 83 | 85 | 89 | 88 | 83 | 80 | 82 | 82 |
| HOUR 07 LST | 30 | 81 | 81 | 81 | 79 | 80 | 82 | 86 | 91 | 92 | 87 | 83 | 83 | 84 |
| HOUR 13 LST | 30 | 71 | 67 | 60 | 53 | 53 | 55 | 56 | 59 | 58 | 58 | 66 | 73 | 61 |
| HOUR 19 LST | 30 | 76 | 72 | 65 | 58 | 57 | 59 | 61 | 68 | 71 | 71 | 74 | 78 | 68 |
| **S** PERCENT POSSIBLE SUNSHINE | 40 | 41 | 46 | 50 | 52 | 60 | 64 | 65 | 63 | 61 | 54 | 37 | 33 | 52 |
| **W/O** MEAN NO. DAYS WITH: | | | | | | | | | | | | | | |
| HEAVY FOG(VISBY <= 1/4 MI) | 46 | 1.8 | 1.7 | 1.8 | 0.7 | 0.8 | 1.0 | 0.8 | 1.5 | 1.6 | 1.7 | 1.5 | 2.3 | 17.2 |
| THUNDERSTORMS | 55 | 0.1 | 0.5 | 1.6 | 3.2 | 4.4 | 6.0 | 6.1 | 5.1 | 3.0 | 1.1 | 0.7 | 0.1 | 31.9 |
| **CLOUDNESS** MEAN: | | | | | | | | | | | | | | |
| SUNRISE-SUNSET (OKTAS) | | | | | | | | | | | | | | |
| MIDNIGHT-MIDNIGHT (OKTAS) | | | | | | | | | | | | | | |
| MEAN NO. DAYS WITH: | | | | | | | | | | | | | | |
| CLEAR | | | | 2.0 | | 2.0 | | | | | | | | |
| PARTLY CLOUDY | | | | 1.0 | | | | | | | | | | |
| CLOUDY | 1 | 1.0 | 1.0 | 2.0 | | | | | | | | | | |
| **PR** MEAN STATION PRESSURE(IN) | 26 | 29.31 | 29.32 | 29.30 | 29.23 | 29.24 | 29.23 | 29.26 | 29.30 | 29.32 | 29.32 | 29.32 | 29.33 | 29.29 |
| MEAN SEA-LEVEL PRES. (IN) | 26 | 30.09 | 30.09 | 30.06 | 29.98 | 29.98 | 29.96 | 29.99 | 30.03 | 30.06 | 30.07 | 30.07 | 30.10 | 30.04 |
| **WINDS** MEAN SPEED (MPH) | 26 | 10.7 | 10.2 | 10.5 | 10.5 | 9.0 | 7.7 | 7.1 | 6.5 | 6.9 | 8.3 | 9.8 | 10.0 | 8.9 |
| PREVAIL.DIR(TENS OF DEGS) | 35 | 25 | 25 | 07 | 07 | 24 | 24 | 24 | 24 | 25 | 24 | 25 | 24 | 25 |
| MAXIMUM 2-MINUTE: | | | | | | | | | | | | | | |
| SPEED (MPH) | 14 | 47 | 46 | 46 | 48 | 46 | 53 | 44 | 43 | 38 | 45 | 51 | 48 | 53 |
| DIR. (TENS OF DEGS) | | 26 | 26 | 24 | 25 | 25 | 25 | 36 | 26 | 24 | 24 | 21 | 30 | 25 |
| YEAR OF OCCURRENCE | | 2008 | 2001 | 2002 | 1997 | 2000 | 2007 | 2008 | 1998 | 2001 | 1996 | 2005 | 1998 | JUN 2007 |
| MAXIMUM 3-SECOND | | | | | | | | | | | | | | |
| SPEED (MPH) | 14 | 56 | 56 | 69 | 61 | 68 | 62 | 54 | 54 | 47 | 59 | 66 | 66 | 69 |
| DIR. (TENS OF DEGS) | | 25 | 26 | 23 | 27 | 27 | 26 | 35 | 26 | 23 | 25 | 24 | 25 | 23 |
| YEAR OF OCCURRENCE | | 2008 | 2001 | 2002 | 2003 | 1999 | 2007 | 2008 | 1998 | 2001 | 1996 | 1998 | 2008 | MAR 2002 |
| **PRECIPITATION** NORMAL (IN) | 30 | 1.93 | 1.88 | 2.62 | 3.24 | 3.14 | 3.80 | 2.80 | 3.19 | 2.84 | 2.35 | 2.78 | 2.64 | 33.21 |
| MAXIMUM MONTHLY (IN) | 54 | 4.61 | 5.50 | 5.70 | 6.10 | 6.80 | 8.48 | 9.19 | 8.47 | 8.10 | 6.26 | 6.86 | 6.81 | 9.19 |
| YEAR OF OCCURRENCE | | 1965 | 2008 | 1985 | 1977 | 2000 | 1981 | 2006 | 1965 | 1972 | 2001 | 1982 | 1967 | JUL 2006 |
| MINIMUM MONTHLY (IN) | 54 | 0.27 | 0.27 | 0.58 | 0.88 | 0.96 | 0.27 | 0.34 | 0.40 | 0.58 | .27 | 0.55 | 0.54 | 0.27 |
| YEAR OF OCCURRENCE | | 1961 | 1969 | 1958 | 1962 | 1964 | 1988 | 1995 | 1976 | 1963 | 2005 | 1976 | 1958 | OCT 2005 |
| MAXIMUM IN 24 HOURS (IN) | 54 | 1.78 | 2.59 | 2.60 | 3.43 | 2.34 | 3.21 | 4.39 | 2.42 | 3.97 | 3.21 | 3.17 | 3.53 | 4.39 |
| YEAR OF OCCURRENCE | | 1959 | 1990 | 1985 | 1977 | 1991 | 1978 | 1969 | 1972 | 1972 | 1988 | 1982 | 1967 | JUL 1969 |
| NORMAL NO. DAYS WITH: | | | | | | | | | | | | | | |
| PRECIPITATION >= 0.01 | 30 | 13.6 | 10.6 | 12.5 | 12.7 | 11.9 | 10.6 | 9.4 | 9.6 | 9.9 | 9.9 | 12.0 | 13.6 | 136.3 |
| PRECIPITATION >= 1.00 | 30 | 0.1 | 0.2 | 0.2 | 0.3 | 0.6 | 0.7 | 0.6 | 0.6 | 0.6 | 0.3 | 0.4 | 0.3 | 4.9 |
| **SNOWFALL** NORMAL (IN) | 30 | 10.8 | 8.5 | 5.6 | 1.3 | 0.1 | 0.0 | 0.0 | 0.0 | 0.0 | 0.2 | 2.6 | 8.3 | 37.4 |
| MAXIMUM MONTHLY (IN) | 48 | 30.8 | 23.6 | 17.7 | 12.0 | 1.3 | T | T | T | T | 2.0 | 17.9 | 24.2 | 30.8 |
| YEAR OF OCCURRENCE | | 1978 | 2008 | 1993 | 1957 | 1989 | 1995 | 1992 | 1994 | 1993 | 1989 | 1966 | 1977 | JAN 1978 |
| MAXIMUM IN 24 HOURS (IN) | 48 | 12.0 | 7.7 | 9.7 | 9.8 | 1.3 | T | T | T | T | 1.8 | 8.3 | 13.9 | 13.9 |
| YEAR OF OCCURRENCE' | | 2005 | 1981 | 1993 | 1957 | 1989 | 1995 | 1992 | 1994 | 1993 | 1989 | 1966 | 1974 | DEC 1974 |
| MAXIMUM SNOW DEPTH (IN) | 46 | 17 | 19 | 8 | 10 | 1 | 0 | 0 | 0 | 0 | 1 | 8 | 16 | 19 |
| YEAR OF OCCURRENCE | | 1978 | 1978 | 2002 | 1957 | 1989 | | | | | 1989 | 1966 | 1977 | FEB 1978 |
| NORMAL NO. DAYS WITH: | | | | | | | | | | | | | | |
| SNOWFALL >= 1.0 | 30 | 3.3 | 2.8 | 1.7 | 0.4 | 0.0 | 0.0 | 0.0 | 0.0 | 0.0 | 0.1 | 1.0 | 2.5 | 11.8 |

## PRECIPITATION (inches) 2009  TOLEDO (KTOL)

| YEAR | JAN | FEB | MAR | APR | MAY | JUN | JUL | AUG | SEP | OCT | NOV | DEC | ANNUAL |
|------|-----|-----|-----|-----|-----|-----|-----|-----|-----|-----|-----|-----|--------|
| 1980 | 0.74 | 0.96 | 3.65 | 3.13 | 2.93 | 3.26 | 4.49 | 5.89 | 1.63 | 1.79 | 0.97 | 2.48 | 31.92 |
| 1981 | 0.48 | 3.27 | 0.63 | 3.54 | 2.38 | 8.48 | 3.72 | 2.28 | 6.05 | 3.79 | 0.84 | 2.93 | 38.39 |
| 1982 | 3.61 | 1.15 | 3.74 | 1.53 | 2.61 | 2.01 | 1.97 | 1.38 | 2.03 | 1.14 | 6.86 | 3.48 | 31.51 |
| 1983 | 0.88 | 0.59 | 1.86 | 4.28 | 3.98 | 4.06 | 3.39 | 2.15 | 1.42 | 3.59 | 5.56 | 3.91 | 35.67 |
| 1984 | 0.99 | 1.18 | 2.95 | 5.15 | 3.48 | 1.49 | 2.30 | 3.87 | 2.02 | 1.75 | 2.74 | 3.22 | 31.14 |
| 1985 | 2.02 | 3.23 | 5.70 | 1.40 | 1.85 | 2.90 | 3.86 | 4.30 | 2.53 | 3.05 | 5.89 | 1.62 | 38.35 |
| 1986 | 0.99 | 2.46 | 2.16 | 2.81 | 2.72 | 5.32 | 3.37 | 5.93 | 4.75 | 4.78 | 1.66 | 1.87 | 38.82 |
| 1987 | 1.87 | 0.53 | 1.78 | 1.72 | 2.32 | 5.62 | 1.51 | 4.45 | 2.31 | 2.21 | 2.59 | 3.80 | 30.71 |
| 1988 | 1.17 | 1.33 | 1.69 | 1.45 | 1.37 | 0.27 | 3.76 | 5.11 | 1.80 | 4.37 | 4.27 | 1.96 | 28.55 |
| 1989 | 1.80 | 0.74 | 2.03 | 3.50 | 4.87 | 6.74 | 6.31 | 3.59 | 3.30 | 1.36 | 1.89 | 1.29 | 37.42 |
| 1990 | 2.18 | 5.39 | 3.46 | 2.09 | 4.63 | 3.14 | 1.89 | 3.32 | 1.72 | 2.63 | 2.27 | 5.69 | 38.41 |
| 1991 | 1.41 | 1.42 | 1.42 | 4.29 | 4.82 | 1.51 | 0.52 | 1.94 | 0.73 | 5.53 | 2.15 | 1.51 | 27.25 |
| 1992 | 1.70 | 1.68 | 3.05 | 3.41 | 3.18 | 1.28 | 6.51 | 2.40 | 4.01 | 1.77 | 4.45 | 3.60 | 37.04 |
| 1993 | 3.17 | 1.71 | 3.46 | 3.06 | 1.13 | 4.60 | 1.60 | 1.15 | 4.50 | 1.51 | 2.73 | 1.25 | 29.87 |
| 1994 | 2.83 | 1.88 | 2.06 | 4.86 | 1.11 | 3.63 | 2.14 | 3.05 | 0.93 | 1.00 | 2.69 | 3.01 | 29.19 |
| 1995 | 3.07 | 0.57 | 1.59 | 4.52 | 2.96 | 4.46 | 0.34 | 2.72 | 1.41 | 3.71 | 2.72 | 0.89 | 28.96 |
| 1996 | 2.22 | 0.95 | 2.67 | 3.85 | 2.62 | 4.91 | 1.81 | .74 | 2.74 | 1.75 | 2.79 | 2.92 | 29.97 |
| 1997 | 2.35 | 4.27 | 2.53 | 1.55 | 6.76 | 3.70 | 2.63 | 4.07 | 4.74 | 1.24 | 2.16 | 2.07 | 38.07 |
| 1998 | 2.96 | 3.77 | 3.32 | 4.54 | 2.07 | 1.73 | 2.70 | 5.44 | 0.96 | 2.13 | 1.63 | 0.61 | 31.86 |
| 1999 | 3.17 | 1.67 | 1.42 | 4.89 | 4.93 | 1.86 | 2.87 | 1.40 | 1.50 | 1.92 | 1.46 | 1.71 | 28.80 |
| 2000 | 1.19 | 1.08 | 1.84 | 3.55 | 6.80 | 5.52 | 2.29 | 4.15 | 4.98 | 2.83 | 1.36 | 2.53 | 38.12 |
| 2001 | 0.52 | 2.45 | 0.64 | 2.47 | 5.06 | 2.87 | 1.87 | 2.48 | 6.26 | 2.11 | 1.96 | 1.96 | 33.41 |
| 2002 | 2.67 | 1.67 | 3.07 | 4.14 | 3.31 | 2.00 | 1.94 | 1.22 | 2.10 | 1.70 | 2.60 | 2.67 | 29.09 |
| 2003 | 1.29 | 1.87 | 2.11 | 2.57 | 5.69 | 3.12 | 4.04 | 3.32 | 5.27 | 2.75 | 1.99 | 3.25 | 37.27 |
| 2004 | 1.29 | 0.44 | 2.36 | 0.97 | 4.67 | 3.89 | 2.57 | 4.10 | 1.41 | 2.36 | 3.33 | 2.08 | 29.47 |
| 2005 | 4.52 | 2.73 | 0.80 | 2.71 | 2.08 | 1.66 | 5.03 | 1.76 | 2.82 | 0.27 | 4.02 | 3.17 | 31.57 |
| 2006 | 2.93 | 1.86 | 2.48 | 1.35 | 6.60 | 3.91 | 9.19 | 3.23 | 2.35 | 4.29 | 3.03 | 4.49 | 45.71 |
| 2007 | 3.56 | 0.96 | 1.91 | 3.77 | 2.23 | 2.95 | 3.40 | 8.26 | 1.45 | 1.81 | 2.79 | 3.86 | 36.95 |
| 2008 | 2.20 | 5.50 | 4.34 | 2.13 | 2.51 | 5.55 | 5.37 | 1.12 | 4.14 | 1.50 | 3.10 | 4.40 | 41.86 |
| 2009 | 1.59 | 3.72 | 4.82 | 4.76 | 2.77 | 3.82 | 2.98 | 2.96 | 2.96 | 3.94 | 0.67 | 3.03 | 38.02 |
| POR= 55 YRS | 1.98 | 1.91 | 2.49 | 3.09 | 3.20 | 3.56 | 3.25 | 3.21 | 2.73 | 2.27 | 2.67 | 2.68 | 33.04 |

WBAN : 94830

## AVERAGE TEMPERATURE (°F) 2009  TOLEDO (KTOL)

| YEAR | JAN | FEB | MAR | APR | MAY | JUN | JUL | AUG | SEP | OCT | NOV | DEC | ANNUAL |
|------|-----|-----|-----|-----|-----|-----|-----|-----|-----|-----|-----|-----|--------|
| 1980 | 24.3 | 21.4 | 32.4 | 46.8 | 59.5 | 65.5 | 73.6 | 73.3 | 63.8 | 46.8 | 37.4 | 26.0 | 47.6 |
| 1981 | 17.6 | 28.5 | 36.5 | 49.9 | 55.4 | 68.4 | 71.7 | 69.8 | 61.3 | 47.7 | 39.6 | 27.4 | 47.8 |
| 1982 | 15.8 | 20.2 | 33.4 | 42.7 | 64.4 | 64.3 | 72.6 | 67.5 | 61.9 | 52.7 | 41.8 | 36.6 | 47.8 |
| 1983 | 27.6 | 30.5 | 37.9 | 44.2 | 54.8 | 67.9 | 74.7 | 73.8 | 64.2 | 51.9 | 41.3 | 20.0 | 49.1 |
| 1984 | 16.6 | 33.0 | 27.6 | 46.8 | 54.4 | 71.2 | 69.8 | 71.2 | 60.8 | 55.2 | 38.7 | 34.0 | 48.3 |
| 1985 | 19.5 | 22.6 | 39.3 | 53.5 | 61.6 | 64.8 | 73.2 | 69.1 | 64.0 | 53.3 | 43.9 | 22.3 | 48.9 |
| 1986 | 25.6 | 25.0 | 39.2 | 50.0 | 60.3 | 66.8 | 73.8 | 67.0 | 65.3 | 53.2 | 37.2 | 31.6 | 49.6 |
| 1987 | 25.8 | 30.0 | 39.7 | 50.3 | 62.5 | 70.8 | 74.9 | 71.0 | 63.8 | 45.4 | 44.4 | 33.0 | 51.0 |
| 1988 | 23.8 | 23.3 | 37.5 | 48.1 | 61.0 | 69.3 | 75.9 | 73.9 | 62.5 | 45.2 | 41.8 | 28.0 | 49.2 |
| 1989 | 33.1 | 24.5 | 36.7 | 45.5 | 57.2 | 68.2 | 73.2 | 69.8 | 61.8 | 52.2 | 38.5 | 16.8 | 48.1 |
| 1990 | 34.3 | 32.4 | 41.1 | 49.4 | 56.6 | 69.1 | 71.8 | 70.0 | 63.7 | 51.8 | 44.3 | 33.1 | 51.5 |
| 1991 | 25.2 | 31.6 | 40.3 | 52.6 | 67.0 | 72.6 | 74.6 | 73.0 | 62.9 | 55.0 | 37.9 | 33.0 | 52.1 |
| 1992 | 28.8 | 31.9 | 36.1 | 47.4 | 57.9 | 65.1 | 70.1 | 67.8 | 61.9 | 49.6 | 40.8 | 32.9 | 49.2 |
| 1993 | 30.2 | 24.7 | 34.3 | 48.3 | 60.6 | 68.1 | 76.1 | 74.3 | 61.1 | 49.8 | 39.7 | 29.5 | 49.7 |
| 1994 | 17.1 | 23.0 | 36.7 | 50.8 | 57.2 | 71.0 | 72.6 | 67.1 | 64.1 | 53.6 | 45.9 | 35.7 | 49.6 |
| 1995 | 28.2 | 25.7 | 40.5 | 46.8 | 60.1 | 72.3 | 76.5 | 78.5 | 62.8 | 55.8 | 36.6 | 25.4 | 50.8 |
| 1996 | 23.7 | 26.9 | 31.7 | 46.1 | 57.5 | 70.8 | 70.5 | 72.3 | 64.3 | 53.1 | 34.4 | 31.6 | 48.6 |
| 1997 | 21.9 | 31.4 | 38.6 | 46.2 | 52.5 | 68.8 | 71.7 | 67.3 | 62.4 | 52.1 | 36.6 | 31.3 | 48.4 |
| 1998 | 33.0 | 36.3 | 40.0 | 50.0 | 66.0 | 70.2 | 73.3 | 72.2 | 67.2 | 53.5 | 42.9 | 35.1 | 53.3 |
| 1999 | 24.4 | 33.6 | 34.4 | 50.7 | 62.6 | 70.8 | 77.3 | 69.7 | 65.1 | 51.5 | 44.8 | 31.0 | 51.3 |
| 2000 | 23.8 | 33.3 | 43.8 | 48.2 | 61.8 | 69.0 | 70.3 | 70.2 | 62.5 | 55.4 | 39.9 | 18.3 | 49.7 |
| 2001 | 26.0 | 30.4 | 35.2 | 51.5 | 61.4 | 68.7 | 72.7 | 73.4 | 63.1 | 53.6 | 48.5 | 37.0 | 51.8 |
| 2002 | 35.2 | 34.9 | 36.6 | 52.4 | 56.3 | 72.2 | 77.6 | 74.2 | 69.0 | 50.6 | 40.1 | 29.3 | 52.4 |
| 2003 | 20.8 | 23.9 | 36.5 | 48.8 | 56.9 | 66.3 | 71.6 | 72.8 | 62.9 | 51.8 | 45.4 | 31.8 | 49.1 |
| 2004 | 20.5 | 28.0 | 40.6 | 51.1 | 62.0 | 67.4 | 72.3 | 68.1 | 66.2 | 52.5 | 42.9 | 29.2 | 50.1 |
| 2005 | 24.2 | 29.3 | 34.0 | 50.2 | 56.4 | 74.3 | 74.9 | 73.9 | 67.3 | 54.3 | 42.9 | 25.2 | 50.6 |
| 2006 | 36.6 | 30.3 | 37.7 | 52.8 | 59.9 | 68.9 | 74.8 | 73.0 | 62.3 | 49.8 | 42.6 | 37.2 | 52.2 |
| 2007 | 29.9 | 17.3 | 40.9 | 47.8 | 63.0 | 70.7 | 71.1 | 73.2 | 66.5 | 59.0 | 39.8 | 29.5 | 50.7 |
| 2008 | 27.4 | 25.0 | 33.8 | 50.8 | 57.6 | 70.1 | 73.6 | 71.7 | 66.3 | 50.9 | 38.4 | 27.7 | 49.4 |
| 2009 | 16.5 | 28.6 | 39.7 | 49.7 | 59.4 | 68.7 | 68.9 | 70.9 | 65.0 | 49.2 | 45.0 | 28.6 | 49.2 |
| POR= 55 YRS | 23.8 | 26.7 | 36.4 | 48.5 | 59.1 | 68.5 | 72.5 | 70.8 | 63.5 | 51.9 | 40.4 | 28.9 | 49.2 |

## HEATING DEGREE DAYS (base 65°F) 2009 TOLEDO (KTOL)

| YEAR | JUL | AUG | SEP | OCT | NOV | DEC | JAN | FEB | MAR | APR | MAY | JUN | TOTAL |
|---|---|---|---|---|---|---|---|---|---|---|---|---|---|
| 1980-81 | 0 | 3 | 113 | 560 | 822 | 1206 | 1464 | 1015 | 879 | 450 | 309 | 24 | 6845 |
| 1981-82 | 7 | 15 | 169 | 529 | 754 | 1160 | 1522 | 1250 | 972 | 665 | 81 | 76 | 7200 |
| 1982-83 | 3 | 47 | 148 | 386 | 690 | 871 | 1154 | 958 | 833 | 624 | 311 | 55 | 6080 |
| 1983-84 | 8 | 0 | 127 | 407 | 705 | 1389 | 1494 | 920 | 1151 | 545 | 341 | 9 | 7096 |
| 1984-85 | 11 | 15 | 173 | 297 | 782 | 951 | 1404 | 1182 | 791 | 368 | 158 | 58 | 6190 |
| 1985-86 | 0 | 16 | 138 | 356 | 626 | 1316 | 1216 | 1113 | 793 | 449 | 185 | 54 | 6262 |
| 1986-87 | 2 | 54 | 87 | 365 | 828 | 1027 | 1209 | 972 | 778 | 439 | 173 | 20 | 5954 |
| 1987-88 | 5 | 34 | 89 | 601 | 611 | 986 | 1269 | 1202 | 845 | 498 | 159 | 53 | 6352 |
| 1988-89 | 4 | 5 | 104 | 613 | 691 | 1141 | 979 | 1127 | 869 | 578 | 270 | 29 | 6410 |
| 1989-90 | 0 | 14 | 159 | 396 | 789 | 1488 | 947 | 907 | 742 | 492 | 262 | 31 | 6227 |
| 1990-91 | 4 | 3 | 125 | 415 | 612 | 981 | 1228 | 928 | 758 | 377 | 115 | 7 | 5553 |
| 1991-92 | 0 | 0 | 167 | 315 | 806 | 986 | 1116 | 953 | 889 | 525 | 245 | 62 | 6064 |
| 1992-93 | 7 | 25 | 146 | 473 | 719 | 987 | 1072 | 1123 | 943 | 493 | 156 | 48 | 6192 |
| 1993-94 | 0 | 3 | 151 | 465 | 756 | 1095 | 1479 | 1170 | 868 | 442 | 272 | 34 | 6735 |
| 1994-95 | 0 | 34 | 87 | 344 | 566 | 897 | 1137 | 1091 | 753 | 537 | 160 | 6 | 5612 |
| 1995-96 | 3 | 0 | 124 | 287 | 846 | 1221 | 1272 | 1099 | 1027 | 559 | 279 | 11 | 6728 |
| 1996-97 | 6 | 0 | 100 | 365 | 911 | 1027 | 1329 | 937 | 813 | 557 | 382 | 44 | 6471 |
| 1997-98 | 4 | 22 | 112 | 430 | 848 | 1037 | 985 | 795 | 783 | 447 | 63 | 72 | 5598 |
| 1998-99 | 0 | 3 | 52 | 363 | 655 | 920 | 1255 | 871 | 942 | 425 | 113 | 39 | 5638 |
| 1999-00 | 0 | 7 | 85 | 411 | 599 | 1047 | 1272 | 911 | 652 | 495 | 163 | 43 | 5685 |
| 2000-01 | 1 | 12 | 149 | 296 | 746 | 1445 | 1201 | 963 | 919 | 409 | 144 | 54 | 6339 |
| 2001-02 | 7 | 0 | 117 | 352 | 490 | 860 | 915 | 833 | 874 | 420 | 293 | 16 | 5177 |
| 2002-03 | 0 | 0 | 39 | 469 | 740 | 1099 | 1362 | 1143 | 877 | 491 | 242 | 55 | 6517 |
| 2003-04 | 0 | 1 | 102 | 409 | 582 | 1019 | 1374 | 1069 | 748 | 429 | 149 | 34 | 5916 |
| 2004-05 | 3 | 26 | 54 | 380 | 656 | 1102 | 1258 | 993 | 954 | 436 | 268 | 7 | 6137 |
| 2005-06 | 0 | 0 | 37 | 352 | 655 | 1229 | 875 | 964 | 840 | 360 | 216 | 15 | 5543 |
| 2006-07 | 0 | 0 | 114 | 469 | 667 | 852 | 1048 | 1328 | 739 | 515 | 140 | 18 | 5890 |
| 2007-08 | 3 | 9 | 65 | 238 | 753 | 1093 | 1159 | 1152 | 959 | 421 | 244 | 12 | 6108 |
| 2008-09 | 1 | 1 | 30 | 431 | 790 | 1151 | 1496 | 1011 | 779 | 482 | 179 | 34 | 6385 |
| 2009- | 4 | 10 | 62 | 484 | 595 | 1124 | | | | | | | |

WBAN : 94830

## COOLING DEGREE DAYS (base 65°F) 2009 TOLEDO (KTOL)

| YEAR | JAN | FEB | MAR | APR | MAY | JUN | JUL | AUG | SEP | OCT | NOV | DEC | TOTAL |
|---|---|---|---|---|---|---|---|---|---|---|---|---|---|
| 1980 | 0 | 0 | 0 | 3 | 35 | 106 | 275 | 265 | 84 | 4 | 0 | 0 | 772 |
| 1981 | 0 | 0 | 1 | 2 | 17 | 132 | 220 | 170 | 64 | 0 | 0 | 0 | 606 |
| 1982 | 0 | 0 | 0 | 0 | 68 | 61 | 245 | 132 | 62 | 11 | 0 | 0 | 579 |
| 1983 | 0 | 0 | 0 | 4 | 2 | 148 | 311 | 279 | 109 | 11 | 0 | 0 | 864 |
| 1984 | 0 | 0 | 0 | 5 | 17 | 203 | 168 | 214 | 51 | 1 | 0 | 0 | 659 |
| 1985 | 0 | 0 | 0 | 29 | 60 | 58 | 263 | 147 | 116 | 0 | 0 | 0 | 673 |
| 1986 | 0 | 0 | 1 | 4 | 48 | 113 | 282 | 125 | 103 | 4 | 0 | 0 | 680 |
| 1987 | 0 | 0 | 0 | 5 | 105 | 202 | 318 | 225 | 59 | 0 | 4 | 0 | 918 |
| 1988 | 0 | 0 | 0 | 0 | 43 | 190 | 350 | 286 | 39 | 5 | 0 | 0 | 913 |
| 1989 | 0 | 0 | 2 | 0 | 34 | 132 | 259 | 168 | 69 | 5 | 0 | 0 | 669 |
| 1990 | 0 | 0 | 7 | 32 | 11 | 164 | 222 | 164 | 91 | 14 | 0 | 0 | 705 |
| 1991 | 0 | 0 | 0 | 14 | 185 | 244 | 305 | 256 | 111 | 13 | 0 | 0 | 1128 |
| 1992 | 0 | 0 | 0 | 3 | 32 | 66 | 170 | 120 | 59 | 2 | 0 | 0 | 452 |
| 1993 | 0 | 0 | 0 | 0 | 26 | 148 | 351 | 297 | 41 | 1 | 0 | 0 | 864 |
| 1994 | 0 | 0 | 0 | 22 | 39 | 222 | 245 | 104 | 66 | 1 | 0 | 0 | 699 |
| 1995 | 0 | 0 | 0 | 0 | 16 | 230 | 367 | 426 | 64 | 8 | 0 | 0 | 1111 |
| 1996 | 0 | 0 | 0 | 2 | 53 | 191 | 184 | 234 | 85 | 2 | 0 | 0 | 751 |
| 1997 | 0 | 0 | 0 | 0 | 0 | 163 | 215 | 101 | 41 | 34 | 0 | 0 | 554 |
| 1998 | 0 | 0 | 13 | 0 | 100 | 233 | 263 | 236 | 126 | 11 | 0 | 0 | 982 |
| 1999 | 0 | 0 | 0 | 2 | 46 | 220 | 386 | 161 | 95 | 2 | 0 | 0 | 912 |
| 2000 | 0 | 0 | 2 | 0 | 69 | 168 | 170 | 182 | 82 | 6 | 0 | 0 | 679 |
| 2001 | 0 | 0 | 0 | 10 | 40 | 171 | 254 | 268 | 64 | 6 | 0 | 0 | 813 |
| 2002 | 0 | 0 | 0 | 50 | 32 | 240 | 398 | 290 | 165 | 29 | 0 | 0 | 1204 |
| 2003 | 0 | 0 | 0 | 9 | 2 | 101 | 216 | 250 | 47 | 7 | 1 | 0 | 633 |
| 2004 | 0 | 0 | 0 | 18 | 63 | 114 | 234 | 129 | 96 | 1 | 0 | 0 | 655 |
| 2005 | 0 | 0 | 0 | 0 | 10 | 294 | 317 | 281 | 114 | 28 | 0 | 0 | 1044 |
| 2006 | 0 | 0 | 0 | 4 | 64 | 141 | 310 | 256 | 42 | 4 | 0 | 0 | 821 |
| 2007 | 0 | 0 | 0 | 4 | 84 | 198 | 200 | 270 | 118 | 58 | 0 | 0 | 932 |
| 2008 | 0 | 0 | 0 | 2 | 21 | 172 | 275 | 213 | 78 | 1 | 0 | 0 | 762 |
| 2009 | 0 | 0 | 0 | 31 | 13 | 152 | 134 | 200 | 67 | 0 | 0 | 0 | 597 |

## SNOWFALL (inches) 2009 TOLEDO (KTOL)

| YEAR | JUL | AUG | SEP | OCT | NOV | DEC | JAN | FEB | MAR | APR | MAY | JUN | TOTAL |
|------|-----|-----|-----|-----|-----|-----|-----|-----|-----|-----|-----|-----|-------|
| 1980-81 | 0.0 | 0.0 | 0.0 | 0.9 | 3.5 | 11.6 | 6.9 | 11.2 | 3.6 | 0.0 | 0.0 | 0.0 | 37.7 |
| 1981-82 | 0.0 | 0.0 | 0.0 | T | 0.8 | 14.9 | 18.4 | 14.3 | 10.7 | 9.1 | 0.0 | 0.0 | 68.2 |
| 1982-83 | 0.0 | 0.0 | 0.0 | T | 2.2 | 1.2 | 0.7 | 4.1 | 3.6 | 0.7 | 0.0 | 0.0 | 12.5 |
| 1983-84 | 0.0 | 0.0 | 0.0 | 0.0 | 3.4 | 13.4 | 12.2 | 6.3 | 9.8 | T | T | 0.0 | 45.1 |
| 1984-85 | 0.0 | 0.0 | 0.0 | 0.0 | 2.4 | 5.1 | 14.0 | 12.4 | 2.6 | 2.0 | 0.0 | 0.0 | 38.5 |
| 1985-86 | 0.0 | 0.0 | 0.0 | 0.0 | 2.5 | 8.7 | 6.6 | 10.2 | 2.2 | 0.2 | 0.0 | 0.0 | 30.4 |
| 1986-87 | 0.0 | 0.0 | 0.0 | T | 4.5 | 1.3 | 20.5 | 0.5 | 10.0 | 2.4 | 0.0 | 0.0 | 39.2 |
| 1987-88 | 0.0 | 0.0 | 0.0 | T | 0.1 | 11.1 | 8.3 | 14.3 | 4.2 | T | 0.0 | 0.0 | 38.0 |
| 1988-89 | 0.0 | 0.0 | 0.0 | T | 2.3 | 6.6 | 2.4 | 4.8 | 2.6 | 0.7 | 1.3 | 0.0 | 20.7 |
| 1989-90 | 0.0 | 0.0 | 0.0 | 2.0 | 2.3 | 6.5 | 2.5 | 10.4 | 3.5 | 0.3 | 0.0 | 0.0 | 27.5 |
| 1990-91 | 0.0 | 0.0 | 0.0 | 0.0 | T | 8.2 | 5.0 | 10.1 | T | T | 0.0 | 0.0 | 23.3 |
| 1991-92 | 0.0 | 0.0 | 0.0 | T | 1.7 | 2.5 | 10.5 | 3.0 | 12.5 | 0.1 | 0.0 | 0.0 | 30.3 |
| 1992-93 | T | 0.0 | 0.0 | 1.0 | 0.2 | 5.2 | 6.3 | 10.2 | 17.7 | 0.8 | 0.0 | 0.0 | 41.4 |
| 1993-94 | 0.0 | T | T | 0.8 | 1.1 | 6.9 | 20.2 | 16.6 | 4.2 | 7.0 | 0.0 | T | 56.8 |
| 1994-95 | 0.0 | T | 0.0 | 0.0 | T | 4.8 | 13.6 | 1.2 | 2.6 | 0.1 | 0.0 | T | 22.3 |
| 1995-96 | 0.0 | 0.0 | 0.0 | 0.0 | 6.8 | 7.0 | 11.0 | 2.5 | 4.4 | T | 0.0 | 0.0 | 31.7 |
| 1996-97 | 0.0 | 0.0 | 0.0 | 0.0 | 8.1 | 15.6 | 15.1 | 3.3 | 3.2 | T | 0.0 | 0.0 | 45.3 |
| 1997-98 | 0.0 | 0.0 | 0.0 | T | 5.2 | 5.2 | 2.6 | 0.9 | 2.1 | 0.0 | 0.0 | 0.0 | 16.0 |
| 1998-99 | 0.0 | 0.0 | 0.0 | 0.0 | 0.0 | 0.5 | 21.9 | 7.3 | 12.7 | 0.0 | 0.0 | 0.0 | 42.4 |
| 1999-00 | 0.0 | 0.0 | 0.0 | 0.0 | T | 3.1 | 12.2 | 4.7 | 2.5 | 0.2 | 0.0 | 0.0 | 22.7 |
| 2000-01 | 0.0 | 0.0 | 0.0 | 0.0 | 1.2 | 26.0 | 3.2 | 1.9 | 5.2 | T | 0.0 | 0.0 | 37.5 |
| 2001-02 | 0.0 | 0.0 | 0.0 | T | 0.0 | 1.2 | 2.5 | 3.9 | 11.4 | T | 0.0 | 0.0 | 19.0 |
| 2002-03 | 0.0 | 0.0 | 0.0 | 0.0 | 5.7 | 13.6 | 14.1 | 18.8 | 3.7 | 0.5 | 0.0 | 0.0 | 56.4 |
| 2003-04 | 0.0 | 0.0 | 0.0 | 0.0 | 0.7 | 9.3 | 14.0 | 1.1 | 3.9 | T | 0.0 | 0.0 | 29.0 |
| 2004-05 | 0.0 | 0.0 | 0.0 | 0.0 | 0.5 | 10.0 | 27.6 | 6.4 | 7.5 | 4.0 | 0.0 | 0.0 | 56.0 |
| 2005-06 | 0.0 | 0.0 | 0.0 | 0.0 | 5.0 | 21.5 | 1.9 | 1.5 | 1.6 | 0.0 | 0.0 | 0.0 | 31.5 |
| 2006-07 | 0.0 | 0.0 | 0.0 | T | T | 1.6 | 6.7 | 14.2 | 3.6 | 2.4 | 0.0 | 0.0 | 28.5 |
| 2007-08 | 0.0 | 0.0 | 0.0 | 0.0 | 0.8 | 10.1 | 6.2 | 23.6 | 17.4 | T | 0.0 | 0.0 | 58.1 |
| 2008-09 | 0.0 | 0.0 | 0.0 | T | 2.4 | 7.0 | 30.7 | 5.2 | 0.1 | 0.5 | 0.0 | 0.0 | 45.9 |
| 2009- | 0.0 | 0.0 | 0.0 | 0.0 | T | 7.0 | | | | | | | |
| POR= 55 YRS | T | T | T | 0.1 | 2.7 | 8.5 | 10.2 | 7.8 | 6.0 | 1.3 | T | T | 36.6 |

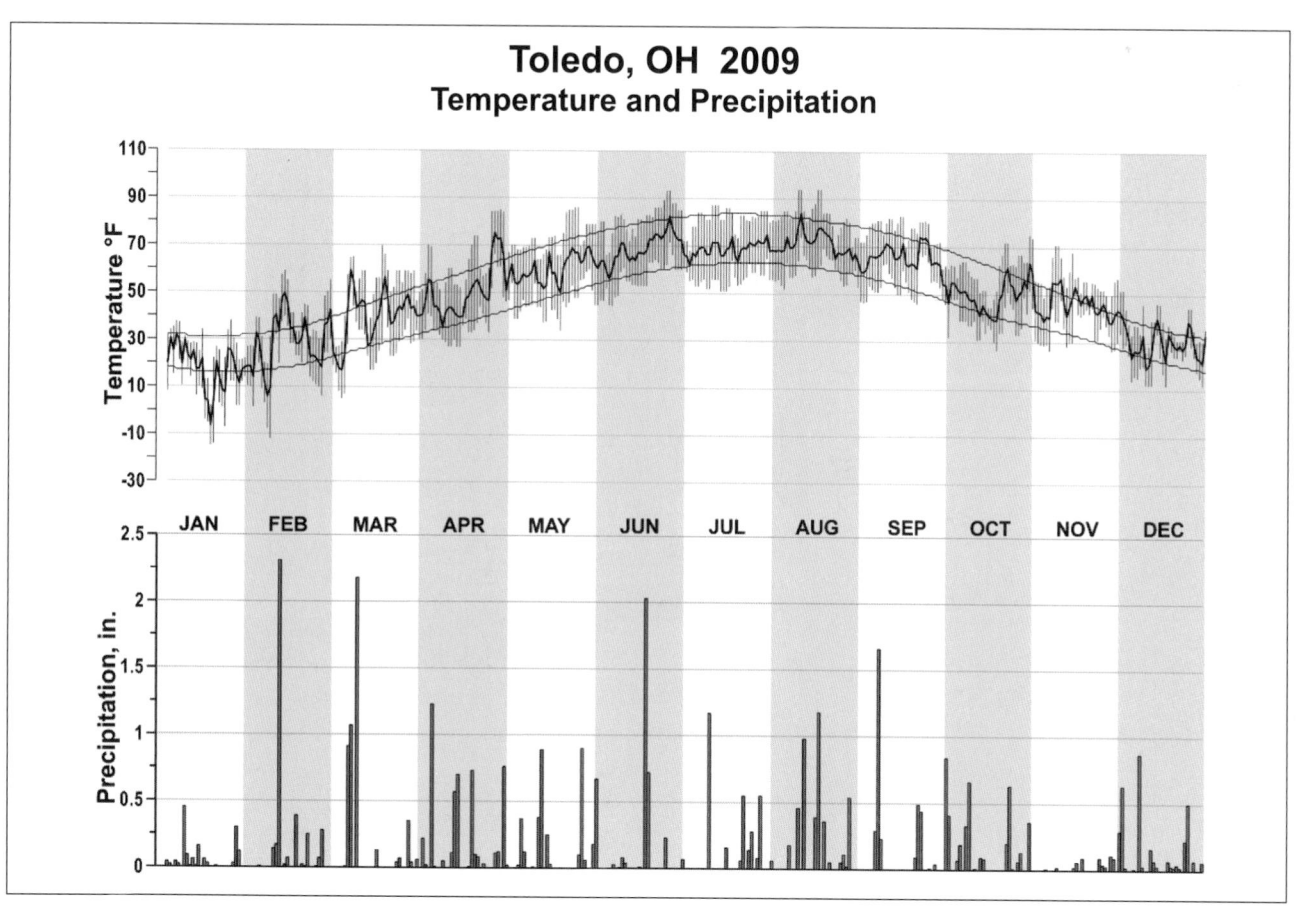

## Toledo, OH 2009
### Temperature and Precipitation

# 2009
# OKLAHOMA CITY
# OKLAHOMA (KOKC)

Oklahoma City is located along the North Canadian River, a frequently nearly-dry stream, at the geographic center of the state. It is not quite 1,000 miles south of the Canadian Border and a little less than 500 miles north of the Gulf of Mexico. The surrounding country is gently rolling with the nearest hills or low mountains, the Arbuckles, 80 miles south. The elevation ranges around 1,250 feet above sea level.

Although some influence is exerted at times by warm, moist air currents from the Gulf of Mexico, the climate of Oklahoma City falls mainly under continental controls characteristic of the Great Plains Region. The continental effect produces pronounced daily and seasonal temperature changes and considerable variation in seasonal and annual precipitation. Summers are long and usually hot. Winters are comparatively mild and short.

During the year, temperatures of 100 degrees or more occur on an average of 10 days, but have occurred on as many as 50 days or more. While summers are usually hot, the discomforting effect of extreme heat is considerably mitigated by low humidity and the prevalence of a moderate southerly breeze. Approximately one winter in three has temperatures of zero or lower.

The length of the growing season varies from 180 to 251 days. Average date of last freeze is early April and average date of first freeze is early November. Freezes have occurred in early October.

During an average year, skies are clear approximately 40 percent of the time, partly cloudy 25 percent, and cloudy 35 percent of the time. The city is almost smoke-free as a result of favorable atmospheric conditions and the almost exclusive use of natural gas for heating. Flying conditions are generally very good with flight by visual flight rules possible about 96 percent of the time.

Summer rainfall comes mainly from showers and thunderstorms. Winter precipitation is generally associated with frontal passages. Measurable precipitation has occurred on as many as 122 days and as few as 55 days during the year. The seasonal distribution of precipitation is normally 12 percent in winter, 34 percent in spring, 30 percent in summer, and 24 percent in fall. The The period with the least number of days with precipitation is November through January, and the month with the most rainy days is May. Thunderstorms occur most often in late spring and early summer. Large hail and/or destructive winds on occasion accompany these thunderstorms.

Snowfall averages less than 10 inches per year and seldom remains on the ground very long. Occasional brief periods of freezing rain and sleet storms occur.

Heavy fogs are infrequent. Prevailing winds are southerly except in January and February when northerly breezes predominate.

# NORMALS, MEANS, AND EXTREMES
## OKLAHOMA CITY (KOKC)

LATITUDE: 35° 23'N　　LONGITUDE: -97° 36'W　　ELEVATION (FT): GRND: 1274　BARO: 1284　　TIME ZONE: CENTRAL (UTC -6)　　WBAN: 13967

| ELEMENT | POR | JAN | FEB | MAR | APR | MAY | JUN | JUL | AUG | SEP | OCT | NOV | DEC | YEAR |
|---|---|---|---|---|---|---|---|---|---|---|---|---|---|---|
| **TEMPERATURE °F** | | | | | | | | | | | | | | |
| NORMAL DAILY MAXIMUM | 30 | 47.1 | 53.5 | 62.5 | 71.2 | 78.9 | 87.2 | 93.1 | 92.5 | 84.1 | 73.4 | 59.6 | 49.8 | 71.1 |
| MEAN DAILY MAXIMUM | 62 | 47.5 | 53.0 | 61.6 | 71.6 | 79.1 | 87.3 | 93.1 | 92.6 | 84.2 | 73.5 | 60.3 | 50.4 | 71.2 |
| HIGHEST DAILY MAXIMUM | 55 | 80 | 92 | 93 | 100 | 104 | 105 | 110 | 110 | 108 | 96 | 87 | 86 | 110 |
| YEAR OF OCCURRENCE | | 1986 | 1996 | 1967 | 1972 | 1985 | 1998 | 1996 | 1980 | 2000 | 1972 | 1980 | 1955 | JUL 1996 |
| MEAN OF EXTREME MAXS. | 62 | 69.9 | 75.2 | 82.6 | 87.5 | 91.9 | 96.5 | 101.5 | 101.5 | 96.5 | 89.0 | 78.3 | 70.8 | 86.8 |
| NORMAL DAILY MINIMUM | 30 | 26.2 | 31.1 | 39.4 | 48.1 | 57.9 | 66.4 | 70.8 | 69.8 | 62.2 | 50.6 | 38.2 | 29.2 | 49.2 |
| MEAN DAILY MINIMUM | 62 | 26.4 | 30.8 | 38.5 | 48.6 | 58.2 | 66.4 | 70.8 | 69.8 | 62.0 | 50.5 | 38.2 | 29.4 | 49.1 |
| LOWEST DAILY MINIMUM | 55 | -4 | -3 | 3 | 20 | 37 | 47 | 53 | 51 | 36 | 16 | 11 | -8 | -8 |
| YEAR OF OCCURRENCE | | 1988 | 1996 | 1960 | 1957 | 1981 | 1954 | 1971 | 1956 | 1989 | 1993 | 1991 | 1989 | DEC 1989 |
| MEAN OF EXTREME MINS. | 62 | 8.3 | 13.3 | 20.2 | 32.1 | 44.5 | 55.7 | 62.4 | 60.6 | 46.7 | 34.4 | null | 12.3 | 32.5 |
| NORMAL DRY BULB | 30 | 36.7 | 42.3 | 51.0 | 59.7 | 68.4 | 76.8 | 82.0 | 81.2 | 73.2 | 62.0 | 48.9 | 39.5 | 60.1 |
| MEAN DRY BULB | 62 | 37.0 | 41.9 | 50.1 | 60.1 | 68.7 | 77.0 | 81.9 | 81.2 | 73.1 | 62.0 | 49.3 | 39.9 | 60.2 |
| MEAN WET BULB | 26 | 32.0 | 36.1 | 43.3 | 51.4 | 61.4 | 68.2 | 70.5 | 69.9 | 63.6 | 53.6 | 43.0 | 34.2 | 52.3 |
| MEAN DEW POINT | 26 | 27.9 | 31.5 | 38.7 | 47.0 | 58.4 | 65.7 | 67.3 | 66.5 | 60.3 | 49.9 | 38.6 | 30.1 | 48.5 |
| NORMAL NO. DAYS WITH: | | | | | | | | | | | | | | |
| MAXIMUM >= 90 | 30 | 0.0 | * | 0.1 | 0.4 | 2.1 | 10.9 | 22.3 | 22.6 | 9.6 | 0.9 | 0.0 | 0.0 | 68.9 |
| MAXIMUM <= 32 | 30 | 4.7 | 2.6 | 0.3 | 0.0 | 0.0 | 0.0 | 0.0 | 0.0 | 0.0 | 0.0 | 0.2 | 2.6 | 10.4 |
| MINIMUM <= 32 | 30 | 22.5 | 14.9 | 7.3 | 1.1 | 0.0 | 0.0 | 0.0 | 0.0 | 0.0 | 0.6 | 8.5 | 19.6 | 74.5 |
| MINIMUM <= 0 | 30 | 0.3 | 0.2 | 0.0 | 0.0 | 0.0 | 0.0 | 0.0 | 0.0 | 0.0 | 0.0 | 0.0 | 0.2 | 0.7 |
| **H/C** | | | | | | | | | | | | | | |
| NORMAL HEATING DEG. DAYS | 30 | 884 | 648 | 446 | 197 | 43 | 1 | 0 | 0 | 30 | 152 | 482 | 780 | 3663 |
| NORMAL COOLING DEG. DAYS | 30 | 0 | 1 | 7 | 38 | 145 | 360 | 527 | 497 | 271 | 58 | 3 | 0 | 1907 |
| **RH** | | | | | | | | | | | | | | |
| NORMAL (PERCENT) | 30 | 67 | 65 | 63 | 63 | 70 | 69 | 63 | 63 | 67 | 67 | 69 | 70 | 66 |
| HOUR 00 LST | 30 | 73 | 71 | 69 | 70 | 78 | 78 | 72 | 71 | 76 | 74 | 75 | 75 | 74 |
| HOUR 06 LST | 30 | 78 | 77 | 77 | 78 | 84 | 85 | 81 | 81 | 83 | 81 | 81 | 79 | 80 |
| HOUR 12 LST | 30 | 58 | 57 | 53 | 52 | 58 | 57 | 50 | 50 | 54 | 53 | 58 | 60 | 55 |
| HOUR 18 LST | 30 | 59 | 54 | 50 | 49 | 56 | 55 | 47 | 47 | 53 | 56 | 62 | 63 | 54 |
| **S** — PERCENT POSSIBLE SUNSHINE | 41 | 60 | 61 | 65 | 66 | 66 | 76 | 80 | 80 | 74 | 70 | 62 | 59 | 68 |
| **W/O** MEAN NO. DAYS WITH: | | | | | | | | | | | | | | |
| HEAVY FOG(VISBY <= 1/4 MI) | 46 | 3.5 | 2.7 | 1.7 | 1.0 | 0.8 | 0.6 | 0.1 | 0.4 | 0.7 | 1.5 | 2.2 | 3.3 | 18.5 |
| THUNDERSTORMS | 62 | 0.7 | 1.3 | 3.5 | 5.5 | 8.7 | 8.8 | 6.0 | 6.4 | 4.8 | 3.4 | 1.4 | 0.8 | 51.3 |
| **CLOUDNESS** MEAN: | | | | | | | | | | | | | | |
| SUNRISE-SUNSET (OKTAS) | 1 | 4.8 | 5.1 | 5.3 | 4.8 | 5.2 | 2.4 | 2.0 | 2.7 | 4.8 | 3.6 | 3.2 | 4.8 | 4.1 |
| MIDNIGHT-MIDNIGHT (OKTAS) | 1 | 4.8 | 5.2 | 5.1 | 4.8 | 5.2 | 2.4 | 2.0 | 2.4 | 2.4 | 3.2 | 3.6 | 4.0 | 3.8 |
| MEAN NO. DAYS WITH: | | | | | | | | | | | | | | |
| CLEAR | 3 | 4.3 | 10.0 | 8.3 | 10.5 | 8.7 | 13.0 | 16.0 | 15.0 | 7.0 | 12.5 | 5.5 | 8.5 | 119.3 |
| PARTLY CLOUDY | 3 | 4.7 | 2.7 | 2.7 | 2.0 | 6.0 | 5.3 | 4.5 | 4.5 | 3.0 | 4.0 | 4.5 | 5.0 | 48.9 |
| CLOUDY | 3 | 7.3 | 10.0 | 7.0 | 8.0 | 6.3 | 3.3 | 2.5 | 3.0 | 2.0 | 5.0 | 6.5 | 7.5 | 68.4 |
| **PR** | | | | | | | | | | | | | | |
| MEAN STATION PRESSURE(IN) | 26 | 28.73 | 28.70 | 28.62 | 28.56 | 28.55 | 28.57 | 28.61 | 28.63 | 28.64 | 28.66 | 28.68 | 28.72 | 28.64 |
| MEAN SEA-LEVEL PRES. (IN) | 26 | 30.14 | 30.09 | 30.00 | 29.93 | 29.90 | 29.90 | 29.95 | 29.96 | 29.99 | 30.03 | 30.07 | 30.13 | 30.01 |
| **WINDS** | | | | | | | | | | | | | | |
| MEAN SPEED (MPH) | 26 | 11.7 | 12.2 | 13.3 | 13.3 | 11.7 | 10.7 | 10.1 | 9.4 | 9.9 | 11.1 | 11.8 | 11.5 | 11.4 |
| PREVAIL.DIR(TENS OF DEGS) | 29 | 36 | 36 | 17 | 17 | 17 | 17 | 17 | 17 | 16 | 17 | 17 | 17 | 17 |
| MAXIMUM 2-MINUTE: | | | | | | | | | | | | | | |
| SPEED (MPH) | 16 | 47 | 45 | 52 | 46 | 53 | 49 | 48 | 46 | 40 | 43 | 48 | 49 | 53 |
| DIR. (TENS OF DEGS) | | 31 | 32 | 24 | 32 | 23 | 03 | 34 | 05 | 30 | 03 | 23 | 34 | 23 |
| YEAR OF OCCURRENCE | | 2008 | 1997 | 1996 | 1999 | 2002 | 2004 | 2000 | 1996 | 2002 | 1994 | 2005 | 2009 | MAY 2002 |
| MAXIMUM 3-SECOND | | | | | | | | | | | | | | |
| SPEED (MPH) | 16 | 53 | 54 | 62 | 66 | 74 | 63 | 64 | 56 | 48 | 51 | 56 | 62 | 74 |
| DIR. (TENS OF DEGS) | | 31 | 33 | 23 | 25 | 27 | 22 | 33 | 14 | 29 | 03 | 20 | 33 | 27 |
| YEAR OF OCCURRENCE | | 2008 | 1997 | 1996 | 2009 | 2008 | 1998 | 2000 | 2007 | 1998 | 1994 | 1994 | 2009 | MAY 2008 |
| **PRECIPITATION** | | | | | | | | | | | | | | |
| NORMAL (IN) | 30 | 1.28 | 1.56 | 2.90 | 3.00 | 5.44 | 4.63 | 2.94 | 2.48 | 3.98 | 3.64 | 2.11 | 1.89 | 35.85 |
| MAXIMUM MONTHLY (IN) | 69 | 5.68 | 4.63 | 8.02 | 10.78 | 12.07 | 14.66 | 11.90 | 9.95 | 11.85 | 13.18 | 5.72 | 8.14 | 14.66 |
| YEAR OF OCCURRENCE | | 1949 | 1990 | 2007 | 1947 | 1982 | 1989 | 1996 | 2008 | 1991 | 1983 | 1994 | 1984 | JUN 1989 |
| MINIMUM MONTHLY (IN) | 69 | 0.00 | T | T | 0.17 | 0.33 | 0.55 | T | 0.00 | T | T | T | 0.03 | 0.00 |
| YEAR OF OCCURRENCE | | 1985 | 1947 | 1940 | 1989 | 1942 | 2001 | 1983 | 2000 | 1948 | 1958 | 1949 | 1996 | AUG 2000 |
| MAXIMUM IN 24 HOURS (IN) | 69 | 3.10 | 2.21 | 3.51 | 4.48 | 7.56 | 4.56 | 5.75 | 5.39 | 7.68 | 8.95 | 2.89 | 2.89 | 8.95 |
| YEAR OF OCCURRENCE | | 1982 | 2008 | 2007 | 1999 | 1993 | 1989 | 1981 | 2007 | 1970 | 1983 | 1994 | 1991 | OCT 1983 |
| NORMAL NO. DAYS WITH: | | | | | | | | | | | | | | |
| PRECIPITATION >= 0.01 | 30 | 5.6 | 5.6 | 7.4 | 7.6 | 10.3 | 8.5 | 5.8 | 6.2 | 7.6 | 7.3 | 6.3 | 5.6 | 83.8 |
| PRECIPITATION >= 1.00 | 30 | 0.3 | 0.3 | 1.0 | 0.8 | 1.5 | 1.6 | 1.0 | 0.8 | 1.2 | 1.1 | 0.7 | 0.6 | 10.9 |
| **SNOWFALL** | | | | | | | | | | | | | | |
| NORMAL (IN) | 30 | 3.1 | 2.1 | 0.7 | 0.* | 0.0 | 0.0 | 0.0 | 0.0 | 0.0 | 0.* | 0.6 | 2.1 | 8.6 |
| MAXIMUM MONTHLY (IN) | 69 | 17.3 | 12.0 | 13.9 | 0.7 | T | T | T | T | T | 0.1 | 7.5 | 14.0 | 17.3 |
| YEAR OF OCCURRENCE | | 1949 | 1978 | 1968 | 1957 | 2009 | 1992 | 1997 | 1997 | 1992 | 1993 | 1972 | 2009 | JAN 1949 |
| MAXIMUM IN 24 HOURS (IN) | 69 | 8.9 | 6.5 | 8.4 | 0.7 | T | T | T | T | T | 0.1 | 5.5 | 13.5 | 13.5 |
| YEAR OF OCCURRENCE' | | 1988 | 1986 | 1948 | 1957 | 2009 | 1992 | 1997 | 1997 | 1992 | 1993 | 1972 | 2009 | DEC 2009 |
| MAXIMUM SNOW DEPTH (IN) | 61 | 12 | 8 | 8 | T | 0 | 0 | 0 | 0 | 0 | T | 3 | 14 | 14 |
| YEAR OF OCCURRENCE | | 1988 | 1951 | 1948 | 1973 | | | | | | 1993 | 1980 | 2009 | DEC 2009 |
| NORMAL NO. DAYS WITH: | | | | | | | | | | | | | | |
| SNOWFALL >= 1.0 | 30 | 1.0 | 0.8 | 0.2 | 0.0 | 0.0 | 0.0 | 0.0 | 0.0 | 0.0 | 0.0 | 0.2 | 0.7 | 2.9 |

## PRECIPITATION (inches) 2009 OKLAHOMA CITY (KOKC)

| YEAR | JAN | FEB | MAR | APR | MAY | JUN | JUL | AUG | SEP | OCT | NOV | DEC | ANNUAL |
|------|-----|-----|-----|-----|-----|-----|-----|-----|-----|-----|-----|-----|--------|
| 1980 | 1.69 | 1.29 | 1.38 | 2.16 | 9.00 | 2.52 | 0.42 | 0.60 | 2.21 | 0.99 | 0.51 | 1.58 | 24.35 |
| 1981 | 0.19 | 1.15 | 2.87 | 2.97 | 2.73 | 7.49 | 6.45 | 3.61 | 1.48 | 7.70 | 2.11 | 0.20 | 38.95 |
| 1982 | 3.68 | 0.98 | 1.63 | 1.92 | 12.07 | 4.06 | 2.11 | 1.13 | 2.86 | 1.03 | 2.78 | 1.94 | 36.19 |
| 1983 | 2.62 | 1.71 | 2.51 | 2.34 | 6.88 | 3.18 | T | 3.18 | 0.90 | 13.18 | 1.90 | 0.70 | 39.10 |
| 1984 | 0.35 | 1.16 | 4.70 | 1.79 | 1.62 | 3.48 | 0.30 | 2.35 | 1.01 | 6.64 | 2.05 | 8.14 | 33.59 |
| 1985 | 0.92 | 3.71 | 6.60 | 5.35 | 1.49 | 8.34 | 1.33 | 2.63 | 4.59 | 5.23 | 3.73 | 0.26 | 44.18 |
| 1986 | 0.00 | 0.68 | 1.75 | 4.42 | 8.21 | 3.11 | 0.38 | 3.29 | 9.54 | 8.00 | 4.63 | 1.16 | 45.17 |
| 1987 | 2.45 | 4.05 | 2.33 | 0.41 | 11.86 | 6.50 | 2.99 | 1.83 | 4.58 | 1.82 | 1.92 | 3.75 | 44.49 |
| 1988 | 1.24 | 0.41 | 7.85 | 3.19 | 1.07 | 3.59 | 1.92 | 1.60 | 5.19 | 2.04 | 2.45 | 1.39 | 31.94 |
| 1989 | 1.17 | 2.20 | 2.72 | 0.17 | 4.33 | 14.66 | 1.91 | 5.55 | 4.51 | 3.26 | 0.09 | 0.32 | 40.89 |
| 1990 | 1.85 | 4.63 | 4.43 | 5.11 | 5.79 | 1.25 | 2.65 | 3.16 | 7.35 | 1.27 | 1.59 | 1.46 | 40.54 |
| 1991 | 0.89 | 0.03 | 1.59 | 2.10 | 6.39 | 3.85 | 1.98 | 3.24 | 11.85 | 3.98 | 1.94 | 5.90 | 43.74 |
| 1992 | 1.15 | 1.28 | 1.08 | 3.64 | 4.88 | 6.35 | 4.01 | 5.82 | 2.92 | 1.13 | 4.51 | 3.08 | 39.85 |
| 1993 | 1.90 | 3.21 | 2.82 | 2.50 | 10.90 | 2.65 | 1.24 | 1.86 | 7.05 | 0.47 | 1.34 | 1.27 | 37.21 |
| 1994 | 0.21 | 2.56 | 3.18 | 3.38 | 2.69 | 1.70 | 2.17 | 1.81 | 2.17 | 1.88 | 5.72 | 1.63 | 29.10 |
| 1995 | 1.28 | 0.04 | 2.21 | 3.76 | 7.39 | 6.06 | 1.94 | 3.15 | 6.66 | 1.54 | 0.39 | 2.35 | 36.77 |
| 1996 | 0.08 | 0.02 | 2.17 | 2.00 | 1.90 | 1.16 | 11.90 | 5.85 | 5.88 | 2.53 | 3.36 | T | 36.85 |
| 1997 | 0.52 | 2.59 | 0.60 | 4.39 | 3.68 | 3.01 | 4.60 | 4.04 | 1.66 | 3.93 | 1.11 | 2.96 | 33.09 |
| 1998 | 4.09 | 0.32 | 6.45 | 3.34 | 2.12 | 2.67 | 0.02 | 0.48 | 4.39 | 6.76 | 3.09 | 1.62 | 35.35 |
| 1999 | 1.81 | 1.20 | 3.45 | 6.92 | 3.10 | 8.61 | 1.94 | 1.35 | 4.88 | 2.22 | 0.06 | 3.71 | 39.25 |
| 2000 | 0.75 | 1.47 | 3.12 | 5.17 | 1.36 | 6.71 | 5.25 | 0.00 | 1.73 | 8.39 | 2.79 | 2.30 | 39.04 |
| 2001 | 2.23 | 2.25 | 1.01 | 1.04 | 7.70 | 0.55 | 1.27 | 1.95 | 5.55 | 3.56 | 1.08 | 0.91 | 29.10 |
| 2002 | 2.62 | 0.47 | 2.24 | 5.10 | 2.48 | 4.56 | 4.94 | 1.58 | 2.94 | 4.64 | 0.74 | 1.84 | 34.15 |
| 2003 | 0.02 | 0.87 | 2.30 | 1.56 | 2.41 | 4.70 | 0.65 | 4.79 | 1.98 | 1.01 | 1.23 | 1.11 | 22.63 |
| 2004 | 1.45 | 1.45 | 3.98 | 1.35 | 1.20 | 7.03 | 3.65 | 5.01 | 0.64 | 4.86 | 5.66 | 0.50 | 36.78 |
| 2005 | 2.05 | 2.69 | 0.44 | 0.29 | 2.23 | 4.89 | 3.22 | 4.45 | 1.89 | 1.17 | T | 0.28 | 23.60 |
| 2006 | 0.27 | 0.08 | 2.78 | 3.18 | 3.01 | 2.32 | 3.42 | 4.01 | 3.76 | 1.56 | 1.43 | 2.02 | 27.84 |
| 2007 | 2.08 | 0.62 | 8.02 | 2.57 | 8.49 | 10.06 | 6.31 | 5.39 | 5.73 | 3.72 | 0.53 | 3.43 | 56.95 |
| 2008 | 0.65 | 2.88 | 3.29 | 4.17 | 4.54 | 5.83 | 1.07 | 9.95 | 0.59 | 1.63 | 0.70 | 0.52 | 35.82 |
| 2009 | 0.43 | 0.98 | 2.53 | 4.80 | 4.54 | 1.13 | 3.53 | 5.74 | 4.62 | 5.63 | 0.29 | 1.47 | 35.69 |
| POR= 62 YRS | 1.22 | 1.43 | 2.60 | 2.91 | 5.19 | 4.38 | 2.98 | 2.88 | 3.65 | 3.17 | 1.72 | 1.51 | 33.64 |

WBAN : 13967

## AVERAGE TEMPERATURE (°F) 2009 OKLAHOMA CITY (KOKC)

| YEAR | JAN | FEB | MAR | APR | MAY | JUN | JUL | AUG | SEP | OCT | NOV | DEC | ANNUAL |
|------|-----|-----|-----|-----|-----|-----|-----|-----|-----|-----|-----|-----|--------|
| 1980 | 38.2 | 38.2 | 46.3 | 56.7 | 69.0 | 81.4 | 88.3 | 88.0 | 76.3 | 61.1 | 50.3 | 41.9 | 61.3 |
| 1981 | 37.7 | 43.9 | 51.9 | 65.6 | 65.7 | 78.4 | 84.2 | 78.8 | 74.1 | 60.1 | 50.3 | 39.1 | 60.8 |
| 1982 | 35.3 | 37.7 | 52.7 | 57.5 | 68.2 | 72.2 | 81.0 | 84.1 | 74.5 | 62.7 | 48.6 | 43.2 | 59.8 |
| 1983 | 38.6 | 42.6 | 48.8 | 54.0 | 64.6 | 73.4 | 81.6 | 84.0 | 74.9 | 62.7 | 50.4 | 25.8 | 58.5 |
| 1984 | 34.0 | 45.4 | 46.4 | 56.5 | 68.4 | 78.6 | 81.6 | 82.6 | 71.5 | 61.6 | 49.7 | 43.0 | 59.9 |
| 1985 | 30.6 | 37.2 | 53.0 | 62.7 | 70.0 | 76.0 | 80.9 | 81.3 | 73.1 | 61.2 | 46.1 | 35.1 | 58.9 |
| 1986 | 43.6 | 44.8 | 55.5 | 62.8 | 69.0 | 79.0 | 85.9 | 80.0 | 74.8 | 61.6 | 44.8 | 40.8 | 61.9 |
| 1987 | 35.1 | 45.9 | 50.3 | 61.8 | 72.6 | 77.1 | 80.1 | 82.2 | 72.4 | 60.0 | 50.5 | 40.6 | 60.7 |
| 1988 | 34.2 | 40.3 | 49.5 | 58.9 | 70.3 | 78.4 | 81.6 | 82.8 | 73.5 | 59.3 | 51.2 | 43.9 | 60.3 |
| 1989 | 42.8 | 33.1 | 51.1 | 63.4 | 69.4 | 74.3 | 79.6 | 78.3 | 67.8 | 63.1 | 52.2 | 32.7 | 59.0 |
| 1990 | 45.9 | 46.0 | 52.6 | 59.2 | 68.6 | 82.0 | 80.7 | 81.6 | 77.0 | 60.9 | 54.9 | 37.1 | 62.2 |
| 1991 | 34.9 | 49.0 | 54.3 | 62.5 | 72.3 | 78.0 | 82.2 | 81.2 | 70.9 | 62.6 | 45.0 | 44.1 | 61.4 |
| 1992 | 42.0 | 49.9 | 54.1 | 61.3 | 66.5 | 74.1 | 81.1 | 74.8 | 72.5 | 62.3 | 45.9 | 39.8 | 60.4 |
| 1993 | 36.5 | 38.8 | 48.0 | 56.2 | 66.0 | 76.8 | 83.6 | 82.3 | 69.8 | 57.1 | 44.2 | 42.0 | 58.4 |
| 1994 | 36.0 | 37.4 | 52.7 | 59.4 | 66.8 | 79.6 | 79.9 | 79.7 | 70.7 | 62.6 | 50.1 | 42.5 | 59.8 |
| 1995 | 38.6 | 44.9 | 49.8 | 56.8 | 64.3 | 73.0 | 81.0 | 81.5 | 70.9 | 61.6 | 49.3 | 39.9 | 59.3 |
| 1996 | 35.8 | 44.3 | 46.0 | 58.7 | 73.8 | 77.9 | 81.3 | 78.0 | 69.4 | 61.2 | 46.3 | 42.3 | 59.6 |
| 1997 | 37.8 | 44.0 | 52.5 | 54.7 | 66.7 | 75.4 | 81.6 | 78.6 | 75.3 | 62.3 | 46.5 | 39.1 | 59.5 |
| 1998 | 40.6 | 45.5 | 47.4 | 57.4 | 72.5 | 81.1 | 88.0 | 85.0 | 81.2 | 64.4 | 53.2 | 41.6 | 63.2 |
| 1999 | 40.5 | 50.7 | 49.8 | 61.3 | 68.1 | 75.7 | 82.2 | 84.8 | 71.1 | 62.7 | 56.8 | 43.3 | 62.3 |
| 2000 | 40.7 | 49.1 | 53.4 | 59.0 | 71.0 | 74.6 | 80.8 | 85.4 | 76.1 | 64.1 | 43.3 | 30.5 | 60.7 |
| 2001 | 36.3 | 40.8 | 46.6 | 63.6 | 69.5 | 76.4 | 85.7 | 82.9 | 70.7 | 60.2 | 53.9 | 42.2 | 60.7 |
| 2002 | 40.0 | 41.1 | 46.0 | 61.0 | 65.9 | 76.3 | 79.8 | 81.3 | 74.2 | 56.2 | 47.4 | 41.0 | 59.2 |
| 2003 | 36.8 | 37.7 | 49.5 | 60.4 | 69.1 | 73.9 | 84.3 | 82.8 | 69.4 | 63.6 | 50.5 | 43.1 | 60.1 |
| 2004 | 39.9 | 39.9 | 55.3 | 61.2 | 71.9 | 75.4 | 78.9 | 76.6 | 75.1 | 64.6 | 50.6 | 43.5 | 61.1 |
| 2005 | 39.6 | 46.8 | 51.7 | 61.3 | 69.2 | 78.0 | 80.4 | 81.0 | 77.1 | 63.5 | 53.8 | 38.9 | 61.8 |
| 2006 | 47.7 | 41.7 | 55.1 | 67.3 | 72.6 | 80.0 | 86.2 | 85.9 | 71.3 | 62.8 | 52.9 | 43.5 | 63.9 |
| 2007 | 36.8 | 42.1 | 60.2 | 57.4 | 71.0 | 77.1 | 80.7 | 84.2 | 76.1 | 65.5 | 52.9 | 39.1 | 61.9 |
| 2008 | 40.1 | 42.0 | 52.7 | 59.6 | 71.4 | 80.1 | 83.5 | 81.1 | 71.4 | 61.3 | 51.0 | 39.5 | 61.1 |
| 2009 | 37.3 | 48.1 | 53.9 | 59.6 | 66.6 | 80.2 | 80.2 | 82.4 | 79.3 | 70.8 | 55.7 | 54.1 | 60.3 |
| POR= 62 YRS | 37.0 | 41.9 | 50.1 | 60.1 | 68.7 | 77.0 | 81.9 | 81.2 | 73.1 | 62.0 | 49.3 | 39.9 | 60.2 |

## HEATING DEGREE DAYS (base 65°F) 2009  OKLAHOMA CITY (KOKC)

| YEAR | JUL | AUG | SEP | OCT | NOV | DEC | JAN | FEB | MAR | APR | MAY | JUN | TOTAL |
|---|---|---|---|---|---|---|---|---|---|---|---|---|---|
| 1980-81 | 0 | 0 | 23 | 180 | 444 | 710 | 839 | 587 | 400 | 69 | 69 | 0 | 3321 |
| 1981-82 | 0 | 0 | 22 | 189 | 434 | 797 | 913 | 759 | 382 | 248 | 25 | 13 | 3782 |
| 1982-83 | 0 | 0 | 14 | 156 | 490 | 671 | 809 | 622 | 496 | 345 | 96 | 9 | 3708 |
| 1983-84 | 0 | 0 | 25 | 117 | 439 | 1207 | 955 | 561 | 572 | 263 | 45 | 0 | 4184 |
| 1984-85 | 0 | 0 | 75 | 162 | 462 | 676 | 1059 | 773 | 377 | 108 | 10 | 0 | 3702 |
| 1985-86 | 0 | 0 | 63 | 146 | 562 | 921 | 656 | 562 | 308 | 122 | 17 | 0 | 3357 |
| 1986-87 | 0 | 0 | 2 | 137 | 599 | 742 | 918 | 528 | 450 | 177 | 3 | 0 | 3556 |
| 1987-88 | 0 | 0 | 1 | 165 | 442 | 748 | 948 | 712 | 473 | 204 | 14 | 0 | 3707 |
| 1988-89 | 0 | 0 | 8 | 196 | 408 | 644 | 679 | 887 | 441 | 140 | 38 | 0 | 3441 |
| 1989-90 | 0 | 0 | 78 | 135 | 386 | 993 | 583 | 525 | 387 | 202 | 52 | 0 | 3341 |
| 1990-91 | 0 | 0 | 9 | 169 | 307 | 860 | 925 | 444 | 339 | 110 | 25 | 0 | 3188 |
| 1991-92 | 0 | 0 | 37 | 150 | 594 | 642 | 704 | 430 | 332 | 154 | 59 | 2 | 3104 |
| 1992-93 | 0 | 1 | 5 | 115 | 563 | 774 | 878 | 725 | 525 | 265 | 53 | 0 | 3904 |
| 1993-94 | 0 | 1 | 27 | 269 | 619 | 706 | 896 | 767 | 394 | 204 | 53 | 0 | 3936 |
| 1994-95 | 0 | 0 | 31 | 138 | 451 | 690 | 810 | 554 | 477 | 253 | 84 | 0 | 3488 |
| 1995-96 | 0 | 0 | 75 | 129 | 465 | 767 | 898 | 602 | 584 | 209 | 10 | 0 | 3739 |
| 1996-97 | 0 | 0 | 29 | 151 | 556 | 697 | 839 | 583 | 385 | 310 | 43 | 0 | 3593 |
| 1997-98 | 0 | 1 | 2 | 188 | 549 | 798 | 750 | 542 | 554 | 238 | 9 | 3 | 3634 |
| 1998-99 | 0 | 0 | 0 | 75 | 347 | 719 | 752 | 398 | 463 | 144 | 28 | 0 | 2926 |
| 1999-00 | 0 | 0 | 34 | 115 | 249 | 669 | 746 | 457 | 354 | 192 | 40 | 1 | 2857 |
| 2000-01 | 0 | 0 | 35 | 113 | 648 | 1063 | 882 | 672 | 561 | 105 | 22 | 0 | 4101 |
| 2001-02 | 0 | 0 | 18 | 169 | 338 | 698 | 768 | 661 | 582 | 163 | 65 | 0 | 3462 |
| 2002-03 | 0 | 0 | 2 | 307 | 527 | 736 | 868 | 757 | 476 | 169 | 19 | 1 | 3862 |
| 2003-04 | 0 | 0 | 24 | 112 | 443 | 672 | 768 | 721 | 313 | 145 | 39 | 0 | 3237 |
| 2004-05 | 0 | 0 | 0 | 79 | 428 | 658 | 780 | 505 | 405 | 139 | 68 | 0 | 3062 |
| 2005-06 | 0 | 0 | 3 | 140 | 356 | 802 | 531 | 645 | 330 | 63 | 28 | 0 | 2898 |
| 2006-07 | 0 | 0 | 9 | 150 | 362 | 658 | 867 | 636 | 179 | 257 | 3 | 0 | 3121 |
| 2007-08 | 0 | 0 | 0 | 109 | 365 | 798 | 766 | 660 | 385 | 192 | 33 | 0 | 3308 |
| 2008-09 | 0 | 0 | 2 | 157 | 420 | 787 | 853 | 470 | 361 | 202 | 57 | 0 | 3309 |
| 2009- | 0 | 0 | 19 | 290 | 324 | 925 |  |  |  |  |  |  |  |

WBAN : 13967

## COOLING DEGREE DAYS (base 65°F) 2009  OKLAHOMA CITY (KOKC)

| YEAR | JAN | FEB | MAR | APR | MAY | JUN | JUL | AUG | SEP | OCT | NOV | DEC | TOTAL |
|---|---|---|---|---|---|---|---|---|---|---|---|---|---|
| 1980 | 0 | 0 | 0 | 7 | 155 | 498 | 729 | 721 | 366 | 65 | 11 | 2 | 2554 |
| 1981 | 0 | 4 | 0 | 94 | 98 | 409 | 603 | 435 | 304 | 47 | 0 | 0 | 1994 |
| 1982 | 0 | 0 | 9 | 28 | 130 | 234 | 503 | 598 | 305 | 90 | 3 | 1 | 1901 |
| 1983 | 0 | 0 | 0 | 20 | 91 | 266 | 523 | 599 | 329 | 54 | 8 | 0 | 1890 |
| 1984 | 0 | 0 | 0 | 16 | 159 | 414 | 521 | 551 | 279 | 64 | 5 | 0 | 2009 |
| 1985 | 0 | 0 | 12 | 43 | 172 | 336 | 501 | 512 | 313 | 38 | 0 | 0 | 1927 |
| 1986 | 0 | 2 | 21 | 63 | 147 | 425 | 653 | 473 | 301 | 40 | 0 | 0 | 2125 |
| 1987 | 0 | 0 | 0 | 88 | 242 | 371 | 475 | 543 | 230 | 18 | 12 | 0 | 1979 |
| 1988 | 0 | 0 | 1 | 29 | 186 | 410 | 525 | 558 | 270 | 25 | 1 | 0 | 2005 |
| 1989 | 0 | 0 | 16 | 100 | 179 | 285 | 459 | 419 | 170 | 83 | 8 | 0 | 1719 |
| 1990 | 0 | 0 | 12 | 33 | 169 | 517 | 495 | 522 | 378 | 48 | 13 | 0 | 2187 |
| 1991 | 0 | 0 | 15 | 45 | 257 | 398 | 542 | 507 | 219 | 85 | 1 | 0 | 2069 |
| 1992 | 0 | 0 | 3 | 51 | 114 | 283 | 508 | 312 | 239 | 36 | 0 | 1 | 1547 |
| 1993 | 0 | 0 | 4 | 9 | 89 | 362 | 584 | 545 | 177 | 32 | 0 | 0 | 1802 |
| 1994 | 0 | 0 | 20 | 44 | 116 | 446 | 470 | 464 | 208 | 70 | 9 | 0 | 1847 |
| 1995 | 0 | 0 | 11 | 14 | 72 | 250 | 506 | 521 | 262 | 33 | 0 | 0 | 1669 |
| 1996 | 0 | 7 | 4 | 28 | 288 | 382 | 514 | 408 | 170 | 42 | 0 | 0 | 1843 |
| 1997 | 2 | 0 | 5 | 7 | 104 | 316 | 520 | 429 | 317 | 112 | 0 | 0 | 1812 |
| 1998 | 0 | 0 | 13 | 16 | 252 | 496 | 719 | 627 | 497 | 62 | 2 | 0 | 2684 |
| 1999 | 0 | 4 | 0 | 40 | 131 | 327 | 540 | 619 | 225 | 51 | 10 | 0 | 1947 |
| 2000 | 0 | 0 | 3 | 20 | 232 | 295 | 498 | 639 | 372 | 94 | 0 | 0 | 2153 |
| 2001 | 0 | 0 | 0 | 70 | 168 | 348 | 650 | 563 | 196 | 27 | 8 | 1 | 2031 |
| 2002 | 0 | 0 | 1 | 50 | 102 | 344 | 468 | 509 | 285 | 41 | 5 | 0 | 1805 |
| 2003 | 0 | 0 | 4 | 36 | 152 | 275 | 603 | 559 | 162 | 75 | 17 | 0 | 1883 |
| 2004 | 0 | 0 | 18 | 39 | 260 | 318 | 439 | 368 | 308 | 73 | 0 | 0 | 1823 |
| 2005 | 0 | 0 | 1 | 37 | 206 | 399 | 484 | 502 | 373 | 102 | 25 | 0 | 2129 |
| 2006 | 0 | 0 | 32 | 139 | 270 | 456 | 664 | 656 | 203 | 90 | 6 | 0 | 2516 |
| 2007 | 0 | 2 | 38 | 35 | 194 | 372 | 494 | 599 | 342 | 130 | 8 | 0 | 2214 |
| 2008 | 1 | 0 | 10 | 37 | 240 | 460 | 581 | 507 | 200 | 50 | 5 | 3 | 2094 |
| 2009 | 0 | 1 | 23 | 48 | 117 | 464 | 545 | 447 | 199 | 8 | 2 | 0 | 1854 |

**SNOWFALL (inches) 2009 OKLAHOMA CITY (KOKC)**

| YEAR | JUL | AUG | SEP | OCT | NOV | DEC | JAN | FEB | MAR | APR | MAY | JUN | TOTAL |
|---|---|---|---|---|---|---|---|---|---|---|---|---|---|
| 1980-81 | 0.0 | 0.0 | 0.0 | 0.0 | 4.0 | 0.0 | T | T | 0.0 | 0.0 | 0.0 | 0.0 | 4.0 |
| 1981-82 | 0.0 | 0.0 | 0.0 | 0.0 | 0.0 | T | 1.0 | 3.9 | 2.5 | 0.0 | 0.0 | 0.0 | 7.4 |
| 1982-83 | 0.0 | 0.0 | 0.0 | 0.0 | T | T | 5.1 | 4.3 | T | 0.0 | 0.0 | 0.0 | 9.4 |
| 1983-84 | 0.0 | 0.0 | 0.0 | 0.0 | T | 1.9 | 5.6 | 2.0 | T | 0.0 | 0.0 | 0.0 | 9.5 |
| 1984-85 | 0.0 | 0.0 | 0.0 | 0.0 | T | 6.1 | 1.5 | 2.3 | 0.0 | 0.0 | 0.0 | 0.0 | 9.9 |
| 1985-86 | 0.0 | 0.0 | 0.0 | 0.0 | T | 2.9 | 0.0 | 10.9 | 0.0 | 0.0 | 0.0 | 0.0 | 13.8 |
| 1986-87 | 0.0 | 0.0 | 0.0 | 0.0 | 0.0 | T | 10.0 | 1.0 | T | 0.0 | 0.0 | 0.0 | 11.0 |
| 1987-88 | 0.0 | 0.0 | 0.0 | 0.0 | 2.0 | 8.3 | 12.1 | 0.2 | 0.9 | 0.0 | 0.0 | 0.0 | 23.5 |
| 1988-89 | 0.0 | 0.0 | 0.0 | 0.0 | 0.6 | 2.0 | 4.8 | T | 4.0 | 0.6 | T | 0.0 | 12.0 |
| 1989-90 | 0.0 | 0.0 | 0.0 | 0.0 | T | 1.7 | 0.0 | 1.7 | 0.1 | 0.0 | 0.0 | 0.0 | 3.5 |
| 1990-91 | 0.0 | 0.0 | 0.0 | 0.0 | 0.0 | 4.2 | T | 0.0 | T | 0.0 | 0.0 | 0.0 | 4.2 |
| 1991-92 | 0.0 | 0.0 | 0.0 | T | 2.1 | 1.0 | 5.0 | 0.0 | T | 0.0 | T | T | 8.1 |
| 1992-93 | 0.0 | 0.0 | T |  | T | 3.3 | 0.4 | 1.8 | T | T |  | 0.0 |  |
| 1993-94 | 0.0 | 0.0 | 0.0 | 0.1 | 0.0 | T | 0.5 | T | 6.0 | 0.0 | 0.0 |  |  |
| 1994-95 | 0.0 | 0.0 | 0.0 | 0.0 | 0.0 | 0.0 | 4.9 | T | 4.5 | T | T | T | 9.4 |
| 1995-96 | 0.0 | 0.0 |  | 0.0 | 0.5 | 4.1 | 1.0 | 0.3 | T | 0.0 | T |  |  |
| 1996-97 |  |  |  |  | T |  | 6.5 |  |  | T | 0.0 | T |  |
| 1997-98 | T | T | 0.0 | 0.0 | 0.1 | 2.0 | T | T | T | 0.0 | T | T | 2.1 |
| 1998-99 | 0.0 | 0.0 | 0.0 | T | 0.0 | 1.0 | T | 0.0 | 1.3 | T | T | 0.0 | 2.3 |
| 1999-00 | 0.0 | 0.0 | T | 0.0 | 0.0 | T | 9.1 | 0.0 | T | T | 0.0 | 0.0 | 9.1 |
| 2000-01 | 0.0 | 0.0 | 0.0 | T | T | 8.2 | 3.4 | T | T | 0.0 | T | 0.0 | 11.6 |
| 2001-02 | 0.0 | 0.0 | T | T | 3.2 | 1.5 | T | 2.9 | 1.3 | 0.0 | 0.0 | 0.0 | 8.9 |
| 2002-03 | 0.0 | 0.0 | 0.0 | 0.0 | 0.0 | 2.0 | T | 5.0 | T | 0.0 | T | 0.0 | 7.0 |
| 2003-04 | 0.0 | 0.0 | 0.0 | 0.0 | 0.0 | 1.9 | 0.3 | 0.3 | 0.0 | 0.0 | T | 0.0 | 2.5 |
| 2004-05 | 0.0 | 0.0 | 0.0 | 0.0 | T | T | 2.8 | 0.1 | 0.0 | 0.0 | 0.0 | 0.0 | 2.9 |
| 2005-06 | 0.0 | 0.0 | 0.0 | 0.0 | 0.0 | 2.2 | 0.5 | 0.2 | 1.6 | 0.0 | 0.0 | 0.0 | 4.5 |
| 2006-07 | 0.0 | 0.0 | 0.0 | 0.0 | 4.1 | T | 2.9 | 2.2 | 0.0 | T | 0.0 | 0.0 | 9.2 |
| 2007-08 | 0.0 | 0.0 | 0.0 | T | T | 2.1 | 0.6 | 0.0 | 0.2 | 0.0 | T | 0.0 | 2.9 |
| 2008-09 | 0.0 | 0.0 | 0.0 | 0.0 | 0.0 | T | 1.4 | T | 2.5 | 0.0 | T | 0.0 | 3.9 |
| 2009- | 0.0 | 0.0 | 0.0 | 0.0 | 0.0 | 14.0 |  |  |  |  |  |  |  |
| POR= 61 YRS | T | T | T | T | 0.6 | 2.0 | 2.8 | 2.0 | 1.3 | T | T | T | 8.7 |

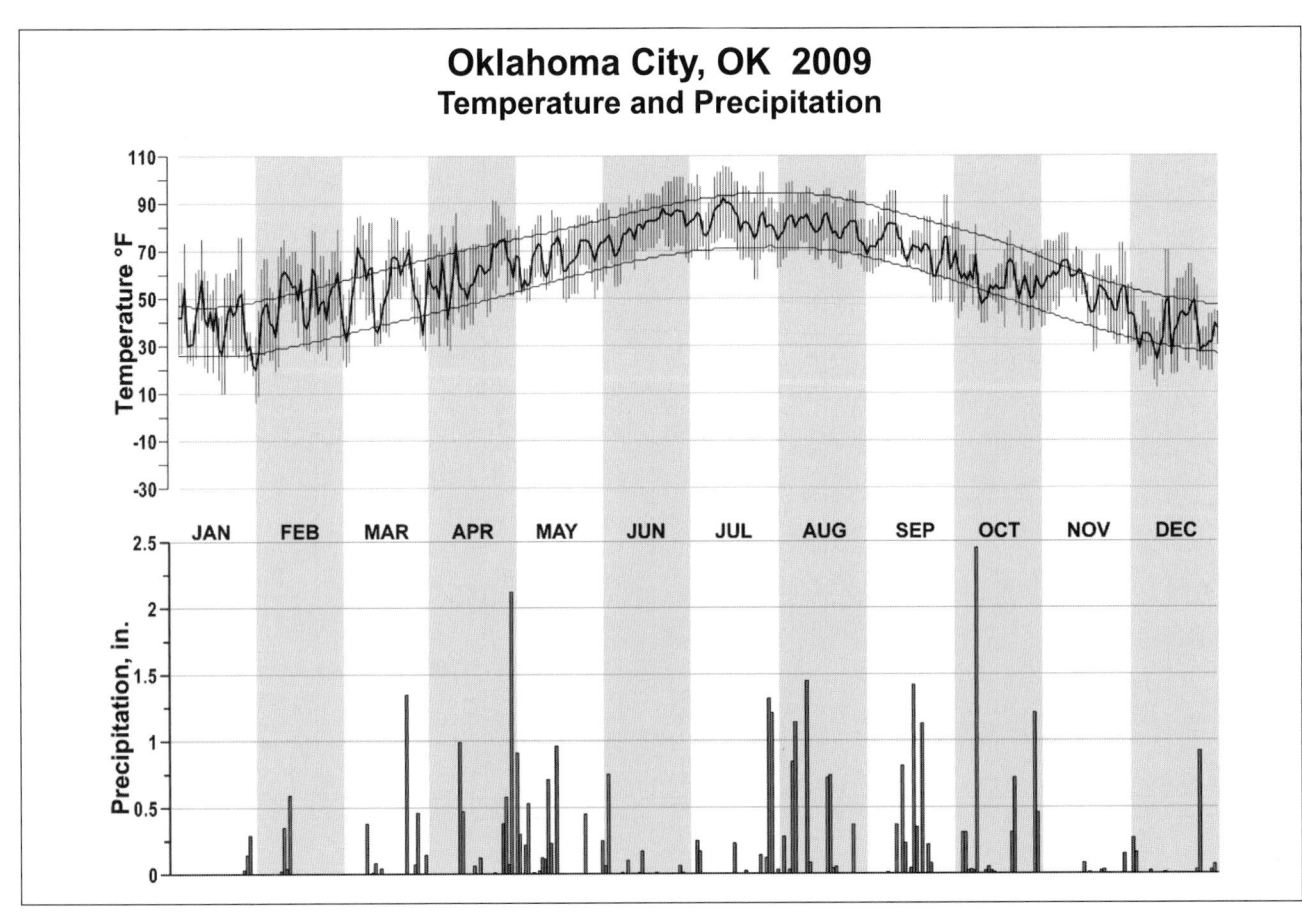

# Oklahoma City, OK  2009
## Temperature and Precipitation

# 2009
# TULSA
# OKLAHOMA (KTUL)

The city of Tulsa lies along the Arkansas River at an elevation of 700 feet above sea level. The surrounding terrain is gently rolling.

At latitude 36 degrees, Tulsa is far enough north to escape the long periods of heat in summer, yet far enough south to miss the extreme cold of winter. The influence of warm moist air from the Gulf of Mexico is often noted, due to the high humidity, but the climate is essentially continental characterized by rapid changes in temperature. Generally the winter months are mild. Temperatures occasionally fall below zero but only last a very short time. Temperatures of 100 degrees or higher are often experienced from late July to early September, but are usually accompanied by low relative humidity and a good southerly breeze. The fall season is long with a great number of pleasant, sunny days and cool, bracing nights.

Rainfall is ample for most agricultural pursuits and is distributed favorably throughout the year. Spring is the wettest season, having an abundance of rain in the form of showers and thunderstorms.

The steady rains of fall are a contrast to the spring and summer showers and provide a good supply of moisture and more ideal conditions for the growth of winter grains and pastures. The greatest amounts of snow are received in January and early March. The snow is usually light and only remains on the ground for brief periods.

The average date of the last 32 degree temperature occurrence is late March and the average date of the first 32 degree occurrence is early November. The average growing season is 216 days.

The Tulsa area is occasionally subjected to large hail and violent windstorms which occur mostly during spring and early summer, although occurrences have been noted throughout the year.

Prevailing surface winds are southerly during most of the year. Heavy fogs are infrequent. Sunshine is abundant. The prevalence of good flying weather throughout the year has contributed to the development of Tulsa as an aviation center.

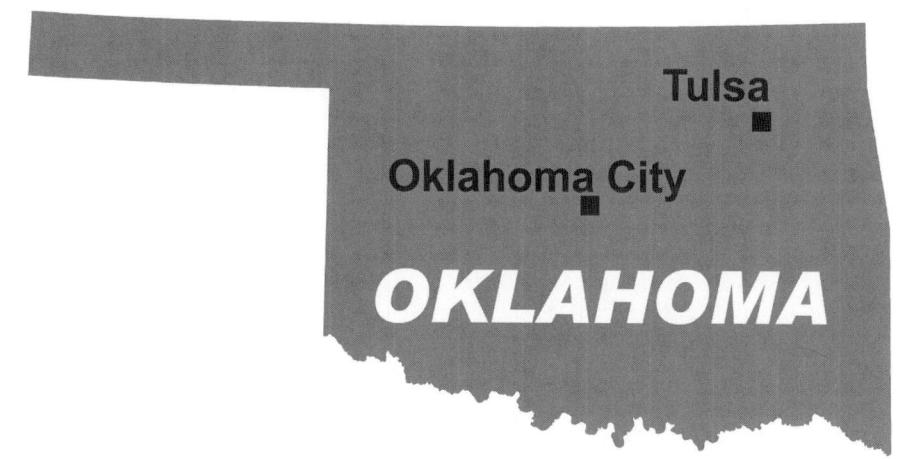

# NORMALS, MEANS, AND EXTREMES
## TULSA (KTUL)

LATITUDE: 36 ° 11'N  LONGITUDE: -95 ° 53'W  ELEVATION (FT): GRND: 640  BARO: 742  TIME ZONE: CENTRAL (UTC -6)  WBAN: 13968

| ELEMENT | POR | JAN | FEB | MAR | APR | MAY | JUN | JUL | AUG | SEP | OCT | NOV | DEC | YEAR |
|---|---|---|---|---|---|---|---|---|---|---|---|---|---|---|
| **TEMPERATURE °F** | | | | | | | | | | | | | | |
| NORMAL DAILY MAXIMUM | 30 | 46.5 | 52.9 | 62.4 | 72.1 | 79.6 | 88.0 | 93.8 | 93.2 | 84.1 | 74.0 | 60.0 | 49.6 | 71.4 |
| MEAN DAILY MAXIMUM | 68 | 46.7 | 51.9 | 61.4 | 71.8 | 79.4 | 87.2 | 93.2 | 92.7 | 84.1 | 74.1 | 60.3 | 50.0 | 71.1 |
| HIGHEST DAILY MAXIMUM | 70 | 79 | 90 | 96 | 102 | 97 | 103 | 112 | 110 | 109 | 98 | 87 | 80 | 112 |
| YEAR OF OCCURRENCE | | 1950 | 1996 | 1974 | 1972 | 2006 | 1953 | 1954 | 1970 | 1939 | 1979 | 1945 | 1966 | JUL 1954 |
| MEAN OF EXTREME MAXS. | 68 | 69.5 | 75.0 | 83.3 | 87.9 | 90.9 | 96.0 | 101.9 | 102.1 | 96.5 | 89.2 | 79.0 | 70.9 | 86.9 |
| NORMAL DAILY MINIMUM | 30 | 26.3 | 31.1 | 40.3 | 49.5 | 59.0 | 67.9 | 73.1 | 71.2 | 62.9 | 51.1 | 39.3 | 29.8 | 50.1 |
| MEAN DAILY MINIMUM | 68 | 26.3 | 30.5 | 38.8 | 49.4 | 59.0 | 67.3 | 72.4 | 70.8 | 62.1 | 50.8 | 38.7 | 29.8 | 49.7 |
| LOWEST DAILY MINIMUM | 70 | -8 | -11 | -3 | 22 | 35 | 49 | 51 | 50 | 35 | 18 | 10 | -8 | -11 |
| YEAR OF OCCURRENCE | | 1947 | 1996 | 1948 | 1957 | 2005 | 1954 | 1971 | 2009 | 1984 | 1993 | 1976 | 1989 | FEB 1996 |
| MEAN OF EXTREME MINS. | 68 | 7.5 | 12.9 | 20.4 | 33.0 | 44.2 | 55.7 | 62.3 | 59.9 | 46.6 | 34.3 | 22.3 | 11.4 | 34.2 |
| NORMAL DRY BULB | 30 | 36.4 | 42.0 | 51.4 | 60.8 | 69.3 | 78.0 | 83.5 | 82.2 | 73.5 | 62.6 | 49.7 | 39.7 | 60.8 |
| MEAN DRY BULB | 68 | 36.5 | 41.2 | 50.1 | 60.6 | 69.2 | 77.4 | 82.8 | 81.8 | 73.1 | 62.5 | 49.5 | 39.9 | 60.4 |
| MEAN WET BULB | 26 | 32.3 | 36.0 | 43.5 | 52.0 | 62.1 | 69.4 | 72.1 | 71.3 | 64.4 | 54.2 | 43.5 | 34.6 | 53.0 |
| MEAN DEW POINT | 26 | 28.1 | 31.3 | 38.5 | 47.3 | 59.1 | 66.8 | 69.0 | 68.1 | 61.1 | 50.4 | 39.2 | 30.2 | 49.1 |
| NORMAL NO. DAYS WITH: | | | | | | | | | | | | | | |
| MAXIMUM >= 90 | 30 | 0.0 | * | 0.1 | 0.4 | 2.1 | 12.7 | 23.8 | 22.4 | 9.2 | 1.0 | 0.0 | 0.0 | 71.7 |
| MAXIMUM <= 32 | 30 | 5.3 | 2.6 | 0.4 | 0.0 | 0.0 | 0.0 | 0.0 | 0.0 | 0.0 | 0.0 | 0.2 | 2.9 | 11.4 |
| MINIMUM <= 32 | 30 | 22.5 | 15.2 | 6.8 | 0.6 | 0.0 | 0.0 | 0.0 | 0.0 | 0.0 | 0.5 | 7.7 | 18.9 | 72.2 |
| MINIMUM <= 0 | 30 | 0.4 | 0.2 | 0.0 | 0.0 | 0.0 | 0.0 | 0.0 | 0.0 | 0.0 | 0.0 | 0.0 | 0.3 | 0.9 |
| **H/C** | | | | | | | | | | | | | | |
| NORMAL HEATING DEG. DAYS | 30 | 898 | 658 | 437 | 179 | 38 | 1 | 0 | 0 | 29 | 152 | 468 | 782 | 3642 |
| NORMAL COOLING DEG. DAYS | 30 | 0 | 1 | 10 | 50 | 163 | 385 | 568 | 524 | 277 | 64 | 6 | 1 | 2049 |
| **RH** | | | | | | | | | | | | | | |
| NORMAL (PERCENT) | 30 | 68 | 65 | 63 | 62 | 70 | 70 | 64 | 65 | 69 | 68 | 68 | 70 | 67 |
| HOUR 00 LST | 30 | 73 | 70 | 68 | 69 | 78 | 79 | 72 | 73 | 77 | 76 | 75 | 75 | 74 |
| HOUR 06 LST | 30 | 79 | 78 | 76 | 78 | 86 | 86 | 82 | 83 | 86 | 83 | 81 | 80 | 82 |
| HOUR 12 LST | 30 | 61 | 57 | 54 | 52 | 59 | 59 | 54 | 52 | 58 | 55 | 59 | 61 | 57 |
| HOUR 18 LST | 30 | 60 | 55 | 51 | 49 | 57 | 57 | 50 | 49 | 56 | 56 | 61 | 63 | 55 |
| **S** PERCENT POSSIBLE SUNSHINE | 53 | 54 | 55 | 58 | 59 | 59 | 69 | 76 | 74 | 68 | 65 | 56 | 53 | 62 |
| **W/O** MEAN NO. DAYS WITH: | | | | | | | | | | | | | | |
| HEAVY FOG(VISBY <= 1/4 MI) | 46 | 1.9 | 1.2 | 0.8 | 0.4 | 0.5 | 0.4 | 0.2 | 0.2 | 0.6 | 1.2 | 1.2 | 1.5 | 10.1 |
| THUNDERSTORMS | 62 | 0.8 | 1.3 | 3.5 | 5.7 | 8.3 | 7.7 | 6.0 | 5.7 | 4.8 | 2.8 | 1.6 | 1.0 | 49.2 |
| **CLOUDNESS** MEAN: | | | | | | | | | | | | | | |
| SUNRISE-SUNSET (OKTAS) | 1 | 6.4 | 5.6 | 5.2 | 6.0 | 5.2 | 2.8 | 3.2 | 2.7 | 5.6 | 3.5 | 4.8 | 4.0 | 4.6 |
| MIDNIGHT-MIDNIGHT (OKTAS) | 1 | 6.4 | 5.6 | 6.4 | 6.0 | 4.8 | 2.4 | 2.8 | 2.8 | 4.0 | 3.5 | 4.8 | 4.8 | 4.5 |
| MEAN NO. DAYS WITH: | | | | | | | | | | | | | | |
| CLEAR | 3 | 5.0 | 8.3 | 8.0 | 9.5 | 9.7 | 10.7 | 13.5 | 15.0 | 7.5 | 10.0 | 7.0 | 9.0 | 113.2 |
| PARTLY CLOUDY | 3 | 3.7 | 3.3 | 1.0 | 1.0 | 3.7 | 8.0 | 6.5 | 5.0 | 3.5 | 3.5 | 1.0 | 5.0 | 45.2 |
| CLOUDY | 3 | 8.3 | 7.3 | 8.7 | 10.5 | 8.7 | 3.7 | 4.0 | 4.5 | 4.5 | 6.5 | 9.5 | 7.0 | 83.2 |
| **PR** MEAN STATION PRESSURE(IN) | 26 | 29.41 | 29.38 | 29.30 | 29.22 | 29.21 | 29.22 | 29.26 | 29.27 | 29.29 | 29.33 | 29.35 | 29.41 | 29.30 |
| MEAN SEA-LEVEL PRES. (IN) | 26 | 30.15 | 30.11 | 30.02 | 29.94 | 29.92 | 29.92 | 29.95 | 29.97 | 30.00 | 30.04 | 30.08 | 30.14 | 30.02 |
| **WINDS** MEAN SPEED (MPH) | 26 | 9.4 | 9.7 | 10.8 | 10.9 | 9.9 | 9.2 | 8.9 | 8.0 | 8.0 | 8.8 | 9.7 | 9.2 | 9.4 |
| PREVAIL.DIR(TENS OF DEGS) | 36 | 19 | 19 | 19 | 19 | 19 | 19 | 19 | 19 | 19 | 19 | 19 | 19 | 19 |
| MAXIMUM 2-MINUTE: | | | | | | | | | | | | | | |
| SPEED (MPH) | 17 | 37 | 41 | 46 | 55 | 41 | 54 | 51 | 46 | 39 | 43 | 44 | 38 | 55 |
| DIR. (TENS OF DEGS) | | 30 | 20 | 18 | 34 | 30 | 32 | 19 | 25 | 27 | 24 | 29 | 34 | 34 |
| YEAR OF OCCURRENCE | | 2008 | 2000 | 1997 | 1993 | 2001 | 2008 | 1993 | 2006 | 1999 | 2007 | 1998 | 2009 | APR 1993 |
| MAXIMUM 3-SECOND | | | | | | | | | | | | | | |
| SPEED (MPH) | 17 | 54 | 53 | 55 | 63 | 48 | 61 | 55 | 56 | 49 | 63 | 55 | 48 | 63 |
| DIR. (TENS OF DEGS) | | 31 | 21 | 16 | 34 | 25 | 31 | 19 | 32 | 27 | 24 | 17 | 34 | 24 |
| YEAR OF OCCURRENCE | | 2008 | 2000 | 1997 | 1993 | 2008 | 2008 | 1993 | 2009 | 1999 | 2007 | 2005 | 2009 | OCT 2007 |
| **PRECIPITATION** NORMAL (IN) | 30 | 1.60 | 1.95 | 3.57 | 3.95 | 6.11 | 4.72 | 2.96 | 2.85 | 4.76 | 4.05 | 3.47 | 2.43 | 42.42 |
| MAXIMUM MONTHLY (IN) | 70 | 6.65 | 5.73 | 11.94 | 9.33 | 18.00 | 11.17 | 11.39 | 8.78 | 18.81 | 16.51 | 7.57 | 8.70 | 18.81 |
| YEAR OF OCCURRENCE | | 1949 | 1985 | 1973 | 2008 | 1943 | 1948 | 1994 | 2003 | 1971 | 1941 | 1946 | 1984 | SEP 1971 |
| MINIMUM MONTHLY (IN) | 70 | 0.00 | 0.16 | 0.08 | 0.34 | 1.17 | 0.53 | 0.03 | 0.01 | T | T | 0.01 | 0.10 | 0.00 |
| YEAR OF OCCURRENCE | | 1993 | 1996 | 1971 | 1989 | 1988 | 1963 | 1954 | 2000 | 1948 | 1952 | 1949 | 1996 | JAN 1993 |
| MAXIMUM IN 24 HOURS (IN) | 70 | 2.25 | 4.34 | 3.17 | 4.58 | 9.27 | 5.01 | 7.54 | 5.37 | 7.25 | 5.80 | 5.14 | 3.27 | 9.27 |
| YEAR OF OCCURRENCE | | 1946 | 1985 | 2004 | 1964 | 1984 | 1941 | 1963 | 1989 | 2007 | 1983 | 1974 | 1984 | MAY 1984 |
| NORMAL NO. DAYS WITH: | | | | | | | | | | | | | | |
| PRECIPITATION >= 0.01 | 30 | 6.5 | 6.5 | 9.4 | 8.9 | 10.4 | 9.2 | 6.0 | 6.8 | 8.2 | 7.3 | 7.0 | 6.4 | 92.6 |
| PRECIPITATION >= 1.00 | 30 | 0.3 | 0.4 | 1.1 | 1.2 | 1.8 | 1.4 | 1.0 | 0.9 | 1.6 | 1.4 | 1.1 | 0.6 | 12.8 |
| **SNOWFALL** NORMAL (IN) | 30 | 3.0 | 2.3 | 1.4 | 0.* | 0.0 | 0.0 | 0.0 | 0.0 | 0.0 | 0.* | 0.6 | 1.9 | 9.2 |
| MAXIMUM MONTHLY (IN) | 68 | 12.7 | 10.5 | 14.1 | 1.7 | T | T | T | 0.0 | T | 0.3 | 10.5 | 11.4 | 14.1 |
| YEAR OF OCCURRENCE | | 1979 | 2003 | 1994 | 1957 | 2009 | 2009 | 1994 | | 1990 | 1993 | 2006 | 2000 | MAR 1994 |
| MAXIMUM IN 24 HOURS (IN) | 68 | 9.0 | 7.0 | 12.9 | 1.7 | T | T | T | 0.0 | T | 0.3 | 4.0 | 8.8 | 12.9 |
| YEAR OF OCCURRENCE' | | 1944 | 2003 | 1994 | 1957 | 2009 | 2009 | 1994 | 1994 | 1990 | 1993 | 1972 | 1954 | MAR 1994 |
| MAXIMUM SNOW DEPTH (IN) | 61 | 11 | 6 | 10 | T | 0 | 0 | 0 | 0 | 0 | 0 | 3 | 8 | 11 |
| YEAR OF OCCURRENCE | | 1988 | 1949 | 1968 | 1993 | | | | | | | 2001 | 1954 | JAN 1988 |
| NORMAL NO. DAYS WITH: | | | | | | | | | | | | | | |
| SNOWFALL >= 1.0 | 30 | 1.0 | 0.7 | 0.3 | 0.0 | 0.0 | 0.0 | 0.0 | 0.0 | 0.0 | 0.0 | 0.3 | 0.8 | 3.1 |

## PRECIPITATION (inches) 2009 TULSA (KTUL)

| YEAR | JAN | FEB | MAR | APR | MAY | JUN | JUL | AUG | SEP | OCT | NOV | DEC | ANNUAL |
|------|------|------|------|------|------|------|------|------|------|------|------|------|--------|
| 1980 | 2.07 | 1.32 | 3.59 | 3.44 | 7.23 | 5.57 | 0.09 | 2.34 | 3.47 | 2.05 | 0.79 | 1.37 | 33.33 |
| 1981 | 0.69 | 1.63 | 1.67 | 1.90 | 6.70 | 3.31 | 6.22 | 2.47 | 3.11 | 6.73 | 2.25 | 0.20 | 36.88 |
| 1982 | 3.58 | 0.67 | 1.04 | 1.28 | 9.30 | 4.13 | 1.65 | 1.42 | 2.95 | 1.22 | 4.61 | 3.39 | 35.24 |
| 1983 | 2.95 | 1.98 | 2.19 | 3.88 | 6.85 | 1.47 | 0.58 | 0.65 | 2.11 | 9.33 | 2.14 | 0.61 | 34.74 |
| 1984 | 1.00 | 1.95 | 6.72 | 2.44 | 11.25 | 1.72 | 0.48 | 1.96 | 2.77 | 6.98 | 2.80 | 8.70 | 48.77 |
| 1985 | 1.24 | 5.74 | 5.39 | 5.62 | 4.19 | 7.63 | 2.38 | 1.91 | 3.29 | 6.26 | 6.27 | 1.39 | 51.31 |
| 1986 | 0.00 | 1.22 | 2.28 | 5.10 | 6.97 | 4.23 | 1.15 | 3.96 | 8.36 | 5.53 | 2.99 | 0.97 | 42.76 |
| 1987 | 2.21 | 4.72 | 2.20 | 0.70 | 10.02 | 2.31 | 4.20 | 3.72 | 3.52 | 1.27 | 5.17 | 5.87 | 45.91 |
| 1988 | 1.11 | 1.03 | 6.52 | 3.18 | 1.17 | 0.58 | 4.20 | 2.43 | 5.37 | 1.43 | 4.38 | 1.82 | 33.22 |
| 1989 | 2.94 | 2.26 | 3.14 | 0.34 | 3.95 | 5.16 | 4.09 | 6.69 | 3.32 | 2.80 | 0.15 | 0.26 | 35.10 |
| 1990 | 2.93 | 4.14 | 6.51 | 5.31 | 5.21 | 1.08 | 0.24 | 1.83 | 4.19 | 2.15 | 2.41 | 2.94 | 38.94 |
| 1991 | 1.47 | 0.38 | 1.02 | 2.58 | 5.11 | 3.64 | 0.35 | 1.17 | 6.15 | 5.12 | 1.98 | 4.57 | 33.54 |
| 1992 | 0.48 | 1.32 | 1.37 | 4.75 | 5.65 | 8.41 | 2.12 | 3.09 | 2.66 | 3.53 | 4.83 | 5.21 | 43.42 |
| 1993 |      | 2.86 | 2.76 | 4.59 | 6.86 | 3.79 | 2.42 | 2.29 | 6.90 | 1.13 | 1.69 | 1.76 |        |
| 1994 | 0.68 | 2.21 | 3.35 | 6.57 | 2.81 | 2.73 | 11.39 | 4.12 | 3.60 | 3.68 | 7.10 | 1.21 | 49.45 |
| 1995 | 0.93 | 0.57 | 1.83 | 5.92 | 10.73 | 9.84 | 2.55 | 1.44 | 4.96 | 1.05 | 0.25 | 1.77 | 41.84 |
| 1996 | 0.47 | 0.16 | 2.07 | 1.40 | 2.14 | 3.64 | 3.22 | 1.34 | 5.04 | 5.60 | 7.16 | 0.10 | 32.34 |
| 1997 | 0.27 | 3.41 | 1.39 | 4.09 | 1.66 | 5.77 | 5.64 | 7.89 | 3.06 | 2.07 | 1.63 | 4.32 | 41.20 |
| 1998 | 3.49 | 0.30 | 7.30 | 4.54 | 2.52 | 3.36 | 4.31 | 1.67 | 5.13 | 9.14 | 3.26 | 1.58 | 46.60 |
| 1999 | 3.03 | 1.25 | 3.55 | 7.20 | 9.55 | 5.21 | 0.40 | 0.42 | 9.70 | 1.75 | 1.32 | 5.10 | 48.48 |
| 2000 | 0.89 | 1.33 | 3.77 | 2.71 | 7.01 | 6.25 | 6.58 | 0.01 | 1.10 | 6.32 | 3.51 | 1.62 | 41.10 |
| 2001 | 2.09 | 2.62 | 0.77 | 1.19 | 6.32 | 3.04 | 0.51 | 2.26 | 1.95 | 2.81 | 3.33 | 2.25 | 29.14 |
| 2002 | 2.67 | 0.90 | 2.39 | 3.71 | 5.21 | 2.86 | 2.18 | 3.55 | 1.24 | 3.33 | 0.45 | 2.74 | 31.23 |
| 2003 | 0.14 | 1.76 | 3.25 | 2.17 | 5.25 | 5.96 | 0.89 | 8.78 | 4.94 | 3.95 | 1.73 | 2.46 | 41.28 |
| 2004 | 2.36 | 1.20 | 6.16 | 5.97 | 3.07 | 6.41 | 8.62 | 1.62 | 0.82 | 8.51 | 3.92 | 0.84 | 49.50 |
| 2005 | 3.69 | 1.93 | 1.21 | 2.80 | 1.61 | 3.94 | 1.62 | 5.91 | 3.09 | 1.58 | 0.31 | 0.52 | 28.21 |
| 2006 | 0.72 | 0.35 | 2.80 | 5.83 | 3.04 | 5.85 | 4.41 | 4.06 | 2.01 | 1.31 | 3.58 | 4.27 | 38.23 |
| 2007 | 2.27 | 1.23 | 3.07 | 2.25 | 10.03 | 9.17 | 6.10 | 0.69 | 10.82 | 3.04 | 0.54 | 3.88 | 53.09 |
| 2008 | 0.88 | 2.01 | 4.73 | 9.33 | 9.61 | 9.43 | 4.64 | 4.59 | 4.40 | 2.75 | 1.96 | 1.77 | 56.10 |
| 2009 | 0.68 | 2.28 | 5.02 | 4.34 | 6.80 | 3.51 | 2.84 | 3.76 | 8.29 | 6.14 | 0.58 | 1.88 | 46.12 |
| POR= 67 YRS | 1.50 | 1.80 | 3.00 | 3.95 | 5.67 | 4.85 | 3.34 | 3.13 | 4.33 | 3.72 | 2.56 | 2.08 | 39.93 |

WBAN : 13968

## AVERAGE TEMPERATURE (°F) 2009 TULSA (KTUL)

| YEAR | JAN | FEB | MAR | APR | MAY | JUN | JUL | AUG | SEP | OCT | NOV | DEC | ANNUAL |
|------|------|------|------|------|------|------|------|------|------|------|------|------|--------|
| 1980 | 38.6 | 37.1 | 48.3 | 61.1 | 70.6 | 82.5 | 91.7 | 89.7 | 78.3 | 61.5 | 50.5 | 42.3 | 62.7 |
| 1981 | 37.6 | 43.6 | 53.3 | 68.0 | 65.9 | 80.0 | 85.9 | 79.4 | 73.9 | 60.9 | 51.4 | 38.5 | 61.5 |
| 1982 | 33.6 | 38.2 | 55.3 | 59.3 | 72.9 | 74.7 | 84.2 | 85.3 | 74.6 | 63.4 | 50.6 | 44.4 | 61.4 |
| 1983 | 39.1 | 42.9 | 49.0 | 55.4 | 67.0 | 76.6 | 84.7 | 88.1 | 77.4 | 64.5 | 52.9 | 26.7 | 60.4 |
| 1984 | 34.4 | 46.4 | 48.3 | 58.0 | 67.5 | 80.1 | 82.0 | 82.7 | 71.5 | 63.8 | 50.4 | 44.7 | 60.8 |
| 1985 | 30.2 | 35.9 | 54.7 | 63.3 | 70.6 | 75.8 | 82.9 | 81.7 | 74.6 | 63.1 | 47.8 | 34.5 | 59.6 |
| 1986 | 42.8 | 43.2 | 55.0 | 62.6 | 69.4 | 79.7 | 86.6 | 78.2 | 74.7 | 61.0 | 43.6 | 40.0 | 61.4 |
| 1987 | 36.0 | 45.4 | 51.5 | 63.2 | 74.1 | 78.9 | 81.9 | 83.1 | 72.4 | 59.3 | 51.6 | 41.4 | 61.6 |
| 1988 | 34.8 | 39.3 | 49.3 | 59.5 | 71.0 | 79.9 | 82.6 | 83.0 | 73.2 | 58.5 | 51.7 | 43.4 | 60.5 |
| 1989 | 43.4 | 31.9 | 49.3 | 63.3 | 69.2 | 74.8 | 80.2 | 80.4 | 68.7 | 64.0 | 52.7 | 31.6 | 59.1 |
| 1990 | 46.1 | 46.1 | 53.2 | 59.6 | 67.4 | 82.1 | 83.2 | 83.5 | 78.3 | 61.2 | 56.4 | 38.5 | 63.0 |
| 1991 | 34.7 | 48.3 | 55.1 | 63.8 | 73.7 | 80.0 | 84.9 | 82.9 | 72.7 | 64.3 | 45.8 | 44.6 | 62.6 |
| 1992 | 42.8 | 50.1 | 54.9 | 61.6 | 67.6 | 74.7 | 81.8 | 76.6 | 72.8 | 60.8 | 45.9 | 38.6 | 60.7 |
| 1993 | 35.7 | 37.8 | 46.8 | 55.8 | 66.0 | 76.8 | 84.4 | 83.5 | 68.6 | 56.2 | 44.5 | 42.3 | 58.2 |
| 1994 | 35.2 | 39.0 | 52.9 | 60.4 | 67.4 | 80.5 | 79.3 | 78.4 | 70.9 | 63.4 | 52.0 | 42.7 | 60.2 |
| 1995 | 39.4 | 44.2 | 51.5 | 58.3 | 65.5 | 74.1 | 82.3 | 84.6 | 70.5 | 62.8 | 48.8 | 39.4 | 60.1 |
| 1996 | 35.4 | 43.0 | 45.4 | 59.2 | 72.9 | 78.5 | 81.6 | 79.8 | 70.0 | 61.4 | 44.9 | 42.1 | 59.5 |
| 1997 | 35.9 | 44.1 | 52.4 | 55.9 | 66.9 | 75.5 | 81.7 | 78.3 | 73.9 | 62.0 | 46.0 | 39.3 | 59.3 |
| 1998 | 40.1 | 44.8 | 46.7 | 57.8 | 72.9 | 79.9 | 85.4 | 84.1 | 80.8 | 63.0 | 53.4 | 40.8 | 62.5 |
| 1999 | 39.0 | 50.0 | 48.8 | 61.5 | 68.1 | 75.4 | 84.4 | 84.6 | 69.8 | 62.3 | 57.7 | 43.4 | 62.1 |
| 2000 | 40.0 | 48.0 | 53.2 | 59.5 | 70.8 | 74.4 | 81.4 | 86.8 | 75.7 | 66.2 | 43.8 | 28.6 | 60.7 |
| 2001 | 35.3 | 41.3 | 47.1 | 66.4 | 70.6 | 78.2 | 87.4 | 85.2 | 71.9 | 62.0 | 55.2 | 43.6 | 62.0 |
| 2002 | 40.3 | 42.1 | 47.2 | 62.4 | 67.2 | 78.3 | 82.7 | 83.0 | 76.3 | 58.0 | 48.7 | 42.7 | 60.7 |
| 2003 | 37.0 | 37.7 | 49.7 | 61.9 | 69.9 | 75.4 | 84.4 | 84.4 | 69.3 | 63.0 | 51.4 | 43.4 | 60.7 |
| 2004 | 38.6 | 39.9 | 54.8 | 60.3 | 71.8 | 75.1 | 79.1 | 77.0 | 74.9 | 64.9 | 51.6 | 42.4 | 60.9 |
| 2005 | 38.9 | 45.8 | 50.7 | 61.1 | 69.3 | 79.6 | 82.5 | 83.0 | 77.2 | 63.0 | 54.3 | 38.7 | 62.0 |
| 2006 | 48.5 | 41.8 | 54.5 | 66.8 | 71.5 | 78.2 | 84.7 | 85.9 | 71.1 | 61.2 | 51.5 | 43.2 | 63.2 |
| 2007 | 36.5 | 40.6 | 60.7 | 57.9 | 70.6 | 77.0 | 81.3 | 85.4 | 75.2 | 64.8 | 52.2 | 38.6 | 61.7 |
| 2008 | 38.5 | 41.1 | 51.4 | 59.0 | 70.4 | 78.3 | 83.7 | 80.8 | 71.5 | 61.7 | 50.8 | 39.4 | 60.6 |
| 2009 | 36.9 | 46.4 | 52.9 | 59.2 | 67.4 | 81.2 | 81.0 | 79.4 | 70.7 | 55.9 | 54.9 | 34.6 | 60.0 |
| POR= 68 YRS | 36.5 | 41.2 | 50.1 | 60.6 | 69.2 | 77.4 | 82.8 | 81.8 | 73.1 | 62.5 | 49.5 | 39.9 | 60.4 |

## HEATING DEGREE DAYS (base 65°F) 2009 TULSA (KTUL)

| YEAR | JUL | AUG | SEP | OCT | NOV | DEC | JAN | FEB | MAR | APR | MAY | JUN | TOTAL |
|---|---|---|---|---|---|---|---|---|---|---|---|---|---|
| 1980-81 | 0 | 0 | 13 | 172 | 438 | 703 | 843 | 598 | 360 | 48 | 58 | 0 | 3233 |
| 1981-82 | 0 | 0 | 23 | 178 | 402 | 817 | 967 | 747 | 322 | 208 | 11 | 5 | 3680 |
| 1982-83 | 0 | 0 | 23 | 146 | 437 | 635 | 794 | 611 | 492 | 321 | 50 | 0 | 3509 |
| 1983-84 | 0 | 0 | 19 | 89 | 378 | 1179 | 941 | 533 | 509 | 229 | 47 | 0 | 3924 |
| 1984-85 | 0 | 0 | 73 | 130 | 438 | 628 | 1073 | 809 | 330 | 103 | 7 | 0 | 3591 |
| 1985-86 | 0 | 0 | 46 | 111 | 510 | 936 | 680 | 602 | 322 | 127 | 13 | 0 | 3347 |
| 1986-87 | 0 | 0 | 5 | 148 | 632 | 771 | 893 | 544 | 413 | 149 | 0 | 0 | 3555 |
| 1987-88 | 0 | 0 | 1 | 189 | 416 | 727 | 928 | 739 | 483 | 187 | 9 | 0 | 3679 |
| 1988-89 | 0 | 0 | 8 | 218 | 393 | 662 | 663 | 921 | 487 | 155 | 53 | 0 | 3560 |
| 1989-90 | 0 | 0 | 67 | 126 | 375 | 1029 | 580 | 527 | 376 | 194 | 54 | 0 | 3328 |
| 1990-91 | 0 | 0 | 8 | 172 | 271 | 813 | 933 | 459 | 327 | 83 | 17 | 0 | 3083 |
| 1991-92 | 0 | 0 | 35 | 121 | 570 | 628 | 682 | 423 | 311 | 156 | 53 | 0 | 2979 |
| 1992-93 | 0 | 0 | 9 | 151 | 565 | 812 | 903 | 755 | 556 | 280 | 57 | 0 | 4088 |
| 1993-94 | 0 | 0 | 40 | 294 | 611 | 695 | 917 | 721 | 387 | 186 | 56 | 0 | 3907 |
| 1994-95 | 0 | 0 | 25 | 126 | 390 | 683 | 783 | 574 | 436 | 219 | 70 | 0 | 3306 |
| 1995-96 | 0 | 0 | 79 | 112 | 480 | 786 | 911 | 640 | 604 | 204 | 19 | 0 | 3835 |
| 1996-97 | 0 | 0 | 26 | 152 | 594 | 701 | 896 | 579 | 393 | 275 | 40 | 0 | 3656 |
| 1997-98 | 0 | 0 | 3 | 195 | 563 | 791 | 764 | 558 | 581 | 225 | 12 | 6 | 3698 |
| 1998-99 | 0 | 0 | 0 | 109 | 344 | 747 | 798 | 415 | 496 | 134 | 18 | 0 | 3061 |
| 1999-00 | 0 | 0 | 38 | 130 | 238 | 664 | 770 | 487 | 363 | 183 | 23 | 2 | 2898 |
| 2000-01 | 0 | 0 | 26 | 100 | 628 | 1121 | 914 | 657 | 546 | 93 | 18 | 0 | 4103 |
| 2001-02 | 0 | 0 | 20 | 142 | 298 | 660 | 760 | 634 | 544 | 146 | 57 | 0 | 3261 |
| 2002-03 | 0 | 0 | 0 | 268 | 491 | 686 | 864 | 758 | 474 | 155 | 11 | 0 | 3707 |
| 2003-04 | 0 | 0 | 32 | 113 | 414 | 661 | 814 | 722 | 322 | 178 | 49 | 0 | 3305 |
| 2004-05 | 0 | 0 | 0 | 74 | 395 | 695 | 804 | 531 | 442 | 151 | 59 | 0 | 3151 |
| 2005-06 | 0 | 0 | 7 | 164 | 342 | 808 | 505 | 646 | 354 | 78 | 32 | 0 | 2936 |
| 2006-07 | 0 | 0 | 12 | 191 | 397 | 669 | 878 | 676 | 186 | 262 | 6 | 0 | 3277 |
| 2007-08 | 0 | 0 | 0 | 111 | 394 | 807 | 818 | 693 | 419 | 205 | 34 | 0 | 3481 |
| 2008-09 | 0 | 0 | 4 | 145 | 431 | 787 | 867 | 516 | 397 | 224 | 54 | 0 | 3425 |
| 2009- | 0 | 1 | 12 | 284 | 304 | 934 | | | | | | | |

WBAN : 13968

## COOLING DEGREE DAYS (base 65°F) 2009 TULSA (KTUL)

| YEAR | JAN | FEB | MAR | APR | MAY | JUN | JUL | AUG | SEP | OCT | NOV | DEC | TOTAL |
|---|---|---|---|---|---|---|---|---|---|---|---|---|---|
| 1980 | 0 | 0 | 0 | 43 | 200 | 533 | 833 | 774 | 419 | 69 | 6 | 4 | 2881 |
| 1981 | 0 | 5 | 4 | 145 | 96 | 456 | 658 | 452 | 296 | 57 | 1 | 0 | 2170 |
| 1982 | 0 | 0 | 28 | 44 | 266 | 300 | 601 | 637 | 319 | 106 | 10 | 5 | 2316 |
| 1983 | 0 | 0 | 3 | 40 | 120 | 353 | 615 | 725 | 396 | 80 | 20 | 0 | 2352 |
| 1984 | 0 | 0 | 0 | 25 | 132 | 464 | 534 | 556 | 272 | 100 | 9 | 2 | 2094 |
| 1985 | 0 | 0 | 19 | 59 | 185 | 333 | 564 | 523 | 340 | 57 | 0 | 0 | 2080 |
| 1986 | 0 | 0 | 20 | 60 | 157 | 448 | 676 | 415 | 303 | 31 | 0 | 0 | 2110 |
| 1987 | 0 | 0 | 2 | 102 | 290 | 421 | 532 | 567 | 230 | 18 | 19 | 0 | 2181 |
| 1988 | 0 | 0 | 2 | 30 | 200 | 454 | 555 | 564 | 262 | 23 | 1 | 0 | 2091 |
| 1989 | 0 | 0 | 6 | 107 | 191 | 300 | 475 | 483 | 183 | 105 | 14 | 0 | 1864 |
| 1990 | 0 | 0 | 17 | 38 | 137 | 521 | 571 | 581 | 416 | 63 | 21 | 0 | 2365 |
| 1991 | 0 | 0 | 29 | 53 | 293 | 458 | 622 | 562 | 274 | 108 | 3 | 0 | 2402 |
| 1992 | 0 | 0 | 5 | 63 | 140 | 298 | 526 | 369 | 251 | 29 | 0 | 0 | 1681 |
| 1993 | 0 | 0 | 1 | 7 | 95 | 360 | 609 | 579 | 153 | 27 | 0 | 0 | 1831 |
| 1994 | 0 | 0 | 21 | 55 | 135 | 470 | 452 | 424 | 212 | 82 | 8 | 0 | 1859 |
| 1995 | 0 | 0 | 24 | 26 | 97 | 282 | 545 | 618 | 252 | 50 | 0 | 0 | 1894 |
| 1996 | 0 | 6 | 0 | 37 | 273 | 410 | 522 | 463 | 183 | 49 | 0 | 0 | 1943 |
| 1997 | 3 | 0 | 9 | 10 | 106 | 321 | 524 | 420 | 278 | 108 | 0 | 0 | 1779 |
| 1998 | 0 | 0 | 18 | 16 | 264 | 459 | 640 | 598 | 480 | 53 | 5 | 2 | 2535 |
| 1999 | 0 | 2 | 0 | 35 | 118 | 318 | 607 | 616 | 190 | 57 | 25 | 0 | 1968 |
| 2000 | 0 | 0 | 2 | 26 | 211 | 290 | 517 | 684 | 353 | 143 | 1 | 0 | 2227 |
| 2001 | 0 | 2 | 0 | 139 | 197 | 401 | 700 | 630 | 236 | 56 | 10 | 3 | 2374 |
| 2002 | 0 | 0 | 1 | 75 | 132 | 405 | 557 | 563 | 348 | 60 | 8 | 0 | 2149 |
| 2003 | 0 | 0 | 5 | 71 | 170 | 316 | 645 | 608 | 165 | 57 | 13 | 0 | 2050 |
| 2004 | 3 | 0 | 14 | 42 | 269 | 309 | 444 | 381 | 302 | 81 | 0 | 0 | 1845 |
| 2005 | 0 | 0 | 6 | 40 | 198 | 443 | 552 | 564 | 381 | 108 | 28 | 0 | 2320 |
| 2006 | 0 | 1 | 35 | 142 | 241 | 404 | 617 | 655 | 201 | 80 | 4 | 1 | 2381 |
| 2007 | 0 | 0 | 62 | 54 | 188 | 367 | 511 | 639 | 314 | 111 | 15 | 0 | 2261 |
| 2008 | 4 | 6 | 3 | 32 | 207 | 405 | 591 | 496 | 205 | 51 | 12 | 0 | 2012 |
| 2009 | 0 | 2 | 28 | 59 | 136 | 490 | 503 | 453 | 194 | 9 | 8 | 0 | 1882 |

### SNOWFALL (inches)  2009  TULSA (KTUL)

| YEAR | JUL | AUG | SEP | OCT | NOV | DEC | JAN | FEB | MAR | APR | MAY | JUN | TOTAL |
|------|-----|-----|-----|-----|-----|-----|-----|-----|-----|-----|-----|-----|-------|
| 1980-81 | 0.0 | 0.0 | 0.0 | 0.0 | T | 0.0 | T | 0.9 | T | 0.0 | 0.0 | 0.0 | 0.9 |
| 1981-82 | 0.0 | 0.0 | 0.0 | 0.0 | 0.0 | T | 0.3 | 5.6 | T | 0.0 | 0.0 | 0.0 | 5.9 |
| 1982-83 | 0.0 | 0.0 | 0.0 | 0.0 | T | T | 3.8 | 1.4 | T | 0.0 | 0.0 | 0.0 | 5.2 |
| 1983-84 | 0.0 | 0.0 | 0.0 | 0.0 | T | 3.0 | 4.6 | 0.2 | T | 0.0 | 0.0 | 0.0 | 7.8 |
| 1984-85 | 0.0 | 0.0 | 0.0 | 0.0 | 0.0 | 6.6 | 3.3 | 4.3 | 0.0 | 0.0 | 0.0 | 0.0 | 14.2 |
| 1985-86 | 0.0 | 0.0 | 0.0 | 0.0 | T | 2.5 | 0.0 | 4.9 | 0.0 | 0.0 | 0.0 | 0.0 | 7.4 |
| 1986-87 | 0.0 | 0.0 | 0.0 | 0.0 | 0.0 | 0.0 | 8.7 | 4.6 | 0.0 | 0.0 | 0.0 | 0.0 | 13.3 |
| 1987-88 | 0.0 | 0.0 | 0.0 | 0.0 | T | 6.7 | 11.0 | T | 0.5 | 0.0 | 0.0 | 0.0 | 18.2 |
| 1988-89 | 0.0 | 0.0 | 0.0 | 0.0 | 0.4 | 2.7 | 3.4 | 0.3 | 9.7 | 0.0 | 0.0 | 0.0 | 16.5 |
| 1989-90 | 0.0 | 0.0 | 0.0 | 0.0 | 0.0 | 2.0 | 0.0 | 0.0 | 0.2 | 0.0 | 0.0 | 0.0 | 2.2 |
| 1990-91 | 0.0 | 0.0 | 0.0 | 0.0 | 0.0 | 4.6 | T | 1.4 | T | 0.0 | T | 0.0 | 6.0 |
| 1991-92 | 0.0 | 0.0 | 0.0 | 0.0 | 0.2 | 0.1 | 0.8 | 0.0 | T | T | 0.0 | 0.0 | 1.1 |
| 1992-93 | 0.0 | 0.0 | 0.0 | 0.0 | 3.5 | 1.1 | 0.8 | 6.7 | T | T | 0.0 | 0.0 | 12.1 |
| 1993-94 | 0.0 | 0.0 | 0.0 | 0.3 | 0.0 | 0.0 | T | T | 14.1 | T | 0.0 | T | 14.4 |
| 1994-95 | T | 0.0 | 0.0 | 0.0 | T | 0.0 | 1.8 | T | 6.3 | 0.0 | 0.0 | 0.0 | 8.1 |
| 1995-96 | 0.0 | 0.0 | 0.0 | 0.0 | 1.8 | T | 1.0 | 5.0 | 0.0 | 0.0 | 0.0 | 0.0 | 7.8 |
| 1996-97 | 0.0 | 0.0 | 0.0 | T | T | T | 5.9 | 0.3 | 0.0 | T | 0.0 | 0.0 | 6.2 |
| 1997-98 | T | 0.0 | 0.0 | 0.0 | T | 0.6 | 4.0 | 0.0 | T | 0.0 | 0.0 | 0.0 | 4.6 |
| 1998-99 | 0.0 | 0.0 | 0.0 | 0.0 | 0.0 | T | 3.3 | 0.0 | 5.9 | 0.0 | 0.0 | 0.0 | 9.2 |
| 1999-00 | 0.0 | 0.0 | 0.0 | 0.0 | 0.0 | T | 7.1 | 0.0 | 2.2 | 0.0 | 0.0 | 0.0 | 9.3 |
| 2000-01 | 0.0 | 0.0 | 0.0 | T | 2.1 | 11.4 | 1.4 | T | 0.0 | 0.0 | 0.0 | 0.0 | 14.9 |
| 2001-02 | 0.0 | 0.0 | 0.0 | 0.0 | 3.0 | T | T | 1.0 | 6.4 | 0.0 | 0.0 | 0.0 | 10.4 |
| 2002-03 | 0.0 | 0.0 | 0.0 | 0.0 | 0.0 | 8.9 | 1.0 | 10.5 | 0.0 | 0.0 | 0.0 | 0.0 | 20.4 |
| 2003-04 | 0.0 | 0.0 | 0.0 | 0.0 | T | 5.2 | 1.0 | 1.6 | 0.0 | T | 0.0 | 0.0 | 7.8 |
| 2004-05 | 0.0 | 0.0 | 0.0 | 0.0 | T | T | 2.0 | T | 0.0 | T | 0.0 | 0.0 | 2.0 |
| 2005-06 | 0.0 | 0.0 | 0.0 | 0.0 | T | 2.2 | 1.2 | 2.1 | 1.5 | T | T | 0.0 | 7.0 |
| 2006-07 | 0.0 | 0.0 | 0.0 | 0.0 | 10.5 | 0.3 | 4.3 | 0.1 | T | 0.4 | 0.0 | 0.0 | 15.6 |
| 2007-08 | 0.0 | 0.0 | 0.0 | T | T | 0.5 | 2.4 | 0.3 | 0.3 | 0.0 | 0.0 | 0.0 | 3.5 |
| 2008-09 | 0.0 | 0.0 | 0.0 | 0.0 | T | 0.4 | 1.6 | 0.0 | 10.4 | T | T | T | 12.4 |
| 2009- | 0.0 | 0.0 | 0.0 | 0.0 | T | 7.4 | | | | | | | |
| POR= 66 YRS | T | 0.0 | 0.0 | T | 0.6 | 1.8 | 3.0 | 2.2 | 1.8 | T | T | T | 9.4 |

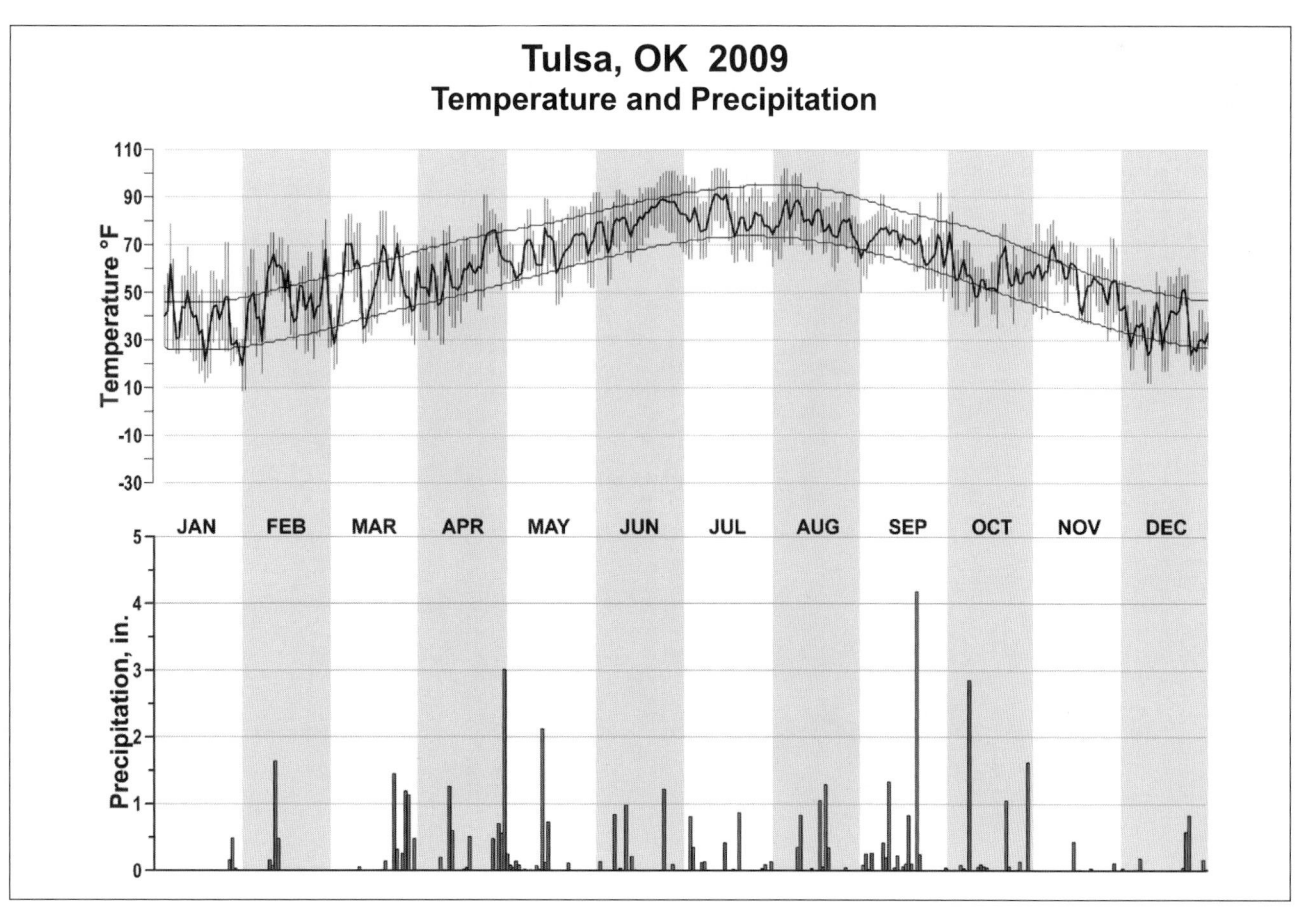

## Tulsa, OK  2009
### Temperature and Precipitation

# 2009
# BURNS
# OREGON (KBNO)

Burns is located near the center of the high plateau area that comprises much of central Oregon. The crest of the Cascade Range is some 135 miles to the west, the Steens Mountains about 45 miles to the southeast. A prong of the Blue Mountains to the north approaches within 40 miles of Burns with lower hills reaching the city limits. Approximately 30 miles east, a number of low hills separate this area from the Malheur Valley. From the Blue Mountains in the north to the southern border of Oregon and between the foothills of the Cascades in the west to the chain of lesser mountains to the east is a series of shallow valleys, each with its own small creek or creeks. These are separated by slightly higher, gently rolling bench lands which in turn are cut here and there by rough, rocky canyons, and with numerous buttes or small mesas interspersed through the area, rising from 500 to 1,500 feet above the general terrain. Well distributed over much of central Oregon south of Burns are a number of large, relatively shallow landlocked lakes and marshes.

Burns, in common with most of the Great Basin area of the Western Plateau, has a semi-arid climate. Maritime air moving in from the Pacific Ocean is greatly modified by its passage over the Coastal and Cascade Mountain Ranges that lie between Burns and the coast. In its ascent over them much of its precipitable moisture has been given up and annual precipitation totals here are small and humidities generally low. This makes for an abundance of sunshine and a rather wide range between daily maximum and minimum temperatures. Nighttime frosts may occur any month in the year. An occasional continental air mass moving from northern Canada southward through this area,

instead of in its usual trajectory along the east slope of the Rocky Mountains, will produce even lower humidities along with fairly strong winds and more extreme temperatures. The average total precipitation for the entire high plateau region of Oregon is between 9 and 12 inches. About one-third of this falls in the form of snow. There are several thunderstorms each year. Occasionally these are accompanied by hail, but seldom do either cause appreciable damage. No tornadoes have ever been reported in the immediate area of Burns, but funnel clouds have been sighted.

Approximately 90 percent of the agricultural income in Harney County, of which Burns is the county seat, is derived from beef cattle and sheep, with far the greater portion coming from cattle. The very low nightly temperatures during much of the year necessitate the raising of only very hardy, frost-resistant crops with a short growing season. The low annual precipitation total limits cultivation to valley bottom lands. Through the use of spring runoff water and some reservoir storage for summer irrigation, about 120,000 acres of native grasslands have been developed into native grass hay meadows and about 40,000 additional acres are in cultivated crops, principally the small hardy grains, with some 6,000 acres of alfalfa. The rest of the more than 6,000,000 acres in Harney County is in range with about 400,000 acres of the rangeland covered by pine forests along the northern border.

# NORMALS, MEANS, AND EXTREMES
## BURNS (KBNO)

| LATITUDE: | LONGITUDE: | ELEVATION (FT): | TIME ZONE: | WBAN: 94185 |
| --- | --- | --- | --- | --- |
| 43 ° 35'N | -118° 57'W | GRND: 4140   BARO: 4148 | PACIFIC    (UTC -8) | |

| ELEMENT | POR | JAN | FEB | MAR | APR | MAY | JUN | JUL | AUG | SEP | OCT | NOV | DEC | YEAR |
| --- | --- | --- | --- | --- | --- | --- | --- | --- | --- | --- | --- | --- | --- | --- |
| NORMAL DAILY MAXIMUM | 30 | 34.7 | 40.5 | 49.0 | 57.4 | 66.1 | 75.1 | 85.4 | 84.5 | 75.0 | 62.4 | 44.8 | 35.1 | 59.2 |
| MEAN DAILY MAXIMUM | 26 | 34.4 | 39.1 | 49.2 | 56.9 | 66.5 | 74.7 | 85.8 | 84.7 | 75.0 | 61.6 | 44.8 | 34.5 | 58.9 |
| HIGHEST DAILY MAXIMUM | 25 | 57 | 67 | 74 | 84 | 94 | 100 | 107 | 100 | 97 | 86 | 70 | 57 | 107 |
| YEAR OF OCCURRENCE | | 2003 | 1995 | 2004 | 1987 | 1986 | 2008 | 2002 | 2009 | 1998 | 2001 | 1999 | 1993 | JUL 2002 |
| MEAN OF EXTREME MAXS. | 40 | 47.0 | 51.2 | 64.4 | 75.6 | 85.3 | 90.5 | 97.2 | 96.3 | 90.0 | 79.1 | 61.6 | 48.0 | 73.9 |
| NORMAL DAILY MINIMUM | 30 | 14.0 | 19.4 | 24.9 | 28.6 | 35.6 | 41.1 | 46.4 | 43.9 | 35.0 | 26.4 | 20.6 | 14.6 | 29.2 |
| MEAN DAILY MINIMUM | 26 | 14.8 | 18.3 | 25.5 | 29.2 | 36.7 | 41.3 | 47.4 | 44.7 | 36.1 | .27.4 | 21.8 | 14.5 | 29.8 |
| LOWEST DAILY MINIMUM | 25 | -27 | -28 | -14 | 10 | 13 | 21 | 25 | 22 | 17 | -7 | -15 | -22 | -28 |
| YEAR OF OCCURRENCE | | 1982 | 1985 | 1993 | 1997 | 2008 | 1996 | 1986 | 1999 | 1999 | 2002 | 2006 | 2008 | FEB 1985 |
| MEAN OF EXTREME MINS. | 40 | -5.7 | 0.8 | 12.4 | 18.2 | 22.6 | 28.0 | 35.6 | 32.6 | 22.9 | 12.7 | 4.7 | -4.7 | 15.0 |
| NORMAL DRY BULB | 30 | 24.4 | 30.0 | 37.0 | 43.0 | 50.9 | 58.1 | 65.9 | 64.2 | 55.0 | 44.4 | 32.7 | 24.9 | 44.2 |
| MEAN DRY BULB | 40 | 25.0 | 30.0 | 37.2 | 42.7 | 51.8 | 59.1 | 67.5 | 65.5 | 56.3 | 45.3 | 34.1 | 25.5 | 45.0 |
| MEAN WET BULB | 22 | 23.5 | 26.7 | 32.7 | 36.4 | 43.1 | 47.9 | 51.5 | 49.5 | 43.0 | 36.5 | 30.0 | 23.5 | 37.0 |
| MEAN DEW POINT | 22 | 21.4 | 23.7 | 27.7 | 30.3 | 36.1 | 40.0 | 42.4 | 39.4 | 34.4 | 29.4 | 26.4 | 21.2 | 31.0 |
| NORMAL NO. DAYS WITH: | | | | | | | | | | | | | | |
| MAXIMUM >= 90 | 30 | 0.0 | 0.0 | 0.0 | 0.0 | 0.2 | 1.8 | 9.2 | 8.3 | 1.3 | 0.0 | 0.0 | 0.0 | 20.8 |
| MAXIMUM <= 32 | 30 | 10.1 | 5.4 | 0.5 | 0.0 | 0.0 | 0.0 | 0.0 | 0.0 | 0.0 | 0.1 | 3.0 | 10.8 | 29.9 |
| MINIMUM <= 32 | 30 | 29.9 | 26.2 | 27.1 | 19.8 | 8.6 | 2.7 | 0.6 | 1.0 | 7.8 | 23.3 | 27.0 | 30.1 | 204.1 |
| MINIMUM <= 0 | 30 | 3.9 | 1.9 | 0.1 | 0.0 | 0.0 | 0.0 | 0.0 | 0.0 | 0.0 | 0.1 | 0.9 | 3.7 | 10.6 |
| NORMAL HEATING DEG. DAYS | 30 | 1259 | 982 | 869 | 661 | 439 | 226 | 89 | 96 | 313 | 638 | 968 | 1245 | 7785 |
| NORMAL COOLING DEG. DAYS | 30 | 0 | 0 | 0 | 0 | 0 | 19 | 117 | 70 | 12 | 0 | 0 | 0 | 218 |
| NORMAL (PERCENT) | 30 | | | | | | | | | | | | | |
| HOUR 04 LST | 30 | | | | | | | | | | | | | |
| HOUR 10 LST | 30 | | 77 | 62 | 50 | 46 | 44 | 35 | 33 | 40 | 51 | 73 | | |
| HOUR 16 LST | 30 | | 65 | 48 | 38 | 35 | | | | 27 | | | | |
| HOUR 22 LST | 30 | | | | | | | | | | | | | |
| PERCENT POSSIBLE SUNSHINE | | | | | | | | | | | | | | |
| MEAN NO. DAYS WITH: | | | | | | | | | | | | | | |
| HEAVY FOG(VISBY <= 1/4 MI) | 22 | 6.1 | 2.9 | 0.7 | 0.6 | 0.3 | 0.2 | 0.1 | 0.1 | 0.2 | 1.1 | 2.5 | 4.8 | 19.6 |
| THUNDERSTORMS | 22 | 0.0 | 0.0 | 0.1 | 0.3 | 0.6 | 0.8 | 0.4 | 0.7 | 0.1 | 0.0 | 0.0 | 0.0 | 3.0 |
| MEAN: | | | | | | | | | | | | | | |
| SUNRISE-SUNSET (OKTAS) | | | | | | | | | | | | | | |
| MIDNIGHT-MIDNIGHT (OKTAS) | | | | | | | | | | | | | | |
| MEAN NO. DAYS WITH: | | | | | | | | | | | | | | |
| CLEAR | | | | | | | | | | | | | | |
| PARTLY CLOUDY | | | | | | | | | | | | | | |
| CLOUDY | | | | | | | | | | | | | | |
| MEAN STATION PRESSURE(IN) | 22 | 26.00 | 25.78 | 25.77 | 25.76 | 25.76 | 25.77 | 25.81 | 25.81 | 25.81 | 25.83 | 25.83 | 25.83 | 25.81 |
| MEAN SEA-LEVEL PRES. (IN) | 22 | 30.22 | 30.11 | 30.05 | 30.01 | 29.97 | 29.96 | 29.97 | 29.98 | 30.02 | 30.09 | 30.15 | 30.19 | 30.06 |
| MEAN SPEED (MPH) | 22 | 4.9 | 5.8 | 7.0 | 8.1 | 7.7 | 7.1 | 6.5 | 6.3 | 6.2 | 6.0 | 5.6 | 5.0 | 6.4 |
| PREVAIL.DIR(TENS OF DEGS) | 16 | 08 | 08 | 31 | 31 | 31 | 31 | 30 | 30 | 31 | 31 | 31 | 08 | 31 |
| MAXIMUM 2-MINUTE: | | | | | | | | | | | | | | |
| SPEED (MPH) | 14 | 44 | 40 | 40 | 45 | 41 | 35 | 44 | 37 | 41 | 41 | 39 | 40 | 45 |
| DIR. (TENS OF DEGS) | | 26 | 24 | 25 | 31 | 20 | 24 | 18 | 22 | 32 | 27 | 20 | 25 | 31 |
| YEAR OF OCCURRENCE | | 2004 | 2000 | 2009 | 2009 | 2000 | 2006 | 2002 | 2007 | 2003 | 2004 | 2009 | 2000 | APR 2009 |
| MAXIMUM 3-SECOND | | | | | | | | | | | | | | |
| SPEED (MPH) | 14 | 51 | 47 | 49 | 55 | 48 | 45 | 52 | 59 | 52 | 54 | 51 | 48 | 59 |
| DIR. (TENS OF DEGS) | | 26 | 24 | 26 | 31 | 30 | 22 | 13 | 11 | 31 | 28 | 19 | 24 | 11 |
| YEAR OF OCCURRENCE | | 2004 | 2000 | 2009 | 2009 | 2008 | 1997 | 2006 | 2006 | 2003 | 2009 | 2009 | 2000 | AUG 2006 |
| NORMAL (IN) | 30 | 1.18 | 1.11 | 1.24 | 0.85 | 1.05 | 0.66 | 0.40 | 0.45 | 0.50 | 0.72 | 1.11 | 1.30 | 10.57 |
| MAXIMUM MONTHLY (IN) | 25 | 2.84 | 3.50 | 3.66 | 2.21 | 3.26 | 3.05 | 1.22 | 1.16 | 2.13 | 1.72 | 2.73 | 4.45 | 4.45 |
| YEAR OF OCCURRENCE | | 1995 | 1986 | 1983 | 1993 | 2005 | 1992 | 1987 | 1984 | 1986 | 2000 | 1984 | 2005 | DEC 2005 |
| MINIMUM MONTHLY (IN) | 25 | 0.09 | 0.12 | 0.23 | 0.12 | 0.10 | 0.11 | T | 0.00 | 0.00 | 0.03 | 0.23 | 0.28 | 0.00 |
| YEAR OF OCCURRENCE | | 1985 | 1997 | 1997 | 1986 | 1992 | 1985 | 1994 | 1994 | 1999 | 1998 | 1993 | 1991 | SEP 1999 |
| MAXIMUM IN 24 HOURS (IN) | 25 | 0.68 | 1.04 | 0.99 | 0.55 | 1.07 | 1.15 | 0.89 | 1.01 | 0.92 | 0.87 | 0.69 | 5.31 | 5.31 |
| YEAR OF OCCURRENCE | | 1997 | 1986 | 1996 | 1993 | 2009 | 1992 | 1985 | 1984 | 2000 | 1982 | 1981 | 1995 | DEC 1995 |
| NORMAL NO. DAYS WITH: | | | | | | | | | | | | | | |
| PRECIPITATION >= 0.01 | 30 | 10.3 | 10.2 | 11.6 | 8.8 | 9.4 | 6.2 | 2.9 | 3.1 | 3.8 | 5.9 | 11.5 | 10.1 | 93.8 |
| PRECIPITATION >= 1.00 | 30 | 0.0 | 0.0 | 0.0 | 0.0 | 0.0 | 0.1 | 0.0 | 0.0 | 0.0 | 0.0 | 0.0 | 0.0 | 0.1 |
| NORMAL (IN) | 30 | 7.3 | 6.7 | 4.3 | 0.7 | 0.2 | 0.* | 0.0 | 0.0 | 0.* | 0.7 | 5.9 | 9.6 | 35.4 |
| MAXIMUM MONTHLY (IN) | 11 | 23.7 | 27.6 | 13.5 | 3.2 | 1.4 | 0.6 | 0.0 | 0.0 | T | 3.6 | 17.4 | 26.8 | 27.6 |
| YEAR OF OCCURRENCE | | 1993 | 1986 | 1985 | 1982 | 1983 | 1981 | | | 1986 | 1984 | 1983 | 1983 | FEB 1986 |
| MAXIMUM IN 24 HOURS (IN) | 11 | 6.2 | 8.5 | 6.4 | 1.7 | 0.9 | 0.6 | 0.0 | 0.0 | T | 5.0 | 7.0 | 6.5 | 8.5 |
| YEAR OF OCCURRENCE' | | 1987 | 1986 | 1985 | 1980 | 1986 | 1981 | | | 1986 | 1991 | 1983 | 1981 | FEB 1986 |
| MAXIMUM SNOW DEPTH (IN) | 22 | 23 | 26 | 26 | 4 | 0 | 0 | 0 | 0 | 0 | 5 | 8 | 18 | 26 |
| YEAR OF OCCURRENCE | | 1993 | 1993 | 1993 | 1999 | | | | | | 1991 | 1985 | 2008 | MAR 1993 |
| NORMAL NO. DAYS WITH: | | | | | | | | | | | | | | |
| SNOWFALL >= 1.0 | 30 | 2.5 | 2.2 | 1.1 | 0.2 | 0.0 | 0.0 | 0.0 | 0.0 | 0.0 | 0.1 | 2.0 | 3.4 | 11.5 |

## PRECIPITATION (inches) 2009 BURNS (KBNO)

| YEAR | JAN | FEB | MAR | APR | MAY | JUN | JUL | AUG | SEP | OCT | NOV | DEC | ANNUAL |
|------|-----|-----|-----|-----|-----|-----|-----|-----|-----|-----|-----|-----|--------|
| 1979 | 2.96 | 1.74 | 0.99 | 0.82 | 0.81 | 0.21 | 0.03 | 1.99 | 0.56 | 1.03 | 1.93 | 1.62 | 14.69 |
| 1980 | 1.58 | 2.01 | 0.61 | 0.57 | 1.08 | 0.85 | 0.26 | 0.03 | 0.56 | 0.88 | 0.59 | 1.36 | 10.38 |
| 1981 | 0.79 | 0.91 | 1.62 | 0.84 | 2.14 | 0.47 | 0.23 | 0.18 | 0.54 | 1.34 | 2.62 | 3.88 | 15.56 |
| 1982 | 1.15 | 1.70 | 0.78 | 0.90 | 0.64 | 0.90 | 0.41 | 0.35 | 1.06 | 1.43 | 1.09 | 2.56 | 12.97 |
| 1983 | 1.04 | 2.13 | 3.66 | 1.10 | 1.54 | 0.31 | 1.09 | 0.70 | 0.06 | 1.08 | 2.44 | 3.09 | 18.24 |
| 1984 | 0.21 | 0.76 | 1.71 | 0.80 | 1.04 | 0.88 | 0.49 | 1.16 | 0.05 | 1.02 | 2.73 | 0.87 | 11.72 |
| 1985 | 0.09 | 0.49 | 1.03 | 0.34 | 1.37 | 0.11 | 0.92 | 0.09 | 1.12 | 0.81 | 0.97 | 0.73 | 8.07 |
| 1986 | 1.41 | 3.50 | 1.33 | 0.12 | 0.53 | 0.26 | 0.37 | 0.04 | 2.13 | 0.34 | 0.30 | 0.28 | 10.61 |
| 1987 | 1.30 | 1.00 | 1.54 | 0.75 | 0.56 | 1.50 | 1.22 | 0.16 | | | | | |
| 1988 | | | | | | | | | | | | | |
| 1989 | | | | | | | | | | | | | |
| 1991 | | | | | 1.91 | 0.73 | T | 0.38 | 0.06 | 1.10 | 1.07 | 0.28 | |
| 1992 | 0.18 | 1.03 | 0.49 | 0.66 | 0.10 | 3.05 | 0.30 | 0.34 | 0.09 | 1.39 | 0.90 | 1.71 | 10.24 |
| 1993 | 2.08 | 1.17 | 1.41 | 2.21 | 0.93 | 0.70 | 0.55 | 0.51 | 0.03 | 0.87 | 0.23 | 0.95 | 11.64 |
| 1994 | 0.39 | 0.74 | 0.37 | 0.51 | 1.23 | 0.52 | T | 0.00 | 0.23 | 0.34 | 1.50 | 0.68 | 6.51 |
| 1995 | 2.84 | 0.47 | 1.68 | 1.39 | | | | 0.83 | T | 0.21 | 0.65 | 2.53 | |
| 1996 | 1.85 | | 1.50 | | 1.69 | 0.66 | | 0.34 | 0.72 | 0.92 | 1.13 | 3.07 | |
| 1997 | 2.46 | 0.12 | 0.23 | 1.03 | 1.47 | 0.29 | 0.68 | 0.17 | 0.61 | 0.68 | 1.08 | 0.63 | 9.45 |
| 1998 | 2.29 | 2.89 | 1.35 | 1.03 | 2.84 | 1.04 | 0.31 | 0.01 | 1.03 | 0.03 | 2.28 | 1.10 | 16.20 |
| 1999 | 1.65 | 1.94 | 0.72 | 0.32 | 0.42 | 0.20 | 0.08 | 0.56 | 0.00 | 0.37 | 0.44 | 0.57 | 7.27 |
| 2000 | 1.63 | 1.89 | 0.77 | 0.80 | 0.28 | 0.18 | 0.96 | T | 1.16 | 1.72 | 0.63 | 0.47 | 10.49 |
| 2001 | 0.33 | 0.39 | 0.64 | 0.53 | 0.63 | 0.71 | 1.07 | T | 0.95 | 0.61 | 1.42 | 1.05 | 8.33 |
| 2002 | 0.88 | 0.39 | 0.32 | 1.00 | 0.46 | 0.40 | 0.07 | 0.07 | 0.05 | 0.05 | 0.27 | 1.97 | 5.93 |
| 2003 | 1.24 | 0.14 | 1.45 | 1.55 | 1.16 | 0.11 | 0.50 | 0.26 | 0.60 | 0.12 | 1.06 | 1.56 | 9.75 |
| 2004 | 1.46 | 1.26 | 0.37 | 0.52 | 1.22 | 0.47 | 0.16 | 0.91 | 0.27 | 1.67 | 0.38 | 1.80 | 10.49 |
| 2005 | 0.60 | 0.39 | 1.24 | 1.69 | 3.26 | 0.73 | 0.65 | T | 0.17 | 1.04 | 2.16 | 4.45 | 16.38 |
| 2006 | 2.13 | 0.35 | 1.05 | 1.32 | 2.29 | 0.72 | 0.66 | 0.11 | 0.33 | 0.77 | 1.31 | 1.38 | 12.42 |
| 2007 | 0.14 | 1.58 | 0.41 | 1.27 | 0.30 | 0.80 | 0.01 | 0.57 | 0.25 | 0.91 | 1.18 | 1.32 | 8.74 |
| 2008 | 1.52 | 0.83 | 0.55 | 0.31 | 1.07 | 0.34 | 0.19 | 0.16 | 0.09 | 0.67 | 0.79 | 1.63 | 8.15 |
| 2009 | 0.41 | 0.67 | 0.57 | 0.56 | 2.55 | 2.70 | 0.08 | 0.32 | 0.07 | 1.07 | 0.52 | 1.46 | 10.98 |
| POR= 40 YRS | 1.49 | 1.09 | 1.13 | 0.81 | 1.07 | 0.67 | 0.39 | 0.39 | 0.45 | 0.76 | 1.28 | 1.62 | 11.15 |

WBAN : 94185

## AVERAGE TEMPERATURE (°F) 2009 BURNS (KBNO)

| YEAR | JAN | FEB | MAR | APR | MAY | JUN | JUL | AUG | SEP | OCT | NOV | DEC | ANNUAL |
|------|-----|-----|-----|-----|-----|-----|-----|-----|-----|-----|-----|-----|--------|
| 1979 | 15.8 | 31.2 | 39.2 | 44.2 | 53.6 | 61.4 | 69.3 | 66.8 | 62.3 | 51.0 | 32.0 | 29.8 | 46.4 |
| 1980 | 25.2 | 35.1 | 35.1 | 45.2 | 50.9 | 54.2 | 65.8 | 61.0 | 55.6 | 44.5 | 35.8 | 29.0 | 44.8 |
| 1981 | 30.1 | 29.2 | 37.6 | 44.6 | 49.2 | 56.6 | 64.1 | 66.5 | 55.5 | 42.3 | 35.6 | 29.1 | 45.0 |
| 1982 | 17.6 | 24.7 | 35.3 | 39.1 | 49.1 | 57.8 | 64.9 | 64.3 | 52.9 | 43.5 | 29.0 | 22.8 | 41.8 |
| 1983 | 28.7 | 33.4 | 39.1 | 40.9 | 51.6 | 55.0 | 61.1 | 64.6 | 53.9 | 47.2 | 35.7 | 21.7 | 44.4 |
| 1984 | 17.0 | 23.1 | 35.8 | 40.7 | 50.0 | 54.5 | 67.0 | 64.9 | 52.7 | 39.7 | 33.6 | 19.0 | 41.5 |
| 1985 | 16.9 | 23.0 | 31.3 | 46.3 | 51.1 | 59.1 | 70.7 | 61.4 | 48.9 | 41.8 | 20.9 | 11.9 | 40.3 |
| 1986 | 27.4 | 33.5 | 42.3 | 42.6 | 51.8 | 63.1 | 61.6 | 69.3 | 50.8 | 46.9 | 36.3 | 26.7 | 46.0 |
| 1987 | 19.8 | 32.7 | 38.8 | 49.0 | 54.1 | 61.1 | 63.2 | 65.1 | | | | | |
| 1988 | | | | | | | | | | | | | |
| 1989 | | | | | | | | | | | | | |
| 1991 | | | | | 47.8 | 54.5 | 68.4 | 67.0 | 59.2 | 45.0 | 35.5 | 28.8 | |
| 1992 | 27.7 | 37.1 | 41.8 | 46.6 | 57.3 | 62.4 | 64.7 | 66.0 | 55.9 | 47.4 | 30.1 | 19.8 | 46.4 |
| 1993 | 14.8 | 18.4 | 33.4 | 41.7 | 54.3 | 54.4 | 56.7 | 60.3 | 56.3 | 46.6 | 27.2 | 28.5 | 41.1 |
| 1994 | 31.8 | 29.0 | 39.8 | 44.4 | 53.4 | 58.6 | 69.9 | 66.2 | 58.7 | 43.2 | 26.1 | 22.8 | 45.3 |
| 1995 | 30.1 | 39.2 | 37.8 | 41.5 | | | | 60.5 | 56.4 | 41.4 | 38.6 | 28.4 | |
| 1996 | 26.9 | | 39.0 | | 48.4 | 57.9 | | 64.6 | 52.8 | | 35.9 | 29.7 | |
| 1997 | 27.6 | 31.6 | 39.0 | 41.8 | 54.8 | 57.3 | 64.0 | 64.7 | 56.8 | 43.3 | 36.3 | 24.6 | 45.2 |
| 1998 | 31.5 | 29.7 | 37.8 | 42.9 | 48.1 | 55.8 | 68.9 | 66.8 | 60.1 | 42.5 | 34.9 | 22.9 | 45.2 |
| 1999 | 28.0 | 25.9 | 34.0 | 41.0 | 48.2 | 57.2 | 62.5 | 63.5 | 54.2 | 45.2 | 39.3 | 26.4 | 43.8 |
| 2000 | 27.2 | 35.0 | 36.9 | 47.4 | 51.5 | 59.8 | 65.8 | 65.3 | 54.3 | 44.5 | 28.7 | 24.2 | 45.1 |
| 2001 | 19.7 | 24.0 | 38.2 | 39.9 | 53.5 | 57.0 | 65.2 | 68.4 | 58.4 | 45.4 | 34.4 | 22.5 | 43.9 |
| 2002 | 22.9 | 23.8 | 33.3 | 43.6 | 49.9 | 60.5 | 70.2 | 61.3 | 55.6 | 41.0 | 34.3 | 29.7 | 43.8 |
| 2003 | 33.2 | 30.8 | 40.1 | 40.1 | 51.6 | 60.7 | 71.5 | 66.3 | 58.5 | 48.9 | 31.1 | 30.2 | 46.9 |
| 2004 | 22.6 | 23.7 | 40.3 | 46.5 | 50.4 | 60.3 | 68.3 | 66.2 | 54.3 | 45.7 | 35.0 | 30.4 | 45.3 |
| 2005 | 23.3 | 29.9 | 40.3 | 43.5 | 52.9 | 55.4 | 68.3 | 66.9 | 54.5 | 45.9 | 34.1 | 21.8 | 44.7 |
| 2006 | 26.5 | 25.8 | 34.8 | 44.6 | 54.5 | 62.5 | 72.6 | 64.6 | 56.9 | 44.1 | 35.7 | 25.4 | 45.7 |
| 2007 | 24.3 | 32.8 | 41.0 | 43.1 | 52.9 | 59.7 | 72.2 | 65.3 | 54.3 | 42.5 | 34.4 | 20.4 | 45.2 |
| 2008 | 16.8 | 24.3 | 31.7 | 37.9 | 51.5 | 57.5 | 67.7 | 65.3 | 56.5 | 45.6 | 37.6 | 22.5 | 42.9 |
| 2009 | 28.4 | 28.5 | 35.6 | 41.3 | 53.3 | 58.8 | 68.4 | 65.0 | 59.8 | 41.2 | 31.8 | 20.4 | 44.4 |
| POR= 40 YRS | 25.0 | 30.0 | 37.2 | 42.7 | 51.8 | 59.1 | 67.5 | 65.5 | 56.3 | 45.3 | 34.1 | 25.5 | 45.0 |

## HEATING DEGREE DAYS (base 65°F) 2009 BURNS (KBNO)

| YEAR | JUL | AUG | SEP | OCT | NOV | DEC | JAN | FEB | MAR | APR | MAY | JUN | TOTAL |
|------|-----|-----|-----|-----|-----|-----|-----|-----|-----|-----|-----|-----|-------|
| 1980-81 | 52 | 130 | 274 | 627 | 868 | 1110 | 1074 | 996 | 844 | 607 | 487 | 255 | 7324 |
| 1981-82 | 80 | 46 | 287 | 696 | 875 | 1106 | 1461 | 1123 | 914 | 770 | 484 | 224 | 8066 |
| 1982-83 | 86 | 76 | 360 | 659 | 1075 | 1303 | 1116 | 882 | 793 | 720 | 424 | 294 | 7788 |
| 1983-84 | 154 | 76 | 328 | 544 | 872 | 1339 | 1481 | 1210 | 898 | 719 | 464 | 324 | 8409 |
| 1984-85 | 21 | 45 | 363 | 778 | 937 | 1421 | 1487 | 1170 | 1037 | 555 | 421 | 190 | 8425 |
| 1985-86 | 14 | 131 | 475 | 715 | 1314 | 1639 | 1155 | 873 | 697 | 665 | 448 | 96 | 8222 |
| 1986-87 | 124 | 19 | 431 | 552 | 853 | 1181 | 1396 | 900 | 806 | 473 | 332 | 158 | 7225 |
| 1987-88 | 93 | 56 | | | | | | | | | | | |
| 1988-89 | | | | | | | | | | | | | |
| 1989-90 | | | | | | | | | | | | | |
| 1990-91 | | | | | | | | | | | 529 | 312 | |
| 1991-92 | 16 | 25 | 177 | 615 | 879 | 1114 | 1149 | 802 | 712 | 547 | 246 | 136 | 6418 |
| 1992-93 | 80 | 88 | 270 | 539 | 1040 | 1395 | 1544 | 1301 | 972 | 693 | 329 | 317 | 8568 |
| 1993-94 | 252 | 166 | 262 | 564 | 1126 | 1123 | 1019 | 1002 | 776 | 611 | 353 | 211 | 7465 |
| 1994-95 | 33 | 31 | 184 | 667 | 1161 | 1302 | 1071 | 718 | 839 | 703 | | | |
| 1995-96 | | 163 | 263 | 727 | 788 | 1127 | 1174 | | 800 | | 507 | 214 | |
| 1996-97 | | 96 | 361 | | 868 | 1090 | 1155 | 926 | 799 | 688 | 313 | 225 | |
| 1997-98 | 87 | 56 | 245 | 662 | 853 | 1246 | 1031 | 982 | 832 | 657 | 516 | 269 | 7436 |
| 1998-99 | 3 | 40 | 182 | 692 | 896 | 1297 | 1140 | 1088 | 953 | 714 | 512 | 240 | 7757 |
| 1999-00 | 125 | 102 | 315 | 608 | 763 | 1190 | 1163 | 865 | 864 | 519 | 411 | 165 | 7090 |
| 2000-01 | 59 | 65 | 323 | 628 | 1082 | 1254 | 1396 | 1140 | 824 | 747 | 360 | 257 | 8135 |
| 2001-02 | 76 | 15 | 194 | 598 | 912 | 1310 | 1296 | 1149 | 974 | 636 | 468 | 178 | 7806 |
| 2002-03 | 9 | 128 | 278 | 736 | 913 | 1084 | 979 | 952 | 765 | 741 | 421 | 147 | 7153 |
| 2003-04 | 9 | 28 | 213 | 491 | 1010 | 1072 | 1310 | 1193 | 760 | 548 | 446 | 167 | 7247 |
| 2004-05 | 17 | 70 | 315 | 590 | 894 | 1066 | 1286 | 978 | 759 | 640 | 367 | 287 | 7269 |
| 2005-06 | 26 | 44 | 312 | 582 | 917 | 1331 | 1187 | 1093 | 930 | 604 | 332 | 111 | 7469 |
| 2006-07 | 5 | 73 | 257 | 642 | 873 | 1220 | 1254 | 896 | 735 | 652 | 367 | 170 | 7144 |
| 2007-08 | 8 | 46 | 320 | 691 | 913 | 1377 | 1488 | 1172 | 1025 | 807 | 420 | 241 | 8508 |
| 2008-09 | 15 | 78 | 247 | 595 | 812 | 1310 | 1129 | 1018 | 904 | 703 | 365 | 186 | 7362 |
| 2009- | 32 | 77 | 161 | 731 | 989 | 1376 | | | | | | | |

WBAN : 94185

## COOLING DEGREE DAYS (base 65°F) 2009 BURNS (KBNO)

| YEAR | JAN | FEB | MAR | APR | MAY | JUN | JUL | AUG | SEP | OCT | NOV | DEC | TOTAL |
|------|-----|-----|-----|-----|-----|-----|-----|-----|-----|-----|-----|-----|-------|
| 1979 | 0 | 0 | 0 | 0 | 3 | 48 | 172 | 111 | 27 | 6 | 0 | 0 | 367 |
| 1980 | 0 | 0 | 0 | 0 | 0 | 2 | 81 | 11 | 0 | 0 | 0 | 0 | 94 |
| 1981 | 0 | 0 | 0 | 0 | 0 | 11 | 61 | 102 | 13 | 0 | 0 | 0 | 187 |
| 1982 | 0 | 0 | 0 | 0 | 0 | 13 | 88 | 62 | 3 | 0 | 0 | 0 | 166 |
| 1983 | 0 | 0 | 0 | 0 | 16 | 0 | 41 | 71 | 0 | 0 | 0 | 0 | 128 |
| 1984 | 0 | 0 | 0 | 0 | 3 | 14 | 89 | 48 | 4 | 0 | 0 | 0 | 158 |
| 1985 | 0 | 0 | 0 | 0 | 0 | 18 | 199 | 27 | 0 | 0 | 0 | 0 | 244 |
| 1986 | 0 | 0 | 0 | 0 | 42 | 48 | 25 | 158 | 12 | 0 | 0 | 0 | 285 |
| 1987 | 0 | 0 | 0 | 0 | 2 | 45 | 44 | 70 | | | | | |
| 1988 | | | | | | | | | | | | | |
| 1989 | | | | | | | | | | | | | |
| 1991 | | | | | 0 | 1 | 129 | 95 | 9 | 0 | 0 | 0 | |
| 1992 | 0 | 0 | 0 | 0 | 13 | 66 | 77 | 128 | 4 | 0 | 0 | 0 | 288 |
| 1993 | 0 | 0 | 0 | 0 | 4 | 4 | 2 | 28 | 10 | 0 | 0 | 0 | 48 |
| 1994 | 0 | 0 | 0 | 0 | 0 | 24 | 194 | 74 | 0 | 0 | 0 | 0 | 292 |
| 1995 | 0 | 0 | 0 | 0 | | | | 29 | 10 | 0 | 0 | 0 | |
| 1996 | 0 | 0 | 0 | 0 | 0 | 9 | | 88 | 2 | 0 | 0 | 0 | |
| 1997 | 0 | 0 | 0 | 0 | 3 | 4 | 64 | 52 | 6 | 0 | 0 | 0 | 129 |
| 1998 | 0 | 0 | 0 | 0 | 0 | 0 | 131 | 103 | 42 | 0 | 0 | 0 | 276 |
| 1999 | 0 | 0 | 0 | 0 | 0 | 14 | 51 | 64 | 0 | 0 | 0 | 0 | 129 |
| 2000 | 0 | 0 | 0 | 0 | 0 | 17 | 91 | 83 | 9 | 0 | 0 | 0 | 200 |
| 2001 | 0 | 0 | 0 | 0 | 10 | 23 | 90 | 127 | 2 | 0 | 0 | 0 | 252 |
| 2002 | 0 | 0 | 0 | 0 | 5 | 49 | 175 | 23 | 3 | 0 | 0 | 0 | 255 |
| 2003 | 0 | 0 | 0 | 0 | 11 | 24 | 217 | 74 | 24 | 0 | 0 | 0 | 350 |
| 2004 | 0 | 0 | 0 | 0 | 0 | 34 | 128 | 116 | 0 | 0 | 0 | 0 | 278 |
| 2005 | 0 | 0 | 0 | 0 | 0 | 4 | 135 | 112 | 6 | 0 | 0 | 0 | 257 |
| 2006 | 0 | 0 | 0 | 0 | 12 | 44 | 247 | 67 | 21 | 0 | 0 | 0 | 391 |
| 2007 | 0 | 0 | 0 | 0 | 2 | 21 | 237 | 61 | 6 | 0 | 0 | 0 | 327 |
| 2008 | 0 | 0 | 0 | 0 | 8 | 25 | 105 | 92 | 0 | 0 | 0 | 0 | 230 |
| 2009 | 0 | 0 | 0 | 0 | 10 | 6 | 145 | 85 | 11 | 0 | 0 | 0 | 257 |

### SNOWFALL (inches)  2009  BURNS (KBNO)

| YEAR | JUL | AUG | SEP | OCT | NOV | DEC | JAN | FEB | MAR | APR | MAY | JUN | TOTAL |
|------|-----|-----|-----|-----|-----|-----|-----|-----|-----|-----|-----|-----|-------|
| 1977-78 | 0.0 | 0.0 | 0.0 | T | 14.6 | 12.7 | 13.3 | 9.4 | 3.6 | 5.8 | T | 0.0 | 59.4 |
| 1978-79 | 0.0 | 0.0 | T | 0.0 | 1.2 | 6.0 | 22.0 | 16.8 | 2.5 | 1.2 | 0.9 | 0.0 | 50.6 |
| 1979-80 | 0.0 | 0.0 | 0.0 | T | 11.0 | 13.7 | 5.5 | 7.0 | 4.9 | 1.9 | T | 0.0 | 44.0 |
| 1980-81 | 0.0 | 0.0 | 0.0 | 0.3 | 1.0 | 5.2 | 4.8 | 1.9 | 1.6 | T | T | 0.6 | 15.4 |
| 1981-82 | 0.0 | 0.0 | 0.0 | T | 4.5 | 24.4 | 13.8 | 2.3 | 3.4 | 3.2 | 0.2 | T | 51.8 |
| 1982-83 | 0.0 | 0.0 | 0.0 | T | 2.1 | 16.0 | 7.1 | 7.2 | 8.3 | 0.3 | 1.4 | 0.0 | 42.4 |
| 1983-84 | 0.0 | 0.0 | 0.0 | 0.0 | 17.4 | 26.8 | 4.2 | 9.4 | 2.7 | 1.5 | T | 0.0 | 62.0 |
| 1984-85 | 0.0 | 0.0 | 0.0 | 3.6 | 9.2 | 7.2 | 1.1 | 5.8 | 13.5 | 1.6 | 0.0 | 0.0 | 42.0 |
| 1985-86 | 0.0 | 0.0 | 0.0 | 1.7 | 13.8 | 7.4 | 5.3 | 27.6 | 5.2 | T | 1.3 | 0.0 | 62.3 |
| 1986-87 | 0.0 | 0.0 | T | 0.0 | 1.4 | 4.2 | 12.8 | 6.9 | 3.6 | 0.0 | 0.0 | 0.0 | 28.9 |
| 1987-88 | 0.0 | 0.0 | | | | | | | | | | | |
| 1988-89 | | | | | | | | | | | | | |
| 1989-90 | | | | | | | | | | | | | |
| 1990-91 | | | | | | | | | | | 0.0 | 0.0 | |
| 1991-92 | 0.0 | 0.0 | 0.0 | | 0.6 | 0.4 | 1.9 | 3.2 | T | 0.0 | 0.0 | 0.0 | |
| 1992-93 | 0.0 | 0.0 | 0.0 | 0.0 | 7.7 | 19.9 | 23.7 | 15.3 | 3.7 | 0.6 | 0.0 | 0.0 | 70.9 |
| 1993-94 | 0.0 | 0.0 | 0.0 | 0.0 | 0.7 | 5.5 | 1.4 | 6.5 | T | T | 0.0 | 0.0 | 14.1 |
| 1994-95 | 0.0 | 0.0 | 0.0 | 0.0 | 17.0 | 13.4 | 10.1 | 1.6 | 1.2 | T | | | |
| 1995-96 | | | | | | | | | | | | | |
| 1996-97 | | | | | | | | | | | | | |
| 1997-98 | | | | | | | | | | | | | |
| 1998-99 | | | | | | | | | | | | | |
| 1999-00 | | | | | | | | | | | | | |
| 2000-01 | | | | | | | | | | | | | |
| 2001-02 | | | | | | | | | | | | | |
| 2002-03 | | | | | | | | | | | | | |
| 2003-04 | | | | | | | | | | | | | |
| 2004-05 | | | | | | | | | | | | | |
| 2005- | | | | | | | | | | | | | |
| 2006- | | | | | | | | | | | | | |
| POR= 26 YRS | 0.0 | 0.0 | T | 0.7 | 4.2 | 7.9 | 6.6 | 5.2 | 3.0 | 0.8 | 0.2 | T | 28.6 |

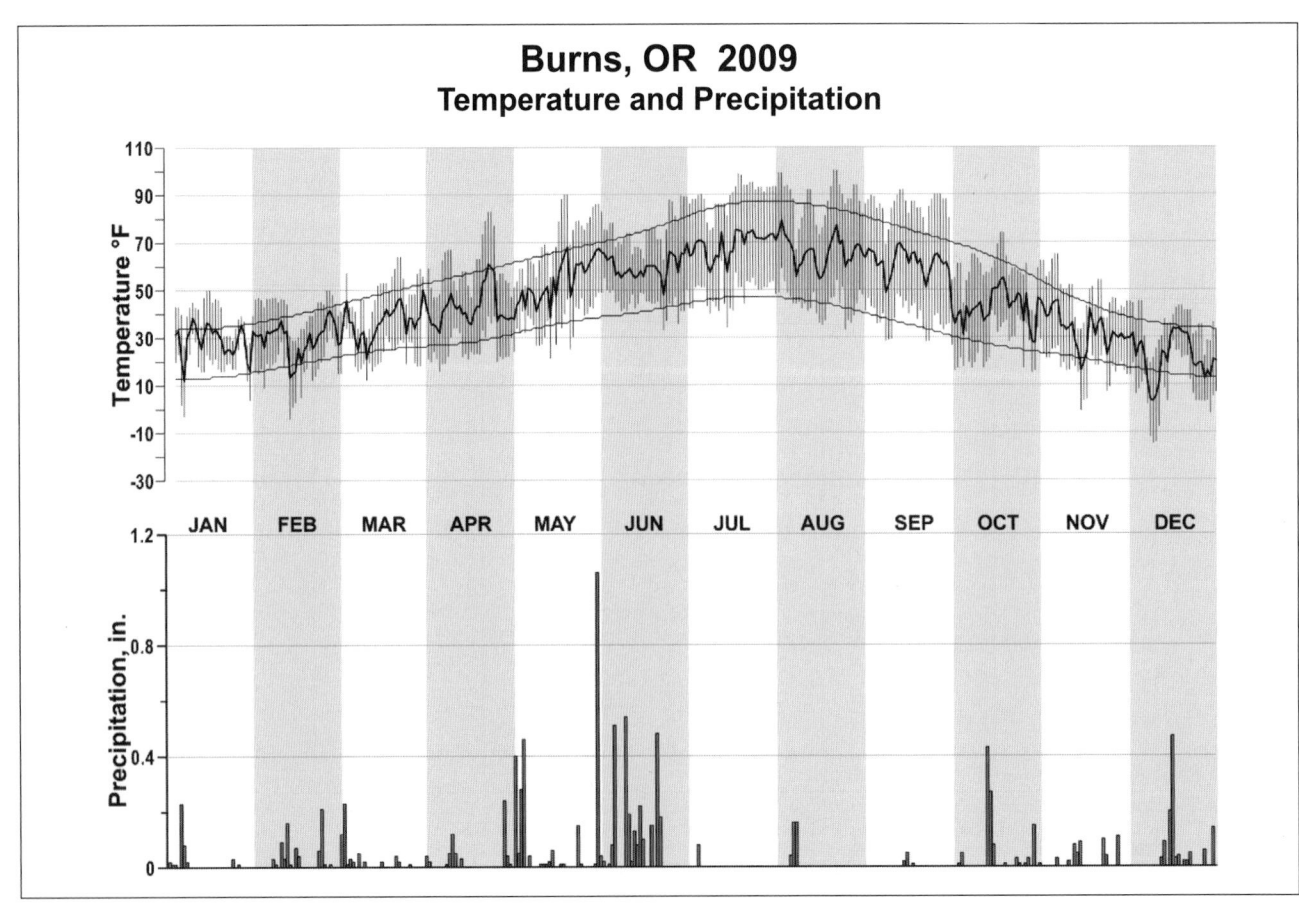

## Burns, OR  2009
### Temperature and Precipitation

# 2009
# PENDLETON
# OREGON (KPDT)

Pendleton is located in the southeastern part of the Columbia Basin, that low country of northern Oregon and central and eastern Washington which is almost entirely surrounded by mountains. This Basin is bounded on the south by the high country of central Oregon, on the north by the mountains of western Canada, on the west by the Cascade Range and on the east by the Blue Mountains and the north Idaho plateau. The gorge in the Cascades through which the Columbia River reaches the Pacific is the most important break in the barriers surrounding this basin. These physical features have important influences on the general climate of Pendleton and the surrounding territory.

The Weather Service Office at Pendleton Airport is located in rolling country which slopes generally upward toward the Blue Mountains about 15 miles to the east and southeast. The Columbia River approaches the area from the northwest to its junction with the Walla Walla River at an elevation of 351 feet and some 25 miles north of Pendleton, then turns southwestward to be joined a few miles below by the Umatilla River. Both the Walla Walla and Umatilla Rivers have their sources in the Blue Mountains and flow westward to the Columbia. The observation station is at an elevation of nearly 1,500 feet, about 3 miles northwest of downtown Pendleton. The city of Pendleton lies in the shallow east-west valley of the Umatilla River, approximately 400 feet lower than the airport.

Precipitation in the Pendleton area is definitely seasonal in occurrence with an average of only 10 percent of the annual total occurring in the three-month period, July-September. Most precipitation reaching this area accompanies cyclonic storms moving in from the Pacific Ocean. These storms reach their greatest intensity and frequency from October through April. The Cascade Range west of the Columbia Basin reduces the amount of precipitation received from the Pacific cyclonic storms. This influence is felt, particularly, in the desert area of the central part of the Basin. A gradual rise in elevation from the Columbia River to the foothills of the Blue Mountains again results in increased precipitation. This increase supplies sufficient moisture for productive wheat, pea, and stock raising activity in the area surrounding Pendleton.

The lighter summertime precipitation usually accompanies thunderstorms which often move into the area from the south or southwest. On occasion, these storms are quite intense, causing flash flooding with resultant heavy property damage and even loss of life.

Seasonal temperature extremes are usually quite moderate for the latitude. The last occurrence in spring of temperatures as low as 32 degrees is mid-April, and the average last occurrence in the fall of 32 degrees is late October. At the city station, where cool air settles in the valley on still nights, temperatures of 32 degrees have been recorded later in the spring and earlier in the fall. Under usual atmospheric conditions, air from the Pacific, with moderate temperature characteristics, moves across the Cascades or through the Columbia Gorge resulting in mild temperatures in the Pendleton area. When this flow of air from the west is impeded by slow-moving high pressure systems over the interior of the continent, temperature conditions sometimes become rather severe, hot in summer and cold in winter. During the summer or early fall, if a stagnant high predominates to the north or east of Pendleton, the hot, dry conditions may prove detrimental to crops during late May and June, and cause fire danger in the forest and grassland areas during late summer and early fall. During winter, coldest temperatures occur when air from a cold high pressure system in central Canada moves southwestward across the Rockies and flows down into the Columbia Basin. Under this condition the heavy cold air sometimes remains at low levels in the Basin for several days while warmer air from the Pacific flows above it, causing comparatively mild temperatures at higher elevations. Extreme winter temperatures are not particularly common in the Pendleton area. Below zero readings are recorded in approximately 60 percent of winters. Maximum temperatures usually reach 100 degrees or slightly higher on a few days during the summer.

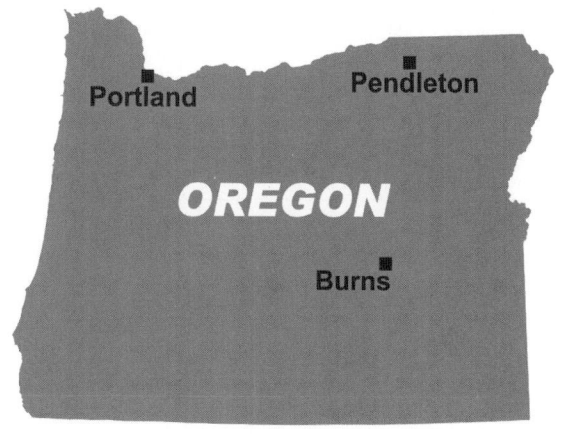

# NORMALS, MEANS, AND EXTREMES
## PENDLETON (KPDT)

| LATITUDE: | LONGITUDE: | | ELEVATION (FT): | | | | TIME ZONE: | | WBAN: 24155 |
|---|---|---|---|---|---|---|---|---|---|
| 45 ° 41'N | -118° 50'W | | GRND: 1481 BARO: 1516 | | | | PACIFIC (UTC -8) | | |

| | ELEMENT | POR | JAN | FEB | MAR | APR | MAY | JUN | JUL | AUG | SEP | OCT | NOV | DEC | YEAR |
|---|---|---|---|---|---|---|---|---|---|---|---|---|---|---|---|
| **TEMPERATURE °F** | NORMAL DAILY MAXIMUM | 30 | 40.1 | 46.5 | 54.8 | 62.2 | 70.2 | 78.7 | 87.7 | 86.6 | 77.1 | 63.8 | 48.5 | 40.0 | 63.0 |
| | MEAN DAILY MAXIMUM | 82 | 39.4 | 45.1 | 54.5 | 61.8 | 70.9 | 78.1 | 88.5 | 86.5 | 76.8 | 63.9 | 48.5 | 41.3 | 62.9 |
| | HIGHEST DAILY MAXIMUM | 74 | 70 | 75 | 80 | 91 | 100 | 108 | 110 | 113 | 102 | 92 | 80 | 67 | 113 |
| | YEAR OF OCCURRENCE | | 1995 | 1996 | 2004 | 1977 | 1986 | 1961 | 1939 | 1961 | 1955 | 1980 | 1999 | 1980 | AUG 1961 |
| | MEAN OF EXTREME MAXS. | 82 | 58.5 | 62.0 | 68.4 | 77.7 | 88.2 | 95.0 | 101.7 | 100.0 | 92.3 | 79.9 | 66.1 | 59.4 | 79.1 |
| | NORMAL DAILY MINIMUM | 30 | 27.4 | 30.9 | 35.4 | 39.7 | 45.9 | 52.0 | 57.5 | 57.3 | 49.7 | 40.7 | 33.8 | 27.7 | 41.5 |
| | MEAN DAILY MINIMUM | 82 | 26.3 | 29.6 | 34.9 | 39.1 | 46.0 | 51.7 | 57.7 | 56.6 | 49.3 | 40.7 | 33.0 | 28.8 | 41.1 |
| | LOWEST DAILY MINIMUM | 74 | -22 | -18 | 1 | 18 | 25 | 35 | 42 | 40 | 30 | 11 | -12 | -19 | -22 |
| | YEAR OF OCCURRENCE | | 1957 | 1950 | 1993 | 1936 | 1954 | 1991 | 2008 | 1980 | 2000 | 1935 | 1985 | 1983 | JAN 1957 |
| | MEAN OF EXTREME MINS. | 82 | 7.7 | 15.3 | 24.0 | 29.6 | 34.9 | 42.2 | 47.7 | 47.3 | 38.6 | 28.0 | 20.6 | 11.8 | 29.0 |
| | NORMAL DRY BULB | 30 | 33.8 | 38.7 | 45.1 | 51.0 | 58.1 | 65.4 | 72.6 | 72.0 | 63.4 | 52.3 | 41.2 | 33.9 | 52.3 |
| | MEAN DRY BULB | 82 | 32.9 | 37.3 | 44.8 | 50.5 | 58.4 | 64.9 | 73.1 | 71.6 | 63.0 | 52.3 | 40.7 | 35.0 | 52.0 |
| | MEAN WET BULB | 26 | 31.4 | 33.6 | 38.6 | 42.6 | 48.0 | 52.0 | 55.0 | 54.4 | 49.7 | 43.3 | 36.7 | 30.2 | 43.0 |
| | MEAN DEW POINT | 26 | 29.0 | 30.3 | 34.0 | 37.4 | 42.5 | 45.0 | 46.1 | 45.3 | 42.1 | 37.6 | 33.7 | 27.8 | 37.6 |
| | NORMAL NO. DAYS WITH: | | | | | | | | | | | | | | |
| | MAXIMUM >= 90 | 30 | 0.0 | 0.0 | 0.0 | * | 0.7 | 4.0 | 13.8 | 11.2 | 2.6 | 0.1 | 0.0 | 0.0 | 32.4 |
| | MAXIMUM <= 32 | 30 | 8.1 | 3.8 | 0.2 | 0.0 | 0.0 | 0.0 | 0.0 | 0.0 | 0.0 | * | 2.2 | 8.9 | 23.2 |
| | MINIMUM <= 32 | 30 | 20.2 | 15.2 | 9.1 | 2.9 | 0.2 | 0.0 | 0.0 | 0.0 | 0.1 | 3.4 | 11.6 | 21.1 | 83.8 |
| | MINIMUM <= 0 | 30 | 1.1 | 0.6 | 0.0 | 0.0 | 0.0 | 0.0 | 0.0 | 0.0 | 0.0 | 0.0 | 0.1 | 1.0 | 2.8 |
| **H/C** | NORMAL HEATING DEG. DAYS | 30 | 971 | 747 | 623 | 433 | 247 | 83 | 14 | 15 | 115 | 400 | 711 | 962 | 5321 |
| | NORMAL COOLING DEG. DAYS | 30 | 0 | 0 | 0 | 2 | 23 | 86 | 243 | 224 | 63 | 3 | 0 | 0 | 644 |
| **RH** | NORMAL (PERCENT) | 30 | 78 | 73 | 66 | 59 | 55 | 48 | 39 | 39 | 48 | 59 | 75 | 80 | 60 |
| | HOUR 04 LST | 30 | 80 | 78 | 75 | 72 | 71 | 67 | 56 | 54 | 62 | 70 | 79 | 81 | 70 |
| | HOUR 10 LST | 30 | 78 | 71 | 61 | 53 | 49 | 43 | 34 | 36 | 43 | 54 | 73 | 78 | 56 |
| | HOUR 16 LST | 30 | 74 | 64 | 51 | 43 | 39 | 33 | 25 | 25 | 26 | 32 | 43 | 69 | 77 | 48 |
| | HOUR 22 LST | 30 | 80 | 77 | 71 | 65 | 60 | 53 | 41 | 41 | 52 | 64 | 78 | 81 | 64 |
| **S** | PERCENT POSSIBLE SUNSHINE | | | | | | | | | | | | | | |
| **W/O** | MEAN NO. DAYS WITH: | | | | | | | | | | | | | | |
| | HEAVY FOG(VISBY <= 1/4 MI) | 46 | 7.7 | 5.1 | 1.6 | 0.4 | 0.1 | 0.1 | 0.0 | 0.0 | 0.3 | 0.8 | 6.0 | 8.2 | 30.3 |
| | THUNDERSTORMS | 62 | 0.0 | 0.0 | 0.2 | 0.9 | 1.5 | 1.9 | 1.8 | 1.9 | 0.9 | 0.2 | 0.1 | 0.0 | 9.4 |
| **CLOUDNESS** | MEAN: | | | | | | | | | | | | | | |
| | SUNRISE-SUNSET (OKTAS) | | | | | | | | | | | | | | |
| | MIDNIGHT-MIDNIGHT (OKTAS) | | | | | | | | | | | | | | |
| | MEAN NO. DAYS WITH: | | | | | | | | | | | | | | |
| | CLEAR | 1 | 1.0 | 3.0 | 4.0 | | 5.0 | 7.0 | | | | | | | |
| | PARTLY CLOUDY | | | 3.0 | 2.0 | | 5.0 | 3.0 | | | | | | | |
| | CLOUDY | 1 | 3.0 | 2.0 | 9.0 | | 10.0 | 3.0 | | | | | | | |
| **PR** | MEAN STATION PRESSURE(IN) | 26 | 28.51 | 28.47 | 28.42 | 28.41 | 28.38 | 28.38 | 28.38 | 28.38 | 28.41 | 28.46 | 28.48 | 28.52 | 28.43 |
| | MEAN SEA-LEVEL PRES. (IN) | 26 | 30.14 | 30.09 | 30.02 | 30.00 | 29.96 | 29.94 | 29.94 | 29.93 | 29.98 | 30.05 | 30.09 | 30.14 | 30.02 |
| **WINDS** | MEAN SPEED (MPH) | 26 | 7.0 | 7.1 | 8.3 | 8.7 | 8.7 | 8.6 | 8.3 | 8.1 | 7.7 | 7.3 | 7.4 | 6.8 | 7.8 |
| | PREVAIL.DIR(TENS OF DEGS) | 31 | 16 | 15 | 26 | 27 | 27 | 27 | 27 | 27 | 15 | 15 | 16 | 17 | 27 |
| | MAXIMUM 2-MINUTE: | | | | | | | | | | | | | | |
| | SPEED (MPH) | 14 | 53 | 52 | 55 | 48 | 44 | 47 | 49 | 43 | 45 | 45 | 44 | 54 | 55 |
| | DIR. (TENS OF DEGS) | | 26 | 22 | 25 | 25 | 26 | 17 | 31 | 23 | 27 | 25 | 27 | 22 | 25 |
| | YEAR OF OCCURRENCE | | 2004 | 2002 | 1997 | 1997 | 2009 | 2008 | 2002 | 1997 | 1999 | 2003 | 2003 | 1998 | MAR 1997 |
| | MAXIMUM 3-SECOND | | | | | | | | | | | | | | |
| | SPEED (MPH) | 14 | 64 | 61 | 63 | 55 | 58 | 62 | 56 | 59 | 53 | 53 | 56 | 66 | 66 |
| | DIR. (TENS OF DEGS) | | 25 | 23 | 25 | 21 | 09 | 17 | 31 | 23 | 27 | 27 | 19 | 22 | 22 |
| | YEAR OF OCCURRENCE | | 2004 | 2002 | 1997 | 1998 | 2007 | 2008 | 2002 | 1997 | 1999 | 2004 | 2007 | 1998 | DEC 1998 |
| **PRECIPITATION** | NORMAL (IN) | 30 | 1.45 | 1.22 | 1.26 | 1.13 | 1.22 | 0.78 | 0.41 | 0.56 | 0.63 | 0.99 | 1.63 | 1.48 | 12.76 |
| | MAXIMUM MONTHLY (IN) | 74 | 3.92 | 3.03 | 2.82 | 2.78 | 3.18 | 2.70 | 1.45 | 2.58 | 2.34 | 2.79 | 3.76 | 4.68 | 4.68 |
| | YEAR OF OCCURRENCE | | 1970 | 1940 | 1983 | 1978 | 1991 | 1947 | 1993 | 1977 | 1941 | 1947 | 1973 | 1973 | DEC 1973 |
| | MINIMUM MONTHLY (IN) | 74 | 0.21 | 0.07 | 0.24 | 0.01 | 0.03 | 0.03 | T | 0.00 | T | T | 0.04 | 0.21 | 0.00 |
| | YEAR OF OCCURRENCE | | 1949 | 1964 | 1941 | 1956 | 1964 | 2003 | 2009 | 1969 | 1993 | 1987 | 1939 | 1989 | AUG 1969 |
| | MAXIMUM IN 24 HOURS (IN) | 74 | 4.97 | 1.41 | 1.33 | 1.24 | 1.52 | 1.49 | 1.19 | 2.19 | 1.23 | 1.88 | 1.35 | 1.25 | 4.97 |
| | YEAR OF OCCURRENCE | | 2009 | 1994 | 1983 | 1990 | 1972 | 1947 | 1948 | 1993 | 1981 | 1982 | 1971 | 1978 | JAN 2009 |
| | NORMAL NO. DAYS WITH: | | | | | | | | | | | | | | |
| | PRECIPITATION >= 0.01 | 30 | 11.7 | 10.5 | 11.0 | 8.6 | 8.1 | 6.0 | 3.2 | 3.0 | 4.4 | 6.0 | 12.2 | 11.3 | 96.0 |
| | PRECIPITATION >= 1.00 | 30 | * | * | * | 0.1 | 0.1 | * | * | 0.1 | * | * | 0.1 | * | 0.4 |
| **SNOWFALL** | NORMAL (IN) | 30 | 4.9 | 3.5 | 1.0 | 0.1 | 0.* | 0.0 | 0.0 | 0.0 | 0.0 | 0.3 | 2.2 | 5.2 | 17.2 |
| | MAXIMUM MONTHLY (IN) | 72 | 41.6 | 16.8 | 4.9 | 2.2 | T | T | T | 0.0 | 0.0 | 3.2 | 14.9 | 32.5 | 41.6 |
| | YEAR OF OCCURRENCE | | 1950 | 1950 | 1971 | 1975 | 2009 | 1994 | 2009 | | | 1973 | 1985 | 2008 | JAN 1950 |
| | MAXIMUM IN 24 HOURS (IN) | 72 | 13.3 | 16.1 | 4.0 | 2.2 | T | T | T | 0.0 | 0.0 | 3.2 | 8.0 | 9.9 | 16.1 |
| | YEAR OF OCCURRENCE' | | 1950 | 1994 | 1970 | 1975 | 2009 | 1994 | 1993 | | | 1973 | 1977 | 1948 | FEB 1994 |
| | MAXIMUM SNOW DEPTH (IN) | 60 | 16 | 12 | 6 | 0 | 0 | 0 | 0 | 0 | 0 | 2 | 8 | 15 | 16 |
| | YEAR OF OCCURRENCE | | 1957 | 1994 | 1993 | | | | | | | 1971 | 1978 | 2008 | JAN 1957 |
| | NORMAL NO. DAYS WITH: | | | | | | | | | | | | | | |
| | SNOWFALL >= 1.0 | 30 | 1.6 | 0.9 | 0.4 | 0.1 | 0.0 | 0.0 | 0.0 | 0.0 | 0.0 | 0.1 | 0.6 | 1.9 | 5.6 |

## PRECIPITATION (inches) 2009  PENDLETON (KPDT)

| YEAR | JAN | FEB | MAR | APR | MAY | JUN | JUL | AUG | SEP | OCT | NOV | DEC | ANNUAL |
|------|-----|-----|-----|-----|-----|-----|-----|-----|-----|-----|-----|-----|--------|
| 1980 | 2.48 | 1.39 | 1.60 | 0.59 | 2.14 | 1.12 | 0.77 | 0.03 | 0.59 | 1.22 | 0.84 | 1.20 | 13.97 |
| 1981 | 0.89 | 1.35 | 1.43 | 1.20 | 1.59 | 1.53 | 0.94 | 0.03 | 1.31 | 0.86 | 1.91 | 2.31 | 15.35 |
| 1982 | 1.54 | 0.77 | 1.22 | 0.84 | 0.31 | 0.63 | 0.51 | 0.24 | 1.47 | 2.67 | 0.34 | 2.20 | 12.74 |
| 1983 | 0.86 | 1.57 | 2.82 | 0.70 | 0.73 | 1.44 | 0.52 | 0.56 | 0.46 | 0.84 | 1.67 | 3.42 | 15.59 |
| 1984 | 0.53 | 1.74 | 1.83 | 1.70 | 1.02 | 1.13 | 0.06 | 0.44 | 0.39 | 1.02 | 2.14 | 0.92 | 12.92 |
| 1985 | 0.44 | 1.33 | 1.13 | 0.37 | 0.44 | 0.69 | 0.34 | 0.26 | 2.10 | 0.89 | 2.11 | 1.27 | 11.37 |
| 1986 | 1.66 | 2.58 | 1.13 | 0.43 | 1.18 | 0.03 | 0.48 | 0.02 | 1.28 | 0.80 | 2.12 | 0.82 | 12.53 |
| 1987 | 1.48 | 0.64 | 1.39 | 0.47 | 0.85 | 0.38 | 0.34 | 0.05 | 0.03 | T | 0.76 | 1.23 | 7.62 |
| 1988 | 1.86 | 0.12 | 0.95 | 2.47 | 1.56 | 0.31 | 0.01 | T | 0.31 | 0.10 | 2.16 | 0.37 | 10.22 |
| 1989 | 1.86 | 1.36 | 1.72 | 1.57 | 1.47 | 0.57 | 0.09 | 1.25 | 0.12 | 0.84 | 1.27 | 0.21 | 12.33 |
| 1990 | 0.77 | 0.28 | 1.14 | 1.54 | 1.83 | 0.58 | 0.18 | 0.62 | T | 0.78 | 0.87 | 0.84 | 9.43 |
| 1991 | 0.98 | 0.57 | 1.00 | 0.71 | 3.18 | 2.14 | 0.24 | 0.42 | T | 0.92 | 2.68 | 0.67 | 13.51 |
| 1992 | 0.41 | 1.04 | 0.26 | 1.21 | 0.07 | 0.94 | 0.70 | 0.43 | 0.42 | 1.32 | 1.15 | 0.73 | 8.68 |
| 1993 | 1.79 | 0.80 | 1.49 | 1.85 | 1.51 | 0.71 | 1.45 | 2.19 | T | 0.22 | 0.93 | 0.92 | 13.86 |
| 1994 | 1.57 | 1.71 | 0.56 | 0.45 | 2.55 | 0.77 | 0.38 | T | 0.36 | 1.28 | 1.98 | 0.85 | 12.46 |
| 1995 | 2.53 | 1.07 | 1.93 | 2.28 | 0.97 | 2.30 | 0.24 | 0.29 | 0.55 | 1.21 | 2.18 | 1.73 | 17.28 |
| 1996 | 1.88 | 1.80 | 1.00 | 1.08 | 2.00 | 0.47 | .06 | .05 | .61 | 1.22 | 1.96 | 2.32 | 14.45 |
| 1997 | 1.84 | 0.39 | 1.16 | 1.56 | 0.33 | 0.76 | 0.66 | 0.07 | 0.77 | 1.43 | 1.64 | 1.05 | 11.66 |
| 1998 | 2.61 | 1.19 | 1.01 | 1.28 | 1.53 | 0.76 | 0.68 | T | 1.11 | 0.60 | 2.31 | 1.37 | 14.45 |
| 1999 | 0.81 | 1.22 | 0.74 | 0.50 | 1.27 | 0.51 | T | 0.54 | 0.01 | 1.51 | 1.23 | 1.01 | 9.35 |
| 2000 | 1.99 | 2.98 | 2.42 | 0.69 | 1.60 | 0.72 | 0.07 | T | 2.01 | 2.06 | 1.22 | 0.57 | 16.33 |
| 2001 | 0.95 | 0.62 | 1.31 | 1.89 | 0.45 | 1.12 | 0.52 | 0.08 | 0.09 | 1.54 | 1.15 | 0.70 | 10.42 |
| 2002 | 0.58 | 0.76 | 0.81 | 1.48 | 1.10 | 1.30 | 0.02 | 0.03 | 0.20 | 0.73 | 0.55 | 2.23 | 9.79 |
| 2003 | 2.92 | 1.07 | 1.51 | 1.62 | 0.78 | T | 0.01 | 0.09 | 0.63 | 0.41 | 1.00 | 2.72 | 12.76 |
| 2004 | 2.32 | 1.76 | 0.72 | 1.29 | 1.81 | 1.47 | 0.48 | 0.97 | 0.45 | 0.78 | 0.88 | 0.65 | 13.58 |
| 2005 | 0.48 | 0.27 | 1.07 | 0.93 | 2.61 | 0.81 | 0.20 | 0.02 | 0.27 | 1.33 | 1.10 | 2.57 | 11.66 |
| 2006 | 2.18 | 0.37 | 1.87 | 1.57 | 1.17 | 1.96 | 0.07 | 0.02 | 0.37 | 0.50 | 2.06 | 1.67 | 13.81 |
| 2007 | 0.45 | 1.63 | 1.21 | 0.87 | 0.64 | 0.88 | 0.21 | 0.44 | 0.60 | 0.96 | 2.05 | 1.56 | 11.50 |
| 2008 | 1.82 | 0.44 | 1.12 | 0.23 | 1.61 | 1.37 | T | 0.76 | 0.13 | 0.32 | 1.39 | 2.58 | 11.77 |
| 2009 | 1.39 | 0.99 | 2.62 | 0.97 | 1.16 | 1.05 | T | 1.04 | 0.04 | 1.50 | 0.91 | 1.52 | 13.19 |
| POR= 82 YRS | 1.50 | 1.12 | 1.20 | 1.06 | 1.14 | 0.93 | 0.30 | 0.41 | 0.57 | 1.04 | 1.45 | 1.55 | 12.27 |

WBAN : 24155

## AVERAGE TEMPERATURE (°F) 2009  PENDLETON (KPDT)

| YEAR | JAN | FEB | MAR | APR | MAY | JUN | JUL | AUG | SEP | OCT | NOV | DEC | ANNUAL |
|------|-----|-----|-----|-----|-----|-----|-----|-----|-----|-----|-----|-----|--------|
| 1980 | 25.6 | 36.1 | 41.3 | 51.9 | 56.4 | 60.4 | 72.1 | 66.9 | 63.3 | 51.3 | 42.0 | 39.2 | 50.5 |
| 1981 | 36.2 | 38.9 | 45.7 | 50.4 | 56.0 | 61.6 | 69.2 | 74.3 | 63.8 | 50.6 | 44.2 | 37.2 | 52.3 |
| 1982 | 35.0 | 38.1 | 43.5 | 47.6 | 56.8 | 67.6 | 71.1 | 71.5 | 60.7 | 50.7 | 37.3 | 35.7 | 51.3 |
| 1983 | 40.8 | 43.8 | 47.8 | 49.0 | 58.9 | 62.7 | 68.4 | 72.7 | 58.9 | 52.5 | 45.9 | 23.2 | 52.1 |
| 1984 | 34.6 | 39.7 | 46.8 | 48.2 | 54.7 | 62.1 | 72.9 | 72.2 | 60.4 | 49.1 | 41.8 | 30.4 | 51.1 |
| 1985 | 26.3 | 33.5 | 43.2 | 53.1 | 58.5 | 65.6 | 77.4 | 68.1 | 57.0 | 50.3 | 26.5 | 19.5 | 48.3 |
| 1986 | 35.9 | 39.0 | 48.8 | 50.0 | 58.6 | 70.0 | 67.6 | 75.8 | 58.9 | 54.0 | 42.2 | 31.5 | 52.7 |
| 1987 | 30.4 | 39.1 | 46.4 | 53.9 | 59.7 | 67.2 | 68.9 | 70.6 | 66.2 | 54.1 | 42.6 | 32.7 | 52.7 |
| 1988 | 32.4 | 41.1 | 44.1 | 51.9 | 56.8 | 63.9 | 72.0 | 70.0 | 63.4 | 58.4 | 44.3 | 33.9 | 52.7 |
| 1989 | 38.3 | 25.1 | 42.5 | 52.9 | 55.9 | 65.9 | 70.3 | 68.8 | 63.6 | 51.8 | 44.6 | 33.2 | 51.1 |
| 1990 | 39.6 | 37.9 | 45.7 | 54.8 | 56.4 | 64.7 | 75.2 | 72.2 | 68.2 | 51.3 | 45.4 | 25.8 | 53.1 |
| 1991 | 31.2 | 44.7 | 42.6 | 49.6 | 53.9 | 59.6 | 71.5 | 73.3 | 64.9 | 51.6 | 41.2 | 36.9 | 51.8 |
| 1992 | 38.7 | 42.4 | 47.7 | 52.8 | 62.4 | 70.4 | 71.9 | 72.6 | 61.9 | 54.2 | 39.8 | 32.0 | 53.9 |
| 1993 | 25.1 | 28.2 | 41.9 | 50.3 | 61.8 | 62.9 | 65.4 | 68.0 | 64.3 | 54.6 | 34.6 | 35.6 | 49.4 |
| 1994 | 41.0 | 34.9 | 46.7 | 54.4 | 60.1 | 64.4 | 75.3 | 71.7 | 67.2 | 51.6 | 40.4 | 35.3 | 53.6 |
| 1995 | 34.9 | 41.7 | 45.3 | 49.0 | 57.7 | 62.1 | 72.0 | 67.4 | 66.0 | 50.2 | 46.1 | 34.5 | 52.2 |
| 1996 | 34.4 | 35.7 | 43.3 | 52.0 | 54.5 | 64.0 | 74.7 | 72.1 | 61.0 | 51.1 | 40.3 | 35.4 | 51.5 |
| 1997 | 32.6 | 38.6 | 45.4 | 48.6 | 59.8 | 63.1 | 70.4 | 73.1 | 64.9 | 51.4 | 42.7 | 34.9 | 52.1 |
| 1998 | 37.9 | 42.9 | 45.9 | 50.3 | 56.2 | 65.4 | 77.4 | 74.6 | 67.8 | 50.9 | 45.5 | 36.2 | 54.3 |
| 1999 | 41.0 | 42.3 | 43.4 | 47.3 | 54.1 | 63.7 | 70.0 | 72.5 | 62.5 | 50.8 | 46.0 | 38.1 | 52.6 |
| 2000 | 35.5 | 39.2 | 44.1 | 54.0 | 58.3 | 65.9 | 71.6 | 70.7 | 61.6 | 49.7 | 35.1 | 31.1 | 51.4 |
| 2001 | 32.7 | 35.5 | 44.4 | 47.3 | 58.7 | 62.1 | 70.5 | 73.2 | 66.2 | 51.2 | 42.2 | 36.6 | 51.7 |
| 2002 | 37.0 | 38.8 | 41.0 | 49.9 | 56.3 | 66.9 | 75.1 | 70.2 | 63.5 | 48.6 | 41.0 | 39.4 | 52.3 |
| 2003 | 40.6 | 38.7 | 48.6 | 49.9 | 57.3 | 67.7 | 76.2 | 73.0 | 65.8 | 55.6 | 38.8 | 35.4 | 54.0 |
| 2004 | 28.5 | 39.6 | 49.2 | 52.5 | 58.6 | 67.0 | 75.6 | 74.4 | 62.5 | 55.4 | 42.8 | 39.5 | 53.8 |
| 2005 | 35.6 | 37.5 | 46.8 | 50.4 | 59.0 | 63.0 | 73.5 | 72.6 | 61.1 | 52.9 | 38.7 | 30.0 | 51.8 |
| 2006 | 42.2 | 36.0 | 43.9 | 50.5 | 59.0 | 65.5 | 75.5 | 71.0 | 63.9 | 50.1 | 42.4 | 32.9 | 52.7 |
| 2007 | 30.7 | 38.4 | 46.4 | 49.4 | 58.2 | 64.8 | 75.9 | 69.4 | 61.0 | 49.6 | 39.6 | 35.6 | 51.6 |
| 2008 | 31.5 | 40.8 | 41.7 | 46.2 | 59.0 | 63.4 | 71.7 | 70.8 | 62.3 | 50.7 | 43.7 | 28.8 | 50.9 |
| 2009 | 33.5 | 35.9 | 40.3 | 49.0 | 58.7 | 65.7 | 74.8 | 72.2 | 64.9 | 47.9 | 41.9 | 27.1 | 51.0 |
| POR= 82 YRS | 32.9 | 37.4 | 44.8 | 50.5 | 58.4 | 64.9 | 73.1 | 71.6 | 63.0 | 52.3 | 40.7 | 35.0 | 52.0 |

## HEATING DEGREE DAYS (base 65°F) 2009 PENDLETON (KPDT)

| YEAR | JUL | AUG | SEP | OCT | NOV | DEC | JAN | FEB | MAR | APR | MAY | JUN | TOTAL |
|------|-----|-----|-----|-----|-----|-----|-----|-----|-----|-----|-----|-----|-------|
| 1980-81 | 4 | 33 | 88 | 438 | 681 | 794 | 886 | 724 | 593 | 435 | 275 | 126 | 5077 |
| 1981-82 | 20 | 1 | 128 | 440 | 617 | 855 | 919 | 747 | 662 | 515 | 256 | 72 | 5232 |
| 1982-83 | 22 | 7 | 171 | 435 | 825 | 901 | 741 | 588 | 528 | 470 | 242 | 95 | 5025 |
| 1983-84 | 42 | 1 | 180 | 381 | 569 | 1292 | 935 | 729 | 558 | 496 | 316 | 134 | 5633 |
| 1984-85 | 4 | 0 | 182 | 490 | 692 | 1065 | 1196 | 876 | 665 | 351 | 224 | 65 | 5810 |
| 1985-86 | 4 | 22 | 242 | 452 | 1149 | 1402 | 898 | 722 | 497 | 446 | 277 | 25 | 6136 |
| 1986-87 | 33 | 0 | 213 | 335 | 675 | 1031 | 1065 | 717 | 571 | 332 | 201 | 71 | 5244 |
| 1987-88 | 25 | 12 | 65 | 334 | 668 | 995 | 1004 | 689 | 637 | 387 | 264 | 126 | 5206 |
| 1988-89 | 22 | 4 | 120 | 208 | 616 | 957 | 821 | 1113 | 691 | 354 | 279 | 42 | 5227 |
| 1989-90 | 11 | 17 | 76 | 403 | 607 | 978 | 781 | 752 | 591 | 299 | 262 | 89 | 4866 |
| 1990-91 | 9 | 13 | 11 | 419 | 583 | 1211 | 1039 | 564 | 689 | 454 | 338 | 162 | 5492 |
| 1991-92 | 4 | 2 | 52 | 418 | 707 | 865 | 810 | 649 | 527 | 362 | 127 | 36 | 4559 |
| 1992-93 | 11 | 28 | 129 | 333 | 752 | 1015 | 1231 | 1025 | 709 | 432 | 153 | 98 | 5916 |
| 1993-94 | 27 | 35 | 114 | 318 | 908 | 903 | 736 | 838 | 559 | 321 | 174 | 83 | 5016 |
| 1994-95 | 15 | 0 | 30 | 406 | 731 | 915 | 928 | 644 | 602 | 473 | 237 | 126 | 5107 |
| 1995-96 | 1 | 26 | 51 | 452 | 560 | 941 | 938 | 840 | 664 | 387 | 321 | 64 | 5245 |
| 1996-97 | 8 | 4 | 153 | 423 | 733 | 909 | 998 | 733 | 599 | 488 | 185 | 84 | 5317 |
| 1997-98 | 13 | 0 | 73 | 415 | 662 | 927 | 832 | 611 | 585 | 442 | 273 | 37 | 4870 |
| 1998-99 | 0 | 3 | 56 | 429 | 578 | 887 | 738 | 628 | 662 | 522 | 336 | 108 | 4947 |
| 1999-00 | 17 | 20 | 117 | 433 | 565 | 830 | 905 | 740 | 640 | 324 | 219 | 57 | 4867 |
| 2000-01 | 7 | 12 | 137 | 468 | 892 | 1046 | 995 | 819 | 631 | 523 | 238 | 121 | 5889 |
| 2001-02 | 14 | 5 | 48 | 419 | 676 | 874 | 863 | 727 | 738 | 445 | 270 | 66 | 5145 |
| 2002-03 | 6 | 10 | 110 | 501 | 714 | 788 | 748 | 729 | 499 | 444 | 259 | 32 | 4840 |
| 2003-04 | 1 | 0 | 74 | 300 | 779 | 913 | 1125 | 729 | 483 | 368 | 195 | 69 | 5036 |
| 2004-05 | 0 | 3 | 87 | 293 | 656 | 785 | 906 | 764 | 556 | 429 | 196 | 98 | 4773 |
| 2005-06 | 1 | 3 | 133 | 370 | 782 | 1077 | 701 | 804 | 644 | 429 | 240 | 61 | 5245 |
| 2006-07 | 5 | 10 | 105 | 458 | 675 | 989 | 1058 | 739 | 568 | 462 | 212 | 74 | 5355 |
| 2007-08 | 0 | 14 | 145 | 470 | 757 | 903 | 1030 | 694 | 714 | 556 | 214 | 115 | 5612 |
| 2008-09 | 3 | 19 | 104 | 444 | 634 | 1116 | 969 | 809 | 760 | 473 | 228 | 33 | 5592 |
| 2009- | 0 | 5 | 68 | 526 | 684 | 1170 | | | | | | | |

WBAN : 24155

## COOLING DEGREE DAYS (base 65°F) 2009 PENDLETON (KPDT)

| YEAR | JAN | FEB | MAR | APR | MAY | JUN | JUL | AUG | SEP | OCT | NOV | DEC | TOTAL |
|------|-----|-----|-----|-----|-----|-----|-----|-----|-----|-----|-----|-----|-------|
| 1980 | 0 | 0 | 0 | 2 | 5 | 13 | 232 | 101 | 44 | 20 | 0 | 0 | 417 |
| 1981 | 0 | 0 | 0 | 4 | 2 | 28 | 155 | 297 | 101 | 0 | 0 | 0 | 587 |
| 1982 | 0 | 0 | 0 | 0 | 7 | 158 | 219 | 215 | 47 | 0 | 0 | 0 | 646 |
| 1983 | 0 | 0 | 0 | 0 | 60 | 32 | 155 | 246 | 6 | 0 | 0 | 0 | 499 |
| 1984 | 0 | 0 | 0 | 0 | 7 | 55 | 256 | 231 | 51 | 3 | 0 | 0 | 603 |
| 1985 | 0 | 0 | 0 | 0 | 28 | 91 | 394 | 127 | 7 | 0 | 0 | 0 | 647 |
| 1986 | 0 | 0 | 0 | 2 | 88 | 184 | 121 | 341 | 35 | 1 | 0 | 0 | 772 |
| 1987 | 0 | 0 | 0 | 8 | 41 | 145 | 152 | 194 | 108 | 4 | 0 | 0 | 652 |
| 1988 | 0 | 0 | 0 | 0 | 16 | 98 | 246 | 164 | 78 | 9 | 0 | 0 | 611 |
| 1989 | 0 | 0 | 0 | 0 | 5 | 76 | 182 | 143 | 41 | 0 | 0 | 0 | 447 |
| 1990 | 0 | 0 | 0 | 0 | 4 | 92 | 330 | 245 | 114 | 3 | 0 | 0 | 788 |
| 1991 | 0 | 0 | 0 | 0 | 0 | 8 | 214 | 267 | 56 | 9 | 0 | 0 | 554 |
| 1992 | 0 | 0 | 0 | 1 | 52 | 204 | 229 | 275 | 45 | 4 | 0 | 0 | 810 |
| 1993 | 0 | 0 | 0 | 0 | 59 | 42 | 47 | 136 | 99 | 4 | 0 | 0 | 387 |
| 1994 | 0 | 0 | 0 | 8 | 29 | 72 | 341 | 214 | 103 | 0 | 0 | 0 | 767 |
| 1995 | 0 | 0 | 0 | 0 | 17 | 44 | 230 | 108 | 89 | 0 | 0 | 0 | 488 |
| 1996 | 0 | 0 | 0 | 0 | 0 | 42 | 312 | 230 | 39 | 0 | 0 | 0 | 623 |
| 1997 | 0 | 0 | 0 | 0 | 31 | 34 | 191 | 258 | 79 | 5 | 0 | 0 | 598 |
| 1998 | 0 | 0 | 0 | 6 | 5 | 56 | 390 | 309 | 144 | 0 | 0 | 0 | 910 |
| 1999 | 0 | 0 | 0 | 0 | 7 | 78 | 180 | 260 | 50 | 0 | 5 | 0 | 580 |
| 2000 | 0 | 0 | 0 | 0 | 17 | 89 | 215 | 196 | 40 | 0 | 0 | 0 | 557 |
| 2001 | 0 | 0 | 0 | 0 | 48 | 43 | 193 | 265 | 93 | 0 | 0 | 0 | 642 |
| 2002 | 0 | 0 | 0 | 0 | 9 | 129 | 323 | 178 | 70 | 0 | 0 | 0 | 709 |
| 2003 | 0 | 0 | 0 | 0 | 27 | 118 | 355 | 255 | 105 | 17 | 0 | 0 | 877 |
| 2004 | 0 | 0 | 0 | 0 | 4 | 135 | 337 | 300 | 19 | 4 | 0 | 0 | 799 |
| 2005 | 0 | 0 | 0 | 0 | 17 | 45 | 271 | 246 | 21 | 0 | 0 | 0 | 600 |
| 2006 | 0 | 0 | 0 | 0 | 58 | 79 | 338 | 202 | 79 | 0 | 3 | 0 | 759 |
| 2007 | 0 | 0 | 0 | 0 | 8 | 77 | 345 | 158 | 31 | 0 | 0 | 0 | 619 |
| 2008 | 0 | 0 | 0 | 0 | 37 | 75 | 218 | 207 | 29 | 7 | 0 | 0 | 573 |
| 2009 | 0 | 0 | 0 | 0 | 38 | 60 | 307 | 238 | 69 | 0 | 0 | 0 | 712 |

## SNOWFALL (inches)  2009  PENDLETON (KPDT)

| YEAR | JUL | AUG | SEP | OCT | NOV | DEC | JAN | FEB | MAR | APR | MAY | JUN | TOTAL |
|---|---|---|---|---|---|---|---|---|---|---|---|---|---|
| 1980-81 | 0.0 | 0.0 | 0.0 | 0.0 | 2.0 | 2.7 | 3.6 | 1.2 | 0.0 | 0.0 | 0.0 | 0.0 | 9.5 |
| 1981-82 | 0.0 | 0.0 | 0.0 | 0.0 | 0.6 | 5.1 | 5.7 | 1.5 | 1.9 | T | 0.0 | 0.0 | 14.8 |
| 1982-83 | 0.0 | 0.0 | 0.0 | 0.0 | T | 1.6 | 0.2 | 0.9 | 0.0 | 0.0 | 0.0 | 0.0 | 2.7 |
| 1983-84 | 0.0 | 0.0 | 0.0 | 0.0 | T | 26.6 | 1.0 | 1.2 | T | T | 0.0 | 0.0 | 28.8 |
| 1984-85 | 0.0 | 0.0 | 0.0 | 0.0 | T | 6.2 | 0.8 | 12.7 | 0.6 | T | 0.0 | 0.0 | 20.3 |
| 1985-86 | 0.0 | 0.0 | 0.0 | 0.0 | 14.9 | 9.1 | T | 7.6 | 0.0 | 0.0 | T | 0.0 | 31.6 |
| 1986-87 | 0.0 | 0.0 | 0.0 | 0.0 | 1.2 | 6.8 | 5.8 | 0.0 | T | 0.0 | 0.0 | 0.0 | 13.8 |
| 1987-88 | 0.0 | 0.0 | 0.0 | 0.0 | 0.3 | 2.3 | 10.6 | 0.0 | 1.5 | 0.0 | 0.0 | 0.0 | 14.7 |
| 1988-89 | 0.0 | 0.0 | 0.0 | 0.0 | T | T | 4.3 | 4.9 | 4.0 | 0.0 | T | 0.0 | 13.2 |
| 1989-90 | 0.0 | 0.0 | 0.0 | 0.0 | 0.0 | 1.0 | T | 2.0 | 1.3 | 0.0 | 0.0 | 0.0 | 4.3 |
| 1990-91 | 0.0 | 0.0 | 0.0 | 0.0 | T | 6.4 | 1.6 | 0.0 | 0.6 | T | 0.0 | 0.0 | 8.6 |
| 1991-92 | 0.0 | 0.0 | 0.0 | 2.3 | 1.0 | T | 0.8 | T | 0.0 | 0.0 | 0.0 | 0.0 | 4.1 |
| 1992-93 | 0.0 | 0.0 | 0.0 | 0.0 | 0.2 | 7.6 | 25.1 | 14.8 | 1.8 | T | T | 0.0 | 49.5 |
| 1993-94 | T | 0.0 | 0.0 | 0.0 | 0.7 | 0.4 | T | 16.8 | 0.2 | 0.0 | 0.0 | T | 18.1 |
| 1994-95 | 0.0 | 0.0 | 0.0 | 0.0 | 0.6 | 3.8 | 2.0 | 7.2 | T | 0.0 | 0.0 | 0.0 | 13.6 |
| 1995-96 | 0.0 | 0.0 | 0.0 |  | 0.0 |  |  |  |  |  |  |  |  |
| 1996-97 |  |  |  |  |  | 10.1 | 3.2 |  |  | T |  |  |  |
| 1997-98 |  |  |  |  |  | 4.2 |  | 0.0 |  |  |  |  |  |
| 1998-99 |  |  | 0.0 | 0.0 |  | 0.8 | 0.0 | 2.4 | T | T | T | 0.0 |  |
| 1999-00 | 0.0 | 0.0 | 0.0 | 0.0 | 0.0 | 0.5 | 5.7 | 4.5 | 1.0 | 0.0 | T | 0.0 | 11.7 |
| 2000-01 | 0.0 | 0.0 | 0.0 | 0.0 | 0.7 | 2.5 | 3.1 | 0.5 | T | T | T | T | 6.8 |
| 2001-02 | 0.0 | 0.0 | 0.0 | 0.0 | T | 3.0 | 1.5 | T | 0.3 | 0.0 | 0.0 | T | 4.8 |
| 2002-03 | T | 0.0 | 0.0 | T | 0.0 | 2.1 | 0.3 | T | T | T | 0.0 | 0.0 | 2.4 |
| 2003-04 | 0.0 | 0.0 | 0.0 | T | 0.1 | 13.9 | 10.0 | 2.2 | T | 0.0 | T | 0.0 | 26.2 |
| 2004-05 | 0.0 | 0.0 | 0.0 | 0.0 | 0.4 | 0.8 | 2.7 | 1.2 | T | 0.0 | 0.0 | 0.0 | 5.1 |
| 2005-06 | 0.0 | 0.0 | 0.0 | 0.0 | 0.2 | 1.6 | T | 0.0 | T | 0.0 | 0.0 | 0.0 | 1.8 |
| 2006-07 | 0.0 | 0.0 | 0.0 | 0.0 | 0.8 | T | 1.2 | 2.5 | 0.3 | 0.0 | 0.0 | 0.0 | 4.8 |
| 2007-08 | 0.0 | 0.0 | 0.0 | 0.0 | 8.7 | 8.1 | 10.7 | 4.3 | 2.8 | T | 0.0 | 0.0 | 34.6 |
| 2008-09 | 0.0 | 0.0 | 0.0 | 0.0 | 0.0 | 32.5 | 6.6 | 4.7 | 1.5 | 0.5 | T | 0.0 | 45.8 |
| 2009- | 0.0 | 0.0 | 0.0 | 0.0 | T | 6.7 |  |  |  |  |  |  |  |
| POR=<br>78 YRS | T | 0.0 | 0.0 | 0.1 | 1.5 | 4.3 | 6.7 | 3.4 | 0.8 | 0.1 | T | T | 16.9 |

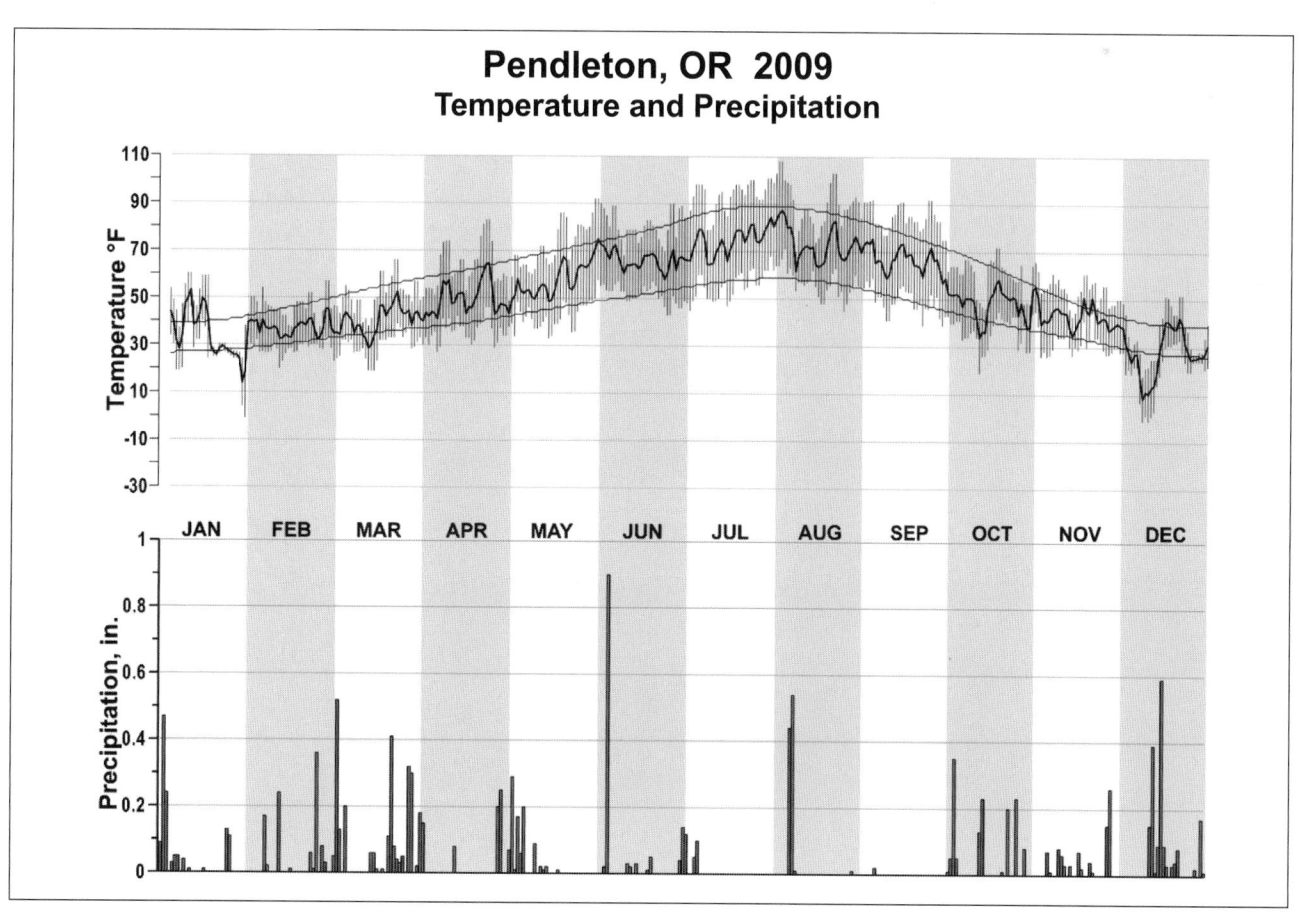

### Pendleton, OR  2009
### Temperature and Precipitation

# 2009
# PORTLAND
# OREGON (KPDX)

The Portland Weather Service Office is located 6 miles north-northeast of downtown Portland. Portland is situated about 65 miles inland from the Pacific Coast and midway between the northerly oriented low coast range on the west and the higher Cascade range on the east, each about 30 miles distant. The airport lies on the south bank of the Columbia River. The coast range provides limited shielding from the Pacific Ocean. The Cascade range provides a steep slope for orographic lift of moisture-laden westerly winds and consequent moderate rainfall, and also forms a barrier from continental air masses originating over the interior Columbia Basin. Airflow is usually northwesterly in Portland in spring and summer and southeasterly in fall and winter. The Portland Airport location is drier than most surrounding localities.

Portland has a very definite winter rainfall climate. Approximately 88 percent of the annual total occurs in the months of October through May, 9 percent in June and September, while only 3 percent comes in July and August. Precipitation is mostly rain, as on the average there are only five days each year with measurable snow. Snowfalls are seldom more than a couple of inches, and generally last only a few days.

The winter season is marked by relatively mild temperatures, cloudy skies and rain with southeasterly surface winds predominating. Summer produces pleasantly mild temperatures, northwesterly winds and very little precipitation. Fall and spring are transitional in nature. Fall and early winter are times with most frequent fog.

At all times, incursions of marine air are a frequent moderating influence. Outbreaks of continental high pressure from east of the Cascade Mountains produce strong easterly flow through the Columbia Gorge into the Portland area. In winter this brings the coldest weather with the extremes of low temperature registered in the cold air mass. Freezing rain and ice glaze are sometimes transitional effects. Temperatures below zero are very infrequent. In summer, hot, dry continental air brings the highest temperatures. Temperatures above 100 degrees are infrequent, but 90 degrees or higher are reached every year, but seldom persist for more than two or three days.

Destructive storms are infrequent in the Portland area. Surface winds seldom exceed gale force and rarely in the period of record have winds reached higher than 75 mph. Thunderstorms occur about once a month through the spring and summer months. Heavy downpours are infrequent but gentle rains occur almost daily during winter months.

Most rural areas around Portland are farmed for berries, green beans, and vegetables for fresh market and processing. The long growing season with mild temperatures and ample moisture favors local nursery and seed industries.

Based on the 1951-1980 period, the average first occurrence of 32 degrees Fahrenheit in the fall is November 7 and the average last occurrence in the spring is April 3.

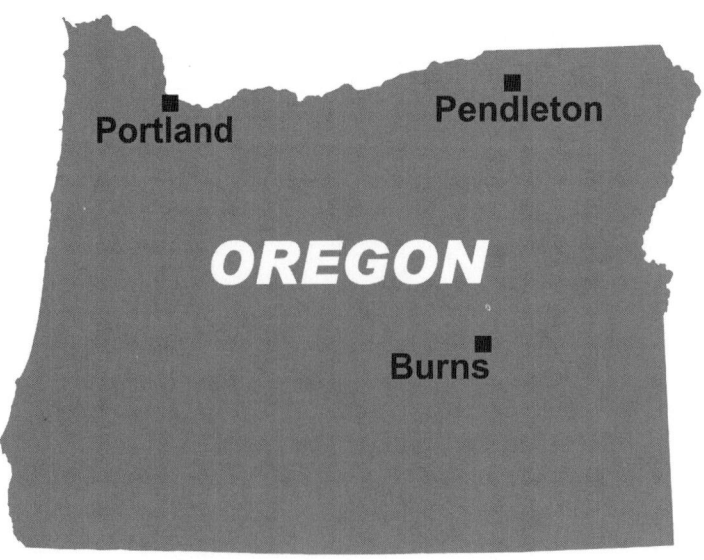

# NORMALS, MEANS, AND EXTREMES
## PORTLAND (KPDX)

LATITUDE: 45° 35'N　LONGITUDE: -122° 36'W　ELEVATION (FT): GRND: 20　BARO: 223　TIME ZONE: PACIFIC (UTC -8)　WBAN: 24229

| | ELEMENT | POR | JAN | FEB | MAR | APR | MAY | JUN | JUL | AUG | SEP | OCT | NOV | DEC | YEAR |
|---|---|---|---|---|---|---|---|---|---|---|---|---|---|---|---|
| TEMPERATURE °F | NORMAL DAILY MAXIMUM | 30 | 45.6 | 50.3 | 55.7 | 60.5 | 66.7 | 72.7 | 79.3 | 79.7 | 74.6 | 63.3 | 51.8 | 45.4 | 62.1 |
| | MEAN DAILY MAXIMUM | 79 | 45.0 | 50.3 | 55.7 | 61.3 | 67.7 | 73.3 | 79.9 | 79.7 | 74.6 | 63.8 | 52.6 | 46.2 | 62.5 |
| | HIGHEST DAILY MAXIMUM | 69 | 66 | 71 | 80 | 90 | 100 | 102 | 107 | 107 | 105 | 92 | 73 | 65 | 107 |
| | YEAR OF OCCURRENCE | | 2005 | 1988 | 1947 | 1998 | 1983 | 2006 | 1965 | 1981 | 1988 | 1987 | 1975 | 1993 | AUG 1981 |
| | MEAN OF EXTREME MAXS. | 79 | 57.1 | 60.7 | 68.8 | 77.7 | 86.1 | 90.8 | 96.0 | 94.9 | 90.3 | 78.3 | 63.5 | 57.8 | 76.8 |
| | NORMAL DAILY MINIMUM | 30 | 34.2 | 35.9 | 38.6 | 41.9 | 47.5 | 52.6 | 56.9 | 57.3 | 52.5 | 45.2 | 39.8 | 35.0 | 44.8 |
| | MEAN DAILY MINIMUM | 79 | 34.2 | 36.3 | 39.1 | 42.5 | 47.9 | 53.2 | 57.0 | 57.1 | 52.5 | 46.0 | 39.9 | 35.9 | 45.1 |
| | LOWEST DAILY MINIMUM | 69 | -2 | -3 | 19 | 29 | 29 | 39 | 43 | 44 | 34 | 26 | 13 | 6 | -3 |
| | YEAR OF OCCURRENCE | | 1950 | 1950 | 1989 | 1955 | 1954 | 1966 | 1955 | 1980 | 1965 | 1971 | 1985 | 1964 | FEB 1950 |
| | MEAN OF EXTREME MINS. | 79 | 21.7 | 24.8 | 29.7 | 33.8 | 38.6 | 45.8 | 49.9 | 49.5 | 43.0 | 34.9 | 28.6 | 24.1 | 35.4 |
| | NORMAL DRY BULB | 30 | 39.9 | 43.1 | 47.2 | 51.2 | 57.1 | 62.7 | 68.1 | 68.5 | 63.6 | 54.3 | 45.8 | 40.2 | 53.5 |
| | MEAN DRY BULB | 79 | 39.6 | 43.3 | 47.4 | 51.9 | 57.8 | 63.3 | 68.5 | 68.4 | 63.6 | 54.9 | 46.3 | 41.1 | 53.8 |
| | MEAN WET BULB | 26 | 38.2 | 39.3 | 43.1 | 46.4 | 51.1 | 55.2 | 59.1 | 59.4 | 56.0 | 49.9 | 43.5 | 37.6 | 48.2 |
| | MEAN DEW POINT | 26 | 35.9 | 36.3 | 40.1 | 43.4 | 48.0 | 51.6 | 55.2 | 55.5 | 52.4 | 47.4 | 41.5 | 35.4 | 45.2 |
| | NORMAL NO. DAYS WITH: | | | | | | | | | | | | | | |
| | MAXIMUM >= 90 | 30 | 0.0 | 0.0 | 0.0 | * | 0.4 | 1.1 | 4.5 | 4.6 | 2.1 | 0.1 | 0.0 | 0.0 | 12.8 |
| | MAXIMUM <= 32 | 30 | 1.3 | 0.4 | 0.0 | 0.0 | 0.0 | 0.0 | 0.0 | 0.0 | 0.0 | 0.0 | 0.2 | 1.2 | 3.1 |
| | MINIMUM <= 32 | 30 | 11.6 | 7.2 | 3.0 | 0.7 | 0.0 | 0.0 | 0.0 | 0.0 | 0.0 | 0.4 | 4.4 | 9.8 | 37.1 |
| | MINIMUM <= 0 | 30 | 0.0 | 0.0 | 0.0 | 0.0 | 0.0 | 0.0 | 0.0 | 0.0 | 0.0 | 0.0 | 0.0 | 0.0 | 0.0 |
| H/C | NORMAL HEATING DEG. DAYS | 30 | 765 | 605 | 536 | 400 | 243 | 96 | 21 | 21 | 78 | 319 | 560 | 756 | 4400 |
| | NORMAL COOLING DEG. DAYS | 30 | 0 | 0 | 0 | 1 | 14 | 43 | 133 | 145 | 52 | 2 | 0 | 0 | 390 |
| RH | NORMAL (PERCENT) | 30 | 81 | 78 | 75 | 73 | 70 | 67 | 64 | 65 | 69 | 78 | 82 | 83 | 74 |
| | HOUR 04 LST | 30 | 85 | 85 | 86 | 86 | 85 | 83 | 81 | 82 | 86 | 90 | 88 | 86 | 85 |
| | HOUR 10 LST | 30 | 82 | 80 | 76 | 70 | 67 | 64 | 62 | 64 | 67 | 77 | 83 | 83 | 73 |
| | HOUR 16 LST | 30 | 74 | 66 | 60 | 56 | 53 | 50 | 45 | 43 | 47 | 59 | 74 | 77 | 59 |
| | HOUR 22 LST | 30 | 83 | 80 | 78 | 75 | 73 | 70 | 67 | 68 | 73 | 82 | 84 | 84 | 76 |
| S | PERCENT POSSIBLE SUNSHINE | 46 | 28 | 38 | 48 | 52 | 57 | 56 | 69 | 66 | 62 | 44 | 28 | 23 | 48 |
| W/O | MEAN NO. DAYS WITH: | | | | | | | | | | | | | | |
| | HEAVY FOG(VISBY <= 1/4 MI) | 46 | 4.7 | 3.9 | 2.1 | 0.9 | 0.2 | 0.1 | 0.0 | 0.1 | 1.5 | 6.0 | 5.7 | 4.3 | 29.5 |
| | THUNDERSTORMS | 62 | 0.0 | 0.1 | 0.5 | 1.0 | 1.3 | 0.8 | 0.8 | 0.9 | 0.7 | 0.4 | 0.3 | 0.0 | 6.8 |
| CLOUDNESS | MEAN: | | | | | | | | | | | | | | |
| | SUNRISE-SUNSET (OKTAS) | 47 | 6.7 | 6.6 | 6.4 | 6.2 | 5.8 | 5.4 | 3.8 | 4.0 | 4.3 | 5.6 | 6.4 | 6.8 | 5.7 |
| | MIDNIGHT-MIDNIGHT (OKTAS) | 32 | 6.3 | 6.0 | 5.7 | 5.7 | 5.3 | 4.9 | 3.6 | 3.6 | 3.9 | 5.0 | 6.3 | 6.4 | 5.2 |
| | MEAN NO. DAYS WITH: | | | | | | | | | | | | | | |
| | CLEAR | 47 | 2.9 | 2.9 | 3.3 | 3.5 | 5.0 | 6.2 | 12.6 | 11.4 | 10.3 | 5.4 | 2.8 | 2.1 | 68.4 |
| | PARTLY CLOUDY | 47 | 3.6 | 3.8 | 4.9 | 5.8 | 7.2 | 7.7 | 8.5 | 9.6 | 8.1 | 7.6 | 4.3 | 3.3 | 74.4 |
| | CLOUDY | 47 | 24.4 | 21.5 | 22.8 | 20.7 | 18.9 | 16.1 | 9.9 | 10.1 | 11.6 | 18.0 | 22.9 | 25.5 | 222.4 |
| PR | MEAN STATION PRESSURE(IN) | 26 | 30.04 | 30.01 | 29.99 | 30.00 | 29.98 | 29.98 | 29.98 | 29.95 | 29.96 | 30.01 | 30.02 | 30.04 | 30.00 |
| | MEAN SEA-LEVEL PRES. (IN) | 26 | 30.10 | 30.07 | 30.06 | 30.06 | 30.04 | 30.04 | 30.04 | 30.02 | 30.02 | 30.08 | 30.08 | 30.10 | 30.06 |
| WINDS | MEAN SPEED (MPH) | 26 | 9.5 | 8.8 | 7.7 | 7.1 | 6.8 | 7.2 | 7.4 | 6.9 | 6.3 | 6.4 | 8.1 | 9.4 | 7.6 |
| | PREVAIL.DIR(TENS OF DEGS) | 39 | 12 | 12 | 12 | 33 | 33 | 33 | 34 | 33 | 33 | 12 | 12 | 12 | 34 |
| | MAXIMUM 2-MINUTE: | | | | | | | | | | | | | | |
| | SPEED (MPH) | 14 | 43 | 43 | 43 | 37 | 33 | 39 | 24 | 26 | 31 | 43 | 43 | 43 | 43 |
| | DIR. (TENS OF DEGS) | | 20 | 08 | 21 | 18 | 24 | 19 | 32 | 32 | 07 | 08 | 20 | 22 | 22 |
| | YEAR OF OCCURRENCE | | 2000 | 2006 | 1999 | 1996 | 2000 | 2009 | 2007 | 1999 | 2000 | 2002 | 1998 | 2006 | DEC 2006 |
| | MAXIMUM 3-SECOND: | | | | | | | | | | | | | | |
| | SPEED (MPH) | 14 | 59 | 52 | 51 | 46 | 41 | 49 | 36 | 33 | 44 | 49 | 53 | 53 | 59 |
| | DIR. (TENS OF DEGS) | | 19 | 10 | 20 | 09 | 24 | 19 | 28 | 17 | 31 | 09 | 19 | 22 | 19 |
| | YEAR OF OCCURRENCE | | 2000 | 2006 | 1999 | 1999 | 2009 | 2009 | 2007 | 1997 | 2004 | 2002 | 2006 | 2006 | JAN 2000 |
| PRECIPITATION | NORMAL (IN) | 30 | 5.07 | 4.18 | 3.71 | 2.64 | 2.38 | 1.59 | 0.72 | 0.93 | 1.65 | 2.88 | 5.61 | 5.71 | 37.07 |
| | MAXIMUM MONTHLY (IN) | 69 | 12.83 | 10.03 | 7.52 | 5.26 | 5.55 | 4.06 | 2.68 | 4.53 | 4.30 | 8.41 | 11.92 | 13.35 | 13.35 |
| | YEAR OF OCCURRENCE | | 1953 | 1996 | 1957 | 1993 | 1998 | 1984 | 1983 | 1968 | 1986 | 2006 | 1996 | | DEC 1996 |
| | MINIMUM MONTHLY (IN) | 69 | 0.06 | 0.72 | 1.10 | 0.53 | 0.10 | 0.03 | 0.00 | T | T | 0.19 | 0.77 | 1.38 | 0.00 |
| | YEAR OF OCCURRENCE | | 1985 | 1993 | 1965 | 1956 | 1992 | 1951 | 1967 | 1970 | 1993 | 1988 | 1976 | 1976 | JUL 1967 |
| | MAXIMUM IN 24 HOURS (IN) | 69 | 3.13 | 2.46 | 1.83 | 1.47 | 1.47 | 1.82 | 1.09 | 1.54 | 2.38 | 4.44 | 4.10 | 2.59 | 4.44 |
| | YEAR OF OCCURRENCE | | 2009 | 1994 | 1943 | 1962 | 1968 | 1958 | 1978 | 1977 | 1982 | 1994 | 1996 | 1977 | OCT 1994 |
| | NORMAL NO. DAYS WITH: | | | | | | | | | | | | | | |
| | PRECIPITATION >= 0.01 | 30 | 17.2 | 15.8 | 17.2 | 15.3 | 12.8 | 8.8 | 4.4 | 4.8 | 7.5 | 11.4 | 18.9 | 18.3 | 152.4 |
| | PRECIPITATION >= 1.00 | 30 | 0.8 | 0.5 | 0.1 | 0.2 | 0.1 | 0.1 | * | 0.1 | 0.2 | 0.3 | 0.9 | 1.1 | 4.4 |
| SNOWFALL | NORMAL (IN) | 30 | 1.6 | 1.6 | 0.1 | 0.* | 0.* | 0.0 | 0.0 | 0.0 | 0.0 | 0.0 | 0.6 | 1.3 | 5.2 |
| | MAXIMUM MONTHLY (IN) | 55 | 41.4 | 13.2 | 12.9 | T | 0.6 | T | 0.0 | T | T | 0.2 | 8.2 | 15.7 | 41.4 |
| | YEAR OF OCCURRENCE | | 1950 | 1949 | 1951 | 1995 | 1953 | 1995 | | 1989 | 1949 | 1950 | 1955 | 1968 | JAN 1950 |
| | MAXIMUM IN 24 HOURS (IN) | 55 | 10.6 | 6.4 | 7.7 | T | 0.5 | T | 0.0 | T | T | 0.2 | 7.4 | 8.0 | 10.6 |
| | YEAR OF OCCURRENCE' | | 1950 | 1993 | 1951 | 1995 | 1953 | 1995 | | 1989 | 1949 | 1950 | 1977 | 1964 | JAN 1950 |
| | MAXIMUM SNOW DEPTH (IN) | 48 | 15 | 14 | 5 | 0 | 0 | 0 | 0 | 0 | 0 | 0 | 5 | 11 | 15 |
| | YEAR OF OCCURRENCE | | 1950 | 1950 | 1951 | | | | | | | | 1977 | 1964 | JAN 1950 |
| | NORMAL NO. DAYS WITH: | | | | | | | | | | | | | | |
| | SNOWFALL >= 1.0 | 30 | 0.4 | 0.5 | 0.0 | 0.0 | 0.0 | 0.0 | 0.0 | 0.0 | 0.0 | 0.0 | 0.2 | 0.6 | 1.7 |

## PRECIPITATION (inches) 2009 PORTLAND (KPDX)

| YEAR | JAN | FEB | MAR | APR | MAY | JUN | JUL | AUG | SEP | OCT | NOV | DEC | ANNUAL |
|------|-----|-----|-----|-----|-----|-----|-----|-----|-----|-----|-----|-----|--------|
| 1980 | 8.51 | 4.01 | 3.11 | 2.58 | 2.19 | 2.50 | 0.19 | 0.39 | 1.56 | 1.18 | 6.47 | 9.72 | 42.41 |
| 1981 | 1.47 | 3.86 | 2.33 | 1.79 | 2.25 | 3.23 | 0.24 | 0.15 | 1.86 | 4.12 | 4.62 | 8.37 | 34.29 |
| 1982 | 6.31 | 5.98 | 2.38 | 3.56 | 0.46 | 1.66 | 0.94 | 1.66 | 3.98 | 4.44 | 3.51 | 8.16 | 43.04 |
| 1983 | 6.23 | 7.78 | 6.80 | 1.87 | 1.30 | 1.95 | 2.68 | 2.29 | 0.39 | 1.95 | 8.65 | 5.30 | 47.19 |
| 1984 | 2.01 | 3.93 | 3.19 | 3.20 | 3.41 | 4.06 | T | 0.09 | 1.46 | 3.85 | 9.74 | 2.56 | 37.50 |
| 1985 | 0.06 | 1.79 | 3.08 | 1.07 | 1.52 | 2.34 | 0.55 | 0.48 | 2.76 | 2.75 | 3.89 | 2.19 | 22.48 |
| 1986 | 4.65 | 5.31 | 2.60 | 1.91 | 2.19 | 0.23 | 1.20 | 0.10 | 4.30 | 1.99 | 6.26 | 4.30 | 35.04 |
| 1987 | 6.93 | 2.45 | 4.91 | 1.94 | 1.63 | 0.14 | 1.03 | 0.35 | 0.30 | 0.27 | 1.96 | 8.00 | 29.91 |
| 1988 | 4.95 | 1.17 | 3.13 | 4.57 | 2.53 | 2.34 | 0.69 | 0.10 | 1.76 | 0.19 | 7.92 | 2.37 | 31.72 |
| 1989 | 3.30 | 2.84 | 6.73 | 2.08 | 2.87 | 0.78 | 0.91 | 1.07 | 1.48 | 1.73 | 3.18 | 3.08 | 30.05 |
| 1990 | 7.95 | 3.43 | 2.52 | 2.31 | 2.37 | 1.94 | 0.32 | 0.95 | 0.34 | 4.65 | 3.68 | 2.40 | 32.86 |
| 1991 | 2.56 | 3.65 | 4.64 | 4.05 | 3.34 | 2.31 | 0.07 | 0.70 | 0.02 | 1.51 | 6.36 | 4.34 | 33.55 |
| 1992 | 4.31 | 4.12 | 1.87 | 3.82 | 0.10 | 0.60 | 0.67 | 0.49 | 1.12 | 2.87 | 4.55 | 4.98 | 29.50 |
| 1993 | 3.06 | 0.72 | 4.39 | 5.26 | 4.36 | 1.69 | 2.41 | 0.37 | T | 1.59 | 1.50 | 5.01 | 30.36 |
| 1994 | 3.56 | 4.92 | 1.84 | 1.91 | 0.56 | 1.67 | 0.07 | 0.13 | 1.13 | 8.41 | 5.91 | 4.85 | 34.96 |
| 1995 | 5.56 | 3.19 | 3.82 | 3.49 | 1.65 | 2.62 | 1.23 | 0.81 | 1.31 | 3.15 | 11.15 | 5.35 | 43.33 |
| 1996 | 7.15 | 10.03 | 3.24 | 5.12 | 4.88 | 0.44 | .73 | .25 | 3.05 | 5.39 | 9.58 | 13.35 | 63.21 |
| 1997 | 7.32 | 1.63 | 7.14 | 3.73 | 3.63 | 2.83 | 0.52 | 1.58 | 1.98 | 6.40 | 4.02 | 3.03 | 43.81 |
| 1998 | 6.77 | 5.27 | 4.06 | 1.04 | 5.55 | 1.73 | 0.59 | T | 1.09 | 2.16 | 11.02 | 6.74 | 46.02 |
| 1999 | 6.63 | 8.73 | 4.03 | 1.56 | 1.97 | 1.73 | 0.51 | 0.75 | 0.10 | 2.44 | 6.81 | 3.62 | 38.88 |
| 2000 | 5.66 | 4.50 | 3.21 | 1.82 | 2.70 | 1.19 | 0.15 | 0.12 | 1.67 | 3.25 | 2.46 | 3.47 | 30.20 |
| 2001 | 1.47 | 1.29 | 3.11 | 2.85 | 0.91 | 1.79 | 0.95 | 0.74 | 0.70 | 3.12 | 6.89 | 6.62 | 30.44 |
| 2002 | 6.22 | 3.55 | 3.40 | 2.34 | 1.86 | 1.57 | 0.19 | 0.04 | 1.54 | 0.63 | 1.91 | 8.00 | 31.25 |
| 2003 | 7.64 | 2.37 | 5.75 | 4.37 | 1.49 | 0.31 | T | 0.19 | 0.85 | 3.01 | 4.09 | 7.45 | 37.52 |
| 2004 | 4.86 | 3.95 | 1.53 | 1.01 | 1.78 | 1.12 | 0.04 | 2.68 | 1.03 | 3.36 | 2.38 | 3.91 | 27.65 |
| 2005 | 1.94 | 1.30 | 3.77 | 3.49 | 4.34 | 2.21 | 0.41 | 1.05 | 1.70 | 3.39 | 4.98 | 7.52 | 36.10 |
| 2006 | 10.92 | 2.15 | 2.96 | 2.46 | 3.00 | 0.92 | 0.47 | 0.10 | 0.86 | 1.39 | 11.92 | 5.85 | 43.00 |
| 2007 | 2.72 | 3.47 | 3.20 | 2.01 | 1.45 | 1.08 | 0.55 | 0.46 | 2.04 | 3.26 | 4.25 | 7.57 | 32.06 |
| 2008 | 4.71 | 2.19 | 3.71 | 2.08 | 2.02 | 1.00 | 0.29 | 1.23 | 0.48 | 1.74 | 4.15 | 3.52 | 27.12 |
| 2009 | 4.50 | 1.36 | 3.36 | 2.31 | 3.26 | 1.30 | 0.34 | 0.76 | 1.40 | 3.02 | 5.13 | 3.76 | 30.50 |
| POR= 79 YRS | 5.44 | 4.07 | 3.73 | 2.50 | 2.17 | 1.53 | 0.50 | 0.84 | 1.59 | 3.06 | 5.53 | 6.42 | 37.38 |

WBAN : 24229

## AVERAGE TEMPERATURE (°F) 2009 PORTLAND (KPDX)

| YEAR | JAN | FEB | MAR | APR | MAY | JUN | JUL | AUG | SEP | OCT | NOV | DEC | ANNUAL |
|------|-----|-----|-----|-----|-----|-----|-----|-----|-----|-----|-----|-----|--------|
| 1980 | 35.1 | 42.5 | 46.3 | 53.8 | 57.3 | 60.7 | 68.9 | 66.4 | 63.8 | 56.0 | 48.5 | 44.0 | 53.6 |
| 1981 | 43.9 | 44.0 | 48.8 | 52.5 | 57.5 | 61.8 | 67.5 | 72.2 | 64.9 | 53.3 | 48.8 | 42.7 | 54.8 |
| 1982 | 39.7 | 43.6 | 48.5 | 49.0 | 57.6 | 66.0 | 67.5 | 68.6 | 63.2 | 54.9 | 44.4 | 41.7 | 53.7 |
| 1983 | 44.4 | 47.3 | 50.7 | 52.7 | 60.4 | 62.8 | 66.5 | 69.1 | 61.5 | 54.2 | 49.3 | 36.4 | 54.6 |
| 1984 | 42.2 | 45.9 | 51.1 | 50.4 | 56.4 | 62.2 | 69.1 | 69.4 | 63.7 | 52.9 | 46.7 | 38.3 | 54.0 |
| 1985 | 36.1 | 41.1 | 45.8 | 53.9 | 58.3 | 64.4 | 74.1 | 69.3 | 60.8 | 52.7 | 37.3 | 33.0 | 52.2 |
| 1986 | 42.5 | 43.7 | 51.3 | 50.2 | 57.6 | 66.3 | 65.3 | 72.3 | 61.5 | 57.0 | 47.7 | 40.6 | 54.7 |
| 1987 | 39.6 | 45.2 | 48.7 | 54.2 | 60.4 | 66.5 | 67.2 | 70.5 | 65.5 | 58.2 | 48.8 | 39.1 | 55.3 |
| 1988 | 39.0 | 44.7 | 47.2 | 52.2 | 56.4 | 62.4 | 68.4 | 68.0 | 64.0 | 58.3 | 47.5 | 42.0 | 54.2 |
| 1989 | 42.2 | 36.0 | 45.6 | 56.0 | 58.0 | 64.3 | 65.5 | 66.1 | 65.3 | 54.9 | 48.6 | 40.3 | 53.6 |
| 1990 | 43.4 | 41.9 | 49.4 | 54.5 | 56.7 | 63.6 | 71.2 | 70.9 | 67.0 | 53.9 | 48.4 | 34.7 | 54.6 |
| 1991 | 38.9 | 48.8 | 46.2 | 50.8 | 54.7 | 59.9 | 69.9 | 70.1 | 67.4 | 55.4 | 47.5 | 42.7 | 54.4 |
| 1992 | 44.5 | 48.1 | 52.3 | 55.4 | 63.1 | 67.4 | 70.2 | 69.8 | 62.3 | 55.9 | 46.3 | 39.3 | 56.2 |
| 1993 | 36.5 | 40.1 | 47.9 | 52.6 | 61.1 | 62.5 | 64.3 | 68.5 | 65.7 | 57.4 | 41.5 | 41.4 | 53.3 |
| 1994 | 44.5 | 40.9 | 50.2 | 54.2 | 60.7 | 62.9 | 71.1 | 70.2 | 67.6 | 54.1 | 42.7 | 42.3 | 55.1 |
| 1995 | 43.4 | 46.5 | 48.3 | 51.3 | 60.7 | 63.8 | 70.6 | 67.3 | 67.1 | 54.5 | 51.7 | 42.4 | 55.6 |
| 1996 | 40.7 | 41.6 | 47.7 | 53.5 | 55.3 | 63.2 | 72.0 | 70.4 | 61.8 | 54.0 | 45.5 | 41.6 | 53.9 |
| 1997 | 41.0 | 42.9 | 46.7 | 51.0 | 62.2 | 62.7 | 68.9 | 71.6 | 65.9 | 53.7 | 49.8 | 41.1 | 54.8 |
| 1998 | 43.0 | 46.1 | 48.8 | 52.8 | 56.3 | 63.4 | 71.0 | 71.0 | 66.9 | 54.3 | 48.9 | 40.5 | 55.3 |
| 1999 | 42.8 | 43.7 | 45.8 | 50.7 | 54.9 | 61.2 | 67.2 | 69.0 | 64.6 | 55.1 | 50.1 | 43.9 | 54.1 |
| 2000 | 39.8 | 43.7 | 45.6 | 54.0 | 57.6 | 65.3 | 67.8 | 67.8 | 63.9 | 55.2 | 42.8 | 40.4 | 53.7 |
| 2001 | 41.2 | 42.0 | 48.0 | 49.6 | 59.4 | 61.0 | 66.5 | 69.2 | 65.1 | 53.6 | 48.0 | 42.0 | 53.8 |
| 2002 | 41.2 | 44.4 | 44.9 | 52.3 | 55.7 | 64.7 | 69.8 | 69.3 | 64.7 | 54.1 | 47.6 | 43.4 | 54.3 |
| 2003 | 44.8 | 44.3 | 49.0 | 50.8 | 57.3 | 66.1 | 71.6 | 70.1 | 66.8 | 58.2 | 45.2 | 42.5 | 55.6 |
| 2004 | 38.5 | 45.3 | 51.2 | 56.3 | 60.1 | 65.5 | 71.5 | 71.5 | 62.8 | 56.3 | 45.9 | 43.3 | 55.7 |
| 2005 | 41.8 | 43.6 | 50.0 | 52.3 | 60.2 | 62.0 | 70.3 | 70.7 | 62.5 | 56.3 | 44.1 | 39.8 | 54.5 |
| 2006 | 45.5 | 42.0 | 46.1 | 53.1 | 59.8 | 66.3 | 71.0 | 69.2 | 65.1 | 54.0 | 47.4 | 40.0 | 55.0 |
| 2007 | 38.1 | 44.2 | 50.1 | 51.7 | 58.6 | 62.9 | 70.7 | 68.3 | 62.4 | 53.1 | 44.8 | 41.0 | 53.8 |
| 2008 | 38.8 | 44.9 | 45.4 | 48.5 | 58.9 | 61.9 | 68.8 | 69.6 | 65.2 | 53.6 | 49.2 | 37.5 | 53.5 |
| 2009 | 40.1 | 41.4 | 45.3 | 52.3 | 60.1 | 65.7 | 73.6 | 69.9 | 66.2 | 54.7 | 47.8 | 35.6 | 54.4 |
| POR= 79 YRS | 39.6 | 43.3 | 47.4 | 51.9 | 57.8 | 63.3 | 68.5 | 68.4 | 63.6 | 54.9 | 46.3 | 41.1 | 53.8 |

## HEATING DEGREE DAYS (base 65°F) 2009 PORTLAND (KPDX)

| YEAR | JUL | AUG | SEP | OCT | NOV | DEC | JAN | FEB | MAR | APR | MAY | JUN | TOTAL |
|---|---|---|---|---|---|---|---|---|---|---|---|---|---|
| 1980-81 | 15 | 25 | 64 | 284 | 485 | 644 | 650 | 583 | 494 | 372 | 229 | 108 | 3953 |
| 1981-82 | 23 | 5 | 76 | 355 | 478 | 687 | 780 | 596 | 502 | 472 | 229 | 71 | 4274 |
| 1982-83 | 22 | 10 | 99 | 307 | 614 | 715 | 635 | 492 | 435 | 363 | 184 | 81 | 3957 |
| 1983-84 | 27 | 2 | 109 | 325 | 463 | 880 | 701 | 546 | 425 | 430 | 269 | 115 | 4292 |
| 1984-85 | 9 | 2 | 80 | 377 | 539 | 820 | 893 | 664 | 588 | 327 | 213 | 62 | 4574 |
| 1985-86 | 0 | 7 | 124 | 373 | 826 | 982 | 691 | 591 | 417 | 437 | 265 | 43 | 4756 |
| 1986-87 | 37 | 0 | 148 | 242 | 510 | 750 | 780 | 550 | 495 | 321 | 173 | 51 | 4057 |
| 1987-88 | 22 | 2 | 54 | 214 | 479 | 798 | 801 | 581 | 544 | 380 | 272 | 109 | 4256 |
| 1988-89 | 33 | 15 | 91 | 208 | 518 | 705 | 699 | 805 | 594 | 263 | 219 | 77 | 4227 |
| 1989-90 | 32 | 27 | 44 | 306 | 486 | 759 | 664 | 641 | 476 | 308 | 251 | 78 | 4072 |
| 1990-91 | 10 | 5 | 14 | 336 | 492 | 933 | 802 | 446 | 575 | 420 | 310 | 156 | 4499 |
| 1991-92 | 4 | 13 | 25 | 295 | 517 | 685 | 627 | 483 | 387 | 282 | 108 | 35 | 3461 |
| 1992-93 | 5 | 9 | 107 | 274 | 556 | 789 | 877 | 692 | 522 | 366 | 135 | 94 | 4426 |
| 1993-94 | 37 | 17 | 57 | 231 | 698 | 726 | 630 | 668 | 451 | 316 | 143 | 85 | 4059 |
| 1994-95 | 16 | 0 | 23 | 330 | 662 | 696 | 666 | 514 | 510 | 405 | 149 | 95 | 4066 |
| 1995-96 | 3 | 15 | 20 | 318 | 392 | 693 | 747 | 671 | 530 | 340 | 294 | 70 | 4093 |
| 1996-97 | 16 | 8 | 113 | 340 | 581 | 717 | 737 | 612 | 563 | 413 | 116 | 79 | 4295 |
| 1997-98 | 7 | 0 | 42 | 341 | 452 | 734 | 676 | 523 | 498 | 376 | 268 | 72 | 3989 |
| 1998-99 | 12 | 4 | 32 | 323 | 475 | 753 | 677 | 592 | 587 | 424 | 312 | 150 | 4341 |
| 1999-00 | 38 | 16 | 73 | 298 | 441 | 648 | 774 | 612 | 593 | 323 | 229 | 62 | 4107 |
| 2000-01 | 19 | 13 | 75 | 296 | 661 | 759 | 733 | 640 | 517 | 455 | 201 | 140 | 4509 |
| 2001-02 | 22 | 5 | 55 | 347 | 504 | 706 | 730 | 572 | 618 | 375 | 282 | 72 | 4288 |
| 2002-03 | 8 | 9 | 58 | 330 | 516 | 662 | 620 | 573 | 490 | 418 | 245 | 67 | 3996 |
| 2003-04 | 1 | 1 | 41 | 219 | 586 | 691 | 815 | 568 | 420 | 258 | 151 | 60 | 3811 |
| 2004-05 | 1 | 3 | 72 | 265 | 569 | 667 | 714 | 595 | 460 | 376 | 171 | 105 | 3998 |
| 2005-06 | 5 | 3 | 93 | 262 | 620 | 774 | 594 | 638 | 579 | 349 | 187 | 36 | 4140 |
| 2006-07 | 5 | 10 | 64 | 334 | 523 | 767 | 826 | 574 | 453 | 393 | 217 | 88 | 4254 |
| 2007-08 | 0 | 7 | 116 | 361 | 600 | 740 | 805 | 578 | 598 | 490 | 223 | 147 | 4665 |
| 2008-09 | 6 | 17 | 57 | 349 | 465 | 847 | 767 | 656 | 599 | 374 | 175 | 29 | 4341 |
| 2009- | 7 | 11 | 45 | 312 | 510 | 906 | | | | | | | |

WBAN : 24229

## COOLING DEGREE DAYS (base 65°F) 2009 PORTLAND (KPDX)

| YEAR | JAN | FEB | MAR | APR | MAY | JUN | JUL | AUG | SEP | OCT | NOV | DEC | TOTAL |
|---|---|---|---|---|---|---|---|---|---|---|---|---|---|
| 1980 | 0 | 0 | 0 | 1 | 0 | 2 | 141 | 75 | 35 | 12 | 0 | 0 | 266 |
| 1981 | 0 | 0 | 0 | 3 | 4 | 16 | 109 | 232 | 82 | 0 | 0 | 0 | 446 |
| 1982 | 0 | 0 | 0 | 0 | 4 | 107 | 103 | 127 | 50 | 0 | 0 | 0 | 391 |
| 1983 | 0 | 0 | 0 | 0 | 48 | 23 | 80 | 137 | 12 | 0 | 0 | 0 | 300 |
| 1984 | 0 | 0 | 0 | 0 | 10 | 34 | 140 | 144 | 47 | 6 | 0 | 0 | 381 |
| 1985 | 0 | 0 | 0 | 0 | 11 | 53 | 291 | 145 | 5 | 0 | 0 | 0 | 505 |
| 1986 | 0 | 0 | 0 | 0 | 0 | 40 | 87 | 52 | 235 | 50 | 0 | 0 | 464 |
| 1987 | 0 | 0 | 0 | 4 | 37 | 102 | 95 | 177 | 77 | 12 | 0 | 0 | 504 |
| 1988 | 0 | 0 | 0 | 0 | 10 | 39 | 147 | 115 | 67 | 8 | 0 | 0 | 386 |
| 1989 | 0 | 0 | 0 | 0 | 9 | 62 | 53 | 66 | 60 | 0 | 0 | 0 | 250 |
| 1990 | 0 | 0 | 0 | 2 | 3 | 45 | 206 | 193 | 83 | 0 | 0 | 0 | 532 |
| 1991 | 0 | 0 | 0 | 0 | 0 | 7 | 164 | 176 | 102 | 2 | 0 | 0 | 451 |
| 1992 | 0 | 0 | 0 | 1 | 57 | 114 | 174 | 164 | 31 | 0 | 0 | 0 | 541 |
| 1993 | 0 | 0 | 0 | 0 | 21 | 24 | 22 | 132 | 80 | 3 | 0 | 0 | 282 |
| 1994 | 0 | 0 | 0 | 0 | 15 | 31 | 210 | 168 | 109 | 0 | 0 | 0 | 533 |
| 1995 | 0 | 0 | 0 | 0 | 23 | 63 | 184 | 92 | 91 | 0 | 0 | 0 | 453 |
| 1996 | 0 | 0 | 0 | 0 | 1 | 36 | 240 | 181 | 25 | 7 | 0 | 0 | 490 |
| 1997 | 0 | 0 | 0 | 0 | 0 | 37 | 16 | 134 | 210 | 75 | 0 | 0 | 472 |
| 1998 | 0 | 0 | 0 | 14 | 4 | 30 | 206 | 197 | 96 | 0 | 0 | 0 | 547 |
| 1999 | 0 | 0 | 0 | 0 | 9 | 41 | 111 | 149 | 68 | 0 | 0 | 0 | 378 |
| 2000 | 0 | 0 | 0 | 0 | 3 | 78 | 113 | 107 | 50 | 0 | 0 | 0 | 351 |
| 2001 | 0 | 0 | 0 | 1 | 35 | 28 | 75 | 144 | 64 | 0 | 0 | 0 | 347 |
| 2002 | 0 | 0 | 0 | 0 | 1 | 69 | 165 | 149 | 57 | 0 | 0 | 0 | 441 |
| 2003 | 0 | 0 | 0 | 0 | 16 | 108 | 212 | 165 | 99 | 14 | 0 | 0 | 614 |
| 2004 | 0 | 0 | 0 | 6 | 8 | 82 | 210 | 211 | 15 | 3 | 0 | 0 | 535 |
| 2005 | 0 | 0 | 0 | 0 | 26 | 22 | 176 | 190 | 25 | 0 | 0 | 0 | 439 |
| 2006 | 0 | 0 | 0 | 1 | 31 | 82 | 199 | 148 | 74 | 0 | 1 | 0 | 536 |
| 2007 | 0 | 0 | 0 | 0 | 22 | 34 | 184 | 114 | 46 | 0 | 0 | 0 | 400 |
| 2008 | 0 | 0 | 0 | 0 | 40 | 58 | 134 | 169 | 73 | 0 | 0 | 0 | 474 |
| 2009 | 0 | 0 | 0 | 0 | 34 | 56 | 281 | 170 | 86 | 0 | 0 | 0 | 627 |

## SNOWFALL (inches) 2009 PORTLAND (KPDX)

| YEAR | JUL | AUG | SEP | OCT | NOV | DEC | JAN | FEB | MAR | APR | MAY | JUN | TOTAL |
|---|---|---|---|---|---|---|---|---|---|---|---|---|---|
| 1976-77 | 0.0 | 0.0 | 0.0 | 0.0 | 0.0 | 0.0 | T | 0.0 | T | 0.0 | T | 0.0 | T |
| 1977-78 | 0.0 | 0.0 | 0.0 | 0.0 | 7.6 | T | 0.0 | 0.0 | 0.1 | T | T | 0.0 | 7.7 |
| 1978-79 | 0.0 | 0.0 | 0.0 | 0.0 | 3.0 | 2.4 | 1.9 | 1.1 | T | T | 0.0 | 0.0 | 8.4 |
| 1979-80 | 0.0 | 0.0 | 0.0 | 0.0 | 0.0 | T | 12.4 | T | T | T | 0.0 | T | 12.4 |
| 1980-81 | 0.0 | 0.0 | 0.0 | 0.0 | T | T | 0.0 | T | T | T | T | T | T |
| 1981-82 | 0.0 | 0.0 | 0.0 | 0.0 | 0.0 | 2.0 | 2.1 | T | T | T | 0.0 | 0.0 | 4.1 |
| 1982-83 | 0.0 | 0.0 | 0.0 | T | T | 0.0 | 0.0 | 0.0 | T | T | T | 0.0 | T |
| 1983-84 | 0.0 | 0.0 | 0.0 | 0.0 | 0.0 | 2.3 | 0.1 | T | T | 0.0 | T | 0.0 | 2.4 |
| 1984-85 | 0.0 | 0.0 | 0.0 | T | 0.0 | 2.8 | T | 4.8 | T | T | 0.0 | 0.0 | 7.6 |
| 1985-86 | 0.0 | 0.0 | 0.0 | 0.0 | 3.4 | 1.6 | T | 5.8 | T | T | T | 0.0 | 10.8 |
| 1986-87 | 0.0 | 0.0 | 0.0 | 0.0 | T | 0.1 | 0.0 | T | T | T | T | 0.0 | 0.1 |
| 1987-88 | 0.0 | 0.0 | 0.0 | 0.0 | 0.0 | 2.9 | 0.6 | 0.0 | 0.0 | T | T | 0.0 | 3.5 |
| 1988-89 | 0.0 | 0.0 | 0.0 | 0.0 | T | T | 0.9 | 0.3 | 2.0 | T | 0.0 | 0.0 | 3.2 |
| 1989-90 | 0.0 | T | 0.0 | T | 0.0 | 0.0 | T | 8.3 | T | 0.0 | 0.0 | 0.0 | 8.3 |
| 1990-91 | 0.0 | 0.0 | 0.0 | 0.0 | 0.0 | 1.3 | 0.6 | 0.0 | T | T | 0.0 | 0.0 | 1.9 |
| 1991-92 | 0.0 | 0.0 | 0.0 | 0.0 | T | 0.0 | 0.0 | 0.0 | 0.0 | T | 0.0 | 0.0 | T |
| 1992-93 | 0.0 | 0.0 | 0.0 | 0.0 | T | 4.6 | 2.9 | 6.6 | 0.0 | T | T | 0.0 | 14.1 |
| 1993-94 | 0.0 | 0.0 | 0.0 | 0.0 | T | T | 0.0 | 2.6 | T | T | 0.0 | T | 2.6 |
| 1994-95 | 0.0 | 0.0 | 0.0 | 0.0 | 0.3 | 1.1 | T | 3.6 | 0.4 | T | 0.0 | T | 5.4 |
| 1995-96 | 0.0 | 0.0 | 0.0 | 0.0 | 0.0 | | | | | | | | |
| 1996-97 | | | | | | | | | | | | | |
| 1997-98 | | | | | | | | | | | | | |
| 1998-99 | | | | | | | | | | | | | |
| 1999-00 | | | | | | | | | | | | | |
| 2000-01 | | | | | | | | | | | | | |
| 2001-02 | | | | | | | | | | | | | |
| 2002-03 | | | | | | | | | | | | | |
| 2003-04 | | | | | | | | | | | | | |
| 2004-05 | | | | | | | | | | | | | |
| 2005- | | | | | | | | | | | | | |
| POR=<br>48 YRS | 0.0 | T | 0.0 | T | 0.5 | 1.6 | 3.3 | 1.3 | 0.5 | T | T | T | 7.2 |

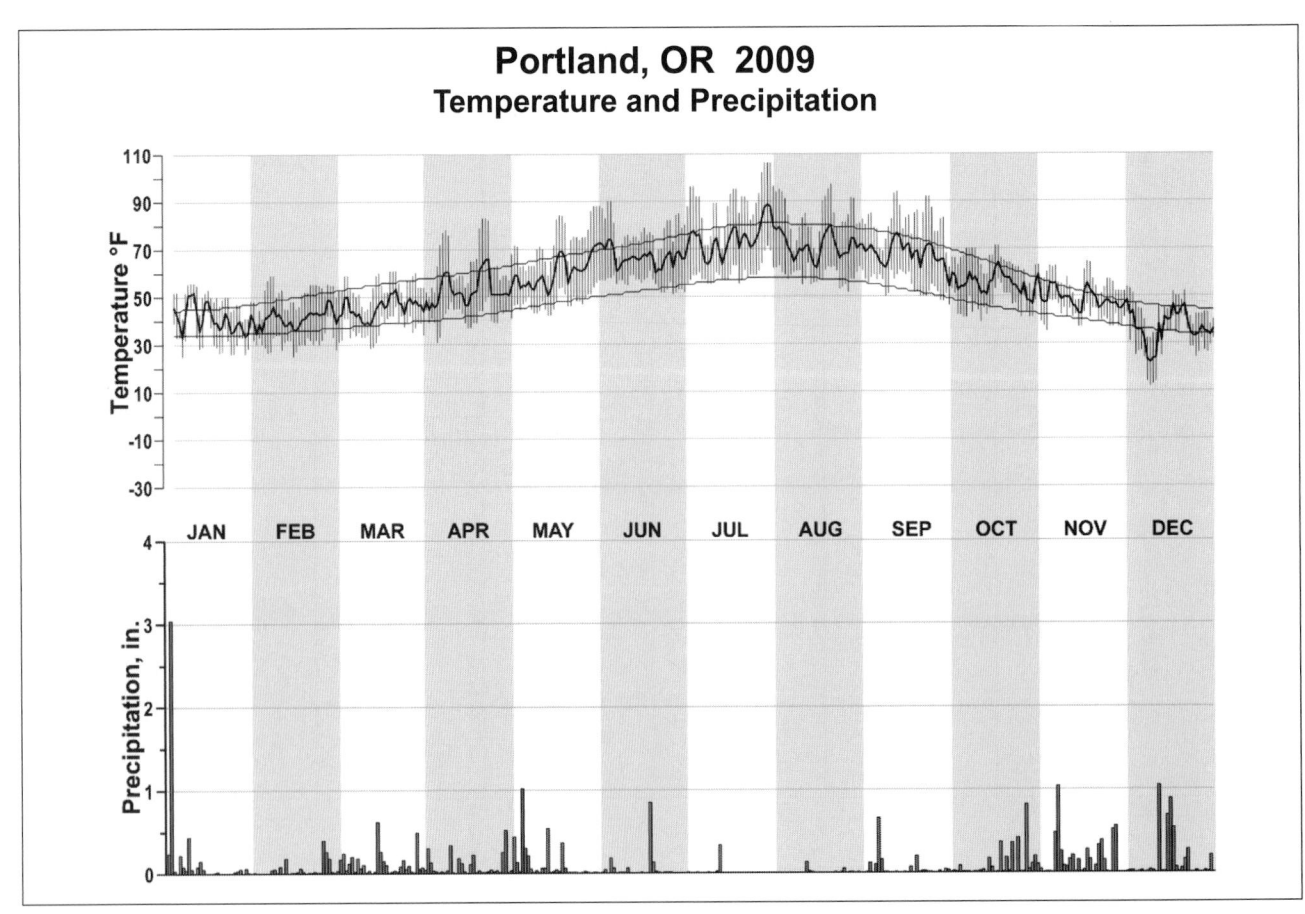

# Portland, OR 2009
## Temperature and Precipitation

# 2009
# AVOCA/WILKES-BARRE/SCRANTON PENNSYLVANIA
# (KAVP)

The Wilkes-Barre Scranton National Weather Service Office is located about midway between the two cities, at the southwest end of the crescent-shaped Lackawanna River Valley. The river flows through this valley and empties into the Susquehanna River and the Wyoming Valley a few miles west of the airport. The surrounding mountains protect both cities and the airport from high winds. They influence the temperature and precipitation during both summer and winter, causing wide departures in both within a few miles of the station. Because of the proximity of the mountains, the climate is relatively cool in summer with frequent shower and thunderstorm activity, usually of brief duration. The winter temperatures in the valley are not severe. The occurrence of sub-zero temperatures and severe snowstorms is infrequent. A high percentage of the winter precipitation occurs as rain.

Although severe snowstorms are infrequent, when they do occur they approach blizzard conditions. High winds cause huge drifts and normal routines are disrupted for several days.

While the incidence of tornadoes is very low, Wilkes-Barre has occasionally been hit with these storms which caused loss of life and great property damage.

The area has felt the effects of tropical storms. Considerable wind damage has occasionally occurred, but the most devastating damage has come from flooding caused by the large amounts of precipitation deposited by the storms. The worst natural disaster to hit the region was the result of the flooding caused by a hurricane.

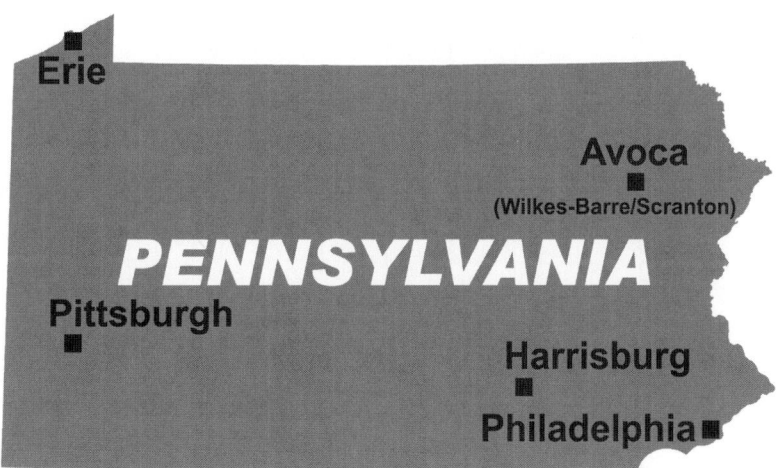

# NORMALS, MEANS, AND EXTREMES
## WILKES-BARRE/SCRANTON (KAVP)

LATITUDE: 41 ° 20'N  LONGITUDE: -75 ° 43'W  ELEVATION (FT): GRND: 953  BARO: 958  TIME ZONE: EASTERN  (UTC -5)  WBAN: 14777

| | ELEMENT | POR | JAN | FEB | MAR | APR | MAY | JUN | JUL | AUG | SEP | OCT | NOV | DEC | YEAR |
|---|---|---|---|---|---|---|---|---|---|---|---|---|---|---|---|
| **TEMPERATURE °F** | NORMAL DAILY MAXIMUM | 30 | 34.1 | 37.3 | 47.3 | 59.2 | 70.8 | 78.2 | 82.6 | 80.5 | 72.4 | 61.2 | 49.3 | 38.6 | 59.3 |
| | MEAN DAILY MAXIMUM | 61 | 33.2 | 35.8 | 45.1 | 58.5 | 69.6 | 77.8 | 82.2 | 80.3 | 72.5 | 61.3 | 48.9 | 37.1 | 58.5 |
| | HIGHEST DAILY MAXIMUM | 54 | 67 | 71 | 85 | 93 | 93 | 97 | 101 | 98 | 95 | 87 | 80 | 71 | 101 |
| | YEAR OF OCCURRENCE | | 2008 | 1985 | 1998 | 2009 | 2006 | 1964 | 1988 | 2001 | 1983 | 2007 | 1982 | 2006 | JUL 1988 |
| | MEAN OF EXTREME MAXS. | 61 | 54.6 | 55.1 | 67.9 | 80.9 | 86.1 | 90.2 | 91.9 | 90.0 | 86.4 | 78.2 | 68.2 | 57.3 | 75.6 |
| | NORMAL DAILY MINIMUM | 30 | 18.5 | 20.4 | 28.4 | 38.1 | 48.4 | 56.7 | 61.5 | 60.1 | 52.6 | 41.7 | 33.7 | 24.2 | 40.4 |
| | MEAN DAILY MINIMUM | 61 | 18.8 | 20.1 | 27.8 | 38.3 | 48.0 | 56.7 | 61.4 | 60.0 | 52.5 | 41.9 | 33.6 | 23.6 | 40.2 |
| | LOWEST DAILY MINIMUM | 54 | -21 | -16 | -4 | 14 | 27 | 34 | 43 | 38 | 29 | 19 | 9 | -9 | -21 |
| | YEAR OF OCCURRENCE | | 1994 | 1979 | 1967 | 1982 | 1974 | 1972 | 1979 | 1982 | 2000 | 1972 | 2000 | 1989 | JAN 1994 |
| | MEAN OF EXTREME MINS. | 61 | 0.1 | 1.6 | 9.8 | 23.6 | 34.0 | 43.0 | 49.4 | 46.9 | 37.2 | 27.7 | 18.1 | 4.9 | 24.7 |
| | NORMAL DRY BULB | 30 | 26.3 | 28.9 | 37.9 | 48.7 | 59.6 | 67.5 | 72.1 | 70.3 | 62.5 | 51.5 | 41.5 | 31.4 | 49.9 |
| | MEAN DRY BULB | 61 | 26.0 | 28.0 | 36.5 | 48.4 | 58.8 | 67.4 | 71.8 | 70.2 | 62.5 | 51.6 | 41.3 | 30.4 | 49.4 |
| | MEAN WET BULB | 26 | 23.4 | 24.6 | 31.2 | 41.1 | 51.1 | 60.5 | 64.2 | 63.3 | 56.8 | 46.0 | 37.0 | 27.5 | 43.9 |
| | MEAN DEW POINT | 26 | 19.5 | 19.7 | 26.0 | 35.7 | 47.0 | 57.3 | 61.4 | 60.6 | 54.2 | 42.5 | 33.0 | 23.6 | 40.0 |
| | NORMAL NO. DAYS WITH: | | | | | | | | | | | | | | |
| | MAXIMUM >= 90 | 30 | 0.0 | 0.0 | 0.0 | 0.1 | 0.3 | 1.0 | 3.6 | 2.0 | 0.4 | 0.0 | 0.0 | 0.0 | 7.4 |
| | MAXIMUM <= 32 | 30 | 14.6 | 10.9 | 3.3 | 0.2 | 0.0 | 0.0 | 0.0 | 0.0 | 0.0 | 0.0 | 1.4 | 8.7 | 39.1 |
| | MINIMUM <= 32 | 30 | 27.7 | 24.0 | 20.5 | 8.4 | 0.5 | 0.0 | 0.0 | 0.0 | 0.1 | 4.4 | 13.7 | 24.6 | 123.9 |
| | MINIMUM <= 0 | 30 | 1.8 | 1.1 | 0.1 | 0.0 | 0.0 | 0.0 | 0.0 | 0.0 | 0.0 | 0.0 | 0.0 | 0.5 | 3.5 |
| **H/C** | NORMAL HEATING DEG. DAYS | 30 | 1214 | 1027 | 857 | 510 | 219 | 53 | 9 | 18 | 138 | 431 | 711 | 1047 | 6234 |
| | NORMAL COOLING DEG. DAYS | 30 | 0 | 0 | 1 | 5 | 36 | 114 | 220 | 174 | 57 | 4 | 0 | 0 | 611 |
| **RH** | NORMAL (PERCENT) | 30 | 71 | 67 | 63 | 61 | 65 | 70 | 71 | 73 | 75 | 72 | 71 | 72 | 69 |
| | HOUR 01 LST | 30 | 73 | 70 | 68 | 66 | 74 | 81 | 82 | 84 | 84 | 80 | 75 | 75 | 76 |
| | HOUR 07 LST | 30 | 76 | 75 | 74 | 72 | 77 | 83 | 84 | 87 | 88 | 84 | 79 | 77 | 80 |
| | HOUR 13 LST | 30 | 66 | 61 | 56 | 52 | 54 | 57 | 57 | 59 | 62 | 59 | 64 | 66 | 59 |
| | HOUR 19 LST | 30 | 68 | 63 | 58 | 54 | 57 | 62 | 63 | 66 | 71 | 67 | 68 | 69 | 64 |
| **S** | PERCENT POSSIBLE SUNSHINE | 41 | 41 | 47 | 50 | 53 | 57 | 61 | 62 | 61 | 55 | 52 | 36 | 34 | 51 |
| **W/O** | MEAN NO. DAYS WITH: | | | | | | | | | | | | | | |
| | HEAVY FOG(VISBY <= 1/4 MI) | 46 | 1.9 | 1.9 | 1.7 | 1.1 | 0.9 | 1.1 | 1.5 | 1.8 | 2.4 | 1.7 | 1.4 | 2.3 | 19.7 |
| | THUNDERSTORMS | 59 | 0.2 | 0.2 | 0.6 | 1.8 | 3.4 | 5.1 | 6.2 | 4.4 | 2.1 | 0.8 | 0.4 | 0.2 | 25.4 |
| **CLOUDNESS** | MEAN: | | | | | | | | | | | | | | |
| | SUNRISE-SUNSET (OKTAS) | 41 | 6.0 | 5.8 | 5.7 | 5.4 | 5.4 | 5.0 | 4.8 | 4.8 | 4.8 | 4.8 | 5.9 | 6.0 | 5.4 |
| | MIDNIGHT-MIDNIGHT (OKTAS) | 32 | 5.7 | 5.4 | 5.4 | 5.0 | 5.0 | 4.6 | 4.5 | 4.4 | 4.6 | 4.6 | 5.7 | 5.9 | 5.1 |
| | MEAN NO. DAYS WITH: | | | | | | | | | | | | | | |
| | CLEAR | 41 | 4.3 | 4.5 | 5.6 | 6.2 | 6.0 | 6.8 | 6.1 | 7.1 | 7.0 | 8.2 | 3.8 | 3.7 | 69.3 |
| | PARTLY CLOUDY | 41 | 7.1 | 7.0 | 7.7 | 7.5 | 9.5 | 10.9 | 12.5 | 11.3 | 9.4 | 8.2 | 6.7 | 6.5 | 104.3 |
| | CLOUDY | 41 | 19.6 | 16.7 | 17.7 | 16.3 | 15.5 | 12.3 | 11.7 | 11.8 | 13.0 | 13.9 | 18.7 | 19.9 | 187.1 |
| **PR** | MEAN STATION PRESSURE(IN) | 26 | 29.01 | 29.01 | 29.00 | 28.95 | 28.97 | 28.97 | 28.99 | 29.03 | 29.07 | 29.06 | 29.05 | 29.03 | 29.01 |
| | MEAN SEA-LEVEL PRES. (IN) | 26 | 30.06 | 30.05 | 30.03 | 29.96 | 29.98 | 29.97 | 29.98 | 30.03 | 30.03 | 30.07 | 30.08 | 30.07 | 30.03 |
| **WINDS** | MEAN SPEED (MPH) | 26 | 7.9 | 8.2 | 8.5 | 8.2 | 7.4 | 6.6 | 6.3 | 5.9 | 6.3 | 6.7 | 7.5 | 7.7 | 7.3 |
| | PREVAIL.DIR(TENS OF DEGS) | 34 | 24 | 25 | 33 | 35 | 23 | 24 | 25 | 11 | 23 | 24 | 24 | 24 | 24 |
| | MAXIMUM 2-MINUTE: | | | | | | | | | | | | | | |
| | SPEED (MPH) | 13 | 36 | 38 | 39 | 34 | 45 | 36 | 39 | 46 | 45 | 36 | 40 | 43 | 46 |
| | DIR. (TENS OF DEGS) | | 23 | 26 | 28 | 26 | 31 | 29 | 36 | 25 | 32 | 28 | 27 | 26 | 25 |
| | YEAR OF OCCURRENCE | | 1999 | 1997 | 2000 | 2000 | 1998 | 2000 | 2001 | 1997 | 1998 | 2003 | 2005 | 2000 | AUG 1997 |
| | MAXIMUM 3-SECOND | | | | | | | | | | | | | | |
| | SPEED (MPH) | 13 | 49 | 48 | 53 | 46 | 55 | 49 | 47 | 55 | 51 | 48 | 52 | 55 | 55 |
| | DIR. (TENS OF DEGS) | | 25 | 26 | 25 | 25 | 31 | 32 | 01 | 23 | 35 | 28 | 26 | 20 | 20 |
| | YEAR OF OCCURRENCE | | 2008 | 2009 | 1997 | 2009 | 1998 | 2008 | 2001 | 1997 | 2001 | 2003 | 2004 | 2006 | DEC 2006 |
| **PRECIPITATION** | NORMAL (IN) | 30 | 2.46 | 2.08 | 2.69 | 3.28 | 3.69 | 3.97 | 3.74 | 3.10 | 3.86 | 3.02 | 3.12 | 2.55 | 37.56 |
| | MAXIMUM MONTHLY (IN) | 54 | 6.48 | 8.06 | 5.21 | 9.56 | 8.02 | 9.00 | 7.25 | 6.78 | 9.76 | 8.12 | 7.69 | 6.58 | 9.76 |
| | YEAR OF OCCURRENCE | | 1979 | 1981 | 2008 | 1983 | 1989 | 2006 | 1986 | 1994 | 1999 | 1976 | 1972 | 1983 | SEP 1999 |
| | MINIMUM MONTHLY (IN) | 54 | 0.39 | 0.30 | 0.49 | 0.96 | 0.77 | 0.27 | 1.04 | 0.95 | 0.80 | 0.03 | 0.80 | 0.35 | 0.03 |
| | YEAR OF OCCURRENCE | | 1980 | 1968 | 1981 | 1997 | 1959 | 1966 | 1993 | 1995 | 2005 | 1963 | 1976 | 1958 | OCT 1963 |
| | MAXIMUM IN 24 HOURS (IN) | 54 | 3.49 | 3.11 | 3.02 | 3.80 | 2.58 | 3.61 | 2.83 | 3.69 | 6.52 | 4.28 | 3.60 | 2.86 | 6.52 |
| | YEAR OF OCCURRENCE | | 2008 | 1981 | 1986 | 1983 | 1972 | 1973 | 2000 | 2003 | 1985 | 1995 | 1996 | 1983 | SEP 1985 |
| | NORMAL NO. DAYS WITH: | | | | | | | | | | | | | | |
| | PRECIPITATION >= 0.01 | 30 | 12.5 | 10.8 | 12.4 | 12.4 | 13.0 | 12.7 | 11.0 | 10.9 | 10.3 | 10.1 | 11.6 | 12.2 | 139.9 |
| | PRECIPITATION >= 1.00 | 30 | 0.3 | 0.3 | 0.2 | 0.6 | 0.8 | 0.6 | 0.8 | 0.6 | 1.1 | 0.7 | 0.6 | 0.3 | 6.9 |
| **SNOWFALL** | NORMAL (IN) | 30 | 13.5 | 10.2 | 8.7 | 2.8 | 0.1 | 0.0 | 0.0 | 0.0 | 0.0 | 0.1 | 4.3 | 7.3 | 47.0 |
| | MAXIMUM MONTHLY (IN) | 43 | 42.3 | 22.0 | 32.0 | 26.7 | 2.4 | T | T | T | T | 4.4 | 22.5 | 33.9 | 42.3 |
| | YEAR OF OCCURRENCE | | 1994 | 1964 | 1993 | 1983 | 1977 | 2008 | 1995 | 1993 | 1993 | 1962 | 1971 | 1969 | JAN 1994 |
| | MAXIMUM IN 24 HOURS (IN) | 43 | 20.6 | 13.3 | 20.4 | 12.2 | 2.4 | T | T | T | T | 4.4 | 20.5 | 12.4 | 20.6 |
| | YEAR OF OCCURRENCE' | | 1996 | 1961 | 1993 | 1983 | 1977 | 2008 | 1995 | 1993 | 1993 | 1962 | 1971 | 1969 | JAN 1996 |
| | MAXIMUM SNOW DEPTH (IN) | 46 | 27 | 25 | 21 | 14 | 0 | 0 | 0 | 0 | 0 | 2 | 17 | 22 | 27 |
| | YEAR OF OCCURRENCE | | 1994 | 1961 | 1993 | 1983 | | | | | | 1962 | 1971 | 1969 | JAN 1994 |
| | NORMAL NO. DAYS WITH: | | | | | | | | | | | | | | |
| | SNOWFALL >= 1.0 | 30 | 3.6 | 2.9 | 2.7 | 0.6 | 0.0 | 0.0 | 0.0 | 0.0 | 0.0 | 0.0 | 1.0 | 2.2 | 13.0 |

## PRECIPITATION (inches) 2009  WILKES-BARRE/SCRANTON (KAVP)

| YEAR | JAN | FEB | MAR | APR | MAY | JUN | JUL | AUG | SEP | OCT | NOV | DEC | ANNUAL |
|------|-----|-----|-----|-----|-----|-----|-----|-----|-----|-----|-----|-----|--------|
| 1980 | 0.39 | 0.69 | 3.72 | 2.35 | 2.37 | 4.36 | 3.76 | 1.23 | 1.43 | 2.17 | 2.83 | 1.24 | 26.54 |
| 1981 | 0.63 | 8.06 | 0.49 | 3.54 | 3.00 | 3.45 | 4.27 | 1.75 | 2.74 | 3.50 | 1.84 | 2.13 | 35.40 |
| 1982 | 2.71 | 2.28 | 2.55 | 3.48 | 3.52 | 7.22 | 3.32 | 3.42 | 1.10 | 0.84 | 3.44 | 1.52 | 35.40 |
| 1983 | 1.17 | 1.46 | 3.28 | 9.56 | 3.28 | 4.81 | 2.76 | 1.77 | 2.12 | 2.73 | 3.71 | 6.58 | 43.23 |
| 1984 | 1.11 | 2.92 | 2.42 | 4.09 | 6.70 | 4.75 | 5.12 | 2.81 | 1.36 | 2.30 | 2.63 | 2.36 | 38.57 |
| 1985 | 0.61 | 1.58 | 2.24 | 2.00 | 6.10 | 3.00 | 6.09 | 2.62 | 7.83 | 1.92 | 4.47 | 1.96 | 40.42 |
| 1986 | 2.59 | 2.58 | 4.25 | 2.98 | 2.24 | 6.77 | 7.25 | 3.94 | 3.07 | 2.61 | 3.94 | 2.04 | 44.26 |
| 1987 | 2.60 | 0.68 | 1.18 | 4.38 | 2.22 | 4.35 | 5.80 | 4.16 | 8.15 | 2.77 | 2.24 | 0.99 | 39.52 |
| 1988 | 1.41 | 2.32 | 1.97 | 2.65 | 4.24 | 0.82 | 6.26 | 5.03 | 1.89 | 1.93 | 3.33 | 1.08 | 32.93 |
| 1989 | 1.02 | 1.73 | 2.23 | 0.97 | 8.02 | 6.10 | 2.76 | 2.92 | 3.92 | 4.73 | 3.57 | 0.96 | 38.93 |
| 1990 | 3.81 | 2.70 | 1.88 | 2.48 | 5.27 | 4.78 | 4.36 | 5.69 | 3.16 | 4.33 | 3.33 | 4.30 | 46.09 |
| 1991 | 1.54 | 1.35 | 2.91 | 2.69 | 2.84 | 1.72 | 2.45 | 3.28 | 2.55 | 2.41 | 3.81 | 2.73 | 30.28 |
| 1992 | 1.23 | 1.41 | 2.65 | 1.67 | 4.21 | 1.45 | 3.83 | 2.13 | 2.65 | 1.91 | 3.47 | 2.91 | 29.52 |
| 1993 | 1.97 | 1.24 | 4.02 | 7.47 | 1.38 | 1.85 | 1.04 | 3.82 | 6.55 | 3.99 | 3.32 | 3.09 | 39.74 |
| 1994 | 3.29 | 1.00 | 3.80 | 4.90 | 2.68 | 3.54 | 3.22 | 6.78 | 5.12 | 0.76 | 4.54 | 2.29 | 41.92 |
| 1995 | 2.73 | 1.45 | 1.34 | 2.15 | 1.40 | 1.45 | 3.16 | 0.95 | 2.31 | 7.15 | 3.65 | 1.22 | 28.96 |
| 1996 | 6.40 | 1.46 | 2.55 | 5.26 | 4.03 | 3.90 | 6.01 | 1.24 | 3.92 | 4.40 | 4.45 | 5.37 | 48.99 |
| 1997 | 1.57 | 1.13 | 3.20 | 0.96 | 2.46 | 2.96 | 1.33 | 5.30 | 1.91 | 1.17 | 3.48 | 2.25 | 27.72 |
| 1998 | 2.96 | 2.91 | 2.54 | 5.41 | 4.38 | 4.16 | 2.43 | 2.69 | 2.46 | 2.99 | 0.95 | 0.86 | 34.74 |
| 1999 | 4.85 | 1.41 | 2.73 | 2.16 | 2.62 | 2.89 | 1.63 | 2.12 | 9.76 | 1.52 | 2.31 | 1.24 | 35.24 |
| 2000 | 2.08 | 2.40 | 2.84 | 2.90 | 2.86 | 6.09 | 6.21 | 1.93 | 3.07 | 1.46 | 1.44 | 2.77 | 36.05 |
| 2001 | 1.13 | 1.14 | 2.54 | 1.97 | 2.92 | 3.04 | 3.67 | 3.02 | 3.95 | 1.01 | 1.89 | 1.11 | 27.39 |
| 2002 | 1.71 | 1.02 | 2.89 | 3.30 | 4.39 | 3.95 | 2.11 | 3.84 | 5.33 | 5.49 | 2.65 | 3.47 | 40.15 |
| 2003 | 1.55 | 1.50 | 1.86 | 2.17 | 3.79 | 7.54 | 3.76 | 6.17 | 8.35 | 5.32 | 3.47 | 3.97 | 49.45 |
| 2004 | 2.08 | 1.92 | 1.44 | 3.22 | 4.64 | 4.31 | 4.20 | 4.90 | 9.38 | 1.96 | 3.60 | 3.39 | 45.04 |
| 2005 | 5.35 | 1.70 | 3.28 | 4.02 | 1.26 | 2.07 | 2.21 | 2.17 | 0.80 | 7.66 | 3.40 | 2.76 | 36.68 |
| 2006 | 4.26 | 1.22 | 1.31 | 3.15 | 2.16 | 9.00 | 3.02 | 4.40 | 5.76 | 3.83 | 6.06 | 1.39 | 45.56 |
| 2007 | 2.65 | 3.41 | 2.33 | 4.06 | 1.72 | 3.03 | 4.15 | 5.04 | 1.33 | 7.87 | 4.13 | 4.06 | 43.78 |
| 2008 | 2.51 | 5.90 | 5.21 | 1.99 | 3.85 | 3.73 | 4.33 | 2.05 | 4.81 | 2.52 | 1.26 | 5.09 | 43.25 |
| 2009 | 1.98 | 0.88 | 1.13 | 1.53 | 5.47 | 4.60 | 4.46 | 5.12 | 2.24 | 4.18 | 1.16 | 2.71 | 35.46 |
| POR= 61 YRS | 2.38 | 2.08 | 2.59 | 3.22 | 3.47 | 3.86 | 3.72 | 3.46 | 3.59 | 3.04 | 3.15 | 2.69 | 37.25 |

WBAN : 14777

## AVERAGE TEMPERATURE (°F) 2009  WILKES-BARRE/SCRANTON (KAVP)

| YEAR | JAN | FEB | MAR | APR | MAY | JUN | JUL | AUG | SEP | OCT | NOV | DEC | ANNUAL |
|------|-----|-----|-----|-----|-----|-----|-----|-----|-----|-----|-----|-----|--------|
| 1980 | 27.8 | 24.2 | 35.9 | 51.0 | 61.8 | 65.3 | 73.2 | 75.2 | 66.1 | 49.8 | 37.6 | 25.7 | 49.5 |
| 1981 | 19.5 | 34.9 | 36.2 | 51.0 | 59.8 | 68.5 | 72.2 | 70.1 | 62.1 | 49.1 | 41.1 | 29.1 | 49.5 |
| 1982 | 18.7 | 27.8 | 36.0 | 46.3 | 61.2 | 64.3 | 71.0 | 66.2 | 62.4 | 52.3 | 44.3 | 36.7 | 48.9 |
| 1983 | 27.3 | 29.4 | 38.8 | 45.9 | 55.7 | 67.6 | 72.4 | 71.6 | 64.8 | 52.5 | 42.8 | 27.1 | 49.7 |
| 1984 | 23.0 | 35.9 | 30.9 | 48.3 | 57.5 | 69.3 | 71.6 | 72.6 | 61.7 | 58.0 | 40.8 | 37.3 | 50.6 |
| 1985 | 21.5 | 29.8 | 39.1 | 51.4 | 60.6 | 63.8 | 70.1 | 69.0 | 64.0 | 52.7 | 44.5 | 26.7 | 49.4 |
| 1986 | 27.2 | 26.1 | 39.8 | 49.3 | 62.7 | 66.2 | 71.0 | 67.4 | 61.6 | 51.2 | 37.3 | 32.8 | 49.4 |
| 1987 | 24.7 | 25.1 | 39.7 | 50.7 | 60.0 | 68.7 | 73.5 | 68.3 | 61.8 | 47.4 | 41.3 | 32.7 | 49.5 |
| 1988 | 22.0 | 27.4 | 38.1 | 46.6 | 60.1 | 65.5 | 75.8 | 72.9 | 60.4 | 46.3 | 43.3 | 29.8 | 49.0 |
| 1989 | 31.3 | 28.0 | 37.2 | 45.6 | 57.6 | 67.0 | 70.4 | 68.5 | 62.2 | 52.8 | 39.7 | 18.6 | 48.2 |
| 1990 | 35.3 | 33.3 | 40.6 | 50.2 | 56.1 | 67.6 | 71.7 | 69.5 | 61.6 | 55.3 | 44.2 | 36.0 | 51.8 |
| 1991 | 27.5 | 33.4 | 40.4 | 51.7 | 65.7 | 69.5 | 73.4 | 72.2 | 61.8 | 53.3 | 41.4 | 32.9 | 51.9 |
| 1992 | 28.6 | 30.3 | 34.4 | 47.4 | 58.7 | 66.3 | 71.1 | 68.4 | 62.8 | 49.0 | 42.2 | 32.8 | 49.3 |
| 1993 | 32.0 | 23.8 | 33.3 | 49.9 | 61.7 | 68.8 | 75.5 | 72.7 | 61.2 | 50.0 | 41.4 | 30.0 | 50.0 |
| 1994 | 18.8 | 24.4 | 35.4 | 52.2 | 56.6 | 69.9 | 73.8 | 67.3 | 60.9 | 50.9 | 47.4 | 36.6 | 49.5 |
| 1995 | 32.9 | 26.5 | 40.5 | 46.4 | 59.2 | 69.9 | 75.2 | 74.7 | 62.4 | 56.5 | 35.7 | 25.7 | 50.5 |
| 1996 | 24.1 | 27.6 | 33.2 | 47.4 | 55.9 | 68.2 | 68.9 | 69.6 | 62.5 | 50.8 | 36.4 | 34.5 | 48.3 |
| 1997 | 25.3 | 32.9 | 36.8 | 45.5 | 54.3 | 65.9 | 70.4 | 67.5 | 60.3 | 50.2 | 37.1 | 31.3 | 48.1 |
| 1998 | 34.1 | 35.3 | 41.1 | 49.5 | 63.3 | 65.7 | 69.8 | 70.4 | 64.1 | 51.6 | 41.8 | 35.9 | 51.9 |
| 1999 | 27.9 | 31.9 | 35.5 | 48.1 | 60.2 | 68.5 | 74.7 | 68.9 | 64.5 | 49.0 | 45.3 | 33.1 | 50.6 |
| 2000 | 24.7 | 30.5 | 42.8 | 47.8 | 59.9 | 66.8 | 67.2 | 67.7 | 60.0 | 51.6 | 38.3 | 23.0 | 48.4 |
| 2001 | 26.6 | 30.1 | 33.1 | 48.6 | 59.6 | 68.5 | 68.3 | 73.1 | 60.4 | 52.9 | 46.9 | 36.9 | 50.4 |
| 2002 | 33.4 | 34.8 | 39.8 | 50.9 | 55.7 | 68.3 | 73.5 | 72.8 | 65.0 | 49.5 | 39.7 | 29.0 | 51.0 |
| 2003 | 20.8 | 24.6 | 37.3 | 47.5 | 56.9 | 64.6 | 71.3 | 71.4 | 63.1 | 48.8 | 44.4 | 31.6 | 48.5 |
| 2004 | 18.7 | 26.8 | 39.3 | 49.3 | 63.8 | 65.1 | 70.0 | 68.8 | 64.1 | 50.4 | 42.2 | 30.7 | 49.1 |
| 2005 | 23.0 | 29.8 | 31.8 | 51.4 | 55.6 | 72.6 | 75.3 | 75.6 | 67.6 | 52.8 | 43.6 | 27.0 | 50.5 |
| 2006 | 34.9 | 29.8 | 37.4 | 50.4 | 59.0 | 66.5 | 73.5 | 70.5 | 60.8 | 50.1 | 45.7 | 38.0 | 51.4 |
| 2007 | 31.4 | 21.5 | 36.0 | 45.1 | 60.9 | 68.8 | 70.2 | 70.7 | 65.6 | 59.0 | 39.4 | 30.7 | 49.9 |
| 2008 | 29.4 | 28.3 | 36.8 | 51.9 | 54.9 | 69.6 | 72.2 | 67.5 | 64.4 | 49.1 | 40.5 | 31.5 | 49.7 |
| 2009 | 20.8 | 30.6 | 39.5 | 51.6 | 59.3 | 66.4 | 68.3 | 70.3 | 60.9 | 49.0 | 45.1 | 29.3 | 49.3 |
| POR= 61 YRS | 26.0 | 28.0 | 36.5 | 48.4 | 58.8 | 67.4 | 71.8 | 70.2 | 62.5 | 51.6 | 41.3 | 30.4 | 49.4 |

## HEATING DEGREE DAYS (base 65°F) 2009  WILKES-BARRE/SCRANTON (KAVP)

| YEAR | JUL | AUG | SEP | OCT | NOV | DEC | JAN | FEB | MAR | APR | MAY | JUN | TOTAL |
|------|-----|-----|-----|-----|-----|-----|-----|-----|-----|-----|-----|-----|-------|
| 1980-81 | 1 | 0 | 82 | 466 | 813 | 1211 | 1407 | 835 | 886 | 416 | 195 | 19 | 6331 |
| 1981-82 | 2 | 5 | 132 | 485 | 706 | 1105 | 1426 | 1034 | 896 | 554 | 147 | 68 | 6560 |
| 1982-83 | 17 | 55 | 112 | 390 | 619 | 870 | 1158 | 992 | 805 | 569 | 292 | 41 | 5920 |
| 1983-84 | 7 | 11 | 119 | 392 | 659 | 1169 | 1297 | 837 | 1052 | 493 | 247 | 34 | 6317 |
| 1984-85 | 7 | 6 | 148 | 219 | 719 | 852 | 1342 | 981 | 799 | 421 | 162 | 78 | 5734 |
| 1985-86 | 4 | 11 | 127 | 376 | 610 | 1181 | 1163 | 1083 | 777 | 467 | 140 | 61 | 6000 |
| 1986-87 | 16 | 50 | 139 | 428 | 823 | 990 | 1243 | 1111 | 779 | 425 | 208 | 20 | 6232 |
| 1987-88 | 2 | 34 | 119 | 539 | 706 | 995 | 1326 | 1082 | 823 | 546 | 176 | 91 | 6439 |
| 1988-89 | 13 | 12 | 156 | 574 | 643 | 1083 | 1037 | 1031 | 853 | 575 | 251 | 39 | 6267 |
| 1989-90 | 6 | 31 | 133 | 377 | 750 | 1433 | 915 | 881 | 757 | 465 | 269 | 44 | 6061 |
| 1990-91 | 10 | 13 | 152 | 320 | 619 | 894 | 1153 | 877 | 757 | 402 | 113 | 23 | 5333 |
| 1991-92 | 0 | 0 | 160 | 370 | 699 | 989 | 1121 | 1001 | 942 | 521 | 209 | 40 | 6052 |
| 1992-93 | 4 | 9 | 132 | 488 | 676 | 993 | 1017 | 1148 | 975 | 447 | 136 | 39 | 6064 |
| 1993-94 | 0 | 4 | 174 | 460 | 703 | 1077 | 1428 | 1131 | 910 | 390 | 270 | 33 | 6580 |
| 1994-95 | 0 | 31 | 143 | 427 | 522 | 875 | 986 | 1072 | 755 | 552 | 189 | 15 | 5567 |
| 1995-96 | 0 | 0 | 123 | 264 | 872 | 1210 | 1260 | 1075 | 979 | 525 | 312 | 16 | 6636 |
| 1996-97 | 9 | 5 | 140 | 432 | 852 | 938 | 1222 | 893 | 866 | 577 | 327 | 60 | 6321 |
| 1997-98 | 7 | 23 | 150 | 466 | 828 | 1036 | 949 | 828 | 754 | 460 | 104 | 97 | 5702 |
| 1998-99 | 2 | 11 | 90 | 408 | 689 | 896 | 1143 | 919 | 910 | 500 | 159 | 32 | 5759 |
| 1999-00 | 4 | 18 | 104 | 491 | 586 | 982 | 1243 | 995 | 683 | 510 | 196 | 59 | 5871 |
| 2000-01 | 26 | 32 | 202 | 410 | 794 | 1289 | 1181 | 973 | 982 | 491 | 187 | 45 | 6612 |
| 2001-02 | 28 | 0 | 156 | 370 | 537 | 863 | 975 | 840 | 775 | 459 | 309 | 35 | 5347 |
| 2002-03 | 4 | 7 | 68 | 483 | 751 | 1110 | 1363 | 1126 | 852 | 520 | 247 | 88 | 6619 |
| 2003-04 | 2 | 10 | 80 | 496 | 611 | 1031 | 1429 | 1103 | 787 | 472 | 119 | 72 | 6212 |
| 2004-05 | 1 | 26 | 63 | 445 | 677 | 1058 | 1294 | 978 | 1021 | 406 | 296 | 14 | 6279 |
| 2005-06 | 0 | 0 | 44 | 377 | 636 | 1172 | 927 | 980 | 849 | 431 | 211 | 54 | 5681 |
| 2006-07 | 0 | 6 | 134 | 454 | 573 | 829 | 1033 | 1212 | 893 | 590 | 177 | 28 | 5929 |
| 2007-08 | 19 | 22 | 72 | 231 | 759 | 1055 | 1095 | 1060 | 868 | 390 | 314 | 25 | 5910 |
| 2008-09 | 0 | 13 | 90 | 487 | 732 | 1031 | 1364 | 958 | 783 | 429 | 189 | 43 | 6119 |
| 2009- | 5 | 11 | 132 | 489 | 590 | 1099 | | | | | | | |

WBAN : 14777

## COOLING DEGREE DAYS (base 65°F) 2009  WILKES-BARRE/SCRANTON (KAVP)

| YEAR | JAN | FEB | MAR | APR | MAY | JUN | JUL | AUG | SEP | OCT | NOV | DEC | TOTAL |
|------|-----|-----|-----|-----|-----|-----|-----|-----|-----|-----|-----|-----|-------|
| 1980 | 0 | 0 | 0 | 0 | 42 | 107 | 263 | 322 | 122 | 3 | 0 | 0 | 859 |
| 1981 | 0 | 0 | 0 | 4 | 42 | 131 | 231 | 172 | 55 | 0 | 0 | 0 | 635 |
| 1982 | 0 | 0 | 0 | 1 | 34 | 55 | 208 | 98 | 41 | 3 | 5 | 0 | 445 |
| 1983 | 0 | 0 | 0 | 4 | 12 | 125 | 243 | 224 | 118 | 9 | 0 | 0 | 735 |
| 1984 | 0 | 0 | 0 | 0 | 20 | 165 | 218 | 248 | 58 | 12 | 0 | 0 | 721 |
| 1985 | 0 | 0 | 0 | 20 | 32 | 47 | 169 | 142 | 104 | 0 | 0 | 0 | 514 |
| 1986 | 0 | 0 | 2 | 0 | 76 | 104 | 205 | 131 | 44 | 8 | 0 | 0 | 570 |
| 1987 | 0 | 0 | 0 | 3 | 62 | 138 | 273 | 141 | 30 | 0 | 0 | 0 | 647 |
| 1988 | 0 | 0 | 0 | 0 | 34 | 111 | 356 | 266 | 23 | 3 | 0 | 0 | 793 |
| 1989 | 0 | 0 | 0 | 0 | 31 | 109 | 179 | 148 | 57 | 4 | 0 | 0 | 528 |
| 1990 | 0 | 0 | 8 | 29 | 2 | 127 | 225 | 160 | 56 | 24 | 0 | 0 | 631 |
| 1991 | 0 | 0 | 0 | 12 | 142 | 164 | 264 | 230 | 69 | 14 | 0 | 0 | 895 |
| 1992 | 0 | 0 | 0 | 0 | 21 | 85 | 201 | 123 | 71 | 0 | 0 | 0 | 501 |
| 1993 | 0 | 0 | 0 | 5 | 41 | 158 | 334 | 249 | 66 | 0 | 0 | 0 | 853 |
| 1994 | 0 | 0 | 0 | 13 | 18 | 188 | 280 | 109 | 27 | 0 | 1 | 0 | 636 |
| 1995 | 0 | 0 | 0 | 0 | 16 | 167 | 323 | 308 | 54 | 6 | 0 | 0 | 874 |
| 1996 | 0 | 0 | 0 | 0 | 38 | 117 | 132 | 155 | 71 | 0 | 0 | 0 | 513 |
| 1997 | 0 | 0 | 0 | 0 | 4 | 97 | 178 | 105 | 16 | 11 | 0 | 0 | 411 |
| 1998 | 0 | 0 | 17 | 0 | 59 | 128 | 157 | 184 | 71 | 0 | 0 | 0 | 616 |
| 1999 | 0 | 0 | 0 | 0 | 19 | 145 | 312 | 146 | 95 | 0 | 0 | 0 | 717 |
| 2000 | 0 | 0 | 0 | 0 | 45 | 117 | 101 | 126 | 60 | 0 | 0 | 0 | 449 |
| 2001 | 0 | 0 | 0 | 6 | 25 | 156 | 137 | 258 | 24 | 2 | 0 | 0 | 608 |
| 2002 | 0 | 0 | 0 | 44 | 26 | 139 | 276 | 254 | 74 | 8 | 0 | 0 | 821 |
| 2003 | 0 | 0 | 0 | 0 | 2 | 84 | 209 | 214 | 31 | 0 | 0 | 0 | 540 |
| 2004 | 0 | 0 | 0 | 7 | 91 | 82 | 164 | 153 | 41 | 0 | 0 | 0 | 538 |
| 2005 | 0 | 0 | 0 | 5 | 9 | 253 | 329 | 331 | 129 | 10 | 0 | 0 | 1066 |
| 2006 | 0 | 0 | 0 | 0 | 32 | 104 | 271 | 184 | 13 | 0 | 0 | 0 | 604 |
| 2007 | 0 | 0 | 0 | 1 | 56 | 147 | 187 | 205 | 98 | 52 | 0 | 0 | 746 |
| 2008 | 0 | 0 | 0 | 6 | 6 | 168 | 230 | 98 | 78 | 0 | 0 | 0 | 586 |
| 2009 | 0 | 0 | 0 | 32 | 20 | 92 | 116 | 183 | 13 | 0 | 0 | 0 | 456 |

**SNOWFALL (inches)  2009  WILKES-BARRE/SCRANTON (KAVP)**

| YEAR | JUL | AUG | SEP | OCT | NOV | DEC | JAN | FEB | MAR | APR | MAY | JUN | TOTAL |
|---|---|---|---|---|---|---|---|---|---|---|---|---|---|
| 1980-81 | 0.0 | 0.0 | 0.0 | T | 8.6 | 8.0 | 11.1 | 7.0 | 5.8 | T | 0.0 | 0.0 | 40.5 |
| 1981-82 | 0.0 | 0.0 | 0.0 | T | 1.0 | 14.2 | 14.1 | 13.5 | 8.7 | 8.1 | 0.0 | 0.0 | 59.6 |
| 1982-83 | 0.0 | 0.0 | 0.0 | 0.0 | 0.5 | 7.4 | 8.4 | 12.3 | 3.8 | 26.7 | 0.0 | 0.0 | 59.1 |
| 1983-84 | 0.0 | 0.0 | 0.0 | 0.0 | 3.1 | 2.7 | 11.2 | 4.0 | 18.4 | 0.0 | 0.0 | 0.0 | 39.4 |
| 1984-85 | 0.0 | 0.0 | 0.0 | 0.0 | 3.0 | 9.2 | 10.8 | 9.1 | 1.4 | 1.8 | 0.0 | 0.0 | 35.3 |
| 1985-86 | 0.0 | 0.0 | 0.0 | 0.0 | 1.7 | 13.4 | 12.9 | 11.6 | 1.1 | 8.6 | 0.0 | 0.0 | 49.3 |
| 1986-87 | 0.0 | 0.0 | 0.0 | 0.0 | 8.6 | 1.4 | 29.6 | 6.4 | 0.9 | 0.6 | 0.0 | 0.0 | 47.5 |
| 1987-88 | 0.0 | 0.0 | 0.0 | T | 6.4 | 6.4 | 13.0 | 14.9 | 4.3 | 0.7 | 0.0 | 0.0 | 45.7 |
| 1988-89 | 0.0 | 0.0 | 0.0 | T | T | 1.1 | 2.1 | 3.0 | 1.1 | T | 0.0 | 0.0 | 7.3 |
| 1989-90 | 0.0 | 0.0 | 0.0 | T | 2.6 | 8.3 | 10.8 | 7.3 | 6.2 | 2.1 | 0.0 | 0.0 | 37.3 |
| 1990-91 | 0.0 | 0.0 | T | T | 0.4 | 8.4 | 6.6 | 7.6 | 7.2 | 1.1 | 0.0 | 0.0 | 31.3 |
| 1991-92 | 0.0 | 0.0 | 0.0 | 0.0 | 0.1 | 3.5 | 5.4 | 6.3 | 8.5 | 0.7 | 0.0 | 0.0 | 24.5 |
| 1992-93 | 0.0 | 0.0 | T | T | 0.5 | 10.0 | 3.8 | 12.1 | 32.0 | 1.9 | 0.0 | 0.0 | 60.3 |
| 1993-94 | 0.0 | T | T | 1.9 | 0.7 | 8.0 | 42.3 | 16.9 | 20.6 | T | T | 0.0 | 90.4 |
| 1994-95 | 0.0 | 0.0 | 0.0 | 0.0 | 1.3 | 0.4 | 7.6 | 9.7 | 5.2 | 0.8 | 0.0 | 0.0 | 25.0 |
| 1995-96 | T | 0.0 | 0.0 | T | 18.6 | 16.0 | 37.5 | 6.5 | 14.2 | 4.9 | 0.0 | 0.0 | 97.7 |
| 1996-97 | 0.0 | 0.0 | 0.0 | 0.0 | 1.4 | 11.7 | 6.5 | 5.2 | 12.7 | T | 0.0 | 0.0 | 37.5 |
| 1997-98 | 0.0 | 0.0 | 0.0 | 0.0 | 4.8 | 16.6 | 5.7 | 7.2 | 6.3 | 0.0 | 0.0 | 0.0 | 40.6 |
| 1998-99 | 0.0 | 0.0 | 0.0 | 0.0 | 0.0 | 1.7 | 10.1 | 6.0 | 9.8 | 0.0 | 0.0 | 0.0 | 27.6 |
| 1999-00 | 0.0 | 0.0 | 0.0 | 0.0 | 0.9 | 1.3 | 21.6 | 5.9 | 1.3 | 5.1 | 0.0 | 0.0 | 36.1 |
| 2000-01 | 0.0 | 0.0 | 0.0 | 0.2 | 1.8 | 14.0 | 11.2 | 9.2 | 13.4 | 0.0 | 0.0 | 0.0 | 49.8 |
| 2001-02 | 0.0 | 0.0 | 0.0 | 0.0 | 0.0 | 1.2 | 22.6 | 2.2 | 4.0 | 0.1 | 0.0 | 0.0 | 30.1 |
| 2002-03 | 0.0 | 0.0 | 0.0 | 0.8 | 6.8 | 28.5 | 14.0 | 19.5 | 8.7 | 3.0 | 0.0 | 0.0 | 81.3 |
| 2003-04 | 0.0 | 0.0 | 0.0 | 0.5 | 1.2 | 19.5 | 15.9 | 5.0 | 10.8 | 0.5 | 0.0 | 0.0 | 53.4 |
| 2004-05 | 0.0 | 0.0 | 0.0 | 0.0 | 0.1 | 3.9 | 23.9 | 11.6 | 11.9 | 0.0 | 0.0 | 0.0 | 51.4 |
| 2005-06 | 0.0 | 0.0 | 0.0 | 0.0 | 1.3 | 10.3 | 4.8 | 3.7 | 6.4 | 0.5 | 0.0 | 0.0 | 27.0 |
| 2006-07 | 0.0 | 0.0 | 0.0 | 0.0 | T | 0.2 | 5.9 | 19.5 | 12.9 | 10.6 | 0.0 | 0.0 | 49.1 |
| 2007-08 |  |  | 0.0 | 0.0 | 5.9 | 16.4 | 5.2 | 13.3 | 0.7 | T | 0.0 | T |  |
| 2008-09 | 0.0 | 0.0 | 0.0 | 1.0 | 0.2 | 12.0 | 11.4 | 3.9 | 2.9 | T | 0.0 | 0.0 | 31.4 |
| 2009- | 0.0 | 0.0 | 0.0 | 0.2 | T | 15.6 |  |  |  |  |  |  |  |
| POR=<br>62 YRS | T | T | T | 0.2 | 3.2 | 9.0 | 11.2 | 9.4 | 8.5 | 2.1 | 0.1 | T | 43.7 |

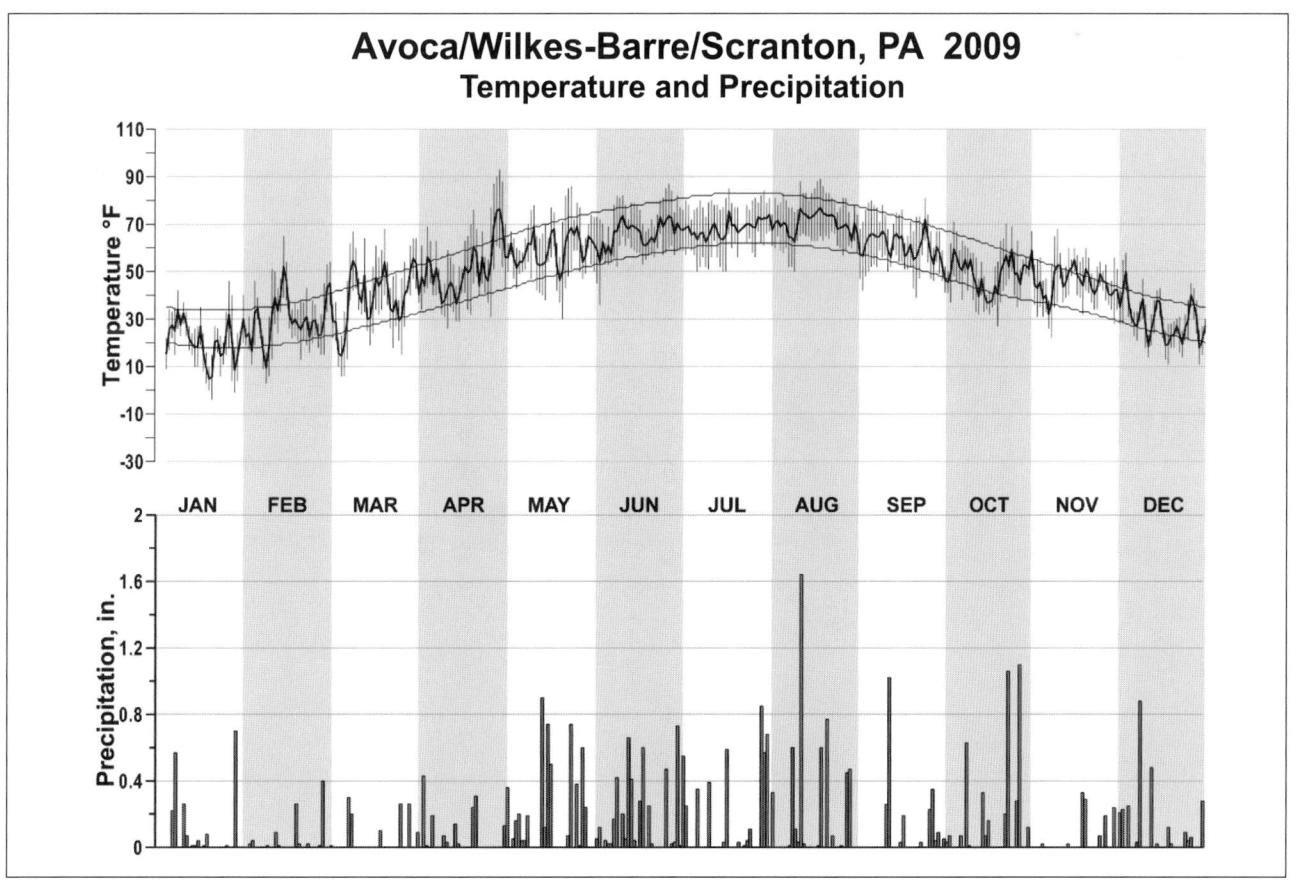

# Avoca/Wilkes-Barre/Scranton, PA  2009
## Temperature and Precipitation

# 2009
# ERIE
# PENNSYLVANIA (KERI)

Erie is located on the southeast shore of Lake Erie and observations are made at Erie International Airport, which is 6 miles southwest of the center of the city and about 1 mile from the lake shore. The terrain rises gradually in a series of ridges paralleling the shoreline to 500 feet above the lake level 3 to 4 miles inland and to 1,000 feet about 15 miles inland. Snowfall from instability showers moving southward off the lake usually increases due to the upslope terrain. Snowfall is somewhat higher south of the city than along the lake shore.

During the winter months, the many cold air masses moving south from Canada are modified by the relatively warm waters of Lake Erie. However, the temperature difference between air and water produces an excess of cloudiness and frequent snow from November through March.

Spring weather is quite variable in Erie, but generally cloudy and cool. Proximity to the lake frequently prevents killing frosts that occur inland. This has led to the establishment of numerous vineyards and orchards in a narrow belt along the shore. Summer heat waves are tempered by cool lake breezes that may reach several miles inland, and days with temperatures above 90 degrees are infrequent. Summer thunderstorms are usually less destructive in Erie than inland areas because of the stabilizing effects of Lake Erie.

Autumn, with long dry periods and an abundance of sunshine, is usually the most pleasant period of the year in Erie. The growing season is extended by the influence of the warmer waters of the lake. Precipitation is well distributed throughout the year, although the number of days with measurable amounts varies considerably from a low average of about one day in three for the period June through September to about one-half of the days from November through March, when snow flurries and squalls move in from the lake.

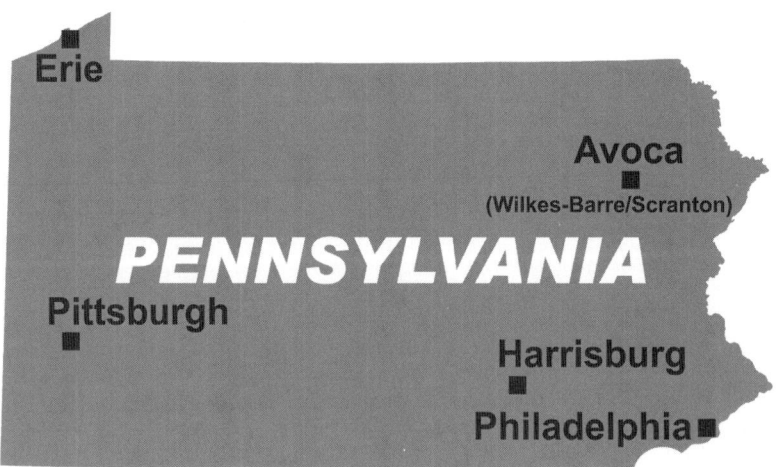

# NORMALS, MEANS, AND EXTREMES
## ERIE (KERI)

| LATITUDE: 42 ° 4 'N | LONGITUDE: -80 ° 10'W | ELEVATION (FT): GRND: 728  BARO: 756 | TIME ZONE: EASTERN  (UTC -5) | WBAN: 14860 |
|---|---|---|---|---|

| | ELEMENT | POR | JAN | FEB | MAR | APR | MAY | JUN | JUL | AUG | SEP | OCT | NOV | DEC | YEAR |
|---|---|---|---|---|---|---|---|---|---|---|---|---|---|---|---|
| **TEMPERATURE °F** | NORMAL DAILY MAXIMUM | 30 | 33.5 | 35.4 | 44.7 | 55.6 | 67.4 | 76.2 | 80.4 | 79.0 | 72.0 | 61.0 | 49.3 | 38.6 | 57.8 |
| | MEAN DAILY MAXIMUM | 84 | 33.4 | 33.3 | 42.5 | 54.0 | 65.7 | 74.4 | 79.5 | 78.2 | 71.2 | 60.9 | 48.3 | 37.6 | 56.6 |
| | HIGHEST DAILY MAXIMUM | 56 | 70 | 75 | 82 | 89 | 90 | 100 | 99 | 94 | 94 | 88 | 80 | 75 | 100 |
| | YEAR OF OCCURRENCE | | 2005 | 2000 | 1998 | 1990 | 1996 | 1988 | 1990 | 2002 | 1959 | 1963 | 1961 | 1982 | JUN 1988 |
| | MEAN OF EXTREME MAXS. | 84 | 55.6 | 56.5 | 70.3 | 77.9 | 83.2 | 88.9 | 89.8 | 88.9 | 85.7 | 78.6 | 69.0 | 58.8 | 75.3 |
| | NORMAL DAILY MINIMUM | 30 | 20.3 | 20.9 | 28.2 | 37.9 | 48.7 | 58.5 | 63.7 | 62.7 | 55.9 | 45.5 | 36.4 | 26.8 | 42.1 |
| | MEAN DAILY MINIMUM | 84 | 20.6 | 19.5 | 27.2 | 36.9 | 47.6 | 57.0 | 62.8 | 61.9 | 54.9 | 45.4 | 35.7 | 26.1 | 41.3 |
| | LOWEST DAILY MINIMUM | 56 | -18 | -17 | -9 | 12 | 26 | 32 | 44 | 37 | 33 | 24 | 7 | -6 | -18 |
| | YEAR OF OCCURRENCE | | 1994 | 1979 | 1980 | 1982 | 1970 | 1972 | 1963 | 1982 | 1974 | 1975 | 1976 | 1983 | JAN 1994 |
| | MEAN OF EXTREME MINS. | 84 | 1.8 | 0.8 | 9.4 | 23.2 | 33.5 | 43.0 | 50.6 | 49.4 | 41.3 | 31.9 | 22.6 | 9.4 | 26.4 |
| | NORMAL DRY BULB | 30 | 26.9 | 28.2 | 36.5 | 46.8 | 58.1 | 67.4 | 72.1 | 70.9 | 64.0 | 53.3 | 42.9 | 32.7 | 50.0 |
| | MEAN DRY BULB | 84 | 27.0 | 26.4 | 34.9 | 45.5 | 56.7 | 65.7 | 71.2 | 70.1 | 63.1 | 53.2 | 42.0 | 31.9 | 49.0 |
| | MEAN WET BULB | 26 | 24.8 | 25.4 | 31.2 | 40.9 | 50.9 | 60.7 | 64.9 | 64.2 | 58.0 | 47.4 | 38.4 | 29.2 | 44.7 |
| | MEAN DEW POINT | 26 | 21.5 | 21.3 | 27.0 | 36.8 | 47.5 | 57.6 | 62.2 | 61.7 | 55.2 | 44.0 | 34.4 | 25.8 | 41.3 |
| | NORMAL NO. DAYS WITH: | | | | | | | | | | | | | | |
| | MAXIMUM >= 90 | 30 | 0.0 | 0.0 | 0.0 | 0.0 | 0.1 | 0.5 | 1.1 | 0.7 | 0.1 | 0.0 | 0.0 | 0.0 | 2.5 |
| | MAXIMUM <= 32 | 30 | 15.5 | 13.1 | 6.4 | 0.4 | 0.0 | 0.0 | 0.0 | 0.0 | 0.0 | 0.0 | 1.2 | 9.0 | 45.6 |
| | MINIMUM <= 32 | 30 | 27.4 | 24.4 | 22.6 | 10.4 | 0.6 | * | 0.0 | 0.0 | 0.0 | 1.4 | 11.0 | 23.5 | 121.3 |
| | MINIMUM <= 0 | 30 | 1.7 | 1.4 | 0.2 | 0.0 | 0.0 | 0.0 | 0.0 | 0.0 | 0.0 | 0.0 | 0.0 | 0.3 | 3.6 |
| **H/C** | NORMAL HEATING DEG. DAYS | 30 | 1196 | 1046 | 900 | 567 | 260 | 58 | 4 | 15 | 116 | 386 | 679 | 1016 | 6243 |
| | NORMAL COOLING DEG. DAYS | 30 | 0 | 0 | 1 | 5 | 30 | 115 | 208 | 183 | 71 | 7 | 0 | 0 | 620 |
| **RH** | NORMAL (PERCENT) | 30 | 75 | 75 | 71 | 68 | 70 | 72 | 72 | 74 | 74 | 71 | 72 | 75 | 72 |
| | HOUR 01 LST | 30 | 76 | 77 | 75 | 73 | 77 | 80 | 80 | 82 | 81 | 75 | 74 | 76 | 77 |
| | HOUR 07 LST | 30 | 78 | 78 | 77 | 75 | 76 | 79 | 80 | 82 | 82 | 78 | 76 | 77 | 78 |
| | HOUR 13 LST | 30 | 72 | 71 | 66 | 62 | 62 | 64 | 64 | 65 | 65 | 63 | 68 | 72 | 66 |
| | HOUR 19 LST | 30 | 75 | 76 | 71 | 65 | 63 | 65 | 65 | 70 | 74 | 73 | 73 | 75 | 70 |
| **S** | PERCENT POSSIBLE SUNSHINE | | | | | | | | | | | | | | |
| **W/O** | MEAN NO. DAYS WITH: | | | | | | | | | | | | | | |
| | HEAVY FOG(VISBY <= 1/4 MI) | 46 | 1.7 | 2.0 | 2.7 | 1.7 | 1.6 | 0.8 | 0.4 | 0.6 | 0.3 | 0.4 | 1.2 | 1.6 | 15.0 |
| | THUNDERSTORMS | 55 | 0.2 | 0.3 | 1.3 | 2.7 | 3.9 | 5.8 | 6.2 | 5.8 | 4.1 | 1.9 | 1.2 | 0.3 | 33.7 |
| **CLOUDNESS** | MEAN: | | | | | | | | | | | | | | |
| | SUNRISE-SUNSET (OKTAS) | | | | | | | | | | | | | | |
| | MIDNIGHT-MIDNIGHT (OKTAS) | | | | | | | | | | | | | | |
| | MEAN NO. DAYS WITH: | | | | | | | | | | | | | | |
| | CLEAR | | | | | | | | | | | | | | |
| | PARTLY CLOUDY | | | | | | | | | | | | | | |
| | CLOUDY | | | | | | | | | | | | | | |
| **PR** | MEAN STATION PRESSURE(IN) | 26 | 29.24 | 29.25 | 29.24 | 29.18 | 29.19 | 29.19 | 29.21 | 29.25 | 29.27 | 29.27 | 29.26 | 29.26 | 29.23 |
| | MEAN SEA-LEVEL PRES. (IN) | 26 | 30.06 | 30.07 | 30.05 | 29.98 | 29.98 | 29.97 | 29.98 | 30.03 | 30.06 | 30.06 | 30.06 | 30.07 | 30.03 |
| **WINDS** | MEAN SPEED (MPH) | 26 | 12.2 | 11.2 | 11.0 | 10.3 | 9.5 | 8.8 | 8.4 | 8.2 | 9.2 | 10.3 | 11.9 | 12.1 | 10.3 |
| | PREVAIL.DIR(TENS OF DEGS) | 35 | 21 | 27 | 27 | 05 | 19 | 19 | 19 | 19 | 19 | 19 | 19 | 21 | 19 |
| | MAXIMUM 2-MINUTE: | | | | | | | | | | | | | | |
| | SPEED (MPH) | 14 | 39 | 43 | 41 | 41 | 36 | 36 | 43 | 36 | 39 | 36 | 40 | 41 | 43 |
| | DIR. (TENS OF DEGS) | | 26 | 25 | 24 | 14 | 15 | 21 | 26 | 25 | 23 | 28 | 26 | 16 | 25 |
| | YEAR OF OCCURRENCE | | 2008 | 2009 | 2006 | 2009 | 2003 | 1998 | 1999 | 2000 | 2008 | 2006 | 2003 | 2009 | FEB 2009 |
| | MAXIMUM 3-SECOND | | | | | | | | | | | | | | |
| | SPEED (MPH) | 14 | 54 | 61 | 66 | 63 | 47 | 45 | 61 | 51 | 58 | 54 | 63 | 62 | 66 |
| | DIR. (TENS OF DEGS) | | 25 | 24 | 25 | 29 | 15 | 21 | 27 | 26 | 25 | 26 | 27 | 15 | 25 |
| | YEAR OF OCCURRENCE | | 2008 | 2009 | 2006 | 2002 | 2003 | 1998 | 1999 | 2009 | 2005 | 2009 | 2003 | 2009 | MAR 2006 |
| **PRECIPITATION** | NORMAL (IN) | 30 | 2.53 | 2.28 | 3.13 | 3.38 | 3.34 | 4.28 | 3.28 | 4.21 | 4.73 | 3.92 | 3.96 | 3.73 | 42.77 |
| | MAXIMUM MONTHLY (IN) | 56 | 6.23 | 5.73 | 6.78 | 7.11 | 6.38 | 8.35 | 7.70 | 11.06 | 10.65 | 9.87 | 10.40 | 7.36 | 11.06 |
| | YEAR OF OCCURRENCE | | 2007 | 1990 | 1976 | 1961 | 2004 | 1996 | 1970 | 1977 | 1977 | 1954 | 1985 | 2008 | AUG 1977 |
| | MINIMUM MONTHLY (IN) | 56 | 0.87 | 0.57 | 0.63 | 1.63 | 1.00 | 0.75 | 0.52 | 0.50 | 1.33 | 1.13 | 1.35 | 1.38 | 0.50 |
| | YEAR OF OCCURRENCE | | 1981 | 1978 | 1960 | 1975 | 1991 | 1991 | 2001 | 2002 | 1995 | 1963 | 2009 | 1960 | AUG 2002 |
| | MAXIMUM IN 24 HOURS (IN) | 56 | 1.63 | 2.16 | 2.38 | 2.53 | 2.23 | 4.66 | 3.22 | 3.91 | 6.11 | 4.35 | 3.67 | 2.39 | 6.11 |
| | YEAR OF OCCURRENCE | | 1998 | 1961 | 1987 | 1977 | 1969 | 1996 | 1970 | 1994 | 1979 | 1954 | 1985 | 1979 | SEP 1979 |
| | NORMAL NO. DAYS WITH: | | | | | | | | | | | | | | |
| | PRECIPITATION >= 0.01 | 30 | 19.4 | 14.9 | 14.9 | 13.9 | 12.5 | 11.0 | 9.9 | 10.7 | 11.6 | 13.0 | 16.1 | 19.2 | 167.1 |
| | PRECIPITATION >= 1.00 | 30 | 0.2 | 0.2 | 0.4 | 0.3 | 0.5 | 1.3 | 0.8 | 1.3 | 1.3 | 0.6 | 0.3 | 0.4 | 7.6 |
| **SNOWFALL** | NORMAL (IN) | 30 | 26.3 | 17.3 | 11.2 | 2.3 | 0.* | 0.0 | 0.0 | 0.0 | 0.0 | 0.3 | 9.0 | 25.3 | 91.7 |
| | MAXIMUM MONTHLY (IN) | 54 | 62.4 | 35.4 | 33.9 | 17.2 | 0.4 | T | T | T | T | 4.0 | 42.2 | 66.9 | 66.9 |
| | YEAR OF OCCURRENCE | | 1978 | 2007 | 2008 | 1957 | 1989 | 1990 | 1999 | 1992 | 1993 | 1954 | 2000 | 1989 | DEC 1989 |
| | MAXIMUM IN 24 HOURS (IN) | 54 | 12.9 | 17.8 | 16.1 | 11.8 | 0.4 | T | T | T | T | 2.4 | 23.0 | 19.2 | 23.0 |
| | YEAR OF OCCURRENCE' | | 1986 | 1979 | 2004 | 2005 | 1989 | 1990 | 1999 | 1992 | 1993 | 2001 | 1956 | 1989 | NOV 1956 |
| | MAXIMUM SNOW DEPTH (IN) | 52 | 28 | 25 | 20 | 9 | 0 | 0 | 0 | 0 | 0 | 2 | 27 | 39 | 39 |
| | YEAR OF OCCURRENCE | | 1985 | 1977 | 1984 | 1987 | | | | | | 1974 | 1950 | 1989 | DEC 1989 |
| | NORMAL NO. DAYS WITH: | | | | | | | | | | | | | | |
| | SNOWFALL >= 1.0 | 30 | 6.6 | 4.7 | 3.2 | 0.8 | 0.0 | 0.0 | 0.0 | 0.0 | 0.0 | 0.1 | 2.4 | 6.6 | 24.4 |

## PRECIPITATION (inches) 2009 ERIE (KERI)

| YEAR | JAN | FEB | MAR | APR | MAY | JUN | JUL | AUG | SEP | OCT | NOV | DEC | ANNUAL |
|---|---|---|---|---|---|---|---|---|---|---|---|---|---|
| 1980 | 1.57 | 1.29 | 4.11 | 3.79 | 2.23 | 4.83 | 5.42 | 6.76 | 5.48 | 6.51 | 2.56 | 2.49 | 47.04 |
| 1981 | 0.87 | 5.21 | 1.58 | 6.09 | 2.13 | 4.84 | 3.04 | 3.85 | 4.26 | 5.04 | 2.22 | 2.84 | 41.97 |
| 1982 | 3.85 | 1.24 | 3.50 | 1.81 | 3.06 | 6.02 | 4.40 | 2.20 | 4.07 | 2.74 | 5.33 | 3.34 | 41.56 |
| 1983 | 1.49 | 1.07 | 3.63 | 2.93 | 3.91 | 3.84 | 5.52 | 4.74 | 5.27 | 3.77 | 6.11 | 3.97 | 46.25 |
| 1984 | 1.65 | 2.42 | 1.91 | 2.63 | 5.83 | 4.49 | 1.94 | 2.09 | 5.29 | 1.82 | 3.62 | 4.10 | 37.79 |
| 1985 | 2.56 | 2.75 | 5.08 | 1.76 | 2.94 | 3.50 | 4.97 | 1.66 | 2.22 | 5.20 | 10.40 | 2.83 | 45.87 |
| 1986 | 2.33 | 2.72 | 2.10 | 2.88 | 5.24 | 7.71 | 2.54 | 1.83 | 7.97 | 4.86 | 2.99 | 4.13 | 47.30 |
| 1987 | 2.15 | 1.05 | 4.28 | 1.87 | 1.78 | 5.15 | 3.91 | 7.82 | 5.45 | 5.76 | 2.25 | 3.39 | 44.86 |
| 1988 | 1.50 | 2.47 | 2.44 | 3.00 | 3.21 | 1.26 | 4.14 | 3.78 | 3.21 | 8.25 | 2.99 | 2.62 | 38.87 |
| 1989 | 1.95 | 2.41 | 4.70 | 2.02 | 6.14 | 5.14 | 1.35 | 3.96 | 3.76 | 3.33 | 3.87 | 3.25 | 41.88 |
| 1990 | 2.30 | 5.73 | 1.29 | 3.52 | 5.74 | 2.84 | 2.53 | 6.49 | 7.74 | 4.15 | 2.69 | 6.94 | 51.96 |
| 1991 | 2.16 | 1.62 | 3.38 | 3.64 | 1.00 | 0.75 | 3.49 | 3.07 | 3.25 | 3.00 | 3.18 | 3.17 | 31.71 |
| 1992 | 2.60 | 1.91 | 2.11 | 4.04 | 1.78 | 1.95 | 6.06 | 4.11 | 6.81 | 4.01 | 5.37 | 3.34 | 44.09 |
| 1993 | 3.36 | 2.03 | 3.59 | 2.34 | 1.28 | 3.94 | 2.80 | 2.85 | 4.52 | 4.41 | 4.10 | 2.88 | 38.10 |
| 1994 | 2.58 | 1.35 | 2.99 | 4.87 | 2.02 | 5.80 | 1.12 | 7.72 | 3.39 | 2.46 | 3.01 | 3.19 | 40.50 |
| 1995 | 3.37 | 1.66 | 1.29 | 3.08 | 2.69 | 1.45 | 2.20 | 3.54 | 1.33 | 4.51 | 4.99 | 3.25 | 33.36 |
| 1996 | 3.26 | 2.03 | 2.04 | 6.07 | 3.37 | 8.35 | 2.99 | 1.43 | 9.63 | 3.28 | 3.26 | 2.96 | 48.67 |
| 1997 | 1.66 | 3.00 | 4.80 | 2.25 | 4.36 | 4.30 | 2.90 | 3.07 | 3.16 | 2.43 | 3.23 | 4.85 | 40.01 |
| 1998 | 5.35 | 1.34 | 2.99 | 4.86 | 2.67 | 2.64 | 2.33 | 2.54 | 1.63 | 1.92 | 1.80 | 3.59 | 33.66 |
| 1999 | 4.98 | 1.88 | 1.86 | 4.09 | 3.20 | 3.00 | 2.42 | 2.77 | 5.15 | 2.94 | 4.48 | 3.84 | 40.61 |
| 2000 | 2.48 | 1.95 | 2.05 | 5.09 | 4.29 | 5.62 | 4.86 | 5.52 | 2.55 | 3.38 | 5.67 | 4.86 | 48.32 |
| 2001 | 1.69 | 2.36 | 2.96 | 2.54 | 3.75 | 2.96 | 0.52 | 4.29 | 2.38 | 4.10 | 2.36 | 4.46 | 34.37 |
| 2002 | 3.54 | 3.64 | 4.40 | 4.74 | 5.65 | 2.81 | 2.43 | 0.50 | 7.77 | 4.37 | 4.90 | 3.98 | 48.73 |
| 2003 | 2.97 | 2.92 | 2.95 | 1.96 | 5.12 | 2.52 | 4.89 | 1.55 | 6.77 | 3.72 | 2.66 | 2.97 | 41.00 |
| 2004 | 3.86 | 0.96 | 3.91 | 3.53 | 6.38 | 1.82 | 5.82 | 2.42 | 5.05 | 4.23 | 2.95 | 5.68 | 46.61 |
| 2005 | 5.35 | 2.01 | 1.71 | 4.79 | 1.27 | 1.73 | 3.89 | 4.06 | 4.42 | 3.00 | 4.98 | 2.96 | 40.17 |
| 2006 | 2.45 | 2.52 | 2.03 | 3.13 | 3.50 | 2.99 | 3.44 | 3.30 | 7.53 | 6.58 | 3.16 | 3.68 | 44.31 |
| 2007 | 6.23 | 1.86 | 2.62 | 2.52 | 1.87 | 1.66 | 4.24 | 6.20 | 2.03 | 2.69 | 5.33 | 4.93 | 42.18 |
| 2008 | 2.88 | 5.07 | 4.96 | 2.25 | 2.65 | 4.33 | 4.72 | 2.58 | 2.94 | 4.81 | 4.88 | 7.36 | 49.43 |
| 2009 | 3.97 | 2.15 | 4.01 | 3.08 | 2.32 | 5.59 | 5.44 | 2.07 | 3.35 | 3.93 | 1.35 | 3.28 | 40.54 |
| POR= 84 YRS | 2.65 | 2.25 | 2.93 | 3.44 | 3.36 | 3.54 | 3.47 | 3.36 | 4.04 | 3.55 | 3.76 | 3.27 | 39.62 |

WBAN : 14860

## AVERAGE TEMPERATURE (°F) 2009 ERIE (KERI)

| YEAR | JAN | FEB | MAR | APR | MAY | JUN | JUL | AUG | SEP | OCT | NOV | DEC | ANNUAL |
|---|---|---|---|---|---|---|---|---|---|---|---|---|---|
| 1980 | 26.6 | 20.9 | 31.0 | 43.3 | 55.7 | 61.3 | 69.6 | 72.2 | 63.7 | 48.1 | 38.9 | 27.5 | 46.6 |
| 1981 | 19.6 | 30.3 | 33.1 | 47.4 | 53.9 | 66.0 | 72.1 | 70.6 | 61.9 | 49.7 | 41.6 | 31.4 | 48.1 |
| 1982 | 19.2 | 23.4 | 34.4 | 42.5 | 60.9 | 61.3 | 70.9 | 66.2 | 62.5 | 54.2 | 44.8 | 40.3 | 48.4 |
| 1983 | 30.6 | 31.6 | 38.5 | 45.3 | 55.0 | 67.0 | 73.0 | 72.4 | 65.7 | 54.4 | 45.2 | 25.9 | 50.4 |
| 1984 | 21.6 | 35.5 | 28.5 | 47.0 | 53.8 | 68.6 | 69.6 | 72.0 | 61.8 | 56.4 | 42.4 | 37.3 | 49.5 |
| 1985 | 21.7 | 25.9 | 37.4 | 51.4 | 59.8 | 63.4 | 70.2 | 70.7 | 66.9 | 54.6 | 46.3 | 27.2 | 49.6 |
| 1986 | 27.5 | 27.0 | 38.7 | 48.2 | 59.6 | 65.2 | 72.0 | 69.2 | 65.0 | 53.6 | 40.0 | 32.9 | 49.9 |
| 1987 | 28.0 | 27.3 | 37.5 | 48.3 | 60.5 | 69.2 | 74.6 | 70.0 | 64.5 | 48.2 | 45.1 | 35.8 | 50.8 |
| 1988 | 27.2 | 25.9 | 36.3 | 46.0 | 58.2 | 65.6 | 74.5 | 72.9 | 63.0 | 48.2 | 44.6 | 32.4 | 49.6 |
| 1989 | 33.4 | 24.8 | 35.6 | 43.1 | 56.1 | 66.4 | 71.9 | 69.4 | 63.2 | 54.8 | 41.2 | 21.7 | 48.5 |
| 1990 | 36.0 | 33.3 | 40.4 | 49.4 | 55.2 | 67.0 | 70.9 | 69.9 | 63.4 | 54.8 | 46.0 | 36.7 | 51.9 |
| 1991 | 28.1 | 32.7 | 39.0 | 51.5 | 64.8 | 70.3 | 73.2 | 72.6 | 63.8 | 55.8 | 41.3 | 34.6 | 52.3 |
| 1992 | 30.4 | 30.4 | 34.7 | 46.6 | 57.5 | 63.9 | 69.4 | 67.8 | 64.0 | 51.1 | 42.4 | 34.7 | 49.4 |
| 1993 | 32.5 | 24.1 | 31.7 | 48.1 | 57.8 | 68.4 | 75.4 | 73.4 | 61.2 | 51.1 | 41.7 | 30.5 | 49.7 |
| 1994 | 18.4 | 24.2 | 33.2 | 49.1 | 54.2 | 68.5 | 73.3 | 69.1 | 63.4 | 53.7 | 48.1 | 36.8 | 49.3 |
| 1995 | 31.4 | 24.4 | 37.8 | 43.0 | 57.5 | 70.1 | 73.7 | 74.7 | 62.1 | 55.8 | 38.1 | 27.6 | 49.7 |
| 1996 | 24.5 | 25.9 | 30.0 | 44.3 | 55.8 | 68.2 | 69.2 | 70.6 | 63.2 | 53.5 | 36.8 | 34.6 | 48.1 |
| 1997 | 25.9 | 31.9 | 36.7 | 43.6 | 50.8 | 66.9 | 69.4 | 67.5 | 62.2 | 53.3 | 39.8 | 33.9 | 48.5 |
| 1998 | 34.5 | 35.7 | 39.8 | 47.9 | 63.5 | 67.6 | 71.5 | 72.1 | 66.9 | 54.5 | 45.2 | 38.2 | 53.1 |
| 1999 | 26.5 | 34.0 | 33.2 | 47.6 | 60.2 | 69.5 | 75.7 | 68.9 | 64.7 | 52.5 | 46.5 | 35.0 | 51.2 |
| 2000 | 27.6 | 32.9 | 41.9 | 46.0 | 59.8 | 67.5 | 67.8 | 68.4 | 63.3 | 54.0 | 40.4 | 24.0 | 49.5 |
| 2001 | 27.6 | 30.6 | 32.4 | 49.0 | 59.6 | 67.0 | 70.6 | 71.9 | 61.3 | 54.3 | 49.2 | 37.5 | 50.9 |
| 2002 | 34.1 | 33.8 | 36.4 | 48.8 | 53.6 | 68.5 | 73.7 | 72.1 | 67.8 | 51.3 | 41.0 | 30.3 | 51.0 |
| 2003 | 20.7 | 21.7 | 36.2 | 44.6 | 55.0 | 64.8 | 70.0 | 71.4 | 63.2 | 51.1 | 46.3 | 34.4 | 48.3 |
| 2004 | 20.8 | 26.8 | 38.5 | 47.5 | 60.3 | 65.2 | 69.7 | 67.3 | 65.5 | 52.9 | 44.2 | 32.2 | 49.2 |
| 2005 | 26.3 | 28.7 | 30.5 | 46.5 | 52.4 | 72.5 | 73.9 | 73.3 | 66.7 | 54.5 | 45.5 | 28.7 | 50.0 |
| 2006 | 37.5 | 30.4 | 35.8 | 48.2 | 57.9 | 65.5 | 73.5 | 71.1 | 61.7 | 49.8 | 45.2 | 38.2 | 51.2 |
| 2007 | 31.8 | 18.9 | 37.7 | 43.9 | 59.4 | 68.4 | 70.1 | 71.5 | 66.9 | 61.1 | 41.7 | 32.8 | 50.4 |
| 2008 | 31.8 | 27.2 | 32.6 | 51.2 | 54.7 | 70.4 | 73.4 | 70.1 | 66.5 | 52.4 | 42.2 | 32.5 | 50.4 |
| 2009 | 21.5 | 30.4 | 36.6 | 48.1 | 57.7 | 64.4 | 67.2 | 70.7 | 63.8 | 50.4 | 46.3 | 30.9 | 49.0 |
| POR= 84 YRS | 27.0 | 26.4 | 34.9 | 45.5 | 56.7 | 65.7 | 71.2 | 70.1 | 63.1 | 53.2 | 42.0 | 31.9 | 49.0 |

## HEATING DEGREE DAYS (base 65°F) 2009 ERIE (KERI)

| YEAR | JUL | AUG | SEP | OCT | NOV | DEC | JAN | FEB | MAR | APR | MAY | JUN | TOTAL |
|------|-----|-----|-----|-----|-----|-----|-----|-----|-----|-----|-----|-----|-------|
| 1980-81 | 12 | 2 | 108 | 520 | 775 | 1156 | 1400 | 967 | 982 | 523 | 345 | 42 | 6832 |
| 1981-82 | 3 | 6 | 153 | 467 | 694 | 1034 | 1413 | 1159 | 942 | 672 | 158 | 123 | 6824 |
| 1982-83 | 11 | 47 | 126 | 336 | 605 | 762 | 1062 | 933 | 814 | 585 | 312 | 67 | 5660 |
| 1983-84 | 10 | 0 | 89 | 336 | 591 | 1207 | 1338 | 847 | 1123 | 539 | 360 | 25 | 6465 |
| 1984-85 | 11 | 5 | 141 | 262 | 673 | 853 | 1337 | 1088 | 846 | 423 | 202 | 87 | 5928 |
| 1985-86 | 5 | 3 | 75 | 316 | 558 | 1164 | 1152 | 1056 | 811 | 502 | 207 | 74 | 5923 |
| 1986-87 | 5 | 32 | 84 | 350 | 742 | 989 | 1138 | 1048 | 844 | 495 | 225 | 31 | 5983 |
| 1987-88 | 0 | 21 | 67 | 513 | 592 | 902 | 1166 | 1128 | 883 | 565 | 236 | 96 | 6169 |
| 1988-89 | 5 | 10 | 102 | 528 | 605 | 1006 | 972 | 1121 | 905 | 651 | 301 | 45 | 6251 |
| 1989-90 | 2 | 16 | 128 | 320 | 706 | 1335 | 892 | 879 | 773 | 500 | 309 | 58 | 5918 |
| 1990-91 | 9 | 0 | 113 | 323 | 564 | 866 | 1137 | 898 | 800 | 412 | 148 | 15 | 5285 |
| 1991-92 | 0 | 0 | 130 | 304 | 704 | 934 | 1065 | 996 | 934 | 556 | 258 | 94 | 5975 |
| 1992-93 | 5 | 23 | 110 | 423 | 673 | 928 | 1002 | 1137 | 1025 | 507 | 227 | 51 | 6111 |
| 1993-94 | 0 | 0 | 162 | 430 | 692 | 1061 | 1438 | 1137 | 979 | 480 | 346 | 61 | 6786 |
| 1994-95 | 0 | 12 | 90 | 344 | 503 | 868 | 1034 | 1130 | 835 | 654 | 237 | 28 | 5735 |
| 1995-96 | 14 | 0 | 127 | 285 | 803 | 1153 | 1245 | 1126 | 1079 | 614 | 321 | 19 | 6786 |
| 1996-97 | 10 | 0 | 109 | 353 | 840 | 934 | 1206 | 920 | 870 | 638 | 430 | 57 | 6367 |
| 1997-98 | 16 | 20 | 111 | 389 | 750 | 957 | 940 | 813 | 794 | 506 | 107 | 82 | 5485 |
| 1998-99 | 0 | 4 | 47 | 323 | 585 | 824 | 1186 | 860 | 980 | 516 | 187 | 52 | 5564 |
| 1999-00 | 0 | 5 | 89 | 380 | 549 | 925 | 1153 | 926 | 710 | 565 | 206 | 64 | 5572 |
| 2000-01 | 23 | 19 | 137 | 342 | 729 | 1261 | 1152 | 956 | 1004 | 486 | 185 | 68 | 6362 |
| 2001-02 | 12 | 1 | 155 | 337 | 469 | 844 | 953 | 866 | 879 | 506 | 368 | 53 | 5443 |
| 2002-03 | 2 | 4 | 45 | 444 | 713 | 1070 | 1365 | 1206 | 888 | 612 | 306 | 69 | 6724 |
| 2003-04 | 4 | 5 | 87 | 425 | 554 | 943 | 1362 | 1105 | 816 | 529 | 186 | 71 | 6087 |
| 2004-05 | 2 | 34 | 59 | 368 | 620 | 1009 | 1192 | 1010 | 1064 | 546 | 387 | 19 | 6310 |
| 2005-06 | 0 | 1 | 31 | 349 | 578 | 1121 | 847 | 961 | 897 | 497 | 261 | 54 | 5597 |
| 2006-07 | 1 | 3 | 119 | 466 | 587 | 825 | 1023 | 1287 | 843 | 627 | 231 | 42 | 6054 |
| 2007-08 | 10 | 11 | 56 | 190 | 690 | 991 | 1019 | 1088 | 997 | 417 | 323 | 22 | 5814 |
| 2008-09 | 1 | 4 | 37 | 388 | 677 | 1002 | 1341 | 961 | 874 | 513 | 239 | 69 | 6106 |
| 2009- | 17 | 12 | 83 | 444 | 556 | 1048 | | | | | | | |

WBAN : 14860

## COOLING DEGREE DAYS (base 65°F) 2009 ERIE (KERI)

| YEAR | JAN | FEB | MAR | APR | MAY | JUN | JUL | AUG | SEP | OCT | NOV | DEC | TOTAL |
|------|-----|-----|-----|-----|-----|-----|-----|-----|-----|-----|-----|-----|-------|
| 1980 | 0 | 0 | 0 | 0 | 20 | 56 | 162 | 234 | 75 | 4 | 0 | 0 | 551 |
| 1981 | 0 | 0 | 0 | 2 | 8 | 81 | 230 | 187 | 65 | 0 | 0 | 0 | 573 |
| 1982 | 0 | 0 | 0 | 4 | 35 | 19 | 201 | 90 | 59 | 8 | 5 | 4 | 425 |
| 1983 | 0 | 0 | 0 | 0 | 7 | 135 | 261 | 235 | 117 | 15 | 0 | 0 | 770 |
| 1984 | 0 | 0 | 0 | 7 | 18 | 145 | 161 | 227 | 52 | 3 | 0 | 0 | 613 |
| 1985 | 0 | 0 | 0 | 21 | 45 | 45 | 172 | 187 | 140 | 4 | 3 | 0 | 617 |
| 1986 | 0 | 0 | 2 | 6 | 46 | 86 | 228 | 169 | 88 | 3 | 0 | 0 | 628 |
| 1987 | 0 | 0 | 0 | 0 | 92 | 164 | 304 | 182 | 61 | 0 | 2 | 0 | 805 |
| 1988 | 0 | 0 | 0 | 0 | 29 | 119 | 305 | 263 | 52 | 10 | 0 | 0 | 778 |
| 1989 | 0 | 0 | 0 | 0 | 35 | 94 | 225 | 159 | 80 | 7 | 0 | 0 | 600 |
| 1990 | 0 | 0 | 16 | 40 | 13 | 123 | 198 | 159 | 73 | 13 | 2 | 0 | 637 |
| 1991 | 0 | 0 | 0 | 14 | 148 | 181 | 259 | 244 | 102 | 25 | 0 | 0 | 973 |
| 1992 | 0 | 0 | 0 | 9 | 30 | 70 | 148 | 117 | 85 | 0 | 0 | 0 | 459 |
| 1993 | 0 | 0 | 0 | 4 | 13 | 161 | 331 | 269 | 55 | 7 | 0 | 0 | 840 |
| 1994 | 0 | 0 | 0 | 9 | 18 | 170 | 266 | 147 | 48 | 3 | 0 | 0 | 661 |
| 1995 | 0 | 0 | 0 | 0 | 8 | 189 | 287 | 308 | 45 | 6 | 0 | 0 | 843 |
| 1996 | 0 | 0 | 0 | 0 | 43 | 122 | 148 | 181 | 61 | 3 | 0 | 0 | 558 |
| 1997 | 0 | 0 | 0 | 0 | 0 | 119 | 157 | 107 | 30 | 33 | 0 | 0 | 446 |
| 1998 | 0 | 0 | 23 | 0 | 70 | 167 | 208 | 232 | 108 | 5 | 0 | 0 | 813 |
| 1999 | 0 | 0 | 0 | 0 | 42 | 195 | 341 | 132 | 87 | 2 | 0 | 0 | 799 |
| 2000 | 0 | 0 | 1 | 0 | 56 | 147 | 118 | 128 | 91 | 6 | 0 | 0 | 547 |
| 2001 | 0 | 0 | 0 | 13 | 24 | 135 | 194 | 224 | 52 | 11 | 0 | 0 | 653 |
| 2002 | 0 | 0 | 0 | 27 | 18 | 163 | 277 | 230 | 135 | 26 | 0 | 0 | 876 |
| 2003 | 0 | 0 | 0 | 9 | 3 | 66 | 168 | 211 | 38 | 3 | 0 | 0 | 498 |
| 2004 | 0 | 0 | 0 | 11 | 51 | 84 | 157 | 112 | 80 | 1 | 0 | 0 | 496 |
| 2005 | 0 | 0 | 0 | 0 | 1 | 253 | 284 | 262 | 89 | 30 | 0 | 0 | 919 |
| 2006 | 0 | 0 | 0 | 0 | 47 | 79 | 271 | 200 | 25 | 0 | 0 | 0 | 622 |
| 2007 | 0 | 0 | 4 | 0 | 64 | 155 | 174 | 218 | 118 | 75 | 0 | 0 | 808 |
| 2008 | 0 | 0 | 0 | 10 | 10 | 188 | 265 | 172 | 87 | 3 | 0 | 0 | 735 |
| 2009 | 0 | 0 | 0 | 15 | 18 | 56 | 92 | 196 | 52 | 0 | 0 | 0 | 429 |

## SNOWFALL (inches) 2009 ERIE (KERI)

| YEAR | JUL | AUG | SEP | OCT | NOV | DEC | JAN | FEB | MAR | APR | MAY | JUN | TOTAL |
|------|-----|-----|-----|-----|-----|-----|-----|-----|-----|-----|-----|-----|-------|
| 1980-81 | 0.0 | 0.0 | 0.0 | T | 4.9 | 21.0 | 27.9 | 22.5 | 13.1 | T | 0.0 | 0.0 | 89.4 |
| 1981-82 | 0.0 | 0.0 | 0.0 | T | 0.4 | 17.3 | 27.1 | 9.0 | 7.3 | 10.2 | 0.0 | 0.0 | 71.3 |
| 1982-83 | 0.0 | 0.0 | 0.0 | T | 9.2 | 8.9 | 6.7 | 9.2 | 5.6 | 1.6 | 0.0 | 0.0 | 41.2 |
| 1983-84 | 0.0 | 0.0 | 0.0 | 0.0 | 1.4 | 41.1 | 18.7 | 27.2 | 21.6 | T | 0.0 | 0.0 | 110.0 |
| 1984-85 | 0.0 | 0.0 | 0.0 | 0.0 | 4.7 | 16.7 | 57.2 | 19.0 | 3.5 | 5.2 | 0.0 | 0.0 | 106.3 |
| 1985-86 | 0.0 | 0.0 | 0.0 | 0.0 | 1.9 | 59.9 | 30.6 | 22.9 | 4.2 | 5.4 | 0.0 | 0.0 | 124.9 |
| 1986-87 | 0.0 | 0.0 | 0.0 | 0.0 | 4.6 | 8.0 | 31.3 | 10.1 | 11.6 | 2.6 | 0.0 | 0.0 | 68.2 |
| 1987-88 | 0.0 | 0.0 | 0.0 | T | T | 24.3 | 30.8 | 31.2 | 16.8 | 0.4 | 0.0 | 0.0 | 103.5 |
| 1988-89 | 0.0 | 0.0 | 0.0 | 1.8 | 0.5 | 28.1 | 10.2 | 21.5 | 10.5 | 3.5 | 0.4 | 0.0 | 76.5 |
| 1989-90 | 0.0 | 0.0 | 0.0 | T | 19.6 | 66.9 | 13.7 | 8.3 | 2.3 | 4.1 | 0.0 | T | 114.9 |
| 1990-91 | 0.0 | T | T | T | 2.0 | 15.4 | 24.3 | 15.4 | 2.1 | 0.4 | 0.0 | 0.0 | 59.6 |
| 1991-92 | 0.0 | 0.0 | 0.0 | T | 13.7 | 30.0 | 32.6 | 8.0 | 12.8 | 7.7 | 0.0 | 0.0 | 104.8 |
| 1992-93 | 0.0 | T | 0.0 | T | 23.0 | 15.6 | 10.0 | 31.7 | 27.3 | 0.9 | 0.0 | 0.0 | 108.5 |
| 1993-94 | 0.0 | 0.0 | T | 0.3 | 3.5 | 25.7 | 46.9 | 27.8 | 22.2 | 4.9 | 0.0 | 0.0 | 131.3 |
| 1994-95 | 0.0 | 0.0 | 0.0 | 0.0 | 1.1 | 1.0 | 27.0 | 15.3 | 5.2 | 4.1 | 0.0 | 0.0 | 53.7 |
| 1995-96 | 0.0 | 0.0 | 0.0 | 0.0 | 20.4 | 39.6 | 23.3 | 11.8 | 31.8 | | | | |
| 1996-97 | | | | | | | | | | | | | |
| 1997-98 | | | | | | | 9.8 | 0.7 | 16.6 | 0.0 | 0.0 | 0.0 | |
| 1998-99 | 0.0 | 0.0 | 0.0 | 0.0 | 0.4 | 28.3 | 57.9 | 9.7 | 14.8 | T | 0.0 | 0.0 | 111.1 |
| 1999-00 | T | 0.0 | 0.0 | T | 0.6 | 29.7 | 24.3 | 10.7 | 6.5 | 0.7 | 0.0 | 0.0 | 72.5 |
| 2000-01 | 0.0 | 0.0 | 0.0 | T | 42.2 | 49.7 | 16.3 | 8.5 | 28.9 | 3.5 | 0.0 | 0.0 | 149.1 |
| 2001-02 | 0.0 | 0.0 | 0.0 | 2.4 | T | 37.1 | 16.1 | 17.4 | 31.1 | 0.9 | 0.0 | 0.0 | 105.0 |
| 2002-03 | 0.0 | 0.0 | 0.0 | T | 21.1 | 26.9 | 51.8 | 32.6 | 8.1 | 2.5 | 0.0 | 0.0 | 143.0 |
| 2003-04 | 0.0 | 0.0 | 0.0 | 0.2 | 1.0 | 17.7 | 59.9 | 5.6 | 22.4 | 5.2 | 0.0 | 0.0 | 112.0 |
| 2004-05 | 0.0 | 0.0 | 0.0 | 0.0 | 2.0 | 30.1 | 38.8 | 16.8 | 20.1 | 14.8 | T | 0.0 | 122.6 |
| 2005-06 | 0.0 | 0.0 | 0.0 | 0.0 | 11.3 | 32.4 | 9.0 | 20.1 | 4.4 | 5.7 | 0.0 | 0.0 | 82.9 |
| 2006-07 | 0.0 | 0.0 | 0.0 | 0.2 | 6.4 | 12.5 | 37.7 | 35.4 | 12.5 | 3.8 | 0.0 | 0.0 | 108.5 |
| 2007-08 | 0.0 | 0.0 | 0.0 | 0.0 | 4.0 | 19.3 | 32.9 | 28.6 | 33.9 | T | 0.0 | 0.0 | 118.7 |
| 2008-09 | 0.0 | 0.0 | 0.0 | 0.4 | 21.1 | 48.8 | 59.2 | 12.0 | 0.7 | 3.6 | 0.0 | 0.0 | 145.8 |
| 2009- | 0.0 | 0.0 | 0.0 | T | T | 28.7 | | | | | | | |
| POR=<br>83 YRS | T | T | T | 0.4 | 9.2 | 20.1 | 20.5 | 14.0 | 11.8 | 2.7 | T | T | 78.7 |

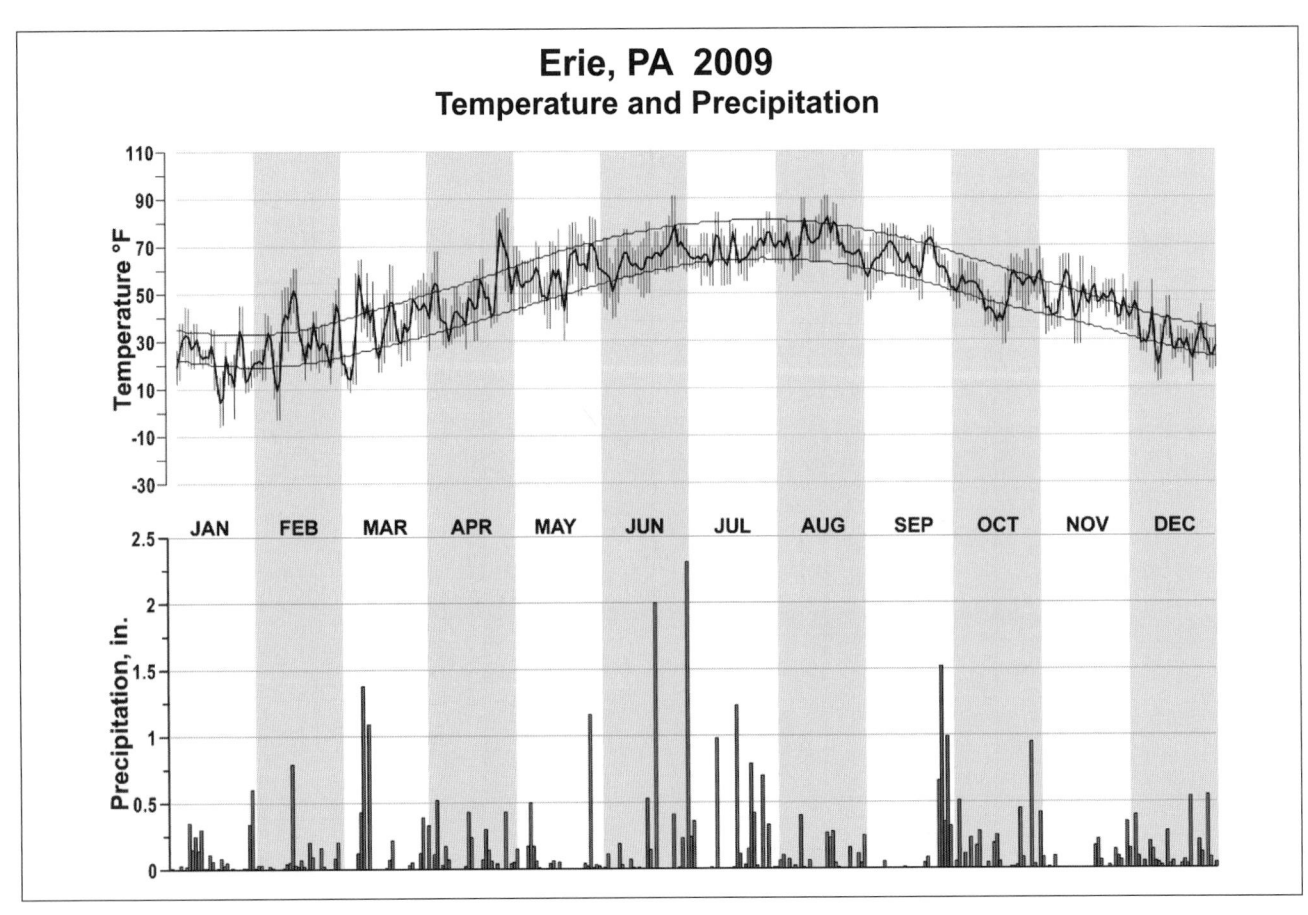

# Erie, PA  2009
## Temperature and Precipitation

# 2009
# HARRISBURG,
# PENNSYLVANIA (KMDT)

Harrisburg, the capital of Pennsylvania, is situated on the east bank of the Susquehanna River. It is in the Great Valley formed by the eastern foothills of the Appalachian Chain, and about 60 miles southeast of the Commonwealths geographic center. It is nested in a saucer-like bowl, 10 miles south of Blue Mountain, which serves as a barrier to provide a modifying influence upon the severe winter climate experienced 50 to 100 miles to the north and west. Although the severity of the winter climate is lessened, the city lies a little too far inland to derive the full benefits of the coastal Climate.

Air masses change with some regularity, and any one condition does not persist for many days in succession. The mountain barrier occasionally prevents cold waves from reaching the Great Valley. The city is favorably located to receive precipitation produced when warm, maritime air from the Atlantic Ocean is forced upslope to cross the Blue Ridge Mountains.

The Growing Season is 192 days.

During June 1992, Hurricane Agnes produced 15.11 inches of rain from the 20th to the 23rd. Prolonged dry spells occur occasionally. During September and October 1947 there were 35 consecutive days with less then .01 inches of precipitation.

Flood stage on the Susquehanna River occurs on the average of about every three years in Harrisburg, but serious flooding is much less frequent. About one-third of all floods have occurred during March. Tropical hurricanes rarely reach Harrisburg with destructive winds.

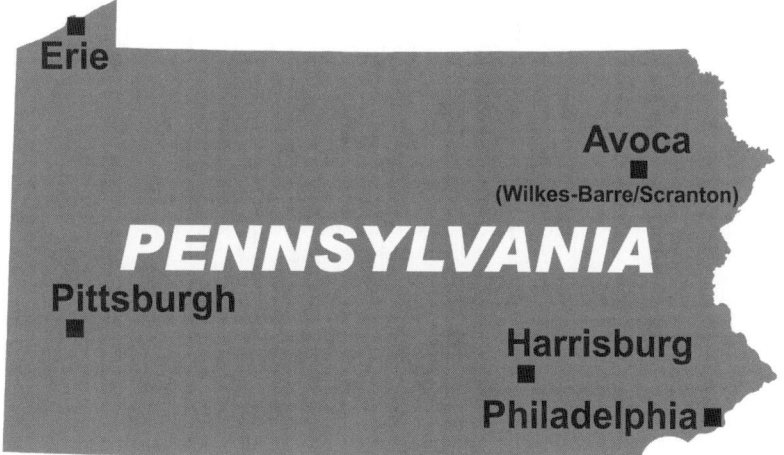

# NORMALS, MEANS, AND EXTREMES
## HARRISBURG (KMDT)

LATITUDE: 40 ° 11'N  LONGITUDE: -76 ° 45'W  ELEVATION (FT): GRND: 300  BARO: 314  TIME ZONE: EASTERN  (UTC -5)  WBAN: 14711

| | ELEMENT | POR | JAN | FEB | MAR | APR | MAY | JUN | JUL | AUG | SEP | OCT | NOV | DEC | YEAR |
|---|---|---|---|---|---|---|---|---|---|---|---|---|---|---|---|
| **TEMPERATURE °F** | NORMAL DAILY MAXIMUM | 30 | 37.5 | 40.9 | 50.9 | 62.6 | 72.6 | 80.8 | 85.7 | 83.7 | 75.7 | 64.3 | 52.5 | 41.7 | 62.4 |
| | MEAN DAILY MAXIMUM | 45 | 37.9 | 39.9 | 49.8 | 61.6 | 71.8 | 78.7 | 84.9 | 82.9 | 75.4 | 65.4 | 52.5 | 41.0 | 61.8 |
| | HIGHEST DAILY MAXIMUM | 71 | 73 | 78 | 87 | 93 | 97 | 100 | 107 | 101 | 102 | 97 | 84 | 75 | 107 |
| | YEAR OF OCCURRENCE | | 1950 | 1997 | 1998 | 1985 | 1942 | 1966 | 1966 | 1944 | 1953 | 1941 | 1950 | 1998 | JUL 1966 |
| | MEAN OF EXTREME MAXS. | 70 | 58.3 | 59.2 | 71.0 | 83.0 | 88.0 | 92.7 | 95.5 | 94.0 | 89.5 | 81.1 | 70.6 | 60.6 | 78.6 |
| | NORMAL DAILY MINIMUM | 30 | 23.1 | 24.7 | 32.5 | 41.5 | 51.4 | 60.6 | 66.0 | 64.2 | 56.7 | 44.6 | 36.1 | 27.8 | 44.1 |
| | MEAN DAILY MINIMUM | 45 | 23.6 | 24.1 | 31.8 | 41.3 | 50.9 | 59.5 | 65.3 | 63.4 | 56.0 | 45.0 | 35.4 | 26.6 | 43.6 |
| | LOWEST DAILY MINIMUM | 71 | -22 | -5 | 5 | 19 | 31 | 40 | 49 | 45 | 30 | 23 | 13 | -8 | -22 |
| | YEAR OF OCCURRENCE | | 1994 | 1979 | 1984 | 1982 | 1966 | 1980 | 1945 | 1976 | 1963 | 2000 | 1955 | 1960 | JAN 1994 |
| | MEAN OF EXTREME MINS. | 70 | 8.1 | 9.4 | 18.2 | 28.8 | 38.4 | 49.0 | 56.7 | 53.7 | 42.2 | 31.5 | 22.2 | 11.6 | 30.8 |
| | NORMAL DRY BULB | 30 | 28.6 | 31.3 | 41.2 | 51.6 | 61.8 | 70.9 | 75.7 | 74.1 | 66.4 | 54.7 | 44.4 | 33.6 | 52.9 |
| | MEAN DRY BULB | 69 | 30.2 | 31.8 | 41.0 | 51.7 | 61.8 | 70.3 | 75.3 | 73.6 | 65.9 | 55.0 | 44.1 | 34.0 | 52.9 |
| | MEAN WET BULB | 18 | 34.8 | 37.2 | 40.7 | 49.2 | 56.4 | 64.6 | 67.5 | 66.8 | 62.1 | 53.5 | 46.8 | 37.5 | 51.4 |
| | MEAN DEW POINT | 18 | 32.2 | 32.8 | 38.2 | 46.1 | 53.9 | 62.6 | 65.7 | 65.2 | 60.2 | 51.3 | 44.1 | 34.6 | 48.9 |
| | NORMAL NO. DAYS WITH: | | | | | | | | | | | | | | |
| | MAXIMUM >= 90 | 30 | 0.0 | 0.0 | 0.0 | 0.0 | 1.9 | 5.2 | 9.1 | 5.1 | 1.1 | 0.0 | 0.0 | 4.6 | 22.4 |
| | MAXIMUM <= 32 | 30 | 8.7 | 5.3 | 1.0 | 0.0 | 0.0 | 0.0 | 0.0 | 0.0 | 0.0 | 0.0 | 0.1 | 0.0 | 19.7 |
| | MINIMUM <= 32 | 30 | 25.3 | 22.4 | 16.0 | 2.6 | 0.0 | 0.0 | 0.0 | 0.0 | 0.0 | 0.6 | 11.8 | 23.0 | 101.7 |
| | MINIMUM <= 0 | 30 | 0.8 | 0.1 | 0.0 | 0.0 | 0.0 | 0.0 | 0.0 | 0.0 | 0.0 | 0.0 | 0.0 | 0.0 | 0.9 |
| **H/C** | NORMAL HEATING DEG. DAYS | 30 | 1076 | 901 | 723 | 390 | 148 | 14 | 0 | 1 | 52 | 338 | 621 | 937 | 5201 |
| | NORMAL COOLING DEG. DAYS | 30 | 0 | 0 | 0 | 1 | 54 | 186 | 337 | 279 | 87 | 11 | 0 | 0 | 955 |
| **RH** | NORMAL (PERCENT) | 30 | | | | | | | | | | | | | |
| | HOUR 01 LST | 30 | | | | | | | | | | | | | |
| | HOUR 07 LST | 30 | | | | | | | | | | | | | |
| | HOUR 13 LST | 30 | | | | | | | | | | | | | |
| | HOUR 19 LST | 30 | | | | | | | | | | | | | |
| **S** | PERCENT POSSIBLE SUNSHINE | 62 | 48 | 54 | 57 | 58 | 59 | 64 | 67 | 66 | 61 | 58 | 47 | 44 | 57 |
| **W/O** | MEAN NO. DAYS WITH: | | | | | | | | | | | | | | |
| | HEAVY FOG(VISBY <= 1/4 MI) | 21 | 2.3 | 2.3 | 2.0 | 1.1 | 0.8 | 0.4 | 0.8 | 0.7 | 1.2 | 1.9 | 1.6 | 2.0 | 17.1 |
| | THUNDERSTORMS | 36 | 0.2 | 0.2 | 0.6 | 2.2 | 3.7 | 4.4 | 4.8 | 4.1 | 2.1 | 0.6 | 0.2 | 0.1 | 23.2 |
| **CLOUDNESS** | MEAN: | | | | | | | | | | | | | | |
| | SUNRISE-SUNSET (OKTAS) | 50 | 5.4 | 5.3 | 5.3 | 5.2 | 5.2 | 4.9 | 4.9 | 4.7 | 4.7 | 4.6 | 5.2 | 5.6 | 5.1 |
| | MIDNIGHT-MIDNIGHT (OKTAS) | 8 | 5.4 | 5.1 | 5.2 | 5.0 | 4.8 | 5.3 | 4.9 | 4.6 | 4.6 | 3.5 | 4.6 | 5.3 | 4.9 |
| | MEAN NO. DAYS WITH: | | | | | | | | | | | | | | |
| | CLEAR | 62 | 6.5 | 6.6 | 7.1 | 6.5 | 5.9 | 6.4 | 7.0 | 7.7 | 8.5 | 9.8 | 6.3 | 6.1 | 84.4 |
| | PARTLY CLOUDY | 62 | 7.8 | 7.4 | 8.3 | 8.7 | 10.9 | 11.7 | 11.7 | 11.3 | 9.5 | 8.4 | 8.2 | 7.5 | 111.4 |
| | CLOUDY | 62 | 16.7 | 14.3 | 15.4 | 14.9 | 11.9 | 11.9 | 11.6 | 11.6 | 11.9 | 12.4 | 15.2 | 16.8 | 167.0 |
| **PR** | MEAN STATION PRESSURE(IN) | 18 | 29.72 | 29.72 | 29.70 | 29.66 | 29.66 | 29.64 | 29.65 | 29.68 | 29.73 | 29.72 | 29.74 | 29.74 | 29.70 |
| | MEAN SEA-LEVEL PRES. (IN) | 18 | 30.09 | 30.08 | 30.05 | 29.99 | 29.99 | 29.97 | 29.97 | 30.02 | 30.06 | 30.09 | 30.10 | 30.11 | 30.04 |
| **WINDS** | MEAN SPEED (MPH) | 18 | 9.1 | 9.2 | 9.8 | 9.2 | 7.8 | 6.9 | 6.3 | 5.8 | 6.2 | 6.4 | 7.8 | 8.5 | 7.8 |
| | PREVAIL.DIR(TENS OF DEGS) | 22 | 31 | 31 | 32 | 32 | 31 | 31 | 32 | 31 | 32 | 31 | 31 | 31 | 31 |
| | MAXIMUM 2-MINUTE: | 9 | | | | | | | | | | | | | |
| | SPEED (MPH) | | 45 | 45 | 40 | 38 | 35 | 46 | 36 | 35 | 48 | 39 | 44 | 41 | 48 |
| | DIR. (TENS OF DEGS) | | 30 | 28 | 27 | 32 | 22 | 30 | 27 | 32 | 13 | 30 | 31 | 32 | 13 |
| | YEAR OF OCCURRENCE | | 2006 | 2008 | 2008 | 2007 | 2007 | 2006 | 2006 | 2008 | 2003 | 2008 | 2003 | 2008 | SEP 2003 |
| | MAXIMUM 3-SECOND | | | | | | | | | | | | | | |
| | SPEED (MPH) | 9 | 54 | 56 | 55 | 48 | 46 | 53 | 46 | 48 | 60 | 54 | 56 | 56 | 60 |
| | DIR. (TENS OF DEGS) | | 30 | 26 | 30 | 32 | 26 | 31 | 32 | 31 | 12 | 29 | 31 | 23 | 12 |
| | YEAR OF OCCURRENCE | | 2006 | 2009 | 2002 | 2007 | 2008 | 2006 | 2004 | 2004 | 2003 | 2008 | 2003 | 2009 | SEP 2003 |
| **PRECIPITATION** | NORMAL (IN) | 30 | 3.18 | 2.88 | 3.58 | 3.31 | 4.60 | 3.99 | 3.21 | 3.24 | 3.65 | 3.06 | 3.53 | 3.22 | 41.45 |
| | MAXIMUM MONTHLY (IN) | 31 | 8.01 | 5.93 | 6.32 | 7.96 | 9.71 | 8.62 | 8.09 | 7.08 | 10.50 | 7.78 | 6.23 | 7.57 | 10.50 |
| | YEAR OF OCCURRENCE | | 1979 | 1981 | 1993 | 1983 | 1989 | 2006 | 1994 | 2004 | 2004 | 2005 | 1985 | 1983 | SEP 2004 |
| | MINIMUM MONTHLY (IN) | 31 | 0.43 | 0.39 | 0.68 | 0.45 | .88 | 1.00 | 0.97 | 0.53 | 0.65 | 0.47 | 0.92 | 0.31 | 0.31 |
| | YEAR OF OCCURRENCE | | 1981 | 2002 | 2006 | 1985 | 2005 | 1988 | 1983 | 1995 | 1986 | 2000 | 1998 | 1998 | DEC 1998 |
| | MAXIMUM IN 24 HOURS (IN) | 31 | 2.09 | 2.40 | 2.69 | 2.06 | 2.91 | 4.56 | 3.84 | 3.04 | 5.82 | 3.97 | 3.27 | 2.58 | 5.82 |
| | YEAR OF OCCURRENCE | | 1979 | 2007 | 2008 | 1992 | 1984 | 2006 | 1994 | 2004 | 2004 | 2005 | 1993 | 1993 | SEP 2004 |
| | NORMAL NO. DAYS WITH: | | | | | | | | | | | | | | |
| | PRECIPITATION >= 0.01 | 30 | 11.4 | 9.8 | 11.1 | 10.2 | 12.1 | 10.1 | 10.9 | 8.4 | 10.1 | 7.3 | 8.3 | 9.5 | 119.2 |
| | PRECIPITATION >= 1.00 | 30 | 0.6 | 0.0 | 1.1 | 0.9 | 0.6 | 0.8 | 0.9 | 1.1 | 1.3 | 0.9 | 0.8 | 0.8 | 9.8 |
| **SNOWFALL** | NORMAL (IN) | 30 | 15.7 | 15.2 | 9.6 | 0.3 | T | 0.0 | 0.0 | 0.0 | 0.0 | T | 2.0 | 11.0 | 53.8 |
| | MAXIMUM MONTHLY (IN) | 31 | 38.9 | 29.8 | 22.8 | 10.2 | 0.0 | T | 0.0 | T | 0.0 | T | 9.7 | 17.4 | 38.9 |
| | YEAR OF OCCURRENCE | | 1996 | 2003 | 1993 | 1982 | | 1993 | | 1993 | | 1982 | 1987 | 1995 | JAN 1996 |
| | MAXIMUM IN 24 HOURS (IN) | 29 | 21.7 | 14.2 | 20.4 | 4.0 | 0.0 | T | 0.0 | T | 0.0 | T | 7.9 | 9.1 | 21.7 |
| | YEAR OF OCCURRENCE' | | 1996 | 1979 | 1993 | 1996 | | 1993 | | 1993 | | 1979 | 1987 | 1990 | JAN 1996 |
| | MAXIMUM SNOW DEPTH (IN) | 36 | 20 | 23 | 20 | 1 | 0 | 0 | 0 | 0 | 0 | 0 | 12 | 13 | 23 |
| | YEAR OF OCCURRENCE | | 1961 | 1961 | 1993 | 1959 | | | | | | | 1953 | 1951 | FEB 1961 |
| | NORMAL NO. DAYS WITH: | | | | | | | | | | | | | | |
| | SNOWFALL >= 1.0 | 30 | 3.0 | 4.2 | 2.3 | 0.2 | 0.0 | 0.0 | 0.0 | 0.0 | 0.0 | 0.0 | 0.4 | 2.7 | 12.8 |

## PRECIPITATION (inches) 2009 HARRISBURG (KMDT)

| YEAR | JAN | FEB | MAR | APR | MAY | JUN | JUL | AUG | SEP | OCT | NOV | DEC | ANNUAL |
|------|-----|-----|-----|-----|-----|-----|-----|-----|-----|-----|-----|-----|--------|
| 1980 | 0.90 | 0.82 | 5.47 | 4.27 | 4.58 | 2.50 | 1.59 | 1.51 | 1.06 | 2.94 | 3.65 | 0.77 | 30.06 |
| 1981 | 0.43 | 5.93 | 1.02 | 2.77 | 1.86 | 4.66 | 4.67 | 4.11 | 2.20 | 3.76 | 0.96 | 2.41 | 34.78 |
| 1982 | 3.63 | 1.92 | 2.20 | 4.17 | 4.89 | 8.12 | 2.90 | 2.47 | 2.87 | 1.82 | 3.37 | 1.56 | 39.92 |
| 1983 | 2.26 | 3.38 | 4.86 | 7.96 | 5.36 | 2.81 | 0.97 | 2.50 | 1.40 | 4.21 | 5.29 | 7.57 | 48.57 |
| 1984 | 1.12 | 4.51 | 5.36 | 4.46 | 6.20 | 6.36 | 3.76 | 2.75 | 1.49 | 1.98 | 3.78 | 2.28 | 44.05 |
| 1985 | 1.06 | 2.91 | 2.78 | 0.45 | 6.29 | 3.07 | 2.50 | 2.14 | 3.76 | 1.34 | 6.23 | 1.28 | 33.81 |
| 1986 | 2.24 | 4.50 | 3.16 | 4.10 | 2.29 | 1.48 | 5.17 | 6.26 | 0.65 | 2.59 | 4.58 | 4.90 | 41.92 |
| 1987 | 3.69 | 1.59 | 1.43 | 2.93 | 3.73 | 3.46 | 1.96 | 2.89 | 8.41 | 2.63 | 4.96 | 1.84 | 39.52 |
| 1988 | 2.18 | 3.28 | 1.98 | 2.65 | 5.79 | 1.00 | 4.40 | 2.67 | 2.42 | 1.81 | 3.67 | 0.90 | 32.75 |
| 1989 | 2.29 | 1.90 | 3.60 | 1.10 | 9.71 | 6.02 | 7.20 | 3.03 | 2.63 | 5.59 | 2.17 | 1.27 | 46.51 |
| 1990 | 3.77 | 2.73 | 1.76 | 2.60 | 7.20 | 1.10 | 3.62 | 6.14 | 1.65 | 4.92 | 2.58 | 6.05 | 44.12 |
| 1991 | 2.61 | 1.39 | 3.54 | 2.00 | 3.15 | 1.08 | 1.99 | 5.29 | 1.35 | 3.15 | 2.08 | 3.49 | 31.12 |
| 1992 | 1.62 | 1.56 | 5.13 | 2.62 | 3.17 | 1.90 | 3.54 | 1.45 | 5.65 | 1.64 | 4.82 | 2.42 | 35.52 |
| 1993 | 2.39 | 2.32 | 6.32 | 6.49 | 1.96 | 3.20 | 3.65 | 3.45 | 7.84 | 2.66 | 4.17 | 3.95 | 48.40 |
| 1994 | 5.00 | 3.24 | 6.22 | 2.96 | 2.73 | 1.81 | 8.09 | 4.94 | 2.33 | 0.74 | 5.18 | 2.92 | 46.16 |
| 1995 | 3.52 | 1.52 | 0.95 | 2.22 | 3.52 | 4.16 | 5.81 | 0.53 | 1.95 | 5.43 | 3.67 | 2.53 | 35.81 |
| 1996 | 5.87 | 1.60 | 3.26 | 4.69 | 3.93 | 6.33 | 7.06 | 2.22 | 3.35 | 4.26 | 3.88 | 5.98 | 52.43 |
| 1997 | 2.00 | 1.47 | 3.44 | 0.92 | 3.66 | 2.41 | 4.82 | 3.60 | 2.52 | 1.50 | 3.97 | 2.01 | 32.32 |
| 1998 | 4.49 | 5.05 | 4.36 | 5.17 | 6.37 | 5.82 | 4.86 | 3.88 | 1.81 | 2.92 | 0.92 | 0.31 | 45.96 |
| 1999 | 4.94 | 1.90 | 2.55 | 2.80 | 1.39 | 1.87 | 2.96 | 3.92 | 9.11 | 2.65 | 1.66 | 2.57 | 38.32 |
| 2000 | 2.01 | 2.33 | 6.06 | 2.63 | 4.03 | 4.09 | 2.32 | 4.14 | 8.61 | 0.47 | 1.55 | 3.99 | 42.23 |
| 2001 | 2.44 | 1.48 | 4.19 | 1.72 | 1.66 | 2.01 | 1.90 | 3.79 | 2.18 | 1.01 | 1.51 | 1.87 | 25.76 |
| 2002 | 2.45 | 0.39 | 5.05 | 3.79 | 4.40 | 2.38 | 1.27 | 2.69 | 3.68 | 6.37 | 3.77 | 4.60 | 40.84 |
| 2003 | 2.29 | 4.25 | 4.40 | 2.50 | 5.40 | 7.31 | 5.09 | 6.19 | 6.10 | 4.73 | 2.97 | 3.40 | 54.63 |
| 2004 | 2.07 | 2.27 | 2.03 | 5.12 | 3.47 | 4.31 | 7.97 | 7.08 | 10.50 | 2.30 | 2.85 | 3.37 | 53.34 |
| 2005 | 4.41 | 1.91 | 3.82 | 4.93 | 0.88 | 1.38 | 6.31 | 1.74 | 0.80 | 7.78 | 2.09 | 2.72 | 38.77 |
| 2006 | 4.20 | 2.40 | 0.68 | 3.16 | 2.16 | 8.62 | 5.13 | 1.41 | 5.53 | 5.01 | 5.48 | 2.30 | 46.08 |
| 2007 | 2.37 | 2.99 | 3.63 | 3.31 | 1.07 | 3.70 | 4.22 | 5.84 | 4.42 | 3.40 | 2.58 | 5.11 | 42.64 |
| 2008 | 1.09 | 5.77 | 4.68 | 3.86 | 5.41 | 2.54 | 3.53 | 1.09 | 6.61 | 1.91 | 2.93 | 6.83 | 46.25 |
| 2009 | 1.92 | 0.65 | 1.25 | 3.69 | 6.99 | 6.14 | 3.20 | 4.84 | 3.95 | 6.10 | 1.61 | 4.99 | 45.33 |
| POR= 69 YRS | 2.91 | 2.57 | 3.48 | 3.30 | 4.12 | 3.76 | 3.83 | 3.40 | 3.60 | 3.03 | 3.38 | 3.28 | 40.66 |

WBAN : 14711

## AVERAGE TEMPERATURE (°F) 2009 HARRISBURG (KMDT)

| YEAR | JAN | FEB | MAR | APR | MAY | JUN | JUL | AUG | SEP | OCT | NOV | DEC | ANNUAL |
|------|-----|-----|-----|-----|-----|-----|-----|-----|-----|-----|-----|-----|--------|
| 1980 | 30.3 | 29.1 | 38.9 | 52.8 | 63.3 | 67.8 | 76.3 | 76.1 | 67.7 | 51.5 | 39.4 | 29.6 | 51.9 |
| 1981 | 23.7 | 34.6 | 38.7 | 53.7 | 61.9 | 71.7 | 75.7 | 72.2 | 63.9 | 50.7 | 44.7 | 31.9 | 52.0 |
| 1982 | 22.8 | 30.9 | 38.6 | 47.6 | 62.2 | 65.1 | 74.4 | 70.5 | 65.3 | 55.1 | 47.6 | 41.4 | 51.8 |
| 1983 | 33.0 | 33.4 | 42.7 | 49.3 | 58.4 | 69.1 | 75.9 | 74.9 | 66.2 | 53.6 | 43.9 | 28.7 | 52.4 |
| 1984 | 24.8 | 36.6 | 33.7 | 48.0 | 58.2 | 72.9 | 74.3 | 75.8 | 64.4 | 61.5 | 43.9 | 41.5 | 53.0 |
| 1985 | 27.9 | 34.4 | 44.5 | 56.9 | 65.2 | 69.4 | 75.9 | 74.1 | 69.2 | 57.2 | 47.9 | 31.0 | 54.5 |
| 1986 | 31.4 | 30.0 | 43.5 | 53.5 | 65.6 | 71.4 | 76.3 | 72.0 | 66.2 | 56.0 | 41.2 | 36.1 | 53.6 |
| 1987 | 30.0 | 32.3 | 44.1 | 52.3 | 63.2 | 72.6 | 78.2 | 73.1 | 65.7 | 49.4 | 43.9 | 36.6 | 53.5 |
| 1988 | 24.4 | 31.8 | 42.4 | 50.0 | 62.3 | 70.8 | 78.8 | 76.2 | 63.5 | 49.5 | 43.7 | 33.3 | 52.2 |
| 1989 | 34.8 | 32.4 | 40.4 | 50.4 | 60.0 | 70.6 | 73.7 | 72.6 | 66.0 | 55.6 | 42.7 | 22.6 | 51.8 |
| 1990 | 38.2 | 38.2 | 44.9 | 53.1 | 59.4 | 71.2 | 75.2 | 72.7 | 65.0 | 58.2 | 46.9 | 38.4 | 55.1 |
| 1991 | 31.7 | 37.4 | 43.7 | 53.7 | 69.1 | 74.2 | 78.7 | 76.2 | 65.4 | 56.9 | 43.7 | 36.5 | 55.6 |
| 1992 | 33.0 | 35.7 | 40.3 | 51.8 | 60.2 | 69.3 | 75.0 | 71.6 | 65.5 | 51.3 | 44.3 | 35.3 | 52.8 |
| 1993 | 34.8 | 28.8 | 36.6 | 51.3 | 64.6 | 72.1 | 78.0 | 75.9 | 65.6 | 52.8 | 43.9 | 34.0 | 53.2 |
| 1994 | 20.3 | 26.7 | 38.1 | 56.8 | 59.8 | 76.2 | 78.7 | 71.6 | 65.9 | 53.8 | 49.0 | 38.5 | 53.0 |
| 1995 | 34.8 | 29.1 | 44.7 | 50.9 | 61.9 | 71.6 | 77.9 | 77.4 | 66.9 | 59.0 | 39.6 | 29.6 | 53.6 |
| 1996 | 26.2 | 31.8 | 37.7 | 52.5 | 60.0 | 73.2 | 74.1 | 74.0 | 67.2 | 54.9 | 39.6 | 37.3 | 52.4 |
| 1997 | 29.8 | 38.3 | 42.5 | 51.1 | 59.6 | 71.4 | 77.4 | 74.0 | 67.1 | 56.2 | 41.6 | 36.0 | 53.8 |
| 1998 | 38.2 | 40.4 | 44.5 | 54.6 | 66.8 | 70.3 | 75.3 | 75.1 | 70.6 | 55.9 | 45.9 | 40.7 | 56.6 |
| 1999 | 30.7 | 36.5 | 41.2 | 53.9 | 65.0 | 73.2 | 81.9 | 75.2 | 68.3 | 54.0 | 49.3 | 37.4 | 55.6 |
| 2000 | 30.2 | 34.7 | 47.5 | 52.4 | 65.1 | 72.2 | 73.2 | 73.7 | 65.3 | 55.6 | 43.5 | 26.5 | 53.3 |
| 2001 | 28.4 | 34.8 | 38.1 | 51.5 | 62.4 | 72.6 | 73.2 | 76.8 | 65.1 | 55.7 | 48.8 | 39.8 | 53.9 |
| 2002 | 35.7 | 37.4 | 42.2 | 54.8 | 59.9 | 72.1 | 78.2 | 77.8 | 68.9 | 52.9 | 42.7 | 31.0 | 54.5 |
| 2003 | 25.7 | 26.7 | 39.8 | 50.6 | 58.9 | 68.2 | 75.1 | 75.4 | 66.2 | 52.3 | 47.7 | 34.2 | 51.7 |
| 2004 | 23.9 | 30.9 | 43.5 | 53.0 | 68.3 | 70.0 | 74.3 | 72.8 | 68.0 | 54.1 | 47.7 | 35.3 | 53.5 |
| 2005 | 31.1 | 34.4 | 38.1 | 54.5 | 59.5 | 75.0 | 77.6 | 77.1 | 71.2 | 56.8 | 46.1 | 30.3 | 54.3 |
| 2006 | 38.3 | 33.5 | 43.3 | 54.6 | 62.3 | 71.7 | 77.9 | 76.1 | 64.1 | 53.7 | 47.4 | 40.2 | 55.3 |
| 2007 | 35.2 | 25.7 | 41.1 | 49.2 | 65.2 | 73.5 | 75.0 | 75.0 | 69.7 | 62.0 | 42.9 | 34.7 | 54.1 |
| 2008 | 32.9 | 32.0 | 41.6 | 54.4 | 58.6 | 74.5 | 76.9 | 72.9 | 68.0 | 53.1 | 43.3 | 34.7 | 53.6 |
| 2009 | 25.9 | 34.2 | 41.9 | 53.5 | 62.5 | 70.2 | 73.0 | 75.4 | 65.7 | 53.4 | 48.2 | 33.5 | 53.1 |
| POR= 69 YRS | 30.2 | 31.8 | 41.0 | 51.7 | 61.8 | 70.3 | 75.3 | 73.6 | 65.9 | 55.0 | 44.1 | 34.0 | 52.9 |

## HEATING DEGREE DAYS (base 65°F) 2009  HARRISBURG (KMDT)

| YEAR | JUL | AUG | SEP | OCT | NOV | DEC | JAN | FEB | MAR | APR | MAY | JUN | TOTAL |
|------|-----|-----|-----|-----|-----|-----|-----|-----|-----|-----|-----|-----|-------|
| 1980-81 | 0 | 0 | 57 | 411 | 761 | 1091 | 1277 | 844 | 809 | 339 | 147 | 6 | 5742 |
| 1981-82 | 0 | 1 | 94 | 437 | 599 | 1021 | 1304 | 948 | 812 | 518 | 128 | 61 | 5923 |
| 1982-83 | 7 | 12 | 67 | 318 | 520 | 725 | 985 | 876 | 686 | 468 | 221 | 25 | 4910 |
| 1983-84 | 0 | 2 | 103 | 362 | 628 | 1117 | 1238 | 817 | 962 | 502 | 240 | 11 | 5982 |
| 1984-85 | 0 | 0 | 105 | 131 | 627 | 724 | 1143 | 849 | 627 | 292 | 87 | 16 | 4601 |
| 1985-86 | 0 | 0 | 41 | 237 | 508 | 1049 | 1038 | 974 | 664 | 349 | 89 | 9 | 4958 |
| 1986-87 | 2 | 17 | 46 | 300 | 705 | 890 | 1080 | 907 | 643 | 380 | 142 | 2 | 5114 |
| 1987-88 | 0 | 8 | 51 | 477 | 627 | 873 | 1252 | 961 | 693 | 445 | 131 | 41 | 5559 |
| 1988-89 | 4 | 5 | 88 | 475 | 633 | 975 | 931 | 912 | 760 | 433 | 196 | 9 | 5421 |
| 1989-90 | 1 | 6 | 81 | 292 | 663 | 1306 | 824 | 744 | 629 | 385 | 175 | 13 | 5119 |
| 1990-91 | 5 | 8 | 96 | 248 | 535 | 816 | 1026 | 769 | 651 | 345 | 69 | 2 | 4570 |
| 1991-92 | 0 | 0 | 103 | 279 | 634 | 877 | 986 | 842 | 761 | 392 | 170 | 16 | 5060 |
| 1992-93 | 0 | 3 | 86 | 420 | 615 | 914 | 931 | 1005 | 875 | 405 | 67 | 15 | 5336 |
| 1993-94 | 0 | 2 | 99 | 371 | 628 | 954 | 1382 | 1067 | 828 | 260 | 195 | 2 | 5788 |
| 1994-95 | 0 | 3 | 41 | 340 | 475 | 815 | 928 | 999 | 623 | 420 | 126 | 3 | 4773 |
| 1995-96 | 0 | 0 | 64 | 218 | 755 | 1090 | 1196 | 960 | 840 | 376 | 208 | 3 | 5710 |
| 1996-97 | 0 | 0 | 54 | 307 | 756 | 852 | 1083 | 741 | 688 | 410 | 175 | 32 | 5098 |
| 1997-98 | 1 | 1 | 46 | 318 | 696 | 892 | 824 | 680 | 653 | 305 | 67 | 29 | 4512 |
| 1998-99 | 0 | 1 | 24 | 278 | 565 | 748 | 1059 | 792 | 733 | 329 | 61 | 6 | 4596 |
| 1999-00 | 0 | 0 | 36 | 334 | 464 | 850 | 1075 | 872 | 534 | 375 | 101 | 16 | 4657 |
| 2000-01 | 0 | 4 | 99 | 293 | 638 | 1187 | 1127 | 836 | 826 | 402 | 112 | 16 | 5540 |
| 2001-02 | 1 | 0 | 82 | 288 | 477 | 774 | 900 | 766 | 699 | 343 | 189 | 5 | 4524 |
| 2002-03 | 0 | 1 | 22 | 395 | 663 | 1047 | 1213 | 1067 | 774 | 427 | 193 | 51 | 5853 |
| 2003-04 | 0 | 0 | 28 | 384 | 513 | 950 | 1267 | 985 | 657 | 361 | 54 | 20 | 5219 |
| 2004-05 | 0 | 4 | 17 | 330 | 514 | 913 | 1045 | 849 | 827 | 309 | 179 | 5 | 4992 |
| 2005-06 | 0 | 0 | 17 | 267 | 560 | 1071 | 818 | 878 | 667 | 306 | 133 | 2 | 4719 |
| 2006-07 | 0 | 0 | 73 | 350 | 522 | 763 | 915 | 1089 | 735 | 465 | 97 | 7 | 5016 |
| 2007-08 | 0 | 11 | 32 | 165 | 655 | 936 | 986 | 952 | 718 | 314 | 210 | 2 | 4981 |
| 2008-09 | 0 | 0 | 34 | 368 | 641 | 932 | 1205 | 853 | 709 | 371 | 120 | 25 | 5258 |
| 2009- | 0 | 2 | 48 | 353 | 499 | 966 | | | | | | | |

WBAN : 14711

## COOLING DEGREE DAYS (base 65°F) 2009  HARRISBURG (KMDT)

| YEAR | JAN | FEB | MAR | APR | MAY | JUN | JUL | AUG | SEP | OCT | NOV | DEC | TOTAL |
|------|-----|-----|-----|-----|-----|-----|-----|-----|-----|-----|-----|-----|-------|
| 1980 | 0 | 0 | 0 | 0 | 57 | 138 | 355 | 350 | 145 | 0 | 0 | 0 | 1045 |
| 1981 | 0 | 0 | 0 | 6 | 60 | 213 | 339 | 232 | 66 | 0 | 0 | 0 | 916 |
| 1982 | 0 | 0 | 0 | 2 | 48 | 70 | 307 | 191 | 83 | 19 | 4 | 0 | 724 |
| 1983 | 0 | 0 | 0 | 6 | 22 | 154 | 343 | 315 | 146 | 13 | 0 | 0 | 999 |
| 1984 | 0 | 0 | 0 | 0 | 34 | 256 | 292 | 342 | 95 | 35 | 0 | 1 | 1055 |
| 1985 | 0 | 0 | 0 | 55 | 99 | 157 | 345 | 290 | 173 | 4 | 0 | 0 | 1123 |
| 1986 | 0 | 0 | 5 | 12 | 116 | 205 | 360 | 238 | 88 | 27 | 0 | 0 | 1051 |
| 1987 | 0 | 0 | 0 | 9 | 94 | 237 | 418 | 266 | 76 | 0 | 0 | 0 | 1100 |
| 1988 | 0 | 0 | 0 | 0 | 53 | 219 | 439 | 355 | 52 | 4 | 0 | 0 | 1122 |
| 1989 | 0 | 0 | 5 | 0 | 49 | 182 | 279 | 249 | 114 | 9 | 1 | 0 | 888 |
| 1990 | 0 | 0 | 14 | 34 | 12 | 205 | 330 | 254 | 102 | 43 | 0 | 0 | 994 |
| 1991 | 0 | 0 | 0 | 16 | 202 | 284 | 429 | 355 | 118 | 31 | 0 | 0 | 1435 |
| 1992 | 0 | 0 | 0 | 1 | 29 | 151 | 314 | 214 | 108 | 0 | 0 | 0 | 817 |
| 1993 | 0 | 0 | 0 | 0 | 63 | 236 | 410 | 348 | 127 | 0 | 0 | 0 | 1184 |
| 1994 | 0 | 0 | 0 | 20 | 40 | 343 | 428 | 213 | 76 | 0 | 1 | 0 | 1121 |
| 1995 | 0 | 0 | 0 | 6 | 41 | 208 | 409 | 392 | 124 | 36 | 0 | 0 | 1216 |
| 1996 | 0 | 0 | 0 | 9 | 61 | 253 | 286 | 288 | 126 | 1 | 0 | 0 | 1024 |
| 1997 | 0 | 0 | 0 | 0 | 14 | 230 | 392 | 283 | 117 | 50 | 0 | 0 | 1086 |
| 1998 | 0 | 0 | 24 | 1 | 130 | 209 | 326 | 323 | 199 | 0 | 0 | 0 | 1212 |
| 1999 | 0 | 0 | 0 | 0 | 72 | 260 | 530 | 323 | 142 | 1 | 0 | 0 | 1328 |
| 2000 | 0 | 0 | 0 | 5 | 110 | 235 | 262 | 280 | 115 | 8 | 0 | 0 | 1015 |
| 2001 | 0 | 0 | 0 | 5 | 39 | 250 | 260 | 371 | 94 | 8 | 0 | 0 | 1027 |
| 2002 | 0 | 0 | 0 | 45 | 38 | 227 | 417 | 402 | 145 | 26 | 0 | 0 | 1300 |
| 2003 | 0 | 0 | 0 | 2 | 10 | 153 | 319 | 328 | 70 | 0 | 0 | 0 | 882 |
| 2004 | 0 | 0 | 0 | 7 | 165 | 176 | 298 | 253 | 118 | 1 | 0 | 0 | 1018 |
| 2005 | 0 | 0 | 0 | 3 | 19 | 311 | 398 | 382 | 212 | 18 | 0 | 0 | 1343 |
| 2006 | 0 | 0 | 0 | 2 | 54 | 211 | 409 | 350 | 53 | 7 | 0 | 0 | 1086 |
| 2007 | 0 | 0 | 0 | 0 | 107 | 268 | 319 | 328 | 178 | 81 | 0 | 0 | 1281 |
| 2008 | 0 | 0 | 0 | 0 | 18 | 293 | 378 | 252 | 131 | 6 | 0 | 0 | 1078 |
| 2009 | 0 | 0 | 0 | 33 | 49 | 186 | 255 | 334 | 74 | 0 | 0 | 0 | 931 |

## SNOWFALL (inches) 2009 HARRISBURG (KMDT)

| YEAR | JUL | AUG | SEP | OCT | NOV | DEC | JAN | FEB | MAR | APR | MAY | JUN | TOTAL |
|---|---|---|---|---|---|---|---|---|---|---|---|---|---|
| 1980-81 | 0.0 | 0.0 | 0.0 | 0.0 | 4.0 | 4.3 | 5.5 | 4.4 | 6.7 | 0.0 | 0.0 | 0.0 | 24.9 |
| 1981-82 | 0.0 | 0.0 | 0.0 | 0.0 | 0.8 | 12.5 | 18.8 | 8.4 | 7.8 | 10.2 | 0.0 | 0.0 | 58.5 |
| 1982-83 | 0.0 | 0.0 | 0.0 | T | T | 1.1 | 4.4 | 28.8 | 0.4 | 1.3 | 0.0 | 0.0 | 36.0 |
| 1983-84 | 0.0 | 0.0 | 0.0 | 0.0 | T | 4.6 | 9.7 | 2.3 | 14.9 | T | 0.0 | 0.0 | 31.5 |
| 1984-85 | 0.0 | 0.0 | 0.0 | 0.0 | 1.9 | 2.6 | 10.6 | 11.4 | T | 3.6 | 0.0 | 0.0 | 30.1 |
| 1985-86 | 0.0 | 0.0 | 0.0 | 0.0 | T | 5.6 | 7.8 | 23.1 | T | T | 0.0 | 0.0 | 36.5 |
| 1986-87 | 0.0 | 0.0 | 0.0 | 0.0 | T | 1.9 | 31.5 | 10.1 | 1.4 | 1.0 | 0.0 | 0.0 | 45.9 |
| 1987-88 | 0.0 | 0.0 | 0.0 | 0.0 | 9.7 | 3.6 | 9.6 | 2.8 | 1.0 | T | 0.0 | 0.0 | 26.7 |
| 1988-89 | 0.0 | 0.0 | 0.0 | 0.0 | T | T | 6.4 | 2.2 | 11.3 | 0.0 | 0.0 | 0.0 | 19.9 |
| 1989-90 | 0.0 | 0.0 | 0.0 | 0.0 | 1.8 | 6.7 | 4.9 | 1.3 | 3.5 | 1.1 | 0.0 | 0.0 | 19.3 |
| 1990-91 | 0.0 | 0.0 | 0.0 | 0.0 | 0.0 | 9.3 | 5.2 | 0.7 | 5.9 | T | 0.0 | 0.0 | 21.1 |
| 1991-92 | 0.0 | 0.0 | 0.0 | 0.0 | 0.0 | T | 1.7 | 4.6 | 6.6 | 0.0 | 0.0 | T | 12.9 |
| 1992-93 | 0.0 | 0.0 | 0.0 | 0.0 | T | 3.9 | 1.4 | 18.5 | 22.8 | 0.6 | 0.0 | T | 47.2 |
| 1993-94 | 0.0 | T | 0.0 | 0.0 | T | 0.9 | 34.2 | 22.0 | 18.8 | 0.0 | 0.0 | 0.0 | 75.9 |
| 1994-95 | 0.0 | 0.0 | 0.0 | 0.0 | 0.5 | 0.5 | 0.6 | 5.6 | 1.8 | 0.0 | 0.0 | 0.0 | 9.0 |
| 1995-96 | 0.0 | 0.0 | 0.0 | 0.0 | 8.1 | 17.4 | 38.9 | 6.3 | 2.3 | 4.6 | | | 22.8 |
| 1996-97 | 0.0 | 0.0 | 0.0 | 0.0 | .2 | 6.1 | 3.7 | 6.9 | 5.9 | 0.0 | 0.0 | 0.0 | 22.8 |
| 1997-98 | 0.0 | 0.0 | 0.0 | 0.0 | 0.3 | 4.8 | 5.3 | 0.4 | 0.6 | 0.0 | 0.0 | 0.0 | 11.4 |
| 1998-99 | 0.0 | 0.0 | 0.0 | 0.0 | 0.0 | 0.8 | 7.4 | 1.3 | 10.6 | 0.0 | 0.0 | 0.0 | 20.1 |
| 1999-00 | 0.0 | 0.0 | 0.0 | 0.0 | T | 0.1 | 15.0 | 7.2 | 0.0 | 0.6 | 0.0 | 0.0 | 22.9 |
| 2000-01 | 0.0 | 0.0 | 0.0 | 0.0 | T | 5.6 | 9.4 | 9.5 | 3.5 | T | 0.0 | 0.0 | 28.0 |
| 2001-02 | 0.0 | 0.0 | 0.0 | 0.0 | 0.0 | T | 8.8 | 0.3 | 1.5 | 0.0 | 0.0 | 0.0 | 10.6 |
| 2002-03 | 0.0 | 0.0 | 0.0 | 0.0 | 1.0 | 15.8 | 7.3 | 29.8 | 1.8 | 2.0 | 0.0 | 0.0 | 57.7 |
| 2003-04 | 0.0 | 0.0 | 0.0 | 0.0 | 0.0 | 11.6 | 12.4 | 4.0 | 6.6 | 0.0 | 0.0 | 0.0 | 34.6 |
| 2004-05 | 0.0 | 0.0 | 0.0 | 0.0 | 0.0 | T | 9.5 | 17.1 | 1.3 | 0.0 | 0.0 | 0.0 | 27.9 |
| 2005-06 | 0.0 | 0.0 | 0.0 | 0.0 | 0.1 | 12.2 | T | 5.7 | T | T | 0.0 | 0.0 | 18.0 |
| 2006-07 | 0.0 | 0.0 | 0.0 | 0.0 | 0.0 | T | 0.6 | 9.2 | 11.5 | T | 0.0 | 0.0 | 21.3 |
| 2007-08 | 0.0 | 0.0 | 0.0 | 0.0 | 0.5 | 2.8 | 2.0 | 7.9 | 0.8 | 0.0 | 0.0 | 0.0 | 14.0 |
| 2008-09 | 0.0 | 0.0 | 0.0 | 0.0 | 1.4 | 4.3 | 6.8 | 0.7 | 2.3 | T | 0.0 | 0.0 | 15.5 |
| 2009- | 0.0 | 0.0 | 0.0 | 0.0 | 0.0 | 14.1 | | | | | | | |
| POR= 40 YRS | 0.0 | T | 0.0 | T | 1.6 | 6.0 | 9.5 | 8.9 | 5.8 | 0.5 | 0.0 | T | 32.3 |

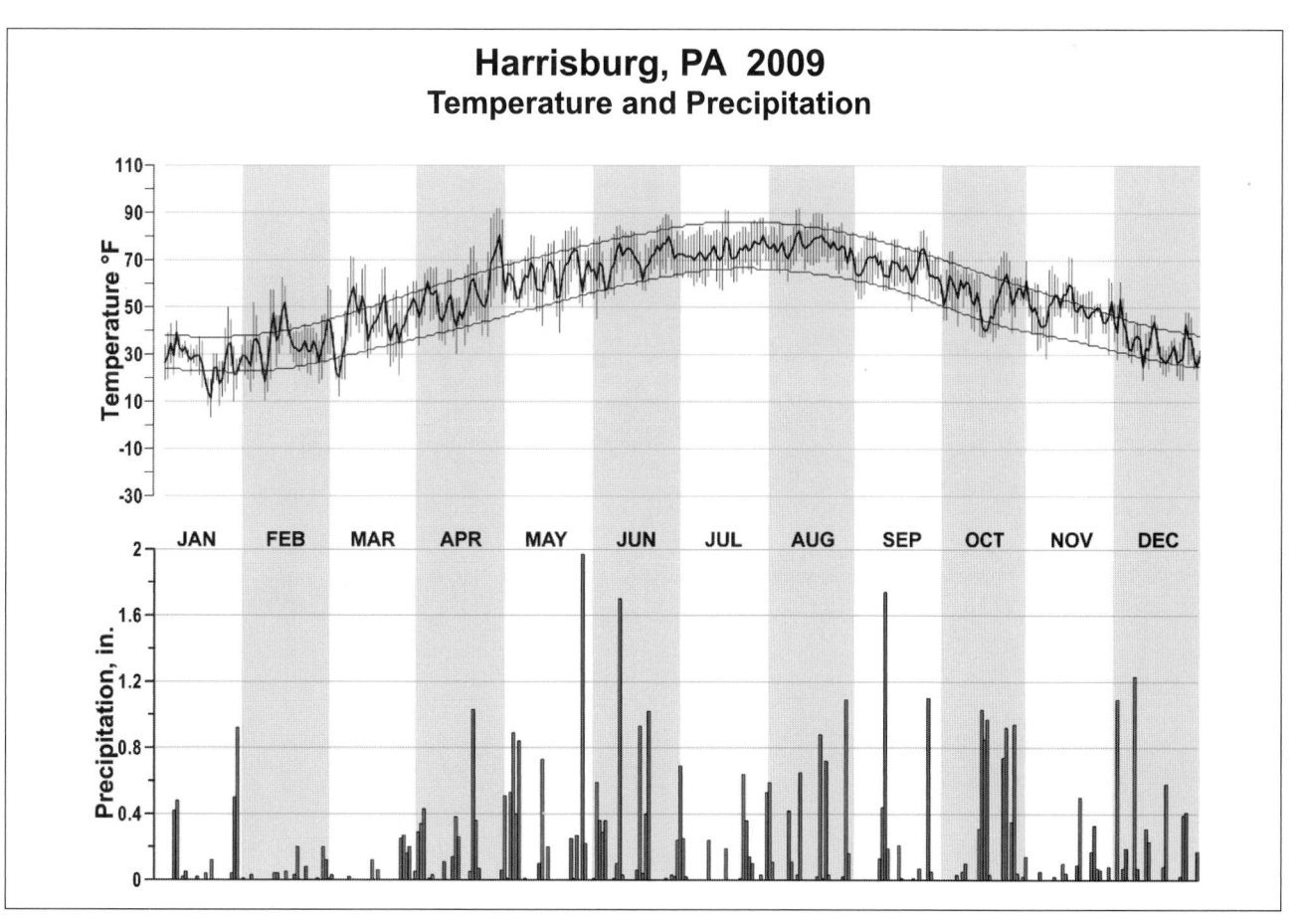

## Harrisburg, PA 2009
### Temperature and Precipitation

# 2009
# PHILADELPHIA
# PENNSYLVANIA (KPHL)

The Appalachian Mountains to the west and the Atlantic Ocean to the east have a moderating effect on climate. Periods of very high or very low temperatures seldom last for more than three or four days. Temperatures below zero or above 100 degrees are a rarity. On occasion, the area becomes engulfed with maritime air during the summer months, and high humidity adds to the discomfort of seasonably warm temperatures.

Precipitation is fairly evenly distributed throughout the year with maximum amounts during the late summer months. Much of the summer rainfall is from local thunderstorms and amounts vary in different areas of the city. This is due, in part, to the higher elevations to the west and north. Snowfall amounts are often considerably larger in the northern suburbs than in the central and southern parts of the city. In many cases, the precipitation will change from snow to rain within the city. Single storms of 10 inches or more occur about every five years.

The prevailing wind direction for the summer months is from the southwest, while northwesterly winds prevail during the winter. The annual prevailing direction is from the west-southwest. Destructive velocities are comparatively rare and occur mostly in gustiness during summer thunderstorms. High winds occurring in the winter months, as a rule, come with the advance of cold air after the passage of a deep low pressure system. Only rarely have hurricanes in the vicinity caused widespread damage, primarily because of flooding.

Flood stages in the Schuylkill River normally occur about twice a year. Flood stages seldom last over 12 hours and usually occur after excessive thunderstorms. Flooding rarely occurs on the Delaware River.

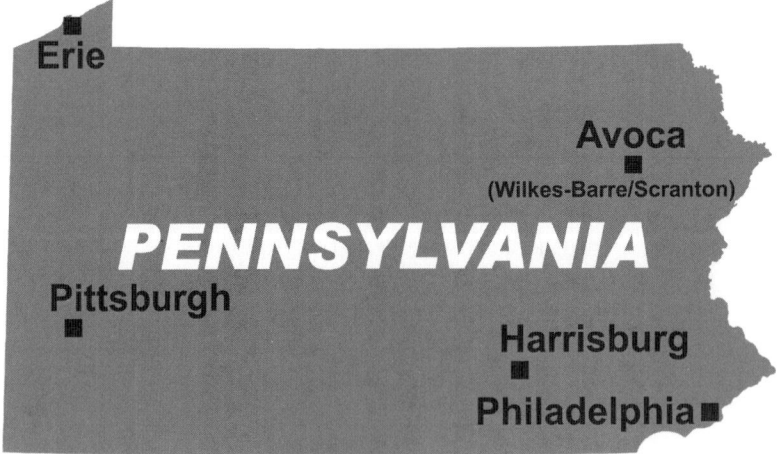

# NORMALS, MEANS, AND EXTREMES
## PHILADELPHIA (KPHL)

LATITUDE: 39°52'N  LONGITUDE: -75°13'W  ELEVATION (FT): GRND: 7  BARO: 62  TIME ZONE: EASTERN (UTC -5)  WBAN: 13739

| | ELEMENT | POR | JAN | FEB | MAR | APR | MAY | JUN | JUL | AUG | SEP | OCT | NOV | DEC | YEAR |
|---|---|---|---|---|---|---|---|---|---|---|---|---|---|---|---|
| **TEMPERATURE °F** | NORMAL DAILY MAXIMUM | 30 | 39.0 | 42.1 | 51.3 | 62.0 | 72.1 | 80.6 | 85.5 | 84.0 | 76.7 | 65.7 | 54.8 | 44.2 | 63.2 |
| | MEAN DAILY MAXIMUM | 62 | 39.7 | 42.6 | 51.5 | 63.4 | 73.3 | 82.0 | 86.5 | 84.8 | 77.7 | 66.6 | 55.3 | 44.0 | 64.0 |
| | HIGHEST DAILY MAXIMUM | 68 | 74 | 74 | 87 | 95 | 97 | 100 | 104 | 101 | 100 | 96 | 81 | 73 | 104 |
| | YEAR OF OCCURRENCE | | 1950 | 1997 | 1945 | 2002 | 1991 | 1994 | 1966 | 2001 | 1953 | 1941 | 1993 | 1998 | JUL 1966 |
| | MEAN OF EXTREME MAXS. | 62 | 60.1 | 61.9 | 73.1 | 83.3 | 88.5 | 93.6 | 95.8 | 94.1 | 90.2 | 81.8 | 72.6 | 63.1 | 79.8 |
| | NORMAL DAILY MINIMUM | 30 | 25.5 | 27.5 | 35.1 | 44.2 | 54.8 | 64.0 | 69.7 | 68.5 | 60.9 | 48.7 | 39.5 | 30.6 | 47.4 |
| | MEAN DAILY MINIMUM | 62 | 24.9 | 26.4 | 33.6 | 43.2 | 53.1 | 62.3 | 68.0 | 66.8 | 59.3 | 47.6 | 38.2 | 29.1 | 46.0 |
| | LOWEST DAILY MINIMUM | 68 | -7 | -4 | 7 | 19 | 28 | 44 | 51 | 44 | 35 | 25 | 15 | 1 | -7 |
| | YEAR OF OCCURRENCE | | 1984 | 1961 | 1984 | 1982 | 1966 | 1984 | 1966 | 1986 | 1963 | 1969 | 1976 | 1983 | JAN 1984 |
| | MEAN OF EXTREME MINS. | 62 | 9.0 | 10.9 | 18.9 | 30.4 | 40.8 | 50.6 | 58.2 | 55.9 | 45.0 | 33.8 | 24.2 | 14.3 | 32.7 |
| | NORMAL DRY BULB | 30 | 32.3 | 34.8 | 43.2 | 53.1 | 63.5 | 72.3 | 77.6 | 76.3 | 68.8 | 57.2 | 47.1 | 37.4 | 55.3 |
| | MEAN DRY BULB | 62 | 32.3 | 34.5 | 42.6 | 53.3 | 63.2 | 72.3 | 77.3 | 75.8 | 68.5 | 57.1 | 46.8 | 36.6 | 55.0 |
| | MEAN WET BULB | 26 | 28.8 | 30.4 | 36.4 | 45.7 | 55.3 | 64.5 | 68.8 | 67.9 | 61.7 | 51.3 | 42.0 | 32.8 | 48.8 |
| | MEAN DEW POINT | 26 | 24.1 | 24.6 | 31.3 | 40.8 | 51.3 | 61.1 | 65.9 | 65.2 | 58.7 | 47.8 | 37.6 | 27.8 | 44.7 |
| | NORMAL NO. DAYS WITH: | | | | | | | | | | | | | | |
| | MAXIMUM >= 90 | 30 | 0.0 | 0.0 | 0.0 | 0.4 | 1.3 | 4.6 | 10.5 | 7.3 | 1.6 | 0.0 | 0.0 | 0.0 | 25.7 |
| | MAXIMUM <= 32 | 30 | 8.0 | 5.3 | 0.7 | 0.0 | 0.0 | 0.0 | 0.0 | 0.0 | 0.0 | 0.0 | 0.1 | 3.6 | 17.7 |
| | MINIMUM <= 32 | 30 | 24.8 | 20.8 | 12.6 | 2.0 | 0.0 | 0.0 | 0.0 | 0.0 | 0.0 | 0.7 | 7.3 | 19.2 | 87.4 |
| | MINIMUM <= 0 | 30 | 0.4 | 0.1 | 0.0 | 0.0 | 0.0 | 0.0 | 0.0 | 0.0 | 0.0 | 0.0 | 0.0 | 0.0 | 0.5 |
| **H/C** | NORMAL HEATING DEG. DAYS | 30 | 1020 | 858 | 681 | 362 | 113 | 12 | 1 | 2 | 39 | 269 | 545 | 857 | 4759 |
| | NORMAL COOLING DEG. DAYS | 30 | 0 | 0 | 2 | 10 | 70 | 234 | 395 | 351 | 152 | 19 | 2 | 0 | 1235 |
| **RH** | NORMAL (PERCENT) | 30 | 67 | 64 | 62 | 61 | 66 | 68 | 69 | 70 | 72 | 71 | 68 | 68 | 67 |
| | HOUR 01 LST | 30 | 71 | 69 | 70 | 70 | 77 | 80 | 80 | 82 | 82 | 82 | 75 | 72 | 76 |
| | HOUR 07 LST | 30 | 74 | 73 | 73 | 71 | 75 | 77 | 78 | 81 | 83 | 83 | 78 | 75 | 77 |
| | HOUR 13 LST | 30 | 60 | 55 | 52 | 50 | 53 | 54 | 54 | 55 | 56 | 55 | 56 | 58 | 55 |
| | HOUR 19 LST | 30 | 65 | 61 | 58 | 55 | 59 | 61 | 62 | 65 | 68 | 69 | 66 | 66 | 63 |
| **S** | PERCENT POSSIBLE SUNSHINE | 59 | 49 | 53 | 55 | 55 | 56 | 62 | 61 | 62 | 59 | 59 | 52 | 49 | 56 |
| **W/O** | MEAN NO. DAYS WITH: | | | | | | | | | | | | | | |
| | HEAVY FOG(VISBY <= 1/4 MI) | 46 | 2.0 | 1.5 | 1.0 | 0.8 | 0.7 | 0.6 | 0.7 | 0.4 | 0.8 | 1.9 | 1.5 | 1.7 | 13.6 |
| | THUNDERSTORMS | 62 | 0.3 | 0.3 | 1.0 | 2.2 | 4.0 | 5.2 | 5.6 | 4.9 | 2.4 | 0.8 | 0.5 | 0.2 | 27.4 |
| **CLOUDNESS** | MEAN: SUNRISE-SUNSET (OKTAS) MIDNIGHT-MIDNIGHT (OKTAS) MEAN NO. DAYS WITH: | | | | | | | | | | | | | | |
| | CLEAR | 1 | 3.0 | 2.0 | 8.0 | | 8.0 | 11.0 | | | | | | | |
| | PARTLY CLOUDY | | | 1.0 | 4.0 | | 5.0 | 3.0 | | | | | | | |
| | CLOUDY | 1 | 3.0 | 6.0 | 8.0 | | 7.0 | 9.0 | | | | | | | |
| **PR** | MEAN STATION PRESSURE(IN) | 26 | 30.06 | 30.05 | 30.02 | 29.95 | 29.96 | 29.94 | 29.95 | 29.99 | 30.03 | 30.06 | 30.03 | 30.08 | 30.01 |
| | MEAN SEA-LEVEL PRES. (IN) | 26 | 30.09 | 30.08 | 30.06 | 29.98 | 29.99 | 29.97 | 29.98 | 30.02 | 30.07 | 30.09 | 30.10 | 30.10 | 30.04 |
| **WINDS** | MEAN SPEED (MPH) | 26 | 10.0 | 10.4 | 10.9 | 10.3 | 9.2 | 8.7 | 8.5 | 8.0 | 8.4 | 8.7 | 9.4 | 10.0 | 9.4 |
| | PREVAIL.DIR(TENS OF DEGS) | 46 | 30 | 31 | 30 | 24 | 24 | 24 | 24 | 24 | 24 | 24 | 24 | 30 | 24 |
| | MAXIMUM 2-MINUTE: | | | | | | | | | | | | | | |
| | SPEED (MPH) | 14 | 44 | 46 | 49 | 39 | 52 | 51 | 41 | 41 | 39 | 45 | 46 | 45 | 52 |
| | DIR. (TENS OF DEGS) | | 27 | 27 | 24 | 28 | 28 | 30 | 33 | 25 | 32 | 24 | 27 | 27 | 28 |
| | YEAR OF OCCURRENCE | | 2008 | 2009 | 2008 | 2009 | 2005 | 1998 | 1999 | 1997 | 1998 | 2003 | 2003 | 2000 | MAY 2005 |
| | MAXIMUM 3-SECOND | | | | | | | | | | | | | | |
| | SPEED (MPH) | 14 | 57 | 56 | 56 | 51 | 69 | 71 | 48 | 52 | 49 | 55 | 58 | 53 | 71 |
| | DIR. (TENS OF DEGS) | | 18 | 28 | 24 | 28 | 27 | 30 | 28 | 24 | 13 | 20 | 28 | 28 | 30 |
| | YEAR OF OCCURRENCE | | 1996 | 2009 | 2008 | 2000 | 2005 | 1998 | 2006 | 1997 | 2003 | 2003 | 2003 | 2000 | JUN 1998 |
| **PRECIPITATION** | NORMAL (IN) | 30 | 3.52 | 2.74 | 3.81 | 3.49 | 3.89 | 3.29 | 4.39 | 3.82 | 3.88 | 2.75 | 3.16 | 3.31 | 42.05 |
| | MAXIMUM MONTHLY (IN) | 67 | 8.86 | 6.44 | 7.01 | 9.05 | 7.41 | 8.08 | 10.42 | 10.29 | 13.07 | 8.68 | 9.06 | 8.86 | 13.07 |
| | YEAR OF OCCURRENCE | | 1978 | 1979 | 1980 | 2007 | 1948 | 2003 | 1994 | 2009 | 1999 | 2005 | 1972 | 2009 | SEP 1999 |
| | MINIMUM MONTHLY (IN) | 67 | 0.45 | 0.55 | 0.68 | 0.52 | 0.47 | 0.11 | 0.64 | 0.49 | .21 | 0.09 | 0.32 | 0.25 | 0.09 |
| | YEAR OF OCCURRENCE | | 1955 | 2002 | 1966 | 1985 | 1964 | 1949 | 1957 | 1964 | 2005 | 1963 | 1976 | 1955 | OCT 1963 |
| | MAXIMUM IN 24 HOURS (IN) | 67 | 2.70 | 3.52 | 3.08 | 4.36 | 3.18 | 4.62 | 4.68 | 5.68 | 6.77 | 5.94 | 3.99 | 3.18 | 6.77 |
| | YEAR OF OCCURRENCE | | 1979 | 2008 | 2000 | 2007 | 1984 | 1973 | 2004 | 1971 | 1999 | 2005 | 1977 | 2008 | SEP 1999 |
| | NORMAL NO. DAYS WITH: | | | | | | | | | | | | | | |
| | PRECIPITATION >= 0.01 | 30 | 10.9 | 9.7 | 10.5 | 10.9 | 11.7 | 10.0 | 9.4 | 8.4 | 9.1 | 8.0 | 9.4 | 10.6 | 118.6 |
| | PRECIPITATION >= 1.00 | 30 | 0.9 | 0.6 | 1.0 | 0.8 | 0.8 | 0.8 | 1.3 | 1.2 | 1.1 | 0.7 | 0.6 | 0.8 | 10.6 |
| **SNOWFALL** | NORMAL (IN) | 30 | 6.4 | 6.6 | 3.2 | 0.6 | 0.0 | 0.0 | 0.0 | 0.0 | 0.0 | 0.1 | 0.4 | 2.0 | 19.3 |
| | MAXIMUM MONTHLY (IN) | 66 | 23.4 | 29.6 | 13.4 | 4.3 | T | T | T | 0.0 | 0.0 | 2.1 | 8.8 | 24.1 | 29.6 |
| | YEAR OF OCCURRENCE | | 1978 | 2003 | 1958 | 1971 | 1963 | 1993 | 2008 | | | 1979 | 1953 | 2009 | FEB 2003 |
| | MAXIMUM IN 24 HOURS (IN) | 66 | 33.8 | 21.3 | 12.0 | 4.3 | T | T | T | 0.0 | 0.0 | 2.1 | 8.7 | 22.5 | 33.8 |
| | YEAR OF OCCURRENCE' | | 1996 | 1983 | 1993 | 1971 | 1963 | 1993 | 2006 | | | 1979 | 1953 | 2009 | JAN 1996 |
| | MAXIMUM SNOW DEPTH (IN) | 61 | 12 | 23 | 12 | 3 | 0 | 0 | 0 | 0 | 0 | 0 | 8 | 21 | 23 |
| | YEAR OF OCCURRENCE | | 1961 | 2003 | 1993 | 1997 | | | | | | | 1953 | 2009 | FEB 2003 |
| | NORMAL NO. DAYS WITH: | | | | | | | | | | | | | | |
| | SNOWFALL >= 1.0 | 30 | 1.9 | 1.5 | 0.8 | 0.2 | 0.0 | 0.0 | 0.0 | 0.0 | 0.0 | 0.0 | 0.2 | 0.5 | 5.1 |

## PRECIPITATION (inches) 2009 PHILADELPHIA (KPHL)

| YEAR | JAN | FEB | MAR | APR | MAY | JUN | JUL | AUG | SEP | OCT | NOV | DEC | ANNUAL |
|------|-----|-----|-----|-----|-----|-----|-----|-----|-----|-----|-----|-----|--------|
| 1980 | 2.27 | 0.96 | 7.01 | 4.79 | 3.22 | 1.73 | 6.58 | 0.80 | 2.79 | 5.03 | 2.85 | 0.77 | 38.80 |
| 1981 | 0.50 | 2.94 | 1.61 | 3.60 | 4.53 | 4.40 | 4.54 | 5.11 | 2.83 | 2.68 | 0.95 | 4.14 | 37.83 |
| 1982 | 4.45 | 3.16 | 2.66 | 6.06 | 4.47 | 5.76 | 1.94 | 2.20 | 2.32 | 1.94 | 3.67 | 1.80 | 40.43 |
| 1983 | 2.81 | 3.53 | 6.70 | 8.12 | 7.03 | 2.75 | 0.68 | 2.57 | 3.45 | 3.69 | 5.71 | 7.37 | 54.41 |
| 1984 | 2.22 | 2.81 | 6.14 | 4.25 | 6.87 | 2.85 | 6.99 | 3.28 | 1.96 | 2.56 | 1.56 | 2.17 | 43.66 |
| 1985 | 1.55 | 2.44 | 1.95 | 0.52 | 4.99 | 1.88 | 4.66 | 2.82 | 5.78 | 1.54 | 6.09 | 0.98 | 35.20 |
| 1986 | 4.13 | 3.38 | 1.25 | 4.46 | 0.70 | 1.99 | 4.10 | 3.70 | 2.33 | 2.22 | 6.27 | 5.89 | 40.42 |
| 1987 | 4.58 | 1.17 | 1.16 | 3.63 | 3.15 | 2.01 | 4.82 | 3.72 | 2.78 | 2.62 | 2.08 | 1.68 | 33.40 |
| 1988 | 2.72 | 4.11 | 2.24 | 2.92 | 3.67 | 0.57 | 8.07 | 3.16 | 2.62 | 2.16 | 5.17 | 1.00 | 38.41 |
| 1989 | 2.41 | 3.25 | 4.41 | 2.27 | 6.76 | 4.73 | 9.44 | 3.92 | 5.03 | 3.44 | 1.79 | 1.21 | 48.66 |
| 1990 | 4.09 | 1.44 | 2.59 | 3.16 | 6.08 | 3.39 | 2.62 | 4.07 | 1.71 | 1.68 | 1.17 | 3.79 | 35.79 |
| 1991 | 4.10 | 0.75 | 4.13 | 2.81 | 1.82 | 3.36 | 4.79 | 3.86 | 3.58 | 1.61 | 1.55 | 3.86 | 36.22 |
| 1992 | 0.88 | 1.31 | 3.19 | 1.26 | 2.74 | 1.84 | 5.05 | 2.00 | 3.04 | 1.23 | 3.26 | 4.61 | 30.41 |
| 1993 | 1.97 | 3.03 | 6.61 | 4.20 | 2.42 | 1.52 | 1.98 | 5.18 | 6.66 | 2.69 | 2.23 | 3.69 | 42.18 |
| 1994 | 4.27 | 3.27 | 6.44 | 2.86 | 3.66 | 1.74 | 10.42 | 4.54 | 1.64 | 0.94 | 3.03 | 2.11 | 44.92 |
| 1995 | 3.10 | 2.41 | 1.67 | 1.96 | 2.67 | 0.62 | 2.92 | 1.15 | 3.55 | 5.99 | 3.34 | 2.15 | 31.53 |
| 1996 | 4.39 | 2.12 | 4.27 | 4.48 | 3.25 | 4.73 | 8.17 | 4.29 | 4.95 | 4.30 | 3.03 | 8.47 | 56.45 |
| 1997 | 2.80 | 2.48 | 3.91 | 2.58 | 2.32 | 1.49 | 2.38 | 4.56 | 1.59 | 1.83 | 3.49 | 3.09 | 32.52 |
| 1998 | 4.24 | 3.25 | 3.93 | 2.70 | 3.87 | 4.91 | 1.79 | 1.26 | 1.86 | 1.84 | 1.18 | 0.82 | 31.65 |
| 1999 | 4.89 | 2.95 | 4.02 | 3.31 | 3.70 | 1.16 | 1.22 | 5.32 | 13.07 | 3.55 | 2.31 | 2.99 | 48.49 |
| 2000 | 3.22 | 2.02 | 6.32 | 3.05 | 3.03 | 3.82 | 5.54 | 2.90 | 8.28 | 1.51 | 2.21 | 2.82 | 44.72 |
| 2001 | 2.77 | 3.04 | 5.44 | 1.49 | 3.99 | 5.93 | 1.30 | 0.97 | 2.58 | 0.83 | 0.56 | 2.11 | 31.01 |
| 2002 | 2.43 | 0.55 | 4.03 | 2.17 | 3.57 | 3.73 | 2.12 | 2.47 | 3.67 | 5.90 | 4.61 | 4.05 | 39.30 |
| 2003 | 1.93 | 5.04 | 4.09 | 2.20 | 4.17 | 8.08 | 2.01 | 3.26 | 4.66 | 4.45 | 2.63 | 5.46 | 47.98 |
| 2004 | 1.70 | 2.50 | 3.54 | 6.02 | 3.62 | 4.57 | 7.91 | 4.17 | 5.19 | 2.24 | 4.55 | 3.17 | 49.18 |
| 2005 | 4.45 | 2.61 | 3.66 | 5.32 | 1.27 | 3.31 | 4.31 | 2.57 | 0.21 | 8.68 | 2.86 | 2.97 | 42.22 |
| 2006 | 4.34 | 1.51 | 0.91 | 3.71 | 2.16 | 7.95 | 4.27 | 3.93 | 5.97 | 6.42 | 4.88 | 2.15 | 48.20 |
| 2007 | 3.35 | 1.73 | 3.82 | 9.05 | 2.68 | 4.02 | 3.44 | 2.94 | 0.58 | 4.66 | 1.45 | 4.41 | 42.13 |
| 2008 | 1.74 | 3.93 | 3.67 | 2.19 | 4.55 | 2.87 | 3.45 | 2.44 | 4.31 | 1.59 | 4.02 | 5.57 | 40.33 |
| 2009 | 2.70 | 0.84 | 1.62 | 3.99 | 4.84 | 4.79 | 3.35 | 10.29 | 3.65 | 5.51 | 2.06 | 8.86 | 52.50 |
| POR= 62 YRS | 3.17 | 2.70 | 3.75 | 3.50 | 3.57 | 3.64 | 4.07 | 3.95 | 3.57 | 2.93 | 3.20 | 3.52 | 41.57 |

WBAN : 13739

## AVERAGE TEMPERATURE (°F) 2009 PHILADELPHIA (KPHL)

| YEAR | JAN | FEB | MAR | APR | MAY | JUN | JUL | AUG | SEP | OCT | NOV | DEC | ANNUAL |
|------|-----|-----|-----|-----|-----|-----|-----|-----|-----|-----|-----|-----|--------|
| 1980 | 31.8 | 29.7 | 40.2 | 54.7 | 65.4 | 70.6 | 78.5 | 80.0 | 72.2 | 54.9 | 43.2 | 32.5 | 54.5 |
| 1981 | 25.3 | 37.9 | 40.0 | 54.7 | 62.6 | 72.0 | 76.9 | 74.9 | 66.8 | 53.1 | 45.6 | 34.6 | 53.7 |
| 1982 | 24.7 | 34.4 | 41.7 | 50.2 | 65.9 | 68.7 | 76.9 | 73.5 | 67.6 | 56.9 | 48.4 | 41.3 | 54.2 |
| 1983 | 34.1 | 34.0 | 43.7 | 51.0 | 62.1 | 72.0 | 77.9 | 77.1 | 69.0 | 56.6 | 46.7 | 33.2 | 54.8 |
| 1984 | 26.2 | 38.7 | 35.5 | 50.2 | 60.2 | 73.0 | 73.9 | 75.2 | 64.7 | 61.2 | 44.4 | 41.9 | 53.8 |
| 1985 | 27.3 | 35.3 | 44.6 | 55.5 | 64.5 | 68.8 | 75.4 | 74.1 | 69.1 | 59.3 | 51.3 | 33.3 | 54.9 |
| 1986 | 32.8 | 32.1 | 44.5 | 53.3 | 66.8 | 73.8 | 78.1 | 74.0 | 68.3 | 57.8 | 44.5 | 37.9 | 55.3 |
| 1987 | 31.9 | 32.5 | 45.7 | 53.1 | 63.9 | 74.6 | 79.5 | 75.4 | 68.8 | 52.5 | 48.0 | 39.2 | 55.4 |
| 1988 | 27.3 | 34.6 | 44.7 | 51.3 | 63.6 | 72.3 | 80.7 | 78.3 | 66.7 | 51.8 | 47.7 | 35.4 | 54.5 |
| 1989 | 36.5 | 34.8 | 42.3 | 52.4 | 62.4 | 74.7 | 76.3 | 75.6 | 69.7 | 58.3 | 44.9 | 25.5 | 54.5 |
| 1990 | 40.3 | 41.2 | 46.1 | 53.3 | 61.3 | 72.2 | 78.0 | 75.8 | 68.0 | 61.9 | 49.7 | 42.1 | 57.5 |
| 1991 | 35.2 | 40.0 | 46.1 | 55.5 | 70.8 | 75.7 | 79.0 | 79.0 | 69.5 | 58.9 | 47.3 | 39.6 | 58.1 |
| 1992 | 35.7 | 37.5 | 41.6 | 53.2 | 62.5 | 71.3 | 77.1 | 73.5 | 68.4 | 55.2 | 48.0 | 38.8 | 55.2 |
| 1993 | 38.2 | 31.9 | 39.8 | 54.2 | 66.4 | 74.4 | 81.4 | 78.9 | 69.7 | 58.0 | 49.0 | 38.4 | 56.7 |
| 1994 | 27.4 | 33.2 | 42.8 | 59.5 | 62.5 | 78.1 | 82.1 | 74.9 | 68.3 | 57.5 | 51.7 | 41.9 | 56.7 |
| 1995 | 38.2 | 31.5 | 47.3 | 54.5 | 64.3 | 74.3 | 81.5 | 79.9 | 70.4 | 61.4 | 43.0 | 31.6 | 56.5 |
| 1996 | 30.2 | 33.9 | 38.7 | 52.7 | 60.6 | 73.0 | 74.4 | 74.5 | 68.7 | 56.3 | 41.3 | 40.2 | 53.7 |
| 1997 | 32.5 | 40.0 | 44.1 | 51.4 | 59.5 | 71.1 | 77.5 | 73.9 | 67.1 | 57.3 | 44.4 | 38.5 | 54.8 |
| 1998 | 41.0 | 41.8 | 45.5 | 55.3 | 66.2 | 71.5 | 77.5 | 78.0 | 71.8 | 58.3 | 48.2 | 42.0 | 58.1 |
| 1999 | 35.0 | 38.0 | 42.4 | 53.5 | 64.0 | 72.9 | 81.2 | 77.4 | 69.9 | 56.1 | 50.9 | 39.9 | 56.8 |
| 2000 | 32.1 | 37.5 | 48.0 | 52.6 | 64.2 | 72.6 | 74.1 | 74.1 | 66.4 | 57.7 | 45.4 | 31.3 | 54.7 |
| 2001 | 32.5 | 37.4 | 41.1 | 54.9 | 64.7 | 75.2 | 75.4 | 79.9 | 68.4 | 59.1 | 52.9 | 43.7 | 57.1 |
| 2002 | 39.2 | 41.1 | 45.5 | 57.2 | 63.3 | 73.9 | 79.6 | 79.5 | 72.2 | 57.6 | 45.9 | 35.4 | 57.5 |
| 2003 | 28.6 | 29.9 | 44.9 | 52.6 | 60.6 | 71.3 | 78.8 | 78.3 | 70.1 | 56.0 | 50.8 | 36.8 | 54.9 |
| 2004 | 26.1 | 35.5 | 45.1 | 54.4 | 69.3 | 71.8 | 76.3 | 75.0 | 70.5 | 56.2 | 47.9 | 37.8 | 55.5 |
| 2005 | 31.8 | 36.1 | 39.5 | 55.3 | 59.6 | 74.8 | 78.9 | 79.7 | 73.8 | 58.8 | 49.3 | 34.4 | 56.0 |
| 2006 | 40.8 | 35.6 | 44.5 | 56.7 | 64.6 | 72.8 | 79.5 | 78.1 | 67.2 | 56.2 | 51.3 | 42.7 | 57.5 |
| 2007 | 38.2 | 28.0 | 43.7 | 50.8 | 66.1 | 73.8 | 77.3 | 77.0 | 72.0 | 64.5 | 45.7 | 37.7 | 56.2 |
| 2008 | 35.9 | 36.9 | 44.8 | 56.3 | 61.2 | 76.4 | 79.3 | 74.6 | 70.4 | 56.6 | 45.6 | 38.7 | 56.4 |
| 2009 | 29.1 | 37.1 | 43.7 | 55.6 | 65.0 | 71.1 | 75.9 | 78.2 | 68.0 | 56.4 | 50.9 | 36.4 | 55.6 |
| POR= 62 YRS | 32.3 | 34.5 | 42.6 | 53.3 | 63.2 | 72.3 | 77.3 | 75.8 | 68.5 | 57.1 | 46.8 | 36.6 | 55.0 |

## HEATING DEGREE DAYS (base 65°F) 2009 PHILADELPHIA (KPHL)

| YEAR | JUL | AUG | SEP | OCT | NOV | DEC | JAN | FEB | MAR | APR | MAY | JUN | TOTAL |
|---|---|---|---|---|---|---|---|---|---|---|---|---|---|
| 1980-81 | 0 | 0 | 22 | 320 | 646 | 999 | 1222 | 752 | 768 | 309 | 129 | 4 | 5171 |
| 1981-82 | 0 | 0 | 58 | 364 | 576 | 936 | 1243 | 850 | 714 | 440 | 50 | 25 | 5256 |
| 1982-83 | 0 | 8 | 31 | 277 | 497 | 730 | 951 | 861 | 653 | 423 | 128 | 2 | 4561 |
| 1983-84 | 0 | 0 | 70 | 283 | 540 | 981 | 1196 | 756 | 911 | 438 | 181 | 13 | 5369 |
| 1984-85 | 0 | 0 | 92 | 138 | 613 | 709 | 1161 | 824 | 627 | 306 | 89 | 9 | 4568 |
| 1985-86 | 0 | 0 | 38 | 187 | 407 | 975 | 990 | 914 | 628 | 345 | 77 | 6 | 4567 |
| 1986-87 | 0 | 21 | 23 | 255 | 609 | 838 | 1017 | 904 | 591 | 359 | 129 | 1 | 4747 |
| 1987-88 | 0 | 0 | 20 | 379 | 504 | 796 | 1162 | 876 | 624 | 404 | 105 | 32 | 4902 |
| 1988-89 | 0 | 0 | 35 | 408 | 513 | 908 | 876 | 840 | 700 | 371 | 138 | 0 | 4789 |
| 1989-90 | 0 | 0 | 43 | 220 | 594 | 1219 | 757 | 662 | 588 | 375 | 127 | 6 | 4591 |
| 1990-91 | 2 | 1 | 55 | 171 | 453 | 701 | 920 | 694 | 576 | 296 | 44 | 0 | 3913 |
| 1991-92 | 0 | 0 | 42 | 215 | 527 | 778 | 903 | 789 | 720 | 356 | 124 | 9 | 4463 |
| 1992-93 | 0 | 1 | 49 | 310 | 504 | 804 | 824 | 920 | 773 | 314 | 47 | 6 | 4552 |
| 1993-94 | 0 | 0 | 34 | 219 | 479 | 820 | 1156 | 881 | 681 | 193 | 132 | 0 | 4595 |
| 1994-95 | 0 | 0 | 20 | 231 | 392 | 710 | 824 | 932 | 542 | 318 | 82 | 0 | 4051 |
| 1995-96 | 0 | 0 | 21 | 159 | 656 | 1029 | 1072 | 894 | 809 | 370 | 191 | 7 | 5208 |
| 1996-97 | 0 | 0 | 34 | 265 | 704 | 761 | 999 | 695 | 643 | 401 | 172 | 42 | 4716 |
| 1997-98 | 2 | 0 | 45 | 278 | 612 | 811 | 738 | 644 | 619 | 287 | 67 | 13 | 4116 |
| 1998-99 | 0 | 0 | 14 | 207 | 498 | 706 | 920 | 751 | 691 | 336 | 77 | 3 | 4203 |
| 1999-00 | 0 | 0 | 22 | 275 | 415 | 772 | 1012 | 792 | 519 | 371 | 107 | 15 | 4300 |
| 2000-01 | 0 | 0 | 73 | 230 | 584 | 1039 | 999 | 768 | 734 | 320 | 74 | 4 | 4825 |
| 2001-02 | 0 | 0 | 48 | 211 | 358 | 652 | 791 | 662 | 595 | 290 | 120 | 4 | 3731 |
| 2002-03 | 0 | 1 | 4 | 277 | 567 | 910 | 1123 | 975 | 618 | 373 | 157 | 25 | 5030 |
| 2003-04 | 0 | 0 | 11 | 284 | 421 | 870 | 1199 | 848 | 609 | 323 | 47 | 11 | 4623 |
| 2004-05 | 0 | 0 | 10 | 268 | 507 | 839 | 1021 | 803 | 780 | 297 | 170 | 5 | 4700 |
| 2005-06 | 0 | 0 | 7 | 221 | 463 | 941 | 743 | 816 | 628 | 251 | 82 | 2 | 4154 |
| 2006-07 | 0 | 0 | 34 | 281 | 404 | 683 | 824 | 1030 | 651 | 427 | 76 | 4 | 4414 |
| 2007-08 | 0 | 6 | 13 | 117 | 570 | 842 | 895 | 810 | 617 | 261 | 149 | 0 | 4280 |
| 2008-09 | 0 | 0 | 12 | 274 | 576 | 809 | 1107 | 774 | 653 | 320 | 79 | 15 | 4619 |
| 2009- | 0 | 0 | 19 | 269 | 417 | 881 | | | | | | | |

WBAN : 13739

## COOLING DEGREE DAYS (base 65°F) 2009 PHILADELPHIA (KPHL)

| YEAR | JAN | FEB | MAR | APR | MAY | JUN | JUL | AUG | SEP | OCT | NOV | DEC | TOTAL |
|---|---|---|---|---|---|---|---|---|---|---|---|---|---|
| 1980 | 0 | 0 | 0 | 0 | 89 | 194 | 428 | 470 | 244 | 10 | 0 | 0 | 1435 |
| 1981 | 0 | 0 | 0 | 9 | 62 | 224 | 373 | 315 | 119 | 1 | 0 | 0 | 1103 |
| 1982 | 0 | 0 | 0 | 3 | 85 | 142 | 376 | 280 | 115 | 31 | 5 | 0 | 1037 |
| 1983 | 0 | 0 | 0 | 11 | 43 | 217 | 409 | 380 | 199 | 27 | 0 | 0 | 1286 |
| 1984 | 0 | 0 | 0 | 0 | 39 | 260 | 283 | 324 | 90 | 30 | 0 | 0 | 1026 |
| 1985 | 0 | 0 | 0 | 27 | 81 | 133 | 330 | 291 | 166 | 19 | 0 | 0 | 1047 |
| 1986 | 0 | 0 | 0 | 0 | 139 | 278 | 413 | 307 | 129 | 40 | 0 | 0 | 1306 |
| 1987 | 0 | 0 | 0 | 7 | 101 | 295 | 456 | 332 | 142 | 0 | 0 | 0 | 1333 |
| 1988 | 0 | 0 | 0 | 1 | 70 | 259 | 495 | 418 | 93 | 7 | 0 | 0 | 1343 |
| 1989 | 0 | 0 | 1 | 1 | 62 | 298 | 357 | 332 | 192 | 18 | 1 | 0 | 1262 |
| 1990 | 0 | 0 | 9 | 29 | 20 | 226 | 413 | 341 | 152 | 83 | 1 | 0 | 1274 |
| 1991 | 0 | 0 | 1 | 21 | 230 | 327 | 443 | 437 | 183 | 30 | 0 | 0 | 1672 |
| 1992 | 0 | 0 | 0 | 10 | 54 | 205 | 380 | 268 | 158 | 13 | 0 | 0 | 1088 |
| 1993 | 0 | 0 | 0 | 0 | 95 | 298 | 517 | 438 | 181 | 11 | 6 | 0 | 1546 |
| 1994 | 0 | 0 | 0 | 34 | 63 | 401 | 540 | 313 | 125 | 4 | 0 | 0 | 1480 |
| 1995 | 0 | 0 | 0 | 8 | 66 | 284 | 519 | 468 | 189 | 53 | 1 | 0 | 1588 |
| 1996 | 0 | 0 | 0 | 9 | 61 | 257 | 301 | 305 | 151 | 2 | 1 | 0 | 1087 |
| 1997 | 0 | 0 | 0 | 0 | 9 | 230 | 397 | 280 | 113 | 49 | 0 | 0 | 1078 |
| 1998 | 0 | 0 | 20 | 3 | 114 | 216 | 397 | 410 | 223 | 3 | 0 | 0 | 1386 |
| 1999 | 0 | 0 | 0 | 0 | 51 | 246 | 508 | 394 | 177 | 4 | 0 | 0 | 1380 |
| 2000 | 0 | 0 | 0 | 3 | 89 | 251 | 286 | 286 | 123 | 12 | 0 | 0 | 1050 |
| 2001 | 0 | 0 | 0 | 22 | 72 | 316 | 330 | 468 | 158 | 37 | 1 | 0 | 1404 |
| 2002 | 0 | 0 | 0 | 63 | 76 | 277 | 459 | 461 | 224 | 54 | 0 | 0 | 1614 |
| 2003 | 0 | 0 | 0 | 7 | 25 | 219 | 432 | 418 | 169 | 7 | 1 | 0 | 1278 |
| 2004 | 0 | 0 | 0 | 11 | 186 | 221 | 357 | 318 | 179 | 2 | 0 | 0 | 1274 |
| 2005 | 0 | 0 | 0 | 11 | 11 | 305 | 437 | 462 | 278 | 34 | 0 | 0 | 1538 |
| 2006 | 0 | 0 | 1 | 11 | 76 | 240 | 454 | 413 | 102 | 17 | 0 | 0 | 1314 |
| 2007 | 0 | 0 | 0 | 8 | 115 | 275 | 391 | 384 | 230 | 110 | 0 | 0 | 1513 |
| 2008 | 0 | 0 | 0 | 9 | 37 | 347 | 452 | 310 | 180 | 19 | 0 | 0 | 1354 |
| 2009 | 0 | 0 | 0 | 42 | 86 | 205 | 347 | 418 | 119 | 10 | 0 | 0 | 1227 |

## SNOWFALL (inches) 2009 PHILADELPHIA (KPHL)

| YEAR | JUL | AUG | SEP | OCT | NOV | DEC | JAN | FEB | MAR | APR | MAY | JUN | TOTAL |
|------|-----|-----|-----|-----|-----|-----|-----|-----|-----|-----|-----|-----|-------|
| 1980-81 | 0.0 | 0.0 | 0.0 | 0.0 | 0.2 | 1.4 | 5.0 | T | 8.8 | 0.0 | 0.0 | 0.0 | 15.4 |
| 1981-82 | 0.0 | 0.0 | 0.0 | 0.0 | T | 2.8 | 14.0 | 3.5 | 1.1 | 4.0 | 0.0 | 0.0 | 25.4 |
| 1982-83 | 0.0 | 0.0 | 0.0 | 0.0 | 0.0 | 6.8 | 0.2 | 26.1 | 0.9 | 1.9 | 0.0 | 0.0 | 35.9 |
| 1983-84 | 0.0 | 0.0 | 0.0 | 0.0 | 0.8 | T | 10.5 | T | 10.3 | T | 0.0 | 0.0 | 21.6 |
| 1984-85 | 0.0 | 0.0 | 0.0 | 0.0 | T | 0.2 | 11.9 | 4.4 | T | T | 0.0 | 0.0 | 16.5 |
| 1985-86 | 0.0 | 0.0 | 0.0 | 0.0 | 0.0 | 1.5 | 3.4 | 11.5 | T | T | 0.0 | 0.0 | 16.4 |
| 1986-87 | 0.0 | 0.0 | 0.0 | 0.0 | T | 0.4 | 15.2 | 10.1 | T | T | 0.0 | 0.0 | 25.7 |
| 1987-88 | 0.0 | 0.0 | 0.0 | 0.0 | 1.4 | 1.5 | 10.6 | 1.5 | T | 0.0 | 0.0 | 0.0 | 15.0 |
| 1988-89 | 0.0 | 0.0 | 0.0 | 0.0 | 0.0 | 0.4 | 6.0 | 2.4 | 2.4 | 0.0 | 0.0 | 0.0 | 11.2 |
| 1989-90 | 0.0 | 0.0 | 0.0 | 0.0 | 4.6 | 5.3 | 1.4 | 0.9 | 2.4 | 2.4 | 0.0 | 0.0 | 17.0 |
| 1990-91 | 0.0 | 0.0 | 0.0 | 0.0 | 0.0 | 6.4 | 6.5 | 1.0 | 0.7 | 0.0 | 0.0 | 0.0 | 14.6 |
| 1991-92 | 0.0 | 0.0 | 0.0 | 0.0 | T | T | 1.2 | 1.0 | 2.5 | T | 0.0 | 0.0 | 4.7 |
| 1992-93 | 0.0 | 0.0 | 0.0 | 0.0 | T | T | 1.0 | 10.9 | 12.4 | T | 0.0 | T | 24.3 |
| 1993-94 | 0.0 | 0.0 | 0.0 | 0.0 | T | 0.9 | 4.1 | 13.2 | 4.9 | 0.0 | 0.0 | 0.0 | 23.1 |
| 1994-95 | 0.0 | 0.0 | 0.0 | 0.0 | T | 0.0 | T | 9.8 | T | 0.0 | 0.0 | 0.0 | 9.8 |
| 1995-96 | 0.0 | 0.0 | 0.0 | 0.0 | 1.9 | 7.3 | 33.8 | 12.9 | 7.2 | 2.4 | 0.0 | 0.0 | 65.5 |
| 1996-97 | 0.0 | 0.0 | 0.0 | 0.0 | T | T | 1.7 | 4.1 | 5.5 | 1.6 | 0.0 | 0.0 | 12.9 |
| 1997-98 | 0.0 | 0.0 | 0.0 | 0.0 | T | 0.2 | 0.5 | T | 0.1 | 0.0 | 0.0 | T | 0.8 |
| 1998-99 | 0.0 | 0.0 | 0.0 | 0.0 | 0.0 | 2.0 | 4.9 | 0.7 | 4.9 | 0.0 | 0.0 | 0.0 | 12.5 |
| 1999-00 | 0.0 | 0.0 | 0.0 | 0.0 | 0.0 | T | 13.7 | 5.7 | 0.0 | 1.6 | T | 0.0 | 21.0 |
| 2000-01 | 0.0 | 0.0 | 0.0 | 0.0 | 0.0 | 10.5 | 3.8 | 10.0 | 1.8 | T | 0.0 | 0.0 | 26.1 |
| 2001-02 | 0.0 | 0.0 | 0.0 | 0.0 | 0.0 | 0.0 | 4.0 | T | T | T | 0.0 | 0.0 | 4.0 |
| 2002-03 | 0.0 | 0.0 | 0.0 | T | T | 8.4 | 6.3 | 29.6 | 0.2 | 1.8 | 0.0 | 0.0 | 46.3 |
| 2003-04 | 0.0 | 0.0 | 0.0 | 0.0 | T | 6.0 | 7.6 | 0.4 | 3.7 | 0.3 | 0.0 | 0.0 | 18.0 |
| 2004-05 | 0.0 | 0.0 | 0.0 | 0.0 | 0.0 | 0.4 | 15.6 | 12.9 | 1.5 | 0.0 | 0.0 | 0.0 | 30.4 |
| 2005-06 | 0.0 | 0.0 | 0.0 | 0.0 | T | 7.0 | 0.4 | 12.0 | T | 0.1 | 0.0 | 0.0 | 19.5 |
| 2006-07 | T | 0.0 | 0.0 | 0.0 | 0.0 | T | 2.2 | 6.5 | 4.7 | T | 0.0 | 0.0 | 13.4 |
| 2007-08 | 0.0 | 0.0 | 0.0 | 0.0 | 0.0 | 1.6 | 1.0 | 3.7 | T | 0.0 | 0.0 | 0.0 | 6.3 |
| 2008-09 | T | 0.0 | 0.0 | 0.0 | 1.0 | 0.4 | 4.1 | 8.4 | 9.0 | 0.0 | 0.0 | 0.0 | 22.9 |
| 2009- | 0.0 | 0.0 | 0.0 | 0.0 | 0.0 | 24.1 | | | | | | | |
| POR= 62 YRS | T | 0.0 | 0.0 | T | 0.6 | 3.5 | 6.5 | 6.8 | 3.5 | 0.5 | T | T | 21.4 |

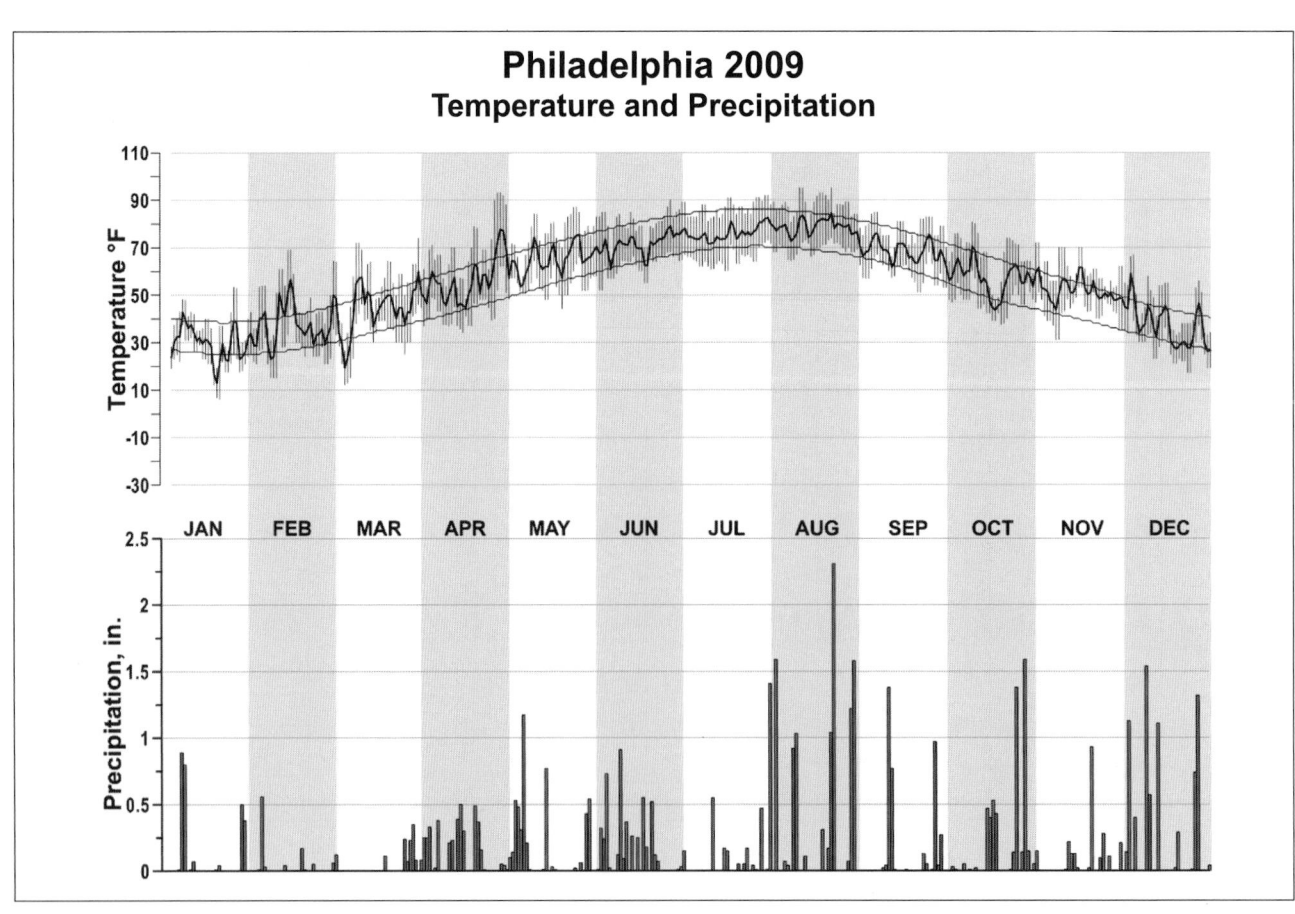

## Philadelphia 2009
### Temperature and Precipitation

# 2009
# PITTSBURGH
# PENNSYLVANIA (KPIT)

Pittsburgh lies at the foothills of the Allegheny Mountains at the confluence of the Allegheny and Monongahela Rivers which form the Ohio. The city is a little over 100 miles southeast of Lake Erie. It has a humid continental type of climate modified only slightly by its nearness to the Atlantic Seaboard and the Great Lakes.

The predominant winter air masses influencing the climate of Pittsburgh have a polar continental source in Canada and move in from the Hudson Bay region or the Canadian Rockies. During the summer, frequent invasions of air from the Gulf of Mexico bring warm humid weather. Occasionally, Gulf air reaches as far north as Pittsburgh during the winter and produces intermittent periods of thawing. The last spring temperature of 32 degrees usually occurs in late April and the first in late October. The average growing season is about 180 days. There is a wide variation in the time of the first and last frosts over a radius of 25 miles from the center of Pittsburgh due to terrain differences.

Precipitation is distributed well throughout the year. During the winter months about a fourth of the precipitation occurs as snow and there is about a 50 percent chance of measurable precipitation on any day. Thunderstorms occur normally during all months, except midwinter, and have a maximum frequency in midsummer. The first appreciable snowfall generally occurs in late November and usually the last occurs early in April. Snow lies on the ground in the suburbs on an average of about 33 days during the year.

Seven months of the year, April through October, have sunshine more than 50 percent of the possible time. During the remaining five months cloudiness is heavier because the track of migratory storms from west to east is closer to the area and because of the frequent periods of cloudy, showery weather associated with northwest winds from across the Great Lakes. Cold air drainage induced by the many hills leads to the frequent formation of early morning fog which may be quite persistent in the river valleys during the colder months.

The Allegheny River flowing south and the Monongahela River flowing north meet to form the Ohio River at Pittsburgh. Heavier rainfall and steeper topography cause the Monongahela River to flood more frequently than the Allegheny River.

Both rivers combine to cause the Ohio River at Pittsburgh to reach the 25 foot flood stage approximately once every four years. The serious flood level of 30 feet is reached much less frequently.

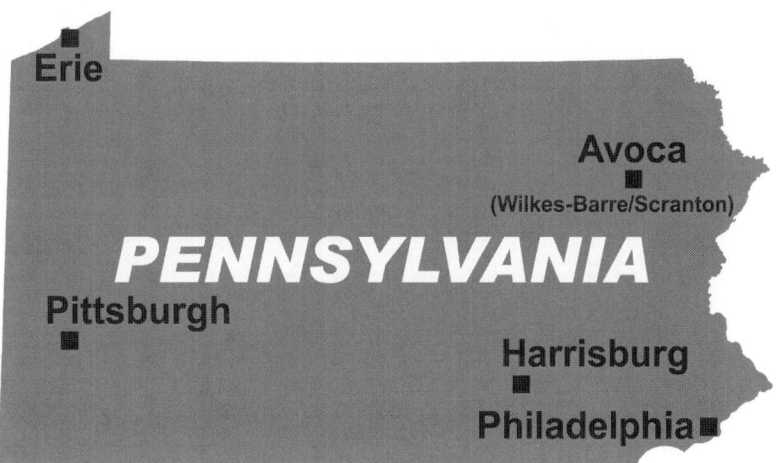

# NORMALS, MEANS, AND EXTREMES
## PITTSBURGH (KPIT)

LATITUDE: 40 ° 30'N    LONGITUDE: -80 ° 13'W    ELEVATION (FT): GRND: 1118   BARO: 1175    TIME ZONE: EASTERN (UTC -5)    WBAN: 94823

| ELEMENT | POR | JAN | FEB | MAR | APR | MAY | JUN | JUL | AUG | SEP | OCT | NOV | DEC | YEAR |
|---|---|---|---|---|---|---|---|---|---|---|---|---|---|---|
| **TEMPERATURE °F** | | | | | | | | | | | | | | |
| NORMAL DAILY MAXIMUM | 30 | 35.1 | 38.8 | 49.5 | 60.7 | 70.8 | 79.1 | 82.7 | 81.1 | 74.2 | 62.5 | 50.5 | 39.8 | 60.4 |
| MEAN DAILY MAXIMUM | 62 | 35.6 | 38.6 | 48.6 | 61.2 | 70.9 | 79.1 | 82.6 | 81.3 | 74.5 | 63.0 | 50.6 | 39.3 | 60.4 |
| HIGHEST DAILY MAXIMUM | 57 | 72 | 76 | 82 | 89 | 91 | 98 | 103 | 100 | 97 | 87 | 82 | 74 | 103 |
| YEAR OF OCCURRENCE | | 2002 | 2000 | 1998 | 1990 | 2006 | 1988 | 1988 | 1988 | 1954 | 2007 | 1961 | 1982 | JUL 1988 |
| MEAN OF EXTREME MAXS. | 62 | 58.9 | 60.8 | 73.5 | 80.9 | 85.5 | 90.0 | 91.5 | 90.5 | 87.5 | 79.3 | 71.2 | 61.3 | 77.6 |
| NORMAL DAILY MINIMUM | 30 | 19.9 | 22.3 | 30.1 | 39.1 | 49.2 | 57.7 | 62.4 | 61.0 | 53.9 | 42.5 | 34.2 | 25.3 | 41.5 |
| MEAN DAILY MINIMUM | 62 | 20.6 | 22.1 | 29.7 | 39.8 | 49.0 | 57.7 | 62.3 | 61.0 | 53.9 | 42.9 | 34.3 | 25.0 | 41.5 |
| LOWEST DAILY MINIMUM | 57 | -22 | -12 | -1 | 14 | 26 | 34 | 42 | 39 | 31 | 16 | -1 | -12 | -22 |
| YEAR OF OCCURRENCE | | 1994 | 1979 | 1980 | 1982 | 1970 | 1972 | 1963 | 1982 | 1959 | 1965 | 1958 | 1989 | JAN 1994 |
| MEAN OF EXTREME MINS. | 62 | 0.1 | 2.0 | 11.3 | 23.6 | 34.4 | 43.8 | 50.6 | 49.0 | 38.5 | 28.3 | 17.6 | 6.3 | 25.5 |
| NORMAL DRY BULB | 30 | 27.5 | 30.5 | 39.8 | 49.9 | 60.0 | 68.4 | 72.6 | 71.0 | 64.0 | 52.5 | 42.3 | 32.5 | 50.9 |
| MEAN DRY BULB | 62 | 28.1 | 30.4 | 39.1 | 50.5 | 60.0 | 68.6 | 72.5 | 71.2 | 64.2 | 52.9 | 42.4 | 32.2 | 51.0 |
| MEAN WET BULB | 26 | 25.6 | 27.3 | 33.8 | 43.3 | 52.6 | 61.5 | 65.0 | 64.2 | 58.0 | 46.9 | 38.0 | 29.0 | 45.4 |
| MEAN DEW POINT | 26 | 22.0 | 22.8 | 29.1 | 38.0 | 48.8 | 58.2 | 62.2 | 61.4 | 55.1 | 43.4 | 34.2 | 25.7 | 41.7 |
| NORMAL NO. DAYS WITH: | | | | | | | | | | | | | | |
| MAXIMUM >= 90 | 30 | 0.0 | 0.0 | 0.0 | 0.0 | 0.2 | 1.5 | 3.9 | 2.2 | 0.6 | 0.0 | 0.0 | 0.0 | 8.4 |
| MAXIMUM <= 32 | 30 | 12.9 | 9.4 | 3.0 | 0.1 | 0.0 | 0.0 | 0.0 | 0.0 | 0.0 | 0.0 | 1.2 | 8.4 | 35.0 |
| MINIMUM <= 32 | 30 | 26.7 | 22.9 | 19.2 | 7.8 | 0.5 | 0.0 | 0.0 | 0.0 | 0.0 | 3.8 | 14.0 | 23.9 | 118.8 |
| MINIMUM <= 0 | 30 | 2.0 | 1.2 | 0.1 | 0.0 | 0.0 | 0.0 | 0.0 | 0.0 | 0.0 | 0.0 | 0.0 | * | 3.9 |
| **H/C** | | | | | | | | | | | | | | |
| NORMAL HEATING DEG. DAYS | 30 | 1163 | 979 | 788 | 462 | 200 | 43 | 6 | 13 | 105 | 397 | 677 | 996 | 5829 |
| NORMAL COOLING DEG. DAYS | 30 | 0 | 0 | 2 | 8 | 41 | 143 | 244 | 203 | 78 | 6 | 1 | 0 | 726 |
| **RH** | | | | | | | | | | | | | | |
| NORMAL (PERCENT) | 30 | 71 | 68 | 64 | 60 | 64 | 67 | 69 | 71 | 73 | 70 | 70 | 72 | 68 |
| HOUR 01 LST | 30 | 74 | 71 | 69 | 67 | 74 | 79 | 80 | 82 | 83 | 78 | 74 | 75 | 76 |
| HOUR 07 LST | 30 | 77 | 75 | 75 | 73 | 77 | 80 | 83 | 86 | 86 | 83 | 78 | 78 | 79 |
| HOUR 13 LST | 30 | 66 | 62 | 56 | 50 | 52 | 53 | 55 | 57 | 58 | 56 | 61 | 66 | 58 |
| HOUR 19 LST | 30 | 69 | 63 | 58 | 52 | 55 | 59 | 60 | 63 | 67 | 64 | 67 | 69 | 62 |
| **S** PERCENT POSSIBLE SUNSHINE | 49 | 32 | 36 | 43 | 46 | 50 | 55 | 57 | 55 | 55 | 51 | 36 | 29 | 45 |
| **W/O** MEAN NO. DAYS WITH: | | | | | | | | | | | | | | |
| HEAVY FOG(VISBY <= 1/4 MI) | 46 | 1.2 | 0.9 | 0.8 | 0.5 | 1.1 | 1.1 | 1.5 | 1.9 | 1.8 | 1.7 | 1.4 | 1.6 | 15.5 |
| THUNDERSTORMS | 62 | 0.2 | 0.5 | 1.7 | 3.4 | 5.5 | 7.2 | 7.0 | 5.6 | 3.0 | 1.3 | 0.7 | 0.3 | 36.4 |
| **CLOUDNESS** MEAN: | | | | | | | | | | | | | | |
| SUNRISE-SUNSET (OKTAS) | 44 | 6.5 | 6.2 | 6.0 | 5.8 | 5.5 | 5.2 | 4.9 | 4.9 | 4.8 | 4.9 | 6.0 | 6.4 | 5.6 |
| MIDNIGHT-MIDNIGHT (OKTAS) | 32 | 6.3 | 5.9 | 5.7 | 5.4 | 5.1 | 4.9 | 4.6 | 4.4 | 4.5 | 4.6 | 5.8 | 6.2 | 5.3 |
| MEAN NO. DAYS WITH: | | | | | | | | | | | | | | |
| CLEAR | 44 | 2.9 | 3.3 | 4.5 | 4.4 | 5.3 | 4.8 | 5.4 | 6.6 | 7.3 | 7.8 | 3.7 | 2.7 | 58.7 |
| PARTLY CLOUDY | 44 | 6.0 | 5.9 | 6.8 | 8.3 | 9.0 | 11.8 | 12.8 | 11.6 | 10.1 | 8.8 | 6.3 | 5.7 | 103.1 |
| CLOUDY | 44 | 22.2 | 19.1 | 19.8 | 17.3 | 16.7 | 13.4 | 12.7 | 12.9 | 12.5 | 14.4 | 20.0 | 22.6 | 203.6 |
| **PR** MEAN STATION PRESSURE(IN) | 26 | 28.75 | 28.75 | 28.73 | 28.68 | 28.70 | 28.71 | 28.73 | 28.76 | 28.79 | 28.79 | 28.78 | 28.77 | 28.75 |
| MEAN SEA-LEVEL PRES. (IN) | 26 | 30.10 | 30.09 | 30.06 | 29.98 | 30.00 | 29.99 | 30.01 | 30.05 | 30.08 | 30.10 | 30.10 | 30.11 | 30.06 |
| **WINDS** MEAN SPEED (MPH) | 26 | 9.6 | 9.4 | 9.6 | 9.3 | 7.9 | 7.2 | 6.5 | 6.2 | 6.5 | 7.2 | 8.6 | 9.0 | 8.1 |
| PREVAIL.DIR(TENS OF DEGS) | 41 | 27 | 27 | 27 | 28 | 24 | 24 | 24 | 24 | 24 | 27 | 27 | 27 | 27 |
| MAXIMUM 2-MINUTE: | | | | | | | | | | | | | | |
| SPEED (MPH) | 13 | 47 | 44 | 39 | 51 | 48 | 53 | 45 | 46 | 40 | 39 | 44 | 40 | 53 |
| DIR. (TENS OF DEGS) | | 21 | 28 | 31 | 30 | 30 | 34 | 29 | 33 | 23 | 31 | 28 | 27 | 34 |
| YEAR OF OCCURRENCE | | 2008 | 2006 | 2002 | 1998 | 2002 | 1998 | 2001 | 2001 | 2008 | 1999 | 2002 | 2008 | JUN 1998 |
| MAXIMUM 3-SECOND | | | | | | | | | | | | | | |
| SPEED (MPH) | 13 | 58 | 56 | 51 | 61 | 59 | 62 | 61 | 60 | 52 | 47 | 55 | 53 | 62 |
| DIR. (TENS OF DEGS) | | 20 | 23 | 28 | 31 | 22 | 34 | 33 | 34 | 22 | 31 | 27 | 23 | 34 |
| YEAR OF OCCURRENCE | | 2008 | 2009 | 2007 | 1998 | 1997 | 1998 | 1997 | 2001 | 2008 | 1999 | 2003 | 2006 | JUN 1998 |
| **PRECIPITATION** NORMAL (IN) | 30 | 2.70 | 2.37 | 3.17 | 3.01 | 3.80 | 4.12 | 3.96 | 3.38 | 3.21 | 2.25 | 3.02 | 2.86 | 37.85 |
| MAXIMUM MONTHLY (IN) | 57 | 6.25 | 5.98 | 6.10 | 7.61 | 6.56 | 10.29 | 8.71 | 7.86 | 10.06 | 8.20 | 11.05 | 8.51 | 11.05 |
| YEAR OF OCCURRENCE | | 1978 | 1956 | 1967 | 1964 | 1989 | 1989 | 1992 | 1987 | 2004 | 1954 | 1985 | 1990 | NOV 1985 |
| MINIMUM MONTHLY (IN) | 57 | 0.77 | 0.51 | 1.14 | 0.48 | 1.21 | 0.64 | 1.62 | 0.78 | 0.28 | 0.16 | 0.90 | 0.40 | 0.16 |
| YEAR OF OCCURRENCE | | 1981 | 1969 | 1969 | 1971 | 1965 | 1992 | 1989 | 1957 | 1985 | 1963 | 1976 | 1955 | OCT 1963 |
| MAXIMUM IN 24 HOURS (IN) | 57 | 1.99 | 2.30 | 2.00 | 2.15 | 2.90 | 3.11 | 4.41 | 3.06 | 5.95 | 3.56 | 1.97 | 2.76 | 5.95 |
| YEAR OF OCCURRENCE | | 2004 | 1975 | 1964 | 1964 | 1997 | 1996 | 1999 | 1956 | 2004 | 1954 | 1985 | 1990 | SEP 2004 |
| NORMAL NO. DAYS WITH: | | | | | | | | | | | | | | |
| PRECIPITATION >= 0.01 | 30 | 16.5 | 13.6 | 14.9 | 13.6 | 13.1 | 12.0 | 10.4 | 10.0 | 10.7 | 10.2 | 12.9 | 15.4 | 153.3 |
| PRECIPITATION >= 1.00 | 30 | 0.3 | 0.2 | 0.2 | 0.3 | 0.7 | 1.0 | 1.0 | 0.8 | 0.5 | 0.2 | 0.4 | 0.4 | 6.0 |
| **SNOWFALL** NORMAL (IN) | 30 | 12.3 | 8.5 | 7.9 | 1.5 | 0.* | 0.0 | 0.0 | 0.0 | 0.0 | 0.4 | 3.1 | 6.9 | 40.6 |
| MAXIMUM MONTHLY (IN) | 56 | 40.2 | 25.3 | 34.1 | 8.1 | 3.1 | T | T | T | T | 8.5 | 13.9 | 21.3 | 40.2 |
| YEAR OF OCCURRENCE | | 1978 | 2003 | 1993 | 1987 | 1966 | 2009 | 2007 | 1994 | 1989 | 1993 | 1995 | 2003 | JAN 1978 |
| MAXIMUM IN 24 HOURS (IN) | 56 | 14.0 | 12.3 | 23.8 | 7.7 | 3.1 | T | T | T | T | 6.6 | 10.5 | 12.5 | 23.8 |
| YEAR OF OCCURRENCE' | | 1966 | 1960 | 1993 | 1987 | 1966 | 2009 | 1991 | 1994 | 1989 | 1993 | 1958 | 1974 | MAR 1993 |
| MAXIMUM SNOW DEPTH (IN) | 61 | 26 | 16 | 25 | 7 | 1 | 0 | 0 | 0 | 0 | 1 | 22 | 13 | 26 |
| YEAR OF OCCURRENCE | | 1978 | 1961 | 1993 | 1987 | 1963 | | | | | 1992 | 1950 | 1974 | JAN 1978 |
| NORMAL NO. DAYS WITH: | | | | | | | | | | | | | | |
| SNOWFALL >= 1.0 | 30 | 3.8 | 2.5 | 2.4 | 0.5 | 0.0 | 0.0 | 0.0 | 0.0 | 0.0 | 0.1 | 0.8 | 1.9 | 12.0 |

## PRECIPITATION (inches) 2009 PITTSBURGH (KPIT)

| YEAR | JAN | FEB | MAR | APR | MAY | JUN | JUL | AUG | SEP | OCT | NOV | DEC | ANNUAL |
|------|-----|-----|-----|-----|-----|-----|-----|-----|-----|-----|-----|-----|--------|
| 1980 | 1.56 | 1.32 | 5.65 | 2.94 | 4.32 | 4.34 | 6.76 | 5.10 | 1.29 | 2.42 | 2.38 | 1.38 | 39.46 |
| 1981 | 0.77 | 4.20 | 2.12 | 4.92 | 2.04 | 8.20 | 3.82 | 0.98 | 4.13 | 1.82 | 1.50 | 3.00 | 37.50 |
| 1982 | 4.44 | 1.93 | 3.52 | 1.44 | 3.98 | 3.05 | 2.36 | 1.97 | 2.80 | 0.40 | 3.33 | 2.79 | 32.01 |
| 1983 | 1.19 | 1.58 | 3.50 | 4.33 | 5.24 | 4.82 | 3.32 | 3.13 | 2.42 | 3.67 | 3.94 | 4.27 | 41.41 |
| 1984 | 1.40 | 2.05 | 2.32 | 3.72 | 5.22 | 1.98 | 3.01 | 5.15 | 0.84 | 3.45 | 3.14 | 3.04 | 35.32 |
| 1985 | 1.43 | 1.45 | 3.37 | 1.64 | 5.80 | 2.26 | 4.06 | 2.64 | 0.28 | 2.27 | 11.05 | 2.26 | 38.51 |
| 1986 | 2.49 | 3.43 | 1.38 | 1.94 | 1.67 | 5.24 | 5.66 | 3.04 | 2.33 | 2.83 | 3.92 | 3.47 | 37.40 |
| 1987 | 2.23 | 0.71 | 2.65 | 5.30 | 2.41 | 6.30 | 2.42 | 7.86 | 3.97 | 0.92 | 2.02 | 2.41 | 39.20 |
| 1988 | 1.49 | 3.46 | 2.56 | 1.97 | 2.78 | 1.26 | 2.82 | 2.04 | 2.34 | 1.40 | 2.80 | 2.17 | 27.09 |
| 1989 | 1.99 | 3.42 | 5.52 | 1.43 | 6.56 | 10.29 | 1.62 | 1.12 | 4.57 | 2.04 | 1.56 | 2.39 | 42.51 |
| 1990 | 3.30 | 3.31 | 1.47 | 3.48 | 6.19 | 4.24 | 6.59 | 3.59 | 6.00 | 3.51 | 2.05 | 8.51 | 52.24 |
| 1991 | 2.55 | 1.88 | 2.92 | 2.56 | 3.29 | 3.82 | 3.74 | 1.63 | 3.45 | 0.55 | 1.97 | 3.66 | 32.02 |
| 1992 | 2.13 | 1.73 | 3.54 | 2.30 | 2.31 | 0.64 | 8.71 | 4.77 | 2.91 | 1.47 | 3.31 | 2.83 | 36.65 |
| 1993 | 2.99 | 2.92 | 4.14 | 3.66 | 2.85 | 3.35 | 2.85 | 2.44 | 3.87 | 2.77 | 4.30 | 2.12 | 38.26 |
| 1994 | 3.90 | 2.13 | 5.00 | 3.72 | 2.54 | 2.91 | 3.27 | 7.75 | 3.59 | 0.88 | 3.64 | 2.01 | 41.34 |
| 1995 | 2.23 | 1.73 | 1.56 | 1.70 | 3.72 | 3.74 | 3.06 | 1.75 | 1.80 | 3.24 | 2.74 | 1.62 | 28.89 |
| 1996 | 3.68 | 2.54 | 4.54 | 4.42 | 2.95 | 7.95 | 4.01 | 1.99 | 5.63 | 2.96 | 2.83 | 1.98 | 45.48 |
| 1997 | 1.58 | 1.15 | 3.22 | 1.57 | 6.33 | 3.95 | 1.82 | 3.80 | 2.90 | 0.95 | 5.98 | 1.29 | 34.54 |
| 1998 | 3.63 | 2.57 | 1.91 | 5.00 | 2.39 | 6.71 | 2.02 | 3.32 | 1.09 | 2.27 | 1.50 | 1.81 | 34.22 |
| 1999 | 4.88 | 2.43 | 1.24 | 4.19 | 4.12 | 1.67 | 6.25 | 2.21 | 1.97 | 1.55 | 3.46 | 2.24 | 36.21 |
| 2000 | 1.70 | 2.53 | 2.26 | 3.15 | 5.69 | 5.64 | 6.28 | 3.66 | 3.07 | 2.09 | 1.38 | 2.64 | 40.09 |
| 2001 | 1.35 | 1.09 | 3.28 | 3.75 | 2.11 | 3.43 | 3.15 | 7.12 | 2.23 | 2.33 | 3.47 | 2.43 | 35.74 |
| 2002 | 1.76 | 1.17 | 3.67 | 3.05 | 4.70 | 2.63 | 1.66 | 2.89 | 3.24 | 2.99 | 2.00 | 2.57 | 32.33 |
| 2003 | 2.18 | 2.86 | 1.55 | 2.45 | 6.14 | 3.87 | 6.01 | 3.17 | 3.48 | 2.57 | 3.42 | 3.34 | 41.04 |
| 2004 | 4.78 | 2.44 | 3.60 | 4.49 | 6.08 | 5.01 | 5.67 | 6.13 | 10.06 | 3.35 | 3.19 | 2.61 | 57.41 |
| 2005 | 6.12 | 3.02 | 2.31 | 3.72 | 4.09 | 3.35 | 4.33 | 3.72 | 1.32 | 3.47 | 4.05 | 1.73 | 41.23 |
| 2006 | 3.74 | 1.74 | 2.12 | 3.00 | 2.96 | 4.37 | 3.86 | 1.60 | 3.26 | 4.84 | 1.40 | 2.01 | 34.90 |
| 2007 | 3.28 | 1.97 | 5.28 | 4.31 | 1.93 | 2.53 | 3.01 | 6.15 | 2.30 | 1.70 | 3.96 | 4.28 | 40.70 |
| 2008 | 1.63 | 5.45 | 3.92 | 2.02 | 3.20 | 6.17 | 2.58 | 3.36 | 2.65 | 2.01 | 2.00 | 4.70 | 39.69 |
| 2009 | 2.98 | 1.56 | 1.69 | 2.36 | 3.83 | 4.42 | 4.12 | 3.55 | 1.55 | 2.29 | 0.96 | 3.53 | 32.84 |
| POR= 62 YRS | 2.82 | 2.39 | 3.30 | 3.26 | 3.68 | 3.94 | 3.84 | 3.37 | 2.89 | 2.38 | 2.80 | 2.75 | 37.42 |

WBAN : 94823

## AVERAGE TEMPERATURE (°F) 2009 PITTSBURGH (KPIT)

| YEAR | JAN | FEB | MAR | APR | MAY | JUN | JUL | AUG | SEP | OCT | NOV | DEC | ANNUAL |
|------|-----|-----|-----|-----|-----|-----|-----|-----|-----|-----|-----|-----|--------|
| 1980 | 26.9 | 24.2 | 35.6 | 48.1 | 60.3 | 66.2 | 75.0 | 74.5 | 67.1 | 49.5 | 38.6 | 28.6 | 49.6 |
| 1981 | 20.5 | 31.4 | 35.6 | 51.9 | 58.4 | 68.8 | 72.1 | 69.7 | 61.9 | 49.4 | 40.3 | 29.4 | 49.1 |
| 1982 | 20.9 | 28.4 | 38.4 | 45.3 | 64.7 | 63.7 | 72.4 | 68.2 | 63.4 | 54.4 | 44.7 | 39.9 | 50.4 |
| 1983 | 30.0 | 32.6 | 40.7 | 47.1 | 55.8 | 67.8 | 73.0 | 72.8 | 64.4 | 53.0 | 43.5 | 25.4 | 50.5 |
| 1984 | 23.2 | 36.4 | 32.2 | 49.2 | 55.3 | 69.7 | 68.5 | 70.8 | 61.4 | 58.3 | 40.2 | 39.3 | 50.4 |
| 1985 | 22.1 | 27.7 | 42.1 | 55.0 | 60.6 | 64.2 | 70.5 | 69.6 | 65.3 | 55.2 | 47.1 | 27.4 | 50.6 |
| 1986 | 28.3 | 31.3 | 41.1 | 53.1 | 62.0 | 68.3 | 73.3 | 68.6 | 66.6 | 54.2 | 40.4 | 33.1 | 51.7 |
| 1987 | 28.0 | 32.6 | 41.9 | 50.0 | 63.0 | 70.9 | 75.7 | 71.8 | 65.1 | 47.8 | 46.2 | 35.1 | 52.3 |
| 1988 | 26.6 | 29.0 | 39.3 | 49.4 | 61.4 | 68.5 | 76.9 | 75.1 | 63.5 | 46.6 | 44.2 | 31.9 | 51.0 |
| 1989 | 35.5 | 27.8 | 41.1 | 47.0 | 58.0 | 69.0 | 74.1 | 71.6 | 64.8 | 53.3 | 40.6 | 19.2 | 50.2 |
| 1990 | 36.8 | 36.9 | 44.0 | 51.3 | 57.7 | 68.3 | 71.7 | 70.5 | 63.7 | 55.0 | 45.5 | 38.0 | 53.3 |
| 1991 | 29.7 | 35.4 | 43.0 | 54.5 | 68.7 | 72.6 | 75.4 | 74.4 | 64.6 | 55.7 | 41.6 | 35.3 | 54.2 |
| 1992 | 30.5 | 34.3 | 38.8 | 51.3 | 59.1 | 66.0 | 72.4 | 67.9 | 63.7 | 50.1 | 42.9 | 33.9 | 50.9 |
| 1993 | 35.1 | 27.8 | 37.6 | 50.0 | 61.9 | 68.9 | 75.7 | 75.4 | 63.3 | 51.7 | 43.0 | 31.7 | 51.8 |
| 1994 | 21.1 | 29.5 | 37.8 | 53.9 | 56.7 | 72.9 | 74.5 | 70.0 | 63.8 | 53.4 | 47.7 | 38.2 | 51.6 |
| 1995 | 31.0 | 26.5 | 42.8 | 48.6 | 60.4 | 72.0 | 75.8 | 77.8 | 64.3 | 56.3 | 38.1 | 27.7 | 51.8 |
| 1996 | 27.5 | 30.1 | 34.7 | 50.7 | 60.0 | 72.3 | 69.6 | 70.6 | 63.7 | 52.0 | 36.3 | 36.5 | 50.3 |
| 1997 | 28.1 | 35.1 | 40.5 | 46.4 | 54.3 | 68.1 | 71.7 | 68.2 | 62.1 | 52.6 | 39.0 | 33.5 | 50.0 |
| 1998 | 37.1 | 38.5 | 42.4 | 51.0 | 64.4 | 67.1 | 71.2 | 73.1 | 67.2 | 53.1 | 44.4 | 37.8 | 53.9 |
| 1999 | 30.1 | 34.0 | 35.1 | 51.8 | 61.0 | 69.4 | 76.0 | 69.0 | 64.2 | 52.3 | 46.7 | 34.6 | 52.0 |
| 2000 | 27.7 | 36.0 | 44.9 | 50.2 | 62.5 | 69.7 | 68.8 | 69.1 | 62.3 | 54.7 | 39.5 | 23.1 | 50.7 |
| 2001 | 28.4 | 35.1 | 35.3 | 54.3 | 60.0 | 68.4 | 70.2 | 73.1 | 62.1 | 54.2 | 48.2 | 37.5 | 52.2 |
| 2002 | 35.5 | 35.0 | 40.9 | 52.7 | 56.7 | 70.6 | 76.0 | 74.1 | 67.4 | 51.0 | 40.3 | 30.7 | 52.6 |
| 2003 | 21.5 | 26.5 | 41.0 | 53.4 | 59.1 | 65.8 | 71.3 | 72.6 | 63.1 | 50.5 | 46.1 | 32.6 | 50.3 |
| 2004 | 22.2 | 31.7 | 42.7 | 51.2 | 65.5 | 67.6 | 71.3 | 68.7 | 65.8 | 53.2 | 46.0 | 33.3 | 51.6 |
| 2005 | 29.7 | 32.4 | 35.3 | 52.3 | 56.3 | 71.7 | 75.1 | 73.7 | 67.3 | 53.4 | 44.0 | 27.6 | 51.6 |
| 2006 | 38.1 | 30.4 | 38.2 | 53.8 | 58.8 | 66.2 | 73.4 | 72.9 | 61.0 | 50.8 | 45.2 | 38.8 | 52.3 |
| 2007 | 32.5 | 20.9 | 43.3 | 47.5 | 63.8 | 69.5 | 70.6 | 74.0 | 66.7 | 59.3 | 41.8 | 34.5 | 52.0 |
| 2008 | 31.4 | 29.4 | 37.1 | 53.9 | 57.0 | 70.0 | 72.6 | 69.7 | 65.9 | 51.1 | 40.1 | 33.0 | 50.9 |
| 2009 | 22.0 | 31.2 | 42.3 | 52.1 | 61.2 | 68.2 | 69.4 | 71.8 | 64.9 | 50.7 | 47.2 | 31.1 | 51.0 |
| POR= 62 YRS | 28.1 | 30.4 | 39.1 | 50.5 | 60.0 | 68.6 | 72.5 | 71.2 | 64.2 | 52.9 | 42.4 | 32.2 | 51.0 |

## HEATING DEGREE DAYS (base 65°F) 2009 PITTSBURGH (KPIT)

| YEAR | JUL | AUG | SEP | OCT | NOV | DEC | JAN | FEB | MAR | APR | MAY | JUN | TOTAL |
|---|---|---|---|---|---|---|---|---|---|---|---|---|---|
| 1980-81 | 0 | 5 | 48 | 476 | 787 | 1117 | 1372 | 936 | 904 | 391 | 223 | 18 | 6277 |
| 1981-82 | 3 | 10 | 159 | 475 | 736 | 1098 | 1361 | 1017 | 819 | 586 | 82 | 67 | 6413 |
| 1982-83 | 9 | 23 | 119 | 336 | 605 | 770 | 1080 | 904 | 746 | 535 | 280 | 44 | 5451 |
| 1983-84 | 10 | 2 | 126 | 365 | 639 | 1223 | 1293 | 823 | 1008 | 471 | 305 | 16 | 6281 |
| 1984-85 | 12 | 7 | 165 | 214 | 734 | 790 | 1322 | 1038 | 701 | 334 | 163 | 65 | 5545 |
| 1985-86 | 3 | 9 | 116 | 300 | 531 | 1160 | 1131 | 936 | 737 | 368 | 148 | 37 | 5476 |
| 1986-87 | 1 | 40 | 65 | 346 | 733 | 983 | 1139 | 904 | 710 | 451 | 145 | 22 | 5539 |
| 1987-88 | 4 | 20 | 61 | 529 | 560 | 920 | 1181 | 1040 | 792 | 461 | 149 | 64 | 5781 |
| 1988-89 | 5 | 3 | 83 | 570 | 619 | 1018 | 905 | 1033 | 739 | 532 | 260 | 25 | 5792 |
| 1989-90 | 1 | 14 | 102 | 364 | 723 | 1414 | 869 | 781 | 657 | 439 | 229 | 49 | 5642 |
| 1990-91 | 4 | 1 | 116 | 314 | 577 | 829 | 1085 | 820 | 674 | 337 | 63 | 5 | 4825 |
| 1991-92 | 0 | 0 | 127 | 308 | 698 | 913 | 1063 | 880 | 805 | 417 | 210 | 50 | 5471 |
| 1992-93 | 1 | 17 | 116 | 457 | 657 | 956 | 920 | 1037 | 841 | 445 | 135 | 43 | 5625 |
| 1993-94 | 0 | 0 | 118 | 407 | 654 | 1028 | 1357 | 988 | 836 | 345 | 278 | 21 | 6032 |
| 1994-95 | 0 | 10 | 80 | 351 | 516 | 824 | 1047 | 1071 | 678 | 487 | 163 | 5 | 5232 |
| 1995-96 | 0 | 0 | 79 | 269 | 799 | 1150 | 1155 | 1005 | 930 | 436 | 215 | 5 | 6043 |
| 1996-97 | 14 | 1 | 104 | 397 | 853 | 873 | 1136 | 832 | 755 | 554 | 329 | 35 | 5883 |
| 1997-98 | 2 | 21 | 109 | 401 | 772 | 973 | 858 | 736 | 716 | 415 | 78 | 75 | 5156 |
| 1998-99 | 0 | 0 | 52 | 369 | 616 | 837 | 1076 | 861 | 917 | 387 | 140 | 40 | 5295 |
| 1999-00 | 1 | 6 | 101 | 386 | 541 | 935 | 1148 | 833 | 620 | 442 | 135 | 33 | 5181 |
| 2000-01 | 8 | 14 | 158 | 317 | 761 | 1291 | 1123 | 833 | 912 | 350 | 171 | 43 | 5981 |
| 2001-02 | 17 | 2 | 125 | 341 | 497 | 846 | 910 | 834 | 739 | 399 | 275 | 15 | 5000 |
| 2002-03 | 0 | 0 | 38 | 455 | 736 | 1055 | 1343 | 1073 | 737 | 348 | 186 | 58 | 6029 |
| 2003-04 | 0 | 0 | 96 | 442 | 558 | 997 | 1319 | 959 | 684 | 417 | 94 | 30 | 5596 |
| 2004-05 | 4 | 24 | 44 | 359 | 565 | 975 | 1087 | 907 | 916 | 376 | 272 | 15 | 5544 |
| 2005-06 | 0 | 2 | 27 | 368 | 623 | 1153 | 825 | 962 | 824 | 330 | 228 | 43 | 5385 |
| 2006-07 | 3 | 0 | 127 | 433 | 586 | 805 | 1000 | 1229 | 674 | 521 | 104 | 21 | 5503 |
| 2007-08 | 3 | 7 | 61 | 222 | 689 | 938 | 1036 | 1025 | 860 | 327 | 250 | 18 | 5436 |
| 2008-09 | 0 | 2 | 42 | 426 | 741 | 984 | 1326 | 940 | 699 | 408 | 151 | 22 | 5741 |
| 2009- | 5 | 9 | 55 | 437 | 529 | 1046 | | | | | | | |

WBAN : 94823

## COOLING DEGREE DAYS (base 65°F) 2009 PITTSBURGH (KPIT)

| YEAR | JAN | FEB | MAR | APR | MAY | JUN | JUL | AUG | SEP | OCT | NOV | DEC | TOTAL |
|---|---|---|---|---|---|---|---|---|---|---|---|---|---|
| 1980 | 0 | 0 | 0 | 0 | 34 | 115 | 317 | 306 | 118 | 0 | 0 | 0 | 890 |
| 1981 | 0 | 0 | 0 | 5 | 25 | 139 | 230 | 160 | 72 | 0 | 0 | 0 | 631 |
| 1982 | 0 | 0 | 0 | 0 | 79 | 33 | 246 | 127 | 77 | 15 | 3 | 0 | 580 |
| 1983 | 0 | 0 | 0 | 3 | 3 | 135 | 263 | 251 | 115 | 0 | 0 | 0 | 770 |
| 1984 | 0 | 0 | 0 | 3 | 12 | 165 | 127 | 194 | 63 | 13 | 0 | 0 | 577 |
| 1985 | 0 | 0 | 0 | 41 | 33 | 49 | 181 | 160 | 130 | 4 | 0 | 0 | 598 |
| 1986 | 0 | 0 | 3 | 20 | 65 | 144 | 265 | 157 | 121 | 20 | 0 | 0 | 795 |
| 1987 | 0 | 0 | 0 | 6 | 93 | 204 | 342 | 240 | 72 | 0 | 1 | 0 | 958 |
| 1988 | 0 | 0 | 0 | 0 | 44 | 174 | 381 | 322 | 47 | 7 | 0 | 0 | 975 |
| 1989 | 0 | 0 | 5 | 0 | 49 | 154 | 291 | 225 | 100 | 9 | 0 | 0 | 833 |
| 1990 | 0 | 0 | 14 | 37 | 9 | 153 | 218 | 179 | 83 | 13 | 0 | 0 | 706 |
| 1991 | 0 | 0 | 0 | 27 | 184 | 239 | 333 | 298 | 124 | 27 | 0 | 0 | 1232 |
| 1992 | 0 | 0 | 0 | 9 | 33 | 87 | 233 | 112 | 86 | 3 | 0 | 0 | 563 |
| 1993 | 0 | 0 | 0 | 2 | 45 | 166 | 337 | 327 | 73 | 1 | 1 | 0 | 952 |
| 1994 | 0 | 0 | 0 | 21 | 28 | 264 | 302 | 175 | 51 | 0 | 2 | 0 | 843 |
| 1995 | 0 | 0 | 0 | 0 | 27 | 220 | 342 | 402 | 65 | 7 | 0 | 0 | 1063 |
| 1996 | 0 | 0 | 0 | 11 | 67 | 230 | 162 | 186 | 71 | 0 | 0 | 0 | 727 |
| 1997 | 0 | 0 | 0 | 2 | 4 | 134 | 217 | 128 | 29 | 24 | 0 | 0 | 538 |
| 1998 | 0 | 0 | 22 | 0 | 67 | 147 | 199 | 254 | 125 | 6 | 0 | 0 | 820 |
| 1999 | 0 | 0 | 0 | 0 | 24 | 178 | 350 | 136 | 83 | 0 | 0 | 0 | 771 |
| 2000 | 0 | 0 | 5 | 2 | 67 | 181 | 131 | 147 | 83 | 6 | 0 | 0 | 622 |
| 2001 | 0 | 0 | 0 | 34 | 20 | 152 | 184 | 258 | 44 | 12 | 0 | 0 | 704 |
| 2002 | 0 | 0 | 0 | 38 | 26 | 191 | 350 | 289 | 115 | 29 | 0 | 0 | 1038 |
| 2003 | 0 | 0 | 0 | 6 | 9 | 87 | 204 | 243 | 47 | 0 | 0 | 0 | 596 |
| 2004 | 0 | 0 | 0 | 9 | 114 | 113 | 207 | 144 | 75 | 0 | 0 | 0 | 662 |
| 2005 | 0 | 0 | 0 | 1 | 8 | 221 | 319 | 277 | 103 | 17 | 0 | 0 | 946 |
| 2006 | 0 | 0 | 0 | 0 | 43 | 89 | 269 | 250 | 12 | 0 | 0 | 0 | 663 |
| 2007 | 0 | 0 | 1 | 0 | 75 | 165 | 184 | 294 | 118 | 56 | 0 | 0 | 893 |
| 2008 | 0 | 0 | 0 | 2 | 12 | 176 | 241 | 156 | 75 | 1 | 0 | 0 | 663 |
| 2009 | 0 | 0 | 0 | 31 | 42 | 126 | 147 | 227 | 58 | 0 | 0 | 0 | 631 |

## SNOWFALL (inches)  2009  PITTSBURGH (KPIT)

| YEAR | JUL | AUG | SEP | OCT | NOV | DEC | JAN | FEB | MAR | APR | MAY | JUN | TOTAL |
|---|---|---|---|---|---|---|---|---|---|---|---|---|---|
| 1980-81 | 0.0 | 0.0 | 0.0 | T | 9.7 | 6.3 | 12.5 | 11.9 | 7.6 | T | 0.0 | 0.0 | 48.0 |
| 1981-82 | 0.0 | 0.0 | 0.0 | T | 0.6 | 11.5 | 13.4 | 3.6 | 12.2 | 3.8 | 0.0 | 0.0 | 45.1 |
| 1982-83 | 0.0 | 0.0 | 0.0 | T | 0.1 | 8.8 | 3.9 | 12.0 | 4.3 | 1.0 | 0.0 | 0.0 | 30.1 |
| 1983-84 | 0.0 | 0.0 | 0.0 | 0.0 | 6.1 | 10.5 | 10.8 | 11.4 | 10.4 | T | 0.0 | 0.0 | 49.2 |
| 1984-85 | 0.0 | 0.0 | 0.0 | 0.0 | 1.5 | 4.8 | 14.6 | 8.1 | 0.2 | 7.2 | 0.0 | 0.0 | 36.4 |
| 1985-86 | 0.0 | 0.0 | 0.0 | 0.0 | T | 15.3 | 11.1 | 12.4 | 4.8 | 2.7 | 0.0 | 0.0 | 46.3 |
| 1986-87 | 0.0 | 0.0 | 0.0 | 0.0 | 1.0 | 0.9 | 11.6 | 1.1 | 7.3 | 8.1 | 0.0 | 0.0 | 30.0 |
| 1987-88 | 0.0 | 0.0 | 0.0 | T | 4.1 | 7.9 | 5.5 | 6.9 | 9.8 | 0.9 | 0.0 | 0.0 | 35.1 |
| 1988-89 | 0.0 | 0.0 | 0.0 | 0.2 | 1.1 | 4.0 | 4.2 | 4.1 | 7.5 | 0.6 | T | 0.0 | 21.7 |
| 1989-90 | 0.0 | 0.0 | T | 0.2 | 1.6 | 12.5 | 7.7 | 2.5 | 0.6 | 3.3 | 0.0 | T | 28.4 |
| 1990-91 | 0.0 | 0.0 | 0.0 | 0.0 | T | 4.6 | 4.8 | 3.6 | 4.2 | T | 0.0 | 0.0 | 17.2 |
| 1991-92 | T | 0.0 | 0.0 | 0.0 | 2.1 | 3.1 | 12.9 | 2.4 | 10.6 | 2.8 | T | 0.0 | 33.9 |
| 1992-93 | 0.0 | 0.0 | 0.0 | 1.3 | 1.5 | 14.1 | 2.1 | 18.5 | 34.1 | 0.5 | 0.0 | 0.0 | 72.1 |
| 1993-94 | 0.0 | 0.0 | 0.0 | 0.0 | 8.5 | 2.6 | 10.4 | 30.1 | 11.0 | 13.6 | 0.6 | 0.0 | 0.0 | 76.8 |
| 1994-95 | 0.0 | T | 0.0 | 0.0 | 0.2 | T | 7.6 | 9.0 | 6.4 | 0.2 | 0.0 | 0.0 | 23.4 |
| 1995-96 | 0.0 | 0.0 | 0.0 | 0.0 | 13.9 | 12.6 | 23.4 | 9.6 | 13.4 | 1.6 | 0.0 | 0.0 | 74.5 |
| 1996-97 | 0.0 | 0.0 | 0.0 | 0.0 | 4.4 | 6.1 | | | | | 0.0 | 0.0 | |
| 1997-98 | 0.0 | 0.0 | 0.0 | 0.0 | | 12.6 | 1.7 | 2.5 | 4.9 | 0.0 | 0.0 | 0.0 | |
| 1998-99 | 0.0 | 0.0 | 0.0 | T | T | 1.8 | 17.1 | 4.7 | 15.6 | T | 0.0 | 0.0 | 39.2 |
| 1999-00 | 0.0 | 0.0 | 0.0 | T | 2.1 | 3.2 | 12.0 | 8.4 | 1.1 | 0.3 | T | 0.0 | 27.1 |
| 2000-01 | 0.0 | 0.0 | 0.0 | T | 2.1 | 8.3 | 13.7 | 2.7 | 8.0 | 1.1 | 0.0 | 0.0 | 35.9 |
| 2001-02 | T | 0.0 | 0.0 | 0.2 | T | 5.0 | 10.0 | 6.4 | 4.0 | 0.1 | T | 0.0 | 25.7 |
| 2002-03 | 0.0 | 0.0 | 0.0 | T | 3.1 | 11.0 | 18.3 | 25.3 | 4.1 | T | 0.0 | 0.0 | 61.8 |
| 2003-04 | 0.0 | 0.0 | 0.0 | T | 1.2 | 21.3 | 18.3 | 4.1 | 8.6 | 0.7 | 0.0 | 0.0 | 54.2 |
| 2004-05 | 0.0 | 0.0 | 0.0 | T | 0.0 | 7.3 | 11.3 | 14.7 | 9.8 | 6.3 | 0.0 | T | 49.4 |
| 2005-06 | 0.0 | 0.0 | 0.0 | T | 3.2 | 12.2 | 5.5 | 9.7 | 1.4 | T | 0.0 | T | 32.0 |
| 2006-07 | 0.0 | 0.0 | 0.0 | T | T | 0.7 | 11.3 | 14.0 | 9.3 | 0.4 | 0.0 | 0.0 | 35.7 |
| 2007-08 | T | 0.0 | 0.0 | 0.0 | 0.7 | 8.2 | 6.9 | 20.7 | 4.7 | 0.0 | 0.0 | 0.0 | 41.2 |
| 2008-09 | 0.0 | 0.0 | 0.0 | 0.7 | 5.6 | 5.3 | 20.8 | 7.1 | 0.2 | 1.3 | 0.0 | T | 41.0 |
| 2009- | 0.0 | 0.0 | 0.0 | T | T | 10.8 | | | | | | | |
| POR=<br>62 YRS | T | T | T | 0.3 | 3.6 | 8.3 | 11.6 | 9.1 | 8.2 | 1.5 | 0.1 | T | 42.7 |

# Pittsburgh, PA  2009
## Temperature and Precipitation

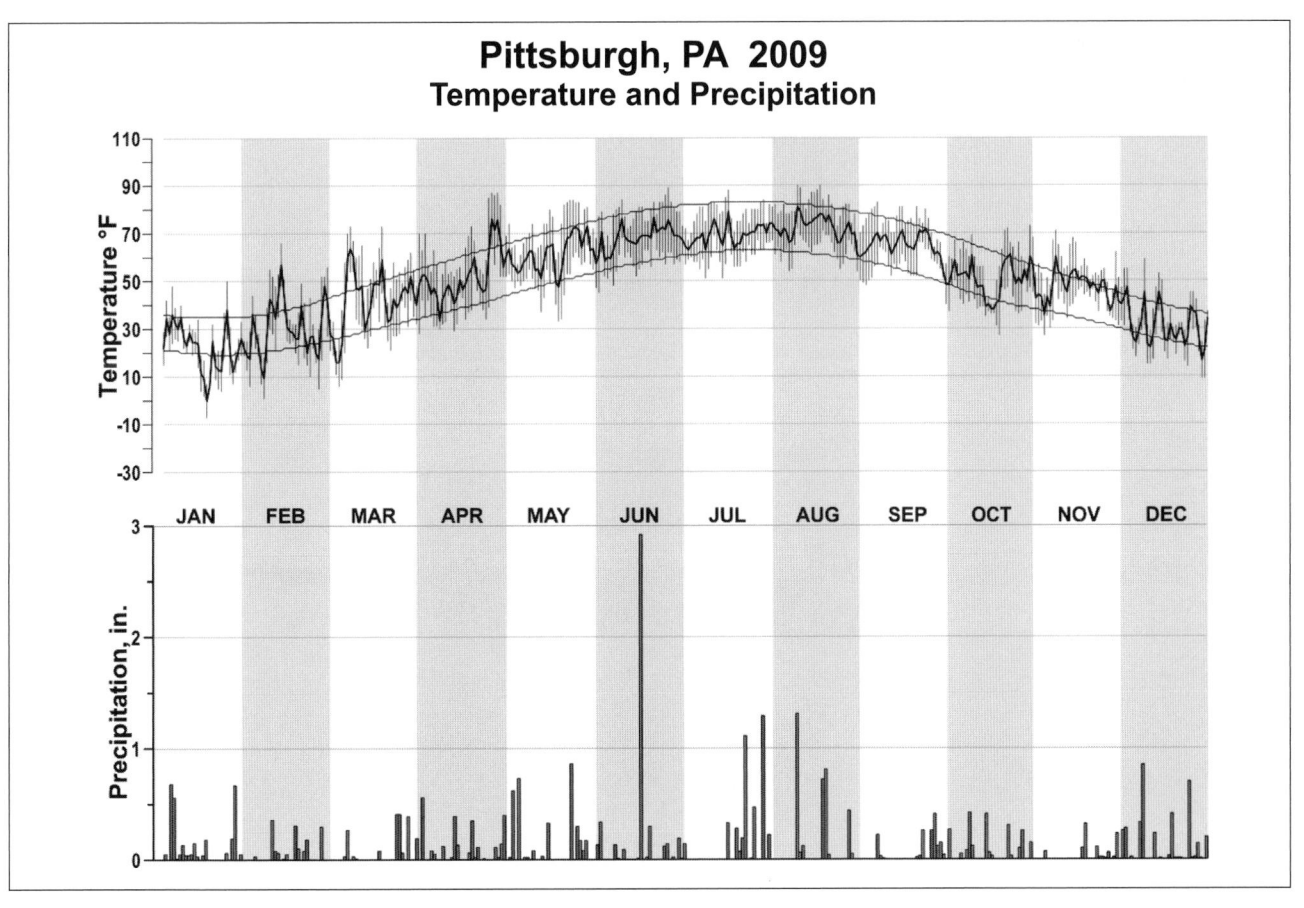

# 2009
# SAN JUAN
# PUERTO RICO (TJSJ)

San Juan, located on the north coast of the island of Puerto Rico, is surrounded by the waters of the Atlantic Ocean and San Juan Bay. Local custom assigns the name San Juan to the old city which lies right on the coast, but the modern metropolitan area extends inland about 12 miles. These inland sections have a temperature and rainfall regime significantly different from the coastal area. Isla Verde Airport, where weather observations are made, lies on the coast about 7 miles east of old San Juan. The surrounding terrain is level with a gradual upslope inland. Mountain ranges, with peak elevations of 4,000 feet, extend east and west through the central portion of Puerto Rico, and are located 15 to 20 miles east and south of San Juan. These mountain ranges have a decided influence on the rainfall of the San Juan metropolitan area, and on the entire island in general.

The climate is tropical maritime, characteristic of all tropical islands. The predominant easterly trade winds, modified by local effects such as the land and sea breeze and the particular island topography, are a primary feature of the climate of San Juan and have a significant influence on the temperature and rainfall. During daylight hours the wind blows almost constantly off the ocean. Usually, after sunset the wind shifts to the south or southeast, off land. This daily wind variation is a contributing factor to the delightful climate of the city. The annual temperature range is small with about a 5-6 degree difference between the temperatures of the warmest and coldest months. The inland sectors have warmer afternoons and cooler nights. In the interior mountain and valley regions even greater daily and annual ranges of temperature occur. The highest temperatures recorded in Puerto Rico have exceeded 105 degrees and the lowest have been near 40. Sea water temperatures range from 78 degrees in March to about 83 degrees in September.

Although rainfall in San Juan is nearly 60 inches, the geographical distribution of rainfall over the island shows the heaviest rainfall, of about 180 inches per year, in the Luquillo Range, only 23 miles distant from San Juan. The driest area, with annual rainfall of 30 to 35 inches, is located in the southwest corner of the island. Rain showers occur mostly in the afternoon and at night. The nocturnal showers, usually light, are a characteristic feature of the San Juan rainfall pattern. Rainfall is generally of the brief showery type except for the continuous rains occuring with the passage of tropical disturbances, or when the trailing edge of a cold front out of the United States reaches Puerto Rico. This normally occurs from about November to April.

Puerto Rico is in the tropical hurricane region of the eastern Caribbean. The hurricane season begins June 1 and ends November 30. Only a few hurricanes have passed close enough to San Juan to produce hurricane force winds or damage.

Mild temperatures, refreshing sea breezes in the daytime, plenty of sunshine, and adequate rainfall make the climate of San Juan most enjoyable for tourists and residents alike.

# NORMALS, MEANS, AND EXTREMES
## SAN JUAN (TJSJ)

| LATITUDE: 18°26'N | LONGITUDE: -66°0'W | ELEVATION (FT): GRND: 6  BARO: 10 | TIME ZONE: ATLANTIC  (UTC -4) | WBAN: 11641 |
|---|---|---|---|---|

| | ELEMENT | POR | JAN | FEB | MAR | APR | MAY | JUN | JUL | AUG | SEP | OCT | NOV | DEC | YEAR |
|---|---|---|---|---|---|---|---|---|---|---|---|---|---|---|---|
| **TEMPERATURE °F** | NORMAL DAILY MAXIMUM | 30 | 82.4 | 82.8 | 83.4 | 84.9 | 86.3 | 87.6 | 87.4 | 87.8 | 87.8 | 87.5 | 85.1 | 83.2 | 85.5 |
| | MEAN DAILY MAXIMUM | 54 | 83.0 | 83.5 | 84.6 | 85.8 | 87.2 | 88.6 | 88.4 | 88.8 | 88.9 | 88.3 | 85.9 | 83.9 | 86.4 |
| | HIGHEST DAILY MAXIMUM | 55 | 92 | 96 | 96 | 97 | 96 | 97 | 95 | 97 | 97 | 98 | 96 | 94 | 98 |
| | YEAR OF OCCURRENCE | | 1983 | 1983 | 1983 | 1983 | 1980 | 1988 | 1981 | 1980 | 1981 | 1981 | 1981 | 1989 | OCT 1981 |
| | MEAN OF EXTREME MAXS. | 54 | 87.1 | 87.8 | 89.7 | 91.2 | 92.2 | 92.6 | 91.8 | 92.2 | 92.8 | 92.5 | 90.1 | 87.4 | 90.6 |
| | NORMAL DAILY MINIMUM | 30 | 70.8 | 70.9 | 71.7 | 73.2 | 74.9 | 76.6 | 76.9 | 77.0 | 76.5 | 75.6 | 74.0 | 72.1 | 74.2 |
| | MEAN DAILY MINIMUM | 54 | 70.7 | 70.5 | 71.4 | 72.9 | 74.6 | 76.1 | 76.6 | 76.7 | 76.3 | 75.5 | 73.9 | 72.2 | 74.0 |
| | LOWEST DAILY MINIMUM | 55 | 61 | 62 | 60 | 64 | 66 | 69 | 69 | 70 | 69 | 67 | 66 | 63 | 60 |
| | YEAR OF OCCURRENCE | | 1962 | 1968 | 1957 | 1968 | 1962 | 1957 | 1959 | 1956 | 1960 | 1959 | 1969 | 1964 | MAR 1957 |
| | MEAN OF EXTREME MINS. | 54 | 66.2 | 66.6 | 67.3 | 69.2 | 71.3 | 73.0 | 73.3 | 73.6 | 73.3 | 72.4 | 70.3 | 68.3 | 70.4 |
| | NORMAL DRY BULB | 30 | 76.6 | 76.9 | 77.6 | 79.1 | 80.6 | 82.1 | 82.2 | 82.2 | 82.2 | 81.6 | 79.6 | 77.7 | 79.9 |
| | MEAN DRY BULB | 54 | 76.9 | 77.0 | 78.0 | 79.4 | 80.9 | 82.4 | 82.6 | 82.8 | 82.6 | 81.9 | 79.9 | 78.1 | 80.2 |
| | MEAN WET BULB | 26 | 70.9 | 70.3 | 70.7 | 72.4 | 74.5 | 75.8 | 76.3 | 76.6 | 76.3 | 75.7 | 74.2 | 72.3 | 73.8 |
| | MEAN DEW POINT | 26 | 69.1 | 68.3 | 68.6 | 70.4 | 72.8 | 74.2 | 74.8 | 75.1 | 74.8 | 74.3 | 72.7 | 70.6 | 72.1 |
| | NORMAL NO. DAYS WITH: | | | | | | | | | | | | | | |
| | MAXIMUM >= 90 | 30 | 0.4 | 1.0 | 2.0 | 4.3 | 7.1 | 11.7 | 11.1 | 13.5 | 13.1 | 10.5 | 2.0 | 0.8 | 77.5 |
| | MAXIMUM <= 32 | 30 | 0.0 | 0.0 | 0.0 | 0.0 | 0.0 | 0.0 | 0.0 | 0.0 | 0.0 | 0.0 | 0.0 | 0.0 | 0.0 |
| | MINIMUM <= 32 | 30 | 0.0 | 0.0 | 0.0 | 0.0 | 0.0 | 0.0 | 0.0 | 0.0 | 0.0 | 0.0 | 0.0 | 0.0 | 0.0 |
| | MINIMUM <= 0 | 30 | 0.0 | 0.0 | 0.0 | 0.0 | 0.0 | 0.0 | 0.0 | 0.0 | 0.0 | 0.0 | 0.0 | 0.0 | 0.0 |
| **H/C** | NORMAL HEATING DEG. DAYS | 30 | 0 | 0 | 0 | 0 | 0 | 0 | 0 | 0 | 0 | 0 | 0 | 0 | 0 |
| | NORMAL COOLING DEG. DAYS | 30 | 360 | 332 | 388 | 421 | 484 | 513 | 533 | 539 | 515 | 513 | 436 | 392 | 5426 |
| **RH** | NORMAL (PERCENT) | 30 | 76 | 74 | 73 | 73 | 76 | 77 | 77 | 78 | 78 | 79 | 79 | 77 | 76 |
| | HOUR 02 LST | 30 | 81 | 80 | 79 | 80 | 83 | 83 | 83 | 84 | 84 | 85 | 84 | 82 | 82 |
| | HOUR 08 LST | 30 | 81 | 79 | 76 | 74 | 76 | 77 | 78 | 79 | 78 | 79 | 80 | 80 | 78 |
| | HOUR 14 LST | 30 | 64 | 62 | 61 | 61 | 65 | 65 | 66 | 67 | 67 | 67 | 68 | 66 | 65 |
| | HOUR 20 LST | 30 | 76 | 74 | 73 | 74 | 76 | 77 | 78 | 78 | 78 | 79 | 79 | 77 | 77 |
| **S** | PERCENT POSSIBLE SUNSHINE | 50 | 68 | 70 | 76 | 71 | 64 | 65 | 69 | 68 | 62 | 64 | 59 | 62 | 67 |
| **W/O** | MEAN NO. DAYS WITH: | | | | | | | | | | | | | | |
| | HEAVY FOG(VISBY <= 1/4 MI) | 46 | 0.0 | 0.0 | 0.0 | 0.0 | 0.1 | 0.0 | 0.0 | 0.1 | 0.0 | 0.0 | 0.0 | 0.0 | 0.2 |
| | THUNDERSTORMS | 54 | 0.3 | 0.3 | 0.3 | 1.3 | 4.2 | 5.0 | 5.1 | 6.2 | 8.4 | 7.8 | 3.1 | 0.9 | 42.9 |
| **CLOUDNESS** | MEAN: | | | | | | | | | | | | | | |
| | SUNRISE-SUNSET (OKTAS) | 41 | 3.9 | 4.0 | 3.9 | 4.3 | 5.1 | 4.8 | 4.6 | 4.5 | 4.8 | 4.7 | 4.4 | 4.2 | 4.4 |
| | MIDNIGHT-MIDNIGHT (OKTAS) | 32 | 3.5 | 3.6 | 3.6 | 3.7 | 4.5 | 4.1 | 4.2 | 4.0 | 4.2 | 4.0 | 4.0 | 3.8 | 3.9 |
| | MEAN NO. DAYS WITH: | | | | | | | | | | | | | | |
| | CLEAR | 41 | 9.0 | 7.5 | 9.4 | 7.5 | 3.7 | 4.4 | 5.2 | 5.5 | 4.0 | 4.7 | 5.4 | 6.7 | 73.0 |
| | PARTLY CLOUDY | 41 | 18.0 | 16.3 | 17.2 | 16.4 | 15.5 | 15.8 | 17.2 | 17.4 | 16.6 | 17.0 | 17.6 | 18.2 | 203.2 |
| | CLOUDY | 41 | 4.0 | 4.4 | 4.4 | 6.1 | 11.8 | 9.8 | 8.6 | 8.1 | 9.4 | 9.4 | 6.9 | 6.1 | 89.0 |
| **PR** | MEAN STATION PRESSURE(IN) | 26 | 29.99 | 29.99 | 29.97 | 29.86 | 29.94 | 29.98 | 29.96 | 29.92 | 29.90 | 29.85 | 29.89 | 29.89 | 29.93 |
| | MEAN SEA-LEVEL PRES. (IN) | 26 | 30.03 | 30.03 | 30.01 | 29.98 | 29.98 | 30.02 | 30.03 | 29.99 | 29.94 | 29.92 | 29.93 | 30.00 | 29.99 |
| **WINDS** | MEAN SPEED (MPH) | 26 | 7.7 | 8.1 | 8.3 | 8.1 | 8.1 | 8.7 | 9.4 | 8.1 | 6.9 | 6.1 | 6.9 | 7.5 | 7.8 |
| | PREVAIL.DIR(TENS OF DEGS) | 33 | 08 | 09 | 07 | 07 | 09 | 09 | 09 | 09 | 07 | 09 | 07 | 08 | 09 |
| | MAXIMUM 2-MINUTE: | | | | | | | | | | | | | | |
| | SPEED (MPH) | 13 | 31 | 30 | 32 | 31 | 30 | 32 | 33 | 34 | 79 | 37 | 33 | 31 | 79 |
| | DIR. (TENS OF DEGS) | | 06 | 06 | 07 | 06 | 06 | 07 | 10 | 04 | 05 | 04 | 36 | 09 | 05 |
| | YEAR OF OCCURRENCE | | 2006 | 2009 | 1997 | 2002 | 2009 | 1998 | 2009 | 1998 | 1998 | 2001 | 1999 | 2006 | SEP 1998 |
| | MAXIMUM 3-SECOND | | | | | | | | | | | | | | |
| | SPEED (MPH) | 13 | 41 | 43 | 37 | 39 | 46 | 45 | 45 | 47 | 93 | 43 | 39 | 39 | 93 |
| | DIR. (TENS OF DEGS) | | 32 | 06 | 07 | 08 | 07 | 10 | 11 | 07 | 07 | 04 | 36 | 02 | 07 |
| | YEAR OF OCCURRENCE | | 2008 | 2009 | 2004 | 2008 | 2009 | 2003 | 2009 | 1998 | 1998 | 2001 | 1999 | 2003 | SEP 1998 |
| **PRECIPITATION** | NORMAL (IN) | 30 | 3.02 | 2.30 | 2.14 | 3.71 | 5.29 | 3.52 | 4.16 | 5.22 | 5.60 | 5.06 | 6.17 | 4.57 | 50.76 |
| | MAXIMUM MONTHLY (IN) | 55 | 7.60 | 6.69 | 5.41 | 15.00 | 14.99 | 10.96 | 9.35 | 11.76 | 15.15 | 15.06 | 15.96 | 16.81 | 16.81 |
| | YEAR OF OCCURRENCE | | 1977 | 1982 | 1958 | 2005 | 1965 | 1965 | 1961 | 1955 | 1996 | 1970 | 1979 | 1981 | DEC 1981 |
| | MINIMUM MONTHLY (IN) | 55 | 0.61 | 0.20 | 0.58 | 0.08 | 0.44 | 0.29 | 1.12 | 1.83 | 1.73 | 1.17 | 1.91 | 0.68 | 0.08 |
| | YEAR OF OCCURRENCE | | 1978 | 1983 | 2008 | 1997 | 1972 | 1985 | 1974 | 1994 | 1987 | 1979 | 1980 | 1963 | APR 1997 |
| | MAXIMUM IN 24 HOURS (IN) | 55 | 5.08 | 2.75 | 3.91 | 7.20 | 4.74 | 3.55 | 2.91 | 5.30 | 8.84 | 5.04 | 7.07 | 8.40 | 8.84 |
| | YEAR OF OCCURRENCE | | 1969 | 1989 | 1969 | 1988 | 1986 | 1965 | 1993 | 2000 | 1989 | 1985 | 1979 | 1981 | SEP 1989 |
| | NORMAL NO. DAYS WITH: | | | | | | | | | | | | | | |
| | PRECIPITATION >= 0.01 | 30 | 17.6 | 13.6 | 12.8 | 12.0 | 14.6 | 14.0 | 17.9 | 18.3 | 17.3 | 17.3 | 19.4 | 19.2 | 194.0 |
| | PRECIPITATION >= 1.00 | 30 | 0.3 | 0.2 | 0.3 | 0.9 | 1.6 | 0.6 | 0.8 | 1.0 | 1.1 | 1.0 | 1.1 | 0.9 | 9.8 |
| **SNOWFALL** | NORMAL (IN) | 30 | 0.0 | 0.0 | 0.0 | 0.0 | 0.0 | 0.0 | 0.0 | 0.0 | 0.0 | 0.0 | 0.0 | 0.0 | 0.0 |
| | MAXIMUM MONTHLY (IN) | 55 | 0.0 | 0.0 | 0.0 | 0.0 | 0.0 | 0.0 | 0.0 | 0.0 | T | 0.0 | 0.0 | 0.0 | T |
| | YEAR OF OCCURRENCE | | | | | | | | | | 1989 | | | | SEP 1989 |
| | MAXIMUM IN 24 HOURS (IN) | 55 | 0.0 | 0.0 | 0.0 | 0.0 | 0.0 | 0.0 | 0.0 | 0.0 | T | 0.0 | 0.0 | 0.0 | T |
| | YEAR OF OCCURRENCE' | | | | | | | | | | 1989 | | | | SEP 1989 |
| | MAXIMUM SNOW DEPTH (IN) | 53 | 0 | 0 | 0 | 0 | 0 | 0 | 0 | 0 | 0 | 0 | 0 | 0 | 0 |
| | YEAR OF OCCURRENCE | | | | | | | | | | | | | | |
| | NORMAL NO. DAYS WITH: | | | | | | | | | | | | | | |
| | SNOWFALL >= 1.0 | 30 | 0.0 | 0.0 | 0.0 | 0.0 | 0.0 | 0.0 | 0.0 | 0.0 | 0.0 | 0.0 | 0.0 | 0.0 | 0.0 |

## PRECIPITATION (inches) 2009 SAN JUAN (TJSJ)

| YEAR | JAN | FEB | MAR | APR | MAY | JUN | JUL | AUG | SEP | OCT | NOV | DEC | ANNUAL |
|---|---|---|---|---|---|---|---|---|---|---|---|---|---|
| 1980 | 1.75 | 1.67 | 1.47 | 2.55 | 5.19 | 1.31 | 2.19 | 3.17 | 4.85 | 6.71 | 1.91 | 3.18 | 35.95 |
| 1981 | 2.55 | 2.72 | 4.39 | 2.89 | 11.02 | 5.48 | 7.04 | 3.32 | 2.98 | 9.32 | 4.94 | 16.81 | 73.46 |
| 1982 | 2.53 | 6.69 | 0.98 | 1.01 | 10.26 | 5.24 | 2.33 | 1.93 | 2.87 | 2.06 | 4.34 | 4.76 | 45.00 |
| 1983 | 0.69 | 0.20 | 1.47 | 8.54 | 3.85 | 1.91 | 6.53 | 5.15 | 2.75 | 4.06 | 3.25 | 3.50 | 41.90 |
| 1984 | 1.96 | 3.13 | 0.82 | 0.28 | 3.75 | 6.85 | 2.66 | 6.04 | 3.16 | 5.10 | 5.65 | 4.69 | 44.09 |
| 1985 | 2.80 | 2.40 | 1.84 | 1.02 | 5.95 | 0.29 | 2.85 | 4.33 | 5.44 | 11.10 | 4.54 | 2.80 | 45.36 |
| 1986 | 2.18 | 1.13 | 1.61 | 8.93 | 12.80 | 1.52 | 1.94 | 5.19 | 1.98 | 8.54 | 5.87 | 3.59 | 55.28 |
| 1987 | 2.16 | 1.20 | 5.17 | 8.88 | 12.17 | 7.07 | 3.26 | 2.48 | 1.73 | 2.70 | 7.49 | 7.69 | 62.00 |
| 1988 | 3.83 | 2.27 | 1.76 | 10.37 | 6.06 | 1.45 | 4.02 | 11.31 | 5.49 | 4.12 | 5.68 | 4.07 | 60.43 |
| 1989 | 2.96 | 6.05 | 3.39 | 2.63 | 4.88 | 2.97 | 5.54 | 7.88 | 14.83 | 2.09 | 4.95 | 2.50 | 60.67 |
| 1990 | 4.56 | 3.02 | 3.14 | 1.05 | 2.44 | 4.32 | 5.76 | 3.42 | 2.23 | 8.65 | 5.33 | 5.03 | 48.95 |
| 1991 | 2.57 | 2.26 | 1.99 | 1.76 | 3.23 | 2.77 | 3.30 | 1.94 | 5.00 | 1.84 | 6.16 | 2.71 | 35.53 |
| 1992 | 4.03 | 1.19 | 1.47 | 2.12 | 8.76 | 5.55 | 4.38 | 4.00 | 5.35 | 1.74 | 11.98 | 4.72 | 55.29 |
| 1993 | 2.35 | 0.51 | 0.78 | 6.55 | 4.48 | 5.46 | 7.34 | 3.01 | 4.36 | 2.78 | 4.34 | 3.00 | 44.96 |
| 1994 | 3.39 | 1.69 | 1.38 | 2.75 | 1.69 | 4.49 | 3.58 | 1.83 | 5.30 | 3.52 | 8.32 | 3.04 | 40.98 |
| 1995 | 3.69 | 3.57 | 1.29 | 2.98 | 9.47 | 4.66 | 4.93 | 5.15 | 7.17 | 6.11 | 3.09 | 3.84 | 55.95 |
| 1996 | 5.74 | 2.44 | 2.06 | 5.04 | 2.39 | 6.02 | 7.64 | 6.94 | 15.15 | 2.55 | 7.98 | 3.29 | 67.24 |
| 1997 | 4.01 | 4.58 | 1.83 | 0.08 | 3.38 | 0.98 | 4.75 | 6.86 | 3.56 | | 6.82 | 1.02 | |
| 1998 | 7.29 | 3.72 | 3.48 | 4.14 | 3.67 | 3.25 | 7.18 | 7.80 | 10.47 | 6.82 | 6.74 | 7.99 | 72.55 |
| 1999 | 3.24 | 2.74 | 0.75 | 2.53 | 3.99 | 5.53 | 5.81 | 7.34 | 5.82 | 7.38 | 9.84 | 6.15 | 61.12 |
| 2000 | 2.37 | 1.01 | 0.78 | 1.20 | 4.72 | 2.97 | 3.67 | 9.52 | 2.96 | 3.73 | 3.76 | 3.08 | 39.77 |
| 2001 | 2.91 | 3.27 | 1.59 | 2.27 | 5.90 | 2.01 | 3.57 | 6.44 | 5.05 | 5.00 | 6.25 | 11.79 | 56.05 |
| 2002 | 3.33 | 0.96 | 1.13 | 4.85 | 3.24 | 1.82 | 5.16 | 6.70 | 7.45 | 4.81 | 2.65 | 4.59 | 46.69 |
| 2003 | 4.59 | 2.22 | 1.48 | 9.37 | 1.67 | 3.43 | 4.70 | 5.65 | 4.50 | 4.83 | 11.26 | 5.25 | 58.95 |
| 2004 | 1.96 | 2.60 | 3.17 | 3.07 | 9.38 | 7.02 | 6.78 | 3.49 | 9.81 | 6.59 | 5.82 | 4.66 | 64.35 |
| 2005 | 5.31 | 0.87 | T | 15.00 | 8.58 | 5.13 | 8.51 | 7.14 | 3.85 | 11.94 | 7.31 | 3.64 | 77.28 |
| 2006 | 6.49 | 1.74 | 1.57 | 10.06 | 4.86 | 4.74 | 8.92 | 4.26 | 1.78 | 11.25 | 5.71 | 4.50 | 65.88 |
| 2007 | 2.55 | 1.10 | 2.23 | 9.77 | 1.52 | 4.04 | 4.58 | 3.44 | 6.71 | 4.96 | 6.00 | 7.95 | 54.85 |
| 2008 | 6.96 | 1.98 | 0.58 | 4.84 | 5.00 | 5.52 | 1.75 | 3.99 | 10.02 | 5.30 | 4.18 | 4.58 | 54.70 |
| 2009 | 2.49 | 3.32 | 3.21 | 2.80 | 5.28 | 10.37 | 4.80 | 8.57 | 6.43 | 3.62 | 12.14 | 1.83 | 64.86 |
| POR= 54 YRS | 3.30 | 2.23 | 2.16 | 4.16 | 5.72 | 4.44 | 4.79 | 5.52 | 5.70 | 5.60 | 6.11 | 4.75 | 54.48 |

WBAN : 11641

## AVERAGE TEMPERATURE (°F) 2009 SAN JUAN (TJSJ)

| YEAR | JAN | FEB | MAR | APR | MAY | JUN | JUL | AUG | SEP | OCT | NOV | DEC | ANNUAL |
|---|---|---|---|---|---|---|---|---|---|---|---|---|---|
| 1980 | 78.2 | 78.6 | 79.2 | 80.8 | 83.8 | 85.1 | 85.2 | 85.1 | 84.5 | 84.4 | 82.2 | 80.7 | 82.3 |
| 1981 | 79.8 | 79.4 | 80.9 | 80.3 | 83.4 | 84.2 | 85.0 | 83.8 | 84.3 | 83.3 | 81.6 | 78.7 | 82.1 |
| 1982 | 78.5 | 77.8 | 78.0 | 80.4 | 81.0 | 82.9 | 83.3 | 84.6 | 84.6 | 83.8 | 80.2 | 78.0 | 81.1 |
| 1983 | 78.5 | 79.9 | 82.2 | 81.8 | 83.2 | 85.4 | 84.2 | 84.0 | 84.3 | 83.6 | 81.4 | 79.9 | 82.4 |
| 1984 | 78.1 | 77.8 | 80.2 | 81.8 | 81.1 | 82.4 | 82.6 | 82.6 | 82.2 | 81.3 | 78.7 | 77.3 | 80.5 |
| 1985 | 76.1 | 77.3 | 76.7 | 78.3 | 80.4 | 83.0 | 83.5 | 83.4 | 81.7 | 79.6 | 79.0 | 76.8 | 79.7 |
| 1986 | 75.5 | 76.0 | 77.4 | 79.0 | 79.1 | 81.6 | 82.0 | 82.2 | 82.5 | 81.4 | 79.1 | 77.7 | 79.5 |
| 1987 | 76.7 | 77.4 | 78.0 | 81.2 | 81.6 | 81.7 | 82.7 | 83.7 | 83.8 | 83.3 | 80.6 | 79.7 | 80.9 |
| 1988 | 76.9 | 76.6 | 77.8 | 80.4 | 82.4 | 84.3 | 83.6 | 82.8 | 82.2 | 81.6 | 79.8 | 76.9 | 80.4 |
| 1989 | 76.2 | 76.0 | 76.1 | 78.8 | 80.3 | 81.2 | 81.8 | 82.3 | 82.3 | 81.7 | 80.0 | 79.2 | 79.7 |
| 1990 | 77.1 | 76.4 | 76.5 | 79.1 | 81.8 | 82.4 | 82.6 | 82.9 | 83.4 | 82.1 | 80.7 | 77.5 | 80.2 |
| 1991 | 77.0 | 77.8 | 78.6 | 79.2 | 81.0 | 83.2 | 83.1 | 83.9 | 83.3 | 82.4 | 79.9 | 77.0 | 80.5 |
| 1992 | 77.1 | 78.2 | 78.9 | 80.8 | 81.3 | 83.4 | 82.4 | 82.7 | 82.3 | 84.0 | 80.8 | 79.3 | 80.9 |
| 1993 | 77.3 | 78.5 | 79.3 | 80.3 | 82.1 | 83.3 | 82.8 | 84.2 | 83.0 | 82.7 | 80.9 | 78.9 | 81.1 |
| 1994 | 77.9 | 78.0 | 79.3 | 79.7 | 82.8 | 82.8 | 82.9 | 83.8 | 83.3 | 82.8 | 81.1 | 79.7 | 81.2 |
| 1995 | 78.0 | 78.7 | 77.2 | 80.6 | 81.5 | 83.3 | 84.2 | 84.5 | 84.9 | 82.3 | 80.9 | 80.0 | 81.3 |
| 1996 | 78.2 | 77.9 | 79.1 | 79.1 | 80.3 | 81.3 | 81.6 | 81.8 | 81.8 | 81.6 | 78.8 | 76.7 | 79.9 |
| 1997 | 76.2 | 76.0 | 76.6 | 80.3 | 81.5 | 83.4 | 83.1 | 83.1 | 83.2 | 82.4 | 80.9 | 80.1 | 80.6 |
| 1998 | 78.7 | 78.7 | 78.2 | 79.6 | 81.7 | 83.1 | 82.6 | 83.7 | 82.8 | 82.2 | 80.2 | 78.6 | 80.8 |
| 1999 | 76.9 | 75.4 | 77.7 | 79.5 | 82.4 | 82.1 | 81.5 | 82.7 | 83.1 | 81.5 | 79.8 | 76.9 | 80.0 |
| 2000 | 75.7 | 76.7 | 76.6 | 79.4 | 80.6 | 82.0 | 82.9 | 82.8 | 82.0 | 81.8 | 79.5 | 78.4 | 79.9 |
| 2001 | 77.3 | 76.9 | 79.2 | 80.2 | 82.7 | 82.8 | 82.9 | 83.2 | 82.9 | 82.1 | 79.4 | 78.3 | 80.7 |
| 2002 | 78.1 | 77.6 | 79.1 | 79.3 | 81.1 | 83.1 | 82.9 | 83.1 | 83.0 | 81.9 | 80.8 | 78.6 | 80.7 |
| 2003 | 77.9 | 77.9 | 79.8 | 79.6 | 81.4 | 82.1 | 82.2 | 82.3 | 83.4 | 81.9 | 79.8 | 78.0 | 80.5 |
| 2004 | 76.2 | 76.7 | 77.6 | 78.9 | 79.5 | 82.8 | 82.5 | 83.9 | 83.2 | 82.1 | 79.5 | 78.4 | 80.1 |
| 2005 | 76.1 | 75.3 | 79.7 | 81.5 | 81.9 | 83.8 | 83.8 | 83.6 | 83.9 | 81.8 | 80.4 | 76.6 | 80.7 |
| 2006 | 75.5 | 76.6 | 78.7 | 79.2 | 82.2 | 83.0 | 82.2 | 82.6 | 83.8 | 83.2 | 81.2 | 78.7 | 80.6 |
| 2007 | 77.2 | 79.0 | 79.9 | 80.1 | 83.3 | 84.0 | 83.7 | 83.2 | 82.9 | 82.1 | 79.8 | 77.7 | 81.1 |
| 2008 | 76.2 | 76.6 | 77.8 | 78.9 | 81.7 | 82.8 | 83.4 | 84.1 | 82.6 | 81.6 | 80.0 | 78.0 | 80.3 |
| 2009 | 78.4 | 77.7 | 77.6 | 79.6 | 80.6 | 83.3 | 84.8 | 84.4 | 84.2 | 84.5 | 81.3 | 80.3 | 81.4 |
| POR= 54 YRS | 76.9 | 77.0 | 78.0 | 79.4 | 80.9 | 82.4 | 82.6 | 82.8 | 82.6 | 81.9 | 79.9 | 78.1 | 80.2 |

## HEATING DEGREE DAYS (base 65°F) 2009  SAN JUAN (TJSJ)

| YEAR | JUL | AUG | SEP | OCT | NOV | DEC | JAN | FEB | MAR | APR | MAY | JUN | TOTAL |
|------|-----|-----|-----|-----|-----|-----|-----|-----|-----|-----|-----|-----|-------|
| 1983-84 | 0 | 0 | 0 | 0 | 0 | 0 | 0 | 0 | 0 | 0 | 0 | 0 | 0 |
| 1984-85 | 0 | 0 | 0 | 0 | 0 | 0 | 0 | 0 | 0 | 0 | 0 | 0 | 0 |
| 1985-86 | 0 | 0 | 0 | 0 | 0 | 0 | 0 | 0 | 0 | 0 | 0 | 0 | 0 |
| 1986-87 | 0 | 0 | 0 | 0 | 0 | 0 | 0 | 0 | 0 | 0 | 0 | 0 | 0 |
| 1987-88 | 0 | 0 | 0 | 0 | 0 | 0 | 0 | 0 | 0 | 0 | 0 | 0 | 0 |
| 1988-89 | 0 | 0 | 0 | 0 | 0 | 0 | 0 | 0 | 0 | 0 | 0 | 0 | 0 |
| 1989-90 | 0 | 0 | 0 | 0 | 0 | 0 | 0 | 0 | 0 | 0 | 0 | 0 | 0 |
| 1990-91 | 0 | 0 | 0 | 0 | 0 | 0 | 0 | 0 | 0 | 0 | 0 | 0 | 0 |
| 1991-92 | 0 | 0 | 0 | 0 | 0 | 0 | 0 | 0 | 0 | 0 | 0 | 0 | 0 |
| 1992-93 | 0 | 0 | 0 | 0 | 0 | 0 | 0 | 0 | 0 | 0 | 0 | 0 | 0 |
| 1993-94 | 0 | 0 | 0 | 0 | 0 | 0 | 0 | 0 | 0 | 0 | 0 | 0 | 0 |
| 1994-95 | 0 | 0 | 0 | 0 | 0 | 0 | 0 | 0 | 0 | 0 | 0 | 0 | 0 |
| 1995-96 | 0 | 0 | 0 | 0 | 0 | 0 | 0 | 0 | 0 | 0 | 0 | 0 | 0 |
| 1996-97 | 0 | 0 | 0 | 0 | 0 | 0 | 0 | 0 | 0 | 0 | 0 | 0 | 0 |
| 1997-98 | 0 | 0 | 0 | 0 | 0 | 0 | 0 | 0 | 0 | 0 | 0 | 0 | 0 |
| 1998-99 | 0 | 0 | 0 | 0 | 0 | 0 | 0 | 0 | 0 | 0 | 0 | 0 | 0 |
| 1999-00 | 0 | 0 | 0 | 0 | 0 | 0 | 0 | 0 | 0 | 0 | 0 | 0 | 0 |
| 2000-01 | 0 | 0 | 0 | 0 | 0 | 0 | 0 | 0 | 0 | 0 | 0 | 0 | 0 |
| 2001-02 | 0 | 0 | 0 | 0 | 0 | 0 | 0 | 0 | 0 | 0 | 0 | 0 | 0 |
| 2002-03 | 0 | 0 | 0 |   | 0 | 0 | 0 | 0 | 0 | 0 | 0 | 0 |   |
| 2003-04 | 0 | 0 | 0 | 0 | 0 | 0 | 0 | 0 | 0 | 0 | 0 | 0 | 0 |
| 2004-05 | 0 | 0 | 0 | 0 | 0 | 0 | 0 | 0 | 0 | 0 | 0 | 0 | 0 |
| 2005-06 | 0 | 0 | 0 | 0 | 0 | 0 | 0 | 0 | 0 | 0 | 0 | 0 | 0 |
| 2006-07 | 0 | 0 | 0 | 0 | 0 | 0 | 0 | 0 | 0 | 0 | 0 | 0 | 0 |
| 2007-08 | 0 | 0 | 0 | 0 | 0 | 0 | 0 | 0 | 0 | 0 | 0 | 0 | 0 |
| 2008-09 | 0 | 0 | 0 | 0 | 0 | 0 | 0 | 0 | 0 | 0 | 0 | 0 | 0 |
| 2009- | 0 | 0 | 0 | 0 | 0 | 0 |   |   |   |   |   |   |   |

WBAN : 11641

## COOLING DEGREE DAYS (base 65°F) 2009  SAN JUAN (TJSJ)

| YEAR | JAN | FEB | MAR | APR | MAY | JUN | JUL | AUG | SEP | OCT | NOV | DEC | TOTAL |
|------|-----|-----|-----|-----|-----|-----|-----|-----|-----|-----|-----|-----|-------|
| 1980 | 414 | 402 | 446 | 479 | 590 | 610 | 633 | 628 | 591 | 608 | 522 | 494 | 6417 |
| 1981 | 467 | 407 | 499 | 468 | 578 | 584 | 626 | 587 | 588 | 574 | 505 | 429 | 6312 |
| 1982 | 423 | 364 | 412 | 470 | 504 | 543 | 573 | 615 | 593 | 590 | 466 | 405 | 5958 |
| 1983 | 426 | 424 | 541 | 509 | 573 | 621 | 604 | 593 | 583 | 582 | 502 | 468 | 6426 |
| 1984 | 415 | 377 | 479 | 508 | 503 | 528 | 553 | 553 | 522 | 517 | 417 | 389 | 5761 |
| 1985 | 352 | 349 | 368 | 403 | 483 | 547 | 579 | 577 | 508 | 459 | 429 | 371 | 5425 |
| 1986 | 329 | 315 | 390 | 428 | 444 | 501 | 535 | 538 | 534 | 517 | 429 | 401 | 5361 |
| 1987 | 366 | 354 | 407 | 490 | 521 | 508 | 556 | 588 | 571 | 575 | 477 | 464 | 5877 |
| 1988 | 377 | 341 | 403 | 470 | 547 | 587 | 581 | 558 | 523 | 522 | 453 | 375 | 5737 |
| 1989 | 354 | 313 | 353 | 421 | 482 | 495 | 529 | 543 | 525 | 526 | 456 | 443 | 5440 |
| 1990 | 383 | 325 | 362 | 430 | 526 | 526 | 551 | 561 | 559 | 539 | 478 | 392 | 5632 |
| 1991 | 379 | 363 | 428 | 430 | 503 | 551 | 571 | 593 | 558 | 547 | 455 | 380 | 5758 |
| 1992 | 383 | 393 | 435 | 479 | 512 | 556 | 547 | 555 | 524 | 596 | 482 | 454 | 5916 |
| 1993 | 391 | 383 | 450 | 469 | 535 | 557 | 561 | 603 | 550 | 557 | 483 | 440 | 5979 |
| 1994 | 404 | 370 | 449 | 448 | 559 | 541 | 563 | 592 | 556 | 562 | 493 | 461 | 5998 |
| 1995 | 412 | 389 | 387 | 475 | 518 | 557 | 602 | 611 | 603 | 545 | 486 | 472 | 6057 |
| 1996 | 414 | 381 | 444 | 430 | 480 | 495 | 519 | 529 | 510 | 524 | 421 | 372 | 5519 |
| 1997 | 353 | 316 | 369 | 467 | 518 | 561 | 569 | 567 | 555 | 545 | 485 | 473 | 5778 |
| 1998 | 430 | 391 | 415 | 446 | 523 | 550 | 555 | 589 | 538 | 541 | 462 | 428 | 5868 |
| 1999 | 374 | 297 | 402 | 443 | 547 | 515 | 519 | 558 | 549 | 516 | 449 | 379 | 5548 |
| 2000 | 339 | 345 | 366 | 443 | 492 | 517 | 560 | 557 | 516 | 530 | 443 | 424 | 5532 |
| 2001 | 388 | 342 | 448 | 464 | 555 | 540 | 563 | 569 | 544 | 535 | 437 | 421 | 5806 |
| 2002 | 410 | 359 | 445 | 435 | 503 | 548 | 562 | 566 | 549 | 530 | 483 | 429 | 5819 |
| 2003 | 405 | 370 | 465 | 446 | 517 | 517 | 542 | 543 | 560 | 532 | 453 | 410 | 5760 |
| 2004 | 351 | 348 | 397 | 421 | 457 | 541 | 553 | 593 | 558 | 538 | 442 | 423 | 5622 |
| 2005 | 353 | 295 | 463 | 503 | 529 | 571 | 590 | 583 | 571 | 529 | 468 | 366 | 5821 |
| 2006 | 332 | 329 | 430 | 433 | 541 | 549 | 541 | 548 | 571 | 568 | 492 | 436 | 5770 |
| 2007 | 385 | 397 | 473 | 461 | 576 | 574 | 582 | 573 | 544 | 542 | 451 | 403 | 5961 |
| 2008 | 355 | 343 | 402 | 423 | 523 | 542 | 578 | 603 | 532 | 520 | 456 | 408 | 5685 |
| 2009 | 420 | 361 | 398 | 448 | 493 | 556 | 620 | 610 | 583 | 610 | 492 | 479 | 6070 |

### SNOWFALL (inches) 2009 SAN JUAN (TJSJ)

| YEAR | JUL | AUG | SEP | OCT | NOV | DEC | JAN | FEB | MAR | APR | MAY | JUN | TOTAL |
|------|-----|-----|-----|-----|-----|-----|-----|-----|-----|-----|-----|-----|-------|
| 1980-81 | 0.0 | 0.0 | 0.0 | 0.0 | 0.0 | 0.0 | 0.0 | 0.0 | 0.0 | 0.0 | 0.0 | 0.0 | 0.0 |
| 1981-82 | 0.0 | 0.0 | 0.0 | 0.0 | 0.0 | 0.0 | 0.0 | 0.0 | 0.0 | 0.0 | 0.0 | 0.0 | 0.0 |
| 1982-83 | 0.0 | 0.0 | 0.0 | 0.0 | 0.0 | 0.0 | 0.0 | 0.0 | 0.0 | 0.0 | 0.0 | 0.0 | 0.0 |
| 1983-84 | 0.0 | 0.0 | 0.0 | 0.0 | 0.0 | 0.0 | 0.0 | 0.0 | 0.0 | 0.0 | 0.0 | 0.0 | 0.0 |
| 1984-85 | 0.0 | 0.0 | 0.0 | 0.0 | 0.0 | 0.0 | 0.0 | 0.0 | 0.0 | 0.0 | 0.0 | 0.0 | 0.0 |
| 1985-86 | 0.0 | 0.0 | 0.0 | 0.0 | 0.0 | 0.0 | 0.0 | 0.0 | 0.0 | 0.0 | 0.0 | 0.0 | 0.0 |
| 1986-87 | 0.0 | 0.0 | 0.0 | 0.0 | 0.0 | 0.0 | 0.0 | 0.0 | 0.0 | 0.0 | 0.0 | 0.0 | 0.0 |
| 1987-88 | 0.0 | 0.0 | 0.0 | 0.0 | 0.0 | 0.0 | 0.0 | 0.0 | 0.0 | 0.0 | 0.0 | 0.0 | 0.0 |
| 1988-89 | 0.0 | 0.0 | 0.0 | 0.0 | 0.0 | 0.0 | 0.0 | 0.0 | 0.0 | 0.0 | 0.0 | 0.0 | 0.0 |
| 1989-90 | 0.0 | 0.0 | T | 0.0 | 0.0 | 0.0 | 0.0 | 0.0 | 0.0 | 0.0 | 0.0 | 0.0 | T |
| 1990-91 | 0.0 | 0.0 | 0.0 | 0.0 | 0.0 | 0.0 | 0.0 | 0.0 | 0.0 | 0.0 | 0.0 | 0.0 | 0.0 |
| 1991-92 | 0.0 | 0.0 | 0.0 | 0.0 | 0.0 | 0.0 | 0.0 | 0.0 | 0.0 | 0.0 | 0.0 | 0.0 | 0.0 |
| 1992-93 | 0.0 | 0.0 | 0.0 | 0.0 | 0.0 | 0.0 | 0.0 | 0.0 | 0.0 | 0.0 | 0.0 | 0.0 | 0.0 |
| 1993-94 | 0.0 | 0.0 | 0.0 | 0.0 | 0.0 | 0.0 | 0.0 | 0.0 | 0.0 | 0.0 | 0.0 | 0.0 | 0.0 |
| 1994-95 | 0.0 | 0.0 | 0.0 | 0.0 | 0.0 | 0.0 | 0.0 | 0.0 | 0.0 | 0.0 | 0.0 | 0.0 | 0.0 |
| 1995-96 | 0.0 | 0.0 | 0.0 | 0.0 | 0.0 | 0.0 | 0.0 | 0.0 | 0.0 | 0.0 | 0.0 | 0.0 | 0.0 |
| 1996-97 | 0.0 | 0.0 | 0.0 | 0.0 | 0.0 | 0.0 | 0.0 | 0.0 | 0.0 | 0.0 | 0.0 | 0.0 | 0.0 |
| 1997-98 | 0.0 | 0.0 | 0.0 | 0.0 | 0.0 | 0.0 | 0.0 | 0.0 | 0.0 | 0.0 | 0.0 | 0.0 | 0.0 |
| 1998-99 | 0.0 | 0.0 | 0.0 | 0.0 | 0.0 | 0.0 | 0.0 | 0.0 | 0.0 | 0.0 | 0.0 | 0.0 | 0.0 |
| 1999-00 | 0.0 | 0.0 | 0.0 | 0.0 | 0.0 | 0.0 | 0.0 | 0.0 | 0.0 | 0.0 | 0.0 | 0.0 | 0.0 |
| 2000-01 | 0.0 | 0.0 | 0.0 | 0.0 | 0.0 | 0.0 | 0.0 | 0.0 | 0.0 | 0.0 | 0.0 | 0.0 | 0.0 |
| 2001-02 | 0.0 | 0.0 | 0.0 | 0.0 | 0.0 | 0.0 | 0.0 | 0.0 | 0.0 | 0.0 | 0.0 | 0.0 | 0.0 |
| 2002-03 | 0.0 | 0.0 | 0.0 | 0.0 | 0.0 | 0.0 | 0.0 | 0.0 | 0.0 | 0.0 | 0.0 | 0.0 | 0.0 |
| 2003-04 | 0.0 | 0.0 | 0.0 | 0.0 | 0.0 | 0.0 | 0.0 | 0.0 | 0.0 | 0.0 | 0.0 | 0.0 | 0.0 |
| 2004-05 | 0.0 | 0.0 | 0.0 | 0.0 | 0.0 | 0.0 | 0.0 | 0.0 | 0.0 | 0.0 | 0.0 | 0.0 | 0.0 |
| 2005-06 | 0.0 | 0.0 | 0.0 | 0.0 | 0.0 | 0.0 | 0.0 | 0.0 | 0.0 | 0.0 | 0.0 | 0.0 | 0.0 |
| 2006-07 | 0.0 | 0.0 | 0.0 | 0.0 | 0.0 | 0.0 | 0.0 | 0.0 | 0.0 | 0.0 | 0.0 | 0.0 | 0.0 |
| 2007-08 | 0.0 | 0.0 | 0.0 | 0.0 | 0.0 | 0.0 | 0.0 | 0.0 | 0.0 | 0.0 | 0.0 | 0.0 | 0.0 |
| 2008-09 | 0.0 | 0.0 | 0.0 | 0.0 | 0.0 | 0.0 | 0.0 | 0.0 | 0.0 | 0.0 | 0.0 | 0.0 | 0.0 |
| 2009- | 0.0 | 0.0 | 0.0 | 0.0 | 0.0 | 0.0 | | | | | | | |
| POR= 54 YRS | 0.0 | 0.0 | T | 0.0 | 0.0 | 0.0 | 0.0 | 0.0 | 0.0 | 0.0 | 0.0 | 0.0 | T |

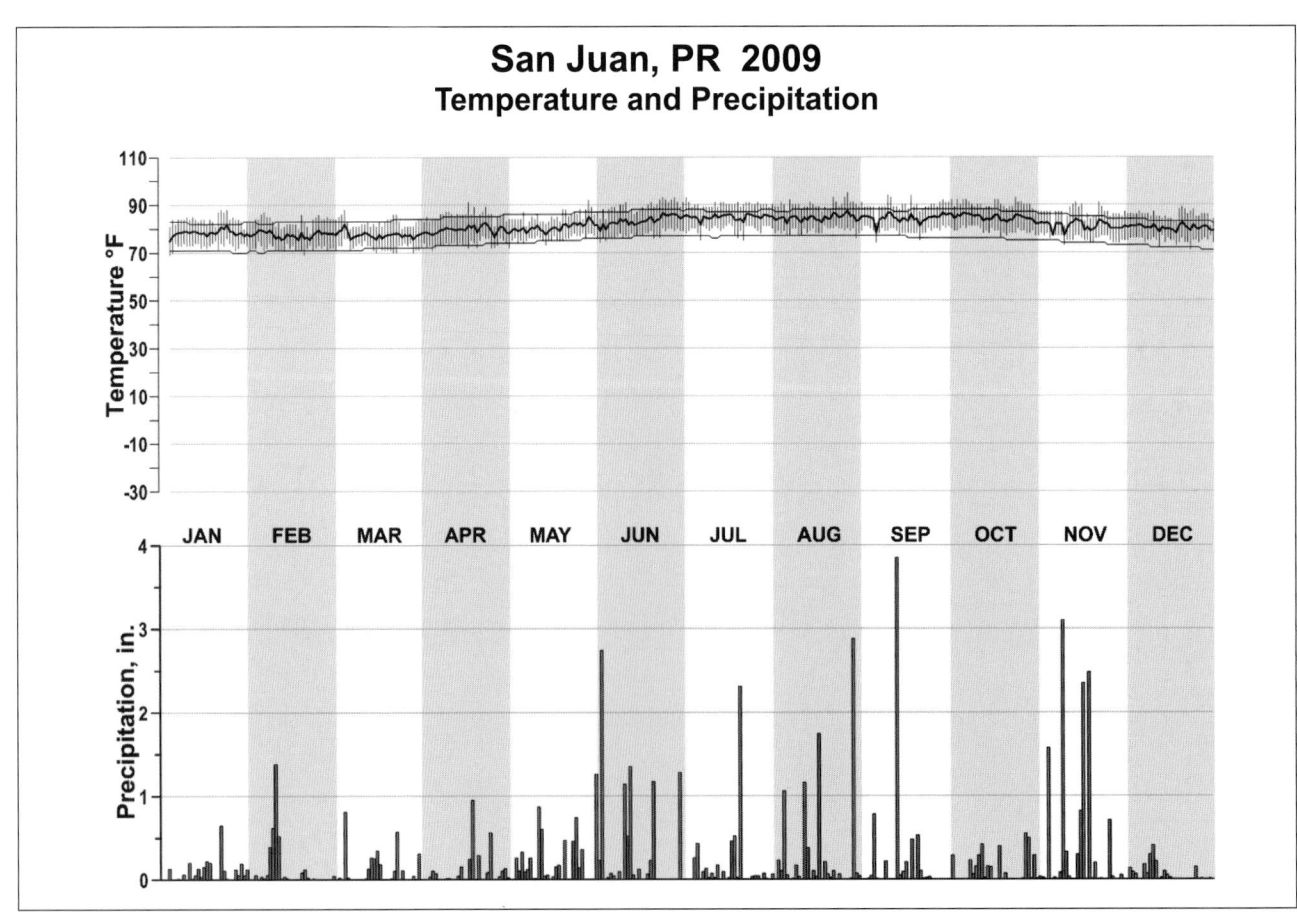

**San Juan, PR 2009**
**Temperature and Precipitation**

# 2009
# PROVIDENCE
# RHODE ISLAND (KPVD)

The proximity to Narragansett Bay and the Atlantic Ocean plays an important part in determining the climate for Providence and vicinity. In winter, the temperatures are modified considerably, and many major snowstorms change to rain before reaching the area. In summer, many days that could be uncomfortably warm are cooled by refreshing sea breezes. At other times of the year, sea fog may be advected in over land by onshore winds. In fact, most cases of dense fog are produced this way, but the number of such days is few, averaging two or three days per month. In early fall, severe coastal storms of tropical origin sometimes bring destructive winds to this area. Even at other times of the year, it is usually coastal storms which produce the severest weather.

The temperature for the entire year averages around 50 degrees with 70 degree temperatures common from near the end of May to the latter part of September. During this period, there may be several days reaching 90 degrees or more. Temperatures of 100 degrees and more are rare.

Freezing temperatures occur on the average about 125 days per year. They become a common daily occurrence in the latter part of November, and become less frequent near the end of March. The average date for the last freeze in spring is mid-April, while the average date for the first freeze in fall is late October, making the growing season about 195 days in length. Sub-zero weather in winter seldom occurs, averaging less than one day for December and one or two days each for January and February.

Measurable precipitation occurs on about one day out of every three, and is fairly evenly distributed throughout the year. There is usually no definite dry season, but occasionally droughts do occur.

Thunderstorms are responsible for much of the rainfall from May through August. They usually produce heavy, and sometimes even excessive amounts of rainfall. However, since their duration is relatively short, damage is ordinarily light. The thunderstorms of summer are frequently accompanied by extremely gusty winds, which may result in some damage to property.

The first measurable snowfall of winter usually comes toward the end of November, and the last in spring is about the middle of March. Winters with over 50 inches of snow are not common. The area normally receives less than 25 inches. The month of greatest snowfall is usually February, but January and March are close seconds. It is unusual for the ground to remain well covered with snow for any long period of time.

# NORMALS, MEANS, AND EXTREMES
## PROVIDENCE (KPVD)

LATITUDE: 41 ° 43'N  LONGITUDE: -71 ° 25'W  ELEVATION (FT): GRND: 52  BARO: 53  TIME ZONE: EASTERN (UTC -5)  WBAN: 14765

| | ELEMENT | POR | JAN | FEB | MAR | APR | MAY | JUN | JUL | AUG | SEP | OCT | NOV | DEC | YEAR |
|---|---|---|---|---|---|---|---|---|---|---|---|---|---|---|---|
| **TEMPERATURE °F** | NORMAL DAILY MAXIMUM | 30 | 37.1 | 39.3 | 47.7 | 58.1 | 68.5 | 77.3 | 82.6 | 80.9 | 73.4 | 62.9 | 52.4 | 42.1 | 60.2 |
| | MEAN DAILY MAXIMUM | 62 | 37.0 | 38.9 | 46.4 | 57.9 | 67.7 | 76.8 | 82.2 | 80.8 | 73.5 | 63.3 | 52.6 | 41.5 | 59.9 |
| | HIGHEST DAILY MAXIMUM | 56 | 69 | 72 | 85 | 98 | 95 | 97 | 102 | 104 | 100 | 86 | 78 | 77 | 104 |
| | YEAR OF OCCURRENCE | | 2002 | 1985 | 1998 | 1976 | 1996 | 2008 | 1991 | 1975 | 1983 | 1979 | 1993 | 1998 | AUG 1975 |
| | MEAN OF EXTREME MAXS. | 62 | 56.4 | 56.2 | 66.0 | 78.8 | 85.5 | 91.2 | 93.6 | 91.7 | 87.0 | 79.1 | 69.5 | 60.2 | 76.3 |
| | NORMAL DAILY MINIMUM | 30 | 20.3 | 22.5 | 30.0 | 39.1 | 48.8 | 57.9 | 64.1 | 62.8 | 54.5 | 43.1 | 35.1 | 25.6 | 42.0 |
| | MEAN DAILY MINIMUM | 62 | 20.7 | 22.2 | 29.4 | 38.9 | 48.0 | 57.4 | 63.8 | 62.5 | 54.4 | 43.5 | 35.4 | 25.4 | 41.8 |
| | LOWEST DAILY MINIMUM | 56 | -13 | -7 | 1 | 14 | 29 | 41 | 0 | 40 | 33 | 20 | 6 | -10 | -13 |
| | YEAR OF OCCURRENCE | | 1976 | 1979 | 1967 | 1954 | 1956 | 1980 | 1996 | 1965 | 1980 | 1976 | 1989 | 1980 | JAN 1976 |
| | MEAN OF EXTREME MINS. | 62 | 2.5 | 5.0 | 14.0 | 27.4 | 36.6 | 46.2 | 53.9 | 50.9 | 39.9 | 29.7 | 21.0 | 8.0 | 27.9 |
| | NORMAL DRY BULB | 30 | 28.7 | 30.9 | 38.8 | 48.6 | 58.7 | 67.6 | 73.3 | 71.9 | 64.0 | 53.0 | 43.8 | 33.8 | 51.1 |
| | MEAN DRY BULB | 62 | 28.9 | 30.6 | 37.9 | 48.4 | 57.9 | 67.3 | 73.0 | 71.7 | 64.0 | 53.4 | 44.0 | 33.4 | 50.9 |
| | MEAN WET BULB | 26 | 25.4 | 26.6 | 32.3 | 41.3 | 51.0 | 60.8 | 65.9 | 65.3 | 58.7 | 48.3 | 39.5 | 30.0 | 45.4 |
| | MEAN DEW POINT | 26 | 20.1 | 21.0 | 27.1 | 36.3 | 47.2 | 57.6 | 63.2 | 62.9 | 56.1 | 44.7 | 35.0 | 25.1 | 41.4 |
| | NORMAL NO. DAYS WITH: | | | | | | | | | | | | | | |
| | MAXIMUM >= 90 | 30 | 0.0 | 0.0 | 0.0 | 0.1 | 0.7 | 2.0 | 4.0 | 2.7 | 0.7 | 0.0 | 0.0 | 0.0 | 10.2 |
| | MAXIMUM <= 32 | 30 | 10.4 | 7.3 | 1.3 | * | 0.0 | 0.0 | 0.0 | 0.0 | 0.0 | 0.0 | 0.2 | 5.1 | 24.3 |
| | MINIMUM <= 32 | 30 | 27.3 | 23.6 | 18.8 | 4.6 | 0.1 | 0.0 | 0.0 | 0.0 | 0.0 | 2.8 | 12.6 | 24.1 | 113.9 |
| | MINIMUM <= 0 | 30 | 1.0 | 0.5 | 0.0 | 0.0 | 0.0 | 0.0 | 0.0 | 0.0 | 0.0 | 0.0 | 0.0 | 0.2 | 1.7 |
| **H/C** | NORMAL HEATING DEG. DAYS | 30 | 1125 | 965 | 817 | 494 | 221 | 44 | 3 | 9 | 101 | 377 | 637 | 961 | 5754 |
| | NORMAL COOLING DEG. DAYS | 30 | 0 | 0 | 0 | 3 | 25 | 122 | 265 | 223 | 71 | 5 | 0 | 0 | 714 |
| **RH** | NORMAL (PERCENT) | 30 | 65 | 63 | 63 | 63 | 69 | 71 | 72 | 73 | 74 | 71 | 69 | 67 | 68 |
| | HOUR 01 LST | 30 | 70 | 69 | 70 | 71 | 79 | 83 | 84 | 86 | 85 | 81 | 75 | 72 | 77 |
| | HOUR 07 LST | 30 | 72 | 71 | 72 | 70 | 74 | 76 | 77 | 81 | 83 | 81 | 78 | 74 | 76 |
| | HOUR 13 LST | 30 | 57 | 54 | 52 | 49 | 54 | 57 | 56 | 57 | 57 | 54 | 56 | 57 | 55 |
| | HOUR 19 LST | 30 | 65 | 62 | 62 | 61 | 65 | 68 | 68 | 73 | 75 | 72 | 69 | 67 | 67 |
| **S** | PERCENT POSSIBLE SUNSHINE | 42 | 56 | 58 | 58 | 57 | 58 | 61 | 63 | 62 | 62 | 61 | 50 | 52 | 58 |
| **W/O** | MEAN NO. DAYS WITH: | | | | | | | | | | | | | | |
| | HEAVY FOG(VISBY <= 1/4 MI) | 46 | 2.0 | 2.0 | 2.3 | 2.0 | 1.9 | 2.2 | 1.5 | 1.3 | 1.4 | 2.6 | 1.7 | 2.2 | 23.1 |
| | THUNDERSTORMS | 62 | 0.2 | 0.3 | 0.7 | 1.4 | 2.6 | 3.7 | 4.4 | 3.7 | 1.7 | 1.0 | 0.7 | 0.2 | 20.6 |
| **CLOUDNESS** | MEAN: | | | | | | | | | | | | | | |
| | SUNRISE-SUNSET (OKTAS) | 42 | 5.0 | 5.0 | 5.3 | 5.3 | 5.4 | 5.1 | 5.0 | 4.8 | 4.6 | 4.3 | 5.0 | 4.9 | 5.0 |
| | MIDNIGHT-MIDNIGHT (OKTAS) | 31 | 4.6 | 4.5 | 4.7 | 4.8 | 5.0 | 4.9 | 4.9 | 4.6 | 4.5 | 4.2 | 4.8 | 4.7 | 4.7 |
| | MEAN NO. DAYS WITH: | | | | | | | | | | | | | | |
| | CLEAR | 42 | 9.5 | 8.0 | 8.4 | 7.3 | 6.5 | 6.6 | 6.7 | 8.4 | 9.4 | 11.0 | 8.3 | 8.3 | 98.4 |
| | PARTLY CLOUDY | 42 | 6.8 | 7.1 | 7.7 | 8.2 | 9.9 | 10.3 | 11.9 | 10.3 | 8.1 | 7.9 | 6.9 | 7.8 | 102.9 |
| | CLOUDY | 42 | 14.7 | 13.2 | 14.9 | 14.5 | 14.6 | 13.1 | 12.4 | 12.3 | 12.4 | 12.1 | 14.9 | 14.9 | 164.0 |
| **PR** | MEAN STATION PRESSURE(IN) | 26 | 29.96 | 29.96 | 29.91 | 29.90 | 29.91 | 29.89 | 29.91 | 29.95 | 30.00 | 30.00 | 29.96 | 29.98 | 29.94 |
| | MEAN SEA-LEVEL PRES. (IN) | 26 | 30.03 | 30.03 | 30.02 | 29.97 | 29.98 | 29.96 | 29.97 | 30.02 | 30.07 | 30.07 | 30.06 | 30.04 | 30.02 |
| **WINDS** | MEAN SPEED (MPH) | 26 | 9.9 | 10.3 | 10.9 | 10.6 | 9.8 | 9.0 | 8.7 | 8.3 | 8.3 | 8.8 | 9.3 | 9.6 | 9.5 |
| | PREVAIL.DIR(TENS OF DEGS) | 37 | 30 | 31 | 31 | 17 | 17 | 19 | 23 | 22 | 23 | 36 | 31 | 31 | 30 |
| | MAXIMUM 2-MINUTE: | | | | | | | | | | | | | | |
| | SPEED (MPH) | 14 | 45 | 43 | 46 | 41 | 38 | 38 | 39 | 35 | 39 | 40 | 44 | 44 | 46 |
| | DIR. (TENS OF DEGS) | | 02 | 21 | 29 | 13 | 22 | 36 | 32 | 34 | 32 | 00 | 36 | 20 | 29 |
| | YEAR OF OCCURRENCE | | 2005 | 2008 | 1997 | 2007 | 2000 | 1997 | 2006 | 2002 | 2002 | 1996 | 2006 | 2009 | MAR 1997 |
| | MAXIMUM 3-SECOND | | | | | | | | | | | | | | |
| | SPEED (MPH) | 14 | 60 | 54 | 59 | 61 | 48 | 51 | 51 | 49 | 47 | 51 | 53 | 56 | 61 |
| | DIR. (TENS OF DEGS) | | 02 | 20 | 28 | 13 | 24 | 36 | 00 | 26 | 19 | 24 | 01 | 20 | 13 |
| | YEAR OF OCCURRENCE | | 2005 | 2000 | 1997 | 2007 | 2000 | 1997 | 1996 | 2007 | 2005 | 2006 | 2006 | 2009 | APR 2007 |
| **PRECIPITATION** | NORMAL (IN) | 30 | 4.37 | 3.45 | 4.43 | 4.16 | 3.66 | 3.38 | 3.17 | 3.90 | 3.70 | 3.69 | 4.40 | 4.14 | 46.45 |
| | MAXIMUM MONTHLY (IN) | 56 | 11.66 | 7.20 | 8.84 | 12.74 | 8.38 | 11.08 | 10.52 | 11.12 | 10.99 | 15.38 | 11.01 | 10.75 | 15.38 |
| | YEAR OF OCCURRENCE | | 1979 | 1984 | 1983 | 1983 | 1984 | 1982 | 2009 | 1955 | 2008 | 2005 | 1983 | 1969 | OCT 2005 |
| | MINIMUM MONTHLY (IN) | 56 | 0.50 | 0.39 | 0.56 | 1.48 | 0.71 | 0.17 | 0.39 | 0.71 | 0.77 | 0.40 | 0.41 | 0.58 | 0.17 |
| | YEAR OF OCCURRENCE | | 1970 | 1987 | 1981 | 1966 | 1964 | 1999 | 2002 | 1984 | 1959 | 1994 | 2001 | 1955 | JUN 1999 |
| | MAXIMUM IN 24 HOURS (IN) | 56 | 3.34 | 3.14 | 4.53 | 4.45 | 5.17 | 5.03 | 4.83 | 6.71 | 4.89 | 6.63 | 4.18 | 3.85 | 6.71 |
| | YEAR OF OCCURRENCE | | 1962 | 1978 | 1968 | 1983 | 1984 | 1984 | 1976 | 1979 | 1961 | 1962 | 1983 | 1969 | AUG 1979 |
| | NORMAL NO. DAYS WITH: | | | | | | | | | | | | | | |
| | PRECIPITATION >= 0.01 | 30 | 11.2 | 9.8 | 12.3 | 11.5 | 11.8 | 10.5 | 8.8 | 9.2 | 8.9 | 8.9 | 10.0 | 12.0 | 124.9 |
| | PRECIPITATION >= 1.00 | 30 | 1.3 | 0.9 | 1.3 | 1.1 | 0.7 | 0.9 | 0.8 | 0.9 | 1.0 | 1.0 | 1.4 | 1.1 | 12.4 |
| **SNOWFALL** | NORMAL (IN) | 30 | 10.6 | 9.0 | 5.2 | 0.7 | 0.3 | 0.0 | 0.0 | 0.0 | 0.0 | 0.1 | 1.4 | 5.6 | 32.9 |
| | MAXIMUM MONTHLY (IN) | 55 | 37.2 | 30.9 | 31.6 | 7.6 | 7.0 | T | 0.0 | 0.0 | 0.0 | 2.5 | 8.0 | 21.7 | 37.2 |
| | YEAR OF OCCURRENCE | | 1996 | 1961 | 1956 | 1982 | 1977 | 2007 | | | | 1979 | 1989 | 2009 | JAN 1996 |
| | MAXIMUM IN 24 HOURS (IN) | 55 | 20.8 | 27.6 | 16.9 | 7.6 | 7.0 | T | 0.0 | 0.0 | 0.0 | 2.5 | 8.0 | 14.3 | 27.6 |
| | YEAR OF OCCURRENCE' | | 1996 | 1978 | 1960 | 1982 | 1977 | 2007 | | | | 1979 | 1989 | 2009 | FEB 1978 |
| | MAXIMUM SNOW DEPTH (IN) | 60 | 20 | 30 | 20 | 10 | 2 | 0 | 0 | 0 | 0 | 1 | 7 | 14 | 30 |
| | YEAR OF OCCURRENCE | | 2005 | 1961 | 1956 | 1970 | 1977 | | | | | 1979 | 1989 | 2009 | FEB 1961 |
| | NORMAL NO. DAYS WITH: | | | | | | | | | | | | | | |
| | SNOWFALL >= 1.0 | 30 | 3.0 | 2.6 | 1.4 | 0.2 | 0.0 | 0.0 | 0.0 | 0.0 | 0.0 | 0.0 | 0.4 | 1.7 | 9.3 |

## PRECIPITATION (inches) 2009 PROVIDENCE (KPVD)

| YEAR | JAN | FEB | MAR | APR | MAY | JUN | JUL | AUG | SEP | OCT | NOV | DEC | ANNUAL |
|------|-----|-----|-----|-----|-----|-----|-----|-----|-----|-----|-----|-----|--------|
| 1980 | 1.40 | 1.16 | 8.11 | 6.18 | 1.78 | 3.85 | 2.03 | 1.99 | 0.90 | 3.41 | 3.73 | 1.57 | 36.11 |
| 1981 | 0.77 | 4.79 | 0.56 | 4.10 | 1.92 | 2.31 | 3.75 | 2.65 | 2.58 | 3.38 | 3.20 | 6.36 | 36.37 |
| 1982 | 6.09 | 3.08 | 3.76 | 3.64 | 1.61 | 11.08 | 3.51 | 3.67 | 3.61 | 3.08 | 4.32 | 1.81 | 49.26 |
| 1983 | 4.32 | 4.81 | 8.84 | 12.74 | 4.67 | 1.91 | 2.14 | 2.71 | 2.16 | 4.50 | 11.01 | 7.71 | 67.52 |
| 1984 | 2.00 | 7.20 | 5.77 | 4.30 | 8.38 | 4.09 | 5.16 | 0.71 | 1.77 | 4.25 | 1.95 | 3.16 | 48.74 |
| 1985 | 1.18 | 1.57 | 3.08 | 1.65 | 4.76 | 4.70 | 2.88 | 8.57 | 1.69 | 1.78 | 7.14 | 1.42 | 40.42 |
| 1986 | 5.88 | 3.18 | 2.86 | 2.10 | 2.29 | 3.27 | 5.95 | 3.29 | 0.97 | 2.48 | 5.77 | 8.09 | 46.13 |
| 1987 | 4.73 | 0.39 | 5.62 | 6.91 | 1.80 | 2.00 | 1.20 | 2.58 | 7.47 | 2.28 | 3.40 | 2.29 | 40.67 |
| 1988 | 2.69 | 5.29 | 4.09 | 3.11 | 2.83 | 0.91 | 5.73 | 0.94 | 2.38 | 1.77 | 7.60 | 1.03 | 38.37 |
| 1989 | 1.17 | 2.69 | 4.13 | 5.30 | 6.07 | 5.84 | 5.59 | 6.14 | 4.75 | 8.37 | 4.35 | 1.66 | 56.06 |
| 1990 | 5.01 | 2.93 | 2.01 | 5.57 | 5.70 | 1.13 | 3.52 | 3.74 | 2.28 | 4.96 | 2.45 | 5.48 | 44.78 |
| 1991 | 3.44 | 2.31 | 6.61 | 4.80 | 3.30 | 0.93 | 2.76 | 5.98 | 5.09 | 2.65 | 4.65 | 3.17 | 45.69 |
| 1992 | 4.82 | 2.10 | 4.04 | 2.34 | 1.42 | 4.61 | 3.59 | 6.06 | 5.09 | 1.53 | 5.05 | 6.83 | 47.48 |
| 1993 | 2.42 | 5.06 | 6.99 | 5.02 | 1.12 | 1.40 | 2.18 | 1.23 | 4.08 | 3.55 | 3.35 | 5.76 | 42.16 |
| 1994 | 5.53 | 2.10 | 7.19 | 2.07 | 2.98 | 2.70 | 1.34 | 6.34 | 4.12 | 0.40 | 5.34 | 4.58 | 44.69 |
| 1995 | 3.67 | 3.14 | 2.03 | 3.34 | 2.83 | 2.89 | 1.17 | 1.80 | 4.06 | 6.37 | 4.76 | 2.18 | 38.24 |
| 1996 | 5.02 | 2.19 | 2.71 | 4.88 | 2.44 | 2.17 | 5.57 | 2.19 | 5.72 | 6.20 | 2.38 | 6.59 | 48.06 |
| 1997 | 4.27 | 1.89 | 4.68 | 3.25 | 2.68 | 2.23 | 0.96 | 6.32 | 0.99 | 1.80 | 6.06 | 2.84 | 37.97 |
| 1998 | 6.55 | 5.85 | 5.86 | 4.91 | 6.05 | 9.61 | 1.37 | 2.39 | 2.30 | 3.78 | 2.76 | 1.27 | 52.70 |
| 1999 | 6.70 | 5.45 | 3.33 | 1.54 | 4.25 | 0.17 | 0.82 | 3.25 | 7.00 | 4.51 | 2.85 | 2.39 | 42.26 |
| 2000 | 4.19 | 2.74 | 5.37 | 5.06 | 3.72 | 4.78 | 3.64 | 2.41 | 3.79 | 1.31 | 4.73 | 4.26 | 46.00 |
| 2001 | 2.40 | 1.96 | 8.78 | 2.04 | 3.96 | 6.72 | 1.92 | 4.50 | 4.40 | 0.64 | 0.41 | 2.46 | 40.19 |
| 2002 | 2.76 | 1.72 | 4.84 | 3.08 | 4.97 | 3.32 | 0.39 | 1.91 | 5.26 | 3.49 | 5.66 | 4.94 | 42.34 |
| 2003 | 2.04 | 3.75 | 5.18 | 4.35 | 3.13 | 5.51 | 3.62 | 5.61 | 3.38 | 5.51 | 1.76 | 6.43 | 50.27 |
| 2004 | 1.52 | 2.10 | 3.50 | 6.58 | 2.45 | 1.44 | 3.23 | 6.39 | 6.95 | 2.13 | 4.14 | 4.90 | 45.33 |
| 2005 | 4.69 | 3.28 | 5.60 | 4.92 | 3.59 | 0.64 | 1.03 | 4.57 | 4.28 | 15.38 | 5.60 | 4.34 | 57.92 |
| 2006 | 5.11 | 2.75 | 0.57 | 3.19 | 7.26 | 9.24 | 2.05 | 3.74 | 3.18 | 7.12 | 7.69 | 2.40 | 54.30 |
| 2007 | 3.51 | 2.33 | 6.48 | 7.92 | 2.42 | 3.23 | 3.96 | 1.08 | 2.55 | 1.81 | 2.89 | 4.63 | 42.81 |
| 2008 | 2.93 | 7.04 | 6.47 | 4.05 | 1.95 | 2.48 | 5.28 | 1.85 | 10.99 | 1.49 | 5.33 | 7.26 | 57.12 |
| 2009 | 3.94 | 1.99 | 2.86 | 5.87 | 3.29 | 3.61 | 10.52 | 2.80 | 2.27 | 7.13 | 4.42 | 6.15 | 54.85 |
| POR= 62 YRS | 3.94 | 3.49 | 4.38 | 4.16 | 3.62 | 3.13 | 3.09 | 3.82 | 3.72 | 3.77 | 4.38 | 4.27 | 45.77 |

WBAN : 14765

## AVERAGE TEMPERATURE (°F) 2009 PROVIDENCE (KPVD)

| YEAR | JAN | FEB | MAR | APR | MAY | JUN | JUL | AUG | SEP | OCT | NOV | DEC | ANNUAL |
|------|-----|-----|-----|-----|-----|-----|-----|-----|-----|-----|-----|-----|--------|
| 1980 | 29.7 | 26.8 | 37.1 | 49.5 | 60.3 | 64.5 | 74.8 | 73.4 | 64.9 | 49.8 | 41.0 | 28.5 | 50.0 |
| 1981 | 20.3 | 37.4 | 38.7 | 51.4 | 58.5 | 69.4 | 75.6 | 70.0 | 62.4 | 49.1 | 43.0 | 31.1 | 50.6 |
| 1982 | 21.5 | 31.5 | 38.8 | 47.8 | 58.9 | 63.9 | 73.6 | 69.2 | 64.1 | 53.2 | 47.5 | 38.6 | 50.7 |
| 1983 | 31.4 | 32.9 | 40.4 | 49.9 | 56.9 | 70.2 | 76.6 | 74.3 | 69.6 | 55.3 | 46.0 | 32.5 | 53.0 |
| 1984 | 26.4 | 37.1 | 33.8 | 47.6 | 57.4 | 69.1 | 71.5 | 73.5 | 62.1 | 56.3 | 43.6 | 37.9 | 51.4 |
| 1985 | 22.5 | 32.1 | 40.8 | 51.0 | 60.2 | 64.8 | 73.0 | 71.1 | 65.2 | 54.6 | 45.9 | 30.4 | 51.0 |
| 1986 | 31.1 | 29.0 | 39.9 | 49.4 | 59.4 | 66.4 | 71.0 | 69.3 | 62.3 | 53.0 | 41.6 | 35.4 | 50.7 |
| 1987 | 29.0 | 28.6 | 39.8 | 48.4 | 59.3 | 68.4 | 72.2 | 69.6 | 64.3 | 51.4 | 43.0 | 35.1 | 50.8 |
| 1988 | 26.8 | 31.8 | 39.4 | 47.0 | 58.0 | 66.9 | 74.3 | 75.3 | 63.0 | 48.9 | 45.2 | 32.4 | 50.8 |
| 1989 | 33.8 | 29.9 | 37.5 | 46.2 | 59.3 | 68.7 | 72.3 | 72.1 | 65.3 | 54.1 | 42.5 | 21.8 | 50.3 |
| 1990 | 36.3 | 34.3 | 40.1 | 48.1 | 56.0 | 67.7 | 73.0 | 73.5 | 63.7 | 58.6 | 46.5 | 39.5 | 53.1 |
| 1991 | 29.6 | 35.1 | 41.3 | 51.8 | 63.9 | 69.3 | 74.2 | 73.6 | 63.1 | 56.1 | 45.4 | 36.3 | 53.3 |
| 1992 | 31.4 | 33.0 | 36.6 | 46.5 | 57.6 | 67.3 | 70.3 | 70.1 | 64.0 | 51.7 | 43.0 | 34.2 | 50.5 |
| 1993 | 31.4 | 26.3 | 35.8 | 49.6 | 61.8 | 69.3 | 74.5 | 73.8 | 65.1 | 51.5 | 44.0 | 33.7 | 51.4 |
| 1994 | 22.7 | 25.8 | 38.8 | 51.4 | 56.5 | 69.4 | 76.2 | 69.9 | 63.0 | 54.2 | 48.5 | 38.4 | 51.2 |
| 1995 | 36.0 | 29.4 | 41.2 | 48.4 | 57.3 | 68.3 | 75.8 | 73.5 | 62.8 | 57.0 | 40.7 | 29.8 | 51.7 |
| 1996 | 28.7 | 29.6 | 35.0 | 48.3 | 57.4 | 68.1 |  | 71.0 | 64.1 | 52.1 | 40.0 | 38.6 |  |
| 1997 | 29.3 | 36.5 | 37.8 | 47.1 | 55.4 | 68.1 | 73.7 | 70.6 | 64.0 | 51.8 | 41.0 | 34.2 | 50.8 |
| 1998 | 35.1 | 37.0 | 41.2 | 49.8 | 61.1 | 66.0 | 73.5 | 73.3 | 66.3 | 54.1 | 43.5 | 38.5 | 53.3 |
| 1999 | 30.2 | 34.8 | 39.9 | 50.0 | 59.8 | 70.5 | 76.6 | 72.0 | 66.5 | 52.4 | 47.4 | 36.8 | 53.1 |
| 2000 | 27.7 | 34.0 | 43.5 | 47.3 | 58.5 | 67.8 | 70.5 | 70.2 | 63.4 | 53.4 | 43.4 | 28.9 | 50.7 |
| 2001 | 29.0 | 31.9 | 36.3 | 49.1 | 60.1 | 70.5 | 69.9 | 74.3 | 64.9 | 54.2 | 47.6 | 39.3 | 52.3 |
| 2002 | 35.2 | 35.6 | 40.2 | 51.0 | 56.6 | 66.3 | 75.6 | 75.3 | 67.8 | 52.9 | 42.9 | 33.0 | 52.7 |
| 2003 | 25.1 | 26.2 | 38.3 | 45.4 | 55.2 | 65.3 | 73.5 | 74.9 | 66.3 | 51.8 | 46.6 | 36.1 | 50.4 |
| 2004 | 21.4 | 33.0 | 39.2 | 49.6 | 59.3 | 66.7 | 71.5 | 71.3 | 65.4 | 53.2 | 43.5 | 34.4 | 50.7 |
| 2005 | 27.5 | 31.3 | 34.8 | 50.0 | 53.4 | 69.7 | 74.3 | 76.5 | 68.3 | 55.3 | 45.8 | 32.4 | 51.6 |
| 2006 | 37.2 | 31.7 | 39.0 | 50.5 | 58.6 | 68.7 | 76.2 | 72.7 | 63.8 | 53.9 | 49.3 | 40.7 | 53.5 |
| 2007 | 34.1 | 27.1 | 38.1 | 46.8 | 60.8 | 68.3 | 73.8 | 73.2 | 67.5 | 59.6 | 42.8 | 32.8 | 52.1 |
| 2008 | 33.3 | 32.9 | 39.5 | 51.4 | 57.7 | 71.3 | 76.4 | 70.2 | 65.4 | 52.2 | 42.6 | 35.3 | 52.4 |
| 2009 | 24.2 | 32.9 | 37.7 | 49.9 | 58.7 | 64.5 | 70.4 | 73.9 | 63.1 | 52.0 | 49.0 | 33.0 | 50.8 |
| POR= 62 YRS | 28.9 | 30.6 | 37.9 | 48.4 | 57.9 | 67.3 | 73.0 | 71.7 | 64.0 | 53.4 | 44.0 | 33.4 | 50.9 |

## HEATING DEGREE DAYS (base 65°F) 2009 PROVIDENCE (KPVD)

| YEAR | JUL | AUG | SEP | OCT | NOV | DEC | JAN | FEB | MAR | APR | MAY | JUN | TOTAL |
|------|-----|-----|-----|-----|-----|-----|-----|-----|-----|-----|-----|-----|-------|
| 1980-81 | 0 | 1 | 120 | 465 | 715 | 1125 | 1379 | 769 | 808 | 405 | 228 | 13 | 6028 |
| 1981-82 | 0 | 20 | 119 | 486 | 651 | 1044 | 1343 | 932 | 802 | 510 | 190 | 91 | 6188 |
| 1982-83 | 1 | 26 | 78 | 363 | 518 | 809 | 1038 | 892 | 755 | 449 | 254 | 13 | 5196 |
| 1983-84 | 0 | 4 | 62 | 323 | 563 | 1001 | 1190 | 802 | 961 | 513 | 236 | 36 | 5691 |
| 1984-85 | 1 | 0 | 125 | 270 | 637 | 832 | 1309 | 914 | 743 | 417 | 177 | 63 | 5488 |
| 1985-86 | 0 | 6 | 78 | 321 | 567 | 1065 | 1045 | 999 | 772 | 460 | 216 | 57 | 5586 |
| 1986-87 | 14 | 25 | 113 | 380 | 697 | 911 | 1111 | 1014 | 772 | 494 | 228 | 23 | 5782 |
| 1987-88 | 2 | 25 | 70 | 414 | 653 | 921 | 1177 | 954 | 787 | 532 | 238 | 67 | 5840 |
| 1988-89 | 8 | 10 | 89 | 491 | 587 | 1003 | 960 | 975 | 847 | 557 | 181 | 22 | 5730 |
| 1989-90 | 2 | 9 | 89 | 332 | 668 | 1329 | 882 | 854 | 761 | 511 | 275 | 24 | 5736 |
| 1990-91 | 6 | 0 | 107 | 242 | 549 | 781 | 1090 | 829 | 726 | 400 | 121 | 29 | 4880 |
| 1991-92 | 1 | 0 | 125 | 275 | 581 | 884 | 1034 | 919 | 876 | 549 | 246 | 27 | 5517 |
| 1992-93 | 11 | 4 | 100 | 404 | 652 | 951 | 1036 | 1077 | 901 | 455 | 118 | 32 | 5741 |
| 1993-94 | 1 | 0 | 102 | 413 | 623 | 966 | 1307 | 1092 | 805 | 401 | 263 | 17 | 5990 |
| 1994-95 | 0 | 5 | 85 | 326 | 487 | 815 | 892 | 990 | 731 | 493 | 244 | 22 | 5090 |
| 1995-96 | 0 | 0 | 111 | 245 | 721 | 1082 | 1116 | 1018 | 921 | 494 | 265 | 18 | 5991 |
| 1996-97 | 0 | 4 | 83 | 389 | 743 | 814 | 1098 | 794 | 836 | 529 | 289 | 77 | 5656 |
| 1997-98 | 3 | 2 | 83 | 410 | 713 | 948 | 919 | 776 | 738 | 451 | 149 | 61 | 5253 |
| 1998-99 | 0 | 0 | 45 | 329 | 636 | 816 | 1072 | 837 | 773 | 441 | 172 | 14 | 5135 |
| 1999-00 | 1 | 5 | 43 | 384 | 519 | 866 | 1148 | 892 | 659 | 524 | 219 | 59 | 5319 |
| 2000-01 | 2 | 7 | 120 | 355 | 642 | 1115 | 1111 | 922 | 883 | 473 | 193 | 15 | 5838 |
| 2001-02 | 3 | 0 | 72 | 334 | 515 | 790 | 918 | 819 | 762 | 424 | 268 | 69 | 4974 |
| 2002-03 | 0 | 2 | 23 | 392 | 659 | 984 | 1231 | 1080 | 818 | 580 | 297 | 77 | 6143 |
| 2003-04 | 0 | 1 | 31 | 403 | 549 | 893 | 1345 | 921 | 796 | 457 | 197 | 52 | 5645 |
| 2004-05 | 2 | 4 | 57 | 358 | 639 | 943 | 1157 | 936 | 928 | 448 | 352 | 42 | 5866 |
| 2005-06 | 6 | 0 | 39 | 321 | 567 | 1004 | 856 | 928 | 797 | 427 | 217 | 38 | 5200 |
| 2006-07 | 0 | 5 | 77 | 342 | 467 | 748 | 948 | 1055 | 823 | 541 | 178 | 33 | 5217 |
| 2007-08 | 0 | 8 | 38 | 203 | 662 | 993 | 972 | 924 | 782 | 404 | 227 | 19 | 5232 |
| 2008-09 | 0 | 2 | 72 | 391 | 665 | 914 | 1256 | 890 | 840 | 457 | 208 | 63 | 5758 |
| 2009- | 6 | 3 | 91 | 396 | 474 | 988 | | | | | | | |

WBAN : 14765

## COOLING DEGREE DAYS (base 65°F) 2009 PROVIDENCE (KPVD)

| YEAR | JAN | FEB | MAR | APR | MAY | JUN | JUL | AUG | SEP | OCT | NOV | DEC | TOTAL |
|------|-----|-----|-----|-----|-----|-----|-----|-----|-----|-----|-----|-----|-------|
| 1980 | 0 | 0 | 0 | 0 | 21 | 84 | 312 | 272 | 122 | 0 | 0 | 0 | 811 |
| 1981 | 0 | 0 | 0 | 2 | 33 | 152 | 335 | 183 | 47 | 0 | 0 | 0 | 752 |
| 1982 | 0 | 0 | 0 | 0 | 11 | 64 | 276 | 165 | 59 | 3 | 2 | 0 | 580 |
| 1983 | 0 | 0 | 0 | 1 | 8 | 177 | 367 | 298 | 206 | 30 | 0 | 0 | 1087 |
| 1984 | 0 | 0 | 0 | 0 | 6 | 164 | 206 | 272 | 47 | 7 | 0 | 0 | 702 |
| 1985 | 0 | 0 | 0 | 5 | 34 | 65 | 256 | 203 | 90 | 5 | 0 | 0 | 658 |
| 1986 | 0 | 0 | 0 | 0 | 51 | 105 | 207 | 164 | 38 | 14 | 0 | 0 | 579 |
| 1987 | 0 | 0 | 0 | 0 | 57 | 130 | 231 | 177 | 53 | 0 | 0 | 0 | 648 |
| 1988 | 0 | 0 | 0 | 0 | 26 | 131 | 302 | 336 | 37 | 2 | 0 | 0 | 834 |
| 1989 | 0 | 0 | 0 | 0 | 10 | 141 | 237 | 237 | 103 | 0 | 0 | 0 | 728 |
| 1990 | 0 | 0 | 0 | 8 | 1 | 114 | 262 | 272 | 74 | 49 | 2 | 0 | 782 |
| 1991 | 0 | 0 | 0 | 12 | 96 | 166 | 295 | 276 | 73 | 8 | 0 | 0 | 926 |
| 1992 | 0 | 0 | 0 | 0 | 27 | 103 | 183 | 169 | 75 | 1 | 0 | 0 | 558 |
| 1993 | 0 | 0 | 0 | 0 | 26 | 167 | 303 | 281 | 110 | 1 | 1 | 0 | 889 |
| 1994 | 0 | 0 | 0 | 0 | 7 | 155 | 352 | 163 | 30 | 0 | 0 | 0 | 707 |
| 1995 | 0 | 0 | 0 | 0 | 12 | 128 | 344 | 272 | 53 | 4 | 0 | 0 | 813 |
| 1996 | 0 | 0 | 0 | 3 | 36 | 117 | | 200 | 65 | 0 | 0 | 0 | |
| 1997 | 0 | 0 | 0 | 0 | 0 | 177 | 281 | 184 | 63 | 10 | 0 | 0 | 715 |
| 1998 | 0 | 0 | 8 | 0 | 36 | 96 | 270 | 266 | 93 | 0 | 0 | 0 | 769 |
| 1999 | 0 | 0 | 0 | 0 | 18 | 185 | 366 | 229 | 96 | 0 | 0 | 0 | 894 |
| 2000 | 0 | 0 | 0 | 0 | 27 | 147 | 179 | 173 | 77 | 1 | 0 | 0 | 604 |
| 2001 | 0 | 0 | 0 | 1 | 48 | 190 | 161 | 293 | 74 | 10 | 0 | 0 | 777 |
| 2002 | 0 | 0 | 0 | 14 | 16 | 118 | 335 | 327 | 112 | 24 | 0 | 0 | 946 |
| 2003 | 0 | 0 | 0 | 0 | 0 | 91 | 272 | 310 | 76 | 0 | 0 | 0 | 749 |
| 2004 | 0 | 0 | 0 | 0 | 24 | 111 | 212 | 205 | 76 | 0 | 0 | 0 | 628 |
| 2005 | 0 | 0 | 0 | 2 | 0 | 189 | 300 | 365 | 145 | 24 | 0 | 0 | 1025 |
| 2006 | 0 | 0 | 0 | 0 | 25 | 156 | 355 | 251 | 49 | 3 | 0 | 0 | 839 |
| 2007 | 0 | 0 | 0 | 0 | 54 | 140 | 280 | 272 | 120 | 42 | 0 | 0 | 908 |
| 2008 | 0 | 0 | 0 | 1 | 6 | 214 | 363 | 172 | 92 | 3 | 0 | 0 | 851 |
| 2009 | 0 | 0 | 0 | 10 | 21 | 55 | 181 | 286 | 38 | 0 | 0 | 0 | 591 |

**SNOWFALL (inches) 2009 PROVIDENCE (KPVD)**

| YEAR | JUL | AUG | SEP | OCT | NOV | DEC | JAN | FEB | MAR | APR | MAY | JUN | TOTAL |
|---|---|---|---|---|---|---|---|---|---|---|---|---|---|
| 1980-81 | 0.0 | 0.0 | 0.0 | 0.0 | 4.1 | 3.6 | 12.9 | 0.6 | 0.3 | 0.0 | 0.0 | 0.0 | 21.5 |
| 1981-82 | 0.0 | 0.0 | 0.0 | T | T | 16.4 | 13.4 | 4.3 | 5.7 | 7.6 | 0.0 | 0.0 | 47.4 |
| 1982-83 | 0.0 | 0.0 | 0.0 | 0.0 | 0.0 | 7.3 | 3.8 | 21.3 | T | T | 0.0 | 0.0 | 32.4 |
| 1983-84 | 0.0 | 0.0 | 0.0 | 0.0 | T | 4.5 | 17.9 | T | 13.7 | T | 0.0 | 0.0 | 36.1 |
| 1984-85 | 0.0 | 0.0 | 0.0 | 0.0 | T | 2.0 | 9.8 | 10.0 | 0.6 | T | 0.0 | 0.0 | 22.4 |
| 1985-86 | 0.0 | 0.0 | 0.0 | 0.0 | 1.8 | 2.6 | 0.7 | 13.0 | 0.5 | T | T | 0.0 | 18.6 |
| 1986-87 | 0.0 | 0.0 | 0.0 | 0.0 | 4.4 | 8.0 | 21.5 | 4.7 | 1.6 | 1.1 | 0.0 | 0.0 | 41.3 |
| 1987-88 | 0.0 | 0.0 | 0.0 | 0.0 | 8.0 | 7.8 | 13.5 | 6.7 | 2.7 | T | 0.0 | 0.0 | 38.7 |
| 1988-89 | 0.0 | 0.0 | 0.0 | 0.0 | T | 1.2 | 0.2 | 7.3 | 1.9 | 0.3 | 0.0 | 0.0 | 10.9 |
| 1989-90 | 0.0 | 0.0 | 0.0 | 0.0 | 8.0 | 15.8 | 10.8 | 10.5 | 9.3 | 1.8 | 0.0 | 0.0 | 56.2 |
| 1990-91 | 0.0 | 0.0 | 0.0 | 0.0 | T | 6.9 | 6.4 | 6.0 | 5.3 | 0.0 | 0.0 | 0.0 | 24.6 |
| 1991-92 | 0.0 | 0.0 | 0.0 | 0.0 | T | 4.8 | 2.4 | 4.9 | 8.2 | 2.0 | 0.0 | 0.0 | 22.3 |
| 1992-93 | 0.0 | 0.0 | 0.0 | T | T | 3.6 | 5.4 | 12.7 | 17.8 | 0.2 | 0.0 | 0.0 | 39.7 |
| 1993-94 | 0.0 | 0.0 | 0.0 | 0.0 | T | 10.1 | 18.0 | 25.8 | 9.6 | 0.0 | T | 0.0 | 63.5 |
| 1994-95 | 0.0 | 0.0 | 0.0 | 0.0 | 0.7 | 0.3 | 3.0 | 8.4 | T | 0.1 | 0.0 | 0.0 | 12.5 |
| 1995-96 | 0.0 | 0.0 | 0.0 | 0.0 | 4.0 | | 37.2 | | | | | | |
| 1996-97 | | | | | | | | | | | | | |
| 1997-98 | | | | | | 0.2 | 1.6 | 0.1 | 0.3 | T | | | |
| 1998-99 | | | | | | | 6.5 | 12.8 | 12.2 | T | | | |
| 1999-00 | | | | | | | 6.9 | 6.8 | 2.6 | | | | |
| 2000-01 | | | | | | | 9.7 | 10.8 | 10.3 | T | 0.0 | 0.0 | |
| 2001-02 | 0.0 | 0.0 | 0.0 | 0.0 | 0.0 | 1.5 | 5.6 | 0.6 | 2.5 | T | 0.0 | 0.0 | 10.2 |
| 2002-03 | 0.0 | 0.0 | 0.0 | 0.0 | 5.8 | 12.0 | 3.9 | 24.7 | 7.8 | 2.1 | 0.0 | 0.0 | 56.3 |
| 2003-04 | 0.0 | 0.0 | 0.0 | T | 0.0 | 20.4 | 11.0 | 3.1 | 6.5 | T | 0.0 | 0.0 | 41.0 |
| 2004-05 | 0.0 | 0.0 | 0.0 | 0.0 | 3.8 | 7.9 | 36.7 | 13.9 | 9.9 | 0.0 | 0.0 | 0.0 | 72.2 |
| 2005-06 | 0.0 | 0.0 | 0.0 | T | 2.0 | 8.8 | 6.9 | 9.8 | 6.2 | 0.2 | 0.0 | 0.0 | 33.9 |
| 2006-07 | 0.0 | 0.0 | 0.0 | 0.0 | 0.0 | 0.8 | 1.2 | 7.3 | 5.8 | T | 0.0 | T | 15.1 |
| 2007-08 | 0.0 | 0.0 | 0.0 | 0.0 | T | 14.4 | 2.7 | 6.2 | 1.2 | 0.0 | 0.0 | 0.0 | 24.5 |
| 2008-09 | 0.0 | 0.0 | 0.0 | 0.0 | T | 20.6 | 14.9 | 3.9 | 11.6 | 0.0 | 0.0 | 0.0 | 51.0 |
| 2009- | 0.0 | 0.0 | 0.0 | 0.0 | 0.0 | 21.7 | | | | | | | |
| POR= 57 YRS | 0.0 | 0.0 | 0.0 | 0.1 | 1.0 | 7.3 | 10.1 | 9.8 | 7.0 | 0.6 | 0.1 | T | 36.0 |

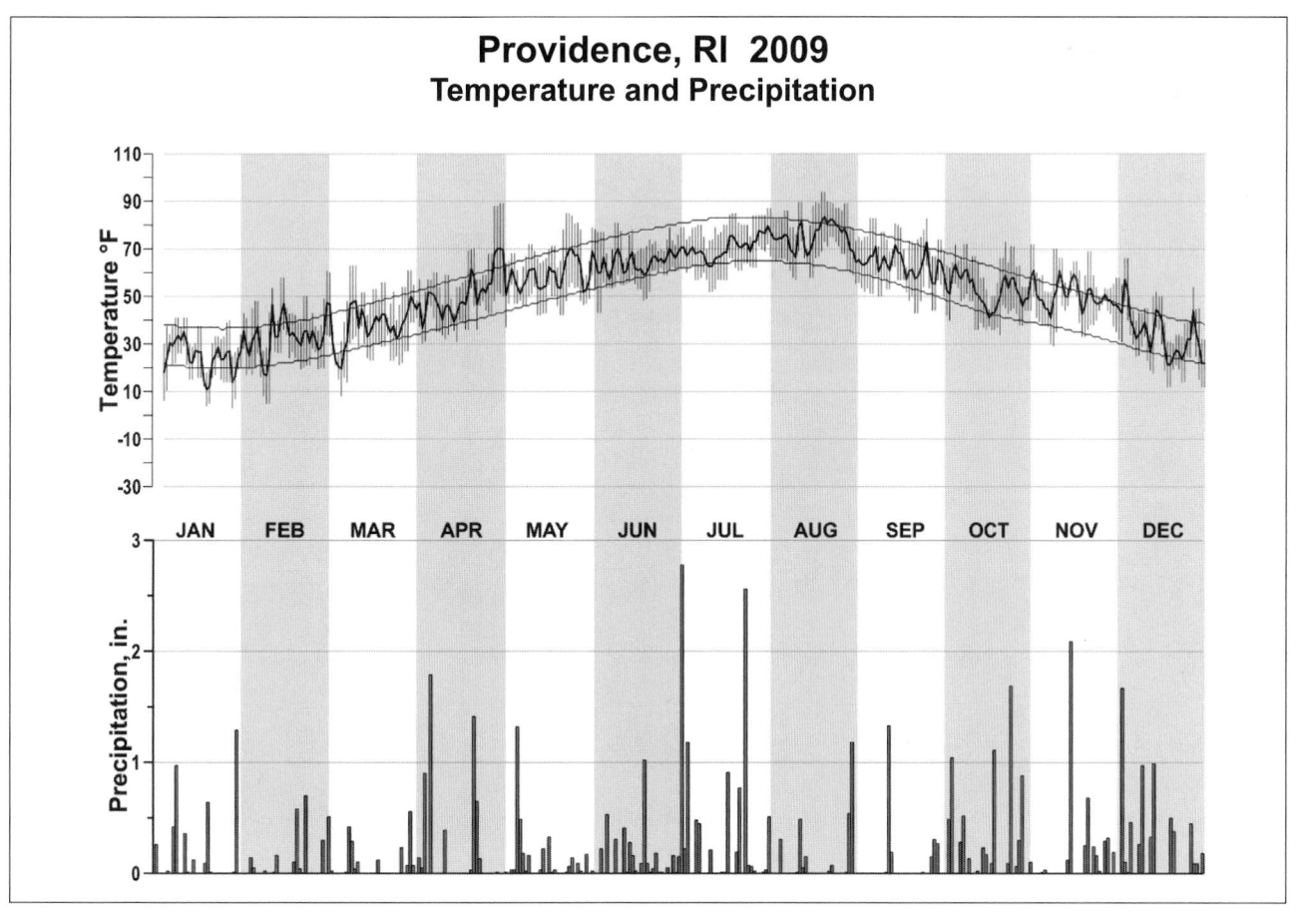

# 2009
# CHARLESTON
# SOUTH CAROLINA (KCHS)

Charleston is a peninsula city bounded on the west and south by the Ashley River, on the east by the Cooper River, and on the southeast by a spacious harbor. Weather records for the airport are from a site some 10 miles inland. The terrain is generally level, ranging in elevation from sea level to 20 feet on the peninsula, with gradual increases in elevation toward inland areas. The soil is sandy to sandy loam with lesser amounts of loam. The drainage varies from good to poor. Because of the very low elevation, a considerable portion of this community and the nearby coastal islands are vulnerable to tidal flooding.

The climate is temperate, modified considerably by the nearness to the ocean. The marine influence is noticeable during winter when the low temperatures are sometimes 10-15 degrees higher on the peninsula than at the airport. By the same token, high temperatures are generally a few degrees lower on the peninsula. The prevailing winds are northerly in the fall and winter, southerly in the spring and summer.

Summer is warm and humid. Temperatures of 100 degrees or more are infrequent. High temperatures are generally several degrees lower along the coast than inland due to the cooling effect of the sea breeze. Summer is the rainiest season with 41 percent of the annual total. The rain, except during occasional tropical storms, generally occurs as showers or thunderstorms.

The fall season passes through the warm Indian Summer period to the pre-winter cold spells which begin late in November. From late September to early November the weather is mostly sunny and temperature extremes are rare. Late summer and early fall is the period of maximum threat to the South Carolina coast from hurricanes.

The winter months, December through February, are mild with periods of rain. However, the winter rainfall is generally of a more uniform type. There is some chance of a snow flurry, with the best probability of its occurrence in January, but a significant amount is rarely measured. An average winter would experience less than one cold wave and severe freeze. Temperatures of 20 degrees or less on the peninsula and along the coast are very unusual.

The most spectacular time of the year, weatherwise, is spring with its rapid changes from windy and cold in March to warm and pleasant in May. Severe local storms are more likely to occur in spring than in summer.

The average occurrence of the first freeze in the fall is early December, and the average last freeze is late February, giving an average growing season of about 294 days.

# NORMALS, MEANS, AND EXTREMES
## CHARLESTON (KCHS)

| LATITUDE: 32 ° 53'N | LONGITUDE: -80 ° 2 'W | ELEVATION (FT): GRND: 39　BARO: 48 | TIME ZONE: EASTERN　(UTC -5) | WBAN: 13880 |
|---|---|---|---|---|

| | ELEMENT | POR | JAN | FEB | MAR | APR | MAY | JUN | JUL | AUG | SEP | OCT | NOV | DEC | YEAR |
|---|---|---|---|---|---|---|---|---|---|---|---|---|---|---|---|
| **TEMPERATURE °F** | NORMAL DAILY MAXIMUM | 30 | 58.9 | 62.3 | 69.3 | 76.1 | 82.9 | 87.9 | 90.9 | 89.4 | 85.0 | 77.0 | 69.6 | 61.6 | 75.9 |
| | MEAN DAILY MAXIMUM | 80 | 59.3 | 60.7 | 68.2 | 75.1 | 82.4 | 86.9 | 89.8 | 88.8 | 84.1 | 76.7 | 68.2 | 60.9 | 75.1 |
| | HIGHEST DAILY MAXIMUM | 67 | 83 | 87 | 90 | 95 | 98 | 103 | 104 | 105 | 99 | 94 | 88 | 83 | 105 |
| | YEAR OF OCCURRENCE | | 1950 | 1989 | 1974 | 2002 | 1989 | 1944 | 1986 | 1999 | 1944 | 1986 | 1961 | 1972 | AUG 1999 |
| | MEAN OF EXTREME MAXS. | 80 | 75.8 | 78.0 | 83.0 | 88.3 | 92.6 | 96.4 | 96.8 | 96.0 | 92.3 | 87.1 | 81.7 | 76.7 | 87.1 |
| | NORMAL DAILY MINIMUM | 30 | 36.9 | 39.1 | 46.0 | 52.2 | 61.3 | 68.5 | 72.5 | 71.6 | 67.1 | 55.3 | 46.4 | 39.3 | 54.7 |
| | MEAN DAILY MINIMUM | 80 | 39.1 | 40.0 | 47.0 | 53.6 | 62.7 | 69.2 | 72.9 | 72.3 | 67.5 | 56.9 | 46.8 | 40.3 | 55.7 |
| | LOWEST DAILY MINIMUM | 67 | 6 | 12 | 15 | 29 | 36 | 50 | 58 | 56 | 42 | 27 | 15 | 8 | 6 |
| | YEAR OF OCCURRENCE | | 1985 | 1973 | 1980 | 1944 | 1963 | 1972 | 1952 | 1979 | 1967 | 1976 | 1950 | 1962 | JAN 1985 |
| | MEAN OF EXTREME MINS. | 80 | 21.1 | 24.1 | 29.4 | 37.6 | 48.4 | 58.9 | 65.8 | 64.7 | 55.5 | 39.4 | 29.7 | 22.9 | 41.5 |
| | NORMAL DRY BULB | 30 | 47.9 | 50.7 | 57.7 | 64.2 | 72.1 | 78.2 | 81.7 | 80.5 | 76.1 | 66.2 | 58.0 | 50.5 | 65.3 |
| | MEAN DRY BULB | 80 | 49.2 | 50.4 | 57.6 | 64.4 | 72.6 | 78.1 | 81.4 | 80.6 | 75.8 | 66.8 | 57.5 | 50.6 | 65.4 |
| | MEAN WET BULB | 26 | 43.5 | 45.8 | 51.3 | 57.1 | 65.1 | 71.9 | 74.9 | 74.4 | 69.9 | 61.1 | 53.1 | 45.9 | 59.5 |
| | MEAN DEW POINT | 26 | 39.5 | 42.1 | 47.2 | 53.3 | 62.4 | 70.0 | 73.3 | 72.9 | 68.0 | 58.7 | 50.1 | 42.1 | 56.6 |
| | NORMAL NO. DAYS WITH: | | | | | | | | | | | | | | |
| | MAXIMUM >= 90 | 30 | 0.0 | 0.0 | * | 0.9 | 3.8 | 11.5 | 19.8 | 15.6 | 6.1 | 0.4 | 0.0 | 0.0 | 58.1 |
| | MAXIMUM <= 32 | 30 | 0.2 | 0.1 | * | 0.0 | 0.0 | 0.0 | 0.0 | 0.0 | 0.0 | 0.0 | 0.0 | 0.1 | 0.4 |
| | MINIMUM <= 32 | 30 | 9.8 | 6.5 | 1.8 | 0.1 | 0.0 | 0.0 | 0.0 | 0.0 | 0.0 | 0.1 | 1.7 | 7.5 | 27.5 |
| | MINIMUM <= 0 | 30 | 0.0 | 0.0 | 0.0 | 0.0 | 0.0 | 0.0 | 0.0 | 0.0 | 0.0 | 0.0 | 0.0 | 0.0 | 0.0 |
| **H/C** | NORMAL HEATING DEG. DAYS | 30 | 523 | 394 | 242 | 95 | 11 | 1 | 0 | 0 | 2 | 69 | 229 | 439 | 2005 |
| | NORMAL COOLING DEG. DAYS | 30 | 3 | 7 | 29 | 84 | 242 | 408 | 532 | 494 | 348 | 121 | 34 | 4 | 2306 |
| **RH** | NORMAL (PERCENT) | 30 | 71 | 69 | 68 | 68 | 73 | 76 | 77 | 79 | 79 | 75 | 74 | 73 | 74 |
| | HOUR 01 LST | 30 | 79 | 78 | 80 | 82 | 87 | 88 | 89 | 90 | 89 | 87 | 85 | 81 | 85 |
| | HOUR 07 LST | 30 | 81 | 81 | 83 | 84 | 85 | 86 | 87 | 90 | 90 | 88 | 86 | 84 | 85 |
| | HOUR 13 LST | 30 | 57 | 53 | 50 | 47 | 53 | 58 | 60 | 62 | 62 | 56 | 54 | 57 | 56 |
| | HOUR 19 LST | 30 | 70 | 66 | 64 | 63 | 68 | 72 | 74 | 77 | 78 | 76 | 75 | 72 | 71 |
| **S** | PERCENT POSSIBLE SUNSHINE | 39 | 56 | 60 | 66 | 71 | 70 | 66 | 67 | 64 | 61 | 63 | 59 | 56 | 63 |
| **W/O** | MEAN NO. DAYS WITH: | | | | | | | | | | | | | | |
| | HEAVY FOG(VISBY <= 1/4 MI) | 46 | 4.1 | 2.0 | 2.3 | 1.8 | 2.0 | 1.2 | 0.7 | 1.0 | 1.7 | 2.3 | 3.4 | 3.8 | 26.3 |
| | THUNDERSTORMS | 64 | 0.8 | 1.0 | 2.5 | 3.2 | 6.3 | 10.1 | 13.4 | 11.5 | 5.2 | 1.6 | 0.8 | 0.6 | 57.0 |
| **CLOUDNESS** | MEAN: | | | | | | | | | | | | | | |
| | SUNRISE-SUNSET (OKTAS) | 1 | | | 5.6 | | 2.4 | 4.0 | | | | | | | |
| | MIDNIGHT-MIDNIGHT (OKTAS) | 1 | | | 5.6 | | | 4.0 | | | | | | | |
| | MEAN NO. DAYS WITH: | | | | | | | | | | | | | | |
| | CLEAR | 1 | 3.0 | 4.0 | 11.0 | | 13.0 | 9.0 | | | | | | | |
| | PARTLY CLOUDY | 1 | 1.0 | 2.0 | 3.0 | | 4.0 | 11.0 | | | | | | | |
| | CLOUDY | 1 | 4.0 | 3.0 | 9.0 | | 3.0 | 2.0 | | | | | | | |
| **PR** | MEAN STATION PRESSURE(IN) | 26 | 30.09 | 30.06 | 30.02 | 29.97 | 29.97 | 29.96 | 29.98 | 29.98 | 29.98 | 30.03 | 30.07 | 30.10 | 30.02 |
| | MEAN SEA-LEVEL PRES. (IN) | 26 | 30.15 | 30.12 | 30.07 | 30.02 | 30.02 | 30.01 | 30.03 | 30.02 | 30.03 | 30.08 | 30.12 | 30.15 | 30.07 |
| **WINDS** | MEAN SPEED (MPH) | 26 | 8.4 | 8.7 | 9.3 | 9.2 | 8.4 | 7.8 | 7.4 | 7.0 | 7.5 | 7.3 | 7.4 | 7.7 | 8.0 |
| | PREVAIL.DIR(TENS OF DEGS) | 38 | 28 | 21 | 21 | 21 | 21 | 21 | 21 | 20 | 03 | 03 | 02 | 02 | 03 |
| | MAXIMUM 2-MINUTE: | | | | | | | | | | | | | | |
| | SPEED (MPH) | 14 | 36 | 38 | 40 | 43 | 47 | 44 | 40 | 43 | 51 | 39 | 36 | 38 | 51 |
| | DIR. (TENS OF DEGS) | | 26 | 19 | 24 | 20 | 30 | 17 | 28 | 29 | 36 | 21 | 22 | 26 | 36 |
| | YEAR OF OCCURRENCE | | 2009 | 2007 | 2008 | 2007 | 2006 | 1997 | 2002 | 2004 | 1999 | 1996 | 2006 | 2000 | SEP 1999 |
| | MAXIMUM 3-SECOND | | | | | | | | | | | | | | |
| | SPEED (MPH) | 14 | 46 | 47 | 52 | 53 | 54 | 56 | 55 | 55 | 67 | 56 | 44 | 49 | 67 |
| | DIR. (TENS OF DEGS) | | 26 | 19 | 24 | 19 | 30 | 19 | 11 | 29 | 36 | 21 | 19 | 23 | 36 |
| | YEAR OF OCCURRENCE | | 2009 | 2007 | 1997 | 2007 | 2006 | 2003 | 1999 | 2004 | 1999 | 1996 | 2006 | 2002 | SEP 1999 |
| **PRECIPITATION** | NORMAL (IN) | 30 | 4.08 | 3.08 | 4.00 | 2.77 | 3.67 | 5.92 | 6.13 | 6.91 | 5.98 | 3.09 | 2.66 | 3.24 | 51.53 |
| | MAXIMUM MONTHLY (IN) | 67 | 8.92 | 10.17 | 11.11 | 9.50 | 9.28 | 27.24 | 18.46 | 16.99 | 17.31 | 12.11 | 7.35 | 10.06 | 27.24 |
| | YEAR OF OCCURRENCE | | 1993 | 1998 | 1983 | 1958 | 1957 | 1973 | 1964 | 1974 | 1945 | 1994 | 1972 | 2009 | JUN 1973 |
| | MINIMUM MONTHLY (IN) | 67 | 0.63 | 0.33 | 0.36 | 0.01 | 0.57 | 0.96 | 1.76 | 0.73 | 0.18 | 0.08 | 0.03 | 0.35 | 0.01 |
| | YEAR OF OCCURRENCE | | 1950 | 1947 | 2004 | 1972 | 2000 | 1970 | 1972 | 1980 | 1990 | 2000 | 2007 | 2008 | APR 1972 |
| | MAXIMUM IN 24 HOURS (IN) | 67 | 3.90 | 5.93 | 6.63 | 4.10 | 6.23 | 10.10 | 5.81 | 5.77 | 10.52 | 6.57 | 5.24 | 3.40 | 10.52 |
| | YEAR OF OCCURRENCE | | 1993 | 1998 | 1959 | 1958 | 1967 | 1973 | 1960 | 1964 | 1998 | 2008 | 1969 | 1978 | SEP 1998 |
| | NORMAL NO. DAYS WITH: | | | | | | | | | | | | | | |
| | PRECIPITATION >= 0.01 | 30 | 10.4 | 8.6 | 8.9 | 7.2 | 8.5 | 11.2 | 12.3 | 13.0 | 10.3 | 6.2 | 7.2 | 8.7 | 112.5 |
| | PRECIPITATION >= 1.00 | 30 | 0.9 | 0.6 | 1.2 | 0.7 | 0.9 | 1.7 | 1.8 | 2.1 | 1.9 | 0.9 | 0.8 | 0.8 | 14.3 |
| **SNOWFALL** | NORMAL (IN) | 30 | 0.1 | 0.4 | 0.1 | 0.0 | 0.0 | 0.0 | 0.0 | 0.0 | 0.0 | 0.0 | 0.* | 0.5 | 1.1 |
| | MAXIMUM MONTHLY (IN) | 62 | 1.0 | 7.1 | 2.0 | T | T | T | T | 0.0 | 0.0 | 0.0 | T | 8.0 | 8.0 |
| | YEAR OF OCCURRENCE | | 1977 | 1973 | 1969 | 2008 | 2006 | 2006 | 1993 | | | | 2006 | 1989 | DEC 1989 |
| | MAXIMUM IN 24 HOURS (IN) | 62 | 0.8 | 5.9 | 2.0 | T | T | T | T | 0.0 | 0.0 | 0.0 | T | 6.6 | 6.6 |
| | YEAR OF OCCURRENCE' | | 1966 | 1973 | 1969 | 1985 | 2006 | 1995 | 1993 | | | | 1995 | 1989 | DEC 1989 |
| | MAXIMUM SNOW DEPTH (IN) | 56 | 1 | 7 | 1 | 0 | 0 | 0 | 0 | 0 | 0 | 0 | 0 | 8 | 8 |
| | YEAR OF OCCURRENCE | | 1966 | 1973 | 1980 | | | | | | | | | 1989 | DEC 1989 |
| | NORMAL NO. DAYS WITH: | | | | | | | | | | | | | | |
| | SNOWFALL >= 1.0 | 30 | 0.0 | 0.1 | 0.0 | 0.0 | 0.0 | 0.0 | 0.0 | 0.0 | 0.0 | 0.0 | 0.0 | 0.1 | 0.2 |

## PRECIPITATION (inches) 2009 CHARLESTON (KCHS)

| YEAR | JAN | FEB | MAR | APR | MAY | JUN | JUL | AUG | SEP | OCT | NOV | DEC | ANNUAL |
|---|---|---|---|---|---|---|---|---|---|---|---|---|---|
| 1980 | 3.99 | 1.25 | 7.99 | 3.43 | 5.85 | 3.15 | 6.97 | 0.73 | 2.60 | 1.52 | 2.19 | 1.25 | 40.92 |
| 1981 | 0.93 | 2.23 | 2.38 | 1.87 | 4.02 | 6.04 | 12.66 | 9.30 | 1.27 | 1.95 | 1.06 | 5.73 | 49.44 |
| 1982 | 2.18 | 3.64 | 1.26 | 6.51 | 3.04 | 9.16 | 5.40 | 4.10 | 3.92 | 2.42 | 1.19 | 4.20 | 47.02 |
| 1983 | 4.86 | 6.35 | 11.11 | 3.57 | 0.75 | 2.37 | 8.89 | 2.90 | 3.50 | 2.36 | 3.08 | 4.35 | 54.09 |
| 1984 | 5.12 | 3.51 | 5.63 | 6.30 | 6.89 | 2.96 | 4.87 | 1.96 | 5.27 | 1.67 | 1.39 | 0.66 | 46.23 |
| 1985 | 0.87 | 2.70 | 1.50 | 1.12 | 2.79 | 7.02 | 12.06 | 8.48 | 2.53 | 4.58 | 5.49 | 1.21 | 50.35 |
| 1986 | 2.05 | 4.17 | 2.67 | 0.83 | 0.93 | 2.51 | 5.07 | 13.41 | 4.60 | 2.95 | 4.03 | 5.21 | 48.43 |
| 1987 | 7.17 | 4.58 | 5.55 | 1.31 | 2.29 | 5.64 | 2.92 | 6.97 | 14.49 | 0.56 | 3.65 | 1.57 | 56.70 |
| 1988 | 2.76 | 2.38 | 1.78 | 3.21 | 1.86 | 2.32 | 4.13 | 11.88 | 9.72 | 0.73 | 1.08 | 0.72 | 42.57 |
| 1989 | 2.31 | 1.17 | 2.87 | 4.84 | 2.14 | 7.26 | 1.93 | 9.18 | 13.35 | 4.08 | 1.85 | 4.74 | 55.72 |
| 1990 | 3.96 | 1.68 | 6.63 | 1.65 | 1.91 | 3.12 | 5.95 | 6.32 | 0.18 | 7.29 | 3.75 | 2.69 | 45.13 |
| 1991 | 7.78 | 0.94 | 4.66 | 4.59 | 5.37 | 4.54 | 7.38 | 8.09 | 2.29 | 0.77 | 1.64 | 1.62 | 49.67 |
| 1992 | 4.93 | 2.23 | 3.59 | 2.75 | 5.07 | 6.22 | 4.36 | 9.55 | 3.04 | 4.87 | 5.76 | 1.50 | 53.87 |
| 1993 | 8.92 | 3.08 | 5.80 | 2.72 | 2.67 | 3.70 | 4.21 | 7.69 | 5.01 | 3.00 | 3.59 | 2.30 | 52.69 |
| 1994 | 7.50 | 1.23 | 4.44 | 0.39 | 2.35 | 11.71 | 8.07 | 5.39 | 8.08 | 12.11 | 2.92 | 6.35 | 70.54 |
| 1995 | 3.94 | 3.73 | 0.70 | 1.77 | 1.31 | 6.72 | 5.81 | 11.07 | 7.98 | 3.52 | 2.02 | 1.02 | 49.59 |
| 1996 | 1.05 | 1.36 | 4.04 | 2.70 | 1.72 | 4.04 | 7.34 | 5.73 | 8.77 | 5.07 | 1.74 | 2.14 | 45.70 |
| 1997 | 2.68 | 2.86 | 1.81 | 6.61 | 2.04 | 13.76 | 8.51 | 2.15 | 9.58 | 4.12 | 3.26 | 5.19 | 62.57 |
| 1998 | 7.58 | 10.17 | 5.51 | 4.01 | 4.63 | 3.41 | 6.74 | 4.44 | 14.74 | 1.99 | 0.16 | 4.34 | 67.72 |
| 1999 | 4.96 | 2.01 | 2.15 | 2.90 | 3.95 | 2.32 | 3.19 | 3.68 | 10.81 | 6.20 | 1.70 | 2.54 | 46.41 |
| 2000 | 4.04 | 2.01 | 3.66 | 1.78 | 0.57 | 4.35 | 10.81 | 4.47 | 8.88 | T | 2.72 | 2.65 | 45.94 |
| 2001 | 1.07 | 2.31 | 6.30 | 0.94 | 1.36 | 8.83 | 9.73 | 1.65 | 4.90 | 0.65 | 0.53 | 1.71 | 39.98 |
| 2002 | 2.42 | 2.19 | 4.48 | 1.57 | 2.39 | 5.89 | 6.55 | 11.37 | 4.92 | 8.44 | 5.66 | 5.09 | 60.97 |
| 2003 | 1.04 | 2.46 | 7.20 | 5.17 | 4.79 | 6.58 | 9.10 | 3.83 | 4.46 | 3.68 | 0.82 | 1.86 | 50.99 |
| 2004 | 1.57 | 4.40 | 0.36 | 4.10 | 3.04 | 3.50 | 3.77 | 10.99 | 3.71 | 1.38 | 1.37 | 1.04 | 39.23 |
| 2005 | 1.70 | 3.07 | 4.31 | 1.73 | 5.17 | 4.68 | 3.22 | 9.30 | 0.97 | 5.42 | 3.34 | 3.25 | 46.16 |
| 2006 | 2.93 | 3.30 | 0.44 | 2.99 | 2.72 | 10.72 | 3.87 | 9.11 | 4.31 | 2.39 | 4.18 | 2.33 | 49.29 |
| 2007 | 3.83 | 2.47 | 0.79 | 0.88 | 0.93 | 4.33 | 6.17 | 5.28 | 7.29 | 5.66 | 0.03 | 4.39 | 42.05 |
| 2008 | 3.07 | 2.72 | 2.41 | 2.49 | 3.66 | 2.61 | 4.97 | 5.99 | 5.68 | 11.10 | 2.27 | 0.35 | 47.32 |
| 2009 | 1.29 | 1.33 | 2.84 | 5.24 | 6.45 | 5.04 | 8.16 | 10.12 | 0.48 | 2.71 | 2.25 | 10.06 | 55.97 |
| POR= 80 YRS | 3.18 | 3.15 | 3.94 | 2.74 | 3.61 | 5.67 | 7.20 | 6.73 | 5.46 | 3.17 | 2.29 | 2.99 | 50.13 |

WBAN : 13880

## AVERAGE TEMPERATURE (°F) 2009 CHARLESTON (KCHS)

| YEAR | JAN | FEB | MAR | APR | MAY | JUN | JUL | AUG | SEP | OCT | NOV | DEC | ANNUAL |
|---|---|---|---|---|---|---|---|---|---|---|---|---|---|
| 1980 | 48.7 | 45.9 | 54.6 | 64.3 | 71.4 | 78.4 | 82.4 | 82.1 | 79.8 | 65.0 | 55.4 | 47.5 | 64.6 |
| 1981 | 41.6 | 50.8 | 54.3 | 67.5 | 70.8 | 82.7 | 83.5 | 80.3 | 74.8 | 64.1 | 55.4 | 46.2 | 64.3 |
| 1982 | 45.1 | 51.5 | 59.2 | 61.8 | 72.2 | 78.8 | 81.2 | 80.0 | 74.5 | 65.1 | 60.9 | 57.0 | 65.6 |
| 1983 | 45.6 | 49.0 | 56.4 | 61.0 | 71.7 | 76.9 | 82.8 | 82.9 | 75.5 | 68.9 | 57.4 | 48.8 | 64.7 |
| 1984 | 46.1 | 52.8 | 57.6 | 64.0 | 71.7 | 78.8 | 79.9 | 81.1 | 73.0 | 71.3 | 54.2 | 57.2 | 65.6 |
| 1985 | 42.6 | 50.5 | 60.7 | 67.8 | 73.6 | 79.6 | 80.9 | 79.9 | 75.8 | 72.2 | 67.3 | 47.9 | 66.6 |
| 1986 | 45.8 | 55.5 | 58.0 | 66.1 | 74.3 | 81.4 | 86.1 | 79.9 | 78.6 | 68.8 | 63.1 | 52.8 | 67.5 |
| 1987 | 47.2 | 48.8 | 56.8 | 62.6 | 73.3 | 80.1 | 83.0 | 83.5 | 77.8 | 61.0 | 60.1 | 53.5 | 65.6 |
| 1988 | 43.2 | 49.2 | 57.4 | 64.4 | 71.9 | 76.7 | 81.8 | 82.0 | 76.3 | 62.7 | 61.0 | 50.6 | 64.8 |
| 1989 | 55.6 | 55.0 | 59.7 | 65.3 | 72.3 | 80.4 | 82.8 | 80.7 | 76.6 | 68.7 | 60.6 | 43.2 | 66.7 |
| 1990 | 55.4 | 59.2 | 62.5 | 66.0 | 74.4 | 81.0 | 83.6 | 82.5 | 79.2 | 70.5 | 60.4 | 56.4 | 69.3 |
| 1991 | 50.8 | 54.9 | 60.5 | 67.6 | 76.3 | 78.9 | 83.6 | 82.0 | 77.4 | 67.6 | 56.0 | 54.3 | 67.5 |
| 1992 | 49.5 | 55.0 | 57.9 | 63.6 | 70.3 | 77.5 | 83.9 | 80.6 | 76.2 | 65.1 | 60.4 | 51.0 | 65.9 |
| 1993 | 53.7 | 49.7 | 55.9 | 61.6 | 72.5 | 79.5 | 85.5 | 82.0 | 78.3 | 67.3 | 59.3 | 48.5 | 66.2 |
| 1994 | 46.5 | 53.4 | 61.3 | 68.3 | 71.1 | 79.8 | 82.0 | 80.0 | 75.7 | 66.4 | 62.5 | 54.4 | 66.8 |
| 1995 | 49.7 | 50.2 | 60.4 | 68.5 | 76.0 | 78.5 | 83.4 | 82.5 | 75.6 | 69.8 | 53.8 | 47.5 | 66.3 |
| 1996 | 48.0 | 51.8 | 53.7 | 64.2 | 74.4 | 78.3 | 81.9 | 78.7 | 75.4 | 65.8 | 54.6 | 52.3 | 64.9 |
| 1997 | 49.7 | 55.1 | 64.3 | 63.1 | 69.2 | 74.8 | 80.8 | 79.0 | 76.0 | 66.0 | 55.2 | 49.9 | 65.3 |
| 1998 | 52.5 | 53.7 | 56.2 | 65.2 | 75.1 | 82.8 | 83.8 | 81.2 | 77.3 | 68.6 | 62.4 | 55.7 | 67.9 |
| 1999 | 53.6 | 53.6 | 54.9 | 67.8 | 70.4 | 76.4 | 82.5 | 83.2 | 74.7 | 66.4 | 59.8 | 50.5 | 66.2 |
| 2000 | 47.4 | 53.2 | 60.9 | 62.8 | 75.6 | 79.7 | 81.5 | 80.3 | 75.2 | 64.9 | 55.3 | 42.6 | 65.0 |
| 2001 | 46.4 | 54.6 | 56.6 | 65.2 | 72.6 | 79.0 | 79.8 | 80.9 | 73.6 | 64.8 | 63.6 | 56.4 | 66.1 |
| 2002 | 51.3 | 51.8 | 60.5 | 69.6 | 71.4 | 78.7 | 82.7 | 79.6 | 79.0 | 70.7 | 56.0 | 48.3 | 66.6 |
| 2003 | 44.7 | 50.9 | 61.2 | 64.7 | 74.0 | 78.7 | 81.3 | 81.6 | 74.6 | 66.9 | 62.1 | 47.1 | 65.7 |
| 2004 | 46.3 | 47.8 | 59.7 | 64.9 | 75.8 | 80.4 | 82.7 | 79.3 | 76.4 | 69.0 | 59.9 | 49.7 | 66.0 |
| 2005 | 50.8 | 51.7 | 55.9 | 63.2 | 69.9 | 78.6 | 83.5 | 82.1 | 79.2 | 67.8 | 60.0 | 49.2 | 66.0 |
| 2006 | 54.1 | 50.0 | 57.2 | 68.1 | 72.2 | 78.5 | 82.2 | 82.6 | 75.5 | 65.0 | 57.6 | 56.4 | 66.6 |
| 2007 | 52.4 | 49.8 | 60.9 | 64.2 | 71.4 | 78.5 | 80.7 | 83.8 | 77.5 | 71.6 | 57.5 | 57.0 | 67.1 |
| 2008 | 48.9 | 55.3 | 58.6 | 65.0 | 71.8 | 81.2 | 81.2 | 81.3 | 76.2 | 64.5 | 54.5 | 56.0 | 66.2 |
| 2009 | 48.2 | 49.8 | 58.0 | 64.5 | 73.4 | 80.9 | 80.9 | 81.6 | 76.6 | 67.5 | 58.8 | 51.0 | 65.9 |
| POR= 80 YRS | 49.2 | 50.4 | 57.6 | 64.4 | 72.6 | 78.1 | 81.4 | 80.6 | 75.8 | 66.8 | 57.5 | 50.6 | 65.4 |

## HEATING DEGREE DAYS (base 65°F) 2009  CHARLESTON (KCHS)

| YEAR | JUL | AUG | SEP | OCT | NOV | DEC | JAN | FEB | MAR | APR | MAY | JUN | TOTAL |
|------|-----|-----|-----|-----|-----|-----|-----|-----|-----|-----|-----|-----|-------|
| 1980-81 | 0 | 0 | 0 | 80 | 287 | 537 | 719 | 393 | 333 | 55 | 16 | 0 | 2420 |
| 1981-82 | 0 | 0 | 3 | 88 | 291 | 577 | 611 | 372 | 214 | 132 | 3 | 0 | 2291 |
| 1982-83 | 0 | 0 | 0 | 102 | 154 | 276 | 596 | 440 | 264 | 146 | 2 | 0 | 1980 |
| 1983-84 | 0 | 0 | 4 | 24 | 230 | 500 | 578 | 347 | 240 | 92 | 16 | 0 | 2031 |
| 1984-85 | 0 | 0 | 9 | 13 | 337 | 249 | 692 | 418 | 183 | 47 | 4 | 0 | 1952 |
| 1985-86 | 0 | 0 | 2 | 16 | 54 | 526 | 586 | 261 | 244 | 74 | 4 | 0 | 1767 |
| 1986-87 | 0 | 6 | 0 | 56 | 128 | 376 | 545 | 446 | 272 | 131 | 7 | 0 | 1967 |
| 1987-88 | 0 | 0 | 0 | 135 | 188 | 358 | 669 | 458 | 239 | 85 | 7 | 2 | 2141 |
| 1988-89 | 0 | 0 | 0 | 107 | 145 | 442 | 286 | 312 | 220 | 121 | 14 | 0 | 1647 |
| 1989-90 | 0 | 0 | 1 | 50 | 169 | 669 | 294 | 189 | 137 | 67 | 0 | 0 | 1576 |
| 1990-91 | 0 | 0 | 0 | 65 | 152 | 280 | 432 | 293 | 185 | 34 | 0 | 0 | 1441 |
| 1991-92 | 0 | 0 | 0 | 41 | 281 | 349 | 472 | 293 | 234 | 119 | 28 | 0 | 1817 |
| 1992-93 | 0 | 0 | 3 | 70 | 197 | 430 | 353 | 421 | 279 | 132 | 1 | 0 | 1886 |
| 1993-94 | 0 | 0 | 3 | 49 | 206 | 504 | 568 | 327 | 152 | 30 | 11 | 0 | 1850 |
| 1994-95 | 0 | 0 | 0 | 46 | 119 | 338 | 468 | 414 | 167 | 35 | 1 | 0 | 1588 |
| 1995-96 | 0 | 0 | 8 | 44 | 357 | 535 | 518 | 383 | 350 | 107 | 15 | 0 | 2317 |
| 1996-97 | 0 | 0 | 0 | 54 | 323 | 391 | 467 | 288 | 94 | 106 | 20 | 9 | 1752 |
| 1997-98 | 0 | 0 | 0 | 87 | 293 | 463 | 391 | 318 | 288 | 66 | 0 | 0 | 1906 |
| 1998-99 | 0 | 0 | 0 | 38 | 113 | 306 | 358 | 315 | 314 | 61 | 20 | 0 | 1525 |
| 1999-00 | 0 | 0 | 4 | 71 | 175 | 446 | 543 | 339 | 148 | 96 | 0 | 0 | 1822 |
| 2000-01 | 0 | 0 | 1 | 80 | 306 | 687 | 571 | 300 | 272 | 94 | 1 | 0 | 2312 |
| 2001-02 | 0 | 0 | 8 | 110 | 94 | 277 | 428 | 368 | 195 | 38 | 21 | 0 | 1539 |
| 2002-03 | 0 | 0 | 0 | 17 | 276 | 510 | 619 | 393 | 145 | 69 | 0 | 0 | 2029 |
| 2003-04 | 0 | 0 | 4 | 26 | 151 | 551 | 576 | 493 | 186 | 83 | 5 | 0 | 2075 |
| 2004-05 | 0 | 0 | 0 | 25 | 205 | 482 | 442 | 369 | 290 | 92 | 25 | 0 | 1930 |
| 2005-06 | 0 | 0 | 0 | 89 | 179 | 485 | 334 | 409 | 261 | 46 | 8 | 0 | 1811 |
| 2006-07 | 0 | 0 | 0 | 98 | 228 | 269 | 390 | 421 | 157 | 109 | 3 | 0 | 1675 |
| 2007-08 | 0 | 0 | 0 | 18 | 231 | 253 | 497 | 285 | 212 | 80 | 5 | 0 | 1581 |
| 2008-09 | 0 | 0 | 2 | 109 | 320 | 292 | 513 | 418 | 236 | 75 | 16 | 0 | 1981 |
| 2009- | 0 | 0 | 0 | 59 | 189 | 436 | | | | | | | |

WBAN : 13880

## COOLING DEGREE DAYS (base 65°F) 2009  CHARLESTON (KCHS)

| YEAR | JAN | FEB | MAR | APR | MAY | JUN | JUL | AUG | SEP | OCT | NOV | DEC | TOTAL |
|------|-----|-----|-----|-----|-----|-----|-----|-----|-----|-----|-----|-----|-------|
| 1980 | 0 | 9 | 7 | 69 | 221 | 407 | 549 | 539 | 451 | 87 | 5 | 1 | 2345 |
| 1981 | 0 | 0 | 9 | 138 | 199 | 539 | 582 | 481 | 307 | 66 | 9 | 0 | 2330 |
| 1982 | 0 | 2 | 42 | 42 | 232 | 420 | 510 | 475 | 293 | 111 | 36 | 34 | 2197 |
| 1983 | 0 | 0 | 6 | 32 | 217 | 362 | 559 | 567 | 322 | 149 | 10 | 4 | 2228 |
| 1984 | 0 | 1 | 19 | 67 | 228 | 420 | 471 | 509 | 254 | 212 | 21 | 10 | 2212 |
| 1985 | 7 | 17 | 57 | 136 | 276 | 445 | 501 | 470 | 332 | 245 | 129 | 5 | 2620 |
| 1986 | 0 | 2 | 36 | 114 | 300 | 499 | 662 | 474 | 414 | 182 | 78 | 5 | 2766 |
| 1987 | 0 | 0 | 26 | 62 | 269 | 459 | 567 | 580 | 389 | 18 | 48 | 9 | 2427 |
| 1988 | 0 | 2 | 12 | 74 | 229 | 359 | 529 | 534 | 349 | 43 | 30 | 4 | 2165 |
| 1989 | 5 | 37 | 64 | 136 | 246 | 470 | 561 | 493 | 358 | 173 | 44 | 0 | 2587 |
| 1990 | 4 | 34 | 65 | 105 | 302 | 487 | 583 | 548 | 430 | 238 | 24 | 21 | 2841 |
| 1991 | 0 | 20 | 52 | 118 | 356 | 426 | 584 | 531 | 379 | 128 | 18 | 25 | 2637 |
| 1992 | 0 | 7 | 20 | 83 | 199 | 381 | 592 | 495 | 345 | 79 | 68 | 5 | 2274 |
| 1993 | 7 | 0 | 4 | 37 | 241 | 440 | 644 | 536 | 410 | 127 | 42 | 0 | 2488 |
| 1994 | 0 | 7 | 42 | 133 | 210 | 449 | 537 | 473 | 327 | 93 | 53 | 17 | 2341 |
| 1995 | 1 | 6 | 29 | 146 | 349 | 410 | 577 | 552 | 334 | 198 | 26 | 0 | 2628 |
| 1996 | 0 | 8 | 6 | 88 | 313 | 405 | 531 | 432 | 320 | 87 | 18 | 6 | 2214 |
| 1997 | 0 | 17 | 77 | 56 | 158 | 309 | 496 | 444 | 336 | 124 | 7 | 0 | 2024 |
| 1998 | 10 | 6 | 23 | 77 | 320 | 542 | 592 | 509 | 378 | 158 | 43 | 26 | 2684 |
| 1999 | 10 | 1 | 6 | 153 | 193 | 351 | 548 | 571 | 303 | 122 | 25 | 1 | 2284 |
| 2000 | 1 | 2 | 29 | 36 | 336 | 448 | 522 | 481 | 315 | 84 | 25 | 0 | 2279 |
| 2001 | 0 | 13 | 16 | 107 | 242 | 423 | 467 | 499 | 269 | 114 | 58 | 17 | 2225 |
| 2002 | 10 | 7 | 62 | 182 | 228 | 419 | 555 | 461 | 428 | 199 | 16 | 0 | 2567 |
| 2003 | 0 | 1 | 35 | 67 | 284 | 417 | 513 | 520 | 300 | 89 | 67 | 0 | 2293 |
| 2004 | 6 | 3 | 32 | 86 | 348 | 470 | 557 | 450 | 351 | 155 | 57 | 14 | 2529 |
| 2005 | 8 | 3 | 13 | 47 | 184 | 416 | 583 | 535 | 433 | 183 | 36 | 0 | 2441 |
| 2006 | 0 | 0 | 24 | 148 | 240 | 411 | 539 | 554 | 323 | 108 | 13 | 10 | 2370 |
| 2007 | 6 | 0 | 34 | 92 | 206 | 414 | 492 | 587 | 381 | 231 | 14 | 12 | 2469 |
| 2008 | 6 | 11 | 19 | 85 | 221 | 491 | 508 | 510 | 346 | 102 | 13 | 20 | 2332 |
| 2009 | 0 | 0 | 26 | 64 | 282 | 485 | 501 | 523 | 357 | 144 | 14 | 7 | 2403 |

## SNOWFALL (inches) 2009 CHARLESTON (KCHS)

| YEAR | JUL | AUG | SEP | OCT | NOV | DEC | JAN | FEB | MAR | APR | MAY | JUN | TOTAL |
|------|-----|-----|-----|-----|-----|-----|-----|-----|-----|-----|-----|-----|-------|
| 1980-81 | 0.0 | 0.0 | 0.0 | 0.0 | 0.0 | 3.8 | 0.0 | 0.0 | 0.0 | 0.0 | 0.0 | 0.0 | 3.8 |
| 1981-82 | 0.0 | 0.0 | 0.0 | 0.0 | 0.0 | 0.0 | T | 0.0 | 0.0 | 0.0 | 0.0 | 0.0 | T |
| 1982-83 | 0.0 | 0.0 | 0.0 | 0.0 | 0.0 | 0.0 | 0.0 | 0.0 | T | 0.0 | 0.0 | 0.0 | T |
| 1983-84 | 0.0 | 0.0 | 0.0 | 0.0 | 0.0 | 0.0 | T | 0.0 | 0.0 | 0.0 | 0.0 | 0.0 | T |
| 1984-85 | 0.0 | 0.0 | 0.0 | 0.0 | 0.0 | 0.0 | T | 0.0 | 0.0 | T | 0.0 | 0.0 | T |
| 1985-86 | 0.0 | 0.0 | 0.0 | 0.0 | 0.0 | 0.0 | 0.5 | 0.0 | 0.0 | 0.0 | 0.0 | 0.0 | 0.5 |
| 1986-87 | 0.0 | 0.0 | 0.0 | 0.0 | 0.0 | 0.0 | T | 0.0 | T | 0.0 | 0.0 | 0.0 | T |
| 1987-88 | 0.0 | 0.0 | 0.0 | 0.0 | 0.0 | 0.0 | 0.4 | 0.0 | 0.0 | 0.0 | 0.0 | 0.0 | 0.4 |
| 1988-89 | 0.0 | 0.0 | 0.0 | 0.0 | 0.0 | T | 0.0 | 0.9 | T | 0.0 | 0.0 | T | 0.9 |
| 1989-90 | 0.0 | 0.0 | 0.0 | 0.0 | 0.0 | 8.0 | 0.0 | 0.0 | 0.0 | 0.0 | 0.0 | 0.0 | 8.0 |
| 1990-91 | 0.0 | 0.0 | 0.0 | 0.0 | 0.0 | 0.0 | 0.0 | T | 0.0 | 0.0 | 0.0 | 0.0 | T |
| 1991-92 | 0.0 | 0.0 | 0.0 | 0.0 | 0.0 | 0.0 | 0.0 | 0.0 | 0.0 | 0.0 | 0.0 | 0.0 | 0.0 |
| 1992-93 | 0.0 | 0.0 | 0.0 | 0.0 | 0.0 | 0.0 | 0.0 | 0.0 | T | 0.0 | 0.0 | 0.0 | T |
| 1993-94 | T | 0.0 | 0.0 | 0.0 | 0.0 | 0.0 | T | 0.0 | 0.0 | 0.0 | 0.0 | 0.0 | T |
| 1994-95 | 0.0 | 0.0 | 0.0 | 0.0 | 0.0 | 0.0 | 0.0 | 0.0 | 0.0 | 0.0 | 0.0 | T | T |
| 1995-96 | 0.0 | 0.0 | 0.0 | 0.0 | T | 0.0 | 0.0 | | | | | | |
| 1996-97 | | | | | | | | | | | | | |
| 1997-98 | | | | | | | | | | | | | |
| 1998-99 | | | | | | | | | | | | | |
| 1999-00 | | | | | | | | | | | | | |
| 2000-01 | | | | | | | | | | | | | |
| 2001-02 | | | | 0.0 | 0.0 | 0.0 | T | 0.0 | 0.0 | 0.0 | 0.0 | 0.0 | |
| 2002-03 | 0.0 | 0.0 | 0.0 | 0.0 | | 0.0 | T | 0.0 | 0.0 | 0.0 | 0.0 | 0.0 | |
| 2003-04 | 0.0 | 0.0 | 0.0 | 0.0 | 0.0 | 0.0 | T | T | 0.0 | 0.0 | 0.0 | 0.0 | T |
| 2004-05 | 0.0 | 0.0 | 0.0 | 0.0 | 0.0 | T | T | 0.0 | 0.0 | 0.0 | 0.0 | 0.0 | T |
| 2005-06 | 0.0 | 0.0 | 0.0 | 0.0 | 0.0 | 0.0 | 0.0 | 0.0 | 0.0 | 0.0 | T | T | T |
| 2006-07 | 0.0 | 0.0 | 0.0 | 0.0 | T | 0.0 | 0.0 | 0.0 | 0.0 | 0.0 | 0.0 | 0.0 | T |
| 2007-08 | 0.0 | 0.0 | 0.0 | 0.0 | 0.0 | 0.0 | 0.0 | 0.0 | 0.0 | T | 0.0 | 0.0 | T |
| 2008-09 | 0.0 | 0.0 | 0.0 | 0.0 | 0.0 | 0.0 | T | 0.0 | 0.0 | 0.0 | 0.0 | 0.0 | T |
| 2009- | 0.0 | 0.0 | 0.0 | 0.0 | 0.0 | 0.0 | | | | | | | |
| POR= 74 YRS | T | 0.0 | 0.0 | 0.0 | T | 0.2 | T | 0.2 | 0.1 | T | T | T | 0.5 |

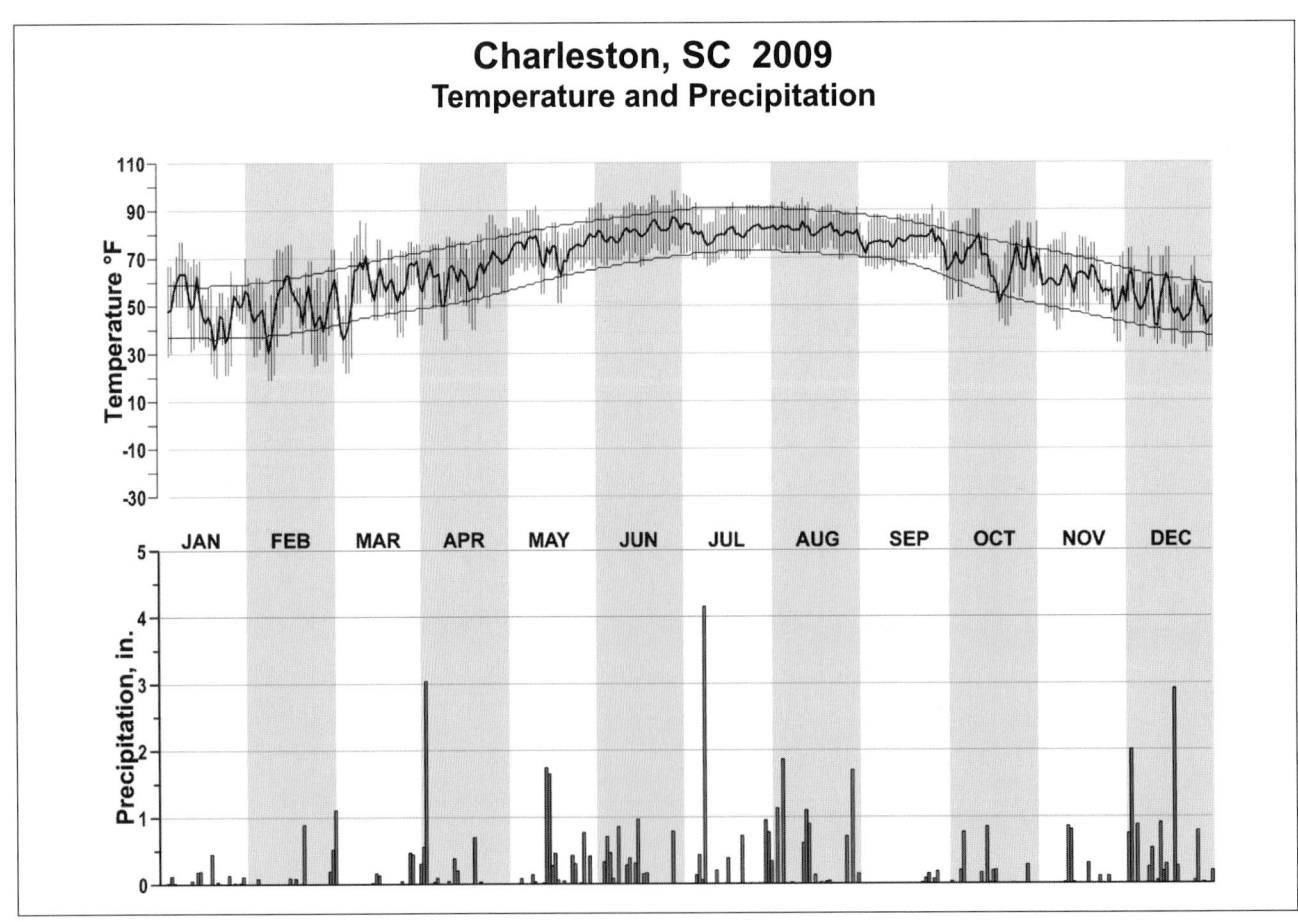

## Charleston, SC 2009
### Temperature and Precipitation

# 2009
# ABERDEEN
# SOUTH DAKOTA (KABR)

Aberdeen is located in the northeast quarter of South Dakota, approximately 200 miles south of the geographical center of the North American continent. The surrounding area, extensively cultivated, is the bed of glacial Lake Dakota, which is by far the largest flat area in South Dakota. The lake bed slopes gently to the south. The elevation of Aberdeen at the northern end of the lake bed is 1,300 feet. The elevation at the southern end, some 30 miles distant is 1,280 feet. Low hills rim the area on the east and west. These hills effect ceilings, visibility, and precipitation, which are a hazard to private aircraft operating in the area during periods of marginal weather. Principal drainage for the area is through the southward flowing, meandering James River with its associated meandering rivers and creeks.

Located near the center of the North American land mass, the climate is continental with distinct seasons. Frequent and rapid weather changes occur during all seasons of the year as migratory storms sweep through the area. The winters are cold and dry. Sub-zero minimum temperatures may set in as early as late November, although temperatures of zero and below are generally not recorded until mid-December. Lowest temperatures of the winter generally occur in the period from mid-January to mid-February. During the coldest periods the days are generally sunny with light winds, and these conditions partially moderate the discomfort experienced at such low temperatures. Some days of the winter will be extremely unpleasant with temperatures near or below zero and brisk winds. Heavy snowfalls rarely occur during the first two-thirds of the winter season, with heaviest snowfalls developing during late February and early March as temperatures moderate.

Blizzards are infrequent, many winters will pass without a single occurrence of this type of weather phenomenon. However, difficult driving conditions occur several times during most winters during periods of weather termed ground blizzards.

Spring is a very short and transitional period, but marked by very rapid weather changes. Cool to quite cold nights prevail into mid-May, although afternoon temperatures may be quite warm, as high as the mid-80s. Frost is rarely experienced after the end of May. Precipitation increases markedly during the spring, with 42 percent of the total annual precipitation normally being recorded in the three month period from April through June.

Summers are pleasant with a maximum of sunshine, warm days, and generally cool and comfortable nights. Temperatures of 100 degrees or above may occur several times during the summer season, but low humidities, brisk winds during the heat of the day, and rapid cooling during the evening hours, which generally occur during the periods of elevated temperatures, markedly moderate the physical discomfort normally experienced at these high temperatures. In June and August, thunderstorms are more likely to occur during the early evening and nighttime hours. During July, thunderstorms are approximately equally distributed throughout the 24 hours of the day. Hail is most likely in late May and early June.

Autumn is most pleasant with mild days, cool nights, ample sunshine, and declining occurrences and amounts of precipitation. The first frost may be expected by late September, although it may occur as early as late August. By mid-October, the temperatures during the night will be near or below freezing. The growing season is about 132 days.

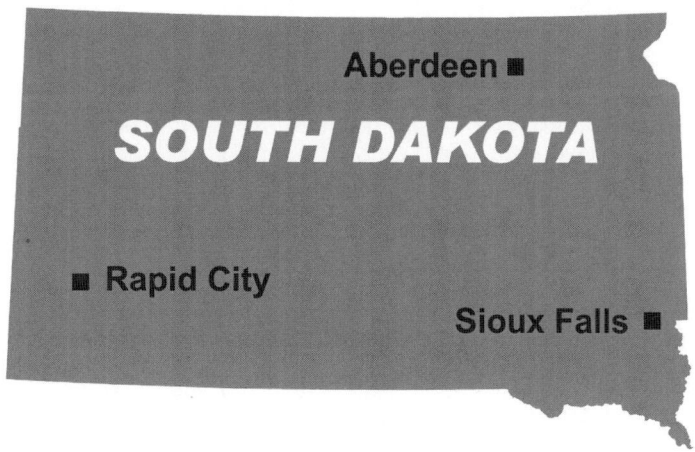

# NORMALS, MEANS, AND EXTREMES
## ABERDEEN (KABR)

LATITUDE: 45°26'N  LONGITUDE: -98°25'W  ELEVATION (FT): GRND: 1295  BARO: 1306  TIME ZONE: CENTRAL (UTC -6)  WBAN: 14929

| | ELEMENT | POR | JAN | FEB | MAR | APR | MAY | JUN | JUL | AUG | SEP | OCT | NOV | DEC | YEAR |
|---|---|---|---|---|---|---|---|---|---|---|---|---|---|---|---|
| TEMPERATURE °F | NORMAL DAILY MAXIMUM | 30 | 21.4 | 28.5 | 40.2 | 57.4 | 70.2 | 78.7 | 84.7 | 83.5 | 73.0 | 59.2 | 38.8 | 25.7 | 55.1 |
| | MEAN DAILY MAXIMUM | 115 | 21.2 | 25.0 | 39.3 | 56.1 | 69.8 | 77.0 | 85.4 | 83.7 | 72.3 | 60.0 | 39.7 | 26.6 | 54.7 |
| | HIGHEST DAILY MAXIMUM | 48 | 60 | 62 | 82 | 98 | 96 | 108 | 110 | 112 | 103 | 96 | 78 | 62 | 112 |
| | YEAR OF OCCURRENCE | | 1981 | 2000 | 1963 | 1992 | 2009 | 1988 | 1966 | 1965 | 1970 | 1963 | 1975 | 1969 | AUG 1965 |
| | MEAN OF EXTREME MAXS. | 116 | 43.4 | 48.1 | 63.2 | 81.1 | 88.0 | 93.0 | 98.1 | 96.7 | 92.1 | 81.8 | 64.2 | 46.4 | 74.7 |
| | NORMAL DAILY MINIMUM | 30 | 0.6 | 8.8 | 21.2 | 33.4 | 45.6 | 54.8 | 59.7 | 57.4 | 46.5 | 34.4 | 19.7 | 6.3 | 32.4 |
| | MEAN DAILY MINIMUM | 115 | 0.3 | 4.9 | 18.4 | 31.7 | 43.8 | 53.0 | 58.8 | 56.2 | 45.3 | 33.8 | 19.3 | 7.1 | 31.1 |
| | LOWEST DAILY MINIMUM | 48 | -42 | -45 | -32 | -2 | 13 | 33 | 39 | 32 | 20 | 8 | -27 | -39 | -45 |
| | YEAR OF OCCURRENCE | | 2009 | 1994 | 1995 | 1975 | 2005 | 1964 | 1971 | 1987 | 1995 | 2006 | 1964 | 1967 | FEB 1994 |
| | MEAN OF EXTREME MINS. | 116 | -23.8 | -17.6 | -5.6 | 16.3 | 28.2 | 40.8 | 46.5 | 42.5 | 29.2 | 17.4 | -0.4 | -15.7 | 13.2 |
| | NORMAL DRY BULB | 30 | 11.0 | 18.7 | 30.7 | 45.4 | 57.9 | 66.8 | 72.2 | 70.5 | 59.8 | 46.8 | 29.3 | 16.0 | 43.8 |
| | MEAN DRY BULB | 115 | 10.8 | 15.0 | 28.9 | 43.9 | 56.8 | 65.1 | 72.1 | 70.0 | 58.8 | 46.9 | 29.5 | 16.9 | 42.9 |
| | MEAN WET BULB | 26 | 12.3 | 16.6 | 27.1 | 38.5 | 50.1 | 59.4 | 64.2 | 62.4 | 53.1 | 40.3 | 26.4 | 15.6 | 38.8 |
| | MEAN DEW POINT | 26 | 9.4 | 13.6 | 23.4 | 32.9 | 45.5 | 56.0 | 61.2 | 59.1 | 49.1 | 35.6 | 23.3 | 12.8 | 35.2 |
| | NORMAL NO. DAYS WITH: | | | | | | | | | | | | | | |
| | MAXIMUM >= 90 | 30 | 0.0 | 0.0 | 0.0 | 0.2 | 0.4 | 2.7 | 8.3 | 6.7 | 2.1 | 0.1 | 0.0 | 0.0 | 20.5 |
| | MAXIMUM <= 32 | 30 | 23.0 | 15.9 | 8.4 | 0.6 | 0.0 | 0.0 | 0.0 | 0.0 | 0.0 | 0.3 | 9.7 | 20.3 | 78.2 |
| | MINIMUM <= 32 | 30 | 30.9 | 27.5 | 26.4 | 14.5 | 2.2 | 0.0 | 0.0 | * | 2.1 | 13.1 | 27.4 | 30.9 | 175.0 |
| | MINIMUM <= 0 | 30 | 15.2 | 8.9 | 2.5 | * | 0.0 | 0.0 | 0.0 | 0.0 | 0.0 | 0.0 | 2.1 | 10.2 | 38.9 |
| H/C | NORMAL HEATING DEG. DAYS | 30 | 1678 | 1312 | 1072 | 591 | 251 | 59 | 11 | 27 | 206 | 569 | 1066 | 1506 | 8348 |
| | NORMAL COOLING DEG. DAYS | 30 | 0 | 0 | 0 | 3 | 29 | 112 | 235 | 196 | 49 | 2 | 0 | 0 | 626 |
| RH | NORMAL (PERCENT) | 30 | 77 | 78 | 75 | 64 | 64 | 68 | | | | | 77 | 78 | |
| | HOUR 00 LST | 30 | 78 | 81 | 82 | 77 | 76 | 80 | | | | | 82 | 81 | |
| | HOUR 06 LST | 30 | 79 | 81 | 84 | 82 | 82 | 86 | 88 | 89 | 86 | 82 | 83 | 82 | 84 |
| | HOUR 12 LST | 30 | 72 | 72 | 68 | 54 | 52 | 57 | 55 | 54 | 53 | 56 | 69 | 73 | 61 |
| | HOUR 18 LST | 30 | 75 | 73 | 66 | 50 | 48 | 53 | 51 | 49 | 51 | 57 | 72 | 77 | 60 |
| S | PERCENT POSSIBLE SUNSHINE | 5 | 39 | 46 | 65 | 56 | 61 | 66 | 67 | 66 | 62 | 42 | 33 | 35 | 53 |
| W/O | MEAN NO. DAYS WITH: | | | | | | | | | | | | | | |
| | HEAVY FOG(VISBY <= 1/4 MI) | 42 | 2.3 | 2.5 | 2.6 | 1.2 | 0.8 | 0.9 | 1.1 | 1.3 | 1.4 | 1.4 | 2.3 | 2.7 | 20.5 |
| | THUNDERSTORMS | 50 | 0.0 | 0.0 | 0.3 | 1.5 | 4.0 | 8.1 | 8.0 | 6.6 | 3.0 | 1.1 | 0.0 | 0.0 | 32.6 |
| CLOUDNESS | MEAN: | | | | | | | | | | | | | | |
| | SUNRISE-SUNSET (OKTAS) | | | | 6.4 | | | 4.0 | | | | | | | |
| | MIDNIGHT-MIDNIGHT (OKTAS) | | | | 6.4 | | | | | | | | | | |
| | MEAN NO. DAYS WITH: | | | | | | | | | | | | | | |
| | CLEAR | 1 | 2.0 | 3.0 | 9.0 | | 2.0 | 9.0 | 3.0 | 4.0 | 5.0 | 2.0 | | 2.0 | |
| | PARTLY CLOUDY | 1 | 1.0 | 1.0 | 4.0 | | | 8.0 | 1.0 | 3.0 | | 3.0 | 1.0 | 1.0 | |
| | CLOUDY | 2 | 2.5 | 3.0 | 11.0 | | 13.0 | 3.0 | | 3.0 | 2.0 | 9.0 | 4.0 | 12.0 | |
| PR | MEAN STATION PRESSURE(IN) | 26 | 28.66 | 28.67 | 28.62 | 28.56 | 28.53 | 28.52 | 28.56 | 28.59 | 28.59 | 28.60 | 28.61 | 28.64 | 28.60 |
| | MEAN SEA-LEVEL PRES. (IN) | 26 | 30.12 | 30.13 | 30.05 | 29.97 | 29.91 | 29.88 | 29.93 | 29.96 | 29.97 | 30.01 | 30.04 | 30.10 | 30.01 |
| WINDS | MEAN SPEED (MPH) | 26 | 10.5 | 10.9 | 11.7 | 12.5 | 12.3 | 10.5 | 9.3 | 9.4 | 10.4 | 10.7 | 10.6 | 10.5 | 10.8 |
| | PREVAIL.DIR(TENS OF DEGS) | 22 | 19 | 19 | 36 | 36 | 17 | 17 | 17 | 17 | 17 | 19 | 19 | 19 | 17 |
| | MAXIMUM 2-MINUTE: | | | | | | | | | | | | | | |
| | SPEED (MPH) | 15 | 45 | 47 | 48 | 55 | 43 | 47 | 63 | 40 | 43 | 49 | 46 | 46 | 63 |
| | DIR. (TENS OF DEGS) | | 33 | 30 | 31 | 31 | 02 | 16 | 33 | 34 | 16 | 33 | 34 | 32 | 33 |
| | YEAR OF OCCURRENCE | | 2000 | 2002 | 2004 | 2000 | 2008 | 2005 | 2008 | 2000 | 2006 | 1996 | 1997 | 2004 | JUL 2008 |
| | MAXIMUM 3-SECOND | | | | | | | | | | | | | | |
| | SPEED (MPH) | 15 | 61 | 66 | 59 | 64 | 58 | 60 | 79 | 52 | 54 | 60 | 58 | 59 | 79 |
| | DIR. (TENS OF DEGS) | | 35 | 33 | 17 | 30 | 28 | 32 | 34 | 33 | 26 | 33 | 32 | 31 | 34 |
| | YEAR OF OCCURRENCE | | 1996 | 1996 | 2004 | 2000 | 2004 | 2008 | 2008 | 2000 | 2005 | 1996 | 2005 | 2004 | JUL 2008 |
| PRECIPITATION | NORMAL (IN) | 30 | 0.48 | 0.48 | 1.34 | 1.83 | 2.69 | 3.49 | 2.92 | 2.42 | 1.81 | 1.63 | 0.75 | 0.38 | 20.22 |
| | MAXIMUM MONTHLY (IN) | 78 | 2.23 | 2.06 | 3.45 | 7.88 | 12.23 | 8.88 | 7.71 | 6.62 | 5.32 | 7.29 | 2.87 | 1.86 | 12.23 |
| | YEAR OF OCCURRENCE | | 1937 | 1952 | 1977 | 1986 | 2007 | 1939 | 1972 | 1942 | 1996 | 1998 | 2000 | 1935 | MAY 2007 |
| | MINIMUM MONTHLY (IN) | 78 | 0.01 | 0.00 | 0.04 | 0.13 | 0.28 | 0.37 | 0.30 | 0.06 | 0.05 | 0.00 | T | T | 0.00 |
| | YEAR OF OCCURRENCE | | 1961 | 1932 | 1971 | 1988 | 1948 | 1974 | 1975 | 1947 | 1979 | 1952 | 1980 | 1986 | OCT 1952 |
| | MAXIMUM IN 24 HOURS (IN) | 78 | 3.41 | 1.02 | 3.00 | 2.28 | 7.75 | 5.20 | 3.46 | 3.50 | 3.49 | 3.75 | 1.30 | 0.91 | 7.75 |
| | YEAR OF OCCURRENCE | | 2008 | 1958 | 1937 | 1938 | 2007 | 1978 | 1983 | 1990 | 1988 | 2000 | 1977 | 1988 | MAY 2007 |
| | NORMAL NO. DAYS WITH: | | | | | | | | | | | | | | |
| | PRECIPITATION >= 0.01 | 30 | 6.0 | 6.6 | 7.4 | 8.2 | 9.9 | 10.4 | 9.7 | 8.1 | 6.4 | 6.0 | 7.0 | 6.3 | 92.0 |
| | PRECIPITATION >= 1.00 | 30 | 0.0 | 0.0 | 0.2 | 0.2 | 0.4 | 0.8 | 0.6 | 0.5 | 0.5 | 0.5 | * | 0.0 | 3.7 |
| SNOWFALL | NORMAL (IN) | 30 | 7.3 | 6.4 | 8.0 | 2.7 | 0.* | 0.0 | 0.0 | 0.0 | 0.* | 0.8 | 7.6 | 5.8 | 38.6 |
| | MAXIMUM MONTHLY (IN) | 78 | 26.2 | 25.1 | 27.9 | 24.4 | 2.0 | T | T | T | 0.2 | 5.5 | 30.5 | 18.5 | 30.5 |
| | YEAR OF OCCURRENCE | | 1937 | 1969 | 1975 | 1970 | 1943 | 1993 | 1994 | 1993 | 1995 | 1970 | 2000 | 1935 | NOV 2000 |
| | MAXIMUM IN 24 HOURS (IN) | 78 | 10.8 | 14.3 | 13.0 | 15.0 | 2.0 | T | T | T | 0.2 | 5.0 | 12.6 | 9.1 | 15.0 |
| | YEAR OF OCCURRENCE' | | 1997 | 1951 | 1937 | 1970 | 1943 | 1993 | 1994 | 1993 | 1995 | 1932 | 1993 | 1988 | APR 1970 |
| | MAXIMUM SNOW DEPTH (IN) | 60 | 30 | 25 | 24 | 15 | 7 | 0 | 0 | 0 | T | 3 | 17 | 21 | 30 |
| | YEAR OF OCCURRENCE | | 1997 | 1969 | 1969 | 1975 | 2005 | | | | 1965 | 2002 | 1996 | 1996 | JAN 1997 |
| | NORMAL NO. DAYS WITH: | | | | | | | | | | | | | | |
| | SNOWFALL >= 1.0 | 30 | 2.1 | 2.0 | 2.1 | 0.9 | 0.0 | 0.0 | 0.0 | 0.0 | 0.0 | 0.3 | 2.2 | 1.8 | 11.4 |

## PRECIPITATION (inches) 2009  ABERDEEN (KABR)

| YEAR | JAN | FEB | MAR | APR | MAY | JUN | JUL | AUG | SEP | OCT | NOV | DEC | ANNUAL |
|------|-----|-----|-----|-----|-----|-----|-----|-----|-----|-----|-----|-----|--------|
| 1980 | 0.51 | 0.44 | 0.88 | 1.15 | 1.64 | 2.53 | 0.80 | 5.93 | 0.92 | 1.44 | T | 0.14 | 16.38 |
| 1981 | 0.12 | 0.20 | 2.00 | 0.12 | 1.60 | 2.10 | 3.97 | 2.91 | | | | | |
| 1982 | | | | | | | | | | 5.14 | 0.59 | 0.09 | |
| 1983 | 0.16 | 0.26 | 2.65 | 0.69 | 1.66 | 3.47 | 6.46 | 2.21 | 1.55 | 0.81 | 0.60 | 0.55 | 21.07 |
| 1984 | 0.47 | 0.70 | 1.94 | 2.39 | 1.13 | 5.65 | 2.64 | 2.23 | 0.84 | 2.93 | 0.06 | 0.61 | 21.59 |
| 1985 | 0.23 | 0.08 | 1.82 | 0.63 | 3.41 | 1.76 | 2.38 | 2.71 | 2.71 | 0.87 | 1.60 | 0.57 | 18.77 |
| 1986 | 0.43 | 0.71 | 0.58 | 7.88 | 3.32 | 2.48 | 3.78 | 2.85 | 2.82 | 0.19 | 0.77 | T | 25.81 |
| 1987 | 0.09 | 1.12 | 1.91 | 0.41 | 2.01 | 0.77 | 2.13 | 1.87 | 1.33 | 0.20 | 0.79 | 0.09 | 12.72 |
| 1988 | 0.35 | 0.31 | 0.37 | 0.13 | 3.43 | 0.93 | 3.14 | 2.80 | 5.31 | 0.11 | 0.73 | 1.37 | 18.98 |
| 1989 | 0.52 | 0.37 | 1.46 | 3.42 | 1.20 | 2.05 | 2.00 | 3.83 | 2.23 | 0.58 | 0.74 | 0.15 | 18.55 |
| 1990 | 0.13 | 0.39 | 0.81 | 1.87 | 1.41 | 7.72 | 1.98 | 4.85 | 3.01 | 0.44 | 0.11 | 0.37 | 23.09 |
| 1991 | 0.11 | 0.70 | 0.77 | 3.70 | 7.36 | 4.76 | 1.32 | 2.28 | 0.57 | 1.06 | 0.32 | 0.13 | 23.08 |
| 1992 | 0.66 | 0.47 | 0.54 | 0.40 | 0.78 | 5.61 | 2.97 | 1.55 | 1.63 | 0.83 | 1.19 | 0.20 | 16.83 |
| 1993 | 0.61 | 0.49 | 0.42 | 1.51 | 3.11 | 6.20 | 7.37 | 4.42 | 1.21 | 0.35 | 1.88 | 0.56 | 28.13 |
| 1994 | 0.80 | 0.42 | 0.43 | 2.28 | 0.30 | 1.10 | 5.37 | 3.87 | 1.63 | 3.36 | 0.77 | 0.38 | 20.71 |
| 1995 | 0.62 | 0.50 | 2.34 | 2.26 | 5.98 | 1.34 | 3.51 | 2.36 | 1.50 | 3.16 | 0.20 | 0.47 | 24.24 |
| 1996 | 1.32 | 0.78 | 0.87 | 0.15 | 4.46 | 4.12 | 1.91 | 0.24 | 5.32 | 3.55 | 1.40 | 0.87 | 24.99 |
| 1997 | 1.34 | 0.88 | 0.79 | 2.01 | 1.72 | 2.65 | 1.41 | 3.75 | 0.82 | 3.37 | 0.46 | 0.18 | 19.38 |
| 1998 | 0.63 | 0.75 | 1.54 | 1.81 | 4.29 | 6.47 | 1.12 | 2.95 | 0.06 | 7.29 | 1.41 | 0.18 | 28.50 |
| 1999 | 0.62 | 0.21 | 0.92 | 1.76 | 2.97 | 5.23 | 2.80 | 3.19 | 4.27 | 0.15 | T | 0.15 | 22.27 |
| 2000 | 0.27 | 0.69 | 1.21 | 2.47 | 2.93 | 4.94 | 4.51 | 1.69 | 0.51 | 4.75 | 2.87 | 0.38 | 27.22 |
| 2001 | 0.28 | 1.01 | 0.30 | 3.43 | 2.67 | 3.31 | 4.81 | 0.79 | 2.61 | 1.96 | 1.39 | 0.06 | 22.62 |
| 2002 | 0.27 | 0.03 | 0.56 | 1.15 | 1.82 | 1.22 | 3.96 | 3.82 | 0.93 | 1.40 | 0.09 | 0.30 | 15.55 |
| 2003 | 0.24 | 0.45 | 0.57 | 2.01 | 4.33 | 6.94 | 1.98 | 1.59 | 1.26 | 0.88 | 0.70 | 0.33 | 21.28 |
| 2004 | 0.56 | 0.72 | 1.27 | 0.62 | 5.10 | 3.68 | 3.02 | 0.96 | 3.53 | 1.75 | 0.23 | 0.33 | 21.77 |
| 2005 | 0.41 | 0.63 | 0.16 | 0.35 | 2.64 | 6.21 | 0.80 | 2.90 | 1.19 | 0.95 | 1.35 | 1.11 | 18.70 |
| 2006 | 0.33 | 0.20 | 0.65 | 2.41 | 2.16 | 3.21 | 0.71 | 2.47 | 2.67 | 0.13 | 0.12 | 0.88 | 15.94 |
| 2007 | 0.09 | 1.16 | 1.88 | 3.42 | 12.23 | 2.43 | 0.79 | 2.19 | 1.64 | 1.48 | 0.02 | 0.90 | 28.23 |
| 2008 | 0.07 | 0.24 | 1.76 | 0.81 | 1.32 | 3.21 | 6.26 | 1.24 | 3.61 | 4.75 | 0.30 | 0.88 | 24.45 |
| 2009 | 0.68 | 0.87 | 1.41 | 1.76 | 0.47 | 3.87 | 2.47 | 2.82 | 4.41 | 4.33 | 0.22 | 0.96 | 24.27 |
| POR= 115 YRS | 0.70 | 0.69 | 1.30 | 2.28 | 2.96 | 3.71 | 2.87 | 2.51 | 1.88 | 1.52 | 0.83 | 0.60 | 21.85 |

WBAN : 14929

## AVERAGE TEMPERATURE (°F) 2009  ABERDEEN (KABR)

| YEAR | JAN | FEB | MAR | APR | MAY | JUN | JUL | AUG | SEP | OCT | NOV | DEC | ANNUAL |
|------|-----|-----|-----|-----|-----|-----|-----|-----|-----|-----|-----|-----|--------|
| 1980 | 13.0 | 14.7 | 25.6 | 49.9 | 59.5 | 67.0 | 72.9 | 68.9 | 59.5 | 45.3 | 36.2 | 20.0 | 44.4 |
| 1981 | 19.4 | 23.9 | 35.6 | 49.9 | 56.3 | 65.3 | 74.4 | 72.1 | | | | | |
| 1982 | | | | | | | | | | 46.6 | 27.7 | 25.0 | |
| 1983 | 23.1 | 28.0 | 33.3 | 41.5 | 53.6 | 65.2 | 75.2 | 76.5 | 60.8 | 47.3 | 32.2 | -0.6 | 44.7 |
| 1984 | 15.7 | 30.3 | 27.9 | 47.0 | 54.4 | 65.6 | 71.3 | 72.7 | 55.4 | 48.6 | 32.5 | 13.0 | 44.5 |
| 1985 | 8.5 | 16.3 | 36.0 | 50.3 | 61.2 | 61.3 | 72.6 | 67.3 | 56.5 | 45.3 | 16.4 | 7.8 | 41.6 |
| 1986 | 17.3 | 12.1 | 35.8 | 44.1 | 57.6 | 68.0 | 72.5 | 66.2 | 56.7 | 47.3 | 26.3 | 24.2 | 44.0 |
| 1987 | 22.7 | 30.5 | 33.3 | 51.5 | 62.3 | 69.3 | 75.7 | 67.8 | 61.1 | 43.3 | 35.1 | 24.5 | 48.1 |
| 1988 | 9.9 | 14.6 | 33.0 | 45.4 | 63.5 | 75.0 | 76.4 | 72.9 | 59.2 | 44.4 | 29.1 | 18.8 | 45.2 |
| 1989 | 15.6 | 5.0 | 22.9 | 45.1 | 57.4 | 64.7 | 75.3 | 72.0 | 59.8 | 46.8 | 28.6 | 9.3 | 41.9 |
| 1990 | 25.3 | 21.3 | 35.8 | 44.0 | 54.9 | 66.2 | 69.7 | 70.9 | 63.6 | 45.8 | 34.7 | 12.2 | 45.4 |
| 1991 | 11.3 | 27.8 | 33.8 | 48.3 | 60.8 | 70.8 | 72.2 | 72.7 | 60.3 | 44.8 | 24.9 | 24.1 | 46.0 |
| 1992 | 21.6 | 28.8 | 36.2 | 44.1 | 59.5 | 63.7 | 64.3 | 64.3 | 58.5 | 46.2 | 29.6 | 14.7 | 44.3 |
| 1993 | 9.7 | 12.0 | 31.1 | 44.7 | 57.6 | 63.0 | 68.6 | 70.2 | 56.1 | 45.6 | 27.9 | 18.3 | 42.1 |
| 1994 | 0.4 | 7.5 | 33.5 | 46.4 | 63.5 | 70.3 | 69.3 | 65.7 | 61.5 | 49.5 | 32.8 | 16.5 | 43.1 |
| 1995 | 11.7 | 15.4 | 26.6 | 38.9 | 53.8 | 68.1 | 70.7 | 71.6 | 57.6 | 44.8 | 26.2 | 16.7 | 41.8 |
| 1996 | 4.1 | 17.5 | 21.3 | 40.7 | 53.5 | 66.9 | 67.8 | 70.1 | 58.6 | 45.2 | 17.3 | 6.0 | 39.1 |
| 1997 | 3.5 | 14.5 | 23.9 | 39.0 | 53.2 | 68.1 | 71.1 | 68.5 | 61.8 | 48.0 | 25.0 | 25.7 | 41.9 |
| 1998 | 13.8 | 28.2 | 25.1 | 48.0 | 60.0 | 62.9 | 72.9 | 71.3 | 64.3 | 48.3 | 33.3 | 22.8 | 45.9 |
| 1999 | 8.8 | 28.8 | 35.3 | 45.8 | 57.8 | 66.1 | 72.4 | 69.6 | 56.1 | 45.4 | 38.7 | 24.6 | 45.8 |
| 2000 | 13.7 | 26.3 | 36.9 | 43.6 | 58.1 | 65.3 | 71.1 | 69.6 | 59.4 | 48.9 | 22.5 | 3.1 | 43.2 |
| 2001 | 15.7 | 4.2 | 23.0 | 44.6 | 59.1 | 66.9 | 74.2 | 71.4 | 60.9 | 45.3 | 39.5 | 19.4 | 43.7 |
| 2002 | 19.5 | 27.5 | 21.6 | 42.4 | 51.3 | 70.8 | 74.8 | 67.9 | 61.0 | 37.7 | 30.5 | 22.5 | 44.0 |
| 2003 | 15.1 | 12.5 | 28.4 | 46.2 | 54.7 | 65.0 | 71.5 | 71.9 | 58.3 | 48.9 | 25.6 | 23.2 | 43.4 |
| 2004 | 9.4 | 18.1 | 36.0 | 45.8 | 54.3 | 61.9 | 68.9 | 63.4 | 62.2 | 47.2 | 34.4 | 22.9 | 43.7 |
| 2005 | 11.4 | 22.2 | 31.1 | 48.1 | 54.5 | 69.0 | 72.7 | 68.2 | 63.7 | 47.5 | 33.1 | 17.5 | 44.9 |
| 2006 | 27.2 | 19.9 | 32.8 | 49.6 | 57.5 | 67.7 | 75.4 | 70.9 | 56.3 | 42.0 | 29.8 | 24.9 | 46.2 |
| 2007 | 11.6 | 8.2 | 34.6 | 41.3 | 60.4 | 68.9 | 73.9 | 68.4 | 60.4 | 49.1 | 31.1 | 11.4 | 43.3 |
| 2008 | 7.3 | 12.1 | 28.5 | 41.9 | 54.4 | 64.2 | 72.3 | 70.0 | 59.7 | 46.0 | 32.0 | 8.8 | 41.4 |
| 2009 | 6.3 | 15.3 | 26.7 | 42.8 | 56.1 | 63.3 | 67.4 | 65.9 | 62.7 | 40.2 | 38.2 | 11.0 | 41.3 |
| POR= 115 YRS | 10.8 | 15.0 | 28.9 | 43.9 | 56.8 | 65.1 | 72.1 | 70.0 | 58.8 | 46.9 | 29.5 | 16.9 | 42.9 |

## HEATING DEGREE DAYS (base 65°F) 2009 ABERDEEN (KABR)

| YEAR | JUL | AUG | SEP | OCT | NOV | DEC | JAN | FEB | MAR | APR | MAY | JUN | TOTAL |
|------|-----|-----|-----|-----|-----|-----|-----|-----|-----|-----|-----|-----|-------|
| 1980-81 | 1 | 24 | 192 | 601 | 860 | 1388 | 1408 | 1146 | 907 | 447 | 273 | 43 | 7290 |
| 1981-82 | 13 | 1 | 0 | 0 | 0 | 0 | | | | | | | |
| 1982-83 | | | | 562 | 1111 | 1232 | 1291 | 1032 | 976 | 697 | 351 | 89 | |
| 1983-84 | 4 | 0 | 198 | 540 | 975 | 2030 | 1522 | 1001 | 1147 | 533 | 332 | 44 | 8326 |
| 1984-85 | 12 | 15 | 312 | 514 | 967 | 1608 | 1747 | 1361 | 891 | 444 | 158 | 145 | 8174 |
| 1985-86 | 2 | 42 | 292 | 608 | 1453 | 1770 | 1473 | 1477 | 897 | 622 | 245 | 32 | 8913 |
| 1986-87 | 0 | 59 | 246 | 540 | 1152 | 1258 | 1303 | 961 | 978 | 409 | 134 | 31 | 7071 |
| 1987-88 | 4 | 60 | 140 | 666 | 892 | 1247 | 1707 | 1461 | 983 | 584 | 120 | 0 | 7864 |
| 1988-89 | 3 | 30 | 193 | 631 | 1069 | 1426 | 1527 | 1682 | 1298 | 597 | 244 | 81 | 8781 |
| 1989-90 | 0 | 5 | 203 | 561 | 1085 | 1724 | 1222 | 1221 | 900 | 639 | 308 | 68 | 7936 |
| 1990-91 | 7 | 12 | 155 | 586 | 901 | 1633 | 1661 | 1035 | 961 | 491 | 205 | 6 | 7653 |
| 1991-92 | 5 | 2 | 206 | 619 | 1195 | 1264 | 1338 | 1041 | 884 | 631 | 224 | 97 | 7506 |
| 1992-93 | 61 | 97 | 206 | 582 | 1054 | 1556 | 1713 | 1481 | 1043 | 602 | 240 | 110 | 8745 |
| 1993-94 | 17 | 18 | 277 | 599 | 1109 | 1441 | 2002 | 1609 | 967 | 560 | 147 | 15 | 8761 |
| 1994-95 | 15 | 65 | 165 | 475 | 958 | 1496 | 1645 | 1385 | 1184 | 773 | 341 | 57 | 8559 |
| 1995-96 | 11 | 3 | 260 | 620 | 1154 | 1493 | 1890 | 1372 | 1348 | 723 | 369 | 59 | 9302 |
| 1996-97 | 12 | 5 | 241 | 606 | 1423 | 1827 | 1899 | 1408 | 1267 | 772 | 360 | 10 | 9830 |
| 1997-98 | 26 | 37 | 135 | 531 | 1196 | 1213 | 1585 | 1025 | 1228 | 503 | 182 | 114 | 7775 |
| 1998-99 | 1 | 0 | 107 | 511 | 947 | 1299 | 1734 | 1008 | 914 | 569 | 222 | 55 | 7367 |
| 1999-00 | 2 | 8 | 264 | 600 | 781 | 1244 | 1583 | 1116 | 866 | 636 | 227 | 76 | 7403 |
| 2000-01 | 28 | 19 | 210 | 491 | 1270 | 1914 | 1520 | 1697 | 1296 | 613 | 198 | 66 | 9322 |
| 2001-02 | 7 | 9 | 172 | 602 | 761 | 1407 | 1403 | 1045 | 1338 | 671 | 448 | 32 | 7895 |
| 2002-03 | 5 | 43 | 207 | 841 | 1028 | 1308 | 1539 | 1463 | 1130 | 565 | 319 | 70 | 8518 |
| 2003-04 | 0 | 24 | 244 | 502 | 1175 | 1288 | 1721 | 1353 | 892 | 570 | 337 | 115 | 8221 |
| 2004-05 | 37 | 99 | 157 | 544 | 911 | 1299 | 1652 | 1192 | 1044 | 502 | 331 | 26 | 7794 |
| 2005-06 | 11 | 30 | 125 | 540 | 952 | 1466 | 1166 | 1258 | 989 | 455 | 259 | 28 | 7279 |
| 2006-07 | 0 | 2 | 277 | 715 | 1052 | 1236 | 1648 | 1584 | 933 | 702 | 168 | 32 | 8349 |
| 2007-08 | 1 | 25 | 212 | 496 | 1009 | 1655 | 1782 | 1527 | 1124 | 689 | 324 | 69 | 8913 |
| 2008-09 | 4 | 15 | 187 | 584 | 983 | 1734 | 1813 | 1385 | 1183 | 662 | 280 | 129 | 8959 |
| 2009- | 21 | 52 | 99 | 760 | 796 | 1667 | | | | | | | |

WBAN : 14929

## COOLING DEGREE DAYS (base 65°F) 2009 ABERDEEN (KABR)

| YEAR | JAN | FEB | MAR | APR | MAY | JUN | JUL | AUG | SEP | OCT | NOV | DEC | TOTAL |
|------|-----|-----|-----|-----|-----|-----|-----|-----|-----|-----|-----|-----|-------|
| 1980 | 0 | 0 | 0 | 16 | 55 | 98 | 251 | 150 | 33 | 0 | 0 | 0 | 603 |
| 1981 | 0 | 0 | 0 | 2 | 11 | 57 | 314 | 230 | | 0 | 0 | | |
| 1982 | | | | | | | | | 0 | 0 | 0 | | |
| 1983 | 0 | 0 | 0 | 0 | 3 | 100 | 329 | 363 | 79 | 0 | 0 | 0 | 874 |
| 1984 | 0 | 0 | 0 | 0 | 12 | 70 | 210 | 259 | 31 | 15 | 0 | 0 | 597 |
| 1985 | 0 | 0 | 0 | 9 | 48 | 41 | 244 | 121 | 46 | 0 | 0 | 0 | 509 |
| 1986 | 0 | 0 | 0 | 1 | 22 | 126 | 240 | 105 | 6 | 0 | 0 | 0 | 500 |
| 1987 | 0 | 0 | 0 | 11 | 57 | 166 | 341 | 153 | 30 | 0 | 0 | 0 | 758 |
| 1988 | 0 | 0 | 0 | 3 | 81 | 308 | 362 | 279 | 28 | 0 | 0 | 0 | 1061 |
| 1989 | 0 | 0 | 0 | 5 | 14 | 77 | 326 | 228 | 54 | 2 | 0 | 0 | 706 |
| 1990 | 0 | 0 | 0 | 16 | 3 | 108 | 159 | 204 | 120 | 0 | 0 | 0 | 610 |
| 1991 | 0 | 0 | 0 | 0 | 82 | 187 | 238 | 244 | 72 | 3 | 0 | 0 | 826 |
| 1992 | 0 | 0 | 0 | 11 | 59 | 65 | 47 | 85 | 17 | 5 | 0 | 0 | 289 |
| 1993 | 0 | 0 | 0 | 0 | 19 | 54 | 137 | 187 | 16 | 2 | 0 | 0 | 415 |
| 1994 | 0 | 0 | 0 | 6 | 108 | 180 | 153 | 96 | 69 | 0 | 0 | 0 | 612 |
| 1995 | 0 | 0 | 0 | 0 | 0 | 161 | 197 | 212 | 49 | 3 | 0 | 0 | 622 |
| 1996 | 0 | 0 | 0 | 0 | 18 | 122 | 106 | 173 | 55 | 1 | 0 | 0 | 475 |
| 1997 | 0 | 0 | 0 | 0 | 1 | 108 | 224 | 149 | 46 | 12 | 0 | 0 | 540 |
| 1998 | 0 | 0 | 0 | 0 | 30 | 58 | 258 | 203 | 96 | 0 | 0 | 0 | 645 |
| 1999 | 0 | 0 | 0 | 0 | 9 | 98 | 238 | 160 | 2 | 0 | 0 | 0 | 507 |
| 2000 | 0 | 0 | 0 | 0 | 21 | 92 | 225 | 168 | 48 | 0 | 0 | 0 | 554 |
| 2001 | 0 | 0 | 0 | 8 | 21 | 129 | 300 | 214 | 55 | 0 | 0 | 0 | 727 |
| 2002 | 0 | 0 | 0 | 0 | 27 | 213 | 317 | 137 | 94 | 0 | 0 | 0 | 788 |
| 2003 | 0 | 0 | 0 | 8 | 5 | 79 | 209 | 243 | 51 | 6 | 0 | 0 | 601 |
| 2004 | 0 | 0 | 0 | 0 | 13 | 29 | 164 | 57 | 78 | 0 | 0 | 0 | 341 |
| 2005 | 0 | 0 | 0 | 3 | 14 | 151 | 258 | 135 | 94 | 4 | 0 | 0 | 659 |
| 2006 | 0 | 0 | 0 | 0 | 36 | 115 | 330 | 193 | 20 | 10 | 0 | 0 | 704 |
| 2007 | 0 | 0 | 0 | 0 | 33 | 155 | 284 | 138 | 81 | 7 | 0 | 0 | 698 |
| 2008 | 0 | 0 | 0 | 0 | 2 | 50 | 238 | 176 | 33 | 0 | 0 | 0 | 499 |
| 2009 | 0 | 0 | 0 | 2 | 14 | 83 | 103 | 88 | 37 | 0 | 0 | 0 | 327 |

## SNOWFALL (inches) 2009 ABERDEEN (KABR)

| YEAR | JUL | AUG | SEP | OCT | NOV | DEC | JAN | FEB | MAR | APR | MAY | JUN | TOTAL |
|------|-----|-----|-----|-----|-----|-----|-----|-----|-----|-----|-----|-----|-------|
| 1980-81 | 0.0 | 0.0 | 0.0 | 0.4 | 0.0 | 2.3 | 1.8 | 2.2 | 1.6 | T | 0.0 | 0.0 | 8.3 |
| 1981-82 | 0.0 | 0.0 | 0.0 | 0.0 | 0.0 | 0.0 | | | | | | | |
| 1982-83 | | | | 0.0 | 0.0 | 0.0 | 0.9 | 2.2 | 7.9 | 2.3 | 0.0 | 0.0 | |
| 1983-84 | 0.0 | 0.0 | 0.0 | T | 5.5 | 5.2 | 2.0 | 3.8 | 17.3 | 0.1 | 0.0 | 0.0 | 33.9 |
| 1984-85 | 0.0 | 0.0 | T | 1.0 | 0.2 | 5.6 | 3.7 | 1.0 | 18.4 | 0.1 | 0.0 | 0.0 | 30.0 |
| 1985-86 | 0.0 | 0.0 | 0.0 | T | 22.4 | 9.6 | 4.0 | 10.5 | 1.9 | 9.2 | 0.0 | 0.0 | 57.6 |
| 1986-87 | 0.0 | 0.0 | 0.0 | 0.0 | 3.4 | 0.1 | 1.1 | 10.5 | 5.7 | T | 0.0 | 0.0 | 20.8 |
| 1987-88 | 0.0 | 0.0 | 0.0 | T | 5.6 | 0.9 | 4.4 | 4.2 | 2.5 | 1.0 | 0.0 | 0.0 | 18.6 |
| 1988-89 | | | | | 5.0 | 16.4 | 5.1 | 5.8 | 13.1 | 1.6 | T | 0.0 | |
| 1989-90 | T | T | 0.0 | 0.1 | 6.4 | 1.5 | 0.8 | 9.2 | 4.8 | 4.7 | T | T | 27.5 |
| 1990-91 | 0.0 | 0.0 | 0.0 | 1.0 | T | 7.0 | 2.2 | 13.0 | 5.1 | 5.5 | 0.0 | 0.0 | 33.8 |
| 1991-92 | 0.0 | 0.0 | 0.0 | 0.4 | 4.0 | 0.3 | 7.9 | 2.4 | 0.8 | 1.1 | 0.0 | 0.0 | 16.9 |
| 1992-93 | T | 0.0 | T | 3.6 | 9.3 | 2.7 | 10.5 | 7.1 | 2.5 | 3.2 | 0.0 | T | 38.9 |
| 1993-94 | 0.0 | T | 0.0 | T | 30.1 | 9.7 | 16.0 | 9.1 | 5.8 | 6.1 | 0.0 | 0.0 | 76.8 |
| 1994-95 | T | 0.0 | 0.0 | 0.0 | 7.6 | 6.6 | 5.8 | 8.0 | 22.3 | 0.0 | 0.0 | 0.0 | 50.3 |
| 1995-96 | 0.0 | 0.0 | 0.2 | 3.5 | 2.4 | 5.3 | 17.7 | 6.0 | 9.0 | 0.4 | T | 0.0 | 44.5 |
| 1996-97 | 0.0 | 0.0 | 0.0 | 0.0 | 20.3 | 13.9 | 19.2 | 9.2 | 8.5 | 4.6 | 0.0 | T | 75.7 |
| 1997-98 | 0.0 | 0.0 | 0.0 | 0.2 | 7.7 | 3.9 | 11.0 | 6.4 | 4.5 | 0.0 | 0.0 | 0.0 | 33.7 |
| 1998-99 | 0.0 | 0.0 | 0.0 | 0.0 | 7.6 | 2.7 | 10.0 | 2.7 | 1.9 | 1.8 | 0.0 | T | 26.7 |
| 1999-00 | 0.0 | T | 0.0 | 0.2 | 0.0 | 2.2 | 4.8 | 7.3 | 4.5 | 3.9 | 0.0 | 0.0 | 22.9 |
| 2000-01 | T | 0.0 | 0.0 | 0.0 | 30.5 | 8.9 | 4.3 | 21.0 | 2.3 | 7.7 | 0.0 | T | 74.7 |
| 2001-02 | 0.0 | 0.0 | 0.0 | 0.1 | 10.5 | 1.4 | 2.5 | 0.6 | 9.3 | 1.9 | 0.4 | 0.0 | 26.7 |
| 2002-03 | 0.0 | T | 0.0 | 4.8 | 1.1 | 2.9 | 4.8 | 5.0 | 4.2 | 1.0 | 0.0 | 0.0 | 23.8 |
| 2003-04 | 0.0 | 0.0 | 0.0 | 1.6 | 9.9 | 3.8 | 10.5 | 3.6 | 4.1 | 0.1 | 0.0 | 0.0 | 33.6 |
| 2004-05 | 0.0 | 0.0 | 0.0 | 0.0 | T | 1.5 | 4.5 | 5.6 | 1.7 | T | T | 0.0 | 13.3 |
| 2005-06 | 0.0 | 0.0 | 0.0 | T | 4.0 | 13.2 | 1.4 | 3.5 | 4.7 | 0.3 | 0.0 | 0.0 | 27.1 |
| 2006-07 | 0.0 | 0.0 | 0.0 | 0.2 | T | 7.3 | 1.2 | 19.1 | 4.7 | 10.5 | 0.0 | 0.0 | 43.0 |
| 2007-08 | 0.0 | 0.0 | 0.0 | 0.0 | 0.2 | 11.6 | 1.2 | 4.4 | 13.0 | 8.2 | 0.0 | 0.0 | 38.6 |
| 2008-09 | 0.0 | 0.0 | 0.0 | 0.0 | 2.7 | 16.9 | 10.5 | 10.2 | 16.0 | T | 0.0 | 0.0 | 56.3 |
| 2009- | 0.0 | 0.0 | 0.0 | 1.6 | T | 14.1 | | | | | | | |
| POR= 83 YRS | T | T | T | 1.1 | 5.4 | 6.1 | 7.1 | 7.3 | 7.9 | 4.3 | 0.3 | T | 39.5 |

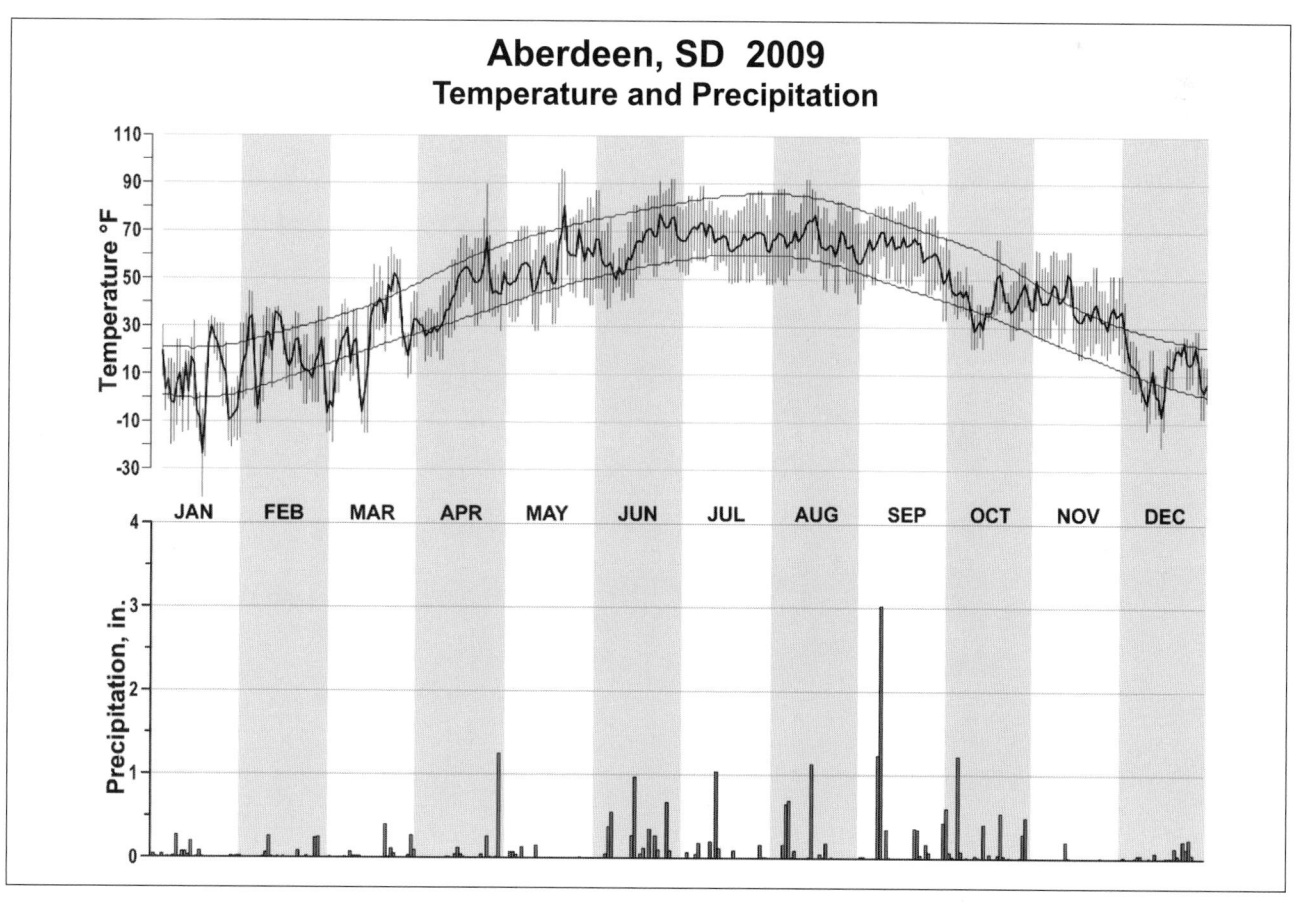

# 2009
# RAPID CITY
# SOUTH DAKOTA (KRAP)

Rapid City, which is not far from the geographical center of North America, experiences the large temperature ranges, both daily and seasonal, that are typical of semi-arid continental climates.

The city is surrounded by contrasting landforms, with the forested Black Hills rising immediately west of the city, and rolling prairie extending out in the other directions. From 40 to 70 miles southeast lie the eroded Badlands. The Black Hills, many of which are more than 5,000 feet above sea level, with a number of peaks above 7,000 feet, exert a pronounced influence on the climate of this area. The rolling land to the east of the city is cut by the valleys of the Box Elder and Rapid Creeks, which flow generally east-southeastward. The station is located on the north slope of the irrigated Rapid Valley. An east-west ridge 200 to 300 feet higher than the airport separates the station from the Box Elder Creek Valley.

The principal agricultural products in the area are cattle and wheat, and ranchers and farmers are dependent on the current weather forecasts, which are at times of vital interest in the protection of livestock.

Although the annual precipitation is light at lower elevations, the distribution is beneficial to agriculture with the greatest amounts occurring during the growing season. The heaviest snows are expected in the spring, which helps to furnish moisture for the early maturing crops such as wheat, while heavy winter snows at the higher elevations provide irrigation water for the fertile valleys.

Summer days are normally warm with cool, comfortable nights. Nearly all of the summer precipitation occurs as thunderstorms. Hail is often associated with the more severe thunderstorms, with resultant damage to vegetation as well as other fragile material in the path of the storms. Autumn, which begins soon after the first of September, is characterized by mild, balmy days, and cool, invigorating mornings and evenings. Autumn weather usually extends into November and often into December.

Temperatures for the winter months of December, January, and February are among the warmest in South Dakota due to the protection of the Black Hills, the frequent occurrence of Chinook winds, and the fact that the winter tracks of arctic air masses usually pass east of Rapid City. Rapid City has become the retirement home for many farmers and ranchers from the western half of the state because of the cool summer nights and the relatively mild winters.

Snowfall is normally light with the greatest monthly average of about 8 inches occurring in March. Cold waves can be expected occasionally, and one or more blizzards may occur each winter.

Spring is characterized by unsettled conditions. Wide variations usually occur in temperatures, and snows may fall as late as May.

Based on the 1951-1980 period, the average first occurrence of 32 degrees Fahrenheit in the fall is September 29 and the average last occurrence in the spring is May 7.

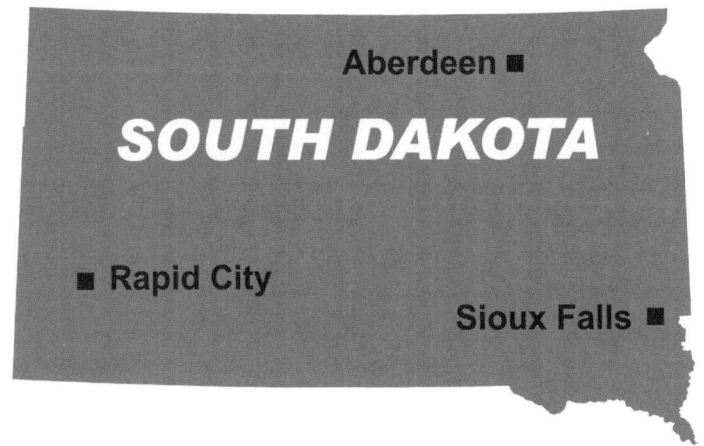

# NORMALS, MEANS, AND EXTREMES
## RAPID CITY (KRAP)

| LATITUDE:<br>44 ° 2 'N | LONGITUDE:<br>-103° 3 'W | ELEVATION (FT):<br>GRND: 3160　BARO: 3153 | TIME ZONE:<br>MOUNTAIN　(UTC -7) | WBAN: 24090 |
|---|---|---|---|---|

| ELEMENT | POR | JAN | FEB | MAR | APR | MAY | JUN | JUL | AUG | SEP | OCT | NOV | DEC | YEAR |
|---|---|---|---|---|---|---|---|---|---|---|---|---|---|---|
| NORMAL DAILY MAXIMUM | 30 | 33.6 | 38.6 | 46.6 | 57.1 | 67.2 | 77.4 | 85.5 | 85.5 | 75.2 | 61.7 | 44.8 | 36.1 | 59.1 |
| MEAN DAILY MAXIMUM | 61 | 34.3 | 38.6 | 45.7 | 57.6 | 68.0 | 77.7 | 86.8 | 86.0 | 75.4 | 62.3 | 47.0 | 37.3 | 59.7 |
| HIGHEST DAILY MAXIMUM | 67 | 76 | 75 | 83 | 93 | 98 | 109 | 111 | 107 | 104 | 94 | 83 | 75 | 111 |
| YEAR OF OCCURRENCE | | 1987 | 1995 | 2007 | 1989 | 1969 | 2002 | 2006 | 2007 | 1978 | 2005 | 1999 | 1965 | JUL 2006 |
| MEAN OF EXTREME MAXS. | 61 | 60.7 | 63.3 | 72.2 | 82.0 | 87.9 | 94.7 | 101.2 | 100.2 | 95.2 | 85.5 | 71.7 | 62.4 | 81.4 |
| NORMAL DAILY MINIMUM | 30 | 11.3 | 15.9 | 23.2 | 32.3 | 42.7 | 51.8 | 57.9 | 56.6 | 46.0 | 34.7 | 22.1 | 13.3 | 34.0 |
| MEAN DAILY MINIMUM | 61 | 10.9 | 15.1 | 21.9 | 32.1 | 42.8 | 51.8 | 58.6 | 57.0 | 46.4 | 35.3 | 22.9 | 14.3 | 34.1 |
| LOWEST DAILY MINIMUM | 67 | -27 | -31 | -21 | 1 | 18 | 31 | 39 | 38 | 18 | -2 | -19 | -30 | -31 |
| YEAR OF OCCURRENCE | | 1950 | 1996 | 1996 | 1975 | 1950 | 1951 | 1987 | 1992 | 1985 | 1991 | 1959 | 1990 | FEB 1996 |
| MEAN OF EXTREME MINS. | 61 | -11.5 | -6.8 | 1.3 | 16.6 | 29.0 | 39.9 | 48.3 | 45.9 | 30.7 | 18.9 | 3.3 | -8.3 | 17.3 |
| NORMAL DRY BULB | 30 | 22.4 | 27.3 | 34.9 | 44.7 | 55.0 | 64.6 | 71.7 | 71.1 | 60.6 | 48.2 | 33.4 | 24.7 | 46.6 |
| MEAN DRY BULB | 61 | 22.6 | 26.9 | 33.8 | 44.8 | 55.4 | 64.9 | 72.7 | 71.5 | 60.9 | 48.8 | 35.0 | 25.8 | 46.9 |
| MEAN WET BULB | 26 | 20.0 | 22.3 | 28.9 | 37.3 | 47.5 | 55.7 | 59.9 | 58.4 | 49.4 | 38.8 | 27.7 | 20.6 | 38.9 |
| MEAN DEW POINT | 26 | 15.4 | 17.0 | 23.9 | 30.7 | 42.7 | 51.6 | 55.1 | 52.7 | 43.0 | 32.6 | 22.9 | 15.4 | 33.6 |
| NORMAL NO. DAYS WITH: | | | | | | | | | | | | | | |
| MAXIMUM >= 90 | 30 | 0.0 | 0.0 | 0.0 | 0.1 | 0.3 | 3.1 | 10.7 | 11.0 | 3.9 | 0.2 | 0.0 | 0.0 | 29.3 |
| MAXIMUM <= 32 | 30 | 12.9 | 9.1 | 5.1 | 0.8 | 0.0 | 0.0 | 0.0 | 0.0 | 0.0 | 0.4 | 6.3 | 10.7 | 45.3 |
| MINIMUM <= 32 | 30 | 30.0 | 26.6 | 26.4 | 15.6 | 2.7 | * | 0.0 | 0.0 | 2.0 | 11.1 | 25.7 | 30.0 | 170.1 |
| MINIMUM <= 0 | 30 | 7.5 | 4.2 | 1.2 | 0.0 | 0.0 | 0.0 | 0.0 | 0.0 | 0.0 | * | 1.2 | 4.6 | 18.7 |
| NORMAL HEATING DEG. DAYS | 30 | 1314 | 1061 | 925 | 595 | 313 | 88 | 16 | 21 | 190 | 521 | 934 | 1233 | 7211 |
| NORMAL COOLING DEG. DAYS | 30 | 0 | 0 | 0 | 2 | 13 | 86 | 227 | 208 | 59 | 3 | 0 | 0 | 598 |
| NORMAL (PERCENT) | 30 | 66 | 65 | 65 | 61 | 64 | 65 | 58 | 56 | 55 | 59 | 66 | 66 | 62 |
| HOUR 05 LST | 30 | 69 | 72 | 75 | 75 | 78 | 80 | 76 | 75 | 70 | 70 | 72 | 71 | 74 |
| HOUR 11 LST | 30 | 60 | 57 | 55 | 49 | 51 | 52 | 46 | 43 | 41 | 45 | 55 | 59 | 51 |
| HOUR 17 LST | 30 | 65 | 59 | 53 | 47 | 49 | 50 | 42 | 39 | 39 | 48 | 62 | 65 | 52 |
| HOUR 23 LST | 30 | 70 | 71 | 72 | 70 | 73 | 75 | 69 | 66 | 63 | 66 | 71 | 69 | 70 |
| PERCENT POSSIBLE SUNSHINE | 54 | 57 | 60 | 63 | 62 | 60 | 65 | 73 | 74 | 70 | 66 | 55 | 55 | 63 |
| MEAN NO. DAYS WITH: | | | | | | | | | | | | | | |
| HEAVY FOG(VISBY <= 1/4 MI) | 46 | 1.8 | 2.7 | 2.8 | 2.0 | 1.4 | 0.9 | 0.7 | 0.7 | 0.7 | 1.0 | 2.2 | 2.0 | 18.9 |
| THUNDERSTORMS | 61 | 0.0 | 0.0 | 0.1 | 1.1 | 5.2 | 9.6 | 10.7 | 7.8 | 2.9 | 0.4 | 0.0 | 0.0 | 37.8 |
| MEAN: | | | | | | | | | | | | | | |
| SUNRISE-SUNSET (OKTAS) | | | | | | | | | | | | | | |
| MIDNIGHT-MIDNIGHT (OKTAS) | | | | | | | | | | | | | | |
| MEAN NO. DAYS WITH: | | | | | | | | | | | | | | |
| CLEAR | | | | | | | | | | | | | | |
| PARTLY CLOUDY | | | | | | | | | | | | | | |
| CLOUDY | | | | | | | | | | | | | | |
| MEAN STATION PRESSURE(IN) | 26 | 26.70 | 26.82 | 26.68 | 26.66 | 26.66 | 26.68 | 26.73 | 26.74 | 26.73 | 26.72 | 26.70 | 26.70 | 26.71 |
| MEAN SEA-LEVEL PRES. (IN) | 26 | 30.08 | 30.09 | 30.02 | 29.96 | 29.92 | 29.90 | 29.93 | 29.95 | 29.98 | 30.02 | 30.05 | 30.09 | 30.00 |
| MEAN SPEED (MPH) | 26 | 10.4 | 11.0 | 11.9 | 12.6 | 11.8 | 10.4 | 9.8 | 9.8 | 10.3 | 10.9 | 10.5 | 10.4 | 10.8 |
| PREVAIL.DIR(TENS OF DEGS) | 30 | 34 | 34 | 34 | 34 | 34 | 34 | 34 | 16 | 34 | 34 | 34 | 34 | 34 |
| MAXIMUM 2-MINUTE: | | | | | | | | | | | | | | |
| SPEED (MPH) | 14 | 59 | 59 | 64 | 61 | 57 | 54 | 69 | 54 | 47 | 58 | 62 | 58 | 69 |
| DIR. (TENS OF DEGS) | | 35 | 33 | 33 | 32 | 32 | 32 | 21 | 32 | 33 | 33 | 31 | 33 | 21 |
| YEAR OF OCCURRENCE | | 2008 | 1998 | 2009 | 1997 | 1999 | 2001 | 2002 | 2002 | 2007 | 2008 | 2008 | 2009 | JUL 2002 |
| MAXIMUM 3-SECOND | | | | | | | | | | | | | | |
| SPEED (MPH) | 14 | 70 | 70 | 74 | 69 | 68 | 83 | 89 | 69 | 62 | 69 | 77 | 71 | 89 |
| DIR. (TENS OF DEGS) | | 35 | 31 | 34 | 31 | 32 | 28 | 21 | 32 | 33 | 33 | 33 | 33 | 21 |
| YEAR OF OCCURRENCE | | 2008 | 1996 | 2009 | 1997 | 1999 | 2009 | 2002 | 2008 | 2007 | 2008 | 2008 | 2009 | JUL 2002 |
| NORMAL (IN) | 30 | 0.37 | 0.46 | 1.03 | 1.86 | 2.96 | 2.83 | 2.03 | 1.61 | 1.10 | 1.37 | 0.61 | 0.41 | 16.64 |
| MAXIMUM MONTHLY (IN) | 67 | 1.77 | 2.46 | 3.02 | 5.16 | 8.18 | 7.00 | 6.13 | 4.83 | 3.94 | 5.60 | 2.22 | 1.65 | 8.18 |
| YEAR OF OCCURRENCE | | 1944 | 1953 | 1945 | 1967 | 1996 | 1968 | 1969 | 1982 | 1946 | 1998 | 1985 | 1975 | MAY 1996 |
| MINIMUM MONTHLY (IN) | 67 | 0.01 | 0.02 | 0.12 | 0.02 | 0.33 | 0.42 | 0.38 | 0.10 | 0.03 | T | 0.01 | 0.01 | T |
| YEAR OF OCCURRENCE | | 1952 | 1999 | 1981 | 1987 | 1966 | 2002 | 1988 | 1943 | 1975 | 1960 | 2004 | 2006 | OCT 1960 |
| MAXIMUM IN 24 HOURS (IN) | 67 | 1.26 | 1.00 | 2.19 | 3.19 | 3.40 | 4.01 | 2.51 | 2.60 | 2.13 | 2.49 | 1.09 | 1.04 | 4.01 |
| YEAR OF OCCURRENCE | | 1944 | 1953 | 1945 | 1997 | 1965 | 1963 | 1944 | 1982 | 1966 | 1982 | 1944 | 1975 | JUN 1963 |
| NORMAL NO. DAYS WITH: | | | | | | | | | | | | | | |
| PRECIPITATION >= 0.01 | 30 | 6.4 | 6.4 | 8.2 | 9.7 | 12.0 | 12.1 | 9.8 | 7.8 | 6.5 | 6.4 | 5.8 | 5.5 | 96.6 |
| PRECIPITATION >= 1.00 | 30 | 0.0 | 0.0 | 0.1 | 0.2 | 0.5 | 0.6 | 0.3 | 0.2 | 0.1 | 0.2 | 0.0 | * | 2.2 |
| NORMAL (IN) | 30 | 5.2 | 6.3 | 9.1 | 6.2 | 0.5 | 0.0 | 0.0 | 0.0 | 0.2 | 1.8 | 6.3 | 5.3 | 40.9 |
| MAXIMUM MONTHLY (IN) | 61 | 24.0 | 23.7 | 30.7 | 30.6 | 11.6 | 3.6 | T | T | 2.0 | 10.2 | 33.6 | 17.9 | 33.6 |
| YEAR OF OCCURRENCE | | 1949 | 1953 | 1950 | 1970 | 1950 | 1951 | 1993 | 1994 | 1970 | 1995 | 1985 | 1975 | NOV 1985 |
| MAXIMUM IN 24 HOURS (IN) | 61 | 16.3 | 10.0 | 14.9 | 16.0 | 13.4 | 3.6 | T | T | 2.0 | 7.8 | 9.4 | 9.8 | 16.3 |
| YEAR OF OCCURRENCE' | | 1944 | 1953 | 1973 | 1970 | 1967 | 1951 | 1993 | 1994 | 1970 | 1995 | 1977 | 1975 | JAN 1944 |
| MAXIMUM SNOW DEPTH (IN) | 53 | 16 | 14 | 16 | 17 | 13 | 2 | 0 | 0 | 1 | 6 | 15 | 11 | 17 |
| YEAR OF OCCURRENCE | | 1993 | 1987 | 1977 | 1970 | 1967 | 1951 | | | 1965 | 1954 | 1985 | 1985 | APR 1970 |
| NORMAL NO. DAYS WITH: | | | | | | | | | | | | | | |
| SNOWFALL >= 1.0 | 30 | 1.5 | 2.4 | 2.5 | 1.9 | 0.1 | 0.0 | 0.0 | 0.0 | 0.0 | 0.5 | 2.2 | 1.6 | 12.7 |

## PRECIPITATION (inches) 2009 RAPID CITY (KRAP)

| YEAR | JAN | FEB | MAR | APR | MAY | JUN | JUL | AUG | SEP | OCT | NOV | DEC | ANNUAL |
|------|-----|-----|-----|-----|-----|-----|-----|-----|-----|-----|-----|-----|--------|
| 1980 | 0.20 | 0.51 | 0.86 | 1.13 | 1.58 | 4.75 | 1.78 | 2.38 | 0.48 | 2.28 | 0.57 | 0.66 | 17.18 |
| 1981 | 0.14 | 0.09 | 0.12 | 0.32 | 2.81 | 1.89 | 4.47 | 1.74 | 0.16 | 1.81 | 0.23 | 0.35 | 14.13 |
| 1982 | 0.39 | 0.37 | 1.35 | 0.69 | 6.50 | 2.89 | 1.81 | 4.83 | 2.69 | 3.82 | 0.27 | 0.36 | 25.97 |
| 1983 | 0.34 | 0.18 | 0.84 | 1.00 | 2.18 | 3.01 | 1.94 | 2.39 | 0.33 | 1.74 | 1.07 | 0.47 | 15.49 |
| 1984 | 0.10 | 0.18 | 0.69 | 3.10 | 1.57 | 4.72 | 1.57 | 1.00 | 0.74 | 0.67 | 0.51 | 0.38 | 15.23 |
| 1985 | 0.46 | 0.06 | 1.55 | 0.32 | 1.24 | 1.58 | 1.03 | 1.86 | 1.57 | 0.98 | 2.22 | 0.77 | 13.64 |
| 1986 | 0.49 | 0.92 | 0.88 | 4.74 | 1.43 | 4.56 | 0.91 | 1.32 | 3.14 | 1.64 | 1.40 | 0.01 | 21.44 |
| 1987 | 0.04 | 1.71 | 1.14 | 0.02 | 3.39 | 1.37 | 0.83 | 2.37 | 0.68 | 0.26 | 0.30 | 0.31 | 12.42 |
| 1988 | 0.17 | 0.34 | 0.52 | 0.60 | 3.25 | 1.09 | 0.38 | 1.98 | 0.56 | 0.76 | 0.81 | 0.46 | 10.92 |
| 1989 | 0.02 | 0.34 | 0.96 | 1.46 | 1.40 | 1.04 | 0.82 | 1.70 | 3.09 | 1.49 | 0.43 | 0.82 | 13.57 |
| 1990 | 0.22 | 0.37 | 1.17 | 0.77 | 4.87 | 1.42 | 1.94 | 1.87 | 2.44 | 0.61 | 0.44 | 0.33 | 16.45 |
| 1991 | 0.32 | 0.77 | 0.63 | 2.99 | 4.40 | 3.27 | 1.97 | 0.58 | 0.59 | 1.00 | 0.73 | 0.04 | 17.29 |
| 1992 | 0.29 | 0.16 | 1.92 | 0.71 | 2.47 | 2.17 | 3.25 | 0.47 | 0.42 | 0.68 | 0.39 | 0.57 | 13.50 |
| 1993 | 0.68 | 0.61 | 0.82 | 3.05 | 2.16 | 3.39 | 4.31 | 1.18 | 1.46 | 0.90 | 0.70 | 0.53 | 19.79 |
| 1994 | 0.45 | 0.66 | 0.37 | 1.20 | 1.47 | 0.67 | 0.64 | 0.92 | 0.27 | 2.84 | 0.66 | 0.35 | 10.50 |
| 1995 | 0.09 | 0.55 | 0.79 | 2.57 | 4.03 | 4.50 | 2.87 | 0.46 | 0.82 | 2.42 | 0.42 | 0.13 | 19.65 |
| 1996 | 0.85 | 0.10 | 1.06 | 1.63 | 8.18 | 1.24 | .52 | 1.85 | 1.55 |  | .07 |  | 23.83 |
| 1997 | 0.65 | 0.28 | 0.20 | 4.80 | 5.35 | 3.43 | 3.67 | 3.93 | 0.78 | 0.47 | 0.19 | 0.08 | 21.89 |
| 1998 | 0.15 | 1.24 | 1.32 | 0.28 | 2.34 | 5.59 | 1.26 | 1.42 | 1.50 | 5.60 | 1.13 | 0.06 | 19.48 |
| 1999 | 0.21 | 0.02 | 1.32 | 2.45 | 4.49 | 5.24 | 3.68 | 0.47 | 0.85 | 0.11 | 0.43 | 0.21 | 15.03 |
| 2000 | 0.23 | 0.15 | 1.37 | 3.95 | 2.40 | 1.60 | 2.06 | 0.70 | 0.45 | 1.54 | 0.47 | 0.11 | 14.29 |
| 2001 | 0.24 | 0.17 | 0.42 | 2.16 | 1.73 | 3.57 | 2.46 | 1.59 | 0.91 | 0.97 | 0.07 | T | 10.27 |
| 2002 | 0.05 | 0.20 | 0.69 | 2.24 | 1.61 | 0.42 | 1.13 | 0.76 | 2.45 | 0.65 | 0.04 | 0.03 | 10.97 |
| 2003 | 0.33 | 0.22 | 1.05 | 2.04 | 1.32 | 2.45 | 0.48 | 0.47 | 1.37 | 0.49 | 0.50 | 0.25 | 13.16 |
| 2004 | 0.03 | 0.93 | 0.98 | 0.50 | 2.55 | 1.02 | 2.96 | 0.80 | 2.04 | 1.28 | 0.01 | 0.06 | 14.41 |
| 2005 | 0.54 | 0.16 | 1.09 | 1.46 | 5.30 | 1.23 | 0.92 | 1.80 | 0.57 | 0.84 | 0.25 | 0.25 | 11.72 |
| 2006 | 0.14 | 0.20 | 1.20 | 2.09 | 1.82 | 0.93 | 0.76 | 1.76 | 2.00 | 0.34 | 0.47 | 0.01 | 12.59 |
| 2007 | 0.12 | 0.79 | 0.49 | 1.32 | 2.93 | 0.96 | 1.20 | 2.82 | 0.83 | 0.59 | 0.03 | 0.51 | 20.55 |
| 2008 | 0.34 | 0.51 | 0.60 | 1.38 | 7.24 | 3.12 | 2.21 | 1.42 | 0.87 | 1.47 | 1.02 | 0.37 | 18.64 |
| 2009 | 0.37 | 0.77 | 2.07 | 3.65 | 0.94 | 2.83 | 1.77 | 1.48 | 1.90 | 1.97 | 0.17 | 0.72 | 16.13 |
| POR= 61 YRS | 0.38 | 0.52 | 1.00 | 1.90 | 2.86 | 2.84 | 1.96 | 1.53 | 1.18 | 1.06 | 0.51 | 0.39 | 16.13 |

WBAN : 24090

## AVERAGE TEMPERATURE (°F) 2009 RAPID CITY (KRAP)

| YEAR | JAN | FEB | MAR | APR | MAY | JUN | JUL | AUG | SEP | OCT | NOV | DEC | ANNUAL |
|------|-----|-----|-----|-----|-----|-----|-----|-----|-----|-----|-----|-----|--------|
| 1980 | 21.0 | 27.1 | 31.5 | 48.9 | 57.4 | 67.1 | 74.9 | 68.6 | 61.6 | 48.9 | 39.3 | 30.3 | 48.1 |
| 1981 | 32.6 | 29.4 | 40.0 | 51.5 | 54.9 | 64.9 | 72.0 | 70.0 | 63.5 | 47.7 | 40.4 | 25.8 | 49.4 |
| 1982 | 11.9 | 23.9 | 33.0 | 42.1 | 53.3 | 59.7 | 70.7 | 70.2 | 58.7 | 47.2 | 32.7 | 28.6 | 44.3 |
| 1983 | 32.1 | 37.3 | 36.4 | 40.7 | 52.0 | 63.1 | 73.6 | 78.0 | 60.8 | 49.4 | 34.9 | 8.1 | 47.2 |
| 1984 | 28.0 | 36.1 | 34.3 | 43.8 | 53.6 | 62.8 | 72.2 | 74.8 | 57.2 | 47.2 | 37.5 | 21.4 | 47.4 |
| 1985 | 21.6 | 23.8 | 35.9 | 52.0 | 61.8 | 62.1 | 74.6 | 69.0 | 55.6 | 47.3 | 16.0 | 21.0 | 45.1 |
| 1986 | 29.8 | 21.5 | 43.0 | 44.2 | 54.9 | 67.6 | 70.9 | 69.6 | 55.0 | 48.7 | 30.6 | 30.5 | 47.2 |
| 1987 | 31.1 | 32.5 | 32.6 | 51.6 | 59.5 | 67.1 | 75.4 | 68.1 | 61.4 | 47.1 | 40.3 | 28.9 | 49.6 |
| 1988 | 21.7 | 26.9 | 35.6 | 47.1 | 60.0 | 75.6 | 76.1 | 72.5 | 60.4 | 49.7 | 36.4 | 28.4 | 49.2 |
| 1989 | 28.7 | 14.4 | 31.1 | 45.8 | 55.4 | 64.0 | 77.0 | 73.1 | 61.0 | 48.8 | 36.4 | 19.4 | 46.3 |
| 1990 | 32.4 | 28.9 | 36.4 | 45.0 | 53.2 | 66.6 | 71.7 | 73.8 | 65.9 | 48.2 | 40.5 | 17.8 | 48.4 |
| 1991 | 18.4 | 36.0 | 38.5 | 46.3 | 56.6 | 67.2 | 72.9 | 74.0 | 61.1 | 46.4 | 30.0 | 32.7 | 48.3 |
| 1992 | 33.6 | 36.3 | 40.9 | 47.1 | 57.4 | 62.6 | 64.3 | 65.9 | 62.6 | 49.3 | 32.4 | 19.0 | 47.6 |
| 1993 | 16.0 | 15.2 | 37.6 | 43.5 | 56.1 | 60.3 | 65.1 | 68.6 | 56.6 | 48.2 | 32.4 | 31.5 | 44.3 |
| 1994 | 22.4 | 19.8 | 40.5 | 46.0 | 60.2 | 68.5 | 70.7 | 73.3 | 65.5 | 49.7 | 35.5 | 30.3 | 48.5 |
| 1995 | 29.2 | 31.9 | 33.2 | 40.3 | 51.8 | 62.9 | 70.6 | 73.6 | 60.1 | 46.5 | 33.7 | 25.4 | 46.6 |
| 1996 | 14.9 | 27.3 | 26.2 | 43.3 | 50.6 | 65.5 | 70.8 | 73.2 | 59.8 |  | 24.3 | 17.9 |  |
| 1997 | 18.1 | 28.8 | 36.7 | 38.7 | 52.8 | 66.3 | 70.9 | 68.7 | 62.5 | 48.6 | 33.2 | 30.3 | 46.3 |
| 1998 | 24.3 | 34.9 | 28.0 | 46.2 | 56.2 | 58.7 | 72.7 | 72.3 | 67.1 | 47.8 | 36.9 | 26.8 | 47.7 |
| 1999 | 25.6 | 36.6 | 36.3 | 42.6 | 53.0 | 62.6 | 71.3 | 71.4 | 55.7 | 49.9 | 44.4 | 33.3 | 48.6 |
| 2000 | 26.2 | 33.2 | 38.7 | 43.1 | 56.0 | 62.5 | 73.6 | 74.4 | 63.4 | 49.8 | 27.0 | 16.7 | 47.1 |
| 2001 | 29.7 | 16.9 | 34.4 | 46.0 | 56.2 | 64.1 | 74.8 | 74.4 | 62.5 | 47.9 | 39.9 | 28.9 | 48.0 |
| 2002 | 27.7 | 30.7 | 22.4 | 44.3 | 51.8 | 69.7 | 78.3 | 71.2 | 61.9 | 39.0 | 37.4 | 31.7 | 47.2 |
| 2003 | 26.1 | 21.5 | 35.0 | 47.9 | 55.0 | 62.2 | 77.1 | 75.9 | 59.6 | 52.5 | 28.6 | 30.5 | 47.7 |
| 2004 | 22.8 | 28.5 | 40.4 | 48.0 | 55.6 | 61.6 | 72.1 | 68.0 | 62.5 | 50.1 | 37.4 | 31.8 | 48.2 |
| 2005 | 21.9 | 32.8 | 38.0 | 48.0 | 53.6 | 66.9 | 76.0 | 70.8 | 65.3 | 50.0 | 39.5 | 25.5 | 49.0 |
| 2006 | 36.4 | 26.3 | 32.9 | 50.0 | 57.6 | 70.2 | 79.3 | 73.1 | 57.8 | 45.0 | 36.5 | 29.3 | 49.5 |
| 2007 | 26.1 | 21.1 | 44.3 | 43.9 | 59.3 | 69.4 | 79.7 | 74.2 | 63.7 | 50.3 | 37.8 | 22.3 | 49.3 |
| 2008 | 21.5 | 26.1 | 34.9 | 41.8 | 51.1 | 61.9 | 71.6 | 70.9 | 59.0 | 46.9 | 36.3 | 17.8 | 45.0 |
| 2009 | 24.9 | 28.3 | 32.7 | 40.0 | 54.1 | 59.9 | 68.2 | 67.6 | 62.6 | 38.7 | 40.4 | 16.2 | 44.5 |
| POR= 61 YRS | 22.6 | 26.9 | 33.8 | 44.8 | 55.4 | 64.9 | 72.7 | 71.5 | 60.9 | 48.8 | 35.0 | 25.8 | 46.9 |

## HEATING DEGREE DAYS (base 65°F) 2009  RAPID CITY (KRAP)

| YEAR | JUL | AUG | SEP | OCT | NOV | DEC | JAN | FEB | MAR | APR | MAY | JUN | TOTAL |
|---|---|---|---|---|---|---|---|---|---|---|---|---|---|
| 1980-81 | 1 | 18 | 144 | 510 | 763 | 1070 | 998 | 993 | 765 | 402 | 311 | 65 | 6040 |
| 1981-82 | 21 | 7 | 108 | 531 | 730 | 1209 | 1646 | 1146 | 985 | 682 | 358 | 170 | 7593 |
| 1982-83 | 7 | 21 | 226 | 545 | 962 | 1119 | 1012 | 772 | 880 | 723 | 407 | 113 | 6787 |
| 1983-84 | 8 | 0 | 208 | 474 | 896 | 1762 | 1139 | 832 | 948 | 626 | 366 | 101 | 7360 |
| 1984-85 | 0 | 0 | 268 | 546 | 820 | 1344 | 1341 | 1148 | 895 | 393 | 146 | 144 | 7045 |
| 1985-86 | 8 | 27 | 327 | 544 | 1466 | 1358 | 1083 | 1211 | 672 | 617 | 317 | 35 | 7665 |
| 1986-87 | 5 | 12 | 296 | 497 | 1025 | 1059 | 1045 | 907 | 997 | 408 | 199 | 46 | 6496 |
| 1987-88 | 10 | 49 | 147 | 545 | 736 | 1111 | 1340 | 1103 | 905 | 533 | 195 | 17 | 6691 |
| 1988-89 | 3 | 18 | 163 | 470 | 850 | 1127 | 1120 | 1414 | 1047 | 586 | 303 | 116 | 7217 |
| 1989-90 | 3 | 6 | 182 | 495 | 847 | 1410 | 1004 | 1004 | 880 | 597 | 363 | 68 | 6859 |
| 1990-91 | 10 | 5 | 112 | 514 | 730 | 1462 | 1440 | 807 | 815 | 556 | 269 | 16 | 6736 |
| 1991-92 | 2 | 6 | 183 | 581 | 1045 | 997 | 964 | 828 | 742 | 539 | 262 | 107 | 6256 |
| 1992-93 | 67 | 83 | 138 | 493 | 973 | 1418 | 1515 | 1390 | 841 | 636 | 269 | 158 | 7981 |
| 1993-94 | 56 | 26 | 256 | 522 | 972 | 1035 | 1312 | 1260 | 749 | 565 | 171 | 27 | 6951 |
| 1994-95 | 9 | 17 | 79 | 468 | 876 | 1068 | 1104 | 919 | 983 | 734 | 400 | 130 | 6787 |
| 1995-96 | 11 | 5 | 215 | 568 | 933 | 1221 | 1551 | 1087 | 1200 | 643 | 447 | 66 | 7947 |
| 1996-97 | 6 | 0 | 203 | | 1215 | 1453 | 1446 | 1008 | 872 | 783 | 373 | 30 | |
| 1997-98 | 23 | 26 | 127 | 510 | 947 | 1067 | 1254 | 837 | 1143 | 558 | 272 | 198 | 6962 |
| 1998-99 | 3 | 0 | 91 | 522 | 840 | 1178 | 1214 | 790 | 885 | 666 | 366 | 104 | 6659 |
| 1999-00 | 22 | 6 | 284 | 460 | 609 | 976 | 1196 | 914 | 808 | 652 | 281 | 139 | 6347 |
| 2000-01 | 10 | 1 | 152 | 462 | 1134 | 1490 | 1086 | 1339 | 944 | 565 | 278 | 127 | 7588 |
| 2001-02 | 9 | 0 | 138 | 523 | 746 | 1111 | 1149 | 954 | 1313 | 614 | 417 | 44 | 7018 |
| 2002-03 | 0 | 17 | 189 | 797 | 817 | 1027 | 1198 | 1212 | 924 | 505 | 320 | 134 | 7140 |
| 2003-04 | 0 | 17 | 220 | 385 | 1087 | 1063 | 1300 | 1053 | 758 | 506 | 298 | 136 | 6823 |
| 2004-05 | 17 | 45 | 136 | 454 | 822 | 1021 | 1331 | 896 | 833 | 510 | 354 | 69 | 6488 |
| 2005-06 | 8 | 15 | 97 | 468 | 759 | 1219 | 880 | 1079 | 990 | 444 | 260 | 8 | 6227 |
| 2006-07 | 0 | 11 | 238 | 619 | 847 | 1100 | 1198 | 1221 | 636 | 627 | 200 | 32 | 6729 |
| 2007-08 | 0 | 18 | 144 | 450 | 808 | 1318 | 1343 | 1122 | 926 | 688 | 427 | 115 | 7359 |
| 2008-09 | 9 | 15 | 199 | 554 | 853 | 1455 | 1239 | 1020 | 995 | 745 | 340 | 188 | 7612 |
| 2009- | 14 | 37 | 127 | 809 | 731 | 1507 | | | | | | | |

WBAN : 24090

## COOLING DEGREE DAYS (base 65°F) 2009  RAPID CITY (KRAP)

| YEAR | JAN | FEB | MAR | APR | MAY | JUN | JUL | AUG | SEP | OCT | NOV | DEC | TOTAL |
|---|---|---|---|---|---|---|---|---|---|---|---|---|---|
| 1980 | 0 | 0 | 0 | 6 | 25 | 123 | 315 | 136 | 48 | 14 | 0 | 0 | 667 |
| 1981 | 0 | 0 | 0 | 3 | 5 | 67 | 243 | 170 | 74 | 0 | 0 | 0 | 562 |
| 1982 | 0 | 0 | 0 | 0 | 3 | 18 | 189 | 190 | 41 | 0 | 0 | 0 | 441 |
| 1983 | 0 | 0 | 0 | 0 | 9 | 62 | 282 | 407 | 88 | 0 | 0 | 0 | 848 |
| 1984 | 0 | 0 | 0 | 0 | 18 | 41 | 234 | 309 | 42 | 2 | 0 | 0 | 646 |
| 1985 | 0 | 0 | 0 | 10 | 53 | 64 | 312 | 158 | 51 | 0 | 0 | 0 | 648 |
| 1986 | 0 | 0 | 0 | 0 | 11 | 124 | 192 | 164 | 0 | 0 | 0 | 0 | 491 |
| 1987 | 0 | 0 | 0 | 13 | 33 | 118 | 341 | 152 | 45 | 0 | 0 | 0 | 702 |
| 1988 | 0 | 0 | 0 | 2 | 46 | 341 | 355 | 255 | 33 | 2 | 0 | 0 | 1034 |
| 1989 | 0 | 0 | 0 | 15 | 9 | 95 | 380 | 265 | 70 | 0 | 0 | 0 | 834 |
| 1990 | 0 | 0 | 0 | 2 | 4 | 120 | 226 | 282 | 147 | 3 | 0 | 0 | 784 |
| 1991 | 0 | 0 | 0 | 3 | 16 | 89 | 256 | 294 | 76 | 15 | 0 | 0 | 749 |
| 1992 | 0 | 0 | 0 | 9 | 36 | 41 | 51 | 118 | 72 | 11 | 0 | 0 | 338 |
| 1993 | 0 | 0 | 0 | 0 | 0 | 25 | 66 | 146 | 9 | 7 | 0 | 0 | 253 |
| 1994 | 0 | 0 | 0 | 0 | 28 | 141 | 193 | 280 | 97 | 0 | 0 | 0 | 739 |
| 1995 | 0 | 0 | 0 | 0 | 0 | 72 | 192 | 280 | 72 | 1 | 0 | 0 | 617 |
| 1996 | 0 | 0 | 0 | 0 | 5 | 86 | 194 | 262 | 52 | | 0 | 0 | |
| 1997 | 0 | 0 | 0 | 0 | 4 | 76 | 211 | 148 | 58 | 7 | 0 | 0 | 504 |
| 1998 | 0 | 0 | 0 | 0 | 8 | 15 | 250 | 232 | 160 | 0 | 0 | 0 | 665 |
| 1999 | 0 | 0 | 0 | 0 | 2 | 40 | 225 | 212 | 9 | 0 | 0 | 0 | 488 |
| 2000 | 0 | 0 | 0 | 0 | 10 | 71 | 282 | 299 | 108 | 0 | 0 | 0 | 770 |
| 2001 | 0 | 0 | 0 | 4 | 11 | 107 | 320 | 300 | 69 | 1 | 0 | 0 | 812 |
| 2002 | 0 | 0 | 0 | 0 | 15 | 193 | 422 | 216 | 103 | 0 | 0 | 0 | 949 |
| 2003 | 0 | 0 | 0 | 0 | 20 | 54 | 381 | 364 | 65 | 7 | 0 | 0 | 891 |
| 2004 | 0 | 0 | 0 | 0 | 16 | 42 | 246 | 148 | 69 | 0 | 0 | 0 | 521 |
| 2005 | 0 | 0 | 0 | 8 | 11 | 129 | 356 | 204 | 113 | 9 | 0 | 0 | 830 |
| 2006 | 0 | 0 | 0 | 1 | 36 | 169 | 452 | 269 | 27 | 6 | 0 | 0 | 960 |
| 2007 | 0 | 0 | 0 | 0 | 30 | 170 | 464 | 310 | 109 | 0 | 0 | 0 | 1083 |
| 2008 | 0 | 0 | 0 | 0 | 0 | 29 | 221 | 206 | 26 | 0 | 0 | 0 | 482 |
| 2009 | 0 | 0 | 0 | 0 | 9 | 42 | 118 | 123 | 62 | 0 | 0 | 0 | 354 |

**SNOWFALL (inches) 2009 RAPID CITY (KRAP)**

| YEAR | JUL | AUG | SEP | OCT | NOV | DEC | JAN | FEB | MAR | APR | MAY | JUN | TOTAL |
|------|-----|-----|-----|-----|-----|-----|-----|-----|-----|-----|-----|-----|-------|
| 1980-81 | 0.0 | 0.0 | 0.0 | 1.4 | 6.9 | 6.1 | 1.2 | 1.3 | T | T | 0.0 | 0.0 | 16.9 |
| 1981-82 | 0.0 | 0.0 | 0.0 | 1.6 | 1.2 | 3.8 | 6.2 | 5.0 | 11.5 | 5.5 | 0.0 | 0.0 | 34.8 |
| 1982-83 | 0.0 | 0.0 | 0.0 | 1.4 | 1.2 | 4.0 | 2.9 | 0.3 | 6.5 | 4.3 | 4.3 | 0.0 | 24.9 |
| 1983-84 | 0.0 | 0.0 | 0.3 | 0.9 | 6.9 | 7.1 | 1.9 | 2.5 | 6.1 | 22.1 | 0.2 | 0.0 | 48.0 |
| 1984-85 | 0.0 | 0.0 | 1.3 | 0.7 | 2.0 | 4.9 | 3.8 | 0.7 | 16.2 | 0.4 | 0.0 | T | 30.0 |
| 1985-86 | 0.0 | 0.0 | 1.4 | 0.6 | 33.6 | 10.2 | 5.7 | 10.7 | 6.0 | 12.7 | 0.0 | 0.0 | 80.9 |
| 1986-87 | 0.0 | 0.0 | 0.0 | T | 12.6 | 0.1 | 0.5 | 21.5 | 10.9 | 0.3 | 0.0 | 0.0 | 45.9 |
| 1987-88 | 0.0 | 0.0 | 0.0 | 1.7 | T | 4.7 | 2.7 | 3.6 | 10.6 | 6.1 | 0.0 | 0.0 | 29.4 |
| 1988-89 | 0.0 | 0.0 | 0.0 | 0.0 | 2.2 | 9.0 | 0.4 | 7.3 | 10.7 | 6.4 | 0.0 | 0.0 | 36.0 |
| 1989-90 | 0.0 | T | 0.0 | 3.9 | 4.6 | 10.9 | 3.1 | 5.0 | 9.2 | 3.0 | 0.5 | T | 40.2 |
| 1990-91 | 0.0 | T | 0.0 | 1.5 | 1.0 | 6.2 | 6.3 | 8.2 | 3.6 | 11.1 | 4.9 | T | 42.8 |
| 1991-92 | T | 0.0 | T | 4.0 | 6.7 | 0.8 | 5.2 | 1.2 | 8.2 | 3.5 | 0.0 | 0.0 | 29.6 |
| 1992-93 | T | 0.0 | 0.0 | 0.9 | 5.0 | 12.4 | 13.6 | 9.0 | 1.0 | 12.2 | 0.0 | 0.0 | 54.1 |
| 1993-94 | T | 0.0 | T | 2.4 | 11.4 | 5.5 | 7.5 | 11.1 | 5.6 | 12.8 | 0.0 | T | 56.3 |
| 1994-95 | 0.0 | T | 0.0 | 0.0 | 6.8 | 4.6 | 0.9 | 6.3 | 9.2 | 10.1 | T | T | 37.9 |
| 1995-96 | 0.0 | 0.0 | T | 10.2 | 3.1 | 1.4 | 19.4 | 4.4 | 16.2 | 10.3 | 0.9 | T | 65.9 |
| 1996-97 | 0.0 | 0.0 | T | | | | | | | | 0.0 | T | |
| 1997-98 | T | T | 0.0 | 3.1 | | | | | | | | | |
| 1998-99 | | | 0.0 | T | 15.9 | 0.5 | 2.0 | 0.2 | 18.5 | 6.3 | 0.0 | T | |
| 1999-00 | 0.0 | 0.0 | 0.0 | T | 0.3 | 3.7 | 2.9 | 1.5 | 5.5 | 19.0 | 0.0 | 0.0 | 32.9 |
| 2000-01 | 0.0 | 0.0 | 1.4 | 0.0 | 2.9 | 3.0 | 2.7 | | | | | | |
| 2001-02 | | | | | | | | | | | | | |
| 2002-03 | | | | | | | | | | | | | |
| 2003-04 | | | | | | | | | | | | | |
| 2004-05 | | | | | | | | | | | | | |
| 2005-06 | | | | | | 4.9 | T | 4.8 | 19.0 | 7.5 | 0.0 | 0.0 | |
| 2006-07 | 0.0 | 0.0 | 0.0 | 3.0 | 6.8 | 2.4 | 0.4 | 8.1 | 2.3 | 1.6 | T | 0.0 | 24.6 |
| 2007-08 | 0.0 | 0.0 | 0.0 | 0.0 | T | 10.7 | 4.3 | 6.2 | 8.7 | 5.2 | 11.0 | 0.0 | 46.1 |
| 2008-09 | 0.0 | 0.0 | 0.0 | 0.0 | 11.3 | 7.9 | 7.7 | 4.7 | 26.8 | 21.9 | 0.0 | 0.0 | 80.3 |
| 2009- | 0.0 | 0.0 | 0.0 | 8.5 | T | 13.1 | | | | | | | |
| POR=<br>55 YRS | T | T | 0.1 | 1.7 | 5.2 | 5.2 | 4.7 | 6.4 | 9.6 | 7.3 | 0.9 | 0.1 | 41.2 |

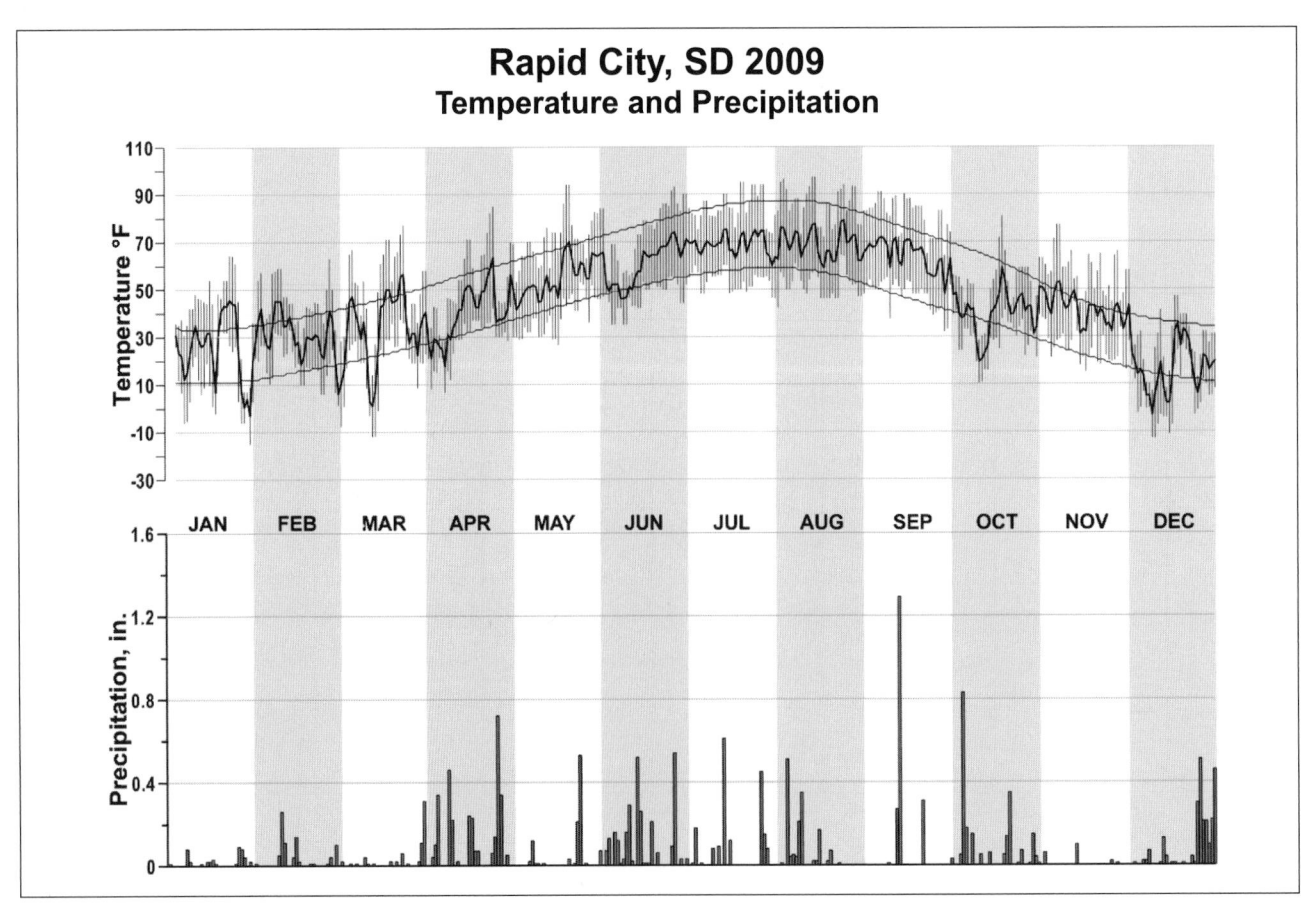

# Rapid City, SD 2009
## Temperature and Precipitation

# 2009
# SIOUX FALLS
# SOUTH DAKOTA (KFSD)

Sioux Falls is located in the Big Sioux River Valley in southeast South Dakota. The surrounding terrain is gently rolling. The land slopes upward for about 100 miles north and northwest to an elevation about 400 feet higher than the city. To the southeast, the land slopes downward 200 to 300 feet over the same distance. Little change in elevation occurs in the remaining directions.

The climate is of the continental type. There are frequent weather changes from day to day or week to week as the locality is visited by differing air masses. Cold air masses arrive from the interior of Canada, cool, dry air from the northern Pacific, warm, moist air from the Gulf of Mexico, or hot, dry air from the southwest.

Temperatures fluctuate frequently as cold air masses move in very rapidly. During the late fall and winter, cold fronts accompanied by strong, gusty winds drop temperatures by 20 to 30 degrees in a 24-hour period. Severe cold spells usually last only a few days. The winter months of December through February have experienced cold spells with average temperatures under 8 degrees and more than 60 consecutive days below 32 degrees.

Temperatures of 100 degrees and above occur about one in every three years, and will most likely happen in July. Summer nights are usually comfortable with temperatures below 70 degrees.

Rainfall is heavier during the spring and summer with lighter amounts in winter. Nearly 64 percent of the normal yearly precipitation falls during the growing season of April through August.

One or two very heavy snows usually fall each winter. Eight to 12 inches of snow may fall in 24 hours. There have been a few snows in excess of 15 inches and almost 30 inches have fallen during a severe winter storm. Strong winds often cause drifting snow, and blizzard conditions may block highways for a day or so.

Southerly winds prevail from late spring to early fall with northwest winds the remainder of the year. Strong winds of 70 mph with gusts to 90 mph have occurred.

Thunderstorms are frequent during the late spring and summer with June and July the most active months. The thunderstorms usually occur during the late afternoon and evening with a secondary peak of activity between 2 and 5 in the morning. Some of the most severe thunderstorms with damaging winds, hail and an occasional tornado, occur most frequently June.

There is occasional flooding in the lower areas of Sioux Falls along the Big Sioux River and Skunk Creek. Runoff from the melting snow in the spring often causes substantial rises in the rivers. A diversion canal around Sioux Falls has reduced the threat of damaging floods.

Based on the 1951-1980 period, the average first occurrence of 32 degrees Fahrenheit in the fall is October 1 and the average last occurrence in the spring is May 10.

# NORMALS, MEANS, AND EXTREMES
## SIOUX FALLS (KFSD)

| LATITUDE: 43° 34'N | LONGITUDE: -96° 45'W | ELEVATION (FT): GRND: 1422  BARO: 1428 | TIME ZONE: CENTRAL  (UTC -6) | WBAN: 14944 |
|---|---|---|---|---|

| | ELEMENT | POR | JAN | FEB | MAR | APR | MAY | JUN | JUL | AUG | SEP | OCT | NOV | DEC | YEAR |
|---|---|---|---|---|---|---|---|---|---|---|---|---|---|---|---|
| **TEMPERATURE °F** | NORMAL DAILY MAXIMUM | 30 | 25.2 | 31.6 | 43.8 | 58.8 | 71.0 | 80.6 | 85.6 | 83.2 | 74.2 | 61.1 | 41.9 | 28.8 | 57.2 |
| | MEAN DAILY MAXIMUM | 78 | 25.1 | 30.3 | 41.8 | 58.4 | 70.7 | 79.9 | 85.8 | 83.2 | 73.9 | 61.5 | 43.2 | 29.6 | 57.0 |
| | HIGHEST DAILY MAXIMUM | 64 | 66 | 70 | 87 | 94 | 100 | 110 | 108 | 108 | 104 | 94 | 81 | 63 | 110 |
| | YEAR OF OCCURRENCE | | 1981 | 1982 | 1968 | 1962 | 1967 | 1988 | 1989 | 1973 | 1976 | 1963 | 1999 | 1998 | JUN 1988 |
| | MEAN OF EXTREME MAXS. | 78 | 46.1 | 51.5 | 67.4 | 82.3 | 88.5 | 94.5 | 98.2 | 96.3 | 90.5 | 81.7 | 65.7 | 49.8 | 76.0 |
| | NORMAL DAILY MINIMUM | 30 | 2.9 | 10.1 | 21.3 | 32.5 | 44.6 | 54.5 | 60.3 | 58.4 | 47.6 | 34.8 | 20.7 | 7.8 | 33.0 |
| | MEAN DAILY MINIMUM | 78 | 5.4 | 10.5 | 22.1 | 34.9 | 46.6 | 56.6 | 62.3 | 59.9 | 49.6 | 37.4 | 23.1 | 11.1 | 35.0 |
| | LOWEST DAILY MINIMUM | 64 | -36 | -31 | -23 | 5 | 17 | 33 | 38 | 34 | 22 | 9 | -17 | -28 | -36 |
| | YEAR OF OCCURRENCE | | 1970 | 1962 | 1948 | 1982 | 1967 | 1969 | 1971 | 1950 | 1974 | 1972 | 1964 | 1990 | JAN 1970 |
| | MEAN OF EXTREME MINS. | 78 | -17.8 | -13.8 | -0.5 | 18.7 | 30.4 | 41.9 | 49.2 | 45.9 | 32.1 | 20.0 | 3.0 | -11.2 | 16.5 |
| | NORMAL DRY BULB | 30 | 14.0 | 20.8 | 32.6 | 45.7 | 57.8 | 67.5 | 73.0 | 70.8 | 60.9 | 48.0 | 31.3 | 18.3 | 45.1 |
| | MEAN DRY BULB | 78 | 15.3 | 20.4 | 32.0 | 46.7 | 58.7 | 68.4 | 74.1 | 71.6 | 61.8 | 49.4 | 33.2 | 20.4 | 46.0 |
| | MEAN WET BULB | 26 | 15.6 | 19.4 | 29.1 | 40.1 | 51.1 | 60.6 | 65.5 | 64.1 | 55.0 | 42.1 | 28.9 | 18.6 | 40.8 |
| | MEAN DEW POINT | 26 | 12.6 | 16.3 | 25.2 | 35.1 | 47.1 | 57.1 | 62.8 | 61.6 | 51.5 | 37.9 | 25.7 | 15.7 | 37.4 |
| | NORMAL NO. DAYS WITH: | | | | | | | | | | | | | | |
| | MAXIMUM >= 90 | 30 | 0.0 | 0.0 | 0.0 | 0.2 | 0.5 | 3.9 | 9.0 | 6.2 | 1.8 | 0.0 | 0.0 | 0.0 | 21.6 |
| | MAXIMUM <= 32 | 30 | 21.2 | 14.8 | 6.5 | 0.7 | 0.0 | 0.0 | 0.0 | 0.0 | 0.0 | 0.2 | 7.6 | 18.5 | 69.5 |
| | MINIMUM <= 32 | 30 | 30.9 | 27.4 | 25.0 | 12.7 | 1.7 | 0.0 | 0.0 | 0.0 | 1.3 | 10.9 | 25.7 | 30.7 | 166.3 |
| | MINIMUM <= 0 | 30 | 12.3 | 6.5 | 1.5 | 0.0 | 0.0 | 0.0 | 0.0 | 0.0 | 0.0 | 0.0 | 1.3 | 8.0 | 29.6 |
| **H/C** | NORMAL HEATING DEG. DAYS | 30 | 1566 | 1236 | 989 | 568 | 242 | 58 | 10 | 20 | 176 | 519 | 995 | 1433 | 7812 |
| | NORMAL COOLING DEG. DAYS | 30 | 0 | 0 | 0 | 5 | 35 | 149 | 274 | 216 | 64 | 4 | 0 | 0 | 747 |
| **RH** | NORMAL (PERCENT) | 30 | 76 | 76 | 73 | 66 | 65 | 66 | 68 | 72 | 70 | 68 | 76 | 78 | 71 |
| | HOUR 00 LST | 30 | 78 | 80 | 80 | 74 | 73 | 76 | 78 | 81 | 79 | 76 | 80 | 81 | 78 |
| | HOUR 06 LST | 30 | 79 | 81 | 83 | 81 | 82 | 82 | 85 | 88 | 86 | 82 | 83 | 82 | 83 |
| | HOUR 12 LST | 30 | 70 | 70 | 64 | 54 | 54 | 55 | 55 | 58 | 56 | 56 | 67 | 72 | 61 |
| | HOUR 18 LST | 30 | 74 | 72 | 64 | 52 | 51 | 52 | 53 | 57 | 56 | 58 | 71 | 77 | 61 |
| **S** | PERCENT POSSIBLE SUNSHINE | 12 | 54 | 52 | 51 | 55 | 51 | 60 | 71 | 67 | 66 | 67 | 54 | 51 | 58 |
| **W/O** | MEAN NO. DAYS WITH: | | | | | | | | | | | | | | |
| | HEAVY FOG(VISBY <= 1/4 MI) | 46 | 2.1 | 2.6 | 2.6 | 1.1 | 0.8 | 0.4 | 0.9 | 1.3 | 1.4 | 1.3 | 2.5 | 3.6 | 20.6 |
| | THUNDERSTORMS | 78 | 0.1 | 0.1 | 0.8 | 2.7 | 5.6 | 7.9 | 7.9 | 6.9 | 4.5 | 1.7 | 0.4 | 0.1 | 38.7 |
| **CLOUDNESS** | MEAN: | | | | | | | | | | | | | | |
| | SUNRISE-SUNSET (OKTAS) | | | | | | | | | | | | | | |
| | MIDNIGHT-MIDNIGHT (OKTAS) | | | | | | | | | | | | | | |
| | MEAN NO. DAYS WITH: | | | | | | | | | | | | | | |
| | CLEAR | | | | | | | | | | | | | | |
| | PARTLY CLOUDY | | | | | | | | | | | | | | |
| | CLOUDY | | | | | | | | | | | | | | |
| **PR** | MEAN STATION PRESSURE(IN) | 26 | 28.52 | 28.53 | 28.49 | 28.42 | 28.41 | 28.40 | 28.45 | 28.48 | 28.48 | 28.49 | 28.48 | 28.52 | 28.47 |
| | MEAN SEA-LEVEL PRES. (IN) | 26 | 30.12 | 30.12 | 30.05 | 29.95 | 29.92 | 29.90 | 29.94 | 29.98 | 29.99 | 30.02 | 30.05 | 30.10 | 30.01 |
| **WINDS** | MEAN SPEED (MPH) | 26 | 10.5 | 10.5 | 11.5 | 12.1 | 11.3 | 10.0 | 9.3 | 9.0 | 9.8 | 10.0 | 10.3 | 10.2 | 10.4 |
| | PREVAIL.DIR(TENS OF DEGS) | 35 | 32 | 32 | 35 | 36 | 19 | 19 | 19 | 19 | 19 | 19 | 32 | 31 | 32 |
| | MAXIMUM 2-MINUTE: | | | | | | | | | | | | | | |
| | SPEED (MPH) | 13 | 44 | 45 | 46 | 51 | 46 | 52 | 51 | 58 | 43 | 44 | 43 | 43 | 58 |
| | DIR. (TENS OF DEGS) | | 33 | 31 | 30 | 32 | 08 | 08 | 32 | 31 | 34 | 31 | 30 | 32 | 31 |
| | YEAR OF OCCURRENCE | | 2005 | 2002 | 2005 | 2000 | 2008 | 2001 | 1998 | 2002 | 1997 | 2008 | 1998 | 2000 | AUG 2002 |
| | MAXIMUM 3-SECOND | | | | | | | | | | | | | | |
| | SPEED (MPH) | 13 | 54 | 55 | 60 | 62 | 60 | 56 | 64 | 77 | 53 | 59 | 55 | 52 | 77 |
| | DIR. (TENS OF DEGS) | | 33 | 30 | 30 | 32 | 09 | 08 | 32 | 31 | 20 | 33 | 30 | 32 | 31 |
| | YEAR OF OCCURRENCE | | 2005 | 2002 | 2005 | 2000 | 2008 | 2001 | 1998 | 2002 | 2007 | 2008 | 2003 | 2000 | AUG 2002 |
| **PRECIPITATION** | NORMAL (IN) | 30 | 0.51 | 0.51 | 1.81 | 2.65 | 3.39 | 3.49 | 2.93 | 3.01 | 2.58 | 1.93 | 1.36 | 0.52 | 24.69 |
| | MAXIMUM MONTHLY (IN) | 64 | 1.71 | 4.05 | 4.97 | 6.97 | 8.26 | 8.43 | 8.41 | 9.09 | 9.26 | 6.28 | 4.76 | 2.62 | 9.26 |
| | YEAR OF OCCURRENCE | | 1969 | 1962 | 2007 | 2001 | 1993 | 1984 | 1992 | 1975 | 1986 | 1998 | 2001 | 1968 | SEP 1986 |
| | MINIMUM MONTHLY (IN) | 64 | 0.05 | 0.05 | 0.14 | 0.17 | 0.61 | 0.91 | 0.25 | 0.53 | 0.29 | T | 0.02 | T | T |
| | YEAR OF OCCURRENCE | | 1958 | 1986 | 1967 | 1969 | 1981 | 1988 | 1947 | 1970 | 1956 | 1952 | 1980 | 1986 | DEC 1986 |
| | MAXIMUM IN 24 HOURS (IN) | 64 | 1.61 | 2.00 | 3.27 | 3.72 | 4.22 | 4.32 | 3.39 | 4.59 | 4.02 | 4.54 | 2.68 | 1.44 | 4.59 |
| | YEAR OF OCCURRENCE | | 1960 | 1962 | 2007 | 2001 | 2004 | 1957 | 1992 | 1975 | 1966 | 1973 | 2001 | 1955 | AUG 1975 |
| | NORMAL NO. DAYS WITH: | | | | | | | | | | | | | | |
| | PRECIPITATION >= 0.01 | 30 | 7.0 | 6.9 | 9.0 | 10.6 | 11.2 | 10.3 | 10.1 | 9.4 | 8.0 | 6.8 | 7.9 | 6.1 | 103.3 |
| | PRECIPITATION >= 1.00 | 30 | * | 0.0 | 0.3 | 0.5 | 0.8 | 0.7 | 0.6 | 0.7 | 0.8 | 0.4 | 0.2 | 0.0 | 5.0 |
| **SNOWFALL** | NORMAL (IN) | 30 | 7.4 | 5.9 | 8.1 | 3.5 | 0.* | 0.* | 0.0 | 0.0 | 0.* | 1.1 | 7.6 | 7.0 | 40.6 |
| | MAXIMUM MONTHLY (IN) | 64 | 19.6 | 48.4 | 31.5 | 18.4 | 0.2 | T | T | T | 0.9 | 10.0 | 21.9 | 41.1 | 48.4 |
| | YEAR OF OCCURRENCE | | 1969 | 1962 | 1951 | 1983 | 1954 | 2006 | 1995 | 2009 | 1985 | 1991 | 1985 | 1968 | FEB 1962 |
| | MAXIMUM IN 24 HOURS (IN) | 64 | 11.8 | 26.0 | 18.9 | 10.5 | 0.2 | T | T | T | 0.9 | 8.8 | 12.6 | 16.6 | 26.0 |
| | YEAR OF OCCURRENCE' | | 1960 | 1962 | 1956 | 1994 | 1954 | 1994 | 1995 | 2009 | 1985 | 1991 | 1998 | 1968 | FEB 1962 |
| | MAXIMUM SNOW DEPTH (IN) | 66 | 33 | 34 | 33 | 10 | 1 | 0 | 0 | 0 | 0 | 3 | 13 | 34 | 34 |
| | YEAR OF OCCURRENCE | | 1969 | 1969 | 1969 | 1969 | 1947 | | | | | 1982 | 1983 | 1968 | FEB 1969 |
| | NORMAL NO. DAYS WITH: | | | | | | | | | | | | | | |
| | SNOWFALL >= 1.0 | 30 | 2.0 | 1.7 | 2.2 | 1.0 | 0.0 | 0.0 | 0.0 | 0.0 | 0.0 | 0.3 | 2.1 | 1.8 | 11.1 |

## PRECIPITATION (inches) 2009  SIOUX FALLS (KFSD)

| YEAR | JAN | FEB | MAR | APR | MAY | JUN | JUL | AUG | SEP | OCT | NOV | DEC | ANNUAL |
|------|-----|-----|-----|-----|-----|-----|-----|-----|-----|-----|-----|-----|--------|
| 1980 | 0.18 | 0.47 | 0.70 | 0.77 | 2.52 | 2.17 | 1.63 | 2.92 | 0.79 | 1.36 | 0.02 | 0.29 | 13.82 |
| 1981 | 0.12 | 0.33 | 1.86 | 0.58 | 0.61 | 3.90 | 3.89 | 2.28 | 0.50 | 2.45 | 1.21 | 0.38 | 18.11 |
| 1982 | 0.76 | 0.13 | 1.17 | 1.87 | 4.72 | 1.18 | 4.60 | 5.23 | 3.49 | 5.18 | 2.94 | 1.99 | 33.26 |
| 1983 | 0.52 | 0.22 | 3.35 | 2.88 | 2.92 | 6.75 | 1.82 | 2.00 | 1.92 | 0.71 | 2.95 | 0.73 | 26.77 |
| 1984 | 0.37 | 1.10 | 1.83 | 5.79 | 2.95 | 8.43 | 1.63 | 0.76 | 1.62 | 4.11 | 0.03 | 1.02 | 29.64 |
| 1985 | 0.45 | 0.05 | 2.37 | 5.18 | 3.29 | 2.52 | 2.70 | 4.07 | 3.34 | 0.75 | 1.97 | 0.47 | 27.16 |
| 1986 | 0.72 | 0.05 | 1.50 | 5.15 | 2.42 | 3.93 | 2.59 | 2.77 | 9.26 | 1.22 | 0.89 | T | 30.50 |
| 1987 | 0.19 | 0.26 | 3.27 | 0.28 | 2.94 | 1.78 | 3.16 | 1.36 | 2.05 | 0.31 | 1.66 | 1.40 | 18.66 |
| 1988 | 1.54 | 0.25 | 0.63 | 3.00 | 1.54 | 0.91 | 0.49 | 4.02 | 4.39 | 0.02 | 1.98 | 0.37 | 19.14 |
| 1989 | 0.23 | 0.51 | 1.07 | 1.59 | 1.42 | 2.50 | 1.37 | 2.46 | 3.38 | 0.10 | 0.91 | 0.25 | 15.79 |
| 1990 | 0.08 | 0.31 | 1.57 | 1.86 | 4.07 | 4.86 | 1.77 | 1.17 | 0.47 | 1.82 | 0.61 | 0.61 | 19.20 |
| 1991 | 0.22 | 0.34 | 0.86 | 2.21 | 6.20 | 6.36 | 2.26 | 1.41 | 3.95 | 1.65 | 1.78 | 0.20 | 27.44 |
| 1992 | 0.75 | 1.76 | 2.36 | 2.01 | 1.80 | 2.44 | 8.41 | 5.29 | 3.06 | 2.72 | 1.04 | 0.83 | 32.47 |
| 1993 | 0.70 | 0.81 | 2.04 | 2.61 | 8.26 | 6.43 | 7.86 | 3.10 | 1.88 | 0.62 | 1.50 | 0.30 | 36.11 |
| 1994 | 0.97 | 0.63 | 0.20 | 3.34 | 1.26 | 6.03 | 1.70 | 2.66 | 2.36 | 2.36 | 1.03 | 0.33 | 22.87 |
| 1995 | 0.18 | 0.13 | 4.06 | 5.83 | 4.76 | 2.70 | 2.55 | 5.11 | 1.86 | 2.76 | 0.38 | 0.10 | 30.42 |
| 1996 | 0.99 | 0.16 | 0.82 | 0.55 | 5.27 | 1.14 | 0.98 | 1.79 | 2.82 | 1.63 | 2.91 | .78 | 19.84 |
| 1997 | 0.41 | 1.39 | 0.23 | 2.43 | 3.58 | 3.77 | 2.94 | 1.58 | 1.59 | 1.75 | 0.35 | 0.24 | 20.26 |
| 1998 | 0.50 | 0.67 | 4.08 | 3.57 | 1.92 | 4.52 | 2.66 | 3.29 | 1.19 | 6.28 | 2.20 | 0.24 | 31.12 |
| 1999 | 0.35 | 0.28 | 1.15 | 4.32 | 6.20 | 2.57 | 4.81 | 0.80 | 0.84 | 0.37 | 0.05 | 0.17 | 21.91 |
| 2000 | 0.68 | 1.04 | 0.91 | 2.27 | 5.56 | 3.26 | 3.22 | 3.17 | 1.34 | 1.79 | 2.52 | 0.35 | 26.11 |
| 2001 | 1.50 | 0.65 | 0.78 | 6.97 | 1.92 | 3.13 | 5.88 | 1.37 | 2.25 | 0.86 | 4.76 | 0.11 | 30.18 |
| 2002 | 0.17 | 0.27 | 1.41 | 2.29 | 1.82 | 2.57 | 1.80 | 8.26 | 1.39 | 3.85 | 0.09 | 0.15 | 24.07 |
| 2003 | 0.30 | 0.64 | 0.22 | 3.69 | 2.64 | 3.54 | 1.64 | 1.82 | 4.74 | 0.92 | 0.59 | 1.07 | 21.81 |
| 2004 | 0.52 | 1.11 | 2.03 | 1.28 | 8.10 | 6.00 | 1.40 | 3.58 | 5.12 | 0.86 | 0.81 | 0.11 | 30.92 |
| 2005 | 0.44 | 1.12 | 1.53 | 3.33 | 5.22 | 3.72 | 4.59 | 1.36 | 4.76 | 1.66 | 2.95 | 1.03 | 31.71 |
| 2006 | 0.76 | 0.14 | 2.67 | 6.17 | 1.02 | 3.81 | 0.68 | 4.33 | 3.88 | 0.33 | 1.00 | 1.95 | 26.74 |
| 2007 | 0.45 | 1.29 | 4.97 | 1.93 | 2.63 | 3.98 | 0.32 | 6.18 | 2.27 | 5.98 | 0.04 | 1.30 | 31.34 |
| 2008 | 0.23 | 0.57 | 1.34 | 2.68 | 3.34 | 3.95 | 2.52 | 1.91 | 1.78 | 5.44 | 1.01 | 0.70 | 25.47 |
| 2009 | 0.32 | 0.41 | 1.31 | 1.95 | 1.43 | 3.07 | 3.71 | 1.93 | 1.21 | 5.52 | 0.17 | 2.03 | 23.06 |
| POR= 78 YRS | 0.57 | 0.77 | 1.63 | 2.61 | 3.39 | 3.88 | 2.88 | 3.20 | 2.73 | 1.76 | 1.09 | 0.68 | 25.19 |

WBAN : 14944

## AVERAGE TEMPERATURE (°F) 2009  SIOUX FALLS (KFSD)

| YEAR | JAN | FEB | MAR | APR | MAY | JUN | JUL | AUG | SEP | OCT | NOV | DEC | ANNUAL |
|------|-----|-----|-----|-----|-----|-----|-----|-----|-----|-----|-----|-----|--------|
| 1980 | 18.1 | 18.1 | 31.7 | 50.1 | 58.4 | 68.8 | 74.2 | 71.1 | 62.0 | 45.5 | 36.7 | 21.2 | 46.3 |
| 1981 | 22.3 | 26.0 | 38.4 | 53.5 | 58.3 | 70.4 | 75.5 | 71.1 | 63.0 | 48.7 | 39.5 | 18.0 | 48.7 |
| 1982 | 3.8 | 20.2 | 32.8 | 43.9 | 59.6 | 62.6 | 74.4 | 71.3 | 60.0 | 48.8 | 30.3 | 24.5 | 44.4 |
| 1983 | 20.1 | 26.2 | 33.1 | 41.5 | 55.1 | 66.7 | 77.1 | 78.3 | 63.5 | 49.6 | 34.9 | 2.1 | 45.7 |
| 1984 | 17.4 | 27.8 | 24.6 | 45.8 | 55.9 | 68.1 | 73.6 | 74.2 | 57.2 | 50.4 | 36.2 | 20.4 | 46.0 |
| 1985 | 13.4 | 19.9 | 37.8 | 52.8 | 62.7 | 64.2 | 71.5 | 66.3 | 58.2 | 46.6 | 20.7 | 9.5 | 43.6 |
| 1986 | 20.7 | 16.9 | 36.9 | 48.2 | 58.8 | 69.6 | 75.0 | 66.5 | 59.6 | 48.6 | 28.7 | 25.2 | 46.2 |
| 1987 | 24.0 | 32.8 | 37.8 | 52.6 | 64.9 | 71.4 | 77.0 | 68.8 | 62.8 | 43.9 | 38.1 | 24.4 | 49.9 |
| 1988 | 9.6 | 15.0 | 36.4 | 46.4 | 65.1 | 76.3 | 77.4 | 74.8 | 62.6 | 44.6 | 34.2 | 22.1 | 47.0 |
| 1989 | 25.5 | 9.2 | 29.8 | 47.8 | 58.0 | 66.9 | 77.3 | 72.0 | 60.2 | 49.5 | 29.5 | 11.5 | 44.8 |
| 1990 | 28.2 | 24.9 | 36.7 | 46.4 | 56.4 | 70.1 | 71.2 | 72.9 | 66.5 | 48.0 | 35.6 | 15.3 | 47.7 |
| 1991 | 13.7 | 29.8 | 37.0 | 50.0 | 62.5 | 73.5 | 73.1 | 72.8 | 62.0 | 46.1 | 25.7 | 26.0 | 47.7 |
| 1992 | 26.5 | 30.0 | 36.7 | 43.7 | 60.5 | 67.0 | 65.5 | 66.1 | 60.3 | 48.7 | 30.4 | 19.2 | 46.2 |
| 1993 | 14.0 | 15.0 | 29.0 | 44.1 | 57.1 | 65.1 | 71.4 | 71.4 | 56.8 | 47.1 | 29.9 | 21.9 | 43.6 |
| 1994 | 6.2 | 13.3 | 36.6 | 46.5 | 63.9 | 71.6 | 70.8 | 69.5 | 65.2 | 52.4 | 36.5 | 21.8 | 46.2 |
| 1995 | 17.7 | 23.5 | 34.8 | 41.8 | 55.7 | 70.0 | 75.1 | 76.9 | 60.4 | 48.3 | 29.2 | 23.7 | 46.4 |
| 1996 | 10.9 | 22.6 | 27.4 | 42.4 | 54.9 | 68.8 | 69.1 | 70.0 | 59.4 | 47.7 | 23.5 | 10.3 | 42.3 |
| 1997 | 8.6 | 17.9 | 30.6 | 40.6 | 51.5 | 68.8 | 72.4 | 68.9 | 62.4 | 49.4 | 27.9 | 26.1 | 43.8 |
| 1998 | 19.6 | 31.8 | 28.0 | 46.5 | 61.7 | 63.1 | 71.9 | 71.0 | 66.3 | 50.6 | 35.5 | 24.0 | 47.5 |
| 1999 | 14.0 | 29.9 | 33.8 | 45.7 | 58.5 | 66.9 | 75.1 | 70.6 | 59.7 | 48.5 | 42.1 | 25.1 | 47.5 |
| 2000 | 17.3 | 30.4 | 39.4 | 46.0 | 59.2 | 65.7 | 71.7 | 71.6 | 61.4 | 51.1 | 25.0 | 7.7 | 45.5 |
| 2001 | 18.8 | 11.1 | 26.7 | 48.1 | 59.1 | 68.7 | 75.7 | 72.2 | 61.3 | 48.2 | 43.6 | 25.0 | 46.5 |
| 2002 | 24.8 | 29.6 | 23.4 | 47.3 | 54.1 | 72.2 | 77.7 | 70.6 | 63.3 | 40.9 | 33.0 | 26.6 | 47.0 |
| 2003 | 17.6 | 18.8 | 33.0 | 47.9 | 55.5 | 66.3 | 73.0 | 73.3 | 60.0 | 50.8 | 31.9 | 25.6 | 46.1 |
| 2004 | 14.9 | 20.3 | 38.5 | 49.0 | 57.8 | 64.8 | 70.5 | 65.5 | 65.8 | 49.8 | 37.5 | 25.7 | 46.7 |
| 2005 | 16.0 | 28.6 | 33.4 | 52.1 | 56.3 | 70.4 | 75.0 | 70.0 | 67.4 | 50.5 | 36.8 | 17.9 | 47.9 |
| 2006 | 31.1 | 23.3 | 33.3 | 51.8 | 59.6 | 69.0 | 76.0 | 71.3 | 58.2 | 46.4 | 34.4 | 28.2 | 48.6 |
| 2007 | 18.5 | 13.6 | 39.0 | 45.5 | 62.4 | 69.9 | 75.0 | 72.6 | 63.5 | 53.5 | 35.1 | 17.5 | 47.2 |
| 2008 | 13.7 | 14.9 | 29.2 | 43.3 | 56.4 | 68.2 | 74.6 | 71.6 | 62.9 | 49.5 | 34.6 | 15.3 | 44.5 |
| 2009 | 14.3 | 23.7 | 33.4 | 45.3 | 58.7 | 65.9 | 68.6 | 68.0 | 63.5 | 42.5 | 41.3 | 15.1 | 45.0 |
| POR= 78 YRS | 15.3 | 20.4 | 32.0 | 46.7 | 58.7 | 68.4 | 74.1 | 71.6 | 61.8 | 49.4 | 33.2 | 20.4 | 46.0 |

**HEATING DEGREE DAYS (base 65°F) 2009 SIOUX FALLS (KFSD)**

| YEAR | JUL | AUG | SEP | OCT | NOV | DEC | JAN | FEB | MAR | APR | MAY | JUN | TOTAL |
|---|---|---|---|---|---|---|---|---|---|---|---|---|---|
| 1980-81 | 1 | 14 | 157 | 602 | 841 | 1354 | 1318 | 1087 | 816 | 353 | 230 | 5 | 6778 |
| 1981-82 | 5 | 5 | 114 | 497 | 758 | 1452 | 1899 | 1247 | 991 | 632 | 174 | 105 | 7879 |
| 1982-83 | 0 | 28 | 188 | 494 | 1035 | 1249 | 1385 | 1082 | 982 | 699 | 317 | 67 | 7526 |
| 1983-84 | 1 | 0 | 160 | 476 | 894 | 1947 | 1468 | 1072 | 1247 | 569 | 285 | 23 | 8142 |
| 1984-85 | 2 | 12 | 265 | 451 | 857 | 1376 | 1594 | 1260 | 836 | 378 | 124 | 101 | 7256 |
| 1985-86 | 5 | 44 | 269 | 562 | 1327 | 1721 | 1363 | 1341 | 865 | 498 | 204 | 17 | 8216 |
| 1986-87 | 0 | 54 | 180 | 504 | 1082 | 1227 | 1265 | 896 | 835 | 387 | 96 | 23 | 6549 |
| 1987-88 | 5 | 54 | 110 | 649 | 801 | 1252 | 1715 | 1448 | 878 | 554 | 83 | 1 | 7550 |
| 1988-89 | 0 | 22 | 126 | 628 | 916 | 1321 | 1219 | 1559 | 1083 | 514 | 235 | 61 | 7684 |
| 1989-90 | 0 | 5 | 192 | 481 | 1056 | 1655 | 1131 | 1117 | 871 | 586 | 268 | 39 | 7401 |
| 1990-91 | 11 | 7 | 109 | 527 | 875 | 1538 | 1587 | 977 | 859 | 455 | 192 | 7 | 7144 |
| 1991-92 | 6 | 3 | 180 | 583 | 1171 | 1200 | 1186 | 1009 | 871 | 636 | 187 | 35 | 7067 |
| 1992-93 | 35 | 63 | 159 | 500 | 1030 | 1414 | 1576 | 1397 | 1109 | 621 | 244 | 74 | 8222 |
| 1993-94 | 2 | 19 | 257 | 557 | 1046 | 1331 | 1821 | 1443 | 872 | 563 | 140 | 8 | 8059 |
| 1994-95 | 5 | 15 | 100 | 386 | 849 | 1332 | 1461 | 1157 | 931 | 689 | 281 | 55 | 7261 |
| 1995-96 | 3 | 0 | 198 | 523 | 1068 | 1275 | 1674 | 1224 | 1159 | 672 | 334 | 44 | 8174 |
| 1996-97 | 8 | 4 | 228 | 535 | 1237 | 1691 | 1738 | 1313 | 1060 | 727 | 413 | 7 | 8961 |
| 1997-98 | 20 | 26 | 130 | 497 | 1105 | 1202 | 1402 | 922 | 1138 | 549 | 147 | 117 | 7255 |
| 1998-99 | 4 | 2 | 79 | 440 | 881 | 1262 | 1574 | 976 | 961 | 572 | 208 | 72 | 7031 |
| 1999-00 | 0 | 4 | 199 | 505 | 686 | 1229 | 1471 | 997 | 789 | 565 | 205 | 80 | 6730 |
| 2000-01 | 23 | 8 | 167 | 425 | 1195 | 1771 | 1424 | 1501 | 1185 | 516 | 203 | 43 | 8461 |
| 2001-02 | 7 | 7 | 161 | 522 | 637 | 1234 | 1239 | 986 | 1283 | 540 | 368 | 27 | 7011 |
| 2002-03 | 2 | 18 | 148 | 739 | 953 | 1184 | 1462 | 1285 | 986 | 516 | 293 | 63 | 7649 |
| 2003-04 | 1 | 12 | 201 | 443 | 987 | 1216 | 1546 | 1291 | 813 | 483 | 250 | 71 | 7314 |
| 2004-05 | 18 | 67 | 85 | 464 | 820 | 1211 | 1511 | 1015 | 971 | 386 | 283 | 2 | 6833 |
| 2005-06 | 7 | 15 | 65 | 462 | 838 | 1453 | 1047 | 1164 | 978 | 396 | 217 | 26 | 6668 |
| 2006-07 | 1 | 4 | 222 | 594 | 912 | 1134 | 1433 | 1431 | 796 | 597 | 132 | 23 | 7279 |
| 2007-08 | 0 | 1 | 135 | 377 | 892 | 1467 | 1582 | 1448 | 1102 | 643 | 273 | 5 | 7925 |
| 2008-09 | 2 | 1 | 122 | 475 | 906 | 1535 | 1565 | 1150 | 970 | 593 | 225 | 85 | 7629 |
| 2009- | 11 | 35 | 89 | 691 | 704 | 1541 | | | | | | | |

WBAN : 14944

**COOLING DEGREE DAYS (base 65°F) 2009 SIOUX FALLS (KFSD)**

| YEAR | JAN | FEB | MAR | APR | MAY | JUN | JUL | AUG | SEP | OCT | NOV | DEC | TOTAL |
|---|---|---|---|---|---|---|---|---|---|---|---|---|---|
| 1980 | 0 | 0 | 0 | 23 | 46 | 153 | 290 | 207 | 74 | 6 | 0 | 0 | 799 |
| 1981 | 0 | 0 | 0 | 14 | 27 | 175 | 342 | 203 | 61 | 0 | 0 | 0 | 822 |
| 1982 | 0 | 0 | 0 | 4 | 12 | 42 | 298 | 230 | 45 | 0 | 0 | 0 | 631 |
| 1983 | 0 | 0 | 0 | 0 | 15 | 122 | 381 | 420 | 120 | 2 | 0 | 0 | 1060 |
| 1984 | 0 | 0 | 0 | 0 | 10 | 120 | 276 | 305 | 36 | 4 | 0 | 0 | 751 |
| 1985 | 0 | 0 | 0 | 19 | 59 | 85 | 212 | 93 | 72 | 0 | 0 | 0 | 540 |
| 1986 | 0 | 0 | 0 | 0 | 20 | 163 | 318 | 108 | 25 | 0 | 0 | 0 | 634 |
| 1987 | 0 | 0 | 0 | 20 | 100 | 219 | 381 | 178 | 53 | 0 | 0 | 0 | 951 |
| 1988 | 0 | 0 | 0 | 1 | 94 | 349 | 393 | 330 | 61 | 0 | 0 | 0 | 1228 |
| 1989 | 0 | 0 | 0 | 6 | 25 | 125 | 387 | 228 | 56 | 7 | 0 | 0 | 834 |
| 1990 | 0 | 0 | 0 | 31 | 11 | 198 | 209 | 258 | 160 | 7 | 0 | 0 | 874 |
| 1991 | 0 | 0 | 0 | 11 | 119 | 265 | 264 | 249 | 100 | 5 | 0 | 0 | 1013 |
| 1992 | 0 | 0 | 0 | 4 | 53 | 103 | 58 | 103 | 25 | 5 | 0 | 0 | 351 |
| 1993 | 0 | 0 | 0 | 0 | 5 | 85 | 208 | 223 | 16 | 9 | 0 | 0 | 546 |
| 1994 | 0 | 0 | 0 | 13 | 113 | 212 | 191 | 160 | 113 | 2 | 0 | 0 | 804 |
| 1995 | 0 | 0 | 0 | 0 | 0 | 214 | 324 | 375 | 68 | 11 | 0 | 0 | 992 |
| 1996 | 0 | 0 | 0 | 0 | 27 | 163 | 140 | 167 | 63 | 4 | 0 | 0 | 564 |
| 1997 | 0 | 0 | 0 | 0 | 1 | 128 | 255 | 152 | 62 | 20 | 0 | 0 | 618 |
| 1998 | 0 | 0 | 0 | 0 | 54 | 69 | 225 | 196 | 123 | 0 | 0 | 0 | 667 |
| 1999 | 0 | 0 | 0 | 0 | 14 | 136 | 319 | 185 | 48 | 3 | 0 | 0 | 705 |
| 2000 | 0 | 0 | 0 | 0 | 31 | 106 | 240 | 219 | 64 | 4 | 0 | 0 | 664 |
| 2001 | 0 | 0 | 0 | 14 | 29 | 163 | 345 | 237 | 58 | 8 | 0 | 0 | 854 |
| 2002 | 0 | 0 | 0 | 16 | 36 | 247 | 403 | 197 | 105 | 0 | 0 | 0 | 1004 |
| 2003 | 0 | 0 | 0 | 9 | 5 | 109 | 254 | 276 | 60 | 11 | 0 | 0 | 724 |
| 2004 | 0 | 0 | 0 | 10 | 33 | 72 | 196 | 90 | 114 | 0 | 0 | 0 | 515 |
| 2005 | 0 | 0 | 0 | 5 | 21 | 171 | 323 | 178 | 145 | 19 | 0 | 0 | 862 |
| 2006 | 0 | 0 | 0 | 6 | 57 | 154 | 350 | 207 | 23 | 22 | 0 | 0 | 819 |
| 2007 | 0 | 0 | 0 | 22 | 58 | 177 | 318 | 244 | 100 | 23 | 0 | 0 | 942 |
| 2008 | 0 | 0 | 0 | 0 | 15 | 107 | 306 | 214 | 67 | 0 | 0 | 0 | 709 |
| 2009 | 0 | 0 | 0 | 9 | 35 | 120 | 134 | 133 | 50 | 0 | 0 | 0 | 481 |

## SNOWFALL (inches)  2009  SIOUX FALLS (KFSD)

| YEAR | JUL | AUG | SEP | OCT | NOV | DEC | JAN | FEB | MAR | APR | MAY | JUN | TOTAL |
|------|-----|-----|-----|-----|-----|-----|-----|-----|-----|-----|-----|-----|-------|
| 1980-81 | 0.0 | 0.0 | 0.0 | 0.8 | 0.4 | 4.6 | 1.4 | 3.1 | T | 0.5 | 0.0 | 0.0 | 10.8 |
| 1981-82 | 0.0 | 0.0 | 0.0 | 0.8 | 8.9 | 5.5 | 16.9 | 1.0 | 1.8 | 7.5 | 0.0 | 0.0 | 42.4 |
| 1982-83 | 0.0 | 0.0 | 0.0 | 3.3 | 4.1 | 17.6 | 4.7 | 3.6 | 18.8 | 18.4 | 0.0 | 0.0 | 70.5 |
| 1983-84 | 0.0 | 0.0 | 0.0 | T | 19.0 | 13.7 | 5.0 | 11.9 | 19.4 | 6.0 | 0.0 | 0.0 | 75.0 |
| 1984-85 | 0.0 | 0.0 | T | T | T | 4.7 | 7.4 | 0.7 | 16.1 | 2.5 | 0.0 | 0.0 | 31.4 |
| 1985-86 | 0.0 | 0.0 | 0.9 | 0.2 | 21.9 | 9.1 | 9.1 | 0.9 | 8.2 | 0.3 | 0.0 | 0.0 | 50.6 |
| 1986-87 | 0.0 | 0.0 | 0.0 | T | 5.4 | T | 2.4 | 0.3 | T | T | 0.0 | 0.0 | 8.1 |
| 1987-88 | 0.0 | 0.0 | 0.0 | 0.1 | 7.5 | 13.3 | 17.9 | 7.3 | 2.5 | 11.3 | 0.0 | 0.0 | 59.9 |
| 1988-89 | 0.0 | 0.0 | 0.0 | T | 10.9 | 2.2 | 2.0 | 10.0 | 16.0 | 0.7 | T | 0.0 | 41.8 |
| 1989-90 | 0.0 | 0.0 | 0.0 | 0.0 | 2.4 | 4.0 | 0.2 | 5.8 | 0.7 | T | 0.0 | T | 13.1 |
| 1990-91 | 0.0 | 0.0 | 0.0 | 1.2 | 8.4 | 8.8 | 5.4 | 5.4 | 3.2 | T | T | T | 32.4 |
| 1991-92 | 0.0 | 0.0 | 0.0 | 10.0 | 10.8 | 0.2 | 3.8 | 11.2 | 8.4 | 3.5 | T | T | 47.9 |
| 1992-93 | 0.0 | 0.0 | 0.0 | T | 4.7 | 8.4 | 8.6 | 13.1 | 14.9 | 2.2 | T | T | 51.9 |
| 1993-94 | T | 0.0 | T | T | 8.8 | 4.6 | 16.8 | 13.0 | 1.1 | 14.9 | 0.0 | T | 59.2 |
| 1994-95 | 0.0 | 0.0 | 0.0 | 0.0 | 9.7 | 7.0 | 0.5 | 2.4 | 8.8 | 12.5 | 0.0 | 0.0 | 40.9 |
| 1995-96 | T | 0.0 | 0.0 | 7.4 | 5.1 | 1.9 | 15.4 | 1.7 | 6.9 | | | | |
| 1996-97 | | | 0.0 | T | 11.3 | 19.8 | 8.8 | 16.5 | 1.3 | 6.0 | T | 0.0 | |
| 1997-98 | 0.0 | 0.0 | 0.0 | 0.4 | 5.0 | 3.3 | 11.7 | 6.9 | 21.4 | 0.3 | 0.0 | T | 49.0 |
| 1998-99 | 0.0 | 0.0 | 0.0 | 0.0 | 14.8 | 9.0 | 9.9 | 7.9 | 11.2 | 2.4 | T | 0.0 | 55.2 |
| 1999-00 | T | 0.0 | T | 2.7 | T | 3.2 | 6.0 | 1.2 | 4.5 | 5.6 | T | 0.0 | 23.2 |
| 2000-01 | T | 0.0 | 0.0 | 0.0 | 19.6 | 11.4 | 13.9 | 8.4 | 3.2 | 1.4 | 0.0 | T | 57.9 |
| 2001-02 | 0.0 | T | 0.0 | T | 13.8 | 2.9 | 3.2 | 2.8 | 17.1 | 4.9 | T | 0.0 | 44.7 |
| 2002-03 | 0.0 | T | 0.0 | 6.3 | 1.0 | 0.8 | 4.1 | 6.3 | 4.6 | 9.0 | 0.0 | 0.0 | 32.1 |
| 2003-04 | 0.0 | 0.0 | 0.0 | 0.1 | 7.6 | 13.4 | 12.3 | 15.8 | 7.4 | T | T | 0.0 | 56.6 |
| 2004-05 | 0.0 | 0.0 | 0.0 | 0.0 | T | 0.5 | 9.5 | 4.0 | 13.4 | T | T | 0.0 | 27.4 |
| 2005-06 | T | T | 0.0 | T | 14.3 | 8.5 | 0.7 | 2.3 | 9.3 | T | 0.0 | T | 35.1 |
| 2006-07 | 0.0 | 0.0 | 0.0 | 1.6 | 0.8 | 1.7 | 8.1 | 15.9 | 8.0 | 8.2 | 0.0 | 0.0 | 44.3 |
| 2007-08 | 0.0 | 0.0 | 0.0 | 0.0 | 0.6 | 11.4 | 4.1 | 7.5 | 14.5 | 7.6 | 0.0 | 0.0 | 45.7 |
| 2008-09 | 0.0 | 0.0 | 0.0 | T | 3.4 | 12.5 | 6.5 | 2.6 | 3.2 | 6.6 | 0.0 | 0.0 | 34.8 |
| 2009- | 0.0 | T | 0.0 | 3.5 | T | 27.8 | | | | | | | |
| POR=<br>77 YRS | T | T | T | 0.7 | 5.1 | 7.1 | 6.8 | 8.1 | 8.9 | 2.9 | 0.1 | T | 39.7 |

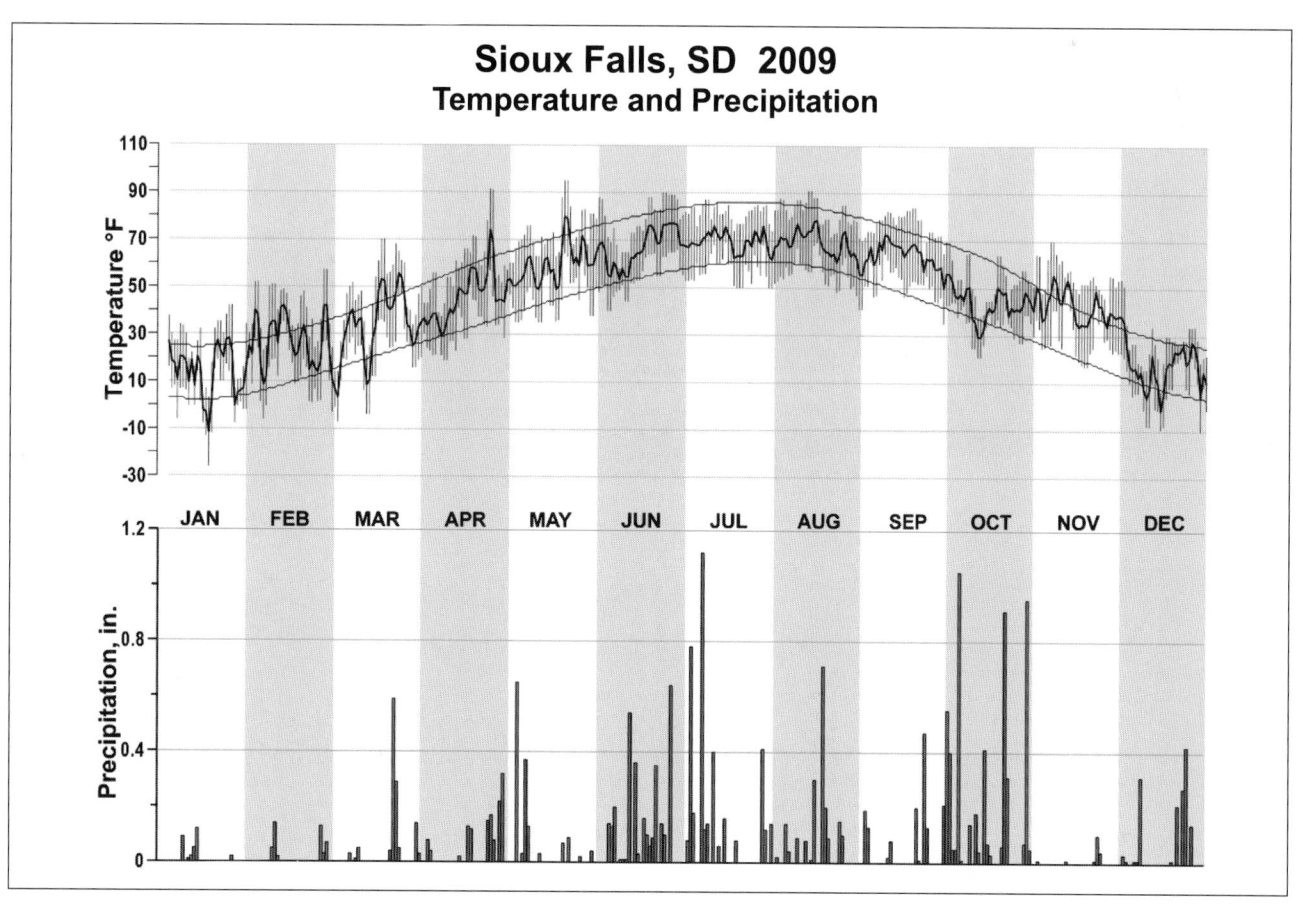

## Sioux Falls, SD  2009
### Temperature and Precipitation

# 2009
# KNOXVILLE
# TENNESSEE (KTYS)

Knoxville is located in a broad valley between the Cumberland Mountains, which lie northwest of the city, and the Great Smoky Mountains, which lie southeast of the city. These two mountain ranges exercise a marked influence upon the climate of the valley. The Cumberland Mountains, to the northwest, serve to retard and weaken the force of the cold winter air which frequently penetrates far south of the latitude of Knoxville over the plains areas to the west of the mountains.

The mountains also serve to modify the hot summer winds which are common to the plains to the west. In addition, they serve as a fixed incline plane which lifts the warm, moist air flowing northward from the Gulf of Mexico and thereby increases the frequency of afternoon thunderstorms. Relief from extremely high temperatures which such thunderstorms produce serves to reduce the number of extremely warm days in the valley.

July is usually the warmest month of the year. The coldest weather usually occurs during the month of January. Sudden great temperature changes occur infrequently. This again is due mainly to the retarding effect of the mountains. Summer nights are nearly always comfortable.

Rainfall is ample for agricultural purposes and is favorably distributed during the year for most crops. Precipitation is greatest in the wintertime. Another peak period occurs during the late spring and summer months. The period of lowest rainfall occurs during the fall. A cumulative total of approximately 12 inches of snow falls annually. However, this usually comes in amounts of less than 4 inches at one time. It is unusual for snow to remain on the ground in measurable amounts longer than one week.

The topography also has a pronounced effect upon the prevailing wind direction. Daytime winds usually have a southwesterly component, while nighttime winds usually move from the northeast. The winds are relatively light and tornadoes are extremely rare.

# NORMALS, MEANS, AND EXTREMES
## KNOXVILLE (KTYS)

| LATITUDE: 35 ° 49'N | LONGITUDE: -83 ° 59'W | ELEVATION (FT): GRND: 962  BARO: 982 | TIME ZONE: EASTERN  (UTC -5) | WBAN: 13891 |
|---|---|---|---|---|

| | ELEMENT | POR | JAN | FEB | MAR | APR | MAY | JUN | JUL | AUG | SEP | OCT | NOV | DEC | YEAR |
|---|---|---|---|---|---|---|---|---|---|---|---|---|---|---|---|
| **TEMPERATURE °F** | NORMAL DAILY MAXIMUM | 30 | 46.3 | 51.7 | 60.3 | 69.0 | 76.3 | 83.6 | 86.9 | 86.4 | 80.7 | 69.9 | 59.0 | 49.8 | 68.3 |
| | MEAN DAILY MAXIMUM | 100 | 48.0 | 50.0 | 60.4 | 69.5 | 78.3 | 84.3 | 87.8 | 87.0 | 81.1 | 71.4 | 58.4 | 49.8 | 68.8 |
| | HIGHEST DAILY MAXIMUM | 68 | 77 | 83 | 86 | 92 | 94 | 102 | 103 | 102 | 103 | 91 | 84 | 80 | 103 |
| | YEAR OF OCCURRENCE | | 1950 | 1996 | 1963 | 1942 | 1962 | 1988 | 1952 | 2007 | 1954 | 1953 | 1948 | 1982 | SEP 1954 |
| | MEAN OF EXTREME MAXS. | 100 | 67.5 | 70.7 | 78.6 | 84.9 | 88.2 | 92.9 | 94.7 | 94.2 | 91.5 | 83.5 | 76.0 | 68.6 | 82.6 |
| | NORMAL DAILY MINIMUM | 30 | 28.9 | 31.8 | 39.1 | 46.6 | 55.6 | 63.9 | 68.5 | 67.3 | 60.8 | 47.7 | 38.9 | 31.9 | 48.4 |
| | MEAN DAILY MINIMUM | 100 | 30.4 | 31.3 | 39.1 | 47.2 | 56.3 | 63.6 | 68.1 | 67.1 | 60.4 | 48.9 | 38.0 | 32.2 | 48.6 |
| | LOWEST DAILY MINIMUM | 68 | -24 | -8 | 1 | 22 | 32 | 43 | 49 | 49 | 36 | 25 | 5 | -6 | -24 |
| | YEAR OF OCCURRENCE | | 1985 | 1996 | 1980 | 1987 | 1986 | 1956 | 1988 | 1946 | 1967 | 1987 | 1950 | 1983 | JAN 1985 |
| | MEAN OF EXTREME MINS. | 100 | 9.9 | 13.5 | 22.2 | 31.6 | 41.3 | 52.8 | 59.8 | 59.3 | 47.3 | 33.0 | 22.8 | 14.8 | 34.0 |
| | NORMAL DRY BULB | 30 | 37.6 | 41.8 | 49.7 | 57.8 | 66.0 | 73.8 | 77.7 | 76.9 | 70.8 | 58.8 | 49.0 | 40.9 | 58.4 |
| | MEAN DRY BULB | 100 | 39.2 | 40.7 | 49.8 | 58.4 | 67.3 | 74.0 | 78.0 | 77.0 | 70.8 | 60.1 | 48.2 | 41.0 | 58.7 |
| | MEAN WET BULB | 26 | 34.1 | 36.8 | 43.7 | 51.0 | 59.9 | 66.9 | 70.1 | 69.5 | 63.6 | 53.3 | 43.8 | 36.6 | 52.4 |
| | MEAN DEW POINT | 26 | 30.5 | 33.0 | 38.7 | 46.7 | 57.2 | 64.8 | 68.2 | 67.6 | 61.1 | 50.5 | 40.5 | 33.3 | 49.3 |
| | NORMAL NO. DAYS WITH: | | | | | | | | | | | | | | |
| | MAXIMUM >= 90 | 3 | 0.0 | 0.0 | 0.0 | * | 0.3 | 5.4 | 12.4 | 9.7 | 3.4 | 0.0 | 0.0 | 0.0 | 31.2 |
| | MAXIMUM <= 32 | 3 | 3.1 | 1.5 | 0.2 | 0.0 | 0.0 | 0.0 | 0.0 | 0.0 | 0.0 | 0.0 | * | 1.3 | 6.1 |
| | MINIMUM <= 32 | 30 | 19.3 | 15.2 | 8.3 | 1.9 | * | 0.0 | 0.0 | 0.0 | 0.0 | 0.9 | 8.2 | 17.1 | 70.9 |
| | MINIMUM <= 0 | 30 | 0.3 | * | 0.0 | 0.0 | 0.0 | 0.0 | 0.0 | 0.0 | 0.0 | 0.0 | 0.0 | 0.1 | 0.4 |
| **H/C** | NORMAL HEATING DEG. DAYS | 30 | 841 | 652 | 467 | 227 | 65 | 3 | 0 | 0 | 22 | 210 | 470 | 733 | 3690 |
| | NORMAL COOLING DEG. DAYS | 30 | 0 | 1 | 5 | 27 | 110 | 282 | 408 | 381 | 205 | 28 | 3 | 0 | 1450 |
| **RH** | NORMAL (PERCENT) | 30 | 74 | 70 | 66 | 65 | 73 | 75 | 75 | 76 | 75 | 75 | 74 | 75 | 73 |
| | HOUR 01 LST | 30 | 79 | 75 | 73 | 73 | 84 | 87 | 88 | 88 | 87 | 87 | 81 | 80 | 82 |
| | HOUR 07 LST | 30 | 82 | 81 | 81 | 82 | 88 | 89 | 90 | 92 | 92 | 91 | 86 | 84 | 87 |
| | HOUR 13 LST | 30 | 64 | 60 | 56 | 52 | 58 | 59 | 61 | 60 | 59 | 56 | 59 | 65 | 59 |
| | HOUR 19 LST | 30 | 69 | 61 | 57 | 54 | 63 | 64 | 65 | 66 | 68 | 66 | 67 | 70 | 64 |
| **S** | PERCENT POSSIBLE SUNSHINE | 57 | 40 | 47 | 53 | 63 | 64 | 65 | 64 | 63 | 61 | 61 | 49 | 40 | 56 |
| **W/O** | MEAN NO. DAYS WITH: | | | | | | | | | | | | | | |
| | HEAVY FOG(VISBY <= 1/4 MI) | 46 | 2.6 | 1.8 | 1.7 | 1.3 | 2.2 | 1.8 | 2.1 | 3.5 | 3.8 | 4.3 | 2.9 | 2.4 | 30.4 |
| | THUNDERSTORMS | 62 | 0.8 | 1.4 | 3.2 | 4.5 | 6.9 | 8.5 | 9.9 | 6.9 | 3.0 | 1.3 | 1.1 | 0.7 | 48.2 |
| **CLOUDNESS** | MEAN: | | | | | | | | | | | | | | |
| | SUNRISE-SUNSET (OKTAS) | | | | | | 4.0 | 3.2 | | | | | | | |
| | MIDNIGHT-MIDNIGHT (OKTAS) | | | | | | | 3.2 | | | | | | | |
| | MEAN NO. DAYS WITH: | | | | | | | | | | | | | | |
| | CLEAR | 1 | 4.0 | 1.0 | 4.0 | | 13.0 | 11.0 | | | | | | | |
| | PARTLY CLOUDY | | | | 1.0 | | 3.0 | 5.0 | | | | | | | |
| | CLOUDY | 1 | 2.0 | 3.0 | 10.0 | | 4.0 | 4.0 | | | | | | | |
| **PR** | MEAN STATION PRESSURE(IN) | 26 | 29.08 | 29.05 | 29.01 | 28.98 | 28.97 | 28.97 | 29.00 | 29.01 | 29.02 | 29.06 | 29.07 | 29.09 | 29.03 |
| | MEAN SEA-LEVEL PRES. (IN) | 26 | 30.14 | 30.11 | 30.05 | 30.01 | 30.00 | 29.99 | 30.01 | 30.02 | 30.04 | 30.09 | 30.12 | 30.15 | 30.06 |
| **WINDS** | MEAN SPEED (MPH) | 26 | 6.7 | 7.0 | 7.4 | 7.1 | 6.3 | 5.6 | 5.5 | 4.9 | 5.0 | 5.0 | 5.7 | 6.3 | 6.0 |
| | PREVAIL.DIR(TENS OF DEGS) | 36 | 25 | 25 | 24 | 24 | 24 | 24 | 24 | 24 | 05 | 05 | 25 | 24 | 24 |
| | MAXIMUM 2-MINUTE: | | | | | | | | | | | | | | |
| | SPEED (MPH) | 14 | 43 | 46 | 43 | 64 | 41 | 44 | 46 | 39 | 31 | 43 | 49 | 44 | 64 |
| | DIR. (TENS OF DEGS) | | 27 | 27 | 24 | 28 | 21 | 23 | 25 | 01 | 23 | 26 | 25 | 24 | 28 |
| | YEAR OF OCCURRENCE | | 1996 | 2009 | 1996 | 1996 | 2008 | 2003 | 2005 | 2003 | 2008 | 2001 | 2000 | 2006 | APR 1996 |
| | MAXIMUM 3-SECOND | | | | | | | | | | | | | | |
| | SPEED (MPH) | 14 | 54 | 56 | 51 | 76 | 55 | 52 | 68 | 48 | 45 | 56 | 61 | 57 | 76 |
| | DIR. (TENS OF DEGS) | | 28 | 27 | 25 | 27 | 12 | 24 | 25 | 03 | 19 | 26 | 24 | 28 | 27 |
| | YEAR OF OCCURRENCE | | 2008 | 2009 | 2008 | 1996 | 2009 | 2003 | 1997 | 2000 | 2008 | 2001 | 2000 | 1996 | APR 1996 |
| **PRECIPITATION** | NORMAL (IN) | 30 | 4.57 | 4.01 | 5.17 | 3.99 | 4.68 | 4.04 | 4.71 | 2.89 | 3.04 | 2.65 | 3.98 | 4.49 | 48.22 |
| | MAXIMUM MONTHLY (IN) | 68 | 11.74 | 9.38 | 11.81 | 11.07 | 10.98 | 8.21 | 12.66 | 8.88 | 9.19 | 6.67 | 10.36 | 11.63 | 12.66 |
| | YEAR OF OCCURRENCE | | 1954 | 1944 | 1994 | 1998 | 1974 | 1989 | 1999 | 1942 | 1989 | 1949 | 1948 | 1961 | JUL 1999 |
| | MINIMUM MONTHLY (IN) | 68 | 0.95 | 0.74 | 1.69 | 0.39 | 0.74 | 0.20 | 0.33 | 0.77 | 0.42 | T | 0.97 | 0.45 | T |
| | YEAR OF OCCURRENCE | | 1986 | 1968 | 1986 | 1976 | 1970 | 1944 | 1995 | 1954 | 1985 | 1963 | 1942 | 1965 | OCT 1963 |
| | MAXIMUM IN 24 HOURS (IN) | 68 | 3.89 | 3.42 | 5.77 | 3.71 | 4.35 | 3.57 | 4.69 | 3.25 | 5.08 | 2.44 | 4.06 | 4.89 | 5.77 |
| | YEAR OF OCCURRENCE | | 1946 | 1991 | 1994 | 2006 | 2003 | 1972 | 1942 | 1959 | 1944 | 1961 | 1948 | 1969 | MAR 1994 |
| | NORMAL NO. DAYS WITH: | | | | | | | | | | | | | | |
| | PRECIPITATION >= 0.01 | 30 | 12.2 | 11.1 | 13.2 | 10.8 | 11.6 | 10.9 | 10.9 | 8.9 | 8.5 | 7.9 | 10.1 | 11.5 | 127.6 |
| | PRECIPITATION >= 1.00 | 30 | 1.1 | 1.0 | 1.3 | 0.9 | 1.2 | 1.0 | 1.5 | 0.7 | 0.9 | 0.7 | 0.7 | 1.3 | 12.3 |
| **SNOWFALL** | NORMAL (IN) | 30 | 3.7 | 3.0 | 1.6 | 0.8 | 0.0 | 0.0 | 0.0 | 0.0 | 0.0 | 0.* | 0.1 | 0.7 | 9.9 |
| | MAXIMUM MONTHLY (IN) | 65 | 15.1 | 23.3 | 20.2 | 10.7 | T | T | 0.0 | T | T | T | 18.2 | 12.2 | 23.3 |
| | YEAR OF OCCURRENCE | | 1962 | 1960 | 1960 | 1987 | 2008 | 2008 | | 1995 | 2006 | 1993 | 1952 | 1963 | FEB 1960 |
| | MAXIMUM IN 24 HOURS (IN) | 65 | 12.0 | 17.5 | 14.1 | 10.7 | T | T | 0.0 | T | T | T | 18.2 | 8.9 | 18.2 |
| | YEAR OF OCCURRENCE' | | 1962 | 1960 | 1993 | 1987 | 1945 | 1998 | | 1995 | 2006 | 1993 | 1952 | 1969 | NOV 1952 |
| | MAXIMUM SNOW DEPTH (IN) | 58 | 10 | 15 | 15 | 7 | 0 | 0 | 0 | 0 | 0 | 0 | 10 | 6 | 15 |
| | YEAR OF OCCURRENCE | | 1966 | 1960 | 1993 | 1987 | | | | | | | 1952 | 1963 | MAR 1993 |
| | NORMAL NO. DAYS WITH: | | | | | | | | | | | | | | |
| | SNOWFALL >= 1.0 | 30 | 1.3 | 1.0 | 0.5 | 0.2 | 0.0 | 0.0 | 0.0 | 0.0 | 0.0 | 0.0 | 0.0 | 0.2 | 3.2 |

## PRECIPITATION (inches) 2009 KNOXVILLE (KTYS)

| YEAR | JAN | FEB | MAR | APR | MAY | JUN | JUL | AUG | SEP | OCT | NOV | DEC | ANNUAL |
|------|-----|-----|-----|-----|-----|-----|-----|-----|-----|-----|-----|-----|--------|
| 1980 | 5.54 | 1.78 | 8.72 | 3.30 | 3.80 | 1.94 | 3.57 | 2.34 | 2.38 | 1.53 | 3.78 | 1.78 | 40.46 |
| 1981 | 1.05 | 3.62 | 2.83 | 4.84 | 3.02 | 5.53 | 2.03 | 3.48 | 6.09 | 4.15 | 3.01 | 4.14 | 43.79 |
| 1982 | 6.03 | 4.88 | 6.36 | 3.26 | 5.52 | 3.93 | 6.60 | 2.68 | 2.68 | 2.66 | 5.21 | 4.89 | 54.70 |
| 1983 | 1.58 | 2.90 | 1.99 | 5.88 | 5.42 | 3.26 | 3.18 | 3.89 | 0.95 | 3.34 | 4.40 | 5.69 | 42.48 |
| 1984 | 2.26 | 4.42 | 3.79 | 3.37 | 10.14 | 4.34 | 9.03 | 1.72 | 0.85 | 3.26 | 2.87 | 2.49 | 48.54 |
| 1985 | 3.17 | 4.11 | 1.98 | 2.86 | 1.60 | 4.77 | 2.63 | 4.07 | 0.42 | 3.04 | 5.39 | 2.36 | 36.40 |
| 1986 | 0.95 | 3.90 | 1.69 | 2.25 | 2.40 | 0.69 | 1.89 | 3.37 | 3.59 | 3.84 | 3.83 | 4.08 | 32.48 |
| 1987 | 4.68 | 4.63 | 2.91 | 2.18 | 4.62 | 2.66 | 4.67 | 1.08 | 1.93 | 0.60 | 1.21 | 3.49 | 34.66 |
| 1988 | 4.29 | 2.94 | 2.42 | 2.34 | 2.35 | 0.51 | 3.60 | 3.20 | 2.68 | 1.52 | 4.82 | 3.99 | 34.66 |
| 1989 | 4.96 | 6.26 | 3.82 | 3.50 | 5.31 | 8.21 | 2.68 | 3.16 | 9.19 | 1.47 | 4.92 | 2.74 | 56.22 |
| 1990 | 5.88 | 6.90 | 5.73 | 2.56 | 4.71 | 1.72 | 7.56 | 3.02 | 2.68 | 3.64 | 1.77 | 8.99 | 55.16 |
| 1991 | 2.53 | 6.99 | 6.36 | 3.97 | 3.10 | 6.02 | 3.45 | 6.13 | 3.14 | 1.10 | 5.24 | 10.23 | 58.26 |
| 1992 | 3.87 | 3.36 | 3.99 | 2.11 | 3.20 | 3.07 | 4.62 | 4.22 | 1.77 | 3.05 | 4.29 | 6.68 | 44.23 |
| 1993 | 4.09 | 2.20 | 6.16 | 2.50 | 3.78 | 0.90 | 2.03 | 6.04 | 4.77 | 2.25 | 3.33 | 7.04 | 45.09 |
| 1994 | 7.08 | 8.81 | 11.81 | 7.90 | 3.04 | 5.97 | 5.94 | 4.19 | 1.20 | 2.38 | 2.93 | 2.02 | 63.27 |
| 1995 | 4.90 | 4.54 | 3.54 | 1.74 | 7.05 | 3.90 | 0.33 | 2.08 | 2.46 | 3.58 | 6.21 | 2.50 | 42.83 |
| 1996 | 7.53 | 2.74 | 4.72 | 4.50 | 4.57 | 2.96 | 4.77 | 1.33 | 3.45 | 1.17 | 7.71 | 5.44 | 50.89 |
| 1997 | 5.20 | 5.17 | 6.34 | 3.82 | 6.76 | 5.70 | 3.75 | 2.02 | 3.32 | 2.90 | 2.78 | 2.37 | 50.13 |
| 1998 | 4.62 | 2.71 | 4.75 | 11.07 | 3.84 | 7.96 | 5.69 | 2.07 | 1.31 | 1.42 | 2.51 | 5.95 | 53.90 |
| 1999 | 6.62 | 3.50 | 4.85 | 3.40 | 4.92 | 5.58 | 12.66 | 0.85 | 0.82 | 2.84 | 2.40 | 1.70 | 50.14 |
| 2000 | 5.14 | 3.42 | 4.37 | 6.69 | 5.90 | 3.36 | 6.12 | 1.40 | 3.82 | T | 4.35 | 2.45 | 47.02 |
| 2001 | 4.74 | 6.46 | 2.63 | 2.82 | 3.46 | 3.88 | 4.14 | 3.74 | 3.56 | 0.86 | 1.55 | 4.66 | 42.50 |
| 2002 | 8.06 | 1.69 | 10.59 | 1.44 | 4.66 | 3.91 | 5.65 | 1.41 | 5.17 | 5.59 | 4.58 | 5.35 | 58.10 |
| 2003 | 3.17 | 8.69 | 2.07 | 5.95 | 8.09 | 2.45 | 8.27 | 5.48 | 5.14 | 1.24 | 4.71 | 3.40 | 58.66 |
| 2004 | 2.74 | 3.64 | 4.61 | 2.73 | 5.35 | 7.90 | 7.62 | 2.18 | 4.75 | 3.10 | 6.01 | 5.56 | 56.19 |
| 2005 | 2.33 | 3.77 | 3.47 | 4.76 | 3.18 | 3.09 | 4.98 | 3.77 | 1.74 | 1.08 | 3.43 | 2.83 | 38.43 |
| 2006 | 4.00 | 2.55 | 3.22 | 7.57 | 2.13 | 2.38 | 3.94 | 6.39 | 6.17 | 4.11 | 3.24 | 2.09 | 47.79 |
| 2007 | 2.12 | 1.54 | 2.62 | 5.05 | 1.48 | 1.34 | 4.71 | 4.13 | 1.49 | 1.23 | 3.98 | 4.20 | 33.89 |
| 2008 | 3.13 | 4.81 | 4.33 | 3.74 | 3.39 | 2.19 | 7.04 | 1.91 | 3.16 | 1.66 | 3.41 | 8.98 | 47.75 |
| 2009 | 6.60 | 2.73 | 3.50 | 2.90 | 8.10 | 4.85 | 6.97 | 6.22 | 4.64 | 4.28 | 3.58 | 6.29 | 60.66 |
| POR= 100 YRS | 4.49 | 4.30 | 4.88 | 3.96 | 4.00 | 3.93 | 4.81 | 3.41 | 2.91 | 2.65 | 3.48 | 4.53 | 47.35 |

WBAN : 13891

## AVERAGE TEMPERATURE (°F) 2009 KNOXVILLE (KTYS)

| YEAR | JAN | FEB | MAR | APR | MAY | JUN | JUL | AUG | SEP | OCT | NOV | DEC | ANNUAL |
|------|-----|-----|-----|-----|-----|-----|-----|-----|-----|-----|-----|-----|--------|
| 1980 | 41.6 | 36.6 | 47.6 | 59.1 | 68.1 | 75.2 | 82.1 | 81.7 | 73.7 | 56.0 | 47.3 | 40.1 | 59.1 |
| 1981 | 33.4 | 41.9 | 46.8 | 63.5 | 63.9 | 77.1 | 81.4 | 78.3 | 70.2 | 58.9 | 49.4 | 38.7 | 58.6 |
| 1982 | 33.8 | 43.1 | 53.0 | 55.2 | 69.3 | 72.6 | 77.6 | 76.7 | 70.0 | 59.5 | 50.9 | 46.4 | 59.0 |
| 1983 | 38.0 | 41.1 | 49.7 | 53.2 | 65.4 | 72.5 | 78.5 | 79.6 | 70.3 | 60.2 | 47.4 | 36.4 | 57.7 |
| 1984 | 35.3 | 43.4 | 47.7 | 57.3 | 62.8 | 74.8 | 73.3 | 74.1 | 67.1 | 67.6 | 46.3 | 47.5 | 58.1 |
| 1985 | 29.4 | 37.1 | 50.2 | 58.6 | 65.7 | 72.5 | 76.5 | 75.1 | 68.5 | 63.7 | 56.8 | 34.8 | 57.4 |
| 1986 | 35.0 | 43.9 | 48.7 | 58.6 | 68.1 | 77.2 | 81.4 | 76.7 | 72.8 | 60.2 | 52.4 | 39.1 | 59.5 |
| 1987 | 36.8 | 42.2 | 49.2 | 55.0 | 71.3 | 75.5 | 78.5 | 79.6 | 70.1 | 52.8 | 50.2 | 43.5 | 58.7 |
| 1988 | 33.9 | 39.4 | 50.1 | 57.2 | 65.0 | 74.4 | 78.7 | 79.6 | 71.3 | 52.7 | 49.7 | 40.4 | 57.7 |
| 1989 | 42.8 | 41.3 | 53.5 | 57.8 | 62.4 | 73.3 | 77.5 | 76.8 | 70.9 | 59.4 | 48.1 | 32.2 | 58.0 |
| 1990 | 44.7 | 48.9 | 53.6 | 57.9 | 65.9 | 75.4 | 78.0 | 77.9 | 72.1 | 59.9 | 52.1 | 45.1 | 61.0 |
| 1991 | 41.2 | 43.8 | 52.0 | 62.8 | 72.0 | 74.9 | 79.3 | 76.5 | 71.4 | 61.0 | 47.1 | 43.5 | 60.5 |
| 1992 | 40.5 | 45.8 | 48.3 | 59.0 | 64.0 | 72.2 | 77.9 | 73.7 | 71.3 | 57.4 | 48.3 | 40.0 | 58.2 |
| 1993 | 43.5 | 39.5 | 45.7 | 56.2 | 67.0 | 76.0 | 83.0 | 78.7 | 70.6 | 57.9 | 48.4 | 40.2 | 58.9 |
| 1994 | 32.0 | 43.4 | 49.5 | 61.9 | 63.9 | 76.5 | 76.8 | 75.8 | 69.0 | 60.1 | 53.6 | 44.8 | 58.9 |
| 1995 | 39.2 | 40.7 | 53.1 | 61.0 | 67.3 | 73.8 | 80.2 | 81.5 | 70.5 | 58.6 | 43.6 | 37.1 | 58.9 |
| 1996 | 36.4 | 39.6 | 44.8 | 55.1 | 69.4 | 74.2 | 76.8 | 76.2 | 68.9 | 59.2 | 44.1 | 43.3 | 57.3 |
| 1997 | 39.4 | 46.6 | 54.2 | 54.0 | 62.2 | 71.9 | 78.4 | 75.7 | 71.7 | 58.9 | 43.9 | 39.2 | 58.0 |
| 1998 | 42.8 | 46.0 | 49.4 | 57.3 | 70.7 | 75.9 | 79.4 | 77.7 | 76.0 | 61.9 | 50.1 | 44.4 | 61.0 |
| 1999 | 43.6 | 44.1 | 45.3 | 61.2 | 66.2 | 75.0 | 78.6 | 77.7 | 69.9 | 58.5 | 52.7 | 42.4 | 59.6 |
| 2000 | 37.5 | 46.1 | 53.2 | 55.9 | 69.9 | 74.8 | 77.3 | 76.9 | 70.7 | 61.0 | 47.4 | 34.0 | 58.7 |
| 2001 | 35.5 | 45.8 | 46.3 | 61.7 | 68.4 | 72.9 | 78.0 | 77.5 | 68.2 | 56.9 | 54.2 | 45.6 | 59.3 |
| 2002 | 41.5 | 41.3 | 50.3 | 62.7 | 65.3 | 76.6 | 79.1 | 78.9 | 74.9 | 62.9 | 46.4 | 40.7 | 60.1 |
| 2003 | 33.9 | 40.9 | 53.3 | 60.4 | 67.5 | 72.1 | 76.3 | 77.8 | 69.4 | 58.5 | 53.0 | 38.5 | 58.5 |
| 2004 | 37.3 | 40.4 | 53.3 | 58.5 | 70.8 | 74.6 | 76.7 | 76.7 | 73.8 | 70.2 | 53.1 | 39.9 | 59.4 |
| 2005 | 43.0 | 44.1 | 48.5 | 58.6 | 64.1 | 75.0 | 78.4 | 79.3 | 74.1 | 61.1 | 51.3 | 37.6 | 59.6 |
| 2006 | 45.7 | 40.7 | 50.8 | 63.6 | 65.5 | 75.4 | 79.4 | 80.4 | 68.2 | 56.8 | 48.9 | 44.3 | 60.0 |
| 2007 | 41.2 | 38.2 | 57.0 | 56.1 | 70.0 | 76.7 | 77.5 | 83.2 | 74.7 | 65.1 | 49.3 | 45.5 | 61.2 |
| 2008 | 37.0 | 43.2 | 49.4 | 59.3 | 66.4 | 76.7 | 77.9 | 77.1 | 73.0 | 59.1 | 45.8 | 42.1 | 58.9 |
| 2009 | 36.6 | 42.8 | 51.3 | 58.4 | 67.5 | 76.0 | 74.5 | 76.4 | 71.8 | 58.0 | 50.9 | 39.0 | 58.6 |
| POR= 100 YRS | 39.2 | 40.7 | 49.8 | 58.4 | 67.3 | 74.0 | 78.0 | 77.0 | 70.8 | 60.1 | 48.2 | 41.0 | 58.7 |

## HEATING DEGREE DAYS (base 65°F) 2009  KNOXVILLE (KTYS)

| YEAR | JUL | AUG | SEP | OCT | NOV | DEC | JAN | FEB | MAR | APR | MAY | JUN | TOTAL |
|---|---|---|---|---|---|---|---|---|---|---|---|---|---|
| 1980-81 | 0 | 0 | 23 | 284 | 523 | 761 | 974 | 641 | 556 | 104 | 94 | 0 | 3960 |
| 1981-82 | 0 | 0 | 32 | 196 | 461 | 809 | 959 | 606 | 384 | 295 | 20 | 0 | 3762 |
| 1982-83 | 0 | 0 | 30 | 228 | 416 | 577 | 829 | 662 | 472 | 356 | 50 | 3 | 3623 |
| 1983-84 | 0 | 0 | 51 | 163 | 523 | 878 | 912 | 619 | 530 | 240 | 139 | 5 | 4060 |
| 1984-85 | 0 | 0 | 44 | 30 | 556 | 536 | 1095 | 776 | 459 | 208 | 51 | 13 | 3768 |
| 1985-86 | 0 | 0 | 44 | 86 | 250 | 927 | 922 | 582 | 499 | 206 | 51 | 0 | 3567 |
| 1986-87 | 0 | 2 | 0 | 197 | 377 | 797 | 863 | 631 | 483 | 313 | 13 | 0 | 3676 |
| 1987-88 | 0 | 0 | 15 | 370 | 437 | 660 | 956 | 734 | 458 | 247 | 66 | 9 | 3952 |
| 1988-89 | 0 | 0 | 3 | 384 | 454 | 755 | 681 | 660 | 360 | 258 | 148 | 1 | 3704 |
| 1989-90 | 0 | 1 | 36 | 204 | 499 | 1011 | 622 | 443 | 358 | 239 | 68 | 0 | 3481 |
| 1990-91 | 0 | 0 | 26 | 182 | 382 | 612 | 730 | 589 | 405 | 93 | 16 | 0 | 3035 |
| 1991-92 | 0 | 0 | 46 | 159 | 538 | 660 | 753 | 549 | 512 | 225 | 113 | 6 | 3561 |
| 1992-93 | 0 | 0 | 13 | 229 | 494 | 769 | 659 | 708 | 592 | 270 | 38 | 2 | 3774 |
| 1993-94 | 0 | 0 | 37 | 232 | 495 | 761 | 1016 | 597 | 473 | 139 | 93 | 0 | 3843 |
| 1994-95 | 0 | 0 | 11 | 158 | 336 | 620 | 794 | 674 | 364 | 162 | 54 | 0 | 3173 |
| 1995-96 | 0 | 0 | 22 | 216 | 638 | 857 | 877 | 732 | 622 | 302 | 45 | 0 | 4311 |
| 1996-97 | 0 | 0 | 30 | 196 | 623 | 663 | 786 | 512 | 332 | 333 | 124 | 8 | 3607 |
| 1997-98 | 0 | 0 | 1 | 226 | 625 | 794 | 681 | 528 | 500 | 234 | 30 | 12 | 3631 |
| 1998-99 | 0 | 0 | 1 | 133 | 438 | 632 | 656 | 578 | 604 | 149 | 24 | 0 | 3215 |
| 1999-00 | 0 | 0 | 28 | 206 | 359 | 694 | 846 | 542 | 354 | 265 | 9 | 6 | 3309 |
| 2000-01 | 0 | 0 | 34 | 149 | 522 | 955 | 908 | 530 | 571 | 155 | 25 | 1 | 3850 |
| 2001-02 | 0 | 0 | 57 | 263 | 320 | 594 | 719 | 657 | 451 | 152 | 99 | 0 | 3312 |
| 2002-03 | 0 | 0 | 2 | 126 | 555 | 749 | 957 | 669 | 358 | 148 | 11 | 4 | 3579 |
| 2003-04 | 0 | 0 | 29 | 206 | 358 | 818 | 853 | 708 | 365 | 225 | 42 | 0 | 3604 |
| 2004-05 | 0 | 0 | 4 | 75 | 362 | 772 | 675 | 580 | 506 | 197 | 90 | 0 | 3261 |
| 2005-06 | 0 | 0 | 1 | 187 | 416 | 844 | 593 | 674 | 444 | 102 | 77 | 0 | 3338 |
| 2006-07 | 0 | 0 | 32 | 271 | 476 | 633 | 732 | 745 | 277 | 288 | 23 | 0 | 3477 |
| 2007-08 | 0 | 0 | 0 | 106 | 464 | 597 | 861 | 624 | 476 | 194 | 53 | 0 | 3375 |
| 2008-09 | 0 | 0 | 0 | 223 | 569 | 702 | 870 | 615 | 418 | 231 | 39 | 0 | 3667 |
| 2009- | 0 | 0 | 13 | 221 | 412 | 800 | | | | | | | |

WBAN : 13891

## COOLING DEGREE DAYS (base 65°F) 2009  KNOXVILLE (KTYS)

| YEAR | JAN | FEB | MAR | APR | MAY | JUN | JUL | AUG | SEP | OCT | NOV | DEC | TOTAL |
|---|---|---|---|---|---|---|---|---|---|---|---|---|---|
| 1980 | 0 | 0 | 0 | 16 | 136 | 315 | 538 | 525 | 290 | 12 | 0 | 0 | 1832 |
| 1981 | 0 | 0 | 0 | 62 | 65 | 373 | 512 | 421 | 193 | 14 | 1 | 0 | 1641 |
| 1982 | 0 | 0 | 17 | 6 | 157 | 233 | 396 | 372 | 187 | 65 | 2 | 8 | 1443 |
| 1983 | 0 | 0 | 4 | 6 | 71 | 235 | 425 | 462 | 217 | 23 | 0 | 0 | 1443 |
| 1984 | 0 | 0 | 0 | 12 | 77 | 306 | 263 | 290 | 116 | 120 | 2 | 0 | 1186 |
| 1985 | 0 | 0 | 11 | 20 | 81 | 247 | 363 | 322 | 155 | 53 | 11 | 0 | 1263 |
| 1986 | 0 | 0 | 0 | 21 | 156 | 374 | 517 | 373 | 241 | 55 | 6 | 0 | 1743 |
| 1987 | 0 | 0 | 0 | 18 | 215 | 321 | 427 | 460 | 172 | 0 | 0 | 0 | 1613 |
| 1988 | 0 | 0 | 3 | 19 | 74 | 297 | 431 | 458 | 200 | 9 | 0 | 0 | 1491 |
| 1989 | 0 | 2 | 10 | 47 | 76 | 257 | 395 | 374 | 219 | 35 | 0 | 0 | 1415 |
| 1990 | 0 | 1 | 13 | 32 | 101 | 316 | 410 | 406 | 245 | 32 | 1 | 0 | 1557 |
| 1991 | 0 | 0 | 9 | 32 | 239 | 306 | 453 | 364 | 244 | 42 | 6 | 2 | 1697 |
| 1992 | 0 | 0 | 0 | 49 | 91 | 230 | 410 | 279 | 206 | 0 | 0 | 0 | 1265 |
| 1993 | 0 | 0 | 0 | 10 | 106 | 339 | 566 | 432 | 212 | 21 | 3 | 0 | 1689 |
| 1994 | 0 | 0 | 0 | 54 | 67 | 349 | 372 | 342 | 137 | 15 | 0 | 0 | 1336 |
| 1995 | 0 | 0 | 3 | 50 | 134 | 268 | 480 | 521 | 192 | 26 | 1 | 0 | 1675 |
| 1996 | 0 | 4 | 0 | 10 | 187 | 285 | 371 | 353 | 152 | 22 | 3 | 0 | 1387 |
| 1997 | 0 | 1 | 5 | 9 | 44 | 222 | 423 | 338 | 209 | 46 | 0 | 0 | 1297 |
| 1998 | 0 | 0 | 24 | 8 | 212 | 346 | 454 | 402 | 338 | 43 | 0 | 1 | 1828 |
| 1999 | 0 | 0 | 0 | 41 | 67 | 304 | 429 | 403 | 181 | 13 | 0 | 0 | 1438 |
| 2000 | 0 | 0 | 0 | 2 | 167 | 309 | 390 | 376 | 211 | 33 | 4 | 0 | 1492 |
| 2001 | 0 | 0 | 0 | 62 | 137 | 247 | 411 | 393 | 160 | 19 | 1 | 0 | 1430 |
| 2002 | 0 | 0 | 2 | 89 | 113 | 355 | 445 | 439 | 304 | 71 | 5 | 0 | 1823 |
| 2003 | 0 | 0 | 0 | 21 | 93 | 226 | 358 | 407 | 168 | 14 | 8 | 0 | 1295 |
| 2004 | 2 | 0 | 8 | 39 | 229 | 297 | 371 | 278 | 166 | 46 | 12 | 0 | 1448 |
| 2005 | 0 | 0 | 1 | 12 | 69 | 304 | 425 | 451 | 278 | 73 | 9 | 0 | 1622 |
| 2006 | 0 | 0 | 11 | 66 | 100 | 318 | 452 | 485 | 138 | 23 | 0 | 0 | 1593 |
| 2007 | 0 | 0 | 35 | 27 | 187 | 359 | 393 | 572 | 300 | 114 | 0 | 0 | 1987 |
| 2008 | 0 | 0 | 0 | 29 | 105 | 355 | 408 | 382 | 246 | 46 | 0 | 0 | 1571 |
| 2009 | 0 | 0 | 2 | 40 | 126 | 339 | 302 | 360 | 222 | 11 | 0 | 0 | 1402 |

## SNOWFALL (inches) 2009 KNOXVILLE (KTYS)

| YEAR | JUL | AUG | SEP | OCT | NOV | DEC | JAN | FEB | MAR | APR | MAY | JUN | TOTAL |
|------|-----|-----|-----|-----|-----|-----|-----|-----|-----|-----|-----|-----|-------|
| 1980-81 | 0.0 | 0.0 | 0.0 | 0.0 | T | T | 5.0 | 2.5 | T | 0.0 | 0.0 | 0.0 | 7.5 |
| 1981-82 | 0.0 | 0.0 | 0.0 | 0.0 | T | 0.1 | 4.4 | 0.1 | 1.1 | T | 0.0 | 0.0 | 5.7 |
| 1982-83 | 0.0 | 0.0 | 0.0 | 0.0 | 0.0 | 3.0 | 1.1 | 3.4 | 0.7 | 2.0 | 0.0 | 0.0 | 10.2 |
| 1983-84 | 0.0 | 0.0 | 0.0 | 0.0 | 0.0 | 0.7 | 2.8 | 3.5 | T | 0.0 | 0.0 | 0.0 | 7.0 |
| 1984-85 | 0.0 | 0.0 | 0.0 | 0.0 | 0.0 | T | 14.2 | 8.3 | 0.0 | 0.0 | 0.0 | 0.0 | 22.5 |
| 1985-86 | 0.0 | 0.0 | 0.0 | 0.0 | 0.0 | 0.4 | 3.6 | 5.0 | T | 0.0 | 0.0 | 0.0 | 9.0 |
| 1986-87 | 0.0 | 0.0 | 0.0 | 0.0 | T | T | 7.5 | T | 1.6 | 10.7 | 0.0 | 0.0 | 19.8 |
| 1987-88 | 0.0 | 0.0 | 0.0 | 0.0 | T | T | 9.2 | 0.9 | T | 0.0 | 0.0 | 0.0 | 10.1 |
| 1988-89 | 0.0 | 0.0 | 0.0 | 0.0 | T | 2.6 | 1.8 | 2.8 | 0.0 | 0.0 | 0.0 | 0.0 | 7.2 |
| 1989-90 | 0.0 | 0.0 | 0.0 | T | T | 1.1 | T | T | T | 0.0 | 0.0 | 0.0 | 1.1 |
| 1990-91 | 0.0 | 0.0 | 0.0 | 0.0 | 0.0 | T | T | 1.2 | 0.7 | T | 0.0 | 0.0 | 1.9 |
| 1991-92 | 0.0 | 0.0 | 0.0 | 0.0 | 0.2 | T | T | 0.0 | T | T | 0.0 | 0.0 | 0.2 |
| 1992-93 | 0.0 | 0.0 | 0.0 | 0.0 | 0.0 | 0.3 | T | 1.0 | 15.1 | T | 0.0 | 0.0 | 16.4 |
| 1993-94 | 0.0 | 0.0 | 0.0 | T | 0.0 | 0.3 | 3.3 | T | 0.6 | 0.0 | 0.0 | 0.0 | 4.2 |
| 1994-95 | 0.0 | 0.0 | 0.0 | 0.0 | 0.0 | 0.0 | 2.0 | 1.1 | 0.6 | 0.0 | 0.0 | 0.0 | 3.7 |
| 1995-96 | 0.0 | T | 0.0 | 0.0 |  | T | 10.0 |  |  |  |  |  |  |
| 1996-97 |  |  |  |  |  |  | 1.5 |  |  |  |  |  |  |
| 1997-98 |  |  |  |  | T | 3.3 | 0.4 | 0.1 | T | 0.0 | T | T |  |
| 1998-99 | 0.0 | 0.0 | 0.0 | 0.0 | 0.0 | T | T | 2.1 | 2.0 | 0.0 | T | 0.0 | 4.1 |
| 1999-00 | 0.0 | 0.0 | 0.0 | 0.0 | 0.0 | T | 2.0 | T | T | 0.0 | T | 0.0 | 2.0 |
| 2000-01 |  |  |  |  |  |  |  |  |  |  |  |  |  |
| 2001-02 |  |  |  |  |  |  |  |  |  |  |  |  |  |
| 2002-03 |  |  |  |  |  |  |  |  |  |  |  |  |  |
| 2003-04 |  |  |  |  |  |  | 1.0 | T | 0.0 | 0.0 | 0.0 | 0.0 |  |
| 2004-05 | 0.0 | 0.0 | 0.0 | 0.0 | 0.0 | 0.1 | T | T | 0.4 | T | 0.0 | 0.0 | 0.5 |
| 2005-06 | 0.0 | 0.0 | 0.0 | 0.0 | 0.0 | T | T | 2.7 | T | T | T | 0.0 | 2.7 |
| 2006-07 | 0.0 | 0.0 | T | 0.0 | T | T | T | 0.3 | T | T | 0.0 | 0.0 | 0.3 |
| 2007-08 | 0.0 | 0.0 | 0.0 | 0.0 | 0.0 | T | T | T | T | T | T | T | T |
| 2008-09 | 0.0 | 0.0 | 0.0 | 0.0 | T | 0.0 | 1.0 | 5.0 | T | T | 0.0 | 0.0 | 6.0 |
| 2009- | 0.0 | 0.0 | 0.0 | 0.0 | 0.0 | 2.0 |  |  |  |  |  |  |  |
| POR= 99 YRS | 0.0 | T | T | T | 0.4 | 1.4 | 3.0 | 2.6 | 1.6 | 0.2 | T | T | 9.2 |

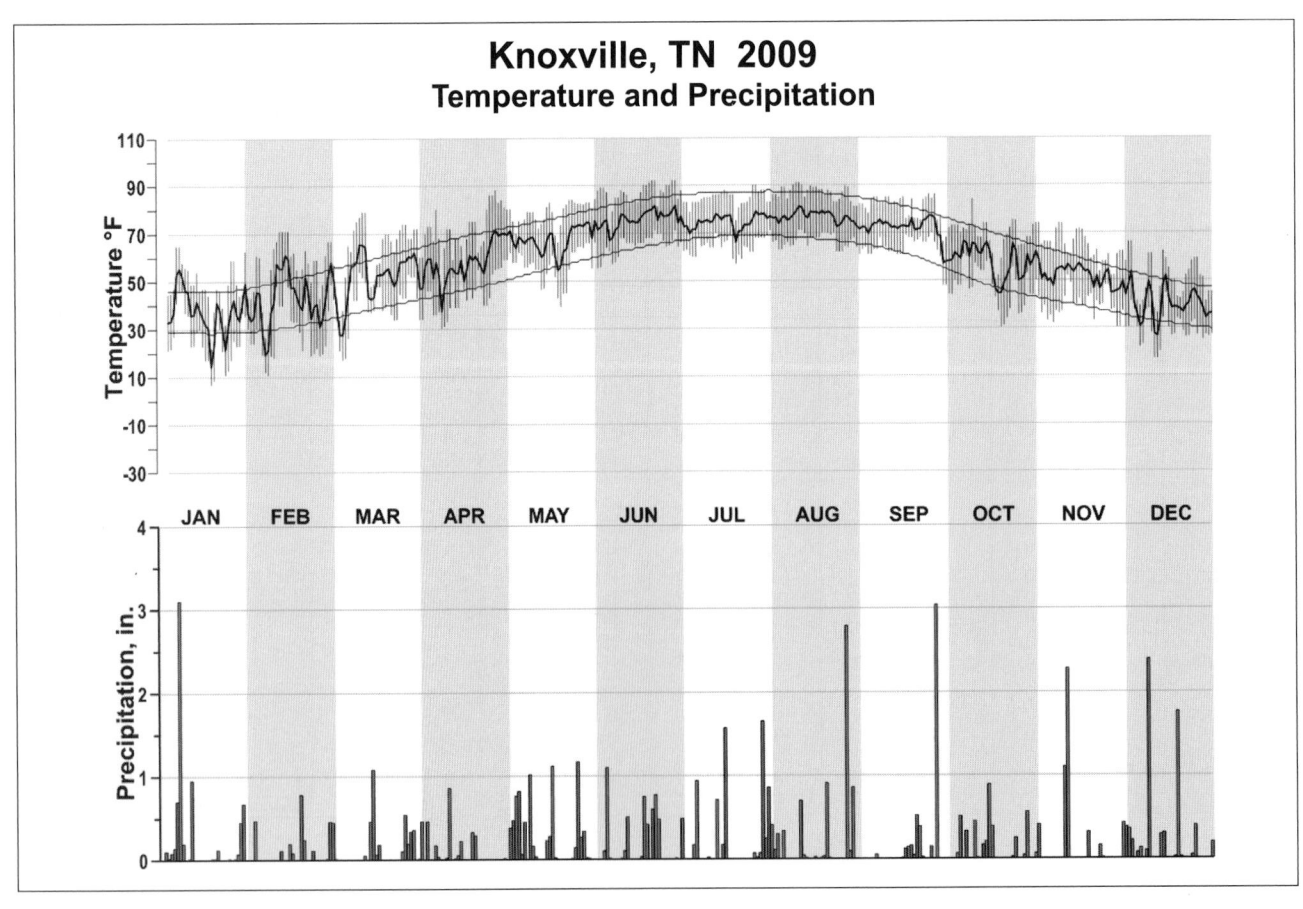

# Knoxville, TN 2009
## Temperature and Precipitation

# 2009
# MEMPHIS
# TENNESSEE (KMEM)

Topography varies from the level alluvial area in east-central Arkansas to the slightly rolling area in northwestern Mississippi and southwestern Tennessee.

Agricultural interests are varied, with major crops being cotton, corn, hay, soybeans, peaches, apples, and a considerable number of vegetables. The climate is quite favorable for dairy interests, and for the raising of cattle and hogs.

The growing season is about 230 days in length. The average date for the last occurrence of temperatures as low as 32 degrees is late March. The average date of the first temperature of 32 degrees or below is early November.

Precipitation of nearly 50 inches per year is fairly well distributed. Crops and pastures receive, on the average, an adequate supply of moisture during the growing season, with lesser amounts during the fall harvesting period.

Sunshine averages slightly over 70 percent of the possible amount during the growing season. Relative humidity averages about 70 percent for the year.

Memphis, although not in the normal paths of storms coming from the Gulf or from western Canada, is affected by both, and thereby has comparatively frequent changes in weather. Extremely high or low temperatures, however, are relatively rare.

# NORMALS, MEANS, AND EXTREMES
## MEMPHIS (KMEM)

LATITUDE: 35 ° 3 'N  LONGITUDE: -89 ° 59'W  ELEVATION (FT): GRND: 305  BARO: 286  TIME ZONE: CENTRAL  (UTC -6)  WBAN: 13893

| ELEMENT | POR | JAN | FEB | MAR | APR | MAY | JUN | JUL | AUG | SEP | OCT | NOV | DEC | YEAR |
|---|---|---|---|---|---|---|---|---|---|---|---|---|---|---|
| **TEMPERATURE °F** | | | | | | | | | | | | | | |
| NORMAL DAILY MAXIMUM | 30 | 48.6 | 54.4 | 63.3 | 72.4 | 80.4 | 88.5 | 92.1 | 91.2 | 85.3 | 75.1 | 62.1 | 52.2 | 72.1 |
| MEAN DAILY MAXIMUM | 70 | 49.1 | 53.3 | 62.5 | 72.5 | 81.0 | 88.2 | 91.5 | 90.7 | 84.2 | 74.7 | 61.7 | 52.2 | 71.8 |
| HIGHEST DAILY MAXIMUM | 68 | 79 | 81 | 86 | 94 | 99 | 104 | 108 | 107 | 103 | 95 | 86 | 81 | 108 |
| YEAR OF OCCURRENCE | | 2002 | 1962 | 2007 | 1987 | 1977 | 1954 | 1980 | 2000 | 1954 | 2007 | 2000 | 1982 | JUL 1980 |
| MEAN OF EXTREME MAXS. | 70 | 70.4 | 73.3 | 79.7 | 85.8 | 90.9 | 96.0 | 98.1 | 97.5 | 94.0 | 87.6 | 79.2 | 71.2 | 85.3 |
| NORMAL DAILY MINIMUM | 30 | 31.3 | 35.5 | 43.7 | 51.9 | 60.8 | 68.8 | 72.9 | 71.2 | 64.3 | 52.5 | 42.6 | 34.5 | 52.5 |
| MEAN DAILY MINIMUM | 70 | 31.8 | 34.9 | 42.8 | 52.3 | 61.3 | 69.0 | 72.8 | 71.4 | 64.0 | 52.5 | 42.0 | 34.6 | 52.5 |
| LOWEST DAILY MINIMUM | 68 | -4 | -11 | 12 | 28 | 38 | 48 | 52 | 48 | 36 | 25 | 9 | -13 | -13 |
| YEAR OF OCCURRENCE | | 1985 | 1951 | 1943 | 2007 | 1944 | 1966 | 1947 | 1946 | 1949 | 1952 | 1950 | 1963 | DEC 1963 |
| MEAN OF EXTREME MINS. | 70 | 13.9 | 18.2 | 26.2 | 35.9 | 47.7 | 58.0 | 64.6 | 62.5 | 49.7 | 36.8 | 25.7 | 17.5 | 38.1 |
| NORMAL DRY BULB | 30 | 39.9 | 44.9 | 53.5 | 62.1 | 70.6 | 78.7 | 82.5 | 81.2 | 74.8 | 63.8 | 52.3 | 43.3 | 62.3 |
| MEAN DRY BULB | 70 | 40.5 | 44.1 | 52.6 | 62.4 | 71.1 | 78.7 | 82.2 | 81.1 | 74.1 | 63.6 | 51.8 | 43.4 | 62.1 |
| MEAN WET BULB | 26 | 36.3 | 39.5 | 46.1 | 54.2 | 63.1 | 69.8 | 73.1 | 72.0 | 65.9 | 56.0 | 46.8 | 38.8 | 55.1 |
| MEAN DEW POINT | 26 | 32.5 | 35.6 | 41.5 | 50.0 | 59.8 | 67.2 | 70.9 | 69.7 | 62.9 | 52.3 | 43.0 | 34.9 | 51.7 |
| NORMAL NO. DAYS WITH: | | | | | | | | | | | | | | |
| MAXIMUM >= 90 | 30 | 0.0 | 0.0 | 0.0 | 0.2 | 2.6 | 14.3 | 22.0 | 19.1 | 8.7 | 0.5 | 0.0 | 0.0 | 67.4 |
| MAXIMUM <= 32 | 30 | 3.1 | 1.4 | 0.1 | 0.0 | 0.0 | 0.0 | 0.0 | 0.0 | 0.0 | 0.0 | * | 1.6 | 6.2 |
| MINIMUM <= 32 | 30 | 16.6 | 11.1 | 3.9 | 0.3 | 0.0 | 0.0 | 0.0 | 0.0 | 0.0 | 0.1 | 3.9 | 12.8 | 48.7 |
| MINIMUM <= 0 | 30 | 0.1 | 0.0 | 0.0 | 0.0 | 0.0 | 0.0 | 0.0 | 0.0 | 0.0 | 0.0 | 0.0 | 0.1 | 0.2 |
| **H/C** | | | | | | | | | | | | | | |
| NORMAL HEATING DEG. DAYS | 30 | 770 | 565 | 366 | 144 | 22 | 0 | 0 | 0 | 13 | 121 | 381 | 659 | 3041 |
| NORMAL COOLING DEG. DAYS | 30 | 1 | 2 | 15 | 72 | 210 | 426 | 554 | 504 | 307 | 84 | 11 | 1 | 2187 |
| **RH** | | | | | | | | | | | | | | |
| NORMAL (PERCENT) | 30 | 70 | 67 | 63 | 63 | 68 | 68 | 69 | 69 | 70 | 67 | 69 | 70 | 68 |
| HOUR 00 LST | 30 | 74 | 71 | 68 | 70 | 75 | 76 | 78 | 78 | 78 | 75 | 73 | 73 | 74 |
| HOUR 06 LST | 30 | 77 | 76 | 75 | 78 | 82 | 82 | 84 | 85 | 86 | 82 | 79 | 78 | 80 |
| HOUR 12 LST | 30 | 63 | 58 | 55 | 53 | 57 | 57 | 58 | 57 | 56 | 53 | 58 | 62 | 57 |
| HOUR 18 LST | 30 | 66 | 60 | 55 | 52 | 57 | 57 | 58 | 59 | 61 | 59 | 63 | 66 | 59 |
| **S** PERCENT POSSIBLE SUNSHINE | 35 | 50 | 54 | 56 | 64 | 69 | 74 | 74 | 75 | 69 | 70 | 58 | 50 | 64 |
| **W/O** MEAN NO. DAYS WITH: | | | | | | | | | | | | | | |
| HEAVY FOG(VISBY <= 1/4 MI) | 46 | 1.6 | 1.2 | 0.7 | 0.2 | 0.3 | 0.3 | 0.3 | 0.3 | 0.6 | 0.7 | 1.2 | 1.4 | 8.8 |
| THUNDERSTORMS | 62 | 1.9 | 2.4 | 4.6 | 6.2 | 7.0 | 7.6 | 8.6 | 6.3 | 3.4 | 2.2 | 2.5 | 1.8 | 54.5 |
| **CLOUDNESS** MEAN: | | | | | | | | | | | | | | |
| SUNRISE-SUNSET (OKTAS) | 45 | 5.4 | 5.1 | 5.2 | 4.8 | 4.7 | 4.3 | 4.2 | 3.8 | 3.9 | 3.6 | 4.5 | 5.0 | 4.5 |
| MIDNIGHT-MIDNIGHT (OKTAS) | 30 | 4.9 | 4.6 | 4.9 | 4.3 | 4.3 | 3.9 | 3.6 | 3.4 | 3.6 | 3.3 | 4.2 | 4.7 | 4.1 |
| MEAN NO. DAYS WITH: | | | | | | | | | | | | | | |
| CLEAR | 43 | 7.7 | 7.9 | 7.9 | 8.6 | 8.4 | 9.6 | 10.2 | 11.8 | 12.5 | 14.4 | 10.2 | 8.9 | 118.1 |
| PARTLY CLOUDY | 43 | 5.8 | 6.5 | 6.5 | 7.3 | 9.6 | 11.1 | 11.9 | 11.7 | 7.5 | 7.0 | 6.2 | 5.6 | 95.8 |
| CLOUDY | 43 | 17.5 | 14.8 | 16.6 | 14.1 | 13.0 | 9.3 | 8.9 | 7.4 | 10.0 | 9.7 | 13.6 | 16.5 | 151.4 |
| **PR** MEAN STATION PRESSURE(IN) | 26 | 29.82 | 29.78 | 29.71 | 29.66 | 29.65 | 29.65 | 29.68 | 29.67 | 29.70 | 29.73 | 29.77 | 29.80 | 29.72 |
| MEAN SEA-LEVEL PRES. (IN) | 26 | 30.17 | 30.12 | 30.06 | 29.99 | 29.98 | 29.97 | 30.00 | 30.01 | 30.03 | 30.08 | 30.12 | 30.16 | 30.06 |
| **WINDS** MEAN SPEED (MPH) | 26 | 9.0 | 9.4 | 9.6 | 9.2 | 8.4 | 7.4 | 7.0 | 6.6 | 7.3 | 7.3 | 8.2 | 8.8 | 8.2 |
| PREVAIL.DIR(TENS OF DEGS) | 35 | 21 | 20 | 19 | 20 | 21 | 20 | 22 | 22 | 04 | 16 | 19 | 19 | 22 |
| MAXIMUM 2-MINUTE: | | | | | | | | | | | | | | |
| SPEED (MPH) | 10 | 41 | 48 | 43 | 41 | 41 | 40 | 48 | 45 | 36 | 36 | 35 | 38 | 48 |
| DIR. (TENS OF DEGS) | | 27 | 30 | 02 | 29 | 29 | 26 | 31 | 07 | 14 | 32 | 30 | 19 | 30 |
| YEAR OF OCCURRENCE | | 2008 | 2008 | 2004 | 2005 | 2008 | 2009 | 2003 | 2003 | 2005 | 2003 | 2005 | 2008 | FEB 2008 |
| MAXIMUM 3-SECOND | | | | | | | | | | | | | | |
| SPEED (MPH) | 10 | 59 | 55 | 48 | 49 | 54 | 53 | 61 | 59 | 44 | 46 | 51 | 48 | 61 |
| DIR. (TENS OF DEGS) | | 29 | 30 | 02 | 31 | 30 | 26 | 30 | 30 | 13 | 32 | 29 | 19 | 30 |
| YEAR OF OCCURRENCE | | 2008 | 2008 | 2004 | 2005 | 2008 | 2009 | 2003 | 2002 | 2005 | 2003 | 2002 | 2008 | JUL 2003 |
| **PRECIPITATION** NORMAL (IN) | 30 | 4.24 | 4.31 | 5.58 | 5.79 | 5.15 | 4.30 | 4.22 | 3.00 | 3.31 | 3.31 | 5.76 | 5.68 | 54.65 |
| MAXIMUM MONTHLY (IN) | 59 | 12.21 | 10.51 | 12.08 | 17.13 | 11.58 | 10.17 | 9.96 | 9.65 | 12.34 | 10.56 | 11.60 | 13.81 | 17.13 |
| YEAR OF OCCURRENCE | | 1951 | 1989 | 1975 | 1991 | 1953 | 1996 | 1998 | 1978 | 2002 | 2009 | 2001 | 1982 | APR 1991 |
| MINIMUM MONTHLY (IN) | 59 | 0.57 | 1.12 | 1.50 | 1.39 | 0.83 | 0.04 | 0.43 | 0.43 | 0.19 | T | 0.75 | 1.05 | T |
| YEAR OF OCCURRENCE | | 1986 | 1980 | 1966 | 1992 | 1977 | 1953 | 1954 | 1953 | 1953 | 1963 | 1965 | 1955 | OCT 1963 |
| MAXIMUM IN 24 HOURS (IN) | 59 | 3.89 | 4.24 | 5.95 | 4.35 | 4.94 | 4.76 | 4.71 | 4.04 | 7.11 | 5.36 | 7.30 | 5.42 | 7.30 |
| YEAR OF OCCURRENCE | | 1974 | 1989 | 1975 | 1985 | 1958 | 1980 | 1980 | 1978 | 2002 | 2002 | 2001 | 1978 | NOV 2001 |
| NORMAL NO. DAYS WITH: | | | | | | | | | | | | | | |
| PRECIPITATION >= 0.01 | 30 | 10.3 | 8.8 | 11.1 | 10.0 | 10.3 | 9.0 | 8.5 | 6.9 | 7.7 | 6.9 | 9.4 | 9.7 | 108.6 |
| PRECIPITATION >= 1.00 | 30 | 1.1 | 1.2 | 1.8 | 2.0 | 1.6 | 1.4 | 1.4 | 0.8 | 1.0 | 1.0 | 2.0 | 1.9 | 17.2 |
| **SNOWFALL** NORMAL (IN) | 30 | 2.2 | 1.9 | 0.3 | 0.* | 0.0 | 0.0 | 0.0 | 0.0 | 0.0 | 0.* | 0.1 | 0.1 | 4.6 |
| MAXIMUM MONTHLY (IN) | 52 | 12.4 | 8.3 | 17.3 | T | T | T | 0.0 | 0.0 | 0.0 | T | 1.5 | 14.3 | 17.3 |
| YEAR OF OCCURRENCE | | 1985 | 1985 | 1968 | 2008 | 2008 | 1995 | | | | 1993 | 1976 | 1963 | MAR 1968 |
| MAXIMUM IN 24 HOURS (IN) | 52 | 8.1 | 5.8 | 16.1 | T | T | T | 0.0 | 0.0 | 0.0 | T | 1.2 | 14.3 | 16.1 |
| YEAR OF OCCURRENCE' | | 1985 | 1960 | 1968 | 1993 | 1995 | 1995 | | | | 1993 | 1976 | 1963 | MAR 1968 |
| MAXIMUM SNOW DEPTH (IN) | 54 | 12 | 6 | 12 | 0 | 0 | 0 | 0 | 0 | 0 | 0 | 1 | 13 | 13 |
| YEAR OF OCCURRENCE | | 1948 | 1985 | 1968 | | | | | | | | 1991 | 1963 | DEC 1963 |
| NORMAL NO. DAYS WITH: | | | | | | | | | | | | | | |
| SNOWFALL >= 1.0 | 30 | 0.7 | 0.6 | 0.1 | 0.0 | 0.0 | 0.0 | 0.0 | 0.0 | 0.0 | 0.0 | 0.0 | 0.0 | 1.4 |

## PRECIPITATION (inches) 2009  MEMPHIS (KMEM)

| YEAR | JAN | FEB | MAR | APR | MAY | JUN | JUL | AUG | SEP | OCT | NOV | DEC | ANNUAL |
|------|-----|-----|-----|-----|-----|-----|-----|-----|-----|-----|-----|-----|--------|
| 1980 | 3.23 | 1.12 | 10.86 | 7.53 | 4.43 | 5.75 | 4.73 | 1.23 | 5.32 | 3.14 | 5.23 | 1.86 | 54.43 |
| 1981 | 1.38 | 3.66 | 4.98 | 3.67 | 7.06 | 2.93 | 1.71 | 4.21 | 0.61 | 5.83 | 2.12 | 1.84 | 40.00 |
| 1982 | 6.61 | 4.16 | 4.47 | 6.76 | 5.50 | 6.68 | 4.13 | 3.11 | 1.92 | 5.23 | 6.43 | 13.81 | 68.81 |
| 1983 | 2.32 | 2.61 | 3.66 | 8.84 | 9.58 | 3.50 | 3.83 | 0.61 | 1.52 | 2.94 | 9.56 | 8.68 | 57.65 |
| 1984 | 1.88 | 4.37 | 6.07 | 5.24 | 9.06 | 1.12 | 4.59 | 5.00 | 1.96 | 7.75 | 5.85 | 4.35 | 57.24 |
| 1985 | 3.78 | 4.10 | 4.96 | 6.51 | 2.23 | 4.55 | 3.50 | 3.50 | 4.03 | 3.36 | 3.87 | 3.27 | 47.66 |
| 1986 | 0.57 | 2.50 | 1.90 | 3.72 | 4.63 | 3.80 | 1.21 | 2.74 | 1.21 | 3.75 | 8.67 | 3.92 | 38.62 |
| 1987 | 1.76 | 5.81 | 3.38 | 3.78 | 2.96 | 3.66 | 2.06 | 4.12 | 2.01 | 1.96 | 10.45 | 11.39 | 53.34 |
| 1988 | 4.25 | 3.49 | 4.20 | 2.85 | 2.38 | 2.15 | 5.21 | 0.85 | 4.73 | 3.62 | 10.52 | 5.99 | 50.24 |
| 1989 | 7.91 | 10.51 | 5.50 | 2.13 | 2.36 | 7.20 | 7.55 | 1.43 | 6.08 | 2.37 | 3.65 | 2.20 | 58.89 |
| 1990 | 3.97 | 8.99 | 5.65 | 6.93 | 4.55 | 2.68 | 2.21 | 1.18 | 5.21 | 4.37 | 3.44 | 10.61 | 59.79 |
| 1991 | 2.90 | 6.46 | 3.68 | 17.13 | 5.10 | 1.42 | 1.92 | 2.06 | 1.47 | 4.39 | 5.54 | 7.04 | 59.11 |
| 1992 | 1.78 | 2.18 | 7.07 | 1.39 | 3.68 | 7.50 | 5.38 | 2.44 | 3.62 | 4.01 | 4.82 | 3.28 | 47.15 |
| 1993 | 3.59 | 2.46 | 3.14 | 6.20 | 4.56 | 4.20 | 0.86 | 3.69 | 3.73 | 1.91 | 4.07 | 5.59 | 44.00 |
| 1994 | 5.53 | 4.67 | 6.65 | 4.04 | 3.00 | 5.03 | 4.03 | 2.00 | 1.44 | 2.72 | 4.34 | 6.04 | 49.49 |
| 1995 | 7.07 | 2.12 | 3.35 | 4.38 | 5.90 | 6.66 | 8.42 | 3.63 | 0.39 | 1.57 | 7.62 | 5.79 | 56.90 |
| 1996 | 5.53 | 2.77 | 5.31 | 3.37 | 6.28 | 10.17 | 9.89 | 2.07 | 5.99 | 7.12 | 11.51 | 6.18 | 76.19 |
| 1997 | 4.37 | 7.63 | 11.28 | 8.91 | 8.00 | 6.75 | 4.35 | 4.01 | 6.31 | 4.06 | 1.69 | 4.52 | 71.88 |
| 1998 | 6.73 | 5.55 | 5.14 | 8.24 | 1.71 | 1.09 | 9.96 | 2.54 | 1.70 | 2.34 | 2.56 | 4.25 | 51.81 |
| 1999 | 6.15 | 2.10 | 6.59 | 8.92 | 4.79 | 2.42 | 3.63 | 1.18 | 1.11 | 1.53 | 2.37 | 4.73 | 45.52 |
| 2000 | 1.37 | 5.37 | 4.42 | 4.41 | 4.02 | 3.79 | 2.23 | 0.77 | 1.08 | 0.45 | 6.89 | 2.47 | 37.27 |
| 2001 | 3.32 | 6.89 | 3.54 | 3.94 | 6.35 | 2.12 | 5.53 | 2.25 | 3.35 | 6.93 | 11.60 | 10.19 | 66.01 |
| 2002 | 4.03 | 1.68 | 11.83 | 2.36 | 5.33 | 2.22 | 8.45 | 5.26 | 12.34 | 8.28 | 3.40 | 9.66 | 74.84 |
| 2003 | 0.87 | 8.23 | 2.95 | 3.52 | 11.40 | 4.82 | 3.29 | 2.22 | 3.54 | 3.01 | 5.05 | 3.08 | 51.98 |
| 2004 | 3.14 | 4.44 | 3.88 | 6.77 | 6.04 | 3.22 | 3.10 | 2.97 | 0.25 | 6.16 | 9.38 | 4.36 | 53.71 |
| 2005 | 4.98 | 3.19 | 3.52 | 5.30 | 1.10 | 1.40 | 8.21 | 5.68 | 1.60 | 0.79 | 2.33 | 1.91 | 40.01 |
| 2006 | 7.17 | 3.77 | 3.79 | 3.72 | 3.83 | 1.73 | 1.20 | 2.99 | 2.87 | 1.95 | 3.08 | 6.10 | 42.20 |
| 2007 | 4.73 | 2.17 | 2.82 | 2.61 | 1.30 | 0.72 | 4.05 | 2.81 | 1.39 | 3.96 | 3.52 | 4.73 | 34.81 |
| 2008 | 4.68 | 2.51 | 10.00 | 8.66 | 7.61 | 2.97 | 2.49 | 7.45 | 2.49 | 4.34 | 2.36 | 8.65 | 64.21 |
| 2009 | 3.07 | 3.26 | 6.12 | 3.63 | 7.73 | 2.13 | 8.46 | 1.21 | 8.59 | 10.56 | 1.37 | 5.13 | 61.26 |
| POR= 70 YRS | 4.47 | 4.50 | 5.35 | 5.44 | 4.90 | 3.71 | 4.03 | 3.36 | 3.26 | 3.22 | 4.82 | 5.33 | 52.39 |

WBAN : 13893

## AVERAGE TEMPERATURE (°F) 2009  MEMPHIS (KMEM)

| YEAR | JAN | FEB | MAR | APR | MAY | JUN | JUL | AUG | SEP | OCT | NOV | DEC | ANNUAL |
|------|-----|-----|-----|-----|-----|-----|-----|-----|-----|-----|-----|-----|--------|
| 1980 | 43.2 | 39.5 | 49.4 | 60.9 | 72.5 | 80.9 | 88.8 | 87.2 | 80.5 | 62.7 | 53.3 | 45.9 | 63.7 |
| 1981 | 40.9 | 47.3 | 54.3 | 70.2 | 70.0 | 82.5 | 84.6 | 81.8 | 74.0 | 62.5 | 53.8 | 40.9 | 63.6 |
| 1982 | 36.6 | 40.5 | 55.5 | 58.5 | 74.5 | 78.0 | 85.0 | 82.9 | 73.3 | 63.7 | 53.4 | 49.5 | 62.6 |
| 1983 | 40.4 | 45.3 | 51.9 | 56.8 | 68.2 | 77.2 | 83.6 | 84.9 | 75.2 | 66.2 | 53.1 | 34.7 | 61.5 |
| 1984 | 35.9 | 47.6 | 51.1 | 61.0 | 69.5 | 80.9 | 80.9 | 79.8 | 71.9 | 68.4 | 50.9 | 53.8 | 62.6 |
| 1985 | 32.4 | 40.6 | 57.7 | 65.0 | 71.8 | 78.9 | 82.2 | 80.2 | 73.6 | 67.2 | 57.6 | 36.9 | 62.0 |
| 1986 | 41.9 | 48.0 | 55.2 | 64.4 | 72.5 | 81.0 | 86.5 | 79.2 | 79.1 | 64.2 | 51.1 | 42.6 | 63.8 |
| 1987 | 39.6 | 47.1 | 54.7 | 62.2 | 76.5 | 80.0 | 82.5 | 83.6 | 75.3 | 59.2 | 54.4 | 46.7 | 63.5 |
| 1988 | 36.8 | 42.2 | 52.3 | 62.8 | 71.8 | 80.3 | 81.6 | 83.7 | 76.3 | 59.1 | 54.6 | 44.9 | 62.2 |
| 1989 | 47.3 | 40.0 | 53.8 | 62.6 | 69.5 | 77.6 | 80.7 | 81.2 | 72.4 | 64.2 | 54.2 | 33.6 | 61.4 |
| 1990 | 48.5 | 52.0 | 55.3 | 61.4 | 68.3 | 80.7 | 82.5 | 82.0 | 77.9 | 61.3 | 57.1 | 45.8 | 64.4 |
| 1991 | 41.2 | 48.0 | 55.4 | 65.0 | 74.6 | 80.3 | 83.3 | 80.9 | 75.3 | 65.5 | 49.2 | 46.8 | 63.8 |
| 1992 | 42.2 | 49.9 | 53.7 | 63.0 | 70.4 | 76.9 | 82.5 | 76.8 | 73.2 | 63.8 | 50.7 | 43.8 | 62.2 |
| 1993 | 42.7 | 42.6 | 51.1 | 59.3 | 69.9 | 79.2 | 86.4 | 83.6 | 73.5 | 62.2 | 50.2 | 44.8 | 62.1 |
| 1994 | 37.8 | 46.3 | 54.2 | 67.0 | 69.4 | 82.3 | 81.1 | 79.4 | 72.5 | 63.6 | 56.6 | 46.7 | 63.1 |
| 1995 | 42.2 | 44.2 | 56.2 | 62.9 | 71.1 | 77.0 | 81.1 | 84.5 | 73.0 | 65.2 | 48.3 | 43.2 | 62.4 |
| 1996 | 39.2 | 45.0 | 48.0 | 59.7 | 74.2 | 78.4 | 80.5 | 79.6 | 72.3 | 63.1 | 49.1 | 46.8 | 61.3 |
| 1997 | 39.7 | 47.6 | 56.7 | 57.9 | 67.9 | 75.7 | 82.5 | 79.3 | 74.7 | 62.6 | 48.5 | 42.3 | 61.3 |
| 1998 | 45.8 | 48.5 | 51.9 | 61.4 | 75.3 | 83.2 | 84.1 | 81.8 | 80.5 | 66.6 | 55.9 | 45.1 | 65.0 |
| 1999 | 46.0 | 51.4 | 50.9 | 67.0 | 71.2 | 79.7 | 83.7 | 82.7 | 75.2 | 64.3 | 57.0 | 45.8 | 64.6 |
| 2000 | 42.8 | 50.2 | 56.4 | 60.6 | 73.9 | 78.6 | 83.4 | 86.3 | 76.7 | 68.1 | 50.2 | 32.7 | 63.3 |
| 2001 | 38.0 | 47.7 | 49.3 | 67.7 | 72.1 | 77.7 | 83.2 | 81.8 | 73.6 | 61.7 | 57.3 | 48.2 | 63.2 |
| 2002 | 45.7 | 43.5 | 50.6 | 66.2 | 69.1 | 80.1 | 82.7 | 82.0 | 77.7 | 63.0 | 49.0 | 44.2 | 62.8 |
| 2003 | 37.7 | 42.0 | 54.3 | 64.2 | 71.7 | 75.1 | 81.1 | 81.4 | 72.2 | 64.2 | 55.5 | 43.9 | 61.9 |
| 2004 | 41.9 | 43.1 | 57.3 | 61.9 | 74.1 | 78.6 | 80.3 | 77.9 | 76.4 | 68.7 | 57.1 | 44.2 | 63.5 |
| 2005 | 46.5 | 49.6 | 53.5 | 63.6 | 71.7 | 81.7 | 82.5 | 84.8 | 78.9 | 64.4 | 55.7 | 42.0 | 64.6 |
| 2006 | 49.7 | 42.2 | 55.2 | 68.2 | 72.3 | 79.8 | 84.1 | 84.9 | 73.1 | 62.5 | 52.1 | 47.5 | 64.3 |
| 2007 | 42.7 | 42.6 | 62.5 | 60.3 | 74.7 | 82.2 | 82.1 | 88.5 | 79.0 | 68.6 | 54.5 | 47.8 | 65.5 |
| 2008 | 40.3 | 46.6 | 54.1 | 61.2 | 70.8 | 80.4 | 83.4 | 80.4 | 74.9 | 63.5 | 50.3 | 43.6 | 62.5 |
| 2009 | 39.6 | 48.0 | 55.1 | 62.0 | 71.7 | 81.3 | 79.3 | 79.9 | 75.7 | 60.8 | 55.1 | 40.9 | 62.5 |
| POR= 70 YRS | 40.5 | 44.1 | 52.6 | 62.4 | 71.1 | 78.7 | 82.2 | 81.1 | 74.1 | 63.6 | 51.8 | 43.4 | 62.1 |

## HEATING DEGREE DAYS (base 65°F) 2009 MEMPHIS (KMEM)

| YEAR | JUL | AUG | SEP | OCT | NOV | DEC | JAN | FEB | MAR | APR | MAY | JUN | TOTAL |
|------|-----|-----|-----|-----|-----|-----|-----|-----|-----|-----|-----|-----|-------|
| 1980-81 | 0 | 0 | 5 | 146 | 362 | 586 | 739 | 492 | 342 | 18 | 23 | 0 | 2713 |
| 1981-82 | 0 | 0 | 9 | 153 | 331 | 739 | 873 | 680 | 324 | 215 | 2 | 0 | 3326 |
| 1982-83 | 0 | 0 | 20 | 134 | 352 | 500 | 759 | 543 | 406 | 273 | 25 | 0 | 3012 |
| 1983-84 | 0 | 0 | 27 | 73 | 368 | 935 | 894 | 499 | 426 | 162 | 24 | 0 | 3408 |
| 1984-85 | 0 | 0 | 37 | 48 | 423 | 367 | 1004 | 683 | 254 | 100 | 6 | 0 | 2922 |
| 1985-86 | 0 | 0 | 17 | 54 | 257 | 864 | 708 | 475 | 307 | 102 | 8 | 0 | 2792 |
| 1986-87 | 0 | 0 | 0 | 102 | 413 | 687 | 782 | 492 | 322 | 154 | 0 | 0 | 2952 |
| 1987-88 | 0 | 0 | 0 | 186 | 324 | 559 | 867 | 657 | 393 | 108 | 0 | 0 | 3094 |
| 1988-89 | 0 | 0 | 1 | 202 | 314 | 619 | 544 | 694 | 369 | 174 | 41 | 0 | 2958 |
| 1989-90 | 0 | 0 | 24 | 102 | 337 | 966 | 503 | 363 | 321 | 181 | 38 | 0 | 2835 |
| 1990-91 | 0 | 0 | 11 | 182 | 249 | 593 | 728 | 471 | 329 | 57 | 7 | 0 | 2627 |
| 1991-92 | 0 | 0 | 23 | 90 | 481 | 556 | 699 | 433 | 347 | 144 | 30 | 0 | 2803 |
| 1992-93 | 0 | 0 | 15 | 87 | 428 | 650 | 685 | 617 | 433 | 200 | 15 | 2 | 3132 |
| 1993-94 | 0 | 0 | 14 | 164 | 447 | 618 | 839 | 524 | 337 | 78 | 38 | 0 | 3059 |
| 1994-95 | 0 | 0 | 16 | 117 | 261 | 559 | 703 | 574 | 292 | 123 | 32 | 0 | 2677 |
| 1995-96 | 0 | 0 | 18 | 79 | 495 | 666 | 792 | 577 | 525 | 190 | 9 | 0 | 3351 |
| 1996-97 | 0 | 0 | 12 | 115 | 475 | 561 | 776 | 479 | 267 | 214 | 36 | 0 | 2935 |
| 1997-98 | 0 | 0 | 0 | 181 | 490 | 696 | 587 | 457 | 439 | 133 | 4 | 0 | 2987 |
| 1998-99 | 0 | 0 | 0 | 58 | 275 | 622 | 586 | 381 | 432 | 68 | 0 | 0 | 2422 |
| 1999-00 | 0 | 0 | 6 | 112 | 241 | 591 | 679 | 430 | 269 | 150 | 0 | 0 | 2478 |
| 2000-01 | 0 | 0 | 11 | 81 | 455 | 996 | 829 | 479 | 479 | 72 | 5 | 0 | 3407 |
| 2001-02 | 0 | 0 | 20 | 150 | 235 | 513 | 601 | 598 | 444 | 94 | 41 | 0 | 2696 |
| 2002-03 | 0 | 0 | 0 | 125 | 480 | 639 | 839 | 639 | 326 | 97 | 0 | 0 | 3145 |
| 2003-04 | 0 | 0 | 16 | 87 | 304 | 645 | 712 | 627 | 264 | 139 | 24 | 0 | 2818 |
| 2004-05 | 0 | 0 | 0 | 43 | 240 | 639 | 573 | 425 | 357 | 94 | 32 | 0 | 2403 |
| 2005-06 | 0 | 0 | 1 | 137 | 304 | 708 | 467 | 629 | 330 | 54 | 21 | 0 | 2651 |
| 2006-07 | 0 | 0 | 8 | 162 | 387 | 539 | 683 | 615 | 172 | 198 | 2 | 0 | 2766 |
| 2007-08 | 0 | 0 | 0 | 84 | 318 | 525 | 764 | 536 | 348 | 163 | 20 | 0 | 2758 |
| 2008-09 | 0 | 0 | 0 | 125 | 438 | 657 | 783 | 475 | 325 | 155 | 8 | 0 | 2966 |
| 2009- | 0 | 0 | 1 | 149 | 291 | 739 | | | | | | | |

WBAN : 13893

## COOLING DEGREE DAYS (base 65°F) 2009 MEMPHIS (KMEM)

| YEAR | JAN | FEB | MAR | APR | MAY | JUN | JUL | AUG | SEP | OCT | NOV | DEC | TOTAL |
|------|-----|-----|-----|-----|-----|-----|-----|-----|-----|-----|-----|-----|-------|
| 1980 | 0 | 0 | 0 | 40 | 249 | 480 | 744 | 695 | 476 | 80 | 18 | 2 | 2784 |
| 1981 | 0 | 5 | 20 | 181 | 184 | 532 | 614 | 527 | 285 | 80 | 2 | 0 | 2430 |
| 1982 | 0 | 1 | 36 | 26 | 305 | 399 | 623 | 563 | 275 | 100 | 14 | 25 | 2367 |
| 1983 | 0 | 0 | 7 | 32 | 131 | 373 | 584 | 622 | 338 | 116 | 17 | 0 | 2220 |
| 1984 | 0 | 1 | 4 | 51 | 169 | 482 | 502 | 462 | 249 | 162 | 5 | 23 | 2110 |
| 1985 | 0 | 6 | 30 | 107 | 224 | 425 | 540 | 478 | 285 | 129 | 42 | 0 | 2266 |
| 1986 | 0 | 3 | 12 | 91 | 247 | 487 | 673 | 448 | 427 | 81 | 4 | 0 | 2473 |
| 1987 | 0 | 0 | 9 | 78 | 366 | 458 | 549 | 584 | 315 | 12 | 13 | 0 | 2384 |
| 1988 | 0 | 0 | 7 | 52 | 221 | 469 | 518 | 586 | 347 | 24 | 8 | 0 | 2232 |
| 1989 | 0 | 2 | 27 | 109 | 186 | 386 | 496 | 510 | 254 | 84 | 20 | 0 | 2074 |
| 1990 | 0 | 7 | 27 | 83 | 146 | 477 | 550 | 534 | 404 | 71 | 18 | 2 | 2319 |
| 1991 | 0 | 0 | 34 | 62 | 309 | 464 | 576 | 500 | 339 | 113 | 13 | 0 | 2410 |
| 1992 | 0 | 0 | 5 | 92 | 204 | 363 | 551 | 375 | 268 | 55 | 4 | 0 | 1917 |
| 1993 | 0 | 0 | 8 | 38 | 173 | 436 | 669 | 584 | 275 | 84 | 13 | 0 | 2280 |
| 1994 | 0 | 7 | 8 | 145 | 179 | 525 | 502 | 451 | 247 | 80 | 15 | 0 | 2159 |
| 1995 | 4 | 0 | 26 | 66 | 229 | 365 | 507 | 612 | 264 | 91 | 2 | 1 | 2167 |
| 1996 | 0 | 6 | 5 | 40 | 302 | 408 | 486 | 458 | 237 | 62 | 4 | 3 | 2011 |
| 1997 | 1 | 1 | 14 | 7 | 132 | 327 | 550 | 446 | 297 | 114 | 0 | 0 | 1889 |
| 1998 | 0 | 0 | 39 | 33 | 330 | 550 | 598 | 528 | 471 | 114 | 9 | 13 | 2685 |
| 1999 | 5 | 6 | 0 | 136 | 199 | 447 | 587 | 555 | 319 | 99 | 10 | 1 | 2364 |
| 2000 | 0 | 7 | 12 | 24 | 284 | 416 | 579 | 667 | 371 | 184 | 18 | 0 | 2562 |
| 2001 | 0 | 2 | 0 | 159 | 232 | 387 | 569 | 526 | 285 | 56 | 8 | 2 | 2226 |
| 2002 | 11 | 0 | 4 | 136 | 175 | 460 | 555 | 531 | 385 | 71 | 6 | 0 | 2334 |
| 2003 | 0 | 0 | 1 | 79 | 216 | 309 | 507 | 517 | 235 | 68 | 28 | 0 | 1960 |
| 2004 | 3 | 0 | 34 | 51 | 311 | 417 | 484 | 405 | 347 | 164 | 9 | 3 | 2228 |
| 2005 | 6 | 0 | 10 | 58 | 247 | 508 | 549 | 619 | 427 | 126 | 31 | 0 | 2581 |
| 2006 | 0 | 0 | 34 | 156 | 253 | 453 | 599 | 621 | 259 | 87 | 6 | 0 | 2468 |
| 2007 | 0 | 0 | 96 | 62 | 310 | 524 | 533 | 733 | 429 | 202 | 11 | 0 | 2900 |
| 2008 | 4 | 10 | 17 | 57 | 209 | 469 | 578 | 485 | 302 | 87 | 3 | 0 | 2221 |
| 2009 | 0 | 2 | 21 | 75 | 221 | 496 | 450 | 471 | 328 | 21 | 3 | 0 | 2088 |

## SNOWFALL (inches)  2009  MEMPHIS (KMEM)

| YEAR | JUL | AUG | SEP | OCT | NOV | DEC | JAN | FEB | MAR | APR | MAY | JUN | TOTAL |
|------|-----|-----|-----|-----|-----|-----|-----|-----|-----|-----|-----|-----|-------|
| 1980-81 | 0.0 | 0.0 | 0.0 | 0.0 | T | T | T | T | 0.0 | 0.0 | 0.0 | 0.0 | T |
| 1981-82 | 0.0 | 0.0 | 0.0 | 0.0 | 0.0 | T | 4.5 | 0.7 | 1.2 | 0.0 | 0.0 | 0.0 | 6.4 |
| 1982-83 | 0.0 | 0.0 | 0.0 | 0.0 | 0.0 | T | 7.3 | T | 0.2 | 0.0 | 0.0 | 0.0 | 7.5 |
| 1983-84 | 0.0 | 0.0 | 0.0 | 0.0 | 0.0 | 0.8 | 2.0 | T | 0.5 | 0.0 | 0.0 | 0.0 | 3.3 |
| 1984-85 | 0.0 | 0.0 | 0.0 | 0.0 | 0.0 | T | 12.4 | 8.3 | 0.0 | 0.0 | 0.0 | 0.0 | 20.7 |
| 1985-86 | 0.0 | 0.0 | 0.0 | 0.0 | 0.0 | T | T | 2.0 | 0.0 | 0.0 | 0.0 | 0.0 | 2.0 |
| 1986-87 | 0.0 | 0.0 | 0.0 | 0.0 | 0.0 |  | T | T | 0.4 | 0.0 | 0.0 | 0.0 |  |
| 1987-88 | 0.0 | 0.0 | 0.0 | 0.0 | 0.0 | T | 8.2 | 3.0 | T | 0.0 | 0.0 | 0.0 | 11.2 |
| 1988-89 | 0.0 | 0.0 | 0.0 | 0.0 | 0.0 | T | T | 0.3 | T | 0.0 | T | 0.0 | 0.3 |
| 1989-90 | 0.0 | 0.0 | 0.0 | T | T | 0.4 | 0.0 | 0.0 | 0.0 | T | 0.0 | 0.0 | 0.4 |
| 1990-91 | 0.0 | 0.0 | 0.0 | 0.0 | 0.0 | 0.4 | T | T | 1.0 | 0.0 | 0.0 | 0.0 | 1.4 |
| 1991-92 | 0.0 | 0.0 | 0.0 | 0.0 | 0.6 | 0.0 | T | 0.0 | T | 0.0 | 0.0 | 0.0 | 0.6 |
| 1992-93 | 0.0 | 0.0 | 0.0 | 0.0 | T | T | T | T | T | T | T | 0.0 | T |
| 1993-94 | 0.0 | 0.0 | 0.0 | T | 0.0 | 0.0 | 0.7 | 2.4 | T | 0.0 | 0.0 | 0.0 | 3.1 |
| 1994-95 | 0.0 | 0.0 | 0.0 | 0.0 | 0.0 | 0.0 | 0.5 | 2.5 | 0.7 | 0.0 | T | T | 3.7 |
| 1995-96 | 0.0 | 0.0 | 0.0 | 0.0 | T | T | 0.4 | 0.2 | T | 0.0 | 0.0 | 0.0 | 0.6 |
| 1996-97 | 0.0 | 0.0 | 0.0 | 0.0 | 0.0 | T | 3.8 | 0.7 | 0.0 | 0.0 | T | 0.0 | 4.5 |
| 1997-98 | 0.0 | 0.0 | 0.0 | 0.0 | 0.0 | T | 3.0 | 0.0 | T | 0.0 | 0.0 | 0.0 | 3.0 |
| 1998-99 | 0.0 | 0.0 | 0.0 | 0.0 | 0.0 | 0.8 | T | T | T | 0.0 |  |  |  |
| 1999-00 |  |  |  |  |  |  |  |  |  |  |  |  |  |
| 2000-01 |  |  |  |  |  |  |  |  |  |  |  |  |  |
| 2001-02 |  |  |  |  |  |  |  |  |  |  |  |  |  |
| 2002-03 |  |  |  |  |  |  |  |  |  |  |  |  |  |
| 2003-04 |  |  |  |  |  |  |  |  |  |  |  |  |  |
| 2004-05 |  |  |  |  |  |  |  |  |  |  |  |  |  |
| 2005-06 |  |  |  |  |  |  |  |  |  |  |  |  |  |
| 2006-07 |  |  |  |  |  | T | T | 2.2 | T | T | 0.0 | 0.0 |  |
| 2007-08 | 0.0 | 0.0 | 0.0 | 0.0 | T | 0.0 | T | T | 5.4 | T | T | 0.0 | 5.4 |
| 2008-09 | 0.0 | 0.0 | 0.0 | 0.0 | 0.0 | 0.0 | 0.8 | 3.0 | 0.3 | 0.0 | 0.0 | 0.0 | 4.1 |
| 2009- | 0.0 | 0.0 | 0.0 | 0.0 | 0.0 | T |  |  |  |  |  |  |  |
| POR=<br>69 YRS | 0.0 | 0.0 | 0.0 | T | 0.1 | 0.4 | 2.0 | 1.1 | 0.6 | T | T | T | 4.2 |

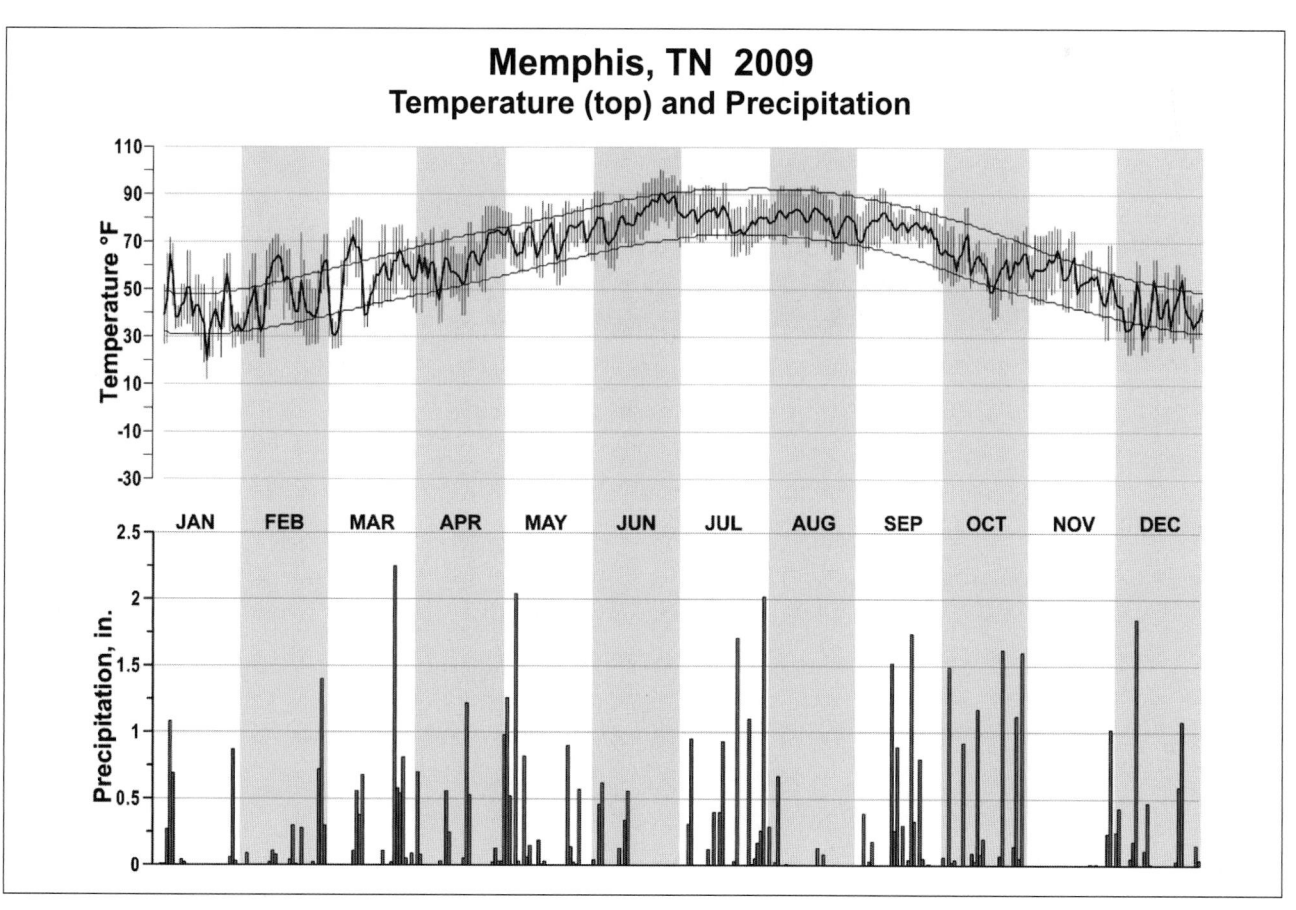

## Memphis, TN  2009
### Temperature (top) and Precipitation

# 2009
# NASHVILLE
# TENNESSEE (KBNA)

The city of Nashville is located on the Cumberland River, in the northwestern corner of the Central Basin of middle Tennessee near the escarpment of the Highland Rim. The Rim, as it is called, rises to the height of 300 to 400 feet above the mean elevation of the basin, forming an amphitheater about the city from the southwest to the southeast, with the south being more or less open but undulating.

Temperatures are moderate, with great extremes of either heat or cold rarely occurring, yet there are changes of sufficient amplitude and frequency to give variety.

Based on the 1951-1980 period, the average first occurrence of 32 degrees Fahrenheit in the fall is October 29 and the average last occurrence in the spring is April 5.

Humidity is an important phase of climate in relation to bodily health and comfort. The Nashville records show that the average relative humidity is moderate as compared with the general conditions east of the Mississippi River and south of the Ohio.

Nashville is not in the most frequented path of general storms that cross the country, however, it is in the zone of moderate frequency of thunderstorms. The thunderstorm season usually begins in the latter part of March and continues through September.

# NORMALS, MEANS, AND EXTREMES
## NASHVILLE (KBNA)

| LATITUDE: 36 ° 7 'N | LONGITUDE: -86 ° 41'W | ELEVATION (FT): GRND: 600   BARO: 574 | TIME ZONE: CENTRAL   (UTC -6) | WBAN: 13897 |
|---|---|---|---|---|

| | ELEMENT | POR | JAN | FEB | MAR | APR | MAY | JUN | JUL | AUG | SEP | OCT | NOV | DEC | YEAR |
|---|---|---|---|---|---|---|---|---|---|---|---|---|---|---|---|
| **TEMPERATURE °F** | NORMAL DAILY MAXIMUM | 30 | 45.6 | 51.4 | 60.7 | 69.8 | 77.5 | 85.1 | 88.7 | 87.8 | 81.5 | 71.1 | 59.0 | 49.4 | 69.0 |
| | MEAN DAILY MAXIMUM | 62 | 47.2 | 51.7 | 60.6 | 70.9 | 79.0 | 86.5 | 89.7 | 89.0 | 82.8 | 72.2 | 59.8 | 50.3 | 70.0 |
| | HIGHEST DAILY MAXIMUM | 70 | 78 | 84 | 86 | 91 | 97 | 106 | 107 | 106 | 105 | 94 | 84 | 79 | 107 |
| | YEAR OF OCCURRENCE | | 1972 | 1962 | 2007 | 1989 | 1941 | 1952 | 1952 | 2007 | 1954 | 1953 | 2000 | 1982 | JUL 1952 |
| | MEAN OF EXTREME MAXS. | 62 | 68.3 | 72.0 | 79.0 | 85.3 | 89.3 | 94.7 | 96.9 | 96.4 | 93.3 | 85.9 | 77.4 | 69.0 | 84.0 |
| | NORMAL DAILY MINIMUM | 30 | 27.9 | 31.2 | 39.4 | 47.1 | 56.7 | 65.0 | 69.5 | 68.0 | 61.0 | 48.6 | 39.5 | 31.5 | 48.8 |
| | MEAN DAILY MINIMUM | 62 | 28.6 | 31.3 | 38.9 | 48.0 | 57.1 | 65.3 | 69.5 | 68.2 | 61.1 | 48.8 | 38.9 | 31.6 | 48.9 |
| | LOWEST DAILY MINIMUM | 70 | -17 | -13 | 2 | 23 | 34 | 42 | 51 | 47 | 36 | 26 | -1 | -10 | -17 |
| | YEAR OF OCCURRENCE | | 1985 | 1951 | 1980 | 2007 | 1976 | 1966 | 1947 | 1946 | 1983 | 1987 | 1950 | 1989 | JAN 1985 |
| | MEAN OF EXTREME MINS. | 62 | 8.0 | 11.9 | 21.9 | 31.3 | 42.3 | 53.2 | 60.3 | 58.6 | 46.0 | 32.7 | 21.4 | 13.2 | 33.4 |
| | NORMAL DRY BULB | 30 | 36.8 | 41.3 | 50.1 | 58.5 | 67.1 | 75.1 | 79.1 | 77.9 | 71.3 | 59.9 | 49.3 | 40.5 | 58.9 |
| | MEAN DRY BULB | 62 | 37.9 | 41.5 | 49.7 | 59.5 | 68.1 | 75.9 | 79.6 | 78.6 | 72.0 | 60.6 | 49.3 | 41.0 | 59.5 |
| | MEAN WET BULB | 26 | 33.8 | 36.6 | 43.2 | 51.3 | 60.4 | 67.6 | 70.8 | 69.8 | 63.9 | 53.6 | 44.3 | 36.6 | 52.7 |
| | MEAN DEW POINT | 26 | 30.0 | 32.5 | 38.1 | 46.5 | 57.2 | 65.2 | 68.5 | 67.4 | 61.0 | 50.0 | 40.4 | 32.9 | 49.1 |
| | NORMAL NO. DAYS WITH: | | | | | | | | | | | | | | |
| | MAXIMUM >= 90 | 30 | 0.0 | 0.0 | 0.0 | 0.1 | 1.0 | 8.8 | 17.1 | 13.1 | 5.7 | 0.1 | 0.0 | 0.0 | 45.9 |
| | MAXIMUM <= 32 | 30 | 4.5 | 2.3 | 0.2 | 0.0 | 0.0 | 0.0 | 0.0 | 0.0 | 0.0 | 0.0 | 0.1 | 2.2 | 9.3 |
| | MINIMUM <= 32 | 30 | 20.6 | 15.6 | 8.4 | 1.9 | 0.0 | 0.0 | 0.0 | 0.0 | 0.0 | 1.1 | 8.1 | 17.0 | 72.7 |
| | MINIMUM <= 0 | 30 | 0.5 | 0.2 | 0.0 | 0.0 | 0.0 | 0.0 | 0.0 | 0.0 | 0.0 | 0.0 | 0.0 | 0.2 | 0.9 |
| **H/C** | NORMAL HEATING DEG. DAYS | 30 | 859 | 664 | 462 | 217 | 56 | 2 | 0 | 0 | 24 | 189 | 460 | 744 | 3677 |
| | NORMAL COOLING DEG. DAYS | 30 | 0 | 0 | 9 | 37 | 136 | 321 | 453 | 416 | 229 | 46 | 5 | 0 | 1652 |
| **RH** | NORMAL (PERCENT) | 30 | 71 | 68 | 64 | 63 | 70 | 71 | 72 | 73 | 73 | 70 | 70 | 72 | 70 |
| | HOUR 00 LST | 30 | 76 | 74 | 71 | 72 | 82 | 84 | 84 | 84 | 84 | 81 | 77 | 77 | 79 |
| | HOUR 06 LST | 30 | 80 | 80 | 78 | 81 | 86 | 87 | 89 | 90 | 90 | 87 | 82 | 81 | 84 |
| | HOUR 12 LST | 30 | 64 | 59 | 54 | 51 | 56 | 56 | 57 | 56 | 57 | 53 | 59 | 64 | 57 |
| | HOUR 18 LST | 30 | 67 | 60 | 55 | 52 | 58 | 60 | 62 | 61 | 63 | 60 | 64 | 67 | 61 |
| **S** | PERCENT POSSIBLE SUNSHINE | 55 | 41 | 47 | 52 | 59 | 60 | 65 | 63 | 63 | 62 | 62 | 50 | 42 | 56 |
| **W/O** | MEAN NO. DAYS WITH: | | | | | | | | | | | | | | |
| | HEAVY FOG(VISBY <= 1/4 MI) | 46 | 2.0 | 1.3 | 1.0 | 0.5 | 1.2 | 1.0 | 0.7 | 1.4 | 1.9 | 2.2 | 1.7 | 1.5 | 16.4 |
| | THUNDERSTORMS | 62 | 1.3 | 1.6 | 3.9 | 5.0 | 7.1 | 8.3 | 9.1 | 7.3 | 3.5 | 1.5 | 1.7 | 0.9 | 51.2 |
| **CLOUDNESS** | MEAN: | | | | | | | | | | | | | | |
| | SUNRISE-SUNSET (OKTAS) | 56 | 5.7 | 5.4 | 5.3 | 4.9 | 4.8 | 4.5 | 4.4 | 4.2 | 4.2 | 3.9 | 4.8 | 5.3 | 4.8 |
| | MIDNIGHT-MIDNIGHT (OKTAS) | 32 | 5.3 | 5.1 | 5.0 | 4.6 | 4.6 | 4.1 | 4.0 | 3.8 | 4.0 | 3.6 | 4.7 | 5.0 | 4.5 |
| | MEAN NO. DAYS WITH: | | | | | | | | | | | | | | |
| | CLEAR | 55 | 6.3 | 6.9 | 7.5 | 8.3 | 8.0 | 8.0 | 8.1 | 9.9 | 10.5 | 12.8 | 8.8 | 7.1 | 102.2 |
| | PARTLY CLOUDY | 55 | 6.1 | 5.9 | 7.2 | 8.5 | 10.1 | 12.5 | 13.2 | 12.2 | 9.1 | 8.0 | 6.8 | 7.0 | 106.6 |
| | CLOUDY | 55 | 18.5 | 15.5 | 16.3 | 13.2 | 13.0 | 9.5 | 9.7 | 8.9 | 10.4 | 10.2 | 14.4 | 17.0 | 156.6 |
| **PR** | MEAN STATION PRESSURE(IN) | 26 | 29.49 | 29.46 | 29.41 | 29.35 | 29.36 | 29.35 | 29.38 | 29.39 | 29.41 | 29.45 | 29.47 | 29.50 | 29.42 |
| | MEAN SEA-LEVEL PRES. (IN) | 26 | 30.16 | 30.13 | 30.07 | 30.00 | 30.00 | 29.99 | 30.01 | 30.02 | 30.05 | 30.10 | 30.13 | 30.16 | 30.07 |
| **WINDS** | MEAN SPEED (MPH) | 26 | 8.3 | 8.5 | 8.8 | 8.2 | 7.1 | 6.4 | 6.2 | 5.9 | 6.1 | 6.5 | 7.5 | 8.1 | 7.3 |
| | PREVAIL.DIR(TENS OF DEGS) | 41 | 19 | 19 | 19 | 19 | 19 | 19 | 19 | 19 | 19 | 19 | 19 | 19 | 19 |
| | MAXIMUM 2-MINUTE: | | | | | | | | | | | | | | |
| | SPEED (MPH) | 13 | 40 | 37 | 39 | 40 | 33 | 38 | 58 | 43 | 31 | 37 | 37 | 38 | 58 |
| | DIR. (TENS OF DEGS) | | 13 | 17 | 17 | 25 | 33 | 02 | 03 | 27 | 03 | 26 | 16 | 16 | 03 |
| | YEAR OF OCCURRENCE | | 2008 | 2008 | 1999 | 1998 | 1998 | 2009 | 2004 | 2003 | 2004 | 2007 | 2005 | 2009 | JUL 2004 |
| | MAXIMUM 3-SECOND | | | | | | | | | | | | | | |
| | SPEED (MPH) | 13 | 61 | 51 | 48 | 59 | 47 | 47 | 67 | 59 | 41 | 55 | 48 | 52 | 67 |
| | DIR. (TENS OF DEGS) | | 26 | 25 | 16 | 23 | 32 | 01 | 02 | 33 | 17 | 27 | 33 | 16 | 02 |
| | YEAR OF OCCURRENCE | | 1999 | 2009 | 1999 | 1998 | 2000 | 1999 | 2004 | 2003 | 2008 | 2007 | 2007 | 2009 | JUL 2004 |
| **PRECIPITATION** | NORMAL (IN) | 30 | 3.97 | 3.69 | 4.87 | 3.93 | 5.07 | 4.08 | 3.77 | 3.28 | 3.59 | 2.87 | 4.45 | 4.54 | 48.11 |
| | MAXIMUM MONTHLY (IN) | 70 | 13.92 | 10.31 | 12.35 | 8.41 | 11.04 | 11.95 | 7.75 | 8.31 | 11.44 | 6.49 | 9.04 | 13.63 | 13.92 |
| | YEAR OF OCCURRENCE | | 1950 | 1956 | 1975 | 1984 | 1983 | 1998 | 1950 | 1942 | 1979 | 2009 | 1945 | 1978 | JAN 1950 |
| | MINIMUM MONTHLY (IN) | 70 | 0.19 | 0.64 | 1.18 | 0.52 | 0.69 | 0.45 | 0.71 | 0.69 | 0.28 | T | 0.54 | 0.98 | T |
| | YEAR OF OCCURRENCE | | 1986 | 1968 | 1987 | 1986 | 1941 | 1988 | 1954 | 1968 | 1956 | 1963 | 1949 | 1985 | OCT 1963 |
| | MAXIMUM IN 24 HOURS (IN) | 70 | 4.40 | 4.73 | 4.66 | 3.29 | 4.63 | 5.24 | 4.32 | 5.34 | 6.68 | 3.75 | 4.20 | 5.12 | 6.68 |
| | YEAR OF OCCURRENCE | | 1946 | 1989 | 1975 | 1979 | 2003 | 1998 | 1992 | 1963 | 1979 | 1975 | 1997 | 1978 | SEP 1979 |
| | NORMAL NO. DAYS WITH: | | | | | | | | | | | | | | |
| | PRECIPITATION >= 0.01 | 30 | 11.1 | 10.1 | 12.1 | 10.7 | 11.3 | 9.7 | 10.0 | 8.4 | 8.3 | 7.4 | 10.1 | 11.0 | 120.2 |
| | PRECIPITATION >= 1.00 | 30 | 0.9 | 0.7 | 1.0 | 1.2 | 1.5 | 1.1 | 1.1 | 1.0 | 0.9 | 0.8 | 1.2 | 1.5 | 12.9 |
| **SNOWFALL** | NORMAL (IN) | 30 | 3.9 | 3.4 | 1.1 | 0.1 | 0.0 | 0.0 | 0.0 | 0.0 | 0.0 | 0.* | 0.1 | 0.5 | 9.1 |
| | MAXIMUM MONTHLY (IN) | 62 | 18.8 | 18.9 | 16.1 | 1.1 | 0.0 | T | T | T | 0.0 | 0.4 | 9.2 | 13.2 | 18.9 |
| | YEAR OF OCCURRENCE | | 1948 | 1979 | 1960 | 1971 | | 2009 | 2007 | 1989 | | 1993 | 1950 | 1963 | FEB 1979 |
| | MAXIMUM IN 24 HOURS (IN) | 62 | 8.1 | 8.3 | 8.8 | 1.1 | 0.0 | T | T | T | 0.0 | 0.4 | 9.2 | 10.2 | 10.2 |
| | YEAR OF OCCURRENCE' | | 1988 | 1979 | 1951 | 1971 | | 2009 | 2007 | 1989 | | 1993 | 1950 | 1963 | DEC 1963 |
| | MAXIMUM SNOW DEPTH (IN) | 53 | 70 | 8 | 7 | 0 | 0 | 0 | 0 | 0 | 0 | 0 | 5 | 7 | 70 |
| | YEAR OF OCCURRENCE | | 1948 | 1979 | 1968 | | | | | | | | 1966 | 1963 | JAN 1948 |
| | NORMAL NO. DAYS WITH: | | | | | | | | | | | | | | |
| | SNOWFALL >= 1.0 | 30 | 1.0 | 1.2 | 0.4 | 0.0 | 0.0 | 0.0 | 0.0 | 0.0 | 0.0 | 0.0 | 0.0 | 0.2 | 2.8 |

## PRECIPITATION (inches) 2009  NASHVILLE (KBNA)

| YEAR | JAN | FEB | MAR | APR | MAY | JUN | JUL | AUG | SEP | OCT | NOV | DEC | ANNUAL |
|------|-----|-----|-----|-----|-----|-----|-----|-----|-----|-----|-----|-----|--------|
| 1980 | 2.59 | 1.38 | 7.27 | 3.67 | 6.14 | 2.89 | 3.53 | 1.24 | 1.09 | 1.17 | 2.55 | 1.40 | 34.92 |
| 1981 | 1.60 | 3.83 | 3.38 | 4.78 | 3.05 | 8.05 | 3.49 | 3.10 | 1.37 | 2.82 | 3.83 | 2.38 | 41.68 |
| 1982 | 6.50 | 4.80 | 3.00 | 4.36 | 4.19 | 2.28 | 5.47 | 3.46 | 3.23 | 1.91 | 3.87 | 6.36 | 49.43 |
| 1983 | 2.56 | 2.93 | 3.44 | 6.80 | 11.04 | 3.93 | 1.71 | 1.36 | 0.45 | 2.77 | 6.98 | 7.75 | 51.72 |
| 1984 | 1.79 | 2.38 | 5.14 | 8.41 | 9.68 | 4.49 | 6.63 | 2.42 | 0.97 | 6.00 | 6.20 | 2.38 | 56.49 |
| 1985 | 3.02 | 3.30 | 2.70 | 2.91 | 2.65 | 1.53 | 2.00 | 3.91 | 2.52 | 1.59 | 3.81 | 0.98 | 30.92 |
| 1986 | 0.19 | 3.59 | 2.29 | 0.52 | 3.36 | 2.38 | 0.77 | 3.38 | 2.19 | 2.19 | 7.43 | 3.31 | 31.60 |
| 1987 | 1.61 | 4.87 | 1.18 | 1.03 | 4.41 | 2.82 | 2.56 | 0.73 | 1.95 | 0.21 | 3.40 | 5.46 | 30.23 |
| 1988 | 3.73 | 2.02 | 2.18 | 2.09 | 1.86 | 0.45 | 3.26 | 2.39 | 2.45 | 1.54 | 5.49 | 3.95 | 31.41 |
| 1989 | 4.52 | 9.36 | 5.31 | 2.68 | 4.61 | 7.87 | 3.18 | 3.67 | 6.30 | 3.62 | 3.94 | 1.97 | 57.03 |
| 1990 | 2.76 | 4.73 | 3.26 | 1.60 | 2.80 | 2.37 | 4.86 | 3.12 | 2.13 | 4.41 | 4.29 | 10.76 | 47.09 |
| 1991 | 2.92 | 5.44 | 4.25 | 3.35 | 5.63 | 1.25 | 2.82 | 1.79 | 5.47 | 3.88 | 2.87 | 7.27 | 46.94 |
| 1992 | 2.97 | 2.60 | 4.50 | 0.77 | 3.12 | 4.31 | 5.89 | 3.25 | 3.45 | 1.62 | 4.48 | 2.88 | 39.84 |
| 1993 | 2.76 | 3.33 | 5.50 | 3.33 | 4.50 | 5.31 | 3.64 | 1.76 | 2.90 | 2.20 | 2.53 | 6.62 | 44.38 |
| 1994 | 4.36 | 6.18 | 7.56 | 5.72 | 3.76 | 8.08 | 4.82 | 5.05 | 4.20 | 3.31 | 4.04 | 2.69 | 59.77 |
| 1995 | 5.61 | 1.81 | 3.87 | 3.95 | 7.66 | 3.69 | 1.95 | 3.40 | 5.00 | 5.60 | 3.98 | 2.32 | 48.84 |
| 1996 | 3.82 | 2.46 | 5.15 | 3.68 | 4.48 | 3.68 | 5.45 | 1.09 | 4.89 | 3.16 | 6.00 | 4.77 | 48.63 |
| 1997 | 4.19 | 3.10 | 9.64 | 2.42 | 4.92 | 6.66 | 3.26 | 3.52 | 5.75 | 2.71 | 6.59 | 2.19 | 54.95 |
| 1998 | 3.68 | 4.11 | 3.13 | 6.31 | 4.46 | 11.95 | 4.63 | 2.93 | 1.39 | 1.59 | 1.30 | 6.53 | 52.01 |
| 1999 | 9.28 | 2.33 | 4.27 | 2.29 | 4.35 | 3.56 | 3.19 | 3.05 | 1.97 | 2.04 | 2.99 | 2.50 | 41.82 |
| 2000 | 3.52 | 3.75 | 3.34 | 6.23 | 7.66 | 1.74 | 2.25 | 1.95 | 1.90 | 0.26 | 6.39 | 3.44 | 42.43 |
| 2001 | 3.21 | 8.54 | 2.73 | 2.42 | 5.54 | 4.47 | 2.77 | 4.07 | 1.79 | 4.61 | 5.09 | 3.32 | 48.56 |
| 2002 | 4.93 | 1.99 | 9.40 | 4.31 | 3.98 | 3.76 | 5.64 | 3.13 | 6.29 | 4.48 | 2.91 | 5.81 | 56.63 |
| 2003 | 1.59 | 8.47 | 2.30 | 4.69 | 10.73 | 7.08 | 2.87 | 3.88 | 8.70 | 1.80 | 4.17 | 3.19 | 59.47 |
| 2004 | 3.60 | 5.77 | 4.81 | 6.69 | 6.90 | 3.39 | 3.19 | 4.24 | 4.55 | 4.90 | 5.21 | 5.93 | 59.18 |
| 2005 | 4.42 | 3.84 | 3.90 | 6.93 | 1.03 | 2.70 | 2.39 | 6.89 | 1.44 | 0.02 | 3.29 | 2.46 | 39.31 |
| 2006 | 6.57 | 2.69 | 2.90 | 4.14 | 4.95 | 2.19 | 2.64 | 5.20 | 4.00 | 2.98 | 4.05 | 3.41 | 45.72 |
| 2007 | 3.32 | 1.84 | 2.26 | 2.75 | 3.30 | 2.37 | 1.47 | 1.38 | 1.99 | 4.95 | 6.20 | 3.83 | 35.66 |
| 2008 | 4.76 | 2.53 | 5.56 | 7.20 | 5.54 | 2.21 | 4.32 | 1.67 | 0.88 | 5.03 | 1.75 | 6.72 | 48.17 |
| 2009 | 4.59 | 2.85 | 2.92 | 4.13 | 8.45 | 4.53 | 6.03 | 2.14 | 11.08 | 6.49 | 0.67 | 3.99 | 57.87 |
| POR= 62 YRS | 4.35 | 4.11 | 4.86 | 4.17 | 4.79 | 3.94 | 3.73 | 3.25 | 3.55 | 2.80 | 3.94 | 4.47 | 47.96 |

WBAN : 13897

## AVERAGE TEMPERATURE (°F) 2009  NASHVILLE (KBNA)

| YEAR | JAN | FEB | MAR | APR | MAY | JUN | JUL | AUG | SEP | OCT | NOV | DEC | ANNUAL |
|------|-----|-----|-----|-----|-----|-----|-----|-----|-----|-----|-----|-----|--------|
| 1980 | 39.7 | 35.7 | 46.3 | 57.3 | 67.8 | 75.5 | 82.8 | 81.7 | 76.0 | 57.8 | 48.5 | 41.0 | 59.2 |
| 1981 | 35.5 | 42.6 | 47.5 | 64.0 | 64.2 | 77.5 | 79.8 | 76.6 | 68.1 | 60.4 | 49.9 | 38.3 | 58.7 |
| 1982 | 34.0 | 39.5 | 52.5 | 54.6 | 71.0 | 73.3 | 79.8 | 76.1 | 69.6 | 61.1 | 51.4 | 48.2 | 59.3 |
| 1983 | 38.8 | 42.7 | 50.3 | 54.5 | 64.8 | 75.5 | 80.5 | 83.2 | 73.7 | 62.4 | 49.9 | 34.0 | 59.2 |
| 1984 | 32.2 | 43.4 | 46.1 | 58.2 | 64.2 | 77.4 | 76.1 | 76.5 | 68.6 | 66.7 | 46.0 | 49.6 | 58.8 |
| 1985 | 27.8 | 36.5 | 53.2 | 61.9 | 68.4 | 75.7 | 80.2 | 77.2 | 70.8 | 64.4 | 56.9 | 34.2 | 58.9 |
| 1986 | 37.2 | 44.7 | 50.8 | 60.8 | 68.6 | 76.5 | 82.4 | 76.7 | 74.9 | 61.0 | 49.9 | 39.9 | 60.3 |
| 1987 | 36.1 | 43.1 | 51.8 | 57.7 | 73.4 | 77.5 | 80.2 | 81.1 | 72.2 | 54.6 | 52.4 | 44.1 | 60.4 |
| 1988 | 34.4 | 38.7 | 49.3 | 57.1 | 67.3 | 77.3 | 81.4 | 81.9 | 72.8 | 54.2 | 51.1 | 42.4 | 59.0 |
| 1989 | 44.9 | 39.0 | 52.6 | 59.3 | 65.7 | 74.7 | 79.1 | 78.0 | 70.5 | 61.0 | 51.4 | 29.5 | 58.8 |
| 1990 | 45.8 | 49.9 | 53.6 | 58.4 | 66.4 | 78.2 | 80.4 | 79.6 | 74.7 | 60.1 | 54.3 | 43.7 | 62.1 |
| 1991 | 39.2 | 43.9 | 52.5 | 63.8 | 74.2 | 78.2 | 81.1 | 78.3 | 72.3 | 61.2 | 47.2 | 44.5 | 61.4 |
| 1992 | 40.0 | 45.9 | 50.1 | 59.6 | 65.8 | 72.4 | 79.9 | 74.9 | 70.9 | 59.4 | 49.5 | 41.2 | 59.1 |
| 1993 | 41.6 | 39.3 | 47.1 | 56.7 | 67.6 | 75.9 | 83.3 | 81.0 | 71.0 | 58.6 | 47.4 | 40.3 | 59.2 |
| 1994 | 33.4 | 44.0 | 50.7 | 62.5 | 64.1 | 78.1 | 78.5 | 77.1 | 69.1 | 61.0 | 54.5 | 45.3 | 59.9 |
| 1995 | 38.6 | 40.4 |  | 60.9 | 68.5 | 74.7 | 80.8 | 83.3 | 70.7 | 60.0 | 44.0 | 39.5 |  |
| 1996 | 36.3 | 40.6 | 44.6 | 55.8 | 71.5 | 75.6 | 77.6 | 77.5 | 69.5 | 60.8 | 45.7 | 44.4 | 58.3 |
| 1997 | 37.3 | 45.6 | 53.6 | 54.4 | 63.2 | 72.0 | 79.8 | 76.9 | 71.9 | 59.8 | 45.5 | 39.5 | 58.3 |
| 1998 | 44.7 | 46.0 | 49.4 | 57.9 | 71.5 | 77.6 | 79.6 | 79.2 | 77.1 | 63.5 | 52.3 | 43.1 | 61.8 |
| 1999 | 42.5 | 45.8 | 45.6 | 62.6 | 67.6 | 76.5 | 81.9 | 79.4 | 71.9 | 59.9 | 54.0 | 43.7 | 61.0 |
| 2000 | 39.2 | 46.7 | 53.1 | 56.7 | 69.9 | 76.4 | 80.3 | 79.8 | 71.3 | 63.8 | 48.0 | 30.8 | 59.7 |
| 2001 | 35.4 | 44.7 | 45.3 | 63.7 | 68.8 | 73.4 | 79.8 | 78.8 | 70.2 | 58.5 | 54.3 | 44.9 | 59.8 |
| 2002 | 42.2 | 40.6 | 49.6 | 62.0 | 65.7 | 76.9 | 80.1 | 80.0 | 75.0 | 61.4 | 46.4 | 40.5 | 60.0 |
| 2003 | 32.9 | 38.0 | 52.0 | 61.2 | 67.2 | 71.9 | 78.3 | 79.3 | 70.2 | 60.6 | 53.5 | 40.3 | 58.8 |
| 2004 | 38.6 | 40.7 | 53.3 | 59.4 | 71.7 | 75.6 | 77.7 | 74.3 | 72.1 | 64.5 | 53.6 | 40.0 | 60.1 |
| 2005 | 43.4 | 45.0 | 47.6 | 59.7 | 66.1 | 76.9 | 81.1 | 81.8 | 75.1 | 61.1 | 51.3 | 37.6 | 60.6 |
| 2006 | 46.2 | 40.7 | 51.1 | 65.2 | 66.7 | 75.7 | 80.6 | 82.2 | 71.4 | 59.5 | 51.5 | 46.4 | 61.4 |
| 2007 | 41.7 | 37.2 | 58.3 | 57.4 | 71.2 | 78.0 | 79.9 | 86.9 | 75.8 | 65.5 | 49.8 | 46.3 | 62.3 |
| 2008 | 37.0 | 42.9 | 50.2 | 58.1 | 66.8 | 78.1 | 79.6 | 79.0 | 73.9 | 60.2 | 46.4 | 40.7 | 59.4 |
| 2009 | 35.2 | 43.8 | 52.1 | 59.0 | 67.7 | 78.0 | 75.8 | 77.1 | 72.3 | 56.9 | 51.3 | 39.3 | 59.0 |
| POR= 62 YRS | 37.9 | 41.5 | 49.7 | 59.5 | 68.1 | 75.9 | 79.6 | 78.6 | 72.0 | 60.6 | 49.3 | 41.0 | 59.5 |

## HEATING DEGREE DAYS (base 65°F) 2009 NASHVILLE (KBNA)

| YEAR | JUL | AUG | SEP | OCT | NOV | DEC | JAN | FEB | MAR | APR | MAY | JUN | TOTAL |
|------|-----|-----|-----|-----|-----|-----|-----|-----|-----|-----|-----|-----|-------|
| 1980-81 | 0 | 0 | 9 | 259 | 487 | 739 | 909 | 621 | 537 | 97 | 96 | 0 | 3754 |
| 1981-82 | 0 | 0 | 42 | 175 | 445 | 820 | 956 | 707 | 416 | 309 | 8 | 0 | 3878 |
| 1982-83 | 0 | 0 | 30 | 194 | 413 | 537 | 806 | 620 | 458 | 322 | 71 | 0 | 3451 |
| 1983-84 | 0 | 0 | 45 | 121 | 447 | 956 | 1009 | 621 | 578 | 220 | 106 | 0 | 4103 |
| 1984-85 | 0 | 0 | 59 | 63 | 564 | 473 | 1146 | 794 | 383 | 145 | 25 | 6 | 3658 |
| 1985-86 | 0 | 0 | 30 | 91 | 264 | 948 | 854 | 561 | 432 | 171 | 55 | 0 | 3406 |
| 1986-87 | 0 | 3 | 0 | 175 | 447 | 773 | 889 | 608 | 401 | 242 | 6 | 0 | 3544 |
| 1987-88 | 0 | 0 | 7 | 317 | 376 | 640 | 941 | 756 | 485 | 242 | 43 | 2 | 3809 |
| 1988-89 | 0 | 0 | 5 | 343 | 408 | 693 | 618 | 721 | 397 | 258 | 90 | 0 | 3533 |
| 1989-90 | 0 | 0 | 36 | 158 | 408 | 1095 | 590 | 422 | 373 | 245 | 65 | 1 | 3393 |
| 1990-91 | 0 | 0 | 21 | 195 | 323 | 654 | 791 | 586 | 402 | 80 | 9 | 0 | 3061 |
| 1991-92 | 0 | 0 | 42 | 166 | 535 | 628 | 768 | 544 | 456 | 217 | 85 | 2 | 3443 |
| 1992-93 | 0 | 0 | 26 | 181 | 461 | 731 | 717 | 713 | 552 | 252 | 32 | 4 | 3669 |
| 1993-94 | 0 | 0 | 27 | 227 | 528 | 759 | 974 | 585 | 437 | 134 | 90 | 0 | 3761 |
| 1994-95 | 0 | 0 | 21 | 144 | 316 | 605 | 814 | 683 | | 175 | 42 | 0 | |
| 1995-96 | 0 | 0 | 31 | 184 | 624 | 783 | 886 | 702 | 626 | 292 | 32 | 0 | 4160 |
| 1996-97 | 0 | 0 | 25 | 160 | 572 | 634 | 851 | 538 | 357 | 319 | 107 | 4 | 3567 |
| 1997-98 | 0 | 0 | 0 | 227 | 576 | 785 | 622 | 527 | 505 | 209 | 19 | 5 | 3475 |
| 1998-99 | 0 | 0 | 0 | 111 | 375 | 671 | 686 | 531 | 595 | 125 | 10 | 0 | 3104 |
| 1999-00 | 0 | 0 | 18 | 185 | 322 | 654 | 796 | 523 | 364 | 248 | 14 | 0 | 3124 |
| 2000-01 | 0 | 0 | 32 | 123 | 518 | 1051 | 912 | 565 | 602 | 124 | 17 | 7 | 3951 |
| 2001-02 | 0 | 0 | 47 | 218 | 315 | 617 | 704 | 677 | 477 | 171 | 89 | 0 | 3315 |
| 2002-03 | 0 | 0 | 2 | 161 | 554 | 752 | 991 | 748 | 393 | 147 | 24 | 8 | 3780 |
| 2003-04 | 0 | 0 | 26 | 150 | 341 | 761 | 814 | 698 | 365 | 201 | 39 | 0 | 3395 |
| 2004-05 | 0 | 3 | 4 | 63 | 346 | 767 | 664 | 554 | 531 | 179 | 75 | 0 | 3186 |
| 2005-06 | 0 | 0 | 2 | 196 | 419 | 845 | 578 | 677 | 439 | 97 | 69 | 0 | 3322 |
| 2006-07 | 0 | 0 | 15 | 222 | 407 | 568 | 714 | 774 | 255 | 269 | 17 | 0 | 3241 |
| 2007-08 | 0 | 0 | 0 | 102 | 451 | 571 | 861 | 636 | 453 | 224 | 46 | 0 | 3344 |
| 2008-09 | 0 | 0 | 0 | 202 | 552 | 747 | 918 | 583 | 399 | 229 | 41 | 0 | 3671 |
| 2009- | 0 | 0 | 11 | 253 | 406 | 791 | | | | | | | |

WBAN : 13897

## COOLING DEGREE DAYS (base 65°F) 2009 NASHVILLE (KBNA)

| YEAR | JAN | FEB | MAR | APR | MAY | JUN | JUL | AUG | SEP | OCT | NOV | DEC | TOTAL |
|------|-----|-----|-----|-----|-----|-----|-----|-----|-----|-----|-----|-----|-------|
| 1980 | 0 | 0 | 0 | 17 | 131 | 322 | 562 | 527 | 344 | 44 | 1 | 0 | 1948 |
| 1981 | 0 | 0 | 1 | 71 | 81 | 383 | 464 | 366 | 145 | 42 | 0 | 0 | 1553 |
| 1982 | 0 | 0 | 37 | 4 | 199 | 256 | 470 | 352 | 177 | 84 | 12 | 21 | 1612 |
| 1983 | 0 | 0 | 9 | 12 | 69 | 320 | 488 | 568 | 315 | 49 | 2 | 0 | 1832 |
| 1984 | 0 | 0 | 0 | 21 | 87 | 382 | 352 | 364 | 173 | 121 | 0 | 1 | 1501 |
| 1985 | 0 | 2 | 24 | 59 | 137 | 335 | 479 | 386 | 206 | 79 | 29 | 0 | 1736 |
| 1986 | 0 | 0 | 1 | 52 | 174 | 352 | 551 | 371 | 304 | 59 | 0 | 0 | 1864 |
| 1987 | 0 | 0 | 0 | 31 | 272 | 381 | 479 | 507 | 227 | 3 | 7 | 0 | 1907 |
| 1988 | 0 | 0 | 5 | 17 | 120 | 380 | 515 | 531 | 246 | 17 | 0 | 0 | 1831 |
| 1989 | 0 | 0 | 21 | 93 | 120 | 298 | 446 | 408 | 208 | 39 | 8 | 0 | 1641 |
| 1990 | 0 | 4 | 26 | 51 | 115 | 401 | 485 | 458 | 315 | 52 | 10 | 0 | 1917 |
| 1991 | 0 | 0 | 22 | 50 | 300 | 403 | 507 | 419 | 268 | 57 | 4 | 0 | 2030 |
| 1992 | 0 | 0 | 0 | 60 | 115 | 233 | 471 | 311 | 208 | 15 | 2 | 0 | 1415 |
| 1993 | 0 | 0 | 1 | 9 | 121 | 336 | 573 | 506 | 215 | 33 | 7 | 0 | 1801 |
| 1994 | 0 | 3 | 0 | 66 | 67 | 400 | 426 | 381 | 152 | 26 | 8 | 0 | 1529 |
| 1995 | 0 | 0 | | 60 | 158 | 297 | 496 | 573 | 210 | 34 | 1 | 0 | |
| 1996 | 0 | 4 | 0 | 22 | 240 | 324 | 397 | 395 | 166 | 35 | 0 | 2 | 1585 |
| 1997 | 0 | 0 | 13 | 6 | 61 | 222 | 465 | 374 | 211 | 72 | 0 | 0 | 1424 |
| 1998 | 0 | 0 | 27 | 5 | 227 | 391 | 459 | 447 | 370 | 69 | 1 | 1 | 1997 |
| 1999 | 0 | 0 | 0 | 60 | 100 | 354 | 533 | 453 | 229 | 35 | 1 | 0 | 1765 |
| 2000 | 2 | 0 | 6 | 6 | 172 | 346 | 483 | 467 | 227 | 95 | 18 | 0 | 1822 |
| 2001 | 0 | 0 | 0 | 93 | 146 | 264 | 468 | 437 | 209 | 27 | 2 | 0 | 1646 |
| 2002 | 3 | 0 | 6 | 87 | 118 | 364 | 474 | 471 | 310 | 55 | 4 | 0 | 1892 |
| 2003 | 0 | 0 | 0 | 40 | 99 | 221 | 417 | 451 | 187 | 22 | 5 | 0 | 1442 |
| 2004 | 4 | 0 | 10 | 38 | 252 | 323 | 398 | 297 | 226 | 52 | 11 | 0 | 1611 |
| 2005 | 0 | 0 | 0 | 26 | 118 | 363 | 502 | 530 | 313 | 83 | 15 | 0 | 1950 |
| 2006 | 0 | 0 | 14 | 112 | 126 | 327 | 491 | 540 | 216 | 58 | 8 | 0 | 1892 |
| 2007 | 0 | 0 | 53 | 49 | 214 | 395 | 473 | 688 | 330 | 122 | 0 | 1 | 2325 |
| 2008 | 0 | 3 | 2 | 25 | 109 | 401 | 458 | 439 | 273 | 62 | 0 | 1 | 1773 |
| 2009 | 0 | 0 | 7 | 56 | 132 | 395 | 340 | 383 | 238 | 8 | 0 | 0 | 1559 |

**SNOWFALL (inches) 2009 NASHVILLE (KBNA)**

| YEAR | JUL | AUG | SEP | OCT | NOV | DEC | JAN | FEB | MAR | APR | MAY | JUN | TOTAL |
|------|-----|-----|-----|-----|-----|-----|-----|-----|-----|-----|-----|-----|-------|
| 1980-81 | 0.0 | 0.0 | 0.0 | 0.0 | T | T | 1.2 | 1.7 | T | 0.0 | 0.0 | 0.0 | 2.9 |
| 1981-82 | 0.0 | 0.0 | 0.0 | 0.0 | 0.0 | 0.2 | 4.8 | 3.7 | 1.0 | 0.0 | 0.0 | 0.0 | 9.7 |
| 1982-83 | 0.0 | 0.0 | 0.0 | 0.0 | 0.0 | 0.4 | 0.3 | 0.8 | T | 0.0 | 0.0 | 0.0 | 1.5 |
| 1983-84 | 0.0 | 0.0 | 0.0 | 0.0 | 0.0 | 0.7 | 5.3 | 3.7 | T | 0.0 | 0.0 | 0.0 | 9.7 |
| 1984-85 | 0.0 | 0.0 | 0.0 | 0.0 | 0.0 | 0.8 | 9.8 | 8.0 | 0.0 | 0.0 | 0.0 | 0.0 | 18.6 |
| 1985-86 | 0.0 | 0.0 | 0.0 | 0.0 | 0.0 | 0.5 | 0.4 | 2.1 | T | 0.0 | 0.0 | 0.0 | 3.0 |
| 1986-87 | 0.0 | 0.0 | 0.0 | 0.0 | 0.0 | T | 1.4 | 1.3 | 1.6 | T | 0.0 | 0.0 | 4.3 |
| 1987-88 | 0.0 | 0.0 | 0.0 | 0.0 | T | T | 8.6 | 1.4 | T | 0.0 | 0.0 | 0.0 | 10.0 |
| 1988-89 | 0.0 | 0.0 | 0.0 | 0.0 | 0.0 | 1.6 | T | 5.2 | 0.0 | 0.0 | 0.0 | 0.0 | 6.8 |
| 1989-90 | 0.0 | T | 0.0 | T | T | 0.4 | T | T | 0.4 | 0.0 | 0.0 | 0.0 | 0.8 |
| 1990-91 | 0.0 | 0.0 | 0.0 | 0.0 | 0.0 | 0.3 | T | 0.6 | 1.1 | 0.0 | 0.0 | 0.0 | 2.0 |
| 1991-92 | 0.0 | 0.0 | 0.0 | 0.0 | T | 0.0 | T | 0.0 | 1.0 | 0.0 | 0.0 | 0.0 | 1.0 |
| 1992-93 | 0.0 | 0.0 | 0.0 | 0.0 | T | 0.3 | T | 5.9 | 2.8 | T | 0.0 | T | 9.0 |
| 1993-94 | 0.0 | 0.0 | 0.0 | 0.4 | T | 0.3 | 2.3 | 1.0 | T | T | 0.0 | T | 4.0 |
| 1994-95 | 0.0 | 0.0 | 0.0 | 0.0 | 0.0 | 0.0 | 1.5 | 1.3 | 0.1 | 0.0 | 0.0 | 0.0 | 2.9 |
| 1995-96 | 0.0 | 0.0 | 0.0 | 0.0 | T | 0.4 | 6.2 | 7.8 | 9.3 | 0.0 | 0.0 | | |
| 1996-97 | | | | | | | | 0.2 | | | | | |
| 1997-98 | | | | | | | | | | | | | |
| 1998-99 | | | | | | | | | 1.0 | | | | |
| 1999-00 | | | | | | | | | | | | | |
| 2000-01 | | | | | | | | | | | | | |
| 2001-02 | | | | | | | | | | | | | |
| 2002-03 | | | | | | | 8.4 | | | | | | |
| 2003-04 | | | | | | | | | | | | | |
| 2004-05 | | | | | | | | | | | | | |
| 2005-06 | | | | | | T | 0.5 | 1.2 | T | T | 0.0 | T | |
| 2006-07 | 0.0 | 0.0 | 0.0 | 0.0 | 0.0 | T | T | 2.2 | 0.0 | T | 0.0 | 0.0 | 2.2 |
| 2007-08 | T | 0.0 | 0.0 | 0.0 | T | T | T | 1.2 | 0.8 | 0.0 | 0.0 | 0.0 | 2.0 |
| 2008-09 | 0.0 | 0.0 | 0.0 | 0.0 | T | 1.0 | 0.4 | T | 0.3 | 0.0 | 0.0 | T | 1.7 |
| 2009- | 0.0 | 0.0 | 0.0 | 0.0 | 0.0 | T | | | | | | | |
| POR= 61 YRS | T | T | 0.0 | T | 0.4 | 1.0 | 3.3 | 2.5 | 1.3 | T | 0.0 | T | 8.5 |

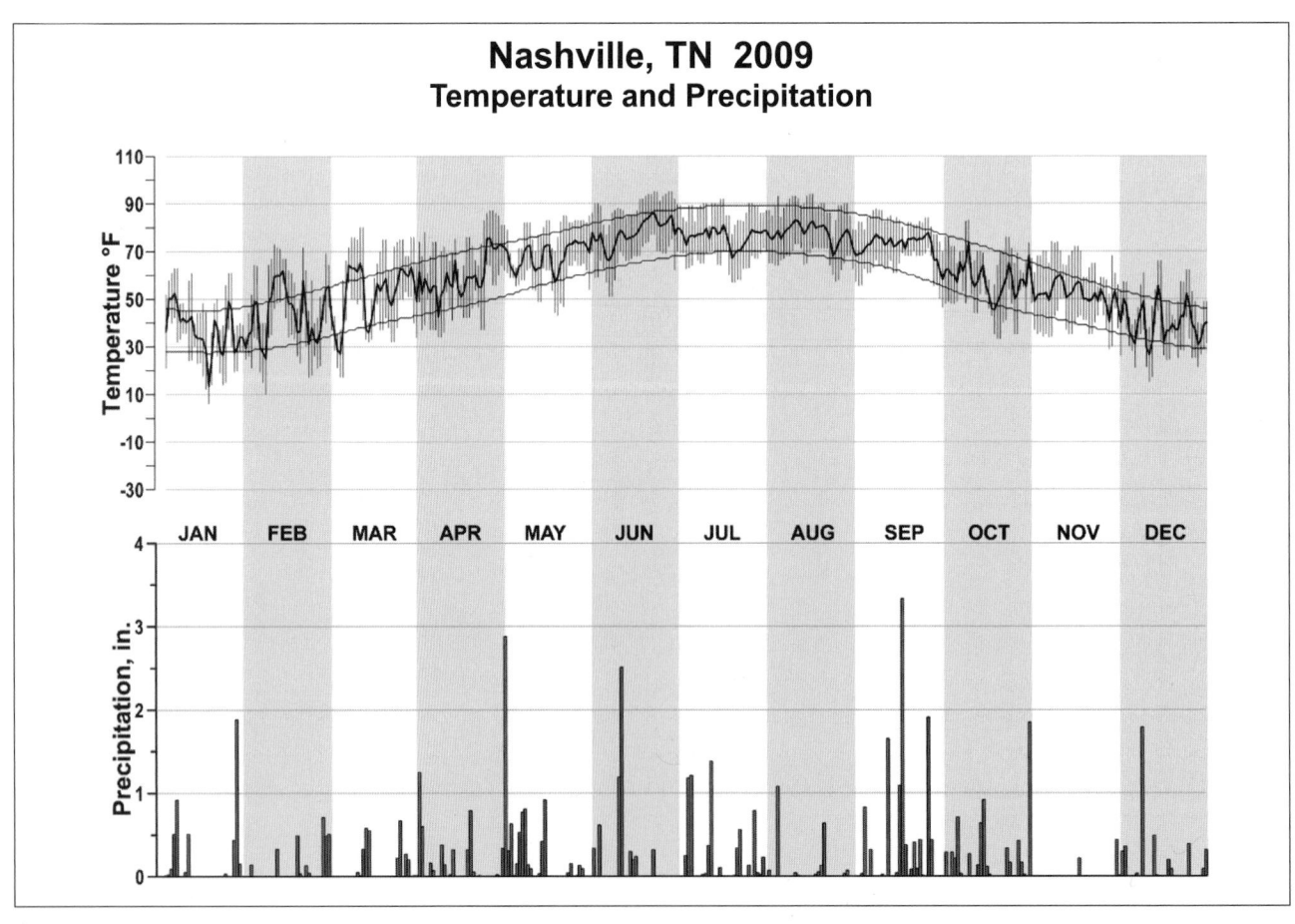

# Nashville, TN 2009
## Temperature and Precipitation

# 2009
# AMARILLO
# TEXAS (KAMA)

The station is located 7 statute miles east northeast of the downtown post office in a region of rather flat topography.   The Canadian River flows eastward 18 miles north of the station, with its bed about 800 feet below the plains.   The Prairie Dog Town Fork of the Red River flows southeastward about 15 miles south of the station where it enters the Palo Duro Canyon, which is about 1,000 feet deep.   There are numerous shallow Playa lakes, often dry, over the area, and the nearly treeless grasslands slope downward to the east.   The terrain gradually rises to the west and northwest.

Three-fourths of the total annual precipitation falls from April through September, occurring from thunderstorm activity.   Snow usually melts within a few days after it falls.   Heavier snowfalls of 10 inches or more, usually with near blizzard conditions, average once every 5 years and last 2 to 3 days.

The Amarillo area is subject to rapid and large temperature changes, especially during the winter months when cold fronts from the northern Rocky Mountain and Plains states sweep across the area. Temperature drops of 50 to 60 degrees within a 12-hour period are not uncommon.   Temperature drops of 40 degrees have occurred within a few minutes.

Humidity averages are low,   occasionally dropping below 20 percent in the spring.   Low humidity moderates the effect of high summer afternoon temperatures, permits evaporative cooling systems to be very effective, and provides many pleasant evenings and nights.

Severe local storms are infrequent, although a few thunderstorms with damaging hail, lightning, and wind in a very localized area occur most years, usually in spring and summer.   These storms are often accompanied by very heavy rain,   which produces local flooding, particularly of roads and streets.   Tornadoes are rare.

Based on the 1951-1980 period, the average first occurrence of 32 degrees Fahrenheit in the fall is October 29 and the average last occurrence in the spring is April 14.

# NORMALS, MEANS, AND EXTREMES
## AMARILLO (KAMA)

LATITUDE: 35 ° 13'N   LONGITUDE: -101° 42'W   ELEVATION (FT): GRND: 3592   BARO: 3589   TIME ZONE: CENTRAL (UTC -6)   WBAN: 23047

| | ELEMENT | POR | JAN | FEB | MAR | APR | MAY | JUN | JUL | AUG | SEP | OCT | NOV | DEC | YEAR |
|---|---|---|---|---|---|---|---|---|---|---|---|---|---|---|---|
| **TEMPERATURE °F** | NORMAL DAILY MAXIMUM | 30 | 48.9 | 54.1 | 62.2 | 70.6 | 78.6 | 87.4 | 91.0 | 88.7 | 81.8 | 71.8 | 58.4 | 49.8 | 70.3 |
| | MEAN DAILY MAXIMUM | 62 | 49.6 | 53.8 | 61.5 | 71.0 | 79.3 | 87.7 | 91.3 | 89.4 | 82.2 | 72.2 | 59.4 | 50.9 | 70.7 |
| | HIGHEST DAILY MAXIMUM | 68 | 81 | 88 | 94 | 98 | 103 | 108 | 106 | 106 | 102 | 99 | 87 | 81 | 108 |
| | YEAR OF OCCURRENCE | | 1950 | 1963 | 1971 | 1989 | 1996 | 1998 | 2009 | 1944 | 2000 | 2000 | 1980 | 1955 | JUN 1998 |
| | MEAN OF EXTREME MAXS. | 62 | 71.6 | 76.2 | 82.5 | 88.8 | 94.3 | 99.7 | 100.0 | 98.5 | 94.8 | 89.1 | 79.0 | 72.2 | 87.2 |
| | NORMAL DAILY MINIMUM | 30 | 22.6 | 27.0 | 33.6 | 41.7 | 51.7 | 61.1 | 65.3 | 63.8 | 56.3 | 44.6 | 31.8 | 24.1 | 43.6 |
| | MEAN DAILY MINIMUM | 62 | 22.4 | 26.3 | 32.5 | 41.8 | 51.9 | 61.0 | 65.7 | 64.3 | 56.5 | 45.0 | 32.3 | 24.4 | 43.7 |
| | LOWEST DAILY MINIMUM | 68 | -11 | -14 | -3 | 14 | 28 | 41 | 51 | 49 | 30 | 12 | 0 | -8 | -14 |
| | YEAR OF OCCURRENCE | | 1984 | 1951 | 1948 | 1945 | 1954 | 1998 | 1990 | 1956 | 1984 | 1993 | 1976 | 1989 | FEB 1951 |
| | MEAN OF EXTREME MINS. | 62 | 4.6 | 9.0 | 15.3 | 26.9 | 37.8 | 50.0 | 57.9 | 56.7 | 42.6 | 30.0 | 15.9 | 7.6 | 29.5 |
| | NORMAL DRY BULB | 30 | 35.8 | 40.6 | 47.9 | 56.2 | 65.2 | 74.3 | 78.2 | 76.3 | 69.1 | 58.2 | 45.1 | 37.0 | 57.0 |
| | MEAN DRY BULB | 62 | 36.0 | 40.1 | 47.0 | 56.4 | 65.6 | 74.6 | 78.5 | 76.8 | 69.4 | 58.6 | 45.9 | 37.6 | 57.2 |
| | MEAN WET BULB | 26 | 28.1 | 31.1 | 36.9 | 43.5 | 53.6 | 61.1 | 64.1 | 64.2 | 57.7 | 47.6 | 36.0 | 29.0 | 46.1 |
| | MEAN DEW POINT | 26 | 22.5 | 25.2 | 30.0 | 36.3 | 47.9 | 56.7 | 59.1 | 59.8 | 52.8 | 42.0 | 30.1 | 23.1 | 40.5 |
| | NORMAL NO. DAYS WITH: | | | | | | | | | | | | | | |
| | MAXIMUM >= 90 | 30 | 0.0 | 0.0 | 0.1 | 0.7 | 4.0 | 12.8 | 19.9 | 16.5 | 7.0 | 0.8 | 0.0 | 0.0 | 61.8 |
| | MAXIMUM <= 32 | 30 | 4.6 | 2.7 | 0.7 | 0.1 | 0.0 | 0.0 | 0.0 | 0.0 | 0.0 | 0.1 | 0.9 | 3.6 | 12.7 |
| | MINIMUM <= 32 | 30 | 27.5 | 21.4 | 14.4 | 4.3 | 0.1 | 0.0 | 0.0 | 0.0 | 0.2 | 2.3 | 15.9 | 27.0 | 113.1 |
| | MINIMUM <= 0 | 30 | 0.5 | 0.5 | 0.0 | 0.0 | 0.0 | 0.0 | 0.0 | 0.0 | 0.0 | 0.0 | * | 0.7 | 1.7 |
| **H/C** | NORMAL HEATING DEG. DAYS | 30 | 920 | 699 | 542 | 291 | 94 | 7 | 1 | 1 | 56 | 239 | 594 | 874 | 4318 |
| | NORMAL COOLING DEG. DAYS | 30 | 0 | 0 | 2 | 18 | 90 | 285 | 405 | 345 | 173 | 26 | 0 | 0 | 1344 |
| **RH** | NORMAL (PERCENT) | 30 | 60 | 59 | 54 | 51 | 58 | 59 | 55 | 60 | 61 | 59 | 59 | 62 | 58 |
| | HOUR 00 LST | 30 | 67 | 65 | 61 | 59 | 68 | 67 | 63 | 69 | 69 | 67 | 68 | 68 | 66 |
| | HOUR 06 LST | 30 | 72 | 72 | 70 | 71 | 78 | 78 | 76 | 80 | 79 | 75 | 74 | 72 | 75 |
| | HOUR 12 LST | 30 | 51 | 49 | 43 | 39 | 45 | 45 | 42 | 47 | 48 | 45 | 47 | 50 | 46 |
| | HOUR 18 LST | 30 | 49 | 43 | 37 | 33 | 40 | 40 | 38 | 44 | 44 | 44 | 50 | 52 | 43 |
| **S** | PERCENT POSSIBLE SUNSHINE | 60 | 69 | 69 | 74 | 76 | 72 | 78 | 79 | 77 | 74 | 76 | 72 | 68 | 74 |
| **W/O** | MEAN NO. DAYS WITH: | | | | | | | | | | | | | | |
| | HEAVY FOG(VISBY <= 1/4 MI) | 46 | 3.3 | 3.8 | 3.4 | 2.2 | 1.8 | 0.5 | 0.5 | 1.0 | 1.7 | 2.6 | 2.8 | 3.0 | 26.6 |
| | THUNDERSTORMS | 62 | 0.2 | 0.5 | 1.6 | 3.4 | 7.9 | 9.8 | 9.0 | 9.1 | 4.0 | 2.5 | 0.6 | 0.2 | 48.8 |
| **CLOUDNESS** | MEAN: | | | | | | | | | | | | | | |
| | SUNRISE-SUNSET (OKTAS) | 1 | 3.2 | 4.4 | 4.0 | 3.7 | 4.0 | 2.0 | 2.0 | 2.9 | 4.0 | 3.6 | 3.2 | 2.4 | 3.3 |
| | MIDNIGHT-MIDNIGHT (OKTAS) | 1 | 3.2 | 4.4 | 4.0 | 3.6 | 4.0 | 1.6 | 2.0 | 3.2 | 1.6 | 3.2 | 2.8 | 2.4 | 3.0 |
| | MEAN NO. DAYS WITH: | | | | | | | | | | | | | | |
| | CLEAR | 3 | 7.7 | 9.7 | 11.0 | 12.0 | 11.0 | 17.8 | 17.0 | 15.5 | 9.5 | 12.0 | 10.0 | 13.0 | 146.2 |
| | PARTLY CLOUDY | 3 | 5.0 | 3.7 | 3.3 | 4.3 | 5.3 | 4.7 | 3.0 | 2.5 | 1.0 | 2.5 | 3.0 | 5.0 | 43.3 |
| | CLOUDY | 3 | 4.7 | 6.0 | 3.3 | 4.7 | 5.7 | 1.5 | 2.0 | 3.5 | 1.5 | 4.0 | 3.5 | 3.5 | 43.9 |
| **PR** | MEAN STATION PRESSURE(IN) | 26 | 26.36 | 26.33 | 26.29 | 26.26 | 26.27 | 26.30 | 26.36 | 26.38 | 26.37 | 26.36 | 26.35 | 26.36 | 26.33 |
| | MEAN SEA-LEVEL PRES. (IN) | 26 | 30.09 | 30.04 | 29.95 | 29.88 | 29.85 | 29.85 | 29.91 | 29.94 | 29.96 | 29.99 | 30.03 | 30.09 | 29.97 |
| **WINDS** | MEAN SPEED (MPH) | 26 | 12.1 | 12.8 | 13.9 | 14.5 | 13.7 | 13.6 | 12.4 | 11.4 | 12.1 | 12.4 | 12.5 | 12.1 | 12.8 |
| | PREVAIL.DIR(TENS OF DEGS) | 31 | 24 | 36 | 22 | 19 | 19 | 19 | 19 | 20 | 19 | 21 | 23 | 36 | 19 |
| | MAXIMUM 2-MINUTE: | 17 | | | | | | | | | | | | | |
| | SPEED (MPH) | | 46 | 51 | 48 | 53 | 58 | 68 | 54 | 46 | 41 | 44 | 45 | 54 | 68 |
| | DIR. (TENS OF DEGS) | | 29 | 33 | 36 | 25 | 23 | 02 | 01 | 35 | 20 | 33 | 32 | 25 | 02 |
| | YEAR OF OCCURRENCE | | 2008 | 2007 | 2008 | 2001 | 2008 | 2008 | 2009 | 2006 | 2003 | 2006 | 2006 | 2009 | JUN 2008 |
| | MAXIMUM 3-SECOND: | | | | | | | | | | | | | | |
| | SPEED (MPH) | 17 | 59 | 64 | 63 | 74 | 69 | 81 | 70 | 59 | 52 | 61 | 59 | 75 | 81 |
| | DIR. (TENS OF DEGS) | | 31 | 32 | 28 | 24 | 23 | 01 | 19 | 21 | 23 | 23 | 32 | 03 | 01 |
| | YEAR OF OCCURRENCE | | 2008 | 2007 | 2009 | 2001 | 2008 | 2008 | 2008 | 2000 | 2006 | 1996 | 2006 | 2006 | JUN 2008 |
| **PRECIPITATION** | NORMAL (IN) | 30 | 0.63 | 0.55 | 1.13 | 1.33 | 2.50 | 3.28 | 2.68 | 2.94 | 1.88 | 1.50 | 0.68 | 0.61 | 19.71 |
| | MAXIMUM MONTHLY (IN) | 68 | 2.67 | 2.08 | 4.14 | 6.45 | 9.81 | 10.73 | 7.59 | 8.07 | 5.02 | 7.64 | 4.06 | 4.52 | 10.73 |
| | YEAR OF OCCURRENCE | | 1999 | 1998 | 2000 | 1997 | 1951 | 1965 | 1960 | 2009 | 1950 | 1941 | 2004 | 1959 | JUN 1965 |
| | MINIMUM MONTHLY (IN) | 68 | 0.00 | T | T | T | 0.04 | 0.01 | 0.04 | 0.28 | 0.03 | 0.00 | 0.00 | T | 0.00 |
| | YEAR OF OCCURRENCE | | 1986 | 1991 | 1950 | 1964 | 1984 | 1953 | 2001 | 1983 | 1977 | 1952 | 1989 | 1976 | NOV 1989 |
| | MAXIMUM IN 24 HOURS (IN) | 68 | 3.73 | 1.28 | 2.27 | 2.65 | 6.75 | 6.15 | 4.74 | 4.26 | 3.42 | 3.45 | 1.53 | 3.11 | 6.75 |
| | YEAR OF OCCURRENCE | | 2007 | 1971 | 1973 | 1999 | 1951 | 1960 | 1982 | 1945 | 1941 | 1948 | 1980 | 1943 | MAY 1951 |
| | NORMAL NO. DAYS WITH: | | | | | | | | | | | | | | |
| | PRECIPITATION >= 0.01 | 30 | 4.4 | 4.4 | 5.4 | 5.4 | 8.3 | 8.3 | 7.8 | 8.4 | 6.4 | 5.0 | 4.1 | 4.2 | 72.1 |
| | PRECIPITATION >= 1.00 | 30 | * | 0.1 | 0.2 | 0.2 | 0.6 | 0.8 | 0.7 | 0.7 | 0.5 | 0.2 | * | * | 4.0 |
| **SNOWFALL** | NORMAL (IN) | 30 | 4.8 | 3.8 | 1.9 | 0.8 | 0.* | 0.0 | 0.0 | 0.0 | 0.* | 0.4 | 2.4 | 3.7 | 17.8 |
| | MAXIMUM MONTHLY (IN) | 68 | 14.5 | 17.3 | 14.7 | 6.5 | 4.7 | T | T | T | 0.3 | 3.9 | 13.8 | 21.2 | 21.2 |
| | YEAR OF OCCURRENCE | | 1983 | 1971 | 1961 | 1997 | 1978 | 2009 | 2009 | 2009 | 1984 | 1976 | 2004 | 2000 | DEC 2000 |
| | MAXIMUM IN 24 HOURS (IN) | 68 | 10.2 | 13.5 | 11.2 | 6.5 | 4.7 | T | T | T | 0.3 | 3.2 | 12.2 | 16.8 | 16.8 |
| | YEAR OF OCCURRENCE' | | 1994 | 1971 | 2005 | 1997 | 2005 | 2009 | 2009 | 2009 | 1984 | 1976 | 1952 | 2000 | DEC 2000 |
| | MAXIMUM SNOW DEPTH (IN) | 64 | 10 | 14 | 12 | 6 | T | T | 0 | 0 | T | 1 | 9 | 15 | 15 |
| | YEAR OF OCCURRENCE | | 1987 | 1983 | 2005 | 1973 | 1988 | 1949 | | | 1984 | 1970 | 1952 | 2000 | DEC 2000 |
| | NORMAL NO. DAYS WITH: | | | | | | | | | | | | | | |
| | SNOWFALL >= 1.0 | 30 | 1.7 | 1.2 | 0.6 | 0.3 | 0.0 | 0.0 | 0.0 | 0.0 | 0.0 | 0.1 | 0.8 | 1.0 | 5.7 |

## PRECIPITATION (inches) 2009  AMARILLO (KAMA)

| YEAR | JAN | FEB | MAR | APR | MAY | JUN | JUL | AUG | SEP | OCT | NOV | DEC | ANNUAL |
|---|---|---|---|---|---|---|---|---|---|---|---|---|---|
| 1980 | 0.85 | 0.55 | 1.38 | 0.82 | 2.88 | 1.30 | 0.65 | 1.80 | 1.55 | 0.42 | 0.84 | 0.35 | 13.39 |
| 1981 | 0.11 | 0.23 | 1.87 | 0.90 | 2.11 | 1.04 | 2.73 | 5.22 | 3.47 | 1.79 | 1.50 | 0.03 | 21.00 |
| 1982 | 0.15 | 0.39 | 0.52 | 0.43 | 1.96 | 4.75 | 6.23 | 0.55 | 1.37 | 0.71 | 0.75 | 0.79 | 18.60 |
| 1983 | 1.78 | 1.19 | 0.98 | 0.83 | 2.85 | 1.76 | 0.74 | 0.28 | 0.37 | 3.23 | 0.33 | 0.64 | 14.98 |
| 1984 | 0.56 | 0.37 | 0.98 | 1.18 | 0.04 | 6.76 | 0.83 | 2.28 | 0.95 | 3.19 | 1.09 | 1.00 | 19.23 |
| 1985 | 0.99 | 0.77 | 1.49 | 2.79 | 0.86 | 3.08 | 2.07 | 1.67 | 4.96 | 3.07 | 0.39 | 0.26 | 22.40 |
| 1986 | 0.00 | 1.02 | 0.60 | 0.30 | 3.28 | 3.70 | 3.52 | 7.04 | 1.45 | 1.94 | 1.82 | 0.66 | 25.33 |
| 1987 | 1.26 | 0.84 | 0.92 | 0.57 | 4.28 | 3.29 | 0.83 | 3.28 | 3.40 | 1.17 | 0.43 | 1.75 | 22.02 |
| 1988 | 0.33 | 0.04 | 1.19 | 2.22 | 6.02 | 3.68 | 3.30 | 3.59 | 3.15 | 0.71 | 0.29 | 0.17 | 24.69 |
| 1989 | 0.16 | 0.55 | 0.52 | 0.75 | 2.51 | 6.07 | 2.74 | 3.22 | 1.80 | 0.74 | 0.00 | 0.49 | 19.55 |
| 1990 | 1.22 | 1.61 | 2.56 | 1.10 | 0.90 | 0.14 | 3.28 | 2.79 | 2.72 | 0.46 | 0.50 | 0.23 | 17.51 |
| 1991 | 0.86 | T | 0.41 | 0.04 | 3.08 | 2.47 | 2.20 | 1.28 | 2.04 | 0.64 | 0.66 | 2.24 | 15.92 |
| 1992 | 0.50 | 0.30 | 1.11 | 1.60 | 3.10 | 7.57 | 2.36 | 2.27 | 0.16 | 0.31 | 0.80 | 0.55 | 20.63 |
| 1993 | 0.76 | 0.36 | 1.29 | 0.35 | 1.92 | 2.76 | 3.36 | 4.64 | 1.00 | 0.53 | 0.51 | 0.95 | 18.43 |
| 1994 | 1.01 | 0.07 | 1.22 | 1.15 | 1.41 | 1.26 | 5.01 | 2.86 | 2.02 | 0.46 | 0.58 | 0.30 | 17.35 |
| 1995 | 0.32 | T | 0.80 | 0.83 | 4.94 | 2.71 | 2.85 | 2.18 | 2.62 | 0.45 | 0.06 | 0.59 | 18.35 |
| 1996 | 0.07 | 0.26 | 0.24 | T | 1.67 | 2.92 | 4.95 | 6.43 | 1.86 | 0.84 | 1.30 | 0.05 | 20.59 |
| 1997 | 0.64 | 0.47 | 0.01 | 6.45 | 2.16 | 2.93 | 5.51 | 1.40 | 1.35 | 0.74 | 1.17 | 2.12 | 24.95 |
| 1998 | 0.68 | 2.08 | 2.46 | 0.97 | 0.53 | 0.12 | 1.09 | 0.76 | 1.23 | 6.48 | 0.34 | 0.41 | 17.15 |
| 1999 | 2.67 | T | 1.35 | 6.30 | 4.29 | 3.61 | 2.87 | 2.04 | 2.54 | 0.38 | 0.00 | 0.52 | 26.57 |
| 2000 | 0.24 | 0.04 | 4.14 | 0.43 | 1.14 | 5.54 | 0.16 | 0.29 | 0.03 | 3.95 | 0.96 | 1.47 | 18.39 |
| 2001 | 1.67 | 0.93 | 3.96 | 0.49 | 3.05 | 1.99 | 0.04 | 1.39 | 3.03 | 0.05 | 1.86 | 0.23 | 18.69 |
| 2002 | 1.17 | 0.19 | 0.45 | 2.24 | 1.05 | 1.54 | 1.66 | 3.83 | 1.64 | 3.34 | 0.04 | 1.10 | 18.25 |
| 2003 | T | 0.20 | 0.88 | 0.28 | 1.46 | 6.42 | 0.09 | 0.80 | 1.81 | 0.95 | 0.44 | 0.09 | 13.42 |
| 2004 | 0.70 | 1.42 | 1.50 | 2.65 | 0.09 | 5.38 | 3.01 | 0.85 | 3.80 | 3.37 | 4.06 | 0.48 | 27.31 |
| 2005 | 1.43 | 0.67 | 1.92 | 0.60 | 2.65 | 1.87 | 1.15 | 3.93 | 0.14 | 0.42 | 0.19 | 0.03 | 15.00 |
| 2006 | 0.03 | 0.05 | 1.56 | 0.23 | 1.26 | 1.02 | 4.40 | 6.67 | 1.10 | 2.71 | 0.37 | 2.48 | 21.88 |
| 2007 | 0.95 | 0.29 | 4.00 | 0.65 | 5.40 | 2.71 | 1.83 | 0.88 | 3.55 | 0.95 | 0.08 | 1.21 | 22.50 |
| 2008 | 0.24 | 0.59 | 0.30 | 0.38 | 2.08 | 4.03 | 4.96 | 4.43 | 1.30 | 3.87 | 0.19 | 0.05 | 22.42 |
| 2009 | 0.03 | 0.45 | 1.01 | 1.84 | 0.43 | 2.79 | 3.78 | 8.07 | 0.83 | 1.41 | 0.26 | 0.25 | 21.15 |
| POR= 62 YRS | 0.60 | 0.57 | 1.09 | 1.20 | 2.60 | 3.39 | 2.76 | 3.01 | 1.86 | 1.52 | 0.65 | 0.60 | 19.85 |

WBAN : 23047

## AVERAGE TEMPERATURE (°F) 2009  AMARILLO (KAMA)

| YEAR | JAN | FEB | MAR | APR | MAY | JUN | JUL | AUG | SEP | OCT | NOV | DEC | ANNUAL |
|---|---|---|---|---|---|---|---|---|---|---|---|---|---|
| 1980 | 34.9 | 37.7 | 43.9 | 52.4 | 61.9 | 78.3 | 82.9 | 78.5 | 70.7 | 56.6 | 42.7 | 41.4 | 56.8 |
| 1981 | 37.9 | 42.2 | 48.8 | 63.1 | 65.6 | 78.5 | 81.3 | 74.4 | 69.0 | 56.6 | 49.1 | 40.1 | 58.9 |
| 1982 | 37.1 | 35.9 | 47.3 | 53.8 | 63.3 | 72.2 | 78.8 | 78.7 | 71.1 | 57.3 | 45.6 | 36.1 | 56.4 |
| 1983 | 33.4 | 36.2 | 45.3 | 50.9 | 60.3 | 70.4 | 80.0 | 81.0 | 73.6 | 60.5 | 47.9 | 24.7 | 55.4 |
| 1984 | 31.6 | 40.3 | 44.0 | 51.8 | 66.9 | 74.6 | 75.6 | 75.3 | 65.7 | 56.9 | 47.0 | 40.5 | 55.9 |
| 1985 | 31.5 | 37.2 | 49.5 | 60.0 | 68.1 | 75.1 | 80.1 | 79.6 | 68.9 | 56.9 | 43.5 | 34.0 | 57.0 |
| 1986 | 42.5 | 40.4 | 52.7 | 59.4 | 65.1 | 73.3 | 80.6 | 74.4 | 68.8 | 55.7 | 42.7 | 37.1 | 57.7 |
| 1987 | 34.0 | 42.0 | 44.6 | 54.8 | 64.5 | 72.1 | 77.2 | 75.3 | 67.6 | 57.9 | 45.4 | 34.6 | 55.8 |
| 1988 | 32.2 | 38.3 | 44.3 | 54.1 | 63.6 | 73.3 | 75.8 | 76.4 | 67.7 | 59.0 | 47.6 | 38.5 | 55.9 |
| 1989 | 40.7 | 32.3 | 51.2 | 59.9 | 67.3 | 69.4 | 76.2 | 76.0 | 66.3 | 60.4 | 47.9 | 31.4 | 56.6 |
| 1990 | 39.2 | 40.8 | 47.0 | 55.9 | 63.6 | 81.3 | 76.4 | 76.3 | 72.0 | 57.7 | 49.7 | 33.2 | 57.8 |
| 1991 | 32.6 | 45.9 | 49.7 | 58.2 | 68.5 | 74.7 | 76.5 | 76.2 | 67.1 | 58.5 | 41.2 | 39.0 | 57.3 |
| 1992 | 38.0 | 45.0 | 50.6 | 58.7 | 63.3 | 70.8 | 76.6 | 73.5 | 70.2 | 60.1 | 39.3 | 34.2 | 56.7 |
| 1993 | 33.1 | 36.0 | 46.1 | 54.4 | 64.5 | 74.0 | 79.1 | 74.9 | 67.9 | 54.6 | 40.9 | 38.7 | 55.4 |
| 1994 | 36.5 | 37.6 | 48.9 | 55.0 | 65.2 | 79.4 | 77.4 | 75.9 | 69.1 | 58.2 | 46.5 | 41.7 | 57.6 |
| 1995 | 39.2 | 43.9 | 47.4 | 53.9 | 61.6 | 70.8 | 77.2 | 78.2 | 67.9 | 58.2 | 48.2 | 38.4 | 57.1 |
| 1996 | 34.9 | 43.1 | 44.2 | 56.9 | 72.5 | 75.6 | 77.1 | 74.4 | 65.9 | 57.6 | 45.7 | 39.7 | 57.3 |
| 1997 | 35.3 | 38.8 | 50.3 | 48.9 | 63.2 | 71.6 | 77.9 | 75.0 | 71.9 | 58.2 | 42.8 | 33.5 | 55.6 |
| 1998 | 39.8 | 41.0 | 43.5 | 53.5 | 69.2 | 77.4 | 82.4 | 78.0 | 75.8 | 60.4 | 49.2 | 38.1 | 59.0 |
| 1999 | 40.1 | 47.3 | 46.3 | 54.8 | 62.3 | 71.8 | 77.5 | 78.0 | 66.2 | 58.1 | 52.3 | 37.6 | 57.7 |
| 2000 | 39.5 | 47.1 | 49.2 | 57.8 | 69.7 | 72.1 | 80.1 | 81.8 | 73.7 | 59.6 | 39.6 | 33.1 | 58.6 |
| 2001 | 34.8 | 41.2 | 45.0 | 61.0 | 65.2 | 76.5 | 83.9 | 78.7 | 70.4 | 60.2 | 50.7 | 40.4 | 59.0 |
| 2002 | 39.5 | 39.7 | 45.6 | 59.8 | 66.4 | 77.7 | 79.0 | 78.1 | 69.2 | 52.9 | 44.1 | 35.6 | 57.3 |
| 2003 | 39.6 | 37.6 | 48.0 | 58.6 | 66.9 | 69.8 | 80.1 | 79.3 | 67.7 | 61.9 | 47.5 | 41.0 | 58.2 |
| 2004 | 39.3 | 37.6 | 52.4 | 55.2 | 69.4 | 73.4 | 75.9 | 73.5 | 70.0 | 58.8 | 43.1 | 39.9 | 57.4 |
| 2005 | 39.7 | 43.2 | 45.6 | 55.0 | 64.0 | 74.8 | 77.9 | 75.7 | 72.6 | 57.5 | 48.3 | 36.5 | 57.6 |
| 2006 | 42.7 | 39.3 | 48.2 | 60.7 | 69.3 | 76.6 | 78.3 | 74.6 | 64.3 | 57.9 | 48.8 | 38.1 | 58.2 |
| 2007 | 31.5 | 39.5 | 53.4 | 52.9 | 64.4 | 72.3 | 77.1 | 79.7 | 72.4 | 62.0 | 48.3 | 37.9 | 57.6 |
| 2008 | 37.9 | 41.9 | 48.2 | 56.7 | 66.3 | 77.2 | 77.5 | 74.6 | 66.9 | 58.2 | 48.2 | 39.3 | 57.7 |
| 2009 | 38.6 | 45.7 | 51.2 | 56.0 | 63.9 | 75.5 | 78.2 | 76.0 | 67.1 | 53.4 | 50.4 | 32.6 | 57.4 |
| POR= 62 YRS | 36.0 | 40.1 | 47.0 | 56.4 | 65.6 | 74.6 | 78.5 | 76.8 | 69.4 | 58.6 | 45.9 | 37.6 | 57.2 |

## HEATING DEGREE DAYS (base 65°F) 2009  AMARILLO (KAMA)

| YEAR | JUL | AUG | SEP | OCT | NOV | DEC | JAN | FEB | MAR | APR | MAY | JUN | TOTAL |
|------|-----|-----|-----|-----|-----|-----|-----|-----|-----|-----|-----|-----|-------|
| 1980-81 | 0 | 0 | 35 | 280 | 662 | 723 | 832 | 630 | 496 | 111 | 65 | 0 | 3834 |
| 1981-82 | 0 | 0 | 26 | 271 | 469 | 765 | 849 | 803 | 534 | 340 | 96 | 2 | 4155 |
| 1982-83 | 0 | 0 | 23 | 252 | 575 | 888 | 972 | 800 | 603 | 421 | 171 | 32 | 4737 |
| 1983-84 | 0 | 0 | 40 | 175 | 506 | 1241 | 1028 | 709 | 642 | 390 | 56 | 0 | 4787 |
| 1984-85 | 0 | 0 | 125 | 262 | 531 | 752 | 1034 | 769 | 474 | 169 | 37 | 5 | 4158 |
| 1985-86 | 0 | 0 | 111 | 249 | 640 | 957 | 691 | 681 | 379 | 203 | 72 | 0 | 3983 |
| 1986-87 | 0 | 2 | 26 | 290 | 665 | 858 | 954 | 634 | 624 | 315 | 70 | 9 | 4447 |
| 1987-88 | 0 | 8 | 18 | 226 | 584 | 936 | 1010 | 765 | 633 | 323 | 102 | 4 | 4609 |
| 1988-89 | 0 | 7 | 32 | 197 | 517 | 815 | 747 | 909 | 429 | 219 | 59 | 27 | 3958 |
| 1989-90 | 0 | 0 | 91 | 185 | 507 | 1037 | 795 | 672 | 551 | 276 | 140 | 0 | 4254 |
| 1990-91 | 0 | 0 | 11 | 234 | 454 | 981 | 994 | 529 | 467 | 208 | 53 | 1 | 3932 |
| 1991-92 | 0 | 0 | 71 | 258 | 709 | 798 | 833 | 576 | 437 | 203 | 98 | 10 | 3993 |
| 1992-93 | 0 | 0 | 21 | 171 | 765 | 949 | 984 | 808 | 577 | 317 | 82 | 5 | 4679 |
| 1993-94 | 0 | 2 | 39 | 332 | 717 | 810 | 875 | 760 | 499 | 309 | 78 | 0 | 4421 |
| 1994-95 | 0 | 0 | 31 | 228 | 553 | 716 | 792 | 583 | 547 | 333 | 146 | 13 | 3942 |
| 1995-96 | 0 | 1 | 89 | 217 | 497 | 813 | 926 | 634 | 636 | 250 | 21 | 0 | 4084 |
| 1996-97 | 3 | 0 | 70 | 251 | 571 | 775 | 917 | 728 | 447 | 476 | 99 | 2 | 4339 |
| 1997-98 | 0 | 4 | 35 | 259 | 661 | 967 | 774 | 667 | 664 | 343 | 39 | 21 | 4434 |
| 1998-99 | 0 | 0 | 13 | 180 | 469 | 828 | 764 | 490 | 573 | 305 | 132 | 9 | 3763 |
| 1999-00 | 0 | 0 | 93 | 226 | 373 | 844 | 785 | 510 | 482 | 220 | 60 | 9 | 3602 |
| 2000-01 | 0 | 0 | 50 | 195 | 755 | 981 | 929 | 660 | 611 | 139 | 75 | 0 | 4395 |
| 2001-02 | 0 | 0 | 12 | 180 | 418 | 755 | 782 | 702 | 597 | 205 | 79 | 1 | 3731 |
| 2002-03 | 0 | 0 | 21 | 377 | 623 | 906 | 784 | 761 | 519 | 205 | 51 | 13 | 4260 |
| 2003-04 | 0 | 6 | 32 | 148 | 524 | 737 | 789 | 787 | 395 | 295 | 54 | 0 | 3767 |
| 2004-05 | 4 | 0 | 18 | 190 | 654 | 773 | 778 | 602 | 595 | 298 | 128 | 0 | 4040 |
| 2005-06 | 0 | 2 | 8 | 263 | 495 | 876 | 682 | 711 | 514 | 178 | 60 | 0 | 3789 |
| 2006-07 | 0 | 0 | 62 | 254 | 481 | 829 | 1030 | 711 | 356 | 356 | 66 | 5 | 4150 |
| 2007-08 | 0 | 0 | 2 | 163 | 496 | 836 | 835 | 665 | 517 | 268 | 82 | 0 | 3864 |
| 2008-09 | 0 | 0 | 21 | 219 | 497 | 791 | 812 | 536 | 420 | 286 | 113 | 0 | 3695 |
| 2009- | 0 | 0 | 52 | 370 | 432 | 996 | | | | | | | |

WBAN : 23047

## COOLING DEGREE DAYS (base 65°F) 2009  AMARILLO (KAMA)

| YEAR | JAN | FEB | MAR | APR | MAY | JUN | JUL | AUG | SEP | OCT | NOV | DEC | TOTAL |
|------|-----|-----|-----|-----|-----|-----|-----|-----|-----|-----|-----|-----|-------|
| 1980 | 0 | 0 | 0 | 4 | 58 | 408 | 562 | 429 | 211 | 26 | 0 | 0 | 1698 |
| 1981 | 0 | 0 | 1 | 59 | 87 | 410 | 512 | 299 | 154 | 16 | 0 | 0 | 1538 |
| 1982 | 0 | 0 | 0 | 9 | 52 | 225 | 437 | 432 | 213 | 22 | 0 | 0 | 1390 |
| 1983 | 0 | 0 | 0 | 2 | 31 | 201 | 473 | 502 | 306 | 41 | 3 | 0 | 1559 |
| 1984 | 0 | 0 | 0 | 1 | 121 | 298 | 331 | 325 | 151 | 17 | 0 | 0 | 1244 |
| 1985 | 0 | 0 | 2 | 27 | 143 | 315 | 473 | 458 | 235 | 6 | 0 | 0 | 1659 |
| 1986 | 0 | 0 | 3 | 45 | 87 | 256 | 489 | 299 | 144 | 10 | 0 | 0 | 1333 |
| 1987 | 0 | 0 | 0 | 19 | 64 | 227 | 386 | 334 | 105 | 11 | 0 | 0 | 1146 |
| 1988 | 0 | 0 | 0 | 3 | 67 | 263 | 340 | 366 | 120 | 19 | 0 | 0 | 1178 |
| 1989 | 0 | 0 | 10 | 75 | 136 | 164 | 354 | 347 | 138 | 51 | 0 | 0 | 1275 |
| 1990 | 0 | 0 | 0 | 9 | 101 | 494 | 359 | 358 | 229 | 17 | 1 | 0 | 1568 |
| 1991 | 0 | 0 | 0 | 12 | 166 | 298 | 364 | 358 | 144 | 66 | 0 | 0 | 1408 |
| 1992 | 0 | 0 | 0 | 19 | 52 | 188 | 366 | 271 | 186 | 25 | 0 | 0 | 1107 |
| 1993 | 0 | 0 | 0 | 6 | 77 | 283 | 443 | 315 | 134 | 15 | 0 | 0 | 1273 |
| 1994 | 0 | 0 | 6 | 18 | 91 | 439 | 392 | 345 | 159 | 24 | 4 | 0 | 1478 |
| 1995 | 0 | 0 | 9 | 7 | 47 | 194 | 388 | 419 | 182 | 13 | 0 | 0 | 1259 |
| 1996 | 0 | 5 | 0 | 14 | 260 | 325 | 386 | 297 | 103 | 27 | 0 | 0 | 1417 |
| 1997 | 0 | 0 | 0 | 0 | 49 | 205 | 407 | 320 | 247 | 55 | 0 | 0 | 1283 |
| 1998 | 0 | 0 | 1 | 6 | 177 | 399 | 547 | 407 | 342 | 43 | 0 | 0 | 1922 |
| 1999 | 0 | 0 | 0 | 5 | 56 | 221 | 396 | 410 | 135 | 16 | 0 | 0 | 1239 |
| 2000 | 0 | 0 | 0 | 14 | 214 | 225 | 475 | 529 | 320 | 39 | 0 | 0 | 1816 |
| 2001 | 0 | 0 | 0 | 27 | 88 | 353 | 594 | 433 | 180 | 38 | 0 | 0 | 1713 |
| 2002 | 0 | 0 | 2 | 56 | 129 | 390 | 442 | 412 | 156 | 10 | 0 | 0 | 1597 |
| 2003 | 0 | 0 | 0 | 21 | 115 | 165 | 473 | 456 | 120 | 60 | 6 | 0 | 1416 |
| 2004 | 0 | 0 | 12 | 6 | 194 | 257 | 345 | 273 | 175 | 6 | 0 | 0 | 1268 |
| 2005 | 0 | 0 | 0 | 4 | 104 | 297 | 404 | 342 | 244 | 34 | 3 | 0 | 1432 |
| 2006 | 0 | 0 | 2 | 54 | 200 | 356 | 418 | 306 | 49 | 41 | 0 | 0 | 1426 |
| 2007 | 0 | 0 | 7 | 1 | 57 | 233 | 380 | 464 | 231 | 77 | 0 | 0 | 1450 |
| 2008 | 0 | 0 | 0 | 25 | 130 | 372 | 391 | 305 | 88 | 19 | 1 | 0 | 1331 |
| 2009 | 0 | 0 | 0 | 23 | 84 | 323 | 416 | 347 | 123 | 20 | 2 | 0 | 1338 |

## SNOWFALL (inches) 2009 AMARILLO (KAMA)

| YEAR | JUL | AUG | SEP | OCT | NOV | DEC | JAN | FEB | MAR | APR | MAY | JUN | TOTAL |
|------|-----|-----|-----|-----|-----|-----|-----|-----|-----|-----|-----|-----|-------|
| 1980-81 | 0.0 | 0.0 | 0.0 | 0.0 | 8.6 | T | 1.1 | 0.1 | 0.1 | 0.0 | 0.0 | 0.0 | 9.9 |
| 1981-82 | 0.0 | 0.0 | 0.0 | 0.0 | T | 0.3 | 1.1 | 4.7 | 1.7 | T | 0.0 | 0.0 | 7.8 |
| 1982-83 | 0.0 | 0.0 | 0.0 | 0.0 | 4.5 | 1.9 | 14.5 | 13.0 | 8.1 | 5.9 | 0.0 | 0.0 | 47.9 |
| 1983-84 | 0.0 | 0.0 | 0.0 | 0.0 | T | 5.2 | 7.0 | 3.5 | 2.5 | 0.0 | 0.0 | 0.0 | 18.2 |
| 1984-85 | 0.0 | 0.0 | 0.3 | 0.0 | 0.1 | 0.3 | 7.4 | 0.3 | 0.2 | 0.0 | 0.0 | 0.0 | 8.6 |
| 1985-86 | 0.0 | 0.0 | T | 0.0 | T | 2.8 | 0.0 | 10.9 | 1.0 | 0.0 | 0.0 | 0.0 | 14.7 |
| 1986-87 | 0.0 | 0.0 | 0.0 | T | 0.2 | 3.3 | 12.1 | 3.1 | 6.4 | 0.1 | 0.0 | 0.0 | 25.2 |
| 1987-88 | 0.0 | 0.0 | 0.0 | 0.0 | 0.7 | 15.3 | 4.3 | 0.5 | 8.5 | 4.2 | 0.0 | 0.0 | 33.5 |
| 1988-89 | 0.0 | 0.0 | 0.0 | 0.0 | 2.4 | 2.2 | T | 0.1 | 4.2 | 0.1 | T | T | 9.0 |
| 1989-90 | 0.0 | 0.0 | T | 0.0 | 0.0 | 5.4 | 8.5 | 3.0 | 0.0 | 0.0 | 0.0 | 0.0 | 16.9 |
| 1990-91 | 0.0 | 0.0 | 0.0 | 0.0 | 2.2 | 3.8 | 5.0 | T | 1.0 | 0.0 | T | T | 12.0 |
| 1991-92 | 0.0 | 0.0 | T | 3.2 | T | 2.0 | 1.0 | 0.2 | T | 0.3 | T | T | 6.7 |
| 1992-93 | 0.0 | 0.0 | 0.0 | T | 9.0 | 10.0 | 2.5 | 2.2 | 2.1 | T | T | 0.0 | 25.8 |
| 1993-94 | 0.0 | 0.0 | 0.0 | T | T | 0.5 | 10.2 | T | 3.0 | T | 0.0 | 0.0 | 13.7 |
| 1994-95 | T | 0.0 | 0.0 | 0.0 | T | 0.0 | 3.7 | T | 2.3 | 0.0 | 0.0 | 0.0 | 6.0 |
| 1995-96 | T | 0.0 | T | 0.0 | 0.5 | 2.7 | 1.1 | 4.6 | 0.1 | 0.0 | T | 0.0 | 9.0 |
| 1996-97 | T | 0.0 | 0.0 | 1.5 | 8.0 | 1.1 | 10.8 | 2.0 | 0.0 | 6.5 | 0.0 | T | 29.9 |
| 1997-98 | T | 0.0 | 0.0 | T | 5.4 | 10.9 | T | 10.4 | 4.4 | T | 0.0 | 0.0 | 31.1 |
| 1998-99 | 0.0 | 0.0 | 0.0 | 0.0 | 0.0 | 1.0 | 13.1 | T | 1.3 | T | 0.0 | T | 15.4 |
| 1999-00 | T | T | 0.0 | T | 0.0 | 8.6 | 2.8 | T | 0.5 | 0.5 | T | 0.0 | 12.4 |
| 2000-01 | 0.0 | 0.0 | 0.0 | T | 8.9 | 21.2 | 13.0 | 1.4 | 1.8 | T | T | 0.0 | 46.3 |
| 2001-02 | 0.0 | 0.0 | T | 0.0 | 2.3 | 2.0 | 3.3 | 1.1 | 0.3 | T | T | T | 9.0 |
| 2002-03 | 0.0 | 0.0 | 0.0 | T | 0.0 | 10.6 | T | 1.1 | 0.3 | 0.0 | 0.0 | 0.0 | 12.0 |
| 2003-04 | 0.0 | 0.0 | 0.0 | 0.0 | T | 0.5 | 4.7 | 6.0 | T | 0.7 | 0.0 | T | 11.9 |
| 2004-05 | 0.0 | 0.0 | 0.0 | 0.0 | 13.8 | 5.3 | 2.3 | 0.3 | 13.8 | 0.0 | 4.7 | 0.0 | 40.2 |
| 2005-06 | 0.0 | T | 0.0 | 0.0 | 1.3 | T | 0.3 | T | 0.9 | 0.0 | 0.0 | 0.0 | 2.5 |
| 2006-07 | 0.0 | 0.0 | 0.0 | 0.0 | 7.2 | 1.7 | 9.1 | 5.7 | T | 1.1 | T | T | 24.8 |
| 2007-08 | 0.0 | 0.0 | T | T | 0.8 | 4.7 | 2.7 | 0.3 | 3.5 | T | T | T | 12.0 |
| 2008-09 | T | 0.0 | 0.0 | 0.0 | 0.0 | 0.6 | 0.7 | 0.0 | 12.6 | T | 0.0 | T | 13.9 |
| 2009- | T | T | 0.0 | T | 2.1 | 3.0 | | | | | | | |
| POR= 62 YRS | T | T | T | 0.2 | 2.2 | 3.1 | 4.0 | 3.3 | 2.6 | 0.5 | 0.1 | T | 16.0 |

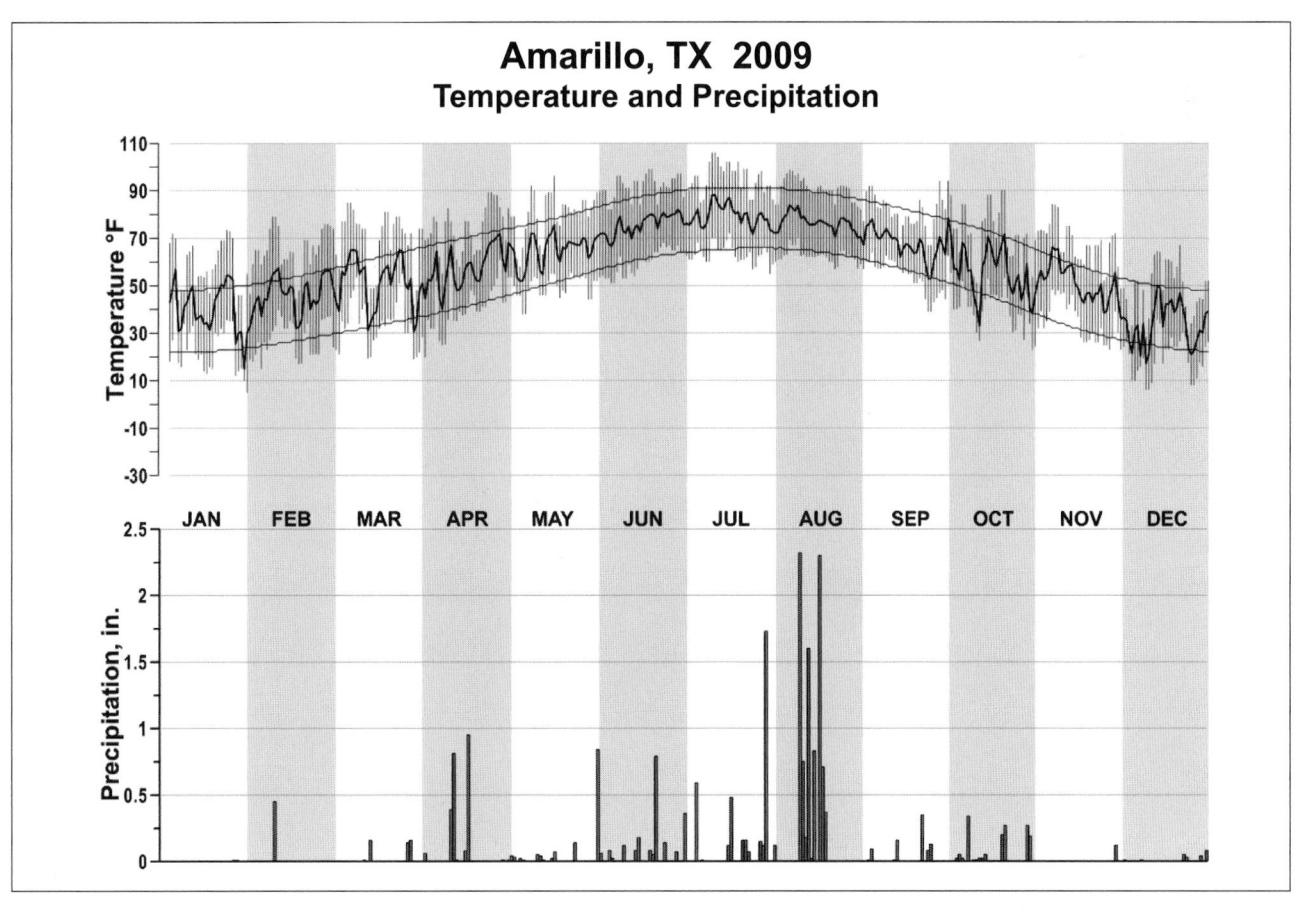

# Amarillo, TX  2009
## Temperature and Precipitation

# 2009
# DALLAS - FORT WORTH
# TEXAS (KDFW)

The Dallas-Fort Worth Metroplex is located in North Central Texas, approximately 250 miles north of the Gulf of Mexico. It is near the headwaters of the Trinity River, which lie in the upper margins of the Coastal Plain. The rolling hills in the area range from 500 to 800 feet in elevation.

The Dallas-Fort Worth climate is humid subtropical with hot summers. It is also continental, characterized by a wide annual temperature range. Precipitation also varies considerably, ranging from less than 20 to more than 50 inches.

Winters are mild, but northers occur about three times each month, and often are accompanied by sudden drops in temperature. Periods of extreme cold that occasionally occur are short-lived, so that even in January mild weather occurs frequently.

The highest temperatures of summer are associated with fair skies, westerly winds and low humidities. Characteristically, hot spells in summer are broken into three-to-five day periods by thunderstorm activity. There are only a few nights each summer when the low temperature exceeds 80 degrees. Summer daytime temperatures frequently exceed 100 degrees. Air conditioners are recommended for maximum comfort indoors and while traveling via automobile.

Throughout the year, rainfall occurs more frequently during the night. Usually, periods of rainy weather last for only a day or two, and are followed by several days with fair skies. A large part of the annual precipitation results from thunderstorm activity, with occasional heavy rainfall over brief periods of time. Thunderstorms occur throughout the year, but are most frequent in the spring. Hail falls on about two or three days a year, ordinarily with only slight and scattered damage. Windstorms occurring during thunderstorm activity are sometimes destructive. Snowfall is rare.

The average length of the warm season (freeze-free period) in the Dallas-Fort Worth Metroplex is about 249 days. The average last occurrence of 32 degrees or below is mid March and the average first occurrence of 32 degrees or below is in late November.

# NORMALS, MEANS, AND EXTREMES
## DALLAS-FORT WORTH (KDFW)

| LATITUDE: 32° 53'N | LONGITUDE: -97° 2 'W | ELEVATION (FT): GRND: 547   BARO: 562 | TIME ZONE: CENTRAL   (UTC -6) | WBAN: 03927 |
|---|---|---|---|---|

| | ELEMENT | POR | JAN | FEB | MAR | APR | MAY | JUN | JUL | AUG | SEP | OCT | NOV | DEC | YEAR |
|---|---|---|---|---|---|---|---|---|---|---|---|---|---|---|---|
| **TEMPERATURE °F** | NORMAL DAILY MAXIMUM | 30 | 54.1 | 60.1 | 68.3 | 75.9 | 83.2 | 91.1 | 95.4 | 94.8 | 87.7 | 77.9 | 65.1 | 56.5 | 75.8 |
| | MEAN DAILY MAXIMUM | 67 | 55.0 | 59.6 | 67.8 | 75.9 | 83.3 | 91.2 | 95.7 | 95.9 | 88.3 | 78.7 | 66.6 | 58.0 | 76.3 |
| | HIGHEST DAILY MAXIMUM | 56 | 88 | 95 | 96 | 101 | 103 | 113 | 110 | 109 | 111 | 102 | 89 | 89 | 113 |
| | YEAR OF OCCURRENCE | | 1969 | 1996 | 1991 | 2006 | 1985 | 1980 | 1998 | 2003 | 2000 | 1979 | 2005 | 2005 | JUN 1980 |
| | MEAN OF EXTREME MAXS. | 67 | 76.5 | 80.2 | 85.2 | 89.2 | 94.3 | 98.5 | 102.6 | 103.2 | 98.5 | 92.4 | 83.0 | 77.5 | 90.1 |
| | NORMAL DAILY MINIMUM | 30 | 34.0 | 38.7 | 46.4 | 54.0 | 63.0 | 70.7 | 74.6 | 74.0 | 67.2 | 56.4 | 45.1 | 36.8 | 55.1 |
| | MEAN DAILY MINIMUM | 67 | 34.2 | 38.0 | 45.6 | 54.3 | 63.2 | 70.7 | 74.6 | 74.2 | 67.0 | 56.4 | 45.1 | 37.1 | 55.0 |
| | LOWEST DAILY MINIMUM | 56 | 4 | 7 | 15 | 29 | 41 | 51 | 59 | 56 | 43 | 29 | 20 | -1 | -1 |
| | YEAR OF OCCURRENCE | | 1964 | 1985 | 2002 | 1989 | 1978 | 1964 | 1972 | 1967 | 1984 | 1993 | 1959 | 1989 | DEC 1989 |
| | MEAN OF EXTREME MINS. | 67 | 16.9 | 21.8 | 28.0 | 37.9 | 50.0 | 60.7 | 67.9 | 66.3 | 53.2 | 40.6 | 29.0 | 20.6 | 41.1 |
| | NORMAL DRY BULB | 30 | 44.1 | 49.4 | 57.4 | 65.0 | 73.1 | 80.9 | 85.0 | 84.4 | 77.5 | 67.2 | 55.1 | 46.7 | 65.5 |
| | MEAN DRY BULB | 67 | 44.6 | 48.8 | 56.7 | 65.1 | 73.3 | 81.0 | 85.2 | 85.1 | 77.6 | 67.6 | 55.9 | 47.6 | 65.7 |
| | MEAN WET BULB | 26 | 39.2 | 43.1 | 49.6 | 56.8 | 65.6 | 71.2 | 72.8 | 72.2 | 67.1 | 58.6 | 49.2 | 41.0 | 57.2 |
| | MEAN DEW POINT | 26 | 35.4 | 39.0 | 45.2 | 52.6 | 62.8 | 68.6 | 69.3 | 68.4 | 63.4 | 54.9 | 45.4 | 37.0 | 53.5 |
| | NORMAL NO. DAYS WITH: | | | | | | | | | | | | | | |
| | MAXIMUM >= 90 | 30 | 0.0 | 0.1 | 0.2 | 0.8 | 5.6 | 20.2 | 28.2 | 27.1 | 15.7 | 3.0 | 0.0 | 0.0 | 100.9 |
| | MAXIMUM <= 32 | 30 | 1.5 | 0.9 | 0.1 | 0.0 | 0.0 | 0.0 | 0.0 | 0.0 | 0.0 | 0.0 | 0.0 | 1.1 | 3.6 |
| | MINIMUM <= 32 | 30 | 13.5 | 6.9 | 2.0 | 0.2 | 0.0 | 0.0 | 0.0 | 0.0 | 0.0 | 0.1 | 2.9 | 10.0 | 35.6 |
| | MINIMUM <= 0 | 30 | 0.0 | 0.0 | 0.0 | 0.0 | 0.0 | 0.0 | 0.0 | 0.0 | 0.0 | 0.0 | 0.0 | * | 0.0 |
| **H/C** | NORMAL HEATING DEG. DAYS | 30 | 650 | 448 | 248 | 74 | 13 | 0 | 0 | 0 | 2 | 52 | 312 | 571 | 2370 |
| | NORMAL COOLING DEG. DAYS | 30 | 3 | 7 | 10 | 72 | 265 | 478 | 621 | 601 | 376 | 118 | 15 | 2 | 2568 |
| **RH** | NORMAL (PERCENT) | 30 | 68 | 67 | 65 | 64 | 70 | 67 | 61 | 59 | 64 | 66 | 69 | 70 | 66 |
| | HOUR 00 LST | 30 | 74 | 71 | 69 | 71 | 78 | 74 | 67 | 66 | 71 | 73 | 74 | 73 | 72 |
| | HOUR 06 LST | 30 | 80 | 80 | 79 | 81 | 87 | 86 | 81 | 80 | 83 | 83 | 82 | 80 | 82 |
| | HOUR 12 LST | 30 | 61 | 58 | 56 | 55 | 59 | 55 | 49 | 49 | 54 | 55 | 58 | 60 | 56 |
| | HOUR 18 LST | 30 | 59 | 55 | 51 | 51 | 57 | 52 | 45 | 45 | 51 | 55 | 60 | 60 | 53 |
| **S** | PERCENT POSSIBLE SUNSHINE | 17 | 52 | 54 | 58 | 61 | 57 | 67 | 75 | 73 | 67 | 63 | 57 | 52 | 61 |
| **W/O** | MEAN NO. DAYS WITH: | | | | | | | | | | | | | | |
| | HEAVY FOG(VISBY <= 1/4 MI) | 46 | 2.2 | 1.2 | 0.9 | 0.5 | 0.3 | 0.1 | 0.1 | 0.0 | 0.1 | 0.7 | 1.5 | 2.2 | 9.8 |
| | THUNDERSTORMS | 62 | 1.3 | 2.0 | 4.1 | 5.7 | 7.5 | 6.4 | 4.7 | 4.7 | 3.3 | 3.2 | 1.9 | 1.2 | 46.0 |
| **CLOUDNESS** | MEAN: | | | | | | | | | | | | | | |
| | SUNRISE-SUNSET (OKTAS) | | | | | | | | | | | | | | |
| | MIDNIGHT-MIDNIGHT (OKTAS) | | | | | | | | | | | | | | |
| | MEAN NO. DAYS WITH: | | | | | | | | | | | | | | |
| | CLEAR | | | | | | | | | | | | | | |
| | PARTLY CLOUDY | | | | | | | | | | | | | | |
| | CLOUDY | | | | | | | | | | | | | | |
| **PR** | MEAN STATION PRESSURE(IN) | 26 | 29.50 | 29.45 | 29.38 | 29.32 | 29.29 | 29.30 | 29.34 | 29.31 | 29.36 | 29.40 | 29.45 | 29.49 | 29.38 |
| | MEAN SEA-LEVEL PRES. (IN) | 26 | 30.14 | 30.09 | 30.01 | 29.94 | 29.91 | 29.91 | 29.96 | 29.95 | 29.97 | 30.03 | 30.08 | 30.13 | 30.01 |
| **WINDS** | MEAN SPEED (MPH) | 26 | 10.8 | 11.3 | 12.1 | 12.1 | 11.2 | 10.4 | 10.0 | 8.6 | 8.7 | 9.7 | 10.6 | 10.4 | 10.5 |
| | PREVAIL.DIR(TENS OF DEGS) | 13 | 19 | 19 | 17 | 17 | 17 | 17 | 19 | 17 | 17 | 17 | 19 | 19 | 17 |
| | MAXIMUM 2-MINUTE: | 14 | | | | | | | | | | | | | |
| | SPEED (MPH) | | 47 | 51 | 48 | 61 | 52 | 53 | 43 | 47 | 43 | 46 | 43 | 44 | 61 |
| | DIR. (TENS OF DEGS) | | 31 | 23 | 30 | 25 | 17 | 32 | 32 | 33 | 35 | 23 | 31 | 18 | 25 |
| | YEAR OF OCCURRENCE | | 2008 | 2000 | 2000 | 2008 | 2006 | 2007 | 2009 | 1996 | 2005 | 2001 | 2006 | 2008 | APR 2008 |
| | MAXIMUM 3-SECOND | 14 | | | | | | | | | | | | | |
| | SPEED (MPH) | | 54 | 78 | 74 | 75 | 67 | 64 | 54 | 47 | 49 | 54 | 54 | 48 | 78 |
| | DIR. (TENS OF DEGS) | | 28 | 23 | 27 | 24 | 15 | 30 | 33 | 34 | 35 | 27 | 31 | 18 | 23 |
| | YEAR OF OCCURRENCE | | 2008 | 2000 | 2000 | 2008 | 2006 | 2009 | 2009 | 2002 | 2005 | 2004 | 2006 | 2008 | FEB 2000 |
| **PRECIPITATION** | NORMAL (IN) | 30 | 1.90 | 2.37 | 3.06 | 3.20 | 5.15 | 3.23 | 2.12 | 2.03 | 2.42 | 4.11 | 2.57 | 2.57 | 34.73 |
| | MAXIMUM MONTHLY (IN) | 56 | 5.58 | 7.40 | 7.39 | 12.19 | 13.66 | 11.10 | 11.13 | 6.85 | 9.52 | 14.18 | 6.95 | 8.75 | 14.18 |
| | YEAR OF OCCURRENCE | | 2007 | 1997 | 2002 | 1957 | 1982 | 2007 | 1973 | 1970 | 1964 | 1981 | 2000 | 1991 | OCT 1981 |
| | MINIMUM MONTHLY (IN) | 56 | T | 0.15 | 0.10 | 0.11 | 0.95 | 0.34 | 0.00 | 0.00 | 0.09 | T | .02 | 0.17 | 0.00 |
| | YEAR OF OCCURRENCE | | 1986 | 1963 | 1972 | 1987 | 1996 | 2006 | 1993 | 2000 | 1984 | 1975 | 2005 | 1981 | AUG 2000 |
| | MAXIMUM IN 24 HOURS (IN) | 56 | 3.46 | 4.06 | 4.39 | 4.55 | 5.34 | 3.60 | 4.01 | 4.05 | 4.76 | 5.91 | 3.89 | 4.22 | 5.91 |
| | YEAR OF OCCURRENCE | | 2002 | 1965 | 1977 | 1957 | 1989 | 2009 | 2004 | 1976 | 1965 | 1959 | 2008 | 1991 | OCT 1959 |
| | NORMAL NO. DAYS WITH: | | | | | | | | | | | | | | |
| | PRECIPITATION >= 0.01 | 30 | 7.2 | 6.3 | 7.8 | 7.1 | 9.3 | 7.2 | 4.3 | 4.5 | 5.9 | 6.7 | 6.4 | 6.5 | 79.2 |
| | PRECIPITATION >= 1.00 | 30 | 0.3 | 0.7 | 0.8 | 1.1 | 1.8 | 0.8 | 0.6 | 0.6 | 0.8 | 1.4 | 0.7 | 0.6 | 10.2 |
| **SNOWFALL** | NORMAL (IN) | 30 | 0.8 | 1.2 | 0.1 | 0.0 | 0.0 | 0.0 | 0.0 | 0.0 | 0.0 | 0.* | 0.2 | 0.2 | 2.5 |
| | MAXIMUM MONTHLY (IN) | 48 | 12.1 | 13.5 | 2.5 | T | T | T | T | 0.0 | 0.0 | T | 5.0 | 3.2 | 13.5 |
| | YEAR OF OCCURRENCE | | 1964 | 1978 | 1962 | 2007 | 2008 | 2007 | 2008 | | | 1993 | 1976 | 2009 | FEB 1978 |
| | MAXIMUM IN 24 HOURS (IN) | 48 | 12.1 | 7.5 | 2.5 | T | T | T | T | 0.0 | 0.0 | T | 4.8 | 3.0 | 12.1 |
| | YEAR OF OCCURRENCE' | | 1964 | 1978 | 1962 | 1995 | 1995 | 2007 | 2008 | | | 1993 | 1976 | 2009 | JAN 1964 |
| | MAXIMUM SNOW DEPTH (IN) | 52 | 6 | 8 | 2 | 0 | 0 | 0 | 0 | 0 | 0 | 0 | 3 | 2 | 8 |
| | YEAR OF OCCURRENCE | | 1964 | 1978 | 1971 | | | | | | | | 1976 | 2009 | FEB 1978 |
| | NORMAL NO. DAYS WITH: | | | | | | | | | | | | | | |
| | SNOWFALL >= 1.0 | 30 | 0.4 | 0.4 | 0.0 | 0.0 | 0.0 | 0.0 | 0.0 | 0.0 | 0.0 | 0.0 | 0.1 | 0.1 | 1.0 |

## PRECIPITATION (inches) 2009 DALLAS-FORT WORTH (KDFW)

| YEAR | JAN | FEB | MAR | APR | MAY | JUN | JUL | AUG | SEP | OCT | NOV | DEC | ANNUAL |
|------|------|------|------|------|------|------|------|------|------|------|------|------|--------|
| 1980 | 2.52 | 0.84 | 1.24 | 2.23 | 3.01 | 1.25 | 0.71 | T | 6.54 | 1.08 | 1.23 | 1.43 | 22.08 |
| 1981 | 0.58 | 1.44 | 3.39 | 2.69 | 6.24 | 7.85 | 1.81 | 2.32 | 2.40 | 14.18 | 1.53 | 0.17 | 44.60 |
| 1982 | 2.33 | 1.89 | 1.71 | 2.71 | 13.66 | 4.28 | 2.73 | 0.52 | 0.58 | 3.36 | 4.22 | 2.76 | 40.75 |
| 1983 | 2.55 | 1.25 | 4.36 | 0.59 | 5.83 | 2.07 | 1.56 | 5.55 | 0.22 | 4.04 | 2.22 | 0.83 | 31.07 |
| 1984 | 1.07 | 3.11 | 4.92 | 1.41 | 3.04 | 2.79 | 0.43 | 1.47 | 0.09 | 6.50 | 2.97 | 6.09 | 33.89 |
| 1985 | 0.81 | 2.62 | 3.70 | 3.75 | 2.13 | 3.78 | 2.40 | 0.53 | 3.35 | 3.91 | 3.11 | 0.61 | 30.70 |
| 1986 | T | 2.49 | 1.08 | 5.30 | 5.52 | 3.92 | 0.41 | 1.63 | 4.60 | 1.81 | 3.25 | 2.44 | 32.45 |
| 1987 | 1.22 | 3.67 | 1.70 | 0.11 | 5.95 | 3.45 | 1.77 | 0.81 | 1.38 | 0.12 | 4.17 | 2.90 | 27.25 |
| 1988 | 0.88 | 1.23 | 2.03 | 2.21 | 2.11 | 3.23 | 2.47 | 0.44 | 4.04 | 1.64 | 2.28 | 2.48 | 25.04 |
| 1989 | 2.56 | 3.70 | 3.72 | 1.86 | 9.62 | 8.75 | 2.61 | 1.89 | 2.40 | 2.02 | 0.49 | 0.33 | 39.95 |
| 1990 | 4.54 | 4.72 | 5.89 | 6.90 | 7.16 | 1.89 | 2.60 | 2.37 | 1.12 | 2.81 | 3.81 | 1.46 | 45.27 |
| 1991 | 2.72 | 2.60 | 1.35 | 3.63 | 6.97 | 4.26 | 3.99 | 4.30 | 4.61 | 9.32 | 1.04 | 8.75 | 53.54 |
| 1992 | 3.25 | 2.40 | 3.24 | 2.46 | 6.93 | 5.23 | 2.48 | 2.08 | 3.25 | 3.05 | 3.56 | 4.26 | 42.19 |
| 1993 | 1.74 | 5.78 | 3.03 | 3.49 | 1.75 | 3.75 | 0.00 | 0.75 | 3.28 | 5.10 | 1.62 | 2.54 | 32.83 |
| 1994 | 1.43 | 2.01 | 1.69 | 3.62 | 5.80 | 2.05 | 4.58 | 4.89 | 1.39 | 8.19 | 6.03 | 2.42 | 44.10 |
| 1995 | 2.11 | 0.44 | 6.69 | 6.83 | 7.50 | 2.41 | 3.45 | 0.86 | 1.54 | 0.75 | 0.74 | 2.08 | 35.40 |
| 1996 | 0.97 | 0.35 | 2.36 | 2.14 | 0.95 | 3.42 | 3.85 | 5.02 | 1.51 | 6.56 | 5.54 | 0.47 | 33.14 |
| 1997 | 0.33 | 7.40 | 2.21 | 6.73 | 3.92 | 3.99 | 1.68 | 3.13 | 2.01 | 5.66 | 1.01 | 6.93 | 45.00 |
| 1998 | 5.07 | 3.22 | 4.45 | 1.25 | 2.38 | 1.75 | 0.11 | 0.35 | 0.68 | 5.64 | 4.91 | 4.43 | 34.24 |
| 1999 | 1.44 | 0.48 | 2.84 | 2.74 | 6.91 | 0.99 | 0.77 | T | 2.30 | 2.26 | 0.31 | 2.55 | 23.59 |
| 2000 | 1.59 | 3.30 | 2.92 | 4.28 | 3.17 | 5.93 | T | 0.00 | 0.16 | 4.38 | 6.95 | 3.57 | 36.25 |
| 2001 | 2.44 | 6.17 | 5.27 | 0.89 | 5.58 | 1.28 | 3.85 | 2.72 | 3.72 | 1.86 | 1.11 | 3.24 | 38.13 |
| 2002 | 4.90 | 0.94 | 7.39 | 5.68 | 5.40 | 5.23 | 3.07 | 1.47 | 1.38 | 6.44 | 0.52 | 4.13 | 44.42 |
| 2003 | 0.22 | 3.07 | 0.85 | 1.90 | 2.53 | 5.17 | 0.08 | 1.85 | 3.99 | 0.78 | 3.15 | 0.96 | 24.55 |
| 2004 | 3.04 | 3.84 | 1.71 | 2.96 | 4.73 | 10.49 | 4.16 | 4.24 | 1.02 | 5.72 | 5.01 | 0.65 | 47.57 |
| 2005 | 4.33 | 1.62 | 2.17 | 0.56 | 3.35 | 1.14 | 0.74 | 2.46 | 1.36 | 0.89 | 0.02 | 0.33 | 18.97 |
| 2006 | 2.25 | 3.85 | 4.40 | 1.86 | 1.90 | 0.34 | 1.78 | 0.52 | 2.60 | 4.34 | 2.58 | 3.33 | 29.75 |
| 2007 | 5.58 | 0.43 | 3.81 | 2.82 | 8.34 | 11.10 | 5.54 | 0.35 | 4.99 | 3.53 | 1.22 | 2.34 | 50.05 |
| 2008 | 0.27 | 2.30 | 6.07 | 3.85 | 2.21 | 0.84 | 0.81 | 2.82 | 0.84 | 2.29 | 4.53 | 0.27 | 27.10 |
| 2009 | 0.82 | 0.72 | 5.56 | 3.54 | 4.36 | 3.98 | 2.09 | 1.64 | 6.52 | 8.05 | 1.76 | 1.85 | 40.89 |
| POR= 67 YRS | 2.01 | 2.25 | 2.87 | 3.61 | 4.84 | 3.16 | 2.14 | 2.01 | 2.80 | 3.58 | 2.32 | 2.03 | 33.62 |

WBAN : 03927

## AVERAGE TEMPERATURE (°F) 2009 DALLAS-FORT WORTH (KDFW)

| YEAR | JAN | FEB | MAR | APR | MAY | JUN | JUL | AUG | SEP | OCT | NOV | DEC | ANNUAL |
|------|------|------|------|------|------|------|------|------|------|------|------|------|--------|
| 1980 | 45.5 | 46.6 | 54.2 | 63.1 | 75.0 | 87.0 | 92.0 | 88.5 | 80.3 | 65.4 | 54.9 | 49.4 | 66.8 |
| 1981 | 44.6 | 48.9 | 55.7 | 69.2 | 70.5 | 80.3 | 85.9 | 83.4 | 76.2 | 66.1 | 57.5 | 47.3 | 65.5 |
| 1982 | 44.6 | 44.5 | 59.8 | 62.5 | 72.5 | 79.2 | 84.6 | 86.7 | 78.1 | 67.0 | 55.6 | 49.2 | 65.4 |
| 1983 | 43.4 | 48.5 | 54.5 | 60.6 | 69.5 | 77.3 | 83.6 | 84.9 | 77.1 | 67.8 | 57.3 | 34.8 | 63.3 |
| 1984 | 39.3 | 50.9 | 56.3 | 63.7 | 73.7 | 82.5 | 85.5 | 85.8 | 76.1 | 67.0 | 54.6 | 52.6 | 65.7 |
| 1985 | 37.8 | 45.0 | 60.8 | 67.2 | 74.0 | 80.2 | 84.4 | 87.6 | 77.7 | 67.6 | 56.3 | 42.3 | 65.1 |
| 1986 | 48.8 | 51.2 | 60.2 | 67.2 | 71.5 | 80.8 | 86.4 | 83.4 | 80.2 | 65.7 | 52.4 | 46.1 | 66.2 |
| 1987 | 44.5 | 50.8 | 53.9 | 65.0 | 75.1 | 79.6 | 83.4 | 86.5 | 77.1 | 66.5 | 55.7 | 47.3 | 65.5 |
| 1988 | 42.2 | 47.1 | 56.0 | 64.5 | 72.8 | 80.4 | 85.3 | 87.9 | 79.2 | 65.7 | 58.1 | 49.1 | 65.7 |
| 1989 | 50.0 | 42.2 | 56.7 | 66.4 | 74.3 | 77.9 | 82.8 | 82.3 | 74.7 | 69.0 | 58.2 | 39.0 | 64.5 |
| 1990 | 51.8 | 53.9 | 57.7 | 64.0 | 73.4 | 84.0 | 82.5 | 84.6 | 80.0 | 66.4 | 59.8 | 44.0 | 66.8 |
| 1991 | 42.8 | 53.7 | 59.8 | 67.4 | 75.4 | 81.0 | 85.0 | 82.5 | 75.2 | 68.1 | 51.7 | 50.3 | 66.1 |
| 1992 | 46.9 | 54.4 | 59.1 | 66.0 | 71.1 | 79.4 | 84.3 | 80.2 | 77.7 | 69.5 | 52.7 | 49.9 | 65.9 |
| 1993 | 45.1 | 49.0 | 56.1 | 63.3 | 71.9 | 81.7 | 87.3 | 87.5 | 78.2 | 63.8 | 51.6 | 49.5 | 65.4 |
| 1994 | 45.7 | 48.6 | 59.0 | 65.9 | 71.5 | 84.1 | 83.9 | 84.0 | 76.3 | 67.3 | 57.9 | 49.0 | 66.1 |
| 1995 | 48.2 | 52.5 | 56.9 | 64.2 | 73.2 | 79.9 | 85.5 | 85.5 | 76.6 | 67.8 | 54.8 | 47.5 | 66.1 |
| 1996 | 43.0 | 52.1 | 53.3 | 64.2 | 79.7 | 82.5 | 86.2 | 82.5 | 74.5 | 66.9 | 54.9 | 49.2 | 65.8 |
| 1997 | 44.0 | 49.5 | 58.3 | 60.4 | 70.1 | 78.8 | 84.9 | 83.1 | 80.2 | 67.2 | 52.1 | 45.6 | 64.5 |
| 1998 | 48.4 | 51.1 | 54.9 | 63.8 | 78.5 | 85.5 | 91.6 | 87.8 | 83.6 | 69.5 | 57.6 | 47.0 | 68.3 |
| 1999 | 48.6 | 55.5 | 56.4 | 67.9 | 73.8 | 82.1 | 86.1 | 90.2 | 79.0 | 69.2 | 62.9 | 51.5 | 68.6 |
| 2000 | 50.6 | 57.3 | 61.0 | 65.3 | 76.6 | 80.7 | 87.3 | 90.2 | 80.4 | 69.7 | 49.8 | 39.4 | 67.4 |
| 2001 | 42.7 | 50.2 | 51.8 | 67.8 | 74.2 | 80.3 | 86.7 | 84.9 | 74.6 | 65.0 | 59.7 | 49.5 | 65.6 |
| 2002 | 47.6 | 47.0 | 54.8 | 68.4 | 72.0 | 80.1 | 83.2 | 84.7 | 79.6 | 64.9 | 53.8 | 47.7 | 65.3 |
| 2003 | 43.7 | 45.6 | 56.2 | 67.0 | 75.5 | 79.0 | 86.3 | 86.7 | 74.5 | 68.8 | 59.4 | 49.7 | 66.0 |
| 2004 | 48.5 | 45.8 | 61.8 | 66.2 | 74.4 | 79.4 | 83.5 | 81.5 | 78.4 | 72.1 | 57.1 | 48.9 | 66.5 |
| 2005 | 49.6 | 52.6 | 56.8 | 65.5 | 73.5 | 83.6 | 85.1 | 86.8 | 83.7 | 68.8 | 60.8 | 48.2 | 67.9 |
| 2006 | 55.1 | 49.4 | 62.2 | 72.0 | 78.0 | 83.7 | 87.7 | 89.8 | 77.6 | 68.1 | 57.5 | 50.0 | 69.3 |
| 2007 | 42.3 | 49.4 | 64.0 | 62.4 | 74.6 | 81.4 | 83.8 | 87.8 | 81.6 | 72.1 | 61.3 | 49.8 | 67.5 |
| 2008 | 46.9 | 54.0 | 61.0 | 66.5 | 77.0 | 86.5 | 89.0 | 86.7 | 78.5 | 69.6 | 59.7 | 49.0 | 68.7 |
| 2009 | 48.1 | 55.8 | 58.9 | 64.8 | 73.3 | 83.8 | 86.6 | 85.3 | 76.2 | 62.7 | 59.5 | 42.7 | 66.5 |
| POR= 67 YRS | 44.6 | 48.8 | 56.7 | 65.1 | 73.3 | 81.0 | 85.2 | 85.1 | 77.6 | 67.6 | 55.9 | 47.6 | 65.7 |

## HEATING DEGREE DAYS (base 65°F) 2009 DALLAS-FORT WORTH (KDFW)

| YEAR | JUL | AUG | SEP | OCT | NOV | DEC | JAN | FEB | MAR | APR | MAY | JUN | TOTAL |
|------|-----|-----|-----|-----|-----|-----|-----|-----|-----|-----|-----|-----|-------|
| 1980-81 | 0 | 0 | 18 | 99 | 330 | 486 | 625 | 448 | 284 | 26 | 23 | 0 | 2339 |
| 1981-82 | 0 | 0 | 10 | 116 | 228 | 541 | 625 | 569 | 232 | 140 | 9 | 0 | 2470 |
| 1982-83 | 0 | 0 | 1 | 94 | 316 | 495 | 663 | 454 | 324 | 186 | 21 | 2 | 2556 |
| 1983-84 | 0 | 0 | 12 | 52 | 269 | 933 | 789 | 401 | 281 | 89 | 11 | 0 | 2837 |
| 1984-85 | 0 | 0 | 38 | 66 | 322 | 389 | 837 | 558 | 171 | 37 | 0 | 0 | 2418 |
| 1985-86 | 0 | 0 | 19 | 53 | 285 | 696 | 495 | 400 | 164 | 41 | 5 | 0 | 2158 |
| 1986-87 | 0 | 0 | 0 | 61 | 376 | 580 | 632 | 387 | 342 | 109 | 0 | 0 | 2487 |
| 1987-88 | 0 | 0 | 0 | 55 | 297 | 540 | 703 | 512 | 301 | 70 | 0 | 0 | 2478 |
| 1988-89 | 0 | 0 | 0 | 51 | 240 | 487 | 460 | 630 | 294 | 102 | 4 | 0 | 2268 |
| 1989-90 | 0 | 0 | 14 | 80 | 251 | 799 | 401 | 306 | 251 | 102 | 19 | 0 | 2223 |
| 1990-91 | 0 | 0 | 0 | 100 | 190 | 646 | 681 | 314 | 198 | 37 | 7 | 0 | 2173 |
| 1991-92 | 0 | 0 | 12 | 69 | 405 | 456 | 555 | 302 | 182 | 73 | 14 | 0 | 2068 |
| 1992-93 | 0 | 0 | 0 | 14 | 366 | 474 | 612 | 445 | 290 | 120 | 4 | 0 | 2325 |
| 1993-94 | 0 | 0 | 2 | 145 | 414 | 476 | 594 | 451 | 229 | 83 | 27 | 0 | 2421 |
| 1994-95 | 0 | 0 | 6 | 71 | 226 | 487 | 521 | 348 | 289 | 86 | 13 | 0 | 2047 |
| 1995-96 | 0 | 0 | 22 | 41 | 304 | 549 | 671 | 403 | 384 | 104 | 0 | 0 | 2478 |
| 1996-97 | 0 | 0 | 14 | 51 | 308 | 483 | 650 | 429 | 213 | 155 | 16 | 0 | 2319 |
| 1997-98 | 0 | 0 | 0 | 89 | 383 | 593 | 512 | 383 | 347 | 86 | 0 | 0 | 2393 |
| 1998-99 | 0 | 0 | 0 | 31 | 226 | 559 | 505 | 266 | 263 | 39 | 3 | 0 | 1892 |
| 1999-00 | 0 | 0 | 0 | 44 | 118 | 421 | 447 | 239 | 169 | 77 | 0 | 0 | 1515 |
| 2000-01 | 0 | 0 | 12 | 64 | 458 | 785 | 685 | 417 | 402 | 41 | 0 | 0 | 2864 |
| 2001-02 | 0 | 0 | 3 | 77 | 191 | 489 | 540 | 501 | 334 | 57 | 12 | 0 | 2204 |
| 2002-03 | 0 | 0 | 0 | 101 | 335 | 528 | 652 | 541 | 280 | 51 | 0 | 0 | 2488 |
| 2003-04 | 0 | 0 | 0 | 30 | 226 | 468 | 515 | 549 | 140 | 57 | 20 | 0 | 2005 |
| 2004-05 | 0 | 0 | 0 | 5 | 230 | 494 | 483 | 357 | 259 | 52 | 24 | 0 | 1904 |
| 2005-06 | 0 | 0 | 0 | 65 | 194 | 521 | 304 | 437 | 170 | 6 | 0 | 0 | 1697 |
| 2006-07 | 0 | 0 | 0 | 63 | 243 | 466 | 699 | 438 | 109 | 141 | 0 | 0 | 2159 |
| 2007-08 | 0 | 0 | 0 | 35 | 188 | 474 | 569 | 331 | 182 | 59 | 3 | 0 | 1841 |
| 2008-09 | 0 | 0 | 0 | 38 | 197 | 499 | 526 | 273 | 239 | 95 | 3 | 0 | 1870 |
| 2009- | 0 | 0 | 0 | 115 | 176 | 681 | | | | | | | |

WBAN : 03927

## COOLING DEGREE DAYS (base 65°F) 2009 DALLAS-FORT WORTH (KDFW)

| YEAR | JAN | FEB | MAR | APR | MAY | JUN | JUL | AUG | SEP | OCT | NOV | DEC | TOTAL |
|------|-----|-----|-----|-----|-----|-----|-----|-----|-----|-----|-----|-----|-------|
| 1980 | 0 | 0 | 11 | 52 | 320 | 668 | 844 | 737 | 485 | 117 | 35 | 10 | 3279 |
| 1981 | 0 | 5 | 5 | 158 | 200 | 467 | 654 | 577 | 352 | 155 | 8 | 0 | 2581 |
| 1982 | 1 | 2 | 77 | 71 | 252 | 433 | 614 | 679 | 403 | 160 | 40 | 10 | 2742 |
| 1983 | 0 | 0 | 7 | 61 | 171 | 382 | 582 | 626 | 381 | 145 | 46 | 0 | 2401 |
| 1984 | 0 | 0 | 20 | 60 | 288 | 531 | 644 | 652 | 376 | 135 | 16 | 12 | 2734 |
| 1985 | 0 | 5 | 51 | 108 | 287 | 460 | 608 | 706 | 408 | 139 | 29 | 0 | 2801 |
| 1986 | 0 | 19 | 24 | 112 | 212 | 480 | 673 | 578 | 464 | 91 | 3 | 0 | 2656 |
| 1987 | 0 | 0 | 6 | 114 | 318 | 442 | 576 | 674 | 370 | 111 | 23 | 0 | 2634 |
| 1988 | 4 | 0 | 28 | 61 | 247 | 467 | 639 | 714 | 433 | 78 | 39 | 1 | 2711 |
| 1989 | 1 | 0 | 45 | 154 | 297 | 393 | 561 | 542 | 314 | 208 | 52 | 0 | 2567 |
| 1990 | 1 | 2 | 30 | 79 | 286 | 575 | 551 | 617 | 457 | 152 | 41 | 2 | 2793 |
| 1991 | 0 | 3 | 42 | 115 | 335 | 484 | 624 | 550 | 324 | 174 | 14 | 5 | 2670 |
| 1992 | 0 | 2 | 4 | 109 | 213 | 437 | 606 | 480 | 386 | 161 | 4 | 13 | 2415 |
| 1993 | 0 | 0 | 19 | 75 | 223 | 507 | 697 | 701 | 406 | 115 | 20 | 4 | 2767 |
| 1994 | 3 | 0 | 50 | 119 | 238 | 578 | 591 | 596 | 350 | 149 | 19 | 1 | 2694 |
| 1995 | 5 | 5 | 44 | 67 | 271 | 456 | 641 | 643 | 374 | 132 | 6 | 12 | 2656 |
| 1996 | 0 | 37 | 27 | 87 | 464 | 531 | 663 | 549 | 302 | 120 | 10 | 2 | 2792 |
| 1997 | 6 | 0 | 12 | 24 | 181 | 423 | 622 | 566 | 461 | 163 | 3 | 0 | 2461 |
| 1998 | 0 | 0 | 45 | 56 | 424 | 623 | 831 | 710 | 563 | 177 | 9 | 9 | 3447 |
| 1999 | 4 | 5 | 4 | 134 | 282 | 520 | 662 | 785 | 425 | 181 | 63 | 9 | 3074 |
| 2000 | 7 | 23 | 49 | 93 | 364 | 475 | 698 | 787 | 482 | 215 | 7 | 0 | 3200 |
| 2001 | 0 | 6 | 0 | 134 | 293 | 465 | 682 | 624 | 300 | 86 | 39 | 13 | 2642 |
| 2002 | 5 | 0 | 24 | 166 | 237 | 460 | 570 | 617 | 444 | 103 | 8 | 0 | 2634 |
| 2003 | 0 | 4 | 13 | 115 | 334 | 428 | 666 | 679 | 294 | 152 | 68 | 0 | 2753 |
| 2004 | 11 | 0 | 47 | 100 | 319 | 438 | 577 | 518 | 407 | 232 | 2 | 2 | 2653 |
| 2005 | 10 | 17 | 10 | 75 | 294 | 564 | 628 | 684 | 570 | 189 | 74 | 5 | 3120 |
| 2006 | 5 | 7 | 89 | 223 | 406 | 569 | 711 | 775 | 386 | 165 | 22 | 8 | 3366 |
| 2007 | 0 | 6 | 81 | 68 | 304 | 497 | 588 | 714 | 506 | 261 | 88 | 11 | 3124 |
| 2008 | 15 | 18 | 66 | 111 | 385 | 652 | 753 | 679 | 410 | 190 | 44 | 9 | 3332 |
| 2009 | 10 | 24 | 58 | 95 | 267 | 569 | 674 | 636 | 340 | 52 | 15 | 0 | 2740 |

**SNOWFALL (inches) 2009 DALLAS-FORT WORTH (KDFW)**

| YEAR | JUL | AUG | SEP | OCT | NOV | DEC | JAN | FEB | MAR | APR | MAY | JUN | TOTAL |
|------|-----|-----|-----|-----|-----|-----|-----|-----|-----|-----|-----|-----|-------|
| 1980-81 | 0.0 | 0.0 | 0.0 | 0.0 | 0.0 | 0.0 | T | T | 0.0 | 0.0 | 0.0 | 0.0 | T |
| 1981-82 | 0.0 | 0.0 | 0.0 | 0.0 | 0.0 | 0.0 | 0.8 | T | 0.0 | 0.0 | 0.0 | 0.0 | 0.8 |
| 1982-83 | 0.0 | 0.0 | 0.0 | 0.0 | 0.0 | T | T | T | 0.0 | 0.0 | 0.0 | 0.0 | T |
| 1983-84 | 0.0 | 0.0 | 0.0 | 0.0 | 0.0 | 2.0 | 0.0 | 0.0 | 0.0 | 0.0 | 0.0 | 0.0 | 2.0 |
| 1984-85 | 0.0 | 0.0 | 0.0 | 0.0 | 0.0 | 0.0 | 3.4 | 1.7 | 0.0 | 0.0 | 0.0 | 0.0 | 5.1 |
| 1985-86 | 0.0 | 0.0 | 0.0 | 0.0 | 0.0 | T | 0.0 | 0.8 | 0.0 | 0.0 | 0.0 | 0.0 | 0.8 |
| 1986-87 | 0.0 | 0.0 | 0.0 | 0.0 | 0.0 | 1.7 | T | T | 0.5 | 0.0 | 0.0 | 0.0 | 2.2 |
| 1987-88 | 0.0 | 0.0 | 0.0 | 0.0 | 0.0 | T | 0.8 | 2.7 | 0.0 | 0.0 | 0.0 | 0.0 | 3.5 |
| 1988-89 | 0.0 | 0.0 | 0.0 | 0.0 | 0.0 | T | T | 0.7 | 1.1 | 0.0 | T | 0.0 | 1.8 |
| 1989-90 | 0.0 | 0.0 | 0.0 | 0.0 | 0.0 | T | 0.0 | 0.0 | 0.0 | T | T | 0.0 | T |
| 1990-91 | 0.0 | 0.0 | 0.0 | 0.0 | 0.0 | 0.3 | 0.3 | 0.0 | 0.0 | 0.0 | 0.0 | 0.0 | 0.6 |
| 1991-92 | 0.0 | 0.0 | 0.0 | 0.0 | 0.0 | 0.0 | T | 0.0 | 0.0 | 0.0 | 0.0 | 0.0 | T |
| 1992-93 | 0.0 | 0.0 | 0.0 | 0.0 | 0.0 | T | 0.0 | 0.0 | 0.0 | 0.0 | 0.0 | 0.0 | T |
| 1993-94 | 0.0 | 0.0 | 0.0 | T | 0.3 | T | T | 0.1 | T | T | T | 0.0 | 0.4 |
| 1994-95 | 0.0 | 0.0 | 0.0 | 0.0 | 0.0 | 0.0 | T | 0.0 | T | T | T | 0.0 | T |
| 1995-96 | 0.0 | 0.0 | 0.0 | 0.0 | T | 0.0 | 0.9 | 1.5 | T | 0.0 | 0.0 | 0.0 | 2.4 |
| 1996-97 | 0.0 | 0.0 | 0.0 | 0.0 | T | T | T | 0.0 | 0.0 | 0.0 | 0.0 | 0.0 | T |
| 1997-98 | 0.0 | 0.0 | 0.0 | 0.0 | T | T | T | 0.0 | 0.5 | 0.0 | 0.0 | 0.0 | 0.5 |
| 1998-99 | 0.0 | 0.0 | 0.0 | 0.0 | 0.0 | T | 0.0 | 0.0 | 0.0 | 0.0 | 0.0 | 0.0 | T |
| 1999-00 | 0.0 | 0.0 | 0.0 | 0.0 | 0.0 | 0.0 | T | 0.0 | 0.0 | 0.0 | 0.0 | 0.0 | T |
| 2000-01 | 0.0 | 0.0 | 0.0 | 0.0 | 0.0 | 1.5 | 0.0 | T | 0.0 | 0.0 | 0.0 | 0.0 | 1.5 |
| 2001-02 | 0.0 | 0.0 | 0.0 | 0.0 | T | 0.0 | T | 3.5 | 0.3 | T | 0.0 | 0.0 | 3.8 |
| 2002-03 | 0.0 | 0.0 | 0.0 | 0.0 | 0.0 | T | 1.2 | 2.0 | 0.0 | 0.0 | 0.0 | 0.0 | 3.2 |
| 2003-04 | 0.0 | 0.0 | 0.0 | 0.0 | 0.0 | 0.0 | 0.0 | 2.6 | 0.0 | 0.0 | 0.0 | 0.0 | 2.6 |
| 2004-05 | 0.0 | 0.0 | 0.0 | 0.0 | 0.0 | 0.3 | 0.0 | 0.0 | 0.0 | 0.0 | 0.0 | 0.0 | 0.3 |
| 2005-06 | 0.0 | 0.0 | 0.0 | 0.0 | 0.0 | T | 0.0 | 0.0 | 0.0 | 0.0 | 0.0 | 0.0 | T |
| 2006-07 | 0.0 | 0.0 | 0.0 | 0.0 | T | 0.0 | 0.3 | T | 0.0 | T | 0.0 | T | 0.3 |
| 2007-08 | 0.0 | 0.0 | 0.0 | 0.0 | T | 0.0 | T | 0.0 | 2.1 | 0.0 | T | 0.0 | 2.1 |
| 2008-09 | T | 0.0 | 0.0 | 0.0 | 0.0 | T | 0.2 | T | 0.0 | 0.0 | 0.0 | 0.0 | 0.2 |
| 2009- | 0.0 | 0.0 | 0.0 | 0.0 | 0.0 | 3.2 | | | | | | | |
| POR= 66 YRS | T | 0.0 | 0.0 | T | 0.1 | 0.2 | 1.2 | 0.8 | 0.2 | T | T | T | 2.5 |

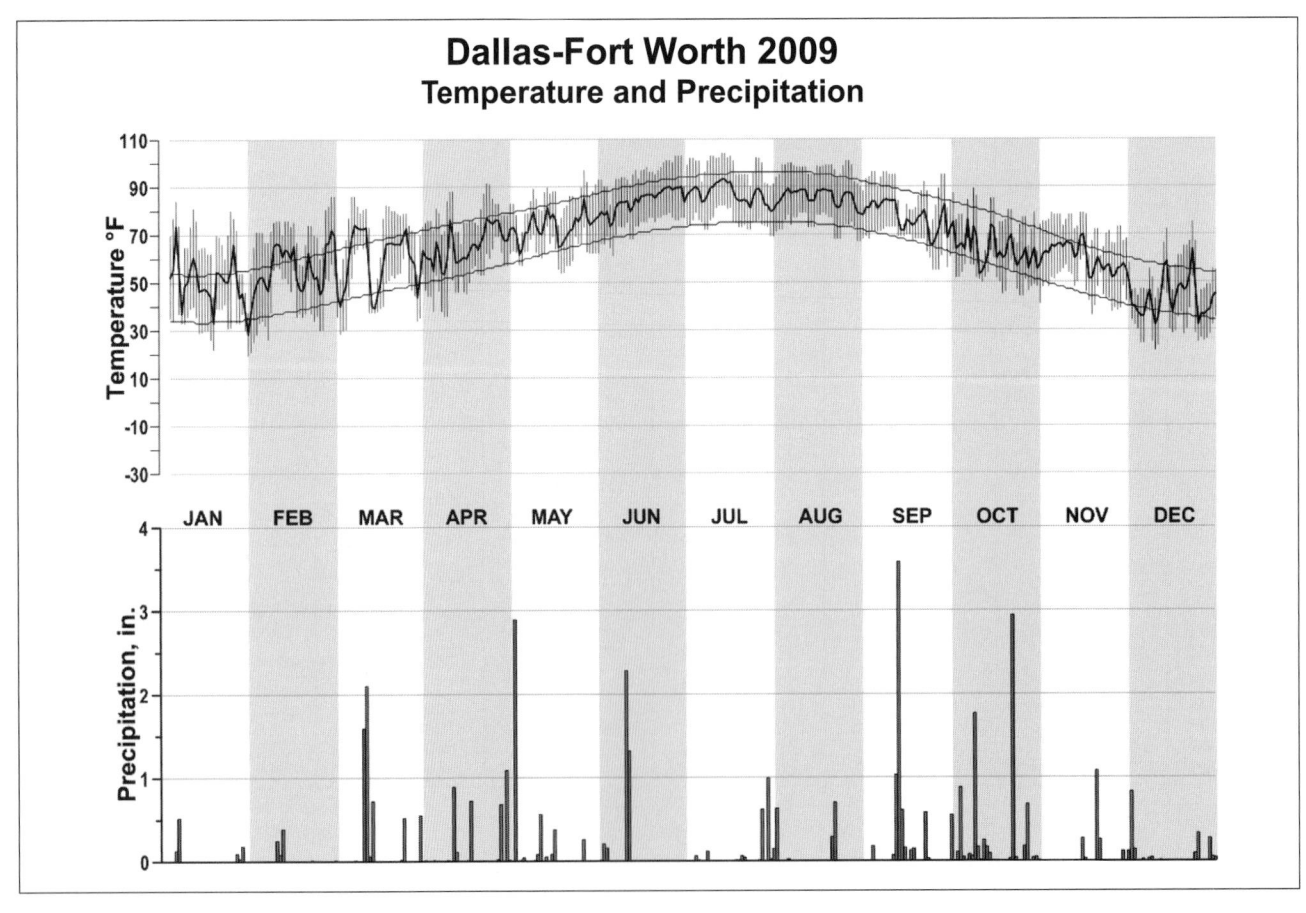

# Dallas-Fort Worth 2009
## Temperature and Precipitation

# 2009
# EL PASO
# TEXAS (KELP)

The city of El Paso is located in the extreme west point of Texas at an elevation of about 3,700 feet . The National Weather Service station is located on a mesa about 200 feet higher than the city. The climate of the region is characterized by an abundance of sunshine throughout the year, high daytime summer temperatures, very low humidity, scanty rainfall, and a relatively mild winter season. The Franklin Mountains begin within the city limits and extend northward for about 16 miles. Peaks of these mountains range from 4,687 to 7,152 feet above sea level.

Rainfall throughout the year is light, insufficient for any growth except desert vegetation. Irrigation is necessary for crops, gardens, and lawns. Dry periods lasting several months are not unusual. Almost half of the precipitation occurs in the three-month period, July through September, from brief but often heavy thunderstorms. Small amounts of snow fall nearly every winter, but snow cover rarely amounts to more than an inch and seldom remains on the ground for more than a few hours.

Daytime summer temperatures are high, frequently above 90 degrees and occasionally above 100 degrees. Summer nights are usually comfortable, with temperatures in the 60s. It should be noted that when temperatures are high the relative humidity is generally quite low. A 20-year tabulation of observations with temperatures above 90 degrees shows that in April, May, and June the humidity averaged from 10 to 14 percent, while in July, August, and September it averaged 22 to 24 percent. This low humidity aids the efficiency of evaporative air coolers, which are widely used in homes and public buildings and are quite effective in cooling the air to comfortable temperatures.

Winter daytime temperatures are mild. At night they drop below freezing about half the time in December and January. The flat, irrigated land of the Rio Grande Valley in the vicinity of El Paso is noticeably cooler, particularly at night, than the airport or the city proper, both in summer and winter. This results in more comfortable temperatures in summer but increases the severity of freezes in winter. The cooler air in the Valley also causes marked short-period fluctuations of temperature and dewpoint at the airport with changes in wind direction, especially during the early morning hours.

Dust and sandstorms are the most unpleasant features of the weather in El Paso. While wind velocities are not excessively high, the soil surface is dry and loose and natural vegetation is sparse, so moderately strong winds raise considerable dust and sand. A tabulation of duststorms for a period of 20 years shows that they are most frequent in March and April, and comparatively rare in the period July through December. prevailing winds are from the north in winter and the south in summer.

# NORMALS, MEANS, AND EXTREMES
## EL PASO (KELP)

LATITUDE: 31° 48'N  LONGITUDE: -106° 22'W  ELEVATION (FT): GRND: 3937  BARO: 3945  TIME ZONE: MOUNTAIN (UTC -7)  WBAN: 23044

| | ELEMENT | POR | JAN | FEB | MAR | APR | MAY | JUN | JUL | AUG | SEP | OCT | NOV | DEC | YEAR |
|---|---|---|---|---|---|---|---|---|---|---|---|---|---|---|---|
| **TEMPERATURE °F** | NORMAL DAILY MAXIMUM | 30 | 57.2 | 63.4 | 70.2 | 78.1 | 86.7 | 95.3 | 94.5 | 92.0 | 87.1 | 77.9 | 65.5 | 57.4 | 77.1 |
| | MEAN DAILY MAXIMUM | 62 | 57.8 | 63.3 | 70.1 | 78.9 | 87.6 | 95.8 | 95.3 | 93.0 | 87.8 | 78.6 | 66.2 | 57.9 | 77.7 |
| | HIGHEST DAILY MAXIMUM | 70 | 80 | 83 | 89 | 98 | 105 | 114 | 112 | 108 | 104 | 96 | 87 | 80 | 114 |
| | YEAR OF OCCURRENCE | | 1970 | 2009 | 1989 | 1989 | 2005 | 1994 | 1979 | 1980 | 1982 | 1994 | 1983 | 1973 | JUN 1994 |
| | MEAN OF EXTREME MAXS. | 62 | 70.8 | 76.5 | 82.7 | 90.4 | 97.9 | 104.3 | 103.4 | 100.4 | 96.5 | 89.7 | 78.8 | 70.7 | 88.5 |
| | NORMAL DAILY MINIMUM | 30 | 32.9 | 37.5 | 43.7 | 51.1 | 60.6 | 68.8 | 72.0 | 70.2 | 63.7 | 51.8 | 39.8 | 33.4 | 52.1 |
| | MEAN DAILY MINIMUM | 62 | 31.5 | 35.6 | 41.7 | 49.7 | 58.6 | 66.8 | 70.1 | 68.6 | 62.3 | 50.7 | 38.4 | 31.8 | 50.5 |
| | LOWEST DAILY MINIMUM | 70 | -8 | 8 | 14 | 23 | 31 | 46 | 57 | 56 | 41 | 25 | 0 | 5 | -8 |
| | YEAR OF OCCURRENCE | | 1962 | 1985 | 1971 | 1983 | 1967 | 1988 | 1988 | 1973 | 1945 | 1970 | 1996 | 1953 | JAN 1962 |
| | MEAN OF EXTREME MINS. | 62 | 17.3 | 21.3 | 26.5 | 35.2 | 45.3 | 55.9 | 63.6 | 62.2 | 51.8 | 37.2 | 24.3 | 18.0 | 38.2 |
| | NORMAL DRY BULB | 30 | 45.1 | 50.5 | 57.0 | 64.6 | 73.7 | 82.1 | 83.3 | 81.1 | 75.4 | 64.9 | 52.7 | 45.4 | 64.7 |
| | MEAN DRY BULB | 62 | 44.7 | 49.4 | 55.9 | 64.3 | 73.1 | 81.5 | 82.7 | 80.8 | 75.0 | 64.7 | 52.3 | 44.9 | 64.1 |
| | MEAN WET BULB | 26 | 34.1 | 36.4 | 39.5 | 43.4 | 50.8 | 57.6 | 63.9 | 64.5 | 59.6 | 50.1 | 40.2 | 34.6 | 47.9 |
| | MEAN DEW POINT | 26 | 26.5 | 27.7 | 28.3 | 30.6 | 38.0 | 47.3 | 56.7 | 58.7 | 52.7 | 42.6 | 32.0 | 27.1 | 39.0 |
| | NORMAL NO. DAYS WITH: | | | | | | | | | | | | | | |
| | MAXIMUM >= 90 | 30 | 0.0 | 0.0 | 0.0 | 2.3 | 13.2 | 25.9 | 26.8 | 24.0 | 14.3 | 2.4 | 0.0 | 0.0 | 108.9 |
| | MAXIMUM <= 32 | 30 | 0.3 | 0.2 | 0.0 | 0.0 | 0.0 | 0.0 | 0.0 | 0.0 | 0.0 | 0.0 | 0.1 | 0.2 | 0.8 |
| | MINIMUM <= 32 | 30 | 17.8 | 10.7 | 4.4 | 0.9 | 0.0 | 0.0 | 0.0 | 0.0 | 0.0 | 0.5 | 8.1 | 17.6 | 60.0 |
| | MINIMUM <= 0 | 30 | 0.0 | 0.0 | 0.0 | 0.0 | 0.0 | 0.0 | 0.0 | 0.0 | 0.0 | 0.0 | 0.0 | 0.0 | 0.0 |
| **H/C** | NORMAL HEATING DEG. DAYS | 30 | 632 | 424 | 272 | 100 | 8 | 0 | 0 | 0 | 9 | 89 | 386 | 623 | 2543 |
| | NORMAL COOLING DEG. DAYS | 30 | 0 | 2 | 8 | 72 | 262 | 497 | 552 | 483 | 305 | 71 | 2 | 0 | 2254 |
| **RH** | NORMAL (PERCENT) | 30 | 52 | 43 | 34 | 28 | 29 | 31 | 45 | 49 | 49 | 48 | 47 | 53 | 42 |
| | HOUR 05 LST | 30 | 66 | 58 | 48 | 42 | 43 | 47 | 64 | 69 | 68 | 65 | 64 | 67 | 58 |
| | HOUR 11 LST | 30 | 44 | 37 | 29 | 23 | 24 | 26 | 37 | 41 | 41 | 39 | 39 | 45 | 35 |
| | HOUR 17 LST | 30 | 35 | 28 | 21 | 18 | 18 | 19 | 29 | 33 | 33 | 31 | 33 | 38 | 28 |
| | HOUR 23 LST | 30 | 56 | 46 | 35 | 30 | 30 | 33 | 48 | 54 | 55 | 53 | 53 | 57 | 46 |
| **S** | PERCENT POSSIBLE SUNSHINE | 54 | 78 | 82 | 86 | 89 | 90 | 90 | 82 | 81 | 83 | 84 | 83 | 77 | 84 |
| **W/O** | MEAN NO. DAYS WITH: | | | | | | | | | | | | | | |
| | HEAVY FOG(VISBY <= 1/4 MI) | 46 | 0.5 | 0.1 | 0.0 | 0.0 | 0.0 | 0.0 | 0.0 | 0.0 | 0.0 | 0.1 | 0.2 | 0.6 | 1.5 |
| | THUNDERSTORMS | 62 | 0.2 | 0.4 | 0.6 | 1.0 | 2.9 | 4.6 | 10.6 | 9.8 | 4.2 | 1.9 | 0.3 | 0.3 | 36.8 |
| **CLOUDNESS** | MEAN: | | | | | | | | | | | | | | |
| | SUNRISE-SUNSET (OKTAS) | 53 | 3.7 | 3.4 | 3.4 | 2.8 | 2.6 | 2.2 | 3.5 | 3.4 | 2.7 | 2.5 | 2.7 | 3.4 | 3.0 |
| | MIDNIGHT-MIDNIGHT (OKTAS) | 32 | 3.3 | 3.1 | 2.8 | 2.4 | 2.4 | 2.2 | 3.6 | 3.6 | 2.9 | 2.2 | 2.5 | 3.0 | 2.8 |
| | MEAN NO. DAYS WITH: | | | | | | | | | | | | | | |
| | CLEAR | 53 | 13.9 | 13.6 | 15.1 | 16.7 | 18.5 | 19.8 | 12.0 | 13.6 | 17.4 | 18.5 | 17.2 | 14.8 | 191.1 |
| | PARTLY CLOUDY | 53 | 7.5 | 7.3 | 8.4 | 8.0 | 8.2 | 7.4 | 13.1 | 11.9 | 7.1 | 6.7 | 6.1 | 7.2 | 98.9 |
| | CLOUDY | 53 | 9.7 | 7.3 | 7.5 | 5.2 | 4.2 | 2.7 | 5.4 | 4.9 | 5.0 | 5.2 | 6.2 | 8.5 | 71.8 |
| **PR** | MEAN STATION PRESSURE(IN) | 26 | 26.10 | 26.06 | 26.01 | 25.98 | 25.97 | 25.98 | 26.05 | 26.07 | 26.06 | 26.07 | 26.09 | 26.08 | 26.04 |
| | MEAN SEA-LEVEL PRES. (IN) | 26 | 30.06 | 29.99 | 29.89 | 29.82 | 29.77 | 29.76 | 29.84 | 29.86 | 29.88 | 29.93 | 30.01 | 30.06 | 29.91 |
| **WINDS** | MEAN SPEED (MPH) | 26 | 7.5 | 8.4 | 9.5 | 10.2 | 9.6 | 8.5 | 7.6 | 6.9 | 6.9 | 7.1 | 7.2 | 7.1 | 8.0 |
| | PREVAIL.DIR(TENS OF DEGS) | 38 | 36 | 28 | 26 | 26 | 26 | 15 | 15 | 15 | 36 | 36 | 01 | 36 | 36 |
| | MAXIMUM 2-MINUTE: | | | | | | | | | | | | | | |
| | SPEED (MPH) | 14 | 64 | 51 | 52 | 56 | 45 | 51 | 47 | 54 | 41 | 45 | 53 | 56 | 64 |
| | DIR. (TENS OF DEGS) | | 26 | 24 | 28 | 26 | 25 | 32 | 03 | 26 | 35 | 26 | 27 | 24 | 26 |
| | YEAR OF OCCURRENCE | | 1996 | 1998 | 2001 | 2001 | 1999 | 2000 | 2008 | 2002 | 1996 | 2009 | 2005 | 2009 | JAN 1996 |
| | MAXIMUM 3-SECOND | | | | | | | | | | | | | | |
| | SPEED (MPH) | 14 | 75 | 60 | 63 | 70 | 58 | 61 | 68 | 69 | 51 | 70 | 64 | 68 | 75 |
| | DIR. (TENS OF DEGS) | | 26 | 26 | 26 | 24 | 24 | 33 | 09 | 26 | 33 | 23 | 24 | 25 | 26 |
| | YEAR OF OCCURRENCE | | 1996 | 1998 | 1999 | 2001 | 1999 | 2006 | 2005 | 2002 | 2009 | 2009 | 1998 | 2009 | JAN 1996 |
| **PRECIPITATION** | NORMAL (IN) | 30 | 0.45 | 0.39 | 0.26 | 0.23 | 0.38 | 0.87 | 1.49 | 1.75 | 1.61 | 0.81 | 0.42 | 0.77 | 9.43 |
| | MAXIMUM MONTHLY (IN) | 70 | 1.84 | 1.92 | 2.26 | 1.42 | 4.22 | 3.18 | 5.53 | 6.85 | 6.68 | 4.31 | 2.01 | 3.29 | 6.85 |
| | YEAR OF OCCURRENCE | | 1949 | 2005 | 1958 | 1983 | 1992 | 1984 | 1968 | 2006 | 1974 | 1945 | 2004 | 1991 | AUG 2006 |
| | MINIMUM MONTHLY (IN) | 70 | 0.00 | 0.00 | 0.00 | 0.00 | 0.00 | T | 0.04 | T | T | 0.00 | 0.00 | 0.00 | 0.00 |
| | YEAR OF OCCURRENCE | | 1967 | 1943 | 2002 | 1978 | 1962 | 1990 | 1978 | 1962 | 1959 | 1952 | 1964 | 1955 | MAR 2002 |
| | MAXIMUM IN 24 HOURS (IN) | 70 | 0.80 | 1.02 | 1.72 | 1.08 | 2.40 | 1.91 | 2.63 | 2.89 | 2.52 | 1.77 | 1.19 | 1.76 | 2.89 |
| | YEAR OF OCCURRENCE | | 2007 | 2003 | 1941 | 1966 | 1992 | 2009 | 1968 | 2005 | 1958 | 1945 | 1943 | 1987 | AUG 2005 |
| | NORMAL NO. DAYS WITH: | | | | | | | | | | | | | | |
| | PRECIPITATION >= 0.01 | 30 | 4.7 | 3.1 | 2.2 | 1.7 | 2.8 | 3.5 | 8.2 | 8.8 | 6.5 | 4.9 | 3.1 | 4.3 | 53.8 |
| | PRECIPITATION >= 1.00 | 30 | 0.0 | 0.0 | 0.0 | 0.0 | 0.1 | 0.1 | 0.3 | 0.3 | 0.4 | * | 0.0 | 0.1 | 1.3 |
| **SNOWFALL** | NORMAL (IN) | 30 | 1.5 | 0.8 | 0.3 | 0.7 | 0.0 | 0.0 | 0.0 | 0.0 | 0.0 | 0.1 | 0.9 | 2.4 | 6.7 |
| | MAXIMUM MONTHLY (IN) | 60 | 8.3 | 8.9 | 7.3 | 16.5 | T | T | T | T | T | 1.0 | 12.7 | 25.9 | 25.9 |
| | YEAR OF OCCURRENCE | | 1949 | 1956 | 1958 | 1983 | 2007 | 1992 | 2007 | 2007 | 2009 | 1980 | 1976 | 1987 | DEC 1987 |
| | MAXIMUM IN 24 HOURS (IN) | 60 | 5.2 | 7.2 | 7.3 | 8.8 | T | T | T | T | T | 1.0 | 7.8 | 16.8 | 16.8 |
| | YEAR OF OCCURRENCE' | | 1992 | 1956 | 1958 | 1983 | 1992 | 1992 | 1990 | 2007 | 2009 | 1980 | 1961 | 1987 | DEC 1987 |
| | MAXIMUM SNOW DEPTH (IN) | 51 | 6 | 7 | 7 | 9 | 0 | 0 | 0 | 0 | 0 | 1 | 7 | 14 | 14 |
| | YEAR OF OCCURRENCE | | 1983 | 1956 | 1958 | 1983 | | | | | | 1993 | 1968 | 1987 | DEC 1987 |
| | NORMAL NO. DAYS WITH: | | | | | | | | | | | | | | |
| | SNOWFALL >= 1.0 | 30 | 0.5 | 0.3 | 0.1 | 0.2 | 0.0 | 0.0 | 0.0 | 0.0 | 0.0 | 0.0 | 0.3 | 0.5 | 1.9 |

## PRECIPITATION (inches) 2009 EL PASO (KELP)

| YEAR | JAN | FEB | MAR | APR | MAY | JUN | JUL | AUG | SEP | OCT | NOV | DEC | ANNUAL |
|------|-----|-----|-----|-----|-----|-----|-----|-----|-----|-----|-----|-----|--------|
| 1980 | 0.54 | 0.73 | 0.25 | 0.31 | 0.08 | T | 0.21 | 1.76 | 1.90 | 0.95 | 0.54 | 0.04 | 7.31 |
| 1981 | 1.10 | 0.36 | 0.39 | 0.65 | 0.72 | 0.64 | 2.08 | 5.26 | 0.52 | 0.53 | 0.30 | 0.08 | 12.63 |
| 1982 | 0.34 | 0.55 | T | 0.05 | 0.19 | 0.18 | 1.00 | 0.48 | 5.28 | T | 0.29 | 2.61 | 10.97 |
| 1983 | 0.35 | 0.60 | 0.45 | 1.42 | 0.05 | 0.23 | 0.43 | 0.97 | 1.51 | 1.48 | 0.34 | 0.16 | 7.99 |
| 1984 | 0.31 | 0.00 | 0.44 | 0.01 | 0.59 | 3.18 | 0.69 | 5.57 | 0.58 | 3.12 | 0.51 | 1.17 | 16.17 |
| 1985 | 0.95 | 0.19 | 0.59 | 0.07 | 0.01 | 0.10 | 1.32 | 1.46 | 1.47 | 1.82 | 0.13 | 0.05 | 8.16 |
| 1986 | 0.01 | 0.39 | 0.39 | T | 0.83 | 3.05 | 2.66 | 0.70 | 0.85 | 0.45 | 1.42 | 1.42 | 12.17 |
| 1987 | 0.29 | 0.30 | 0.49 | 0.32 | 0.24 | 2.24 | 0.64 | 2.22 | 0.89 | 0.15 | 0.29 | 2.87 | 10.94 |
| 1988 | 0.25 | 0.70 | 0.10 | 0.23 | 0.15 | 0.03 | 3.35 | 3.46 | 1.52 | 0.59 | 0.24 | 0.44 | 11.06 |
| 1989 | 0.11 | 0.72 | 0.62 | T | 0.65 | T | 1.23 | 3.06 | 0.48 | 0.23 | T | 0.16 | 7.26 |
| 1990 | 0.29 | 0.14 | 0.41 | 0.25 | 0.10 | T | 3.96 | 1.98 | 3.46 | 0.58 | 1.34 | 0.34 | 12.85 |
| 1991 | 0.82 | 0.66 | 0.10 | T | 0.23 | 0.01 | 2.69 | 2.06 | 1.82 | 0.20 | 0.50 | 3.29 | 12.38 |
| 1992 | 1.14 | 0.16 | 0.50 | 0.30 | 4.22 | 0.27 | 0.65 | 2.11 | 0.15 | 0.27 | 0.28 | 1.35 | 11.40 |
| 1993 | 1.34 | 0.32 | 0.01 | 0.12 | T | 1.47 | 0.95 | 2.73 | 1.32 | 0.17 | 0.49 | 0.71 | 9.63 |
| 1994 | 0.03 | 0.23 | 0.37 | 0.65 | 0.80 | 0.67 | 0.18 | 0.02 | 0.03 | 0.35 | 0.54 | 1.61 | 5.48 |
| 1995 | 0.26 | 0.88 | 0.42 | 0.04 | 0.01 | 1.74 | 0.28 | 0.76 | 3.18 | T | 0.26 | 0.03 | 7.86 |
| 1996 | 0.11 | 0.19 | T | 0.49 | 0.00 | 2.36 | 1.97 | 1.87 | 1.24 | T | 0.16 | .00 | 8.39 |
| 1997 | 0.38 | 0.29 | 0.64 | 0.43 | 0.52 | 1.11 | 0.91 | 1.41 | 1.55 | 0.19 | 0.79 | 1.41 | 9.63 |
| 1998 | 0.05 | 0.15 | 0.18 | 0.04 | T | 0.27 | 2.07 | 0.53 | 0.66 | 2.14 | 0.34 | 0.34 | 6.77 |
| 1999 | 0.10 | 0.00 | 0.04 | T | 0.02 | 1.44 | 2.00 | 1.43 | 1.94 | 0.56 | 0.00 | 0.63 | 8.16 |
| 2000 | 0.00 | 0.03 | 0.06 | 0.28 | T | 2.45 | 1.59 | 0.70 | T | 0.82 | 1.06 | 0.42 | 7.41 |
| 2001 | 0.06 | 0.24 | 0.40 | T | 0.18 | 0.30 | 0.36 | 1.72 | 0.30 | T | 0.60 | 0.13 | 4.29 |
| 2002 | T | 1.22 | 0.00 | 0.00 | T | 0.35 | 1.34 | 0.76 | 0.48 | 1.09 | T | 1.65 | 6.89 |
| 2003 | T | 1.37 | 0.18 | 0.02 | T | 0.49 | 0.55 | 0.66 | 0.08 | 0.33 | 0.52 | 0.01 | 4.21 |
| 2004 | 0.36 | 0.05 | 0.80 | 1.06 | 0.50 | 0.93 | 1.70 | 3.04 | 0.89 | 0.39 | 2.01 | 0.36 | 12.09 |
| 2005 | 0.66 | 1.92 | 0.08 | 0.14 | 0.93 | T | 0.66 | 4.35 | 2.77 | 1.36 | 0.00 | T | 12.87 |
| 2006 | 0.02 | 0.28 | T | 0.01 | 0.89 | 0.27 | 3.17 | 6.85 | 4.99 | 0.92 | 0.06 | 0.05 | 17.51 |
| 2007 | 1.81 | 0.19 | 0.02 | 0.31 | 1.30 | 0.51 | 2.08 | 0.57 | 1.71 | 0.09 | 1.07 | 0.46 | 10.12 |
| 2008 | 0.14 | 0.15 | T | T | 0.03 | 0.48 | 4.34 | 2.61 | 1.52 | 0.15 | 0.17 | 0.27 | 9.86 |
| 2009 | 0.01 | T | 0.06 | 0.01 | 0.77 | 2.24 | 0.49 | 0.59 | 2.50 | 0.21 | 0.96 | 0.84 | 8.68 |
| POR= 62 YRS | 0.40 | 0.44 | 0.28 | 0.21 | 0.35 | 0.73 | 1.59 | 1.59 | 1.42 | 0.68 | 0.40 | 0.59 | 8.68 |

WBAN : 23044

## AVERAGE TEMPERATURE (°F) 2009 EL PASO (KELP)

| YEAR | JAN | FEB | MAR | APR | MAY | JUN | JUL | AUG | SEP | OCT | NOV | DEC | ANNUAL |
|------|-----|-----|-----|-----|-----|-----|-----|-----|-----|-----|-----|-----|--------|
| 1980 | 46.8 | 50.6 | 54.1 | 60.6 | 70.5 | 86.3 | 87.2 | 82.4 | 75.6 | 60.3 | 49.2 | 48.5 | 64.3 |
| 1981 | 45.2 | 50.3 | 57.2 | 64.8 | 73.6 | 82.6 | 83.6 | 79.5 | 75.9 | 64.6 | 54.3 | 48.7 | 65.0 |
| 1982 | 42.3 | 48.7 | 57.7 | 64.4 | 69.6 | 80.9 | 84.2 | 83.5 | 77.1 | 64.6 | 53.2 | 43.3 | 64.1 |
| 1983 | 41.6 | 49.5 | 54.6 | 56.3 | 68.9 | 77.3 | 82.9 | 81.8 | 78.7 | 66.6 | 54.3 | 45.5 | 63.2 |
| 1984 | 44.4 | 47.0 | 55.7 | 62.0 | 75.0 | 79.5 | 81.1 | 80.4 | 72.9 | 61.4 | 51.6 | 45.7 | 63.1 |
| 1985 | 40.0 | 45.6 | 55.2 | 64.2 | 72.1 | 79.0 | 79.4 | 80.6 | 72.8 | 61.4 | 52.9 | 43.1 | 62.2 |
| 1986 | 44.7 | 52.1 | 55.7 | 67.3 | 71.5 | 77.7 | 80.0 | 80.4 | 74.1 | 62.4 | 49.4 | 42.6 | 63.2 |
| 1987 | 41.3 | 46.2 | 51.2 | 59.7 | 68.6 | 78.1 | 81.6 | 79.1 | 72.1 | 67.0 | 51.4 | 40.5 | 61.4 |
| 1988 | 42.6 | 48.4 | 53.4 | 61.1 | 70.3 | 79.0 | 80.3 | 77.6 | 72.4 | 66.7 | 54.0 | 42.6 | 62.4 |
| 1989 | 43.6 | 51.4 | 58.7 | 67.4 | 74.2 | 81.3 | 81.8 | 79.1 | 73.4 | 63.2 | 53.3 | 41.9 | 64.1 |
| 1990 | 44.1 | 49.1 | 56.1 | 66.4 | 73.1 | 87.1 | 80.2 | 76.8 | 73.9 | 63.4 | 52.9 | 44.8 | 64.0 |
| 1991 | 44.3 | 51.1 | 53.9 | 64.0 | 72.6 | 79.0 | 78.6 | 79.5 | 70.7 | 65.4 | 50.3 | 46.3 | 63.0 |
| 1992 | 44.0 | 50.2 | 58.1 | 67.9 | 70.6 | 81.5 | 85.1 | 81.3 | 77.8 | 66.5 | 48.0 | 44.7 | 64.6 |
| 1993 | 48.5 | 51.4 | 57.9 | 66.7 | 74.6 | 82.7 | 85.2 | 82.5 | 75.8 | 64.4 | 52.3 | 46.9 | 65.7 |
| 1994 | 45.6 | 50.4 | 58.3 | 66.7 | 75.7 | 89.0 | 88.1 | 86.2 | 78.3 | 66.6 | 54.4 | 49.4 | 67.4 |
| 1995 | 47.1 | 56.3 | 59.7 | 65.2 | 74.3 | 80.2 | 83.1 | 82.8 | 74.1 | 65.6 | 55.0 | 47.6 | 65.9 |
| 1996 | 46.5 | 54.5 | 54.6 | 65.5 | 80.1 | 83.5 | 82.9 | 79.6 | 73.5 | 65.5 |  | 46.3 |  |
| 1997 | 44.2 | 48.2 | 58.6 | 61.5 | 74.3 | 81.0 | 82.7 | 81.6 | 78.4 | 63.6 | 52.2 | 40.6 | 63.9 |
| 1998 | 47.1 | 47.6 | 54.0 | 59.5 | 74.4 | 82.2 | 83.5 | 80.6 | 79.2 | 66.0 | 55.2 | 45.8 | 64.6 |
| 1999 | 48.1 | 53.6 | 58.0 | 63.2 | 74.2 | 80.0 | 81.2 | 81.2 | 75.4 | 63.9 | 56.5 | 43.4 | 64.9 |
| 2000 | 49.7 | 54.0 | 57.2 | 68.3 | 79.0 | 80.3 | 84.1 | 81.5 | 78.6 | 62.5 | 47.6 | 44.0 | 65.6 |
| 2001 | 41.8 | 50.1 | 56.3 | 65.5 | 76.0 | 83.3 | 84.3 | 81.0 | 76.9 | 67.6 | 54.7 | 42.7 | 65.0 |
| 2002 | 46.2 | 47.1 | 56.8 | 69.9 | 75.0 | 85.2 | 82.3 | 84.5 | 77.0 | 64.4 | 52.2 | 43.9 | 65.4 |
| 2003 | 48.7 | 50.2 | 55.5 | 65.4 | 76.0 | 81.3 | 84.9 | 83.8 | 77.6 | 68.5 | 55.2 | 44.6 | 66.0 |
| 2004 | 46.4 | 45.8 | 59.9 | 63.8 | 76.3 | 81.5 | 82.8 | 79.2 | 73.8 | 65.8 | 50.3 | 43.8 | 64.1 |
| 2005 | 48.7 | 50.0 | 55.4 | 64.8 | 74.0 | 83.8 | 85.6 | 80.0 | 78.1 | 65.0 | 55.4 | 47.9 | 65.7 |
| 2006 | 49.1 | 52.3 | 59.1 | 69.8 | 78.2 | 83.6 | 83.7 | 78.8 | 71.5 | 64.5 | 56.5 | 43.6 | 65.9 |
| 2007 | 42.6 | 50.6 | 59.3 | 64.3 | 72.6 | 81.7 | 82.2 | 83.4 | 77.6 | 68.0 | 55.5 | 46.0 | 65.3 |
| 2008 | 45.3 | 52.8 | 57.9 | 65.4 | 74.2 | 85.3 | 80.5 | 79.2 | 71.9 | 65.7 | 54.1 | 47.8 | 65.0 |
| 2009 | 48.1 | 53.4 | 59.6 | 65.7 | 77.4 | 81.8 | 86.7 | 84.1 | 75.6 | 65.4 | 55.2 | 44.0 | 66.4 |
| POR= 62 YRS | 44.7 | 49.4 | 55.9 | 64.3 | 73.1 | 81.5 | 82.7 | 80.8 | 75.0 | 64.7 | 52.3 | 44.9 | 64.1 |

## HEATING DEGREE DAYS (base 65°F) 2009 EL PASO (KELP)

| YEAR | JUL | AUG | SEP | OCT | NOV | DEC | JAN | FEB | MAR | APR | MAY | JUN | TOTAL |
|------|-----|-----|-----|-----|-----|-----|-----|-----|-----|-----|-----|-----|-------|
| 1980-81 | 0 | 0 | 2 | 203 | 467 | 503 | 607 | 405 | 233 | 82 | 2 | 0 | 2504 |
| 1981-82 | 0 | 0 | 0 | 93 | 313 | 499 | 697 | 449 | 237 | 82 | 17 | 0 | 2387 |
| 1982-83 | 0 | 0 | 0 | 88 | 344 | 668 | 720 | 430 | 316 | 284 | 23 | 0 | 2873 |
| 1983-84 | 0 | 0 | 0 | 52 | 317 | 599 | 633 | 514 | 285 | 126 | 8 | 0 | 2534 |
| 1984-85 | 0 | 0 | 18 | 144 | 404 | 592 | 768 | 537 | 302 | 71 | 5 | 0 | 2841 |
| 1985-86 | 0 | 0 | 10 | 125 | 358 | 670 | 621 | 356 | 283 | 47 | 22 | 0 | 2492 |
| 1986-87 | 0 | 0 | 1 | 116 | 460 | 687 | 725 | 521 | 420 | 173 | 5 | 0 | 3108 |
| 1987-88 | 0 | 0 | 0 | 13 | 405 | 750 | 686 | 474 | 360 | 121 | 16 | 0 | 2825 |
| 1988-89 | 0 | 2 | 2 | 17 | 337 | 684 | 661 | 377 | 208 | 52 | 9 | 0 | 2349 |
| 1989-90 | 0 | 0 | 2 | 108 | 344 | 708 | 640 | 439 | 271 | 45 | 20 | 0 | 2577 |
| 1990-91 | 0 | 2 | 8 | 88 | 354 | 617 | 636 | 383 | 336 | 78 | 3 | 0 | 2505 |
| 1991-92 | 0 | 0 | 26 | 66 | 434 | 576 | 644 | 423 | 209 | 49 | 3 | 0 | 2430 |
| 1992-93 | 0 | 0 | 0 | 21 | 502 | 624 | 504 | 373 | 222 | 53 | 4 | 0 | 2303 |
| 1993-94 | 0 | 0 | 0 | 145 | 375 | 551 | 596 | 401 | 218 | 38 | 0 | 0 | 2324 |
| 1994-95 | 0 | 0 | 0 | 41 | 313 | 475 | 548 | 240 | 185 | 84 | 6 | 0 | 1892 |
| 1995-96 | 0 | 0 | 15 | 31 | 292 | 533 | 562 | 305 | 322 | 89 | 0 | 0 | 2149 |
| 1996-97 | 0 | 0 | 0 | 103 | | 571 | 635 | 467 | 202 | 155 | 0 | 0 | |
| 1997-98 | 0 | 0 | 0 | 150 | 378 | 746 | 550 | 481 | 342 | 187 | 0 | 0 | 2834 |
| 1998-99 | 0 | 0 | 0 | 80 | 289 | 590 | 518 | 315 | 218 | 114 | 1 | 0 | 2125 |
| 1999-00 | 0 | 0 | 4 | 108 | 253 | 662 | 466 | 314 | 242 | 49 | 2 | 0 | 2100 |
| 2000-01 | 0 | 0 | 3 | 145 | 513 | 644 | 711 | 412 | 265 | 71 | 0 | 0 | 2764 |
| 2001-02 | 0 | 0 | 0 | 18 | 315 | 684 | 578 | 496 | 264 | 12 | 0 | 0 | 2367 |
| 2002-03 | 0 | 0 | 0 | 87 | 377 | 647 | 499 | 409 | 291 | 65 | 1 | 0 | 2376 |
| 2003-04 | 0 | 0 | 0 | 41 | 302 | 625 | 568 | 549 | 188 | 100 | 6 | 0 | 2379 |
| 2004-05 | 0 | 0 | 1 | 44 | 434 | 654 | 498 | 416 | 293 | 75 | 9 | 0 | 2424 |
| 2005-06 | 0 | 0 | 0 | 69 | 293 | 523 | 486 | 349 | 189 | 16 | 0 | 0 | 1925 |
| 2006-07 | 0 | 0 | 0 | 75 | 251 | 654 | 689 | 397 | 202 | 82 | 5 | 0 | 2355 |
| 2007-08 | 0 | 0 | 0 | 51 | 279 | 581 | 607 | 349 | 235 | 73 | 10 | 0 | 2185 |
| 2008-09 | 0 | 0 | 1 | 56 | 330 | 527 | 515 | 332 | 183 | 82 | 0 | 0 | 2026 |
| 2009- | 0 | 0 | 2 | 98 | 289 | 643 | | | | | | | |

WBAN : 23044

## COOLING DEGREE DAYS (base 65°F) 2009 EL PASO (KELP)

| YEAR | JAN | FEB | MAR | APR | MAY | JUN | JUL | AUG | SEP | OCT | NOV | DEC | TOTAL |
|------|-----|-----|-----|-----|-----|-----|-----|-----|-----|-----|-----|-----|-------|
| 1980 | 0 | 0 | 0 | 34 | 198 | 646 | 693 | 546 | 329 | 63 | 0 | 0 | 2509 |
| 1981 | 0 | 2 | 2 | 84 | 275 | 534 | 586 | 455 | 333 | 89 | 0 | 2 | 2362 |
| 1982 | 0 | 0 | 14 | 70 | 167 | 484 | 602 | 583 | 371 | 82 | 0 | 0 | 2373 |
| 1983 | 0 | 0 | 0 | 32 | 151 | 374 | 564 | 527 | 417 | 108 | 6 | 0 | 2179 |
| 1984 | 0 | 0 | 2 | 44 | 324 | 441 | 507 | 482 | 260 | 38 | 8 | 0 | 2106 |
| 1985 | 0 | 0 | 9 | 55 | 233 | 428 | 457 | 488 | 252 | 18 | 0 | 0 | 1940 |
| 1986 | 0 | 3 | 0 | 122 | 228 | 391 | 474 | 485 | 281 | 41 | 0 | 0 | 2025 |
| 1987 | 0 | 0 | 0 | 22 | 121 | 399 | 521 | 446 | 222 | 83 | 2 | 0 | 1816 |
| 1988 | 0 | 0 | 9 | 15 | 186 | 429 | 478 | 399 | 226 | 76 | 16 | 0 | 1834 |
| 1989 | 0 | 2 | 20 | 130 | 300 | 494 | 530 | 442 | 261 | 60 | 0 | 0 | 2239 |
| 1990 | 0 | 1 | 0 | 96 | 278 | 667 | 480 | 374 | 282 | 43 | 0 | 0 | 2221 |
| 1991 | 0 | 0 | 0 | 55 | 245 | 424 | 428 | 455 | 204 | 88 | 0 | 0 | 1899 |
| 1992 | 0 | 0 | 3 | 144 | 182 | 500 | 628 | 511 | 389 | 75 | 0 | 0 | 2432 |
| 1993 | 0 | 0 | 11 | 110 | 306 | 538 | 632 | 550 | 331 | 132 | 2 | 0 | 2612 |
| 1994 | 0 | 0 | 17 | 92 | 338 | 725 | 722 | 666 | 404 | 100 | 3 | 0 | 3067 |
| 1995 | 0 | 0 | 27 | 97 | 304 | 462 | 567 | 559 | 292 | 58 | 1 | 0 | 2367 |
| 1996 | 0 | 6 | 7 | 112 | 475 | 560 | 561 | 462 | 263 | 125 | | 0 | |
| 1997 | 0 | 0 | 10 | 53 | 298 | 489 | 556 | 518 | 408 | 116 | 0 | 0 | 2448 |
| 1998 | 0 | 0 | 8 | 30 | 299 | 527 | 581 | 491 | 434 | 116 | 0 | 0 | 2486 |
| 1999 | 0 | 0 | 8 | 67 | 291 | 457 | 506 | 511 | 324 | 81 | 5 | 0 | 2250 |
| 2000 | 0 | 0 | 4 | 159 | 443 | 463 | 600 | 517 | 417 | 77 | 0 | 0 | 2680 |
| 2001 | 0 | 0 | 4 | 95 | 351 | 554 | 606 | 506 | 365 | 108 | 13 | 0 | 2602 |
| 2002 | 0 | 0 | 16 | 166 | 316 | 612 | 543 | 609 | 365 | 74 | 0 | 0 | 2701 |
| 2003 | 0 | 0 | 3 | 83 | 346 | 494 | 626 | 591 | 386 | 155 | 13 | 0 | 2697 |
| 2004 | 0 | 0 | 37 | 72 | 363 | 501 | 559 | 448 | 274 | 75 | 0 | 0 | 2329 |
| 2005 | 0 | 0 | 2 | 75 | 298 | 569 | 647 | 473 | 398 | 75 | 9 | 0 | 2546 |
| 2006 | 0 | 1 | 12 | 167 | 416 | 566 | 585 | 437 | 201 | 68 | 4 | 0 | 2457 |
| 2007 | 0 | 0 | 33 | 65 | 251 | 508 | 543 | 577 | 384 | 150 | 1 | 0 | 2512 |
| 2008 | 0 | 0 | 24 | 90 | 301 | 617 | 487 | 448 | 212 | 85 | 8 | 0 | 2272 |
| 2009 | 0 | 11 | 26 | 109 | 393 | 511 | 679 | 597 | 325 | 117 | 0 | 0 | 2768 |

## SNOWFALL (inches)  2009  EL PASO (KELP)

| YEAR | JUL | AUG | SEP | OCT | NOV | DEC | JAN | FEB | MAR | APR | MAY | JUN | TOTAL |
|---|---|---|---|---|---|---|---|---|---|---|---|---|---|
| 1980-81 | 0.0 | 0.0 | 0.0 | 1.0 | 4.0 | 0.0 | 4.8 | 0.0 | 0.0 | 0.0 | 0.0 | 0.0 | 9.8 |
| 1981-82 | 0.0 | 0.0 | 0.0 | 0.0 | 0.0 | 0.0 | T | T | 0.0 | 0.0 | 0.0 | 0.0 | T |
| 1982-83 | 0.0 | 0.0 | 0.0 | 0.0 | 0.3 | 18.2 | T | 0.0 | T | 16.5 | 0.0 | 0.0 | 35.0 |
| 1983-84 | 0.0 | 0.0 | 0.0 | 0.0 | T | T | T | 0.0 | 6.1 | 0.0 | 0.0 | 0.0 | 6.1 |
| 1984-85 | 0.0 | 0.0 | 0.0 | 0.0 | T | 2.9 | 5.4 | 1.1 | 0.0 | 0.0 | 0.0 | 0.0 | 9.4 |
| 1985-86 | 0.0 | 0.0 | 0.0 | 0.0 | 0.0 | 0.9 | T | 0.9 | T | 0.0 | 0.0 | 0.0 | 1.8 |
| 1986-87 | 0.0 | 0.0 | 0.0 | 0.0 | 0.0 | 0.6 | 3.4 | 2.8 | 0.6 | 0.0 | 0.0 | 0.0 | 7.4 |
| 1987-88 | 0.0 | 0.0 | 0.0 | 0.0 | 0.0 | 25.9 | T | 6.6 | 0.0 | 0.0 | 0.0 | 0.0 | 32.5 |
| 1988-89 | 0.0 | 0.0 | 0.0 | 0.0 | T | 0.3 | 0.0 | 0.0 | T | 0.0 | T | 0.0 | 0.3 |
| 1989-90 | 0.0 | 0.0 | 0.0 | 0.0 | 0.0 | T | T | T | 0.0 | T | T | 0.0 | T |
| 1990-91 | T | 0.0 | 0.0 | 0.0 | 4.3 | T | 1.5 | T | T | 0.0 | T | 0.0 | 5.8 |
| 1991-92 | 0.0 | 0.0 | 0.0 | 0.0 | 0.0 | 0.5 | 6.9 | 0.0 | T | 0.0 | T | T | 7.4 |
| 1992-93 | 0.0 | 0.0 | 0.0 | 0.0 | 0.0 | 1.6 | 0.0 | 0.0 | 0.0 | 0.0 | 0.0 | 0.0 | 1.6 |
| 1993-94 | 0.0 | 0.0 | T | 0.8 | 0.0 | 1.6 | 0.3 | 0.0 | 0.0 | 0.0 | 0.0 | 0.0 | 2.7 |
| 1994-95 | 0.0 | 0.0 | 0.0 | 0.0 | 0.0 | 0.0 | 0.0 | 0.0 | 0.0 | 0.0 | 0.0 | 0.0 | 0.0 |
| 1995-96 | 0.0 | 0.0 | 0.0 | 0.0 | 0.0 | T | 1.6 | 0.0 |  | T | 0.0 |  |  |
| 1996-97 |  |  |  |  |  |  |  |  |  |  |  |  |  |
| 1997-98 |  |  |  |  |  |  |  |  |  |  |  |  |  |
| 1998-99 |  |  |  |  |  |  |  |  |  |  |  |  |  |
| 1999-00 |  |  |  |  |  |  |  |  |  |  |  |  |  |
| 2000-01 |  |  |  |  |  |  |  |  |  |  |  |  |  |
| 2001-02 |  |  |  |  |  |  |  |  |  |  |  |  |  |
| 2002-03 |  |  |  |  |  |  |  |  |  |  |  |  |  |
| 2003-04 |  |  |  |  |  |  |  |  |  |  |  |  |  |
| 2004-05 |  |  |  |  |  |  |  |  |  |  |  |  |  |
| 2005- |  |  |  |  |  |  |  |  |  |  |  |  |  |
| 2006-07 |  |  |  |  |  |  | 2.1 | T | 0.0 | 0.0 | T | 0.0 |  |
| 2007-08 | T | T | T | 0.0 | 7.7 | 0.0 | 0.0 | T | T | 0.0 | 0.0 | 0.0 | 7.7 |
| 2008-09 | 0.0 | 0.0 | 0.0 | 0.0 | 0.0 | T | T | T | 0.0 | 0.0 | 0.0 | 0.0 | T |
| 2009- | 0.0 | 0.0 | T | T | 2.5 | 5.8 |  |  |  |  |  |  |  |
| POR= 61 YRS | T | T | T | T | 0.9 | 1.6 | 1.0 | 0.7 | 0.3 | 0.3 | T | T | 4.8 |

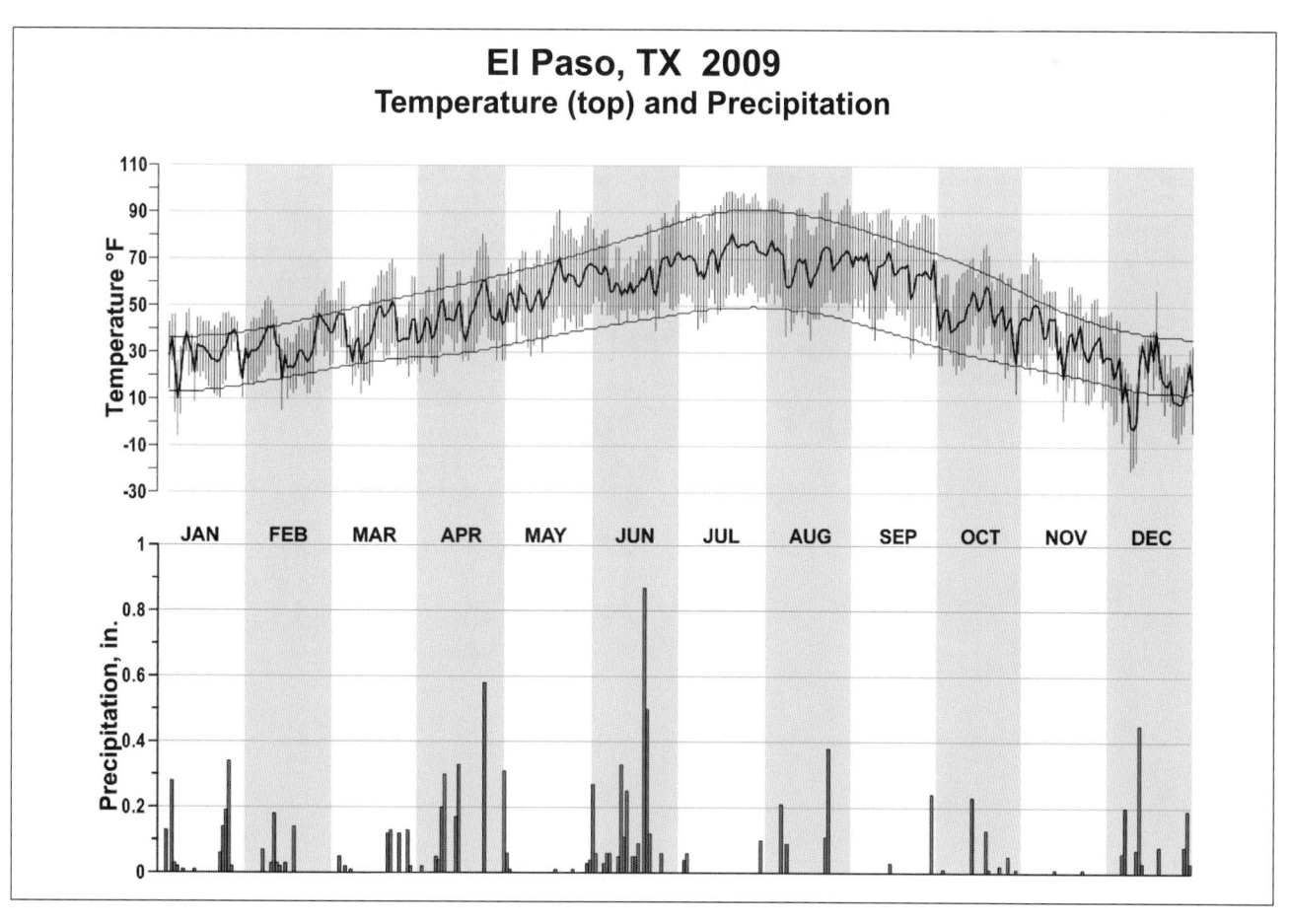

# El Paso, TX  2009
## Temperature (top) and Precipitation

# 2009
# HOUSTON
# TEXAS (KIAH)

Houston, the largest city in Texas, is located in the flat Coastal Plains, about 50 miles from the Gulf of Mexico and about 25 miles from Galveston Bay. The climate is predominantly marine. The terrain includes numerous small streams and bayous which, together with the nearness to Galveston Bay, favor the development of both ground and advective fogs. Prevailing winds are from the southeast and south, except in January, when frequent passages of high pressure areas bring invasions of polar air and prevailing northerly winds.

Temperatures are moderated by the influence of winds from the Gulf, which result in mild winters. Another effect of the nearness of the Gulf is abundant rainfall, except for rare extended dry periods. Polar air penetrates the area frequently enough to provide variability in the weather.

Records of sky cover for daylight hours indicate about one-fourth of the days per year as clear, with a high number of clear days in October and November. Cloudy days are relatively frequent from December to May and partly cloudy days are the more frequent for June through September. Sunshine averages nearly 60 percent of the possible amount for the year ranging from 42 percent in January to 67 percent in June.

Heavy fog occurs on an average of 16 days a year and light fog occurs about 62 days a year in the city. The frequency of heavy fog is considerably higher at William P. Hobby Airport and at Intercontinental Airport.

Destructive windstorms are fairly infrequent, but both thundersqualls and tropical storms occasionally pass through the area.

# NORMALS, MEANS, AND EXTREMES
## HOUSTON (KIAH)

**LATITUDE:** 29 ° 59'N  **LONGITUDE:** -95 ° 21'W   **ELEVATION (FT):** GRND: 94  BARO: 107   **TIME ZONE:** CENTRAL  (UTC -6)   **WBAN: 12960**

| | ELEMENT | POR | JAN | FEB | MAR | APR | MAY | JUN | JUL | AUG | SEP | OCT | NOV | DEC | YEAR |
|---|---|---|---|---|---|---|---|---|---|---|---|---|---|---|---|
| **TEMPERATURE °F** | NORMAL DAILY MAXIMUM | 30 | 62.3 | 66.5 | 73.3 | 79.1 | 85.5 | 90.7 | 93.6 | 93.5 | 89.3 | 82.0 | 72.0 | 64.6 | 79.4 |
| | MEAN DAILY MAXIMUM | 40 | 62.2 | 66.3 | 72.9 | 79.2 | 85.7 | 91.0 | 93.7 | 93.7 | 89.1 | 81.6 | 71.9 | 64.6 | 79.3 |
| | HIGHEST DAILY MAXIMUM | 40 | 84 | 91 | 91 | 95 | 99 | 104 | 104 | 107 | 109 | 96 | 89 | 85 | 109 |
| | YEAR OF OCCURRENCE | | 1975 | 1986 | 1989 | 1987 | 1996 | 2009 | 1980 | 2000 | 2000 | 1991 | 1989 | 1995 | SEP 2000 |
| | MEAN OF EXTREME MAXS. | 43 | 78.5 | 81.2 | 85.2 | 88.4 | 92.9 | 97.0 | 98.6 | 99.4 | 96.4 | 91.3 | 85.0 | 80.1 | 89.5 |
| | NORMAL DAILY MINIMUM | 30 | 41.2 | 44.3 | 51.3 | 57.9 | 66.1 | 71.8 | 73.5 | 73.0 | 68.4 | 58.8 | 49.8 | 42.8 | 58.2 |
| | MEAN DAILY MINIMUM | 40 | 41.9 | 44.9 | 51.5 | 58.3 | 66.1 | 71.5 | 73.6 | 73.3 | 68.8 | 59.3 | 50.1 | 43.5 | 58.6 |
| | LOWEST DAILY MINIMUM | 40 | 12 | 20 | 22 | 31 | 44 | 52 | 62 | 60 | 48 | 29 | 19 | 7 | 7 |
| | YEAR OF OCCURRENCE | | 1982 | 1997 | 2002 | 1987 | 1978 | 1970 | 1990 | 1992 | 1975 | 1993 | 1976 | 1989 | DEC 1989 |
| | MEAN OF EXTREME MINS. | 43 | 25.6 | 28.4 | 33.4 | 40.8 | 52.6 | 63.0 | 68.8 | 67.9 | 55.7 | 42.6 | 32.9 | 27.1 | 44.9 |
| | NORMAL DRY BULB | 30 | 51.8 | 55.4 | 62.3 | 68.5 | 75.8 | 81.3 | 83.6 | 83.3 | 78.9 | 70.4 | 60.9 | 53.7 | 68.8 |
| | MEAN DRY BULB | 43 | 52.3 | 55.5 | 62.1 | 69.0 | 75.9 | 81.4 | 83.6 | 83.5 | 78.9 | 70.5 | 61.1 | 54.2 | 69.0 |
| | MEAN WET BULB | 26 | 47.4 | 50.6 | 56.0 | 62.0 | 69.3 | 73.8 | 75.4 | 75.2 | 71.1 | 63.9 | 55.8 | 49.1 | 62.5 |
| | MEAN DEW POINT | 26 | 44.1 | 47.5 | 52.7 | 59.1 | 67.1 | 72.0 | 73.3 | 72.9 | 68.6 | 61.3 | 53.0 | 46.1 | 59.8 |
| | NORMAL NO. DAYS WITH: | | | | | | | | | | | | | | |
| | MAXIMUM >= 90 | 30 | 0.0 | 0.1 | 0.1 | 0.9 | 6.4 | 20.1 | 27.3 | 25.7 | 15.9 | 3.1 | 0.0 | 0.0 | 99.6 |
| | MAXIMUM <= 32 | 30 | 0.1 | 0.2 | 0.0 | 0.0 | 0.0 | 0.0 | 0.0 | 0.0 | 0.0 | 0.0 | 0.0 | 0.2 | 0.5 |
| | MINIMUM <= 32 | 30 | 6.3 | 3.8 | 1.3 | 0.1 | 0.0 | 0.0 | 0.0 | 0.0 | 0.0 | 0.1 | 1.3 | 5.1 | 18.0 |
| | MINIMUM <= 0 | 30 | 0.0 | 0.0 | 0.0 | 0.0 | 0.0 | 0.0 | 0.0 | 0.0 | 0.0 | 0.0 | 0.0 | 0.0 | 0.0 |
| **H/C** | NORMAL HEATING DEG. DAYS | 30 | 427 | 298 | 156 | 48 | 2 | 0 | 0 | 0 | 1 | 37 | 189 | 367 | 1525 |
| | NORMAL COOLING DEG. DAYS | 30 | 15 | 21 | 63 | 147 | 328 | 485 | 573 | 563 | 412 | 196 | 65 | 25 | 2893 |
| **RH** | NORMAL (PERCENT) | 30 | 75 | 74 | 73 | 74 | 76 | 76 | 74 | 75 | 76 | 75 | 77 | 76 | 75 |
| | HOUR 00 LST | 30 | 82 | 83 | 83 | 85 | 87 | 88 | 86 | 87 | 88 | 88 | 86 | 84 | 86 |
| | HOUR 06 LST | 30 | 86 | 87 | 88 | 89 | 92 | 93 | 93 | 94 | 93 | 91 | 89 | 87 | 90 |
| | HOUR 12 LST | 30 | 64 | 61 | 59 | 58 | 60 | 60 | 57 | 57 | 59 | 57 | 61 | 62 | 60 |
| | HOUR 18 LST | 30 | 67 | 62 | 60 | 60 | 64 | 64 | 62 | 62 | 66 | 68 | 72 | 71 | 65 |
| **S** | PERCENT POSSIBLE SUNSHINE | 27 | 45 | 50 | 54 | 58 | 62 | 68 | 70 | 68 | 66 | 64 | 52 | 51 | 59 |
| **W/O** | MEAN NO. DAYS WITH: | | | | | | | | | | | | | | |
| | HEAVY FOG(VISBY <= 1/4 MI) | 40 | 4.2 | 3.2 | 2.9 | 2.1 | 1.5 | 0.6 | 0.3 | 0.4 | 1.1 | 2.5 | 3.4 | 3.9 | 26.1 |
| | THUNDERSTORMS | 40 | 2.0 | 2.5 | 3.6 | 3.9 | 6.6 | 8.7 | 10.6 | 10.6 | 7.0 | 4.1 | 3.0 | 2.1 | 64.7 |
| **CLOUDNESS** | MEAN: | | | | | | | | | | | | | | |
| | SUNRISE-SUNSET (OKTAS) | 27 | 5.5 | 5.3 | 5.4 | 5.3 | 5.2 | 4.6 | 4.4 | 4.4 | 4.2 | 4.0 | 4.6 | 5.3 | 4.9 |
| | MIDNIGHT-MIDNIGHT (OKTAS) | 26 | 5.3 | 5.0 | 5.1 | 5.0 | 4.9 | 3.9 | 3.8 | 3.7 | 3.6 | 3.5 | 4.3 | 5.0 | 4.4 |
| | MEAN NO. DAYS WITH: | | | | | | | | | | | | | | |
| | CLEAR | 27 | 7.2 | 7.0 | 7.0 | 7.2 | 5.6 | 7.3 | 6.9 | 6.1 | 8.6 | 11.2 | 9.0 | 7.2 | 90.3 |
| | PARTLY CLOUDY | 27 | 5.4 | 5.6 | 6.4 | 7.1 | 11.0 | 13.3 | 15.9 | 16.9 | 11.5 | 8.9 | 7.0 | 5.5 | 114.5 |
| | CLOUDY | 27 | 18.4 | 15.6 | 17.6 | 15.7 | 14.3 | 9.3 | 8.2 | 8.0 | 10.0 | 10.9 | 14.0 | 18.3 | 160.3 |
| **PR** | MEAN STATION PRESSURE(IN) | 26 | 30.03 | 29.98 | 29.91 | 29.86 | 29.83 | 29.83 | 29.88 | 29.87 | 29.85 | 29.92 | 29.98 | 30.02 | 29.91 |
| | MEAN SEA-LEVEL PRES. (IN) | 26 | 30.15 | 30.10 | 30.02 | 29.97 | 29.94 | 29.94 | 29.99 | 29.98 | 29.97 | 30.03 | 30.09 | 30.14 | 30.03 |
| **WINDS** | MEAN SPEED (MPH) | 26 | 8.0 | 8.4 | 8.7 | 8.8 | 8.1 | 7.1 | 6.3 | 5.9 | 6.4 | 7.0 | 7.4 | 7.6 | 7.5 |
| | PREVAIL.DIR(TENS OF DEGS) | 37 | 36 | 17 | 17 | 17 | 17 | 17 | 17 | 17 | 01 | 14 | 36 | 36 | 17 |
| | MAXIMUM 2-MINUTE: | | | | | | | | | | | | | | |
| | SPEED (MPH) | 13 | 43 | 40 | 35 | 39 | 47 | 45 | 39 | 48 | 47 | 36 | 36 | 46 | 48 |
| | DIR. (TENS OF DEGS) | | 34 | 10 | 14 | 14 | 31 | 13 | 03 | 31 | 01 | 35 | 29 | 14 | 31 |
| | YEAR OF OCCURRENCE | | 2006 | 2000 | 2004 | 2007 | 2006 | 2004 | 2005 | 2009 | 2008 | 2007 | 2006 | 2002 | AUG 2009 |
| | MAXIMUM 3-SECOND | | | | | | | | | | | | | | |
| | SPEED (MPH) | 13 | 49 | 44 | 43 | 48 | 58 | 66 | 49 | 70 | 61 | 47 | 47 | 52 | 70 |
| | DIR. (TENS OF DEGS) | | 33 | 10 | 34 | 31 | 31 | 15 | 33 | 32 | 34 | 13 | 29 | 13 | 32 |
| | YEAR OF OCCURRENCE | | 2006 | 2000 | 2008 | 2009 | 2006 | 2000 | 1998 | 2009 | 2005 | 1998 | 2006 | 2002 | AUG 2009 |
| **PRECIPITATION** | NORMAL (IN) | 30 | 3.68 | 2.98 | 3.36 | 3.60 | 5.15 | 5.35 | 3.18 | 3.83 | 4.33 | 4.50 | 4.19 | 3.69 | 47.84 |
| | MAXIMUM MONTHLY (IN) | 40 | 9.78 | 6.10 | 8.52 | 10.92 | 14.39 | 19.21 | 9.94 | 10.58 | 12.07 | 16.05 | 11.73 | 9.34 | 19.21 |
| | YEAR OF OCCURRENCE | | 1991 | 2005 | 1972 | 1976 | 1970 | 2001 | 2007 | 1996 | 2008 | 1984 | 2004 | 1991 | JUN 2001 |
| | MINIMUM MONTHLY (IN) | 40 | 0.36 | 0.38 | 0.12 | 0.43 | 0.04 | .08 | 0.47 | 0.31 | 0.80 | 0.05 | 0.41 | 0.64 | 0.04 |
| | YEAR OF OCCURRENCE | | 1971 | 1976 | 1996 | 1983 | 1998 | 2005 | 1993 | 1990 | 1975 | 1978 | 1988 | 1973 | MAY 1998 |
| | MAXIMUM IN 24 HOURS (IN) | 40 | 2.73 | 2.22 | 7.47 | 8.16 | 10.36 | 11.02 | 5.40 | 6.83 | 7.98 | 9.31 | 6.33 | 5.64 | 11.02 |
| | YEAR OF OCCURRENCE | | 1995 | 1985 | 1972 | 1976 | 1989 | 2001 | 2002 | 1981 | 1976 | 1984 | 1998 | 2005 | JUN 2001 |
| | NORMAL NO. DAYS WITH: | | | | | | | | | | | | | | |
| | PRECIPITATION >= 0.01 | 30 | 10.5 | 8.5 | 9.1 | 7.0 | 8.2 | 9.7 | 8.9 | 8.9 | 8.9 | 7.5 | 8.6 | 9.4 | 105.2 |
| | PRECIPITATION >= 1.00 | 30 | 1.1 | 0.9 | 0.8 | 1.2 | 1.7 | 1.7 | 0.9 | 1.1 | 1.2 | 1.5 | 1.2 | 1.2 | 14.5 |
| **SNOWFALL** | NORMAL (IN) | 30 | 0.1 | 0.2 | 0.0 | 0.0 | 0.0 | 0.0 | 0.0 | 0.0 | 0.0 | 0.0 | 0.* | 0.1 | 0.4 |
| | MAXIMUM MONTHLY (IN) | 40 | 2.0 | 2.8 | T | T | T | T | 0.0 | 0.0 | 0.0 | 0.0 | T | 1.7 | 2.8 |
| | YEAR OF OCCURRENCE | | 1973 | 1973 | 2009 | 1993 | 1993 | 1996 | | | | | 1979 | 1989 | FEB 1973 |
| | MAXIMUM IN 24 HOURS (IN) | 40 | 2.0 | 1.4 | T | T | T | T | 0.0 | 0.0 | 0.0 | 0.0 | T | 1.7 | 2.0 |
| | YEAR OF OCCURRENCE' | | 1973 | 1980 | 2009 | 1993 | 1993 | 1996 | | | | | 1979 | 1989 | JAN 1973 |
| | MAXIMUM SNOW DEPTH (IN) | 39 | 1 | 1 | 0 | 0 | 0 | 0 | 0 | 0 | 0 | 0 | 0 | 0 | 1 |
| | YEAR OF OCCURRENCE | | 1973 | 1973 | | | | | | | | | | | FEB 1973 |
| | NORMAL NO. DAYS WITH: | | | | | | | | | | | | | | |
| | SNOWFALL >= 1.0 | 30 | 0.1 | 0.1 | 0.0 | 0.0 | 0.0 | 0.0 | 0.0 | 0.0 | 0.0 | 0.0 | 0.0 | 0.0 | 0.2 |

## PRECIPITATION (inches) 2009 HOUSTON (KIAH)

| YEAR | JAN | FEB | MAR | APR | MAY | JUN | JUL | AUG | SEP | OCT | NOV | DEC | ANNUAL |
|------|-----|-----|-----|-----|-----|-----|-----|-----|-----|-----|-----|-----|--------|
| 1980 | 6.09 | 2.54 | 5.39 | 2.05 | 5.63 | 0.92 | 1.57 | 1.40 | 6.00 | 4.03 | 2.12 | 1.25 | 38.99 |
| 1981 | 2.32 | 2.21 | 1.74 | 2.69 | 8.75 | 9.65 | 4.43 | 7.01 | 2.91 | 6.96 | 5.26 | 2.05 | 55.98 |
| 1982 | 1.82 | 1.59 | 1.55 | 2.28 | 6.87 | 1.10 | 4.32 | 1.90 | 0.98 | 6.64 | 8.91 | 4.91 | 42.87 |
| 1983 | 2.00 | 3.97 | 3.85 | 0.43 | 7.29 | 5.37 | 5.23 | 9.42 | 7.23 | 1.56 | 3.17 | 3.69 | 53.21 |
| 1984 | 3.99 | 4.37 | 2.41 | 0.56 | 3.13 | 1.99 | 3.43 | 3.52 | 3.87 | 16.05 | 2.28 | 2.59 | 48.19 |
| 1985 | 2.10 | 5.38 | 4.52 | 4.31 | 1.57 | 5.29 | 4.93 | 1.14 | 4.67 | 6.54 | 4.84 | 3.85 | 49.14 |
| 1986 | 0.71 | 2.74 | 1.44 | 2.63 | 4.29 | 6.34 | 0.61 | 3.27 | 3.70 | 6.83 | 6.66 | 5.71 | 44.93 |
| 1987 | 2.42 | 4.26 | 0.88 | 0.47 | 5.39 | 9.31 | 4.79 | 1.48 | 3.46 | 0.17 | 3.41 | 4.56 | 40.60 |
| 1988 | 1.27 | 1.29 | 4.88 | 1.26 | 1.32 | 2.00 | 3.23 | 3.52 | 1.20 | 1.29 | 0.41 | 1.26 | 22.93 |
| 1989 | 4.80 | 0.90 | 3.96 | 1.48 | 13.56 | 16.28 | 1.92 | 2.74 | 2.69 | 1.76 | 1.84 | 0.80 | 52.73 |
| 1990 | 3.96 | 4.54 | 5.11 | 6.21 | 2.23 | 2.98 | 4.85 | 0.31 | 1.57 | 3.79 | 3.01 | 1.81 | 40.37 |
| 1991 | 9.78 | 5.79 | 1.77 | 8.06 | 4.02 | 7.69 | 1.31 | 2.97 | 2.76 | 2.57 | 5.03 | 9.34 | 61.09 |
| 1992 | 7.70 | 5.99 | 6.28 | 3.74 | 7.05 | 3.38 | 3.85 | 2.78 | 1.08 | 1.03 | 5.99 | 3.46 | 52.33 |
| 1993 | 5.79 | 2.67 | 6.41 | 7.88 | 8.50 | 12.08 | 0.47 | 1.82 | 1.10 | 5.32 | 3.27 | 2.68 | 57.99 |
| 1994 | 2.08 | 2.79 | 2.39 | 2.11 | 5.02 | 3.40 | 1.60 | 5.45 | 1.12 | 10.62 | 1.67 | 4.90 | 43.15 |
| 1995 | 5.95 | 2.55 | 4.11 | 2.59 | 3.83 | 4.11 | 2.68 | 4.90 | 2.52 | 2.77 | 3.63 | 4.99 | 44.63 |
| 1996 | 0.88 | 1.29 | 0.12 | 2.05 | 0.56 | 8.37 | 1.11 | 10.58 | 6.96 | 2.60 | 4.55 | 3.74 | 42.81 |
| 1997 | 3.26 | 5.35 | 7.96 | 7.17 | 6.69 | 4.46 | 2.30 | 2.26 | 4.86 | 7.11 | 3.38 | 5.42 | 60.22 |
| 1998 | 4.35 | 5.85 | 2.32 | 1.21 | 0.04 | 2.87 | 1.65 | 4.38 | 10.16 | 7.79 | 10.21 | 4.01 | 54.84 |
| 1999 | 2.12 | 0.80 | 3.44 | 1.06 | 4.10 | 5.26 | 5.11 | 0.50 | 1.36 | 0.56 | 1.53 | 2.20 | 28.04 |
| 2000 | 1.25 | 2.32 | 1.35 | 5.52 | 12.35 | 3.29 | 0.64 | 2.11 | 4.34 | 3.27 | 8.50 | 2.69 | 47.63 |
| 2001 | 4.25 | 0.82 | 7.97 | 2.00 | 3.53 | 19.21 | 2.05 | 4.83 | 8.82 | 8.95 | 2.58 | 6.18 | 71.19 |
| 2002 | 1.24 | 0.89 | 2.36 | 3.79 | 1.79 | 4.54 | 7.11 | 5.47 | 8.02 | 14.65 | 4.20 | 5.65 | 59.71 |
| 2003 | 2.09 | 4.08 | 2.04 | 1.46 | 0.06 | 3.62 | 5.35 | 4.47 | 6.79 | 4.99 | 7.80 | 2.99 | 45.74 |
| 2004 | 6.01 | 5.58 | 2.23 | 5.56 | 7.33 | 18.33 | 0.79 | 2.49 | 1.01 | 2.05 | 11.73 | 1.95 | 65.06 |
| 2005 | 3.41 | 6.10 | 4.05 | 1.28 | 6.06 | 0.08 | 5.30 | 1.52 | 2.63 | 1.69 | 2.72 | 6.37 | 41.21 |
| 2006 | 2.50 | 1.46 | 2.36 | 2.93 | 8.78 | 7.84 | 7.85 | 3.40 | 3.22 | 14.53 | 0.92 | 2.07 | 57.86 |
| 2007 | 5.72 | 1.15 | 6.40 | 3.86 | 9.88 | 3.07 | 9.94 | 8.05 | 3.05 | 6.85 | 5.49 | 2.06 | 65.52 |
| 2008 | 4.62 | 4.00 | 2.41 | 1.46 | 4.57 | 2.06 | 1.09 | 7.45 | 12.07 | 8.67 | 2.92 | 1.68 | 53.00 |
| 2009 | 0.49 | 1.52 | 4.08 | 10.38 | 0.38 | 0.27 | 2.84 | 2.11 | 4.68 | 13.16 | 1.66 | 5.44 | 47.01 |
| POR= 43 YRS | 3.63 | 2.95 | 3.43 | 3.57 | 5.39 | 5.41 | 3.68 | 3.81 | 4.68 | 5.35 | 4.01 | 3.63 | 49.54 |

WBAN : 12960

## AVERAGE TEMPERATURE (°F) 2009 HOUSTON (KIAH)

| YEAR | JAN | FEB | MAR | APR | MAY | JUN | JUL | AUG | SEP | OCT | NOV | DEC | ANNUAL |
|------|-----|-----|-----|-----|-----|-----|-----|-----|-----|-----|-----|-----|--------|
| 1980 | 55.0 | 53.7 | 60.9 | 66.2 | 77.3 | 85.1 | 87.5 | 86.6 | 83.2 | 67.8 | 58.0 | 55.2 | 69.7 |
| 1981 | 51.4 | 55.4 | 60.9 | 74.3 | 75.3 | 82.7 | 84.4 | 84.4 | 78.6 | 72.3 | 64.4 | 54.5 | 69.9 |
| 1982 | 52.9 | 52.1 | 64.9 | 67.8 | 75.3 | 83.0 | 85.4 | 84.1 | 79.3 | 69.5 | 60.9 | 55.4 | 69.2 |
| 1983 | 50.1 | 52.5 | 58.3 | 64.0 | 73.4 | 79.0 | 82.2 | 82.6 | 76.6 | 70.1 | 63.1 | 45.7 | 66.5 |
| 1984 | 47.0 | 54.0 | 61.9 | 67.8 | 74.9 | 78.6 | 81.8 | 82.9 | 77.4 | 74.2 | 60.0 | 63.4 | 68.7 |
| 1985 | 45.7 | 49.6 | 64.7 | 70.0 | 75.6 | 81.0 | 81.6 | 84.2 | 79.8 | 72.5 | 67.0 | 51.0 | 68.6 |
| 1986 | 54.4 | 59.9 | 63.3 | 71.7 | 75.8 | 82.0 | 85.9 | 82.6 | 81.8 | 68.9 | 62.0 | 51.7 | 70.0 |
| 1987 | 51.4 | 56.1 | 58.9 | 67.2 | 77.1 | 81.3 | 83.5 | 86.2 | 78.9 | 68.7 | 60.5 | 55.6 | 68.8 |
| 1988 | 48.1 | 54.1 | 61.3 | 67.6 | 73.6 | 80.5 | 84.4 | 85.3 | 80.8 | 72.0 | 65.7 | 55.4 | 69.1 |
| 1989 | 57.5 | 52.7 | 61.3 | 69.4 | 77.8 | 79.9 | 82.4 | 81.7 | 77.0 | 70.2 | 62.9 | 44.4 | 68.1 |
| 1990 | 57.0 | 59.1 | 62.9 | 69.4 | 78.1 | 84.8 | 82.1 | 85.1 | 80.1 | 68.7 | 63.4 | 53.6 | 70.4 |
| 1991 | 50.4 | 57.4 | 63.5 | 72.2 | 78.0 | 82.0 | 84.0 | 83.0 | 77.4 | 72.3 | 56.7 | 56.2 | 69.4 |
| 1992 | 51.0 | 58.5 | 64.0 | 68.7 | 73.7 | 81.7 | 83.6 | 80.1 | 79.3 | 71.4 | 56.8 | 56.7 | 68.8 |
| 1993 | 53.6 | 56.7 | 61.1 | 65.9 | 73.4 | 81.6 | 85.8 | 86.5 | 80.2 | 69.5 | 56.9 | 54.6 | 68.8 |
| 1994 | 52.6 | 55.2 | 62.7 | 69.6 | 76.0 | 83.5 | 85.5 | 83.1 | 78.3 | 71.9 | 65.7 | 57.2 | 70.1 |
| 1995 | 54.3 | 58.7 | 62.9 | 68.6 | 77.9 | 80.6 | 84.8 | 84.9 | 81.6 | 70.4 | 61.4 | 57.1 | 70.3 |
| 1996 | 52.0 | 58.7 | 58.1 | 69.4 | 81.4 | 80.7 | 83.8 | 82.0 | 77.5 | 70.6 | 62.0 | 57.4 | 69.5 |
| 1997 | 50.8 | 55.3 | 65.3 | 64.2 | 73.6 | 79.2 | 83.1 | 83.2 | 79.1 | 68.9 | 55.7 | 50.1 | 67.4 |
| 1998 | 57.1 | 55.1 | 60.1 | 65.9 | 78.7 | 85.5 | 86.6 | 84.7 | 82.2 | 72.6 | 64.3 | 55.1 | 70.7 |
| 1999 | 57.1 | 61.5 | 63.7 | 73.0 | 76.6 | 81.9 | 83.1 | 86.8 | 78.0 | 69.0 | 62.2 | 53.7 | 70.6 |
| 2000 | 56.5 | 61.7 | 66.4 | 67.9 | 78.1 | 81.4 | 85.2 | 84.8 | 79.4 | 70.9 | 57.6 | 47.6 | 69.8 |
| 2001 | 49.3 | 59.3 | 56.4 | 71.7 | 75.9 | 80.5 | 83.6 | 83.5 | 77.0 | 66.9 | 63.4 | 56.0 | 68.6 |
| 2002 | 54.6 | 50.7 | 61.3 | 73.5 | 77.0 | 81.7 | 84.5 | 83.9 | 79.7 | 71.6 | 58.8 | 54.6 | 69.3 |
| 2003 | 50.1 | 53.8 | 61.4 | 70.5 | 80.8 | 82.7 | 83.4 | 84.7 | 77.7 | 71.7 | 65.0 | 53.8 | 69.6 |
| 2004 | 54.7 | 53.5 | 67.3 | 69.5 | 77.0 | 81.1 | 84.6 | 83.2 | 81.2 | 77.5 | 62.0 | 53.8 | 70.5 |
| 2005 | 56.3 | 58.8 | 61.7 | 67.7 | 75.3 | 83.3 | 84.5 | 84.6 | 83.4 | 71.1 | 64.5 | 53.1 | 70.4 |
| 2006 | 58.9 | 55.5 | 65.6 | 73.7 | 76.7 | 81.4 | 83.3 | 84.9 | 79.4 | 72.4 | 62.7 | 55.7 | 70.9 |
| 2007 | 50.5 | 55.2 | 65.9 | 66.7 | 76.4 | 82.5 | 82.2 | 85.4 | 81.9 | 73.0 | 63.7 | 57.7 | 70.1 |
| 2008 | 52.2 | 60.1 | 63.6 | 69.4 | 77.8 | 84.6 | 84.9 | 84.0 | 78.2 | 69.6 | 62.5 | 55.7 | 70.2 |
| 2009 | 54.0 | 61.2 | 62.9 | 68.6 | 78.2 | 85.7 | 87.4 | 86.3 | 79.0 | 71.0 | 62.5 | 50.2 | 70.6 |
| POR= 43 YRS | 52.3 | 55.5 | 62.1 | 69.0 | 75.9 | 81.4 | 83.6 | 83.5 | 78.9 | 70.5 | 61.1 | 54.2 | 69.0 |

## HEATING DEGREE DAYS (base 65°F) 2009  HOUSTON (KIAH)

| YEAR | JUL | AUG | SEP | OCT | NOV | DEC | JAN | FEB | MAR | APR | MAY | JUN | TOTAL |
|------|-----|-----|-----|-----|-----|-----|-----|-----|-----|-----|-----|-----|-------|
| 1980-81 | 0 | 0 | 0 | 67 | 255 | 323 | 416 | 291 | 144 | 6 | 1 | 0 | 1503 |
| 1981-82 | 0 | 0 | 0 | 50 | 82 | 326 | 409 | 363 | 143 | 79 | 1 | 0 | 1453 |
| 1982-83 | 0 | 0 | 0 | 53 | 175 | 328 | 457 | 346 | 219 | 96 | 0 | 0 | 1674 |
| 1983-84 | 0 | 0 | 6 | 27 | 138 | 606 | 549 | 325 | 150 | 45 | 2 | 0 | 1848 |
| 1984-85 | 0 | 0 | 6 | 12 | 204 | 144 | 591 | 432 | 91 | 22 | 0 | 0 | 1502 |
| 1985-86 | 0 | 0 | 5 | 17 | 76 | 434 | 326 | 209 | 99 | 11 | 0 | 0 | 1177 |
| 1986-87 | 0 | 0 | 0 | 28 | 175 | 411 | 421 | 245 | 196 | 82 | 0 | 0 | 1558 |
| 1987-88 | 0 | 0 | 0 | 16 | 185 | 301 | 525 | 331 | 171 | 35 | 0 | 0 | 1564 |
| 1988-89 | 0 | 0 | 0 | 5 | 120 | 309 | 260 | 379 | 210 | 56 | 0 | 0 | 1339 |
| 1989-90 | 0 | 0 | 0 | 47 | 160 | 637 | 264 | 177 | 122 | 34 | 0 | 0 | 1441 |
| 1990-91 | 0 | 0 | 0 | 61 | 129 | 395 | 448 | 222 | 115 | 8 | 0 | 0 | 1378 |
| 1991-92 | 0 | 0 | 0 | 15 | 289 | 303 | 428 | 197 | 95 | 37 | 4 | 0 | 1368 |
| 1992-93 | 0 | 0 | 0 | 1 | 270 | 268 | 351 | 235 | 157 | 62 | 0 | 0 | 1344 |
| 1993-94 | 0 | 0 | 0 | 76 | 269 | 343 | 391 | 291 | 136 | 40 | 1 | 0 | 1547 |
| 1994-95 | 0 | 0 | 0 | 21 | 75 | 268 | 347 | 192 | 155 | 28 | 0 | 0 | 1086 |
| 1995-96 | 0 | 0 | 4 | 8 | 145 | 303 | 408 | 267 | 259 | 54 | 0 | 0 | 1448 |
| 1996-97 | 0 | 0 | 1 | 29 | 159 | 280 | 458 | 287 | 77 | 70 | 0 | 0 | 1361 |
| 1997-98 | 0 | 0 | 0 | 58 | 282 | 454 | 254 | 276 | 212 | 57 | 0 | 0 | 1593 |
| 1998-99 | 0 | 0 | 0 | 8 | 92 | 349 | 276 | 153 | 106 | 25 | 0 | 0 | 1009 |
| 1999-00 | 0 | 0 | 0 | 51 | 114 | 355 | 298 | 152 | 72 | 50 | 0 | 0 | 1092 |
| 2000-01 | 0 | 0 | 5 | 51 | 267 | 532 | 480 | 207 | 262 | 19 | 0 | 0 | 1823 |
| 2001-02 | 0 | 0 | 0 | 57 | 111 | 311 | 343 | 395 | 181 | 11 | 1 | 0 | 1410 |
| 2002-03 | 0 | 0 | 0 | 13 | 201 | 332 | 461 | 314 | 135 | 24 | 0 | 0 | 1480 |
| 2003-04 | 0 | 0 | 0 | 12 | 105 | 342 | 328 | 328 | 29 | 33 | 0 | 0 | 1177 |
| 2004-05 | 0 | 0 | 0 | 3 | 121 | 350 | 303 | 210 | 144 | 19 | 3 | 0 | 1153 |
| 2005-06 | 0 | 0 | 0 | 36 | 138 | 375 | 203 | 278 | 90 | 1 | 0 | 0 | 1121 |
| 2006-07 | 0 | 0 | 0 | 14 | 132 | 316 | 451 | 288 | 75 | 65 | 0 | 0 | 1341 |
| 2007-08 | 0 | 0 | 0 | 23 | 137 | 267 | 407 | 170 | 129 | 28 | 0 | 0 | 1161 |
| 2008-09 | 0 | 0 | 0 | 35 | 122 | 319 | 355 | 145 | 152 | 31 | 0 | 0 | 1159 |
| 2009- | 0 | 0 | 0 | 32 | 108 | 459 |  |  |  |  |  |  |  |

WBAN : 12960

## COOLING DEGREE DAYS (base 65°F) 2009  HOUSTON (KIAH)

| YEAR | JAN | FEB | MAR | APR | MAY | JUN | JUL | AUG | SEP | OCT | NOV | DEC | TOTAL |
|------|-----|-----|-----|-----|-----|-----|-----|-----|-----|-----|-----|-----|-------|
| 1980 | 4 | 31 | 49 | 86 | 388 | 610 | 705 | 677 | 553 | 162 | 52 | 26 | 3343 |
| 1981 | 1 | 28 | 23 | 295 | 330 | 538 | 606 | 609 | 413 | 285 | 71 | 7 | 3206 |
| 1982 | 39 | 11 | 147 | 170 | 329 | 547 | 641 | 599 | 437 | 199 | 60 | 40 | 3219 |
| 1983 | 0 | 0 | 18 | 76 | 268 | 427 | 541 | 554 | 362 | 196 | 87 | 18 | 2547 |
| 1984 | 0 | 13 | 64 | 135 | 315 | 415 | 527 | 562 | 384 | 302 | 62 | 100 | 2879 |
| 1985 | 0 | 6 | 87 | 180 | 335 | 487 | 521 | 602 | 456 | 257 | 143 | 9 | 3083 |
| 1986 | 4 | 71 | 52 | 220 | 341 | 518 | 654 | 553 | 510 | 157 | 92 | 4 | 3176 |
| 1987 | 4 | 4 | 14 | 154 | 383 | 497 | 580 | 661 | 423 | 137 | 54 | 15 | 2926 |
| 1988 | 7 | 20 | 65 | 121 | 274 | 472 | 609 | 637 | 478 | 229 | 144 | 20 | 3076 |
| 1989 | 33 | 44 | 105 | 194 | 405 | 454 | 547 | 526 | 363 | 218 | 105 | 5 | 2999 |
| 1990 | 20 | 19 | 65 | 174 | 413 | 603 | 536 | 630 | 456 | 181 | 87 | 47 | 3231 |
| 1991 | 0 | 14 | 76 | 231 | 408 | 514 | 593 | 565 | 376 | 248 | 47 | 37 | 3109 |
| 1992 | 0 | 17 | 74 | 155 | 281 | 508 | 584 | 476 | 437 | 210 | 32 | 20 | 2794 |
| 1993 | 4 | 10 | 42 | 92 | 267 | 506 | 652 | 674 | 463 | 221 | 31 | 27 | 2989 |
| 1994 | 15 | 25 | 74 | 186 | 347 | 561 | 644 | 569 | 406 | 243 | 102 | 33 | 3205 |
| 1995 | 24 | 22 | 98 | 142 | 406 | 476 | 622 | 622 | 508 | 183 | 42 | 65 | 3210 |
| 1996 | 15 | 89 | 53 | 192 | 513 | 475 | 587 | 532 | 384 | 205 | 78 | 51 | 3174 |
| 1997 | 23 | 21 | 94 | 52 | 274 | 432 | 567 | 570 | 426 | 187 | 9 | 0 | 2655 |
| 1998 | 16 | 6 | 64 | 90 | 429 | 621 | 678 | 616 | 524 | 250 | 75 | 50 | 3419 |
| 1999 | 37 | 60 | 73 | 270 | 365 | 515 | 568 | 685 | 395 | 182 | 35 | 11 | 3196 |
| 2000 | 41 | 64 | 126 | 143 | 413 | 495 | 632 | 622 | 443 | 242 | 51 | 0 | 3272 |
| 2001 | 2 | 51 | 3 | 230 | 344 | 474 | 585 | 581 | 370 | 123 | 70 | 40 | 2873 |
| 2002 | 27 | 2 | 72 | 272 | 380 | 507 | 611 | 590 | 449 | 223 | 24 | 17 | 3174 |
| 2003 | 6 | 6 | 31 | 196 | 495 | 538 | 575 | 617 | 387 | 228 | 111 | 3 | 3193 |
| 2004 | 17 | 1 | 108 | 176 | 379 | 491 | 616 | 570 | 495 | 398 | 37 | 11 | 3299 |
| 2005 | 41 | 42 | 51 | 109 | 329 | 558 | 611 | 615 | 558 | 233 | 131 | 14 | 3292 |
| 2006 | 23 | 17 | 116 | 265 | 370 | 499 | 574 | 625 | 441 | 252 | 70 | 34 | 3286 |
| 2007 | 9 | 19 | 108 | 125 | 360 | 530 | 539 | 640 | 513 | 280 | 106 | 48 | 3277 |
| 2008 | 17 | 32 | 93 | 166 | 403 | 594 | 625 | 594 | 403 | 184 | 54 | 38 | 3203 |
| 2009 | 19 | 44 | 92 | 150 | 418 | 627 | 699 | 665 | 430 | 227 | 40 | 8 | 3419 |

## SNOWFALL (inches)  2009  HOUSTON (KIAH)

| YEAR | JUL | AUG | SEP | OCT | NOV | DEC | JAN | FEB | MAR | APR | MAY | JUN | TOTAL |
|------|-----|-----|-----|-----|-----|-----|-----|-----|-----|-----|-----|-----|-------|
| 1980-81 | 0.0 | 0.0 | 0.0 | 0.0 | 0.0 | 0.0 | T | T | 0.0 | 0.0 | 0.0 | 0.0 | T |
| 1981-82 | 0.0 | 0.0 | 0.0 | 0.0 | 0.0 | 0.0 | T | 0.0 | 0.0 | 0.0 | 0.0 | 0.0 | T |
| 1982-83 | 0.0 | 0.0 | 0.0 | 0.0 | 0.0 | 0.0 | 0.0 | 0.0 | 0.0 | 0.0 | 0.0 | 0.0 | 0.0 |
| 1983-84 | 0.0 | 0.0 | 0.0 | 0.0 | 0.0 | 0.0 | 0.0 | 0.0 | 0.0 | 0.0 | 0.0 | 0.0 | 0.0 |
| 1984-85 | 0.0 | 0.0 | 0.0 | 0.0 | 0.0 | 0.0 | 1.4 | 0.3 | 0.0 | 0.0 | 0.0 | 0.0 | 1.7 |
| 1985-86 | 0.0 | 0.0 | 0.0 | 0.0 | 0.0 | 0.0 | 0.0 | 0.0 | 0.0 | 0.0 | 0.0 | 0.0 | 0.0 |
| 1986-87 | 0.0 | 0.0 | 0.0 | 0.0 | 0.0 | 0.0 | 0.0 | 0.0 | 0.0 | 0.0 | 0.0 | 0.0 | 0.0 |
| 1987-88 | 0.0 | 0.0 | 0.0 | 0.0 | 0.0 | 0.0 | 0.0 | T | 0.0 | 0.0 | 0.0 | 0.0 | T |
| 1988-89 | 0.0 | 0.0 | 0.0 | 0.0 | 0.0 | 0.0 | 0.0 | T | 0.0 | 0.0 | 0.0 | 0.0 | T |
| 1989-90 | 0.0 | 0.0 | 0.0 | 0.0 | 0.0 | 1.7 | 0.0 | 0.0 | 0.0 | 0.0 | 0.0 | T | 1.7 |
| 1990-91 | 0.0 | 0.0 | 0.0 | 0.0 | 0.0 | 0.0 | 0.0 | 0.0 | 0.0 | 0.0 | 0.0 | 0.0 | 0.0 |
| 1991-92 | 0.0 | 0.0 | 0.0 | 0.0 | 0.0 | 0.0 | 0.0 | 0.0 | T | 0.0 | T | 0.0 | T |
| 1992-93 | 0.0 | 0.0 | 0.0 | 0.0 | 0.0 | 0.0 | 0.0 | 0.0 | 0.0 | T | T | 0.0 | T |
| 1993-94 | 0.0 | 0.0 | 0.0 | 0.0 | 0.0 | 0.0 | 0.0 | 0.1 | 0.0 | 0.0 | 0.0 | 0.0 | 0.1 |
| 1994-95 | 0.0 | 0.0 | 0.0 | 0.0 | 0.0 | 0.0 | T | 0.0 | 0.0 | 0.0 | 0.0 | 0.0 | T |
| 1995-96 | 0.0 | 0.0 | 0.0 | 0.0 | 0.0 | 0.0 | 0.0 | T | 0.0 | 0.0 | 0.0 | T | T |
| 1996-97 | 0.0 | 0.0 | 0.0 | 0.0 | 0.0 | T | 0.0 | 0.0 | 0.0 | 0.0 | 0.0 | 0.0 | T |
| 1997-98 | 0.0 | 0.0 | 0.0 | 0.0 | 0.0 | 0.0 | 0.0 | 0.0 | 0.0 | 0.0 | 0.0 | 0.0 | 0.0 |
| 1998-99 | 0.0 | 0.0 | 0.0 | 0.0 | 0.0 | 0.0 | 0.0 | 0.0 | T | 0.0 | 0.0 | 0.0 | T |
| 1999-00 | 0.0 | 0.0 | 0.0 | 0.0 | 0.0 | 0.0 | 0.0 | 0.0 | 0.0 | 0.0 | T | 0.0 | T |
| 2000-01 | 0.0 | 0.0 | 0.0 | 0.0 | 0.0 | 0.0 | 0.0 | 0.0 | T | 0.0 | 0.0 | 0.0 | T |
| 2001-02 | 0.0 | 0.0 | 0.0 | 0.0 | 0.0 | 0.0 | 0.0 | 0.0 | 0.0 | 0.0 | 0.0 | 0.0 | 0.0 |
| 2002-03 | 0.0 | 0.0 | 0.0 | 0.0 | 0.0 | 0.0 | 0.0 | 0.0 | 0.0 | 0.0 | 0.0 | 0.0 | 0.0 |
| 2003-04 | 0.0 | 0.0 | 0.0 | 0.0 | 0.0 | 0.0 | 0.0 | 0.0 | 0.0 | 0.0 | 0.0 | 0.0 | 0.0 |
| 2004-05 | 0.0 | 0.0 | 0.0 | 0.0 | 0.0 | T | T | 0.0 | 0.0 | 0.0 | 0.0 | 0.0 | T |
| 2005-06 | 0.0 | 0.0 | 0.0 | 0.0 | 0.0 | 0.0 | 0.0 | 0.0 | T | 0.0 | 0.0 | 0.0 | T |
| 2006-07 | 0.0 | 0.0 | 0.0 | 0.0 | 0.0 | 0.0 | 0.0 | 0.0 | 0.0 | 0.0 | 0.0 | 0.0 | 0.0 |
| 2007-08 | 0.0 | 0.0 | 0.0 | 0.0 | 0.0 | 0.0 | 0.0 | 0.0 | 0.0 | 0.0 | 0.0 | 0.0 | 0.0 |
| 2008-09 | 0.0 | 0.0 | 0.0 | 0.0 | 0.0 | 1.4 | 0.0 | 0.0 | T | 0.0 | 0.0 | 0.0 | 1.4 |
| 2009- | 0.0 | 0.0 | 0.0 | 0.0 | 0.0 | 1.0 | | | | | | | |
| POR=41 YRS | 0.0 | 0.0 | 0.0 | 0.0 | 0.0 | 0.1 | 0.1 | 0.1 | T | T | T | T | 0.3 |

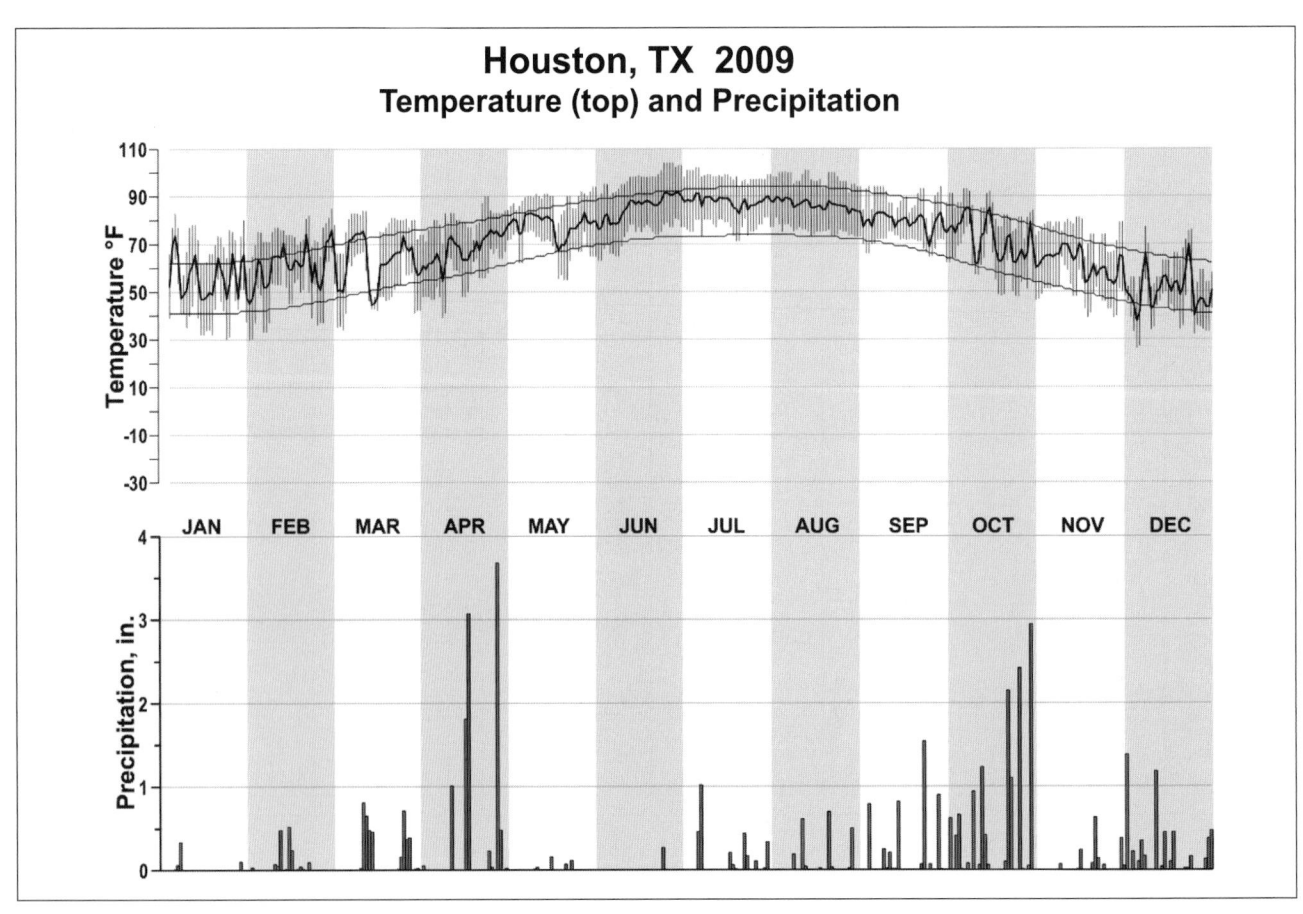

**Houston, TX  2009**
**Temperature (top) and Precipitation**

# 2009
# SAN ANTONIO
# TEXAS (KSAT)

The city of San Antonio is located in the south-central portion of Texas on the Balcones escarpment. Northwest of the city, the terrain slopes upward to the Edwards Plateau and to the southeast it slopes downward to the Gulf Coastal Plains. Soils are blackland clay and silty loam on the Plains and thin limestone soils on the Edwards Plateau.

The location of San Antonio on the edge of the Gulf Coastal Plains is influenced by a modified subtropical climate, predominantly continental during the winter months and marine during the summer months. Temperatures range from 50 degrees in January to the middle 80s in July and August. While the summer is hot, with daily temperatures above 90 degrees over 80 percent of the time, extremely high temperatures are rare. Mild weather prevails during much of the winter months, with below-freezing temperatures occurring on an average of about 20 days each year.

San Antonio is situated between a semi-arid area to the west and the coastal area of heavy precipitation to the east. The normal annual rainfall of nearly 28 inches is sufficient for the production of most crops. Precipitation is fairly well distributed throughout the year with the heaviest amounts occurring during May and September. The precipitation from April through September usually occurs from thunderstorms. Large amounts of precipitation may fall during short periods of time. Most of the winter precipitation occurs as light rain or drizzle. Thunderstorms and heavy rains have occurred in all months of the year. Hail of damaging intensity seldom occurs but light hail is frequent with the springtime thunderstorms. Measurable snow occurs only once in three or four years. Snowfall of 2 to 4 inches occurs about every ten years.

Northerly winds prevail during most of the winter, and strong northerly winds occasionally occur during storms called northers. Southeasterly winds from the Gulf of Mexico also occur frequently during winter and are predominant in summer.

Since San Antonio is located only 140 miles from the Gulf of Mexico, tropical storms occasionally affect the city with strong winds and heavy rains. One of the fastest winds recorded, 74 mph, occurred as a tropical storm moved inland east of the city in August 1942.

Relative humidity is above 80 percent during the early morning hours most of the year, dropping to near 50 percent in the late afternoon.

San Antonio has about 50 percent of the possible amount of sunshine during the winter months and more than 70 percent during the summer months. Skies are clear to partly cloudy more than 60 percent of the time and cloudy less than 40 percent. Air carried over San Antonio by southeasterly winds is lifted orographically, causing low stratus clouds to develop frequently during the later part of the night. These clouds usually dissipate around noon, and clear skies prevail a high percentage of the time during the afternoon.

The first occurrence of 32 degrees Fahrenheit is in late November and the average last occurrence is in early March.

# NORMALS, MEANS, AND EXTREMES
## SAN ANTONIO (KSAT)

LATITUDE: 29°31'N  LONGITUDE: -98°27'W  ELEVATION (FT): GRND: 774  BARO: 821  TIME ZONE: CENTRAL (UTC -6)  WBAN: 12921

| | ELEMENT | POR | JAN | FEB | MAR | APR | MAY | JUN | JUL | AUG | SEP | OCT | NOV | DEC | YEAR |
|---|---|---|---|---|---|---|---|---|---|---|---|---|---|---|---|
| **TEMPERATURE °F** | NORMAL DAILY MAXIMUM | 30 | 62.1 | 67.1 | 74.3 | 80.4 | 86.0 | 91.4 | 94.6 | 94.7 | 90.0 | 82.0 | 71.4 | 64.0 | 79.8 |
| | MEAN DAILY MAXIMUM | 63 | 62.2 | 66.5 | 73.6 | 80.4 | 86.4 | 92.0 | 94.9 | 95.3 | 89.8 | 82.0 | 71.5 | 64.5 | 79.9 |
| | HIGHEST DAILY MAXIMUM | 68 | 89 | 100 | 100 | 101 | 104 | 107 | 106 | 108 | 111 | 99 | 94 | 90 | 111 |
| | YEAR OF OCCURRENCE | | 1971 | 1996 | 1991 | 1996 | 2004 | 1998 | 1989 | 1986 | 2000 | 1991 | 1988 | 1955 | SEP 2000 |
| | MEAN OF EXTREME MAXS. | 63 | 79.9 | 83.6 | 88.3 | 91.7 | 94.7 | 98.0 | 99.8 | 100.5 | 97.5 | 92.1 | 85.2 | 80.2 | 91.0 |
| | NORMAL DAILY MINIMUM | 30 | 38.6 | 42.4 | 49.9 | 56.9 | 65.5 | 71.6 | 74.0 | 73.6 | 69.1 | 59.4 | 48.6 | 40.8 | 57.5 |
| | MEAN DAILY MINIMUM | 63 | 39.8 | 43.3 | 50.2 | 58.4 | 66.2 | 72.1 | 74.4 | 74.0 | 69.1 | 59.7 | 48.9 | 41.7 | 58.2 |
| | LOWEST DAILY MINIMUM | 68 | 0 | 6 | 19 | 31 | 43 | 53 | 62 | 61 | 41 | 27 | 21 | 6 | 0 |
| | YEAR OF OCCURRENCE | | 1949 | 1951 | 2002 | 1987 | 1984 | 1964 | 1967 | 1992 | 1942 | 1993 | 1976 | 1989 | JAN 1949 |
| | MEAN OF EXTREME MINS. | 63 | 22.9 | 26.5 | 31.3 | 40.7 | 52.5 | 62.7 | 68.9 | 67.8 | 56.0 | 42.2 | 31.2 | 24.9 | 44.0 |
| | NORMAL DRY BULB | 30 | 50.3 | 54.7 | 62.1 | 68.6 | 75.8 | 81.5 | 84.3 | 84.2 | 79.4 | 70.7 | 60.0 | 52.4 | 68.7 |
| | MEAN DRY BULB | 63 | 51.0 | 54.9 | 61.9 | 69.4 | 76.3 | 82.2 | 84.7 | 84.7 | 79.5 | 70.9 | 60.2 | 53.1 | 69.1 |
| | MEAN WET BULB | 26 | 44.7 | 48.2 | 54.0 | 60.4 | 68.0 | 72.1 | 73.3 | 73.2 | 69.3 | 62.5 | 53.7 | 46.3 | 60.5 |
| | MEAN DEW POINT | 26 | 40.5 | 44.1 | 49.6 | 56.4 | 65.3 | 69.6 | 70.2 | 69.9 | 65.9 | 59.2 | 50.0 | 42.0 | 56.9 |
| | NORMAL NO. DAYS WITH: | | | | | | | | | | | | | | |
| | MAXIMUM >= 90 | 30 | 0.0 | 0.3 | 0.9 | 2.2 | 8.3 | 20.6 | 27.7 | 28.0 | 18.4 | 4.9 | 0.1 | 0.0 | 111.4 |
| | MAXIMUM <= 32 | 30 | 0.2 | 0.3 | 0.0 | 0.0 | 0.0 | 0.0 | 0.0 | 0.0 | 0.0 | 0.0 | 0.0 | 0.2 | 0.7 |
| | MINIMUM <= 32 | 30 | 7.8 | 4.6 | 1.5 | * | 0.0 | 0.0 | 0.0 | 0.0 | 0.0 | * | 1.9 | 6.2 | 22.0 |
| | MINIMUM <= 0 | 30 | 0.0 | 0.0 | 0.0 | 0.0 | 0.0 | 0.0 | 0.0 | 0.0 | 0.0 | 0.0 | 0.0 | 0.0 | 0.0 |
| **H/C** | NORMAL HEATING DEG. DAYS | 30 | 455 | 303 | 149 | 42 | 1 | 0 | 0 | 0 | 2 | 33 | 197 | 391 | 1573 |
| | NORMAL COOLING DEG. DAYS | 30 | 7 | 19 | 68 | 161 | 344 | 505 | 607 | 601 | 439 | 215 | 57 | 15 | 3038 |
| **RH** | NORMAL (PERCENT) | 30 | 67 | 66 | 65 | 65 | 71 | 70 | 65 | 64 | 66 | 68 | 69 | 68 | 67 |
| | HOUR 00 LST | 30 | 75 | 74 | 73 | 75 | 81 | 81 | 75 | 75 | 77 | 78 | 79 | 76 | 77 |
| | HOUR 06 LST | 30 | 80 | 80 | 81 | 82 | 88 | 88 | 87 | 87 | 86 | 85 | 83 | 81 | 84 |
| | HOUR 12 LST | 30 | 57 | 56 | 54 | 55 | 61 | 59 | 53 | 52 | 54 | 55 | 57 | 57 | 56 |
| | HOUR 18 LST | 30 | 55 | 50 | 48 | 49 | 55 | 53 | 46 | 46 | 51 | 53 | 58 | 56 | 52 |
| **S** | PERCENT POSSIBLE SUNSHINE | 53 | 47 | 50 | 57 | 56 | 56 | 67 | 74 | 74 | 67 | 64 | 54 | 48 | 60 |
| **W/O** | MEAN NO. DAYS WITH: | | | | | | | | | | | | | | |
| | HEAVY FOG(VISBY <= 1/4 MI) | 46 | 4.2 | 2.4 | 2.0 | 1.1 | 0.6 | 0.0 | 0.1 | 0.0 | 0.2 | 1.1 | 2.6 | 3.7 | 18.0 |
| | THUNDERSTORMS | 62 | 0.9 | 1.5 | 2.6 | 3.9 | 6.3 | 4.9 | 3.8 | 4.2 | 3.9 | 2.8 | 1.6 | 1.0 | 37.4 |
| **CLOUDNESS** | MEAN: | | | | | | | | | | | | | | |
| | SUNRISE-SUNSET (OKTAS) | | | | | | | | | | | | | | |
| | MIDNIGHT-MIDNIGHT (OKTAS) | | | | | | | | | | | | | | |
| | MEAN NO. DAYS WITH: | | | | | | | | | | | | | | |
| | CLEAR | 1 | 4.0 | 5.0 | 8.0 | | 6.0 | 12.0 | | | | | | | |
| | PARTLY CLOUDY | 1 | 1.0 | 4.0 | 2.0 | | 11.0 | 5.0 | | | | | | | |
| | CLOUDY | 1 | 1.0 | | 6.0 | | 3.0 | 2.0 | | | | | | | |
| **PR** | MEAN STATION PRESSURE(IN) | 26 | 29.28 | 29.23 | 29.13 | 29.14 | 29.05 | 29.09 | 29.14 | 29.14 | 29.13 | 29.18 | 29.23 | 29.27 | 29.17 |
| | MEAN SEA-LEVEL PRES. (IN) | 26 | 30.12 | 30.07 | 29.99 | 29.94 | 29.89 | 29.90 | 29.95 | 29.94 | 29.95 | 30.00 | 30.07 | 30.11 | 29.99 |
| **WINDS** | MEAN SPEED (MPH) | 26 | 7.7 | 8.1 | 8.9 | 9.1 | 9.0 | 9.0 | 8.5 | 7.8 | 7.2 | 7.5 | 7.5 | 7.4 | 8.1 |
| | PREVAIL.DIR(TENS OF DEGS) | 32 | 02 | 02 | 15 | 15 | 15 | 15 | 17 | 17 | 17 | 17 | 02 | 36 | 17 |
| | MAXIMUM 2-MINUTE: | 67 | | | | | | | | | | | | | |
| | SPEED (MPH) | | 39 | 38 | 41 | 43 | 46 | 38 | 39 | 40 | 43 | 36 | 38 | 39 | 46 |
| | DIR. (TENS OF DEGS) | | 29 | 29 | 29 | 17 | 28 | 35 | 17 | 12 | 25 | 34 | 30 | 31 | 28 |
| | YEAR OF OCCURRENCE | | 1999 | 2009 | 2002 | 2009 | 1997 | 2003 | 2004 | 2006 | 1996 | 2009 | 2003 | 2009 | MAY 1997 |
| | MAXIMUM 3-SECOND | 67 | | | | | | | | | | | | | |
| | SPEED (MPH) | | 46 | 49 | 49 | 55 | 55 | 51 | 54 | 46 | 71 | 46 | 46 | 51 | 71 |
| | DIR. (TENS OF DEGS) | | 28 | 29 | 29 | 17 | 29 | 32 | 12 | 04 | 26 | 34 | 34 | 29 | 26 |
| | YEAR OF OCCURRENCE | | 1999 | 2009 | 2002 | 2009 | 1997 | 2009 | 2002 | 2003 | 1996 | 2009 | 2006 | 2009 | SEP 1996 |
| **PRECIPITATION** | NORMAL (IN) | 30 | 1.66 | 1.75 | 1.89 | 2.60 | 4.72 | 4.30 | 2.03 | 2.57 | 3.00 | 3.86 | 2.58 | 1.96 | 32.92 |
| | MAXIMUM MONTHLY (IN) | 67 | 8.52 | 6.43 | 7.24 | 9.32 | 12.85 | 11.95 | 16.92 | 11.14 | 15.78 | 18.07 | 9.46 | 13.96 | 18.07 |
| | YEAR OF OCCURRENCE | | 1968 | 1965 | 2007 | 1957 | 1987 | 1986 | 2002 | 1974 | 1946 | 1998 | 2004 | 1991 | OCT 1998 |
| | MINIMUM MONTHLY (IN) | 67 | T | 0.01 | 0.03 | .01 | 0.12 | 0.01 | T | 0.00 | 0.05 | T | T | 0.03 | 0.00 |
| | YEAR OF OCCURRENCE | | 1996 | 1999 | 1961 | 2005 | 2003 | 2008 | 1993 | 1952 | 1999 | 1952 | 1966 | 1950 | AUG 1952 |
| | MAXIMUM IN 24 HOURS (IN) | 67 | 3.18 | 2.44 | 3.59 | 4.88 | 6.53 | 6.30 | 9.79 | 5.79 | 7.28 | 13.35 | 4.87 | 6.90 | 13.35 |
| | YEAR OF OCCURRENCE | | 1968 | 1986 | 1992 | 1977 | 1972 | 1986 | 2002 | 2007 | 1973 | 1998 | 1977 | 1991 | OCT 1998 |
| | NORMAL NO. DAYS WITH: | | | | | | | | | | | | | | |
| | PRECIPITATION >= 0.01 | 30 | 7.6 | 6.8 | 7.9 | 7.3 | 9.2 | 7.7 | 4.6 | 5.2 | 6.5 | 6.9 | 7.3 | 7.8 | 84.8 |
| | PRECIPITATION >= 1.00 | 30 | 0.3 | 0.4 | 0.5 | 0.8 | 1.5 | 1.5 | 0.6 | 0.9 | 0.7 | 1.2 | 0.8 | 0.3 | 9.5 |
| **SNOWFALL** | NORMAL (IN) | 30 | 0.7 | 0.1 | 0.* | 0.0 | 0.0 | 0.0 | 0.0 | 0.0 | 0.0 | 0.* | 0.* | 0.* | 0.8 |
| | MAXIMUM MONTHLY (IN) | 64 | 15.9 | 3.5 | T | T | T | T | 0.0 | 0.0 | 0.0 | T | 0.3 | 0.2 | 15.9 |
| | YEAR OF OCCURRENCE | | 1985 | 1966 | 2006 | 1993 | 2006 | 1989 | | | | 1993 | 1957 | 1964 | JAN 1985 |
| | MAXIMUM IN 24 HOURS (IN) | 64 | 13.2 | 3.5 | T | T | T | T | 0.0 | 0.0 | 0.0 | T | 0.3 | 0.2 | 13.2 |
| | YEAR OF OCCURRENCE' | | 1985 | 1966 | 1994 | 1993 | 1993 | 1989 | | | | 1993 | 1957 | 1964 | JAN 1985 |
| | MAXIMUM SNOW DEPTH (IN) | 58 | 9 | 3 | 0 | 0 | 0 | 0 | 0 | 0 | 0 | 0 | 0 | 0 | 9 |
| | YEAR OF OCCURRENCE | | 1985 | 1966 | | | | | | | | | | | JAN 1985 |
| | NORMAL NO. DAYS WITH: | | | | | | | | | | | | | | |
| | SNOWFALL >= 1.0 | 30 | 0.1 | 0.0 | 0.0 | 0.0 | 0.0 | 0.0 | 0.0 | 0.0 | 0.0 | 0.0 | 0.0 | 0.0 | 0.1 |

## PRECIPITATION (inches) 2009 SAN ANTONIO (KSAT)

| YEAR | JAN | FEB | MAR | APR | MAY | JUN | JUL | AUG | SEP | OCT | NOV | DEC | ANNUAL |
|------|-----|-----|-----|-----|-----|-----|-----|-----|-----|-----|-----|-----|--------|
| 1980 | 0.72 | 0.74 | 0.98 | 1.67 | 6.42 | 0.52 | 0.26 | 2.64 | 5.05 | 1.09 | 3.53 | 0.61 | 24.23 |
| 1981 | 2.06 | 0.96 | 1.96 | 2.21 | 6.43 | 8.71 | 0.25 | 2.41 | 1.36 | 8.61 | 0.72 | 0.69 | 36.37 |
| 1982 | 0.72 | 1.28 | 0.69 | 1.23 | 6.42 | 1.37 | 0.14 | 0.55 | 0.87 | 2.84 | 4.54 | 2.31 | 22.96 |
| 1983 | 1.48 | 1.54 | 3.89 | 0.18 | 4.37 | 1.27 | 2.43 | 2.00 | 3.86 | 1.64 | 3.06 | 0.39 | 26.11 |
| 1984 | 1.87 | 0.54 | 1.91 | 0.11 | 3.76 | 1.40 | T | 3.04 | 1.06 | 5.94 | 2.91 | 3.41 | 25.95 |
| 1985 | 2.68 | 1.91 | 2.85 | 3.27 | 2.47 | 8.20 | 5.80 | 0.45 | 4.80 | 3.91 | 5.00 | 0.09 | 41.43 |
| 1986 | 0.76 | 2.52 | 0.35 | 0.60 | 6.29 | 11.95 | 0.05 | 1.86 | 2.83 | 6.58 | 1.83 | 7.11 | 42.73 |
| 1987 | 1.13 | 4.78 | 1.10 | 1.48 | 12.85 | 7.69 | 1.21 | 0.33 | 2.24 | 0.44 | 2.53 | 2.18 | 37.96 |
| 1988 | 0.39 | 0.92 | 0.86 | 1.23 | 0.41 | 5.50 | 5.58 | 1.98 | 0.83 | 0.62 | 0.02 | 0.67 | 19.01 |
| 1989 | 2.96 | 0.29 | 1.24 | 2.55 | 0.33 | 3.96 | 0.69 | 0.48 | 1.54 | 5.81 | 1.93 | 0.36 | 22.14 |
| 1990 | 1.17 | 2.68 | 5.17 | 4.52 | 3.28 | 1.18 | 8.29 | 1.30 | 3.70 | 3.71 | 3.11 | 0.20 | 38.31 |
| 1991 | 5.08 | 2.34 | 1.06 | 4.91 | 5.30 | 2.28 | 2.23 | 2.84 | 1.42 | 0.87 | 0.47 | 13.96 | 42.76 |
| 1992 | 5.64 | 6.37 | 6.12 | 3.03 | 8.15 | 5.67 | 1.28 | 2.56 | 1.12 | 0.92 | 3.47 | 2.16 | 46.49 |
| 1993 | 1.31 | 3.72 | 1.56 | 1.81 | 12.47 | 6.43 | T | 0.01 | 0.52 | 3.07 | 0.66 | 0.44 | 32.00 |
| 1994 | 1.55 | 0.64 | 5.06 | 2.21 | 7.01 | 1.66 | 0.50 | 2.54 | 5.52 | 9.75 | 0.71 | 3.28 | 40.43 |
| 1995 | 0.28 | 1.19 | 1.58 | 1.07 | 5.36 | 4.81 | 0.71 | 2.03 | 4.49 | 0.23 | 0.82 | 0.09 | 22.66 |
| 1996 | T | 0.70 | 0.30 | 0.89 | 1.26 | 2.12 | 1.31 | 2.86 | 3.66 | .36 | 2.79 | 1.56 | 17.81 |
| 1997 | 0.44 | 2.44 | 2.24 | 5.72 | 3.91 | 7.30 | T | 0.62 | 1.86 | 4.08 | 1.76 | 3.55 | 33.92 |
| 1998 | 3.21 | 3.37 | 2.85 | 0.05 | 0.34 | 0.81 | 0.21 | 7.78 | 1.57 | 18.07 | 3.40 | 0.39 | 42.05 |
| 1999 | 0.04 | 0.01 | 3.48 | 0.91 | 2.78 | 3.37 | 1.80 | 2.11 | 0.05 | 1.29 | 0.05 | 0.52 | 16.41 |
| 2000 | 1.40 | 2.20 | 0.91 | 1.22 | 3.59 | 7.61 | 0.34 | 0.16 | 2.65 | 5.62 | 8.58 | 1.57 | 35.85 |
| 2001 | 2.85 | 0.70 | 2.77 | 2.29 | 2.48 | 3.39 | 0.50 | 7.83 | 4.05 | 2.06 | 4.37 | 3.43 | 36.72 |
| 2002 | 0.37 | 0.42 | 1.19 | 3.82 | 2.26 | 1.48 | 16.92 | 0.54 | 7.02 | 7.64 | 2.08 | 2.53 | 46.27 |
| 2003 | 0.99 | 2.15 | 0.77 | 0.17 | 0.12 | 2.90 | 8.12 | 1.65 | 9.21 | 1.94 | 0.32 | 0.11 | 28.45 |
| 2004 | 2.31 | 1.73 | 2.35 | 5.02 | 1.80 | 9.47 | 0.61 | 1.10 | 1.92 | 9.47 | 9.46 | 0.08 | 45.32 |
| 2005 | 2.18 | 2.42 | 2.00 | 0.01 | 2.97 | 0.81 | 2.10 | 1.22 | 1.39 | 1.14 | 0.20 | 0.10 | 16.54 |
| 2006 | 0.35 | 0.62 | 1.36 | 1.40 | 3.80 | 1.63 | 1.41 | 0.03 | 4.11 | 3.44 | 0.75 | 2.44 | 21.34 |
| 2007 | 4.33 | 0.08 | 7.24 | 4.61 | 3.35 | 6.47 | 11.76 | 6.77 | 1.09 | 0.75 | 0.40 | 0.40 | 47.25 |
| 2008 | 0.42 | 0.20 | 1.82 | 0.83 | 0.66 | 0.01 | 3.86 | 4.98 | 0.46 | 0.26 | 0.01 | 0.25 | 13.76 |
| 2009 | 0.27 | 0.65 | 2.51 | 2.05 | 1.57 | 0.45 | 0.48 | 0.45 | 6.35 | 11.90 | 2.09 | 1.92 | 30.69 |
| POR= 63 YRS | 1.60 | 1.77 | 1.77 | 2.47 | 3.72 | 3.55 | 2.27 | 2.56 | 3.37 | 3.46 | 2.22 | 1.64 | 30.40 |

WBAN : 12921

## AVERAGE TEMPERATURE (°F) 2009 SAN ANTONIO (KSAT)

| YEAR | JAN | FEB | MAR | APR | MAY | JUN | JUL | AUG | SEP | OCT | NOV | DEC | ANNUAL |
|------|-----|-----|-----|-----|-----|-----|-----|-----|-----|-----|-----|-----|--------|
| 1980 | 52.6 | 53.7 | 61.5 | 67.6 | 76.1 | 85.1 | 88.1 | 85.3 | 83.7 | 70.7 | 58.3 | 55.0 | 69.8 |
| 1981 | 50.8 | 53.7 | 60.7 | 72.9 | 75.3 | 81.5 | 84.2 | 84.7 | 78.9 | 71.9 | 62.4 | 53.0 | 69.2 |
| 1982 | 50.8 | 49.7 | 63.1 | 66.9 | 74.5 | 81.6 | 85.5 | 86.0 | 80.1 | 69.3 | 59.4 | 52.4 | 68.3 |
| 1983 | 48.9 | 52.1 | 58.7 | 65.2 | 73.6 | 79.2 | 82.9 | 84.5 | 78.5 | 70.9 | 62.5 | 43.0 | 66.7 |
| 1984 | 46.7 | 54.1 | 64.2 | 69.7 | 77.1 | 82.8 | 85.0 | 84.7 | 77.6 | 71.2 | 58.8 | 59.6 | 69.3 |
| 1985 | 44.2 | 50.5 | 64.1 | 69.4 | 76.7 | 80.2 | 82.2 | 85.5 | 79.4 | 71.7 | 64.4 | 49.9 | 68.2 |
| 1986 | 53.4 | 58.0 | 62.9 | 72.6 | 74.6 | 81.5 | 85.8 | 85.7 | 83.7 | 69.7 | 59.4 | 51.6 | 69.9 |
| 1987 | 50.7 | 55.9 | 57.8 | 66.1 | 75.8 | 80.5 | 83.8 | 86.0 | 79.2 | 71.2 | 60.6 | 54.2 | 68.5 |
| 1988 | 47.6 | 54.3 | 61.3 | 69.1 | 76.1 | 81.2 | 84.6 | 86.4 | 80.7 | 73.2 | 65.1 | 56.0 | 69.6 |
| 1989 | 56.2 | 51.6 | 61.9 | 70.4 | 81.7 | 83.3 | 86.6 | 86.0 | 79.1 | 71.3 | 61.8 | 43.4 | 69.4 |
| 1990 | 56.4 | 58.9 | 61.5 | 69.7 | 79.3 | 87.5 | 83.4 | 85.3 | 80.0 | 69.3 | 63.0 | 51.9 | 70.5 |
| 1991 | 48.9 | 56.6 | 64.0 | 72.4 | 77.7 | 82.8 | 84.5 | 85.8 | 77.8 | 73.3 | 57.4 | 55.5 | 69.7 |
| 1992 | 50.8 | 59.1 | 63.3 | 69.0 | 73.7 | 82.5 | 84.7 | 82.2 | 81.7 | 73.4 | 57.3 | 56.3 | 69.5 |
| 1993 | 51.2 | 55.5 | 61.5 | 67.3 | 73.9 | 81.6 | 86.1 | 87.3 | 81.5 | 70.7 | 56.3 | 55.1 | 69.0 |
| 1994 | 52.3 | 56.2 | 63.9 | 69.8 | 76.0 | 84.5 | 87.9 | 86.1 | 78.4 | 72.7 | 64.8 | 57.0 | 70.8 |
| 1995 | 53.5 | 57.4 | 61.9 | 69.8 | 78.6 | 79.3 | 84.3 | 85.5 | 80.1 | 69.8 | 59.5 | 55.6 | 69.6 |
| 1996 | 51.0 | 57.9 | 57.6 | 69.5 | 81.9 | 84.1 | 87.3 | 84.4 | 78.4 | 71.1 | 61.3 | 54.5 | 69.9 |
| 1997 | 49.2 | 53.1 | 63.3 | 63.9 | 74.0 | 79.8 | 85.1 | 86.1 | 82.2 | 70.2 | 57.4 | 50.2 | 67.9 |
| 1998 | 56.4 | 55.3 | 59.8 | 66.7 | 79.8 | 86.3 | 88.1 | 83.6 | 80.5 | 71.4 | 62.4 | 52.7 | 70.3 |
| 1999 | 54.6 | 61.8 | 62.7 | 71.2 | 76.2 | 81.9 | 82.9 | 86.1 | 80.3 | 69.6 | 63.1 | 54.0 | 70.4 |
| 2000 | 55.2 | 62.6 | 67.0 | 70.7 | 78.6 | 81.0 | 85.9 | 86.3 | 81.0 | 71.1 | 56.9 | 46.4 | 70.2 |
| 2001 | 49.2 | 57.5 | 56.6 | 70.8 | 76.3 | 82.6 | 85.4 | 85.6 | 76.9 | 67.9 | 62.9 | 53.8 | 68.8 |
| 2002 | 54.0 | 50.8 | 60.3 | 73.2 | 76.8 | 83.4 | 82.5 | 85.3 | 78.7 | 70.7 | 57.8 | 53.8 | 68.9 |
| 2003 | 49.8 | 53.1 | 60.6 | 71.6 | 80.4 | 81.7 | 82.0 | 83.7 | 76.7 | 70.6 | 63.1 | 53.9 | 68.9 |
| 2004 | 54.5 | 52.7 | 66.0 | 67.2 | 76.2 | 80.9 | 82.9 | 83.4 | 80.5 | 76.9 | 61.1 | 53.2 | 69.6 |
| 2005 | 55.9 | 56.3 | 61.4 | 68.4 | 75.0 | 82.6 | 85.3 | 85.7 | 84.3 | 70.9 | 64.9 | 53.0 | 70.3 |
| 2006 | 58.2 | 56.0 | 67.6 | 76.7 | 78.8 | 83.6 | 85.7 | 88.3 | 79.7 | 72.4 | 63.8 | 54.4 | 72.1 |
| 2007 | 48.3 | 54.9 | 65.0 | 65.2 | 75.5 | 80.7 | 80.4 | 83.7 | 80.3 | 73.1 | 62.7 | 56.1 | 68.8 |
| 2008 | 51.9 | 61.7 | 64.5 | 70.7 | 80.1 | 86.8 | 84.1 | 84.4 | 79.5 | 71.5 | 63.7 | 55.0 | 71.2 |
| 2009 | 54.5 | 62.9 | 65.2 | 69.8 | 79.5 | 86.3 | 88.8 | 88.4 | 78.5 | 69.9 | 60.7 | 48.3 | 71.1 |
| POR= 63 YRS | 51.0 | 54.9 | 61.9 | 69.4 | 76.3 | 82.2 | 84.7 | 84.7 | 79.5 | 70.9 | 60.2 | 53.1 | 69.1 |

## HEATING DEGREE DAYS (base 65°F) 2009 SAN ANTONIO (KSAT)

| YEAR | JUL | AUG | SEP | OCT | NOV | DEC | JAN | FEB | MAR | APR | MAY | JUN | TOTAL |
|------|-----|-----|-----|-----|-----|-----|-----|-----|-----|-----|-----|-----|-------|
| 1980-81 | 0 | 0 | 0 | 62 | 245 | 331 | 437 | 332 | 157 | 10 | 0 | 0 | 1574 |
| 1981-82 | 0 | 0 | 2 | 52 | 112 | 368 | 445 | 430 | 171 | 77 | 2 | 0 | 1659 |
| 1982-83 | 0 | 0 | 0 | 49 | 237 | 404 | 490 | 356 | 208 | 99 | 1 | 0 | 1844 |
| 1983-84 | 0 | 0 | 5 | 20 | 154 | 681 | 563 | 315 | 120 | 21 | 2 | 0 | 1881 |
| 1984-85 | 0 | 0 | 9 | 28 | 228 | 203 | 635 | 406 | 109 | 26 | 0 | 0 | 1644 |
| 1985-86 | 0 | 0 | 10 | 9 | 112 | 467 | 354 | 232 | 106 | 8 | 1 | 0 | 1299 |
| 1986-87 | 0 | 0 | 0 | 14 | 204 | 413 | 443 | 254 | 233 | 98 | 0 | 0 | 1659 |
| 1987-88 | 0 | 0 | 0 | 1 | 194 | 339 | 538 | 323 | 179 | 38 | 0 | 0 | 1612 |
| 1988-89 | 0 | 0 | 0 | 0 | 122 | 291 | 292 | 392 | 187 | 55 | 0 | 0 | 1339 |
| 1989-90 | 0 | 0 | 0 | 42 | 165 | 663 | 283 | 190 | 154 | 32 | 0 | 0 | 1529 |
| 1990-91 | 0 | 0 | 0 | 50 | 142 | 422 | 494 | 240 | 96 | 7 | 0 | 0 | 1451 |
| 1991-92 | 0 | 0 | 5 | 30 | 271 | 306 | 435 | 188 | 91 | 31 | 5 | 0 | 1362 |
| 1992-93 | 0 | 0 | 0 | 0 | 260 | 287 | 421 | 269 | 147 | 42 | 0 | 0 | 1426 |
| 1993-94 | 0 | 0 | 0 | 85 | 287 | 323 | 391 | 273 | 130 | 28 | 10 | 0 | 1527 |
| 1994-95 | 0 | 0 | 0 | 19 | 99 | 267 | 359 | 215 | 180 | 29 | 0 | 0 | 1168 |
| 1995-96 | 0 | 0 | 9 | 12 | 187 | 324 | 435 | 280 | 277 | 63 | 0 | 0 | 1587 |
| 1996-97 | 0 | 0 | 0 | 27 | 165 | 325 | 498 | 338 | 108 | 84 | 0 | 0 | 1545 |
| 1997-98 | 0 | 0 | 0 | 44 | 251 | 449 | 269 | 270 | 214 | 30 | 0 | 0 | 1527 |
| 1998-99 | 0 | 0 | 0 | 22 | 120 | 404 | 328 | 137 | 109 | 32 | 0 | 0 | 1152 |
| 1999-00 | 0 | 0 | 0 | 56 | 98 | 346 | 300 | 135 | 79 | 29 | 0 | 0 | 1043 |
| 2000-01 | 0 | 0 | 2 | 76 | 269 | 570 | 484 | 234 | 255 | 15 | 0 | 0 | 1905 |
| 2001-02 | 0 | 0 | 0 | 38 | 133 | 367 | 357 | 389 | 202 | 14 | 1 | 0 | 1501 |
| 2002-03 | 0 | 0 | 0 | 24 | 217 | 353 | 465 | 339 | 160 | 20 | 0 | 0 | 1578 |
| 2003-04 | 0 | 0 | 0 | 16 | 154 | 345 | 336 | 356 | 38 | 43 | 2 | 0 | 1290 |
| 2004-05 | 0 | 0 | 0 | 2 | 135 | 368 | 312 | 248 | 146 | 28 | 2 | 0 | 1241 |
| 2005-06 | 0 | 0 | 0 | 40 | 132 | 372 | 205 | 263 | 73 | 0 | 0 | 0 | 1085 |
| 2006-07 | 0 | 0 | 0 | 20 | 113 | 341 | 516 | 290 | 81 | 84 | 0 | 0 | 1445 |
| 2007-08 | 0 | 0 | 0 | 22 | 160 | 307 | 414 | 146 | 116 | 18 | 0 | 0 | 1183 |
| 2008-09 | 0 | 0 | 0 | 32 | 107 | 327 | 331 | 117 | 117 | 27 | 0 | 0 | 1058 |
| 2009- | 0 | 0 | 7 | 44 | 143 | 508 | | | | | | | |

WBAN : 12921

## COOLING DEGREE DAYS (base 65°F) 2009 SAN ANTONIO (KSAT)

| YEAR | JAN | FEB | MAR | APR | MAY | JUN | JUL | AUG | SEP | OCT | NOV | DEC | TOTAL |
|------|-----|-----|-----|-----|-----|-----|-----|-----|-----|-----|-----|-----|-------|
| 1980 | 11 | 14 | 61 | 127 | 355 | 614 | 725 | 635 | 567 | 245 | 51 | 26 | 3431 |
| 1981 | 3 | 24 | 30 | 255 | 324 | 502 | 603 | 619 | 424 | 273 | 41 | 3 | 3101 |
| 1982 | 11 | 5 | 117 | 142 | 304 | 504 | 645 | 659 | 459 | 191 | 72 | 19 | 3128 |
| 1983 | 0 | 0 | 21 | 111 | 276 | 435 | 560 | 611 | 417 | 207 | 84 | 8 | 2730 |
| 1984 | 0 | 8 | 101 | 169 | 383 | 541 | 625 | 618 | 394 | 230 | 46 | 44 | 3159 |
| 1985 | 0 | 8 | 85 | 165 | 368 | 462 | 539 | 641 | 450 | 223 | 101 | 5 | 3047 |
| 1986 | 2 | 45 | 49 | 244 | 304 | 500 | 652 | 646 | 568 | 166 | 40 | 4 | 3220 |
| 1987 | 4 | 5 | 17 | 135 | 340 | 471 | 589 | 658 | 434 | 199 | 67 | 12 | 2931 |
| 1988 | 6 | 19 | 71 | 166 | 352 | 492 | 617 | 671 | 480 | 264 | 131 | 20 | 3289 |
| 1989 | 24 | 23 | 99 | 222 | 524 | 557 | 678 | 656 | 429 | 244 | 75 | 0 | 3531 |
| 1990 | 22 | 26 | 53 | 177 | 450 | 681 | 578 | 635 | 459 | 192 | 91 | 23 | 3387 |
| 1991 | 0 | 10 | 70 | 234 | 402 | 541 | 612 | 654 | 396 | 295 | 49 | 20 | 3283 |
| 1992 | 0 | 23 | 47 | 158 | 281 | 531 | 618 | 542 | 508 | 267 | 34 | 24 | 3033 |
| 1993 | 1 | 7 | 45 | 117 | 283 | 503 | 660 | 698 | 503 | 267 | 31 | 22 | 3137 |
| 1994 | 8 | 34 | 102 | 183 | 357 | 593 | 715 | 659 | 410 | 261 | 97 | 27 | 3446 |
| 1995 | 7 | 9 | 90 | 177 | 429 | 436 | 603 | 644 | 467 | 167 | 28 | 40 | 3097 |
| 1996 | 9 | 79 | 57 | 203 | 530 | 578 | 700 | 610 | 411 | 220 | 60 | 8 | 3465 |
| 1997 | 12 | 11 | 61 | 60 | 286 | 449 | 628 | 660 | 522 | 212 | 27 | 0 | 2928 |
| 1998 | 11 | 3 | 56 | 84 | 467 | 645 | 724 | 582 | 470 | 227 | 49 | 32 | 3350 |
| 1999 | 12 | 56 | 45 | 225 | 352 | 512 | 561 | 664 | 467 | 206 | 46 | 11 | 3157 |
| 2000 | 7 | 73 | 147 | 207 | 428 | 487 | 654 | 668 | 487 | 270 | 33 | 0 | 3461 |
| 2001 | 1 | 30 | 1 | 195 | 357 | 530 | 640 | 642 | 362 | 135 | 76 | 26 | 2995 |
| 2002 | 24 | 0 | 64 | 268 | 375 | 559 | 550 | 634 | 416 | 206 | 11 | 11 | 3118 |
| 2003 | 0 | 12 | 29 | 224 | 482 | 510 | 533 | 588 | 357 | 197 | 101 | 7 | 3040 |
| 2004 | 19 | 2 | 74 | 116 | 353 | 483 | 563 | 576 | 472 | 378 | 26 | 7 | 3069 |
| 2005 | 37 | 13 | 41 | 136 | 317 | 534 | 637 | 649 | 585 | 231 | 135 | 6 | 3321 |
| 2006 | 4 | 17 | 160 | 352 | 434 | 567 | 649 | 730 | 447 | 257 | 85 | 21 | 3723 |
| 2007 | 4 | 13 | 89 | 99 | 328 | 478 | 486 | 586 | 465 | 282 | 99 | 41 | 2970 |
| 2008 | 13 | 57 | 108 | 191 | 477 | 661 | 598 | 610 | 439 | 240 | 73 | 26 | 3493 |
| 2009 | 14 | 63 | 128 | 180 | 456 | 647 | 744 | 730 | 416 | 202 | 20 | 0 | 3600 |

### SNOWFALL (inches)  2009  SAN ANTONIO (KSAT)

| YEAR | JUL | AUG | SEP | OCT | NOV | DEC | JAN | FEB | MAR | APR | MAY | JUN | TOTAL |
|---|---|---|---|---|---|---|---|---|---|---|---|---|---|
| 1980-81 | 0.0 | 0.0 | 0.0 | 0.0 | T | 0.0 | T | 0.0 | 0.0 | 0.0 | 0.0 | 0.0 | T |
| 1981-82 | 0.0 | 0.0 | 0.0 | 0.0 | 0.0 | 0.0 | 0.5 | 0.0 | 0.0 | 0.0 | 0.0 | 0.0 | 0.5 |
| 1982-83 | 0.0 | 0.0 | 0.0 | 0.0 | 0.0 | 0.0 | 0.0 | 0.0 | 0.0 | 0.0 | 0.0 | 0.0 | 0.0 |
| 1983-84 | 0.0 | 0.0 | 0.0 | 0.0 | 0.0 | 0.0 | 0.0 | 0.0 | 0.0 | 0.0 | 0.0 | 0.0 | 0.0 |
| 1984-85 | 0.0 | 0.0 | 0.0 | 0.0 | 0.0 | 0.0 | 15.9 | T | 0.0 | 0.0 | 0.0 | 0.0 | 15.9 |
| 1985-86 | 0.0 | 0.0 | 0.0 | 0.0 | 0.0 | 0.0 | T | 0.0 | 0.0 | 0.0 | 0.0 | 0.0 | T |
| 1986-87 | 0.0 | 0.0 | 0.0 | 0.0 | 0.0 | 0.0 | 1.3 | 0.0 | 0.0 | 0.0 | 0.0 | 0.0 | 1.3 |
| 1987-88 | 0.0 | 0.0 | 0.0 | 0.0 | 0.0 | 0.0 | 0.0 | 0.1 | 0.0 | 0.0 | 0.0 | 0.0 | 0.1 |
| 1988-89 | 0.0 | 0.0 | 0.0 | 0.0 | 0.0 | 0.0 | 0.0 | 0.0 | T | 0.0 | 0.0 | T | T |
| 1989-90 | 0.0 | 0.0 | 0.0 | 0.0 | 0.0 | T | 0.0 | T | T | T | 0.0 | 0.0 | T |
| 1990-91 | 0.0 | 0.0 | 0.0 | 0.0 | 0.0 | T | 0.0 | T | 0.0 | T | T | 0.0 | T |
| 1991-92 | 0.0 | 0.0 | 0.0 | 0.0 | 0.0 | 0.0 | 0.0 | T | T | 0.0 | 0.0 | 0.0 | T |
| 1992-93 | 0.0 | 0.0 | 0.0 | 0.0 | T | 0.0 | T | 0.0 | 0.0 | T | T | 0.0 | T |
| 1993-94 | 0.0 | 0.0 | 0.0 | T | 0.0 | 0.0 | T | T | T | 0.0 | 0.0 | 0.0 | T |
| 1994-95 | 0.0 | 0.0 | 0.0 | 0.0 | 0.0 | 0.0 | T | 0.0 | 0.0 | 0.0 | 0.0 | 0.0 | T |
| 1995-96 | 0.0 | 0.0 | 0.0 | 0.0 | 0.0 | 0.0 | 0.0 | | | | | | |
| 1996-97 | | | | | | | | | | | | | |
| 1997-98 | | | | | | | | | | | | | |
| 1998-99 | | | | | | | | | | | | | |
| 1999-00 | | | | | | | 0.0 | T | 0.0 | T | 0.0 | 0.0 | |
| 2000-01 | 0.0 | 0.0 | 0.0 | 0.0 | 0.0 | 0.0 | 0.0 | 0.0 | 0.0 | 0.0 | 0.0 | 0.0 | 0.0 |
| 2001-02 | 0.0 | 0.0 | 0.0 | T | 0.0 | 0.0 | 0.0 | T | T | T | 0.0 | 0.0 | T |
| 2002-03 | 0.0 | 0.0 | 0.0 | 0.0 | T | T | 0.0 | 0.2 | 0.0 | 0.0 | 0.0 | 0.0 | 0.2 |
| 2003-04 | 0.0 | 0.0 | 0.0 | 0.0 | 0.0 | 0.0 | 0.0 | 0.7 | 0.0 | 0.0 | 0.0 | 0.0 | 0.7 |
| 2004-05 | 0.0 | 0.0 | 0.0 | 0.0 | 0.0 | 0.0 | 0.0 | 0.0 | 0.0 | 0.0 | 0.0 | T | T |
| 2005-06 | 0.0 | 0.0 | 0.0 | 0.0 | 0.0 | 0.0 | 0.0 | 0.0 | T | 0.0 | T | 0.0 | T |
| 2006-07 | 0.0 | 0.0 | 0.0 | 0.0 | 0.0 | T | T | 0.0 | 0.0 | 0.0 | 0.0 | 0.0 | T |
| 2007-08 | 0.0 | 0.0 | 0.0 | 0.0 | 0.0 | 0.0 | 0.0 | 0.0 | 0.0 | 0.0 | 0.0 | 0.0 | 0.0 |
| 2008-09 | 0.0 | 0.0 | 0.0 | 0.0 | 0.0 | T | 0.0 | T | 0.0 | 0.0 | 0.0 | 0.0 | T |
| 2009- | 0.0 | 0.0 | 0.0 | 0.0 | 0.0 | T | | | | | | | |
| POR= 62 YRS | 0.0 | 0.0 | 0.0 | T | T | T | 0.4 | 0.2 | T | T | T | T | 0.6 |

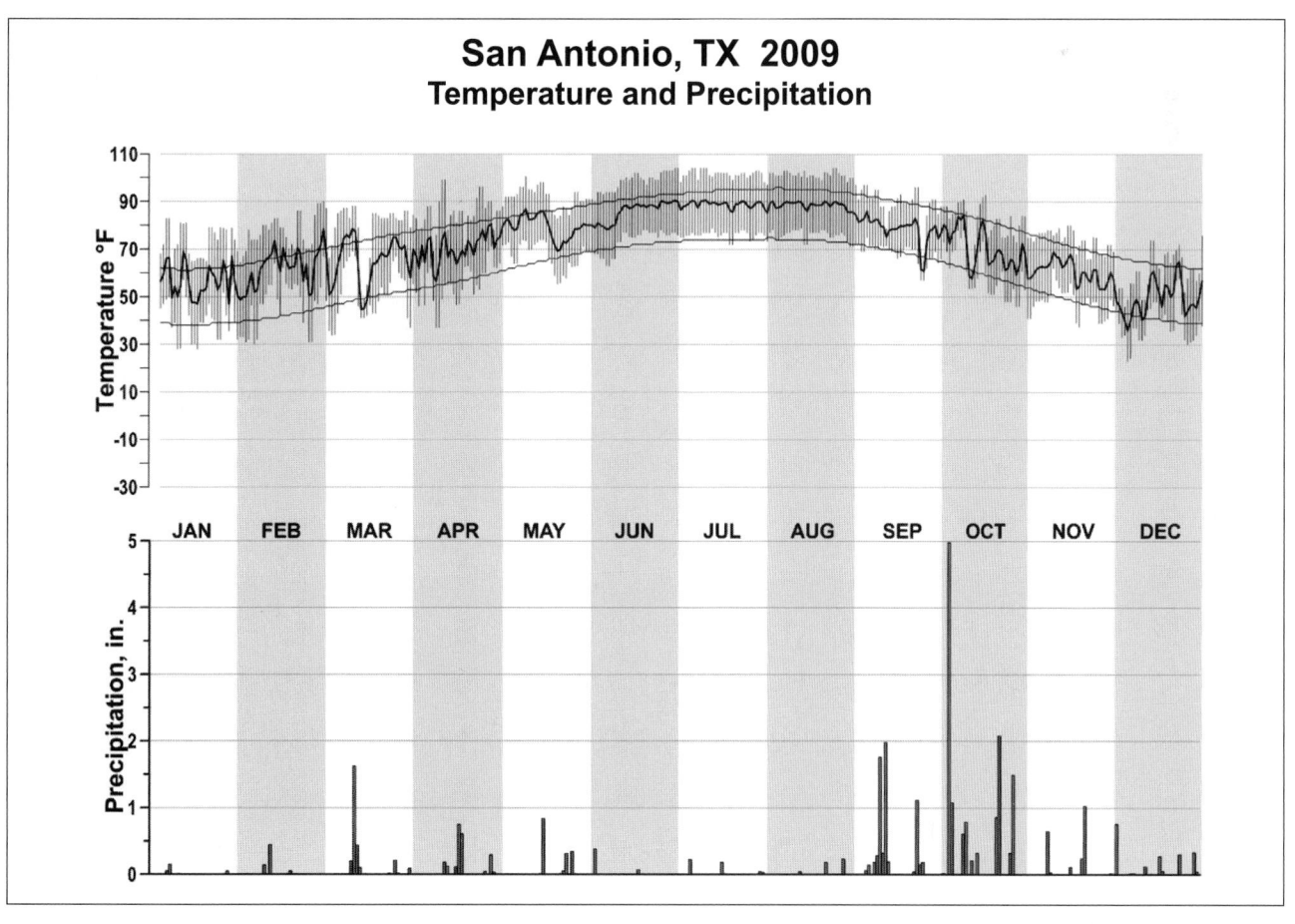

## San Antonio, TX  2009
### Temperature and Precipitation

# 2009
# SALT LAKE CITY
# UTAH (KSLC)

Salt Lake City is located in a northern Utah valley surrounded by mountains on three sides and the Great Salt Lake to the northwest. The city varies in altitude from near 4,200 to 5,000 feet above sea level.

The Wasatch Mountains to the east have peaks to nearly 12,000 feet above sea level. Their orographic effects cause more precipitation in the eastern part of the city than over the western part.

The Oquirrh Mountains to the southwest of the city have several peaks to above 10,000 feet above sea level. The Traverse Mountain Range at the south end of the Salt Lake Valley rises to above 6,000 feet above sea level. These mountain ranges help to shelter the valleys from storms from the southwest in the winter, but are instrumental in developing thunderstorms which can drift over the valley in the summer.

Besides the mountain ranges, the most influential natural condition affecting the climate of Salt Lake City is the Great Salt Lake. This large inland body of water, which never freezes over due to its high salt content, can moderate the temperatures of cold winter winds blowing from the northwest and helps drive a lake/valley wind system. The warmer lake water during the winter and spring also contributes to increased precipitation in the valley downwind from the lake. The combination of the Great Salt Lake and the Wasatch Mountains often enhances storm precipitation in the valley.

Salt Lake City normally has a semi-arid continental climate with four well-defined seasons. Summers are characterized by hot, dry weather, but the high temperatures are usually not oppressive since the relative humidity is generally low and the nights usually cool. July is the hottest month with temperature readings in the 90s.

The mean diurnal temperature range is about 30 degrees in the summer and 18 degrees during the winter. Temperatures above 102 degrees in the summer or colder than -10 degrees in the winter are likely to occur one season out of four.

Winters are cold, but usually not severe. Mountains to the north and east act as a barrier to frequent invasions of cold continental air. The average annual snowfall is under 60 inches at the airport but much higher amounts fall in higher bench locations. Heavy fog can develop under temperature inversions in the winter and persist for several days.

Precipitation, generally light during the summer and early fall, is heavy in the spring when storms from the Pacific Ocean are moving through the area more frequently than at any other season of the year.

Winds are usually light, although occasional high winds have occurred in every month of the year, particularly in March.

The growing season is over five months in length. Yard and garden foilage generally are making good growth by mid-April. The last freezing temperature in the spring averages late April and the first freeze of the fall is mid-October.

# NORMALS, MEANS, AND EXTREMES
## SALT LAKE CITY (KSLC)

LATITUDE: 40° 47'N    LONGITUDE: -111° 58'W    ELEVATION (FT): GRND: 4220  BARO: 4224    TIME ZONE: MOUNTAIN (UTC -7)    WBAN: 24127

| | ELEMENT | POR | JAN | FEB | MAR | APR | MAY | JUN | JUL | AUG | SEP | OCT | NOV | DEC | YEAR |
|---|---|---|---|---|---|---|---|---|---|---|---|---|---|---|---|
| TEMPERATURE °F | NORMAL DAILY MAXIMUM | 30 | 37.0 | 43.4 | 52.8 | 60.9 | 70.6 | 82.2 | 90.6 | 88.7 | 77.6 | 64.0 | 48.7 | 38.0 | 62.9 |
| | MEAN DAILY MAXIMUM | 62 | 37.3 | 43.4 | 52.4 | 61.5 | 72.2 | 82.9 | 92.8 | 90.4 | 79.5 | 65.6 | 50.0 | 38.5 | 63.9 |
| | HIGHEST DAILY MAXIMUM | 81 | 63 | 69 | 78 | 89 | 99 | 104 | 107 | 106 | 100 | 89 | 75 | 69 | 107 |
| | YEAR OF OCCURRENCE | | 2003 | 1972 | 2004 | 2007 | 2003 | 1979 | 2002 | 1994 | 1979 | 1963 | 2009 | 1995 | JUL 2002 |
| | MEAN OF EXTREME MAXS. | 62 | 52.1 | 58.7 | 69.8 | 79.2 | 88.7 | 97.5 | 101.4 | 99.4 | 93.6 | 82.3 | 66.8 | 54.6 | 78.7 |
| | NORMAL DAILY MINIMUM | 30 | 21.3 | 25.5 | 33.4 | 39.0 | 46.9 | 55.8 | 63.4 | 62.4 | 52.4 | 41.0 | 30.4 | 22.4 | 41.2 |
| | MEAN DAILY MINIMUM | 62 | 20.3 | 24.7 | 31.6 | 38.3 | 46.4 | 54.6 | 63.2 | 61.7 | 51.6 | 40.2 | 29.9 | 22.1 | 40.4 |
| | LOWEST DAILY MINIMUM | 81 | -22 | -30 | 2 | 14 | 25 | 35 | 40 | 37 | 27 | 16 | -14 | -21 | -30 |
| | YEAR OF OCCURRENCE | | 1949 | 1933 | 1966 | 1936 | 1965 | 1962 | 1968 | 1965 | 1965 | 1971 | 1955 | 1932 | FEB 1933 |
| | MEAN OF EXTREME MINS. | 62 | 2.8 | 7.3 | 18.4 | 27.5 | 33.6 | 42.0 | 52.1 | 50.1 | 38.0 | 27.9 | 15.6 | 6.5 | 26.8 |
| | NORMAL DRY BULB | 30 | 29.2 | 34.5 | 43.1 | 50.0 | 58.8 | 69.0 | 77.0 | 75.6 | 65.0 | 52.5 | 39.6 | 30.2 | 52.0 |
| | MEAN DRY BULB | 62 | 28.8 | 34.1 | 42.0 | 49.9 | 59.3 | 68.9 | 78.0 | 76.0 | 65.5 | 52.9 | 40.0 | 30.3 | 52.1 |
| | MEAN WET BULB | 26 | 25.7 | 29.4 | 35.6 | 40.7 | 47.0 | 52.5 | 57.6 | 56.6 | 50.5 | 42.7 | 33.7 | 26.8 | 41.6 |
| | MEAN DEW POINT | 26 | 22.8 | 25.5 | 30.0 | 34.2 | 40.1 | 44.1 | 48.6 | 48.1 | 43.0 | 36.0 | 29.6 | 23.2 | 35.4 |
| | NORMAL NO. DAYS WITH: | | | | | | | | | | | | | | |
| | MAXIMUM >= 90 | 30 | 0.0 | 0.0 | 0.0 | 0.0 | 0.3 | 9.3 | 22.1 | 19.5 | 4.1 | 0.0 | 0.0 | 0.0 | 55.3 |
| | MAXIMUM <= 32 | 30 | 9.4 | 3.3 | 0.3 | 0.0 | 0.0 | 0.0 | 0.0 | 0.0 | 0.0 | * | 1.1 | 7.6 | 21.7 |
| | MINIMUM <= 32 | 30 | 27.1 | 21.9 | 13.1 | 5.0 | 0.2 | 0.0 | 0.0 | 0.0 | * | 3.1 | 18.0 | 26.9 | 115.3 |
| | MINIMUM <= 0 | 30 | 1.2 | 0.4 | 0.0 | 0.0 | 0.0 | 0.0 | 0.0 | 0.0 | 0.0 | 0.0 | 0.0 | 0.7 | 2.3 |
| H/C | NORMAL HEATING DEG. DAYS | 30 | 1108 | 857 | 665 | 448 | 215 | 50 | 3 | 3 | 89 | 379 | 747 | 1067 | 5631 |
| | NORMAL COOLING DEG. DAYS | 30 | 0 | 0 | 0 | 4 | 34 | 183 | 387 | 347 | 105 | 6 | 0 | 0 | 1066 |
| RH | NORMAL (PERCENT) | 30 | 76 | 72 | 61 | 54 | 52 | 43 | 38 | 39 | 48 | 58 | 68 | 75 | 57 |
| | HOUR 05 LST | 30 | 81 | 80 | 71 | 67 | 68 | 60 | 54 | 54 | 63 | 70 | 77 | 80 | 69 |
| | HOUR 11 LST | 30 | 71 | 65 | 51 | 44 | 41 | 33 | 29 | 30 | 37 | 45 | 58 | 70 | 48 |
| | HOUR 17 LST | 30 | 71 | 61 | 47 | 40 | 36 | 28 | 24 | 25 | 32 | 43 | 59 | 70 | 45 |
| | HOUR 23 LST | 30 | 80 | 78 | 68 | 61 | 61 | 51 | 45 | 45 | 55 | 66 | 75 | 79 | 64 |
| S | PERCENT POSSIBLE SUNSHINE | 64 | 45 | 54 | 64 | 68 | 72 | 80 | 84 | 83 | 81 | 72 | 53 | 40 | 66 |
| W/O | MEAN NO. DAYS WITH: | | | | | | | | | | | | | | |
| | HEAVY FOG(VISBY <= 1/4 MI) | 46 | 4.6 | 2.5 | 0.6 | 0.2 | 0.0 | 0.0 | 0.0 | 0.0 | 0.0 | 0.0 | 0.8 | 3.7 | 12.4 |
| | THUNDERSTORMS | 62 | 0.3 | 0.5 | 1.3 | 2.0 | 5.2 | 5.3 | 6.2 | 7.3 | 4.4 | 1.9 | 0.6 | 0.3 | 35.3 |
| CLOUDNESS | MEAN: | | | | | | | | | | | | | | |
| | SUNRISE-SUNSET (OKTAS) | 63 | 5.9 | 5.7 | 5.3 | 5.1 | 4.7 | 3.4 | 2.9 | 2.9 | 2.9 | 3.8 | 5.0 | 5.8 | 4.5 |
| | MIDNIGHT-MIDNIGHT (OKTAS) | 34 | 6.0 | 5.4 | 5.3 | 5.1 | 4.6 | 3.5 | 3.1 | 3.1 | 3.0 | 3.7 | 5.2 | 5.8 | 4.5 |
| | MEAN NO. DAYS WITH: | | | | | | | | | | | | | | |
| | CLEAR | 70 | 5.4 | 5.2 | 7.0 | 6.7 | 9.0 | 13.8 | 16.7 | 15.9 | 16.4 | 13.9 | 8.4 | 6.2 | 124.6 |
| | PARTLY CLOUDY | 70 | 6.5 | 6.9 | 8.3 | 9.4 | 10.2 | 9.9 | 9.8 | 10.6 | 8.3 | 7.7 | 7.2 | 6.5 | 101.3 |
| | CLOUDY | 70 | 19.0 | 16.1 | 15.7 | 14.0 | 11.8 | 6.3 | 4.6 | 4.5 | 5.3 | 9.6 | 14.5 | 18.3 | 139.7 |
| PR | MEAN STATION PRESSURE(IN) | 26 | 25.84 | 25.77 | 25.71 | 25.68 | 25.66 | 25.67 | 25.72 | 25.71 | 25.74 | 25.77 | 25.80 | 25.80 | 25.74 |
| | MEAN SEA-LEVEL PRES. (IN) | 26 | 30.22 | 30.11 | 29.99 | 29.91 | 29.86 | 29.84 | 29.85 | 29.88 | 29.93 | 30.02 | 30.12 | 30.20 | 29.99 |
| WINDS | MEAN SPEED (MPH) | 26 | 6.9 | 7.6 | 9.1 | 9.7 | 9.3 | 9.4 | 9.3 | 9.6 | 8.9 | 8.2 | 7.7 | 7.2 | 8.6 |
| | PREVAIL.DIR(TENS OF DEGS) | 45 | 15 | 15 | 15 | 16 | 16 | 17 | 17 | 17 | 15 | 15 | 15 | 15 | 15 |
| | MAXIMUM 2-MINUTE: | | | | | | | | | | | | | | |
| | SPEED (MPH) | 11 | 44 | 47 | 45 | 53 | 63 | 51 | 48 | 43 | 47 | 38 | 40 | 43 | 63 |
| | DIR. (TENS OF DEGS) | | 18 | 18 | 30 | 19 | 20 | 19 | 17 | 20 | 32 | 20 | 17 | 17 | 20 |
| | YEAR OF OCCURRENCE | | 2008 | 1999 | 2006 | 2002 | 2006 | 2002 | 2008 | 2009 | 2007 | 2009 | 2008 | 2008 | MAY 2006 |
| | MAXIMUM 3-SECOND | | | | | | | | | | | | | | |
| | SPEED (MPH) | 11 | 53 | 62 | 53 | 64 | 75 | 60 | 56 | 56 | 58 | 51 | 55 | 49 | 75 |
| | DIR. (TENS OF DEGS) | | 17 | 18 | 19 | 19 | 19 | 24 | 18 | 16 | 32 | 21 | 28 | 17 | 19 |
| | YEAR OF OCCURRENCE | | 2008 | 1999 | 2009 | 2002 | 2006 | 2004 | 2008 | 2006 | 2007 | 2009 | 2000 | 2008 | MAY 2006 |
| PRECIPITATION | NORMAL (IN) | 30 | 1.37 | 1.33 | 1.91 | 2.02 | 2.09 | 0.77 | 0.72 | 0.76 | 1.33 | 1.57 | 1.40 | 1.23 | 16.50 |
| | MAXIMUM MONTHLY (IN) | 81 | 3.23 | 4.89 | 3.97 | 4.90 | 4.76 | 3.84 | 2.57 | 3.66 | 7.04 | 3.91 | 3.34 | 4.37 | 7.04 |
| | YEAR OF OCCURRENCE | | 1993 | 1998 | 1983 | 1944 | 1977 | 1998 | 1982 | 1968 | 1982 | 1981 | 2001 | 1983 | SEP 1982 |
| | MINIMUM MONTHLY (IN) | 81 | 0.09 | 0.12 | 0.10 | 0.45 | T | T | T | T | T | 0.00 | 0.01 | 0.08 | 0.00 |
| | YEAR OF OCCURRENCE | | 1961 | 1946 | 1956 | 1981 | 1934 | 1994 | 1963 | 1944 | 1951 | 1952 | 1939 | 1976 | OCT 1952 |
| | MAXIMUM IN 24 HOURS (IN) | 81 | 1.36 | 1.76 | 1.83 | 2.41 | 2.03 | 1.88 | 2.35 | 1.96 | 2.30 | 1.76 | 1.53 | 1.82 | 2.41 |
| | YEAR OF OCCURRENCE | | 1953 | 1998 | 1944 | 1957 | 1942 | 1948 | 1962 | 1932 | 1982 | 1984 | 2001 | 1972 | APR 1957 |
| | NORMAL NO. DAYS WITH: | | | | | | | | | | | | | | |
| | PRECIPITATION >= 0.01 | 30 | 10.9 | 9.3 | 10.5 | 9.9 | 9.7 | 5.2 | 4.8 | 5.6 | 6.0 | 6.9 | 9.0 | 9.1 | 96.9 |
| | PRECIPITATION >= 1.00 | 30 | 0.0 | * | 0.0 | 0.2 | 0.2 | 0.1 | * | * | 0.3 | 0.1 | * | * | 0.9 |
| SNOWFALL | NORMAL (IN) | 30 | 14.5 | 10.2 | 9.2 | 5.7 | 0.7 | 0.* | 0.0 | 0.0 | 0.2 | 2.1 | 7.9 | 12.2 | 62.7 |
| | MAXIMUM MONTHLY (IN) | 81 | 50.3 | 32.1 | 41.9 | 26.4 | 7.5 | T | T | T | 4.0 | 20.4 | 33.3 | 35.2 | 50.3 |
| | YEAR OF OCCURRENCE | | 1993 | 1998 | 1977 | 1974 | 1975 | 1995 | 1996 | 2006 | 1971 | 1984 | 1994 | 1972 | JAN 1993 |
| | MAXIMUM IN 24 HOURS (IN) | 81 | 16.5 | 18.0 | 15.4 | 16.2 | 6.4 | T | T | T | 4.0 | 18.4 | 11.0 | 18.1 | 18.4 |
| | YEAR OF OCCURRENCE' | | 1996 | 1998 | 1944 | 1974 | 1975 | 1995 | 1991 | 1993 | 1971 | 1984 | 1930 | 1972 | OCT 1984 |
| | MAXIMUM SNOW DEPTH (IN) | 61 | 25 | 17 | 14 | 8 | 4 | 0 | 0 | 0 | 0 | 9 | 11 | 14 | 25 |
| | YEAR OF OCCURRENCE | | 1993 | 1949 | 1998 | 1974 | 1978 | | | | | 1984 | 1985 | 1972 | JAN 1993 |
| | NORMAL NO. DAYS WITH: | | | | | | | | | | | | | | |
| | SNOWFALL >= 1.0 | 30 | 4.1 | 3.1 | 2.8 | 1.5 | 0.2 | 0.0 | 0.0 | 0.0 | 0.1 | 0.6 | 2.3 | 3.5 | 18.2 |

## PRECIPITATION (inches) 2009 SALT LAKE CITY (KSLC)

| YEAR | JAN | FEB | MAR | APR | MAY | JUN | JUL | AUG | SEP | OCT | NOV | DEC | ANNUAL |
|------|-----|-----|-----|-----|-----|-----|-----|-----|-----|-----|-----|-----|--------|
| 1980 | 2.87 | 2.25 | 2.46 | 0.89 | 2.70 | 0.42 | 1.34 | 0.26 | 0.72 | 1.74 | 1.17 | 0.37 | 17.19 |
| 1981 | 0.64 | 0.81 | 2.11 | 0.45 | 3.68 | 1.03 | 0.33 | 0.23 | 0.48 | 3.91 | 1.03 | 1.89 | 16.59 |
| 1982 | 1.08 | 0.53 | 2.39 | 1.63 | 1.86 | 0.66 | 2.57 | 0.56 | 7.04 | 1.87 | 0.75 | 1.92 | 22.86 |
| 1983 | 1.19 | 1.36 | 3.97 | 1.63 | 2.58 | 0.62 | 1.02 | 2.64 | 1.03 | 1.62 | 2.23 | 4.37 | 24.26 |
| 1984 | 0.50 | 0.95 | 1.76 | 4.43 | 1.17 | 1.86 | 1.72 | 1.49 | 1.72 | 3.70 | 1.45 | 0.80 | 21.55 |
| 1985 | 0.91 | 0.85 | 1.80 | 0.64 | 2.95 | 1.30 | 0.85 | 0.03 | 1.98 | 1.61 | 2.63 | 1.42 | 16.97 |
| 1986 | 0.86 | 1.28 | 2.32 | 4.55 | 3.39 | 0.42 | 0.85 | 1.32 | 2.75 | 0.39 | 1.17 | 0.10 | 19.40 |
| 1987 | 1.53 | 1.41 | 1.52 | 0.79 | 2.41 | 0.19 | 0.79 | 0.36 | 0.05 | 1.18 | 1.17 | 1.10 | 12.50 |
| 1988 | 1.06 | 0.13 | 0.94 | 1.84 | 2.16 | 0.03 | 0.04 | 0.22 | 0.07 | 0.01 | 2.17 | 0.62 | 9.29 |
| 1989 | 0.56 | 1.57 | 1.77 | 0.46 | 1.83 | 0.22 | 0.39 | 0.90 | 0.49 | 1.82 | 0.73 | 0.13 | 10.87 |
| 1990 | 0.57 | 0.35 | 2.17 | 1.14 | 1.65 | 0.66 | 0.64 | 0.46 | 0.56 | 0.69 | 1.24 | 0.56 | 10.69 |
| 1991 | 1.11 | 0.61 | 1.11 | 2.71 | 2.76 | 1.09 | 0.32 | 0.86 | 2.55 | 2.10 | 2.17 | 0.40 | 17.79 |
| 1992 | 0.78 | 1.24 | 1.11 | 0.96 | 1.86 | 0.45 | 0.29 | 0.35 | 0.47 | 1.03 | 2.46 | 1.07 | 12.07 |
| 1993 | 3.23 | 1.35 | 1.37 |  | 3.99 | 1.14 | 1.38 | 0.46 | 0.22 | 2.77 | 0.54 | 0.88 |  |
| 1994 | 0.62 | 1.53 | 1.28 | 2.94 | 1.29 | T | 0.06 | 0.61 | 0.32 | 2.24 | 2.96 | 1.43 | 15.28 |
| 1995 | 1.81 | 1.08 | 2.35 | 2.07 | 3.68 | 1.49 | 0.32 | 0.21 | 1.33 | 0.53 | 0.85 | 1.20 | 16.92 |
| 1996 | 3.09 | 1.54 | 2.71 | 2.20 | 1.32 | 0.09 | .41 | .02 | 1.03 | 1.45 | 1.72 | 1.73 | 17.31 |
| 1997 | 2.27 | 1.62 | 0.97 | 2.22 | 1.77 | 1.73 | 0.84 | 0.63 | 1.50 | 1.87 | 0.87 | 0.64 | 16.93 |
| 1998 | 1.63 | 4.89 | 2.97 | 2.09 | 1.04 | 3.84 | 1.57 | 0.46 | 1.53 | 1.25 | 1.27 | 1.27 | 23.81 |
| 1999 | 1.29 | 0.96 | 0.80 | 3.09 | 2.59 | 0.82 | 0.25 | 0.70 | 0.45 | 0.02 | 0.70 | 1.84 | 13.51 |
| 2000 | 2.17 | 1.80 | 0.80 | 0.76 | 1.62 | 0.36 | 0.42 | 2.00 | 1.86 | 2.00 | 1.31 | 1.24 | 16.34 |
| 2001 | 0.78 | 1.50 | 1.55 | 2.46 | 0.22 | 1.12 | 1.13 | 0.53 | 0.05 | 0.92 | 3.34 | 1.44 | 15.04 |
| 2002 | 1.19 | 0.30 | 2.47 | 2.49 | 0.48 | 0.18 | 0.14 | 0.03 | 1.12 | 0.71 | 0.64 | 0.54 | 10.29 |
| 2003 | 0.64 | 1.06 | 1.39 | 1.65 | 1.67 | 0.88 | 0.33 | 0.67 | 1.59 | 0.16 | 1.94 | 3.97 | 15.95 |
| 2004 | 0.46 | 2.25 | 0.88 | 2.38 | 0.95 | 1.70 | 0.34 | 0.19 | 0.50 | 3.48 | 1.24 | 0.52 | 14.89 |
| 2005 | 1.44 | 1.23 | 2.44 | 3.15 | 2.88 | 1.64 | 0.01 | 0.72 | 0.40 | 0.91 | 0.80 | 1.26 | 16.88 |
| 2006 | 1.36 | 1.26 | 2.76 | 3.14 | 0.79 | 0.72 | 0.26 | 0.92 | 1.87 | 1.02 | 1.13 | 0.91 | 16.14 |
| 2007 | 0.73 | 1.53 | 1.11 | 0.53 | 0.57 | 0.80 | 0.53 | 0.10 | 1.74 | 1.88 | 0.49 | 3.35 | 13.36 |
| 2008 | 1.30 | 1.24 | 1.34 | 0.75 | 1.01 | 0.75 | 0.14 | 0.83 | 0.31 | 1.29 | 1.50 | 1.28 | 11.74 |
| 2009 | 2.15 | 0.83 | 1.72 | 2.55 | 1.36 | 2.64 | 0.28 | 0.40 | 1.19 | 1.18 | 0.20 | 1.35 | 15.85 |
| POR= 62 YRS | 1.32 | 1.28 | 1.77 | 2.06 | 1.69 | 0.97 | 0.67 | 0.77 | 1.06 | 1.32 | 1.32 | 1.37 | 15.60 |

WBAN : 24127

## AVERAGE TEMPERATURE (°F) 2009 SALT LAKE CITY (KSLC)

| YEAR | JAN | FEB | MAR | APR | MAY | JUN | JUL | AUG | SEP | OCT | NOV | DEC | ANNUAL |
|------|-----|-----|-----|-----|-----|-----|-----|-----|-----|-----|-----|-----|--------|
| 1980 | 33.7 | 36.0 | 41.5 | 52.7 | 57.0 | 67.5 | 77.6 | 74.1 | 66.3 | 52.6 | 41.3 | 33.6 | 52.8 |
| 1981 | 32.1 | 38.3 | 44.1 | 53.4 | 57.6 | 69.6 | 78.2 | 78.0 | 68.5 | 50.5 | 44.3 | 36.4 | 54.3 |
| 1982 | 29.8 | 32.3 | 43.3 | 46.5 | 56.7 | 68.0 | 75.5 | 78.4 | 64.0 | 48.8 | 38.1 | 29.9 | 50.9 |
| 1983 | 35.2 | 39.4 | 44.6 | 45.9 | 55.8 | 67.7 | 76.6 | 77.8 | 67.8 | 56.0 | 43.0 | 31.9 | 53.5 |
| 1984 | 23.8 | 25.8 | 40.1 | 48.5 | 61.6 | 67.3 | 78.5 | 77.2 | 66.5 | 49.5 | 42.7 | 29.9 | 51.0 |
| 1985 | 24.2 | 25.6 | 40.8 | 55.7 | 48.7 | 63.9 | 72.5 | 76.5 | 62.7 | 53.1 | 37.4 | 27.7 | 51.7 |
| 1986 | 29.0 | 41.4 | 47.7 | 48.8 | 57.2 | 73.5 | 74.2 | 77.9 | 60.2 | 51.3 | 40.9 | 29.8 | 52.7 |
| 1987 | 26.5 | 36.1 | 42.8 | 55.9 | 62.7 | 71.6 | 75.7 | 74.7 | 66.5 | 56.4 | 40.8 | 30.5 | 53.4 |
| 1988 | 25.0 | 34.8 | 41.4 | 52.0 | 59.6 | 75.7 | 80.9 | 76.5 | 63.8 | 60.0 | 41.1 | 28.1 | 53.2 |
| 1989 | 22.3 | 25.3 | 45.8 | 54.8 | 59.9 | 69.2 | 81.1 | 75.1 | 66.4 | 53.4 | 40.5 | 31.4 | 52.1 |
| 1990 | 33.4 | 32.8 | 45.0 | 54.9 | 57.8 | 72.0 | 78.9 | 76.2 | 72.0 | 54.0 | 41.4 | 21.0 | 53.3 |
| 1991 | 24.4 | 36.5 | 42.8 | 47.9 | 55.6 | 68.3 | 79.1 | 78.4 | 64.4 | 53.6 | 40.0 | 30.0 | 51.8 |
| 1992 | 25.6 | 39.3 | 49.3 | 57.1 | 65.6 | 70.4 | 75.4 | 77.3 | 66.5 | 56.0 | 34.1 | 27.1 | 53.6 |
| 1993 | 24.9 | 29.5 | 45.5 | 48.5 | 63.4 | 63.7 | 69.9 | 72.5 | 65.5 | 52.6 | 34.8 | 31.5 | 50.2 |
| 1994 | 36.8 | 35.3 | 45.6 | 52.1 | 63.0 | 74.3 | 80.7 | 80.8 | 70.5 | 51.3 | 32.6 | 31.6 | 54.6 |
| 1995 | 34.9 | 42.3 | 44.1 | 48.6 | 55.5 | 64.3 | 76.0 | 78.0 | 67.5 | 51.5 | 46.1 | 37.1 | 53.8 |
| 1996 | 33.0 | 30.8 | 45.3 | 50.6 | 59.9 | 73.4 | 80.5 | 77.8 | 64.7 | 53.9 | 43.4 | 37.1 | 54.2 |
| 1997 | 32.3 | 35.3 | 46.0 | 48.3 | 63.4 | 70.1 | 75.1 | 78.7 | 67.8 | 52.8 | 42.2 | 27.9 | 53.3 |
| 1998 | 37.9 | 37.5 | 40.4 | 48.1 | 58.6 | 62.6 | 79.7 | 77.6 | 68.1 | 51.0 | 42.9 | 29.6 | 52.8 |
| 1999 | 35.8 | 37.4 | 44.8 | 45.5 | 56.0 | 68.4 | 78.3 | 77.3 | 63.5 | 54.5 | 45.7 | 31.0 | 53.2 |
| 2000 | 35.1 | 39.8 | 42.0 | 54.5 | 61.7 | 72.1 | 80.8 | 78.9 | 64.6 | 52.3 | 31.4 | 30.7 | 53.7 |
| 2001 | 27.3 | 34.4 | 45.4 | 50.1 | 63.6 | 70.9 | 79.4 | 79.1 | 70.2 | 55.0 | 42.6 | 26.3 | 53.7 |
| 2002 | 26.0 | 27.9 | 38.9 | 51.6 | 59.9 | 71.9 | 81.9 | 75.5 | 66.0 | 49.8 | 37.6 | 35.5 | 51.9 |
| 2003 | 38.3 | 34.6 | 44.6 | 50.4 | 61.3 | 71.2 | 83.6 | 80.0 | 65.8 | 57.9 | 37.3 | 33.6 | 54.9 |
| 2004 | 22.4 | 26.9 | 47.8 | 52.4 | 60.3 | 70.2 | 79.0 | 74.3 | 65.3 | 53.9 | 39.0 | 32.7 | 52.0 |
| 2005 | 34.5 | 35.0 | 42.7 | 50.5 | 59.3 | 66.8 | 80.8 | 77.0 | 65.5 | 54.4 | 41.4 | 31.5 | 53.3 |
| 2006 | 34.4 | 33.5 | 41.7 | 53.3 | 63.1 | 73.2 | 83.0 | 76.5 | 63.5 | 50.6 | 41.0 | 30.7 | 53.7 |
| 2007 | 21.1 | 36.8 | 46.3 | 52.5 | 63.0 | 73.2 | 84.0 | 80.7 | 66.7 | 51.9 | 41.9 | 27.0 | 53.8 |
| 2008 | 23.9 | 33.3 | 40.4 | 46.3 | 57.4 | 69.9 | 81.4 | 77.8 | 66.6 | 53.1 | 43.0 | 30.1 | 51.9 |
| 2009 | 30.7 | 36.2 | 42.1 | 48.9 | 61.5 | 66.4 | 79.0 | 75.5 | 70.6 | 49.4 | 41.0 | 23.6 | 52.1 |
| POR= 62 YRS | 28.8 | 34.1 | 42.0 | 49.9 | 59.3 | 68.9 | 78.0 | 76.0 | 65.5 | 52.9 | 40.0 | 30.3 | 52.2 |

## HEATING DEGREE DAYS (base 65°F) 2009  SALT LAKE CITY (KSLC)

| YEAR | JUL | AUG | SEP | OCT | NOV | DEC | JAN | FEB | MAR | APR | MAY | JUN | TOTAL |
|---|---|---|---|---|---|---|---|---|---|---|---|---|---|
| 1980-81 | 0 | 10 | 57 | 379 | 704 | 965 | 1013 | 742 | 641 | 346 | 233 | 46 | 5136 |
| 1981-82 | 0 | 0 | 34 | 444 | 614 | 879 | 1087 | 909 | 668 | 548 | 259 | 62 | 5504 |
| 1982-83 | 7 | 0 | 134 | 495 | 800 | 1080 | 916 | 710 | 624 | 569 | 314 | 36 | 5685 |
| 1983-84 | 6 | 0 | 49 | 276 | 650 | 1018 | 1269 | 1130 | 763 | 493 | 157 | 76 | 5887 |
| 1984-85 | 0 | 0 | 98 | 480 | 662 | 1084 | 1260 | 1097 | 740 | 285 | 109 | 17 | 5832 |
| 1985-86 | 0 | 0 | 140 | 360 | 821 | 1151 | 1110 | 655 | 527 | 477 | 283 | 14 | 5538 |
| 1986-87 | 6 | 0 | 203 | 416 | 720 | 1085 | 1186 | 803 | 679 | 291 | 123 | 17 | 5529 |
| 1987-88 | 0 | 0 | 51 | 260 | 719 | 1060 | 1235 | 870 | 723 | 381 | 222 | 3 | 5524 |
| 1988-89 | 0 | 0 | 142 | 158 | 711 | 1138 | 1318 | 1105 | 587 | 313 | 193 | 35 | 5700 |
| 1989-90 | 0 | 15 | 44 | 355 | 729 | 1036 | 971 | 895 | 612 | 297 | 232 | 30 | 5216 |
| 1990-91 | 0 | 0 | 17 | 347 | 704 | 1359 | 1249 | 792 | 682 | 508 | 289 | 32 | 5979 |
| 1991-92 | 0 | 0 | 78 | 347 | 742 | 1078 | 1220 | 741 | 481 | 247 | 57 | 33 | 5024 |
| 1992-93 | 11 | 14 | 42 | 285 | 921 | 1169 | 1237 | 986 | 596 | 489 | 151 | 108 | 6009 |
| 1993-94 | 19 | 3 | 82 | 391 | 902 | 1031 | 863 | 823 | 590 | 397 | 100 | 12 | 5213 |
| 1994-95 | 3 | 0 | 19 | 416 | 963 | 1028 | 927 | 628 | 642 | 486 | 288 | 104 | 5504 |
| 1995-96 | 0 | 0 | 75 | 415 | 559 | 856 | 986 | 986 | 607 | 428 | 175 | 5 | 5092 |
| 1996-97 | 0 | 0 | 123 | 384 | 642 | 858 | 1009 | 827 | 583 | 495 | 115 | 17 | 5053 |
| 1997-98 | 6 | 0 | 61 | 382 | 677 | 1145 | 835 | 763 | 754 | 504 | 202 | 118 | 5447 |
| 1998-99 | 0 | 0 | 57 | 428 | 657 | 1089 | 898 | 766 | 619 | 578 | 295 | 63 | 5450 |
| 1999-00 | 0 | 0 | 95 | 322 | 573 | 1047 | 920 | 723 | 704 | 311 | 154 | 15 | 4864 |
| 2000-01 | 0 | 0 | 99 | 388 | 1002 | 1056 | 1162 | 850 | 598 | 450 | 123 | 62 | 5790 |
| 2001-02 | 0 | 0 | 27 | 310 | 662 | 1194 | 1201 | 1035 | 800 | 396 | 221 | 47 | 5893 |
| 2002-03 | 0 | 0 | 56 | 466 | 817 | 907 | 821 | 846 | 628 | 434 | 230 | 25 | 5230 |
| 2003-04 | 0 | 0 | 75 | 242 | 827 | 967 | 1311 | 1098 | 526 | 371 | 178 | 18 | 5613 |
| 2004-05 | 0 | 12 | 97 | 344 | 774 | 993 | 940 | 832 | 684 | 429 | 201 | 101 | 5407 |
| 2005-06 | 0 | 3 | 99 | 328 | 701 | 1028 | 943 | 873 | 719 | 344 | 158 | 7 | 5203 |
| 2006-07 | 0 | 2 | 141 | 450 | 713 | 1055 | 1350 | 783 | 574 | 379 | 159 | 35 | 5641 |
| 2007-08 | 0 | 0 | 88 | 399 | 688 | 1170 | 1267 | 914 | 755 | 557 | 258 | 77 | 6173 |
| 2008-09 | 0 | 0 | 41 | 374 | 655 | 1077 | 1056 | 801 | 704 | 477 | 152 | 48 | 5385 |
| 2009- | 0 | 5 | 30 | 480 | 711 | 1275 | | | | | | | |

WBAN : 24127

## COOLING DEGREE DAYS (base 65°F) 2009  SALT LAKE CITY (KSLC)

| YEAR | JAN | FEB | MAR | APR | MAY | JUN | JUL | AUG | SEP | OCT | NOV | DEC | TOTAL |
|---|---|---|---|---|---|---|---|---|---|---|---|---|---|
| 1980 | 0 | 0 | 0 | 9 | 10 | 159 | 399 | 301 | 99 | 1 | 0 | 0 | 978 |
| 1981 | 0 | 0 | 0 | 3 | 12 | 190 | 412 | 409 | 145 | 2 | 0 | 0 | 1173 |
| 1982 | 0 | 0 | 0 | 0 | 11 | 158 | 338 | 423 | 109 | 0 | 0 | 0 | 1039 |
| 1983 | 0 | 0 | 0 | 0 | 37 | 123 | 370 | 405 | 138 | 4 | 0 | 0 | 1077 |
| 1984 | 0 | 0 | 0 | 3 | 58 | 153 | 426 | 383 | 147 | 4 | 0 | 0 | 1174 |
| 1985 | 0 | 0 | 0 | 11 | 78 | 249 | 493 | 364 | 79 | 0 | 0 | 0 | 1274 |
| 1986 | 0 | 0 | 0 | 0 | 47 | 277 | 296 | 407 | 66 | 0 | 0 | 0 | 1093 |
| 1987 | 0 | 0 | 0 | 25 | 60 | 222 | 338 | 309 | 103 | 0 | 0 | 0 | 1057 |
| 1988 | 0 | 0 | 0 | 0 | 61 | 334 | 501 | 363 | 112 | 9 | 0 | 0 | 1380 |
| 1989 | 0 | 0 | 0 | 13 | 43 | 171 | 506 | 337 | 92 | 0 | 0 | 0 | 1162 |
| 1990 | 0 | 0 | 0 | 2 | 19 | 247 | 438 | 351 | 235 | 11 | 0 | 0 | 1303 |
| 1991 | 0 | 0 | 0 | 1 | 3 | 138 | 442 | 425 | 63 | 4 | 0 | 0 | 1076 |
| 1992 | 0 | 0 | 0 | 15 | 81 | 204 | 340 | 401 | 93 | 14 | 0 | 0 | 1148 |
| 1993 | 0 | 0 | 0 | 0 | 108 | 75 | 178 | 246 | 102 | 13 | 0 | 0 | 722 |
| 1994 | 0 | 0 | 0 | 16 | 46 | 300 | 498 | 500 | 189 | 0 | 0 | 0 | 1549 |
| 1995 | 0 | 0 | 0 | 0 | 2 | 89 | 347 | 409 | 158 | 5 | 0 | 0 | 1010 |
| 1996 | 0 | 0 | 0 | 2 | 24 | 264 | 490 | 405 | 125 | 48 | 0 | 0 | 1358 |
| 1997 | 0 | 0 | 0 | 0 | 71 | 178 | 324 | 431 | 151 | 15 | 0 | 0 | 1170 |
| 1998 | 0 | 0 | 0 | 3 | 13 | 53 | 462 | 399 | 156 | 0 | 0 | 0 | 1086 |
| 1999 | 0 | 0 | 0 | 0 | 22 | 172 | 416 | 388 | 58 | 6 | 0 | 0 | 1062 |
| 2000 | 0 | 0 | 0 | 4 | 61 | 233 | 497 | 438 | 91 | 2 | 0 | 0 | 1326 |
| 2001 | 0 | 0 | 0 | 9 | 87 | 249 | 452 | 445 | 191 | 8 | 0 | 0 | 1441 |
| 2002 | 0 | 0 | 0 | 4 | 70 | 259 | 529 | 330 | 95 | 0 | 0 | 0 | 1287 |
| 2003 | 0 | 0 | 0 | 3 | 122 | 219 | 584 | 474 | 110 | 29 | 0 | 0 | 1541 |
| 2004 | 0 | 0 | 0 | 0 | 42 | 181 | 441 | 304 | 114 | 7 | 0 | 0 | 1089 |
| 2005 | 0 | 0 | 0 | 0 | 31 | 161 | 499 | 381 | 116 | 8 | 0 | 0 | 1196 |
| 2006 | 0 | 0 | 0 | 0 | 105 | 260 | 568 | 367 | 101 | 10 | 0 | 0 | 1411 |
| 2007 | 0 | 0 | 0 | 11 | 104 | 288 | 595 | 492 | 145 | 0 | 0 | 0 | 1635 |
| 2008 | 0 | 0 | 0 | 3 | 30 | 230 | 516 | 403 | 97 | 14 | 0 | 0 | 1293 |
| 2009 | 0 | 0 | 0 | 2 | 51 | 99 | 441 | 338 | 204 | 0 | 0 | 0 | 1135 |

**SNOWFALL (inches) 2009 SALT LAKE CITY (KSLC)**

| YEAR | JUL | AUG | SEP | OCT | NOV | DEC | JAN | FEB | MAR | APR | MAY | JUN | TOTAL |
|------|-----|-----|-----|-----|-----|-----|-----|-----|-----|-----|-----|-----|-------|
| 1980-81 | 0.0 | 0.0 | 0.0 | T | 3.9 | 3.3 | 8.9 | 2.7 | 11.1 | 0.3 | T | T | 30.2 |
| 1981-82 | 0.0 | 0.0 | 0.0 | 4.4 | 2.4 | 11.5 | 15.3 | 4.5 | 10.2 | 9.5 | T | T | 57.8 |
| 1982-83 | 0.0 | 0.0 | 0.0 | 0.2 | 1.0 | 20.1 | 6.2 | 1.0 | 13.3 | 9.0 | 5.0 | 0.0 | 55.8 |
| 1983-84 | 0.0 | 0.0 | 0.0 | 0.0 | 5.9 | 34.2 | 7.6 | 18.5 | 6.7 | 25.1 | T | T | 98.0 |
| 1984-85 | 0.0 | 0.0 | T | 20.4 | 6.6 | 12.9 | 12.7 | 11.4 | 8.0 | 0.7 | 0.0 | 0.0 | 72.7 |
| 1985-86 | 0.0 | 0.0 | 0.0 | T | 27.2 | 14.7 | 3.9 | 1.7 | 1.0 | 5.5 | T | 0.0 | 54.0 |
| 1986-87 | 0.0 | 0.0 | T | 0.0 | 4.4 | 1.7 | 16.4 | 9.9 | 3.0 | 2.1 | 0.0 | 0.0 | 37.5 |
| 1987-88 | 0.0 | 0.0 | 0.0 | 0.0 | 0.6 | 11.0 | 16.3 | 0.4 | 6.1 | T | 0.9 | 0.0 | 35.3 |
| 1988-89 | 0.0 | 0.0 | T | 0.0 | 8.5 | 12.5 | 9.4 | 27.5 | 2.1 | T | 0.0 | 0.0 | 60.0 |
| 1989-90 | 0.0 | 0.0 | 0.0 | 2.7 | 2.4 | 1.7 | 8.2 | 8.5 | 11.8 | 0.7 | T | T | 36.0 |
| 1990-91 | 0.0 | 0.0 | T | 0.0 | 4.8 | 14.3 | 10.9 | 1.2 | 1.9 | 13.7 | T | 0.0 | 46.8 |
| 1991-92 | T | T | T | 2.8 | 11.4 | 6.8 | 12.2 | 4.7 | 0.2 | 0.4 | 0.0 | 0.0 | 38.5 |
| 1992-93 | 0.0 | T | 0.0 | 0.0 | 14.2 | 16.9 | 50.3 | 13.2 | T | 2.4 | 1.7 | T | 98.7 |
| 1993-94 | 0.0 | T | 0.0 | 0.0 | 6.1 | 8.3 | 6.2 | 15.0 | 3.1 | 0.1 | 0.0 | 0.0 | 38.8 |
| 1994-95 | 0.0 | 0.0 | 0.0 | T | 33.3 | 13.6 | 11.1 | 13.3 | 7.3 | 6.6 | T | T | 85.2 |
| 1995-96 | 0.0 | 0.0 | 0.0 | 0.5 | 0.1 | 1.4 | 45.0 | 22.6 | 11.2 | 4.9 | T | 0.0 | 85.7 |
| 1996-97 | T | 0.0 | 0.0 | 5.1 | 3.5 | 16.0 | 16.2 | 11.0 | 7.1 | 4.4 | T | T | 63.3 |
| 1997-98 | 0.0 | 0.0 | 0.0 | 1.7 | T | 6.6 | 6.3 | 32.1 | 14.8 | 3.7 | 0.0 | T | 65.2 |
| 1998-99 | 0.0 | 0.0 | T | 0.0 | 2.1 | 6.6 | 8.4 | 8.6 | 3.0 | 3.5 | T | 0.0 | 32.2 |
| 1999-00 | 0.0 | T | 0.0 | 0.0 | 5.5 | 18.0 | 15.0 | 5.1 | 6.2 | T | 0.5 | 0.0 | 50.3 |
| 2000-01 | 0.0 | 0.0 | 0.0 | 0.0 | 17.0 | 12.9 | 6.5 | 13.1 | 5.3 | 9.9 | 0.0 | 0.0 | 64.7 |
| 2001-02 | T | 0.0 | 0.0 | T | 18.8 | 19.9 | 16.1 | 0.7 | 14.4 | 3.6 | 0.0 | 0.0 | 73.5 |
| 2002-03 | 0.0 | 0.0 | 0.0 | T | 0.2 | 3.4 | T | 9.0 | 5.3 | 4.4 | T | 0.0 | 22.3 |
| 2003-04 | 0.0 | 0.0 | 0.0 | 1.0 | 10.7 | 31.2 | 9.9 | 19.8 | 3.9 | 1.0 | 0.0 | T | 77.5 |
| 2004-05 | 0.0 | 0.0 | 0.0 | 0.1 | 3.3 | 3.7 | 6.7 | 11.3 | 2.8 | T | 0.0 | T | 27.9 |
| 2005-06 | 0.0 | 0.0 | T | 0.0 | 1.1 | 6.9 | 9.2 | 13.5 | 13.3 | 2.2 | 0.0 | 0.0 | 46.2 |
| 2006-07 | 0.0 | T | T | T | 8.1 | 9.9 | 10.8 | 13.1 | 2.5 | T | T | 0.0 | 44.4 |
| 2007-08 | 0.0 | 0.0 | T | 0.1 | 0.8 | 29.8 | 17.1 | 17.4 | 8.3 | T | T | 0.0 | 73.5 |
| 2008-09 | 0.0 | 0.0 | 0.0 | 0.5 | 3.0 | 20.8 | 10.3 | 6.3 | 6.0 | 1.4 | 0.0 | 0.0 | 48.3 |
| 2009- | 0.0 | 0.0 | 0.0 | 0.5 | 2.8 | 17.1 | | | | | | | |
| POR= 62 YRS | T | T | 0.1 | 1.4 | 6.7 | 13.3 | 13.5 | 10.5 | 9.6 | 5.2 | 0.6 | T | 60.9 |

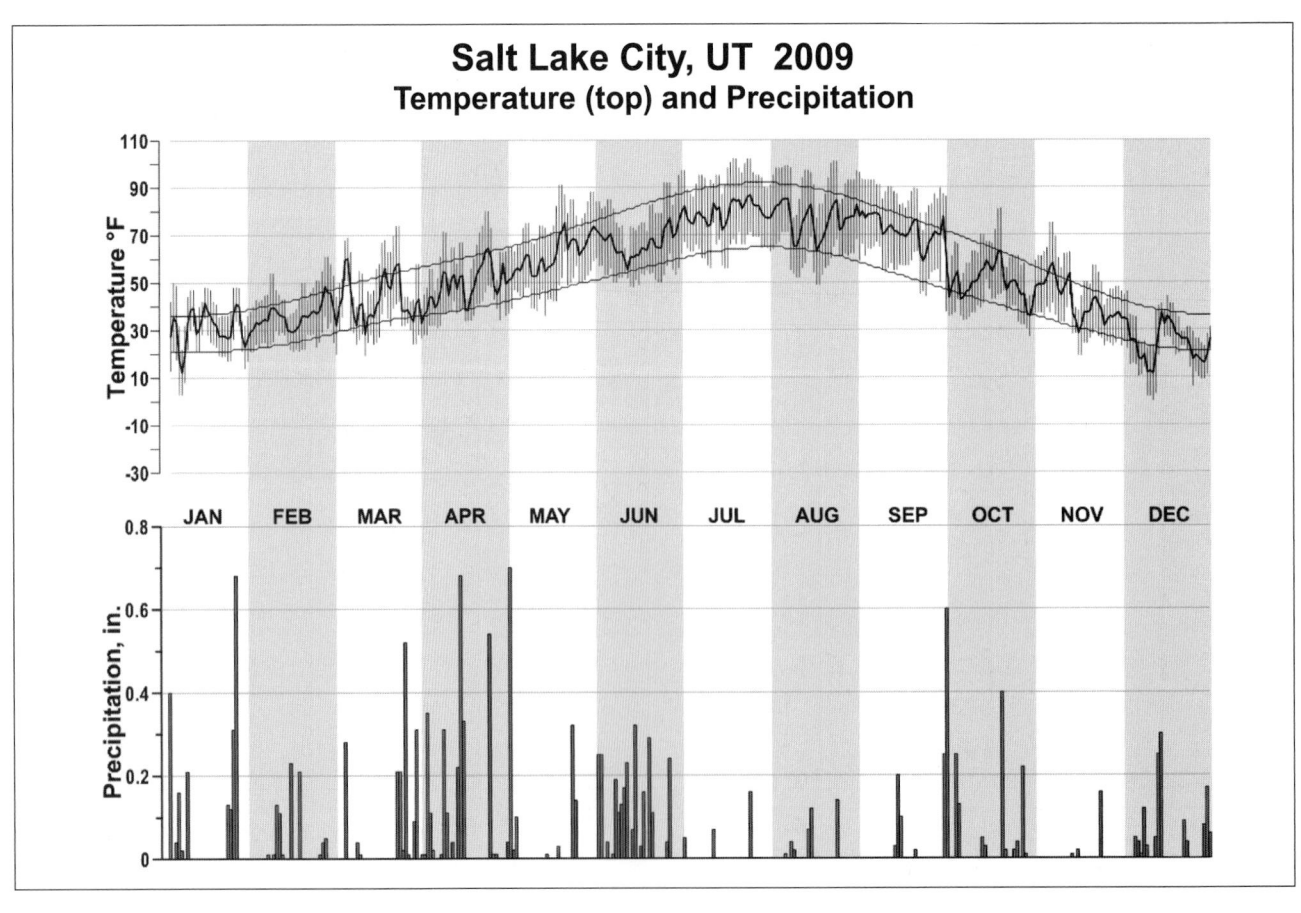

**Salt Lake City, UT 2009**
**Temperature (top) and Precipitation**

# 2009
# BURLINGTON
# VERMONT (KBTV)

Burlington is located on the eastern shore of Lake Champlain at the widest part of the lake. About 35 miles to the west lie the highest peaks of the Adirondacks, while the foothills of the Green Mountains begin 10 miles to the east and southeast.

Its northerly latitude assures the variety and vigor of a true New England climate, while thanks to the modifying influence of the lake, the many rapid and marked weather changes are tempered in severity. Due to its location in the path of the St. Lawrence Valley storm track and the lake effects, the city is one of the cloudiest in the United States.

Lake Champlain exercises a tempering influence on the local temperature. During the winter months and prior to the lake freezing, temperatures along the lake shore are often 5-10 degrees warmer than at the airport 3 1/2 miles inland. At the airport the average occurrence of the last freeze in spring is around May 10th and that of the first in fall is early October, giving a growing season of 145 days. This location is justly proud of its delightful summer weather. On average, there are few days a year with maxima of 90 degrees or higher. This moderate summer heat gives way to a cooler, but none the less pleasant fall period, usually extending well into October. High pressure systems moving down rapidly from central Canada or Hudson Bay produce the coldest temperatures during the winter months, but extended periods of very cold weather are rare.

Precipitation, although generally plentiful and well distributed throughout the year, is less in the Champlain Valley than in other areas of Vermont due to the shielding effect of the mountain barriers to the east and west. The heaviest rainfall usually occurs during summer thunderstorms, but excessively heavy rainfall is quite uncommon. Droughts are infrequent.

Because of the trend of the Champlain Valley between the Adirondack and Green Mountain ranges, most winds have a northerly or southerly component. The prevailing direction most of the year is from the south. Winds of damaging force are very uncommon.

Smoke pollution is nearly non-existent since there is no concentration of heavy industry here, however, haze has been on the increase over the years due to the large increase in industry to the north and south. During the spring and fall months, fog occasionally forms along the Winooski River to the north and east and may drift over the airport with favorable winds. In spite of the high percentage of cloudiness, periods of low aircraft ceilings and visibilities are usually of short duration, allowing this area to have one of the highest percentages of flying weather in New England.

# NORMALS, MEANS, AND EXTREMES
## BURLINGTON (KBTV)

LATITUDE: 44 ° 28'N    LONGITUDE: -73 ° 9 'W    ELEVATION (FT): GRND: 330   BARO: 348    TIME ZONE: EASTERN (UTC -5)    WBAN: 14742

| | ELEMENT | POR | JAN | FEB | MAR | APR | MAY | JUN | JUL | AUG | SEP | OCT | NOV | DEC | YEAR |
|---|---|---|---|---|---|---|---|---|---|---|---|---|---|---|---|
| TEMPERATURE °F | NORMAL DAILY MAXIMUM | 30 | 26.7 | 29.0 | 39.6 | 53.3 | 67.8 | 76.5 | 81.4 | 78.4 | 68.9 | 56.4 | 44.0 | 32.3 | 54.5 |
| | MEAN DAILY MAXIMUM | 104 | 26.4 | 27.3 | 38.5 | 52.1 | 66.3 | 74.5 | 80.0 | 77.7 | 68.5 | 57.2 | 43.3 | 31.4 | 53.6 |
| | HIGHEST DAILY MAXIMUM | 66 | 66 | 62 | 84 | 91 | 93 | 100 | 100 | 101 | 98 | 85 | 75 | 67 | 101 |
| | YEAR OF OCCURRENCE | | 1995 | 1981 | 1998 | 1976 | 1977 | 1995 | 1995 | 1944 | 2002 | 1949 | 1948 | 1998 | AUG 1944 |
| | MEAN OF EXTREME MAXS. | 104 | 48.4 | 47.3 | 61.2 | 75.8 | 84.8 | 90.5 | 92.1 | 90.1 | 84.9 | 75.0 | 64.6 | 52.0 | 72.2 |
| | NORMAL DAILY MINIMUM | 30 | 9.3 | 10.9 | 21.8 | 33.6 | 45.2 | 54.7 | 59.8 | 58.1 | 49.9 | 38.9 | 30.3 | 17.3 | 35.8 |
| | MEAN DAILY MINIMUM | 104 | 9.6 | 10.2 | 21.6 | 33.5 | 45.3 | 54.4 | 59.9 | 58.0 | 49.7 | 39.7 | 29.7 | 16.7 | 35.7 |
| | LOWEST DAILY MINIMUM | 66 | -30 | -30 | -20 | 2 | 24 | 33 | 39 | 35 | 25 | 15 | -2 | -26 | -30 |
| | YEAR OF OCCURRENCE | | 1957 | 1979 | 1948 | 1972 | 1966 | 1986 | 1962 | 1976 | 1963 | 1972 | 1958 | 1980 | FEB 1979 |
| | MEAN OF EXTREME MINS. | 104 | -14.0 | -12.2 | -1.0 | 19.2 | 30.5 | 39.9 | 47.4 | 44.2 | 34.0 | 24.3 | 12.2 | -6.8 | 18.1 |
| | NORMAL DRY BULB | 30 | 18.0 | 19.9 | 30.7 | 43.5 | 56.5 | 65.6 | 70.6 | 68.2 | 59.4 | 47.7 | 37.1 | 24.8 | 45.2 |
| | MEAN DRY BULB | 104 | 18.0 | 18.8 | 30.1 | 42.8 | 55.8 | 64.6 | 70.0 | 67.9 | 59.1 | 48.5 | 36.5 | 24.1 | 44.7 |
| | MEAN WET BULB | 26 | 16.9 | 18.0 | 26.1 | 37.8 | 48.8 | 58.6 | 63.1 | 61.9 | 54.7 | 43.5 | 34.0 | 23.1 | 40.5 |
| | MEAN DEW POINT | 26 | 12.8 | 13.4 | 21.3 | 32.5 | 44.3 | 54.7 | 59.9 | 59.0 | 51.9 | 39.9 | 30.0 | 19.3 | 36.6 |
| | NORMAL NO. DAYS WITH: | | | | | | | | | | | | | | |
| | MAXIMUM >= 90 | 3 0 | 0.0 | 0.0 | 0.0 | 0.1 | 0.6 | 1.4 | 3.1 | 1.1 | 0.2 | 0.0 | 0.0 | 0.0 | 6.5 |
| | MAXIMUM <= 32 | 3 0 | 20.7 | 17.0 | 8.0 | 0.6 | 0.0 | 0.0 | 0.0 | 0.0 | 0.0 | 0.0 | 3.9 | 14.5 | 64.7 |
| | MINIMUM <= 32 | 30 | 29.4 | 26.2 | 25.2 | 13.9 | 1.6 | 0.0 | 0.0 | 0.0 | 0.6 | 8.6 | 18.1 | 27.5 | 151.1 |
| | MINIMUM <= 0 | 30 | 9.2 | 7.5 | 1.6 | 0.0 | 0.0 | 0.0 | 0.0 | 0.0 | 0.0 | 0.0 | 0.0 | 3.8 | 22.1 |
| H/C | NORMAL HEATING DEG. DAYS | 30 | 1457 | 1273 | 1063 | 642 | 283 | 77 | 17 | 38 | 203 | 538 | 834 | 1240 | 7665 |
| | NORMAL COOLING DEG. DAYS | 30 | 0 | 0 | 0 | 3 | 23 | 96 | 192 | 139 | 35 | 1 | 0 | 0 | 489 |
| RH | NORMAL (PERCENT) | 30 | 70 | 69 | 67 | 64 | 65 | 68 | 69 | 73 | 75 | 72 | 71 | 72 | 70 |
| | HOUR 01 LST | 30 | 72 | 73 | 74 | 73 | 77 | 81 | 83 | 85 | 85 | 79 | 75 | 75 | 78 |
| | HOUR 07 LST | 30 | 74 | 75 | 76 | 73 | 74 | 76 | 79 | 83 | 85 | 81 | 77 | 76 | 77 |
| | HOUR 13 LST | 30 | 64 | 61 | 58 | 53 | 51 | 54 | 53 | 57 | 60 | 60 | 64 | 67 | 59 |
| | HOUR 19 LST | 30 | 69 | 66 | 63 | 58 | 57 | 61 | 61 | 67 | 73 | 70 | 71 | 72 | 66 |
| S | PERCENT POSSIBLE SUNSHINE | 65 | 41 | 48 | 51 | 49 | 55 | 59 | 64 | 60 | 54 | 47 | 31 | 33 | 49 |
| W/O | MEAN NO. DAYS WITH: | | | | | | | | | | | | | | |
| | HEAVY FOG(VISBY <= 1/4 MI) | 46 | 1.6 | 1.5 | 1.8 | 1.2 | 0.9 | 1.3 | 1.0 | 1.4 | 2.2 | 1.5 | 1.0 | 1.8 | 17.2 |
| | THUNDERSTORMS | 62 | 0.0 | 0.0 | 0.3 | 1.0 | 2.3 | 5.1 | 5.9 | 5.0 | 1.8 | 0.5 | 0.4 | 0.0 | 22.3 |
| CLOUDNESS | MEAN: | | | | | | | | | | | | | | |
| | SUNRISE-SUNSET (OKTAS) | 53 | 6.0 | 5.8 | 5.7 | 5.8 | 5.6 | 5.4 | 5.0 | 5.0 | 5.1 | 5.4 | 6.4 | 6.3 | 5.6 |
| | MIDNIGHT-MIDNIGHT (OKTAS) | 32 | 5.9 | 5.4 | 5.3 | 5.3 | 5.3 | 5.0 | 4.8 | 4.8 | 5.0 | 5.3 | 6.2 | 6.2 | 5.4 |
| | MEAN NO. DAYS WITH: | | | | | | | | | | | | | | |
| | CLEAR | 53 | 4.6 | 4.5 | 5.8 | 5.0 | 4.9 | 4.8 | 5.1 | 5.9 | 5.9 | 5.9 | 2.5 | 2.9 | 57.8 |
| | PARTLY CLOUDY | 53 | 6.4 | 6.9 | 6.9 | 7.5 | 9.3 | 10.8 | 12.7 | 11.6 | 9.8 | 7.6 | 5.2 | 5.7 | 100.4 |
| | CLOUDY | 53 | 20.1 | 16.9 | 18.3 | 17.5 | 16.7 | 14.4 | 12.7 | 13.1 | 13.8 | 17.1 | 21.7 | 21.7 | 204.0 |
| PR | MEAN STATION PRESSURE(IN) | 26 | 29.65 | 29.66 | 29.66 | 29.60 | 29.59 | 29.57 | 29.58 | 29.64 | 29.68 | 29.68 | 29.67 | 29.67 | 29.64 |
| | MEAN SEA-LEVEL PRES. (IN) | 26 | 30.04 | 30.05 | 30.04 | 29.97 | 29.96 | 29.93 | 29.95 | 30.00 | 30.05 | 30.06 | 30.05 | 30.05 | 30.01 |
| WINDS | MEAN SPEED (MPH) | 26 | 9.5 | 8.9 | 9.0 | 9.1 | 8.5 | 7.9 | 7.4 | 7.2 | 7.8 | 8.4 | 9.2 | 9.3 | 8.5 |
| | PREVAIL.DIR(TENS OF DEGS) | 39 | 19 | 19 | 19 | 19 | 19 | 19 | 19 | 19 | 19 | 19 | 19 | 19 | 19 |
| | MAXIMUM 2-MINUTE: | | | | | | | | | | | | | | |
| | SPEED (MPH) | 13 | 38 | 37 | 35 | 36 | 35 | 39 | 34 | 35 | 36 | 38 | 36 | 35 | 39 |
| | DIR. (TENS OF DEGS) | | 18 | 25 | 24 | 19 | 18 | 29 | 32 | 27 | 33 | 19 | 15 | 18 | 29 |
| | YEAR OF OCCURRENCE | | 1997 | 2001 | 2009 | 2002 | 2009 | 2002 | 1999 | 2000 | 2002 | 2009 | 2004 | 2008 | JUN 2002 |
| | MAXIMUM 3-SECOND | | | | | | | | | | | | | | |
| | SPEED (MPH) | 13 | 51 | 52 | 51 | 53 | 48 | 56 | 56 | 46 | 53 | 51 | 51 | 54 | 56 |
| | DIR. (TENS OF DEGS) | | 19 | 24 | 20 | 26 | 28 | 31 | 34 | 26 | 18 | 18 | 18 | 18 | 34 |
| | YEAR OF OCCURRENCE | | 1997 | 2006 | 2007 | 2007 | 2009 | 2002 | 2007 | 2000 | 2005 | 2009 | 2005 | 2004 | JUL 2007 |
| PRECIPITATION | NORMAL (IN) | 30 | 2.22 | 1.67 | 2.32 | 2.88 | 3.32 | 3.43 | 3.97 | 4.01 | 3.83 | 3.12 | 3.06 | 2.22 | 36.05 |
| | MAXIMUM MONTHLY (IN) | 66 | 5.15 | 5.38 | 4.14 | 6.55 | 7.10 | 8.66 | 9.31 | 11.54 | 10.26 | 6.26 | 6.85 | 5.95 | 11.54 |
| | YEAR OF OCCURRENCE | | 1998 | 1981 | 2001 | 1983 | 2006 | 1998 | 1998 | 1955 | 1999 | 2005 | 1983 | 1973 | AUG 1955 |
| | MINIMUM MONTHLY (IN) | 66 | 0.42 | 0.21 | 0.38 | 0.73 | 0.29 | 0.82 | 0.77 | 0.72 | 0.87 | 0.50 | 0.63 | 0.37 | 0.21 |
| | YEAR OF OCCURRENCE | | 1989 | 1978 | 1965 | 1999 | 1977 | 1995 | 2001 | 1957 | 1948 | 1963 | 1952 | 1998 | FEB 1978 |
| | MAXIMUM IN 24 HOURS (IN) | 66 | 2.11 | 1.94 | 1.62 | 2.16 | 2.36 | 2.83 | 2.69 | 3.62 | 3.96 | 2.17 | 2.48 | 2.60 | 3.96 |
| | YEAR OF OCCURRENCE | | 1998 | 2007 | 1971 | 1968 | 2006 | 1972 | 1985 | 1998 | 1999 | 1983 | 1990 | 1950 | SEP 1999 |
| | NORMAL NO. DAYS WITH: | | | | | | | | | | | | | | |
| | PRECIPITATION >= 0.01 | 30 | 15.8 | 11.2 | 13.8 | 12.7 | 13.9 | 13.3 | 12.2 | 12.9 | 12.4 | 12.4 | 14.0 | 15.2 | 159.8 |
| | PRECIPITATION >= 1.00 | 30 | 0.2 | 0.1 | 0.1 | 0.1 | 0.5 | 0.5 | 1.0 | 0.9 | 0.8 | 0.7 | 0.4 | 0.2 | 5.5 |
| SNOWFALL | NORMAL (IN) | 30 | 21.2 | 15.6 | 15.4 | 6.2 | 0.* | 0.0 | 0.0 | 0.0 | 0.* | 0.3 | 7.2 | 17.2 | 83.1 |
| | MAXIMUM MONTHLY (IN) | 66 | 42.4 | 42.3 | 47.6 | 21.3 | 3.9 | T | T | T | 0.1 | 5.1 | 20.4 | 56.7 | 56.7 |
| | YEAR OF OCCURRENCE | | 1978 | 2008 | 2001 | 1983 | 1966 | 2009 | 2009 | 2001 | 1992 | 1969 | 2002 | 1970 | DEC 1970 |
| | MAXIMUM IN 24 HOURS (IN) | 66 | 17.4 | 17.7 | 22.4 | 15.6 | 3.5 | T | T | T | 0.1 | 5.1 | 10.1 | 17.0 | 22.4 |
| | YEAR OF OCCURRENCE' | | 2003 | 1995 | 1993 | 1983 | 1966 | 2009 | 2009 | 2001 | 1992 | 1969 | 1958 | 1978 | MAR 1993 |
| | MAXIMUM SNOW DEPTH (IN) | 61 | 40 | 33 | 31 | 16 | 1 | 0 | 0 | 0 | 0 | 5 | 10 | 33 | 40 |
| | YEAR OF OCCURRENCE | | 1964 | 1958 | 1993 | 2001 | 1967 | | | | | 1969 | 1958 | 1969 | JAN 1964 |
| | NORMAL NO. DAYS WITH: | | | | | | | | | | | | | | |
| | SNOWFALL >= 1.0 | 30 | 5.8 | 3.8 | 4.0 | 1.6 | 0.0 | 0.0 | 0.0 | 0.0 | 0.0 | 0.1 | 2.1 | 5.3 | 22.7 |

## PRECIPITATION (inches) 2009 BURLINGTON (KBTV)

| YEAR | JAN | FEB | MAR | APR | MAY | JUN | JUL | AUG | SEP | OCT | NOV | DEC | ANNUAL |
|------|------|------|------|------|------|------|------|------|------|------|------|------|--------|
| 1980 | 0.61 | 0.67 | 2.44 | 2.39 | 1.61 | 1.92 | 6.11 | 3.83 | 4.41 | 2.48 | 2.92 | 1.50 | 30.89 |
| 1981 | 0.49 | 5.38 | 1.32 | 3.05 | 3.76 | 3.07 | 3.22 | 5.58 | 6.24 | 5.26 | 2.73 | 2.03 | 42.13 |
| 1982 | 2.74 | 1.43 | 2.31 | 2.63 | 1.95 | 4.95 | 3.07 | 3.55 | 2.12 | 2.31 | 3.59 | 1.69 | 32.34 |
| 1983 | 3.09 | 1.66 | 2.60 | 6.55 | 6.31 | 1.49 | 3.92 | 4.31 | 3.77 | 4.38 | 6.85 | 5.23 | 50.16 |
| 1984 | 0.81 | 2.73 | 1.72 | 4.25 | 5.27 | 1.70 | 5.11 | 3.30 | 2.81 | 1.89 | 3.08 | 3.14 | 35.81 |
| 1985 | 1.46 | 1.26 | 2.46 | 1.90 | 3.53 | 3.76 | 4.42 | 2.67 | 3.30 | 3.31 | 3.68 | 1.59 | 33.34 |
| 1986 | 3.69 | 1.68 | 3.17 | 0.95 | 4.11 | 4.40 | 4.53 | 5.82 | 4.86 | 2.50 | 2.99 | 1.32 | 40.02 |
| 1987 | 1.91 | 0.49 | 1.33 | 1.42 | 2.69 | 4.42 | 2.79 | 2.09 | 3.58 | 3.28 | 2.24 | 1.17 | 27.41 |
| 1988 | 0.69 | 1.69 | 1.55 | 1.91 | 1.80 | 3.26 | 2.55 | 4.27 | 1.50 | 2.05 | 4.51 | 0.90 | 26.68 |
| 1989 | 0.42 | 0.67 | 2.60 | 1.89 | 3.19 | 3.68 | 3.65 | 7.30 | 5.98 | 2.98 | 2.41 | 1.26 | 36.03 |
| 1990 | 2.36 | 2.82 | 1.81 | 2.97 | 3.66 | 3.08 | 5.12 | 4.85 | 2.03 | 5.99 | 3.91 | 3.58 | 42.18 |
| 1991 | 1.65 | 0.51 | 2.55 | 3.41 | 3.15 | 1.28 | 2.83 | 4.00 | 5.14 | 5.07 | 1.58 | 1.35 | 32.52 |
| 1992 | 1.65 | 1.56 | 2.13 | 2.58 | 2.38 | 1.72 | 4.58 | 1.89 | 4.73 | 3.00 | 3.67 | 0.96 | 30.85 |
| 1993 | 2.17 | 1.90 | 1.54 | 3.76 | 2.19 | 3.35 | 3.34 | 4.46 | 3.38 | 2.93 | 2.27 | 1.57 | 32.86 |
| 1994 | 2.19 | 1.21 | 2.93 | 3.37 | 4.58 | 3.65 | 5.30 | 4.50 | 1.74 | 1.25 | 2.48 | 1.66 | 34.86 |
| 1995 | 1.88 | 1.26 | 1.60 | 2.35 | 1.41 | 0.82 | 3.49 | 4.64 | 2.97 | 5.81 | 3.33 | 2.63 | 32.19 |
| 1996 | 3.91 | 0.83 | 0.80 | 6.12 | 5.33 | 4.54 | 4.74 | 1.47 | 2.75 | 3.64 | 3.30 | 0.64 | 38.07 |
| 1997 | 1.71 | 1.38 | 2.59 | 1.54 | 2.24 | 2.62 | 3.89 | 4.63 | 2.98 | 1.23 | 4.16 | 1.65 | 30.62 |
| 1998 | 5.15 | 1.84 | 3.81 | 1.79 | 3.61 | 8.66 | 9.31 | 6.80 | 5.64 | 2.42 | 1.02 | 0.37 | 50.42 |
| 1999 | 3.51 | 1.13 | 2.22 | 0.73 | 2.40 | 1.79 | 1.97 | 2.41 | 10.26 | 3.18 | 1.86 | 1.12 | 32.58 |
| 2000 | 2.30 | 2.67 | 1.63 | 5.01 | 6.13 | 3.55 | 3.16 | 3.67 | 3.02 | 1.80 | 2.96 | 3.36 | 39.26 |
| 2001 | 0.98 | 1.54 | 4.14 | 0.85 | 2.28 | 2.32 | 0.77 | 4.32 | 1.40 | 1.37 | 1.81 | 1.49 | 23.27 |
| 2002 | 1.32 | 1.93 | 1.90 | 3.02 | 3.63 | 6.73 | 3.35 | 1.16 | 6.25 | 3.30 | 3.15 | 1.28 | 37.02 |
| 2003 | 0.99 | 0.99 | 2.06 | 2.09 | 3.32 | 2.98 | 3.48 | 2.24 | 3.29 | 5.54 | 4.23 | 5.00 | 36.21 |
| 2004 | 0.47 | 0.85 | 1.29 | 2.36 | 5.04 | 3.61 | 7.65 | 7.87 | 2.46 | 1.20 | 2.11 | 3.23 | 38.14 |
| 2005 | 1.83 | 1.94 | 1.37 | 3.85 | 1.63 | 3.78 | 5.22 | 4.17 | 2.70 | 6.26 | 4.50 | 2.16 | 39.41 |
| 2006 | 3.83 | 1.51 | 1.62 | 2.92 | 7.10 | 6.77 | 2.95 | 4.36 | 3.22 | 6.25 | 2.62 | 3.84 | 46.99 |
| 2007 | 2.57 | 2.18 | 2.67 | 3.79 | 1.95 | 2.26 | 6.29 | 1.46 | 1.95 | 5.72 | 4.72 | 4.25 | 39.81 |
| 2008 | 1.55 | 3.71 | 3.93 | 2.60 | 1.94 | 5.21 | 7.07 | 3.68 | 1.20 | 4.89 | 1.88 | 2.93 | 40.59 |
| 2009 | 1.76 | 1.81 | 1.90 | 1.86 | 5.25 | 5.25 | 4.62 | 2.32 | 3.67 | 2.98 | 2.98 | 3.02 | 37.42 |
| POR= 104 YRS | 1.86 | 1.67 | 2.24 | 2.68 | 3.07 | 3.58 | 3.86 | 3.64 | 3.38 | 3.00 | 2.85 | 2.19 | 34.02 |

WBAN : 14742

## AVERAGE TEMPERATURE (°F) 2009 BURLINGTON (KBTV)

| YEAR | JAN | FEB | MAR | APR | MAY | JUN | JUL | AUG | SEP | OCT | NOV | DEC | ANNUAL |
|------|------|------|------|------|------|------|------|------|------|------|------|------|--------|
| 1980 | 21.2 | 17.6 | 31.1 | 46.5 | 58.9 | 64.4 | 70.6 | 70.7 | 57.9 | 45.0 | 32.3 | 15.0 | 44.3 |
| 1981 | 8.9 | 32.9 | 33.5 | 46.7 | 58.2 | 66.1 | 71.1 | 67.1 | 59.3 | 44.9 | 36.9 | 25.3 | 45.9 |
| 1982 | 9.6 | 19.1 | 30.3 | 43.4 | 57.3 | 60.7 | 69.5 | 65.9 | 62.3 | 50.1 | 42.3 | 31.9 | 45.2 |
| 1983 | 21.0 | 22.3 | 33.0 | 42.3 | 52.9 | 66.3 | 71.3 | 68.6 | 62.9 | 48.2 | 38.1 | 22.4 | 45.8 |
| 1984 | 16.5 | 28.7 | 21.9 | 44.7 | 52.3 | 66.0 | 70.3 | 71.1 | 57.2 | 50.0 | 38.4 | 30.3 | 45.6 |
| 1985 | 13.4 | 22.5 | 31.6 | 44.3 | 55.8 | 61.7 | 69.6 | 67.5 | 60.3 | 49.1 | 36.9 | 21.3 | 44.5 |
| 1986 | 18.5 | 16.2 | 33.7 | 48.5 | 58.3 | 62.3 | 68.5 | 66.1 | 58.1 | 46.9 | 34.5 | 27.8 | 45.0 |
| 1987 | 18.1 | 15.0 | 33.3 | 48.6 | 55.5 | 66.3 | 71.5 | 66.7 | 59.5 | 45.9 | 37.0 | 28.5 | 45.5 |
| 1988 | 19.9 | 21.4 | 29.7 | 44.3 | 57.9 | 63.4 | 73.2 | 70.7 | 58.2 | 44.4 | 39.6 | 22.9 | 45.5 |
| 1989 | 23.7 | 19.7 | 28.4 | 41.6 | 59.6 | 67.2 | 71.7 | 67.7 | 61.4 | 50.3 | 36.4 | 7.6 | 44.6 |
| 1990 | 29.8 | 23.5 | 33.8 | 46.2 | 52.9 | 65.9 | 70.2 | 69.8 | 59.4 | 49.4 | 39.5 | 30.1 | 47.5 |
| 1991 | 18.9 | 26.5 | 34.0 | 49.1 | 59.3 | 65.9 | 70.4 | 70.5 | 57.8 | 50.4 | 37.6 | 24.0 | 47.0 |
| 1992 | 18.6 | 19.1 | 26.5 | 42.3 | 56.5 | 64.4 | 66.1 | 67.6 | 60.2 | 45.5 | 36.7 | 28.1 | 44.3 |
| 1993 | 21.7 | 10.6 | 27.4 | 45.3 | 56.6 | 64.7 | 72.2 | 70.8 | 59.1 | 46.1 | 36.8 | 24.7 | 44.7 |
| 1994 | 7.1 | 15.4 | 30.2 | 44.5 | 54.8 | 68.6 | 74.2 | 66.5 | 59.6 | 49.8 | 41.1 | 28.7 | 45.0 |
| 1995 | 27.9 | 19.0 | 35.0 | 40.4 | 56.4 | 69.6 | 74.7 | 70.0 | 57.5 | 54.1 | 35.2 | 22.0 | 46.8 |
| 1996 | 17.5 | 21.3 | 28.6 | 42.8 | 54.4 | 66.3 | 68.6 | 68.9 | 62.0 | 47.4 | 32.8 | 32.7 | 45.3 |
| 1997 | 19.1 | 25.0 | 26.9 | 41.4 | 51.4 | 67.3 | 68.7 | 66.8 | 58.6 | 46.6 | 35.1 | 25.8 | 44.4 |
| 1998 | 22.7 | 27.7 | 34.3 | 46.4 | 62.0 | 66.1 | 68.9 | 68.5 | 61.5 | 49.6 | 39.7 | 32.0 | 48.3 |
| 1999 | 19.1 | 24.8 | 30.9 | 44.7 | 59.8 | 70.5 | 74.2 | 68.1 | 64.5 | 46.2 | 42.6 | 28.7 | 47.8 |
| 2000 | 18.1 | 22.1 | 36.5 | 42.7 | 56.7 | 63.6 | 67.5 | 67.7 | 58.8 | 48.8 | 37.2 | 19.6 | 44.9 |
| 2001 | 20.1 | 22.1 | 27.2 | 43.2 | 58.7 | 67.2 | 68.0 | 72.6 | 61.3 | 51.1 | 42.4 | 32.7 | 47.2 |
| 2002 | 27.4 | 26.1 | 33.1 | 46.1 | 53.0 | 64.2 | 70.6 | 71.3 | 64.4 | 46.4 | 36.1 | 25.3 | 47.0 |
| 2003 | 11.8 | 15.6 | 30.1 | 41.4 | 55.6 | 65.6 | 71.4 | 71.5 | 63.2 | 47.4 | 39.8 | 24.9 | 44.9 |
| 2004 | 8.8 | 19.2 | 34.8 | 43.7 | 58.1 | 63.1 | 69.7 | 67.0 | 61.1 | 47.6 | 37.2 | 23.9 | 44.5 |
| 2005 | 15.2 | 21.9 | 28.4 | 46.6 | 52.4 | 70.7 | 72.6 | 71.1 | 63.6 | 50.1 | 40.5 | 24.1 | 46.4 |
| 2006 | 28.2 | 23.5 | 32.0 | 46.0 | 58.2 | 66.3 | 73.4 | 67.3 | 60.2 | 46.9 | 42.8 | 32.7 | 48.1 |
| 2007 | 21.3 | 14.2 | 28.6 | 42.9 | 56.8 | 68.0 | 69.1 | 69.1 | 62.9 | 54.1 | 35.6 | 24.7 | 45.6 |
| 2008 | 25.0 | 22.3 | 28.5 | 49.1 | 54.1 | 67.7 | 70.8 | 67.0 | 61.3 | 46.6 | 38.3 | 24.8 | 46.3 |
| 2009 | 14.1 | 23.2 | 32.4 | 46.1 | 56.5 | 64.4 | 68.3 | 69.9 | 59.3 | 46.0 | 41.8 | 25.5 | 45.6 |
| POR= 104 YRS | 18.0 | 18.8 | 30.1 | 42.8 | 55.8 | 64.6 | 70.0 | 67.9 | 59.1 | 48.5 | 36.5 | 24.1 | 44.7 |

## HEATING DEGREE DAYS (base 65°F) 2009  BURLINGTON (KBTV)

| YEAR | JUL | AUG | SEP | OCT | NOV | DEC | JAN | FEB | MAR | APR | MAY | JUN | TOTAL |
|---|---|---|---|---|---|---|---|---|---|---|---|---|---|
| 1980-81 | 10 | 3 | 240 | 611 | 976 | 1545 | 1738 | 894 | 969 | 544 | 239 | 43 | 7812 |
| 1981-82 | 13 | 36 | 204 | 617 | 837 | 1224 | 1716 | 1277 | 1069 | 643 | 255 | 133 | 8024 |
| 1982-83 | 30 | 54 | 124 | 455 | 676 | 1021 | 1356 | 1188 | 983 | 675 | 367 | 77 | 7006 |
| 1983-84 | 19 | 36 | 148 | 518 | 803 | 1317 | 1500 | 1044 | 1331 | 602 | 395 | 68 | 7781 |
| 1984-85 | 6 | 24 | 241 | 460 | 792 | 1068 | 1592 | 1185 | 1029 | 615 | 296 | 118 | 7426 |
| 1985-86 | 11 | 42 | 169 | 489 | 835 | 1344 | 1436 | 1361 | 966 | 492 | 219 | 113 | 7477 |
| 1986-87 | 40 | 60 | 215 | 553 | 906 | 1144 | 1446 | 1397 | 975 | 488 | 328 | 48 | 7600 |
| 1987-88 | 19 | 66 | 185 | 584 | 833 | 1125 | 1389 | 1260 | 1088 | 614 | 236 | 136 | 7535 |
| 1988-89 | 15 | 52 | 212 | 635 | 755 | 1298 | 1273 | 1265 | 1128 | 691 | 188 | 45 | 7557 |
| 1989-90 | 2 | 43 | 164 | 451 | 849 | 1776 | 1084 | 1156 | 961 | 577 | 370 | 63 | 7496 |
| 1990-91 | 19 | 10 | 180 | 480 | 758 | 1074 | 1424 | 1072 | 954 | 475 | 206 | 59 | 6711 |
| 1991-92 | 7 | 11 | 240 | 451 | 813 | 1266 | 1434 | 1327 | 1187 | 674 | 277 | 83 | 7770 |
| 1992-93 | 49 | 33 | 197 | 597 | 843 | 1137 | 1335 | 1517 | 1159 | 584 | 256 | 80 | 7787 |
| 1993-94 | 3 | 12 | 211 | 579 | 839 | 1243 | 1793 | 1385 | 1073 | 609 | 328 | 48 | 8123 |
| 1994-95 | 1 | 57 | 168 | 467 | 711 | 1118 | 1145 | 1283 | 925 | 733 | 268 | 36 | 6912 |
| 1995-96 | 0 | 21 | 232 | 330 | 885 | 1326 | 1466 | 1262 | 1123 | 659 | 336 | 27 | 7667 |
| 1996-97 | 11 | 10 | 138 | 542 | 960 | 995 | 1416 | 1113 | 1173 | 701 | 412 | 31 | 7502 |
| 1997-98 | 17 | 20 | 205 | 564 | 889 | 1208 | 1306 | 1039 | 947 | 552 | 115 | 81 | 6943 |
| 1998-99 | 8 | 19 | 129 | 473 | 754 | 1017 | 1418 | 1120 | 1053 | 606 | 184 | 31 | 6812 |
| 1999-00 | 1 | 20 | 113 | 578 | 666 | 1119 | 1448 | 1235 | 879 | 663 | 263 | 108 | 7093 |
| 2000-01 | 28 | 28 | 209 | 498 | 828 | 1399 | 1385 | 1193 | 1169 | 649 | 206 | 49 | 7641 |
| 2001-02 | 32 | 10 | 152 | 431 | 670 | 998 | 1159 | 1084 | 980 | 579 | 381 | 93 | 6569 |
| 2002-03 | 16 | 16 | 92 | 581 | 860 | 1221 | 1641 | 1380 | 1075 | 703 | 289 | 72 | 7946 |
| 2003-04 | 1 | 21 | 92 | 540 | 750 | 1234 | 1732 | 1319 | 927 | 644 | 237 | 102 | 7599 |
| 2004-05 | 6 | 44 | 127 | 533 | 830 | 1267 | 1535 | 1200 | 1126 | 543 | 380 | 35 | 7626 |
| 2005-06 | 2 | 10 | 107 | 457 | 727 | 1258 | 1131 | 1154 | 1015 | 566 | 232 | 58 | 6717 |
| 2006-07 | 1 | 41 | 150 | 553 | 655 | 994 | 1349 | 1416 | 1124 | 660 | 286 | 39 | 7268 |
| 2007-08 | 11 | 30 | 126 | 340 | 877 | 1246 | 1234 | 1231 | 1125 | 472 | 330 | 42 | 7064 |
| 2008-09 | 2 | 23 | 155 | 562 | 796 | 1240 | 1575 | 1166 | 1002 | 567 | 266 | 75 | 7429 |
| 2009- | 13 | 31 | 174 | 582 | 685 | 1216 | | | | | | | |

WBAN : 14742

## COOLING DEGREE DAYS (base 65°F) 2009  BURLINGTON (KBTV)

| YEAR | JAN | FEB | MAR | APR | MAY | JUN | JUL | AUG | SEP | OCT | NOV | DEC | TOTAL |
|---|---|---|---|---|---|---|---|---|---|---|---|---|---|
| 1980 | 0 | 0 | 0 | 0 | 24 | 78 | 189 | 184 | 34 | 0 | 0 | 0 | 509 |
| 1981 | 0 | 0 | 0 | 2 | 35 | 85 | 211 | 110 | 39 | 0 | 0 | 0 | 482 |
| 1982 | 0 | 0 | 0 | 1 | 24 | 11 | 179 | 90 | 51 | 0 | 0 | 0 | 356 |
| 1983 | 0 | 0 | 0 | 0 | 0 | 121 | 223 | 155 | 92 | 6 | 0 | 0 | 597 |
| 1984 | 0 | 0 | 0 | 0 | 7 | 106 | 175 | 217 | 15 | 3 | 0 | 0 | 523 |
| 1985 | 0 | 0 | 0 | 0 | 15 | 25 | 160 | 123 | 34 | 0 | 0 | 0 | 357 |
| 1986 | 0 | 0 | 0 | 4 | 19 | 38 | 156 | 104 | 14 | 0 | 0 | 0 | 335 |
| 1987 | 0 | 0 | 0 | 3 | 42 | 92 | 228 | 126 | 30 | 0 | 0 | 0 | 521 |
| 1988 | 0 | 0 | 0 | 0 | 19 | 96 | 274 | 238 | 15 | 3 | 0 | 0 | 645 |
| 1989 | 0 | 0 | 0 | 0 | 28 | 117 | 216 | 134 | 63 | 0 | 0 | 0 | 558 |
| 1990 | 0 | 0 | 0 | 16 | 1 | 95 | 189 | 165 | 18 | 6 | 0 | 0 | 490 |
| 1991 | 0 | 0 | 0 | 5 | 35 | 92 | 182 | 186 | 32 | 6 | 0 | 0 | 538 |
| 1992 | 0 | 0 | 0 | 3 | 21 | 71 | 91 | 121 | 61 | 0 | 0 | 0 | 368 |
| 1993 | 0 | 0 | 0 | 0 | 2 | 79 | 235 | 198 | 39 | 0 | 0 | 0 | 553 |
| 1994 | 0 | 0 | 0 | 2 | 17 | 165 | 293 | 110 | 14 | 0 | 0 | 0 | 601 |
| 1995 | 0 | 0 | 0 | 0 | 8 | 179 | 306 | 182 | 12 | 0 | 0 | 0 | 687 |
| 1996 | 0 | 0 | 0 | 0 | 17 | 72 | 124 | 139 | 56 | 0 | 0 | 0 | 408 |
| 1997 | 0 | 0 | 0 | 0 | 0 | 108 | 142 | 82 | 17 | 0 | 0 | 0 | 349 |
| 1998 | 0 | 0 | 2 | 0 | 27 | 121 | 141 | 134 | 33 | 1 | 0 | 0 | 459 |
| 1999 | 0 | 0 | 0 | 0 | 29 | 207 | 295 | 125 | 106 | 0 | 0 | 0 | 762 |
| 2000 | 0 | 0 | 0 | 0 | 12 | 74 | 112 | 118 | 31 | 1 | 0 | 0 | 348 |
| 2001 | 0 | 0 | 0 | 0 | 17 | 122 | 132 | 251 | 49 | 6 | 0 | 0 | 577 |
| 2002 | 0 | 0 | 0 | 19 | 17 | 76 | 195 | 219 | 83 | 10 | 0 | 0 | 619 |
| 2003 | 0 | 0 | 0 | 3 | 1 | 96 | 208 | 231 | 45 | 0 | 0 | 0 | 584 |
| 2004 | 0 | 0 | 0 | 11 | 31 | 53 | 159 | 112 | 16 | 0 | 0 | 0 | 382 |
| 2005 | 0 | 0 | 0 | 0 | 0 | 212 | 247 | 205 | 72 | 4 | 0 | 0 | 740 |
| 2006 | 0 | 0 | 0 | 0 | 20 | 100 | 269 | 120 | 14 | 0 | 0 | 0 | 523 |
| 2007 | 0 | 0 | 0 | 5 | 36 | 135 | 144 | 163 | 66 | 12 | 0 | 0 | 561 |
| 2008 | 0 | 0 | 0 | 2 | 0 | 130 | 186 | 94 | 51 | 0 | 0 | 0 | 463 |
| 2009 | 0 | 0 | 0 | 7 | 11 | 61 | 121 | 190 | 11 | 0 | 0 | 0 | 401 |

## SNOWFALL (inches)  2009  BURLINGTON (KBTV)

| YEAR | JUL | AUG | SEP | OCT | NOV | DEC | JAN | FEB | MAR | APR | MAY | JUN | TOTAL |
|------|-----|-----|-----|-----|-----|-----|-----|-----|-----|-----|-----|-----|-------|
| 1980-81 | 0.0 | 0.0 | 0.0 | T | 12.2 | 17.5 | 8.7 | 11.9 | 13.3 | 1.1 | 0.0 | 0.0 | 64.7 |
| 1981-82 | 0.0 | 0.0 | 0.0 | T | 3.9 | 32.8 | 19.4 | 8.3 | 13.0 | 4.1 | 0.0 | 0.0 | 81.5 |
| 1982-83 | 0.0 | 0.0 | 0.0 | T | 0.8 | 5.0 | 22.5 | 18.3 | 11.9 | 21.3 | 0.7 | 0.0 | 80.5 |
| 1983-84 | 0.0 | 0.0 | 0.0 | T | 4.7 | 14.4 | 15.2 | 13.7 | 16.1 | 0.4 | T | 0.0 | 64.5 |
| 1984-85 | 0.0 | 0.0 | 0.0 | 0.0 | 6.0 | 29.3 | 25.9 | 10.9 | 16.6 | 2.7 | 0.0 | 0.0 | 91.4 |
| 1985-86 | 0.0 | 0.0 | 0.0 | T | 4.6 | 21.3 | 33.6 | 18.3 | 8.4 | T | T | 0.0 | 86.2 |
| 1986-87 | 0.0 | 0.0 | 0.0 | T | 10.5 | 7.7 | 34.4 | 7.0 | 6.0 | 2.1 | 0.0 | 0.0 | 67.7 |
| 1987-88 | 0.0 | 0.0 | 0.0 | 0.6 | 6.5 | 12.4 | 9.2 | 26.9 | 6.4 | 2.4 | 0.0 | 0.0 | 64.4 |
| 1988-89 | 0.0 | 0.0 | 0.0 | 0.3 | 0.6 | 12.4 | 6.6 | 8.5 | 9.7 | 2.3 | 0.0 | 0.0 | 40.4 |
| 1989-90 | T | 0.0 | 0.0 | 0.0 | 5.6 | 20.7 | 17.6 | 20.5 | 10.2 | 2.1 | 0.0 | 0.0 | 76.7 |
| 1990-91 | 0.0 | 0.0 | T | T | 7.3 | 10.3 | 17.8 | 3.9 | 3.2 | T | 0.0 | 0.0 | 42.5 |
| 1991-92 | 0.0 | 0.0 | T | T | 2.3 | 14.9 | 12.2 | 27.2 | 14.0 | 8.6 | 0.0 | 0.0 | 79.2 |
| 1992-93 | 0.0 | 0.0 | 0.1 | T | 2.9 | 2.6 | 24.8 | 33.8 | 39.9 | 12.8 | 0.0 | 0.0 | 116.9 |
| 1993-94 | 0.0 | 0.0 | 0.0 | 1.3 | 7.9 | 9.1 | 38.6 | 15.9 | 26.6 | 7.8 | T | 0.0 | 107.2 |
| 1994-95 | 0.0 | 0.0 | 0.0 | 0.0 | 5.1 | 4.3 | 8.7 | 26.8 | 10.7 | 4.9 | 0.0 | 0.0 | 60.5 |
| 1995-96 | 0.0 | 0.0 | 0.0 | 0.1 | 7.3 | 44.0 | 19.0 | 4.5 | 11.5 | 12.4 | 0.3 | 0.0 | 99.1 |
| 1996-97 | 0.0 | 0.0 | 0.0 | T | 14.3 | 13.7 | 22.0 | 8.8 | 27.0 | 9.1 | T | 0.0 | 94.9 |
| 1997-98 | 0.0 | 0.0 | 0.0 | 0.6 | 15.5 | 20.6 | 25.1 | 10.3 | 21.5 | 0.3 | 0.0 | 0.0 | 93.9 |
| 1998-99 | 0.0 | 0.0 | 0.0 | T | 2.4 | 4.4 | 30.4 | 10.8 | 22.7 | T | 0.0 | 0.0 | 70.7 |
| 1999-00 | 0.0 | 0.0 | 0.0 | T | 0.7 | 2.3 | 21.9 | 23.1 | 9.3 | 19.1 | 0.0 | 0.0 | 76.4 |
| 2000-01 | T | 0.0 | 0.0 | 3.0 | 8.7 | 32.8 | 15.7 | 14.4 | 47.6 | 0.2 | T | 0.0 | 122.4 |
| 2001-02 | 0.0 | T | 0.0 | T | 1.8 | 13.9 | 16.6 | 8.5 | 15.3 | 0.4 | 0.0 | 0.0 | 56.5 |
| 2002-03 | 0.0 | 0.0 | 0.0 | T | 20.4 | 8.6 | 31.3 | 9.6 | 8.3 | 4.8 | 0.0 | 0.0 | 83.0 |
| 2003-04 | 0.0 | 0.0 | T | T | 0.5 | 53.6 | 14.9 | 12.2 | 9.7 | 3.8 | T | 0.0 | 94.7 |
| 2004-05 | T | 0.0 | 0.0 | 0.0 | 0.1 | 22.8 | 19.1 | 29.7 | 18.0 | 0.0 | 0.0 | 0.0 | 89.7 |
| 2005-06 | 0.0 | 0.0 | 0.0 | 0.9 | 5.5 | 18.4 | 20.0 | 8.3 | 16.3 | 1.0 | 0.0 | T | 70.4 |
| 2006-07 | 0.0 | 0.0 | 0.0 | 2.0 | T | 10.0 | 19.5 | 32.1 | 21.8 | 9.2 | 0.0 | T | 94.6 |
| 2007-08 | 0.0 | 0.0 | 0.0 | 0.0 | 3.7 | 45.3 | 15.8 | 42.3 | 13.0 | 0.1 | 0.0 | 0.0 | 120.2 |
| 2008-09 | 0.0 | 0.0 | 0.0 | 0.3 | 3.5 | 40.3 | 27.8 | 11.3 | 7.8 | 0.4 | 0.0 | T | 91.4 |
| 2009- | T | 0.0 | 0.0 | T | T | 17.7 | | | | | | | |
| POR=<br>78 YRS | 0.7 | 0.2 | T | 0.2 | 6.0 | 16.3 | 18.0 | 16.0 | 13.4 | 4.0 | 0.1 | T | 74.9 |

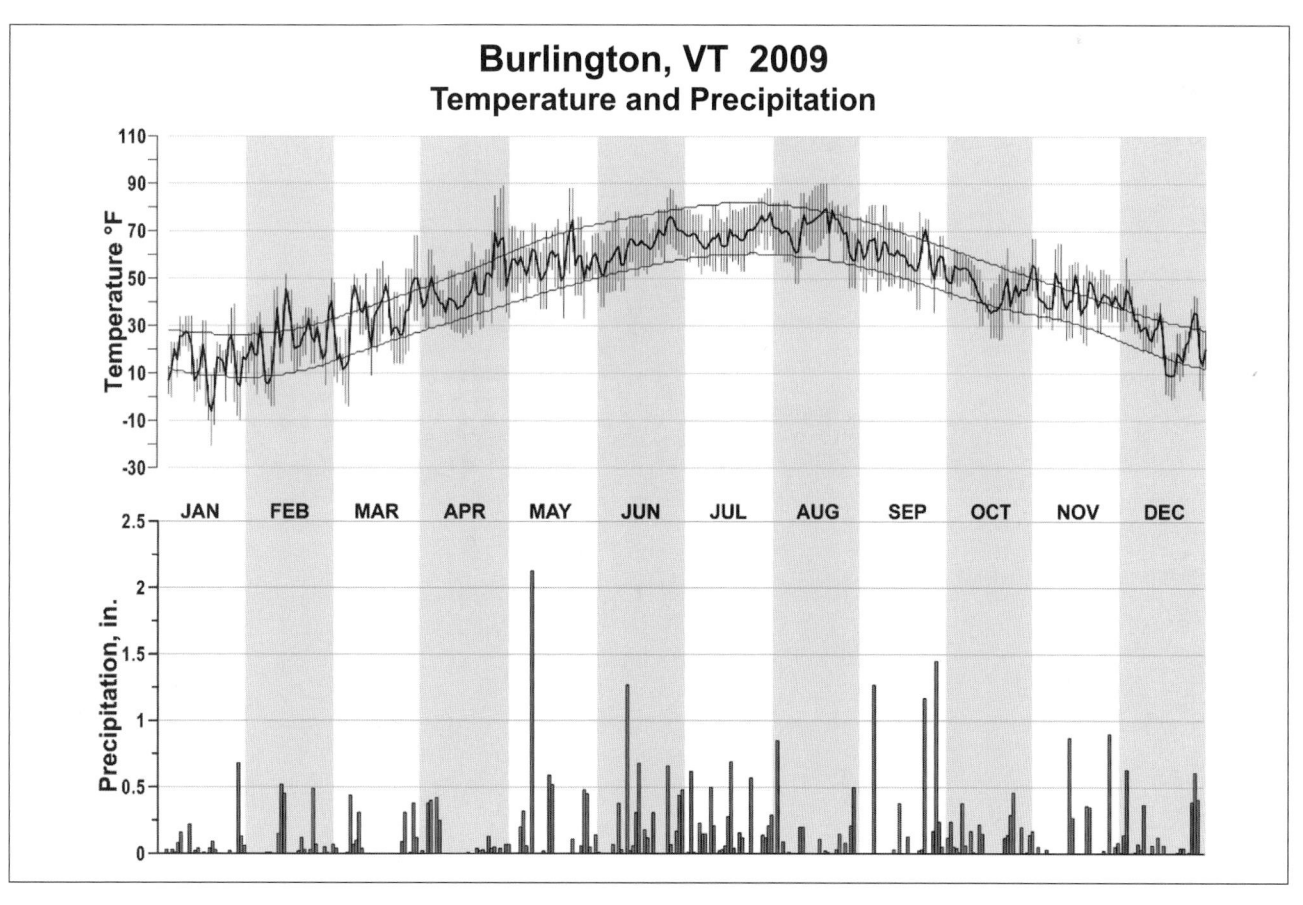

# Burlington, VT  2009
## Temperature and Precipitation

# 2009
# NORFOLK
# VIRGINIA (KORF)

The city of Norfolk, Virginia, is located near the coast and the southern border of the state. It is almost surrounded by water, with the Chesapeake Bay immediately to the north, Hampton Roads to the west, and the Atlantic Ocean only 18 miles to the east. It is traversed by numerous rivers and waterways and its average elevation above sea level is 13 feet. There are no nearby hilly areas and the land is low and level throughout the city. The climate is generally marine. The geographic location of the city with respect to the principal storm tracks, is especially favorable, being south of the average path of storms originating in the higher latitudes and north of the usual tracks of hurricanes and other tropical storms.

The winters are usually mild, while the autumn and spring seasons usually are delightful. Summers, though warm and long, frequently are tempered by cool periods, often associated with northeasterly winds off the Atlantic. Temperatures of 100 degrees or higher occur infrequently. Extreme cold waves seldom penetrate the area and temperatures of zero or below are almost nonexistent. Winters pass, on occasion, without a measurable amount of snowfall. Most of the snowfall in Norfolk is light and generally melts within 24 hours.

Based on the 1951-1980 period, the average first occurrence of 32 degrees Fahrenheit in the fall is November 17 and the average last occurrence in the spring is March 23.

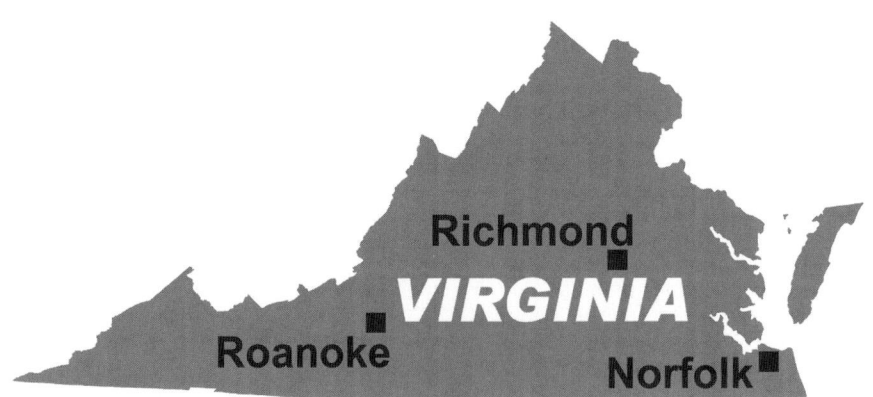

# NORMALS, MEANS, AND EXTREMES
## NORFOLK (KORF)

| LATITUDE: 36 ° 54'N | LONGITUDE: -76 ° 11'W | ELEVATION (FT): GRND: 13  BARO: 69 | TIME ZONE: EASTERN  (UTC -5) | WBAN: 13737 |
|---|---|---|---|---|

| | ELEMENT | POR | JAN | FEB | MAR | APR | MAY | JUN | JUL | AUG | SEP | OCT | NOV | DEC | YEAR |
|---|---|---|---|---|---|---|---|---|---|---|---|---|---|---|---|
| **TEMPERATURE °F** | NORMAL DAILY MAXIMUM | 30 | 47.8 | 50.3 | 57.8 | 67.0 | 74.9 | 82.8 | 86.8 | 84.7 | 79.4 | 69.4 | 60.9 | 52.3 | 67.8 |
| | MEAN DAILY MAXIMUM | 64 | 48.9 | 50.7 | 58.1 | 68.0 | 75.8 | 83.3 | 87.2 | 85.6 | 79.8 | 70.3 | 61.3 | 52.5 | 68.5 |
| | HIGHEST DAILY MAXIMUM | 61 | 80 | 82 | 88 | 97 | 100 | 101 | 103 | 104 | 99 | 95 | 86 | 80 | 104 |
| | YEAR OF OCCURRENCE | | 2002 | 1997 | 1990 | 1960 | 1991 | 2008 | 1993 | 1980 | 1983 | 1954 | 1974 | 1991 | AUG 1980 |
| | MEAN OF EXTREME MAXS. | 64 | 70.9 | 72.5 | 80.0 | 86.7 | 90.7 | 95.1 | 96.9 | 95.5 | 91.8 | 84.7 | 78.6 | 72.1 | 84.6 |
| | NORMAL DAILY MINIMUM | 30 | 32.3 | 33.6 | 40.1 | 47.8 | 57.6 | 66.2 | 71.4 | 70.1 | 64.8 | 52.8 | 43.7 | 36.1 | 51.4 |
| | MEAN DAILY MINIMUM | 64 | 32.6 | 33.4 | 40.0 | 48.3 | 57.5 | 65.8 | 70.8 | 70.0 | 64.6 | 53.5 | 43.7 | 35.7 | 51.3 |
| | LOWEST DAILY MINIMUM | 61 | -3 | 8 | 18 | 28 | 36 | 45 | 54 | 49 | 45 | 27 | 20 | 7 | -3 |
| | YEAR OF OCCURRENCE | | 1985 | 1965 | 2009 | 1982 | 1966 | 1967 | 1979 | 1982 | 1967 | 1976 | 1950 | 1983 | JAN 1985 |
| | MEAN OF EXTREME MINS. | 64 | 17.5 | 20.2 | 26.9 | 35.4 | 44.6 | 54.4 | 62.0 | 60.7 | 52.6 | 38.6 | 29.4 | 21.3 | 38.6 |
| | NORMAL DRY BULB | 30 | 40.1 | 42.0 | 49.0 | 57.4 | 66.3 | 74.5 | 79.1 | 77.4 | 72.1 | 61.1 | 52.3 | 44.2 | 59.6 |
| | MEAN DRY BULB | 64 | 40.8 | 42.1 | 49.1 | 58.2 | 66.7 | 74.7 | 79.0 | 77.8 | 72.2 | 61.9 | 52.5 | 44.1 | 59.9 |
| | MEAN WET BULB | 26 | 36.4 | 37.6 | 43.1 | 51.3 | 59.7 | 68.0 | 72.1 | 71.4 | 66.3 | 56.6 | 48.1 | 40.0 | 54.2 |
| | MEAN DEW POINT | 26 | 32.0 | 32.9 | 38.8 | 46.7 | 56.3 | 65.4 | 70.0 | 69.5 | 63.9 | 53.4 | 43.6 | 35.3 | 50.7 |
| | NORMAL NO. DAYS WITH: | | | | | | | | | | | | | | |
| | MAXIMUM >= 90 | 30 | 0.0 | 0.0 | 0.0 | 0.4 | 1.7 | 6.8 | 12.9 | 9.2 | 2.6 | 0.1 | 0.0 | 0.0 | 33.7 |
| | MAXIMUM <= 32 | 30 | 2.4 | 1.3 | 0.1 | 0.0 | 0.0 | 0.0 | 0.0 | 0.0 | 0.0 | 0.0 | 0.0 | 0.7 | 4.5 |
| | MINIMUM <= 32 | 30 | 15.1 | 13.3 | 5.1 | 0.3 | 0.0 | 0.0 | 0.0 | 0.0 | 0.0 | 0.1 | 2.7 | 11.5 | 48.1 |
| | MINIMUM <= 0 | 30 | * | 0.0 | 0.0 | 0.0 | 0.0 | 0.0 | 0.0 | 0.0 | 0.0 | 0.0 | 0.0 | 0.0 | 0.0 |
| **H/C** | NORMAL HEATING DEG. DAYS | 30 | 759 | 638 | 488 | 247 | 66 | 4 | 0 | 0 | 8 | 152 | 375 | 631 | 3368 |
| | NORMAL COOLING DEG. DAYS | 30 | 1 | 2 | 8 | 35 | 119 | 303 | 453 | 400 | 235 | 45 | 10 | 1 | 1612 |
| **RH** | NORMAL (PERCENT) | 30 | 68 | 66 | 66 | 64 | 70 | 71 | 73 | 75 | 74 | 72 | 70 | 69 | 70 |
| | HOUR 01 LST | 30 | 72 | 72 | 73 | 73 | 80 | 83 | 84 | 85 | 84 | 81 | 76 | 74 | 78 |
| | HOUR 07 LST | 30 | 74 | 75 | 75 | 73 | 77 | 79 | 81 | 84 | 83 | 82 | 78 | 76 | 78 |
| | HOUR 13 LST | 30 | 59 | 56 | 55 | 50 | 56 | 57 | 58 | 61 | 61 | 58 | 57 | 59 | 57 |
| | HOUR 19 LST | 30 | 67 | 65 | 63 | 60 | 65 | 66 | 69 | 73 | 74 | 73 | 70 | 68 | 68 |
| **S** | PERCENT POSSIBLE SUNSHINE | 32 | 53 | 56 | 60 | 63 | 62 | 67 | 62 | 62 | 61 | 58 | 56 | 54 | 60 |
| **W/O** | MEAN NO. DAYS WITH: | | | | | | | | | | | | | | |
| | HEAVY FOG(VISBY <= 1/4 MI) | 46 | 1.9 | 1.9 | 1.8 | 1.3 | 1.2 | 0.5 | 0.4 | 0.5 | 0.7 | 1.3 | 1.5 | 1.8 | 14.8 |
| | THUNDERSTORMS | 62 | 0.4 | 0.6 | 1.8 | 2.7 | 4.8 | 5.6 | 8.3 | 6.8 | 2.8 | 1.1 | 0.5 | 0.4 | 35.8 |
| **CLOUDNESS** | MEAN: | | | | | | | | | | | | | | |
| | SUNRISE-SUNSET (OKTAS) | 48 | 5.0 | 5.0 | 4.9 | 4.6 | 4.9 | 4.7 | 4.7 | 4.6 | 4.5 | 4.2 | 4.4 | 4.8 | 4.7 |
| | MIDNIGHT-MIDNIGHT (OKTAS) | 32 | 4.7 | 4.7 | 4.4 | 4.2 | 4.6 | 4.4 | 4.5 | 4.4 | 4.2 | 4.0 | 4.1 | 4.4 | 4.4 |
| | MEAN NO. DAYS WITH: | | | | | | | | | | | | | | |
| | CLEAR | 48 | 8.8 | 8.1 | 8.9 | 8.8 | 7.7 | 7.5 | 7.4 | 7.9 | 9.1 | 11.6 | 10.3 | 9.4 | 105.5 |
| | PARTLY CLOUDY | 48 | 6.5 | 6.2 | 7.6 | 9.3 | 10.0 | 11.7 | 12.0 | 11.9 | 9.5 | 7.2 | 7.9 | 6.8 | 106.6 |
| | CLOUDY | 48 | 15.7 | 13.9 | 14.5 | 11.9 | 13.3 | 10.8 | 11.6 | 11.3 | 11.5 | 12.2 | 11.7 | 14.8 | 153.2 |
| **PR** | MEAN STATION PRESSURE(IN) | 26 | 30.08 | 30.06 | 30.03 | 29.96 | 29.97 | 29.96 | 29.97 | 29.99 | 30.02 | 30.06 | 30.08 | 30.09 | 30.02 |
| | MEAN SEA-LEVEL PRES. (IN) | 26 | 30.11 | 30.09 | 30.06 | 29.99 | 30.00 | 29.99 | 30.00 | 30.03 | 30.06 | 30.09 | 30.11 | 30.12 | 30.05 |
| **WINDS** | MEAN SPEED (MPH) | 26 | 10.4 | 10.7 | 11.3 | 11.1 | 10.3 | 9.3 | 8.7 | 8.2 | 9.4 | 9.2 | 9.5 | 9.9 | 9.8 |
| | PREVAIL.DIR(TENS OF DEGS) | 41 | 24 | 04 | 05 | 23 | 23 | 24 | 23 | 23 | 05 | 05 | 24 | 24 | 23 |
| | MAXIMUM 2-MINUTE: | | | | | | | | | | | | | | |
| | SPEED (MPH) | 13 | 44 | 41 | 45 | 38 | 41 | 37 | 44 | 46 | 47 | 38 | 58 | 37 | 58 |
| | DIR. (TENS OF DEGS) | | 24 | 07 | 04 | 06 | 04 | 01 | 17 | 11 | 06 | 08 | 05 | 23 | 05 |
| | YEAR OF OCCURRENCE | | 2009 | 1998 | 2009 | 2005 | 2007 | 2006 | 2007 | 1998 | 2003 | 2006 | 2009 | 2009 | NOV 2009 |
| | MAXIMUM 3-SECOND | | | | | | | | | | | | | | |
| | SPEED (MPH) | 13 | 60 | 53 | 69 | 47 | 52 | 60 | 68 | 67 | 74 | 51 | 74 | 49 | 74 |
| | DIR. (TENS OF DEGS) | | 24 | 34 | 30 | 21 | 24 | 02 | 16 | 22 | 11 | 08 | 05 | 02 | 05 |
| | YEAR OF OCCURRENCE | | 2009 | 1999 | 2007 | 2009 | 2008 | 2006 | 2003 | 2000 | 2003 | 2006 | 2009 | 2004 | NOV 2009 |
| **PRECIPITATION** | NORMAL (IN) | 30 | 3.93 | 3.34 | 4.08 | 3.38 | 3.74 | 3.77 | 5.17 | 4.79 | 4.06 | 3.47 | 2.98 | 3.03 | 45.74 |
| | MAXIMUM MONTHLY (IN) | 61 | 9.93 | 8.21 | 10.36 | 7.25 | 10.12 | 10.53 | 14.37 | 14.32 | 13.80 | 10.12 | 9.20 | 7.57 | 14.37 |
| | YEAR OF OCCURRENCE | | 1987 | 1998 | 1994 | 1984 | 1979 | 2006 | 1994 | 1992 | 1979 | 1971 | 2009 | 2009 | JUL 1994 |
| | MINIMUM MONTHLY (IN) | 61 | 1.05 | 0.65 | 0.52 | 0.43 | 0.64 | 0.37 | 0.36 | 0.67 | 0.26 | 0.01 | 0.08 | 0.67 | 0.01 |
| | YEAR OF OCCURRENCE | | 1981 | 2006 | 2006 | 1985 | 1991 | 1954 | 1993 | 2008 | 1986 | 2000 | 2001 | 1988 | OCT 2000 |
| | MAXIMUM IN 24 HOURS (IN) | 61 | 3.80 | 4.78 | 4.02 | 5.90 | 3.41 | 6.85 | 5.64 | 11.40 | 8.93 | 7.29 | 5.01 | 2.80 | 11.40 |
| | YEAR OF OCCURRENCE | | 1967 | 1998 | 1994 | 1991 | 1980 | 1963 | 1969 | 1964 | 2006 | 1999 | 2009 | 2008 | AUG 1964 |
| | NORMAL NO. DAYS WITH: | | | | | | | | | | | | | | |
| | PRECIPITATION >= 0.01 | 30 | 11.4 | 10.1 | 10.9 | 9.5 | 10.5 | 9.5 | 11.0 | 9.8 | 8.4 | 7.5 | 8.3 | 10.2 | 117.1 |
| | PRECIPITATION >= 1.00 | 30 | 0.8 | 0.6 | 1.0 | 0.9 | 0.8 | 1.1 | 1.7 | 1.4 | 1.2 | 0.9 | 0.5 | 0.6 | 11.5 |
| **SNOWFALL** | NORMAL (IN) | 30 | 2.6 | 3.8 | 1.3 | 0.* | 0.0 | 0.0 | 0.0 | 0.0 | 0.0 | 0.0 | 0.* | 0.4 | 8.1 |
| | MAXIMUM MONTHLY (IN) | 57 | 14.2 | 24.4 | 13.7 | 1.2 | T | T | 0.0 | T | 0.0 | 0.0 | 0.6 | 14.7 | 24.4 |
| | YEAR OF OCCURRENCE | | 1966 | 1989 | 1980 | 1964 | 1994 | 1990 | | 1991 | | | 1950 | 1958 | FEB 1958 |
| | MAXIMUM IN 24 HOURS (IN) | 57 | 9.1 | 14.2 | 9.9 | 1.2 | T | T | 0.0 | T | 0.0 | 0.0 | 0.6 | 11.4 | 14.2 |
| | YEAR OF OCCURRENCE' | | 1973 | 1989 | 1980 | 1964 | 1994 | 1990 | | 1991 | | | 1950 | 1958 | FEB 1989 |
| | MAXIMUM SNOW DEPTH (IN) | 49 | 9 | 14 | 14 | 0 | 0 | 0 | 0 | 0 | 0 | 0 | 0 | 11 | 14 |
| | YEAR OF OCCURRENCE | | 1966 | 1980 | 1980 | | | | | | | | | 1958 | MAR 1980 |
| | NORMAL NO. DAYS WITH: | | | | | | | | | | | | | | |
| | SNOWFALL >= 1.0 | 30 | 0.9 | 0.8 | 0.2 | 0.0 | 0.0 | 0.0 | 0.0 | 0.0 | 0.0 | 0.0 | 0.0 | 0.2 | 2.1 |

## PRECIPITATION (inches) 2009 NORFOLK (KORF)

| YEAR | JAN | FEB | MAR | APR | MAY | JUN | JUL | AUG | SEP | OCT | NOV | DEC | ANNUAL |
|------|------|------|------|------|------|------|------|------|------|------|------|------|--------|
| 1980 | 4.54 | 2.91 | 4.40 | 3.25 | 5.17 | 1.39 | 1.85 | 4.54 | 1.47 | 4.21 | 2.01 | 2.64 | 38.38 |
| 1981 | 1.05 | 2.26 | 1.88 | 2.26 | 2.75 | 5.00 | 5.10 | 6.87 | 3.18 | 3.28 | 1.78 | 5.77 | 41.18 |
| 1982 | 3.35 | 5.81 | 3.04 | 1.71 | 3.07 | 4.22 | 5.83 | 6.51 | 3.63 | 4.25 | 3.43 | 4.30 | 49.15 |
| 1983 | 2.21 | 6.23 | 4.55 | 6.13 | 3.52 | 3.84 | 0.77 | 3.07 | 4.52 | 5.29 | 3.24 | 6.10 | 49.47 |
| 1984 | 2.77 | 4.66 | 5.09 | 7.25 | 6.23 | 1.50 | 7.66 | 2.25 | 1.94 | 0.57 | 2.68 | 2.22 | 44.82 |
| 1985 | 3.98 | 3.53 | 2.02 | 0.43 | 3.23 | 6.81 | 6.14 | 1.89 | 6.36 | 3.92 | 5.71 | 0.79 | 44.81 |
| 1986 | 2.52 | 2.71 | 0.75 | 3.31 | 1.41 | 1.51 | 2.59 | 4.80 | 0.26 | 1.67 | 1.21 | 3.74 | 26.48 |
| 1987 | 9.93 | 3.11 | 2.30 | 3.83 | 2.65 | 2.98 | 3.20 | 2.04 | 7.00 | 1.81 | 3.51 | 2.33 | 44.69 |
| 1988 | 3.12 | 2.70 | 2.11 | 3.53 | 5.49 | 3.83 | 2.93 | 5.69 | 1.74 | 2.85 | 4.02 | 0.67 | 38.68 |
| 1989 | 2.70 | 5.80 | 8.50 | 3.62 | 2.97 | 5.10 | 4.86 | 7.49 | 5.10 | 2.94 | 3.69 | 3.86 | 56.63 |
| 1990 | 3.26 | 2.93 | 3.49 | 3.55 | 3.79 | 3.51 | 4.06 | 11.85 | 1.00 | 3.73 | 1.68 | 2.67 | 45.52 |
| 1991 | 4.74 | 0.84 | 4.70 | 6.39 | 0.64 | 4.54 | 6.46 | 3.77 | 2.04 | 4.65 | 1.72 | 2.43 | 42.92 |
| 1992 | 4.48 | 2.07 | 2.63 | 1.26 | 3.46 | 2.22 | 4.52 | 14.32 | 2.06 | 2.85 | 4.26 | 3.15 | 47.28 |
| 1993 | 4.89 | 2.36 | 5.91 | 3.59 | 2.88 | 2.79 | 0.36 | 1.45 | 4.14 | 3.40 | 0.97 | 3.29 | 36.03 |
| 1994 | 4.06 | 3.64 | 10.36 | 0.56 | 3.52 | 1.27 | 14.37 | 3.65 | 2.02 | 2.50 | 5.25 | 1.20 | 52.40 |
| 1995 | 2.40 | 2.80 | 2.95 | 2.76 | 2.42 | 4.54 | 1.78 | 1.50 | 5.05 | 4.82 | 2.94 | 1.86 | 35.82 |
| 1996 | 5.49 | 3.04 | 3.46 | 4.94 | 3.59 | 6.60 | 7.46 | 5.19 | 2.72 | 4.71 | 2.87 | 3.86 | 53.93 |
| 1997 | 2.09 | 2.94 | 3.21 | 3.01 | 1.66 | 1.10 | 7.85 | 1.76 | 1.97 | 4.31 | 4.94 | 2.64 | 37.48 |
| 1998 | 6.02 | 8.21 | 4.15 | 4.31 | 3.99 | 4.56 | 3.91 | 8.47 | 2.25 | 1.73 | 1.83 | 5.33 | 54.76 |
| 1999 | 3.51 | 2.33 | 3.29 | 3.66 | 3.85 | 3.60 | 6.17 | 4.47 | 13.16 | 8.24 | 1.40 | 1.71 | 55.39 |
| 2000 | 5.07 | 1.13 | 2.40 | 3.70 | 4.05 | 8.31 | 7.52 | 8.35 | 6.25 | 0.01 | 1.67 | 0.97 | 49.43 |
| 2001 | 1.46 | 2.16 | 4.73 | 1.48 | 2.89 | 6.91 | 4.40 | 4.21 | 2.46 | 0.75 | 0.08 | 1.83 | 33.36 |
| 2002 | 4.22 | 1.38 | 4.91 | 2.40 | 3.50 | 4.28 | 3.59 | 4.27 | 6.69 | 6.55 | 4.92 | 4.16 | 50.87 |
| 2003 | 2.34 | 5.25 | 2.91 | 6.39 | 4.66 | 3.43 | 8.56 | 6.08 | 9.54 | 3.91 | 2.45 | 6.24 | 61.76 |
| 2004 | 1.59 | 1.82 | 2.09 | 2.82 | 4.67 | 4.86 | 10.89 | 11.11 | 3.30 | 1.88 | 2.65 | 2.41 | 50.09 |
| 2005 | 2.48 | 2.32 | 2.25 | 2.74 | 4.02 | 4.50 | 3.67 | 7.61 | 2.90 | 5.53 | 3.79 | 4.30 | 46.11 |
| 2006 | 2.68 | 0.65 | 0.52 | 3.65 | 2.96 | 10.53 | 1.34 | 3.13 | 11.64 | 3.54 | 6.46 | 2.06 | 49.16 |
| 2007 | 2.71 | 2.09 | 1.84 | 3.19 | 2.06 | 3.80 | 4.77 | 3.71 | 0.38 | 5.39 | 0.31 | 3.50 | 33.75 |
| 2008 | 1.36 | 3.41 | 2.96 | 6.37 | 2.88 | 1.93 | 5.19 | 0.67 | 9.41 | 1.47 | 5.32 | 3.83 | 44.80 |
| 2009 | 1.82 | 1.26 | 5.28 | 2.28 | 4.77 | 5.81 | 2.47 | 13.22 | 7.77 | 3.21 | 9.20 | 7.57 | 64.66 |
| POR= 64 YRS | 3.51 | 3.15 | 3.64 | 3.17 | 3.63 | 3.87 | 5.28 | 5.43 | 4.33 | 3.32 | 3.12 | 3.17 | 45.62 |

WBAN : 13737

## AVERAGE TEMPERATURE (°F) 2009 NORFOLK (KORF)

| YEAR | JAN | FEB | MAR | APR | MAY | JUN | JUL | AUG | SEP | OCT | NOV | DEC | ANNUAL |
|------|------|------|------|------|------|------|------|------|------|------|------|------|--------|
| 1980 | 40.3 | 34.7 | 46.5 | 58.6 | 67.8 | 73.9 | 80.9 | 80.9 | 76.1 | 60.4 | 49.9 | 42.3 | 59.4 |
| 1981 | 32.7 | 43.1 | 45.4 | 61.2 | 65.1 | 78.3 | 79.8 | 75.1 | 70.7 | 59.6 | 50.7 | 41.0 | 58.6 |
| 1982 | 35.4 | 42.0 | 48.8 | 55.0 | 69.4 | 73.4 | 78.6 | 75.3 | 70.0 | 60.2 | 54.4 | 48.8 | 59.3 |
| 1983 | 40.2 | 40.8 | 51.0 | 55.7 | 65.8 | 73.0 | 80.3 | 79.0 | 72.8 | 62.7 | 52.6 | 41.6 | 59.6 |
| 1984 | 35.5 | 46.7 | 45.4 | 55.6 | 67.8 | 76.2 | 76.7 | 78.4 | 70.5 | 66.9 | 49.9 | 50.9 | 60.0 |
| 1985 | 34.9 | 40.4 | 51.8 | 62.0 | 68.8 | 74.2 | 78.2 | 77.2 | 73.4 | 65.9 | 60.3 | 41.2 | 60.7 |
| 1986 | 39.3 | 42.1 | 49.9 | 57.3 | 67.6 | 76.1 | 82.1 | 76.6 | 72.4 | 65.4 | 54.9 | 44.8 | 60.7 |
| 1987 | 39.6 | 38.7 | 47.5 | 54.6 | 68.3 | 77.0 | 82.4 | 79.6 | 74.3 | 56.6 | 54.3 | 46.0 | 59.9 |
| 1988 | 37.3 | 42.5 | 49.5 | 56.5 | 65.8 | 73.6 | 80.1 | 80.8 | 70.5 | 56.9 | 54.2 | 42.4 | 59.2 |
| 1989 | 45.3 | 43.6 | 50.1 | 56.5 | 65.6 | 78.5 | 79.2 | 77.7 | 73.9 | 62.7 | 53.3 | 34.8 | 60.1 |
| 1990 | 47.3 | 50.2 | 53.2 | 58.7 | 66.6 | 75.5 | 80.6 | 78.0 | 71.6 | 65.9 | 55.0 | 50.5 | 62.8 |
| 1991 | 43.5 | 46.0 | 52.7 | 61.6 | 72.8 | 76.2 | 82.0 | 79.5 | 72.3 | 61.9 | 52.7 | 47.6 | 62.4 |
| 1992 | 42.7 | 45.0 | 48.8 | 58.2 | 62.4 | 72.0 | 81.8 | 75.6 | 73.5 | 59.7 | 53.8 | 44.6 | 59.8 |
| 1993 | 44.2 | 39.1 | 46.8 | 57.3 | 68.5 | 76.1 | 83.4 | 78.9 | 74.6 | 61.3 | 53.5 | 41.5 | 60.4 |
| 1994 | 36.5 | 43.4 | 51.7 | 64.7 | 64.9 | 79.6 | 83.0 | 77.6 | 72.4 | 62.0 | 57.5 | 50.1 | 62.0 |
| 1995 | 45.1 | 41.5 | 51.7 | 60.4 | 67.9 | 75.8 | 83.3 | 80.1 | 72.2 | 65.5 | 48.4 | 40.0 | 61.0 |
| 1996 | 38.7 | 41.3 | 45.0 | 58.8 | 65.2 | 75.2 | 77.7 | 74.9 | 72.4 | 61.8 | 46.5 | 46.7 | 58.7 |
| 1997 | 41.3 | 46.0 | 51.5 | 54.6 | 64.0 | 72.0 | 79.1 | 76.4 | 72.0 | 62.1 | 50.3 | 43.5 | 59.4 |
| 1998 | 46.5 | 47.2 | 50.8 | 58.9 | 67.5 | 75.9 | 79.0 | 78.9 | 75.3 | 61.9 | 52.6 | 48.7 | 61.9 |
| 1999 | 46.1 | 44.5 | 47.6 | 58.0 | 66.5 | 73.9 | 81.1 | 79.5 | 72.1 | 61.5 | 55.5 | 45.8 | 61.0 |
| 2000 | 40.2 | 45.1 | 53.0 | 57.5 | 70.3 | 76.1 | 76.3 | 76.6 | 71.5 | 62.1 | 49.3 | 38.2 | 59.7 |
| 2001 | 39.3 | 44.0 | 46.5 | 58.2 | 66.5 | 75.8 | 76.3 | 78.8 | 70.4 | 61.3 | 56.0 | 49.7 | 60.2 |
| 2002 | 44.6 | 45.7 | 52.2 | 62.7 | 66.7 | 76.8 | 80.8 | 79.5 | 74.1 | 64.4 | 51.1 | 42.3 | 61.7 |
| 2003 | 36.3 | 41.0 | 51.6 | 58.0 | 65.7 | 75.0 | 80.2 | 80.3 | 74.1 | 62.0 | 58.1 | 44.4 | 60.6 |
| 2004 | 38.8 | 42.6 | 52.4 | 60.3 | 73.1 | 75.1 | 79.2 | 76.0 | 72.6 | 61.0 | 53.3 | 44.3 | 60.7 |
| 2005 | 41.7 | 41.4 | 45.3 | 57.5 | 62.5 | 75.1 | 81.0 | 80.1 | 75.8 | 63.8 | 54.6 | 42.0 | 60.1 |
| 2006 | 47.4 | 42.6 | 49.7 | 60.5 | 66.2 | 74.4 | 80.8 | 81.3 | 70.4 | 61.1 | 54.5 | 48.2 | 61.4 |
| 2007 | 46.1 | 38.2 | 51.3 | 57.6 | 65.6 | 75.9 | 79.4 | 79.7 | 74.4 | 69.1 | 52.0 | 47.8 | 61.4 |
| 2008 | 42.2 | 46.8 | 52.9 | 59.0 | 65.2 | 79.2 | 79.0 | 77.7 | 73.9 | 60.4 | 49.4 | 48.2 | 61.2 |
| 2009 | 37.8 | 43.1 | 47.5 | 59.5 | 69.0 | 75.6 | 77.6 | 79.6 | 71.1 | 62.5 | 55.3 | 43.9 | 60.2 |
| POR= 64 YRS | 40.8 | 42.1 | 49.1 | 58.2 | 66.7 | 74.7 | 79.0 | 77.8 | 72.2 | 61.9 | 52.5 | 44.1 | 59.9 |

## HEATING DEGREE DAYS (base 65°F) 2009  NORFOLK (KORF)

| YEAR | JUL | AUG | SEP | OCT | NOV | DEC | JAN | FEB | MAR | APR | MAY | JUN | TOTAL |
|------|-----|-----|-----|-----|-----|-----|-----|-----|-----|-----|-----|-----|-------|
| 1980-81 | 0 | 0 | 11 | 181 | 449 | 699 | 994 | 610 | 605 | 159 | 96 | 0 | 3804 |
| 1981-82 | 0 | 0 | 12 | 189 | 423 | 739 | 907 | 636 | 495 | 303 | 21 | 0 | 3725 |
| 1982-83 | 0 | 4 | 6 | 177 | 334 | 498 | 762 | 674 | 426 | 295 | 85 | 3 | 3264 |
| 1983-84 | 0 | 0 | 27 | 126 | 370 | 718 | 908 | 522 | 601 | 281 | 54 | 3 | 3610 |
| 1984-85 | 0 | 0 | 16 | 37 | 450 | 432 | 928 | 686 | 421 | 172 | 21 | 0 | 3163 |
| 1985-86 | 0 | 0 | 6 | 61 | 162 | 731 | 790 | 637 | 465 | 228 | 69 | 1 | 3150 |
| 1986-87 | 0 | 1 | 8 | 88 | 311 | 620 | 779 | 730 | 538 | 306 | 58 | 0 | 3439 |
| 1987-88 | 0 | 0 | 0 | 252 | 320 | 582 | 851 | 646 | 474 | 266 | 86 | 15 | 3492 |
| 1988-89 | 0 | 0 | 2 | 265 | 324 | 692 | 602 | 601 | 486 | 282 | 80 | 0 | 3334 |
| 1989-90 | 0 | 0 | 12 | 134 | 356 | 928 | 541 | 417 | 410 | 234 | 39 | 3 | 3074 |
| 1990-91 | 0 | 0 | 13 | 102 | 301 | 444 | 657 | 527 | 386 | 166 | 19 | 2 | 2617 |
| 1991-92 | 0 | 0 | 22 | 132 | 377 | 542 | 686 | 575 | 496 | 262 | 125 | 1 | 3218 |
| 1992-93 | 0 | 0 | 9 | 179 | 337 | 623 | 638 | 722 | 555 | 246 | 22 | 5 | 3336 |
| 1993-94 | 0 | 0 | 10 | 140 | 353 | 726 | 877 | 599 | 414 | 111 | 89 | 0 | 3319 |
| 1994-95 | 0 | 0 | 0 | 113 | 239 | 463 | 614 | 651 | 411 | 190 | 45 | 0 | 2726 |
| 1995-96 | 0 | 0 | 4 | 80 | 497 | 769 | 810 | 681 | 614 | 237 | 99 | 7 | 3798 |
| 1996-97 | 0 | 0 | 1 | 123 | 559 | 561 | 728 | 532 | 418 | 314 | 82 | 44 | 3362 |
| 1997-98 | 0 | 0 | 1 | 148 | 434 | 663 | 573 | 492 | 466 | 211 | 60 | 0 | 3048 |
| 1998-99 | 0 | 0 | 2 | 127 | 367 | 513 | 583 | 567 | 531 | 226 | 60 | 0 | 2976 |
| 1999-00 | 0 | 0 | 4 | 146 | 282 | 588 | 762 | 570 | 374 | 227 | 28 | 1 | 2982 |
| 2000-01 | 0 | 0 | 13 | 133 | 466 | 825 | 793 | 583 | 571 | 239 | 52 | 0 | 3675 |
| 2001-02 | 0 | 0 | 19 | 163 | 275 | 468 | 629 | 537 | 398 | 158 | 75 | 3 | 2725 |
| 2002-03 | 0 | 0 | 0 | 115 | 418 | 697 | 884 | 665 | 413 | 231 | 71 | 2 | 3496 |
| 2003-04 | 0 | 0 | 5 | 120 | 239 | 634 | 814 | 643 | 393 | 198 | 25 | 0 | 3071 |
| 2004-05 | 0 | 0 | 2 | 155 | 353 | 633 | 719 | 654 | 605 | 241 | 108 | 6 | 3476 |
| 2005-06 | 0 | 0 | 0 | 110 | 318 | 708 | 538 | 620 | 484 | 168 | 56 | 0 | 3002 |
| 2006-07 | 0 | 0 | 11 | 164 | 308 | 521 | 580 | 748 | 436 | 261 | 85 | 2 | 3116 |
| 2007-08 | 0 | 0 | 0 | 58 | 381 | 526 | 702 | 528 | 371 | 206 | 73 | 0 | 2845 |
| 2008-09 | 0 | 0 | 0 | 174 | 463 | 515 | 837 | 609 | 542 | 210 | 46 | 0 | 3396 |
| 2009- | 0 | 0 | 3 | 131 | 286 | 648 | | | | | | | |

WBAN : 13737

## COOLING DEGREE DAYS (base 65°F) 2009  NORFOLK (KORF)

| YEAR | JAN | FEB | MAR | APR | MAY | JUN | JUL | AUG | SEP | OCT | NOV | DEC | TOTAL |
|------|-----|-----|-----|-----|-----|-----|-----|-----|-----|-----|-----|-----|-------|
| 1980 | 0 | 0 | 0 | 11 | 153 | 274 | 499 | 497 | 351 | 45 | 1 | 1 | 1832 |
| 1981 | 0 | 0 | 0 | 51 | 103 | 407 | 468 | 320 | 189 | 29 | 0 | 0 | 1567 |
| 1982 | 0 | 0 | 1 | 8 | 166 | 257 | 428 | 331 | 164 | 39 | 21 | 4 | 1419 |
| 1983 | 0 | 0 | 0 | 21 | 115 | 250 | 481 | 440 | 265 | 62 | 4 | 0 | 1638 |
| 1984 | 0 | 0 | 0 | 5 | 146 | 345 | 368 | 426 | 188 | 102 | 5 | 2 | 1587 |
| 1985 | 0 | 5 | 20 | 91 | 146 | 284 | 419 | 382 | 267 | 97 | 28 | 0 | 1739 |
| 1986 | 0 | 0 | 2 | 2 | 153 | 343 | 537 | 367 | 237 | 109 | 15 | 0 | 1765 |
| 1987 | 0 | 0 | 0 | 2 | 168 | 364 | 544 | 461 | 285 | 0 | 7 | 0 | 1831 |
| 1988 | 0 | 0 | 1 | 18 | 118 | 280 | 477 | 498 | 173 | 17 | 10 | 0 | 1592 |
| 1989 | 0 | 9 | 30 | 31 | 106 | 412 | 447 | 399 | 286 | 69 | 10 | 0 | 1799 |
| 1990 | 0 | 8 | 52 | 51 | 98 | 324 | 489 | 407 | 218 | 137 | 8 | 5 | 1797 |
| 1991 | 0 | 0 | 13 | 71 | 269 | 347 | 534 | 456 | 248 | 43 | 18 | 11 | 2010 |
| 1992 | 0 | 0 | 0 | 63 | 51 | 220 | 527 | 336 | 270 | 23 | 9 | 0 | 1499 |
| 1993 | 0 | 0 | 0 | 23 | 136 | 345 | 579 | 438 | 305 | 31 | 18 | 0 | 1875 |
| 1994 | 0 | 0 | 11 | 111 | 94 | 447 | 562 | 400 | 229 | 30 | 20 | 8 | 1912 |
| 1995 | 2 | 0 | 4 | 59 | 143 | 330 | 574 | 475 | 226 | 102 | 6 | 0 | 1921 |
| 1996 | 0 | 0 | 1 | 55 | 116 | 320 | 401 | 312 | 227 | 31 | 10 | 0 | 1473 |
| 1997 | 0 | 7 | 7 | 7 | 59 | 262 | 443 | 360 | 218 | 63 | 0 | 0 | 1426 |
| 1998 | 7 | 0 | 33 | 37 | 143 | 338 | 440 | 440 | 317 | 39 | 0 | 12 | 1806 |
| 1999 | 4 | 0 | 0 | 25 | 112 | 275 | 509 | 458 | 222 | 42 | 4 | 0 | 1651 |
| 2000 | 0 | 0 | 7 | 10 | 196 | 342 | 358 | 368 | 216 | 51 | 0 | 0 | 1548 |
| 2001 | 0 | 0 | 3 | 40 | 108 | 331 | 357 | 435 | 187 | 57 | 15 | 3 | 1536 |
| 2002 | 5 | 1 | 8 | 97 | 134 | 365 | 495 | 454 | 276 | 100 | 9 | 0 | 1944 |
| 2003 | 0 | 0 | 7 | 28 | 103 | 310 | 482 | 485 | 285 | 35 | 39 | 0 | 1774 |
| 2004 | 5 | 0 | 10 | 65 | 283 | 312 | 449 | 348 | 235 | 39 | 6 | 0 | 1752 |
| 2005 | 4 | 0 | 0 | 22 | 37 | 318 | 500 | 478 | 333 | 82 | 13 | 0 | 1787 |
| 2006 | 0 | 0 | 18 | 41 | 100 | 291 | 498 | 514 | 178 | 50 | 3 | 7 | 1700 |
| 2007 | 2 | 0 | 18 | 43 | 112 | 335 | 451 | 464 | 289 | 194 | 1 | 0 | 1909 |
| 2008 | 0 | 6 | 4 | 31 | 85 | 433 | 442 | 404 | 271 | 39 | 1 | 3 | 1719 |
| 2009 | 0 | 0 | 8 | 50 | 176 | 323 | 397 | 457 | 195 | 59 | 1 | 0 | 1666 |

## SNOWFALL (inches) 2009 NORFOLK (KORF)

| YEAR | JUL | AUG | SEP | OCT | NOV | DEC | JAN | FEB | MAR | APR | MAY | JUN | TOTAL |
|------|-----|-----|-----|-----|-----|-----|-----|-----|-----|-----|-----|-----|-------|
| 1980-81 | 0.0 | 0.0 | 0.0 | 0.0 | 0.0 | T | T | 0.0 | 0.3 | 0.0 | 0.0 | 0.0 | 0.3 |
| 1981-82 | 0.0 | 0.0 | 0.0 | 0.0 | 0.0 | 1.8 | 4.2 | 0.1 | T | T | 0.0 | 0.0 | 6.1 |
| 1982-83 | 0.0 | 0.0 | 0.0 | 0.0 | 0.0 | 0.4 | T | 3.0 | T | T | 0.0 | 0.0 | 3.4 |
| 1983-84 | 0.0 | 0.0 | 0.0 | 0.0 | 0.0 | T | T | 5.2 | T | 0.0 | 0.0 | 0.0 | 5.2 |
| 1984-85 | 0.0 | 0.0 | 0.0 | 0.0 | 0.0 | T | 4.3 | 0.0 | T | 0.0 | 0.0 | 0.0 | 4.3 |
| 1985-86 | 0.0 | 0.0 | 0.0 | 0.0 | 0.0 | T | 3.6 | 1.1 | T | T | 0.0 | 0.0 | 4.7 |
| 1986-87 | 0.0 | 0.0 | 0.0 | 0.0 | 0.0 | T | 1.6 | 1.0 | 1.2 | T | 0.0 | 0.0 | 3.8 |
| 1987-88 | 0.0 | 0.0 | 0.0 | 0.0 | 0.3 | T | 4.4 | T | T | 0.0 | 0.0 | 0.0 | 4.7 |
| 1988-89 | 0.0 | 0.0 | 0.0 | 0.0 | 0.0 | T | T | 24.4 | T | 0.5 | 0.0 | 0.0 | 24.9 |
| 1989-90 | 0.0 | 0.0 | 0.0 | 0.0 | 0.0 | 0.5 | 0.0 | 0.0 | T | 0.0 | 0.0 | T | 0.5 |
| 1990-91 | 0.0 | 0.0 | 0.0 | 0.0 | 0.0 | T | 0.0 | T | T | 0.0 | 0.0 | 0.0 | T |
| 1991-92 | 0.0 | T | 0.0 | 0.0 | 0.0 | T | T | 0.0 | T | T | 0.0 | 0.0 | T |
| 1992-93 | 0.0 | 0.0 | 0.0 | 0.0 | 0.0 | 0.0 | 0.3 | 1.9 | 1.7 | 0.0 | 0.0 | 0.0 | 3.9 |
| 1993-94 | 0.0 | 0.0 | 0.0 | 0.0 | 0.0 | 3.9 | 7.4 | 1.2 | 0.3 | 0.0 | T | 0.0 | 12.8 |
| 1994-95 | 0.0 | 0.0 | 0.0 | 0.0 | 0.0 | 0.0 | T | 0.3 | T | 0.0 | 0.0 | 0.0 | 0.3 |
| 1995-96 | 0.0 | 0.0 | 0.0 | 0.0 | 0.0 | T | 6.2 | 11.0 | | | | | |
| 1996-97 | | | | | | | | | | | | | |
| 1997-98 | | | | | | | | | | | | | |
| 1998-99 | | | | | | | | T | | | | | |
| 1999-00 | | | | | | | 9.0 | | | | | | |
| 2000-01 | | | | | | | | | 0.3 | 0.0 | 0.0 | 0.0 | |
| 2001-02 | 0.0 | 0.0 | 0.0 | 0.0 | 0.0 | 0.0 | 7.2 | 0.2 | 0.0 | 0.0 | 0.0 | T | 7.4 |
| 2002-03 | 0.0 | 0.0 | 0.0 | 0.0 | 0.0 | T | 5.1 | T | T | 0.0 | 0.0 | 0.0 | 5.1 |
| 2003-04 | 0.0 | 0.0 | 0.0 | 0.0 | 0.0 | T | 4.0 | 2.2 | T | 0.0 | 0.0 | 0.0 | 6.2 |
| 2004-05 | 0.0 | 0.0 | 0.0 | 0.0 | 0.0 | 5.5 | 1.3 | 0.5 | 0.9 | 0.0 | 0.0 | 0.0 | 8.2 |
| 2005-06 | 0.0 | 0.0 | 0.0 | 0.0 | 0.0 | T | 0.7 | 0.1 | T | 0.0 | 0.0 | 0.0 | 0.8 |
| 2006-07 | 0.0 | 0.0 | 0.0 | 0.0 | 0.0 | 0.0 | T | T | 0.0 | 0.1 | 0.0 | 0.0 | 0.1 |
| 2007-08 | 0.0 | 0.0 | 0.0 | 0.0 | 0.0 | 0.0 | T | 0.2 | 0.0 | 0.0 | 0.0 | 0.0 | 0.2 |
| 2008-09 | 0.0 | 0.0 | 0.0 | 0.0 | 0.0 | 0.0 | T | T | 0.5 | T | 0.0 | 0.0 | 0.5 |
| 2009- | 0.0 | 0.0 | 0.0 | 0.0 | 0.0 | 0.2 | | | | | | | |
| POR= 62 YRS | 0.0 | T | 0.0 | 0.0 | 0.0 | 0.8 | 2.7 | 2.6 | 0.9 | T | T | T | 7.0 |

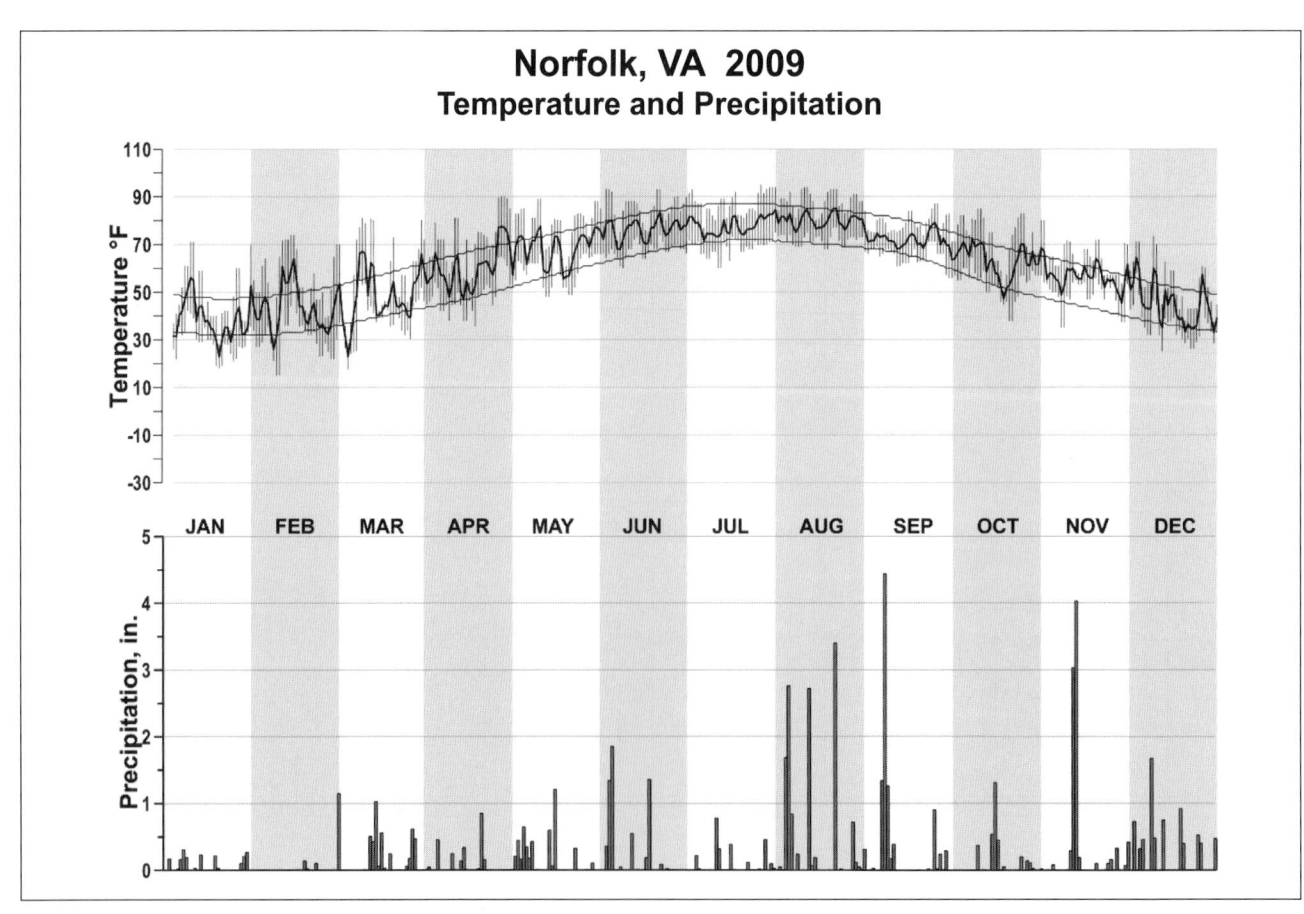

# Norfolk, VA  2009
## Temperature and Precipitation

# 2009
# RICHMOND
# VIRGINIA (KRIC)

Richmond is located in east-central Virginia at the head of navigation on the James River and along a line separating the Coastal Plains (Tidewater Virginia) from the Piedmont. The Blue Ridge Mountains lie about 90 miles to the west and the Chesapeake Bay 60 miles to the east. Elevations range from a few feet above sea level along the river to a little over 300 feet in parts of the western section of the city.

The climate might be classified as modified continental. Summers are warm and humid and winters generally mild. The mountains to the west act as a partial barrier to outbreaks of cold, continental air in winter. The cold winter air is delayed long enough to be modified, then further warmed as it subsides in its approach to Richmond. The open waters of the Chesapeake Bay and Atlantic Ocean contribute to the humid summers and mild winters. The coldest weather normally occurs in late December and January, when low temperatures usually average in the upper 20s, and the high temperatures in the upper 40s. Temperatures seldom lower to zero, but there have been several occurrences of below zero temperatures. Summertime high temperatures above 100 degrees are not uncommon, but do not occur every year.

Precipitation is rather uniformly distributed throughout the year. However, dry periods lasting several weeks do occur, especially in autumn when long periods of pleasant, mild weather are most common. There is considerable variability in total monthly amounts from year to year. Snow usually remains on the ground only one or two days at a time. Ice storms (freezing rain or glaze) are not uncommon, but they are seldom severe enough to do any considerable damage. A notable exception was the spectacular glaze storm of January 27-28, 1943, when nearly 1 inch of ice accumulation caused heavy damage to trees and overhead transmission lines.

The James River reaches tidewater at Richmond where flooding may occur in every month of the year, most frequently in March and least in July. Hurricanes and tropical storms have been responsible for most of the flooding during the summer and early fall months. Hurricanes passing near Richmond have produced record rainfalls. In 1955, three hurricanes brought record rainfall to Richmond within a six-week period. The most noteworthy of these were Hurricanes Connie and Diane that brought heavy rains five days apart.

Damaging storms occur mainly from snow and freezing rain in winter and from hurricanes, tornadoes, and severe thunderstorms in other seasons. Damage may be from wind, flooding, or rain, or from any combination of these. Tornadoes are infrequent but some notable occurrences have been observed within the Richmond area.

Based on the 1951-1980 period, the average first occurrence of 32 degrees Fahrenheit in the fall is October 26 and the average last occurrence in the spring is April 10.

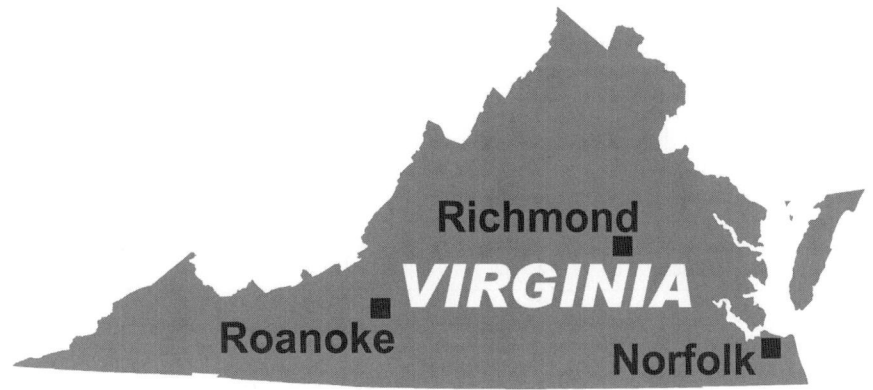

# NORMALS, MEANS, AND EXTREMES
## RICHMOND (KRIC)

| LATITUDE: 37° 30'N | LONGITUDE: -77° 19'W | ELEVATION (FT): GRND: 163  BARO: 167 | TIME ZONE: EASTERN  (UTC -5) | WBAN: 13740 |
|---|---|---|---|---|

| | ELEMENT | POR | JAN | FEB | MAR | APR | MAY | JUN | JUL | AUG | SEP | OCT | NOV | DEC | YEAR |
|---|---|---|---|---|---|---|---|---|---|---|---|---|---|---|---|
| **TEMPERATURE °F** | NORMAL DAILY MAXIMUM | 30 | 45.3 | 49.3 | 58.4 | 68.9 | 76.2 | 83.6 | 87.5 | 85.7 | 79.7 | 69.3 | 59.7 | 49.7 | 67.8 |
| | MEAN DAILY MAXIMUM | 89 | 47.4 | 50.4 | 59.0 | 69.4 | 77.6 | 85.0 | 88.3 | 86.7 | 80.9 | 70.6 | 60.5 | 50.2 | 68.8 |
| | HIGHEST DAILY MAXIMUM | 80 | 81 | 83 | 93 | 96 | 100 | 104 | 105 | 105 | 104 | 103 | 86 | 81 | 105 |
| | YEAR OF OCCURRENCE | | 2002 | 1932 | 1938 | 1990 | 1941 | 1952 | 1977 | 2007 | 1954 | 1941 | 1993 | 1998 | JUL 1977 |
| | MEAN OF EXTREME MAXS. | 89 | 69.6 | 71.3 | 80.3 | 88.0 | 91.0 | 96.0 | 97.5 | 96.1 | 93.2 | 85.5 | 77.8 | 70.1 | 84.7 |
| | NORMAL DAILY MINIMUM | 30 | 27.6 | 29.7 | 37.0 | 45.3 | 54.6 | 63.3 | 68.3 | 66.8 | 59.9 | 47.2 | 38.4 | 31.1 | 47.4 |
| | MEAN DAILY MINIMUM | 89 | 28.5 | 29.9 | 36.8 | 45.5 | 54.9 | 63.5 | 68.1 | 66.9 | 60.2 | 48.0 | 38.7 | 30.9 | 47.7 |
| | LOWEST DAILY MINIMUM | 80 | -12 | -10 | 10 | 23 | 31 | 40 | 51 | 46 | 35 | 21 | 10 | -1 | -12 |
| | YEAR OF OCCURRENCE | | 1940 | 1936 | 2009 | 1985 | 1956 | 1967 | 1965 | 1934 | 1974 | 1962 | 1933 | 1942 | JAN 1940 |
| | MEAN OF EXTREME MINS. | 89 | 10.4 | 14.4 | 21.4 | 30.8 | 41.0 | 51.2 | 58.1 | 56.0 | 45.8 | 32.6 | 23.1 | 14.8 | 33.3 |
| | NORMAL DRY BULB | 30 | 36.4 | 39.5 | 47.7 | 57.1 | 65.4 | 73.5 | 77.9 | 76.3 | 69.8 | 58.3 | 49.0 | 40.4 | 57.6 |
| | MEAN DRY BULB | 89 | 38.0 | 40.2 | 47.9 | 57.5 | 66.2 | 74.4 | 78.2 | 76.8 | 70.6 | 59.3 | 49.6 | 40.6 | 58.3 |
| | MEAN WET BULB | 26 | 32.9 | 34.5 | 40.8 | 49.6 | 58.4 | 66.7 | 70.7 | 69.6 | 63.6 | 53.2 | 44.2 | 36.1 | 51.7 |
| | MEAN DEW POINT | 26 | 28.3 | 29.7 | 35.8 | 44.8 | 55.0 | 64.0 | 68.5 | 67.6 | 61.2 | 50.1 | 40.1 | 31.4 | 48.0 |
| | NORMAL NO. DAYS WITH: | | | | | | | | | | | | | | |
| | MAXIMUM >= 90 | 30 | 0.0 | 0.0 | * | 0.8 | 1.9 | 8.2 | 15.0 | 10.9 | 3.6 | 0.3 | 0.0 | 0.0 | 40.7 |
| | MAXIMUM <= 32 | 30 | 3.5 | 1.9 | 0.2 | 0.0 | 0.0 | 0.0 | 0.0 | 0.0 | 0.0 | 0.0 | 0.0 | 1.6 | 7.2 |
| | MINIMUM <= 32 | 30 | 21.6 | 18.3 | 10.3 | 1.9 | * | 0.0 | 0.0 | 0.0 | 0.0 | 1.3 | 9.2 | 18.3 | 80.9 |
| | MINIMUM <= 0 | 30 | 0.2 | 0.1 | 0.0 | 0.0 | 0.0 | 0.0 | 0.0 | 0.0 | 0.0 | 0.0 | 0.0 | 0.0 | 0.3 |
| **H/C** | NORMAL HEATING DEG. DAYS | 30 | 873 | 705 | 528 | 254 | 80 | 8 | 0 | 1 | 27 | 225 | 470 | 748 | 3919 |
| | NORMAL COOLING DEG. DAYS | 30 | 0 | 1 | 8 | 33 | 107 | 277 | 415 | 367 | 187 | 33 | 6 | 1 | 1435 |
| **RH** | NORMAL (PERCENT) | 30 | 69 | 66 | 63 | 62 | 70 | 72 | 74 | 76 | 76 | 74 | 70 | 69 | 70 |
| | HOUR 01 LST | 30 | 75 | 73 | 72 | 73 | 83 | 86 | 88 | 90 | 89 | 87 | 80 | 76 | 81 |
| | HOUR 07 LST | 30 | 80 | 79 | 78 | 76 | 81 | 83 | 85 | 89 | 90 | 89 | 84 | 81 | 83 |
| | HOUR 13 LST | 30 | 57 | 53 | 50 | 46 | 53 | 54 | 56 | 57 | 57 | 53 | 52 | 56 | 54 |
| | HOUR 19 LST | 30 | 67 | 62 | 57 | 54 | 63 | 66 | 68 | 72 | 75 | 76 | 69 | 68 | 66 |
| **S** | PERCENT POSSIBLE SUNSHINE | 46 | 54 | 58 | 62 | 66 | 65 | 69 | 68 | 66 | 65 | 63 | 59 | 54 | 62 |
| **W/O** | MEAN NO. DAYS WITH: | | | | | | | | | | | | | | |
| | HEAVY FOG(VISBY <= 1/4 MI) | 46 | 2.5 | 2.0 | 1.5 | 1.4 | 1.8 | 1.3 | 1.3 | 1.9 | 2.1 | 2.6 | 2.2 | 2.4 | 23.0 |
| | THUNDERSTORMS | 89 | 0.3 | 0.3 | 1.4 | 2.3 | 5.0 | 6.3 | 8.1 | 6.1 | 2.7 | 0.8 | 0.6 | 0.2 | 34.1 |
| **CLOUDNESS** | MEAN: | | | | | | | | | | | | | | |
| | SUNRISE-SUNSET (OKTAS) | | | | | | | | | | | | | | |
| | MIDNIGHT-MIDNIGHT (OKTAS) | | | | | | | | | | | | | | |
| | MEAN NO. DAYS WITH: | | | | | | | | | | | | | | |
| | CLEAR | 1 | 2.0 | 1.0 | 9.0 | | 9.0 | 10.0 | | | | | | | |
| | PARTLY CLOUDY | | | 1.0 | 5.0 | | 4.0 | 8.0 | | | | | | | |
| | CLOUDY | 1 | 7.0 | 4.0 | 10.0 | | 8.0 | 1.0 | | | | | | | |
| **PR** | MEAN STATION PRESSURE(IN) | 26 | 29.92 | 29.91 | 29.87 | 29.80 | 29.82 | 29.81 | 29.82 | 29.85 | 29.88 | 29.92 | 29.93 | 29.94 | 29.87 |
| | MEAN SEA-LEVEL PRES. (IN) | 26 | 30.12 | 30.10 | 30.07 | 29.99 | 30.01 | 29.99 | 30.01 | 30.04 | 30.07 | 30.11 | 30.12 | 30.14 | 30.06 |
| **WINDS** | MEAN SPEED (MPH) | 26 | 8.4 | 8.7 | 9.3 | 9.1 | 8.1 | 7.5 | 7.1 | 6.5 | 6.9 | 6.9 | 7.6 | 7.8 | 7.8 |
| | PREVAIL.DIR(TENS OF DEGS) | 34 | 01 | 36 | 21 | 20 | 22 | 21 | 20 | 20 | 36 | 01 | 36 | 21 | 20 |
| | MAXIMUM 2-MINUTE: | | | | | | | | | | | | | | |
| | SPEED (MPH) | 14 | 38 | 44 | 48 | 46 | 43 | 45 | 47 | 44 | 46 | 37 | 36 | 40 | 48 |
| | DIR. (TENS OF DEGS) | | 31 | 28 | 24 | 33 | 27 | 26 | 31 | 36 | 09 | 10 | 30 | 15 | 24 |
| | YEAR OF OCCURRENCE | | 2000 | 2008 | 2008 | 1999 | 2009 | 2000 | 2007 | 1996 | 2003 | 1996 | 2003 | 1996 | MAR 2008 |
| | MAXIMUM 3-SECOND | | | | | | | | | | | | | | |
| | SPEED (MPH) | 14 | 48 | 63 | 59 | 56 | 63 | 55 | 58 | 64 | 72 | 46 | 46 | 49 | 72 |
| | DIR. (TENS OF DEGS) | | 31 | 29 | 23 | 25 | 26 | 27 | 21 | 28 | 10 | 10 | 27 | 33 | 10 |
| | YEAR OF OCCURRENCE | | 2000 | 2008 | 2008 | 1998 | 2009 | 2000 | 2008 | 2009 | 2003 | 1996 | 2003 | 2008 | SEP 2003 |
| **PRECIPITATION** | NORMAL (IN) | 30 | 3.55 | 2.98 | 4.09 | 3.18 | 3.96 | 3.54 | 4.67 | 4.18 | 3.98 | 3.60 | 3.06 | 3.12 | 43.91 |
| | MAXIMUM MONTHLY (IN) | 72 | 7.97 | 5.97 | 8.65 | 8.32 | 8.87 | 9.93 | 18.87 | 16.30 | 16.60 | 9.39 | 9.60 | 8.16 | 18.87 |
| | YEAR OF OCCURRENCE | | 1978 | 1979 | 1984 | 2008 | 1972 | 2004 | 1945 | 2004 | 1999 | 1971 | 2009 | 2009 | JUL 1945 |
| | MINIMUM MONTHLY (IN) | 72 | 0.64 | 0.48 | 0.20 | 0.64 | 0.87 | 0.38 | 0.51 | 0.52 | .08 | 0.01 | 0.17 | 0.40 | 0.01 |
| | YEAR OF OCCURRENCE | | 1981 | 1978 | 2006 | 1963 | 1965 | 1980 | 1983 | 1943 | 2005 | 2000 | 2001 | 1980 | OCT 2000 |
| | MAXIMUM IN 24 HOURS (IN) | 72 | 3.31 | 2.67 | 3.43 | 3.54 | 3.40 | 4.61 | 5.73 | 8.79 | 6.52 | 6.50 | 4.07 | 3.16 | 8.79 |
| | YEAR OF OCCURRENCE | | 1962 | 1979 | 1992 | 2008 | 2003 | 1963 | 1969 | 1955 | 1999 | 1961 | 1956 | 1958 | AUG 1955 |
| | NORMAL NO. DAYS WITH: | | | | | | | | | | | | | | |
| | PRECIPITATION >= 0.01 | 30 | 10.7 | 9.5 | 10.5 | 9.2 | 11.0 | 9.6 | 11.1 | 9.0 | 8.7 | 7.5 | 8.1 | 9.7 | 114.6 |
| | PRECIPITATION >= 1.00 | 30 | 1.0 | 0.7 | 1.0 | 0.8 | 1.0 | 0.8 | 1.3 | 1.2 | 1.2 | 1.2 | 0.8 | 0.5 | 11.5 |
| **SNOWFALL** | NORMAL (IN) | 30 | 4.3 | 4.8 | 1.4 | 0.* | 0.0 | 0.0 | 0.0 | 0.0 | 0.0 | 0.* | 0.3 | 1.6 | 12.4 |
| | MAXIMUM MONTHLY (IN) | 70 | 28.5 | 21.4 | 19.7 | 2.0 | T | 0.0 | T | 0.0 | T | T | 7.3 | 12.5 | 28.5 |
| | YEAR OF OCCURRENCE | | 1940 | 1983 | 1960 | 1940 | 2007 | | 2003 | | 2008 | 1979 | 1953 | 1958 | JAN 1940 |
| | MAXIMUM IN 24 HOURS (IN) | 70 | 21.6 | 16.8 | 12.1 | 2.0 | T | 0.0 | T | 0.0 | T | T | 7.3 | 7.5 | 21.6 |
| | YEAR OF OCCURRENCE' | | 1940 | 1983 | 1962 | 1940 | 1994 | | 2003 | | 2008 | 1979 | 1953 | 1966 | JAN 1940 |
| | MAXIMUM SNOW DEPTH (IN) | 79 | 18 | 20 | 13 | 1 | 0 | 0 | 0 | 0 | 0 | 0 | 6 | 9 | 20 |
| | YEAR OF OCCURRENCE | | 1922 | 1922 | 1980 | 1964 | | | | | | | 1938 | 1958 | FEB 1922 |
| | NORMAL NO. DAYS WITH: | | | | | | | | | | | | | | |
| | SNOWFALL >= 1.0 | 30 | 1.1 | 1.3 | 0.4 | 0.0 | 0.0 | 0.0 | 0.0 | 0.0 | 0.0 | 0.0 | 0.1 | 0.6 | 3.5 |

## PRECIPITATION (inches) 2009 RICHMOND (KRIC)

| YEAR | JAN | FEB | MAR | APR | MAY | JUN | JUL | AUG | SEP | OCT | NOV | DEC | ANNUAL |
|------|-----|-----|-----|-----|-----|-----|-----|-----|-----|-----|-----|-----|--------|
| 1980 | 6.05 | 1.01 | 5.49 | 4.28 | 4.68 | 0.38 | 5.18 | 2.15 | 2.37 | 6.96 | 2.18 | 0.40 | 41.13 |
| 1981 | 0.64 | 2.76 | 1.52 | 2.96 | 6.62 | 3.69 | 4.01 | 2.89 | 2.70 | 2.36 | 0.68 | 5.04 | 35.87 |
| 1982 | 2.76 | 4.44 | 3.74 | 2.97 | 3.48 | 3.97 | 9.21 | 4.39 | 2.55 | 2.90 | 2.70 | 3.37 | 46.48 |
| 1983 | 1.59 | 3.95 | 6.04 | 5.21 | 2.50 | 5.46 | 0.51 | 0.97 | 3.05 | 4.02 | 5.63 | 4.50 | 43.43 |
| 1984 | 3.98 | 3.97 | 8.65 | 5.92 | 4.52 | 2.01 | 3.55 | 4.58 | 1.86 | 2.14 | 3.34 | 1.52 | 46.04 |
| 1985 | 3.54 | 3.20 | 1.80 | 0.65 | 2.36 | 4.01 | 5.31 | 10.58 | 4.97 | 5.09 | 6.99 | 0.58 | 49.08 |
| 1986 | 2.69 | 2.67 | 1.16 | 1.16 | 3.15 | 1.30 | 7.01 | 6.75 | 0.63 | 2.43 | 2.46 | 5.15 | 36.56 |
| 1987 | 5.53 | 2.57 | 1.65 | 7.31 | 2.94 | 6.29 | 1.20 | 1.11 | 4.43 | 1.25 | 3.13 | 2.86 | 40.27 |
| 1988 | 2.53 | 3.08 | 1.98 | 2.55 | 4.81 | 2.25 | 7.50 | 2.95 | 1.74 | 2.74 | 4.34 | 0.79 | 37.26 |
| 1989 | 1.88 | 4.34 | 5.00 | 4.27 | 5.02 | 5.85 | 4.00 | 4.89 | 5.33 | 3.54 | 3.00 | 2.62 | 49.74 |
| 1990 | 2.84 | 2.38 | 2.54 | 2.81 | 6.85 | 0.97 | 6.74 | 5.76 | 1.92 | 3.90 | 1.70 | 3.52 | 41.93 |
| 1991 | 3.62 | 1.09 | 5.87 | 0.87 | 0.91 | 6.24 | 3.47 | 3.32 | 2.69 | 2.50 | 0.67 | 4.53 | 35.78 |
| 1992 | 1.57 | 2.89 | 5.87 | 2.21 | 4.95 | 2.28 | 5.68 | 6.40 | 2.35 | 1.94 | 2.62 | 2.79 | 41.55 |
| 1993 | 4.48 | 2.88 | 7.24 | 3.23 | 4.66 | 1.75 | 1.91 | 3.89 | 2.97 | 2.23 | 3.24 | 3.77 | 42.25 |
| 1994 | 3.10 | 4.38 | 7.92 | 2.70 | 2.49 | 1.73 | 7.46 | 2.54 | 3.99 | 2.56 | 3.99 | 0.95 | 43.81 |
| 1995 | 3.15 | 1.14 | 3.00 | 1.98 | 4.33 | 1.85 | 2.89 | 2.59 | 3.82 | 5.11 | 2.87 | 1.71 | 34.44 |
| 1996 | 4.65 | 2.97 | 2.71 | 2.88 | 3.18 | 4.35 | 6.51 | 4.40 | 6.87 | 7.18 | 3.52 | 4.91 | 54.13 |
| 1997 | 1.93 | 3.71 | 2.96 | 3.94 | 1.36 | 2.21 | 4.85 | 1.41 | 0.82 | 3.25 | 5.33 | 2.36 | 34.13 |
| 1998 | 6.85 | 5.76 | 6.72 | 4.32 | 3.72 | 4.41 | 2.37 | 1.89 | 3.94 | 0.47 | 1.30 | 5.00 | 46.75 |
| 1999 | 4.70 | 1.47 | 4.04 | 2.60 | 2.75 | 6.29 | 2.77 | 2.00 | 16.60 | 2.25 | 1.01 | 1.72 | 48.20 |
| 2000 | 3.96 | 1.60 | 3.67 | 4.78 | 3.03 | 6.07 | 4.05 | 8.28 | 3.60 | 0.01 | 1.72 | 2.38 | 43.15 |
| 2001 | 2.06 | 2.55 | 3.77 | 2.14 | 2.03 | 6.53 | 2.73 | 5.08 | 2.14 | 0.65 | 0.17 | 1.67 | 31.52 |
| 2002 | 3.58 | 0.82 | 4.48 | 2.33 | 3.49 | 1.56 | 1.63 | 3.18 | 2.88 | 6.09 | 4.28 | 3.45 | 37.77 |
| 2003 | 2.18 | 4.21 | 5.92 | 4.38 | 8.59 | 3.87 | 9.26 | 4.66 | 10.12 | 2.43 | 3.39 | 4.28 | 63.29 |
| 2004 | 1.55 | 1.87 | 2.22 | 3.42 | 3.06 | 9.93 | 6.42 | 16.30 | 6.14 | 1.95 | 3.27 | 2.37 | 58.50 |
| 2005 | 3.42 | 1.39 | 3.99 | 2.05 | 4.22 | 1.19 | 9.28 | 2.56 | 0.08 | 3.74 | 3.81 | 5.81 | 41.54 |
| 2006 | 2.89 | 1.47 | 0.20 | 2.18 | 3.24 | 7.85 | 4.57 | 5.99 | 9.52 | 6.12 | 6.67 | 1.42 | 52.12 |
| 2007 | 3.46 | 2.06 | 2.66 | 3.62 | 3.69 | 5.22 | 1.69 | 6.81 | 1.11 | 3.54 | 0.80 | 3.24 | 37.90 |
| 2008 | 0.96 | 3.41 | 3.50 | 8.32 | 5.10 | 3.64 | 4.05 | 5.73 | 5.94 | 1.32 | 3.51 | 4.07 | 49.55 |
| 2009 | 1.49 | 0.74 | 4.26 | 2.56 | 3.71 | 4.32 | 3.39 | 4.04 | 2.46 | 3.59 | 9.60 | 8.16 | 48.32 |
| POR= 89 YRS | 3.28 | 2.87 | 3.65 | 3.16 | 3.67 | 3.82 | 5.05 | 4.87 | 3.63 | 3.11 | 2.99 | 3.18 | 43.28 |

WBAN : 13740

## AVERAGE TEMPERATURE (°F) 2009 RICHMOND (KRIC)

| YEAR | JAN | FEB | MAR | APR | MAY | JUN | JUL | AUG | SEP | OCT | NOV | DEC | ANNUAL |
|------|-----|-----|-----|-----|-----|-----|-----|-----|-----|-----|-----|-----|--------|
| 1980 | 38.8 | 36.0 | 47.4 | 61.1 | 68.3 | 72.8 | 80.0 | 80.7 | 74.7 | 56.9 | 46.2 | 38.6 | 58.5 |
| 1981 | 31.2 | 42.2 | 44.6 | 60.6 | 64.1 | 77.9 | 79.6 | 75.1 | 69.4 | 56.4 | 49.1 | 38.0 | 57.4 |
| 1982 | 31.6 | 41.7 | 49.1 | 55.9 | 70.4 | 73.4 | 78.6 | 75.0 | 69.8 | 59.2 | 51.9 | 46.1 | 58.6 |
| 1983 | 37.8 | 39.1 | 50.4 | 56.1 | 66.1 | 75.6 | 79.4 | 77.7 | 68.8 | 58.1 | 49.0 | 36.2 | 57.9 |
| 1984 | 32.6 | 44.5 | 43.6 | 55.8 | 65.4 | 77.7 | 76.0 | 77.0 | 67.5 | 66.1 | 46.6 | 47.7 | 58.4 |
| 1985 | 32.6 | 40.2 | 49.7 | 62.0 | 68.0 | 74.3 | 79.0 | 77.5 | 70.8 | 62.6 | 56.6 | 37.8 | 59.3 |
| 1986 | 36.2 | 39.3 | 50.0 | 59.2 | 66.9 | 76.1 | 80.9 | 74.2 | 70.8 | 61.8 | 49.1 | 40.9 | 58.8 |
| 1987 | 34.7 | 37.0 | 47.1 | 54.3 | 67.3 | 75.8 | 81.3 | 78.5 | 72.3 | 52.9 | 51.3 | 43.0 | 58.0 |
| 1988 | 32.3 | 39.1 | 47.9 | 56.0 | 65.8 | 72.8 | 79.9 | 79.8 | 68.4 | 53.6 | 50.6 | 39.2 | 57.1 |
| 1989 | 42.3 | 39.6 | 47.9 | 55.8 | 64.1 | 76.1 | 77.7 | 75.3 | 70.8 | 60.2 | 49.3 | 31.3 | 57.5 |
| 1990 | 46.3 | 48.0 | 52.1 | 57.9 | 65.8 | 75.0 | 79.9 | 76.3 | 69.5 | 63.1 | 52.6 | 46.3 | 61.1 |
| 1991 | 40.2 | 44.0 | 50.8 | 60.2 | 72.1 | 75.3 | 80.7 | 78.6 | 71.6 | 59.9 | 49.6 | 44.3 | 60.6 |
| 1992 | 40.7 | 42.7 | 47.2 | 57.8 | 61.7 | 70.8 | 79.5 | 73.7 | 70.3 | 56.3 | 50.6 | 40.9 | 57.7 |
| 1993 | 41.4 | 37.8 | 45.9 | 55.9 | 67.7 | 75.0 | 82.5 | 78.5 | 72.5 | 58.6 | 50.9 | 39.0 | 58.8 |
| 1994 | 33.4 | 40.2 | 48.9 | 63.2 | 63.2 | 78.0 | 81.1 | 75.4 | 69.2 | 57.6 | 52.6 | 45.6 | 59.0 |
| 1995 | 40.7 | 38.7 | 50.2 | 58.1 | 65.8 | 74.1 | 80.8 | 78.9 | 69.1 | 61.1 | 44.2 | 35.6 | 58.1 |
| 1996 | 34.1 | 37.8 | 43.2 | 57.9 | 64.8 | 75.1 | 76.4 | 74.3 | 70.1 | 58.9 | 43.2 | 43.8 | 56.6 |
| 1997 | 37.7 | 43.7 | 49.6 | 53.6 | 62.9 | 71.4 | 77.8 | 75.8 | 70.3 | 58.9 | 46.5 | 39.9 | 57.3 |
| 1998 | 43.2 | 44.2 | 48.5 | 58.0 | 67.4 | 74.0 | 78.3 | 78.3 | 74.2 | 59.7 | 49.5 | 44.2 | 60.0 |
| 1999 | 42.0 | 42.2 | 45.2 | 57.7 | 65.5 | 72.3 | 79.6 | 77.9 | 68.5 | 57.2 | 52.9 | 42.4 | 58.6 |
| 2000 | 36.3 | 43.4 | 52.0 | 56.4 | 68.0 | 75.4 | 74.7 | 74.9 | 67.5 | 59.7 | 46.1 | 33.0 | 57.3 |
| 2001 | 37.1 | 42.3 | 45.2 | 58.2 | 65.2 | 75.0 | 74.9 | 78.2 | 67.8 | 59.1 | 54.6 | 46.0 | 58.6 |
| 2002 | 42.1 | 42.3 | 49.1 | 60.5 | 65.4 | 76.0 | 80.2 | 79.3 | 71.8 | 60.0 | 47.7 | 38.5 | 59.4 |
| 2003 | 33.7 | 36.7 | 49.6 | 56.9 | 63.5 | 72.7 | 77.8 | 78.3 | 70.0 | 58.1 | 53.9 | 39.6 | 57.6 |
| 2004 | 33.4 | 38.7 | 50.0 | 59.1 | 73.1 | 74.5 | 78.7 | 75.7 | 71.6 | 60.2 | 52.6 | 42.4 | 59.2 |
| 2005 | 40.1 | 42.1 | 45.2 | 58.5 | 63.4 | 76.2 | 81.7 | 80.6 | 75.6 | 61.9 | 53.1 | 38.6 | 59.8 |
| 2006 | 45.1 | 40.3 | 49.8 | 60.2 | 65.4 | 74.9 | 81.5 | 80.7 | 69.9 | 59.8 | 53.7 | 47.3 | 60.7 |
| 2007 | 43.5 | 35.5 | 51.7 | 57.1 | 67.5 | 75.8 | 78.7 | 80.4 | 72.6 | 66.9 | 49.5 | 44.0 | 60.3 |
| 2008 | 40.0 | 43.5 | 50.9 | 59.5 | 65.5 | 79.0 | 79.1 | 77.2 | 71.9 | 58.8 | 48.9 | 44.5 | 59.9 |
| 2009 | 35.3 | 42.2 | 46.8 | 58.9 | 68.2 | 75.5 | 77.0 | 79.9 | 70.0 | 59.0 | 53.0 | 39.4 | 58.8 |
| POR= 89 YRS | 38.0 | 40.2 | 47.9 | 57.5 | 66.2 | 74.4 | 78.2 | 76.8 | 70.6 | 59.3 | 49.6 | 40.6 | 58.3 |

## HEATING DEGREE DAYS (base 65°F) 2009 RICHMOND (KRIC)

| YEAR | JUL | AUG | SEP | OCT | NOV | DEC | JAN | FEB | MAR | APR | MAY | JUN | TOTAL |
|------|-----|-----|-----|-----|-----|-----|-----|-----|-----|-----|-----|-----|-------|
| 1980-81 | 0 | 0 | 14 | 267 | 557 | 813 | 1042 | 633 | 626 | 171 | 107 | 0 | 4230 |
| 1981-82 | 0 | 1 | 29 | 273 | 473 | 834 | 1029 | 645 | 486 | 280 | 6 | 1 | 4057 |
| 1982-83 | 0 | 6 | 10 | 213 | 399 | 585 | 836 | 718 | 445 | 282 | 69 | 2 | 3565 |
| 1983-84 | 0 | 1 | 86 | 236 | 475 | 887 | 994 | 589 | 657 | 282 | 93 | 3 | 4303 |
| 1984-85 | 0 | 0 | 73 | 57 | 546 | 531 | 997 | 692 | 484 | 177 | 35 | 5 | 3597 |
| 1985-86 | 0 | 0 | 31 | 114 | 257 | 838 | 886 | 713 | 465 | 187 | 78 | 3 | 3572 |
| 1986-87 | 0 | 16 | 24 | 172 | 476 | 741 | 931 | 777 | 550 | 317 | 57 | 0 | 4061 |
| 1987-88 | 0 | 0 | 5 | 370 | 409 | 677 | 1008 | 746 | 527 | 279 | 79 | 32 | 4132 |
| 1988-89 | 0 | 0 | 27 | 361 | 425 | 794 | 696 | 709 | 546 | 293 | 108 | 0 | 3959 |
| 1989-90 | 0 | 3 | 38 | 181 | 468 | 1036 | 574 | 472 | 436 | 258 | 50 | 3 | 3519 |
| 1990-91 | 0 | 0 | 33 | 146 | 365 | 574 | 762 | 582 | 443 | 193 | 27 | 1 | 3126 |
| 1991-92 | 0 | 0 | 25 | 190 | 458 | 637 | 749 | 640 | 545 | 261 | 142 | 10 | 3657 |
| 1992-93 | 0 | 0 | 37 | 275 | 427 | 739 | 724 | 757 | 585 | 280 | 24 | 0 | 3848 |
| 1993-94 | 0 | 0 | 26 | 213 | 436 | 799 | 973 | 687 | 497 | 130 | 120 | 0 | 3881 |
| 1994-95 | 0 | 0 | 11 | 232 | 365 | 595 | 750 | 731 | 453 | 236 | 72 | 1 | 3446 |
| 1995-96 | 0 | 0 | 28 | 166 | 620 | 905 | 952 | 782 | 669 | 247 | 122 | 8 | 4499 |
| 1996-97 | 0 | 0 | 10 | 193 | 652 | 654 | 840 | 589 | 472 | 337 | 117 | 45 | 3909 |
| 1997-98 | 0 | 0 | 19 | 247 | 547 | 771 | 667 | 575 | 536 | 232 | 50 | 7 | 3651 |
| 1998-99 | 0 | 0 | 10 | 173 | 457 | 652 | 710 | 631 | 606 | 237 | 60 | 7 | 3543 |
| 1999-00 | 0 | 0 | 29 | 249 | 362 | 693 | 882 | 620 | 404 | 257 | 45 | 2 | 3543 |
| 2000-01 | 0 | 0 | 61 | 190 | 561 | 983 | 858 | 631 | 605 | 244 | 67 | 0 | 4200 |
| 2001-02 | 0 | 0 | 57 | 213 | 312 | 582 | 704 | 616 | 490 | 210 | 101 | 1 | 3286 |
| 2002-03 | 0 | 0 | 0 | 214 | 512 | 814 | 962 | 785 | 470 | 257 | 98 | 7 | 4119 |
| 2003-04 | 0 | 0 | 13 | 216 | 341 | 779 | 974 | 756 | 461 | 238 | 25 | 1 | 3804 |
| 2004-05 | 0 | 0 | 6 | 178 | 370 | 695 | 770 | 634 | 608 | 214 | 102 | 3 | 3580 |
| 2005-06 | 0 | 0 | 2 | 150 | 359 | 810 | 612 | 686 | 481 | 173 | 84 | 3 | 3360 |
| 2006-07 | 0 | 0 | 16 | 197 | 339 | 543 | 662 | 821 | 421 | 264 | 64 | 0 | 3327 |
| 2007-08 | 0 | 0 | 12 | 80 | 458 | 643 | 767 | 620 | 430 | 193 | 63 | 0 | 3266 |
| 2008-09 | 0 | 0 | 9 | 218 | 476 | 631 | 913 | 633 | 567 | 226 | 56 | 0 | 3729 |
| 2009- | 0 | 0 | 5 | 204 | 351 | 787 | | | | | | | |

WBAN : 13740

## COOLING DEGREE DAYS (base 65°F) 2009 RICHMOND (KRIC)

| YEAR | JAN | FEB | MAR | APR | MAY | JUN | JUL | AUG | SEP | OCT | NOV | DEC | TOTAL |
|------|-----|-----|-----|-----|-----|-----|-----|-----|-----|-----|-----|-----|-------|
| 1980 | 0 | 0 | 1 | 25 | 157 | 243 | 472 | 494 | 313 | 23 | 1 | 0 | 1729 |
| 1981 | 0 | 0 | 1 | 45 | 89 | 395 | 458 | 319 | 169 | 16 | 0 | 0 | 1492 |
| 1982 | 0 | 0 | 0 | 13 | 181 | 259 | 428 | 323 | 157 | 43 | 13 | 7 | 1424 |
| 1983 | 0 | 0 | 0 | 23 | 108 | 325 | 452 | 405 | 207 | 27 | 0 | 0 | 1547 |
| 1984 | 0 | 0 | 0 | 10 | 114 | 392 | 346 | 381 | 154 | 100 | 0 | 2 | 1499 |
| 1985 | 0 | 4 | 20 | 94 | 139 | 290 | 441 | 392 | 213 | 51 | 10 | 0 | 1654 |
| 1986 | 0 | 0 | 8 | 19 | 142 | 344 | 498 | 308 | 205 | 79 | 6 | 0 | 1609 |
| 1987 | 0 | 0 | 0 | 2 | 136 | 329 | 513 | 427 | 227 | 0 | 3 | 0 | 1637 |
| 1988 | 0 | 0 | 3 | 16 | 108 | 269 | 466 | 465 | 137 | 12 | 0 | 0 | 1476 |
| 1989 | 0 | 3 | 24 | 27 | 88 | 341 | 403 | 331 | 218 | 42 | 4 | 0 | 1481 |
| 1990 | 0 | 1 | 43 | 51 | 81 | 312 | 470 | 356 | 177 | 96 | 0 | 2 | 1589 |
| 1991 | 0 | 0 | 10 | 53 | 254 | 316 | 494 | 428 | 227 | 40 | 3 | 0 | 1825 |
| 1992 | 0 | 0 | 0 | 51 | 48 | 192 | 457 | 275 | 202 | 12 | 3 | 0 | 1240 |
| 1993 | 0 | 0 | 0 | 14 | 112 | 307 | 547 | 426 | 261 | 20 | 18 | 0 | 1705 |
| 1994 | 0 | 0 | 5 | 80 | 70 | 396 | 507 | 332 | 142 | 9 | 2 | 2 | 1545 |
| 1995 | 1 | 0 | 0 | 35 | 105 | 281 | 495 | 437 | 157 | 52 | 3 | 0 | 1566 |
| 1996 | 0 | 0 | 0 | 40 | 123 | 315 | 360 | 293 | 173 | 12 | 0 | 0 | 1316 |
| 1997 | 0 | 0 | 4 | 2 | 59 | 244 | 406 | 341 | 182 | 65 | 0 | 0 | 1303 |
| 1998 | 2 | 0 | 32 | 26 | 131 | 282 | 420 | 418 | 292 | 15 | 0 | 12 | 1630 |
| 1999 | 1 | 0 | 0 | 24 | 82 | 231 | 460 | 409 | 140 | 13 | 2 | 0 | 1362 |
| 2000 | 0 | 0 | 5 | 6 | 143 | 320 | 305 | 313 | 141 | 33 | 0 | 0 | 1266 |
| 2001 | 0 | 0 | 0 | 45 | 79 | 307 | 314 | 414 | 149 | 38 | 7 | 0 | 1353 |
| 2002 | 3 | 0 | 6 | 83 | 122 | 340 | 478 | 451 | 213 | 67 | 1 | 0 | 1764 |
| 2003 | 0 | 0 | 0 | 20 | 56 | 246 | 406 | 418 | 172 | 9 | 15 | 0 | 1342 |
| 2004 | 0 | 0 | 3 | 66 | 284 | 293 | 433 | 336 | 210 | 36 | 5 | 0 | 1666 |
| 2005 | 3 | 0 | 0 | 26 | 55 | 346 | 521 | 492 | 328 | 60 | 7 | 0 | 1838 |
| 2006 | 0 | 0 | 17 | 34 | 103 | 308 | 519 | 492 | 171 | 41 | 7 | 3 | 1695 |
| 2007 | 0 | 0 | 16 | 36 | 148 | 332 | 431 | 488 | 247 | 146 | 0 | 0 | 1844 |
| 2008 | 0 | 6 | 1 | 34 | 84 | 425 | 443 | 386 | 224 | 33 | 1 | 1 | 1638 |
| 2009 | 0 | 0 | 8 | 48 | 160 | 322 | 378 | 470 | 160 | 23 | 0 | 0 | 1569 |

## SNOWFALL (inches)  2009  RICHMOND (KRIC)

| YEAR | JUL | AUG | SEP | OCT | NOV | DEC | JAN | FEB | MAR | APR | MAY | JUN | TOTAL |
|------|-----|-----|-----|-----|-----|-----|-----|-----|-----|-----|-----|-----|-------|
| 1980-81 | 0.0 | 0.0 | 0.0 | 0.0 | 0.0 | 0.2 | 0.6 | 0.0 | 0.2 | 0.0 | 0.0 | 0.0 | 1.0 |
| 1981-82 | 0.0 | 0.0 | 0.0 | 0.0 | T | 1.9 | 8.3 | 10.8 | T | 0.2 | 0.0 | 0.0 | 21.2 |
| 1982-83 | 0.0 | 0.0 | 0.0 | 0.0 | 0.0 | 7.9 | 0.1 | 21.4 | 0.0 | T | 0.0 | 0.0 | 29.4 |
| 1983-84 | 0.0 | 0.0 | 0.0 | 0.0 | 0.1 | T | 1.1 | 2.8 | 0.3 | 0.0 | 0.0 | 0.0 | 4.3 |
| 1984-85 | 0.0 | 0.0 | 0.0 | 0.0 | 0.0 | T | 8.3 | T | T | 0.0 | 0.0 | 0.0 | 8.3 |
| 1985-86 | 0.0 | 0.0 | 0.0 | 0.0 | 0.0 | 1.3 | 3.3 | 4.4 | T | T | 0.0 | 0.0 | 9.0 |
| 1986-87 | 0.0 | 0.0 | 0.0 | 0.0 | 0.0 | 0.0 | 15.8 | 5.3 | 0.7 | T | 0.0 | 0.0 | 21.8 |
| 1987-88 | 0.0 | 0.0 | 0.0 | 0.0 | 4.5 | T | 8.1 | T | T | T | 0.0 | 0.0 | 12.6 |
| 1988-89 | 0.0 | 0.0 | 0.0 | 0.0 | 0.0 | 1.8 | T | 13.6 | T | T | T | 0.0 | 15.4 |
| 1989-90 | 0.0 | 0.0 | 0.0 | 0.0 | 1.1 | 9.9 | 0.0 | T | T | 0.2 | 0.0 | 0.0 | 11.2 |
| 1990-91 | 0.0 | 0.0 | 0.0 | 0.0 | 0.0 | T | T | 1.9 | T | 0.0 | 0.0 | 0.0 | 1.9 |
| 1991-92 | 0.0 | 0.0 | 0.0 | 0.0 | 0.1 | T | 0.0 | 0.8 | T | T | 0.0 | 0.0 | 0.9 |
| 1992-93 | 0.0 | 0.0 | 0.0 | 0.0 | 0.0 | 0.0 | 0.5 | 5.3 | 3.5 | T | 0.0 | 0.0 | 9.3 |
| 1993-94 | 0.0 | 0.0 | 0.0 | 0.0 | 0.0 | 5.1 | 2.2 | 1.7 | 0.7 | 0.0 | T | 0.0 | 9.7 |
| 1994-95 | 0.0 | 0.0 | 0.0 | 0.0 | 0.0 | T | 3.1 | T | 0.8 | 0.0 | 0.0 | 0.0 | 3.9 |
| 1995-96 | 0.0 | 0.0 | 0.0 | 0.0 | T | | 12.3 | | | | | | |
| 1996-97 | | | | | T | 1.0 | 0.7 | 1.0 | T | T | 0.0 | 0.0 | |
| 1997-98 | 0.0 | 0.0 | 0.0 | 0.0 | | | | | | | | | |
| 1998-99 | | | | | | | | | | | | | |
| 1999-00 | | | | | | | 15.2 | 0.2 | | | | | |
| 2000-01 | | | | | | | 0.3 | 2.5 | 0.5 | 0.0 | 0.0 | 0.0 | |
| 2001-02 | 0.0 | 0.0 | 0.0 | 0.0 | 0.0 | 0.0 | 8.7 | T | T | 0.0 | 0.0 | 0.0 | 8.7 |
| 2002-03 | 0.0 | 0.0 | 0.0 | 0.0 | 0.0 | 5.0 | 5.9 | 6.0 | 0.1 | 0.0 | 0.0 | 0.0 | 17.0 |
| 2003-04 | T | 0.0 | 0.0 | 0.0 | 0.0 | T | 4.9 | 1.4 | 0.0 | 0.0 | 0.0 | 0.0 | 6.3 |
| 2004-05 | 0.0 | 0.0 | 0.0 | 0.0 | 0.0 | 1.3 | 2.1 | 2.0 | 0.3 | 0.0 | 0.0 | 0.0 | 5.7 |
| 2005-06 | 0.0 | 0.0 | 0.0 | 0.0 | T | 4.4 | T | 3.2 | 0.9 | T | 0.0 | 0.0 | 8.5 |
| 2006-07 | 0.0 | 0.0 | 0.0 | 0.0 | 0.0 | 0.0 | T | 0.3 | 0.0 | 1.0 | T | 0.0 | 1.3 |
| 2007-08 | 0.0 | 0.0 | 0.0 | 0.0 | 0.0 | T | 0.8 | T | T | 0.0 | 0.0 | 0.0 | 0.8 |
| 2008-09 | 0.0 | 0.0 | T | 0.0 | T | T | T | 0.3 | 6.3 | 0.0 | 0.0 | 0.0 | 6.6 |
| 2009- | 0.0 | 0.0 | 0.0 | 0.0 | 0.0 | 7.4 | | | | | | | |
| POR= 85 YRS | T | 0.0 | T | 0.0 | 0.4 | 2.1 | 4.4 | 3.5 | 1.9 | 0.1 | T | 0.0 | 12.4 |

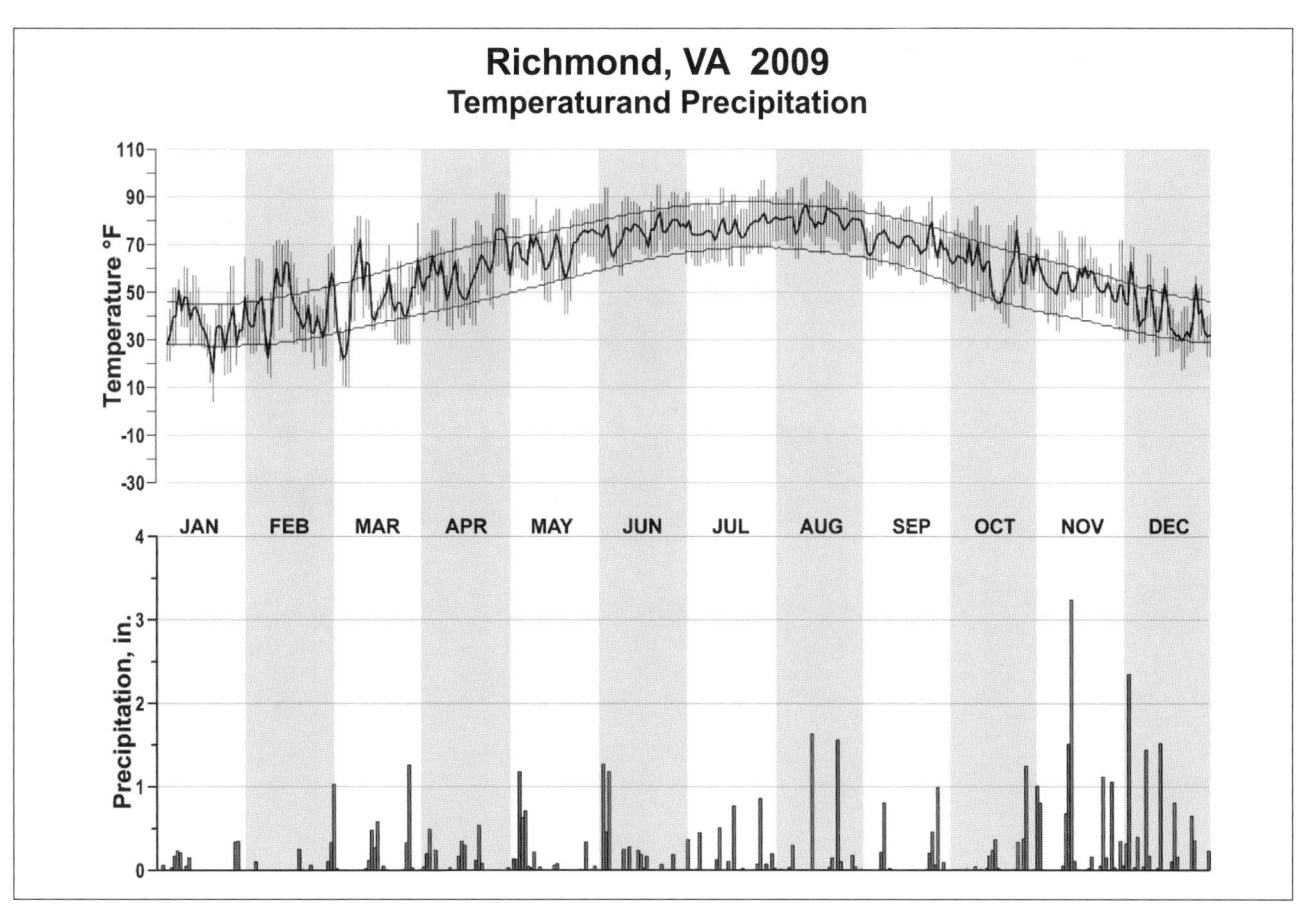

# Richmond, VA  2009
## Temperaturand Precipitation

# 2009
# ROANOKE
# VIRGINIA (KROA)

The climate of Roanoke is relatively mild. Roanoke is nestled among mountains which interrupt the Great Valley, extending from northernmost Virginia southwestward into east Tennessee. This location, at a point where the valley is pinched between the Blue Ridges and the Alleghenies, offers a natural barrier to the winter cold as it moves southward. It is also far enough inland that hurricanes lose much of their destructive force before reaching Roanoke. Finally, the rough terrain is an inhospitable breeding ground for tornadic activity. The elevation in the vicinity usually produces cool summer nights that make a light cover comfortable for sleeping. Although past records show extremes over 100 degrees and below zero, many years pass without either extreme being threatened.

Roanoke is located near the headwaters of the Roanoke River, which flows in a general southeasterly direction. Numerous creeks and small streams from nearby mountainous areas empty into the Roanoke River. The usual low water stage is 1 to 1.5 feet, and flood stage is 10 feet. Some low-lying streets in Roanoke and nearby Salem have to be blocked off during 7 to 8 foot stages, but damage is minor until the river overflows its banks. The highest stage on record exceeds 19 feet. Damage has been widespread on occasion and has amounted to several million dollars in the city of Roanoke alone.

The growing season averages 190 days. The average date of the last freezing temperature in spring is mid-April and the average date of the first freezing date in the fall is late October.

Rainfall is well apportioned throughout the year. Droughts are so infrequent that quoting actual records would be difficult. Snow usually falls each winter, ranging from only a trace to more than 60 inches.

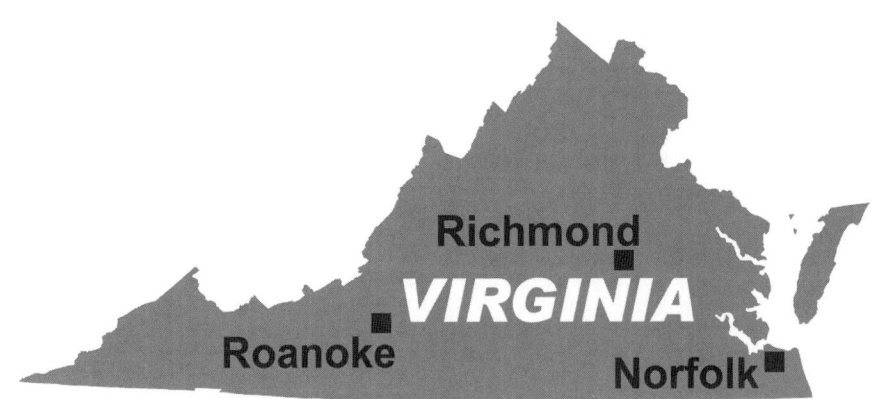

# NORMALS, MEANS, AND EXTREMES
## ROANOKE (KROA)

| LATITUDE: | LONGITUDE: | ELEVATION (FT): | TIME ZONE: | WBAN: 13741 |
|---|---|---|---|---|
| 37 ° 19'N | -79 ° 58'W | GRND: 1135   BARO: 1192 | EASTERN   (UTC -5) | |

| | ELEMENT | POR | JAN | FEB | MAR | APR | MAY | JUN | JUL | AUG | SEP | OCT | NOV | DEC | YEAR |
|---|---|---|---|---|---|---|---|---|---|---|---|---|---|---|---|
| **TEMPERATURE °F** | NORMAL DAILY MAXIMUM | 30 | 45.0 | 49.1 | 57.9 | 68.0 | 75.9 | 83.3 | 87.5 | 86.0 | 78.8 | 68.6 | 58.0 | 48.6 | 67.2 |
| | MEAN DAILY MAXIMUM | 62 | 45.4 | 48.6 | 57.2 | 67.9 | 76.0 | 83.3 | 86.9 | 85.7 | 78.7 | 68.7 | 57.7 | 47.8 | 67.0 |
| | HIGHEST DAILY MAXIMUM | 62 | 79 | 80 | 87 | 95 | 96 | 100 | 104 | 105 | 101 | 93 | 83 | 80 | 105 |
| | YEAR OF OCCURRENCE | | 2002 | 1985 | 1986 | 1957 | 1962 | 1959 | 1954 | 1983 | 1954 | 1951 | 1950 | 1998 | AUG 1983 |
| | MEAN OF EXTREME MAXS. | 62 | 66.4 | 69.3 | 78.0 | 85.7 | 89.1 | 93.4 | 95.5 | 94.6 | 91.1 | 83.4 | 75.3 | 67.5 | 82.4 |
| | NORMAL DAILY MINIMUM | 30 | 26.6 | 29.0 | 36.5 | 44.2 | 52.3 | 60.4 | 64.9 | 63.4 | 56.6 | 44.6 | 36.6 | 29.6 | 45.4 |
| | MEAN DAILY MINIMUM | 62 | 27.2 | 28.9 | 35.9 | 44.9 | 53.3 | 61.1 | 65.6 | 64.5 | 57.2 | 46.0 | 37.1 | 29.7 | 46.0 |
| | LOWEST DAILY MINIMUM | 62 | -11 | -1 | 9 | 20 | 31 | 39 | 47 | 42 | 34 | 22 | 9 | -4 | -11 |
| | YEAR OF OCCURRENCE | | 1985 | 1996 | 1996 | 1985 | 1966 | 1977 | 1988 | 1986 | 1983 | 1976 | 1950 | 1983 | JAN 1985 |
| | MEAN OF EXTREME MINS. | 62 | 9.1 | 12.8 | 20.1 | 29.0 | 38.6 | 48.2 | 55.2 | 54.1 | 42.1 | 30.4 | 21.3 | 12.9 | 31.2 |
| | NORMAL DRY BULB | 30 | 35.8 | 39.1 | 47.2 | 56.1 | 64.1 | 71.9 | 76.2 | 74.7 | 67.7 | 56.6 | 47.3 | 39.1 | 56.3 |
| | MEAN DRY BULB | 62 | 36.3 | 38.8 | 46.5 | 56.4 | 64.7 | 72.3 | 76.3 | 75.1 | 68.0 | 57.3 | 47.4 | 38.8 | 56.5 |
| | MEAN WET BULB | 26 | 30.6 | 32.5 | 38.8 | 47.4 | 56.4 | 64.5 | 67.8 | 67.0 | 61.0 | 50.3 | 41.2 | 33.2 | 49.2 |
| | MEAN DEW POINT | 26 | 25.0 | 26.8 | 32.7 | 41.6 | 52.8 | 61.7 | 65.4 | 64.7 | 58.4 | 46.5 | 36.3 | 27.8 | 45.0 |
| | NORMAL NO. DAYS WITH: | | | | | | | | | | | | | | |
| | MAXIMUM >= 90 | 30 | 0.0 | 0.0 | 0.0 | 0.4 | 0.6 | 5.2 | 10.8 | 8.1 | 2.3 | 0.0 | 0.0 | 0.0 | 27.4 |
| | MAXIMUM <= 32 | 30 | 4.4 | 2.6 | 0.3 | 0.0 | 0.0 | 0.0 | 0.0 | 0.0 | 0.0 | 0.0 | 0.1 | 2.3 | 9.7 |
| | MINIMUM <= 32 | 30 | 22.3 | 19.0 | 11.5 | 2.5 | 0.1 | 0.0 | 0.0 | 0.0 | 0.0 | 2.2 | 10.3 | 19.3 | 87.2 |
| | MINIMUM <= 0 | 30 | 0.4 | * | 0.0 | 0.0 | 0.0 | 0.0 | 0.0 | 0.0 | 0.0 | 0.0 | 0.0 | 0.2 | 0.6 |
| **H/C** | NORMAL HEATING DEG. DAYS | 30 | 911 | 741 | 569 | 290 | 107 | 13 | 1 | 2 | 48 | 276 | 528 | 798 | 4284 |
| | NORMAL COOLING DEG. DAYS | 30 | 0 | 0 | 4 | 20 | 74 | 217 | 355 | 309 | 136 | 17 | 2 | 0 | 1134 |
| **RH** | NORMAL (PERCENT) | 30 | 63 | 61 | 58 | 58 | 67 | 71 | 71 | 73 | 74 | 69 | 66 | 65 | 66 |
| | HOUR 01 LST | 30 | 69 | 66 | 65 | 65 | 78 | 83 | 85 | 86 | 87 | 81 | 73 | 70 | 76 |
| | HOUR 07 LST | 30 | 71 | 71 | 71 | 71 | 80 | 82 | 84 | 88 | 89 | 84 | 77 | 73 | 78 |
| | HOUR 13 LST | 30 | 54 | 51 | 49 | 47 | 53 | 55 | 55 | 56 | 58 | 52 | 53 | 55 | 53 |
| | HOUR 19 LST | 30 | 60 | 56 | 52 | 49 | 59 | 64 | 65 | 66 | 70 | 64 | 61 | 61 | 61 |
| **S** | PERCENT POSSIBLE SUNSHINE | | | | | | | | | | | | | | |
| **W/O** | MEAN NO. DAYS WITH: | | | | | | | | | | | | | | |
| | HEAVY FOG(VISBY <= 1/4 MI) | 46 | 2.0 | 2.3 | 1.6 | 0.8 | 1.3 | 1.0 | 1.3 | 1.2 | 1.6 | 1.5 | 1.6 | 2.0 | 18.2 |
| | THUNDERSTORMS | 62 | 0.2 | 0.3 | 1.0 | 2.9 | 6.0 | 6.8 | 8.0 | 6.3 | 2.4 | 0.8 | 0.4 | 0.1 | 35.2 |
| **CLOUDINESS** | MEAN: | | | | | | | | | | | | | | |
| | SUNRISE-SUNSET (OKTAS) | 48 | 5.1 | 5.0 | 5.0 | 4.8 | 5.0 | 4.8 | 4.7 | 4.6 | 4.5 | 3.9 | 4.6 | 4.8 | 4.7 |
| | MIDNIGHT-MIDNIGHT (OKTAS) | 32 | 4.7 | 4.6 | 4.7 | 4.3 | 4.5 | 4.3 | 4.3 | 4.2 | 4.2 | 3.7 | 4.3 | 4.6 | 4.4 |
| | MEAN NO. DAYS WITH: | | | | | | | | | | | | | | |
| | CLEAR | 49 | 8.1 | 7.6 | 8.1 | 8.5 | 7.4 | 7.1 | 6.9 | 7.8 | 9.6 | 12.8 | 8.8 | 8.7 | 101.4 |
| | PARTLY CLOUDY | 49 | 7.7 | 7.1 | 8.4 | 8.8 | 10.5 | 11.9 | 13.0 | 12.4 | 8.8 | 7.5 | 8.4 | 7.9 | 112.4 |
| | CLOUDY | 49 | 15.2 | 13.6 | 14.4 | 12.7 | 13.0 | 11.1 | 11.1 | 10.8 | 11.6 | 10.7 | 12.8 | 14.4 | 151.4 |
| **PR** | MEAN STATION PRESSURE(IN) | 26 | 28.83 | 28.82 | 28.79 | 28.74 | 28.77 | 28.78 | 28.80 | 28.82 | 28.85 | 28.86 | 28.86 | 28.86 | 28.82 |
| | MEAN SEA-LEVEL PRES. (IN) | 26 | 30.11 | 30.09 | 30.05 | 29.99 | 30.00 | 30.00 | 30.02 | 30.04 | 30.08 | 30.11 | 30.12 | 30.13 | 30.06 |
| **WINDS** | MEAN SPEED (MPH) | 26 | 8.1 | 8.1 | 8.4 | 8.1 | 6.7 | 6.0 | 5.7 | 5.1 | 5.2 | 5.7 | 6.8 | 7.3 | 6.8 |
| | PREVAIL.DIR(TENS OF DEGS) | 34 | 31 | 31 | 31 | 31 | 31 | 28 | 28 | 28 | 14 | 31 | 31 | 31 | 31 |
| | MAXIMUM 2-MINUTE: | 13 | | | | | | | | | | | | | |
| | SPEED (MPH) | | 44 | 52 | 44 | 47 | 39 | 39 | 41 | 37 | 36 | 37 | 40 | 40 | 52 |
| | DIR. (TENS OF DEGS) | | 28 | 28 | 32 | 30 | 31 | 30 | 30 | 32 | 35 | 30 | 30 | 30 | 28 |
| | YEAR OF OCCURRENCE | | 2000 | 2008 | 2004 | 2006 | 2002 | 2008 | 2008 | 2007 | 2003 | 2003 | 2003 | 2008 | FEB 2008 |
| | MAXIMUM 3-SECOND | | | | | | | | | | | | | | |
| | SPEED (MPH) | 13 | 53 | 67 | 53 | 60 | 51 | 47 | 54 | 49 | 45 | 49 | 55 | 59 | 67 |
| | DIR. (TENS OF DEGS) | | 28 | 28 | 31 | 30 | 26 | 30 | 30 | 28 | 29 | 27 | 18 | 30 | 28 |
| | YEAR OF OCCURRENCE | | 2000 | 2008 | 2007 | 2006 | 2001 | 2008 | 2005 | 1997 | 1999 | 2003 | 2003 | 2008 | FEB 2008 |
| **PRECIPITATION** | NORMAL (IN) | 30 | 3.23 | 3.08 | 3.84 | 3.61 | 4.24 | 3.68 | 4.00 | 3.74 | 3.85 | 3.15 | 3.21 | 2.86 | 42.49 |
| | MAXIMUM MONTHLY (IN) | 62 | 7.97 | 8.00 | 7.91 | 11.35 | 10.11 | 10.32 | 10.09 | 9.54 | 11.72 | 9.89 | 12.36 | 8.22 | 12.36 |
| | YEAR OF OCCURRENCE | | 1998 | 1998 | 1993 | 1987 | 2003 | 1995 | 1989 | 1984 | 2004 | 1990 | 1985 | 2009 | NOV 1985 |
| | MINIMUM MONTHLY (IN) | 62 | 0.29 | 0.56 | 0.35 | 0.48 | 1.04 | 0.62 | 0.45 | 0.74 | 0.15 | 0.02 | 0.18 | 0.18 | 0.02 |
| | YEAR OF OCCURRENCE | | 1981 | 1968 | 2006 | 1976 | 1997 | 1986 | 1977 | 1995 | 1991 | 2000 | 2007 | 1965 | OCT 2000 |
| | MAXIMUM IN 24 HOURS (IN) | 62 | 3.64 | 3.32 | 3.09 | 5.57 | 3.99 | 4.78 | 2.74 | 5.22 | 6.60 | 6.41 | 6.63 | 3.40 | 6.63 |
| | YEAR OF OCCURRENCE | | 1995 | 1994 | 1998 | 1978 | 1973 | 2006 | 1989 | 1985 | 1987 | 1968 | 1985 | 1948 | NOV 1985 |
| | NORMAL NO. DAYS WITH: | | | | | | | | | | | | | | |
| | PRECIPITATION >= 0.01 | 30 | 10.1 | 9.8 | 10.8 | 10.1 | 12.2 | 10.4 | 11.5 | 9.6 | 8.9 | 7.8 | 8.3 | 9.4 | 118.9 |
| | PRECIPITATION >= 1.00 | 30 | 0.7 | 0.6 | 0.9 | 0.8 | 1.0 | 0.8 | 1.1 | 1.2 | 1.0 | 1.0 | 0.8 | 0.4 | 10.3 |
| **SNOWFALL** | NORMAL (IN) | 30 | 6.1 | 7.6 | 3.0 | 0.5 | 0.* | 0.0 | 0.0 | 0.0 | 0.0 | 0.* | 0.8 | 2.9 | 20.9 |
| | MAXIMUM MONTHLY (IN) | 55 | 41.2 | 27.6 | 30.3 | 7.3 | T | T | 0.0 | 0.0 | T | 1.0 | 13.8 | 22.6 | 41.2 |
| | YEAR OF OCCURRENCE | | 1966 | 1960 | 1960 | 1971 | 1990 | 1989 | | | 2006 | 1957 | 1968 | 1966 | JAN 1966 |
| | MAXIMUM IN 24 HOURS (IN) | 55 | 22.2 | 18.4 | 17.4 | 7.3 | T | T | 0.0 | 0.0 | T | 1.0 | 10.0 | 16.4 | 22.2 |
| | YEAR OF OCCURRENCE' | | 1996 | 1983 | 1960 | 1971 | 1990 | 1989 | | | 1953 | 1957 | 1968 | 1969 | JAN 1996 |
| | MAXIMUM SNOW DEPTH (IN) | 54 | 23 | 18 | 17 | 3 | 0 | 0 | 0 | 0 | 0 | 0 | 8 | 15 | 23 |
| | YEAR OF OCCURRENCE | | 1987 | 1983 | 1960 | 1987 | | | | | | | 1950 | 1969 | JAN 1987 |
| | NORMAL NO. DAYS WITH: | | | | | | | | | | | | | | |
| | SNOWFALL >= 1.0 | 30 | 1.4 | 1.7 | 0.7 | 0.2 | 0.0 | 0.0 | 0.0 | 0.0 | 0.0 | 0.0 | 0.4 | 0.7 | 5.1 |

## PRECIPITATION (inches) 2009 ROANOKE (KROA)

| YEAR | JAN | FEB | MAR | APR | MAY | JUN | JUL | AUG | SEP | OCT | NOV | DEC | ANNUAL |
|------|-----|-----|-----|-----|-----|-----|-----|-----|-----|-----|-----|-----|--------|
| 1980 | 4.10 | 0.67 | 5.41 | 5.51 | 2.66 | 1.81 | 5.18 | 2.87 | 1.66 | 2.30 | 1.78 | 0.60 | 34.55 |
| 1981 | 0.29 | 2.43 | 2.30 | 1.75 | 4.56 | 2.49 | 2.86 | 1.32 | 4.52 | 3.90 | 0.68 | 3.79 | 30.89 |
| 1982 | 3.76 | 4.75 | 2.33 | 2.01 | 4.83 | 4.99 | 3.98 | 5.20 | 2.67 | 4.13 | 3.65 | 2.53 | 44.83 |
| 1983 | 1.28 | 4.12 | 6.41 | 7.95 | 3.17 | 2.38 | 1.67 | 2.23 | 1.52 | 7.73 | 4.26 | 5.61 | 48.33 |
| 1984 | 1.35 | 4.85 | 4.30 | 3.97 | 4.49 | 2.34 | 4.17 | 9.54 | 2.69 | 1.42 | 2.67 | 1.84 | 43.63 |
| 1985 | 2.45 | 3.64 | 1.80 | 1.75 | 6.89 | 2.08 | 4.18 | 8.67 | 1.26 | 3.77 | 12.36 | 0.85 | 49.70 |
| 1986 | 0.93 | 2.87 | 1.36 | 1.67 | 4.15 | 0.62 | 2.83 | 4.31 | 3.04 | 2.76 | 3.73 | 5.48 | 33.75 |
| 1987 | 4.53 | 4.55 | 4.11 | 11.35 | 2.68 | 0.71 | 3.21 | 1.08 | 11.09 | 1.10 | 5.00 | 2.16 | 51.57 |
| 1988 | 1.87 | 1.07 | 0.88 | 3.40 | 2.76 | 3.66 | 3.75 | 4.30 | 3.01 | 1.26 | 2.42 | 1.28 | 29.66 |
| 1989 | 1.31 | 2.04 | 2.96 | 2.54 | 6.46 | 7.76 | 10.09 | 1.65 | 8.94 | 4.13 | 3.86 | 2.60 | 54.34 |
| 1990 | 2.33 | 2.76 | 3.42 | 2.07 | 7.45 | 0.83 | 3.80 | 4.42 | 1.86 | 9.89 | 1.08 | 3.79 | 43.70 |
| 1991 | 3.55 | 2.10 | 7.58 | 2.49 | 2.88 | 2.42 | 7.22 | 2.31 | 0.15 | 0.04 | 2.51 | 3.81 | 37.06 |
| 1992 | 2.51 | 3.75 | 2.54 | 4.89 | 6.06 | 6.87 | 1.73 | 2.38 | 2.07 | 1.90 | 5.60 | 2.62 | 42.92 |
| 1993 | 3.77 | 3.44 | 7.91 | 2.77 | 2.37 | 2.49 | 1.24 | 1.04 | 4.58 | 2.15 | 2.60 | 5.27 | 39.63 |
| 1994 | 4.52 | 5.34 | 5.67 | 5.59 | 1.69 | 1.99 | 7.07 | 4.22 | 0.76 | 1.84 | 1.53 | 2.41 | 42.63 |
| 1995 | 7.28 | 1.88 | 1.31 | 0.79 | 4.55 | 10.32 | 2.34 | 0.74 | 2.04 | 3.49 | 3.39 | 2.32 | 40.45 |
| 1996 | 6.87 | 2.12 | 3.75 | 1.76 | 4.58 | 7.65 | 2.63 | 6.40 | 10.14 | 1.57 | 4.75 | 2.72 | 54.94 |
| 1997 | 2.18 | 2.21 | 3.12 | 2.51 | 1.04 | 5.43 | 3.38 | 2.22 | 3.43 | 1.50 | 2.65 | 2.37 | 32.04 |
| 1998 | 7.97 | 8.00 | 5.20 | 4.58 | 4.47 | 2.03 | 0.86 | 6.17 | 1.10 | 1.47 | 0.85 | 2.32 | 45.02 |
| 1999 | 3.70 | 2.14 | 2.77 | 2.70 | 2.27 | 0.85 | 5.96 | 2.63 | 7.38 | 1.24 | 2.09 | 2.46 | 36.19 |
| 2000 | 2.08 | 1.69 | 2.86 | 5.71 | 2.67 | 4.57 | 7.16 | 3.16 | 5.88 | 0.02 | 2.08 | 1.67 | 39.55 |
| 2001 | 1.79 | 0.90 | 4.48 | 0.81 | 4.67 | 1.51 | 3.13 | 2.02 | 2.12 | 0.43 | 0.60 | 2.48 | 24.94 |
| 2002 | 1.77 | 0.71 | 4.00 | 1.89 | 2.67 | 1.40 | 3.09 | 2.21 | 3.54 | 5.03 | 4.25 | 3.16 | 33.72 |
| 2003 | 1.44 | 5.80 | 3.52 | 4.99 | 10.11 | 6.31 | 5.16 | 4.03 | 4.32 | 1.49 | 4.14 | 2.95 | 54.26 |
| 2004 | 1.63 | 2.27 | 2.09 | 3.45 | 3.87 | 6.47 | 4.31 | 2.97 | 11.72 | 2.38 | 4.26 | 2.13 | 47.55 |
| 2005 | 2.15 | 2.20 | 3.72 | 2.63 | 2.15 | 5.07 | 4.96 | 3.72 | 0.23 | 5.05 | 3.63 | 2.36 | 37.87 |
| 2006 | 3.56 | 1.56 | 0.35 | 3.00 | 1.46 | 8.51 | 1.92 | 2.35 | 3.21 | 5.33 | 4.22 | 1.98 | 37.45 |
| 2007 | 2.40 | 2.23 | 3.32 | 2.60 | 2.59 | 2.62 | 3.22 | 1.50 | 1.11 | 5.33 | 0.18 | 2.76 | 29.86 |
| 2008 | 0.96 | 1.86 | 2.27 | 4.94 | 2.08 | 4.64 | 3.67 | 4.65 | 2.20 | 1.87 | 1.92 | 2.25 | 33.31 |
| 2009 | 2.72 | 1.22 | 3.47 | 3.20 | 6.87 | 4.54 | 5.84 | 4.43 | 3.14 | 2.69 | 7.44 | 8.22 | 53.78 |
| POR= 62 YRS | 2.88 | 3.06 | 3.56 | 3.27 | 3.86 | 3.69 | 3.81 | 3.87 | 3.48 | 3.19 | 2.99 | 3.01 | 40.67 |

WBAN : 13741

## AVERAGE TEMPERATURE (°F) 2009 ROANOKE (KROA)

| YEAR | JAN | FEB | MAR | APR | MAY | JUN | JUL | AUG | SEP | OCT | NOV | DEC | ANNUAL |
|------|-----|-----|-----|-----|-----|-----|-----|-----|-----|-----|-----|-----|--------|
| 1980 | 37.5 | 34.0 | 43.5 | 56.7 | 64.8 | 70.0 | 78.1 | 76.8 | 71.5 | 55.5 | 45.4 | 38.7 | 56.0 |
| 1981 | 32.0 | 38.2 | 43.1 | 58.6 | 60.9 | 74.0 | 76.1 | 73.4 | 66.1 | 53.7 | 45.2 | 32.8 | 54.5 |
| 1982 | 28.7 | 38.1 | 44.8 | 50.8 | 67.5 | 70.1 | 75.5 | 72.3 | 65.8 | 57.4 | 47.8 | 42.4 | 55.1 |
| 1983 | 35.6 | 36.5 | 47.0 | 52.4 | 61.0 | 70.1 | 77.0 | 77.9 | 68.1 | 57.1 | 47.1 | 34.9 | 55.4 |
| 1984 | 33.6 | 43.6 | 43.3 | 54.2 | 63.1 | 74.2 | 73.2 | 74.4 | 64.1 | 64.3 | 46.0 | 47.1 | 56.8 |
| 1985 | 31.3 | 38.6 | 50.3 | 61.9 | 67.1 | 72.7 | 76.5 | 73.8 | 68.4 | 60.3 | 55.4 | 35.1 | 57.6 |
| 1986 | 35.2 | 38.3 | 47.2 | 58.8 | 63.9 | 74.2 | 78.8 | 72.4 | 68.9 | 59.0 | 47.4 | 38.5 | 56.9 |
| 1987 | 34.2 | 36.9 | 46.6 | 53.2 | 67.1 | 75.0 | 79.3 | 77.8 | 68.9 | 51.5 | 49.6 | 41.4 | 56.8 |
| 1988 | 30.8 | 37.9 | 47.5 | 55.4 | 63.4 | 70.9 | 77.1 | 77.5 | 66.0 | 51.2 | 46.7 | 38.9 | 55.3 |
| 1989 | 41.3 | 38.2 | 47.4 | 54.3 | 61.2 | 73.7 | 75.9 | 74.1 | 68.1 | 58.5 | 46.0 | 29.6 | 55.7 |
| 1990 | 43.7 | 45.9 | 51.5 | 56.0 | 64.3 | 72.6 | 76.8 | 74.6 | 68.4 | 59.2 | 52.2 | 43.8 | 59.1 |
| 1991 | 39.3 | 42.7 | 49.2 | 57.5 | 69.6 | 73.5 | 77.9 | 74.8 | 69.3 | 59.4 | 46.8 | 42.4 | 58.5 |
| 1992 | 39.6 | 41.5 | 46.5 | 55.5 | 60.2 | 68.4 | 77.0 | 72.1 | 68.0 | 55.5 | 46.3 | 38.5 | 55.8 |
| 1993 | 39.7 | 36.0 | 42.1 | 54.5 | 65.6 | 71.9 | 80.3 | 76.8 | 68.6 | 55.3 | 47.3 | 36.2 | 56.2 |
| 1994 | 30.6 | 39.4 | 46.7 | 60.9 | 60.6 | 75.1 | 77.6 | 73.9 | 66.9 | 56.7 | 51.4 | 43.0 | 56.9 |
| 1995 | 38.5 | 36.9 | 50.1 | 56.8 | 64.4 | 72.0 | 76.9 | 78.6 | 67.5 | 58.5 | 43.2 | 35.5 | 56.6 |
| 1996 | 33.7 | 38.2 | 41.7 | 56.3 | 65.2 | 72.8 | 74.4 | 73.3 | 66.1 | 58.0 | 41.0 | 40.2 | 55.1 |
| 1997 | 35.8 | 43.0 | 49.7 | 52.5 | 60.3 | 69.2 | 76.1 | 73.4 | 67.7 | 56.3 | 43.6 | 38.3 | 55.5 |
| 1998 | 41.2 | 42.2 | 45.8 | 56.5 | 67.1 | 73.5 | 77.5 | 75.9 | 73.6 | 59.0 | 49.0 | 43.4 | 58.7 |
| 1999 | 39.8 | 42.0 | 44.0 | 58.2 | 64.3 | 72.3 | 79.2 | 75.5 | 66.4 | 55.7 | 52.8 | 42.5 | 57.7 |
| 2000 | 35.3 | 43.5 | 51.7 | 54.8 | 67.6 | 73.8 | 73.3 | 72.9 | 65.9 | 59.4 | 44.2 | 31.1 | 56.1 |
| 2001 | 37.0 | 42.9 | 43.4 | 59.5 | 64.8 | 73.3 | 74.3 | 76.9 | 66.5 | 56.9 | 53.2 | 44.3 | 57.8 |
| 2002 | 42.0 | 40.9 | 44.2 | 58.8 | 64.0 | 74.5 | 78.0 | 77.0 | 70.7 | 57.6 | 46.3 | 37.7 | 58.0 |
| 2003 | 33.1 | 35.7 | 49.6 | 56.8 | 62.6 | 69.6 | 75.1 | 76.4 | 66.7 | 56.1 | 52.2 | 37.9 | 56.0 |
| 2004 | 34.6 | 38.5 | 50.2 | 58.1 | 70.8 | 72.3 | 76.5 | 73.4 | 69.1 | 59.3 | 51.4 | 39.5 | 57.8 |
| 2005 | 40.5 | 41.7 | 44.1 | 56.8 | 62.2 | 72.8 | 77.8 | 77.6 | 71.2 | 60.2 | 49.9 | 36.9 | 57.6 |
| 2006 | 43.6 | 39.6 | 48.6 | 60.9 | 63.4 | 72.2 | 77.6 | 78.4 | 65.2 | 55.4 | 50.3 | 45.1 | 58.4 |
| 2007 | 41.2 | 34.2 | 53.2 | 56.0 | 67.6 | 74.4 | 75.7 | 82.2 | 71.5 | 63.6 | 48.3 | 43.5 | 59.3 |
| 2008 | 37.5 | 41.6 | 49.4 | 57.0 | 64.2 | 75.9 | 76.1 | 74.2 | 69.7 | 58.2 | 46.0 | 41.8 | 57.6 |
| 2009 | 35.9 | 42.2 | 48.3 | 57.5 | 65.4 | 74.1 | 73.3 | 76.1 | 68.8 | 57.3 | 51.5 | 37.3 | 57.3 |
| POR= 62 YRS | 36.3 | 38.8 | 46.5 | 56.4 | 64.7 | 72.3 | 76.3 | 75.1 | 68.0 | 57.3 | 47.4 | 38.8 | 56.5 |

## HEATING DEGREE DAYS (base 65°F) 2009  ROANOKE (KROA)

| YEAR | JUL | AUG | SEP | OCT | NOV | DEC | JAN | FEB | MAR | APR | MAY | JUN | TOTAL |
|------|-----|-----|-----|-----|-----|-----|-----|-----|-----|-----|-----|-----|-------|
| 1980-81 | 0 | 0 | 30 | 301 | 582 | 807 | 1016 | 744 | 672 | 212 | 158 | 3 | 4525 |
| 1981-82 | 0 | 0 | 58 | 357 | 589 | 991 | 1121 | 746 | 618 | 421 | 51 | 3 | 4955 |
| 1982-83 | 0 | 6 | 61 | 264 | 509 | 695 | 904 | 792 | 551 | 381 | 157 | 14 | 4334 |
| 1983-84 | 1 | 0 | 87 | 246 | 531 | 924 | 966 | 614 | 664 | 336 | 123 | 6 | 4498 |
| 1984-85 | 0 | 0 | 120 | 74 | 565 | 549 | 1041 | 734 | 471 | 168 | 39 | 11 | 3772 |
| 1985-86 | 0 | 0 | 55 | 171 | 282 | 918 | 917 | 739 | 551 | 206 | 115 | 1 | 3955 |
| 1986-87 | 0 | 22 | 22 | 231 | 523 | 813 | 950 | 782 | 562 | 352 | 58 | 0 | 4315 |
| 1987-88 | 0 | 0 | 18 | 412 | 455 | 723 | 1055 | 778 | 535 | 289 | 101 | 46 | 4412 |
| 1988-89 | 1 | 0 | 53 | 423 | 543 | 802 | 726 | 743 | 556 | 340 | 172 | 0 | 4359 |
| 1989-90 | 0 | 3 | 63 | 234 | 560 | 1091 | 654 | 528 | 441 | 297 | 77 | 3 | 3951 |
| 1990-91 | 0 | 0 | 38 | 200 | 381 | 652 | 787 | 618 | 486 | 245 | 36 | 4 | 3447 |
| 1991-92 | 0 | 0 | 49 | 194 | 539 | 695 | 777 | 677 | 567 | 290 | 172 | 18 | 3978 |
| 1992-93 | 0 | 2 | 55 | 290 | 552 | 813 | 777 | 807 | 698 | 310 | 52 | 12 | 4368 |
| 1993-94 | 0 | 0 | 52 | 300 | 529 | 886 | 1061 | 709 | 559 | 172 | 176 | 4 | 4448 |
| 1994-95 | 0 | 0 | 22 | 261 | 401 | 675 | 815 | 779 | 454 | 258 | 93 | 0 | 3758 |
| 1995-96 | 0 | 0 | 54 | 226 | 647 | 908 | 965 | 771 | 717 | 283 | 118 | 6 | 4695 |
| 1996-97 | 0 | 0 | 52 | 221 | 712 | 762 | 898 | 610 | 467 | 369 | 175 | 49 | 4315 |
| 1997-98 | 1 | 0 | 32 | 294 | 634 | 819 | 730 | 629 | 615 | 255 | 46 | 22 | 4077 |
| 1998-99 | 0 | 0 | 10 | 184 | 473 | 662 | 775 | 638 | 646 | 223 | 61 | 8 | 3680 |
| 1999-00 | 5 | 3 | 38 | 289 | 357 | 689 | 913 | 619 | 414 | 301 | 44 | 5 | 3677 |
| 2000-01 | 2 | 0 | 86 | 190 | 617 | 1046 | 863 | 612 | 666 | 221 | 66 | 4 | 4373 |
| 2001-02 | 4 | 0 | 70 | 263 | 345 | 638 | 712 | 667 | 516 | 232 | 129 | 0 | 3576 |
| 2002-03 | 0 | 0 | 8 | 263 | 558 | 838 | 983 | 813 | 468 | 260 | 113 | 16 | 4320 |
| 2003-04 | 0 | 0 | 37 | 272 | 384 | 830 | 935 | 762 | 454 | 243 | 26 | 0 | 3943 |
| 2004-05 | 0 | 1 | 10 | 182 | 410 | 782 | 751 | 647 | 640 | 251 | 113 | 9 | 3796 |
| 2005-06 | 0 | 0 | 11 | 186 | 447 | 864 | 658 | 704 | 506 | 153 | 123 | 3 | 3655 |
| 2006-07 | 0 | 0 | 65 | 296 | 435 | 612 | 731 | 854 | 393 | 297 | 76 | 4 | 3763 |
| 2007-08 | 0 | 0 | 21 | 137 | 494 | 659 | 847 | 671 | 477 | 246 | 79 | 0 | 3631 |
| 2008-09 | 0 | 2 | 12 | 231 | 565 | 712 | 895 | 630 | 527 | 257 | 79 | 0 | 3910 |
| 2009- | 0 | 0 | 18 | 237 | 398 | 852 | | | | | | | |

WBAN : 13741

## COOLING DEGREE DAYS (base 65°F) 2009  ROANOKE (KROA)

| YEAR | JAN | FEB | MAR | APR | MAY | JUN | JUL | AUG | SEP | OCT | NOV | DEC | TOTAL |
|------|-----|-----|-----|-----|-----|-----|-----|-----|-----|-----|-----|-----|-------|
| 1980 | 0 | 0 | 0 | 21 | 78 | 171 | 412 | 374 | 231 | 13 | 0 | 0 | 1300 |
| 1981 | 0 | 0 | 2 | 26 | 39 | 278 | 350 | 267 | 97 | 13 | 0 | 0 | 1072 |
| 1982 | 0 | 0 | 0 | 0 | 137 | 165 | 332 | 242 | 90 | 35 | 0 | 0 | 1001 |
| 1983 | 0 | 0 | 0 | 11 | 41 | 172 | 382 | 407 | 188 | 10 | 0 | 0 | 1211 |
| 1984 | 0 | 0 | 0 | 20 | 71 | 290 | 260 | 301 | 101 | 56 | 0 | 0 | 1099 |
| 1985 | 0 | 1 | 18 | 78 | 112 | 247 | 365 | 280 | 163 | 34 | 3 | 0 | 1301 |
| 1986 | 0 | 0 | 4 | 24 | 84 | 282 | 438 | 257 | 145 | 48 | 0 | 0 | 1282 |
| 1987 | 0 | 0 | 0 | 4 | 130 | 306 | 450 | 403 | 140 | 0 | 0 | 0 | 1433 |
| 1988 | 0 | 0 | 0 | 7 | 58 | 232 | 386 | 395 | 90 | 1 | 0 | 0 | 1169 |
| 1989 | 0 | 0 | 17 | 27 | 61 | 265 | 344 | 293 | 163 | 38 | 0 | 0 | 1208 |
| 1990 | 0 | 0 | 29 | 34 | 61 | 238 | 377 | 305 | 145 | 27 | 3 | 0 | 1219 |
| 1991 | 0 | 0 | 2 | 29 | 188 | 267 | 407 | 311 | 185 | 28 | 1 | 0 | 1418 |
| 1992 | 0 | 0 | 2 | 12 | 29 | 125 | 377 | 230 | 152 | 4 | 0 | 0 | 931 |
| 1993 | 0 | 0 | 0 | 2 | 78 | 226 | 481 | 374 | 166 | 8 | 6 | 0 | 1341 |
| 1994 | 0 | 0 | 0 | 57 | 45 | 317 | 397 | 286 | 84 | 9 | 3 | 0 | 1198 |
| 1995 | 0 | 0 | 0 | 20 | 80 | 220 | 375 | 427 | 134 | 30 | 0 | 0 | 1286 |
| 1996 | 0 | 1 | 0 | 30 | 130 | 247 | 300 | 263 | 93 | 10 | 0 | 0 | 1074 |
| 1997 | 0 | 0 | 0 | 1 | 36 | 181 | 353 | 264 | 121 | 31 | 0 | 0 | 987 |
| 1998 | 0 | 0 | 24 | 5 | 120 | 282 | 394 | 344 | 276 | 7 | 0 | 0 | 1452 |
| 1999 | 0 | 0 | 0 | 26 | 47 | 233 | 451 | 336 | 86 | 9 | 0 | 0 | 1188 |
| 2000 | 0 | 0 | 10 | 4 | 131 | 277 | 269 | 253 | 121 | 25 | 0 | 0 | 1090 |
| 2001 | 0 | 0 | 0 | 64 | 67 | 259 | 301 | 377 | 121 | 20 | 0 | 0 | 1209 |
| 2002 | 3 | 0 | 0 | 53 | 105 | 292 | 409 | 378 | 187 | 44 | 1 | 0 | 1472 |
| 2003 | 0 | 0 | 0 | 22 | 42 | 160 | 321 | 360 | 96 | 5 | 8 | 0 | 1014 |
| 2004 | 0 | 0 | 3 | 44 | 211 | 224 | 365 | 271 | 142 | 14 | 6 | 0 | 1280 |
| 2005 | 0 | 0 | 0 | 11 | 37 | 247 | 403 | 399 | 201 | 44 | 0 | 0 | 1342 |
| 2006 | 0 | 0 | 5 | 35 | 79 | 225 | 398 | 420 | 77 | 16 | 1 | 0 | 1256 |
| 2007 | 0 | 0 | 32 | 34 | 163 | 292 | 341 | 537 | 225 | 101 | 0 | 0 | 1725 |
| 2008 | 0 | 0 | 0 | 14 | 64 | 333 | 352 | 292 | 158 | 27 | 0 | 0 | 1240 |
| 2009 | 0 | 0 | 15 | 37 | 96 | 280 | 266 | 353 | 137 | 6 | 0 | 0 | 1190 |

## SNOWFALL (inches) 2009 ROANOKE (KROA)

| YEAR | JUL | AUG | SEP | OCT | NOV | DEC | JAN | FEB | MAR | APR | MAY | JUN | TOTAL |
|------|-----|-----|-----|-----|-----|-----|-----|-----|-----|-----|-----|-----|-------|
| 1980-81 | 0.0 | 0.0 | 0.0 | T | T | T | 1.4 | T | 10.4 | 0.0 | 0.0 | 0.0 | 11.8 |
| 1981-82 | 0.0 | 0.0 | 0.0 | 0.0 | 4.0 | 3.9 | 8.4 | 12.2 | 0.5 | 1.9 | 0.0 | 0.0 | 30.9 |
| 1982-83 | 0.0 | 0.0 | 0.0 | T | 0.0 | 6.2 | 3.8 | 24.3 | 0.3 | 0.4 | 0.0 | 0.0 | 35.0 |
| 1983-84 | 0.0 | 0.0 | 0.0 | 0.0 | T | 0.6 | 7.2 | 1.5 | 0.3 | 0.2 | 0.0 | 0.0 | 9.8 |
| 1984-85 | 0.0 | 0.0 | 0.0 | 0.0 | T | 0.8 | 2.7 | 1.2 | 1.3 | T | 0.0 | 0.0 | 6.0 |
| 1985-86 | 0.0 | 0.0 | 0.0 | 0.0 | 0.0 | 1.2 | 1.7 | 7.1 | T | T | 0.0 | 0.0 | 10.0 |
| 1986-87 | 0.0 | 0.0 | 0.0 | 0.0 | T | T | 27.9 | 19.1 | 2.7 | 6.3 | 0.0 | 0.0 | 56.0 |
| 1987-88 | 0.0 | 0.0 | 0.0 | 0.0 | 1.7 | T | 6.7 | T | T | T | 0.0 | 0.0 | 8.4 |
| 1988-89 | 0.0 | 0.0 | 0.0 | T | 0.0 | 3.8 | 0.7 | 9.1 | 0.2 | 0.2 | T | T | 14.0 |
| 1989-90 | 0.0 | 0.0 | 0.0 | 0.0 | 3.3 | 11.1 | T | 0.2 | 1.5 | 0.0 | T | 0.0 | 16.1 |
| 1990-91 | 0.0 | 0.0 | 0.0 | 0.0 | 0.0 | 0.8 | T | T | 0.4 | 0.0 | 0.0 | 0.0 | 1.2 |
| 1991-92 | 0.0 | 0.0 | 0.0 | 0.0 | T | T | 0.4 | 2.2 | 0.5 | 1.9 | 0.0 | 0.0 | 5.0 |
| 1992-93 | 0.0 | 0.0 | 0.0 | 0.0 | T | 0.9 | 0.1 | 10.7 | 16.0 | 0.3 | 0.0 | 0.0 | 28.0 |
| 1993-94 | 0.0 | 0.0 | 0.0 | T | T | 6.2 | 5.2 | 1.8 | 3.1 | T | 0.0 | 0.0 | 16.3 |
| 1994-95 | 0.0 | 0.0 | 0.0 | 0.0 | T | 0.0 | 9.6 | 1.2 | 1.0 | 0.0 | 0.0 | 0.0 | 11.8 |
| 1995-96 | 0.0 | 0.0 | 0.0 | 0.0 | 0.4 | 10.2 | 28.2 | 17.1 | 0.1 | T | 0.0 | 0.0 | 56.0 |
| 1996-97 | 0.0 | 0.0 | 0.0 | T | 0.0 | 4.0 | T | 0.0 | 0.0 | 0.0 | 0.0 | 0.0 | 4.0 |
| 1997-98 | 0.0 | 0.0 | 0.0 | 0.0 | 0.0 | 6.5 | 4.4 | T | T | 0.0 | 0.0 | 0.0 | 10.9 |
| 1998-99 | 0.0 | 0.0 | 0.0 | 0.0 | 0.0 | T | 2.5 | T | 4.5 | 0.0 | 0.0 | 0.0 | 7.0 |
| 1999-00 | 0.0 | 0.0 | 0.0 | 0.0 | T | 0.0 | 10.5 | 0.1 | 0.0 | T | 0.0 | 0.0 | 10.6 |
| 2000-01 | 0.0 | 0.0 | 0.0 | 0.0 | T | 3.6 | 0.2 | 4.0 | 0.5 | T | 0.0 | 0.0 | 8.3 |
| 2001-02 | 0.0 | 0.0 | 0.0 | 0.0 | 0.0 | 0.0 | 3.6 | 0.5 | 0.0 | 0.0 | 0.0 | 0.0 | 4.1 |
| 2002-03 | 0.0 | 0.0 | 0.0 | 0.0 | T | 5.8 | 3.9 | 8.8 | 7.0 | 0.0 | 0.0 | 0.0 | 25.5 |
| 2003-04 | 0.0 | 0.0 | 0.0 | T | 0.0 | 9.6 | 7.9 | 4.4 | T | 0.0 | 0.0 | 0.0 | 21.9 |
| 2004-05 | 0.0 | 0.0 | 0.0 | 0.0 | 0.0 | T | 3.0 | 9.9 | 3.2 | 0.0 | 0.0 | 0.0 | 16.1 |
| 2005-06 | 0.0 | 0.0 | 0.0 | 0.0 | 0.4 | 3.7 | T | 5.9 | T | 0.0 | 0.0 | 0.0 | 10.0 |
| 2006-07 | 0.0 | 0.0 | T | 0.0 | 0.0 | T | T | 3.4 | T | T | 0.0 | 0.0 | 3.4 |
| 2007-08 | 0.0 | 0.0 | 0.0 | 0.0 | 0.0 | 0.9 | 4.0 | T | T | 0.0 | 0.0 | 0.0 | 4.9 |
| 2008-09 | 0.0 | 0.0 | 0.0 | 0.0 | 0.4 | T | 0.0 | 0.4 | 3.6 | 0.0 | 0.0 | 0.0 | 4.4 |
| 2009- | 0.0 | 0.0 | 0.0 | 0.0 | T | 19.0 | | | | | | | |
| POR=<br>62 YRS | 0.0 | 0.0 | T | T | 1.1 | 3.9 | 5.8 | 6.4 | 3.2 | 0.3 | T | T | 20.7 |

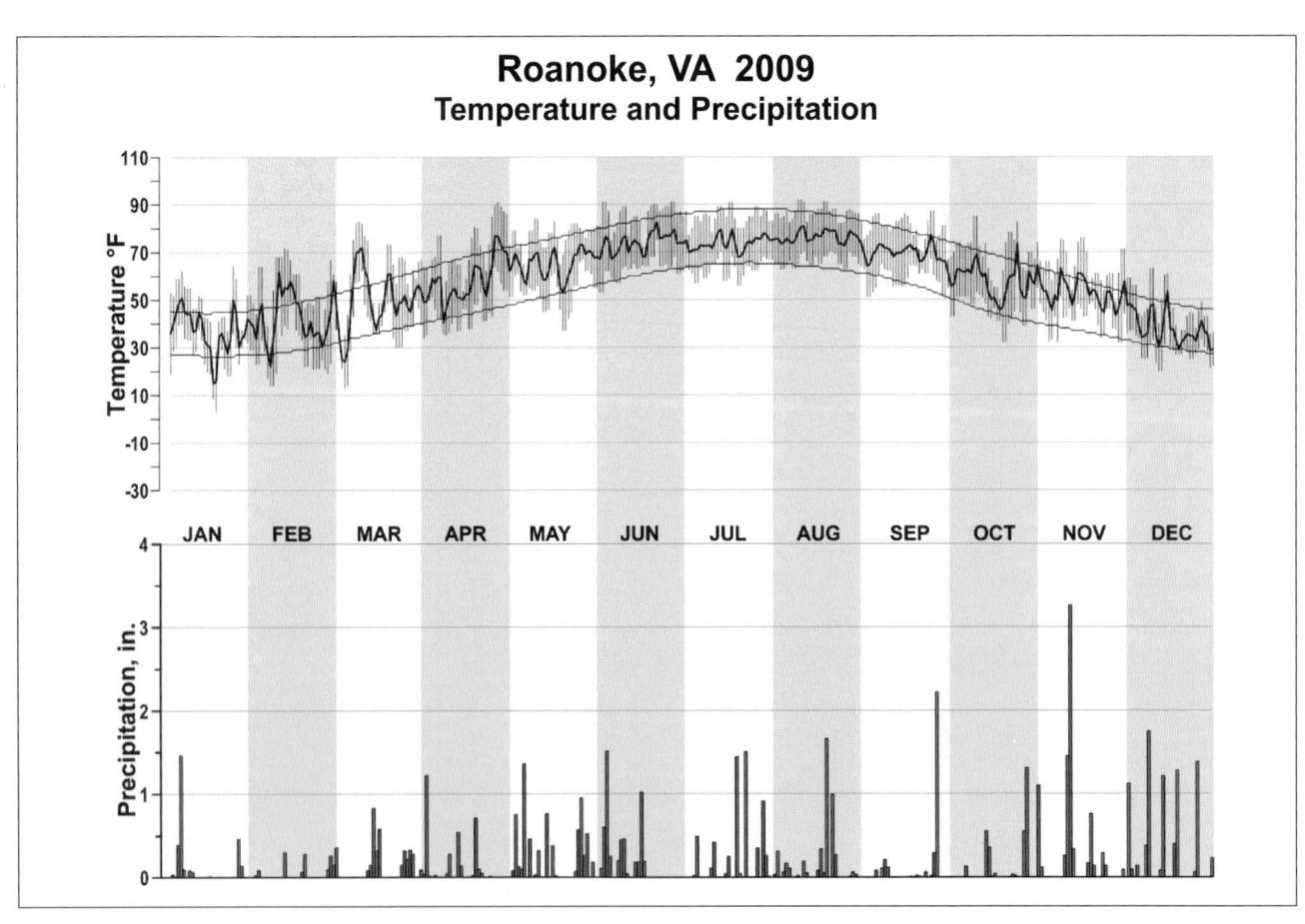

## Roanoke, VA 2009
### Temperature and Precipitation

# 2009
# SEATTLE
# SEATTLE - TACOMA AIRPORT (KSEA)

The Seattle-Tacoma International Airport is located 6 miles south of the Seattle city limits and 14 miles north of Tacoma. It is situated on a low ridge lying between Puget Sound on the west and the Green River valley on the east with terrain sloping moderately to the shores of Puget Sound some 2 miles to the west. The Olympic Mountains, rising sharply from Puget Sound, are about 50 miles to the northwest. Rather steep bluffs border the Green River Valley about 2.5 miles to the east and the foothills of the Cascade Range begin 10 to 15 miles to the east of the airport.

The mild climate of the Pacific Coast is modified by the Cascade Mountains and, to a lesser extent, by the Olympic Mountains. The climate is characterized by mild temperatures, a pronounced though not sharply defined rainy season, and considerable cloudiness, particularly during the winter months. The Cascades are very effective in shielding the Seattle-Tacoma area from the cold, dry continental air during the winter and the hot, dry continental air during the summer months. The extremes of temperature that occur in western Washington are the result of the occasional pressure distributions that force the continental air into the Puget Sound area. But the prevailing southwesterly circulation keeps the average winter daytime temperatures in the 40s and the nighttime readings in the 30s. During the summer, daytime temperatures are usually in the 70s with nighttime lows in the 50s. Extremes of temperatures, both in the winter and summer, are usually of short duration. The dry season is centered around July and early August with July being the driest month of the year. The rainy season extends from October to March with December normally the wettest month, however, precipitation is rather evenly distributed through the winter and early spring months with more than 75 percent of the yearly precipitation falling during the winter wet season. Most of the rainfall in the Seattle area comes from storms common to the middle latitudes. These disturbances are most vigorous during the winter as they move through western Washington. The storm track shifts to the north during the summer and those that reach the State are not the wind and rain producers of the winter months. Local summer afternoon showers and a few thunderstorms occur in the Seattle-Tacoma area but they do not contribute materially to the precipitation.

The occurrence of snow in the Seattle-Tacoma area is extremely variable and usually melts before accumulating measurable depths. There are winters on record with only a trace of snow, but at the other extreme, over 21 inches has fallen in a 24-hour period. Usually, winter storms do not produce snow unless the storm moves in such a way to bring cold air out of Canada directly or with only a short over water trajectory.

The highest winds recorded in the Seattle-Tacoma area were associated with strong storms crossing the state from the southwest. Prevailing winds are from the southwest but occasional severe winter storms will produce strong northerly winds. Winds during the summer months are relatively light with occasional land-sea breeze effects creating afternoon northerly winds of 8 to 15 miles an hour. Fog or low clouds that form over the southern Puget Sound area in the late summer, fall, and early winter months, often dominate the weather conditions during the late night and early morning hours with visibilities occasionally lower for a few hours near sunrise. Most of the summer clouds form along the coast and move into the Seattle area from the southwest.

Based on the 1951-1980 period, the average first occurrence of 32 degrees Fahrenheit in the fall is November 11 and the average last occurrence in the spring is March 24.

# NORMALS, MEANS, AND EXTREMES
## SEATTLE (KSEA)

**LATITUDE:** 47° 27'N  **LONGITUDE:** -122° 18'W  **ELEVATION (FT):** GRND: 370  BARO: 434  **TIME ZONE:** PACIFIC (UTC -8)  **WBAN: 24233**

| | ELEMENT | POR | JAN | FEB | MAR | APR | MAY | JUN | JUL | AUG | SEP | OCT | NOV | DEC | YEAR |
|---|---|---|---|---|---|---|---|---|---|---|---|---|---|---|---|
| **TEMPERATURE °F** | NORMAL DAILY MAXIMUM | 30 | 45.8 | 49.5 | 53.2 | 58.2 | 64.4 | 69.6 | 75.3 | 75.6 | 70.2 | 59.7 | 50.5 | 45.5 | 59.8 |
| | MEAN DAILY MAXIMUM | 62 | 44.9 | 48.9 | 52.1 | 57.4 | 64.2 | 69.5 | 74.1 | 74.8 | 69.4 | 59.3 | 50.5 | 45.4 | 59.2 |
| | HIGHEST DAILY MAXIMUM | 65 | 64 | 70 | 78 | 85 | 93 | 96 | 103 | 99 | 98 | 89 | 74 | 64 | 103 |
| | YEAR OF OCCURRENCE | | 1981 | 1968 | 2004 | 1976 | 1963 | 1995 | 2009 | 1981 | 1988 | 1987 | 1949 | 1993 | JUL 2009 |
| | MEAN OF EXTREME MAXS. | 62 | 55.4 | 59.6 | 64.4 | 73.2 | 81.4 | 85.4 | 89.7 | 88.1 | 84.2 | 72.8 | 60.9 | 55.7 | 72.6 |
| | NORMAL DAILY MINIMUM | 30 | 35.9 | 37.2 | 39.1 | 42.1 | 47.2 | 51.7 | 55.3 | 55.7 | 51.9 | 45.7 | 39.9 | 35.9 | 44.8 |
| | MEAN DAILY MINIMUM | 62 | 35.0 | 36.6 | 38.1 | 41.2 | 46.5 | 51.4 | 54.0 | 55.0 | 51.4 | 45.4 | 39.6 | 35.9 | 44.2 |
| | LOWEST DAILY MINIMUM | 65 | 0 | 1 | 11 | 29 | 28 | 38 | 43 | 44 | 35 | 28 | 6 | 6 | 0 |
| | YEAR OF OCCURRENCE | | 1950 | 1950 | 1955 | 1975 | 1954 | 1952 | 1954 | 1955 | 1972 | 1949 | 1955 | 1968 | JAN 1950 |
| | MEAN OF EXTREME MINS. | 62 | 22.4 | 25.7 | 29.5 | 33.9 | 38.5 | 45.3 | 49.2 | 49.4 | 43.3 | 34.8 | 28.2 | 24.0 | 35.4 |
| | NORMAL DRY BULB | 30 | 40.9 | 43.3 | 46.2 | 50.2 | 55.8 | 60.7 | 65.3 | 65.6 | 61.1 | 52.7 | 45.2 | 40.7 | 52.3 |
| | MEAN DRY BULB | 62 | 39.9 | 42.8 | 45.1 | 49.3 | 55.3 | 60.5 | 64.1 | 64.9 | 60.4 | 52.3 | 45.1 | 40.6 | 51.7 |
| | MEAN WET BULB | 26 | 38.6 | 38.9 | 41.7 | 44.7 | 49.3 | 53.3 | 57.0 | 57.5 | 54.4 | 48.4 | 42.3 | 37.7 | 47.0 |
| | MEAN DEW POINT | 26 | 36.3 | 36.0 | 38.9 | 41.6 | 46.3 | 50.0 | 53.5 | 54.0 | 51.4 | 46.3 | 40.4 | 35.6 | 44.2 |
| | NORMAL NO. DAYS WITH: | | | | | | | | | | | | | | |
| | MAXIMUM >= 90 | 30 | 0.0 | 0.0 | 0.0 | 0.0 | 0.1 | 0.4 | 1.2 | 1.1 | 0.3 | 0.0 | 0.0 | 0.0 | 3.1 |
| | MAXIMUM <= 32 | 30 | 0.7 | 0.3 | 0.0 | 0.0 | 0.0 | 0.0 | 0.0 | 0.0 | 0.0 | 0.0 | 0.2 | 1.2 | 2.4 |
| | MINIMUM <= 32 | 30 | 8.8 | 6.0 | 2.3 | 0.2 | 0.0 | 0.0 | 0.0 | 0.0 | 0.0 | 0.2 | 3.4 | 8.8 | 29.7 |
| | MINIMUM <= 0 | 30 | 0.0 | 0.0 | 0.0 | 0.0 | 0.0 | 0.0 | 0.0 | 0.0 | 0.0 | 0.0 | 0.0 | 0.0 | 0.0 |
| **H/C** | NORMAL HEATING DEG. DAYS | 30 | 747 | 613 | 582 | 447 | 291 | 150 | 55 | 45 | 138 | 383 | 592 | 754 | 4797 |
| | NORMAL COOLING DEG. DAYS | 30 | 0 | 0 | 0 | 0 | 5 | 19 | 65 | 65 | 19 | 0 | 0 | 0 | 173 |
| **RH** | NORMAL (PERCENT) | 30 | 81 | 77 | 76 | 73 | 72 | 70 | 68 | 68 | 73 | 80 | 82 | 82 | 75 |
| | HOUR 04 LST | 30 | 85 | 83 | 86 | 85 | 86 | 85 | 83 | 84 | 88 | 89 | 86 | 85 | 85 |
| | HOUR 10 LST | 30 | 82 | 78 | 76 | 72 | 70 | 69 | 67 | 67 | 72 | 79 | 81 | 83 | 75 |
| | HOUR 16 LST | 30 | 75 | 68 | 62 | 58 | 56 | 55 | 50 | 48 | 54 | 66 | 76 | 79 | 62 |
| | HOUR 22 LST | 30 | 81 | 78 | 78 | 76 | 75 | 74 | 70 | 70 | 76 | 82 | 83 | 82 | 77 |
| **S** | PERCENT POSSIBLE SUNSHINE | 30 | 28 | 40 | 50 | 52 | 56 | 56 | 64 | 65 | 62 | 43 | 28 | 23 | 47 |
| **W/O** | MEAN NO. DAYS WITH: | | | | | | | | | | | | | | |
| | HEAVY FOG(VISBY <= 1/4 MI) | 46 | 4.5 | 2.9 | 1.6 | 0.8 | 0.3 | 0.4 | 0.6 | 1.5 | 3.0 | 5.3 | 3.8 | 4.9 | 29.6 |
| | THUNDERSTORMS | 62 | 0.2 | 0.4 | 0.6 | 0.8 | 0.8 | 0.6 | 0.7 | 0.7 | 0.6 | 0.4 | 0.6 | 0.3 | 6.7 |
| **CLOUDNESS** | MEAN: | | | | | | | | | | | | | | |
| | SUNRISE-SUNSET (OKTAS) | 52 | 6.7 | 6.5 | 6.3 | 6.2 | 5.7 | 5.6 | 4.2 | 4.5 | 4.8 | 5.9 | 6.6 | 6.7 | 5.8 |
| | MIDNIGHT-MIDNIGHT (OKTAS) | 32 | 6.6 | 6.1 | 5.8 | 5.9 | 5.5 | 5.3 | 4.1 | 4.2 | 4.4 | 5.3 | 6.3 | 6.4 | 5.5 |
| | MEAN NO. DAYS WITH: | | | | | | | | | | | | | | |
| | CLEAR | 52 | 2.8 | 3.0 | 3.4 | 2.7 | 4.4 | 5.1 | 10.2 | 9.0 | 8.1 | 4.0 | 2.5 | 2.3 | 57.5 |
| | PARTLY CLOUDY | 52 | 3.9 | 4.2 | 5.8 | 7.3 | 8.8 | 7.9 | 9.5 | 9.7 | 8.5 | 7.4 | 4.2 | 3.9 | 81.1 |
| | CLOUDY | 52 | 24.3 | 21.1 | 21.9 | 20.0 | 17.7 | 16.9 | 10.5 | 11.9 | 13.0 | 19.5 | 23.3 | 24.9 | 225.0 |
| **PR** | MEAN STATION PRESSURE(IN) | 26 | 29.53 | 29.55 | 29.54 | 29.55 | 29.55 | 29.55 | 29.57 | 29.55 | 29.55 | 29.57 | 29.54 | 29.57 | 29.55 |
| | MEAN SEA-LEVEL PRES. (IN) | 26 | 30.06 | 30.04 | 30.03 | 30.04 | 30.03 | 30.04 | 30.05 | 30.03 | 30.04 | 30.06 | 30.04 | 30.06 | 30.04 |
| **WINDS** | MEAN SPEED (MPH) | 26 | 8.7 | 8.6 | 8.6 | 8.2 | 8.0 | 7.9 | 7.5 | 7.2 | 7.1 | 7.5 | 8.6 | 8.5 | 8.0 |
| | PREVAIL.DIR(TENS OF DEGS) | 43 | 20 | 20 | 22 | 22 | 22 | 22 | 23 | 22 | 01 | 20 | 20 | 19 | 22 |
| | MAXIMUM 2-MINUTE: | | | | | | | | | | | | | | |
| | SPEED (MPH) | 13 | 40 | 39 | 44 | 31 | 30 | 36 | 23 | 24 | 36 | 39 | 37 | 52 | 52 |
| | DIR. (TENS OF DEGS) | | 20 | 21 | 19 | 24 | 23 | 23 | 23 | 04 | 20 | 22 | 20 | 22 | 22 |
| | YEAR OF OCCURRENCE | | 2000 | 2008 | 1999 | 2002 | 2006 | 2008 | 2009 | 2008 | 1999 | 2007 | 1998 | 2006 | DEC 2006 |
| | MAXIMUM 3-SECOND | | | | | | | | | | | | | | |
| | SPEED (MPH) | 13 | 52 | 53 | 60 | 44 | 40 | 46 | 31 | 29 | 43 | 53 | 48 | 69 | 69 |
| | DIR. (TENS OF DEGS) | | 20 | 22 | 18 | 24 | 22 | 23 | 23 | 05 | 21 | 22 | 21 | 22 | 22 |
| | YEAR OF OCCURRENCE | | 2000 | 2008 | 1999 | 2002 | 2005 | 2008 | 1999 | 2008 | 1999 | 2007 | 1998 | 2006 | DEC 2006 |
| **PRECIPITATION** | NORMAL (IN) | 30 | 5.13 | 4.18 | 3.75 | 2.59 | 1.78 | 1.49 | 0.79 | 1.02 | 1.63 | 3.19 | 5.90 | 5.62 | 37.07 |
| | MAXIMUM MONTHLY (IN) | 65 | 12.92 | 9.11 | 8.40 | 6.53 | 4.76 | 3.90 | 2.39 | 4.59 | 5.95 | 8.96 | 15.63 | 11.85 | 15.63 |
| | YEAR OF OCCURRENCE | | 1953 | 1961 | 1950 | 1991 | 1948 | 1946 | 1983 | 1975 | 1978 | 2003 | 2006 | 1979 | NOV 2006 |
| | MINIMUM MONTHLY (IN) | 65 | 0.58 | 0.35 | 0.57 | 0.33 | 0.12 | 0.13 | T | 0.01 | T | 0.31 | 0.74 | 1.37 | T |
| | YEAR OF OCCURRENCE | | 1985 | 1993 | 1965 | 1956 | 1992 | 1951 | 1960 | 1974 | 1991 | 1987 | 1976 | 1978 | SEP 1991 |
| | MAXIMUM IN 24 HOURS (IN) | 65 | 3.22 | 3.41 | 2.86 | 3.32 | 1.83 | 2.21 | 0.85 | 1.75 | 2.23 | 5.02 | 3.78 | 4.18 | 5.02 |
| | YEAR OF OCCURRENCE | | 1986 | 1951 | 1972 | 1991 | 1969 | 2008 | 1981 | 1968 | 1978 | 2003 | 2006 | 2007 | OCT 2003 |
| | NORMAL NO. DAYS WITH: | | | | | | | | | | | | | | |
| | PRECIPITATION >= 0.01 | 30 | 17.8 | 15.6 | 16.4 | 13.6 | 11.6 | 8.5 | 5.3 | 5.5 | 8.3 | 11.7 | 17.9 | 17.8 | 150.0 |
| | PRECIPITATION >= 1.00 | 30 | 0.7 | 0.4 | 0.3 | 0.2 | * | 0.1 | 0.0 | 0.1 | 0.1 | 0.4 | 1.0 | 1.2 | 4.5 |
| **SNOWFALL** | NORMAL (IN) | 30 | 2.4 | 1.3 | 0.6 | 0.1 | 0.0 | 0.0 | 0.0 | 0.0 | 0.0 | 0.1 | 1.1 | 2.5 | 8.1 |
| | MAXIMUM MONTHLY (IN) | 53 | 57.2 | 13.1 | 18.2 | 2.3 | T | 0.0 | T | 0.0 | | T | 17.5 | 22.1 | 57.2 |
| | YEAR OF OCCURRENCE | | 1950 | 1949 | 1951 | 1972 | 1993 | | 1980 | | 1972 | 1971 | 1985 | 1968 | JAN 1950 |
| | MAXIMUM IN 24 HOURS (IN) | 53 | 21.4 | 9.8 | 7.4 | 2.3 | T | 0.0 | T | 0.0 | | T | 2.0 | 9.4 | 13.0 | 21.4 |
| | YEAR OF OCCURRENCE' | | 1950 | 1990 | 1989 | 1972 | 1993 | | 1980 | | 1972 | 1971 | 1946 | 1968 | JAN 1950 |
| | MAXIMUM SNOW DEPTH (IN) | 48 | 21 | 11 | 7 | 2 | 0 | 0 | 0 | 0 | | 0 | 2 | 8 | 10 | 21 |
| | YEAR OF OCCURRENCE | | 1969 | 1969 | 1989 | 1972 | | | | | | 1971 | 1985 | 1974 | JAN 1969 |
| | NORMAL NO. DAYS WITH: | | | | | | | | | | | | | | |
| | SNOWFALL >= 1.0 | 30 | 0.9 | 0.4 | 0.2 | 0.1 | 0.0 | 0.0 | 0.0 | 0.0 | 0.0 | 0.0 | 0.3 | 0.7 | 2.6 |

## PRECIPITATION (inches) 2009 SEATTLE (KSEA)

| YEAR | JAN | FEB | MAR | APR | MAY | JUN | JUL | AUG | SEP | OCT | NOV | DEC | ANNUAL |
|------|-----|-----|-----|-----|-----|-----|-----|-----|-----|-----|-----|-----|--------|
| 1980 | 4.09 | 5.04 | 2.10 | 3.23 | 0.97 | 1.77 | 0.46 | 0.64 | 1.43 | 1.32 | 7.16 | 7.39 | 35.60 |
| 1981 | 2.42 | 4.45 | 2.23 | 1.58 | 1.33 | 2.31 | 1.38 | 0.25 | 3.42 | 6.40 | 4.07 | 5.56 | 35.40 |
| 1982 | 5.35 | 7.57 | 3.73 | 2.07 | 0.63 | 1.03 | 0.59 | 0.62 | 1.49 | 4.07 | 5.31 | 6.86 | 39.32 |
| 1983 | 7.07 | 4.57 | 3.81 | 1.06 | 2.10 | 1.85 | 2.39 | 1.90 | 1.85 | 1.34 | 7.97 | 5.02 | 40.93 |
| 1984 | 3.62 | 3.91 | 3.91 | 2.87 | 3.38 | 2.81 | 0.17 | 0.13 | 1.01 | 2.14 | 8.09 | 4.95 | 36.99 |
| 1985 | 0.58 | 2.63 | 2.56 | 1.30 | 0.85 | 2.80 | 0.10 | 0.55 | 1.98 | 5.74 | 4.26 | 1.78 | 25.13 |
| 1986 | 8.54 | 4.41 | 2.67 | 1.38 | 1.71 | 0.68 | 1.10 | 0.10 | 1.89 | 4.21 | 7.98 | 3.67 | 38.34 |
| 1987 | 5.98 | 2.05 | 5.53 | 2.61 | 2.38 | 0.16 | 0.39 | 0.29 | 0.91 | 0.31 | 3.21 | 6.11 | 29.93 |
| 1988 | 4.07 | 0.71 | 3.75 | 3.20 | 3.01 | 1.56 | 0.50 | 0.28 | 1.75 | 2.24 | 8.43 | 3.48 | 32.98 |
| 1989 | 2.78 | 3.43 | 5.79 | 2.80 | 2.78 | 1.14 | 0.64 | 0.89 | 0.54 | 2.98 | 6.13 | 4.79 | 34.69 |
| 1990 | 9.41 | 3.72 | 2.58 | 2.54 | 1.98 | 3.05 | 0.58 | 0.71 | 0.05 | 5.79 | 10.71 | 3.63 | 44.75 |
| 1991 | 4.46 | 4.69 | 4.66 | 6.53 | 1.39 | 1.29 | 0.28 | 2.17 | T | 1.31 | 5.33 | 3.31 | 35.42 |
| 1992 | 7.82 | 3.09 | 1.68 | 4.12 | 0.12 | 1.14 | 0.89 | 0.66 | 1.15 | 2.45 | 5.57 | 4.09 | 32.78 |
| 1993 | 4.09 | 0.35 | 4.80 | 4.54 | 2.86 | 2.48 | 1.27 | 0.16 | 0.03 | 1.54 | 2.20 | 4.48 | 28.80 |
| 1994 | 2.51 | 4.47 | 3.17 | 2.27 | 1.43 | 1.25 | 0.28 | 0.30 | 1.69 | 3.51 | 5.79 | 8.15 | 34.82 |
| 1995 | 4.48 | 4.97 | 4.07 | 2.05 | 0.81 | 1.46 | 1.34 | 1.81 | 0.91 | 3.93 | 10.40 | 6.37 | 42.60 |
| 1996 | 7.34 | 8.35 | 2.06 | 5.37 | 2.07 | 0.59 | .77 | 1.32 | 1.85 | 5.54 | 5.23 | 10.18 | 50.67 |
| 1997 | 7.02 | 1.99 | 8.15 | 4.32 | 1.87 | 1.64 | 1.20 | 1.27 | 3.41 | 5.83 | 3.93 | 2.63 | 43.26 |
| 1998 | 7.15 | 3.31 | 3.96 | 0.99 | 1.98 | 1.11 | 0.41 | 0.35 | 0.72 | 3.48 | 11.62 | 8.98 | 44.06 |
| 1999 | 6.84 | 6.95 | 3.66 | 1.49 | 2.12 | 1.86 | 1.18 | 0.92 | 0.17 | 2.26 | 9.60 | 5.06 | 42.11 |
| 2000 | 3.77 | 5.25 | 2.82 | 1.48 | 3.27 | 1.61 | 0.23 | 0.33 | 1.12 | 3.00 | 3.27 | 2.51 | 28.66 |
| 2001 | 2.70 | 2.07 | 2.73 | 3.16 | 1.39 | 3.05 | 1.03 | 2.32 | 0.83 | 3.13 | 9.26 | 5.89 | 37.56 |
| 2002 | 5.98 | 4.17 | 2.82 | 4.29 | 1.11 | 1.73 | 0.64 | 0.04 | 0.42 | 0.67 | 3.51 | 5.98 | 31.36 |
| 2003 | 8.39 | 1.76 | 6.34 | 2.74 | 1.16 | 0.51 | 0.06 | 0.32 | 0.89 | 8.96 | 6.77 | 3.88 | 41.78 |
| 2004 | 6.36 | 2.44 | 2.14 | 0.65 | 2.51 | 0.71 | 0.16 | 3.00 | 2.80 | 2.80 | 3.16 | 4.37 | 31.10 |
| 2005 | 4.44 | 1.20 | 3.71 | 3.68 | 3.32 | 1.63 | 1.03 | 0.29 | 0.75 | 3.02 | 5.52 | 6.85 | 35.44 |
| 2006 | 11.65 | 2.55 | 2.18 | 2.73 | 1.65 | 1.67 | 0.06 | 0.02 | 1.43 | 1.55 | 15.63 | 7.30 | 48.42 |
| 2007 | 6.22 | 3.38 | 4.42 | 0.69 | 1.46 | 1.34 | 1.44 | 0.73 | 3.16 | 3.32 | 3.71 | 9.08 | 38.95 |
| 2008 | 4.26 | 1.47 | 3.65 | 1.90 | 0.89 | 1.64 | 0.48 | 2.87 | 0.78 | 2.17 | 6.52 | 4.10 | 30.73 |
| 2009 | 5.40 | 1.51 | 4.16 | 3.36 | 3.61 | 0.18 | 0.06 | 1.16 | 1.75 | 5.54 | 8.96 | 2.75 | 38.44 |
| POR= 62 YRS | 5.74 | 3.95 | 3.72 | 2.54 | 1.77 | 1.44 | 0.74 | 1.10 | 1.73 | 3.48 | 6.15 | 5.81 | 38.17 |

WBAN : 24233

## AVERAGE TEMPERATURE (°F) 2009 SEATTLE (KSEA)

| YEAR | JAN | FEB | MAR | APR | MAY | JUN | JUL | AUG | SEP | OCT | NOV | DEC | ANNUAL |
|------|-----|-----|-----|-----|-----|-----|-----|-----|-----|-----|-----|-----|--------|
| 1980 | 34.8 | 43.8 | 44.3 | 51.6 | 54.2 | 57.5 | 63.8 | 61.9 | 59.6 | 53.9 | 46.7 | 44.1 | 51.4 |
| 1981 | 44.4 | 44.2 | 48.8 | 49.6 | 54.7 | 57.5 | 63.3 | 68.1 | 61.1 | 50.9 | 47.2 | 41.7 | 52.6 |
| 1982 | 39.3 | 42.1 | 44.1 | 47.4 | 54.7 | 63.1 | 62.8 | 65.1 | 60.6 | 52.7 | 43.2 | 40.8 | 51.3 |
| 1983 | 45.0 | 46.9 | 49.4 | 50.7 | 57.7 | 59.9 | 63.3 | 65.6 | 58.3 | 51.7 | 47.8 | 36.1 | 52.7 |
| 1984 | 43.2 | 44.8 | 48.5 | 48.7 | 52.9 | 58.8 | 65.0 | 64.9 | 59.9 | 49.7 | 44.6 | 36.8 | 51.5 |
| 1985 | 37.1 | 39.0 | 43.3 | 49.2 | 54.8 | 60.0 | 68.6 | 65.2 | 58.1 | 51.4 | 35.8 | 36.2 | 49.9 |
| 1986 | 44.9 | 42.8 | 49.2 | 48.1 | 55.7 | 62.7 | 61.7 | 68.4 | 59.1 | 54.3 | 45.3 | 42.0 | 52.9 |
| 1987 | 40.5 | 46.3 | 48.9 | 52.0 | 56.9 | 62.6 | 64.2 | 66.1 | 62.6 | 55.8 | 48.5 | 39.2 | 53.6 |
| 1988 | 40.1 | 44.4 | 45.6 | 50.3 | 54.9 | 59.6 | 65.3 | 65.4 | 60.5 | 55.4 | 45.4 | 41.9 | 52.4 |
| 1989 | 40.5 | 35.9 | 43.7 | 53.4 | 56.0 | 63.2 | 64.5 | 65.3 | 64.1 | 53.1 | 47.0 | 42.9 | 52.5 |
| 1990 | 42.5 | 40.0 | 47.1 | 52.1 | 54.7 | 59.8 | 68.0 | 67.3 | 63.4 | 51.2 | 46.6 | 35.3 | 52.3 |
| 1991 | 40.0 | 47.7 | 44.1 | 49.1 | 54.3 | 58.9 | 66.8 | 66.6 | 62.9 | 52.9 | 47.3 | 43.7 | 52.9 |
| 1992 | 43.9 | 47.3 | 50.3 | 53.1 | 59.8 | 65.0 | 66.7 | 66.8 | 60.0 | 54.4 | 45.5 | 38.8 | 54.3 |
| 1993 | 37.9 | 42.3 | 48.1 | 50.6 | 59.6 | 60.6 | 61.2 | 65.5 | 61.9 | 55.4 | 42.0 | 41.4 | 52.2 |
| 1994 | 44.9 | 40.4 | 48.5 | 52.3 | 58.2 | 60.6 | 67.9 | 67.3 | 64.4 | 52.8 | 42.4 | 41.7 | 53.5 |
| 1995 | 46.4 | 45.9 | 47.8 | 51.8 | 60.1 | 62.7 | 67.0 | 63.0 | 64.0 | 52.5 | 49.1 | 42.4 | 54.4 |
| 1996 | 39.7 | 43.2 | 47.2 | 51.5 | 53.1 | 60.5 | 67.9 | 66.5 | 59.2 | 51.1 | 43.1 | 39.3 | 51.9 |
| 1997 | 41.1 | 41.9 | 44.8 | 49.2 | 58.2 | 59.5 | 64.5 | 67.7 | 62.2 | 51.3 | 48.4 | 41.5 | 52.5 |
| 1998 | 42.2 | 45.8 | 46.6 | 50.4 | 54.9 | 59.9 | 67.6 | 66.8 | 63.1 | 52.6 | 46.7 | 39.9 | 53.0 |
| 1999 | 42.0 | 42.5 | 44.3 | 48.5 | 51.8 | 58.2 | 62.4 | 64.8 | 60.8 | 52.2 | 47.9 | 42.0 | 51.5 |
| 2000 | 40.3 | 43.7 | 44.5 | 50.9 | 53.8 | 60.6 | 64.3 | 63.5 | 60.2 | 52.8 | 42.5 | 40.8 | 51.5 |
| 2001 | 42.0 | 40.7 | 45.4 | 48.0 | 55.4 | 57.6 | 62.5 | 64.8 | 59.8 | 50.9 | 46.7 | 41.5 | 51.3 |
| 2002 | 40.7 | 41.9 | 41.9 | 48.8 | 53.1 | 61.1 | 65.0 | 65.2 | 60.6 | 51.4 | 47.1 | 43.0 | 51.7 |
| 2003 | 45.8 | 41.7 | 46.8 | 48.9 | 54.8 | 62.8 | 67.9 | 66.4 | 62.6 | 54.3 | 42.8 | 41.8 | 53.1 |
| 2004 | 40.3 | 44.5 | 47.6 | 53.3 | 57.0 | 63.4 | 68.2 | 67.1 | 58.8 | 53.3 | 45.0 | 42.5 | 53.4 |
| 2005 | 42.1 | 42.6 | 49.1 | 50.8 | 58.2 | 59.9 | 65.6 | 66.8 | 59.5 | 54.0 | 42.9 | 42.7 | 52.9 |
| 2006 | 46.6 | 42.9 | 45.5 | 50.3 | 56.8 | 63.1 | 67.5 | 65.9 | 62.3 | 52.3 | 44.2 | 40.6 | 53.2 |
| 2007 | 38.0 | 43.6 | 47.1 | 50.6 | 56.6 | 60.3 | 67.8 | 65.6 | 60.0 | 50.6 | 44.3 | 39.8 | 52.0 |
| 2008 | 38.7 | 43.9 | 43.5 | 47.0 | 56.1 | 58.4 | 64.9 | 65.9 | 61.0 | 51.7 | 49.2 | 36.9 | 51.4 |
| 2009 | 39.1 | 41.7 | 41.7 | 49.2 | 56.3 | 63.8 | 69.5 | 65.6 | 62.4 | 52.2 | 46.7 | 37.9 | 52.2 |
| POR= 62 YRS | 39.9 | 42.8 | 45.1 | 49.3 | 55.3 | 60.5 | 64.1 | 64.9 | 60.4 | 52.3 | 45.1 | 40.6 | 51.7 |

## HEATING DEGREE DAYS (base 65°F) 2009 SEATTLE (KSEA)

| YEAR | JUL | AUG | SEP | OCT | NOV | DEC | JAN | FEB | MAR | APR | MAY | JUN | TOTAL |
|------|-----|-----|-----|-----|-----|-----|-----|-----|-----|-----|-----|-----|-------|
| 1980-81 | 66 | 104 | 158 | 343 | 543 | 639 | 633 | 577 | 494 | 455 | 316 | 220 | 4548 |
| 1981-82 | 80 | 28 | 138 | 430 | 530 | 715 | 790 | 636 | 640 | 521 | 312 | 103 | 4923 |
| 1982-83 | 93 | 42 | 141 | 373 | 647 | 745 | 613 | 502 | 479 | 422 | 244 | 149 | 4450 |
| 1983-84 | 72 | 19 | 196 | 406 | 511 | 890 | 672 | 577 | 507 | 482 | 372 | 183 | 4887 |
| 1984-85 | 54 | 42 | 159 | 467 | 603 | 867 | 857 | 719 | 666 | 469 | 310 | 160 | 5373 |
| 1985-86 | 8 | 48 | 199 | 413 | 870 | 888 | 618 | 616 | 479 | 502 | 305 | 90 | 5036 |
| 1986-87 | 105 | 12 | 196 | 323 | 586 | 707 | 754 | 522 | 491 | 384 | 253 | 105 | 4438 |
| 1987-88 | 58 | 37 | 102 | 284 | 485 | 792 | 767 | 590 | 593 | 435 | 316 | 165 | 4624 |
| 1988-89 | 60 | 38 | 162 | 291 | 583 | 708 | 749 | 807 | 654 | 340 | 273 | 93 | 4758 |
| 1989-90 | 41 | 29 | 68 | 362 | 534 | 677 | 689 | 696 | 547 | 379 | 312 | 158 | 4492 |
| 1990-91 | 29 | 23 | 61 | 420 | 546 | 913 | 767 | 475 | 639 | 472 | 322 | 179 | 4846 |
| 1991-92 | 24 | 40 | 90 | 368 | 526 | 654 | 645 | 507 | 447 | 351 | 171 | 63 | 3886 |
| 1992-93 | 26 | 22 | 151 | 322 | 580 | 805 | 834 | 629 | 517 | 427 | 170 | 135 | 4618 |
| 1993-94 | 113 | 48 | 120 | 290 | 685 | 723 | 618 | 682 | 503 | 373 | 209 | 132 | 4496 |
| 1994-95 | 23 | 5 | 43 | 369 | 671 | 713 | 570 | 527 | 527 | 390 | 160 | 118 | 4116 |
| 1995-96 | 19 | 77 | 60 | 382 | 471 | 693 | 779 | 630 | 543 | 399 | 362 | 135 | 4550 |
| 1996-97 | 42 | 32 | 167 | 422 | 649 | 793 | 732 | 640 | 622 | 468 | 211 | 159 | 4937 |
| 1997-98 | 38 | 11 | 95 | 416 | 489 | 718 | 700 | 533 | 562 | 433 | 307 | 152 | 4454 |
| 1998-99 | 21 | 21 | 79 | 376 | 543 | 771 | 706 | 623 | 634 | 487 | 400 | 218 | 4879 |
| 1999-00 | 101 | 42 | 141 | 393 | 504 | 706 | 760 | 611 | 628 | 415 | 340 | 154 | 4795 |
| 2000-01 | 60 | 65 | 146 | 373 | 668 | 745 | 707 | 675 | 599 | 504 | 305 | 218 | 5065 |
| 2001-02 | 86 | 44 | 150 | 430 | 547 | 723 | 746 | 640 | 709 | 481 | 363 | 143 | 5062 |
| 2002-03 | 55 | 46 | 134 | 412 | 533 | 675 | 590 | 648 | 556 | 478 | 308 | 117 | 4552 |
| 2003-04 | 16 | 11 | 108 | 326 | 656 | 713 | 757 | 589 | 530 | 344 | 243 | 97 | 4390 |
| 2004-05 | 23 | 19 | 177 | 355 | 596 | 691 | 702 | 622 | 487 | 419 | 219 | 150 | 4460 |
| 2005-06 | 34 | 17 | 161 | 336 | 655 | 687 | 565 | 612 | 596 | 433 | 255 | 91 | 4442 |
| 2006-07 | 33 | 31 | 110 | 388 | 616 | 752 | 833 | 589 | 545 | 423 | 267 | 146 | 4733 |
| 2007-08 | 6 | 24 | 164 | 440 | 613 | 773 | 808 | 601 | 659 | 532 | 281 | 226 | 5127 |
| 2008-09 | 42 | 53 | 123 | 405 | 468 | 864 | 797 | 649 | 714 | 465 | 273 | 70 | 4923 |
| 2009- | 29 | 36 | 99 | 385 | 545 | 835 | | | | | | | |

WBAN : 24233

## COOLING DEGREE DAYS (base 65°F) 2009 SEATTLE (KSEA)

| YEAR | JAN | FEB | MAR | APR | MAY | JUN | JUL | AUG | SEP | OCT | NOV | DEC | TOTAL |
|------|-----|-----|-----|-----|-----|-----|-----|-----|-----|-----|-----|-----|-------|
| 1980 | 0 | 0 | 0 | 0 | 0 | 0 | 34 | 15 | 3 | 2 | 0 | 0 | 54 |
| 1981 | 0 | 0 | 0 | 0 | 1 | 3 | 35 | 131 | 24 | 0 | 0 | 0 | 194 |
| 1982 | 0 | 0 | 0 | 0 | 0 | 53 | 31 | 55 | 15 | 0 | 0 | 0 | 154 |
| 1983 | 0 | 0 | 0 | 0 | 24 | 2 | 24 | 44 | 0 | 0 | 0 | 0 | 94 |
| 1984 | 0 | 0 | 0 | 0 | 1 | 5 | 62 | 45 | 11 | 0 | 0 | 0 | 124 |
| 1985 | 0 | 0 | 0 | 0 | 3 | 17 | 125 | 59 | 0 | 0 | 0 | 0 | 204 |
| 1986 | 0 | 0 | 0 | 0 | 22 | 27 | 10 | 124 | 26 | 0 | 0 | 0 | 209 |
| 1987 | 0 | 0 | 0 | 0 | 11 | 42 | 39 | 80 | 35 | 5 | 0 | 0 | 212 |
| 1988 | 0 | 0 | 0 | 0 | 7 | 10 | 79 | 56 | 36 | 1 | 0 | 0 | 189 |
| 1989 | 0 | 0 | 0 | 0 | 2 | 47 | 32 | 45 | 46 | 0 | 0 | 0 | 172 |
| 1990 | 0 | 0 | 0 | 0 | 0 | 10 | 129 | 100 | 21 | 0 | 0 | 0 | 260 |
| 1991 | 0 | 0 | 0 | 0 | 0 | 0 | 85 | 96 | 34 | 0 | 0 | 0 | 215 |
| 1992 | 0 | 0 | 0 | 0 | 18 | 68 | 83 | 85 | 8 | 0 | 0 | 0 | 262 |
| 1993 | 0 | 0 | 0 | 0 | 10 | 8 | 1 | 72 | 32 | 0 | 0 | 0 | 123 |
| 1994 | 0 | 0 | 0 | 0 | 1 | 5 | 120 | 86 | 29 | 0 | 0 | 0 | 241 |
| 1995 | 0 | 0 | 0 | 0 | 15 | 55 | 88 | 20 | 34 | 0 | 0 | 0 | 212 |
| 1996 | 0 | 0 | 0 | 1 | 0 | 2 | 137 | 81 | 1 | 0 | 0 | 0 | 222 |
| 1997 | 0 | 0 | 0 | 0 | 4 | 0 | 29 | 100 | 17 | 0 | 0 | 0 | 150 |
| 1998 | 0 | 0 | 0 | 2 | 1 | 6 | 108 | 84 | 29 | 0 | 0 | 0 | 230 |
| 1999 | 0 | 0 | 0 | 0 | 0 | 18 | 29 | 43 | 22 | 0 | 0 | 0 | 112 |
| 2000 | 0 | 0 | 0 | 0 | 0 | 28 | 44 | 27 | 8 | 0 | 0 | 0 | 107 |
| 2001 | 0 | 0 | 0 | 0 | 11 | 3 | 14 | 44 | 0 | 0 | 0 | 0 | 72 |
| 2002 | 0 | 0 | 0 | 0 | 0 | 33 | 63 | 59 | 7 | 0 | 0 | 0 | 162 |
| 2003 | 0 | 0 | 0 | 0 | 1 | 56 | 111 | 63 | 42 | 0 | 0 | 0 | 273 |
| 2004 | 0 | 0 | 0 | 0 | 0 | 53 | 129 | 91 | 0 | 0 | 0 | 0 | 273 |
| 2005 | 0 | 0 | 0 | 0 | 17 | 5 | 58 | 78 | 4 | 0 | 0 | 0 | 162 |
| 2006 | 0 | 0 | 0 | 0 | 8 | 42 | 118 | 67 | 35 | 0 | 0 | 0 | 270 |
| 2007 | 0 | 0 | 0 | 0 | 11 | 13 | 99 | 50 | 19 | 0 | 0 | 0 | 192 |
| 2008 | 0 | 0 | 0 | 0 | 14 | 36 | 47 | 88 | 8 | 0 | 0 | 0 | 193 |
| 2009 | 0 | 0 | 0 | 0 | 7 | 41 | 174 | 64 | 28 | 0 | 0 | 0 | 314 |

## SNOWFALL (inches) 2009 SEATTLE (KSEA)

| YEAR | JUL | AUG | SEP | OCT | NOV | DEC | JAN | FEB | MAR | APR | MAY | JUN | TOTAL |
|------|-----|-----|-----|-----|-----|-----|-----|-----|-----|-----|-----|-----|-------|
| 1980-81 | T | 0.0 | 0.0 | 0.0 | T | 0.3 | 0.0 | 1.1 | 0.0 | 0.0 | 0.0 | 0.0 | 1.4 |
| 1981-82 | 0.0 | 0.0 | 0.0 | 0.0 | 0.0 | T | 7.0 | T | 2.0 | T | 0.0 | 0.0 | 9.0 |
| 1982-83 | 0.0 | 0.0 | 0.0 | T | 0.0 | T | 0.0 | 0.0 | 0.0 | 0.0 | 0.0 | 0.0 | T |
| 1983-84 | 0.0 | 0.0 | 0.0 | 0.0 | T | 0.3 | T | 0.0 | 0.0 | T | 0.0 | 0.0 | 0.3 |
| 1984-85 | 0.0 | 0.0 | 0.0 | T | T | 2.4 | T | 5.7 | T | T | 0.0 | 0.0 | 8.1 |
| 1985-86 | 0.0 | 0.0 | 0.0 | T | 17.5 | 1.7 | 0.0 | 1.1 | 0.0 | T | 0.0 | 0.0 | 20.3 |
| 1986-87 | 0.0 | 0.0 | 0.0 | 0.0 | T | 0.0 | 1.4 | 0.0 | 0.0 | 0.0 | 0.0 | 0.0 | 1.4 |
| 1987-88 | 0.0 | 0.0 | 0.0 | 0.0 | 0.0 | T | T | 0.0 | T | 0.0 | 0.0 | 0.0 | T |
| 1988-89 | 0.0 | 0.0 | 0.0 | 0.0 | T | T | 1.0 | 5.8 | 7.4 | T | T | 0.0 | 14.2 |
| 1989-90 | 0.0 | 0.0 | 0.0 | 0.0 | 0.0 | 0.0 | T | 9.8 | T | 0.0 | T | 0.0 | 9.8 |
| 1990-91 | 0.0 | 0.0 | 0.0 | 0.0 | 0.0 | 3.8 | 0.4 | 0.0 | 2.5 | T | 0.0 | 0.0 | 6.7 |
| 1991-92 | 0.0 | 0.0 | 0.0 | 0.0 | 0.0 | 0.0 | 0.0 | 0.0 | 0.0 | 0.0 | 0.0 | 0.0 | 0.0 |
| 1992-93 | 0.0 | 0.0 | 0.0 | 0.0 | T | 6.7 | 1.3 | 1.4 | 0.0 | 0.0 | T | 0.0 | 9.4 |
| 1993-94 | 0.0 | 0.0 | 0.0 | 0.0 | T | 0.0 | 0.0 | 2.1 | 0.2 | 0.0 | 0.0 | 0.0 | 2.3 |
| 1994-95 | 0.0 | 0.0 | 0.0 | 0.0 | T | 1.9 | 0.0 | 0.2 | T | 0.0 | 0.0 | 0.0 | 2.1 |
| 1995-96 | 0.0 | 0.0 | 0.0 | 0.0 | 0.0 | 0.0 | 10.5 | 0.5 | 0.0 | 0.0 | 0.0 | 0.0 | 11.0 |
| 1996-97 | 0.0 | 0.0 | 0.0 | | | | | | | | | | |
| 1997-98 | | | | | | | | | | | | | |
| 1998-99 | | | | | | | | | | | | | |
| 1999-00 | | | | | | | | | | | | | |
| 2000-01 | | | | | | | | | | | | | |
| 2001-02 | | | | | | | | | | | | | |
| 2002-03 | | | | | | | | | | | | | |
| 2003-04 | | | | | | | | | | | | | |
| 2004-05 | | | | | | | | | | | | | |
| 2005-06 | | | | | | | | | | | | | |
| 2006-07 | | | | | | | | | | | | | |
| 2007-08 | | | | | | | | | | | | | |
| 2008-09 | | | | | | 13.9 | 3.9 | 2.5 | 3.0 | T | 0.0 | 0.0 | |
| 2009- | 0.0 | 0.0 | 0.0 | 0.0 | T | T | | | | | | | |
| POR= 63 YRS | T | 0.0 | 0.0 | T | 0.7 | 2.3 | 4.1 | 1.4 | 1.1 | T | T | 0.0 | 9.6 |

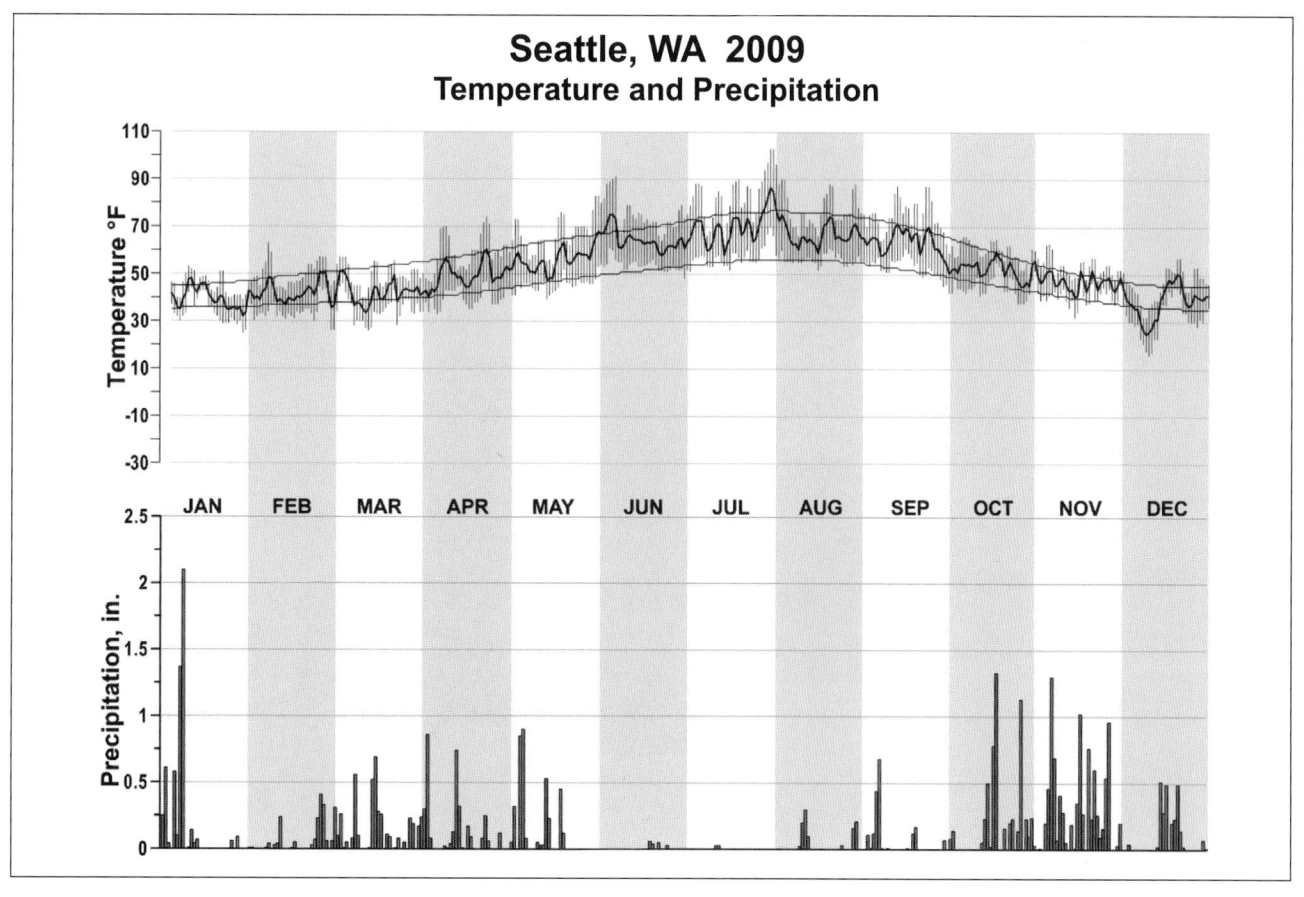

# Seattle, WA 2009
## Temperature and Precipitation

# 2009
# SPOKANE
# WASHINGTON (KGEG)

Spokane lies on the eastern edge of the broad Columbia Basin area of Washington which is bounded by the Cascade Range on the west and the Rocky Mountains on the east. The elevations in eastern Washington vary from less than 400 feet above sea level near Pasco where the Columbia River flows out of Washington to over 5,000 feet in the mountain areas of the extreme eastern edge of the State. Spokane is located on the upper plateau area where the long gradual slope from the Columbia River meets the sharp rise of the Rocky Mountain Ranges.

Much of the urban area of Spokane lies along both sides of the Spokane River at an elevation of approximately 2,000 feet, but the residential areas have spread to the crests of the plateaus on either side of the river with elevations up to 2,500 feet above sea level. Spokane International Airport is situated on the plateau area 6 miles west-southwest and some 400 feet higher than the downtown business district.

The climate of Spokane combines some of the characteristics of damp coastal type weather and arid interior conditions. Most of the air masses which reach Spokane are brought in by the prevailing westerly and southwesterly circulations. Frequently, much of the moisture in the storms that move eastward and southeastward from the Gulf of Alaska and the eastern Pacific Ocean is precipitated out as the storms are lifted across the Coast and Cascade Ranges. Annual precipitation totals in the Spokane area are generally less than 20 inches and less than 50 percent of the amounts received west of the Cascades. However, the precipitation and total cloudiness in the Spokane vicinity is greater than that of the desert areas of south-central Washington. The lifting action of the air masses as they move up the east slope of the Columbia Basin frequently produces the cooling and condensation necessary for formation of clouds and precipitation.

Infrequently, the Spokane area comes under the influence of dry continental air masses from the north or east. On occasions when these air masses penetrate into eastern Washington the result is high temperatures and very low humidity in the summer and sub-zero temperatures in the winter. In the winter most of the severe arctic outbursts of cold air move southward on the east side of the Continental Divide and do not affect Spokane.

In general, Spokane weather has the characteristics of a mild, arid climate during the summer months and a cold, coastal type in the winter. Approximately 70 percent of the total annual precipitation falls between the first of October and the end of March and about half of that falls as snow. The growing season usually extends over nearly six months from mid-April to mid-October. Irrigation is required for all crops except dry-land type grains. The summer weather is ideal for full enjoyment of the many mountain and lake recreational areas in the immediate vicinity. Winter weather includes many cloudy or foggy days and below freezing temperatures with occasional snowfall of several inches in depth. Sub-zero temperatures and traffic-stopping snowfalls are infrequent.

Based on the 1951-1980 period, the average first occurrence of 32 degrees Fahrenheit in the fall is October 6 and the average last occurrence in the spring is May 4.

# NORMALS, MEANS, AND EXTREMES
## SPOKANE (KGEG)

| LATITUDE: 47° 37'N | LONGITUDE: -117° 31'W | ELEVATION (FT): GRND: 2353　BARO: 2384 | TIME ZONE: PACIFIC　(UTC -8) | WBAN: 24157 |
|---|---|---|---|---|

| | ELEMENT | POR | JAN | FEB | MAR | APR | MAY | JUN | JUL | AUG | SEP | OCT | NOV | DEC | YEAR |
|---|---|---|---|---|---|---|---|---|---|---|---|---|---|---|---|
| **TEMPERATURE °F** | NORMAL DAILY MAXIMUM | 30 | 32.8 | 39.3 | 48.6 | 57.5 | 66.2 | 73.9 | 82.5 | 82.6 | 72.5 | 58.5 | 41.1 | 32.8 | 57.4 |
| | MEAN DAILY MAXIMUM | 72 | 32.4 | 38.5 | 47.2 | 56.8 | 66.4 | 73.5 | 83.4 | 82.5 | 72.2 | 58.3 | 41.9 | 33.8 | 57.2 |
| | HIGHEST DAILY MAXIMUM | 62 | 59 | 63 | 71 | 90 | 96 | 101 | 103 | 108 | 98 | 82 | 67 | 56 | 108 |
| | YEAR OF OCCURRENCE | | 1971 | 1995 | 1960 | 1977 | 1986 | 1992 | 1998 | 1961 | 1988 | 2008 | 1999 | 1980 | AUG 1961 |
| | MEAN OF EXTREME MAXS. | 73 | 46.2 | 51.3 | 61.6 | 73.7 | 84.3 | 90.5 | 97.0 | 96.1 | 88.6 | 75.0 | 55.9 | 47.5 | 72.3 |
| | NORMAL DAILY MINIMUM | 30 | 21.7 | 25.7 | 30.4 | 35.5 | 42.6 | 49.2 | 54.6 | 54.5 | 45.9 | 35.8 | 28.7 | 21.6 | 37.2 |
| | MEAN DAILY MINIMUM | 72 | 21.0 | 24.7 | 29.6 | 35.4 | 43.0 | 49.2 | 55.2 | 54.5 | 46.2 | 36.7 | 29.0 | 23.1 | 37.3 |
| | LOWEST DAILY MINIMUM | 62 | -22 | -24 | -7 | 17 | 24 | 33 | 37 | 35 | 22 | 7 | -21 | -25 | -25 |
| | YEAR OF OCCURRENCE | | 2004 | 1996 | 1989 | 1966 | 2002 | 1984 | 1981 | 1965 | 2000 | 2004 | 1985 | 1968 | DEC 1968 |
| | MEAN OF EXTREME MINS. | 73 | 0.0 | 8.1 | 16.6 | 25.7 | 31.2 | 38.9 | 44.6 | 43.8 | 33.7 | 23.4 | 13.7 | 3.0 | 23.6 |
| | NORMAL DRY BULB | 30 | 27.3 | 32.5 | 39.5 | 46.5 | 54.4 | 61.6 | 68.6 | 68.6 | 59.2 | 47.2 | 34.9 | 27.2 | 47.3 |
| | MEAN DRY BULB | 72 | 26.7 | 31.6 | 38.4 | 46.1 | 54.7 | 61.4 | 69.3 | 68.5 | 59.2 | 47.5 | 35.5 | 28.5 | 47.3 |
| | MEAN WET BULB | 26 | 27.2 | 29.2 | 34.5 | 39.3 | 45.3 | 50.6 | 53.9 | 52.9 | 47.6 | 40.1 | 32.8 | 26.0 | 40.0 |
| | MEAN DEW POINT | 26 | 25.7 | 26.6 | 30.5 | 34.2 | 40.1 | 44.7 | 46.7 | 45.2 | 41.2 | 35.7 | 31.1 | 24.8 | 35.5 |
| | NORMAL NO. DAYS WITH: | | | | | | | | | | | | | | |
| | MAXIMUM >= 90 | 30 | 0.0 | 0.0 | 0.0 | * | 0.3 | 1.5 | 7.8 | 7.5 | 1.0 | 0.0 | 0.0 | 0.0 | 18.1 |
| | MAXIMUM <= 32 | 30 | 13.2 | 5.5 | 0.5 | 0.0 | 0.0 | 0.0 | 0.0 | 0.0 | 0.0 | 0.1 | 4.3 | 13.8 | 37.4 |
| | MINIMUM <= 32 | 30 | 25.9 | 22.3 | 20.4 | 10.1 | 1.7 | 0.0 | 0.0 | 0.0 | 0.9 | 9.7 | 19.9 | 26.7 | 137.6 |
| | MINIMUM <= 0 | 30 | 1.8 | 0.8 | 0.0 | 0.0 | 0.0 | 0.0 | 0.0 | 0.0 | 0.0 | 0.0 | 0.4 | 1.7 | 4.7 |
| **H/C** | NORMAL HEATING DEG. DAYS | 30 | 1169 | 916 | 790 | 557 | 338 | 149 | 44 | 42 | 196 | 554 | 897 | 1168 | 6820 |
| | NORMAL COOLING DEG. DAYS | 30 | 0 | 0 | 0 | 1 | 11 | 46 | 155 | 154 | 26 | 1 | 0 | 0 | 394 |
| **RH** | NORMAL (PERCENT) | 30 | 86 | 81 | 72 | 63 | 60 | 56 | 48 | 46 | 54 | 67 | 85 | 88 | 67 |
| | HOUR 04 LST | 30 | 88 | 86 | 83 | 78 | 78 | 76 | 68 | 65 | 73 | 80 | 88 | 89 | 79 |
| | HOUR 10 LST | 30 | 86 | 81 | 70 | 58 | 54 | 50 | 43 | 44 | 52 | 66 | 84 | 87 | 65 |
| | HOUR 16 LST | 30 | 80 | 70 | 55 | 44 | 42 | 37 | 29 | 28 | 35 | 48 | 76 | 83 | 52 |
| | HOUR 22 LST | 30 | 86 | 82 | 75 | 66 | 64 | 60 | 50 | 47 | 57 | 69 | 86 | 88 | 69 |
| **S** | PERCENT POSSIBLE SUNSHINE | 48 | 28 | 41 | 55 | 61 | 65 | 67 | 80 | 78 | 72 | 28 | 29 | 23 | 52 |
| **W/O** | MEAN NO. DAYS WITH: | | | | | | | | | | | | | | |
| | HEAVY FOG(VISBY <= 1/4 MI) | 46 | 9.7 | 6.9 | 3.1 | 1.1 | 0.7 | 0.3 | 0.2 | 0.2 | 0.9 | 3.5 | 8.2 | 9.9 | 44.7 |
| | THUNDERSTORMS | 62 | 0.0 | 0.0 | 0.3 | 0.6 | 1.6 | 2.7 | 2.4 | 2.1 | 0.7 | 0.2 | 0.1 | 0.0 | 10.7 |
| **CLOUDNESS** | MEAN: | | | | | | | | | | | | | | |
| | SUNRISE-SUNSET (OKTAS) | | | | | | | | | | | | | | |
| | MIDNIGHT-MIDNIGHT (OKTAS) | | | | | | | | | | | | | | |
| | MEAN NO. DAYS WITH: | | | | | | | | | | | | | | |
| | CLEAR | | | | | | | | | | | | | | |
| | PARTLY CLOUDY | | | | | | | | | | | | | | |
| | CLOUDY | | | | | | | | | | | | | | |
| **PR** | MEAN STATION PRESSURE(IN) | 26 | 27.54 | 27.54 | 27.50 | 27.50 | 27.48 | 27.49 | 27.51 | 27.51 | 27.53 | 27.56 | 27.55 | 27.58 | 27.52 |
| | MEAN SEA-LEVEL PRES. (IN) | 26 | 30.13 | 30.08 | 30.02 | 29.99 | 29.95 | 29.94 | 29.95 | 29.94 | 29.99 | 30.06 | 30.08 | 30.14 | 30.02 |
| **WINDS** | MEAN SPEED (MPH) | 26 | 8.6 | 8.7 | 9.8 | 9.9 | 9.7 | 9.5 | 8.9 | 8.4 | 8.1 | 8.5 | 9.0 | 8.1 | 8.9 |
| | PREVAIL.DIR(TENS OF DEGS) | 31 | 22 | 05 | 23 | 22 | 22 | 22 | 23 | 23 | 23 | 22 | 05 | 05 | 22 |
| | MAXIMUM 2-MINUTE: | | | | | | | | | | | | | | |
| | SPEED (MPH) | 14 | 48 | 44 | 53 | 46 | 44 | 62 | 41 | 46 | 47 | 48 | 52 | 47 | 62 |
| | DIR. (TENS OF DEGS) | | 22 | 21 | 23 | 25 | 26 | 23 | 25 | 21 | 25 | 22 | 22 | 22 | 23 |
| | YEAR OF OCCURRENCE | | 2007 | 1999 | 2009 | 1997 | 1997 | 2005 | 1998 | 2004 | 2009 | 2007 | 2006 | 2006 | JUN 2005 |
| | MAXIMUM 3-SECOND | | | | | | | | | | | | | | |
| | SPEED (MPH) | 14 | 55 | 53 | 66 | 54 | 48 | 77 | 53 | 54 | 55 | 55 | 63 | 56 | 77 |
| | DIR. (TENS OF DEGS) | | 22 | 22 | 23 | 23 | 26 | 23 | 17 | 20 | 25 | 22 | 22 | 23 | 23 |
| | YEAR OF OCCURRENCE | | 2007 | 1999 | 2009 | 2000 | 1997 | 2005 | 2007 | 2004 | 2009 | 2007 | 2003 | 2006 | JUN 2005 |
| **PRECIPITATION** | NORMAL (IN) | 30 | 1.82 | 1.51 | 1.53 | 1.28 | 1.60 | 1.18 | 0.76 | 0.68 | 0.76 | 1.06 | 2.24 | 2.25 | 16.67 |
| | MAXIMUM MONTHLY (IN) | 62 | 4.96 | 3.94 | 3.81 | 3.08 | 5.71 | 3.09 | 2.33 | 1.88 | 2.05 | 4.96 | 5.10 | 5.13 | 5.71 |
| | YEAR OF OCCURRENCE | | 1959 | 1961 | 1995 | 1948 | 1948 | 2006 | 1990 | 2004 | 1959 | 1959 | 1973 | 1964 | MAY 1948 |
| | MINIMUM MONTHLY (IN) | 62 | 0.38 | .04 | 0.31 | 0.08 | 0.20 | 0.16 | T | T | T | 0.30 | 0.22 | 0.60 | T |
| | YEAR OF OCCURRENCE | | 1985 | 2005 | 1965 | 1956 | 1982 | 1960 | 2008 | 1988 | 1990 | 2008 | 1976 | 1976 | JUL 2008 |
| | MAXIMUM IN 24 HOURS (IN) | 62 | 1.76 | 1.11 | 1.08 | 1.51 | 2.19 | 2.07 | 1.80 | 1.18 | 1.12 | 1.76 | 1.41 | 1.60 | 2.19 |
| | YEAR OF OCCURRENCE | | 2007 | 1963 | 1995 | 2000 | 2004 | 1964 | 1990 | 2002 | 1973 | 2007 | 1960 | 1951 | MAY 2004 |
| | NORMAL NO. DAYS WITH: | | | | | | | | | | | | | | |
| | PRECIPITATION >= 0.01 | 30 | 13.1 | 11.1 | 11.1 | 9.3 | 10.1 | 7.9 | 5.4 | 4.6 | 5.4 | 7.0 | 13.0 | 13.3 | 111.3 |
| | PRECIPITATION >= 1.00 | 30 | * | 0.0 | 0.0 | * | 0.0 | 0.1 | * | 0.0 | 0.1 | 0.0 | 0.1 | * | 0.3 |
| **SNOWFALL** | NORMAL (IN) | 30 | 14.2 | 6.7 | 3.6 | 0.9 | 0.2 | 0.0 | 0.0 | 0.0 | 0.0 | 0.3 | 6.4 | 15.1 | 47.4 |
| | MAXIMUM MONTHLY (IN) | 61 | 56.9 | 28.5 | 15.8 | 6.6 | 3.5 | T | T | 0.0 | T | 6.1 | 24.7 | 61.5 | 61.5 |
| | YEAR OF OCCURRENCE | | 1950 | 1975 | 2008 | 1964 | 1967 | 2009 | 2008 | | 1991 | 1957 | 1955 | 2008 | DEC 2008 |
| | MAXIMUM IN 24 HOURS (IN) | 61 | 13.0 | 11.0 | 6.1 | 4.9 | 3.5 | T | 0.0 | 0.0 | T | 6.1 | 9.0 | 12.1 | 13.0 |
| | YEAR OF OCCURRENCE' | | 1950 | 1993 | 1989 | 1964 | 1967 | 2009 | | | 1991 | 1957 | 1973 | 1951 | JAN 1950 |
| | MAXIMUM SNOW DEPTH (IN) | 60 | 39 | 42 | 16 | 4 | 0 | 0 | 0 | 0 | 0 | 4 | 12 | 23 | 42 |
| | YEAR OF OCCURRENCE | | 1969 | 1969 | 1969 | 2009 | | | | | | 1957 | 1985 | 2008 | FEB 1969 |
| | NORMAL NO. DAYS WITH: | | | | | | | | | | | | | | |
| | SNOWFALL >= 1.0 | 30 | 4.4 | 2.8 | 0.9 | 0.3 | 0.0 | 0.0 | 0.0 | 0.0 | 0.0 | 0.2 | 2.0 | 4.6 | 15.2 |

## PRECIPITATION (inches) 2009 SPOKANE (KGEG)

| YEAR | JAN | FEB | MAR | APR | MAY | JUN | JUL | AUG | SEP | OCT | NOV | DEC | ANNUAL |
|------|------|------|------|------|------|------|------|------|------|------|------|------|--------|
| 1980 | 1.96 | 1.90 | 0.91 | 1.06 | 2.34 | 0.99 | 0.21 | 0.79 | 0.84 | 0.64 | 1.67 | 3.72 | 17.03 |
| 1981 | 1.00 | 1.41 | 1.57 | 0.85 | 2.02 | 1.92 | 0.51 | 0.04 | 0.59 | 1.53 | 0.96 | 2.51 | 14.91 |
| 1982 | 1.61 | 1.67 | 1.49 | 2.23 | 0.20 | 0.85 | 1.05 | 0.25 | 1.77 | 1.48 | 1.86 | 2.79 | 17.25 |
| 1983 | 1.89 | 2.07 | 2.20 | 0.61 | 0.92 | 2.84 | 1.85 | 0.96 | 0.79 | 1.33 | 4.80 | 2.38 | 22.64 |
| 1984 | 0.99 | 1.37 | 1.80 | 1.75 | 2.01 | 1.89 | 0.07 | 0.27 | 0.56 | 0.76 | 4.26 | 2.28 | 18.01 |
| 1985 | 0.38 | 0.93 | 1.39 | 0.28 | 1.13 | 0.67 | 0.26 | 0.19 | 1.64 | 1.40 | 2.23 | 0.71 | 11.21 |
| 1986 | 3.08 | 2.02 | 1.58 | 1.33 | 1.08 | 0.48 | 0.44 | 0.15 | 1.65 | 0.46 | 2.25 | 1.03 | 15.55 |
| 1987 | 1.59 | 0.88 | 2.18 | 1.12 | 0.90 | 0.59 | 2.27 | 1.81 | 0.01 | 0.03 | 1.37 | 4.93 | 17.68 |
| 1988 | 1.76 | 0.35 | 1.57 | 2.15 | 1.50 | 1.12 | 0.23 | T | 1.63 | 0.11 | 4.35 | 1.75 | 16.52 |
| 1989 | 0.82 | 1.34 | 2.87 | 0.72 | 2.17 | 0.41 | 0.40 | 1.61 | 0.18 | 1.58 | 1.66 | 0.95 | 14.71 |
| 1990 | 2.45 | 1.01 | 0.85 | 1.34 | 3.11 | 1.91 | 2.33 | 1.03 | T | 3.05 | 0.84 | 1.69 | 19.61 |
| 1991 | 1.72 | 0.81 | 2.31 | 1.35 | 1.72 | 1.13 | 0.58 | 0.17 | 0.01 | 0.34 | 3.08 | 1.23 | 14.45 |
| 1992 | 2.12 | 1.76 | 0.43 | 0.65 | 0.28 | 1.51 | 1.09 | 0.33 | 0.36 | 0.81 | 3.02 | 2.16 | 14.52 |
| 1993 | 1.40 | 0.86 | 1.13 | 1.90 | 1.36 | 0.48 | 2.08 | 1.24 | 0.28 | 0.42 | 0.68 | 1.80 | 13.63 |
| 1994 | 1.43 | 0.83 | 0.49 | 1.64 | 1.37 | 0.90 | T | 0.10 | 0.45 | 2.79 | 2.24 | 1.57 | 13.81 |
| 1995 | 2.74 | 1.60 | 3.81 | 0.93 | 1.33 | 2.17 | 1.08 | 0.63 | 0.66 | 1.50 | 0.77 | 2.63 | 19.85 |
| 1996 | 2.44 | 2.95 | 1.61 | 2.15 | 1.78 | 1.19 | 0.34 | .80 | .79 | 3.27 | 4.04 | 4.10 | 25.46 |
| 1997 | 1.67 | 1.40 | 2.40 | 2.56 | 2.27 | 0.63 | 0.80 | 0.14 | 0.92 | 1.67 | 1.99 | 1.00 | 17.45 |
| 1998 | 2.08 | 1.59 | 1.21 | 0.89 | 3.09 | 0.84 | 0.26 | 0.27 | 0.21 | 0.27 | 3.78 | 3.28 | 17.77 |
| 1999 | 1.89 | 3.27 | 0.69 | 0.44 | 0.73 | 1.36 | 0.13 | 1.07 | T | 0.89 | 2.06 | 2.26 | 14.79 |
| 2000 | 1.96 | 1.61 | 1.64 | 2.16 | 2.22 | 0.91 | 0.35 | T | 1.12 | 0.64 | 1.13 | 0.93 | 14.67 |
| 2001 | 0.63 | 0.66 | 1.37 | 1.71 | 0.79 | 1.10 | 0.28 | 0.26 | 0.17 | 2.10 | 2.61 | 2.03 | 13.71 |
| 2002 | 1.15 | 1.04 | 1.02 | 0.88 | 1.10 | 1.50 | 0.25 | 1.24 | 0.55 | 0.18 | 1.65 | 3.27 | 13.83 |
| 2003 | 3.40 | 0.52 | 2.13 | 1.41 | 1.49 | 0.22 | T | 0.44 | 0.58 | 0.51 | 1.57 | 2.14 | 14.41 |
| 2004 | 1.42 | 1.46 | 0.67 | 0.57 | 3.67 | 1.05 | 0.08 | 1.88 | 0.69 | 1.06 | 1.13 | 1.34 | 15.02 |
| 2005 | 1.15 | 0.04 | 2.03 | 0.79 | 3.58 | 1.38 | 1.10 | 0.46 | 0.84 | 1.03 | 2.02 | 2.96 | 17.38 |
| 2006 | 4.48 | 1.20 | 1.23 | 1.69 | 1.09 | 3.09 | 0.10 | 0.25 | 0.32 | 0.93 | 4.38 | 2.37 | 21.13 |
| 2007 | 0.67 | 1.81 | 1.00 | 0.50 | 1.60 | 0.59 | 0.43 | 0.57 | 0.37 | 1.18 | 1.53 | 3.72 | 13.97 |
| 2008 | 3.18 | 0.93 | 1.86 | 1.27 | 0.93 | 1.00 | T | 0.57 | 0.54 | 0.30 | 1.76 | 3.94 | 16.28 |
| 2009 | 1.19 | 1.22 | 2.43 | 1.29 | 0.93 | 1.18 | 0.48 | 0.74 | 0.49 | 2.31 | 1.31 | 1.88 | 15.45 |
| POR= 72 YRS | 2.17 | 1.58 | 1.52 | 1.20 | 1.54 | 1.26 | 0.58 | 0.64 | 0.72 | 1.15 | 2.20 | 2.38 | 16.94 |

WBAN : 24157

## AVERAGE TEMPERATURE (°F) 2009 SPOKANE (KGEG)

| YEAR | JAN | FEB | MAR | APR | MAY | JUN | JUL | AUG | SEP | OCT | NOV | DEC | ANNUAL |
|------|------|------|------|------|------|------|------|------|------|------|------|------|--------|
| 1980 | 20.7 | 34.5 | 38.6 | 51.7 | 55.8 | 57.8 | 69.2 | 64.1 | 58.4 | 47.4 | 36.3 | 33.2 | 47.3 |
| 1981 | 32.8 | 33.9 | 40.9 | 45.7 | 52.0 | 57.0 | 65.1 | 71.5 | 59.7 | 45.9 | 39.9 | 29.7 | 47.8 |
| 1982 | 26.0 | 32.1 | 40.3 | 43.5 | 54.2 | 66.5 | 67.6 | 69.8 | 59.5 | 46.1 | 31.7 | 27.3 | 47.1 |
| 1983 | 35.8 | 38.1 | 43.0 | 46.3 | 57.1 | 61.9 | 65.5 | 72.3 | 57.1 | 49.7 | 39.3 | 16.2 | 48.5 |
| 1984 | 30.5 | 34.5 | 41.7 | 44.0 | 50.1 | 59.2 | 69.1 | 70.1 | 56.7 | 43.4 | 35.8 | 20.4 | 46.3 |
| 1985 | 21.4 | 24.9 | 35.9 | 48.0 | 56.2 | 61.8 | 75.0 | 64.9 | 53.3 | 44.7 | 19.5 | 19.3 | 43.7 |
| 1986 | 30.1 | 31.6 | 42.8 | 44.9 | 55.3 | 66.2 | 64.0 | 72.6 | 54.8 | 49.0 | 34.8 | 26.3 | 47.7 |
| 1987 | 26.5 | 35.1 | 41.8 | 51.1 | 57.2 | 65.1 | 66.6 | 66.2 | 62.8 | 49.5 | 38.1 | 25.9 | 48.8 |
| 1988 | 24.7 | 35.4 | 39.7 | 48.9 | 54.6 | 61.1 | 68.7 | 68.4 | 58.9 | 53.3 | 36.3 | 27.0 | 48.1 |
| 1989 | 28.8 | 21.8 | 36.6 | 48.9 | 53.1 | 64.3 | 68.7 | 64.8 | 60.1 | 47.0 | 38.0 | 31.0 | 46.9 |
| 1990 | 33.4 | 30.2 | 40.9 | 49.7 | 52.8 | 60.7 | 70.4 | 68.5 | 65.3 | 45.1 | 39.0 | 21.1 | 48.1 |
| 1991 | 25.7 | 39.2 | 36.8 | 45.8 | 51.6 | 56.6 | 68.7 | 70.2 | 61.8 | 46.2 | 34.2 | 32.8 | 47.5 |
| 1992 | 31.8 | 38.9 | 45.5 | 48.8 | 58.9 | 68.0 | 67.8 | 69.6 | 57.4 | 49.5 | 34.3 | 22.9 | 49.5 |
| 1993 | 21.9 | 25.4 | 37.8 | 45.5 | 59.8 | 60.2 | 60.2 | 64.2 | 58.7 | 50.0 | 29.4 | 30.9 | 45.3 |
| 1994 | 35.6 | 29.1 | 41.8 | 49.1 | 56.7 | 60.8 | 73.0 | 69.4 | 63.4 | 46.8 | 32.4 | 30.3 | 49.0 |
| 1995 | 31.0 | 37.3 | 39.9 | 45.5 | 56.8 | 60.1 | 67.9 | 63.9 | 61.2 | 43.9 | 40.2 | 28.6 | 48.0 |
| 1996 | 25.4 | 28.7 | 36.4 | 46.3 | 49.6 | 60.5 | 70.0 | 68.1 | 56.0 | 45.3 | 33.2 | 24.8 | 45.4 |
| 1997 | 28.4 | 31.7 | 39.2 | 43.3 | 56.7 | 59.9 | 67.6 | 71.0 | 61.9 | 47.3 | 38.6 | 29.3 | 47.9 |
| 1998 | 30.7 | 38.1 | 41.5 | 48.0 | 56.1 | 62.5 | 75.3 | 71.7 | 65.1 | 46.5 | 39.9 | 28.6 | 50.3 |
| 1999 | 32.2 | 34.9 | 39.9 | 44.9 | 50.6 | 59.9 | 66.2 | 70.3 | 59.1 | 47.4 | 41.4 | 31.6 | 48.2 |
| 2000 | 27.9 | 33.5 | 39.0 | 48.2 | 53.0 | 61.0 | 67.9 | 67.6 | 55.8 | 46.3 | 26.9 | 24.7 | 46.0 |
| 2001 | 27.2 | 26.8 | 39.2 | 43.7 | 55.4 | 58.7 | 68.4 | 71.1 | 63.3 | 45.9 | 39.9 | 28.1 | 47.3 |
| 2002 | 30.6 | 31.4 | 34.4 | 45.4 | 51.5 | 62.3 | 71.4 | 66.4 | 58.5 | 42.9 | 36.8 | 33.8 | 47.1 |
| 2003 | 33.9 | 33.1 | 40.8 | 42.9 | 53.4 | 63.6 | 73.0 | 70.3 | 61.9 | 51.4 | 29.5 | 29.8 | 48.8 |
| 2004 | 26.2 | 32.1 | 43.3 | 49.6 | 54.6 | 63.6 | 72.3 | 71.0 | 58.0 | 49.3 | 36.2 | 31.9 | 49.0 |
| 2005 | 28.3 | 34.6 | 41.8 | 48.0 | 56.8 | 60.2 | 70.1 | 69.6 | 57.5 | 49.0 | 34.1 | 24.1 | 47.8 |
| 2006 | 35.5 | 30.9 | 38.6 | 47.3 | 56.3 | 63.6 | 73.7 | 68.9 | 61.2 | 47.0 | 36.0 | 28.5 | 49.0 |
| 2007 | 24.7 | 34.0 | 42.7 | 46.6 | 56.2 | 62.2 | 75.7 | 68.4 | 59.3 | 46.9 | 35.0 | 28.5 | 48.4 |
| 2008 | 24.7 | 31.9 | 36.3 | 42.0 | 56.9 | 60.8 | 70.3 | 68.7 | 61.0 | 47.8 | 38.6 | 21.9 | 46.7 |
| 2009 | 25.9 | 30.6 | 34.6 | 45.3 | 55.7 | 63.2 | 72.2 | 70.4 | 63.9 | 43.3 | 36.9 | 24.4 | 47.2 |
| POR= 72 YRS | 26.7 | 31.6 | 38.4 | 46.1 | 54.7 | 61.4 | 69.3 | 68.5 | 59.2 | 47.5 | 35.5 | 28.5 | 47.3 |

## HEATING DEGREE DAYS (base 65°F) 2009  SPOKANE (KGEG)

| YEAR | JUL | AUG | SEP | OCT | NOV | DEC | JAN | FEB | MAR | APR | MAY | JUN | TOTAL |
|------|-----|-----|-----|-----|-----|-----|-----|-----|-----|-----|-----|-----|-------|
| 1980-81 | 19 | 77 | 195 | 543 | 854 | 977 | 992 | 867 | 741 | 570 | 395 | 243 | 6473 |
| 1981-82 | 73 | 7 | 209 | 584 | 747 | 1088 | 1202 | 912 | 761 | 639 | 328 | 76 | 6626 |
| 1982-83 | 62 | 17 | 193 | 582 | 996 | 1163 | 897 | 747 | 672 | 558 | 285 | 113 | 6285 |
| 1983-84 | 55 | 2 | 230 | 468 | 765 | 1508 | 1065 | 880 | 715 | 621 | 460 | 194 | 6963 |
| 1984-85 | 21 | 18 | 264 | 662 | 870 | 1381 | 1345 | 1117 | 895 | 501 | 280 | 128 | 7482 |
| 1985-86 | 0 | 64 | 343 | 622 | 1363 | 1409 | 1076 | 927 | 680 | 595 | 357 | 67 | 7503 |
| 1986-87 | 81 | 4 | 311 | 488 | 902 | 1193 | 1186 | 831 | 710 | 417 | 253 | 86 | 6462 |
| 1987-88 | 51 | 50 | 116 | 474 | 799 | 1206 | 1240 | 850 | 775 | 477 | 330 | 173 | 6541 |
| 1988-89 | 47 | 16 | 240 | 361 | 856 | 1171 | 1113 | 1205 | 873 | 473 | 364 | 65 | 6784 |
| 1989-90 | 22 | 76 | 149 | 554 | 805 | 1048 | 976 | 968 | 739 | 454 | 373 | 166 | 6330 |
| 1990-91 | 37 | 42 | 54 | 610 | 774 | 1356 | 1212 | 716 | 866 | 568 | 406 | 248 | 6889 |
| 1991-92 | 15 | 16 | 108 | 574 | 918 | 992 | 1024 | 750 | 598 | 477 | 206 | 61 | 5739 |
| 1992-93 | 32 | 60 | 232 | 481 | 916 | 1297 | 1331 | 1102 | 834 | 578 | 192 | 165 | 7220 |
| 1993-94 | 151 | 83 | 217 | 457 | 1063 | 1051 | 904 | 998 | 713 | 469 | 262 | 160 | 6528 |
| 1994-95 | 26 | 13 | 81 | 558 | 970 | 1071 | 1045 | 771 | 771 | 578 | 262 | 170 | 6316 |
| 1995-96 | 21 | 88 | 146 | 648 | 742 | 1120 | 1217 | 1045 | 880 | 556 | 471 | 143 | 7077 |
| 1996-97 | 35 | 49 | 281 | 603 | 949 | 1241 | 1130 | 928 | 794 | 642 | 264 | 154 | 7070 |
| 1997-98 | 35 | 15 | 116 | 549 | 785 | 1098 | 1058 | 747 | 721 | 505 | 276 | 90 | 5995 |
| 1998-99 | 0 | 20 | 101 | 565 | 748 | 1119 | 1010 | 836 | 769 | 594 | 448 | 186 | 6396 |
| 1999-00 | 75 | 36 | 181 | 540 | 703 | 1030 | 1143 | 908 | 799 | 496 | 363 | 142 | 6416 |
| 2000-01 | 51 | 43 | 285 | 572 | 1134 | 1245 | 1168 | 1060 | 795 | 634 | 320 | 201 | 7508 |
| 2001-02 | 33 | 20 | 100 | 588 | 744 | 1136 | 1063 | 934 | 938 | 581 | 412 | 137 | 6686 |
| 2002-03 | 28 | 26 | 219 | 678 | 839 | 962 | 957 | 885 | 745 | 588 | 365 | 90 | 6382 |
| 2003-04 | 9 | 1 | 151 | 418 | 1056 | 1083 | 1193 | 945 | 668 | 455 | 315 | 131 | 6425 |
| 2004-05 | 16 | 34 | 204 | 480 | 857 | 1020 | 1128 | 842 | 711 | 503 | 260 | 166 | 6221 |
| 2005-06 | 11 | 22 | 229 | 489 | 919 | 1258 | 905 | 949 | 812 | 525 | 301 | 104 | 6524 |
| 2006-07 | 8 | 30 | 170 | 552 | 865 | 1122 | 1243 | 864 | 685 | 548 | 270 | 136 | 6493 |
| 2007-08 | 0 | 27 | 194 | 553 | 894 | 1126 | 1243 | 952 | 880 | 683 | 274 | 176 | 7002 |
| 2008-09 | 8 | 52 | 142 | 529 | 785 | 1328 | 1204 | 957 | 936 | 586 | 303 | 93 | 6923 |
| 2009- | 17 | 23 | 103 | 668 | 834 | 1252 | | | | | | | |

WBAN : 24157

## COOLING DEGREE DAYS (base 65°F) 2009  SPOKANE (KGEG)

| YEAR | JAN | FEB | MAR | APR | MAY | JUN | JUL | AUG | SEP | OCT | NOV | DEC | TOTAL |
|------|-----|-----|-----|-----|-----|-----|-----|-----|-----|-----|-----|-----|-------|
| 1980 | 0 | 0 | 0 | 1 | 3 | 2 | 156 | 56 | 6 | 3 | 0 | 0 | 227 |
| 1981 | 0 | 0 | 0 | 0 | 0 | 9 | 82 | 213 | 60 | 0 | 0 | 0 | 364 |
| 1982 | 0 | 0 | 0 | 0 | 2 | 128 | 148 | 171 | 32 | 0 | 0 | 0 | 481 |
| 1983 | 0 | 0 | 0 | 0 | 46 | 26 | 77 | 235 | 1 | 0 | 0 | 0 | 385 |
| 1984 | 0 | 0 | 0 | 0 | 3 | 28 | 155 | 181 | 23 | 1 | 0 | 0 | 391 |
| 1985 | 0 | 0 | 0 | 0 | 15 | 36 | 317 | 68 | 0 | 0 | 0 | 0 | 436 |
| 1986 | 0 | 0 | 0 | 0 | 65 | 109 | 57 | 247 | 8 | 0 | 0 | 0 | 486 |
| 1987 | 0 | 0 | 0 | 8 | 20 | 94 | 110 | 97 | 53 | 1 | 0 | 0 | 383 |
| 1988 | 0 | 0 | 0 | 0 | 12 | 63 | 169 | 128 | 67 | 0 | 0 | 0 | 439 |
| 1989 | 0 | 0 | 0 | 0 | 0 | 49 | 145 | 78 | 9 | 0 | 0 | 0 | 281 |
| 1990 | 0 | 0 | 0 | 0 | 0 | 42 | 213 | 157 | 68 | 0 | 0 | 0 | 480 |
| 1991 | 0 | 0 | 0 | 0 | 0 | 0 | 139 | 187 | 20 | 0 | 0 | 0 | 346 |
| 1992 | 0 | 0 | 0 | 0 | 25 | 159 | 124 | 209 | 11 | 8 | 0 | 0 | 536 |
| 1993 | 0 | 0 | 0 | 0 | 36 | 27 | 11 | 64 | 34 | 0 | 0 | 0 | 172 |
| 1994 | 0 | 0 | 0 | 0 | 9 | 37 | 280 | 159 | 43 | 0 | 0 | 0 | 528 |
| 1995 | 0 | 0 | 0 | 0 | 14 | 29 | 119 | 59 | 38 | 0 | 0 | 0 | 259 |
| 1996 | 0 | 0 | 0 | 0 | 0 | 16 | 198 | 150 | 17 | 0 | 0 | 0 | 381 |
| 1997 | 0 | 0 | 0 | 0 | 14 | 9 | 122 | 209 | 30 | 6 | 0 | 0 | 390 |
| 1998 | 0 | 0 | 0 | 0 | 6 | 22 | 325 | 234 | 110 | 0 | 0 | 0 | 697 |
| 1999 | 0 | 0 | 0 | 0 | 7 | 41 | 118 | 210 | 14 | 0 | 0 | 0 | 390 |
| 2000 | 0 | 0 | 0 | 0 | 0 | 29 | 146 | 129 | 16 | 0 | 0 | 0 | 320 |
| 2001 | 0 | 0 | 0 | 0 | 29 | 19 | 146 | 213 | 54 | 0 | 0 | 0 | 461 |
| 2002 | 0 | 0 | 0 | 0 | 0 | 63 | 231 | 81 | 30 | 0 | 0 | 0 | 405 |
| 2003 | 0 | 0 | 0 | 0 | 12 | 58 | 266 | 174 | 66 | 2 | 0 | 0 | 578 |
| 2004 | 0 | 0 | 0 | 0 | 0 | 96 | 249 | 225 | 1 | 0 | 0 | 0 | 571 |
| 2005 | 0 | 0 | 0 | 0 | 13 | 32 | 179 | 174 | 11 | 0 | 0 | 0 | 409 |
| 2006 | 0 | 0 | 0 | 0 | 41 | 66 | 285 | 161 | 62 | 0 | 0 | 0 | 615 |
| 2007 | 0 | 0 | 0 | 0 | 7 | 56 | 338 | 143 | 32 | 0 | 0 | 0 | 576 |
| 2008 | 0 | 0 | 0 | 0 | 27 | 60 | 182 | 176 | 29 | 4 | 0 | 0 | 478 |
| 2009 | 0 | 0 | 0 | 0 | 23 | 47 | 245 | 196 | 78 | 0 | 0 | 0 | 589 |

## SNOWFALL (inches)  2009  SPOKANE (KGEG)

| YEAR | JUL | AUG | SEP | OCT | NOV | DEC | JAN | FEB | MAR | APR | MAY | JUN | TOTAL |
|---|---|---|---|---|---|---|---|---|---|---|---|---|---|
| 1980-81 | 0.0 | 0.0 | 0.0 | 0.0 | 1.2 | 6.8 | 2.6 | 3.3 | T | T | 0.3 | 0.0 | 14.2 |
| 1981-82 | 0.0 | 0.0 | 0.0 | T | 0.8 | 13.0 | 23.3 | 2.2 | 2.1 | 6.0 | T | 0.0 | 47.4 |
| 1982-83 | 0.0 | 0.0 | 0.0 | T | 5.4 | 17.4 | 8.1 | 5.5 | T | 0.2 | T | 0.0 | 36.6 |
| 1983-84 | 0.0 | 0.0 | 0.0 | 0.0 | 5.7 | 24.8 | 5.3 | 8.0 | 1.9 | 1.3 | 0.8 | 0.0 | 47.8 |
| 1984-85 | 0.0 | 0.0 | 0.0 | 1.1 | 12.0 | 24.7 | 4.6 | 14.8 | 9.6 | T | T | 0.0 | 66.8 |
| 1985-86 | 0.0 | 0.0 | 0.0 | 0.4 | 23.7 | 8.3 | 14.7 | 13.8 | T | 0.2 | T | 0.0 | 61.1 |
| 1986-87 | 0.0 | 0.0 | 0.0 | 0.0 | 5.0 | 7.9 | 11.7 | 1.1 | T | T | T | 0.0 | 25.7 |
| 1987-88 | 0.0 | 0.0 | 0.0 | 0.0 | 1.5 | 20.3 | 9.1 | 1.2 | 1.6 | T | T | 0.0 | 33.7 |
| 1988-89 | 0.0 | 0.0 | 0.0 | 0.0 | 10.9 | 16.3 | 10.5 | 19.0 | 9.4 | T | T | 0.0 | 66.1 |
| 1989-90 | 0.0 | 0.0 | 0.0 | T | 5.2 | 1.1 | 10.3 | 18.0 | 2.6 | 3.5 | T | 0.0 | 40.7 |
| 1990-91 | 0.0 | 0.0 | 0.0 | 0.0 | 1.2 | 14.3 | 15.9 | 1.1 | 9.5 | 0.2 | 0.0 | 0.0 | 42.2 |
| 1991-92 | 0.0 | 0.0 | T | 0.8 | 4.9 | 2.4 | 9.0 | 1.4 | 0.0 | T | 0.0 | 0.0 | 18.5 |
| 1992-93 | 0.0 | 0.0 | 0.0 | 0.0 | 11.1 | 40.2 | 18.8 | 15.1 | 2.1 | T | T | T | 87.3 |
| 1993-94 | 0.0 | 0.0 | 0.0 | T | 3.7 | 6.4 | 0.9 | 8.2 | 0.5 | T | 0.0 | T | 19.7 |
| 1994-95 | 0.0 | 0.0 | 0.0 | 0.8 | 13.7 | 6.3 | 3.9 | 4.4 | 0.7 | T | 0.0 | 0.0 | 29.8 |
| 1995-96 | 0.0 | 0.0 | 0.0 |  |  |  | 22.7 |  |  |  |  |  |  |
| 1996-97 |  |  |  |  |  |  |  |  |  |  |  |  |  |
| 1997-98 |  |  |  |  |  | 6.4 | 8.5 | 1.6 | 1.8 | T | 0.0 | 0.0 |  |
| 1998-99 | 0.0 | 0.0 | 0.0 | T | 0.8 | 11.2 | 8.7 | 14.7 | 3.8 | 2.7 | 0.6 | 0.0 | 42.5 |
| 1999-00 | 0.0 | 0.0 | 0.0 | 0.0 | 2.1 | 9.7 | 21.3 | 6.7 | 1.1 | T | 0.3 |  |  |
| 2000-01 | 0.0 | 0.0 | 0.0 | T | 10.9 | 15.1 | 9.6 | 8.3 | 2.2 | 2.5 | 0.0 | 0.0 | 48.6 |
| 2001-02 | 0.0 | 0.0 | 0.0 | 0.7 | 12.5 | 21.9 | 9.6 | 6.7 | 11.7 | T | 0.9 | 0.0 | 64.0 |
| 2002-03 | 0.0 | 0.0 | 0.0 | T | 0.0 | 11.3 | 6.8 | 0.4 | 2.0 | 0.7 | T | T | 21.2 |
| 2003-04 | 0.0 | 0.0 | 0.0 | 0.0 | 15.4 | 7.2 | 20.3 | 8.7 | 2.7 | T | 0.4 | 0.0 | 54.7 |
| 2004-05 | 0.0 | 0.0 | 0.0 | T | 1.5 | 6.5 | 14.9 | T | 2.9 | T | T | 0.0 | 25.8 |
| 2005-06 | 0.0 | 0.0 | 0.0 | 0.0 | 7.4 | 4.4 | 9.4 | 3.8 | 4.1 | T | 0.0 | 0.0 | 29.1 |
| 2006-07 | 0.0 | 0.0 | 0.0 | T | 8.4 | 5.3 | 11.9 | 9.6 | 0.1 | T | 0.3 | 0.0 | 35.6 |
| 2007-08 | 0.0 | 0.0 | 0.0 | T | 2.9 | 20.1 | 40.0 | 9.0 | 15.8 | 4.8 | 0.0 | T | 92.6 |
| 2008-09 | 0.0 | 0.0 | 0.0 | 0.0 | 1.5 | 61.5 | 17.6 | 3.7 | 9.5 | 3.9 | 0.0 | T | 97.7 |
| 2009- | 0.0 | 0.0 | 0.0 | T | 4.7 | 6.7 |  |  |  |  |  |  |  |
| POR= 67 YRS | 0.0 | 0.0 | T | 0.3 | 6.6 | 14.2 | 15.4 | 7.7 | 4.3 | 0.7 | 0.1 | T | 49.3 |

# Spokane, WA  2009
## Temperature and Precipitation

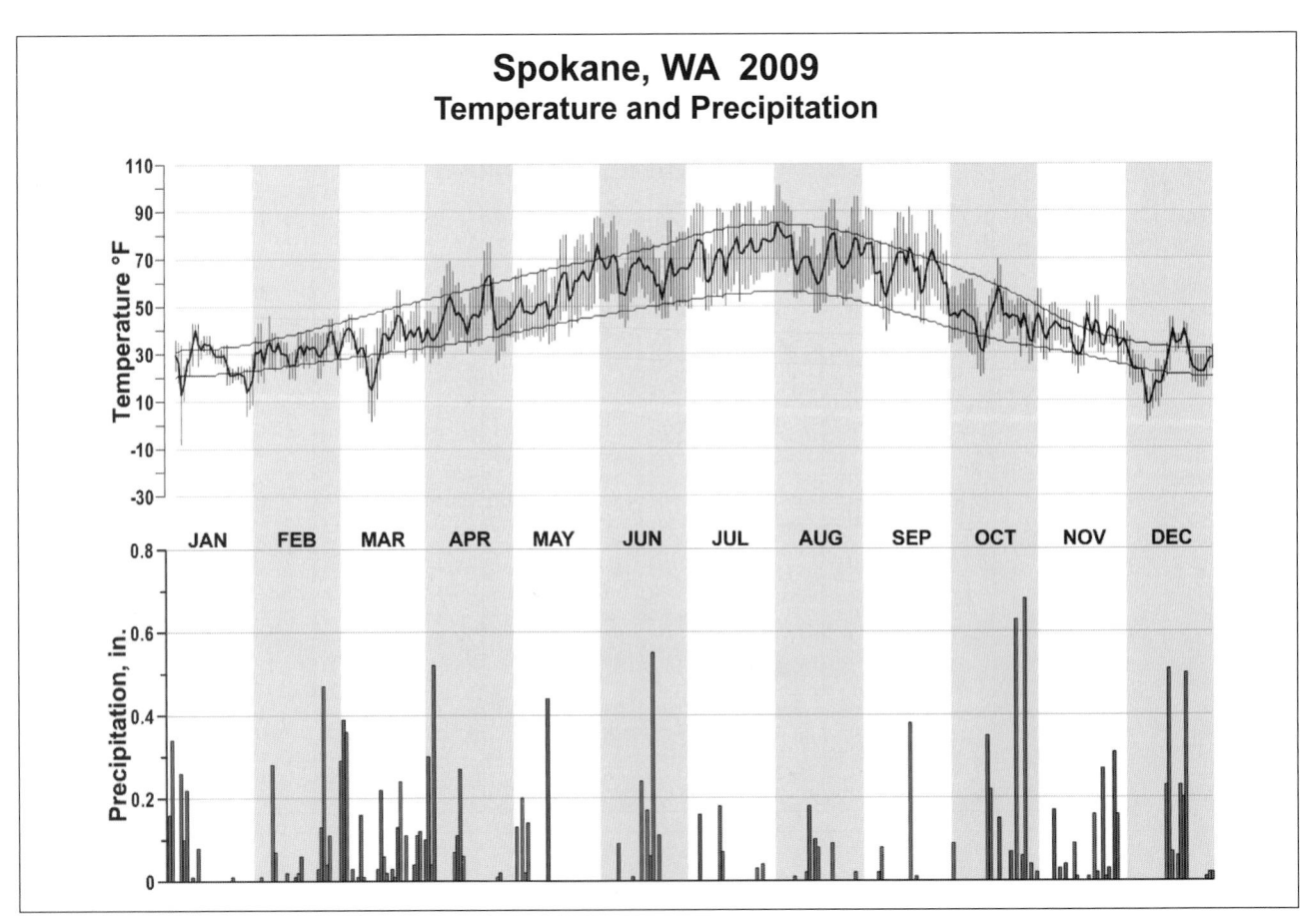

# 2009
# CHARLESTON
# WEST VIRGINIA (KCRW)

Charleston lies at the junction of the Kanawha and Elk Rivers in the western foothills of the Appalachian Mountains. The main urban and business areas have developed along the two river valleys, while some residential areas are in nearby valleys and on the surrounding hills. The hilltops are around 1,100 feet above sea level, about 500 feet higher than the valleys. The Kanawha Airport is just over 2 miles northeast of the center-city area, on an artificial plateau constructed from several hilltops.

Weather records are maintained at the Kanawha Airport by National Weather Service personnel. This site tends to be slightly cooler than the river valleys during the afternoons. Conversely, the valleys can become cooler than the hilltops during clear, calm nights. The weather at Charleston is highly changeable, especially from mid-autumn through the spring.

Winters can vary greatly from one season to the next. Snow does not favor any given winter month, heavy snowstorms are infrequent, and most snowfalls are in the 4-inch or less category. Snow and ice usually do not persist on valley roads, but can linger longer on nearby hills and outlying rural roads.

Afternoon temperatures in the 40s and morning readings in the 20s are common during the winter. Yet, every winter typically has two or three extended cold spells when temperatures stay below freezing for a few consecutive days. Northwesterly winds are associated with the cold weather. Air reaching Charleston from the northwest can cause cloudiness and flurries, even when there is no nearby organized storm system. Winter conditions are much more severe over the higher mountains less than 50 miles to the northeast through the southeast. Temperatures warm rapidly in the spring and are accompanied by low daytime humidities.

Summer and early autumn have more day-to-day consistency in the weather. Sunshine is more abundant than in winter. Summer precipitation falls mostly in brief, but sometimes heavy, showers. Flash flooding can occur along small streams, but flooding is rare on the dam-controlled Kanawha and Elk Rivers.

Afternoon summer temperatures are mostly in the 80s. Readings above 95 degrees are rare. However, during a hot spell, haze and humidity can add to the unpleasantness and indoor air conditioning is recommended. Cooler and less humid air often penetrates the area from the north to end a hot spell.

Early morning fog is common from late June into October. Industrial and vehicular pollutants can contribute to limited visibility any time of the year, especially when cooler air becomes trapped in the valleys. Autumn foliage is generally at its peak during the second and third weeks of October. By the end of October, the first 32 degree temperature has usually arrived.

Ample precipitation is well distributed throughout the year. July is quite often the wettest month of the year, while October averages the least rain. Droughts severe enough to limit water use are scarce. Any dry spells during the spring or autumn can cause conditions favorable for brush fires in outlying areas.

# NORMALS, MEANS, AND EXTREMES
## CHARLESTON (KCRW)

| LATITUDE: 38 ° 22'N | LONGITUDE: -81 ° 35'W | ELEVATION (FT): GRND: 910  BARO: 1026 | TIME ZONE: EASTERN  (UTC -5) | WBAN: 13866 |
|---|---|---|---|---|

| | ELEMENT | POR | JAN | FEB | MAR | APR | MAY | JUN | JUL | AUG | SEP | OCT | NOV | DEC | YEAR |
|---|---|---|---|---|---|---|---|---|---|---|---|---|---|---|---|
| **TEMPERATURE °F** | NORMAL DAILY MAXIMUM | 30 | 42.6 | 47.0 | 56.6 | 66.7 | 74.6 | 81.5 | 84.9 | 83.5 | 77.3 | 67.1 | 56.4 | 46.8 | 65.4 |
| | MEAN DAILY MAXIMUM | 61 | 43.1 | 46.7 | 56.0 | 67.6 | 75.7 | 82.6 | 85.5 | 84.5 | 78.5 | 68.0 | 56.5 | 46.3 | 65.9 |
| | HIGHEST DAILY MAXIMUM | 62 | 79 | 79 | 89 | 94 | 93 | 98 | 104 | 104 | 102 | 93 | 85 | 80 | 104 |
| | YEAR OF OCCURRENCE | | 1950 | 2000 | 1990 | 1990 | 2006 | 1999 | 1988 | 2007 | 1953 | 2007 | 1993 | 1982 | AUG 2007 |
| | MEAN OF EXTREME MAXS. | 61 | 67.2 | 69.3 | 79.3 | 86.2 | 88.9 | 92.0 | 93.8 | 92.9 | 90.3 | 83.3 | 77.0 | 68.6 | 82.4 |
| | NORMAL DAILY MINIMUM | 30 | 24.2 | 26.7 | 34.0 | 41.8 | 50.3 | 58.3 | 62.9 | 61.7 | 55.0 | 43.1 | 35.3 | 28.2 | 43.5 |
| | MEAN DAILY MINIMUM | 61 | 25.2 | 27.2 | 34.4 | 43.5 | 51.8 | 60.0 | 64.4 | 63.4 | 56.3 | 44.6 | 35.8 | 28.4 | 44.6 |
| | LOWEST DAILY MINIMUM | 62 | -16 | -12 | 0 | 19 | 26 | 33 | 46 | 41 | 34 | 17 | 6 | -12 | -16 |
| | YEAR OF OCCURRENCE | | 1994 | 1996 | 1980 | 1982 | 1966 | 1972 | 1963 | 1965 | 1983 | 1962 | 1950 | 1989 | JAN 1994 |
| | MEAN OF EXTREME MINS. | 61 | 3.6 | 7.6 | 17.0 | 26.7 | 35.6 | 46.3 | 53.1 | 52.0 | 41.3 | 28.8 | 18.9 | 9.3 | 28.4 |
| | NORMAL DRY BULB | 30 | 33.4 | 36.9 | 45.3 | 54.3 | 62.4 | 69.9 | 73.9 | 72.6 | 66.2 | 55.1 | 45.9 | 37.5 | 54.5 |
| | MEAN DRY BULB | 61 | 34.2 | 36.9 | 45.2 | 55.6 | 63.8 | 71.4 | 75.0 | 74.0 | 67.4 | 56.3 | 46.1 | 37.4 | 55.3 |
| | MEAN WET BULB | 26 | 30.5 | 32.6 | 38.8 | 47.2 | 56.6 | 64.9 | 68.8 | 67.7 | 61.0 | 50.6 | 41.2 | 33.7 | 49.5 |
| | MEAN DEW POINT | 26 | 26.7 | 27.9 | 33.7 | 42.0 | 53.6 | 62.3 | 66.3 | 65.6 | 58.7 | 47.3 | 36.8 | 28.9 | 45.8 |
| | NORMAL NO. DAYS WITH: | | | | | | | | | | | | | | |
| | MAXIMUM >= 90 | 30 | 0.0 | 0.0 | 0.0 | 0.4 | 0.7 | 4.0 | 8.4 | 5.1 | 1.4 | 0.0 | 0.0 | 0.0 | 20.0 |
| | MAXIMUM <= 32 | 30 | 7.3 | 4.8 | 0.8 | * | 0.0 | 0.0 | 0.0 | 0.0 | 0.0 | 0.0 | 0.2 | 4.3 | 17.4 |
| | MINIMUM <= 32 | 30 | 23.1 | 19.6 | 14.2 | 4.8 | 0.3 | 0.0 | 0.0 | 0.0 | 0.0 | 3.2 | 12.6 | 20.2 | 98.0 |
| | MINIMUM <= 0 | 30 | 1.0 | 0.2 | * | 0.0 | 0.0 | 0.0 | 0.0 | 0.0 | 0.0 | 0.0 | 0.0 | 0.3 | 1.5 |
| **H/C** | NORMAL HEATING DEG. DAYS | 30 | 977 | 794 | 604 | 330 | 141 | 19 | 8 | 3 | 62 | 309 | 560 | 837 | 4644 |
| | NORMAL COOLING DEG. DAYS | 30 | 0 | 0 | 7 | 25 | 76 | 182 | 300 | 254 | 114 | 17 | 3 | 0 | 978 |
| **RH** | NORMAL (PERCENT) | 30 | 72 | 69 | 63 | 60 | 70 | 74 | 77 | 78 | 78 | 74 | 70 | 73 | 72 |
| | HOUR 01 LST | 30 | 76 | 74 | 69 | 68 | 82 | 87 | 90 | 92 | 91 | 86 | 77 | 76 | 81 |
| | HOUR 07 LST | 30 | 79 | 78 | 77 | 77 | 85 | 88 | 91 | 93 | 93 | 89 | 82 | 79 | 84 |
| | HOUR 13 LST | 30 | 65 | 60 | 53 | 47 | 53 | 56 | 59 | 60 | 59 | 55 | 57 | 63 | 57 |
| | HOUR 19 LST | 30 | 67 | 61 | 53 | 48 | 57 | 63 | 67 | 71 | 74 | 68 | 63 | 67 | 63 |
| **S** | PERCENT POSSIBLE SUNSHINE | | | | | | | | | | | | | | |
| **W/O** | MEAN NO. DAYS WITH: | | | | | | | | | | | | | | |
| | HEAVY FOG(VISBY <= 1/4 MI) | 46 | 3.2 | 2.3 | 2.2 | 2.6 | 6.4 | 8.8 | 11.4 | 15.3 | 14.8 | 9.0 | 3.2 | 2.4 | 81.6 |
| | THUNDERSTORMS | 61 | 0.5 | 0.9 | 2.4 | 4.0 | 6.9 | 8.2 | 9.5 | 7.2 | 3.1 | 1.1 | 0.8 | 0.3 | 44.9 |
| **CLOUDNESS** | MEAN: | | | | | | | | | | | | | | |
| | SUNRISE-SUNSET (OKTAS) | | | | | | | | | | | | | | |
| | MIDNIGHT-MIDNIGHT (OKTAS) | | | | | | | | | | | | | | |
| | MEAN NO. DAYS WITH: | | | | | | | | | | | | | | |
| | CLEAR | 1 | 1.0 | | | | 7.0 | 12.0 | 1.0 | 7.0 | 4.0 | 6.0 | | 3.0 | |
| | PARTLY CLOUDY | | | | | | | | | | | | | | |
| | CLOUDY | 1 | 8.0 | 10.0 | 8.0 | | 12.0 | 5.0 | 2.0 | 1.0 | 7.0 | 4.0 | | 14.0 | |
| **PR** | MEAN STATION PRESSURE(IN) | 26 | 29.04 | 29.03 | 28.99 | 28.94 | 28.96 | 28.96 | 28.96 | 29.01 | 29.04 | 29.07 | 29.06 | 29.06 | 29.01 |
| | MEAN SEA-LEVEL PRES. (IN) | 26 | 30.12 | 30.10 | 30.05 | 29.98 | 29.99 | 29.99 | 30.01 | 30.04 | 30.07 | 30.12 | 30.12 | 30.13 | 30.06 |
| **WINDS** | MEAN SPEED (MPH) | 26 | 5.7 | 5.6 | 5.8 | 5.7 | 4.4 | 3.9 | 3.6 | 3.1 | 3.3 | 3.6 | 4.6 | 5.2 | 4.5 |
| | PREVAIL.DIR(TENS OF DEGS) | 30 | 24 | 24 | 24 | 25 | 23 | 23 | 23 | 23 | 06 | 24 | 24 | 24 | 24 |
| | MAXIMUM 2-MINUTE: | | | | | | | | | | | | | | |
| | SPEED (MPH) | 15 | 36 | 38 | 53 | 38 | 36 | 44 | 48 | 38 | 30 | 38 | 38 | 46 | 53 |
| | DIR. (TENS OF DEGS) | | 27 | 23 | 30 | 23 | 28 | 30 | 26 | 32 | 25 | 26 | 26 | 25 | 30 |
| | YEAR OF OCCURRENCE | | 2008 | 2009 | 2008 | 2006 | 2001 | 2004 | 2007 | 2006 | 2008 | 2003 | 2006 | 2009 | MAR 2008 |
| | MAXIMUM 3-SECOND | | | | | | | | | | | | | | |
| | SPEED (MPH) | 15 | 52 | 62 | 84 | 51 | 54 | 67 | 76 | 61 | 44 | 51 | 46 | 74 | 84 |
| | DIR. (TENS OF DEGS) | | 27 | 24 | 29 | 27 | 28 | 30 | 25 | 31 | 24 | 26 | 27 | 25 | 29 |
| | YEAR OF OCCURRENCE | | 2008 | 2009 | 2008 | 2004 | 2001 | 2004 | 2007 | 2000 | 2008 | 2003 | 2003 | 2009 | MAR 2008 |
| **PRECIPITATION** | NORMAL (IN) | 30 | 3.25 | 3.19 | 3.90 | 3.25 | 4.30 | 4.09 | 4.86 | 4.11 | 3.45 | 2.67 | 3.66 | 3.32 | 44.05 |
| | MAXIMUM MONTHLY (IN) | 62 | 9.11 | 7.46 | 8.35 | 6.46 | 8.76 | 10.56 | 13.54 | 10.45 | 7.69 | 6.49 | 9.12 | 8.02 | 13.54 |
| | YEAR OF OCCURRENCE | | 1950 | 2003 | 1997 | 1965 | 2001 | 1998 | 1961 | 1958 | 2004 | 1983 | 2003 | 1978 | JUL 1961 |
| | MINIMUM MONTHLY (IN) | 62 | 1.09 | 0.64 | 1.30 | 0.50 | 0.84 | 0.70 | 1.98 | 0.66 | 0.65 | 0.09 | 0.64 | 0.45 | 0.09 |
| | YEAR OF OCCURRENCE | | 1981 | 1968 | 1987 | 1976 | 1977 | 1966 | 1993 | 1957 | 1959 | 1963 | 1965 | 1965 | OCT 1963 |
| | MAXIMUM IN 24 HOURS (IN) | 62 | 2.45 | 2.71 | 2.86 | 2.72 | 3.31 | 2.73 | 5.60 | 4.17 | 4.17 | 2.48 | 3.66 | 2.47 | 5.60 |
| | YEAR OF OCCURRENCE | | 1994 | 2003 | 1967 | 1948 | 1982 | 2003 | 1961 | 1958 | 2004 | 1961 | 2003 | 1978 | JUL 1961 |
| | NORMAL NO. DAYS WITH: | | | | | | | | | | | | | | |
| | PRECIPITATION >= 0.01 | 30 | 15.8 | 14.0 | 14.7 | 13.6 | 13.8 | 12.4 | 12.4 | 11.1 | 10.0 | 9.5 | 12.4 | 14.5 | 154.2 |
| | PRECIPITATION >= 1.00 | 30 | 0.4 | 0.3 | 0.7 | 0.3 | 0.8 | 1.0 | 1.2 | 1.0 | 0.8 | 0.4 | 0.8 | 0.4 | 8.1 |
| **SNOWFALL** | NORMAL (IN) | 30 | 13.4 | 9.7 | 6.6 | 1.3 | 0.* | 0.0 | 0.0 | 0.0 | 0.0 | 0.1 | 2.0 | 5.3 | 38.4 |
| | MAXIMUM MONTHLY (IN) | 54 | 39.5 | 21.8 | 20.4 | 20.7 | 0.6 | T | T | T | T | 2.8 | 25.8 | 21.9 | 39.5 |
| | YEAR OF OCCURRENCE | | 1978 | 1964 | 1993 | 1987 | 1989 | | 1994 | 1990 | 2008 | 2006 | 1961 | 1950 | JAN 1978 |
| | MAXIMUM IN 24 HOURS (IN) | 54 | 15.8 | 11.2 | 17.1 | 11.3 | 0.6 | T | T | T | T | 2.8 | 15.1 | 11.2 | 17.1 |
| | YEAR OF OCCURRENCE' | | 1978 | 1983 | 1993 | 1987 | 1989 | | 1994 | 1990 | 1989 | 1994 | 1961 | 1950 | MAR 1993 |
| | MAXIMUM SNOW DEPTH (IN) | 49 | 23 | 13 | 9 | 17 | T | 0 | 0 | 0 | 0 | 2 | 19 | 11 | 23 |
| | YEAR OF OCCURRENCE | | 1978 | 1985 | 1980 | 1987 | 1989 | | | | | 1961 | 1950 | 2009 | JAN 1978 |
| | NORMAL NO. DAYS WITH: | | | | | | | | | | | | | | |
| | SNOWFALL >= 1.0 | 30 | 3.6 | 2.9 | 2.0 | 0.2 | 0.0 | 0.0 | 0.0 | 0.0 | 0.0 | 0.0 | 0.8 | 2.1 | 11.6 |

## PRECIPITATION (inches) 2009  CHARLESTON (KCRW)

| YEAR | JAN | FEB | MAR | APR | MAY | JUN | JUL | AUG | SEP | OCT | NOV | DEC | ANNUAL |
|------|-----|-----|-----|-----|-----|-----|-----|-----|-----|-----|-----|-----|--------|
| 1980 | 2.85 | 2.25 | 5.32 | 4.49 | 2.67 | 2.17 | 8.47 | 10.32 | 2.37 | 2.03 | 3.02 | 1.85 | 47.81 |
| 1981 | 1.09 | 4.59 | 1.80 | 4.04 | 3.78 | 6.46 | 3.02 | 2.24 | 2.36 | 2.43 | 1.29 | 2.71 | 35.81 |
| 1982 | 3.74 | 3.23 | 4.96 | 1.14 | 6.19 | 7.00 | 2.68 | 2.65 | 2.58 | 1.65 | 4.65 | 2.71 | 43.18 |
| 1983 | 1.24 | 2.72 | 3.15 | 3.96 | 5.98 | 2.77 | 4.19 | 2.54 | 1.33 | 6.49 | 4.80 | 3.19 | 42.36 |
| 1984 | 1.67 | 2.56 | 2.72 | 4.00 | 3.71 | 2.56 | 4.37 | 4.57 | 2.95 | 3.28 | 4.73 | 3.78 | 40.90 |
| 1985 | 3.07 | 2.32 | 4.23 | 1.84 | 5.88 | 3.07 | 3.22 | 2.02 | 0.71 | 3.65 | 8.45 | 2.71 | 41.17 |
| 1986 | 2.12 | 4.35 | 1.87 | 1.39 | 4.86 | 2.36 | 7.61 | 4.71 | 3.51 | 2.20 | 6.88 | 3.89 | 45.75 |
| 1987 | 3.23 | 3.34 | 1.30 | 4.05 | 2.49 | 3.38 | 4.23 | 3.56 | 3.89 | 1.10 | 2.71 | 4.13 | 37.41 |
| 1988 | 1.62 | 2.50 | 2.71 | 2.17 | 2.59 | 0.94 | 3.00 | 2.86 | 3.46 | 1.87 | 5.02 | 2.66 | 31.40 |
| 1989 | 2.92 | 6.05 | 5.81 | 4.13 | 6.79 | 7.54 | 3.04 | 5.62 | 7.28 | 4.09 | 2.87 | 1.83 | 57.97 |
| 1990 | 2.86 | 3.74 | 1.94 | 2.89 | 4.87 | 3.01 | 5.35 | 2.54 | 4.26 | 3.51 | 2.07 | 7.01 | 44.05 |
| 1991 | 2.68 | 2.98 | 6.07 | 3.49 | 1.47 | 2.49 | 2.84 | 2.95 | 5.51 | 1.10 | 5.00 | 5.89 | 42.47 |
| 1992 | 1.94 | 2.72 | 4.79 | 2.93 | 4.66 | 3.21 | 6.41 | 4.41 | 1.38 | 0.94 | 3.15 | 3.50 | 40.04 |
| 1993 | 1.87 | 2.98 | 6.68 | 1.78 | 1.98 | 5.01 | 1.98 | 2.71 | 5.99 | 3.50 | 3.95 | 3.23 | 41.66 |
| 1994 | 6.42 | 5.56 | 7.73 | 3.78 | 3.98 | 4.43 | 3.71 | 6.20 | 1.95 | 1.13 | 1.95 | 2.52 | 49.36 |
| 1995 | 6.02 | 2.98 | 2.73 | 2.59 | 6.15 | 4.93 | 2.91 | 5.81 | 2.70 | 2.61 | 3.31 | 2.79 | 45.53 |
| 1996 | 5.18 | 2.82 | 4.32 | 3.77 | 7.40 | 3.59 | 8.50 | 2.82 | 7.37 | 2.49 | 4.36 | 2.04 | 54.66 |
| 1997 | 1.76 | 1.76 | 8.35 | 2.77 | 3.60 | 5.24 | 5.83 | 4.14 | 1.94 | 0.84 | 2.96 | 1.57 | 40.76 |
| 1998 | 3.43 | 4.23 | 3.41 | 4.77 | 5.27 | 10.56 | 3.65 | 3.70 | 2.50 | 1.67 | 1.89 | 3.18 | 48.26 |
| 1999 | 4.81 | 2.67 | 3.70 | 2.20 | 1.90 | 1.30 | 5.37 | 2.97 | 1.81 | 3.43 | 4.53 | 2.55 | 37.24 |
| 2000 | 1.41 | 4.25 | 2.26 | 4.67 | 4.75 | 3.38 | 6.06 | 4.35 | 2.87 | 0.87 | 1.27 | 2.10 | 38.24 |
| 2001 | 2.43 | 1.90 | 3.28 | 1.30 | 8.76 | 4.19 | 10.06 | 2.74 | 1.85 | 1.36 | 1.43 | 2.47 | 41.77 |
| 2002 | 3.15 | 0.89 | 5.92 | 4.49 | 4.86 | 3.61 | 3.67 | 1.72 | 3.24 | 6.11 | 4.12 | 2.94 | 44.72 |
| 2003 | 1.79 | 7.46 | 1.78 | 3.40 | 4.99 | 9.93 | 5.89 | 6.53 | 4.77 | 2.46 | 9.12 | 2.89 | 61.01 |
| 2004 | 3.74 | 2.38 | 4.45 | 4.96 | 8.09 | 5.70 | 3.73 | 3.92 | 7.69 | 3.48 | 4.29 | 2.91 | 55.34 |
| 2005 | 3.04 | 3.12 | 3.51 | 4.43 | 3.01 | 3.35 | 4.91 | 5.62 | 1.07 | 3.34 | 3.56 | 2.65 | 41.61 |
| 2006 | 3.85 | 1.20 | 1.90 | 3.85 | 1.68 | 3.72 | 8.44 | 5.78 | 4.56 | 4.37 | 2.07 | 1.99 | 43.41 |
| 2007 | 2.66 | 1.50 | 4.49 | 3.81 | 2.12 | 1.10 | 5.18 | 2.82 | 1.34 | 3.64 | 3.13 | 5.64 | 37.43 |
| 2008 | 2.43 | 4.62 | 4.44 | 3.21 | 6.12 | 5.46 | 5.16 | 2.97 | 1.16 | 1.92 | 2.71 | 5.08 | 45.28 |
| 2009 | 5.20 | 1.13 | 2.90 | 2.97 | 7.92 | 4.65 | 5.32 | 3.25 | 4.42 | 2.65 | 0.74 | 4.86 | 46.01 |
| POR=<br>61 YRS | 3.38 | 3.16 | 3.93 | 3.37 | 4.17 | 3.86 | 5.14 | 3.96 | 3.21 | 2.67 | 3.32 | 3.30 | 43.47 |

WBAN : 13866

## AVERAGE TEMPERATURE (°F) 2009  CHARLESTON (KCRW)

| YEAR | JAN | FEB | MAR | APR | MAY | JUN | JUL | AUG | SEP | OCT | NOV | DEC | ANNUAL |
|------|-----|-----|-----|-----|-----|-----|-----|-----|-----|-----|-----|-----|--------|
| 1980 | 34.1 | 29.7 | 42.0 | 53.3 | 63.6 | 68.8 | 76.6 | 76.3 | 69.8 | 53.6 | 43.2 | 36.3 | 53.9 |
| 1981 | 28.0 | 37.2 | 41.4 | 59.1 | 60.4 | 73.2 | 75.7 | 72.8 | 66.5 | 54.2 | 45.3 | 34.6 | 54.0 |
| 1982 | 29.8 | 36.1 | 47.2 | 51.5 | 68.6 | 68.8 | 76.2 | 71.1 | 65.9 | 57.7 | 49.0 | 44.8 | 55.6 |
| 1983 | 34.0 | 37.7 | 47.0 | 52.1 | 61.1 | 71.6 | 77.0 | 78.0 | 68.4 | 58.1 | 47.4 | 32.0 | 55.4 |
| 1984 | 30.6 | 41.5 | 41.1 | 54.2 | 61.4 | 75.3 | 73.2 | 74.9 | 65.4 | 64.4 | 44.6 | 46.9 | 56.1 |
| 1985 | 27.2 | 34.0 | 49.4 | 60.8 | 66.3 | 71.0 | 75.8 | 74.0 | 69.6 | 62.3 | 55.5 | 33.8 | 56.6 |
| 1986 | 34.1 | 40.5 | 47.1 | 57.9 | 65.3 | 72.2 | 77.2 | 71.9 | 69.5 | 57.9 | 46.4 | 36.4 | 56.4 |
| 1987 | 33.0 | 37.2 | 47.0 | 52.7 | 68.4 | 73.5 | 77.1 | 77.0 | 67.1 | 50.3 | 49.0 | 39.8 | 56.0 |
| 1988 | 31.1 | 35.2 | 46.1 | 54.1 | 63.2 | 71.0 | 78.6 | 77.4 | 66.6 | 49.3 | 47.0 | 37.4 | 54.8 |
| 1989 | 41.1 | 34.9 | 47.8 | 52.8 | 59.4 | 71.7 | 75.7 | 73.2 | 67.0 | 56.7 | 46.4 | 26.0 | 54.4 |
| 1990 | 42.3 | 45.2 | 51.7 | 55.1 | 62.9 | 72.3 | 75.8 | 74.1 | 68.7 | 58.2 | 50.7 | 43.6 | 58.4 |
| 1991 | 36.5 | 40.3 | 47.4 | 60.3 | 71.7 | 73.9 | 77.9 | 75.2 | 68.4 | 59.1 | 45.8 | 41.3 | 58.2 |
| 1992 | 35.6 | 41.8 | 45.4 | 55.7 | 61.2 | 68.8 | 76.0 | 70.9 | 67.4 | 54.1 | 47.4 | 37.9 | 55.2 |
| 1993 | 40.2 | 34.0 | 41.8 | 55.0 | 64.9 | 71.4 | 79.0 | 76.4 | 66.6 | 55.1 | 47.1 | 36.6 | 55.7 |
| 1994 | 28.1 | 38.7 | 45.3 | 60.5 | 59.7 | 74.2 | 76.1 | 72.4 | 65.4 | 54.8 | 50.6 | 41.2 | 55.6 |
| 1995 | 34.7 | 34.0 | 46.9 | 55.2 | 61.9 | 71.4 | 76.4 | 77.6 | 64.6 | 56.9 | 40.8 | 32.5 | 54.4 |
| 1996 | 32.1 | 35.5 | 39.6 | 54.0 | 64.7 | 71.9 | 71.6 | 72.8 | 65.9 | 55.8 | 40.2 | 41.4 | 53.8 |
| 1997 | 35.6 | 43.8 | 47.1 | 50.8 | 58.7 | 70.1 | 74.5 | 70.8 | 65.0 | 55.1 | 42.2 | 36.5 | 54.2 |
| 1998 | 41.0 | 42.0 | 45.9 | 55.4 | 65.8 | 70.3 | 73.7 | 73.9 | 70.2 | 56.1 | 46.7 | 40.4 | 56.8 |
| 1999 | 38.3 | 39.4 | 40.1 | 58.5 | 64.0 | 72.9 | 79.1 | 72.3 | 65.8 | 54.8 | 48.9 | 38.2 | 56.0 |
| 2000 | 32.1 | 43.2 | 49.2 | 54.1 | 66.0 | 72.3 | 71.3 | 71.8 | 65.7 | 57.7 | 43.3 | 28.6 | 54.6 |
| 2001 | 33.2 | 40.6 | 40.3 | 59.1 | 64.0 | 70.6 | 72.4 | 74.8 | 64.6 | 55.1 | 50.4 | 42.3 | 55.6 |
| 2002 | 38.4 | 38.1 | 45.7 | 57.8 | 60.5 | 72.8 | 76.3 | 75.9 | 71.2 | 57.0 | 43.9 | 37.6 | 56.3 |
| 2003 | 28.2 | 33.9 | 48.6 | 57.4 | 62.2 | 67.6 | 73.2 | 74.4 | 65.1 | 54.8 | 49.7 | 36.3 | 54.3 |
| 2004 | 30.9 | 36.8 | 48.1 | 55.8 | 69.0 | 70.8 | 74.5 | 71.7 | 68.4 | 58.9 | 50.0 | 37.8 | 56.1 |
| 2005 | 38.4 | 39.7 | 41.7 | 56.3 | 59.8 | 73.9 | 77.1 | 77.7 | 70.4 | 57.6 | 48.3 | 33.4 | 56.2 |
| 2006 | 43.2 | 36.1 | 45.1 | 59.8 | 61.7 | 70.1 | 75.9 | 76.9 | 64.6 | 54.9 | 48.0 | 42.4 | 56.6 |
| 2007 | 38.8 | 28.2 | 51.5 | 54.1 | 67.1 | 73.8 | 74.0 | 79.4 | 70.9 | 62.6 | 46.1 | 41.1 | 57.3 |
| 2008 | 34.7 | 38.0 | 46.3 | 57.3 | 61.3 | 73.2 | 74.2 | 72.8 | 70.9 | 55.7 | 43.3 | 38.2 | 55.5 |
| 2009 | 31.2 | 38.1 | 47.9 | 56.9 | 64.3 | 71.6 | 71.3 | 73.8 | 68.1 | 54.5 | 49.4 | 35.7 | 55.2 |
| POR=<br>61 YRS | 34.2 | 36.9 | 45.2 | 55.6 | 63.8 | 71.4 | 75.0 | 74.0 | 67.4 | 56.3 | 46.1 | 37.4 | 55.3 |

## HEATING DEGREE DAYS (base 65°F) 2009 CHARLESTON (KCRW)

| YEAR | JUL | AUG | SEP | OCT | NOV | DEC | JAN | FEB | MAR | APR | MAY | JUN | TOTAL |
|---|---|---|---|---|---|---|---|---|---|---|---|---|---|
| 1980-81 | 0 | 0 | 33 | 356 | 650 | 882 | 1138 | 774 | 727 | 207 | 175 | 2 | 4944 |
| 1981-82 | 0 | 1 | 76 | 335 | 585 | 936 | 1086 | 801 | 545 | 405 | 36 | 2 | 4808 |
| 1982-83 | 1 | 2 | 69 | 268 | 480 | 626 | 955 | 757 | 554 | 388 | 153 | 16 | 4269 |
| 1983-84 | 4 | 0 | 66 | 227 | 521 | 1019 | 1059 | 674 | 734 | 346 | 171 | 5 | 4826 |
| 1984-85 | 1 | 0 | 98 | 74 | 613 | 563 | 1164 | 860 | 488 | 192 | 54 | 18 | 4125 |
| 1985-86 | 0 | 0 | 51 | 127 | 294 | 960 | 954 | 679 | 554 | 249 | 83 | 7 | 3958 |
| 1986-87 | 0 | 23 | 23 | 255 | 550 | 880 | 989 | 770 | 549 | 374 | 63 | 4 | 4480 |
| 1987-88 | 0 | 0 | 37 | 447 | 473 | 774 | 1043 | 859 | 577 | 326 | 112 | 38 | 4686 |
| 1988-89 | 2 | 0 | 37 | 484 | 534 | 849 | 735 | 837 | 536 | 367 | 221 | 2 | 4604 |
| 1989-90 | 0 | 7 | 72 | 270 | 553 | 1203 | 697 | 549 | 446 | 323 | 111 | 8 | 4239 |
| 1990-91 | 0 | 0 | 59 | 230 | 428 | 655 | 876 | 685 | 558 | 192 | 21 | 1 | 3705 |
| 1991-92 | 0 | 0 | 80 | 229 | 576 | 729 | 904 | 670 | 602 | 319 | 170 | 24 | 4303 |
| 1992-93 | 0 | 1 | 67 | 335 | 522 | 834 | 760 | 862 | 713 | 306 | 73 | 24 | 4497 |
| 1993-94 | 0 | 0 | 64 | 307 | 540 | 873 | 1136 | 732 | 607 | 191 | 197 | 3 | 4650 |
| 1994-95 | 0 | 4 | 34 | 308 | 424 | 729 | 932 | 861 | 555 | 306 | 138 | 7 | 4298 |
| 1995-96 | 0 | 0 | 72 | 262 | 721 | 999 | 1008 | 848 | 779 | 357 | 105 | 4 | 5155 |
| 1996-97 | 2 | 0 | 59 | 278 | 743 | 727 | 908 | 592 | 547 | 429 | 213 | 23 | 4521 |
| 1997-98 | 0 | 6 | 53 | 324 | 676 | 877 | 733 | 639 | 626 | 283 | 60 | 38 | 4315 |
| 1998-99 | 0 | 0 | 29 | 290 | 541 | 758 | 821 | 711 | 763 | 213 | 74 | 13 | 4213 |
| 1999-00 | 0 | 1 | 56 | 312 | 475 | 821 | 1014 | 627 | 490 | 320 | 51 | 14 | 4181 |
| 2000-01 | 0 | 0 | 94 | 250 | 644 | 1122 | 978 | 681 | 759 | 249 | 82 | 13 | 4872 |
| 2001-02 | 0 | 0 | 95 | 309 | 433 | 699 | 822 | 748 | 591 | 256 | 198 | 0 | 4151 |
| 2002-03 | 0 | 0 | 14 | 280 | 624 | 843 | 1135 | 865 | 503 | 244 | 106 | 34 | 4648 |
| 2003-04 | 0 | 0 | 61 | 312 | 451 | 883 | 1053 | 811 | 523 | 293 | 43 | 6 | 4436 |
| 2004-05 | 0 | 5 | 21 | 187 | 446 | 837 | 820 | 702 | 713 | 259 | 180 | 2 | 4172 |
| 2005-06 | 0 | 0 | 11 | 245 | 502 | 974 | 667 | 806 | 611 | 181 | 161 | 10 | 4168 |
| 2006-07 | 0 | 0 | 74 | 318 | 502 | 695 | 808 | 1022 | 441 | 351 | 64 | 0 | 4275 |
| 2007-08 | 0 | 0 | 28 | 163 | 562 | 735 | 932 | 777 | 572 | 245 | 132 | 0 | 4146 |
| 2008-09 | 0 | 0 | 3 | 295 | 644 | 823 | 1039 | 746 | 533 | 283 | 98 | 10 | 4474 |
| 2009- | 0 | 1 | 26 | 325 | 460 | 898 | | | | | | | |

WBAN : 13866

## COOLING DEGREE DAYS (base 65°F) 2009 CHARLESTON (KCRW)

| YEAR | JAN | FEB | MAR | APR | MAY | JUN | JUL | AUG | SEP | OCT | NOV | DEC | TOTAL |
|---|---|---|---|---|---|---|---|---|---|---|---|---|---|
| 1980 | 0 | 0 | 0 | 6 | 71 | 147 | 370 | 358 | 182 | 9 | 0 | 0 | 1143 |
| 1981 | 0 | 0 | 2 | 38 | 41 | 256 | 340 | 251 | 126 | 5 | 0 | 0 | 1059 |
| 1982 | 0 | 0 | 0 | 6 | 154 | 122 | 355 | 196 | 101 | 47 | 5 | 6 | 992 |
| 1983 | 0 | 0 | 2 | 6 | 39 | 222 | 385 | 407 | 177 | 18 | 0 | 0 | 1256 |
| 1984 | 0 | 0 | 0 | 27 | 64 | 318 | 261 | 312 | 116 | 65 | 7 | 8 | 1178 |
| 1985 | 0 | 0 | 9 | 72 | 105 | 204 | 339 | 285 | 194 | 52 | 14 | 0 | 1274 |
| 1986 | 0 | 0 | 4 | 41 | 100 | 227 | 384 | 244 | 167 | 43 | 0 | 0 | 1210 |
| 1987 | 0 | 0 | 0 | 13 | 177 | 268 | 381 | 379 | 108 | 0 | 2 | 0 | 1328 |
| 1988 | 0 | 0 | 3 | 9 | 64 | 225 | 430 | 392 | 91 | 4 | 3 | 0 | 1221 |
| 1989 | 0 | 0 | 11 | 6 | 55 | 211 | 339 | 273 | 140 | 23 | 2 | 0 | 1060 |
| 1990 | 0 | 0 | 41 | 33 | 54 | 232 | 342 | 286 | 174 | 28 | 7 | 0 | 1197 |
| 1991 | 0 | 0 | 17 | 60 | 236 | 273 | 408 | 324 | 190 | 53 | 5 | 0 | 1566 |
| 1992 | 0 | 0 | 2 | 47 | 57 | 143 | 347 | 192 | 145 | 4 | 0 | 0 | 937 |
| 1993 | 0 | 0 | 0 | 14 | 76 | 222 | 444 | 364 | 119 | 7 | 8 | 0 | 1254 |
| 1994 | 0 | 0 | 4 | 62 | 37 | 288 | 353 | 240 | 51 | 0 | 1 | 0 | 1036 |
| 1995 | 1 | 0 | 1 | 20 | 51 | 209 | 363 | 401 | 69 | 23 | 2 | 0 | 1140 |
| 1996 | 0 | 0 | 0 | 34 | 105 | 214 | 213 | 247 | 93 | 2 | 5 | 0 | 913 |
| 1997 | 1 | 5 | 0 | 11 | 23 | 182 | 302 | 195 | 61 | 23 | 0 | 0 | 803 |
| 1998 | 0 | 0 | 39 | 3 | 93 | 205 | 275 | 283 | 194 | 20 | 0 | 4 | 1116 |
| 1999 | 0 | 0 | 0 | 24 | 48 | 254 | 443 | 236 | 89 | 1 | 0 | 0 | 1095 |
| 2000 | 0 | 0 | 5 | 3 | 93 | 241 | 201 | 218 | 122 | 27 | 0 | 0 | 910 |
| 2001 | 0 | 0 | 0 | 80 | 56 | 185 | 235 | 310 | 93 | 12 | 0 | 0 | 971 |
| 2002 | 0 | 0 | 0 | 48 | 64 | 241 | 360 | 344 | 208 | 42 | 0 | 0 | 1307 |
| 2003 | 0 | 0 | 1 | 22 | 26 | 120 | 260 | 297 | 72 | 1 | 1 | 0 | 800 |
| 2004 | 1 | 0 | 8 | 22 | 173 | 188 | 303 | 219 | 131 | 7 | 3 | 0 | 1055 |
| 2005 | 0 | 0 | 0 | 6 | 23 | 278 | 383 | 400 | 182 | 24 | 8 | 0 | 1304 |
| 2006 | 0 | 0 | 2 | 31 | 65 | 168 | 345 | 373 | 68 | 11 | 0 | 0 | 1063 |
| 2007 | 0 | 0 | 29 | 31 | 134 | 269 | 283 | 454 | 209 | 97 | 0 | 0 | 1506 |
| 2008 | 0 | 0 | 0 | 20 | 25 | 254 | 294 | 252 | 188 | 12 | 0 | 0 | 1045 |
| 2009 | 0 | 0 | 8 | 48 | 82 | 214 | 203 | 284 | 125 | 9 | 0 | 0 | 973 |

## SNOWFALL (inches)  2009  CHARLESTON (KCRW)

| YEAR | JUL | AUG | SEP | OCT | NOV | DEC | JAN | FEB | MAR | APR | MAY | JUN | TOTAL |
|------|-----|-----|-----|-----|-----|-----|-----|-----|-----|-----|-----|-----|-------|
| 1980-81 | 0.0 | 0.0 | 0.0 | T | 0.5 | 2.6 | 9.1 | 6.8 | 7.5 | T | 0.0 | 0.0 | 26.5 |
| 1981-82 | 0.0 | 0.0 | 0.0 | T | 0.5 | 8.2 | 12.1 | 6.4 | 7.4 | 1.0 | 0.0 | 0.0 | 35.6 |
| 1982-83 | 0.0 | 0.0 | 0.0 | 0.0 | T | 2.9 | 5.8 | 15.0 | 5.2 | 0.1 | 0.0 | 0.0 | 29.0 |
| 1983-84 | 0.0 | 0.0 | 0.0 | 0.0 | 0.3 | 3.8 | 12.8 | 9.7 | 2.4 | 0.0 | 0.0 | 0.0 | 29.0 |
| 1984-85 | 0.0 | 0.0 | 0.0 | 0.0 | T | 3.7 | 17.6 | 20.1 | 0.9 | 1.7 | 0.0 | 0.0 | 44.0 |
| 1985-86 | 0.0 | 0.0 | 0.0 | 0.0 | 0.0 | 8.9 | 13.1 | 17.7 | 3.7 | T | 0.0 | 0.0 | 43.4 |
| 1986-87 | 0.0 | 0.0 | 0.0 | 0.0 | 0.2 | 0.1 | 16.3 | 9.7 | 3.9 | 20.7 | 0.0 | 0.0 | 50.9 |
| 1987-88 | 0.0 | 0.0 | 0.0 | T | 2.4 | 5.7 | 8.3 | 7.8 | 4.6 | T | 0.0 | 0.0 | 28.8 |
| 1988-89 | 0.0 | 0.0 | 0.0 | T | T | 6.9 | 1.7 | 4.6 | T | 1.4 | 0.6 | 0.0 | 15.2 |
| 1989-90 | 0.0 | T | T | T | 2.0 | 14.1 | 11.0 | 3.8 | 6.6 | 1.1 | 0.0 | 0.0 | 38.6 |
| 1990-91 | T | 0.0 | 0.0 | 0.0 | 0.0 | 1.2 | 3.5 | 6.5 | 5.3 | T | 0.0 | 0.0 | 16.5 |
| 1991-92 | 0.0 | 0.0 | 0.0 | 0.0 | 4.1 | 0.7 | 5.6 | 1.1 | 8.6 | 3.6 | T | T | 23.7 |
| 1992-93 | 0.0 | 0.0 | 0.0 | T | 2.5 | 3.7 | 0.4 | 12.0 | 20.4 | T | 0.0 | T | 39.0 |
| 1993-94 | 0.0 | 0.0 | 0.0 | 1.5 | 0.4 | 12.4 | 34.2 | 7.0 | 3.1 | 0.0 | 0.0 | T | 58.6 |
| 1994-95 | 0.0 | 0.0 | T | 0.0 | T | T | 9.1 | 7.9 | 8.7 | 0.0 | 0.0 | 0.0 | 25.7 |
| 1995-96 | 0.0 | 0.0 | 0.0 | 0.0 | 13.6 | 21.9 | 35.1 | 14.2 | 20.4 | 0.8 | 0.0 | T | 106.0 |
| 1996-97 | 0.0 | 0.0 | 0.0 | | | | | | | | | | |
| 1997-98 | | | | | | | | | | | | | |
| 1998-99 | | | | | | | | | | | | | |
| 1999-00 | | | | | | | | | | | | | |
| 2000-01 | | | | | | | | | | | | | |
| 2001-02 | | | | | | | | | | | | | |
| 2002-03 | | | | | | | | | | | | | |
| 2003-04 | | | | | | | | | | | | | |
| 2004-05 | | | | | | 4.8 | 6.8 | 6.0 | 8.8 | 0.7 | 0.0 | 0.0 | |
| 2005-06 | 0.0 | 0.0 | 0.0 | 0.0 | 0.9 | 4.7 | 3.1 | 8.5 | 0.5 | T | 0.0 | 0.0 | 17.7 |
| 2006-07 | 0.0 | 0.0 | T | T | T | 0.4 | 4.0 | 9.4 | 2.4 | 0.8 | 0.0 | 0.0 | 17.0 |
| 2007-08 | 0.0 | 0.0 | 0.0 | 0.0 | 0.0 | 3.6 | 7.0 | 7.8 | 3.9 | 0.0 | 0.0 | 0.0 | 22.3 |
| 2008-09 | 0.0 | T | 0.0 | T | 1.3 | 4.7 | 14.3 | 8.0 | 3.4 | T | 0.0 | 0.0 | 31.7 |
| 2009- | 0.0 | 0.0 | 0.0 | 0.0 | T | 17.2 | | | | | | | |
| POR=<br>61 YRS | T | T | T | 0.1 | 1.9 | 4.7 | 9.1 | 7.6 | 4.7 | 0.7 | T | T | 28.8 |

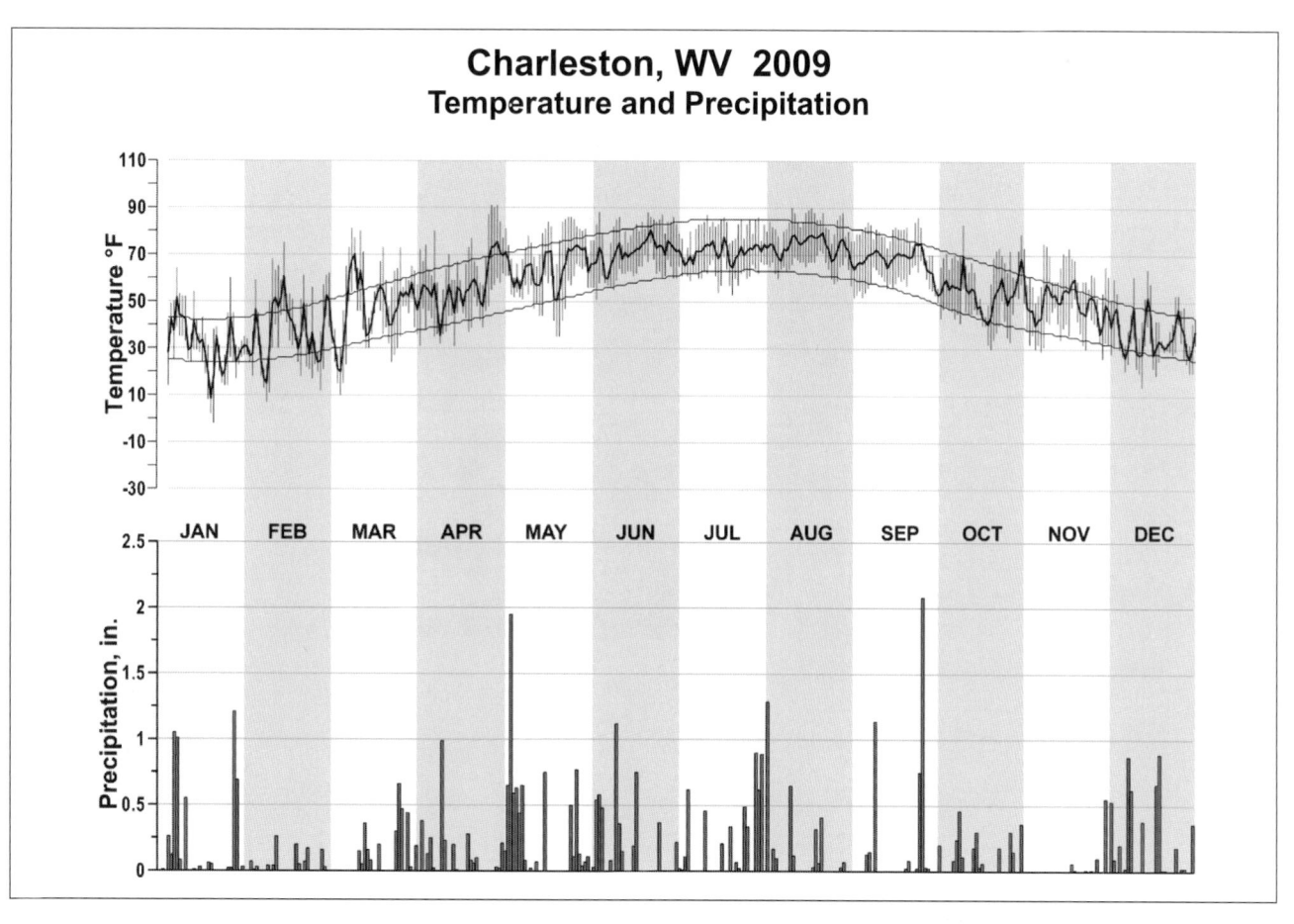

## Charleston, WV  2009
### Temperature and Precipitation

# 2009
# GREEN BAY
# WISCONSIN (KGRB)

The Green Bay climate is modified by surrounding topography. The modification is caused by the Bay of Green Bay, Lakes Michigan, and Superior, and to a lesser extent, the slightly higher surrounding terrain terminating in the Fox River Valley. The city of Green Bay is located at the mouth of the Fox River, one of the largest rivers flowing northward in the United States. It empties into the south end of the Bay.

The modified continental climate of Green Bay is shown by the few occurrences of 90 degree temperatures in the summer season and the few occurrences of sub-zero temperatures in the winter season. The narrow temperature range stems from the lake effects and the limited hours of sunshine caused by cloudiness.

Precipitation normally falls in the five-month period May through September. Three-fifths of the annual total is in the growing season, most often falling during thunderstorms. During the winter months, snowfall is less than in nearby communities where the ground is slightly higher.

The comparatively low range in temperature along with the greater portion of the precipitation falling during the growing season is conducive to the development of the dairy industry. Cherry and apple orchards are important crops in nearby lake communities. The growing of potatoes and canning vegetables are predominant inland. Paper products are the major manufacturing industry.

High winds, excessive precipitation, and electrical storms cause occasional damage. Snowstorms are the principal winter hazard. While the winters are long in Green Bay, the extremes are never as severe as the northern latitude location would indicate.

Based on the 1951-1980 period, the average first occurrence of 32 degrees Fahrenheit in the fall is October 2 and the average last occurrence in the spring is May 12.

# NORMALS, MEANS, AND EXTREMES
## GREEN BAY (KGRB)

LATITUDE: 44 ° 30'N  LONGITUDE: -88 ° 7 'W  ELEVATION (FT): GRND: 688  BARO: 685  TIME ZONE: CENTRAL  (UTC -6)  WBAN: 14898

| | ELEMENT | POR | JAN | FEB | MAR | APR | MAY | JUN | JUL | AUG | SEP | OCT | NOV | DEC | YEAR |
|---|---|---|---|---|---|---|---|---|---|---|---|---|---|---|---|
| **TEMPERATURE °F** | NORMAL DAILY MAXIMUM | 30 | 24.1 | 28.9 | 40.0 | 54.6 | 68.0 | 76.8 | 81.2 | 78.5 | 70.2 | 57.9 | 42.4 | 29.0 | 54.3 |
| | MEAN DAILY MAXIMUM | 60 | 23.9 | 28.0 | 38.4 | 54.1 | 66.7 | 76.2 | 80.8 | 78.5 | 70.3 | 58.1 | 42.5 | 29.0 | 53.9 |
| | HIGHEST DAILY MAXIMUM | 60 | 53 | 61 | 78 | 89 | 91 | 98 | 103 | 99 | 95 | 88 | 74 | 64 | 103 |
| | YEAR OF OCCURRENCE | | 2002 | 2000 | 2007 | 1980 | 1959 | 1988 | 1995 | 1988 | 1955 | 1963 | 2008 | 2001 | JUL 1995 |
| | MEAN OF EXTREME MAXS. | 60 | 41.0 | 44.0 | 59.8 | 77.0 | 84.2 | 90.0 | 91.4 | 90.3 | 86.1 | 77.0 | 62.1 | 46.3 | 70.8 |
| | NORMAL DAILY MINIMUM | 30 | 7.1 | 12.1 | 22.6 | 33.9 | 44.7 | 54.0 | 58.6 | 56.5 | 47.5 | 36.9 | 25.6 | 13.3 | 34.4 |
| | MEAN DAILY MINIMUM | 60 | 7.6 | 10.8 | 21.4 | 33.8 | 44.0 | 53.7 | 58.4 | 56.7 | 48.4 | 38.3 | 26.8 | 14.0 | 34.5 |
| | LOWEST DAILY MINIMUM | 60 | -31 | -28 | -29 | 7 | 21 | 32 | 40 | 38 | 24 | 15 | -9 | -27 | -31 |
| | YEAR OF OCCURRENCE | | 1951 | 1996 | 1962 | 1954 | 1966 | 1958 | 1965 | 1967 | 1949 | 1966 | 1976 | 1983 | JAN 1951 |
| | MEAN OF EXTREME MINS. | 60 | -14.6 | -10.5 | 0.5 | 19.9 | 30.3 | 40.4 | 46.5 | 44.3 | 32.9 | 24.1 | 9.5 | -8.1 | 17.9 |
| | NORMAL DRY BULB | 30 | 15.6 | 20.5 | 31.3 | 44.2 | 56.4 | 65.4 | 69.9 | 67.5 | 58.8 | 47.4 | 34.0 | 21.2 | 44.4 |
| | MEAN DRY BULB | 60 | 15.8 | 19.4 | 29.9 | 44.0 | 55.4 | 65.1 | 69.6 | 67.6 | 59.4 | 48.2 | 34.6 | 21.5 | 44.2 |
| | MEAN WET BULB | 26 | 16.2 | 19.1 | 27.8 | 38.6 | 49.3 | 59.3 | 63.3 | 62.3 | 55.1 | 43.3 | 31.9 | 20.8 | 40.6 |
| | MEAN DEW POINT | 26 | 13.0 | 15.3 | 23.6 | 34.1 | 45.4 | 56.2 | 60.8 | 60.2 | 52.4 | 40.2 | 28.8 | 17.8 | 37.3 |
| | NORMAL NO. DAYS WITH: | | | | | | | | | | | | | | |
| | MAXIMUM >= 90 | 30 | 0.0 | 0.0 | 0.0 | 0.0 | 0.1 | 1.9 | 2.9 | 1.4 | 0.2 | 0.0 | 0.0 | 0.0 | 6.5 |
| | MAXIMUM <= 32 | 30 | 23.0 | 17.3 | 7.5 | 0.4 | 0.0 | 0.0 | 0.0 | 0.0 | 0.0 | 0.0 | 5.3 | 17.8 | 71.3 |
| | MINIMUM <= 32 | 30 | 30.5 | 27.0 | 26.1 | 13.8 | 2.0 | 0.0 | 0.0 | 0.0 | 0.7 | 8.7 | 22.6 | 29.5 | 160.9 |
| | MINIMUM <= 0 | 30 | 10.7 | 6.2 | 1.1 | 0.0 | 0.0 | 0.0 | 0.0 | 0.0 | 0.0 | 0.0 | 0.3 | 5.4 | 23.7 |
| **H/C** | NORMAL HEATING DEG. DAYS | 30 | 1537 | 1262 | 1060 | 638 | 301 | 85 | 19 | 38 | 208 | 540 | 925 | 1350 | 7963 |
| | NORMAL COOLING DEG. DAYS | 30 | 0 | 0 | 0 | 3 | 24 | 95 | 177 | 126 | 36 | 2 | 0 | 0 | 463 |
| **RH** | NORMAL (PERCENT) | 30 | 76 | 75 | 73 | 67 | 67 | 70 | 74 | 77 | 77 | 75 | 76 | 78 | 74 |
| | HOUR 00 LST | 30 | 77 | 78 | 78 | 75 | 76 | 80 | 84 | 88 | 87 | 82 | 81 | 80 | 81 |
| | HOUR 06 LST | 30 | 79 | 80 | 82 | 79 | 79 | 82 | 86 | 90 | 90 | 85 | 83 | 81 | 83 |
| | HOUR 12 LST | 30 | 71 | 68 | 64 | 57 | 55 | 57 | 59 | 63 | 62 | 62 | 68 | 73 | 63 |
| | HOUR 18 LST | 30 | 74 | 71 | 67 | 59 | 57 | 59 | 62 | 67 | 71 | 72 | 75 | 77 | 68 |
| **S** | PERCENT POSSIBLE SUNSHINE | 59 | 49 | 52 | 54 | 56 | 61 | 65 | 66 | 62 | 56 | 48 | 37 | 40 | 54 |
| **W/O** | MEAN NO. DAYS WITH: | | | | | | | | | | | | | | |
| | HEAVY FOG(VISBY <= 1/4 MI) | 46 | 1.6 | 2.0 | 3.0 | 1.6 | 1.4 | 1.3 | 1.3 | 2.5 | 2.6 | 2.2 | 1.9 | 2.5 | 23.9 |
| | THUNDERSTORMS | 60 | 0.1 | 0.1 | 1.2 | 2.1 | 3.8 | 6.2 | 6.3 | 5.6 | 3.6 | 1.8 | 0.4 | 0.1 | 31.3 |
| **CLOUDINESS** | MEAN: | | | | | | | | | | | | | | |
| | SUNRISE-SUNSET (OKTAS) | 47 | 5.3 | 5.2 | 5.4 | 5.4 | 5.1 | 4.8 | 4.5 | 4.6 | 4.8 | 5.1 | 5.8 | 5.6 | 5.1 |
| | MIDNIGHT-MIDNIGHT (OKTAS) | 32 | 5.1 | 4.9 | 5.1 | 5.2 | 4.8 | 4.6 | 4.2 | 4.3 | 4.5 | 4.9 | 5.6 | 5.3 | 4.9 |
| | MEAN NO. DAYS WITH: | | | | | | | | | | | | | | |
| | CLEAR | 47 | 7.7 | 7.2 | 7.0 | 6.2 | 7.0 | 7.5 | 8.0 | 8.4 | 8.1 | 7.1 | 4.8 | 6.3 | 85.3 |
| | PARTLY CLOUDY | 47 | 6.6 | 6.5 | 7.6 | 7.8 | 9.5 | 10.9 | 12.2 | 10.7 | 9.4 | 8.5 | 6.5 | 6.1 | 102.3 |
| | CLOUDY | 47 | 16.7 | 14.6 | 16.4 | 16.0 | 14.5 | 11.6 | 10.8 | 11.9 | 12.5 | 15.4 | 18.8 | 18.6 | 177.8 |
| **PR** | MEAN STATION PRESSURE(IN) | 26 | 29.27 | 29.29 | 29.28 | 29.21 | 29.20 | 29.19 | 29.22 | 29.27 | 29.27 | 29.26 | 29.25 | 29.27 | 29.25 |
| | MEAN SEA-LEVEL PRES. (IN) | 26 | 30.06 | 30.08 | 30.06 | 29.98 | 29.96 | 29.94 | 29.96 | 30.01 | 30.02 | 30.03 | 30.03 | 30.06 | 30.02 |
| **WINDS** | MEAN SPEED (MPH) | 26 | 9.7 | 9.6 | 9.8 | 10.4 | 9.3 | 8.0 | 7.3 | 6.9 | 7.7 | 8.9 | 9.5 | 9.3 | 8.9 |
| | PREVAIL.DIR(TENS OF DEGS) | 33 | 28 | 22 | 22 | 05 | 05 | 22 | 22 | 22 | 22 | 22 | 22 | 28 | 28 |
| | MAXIMUM 2-MINUTE: | | | | | | | | | | | | | | |
| | SPEED (MPH) | 13 | 39 | 37 | 44 | 41 | 39 | 41 | 51 | 32 | 39 | 36 | 45 | 39 | 51 |
| | DIR. (TENS OF DEGS) | | 04 | 30 | 29 | 22 | 33 | 01 | 29 | 32 | 26 | 28 | 20 | 03 | 29 |
| | YEAR OF OCCURRENCE | | 1999 | 2007 | 2002 | 1997 | 2006 | 2005 | 2006 | 2009 | 2005 | 2001 | 1998 | 2009 | JUL 2006 |
| | MAXIMUM 3-SECOND | | | | | | | | | | | | | | |
| | SPEED (MPH) | 13 | 48 | 49 | 54 | 54 | 51 | 66 | 62 | 46 | 54 | 47 | 59 | 52 | 66 |
| | DIR. (TENS OF DEGS) | | 27 | 04 | 29 | 23 | 27 | 09 | 28 | 24 | 26 | 27 | 21 | 27 | 09 |
| | YEAR OF OCCURRENCE | | 2008 | 2006 | 2002 | 1997 | 1997 | 2007 | 2006 | 2004 | 2005 | 2001 | 1998 | 2003 | JUN 2007 |
| **PRECIPITATION** | NORMAL (IN) | 30 | 1.21 | 1.01 | 2.06 | 2.56 | 2.75 | 3.43 | 3.44 | 3.77 | 3.11 | 2.17 | 2.27 | 1.41 | 29.19 |
| | MAXIMUM MONTHLY (IN) | 60 | 3.65 | 3.56 | 4.68 | 5.91 | 8.31 | 10.29 | 7.00 | 9.04 | 7.80 | 5.16 | 5.32 | 3.72 | 10.29 |
| | YEAR OF OCCURRENCE | | 2008 | 1953 | 1977 | 1994 | 2004 | 1990 | 1994 | 1975 | 1965 | 2009 | 1992 | 2008 | JUN 1990 |
| | MINIMUM MONTHLY (IN) | 60 | 0.12 | 0.04 | 0.15 | 0.49 | 0.06 | 0.31 | 0.83 | 0.59 | 0.28 | T | 0.11 | T | T |
| | YEAR OF OCCURRENCE | | 1981 | 1969 | 1999 | 1989 | 1988 | 1976 | 1981 | 2008 | 1976 | 1952 | 2007 | 1952 | DEC 1952 |
| | MAXIMUM IN 24 HOURS (IN) | 60 | 1.14 | 1.78 | 1.83 | 3.24 | 3.28 | 4.90 | 4.65 | 4.60 | 2.99 | 3.68 | 2.30 | 1.55 | 4.90 |
| | YEAR OF OCCURRENCE | | 1980 | 1966 | 1998 | 1994 | 1973 | 1990 | 2000 | 1975 | 1964 | 1954 | 1985 | 1959 | JUN 1990 |
| | NORMAL NO. DAYS WITH: | | | | | | | | | | | | | | |
| | PRECIPITATION >= 0.01 | 30 | 10.7 | 8.5 | 10.8 | 11.1 | 10.1 | 10.1 | 10.4 | 11.3 | 10.0 | 9.7 | 10.3 | 10.6 | 123.6 |
| | PRECIPITATION >= 1.00 | 30 | 0.0 | 0.1 | 0.2 | 0.3 | 0.5 | 0.9 | 0.9 | 0.8 | 0.8 | 0.3 | 0.4 | 0.1 | 5.3 |
| **SNOWFALL** | NORMAL (IN) | 30 | 13.9 | 8.7 | 9.2 | 2.9 | 0.2 | 0.0 | 0.0 | 0.0 | 0.* | 0.2 | 5.4 | 12.6 | 53.1 |
| | MAXIMUM MONTHLY (IN) | 60 | 31.5 | 24.5 | 24.2 | 11.8 | 4.3 | T | T | T | T | 1.8 | 17.1 | 45.6 | 45.6 |
| | YEAR OF OCCURRENCE | | 1996 | 2008 | 1989 | 1977 | 1990 | 1992 | 2006 | 1993 | 2006 | 2002 | 1995 | 2008 | DEC 2008 |
| | MAXIMUM IN 24 HOURS (IN) | 60 | 15.3 | 9.2 | 13.0 | 10.2 | 4.3 | T | T | T | T | 1.6 | 10.1 | 14.4 | 15.3 |
| | YEAR OF OCCURRENCE' | | 1996 | 1959 | 1997 | 1977 | 1990 | 1992 | 2006 | 1993 | 1995 | 1989 | 1995 | 1990 | JAN 1996 |
| | MAXIMUM SNOW DEPTH (IN) | 59 | 25 | 24 | 19 | 11 | 2 | 0 | 0 | 0 | 0 | 1 | 11 | 19 | 25 |
| | YEAR OF OCCURRENCE | | 1979 | 1979 | 1962 | 1977 | 1990 | | | | | 1992 | 1977 | 2008 | JAN 1979 |
| | NORMAL NO. DAYS WITH: | | | | | | | | | | | | | | |
| | SNOWFALL >= 1.0 | 30 | 4.5 | 2.8 | 2.8 | 1.0 | 0.0 | 0.0 | 0.0 | 0.0 | 0.0 | 0.1 | 1.5 | 3.2 | 15.9 |

## PRECIPITATION (inches) 2009  GREEN BAY (KGRB)

| YEAR | JAN | FEB | MAR | APR | MAY | JUN | JUL | AUG | SEP | OCT | NOV | DEC | ANNUAL |
|------|-----|-----|-----|-----|-----|-----|-----|-----|-----|-----|-----|-----|--------|
| 1980 | 1.92 | 0.35 | 1.00 | 2.73 | 1.77 | 3.82 | 1.87 | 7.31 | 3.42 | 1.79 | 1.25 | 1.35 | 28.58 |
| 1981 | 0.12 | 2.76 | 0.42 | 4.22 | 0.56 | 2.63 | 0.83 | 3.37 | 3.25 | 3.44 | 1.08 | 1.10 | 23.78 |
| 1982 | 1.34 | 0.14 | 1.95 | 2.66 | 2.74 | 2.67 | 5.10 | 2.91 | 1.43 | 1.20 | 4.51 | 2.50 | 29.15 |
| 1983 | 0.72 | 1.46 | 1.52 | 1.39 | 4.80 | 1.82 | 3.76 | 5.27 | 3.59 | 2.24 | 2.63 | 1.18 | 30.38 |
| 1984 | 0.59 | 1.59 | 1.64 | 3.33 | 1.65 | 5.60 | 3.17 | 3.78 | 5.66 | 4.92 | 2.55 | 1.72 | 36.20 |
| 1985 | 0.86 | 2.55 | 2.70 | 2.24 | 2.58 | 2.21 | 4.03 | 8.03 | 3.65 | 2.72 | 4.96 | 1.83 | 38.36 |
| 1986 | 0.60 | 0.83 | 2.48 | 2.26 | 1.15 | 4.06 | 4.95 | 3.85 | 7.51 | 1.89 | 1.27 | 0.48 | 31.33 |
| 1987 | 0.47 | 0.39 | 1.53 | 2.33 | 2.58 | 1.83 | 2.18 | 3.41 | 1.57 | 1.76 | 3.07 | 2.04 | 23.16 |
| 1988 | 1.79 | 0.73 | 1.10 | 2.53 | 0.06 | 0.67 | 2.34 | 3.47 | 4.11 | 1.96 | 4.43 | 0.84 | 24.03 |
| 1989 | 0.41 | 0.38 | 2.88 | 0.49 | 4.22 | 1.56 | 2.27 | 1.05 | 0.58 | 4.76 | 1.25 | 0.55 | 20.40 |
| 1990 | 0.64 | 0.58 | 3.25 | 1.28 | 3.99 | 10.29 | 2.93 | 2.51 | 5.13 | 2.34 | 1.61 | 2.10 | 36.65 |
| 1991 | 0.57 | 0.37 | 2.87 | 2.77 | 2.42 | 1.08 | 4.16 | 2.11 | 2.55 | 3.50 | 2.72 | 1.42 | 26.54 |
| 1992 | 0.72 | 0.55 | 2.48 | 3.01 | 1.54 | 1.61 | 4.18 | 2.10 | 5.61 | 0.92 | 5.32 | 2.27 | 30.31 |
| 1993 | 1.42 | 0.34 | 0.76 | 3.99 | 4.28 | 6.82 | 6.83 | 2.30 | 2.78 | 2.29 | 1.56 | 0.44 | 33.81 |
| 1994 | 1.47 | 1.11 | 1.14 | 5.91 | 1.69 | 2.84 | 7.00 | 3.69 | 2.19 | 0.98 | 1.43 | 0.34 | 29.79 |
| 1995 | 0.65 | 0.39 | 1.92 | 2.22 | 2.88 | 1.80 | 1.15 | 7.31 | 2.76 | 4.80 | 3.32 | 1.25 | 30.45 |
| 1996 | 1.77 | 0.76 | 1.16 | 3.85 | 1.40 | 5.57 | 2.49 | 1.40 | 1.40 | 2.93 | .80 | 1.89 | 25.42 |
| 1997 | 1.81 | 1.40 | 1.92 | 1.67 | 2.60 | 5.51 | 2.11 | 5.73 | 2.76 | 0.93 | 0.30 | 0.61 | 27.35 |
| 1998 | 2.21 | 0.80 | 3.66 | 1.85 | 2.21 | 6.17 | 1.86 | 2.93 | 3.54 | 1.56 | 1.67 | 0.30 | 28.76 |
| 1999 | 2.37 | 1.10 | 0.15 | 2.11 | 3.77 | 3.98 | 5.67 | 1.32 | 1.24 | 0.67 | 1.57 | 0.83 | 24.78 |
| 2000 | 0.87 | 1.04 | 0.98 | 2.15 | 4.41 | 5.33 | 6.27 | 3.38 | 3.94 | 0.46 | 1.25 | 1.16 | 31.24 |
| 2001 | 1.19 | 1.26 | 0.42 | 3.66 | 4.74 | 5.17 | 0.85 | 3.42 | 2.35 | 1.71 | 1.70 | 1.23 | 27.70 |
| 2002 | 0.60 | 1.50 | 2.08 | 3.02 | 2.81 | 4.69 | 2.16 | 4.01 | 2.67 | 3.26 | 0.44 | 0.73 | 27.97 |
| 2003 | 0.58 | 0.56 | 2.32 | 2.36 | 3.17 | 3.71 | 4.26 | 4.15 | 3.32 | 1.05 | 3.83 | 1.68 | 30.99 |
| 2004 | 1.24 | 1.62 | 3.58 | 1.56 | 8.31 | 4.87 | 1.78 | 2.00 | 0.47 | 3.70 | 1.80 | 2.26 | 33.19 |
| 2005 | 1.60 | 1.33 | 1.33 | 1.53 | 2.52 | 3.44 | 1.46 | 4.23 | 3.08 | 1.59 | 3.07 | 1.04 | 26.22 |
| 2006 | 1.64 | 1.34 | 1.16 | 1.97 | 5.90 | 2.83 | 3.14 | 2.11 | 3.33 | 3.14 | 1.23 | 2.88 | 30.67 |
| 2007 | 0.63 | 1.39 | 2.74 | 1.72 | 2.39 | 3.71 | 2.41 | 2.72 | 3.16 | 3.62 | 0.11 | 2.54 | 27.14 |
| 2008 | 3.65 | 2.30 | 2.52 | 4.61 | 1.43 | 4.77 | 4.71 | 0.59 | 1.89 | 1.59 | 1.49 | 3.72 | 33.27 |
| 2009 | 0.66 | 1.55 | 2.59 | 2.62 | 3.01 | 2.53 | 1.33 | 3.33 | 1.22 | 5.16 | 1.38 | 2.28 | 27.66 |
| POR=<br>60 YRS | 1.19 | 1.08 | 1.91 | 2.66 | 3.02 | 3.47 | 3.32 | 3.21 | 3.04 | 2.23 | 1.93 | 1.46 | 28.52 |

WBAN : 14898

## AVERAGE TEMPERATURE (°F) 2009  GREEN BAY (KGRB)

| YEAR | JAN | FEB | MAR | APR | MAY | JUN | JUL | AUG | SEP | OCT | NOV | DEC | ANNUAL |
|------|-----|-----|-----|-----|-----|-----|-----|-----|-----|-----|-----|-----|--------|
| 1980 | 17.7 | 17.2 | 27.2 | 45.1 | 58.3 | 62.6 | 70.6 | 69.1 | 59.1 | 43.4 | 35.0 | 19.7 | 43.8 |
| 1981 | 15.3 | 22.9 | 34.7 | 45.7 | 53.6 | 65.6 | 69.3 | 68.2 | 56.8 | 45.0 | 36.9 | 22.9 | 44.7 |
| 1982 | 6.7 | 15.8 | 28.4 | 40.3 | 60.8 | 59.3 | 70.8 | 65.1 | 57.6 | 49.3 | 33.3 | 28.0 | 43.0 |
| 1983 | 21.4 | 26.3 | 31.1 | 40.5 | 48.8 | 64.5 | 72.9 | 70.8 | 59.8 | 48.5 | 36.5 | 10.6 | 44.3 |
| 1984 | 12.7 | 28.3 | 25.3 | 44.5 | 51.7 | 67.4 | 68.6 | 69.5 | 57.5 | 51.0 | 34.8 | 24.0 | 44.6 |
| 1985 | 12.2 | 17.6 | 34.2 | 47.6 | 58.5 | 62.3 | 69.1 | 66.4 | 60.8 | 48.3 | 30.9 | 9.4 | 43.1 |
| 1986 | 16.7 | 18.1 | 32.5 | 48.1 | 57.7 | 63.9 | 71.5 | 64.3 | 59.3 | 48.0 | 29.7 | 24.8 | 44.6 |
| 1987 | 21.6 | 27.9 | 35.1 | 49.1 | 58.8 | 69.0 | 73.0 | 67.6 | 61.1 | 43.2 | 37.6 | 27.1 | 47.6 |
| 1988 | 12.5 | 15.0 | 32.1 | 44.1 | 59.8 | 68.3 | 73.4 | 72.3 | 61.0 | 42.4 | 37.1 | 20.7 | 44.9 |
| 1989 | 25.5 | 13.0 | 25.3 | 41.9 | 54.9 | 63.4 | 70.9 | 68.7 | 59.0 | 49.6 | 31.4 | 11.2 | 42.9 |
| 1990 | 26.5 | 22.3 | 34.0 | 47.6 | 52.5 | 66.0 | 68.6 | 67.6 | 61.8 | 47.2 | 40.0 | 21.1 | 46.3 |
| 1991 | 13.8 | 23.5 | 33.6 | 47.9 | 61.4 | 69.2 | 69.8 | 69.9 | 58.5 | 48.2 | 30.2 | 24.5 | 45.9 |
| 1992 | 22.9 | 27.1 | 30.4 | 42.0 | 56.6 | 62.3 | 64.9 | 64.3 | 58.5 | 46.8 | 32.8 | 23.7 | 44.4 |
| 1993 | 19.6 | 19.0 | 30.2 | 40.4 | 56.2 | 62.8 | 70.3 | 70.3 | 55.4 | 45.9 | 33.7 | 25.3 | 44.1 |
| 1994 | 6.2 | 14.5 | 33.3 | 44.5 | 57.1 | 67.9 | 69.8 | 65.9 | 63.7 | 51.2 | 39.4 | 29.7 | 45.3 |
| 1995 | 21.4 | 20.5 | 34.6 | 40.5 | 56.0 | 70.8 | 73.1 | 73.9 | 58.3 | 49.2 | 27.8 | 18.9 | 45.4 |
| 1996 | 13.7 | 18.1 | 25.8 | 40.7 | 52.9 | 66.2 | 65.8 | 68.0 | 59.5 | 46.7 | 28.5 | 21.7 | 42.3 |
| 1997 | 15.1 | 21.1 | 28.9 | 42.3 | 49.1 | 64.3 | 66.4 | 63.7 | 60.0 | 47.7 | 31.9 | 27.9 | 43.2 |
| 1998 | 22.2 | 31.3 | 33.1 | 46.9 | 60.5 | 64.9 | 69.4 | 69.5 | 63.5 | 50.9 | 39.4 | 27.4 | 48.3 |
| 1999 | 14.9 | 28.3 | 34.0 | 45.9 | 58.1 | 65.4 | 72.2 | 65.4 | 58.6 | 46.1 | 40.0 | 24.1 | 46.1 |
| 2000 | 15.6 | 24.5 | 39.4 | 42.3 | 56.8 | 63.8 | 67.0 | 67.1 | 58.0 | 50.7 | 33.9 | 11.4 | 44.2 |
| 2001 | 20.3 | 17.4 | 29.5 | 47.8 | 57.3 | 65.6 | 70.6 | 70.7 | 58.3 | 47.3 | 43.4 | 30.1 | 46.5 |
| 2002 | 25.9 | 27.0 | 26.7 | 44.8 | 50.5 | 66.7 | 73.2 | 67.7 | 62.6 | 43.7 | 33.9 | 26.3 | 45.8 |
| 2003 | 16.4 | 14.0 | 30.0 | 41.5 | 53.1 | 63.2 | 67.9 | 69.1 | 60.4 | 48.0 | 36.1 | 28.0 | 44.0 |
| 2004 | 12.2 | 21.6 | 35.2 | 45.4 | 53.6 | 63.3 | 67.6 | 64.3 | 64.6 | 50.0 | 39.1 | 23.2 | 45.0 |
| 2005 | 17.0 | 26.1 | 27.2 | 47.4 | 54.1 | 71.0 | 70.5 | 69.0 | 64.4 | 51.6 | 36.9 | 19.8 | 46.3 |
| 2006 | 30.6 | 19.1 | 33.0 | 49.3 | 57.8 | 65.6 | 73.3 | 67.7 | 58.0 | 45.1 | 39.4 | 30.5 | 47.5 |
| 2007 | 23.0 | 15.3 | 35.8 | 44.6 | 59.7 | 67.5 | 69.6 | 69.3 | 62.6 | 54.8 | 34.8 | 21.5 | 46.5 |
| 2008 | 17.0 | 15.9 | 27.6 | 46.2 | 53.9 | 67.2 | 70.8 | 68.8 | 61.2 | 48.3 | 34.7 | 15.0 | 43.9 |
| 2009 | 7.6 | 21.4 | 31.0 | 43.6 | 55.2 | 64.5 | 65.4 | 65.9 | 61.3 | 44.3 | 41.8 | 20.4 | 43.5 |
| POR=<br>60 YRS | 15.8 | 19.4 | 29.9 | 44.0 | 55.4 | 65.1 | 69.6 | 67.6 | 59.4 | 48.2 | 34.6 | 21.5 | 44.2 |

## HEATING DEGREE DAYS (base 65°F) 2009  GREEN BAY (KGRB)

| YEAR | JUL | AUG | SEP | OCT | NOV | DEC | JAN | FEB | MAR | APR | MAY | JUN | TOTAL |
|------|-----|-----|-----|-----|-----|-----|-----|-----|-----|-----|-----|-----|-------|
| 1980-81 | 11 | 11 | 189 | 661 | 893 | 1398 | 1538 | 1175 | 932 | 573 | 350 | 45 | 7776 |
| 1981-82 | 26 | 21 | 248 | 614 | 839 | 1300 | 1805 | 1373 | 1127 | 733 | 152 | 178 | 8416 |
| 1982-83 | 3 | 75 | 250 | 483 | 946 | 1140 | 1344 | 1077 | 1046 | 727 | 495 | 100 | 7686 |
| 1983-84 | 17 | 2 | 210 | 507 | 847 | 1682 | 1617 | 1055 | 1223 | 611 | 406 | 13 | 8190 |
| 1984-85 | 18 | 20 | 237 | 430 | 899 | 1262 | 1632 | 1324 | 949 | 533 | 204 | 114 | 7622 |
| 1985-86 | 9 | 34 | 196 | 508 | 1016 | 1719 | 1491 | 1307 | 1002 | 508 | 244 | 90 | 8124 |
| 1986-87 | 12 | 65 | 191 | 519 | 1052 | 1240 | 1341 | 1033 | 918 | 478 | 240 | 36 | 7125 |
| 1987-88 | 18 | 44 | 132 | 673 | 815 | 1167 | 1623 | 1447 | 1012 | 624 | 201 | 74 | 7830 |
| 1988-89 | 4 | 23 | 146 | 694 | 830 | 1365 | 1216 | 1451 | 1224 | 687 | 315 | 98 | 8053 |
| 1989-90 | 7 | 19 | 200 | 475 | 1000 | 1666 | 1189 | 1191 | 952 | 547 | 380 | 55 | 7681 |
| 1990-91 | 24 | 28 | 157 | 547 | 744 | 1357 | 1579 | 1154 | 967 | 516 | 220 | 23 | 7316 |
| 1991-92 | 17 | 17 | 250 | 515 | 1038 | 1248 | 1296 | 1093 | 1067 | 683 | 275 | 122 | 7621 |
| 1992-93 | 44 | 87 | 209 | 560 | 960 | 1274 | 1398 | 1281 | 1071 | 731 | 277 | 108 | 8000 |
| 1993-94 | 1 | 16 | 292 | 585 | 930 | 1226 | 1818 | 1409 | 974 | 609 | 267 | 59 | 8186 |
| 1994-95 | 15 | 55 | 102 | 423 | 760 | 1087 | 1342 | 1237 | 937 | 730 | 275 | 43 | 7006 |
| 1995-96 | 8 | 0 | 229 | 483 | 1110 | 1419 | 1583 | 1356 | 1208 | 721 | 385 | 73 | 8575 |
| 1996-97 | 32 | 20 | 199 | 561 | 1089 | 1335 | 1541 | 1224 | 1114 | 676 | 486 | 87 | 8364 |
| 1997-98 | 52 | 88 | 163 | 533 | 988 | 1145 | 1319 | 937 | 985 | 535 | 175 | 118 | 7038 |
| 1998-99 | 5 | 5 | 98 | 431 | 760 | 1160 | 1546 | 1022 | 954 | 569 | 223 | 101 | 6874 |
| 1999-00 | 5 | 46 | 226 | 576 | 744 | 1259 | 1526 | 1170 | 785 | 672 | 275 | 94 | 7378 |
| 2000-01 | 39 | 30 | 229 | 435 | 929 | 1657 | 1377 | 1324 | 1094 | 513 | 247 | 87 | 7961 |
| 2001-02 | 19 | 8 | 217 | 540 | 639 | 1074 | 1205 | 1061 | 1180 | 611 | 453 | 77 | 7084 |
| 2002-03 | 2 | 13 | 124 | 661 | 928 | 1191 | 1501 | 1420 | 1078 | 698 | 364 | 93 | 8073 |
| 2003-04 | 16 | 7 | 182 | 522 | 862 | 1140 | 1629 | 1254 | 915 | 579 | 355 | 97 | 7558 |
| 2004-05 | 30 | 89 | 82 | 458 | 770 | 1288 | 1480 | 1083 | 1168 | 521 | 337 | 17 | 7323 |
| 2005-06 | 9 | 15 | 90 | 440 | 833 | 1397 | 1060 | 1279 | 983 | 465 | 263 | 52 | 6886 |
| 2006-07 | 6 | 20 | 218 | 613 | 761 | 1066 | 1295 | 1386 | 897 | 607 | 200 | 42 | 7111 |
| 2007-08 | 19 | 20 | 145 | 341 | 899 | 1341 | 1481 | 1415 | 1155 | 559 | 337 | 31 | 7743 |
| 2008-09 | 3 | 10 | 140 | 513 | 904 | 1542 | 1773 | 1215 | 1049 | 636 | 302 | 116 | 8203 |
| 2009- | 33 | 59 | 126 | 634 | 690 | 1375 | | | | | | | |

WBAN : 14898

## COOLING DEGREE DAYS (base 65°F) 2009  GREEN BAY (KGRB)

| YEAR | JAN | FEB | MAR | APR | MAY | JUN | JUL | AUG | SEP | OCT | NOV | DEC | TOTAL |
|------|-----|-----|-----|-----|-----|-----|-----|-----|-----|-----|-----|-----|-------|
| 1980 | 0 | 0 | 0 | 5 | 27 | 64 | 192 | 146 | 18 | 0 | 0 | 0 | 452 |
| 1981 | 0 | 0 | 0 | 0 | 5 | 71 | 168 | 127 | 9 | 0 | 0 | 0 | 380 |
| 1982 | 0 | 0 | 0 | 0 | 30 | 16 | 187 | 85 | 35 | 4 | 0 | 0 | 357 |
| 1983 | 0 | 0 | 0 | 0 | 0 | 95 | 270 | 188 | 60 | 4 | 0 | 0 | 617 |
| 1984 | 0 | 0 | 0 | 0 | 3 | 94 | 136 | 165 | 17 | 0 | 0 | 0 | 415 |
| 1985 | 0 | 0 | 0 | 16 | 11 | 41 | 141 | 85 | 79 | 0 | 0 | 0 | 373 |
| 1986 | 0 | 0 | 0 | 8 | 25 | 65 | 220 | 48 | 27 | 0 | 0 | 0 | 393 |
| 1987 | 0 | 0 | 0 | 7 | 56 | 161 | 274 | 133 | 23 | 0 | 0 | 0 | 654 |
| 1988 | 0 | 0 | 0 | 0 | 46 | 182 | 270 | 255 | 33 | 0 | 0 | 0 | 786 |
| 1989 | 0 | 0 | 0 | 0 | 7 | 55 | 199 | 141 | 27 | 0 | 0 | 0 | 429 |
| 1990 | 0 | 0 | 0 | 34 | 0 | 92 | 140 | 116 | 70 | 0 | 0 | 0 | 452 |
| 1991 | 0 | 0 | 0 | 8 | 115 | 155 | 171 | 177 | 60 | 0 | 0 | 0 | 686 |
| 1992 | 0 | 0 | 0 | 0 | 22 | 49 | 49 | 69 | 23 | 1 | 0 | 0 | 213 |
| 1993 | 0 | 0 | 0 | 0 | 13 | 48 | 173 | 187 | 11 | 0 | 0 | 0 | 432 |
| 1994 | 0 | 0 | 0 | 0 | 30 | 154 | 172 | 89 | 71 | 3 | 0 | 0 | 519 |
| 1995 | 0 | 0 | 0 | 0 | 3 | 224 | 265 | 282 | 33 | 1 | 0 | 0 | 808 |
| 1996 | 0 | 0 | 0 | 0 | 15 | 115 | 64 | 120 | 38 | 0 | 0 | 0 | 352 |
| 1997 | 0 | 0 | 0 | 0 | 0 | 70 | 103 | 55 | 20 | 7 | 0 | 0 | 255 |
| 1998 | 0 | 0 | 0 | 0 | 44 | 121 | 145 | 150 | 60 | 2 | 0 | 0 | 522 |
| 1999 | 0 | 0 | 0 | 0 | 15 | 122 | 237 | 66 | 40 | 0 | 0 | 0 | 480 |
| 2000 | 0 | 0 | 0 | 0 | 29 | 65 | 107 | 104 | 25 | 0 | 0 | 0 | 330 |
| 2001 | 0 | 0 | 0 | 2 | 16 | 113 | 202 | 190 | 23 | 0 | 0 | 0 | 546 |
| 2002 | 0 | 0 | 0 | 9 | 11 | 135 | 262 | 104 | 58 | 6 | 0 | 0 | 585 |
| 2003 | 0 | 0 | 0 | 0 | 0 | 46 | 113 | 142 | 49 | 1 | 0 | 0 | 351 |
| 2004 | 0 | 0 | 0 | 0 | 9 | 51 | 116 | 73 | 78 | 0 | 0 | 0 | 327 |
| 2005 | 0 | 0 | 0 | 1 | 3 | 203 | 186 | 147 | 77 | 31 | 0 | 0 | 648 |
| 2006 | 0 | 0 | 0 | 0 | 46 | 76 | 271 | 110 | 14 | 3 | 0 | 0 | 520 |
| 2007 | 0 | 0 | 2 | 3 | 46 | 124 | 168 | 160 | 80 | 33 | 0 | 0 | 616 |
| 2008 | 0 | 0 | 0 | 0 | 0 | 104 | 188 | 138 | 33 | 3 | 0 | 0 | 466 |
| 2009 | 0 | 0 | 0 | 0 | 6 | 106 | 49 | 92 | 22 | 0 | 0 | 0 | 275 |

**SNOWFALL (inches) 2009 GREEN BAY (KGRB)**

| YEAR | JUL | AUG | SEP | OCT | NOV | DEC | JAN | FEB | MAR | APR | MAY | JUN | TOTAL |
|------|-----|-----|-----|-----|-----|-----|-----|-----|-----|-----|-----|-----|-------|
| 1980-81 | 0.0 | 0.0 | 0.0 | T | 3.4 | 14.5 | 2.3 | 7.5 | 2.5 | T | 0.0 | 0.0 | 30.2 |
| 1981-82 | 0.0 | 0.0 | 0.0 | 1.0 | 2.7 | 11.4 | 28.0 | 2.0 | 6.0 | 2.9 | 0.0 | 0.0 | 54.0 |
| 1982-83 | 0.0 | 0.0 | 0.0 | T | 2.6 | 1.3 | 8.5 | 15.3 | 11.4 | 0.6 | 0.0 | 0.0 | 39.7 |
| 1983-84 | 0.0 | 0.0 | 0.0 | 0.0 | 7.2 | 12.7 | 10.5 | 1.8 | 6.5 | T | 0.0 | 0.0 | 38.7 |
| 1984-85 | 0.0 | 0.0 | 0.0 | 0.0 | 2.0 | 15.9 | 13.8 | 15.1 | 17.7 | 6.2 | 0.0 | 0.0 | 70.7 |
| 1985-86 | 0.0 | 0.0 | 0.0 | 0.0 | 16.5 | 22.7 | 6.8 | 9.9 | 6.8 | 0.5 | 0.0 | 0.0 | 63.2 |
| 1986-87 | 0.0 | 0.0 | 0.0 | T | 9.1 | 5.2 | 7.8 | 1.6 | 10.8 | 1.6 | 0.0 | 0.0 | 36.1 |
| 1987-88 | 0.0 | 0.0 | 0.0 | T | 1.8 | 15.8 | 20.6 | 8.9 | 1.8 | 0.4 | 0.0 | 0.0 | 49.3 |
| 1988-89 | 0.0 | 0.0 | 0.0 | T | 11.5 | 5.8 | 1.5 | 8.6 | 24.2 | 0.3 | T | 0.0 | 51.9 |
| 1989-90 | 0.0 | 0.0 | 0.0 | 1.6 | 1.5 | 15.3 | 10.7 | 6.9 | 2.7 | 2.7 | 4.3 | 0.0 | 45.7 |
| 1990-91 | 0.0 | 0.0 | 0.0 | T | 1.2 | 26.9 | 8.8 | 6.9 | 5.3 | 7.5 | 0.0 | 0.0 | 56.6 |
| 1991-92 | 0.0 | 0.0 | 0.0 | T | 8.7 | 10.7 | 4.3 | 7.4 | 9.0 | 2.8 | 0.0 | T | 42.9 |
| 1992-93 | 0.0 | T | 0.0 | 1.2 | 8.5 | 15.5 | 11.8 | 5.6 | 7.2 | 7.2 | T | 0.0 | 57.0 |
| 1993-94 | 0.0 | T | 0.0 | T | 1.6 | 1.9 | 30.0 | 16.4 | 8.1 | 3.6 | 0.0 | 0.0 | 61.6 |
| 1994-95 | 0.0 | 0.0 | 0.0 | T | 1.3 | 4.3 | 9.8 | 6.8 | 7.1 | 1.9 | T | 0.0 | 31.2 |
| 1995-96 | 0.0 | 0.0 | T | T | 17.1 | 11.2 | 31.5 | 2.9 | 4.8 | 10.0 | T | 0.0 | 77.5 |
| 1996-97 | 0.0 | 0.0 | 0.0 | T | 3.7 | 19.4 | 17.2 | 15.5 | 20.7 | 0.3 | 0.2 | 0.0 | 77.0 |
| 1997-98 | 0.0 | 0.0 | 0.0 | 0.2 | 0.7 | 6.2 | 24.2 | 1.2 | 11.5 | 2.1 | 0.0 | 0.0 | 46.1 |
| 1998-99 | 0.0 | 0.0 | T | 0.0 | 0.3 | 5.4 | 21.2 | 2.2 | 4.4 | T | 0.0 | 0.0 | 33.5 |
| 1999-00 | 0.0 | 0.0 | T | 0.0 | 0.0 | 5.4 | 5.4 | 15.4 | 11.3 | 2.1 | 2.5 | 0.0 | 36.7 |
| 2000-01 | 0.0 | T | 0.0 | T | 7.8 | 28.9 | 5.6 | 7.8 | 4.7 | 1.9 | 0.0 | 0.0 | 56.7 |
| 2001-02 | 0.0 | 0.0 | 0.0 | T | T | 1.8 | 12.0 | 8.5 | 17.1 | 7.5 | T | T | 46.9 |
| 2002-03 | 0.0 | 0.0 | 0.0 | 1.8 | 0.9 | 3.3 | 9.6 | 9.2 | 9.1 | 5.0 | 0.0 | 0.0 | 38.9 |
| 2003-04 | 0.0 | 0.0 | 0.0 | T | 0.3 | 2.5 | 16.3 | 16.6 | 8.4 | 0.7 | T | 0.0 | 44.8 |
| 2004-05 | 0.0 | 0.0 | 0.0 | T | T | 14.7 | 17.8 | 11.1 | 12.0 | T | T | 0.0 | 55.6 |
| 2005-06 | 0.0 | 0.0 | 0.0 | T | 4.1 | 12.7 | 2.0 | 15.7 | 3.5 | 0.0 | T | 0.0 | 38.0 |
| 2006-07 | T | 0.0 | T | T | 1.5 | 4.3 | 9.8 | 18.0 | 8.9 | 6.8 | T | 0.0 | 49.3 |
| 2007-08 | 0.0 | 0.0 | 0.0 | T | 0.7 | 24.1 | 28.5 | 24.5 | 4.3 | 5.3 | 0.0 | 0.0 | 87.4 |
| 2008-09 | 0.0 | 0.0 | 0.0 | T | 6.7 | 45.6 | 10.2 | 15.7 | 9.1 | 0.4 | 0.0 | 0.0 | 87.7 |
| 2009- | 0.0 | 0.0 | 0.0 | T | T | 20.7 | | | | | | | |
| POR=<br>60 YRS | T | T | T | 0.2 | 4.2 | 11.5 | 12.1 | 9.1 | 8.8 | 2.5 | 0.1 | T | 48.5 |

# Green Bay, WI 2009
## Temperature and Precipitation

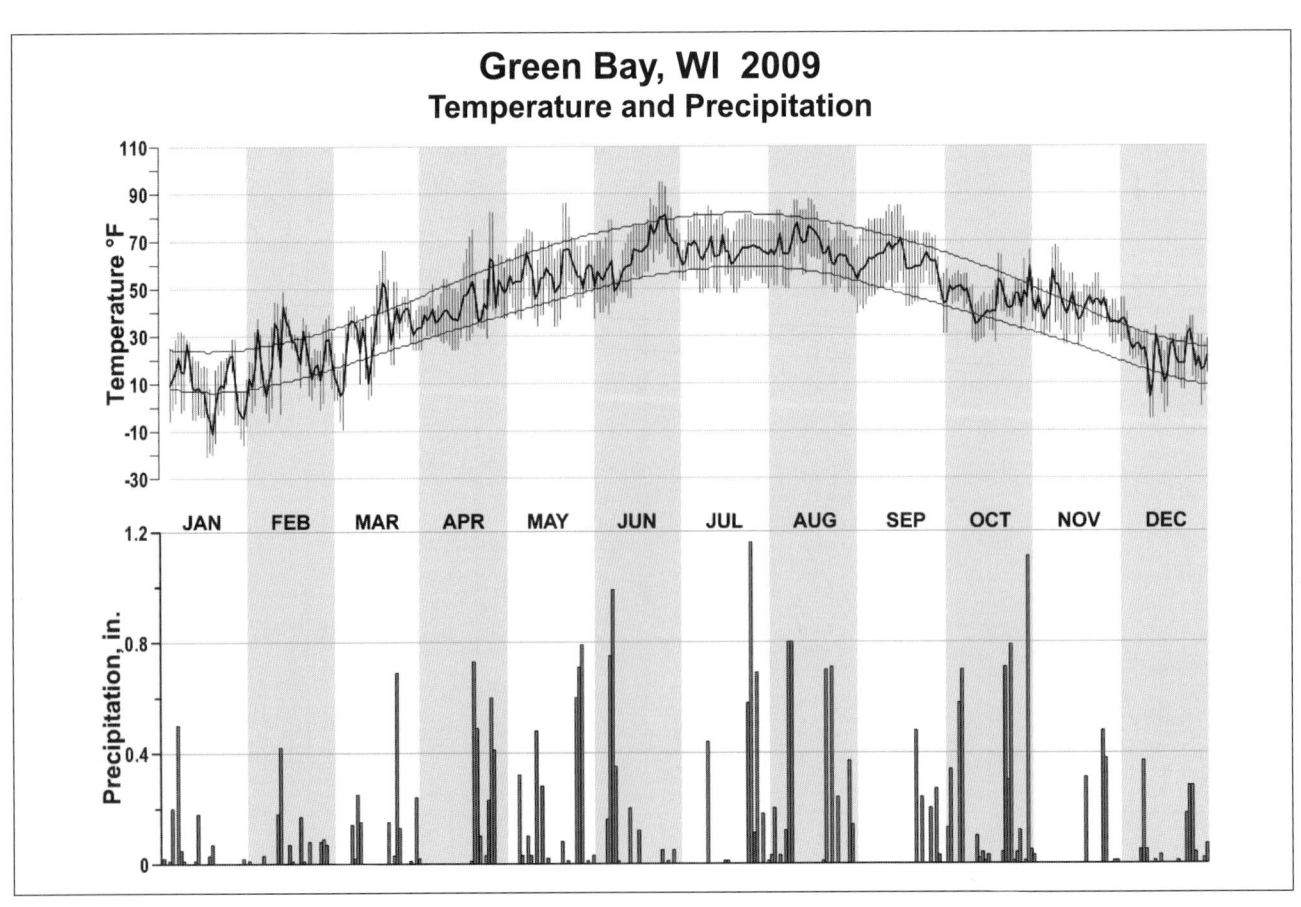

# 2009
# LA CROSSE
# WISCONSIN (KLSE)

The city of La Crosse is situated on the east bank of the Mississippi River at the confluence of the Mississippi, Black, and La Crosse Rivers. The official records are taken at the La Crosse Municipal Airport which is 6 1/2 miles north of the main Post Office, on the north end of French Island. This island is about 6 miles long from north to south and 2 to 4 miles wide with the Mississippi River to the west and the old channel of the Black River to the east. A rather level sandy plain exists on each side of the river extending between the Wisconsin and Minnesota bluffs which rise 450 to 500 feet above the valley floor. The distance from bluff to bluff averages about 5 miles. The Mississippi River bends to the northwest and continues directly southward from the city.

The prevailing winds in the area are from the northwest from January through April and southerly during the remainder of the year. The situation of the city and airport in a natural bowl between the hills results in somewhat colder temperatures at night due to the settling of cooler air. Valley fogs often persist to mid-forenoon. Steepsided hills with narrow valleys are characteristic of most of the surrounding area.

The flow of the Mississippi River is regulated by dams built for the purpose of navigation, but the reservoirs have limited storage capacity. La Crosse is in the area of Pool No. 8 with a mean sea level elevation of 631 feet. When the river reaches an elevation of 639 feet, with open gate operation, there is considerable flooding of land near the river and some industrial sections of the city.

The invigorating continental-type climate results in wide and frequent variations in temperature. General storms moving eastward or northeastward into our area bring warmer weather and supply most of our moisture. These are usually followed by cooler air from Canada. The winters are cold and humid. The summers are warm with moderate humidities, while periods of hot and humid weather occur occasionally, usually lasting from a few days to a week at a time.

Sixty percent of the precipitation falls during the main growing season, extending from May through September. Most of the summer rainfall occurs during scattered thunderstorms. Some damage from heavy rains, high winds, and hail occurs each year, but tornadoes are infrequent and cover very small areas. Snow is frequent and is the predominant form of precipitation in winter. Heavy snow sometimes falls with larger amounts over the ridges. Glaze storms are not numerous since La Crosse is north of the main path of freezing rain.

Farming is diversified with dairying the leading activity. The more important field crops are corn, oats, and hay. Some of the more specialized crops are soybeans, tobacco, small fruits, and cranberries. Commercial apple orchards are numerous across the Mississippi River in Minnesota.

Based on the 1951-1980 period, the average first occurrence of 32 degrees Fahrenheit in the fall is October 13 and the average last occurrence in the spring is April 29.

# NORMALS, MEANS, AND EXTREMES
## LA CROSSE (KLSE)

LATITUDE: 43 ° 45'N  LONGITUDE: -91 ° 15'W  ELEVATION (FT): GRND: 652  BARO: 658  TIME ZONE: CENTRAL  (UTC -6)  WBAN: 14920

| ELEMENT | POR | JAN | FEB | MAR | APR | MAY | JUN | JUL | AUG | SEP | OCT | NOV | DEC | YEAR |
|---|---|---|---|---|---|---|---|---|---|---|---|---|---|---|
| **TEMPERATURE °F** | | | | | | | | | | | | | | |
| NORMAL DAILY MAXIMUM | 30 | 25.5 | 32.4 | 44.6 | 59.7 | 72.5 | 81.3 | 85.2 | 82.5 | 73.7 | 61.1 | 43.6 | 29.9 | 57.7 |
| MEAN DAILY MAXIMUM | 66 | 24.8 | 30.1 | 41.7 | 57.9 | 70.2 | 79.1 | 83.8 | 81.7 | 72.5 | 60.8 | 43.3 | 29.4 | 56.3 |
| HIGHEST DAILY MAXIMUM | 57 | 57 | 64 | 84 | 93 | 95 | 102 | 108 | 105 | 100 | 93 | 75 | 67 | 108 |
| YEAR OF OCCURRENCE | | 1981 | 2000 | 1986 | 1980 | 2006 | 1988 | 1995 | 1988 | 1978 | 1963 | 2008 | 1998 | JUL 1995 |
| MEAN OF EXTREME MAXS. | 68 | 43.9 | 48.7 | 65.5 | 80.8 | 87.8 | 93.1 | 95.1 | 93.8 | 89.1 | 80.2 | 64.2 | 48.9 | 74.3 |
| NORMAL DAILY MINIMUM | 30 | 6.3 | 12.8 | 24.5 | 37.1 | 48.7 | 57.9 | 62.8 | 60.7 | 51.7 | 40.1 | 27.4 | 13.6 | 37.0 |
| MEAN DAILY MINIMUM | 66 | 6.8 | 11.3 | 23.2 | 36.9 | 48.3 | 57.7 | 62.5 | 60.5 | 51.4 | 40.7 | 27.6 | 13.9 | 36.7 |
| LOWEST DAILY MINIMUM | 58 | -37 | -36 | -28 | 7 | 26 | 37 | 33 | 40 | 28 | 14 | -9 | -30 | -37 |
| YEAR OF OCCURRENCE | | 1951 | 1971 | 1962 | 1982 | 1989 | 1978 | 1982 | 2004 | 1967 | 1988 | 1977 | 1983 | JAN 1951 |
| MEAN OF EXTREME MINS. | 68 | -17.1 | -12.9 | 1.6 | 21.6 | 33.6 | 44.4 | 50.9 | 48.1 | 35.9 | 25.0 | 10.1 | -9.1 | 19.3 |
| NORMAL DRY BULB | 30 | 15.9 | 22.6 | 34.6 | 48.4 | 60.6 | 69.6 | 74.0 | 71.6 | 62.7 | 50.6 | 35.5 | 21.8 | 47.3 |
| MEAN DRY BULB | 66 | 15.8 | 20.7 | 32.5 | 47.4 | 59.2 | 68.5 | 73.1 | 71.1 | 61.9 | 50.7 | 35.5 | 21.7 | 46.5 |
| MEAN WET BULB | 25 | 23.4 | 25.2 | 34.1 | 44.2 | 53.0 | 61.4 | 64.9 | 63.9 | 58.3 | 46.9 | 36.8 | 26.9 | 44.9 |
| MEAN DEW POINT | 25 | 19.9 | 21.5 | 29.7 | 38.5 | 48.6 | 58.1 | 62.2 | 61.4 | 55.5 | 43.2 | 33.3 | 23.7 | 41.3 |
| NORMAL NO. DAYS WITH: | | | | | | | | | | | | | | |
| MAXIMUM >= 90 | 30 | 0.0 | 0.0 | 0.0 | 0.1 | 1.0 | 4.0 | 7.3 | 4.1 | 1.3 | 0.1 | 0.0 | 0.0 | 17.9 |
| MAXIMUM <= 32 | 30 | 22.0 | 14.2 | 5.5 | 0.3 | 0.0 | 0.0 | 0.0 | 0.0 | 0.0 | 0.0 | 5.5 | 17.9 | 65.4 |
| MINIMUM <= 32 | 30 | 30.5 | 26.5 | 23.9 | 9.4 | 0.6 | 0.0 | 0.0 | 0.0 | 0.4 | 6.7 | 21.3 | 29.4 | 148.7 |
| MINIMUM <= 0 | 30 | 11.5 | 6.7 | 1.3 | 0.0 | 0.0 | 0.0 | 0.0 | 0.0 | 0.0 | 0.0 | 0.4 | 6.1 | 26.0 |
| **H/C** | | | | | | | | | | | | | | |
| NORMAL HEATING DEG. DAYS | 30 | 1536 | 1202 | 959 | 521 | 202 | 38 | 6 | 17 | 152 | 467 | 893 | 1347 | 7340 |
| NORMAL COOLING DEG. DAYS | 30 | 0 | 0 | 0 | 8 | 49 | 162 | 272 | 208 | 70 | 6 | 0 | 0 | 775 |
| **RH** | | | | | | | | | | | | | | |
| NORMAL (PERCENT) | 30 | 74 | 73 | 70 | 63 | 65 | 69 | 72 | 76 | 76 | 70 | 74 | 77 | 72 |
| HOUR 00 LST | 30 | 77 | 78 | 77 | 72 | 75 | 80 | 85 | 88 | 86 | 78 | 79 | 80 | 80 |
| HOUR 06 LST | 30 | 79 | 80 | 81 | 79 | 81 | 84 | 88 | 91 | 91 | 83 | 82 | 82 | 83 |
| HOUR 12 LST | 30 | 69 | 66 | 61 | 52 | 52 | 55 | 58 | 61 | 62 | 58 | 67 | 72 | 61 |
| HOUR 18 LST | 30 | 72 | 68 | 62 | 52 | 52 | 55 | 58 | 62 | 66 | 63 | 71 | 75 | 63 |
| **S** PERCENT POSSIBLE SUNSHINE | | | | | | | | | | | | | | |
| **W/O** MEAN NO. DAYS WITH: | | | | | | | | | | | | | | |
| HEAVY FOG(VISBY <= 1/4 MI) | 45 | 1.3 | 0.9 | 1.9 | 0.6 | 0.9 | 1.2 | 1.5 | 3.5 | 3.6 | 1.7 | 1.2 | 1.3 | 19.6 |
| THUNDERSTORMS | 61 | 0.0 | 0.2 | 1.2 | 2.9 | 5.1 | 7.4 | 7.3 | 6.7 | 4.4 | 2.1 | 0.5 | 0.2 | 38.0 |
| **CLOUDNESS** MEAN: | | | | | | | | | | | | | | |
| SUNRISE-SUNSET (OKTAS) | 17 | 5.4 | 5.0 | 5.4 | 5.4 | 5.2 | 5.0 | 4.2 | 4.3 | 4.4 | 4.3 | 5.3 | 5.3 | 4.9 |
| MIDNIGHT-MIDNIGHT (OKTAS) | 5 | 3.8 | 3.3 | 4.0 | 4.4 | 3.8 | 3.8 | 2.9 | 3.1 | 3.5 | 3.0 | 4.1 | 3.8 | 3.6 |
| MEAN NO. DAYS WITH: | | | | | | | | | | | | | | |
| CLEAR | 17 | 7.4 | 7.9 | 7.4 | 6.8 | 6.9 | 7.3 | 9.8 | 9.5 | 9.1 | 10.8 | 5.8 | 6.8 | 95.5 |
| PARTLY CLOUDY | 17 | 6.9 | 7.1 | 7.1 | 7.1 | 9.1 | 10.4 | 10.9 | 10.6 | 8.9 | 6.9 | 6.1 | 5.8 | 96.9 |
| CLOUDY | 17 | 16.7 | 13.2 | 16.5 | 16.1 | 15.0 | 12.3 | 10.4 | 10.9 | 11.9 | 13.4 | 18.1 | 18.5 | 173.0 |
| **PR** MEAN STATION PRESSURE(IN) | 25 | 29.28 | 29.30 | 29.27 | 29.22 | 29.21 | 29.21 | 29.23 | 29.25 | 29.26 | 29.27 | 29.26 | 29.28 | 29.25 |
| MEAN SEA-LEVEL PRES. (IN) | 25 | 30.03 | 30.05 | 30.01 | 29.95 | 29.93 | 29.93 | 29.95 | 29.96 | 29.98 | 30.00 | 29.99 | 30.03 | 29.98 |
| **WINDS** MEAN SPEED (MPH) | 25 | 8.7 | 8.9 | 9.3 | 9.9 | 9.1 | 8.3 | 7.6 | 7.3 | 8.1 | 9.1 | 9.3 | 8.8 | 8.7 |
| PREVAIL.DIR(TENS OF DEGS) | 36 | 33 | 33 | 34 | 19 | 19 | 19 | 19 | 19 | 19 | 19 | 19 | 19 | 19 |
| MAXIMUM 2-MINUTE: | | | | | | | | | | | | | | |
| SPEED (MPH) | 9 | 36 | 35 | 39 | 49 | 45 | 44 | 53 | 44 | 46 | 47 | 37 | 38 | 53 |
| DIR. (TENS OF DEGS) | | 18 | 29 | 20 | 22 | 24 | 24 | 28 | 33 | 25 | 25 | 21 | 32 | 28 |
| YEAR OF OCCURRENCE | | 2005 | 2001 | 2005 | 2001 | 2007 | 2003 | 2003 | 2007 | 2007 | 2004 | 2008 | 2004 | JUL 2003 |
| MAXIMUM 3-SECOND | | | | | | | | | | | | | | |
| SPEED (MPH) | 9 | 45 | 46 | 51 | 61 | 59 | 62 | 66 | 64 | 56 | 60 | 47 | 47 | 66 |
| DIR. (TENS OF DEGS) | | 31 | 22 | 20 | 24 | 27 | 23 | 29 | 01 | 25 | 26 | 21 | 32 | 29 |
| YEAR OF OCCURRENCE | | 2006 | 2009 | 2005 | 2001 | 2005 | 2003 | 2003 | 2007 | 2007 | 2004 | 2008 | 2004 | JUL 2003 |
| **PRECIPITATION** NORMAL (IN) | 30 | 1.19 | 0.99 | 2.00 | 3.38 | 3.38 | 4.00 | 4.25 | 4.28 | 3.40 | 2.16 | 2.10 | 1.23 | 32.36 |
| MAXIMUM MONTHLY (IN) | 57 | 3.03 | 2.71 | 3.82 | 7.31 | 9.73 | 10.79 | 9.35 | 13.75 | 10.52 | 5.67 | 6.23 | 3.36 | 13.75 |
| YEAR OF OCCURRENCE | | 1996 | 1998 | 1951 | 1973 | 2004 | 1993 | 1987 | 2007 | 1965 | 2009 | 1991 | 2009 | AUG 2007 |
| MINIMUM MONTHLY (IN) | 57 | 0.14 | 0.05 | 0.30 | 0.60 | 0.94 | 1.33 | 0.16 | 0.54 | 0.42 | 0.02 | T | 0.30 | T |
| YEAR OF OCCURRENCE | | 1981 | 1969 | 1978 | 1966 | 1988 | 1989 | 1967 | 1976 | 1952 | 1952 | 1976 | 1962 | NOV 1976 |
| MAXIMUM IN 24 HOURS (IN) | 57 | 1.31 | 1.11 | 1.64 | 3.84 | 2.87 | 3.94 | 5.24 | 5.98 | 3.15 | 2.28 | 2.80 | 1.42 | 5.98 |
| YEAR OF OCCURRENCE | | 1967 | 2002 | 1966 | 1954 | 2000 | 1967 | 1987 | 2007 | 2004 | 1998 | 1991 | 1990 | AUG 2007 |
| NORMAL NO. DAYS WITH: | | | | | | | | | | | | | | |
| PRECIPITATION >= 0.01 | 30 | 9.5 | 7.6 | 9.6 | 10.8 | 11.0 | 11.0 | 10.8 | 10.4 | 9.6 | 8.3 | 8.9 | 9.3 | 116.8 |
| PRECIPITATION >= 1.00 | 30 | 0.0 | 0.1 | 0.3 | 0.7 | 0.7 | 1.0 | 1.0 | 1.1 | 0.8 | 0.3 | 0.4 | 0.1 | 6.5 |
| **SNOWFALL** NORMAL (IN) | 30 | 12.9 | 8.0 | 7.2 | 2.0 | 0.* | 0.0 | 0.0 | 0.0 | 0.* | 0.2 | 4.4 | 9.6 | 44.3 |
| MAXIMUM MONTHLY (IN) | 57 | 35.0 | 31.0 | 33.5 | 17.0 | 0.8 | T | T | T | T | 1.8 | 30.3 | 32.7 | 35.0 |
| YEAR OF OCCURRENCE | | 1996 | 1959 | 1959 | 1973 | 1960 | 1997 | 2004 | 1995 | 1994 | 1992 | 1991 | 2008 | JAN 1996 |
| MAXIMUM IN 24 HOURS (IN) | 43 | 12.0 | 10.9 | 15.7 | 7.3 | 0.8 | T | T | T | T | 1.8 | 13.0 | 14.4 | 15.7 |
| YEAR OF OCCURRENCE' | | 1996 | 1959 | 1959 | 1952 | 1960 | 1997 | 2004 | 1995 | 1994 | 1992 | 1991 | 1990 | MAR 1959 |
| MAXIMUM SNOW DEPTH (IN) | 60 | 34 | 29 | 31 | 16 | 0 | 0 | 0 | 0 | 0 | 1 | 15 | 20 | 34 |
| YEAR OF OCCURRENCE | | 1979 | 1979 | 1959 | 1973 | | | | | | 1992 | 1991 | 1968 | JAN 1979 |
| NORMAL NO. DAYS WITH: | | | | | | | | | | | | | | |
| SNOWFALL >= 1.0 | 30 | 4.0 | 2.5 | 2.2 | 0.6 | 0.0 | 0.0 | 0.0 | 0.0 | 0.0 | 0.1 | 1.3 | 3.1 | 13.8 |

## PRECIPITATION (inches) 2009  LA CROSSE (KLSE)

| YEAR | JAN | FEB | MAR | APR | MAY | JUN | JUL | AUG | SEP | OCT | NOV | DEC | ANNUAL |
|------|-----|-----|-----|-----|-----|-----|-----|-----|-----|-----|-----|-----|--------|
| 1980 | 1.61 | 0.35 | 0.65 | 1.74 | 2.66 | 3.57 | 2.26 | 9.84 | 8.51 | 2.36 | 0.14 | 0.61 | 34.30 |
| 1981 | 0.14 | 2.10 | 0.60 | 4.37 | 1.88 | 2.60 | 7.66 | 9.56 | 1.77 | 1.89 | 1.08 | 0.86 | 34.51 |
| 1982 | 1.34 | 0.17 |      | 1.65 | 4.71 | 1.35 | 1.89 | 2.91 | 2.67 | 2.83 | 3.64 | 2.03 |       |
| 1983 | 0.89 | 2.27 | 1.60 | 2.37 | 4.50 | 1.67 | 3.16 | 3.06 | 4.94 | 3.35 | 3.72 | 0.68 | 32.21 |
| 1984 | 0.28 | 0.92 | 1.94 | 3.65 | 2.18 | 7.43 | 3.00 | 2.11 | 2.87 | 5.09 | 1.36 | 2.42 | 33.25 |
| 1985 | 0.88 | 1.27 | 2.66 | 2.85 | 1.08 | 2.82 | 2.39 | 3.21 | 5.63 |      |      |      |       |
| 1986 |      | 0.77 | 1.91 | 3.42 | 1.49 | 4.04 | 4.79 | 2.38 | 8.10 | 3.68 | 0.95 | 0.38 |       |
| 1987 | 1.17 | 0.30 | 2.23 | 2.35 | 4.58 | 2.55 | 9.35 | 3.71 | 2.10 | 0.56 | 2.67 | 1.82 | 33.39 |
| 1988 | 1.09 | 0.19 | 1.89 | 2.01 | 0.94 | 3.25 | 2.39 | 4.60 | 5.21 | 0.64 | 3.48 | 0.78 | 26.47 |
| 1989 | 0.41 | 0.40 | 2.36 | 1.78 | 3.02 | 1.33 | 2.59 | 4.60 | 1.95 | 2.71 | 1.47 | 0.49 | 23.11 |
| 1990 | 0.79 | 0.68 | 3.23 | 2.61 | 3.74 | 8.07 | 4.03 | 8.02 | 1.82 | 1.43 | 0.74 | 2.91 | 38.07 |
| 1991 | 0.93 | 0.46 | 2.23 | 5.96 | 6.85 | 1.72 | 7.67 | 2.91 | 3.90 | 1.76 | 6.23 | 1.67 | 42.29 |
| 1992 | 0.87 | 0.81 | 3.64 | 4.38 | 1.49 | 2.32 | 4.02 | 2.67 | 7.01 | 0.62 | 3.64 | 1.58 | 33.05 |
| 1993 | 1.18 | 1.10 | 2.51 | 5.78 | 5.41 | 10.79 | 3.78 | 5.31 | 1.80 | 0.94 | 1.38 | 0.75 | 40.73 |
| 1994 | 2.24 | 1.65 | 0.37 | 6.51 | 1.49 | 2.74 | 6.29 | 4.84 | 6.51 | 1.99 | 2.05 | 0.71 | 37.39 |
| 1995 | 0.73 | 0.38 | 2.78 | 3.99 | 3.76 | 2.71 | 3.84 | 4.20 | 2.23 | 3.36 | 1.45 | 0.82 | 30.25 |
| 1996 | 3.03 | 0.41 | 2.05 | 1.94 | 1.50 | 5.98 | 2.32 | 2.16 | 1.64 | 2.98 | 4.45 | 1.42 | 29.88 |
| 1997 | 1.81 | 1.16 | 3.05 | 1.89 | 2.38 | 3.12 | 5.46 | 5.55 | 2.43 | 2.42 | 0.24 | 0.64 | 30.15 |
| 1998 | 1.76 | 2.71 | 2.43 | 1.74 | 2.95 | 8.22 | 2.78 | 6.21 | 0.85 | 4.61 | 1.26 | 0.30 | 35.82 |
| 1999 | 2.84 | 0.78 | 0.60 | 6.02 | 4.37 | 2.43 | 8.42 | 2.19 | 2.39 | 1.43 | 1.50 | 0.67 | 33.64 |
| 2000 | 1.43 | 0.91 | 1.35 | 1.40 | 5.82 | 7.54 | 4.39 | 2.47 | 1.64 | 1.04 | 2.41 | 1.90 | 32.30 |
| 2001 | 1.19 | 0.99 | 1.16 | 4.61 | 4.96 | 3.88 | 1.12 | 4.45 | 5.58 | 1.44 | 1.92 | 0.83 | 32.13 |
| 2002 | 0.44 | 2.20 | 1.52 | 4.27 | 1.31 | 6.39 | 3.75 | 2.70 | 3.68 | 3.33 | 0.55 | 0.36 | 30.50 |
| 2003 | 0.53 | 0.56 | 2.46 | 2.48 | 3.99 | 2.45 | 2.51 | 1.34 | 2.61 | 0.55 | 2.23 | 0.72 | 22.43 |
| 2004 | 0.62 | 1.63 | 3.38 | 1.50 | 9.73 | 7.37 | 4.93 | 3.92 | 3.48 | 2.09 | 1.49 | 1.29 | 41.43 |
| 2005 | 1.40 | 1.28 | 1.90 | 2.00 | 2.40 | 2.42 | 4.86 | 3.95 | 6.92 | 0.39 | 2.23 | 0.56 | 30.31 |
| 2006 | 0.47 | 0.71 | 2.61 | 4.78 | 4.12 | 3.19 | 1.75 | 4.16 | 3.61 | 0.90 | 1.65 | 2.12 | 30.07 |
| 2007 | 0.67 | 1.87 | 3.18 | 2.17 | 3.91 | 3.03 | 3.54 | 13.75 | 3.25 | 2.88 | 0.21 | 2.64 | 41.10 |
| 2008 | 1.30 | 1.14 | 2.15 | 6.74 | 3.52 | 7.00 | 6.84 | 0.69 | 2.08 | 1.16 | 1.70 | 2.32 | 36.64 |
| 2009 | 0.74 | 0.97 | 1.18 | 2.51 | 3.94 | 2.85 | 2.27 | 5.29 | 1.02 | 5.67 | 0.58 | 3.36 | 30.38 |
| POR= 67 YRS | 1.04 | 0.98 | 2.03 | 3.06 | 3.63 | 4.23 | 3.84 | 3.76 | 3.37 | 2.11 | 1.71 | 1.19 | 30.95 |

WBAN : 14920

## AVERAGE TEMPERATURE (°F) 2009  LA CROSSE (KLSE)

| YEAR | JAN | FEB | MAR | APR | MAY | JUN | JUL | AUG | SEP | OCT | NOV | DEC | ANNUAL |
|------|-----|-----|-----|-----|-----|-----|-----|-----|-----|-----|-----|-----|--------|
| 1980 | 18.5 | 17.9 | 29.7 | 49.4 | 62.8 | 70.3 | 78.3 | 74.1 | 63.7 | 45.2 | 36.5 | 20.8 | 47.3 |
| 1981 | 19.5 | 25.3 | 37.7 | 51.2 | 58.4 | 69.1 | 73.6 | 71.4 | 60.8 | 47.6 | 38.0 | 19.6 | 47.7 |
| 1982 | 4.3  | 17.2 |      | 44.2 | 62.8 | 63.5 | 74.2 | 70.3 | 60.9 | 50.8 | 34.1 | 27.2 |      |
| 1983 | 19.9 | 25.2 | 33.8 | 42.6 | 54.3 | 69.0 | 76.7 | 75.3 | 62.9 | 49.6 | 35.9 | 6.4  | 46.0 |
| 1984 | 14.1 | 30.3 | 27.3 | 48.2 | 55.9 | 69.6 | 71.1 | 72.9 | 59.0 | 52.4 | 35.0 | 21.5 | 46.4 |
| 1985 | 10.9 | 16.5 | 37.2 | 52.5 | 62.0 | 64.5 | 72.1 | 68.4 | 62.2 |      |      |      |      |
| 1986 |      | 18.4 | 36.4 | 52.1 | 61.9 | 69.1 | 75.0 | 67.5 | 62.5 | 50.9 | 30.2 | 25.1 |      |
| 1987 | 21.7 | 31.2 | 39.0 | 52.9 | 63.7 | 72.8 | 77.1 | 70.2 | 63.0 | 45.3 | 40.3 | 27.6 | 50.4 |
| 1988 | 13.4 | 15.8 | 35.0 | 48.0 | 65.2 | 73.4 | 76.6 | 76.0 | 64.6 | 44.6 | 36.6 | 21.8 | 47.6 |
| 1989 | 24.2 | 12.8 | 29.2 | 46.0 | 58.9 | 67.7 | 75.4 | 71.4 | 61.1 | 51.7 | 31.5 | 12.1 | 45.2 |
| 1990 | 28.1 | 25.4 | 38.4 | 49.3 | 56.6 | 69.5 | 73.2 | 72.1 | 65.5 | 49.6 | 41.4 | 19.0 | 49.0 |
| 1991 |      |      |      | 51.0 | 63.9 | 74.3 | 73.2 | 72.7 | 60.5 | 49.0 | 27.3 | 23.8 |      |
| 1992 | 23.9 | 30.3 | 33.9 | 45.6 | 61.2 | 66.9 | 68.0 | 66.6 |      | 48.9 | 33.1 | 23.6 |      |
| 1993 | 16.9 | 18.6 | 30.7 | 44.1 | 59.0 | 66.3 | 71.5 | 71.9 | 56.5 | 48.7 | 33.8 | 25.2 | 45.3 |
| 1994 | 6.0  | 14.9 | 35.7 | 47.7 | 60.7 | 71.8 | 70.8 | 67.5 | 66.0 | 55.2 | 41.6 | 29.0 | 47.2 |
| 1995 | 19.6 | 22.3 | 37.0 | 43.7 | 58.9 | 73.5 | 76.0 | 77.8 | 61.4 | 51.1 | 28.9 | 21.0 | 47.6 |
| 1996 | 11.9 | 19.9 | 28.5 | 44.8 | 57.1 | 69.4 | 71.5 | 72.4 | 64.1 | 51.7 | 29.2 | 19.9 | 45.0 |
| 1997 | 14.7 | 24.8 | 34.0 | 46.4 | 55.0 | 71.2 | 73.0 | 69.2 | 64.0 | 52.9 | 32.8 | 28.9 | 47.2 |
| 1998 | 24.0 | 34.1 | 35.7 | 52.7 | 66.0 | 68.5 | 75.3 | 73.7 | 68.8 | 53.4 | 40.9 | 29.2 | 51.9 |
| 1999 | 14.7 | 31.2 | 37.0 | 51.5 | 63.0 | 70.3 | 77.2 | 70.2 | 61.8 | 49.9 | 43.0 | 27.4 | 49.8 |
| 2000 | 18.1 | 30.6 | 42.5 | 48.7 | 61.7 | 67.9 | 72.5 | 72.1 | 62.7 | 53.5 | 34.0 | 8.7  | 47.8 |
| 2001 | 20.7 | 16.3 | 29.5 | 51.5 | 60.6 | 69.0 | 75.7 | 73.3 | 60.0 | 49.0 | 47.5 | 29.8 | 48.6 |
| 2002 | 27.9 | 30.0 | 28.8 | 47.5 | 56.4 | 71.1 | 76.7 | 70.9 | 65.5 | 45.3 | 34.8 | 27.7 | 48.6 |
| 2003 | 16.6 | 18.0 | 33.5 | 47.8 | 57.9 | 67.7 | 73.3 | 75.3 | 63.3 | 50.8 | 36.1 | 27.5 | 47.3 |
| 2004 | 15.0 | 24.2 | 39.2 | 50.4 | 58.4 | 65.8 | 70.9 | 66.7 | 67.2 | 51.3 | 39.2 | 23.7 | 47.7 |
| 2005 | 17.5 | 28.5 | 32.5 | 52.5 | 57.1 | 74.3 | 75.6 | 71.7 | 67.1 | 53.3 | 38.0 | 19.8 | 49.0 |
| 2006 | 31.1 | 21.9 | 35.5 | 53.6 | 60.4 | 69.5 | 78.0 | 71.9 | 59.9 | 46.6 | 38.9 | 30.8 | 49.8 |
| 2007 | 22.6 | 13.8 | 38.7 | 47.9 | 64.3 | 71.2 | 74.5 | 72.3 | 65.6 | 56.5 | 36.0 | 19.4 | 48.6 |
| 2008 | 14.6 | 15.0 | 28.9 | 46.4 | 56.5 | 68.4 | 73.5 | 70.4 | 64.6 | 50.9 | 36.2 | 14.1 | 45.0 |
| 2009 | 8.8  | 22.8 | 34.5 | 48.1 | 58.9 | 67.8 | 68.1 | 69.2 | 64.6 | 44.6 | 42.7 | 20.1 | 45.9 |
| POR= 66 YRS | 15.8 | 20.7 | 32.5 | 47.4 | 59.2 | 68.5 | 73.1 | 71.1 | 61.9 | 50.7 | 35.5 | 21.7 | 46.5 |

## HEATING DEGREE DAYS (base 65°F) 2009 LA CROSSE (KLSE)

| YEAR | JUL | AUG | SEP | OCT | NOV | DEC | JAN | FEB | MAR | APR | MAY | JUN | TOTAL |
|------|-----|-----|-----|-----|-----|-----|-----|-----|-----|-----|-----|-----|-------|
| 1980-81 | 0 | 3 | 112 | 607 | 849 | 1362 | 1402 | 1107 | 840 | 408 | 214 | 11 | 6915 |
| 1981-82 | 5 | 7 | 157 | 535 | 802 | 1402 | 1878 | 1333 | | 619 | 112 | 77 | |
| 1982-83 | 0 | 41 | 177 | 433 | 919 | 1164 | 1391 | 1109 | 960 | 666 | 325 | 46 | 7231 |
| 1983-84 | 6 | 0 | 160 | 480 | 867 | 1814 | 1572 | 1002 | 1159 | 501 | 285 | 5 | 7851 |
| 1984-85 | 9 | 6 | 219 | 384 | 893 | 1344 | 1673 | 1355 | 854 | 409 | 129 | 92 | 7367 |
| 1985-86 | 0 | 24 | 203 | | | | | 1298 | 880 | 394 | 141 | 25 | |
| 1986-87 | 1 | 35 | 120 | 431 | 1039 | 1230 | 1336 | 942 | 798 | 372 | 137 | 11 | 6452 |
| 1987-88 | 0 | 28 | 102 | 602 | 733 | 1151 | 1595 | 1421 | 924 | 503 | 85 | 10 | 7154 |
| 1988-89 | 0 | 10 | 78 | 627 | 844 | 1332 | 1257 | 1458 | 1100 | 563 | 218 | 52 | 7539 |
| 1989-90 | 1 | 4 | 159 | 417 | 1000 | 1638 | 1138 | 1102 | 818 | 504 | 267 | 38 | 7086 |
| 1990-91 | 3 | 2 | 112 | 474 | 700 | 1421 | | | | 438 | 176 | 2 | |
| 1991-92 | 5 | 3 | 217 | 490 | 1123 | 1275 | 1268 | 1001 | 958 | 576 | 173 | 59 | 7148 |
| 1992-93 | 22 | 46 | | 498 | 952 | 1274 | 1486 | 1295 | 1057 | 620 | 213 | 55 | |
| 1993-94 | 1 | 10 | 265 | 507 | 929 | 1227 | 1829 | 1399 | 899 | 519 | 175 | 23 | 7783 |
| 1994-95 | 7 | 39 | 76 | 304 | 694 | 1110 | 1400 | 1189 | 862 | 632 | 194 | 24 | 6531 |
| 1995-96 | 5 | 0 | 171 | 436 | 1076 | 1360 | 1641 | 1304 | 1125 | 601 | 277 | 35 | 8031 |
| 1996-97 | 2 | 0 | 121 | 410 | 1068 | 1393 | 1552 | 1120 | 951 | 553 | 304 | 7 | 7481 |
| 1997-98 | 11 | 13 | 80 | 414 | 960 | 1108 | 1268 | 859 | 902 | 366 | 57 | 72 | 6110 |
| 1998-99 | 0 | 0 | 42 | 356 | 714 | 1104 | 1553 | 940 | 861 | 396 | 114 | 40 | 6120 |
| 1999-00 | 0 | 4 | 148 | 459 | 653 | 1158 | 1448 | 992 | 695 | 481 | 161 | 42 | 6241 |
| 2000-01 | 9 | 2 | 141 | 359 | 926 | 1740 | 1367 | 1358 | 1094 | 410 | 177 | 59 | 7642 |
| 2001-02 | 5 | 6 | 171 | 491 | 518 | 1085 | 1143 | 972 | 1116 | 545 | 305 | 28 | 6385 |
| 2002-03 | 0 | 3 | 101 | 610 | 899 | 1150 | 1493 | 1310 | 970 | 525 | 217 | 37 | 7315 |
| 2003-04 | 1 | 2 | 151 | 446 | 858 | 1158 | 1545 | 1176 | 791 | 446 | 217 | 60 | 6851 |
| 2004-05 | 9 | 54 | 64 | 422 | 766 | 1274 | 1466 | 1014 | 1002 | 374 | 246 | 0 | 6691 |
| 2005-06 | 0 | 7 | 56 | 391 | 804 | 1397 | 1044 | 1199 | 909 | 336 | 221 | 21 | 6385 |
| 2006-07 | 0 | 1 | 182 | 580 | 775 | 1053 | 1309 | 1428 | 816 | 518 | 94 | 8 | 6764 |
| 2007-08 | 0 | 8 | 97 | 300 | 859 | 1407 | 1553 | 1445 | 1111 | 556 | 260 | 11 | 7607 |
| 2008-09 | 0 | 4 | 75 | 435 | 860 | 1572 | 1731 | 1175 | 940 | 508 | 200 | 66 | 7566 |
| 2009- | 11 | 24 | 68 | 627 | 661 | 1385 | | | | | | | |

WBAN : 14920

## COOLING DEGREE DAYS (base 65°F) 2009 LA CROSSE (KLSE)

| YEAR | JAN | FEB | MAR | APR | MAY | JUN | JUL | AUG | SEP | OCT | NOV | DEC | TOTAL |
|------|-----|-----|-----|-----|-----|-----|-----|-----|-----|-----|-----|-----|-------|
| 1980 | 0 | 0 | 0 | 20 | 96 | 180 | 416 | 293 | 79 | 0 | 0 | 0 | 1084 |
| 1981 | 0 | 0 | 0 | 0 | 19 | 139 | 277 | 211 | 37 | 0 | 0 | 0 | 683 |
| 1982 | | | 0 | | 54 | 38 | 292 | 212 | 60 | 0 | 0 | 0 | |
| 1983 | 0 | 0 | 0 | 1 | 0 | 172 | 375 | 324 | 104 | 12 | 0 | 0 | 988 |
| 1984 | 0 | 0 | 0 | 2 | 8 | 149 | 206 | 257 | 48 | 0 | 0 | 0 | 670 |
| 1985 | 0 | | 0 | 41 | 42 | 84 | 229 | 136 | 126 | | | | |
| 1986 | | 0 | 1 | 13 | 54 | 156 | 319 | 118 | 53 | 0 | 0 | 0 | |
| 1987 | 0 | 0 | 0 | 16 | 103 | 249 | 383 | 196 | 48 | 0 | 0 | 0 | 995 |
| 1988 | 0 | 0 | 0 | 0 | 101 | 269 | 365 | 360 | 73 | 0 | 0 | 0 | 1168 |
| 1989 | 0 | 0 | 1 | 0 | 38 | 138 | 331 | 212 | 48 | 11 | 0 | 0 | 779 |
| 1990 | 0 | 0 | 0 | 39 | 13 | 181 | 264 | 230 | 134 | 2 | 0 | 0 | 863 |
| 1991 | | | | 24 | 148 | 286 | 267 | 251 | 84 | 0 | 0 | 0 | |
| 1992 | 0 | 0 | 0 | 0 | 59 | 124 | 121 | 102 | | 5 | 0 | 0 | |
| 1993 | 0 | 0 | 0 | 0 | 35 | 100 | 209 | 232 | 14 | 10 | 0 | 0 | 600 |
| 1994 | 0 | 0 | 0 | 9 | 50 | 232 | 193 | 124 | 111 | 5 | 0 | 0 | 724 |
| 1995 | 0 | 0 | 0 | 0 | 13 | 286 | 353 | 405 | 71 | 11 | 0 | 0 | 1139 |
| 1996 | 0 | 0 | 0 | 0 | 38 | 174 | 210 | 236 | 101 | 5 | 0 | 0 | 764 |
| 1997 | 0 | 0 | 0 | 1 | 2 | 202 | 268 | 147 | 54 | 48 | 0 | 0 | 722 |
| 1998 | 0 | 0 | 0 | 3 | 96 | 185 | 326 | 277 | 163 | 2 | 0 | 0 | 1052 |
| 1999 | 0 | 0 | 0 | 0 | 55 | 209 | 384 | 169 | 59 | 0 | 0 | 0 | 876 |
| 2000 | 0 | 0 | 4 | 1 | 66 | 135 | 248 | 228 | 77 | 8 | 2 | 0 | 769 |
| 2001 | 0 | 0 | 0 | 10 | 48 | 185 | 344 | 269 | 28 | 1 | 0 | 0 | 885 |
| 2002 | 0 | 0 | 0 | 25 | 44 | 218 | 371 | 192 | 123 | 6 | 0 | 0 | 979 |
| 2003 | 0 | 0 | 0 | 16 | 4 | 124 | 265 | 326 | 104 | 13 | 0 | 0 | 852 |
| 2004 | 0 | 0 | 0 | 14 | 19 | 91 | 198 | 114 | 136 | 3 | 0 | 0 | 575 |
| 2005 | 0 | 0 | 0 | 5 | 7 | 285 | 335 | 222 | 125 | 38 | 0 | 0 | 1017 |
| 2006 | 0 | 0 | 0 | 0 | 83 | 163 | 411 | 222 | 36 | 14 | 0 | 0 | 929 |
| 2007 | 0 | 0 | 9 | 12 | 80 | 199 | 302 | 242 | 122 | 43 | 0 | 0 | 1009 |
| 2008 | 0 | 0 | 0 | 0 | 8 | 122 | 267 | 179 | 70 | 6 | 1 | 0 | 653 |
| 2009 | 0 | 0 | 0 | 6 | 21 | 159 | 113 | 158 | 63 | 0 | 0 | 0 | 520 |

## SNOWFALL (inches) 2009 LA CROSSE (KLSE)

| YEAR | JUL | AUG | SEP | OCT | NOV | DEC | JAN | FEB | MAR | APR | MAY | JUN | TOTAL |
|------|-----|-----|-----|-----|-----|-----|-----|-----|-----|-----|-----|-----|-------|
| 1980-81 | 0.0 | 0.0 | 0.0 | T | 0.4 | 4.6 | 1.9 | 13.9 | 1.0 | T | 0.0 | 0.0 | 21.8 |
| 1981-82 | 0.0 | 0.0 | 0.0 | 0.5 | 3.6 | 11.6 | 14.3 | 2.2 | | 0.0 | 0.0 | 0.0 | |
| 1982-83 | 0.0 | 0.0 | 0.0 | 0.0 | T | 0.0 | 9.8 | 18.5 | 4.2 | 2.1 | 0.0 | 0.0 | 34.6 |
| 1983-84 | 0.0 | 0.0 | 0.0 | 0.0 | 4.0 | 9.2 | 3.5 | 1.7 | 11.0 | T | 0.0 | 0.0 | 29.4 |
| 1984-85 | 0.0 | 0.0 | 0.0 | T | 2.9 | 13.3 | 9.4 | 5.3 | 14.5 | T | 0.0 | 0.0 | 45.4 |
| 1985-86 | 0.0 | 0.0 | T | | | | | 8.2 | T | T | 0.0 | 0.0 | |
| 1986-87 | 0.0 | 0.0 | 0.0 | T | 8.7 | 2.6 | 12.6 | 2.3 | 12.5 | T | 0.0 | 0.0 | 38.7 |
| 1987-88 | 0.0 | 0.0 | 0.0 | T | T | 13.3 | 19.1 | 4.2 | 1.2 | T | 0.0 | 0.0 | 37.8 |
| 1988-89 | 0.0 | 0.0 | 0.0 | T | 7.6 | 5.2 | 3.8 | 7.9 | 19.3 | T | T | 0.0 | 43.8 |
| 1989-90 | 0.0 | 0.0 | 0.0 | T | 8.7 | 5.9 | 10.0 | 7.0 | T | T | 0.0 | 0.0 | 31.6 |
| 1990-91 | 0.0 | 0.0 | 0.0 | T | 0.8 | 30.4 | 12.6 | 2.6 | 0.3 | 2.5 | 0.0 | 0.0 | 49.2 |
| 1991-92 | 0.0 | 0.0 | 0.0 | 1.2 | 30.3 | 8.7 | 5.3 | 7.5 | 11.1 | T | 0.0 | 0.0 | 64.1 |
| 1992-93 | 0.0 | 0.0 | 0.0 | 1.8 | 3.8 | 10.6 | 10.4 | 12.6 | 10.0 | 8.9 | 0.0 | 0.0 | 58.1 |
| 1993-94 | 0.0 | 0.0 | 0.0 | 0.2 | 4.5 | 2.3 | 21.4 | 18.7 | 0.8 | 1.9 | 0.0 | 0.0 | 49.8 |
| 1994-95 | 0.0 | 0.0 | T | 0.0 | 3.2 | 5.7 | 5.9 | 2.2 | 9.6 | 3.8 | T | 0.0 | 30.4 |
| 1995-96 | 0.0 | T | 0.0 | 0.5 | 6.1 | 8.8 | 35.0 | 1.2 | 5.8 | 3.3 | 0.0 | 0.0 | 60.7 |
| 1996-97 | 0.0 | 0.0 | 0.0 | 0.0 | 12.2 | 11.3 | 11.1 | 10.9 | 20.5 | 3.2 | 0.0 | T | 69.2 |
| 1997-98 | 0.0 | 0.0 | 0.0 | 0.0 | 1.8 | 6.0 | 16.1 | 3.8 | 9.7 | 0.1 | 0.0 | T | 37.5 |
| 1998-99 | 0.0 | T | 0.0 | 0.0 | 0.3 | 4.1 | 31.9 | 2.4 | 5.3 | 0.0 | 0.0 | 0.0 | 44.0 |
| 1999-00 | 0.0 | 0.0 | 0.0 | 0.3 | 0.0 | 4.9 | 9.4 | 5.4 | 3.7 | 1.9 | 0.0 | 0.0 | 25.6 |
| 2000-01 | 0.0 | 0.0 | 0.0 | 0.0 | 2.8 | 25.5 | 4.3 | 6.2 | 8.1 | 0.1 | 0.0 | 0.0 | 47.0 |
| 2001-02 | 0.0 | 0.0 | 0.0 | T | T | 1.0 | 6.4 | 6.2 | 8.5 | 10.8 | T | 0.0 | 32.9 |
| 2002-03 | 0.0 | 0.0 | 0.0 | 0.0 | 2.7 | 1.7 | 7.0 | 6.2 | 6.0 | 6.7 | 0.0 | 0.0 | 30.3 |
| 2003-04 | 0.0 | 0.0 | 0.0 | T | 0.5 | 6.0 | 8.2 | 14.5 | 5.5 | 0.0 | T | 0.0 | 34.7 |
| 2004-05 | T | 0.0 | 0.0 | T | 0.2 | 9.1 | 14.3 | 5.6 | 19.7 | T | T | 0.0 | 48.9 |
| 2005-06 | 0.0 | 0.0 | 0.0 | 0.0 | 6.1 | 12.9 | 1.6 | 11.4 | 7.8 | T | 0.0 | 0.0 | 39.8 |
| 2006-07 | 0.0 | 0.0 | 0.0 | T | 2.9 | T | 12.4 | 23.8 | 7.3 | 3.5 | 0.0 | 0.0 | 49.9 |
| 2007-08 | 0.0 | 0.0 | 0.0 | 0.0 | 1.7 | 24.2 | 18.3 | 15.0 | 7.9 | 0.8 | 0.0 | 0.0 | 67.9 |
| 2008-09 | 0.0 | 0.0 | 0.0 | T | 4.0 | 32.7 | 10.1 | 7.7 | 1.2 | T | 0.0 | 0.0 | 55.7 |
| 2009- | 0.0 | 0.0 | 0.0 | 0.4 | T | 24.9 | | | | | | | |
| POR= 61 YRS | T | T | T | 0.1 | 3.8 | 9.7 | 10.7 | 8.1 | 8.5 | 1.9 | T | T | 42.8 |

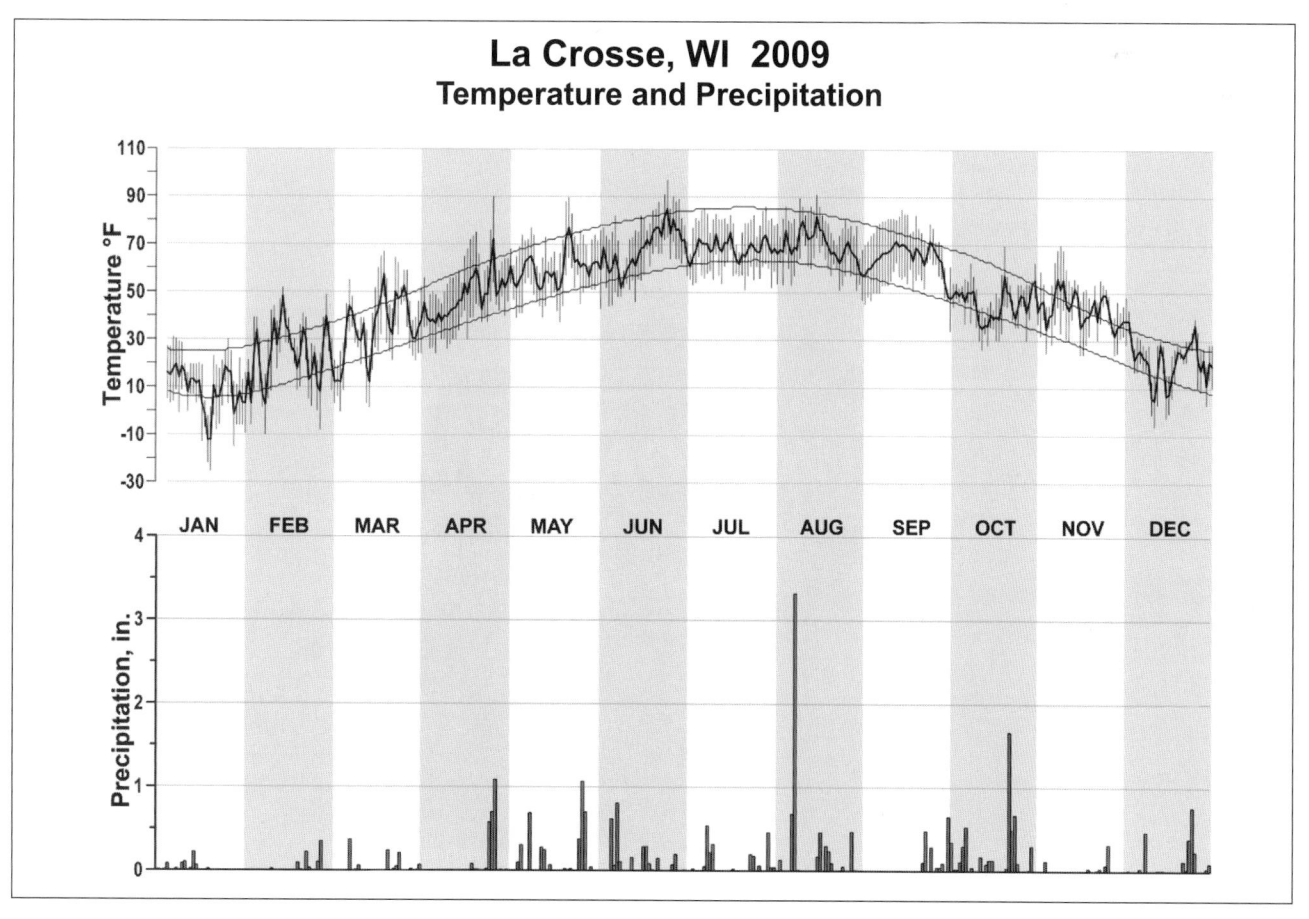

## La Crosse, WI 2009
### Temperature and Precipitation

# 2009
# MADISON
# WISCONSIN (KMSN)

Madison is set on a narrow isthmus of land between Lakes Mendota and Monona. Lake Mendota (15 square miles) lies northwest of Lake Monona (5 square miles) and the lakes are only two-thirds of a mile apart at one point. Drainage at Madison is southeast through two other lakes into the Rock River, which flows south into Illinois, and then west to the Mississippi. The westward flowing Wisconsin River is only 20 miles northwest of Madison. Madison lakes are normally frozen from mid-December to early April.

Madison has the typical continental climate of interior North America with a large annual temperature range and with frequent short period temperature changes. The range of extreme temperatures is from about 110 to -40 degrees. Winter temperatures (December-February) average near 20 degrees and the summer average (June-August) is in the upper 60s. Daily temperatures average below 32 degrees about 120 days and above 40 degrees for about 210 days of the year.

Madison lies in the path of the frequent cyclones and anticyclones which move eastward over this area during fall, winter and spring. In summer, the cyclones have diminished intensity and tend to pass farther north. The most frequent air masses are of polar origin. Occasional outbreaks of arctic air affect this area during the winter months. Although northward moving tropical air masses contribute considerable cloudiness and precipitation, the true Gulf air mass does not reach this area in winter, and only occasionally at other seasons. Summers are pleasant, with only occasional periods of extreme heat or high humidity.

There are no dry and wet seasons, but about 60 percent of the annual precipitation falls in the five months of May through September. Cold season precipitation is lighter, but lasts longer. Soil moisture is usually adequate in the first part of the growing season. During July, August, and September, the crops depend on current rainfall, which is mostly from thunderstorms and tends to be erratic and variable. Average occurrence of thunderstorms is just under 7 days per month during this period.

March and November are the windiest months. Tornadoes are infrequent. Dane County has about one tornado in every three to five years.

The ground is covered with 1 inch or more of snow about 60 percent of the time from about December 10 to near February 25 in an average winter. The soil is usually frozen from the first of December through most of March with an average frost penetration of 25 to 30 inches. The growing season averages 175 days.

Farming is diversified with the main emphasis on dairying. Field crops are mainly corn, oats, clover, and alfalfa, but barley, wheat, rye, and tobacco are also raised. Canning factories pack peas, sweet corn, and lima beans. Fruits are mainly apples, strawberries, and raspberries.

# NORMALS, MEANS, AND EXTREMES
## MADISON (KMSN)

| LATITUDE: 43 ° 8 'N | LONGITUDE: -89 ° 20'W | ELEVATION (FT): GRND: 859  BARO: 860 | TIME ZONE: CENTRAL  (UTC -6) | WBAN: 14837 |
|---|---|---|---|---|

| | ELEMENT | POR | JAN | FEB | MAR | APR | MAY | JUN | JUL | AUG | SEP | OCT | NOV | DEC | YEAR |
|---|---|---|---|---|---|---|---|---|---|---|---|---|---|---|---|
| **TEMPERATURE °F** | NORMAL DAILY MAXIMUM | 30 | 25.2 | 30.8 | 42.8 | 56.6 | 69.4 | 78.3 | 82.1 | 79.4 | 71.4 | 59.6 | 43.3 | 30.2 | 55.8 |
| | MEAN DAILY MAXIMUM | 63 | 26.2 | 30.9 | 42.1 | 57.6 | 69.4 | 78.8 | 82.9 | 80.9 | 72.6 | 60.7 | 44.3 | 30.9 | 56.4 |
| | HIGHEST DAILY MAXIMUM | 71 | 56 | 64 | 82 | 94 | 93 | 101 | 104 | 102 | 99 | 90 | 76 | 64 | 104 |
| | YEAR OF OCCURRENCE | | 1989 | 2000 | 1986 | 1980 | 1975 | 1988 | 1976 | 1988 | 1953 | 1976 | 1964 | 2001 | JUL 1976 |
| | MEAN OF EXTREME MAXS. | 67 | 44.8 | 48.8 | 66.6 | 79.5 | 85.8 | 91.4 | 93.3 | 92.0 | 87.6 | 79.3 | 64.9 | 50.3 | 73.7 |
| | NORMAL DAILY MINIMUM | 30 | 9.3 | 14.3 | 24.6 | 35.2 | 46.0 | 55.7 | 61.0 | 58.7 | 49.9 | 38.9 | 27.7 | 15.8 | 36.4 |
| | MEAN DAILY MINIMUM | 63 | 9.2 | 12.9 | 23.1 | 34.9 | 45.2 | 54.9 | 59.7 | 57.8 | 49.1 | 38.7 | 27.2 | 15.1 | 35.7 |
| | LOWEST DAILY MINIMUM | 71 | -37 | -29 | -29 | 0 | 19 | 31 | 36 | 35 | 25 | 13 | -11 | -25 | -37 |
| | YEAR OF OCCURRENCE | | 1951 | 1996 | 1962 | 1982 | 1978 | 1972 | 1965 | 1968 | 1974 | 1988 | 1947 | 1983 | JAN 1951 |
| | MEAN OF EXTREME MINS. | 67 | -13.5 | -9.5 | 2.7 | 19.7 | 30.0 | 40.1 | 46.7 | 44.2 | 32.7 | 23.2 | 9.7 | -6.7 | 18.3 |
| | NORMAL DRY BULB | 30 | 17.3 | 22.6 | 33.7 | 45.9 | 57.7 | 67.0 | 71.6 | 69.1 | 60.7 | 49.3 | 35.5 | 23.0 | 46.1 |
| | MEAN DRY BULB | 63 | 17.7 | 21.9 | 32.6 | 46.3 | 57.3 | 67.0 | 71.3 | 69.4 | 60.8 | 49.7 | 35.8 | 23.1 | 46.1 |
| | MEAN WET BULB | 26 | 17.7 | 20.9 | 29.9 | 40.3 | 50.9 | 60.5 | 64.7 | 63.5 | 56.0 | 44.2 | 32.7 | 21.9 | 41.9 |
| | MEAN DEW POINT | 26 | 14.2 | 17.2 | 25.4 | 35.3 | 46.8 | 57.3 | 62.1 | 61.1 | 53.2 | 40.5 | 29.2 | 18.6 | 38.4 |
| | NORMAL NO. DAYS WITH: | | | | | | | | | | | | | | |
| | MAXIMUM >= 90 | 30 | 0.0 | 0.0 | 0.0 | * | 0.3 | 2.9 | 5.3 | 2.8 | 0.6 | 0.1 | 0.0 | 0.0 | 12.0 |
| | MAXIMUM <= 32 | 30 | 21.1 | 14.4 | 5.0 | 0.5 | 0.0 | 0.0 | 0.0 | 0.0 | 0.0 | 0.0 | 4.6 | 15.8 | 61.4 |
| | MINIMUM <= 32 | 30 | 30.1 | 26.5 | 24.7 | 12.7 | 2.5 | * | 0.0 | 0.0 | 1.0 | 9.5 | 21.4 | 28.8 | 157.2 |
| | MINIMUM <= 0 | 30 | 10.0 | 5.9 | 0.9 | * | 0.0 | 0.0 | 0.0 | 0.0 | 0.0 | 0.0 | 0.2 | 4.9 | 21.9 |
| **H/C** | NORMAL HEATING DEG. DAYS | 30 | 1490 | 1203 | 978 | 576 | 261 | 63 | 12 | 33 | 183 | 504 | 892 | 1298 | 7493 |
| | NORMAL COOLING DEG. DAYS | 30 | 0 | 0 | 0 | 6 | 33 | 123 | 214 | 154 | 48 | 4 | 0 | 0 | 582 |
| **RH** | NORMAL (PERCENT) | 30 | 76 | 74 | 71 | 66 | 66 | 68 | 72 | 76 | 76 | 73 | 76 | 78 | 73 |
| | HOUR 00 LST | 30 | 79 | 79 | 78 | 76 | 77 | 80 | 83 | 87 | 87 | 81 | 81 | 81 | 81 |
| | HOUR 06 LST | 30 | 80 | 82 | 82 | 81 | 81 | 82 | 85 | 91 | 91 | 86 | 84 | 83 | 84 |
| | HOUR 12 LST | 30 | 70 | 67 | 61 | 54 | 53 | 55 | 57 | 61 | 60 | 59 | 67 | 72 | 61 |
| | HOUR 18 LST | 30 | 75 | 71 | 64 | 56 | 54 | 56 | 59 | 64 | 68 | 68 | 74 | 77 | 66 |
| **S** | PERCENT POSSIBLE SUNSHINE | 50 | 47 | 51 | 52 | 52 | 58 | 64 | 67 | 64 | 60 | 54 | 39 | 40 | 54 |
| **W/O** | MEAN NO. DAYS WITH: | | | | | | | | | | | | | | |
| | HEAVY FOG(VISBY <= 1/4 MI) | 46 | 2.0 | 1.9 | 2.6 | 1.1 | 1.3 | 1.2 | 1.4 | 2.2 | 1.8 | 1.3 | 1.6 | 3.0 | 21.4 |
| | THUNDERSTORMS | 62 | 0.2 | 0.2 | 1.8 | 3.7 | 5.2 | 7.0 | 7.5 | 6.6 | 4.3 | 1.9 | 0.8 | 0.3 | 39.5 |
| **CLOUDINESS** | MEAN: | | | | | | | | | | | | | | |
| | SUNRISE-SUNSET (OKTAS) | | | | | | | | | | | | | | |
| | MIDNIGHT-MIDNIGHT (OKTAS) | | | | | | | | | | | | | | |
| | MEAN NO. DAYS WITH: | | | | | | | | | | | | | | |
| | CLEAR | | | | | | | | | | | | | | |
| | PARTLY CLOUDY | | | | | | | | | | | | | | |
| | CLOUDY | | | | | | | | | | | | | | |
| **PR** | MEAN STATION PRESSURE(IN) | 26 | 29.11 | 29.12 | 29.08 | 29.02 | 29.02 | 29.02 | 29.06 | 29.10 | 29.11 | 29.10 | 29.09 | 29.11 | 29.08 |
| | MEAN SEA-LEVEL PRES. (IN) | 26 | 30.08 | 30.09 | 30.04 | 29.96 | 29.95 | 29.94 | 29.97 | 30.02 | 30.03 | 30.03 | 30.04 | 30.08 | 30.02 |
| **WINDS** | MEAN SPEED (MPH) | 26 | 9.2 | 9.3 | 9.8 | 10.2 | 8.7 | 7.7 | 7.0 | 6.8 | 7.3 | 8.4 | 9.2 | 8.9 | 8.5 |
| | PREVAIL.DIR(TENS OF DEGS) | 38 | 31 | 31 | 32 | 19 | 19 | 19 | 19 | 19 | 19 | 19 | 19 | 31 | 19 |
| | MAXIMUM 2-MINUTE: | | | | | | | | | | | | | | |
| | SPEED (MPH) | 14 | 34 | 30 | 34 | 37 | 39 | 41 | 40 | 35 | 43 | 32 | 38 | 31 | 43 |
| | DIR. (TENS OF DEGS) | | 06 | 10 | 01 | 18 | 16 | 36 | 31 | 33 | 15 | 19 | 18 | 15 | 15 |
| | YEAR OF OCCURRENCE | | 1999 | 2007 | 1998 | 2008 | 2008 | 1998 | 2006 | 2006 | 2000 | 2007 | 1998 | 2002 | SEP 2000 |
| | MAXIMUM 3-SECOND | | | | | | | | | | | | | | |
| | SPEED (MPH) | 14 | 45 | 41 | 45 | 53 | 45 | 53 | 66 | 44 | 74 | 49 | 52 | 40 | 74 |
| | DIR. (TENS OF DEGS) | | 07 | 25 | 09 | 26 | 17 | 31 | 02 | 15 | 15 | 19 | 21 | 25 | 15 |
| | YEAR OF OCCURRENCE | | 1999 | 2001 | 2009 | 1997 | 2008 | 2000 | 2004 | 2007 | 2000 | 2007 | 1998 | 2007 | SEP 2000 |
| **PRECIPITATION** | NORMAL (IN) | 30 | 1.25 | 1.28 | 2.28 | 3.35 | 3.25 | 4.05 | 3.93 | 4.33 | 3.08 | 2.18 | 2.31 | 1.66 | 32.95 |
| | MAXIMUM MONTHLY (IN) | 71 | 2.53 | 3.30 | 6.19 | 7.11 | 10.84 | 10.93 | 10.93 | 15.18 | 9.51 | 5.63 | 7.49 | 4.09 | 15.18 |
| | YEAR OF OCCURRENCE | | 1996 | 2008 | 2009 | 1973 | 2004 | 2008 | 1950 | 2007 | 1941 | 1984 | 2003 | 1987 | AUG 2007 |
| | MINIMUM MONTHLY (IN) | 71 | 0.14 | 0.06 | 0.28 | 0.96 | 0.64 | 0.81 | 1.38 | 0.70 | 0.11 | 0.06 | 0.11 | 0.25 | 0.06 |
| | YEAR OF OCCURRENCE | | 1981 | 1995 | 1978 | 1946 | 1981 | 1973 | 1946 | 1948 | 1979 | 1952 | 1976 | 1960 | FEB 1995 |
| | MAXIMUM IN 24 HOURS (IN) | 71 | 1.27 | 1.59 | 3.01 | 2.83 | 4.37 | 5.28 | 5.25 | 5.00 | 3.67 | 2.78 | 3.43 | 2.19 | 5.28 |
| | YEAR OF OCCURRENCE | | 1960 | 2001 | 1998 | 1975 | 2004 | 2008 | 1950 | 2007 | 2009 | 1984 | 2003 | 1990 | JUN 2008 |
| | NORMAL NO. DAYS WITH: | | | | | | | | | | | | | | |
| | PRECIPITATION >= 0.01 | 30 | 11.1 | 8.7 | 10.6 | 11.8 | 11.5 | 10.7 | 10.5 | 9.9 | 9.6 | 9.3 | 10.9 | 10.3 | 124.9 |
| | PRECIPITATION >= 1.00 | 30 | 0.1 | 0.2 | 0.2 | 0.7 | 0.8 | 1.0 | 1.1 | 1.2 | 0.8 | 0.3 | 0.5 | 0.2 | 7.1 |
| **SNOWFALL** | NORMAL (IN) | 30 | 12.9 | 8.6 | 7.1 | 3.5 | 0.1 | 0.0 | 0.0 | 0.0 | 0.0 | 0.4 | 4.7 | 12.6 | 49.9 |
| | MAXIMUM MONTHLY (IN) | 62 | 27.5 | 9.9E6 | 25.4 | 17.4 | 3.0 | T | T | T | T | 3.9 | 18.3 | 40.4 | 40.4 |
| | YEAR OF OCCURRENCE | | 1995 | 2009 | 1959 | 1973 | 1990 | 2008 | 2009 | 2006 | 2007 | 1997 | 1985 | 2008 | DEC 2008 |
| | MAXIMUM IN 24 HOURS (IN) | 62 | 13.0 | 14.2 | 13.6 | 12.9 | 3.0 | T | T | T | T | 3.8 | 9.0 | 17.3 | 17.3 |
| | YEAR OF OCCURRENCE' | | 1996 | 1994 | 1971 | 1973 | 1990 | 1992 | 2009 | 1994 | 1994 | 1997 | 1985 | 1990 | DEC 1990 |
| | MAXIMUM SNOW DEPTH (IN) | 62 | 32 | 28 | 16 | 14 | 4 | 0 | 0 | 0 | 0 | 4 | 9 | 17 | 32 |
| | YEAR OF OCCURRENCE | | 1979 | 1979 | 1986 | 1973 | 1994 | | | | | 1997 | 1985 | 1990 | JAN 1979 |
| | NORMAL NO. DAYS WITH: | | | | | | | | | | | | | | |
| | SNOWFALL >= 1.0 | 30 | 3.3 | 2.6 | 2.1 | 1.0 | 0.0 | 0.0 | 0.0 | 0.0 | 0.0 | 0.1 | 1.7 | 3.4 | 14.2 |

## PRECIPITATION (inches) 2009 MADISON (KMSN)

| YEAR | JAN | FEB | MAR | APR | MAY | JUN | JUL | AUG | SEP | OCT | NOV | DEC | ANNUAL |
|------|-----|-----|-----|-----|-----|-----|-----|-----|-----|-----|-----|-----|--------|
| 1980 | 1.11 | 0.64 | 0.68 | 2.36 | 2.08 | 3.43 | 2.67 | 9.49 | 7.84 | 1.13 | 1.33 | 1.62 | 34.38 |
| 1981 | 0.14 | 2.47 | 0.33 | 3.42 | 0.64 | 4.99 | 4.81 | 7.06 | 3.10 | 2.68 | 1.71 | 0.75 | 32.10 |
| 1982 | 1.42 | 0.17 | 2.11 | 3.26 | 4.34 | 3.40 | 3.47 | 2.67 | 1.42 | 1.46 | 4.21 | 3.65 | 31.58 |
| 1983 | 0.53 | 2.26 | 2.70 | 2.23 | 4.21 | 1.85 | 1.92 | 5.05 | 2.85 | 2.59 | 3.18 | 2.30 | 31.67 |
| 1984 | 0.36 | 1.26 | 1.15 | 3.86 | 3.32 | 7.01 | 1.96 | 1.89 | 2.79 | 5.63 | 1.83 | 2.66 | 33.72 |
| 1985 | 1.43 | 1.89 | 3.13 | 1.52 | 3.35 | 3.06 | 4.48 | 2.98 | 5.00 | 4.58 | 5.13 | 2.39 | 38.94 |
| 1986 | 1.02 | 2.72 | 1.55 | 2.27 | 1.97 | 3.24 | 4.31 | 4.38 | 6.82 | 1.85 | 1.03 | 0.69 | 31.85 |
| 1987 | 0.68 | 0.62 | 1.99 | 2.46 | 3.90 | 1.17 | 3.26 | 7.16 | 3.61 | 1.24 | 3.24 | 4.09 | 33.42 |
| 1988 | 1.82 | 0.46 | 1.20 | 2.65 | 0.92 | 2.06 | 2.44 | 2.95 | 3.33 | 1.60 | 3.58 | 1.56 | 24.57 |
| 1989 | 0.61 | 0.57 | 1.69 | 1.69 | 1.72 | 1.67 | 4.97 | 6.46 | 0.89 | 1.88 | 0.98 | 0.26 | 23.39 |
| 1990 | 1.60 | 0.99 | 4.18 | 1.90 | 5.35 | 4.88 | 2.61 | 6.03 | 1.64 | 2.25 | 1.65 | 3.46 | 36.54 |
| 1991 | 1.17 | 0.44 | 4.24 | 4.89 | 2.20 | 3.75 | 5.18 | 2.34 | 3.96 | 5.35 | 3.86 | 1.71 | 39.09 |
| 1992 | 0.78 | 1.34 | 1.90 | 3.17 | 1.12 | 1.53 | 5.54 | 2.48 | 5.99 | 1.06 | 4.83 | 2.39 | 32.13 |
| 1993 | 1.60 | 1.18 | 3.29 | 5.33 | 3.81 | 6.67 | 9.34 | 5.57 | 3.74 | 0.91 | 1.55 | 0.35 | 43.34 |
| 1994 | 1.46 | 2.76 | 0.46 | 2.57 | 1.33 | 5.66 | 4.10 | 4.56 | 6.14 | 0.65 | 2.77 | 1.08 | 33.54 |
| 1995 | 2.12 | 0.06 | 2.17 | 4.14 | 3.92 | 1.22 | 4.36 | 5.58 | 1.78 | 4.29 | 3.17 | 0.77 | 33.58 |
| 1996 | 2.53 | 0.53 | 0.80 | 2.76 | 2.95 | 9.69 | 4.08 | 1.84 | 1.07 | 3.14 | 1.01 | 1.27 | 31.67 |
| 1997 | 1.36 | 2.52 | 1.54 | 2.50 | 1.94 | 5.23 | 6.23 | 2.33 | 1.38 | 1.31 | 1.20 | 0.95 | 28.49 |
| 1998 | 0.88 | 1.44 | 5.46 | 4.10 | 4.58 | 7.46 | 2.50 | 4.24 | 2.48 | 3.20 | 1.95 | 0.29 | 38.58 |
| 1999 | 2.10 | 0.91 | 0.47 | 6.91 | 3.72 | 5.57 | 4.49 | 3.26 | 1.55 | 0.88 | 1.21 | 0.86 | 31.93 |
| 2000 | 0.91 | 1.95 | 1.17 | 3.18 | 9.63 | 8.63 | 3.27 | 3.94 | 3.59 | 0.68 | 2.00 | 1.39 | 40.34 |
| 2001 | 0.99 | 2.64 | 0.59 | 3.07 | 4.16 | 5.40 | 3.09 | 7.64 | 5.53 | 2.62 | 1.59 | 1.13 | 38.45 |
| 2002 | 0.63 | 2.17 | 1.70 | 3.45 | 2.92 | 3.70 | 2.06 | 3.04 | 2.74 | 2.10 | 1.01 | 0.67 | 26.19 |
| 2003 | 0.36 | 0.50 | 1.72 | 2.95 | 3.67 | 2.10 | 4.24 | 0.87 | 4.24 | 1.60 | 7.49 | 2.00 | 31.74 |
| 2004 | 0.62 | 1.44 | 3.61 | 1.76 | 10.84 | 3.93 | 6.05 | 3.96 | 1.00 | 3.20 | 1.51 | 1.46 | 39.38 |
| 2005 | 2.20 | 1.45 | 1.56 | 1.68 | 3.96 | 1.65 | 3.92 | 1.22 | 1.95 | 0.76 | 3.36 | 0.99 | 24.70 |
| 2006 | 1.96 | 0.81 | 2.34 | 5.04 | 4.61 | 2.29 | 4.45 | 5.43 | 3.33 | 2.87 | 2.24 | 1.36 | 36.73 |
| 2007 | 0.84 | 1.59 | 3.39 | 4.68 | 1.40 | 4.82 | 2.69 | 15.18 | 2.45 | 3.35 | 0.39 | 3.63 | 44.41 |
| 2008 | 2.17 | 3.30 | 2.47 | 6.43 | 2.55 | 10.93 | 5.62 | 1.41 | 2.23 | 2.20 | 1.46 | 3.29 | 44.06 |
| 2009 | 0.54 | 1.91 | 6.19 | 4.43 | 3.68 | 4.17 | 1.94 | 2.49 | 4.68 | 3.80 | 1.32 | 3.20 | 38.35 |
| POR= 67 YRS | 1.22 | 1.22 | 2.19 | 3.16 | 3.39 | 4.21 | 3.88 | 3.86 | 3.06 | 2.19 | 2.13 | 1.62 | 32.13 |

WBAN : 14837

## AVERAGE TEMPERATURE (°F) 2009 MADISON (KMSN)

| YEAR | JAN | FEB | MAR | APR | MAY | JUN | JUL | AUG | SEP | OCT | NOV | DEC | ANNUAL |
|------|-----|-----|-----|-----|-----|-----|-----|-----|-----|-----|-----|-----|--------|
| 1980 | 17.3 | 15.7 | 28.0 | 45.5 | 57.8 | 65.3 | 73.4 | 70.3 | 59.9 | 43.7 | 35.4 | 22.6 | 44.6 |
| 1981 | 20.5 | 25.3 | 36.9 | 48.7 | 55.3 | 67.4 | 70.6 | 68.7 | 59.1 | 46.6 | 36.7 | 22.0 | 46.5 |
| 1982 | 8.0 | 19.1 | 30.6 | 41.7 | 60.8 | 59.6 | 70.9 | 66.3 | 59.0 | 50.6 | 34.2 | 28.8 | 44.1 |
| 1983 | 21.4 | 26.3 | 33.1 | 41.6 | 51.9 | 67.5 | 75.0 | 72.2 | 60.1 | 48.2 | 37.3 | 10.8 | 45.5 |
| 1984 | 14.8 | 30.2 | 26.7 | 45.6 | 53.4 | 67.5 | 70.2 | 71.3 | 59.3 | 52.0 | 33.9 | 26.4 | 45.9 |
| 1985 | 12.2 | 19.0 | 37.7 | 52.2 | 60.7 | 63.8 | 70.0 | 66.4 | 61.6 | 49.4 | 31.0 | 11.3 | 44.6 |
| 1986 | 18.2 | 19.4 | 36.2 | 49.8 | 58.4 | 65.9 | 73.2 | 64.8 | 61.6 | 49.7 | 31.2 | 25.5 | 46.2 |
| 1987 | 22.6 | 30.5 | 37.2 | 49.9 | 60.8 | 70.4 | 74.5 | 68.7 | 60.6 | 43.4 | 40.0 | 28.4 | 48.9 |
| 1988 | 13.8 | 17.4 | 34.6 | 46.0 | 60.5 | 69.5 | 74.1 | 74.5 | 63.0 | 43.5 | 38.8 | 24.6 | 46.7 |
| 1989 | 27.6 | 14.6 | 30.1 | 44.7 | 56.1 | 65.7 | 72.3 | 68.6 | 58.7 | 50.8 | 33.1 | 14.2 | 44.7 |
| 1990 | 28.6 | 25.8 | 37.7 | 48.5 | 53.6 | 67.6 | 70.6 | 69.9 | 63.7 | 48.3 | 41.0 | 21.4 | 48.1 |
| 1991 | 15.1 | 26.5 | 36.8 | 49.4 | 63.6 | 71.1 | 72.3 | 70.2 | 60.0 | 49.4 | 31.4 | 26.3 | 47.7 |
| 1992 | 25.5 | 29.3 | 34.8 | 43.6 | 58.0 | 64.9 | 67.2 | 65.3 | 60.4 | 48.4 | 34.7 | 24.9 | 46.4 |
| 1993 | 21.8 | 21.2 | 31.5 | 43.7 | 59.6 | 65.9 | 72.0 | 72.1 | 56.7 | 47.9 | 35.2 | 26.6 | 46.2 |
| 1994 | 8.8 | 15.8 | 35.8 | 48.3 | 58.0 | 70.3 | 70.7 | 66.5 | 64.9 | 52.4 | 40.2 | 28.8 | 46.7 |
| 1995 | 20.4 | 22.9 | 36.9 | 43.8 | 57.6 | 72.0 | 74.8 | 76.9 | 58.7 | 50.2 | 28.7 | 20.6 | 47.0 |
| 1996 | 15.2 | 21.6 | 28.6 | 42.2 | 53.2 | 66.7 | 67.5 | 68.6 | 60.3 | 48.5 | 28.7 | 22.6 | 43.6 |
| 1997 | 15.7 | 24.1 | 34.1 | 43.3 | 51.3 | 67.3 | 69.3 | 65.4 | 60.2 | 50.2 | 33.3 | 27.9 | 45.2 |
| 1998 | 23.7 | 33.5 | 34.2 | 48.1 | 62.5 | 64.1 | 71.2 | 70.6 | 64.7 | 51.6 | 40.6 | 31.3 | 49.9 |
| 1999 | 16.5 | 30.8 | 33.8 | 48.1 | 59.9 | 68.4 | 74.9 | 66.9 | 60.3 | 48.9 | 42.0 | 26.4 | 48.1 |
| 2000 | 20.6 | 29.6 | 40.7 | 45.3 | 59.1 | 65.8 | 69.4 | 69.8 | 61.1 | 52.5 | 34.9 | 11.2 | 46.7 |
| 2001 | 21.7 | 19.7 | 31.7 | 51.1 | 58.4 | 66.0 | 72.3 | 70.9 | 58.7 | 48.9 | 46.1 | 30.4 | 48.0 |
| 2002 | 28.2 | 30.5 | 30.1 | 46.8 | 53.4 | 68.8 | 74.1 | 69.5 | 64.2 | 44.8 | 34.3 | 27.4 | 47.7 |
| 2003 | 17.7 | 18.5 | 32.3 | 44.7 | 55.2 | 65.4 | 70.3 | 71.9 | 61.0 | 49.3 | 37.2 | 27.9 | 46.0 |
| 2004 | 15.2 | 23.4 | 38.3 | 47.3 | 57.0 | 65.6 | 69.4 | 65.4 | 65.1 | 51.4 | 40.1 | 26.1 | 47.0 |
| 2005 | 19.6 | 29.3 | 32.2 | 51.0 | 55.9 | 72.3 | 72.2 | 70.6 | 66.8 | 51.6 | 38.3 | 19.9 | 48.3 |
| 2006 | 31.6 | 22.3 | 34.7 | 51.0 | 58.6 | 66.6 | 74.2 | 69.9 | 58.3 | 45.3 | 39.3 | 30.5 | 48.5 |
| 2007 | 23.4 | 14.5 | 38.4 | 45.1 | 62.2 | 68.8 | 71.9 | 71.6 | 63.9 | 56.4 | 35.4 | 21.2 | 47.7 |
| 2008 | 17.4 | 16.8 | 29.7 | 47.6 | 54.6 | 68.1 | 71.8 | 69.2 | 63.4 | 49.2 | 36.7 | 17.0 | 45.1 |
| 2009 | 10.6 | 23.7 | 35.2 | 45.8 | 58.5 | 67.1 | 65.7 | 67.9 | 62.4 | 45.5 | 42.2 | 22.2 | 45.6 |
| POR= 63 YRS | 17.7 | 21.9 | 32.6 | 46.3 | 57.3 | 67.0 | 71.3 | 69.4 | 60.8 | 49.7 | 35.8 | 23.1 | 46.1 |

## HEATING DEGREE DAYS (base 65°F) 2009  MADISON (KMSN)

| YEAR | JUL | AUG | SEP | OCT | NOV | DEC | JAN | FEB | MAR | APR | MAY | JUN | TOTAL |
|------|-----|-----|-----|-----|-----|-----|-----|-----|-----|-----|-----|-----|-------|
| 1980-81 | 2 | 11 | 178 | 651 | 881 | 1303 | 1373 | 1107 | 864 | 482 | 307 | 30 | 7189 |
| 1981-82 | 16 | 27 | 193 | 566 | 842 | 1327 | 1765 | 1281 | 1059 | 688 | 155 | 172 | 8091 |
| 1982-83 | 5 | 66 | 230 | 444 | 918 | 1117 | 1346 | 1078 | 978 | 693 | 400 | 57 | 7332 |
| 1983-84 | 11 | 6 | 193 | 519 | 823 | 1678 | 1550 | 1006 | 1181 | 575 | 358 | 20 | 7920 |
| 1984-85 | 9 | 21 | 215 | 397 | 927 | 1191 | 1632 | 1287 | 839 | 418 | 155 | 96 | 7187 |
| 1985-86 | 12 | 36 | 198 | 475 | 1012 | 1661 | 1444 | 1272 | 888 | 462 | 220 | 73 | 7753 |
| 1986-87 | 7 | 59 | 145 | 471 | 1007 | 1218 | 1309 | 963 | 857 | 452 | 192 | 27 | 6707 |
| 1987-88 | 3 | 45 | 150 | 661 | 743 | 1127 | 1586 | 1377 | 938 | 565 | 176 | 53 | 7424 |
| 1988-89 | 4 | 18 | 107 | 661 | 777 | 1242 | 1153 | 1404 | 1076 | 602 | 290 | 68 | 7402 |
| 1989-90 | 5 | 22 | 207 | 437 | 952 | 1568 | 1122 | 1092 | 835 | 519 | 349 | 46 | 7154 |
| 1990-91 | 7 | 12 | 133 | 511 | 713 | 1349 | 1539 | 1072 | 868 | 467 | 173 | 22 | 6866 |
| 1991-92 | 8 | 11 | 222 | 476 | 1002 | 1195 | 1216 | 1031 | 929 | 634 | 244 | 73 | 7041 |
| 1992-93 | 26 | 68 | 176 | 514 | 903 | 1236 | 1333 | 1220 | 1032 | 633 | 196 | 74 | 7411 |
| 1993-94 | 0 | 9 | 260 | 525 | 887 | 1182 | 1739 | 1375 | 896 | 501 | 241 | 39 | 7654 |
| 1994-95 | 6 | 52 | 80 | 389 | 736 | 1116 | 1376 | 1173 | 867 | 629 | 226 | 20 | 6670 |
| 1995-96 | 9 | 0 | 228 | 452 | 1081 | 1370 | 1537 | 1256 | 1123 | 675 | 385 | 52 | 8168 |
| 1996-97 | 18 | 7 | 174 | 505 | 1083 | 1306 | 1521 | 1138 | 951 | 646 | 416 | 44 | 7809 |
| 1997-98 | 28 | 52 | 157 | 481 | 945 | 1141 | 1273 | 875 | 949 | 496 | 129 | 95 | 6621 |
| 1998-99 | 0 | 1 | 78 | 409 | 724 | 1040 | 1497 | 952 | 962 | 496 | 178 | 53 | 6390 |
| 1999-00 | 0 | 26 | 177 | 491 | 681 | 1190 | 1372 | 1021 | 747 | 587 | 206 | 64 | 6562 |
| 2000-01 | 17 | 10 | 187 | 386 | 895 | 1663 | 1334 | 1260 | 1027 | 417 | 231 | 89 | 7516 |
| 2001-02 | 17 | 8 | 210 | 490 | 560 | 1063 | 1135 | 961 | 1074 | 560 | 371 | 52 | 6501 |
| 2002-03 | 1 | 7 | 102 | 623 | 914 | 1157 | 1459 | 1296 | 1006 | 607 | 298 | 70 | 7540 |
| 2003-04 | 3 | 5 | 177 | 478 | 824 | 1143 | 1537 | 1200 | 823 | 530 | 258 | 64 | 7042 |
| 2004-05 | 11 | 72 | 82 | 418 | 738 | 1200 | 1400 | 995 | 1010 | 419 | 283 | 14 | 6642 |
| 2005-06 | 9 | 12 | 69 | 445 | 795 | 1391 | 1029 | 1192 | 935 | 417 | 257 | 36 | 6587 |
| 2006-07 | 2 | 2 | 215 | 611 | 762 | 1062 | 1283 | 1408 | 825 | 592 | 145 | 21 | 6928 |
| 2007-08 | 2 | 6 | 117 | 299 | 883 | 1350 | 1468 | 1392 | 1089 | 514 | 317 | 16 | 7453 |
| 2008-09 | 2 | 8 | 97 | 487 | 844 | 1480 | 1681 | 1153 | 918 | 569 | 210 | 63 | 7512 |
| 2009- | 34 | 42 | 94 | 597 | 677 | 1318 | | | | | | | |

WBAN : 14837

## COOLING DEGREE DAYS (base 65°F) 2009  MADISON (KMSN)

| YEAR | JAN | FEB | MAR | APR | MAY | JUN | JUL | AUG | SEP | OCT | NOV | DEC | TOTAL |
|------|-----|-----|-----|-----|-----|-----|-----|-----|-----|-----|-----|-----|-------|
| 1980 | 0 | 0 | 0 | 8 | 39 | 100 | 268 | 183 | 31 | 0 | 0 | 0 | 629 |
| 1981 | 0 | 0 | 0 | 0 | 13 | 107 | 198 | 148 | 19 | 0 | 0 | 0 | 485 |
| 1982 | 0 | 0 | 0 | 0 | 29 | 16 | 194 | 114 | 53 | 3 | 0 | 0 | 409 |
| 1983 | 0 | 0 | 0 | 0 | 0 | 138 | 327 | 237 | 52 | 6 | 0 | 0 | 760 |
| 1984 | 0 | 0 | 0 | 1 | 5 | 102 | 177 | 224 | 50 | 0 | 0 | 0 | 559 |
| 1985 | 0 | 0 | 0 | 40 | 29 | 66 | 175 | 84 | 102 | 0 | 0 | 0 | 496 |
| 1986 | 0 | 0 | 0 | 13 | 24 | 105 | 269 | 59 | 49 | 0 | 0 | 0 | 519 |
| 1987 | 0 | 0 | 0 | 8 | 69 | 194 | 304 | 165 | 26 | 0 | 0 | 0 | 766 |
| 1988 | 0 | 0 | 0 | 0 | 43 | 194 | 296 | 315 | 54 | 0 | 0 | 0 | 902 |
| 1989 | 0 | 0 | 0 | 0 | 21 | 97 | 237 | 141 | 25 | 3 | 0 | 0 | 524 |
| 1990 | 0 | 0 | 0 | 32 | 2 | 132 | 191 | 171 | 100 | 1 | 0 | 0 | 629 |
| 1991 | 0 | 0 | 0 | 8 | 136 | 210 | 241 | 180 | 80 | 3 | 0 | 0 | 858 |
| 1992 | 0 | 0 | 0 | 0 | 33 | 76 | 100 | 86 | 42 | 5 | 0 | 0 | 342 |
| 1993 | 0 | 0 | 0 | 0 | 33 | 111 | 223 | 240 | 18 | 5 | 0 | 0 | 630 |
| 1994 | 0 | 0 | 0 | 10 | 32 | 207 | 192 | 108 | 85 | 3 | 0 | 0 | 637 |
| 1995 | 0 | 0 | 0 | 0 | 5 | 237 | 320 | 374 | 45 | 1 | 0 | 0 | 982 |
| 1996 | 0 | 0 | 0 | 0 | 28 | 110 | 103 | 125 | 41 | 0 | 0 | 0 | 407 |
| 1997 | 0 | 0 | 0 | 0 | 0 | 123 | 168 | 69 | 21 | 29 | 0 | 0 | 410 |
| 1998 | 0 | 0 | 0 | 0 | 57 | 140 | 199 | 182 | 76 | 0 | 0 | 0 | 654 |
| 1999 | 0 | 0 | 0 | 0 | 24 | 161 | 315 | 95 | 42 | 0 | 0 | 0 | 637 |
| 2000 | 0 | 0 | 0 | 0 | 32 | 96 | 159 | 165 | 78 | 5 | 0 | 0 | 535 |
| 2001 | 0 | 0 | 0 | 4 | 37 | 125 | 250 | 199 | 27 | 0 | 0 | 0 | 642 |
| 2002 | 0 | 0 | 0 | 21 | 19 | 173 | 292 | 156 | 86 | 5 | 0 | 0 | 752 |
| 2003 | 0 | 0 | 0 | 4 | 0 | 88 | 173 | 225 | 65 | 0 | 0 | 0 | 555 |
| 2004 | 0 | 0 | 0 | 4 | 16 | 89 | 155 | 91 | 93 | 2 | 0 | 0 | 450 |
| 2005 | 0 | 0 | 0 | 2 | 7 | 239 | 239 | 194 | 131 | 34 | 0 | 0 | 846 |
| 2006 | 0 | 0 | 0 | 2 | 63 | 90 | 297 | 161 | 18 | 6 | 0 | 0 | 637 |
| 2007 | 0 | 0 | 5 | 5 | 61 | 139 | 222 | 219 | 91 | 38 | 0 | 0 | 780 |
| 2008 | 0 | 0 | 0 | 0 | 2 | 111 | 220 | 144 | 57 | 4 | 0 | 0 | 538 |
| 2009 | 0 | 0 | 0 | 0 | 16 | 130 | 62 | 139 | 21 | 0 | 0 | 0 | 368 |

### SNOWFALL (inches) 2009 MADISON (KMSN)

| YEAR | JUL | AUG | SEP | OCT | NOV | DEC | JAN | FEB | MAR | APR | MAY | JUN | TOTAL |
|------|-----|-----|-----|-----|-----|-----|-----|-----|-----|-----|-----|-----|-------|
| 1980-81 | 0.0 | 0.0 | 0.0 | T | 3.5 | 9.2 | 2.9 | 9.2 | 1.7 | 0.0 | 0.0 | 0.0 | 26.5 |
| 1981-82 | 0.0 | 0.0 | 0.0 | 0.1 | 2.0 | 7.2 | 19.4 | 2.4 | 8.6 | 10.3 | 0.0 | 0.0 | 50.0 |
| 1982-83 | 0.0 | 0.0 | 0.0 | 0.0 | 0.3 | 3.3 | 6.5 | 13.0 | 14.1 | 4.2 | 0.0 | 0.0 | 41.4 |
| 1983-84 | 0.0 | 0.0 | 0.0 | 0.0 | 2.1 | 22.6 | 6.0 | 0.8 | 6.8 | 3.9 | 0.0 | 0.0 | 42.2 |
| 1984-85 | 0.0 | 0.0 | 0.0 | 0.0 | 0.5 | 15.8 | 19.9 | 7.4 | 8.2 | 1.9 | 0.0 | 0.0 | 53.7 |
| 1985-86 | 0.0 | 0.0 | 0.0 | 0.0 | 18.3 | 24.0 | 13.9 | 13.3 | 2.7 | 0.2 | 0.0 | 0.0 | 72.4 |
| 1986-87 | 0.0 | 0.0 | 0.0 | T | 8.6 | 8.0 | 8.7 | 0.3 | 8.9 | T | 0.0 | 0.0 | 34.5 |
| 1987-88 | 0.0 | 0.0 | 0.0 | 0.4 | 3.9 | 32.8 | 16.3 | 6.4 | 1.1 | 1.3 | 0.0 | 0.0 | 62.2 |
| 1988-89 | 0.0 | 0.0 | 0.0 | 0.2 | 5.5 | 8.2 | 2.6 | 9.7 | 9.3 | 0.2 | 0.5 | T | 36.2 |
| 1989-90 | 0.0 | 0.0 | 0.0 | 0.7 | 4.4 | 4.3 | 10.1 | 11.7 | 0.1 | 0.5 | 3.0 | T | 34.8 |
| 1990-91 | 0.0 | 0.0 | 0.0 | 3.1 | 4.5 | 23.0 | 14.5 | 5.0 | 3.6 | 1.3 | 0.0 | 0.0 | 55.0 |
| 1991-92 | T | 0.0 | 0.0 | 0.5 | 8.0 | 10.2 | 4.5 | 12.3 | 6.9 | 0.1 | 0.0 | T | 42.5 |
| 1992-93 | 0.0 | T | 0.0 | 2.1 | 3.9 | 10.5 | 12.5 | 12.1 | 21.6 | 8.5 | 0.0 | 0.0 | 71.2 |
| 1993-94 | 0.0 | 0.0 | T | 0.2 | 1.2 | 2.5 | 22.5 | 37.0 | 0.6 | 9.7 | 0.0 | 0.0 | 73.7 |
| 1994-95 | T | T | T | 0.0 | 0.3 | 12.5 | 27.5 | 0.7 | 11.2 | 0.6 | 0.0 | 0.0 | 52.8 |
| 1995-96 | 0.0 | 0.0 | 0.0 | 0.1 | 12.8 | 10.3 | 26.4 | 1.3 | 4.7 | 4.9 | 0.0 | 0.0 | 60.5 |
| 1996-97 | 0.0 | 0.0 | 0.0 | 0.0 | 5.9 | 6.7 | 13.1 | 14.4 | 2.7 | 7.1 | 0.1 | 0.0 | 50.0 |
| 1997-98 | 0.0 | 0.0 | 0.0 | 3.9 | 3.0 | 14.3 | 18.9 | 1.8 | 12.0 | T | 0.0 | 0.0 | 53.9 |
| 1998-99 | 0.0 | 0.0 | 0.0 | 0.0 | 0.4 | 2.2 | 23.9 | 3.8 | 7.8 | T | 0.0 | 0.0 | 38.1 |
| 1999-00 | 0.0 | T | 0.0 | 0.0 | T | 3.2 | 11.7 | 11.2 | 3.5 | 4.5 | T | 0.0 | 34.1 |
| 2000-01 | 0.0 | 0.0 | 0.0 | T | 5.3 | 35.0 | 1.6 | 7.8 | 1.9 | 0.6 | 0.0 | 0.0 | 52.2 |
| 2001-02 | 0.0 | 0.0 | 0.0 | T | T | 2.3 | 7.0 | 7.5 | 11.7 | 3.3 | 0.0 | 0.0 | 31.8 |
| 2002-03 | 0.0 | 0.0 | 0.0 | 0.3 | 1.2 | 1.6 | 4.4 | 7.8 | 8.6 | 4.9 | 0.0 | 0.0 | 28.8 |
| 2003-04 | 0.0 | 0.0 | 0.0 | 0.0 | 0.3 | 3.6 | 9.2 | 13.9 | 3.4 | T | 1.2 | 0.0 | 31.6 |
| 2004-05 | T | 0.0 | 0.0 | T | T | 1.9 | 19.0 | 10.9 | 12.1 | T | T | 0.0 | 43.9 |
| 2005-06 | 0.0 | 0.0 | 0.0 | 0.0 | 5.8 | 15.2 | 4.2 | 12.6 | 9.8 | T | 0.3 | 0.0 | 47.9 |
| 2006-07 | 0.0 | T | 0.0 | 2.3 | 2.7 | 2.6 | 15.9 | 22.3 | 3.6 | 5.7 | 0.0 | 0.0 | 55.1 |
| 2007-08 | 0.0 | 0.0 | T | 0.0 | 1.5 | 33.5 | 23.2 | 31.6 | 10.9 | 0.7 | T | T | 101.4 |
| 2008-09 | 0.0 | 0.0 | 0.0 | T | 4.3 | 40.4 | 12.0 | 9.9 | 4.5 | 0.9 | 0.0 | 0.0 | 71.4 |
| 2009- | T | 0.0 | 0.0 | T | 0.5 | 26.8 | | | | | | | |
| POR=<br>62 YRS | T | T | T | 0.3 | 3.4 | 11.5 | 11.0 | 8.5 | 8.1 | 2.3 | 0.1 | T | 45.4 |

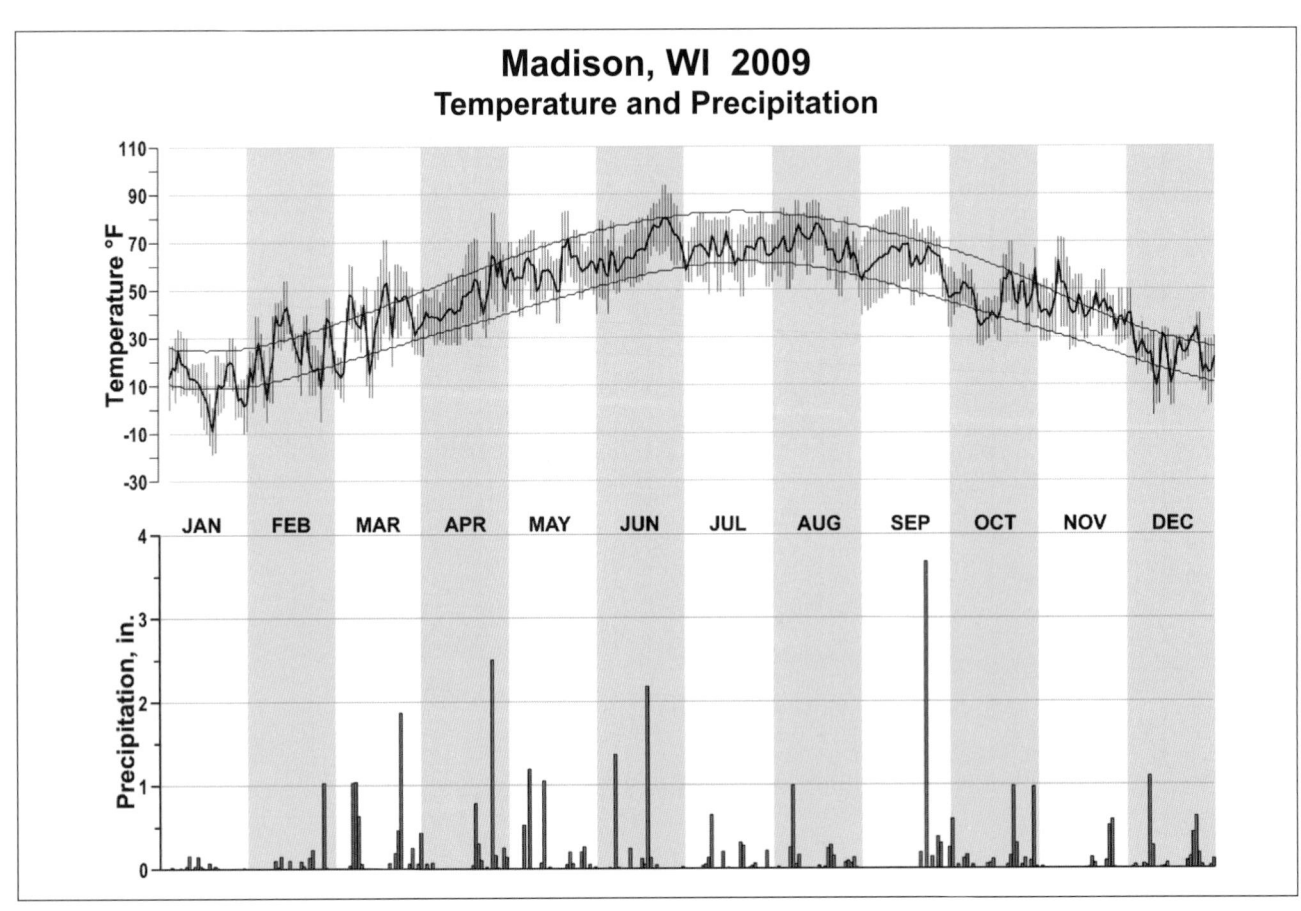

## Madison, WI 2009
### Temperature and Precipitation

# 2009
# MILWAUKEE
# WISCONSIN (KMKE)

Milwaukee possesses a continental climate characterized by a wide range of temperatures between summer and winter. Precipitation is moderate and occurs mostly in the spring, less in the autumn, and very little in the wintertime. Rainfall is well distributed for agricultural purposes, although spring planting is sometimes delayed by wet ground and cold weather.

Milwaukee is in a region of frequently changeable weather and its climate is influenced by general easterly-moving storms which traverse the nations midsection. The most severe winter storms, which produce in excess of 10 inches of snow, develop in the southern Great Plains and move northeast across Illinois and Indiana.

Occasionally during the cold season, frigid air masses from Canada push southeast across the Great Lakes region. These arctic air masses account for the coldest winter temperatures. Very low temperatures, zero degrees or lower, most often occur in air that flows southward to the west of Lake Superior before reaching the Milwaukee area. If northwesterly wind circulation persists, repeated incursions of arctic air will result in a period of bitterly cold weather lasting several days.

Summer temperatures, which reach into the 90s but rarely exceed 100 degrees, occur with brisk southwest winds that carry hot air from the plains and lower Mississippi River Valley across the city. A combination of high temperatures and humidity occasionally develops, usually building up over a period of several days when persistent southerly winds transport moisture from the Gulf of Mexico into the area.

The Gulf is a major source of moisture for Milwaukee in all seasons, but the type of precipitation which results is dependent upon the time of year. Cold-season precipitation (rain, snow, or a mixture) is usually of relatively long duration and low intensity, and occasionally persists for two days or more, whereas in the warm season, relatively short-duration and high-intensity showery rainfall, usually lasting a few hours or less, predominates.

The Great Lakes significantly influence the local climate. Temperature extremes are modified by Lake Michigan and, to a lesser extent, the other Great Lakes. In late autumn and winter, air masses that are initially very cold often reach the city only after being tempered by passage over one or more of the lakes. Similarly, air masses that approach from the northeast in the spring and summer are cooler because of movement over the Great Lakes.

The influence of Lake Michigan is variable and occasionally dramatic, especially when the temperature of the lake water differs strongly from the air temperature. During the spring and early summer, a wind shift from a westerly to an easterly direction frequently causes a sudden 10 to 20 degree temperature drop. When the breeze off the lake is light, this effect reaches inland only a mile or two. With stronger on-shore winds, the entire city is cooled. In the winter the relatively warm water of the lake moderates the temperature during easterly wind situations. Lake-induced snows usually occur a few times each winter, but snow accumulation is rarely heavy.

Topography does not significantly affect air flow, except that lesser frictional drag over Lake Michigan causes winds to be frequently stronger along the lake shore, and often permits air masses approaching from the north to reach shore areas one hour or more before affecting inland portions of the city.

# NORMALS, MEANS, AND EXTREMES
## MILWAUKEE (KMKE)

| LATITUDE: 42 ° 56'N | LONGITUDE: -87 ° 53'W | | ELEVATION (FT): GRND: 676  BARO: 680 | | | | TIME ZONE: CENTRAL  (UTC -6) | | | WBAN: 14839 | | |

| | ELEMENT | POR | JAN | FEB | MAR | APR | MAY | JUN | JUL | AUG | SEP | OCT | NOV | DEC | YEAR |
|---|---|---|---|---|---|---|---|---|---|---|---|---|---|---|---|
| **TEMPERATURE °F** | NORMAL DAILY MAXIMUM | 30 | 28.0 | 32.5 | 42.6 | 53.9 | 66.0 | 76.3 | 81.1 | 79.1 | 71.9 | 60.2 | 45.7 | 33.1 | 55.9 |
| | MEAN DAILY MAXIMUM | 68 | 28.1 | 31.3 | 41.1 | 53.6 | 64.7 | 75.1 | 80.3 | 78.8 | 71.2 | 60.1 | 45.3 | 32.6 | 55.2 |
| | HIGHEST DAILY MAXIMUM | 69 | 63 | 68 | 82 | 91 | 93 | 101 | 103 | 103 | 98 | 89 | 77 | 68 | 103 |
| | YEAR OF OCCURRENCE | | 2008 | 1999 | 1986 | 1980 | 1991 | 1988 | 1995 | 1988 | 1953 | 1963 | 1950 | 2001 | JUL 1995 |
| | MEAN OF EXTREME MAXS. | 82 | 46.8 | 48.7 | 65.6 | 78.5 | 85.3 | 91.6 | 93.1 | 91.9 | 87.6 | 78.8 | 65.0 | 51.7 | 73.7 |
| | NORMAL DAILY MINIMUM | 30 | 13.4 | 18.3 | 27.3 | 36.4 | 46.2 | 56.3 | 62.9 | 62.1 | 54.1 | 42.6 | 31.0 | 19.4 | 39.2 |
| | MEAN DAILY MINIMUM | 68 | 13.7 | 17.2 | 26.3 | 36.3 | 45.4 | 55.5 | 62.2 | 61.6 | 53.4 | 42.7 | 30.8 | 19.0 | 38.7 |
| | LOWEST DAILY MINIMUM | 69 | -26 | -26 | -10 | 12 | 21 | 33 | 40 | 44 | 28 | 18 | -5 | -20 | -26 |
| | YEAR OF OCCURRENCE | | 1982 | 1996 | 1962 | 1982 | 1966 | 1945 | 1965 | 1982 | 1974 | 1981 | 1950 | 1983 | FEB 1996 |
| | MEAN OF EXTREME MINS. | 82 | -7.9 | -2.5 | 8.5 | 23.8 | 33.9 | 43.6 | 51.5 | 50.9 | 39.1 | 28.4 | 14.7 | -1.0 | 23.6 |
| | NORMAL DRY BULB | 30 | 20.7 | 25.4 | 34.9 | 45.2 | 56.1 | 66.3 | 72.0 | 70.6 | 63.0 | 51.4 | 38.4 | 26.2 | 47.5 |
| | MEAN DRY BULB | 68 | 20.9 | 24.3 | 33.7 | 44.9 | 55.1 | 65.5 | 71.2 | 70.3 | 62.3 | 51.4 | 38.0 | 25.8 | 47.0 |
| | MEAN WET BULB | 26 | 20.6 | 23.5 | 31.1 | 39.6 | 49.2 | 59.3 | 64.6 | 64.3 | 57.5 | 45.9 | 35.0 | 24.9 | 43.0 |
| | MEAN DEW POINT | 26 | 17.0 | 19.4 | 26.3 | 35.3 | 45.4 | 56.1 | 62.0 | 61.8 | 54.5 | 42.2 | 31.4 | 20.9 | 39.4 |
| | NORMAL NO. DAYS WITH: | | | | | | | | | | | | | | |
| | MAXIMUM >= 90 | 30 | 0.0 | 0.0 | 0.0 | * | 0.2 | 2.0 | 4.0 | 2.3 | 0.6 | 0.0 | 0.0 | 0.0 | 9.1 |
| | MAXIMUM <= 32 | 30 | 19.6 | 14.1 | 5.5 | 0.6 | 0.0 | 0.0 | 0.0 | 0.0 | 0.0 | 0.0 | 2.8 | 13.8 | 56.4 |
| | MINIMUM <= 32 | 30 | 29.2 | 25.3 | 22.2 | 8.2 | 0.5 | 0.0 | 0.0 | 0.0 | 0.1 | 3.1 | 17.1 | 27.3 | 133.0 |
| | MINIMUM <= 0 | 30 | 5.9 | 2.7 | 0.2 | 0.0 | 0.0 | 0.0 | 0.0 | 0.0 | 0.0 | 0.0 | 0.1 | 2.8 | 11.7 |
| **H/C** | NORMAL HEATING DEG. DAYS | 30 | 1384 | 1124 | 948 | 611 | 318 | 86 | 13 | 18 | 134 | 443 | 808 | 1200 | 7087 |
| | NORMAL COOLING DEG. DAYS | 30 | 0 | 0 | 0 | 5 | 5 | 27 | 114 | 222 | 180 | 63 | 0 | 0 | 616 |
| **RH** | NORMAL (PERCENT) | 30 | 74 | 73 | 70 | 68 | 68 | 70 | 72 | 75 | 74 | 72 | 74 | 75 | 72 |
| | HOUR 00 LST | 30 | 76 | 75 | 75 | 74 | 75 | 78 | 80 | 83 | 81 | 78 | 77 | 78 | 78 |
| | HOUR 06 LST | 30 | 77 | 78 | 78 | 77 | 77 | 79 | 82 | 86 | 85 | 81 | 80 | 80 | 80 |
| | HOUR 12 LST | 30 | 69 | 67 | 63 | 60 | 60 | 61 | 61 | 64 | 62 | 61 | 66 | 71 | 64 |
| | HOUR 18 LST | 30 | 73 | 71 | 67 | 63 | 62 | 63 | 64 | 68 | 70 | 69 | 73 | 74 | 68 |
| **S** | PERCENT POSSIBLE SUNSHINE | 56 | 44 | 47 | 50 | 53 | 60 | 65 | 69 | 66 | 59 | 54 | 39 | 38 | 54 |
| **W/O** | MEAN NO. DAYS WITH: | | | | | | | | | | | | | | |
| | HEAVY FOG(VISBY <= 1/4 MI) | 46 | 1.9 | 2.0 | 2.9 | 2.4 | 3.0 | 1.9 | 1.3 | 1.3 | 1.4 | 1.6 | 1.5 | 2.0 | 23.2 |
| | THUNDERSTORMS | 62 | 0.3 | 0.4 | 1.5 | 3.5 | 4.4 | 6.2 | 6.6 | 5.6 | 3.6 | 1.5 | 0.9 | 0.3 | 34.8 |
| **CLOUDNESS** | MEAN: | | | | | | | | | | | | | | |
| | SUNRISE-SUNSET (OKTAS) | | | | | | | | | | | | | | |
| | MIDNIGHT-MIDNIGHT (OKTAS) | | | | | | | | | | | | | | |
| | MEAN NO. DAYS WITH: | | | | | | | | | | | | | | |
| | CLEAR | | | | | | | | | | | | | | |
| | PARTLY CLOUDY | | | | | | | | | | | | | | |
| | CLOUDY | | | | | | | | | | | | | | |
| **PR** | MEAN STATION PRESSURE(IN) | 26 | 29.29 | 29.31 | 29.28 | 29.22 | 29.19 | 29.21 | 29.24 | 29.28 | 29.29 | 29.29 | 29.28 | 29.30 | 29.27 |
| | MEAN SEA-LEVEL PRES. (IN) | 26 | 30.07 | 30.08 | 30.05 | 29.97 | 29.96 | 29.95 | 29.97 | 30.01 | 30.03 | 30.04 | 30.03 | 30.07 | 30.02 |
| **WINDS** | MEAN SPEED (MPH) | 26 | 11.3 | 11.0 | 11.3 | 11.7 | 10.5 | 9.2 | 9.1 | 8.8 | 9.2 | 10.5 | 11.1 | 10.9 | 10.4 |
| | PREVAIL.DIR(TENS OF DEGS) | 41 | 30 | 31 | 31 | 02 | 02 | 03 | 24 | 23 | 22 | 22 | 31 | 31 | 30 |
| | MAXIMUM 2-MINUTE: | | | | | | | | | | | | | | |
| | SPEED (MPH) | 14 | 44 | 52 | 41 | 48 | 46 | 47 | 39 | 37 | 41 | 43 | 52 | 41 | 52 |
| | DIR. (TENS OF DEGS) | | 08 | 27 | 10 | 24 | 30 | 31 | 30 | 16 | 07 | 24 | 23 | 21 | 27 |
| | YEAR OF OCCURRENCE | | 1999 | 1999 | 2007 | 1997 | 1998 | 2001 | 2003 | 2007 | 2001 | 1996 | 1998 | 2007 | FEB 1999 |
| | MAXIMUM 3-SECOND | | | | | | | | | | | | | | |
| | SPEED (MPH) | 14 | 52 | 70 | 58 | 62 | 61 | 69 | 51 | 53 | 48 | 48 | 68 | 52 | 70 |
| | DIR. (TENS OF DEGS) | | 27 | 25 | 20 | 24 | 30 | 28 | 22 | 30 | 25 | 19 | 23 | 23 | 25 |
| | YEAR OF OCCURRENCE | | 2008 | 1999 | 2005 | 1997 | 1998 | 2001 | 2008 | 2007 | 2005 | 2007 | 1998 | 2007 | FEB 1999 |
| **PRECIPITATION** | NORMAL (IN) | 30 | 1.85 | 1.65 | 2.59 | 3.78 | 3.06 | 3.56 | 3.58 | 4.03 | 3.30 | 2.49 | 2.70 | 2.22 | 34.81 |
| | MAXIMUM MONTHLY (IN) | 69 | 4.38 | 3.94 | 6.93 | 7.31 | 8.42 | 12.27 | 7.66 | 9.05 | 9.87 | 7.03 | 7.11 | 5.42 | 12.27 |
| | YEAR OF OCCURRENCE | | 1999 | 1986 | 1976 | 1973 | 2000 | 2008 | 1964 | 1987 | 1941 | 1991 | 1985 | 1987 | JUN 2008 |
| | MINIMUM MONTHLY (IN) | 69 | 0.31 | 0.05 | 0.31 | 0.81 | 0.50 | 0.70 | 0.71 | 0.46 | 0.02 | 0.15 | 0.36 | 0.29 | 0.02 |
| | YEAR OF OCCURRENCE | | 1981 | 1969 | 1968 | 1942 | 1988 | 1988 | 2009 | 1948 | 1979 | 1956 | 2007 | 1976 | SEP 1979 |
| | MAXIMUM IN 24 HOURS (IN) | 69 | 1.73 | 1.92 | 2.57 | 3.11 | 3.11 | 5.89 | 4.42 | 6.84 | 5.28 | 2.60 | 2.69 | 2.24 | 6.84 |
| | YEAR OF OCCURRENCE | | 1985 | 2001 | 1960 | 1976 | 1978 | 2008 | 2000 | 1986 | 1941 | 1959 | 1998 | 1982 | AUG 1986 |
| | NORMAL NO. DAYS WITH: | | | | | | | | | | | | | | |
| | PRECIPITATION >= 0.01 | 30 | 12.3 | 10.1 | 11.9 | 12.8 | 10.9 | 10.7 | 10.2 | 9.9 | 9.1 | 9.6 | 11.4 | 11.7 | 130.6 |
| | PRECIPITATION >= 1.00 | 30 | 0.2 | 0.2 | 0.3 | 0.9 | 0.7 | 0.8 | 0.9 | 0.8 | 1.0 | 0.4 | 0.5 | 0.4 | 7.1 |
| **SNOWFALL** | NORMAL (IN) | 30 | 15.2 | 11.3 | 7.4 | 2.6 | 0.1 | 0.0 | 0.0 | 0.0 | 0.0 | 0.4 | 3.7 | 11.7 | 52.4 |
| | MAXIMUM MONTHLY (IN) | 69 | 39.0 | 42.0 | 26.7 | 15.8 | 3.2 | .4 | T | T | T | 6.3 | 16.1 | 49.5 | 49.5 |
| | YEAR OF OCCURRENCE | | 1999 | 1974 | 1965 | 1973 | 1990 | 2005 | 1990 | 1989 | 2007 | 1989 | 1977 | 2000 | DEC 2000 |
| | MAXIMUM IN 24 HOURS (IN) | 69 | 13.8 | 16.7 | 14.2 | 11.6 | 3.2 | .4 | T | T | T | 6.3 | 10.6 | 13.6 | 16.7 |
| | YEAR OF OCCURRENCE' | | 1990 | 1960 | 2009 | 1973 | 1990 | 2005 | 1990 | 1989 | 1993 | 1989 | 1977 | 2000 | FEB 1960 |
| | MAXIMUM SNOW DEPTH (IN) | 61 | 33 | 29 | 24 | 13 | 2 | 0 | 0 | 0 | 0 | 6 | 11 | 32 | 33 |
| | YEAR OF OCCURRENCE | | 1979 | 1979 | 1960 | 1973 | 1990 | | | | | 1989 | 1977 | 2000 | JAN 1979 |
| | NORMAL NO. DAYS WITH: | | | | | | | | | | | | | | |
| | SNOWFALL >= 1.0 | 30 | 4.5 | 2.9 | 2.1 | 0.5 | 0.0 | 0.0 | 0.0 | 0.0 | 0.0 | 0.1 | 0.9 | 3.3 | 14.3 |

## PRECIPITATION (inches) 2009  MILWAUKEE (KMKE)

| YEAR | JAN | FEB | MAR | APR | MAY | JUN | JUL | AUG | SEP | OCT | NOV | DEC | ANNUAL |
|---|---|---|---|---|---|---|---|---|---|---|---|---|---|
| 1980 | 1.65 | 1.75 | 0.77 | 4.02 | 1.81 | 4.67 | 3.39 | 5.06 | 3.57 | 1.63 | 1.57 | 3.52 | 33.41 |
| 1981 | 0.31 | 2.88 | 0.51 | 4.87 | 3.05 | 2.39 | 4.35 | 4.26 | 5.47 | 2.71 | 2.05 | 1.03 | 33.88 |
| 1982 | 2.92 | 0.29 | 3.20 | 4.47 | 2.76 | 3.06 | 3.88 | 3.33 | 0.64 | 3.17 | 4.74 | 4.10 | 36.56 |
| 1983 | 0.75 | 2.23 | 4.12 | 4.66 | 5.83 | 1.41 | 1.34 | 4.70 | 2.79 | 2.65 | 4.10 | 2.89 | 37.47 |
| 1984 | 0.79 | 1.20 | 2.17 | 5.04 | 4.21 | 4.07 | 3.39 | 2.93 | 2.51 | 5.30 | 3.74 | 4.25 | 39.60 |
| 1985 | 1.94 | 2.34 | 4.11 | 1.93 | 2.73 | 1.27 | 2.18 | 2.23 | 3.44 | 5.39 | 7.11 | 2.62 | 37.29 |
| 1986 | 0.91 | 3.94 | 1.85 | 1.83 | 2.74 | 4.51 | 6.15 | 8.82 | 7.26 | 2.24 | 0.89 | 1.03 | 42.17 |
| 1987 | 1.22 | 1.22 | 1.74 | 4.26 | 3.76 | 2.23 | 4.20 | 9.05 | 2.22 | 1.09 | 2.73 | 5.42 | 39.14 |
| 1988 | 3.25 | 1.29 | 1.30 | 4.12 | 0.50 | 0.70 | 1.53 | 3.25 | 4.94 | 2.97 | 5.15 | 1.43 | 30.43 |
| 1989 | 0.86 | 0.69 | 3.03 | 1.33 | 2.86 | 1.89 | 6.16 | 5.19 | 3.25 | 2.67 | 1.90 | 0.47 | 30.30 |
| 1990 | 2.57 | 1.90 | 2.75 | 2.67 | 7.56 | 4.97 | 3.02 | 4.68 | 1.89 | 2.65 | 3.54 | 2.66 | 40.86 |
| 1991 | 1.55 | 0.38 | 4.06 | 3.70 | 4.25 | 2.13 | 4.34 | 2.27 | 4.34 | 7.03 | 3.36 | 1.94 | 39.35 |
| 1992 | 1.09 | 1.54 | 2.61 | 2.41 | 0.60 | 3.13 | 5.64 | 3.50 | 4.13 | 1.45 | 5.40 | 2.45 | 33.95 |
| 1993 | 2.63 | 0.98 | 3.19 | 6.64 | 1.56 | 6.39 | 4.22 | 4.20 | 3.91 | 0.44 | 1.98 | 0.70 | 36.84 |
| 1994 | 2.20 | 3.52 | 1.21 | 2.35 | 0.67 | 3.08 | 2.51 | 4.91 | 1.68 | 0.78 | 3.31 | 1.14 | 27.36 |
| 1995 | 2.14 | 0.25 | 1.76 | 3.86 | 3.41 | 1.46 | 2.80 | 5.83 | 1.24 | 4.64 | 3.42 | 0.53 | 31.34 |
| 1996 | 1.66 | 0.52 | 0.76 | 2.99 | 2.89 | 5.47 | 1.61 | 1.24 | 1.82 | 3.00 | .63 | 1.53 | 24.12 |
| 1997 | 1.59 | 2.47 | 0.63 | 2.16 | 1.95 | 9.98 | 3.59 | 3.95 | 2.91 | 1.11 | 1.11 | 1.30 | 32.75 |
| 1998 | 3.60 | 2.19 | 3.18 | 4.18 | 2.48 | 2.82 | 1.78 | 5.98 | 2.17 | 2.47 | 2.91 | 0.88 | 34.64 |
| 1999 | 4.38 | 0.98 | 1.35 | 6.14 | 3.74 | 6.96 | 5.58 | 1.69 | 4.16 | 0.94 | 0.70 | 1.26 | 37.88 |
| 2000 | 1.20 | 1.66 | 1.12 | 3.64 | 8.42 | 3.42 | 7.12 | 5.17 | 7.04 | 0.84 | 2.33 | 2.41 | 44.37 |
| 2001 | 1.11 | 3.48 | 0.67 | 3.45 | 4.68 | 4.13 | 2.70 | 5.41 | 4.76 | 4.29 | 1.19 | 0.86 | 36.73 |
| 2002 | 1.44 | 1.46 | 1.76 | 3.59 | 2.31 | 2.99 | 2.33 | 4.73 | 2.79 | 1.66 | 0.88 | 0.75 | 26.69 |
| 2003 | 0.35 | 0.43 | 1.64 | 2.61 | 3.65 | 1.49 | 2.43 | 0.57 | 1.65 | 1.51 | 3.94 | 2.03 | 22.30 |
| 2004 | 1.43 | 1.10 | 3.99 | 1.87 | 8.18 | 4.07 | 3.25 | 3.43 | 0.24 | 1.47 | 2.38 | 1.53 | 32.94 |
| 2005 | 3.31 | 1.79 | 0.72 | 1.41 | 2.62 | 2.23 | 2.60 | 1.29 | 4.17 | 0.95 | 3.65 | 1.18 | 25.92 |
| 2006 | 2.92 | 0.91 | 3.69 | 4.23 | 3.73 | 2.54 | 5.23 | 2.18 | 3.57 | 3.30 | 2.72 | 2.91 | 37.93 |
| 2007 | 0.86 | 1.36 | 3.21 | 4.02 | 1.99 | 3.64 | 1.40 | 7.92 | 1.93 | 2.96 | 0.36 | 3.41 | 33.06 |
| 2008 | 2.07 | 3.32 | 3.11 | 4.42 | 2.92 | 12.27 | 3.20 | 0.88 | 4.16 | 2.62 | 1.47 | 4.00 | 44.44 |
| 2009 | 0.97 | 2.29 | 3.68 | 4.50 | 2.56 | 5.44 | 0.71 | 4.04 | 1.57 | 5.57 | 1.80 | 2.68 | 35.81 |
| POR= 82 YRS | 1.70 | 1.42 | 2.33 | 3.14 | 3.06 | 3.59 | 3.22 | 3.27 | 3.08 | 2.27 | 2.28 | 1.89 | 31.25 |

WBAN : 14839

## AVERAGE TEMPERATURE (°F) 2009  MILWAUKEE (KMKE)

| YEAR | JAN | FEB | MAR | APR | MAY | JUN | JUL | AUG | SEP | OCT | NOV | DEC | ANNUAL |
|---|---|---|---|---|---|---|---|---|---|---|---|---|---|
| 1980 | 20.7 | 20.3 | 30.5 | 45.3 | 57.2 | 61.3 | 71.2 | 69.7 | 61.1 | 45.7 | 37.7 | 24.4 | 45.4 |
| 1981 | 18.9 | 25.3 | 35.6 | 46.5 | 51.5 | 65.2 | 67.3 | 67.8 | 59.1 | 45.8 | 37.4 | 24.2 | 45.4 |
| 1982 | 9.7 | 19.4 | 31.5 | 41.2 | 58.5 | 59.8 | 71.1 | 67.3 | 60.8 | 52.7 | 38.0 | 33.2 | 45.3 |
| 1983 | 26.4 | 29.4 | 35.0 | 41.7 | 50.2 | 66.3 | 76.2 | 74.4 | 62.9 | 52.0 | 39.9 | 14.4 | 47.4 |
| 1984 | 18.9 | 33.4 | 29.2 | 45.5 | 54.9 | 68.7 | 71.7 | 73.3 | 60.9 | 52.9 | 37.7 | 29.1 | 48.0 |
| 1985 | 15.2 | 21.3 | 37.9 | 50.8 | 58.8 | 63.8 | 72.4 | 68.4 | 64.3 | 50.7 | 36.7 | 15.7 | 46.3 |
| 1986 | 21.9 | 23.3 | 38.1 | 48.5 | 56.1 | 63.3 | 72.5 | 67.1 | 63.8 | 51.6 | 35.1 | 29.2 | 47.5 |
| 1987 | 24.8 | 32.0 | 37.8 | 48.0 | 60.1 | 72.2 | 74.8 | 69.9 | 63.3 | 46.2 | 41.9 | 31.4 | 50.2 |
| 1988 | 18.3 | 20.1 | 34.8 | 45.7 | 58.7 | 70.2 | 75.4 | 75.7 | 63.5 | 45.8 | 40.7 | 26.6 | 48.0 |
| 1989 | 30.4 | 18.0 | 32.4 | 43.2 | 54.9 | 64.4 | 71.6 | 68.8 | 60.2 | 52.7 | 35.1 | 16.7 | 45.7 |
| 1990 | 31.1 | 28.9 | 38.8 | 49.3 | 52.8 | 67.6 | 70.5 | 71.2 | 66.0 | 51.5 | 44.5 | 26.9 | 49.9 |
| 1991 | 20.1 | 29.9 | 38.2 | 49.7 | 63.2 | 70.7 | 74.4 | 73.5 | 63.2 | 52.3 | 35.1 | 29.6 | 50.0 |
| 1992 | 27.9 | 31.9 | 35.4 | 42.9 | 56.7 | 63.0 | 67.7 | 67.4 | 61.6 | 49.8 | 37.3 | 28.2 | 47.5 |
| 1993 | 25.7 | 24.4 | 32.5 | 43.0 | 56.9 | 63.8 | 73.0 | 73.6 | 60.6 | 51.0 | 39.5 | 29.9 | 47.8 |
| 1994 | 14.8 | 21.6 | 35.8 | 48.1 | 57.0 | 70.2 | 73.7 | 70.1 | 67.5 | 55.9 | 43.7 | 34.5 | 49.4 |
| 1995 | 25.3 | 25.5 | 38.7 | 44.0 | 57.8 | 71.4 | 74.4 | 75.7 | 61.1 | 52.8 | 31.1 | 24.1 | 48.5 |
| 1996 | 20.7 | 24.0 | 29.8 | 42.4 | 52.0 | 65.2 | 68.7 | 72.4 | 63.8 | 52.3 | 33.1 | 27.1 | 46.0 |
| 1997 | 20.4 | 28.2 | 35.9 | 43.6 | 49.6 | 64.8 | 69.2 | 66.5 | 61.4 | 51.6 | 34.8 | 30.4 | 46.4 |
| 1998 | 26.9 | 34.6 | 35.9 | 45.8 | 59.9 | 66.7 | 72.4 | 72.5 | 67.0 | 54.0 | 43.0 | 32.0 | 50.9 |
| 1999 | 20.2 | 32.0 | 34.6 | 46.2 | 58.1 | 67.2 | 76.5 | 68.8 | 63.4 | 52.3 | 44.8 | 29.3 | 49.5 |
| 2000 | 23.6 | 32.2 | 41.9 | 44.3 | 58.0 | 65.9 | 68.2 | 70.7 | 62.2 | 54.7 | 36.9 | 16.6 | 47.9 |
| 2001 | 24.8 | 23.5 | 32.5 | 49.0 | 56.8 | 66.2 | 72.4 | 72.9 | 61.2 | 51.2 | 47.5 | 32.9 | 49.2 |
| 2002 | 30.7 | 31.7 | 31.5 | 46.5 | 52.3 | 68.5 | 75.6 | 71.3 | 66.2 | 48.2 | 37.9 | 30.1 | 49.2 |
| 2003 | 20.3 | 21.5 | 33.9 | 43.2 | 52.4 | 63.0 | 70.5 | 73.3 | 62.9 | 51.2 | 40.0 | 30.8 | 46.9 |
| 2004 | 17.9 | 26.5 | 39.0 | 46.9 | 54.8 | 63.7 | 69.1 | 66.5 | 65.7 | 52.9 | 42.3 | 27.4 | 47.7 |
| 2005 | 22.3 | 30.1 | 32.6 | 47.8 | 53.9 | 71.5 | 72.5 | 73.4 | 68.6 | 54.1 | 40.4 | 23.1 | 49.2 |
| 2006 | 34.0 | 25.8 | 36.2 | 49.9 | 57.4 | 66.3 | 75.0 | 71.9 | 61.9 | 47.9 | 41.9 | 32.6 | 50.1 |
| 2007 | 26.8 | 17.7 | 39.8 | 44.5 | 58.8 | 68.0 | 71.6 | 72.3 | 65.3 | 58.1 | 38.0 | 26.3 | 48.9 |
| 2008 | 22.4 | 22.0 | 32.5 | 46.3 | 53.1 | 66.9 | 71.3 | 71.0 | 65.2 | 52.1 | 38.8 | 22.4 | 47.0 |
| 2009 | 15.8 | 27.1 | 36.1 | 45.1 | 57.2 | 65.1 | 68.6 | 68.8 | 63.9 | 48.3 | 44.9 | 26.4 | 47.3 |
| POR= 68 YRS | 20.9 | 24.3 | 33.7 | 44.9 | 55.1 | 65.5 | 71.2 | 70.3 | 62.3 | 51.4 | 38.0 | 25.8 | 47.0 |

## HEATING DEGREE DAYS (base 65°F) 2009 MILWAUKEE (KMKE)

| YEAR | JUL | AUG | SEP | OCT | NOV | DEC | JAN | FEB | MAR | APR | MAY | JUN | TOTAL |
|---|---|---|---|---|---|---|---|---|---|---|---|---|---|
| 1980-81 | 8 | 9 | 140 | 590 | 812 | 1250 | 1423 | 1106 | 905 | 548 | 417 | 69 | 7277 |
| 1981-82 | 44 | 21 | 187 | 590 | 820 | 1257 | 1712 | 1272 | 1032 | 707 | 215 | 172 | 8029 |
| 1982-83 | 3 | 44 | 170 | 381 | 802 | 983 | 1186 | 990 | 925 | 692 | 453 | 81 | 6710 |
| 1983-84 | 10 | 0 | 148 | 405 | 748 | 1565 | 1424 | 910 | 1103 | 579 | 318 | 35 | 7245 |
| 1984-85 | 3 | 7 | 179 | 373 | 812 | 1103 | 1542 | 1215 | 831 | 461 | 222 | 96 | 6844 |
| 1985-86 | 2 | 13 | 139 | 436 | 843 | 1523 | 1328 | 1161 | 827 | 494 | 302 | 128 | 7196 |
| 1986-87 | 13 | 34 | 98 | 407 | 891 | 1106 | 1242 | 917 | 839 | 502 | 236 | 19 | 6304 |
| 1987-88 | 12 | 28 | 91 | 576 | 686 | 1037 | 1442 | 1294 | 930 | 571 | 245 | 55 | 6967 |
| 1988-89 | 3 | 7 | 87 | 587 | 720 | 1183 | 1065 | 1307 | 1006 | 649 | 324 | 85 | 7023 |
| 1989-90 | 0 | 16 | 166 | 381 | 890 | 1493 | 1040 | 1004 | 805 | 502 | 375 | 51 | 6723 |
| 1990-91 | 21 | 9 | 93 | 418 | 612 | 1173 | 1385 | 980 | 822 | 467 | 201 | 26 | 6207 |
| 1991-92 | 0 | 1 | 160 | 394 | 889 | 1091 | 1141 | 955 | 914 | 655 | 278 | 104 | 6582 |
| 1992-93 | 27 | 35 | 145 | 472 | 825 | 1132 | 1211 | 1132 | 1004 | 654 | 245 | 104 | 6986 |
| 1993-94 | 1 | 2 | 158 | 428 | 757 | 1080 | 1553 | 1211 | 899 | 502 | 286 | 55 | 6932 |
| 1994-95 | 0 | 22 | 49 | 284 | 633 | 937 | 1222 | 1096 | 807 | 623 | 230 | 31 | 5934 |
| 1995-96 | 1 | 0 | 171 | 379 | 1013 | 1260 | 1365 | 1180 | 1084 | 672 | 431 | 102 | 7658 |
| 1996-97 | 17 | 0 | 103 | 398 | 950 | 1167 | 1376 | 1026 | 896 | 635 | 470 | 100 | 7138 |
| 1997-98 | 23 | 37 | 129 | 441 | 899 | 1069 | 1176 | 847 | 897 | 571 | 203 | 100 | 6392 |
| 1998-99 | 0 | 0 | 47 | 337 | 653 | 1017 | 1380 | 919 | 936 | 558 | 239 | 78 | 6164 |
| 1999-00 | 0 | 9 | 113 | 388 | 598 | 1100 | 1275 | 945 | 711 | 616 | 252 | 84 | 6091 |
| 2000-01 | 25 | 8 | 151 | 323 | 836 | 1491 | 1239 | 1155 | 1001 | 481 | 267 | 99 | 7076 |
| 2001-02 | 16 | 0 | 146 | 420 | 522 | 992 | 1060 | 927 | 1031 | 580 | 410 | 74 | 6178 |
| 2002-03 | 0 | 0 | 67 | 518 | 806 | 1078 | 1379 | 1212 | 956 | 646 | 387 | 125 | 7174 |
| 2003-04 | 6 | 0 | 128 | 427 | 742 | 1055 | 1453 | 1111 | 798 | 539 | 324 | 100 | 6683 |
| 2004-05 | 19 | 48 | 66 | 369 | 675 | 1161 | 1318 | 973 | 997 | 518 | 344 | 29 | 6517 |
| 2005-06 | 5 | 2 | 46 | 370 | 733 | 1293 | 955 | 1092 | 888 | 451 | 270 | 50 | 6155 |
| 2006-07 | 3 | 0 | 125 | 527 | 686 | 996 | 1175 | 1317 | 779 | 612 | 232 | 36 | 6488 |
| 2007-08 | 7 | 3 | 98 | 252 | 803 | 1194 | 1313 | 1240 | 1000 | 554 | 359 | 49 | 6872 |
| 2008-09 | 11 | 2 | 58 | 396 | 780 | 1311 | 1517 | 1051 | 890 | 589 | 249 | 108 | 6962 |
| 2009- | 15 | 37 | 72 | 512 | 598 | 1187 | | | | | | | |

WBAN : 14839

## COOLING DEGREE DAYS (base 65°F) 2009 MILWAUKEE (KMKE)

| YEAR | JAN | FEB | MAR | APR | MAY | JUN | JUL | AUG | SEP | OCT | NOV | DEC | TOTAL |
|---|---|---|---|---|---|---|---|---|---|---|---|---|---|
| 1980 | 0 | 0 | 0 | 9 | 25 | 50 | 207 | 164 | 29 | 0 | 0 | 0 | 484 |
| 1981 | 0 | 0 | 0 | 2 | 3 | 84 | 121 | 112 | 16 | 0 | 0 | 0 | 338 |
| 1982 | 0 | 0 | 0 | 0 | 21 | 24 | 199 | 121 | 51 | 5 | 0 | 0 | 421 |
| 1983 | 0 | 0 | 0 | 0 | 0 | 127 | 364 | 299 | 92 | 9 | 0 | 0 | 891 |
| 1984 | 0 | 0 | 0 | 1 | 11 | 152 | 216 | 270 | 63 | 3 | 0 | 0 | 716 |
| 1985 | 0 | 0 | 0 | 42 | 35 | 68 | 240 | 127 | 127 | 0 | 0 | 0 | 639 |
| 1986 | 0 | 0 | 3 | 7 | 31 | 84 | 251 | 105 | 70 | 0 | 0 | 0 | 551 |
| 1987 | 0 | 0 | 0 | 2 | 87 | 244 | 323 | 189 | 47 | 0 | 1 | 0 | 893 |
| 1988 | 0 | 0 | 0 | 0 | 57 | 215 | 333 | 344 | 48 | 0 | 0 | 0 | 997 |
| 1989 | 0 | 0 | 2 | 0 | 16 | 76 | 214 | 144 | 29 | 4 | 0 | 0 | 485 |
| 1990 | 0 | 0 | 2 | 38 | 5 | 135 | 198 | 210 | 132 | 7 | 1 | 0 | 728 |
| 1991 | 0 | 0 | 0 | 13 | 150 | 204 | 300 | 268 | 114 | 7 | 0 | 0 | 1056 |
| 1992 | 0 | 0 | 0 | 0 | 25 | 49 | 117 | 119 | 50 | 4 | 0 | 0 | 364 |
| 1993 | 0 | 0 | 0 | 0 | 5 | 73 | 258 | 277 | 35 | 3 | 0 | 0 | 651 |
| 1994 | 0 | 0 | 0 | 5 | 46 | 218 | 278 | 187 | 133 | 10 | 0 | 0 | 877 |
| 1995 | 0 | 0 | 0 | 0 | 16 | 232 | 297 | 338 | 61 | 8 | 0 | 0 | 952 |
| 1996 | 0 | 0 | 0 | 0 | 37 | 112 | 140 | 237 | 75 | 7 | 0 | 0 | 608 |
| 1997 | 0 | 0 | 0 | 0 | 0 | 103 | 158 | 89 | 26 | 31 | 0 | 0 | 407 |
| 1998 | 0 | 0 | 3 | 0 | 51 | 157 | 235 | 238 | 114 | 2 | 0 | 0 | 800 |
| 1999 | 0 | 0 | 0 | 0 | 31 | 151 | 363 | 136 | 71 | 0 | 1 | 0 | 753 |
| 2000 | 0 | 0 | 1 | 0 | 44 | 116 | 131 | 188 | 76 | 10 | 0 | 0 | 566 |
| 2001 | 0 | 0 | 0 | 5 | 20 | 143 | 252 | 250 | 39 | 2 | 0 | 0 | 711 |
| 2002 | 0 | 0 | 0 | 30 | 24 | 187 | 336 | 202 | 109 | 9 | 0 | 0 | 897 |
| 2003 | 0 | 0 | 0 | 1 | 0 | 74 | 187 | 265 | 73 | 6 | 0 | 0 | 606 |
| 2004 | 0 | 0 | 0 | 6 | 15 | 70 | 151 | 105 | 93 | 2 | 0 | 0 | 442 |
| 2005 | 0 | 0 | 0 | 4 | 4 | 229 | 244 | 269 | 162 | 39 | 0 | 0 | 951 |
| 2006 | 0 | 0 | 0 | 3 | 42 | 98 | 316 | 222 | 39 | 3 | 0 | 0 | 723 |
| 2007 | 0 | 0 | 7 | 3 | 46 | 132 | 219 | 235 | 113 | 45 | 0 | 0 | 800 |
| 2008 | 0 | 0 | 0 | 0 | 0 | 109 | 214 | 195 | 69 | 6 | 0 | 0 | 593 |
| 2009 | 0 | 0 | 0 | 0 | 15 | 119 | 134 | 162 | 45 | 0 | 0 | 0 | 475 |

### SNOWFALL (inches) 2009 MILWAUKEE (KMKE)

| YEAR | JUL | AUG | SEP | OCT | NOV | DEC | JAN | FEB | MAR | APR | MAY | JUN | TOTAL |
|------|-----|-----|-----|-----|-----|-----|-----|-----|-----|-----|-----|-----|-------|
| 1980-81 | 0.0 | 0.0 | 0.0 | T | 2.3 | 17.5 | 4.9 | 15.7 | 1.5 | T | 0.0 | 0.0 | 41.9 |
| 1981-82 | 0.0 | 0.0 | 0.0 | T | 2.0 | 8.3 | 29.2 | 3.0 | 13.0 | 11.7 | 0.0 | 0.0 | 67.2 |
| 1982-83 | 0.0 | 0.0 | 0.0 | T | 0.4 | 3.1 | 6.3 | 13.5 | 13.8 | 1.0 | 0.0 | 0.0 | 38.1 |
| 1983-84 | 0.0 | 0.0 | 0.0 | 0.0 | 0.3 | 13.3 | 9.6 | 1.2 | 8.2 | 0.5 | T | 0.0 | 33.1 |
| 1984-85 | 0.0 | 0.0 | 0.0 | 0.0 | T | 19.0 | 20.8 | 15.3 | 9.0 | 2.5 | 0.0 | 0.0 | 66.6 |
| 1985-86 | 0.0 | 0.0 | 0.0 | 0.0 | 3.5 | 13.5 | 10.4 | 14.0 | 0.7 | 0.3 | 0.0 | 0.0 | 42.4 |
| 1986-87 | 0.0 | 0.0 | 0.0 | T | 2.4 | 2.5 | 11.4 | T | 5.2 | 0.4 | 0.0 | 0.0 | 21.9 |
| 1987-88 | 0.0 | 0.0 | 0.0 | 0.6 | 0.4 | 19.9 | 10.2 | 20.7 | 2.9 | T | 0.0 | 0.0 | 54.7 |
| 1988-89 | 0.0 | 0.0 | 0.0 | T | 2.7 | 7.1 | 2.7 | 13.1 | 13.3 | 0.4 | 0.6 | 0.0 | 39.9 |
| 1989-90 | T | T | 0.0 | 6.3 | 11.6 | 7.4 | 19.9 | 17.9 | 0.2 | 1.2 | 3.2 | 0.0 | 67.7 |
| 1990-91 | T | 0.0 | 0.0 | 0.0 | 0.4 | 10.5 | 15.2 | 2.4 | 1.1 | 0.4 | 0.0 | 0.0 | 30.0 |
| 1991-92 | 0.0 | 0.0 | T | T | 4.8 | 14.7 | 4.3 | 5.3 | 11.1 | 0.8 | 0.0 | T | 41.0 |
| 1992-93 | 0.0 | 0.0 | 0.0 | 1.2 | 0.4 | 8.4 | 12.2 | 11.6 | 12.7 | 3.4 | T | 0.0 | 49.9 |
| 1993-94 | 0.0 | 0.0 | T | T | 8.0 | 1.2 | 27.0 | 38.7 | 3.9 | 3.1 | T | 0.0 | 81.9 |
| 1994-95 | 0.0 | 0.0 | 0.0 | 0.0 | 0.2 | 10.4 | 15.2 | 2.1 | 7.6 | 0.2 | 0.0 | 0.0 | 35.7 |
| 1995-96 | 0.0 | 0.0 | 0.0 | T | 14.9 | 6.2 | 22.7 | 0.6 | 2.9 | 4.2 | 0.0 | 0.0 | 51.5 |
| 1996-97 | 0.0 | 0.0 | 0.0 | 0.0 | 1.8 | 9.2 | 23.6 | 10.7 | 0.5 | 5.4 | 0.0 | T | 51.2 |
| 1997-98 | 0.0 | T | 0.0 | T | 1.1 | 10.6 | 23.7 | 0.5 | 3.7 | T | 0.0 | 0.0 | 39.6 |
| 1998-99 | 0.0 | 0.0 | 0.0 | 0.0 | 0.3 | 3.4 | 39.0 | 4.4 | 13.6 | 0.0 | 0.0 | T | 60.7 |
| 1999-00 | 0.0 | 0.0 | 0.0 | 0.0 | 0.0 | 2.3 | 15.2 | 12.1 | 1.0 | 7.0 | T | 0.0 | 37.6 |
| 2000-01 | 0.0 | 0.0 | 0.0 | 0.2 | 2.8 | 49.5 | 1.3 | 4.2 | 0.8 | 0.5 | T | 0.0 | 59.3 |
| 2001-02 | 0.0 | 0.0 | 0.0 | T | 0.0 | 2.6 | 13.1 | 4.2 | 13.0 | 3.6 | 0.0 | T | 36.5 |
| 2002-03 | 0.0 | 0.0 | 0.0 | 0.0 | 1.2 | 4.7 | 3.7 | 6.1 | 14.7 | 4.3 | 0.0 | 0.0 | 34.7 |
| 2003-04 | 0.0 | 0.0 | 0.0 | 0.0 | 0.7 | 3.1 | 22.7 | 9.2 | 3.4 | 0.1 | 0.0 | T | 39.2 |
| 2004-05 | 0.0 | 0.0 | 0.0 | 0.0 | 2.6 | 1.3 | 29.1 | 8.8 | 6.4 | T | 0.0 | 0.4 | 48.6 |
| 2005-06 | 0.0 | 0.0 | 0.0 | 0.0 | 4.6 | 12.5 | 6.6 | 6.7 | 7.5 | T | T | 0.0 | 37.9 |
| 2006-07 | 0.0 | 0.0 | 0.0 | 0.1 | T | 11.0 | 11.9 | 23.7 | 4.4 | 7.0 | T | 0.0 | 58.1 |
| 2007-08 | 0.0 | 0.0 | T | 0.0 | 1.3 | 29.5 | 18.4 | 31.0 | 18.7 | 0.2 | 0.0 | 0.0 | 99.1 |
| 2008-09 | 0.0 | 0.0 | 0.0 | T | 3.0 | 35.3 | 9.9 | 7.9 | 19.8 | 0.1 | 0.0 | 0.0 | 76.0 |
| 2009- | 0.0 | 0.0 | 0.0 | 0.0 | 0.4 | 11.7 | | | | | | | |
| POR=<br>62 YRS | T | T | T | 0.2 | 2.8 | 11.3 | 14.2 | 10.3 | 8.8 | 2.0 | 0.1 | 0.1 | 49.8 |

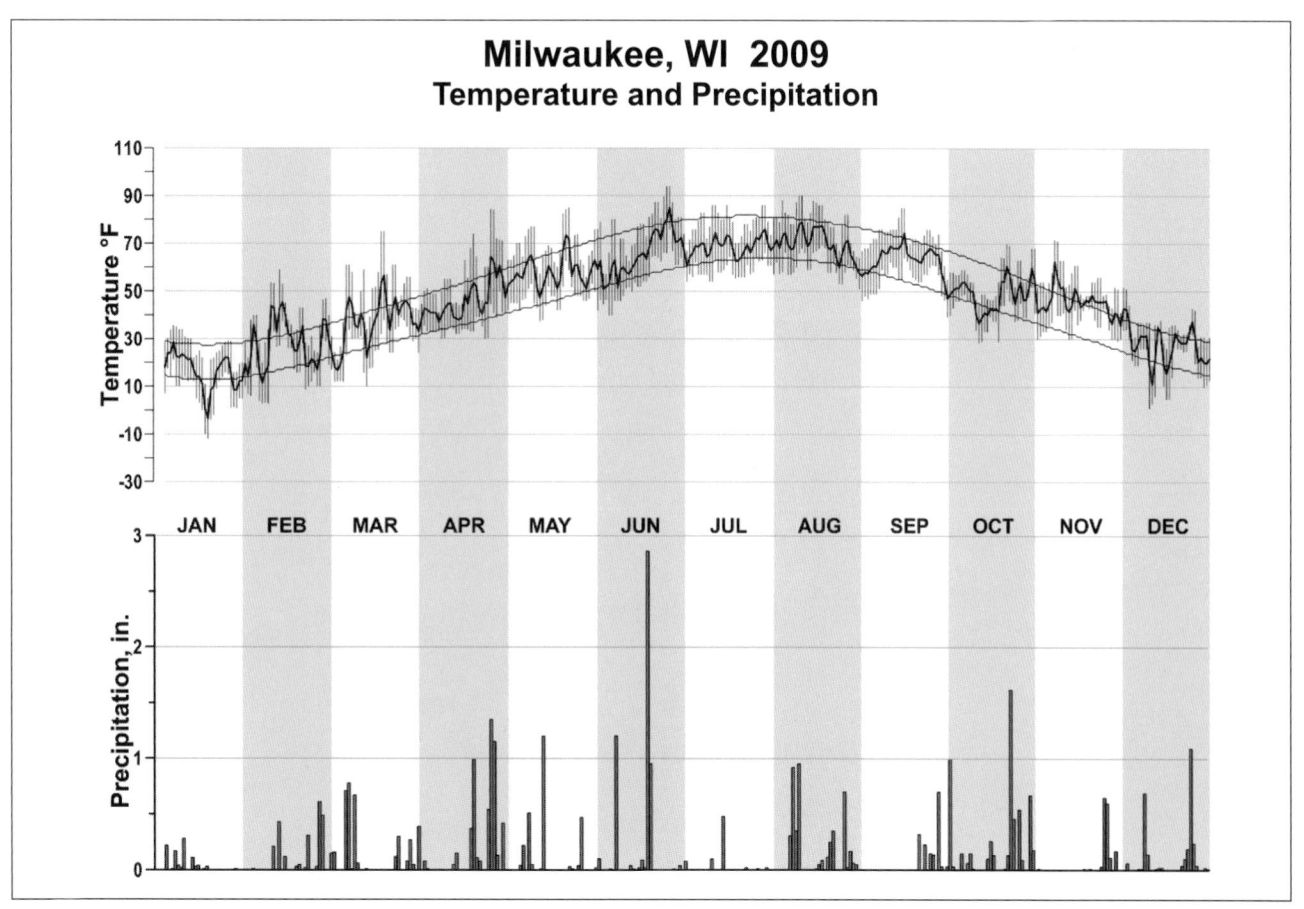

## Milwaukee, WI 2009
### Temperature and Precipitation

# 2009
# CHEYENNE
# WYOMING (KCYS)

The city of Cheyenne is located on a broad plateau between the North and South Platte Rivers in the extreme southeastern corner of Wyoming at an elevation of approximately 6,100 feet. The surrounding country is mostly rolling prairie which is used primarily for grazing. The ground level rises rapidly to a ridge approximately 9,000 feet in elevation about 30 miles west of the city. This ridge is known as the Laramie Mountains, one of the ranges of the Rockies, and extends in a north-south direction. Because of this ridge, winds from the northwest through west to southwest are downslope and produce a marked chinook effect in Cheyenne which is especially noticeable during the winter months. Also, winds from the north through east to south are upslope and may cause fog or low stratus clouds in the Cheyenne area throughout the year. Because of this terrain variation, the wind direction plays an important role in controlling the local temperature and weather.

Cheyenne experiences large diurnal and annual temperature ranges. This is due to the advent of both warm and cold air masses and the relatively high elevation of the city which permits rapid incoming and outgoing radiation. The daily temperature range averages about 30 degrees in the summer and 23 degrees in the winter. Many cold air masses from the north during the winter months miss Cheyenne. Because of the downslope of land to the east and the prevailing westerlies, some of the cold air masses do move over the city, but only about 13 percent of the days in an average January, the coldest month of the year, show temperatures dropping to zero or below. Temperatures during the winter months average a few degrees higher than over the Mississippi and Missouri Valleys at the same latitude.

Windy days are quire frequent during the winter and spring months. Since the wind is usually strongest during the daytime it is a very noticeable weather element. Usually the strong winds are from a westerly direction and this tends to raise the temperature because the air is moving downslope.

Most of the air masses reaching this area move in from the Pacific and since the mountains to the west are quite effective moisture barriers the climate is semi-arid. Fortunately, about 70 percent of normal annual precipitation occurs during the growing season. In the summer months, precipitation is mostly of the shower type and occurs mainly with thunderstorms. Hail is frequent and occasionally destructive in some thunderstorms. Most of the snow falls during the late winter and early spring months. It is not uncommon to have heavy snow in May.

The growing season in Cheyenne averages about 132 days a year and extends from around May 18th to September 27th. Freezing temperatures have occurred as late in the spring as mid-June, and as early in the fall as late August.

Relative humidity averages near 50 percent on an annual basis with large daily variations. Very seldom is the relative humidity above 30 percent when the temperature is above 80 degrees.

# NORMALS, MEANS, AND EXTREMES
## CHEYENNE (KCYS)

**LATITUDE:** 41° 9 'N  **LONGITUDE:** -104° 48'W  **ELEVATION (FT):** GRND: 6115  BARO: 6128  **TIME ZONE:** MOUNTAIN  (UTC -7)  **WBAN: 24018**

| | ELEMENT | POR | JAN | FEB | MAR | APR | MAY | JUN | JUL | AUG | SEP | OCT | NOV | DEC | YEAR |
|---|---|---|---|---|---|---|---|---|---|---|---|---|---|---|---|
| **TEMPERATURE °F** | NORMAL DAILY MAXIMUM | 30 | 37.1 | 40.5 | 46.4 | 54.4 | 64.4 | 75.4 | 81.9 | 79.8 | 70.3 | 58.2 | 44.5 | 38.1 | 57.6 |
| | MEAN DAILY MAXIMUM | 44 | 37.7 | 37.2 | 44.9 | 52.0 | 63.1 | 72.2 | 82.9 | 80.1 | 69.4 | 59.1 | 45.7 | 39.1 | 57.0 |
| | HIGHEST DAILY MAXIMUM | 74 | 66 | 71 | 74 | 83 | 91 | 100 | 100 | 98 | 95 | 83 | 75 | 69 | 100 |
| | YEAR OF OCCURRENCE | | 2005 | 1962 | 2004 | 1992 | 2003 | 1954 | 1939 | 2008 | 1995 | 1992 | 1999 | 1939 | JUN 1954 |
| | MEAN OF EXTREME MAXS. | 44 | 58.1 | 60.5 | 68.1 | 75.3 | 84.4 | 90.1 | 95.4 | 93.9 | 86.6 | 77.5 | 67.8 | 57.2 | 76.2 |
| | NORMAL DAILY MINIMUM | 30 | 14.8 | 17.2 | 22.0 | 28.7 | 38.3 | 47.5 | 53.4 | 52.0 | 42.9 | 32.5 | 22.1 | 16.1 | 32.3 |
| | MEAN DAILY MINIMUM | 44 | 16.1 | 16.3 | 22.2 | 29.1 | 38.7 | 46.5 | 55.0 | 53.2 | 42.7 | 33.6 | 23.5 | 17.8 | 32.9 |
| | LOWEST DAILY MINIMUM | 74 | -29 | -34 | -21 | -8 | 16 | 25 | 38 | 36 | 8 | -1 | -16 | -28 | -34 |
| | YEAR OF OCCURRENCE | | 1984 | 1936 | 1943 | 1975 | 1947 | 1951 | 1952 | 1975 | 1985 | 1991 | 1993 | 1990 | FEB 1936 |
| | MEAN OF EXTREME MINS. | 44 | -3.6 | -2.3 | 8.5 | 16.4 | 26.1 | 36.3 | 48.8 | 45.2 | 31.4 | 18.0 | 3.1 | -3.1 | 18.7 |
| | NORMAL DRY BULB | 30 | 25.9 | 28.8 | 34.2 | 41.6 | 51.3 | 61.5 | 67.7 | 65.9 | 56.6 | 45.4 | 33.3 | 27.1 | 44.9 |
| | MEAN DRY BULB | 44 | 26.9 | 26.8 | 33.6 | 40.6 | 50.9 | 59.6 | 69.0 | 66.7 | 56.1 | 46.4 | 34.6 | 28.5 | 45.0 |
| | MEAN WET BULB | 11 | 19.3 | 20.9 | 24.8 | 31.8 | 39.8 | 46.6 | 52.6 | 52.0 | 43.4 | 34.6 | 24.5 | 19.3 | 34.1 |
| | MEAN DEW POINT | 11 | 17.2 | 16.8 | 22.9 | 28.6 | 37.4 | 44.8 | 49.3 | 48.3 | 40.2 | 30.9 | 22.1 | 16.3 | 31.2 |
| | NORMAL NO. DAYS WITH: | | | | | | | | | | | | | | |
| | MAXIMUM >= 90 | 30 | 0.0 | 0.0 | 0.0 | 0.0 | * | 1.0 | 4.9 | 2.0 | 0.4 | 0.0 | 0.0 | 0.0 | 8.3 |
| | MAXIMUM <= 32 | 30 | 9.1 | 6.7 | 4.1 | 1.6 | * | 0.0 | 0.0 | 0.0 | 0.2 | 0.8 | 5.2 | 8.9 | 36.6 |
| | MINIMUM <= 32 | 30 | 28.8 | 26.2 | 26.8 | 18.4 | 3.8 | * | 0.0 | 0.0 | 2.1 | 12.2 | 24.7 | 28.6 | 171.6 |
| | MINIMUM <= 0 | 30 | 4.2 | 2.5 | 0.6 | 0.1 | 0.0 | 0.0 | 0.0 | 0.0 | 0.0 | 0.1 | 0.8 | 3.2 | 11.5 |
| **H/C** | NORMAL HEATING DEG. DAYS | 30 | 1197 | 1001 | 939 | 686 | 414 | 136 | 29 | 42 | 255 | 593 | 936 | 1160 | 7388 |
| | NORMAL COOLING DEG. DAYS | 30 | 0 | 0 | 0 | 0 | 1 | 41 | 126 | 86 | 19 | 0 | 0 | 0 | 273 |
| **RH** | NORMAL (PERCENT) | 30 | 56 | 57 | 58 | 58 | 59 | 56 | 52 | 55 | 54 | 55 | 57 | 57 | 56 |
| | HOUR 05 LST | 30 | 60 | 63 | 67 | 70 | 75 | 73 | 71 | 72 | 68 | 64 | 62 | 60 | 67 |
| | HOUR 11 LST | 30 | 48 | 46 | 46 | 45 | 45 | 40 | 37 | 37 | 38 | 41 | 46 | 47 | 43 |
| | HOUR 17 LST | 30 | 52 | 49 | 46 | 44 | 46 | 41 | 38 | 40 | 39 | 42 | 53 | 53 | 45 |
| | HOUR 23 LST | 30 | 61 | 63 | 65 | 67 | 70 | 65 | 63 | 66 | 63 | 62 | 63 | 60 | 64 |
| **S** | PERCENT POSSIBLE SUNSHINE | 64 | 64 | 67 | 67 | 63 | 61 | 67 | 69 | 68 | 70 | 69 | 61 | 60 | 66 |
| **W/O** | MEAN NO. DAYS WITH: | | | | | | | | | | | | | | |
| | HEAVY FOG(VISBY <= 1/4 MI) | 11 | 1.6 | 2.5 | 3.9 | 4.1 | 2.0 | 1.4 | 0.9 | 1.2 | 0.8 | 2.7 | 1.5 | 1.8 | 24.4 |
| | THUNDERSTORMS | 11 | 0.0 | 0.1 | 0.2 | 1.5 | 4.3 | 8.0 | 9.5 | 7.4 | 4.0 | 0.7 | 0.2 | 0.0 | 35.9 |
| **CLOUDNESS** | MEAN: | | | | | | | | | | | | | | |
| | SUNRISE-SUNSET (OKTAS) | | | | | | | | | | | | | | |
| | MIDNIGHT-MIDNIGHT (OKTAS) | | | | | | | | | | | | | | |
| | MEAN NO. DAYS WITH: | | | | | | | | | | | | | | |
| | CLEAR | 1 | 4.0 | 7.0 | 5.0 | | 4.0 | 13.0 | | | | | | | |
| | PARTLY CLOUDY | 1 | 3.0 | 6.0 | 5.0 | | 12.0 | 8.0 | | | | | | | |
| | CLOUDY | 1 | 3.0 | 4.0 | 11.0 | | 13.0 | 7.0 | | | | | | | |
| **PR** | MEAN STATION PRESSURE(IN) | 11 | 23.91 | 23.88 | 23.87 | 23.88 | 23.93 | 23.98 | 24.07 | 24.07 | 24.03 | 23.98 | 23.96 | 23.89 | 23.95 |
| | MEAN SEA-LEVEL PRES. (IN) | 11 | 30.06 | 30.02 | 29.95 | 29.90 | 29.87 | 29.87 | 29.92 | 29.95 | 29.96 | 29.99 | 30.06 | 30.05 | 29.97 |
| **WINDS** | MEAN SPEED (MPH) | 11 | 14.0 | 13.6 | 13.6 | 13.2 | 12.0 | 10.8 | 9.5 | 9.8 | 10.2 | 11.6 | 12.7 | 13.7 | 12.1 |
| | PREVAIL.DIR(TENS OF DEGS) | 30 | 29 | 29 | 29 | 29 | 29 | 29 | 29 | 29 | 29 | 29 | 29 | 29 | 29 |
| | MAXIMUM 2-MINUTE: | | | | | | | | | | | | | | |
| | SPEED (MPH) | 14 | 55 | 59 | 59 | 55 | 54 | 52 | 46 | 56 | 48 | 53 | 59 | 63 | 63 |
| | DIR. (TENS OF DEGS) | | 27 | 27 | 28 | 30 | 27 | 28 | 20 | 25 | 28 | 26 | 27 | 27 | 27 |
| | YEAR OF OCCURRENCE | | 2003 | 1999 | 2004 | 1999 | 2005 | 2008 | 2001 | 1996 | 2005 | 2001 | 1999 | 1998 | DEC 1998 |
| | MAXIMUM 3-SECOND | | | | | | | | | | | | | | |
| | SPEED (MPH) | 14 | 69 | 71 | 69 | 69 | 67 | 66 | 59 | 60 | 60 | 63 | 68 | 75 | 75 |
| | DIR. (TENS OF DEGS) | | 27 | 27 | 29 | 32 | 31 | 29 | 32 | 25 | 28 | 26 | 27 | 26 | 26 |
| | YEAR OF OCCURRENCE | | 2003 | 1999 | 2004 | 1999 | 2008 | 2008 | 1996 | 1996 | 2009 | 2001 | 1999 | 1998 | DEC 1998 |
| **PRECIPITATION** | NORMAL (IN) | 30 | 0.45 | 0.44 | 1.05 | 1.55 | 2.48 | 2.12 | 2.26 | 1.82 | 1.43 | 0.75 | 0.64 | 0.46 | 15.45 |
| | MAXIMUM MONTHLY (IN) | 74 | 2.78 | 2.16 | 3.65 | 5.04 | 6.00 | 5.32 | 5.01 | 6.64 | 4.52 | 3.57 | 2.48 | 1.68 | 6.64 |
| | YEAR OF OCCURRENCE | | 1949 | 1953 | 1990 | 1942 | 1995 | 1955 | 1973 | 1985 | 1973 | 1942 | 1979 | 1937 | AUG 1985 |
| | MINIMUM MONTHLY (IN) | 74 | T | T | 0.12 | 0.34 | 0.11 | 0.07 | 0.43 | 0.03 | 0.01 | 0.03 | T | 0.03 | T |
| | YEAR OF OCCURRENCE | | 1952 | 1983 | 1966 | 1992 | 1974 | 1980 | 2008 | 1944 | 1992 | 1964 | 1965 | 1959 | FEB 1983 |
| | MAXIMUM IN 24 HOURS (IN) | 74 | 1.41 | 1.60 | 1.88 | 1.94 | 2.07 | 2.68 | 3.42 | 6.06 | 2.75 | 1.70 | 1.66 | 1.19 | 6.06 |
| | YEAR OF OCCURRENCE | | 1949 | 1953 | 1946 | 1984 | 1991 | 1955 | 1973 | 1985 | 1973 | 1947 | 1979 | 1979 | AUG 1985 |
| | NORMAL NO. DAYS WITH: | | | | | | | | | | | | | | |
| | PRECIPITATION >= 0.01 | 30 | 5.6 | 5.9 | 9.1 | 10.3 | 12.5 | 10.4 | 11.3 | 11.2 | 8.1 | 6.6 | 6.8 | 6.1 | 103.9 |
| | PRECIPITATION >= 1.00 | 30 | * | 0.0 | 0.0 | 0.2 | 0.4 | 0.3 | 0.3 | 0.1 | 0.2 | 0.0 | * | * | 1.5 |
| **SNOWFALL** | NORMAL (IN) | 30 | 7.3 | 6.8 | 12.5 | 9.1 | 2.5 | 0.* | 0.0 | 0.0 | 1.6 | 4.1 | 8.5 | 7.9 | 60.3 |
| | MAXIMUM MONTHLY (IN) | 74 | 35.5 | 23.3 | 39.2 | 31.8 | 30.4 | 8.7 | 1.0 | 0.5 | 11.8 | 28.2 | 31.1 | 24.4 | 39.2 |
| | YEAR OF OCCURRENCE | | 1980 | 1995 | 1990 | 1984 | 1943 | 1947 | 1994 | 1993 | 2000 | 2009 | 1979 | 2006 | MAR 1990 |
| | MAXIMUM IN 24 HOURS (IN) | 74 | 12.7 | 14.0 | 15.9 | 17.4 | 15.0 | 8.7 | 1.0 | 0.5 | 10.1 | 8.6 | 19.8 | 11.7 | 19.8 |
| | YEAR OF OCCURRENCE' | | 1992 | 1953 | 2003 | 1984 | 1942 | 1947 | 1994 | 1993 | 2000 | 1990 | 1979 | 1979 | NOV 1979 |
| | MAXIMUM SNOW DEPTH (IN) | 61 | 23 | 12 | 19 | 15 | 12 | 0 | 0 | 0 | 8 | 11 | 26 | 17 | 26 |
| | YEAR OF OCCURRENCE | | 1980 | 1989 | 1990 | 1984 | 1978 | | | | 2000 | 2009 | 1979 | 1979 | NOV 1979 |
| | NORMAL NO. DAYS WITH: | | | | | | | | | | | | | | |
| | SNOWFALL >= 1.0 | 30 | 2.4 | 2.3 | 3.7 | 2.4 | 0.7 | 0.0 | 0.0 | 0.0 | 0.5 | 1.3 | 2.5 | 2.3 | 18.1 |

## PRECIPITATION (inches) 2009 CHEYENNE (KCYS)

| YEAR | JAN | FEB | MAR | APR | MAY | JUN | JUL | AUG | SEP | OCT | NOV | DEC | ANNUAL |
|------|-----|-----|-----|-----|-----|-----|-----|-----|-----|-----|-----|-----|--------|
| 1980 | 2.71 | 0.73 | 1.36 | 0.93 | 2.39 | 0.07 | 2.00 | 1.55 | 0.97 | 0.51 | 0.46 | 0.08 | 13.76 |
| 1981 | 0.30 | 0.20 | 0.70 | 0.73 | 5.67 | 1.66 | 2.85 | 2.90 | 0.31 | 0.85 | 0.09 | 0.45 | 16.71 |
| 1982 | 0.41 | 0.19 | 0.17 | 0.53 | 3.56 | 4.52 | 2.71 | 1.81 | 2.87 | 1.20 | 0.43 | 0.83 | 19.23 |
| 1983 | 0.02 | T | 2.96 | 4.45 | 2.31 | 2.81 | 2.12 | 1.95 | 0.78 | 0.49 | 2.34 | 0.46 | 20.69 |
| 1984 | 0.54 | 0.84 | 1.28 | 3.71 | 0.78 | 2.43 | 2.57 | 2.84 | 0.65 | 1.55 | 0.11 | 0.34 | 17.64 |
| 1985 | 0.66 | 0.19 | 0.36 | 1.10 | 1.05 | 1.59 | 3.99 | 6.64 | 1.78 | 0.94 | 0.84 | 0.80 | 19.94 |
| 1986 | 0.13 | 0.50 | 0.54 | 2.26 | 1.03 | 2.42 | 1.04 | 1.55 | 2.47 | 1.78 | 0.66 | 0.18 | 14.56 |
| 1987 | 0.09 | 0.90 | 1.25 | 0.68 | 4.43 | 1.80 | 2.04 | 1.23 | 0.93 | 0.33 | 0.76 | 0.85 | 15.29 |
| 1988 | 0.52 | 0.65 | 1.34 | 1.84 | 3.09 | 2.03 | 1.79 | 1.79 | 1.66 | 0.09 | 0.42 | 0.53 | 15.75 |
| 1989 | 0.27 | 1.26 | 0.49 | 0.48 | 1.37 | 2.51 | 1.70 | 1.79 | 1.62 | 0.41 | 0.14 | 0.69 | 12.73 |
| 1990 | 0.35 | 0.69 | 3.65 | 1.66 | 3.37 | 1.03 | 3.64 | 1.98 | 0.80 | 1.35 | 0.72 | 0.39 | 19.63 |
| 1991 | 0.36 | 0.12 | 0.43 | 1.15 | 3.84 | 4.56 | 3.39 | 1.49 | 1.87 | 0.44 | 0.87 | 0.13 | 18.65 |
| 1992 | 1.18 | 0.11 | 1.75 | 0.34 | 1.86 | 2.11 | 1.87 | 1.90 | 0.01 | 0.46 | 1.69 | 0.49 | 13.77 |
| 1993 | 0.35 | 0.81 | 0.63 | 2.49 | 1.92 | 3.32 | 0.64 | 2.21 | 3.16 | 1.85 | 1.23 | 0.30 | 18.91 |
| 1994 | 0.66 | 0.88 | 0.55 | 1.45 | 2.05 | 1.44 | 2.13 | 1.49 | 0.38 | 1.55 | 0.34 | 0.57 | 13.49 |
| 1995 | 0.12 | 1.02 | 0.19 | 1.49 | 6.00 | 4.58 | 1.08 | 0.89 | 3.39 | 0.78 | 0.37 | 0.18 | 20.09 |
| 1996 | 0.50 | 0.07 | 1.50 | 1.51 | 2.24 | 1.79 | 3.10 | 1.79 | 2.34 | .50 | .45 | .01 | 15.80 |
| 1997 | 0.39 | 0.31 | 0.44 | 1.68 | 2.47 | 3.48 | 2.80 | 4.33 | 2.06 | 1.09 | 0.20 | 0.59 | 19.84 |
| 1998 | 0.07 | 0.29 | 0.69 | 1.09 | 2.35 | 1.63 | 1.21 | 0.96 | 0.49 | 1.27 | 0.31 | 0.46 | 10.82 |
| 1999 | 0.35 | 0.13 | 0.46 | 5.02 | 2.04 | 2.13 | 2.38 | 0.78 | 2.11 | 0.27 | 0.25 | 0.19 | 16.11 |
| 2000 | 0.35 | 0.59 | 1.48 | 0.58 | 1.38 | 0.46 | 2.64 | 1.87 | 2.64 | 0.69 | 0.30 | 0.75 | 13.73 |
| 2001 | 0.41 | 0.33 | 0.56 | 2.49 | 2.24 | 1.45 | 3.31 | 0.74 | 1.18 | 0.40 | 0.23 | 0.13 | 13.47 |
| 2002 | 0.17 | 0.64 | 0.85 | 0.68 | 0.75 | 0.68 | 0.48 | 2.21 | 1.73 | 1.10 | 0.46 | 0.11 | 9.86 |
| 2003 | 0.09 | 0.47 | 2.23 | 2.70 | 1.41 | 2.69 | 0.44 | 0.82 | 1.24 | 0.23 | 0.46 | 0.76 | 13.54 |
| 2004 | 0.07 | 0.45 | 0.15 | 1.24 | 1.04 | 3.37 | 1.73 | 0.92 | 3.06 | 0.64 | 1.55 | 0.13 | 14.35 |
| 2005 | 0.41 | 0.34 | 0.88 | 1.65 | 1.29 | 4.33 | 2.59 | 1.61 | 0.28 | 1.87 | 0.30 | 0.28 | 15.83 |
| 2006 | 0.05 | 0.52 | 1.70 | 0.82 | 1.61 | 0.25 | 2.98 | 0.98 | 1.02 | 0.58 | 0.07 | 1.55 | 12.13 |
| 2007 | 0.32 | 0.34 | 1.40 | 1.49 | 1.40 | 0.36 | 3.97 | 1.62 | 1.23 | 1.28 | 0.32 | 1.01 | 14.74 |
| 2008 | 0.02 | 0.17 | 0.74 | 0.54 | 2.50 | 1.91 | 0.43 | 6.55 | 1.12 | 0.57 | 0.40 | 0.31 | 15.26 |
| 2009 | 0.71 | 0.33 | 0.72 | 3.60 | 2.08 | 4.26 | 1.83 | 0.37 | 1.14 | 2.52 | 0.49 | 0.69 | 18.74 |
| POR= 44 YRS | 0.42 | 0.61 | 1.14 | 2.17 | 2.24 | 1.85 | 2.00 | 1.68 | 1.35 | 1.11 | 0.65 | 0.55 | 15.77 |

WBAN : 24018

## AVERAGE TEMPERATURE (°F) 2009 CHEYENNE (KCYS)

| YEAR | JAN | FEB | MAR | APR | MAY | JUN | JUL | AUG | SEP | OCT | NOV | DEC | ANNUAL |
|------|-----|-----|-----|-----|-----|-----|-----|-----|-----|-----|-----|-----|--------|
| 1980 | 22.2 | 29.4 | 33.0 | 42.3 | 51.2 | 65.5 | 71.4 | 66.5 | 60.4 | 46.7 | 36.8 | 38.5 | 47.0 |
| 1981 | 33.4 | 32.2 | 36.8 | 50.1 | 50.5 | 63.7 | 69.0 | 65.2 | 61.2 | 46.1 | 40.7 | 30.7 | 48.3 |
| 1982 | 25.5 | 28.2 | 36.1 | 41.8 | 50.5 | 57.5 | 67.7 | 69.1 | 56.7 | 45.1 | 31.4 | 28.3 | 44.8 |
| 1983 | 32.8 | 33.5 | 32.0 | 34.8 | 47.4 | 57.3 | 67.8 | 70.2 | 59.0 | 48.6 | 32.3 | 15.0 | 44.2 |
| 1984 | 24.3 | 28.4 | 32.4 | 35.7 | 53.6 | 59.4 | 68.4 | 66.6 | 53.0 | 40.0 | 35.0 | 27.0 | 43.7 |
| 1985 | 19.5 | 23.3 | 35.2 | 45.6 | 54.3 | 61.1 | 69.3 | 66.8 | 52.8 | 45.1 | 26.2 | 26.1 | 43.8 |
| 1986 | 37.0 | 30.1 | 42.2 | 43.9 | 50.6 | 64.5 | 68.5 | 67.0 | 55.3 | 44.6 | 34.3 | 28.7 | 47.2 |
| 1987 | 29.1 | 31.6 | 32.4 | 46.7 | 54.5 | 63.3 | 69.2 | 64.9 | 58.0 | 46.9 | 36.7 | 25.7 | 46.6 |
| 1988 | 21.5 | 28.4 | 32.3 | 44.4 | 53.3 | 67.4 | 69.7 | 68.7 | 57.5 | 50.3 | 35.3 | 28.7 | 46.5 |
| 1989 | 30.2 | 17.3 | 37.4 | 44.3 | 54.3 | 60.3 | 70.6 | 67.0 | 57.5 | 46.5 | 38.7 | 24.3 | 45.7 |
| 1990 | 31.5 | 28.7 | 31.9 | 42.8 | 49.3 | 64.1 | 65.0 | 66.4 | 62.2 | 46.7 | 38.7 | 20.8 | 45.7 |
| 1991 | 25.0 | 36.2 | 36.4 | 41.1 | 52.4 | 62.5 | 67.7 | 67.7 | 58.0 | 46.1 | 31.7 | 31.6 | 46.4 |
| 1992 | 30.5 | 36.7 | 38.6 | 48.6 | 55.0 | 60.8 | 64.6 | 63.8 | 60.0 | 49.2 | 30.9 | 25.5 | 47.0 |
| 1993 | 26.1 | 24.9 | 37.2 | 42.1 | 53.4 | 58.7 | 66.1 | 64.8 | 54.3 | 44.1 | 29.2 | 30.7 | 44.3 |
| 1994 | 29.4 | 27.2 | 38.6 | 43.3 | 56.9 | 66.3 | 68.0 | 69.4 | 61.7 | 46.9 | 34.6 | 33.3 | 48.0 |
| 1995 | 30.7 | 32.6 | 36.8 | 39.5 | 45.7 | 59.0 | 67.8 | 71.6 | 57.1 | 45.8 | 38.4 | 30.0 | 46.3 |
| 1996 | 23.0 | 30.2 | 31.6 | 42.7 | 51.7 | 63.6 | 68.3 | 66.3 | 56.6 | 45.8 | 34.7 | 30.1 | 45.4 |
| 1997 | 24.7 | 26.7 | 38.3 | 35.9 | 51.7 | 62.6 | 68.3 | 65.0 | 59.2 | 46.1 | 31.4 | 28.3 | 44.9 |
| 1998 | 30.6 | 30.6 | 32.7 | 41.2 | 53.3 | 56.8 | 70.3 | 68.7 | 64.0 | 45.8 | 39.5 | 26.6 | 46.7 |
| 1999 | 30.8 | 34.6 | 37.6 | 38.4 | 50.5 | 60.3 | 70.0 | 67.7 | 54.0 | 47.6 | 44.0 | 32.0 | 47.3 |
| 2000 | 29.8 | 34.7 | 37.1 | 45.4 | 55.5 | 62.5 | 72.2 | 70.5 | 58.2 | 46.3 | 25.9 | 25.3 | 47.0 |
| 2001 | 27.6 | 26.9 | 35.1 | 44.3 | 52.6 | 64.2 | 72.2 | 69.5 | 61.5 | 47.2 | 38.9 | 29.2 | 47.4 |
| 2002 | 27.5 | 29.2 | 29.5 | 44.4 | 51.2 | 67.7 | 73.6 | 66.6 | 59.1 | 39.2 | 34.4 | 32.4 | 46.2 |
| 2003 | 35.0 | 24.9 | 37.7 | 45.8 | 53.5 | 58.3 | 75.1 | 70.7 | 56.4 | 51.3 | 33.5 | 30.0 | 47.7 |
| 2004 | 30.4 | 28.8 | 42.1 | 44.3 | 55.5 | 60.5 | 67.3 | 65.1 | 58.7 | 48.0 | 34.8 | 31.8 | 47.3 |
| 2005 | 31.9 | 32.7 | 36.1 | 43.1 | 52.1 | 61.7 | 71.9 | 66.5 | 62.1 | 48.2 | 39.2 | 27.6 | 47.8 |
| 2006 | 34.6 | 26.7 | 32.7 | 47.0 | 54.6 | 68.0 | 71.8 | 68.0 | 54.2 | 44.9 | 36.8 | 29.4 | 47.4 |
| 2007 | 21.7 | 29.8 | 41.8 | 42.2 | 53.3 | 64.3 | 72.2 | 70.6 | 59.1 | 49.0 | 37.7 | 23.0 | 47.1 |
| 2008 | 23.5 | 30.0 | 34.2 | 40.6 | 50.4 | 61.2 | 72.2 | 66.2 | 56.4 | 46.8 | 39.2 | 24.5 | 45.4 |
| 2009 | 30.3 | 33.2 | 36.7 | 40.6 | 53.3 | 59.6 | 66.0 | 65.5 | 58.7 | 37.0 | 38.0 | 20.7 | 45.0 |
| POR= 44 YRS | 26.9 | 26.8 | 33.6 | 40.6 | 50.9 | 59.6 | 69.0 | 66.7 | 56.1 | 46.4 | 34.6 | 28.5 | 45.0 |

## HEATING DEGREE DAYS (base 65°F) 2009 CHEYENNE (KCYS)

| YEAR | JUL | AUG | SEP | OCT | NOV | DEC | JAN | FEB | MAR | APR | MAY | JUN | TOTAL |
|---|---|---|---|---|---|---|---|---|---|---|---|---|---|
| 1980-81 | 0 | 41 | 151 | 558 | 840 | 812 | 974 | 913 | 868 | 440 | 445 | 95 | 6137 |
| 1981-82 | 21 | 50 | 120 | 580 | 722 | 1058 | 1216 | 1025 | 892 | 687 | 446 | 227 | 7044 |
| 1982-83 | 29 | 7 | 264 | 608 | 1002 | 1131 | 992 | 875 | 1016 | 898 | 538 | 233 | 7593 |
| 1983-84 | 23 | 0 | 202 | 502 | 974 | 1547 | 1259 | 1058 | 1002 | 870 | 348 | 177 | 7962 |
| 1984-85 | 6 | 15 | 365 | 769 | 892 | 1171 | 1403 | 1163 | 917 | 576 | 323 | 162 | 7762 |
| 1985-86 | 11 | 37 | 364 | 612 | 1158 | 1199 | 862 | 970 | 698 | 629 | 440 | 71 | 7051 |
| 1986-87 | 4 | 20 | 286 | 630 | 914 | 1121 | 1102 | 928 | 1002 | 542 | 321 | 78 | 6948 |
| 1987-88 | 27 | 79 | 214 | 554 | 841 | 1213 | 1343 | 1056 | 1008 | 611 | 361 | 42 | 7349 |
| 1988-89 | 12 | 18 | 236 | 449 | 884 | 1116 | 1075 | 1332 | 851 | 615 | 334 | 180 | 7102 |
| 1989-90 | 4 | 18 | 236 | 566 | 779 | 1257 | 1037 | 1012 | 1021 | 662 | 480 | 106 | 7178 |
| 1990-91 | 75 | 28 | 127 | 558 | 784 | 1364 | 1231 | 799 | 879 | 712 | 385 | 95 | 7037 |
| 1991-92 | 26 | 17 | 217 | 586 | 993 | 1031 | 1062 | 813 | 811 | 488 | 307 | 131 | 6482 |
| 1992-93 | 63 | 92 | 161 | 484 | 1014 | 1216 | 1198 | 1118 | 856 | 680 | 351 | 202 | 7435 |
| 1993-94 | 45 | 58 | 322 | 642 | 1068 | 1059 | 1097 | 1053 | 809 | 644 | 249 | 43 | 7089 |
| 1994-95 | 24 | 16 | 131 | 552 | 904 | 978 | 1057 | 901 | 866 | 758 | 588 | 199 | 6974 |
| 1995-96 | 37 | 6 | 283 | 589 | 793 | 1076 | 1293 | 1002 | 1031 | 663 | 411 | 77 | 7261 |
| 1996-97 | 16 | 29 | 269 | 590 | 905 | 1076 | 1242 | 1065 | 820 | 867 | 400 | 102 | 7381 |
| 1997-98 | 34 | 58 | 181 | 577 | 1001 | 1130 | 1059 | 957 | 994 | 706 | 352 | 256 | 7305 |
| 1998-99 | 7 | 10 | 104 | 587 | 754 | 1183 | 1055 | 845 | 840 | 791 | 444 | 150 | 6770 |
| 1999-00 | 8 | 27 | 322 | 534 | 625 | 1015 | 1083 | 870 | 858 | 581 | 305 | 130 | 6358 |
| 2000-01 | 2 | 8 | 242 | 573 | 1167 | 1223 | 1155 | 1064 | 922 | 614 | 377 | 106 | 7453 |
| 2001-02 | 0 | 11 | 135 | 545 | 776 | 1102 | 1154 | 997 | 1090 | 610 | 432 | 56 | 6908 |
| 2002-03 | 0 | 41 | 205 | 791 | 911 | 1006 | 925 | 1114 | 843 | 570 | 360 | 200 | 6966 |
| 2003-04 | 0 | 26 | 265 | 413 | 939 | 1077 | 1065 | 1043 | 704 | 613 | 295 | 154 | 6594 |
| 2004-05 | 50 | 68 | 203 | 518 | 897 | 1025 | 1020 | 897 | 889 | 647 | 401 | 153 | 6768 |
| 2005-06 | 13 | 41 | 129 | 515 | 767 | 1152 | 935 | 1065 | 995 | 533 | 325 | 19 | 6489 |
| 2006-07 | 8 | 39 | 322 | 613 | 840 | 1099 | 1334 | 980 | 711 | 676 | 358 | 105 | 7085 |
| 2007-08 | 0 | 11 | 194 | 489 | 813 | 1295 | 1281 | 1008 | 947 | 725 | 447 | 134 | 7344 |
| 2008-09 | 4 | 60 | 253 | 558 | 767 | 1251 | 1070 | 885 | 869 | 726 | 360 | 175 | 6978 |
| 2009- | 43 | 51 | 203 | 858 | 804 | 1367 | | | | | | | |

WBAN : 24018

## COOLING DEGREE DAYS (base 65°F) 2009 CHEYENNE (KCYS)

| YEAR | JAN | FEB | MAR | APR | MAY | JUN | JUL | AUG | SEP | OCT | NOV | DEC | TOTAL |
|---|---|---|---|---|---|---|---|---|---|---|---|---|---|
| 1980 | 0 | 0 | 0 | 0 | 0 | 88 | 205 | 94 | 21 | 0 | 0 | 0 | 408 |
| 1981 | 0 | 0 | 0 | 0 | 0 | 59 | 156 | 64 | 15 | 0 | 0 | 0 | 294 |
| 1982 | 0 | 0 | 0 | 0 | 0 | 8 | 120 | 140 | 21 | 0 | 0 | 0 | 289 |
| 1983 | 0 | 0 | 0 | 0 | 0 | 10 | 115 | 169 | 28 | 0 | 0 | 0 | 322 |
| 1984 | 0 | 0 | 0 | 0 | 1 | 14 | 118 | 72 | 8 | 0 | 0 | 0 | 213 |
| 1985 | 0 | 0 | 0 | 0 | 0 | 52 | 150 | 98 | 8 | 0 | 0 | 0 | 308 |
| 1986 | 0 | 0 | 0 | 0 | 0 | 62 | 118 | 89 | 0 | 0 | 0 | 0 | 269 |
| 1987 | 0 | 0 | 0 | 0 | 0 | 35 | 164 | 83 | 9 | 0 | 0 | 0 | 291 |
| 1988 | 0 | 0 | 0 | 0 | 4 | 122 | 166 | 140 | 18 | 0 | 0 | 0 | 450 |
| 1989 | 0 | 0 | 0 | 0 | 7 | 46 | 188 | 86 | 19 | 0 | 0 | 0 | 346 |
| 1990 | 0 | 0 | 0 | 0 | 0 | 86 | 84 | 79 | 49 | 0 | 0 | 0 | 298 |
| 1991 | 0 | 0 | 0 | 0 | 2 | 28 | 119 | 106 | 13 | 4 | 0 | 0 | 272 |
| 1992 | 0 | 0 | 0 | 1 | 2 | 15 | 58 | 63 | 16 | 0 | 0 | 0 | 155 |
| 1993 | 0 | 0 | 0 | 0 | 0 | 19 | 89 | 60 | 6 | 0 | 0 | 0 | 174 |
| 1994 | 0 | 0 | 0 | 0 | 3 | 91 | 125 | 156 | 38 | 0 | 0 | 0 | 413 |
| 1995 | 0 | 0 | 0 | 0 | 0 | 23 | 132 | 218 | 50 | 0 | 0 | 0 | 423 |
| 1996 | 0 | 0 | 0 | 0 | 2 | 40 | 124 | 73 | 24 | 0 | 0 | 0 | 263 |
| 1997 | 0 | 0 | 0 | 0 | 0 | 37 | 144 | 65 | 13 | 0 | 0 | 0 | 259 |
| 1998 | 0 | 0 | 0 | 0 | 0 | 17 | 182 | 131 | 78 | 0 | 0 | 0 | 408 |
| 1999 | 0 | 0 | 0 | 0 | 0 | 18 | 170 | 114 | 0 | 0 | 0 | 0 | 302 |
| 2000 | 0 | 0 | 0 | 0 | 19 | 59 | 231 | 187 | 44 | 0 | 0 | 0 | 540 |
| 2001 | 0 | 0 | 0 | 0 | 0 | 91 | 229 | 158 | 38 | 0 | 0 | 0 | 516 |
| 2002 | 0 | 0 | 0 | 0 | 12 | 145 | 271 | 95 | 36 | 0 | 0 | 0 | 559 |
| 2003 | 0 | 0 | 0 | 0 | 10 | 6 | 322 | 210 | 13 | 0 | 0 | 0 | 561 |
| 2004 | 0 | 0 | 0 | 0 | 8 | 24 | 127 | 80 | 21 | 0 | 0 | 0 | 260 |
| 2005 | 0 | 0 | 0 | 0 | 6 | 59 | 232 | 95 | 49 | 2 | 0 | 0 | 443 |
| 2006 | 0 | 0 | 0 | 0 | 8 | 116 | 223 | 136 | 3 | 0 | 0 | 0 | 486 |
| 2007 | 0 | 0 | 0 | 0 | 2 | 88 | 230 | 191 | 25 | 0 | 0 | 0 | 536 |
| 2008 | 0 | 0 | 0 | 0 | 1 | 26 | 239 | 105 | 1 | 0 | 0 | 0 | 372 |
| 2009 | 0 | 0 | 0 | 0 | 4 | 20 | 82 | 73 | 22 | 0 | 0 | 0 | 201 |

**SNOWFALL (inches) 2009 CHEYENNE (KCYS)**

| YEAR | JUL | AUG | SEP | OCT | NOV | DEC | JAN | FEB | MAR | APR | MAY | JUN | TOTAL |
|---|---|---|---|---|---|---|---|---|---|---|---|---|---|
| 1980-81 | 0.0 | 0.0 | 0.0 | 1.1 | 6.3 | 2.1 | 3.4 | 2.9 | 9.0 | 2.0 | 0.8 | 0.0 | 27.6 |
| 1981-82 | 0.0 | 0.0 | 0.0 | 4.6 | 0.8 | 5.8 | 5.6 | 2.4 | 1.7 | 2.0 | 4.0 | 0.0 | 26.9 |
| 1982-83 | 0.0 | 0.0 | 0.0 | 12.2 | 7.9 | 13.1 | 0.1 | T | 31.9 | 25.7 | 10.1 | 0.0 | 101.0 |
| 1983-84 | 0.0 | 0.0 | T | 0.2 | 27.2 | 7.0 | 7.9 | 12.1 | 13.0 | 31.8 | T | 0.0 | 99.2 |
| 1984-85 | 0.0 | 0.0 | 1.6 | 3.8 | 1.5 | 5.6 | 9.8 | 1.9 | 3.6 | 2.6 | 0.4 | 0.0 | 30.8 |
| 1985-86 | 0.0 | 0.0 | 7.4 | 6.0 | 12.4 | 13.0 | 1.3 | 4.9 | 5.6 | 11.3 | 3.8 | 0.0 | 65.7 |
| 1986-87 | 0.0 | 0.0 | 0.0 | 9.2 | 6.9 | 2.2 | 1.1 | 9.9 | 13.7 | 4.4 | T | 0.0 | 47.4 |
| 1987-88 | 0.0 | 0.0 | 0.0 | 1.3 | 4.5 | 16.1 | 9.0 | 7.5 | 16.0 | 7.5 | 2.2 | 0.0 | 64.1 |
| 1988-89 | 0.0 | 0.0 | 0.2 | 0.0 | 4.5 | 7.2 | 4.4 | 17.6 | 5.0 | 4.3 | T | T | 43.2 |
| 1989-90 | T | T | 2.6 | 3.4 | 1.9 | 9.7 | 5.6 | 11.6 | 39.2 | 5.6 | 1.3 | T | 80.9 |
| 1990-91 | T | T | T | 12.3 | 8.8 | 6.6 | 7.7 | 1.6 | 3.7 | 10.5 | T | T | 51.2 |
| 1991-92 | T | T | T | 7.5 | 9.1 | 2.0 | 18.2 | 0.8 | 11.3 | 0.7 | 4.4 | 0.5 | 54.5 |
| 1992-93 | T | T | 0.0 | 2.1 | 25.5 | 10.9 | 7.1 | 14.6 | 6.6 | 11.3 | T | 0.5 | 78.6 |
| 1993-94 | 0.0 | 0.5 | 5.5 | 3.1 | 18.0 | 5.7 | 11.5 | 16.1 | 4.2 | 9.1 | T | T | 73.7 |
| 1994-95 | 1.0 | 0.0 | 0.9 | 2.3 | 3.0 | 10.0 | 2.7 | 23.3 | 4.4 | 13.2 | 2.6 | T | 63.4 |
| 1995-96 | 0.0 | 0.0 | 3.0 | 6.4 | 6.4 | 3.6 | 11.9 | 2.1 | 21.9 | 8.7 | 0.1 | T | 64.1 |
| 1996-97 | T | T | 1.2 | 2.0 | 7.0 | 0.4 | 10.5 | 7.9 | 8.4 | 23.3 | 2.8 | T | 63.5 |
| 1997-98 | T | T | T | 8.8 | 2.5 | 10.9 | 1.3 | 3.4 | 8.4 | 10.1 | 1.1 | 0.7 | 47.2 |
| 1998-99 | T | 0.0 | 0.0 | 0.3 | 4.0 | 13.5 | 6.4 | 2.4 | 4.9 | 17.0 | 0.9 | T | 49.4 |
| 1999-00 | 0.0 | 0.0 | 3.7 | 3.3 | 4.3 | 2.9 | 2.9 | 10.7 | 16.2 | 1.7 | 0.2 | 0.0 | 45.9 |
| 2000-01 | T | T | 11.8 | 0.7 | 4.3 | 13.5 | 8.6 | 6.8 | 7.0 | 23.2 | 13.1 | T | 89.0 |
| 2001-02 | T | 0.0 | T | 2.6 | 4.3 | 2.1 | 2.7 | 10.3 | 11.4 | 1.7 | 2.5 | T | 37.6 |
| 2002-03 | T | T | 0.0 | 13.1 | 8.3 | 1.6 | 2.3 | 8.6 | 25.2 | 6.2 | 2.4 | T | 67.7 |
| 2003-04 | 0.0 | T | 1.4 | 2.0 | 9.8 | 9.7 | 1.9 | 6.8 | 0.3 | 7.9 | 3.5 | T | 43.3 |
| 2004-05 | T | 0.0 | T | 1.4 | 24.4 | 3.0 | 6.9 | 4.4 | 8.3 | 13.4 | 0.3 | 0.0 | 62.1 |
| 2005-06 | T | 0.0 | 0.0 | 7.5 | 1.2 | 7.1 | 1.0 | 13.1 | 21.6 | 2.3 | 2.0 | 0.2 | 55.8 |
| 2006-07 | T | 0.0 | T | 5.3 | 1.3 | 24.4 | 9.3 | 5.5 | 2.8 | 0.7 | 0.0 | 0.2 | 49.5 |
| 2007-08 | T | T | T | 1.2 | 5.4 | 17.8 | 0.5 | 2.7 | 8.4 | 6.7 | 0.2 | T | 42.9 |
| 2008-09 | 0.0 | T | T | 0.1 | 4.9 | 8.7 | 12.8 | 5.9 | 9.3 | 27.7 | 0.1 | 0.0 | 69.5 |
| 2009- | 0.0 | 0.0 | T | 28.2 | 6.7 | 16.4 | | | | | | | |
| POR=<br>44 YRS | T | T | 0.8 | 4.4 | 7.3 | 7.5 | 5.4 | 7.9 | 11.1 | 12.0 | 3.7 | 0.3 | 60.4 |

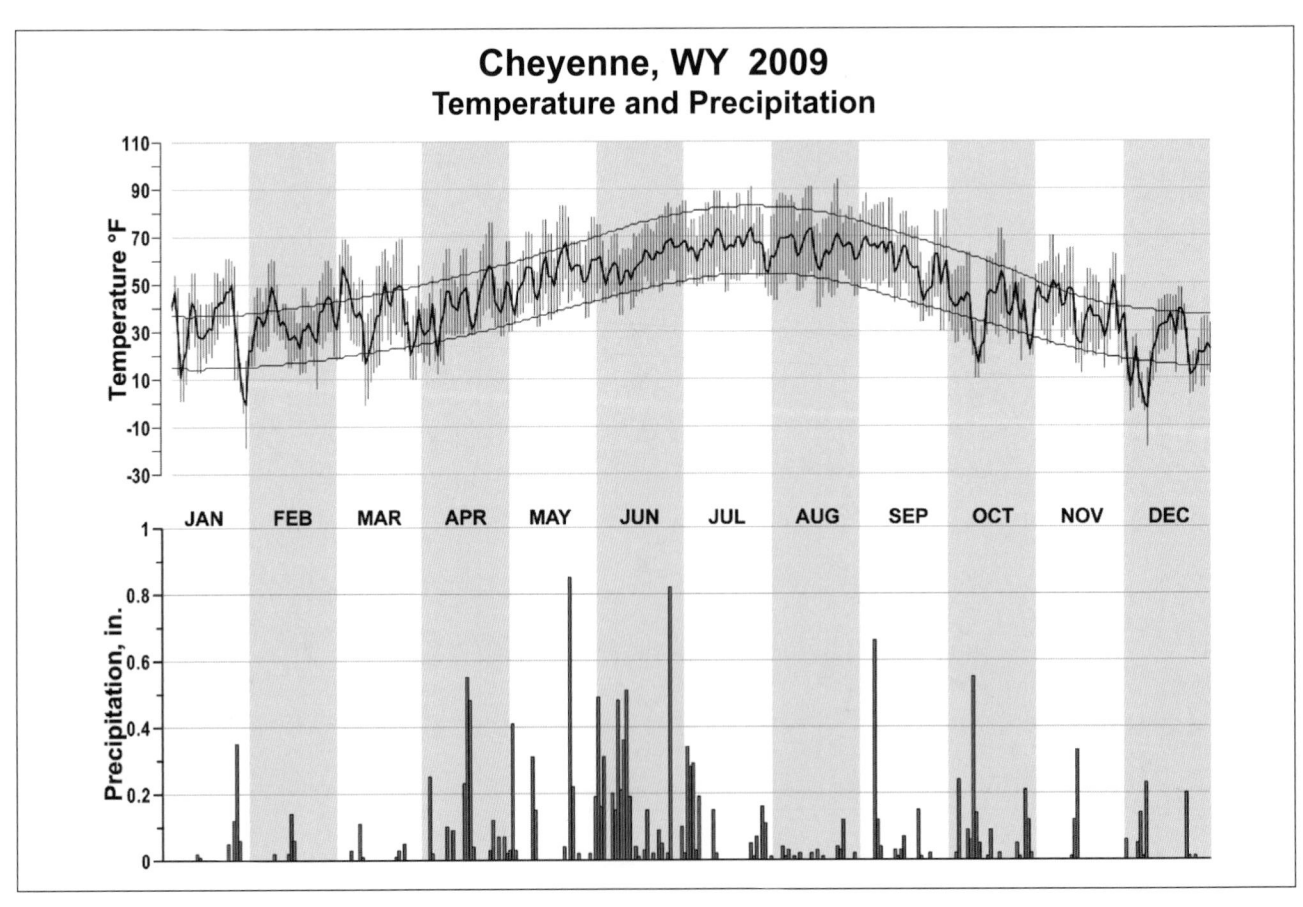

# Cheyenne, WY 2009
## Temperature and Precipitation

# 11

# A Time Line of Meteorology: 9000 B.C.–2000 A.D.

## • INTRODUCTION

There is no universally accepted division of the history of meteorology. In fact, as of this writing, the author is aware of no other extensive chronology in print meant for the general public.

This chapter lists events in chronological order important to our understanding of the meteorology. Work in this area is fairly rare as there is no generally accepted terminology that describes the stages human knowledge has gone through to arrive at the modern science of meteorology.

I have divided the history of meteorology into six time spans having common characteristics. The names of these eras are purely descriptive:

1. The beginning—9000 B.C.–ca. 600 B.C.
2. The period of speculation—600 B.C.–ca. 1500 A.D.
3. The dawn of scientific meteorology—ca. 1450 A.D.–1800 A.D.
4. Prelude to modern meteorology—1800 A.D.–1900 A.D.
5. The birth of fronts and the middle latitude cyclone—1900–1940.
6. The jet stream era—1940s–1950s.

By the middle 1950s the role of the jet stream in energizing weather systems was clear and the scientific foundation of modern meteorology was complete. There is still much to learn, but the framework is essentially complete. Most of what is to come will involve more specific knowledge of all atmospheric processes, extending accurate forecasts to longer time periods and changing climate predictions.

This chapter concludes with listings of six technologies and nine concepts that stand out in the history of meteorology as great leaps forward in our understanding of the atmosphere.

## The Beginning—9000 B.C.–ca. 600 B.C.: Mythology, Astrometeorology and Oral Tradition

There is much more about this time period that we will never know because it is the dawn of civilization, but even at this early stage of human social development, there is evidence for a rich oral history of weather.

### 9000 B.C.

Cultivation of wild wheat and barley begins as humans transition from hunting to gathering to farming in ancient Babylonia. No evidence of permanent settlements exists this far back in time. As agriculture grew, weather assumed greater importance.

### 7000 B.C.

Jarmo—In present-day northern Iraq, in the foothills of the Zagros Mountains, east of present-day Kirkuk, Iraq, is the oldest known permanent agricultural settlement in the world. Humans grew wheat, barley, and lentils and kept domesticated sheep, dogs, and goats.

### 6000 B.C.

Hassuna residents in present-day Iraq, south of Mosul, developed irrigation and practiced dry farming in places.

### 5000 B.C.

Ubadian culture, in southern Iraq, between the Tigris and Euphrates Rivers, built the first known shrines, offering tables niches for religious or cult objects.

### 4000 B.C.

Stone artifacts from Mesopotamia, Sumeria, and Babylon have likenesses of a lightning god.

### 3500 B.C.

Egyptians practiced a sky religion and had rain-making rituals.

*The Weather Almanac: A Reference Guide to Weather, Climate, and Related Issues in the United States and Its Key Cities*, Twelfth Edition. Steve Horstmeyer.
© 2011 John Wiley & Sons, Inc. Published 2011 by John Wiley & Sons, Inc.

### 3000 B.C.–300 B.C.

Babylonians practiced astrometeorology and left clay tablet records in cuneiform of 7000 omens used in politics, military operations, and weather forecasting. Scribes made regular astronomical and meteorological observations. The oldest surviving observation is from 625 B.C. and continues for 600 years. Omens took the following form, "When a cloud grows dark in heaven, a wind will blow." The Babylonian wind rose had eight directions.

### 3000 B.C.

Sumer, present-day southern Iraq—The "bow of the deluge" is part of a hymn and probably the predecessor of the biblical story of the great flood and God's pledge not to destroy the world by flood again.

### 3000 B.C.

India—Cloud formation and seasonal cycles are mentioned in records.

### 1200 B.C.

China—Shang Dynasty in northeastern modern China kept systematic meteorological records.

Rainbows were thought to be visible rain dragons.

Humidity was measured by weighing charcoal after exposure to the atmosphere and determining the increase of weight.

### 1066 B.C.

Chou dynasty in China kept official records including climate descriptions.

## The Period of Speculation—600 B.C.–ca. 1500 A.D.: Natural Philosophers and Sacred Science

This period is characterized by natural philosophers who observed natural phenomena and came to philosophical conclusions as to their origin and the processes by which they operated. Natural philosophers did not necessarily test their ideas, and thus their work was not science. The scientific method was born in the Middle East around 1000 A.D.

### ca. 600 B.C.

Thales of Miletus (ca. 624 B.C.–ca. 547 B.C.) is now known as the "first natural philosopher."

He studied Babylonian writings on astrometeorology and wrote on the hydrologic cycle.

### ca. 570 B.C.

Greece—Anaximander (610 B.C.–545 B.C.) wrote "On Nature," and it is the first known work on natural philosophy.

He correctly wrote that wind is moving air, but his idea was rejected for over 2000 years.

He proposed evolution in the animal kingdom and stated that Earth was spherical because stars disappeared below the horizon.

He attributed Nile River floods to wind changes.

### ca. 550 B.C.

Greece—Anaximander (570 B.C.–502 B.C.) wrote that lightning and thunder were from wind breaking out of clouds, rainbows were from sunlight falling on clouds, and hail was frozen rain.

### 500 B.C.

Greece—The first mention of measuring rainfall in the western world.

Parmenides classified climates by latitude into torrid, temperate, and frigid.

### ca. 500 B.C.

The Book of Job 37:22, "Fair weather cometh out of the north."

### 400 B.C.

India—Rainfall is measured using bowls 1 Aratni (18 inches) in diameter, and taxation of crops is based on rainfall.

### ca. 465 B.C.

Greece—Anaxagoras (499 B.C.–427 B.C.) almost correctly explained hailstorms:

1. Air temperature decreases with height because less reflected light from surface reaches the high atmosphere. Actually, heat is conducted to air from ground and carried aloft in updrafts.
2. Clouds contain moisture (correct).
3. Moisture will freeze at high altitude even in summer (correct).
4. Air warmed by the reflected sunlight rises (almost correct as air is warmed by conduction).
5. Very high up temperature again warms (correct but wrong reason).

Anaxagoras deduced the vertical temperature structure of the atmosphere.

He correctly explained the seasonal cycle of the Nile:

"The Nile comes from the snow in Ethiopia which melts in summer and freezes in winter."

The world would not catch up with him until the 1800s.

### ca. 440 B.C.

Sicily—Empedocles of Agrigentum (ca. 492 B.C.–432 B.C.) wrote about the four basic elements (air, earth, fire, water) and the four basic qualities (hot, cold, dry, wet). This was adopted by Aristotle and dominated meteorological thought for 2000 years.

### 400 B.C.

Greece—Hippocrates of Cos (460 B.C.–375 B.C.), called "The Father of Medicine," wrote about the influence of climate on health.

### 340 B.C.

Greece—Aristotle (384 B.C.–322 B.C.) published "Meteorologica." It dominated western prescientific thought for

2000 years until the Renaissance. In it he stated that air has weight and weather was caused by four elements (air, earth, fire, water) and four contraries (hot, cold, dry, moist). He adopted this from Empedocles.

He considered wind to be exhalations of earth.

Aristotle was mostly wrong.

### ca. 330 B.C.

Greece—Theophrastus (371 B.C.–287 B.C.) wrote the first weather forecasting manual, "The Book of Signs," that contained 80 signs of rain, 45 signs of wind force and direction, 50 signs of storms, 24 signs of fair weather, and 7 signs for weather for a period of less than a year. It also included folklore.

### ca. 300 B.C.

Greece—Theophrastus published "On Winds," which stated that wind is air in motion. He also had a very basic understanding of atmospheric pressure. He discussed the impact of distance from source region on temperature of an air mass and covered the role of mountains in lifting and blocking air. He also wrote about local winds, like the sea breeze.

### 250 B.C.

Greece—Archimedes (ca. 287 B.C.–212/11 B.C.) explained the buoyancy principle.

### 240 B.C.

Greece—Eratosthenes wrote that the Earth is a globe 40,000 km in circumference.

### 200 B.C.–200 A.D.

Palestine—Records of rainfall measurements were being kept.

### 29 B.C.

Rome—Virgil (70 B.C.–19 B.C.) wrote "Georgics," 2000 lines of poetry on agriculture and weather. He also established the tradition of including weather signs in handbooks of animal husbandry.

### 25 A.D.

Spain (Roman Empire)—Pomponius Mela, a geographer, formalized a climate zone system.

### ca. 35 A.D.

Byzantium—Philo (?–50 A.D.), a Jewish philosopher, invented a device that demonstrated the expansion and contraction of air as it was warmed and cooled. He did not develop this as a thermometer.

Alexandria—Heron (10–70 A.D.) drew plans for a basic medical thermometer, but he did not build one as far as is known.

### 61 A.D.

Rome—Lucius Seneca (4 B.C.–65 A.D.) complained of the air pollution in Rome.

### 63 A.D.

Rome—Lucius Seneca made weather observations and wrote "Natural Questions," in which he summarized the works of others and added his views.

### 70 A.D.

Rome—Gaius Pliny Secundus (Pliny the Elder) (23 A.D.–79 A.D.) wrote "Natural History," which he compiled from 2000 works by 146 Roman and 326 Greeks authors. He stated that everyone agrees on the spherical shape of Earth.

### 77 A.D.

The first known record of snow accumulation in Korea.

### ca. 80 A.D.

Matthew 16.2-3. Jesus said to a group of fishermen: "When it is evening, you say, 'it will be fair weather, for the sky is red.' And in the morning, 'It will be stormy today, for the sky is red and threatening.' "

### ca. 200 A.D.

Tunisia (Roman Empire)—Tertullian of Carthage (ca. 160–?225) held that various passages of scripture prove that lightning is identical to "hellfire." This idea was transmitted from generation to generation of churchmen who found support for Tertullian's view in the sulfurous smell experienced during thunderstorms.

Marks the beginning of a temporary end of science based on observation in the west, and the beginning of "sacred science" based on the "authority" of scripture.

Science based on observation would not rebound for more than 1000 years.

### 350 A.D.

Flood levels were recorded in Korea.

### ca. 380 A.D.

St. Jerome (340/2–420) wrote on the basis of scripture that the air is full of devils. This started the doctrine of the diabolical origin of storms, which held that they are the work of the devil.

### ca. 400 A.D.

Algeria (Roman Empire)—St. Augustine, Bishop of Hippo (354–430), who was the most influential philosopher of early Christianity held that scripture should rule the thoughts of man, not observation. He wrote:

> "Nothing is to be accepted save on the authority of Scripture, since greater is that authority than all the powers of the human mind."

Augustine emphasized absolute belief as opposed to understanding.

China—Liu Thien Chun was made comptroller general of Crops and Weather in China.

**410 A.D.**

City of Rome, "the Eternal City," was sacked, and Western civilization disintegrated into a feudal system.

**476 A.D.**

The barbarian Odovacer (434–493) deposed western Roman emperor Romulus Augustulus, and the Middle Ages began. Aristotle ruled science, superstition, and scripture; divine revelation ruled in Western culture; and little progress was made in the sciences.

**541 A.D.**

China—"Qi (life breath) rides the wind (feng) and is scattered, but is retained when encountering the rain (shui)," which is an early statement of biometeorology.

Qi is life breath, but originally it is a meteorological category made up of six phases: cold, warmth, wind, rain, darkness, and light. Qi originally meant water in the form of clouds. It evolved to mean an animating force in the atmosphere manifested by weather phenomena that actively influenced the human body.

**ca. 600 A.D.**

Spain—St. Isidore, Bishop of Seville (560–636), printed the first world map in Europe. Isidore wrote "On the Nature of Things," a manual of elementary physics including meteorology. He discussed frost, rain, hail, and snow.

He wrote of four elements (earth, air, fire, and water) and two pairs of opposing qualities (hot–cold and moist–dry).

Isidore relied on past authority, not observation, so his work is not science.

**703 A.D.**

Bede the Venerable (672/3–735) wrote "On the Nature of Things," which was a translation of extracts of Isidore's work. It had 51 chapters on Earth, heavens, stars, and planets.

**ca. 770 A.D.**

During the times of Charles the Great (Charlemagne) (742–814), church bells become popular and sometimes mandatory. Washing of bells begins and becomes popular. This was later taken to be "baptism," but is more accurately described as a "blessing." The popularity of church bells rises during the late 7th century, and bells had many purposes as revealed in bell inscriptions like:

> "I praise the true God, I call the people, I assemble the clergy;
>
> I bewail the dead, I dispense storm clouds, I do honor to feasts."

Another example is: "At obsequies I mourn, the thunderbolts I scatter, I ring in the sabbaths; I hustle the sluggards, I drive away storms, I proclaim peace after bloodshed."

**825 A.D.**

Irish monk Discuil commented on the lack of ice around Iceland.

**ca. 830 A.D.**

Agobard (770–840), Archbishop of Lyons, denounced that tempests were the result of the magic of "storm-makers," which was considered a form of witchcraft. He wrote,

> "For as soon as they hear thunder and see lightning, they say 'a gale has been raised.' When they are asked how the gale is raised, they answer that the gale has been raised by the incantations of men called 'storm-makers,' and it is called a 'raised gale.' "

He was against the diabolical origin of storms, not for scientific reasons but for religious ones. He also wrote,

> "We also read in the afore-mentioned book (Ecclesiasticus 43:12–25): 'Look upon the rainbow, and bless Him that made it: it is very beautiful in its brightness. It encompasseth the heaven about with the circle of its glory, the hands of the most High have displayed it. By His command. He maketh the snow to fall, and sendeth forth swiftly the lightnings of His judgment ... and at His will the south wind shall blow. The noise of His thunders shall strike the earth, so doth the northern storm, and the whirlwind ... "

**968 A.D.**

Pope John XIII writes that baptism of bells more accurately should be termed the blessing of bells.

**ca. 1000 A.D.**

Persia (presently Iran and Iraq)—According to Ibn al-Haytham (965–1040), "founder of the science of optics," conclusions must be based on experiment and geometric proofs, not on past authority. This is the earliest known statement of the scientific method.

He wrote that light emanates from a luminous surface, not the eye, and is the same thing, no matter what the source.

He correctly explained the atmospheric refraction of light, dispersion into colors, and the apparent increase of the size of the sun and moon near the horizon.

He discussed the density of the atmosphere.

He explained colors of sunset, colors of the rainbow, and the nature of light.

He calculated lens curvature versus focal point and the acceleration of gravity, and showed that twilight begins when sun is 19° below the horizon.

He calculated height of the atmosphere to be 52,000 passuum (paces) = 52 km = 32 mi, which is very close.

He argued the Milky Way was very far away, no matter what Aristotle said.

He was the first to describe and build a camera obscura, 230 years before Roger Bacon did.

He wrote that he would seek "to employ justice, not follow prejudice, and to take care in all that we judge and criticize that we seek the truth and not be swayed by opinions."

Science was alive in the Arab world, while the West struggled through the Middle Ages.

Abu Ali ibn Sina (Avicenna) (980–1037) invented what was possibly the first working thermometer.

**1070 A.D.**
China (Shang Dynasty)—A double rainbow was described and thought to be caused by reflection from suspended raindrops.

**1200 A.D.**
Arab scientist Idrisi classified climate using 7 climatic zones, each with 10 divisions.

**1247 A.D.**
China—Rain and snow were measured in each provincial capital by gauges made of large bamboo segments.

**ca. 1260 A.D.**
St. Thomas Aquinas (1225/7–1274) wrote "Summa Theologica," which reinforced the doctrine of the diabolical origin of storms.

> "Rains and winds, and whatsoever occurs by local impulse alone, can be caused by demons … It is, a dogma of faith that the demons can produce wind, storms, and rain of fire from heaven."

Elsewhere he wrote that bells, "provided they have been duly consecrated and baptized," are the foremost means of "frustrating the atmospheric mischiefs of the devil … for the tones of the consecrated metal repel the demons and avert storm and lightning."

**1269–1270 A.D.**
A weather diary records existing conditions, most likely in Oxford, England.

**ca. 1300 A.D.**
Kamal al-Din Abu'l Hasan Muhammad Al-Farisi (1260–1320) developed the first mathematically correct explanation of the rainbow as two refractions and one reflection of sunlight on the way to the eye. He experimentally verified this with a sphere of water.

He also explained that the colors of the rainbow were from superimposition of different forms of light on a dark background.

**1328 A.D.**
William of Ockham (1288–1348) wrote "Summa Logicae," which contained the basic foundation of the scientific method, and "Occam's Razor," which stated that the simplest of two (or more) equally likely explanations is probably the correct one.

**1337 A.D.**
England—William Merle was the first westerner known to keep a weather diary from 1337 to 1344 (see 1269 A.D.).

**1423 A.D.**
Korea—Soil moisture was measured to determine severity of a drought.

**1437 A.D.**
Pope Eugene IV (1388–1447) wrote a bull encouraging the inquisitors to use greater diligence against human agents of the "Prince of Darkness," especially against those that have the power to produce bad weather.

**1441 A.D.**
Korea—Because of different soil types it was impossible to compare regions and get an accurate estimate of drought severity. Provincial governors were ordered to construct a pedestal to hold an iron vessel of specified dimensions to measure rainfall. The instrument was designed by Prince Munjong and he called it a Chugugi.

The rainfall-measuring network was in operation for more than 700 years until the early 20th century.

## The Dawn of Scientific Meteorology—ca. 1450 A.D.–1800 A.D.: The Development of Weather Instruments and Observations

The scientific method was developed before this period began, but what distinguishes this era is that technology was now giving scientists the tools to test their ideas and measure results. To test ideas, science must have the tools to do so, and those tools were the first weather instruments.

**1450**
Italy—Leon Battista Alberti described a flat plate anemometer to measure wind speed.

**ca. 1460**
Germany—Cardinal Nicholas de Cusa (1400–1464) developed the first hygrometer to measure humidity in the west. He balanced wool against stones, and as the wool absorbed moisture from the air, it became heavier.

**1484**
Pope Innocent VIII (1442–1492) told clergy to leave no means untried to detect sorcerers, especially those who by evil weather destroy vineyards, gardens, meadows, and growing crops. The doctrine of satanic agency in atmospheric phenomena became a moving force in the Inquisition.

**ca. 1485**
As the Renaissance began, weather "science" and forecasting were dominated by astrology and weather signs dating back to the Greeks. There had been little change from the Middle Ages, and there would not be much advancement until the age of instrumentation began in the 17th century.

The term "meteors" is used for anything falling from or crossing the sky, other than planets and stars. "Meteor" is a direct transliteration from the Greek meaning "something raised up," and thus different from the modern use of "meteor" in the astronomical sense.

**1500**
Italy—Leonardo Da Vinci (1452–1519) improved the hygrometer and described a pressure plate anemometer.

## 1535

Chile—Estimates of annual rainfall for the Valparaiso–Santiago area exist.

## 1555

The term "meteorologer," an authority on meteors, was used by Leonard Digges in his book "A Prognostication of Right Good Effect." He credited Aristotle with the theoretical basis for his work.

## 1563

Possibly the earliest use of the term "meteorology" in English writing by William Fulke in his book "A Goodly Gallery with a Most Pleasant Prospect, into the Garden of Naturall Contemplation, To Beholde the Naturall Causes of All Kind of Meteors." He listed many alternative explanations and wrote on the plausibility of each.

## 1580 A.D.

Italy—Giambattista Della Porta reported on oat beard hygrometers.

## ca. 1593

Italy—Galileo Galilei (1564–1642) invented the thermometer, really a thermoscope. A thermoscope registers changes in temperature, while a thermometer has a fixed scale, and temperature, not just changes, can be measured.

## ca. 1600

British physician Robert Fudd drew the first diagram of a true thermometer with sensor and measurement scale.

## ca. 1610

Majoli, Bishop of Voltoraria (now southern Italy), wrote "Dies Canicularii" (Dog Days) referring to the warm days of summer thought to partially be caused by the appearance of Sirius, the Dog Star.

He wrote that thunderbolts are "an exhalation condensed and cooked into stone," and that "it is not to be doubted that, of all instruments of God's vengeance, the thunderbolt is the chief."

## ca. 1620

Cornelius Drebble (1572–1633) invented a thermoscope and also invented the submarine.

Santorio Santorio (1561–1631), a physician, developed a medical thermometer for use at the University of Padua. He claimed to base it on the design of Heron of Alexandria (see ca. 35 A.D.). He was the first to develop the idea of a thermometer and use it.

## 1633

Italy—June 22. After his observations lead him to conclude Earth orbits the sun, 69-year-old Galileo is accused of heresy. In 1992, 359 years later, Pope John Paul II issued an apology to Galileo.

## 1634

Pont-Mousson, France—Church bells were still being rung because, according to the inscription, "They praise God, put to flight the clouds, affright the demons, and call the people." Other bell inscriptions in the same part of France: "It is I who dissipate the thunders."

Germany—"They ward off lightning and malignant demons." In Germany, 400 towers damaged and 120 bell ringers were killed in 33 years.

## 1639

Italy—Benedetto Castelli (1578–1643), first scientific rain measurement in Europe.

## ca. 1640

France—René Descartes (1596–1650) discovered that water vapor is a distinct substance in the air. He also developed analytic geometry and much of calculus.

## 1643

Italy—Evangelista Torricelli (1608–1647) invented the barometer based on studies by recently deceased Galileo to verify "vacuum theory." His first barometer used water. Torricelli was the first to notice weather changes associated with barometric pressure changes.

## 1644

America—The first weather records were made in America by Reverend John Campanius at Swedes' Fort near Wilmington, DE.

## 1648

France—Blaise Pascal (1623–1662) along with René Descartes, carried a barometer, carried up Puy-de-Dôme, and demonstrated the decrease of atmospheric pressure with increasing altitude.

## 1654

Fernando II de Medici developed the first thermometer dependent only on temperature and not on temperature and air pressure. His design was a sealed tube filled partially with alcohol with a reserve in a bulb. A sealed tube was not affected by air pressure like earlier designs.

## 1661

Protestant Swabia (now eastern Switzerland, southwestern Germany and Alsace)—Pastor Georg Nuber wrote "Weather Sermons" and discussed nearly every sort of elemental disturbance, storms, floods, droughts, lightning, and hail, which he said come directly from God because of human sins.

## 1662

England—Robert Boyle (1627–1691) developed the gas law relating pressure, volume, and temperature.

England—Christopher Wren (1632–1723), first recording rain gauge, a tipping bucket.

## 1663

England—Robert Hooke (1635–1703) wrote "Method for Making a History of Weather," in which he treated the issues of standardization and recording observations. The Royal Society of London proposed standardizing thermometer scales using Hooke's proposal, published in 1665.

### 1664

Paris—Formal weather observations begin.

### 1665

England—Robert Hooke fixed 0° as the freezing point of water. Christian Huygens proposed the melting and boiling points of water as extremes for a thermometer scale.

### 1667

England—Robert Hooke invented the anemometer.

### 1670

First mercury in glass thermometer. Barometers were being produced and marketed for use in private homes.

### 1671

Ralph Bohun made one of the earliest attempts to scientifically explain wind.

### 1673

Rome—Father Augustin de Angelis, Clementine College, wrote "Meteorology Lectures" in an attempt to compromise the established, nonscientific, Aristotilian view with the new scientific view; he was using the new methodology to bolster the old, fading ways. He always came to the medieval conclusion but by the modern process of observation.

### 1686

Engand—Edmund Halley (1656–1742) published the first comprehensive map of the trade winds. Halley's comet is named after him, and he was the first to connect Earth's general circulation with the distribution of solar heating.

### 1687

England—Isaac Newton (1643–1727) wrote "Principia" and detailed his three laws of motion. This is perhaps the greatest scientific work ever written.

### 1694

Italy—Carlo Renaldini (1615–1679) used both freezing and boiling points of water for scaling a thermometer.

### 1698

Gottfried Wilhelm Liebniz (1646–1716) proposed the aneroid barometer in a letter to Johann Bernoulli (1667–1748). He wrote, " . . . a small closed bellows which would be compressed and dilated by itself as the weight of the air increases or decreases." In 1702, he proposed a bellows of metal that could be carried like a pocket watch but could find no one that had technology to manufacture it.

### ca. 1700

The doctrine of "storm-makers," that is, the satanic agency, in weather is dying out.

### 1701

Sir Isaac Newton proposed a 12-degree scale as a thermometer standard ranging from the melting point of ice to human body temperature.

### ca. 1701

Ole Christiansen Romer (1644–1710), a Danish Royal mathematician, proposed the temperature of ice and salt mix, the "frigorific" temperature, as the lower limit of the scale and the temperature of the armpit of a healthy male as the upper.

### 1710

Germany—Daniel Gabriel Fahrenheit (1668–1736) developed the Fahrenheit temperature scale based on Romer's idea.

### 1714

Germany—Daniel Gabriel Fahrenheit developed the first mercury thermometer with reliable scales.

### 1716

England—Edmund Halley correctly wrote that the auroras are caused by what he called "magnetic effluvia" moving along Earth's magnetic field lines.

### 1731

France—Réaumur temperature scale developed by René Réaumur (1683–1757).

### 1735

England—George Hadley (1685–1758) wrote "Concerning the Cause of the General Trade Winds." He built on Halley's theory and explained that the relative motion of Earth and atmosphere resulted in an easterly component of trade winds. His entire theory stood until the 1920s. His work was just incomplete based on modern knowledge. The atmosphere's Hadley circulation cells are named for him.

### 1742

Sweden—Centigrade temperature scale was developed by Anders Celsius (1701–1744).

### 1743

Father Vincent of Berg wrote a manual on "storm-makers."

### 1743

America—Benjamin Franklin (1706–1790) deduced the northeastward movement of a hurricane from eclipse observations at Philadelphia and Boston. This was the first recorded instance in which the progressive movement of a storm system as a whole was recognized.

He also proved that lightning was electricity (June, 1752) and charted the Gulf Stream with his cousin, whaler Capt. Timothy Folger in 1770.

### 1745

England—Peter Ahlwardts wrote "Reasonable and Theological Considerations About Thunder and Lightning," and advised his readers to seek refuge from storms anywhere except in or around a church. "Had not lightning struck only the churches ringing bells during the terrific storm in lower Brittany on Good Friday, 1718?"

## 1747

Ben Franklin does electrical experiments and coins "positive" and "negative" for the opposite electric currents instead of the established "vitreous" and "resinous," and conceptualized the "battery."

Ben Franklin described similarities between electricity and lightning, like color, crackling, and a forked path.

## 1749

Scotland—Alexander Wilson (1766–1813) and student Thomas Melville used kites for upper air measurements, and a thermometer was carried to 3000 feet.

## 1750

Franklin writes about his idea for a lightning rod after he observed a sharp iron needle conducting electricity away from a charged metal sphere.

> "May not the knowledge of this power of points be of use to mankind, in preserving houses, churches, ships, etc., from the stroke of lightning, by directing us to fix, on the highest parts of those edifices, upright rods of iron made sharp as a needle ... Would not these pointed rods probably draw the electrical fire silently out of a cloud before it came nigh enough to strike, and thereby secure us from that most sudden and terrible mischief!"

## 1752, June 15

Benjamin Franklin demonstrated that lightning is electricity with his famous and very dangerous kite experiment.

## 1753, February 4

Ben Franklin in a letter to John Perkins sends an illustration on his idea of the airflow in a waterspout. He wrote,

> "In my Paper, I supposed a Whirlwind and a Spout, to be the same Thing, and to proceed from the same Cause; the only Difference between them being, that the one passes over Land, the other over Water ... You agree that the Wind blows every way towards a Whirlwind from a large Space round; ... A Fluid moving from all Points horizontally towards a Center, must at that Center either ascend or descend. Water being in a Tub, if a Hole be open'din the Middle of the Bottom, will flow from all Sides to the Center, and there descend in a Whirl."

## 1755

In Massachusetts, the earthquake of 1755 is blamed on the use of Franklin's lightning rod. Reverend Thomas Prince, Pastor, Old South Church, said earthquakes aredue to "iron points invented by the sagacious Mr. Franklin." He continued, "in Boston are more erected than anywhere else in New England, and Boston seems to be more dreadfully shaken. Oh! there is no getting out of the mighty hand of God."

Afterward, superstition delayed installation of lightning rods across the United States and Europe.

## 1762

England—The first lightning rod in was installed.

## 1766

Venice, St. Mark's—A lightning rod was installed. The church was struck by lightning in 1388, 1417, 1489, 1548, 1565, 1653, 1745, 1761, and 1762 and damaged each time. It has not been damaged since a lightning rod was installed.

## 1770

Prof. John Winthrop of Harvard declared as to the arguments against Franklin's rods:

> "It is as much our duty to secure ourselves against the effects of lightning as against those of rain, snow, and wind by the means God has put into our hands."

## 1770

Ben Franklin and Capt. Timothy Folger chart the Gulf Stream. Franklin did not discover the current, Ponce de Leon described it in 1513, and the first chart was published in 1665 by Kircher.

## 1775

Ben Franklin and Charles Blagden use the thermometer as a navigation instrument and record water temperature across the Atlantic. By measuring water temperature, they could deduce how close to the center of the Gulf Stream they were sailing.

## 1776

New Haven—Formal weather observations begin.

## 1776, July 5

Thomas Jefferson bought his first barometer in Philadelphia.

## 1776–1778

Thomas Jefferson (Monticello) and James Madison (not the president, near Williamsburg) take first known simultaneous observations in America.

## 1783, December 1

France—Jacques Charles rising in a hydrogen balloon recorded a fall of temperature with height and the first measure of an atmospheric lapse rate.

## 1783

Switzerland—Horace-Benedict De Saussure (1740–1799), who wrote "Essay on Hygrometry," invented the human hair hygrometer when he discovered that human hair would change length 2.3% between complete dryness (short) and complete saturation (long). The strands had to be grease free. By linking the hair to a mechanism that scaled the change of length an arm would move indicating the humidity of the air.

## 1783

Paris—An edict, "to make the custom of ringing church bells during storms illegal on account of the many deaths it caused to those pulling the ropes."

## 1785

Rainfall measurements began in Edinburgh, Scotland.

## Prelude to Modern Meteorology—1800 A.D.–1900 A.D.: The Thermal Theory of Cyclones

For nearly 400 years, man was observing with instruments and testing his hypotheses using the scientific method. Now the ideas began to develop into a coherent body of knowledge. The thermal theory of cyclones held that the energy source for storms was the heat released when water condensed. Meteorologists were beginning to discover the basics of storm energy but this was not the final answer.

### 1802
England—Luke Howard (1772–1864) classified clouds by type (form) in a lecture and published a paper in summer 1803 titled "On the Modification of Clouds." By "modifications" he meant "form."

England—John Dalton's (1766–1844) published his work on the Law of Partial Pressures. He also kept a weather diary.

### 1805
Sir Admiral Francis Beaufort (1774–1856) developed the Beaufort Wind Scale because there were with no or very few devices to measure wind speed.

### 1807
Thomas Jefferson founded the Coast Survey.

### 1814
US Surgeon General ordered all Surgeons in the US Army to take weather observations.

### 1816
France—Pierre Simon de Laplace (1749–1827) wrote on adiabatic changes of temperature. Adiabatic changes of temperature occur without the addition or removal of heat and are entirely due to pressure or volume changes.

### 1816
Germany—H. W. Brandes drew the first synoptic maps using pressure and wind observations from a scanty network in central and western Europe for a few days in 1783.

This is the earliest known statement of the synoptic method. Synoptic is from the Greek meaning "a view of the entirety."

> "Even though these charts . . . appear ridiculous to some, I do believe that one should consider to pursue this thought. So much as least is certain: that 365 charts of Europe, depicting blue sky, and thin and dark clouds and rain . . . the direction of the wind . . . [and] a few well selected indications of temperature, would give the audience more pleasure and would teach more than meteorological tables."

### 1823
France—Simon Denise Poisson (1781–1840) developed the adiabatic volume change equation.

### 1828
England—Robert Fitzroy (1805–1865) was placed in charge of H.M.S. Beagle by Sir Francis Beaufort.

### 1830
America—William Redfield (1789–1857), his "Remarks on the Prevailing Storms of the Atlantic Coast of the North American States" appeared in the American Journal of Science and Arts. Redfield noticed that the trees in eastern Connecticut fell one way while trees in western parts fell the other and developed the "centrifugal theory." Where centrifugal force balances pressure gradient force as storms rotate, air converges to the center.

He investigated smaller scale storms than Dove (see 1837). He could easily see evidence of rotation in damage from tornadoes and hurricanes.

His theory was a purely mechanical theory and did not consider the source of energy.

### 1831
America—William Redfield drew the first known weather map of the United States.

England—Robert Fitzroy and the H.M.S. Beagle depart for a second journey. On board is Charles Darwin.

### Early 1830s
America—James Pollard Espy (1785–1860) stated that vertical motions lead to adiabatic temperature changes in storms. He linked adiabatic cooling and thermal convection with the release of latent heat to large-scale cloud formation and precipitation. He was the first to incorporate thermodynamics in meteorology (see 1841).

### 1833
America—The telegraph was being developed by Joseph Henry and others.

### 1835
Paris—Gaspard Gustave de Coriolis (1792–1843) in "Sur les équations du movement relative des systémes de corps," defined the coriolis effect mathematically. He discovered how to modify Newton's equations of motion for a rotating frame of reference.

### 1837
Invention of the pyrheliometer, an instrument that measures sunshine.

Heinrich Wilhelm Dove (1844–1916) published his "Linear Two Current Theory." In it he opposed the idea that storms were vortices except in the tropics.

He stated that midlatitude storms were from the conflict of opposite currents, a "polar current" and an "equatorial current." He also sought a relationship between pressure distribution and wind. Because he investigated large-scale storms like the midlatitude cyclone, it was hard to see evidence of rotation.

Samuel Morse invented the telegraph. He also compiled global climate maps.

### 1839
James Pollard Espy suggests creating rain artificially.

## 1841

James Pollard Espy published "Philosophy of Storms." In his book he took a giant step forward in modernizing the science of meteorology. He discussed the role of water vapor as an energy source for storms through the release of latent heat.

## 1842

Christian Doppler (1803–1853) conceptualized what is now called the "Doppler effect."

James Pollard Espy was appointed the first federal government meteorologist.

## 1843

America—Elias Loomis wrote an article titled "On Two Storms Which Were Experienced Throughout the United States in the Month of February, 1842." This is the first scientific use of synoptic weather maps.

Sunspot cycle was discovered.

America—William Redfield in his article "Observations of the Storm of December 15, 1839," attributed the low pressure at the center of a storm incorrectly to the centrifugal effect.

## 1844

France—Lucien Vidie (1805–1866) invented and patented the first aneroid barometer.

America—The first telegraph line is installed between Baltimore and Washington City. Soon, weather from distant cities would be known instantly.

## 1845

America—The telegraph is first made available to the general public. Soon, weather forecasting would no longer be based on local signs but on weather traveling from one place to another.

## 1848

First telegraphic weather report in the world was published in England.

## 1849

America—Joseph Henry and the Smithsonian Institution established a network of 150 volunteer weather observers in what Henry called a "system of extended meteorological observations for solving the problem of American storm."

By 1853, there are 40,000 miles of telegraph lines in operation; 27,000 miles are in the United States.

## 1851

Weather maps using 22 stations were sold at the Great Exhibition in London.

## 1853

Matthew Fontaine Maury (1806–1873) published "The Physical Geography of the Sea and Its Meteorology," the first textbook of modern oceanography.

## 1855–1875

Many European countries establish meteorological services.

## 1856

A daily weather map was on public display each day at the Smithsonian Institution. Telegraphed weather observations were color coded on the map.

America—Willam Ferrel (1817–1891) developed what is essentially the modern scheme of the general circulation of the atmosphere.

## 1857

America—Joseph Henry envisioned a system for storm warnings.

Christophorus Heinrich Didericus Buys-Ballot developed the wind–pressure relationship known as "Buys-Ballot's Law" which stated that in the northern hemisphere with the wind at your back, high pressure is to the right.

Buys-Ballot also confirmed the Doppler shift by having musicians hold a constant note as they rushed by on a train.

Lorin Blodget published "Climatology of the United States."

## 1859

William Ferrel developed the first mathematical formulation of atmospheric motions on a rotating Earth.

## Post 1850

England—Lord Kelvin developed the absolute temperature scale.

## ca. 1860

Admiral Fitzroy provided weather forecasting scripts to be used with barometers on British ships.

## 1861

Synoptic weather maps were introduced to England by Rear Admiral R. Fitzroy (1805–1865) and used to produce public forecasts.

## 1863

Synoptic weather maps were introduced to France by astronomer Urbain Jean Joseph Le Verrier (1811–1877). When a French–British fleet was destroyed by a storm on the Black Sea in 1854, while reconstructing weather maps, he discovered that the storm was observed on the Mediterranean Sea the previous day and he realized that a network of observers could provide warnings.

England—The term "synoptic charts" was coined by Rear Admiral Robert Fitzroy.

## 1864

James Croll published a paper, "Physical Cause of the Change of Climate During Glacial Epochs."

## 1860s–1870s

Europe—The Thermal Theory of Cyclones ruled. It stated that the only source of energy for cyclones was the latent heat released during condensation.

*1868*

Cleveland Abbe (1838–1916), the Cincinnati Chamber of Commerce and Western Union established a telegraphic system to collect weather observations. Abbe worked with Joseph Henry of the Smithsonian Institution.

*1869, May 7*

America—Cleveland Abbe proposed daily forecasts for newspapers.

*1869, September 1*

Cleveland Abbe's first public forecast, "The Cincinnati Weather Bulletin" is issued. He said, "I have started that which the country will not willingly let die." These were the first regular maps produced in the United States. The project benefited from the telegraph.

*1870*

US Weather Bureau was established.

R. Robinson introduced the 4-cup anemometer to the United States.

*1871, November 1*

The first weather map with isobars and weather probabilities was issued by the US Army Signal Service.

*1872*

William Clement Ley (1840–1896) published "The Laws of Winds Prevailing in Western Europe." He included a diagram that allowed ground-based observers to envision the typical sequence of weather types as a cyclone passed. He also gave a good description of the weather and wind associated with the passage of a cold front.

*1873–1879*

Northern hemisphere charts were published by Cleveland Abbe.

*1875*

First wind charts of the world by James Henry Coffin.

*1879*

The first weather map published in a newspaper appeared in the "New York Graphic."

*1882*

John P. Finley wrote, "The Character of Six Hundred Tornadoes," an exhaustive study of tornado paths, damage, and characteristics.

*1884*

The "American Meteorological Journal" reported on the efforts of John P. Finley to forecast tornadoes. By 1883, he had 120 "tornado reporters" and by 1887 the number had grown to 2403. Finley's was the first scientific effort in the United States to anticipate severe weather. He used isobaric and "thermometric" charts along with climatological data in his work.

*1884, April 26*

The first time a tornado was photographed was on this date in Garnet, KS by A.A. Adams.

*1884, August 28*

Second time a tornado was photographed occurred near Forestburg, Dakota Territory (now South Dakota) by F.N. Robinson. His is often incorrectly listed as the earliest tornado photograph.

"Principals of Forecasting by Means of Weather Charts" by Ralph Abercromby is published.

*1886*

The word "tornado" is banned from US forecasts to avoid panic.

First world maps of annual and monthly cloud cover.

*1887*

Germany—Heinrich Hertz (1857–1894) experimented with radio waves and discovered some things reflected radio waves. He also measured their speed. This led directly to the development of radar.

Scotland—Ralph Abercromby (1842–1897) drew an accurate forerunner to the modern midlatitude cyclone model. He theorized that gunfire could affect rainfall.

Europe—Luke Howard's cloud classification scheme was modified to include altitude as a variable and adopted by committee in 1891. It was published as the "International Cloud Atlas" in 1896.

America—"Tornadoes: What They Are and How to Observe Them; With Practical Suggestions for the Protection of Life and Property" was published by John P. Finley.

*1889*

Mikhail Milankovitch wrote one of the first books on synoptic meteorology.

*Late 1800s*

Kite observation stations sent "meteorographs" aloft that measured temperature and relative humidity on a graph.

*1891*

The US Weather Bureau was moved from the Signal Service to the Department of Agriculture.

*1892*

America—The systematic use of weather balloons began.

*1893*

Hail formation theories by Rollo Russell.

*1894*

Blue Hill Observatory, Massachusetts—Recording barometer and recording thermometer were suspended from a kite.

*1896*

The first "International Cloud Atlas" was published.

"The Air of Towns" by J.B. Cohen, a comprehensive study of urban air pollution was published.

*1896, May 27*

The US Weather Bureau forecast called for "destructive local storms." That day a killer tornado struck St. Louis. The word "tornado" was banned from use in forecasts in 1886 to avoid panic.

*1898*

Regular kite observations were started by the US Weather Bureau and would continue until 1933.

Norway Vilhelm Bjerknes (1862–1962) applied the science of hydrodynamics to the atmosphere and opened the door for forecasting computer models in the future.

### The Discovery of Fronts, Air Masses, and the Middle Latitude Cyclone—1900–1940

This is the first phase of modern meteorology. The concepts the public sees everyday on television, highs, lows, and fronts were developed during this period in Bergen, Norway by Jacob Bjerknes, his son Vilhelm and their collaborators. They were very close to finalizing the framework of modern meteorology but that would have to wait until the role of the jet stream was discovered.

*1900*

Vladimir Köppen introduced the first true climate classification system.

*1902*

The stratosphere was discovered and named by Leon Teisserenc de Bort (1855–1913 and Richard Assmann (1845–1918).

*1904*

V. Bjerknes suggests numerical weather prediction, but there were too few observations and complex mathematical calculations took too long to crank out by hand.

*1906*

R. G. K. Lempfert (1875–1957) and W. Napier Shaw (1854–1945), constructed air trajectories, backward in time using synoptic sea-level pressure maps.

*1909*

America—Free balloon program of meteorological soundings began at USWB.

*1917*

Jacob Bjerknes (1897–1975), son of Vilhelm formulated the polar front theory of cyclone formation.

*1917*

First work on forecasting with probabilities by B. Rolf.

*1917–1918*

Norway—Tor Bergeron (1891–1977) confirmed the existence of different air mass separated by what Bergeron and his colleagues at the Bergen School of Meteorology called fronts.

*1918, August 15*

Norway—J. Bjerknes combined a warm front with and cold front and the modern model of the middle latitude cyclone was born.

*1918*

Vladimir Köppen began work on his climate classification system.

*1919, November 19*

Norway—Tor Bergeron discovers the process of occlusion in the middle latitude cyclone. As the cyclone ages, the cold front catches up with the warm front and the combination of the two is the occluded front.

*1920*

Milutin Milankovitch refined the theory of climate change through the variation of Earth's orbital geometry.

*1922*

Lewis Fry Richardson (1881–1953) published "Weather Prediction by Numerical Process." His first attempt to predict the weather 1 hour into the future took six weeks of hand calculation. He proposed a calculating factory with 64,000 people operating mechanical calculators. The weather would still happen faster than they could get results.

He is famous for his statement of continuity across spatial scales from the largest to the very smallest:

> Big whirls have little whirls, which feed on their velocity.
> Little whirls have lesser whirls, and so on to viscosity.

*1925*

The US Weather Bureau established 20 aircraft sounding stations across the country. Airplanes would collect upper air data for forecasting purposes.

J. Durward noted the jet stream above northwestern Europe.

*1928*

The first radiosondes were put into use. A radiosonde is an instrument package carried aloft by balloon. It radioed back weather conditions.

*1930s*

Tor Bergeron (1891–1977) and Walter Findiesen (1909–1945) developed what is known today as the Bergeron Process that is thought to be the primary mechanism for the formation of rain.

Carl Gustav Rossby (1898–1957) studied atmospheric circulation and formalized the theory of planetary waves, now called Rossby waves.

*1931*

Francis Reichelderfer, future chief of the US Weather Bureau was sent to Norway by the US Navy to study the Norwegian methods of air mass analysis.

### 1932

The US Navy adopts the Norwegian methods of weather map analysis. It is the first organization in the Americas to do so.

### 1933

Meteorological Service of Canada begins using air mass and frontal analysis for instructional purposes.

### 1937, August 17

The first official US Weather Bureau radio meteorograph (radiosonde, see 1928) was made at East Boston Airport.

### Late 1930s

Weather Radar (radio detecting and ranging) is developed.

### 1940

First attempts at 5-day forecasting using upper air patterns called Rossby waves.

US Weather Bureau is moved to the US Department of Commerce.

### 1941

Pilots flying westward over the Pacific Ocean encounter headwinds up to 300 mph. Intensive investigations began into to the winds that became known as jet streams. Eventually, The Chicago School of Meteorology in the early 1950s, using the jet stream, would complete the theoretical foundation of modern meteorology started by the Bergen School (see early 1900s).

### 1941, July 31

The last map of the "Daily Weather Map" series to be published without fronts. Isobaric analysis began on November 1, 1871 and the United States. was about to adopt the technique of frontal analysis developed in Norway at the Bergen School.

### 1941, August 1

The first map of the "Daily Weather Map" series is published using frontal and air mass analysis.

## The Final Piece of the Puzzle: the Jet Stream—1940s and 1950s–Present

Two words summarize the importance of this era; jet streams. What earlier work on fronts, high, lows, and air masses could not explain became understandable and more easily forecast once the role of the jet stream was known, which culminated in numerical weather forecasting.

### 1943

Col. Joseph P. Duckworth and Lt. Ralph O 'Hair of the Army Air Forces accomplished the first known penetration of the eye of a hurricane.

### 1944

The first quantitative map identifying the jet stream.

### 1945, April

Joint Meteorology Committee adopted constant pressure analysis for upper air charts. Until that time the standard was pressure on constant altitude charts.

### 1945, October

V.K. Zworykin, Associate Director of Research, RCA Corporation wrote "Outline of Weather Proposal" proposing using newly developed electronic technologies for gathering, storing, and analyzing weather data. The data would be used to forecast and further understand weather systems, with a goal of weather control.

### 1946, November 13

Irving Langmuir and Vincent Schaefer of General Electric successfully modify clouds by cloud seeding in the Berkshire Mountains.

### 1946

The Weather Radar Research Project at M.I.T. explored and improved applications of weather radar. In Stage II (1956–1966), the goal was to make information into a usable product for real-time forecasting, and in Stage III (1966–1976), they sought to increase usability, develop computerization, and make the system digital.

John von Neumann (1903–1957), helped develop high-speed, digital, and electronic computers, and apply the technology to weather forecasting.

### 1948, March 25

Ernest Fawbush and Robert Miller successfully forecast a tornado at Tinker AFB, OK.

### 1948

Rockets first probed the upper atmosphere.

C. Warren Thornthwaite classified climates using water budgets.

### 1949

Delmar Crowson discussed the uses of cloud pictures taken from space.

### 1950

First computerized weather forecast on ENIAC.

### ca. 1950

The link between the jet stream and the development of the surface low is refined and put into use in forecasting.

### 1953

Numerical models were demonstrated that could forecast development of the surface low.

### 1955

US Weather Bureau began full-time numerical weather prediction under the Joint Numerical Weather Prediction unit.

### 1956

Norman Phillips proposed numerical modeling of the large-scale features of the atmosphere.

## 1957

The WSR-57 (Weather Surveillance Radar, 1957) was developed. It is the first radar network designed as a national warning network. The first was installed at The Hurricane Forecast Center, Miami, FL. in 1959. It was destroyed by Hurricane Andrew in 1992. One hundred and twenty-eight radar sites made up the network with both WSR-57 and WSR-74 (an updated version) radars. The last one was decommissioned on December 2, 1996 in Charleston, SC.

The network was replaced by the NEXRAD WSR88D (Next Generation Radar, Weather Surveillance Radar, 1988, Doppler).

## 1960, April 1

TIROS (Television Infrared Observation Satellite), the first weather satellite was launched from Cape Canaveral, FL.

## 1961

President John F. Kennedy made a speech at the United Nations proposing cooperation amongst the world's nations in weather forecasting and weather modification.

## 1962, August

J.C.R. Licklider wrote a series of memos detailing the design of the "Intergalactic Computer Network." Essentially everything the Internet is today was contained in his plans.

## 1963

Through international cooperation, World Weather Watch was formed to bring a greater global coverage of weather observations. The research arm of was called GARP, Global Atmospheric Research Program.

## 1964

Satellite Nimbus I was launched. The camera always pointed toward Earth.

## 1966, December 7

The first ATS (Applications Technology Satellite) was launched as a multipurpose engineering satellite to test whether gravity could anchor a satellite in a synchronous orbit 22,300 miles in space in the equatorial plane. The satellite would move at the same speed as Earth's surface, thus appear to remain stationary.

## 1969, April 7

BBN Technologies is awarded a contract by ARPA (Advanced Research Projects Agency) of the US Department of Defense to develop a computer-to-computer network to allow sharing files, software, and information. The ARPANET was the direct predecessor of the Internet.

## 1969, November 21

The first permanent ARPANET link was established between UCLA and the Stanford Research Institute.

## 1972

Numerical models of the atmosphere were developed.

## 1975, October 16

The GOES I was launched. GOES satellites are meant for continuous observation of Earth for weather forecasting, storm tracking, and research. The acronym stands for Geostationary Operational Environmental Satellite.

## 1981

CSNET (Computer Science Network) was deployed and extended ARPANET allowing many more college and university science departments to connect and communicate computer-to-computer.

## 1986

NSFNET went on line allowing 5 super computer centers to communicate. This network funded and developed by the US National Science Foundation bridged the gap between the very restrictive ARPANET of the Department of Defense and the explosive commercialization of the public Internet beginning in the mid 1990s.

## 1991, December 9

The High Performance Computing and Communication Act of 1991 (also known as "The Gore Bill") was enacted into law. The law provided for the development of the "National Research and Education Network" and the "National Information Infrastructure." Both were popularly called the Information Super Highway.

The bill funded the development of Mosaic the first World Wide Web browser that is credited with the boom in the Internet in the mid 1990s.

## 1990, Autumn

The first NEXRAD WSR88D radar was installed in Norman, OK.

## 1992, June 12

The first NEXRAD WSR88D doppler radar for daily use was installed at Sterling, VA.

## • 1850–PRESENT: SIX TECHNOLOGICAL INNOVATIONS ON THE ROAD TO MODERN METEOROLOGY

### 1850s Telegraph

Synoptic weather maps became possible with rapid transmission of weather observations. Joseph Henry of the Smithsonian Institution was an innovator in this area. Cleveland Abbe compiled some of the first maps in the United States from his observatory in Cincinnati. Because weather moved, by knowing what was happening to the west primitive forecasts could be made.

### 1928–1931 Radiosonde

This was the next important step in the advance of weather observations. Now a 3-dimensional picture of the atmosphere could emerge. The evaluation and application of static stability to forecasting became possible and the idea of upper winds as "Rossby Waves" took forecasting into new territory.

## Late 1930s RADAR (and Later Doppler RADAR)

Radio Detection and Ranging allowed pinpoint accuracy in tracking of thunderstorms and large-scale events, making warnings possible using the PPI (plan position indicator, or map view). The RHI (range height indicator) revealed details of weather systems never before available by scanning vertically.

## 1950s Electronic Computers

"Faster" is a good way to describe the effect of computers. Gathering, organizing, plotting, and analyzing data for charts and diagrams are 4 tasks that computers do fast and well. In addition, accurately calculating parameters that are very complex and practically impossible to do by hand is another way computers make forecasting more reliable. Numerical modeling of the atmosphere finally puts into operation the dream of Vilhlem Bjerknes (see 1904) and makes forecasting much more accurate than in the past. Finally, analysis of weather information is fast by using the electronic computer.

## 1960s Weather Satellites

Hurricanes that are far out to sea (and even into the 1960s completely unknown until accidentally discovered) are now routinely detected and tracked even before they become hurricanes. But there is more! By using infrared wavelength, energy clouds can be seen at night and temperature can be determined by measuring the wavelength of the radiated energy. By comparing, subtracting, or adding different wavelength bands, very low clouds (i.e., fog) can be separated from the ground and tracked. Precipitation estimates are possible and soon, radiosondes will be obsolete and the inadequate network upon which we now rely will be replaced by nearly continuous observations at any spot on Earth.

## 1990s The Internet

Simply put, never before in history has more information been available to so many, so quickly. The first technological innovation in modern meteorology, the telegraph revolutionized the transmission of weather information. The last innovation in this list pushed the limit of weather data transmission to extremes, not though possible, just 10 years earlier.

## • 1830–PRESENT: NINE CONCEPTUAL ADVANCES ON THE ROAD TO MODERN METEOROLOGY

### 1830s William Redfield—Storms Rotate

It seems too simplistic now but this was a novel idea in the 1830s. Redfield mapped storms and made important contributions to the nature of storms. His theory was purely mechanical emphasizing a balance between pressure change and centrifugal force.

## 1830s James Pollard Espy—Latent Heat Release

He was one of the first to state that the release of heat as water vapor condensed to liquid drove storms. His theory was called the convective theory of storms. Today we know that latent heat is a secondary source of energy for extra tropical storms. It is the primary source of energy driving tropical cyclones. During Espy's lifetime, heat was treated as a mysterious substance called "caloric." Espy's work led to the thermal theory of cyclones popular in the 1860s and 1970s.

### 1840s Elias Loomis—Use of Synoptic Charts

Elias Loomis was the first to use synoptic charts, a fundamental meteorological tool, for investigative purposes. The word synoptic comes from the Greek meaning "a view of the entirety" and a synoptic chart is an overall view of weather using simultaneous observations.

### 1850s William Ferrel—General Circulation

He was the first to explain the large-scale circulation of the global atmosphere mathematically, taking the first steps on the path to numerical weather prediction.

### 1890s Vilhelm Bjerknes—Atmospheric Hydrodynamics

As physicist, he proposed and worked out many of the details of modern meteorology and posed them as a mathematical problem in fluid dynamics.

### 1918–1921 Jacob Bjerknes—Cyclone Model, Air Masses, and Polar Front Theory

He was son of Vilhelm, and he and his collaborators known as the "Bergen School" or "Norwegian School" of meteorology conceptualized fronts, air masses, and the middle latitude cyclone.

### 1930s Carl Gustave Rossby—Long Waves

Rossby waves, or long waves are named after him. He advanced the theory and application of the concepts of upper atmosphere winds flowing through troughs and ridges to forecasting.

### 1945–1955 Rossby, Riehl, Palmen, Fultz, Petterssen—Jet Streams

What the middle latitude cyclone model of the Bergen School could not explain was how quickly a low developed, that is how fast the pressure dropped as the low strengthened. The answer was in the jet stream and the "Chicago School" of meteorology is credited with working out the details.

### 1970s Various—Air–Sea Interaction

A subject of study for years, in modern literature this became a hot topic when Jacob Bjerknes (then at UCLA) published a paper on sea surface temperatures off Ecuador and Peru. Jerome Namias followed, and soon the topic of El Niño/La Niña was everywhere.

# Index of Acronyms and Abbreviations

*The Weather Almanac: A Reference Guide to Weather, Climate, and Related Issues in the United States and Its Key Cities*, Twelfth Edition. Steve Horstmeyer.
© 2011 John Wiley & Sons, Inc. Published 2011 by John Wiley & Sons, Inc.

# General Index

# Index of Persons

## A

Abbe, Cleveland, 1110–1111, 1114
Abercromby, Ralph, 1111
Abu Ali, 1105
Adams, A.A., 1111
Adhémar, Joseph, 335
Agassiz, Louis, 335
Agobard, Archbishop of Lyons, 1104
Ahlwardts, Peter, 1107
Al-Farisi, Kamal al-Din Abu'l Hasan Muhammad, 1105
Alberti, Leon Battista, 1105
Anaximander, 1102
Aquinas, St. Thomas, 1105
Archimedes, 1103
Aristotle, 1102–1104
Arrhenius, Svante, 324
Ashley, Walker, 185, 199, 231–232
Assmann, Richard, 1112
Augustine, 1103
Augustine Bishop of Hippo, St., 1103
Augustulus, Romulus, 1104
Avicenna, 1105

## B

Bacon, Roger, 1104
Beaufort, Admiral Francis, 1109
Beaufort, Sir Francis, 1109
Bede the Venerable, 1104
Benedetto Castelli, 1106
Berg, Father Vincent of, 1107
Bergeron, Tor, 1112
Bernoulii, Johann, 1107
Bjerknes, Jacob, 298, 1112, 1115
Bjerknes, Vilhelm, 1112
Blagden, Charles, 1108
Blodget, Lorin, 1110
Bohun, Ralph, 1107
Boyle, Robert, 1106
Brandes, William Vincent
Buys-Ballot, Christophorus Heinrich Didericus, 1110

## C

Callendar, Guy Stewart, 324
Campanius, Rev. John, 1106
Castelli, Benedetto, 1106
Celsius, Anders, 1107
Charlemagne (Charles the Great), 1104
Charles, Jacques, 1108
Chun, Liu Thien, 1103
Coffin, James Henry, 1111
Cohen, Julius, 353
Coriolis, Gaspard Gustave de, 1109
Croll, James, 1110
Crowson, Delmar, 1113
Cuillandre, Jean-Charles, 101, 115

## D

Da Vinci, Leonardo, 1105
Dalton, John, 1109
Darwin, Charles, 1109
de Angelis, Father Augustin, 1107
de Bort, Leon Teisserenc, 1112
de Coriolis, Gaspard Gustave, 1109
de Cusa, Cardinal Nicholas, 1105
de Medici, Fernando II, 1106
Della Porta, Giambattista, 1106
Descartes, René, 1106
Des Voeux, Dr. Henry Antoine, 353
Discuil, 1104
Doppler, Christian, 1110
Dove, Heinrich Wilhelm, 1109
Duckworth, Joseph P., 1113
Durward, J., 1112
Dvorak, Vernon, 264

## E

Edward I, King, 353
Edzani, Rahim, 290
Elizabeth I, Queen, 353
Eratosthenes, 1103
Espy, James Pollard, 1109–1110, 1115

## F

Fahrenheit, Daniel Gabriel, 1107
Fawbush, Major Ernest J., 179, 1113

Feliz, Fray Isidro, 148
Ferrel, Willam, 1110
Findiesen, Walter, 1112
Finley, John P., 179, 184–185, 187, 1111
Fitzroy, Rear Admiral Robert, 1109–1110
Folger, Capt. Timothy, 1108
Fourier, Jean Baptiste Joseph, 324
Franklin, Benjamin, 167, 1107–1108
Fujita, Theodore, 171, 178–179
Fulke, William, 1106
Fultz, David, 1115

## G

Galilei, Galileo, 1106
Gilson, Christopher, 185, 199, 231–232
Gore, Al, 1114

## H

Hadley, George, 1107
Halley, Edmund, 1107
Harmar, General Josiah, 153
Henry, Joseph, 1109–1111, 1114
Hertz, Heinrich, 1111
Hippocrates of Cos, 1102
Hooke, Robert, 1106–1107
Howard, Luke, 1109, 1111
Huygens, Christian, 1107

## I

Ibn al Haytaham, 1104
Ibn Sina, Ali, 1105
Idrisi, 1105
Imbrie, John, 335
Isidore Bishop of Seville, St., 1104

## J

Jefferson, Thomas, 153, 1108–1109
Jerome, St., 1103
Jesus Christ, 297, 1103

## K

Keeling, David, 324
Keith, David, 231–232

*The Weather Almanac: A Reference Guide to Weather, Climate, and Related Issues in the United States and Its Key Cities*, Twelfth Edition. Steve Horstmeyer.
© 2011 John Wiley & Sons, Inc. Published 2011 by John Wiley & Sons, Inc.

# Index of Select Storms and Events

*The Weather Almanac: A Reference Guide to Weather, Climate, and Related Issues in the United States and Its Key Cities*, Twelfth Edition. Steve Horstmeyer.
© 2011 John Wiley & Sons, Inc. Published 2011 by John Wiley & Sons, Inc.